GENOMICS AND HEALTH IN THE DEVELOPING WORLD

OXFORD MONOGRAPHS ON MEDICAL GENETICS

Editors:

Judith G. Hall Peter S. Harper Louanne Hudgkins Evan Eichler Arno G. Motulsky

1. R.B. McConnell *The Genetics of Gastrointestinal Disorders*
2. A.C. Kopéc: *The Distribution of the Blood Groups in the United Kingdom*
3. E. Slater and V.A. Cowie: *The Genetics of Mental Disorders*
4. C.O. Carter and T.J. Fairbank: *The Genetics of Locomotor Disorders*
5. A.E. Mourant, A.C. Kopéc, and K. Domaniewska-Sobezak: *The Distribution of the Human Blood Groups and Other Polymorphisms*
6. A.E. Mourant, A.C. Kopéc, and K. Domaniewska-Sobezak: *Blood Groups and Diseases*
7. A.G. Steinbert and C.E. Cook: *The Distribution of the Human Immunoglobulin Allotypes*
8. D. Tills, A.C. Kopéc, and R.E. Tills: *The Distribution of the Human Blood Groups and Other Polymorphisms: Supplement I*
10. D.Z. Loesch: *Quantitative Dermatoglyphics: Classification, Genetics, and Pathology*
11. D.J. Bond and A.C. Chandley: *Aneuploidy*
12. P.F. Benson and A.H. Fensom: *Genetic Biochemical Disorders*
13. G.R. Sutherland and F. Hecht: *Fragile Sites on Human Chromosomes*
14. M. d'A Crawfurd: *The Genetics of Renal Tract Disorders*
16. C.R. Scriver and B. Child: *Garrod's Inborn Factors in Disease*
18. M. Baraitser: *The Genetics of Neurological Disorders*
19. R.J. Gorlin, M.M. Cohen, Jr. and L.S. Levin: *Syndromes of the Head and Neck, Third Edition*
21. D. Warburton, J. Byrne, and N. Canki: *Chromosome Anomalies and Prenatal Development: An Atlas*
22. J.J. Nora, K. Berg, and A.H. Nora: *Cardiovascular Disease: Genetics, Epidemiology, and Prevention*
24. A.E.H. Emery: *Duchenne Muscular Dystrophy, Second Edition*
25. E.G.D. Tuddenham and D.N. Cooper: *The Molecular Genetics Of Haemostasis And Its Inherited Disorders*
26. A. Boué: *Foetal Medicine*
27. R.E. Stevenson, J.G. Hall, and R.M. Goodman: *Human Malformations*
28. R.J. Gorlin, H.V. Toriello, and M.M. Cohen, Jr.: *Hereditary Hearing Loss and its Syndromes*
29. R.J. M. Gardner and G. R. Sutherland: *Chromosome Abnormalities and Genetic Counseling, Second Edition*
30. A.S. Teebi and T.I. Farag: *Genetic Disorders among Arab Populations*
31. M.M. Cohen, Jr.: *The Child with Multiple Birth Defects*
32. W.W. Weber: *Pharmacogenetics*
33. V.P. Sybert: *Genetic Skin Disorders*
34. M. Baraitser: *The Genetics of Neurological Disorders, third edition*
35. H. Ostrer: *Non-Mendelian Genetics in Humans*
36. E. Traboulsi: *Genetic Factors in Human Disease*
37. G.L. Semenza: *Transcription Factors and Human Disease*
38. L. Pinsky, R.P. Erickson, and R.N. Schimke: *Genetic Disorders of Human Sexual Development*
39. R.E. Stevenson, C.E. Schwartz, and R. J. Schroer: *X-linked Mental Retardation*
40. M.J. Khoury, W. Burke, and E. Thomson: *Genetics and Public Health in the 21st Century*
41. J. Weil: *Psychosocial Genetic Counseling*
42. R.J. Gorlin, M.M. Cohen, Jr., and R.C.M. Hennekam: *Syndromes of the Head and Neck, Fourth Edition*
43. M.M. Cohen, Jr., G. Neri, and R. Weksberg: *Overgrowth Syndromes*
44. R.A. King, J.I. Rotter, and A.G. Motulsky: *The Genetic Basis of Common Diseases, Second Edition*
45. G.P. Bates, P.S. Harper, and L. Jones: *Huntington's Disease, Third Edition*
46. R.J.M. Gardner and G.R. Sutherland: *Chromosome Abnormalities and Genetic Counseling, Third Edition*
47. I.J. Holt: *Genetics of Mitochondrial Disease*
48. F. Flinter, E. Maher, and A. Saggar-Malik: *The Genetics of Renal Disease*
49. C.J. Epstein, R.P. Erickson, and A. Wynshaw-Boris: *Inborn Errors of Development: The Molecular Basis of Clinical Disorders of Morphogenesis*
50. H.V. Toriello, W. Reardon, and R.J. Gorlin: *Hereditary Hearing Loss and its Syndromes, Second Edition*
51. P. S. Harper: *Landmarks in Medical Genetics*
52. R.E. Stevenson and J.G. Hall: *Human Malformations and Related Anomalies, Second Edition*
53. D. Kumar and D. Weatherall: *Genomics and Clinical Medicine*
54. C.J. Epstein, R.P. Erickson, and A. Wynshaw-Boris: *Inborn Errors of Development: The Molecular Basis of Clinical Disorders of Morphogenesis, Second Edition*
55. W. Weber: *Pharmacogenetics, Second Edition*
56. P.L. Beales, I.S. Farooqi, and S. O'Rahilly: *The Genetics of Obesity Syndromes*
57. P.S. Harper: *A Short History of Medical Genetics*
58. R.C.M. Hennekam, I.D. Krantz, and J.E. Allanson: *Gorlin's Syndromes of the Head and Neck, Fifth Edition*
59. D. Kumar and P. Elliot: *Principles and Practice of Clinical Cardiovascular Genetics*
60. V.P. Sybert: *Genetic Skin Disorders, Second Edition*
61. E. Traboulsi: *Genetic Diseases of the Eye, Second Edition*
62. D. Kumar: *Genomics and Health in the Developing World*

GENOMICS AND HEALTH IN THE DEVELOPING WORLD

EDITED BY

Dhavendra Kumar, MD, FRCP, FRCPCH, FACMG

CONSULTANT IN CLINICAL GENETICS

UNIVERSITY HOSPITAL OF WALES, CARDIFF, WALES;

INSTITUTE OF MOLECULAR AND EXPERIMENTAL MEDICINE, SCHOOL OF MEDICINE

CARDIFF UNIVERSITY & GENOMIC POLICY RESEARCH UNIT, SCHOOL OF HEALTH

SPORT & SCIENCE, THE UNIVERSITY OF GLAMORGAN, PONTYPRIDD, WALES

UNITED KINGDOM

OXFORD

UNIVERSITY PRESS

OXFORD
UNIVERSITY PRESS

Oxford University Press, Inc., publishes works that further
Oxford University's objective of excellence
in research, scholarship, and education.

Oxford New York
Auckland Cape Town Dar es Salaam Hong Kong Karachi
Kuala Lumpur Madrid Melbourne Mexico City Nairobi
New Delhi Shanghai Taipei Toronto

With offices in
Argentina Austria Brazil Chile Czech Republic France Greece
Guatemala Hungary Italy Japan Poland Portugal Singapore
South Korea Switzerland Thailand Turkey Ukraine Vietnam

Published by Oxford University Press, Inc.
198 Madison Avenue, New York, New York 10016
www.oup.com

Library of Congress Cataloging-in-Publication Data
Genomics and health in the developing world / edited by Dhavendra Kumar.
p. ; cm. — (Oxford monographs on medical genetics ; no. 62)
Includes bibliographical references and index.
ISBN-13: 978–0–19–537475–9 (hardcover : alk. paper)
ISBN-10: 0–19–537475–4 (hardcover : alk. paper)
I. Kumar, Dhavendra. II. Series: Oxford monographs on medical genetics ; no. 62.
[DNLM: 1. Genetic Diseases, Inborn—epidemiology. 2. Developing Countries.
3. Genetics, Medical. 4. Genomics. QZ 50]
616'.042091724—dc23 2012002317

1 3 5 7 9 8 6 4 2
Printed in the United States of America
on acid-free paper

"To millions of people in the developing world who live with poverty, hunger and disease"

"Science to be science must afford the fullest scope for satisfying the hunger of body, mind and soul"

– Mahatma Gandhi

PREFACE

Health of any nation depends upon its sound economic base, political stability, healthy agriculture (crops/animal stock), a safe environment (clean water, sanitation and climate control) and the general well-being of its peoples. Among several biological and ecological factors, the role of genetics and genomics is now widely accepted as a major determinant for the health of a nation. There are diverse and far reaching applications of genetics and genomics that we are beginning to utilize in all walks of life, including genetically engineered vaccines and biological products (insulin and growth hormone), new drugs (cancer and heart disease), genetically modified crops (enhanced produce and biofortification), developing bioengineered vectors for controlling malaria and other parasitic diseases, and the preservation of animal and plant life (rare and extinct species).

Most developed nations in the West have invested, and continue to do so, in harnessing the benefits of genetics and genomics. However, large populations in the less developed and least developed countries—comprising more than two-thirds of the world's population—continue to be deprived of potential health and socioeconomic applications (Figure 1). It is estimated that the current global population of around 7 billion will rise to around 10 billion by 2050 (Figure 2). There is concern that currently around 5 billion people are deprived of the benefits of modern science and technology, including genetics and genomics.

The clinical use of medical genetics and human genetic research is concentrated in the developed nations of Europe and North America that cover less than one billion people. In contrast, India alone has over 1 billion people equally at risk for being affected with a wide range of genetic diseases! The volume and burden of genetic disease afflicting those in other, developing or less developed or "low income" countries are beyond both imagination and comprehension. In addition to medicine and health, whole nations and peoples are being deprived of the socioeconomic benefits from investment and infrastructure developments in areas like biomedical and agriculture technology. The current and future applications of genome science and technology in pharmaceutical, bioengineering, and the food and agriculture industries hold great promise for the new emerging economies of the developing world (Juma and Yee-Cheong, 2005). This challenge offers new opportunities for promoting international cooperation in relevant biomedical research in developing countries as recommended by the United Nations' Millennium Project Task Force on Science, Technology and Innovation (UN Millennium Project Report, 2005). These recommendations are included in the UN Millennium Development Goals (Acharya et al., 2003).

Since the completion of mapping/sequencing of the human genome and other genomes, rapid progress has been made in a number of areas of biomedical research and biotechnology. Biotechnology has emerged as one of the methods that can be used to address health and other socioeconomic challenges in developing countries (Acharya et al., 2003). This largely depends upon the development of infrastructure and adopting a diverse set of policies aimed at translating scientific discoveries into goods and services. These technologies include molecular diagnostics, recombinant vaccines, techniques of vaccine and drug delivery, bioremediation (use of living organisms to degrade hazardous matter), sequencing pathogen genomes, female-controlled protection against sexually transmitted infections, bioinformatics, nutritionally enriched genetically modified crops, recombinant therapeutic proteins, and combinatorial chemistry (Daar et al., 2002).

While it is necessary to prioritize the current limited resources in dealing with the most common socioeconomic and health problems faced by developing countries, especially the least developed countries, resulting from poverty, malnutrition, unsafe water supply, poor sanitation, and communicable diseases, it is nevertheless important to invest in new science and technology to bridge the gap and prepare some ground for future developments. Unfortunately, geographic, economic, and political challenges in these countries often restrict investment and improvement in the infrastructure development necessary to sustain progress in any area. There is ample evidence to argue that genetic and genomic factors play an important role in the causation of the common health problems affecting developing and least-developed nations, and in their consequences for these populations. This issue was examined and highlighted in the WHO

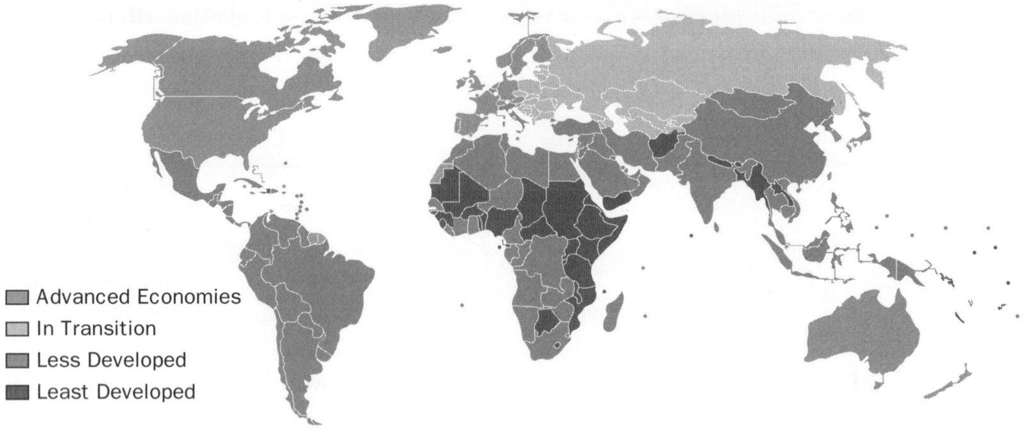

Figure 1 The world map showing least and less developed nations including those in transition.

expert committee report on role of genomics for global health (WHO, 2002). This has been followed by several reports and publications that lend further support to the view that the developing nations in Asia should not be left behind in acquiring relevant genetic and genomic technologies for the betterment of health. A recent editorial commentary (*Nature*, 2007) points to this subject and sets out an optimistic picture of the future of genetics and genomics in developing countries, particularly India, Latin America, the Arab world, and parts of Africa. The importance of genetics and genomics is acknowledged in the recent WHO statement on the global prevention of birth defects endorsed by the Sixty-Third World Assembly (WHO, April 2010). The WHO 'Grand Challenges' project on applications of genomics in the public health in the developing world (July 2011) is expected to identify potential areas and offer strategic guidance to member Nations in the developing world.

The practice of clinical medicine in public health is rapidly changing in the light of recent critical advances in genetics and genomics. This is largely based upon the sequencing of the human genome and a number of other genomes, including agricultural crops and other plants, as well as pathogens and important model organisms (microbes, worms, insects, and mammals). The true extent of the power of genomic science and technology in shaping and influencing medicine and the economy is difficult to describe but is likely to be vast. There are now several dedicated publications and on-line resources specifically reporting and discussing this subject. The publication of this book is timely, and we hope that the reader will find the insights presented throughout the volume to be exciting, challenging, and constructive. In addition, the book contains information that will be helpful beyond the confines of the profession of medicine, including the range of those with responsibility for making decisions on biotechnology development and health planning, and those concerned with related social, ethical, economic, and legal matters.

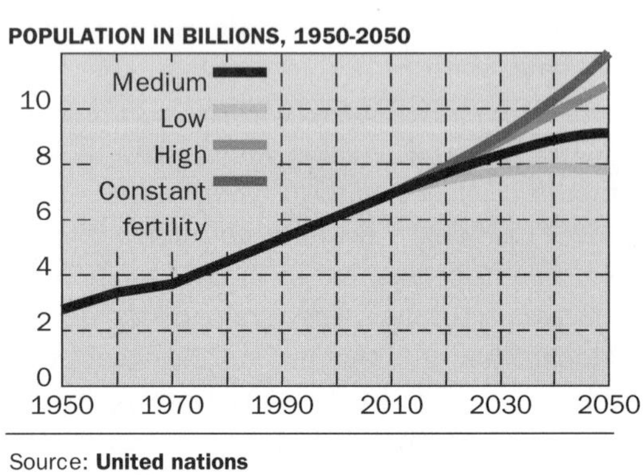

Figure 2 The United Nations projection of the global population.

There is no single book that deals with the issues and challenges for the emerging economies of the developing world in confronting the rapid advances of genome science and technology. This is probably the first book providing a detailed and comprehensive account of this extremely important field, and it is likely to have a major impact on healthcare and the economy in most countries in the developing and least-developed world. There are a few volumes currently available that cover selected population groups only, for example Jewish peoples (Abel, 2001) and peoples of African descent (Bowman and Murray Jr., 1998). There are only two books that cover in some detail the human genetics of the Arab world (Teebi and Farag, 1996) and the populations of the Indian subcontinent (Kumar, 2004). It is anticipated that the publication of this new book will fill a huge gap that currently exists in information and issues about human health in those countries that constitute the newly emerging and other developing economies of the world. It is now widely believed that the rapid development and inclusion of genetic and genomic technologies will be crucial to any nation's socioeconomic well-being, and the health of its population. It is important that this technology and expertise should not remain beyond the reach of the developing world, confined to the global West and North. We hope that the structured presentation of factual information in this book will serve as a useful resource for a wide range of professionals in medicine, population health, the biotechnology industries, media, and public services.

The book is expected to be useful to a broad range of audiences including practicing and trainee medical practitioners, health professionals, human genetic scientists, medical geneticists, genetic counselors, public health physicians and consultants, experts in health economics, bioethics, sociology, anthropology, and others leading various projects at national and international levels. The book includes material that is also applicable to medical and related professionals about the minority migrant or resident ethnic populations working and living in the developed countries of Europe and North America, as well as Australasia.

This "multi-author" edited book includes several expertly written chapters providing a detailed and comprehensive account on most aspects of population and human genetics, with an emphasis on medical and health issues related to a defined geographic region and, as far as possible, covering all major ethnic groups. The book is set in two parts, each part with several chapters contributed by a panel of internationally renowned experts in the designated field or area of interest. The first part covers general aspects of population origin and heterogeneity, an introduction to basic genetics and genomics, and the principles of human genetics, the classification of human disease from genetic and genomic perspectives, the current and future health strategies in dealing with the common human genetic diseases affecting developing countries, social and cultural practices and legislation pertinent to human genetics and genomics, the impact of genomics on human health in the developing world, and the related social, ethical, economic, and legal issues. The second part gives an account of selected genetic disorders and the genetic services relevant to major geographic and population-specific regions. Chapters within each section are written and carefully edited with the advice of experts with specialist knowledge and understanding of the subject and related regional and ethnic issues.

This book is only a small effort by several dedicated professionals, who share the commitment and strong belief that the applications of genome science and technology are crucial for the socio-economic development and provision of modern healthcare to large heterogeneous populations in the developed and least-developed nations of the world. Hopefully, it will be a contribution, however small, to achieving the goals of the World Health Organization and that of emerging low- and middle-income economies across the world.

Dhavendra Kumar
University Hospital of Wales , Institute of Molecular and Experimental Medicine, School of Medicine, Cardiff University; Genomic Policy Research Unit, School of Health, Sport & Science The University of Galmorgan United Kingdom

REFERENCES

Abel EL (2001). *Genetic disease of the Jewish peoples.*

Acharya T, Daar AS, Singer P. (2003). Biotechnology and the UN Millennium Development Goals. *Nat Biotechnology* 21, 1434–1436.

Bowman JE, Murray RF Jr. (1998). *Genetic variation and diseases among peoples of African descent.* Baltimore: Johns Hopkins University Press.

Daar AS, Thorsteinsdóttir H, Martin DK, Smith AC, Nast S, Singer PA. (2002). Top ten biotechnologies for improving health in developing countries. *Nature Genetics*, 32(2), 229–232.

Editorial—Nature (2007). Asia on the rise. *Nature* 447, 885.

Juma C, Yee-Cheong L. (2005). Reinventing global health: the role of science, technology, and innovation. *The Lancet*; 365(9464), 1105–1107.

Kumar D. (2004). *Genetic disorders of the Indian subcontinent.* Dordrecht, The Netherlands: Kluwer Academic/Springer.

Teebi A, Farag S. (1996). *Genetic diseases among Arab populations.* New York: Oxford University Press.

WHO. (2002). *Genomics and World Health.* Report of the Scientific Advisory Committee, Geneva, WHO.

WHO (2010). *Birth defects*—Report of the secretariat, 3rd World Health Assembly, April 1, 2010.

UN Millennium Project Task Force on Science, Technology and Innovation. (2005). *Innovation: applying knowledge in development*, Available at: http://www.unmillenniumproject.org/documents/Sciencecomplete.pdf.

FOREWORD I

The past one hundred years has seen the remarkable rise of science and medicine in the West. The great public investments in biomedical research and technology by the developed economies have not only contributed to the vastly improved health of their citizens, but that knowledge and its practice has, in no small measure, contributed to the improved health of populations in the developing economies as well. It is now an expectation, of people and their governments, that investments in science and medicine should improve health in their lifetimes. For someone like myself, who started academic life merely interested in the mysteries that genes offered, and who has found himself in the rapidly advancing biomedical field of human genetics, this last statement is extraordinary but possibly realizable.

We are likely in the midst of another major revolution in medicine occasioned by the human genome sequence. The rise of genetics and genomics has not only produced a common intellectual (and intelligible) language for the biomedical sciences, but has raised huge expectations that we can rationally understand the molecular causes of any human disease, create efficient diagnostic tools, and invent rational chemical therapies. More importantly, genomics has added a new dimension to medical practice, raising the expectation that we might be able to "personalize" treatment to the individual patient. Will the transfer of this new science and health knowledge, from the developed to the developing countries, be the same as that in the past? There are many reasons to suspect that it will not, and this edited volume by Professor Dhavendra Kumar is an argument for this view. Three major underlying themes in his book deserve mention.

First, genetics explains our individuality and how our recent shared human ancestry affects the variation in our genomes. Consequently, there is no full understanding of a human disease, not even a purely single-gene Mendelian disorder, from studying patients in one geographical locale. Alternatively, a comparative study of the same disease across cultures, ecologies, and ancestries is more informative for all patients. Second, unique cultures often produce unique genetic outcomes not always observed elsewhere. This could include the occurrence of consanguinity, or unique diets and lifestyles, and have effects on the distribution of the types and nature of genetic mutations. If we are to understand gene–environmental interactions it is quite clear that we will need a global study. Third, modern genetic and genomic technologies, ever since the recombinant DNA days, are powerful primarily because they are "species-independent" and can be of great benefit not only to medicine but also to agriculture and veterinary applications. This has not escaped the notice of many governments who are now enthusiastic about their investments in "biotechnology." Consequently, and despite the differing magnitudes of investment, many developing nations think it in their self-interest to develop biotechnology for solving health problems unique to their lives.

Genomics and Health in the Developing World is a potpourri of chapters that attempts to educate us to the nature, magnitude, successes, and opportunities for genetic studies in the developing world. Individually, the chapters inform us about the current state of the art. Collectively, the chapters demonstrate that a true understanding of the problems of genetic disease and personalized genetics will require solid and persistent contributions from the developing world.

We all live in a remarkable time in biomedicine, where each human has something credible to contribute to humanity's understanding of health and disease, and the practice of medicine. Whether you are a student or teacher, geneticist or not, researcher or health professional, you will find this treasure exciting.

Aravinda Chakravarti, PhD
Baltimore, MD
December, 2011

FOREWORD

FOREWORD II

Because of the extremely high childhood mortality in the poorer countries of the world resulting from malnutrition, limited public health services, and the ravages of communicable disease, international health agencies have tended to ignore the growing importance of congenital malformation and genetic disease as an emerging global health problem. In 2006 the American charity, March of Dimes, in its *Global Report on Birth Defects* estimated that more than 7,000,000 babies are born each year with either a congenital abnormality or genetic disease, and that up to 90% of these births occur in low- or middle-income countries. Remarkably, over 25% of them consist of only five conditions: congenital heart defects, neural tube defects, inherited disorders of hemoglobin, Down syndrome, and glucose-6-phosphate dehydrogenase deficiency. Overall, however, over 7,000 birth defects of genetic or partial genetic origin have been identified.

There are several reasons why there is a higher frequency of births with these conditions in the poorer countries. They include natural selection, a high frequency of consanguineous marriages, increased maternal age, and the effects of the epidemiological transition whereby, as public health measures and improved standards of living reduce childhood mortality rates, many babies who would have died with genetic disorders are now surviving long enough to present for diagnosis and treatment.

It is clear, therefore, that genetic disease cannot continue to be ignored by governments of the developing countries or by the international health agencies. One of the major obstacles toward progress in this field is lack of accurate global data about the frequency of these conditions, and current facilities for their diagnosis and management. In this book, many aspects of the current background to the high frequency of genetic disease in the poorer countries are explored, together with representative accounts of the problems raised by genetic disease in many of these countries. It is hoped that information of this type will form the basis for educating governments and health agencies about the increasing problems that genetic disease will pose in the future, particularly in countries that are going through the epidemiological transition with falling childhood mortality rates. This is a critically important topic for the future of global healthcare, and I wish this book all the success it deserves.

D.J. Weatherall, FRS
Oxford
September 2011

CONTENTS

CONTRIBUTORS

Abdelhak, Sonia, PhD
Molecular Investigation of Genetic Orphan Diseases Research Unit
Institut Pasteur de Tunis
Tunis
Tunisia

Abu-Libdeh, Bassam, MD
Department of Pediatrics & Clinical Genetics
Al Makassad Arabic Hospital
East Jerusalem
Israel

Adib, Salim M, MD
Department of Social and Family Medicine
Faculty of Medicine
Saint Joseph University
Beirut
Lebanon

Agarwal, Neeraj
National Eye Institute
National Institutes of Health
Bethesda, MD
USA

Agrawal, Anurag, PhD
Institute of Genomics and Integrative Biology (IGIB-CSIR)
Mall Road
Delhi
India

Akhmetova, Vita L
Institute of Biochemistry and Genetics
Ufa Science Center
Russian Academy of Sciences
Prospekt Oktyabrya, 71
Ufa, Bashkortostan, 450054
Russia

Al-Aqeel, Aida I, MD, FACMG
Senior Consultant Pediatrics, Medical Genetics and Endocrinology
Department of Paediatrics
King Faisal Specialist Hospital and Research Centre
Riyadh Military Hospital
Kingdom of Saudi Arabia

Al Ali, Mahmoud Taleb
Centre for Arab Genomic Studies
Dubai
United Arab Emirates

Al-Arrayed, Shaikha Salim, PhD
Genetics Department
Salmaniya Medical Complex
Ministry of Health
Bahrain

Al-Gazali, Lihadh, MSc, FRCP
Professor of Clinical Genetics
Department of Paediatrics
Faculty of Medicine & Health Sciences
UAE University
Al-Ain
United Arab Emirates

Al Homssi, Mohamad
College of Medicine
University of Sharjah
Sharjah
United Arab Emirates

Al-Saud, Haya
Department of Genomics of Common Disease
School of Public Health
Imperial College London
London
UK

Alessandrini, Marco
ME Health Enterprises CC
Centurion
South Africa

Apte, Bal N
Formerly Head of Human Genetics Laboratory
Bombay Hospital
Mumbai
India

Arunkumar, Ganesh Prasad
Department of Immunology
Madurai Kamaraj University
Madurai-625021
Tamil Nadu
India

Arun, Varatharajan Santhakumari
Dept of Immunology
Madurai Kamaraj University
Madurai-625021
Tamil Nadu
India

Asfaha, Sahle M.
Departments of Immunology
Faculty of Health Science
University of Pretoria
South Africa

Balu, Sam
Department of Human Genetics
National Institute of Mental Health and Neurological Sciences (NIMHANS)
Bangalore, Karnataka
India

Baldo, Guilherme
Gene Therapy Center
Hospital de Clinicas de Porto Alegre
Rua Ramiro Barcelos
Porto Alegre, RS
Brazil

Beighton, Peter, PhD, FRCP
Emeritus Professor
Division of Human Genetics
Faculty of Health Sciences
University of Cape Town
South Africa

Bharadwaj, Dwaipayan, PhD
Genomics and Molecular Medicine Unit
Institute of Genomics and Integrative Biology (IGIB-CSIR)
Mall Road
Delhi
India

Bhardwaj, Minakshi, PhD
Centre for Economic and Social applications of Genomics (CESAGEN)
School of Social Sciences
Cardiff University
UK

Bhatt, Meenakshi, MD, MRCPI
Associate Professor/Consultant in Clinical Genetics
Centre for Human Genetics & Indira Gandhi Child Health Institute
Bangalore, Karnataka
India

Bhatnagar, Deepak, PhD, FRCP
Consultant/Senior Lecturer in Diabetes & Metabolism
The Royal Oldham Hospital
University of Manchester Cardiovascular Research Group/Diabetes Centre
Rochdale Road
Oldham OL1 2JH
UK

Bissar-Tadmouri, Nisrine, MD
Assistant Professor
University Of Sharjah
College of Medicine
Department of Basic Medical Sciences
Sharjah
United Arab Emirates

Bittles, Alan H, PhD
Professor, Centre for Comparative Genomics
Edith Cowan University & Murdoch University
South Street
Perth, WA
Australia

Black, Michael L.
Centre for Comparative Genomics
Murdoch University
South Street
Perth, WA
Australia

Blakemore, Alexandra, PhD
Department of Genomics of Common Disease
School of Public Health
Imperial College London
London
UK

Brahmachari, Samir K, PhD
Director-General & Secretary
Council of Scientific and Industrial Research
Government of India
New Delhi
India

Buxton, Jessica
Department of Genomics of Common Disease
School of Public Health
Imperial College London
London
UK

Campino, Susana G, D Phil
The Genome Campus
Wellcome Trust Sanger Centre
Hinxton
UK

Cerda-Flores, Ricardo M, PhD
School of Nursery and Center for Research
Development in Health Sciences
Universidad Autonoma de Nuevo Leon
Monterrey
Mexico

Chakrapani, Anupam, FRCPCH
Inherited Metabolic Disease Unit
Birmingham Children's Hospital
Birmingham
England
UK

Charfeddine, Cherine
Molecular Investigation of Genetic Orphan Diseases
Research Unit
Institut Pasteur de Tunis
Tunis
Tunisia

Chen, Chien-Hsiun
Institute of Biomedical Sciences
Academia Sinica Taipei
Taiwan
and
Graduate Institute of Chinese Medical Science
China Medical University
Taichung
Taiwan

Chen, Yuan-Tsong, MD, PhD
Institute of Biomedical Sciences
Academia Sinica, Taipei
Taiwan
and
Department of Pediatrics
Duke University Medical Center Durham, NC
USA

Chetan GK
Department of Human Genetics
National Institute of Mental Health and Neurosciences
(NIMHANS)
Bangalore, Karnataka
India

Clark, Taane G
London School of Hygiene and Tropical Medicine
Gower Street
The University of London
London
UK

Clarke, Angus J, DM, FRCPCH
Institute of Medical Genetics
School of Medicine
University Hospital of Wales
Cardiff
UK

Cleary, Maureen, MD, FRCPCH
Inherited Metabolic Disease Unit
Great Ormand Street Hospital for Sick Children
London WC1
UK

Cogliati, Tiziana
National Eye Institute
National Institutes of Health
Bethesda, MD
USA

Cohen, Joel I, PhD
International Food Policy Research Institute
Washington, DC
USA

Cotton, Richard GH, PhD, DSc
Genomic Disorders Research Centre
Howard Florey Institute and Convenor,
Human Variome Project
Faculty of Medicine, Dentistry and
Health Sciences
University of Melbourne
Parkville VIC 3010
Australia

Cutiongco-de la Paz, Eva Maria
Institute of Human Genetics
National Institutes of Health
Manilla
The Philippines

Daar, Abdallah S, MD, FRCS
The McLaughlin–Rotman Centre for
Global Health
Program on Life Sciences, Ethics and Policy
McLaughlin Centre for Molecular Medicine
The Faculty of Medicine
Medical Sciences Building
University of Toronto
Ontario
Canada

Dai, Ying
Department of Public Health and Epidemiology
Capital Medical University
Beijing
Republic of China

Das, Reena, MD
Thalassaemia and Haemoglobinopathy Laboratory
Department of Haematology
Post Graduate Institute of Medical Sciences
Chandigarh
India

de Silva, Deepthi, MRCP
Department of Physiology
Faculty of Medicine
University of Kelaniya
Ragama
Sri Lanka

de Wit, Erika
University of Wittwatersand
Johannesburg
South Africa

Dissanayake, Vajira HW, MBBS, PhD
Senior Lecturer/Medical Geneticist
Human Genetics Unit
Faculty of Medicine
University of Colombo
Kynsey Road, Colombo 00800
Sri Lanka

Dodgen, Tyren M.
Departments of Immunology and Pharmacology
Faculty of Health Science
University of Pretoria
South Africa

Elles, Robert George, PhD
Director of Genetic Laboratory
Department of Genetic Medicine
St Mary's Hospital
Central Manchester Hospitals Foundation
Manchester, England
UK

Eloff, Arinda
Departments of Immunology
Faculty of Health Science
University of Pretoria
South Africa

Faruq, Mohammed, PhD
Functional Genomics Unit
Institute of Genomics and Integrative Biology (IGIB-CSIR)
Mall Road, Delhi 110007
India

Faskhutdinova, G
Institute of Biochemistry and Genetics
Ufa Scientific Center
Russian Academy of Sciences
Prospekt Oktyabrya 71
Ufa 450054
Russia

Fedorova, Sardana A
Department of Molecular Genetics
Yakut Research Center of Complex
Medical Problems
Russian Academy of Medical Sciences
Sergelyakhskoe shosse 4, Yakutsk, 677010
Russia

Ferraz, Victor Evangelista de Faria
Department of Medical Genetics
Instituto Fernandes Figueira
Fundação Oswaldo Cruz
Rio de Janeiro—RJ
Brazil

Frenk, Julio
Dean of the School of Public Health
Harvard University
Boston, MS
USA

Friedman, James
National Eye Institute
National Institutes of Health
Bethesda, MD
USA

Fucharoen, Suthat, MD
Department of Health, Ministry of Public Health and Thalassemia Research Center
Institute of Molecular Biosciences
Mahidol University
Thailand

Gaysina D
Institute of Biochemistry and Genetics
Ufa Scientific Center
Russian Academy of Sciences
Ufa, 450054
Russia

Geerts, Lut, BSc (Hons.), MRCOG
Associate Professor in Obstetrics and Gynaecology
University of Stellenbosch and Tygerberg Hospital
Stellenbosch
South Africa

Ghosh, Balram
Institute of Genomics and Integrative Biology, (IGIB-CSIR)
Mall Road
Delhi
India

Gilliat-Ray, Sophie, PhD
Centre for the Study of Islam in the UK (Islam-UK)
Department of Religious Studies and Theology
School of History, Archaeology and Religion
Cardiff University
Cardiff
UK

Gilyazova, Irina R
Institute of Biochemistry and Genetics
Ufa Science Center
Russian Academy of Sciences
Prospekt Oktyabrya, 71
Ufa, Bashkortostan, 450054
Russia

Ginter, EK
Research Center for Medical Genetics
Russian Academy of Medical Sciences
Moskvorechje St., 1
Moscow
Russia

Giugliani, Roberto, MD
Gene Therapy Center
Hospital de Clinicas de Porto Alegre
Rua Ramiro Barcelos 2350—CEP 90035–903
Porto Alegre, RS
Brazil

Goodsaid, Federico
The Genomics Group
Office of Clinical Pharmacology, Office of Translational Science
Center for Drug Evaluation and Research, US Food and Drug Administration
Silver Spring, MD
USA

Gotoh, Norimoto
National Eye Institute
National Institutes of Health
Bethesda, MD
USA

Gouider-Khouja, Neziha, MD
Department of Neurology
Institut National de Neurologie
La Rabta 1007
Tunis
Tunisa

Greenberg, Jacquie, PhD
Professor, Genetic Counsellor
Department of Human Genetics
University of Cape Town
South Africa

Grimaji SC
Department of Psychiatry
National Institute of Mental Health and neurological Sciences (NIMHANS)
Bangalore, Karnataka
India

Gupta, Amrendra Kumar, PhD
Institute of Genomics and Integrative Biology (IGIB-CSIR)
Mall Road
Delhi 110007
India

Gupta, Neerja, DM
Genetics Division, Department of Pediatrics
All India Institute of Medical Sciences
New Delhi
India

Hamamy, Hanan, MD
Department of Genetic Medicine and Development
Geneva University Hospital
Geneva Foundation for Medical Education and Research
Geneva
Switzerland

Hardy, Billie-Jo, PhD
The McLaughlin–Rotman Centre for Global Health
McLaughlin Centre for Molecular Medicine
University of Toronto
College Street, Toronto, Ontario
Canada

He, Yan
School of Public Health and Family Medicine, Capital Medical University
College of life Sciences, Graduate University
Chinese Academy of Sciences
Beijing
China

Hoal, Eileen, PhD
Associate Professor in Molecular Biology
University of Stellenbosch
Cape Town Area
South Africa

Hochfeld, Warren E
Department of Medical Genetics
School of Clinical Medicine
Cambridge University
UK

Horovitz, Dafne Dain Gandelman, MD, PhD
Department of Medical Genetics
Instituto Fernandes Figueira
Fundação Oswaldo Cruz
Rio de Janeiro—RJ
Brazil

Hurst, Jane A, FRCP
Department of Clinical Genetics and Molecular Genetics
Great Ormond Street Hospital for Sick Children
Institute of Child Health
Guilford Street
London, WC1
UK

Hussain, Rafat, PhD
Associate Professor and Deputy Head of School
School of Rural Medicine
University of New England
Australia

Izhevskaya, VL
Institute of Biochemistry and Genetics of Ufa Science Center
Russian Academy of Sciences
Prospekt Oktyabrya, 71, Ufa, 450054
Russia

Ivanov, VI
Institute of Biochemistry and Genetics of Ufa Science Center
Russian Academy of Sciences
Prospekt Oktyabrya, 71, Ufa, 450054
Russia

Jayalakshmi, Mariakuttikan
Department of Immunology
Madurai Kamaraj University
Madurai-625021
Tamil Nadu
India

Jenkins, Trefor, PhD
Department of Human Genetics
University of the Witwatersrand
National Health Laboratory Service
Johannesburg
South Africa

Jiménez-Sánchez, Gerardo, MD, PhD
Formerly Executive Director-The National Institute of Genomic Medicine
Periferico Sur 4124, Torre Zafiro II
Piso 6 Col. Jardines del Pedregal, Mexico D.F. 01900
Mexico

Jones, Wendy, MRCP
Department of Clinical and Molecular Genetics
Great Ormond Street Hospital for Sick Children
Institute of Child Health
Guilford Street
London, WC1
London
UK

Johnson, Colin A, PhD
International Resource for Autozygosity Mapping
Division of Molecular & Translational Medicine
Leeds Institute of Molecular Medicine
University of Leeds
UK

Jorde, Lynn B, PhD
The Department of Human Genetics
University of Utah Health Sciences Center
Salt Lake City, Utah
USA

Kabra, Madhulika, MD
Genetics Division
Department of Pediatrics
All India Institute of Medical Sciences
New Delhi
India

Kalyanaraman, Narayanan
Dept of Immunology
Madurai Kamaraj University
Madurai-625 021
Tamil Nadu
India

Kaput, Jim, PhD
Division of Personalized Nutrition and Medicine
National Center for Toxicological Research
United States Food and Drug Administration
3900 NCTR Road
Jefferson, AR 72079
USA

Kaur, Gurvinder, PhD
Department of Immunology and Transplantation
All India Institute of Medical Sciences
New Delhi
India

Kavitha, Valampuri John Mary Selvam
Department of Immunology
Madurai Kamaraj University
Madurai-625021
India

Kazantseva, A
Institute of Biochemistry and Genetics
Ufa Scientific Center
Russian Academy of Sciences
Prospekt Oktyabrya 71, Ufa 450054
Russia

Kefi, Rym
Institut Pasteur de Tunis Laboratoire de séquençage et de typage génétique
UR: Exploration Moléculaire des maladies orphelines d'origine génétique
Tunis
Tunisia

Kent, Alastair, MA
Director
Genetic Alliance
London
UK

Khidiyatova, Irina M
Institute of Biochemistry and Genetics
Ufa Science Center
Russian Academy of Sciences
Prospekt Oktyabrya, 71
Ufa, Bashkortostan, 450054
Russia

Khneisser, Issam, MBA
Newborn Screening Laboratory Manager
Saint Joseph University
11–5076 Riad Elsolh, Beirut
Lebanon

Khusnutdinova, Elza K
Institute of Biochemistry and Genetics
Ufa Science Center
Russian Academy of Sciences
Prospekt Oktyabrya, 71, Ufa, 450054
Russia

Kononova, SK
Department of Molecular Genetics
The Yakut Research Center of Complex Medical Problems
Siberian Branch, Russian Academy of Medical Sciences
Sergelyakhskoe shosse 4
Yakutsk 677010
Russia

Kozlova, SI
Russian Medical Academy for Postgraduate study
Moscow
Russia

Krause, Amanda, MB, ChB, PhD
Department of Human Genetics
National Health Laboratory Service
University of Johannesburg
Johannesburg
South Africa

Kromberg, Jennifer GR, PhD
Department of Human Genetics
University of the Witwatersrand
National Health Laboratory Service
Johannesburg
South Africa

Kumar, Dhavendra, MD FRCP FACMG
Institute of Molecular and Experimental Medicine,
Cardiff University School of Medicine
University Hospital of Wales
Cardiff
UK
and
Genomic Policy Research Unit
School of Health, Sport and Science
University of Glamorgan
Pontypridd, Wales
UK

Kumaramanickavel, Govindasamy, PhD
Director of Research
Narayana Nethralaya, Health City
Bangalore
India
and
Aditya Jyot Eye Hospital
Mumbai
India

Kutuev, IA
Institute of Biochemistry and Genetics
Ufa Science Center
Russian Academy of Sciences
Prospekt Oktyabrya, 71, Ufa, 450054
Russia

Laffita-Mesa, José Miguel, BSc
Centre for the Research and Rehabilitation of Hereditary Ataxias
Holguín
Cuba

Lee, Ming Ta Michael
Institute of Biomedical Sciences,
Academia Sinica Taipei
Taiwan
and
Graduate Institute of Chinese Medical Science
China Medical University
Taichung
Taiwan

Li, Peng-Gao
Public Health Genetics Unit
Department of Public Health and Epidemiology
Capital Medical University
Beijing
Republic of China

Liu, Fen
China Capital Medical University
Beijing
China

Loggenberg, Kelly, MSc
Genetic Counsellor
Department of Human Genetics
The University of Cape Town
South Africa

Lok, Chun Yu, DPhil
Weatherall Institute of Molecular Medicine
University of Oxford
John Radcliffe Hospital
Headington
Oxford
UK

Lunt, Peter, MD FRCP
Department of Clinical Genetics
Bristol Children's Hospital
St. Michael's Hill
Bristol, England
UK

Luzzatto, Lucio, MD
Direttore Scientifico, Istituto Toscano Tumori (ITT)
Professor of Haematology
University of Firenze
Via Taddeo Alderotti 26N
50139 Firenze
Italy

Madon, Prochi F, PhD
Dept. of Assisted Reproduction and Genetics
Jaslok Hospital and Research Centre
Mumbai
India

Mahajan, Anubha, PhD
Genomics and Molecular Medicine Unit
Institute of Genomics and Integrative Biology (CSIR)
Mall Road
Delhi
India

Maher, Eamonn R, MD FRCP
Medical and Molecular Genetics
School of Clinical and Experimental Medicine
University of Birmingham
Birmingham B15 2TT
UK

Megarbane, Andre, MD, PhD
Medical Genetics Unit
Saint Joseph University
Beirut
Lebanon

Merryweather-Clarke, Alison T, DPhil
Blood Research Laboratory
Nuffield Department of Clinical Laboratory Sciences
John Radcliffe Hospital
Headington
Oxford OX3 9DU
UK

Manjunatha KR, PhD
Department of Human Genetics
National Institute of Mental Health and Neurological Sciences (NIMHANS)
Bangalore, Karnataka
India

Maksimova, Nadezda R
Yakut Research Center of Complex Medical Problems
Russian Academy of Medical Sciences
Yakutsk, 677019
Russia

Marques de Faria, Antonia Paula, MD, PhD
Department of Medical Genetics
School of Medical Sciences
University of Campinas (Unicamp)
Campinas-SP
Brazil

Matte, Ursula
Gene Therapy Center
Hospital de Clinicas de Porto Alegre
Rua Ramiro Barcelos
Porto Alegre, RS
Brazil

Mehra, Narinder, PhD
Department of Immunology and Transplantation
All India Institute of Medical Sciences
New Delhi
India

Melegh, Béla, MD, PhD, DSc
Department of Medical Genetics
University Pécs
Szigeti
Hungary

Mittal, Uma
Genomics and Molecular medicine
Functional Genomics Unit
Institute of Genomics and Integrative Biology, (IGIB-CSIR)
Mall Road
Delhi
India

Mohan, Viswanathan
Dr. Mohan's Diabetes Specialities Centre
WHO Collaborating Centre for Non Communicable Diseases Prevention & Control
Madras Diabetes Research Foundation
ICMR Advanced Centre for Genomics Of Diabetes
Gopalapuram
Chennai
India

Mokni, Mourad
Dermatology Department—La Rabta Hospital
Hereditary Keratinisation Disorders Research Unit-La Rabta Hospital
Tunis
Tunisia

Möller, Marlo
Department of Human Genetics
University of Wittwatersand
Johannesburg
South Africa

Moustafa, Julia El-Sayed
Department of Genomics of Common Disease
School of Public Health
Imperial College London
London
UK

Mukerji, Mitali, PhD
Senior Scientist-Genomics and Molecular medicine
Functional Genomics Unit
Institute of Genomics and Integrative Biology (IGIB-CSIR)
Mall Road
Delhi
India

Myburgh, Renier
Departments of Immunology
Faculty of Health Science
University of Pretoria
South Africa

Naik, Swati, MD
Kennedy-Galton Centre
Department of Clinical Genetics
Northwick Park Hospital
Harrow
Middlesex
UK

Nair, Pratibha, MSc
Centre for Arab Genomic Studies
Dubai
United Arab Emirates

Nazarenko, Ludmila P
Research Institute of Medical Genetics
Siberian Branch of Russian Academy of Medical Sciences
Tomsk
Russia

Nogovitsina, Anna N
Yakut Research Center of Complex Medical Problems
Russian Academy of Medical Sciences
Yakutsk
Russia

Obeid, Tasneem
Centre for Arab Genomic Studies
Dubai
United Arab Emirates

Old, John, PhD
The Oxford Thalassaemia and Haemoglobinopathy Unit
The Churchill Hospital
John Radcliffe NHS Trust
Headington
Oxford
UK

Olopade, Olufunmilayo I, MD
Center for Clinical Cancer Genetics & Global Health
Section of Hematology/Oncology
Department of Medicine
The University of Chicago
Chicago, IL
USA

Padilla, Carmencita, MD
Department of Pediatrics
College of Medicine and Institute of Human Genetics
National Institutes of Health
University of the Philippines Manila
Manila
The Philippines

Pang, Tikki
Research Policy & Cooperation
World Health Organization
Avenue Appia
Geneva
Switzerland

Pattanapongthorn, Jintana
Department of Health, Ministry of Public Health and
Thalassemia Research Center, Institute of Molecular Biosciences
Mahidol University
Thailand

Patton, Michael A, MD, FRCP
Department of Medical Genetics
St Georges Hospital & Medical School
University of London
London
UK

Paz-y-Miño, César, MD, DB
Decano del Instituto de
Investigaciones Biomédicas
Facultad de Ciencias de la Salud
Quito
Ecuador

Pena, Sérgio DJ, MD, PhD, FRCP(C)
Universidade Federal de Minas Gerais
Av. Afonso Pena
Belo Horizonte
Brazil

Penchaszadeh, Victor B, MD
Mailman School of Public Health
Columbia University, New York
USA
and
Argentine Forum for Health Research, Buenos Aires
República de la India
Buenos Aires
Argentina

Penn, Claire, PhD
Health Communication Project
School of Human and Community Development
University of the Witwatersrand
South Africa

Pepper, Michael S, MD PhD
Professor, Department of Immunology
Faculty of Health Sciences
University of Pretoria
P.O. Box 2034
Pretoria 0001
South Africa

Phadke, Shubha R, MD, DM
Department of Medical Genetics
Sanjay Gandhi Postgraduate Institute of Medical Sciences
Raebarelli Road
Lucknow
India

Pitchappan, Ramasamy M, PhD
Department of Immunology
Madurai Kamaraj University
Madurai-625021
Tamil Nadu
India

Pollitt, Rodney J, PhD
Department of Paediatric Pathology
The Children's Hospital
Sheffield Children's Hospital NHS Foundation
The University of Sheffield
Sheffield
UK

Puri, Ratna D, MD, DM
Center of Genetic Medicine
Sir Ganga Ram Hospital
Rajinnder Nagar
New Delhi
India

Puzyrev, VP
Research Institute of Medical Genetics
Siberian Branch of Russian Academy of Medical Sciences
Tomsk
Russia

Qubbaj, Wafa, PhD
Assisted Reproductive Technology Unit
Department of Pathology and Laboratory Medicine
King Faisal Specialist Hospital and Research Center
Riyadh, 11211
Saudi Arabia

Rajab, Anna, PhD, FRCP
Genetics Unit, Directorate General of Health Affairs
Ministry of Health, Sultanate of Oman
Muscat
Sultanate of Oman

Ramesar, Raj, PhD
MRC Human Genetics Research Unit
Division of Human Genetics
Institute for Infectious Diseases and Molecular Medicine
University of Cape Town
Cape Town
South Africa

Rappuoli, Rino
Chiron Vaccines
Via Fiorentina 1
53100 Siena
Italy

Ray, Jharna, MSc, PhD
S. N. Pradhan Centre for Neurosciences
University of Calcutta
Ballygunge Circular Road
Kolkata
India

Rebai, Ibtihel
National Institute of Neurology.
La Rabta
Tunis
Tunisia

Robson, Kathryn JH, DPhil
Blood Research Laboratory
Nuffield Department of Clinical Laboratory Sciences
John Radcliffe Hospital
Headington
Oxford OX3 9DU
UK

Rodríguez-Labrada, Roberto, BSc
Centre for the Research and Rehabilitation of Hereditary Ataxias
Holguín
Cuba

Rojas-Martínez, Augusto, MD, DSc
Centro de Investigación y Desarrollo en Ciencias de la Salud
Universidad Autónoma de Nuevo León. Calle Carlos Canseco
S.N. Colonia Mitras Centro
Monterrey, C.P. 64460
Mexico

Romdhane, Lilia
Molecular Investigation of Genetic Orphan Diseases Research Unit
Pasteur Institute of Tunis
Tunisia

Rotimi, Charles N, PhD
Center for Research on Genomics and Global Health
National Human Genome Research Institute
The National Institutes of Health
Bethesda, MD 20851–5635
USA

Rozario, Santi, PhD
Department of Religious Studies
and Theology
School of History, Archaelogy and Religion
School of Sociology and Social Work
University of Tasmania
Australia

Saggar, Anand MD, FRCP
Department of Medical Genetics
St Georges University of London
Cranmer Terrace
London SW17 0RE
UK

Schlebusch, Carina M.
Human Genomic Diversity and Disease|Research Unit
Division of Human Genetics
School of Pathology
National Health Laboratory Service & University of the
Witwatersrand
Johannesburg
South Africa

Schuller-Faccini, Lavinia
Brazilian Institute of Population Medical Genetics
INAGEMP, Medical Genetics Service
HCPA and Department of Genetics—UFRGS
Porto Alegre, RS
Brazil

Séguin, Béatrice PhD
The McLaughlin–Rotman Centre for
Global Health
Program on Life Sciences, Ethics and Policy
McLaughlin Centre for Molecular Medicine
University of Toronto
Ontario
Canada

Sekar, Kanthimathi
Madras Diabetes Research Foundation
ICMR Advanced Centre for Genomics of Diabetes
Gopalapuram
Chennai
India

Sellars, Sean, MA, FACS
Emeritus Professor
Division of Otolaryngology
Faculty of Health Sciences
University of Cape Town
South Africa

Coskun, Serdar, PhD
Section Head, Assisted Reproductive Technology
Department of pathology and Laboratory Medicine
King Faisal Specialist Hospital and Research Center
Riyadh, 11211
Saudi Arabia

Shaw, Alison
Ethox Centre, Department of Public Health
and Primary Care
University of Oxford
Badenoch Building, Old Road Campus, Headington
Oxford
UK

Silao, Catherine Lynn
Institute of Human Genetics
National Institutes of Health
Manilla
The Philippines

Singer, Peter A, PhD
The McLaughlin–Rotman Centre for Global Health
Program on Life Sciences, Ethics and Policy
University Health Network/McLaughlin Centre for
Molecular Medicine
The Faculty of Medicine, Medical Sciences Building
University of Toronto
Toronto, Ontario
Canada

Singh, Jai Rup, PhD
Vice Chancellor
Central University of Punjab
Bathinda, Punjab
India

Singh, Arvind Rup, PhD
45 Taft Ave
Providence, RI 02906
USA

Soberón, Guillermo, PhD
Socio y Representante en Sonora at Aita
Infraestructura
Representante regional at Aita S. C., Varios, Campus
Guaymas at Tecnológico de Monterrey, Director
Campus Guaymas at Tecnológico de Monterrey
Universidad Nacional Autónoma de México, Universidad
Nacional Autónoma de México, Instituto Tecnológico y
de Estudios Superiores de Monterrey
Mexico

Soodyall, Himla, PhD
Professor, Human Genomic Diversity and Disease Research Unit
Division of Human Genetics
School of Pathology
National Health Laboratory Service & University of the Witwatersrand
Johannesburg
South Africa

Song, Manshu
China Capital Medical University
Beijing
China

Srivastava, Achal K, MD
Neuroscience Centre
All India Institute of Medical Sciences
New Delhi, 10029
India

Swaroop, Anand, PhD
National Eye Institute
National Institutes of Health
Bethesda, MD
USA

Syama, Adhikarla, PhD
Dept of Immunology
Madurai Kamaraj University
Madurai-625021
Tamil Nadu
India

Tadmouri, Ghazi Omar
Centre for Arab Genomic Studies
Dubai
United Arab Emirates

Tandon, Nikhil, MD, DNB
Department of Endocrinology
All India Institute of Medical Sciences
New Delhi
India

Teebi, Ahmed, MD (Deceased)
The Hospital for Sick Children
Toronto
Canada

Therrell, Bradford L, BD
Department of Pediatrics
University of Texas Health Science Center at San Antonio
San Antonio, Texas
USA
and
National Newborn Screening and Genetics Resource Center
Austin, Texas
USA

Thong, Meow Keong, MD
Genetics & Metabolism Unit
Department of Paediatrics
Faculty of Medicine, University of Malaya
Kuala Lumpur
Malaysia

Tsigankova PG
Research Centre for Medical Genetics
Russian Academy of Medical Sciences
Moscow
Russia

Turnpenny, Peter D, FRCP
Consultant in Clinical Genetics
Royal Devon & Exeter Hospital
Exeter
England
UK

Urban, Michael, FCPaed, MMed(Paed)
Senior specialist in Medical Genetics
University of Cape Town and Groote Schuur Hospital
Capte Town
South Africa

Vasani, Meera, PhD
Professor of Psychiatry
Director—National Drug Dependence Treatment Centre
All India Institute of Medical Sciences
New Delhi
India

Velázquez-Pérez, Luis, MD, PhD
Centre for the Research and Rehabilitation of Hereditary Ataxias
Holguín
Cuba

Venkatesan, Radha
Madras Diabetes Research Foundation
ICMR Advanced Centre for Genomics Of Diabetes
Gopalapuram
Chennai
India

Venkatesh HN
Department of Human Genetics
National Institute of Mental Health and Neurological Sciences (NIMHANS)
Karnataka, Bangalore
India

Verma, Ishwar C, FRCP
Formerly Professor/Head, Genetics Unit
All India Institute of Medical Sciences, New Delhi
Senior Consultant—Center of Medical Genetics
Sir Ganga Ram Hospital
Rajinder Nagar, New Delhi
India

Villems, Richard
Department of Evolutionary Biology
University of Tartu and Estonian Biocentre
Riia 23, Tartu
Estonia

Wanapirak, Chanane
Department of Obstetrics and Gynecology
Faculty of Medicine
Chiangmai University
Thailand

Wang, Youxin
China Capital Medical University
Beijing
China

Wang, Yun
China Capital Medical University
Beijing
China

Wang, Wei, MD, PhD
School of Public Health and Family Medicine
Capital Medical University
College of life Sciences, Graduate University
Chinese Academy of Sciences
Beijing
China

Wasant, Pornswan, MD
Professor of Pediatrics
Division of Medical Genetics
Department of Pediatrics
Faculty of Medicine Siriraj Hospital, Mahidol University
Bangkok
Thailand

Watermeyer, Jennifer, PhD
Health Communication Project
School of Human and Community Development
University of the Witwatersrand
South Africa

Weatherall, Sir David J., DM FRS
Weatherall Institute of Molecular Medicine
University of Oxford
John Radcliffe Hospital
Headington
Oxford
UK

Wessels, Tina-Marié, MSc
Genetic Counsellor
University of the Witwatersrand
Johannesburg
South Africa

Winichagoon, Pranee
Thalassemia Research Center
Institute of Molecular Biosciences
Mahidol University
Thailand

Wonkam, Ambroise, MD
Senior Clinical Geneticist
Department of Human Genetics
The University of Capetown
South Africa

Wu, Lijuan
Capital Medical University
Beijing
China

Yoshimatsu, Toshio F.
Center for Clinical Cancer Genetics and Global Health
Department of Medicine
The University of Chicago, IL
USA

Yun, Hong-Min
Peking University Shenzhen Hospital
Shenzhen
Guangdong Province 518036
China

Yunusbayev, BB
Institute of Biochemistry and Genetics
Ufa Science Center of Russian Academy of Sciences
Prospekt Oktyabrya Ufa
Russia

Zainullina, A
Institute of Biochemistry and Genetics
Ufa Scientific Center, Russian Academy of Sciences
Ufa, 450054
Russia

Zhang, Ling
School of Public Health and Family Medicine
Capital Medical University, Beijing
College of life Sciences, Graduate University, Chinese Academy of Sciences
Beijing
China

Zheng, Yonglan
Center for Clinical Cancer Genetics and Global Health
Department of Medicine
The University of Chicago
USA

Zinchenko, RA
Research Center for Medical Genetics
The Russian Academy of Medical Sciences
Moscow
Russia

PART I.

GENOMICS IN MEDICINE AND HEALTH—GENERAL

1.

GENES AND GENOMES IN HEALTH AND DISEASE: AN INTRODUCTION

Dhavendra Kumar

INTRODUCTION

The concept of heredity dates back to several hundreds and thousands of years. It is evident in all forms of biological life and evolution. Evolutionary scientists, philosophers, and biologists have used "heredity" to put forward their views on procreation, development, adaptation, and transmission of species-specific traits. The popular Darwinian theory of natural selection rests on *transmission of hereditary factors*. For several thousand years, various descriptions and explanations have been put forward to define the physical shape and functional nature of hereditary factors. From ancient times, and in almost every civilization, intense debate and arguments failed to arrive at a consensus. Most arguments focused on whether the hereditary factor was a creation by God, a new product fresh from the soil and water, or something in the blood and in the semen. The symbolic representation of the phallus in ancient sculptures and paintings of the Indian subcontinent is an example of the concept that the phallus, and thus semen, is a key factor in the creation and transmission of individual's (including family) physical traits and behavior characteristics.

In the historical context, the term *gene* was introduced only recently as the most acceptable answer to explain one of the hereditary factors. It is unclear when and by whom this term was first introduced. It does not matter, as the term *gene* is now universally accepted and used in the context of understanding heredity, and probably the single most important biological factor regulating biological life ranging from single-cell organisms to multicellular mammals. Rapid and extraordinary scientific progress made during 19th and 20th centuries has led to the development of *genetics*, the science of heredity. This is now transformed into a broader field of *genomics* that includes not only genes but all possible regulatory factors, whether recent or extending back through several thousand years of evolution.

In biological terms, genes, genetics, and genomics are keys to procreation, development, growth, function, and survival. The health of any living organism is judged by its physical and functional wellbeing. Thus, genes, genetics, and genomics are central to all forms of biological health, including that of humans. The human health depends not only on its own *genetic* or *genomic* constitution, but on that of other organisms whose wellbeing is also essential to human health—for example, food (plant, fish and animals), shelter (homes made of wood from trees), environment (water, trees, and plants), protection (clothes from cotton and animal skin), and transport (animals and vehicles made of wood from trees). From a medical perspective, the science of genetics or genomics offers deep insight and evidence for a number of human diseases, including infectious diseases resulting from either lack of protection and/or failure in controlling the spread of microbial infections or parasitic infestations. This chapter introduces the reader some of the basic facts about genes, genetics, and genomics, and how these impact human health and that of plants, crops, and animals necessary for human health and survival. This is obviously more relevant to millions of people in the developing and less developed countries, where limited resources and lack of infrastructure limit the optimal use of the science of genetics and genomics in applications to eradicate poverty and ensure optimal health. The reader will find cross-references to separate chapters in the book containing detailed information and further discussion of each subject.

GENES, GENETICS AND GENOMICS

A detailed description of the basic principles of genetics and human genetic diseases is beyond the scope of this chapter. These facts are explained in various other texts on basic genetics and medical genetics (see Further Reading). However, some basic principles and relevant information are outlined in this section to assist the reader with limited understanding of the basic genetics.

Living organisms are divided into two large classes—the eukaryotes and prokaryotes. The cells of the eukaryotes have a complex compartmentalized internal structure; these include algae, fungi, plants, and animals. Prokaryotes, on the other hand, are single-celled microorganisms, which include bacteria and other related microorganisms. The other types of living organisms are viruses, which are intracellular parasites living in both eukaryotes and prokaryotes.

Genetic information is transferred from one generation to the next by small sections of the nucleic acid, deoxyribonucleic acid (DNA), which is tightly packaged into subcellular structures called *chromosomes*. Prokaryotes usually have a single circular chromosome, while most eukaryotes have more than two and in some cases up to several hundred. In humans there are 46 chromosomes arranged in 23 pairs, with one of each pair inherited from each parent (Figure 1-4). Twenty-two pairs are called *autosomes* and one pair is called sex chromosomes, designated as X and Y; females have two X chromosomes (46, XX) and males have an X and Y (46, XY).

A chromosome consists of a tightly coiled length of DNA, together with the proteins (e.g., chromatins) that help to define its structure and level of activity. DNA consists of two long strands of nucleotide bases wrapped round each other along a central spine made up of phosphate and sugar (Figure 1-1). There are four bases: adenine (A), guanine (G) cytosine (C) and thymine (T). Pairing of these bases follows strict rules: A always pairs with T, and C with G. Two strands are, therefore, complementary to each other.

Genes are made up of specific lengths of DNA that encode the information to make a protein, or ribonucleic acid (RNA) product. RNA differs from DNA in that the base thymine (T) is replaced by uracil (U), and the sugar is ribose. It acts as a template to take the coded information across to ribosomes for final assembly of amino acids into the protein peptide chain (Figure 1-2). The bases are arranged in sets of three, referred to as *codons*. Each *codon* codes for a specific amino acid; hence the term *genetic code*. Codons are located in *exons*, which contain the coding sequences. A gene may consist of several such coding DNA segments. Exons are separated from each other by non-coding sequences of DNA, called *introns*. Although they are not known to be associated with any specific function, it is likely that some of these introns might be of evolutionary significance, or associated with

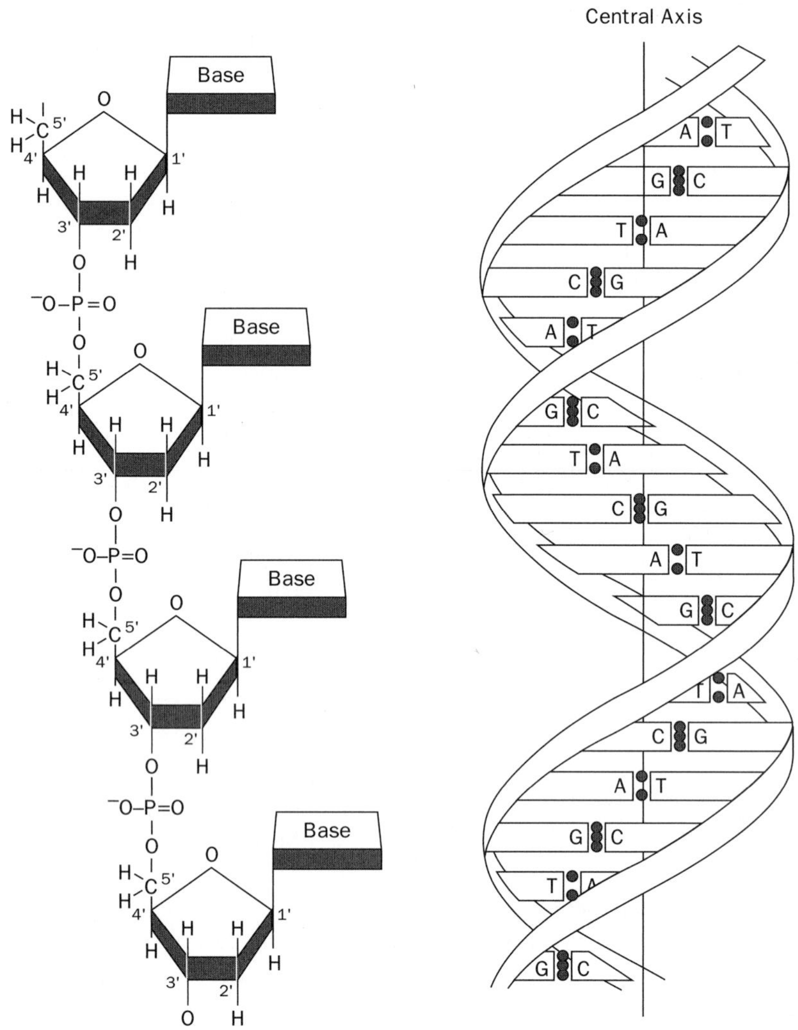

Figure 1-1 The Watson-Crick double-helix structure of the DNA molecule.
SOURCE: with permission, Weatherall, 1991, Oxford University Press, UK

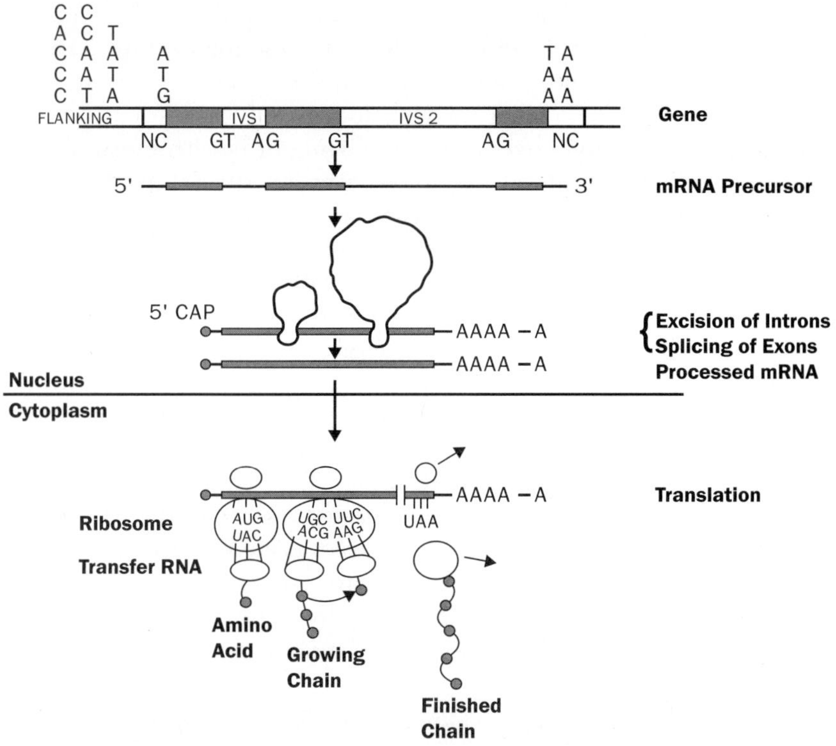

Figure 1-2 The synthesis of a protein (peptide chain) from the gene.

other fundamental biological functions. During the transcription of DNA, the introns are spliced out, and the exons then attach to mRNA to start the process of protein synthesis.

Proteins are one of the major constituents of the body chemistry. These are remarkably variable in their structure, ranging from tough collagen that forms connective tissue and bone, through the fluid hemoglobin that transports oxygen, to thousands of enzymes, hormones, and other biological effectors and their receptors that are essential for the structure and function of the body. Each protein is made up of one or more peptide chains consisting of series of amino acids, of which only 20 occur in living organisms. The different structures and functions of proteins depend on the order of amino acids as determined by the genetic code.

DNA has the remarkable property of self-replication. The two strands of the DNA separate as chromosomes during cell division. There are two types of cell division; *mitosis* in all body cells, and *meiosis*, which is specifically confined to the gonads in making sperms and eggs (Figure 1-3).

During mitosis, no reduction of the number of chromosomes takes place (diploid, or 2n), while meiosis results in half the number of chromosomes (haploid, or 1n). The new pairs of DNA

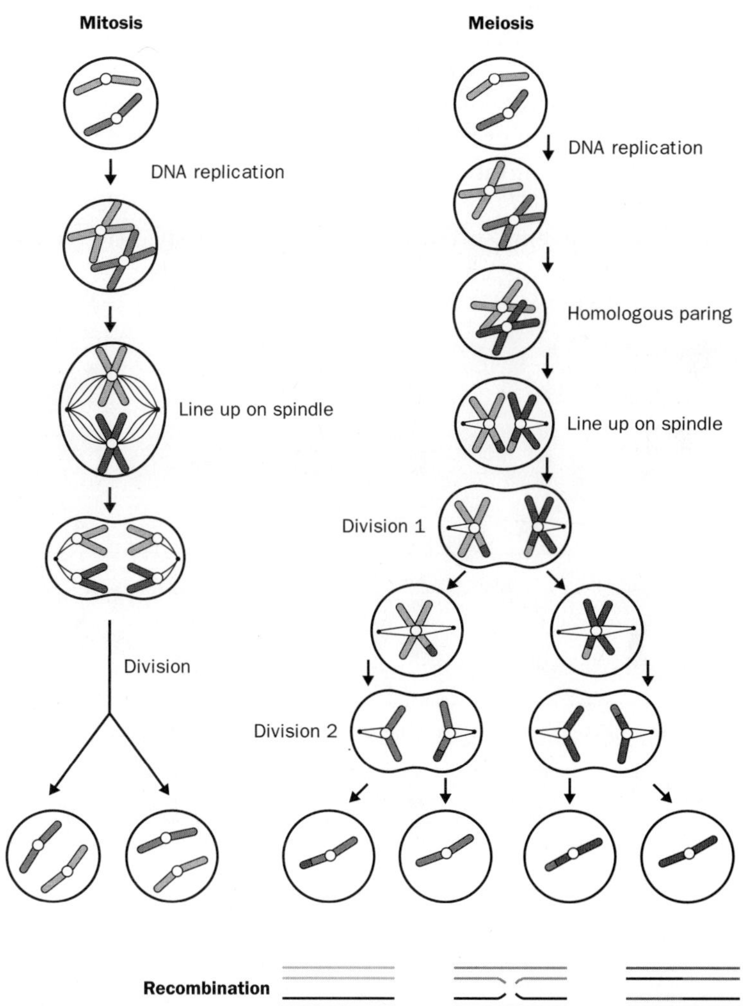

Figure 1-3 Steps in mitosis and meiosis cell divisions; recombination of genetic material during homologous pairing is shown at the bottom of the figure
SOURCE: with permission, Weatherall, 1991, Oxford University Press, U.K.

are identical to those from which they were synthesized. However, sometimes mistakes or mutations occur. These usually result from substitution of a different base, or are due to extensive structural changes to genes. In other words, any spelling mistake in the letters A-T or C-G could result in either absence of coded information (*nonsense mutation*) or a different message (*missense mutation*). However, not all mutations or spelling mistakes have an adverse effect (neutral mutations). Conversely, some changes in the genes might result in a favorable property; for example, resistance to disease or other environmental hazard. This is the basis for the gradual change in species during millions of years of evolution. On the other hand, mutations may result in defective gene function leading to a disease, or susceptibility to disease, due to qualitative or quantitative changes in the gene product, the peptide chain. However, these changes may also result from epigenetic mechanisms, abnormal RNA molecules, and post-translational modifications. A brief introduction to these molecular processes is provided elsewhere in this chapter; interested readers are advised to consult dedicated texts on cell and molecular biology (see Further Reading).

GENETICS AND GENOMICS OF HUMAN DISEASE

Changes in the DNA structure or its organization may result in disruption or function of a number of genes or a single gene. A number of chromosomal abnormalities are now known to be associated with a wide range of clinical problems, often designated with a specific syndrome title. Any change in the diploid number of 46 chromosomes (*aneuploidy)* may result in physical and developmental problems; for example, Down syndrome (Figure 1-4). On the other hand, structural changes, such as loss (deletion), gain (duplication), position effect (inversion, ring, or isochromosome) are also known to result in a broad spectrum of clinical problems. All these types of genetic diseases are called *chromosomal disorders*. The field of *clinical cytogenetics* is now exclusively dedicated to the diagnosis and understanding of chromosomal disorders.

Mutations in a single gene or in the alternate form (allele) are transmitted along the principles of Mendelian inheritance (Mendelian genetics). Mendel's laws of inheritance are discussed in detail in all leading genetics textbooks (see Further Reading). The Mendelian diseases are either *autosomal* (gene located on one of the 22 pairs of autosomes) or *X-linked* (gene located on the X chromosome). A mutation may have an adverse outcome when present only in one copy of the gene pair (heterozygote). This is called a "dominant" mutation. On the other hand, the clinical symptoms are only seen when the mutation is present in both copies (homozygous), also called a "recessive" mutation. Both dominant and recessive mutations could be either autosomal or X-linked (Figure 1-5). The main distinction between an autosomal and X-linked disease lies in the gender differences among the affected persons. No appreciable difference would be expected in the number of males or females affected with an autosomal dominant or recessive disease, while preponderance of males is seen in an X-linked recessive disorder. In an X-linked dominant disorder, females are more likely to be encountered either due to lethality in the male or reduction of the biological fitness of the affected male.

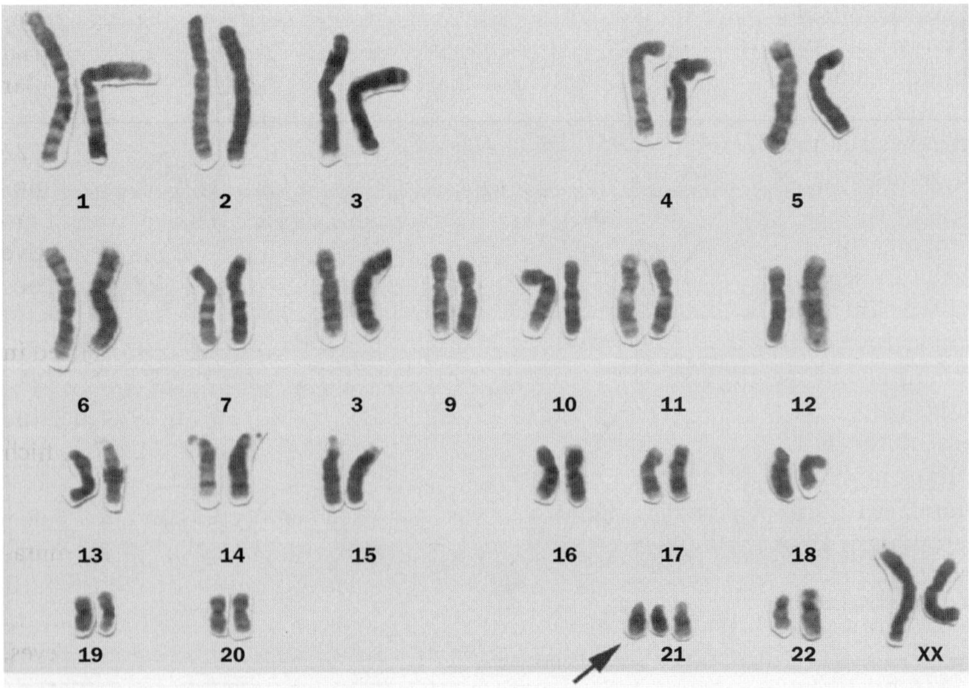

Figure 1-4 Aneuploid karyotype (47,+21,XX) of a girl with Down syndrome due to one extra chromosome 21 (trisomy 21).

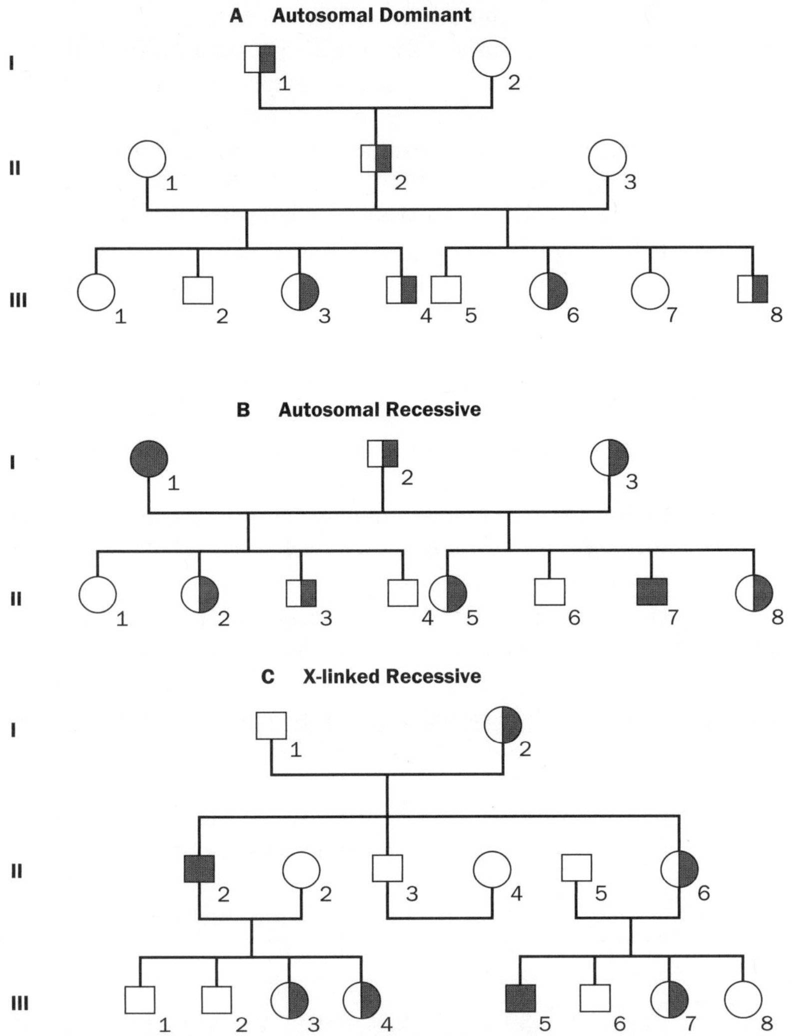

Figure 1-5 Different forms of monogenic (Mendelian) inheritance
SOURCE: with permission, Weatherall, 1991, Oxford University Press, UK

Mutations in several genes might predispose to a disease or structural defect, when combined with one or more environmental factors. This concept is the basis of multifactorial/polygenic inheritance. These comprise probably the largest group of genetic diseases. The mechanisms are often complicated, involving a number of different gene systems and interactions with various environmental factors. The underlying genetic factors confer the "threshold," which is reflected in the general population frequency of a given trait; for example, cleft lip/palate. A lowered threshold would indicate increased genetic liability acting in conjunction with environmental factors. A classic way of demonstrating this concept is by the use of the "Gaussian" curve (Figure 1-6).

A shift of the curve to the right would imply increased genetic liability and a lowered "genetic threshold," manifesting with an increased recurrence risk in a close relative. The term *complex disease* is now widely used in reference to these diseases. The notable examples include cleft lip and/or palate, congenital heart defect, limb anomalies, diabetes, hypertension, coronary heart disease, and epilepsy. The list is long, and more and more disorders are continually added, including susceptibility to infection or an adverse reaction to a drug or toxic substance. The completion of the *Human Genome Project* (see below) has provided several avenues for studying the genetic basis of complex diseases.

Another group of genetic disorders is related to mutations within the mitochondrial genome (mitochondrial genetics). The mitochondria consist of 10 single circular double helices of DNA comprising 16,596 base pairs. In addition, the mitochondrial chromosome contains 22 transfer RNAs, two types of ribosomal RNA for mitochondrial protein synthesis, and 13 peptides, which are subunits of various steps involved in cellular oxidative phosphorylation. The mitochondrial DNA does not contain any introns, and the whole strand is transcribed and translated. As the mitochondria are exclusively located in the cytoplasm, they are transmitted in the egg from the mother to all of her children (maternal inheritance). The mitochondria in the sperm are located in the tail, which is discarded at the time of fertilization; thus, no paternal inheritance of mitochondrial DNA is possible. A number of disorders associated with mitochondrial DNA mutations are now known (see below). These are often multisystem, involving body systems or tissues dependent on high energy turnover; for example, muscle, heart, eyes, and the inner ear. Although the inheritance is maternal,

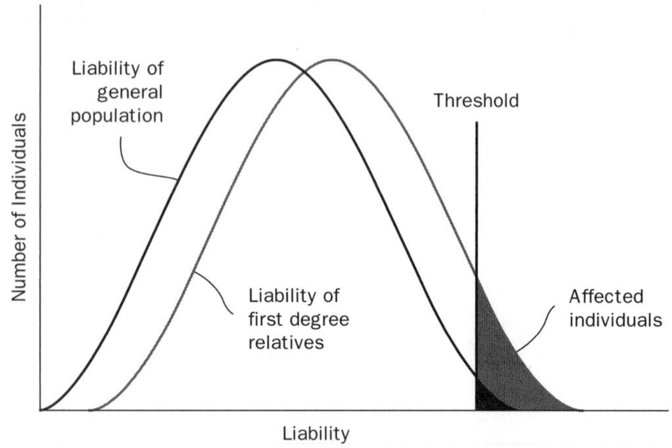

Figure 1-6 The 'Gaussian' bell shaped curve to illustrate 'genetic threshold', indicated by liability in the general population. A shift to the right (in dark grey) indicates increased liability in first degree relatives with an increased risk of recurrence.
SOURCE: with permission, Weatherall, 1991, Oxford University Press, UK. (see colour insert)

approaching a recurrence risk of up to 50%, a prediction of the clinical outcome is often difficult due to variation in the mitochondrial content in different tissues (heteroplasmy) as opposed to uniform distribution (homoplasmy). Some of the mitochondrial function is dependent on autosomal genes, which might present in an autosomal dominant or recessive manner.

GENETICS AND GENOMICS OF COMPLEX DISEASES

All inherited disorders have a genetic abnormality present in the DNA of all nucleated cells in the body, including germ cells (sperm and egg), and can be transmitted to subsequent generations. In contrast, a genetic abnormality that is present only in specific somatic cells cannot be transmitted. The genetic abnormality in a somatic cell can occur at any time from the post-conception stage to late adult life. The paradigm of somatic cell genetic disorder is cancer, where the development of malignancy is often the consequence of mutations in genes that control cellular growth. There are several such genes, and these are designated *oncogenes*. It is now accepted that all human cancer results from mutations in the nuclear DNA of a specific somatic cell, making it the most common genetic disease (see also Chapters 44 & 85).

The clinical course and outcome of treatment in a number of acute and chronic medical conditions depend upon various factors. For instance, there is overwhelming evidence that highly polymorphic cytokine, interferon, and interleukin families of complex proteins influence the host response to acute infection and physical injury. All these proteins are encoded by several genes. Similarly, association of human leucocyte antigens (HLA) in the pathogenesis of a number of acute and chronic medical disorders is well known (see Chapter 75). In addition, interaction of mutations within these genes, and with several other genomic polymorphisms, such as single nucleotide polymorphisms (SNPs), is probably important in several acute medical conditions including trauma. This will have a major impact in critical care and acute medicine (see Chapter 6). The role of SNPs in modulating complex medical disorders such as diabetes mellitus, coronary heart disease, hypertension, and various forms of cancer is unclear. However, the complexity of interaction of SNPs with other genetic traits and loci is probably important in the prognosis of these disorders, in particular the outcome of therapeutic interventions. This argument probably justifies separating some of these disorders under the title of "complex genomic diseases."

Various cancers and degenerative diseases occur with increasing frequency in old age. However, these may also present at a younger age, such as childhood leukemias. The molecular mechanisms in these diseases are not entirely clear, but probably include defects in DNA repair mechanisms, accelerated apoptosis, deregulation of imprinted genomic regions, and *de novo* chromosome rearrangements involving specific genomic regions. Although these disorders arguably can be included under the broad category of multifactorial/polygenic diseases, the pattern of distribution and recurrence does not follow the agreed principles of multifactorial/polygenic inheritance as discussed elsewhere in this chapter.

As discussed in the following section (see Non-Traditional Inheritance), epigenetic changes play a major role in the development of human cancer (Egger et al., 2004). A high percentage of patients with sporadic colorectal cancer (CRC) possess microsatellite instability, and show methylation and silencing of the gene encoding MLH1 (Kane et al., 1997). It is thus likely that epigenetic changes also predispose to genetic instability. In some cases, promoter-associated methylation of *MLH1* is found not only in the tumor, but also in normal somatic tissues, including spermatozoa. These germline "epimutations" predispose individuals carrying abnormal methylation patterns to multiple cancers. Indeed, disruption of pathways that lead to cancer is often caused by the *de novo* methylation of the relevant gene's promoters (Jones & Baylin, 2002). Epigenetic silencing has been recognized as a third pathway satisfying Knudson's "two-hit" hypothesis for the silencing of tumor suppressor genes (Jones & Laird, 1999).

Chromosomal rearrangements have long been associated with human leukemias. These result in formation of fusion proteins including histone acetyltransferases and histone methyltransferases that influence upregulation of target genes. In acute promyelocytic leukemia, the oncogenic fusion protein PML-RARα (promyelocytic leukemia–retinoic acid receptor-α) causes repression of genes that are essential for differentiation of hematopoietic cells. Similarly, in acute myeloid leukemia, AML-ETO fusions recruit the repressive N-COR-Sin3-HDAC1 complex and inhibit myeloid development

(Jones & Saha, 2002). There are further examples of complex genomic arrangements that result in other cancers, and which can modify the therapeutic response. For example, mutations in genes for ATPase complex are associated with poorer prognosis in patients with non-small-cell lung cancer (Roberts & Orkin, 2004).

NONTRADITIONAL INHERITANCE

A number of hereditary disorders present with complex genetic pathologies that do not follow the conventional principles of inheritance as outlined in the previous sections. There is now overwhelming evidence within these disorders indicating unusual mechanisms that suggest *nontraditional inheritance*. The mechanisms involve certain genomic regions that directly or indirectly influence regulation and expression of one or more genes manifesting in complex phenotypes. Currently, some of these disorders are either listed as chromosomal or single-gene disorders. However, there is strong evidence to classify these as "genomic disorders" (Table 1-1).

DISORDERS OF GENOMIC IMPRINTING: EPIGENETIC DISEASES

The term *epigenetics* refers to heritable factors that affect gene expression without any change in the gene coding sequence. These factors could be operational either during meiosis or mitosis, and are often selective and preferential on the basis of "parent of origin." The term *imprinting* is commonly used to describe this important biological mechanism that is recognized to influence wide-ranging physical and molecular phenotypes (Figure 1-7). A number of human diseases have now been confirmed to result from epigenetic changes in various parts of the genome. The terms *epigenetic diseases* or *genomic imprinting disorders* refer to this group of diseases. Basic mechanisms related to the phenomenon of epigenetics or epigenomics are not discussed here.

Epigenetic initiation and silencing is regulated by the complex interaction of three systems, including DNA methylation, RNA-associated silencing, and histone modification (Egger et al., 2004). The relationship between these three components is vital for expression or silencing of genes (Figure 1-6). Disruption of one or another of these interacting systems can lead to inappropriate expression or silencing of genes, leading to "epigenetic diseases." Methylation of the C^5 position of cytosine residues in DNA has long been recognized as an epigenetic silencing mechanism of fundamental importance (Holliday & Pugh, 1975). The methylation of CpG sites within the human genome is maintained by a number of DNA methyltransferases (DNMTs) and has multifaceted roles in the silencing of transportable elements, defense against viral sequences, and transcriptional repression of certain genes. A strong suppression of the CpG methyl acceptor site in human DNA results from mutagenic changes in 5-Methylcytosine, causing C:G to T:A transitions. Normally, CpG islands, which are GC-rich, evolutionary conserved regions of more than 500 base pairs, are kept free of methylation. These stretches of DNA are located within the promoter region of about 40% of mammalian genes and, when methylated, cause stable heritable transcriptional silencing. Aberrant *de novo* methylation of CpG islands is a hallmark of human cancers and is found early during carcinogenesis (Jones & Baylin, 2002).

In addition to DNA methylation, histone modifications have also been found to have epigenetic effects. Acetylation and methylation of conserved lysine residues of the amino-terminal tail domains are the key elements in histone modification. Generally, the acetylation of histones marks active, transcriptionally competent regions, whereas hypoacetylation histones are found in transcriptionally inactive euchromatic and heterochromatic regions. On the other hand, histone methylation can be a marker for both active and inactive regions of chromatin. Methylation of lysine residue 9 on the N terminus of histone 3 (H3-K9) is a hallmark of silent DNA, and is evenly distributed throughout the heterochromatic regions such as centromeres and telomeres, including the inactive X chromosome. In contrast, methylation of lysine 4 of histone 3 (H3-K4) denotes activity and is predominantly found at promoter regions of active genes (Lachner & Jenuwein, 2002). This constitutes a "histone code," which can be read and interpreted by different cellular factors.

TABLE 1-1 THE CLASSIFICATION OF GENOMIC DISORDERS

Disorders of genomic imprinting (Epigenetic diseases)
Disorders of genome architecture
Trinucleotide repeats disorders
Complex genomic diseases

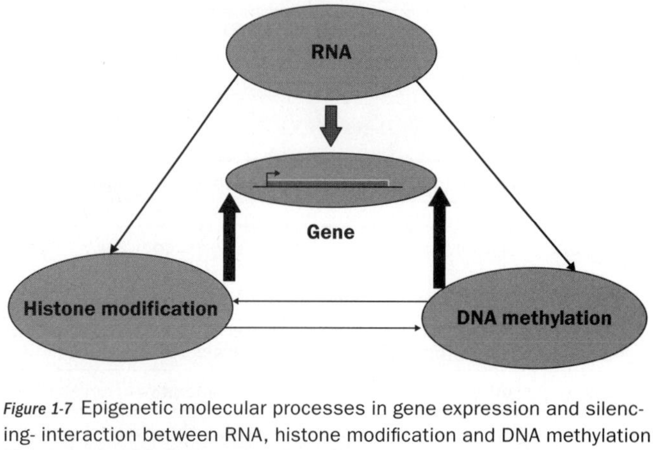

Figure 1-7 Epigenetic molecular processes in gene expression and silencing- interaction between RNA, histone modification and DNA methylation (Egger et al. 2004) (see colour insert)

There is evidence that DNA methylation depends on methylation of H3-K9, and can also be a trigger for its methylation. Recently, evidence has accumulated on the role of RNA in post-transcriptional silencing. In addition, RNA in the form of antisense transcripts (Xist or RNAi) can also lead to mitotically heritable transcriptional silencing by the formation of heterochromatin. For example, transcription of antisense RNA led to gene silencing and to the methylation of the structurally normal α-globin gene in patients with alpha thalassemia. This could be one of the many human diseases resulting from epigenetic silencing due to antisense RNA transcripts (Tufarelli et al., 2003).

Mutations in genes that affect genomic epigenetic profiles can give rise to human diseases that can be inherited or somatically acquired (Table 1-2). These epigenetic mutations can be either due to hypermethylation (silencing) of a regulating gene, or loss of methylation (LOM; activation) of another gene that has a positively modifying effect on the phenotype. The parental imprinting effect can be inferred by demonstrating the parental origin of the mutant allele. Similarly, either a loss or gain of a chromosomal segment can result in the same situation. Confirmation of a specific chromosomal deletion or duplication is usually possible by using the FISH method. The paternal imprinting in this situation is commonly demonstrated by genotyping a set of polymorphic markers located within the chromosomal segment. Inheritance of the whole chromosomal homologue from one parent effectively confirms the imprinting phenomenon, since the regulatory gene sequences for the pathogenic gene would be missing from the other parent. This characteristic abnormality is commonly referred to as *uniparental disomy*, or UPD. This could either be isodisomy (similar parental homologues) or heterodisomy (parental and grandparental homologues; see Figure 1-8). The origin of UPD is believed to result through loss of the additional chromosomal homologue, failing which the conceptus would be trisomic. This mechanism is also called *trisomic rescue*.

For a maternally imprinted disorder, paternal UPD would be confirmatory and maternal UPD diagnostic for the paternally imprinted condition. For example, maternal UPD is diagnostic for Prader-Willi syndrome, and paternal UPD for Angelman syndrome, both conditions being associated with a microdeletion of 15q11 region. The parental origin of the 15q micro deletion follows the expected epigenetic pattern and is in keeping with the clinical diagnosis. Recurrence risk estimates vary dependent on the specific epigenetic pattern. This information is crucial to offer accurate genetic counseling in any genomic imprinting disorder.

Many epigenetic diseases are associated with chromosomal alterations and manifest with physical and learning difficulties. For example, mutations in X-linked mental retardation with the alpha-thalassemia phenotype (ATRX) result in consistent changes in the methylation pattern of ribosomal DNA, Y-specific repeats and subtelomeric repeats. Another X-linked recessive mental retardation syndrome, associated with a visible "fragile site" on the terminal part of the long arm of the X-chromosome (fragile-X syndrome), results from *de novo* silencing of the pathogenic gene *FMR1*. The syndrome is characteristically associated with an abnormal expansion of CGG triplet repeats in the *FMR1* 5' untranslated terminal region. Methylation of the expansion leads to silencing of the *FMR1* gene, and under certain cultural conditions creates the visible "fragile site" on the X-chromosome.

TABLE 1-2 RECOGNIZABLE EPIGENETIC DYSMORPHIC SYNDROMES (EGGER ET AL., 2004)

DISEASE	MAIN FEATURES	EPIGENETIC MECHANISM
ATR-X syndrome	α-thalassemia, facial dysmorphic features, neurodevelopmental disabilities	Mutations in *ATRX* gene; hypomethylation of repeat and satellite sequences
Fragile-X syndrome	Chromosome instability, physical and learning/ behavioral difficulties	Expansion and methylation of CGG repeat in *FMR1* 5' UTR, promoter methylation
ICF syndrome	Chromosome instability, immunodeficiency	*DNMT3* mutations; DNA hypomethylation
Angelman syndrome	seizures and intellectual disabilities	Deregulation of one or more imprinted genes at 15q11-13 (maternal)
Prader-Willi syndrome	Obesity, intellectual disabilities	Deregulation of one or more imprinted genes at 15q11-13 (paternal)
Beckwith-Wiedemann (BWS)	Organ overgrowth, childhood tumors	Deregulation of one or more syndrome imprinted genes at 11p15.5 (*IGF2, CDKN1C, KvDMR1* etc.)
Russel-Silver syndrome	Growth delay, body asymmetry	Deregulation of one or more imprinted genes at 7p (maternal)
Rett syndrome	Seizures, intellectual disabilities	*MeCP2* mutations
Rubinstein-Taybi Syndrome	Facial dysmorphism, intellectual disabilities	Mutation in CREB-binding protein (histone acetylation)
Coffin-Lowry syndrome	Facial dysmorphism, developmental delay	Mutation in *RSk-2* (histone phosphorylation)

NOTES: ATR-X- α-thalassemia, X-linked mental retardation; UT - untranslated region; ICF - immunodeficiency, chromosome instability, facial anomalies; CREB-cAMP-response-element-binding protein

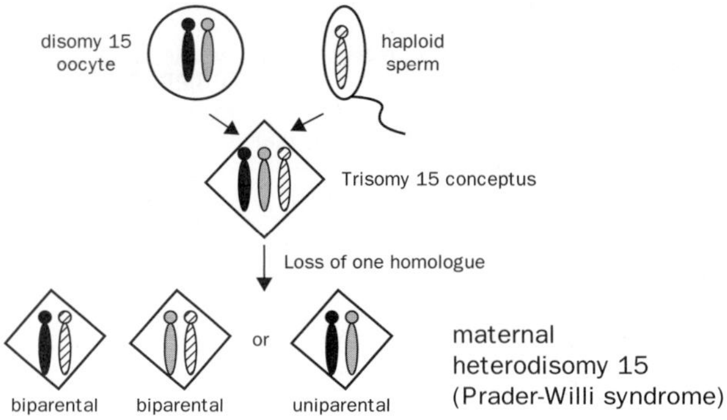

Figure 1-8 The origin of uniparental disomy 15 in Prader-Willi syndrome through trisomic rescue during early embryogenesis - note different homologues (maternal heterodisomy) (see colour insert)

Epigenetic silencing is probably also significant in other neurodevelopmental disorders. For example, in Rett syndrome, a common cause of intellectual disability in young girls, mutations of the *MeCP2* gene are seen in about 80% of cases. The MeCP protein binds to methylcytosine residues and causes de-repression of genes normally suppressed by DNA methylation. Despite lack of firm evidence, it is likely that *MeCP2* might have a key role in the control of neuronal gene activity resulting in the pathology of Rett syndrome (Chen et al., 2003). Interaction with another pathogenic gene (*CTKL5* or *STK9*) in Rett syndrome is likely to be important in the pathogenesis of this neurodevelopmental disorder (Slager et al., 2003). On a wider genomic level, mutations in the *DNMT3b* gene, causing the ICF (immunodeficiency, centromeric region instability, and facial anomalies) syndrome, result in deregulation of DNA methylation patterns. A notable example is that of Beckwith-Wiedemann syndrome, an overgrowth syndrome predisposing to Wilms' tumor and other childhood tumors, which is associated with duplications and rearrangements of a small chromosomal region on the short arm of chromosome 11p15.5. This region contains a cluster of genes, which is susceptible to a number of epigenetic alterations, manifesting with the BWS phenotype and tumorigenesis, particularly Wilms' tumor and other childhood embryonal tumors (Figure 1-8). Loss of methylation in imprinting control regions (such as *KvDMR1*) can cause deregulation of imprinting and either biallelic expression (*IGF2* and *H19*) or silencing (such as *CDKN1C*) of imprinted genes, which is seen in most sporadic cases (Weksberg et al., 1993).

The epigenetic phenomenon is probably significant for the phenotypic manifestations in some other hereditary tumors. For example, transmission of autosomal dominant familial chemodectomas (nonchromaffin paragangliomas or glomus tumors) is exclusively via the paternal line (Figure 1-9; van der Mey et al., 1989). The maternally derived gene is inactivated during oogenesis and can be reactivated only during spermatogenesis (Heutnik et al., 1992). This genetically heterogeneous cancer family

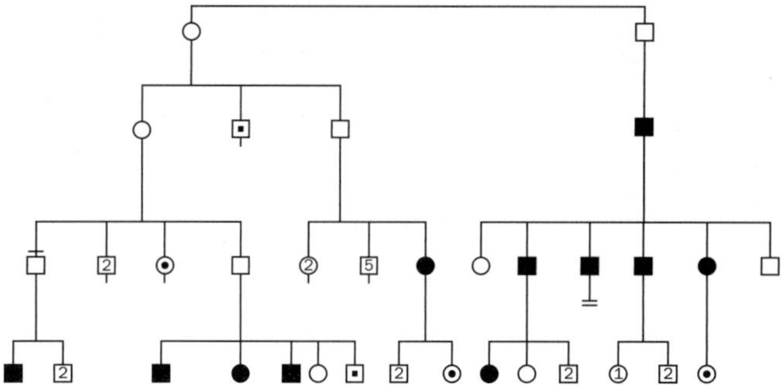

Figure 1-9 Pedigree showing paternal transmission of paraganglioma in a family: note no maternal transmission among 'at-risk' family members

syndrome is associated with germline mutations in succinate dehydrogenase subunits B (SDHB) and D (SDHD; Neumann et al., 2004).

Thus, epigenetic changes are probably significant in a number of other complex phenotypes, particularly those associated with cancer and a number of degenerative diseases (see "Complex Genomic Diseases").

DISORDERS OF GENOME ARCHITECTURE

Recent completion of the human genome project, and sequencing of the total genomes of yeast and other bacterial species, have enabled investigators to view genetic information in the context of the entire genome. As a result, it is now possible to recognize mechanisms of some genetic diseases at the genomic level. The evolution of the mammalian genome has resulted in the duplication of genes, gene segments, and repeat gene clusters (Lupski, 1998). This aspect of genome architecture provides recombination hot spots between regions of chromosomes that are distributed across the whole genome. These genomic regions become susceptible to further DNA rearrangements that may be associated with an abnormal phenotype. Such disorders (Table 1-3) are collectively grouped under the broad category of genome architecture disorders.

The term *genome architecture disorder* refers to a disease that is caused by an alteration of the genome that results in complete loss, gain, or disruption of the structural integrity of a dosage-sensitive gene(s) (Figure 1-10; Shaw & Lupski, 2004; Lupski & Stankiewicz, 2005). Notable examples include a number of chromosome deletion/duplication syndromes (Table 1-3). In these conditions, there is a critical rearranged genomic segment flanked by large (usually >10 kb), highly homologous low-copy repeat (LCR) structures that can act as recombination substrates. Meiotic recombination between non-allelic LCR copies, also known as non-allelic homologous recombination, can result in deletion or duplication of the intervening segment.

Similarly, other chromosomal rearrangements, including reciprocal, Robertsonian, and jumping translocations,

TABLE 1-3 CONTIGUOUS GENE SYNDROMES: EXAMPLES OF DISORDERS OF THE GENOME ARCHITECTURE

DISORDER [OMIM]	INHERITANCE	LOCUS	GENE	REARRANGEMENT	RECOMBINATION Size (kb)	Repeat
William-Beuren syndrome [194050]	AD	7q11.23	*ELN*	del;inv	1600	>320
Prader-Willi syndrome [176270]	AD	5q11.2q13	?	del	3500	>500
Angelman syndrome [105830]	AD	15q11.2q13	*UBE3A*	del	3500	>500
Dup(15)(q11.2q13)		15q11.2q13	?	dup	3500	>500
Triplication 15q11.2q13		15q11.2q13	?	trip		>500
Smith-Magenis syndrome [18290]		17p11.2	*RA13*	del	4000	~250
Dup(17(p11.2p11.2)	AD	17p11.2	*PMP22*	dup	4000	~250
DiGoerge/VCFS [188400]	AD	22q11.2	*TBX1*	del	3000	~225
Male infertility [415000]	YL	Yq11.2	*DBY,*	del	800	~10
AZFa microdeletion			*USP9Y*			
AZFc microdeletion 400024	YL	Yq11.2	*RBMY*	del	3500	~220
			DAZ?			

Abbreviations: del, deletion; dup, duplication; inv, inversion; D, direct; C, complex

inversions, isochromosomes, and small marker chromosomes, may also involve susceptibility to rearrangement related to genome structure or architecture. In several cases, LCRs, AT-rich palindromes, and pericentromeric repeats are located at such rearrangement breakpoints. This susceptibility to genomic rearrangements is not only implicated in disease etiology, but also in primate genome evolution (Shaw & Lupski, 2004).

An increasing number of Mendelian diseases (Table 1-4) are recognized to result from recurrent interchromosomal and intrachromosomal rearrangements, involving unstable genomic regions facilitated by low-copy repeats (LCRs). These genomic regions are predisposed to non-allelic homologous recombination (NAHR) between paralogous genomic segments. LCRs usually span approximately 10–400 kb of genomic DNA, share >or=97% sequence identity, and provide the substrates for NAHR, thus predisposing to rearrangements. LCRs have been shown to facilitate meiotic DNA rearrangements associated with several multiple malformation syndromes and some disease traits (Table 1-3). Seminal examples include microdeletion syndromes (Williams-Beuren syndrome [7q11del],

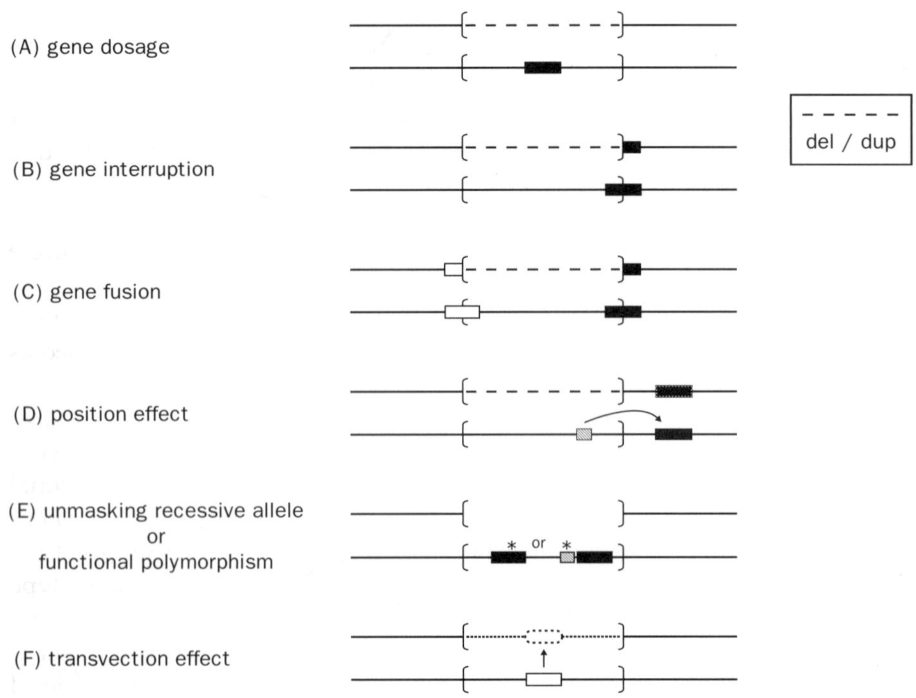

Figure 1-10 Molecular mechanisms for Genomic Disorders (Lupski and Stankiewicz, 2005)- dashed lines indicate either deleted or duplicated region; the rearranged genomic interval is shown in brackets; gene is depicted by filled horizontal rectangle; regulatory gene is shown as horizontal hach-marked rectangle; asterisks denote point mutations.

TABLE 1-4 MENDELIAN DISORDERS WITH GENOMIC REARRANGEMENTS

				REARRANGEMENT		RECOMBINATION SUBSTRATES		
DISORDERS	INHERITANCE	CHROMOSOME LOCATION	GENE(S)	Type	Size(kb)	Size (kb)	Repeat	%Identity Orientation
Barter syndrome type III	AD	1p36	*CLCNKA/8*	del	11		91	D
Gaucher disease	AR	1q21	*GBA*	del	16	14		D
Familial juvenile	AR	2q13	*NPHP1*	del	290	45	>97	D
nephronophthisis								
Facioscapulohumeral muscular dystrophy	AD	4q35	*FRG1?*	del	25-222	3.3		D
Spinal muscular dystrophy	AR	5q13.2	*SMN*	inv/dup	500			I
Congenital adrenal hyperplasia-21 hydroxylase deficiency	AR	6p21.3	*CYP21*	del	30		96–98	D
Glucorticoid remediable aldosteronism	AD	8q21	*CYP11B1/2*	dup	45	10	95	D
β-thalassemia	AR	11p15.5	*β-globin*	del	4,(7)?			D
α-thalassemia	AR	16p13.3	*α-globin*	del	3,7,4.2?	4		D
Polycystic kidney disease type 1	AD	16p13.3	*PKD1*			50	95	
Hereditary neuropathy with liability to pressure palsy(HNPP)	AD	17p12	PMP22	del	1400	24	98.7	D
Neurofibromatosis type 1 (NF1)	AD	17q11.2	*NF1*	del	1500	85		D
Ichthyosis	XL	Xq28	*STS*	del	1900	20		D
Red-green color blindness	XL	Xq28	*RCP/GCP*	del	0	39	98	D
Incontinentia pigmenti	XL	Xq28	*NEMO*	del	10	0.870		D
Hemophilia A	XL	Xq28	*F8*	inv	300-500	9.5	99.9	I
Emery-Dreifuss muscular Dystrophy (EMD)	XL	Xq28	Emerin/FLN1	del/dup/ inv	48	11.3	99.2	
Hunter syndrome	XL	Xq28	*IDS*	inv/del	20	3	>88	

Abbreviations: del, deletion; dup, duplication; inv, inversion; D, direct; C, complex; I, inverted

DiGeorge syndrome [22q11del]), autosomal dominant Charcot-Marie-Tooth disease type 1A (PMP22 gene duplication), hereditary neuropathy of pressure palsy (HNPP: PMP22 gene deletion) mapped to 17p11.2, and Smith-Magenis, a contiguous gene syndrome (CGS) with del (17)(p11.2p11.2). Dominantly inherited male infertility related to AZF gene deletion follows a similar mechanism. In addition, this LCR-based complex genome architecture appears to play a major role in the primate karyotype evolution, pathogenesis of complex traits, and human carcinogenesis.

A notable example includes genetically heterogeneous Charcot-Marie-Tooth disease (CMTD). The disorder is also known as *hereditary motor and sensory neuropathy* (HMSN), by virtue of being a peripheral neuropathy due to either involvement of the axonal or myelinated segments of the peripheral nerve. Genetically autosomal dominant, autosomal recessive, and X-linked dominant types are recognized. The disorder is not uncommon, affecting approximately 1 in 2500 of the adult population. This could be an underestimate, since medically the condition is benign, often not requiring any medical and/or surgical intervention. However, some affected individuals experience increasingly progressive neuromuscular weakness of distal muscles of lower legs, feet, distal forearms, and hands, with onset in early teens and causing severe locomotor restrictions.

An affected person usually presents late, with relative hypertrophy of the upper calf muscles, described as "inverted champagne bottle" associated with pes cavus due to wasting of the small muscles of the feet. Similarly, wasting of the small muscles of the hand leads to "claw hands." Neurophysiological studies remain an essential method of differentiating the two major types of CMTD. A reduced motor nerve conduction velocity of less than 35 m/sec helps in differentiating type 1 CMTD from type 2 CMTD, in which the motor nerve conduction velocity is usually normal but the sensory nerve conduction is often slow. While this distinction is undoubtedly helpful in clinical management, the application for genetic counseling is limited because both types are genetically heterogeneous. For instance, molecular characterization and gene mapping have confirmed the existence of at least four types of type 1 CMTD—autosomal dominant types 1a, 1b, and 1c, and the X-linked type (XCMT). Similarly, there are distinct genetic types within the type 2 CMTD group.

Approximately two-thirds of cases of CMT1 have a detecTable 1-5Mb duplication within a proximal chromosomal segment of the short arm of chromosome 17 (17p12; see Lupski et al., 1991). This duplicated chromosomal segment contains a gene for peripheral myelin protein called *PMP22*. This duplication results in disruption of the gene, leading to abnormal myelination of the peripheral nerves, an essential molecular pathological step resulting in the CMT1 phenotype designated as CMT1A. The CMT1A duplication was visualized by multiple molecular methods (Patel & Lupski, 1994), including fluorescence in-situ hybridization (FISH), pulsed-field gel electrophoresis (PFGE), and dosage differences of heterozygous alleles by restriction fragment length polymorphisms (RFLPs; see Figure 1-11). This finding led to further molecular studies on the origin of the 1.5Mb duplicated 17p12 segment (Lupski, 2003).

Studies by several investigators have revealed a significant variation in the size of marker alleles flanking the duplicated 17p12 region. It soon became apparent that a 500 kb allele co-segregated with 17p duplication in all affected individuals. This suggested a stable mutation, and followed a precise recombination mechanism. However, in *de novo* duplication, the presence of repeated flanking marker alleles indicated the mechanism of unequal crossing-over, leading to duplication. Indeed, this was confirmed when a highly homologous >20 kb size repeat sequence was confirmed flanking the 17 p duplication. It was appropriately termed *CMT1A-REP*. As predicted by the unequal crossing-over model, CMT1A-REP was found to be present in three copies on the CMT1A duplication-bearing chromosome (Pentao et al., 1992). Interestingly, the presence of only one copy was soon demonstrated in another peripheral nervous system disorder known as *hereditary neuropathy with liability to pressure* (HNPP; see) Chance et al., 1994; Reiter et al., 1996). Individuals affected with this disorder present with mild to moderate episodic weakness of the lower limbs, and occasionally of upper limbs, when subjected to prolonged pressure, such as sitting and sleeping. The disorder is dominantly inherited in an autosomal dominant manner. This is generally a clinically mild and benign hereditary neuropathy. The presence of only one copy results from a reciprocal deletion following unequal crossing-over involving the CMT1A-REP repeat.

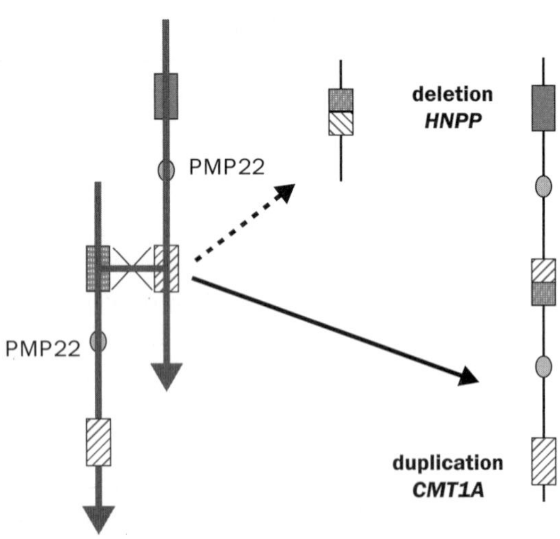

Figure 1-11 The unequal meiotic recombination (crossing-over) resulting in duplication [CMT1A] and deletion [HNPP].

Similar observations were also made in relation to Smith-Magenis syndrome (SMS), a contiguous gene syndrome associated with a microdeletion of 17p11.2 segment (Greenberg et al., 1991). Affected children present with facial dysmorphic features, severe speech delay, and behavioral problems with signs of self-harm. A specific junction fragment was detected by PFGE (SMS-REP) that was involved in recurrent rearrangement, resulting in either SMS or reciprocal 17p11.2 duplication. Pathogenic mutations in *RAI1* gene, mapped to the 17p11.2 chromosomal region, are now shown to be etiologically linked with SMS (Slager et al., 2003). It is also possible to have both duplication and deletion at the same time, resulting from DNA rearrangements on both homologues of chromosome 17. This was demonstrated in a patient with mild delay and a family history of autosomal dominant carpal tunnel syndrome (Potocki et al., 1999). The occurrence of both the 17p11.2 duplication and HNPP deletion in this patient reflects the relatively high rate at which these abnormalities arise, and the underlying molecular characteristics of the genome in this region.

It is perfectly reasonable to accept the argument that similar molecular mechanisms are at work in causing other human disorders. The human genome has evolved an architecture that may make us, as a species, more susceptible to rearrangements causing genomic disorders (Lupski, 2003).

DISORDERS WITH TRINUCLEOTIDE (TRIPLET) REPEATS

Several disorders are recognized to have a phenomenon of earlier age at onset of disease in successive generations. This is known as *anticipation*. This observation failed to secure a valid biological explanation, and had been put aside simply on the assumption of ascertainment bias for probands, or random variations in the age of onset.

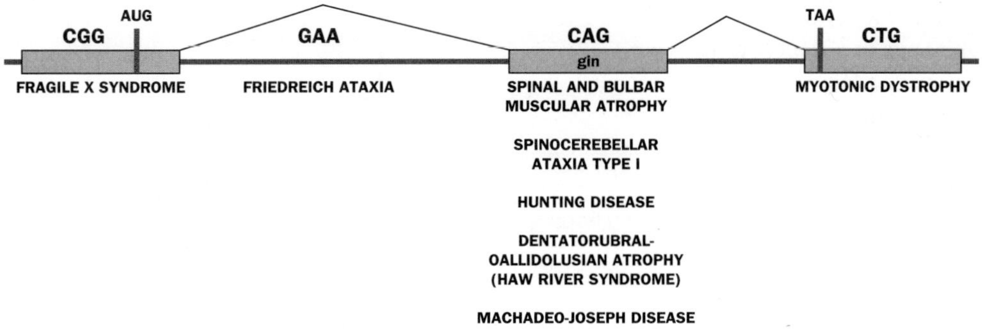

Figure 1-12 Location of four classes of triplet repeats in human diseases. Exons are shown in light pink with intervening introns as a pink solid line. The translation site AUG and termination signal TAA are indicated by red vertical bars.

SOURCE: Adapted with permission from Principles of Medical Genetics by Thomas D. Gelehrter, Francis S. Collins, and David Ginsburg, Williams and Wilkins, Baltimore, 2nd edition, 1998.

With the identification of unstable DNA repeats distributed across the genome, a molecular basis has been found for the phenomenon of anticipation. These unstable DNA repeats tend to increase in size during meiosis over successive generations. The abnormal expansion is correlated with reducing age of onset, and increased severity with further expansion of DNA repeats. The characteristic pattern of the DNA repeat involving a set of three nucleotides is commonly referred to *trinucleotide* or *triplet* repeats. This soon became established as a novel class of mutation and offered a plausible explanation for the phenomenon of anticipation and variable clinical severity in a number of neurodegenerative diseases (Figure 1-12; Table 1-5).

The X-linked recessive spinal bulbar atrophy (SBA) was one of the first hereditary neurological disorders recognized to be associated with CAG triplet repeats (Ashley & Warren, 1995).

The expanded region can occur anywhere in the gene, and thus can disrupt the expression of the gene. In the case of X-linked fragile X syndrome (FRAXA), the CGG repeats are found in the 5' untranslated region of the first exon of *FMR1*, the pathogenic gene for FRAXA. However, in the case of Friedreich's ataxia (FA), an autosomal recessive form of spinocerebellar ataxia (SCA), the expanded triplet repeat allele (GAA) occurs in the first intron of *X25*, the gene encoding frataxin. In Huntington's disease (HD) and other inherited neurodegenerative disorders, the CAG triplet repeats occur within exons and encode an elongated polyglutamine tract. However, the expanded CTG triplet repeats of myotonic dystrophy (DM) are found in the 3' untranslated region of the last exon of the DM protein kinase (myotonin) gene (*DM*).

Each class of trinucleotide repeats exists in normal individuals. A pathogenic expansion is the one that is seen in clinically symptomatic individuals. Carriers for an X-linked disease also have an expanded allele (premutation), which does not usually result in abnormal phenotype. However, it is likely that some carrier females might

TABLE 1-5 DISORDERS WITH TRINUCLEOTIDE (TRIPLET) REPEAT EXPANSION

DISORDER	TRIPLET	LOCATION	NORMAL#	MUTATION#
Fragile X syndrome	CGG	5'UTR	10–50	200–2000
Friedreich's ataxia	GAA	Intronic	17–22	200–900
Kennedy disease [SBMA]	CAG	Coding	17–24	40–55
Spinocerebellar ataxia 1[SCA1]	CAG	Coding	19–36	43–81
Huntington disease	CAG	Coding	9–35	37–100
Dentatorubral-Pallidoluysian Atrophy (DRPLA)	CAG	Coding	7–23	49–>75
Machado-Joseph disease[SCA3]	CAG	Coding	12–36	67–>79
Spinocerebellar ataxia 2[SCA2]	CAG	Coding	15–24	35–39
Spinocerebellar ataxia 6[SCA6]	CAG	Coding	4–16	21–27
Spinocerbellar ataxia 7[SCA7]	CAG	Coding	7–35	37–200
Spinocerebellar ataxia 8[SCA8]	TG	UTR	16–37	100–>500
Myotonic dystrophy	CTG	3'UTR	5–35	50–4000
Fragile site E [FRAXE]	CCG	Promoter	6–25	>200
Fragile site F [FRAXF]	GCC	?	6–29	>500
Fragile site 16 A [FRA16A]	CCG	?	16-49	1000–2000

UTR- untranslated region

exhibit some manifestations, as in fragile X syndrome. An expanded allele in the premutation range in a male would not be associated with any clinical manifestations (normal transmitting male NTM), but this could further expand, resulting in all his daughters being carriers. However, recent studies have provided data on the existence of late-onset gait ataxia in NTMs (Greco et al., 2006). On the other hand, a normal sized CGG repeat in a normal male could undergo further expansion during meiosis, leading to a carrier daughter. This usually comes to light when a symptomatic grandson is confirmed to have pathogenic FRAXA expansion. Prior to availability of the molecular testing in FRAXA, this kind of unusual pedigree pattern in fragile X syndrome was called the "Sherman paradox." Detailed molecular studies in the family are often necessary to offer accurate genetic counseling to "at-risk" carrier females. Carrier females are at an additional risk for developing premature ovarian failure, usually diagnosed when investigated for secondary infertility (Ennis et al., 2006)

Genetic counseling in other neurodegenerative disorders with triplet repeats is often complicated. In particular, the clinical prediction in "borderline" expanded triplet repeats (intermediate allele) in HD is extremely difficult due to lack of reliable data. However, recent studies have produced some data that is likely to be helpful in genetic counseling (Harper, 2003).

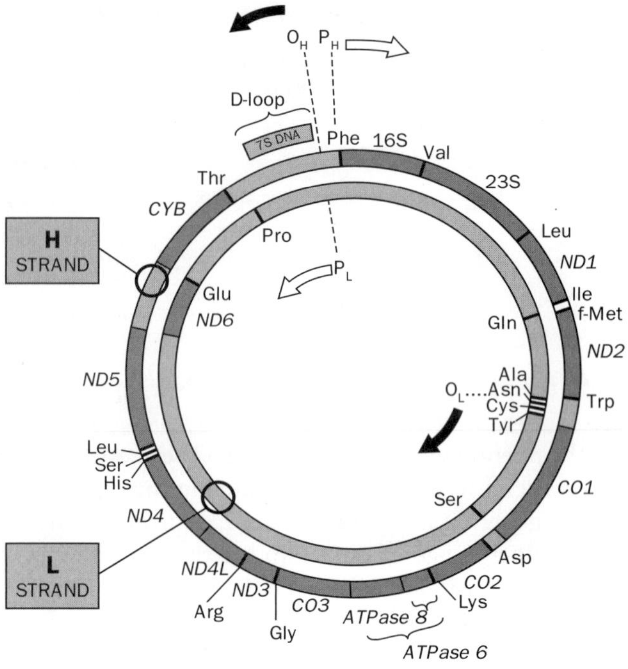

Figure 1-13 The heavy (H) and light (L) strands of the circular 16,659 bp double helix are shown. Protein-coding genes are shaded; transfer RNA genes are shown as short lines with the name of the amino acid. There are no introns. O_R, O_L, and the heavy arrows indicate the origins and directions of replication of the two strands. P_R, P_L, and the light arrows show the promoters and direction of transcription of the two multicistronic transcripts that are subsequently cleaved into individual mRNAs.
SOURCE: Adapted with permission from Figure 9.1 of Strachan & Read, Human Molecular Genetics, 3rd edition, London and New York: Garland, 2004.

MITOCHONDRIAL GENOME AND HUMAN DISEASE

The bulk of the eukaryotic cellular DNA is contained in the nucleus (nuclear DNA, or nDNA). Apart from nDNA, a small amount of DNA is contained in the mitochondria (mtDNA). The mitochondrial genome is very different from the nuclear genome (Figure 1-13, Table 1-6). In many respects, it has more in common with bacterial genomes than the eukaryotic nuclear genome. This is consistent with the idea that mitochondria originated as endosymbiotic bacteria within some ancestral eukaryotic cell. If this theory is correct, then over the years the mitochondria have gradually transferred more and more of their functions to the nucleus. The great majority of mitochondrial proteins are now encoded by nuclear genes. Cells contain many mitochondria (typically 100–1000, and maybe 100,000 in an oocyte) so that mitochondrial DNA (mtDNA) might be formally classified among the repetitive DNA in a cell. Although the mitochondrial genome is very small compared to its nuclear counterpart, because there are many copies mtDNA often makes up 1% or so of total cellular DNA.

As in bacteria, the mitochondrial genome is circular and closely packed with genes. There are no introns, and little intergenic noncoding DNA. Some genes even overlap. In the nuclear genome it is not uncommon for genes on opposite strands to overlap — Nusbaum et al. (2005) recorded 59 such pairs on chromosome 18 — but in this case, genes on the same strand overlap, using the same template but read in different reading frames. Twenty-four of the 37 genes specify functional RNAs (two ribosomal RNAs and 22 transfer RNAs); the other 13 genes encode components of the electron transport pathway. A short segment of the genome is triple-stranded.

This D-loop (displacement loop) is produced by replication forks overlapping as they travel in opposite directions around the circular DNA. The D-loop contains the only significant amount of noncoding DNA in the mitochondrial genome. Perhaps because of this, it is the location of many of the DNA polymorphisms that are such useful tools for anthropologists researching the origins of human populations. Because there is no recombination among mitochondrial DNAs, complete haplotypes of polymorphisms are transmitted through the generations, modified only by recurrent mutation, making mtDNA a highly informative marker of ancestry, at least along the maternal line.

Mitochondrial DNA replication and transcription use nuclear-encoded polymerases. Transcription proceeds in both directions around the circle. The initial products are

TABLE 1-6 COMPARISON OF THE HUMAN NUCLEAR AND MITOCHONDRIAL GENOMES

	NUCLEAR GENOME	MITOCHONDRIAL GENOME
Size	3 x 10^9 bp	16,659 bp
Topology	23 linear molecules	1 circular molecule
No. of genes	Ca. 24,000	37
% coding sequence (incl. genes for functional RNAs)	Ca . 1.4%	93%
Average gene density	Ca. 1 per 125 kb (variable)	1 per 0.45 kb
Introns	Average 8 per gene (variable)	None
Repetitive DNA	Ca. 50%	None

two large multicistronic RNAs, which are subsequently cleaved to make the individual mRNAs. All the protein components of the translation machinery are nuclear encoded, but the transfer RNAs (tRNAs) are exclusively mitochondrially encoded, and these use a coding scheme slightly different from the otherwise universal code. There are four stop codons—UAG, UAA, AGG and AGA; UGA encodes tryptophan, and AUA specifies isoleucine, rather than arginine as normally. Presumably, with only 13 protein-coding genes, the mitochondrial system could tolerate mutations that modified the coding scheme in a way the main genome could not.

Mutations in mtDNA are important causes of disease, and perhaps also of aging (Table 1-7). Phenotypes caused by variations in mtDNA are transmitted exclusively down the maternal line (matrilineal inheritance), but most genetic diseases where there is mitochondrial dysfunction are caused by mutations in nuclear-encoded genes, and so follow normal Mendelian patterns. Since cells contain many copies of the mitochondrial genome, they can be heteroplasmic, containing a mix of different sequences. Unlike mosaicism for nuclear variants, heteroplasmy can be transmitted by a mother to her children.

Several mitochondrial diseases have now been characterized (Table 1-7). One of the best-characterized is Leber's hereditary optic neuropathy (LHON), which exclusively affects males. There is loss of central vision secondary to optic nerve degeneration. The vision loss usually occurs in the 20s, and can progress rapidly in some men. Eleven different missense mtDNA mutations, in three different mitochondrial genes encoding respiratory chain enzyme subunits, have been described. The phenotype in other mitochondrial diseases tends to include a combination of heart, muscle, and central nervous system manifestations with considerable intra/interfamilial variability for the same mtDNA mutation. In addition, mitochondrial dysfunction can be part of the phenotype in some Mendelian diseases where the mutant gene product presumably has pathogenic influence on the mitochondrial mediated metabolic pathway. Examples include autosomal recessive respiratory enzyme disorders. Genetic counseling and decision for prenatal diagnosis can be difficult in mitochondrial disorders, due to difficulty in predicting the phenotype in the affected pregnancy.

Finally, a high degree of sequence variation (polymorphism) is known to occur in the noncoding region of the mitochondrial chromosome (the D-loop). This polymorphism has been used in anthropologic and evolutionary studies to trace back the origins and links of human populations. In addition, this information has been applied in forensic analysis as well, to match a maternal grandparent's mtDNA with an orphaned child whose parents have "disappeared" during war, a natural disaster, or in mysterious circumstances.

So far, no mitochondrial disease has been exclusively associated with any population-specific subgroup. However, it is likely that mtDNA mutations and/or polymorphisms are important in disease causation when they occur in association with either heterozygous or homozygous nuclear DNA mutations. Some of these mutations, in the homozygous combination, manifest as multisystem and multiorgan severe phenotypes. These phenotypes are indistinguishable from autosomal recessive respiratory chain disorders. In addition, it is likely that mtDNA interaction with nongenetic environmental factors, such as malnutrition, infection, toxic exposure, and tropical climate may be clinically important with poor prognosis.

MOLECULAR GENETIC DIAGNOSIS OF HUMAN DISEASE

Since the 1970s, remarkable series of new inventions and developments in DNA recombinant technology have taken place. These have led to the development of the present-day field of clinical molecular genetics. With the help of restriction enzymes, it became possible to fractionate DNA into pieces of predictable size. These could be inserted into small circular plasmids, which could then be inserted into bacteria, allowing multiplication via bacterial cell division. This process was later augmented with the availability of more sophisticated vectors, which would accommodate larger pieces of DNA. It helped to generate and set up DNA libraries of recombinant DNA carriers,

TABLE 1-7 GENETIC CLASSIFICATION OF MITOCHONDRIAL DISORDERS

DISORDER	MAJOR CLINICAL FEATURES	TYPE OF GENE	MITOCHONDRIAL DNA MUTATION
Chronic progressive external ophthalmoplegia [CPEO]	External ophthalmoplegia, bilateral ptosis, mild proximal myopathy	tRNA	A3243G, T8356C Rearrangement (deletion/duplication)
Kearns-Sayre syndrome [KSS]	PEO onset <20 years, pigmentary retinopathy, cerebellar ataxia, heart block, CSF protein >1g/L		Rearrangement (deletion/duplication)
Pearson syndrome	Sideroblastic anemia of childhood, pancytopenia, renal tubular defects, exocrine pancreatic deficiency		Rearrangement (deletion/duplication)
Diabetes and Deafness	Diabetes mellitus, sensori-neural hearing loss	tRNA	A3243G, C12258A Rearrangement (deletion/duplication)
Leber's hereditary optic neuropathy [LHON]	Subacute painless bilateral visual loss, age of onset 24 years, males>females (~4:1), dystonia, cardiac pre-excitation syndromes	Protein encoding	G11778A, T14484C, G3460A
Neurogenic ataxia with retinitis pigmentosa [NARP]	Late-childhood or adult onset peripheral neuropathy, ataxia, pigmentary retinopathy	Protein encoding	T8993G/C
Leigh syndrome [LS]	Subacute relapsing encephalopathy, cerebellar and brainstem signs, infantile onset	Protein encoding	T8993G/C
Exercise intolerance and myoglobulinuria	Exercise induced myoglobulinuria	Protein encoding	Cyt B mutations
Mitochondrial encephalomyopathy with lactic acidosis and stroke-like episodes [MELAS]	Stroke-like episodes before 40 years, seizures and/or dementia, ragged-red fibers and/or lactic acidosis, diabetes mellitus, cardiomyopathy (HCM/DCM), deafness, cerebellar ataxia	tRNA	A32343G, T3271C, A3251G
Myoclonic epilepsy with ragged-red fibers [MERRF]	Myoclonus, seizures, cerebellar ataxia, myopathy, dementia, optic atrophy, bilateral deafness, peripheral neuropathy, spasticity, multiple lipomata	tRNA	A8344G, T8356C
Cardiomyopathy	Hypertrophic cardiomyopathy [HCM] progressing to dilated cardiomyopathy [DCM]	tRNA	A3243G, A4269G
Infantile myopathy/ encephalopathy	Early onset progressive muscle weakness with developmental delay	tRNA	T14709C, A12320G, G1606A, T10010C
Nonsyndromic sensorineural deafness	Early onset progressive bilateral moderate to severe sensorineural hearing loss	rRNA	A7445G
Aminoglycoside-induced nonsyndromic deafness	Early-onset nonprogressive sensorineural deafness secondary to aminoglycoside administration	rRNA	A1555G

which were grown in bacterial cultures. Individual colonies containing genes of interest could then be isolated and grown in sufficient amounts for analysis. This technique is commonly referred to as *cloning*, since each colony would represent an identical fragment of DNA. This could have wider implications, ranging from developing specific DNA probes for molecular genetic diagnosis to the commercial production of recombinant DNA drugs and vaccines. Using hybridization of DNA or RNA, with a cloned library having a sufficiently similar base composition, could isolate individual genes of interest. Once isolated, a gene could be sequenced to confirm the mutation. This would also allow further studies on the gene product.

There are, broadly, two ways of exploring the gene. The first is to characterize the gene from a known protein or the gene product; for example, hemoglobin or blood clotting factors. This is sometimes referred to as *forward genetics*. However, in a large number of cases no information was available on the gene product, protein involved, or pathologic process that resulted from its mutation. In these situations, the tool of genetic linkage was used. In the past, the genetic linkage had enabled localizing certain genes using a number of biochemical or serological markers. With the help of restriction enzymes, it became possible to generate a number of DNA fragments of different sizes. These could be used as genetic markers, commonly referred to as *restriction fragment length polymorphisms*

(RFLP). These served as more specific genetic markers, since there is marked individual variation in the human DNA. Subsequent work provided even more specific markers of individual DNA variation, including *variable number of tandem repeats* (VNTR) and *single nucleotide polymorphisms* (SNPs). These markers have been extensively used in population genetics (see Chapter 3) and in the development of DNA fingerprinting, which has revolutionized forensic science.

Using DNA linkage markers, it became possible to trace genes for diseases through families and determine their approximate location on different chromosomes. It was then possible to explore a particular chromosome (walking along the chromosome) and to isolate the defective gene. With the help of its bases, it became possible to assess the structure of the gene product (protein) and determine its function. This revolutionary procedure was initially called "reverse genetics," later on changing to *positional cloning.*

Direct DNA analysis of a specific gene or a DNA fragment of interest became possible with the availability of specific DNA probes. One of the commonly employed techniques is *Southern blotting*, which uses radioactively labeled DNA probes to identify a particular DNA fragment of interest. The DNA fragments on the gel are denatured with alkali, making them single-stranded and rendering them capable of hybridizing with complementary DNA sequences. A "permanent copy" of this single-stranded DNA fragment is then made by transferring it onto a nitrocellulose filter, which binds the single-stranded DNA. In order to localize and visualize a particular DNA fragment of interest from a collection of DNA fragments, a phosphorus-32 (^{32}P) radioactively labeled DNA probe, which has been made single-stranded, is allowed to hybridize with DNA fragments in the *Southern blot.*

A powerful technique of *polymerase chain reaction* (PCR) has made a tremendous impact on the application of molecular genetics in a number of areas. This method allows amplification up to 30-fold of a small piece of DNA containing a specific sequence. It involves making two oligonucleotide primers of known sequence, up to 20 base pairs in length, complementary to the DNA flanking a particular DNA sequence of interest. The primers are used to amplify that sequence by means of repeated cycles of denaturation through heating of the double-stranded source DNA, annealing of the primer sequences to the resulting single-stranded DNA, and extension of the primer DNA on the single-stranded DNA template using DNA polymerase to produce double-stranded copies of the DNA sequence of interest. The successive cycles of DNA synthesis result in an exponential amplification of the target DNA sequence, leading to a 10^5–10^6-fold increase in the amount of the DNA sequence of interest. This amplification results in sufficient quantities of DNA for direct visualization by ultraviolet fluorescence after ethidium bromide staining, without the need to use indirect detection techniques such as radioactivity or fluorescently labeled probes. Use of PCR allows analysis of DNA from blood, buccal smears, semen, and pathological archival material.

A number of other new molecular techniques are being used in both service and research molecular genetic laboratories, using the next-generation sequencing technology—for example, exome sequencing, whole genome sequencing, and RNA sequencing. Several commercial providers (Illumina, Affymetrix, Roche-Nimblegen etc.) offer these services at highly competitive costs, circumventing the need to incur heavy investments in these rapidly evolving genomic tools and technologies.

GENETIC AND GENOMIC BASIS OF NATURAL SELECTION—DARWINIAN MEDICINE

Despite controversies and disagreements, the Darwinian concept of natural selection offers plausible explanations for any organism's developmental, life-sustaining and reproductive mechanisms. All organisms are products of evolutionary history influenced by the process of natural selection. The whole process of natural selection favors evolutionary mechanisms, and maximizes an organism's ability to get its own genes into future generations of its local population. In biological terms, "Darwinian" and "evolutionary" are synonymous. The process of natural selection is the key to adaptive evolutionary change. An organism's ability to procreate (biological fitness) is fundamental to the whole evolutionary process. Biological fitness is likely to be reduced due to severe life-threatening diseases, infectious diseases of epidemic and pandemic proportions, famine, and natural calamities. However, fitness in the Darwinian sense is different. The Darwinian fitness is related to an organism's ability to adapt and improve with the environment. From a medical perspective, any age-limiting and/or life-threatening disease may negatively influence an individual's ability to procreate, and thus interferes in the evolutionary process. On the other hand, certain biological characteristics may help in the adaptive process; for example, hibernating in adverse weather conditions (amphibians), survival in famines (the thrifty factor), and having protection against infections and parasitic infestations (hemoglobinopathies and malaria). However, it is not entirely clear how physical and physiological variations occur in species within a taxonomic category. To some extent, this can be explained on the basis of gene and genome sequences that are evolutionarily conserved, and can be traced back to a very small organism not necessarily belonging to the same taxonomic family. Several genes and polymorphisms are now confirmed to be present in the Drosophila fruit fly that are homologous to several higher mammalian species, including humans.

Around the time when Charles Darwin put forward his theory of natural selection, there was no clear concept of genes, genetics, and genomics. However, he was far ahead of his time in supporting his arguments on the basis of some innate biological characteristics crucial in adaptive evolutionary process. What Charles Darwin hypothesized in the nineteenth century has become a reality in the twenty-first century. We are now adequately equipped with theoretical and experimental evidence to support Darwin's concept of "innate characteristics." There are number of ways in which genes and genome can influence the process of natural selection. Some notable examples include:

1. Genes conferring positive selection against microbial infections; for example, the chemoreceptor 5 allele in Indian subcontinent populations with human immunodeficiency virus (HIV), associated with low incidence of progression to acquired immune deficiency syndrome (AIDS)
2. Genes and genomic variation conferring protection against parasitic infestations; for example, hemoglobinopathies (beta-thalassemia, sickle cell disease, and hemoglobin C).
3. Genes and other regulatory sequences influencing the internal metabolic environment; for example, *thrifty factor* positively selecting in extreme starvation such as in a famine.
4. Genetic or genomic advantage to people of African origin against catecholamine cardiac complications, such as cardiomyopathy. This is probably part of an adaptive behavior process to possess and maintain a high level of physical prowess for survival in an extremely challenging environment.

It is not entirely clear how the process of natural selection and adaptive evolutionary behavior are linked to medicine and health. However, believers in the Darwinian theory remain convinced. Studies of human genomic variation in different population groups, and the resemblance to several genome sequences among other genomes (comparative genomics) offer wide-ranging evidence to support the followers of Charles Darwin. Apart from reproduction, genes, gene sequence variation, genomic variation, and epigenetic factors are important in growth, development, aging, and senescence. Some of these may be evolutionarily conserved across species, but relevant to human health. Mutations and alterations in several of these genomic elements are linked to a broad range of medical conditions.

THE HUMAN GENOME PROJECT

The advent of recombinant DNA technology in the 1970s revolutionized our ability to characterize the molecular basis of human genetic disease. This laid the foundation of eventual mapping and deciphering the DNA sequence of all the structural and functional genes of the human genome. The human genome project was, thus, a natural progression of all previous developments in the field of human molecular genetics. Such a mammoth task could not have been accomplished without the international collective effort supported by generous funding from governmental and nongovernmental sources.

It has helped to map and provide nucleotide sequences of about 25,000 nuclear genes, which, along with a number of other sequence variations, comprise the whole human genome. Although a large number of the nuclear genes have been assigned with a structural or functional link, the precise roles of other parts of the genome are not yet fully understood. However, the Human Genome Project provides the basis for "functional genomics" to explore further its functional role, and understand the complex mechanisms through which genes and their products interact to effect biological function and influence disease processes. The development of new therapeutic agents is now possible on the basis of genomic arrangement and its designated functional role. It also helps to characterize the genomes of various pathogens and other organisms, an invaluable tool in realizing the full potential of this field to improve human health.

HUMAN GENOME VARIATION AND HUMAN DISEASE

MEASURING HUMAN GENETIC AND GENOMIC VARIATION

The most direct way to measure genetic differences, or genetic variation, is to estimate how often two individuals differ at a specific site in their DNA sequences—that is, whether they have a different nucleotide base pair at a specific location in their DNA. First, DNA sequences are obtained from a sample of individuals. The sequences of all possible pairs of individuals are then compared to see how often each nucleotide differs. When this is done for a sample of humans, the result is that individuals differ, on average, at only about one in 1300 DNA base pairs. In other words, any two humans are about 99.9% identical in terms of their DNA sequences.

During the past several years, a new type of genetic variation has been studied extensively in humans: copy-number variants (CNVs) are DNA sequences of 1000 base pairs or larger are fairly distributed across the genome.

In some instances, CNVs could be deleted, duplicated, or inverted in some individuals with mild phenotypic effects. Several thousand CNVs have been discovered in humans, indicating that at least 4 million nucleotides of the human genome (and perhaps several times more) vary in copy-number among individuals. CNVs thus represent another important class of genetic variation and contribute to at least an additional 0.1% difference, on average, between individuals. Despite significant progress, medical and health implications of CNVs are not entirely clear.

Comparisons of DNA sequences can be done for pairs of individuals from the same population or for pairs of individuals from different populations. Populations can be defined in various ways; one common way is to group individuals into populations according to a continent of origin. Using this definition, individuals from different populations have roughly 10% to 15% more sequence differences than do individuals from the same population (this estimate is approximately the same for both SNPs—see below—and CNVs). In other words, people from different populations are slightly more different at the DNA level than are people from the same population. The slightness of this difference supports the conclusion that all humans are genetically quite similar to one another, irrespective of their geographic ancestry.

Because it is still fairly expensive to assess DNA sequences on a large scale, investigators often study genetic variation at specific sites that are known to vary among individuals. Suppose that a specific site in the DNA sequence harbors an A in some individuals' DNA sequences, and a G in others. This is a *single nucleotide polymorphism* (SNP), where *polymorphism* refers to a genetic site that exists in multiple forms. The proportion of individuals who have an A and the proportion who have a G give the frequency of each form, or allele, and this frequency can be estimated for a sample of individuals from a population. If the frequencies of A in three different populations are .10, .20, and .50, the *genetic distance* between the first two populations is smaller than that between the third population and the first two. On the basis of this assessment, the first two populations are genetically more similar than either is to the third. To get a more accurate picture of genetic differences, hundreds or thousands of SNP frequencies would be assessed to yield the average genetic difference among pairs of populations.

These comparisons can be summarized graphically in a variety of ways. An example is given in Figure 1-1, which portrays population differences measured in approximately 11,000 SNPs throughout the human genome. This diagram shows that populations that are geographically closer together tend to be genetically more similar to one another. This is expected, because geographic neighbors are more likely to have historical connections and to exchange mates.

GENOME VARIATION AND HUMAN DISEASE

Nearly all human diseases are influenced by genes. Because individuals have different variants of genes, it follows that the risk of developing various diseases will also differ among individuals. Consider a simple example. Jim Fixx, a well-known runner and fitness enthusiast, died of a heart attack at the age of 52. Sir Winston Churchill, who was renowned for his abhorrence of exercise and his love of food, drink, and tobacco, lived to the age of 90. It is plausible that genetic differences between Fixx and Churchill were responsible, at least in part, for the paradoxical difference in their life spans. (Indeed, Jim Fixx's father had a heart attack at the age of 35, and died of a second heart attack at the age of 43.)

Because genes are passed down from parents to offspring, diseases tend to "cluster" in families. For example, if an individual has had a heart attack, the risk that his or her close relatives, offspring, or siblings will have a heart attack is two to three times higher than that of the general population. Similar levels of increased risk among family members are seen for colon cancer, breast cancer, prostate cancer, type 2 diabetes, and many other diseases. This clustering in families is partly the result of shared nongenetic factors (e.g., families tend to be similar in terms of their dietary and exercise habits), and partly the result of shared genes. As we have seen, populations differ somewhat in their genetic backgrounds. It is thus possible that genetic differences could be partly responsible for differences in disease prevalence. For many disorders caused by genetic changes in single genes, these differences are readily apparent. Cystic fibrosis, for example, is seen in about one in 2,500 Europeans, but only in one in 90,000 Asians. Sickle-cell disease is much more common in individuals of African and Mediterranean descent than in others, although it is found in lower frequency in many other populations due to migration and intermarriage (Figure 1-14).

These differences in prevalence can be attributed to the evolutionary factors that influence genetic variation in general. *Mutation* is the ultimate source of all genetic variation. In some cases, such as hemochromatosis in Europeans and sickle-cell disease in Africans, the responsible mutations have arisen within the last few thousand years, helping to account for a fairly restricted distribution of the disease. *Natural selection* also plays a role in population differences in some genetic diseases. For sickle-cell disease and related diseases known as the thalassemias, heterozygotes (those who carry a single copy of the disease-causing mutation) are relatively resistant to the malaria parasite. Cystic fibrosis heterozygotes are resistant to typhoid fever, and

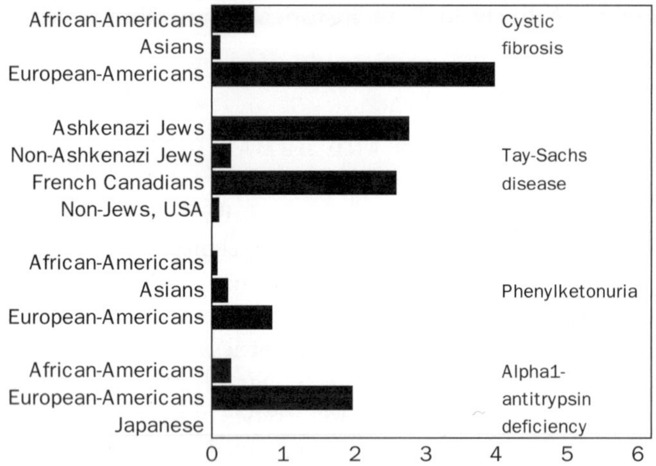

Figure 1-14 Prevalence of several single-gene disorders (per 10,000 births) in a series of human populations. The prevalence of Tay-Sachs disease in the Ashkenazi Jewish population refers to pre-1980 data; widespread genetic screening has greatly reduced the prevalence of this disease in the Ashkenazi population.
SOURCE: Jorde LB. 2007, Human Genetic Variation and Disease, In Meyers RA, ed. Genomics and Genetics: From Molecular Details to Analysis and Techniques, pp. 939–953, Weinheim: Wiley-VCH Publishers, used with permission.

hemochromatosis heterozygotes absorb iron more readily, perhaps protecting them against anemia. Also, the process of *genetic drift*, which is accentuated in small populations, can raise the frequencies of disease-causing mutation quickly just by chance (e.g., Ellis-van Creveld disease, a reduced-stature disorder, is unusually common among the Old Order Amish of Pennsylvania). In contrast to the effects of natural selection and genetic drift, which tend to promote population differences in disease prevalence, *gene flow* (the exchange of DNA among populations) tends to decrease differences among populations. With the enhanced mobility of populations worldwide, gene flow is thought to be increasing steadily.

These same factors can affect common diseases such as cancer, diabetes, hypertension, and heart disease, but the picture is more complex because these diseases are influenced by multiple genetic and nongenetic factors. Common diseases do vary in frequency among populations: hypertension occurs more frequently in African Americans than European Americans, and type 2 diabetes is especially common among Hispanic and Native-American populations. Although genes clearly play a role in causing common diseases, it is less clear that genetic differences between populations play a significant role in causing differences in prevalence rates among populations. Consider another example: the Pima Native-American population in the southwestern United States now has one of the highest known rates of type 2 diabetes in the world. About half of adult Pimas are affected. Yet, this disease was virtually unknown in this population prior to World War II. Obviously, the Pimas' genes have not changed much during the past 50 or so years. Their environment, however, has changed dramatically with the adoption of a "Western" high-calorie, high-fat diet, and a decrease in physical exercise. In this case, it is almost certain that the rapid increase in type 2 diabetes prevalence has much more to do with nongenetic than genetic causes.

But why does a Western diet seem to have a greater effect on some populations than others? Perhaps differences in genetic background, interacting with dietary and other lifestyle changes, help to account for this variation. As additional genes that influence susceptibility to common diseases are discovered, and as the roles of nongenetic factors are also taken into account, it is likely that this picture will become clearer.

FUNCTIONAL GENOMICS AND PROTEOMICS

Functional genomics is a systematic effort to understand the function of genes and gene products by high-throughput analysis of gene transcripts in a biological system (cell, tissue, or organism) with the use of automated procedures that allow scale-up of experiments classically performed for single genes (Yaspo, 2001). Functional genomics can be conceptually divided into gene-driven and phenotype-driven approaches. Gene-driven approaches rely on genomic information to identify, clone, and express genes, as well as to characterize them at the molecular level. Phenotype-driven approaches rely on phenotypes either identified from random mutation screens or associated with naturally occurring gene variants, such as those responsible for mouse mutants or human diseases, to identify and clone the responsible genes without prior knowledge of the underlying molecular mechanisms (Yaspo, 2001). The tools of functional genomics have enabled systematic approaches to obtaining basic information for most genes in a genome, including when and where a gene is expressed and what phenotype results if it is mutated, as well as the localization of the gene product and the identity of other proteins with which it interacts (Steinmetz & Davis, 2004). Functional genomics aspires to answer such questions systematically for all genes in a genome, in contrast to conventional approaches that address one gene at a time.

Analysis and applications of the rapid accumulation of highly sophisticated genome and proteome data necessitated development of powerful computational program and relevant hardware tools. Storage, retrieval, and assimilation of enormous data require fast and accurate computational skills. Bioinformatics deals with these requirements within the broad biomedical and biotechnology sectors. There are several literature and online resources with detailed descriptions of the role and scope of bioinformatics.

A number of biomedical and biotechnology disciplines have emerged during the last two decades, all ending with the suffix *omics. Omics* is derived from *ome (Greek, omoyous),*

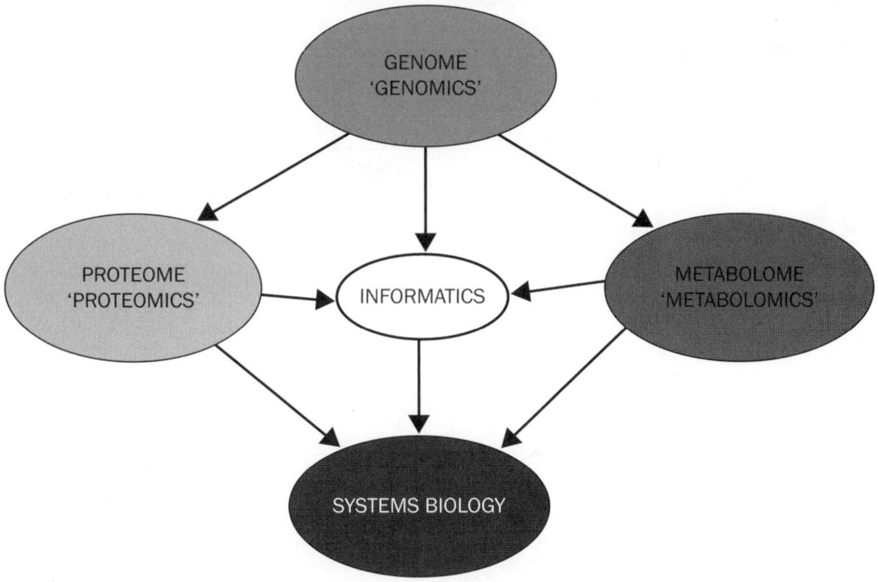

Figure 1-15 Informatics as the central dogma for systems biology and genome sciences.

which refers to *complete knowledge*. The ancient language Sanskrit has a similar phrase, *Ohm*, with similar meaning and expression. A number of these "omics" have direct or indirect links to the fundamentals of genome science and technology. A number of biological models have been developed and tested using genomic, transcriptomic, proteomic and metabolomic approaches (Figure 1-15). Systems biology refers to developing and testing of biological models based on *omic* sciences (Figure 1-16). The central dogma is the computational analysis of complex and enormous data at all biological levels—gene, molecule, cell, tissue, organ, and whole body.

TRANSLATIONAL GENOME RESEARCH AND GENOMIC MEDICINE

The potential of applications of genome science and technology in medicine and health has led to the emergence of *genomic medicine*, a natural outcome of tremendous progress made in medical genetics and genomics. However, final endpoints in genomic medicine will largely depend upon judicious and efficacious application and utilization of the diagnostic and therapeutic potential of genome-based technologies, for example clinical applications of microarray technology (Table 1-8). This process requires multifaceted systematic and analytical research efforts to translate the basic scientific information to practical and pragmatic applications following the principles of good medical practice. There is no disagreement that translational genome research is vital for the successful and efficient delivery of promises made by researchers and physicians behind the genomic medicine movement.

The process for translational genome research includes the participation of several researchers drawn from different disciplines. The multidisciplinary model for translational genome research is widely accepted, and includes key elements (Figure 1-17). Khoury et al. (2007) recommend a framework for the continuum of multidisciplinary translation research, to utilize previous research outcomes in genomics and related areas of health and prevention. The whole process includes four phases and revolves around the development of evidence-based guidelines. Phase 1 translation (T1) research seeks to move a basic genome-based discovery into a candidate health application, such as a genetic test or intervention. Phase 2 translation (T2) research assesses the value of genomic applications for health practice, leading to the development of evidence-based guidelines. Phase 3 translation (T3) research attempts to move evidence-based guidelines into health practice through delivery, dissemination, and diffusion research. Phase 4 translation (T4) research seeks to evaluate the "real world" health outcomes of a genomic application in practice. It is important to appreciate that the whole process of translation research leading to evidence-based guidelines is a dynamic one, with considerable overlap between the different stages. The process should be able to accommodate new knowledge that will inevitably arrive during translation research.

The role of translational genome research, including that of clinical trials, is crucial in developing evidence-based good practice guidelines (Kumar, 2007). The aim

TABLE 1-8 CLINICAL APPLICATIONS OF MICROARRAY TECHNOLOGY

- Simultaneous hybridization of analytes in a single sample to thousands of different specified targets on a DNA microarray
- Evaluation of quantitative changes in mRNA in clinical samples useful in assessing predisposition to disease
- Establishment of laboratory methods for with high sensitivity and specificity
- Evaluation of disease severity and patient prognosis
- Selection of optimal therapies
- Identification of new therapeutic target genes

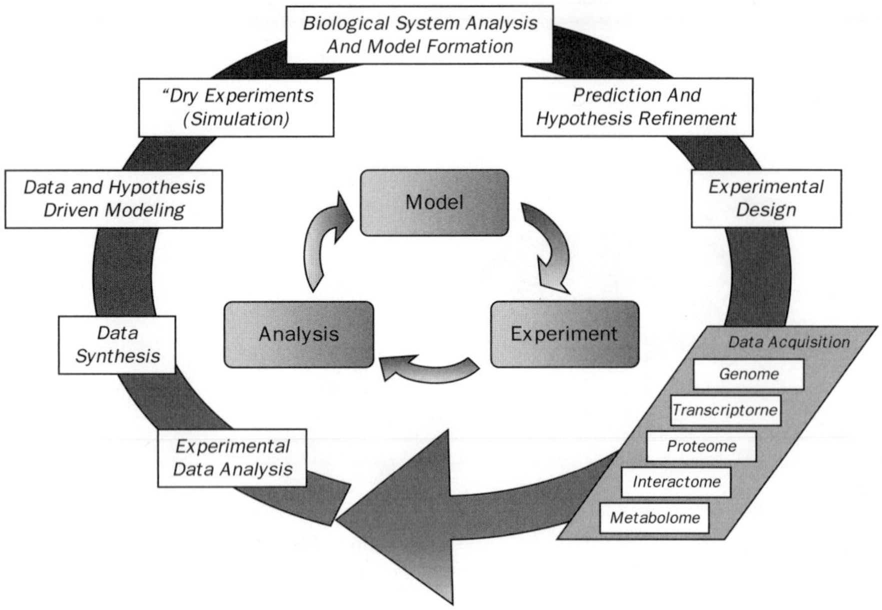

Figure 1-16 A classic model for systems biology.

should be to obtain vital genetic and genomic information, including laboratory material for research, from the patient, family, and community, and then use this scientific data and information for clarification and ratification. The outcomes of translational gene research should be valid and deliverable in the clinic for diagnostic and therapeutic applications.

INTEGRATION OF GENOME SCIENCE AND TECHNOLOGY FOR ECONOMIC DEVELOPMENT

During the last decade rapid progress has been made in harnessing the huge potential of genome science and technology for its economic and health benefits globally, in particular in less and least developed nations. Apart from the World Health Organization (WHO), other international and national institutions engaged in this endeavor include the Human Genome Organization (HUGO), Organization for Economic Cooperation and Development (OECD), the McLaughlin-Rotman Center for Global Health (The University of Toronto, Canada), the Mexican Health Foundation, the Beijing Genomics Institute, the Department of Science & Technology (Government of India), and many more. All these institutions are focused on supporting and exploiting the huge potential of genomic technologies and related bioinformatics developments on the global economy and on health. The impact of genome sciences and technologies will manifest in the following wide-ranging fields:

- Personalized medicine and health approaches that will help people and societies to shift the focus from "sick-care" to "well-care and prevention."
- Biotechnology methods to produce environmentally clean and efficient fuel and chemicals to accelerate transition from petroleum-based economies
- Genome-driven plant and crop growing methods for producing affordable food for less and least developed economies
- Promoting genomic science and technology in animal breeding and livestock improvement
- Supporting genome research for new drug discovery and drug development for enhancing pharmaceutical efficacy
- Applications of genomic biotechnologies in the study and monitoring of environmental health

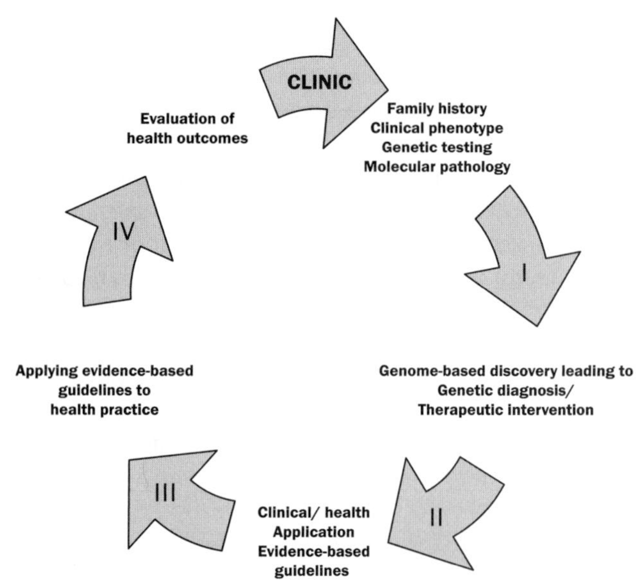

Figure 1-17 A model for multidisciplinary translational genomic research
SOURCE: (Khoury et al, 2007).

CONCLUSIONS

Developments in genetics and subsequent sequencing of the human and other genomes have provided us with an opportunity to review the role of genes and genomes in all aspects of health and disease. Human health, including causation of disease, is not exclusively dependent on human genes and genome. Evolutionary links with other genomes and ecologically relevant and beneficial parts of other genomes play crucial roles in the maintenance of human health and, to some extent, in morbidity and mortality. Understanding genomes of microbes, parasites, animals, plants, and crops are acknowledged priorities of current biomedical and biotechnology research.

Conventionally, the causation of human disease includes malformations, trauma, infection, immune dysfunction, metabolic abnormality, malignancy, and degenerative conditions associated with aging. Genetic factors have long been recognized in all of these disease groups. The traditional genetic categories of diseases include chromosomal disorders, single-gene or Mendelian diseases, and several forms of multifactorial/polygenic conditions. In addition, somatic genetic changes and mutations of the mitochondrial genome probably account for a small, albeit important, number of diseases. These groups of disorders are well recognized and have an established place in the classification of human disease. Recent developments in genome research have provided vast data indicating different genomic mechanisms to explain complex pathogenesis in some disorders. The spectrum of these disorders is wide and includes both acute and chronic medical and surgical diseases. Perhaps it is reasonable to identify these disorders on the basis of underlying molecular pathology, including genomic imprinting, genomic rearrangements, and gene–environment interactions involving multiple genes and genomic polymorphisms.

This chapter has reviewed the genetic and genomic approaches to human health and disease. The genomic approaches to understanding and managing human disease are rapidly being incorporated in the practice of clinical medicine. In addition, applications of genome science and technology are also reforming biotechnologies in a number of industries, including pharmaceutical, agricultural, and ecological bioengineering. The enormous impact of genome sciences and technologies on the economy of the developing world will be judged on applications in a number of areas including biofuels, accelerated breeding of crops and livestock, personalized health products, pharmaceutical efficacy, and genomic monitoring of environmental health.

FURTHER READING

Weatherall DJ. (1991). *The New Genetics and Clinical Practice.* Oxford: Oxford University Press.

Gelehrter TD, Collins FS, Ginsburg D. (1998). *Principles of Medical Genetics, 2nd ed.* Baltimore: Williams and Wilkins.

Strachan T, Read A. (2009). *Human Molecular Genetics, 5th ed.* New York: Garland.

Rimoin DL, Connor JM, Pyeritz RE, Korf, B. (Eds.) (2006). *Emery and Rimoin's Principles and Practice of Medical Genetics*, e-edition (www.elsevier.com).

REFERENCES

Ashley CT Jr, Warren ST. (1995). Trinucleotide repeat expansion and human disease. *Annu Rev Genet 29*, 703–728.

Chen WG, Chang Q, Lin Y, Meissner A, West AE. (2003). Derepression of BDNF transcription involves calcium-dependent phosphorylation of MeCP2. *Science 302*(5646), 885–889.

Chance PF, Abbas N, Lesch MW, et al. (1994). Two autosomal dominant neuropathies result from reciprocal DNA duplication/deletion of a region on chromosome 17. *Hum Mol Genet 3*, 223–228.

Egger G, Liang G, Aparicio A, Jones P. (2004). Epigenetics in human disease and prospects of epigenetic therapy. *Nature 429*, 457–463.

Ennis S, et al. (2006). Women with mid-size FMR1 premutations are at the highest risk for premature ovarian failure. *Eur J Hum Genet 14*, 253.

Greco CM, Berman RF, Martin RM, Tassone F, Schwartz PH, Chang A, Trapp BD, Iwahashi C, Brunberg J, Grigsby J, Hessl D, Becker EJ, Papazian J, Leehey MA, Hagerman RJ, Hagerman PJ. (2006). Neuropathology of fragile X-associated tremor/ataxia syndrome (FXTAS). *Brain 129*(Pt 1), 243–255.

Greenberg F, Guzzeta V, Montes de Oca-Luna R, Magenis RE, Smith ACM et al. (1991). Molecular analysis of the Smith-Magenis syndrome: a possible contiguous gene syndrome associated with del(17)(p11.2). *Am J Hum Genet 49*, 1207–1218.

Harper PS (2003). Practical Table 1-9: Genetic Counselling. Arnold, London, pp 165–166.

Heutnik P, van der Mey AG, Sandkuiji LA, van Gils AP et al. (1992). A gene subject to genomic imprinting and responsible for hereditary paragangliomas maps to chromosome 11q23-qter. *Hum Mol Genet 1*, 7–10.

Holliday R, Pugh JE (1975). DNA modification mechanisms and gene activity during development. *Science 187*, 226–232.

Jones LK, Saha V (2002). Chromatin modification, leukaemia and implications for therapy. *Br J Haematol 118*, 714–727.

Jones P, Baylin SB (2002). The fundamental role of epigenetic events in cancer (2002). *Nature Rev Genet 3*, 415–428.

Jones PA, Laird PW (1999). Cancer epigenetics comes of age. *Nature Genet 21*, 163–167.

Jorde, LB. 2007. Human Genetic Variation and Disease, In Meyers RA (ed.), *Genomics and Genetics: From Molecular Details to Analysis and Techniques*, pp. 939–953, Weinheim: Wiley-VCH Publishers, pp. 939–953.

Kane MF, Loda M, Gaida GM, et al. (1997). Methylation of the hMLH1 promoter correlates with lack of expression of hMLH1 in sporadic colon tumors and mismatch repair-defective human tumor cell lines. *Cancer Res 57*(5), 808–811.

Khoury MJ, Gwinn M, Yoon PW, Dowling N, Moore CA, Bradley L.(2007). The continuum of translation research in genomic medicine: how can we accelerate the appropriate integration of human genome discoveries into health care and disease prevention? *Genetic Med* 9(10), 665–674

Kumar D. (2007). Genomic medicine: a new frontier of medicine in the twenty first century. Genomic Med 2007;1&2, 3–7

Lachner M, Jenuwein T (2002). The many facets of histone lysine methylation. *Curr Opin Cell Biol 14*, 286–298.

Lupski JR (1998). Genomic disorders: structural features of the genome can lead to DNA rearrangements and human disease traits. *Trends Genet 14*, 417–420

Lupski JR and Stankiewicz, P (2005). Genomic disorders: molecular mechanisms for rearrangements and conveyed phenotypes. *PLoS Genet 1*, e49

Lupski JR, Montes de Oca-Luna R, Slaugenhaupt S, Pentao L et al. (1991). DNA duplication associated with Charcot-Marie-Tooth disease type 1A. *Cell 66*, 219–232

Lupski, JR (2003). Genomic disorders: Recombination-based disease resulting from genome architecture. *Am J Hum Genet 72*, 246–252.

Neumann HP, Pawlu C, Peczkowska M, Bausch B, McWhinney SR et al. (2004). Distinct clinical features of paraganglioma syndromes associated with SDHB and SDHD gene mutations. *JAMA 292(8)*, 943–951.

Nusbaum C, et al., (2005). DNA sequence analysis of human chromosome 18. *Nature* 437(7058), 551–555

Patel P, Lupski JR (1994). Charcot-Marie-Tooth disease: a new paradigm for the mechanism of inherited disease. *Trends Genet 10*, 128–133

Pentao L, Wise CA, Chinault AC, Patel PI, Lupski JR (1992). Charcot-Marie-Tooth type 1A duplication appears to arise from recombination at repeat sequences flanking the 1.5 Mb monomer unit. *Nat Genet 2*, 292–300

Potocki L, Chen K-S, Koeuth T, Killian J, Iannaccone ST et al. (1999). DNA rearrangements on both homologues of chromosome 17 in a mildly delayed individual with a family history of autosomal dominant carpal tunnel syndrome. *Am J Hum Genet 64*, 471–478

Reiter LT, Murakami T, Koeuth T, Pentao L et al. (1996). A recombination hotspot responsible for two inherited peripheral neuropathies is located near a *mariner* transposon-like element. *Nat Genet 12*, 288–297

Roberts CW, Orkin SH (2004). The SW1/SNF complex-chromatin and cancer. *Nature Rev Cancer 4,* 133–142.

Shaw CJ and Lupski JR (2004). Implications of human genome architecture for rearrangement-based disorders: the genomic basis of disease. *Hum Mol Genet 13(1)*, R57–64

Slager RE, Newton TL, Vlangos CN, Finucane B, Elsea SH (2003). Mutations in *RAII* associated with Smith-Magenis syndrome. *Nat Genet* 33, 466–468.

Steinmetz LM, Davis RW (2004). Maximising the potential of functional genomics. *Nat Rev Genet* 5(3), 190–201

Tufarelli C et al. (2003). Transcription of antisense RNA leading to gene silencing and methylation as a novel cause of human genetic disease. *Nature Genet 203*, 157–165.

Van der Mey AGL, Maaswinkel-Mooy PD, Cornelisse CJ, Schmidt PH, van de Kemp JJP (1989). Genomic imprinting in hereditary glomus tumors: Evidence for new genetic theory. *Lancet 2(8675)*, 1291–1294.

Weksberg R. *et al.* (1993). Molecular characterization of cytogenetic alterations associated with the Beckwith-Wiedemann syndrome (BWS) phenotype refines the localization and suggests the gene for BWS is imprinted. *Hum molec Genet 2*, 549–556.

Yaspo M-L. (2001). Taking a functional genomics approach in molecular medicine. *Trend Mol Med 7*(11), 494–502.

2.

THE HUMAN VARIOME PROJECT AND THE DEVELOPING WORLD

Richard G.H. Cotton

INTRODUCTION

Individual scientists and researchers started collecting mutations in genes soon after the first globin mutation was defined at the protein level over 60 years ago. This was for both research purposes and, particularly later, for guidance in clinical decisions. Now, access to all cases of specific diseases and the mutations causing them is becoming essential not only for clinical judgment, but also for developing disease/country/ethnic-specific mutation detection and prevention strategies. Also, a patient population for new therapies, especially as mutation-specific strategies are becoming a target of research is needed.

A focus on specific ethnic groups became important when it was clear that thalassemia was a Mediterranean disease, and when specific mutations were found in many diseases that predominated in the Ashkenazi Jews. Similarly, the most common mutation in phenylketonuria (PKU) and cystic fibrosis were found to be not prominent in Asian populations. These few examples highlight the need for specific collections of mutations in homogenous ethnic populations, often confined to specific countries. Increased migration means that, in Europe and Northern America, for example, economical and rapid genetic healthcare can be given to specific ethnic groups among whom the most common mutations are known, particularly in inherited diseases, and can be tested for easily and cheaply.

The understanding of the above reality, and the recognition that much important data is inaccessible in hospital records, led to the obvious need to make publicly available all genetic variation that causes inherited disease. Meeting this need was set as a goal in the creation of the Human Variome Project (HVP; www.humanvariomeproject.org; also seeAxton, 2007; Cotton et al., 2007).

THE HUMAN VARIOME PROJECT

In simple terms, the HVP aims to collect all variation in all genes causing disease in humans. This title could be construed to also include variation not causing disease, or those associated with disease in ubiquitous genome-wide association studies (GWAS). In fact, one day the reference human genome sequence will be annotated with all variations. Currently, though, the data for the SNP, HapMap, CNV, GWAS, and 1,000 Genomes consortia are captured at NCBI/EBI as part of the funding rules for National Human Genome Research Institute (NHGRI) and The Welcome Trust. This data does not include variation *causing* human disease (only *associated* with it). Collection of variation causing human disease (termed *mutation* in this chapter) is relatively poorly funded, often collected by volunteer curators in their spare time, only because it is needed for proper genetic healthcare and research.

The HVP was formally initiated in 2006 at a meeting in Melbourne, Australia (Ring et al., 2006). The meeting was co-sponsored by the World Health Organization (WHO) and the American College of Medical Genetics, and attended by representatives of the European Community, WHO, UNESCO, OECD, major genetic societies, and 34 countries. The topics for the meeting were: The Clinic and Phenotype, The Diagnostic Laboratory, The Research Laboratory, Curation and Collection, Informatics, Relevance to the Emerging World, Integration and Federation, and Funding and Sustainability.

The meeting generated 96 recommendations (Cotton et al., 2007) for the field, indicating the need to address the poor state of affairs arising from poor funding and the division of the field by gene, country, and expertise.

A follow-up meeting occurred in May 2008 in Spain. The meeting groupings included Pilot Projects, Data Collection from Clinics, Data Collection from Laboratories, Data Transfer and Databasing (Gene Specific/or Locus Specific Databases (LSDB)), Overall Data Integration and Access, Assessment of Pathogenicity, Publication Credit and Incentives, Developing/Emerging Countries and Worldwide, Funding Mechanisms and Governance, and Ethics. This was a planning meeting, and there was an attempt to have those present who either had made a major contribution, were promising to do so, or had systems that needed to be assessed for their utility in a final scheme of mutation collection/databasing. The plans, which had to be followed to satisfy the 96 recommendations, were developed in a recent manuscript (Kaput et al., 2009).

PATHWAYS AND DATA PIPELINES FOR VARIATION COLLECTION

To perform the HVP, pilot studies are needed to test software, methodology, and data pipelines (see below). A simple fundamental pathway can be seen in Figure 2-1.

It is essential that all mutations in all genes in all countries are collected, and this requires specific activities in all countries. The data from past patients, referred to as "legacy data," resides in molecular genetics diagnostic laboratories, notes of medical geneticists, and hospital records.

In simple terms, this data needs to be collected, which is easily done and has already started in some countries. One simple way to begin is to use the currently available electronic database software but many have downloaded the LOVD software (www.lovd.nl) and start entering data. Funding can be obtained to do this, and it could be part of training of students/doctors. Once this has been done, the data can be held on dedicated servers in a country (currently being piloted in Australia), sent to the relevant LSDB, and then to NCBI and other central databases.

New data can be handled similarly, but because entering data into databases as it is collected takes valuable time of expert personnel, there is a need for software that both sends data to clinical records in an identified way, *and* to a public database in an anonymous way. The pathway for this activity is shown in Figure 2-2.

All patient-related data from labs and clinics need to be accessible in an anonymous way to clinicians and others within a country. It then needs to be sent to expertly curated databases (LSDB) and, finally, become part of secure long-term databases at NCBI, EBI, and the UCSC Genome Browser.

ENSURING COLLECTION IN A COUNTRY

Diagnostic laboratories in each country that carry out genetic testing (or receiving results from elsewhere) will be generating test result data. When a result is obtained

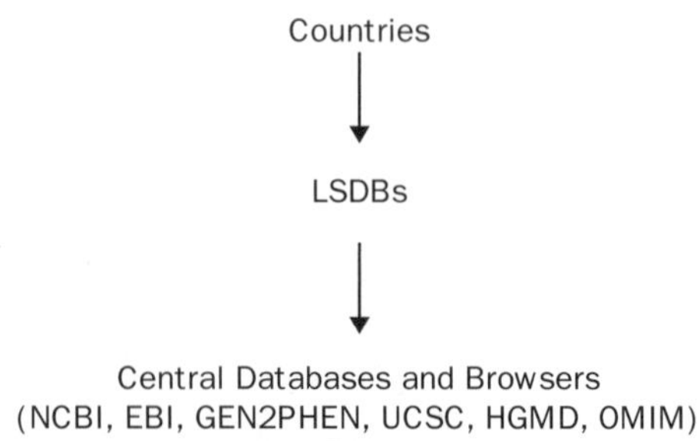

Figure 2-1 Mutation Flow Needed

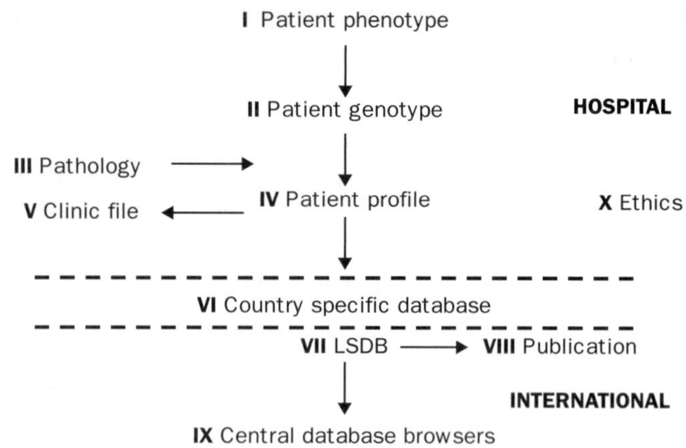

Figure 2-2 Overall Bioinformatic Needs

the pathologist will often consult public databases. It is hoped that these people will enter *their* mutations into databases for use by others. The main obstacles to this at present are lack of time, databases and standard procedures. The main steps need to alleviate these problems are *(a)* development of software to simplify submission, *(b)* changes in procedures and practices, and *(c)* establishment of national databases. Several countries, e.g., UK, Australia, Korea and Thailand have initiated, or are about to initiate HVP Country Nodes (12 countries have already signed up: Australia, Austria, Belgium, China, Cyprus, Greece, Egypt, Malaysia, Kuwait, Spain, Vietnam and Nepal), pilot programs that will be able to guide other countries in making software available. A possible strategy is shown in Figure 2-2. These country-specific databases should deliver mutations to LSDBs, and finally to NCBI/EBI/UCSCC Genome Browser for safekeeping.

Collecting legacy data requires funding, software, and volunteers to enter the data in a national database.

CURRENT COUNTRY ACTIVITIES: SPREAD OF PROTOCOLS AND SOFTWARE

Current country-specific activities are shown in Table 2-1. They are variable in aims, strategy, and coverage, and the software is not portable to other countries.

Once protocols are available to collect all mutations in a country, it could be spread to others. We hope that such software will be available in the near future.

WHAT CAN BE DONE NOW?

Of course, mutations have been collected in countries already *outside* publication through several mechanisms. But a readily available and immediate route could be called "System 1." This would be to download the freely available, open source software (e.g. LOVD or UMD), and initiate an LSDB for each gene in the lab.

TABLE 2-1 CURRENT COUNTRY-SPECIFIC ACTIVITIES

Finnish disease heritage	http://www.findis.org
Israeli population database	http://www.goldenhelix.org/israeli
Singapore Human Mutation and Polymorphism Database	http://shmpd.bii.a-star.edu.sg/
FINDbase	http://www.findbase.org
RAMEDIS – Germany	http://www.ramedis.de/
CETT/caBIG – USA	? https://cabig.nci.nih.gov/
DmuDB – UK	https://secure.dmudb.net/ngrl-rep/Home.do
Hellenic Genetic Database Consortium	Recently initiated
Centre for Arab Genomic Studies	http://www.cags.org.ae/
Australian node of HVP	Recently initiated
China-HVP	http://www.genomed.org/LOVD/hvp-china.php
Korean node of HVP	Recently initiated

WHO WILL PERFORM THE PROJECT?

The project needs to be performed by perhaps tens of thousands of people in, say, 100 countries. These people would mainly be diagnostic laboratory heads and clinical geneticists. They may have to create ways of collecting legacy data. In the case of new data, it is hoped that policy changes and new software will make this a routine, trivial exercise. Informaticists are clearly needed to derive the software initially, some of which may be based on preexisting software. Policy change needs to come through professional societies and those responsible for quality assessment and registration of laboratories and tests.

WHO WILL PAY FOR THE COLLECTION OF NEW DATA?

If submission of new data becomes routine, this should not cost anything. Two major costs are for software and loading-up of legacy data. It is hoped that software will be available from countries that have been funded to produce the software, e.g., Australia. Loading up legacy data can be done by trainees, students, volunteers, and by those funded by government or support groups to do so.

CHALLENGES IN THE DEVELOPING WORLD

While clearly there are differences in funding levels for genetic testing in the developing world versus the developed world, the funding for databasing data in the developed world is far from adequate. In fact, most variation causing disease in the developing world are found by directed research grants. Despite this, there are special issues that are present in the developing world (see Table 2-2).

TABLE 2-2 SPECIAL ISSUES FOR DEVELOPING COUNTRIES

Many higher priorities; e.g., malaria
Budget restrictions
Lack of trained personnel
Lack of equipment
Lack of travel money
Consanguineous marriages
Ethical restraints
Lack of communication and interaction with developed countries
Long-term sustainability

The items in Table 2-2 have been self-evident for decades and need little discussion, but they do need to be addressed. Possible solutions are listed in Table 2-3, many of which are being put into action but perhaps need refocusing in the interests of genetic healthcare. Intervention can relieve enormous suffering for families, as well as lighten the financial burden on society.

A single person can make a huge difference to human genetics practice in a community. There are many examples, but a well-known one is Dr. Pinar Ozand, who developed the field to a high degree in Riyadh. Governments need to find enough funding to repatriate only one person of very high international profile from the developed world, and this can make an enormous difference.

There are some examples of partnering with the developed-world clinics and laboratories, but this could be further developed through policy initiatives in healthcare with funding sought through UNESCO/OECD.

If mutation/phenotype data could be published in high-profile journals covered by PubMed, Google Scholar, etc., local journals could be used, and arrangements might be made to have contents available on selected sites such as the HVP website, so that data with publication can be achieved.

Regional alliances such as the Centre for Arab Genomic Studies (CAGS) in the Arab world, and Latin American Network of Human Genetics (Rede Latino Americana

TABLE 2-3 POSSIBLE SOLUTIONS AND STRATEGIES FOR DEVELOPING COUNTRIES

Attract leaders experienced and educated in the developed world to come back to home country
Partner with developed-world laboratories, clinicians, and projects
Publish in local journals and ensure contents available
Corporate support
Regional alliances
Shared equipment to initiate activities
Establish Nodes of the Human Variome Project and import expertise/software
Lobby human genetics, cancer, and other relevant societies

de Genética Humana/RELAGH) in South and Central America are clearly valuable and exist today—but again, they may need refocusing. Such alliances could develop economies regarding reagents, equipment, databases and assay sharing.

Critical is the development of nodes of the HVP and setting up databasing using open source or free software; e.g., LOVD, UMD. The Australian Node of the HVP has developed open source software which will be available for use by those countries that need it (see www.hvpaustralia.org.au). This should be done in association with the local relevant cancer, human/molecular genetics, and pathology societies.

REFERENCES

Axton M. (2007). What is the Human Variome Project? *Nat Genet 39*(4), 423.

Cotton RG, Appelbe W, Auerbach AD, Becker K, Bodmer W, Boone DJ, Boulyjenkov V, Brahmachari S, Brody L, Brookes A, et al. (2007). Recommendations of the 2006 Human Variome Project meeting. *Nat Genet 39*(4), 433–436.

Kaput J, Cotton RG, Hardman L, Watson M, Al Aqeel AI, Al-Aama JY, Al-Mulla F, Alonso S, Aretz S, Auerbach AD, et al. (2009). Planning the Human Variome Project: The Spain report. *Hum Mutat 30*(4), 496–510.

Ring HZ, Kwok PY, Cotton RG. 2006. Human Variome Project: an international collaboration to catalogue human genetic variation. *Pharmacogenomics* 7(7), 969–972.

3.

HUMAN GENOMIC VARIATION INITIATIVES IN EMERGING ECONOMIES AND DEVELOPING COUNTRIES[1]

Béatrice Séguin, Billie-Jo Hardy, Peter A. Singer, and Abdallah S. Daar

INTRODUCTION

Since the completion of the human genome and other genomes, the opportunities and implications for global health in the developing world are enormous. The population in most developing countries, in particular the emerging economies of Brazil, Russia, India China, and South Africa (BRICS), is vast and heterogeneous. Even before the advent of sophisticated genetic and genomic technologies, a wealth of data and information were available to document population heterogeneity at physical, social, and cultural levels. Leading developing countries including BRIC nations, Asia-Pacific countries, and in other regions (Arab and the Middle East, South Africa and Latin America) initiated and collected data on human genomic variation. This chapter discusses the application of knowledge of human genomic variation in Mexico, Thailand, South Africa, and India. Most observations are relevant to the field of genomics as it intersects with public health and is a potential source of economic development in the developing world.

The investment in, and adoption of, innovative science and technology is crucial to improving the health of populations and stimulating emerging economies and developing countries (UN Millennium Project, 2005) These countries are increasingly building their own infrastructures (Frew et al., 2007; Frew et al., 2008; Masum et al., 2007) and spurring local innovation to capture the value of their investments in science and technology (Séguin et al., 2008). These efforts are crucial for breaking the cycle of dependence on industrialized countries.

Following the completion of the HapMap Project, a number of large-scale genotyping projects have been initiated in developed countries to explore the relationship between SNPs and population variance in disease predisposition, diagnosis, and drug response. These include, but are not limited to, CartaGene in Canada, the UK Biobank in the United Kingdom, the Western Australian Genome Health Project in Australia, the Singapore Tissue Network in Singapore, and KoraGen in Germany. These initiatives, together with recent United States Food and Drug Administration regulatory incentives, increasing biomarker validation, pharmacoeconomic evidence, as well as potential litigation and patient demand, are driving the adoption of genomic medicine in developed countries. Many believe that these projects will result in significant health benefits, including individual diagnostics and personalized therapeutics, which will improve treatment outcomes (Daar et al., 2005). Emerging economies such as India, Mexico, and Thailand have also chosen to publicly fund large-scale national human genotyping initiatives to explore human genomic variation in their respective populations. They suggest that establishing these databases will enable them to access the benefits associated with genomic medicine and pharmacogenomics, which have the potential to provide cost savings in drug development, to reduce healthcare costs, to stimulate growth and investment within the local private sector, and to uncover genetic diversity that is relevant to drug response or disease predisposition (Séguin et al., 2008; Daar et al., 2005). Indeed, we have previously argued that developing countries can least afford to waste precious resources on ineffective diagnostics and therapies; genomic medicine thus has significant relevance to the health of people in developing countries (Séguin et al., 2008; Daar et al., 2005) Emerging knowledge about local human genomic variation in these countries can also be harnessed by the local domestic private sector toward creating cost efficiencies and improving their drug development pipelines (Daar et al., 2005). In order to capitalize on these large-scale genotyping initiatives, and on the potential of genomic medicine, emerging regulatory issues and local barriers to the translation of basic research, as well as commercialization and delivery of the potential products, will all need to be addressed. However, how a developing country

1. Reprinted from Séguin B, Hardy BJ, Singer PA, Daar AS (2008). Human genomic variation initiatives in emerging economies and developing countries. Copyright 2008 by Nature Publishing. Reproduced with permission of Nature Publishing.

chooses to leverage the knowledge output will depend upon the existing research and commercialization infrastructure; its way of protecting the data will depend upon the existing legislative, regulatory, and governance systems in place, and its way of integrating the eventual benefits into its healthcare system will depend upon domestic health needs and local healthcare infrastructure.

Qualitative case study methods were used for this research. The case studies of the human genotyping initiatives in Mexico, India, Thailand, and South Africa were selected for the following reasons: at the time, they each had a head start when compared with the other countries at the same level of economic development. Data were collected between June 2006 and March 2008. In-depth, face-to-face (or by teleconference call) semistructured interviews with key informants, including scientists and managers from the Mexican National Institute of Genomic Medicine (INMEGEN) in Mexico, the Indian Genome Variation (IGV) Database Consortium in India, the Thai SNP Discovery Project, the National Center for Genetic Engineering and Biotechnology (BIOTEC) in Thailand, the Thai Center of Excellence for Life Sciences (TCELS) in Thailand, and the African Genome Education Institute (AGEI) and the Division of Human Genetics at the University of Cape Town in South Africa. The other key informants from diverse backgrounds such as the media, nongovernmental organizations, regulatory agencies, and ministries of health, as well as experts in the area of genomics/pharmacogenetics and ethics from developed and developing countries were also interviewed. Data from other resources, such as publicly available materials, were also included that were relevant to the study questions.

Data were analyzed in several phases. First, theme categories were identified by analyzing the interview transcripts by generative or open coding (that is, analysis of word repetitions, key terms, and key words). The next phase of data analysis consisted of axial coding of the data, which allowed us to build connections within and between theme categories. In the final phase we identified core concepts by using selective coding. The results were cross-validated by different members of our research group.

This study was approved by the committee on use of human subjects of the University of Toronto. Each participant that we interviewed provided written consent.

Emerging genomic technologies, such as sequencing and genotyping, can potentially be harnessed to improve health and reduce the burden of disease. In recognition of this consideration, in 2002 the World Health Organization (WHO) suggested that countries should: "strengthen existing, or establish new centers and institutions engaged in genomics research with a view to strengthening national capacity and accelerating application of the advances in genomics relevant to countries' health problems." Otherwise, the gap in quality of healthcare between developing and developed nations will continue to widen.

Now, four years after this report, is an opportune time to begin to explore how emerging economies and developing countries are responding to this challenge. Wishing to understand how the advances of innovative technology can be harnessed toward global health equity, we examine how it is affected by the emerging trend to apply knowledge of human genomic variation toward the understanding of disease susceptibility and drug response. We also examine the potential for development of life-science health products that can help improve health in the developing world—ideally, products that are developed by scientists in the developing world in order to create value for their own societies. The articles that follow provide an overview of how four countries—Mexico, India, Thailand and South Africa—are beginning to innovate within the emerging life-sciences sector through investments in human genomic diversity studies, a logical step following the sequencing of the human genome. We focused on: exploring the motivations for proposing and undertaking large-scale genotyping initiatives; understanding the mechanisms that developing countries are envisioning or implementing to develop genomic medicine appropriate to their own circumstances, and exploring the potential for commercialization of the results of such studies; understanding the challenges faced and how those challenges are being explored; and investigating the emerging potential ethical, legal, social, and cultural issues that have arisen or might arise as a result of these initiatives. Most developing countries, including Mexico, India, South Africa, and Thailand, focus on six major cross-cutting themes: political will, institutional leadership, local health benefits, genomic sovereignty, knowledge-based economy, and challenges.

Together, these themes represents a "taxonomy" of driving forces toward the adoption of genomic medicine in emerging economies, together with the challenges each initiative has subsequently encountered. The South African case study is distinct because although there have been efforts to launch a national large-scale genotyping initiative to catalog baseline variation in the entire South African population, such a project has yet to be implemented. In the final article, we present a perspective on the challenges and opportunities associated with the adoption of genomic medicine, particularly in the developing world, and the need to understand the interdependent nature of efforts to develop genomic medicine. Our findings should be of significance to those emerging and developing economies that are interested in the evolution of genomic studies, particularly in their intersection with public health and as potential sources of economic activity. In this respect the findings are relevant to: leaders of research institutions; politicians, especially in science and technology, in industry and commerce, and in health ministries; legislators considering research funding; individual scientists; investors and small and medium enterprises in the private sector in both industrialized and developing countries; and international organizations interested in how science and technology, and particularly the life sciences, can be used to achieve and accelerate health security and economic development in general.

ACKNOWLEDGMENTS

This project was funded by Genome Canada through the Ontario Genomics Institute. The McLaughlin–Rotman Centre for Global Health, Program on Life Sciences, Ethics and Policy is primarily supported by Genome Canada through the Ontario Genomics Institute, the Ontario Research Fund, and the Bill and Melinda Gates Foundation. Other matching partners are listed at The McLaughlin–Rotman Centre for Global Health web site. A.S.D. and P.A.S. are supported by the McLaughlin Centre for Molecular Medicine. P.A.S. is supported by a Canadian Institutes of Health Research Distinguished Investigator award.

REFERENCES

Daar AS, Singer PA. (2005). Pharmacogenetics and geographical ancestry: implications for drug development and global health. *Nature Rev Genet 6*, 241–246.

Frew SE, et al. (2007). India's health biotech sector at a crossroads. *Nature Biotechnol 25*, 403–417.

Frew SE, et al. (2008). Chinese health biotech and the billion-patient market. *Nature Biotechnol 26*, 37–53.

Masum H, Daar AD, Al-Bader S, Shah R, Singer PA. (2007). Accelerating health product innovation in sub-Saharan Africa. *Innovations: Technology, Governance, Globalization 2*, 129–149.

Séguin B, Hardy BJ, Singer PA, Daar AS. (2008). Genomic medicine and developing countries: creating a room of their own. *Nature Rev Genet 9*, 487–493.

UN Millennium Project 2005 Task Force on Science, Technology, and Innovation. (2005). *Innovation: Applying Knowledge in Development*. Available at: http://www.unmillenniumproject.org/documents/Sciencecomplete. pdf.

World Health Organization. (2002). Genomics and World Health: Report of the Advisory Committee on Health Research. Available at: http://www.who.int/gb/ebwha/pdf_files/EB111/eeb11112.pdf.

4.

THE GENOMIC MAP OF POPULATION DIVERSITY IN ASIA

Dhavendra Kumar

INTRODUCTION

Anthropologists, ethnographers and linguists have long been lured by and struggled to understand and unravel the mysteries surrounding the population origins, route of migration from Africa, and eventually the population structure of a given geographic area. Population diversity— judged by physical characteristics, linguistics and sociocultural patterns—is the norm for indigenous and new migrant populations. Examples in the West include Europe and the north and south American continents. The Asian continent has always been a huge challenge to explore and delineate the patchwork-quilt diversity spread across the continent. For example, Indonesia alone claims some 300 ethnic groups, and the Philippines has 180 native languages and dialects. Large geographic regions like the Chinese and Indian subcontinents harbor peoples of widely differing physical, linguistic, and sociocultural features. Researchers have always been intrigued to answer one question: "Where did they all come from?" Several paradigms of the continental and regional physical map of population diversity now exist, based on massive efforts throughout history that each represents the available investigative methods of the time and period. Among many such methods, geneticists have attracted and enjoyed great confidence and support, as they were able to offer plausible explanations on the population origin, migration, and structure and relationships between differing ethnic groups. In recent years, following the revolutionary advances in genome science and technology, scientists have applied conventional and sophisticated genomic markers to unravel the mysteries of population heterogeneity. Among these, single nucleotide polymorphisms (SNPs) feature high on the list.

The precise mathematical analysis of haploid sets of SNPs was the basis for the international "HapMap project" (International HapMap Consortium, 2005). The multi-institution SNP Consortium has helped to draw the physical maps of population diversity in Europe, North America and Asia. This chapter summarizes the outcome of investigations by the Pan Asian SNP Consortium, led by the Human Genome Organization (HUGO).

THE MIGRATION OF MAN TO ASIA

There are several theories on the migration of man to Asia (Normille, 2009). Several assumptions and hypotheses have produced two basic theories to explain the initial peopling of the Asian continent. The main theory supports the view that there were two major waves of migration from the Middle East. One wave followed southern coastal route, around the rim of the present-day India (see also Chapter 74). This wave continued further east from island to island, eventually reaching the Australasia and the Pacific. This offers a satisfactory explanation for peopling of a number of small to large island nations like Malaysia, Indonesia, the Philippines, and Polynesia. Peoples in these countries evidently share physical, linguistic, and sociocultural characteristics. A separate and distinct wave of population migration took place across the Eurasian subcontinent, and turned south through the Asian mainland (see Figure 4-1). The second theory favors a predominant coastal "out of Africa" route to Southeast Asia (SEA), and eventually reaching East Asia (EA). This then turned north, from SEA and EA respectively, to central and eastern regions of mainland Asia.

The major initiative undertaken by the HUGO Pan Asian SNP Consortium (see section in this chapter) strongly favors the southern coastal route over the northern. It concludes that the southern route made an important contribution to SEA and EA populations. Nevertheless, the contribution made by the northern route is also acknowledged. The population diversity spanning across the Middle East, Eurasia, and Central Asia is better understood by the northern route of "out of Africa" population migration.

THE POPULATION ADMIXTURE IN ASIA

The largest geopolitical continent of the world includes several regions that can be distinguished on the basis of

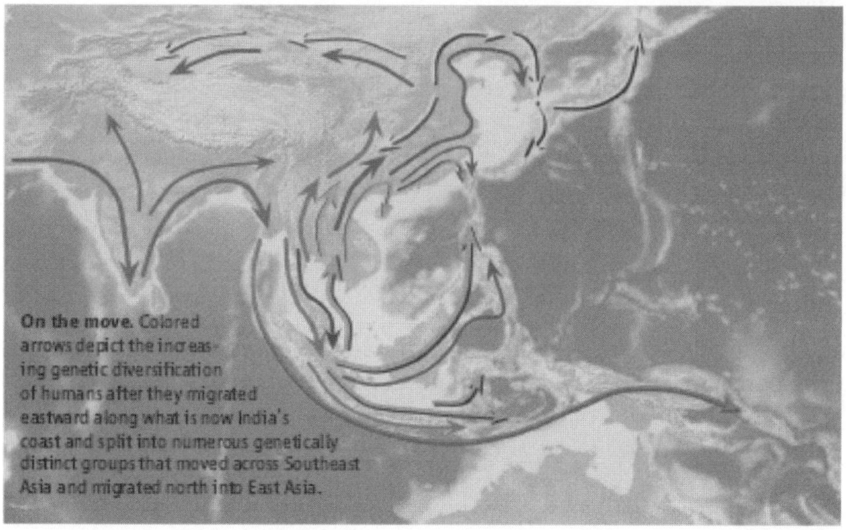

Figure 4-1 The migration of man and peopling in Asia (Adapted with permission from *Science,* December 11, 2009, page 1470). Refer color figure.

physical, linguistic, and sociocultural characteristics of the people. There are several such maps available, and it is beyond the scope of this chapter to elaborate on specific details on peopling of various regions. This section draws attention to the specific region and its people relevant to the genomic variation databases. The interested reader is referred to specific chapters in regional sections in this book for a detailed account.

THE MIDDLE EAST AND THE ARAB WORLD

Geographically, the Arab world and the Middle East span from North Africa in the west to the Indian subcontinent in the east. Many of the most important events during the early evolutionary history of modern humans took place in this vast geographical region. DNA evidence indicates that modern humans originated in East Africa about 100,000–200,000 years ago (Li et al., 2007), then established regional populations throughout the continent. Archaeological artifacts indicate that human inhabitation of today's Maghreb region (modern-day Morocco, Algeria, Tunisia, and Libya) date back to some 82,000 years ago (Taforalt, Morocco). According to the *Recent Out-of-Africa* model, members of one branch of anatomically modern humans left Africa for the Near East some 60,000–70,000 years ago. Phylogenies constructed on the basis of mitochondrial DNA comparisons are indicative for two possible migration routes in this episode of human history. One route laid across Bab-el-Mandeb straits in the Red Sea linking modern-day Eritrea and Djibouti in Africa to Yemen in the Arabian Peninsula (Bailey et al., 2007). This route maintained its important role in influencing gene flow from Africa into the region until nearly 2500 years ago (Cadenas et al., 2008). Another route followed the Nile from East Africa, heading northwards and crossing through Sinai into the Levant between 45,000–50,000 years ago. Human populations in the Near East then branched in several directions, some moving into Europe and others heading east into Asia (Ingman et al., 2000).

Like other population groups, the Arab people share physical, linguistic, and sociocultural characteristics (see Chapter 50). Arab populations encompass a vast geographical region that extends from south of Iran in the east to Morocco in the west, including parts in the southeast of Asia Minor, eastern Africa, and West Africa. However, the political definition of Arab populations is more conservative, as it only includes those populations residing in 23 Arab states, namely: Algeria, Bahrain, Comoros, Djibouti, Egypt, Eritrea, Iraq, Jordan, Kuwait, Lebanon, Libya, Mauritania, Morocco, Oman, Palestine, Qatar, Saudi Arabia, Somalia, Sudan, Syria, Tunisia, United Arab Emirates (UAE), and Yemen.

THE INDIAN SUBCONTINENT AND SOUTHEAST ASIA

The Indian subcontinent (ISC) is a large geopolitical region that encompasses several sovereign nations, including Afghanistan, Pakistan, India, Sri Lanka, Bangladesh, Nepal, Bhutan, and Burma (now Myanmar). The northeastern region overlaps with Tibet, southwest China, and Thailand. The extended map includes Southeast Asian (SEA) countries (Malaysia and Singapore). The whole region is densely populated by diverse peoples with distinct physical, linguistic, and sociocultural characteristics (see also Chapters 73 and 74). In broad terms, major populations groups include Indo-Europeans in the northwest, Indo-Tibetans in the northeast, Tibeto-Burmans in the east, Dravidians in the south, and Indo-Australoids in the central region. The vast plains surrounding the Indus and Ganges rivers are inhabited by a complex admixture of people. Other sections in this book provide details on population origins, migration, and structure of ISC and SEA.

EAST ASIA AND CENTRAL EAST ASIA

East Asia (EA) encompasses China, Mongolia, Korea, and Japan. The fine-detailed phylogenetic status of EA has been achieved by elaborate analyses of the mitochondrial DNA (mtDNA) genomes from 84 populations across China (Kong et al., 2010). To a large extent, the distinction between peoples inhabiting EA and Central East Asia (CEA) is arbitrary. Population groups in this region share physical, linguistic and sociocultural characteristics. The main population groups of CEA include Chinese, Mongols, Hans, Eurasians, and people inhabiting southern Siberia.

Several studies have looked into the complexity of the peopling of the Chinese subcontinent. There is evidence that the earlier hominid populations had been completely replaced by modern humans, as indicated by abundant hominid fossil records depicting a continuity of morphological characteristics between *Homo erectus* and modern East Asians (Wu 2006). This hypothesis is supported by a recent report of the earliest modern human remains (Tianyuan 1) uncovered in Zhoukoudian, China, which shows the mosaic characteristics of both modern and archaic humans and suggests "substantial" gene flow from the archaic hominid populations (Shang et al., 2007). Like in other population studies, both Y-chromosome and mitochondrial DNA studies have helped in unraveling the complexities of EA population origin and admixture. The Y-chromosome study (Ke et al., 2001) has depicted a clear scenario for the recent African origin of modern EA peoples. However, there is evidence to support the debate that there was significant maternal contribution from the archaic hominid female population leading to "completeness of the replacements of modern humans" in the EA region (Hawks, 2001). Large scale mtDNA screening of EA populations has revealed matrilineal complexity in EA (Kong et al., 2010). There is evidence to suggest that uncharacterized mtDNA haplogroups exist in EA, inferring either novel haplogroups or "foreign" lineages introduced by ancient and/or recent migration event(s). Furthermore, these novel basal lineages are sufficiently enriched to contribute to higher genetic diversity in south and southwest China. This observation supports the argument of southern route of ancient population migration from Africa in the late Pleistocene, en route to the Chinese subcontinent.

EURASIA

Different populations inhabit Eurasia, the geographic region of Asia bordering Europe including the Volga-Ural region, Central Asia, and the Caucasus (see Chapter 108). It is common to regard Central Asia and the Volga-Ural region as a crossroads area that witnessed numerous waves of invasions and population migrations in the past. Until the early Medieval Age, the Southern Urals, according to archeological and paleoanthropological data, were inhabited by a Europeoid population that can be associated with the currently living representatives of different divisions of the Europeoid major race (Pontic type of the Indo-Mediterranean race and Suburalic type), which prevails among Volga basin Finnic populations. Ancient populations of the Volga basin could be linked to present-day Baltic and Suburalic types of the Europeoid major race. Alekseev & Gokhman (1984) reviewed paleoanthropological studies and demonstrated that ancient populations of Central Asia from Neolithic times and until Iron Age were predominantly Europeoid. Turkic-speaking nomadic people from southern Siberia and, later, Tataro-Mongol invaders, are considered to be relatively recent newcomers to Eurasia.

GENOMICS, NATURAL SELECTION AND POPULATION DIVERSITY

Population genomics is a rapidly emerging new discipline with roots lying in the field of population genetics. It combines genomic concepts and technologies with the population genetics objective of understanding evolution (Luikart et al., 2003). Population genomics can be broadly defined as the simultaneous study of numerous loci or genome regions to better understand the roles of evolutionary processes (for example, mutation, gene flow, genetic drift, natural selection) that influence variation across genomes and populations. The field of population genomics surveys patterns in the genome within and among different populations to make inferences about evolution and the genome (Nosil & Buerkle, 2010). Commonly, genome scans are used to examine genetic divergence between populations at numerous loci across the genome. The degree of genetic divergence is measured by fixation indices, with larger index values representing greater differentiation between populations. Different genomic regions of the genome (Figure 4-2) are expected to reveal highly variable levels of genetic divergence compared to certain specific genomic regions exhibiting minimal or no divergence (Nosil et al., 2009). This observation is important in studying population divergence across genomic regions. Genomic divergence is expected to be highly heterogeneous, reflecting the process of species formation, evolution, and population differentiation. In evolutionary terms, the genetic differentiation associated with natural selection accumulates in some regions, while the homogenizing effects of gene flow or genetic drift preclude divergence in other regions.

Population genomic analyses require multi-locus datasets from multiple populations, and identify non-neutral or outlier loci by contrasting patterns of population

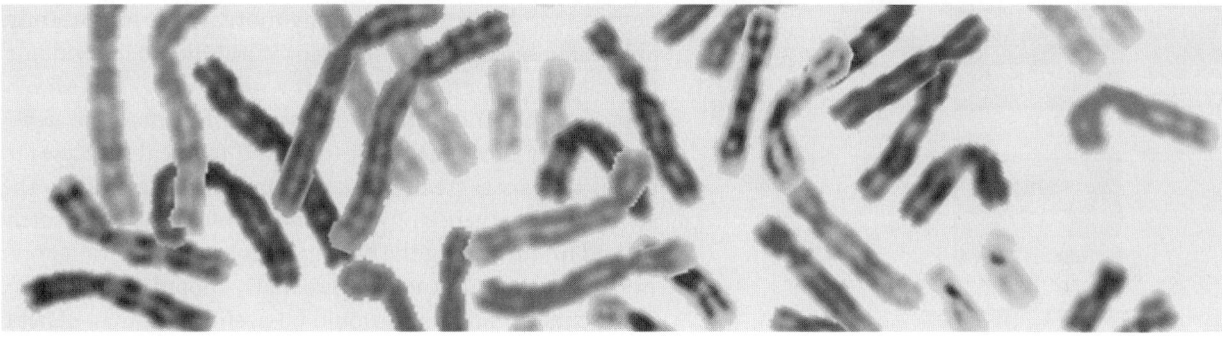

Figure 4-2 Heterogeneous human genomic regions showing minimal or no genetic differentiation. Refer color figure.

divergence among genomic regions (Foll & Gaggiotti, 2008). This approach is often combined with statistical analysis using the Bayesian methods, which provide a probabilistic framework in which to interpret differentiation at individual loci (Hahn, 2006).

As humans expanded into new geographic regions and came to occupy novel environments, our ancestors experienced diverse patterns of natural selection that were recorded in the genomes of divergent populations. For example, humans moved from low elevations to occupy some of the highest plateaus and mountain ranges in the world, including the plateaus of Central Asia and the Andes of South America. These populations exhibit heritable physiological attributes that allow individuals to function at high altitudes (3250–4500 m) with low oxygen concentrations that are challenging to humans from lower elevations. Interestingly, humans from the Tibetan plateau exhibit several physiological attributes that differ from those of Andean highlanders, suggesting that independent evolutionary trajectories led to different adaptations to high altitude. Three studies have used genome scans to identify genes with exceptional allele frequency shifts between populations that were likely targets of divergent natural selection in Tibetan highlanders relative to other human populations (Storz, 2010). Several genes were identified as associated with a history of positive selection in Tibetan highlanders; the gene EPAS1 was identified as one of the most exceptional genes in each study, and was also shown to be associated with presumably adaptive variation in hemoglobin concentration.

A second example of adaptive divergence among human populations is related to the persistence of lactase production in adults. Lactase production in the gut functions in the digestion of the milk sugar lactose, and lactase production in adults is prevalent in humans with ancestry in northern and western Europe, and pastoralist populations in several regions of the world. Adult persistence of lactase is much less common in southern Europe and the Middle East, and rare in nonpastoralist populations in Asia and Africa. Genetic studies have associated adult persistence of lactase with different genes in different populations that exhibit the trait at high frequency, indicating that the trait has arisen independently in multiple populations (Tishkoff et al., 2007). Remarkably, genomic variation surrounding each of the underlying genes is consistent with strong natural selection within the last several thousand years, increasing the frequency of the derived, adaptive alleles (Nosil & Buerkle, 2010).

Recently, evolutionary biologists have used the concept of "genomic islands of divergence" to explain divergence within and between genomic regions (Turner et al., 2005). A genomic island is any gene region—it could be a single nucleotide, copy-number variation, or an entire chromosome—that exhibits significantly greater differentiation than expected under neutrality. This metaphor draws a parallel between genetic differentiation observed along a chromosome, and the topography of oceanic islands and the contiguous sea floor to which they are connected. It is argued that the sea level represents the threshold above which observed differentiation is significantly greater than expected by neutral evolution alone. Thus, a "genomic island of divergence" is composed of both directly selected and tightly linked (potentially neutral) loci.

Population genomics holds great promise for understanding the evolutionary processes affecting genomes of diverse populations. Several different methods and approaches are currently employed in population genomics; however, analyses must be conducted with care and should involve steps that are robust and replicable (Luikart et al., 2003; Figure 4-3). For example, the stochastic nature of next-generation sequencing technologies creates uneven coverage among individuals and genetic regions, which misses data for many individuals and loci, and thus to increased uncertainty in the genotypes of individuals relative to traditional Sanger sequencing. Appropriately modeling and accounting for this uncertainty is important, and preferable to discarding large amounts of sequence data (Gompert et al., 2010). Further advances in molecular and computational biology, and increased computing power itself, will allow more powerful and accurate application of population genomics.

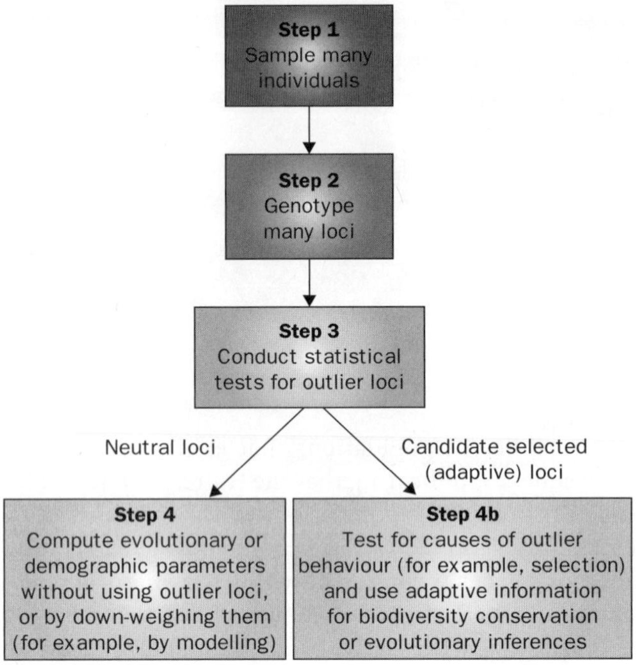

Figure 4-3 The four main steps in the population-genomic analysis (adopted with permission from *Nature Reviews Genetics*, Luikart et al., 2003, vol. 4, pp. 981–987).

THE HUGO PAN-ASIAN SINGLE NUCLEOTIDE POLYMORPHISM (SNP) CONSORTIUM

Single nucleotide polymorphisms (SNPs) are emerging as a critical tool for human genetics in general, and pharmacogenomics in particular. The relevance of common DNA sequence variations in mapping susceptibility loci for common (complex) diseases is now universally acknowledged. Among the many DNA sequence variations, SNPs together constitute a finite collection of genomic variants that underlie complex diseases. There are around 3–4 million SNPs scattered across the human genome, the full set of genetic instructions and regulatory elements encoded in long strands of DNA. The value of SNPs resides in their simplicity, frequency, and distribution throughout the human genome. There are approximately 3 billion nucleotide pairs, also called *base pairs*, 99.9% of which are similar in all humans. A variation in a single base, i.e., SNP, occur roughly every 1000 bases. SNPs are the most common form of genetic variation.

A number of examples now exist where SNPs have been utilized in studying complex disease traits, including ApoE in Alzhiemer's disease, CKR5 in susceptibility to HIV, factor V Leiden in risk for deep venous thrombosis, MTHFR in cardiovascular disease and neural tube defects, p53 in HPV infection, and a number of cytochrome p450s in drug metabolism. It is now established that SNPs should help pinpoint subtle genetic differences that predispose (or protect) some but not others to common diseases such as arthritis, diabetes mellitus, coronary artery disease, cancer, depression, dementia, and many more. More importantly, SNPs offer biological evidence for variability to individuals' favorable or adverse reaction to a given drug. It is widely expected that such knowledge should facilitate the developments of new diagnostic tools and medical therapies to better diagnose, treat, and even prevent illness. Thus the success of the genomic medicine revolution is intricately related to unraveling the biological role of SNPs.

It was soon acknowledged that a population-specific, representative SNP map would be useful in human genetic research in complex disorders directly, via genome-wide association studies (GWAS), and indirectly, by linkage disequilibrium studies. The need for a high-density, high-quality SNP map was recommended by several investigators. It was unanimously agreed that such a map would be made available to academic, government, and independent medical researchers worldwide to enable investigations of genes and variants associated with common and rare diseases. In view of the enormous financial implications and the high quality requirements, a multi-institution collaborative effort was launched in 1998 to create the human SNPs map.

The SNP Consortium (TSC) was established in the spring of 1999 as an innovative collaboration among major pharmaceutical companies, the Wellcome Trust (the world's largest charity supporting bio-medical research), and five leading academic centers (Holden, 2002). The history of TSC goes back to early 1998, when Glaxo Wellcome established a "private entity" to pursue three major genomic activities: (i) identify and map enough SNPs to create a high-quality, genome representative map; (ii) develop alone or in conjunction with other companies a cost-effective technology for genotyping high numbers of SNPs in a reasonable time frame; and (iii) develop "multipoint" software programs to facilitate the analysis of pharmacogenomic studies. The group also agreed to make this entity a "nonprofit" organization. The central concept was to establish TSC as an independent entity and one that would function in the public good. The members of the consortium would not enjoy additional benefits compared to the public. The total initial budget of TSC, of $53 million, was funded by Wellcome Trust ($14 million) and ten other pharmaceutical companies including Bayer Group AG, Bristol-Myers Squibb Company, Glaxo Wellcome PLC, Aventis, Monsanto Company, Novartis AG, Pfizer Inc., Roche Holding Ltd., SmithKline Beecham PLC, and Zeneca Group PLC. Three nonpharmaceutical companies also joined TSC, including Motorola, Amersham Pharmacia Biotech, and IBM.

Objectives of the SNP Consortium are as follows:

1. Identify 300,000 SNPs within two years of launch of the scientific work plan.
2. Map 150,000 of the SNPs over the two-year term of the program.

3. Manage publication of the resulting SNP map in a manner intended to maximize the number of SNPs that enter the public domain.
4. Complete the project for less than $50 million.

The significance of TSC efforts can be judged by its vast scope and costs. Moreover, the group agreed to make this available free for all. There are several achievements of TSC:

a. A high-quality systematic SNP map of the human genome in public domain has facilitated and accelerated basic human genetic research, unraveling the complexity of population origin and structure.
b. It has provided a powerful resource for applied genomic research worldwide in the identification of genetic markers in both common and rare diseases.
c. It has helped in the understanding of disease processes and developing novel diagnostic tools.
d. Discovery and development of new drugs have led the way for personalized therapies for more effective clinical management of the patient.
e. It demonstrates the practicality and benefits of cross-industry research collaboration. This model is acknowledged to be relevant for the life science industry, as the cost and risks of basic research continue to increase.

Since TSC largely concentrated on Caucasian population groups, this was not representative for other populations—specifically, the Asian peoples that comprise almost two-thirds of the world's population. In this context, several Asian countries conceived and developed a genomic program to map genetic variations. It was agreed that a centralized database on human genetic variation was necessary to include Asian population groups. In 2007, the Human Genome Organization (HUGO) led a multicenter initiative to create a physical map of human genetic variation within Asia. A number of investigators from 40 institutions in 11 Asian countries agreed to form the Pan Asian Consortium (Box 4-1). Huge human, technical, and financial resources were invested to achieve the primary objective of the consortium, and its first report was published in 2009 (*Science,* December 11, 2009).

The consortium is working on the second phase of the project that is expected to be released by 2012. This phase intends to resolve the question of the "multi-wave theory" for population migrations across SEA, CEA, and EA. This will be much more ambitious geographically, and is to include mainland central Asia and Pacific islands. It will employ many more genetic markers to map population diversity. The consortium hopes that this project will eventually lead the way forward for genomic applications in health and medicine.

BOX 4-1 THE HUGO PAN ASIAN SNP CONSORTIUM

COUNTRY	INSTITUTION (S)
China	Institute of Medical Biology, Chinese Academy of Medical Science, Kunming; National Engineering Center for Biochip, Shanghai; Chinese National Human Genome Center, Shanghai; Laboratory of Genetic Engineering, Fudan University, Shanghai; Institute of Computational Biology, Shanghai Institutes of Biological Sciences.
Hong Kong	Department of Biochemistry, University of Hong Kong.
India	Institute of Genomics and Integrative Biology, CSIR, Delhi; The Center for Genomic Applications, New Delhi, Indian Statistical Institute, Calcutta, The Indian Genome Variation Consortium.
Indonesia	Eijkman Institute of Molecular Biology, Jakarta.
Japan	Health Sciences University of Hokkaido, Tobetsu; RIKEN Genomic Sciences Center, Yokohama; Centre for Information Biology and DNA Data Bank of Japan, Shizuoka; National Institute of Advance Industrial Science and Technology, Kotoku; Molecular Life Sciences, Tokai University School of Medicine; Department of Medical Genetics, University of Ryukyus Faculty of Medicine, Okinawa; Laboratory of Functional Genomics, University of Tokyo; Research Institute of Personalized Health Sciences, Health Sciences University, Tobetsu.
Korea	Korea National Institute of Health, Seoul; Korea Research Institute of Bioscience and Biotechnology, Deajeon; Seongsil University, Seoul; Eulji University College of Medicine, Deajeon; Department of Human Genetics, University of Tokyo.
Malaysia	Department of Molecular Medicine and Anthropology, University of Malaya, Kuala Lumpur; Human Genome Center, Kelantan; Monash University, Selangor.
Singapore	The Genome Institute of Singapore, Biopolis; Department of Pediatrics, National University Hospital of Singapore; Defense Medical and Environmental Research Institute, Singapore; Informatics Experimental Therapeutic Center, Biopolis; Division of Information Sciences, School of Computer Engineering.

Thailand	Mahidol University, Nakompathom; National Center for Genetic Engineering and Biotechnology, Pathumtani; Department of Biology, Chiang Mai University; National Science and Technology Division, Pathumtani.
Taiwan	Institute of Biomedical Sciences, Academia Sinica, Taepei City.
The Philippines	Institute of Human Genetics, Manila; National Sciences Research Institute, Quezon City.
United States of America	Genomic Laboratories, Affymetrix, Santa Clara, CA; Genomic Research Branch, National Institute of Health, Bethesda, MD.

THE CURRENT GENOMIC MAP OF ASIAN POPULATIONS

The HUGO-Pan Asian SNP Consortium carried out an extensive haplogroup analysis on the data derived from 10 major population categories based on geophysical and linguistic characteristics (Figure 4-4). Previously, similar approaches have led to fine-scale analysis of human genetic diversity among the European populations (Li et al., 2008; Novembre et al., 2008). The consortium targeted Southeast Asian (SEA) and East Asian (EA) populations by using the Affymetrix GeneChip HumanMapping 50K Xba Array. Stringently quality-controlled genotypes were obtained at 54,794 autosomal single-nucleotide polymorphisms (SNPs) in 1928 individuals representing 73 Asian and two non-Asian HapMap populations (International HapMap Consortium, 2003). Apart from developing a general description of Asian population structure and its relation to geography, language, and demographic history, the consortium concentrated on uncovering the geographic source(s) of EA and SEA populations.

The HUGO consortium carried out a Bayesian clustering procedure using the STRUCTURE algorithm (Pritchard et al., 2000) to examine the ancestry of each individual. Each person was posited to derive from an arbitrary number of ancestral populations, denoted by K, assigned K = 2 to K = 14 using both the complete dataset and SNP subsets to exclude those in strong linkage disequilibrium (Figure 4-4). All SEA and EA samples were united at K=2 and K=3 by predominant membership in a common cluster, with the other cluster(s) corresponding largely to Indo-European (IE) and African (AF) ancestries. Negrito populations shared K=4 component with SEA populations, suggesting a common SEA ancestry. Each value of K beyond 4 corresponded with a new component associated with a group of populations united by membership in a linguistic family, by geographic proximity, by a known history of admixture, or by membership in a small population isolate. The results obtained using *frappe* (Tang, Peng and Wang, 2005), a maximum-likelihood– based clustering analysis, showed a general concordance with those of STRUCTURE. These analyses showed that most individuals within a population share very similar ancestry estimates at all Ks, an observation that is consistent with phylogenetically related individuals based on an allele-sharing distance (Mountain & Cavalli-Sforja, 1997). This relationship was also analyzed using the *maximum-likelihood* approach. A maximum-likelihood tree of populations, based on 42,793 SNPs whose ancestral states were known (Figure 4-4), showed that all the SEA and EA populations make up a monophyletic clade that is supported by 100% of bootstrap replicates. This pattern remained even after data from 51 additional populations, and 19,934 commonly typed SNPs from a recent study (Patterson et al., 2006), were integrated into the tree. These observations suggest that SEA and EA populations share a common origin.

Using STRUCTURE/*frappe* and principal components analyses (PCA; Patterson et al., 2006; Figures 4-3 and 4-4), as many as 10 main population components were identified. Each component corresponds largely to one of the five major linguistic groups (Altaic, Sino-Tibetan/Tai-Kadai, Hmong-Mien, Austro-Asiatic, and Austronesian), three ethnic categories (Philippine Negritos, Malaysian Negritos, and East Indonesians/Melanesians) and two small population isolates (the Bidayuh of Borneo and the hunter-gatherer Mlabri population of central and northern Thailand). The STRUCTURE results, population phylogenies (Figure 4-4) and PCA results (Figure 4-5) all show that populations from the same linguistic group tend to cluster together. A Mantel test confirmed the correlation between linguistic and genetic affinities ($R2 = 0.253$; $P < 0.0001$ with 10,000 permutations), even after controlling for geography (partial correlation = 0.136; $P <0.005$with 10,000 permutations). The consortium identified eight population outliers whose linguistic and genetic affinities are inconsistent: Affymetrix-Melanesian (AX-ME), Malaysia-Jehai (MY-JH; Negrito), Malaysia-Kensiu (MY-KS; Negrito), Thailand-Mon (TH-MO), Thailand-Karen (THKA), China-Jinuo (CN-JN), India-Spiti (IN-TB), and China-Uyghur (CN-UG). These linguistic outliers tend to cluster with their geographic neighbors and occupy an intermediate position between their geographic neighbors and the more-distant members of their linguistic group. These patterns are consistent either with substantial recent admixture among the populations (Xu & Jin, 2008, 2009), a history of language replacement, or uncertainties in the linguistic classifications themselves (for example, the controversial Altaic family, which groups Korean and Japanese with Uyghur).

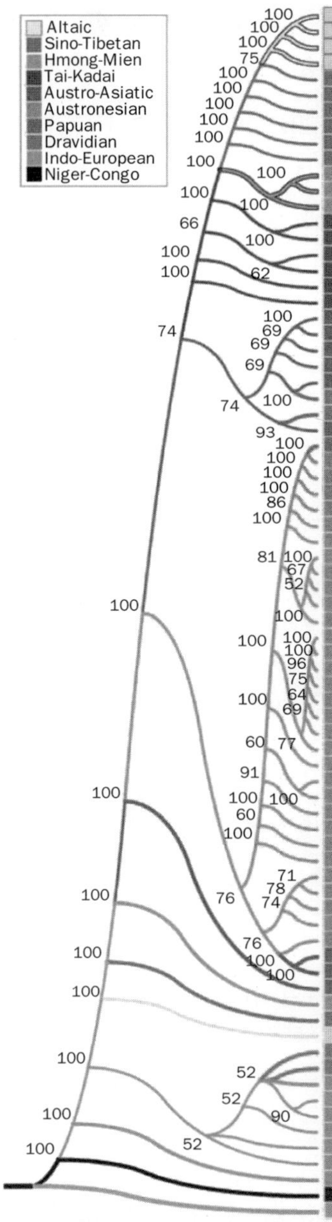

ID	Location	Latitude	Longitude	Ethnicity	Language	size
JP-RK	Japan	26.5	127.9	Ryukyuan	Okinawan	49
JP-ML	Japan	35.7	139.8	Japanese	Japanese	71
JPT	Japan	35.7	139.8	Japanese	Japanese	44
KR-KR	Korea	36.9	127.5	Korean	Korean	90
CHB	China	40.0	116.4	Han	Chinese	45
CN-SH	China	31.2	121.5	Han	Chinese	21
TW-HA	Taiwan	25.0	121.5	Han	MinNan	48
TW-HB	Taiwan	25.0	121.5	Han	Hakka	32
SG-CH	Singapore	1.4	103.8	Han	MinNan	30
CN-GA	China	23.3	113.5	Han	Cantonese	30
CN-HM	China	26.3	108.7	Hmong	Hmong	26
TH-HM	Thailand	18.6	98.1	Hmong	Hmong	20
TH-YA	Thailand	20.0	100.2	Yao	Iu-Mien	19
CN-CC	China	24.4	110.2	Zhuang	Zhuang	26
CN-JI	China	18.9	109.8	Jiamao	Jiamao	31
TH-TL	Thailand	19.2	100.9	Tai Lue	Lue	20
TH-TY	Thailand	18.4	98.9	Tai Yong	Tai Yong	18
TH-TK	Thailand	18.6	98.9	Tai Kern	Tai Kern	18
TH-TU	Thailand	19.0	99.0	Tai Yuan	Tai Yuan	20
TH-MA	Thailand	18.7	100.5	Mlabri	Mlabri	18
TH-TN	Thailand	19.1	100.9	H'Tin	Mal	18
TH-PP	Thailand	20.4	99.9	Plang	Blang	18
CN-WA	China	22.8	100.2	Wa	Wa	56
TH-LW	Thailand	18.4	98.1	Lawa	Lawa	19
TH-KA	Thailand	18.0	98.4	Karen	Karen	20
CN-JN	China	22.0	101.0	Jinuo	Jinuo	29
TH-PL	Thailand	19.9	99.2	Palong	Palong	18
AX-ME	Pacific	-5.8	155.1	Melanesian	Nasioi	5
ID-AL	Indonesia	-8.3	124.7	Alorese	Alor	19
ID-LE	Indonesia	-8.3	124.7	Lembata	Lembata	19
ID-LA	Indonesia	-8.3	123.0	Lamaholot	Lamaholot	20
ID-SO	Indonesia	-8.6	120.1	Manggarai	Manggarai	19
ID-RA	Indonesia	-8.7	120.5	Manggarai	Manggarai	17
ID-SB	Indonesia	-9.8	120.0	Kambera	Kambera	20
PI-AG	Philippines	13.7	123.3	Negrito	Agta	8
PI-AE	Philippines	14.9	120.2	Negrito	Aeta	8
PI-MW	Philippines	9.7	125.6	Negrito	Mamanwa	19
PI-IR	Philippines	13.0	121.1	Negrito	Iraya	9
PI-AT	Philippines	11.9	122.0	Negrito	Ati	23
AX-AM	Taiwan	23.7	121.4	Ami	Ami	10
AX-AT	Taiwan	24.6	121.4	Atayal	Atayal	10
PI-UB	Philippines	17.2	121.9	Urban	Ilocano	20
PI-UN	Philippines	14.6	121.0	Urban	Tagalog	19
PI-UI	Philippines	6.9	122.1	Urban	Visaya	20
PI-MA	Philippines	8.2	125.9	Manobo	Manobo	18
ID-MT	Indonesia	-0.3	98.4	Mentawai	Mentawai	15
ID-TR	Indonesia	-4.7	119.7	Toraja	Toraja	20
ID-ML	Indonesia	-3.0	104.7	Malay	Malay	12
ID-KR	Indonesia	1.5	100.0	Batak Karo	Batak Karo	17
ID-TB	Indonesia	2.3	99.1	Batak	Batak Toba	20
ID-DY	Indonesia	1.2	116.7	Dayak	Benuak	12
MY-MN	Malaysia	2.8	102.2	Malay	Minangkabau	20
SG-MY	Singapore	1.4	103.8	Malay	Malay	30
MY-KN	Malaysia	5.3	102.0	Malay	Malay	18
ID-JA	Indonesia	-6.2	106.7	Javanese	Javanese	34
ID-JV	Indonesia	-7.3	110.4	Javanese	Javanese	19
ID-SU	Indonesia	-6.2	106.7	Sundanese	Sunda	25
MY-BD	Malaysia	1.4	110.2	Bidayuh	Jagoi	50
MY-TM	Malaysia	2.9	102.1	Proto-Malay	Temuan	49
MY-JH	Malaysia	5.4	101.1	Negrito	Jehai	50
MY-KS	Malaysia	5.7	100.9	Negrito	Kensiu	30
TH-MO	Thailand	18.5	98.9	Mon	Mon	19
IN-NI	India	30.4	79.2	Tharu	Pahari	20
IN-TB	India	34.7	76.5	Ladakhi	Spiti	23
CN-UG	China	37.1	86.6	Uyghur	Uyghur	26
IN-DR	India	15.3	77.8	Upper-caste	Telugu	24
SG-ID	Singapore	1.4	103.8	India origin	Tamil	30
IN-WI	India	26.7	74.0	Bhil	Bhili	25
IN-EL	India	23.0	88.2	Upper-caste	Bengali	16
IN-SP	India	29.1	76.5	Upper-caste	Hindi	23
IN-WL	India	19.7	75.9	Upper-caste	Marathi	14
IN-NL	India	26.8	81.4	Upper-caste	Hindi	15
IN-IL	India	26.7	74.0	Upper-caste	Hindi	15
CEU	USA	51.5	-0.1	European	English	60
YRI	Nigeria	7.4	3.9	Yoruba	Yoruba	60

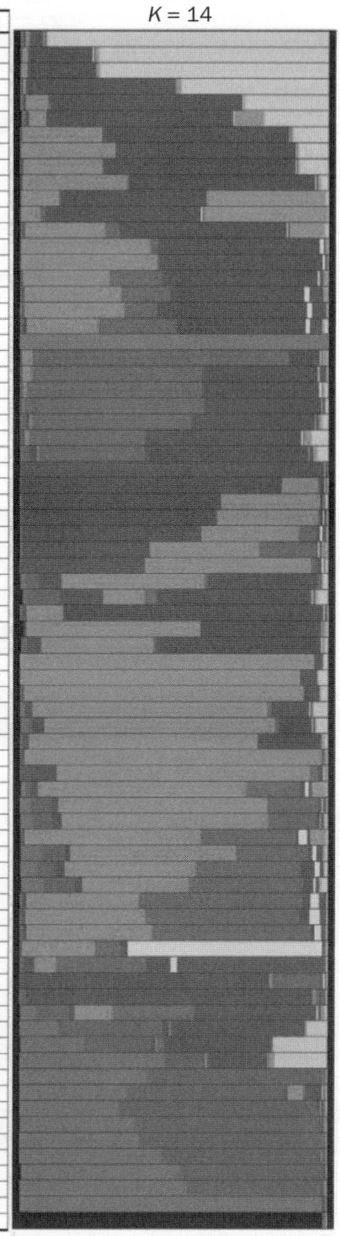

Figure 4-4 The geophysical and linguistic categories of the Asian populations analyzed by the HUGO Pan Asian SNP Consortium (adapted with permission from *Science*, September 17, 2009, page 1542). All population IDs except the four HapMap samples are denoted by four characters. The first two letters indicate the country where the samples were collected or (in the case of Affymetrix) genotyped, according to the following convention: AX, Affymetrix; CN, China; ID, Indonesia; IN, India; JP, Japan; KR, Korea; MY, Malaysia; PI, the Philippines; SG, Singapore; TH, Thailand; and TW, Taiwan. The last two letters are unique IDs for the population. To the right of the table, an averaged graph of results from STRUCTURE is shown for K = 14.

Considerable gene flow among Asian populations was observed among subpopulations in these clusters, including those groups believed to practice endogamy based on linguistic, cultural, and ethnic information. In fact, most populations studied, even at lower Ks, showed evidence of admixture in the STRUCTURE analyses. For example, the Han Chinese have grown to become the largest ethnic group today in a demographic expansion that has occurred mostly within historical times. STRUCTURE revealed that the six Han Chinese population samples showed varying degrees of admixture between a northern Altaic cluster and a Sino-Tibetan/Tai-Kadai cluster, which most frequently appears in the ethnic groups sampled from southern China and northern Thailand. Finally, most of the Indian populations showed evidence of shared ancestry with European populations, which is consistent with recent observations (Indian Genome Variation Consortium, 2008) and our understanding of

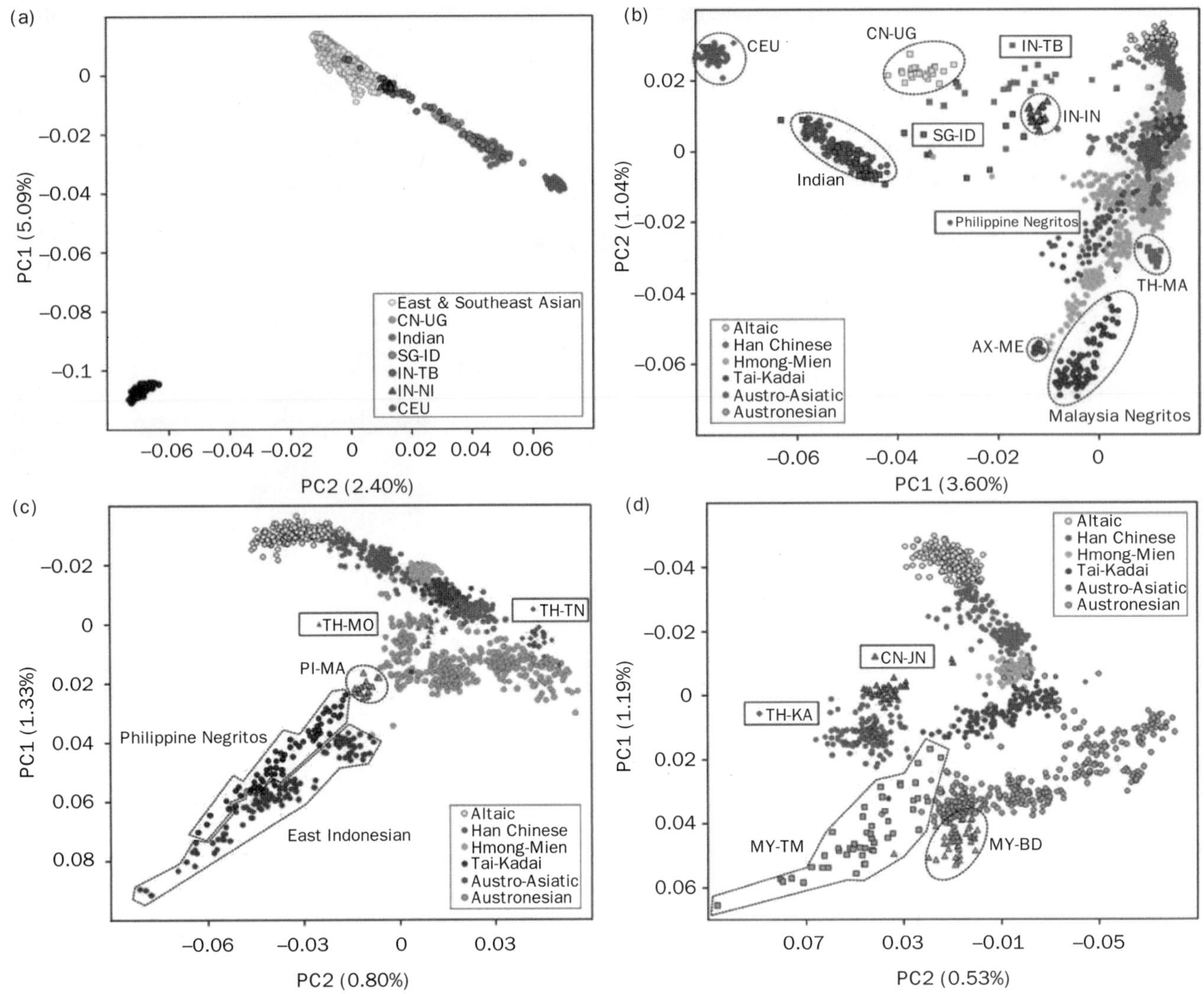

Figure 4-5 The population admixture of Asian population groups based on the principal component analysis (adapted with permission from *Science,* September 17, 2009, page 1543). Refer color figure.

the expansion of Indo-European speaking populations (Figure 4-4).

The geographic source(s) contributing to EA populations have long been debated. One hypothesis suggests that all SEA and EA populations derive primarily from a single initial migration, which entered the continent along a southern, largely coastal route (Su et al., 1999). Another hypothesis argues for at least two independent migrations into East Asia, first along a southern route, followed later by a series of migrations along a more northern route that served to bridge European and EA populations, but with little contribution to populations in Southeast Asia (Su et al., 1999). The topology of a maximum-likelihood tree (Figure 4-4) displays a largely south-to-north ordering of the populations, and a plot of the first two PCs (Figure 4-5) similarly places most populations according to their geographic coordinates. The average value of the first PC is highly correlated with the latitude at which the populations were sampled ($R2 = 0.79$, $P < 0.0001$). Such a pattern could result simply from isolation by distance (IBD), as suggested by Ding et al. (2000), although a recent study failed to detect IBD in East Asia with data from the Human Genome Diversity Project (Manica et al., 2005).

In an effort to distinguish between long-term historical divergence and the effects of IBD, the HUGO Consortium applied partial and multiple Mantel tests to the data (Telles & Diniz-Filho, 2005). The primary approach was to ascertain the differential correlation between genetic distance, geographical distance, and a group indicator matrix as an indication of prehistoric population divergence. The superior association between genetic distance and the group indicator matrix, as measured by the correlation coefficients, suggests that prehistorical population divergence is the favored model over IBD in explaining the data (Cavalli-Sforja et al., 1993).

This conclusion is supported by simulation studies that also suggest that the observed patterns cannot be explained by simple IBD effects alone.

The consortium looked in fine detail at haplotype organization to limit the effect of fluctuations in single-nucleotide determinations, and to increase the resolution around genetic diversity. The IBD model predicts a correlation of genetic distance with geographical distance, but not genetic diversity and geographic distance (Cavalli-Sforja et al., 1993). Haplotype diversity is strongly correlated with latitude ($R2 = 0.91$, $P < 0.0001$), with diversity decreasing from south to north, which is consistent with a loss of diversity as populations moved to higher latitudes (Figure 4-6A). Furthermore, the consortium found that more than 90% of haplotypes in EA populations could be found in SEA and CSA populations, of which about 50% were found in SEA and EA only, and 5% found in CSA only (Figure 4-6B). Phylogenetic analysis of private haplotypes indicates greater similarity between EA and SEA populations relative to EA and CSA populations (Figure 4-6C). These observations suggest that the geographic source(s) contributing to EA populations were mainly from SEA populations, with rather minor contributions from CSA, and that this clinal structure of EA populations arose from prehistoric population divergence rather than IBD or gene flow from CSA populations.

On the basis of increased cultural, linguistic, and genetic diversity, the origins of SEA populations are thought to be more complex than the origins of those to their north. Notably, the Negritos of the Philippines and Malaysia differ from neighboring populations in aspects of their physical appearance, prompting intense speculation about models of human settlement in Southeast Asia. The two-wave hypothesis, which suggests that ancestral Negrito populations settled in Southeast Asia, Australia, and Oceania before a more northerly migration originating in or near the Middle East, and spreading both toward Europe and Northeast Asia via Central Asia, has been supported by phylogenetic trees constructed from data on a limited number of protein markers (Cavalli-Sforja & Feldman, 2003).

The topology of our population trees, both with and without the data from additional European and Asian populations discussed in, is inconsistent with regard to this genetic similarity of European and EA populations (Figures 4-4 and 4-6D). Instead, on the basis of variation at a large number of independent SNPs, there is substantial

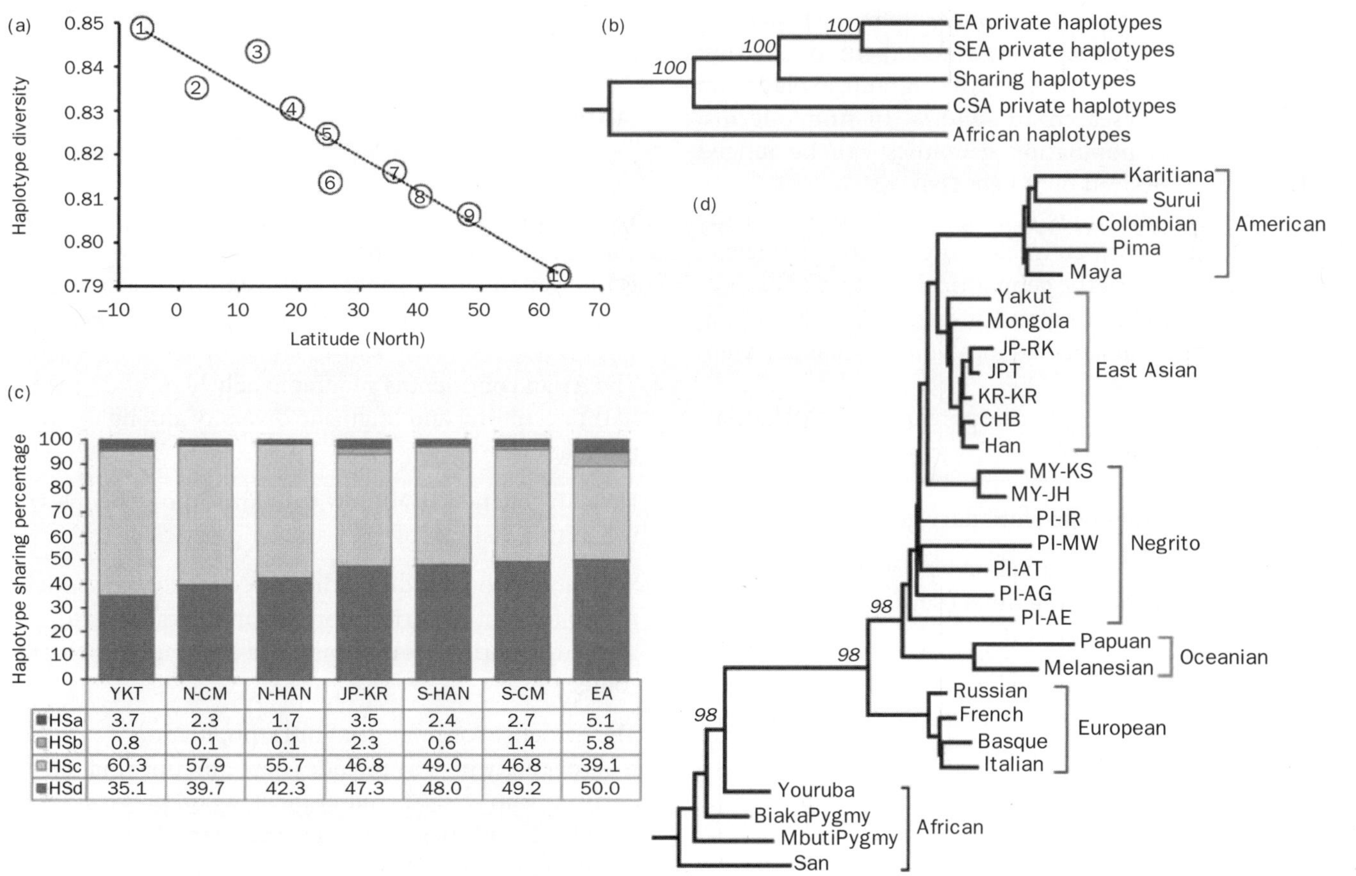

Figure 4-6 The population phylogeny in Asia based on haplotype analysis (adopted with permission from *Science*, December 11, 2009, page 1544).

genetic proximity of SEA and EA populations. An identical pattern is seen in the population tree based on all of their 642,690 SNPs (Li et al., 2008) Our forward-time simulation results under extreme ascertainment scenarios show that the observed phylogeny is not the result of ascertainment bias. Simulation studies also suggest that substantial levels of migration between populations after their initial separation are unlikely to distort the topology of the phylogeny. To unambiguously infer population histories represents a considerable challenge (Hellenthal et al., 2008). Although the HUGO Consortium does not disprove a two-wave model of migration, the evidence from the autosomal data and the accompanying simulation studies point toward a history that unites the Negrito and non-Negrito populations of Southeast and East Asia via a single primary wave of entry of humans into the continent.

THE LONG-TERM APPLICATIONS OF POPULATION GENOMICS

While the importance of anthropological and human genetic studies on population origin, migration, and admixture is not challenged, the question of its relevance to economic and health applications has been raised from all quarters. Of course, employing expensive and sophisticated genomic technology in population studies does require careful thinking and explanation to justify future research and efforts. In simple terms, the relevance of population genomics can be judged against several desired outcomes that may overlap, with far-reaching implications (Luikart et al., 2003). In the past, such studies helped in charting out the population admixture in a strictly geopolitical context. This has been helpful in planning and meeting the educational and social requirements at local and to some extent national levels.

Population genomics has the potential for applications in a number of areas:

- To improve studies of evolutionary genetics.
- Important tool for conservation biology through comparative genome mapping of several living creatures.
- The genomics model applied to population genetics may assist in improving inferences about population demography and evolutionary history.
- Improving the estimation of important parameters such as population size, migration rates, and phylogenetic relationships.
- Identification of adaptive molecular variation using the population genomic approach.
- Enhance understanding of molecular ecology with implications on nutrition and gene-environment interactions, for example nutrigenomics and toxicogenomics.
- Facilitating the identification population-level susceptibility to and prevention of microbial infections and parasitic infestations.
- Population-specific genomic variation databases have an important role in the genetic epidemiology of complex disorders.
- Population genomic variation is essential for developing targeted array, high-throughput diagnostic tools for screening unexplained developmental delay and certain common disorders.
- Population genomic databases appear to be important in the new drug discovery and development in assessing the favorable or adverse drug response.

Finally, with the rapid advances in genomic technologies and bioinformatics, complete personal genomic sequencing would become affordable and feasible. A logical explanation and interpretation of an individual's genome would require population and ethnic specific genomic variation database applicable to the individual concerned (Venter, 2010).

CONCLUSIONS

This chapter provides an overview on the current knowledge and understanding on the genomic perspectives of population origins, population diversity and population-specific genetic divergence and genomic variation. A number of observations can be made:

- The Asian continent is geographically huge and harbors cultural and linguistic diversity among many populations.
- Human genetic variation within and among peoples inhabiting Asia is complex and enigmatic.
- Several previous studies indicate broadly two essential routes for "out of Africa" population migration into Asia—the northern land route and the southern coastal route.
- The evidence for inter- and intra-population heterogeneity in Asia includes data and information derived from several studies focusing on physical, linguistic, and sociocultural aspects, and human genetic analysis.
- Human genetic data and information drawn from several regional populations exist, but do not help in

assembling a comprehensive physical map of human genetic diversity.

- The HUGO Pan Asian SNP Consortium, a multicenter major undertaking by 40 institutions from 11 Asian countries, provides the most up-to-date physical map of human genetic variation in Southeast Asian (SEA), East Asian (EA) and Central South Asian (CSA) populations using the Affymatrix GeneChip Human Mapping 50K Xba Array.
- The findings of the consortium indicate genetic proximity of EA populations with SEA and CSA peoples.
- Autosomal SNP haplotype analysis and accompanying simulation studies unite Negrito and non-Negrito populations of SEA, CSA and EA, indicating a single primary wave of entry with SEA being a major source for EA populations.

ACKNOWLEDGMENTS

This chapter is based on a number of original research papers and review articles cited in references. The author is indebted to HUGO, several prominent investigators and leaders in the field of population genetics, molecular evolution, medical genomics and public/ population health who generated author's interest and understanding of the complexities and challenges of population genomics in the context of health and medicine.

REFERENCES

Alekseev VP, & Gokhman II. (1984). *Antropologiya azi-atskoi chasti SSSR* (Anthropology of the Asian Part of the USSR). Moscow: Nauka.

Bailey GN, Flemming NC, King GCP, et al. (2007). *Coastlines, submerged landscapes, and human evolution: the Red Sea Basin and the Farasan Islands. J Island Coast Archaeol, 2*(2), 127–160.

Cadenas AM, Zhivotovsky LA, Cavalli-Sforza LL et al. (2008). Y-chromosome diversity characterizes the Gulf of Oman. *Eur J Hum Genet 16*, 374–386; doi:10.1038/sj.ejhg.5201934; published online 10 October 2007.

Cavalli-Sforza LL, Menozzi P, & Piazza A. (1993). *The History and Geography of Human Genes.* Princeton, NJ: Princeton Univ. Press.

Cavalli-Sforza LL & Feldman MW. (2003). The application of molecular genetic approaches to the study of human evolution. *Nat Genet* 33, 266–275.

Chu JY et al. (1998). Genetic relationship of populations in China. *Proc Natl Acad Sci USA 95*, 11763.

Ding YC et al. (2000). Population structure and history in East Asia. *Proc Natl Acad Sci USA 97*, 14003.

Foll M & Gaggiotti O. (2008). A genome scan method to identify selected loci appropriate for both dominant and codominant markers: a Bayesian perspective. *Genetics* 180, 977–993.

Gompert Z. *et al.* (2010). Bayesian analysis of molecular variance in pyrosequences quantifies population genetic structure across the genome of Lycaeides butterflies. *Mol Ecol 19*, 2455–2473.

Hahn MW. (2006). Proceedings of the SMBE Tri-National Young Investigators' Workshop 2005: Accurate Inference and Estimation in Population Genomics. *Mol Biol Evol 23*(5), 911–918.

Hawks J. (2001). The Y chromosome and the replacement hypothesis. *Science 293*, 567.

Hellenthal G, Auton A, Falush D. (2008). Inferring human colonization history using a copying model. *PLoS Genet* 4, e1000078.

Holden AL. (2002). The SNP Consortium: Summary of a private consortium effort to develop an applied map of the human genome. *BioTechniques 32,* S22–S26.

Indian Genome Variation Consortium. (2008). Genetic landscape of the people of India: a canvas for disease gene exploration. *J Genet* 87, 3.

Ingman M, Kaessmann H, Pääbo S, Gyllensten U. (2000). Mitochondrial genome variation and the origin of modern humans. *Nature 408*(6813), 708–713.

International HapMap Consortium. (2005). A haplotype map of the human genome. *Nature 437*, 1299–1320.

Ke Y, Su B, Song X, et al. (2001). African origin of modern humans in East Asia: a tale of 12,000 Y chromosomes. *Science* 292, 1151–1153.

Kong QP, et al. (2010). Large-Scale mtDNA screening reveals a surprising matrilineal complexity in East Asia and its implications to the peopling of the Region. *Mol Biol Evol 28*(1), 513–522.

Kumar S, et al. (2009). Reconstructing Indian–Australian phylogenetic link. *BMC Evol Bio* 9, 173.

Li JZ. et al. (2008). Worldwide human relationships inferred from genome-wide patterns of variation. *Science* 319, 1100.

Li H, Cai X, Winograd-Cort ER, et al. (2007). Mitochondrial DNA diversity and population differentiation in southern East Asia. *Am J Phys Anthropol* 134, 481–488.

Luikart et al. (2003). The power and promise of population genomics. *Nat Rev Genet* 4, 981–987.

Manica A, Prugnolle F, Balloux F. (2005). Geography is a better determinant of human genetic differentiation than ethnicity. *Hum Genet* 118, 366.

Mountain JL, Cavalli-Sforza LL. (1997). Multilocus genotypes, a tree of individuals, and human evolutionary history. *Am J Hum Genet* 61, 705.

Normille D. (2009). SNP study supports southern migration route to Asia. *Science* 326, 1470.

Nosil P, & Buerkle A. (2010). Population genomics. *Nat Educ Knowledge 1*(11), 8.

Nosil P. et al. (2009). Divergent selection and heterogeneous genomic divergence. *Mol Ecol 18*, 375–402.

Novembre J. et al. (2008). Genes mirror geography within Europe. *Nature* 456, 98.

Patterson N, Price AL, Reich D. (2006). Population structure and Eigen analysis. *PLoS Genet* 2, e190.

Pritchard JK, Stephens M, Donnelly P. (2000). Inference of population structure using multilocus genotype data. *Genetics 155*, 945.

Reid L. (1994). Unravelling the linguistic histories of Philippine negritos. In Dutton T, & Tryon T. (eds.), *Language Contact and Change in the Austronesian World* (pp. 443–475). Berlin: Mouton de Gruyter.

Shang H, Tong H, Zhang S, Chen F, Trinkaus E. (2007). An early modern human from Tianyuan Cave, Zhoukoudian, China. *Proc Natl Acad Sci USA 104*, 6573–6578.

Storz JF. (2010). Genes for high altitudes. *Science 329*, 40–41.

Su B. et al. (1999). Y-Chromosome evidence for a northward migration of modern humans into Eastern Asia during the last Ice Age. *Am J Hum Genet* 65, 1718.

Tang H, Peng J, Wang P, Risch NJ. (2005). Estimation of individual admixture: analytical and study design considerations. *Genet Epidemiol* 28, 289.

Telles MP, Diniz-Filho JA. (2005). Multiple Mantel tests and isolation-by-distance, taking into account long-term historical divergence. *Genet Mol Res* 4, 742.

The HUGO Pan-Asian SNP Consortium. (2009). Mapping human genetic diversity in Asia. *Science* 326, 1541–1545.

The International HapMap Consortium. (2003). *Nature* 426, 789.

Tishkoff SA. et al. (2007). Convergent adaptation of human lactase persistence in Africa and in Europe. *Nat Genet* 39, 31–40.

Turner TL. et al. (2005). Genomic islands of speciation in *Anopheles gambiae. Publ Library Sci Biol 3*, 1572–1578.

Venter JC. (2010). Multiple personal genomes await. *Nature* 464, 676–677.

Wu X-Z. (2006). New arguments on continuity of human evolution in China. *Acta Anthropol Sin 25*, 17–25.

Xu S. & Jin L. (2008). A genome-wide analysis of admixture in Uyghurs and a high-density admixture map for disease-gene discovery. *Am J Hum Genet 83*, 322.

Xu S, Huang W, Qian J, Jin L. (2008). Analysis of genomic admixture in Uyghur and its implication in mapping strategy. *Am J Hum Genet 82*, 883.

Xu S, Jin W, Jin L. (2009). Haplotype-sharing analysis showing Uyghurs are unlikely genetic donors. *Mol Biol Evol* 26, 2197.

5.

ANCESTRY, DISEASE AND VARIABLE DRUG RESPONSE IN THE GENOMIC ERA

Charles Rotimi and Lynn Jorde

INTRODUCTION

Human genetic data are accumulating at an ever-increasing pace, and whole-genome sequences of individuals from multiple populations are now publicly available (Dramanac et al., 2010; Yngvodottir et al., 2009; Schuster et al., 2010). The growing inventory of human genetic variation is facilitating an understanding of why susceptibility to common diseases varies among individuals and populations.

In addition, we are gaining insights that may improve the efficacy and safety of therapeutic drugs. Such knowledge is relevant to fundamental questions about our origins, differences, and similarities. Here, we provide a brief review of the current knowledge of human genetic variation and how it contributes to our understanding of human evolutionary history, group identity, and health disparities.

GENETIC VARIATION IN INDIVIDUALS AND POPULATIONS

Human genetic variation contributes substantially to the physical variation we observe among individuals. The two most important components of human genetic variation are single-nucleotide polymorphisms (SNPs) and copy-number variants. SNPs are single DNA nucleotide base pairs that differ among individual DNA sequences. (For example, one individual will have an A-T at a specific position in the haploid DNA sequence, whereas another will have a C-G.) Copy-number variants are larger contiguous blocks of DNA sequence (usually exceeding 1000 bp) that vary in copy number among individuals. (For example, a block of DNA could be duplicated in one person but deleted in another.) Millions of SNPs have been discovered in humans, and at the single-nucleotide level any two humans differ at about 1 in 1000 bp (Sachidanandan et al., 2001). Copy-number variants are less common, but because they comprise much larger segments of DNA, they contribute an additional 0.4% difference in DNA sequence between any two individuals (Redon et al., 2006; Sebat et al., 2004). Any two humans would be about 99.5% identical in DNA sequence, reflecting a relatively recent, common origin of our species.

Humans have about half the amount of genetic variation that is observed in central African chimpanzees (Chimpanzee Sequencing and Analysis Consortium, 2005) and gorillas (Yu et al., 2004), and about one-tenth the amount of variation seen in the fruit fly Drosophila pseudoobscura (Li and Sadler, 1991). The fact that we have less variation than many gorilla and chimpanzee populations, despite our much larger current population size, reflects prehistoric bottlenecks in human population size (Amos & Hoffman, 2009). Some mammals, such as cheetahs, have also undergone population bottlenecks and have even less genetic diversity than humans (Menotti-Raymond & O'Brien, 1993).

Most common SNPs (i.e., those for which the prevalence of the less common allele is greater than 5%) are shared among populations from different continents (Xing et al., 2009). This commonality reflects continued migration and gene flow among human populations throughout history, in addition to our recent common origin. Many studies have shown that the great majority of genetic variation (around 85% to 90%) can be found within any human population (e.g., samples of persons from Great Britain and from Ghana); only an additional 10% to 15% of variation is gained when considering the entire human population (Li et al., 2008; Jorde et al., 2000). This is another measure of the genetic similarity among persons of different ancestries.

Occasionally, an SNP or a copy-number variant is relatively common in one population but absent (or nearly so) in another. This is sometimes due to the recent emergence of a variant that has not yet had time to spread to other populations—such as a SNP that causes hereditary hemochromatosis and is common in Europe but very rare elsewhere (Merryweather-Clarke et al., 1997). In other cases, the difference in prevalence is caused by natural selection in a specific local environment. For example, hereditary lactase persistence is unusual in most human populations, but it is prevalent in European and African pastoral populations, where milk consumption beyond childhood has had a selective advantage (Tishkoff et al., 2007). Convincing evidence of positive selection has also been seen in genes that affect skin pigmentation (Sturm, 2009), resistance to malaria (Tishkoff et al., 2001; Ayodo et al., 2007), and resistance to Lassa fever (Sabeti et al., 2007).

In comparisons of the prevalence of an SNP or copy-number variant among populations, a strong correlation between geographic location and genetic similarity is generally observed (Xing et al., 2009; Li et al., 2008; Jakobsson et al., 2008). This is hardly surprising, because populations that are geographically close to one another are likely to share a common ancestry and to exchange migrants. Population-level studies have also shown that African populations have relatively greater genetic diversity than other populations, and that the variation found outside Africa tends to be a subgroup of African genetic variation (Conrad et al., 2006). These patterns, which are observed even when comparing a relatively small number of SNPs or copy-number variants (100 or so), support a common African origin of our species (Fagundes et al., 2007).

If larger numbers of polymorphisms are analyzed, individual ancestry can be approximately inferred. Early studies succeeded in correctly classifying most sampled individuals into their continent of origin (Rosenberg et al., 2002; Bamshad et al., 2003; Tang et al., 2005). More recently, by analyzing up to 1 million SNPs at a time, investigators have been able to infer individual ancestry more precisely, often assigning ancestry to specific geographic regions (Xing et al., 2009; Novembre et al., 2008; Reich et al., 2009). Because the prevalence of most SNPs differs by only a little among populations, fine-scaled inferences about ancestry are possible only when large numbers of SNPs are assayed or when SNPs that are already known to show large frequency differences among populations are selected (Kosoy et al., 2009). Few if any SNP variants are present in all members of one population but absent in all members of another. Thus, it is possible to make approximate conclusions about an individual's ancestry from a panel of SNPs (Bolnick et al., 2007; Lee et al., 2009; Royal et al., 2010), but it is not possible to determine with any certainty an individual SNP genotype on the basis of one's ancestry (Jorde & Wooding 2004).

Although individual ancestry can be inferred approximately from genetic data, our growing understanding of these data shows that population categories are not discrete and readily discernible entities. Many populations, such as African Americans and Hispanic Americans, have complex recent ancestral histories. African Americans, on average, are estimated to have approximately 20% European ancestry (Tishkoff et al., 2009), but this proportion varies substantially among different African-American populations within North America (Parra et al., 1998; Halder et al., 2009; Bryc et al., 2010). More important, genetic analyses of individual ancestry show that some self-identified African Americans have large proportions (more than 50%) of European genetic ancestry, whereas some self-identified European Americans have substantial recent African genetic ancestry (Bryc et al., 2010; Shriver et al., 2003).

Moreover, perception of population differences partly depends on the human populations that have been genetically sampled. The current extent of population sampling is, at best, incomplete. The greater the number of human populations genotyped and the smaller the average level of between-population variation (Watkins et al., 2003), the more difficult it is to delineate population boundaries. Definitions of populations or ethnic groups are complex and sometimes overlap or are mutually contradictory, incorporating cultural, linguistic, biologic, and geopolitical factors (Census Bureau, 2000). These definitions vary within countries, as they have throughout time, with racial or ethnic categories having changed every 10 years in the U.S. Census (Census Bureau, 2000).

Population categories, including race and ethnic group, are clearly inadequate to describe fully the pattern and range of variation among individuals. A more accurate assessment of disease risk may be obtained by genotyping disease-associated genetic variants in individuals, rather than using population affiliation as a surrogate.

GENETIC VARIATION AND DISEASE DISPARITIES

Disease prevalence, severity, and resistance vary considerably among ethnic groups as a consequence of inherited factors and noninherited causes, such as poverty, unequal access to care, lifestyle, and health-related cultural practices (Smedley et al, 2003; Sankar et al., 2004; Molnar, 2001). Well-known examples include the elevated prevalence of Tay-Sachs disease among persons of Ashkenazi Jewish ancestry (Molnar, 2001) and sickle cell disease, thalassemia, and glucose-6-phosphate dehydrogenase deficiency in some populations of African descent (Tishkoff et al., 2001; Molnar, 2001; Luzzatto, 2006; Braun, 2002). However, Tay-Sachs disease is observed in non-Jewish populations and has a relatively high prevalence in parts of French-speaking Canada (Scriver, 2001). Similarly, hemoglobinopathies are found in many global populations (Molnar, 2001). Labeling of these diseases as occurring only in black populations can reduce the likelihood of diagnosis in other populations.

Because diseases vary in prevalence among populations, it is to be expected that the frequencies of genetic variants that contribute to their causation also vary. A comparison of the prevalence of risk variants associated with susceptibility to 26 diseases in 11 populations from across the world in the International HapMap Project showed significant differences in the prevalence of such variants among populations (Adeyemo & Rotimi 2010). Some variants showed a difference in prevalence by factors of 20 to 40 (Table 5-1). This variation is important in the context of genome wide association studies, in which the SNPs in the genomes of thousands of affected case subjects and unaffected control subjects are genotyped, with the aim of identifying SNPs that are associated with disease - and ultimately, identifying variants that directly confer susceptibility to disease. Such studies have identified more than 800 unique SNP-trait associations, significant at P-8, that contribute to common diseases, such as age-related macular degeneration, diabetes, and Crohn's disease (Hindorff et al., 2009). However, nearly 90% of all genome-wide association studies have been carried out in populations of European ancestry. Remarkably, only one genome-wide association study (the MalariaGen Project) has involved a large collection of samples from African subjects (Jallow et al., 2009), and only a handful have involved African Americans and other groups of non-European ancestry (Need & Goldstein, 2009).

Clearly, the catalog of worldwide variation in disease-causing SNPs is incomplete, and it will be important to conduct genome-wide association studies in many additional populations. Such studies may be particularly informative in African populations because of their complex history and high level of genetic diversity. Because of this diversity, most available SNP microarrays may not adequately capture disease-related variation, as shown in two studies describing the complete sequences of the genomes of two groups (Khoisan and Bantu) in southern Africa (Schuster et al., 2010). It has been estimated that a genome-wide association study of an African population would require approximately 1.5 million SNPs to achieve the same resolution as a study of a European population using 0.6 million SNPs (Jallow et al., 2009; Achidi, 2008). Furthermore, high levels of diversity within and among African populations increase the likelihood of false positive associations, owing to population stratification. Perhaps more important, heterogeneity in haplotype structure among various African ethnic groups is likely to lead to reduced statistical power to detect true positive signals by means of a genome-wide association study, especially in multisite projects such as MalariaGen (Jallow et al., 2009).

Whole-genome sequence analysis avoids the difficulties associated with SNP ascertainment in specific populations. Whole-genome sequences will reveal collections of rare variants that may account for substantial disease susceptibility (Freedman, 2003) but are more likely to be population-specific. This further emphasizes the need for whole-genome data collection in a wide variety of human populations, an effort that has been initiated as part of the ongoing 1000 Genomes Project (Kuehn, 2008).

MAPPING OF COMMON DISEASE GENES

The use of ancestry in mapping of genes that contribute to common diseases, such as prostate cancer, chronic kidney disease, and hepatitis C virus (HCV) infection, illustrates how genomic science may inform our understanding of the causes of disease and health disparities. It is well established that Americans of African ancestry have a significantly higher risk of chronic kidney disease and kidney failure than do European Americans (Freedman, 2003; USRDS, 2007). An increase in the risk of kidney failure (USRDS, 2007) by a factor of 3.6 can be attributed in part to socioeconomic factors, such as inadequate access to health care, and clinical factors, such as a higher incidence of hypertension and diabetes (which may, in turn, be influenced by socioeconomic factors). Even after adjustment for these factors, however, substantial risk remains (Klag et al., 1997; Rao & Balakrishnan, 2009), suggesting a genetic influence. Two independent studies have tested this hypothesis with the use of a gene-mapping technique that incorporates genetic admixture in

TABLE 5-1 MEDIAN AND RANGE OF MINOR ALLELE FREQUENCY (MAF) FOR LOCI IDENTIFIED BY GENOME-WIDE ASSOCIATION STUDIES (GWAS) IN ELEVEN HAPMAP POPULATIONS WITH ANCESTRY FROM ACROSS THE WORLD*

DISEASE	NUMBER OF GWAS REPORTED LOCI	MEDIAN OF MAF	RANGE OF MAF†
Type 2 Diabetes	41	0.365	0.099–0.564
Crohn's Disease	45	0.381	0.137–0.769
Systemic Lupus Erythematosus	21	0.338	0.106–0.896
Psoriasis	15	0.336	0.096–0.681
Breast Cancer	17	0.243	0.120–0.707
Prostate Cancer	25	0.362	0.144–0.915

*Eleven HapMap global populations: African ancestry in Southwest USA; Utah residents with Northern and Western European ancestry; Han Chinese in Beijing; Chinese in Metropolitan Denver, Colorado; Gujarati Indians in Houston, Texas; Japanese in Tokyo, Japan; Luhya in Webuye, Kenya; Mexican ancestry in Los Angeles, California; Maasai in Kinyawa, Kenya; Tuscans in Italy; and Yoruba in Ibadan, Nigeria.

†Data show considerable population frequency differences for GWAS identified loci for the six selected diseases.

African Americans. Multiple common SNPs in the gene that encodes nonmuscle myosin heavy chain type II isoform A (MYH9) were associated with an increase in the risk of focal segmental glomerulosclerosis and end-stage renal disease by a factor of 2 to 4 in persons without diabetes (Rao & Balakrishnan, 2009; Kao et al., 2008; Kopp et al., 2008). Genetic variation in MYH9 accounted for a large proportion of the excess risk of both disorders that was observed in African Americans, as compared with European Americans. However, sequence data from the 1000 Genomes Project, which included newly discovered variants in a part of chromosome 22q12 that had poor coverage in HapMap, showed that the major effect gene is the encoding apolipoprotein L1 (APOL1), which is in linkage disequilibrium with MYH9 (Genovese et al., 2010). Two independent sequence variants (GI and G2) located in the last exon of the gene encoding APOL1 explained the associations previous attributed to MYH9 in African Americans (Genovese et al., 2010). G1 and G2 are in perfect negative LD (i.e., never occur together on the same parental chromosome). The two APOL1 variants are common in African chromosomes but absent from European chromosomes. The variants reside within haplotypes that harbor signatures of positive selection. APOL 1 is a serum factor that lyses trypanosomes (Genovese et al., 2010). In vitro assays revealed that only the kidney disease-associated ApoL1 variants lysed *Trypanosoma brucei rhodesiense* (Genovese et al., 2010). It is proposed that the G1 and G2 kidney disease risk variants likely rose to high frequency in Africa because they confer resistance to trypanosomal infection and protect against the lethal form of African sleeping sickness (Genovese et al., 2010; Rosset et al., 2011).

Prostate cancer is a second example of a common disease that varies considerably among populations; its prevalence in the United States is greatest among African-American men (Jemal et al., 2008; Shavers & Brown, 2002). The rate of death from prostate cancer is increased by a factor of 2.4 among African-American men, as compared with European-American men (Jemal et al., 2008). Again, disparities in treatment and access to adequate health care contribute to this difference (Shavers & Brown, 2002), but there is increasing evidence that genetic variation makes a significant contribution. Several independent studies have identified and replicated a locus on chromosome 8q24 that is associated with prostate cancer in men from different ancestral backgrounds, including European, African, Latino, and Japanese men (Cheng et al., 2008; Freedman et al., 2006; Robbins et al., 2007; Witte, 2007; Haiman et al., 2007). The effect of SNPs associated with prostate cancer on chromosome 8q24 varies according to ancestry, with population attributable risks ranging from 32% among European Americans to 46% among Hispanic Americans and 68% among African Americans (Haiman et al., 2007). (The population attributable risk is defined as the proportion of cases that would not occur if the risk factor were eliminated.) These SNPs, if shown in follow-up studies to be causal, may explain a substantial proportion of the observed disparity in prostate-cancer incidence.

On the one hand, these results provide compelling reasons to use information about individual and group ancestry in investigations of the causes of disease. On the other hand, continental ancestry, race, and ethnic group may be inadequate descriptors for the reasons we have discussed. We present two examples in which direct genetic testing may provide a better and more accurate alternative than such traditional descriptors.

The first example concerns population screening for the HLA-B*5701 allele to prevent the abacavir hypersensitivity syndrome (AHS). About 4% to 8% of patients with human immunodeficiency virus infection who are treated with abacavir have serious, potentially life-threatening

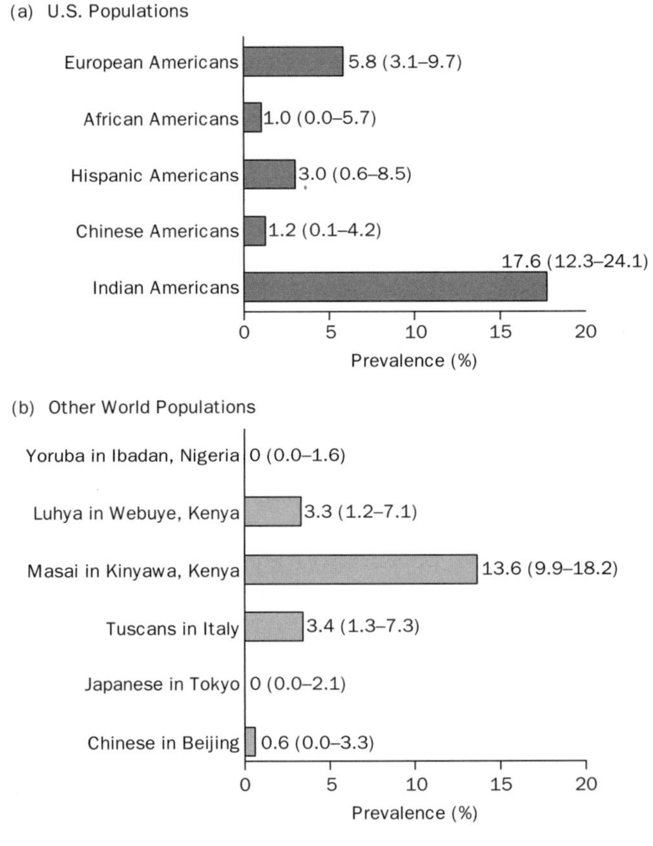

Figure 5-1 Variation in the HLA-B*5701 (rs2395029 (G)) locus that influences response to abacavir treatment for HIV in eleven HapMap samples. Screening for the HLA-B*5701 allele substantially reduces the incidence of abacavir hypersensitive, a potentially life-treating syndrome. Top Panel: frequency and 95% confidence interval of the HLA-B*5701 variant in five US ethnic groups (ASW—African ancestry in Southwest USA; CEU—Utah residents with Northern and Western European ancestry; CHD—Chinese in Metropolitan Denver, Colorado; GIH—Gujarati Indians in Houston, Texas and MEX (Mexican ancestry in Los Angeles, California); Bottom Panel: frequency and 95% confidence interval of the HLA-B*5701 variant in six global ethnic groups from Africa (LWK—Luhya in Webuye, Kenya; MKK—Maasai in Kinyawa, Kenya and YRI—Yoruba in Ibadan, Nigeria), East Asia (CHB—Han Chinese in Beijing, China and JPT—Japanese in Tokyo, Japan) and Europe (TSI—Tuscans in Italy). The frequency of 13.6 in MKK compared to zero in YRI shows how the use of racial or continental labels such as "Black" or "African" can sometimes obscure important, biomedically relevant variation.

AHS (Ingelman-Sundberg 2008; Phillips & Mallal, 2009). A double-blind, prospective, randomized study involving 1956 patients from 19 countries showed that screening for the HLA-B*5701 allele substantially reduced the incidence of AHS (Hetherington et al., 2002; Mallal et al., 2002). However, there is considerable interpopulation variation in the frequency of this allele, ranging from 0% to approximately 20% (Figure 5-1; Nolan, Gaudieri & Mallal, 2003). As a result, the prevalence of AHS varies considerably, a variation that has implications for the utility and cost-effectiveness of screening in different population groups (Ingelman-Sundberg 2008; Nolan, Gaudieri & Mallal, 2003). For example, with an overall AHS incidence of 5% to 8% in populations of European ancestry, only about 14 European patients would have to be screened, on average, to prevent one case of AHS. The incidences of AHS and HLA-B*5701 are much lower in some other populations, leading some observers to recommend screening for HLA-B*5701 in European populations but not in some Asian or African populations (Abel, Paturel, & Cabie 2008; Park et al., 2009).

A better understanding of the global distribution of the HLA-B*5701 variant leads to the conclusion that such recommendations may be problematic. Among five U.S. ethnic groups that participated in the International HapMap Project, the highest prevalence of HLA-B*5701 (17.6%) was seen not among Americans of European ancestry but among Gujarati Indians in Houston (Figure 5-1). A similar estimate of 20% was observed in a sample of persons from India (Nolan, Gaudieri, & Mallal 2003). Substantial differences in allele prevalence were seen in two African HapMap populations: the prevalence in the Kenyan Masai group was 13.6%, a proportion that is more than double that in European samples (3.4% among Tuscans and 5.8% among residents of Utah with European ancestry); the frequency of the allele was zero among the Yoruba in Nigeria. This range in prevalence in the samples from African populations overlaps the ranges in Europeans and for the most part those in other groups. Labels such as "black" or "African" therefore obscure biomedically relevant variation and could lead to less vigilance among physicians for AHS in susceptible patients. For these reasons, the Food and Drug Administration's recommendation to screen all groups, regardless of population identity, for HLA-B*5701 before abacavir administration is prudent and justified (U.S. Food and Drug Administration, 2008).

A second example is provided by population differences in the spontaneous clearance of HCV infection and the response to treatment of the virus. HCV infection may spontaneously resolve (viral clearance), persist without complications, or cause cirrhosis and hepatocellular carcinoma (Seeff, 2002; Thomas et al., 2000). Spontaneous clearance of the infection has been reported to occur in 36.4% of persons of non-African ancestry, as compared with only 9.3% in persons of African ancestry (Thomas et al., 2000). A genome-wide association study of persons chronically infected with hepatitis C and treated with pegylated interferon (peginterferon) alfa and ribavirin showed a strong association between SNP rs12979860 on chromosome 19q13 and a sustained virologic response (Ge et al., 2009). This SNP, close to the gene encoding interferon lambda-3 (IL28B), accounted for more than a doubling in the difference in response to drug treatment for hepatitis, and the effect was similar in patients with European ancestry and in those with African ancestry (Ge et al., 2009). A follow-up study showed that the C/C genotype of SNP rs12979860 was strongly associated with resolution of HCV infection among persons of either European ancestry or African ancestry (Thomas et al., 2009). The frequency of the advantageous allele varies considerably among global populations, reaching more than 90% in East Asia but having an intermediate prevalence in Europe and a low prevalence in Africa (Table 5-2; Ge et al., 2009, Thomas et al., 2009). This difference in allele frequency explains about half the variation in response rates between patients of African-American ancestry and those of European ancestry (Ge et al., 2009; Thomas et al., 2009).

These examples indicate how an understanding of the disease-related effects of specific genetic variants provides the basis for direct genetic testing in individuals, and alleviates reliance on population categories to improve disease diagnosis and treatment. It is therefore anticipated that the future use of drug therapy will not depend on the imprecise indicators of race or ethnicity, but on the individual patient's genotype (Rothstein & Epps, 2001).

REDUCING DISPARITIES IN THE GENOMIC ERA

Genomics, like most human tools, is a double-edged sword. This is especially true in the context of understanding

TABLE 5-2 MEAN AND RANGE OF THE C ALLELE FREQUENCY FOR SNP RS12979860 IN AFRICAN, EUROPEAN AND EAST ASIAN POPULATIONS*

POPULATION GROUPS	# OF INDIVIDUALS/# OF POPULATIONS STUDIED	MEAN FREQUENCY	FREQUENCY RANGE
Africa	428/10	36.2	23.1–54.8
Europe	761/13	68.35	52.8–85.7
East Asia	380/8	94.93	90.0–100.0

*SNP rs12979860 is 3 kb upstream of the IL28B gene which encodes the type III interferon IFN-λ3. The C/C genotype of SNP rs12979860 strongly enhances resolution of hepatitis C infection among individuals of both European and African Ancestry (Ref # 78&79).

and eliminating health disparities (Sankar & Cho, 2002; Braun, 2002; Cooper, Kaufman, & Ward, 2003). Unquestionably, genomics provides novel insights into the causes of and susceptibility to disease and adverse reactions to drugs (Goldstein, 2009; Hirschhorn, 2009). If used carelessly, however, genomics and associated technologies may exacerbate disparities at multiple levels because of unequal application among human populations and because of reinforcement of racial stereotyping, identified by the Institute of Medicine as a primary factor in the unequal treatment of minority patients (Smedley et al., 2003); overshadowing of perhaps more important determinants of disparities, including poverty, racism, unequal access to health care, and cultural practices; exacerbation of existing disparities in access to care, owing to increased costs that are likely to be associated with genetic tests and procedures; and conflation of genetic variation at the population level with that among individuals. However, it is worth mentioning that most of these potential drawbacks do not rest on genomic and related technologies themselves, but on the interpretation of genomic data within the sociopolitical and economic context of societies.

Genomic data present a new array of opportunities and challenges. Data collection should be extended to as many diverse populations as possible. To achieve this goal, innovative funding mechanisms and additional commitments from genome scientists will be required. It is also critical to assess nongenetic factors, which vary substantially among populations and may interact in important ways with genetic risk factors. Analyses of these effects and interactions can be especially powerful in the context of large, long-term prospective studies (Manolio, Bailey-Wilson, & Collins, 2006). Finally, to make the best use of genomic data, it will be essential to educate students, practitioners, and the public about the beneficial applications of genomics, as well as its limitations. Such measures may help to reduce disparities in health.

ACKNOWLEDGMENT

This chapter is based on the authors' article, "Ancestry and Disease in the Age of Genomic Medicine" published in *New England Journal of Medicine* (2010; 363:1551–1558). Editor and Oxford University Press acknowledge permission to use the material for the text in this chapter.

REFERENCES

Abel S, Paturel L, Cabie A. (2008). Abacavir hypersensitivity. *N Engl J Med, 358,* 2515.

Achidi EA. (2008). A global network for investigating the genomic epidemiology of malaria. *Nature 456*, 732–737.

Adeyemo A, Rotimi C. (2010). Genetic variants associated with complex human diseases show wide variation across multiple populations. *Public Health Genomics 13,* 72–79.

Amos W, Hoffman JI. (2009). Evidence that two main bottleneck events shaped modern human genetic diversity. *Proc Biol Sci 277,* 131–137.

Ayodo G, Price AL, Keinan A, et al. (2007). Combining evidence of natural selection with association analysis increases power to detect malaria-resistance variants. *Am J Hum Genet 81*(2), 234–242.

Bamshad MJ, Wooding S, Watkins WS, Ostler CT, Batzer MA, Jorde LB. (2003). Human population genetic structure and inference of group membership. *Am J Hum Genet 72,* 578–589.

Bolnick DA, Fullwiley D, Duster T, et al. (2007). The science and business of genetic ancestry testing. *Science 318,* 399–400.

Braun L. (2002). Race, ethnicity, and health: can genetics explain disparities? Perspect Biol Med 2002;45, 159–174.

Bryc K, Auton A, Nelson MR, et al. (2010). Genome-wide patterns of population structure and admixture in West Africans and African Americans. *Proc Natl Acad Sci U S A 107,* 786–791.

Cheng I, Plummer SJ, Jorgenson E, et al. (2008). 8q24 and prostate cancer: association with advanced disease and meta-analysis. *Eur J Hum Genet 16,* 496–505.

Chimpanzee Sequencing and Analysis Consortium. (2005). Initial sequence of the chimpanzee genome and comparison with the human genome. *Nature 437,* 69–87.

Conrad DF, Jakobsson M, Coop G, et al. (2006). A worldwide survey of haplotype variation and linkage disequilibrium in the human genome. *Nat Genet 38,* 1251–1260.

Cooper RS, Kaufman JS, Ward R. (2003). Race and genomics. *N Engl J Med 348,* 1166–1170.

Drmanac R, Sparks AB, Callow MJ, et al. (2010). Human genome sequencing using unchained base reads on self-assembling DNA nanoarrays. *Science 327,* 78–81.

Fagundes NJ, Ray N, Beaumont M, et al. (2007). Statistical evaluation of alternative models of human evolution. *Proc Natl Acad Sci U S A 104,* 17614–176149.

Freedman BI. (2003). Susceptibility genes for hypertension and renal failure. *J Am Soc Nephrol 14*(Suppl 2), S192–S194.

Freedman ML, Haiman CA, Patterson N, et al. (2006). Admixture mapping identifies 8q24 as a prostate cancer risk locus in African-American men. *Proc Natl Acad Sci U S A 103,* 14068–14073.

Ge D, Fellay J, Thompson AJ, et al. (2009). Genetic variation in IL28B predicts hepatitis C treatment-induced viral clearance. *Nature 461,* 399–401.

Genovese G, Friedman DJ, Ross MD, et al. (2010). Association of trypanolytic ApoL1 variants with kidney disease in African Americans. *Science 329*(5993), 841–845.

Goldstein DB. (2009). Common genetic variation and human traits. *N Engl J Med 360,* 1696–1698.

Haiman CA, Patterson N, Freedman ML, et al. (2007). Multiple regions within 8q24 independently affect risk for prostate cancer. *Nat Genet 39,* 638–644.

Halder I, Yang BZ, Kranzler HR, Stein MB, Shriver MD, Gelernter J. (2009). Measurement of admixture proportions and description of admixture structure in different U.S. populations. *Hum Mutat 30,* 1299–1309.

Hetherington S, Hughes AR, Mosteller M, et al. (2002). Genetic variations in HLA-B region and hypersensitivity reactions to abacavir. *Lancet 359,* 1121–1122.

Hindorff LA, Sethupathy P, Junkins HA, et al. (2009). Potential etiologic and functional implications of genome-wide association loci for human diseases and traits. *Proc Natl Acad Sci U S A106,* 9362–9367.

Hirschhorn JN. (2009). Genomewide association studies—illuminating biologic pathways. *N Engl J Med 360,* 1699–1701.

Hughes DA, Vilar FJ, Ward CC, Alfirevic A, Park BK, Pirmohamed M. (2004). Cost-effectiveness analysis of HLA B*5701 genotyping in preventing abacavir hypersensitivity. *Pharmacogenetics 14,* 335–342.

Ingelman-Sundberg M. (2008). Pharmacogenomic biomarkers for prediction of severe adverse drug reactions. *N Engl J Med 358,* 637–639.

Jakobsson M, Scholz SW, Scheet P, et al. (2008). Genotype, haplotype and copy-number variation in worldwide human populations. *Nature 451,* 998–1003.

Jallow M, Teo YY, Small KS, et al. (2009). Genome-wide and fine-resolution association analysis of malaria in West Africa. *Nat Genet 41,* 657–665.

Jemal A, Siegel R, Ward E, et al. (2008). Cancer statistics, 2008. *CA Cancer J Clin 58,* 71–96.

Jorde LB, Watkins WS, Bamshad MJ, et al. (2000). The distribution of human genetic diversity: a comparison of mitochondrial, autosomal, and Y-chromosome data. *Am J Hum Genet 66,* 979–988.

Jorde LB, Wooding SP. (2004). Genetic variation, classification and "race." *Nat Genet 36*(Suppl), S28–S33.

Kao WH, Klag MJ, Meoni LA, et al. (2008). MYH9 is associated with nondiabetic end-stage renal disease in African Americans. *Nat Genet 40,* 1185–1192.

Klag MJ, Whelton PK, Randall BL, Neaton JD, Brancati FL, Stamler J. (1997). End-stage renal disease in African-American and white men: 16-year MRFIT findings. *JAMA 277,* 1293–1298.

Kopp JB, Smith MW, Nelson GW, et al. (2008). MYH9 is a major-effect risk gene for focal segmental glomerulosclerosis. *Nat Genet 40,* 1175–1184.

Kosoy R, Nassir R, Tian C, et al. (2009). Ancestry informative marker sets for determining continental origin and admixture proportions in common populations in America. *Hum Mutat 30,* 69–78.

Kuehn BM. (2008). 1000 Genomes Project promises closer look at variation in human genome. *JAMA 300,* 2715.

Lee SS, Bolnick DA, Duster T, Ossorio P, Tallbear K. (2009). The illusive gold standard in genetic ancestry testing. *Science 325,* 38–39. [Erratum, *Science 325,* 946.]

Li JZ, Absher DM, Tang H, et al. (2008). Worldwide human relationships inferred from genome-wide patterns of variation. *Science 319,* 1100–1104.

Li W-H, Sadler LA. (1991). Low nucleotide diversity in man. *Genetics 129,* 513–523.

Luzzatto L. (2006). Glucose 6-phosphate dehydrogenase deficiency: from genotype to phenotype. *Haematologica 91,* 1303–1306.

Mallal S, Nolan D, Witt C, et al. (2002). Association between presence of HLA-B*5701, HLA-DR7, and HLA-DQ3 and hypersensitivity to HIV-1 reverse-transcriptase inhibitor abacavir. *Lancet 359,* 727–732.

Manolio TA, Bailey-Wilson JE, Collins FS. (2006). Genes, environment and the value of prospective cohort studies. *Nat Rev Genet 7,* 812–820.

Manolio TA, Collins FS, Cox NJ, et al. (2009). Finding the missing heritability of complex diseases. *Nature 461,* 747–753.

Menotti-Raymond M, O'Brien SJ. (1993). Dating the genetic bottleneck of the African cheetah. *Proc Natl Acad Sci U S A 90,* 3172–3176.

Merryweather-Clarke AT, Pointon JJ, Shearman JD, Robson KJ. (1997). Global prevalence of putative haemochromatosis mutations. *J Med Genet 34,* 275–278.

Molnar S. (2001). *Human variation: race, types and ethnic groups.* 5th ed. Upper Saddle River, NJ: Prentice-Hall.

Need AC, Goldstein DB. (2009). Next generation disparities in human genomics: concerns and remedies. *Trends Genet 25,* 489–494.

Nolan D, Gaudieri S, Mallal S. (2003). Pharmacogenetics: a practical role in predicting antiretroviral drug toxicity? *J HIV Ther 8,* 36–41.

Novembre J, Johnson T, Bryc K, et al. (2008). Genes mirror geography within Europe. *Nature 456,* 98–101. [Erratum, *Nature 456,* 274.]

Park WB, Choe PG, Song KH, et al. (2009). Should HLA-B*5701 screening be performed in every ethnic group before starting abacavir? *Clin Infect Dis 48,* 365–367.

Parra EJ, Marcini A, Akey J, et al. (1998). Estimating African American admixture proportions by use of population-specific alleles. *Am J Hum Genet 63,* 1839–18351.

Phillips E, Mallal S. (2009). Successful translation of pharmacogenetics into the clinic: the abacavir example. *Mol Diagn Ther 13,* 1–9.

Rao M, Balakrishnan VS. (2009). The genetic basis of kidney disease risk in African Americans: MYH9 as a new candidate gene. *Am J Kidney Dis 53,* 579–583.

Redon R, Ishikawa S, Fitch KR, et al. (2006). Global variation in copy number in the human genome. *Nature 444,* 444–454.

Reich D, Thangaraj K, Patterson N, Price AL, Singh L. (2009). Reconstructing Indian population history. *Nature 461,* 489–494.

Robbins C, Torres JB, Hooker S, et al. (2007). Confirmation study of prostate cancer risk variants at 8q24 in African Americans identifies a novel risk locus. *Genome Res 17,* 1717–1722.

Rosenberg NA, Pritchard JK, Weber JL, et al. (2002). Genetic structure of human populations. *Science 298,* 2381–2385.

Rosset S, Tzur S, Behar DM, Wasser WG, Skorecki K. (2011). The population genetics of chronic kidney disease: insights from the MYH9-APOL1 locus. *Nat Rev Nephrol 7*(6), 313–326.

Royal CD, Novembre J, Fullerton SM, et al. (2010). Inferring genetic ancestry: opportunities, challenges, and implications. *Am J Hum Genet 86,* 661–673.

Sabeti PC, Varilly P, Fry B, et al. (2007). Genome-wide detection and characterization of positive selection in human populations. *Nature 449,* 913–918.

Sachidanandam R, Weissman D, Schmidt SC, et al. (2001). A map of human genome sequence variation containing 1.42 million single nucleotide polymorphisms. *Nature 409,* 928–933.

Sankar P, Cho MK, Condit CM, et al. (2004). Genetic research and health disparities. *JAMA 291,* 2985–2989.

Sankar P, Cho MK. (2002). Toward a new vocabulary of human genetic variation. *Science 298,* 1337–1338.

Schuster SC, Miller W, Ratan A, et al. (2010). Complete Khoisan and Bantu genomes from southern Africa. *Nature 463,* 943–947.

Scriver CR. (2001). Human genetics: lessons from Quebec populations. *Annu Rev Genomics Hum Genet 2,* 69–101.

Sebat J, Lakshmi B, Troge J, et al. (2004). Large-scale copy number polymorphism in the human genome. *Science 305,* 525–528.

Seeff LB. (2002). Natural history of chronic hepatitis C. *Hepatology 36*(Suppl 1), S35–S46.

Shavers VL, Brown ML. (2002). Racial and ethnic disparities in the receipt of cancer treatment. *J Natl Cancer Inst 94,* 334–357.

Shriver MD, Parra EJ, Dios S, et al. (2003). Skin pigmentation, biogeographical ancestry and admixture mapping. *Hum Genet 112,* 387–399.

Smedley BD, Stith AY, Nelson AR, eds. (2003). *Unequal treatment: confronting racial and ethnic disparities in health care.* Washington, DC: National Academy Press.

Sturm RA. (2009). Molecular genetics of human pigmentation diversity. *Hum Mol Genet 18,* R9–R17.

Tang H, Quertermous T, Rodriguez B, et al. (2005). Genetic structure, self-identified race/ethnicity, and confounding in case-control association studies. *Am J Hum Genet 76,* 268–275.

Thomas DL, Astemborski J, Rai RM, et al. (2000). The natural history of hepatitis C virus infection: host, viral, and environmental factors. *JAMA 284,* 450–456.

Thomas DL, Thio CL, Martin MP, et al. (2009). Genetic variation in IL28B and spontaneous clearance of hepatitis C virus. *Nature 461* 798–801.

Tishkoff SA, Reed FA, Friedlaender FR, et al. (2009). The genetic structure and history of Africans and African Americans. *Science 324,* 1035–1044.

Tishkoff SA, Reed FA, Ranciaro A, et al. (2007). Convergent adaptation of human lactase persistence in Africa and Europe. *Nat Genet 39,* 31–40.

Tishkoff SA, Varkonyi R, Cahinhinan N, et al. (2001). Haplotype diversity and linkage disequilibrium at human G6PD: recent origin of alleles that confer malarial resistance. *Science 293,* 455–462.

U.S. Census Bureau. (2008). Racial and ethnic classification used in census 2000 and beyond. Washington, DC. Available at: http://

www.census.gov/population/www/socdemo/race/racefactcb.html.
U.S. Food and Drug Administration. (2008). Information for healthcare professionals: abacavir (marketed as Ziagen) and abacavir-containing medications. Silver Spring, MD: Food and Drug Administration. Available at: http://www.fda.gov/Drugs/DrugSafety/PostmarketDrugSafetyInformationforPatientsandProviders/ucm123927.htm.
USRDS [Renal Data System]. (2007). USRDS 2007 annual data report: atlas of end-stage renal disease in the United States. Bethesda, MD: National Institute of Diabetes and Digestive and Kidney Diseases.
Watkins WS, Rogers AR, Ostler CT, et al. (2003). Genetic variation among world populations: inferences from 100 Alu insertion polymorphisms. *Genome Res 13,* 1607–1608.
Witte JS. (2007). Multiple prostate cancer risk variants on 8q24. *Nat Genet 39,* 579–580.
Xing J, Watkins WS, Witherspoon DJ, et al. (2009). Fine-scaled human genetic structure revealed by SNP microarrays. *Genome Res 19,* 815–825.
Yngvadottir B, Macarthur DG, Jin H, Tyler-Smith C. (2009). The promise and reality of personal genomics. *Genome Biol 10,* 237.
Yu N, Jensen-Seaman MI, Chemnick L, Ryder O, Li WH. (2004). Nucleotide diversity in gorillas. *Genetics 166,* 1375–1383.

6.

THE NEXT STEPS FOR GENOMIC MEDICINE: CHALLENGES AND OPPORTUNITIES FOR THE DEVELOPING WORLD[1]

Billie-Jo Hardy, Béatrice Séguin, Federico Goodsaid, Gerardo Jiménez-Sánchez, Peter A. Singer, and Abdallah S. Daar

INTRODUCTION

The genomic revolution is a promising phase in the history of biological sciences. The ultimate goal is to translate the theory into practice—the path to genomic medicine. In previous chapters, the human genome variation initiatives and concerted efforts in major emerging economies of Mexico, India, Thailand, and South Africa are discussed. The political will and serious investments into science and technology in these developing countries will enable them to steer ahead on the path to the medical and health applications of genomics, and to benefit economically. This chapter, previously published in 2008 provides a perspective on the challenges and opportunities facing these and other countries in the developing world as they begin to harness genomics for the benefit of their populations.

Emerging economies in the developing world, such as India, China and Brazil, are investing heavily in innovative science and technology (S&T) and making significant progress in the life-sciences arena, where they are increasingly protecting the subsequent intellectual property (Frew et al., 2007, 2008; Rezaie et al., 2008). The situation in the poorer parts of the developing world, especially in sub-Saharan Africa, has so far been different. The proportion of gross domestic product (GDP) spent on research and development (R&D) has been extremely low in sub-Saharan Africa (Banji Oyelaran-Oyeyinka, 2006). However, in terms of understanding the value of S&T for development, of investing in S&T, and of realizing the need to spend more on health, the situation might be changing. For example, African Union countries have endorsed the call to spend 1% of their GDP on S&T and have undertaken to spend more on health (African Union, 2006; Juma & Serageldin, 2007).

Other chapters in this book discuss applications of genomics in health and medicine in various countries (South Africa, chapter 49; India, chapter 76; Thailand, chapter 60) which demonstrate how emerging economies in the developing world are investing in large scale human genomic variation studies (Séguin et al., 2008a,b; Hardy et al., 2008a,b).

Another comprehensive program is Mexico's National Institute for Genomic Medicine (INMEGEN), which has described in a recent publication a nine-point strategy for the adoption of genomic medicine, including: building an innovative organizational design; establishing the initial infrastructure; initiating nationwide strategic alliances; conducting R&D in genomic medicine; applying genomic technology to common health problems; reaching excellence in teaching and training programs; supporting academic programs in genomic medicine; addressing ethical, social and legal issues; and translating genomic knowledge into products and services (Jimenez-Sanchez et al., 2008). For these countries that have already embarked on major genomics initiatives, establishing research institutes and conducting the research are the first steps on the path to genomic medicine. For them the major question now is how they will go from this early-phase investment toward the hoped-for health-oriented applications, and the economic benefits. Other countries that have not started genomics initiatives are keen to embark and will benefit to receive advice on the potential entry level. And for all developing countries, with or without current genomics initiatives, the question is: what are the challenges and opportunities along the way to the adoption of genomic medicine and to deriving economic benefit from genomics? Some challenges are common to all countries, whether economically developed or developing. Other challenges will be more specific to developing countries. This chapter discusses the challenges and opportunities associated with the adoption of genomic medicine, particularly in the developing world, and the need to understand the

1. Reprinted from Hardy BJ, Séguin B, Goodsaid F, Jimenez-Sanchez G, Singer PA, Daar AS (2008). The next steps for genomic medicine: challenges and opportunities for the developing world. *Nature Reviews*: Genetics, Special issue 'genomic medicine in developing countries' S23–S27.

interdependent nature of efforts to develop genomic medicine globally.

PROJECTS IN THE DEVELOPING WORLD

Apart from the large-scale human genomic variation projects (Chapters 2–5), there are other similar projects taking place in emerging economies and the developing world. China was the only developing country that participated in the sequencing of the human genome. Several years later, it has a number of important initiatives related to genomics. For example, China's Beijing Institute of Genomics plans to sequence the entire genome of 100 Chinese individuals. Some relatively wealthy countries, previously unknown for their involvement in human genomics, are also investing in this field. Researchers in the Al-Mulla molecular pathology laboratory at the University of Kuwait, for example, are using haplotype mapping to study the Arab genome and to identify genomic sites linked to colorectal cancer and type 2 diabetes mellitus. Iran has initiated the Human Genome Diversity Project of Iran (HGDPI), with the main objective of documenting genomic diversity in the populations of Iran for anthropological purposes, and generating a database toward furthering the understanding of disease predisposition. Finally, sub-Saharan African countries other than South Africa are also investing in this field. In the year 2000, a national DNA data bank was initiated in The Gambia containing samples from ~57,000 west Africans (Sirugo et al., 2004; Sgaier et al., 2007). More recently, a biobank and pharmacogenetics database was established in Harare, Zimbabwe, containing 1488 samples from several ethnic sub-Saharan African populations (Nigeria, Kenya, Tanzania, Zimbabwe and South Africa; see Matimba et al., 2008).

Finally, the Pharmacogenetics for Every Nation Initiative (PGENI) is also worth noting in this context as its goals include: enhancing the understanding of pharmacogenetics; building local infrastructure for pharmacogenetic studies; providing guidelines for medical prioritization; and promoting the integration of genetic information in the developing world.

EXPLORING POTENTIAL OPPORTUNITIES

Countries in the developed world, such as the United Kingdom, the United States and Japan have made tremendous investments in R&D toward genomic medicine (Ohnishi & Nakamura, 2005; Palmer, 2007). Emerging economies in the developing world that have made a similar commitment will need to consider how best to identify their next steps into genomic medicine. Thinking of these next steps might also help them identify unique niches that would give them commercial advantages. For other countries that have not yet started genomics initiatives, their entry points will depend, to some degree, on their respective life-sciences innovation infrastructure. Such entry points (see below) would need to be appropriate for their level of investment in genomics R&D and their existing health-care delivery systems. The current trend in terms of both next steps and entry points is for countries in the developing world to collaborate in R&D with more developed nations (north–south collaborations). The Human Genome Organization (HUGO) Pan-Asian SNP Consortium provides an example of a recent north–south R&D collaboration between Asian countries (Normile, 2004; Editorial-Nature, 2007). Lessons learned from such collaborations have contributed to the further development of international ethical guidelines for benefit sharing, ownership and R&D capacity building in human genomic research. However, increasingly there is a trend towards south–south collaborations (Osama, 2008), in which developing countries pool their limited resources, help each other and learn from each other's experience. For example, Mexico's significant investment in genomic research infrastructure provides other Latin-American countries lacking genomic R&D capacity the opportunity to pool their resources with Mexico, as opposed to the United States or Europe, towards the development of genomic medicine and innovative genomic medicine products for this region. Next steps and entry points will need to be cost effective. Pharmacogenomic approaches, including diagnostics, can reduce adverse drug reactions in countries that can least afford to waste money on drugs that might not have the expected therapeutic effect. Diagnostics might be easier to develop than new drugs and vaccines, as they bypass the costly clinical trial stage and tend to have a shorter regulatory review schedule. In this respect, once the cost drops significantly, pharmacogenomic diagnostics might be an early next step or even an entry point for some developing countries. Given their access to large populations exposed to multiple infections (for example, HIV/AIDS, malaria and tuberculosis), another viable option for developing countries could be to focus genomics R&D on the host–pathogen responses for these infections. bioinformatics provides another potential option. The South African National Bioinformatics Institute (SANBI), for example, has developed evOC—a software program that unifies gene expression data by facilitating a link between the genome sequence and expression phenotype information (Kelso et al., 2003). The World Health Organization–based Special Program on Tropical Diseases Research and Training runs training programs on bioinformatics for scientists in the developing world. National genotyping projects are useful for establishing base-line profiles, which might have great benefit for subsequent studies. For example, INMEGEN has revealed significant ancestral components between populations from different regions of Mexico. Moreover, identification of unique SNPs and significant differences of functional variations related to drug metabolism suggest the need for regional approaches for the study and applications of genomic medicine in Mexico. The Indian Genome Variation Consortium (IGVC) recently uncovered high levels of genetic divergence between groups of Indian populations that cluster largely on the basis of ethnicity and language (IGVC, 2008). The study of such population groups will be useful for addressing stratification and complex study design issues (IGVC, 2008). Here, large collaborative efforts on R&D in fields such as pharmacogenomics, vaccinogenomics and toxicogenomics could serve as entry points for developing countries. Examples include pharmacovigilance programs and detailed analyses of data from vaccine trials in which some but not all the human subjects respond to a particular vaccine. A well known example demonstrating that genetic factors can have a strong effect on the immune response to certain vaccines is the response to hepatitis b surface antigen (HbsAg). Up to 10% of people do not respond to HbsAg vaccination. recent evidence suggests that although genes encoded within the major histocompatibility complex are important for this immune unresponsiveness, more than half the heritability is determined outside of this complex. Identification of these genes will help us to understand regulation of immune responses to viral proteins (Höhler et al., 2002). A potential to improve the understanding of genomics and traditional medicines through fields such as nutrigenomics provides another possible entry point that offers these countries an intellectual property advantage. Traditional medicine is well established in China and India, where a memorandum has recently been signed to further the understanding of their respective traditional medicine sectors (Economic Times, India, 2008). Whether genomics can add any value to such endeavors remains to be seen. Although limited, private sector firms in developing countries have also begun to leverage the opportunities in genomic medicine, identifying possible entry points and using unique resources. Avesthagen Ltd., an Indian-owned life-sciences company, has a large-scale genotyping project of the Parsi population in India (Jayaraman, 2007). They predict an initial market in translational medicine, such as genomic medicine, with an eventual foray into early diagnosis, presymptomatic and life long treatment through a combined offering of wellness products (for example, nutrigenomics) and personalized health care. In other countries, such as South Africa and Thailand, a few innovative firms are considering targeting the medical tourism market (Hardy et al., 2008b; Séguin et al., 2008b). Finally, additional possible entry points might involve anthropology or human history and migration studies, as part of establishing a base-line of data for possible health applications. This, for example, was the original impetus for the HUGO Pan-Asian SNP Consortium (Normile, 2004; Editorial–*Nature*, 2007). Other countries might become involved as a result of participating in the HapMap project, as did Nigeria (International HapMap Consortium, 2003, 2005).

CHALLENGES

Genomic research platforms in emerging economies and developing countries will be faced by a number of similar challenges as they become established and proceed toward the adoption of genomic medicine in their respective countries. These challenges are: the current lack of skilled human resources; ensuring sustainable funding and political will; sourcing alternative funding; improving collaboration within the public research sector as well as between the public research sector and the private sector; developing opportunities for south–south and north–south collaboration; improving the commercialization infrastructure in both the public and private sector; developing and improving the existing regulatory infrastructure; developing a health-care infrastructure that can address access and delivery issues of genomic medicine; training health-care workers; and engaging with the public to improve awareness and participation (Séguin et al., 2008a,b; Hardy et al., 2008a,b; Jimenez-Sanchez et al., 2008). A number of these challenges are local and can be addressed as such by and in individual countries. Others, such as establishing international R&D collaborations in genomics research and the need to address the lack of harmonized regulatory infrastructure for genomic medicine, require collaborative efforts on an international scale to address them. Internationally, issues that rapidly need to be addressed for productive and equitable collaborations include data and sample sharing, research capacity building in developing countries, and rules and guidelines for building and using international repositories containing long-term treatment outcomes in both developed and developing nations. These issues are often not straightforward to address. For example, data and sample sharing in many developed countries has traditionally focused upon consent and the concerns associated with privacy and confidentiality (Caulfield et al., 2008). But in international collaborations, in addition to these concerns, considerations will also have to be given to the sovereign nature of the data and of samples sourced in emerging economies and developing countries (Séguin et al., 2008c). The Public Population Project in Genomics (P3G), an international consortium that aims to build the necessary collaborative infrastructure between institutes in order to foster interoperability on the research level (Knoppers et al., 2008), provides one example of how international consortia can contribute to addressing these challenges. It is important to think about such issues early, because the need for large-scale collaborative research is becoming more pressing as R&D in genomic medicine advances. These studies will require data comparisons and validation in large sample sets across different populations (Ginsburg, 2008), which is one reason the European Science Foundation has recommended that European biobanking initiatives harmonize their efforts to achieve maximal benefit (European Science Foundation, 2008).

Collaborative efforts will also be necessary to address policy issues related to translation of genomic research and applications for population health. An example is the Genome-Based Research and Population Health International Network (GRAPH-Int), a global collaborative network that fosters dialog, research, education and training, and communication and stakeholder engagement, with the aim of establishing public policies, programs and services in public health genomics. These types of consortia will be necessary for integrating approaches to genomic medicine and accelerating global consensus-building toward the development of international standards, the regulation of genetic testing and consumer genomics, and accessibility of genetic data and associated information. Broader issues that require consideration include the integration of genetic information into public health decision making, guidelines for medical prioritization, intellectual property regimes, and the association between regulatory bodies and health technology assessment bodies. For instance, forward-looking intellectual property regimes will need to be developed that can be used by countries, depending on the stage of economic and scientific development, to facilitate access to health products, scientific capacity-building, or economic profit. There is also a great need to build capacity in technology transfer in many developing countries.

One of the major challenges in the application of genomic medicine in emerging economies and developing countries involves the limited, or even absent, regulatory infrastructure. Some of these countries may have limited capacity to regulate traditional drugs and diagnostics, and will need to build capacity for these and for emerging genomic medicine products. Furthermore, regulatory capacity in many developing countries will need to encompass the work of ministries of health, science and technology, industry, commerce, natural resources, the judiciary and legislative bodies, as well as the work of drug licensing agencies. Developing countries might benefit from the experience of developed countries that are currently drafting the necessary guidelines to address unique emerging issues associated with the regulation of genomic medicine products (Goodsaid & Frueh, 2007). In the developed world, these issues are being addressed on a national scale by their regulatory agencies and on an international scale through the International Conference on Harmonization (ICH) (International Conference on Harmonization, 2007). The inclusion of emerging or developing countries in the ICH harmonization of guidelines associated with genomic medicine, as well as in other consortia that can help to improve regulatory capacity, will provide a concrete opportunity to improve the application of genomic medicine to global health.

Genomic medicine is redefining how both developed and developing countries need to work together in the application of new knowledge to improve public health. The intrinsic value of information in the human genome is associated less with national boundaries and income than with how comprehensive this information is, and is strongly dependent on the information from many individual genomes and from many individual countries. Data acquired from the genotyping and sequencing projects in emerging and developing countries summarized above help to bridge the gaps in the application of genomic medicine between developing and developed countries. Access to these data throughout the world will be crucial for the identification of novel biomarkers of drug safety and efficacy. Whereas the generation of information from the human genome requires a global effort, its application will require development of therapeutic, diagnostic, and other applications and products aimed at subpopulations and individuals. The most important benefits will probably be in the use of genomics knowledge to prevent diseases and promote health. At an individual level, and perhaps a subpopulation level, the benefits will depend on how genomics is integrated in health systems. To achieve these aims of genomic medicine on a larger scale, public health systems in countries throughout the world will need to integrate pharmacogenomic data from their citizens, and will need to use these data effectively for the optimal allocation of diagnostic and therapeutic resources to improve health education of their publics. This will help their public to change their health-related behavior, to educate health professionals, and to reengineer healthcare systems more toward prevention and health promotion. In terms of the benefits of S&T generally, the current trend to develop knowledge, skills, and products in the economically and scientifically more developed countries, and then struggle to make these available to the less scientifically developed and poorer countries, is not sustainable in the long run. This is why so many developing countries, especially the emerging economies, are focusing more on local innovation, invention, and commercialization to break the cycle of dependency. To go "from the lab to the village" through discovery to product development to commercialization by, and in, developing countries will require that science, business, and capital be brought together creatively. One proposal involves creating "convergence centers," which are an evolution beyond science parks and incubators, aimed specifically at enhancing opportunities for knowledge sharing and rapid innovation, and have a focus on product development and commercialization (Masum et al., 2007).

CONCLUSIONS

The existing initiatives in emerging economies in the developing world, which range from databases run by research networks to comprehensive national institutes with public health mandates, serve as models for how to invest in innovative S&T toward development. Such initiatives can strengthen local research infrastructure and local intellectual property regimes, address local health needs, and reduce healthcare costs, thus improving local health equity. How these countries choose to capture these benefits early on will depend, in part, on the types of entry points they opt for, which in turn will depend on the level of development of their research and health infrastructure. There is much to be learned from the experience and examples of India, Mexico, Thailand, and South Africa. There are, however, significant challenges to be addressed before the adoption of genomic medicine, and although these are for the most part being experienced by all countries, the challenges are more pronounced for developing countries. There is, therefore, a need to strengthen existing collaborative efforts, especially south–south collaborations, which in the long run might be more sustainable. Current and future initiatives and investments in R&D capacity will further enable countries in the developing world to participate as equal R&D partners with more developed countries, instead of merely facilitating access to local biological resources.

In terms of international collaborations that would add value to all countries, at a recent workshop in Mexico City (see Emerging Regulatory Issues in Genomic Medicine conference overview) experts in the emerging issues in genomic medicine from both the developed and developing world discussed a number of ways forward. They included the creation of an information clearinghouse, a virtual network of interested parties, phenotypic and genotypic databases associated with drug development, an international consortium for biomarker development with real involvement of emerging economies and developing countries, and methods of international harmonization of genomic product regulation. Many in the developing world are excited about the prospects of genomic medicine. We are at the crossroads between theory and practice. It is important now that we all think seriously about the challenges and opportunities lying ahead.

ACKNOWLEDGMENTS

This project was funded by Genome Canada through the Ontario Genomics Institute. The Indian Council of Medical Research provided in kind co-funding for this research. The McLaughlin–Rotman Centre for Global Health, Program on Life Sciences, Ethics and Policy is primarily supported by Genome Canada through the Ontario Genomics Institute, the Ontario Research Fund, and the Bill and Melinda Gates Foundation. Other matching partners are listed at The McLaughlin–Rotman Centre for Global Health web site. A.S.D. and P.A.S. are supported by the McLaughlin Centre for Molecular Medicine. P.A.S. is supported by a Canadian Institutes of Health Research Distinguished Investigator award.

REFERENCES

African Union (2006). Decisions of the Executive Council. Eighth Ordinary Session January 16–21, 2006, Khartoum. African Union. Available at: http://www.africa-union.org/root/au/Documents/Decisions/com/AU6th_ord_Council_Decisions_Jan2006_Khartoum.pdf.

Banji Oyelaran-Oyeyinka (2006). Don't make a fetish out of R&D spending. SciDev.net Available at: http://www.scidev.org/en/opinions/dont-make-a-fetish-out-of-rdspending.html.

Caulfield et al. (2008). Research ethics recommendations for whole-genome research: consensus statement. *PLoS Biol 6*, e73.

Cohen R (2008). The world is upside down. *New York Times*. Available at: http://www.nytimes.com/2008/06/02/opinion/l02cohen.html?ref=opinion.

Editorial—Nature (2007). Asia on the rise. *Nature 447*, 885.

European Science Foundation (2008). Populations surveys and biobanking: ESF science policy briefing 32. Available at: http://www.esf.org/publications.html.

Frew SE, et al. (2007). India's health biotech sector at a crossroads. *Nature Biotechnol 25*, 403–417.

Frew SE, et al. (2008). Chinese health biotech and the billion-patient market. *Nature Biotechnol 26*, 37–53.

Ginsburg GS (2008). Genomic medicine: 'grand challenges' in the translation of genomics to human health. *Eur J Hum Genet 16*, 873–874.

Goldsbrough D, Adovor E, Elberger D (2007). What has happened to health spending and fiscal flexibility in low income countries with IMF programs? Center for Global Development. Available at: http://www.cgdev.org/doc/IMF/Health%20Spending.pdf.

Goodsaid F, Frueh FW (2007). Implementing the U.S. FDA guidance on pharmacogenomic data submissions. *Environ Mol Mutagen 48*, 354–358.

Hardy B, Séguin B, Singer PA, Mukerji M, Brahmachari SK, Daar AS (2008a). From diversity to delivery: the case of the Indian Genome Variation initiative. *Nature Rev Genet 9*(Suppl), S9–S14.

Hardy B, Séguin B, Ramesar R, Singer PA, Daar, A. S (2008b). South Africa: from species cradle to genomic application. Nature Rev. Genet. 9 (Suppl), S19–S23.

Höhler T, et al. (2002). Differential genetic determination of immune responsiveness to hepatitis B surface antigen and to hepatitis A virus: a vaccination study in twins. Lancet 360, 991–995.

International Conference on Harmonization (2007). ICH Harmonised Tripartite Guideline Definitions For Genomic Biomarkers, Pharmacogenomics, Pharmacogenetics, Genomic Data And Sample Coding Categories E15. [online], <http://www.ich.org/LOB/media/MEDIA3383.pdf.

Indian Genome Variation Consortium (2008). Genetic landscape of the people of India: a canvas for disease gene exploration. J Genet 87, 3–20.

Jayaraman KS (2007). Database targets Parsi genes. *Nature 446*, 475.

Jimenez-Sanchez G, Silva-Zolezzi I, Hidalgo-Miranda A, March S (2008). Genomic medicine in Mexico: initial steps and the road ahead. *Genome Res 18*, 1191–1198.

Juma C, Serageldin I (2007). *Freedom to Innovate: Biotechnology in Africa's Development*. Report of the High-Level African Panel on Modern Biotechnology. African Ministerial Council on Science and Technology. Available at: http://www.nepadst.org/doclibrary/pdfs/biotech_africarep_2007.pdf.

Kelso J, et al. (2003). eVOC: a controlled vocabulary for unifying gene expression data. *Genome Res 13*, 1222–1230.

Knoppers BM, Fortier I, Legault D, Burton P (2008). The Public Population Project in Genomics (P3G): a proof of concept? *Eur J Hum Genet 16*, 664–665.

Masum H, Daar AD, Al-Bader S, Shah R, Singer PA (2007). Accelerating health product innovation in sub-Saharan Africa. *Innovations: Technology, Governance, Globalization 2*, 129–149.

Matimba A, et al. (2008). Establishment of a biobank and pharmacogenetics database of African populations. *Eur J Hum Genet 16*, 780–783.

Normile D (2004). Genetic diversity. Consortium hopes to map human history in Asia. *Science 306*, 1667.

Ohnishi Y, Nakamura Y (2005). BioBank Japan project. *Nippon Rinsho 63*(Suppl 12), 35–41.

Organization for Economic Cooperation and Development (2008). *How does the United States compare? OECD Health Data 2008*. Available at: http://www.oecd.org/dataoecd/46/2/38980580.pdf.

Osama A (2008). Opportunities and challenges in South–South collaboration. *SciDev.net* Available at: http://www.scidev.net/en/science-and-innovation-policy/southsouth-cooperation/policy-briefs/opportunities-andchallenges-in-south-south-collab.html.

Palmer LJ (2007). UK Biobank: bank on it. *Lancet 369*, 1980–1982.

Rezaie R, et al. (2008). Brazilian health biotech—fostering crosstalk between public and private sectors. *Nature Biotechnol 26*, 627–644.

Séguin B, Hardy B, Singer PA, Daar AS (2008a). Genomics, public health, and developing countries: the case of the Mexican National Institute of Genomic Medicine (INMEGEN). *Nature Rev Genet. 9*(Suppl), S5–S9.

Séguin B, Hardy B, Singer PA, Daar AS (2008b). Universal health care, genomic medicine and Thailand: investing in today and tomorrow. *Nature Rev Genet 9*(Suppl), S14–S19.

Séguin B, Hardy BJ, Singer PA, Daar AS (2008c). Genomic medicine and developing countries: creating a room of their own. *Nature Rev Genet 9*, 487–493.

Sirugo G, et al. (2004). A national DNA bank in The Gambia, West Africa, and genomic research in developing countries. *Nature Genet 36*, 785–786.

Sgaier SK, et al. (2007). Public health. Biobanks in developing countries: needs and feasibility. *Science 318*, 1074–1075.

The Economic Times, "India, China to step up cooperation in traditional medicine," July 8, 2008. http://economictimes.indiatimes.com/News/News_By_Industry/Healthcare_Biotech/Healthcare/India_China_to_step_up_cooperation_in_traditional_medicine/rssarticleshow/3210228.cms.

The International HapMap Consortium (2003). The International HapMap Project. *Nature 426*, 789–796.

The International HapMap Consortium (2005). A haplotype map of the human genome. *Nature 437*, 1299–1320.

FURTHER INFORMATION

Al-Mulla molecular pathology laboratory, University of Kuwait: http://www.al-mulla.org

Avesthagen Ltd: http://www.avesthagen.com

Beijing Institute of Genomics, china: http://www.big.ac.cn/big/english/index.jsp

Emerging Regulatory Issues in Genomic Medicine conference overview: http://regulatory.inmegen.gob.mx/english/home.html

European Science Foundation: http://www.esf.org

EevOc software program: http://www.evocontology.org

Genome-based research and Population Health International Network (GraPH-int): http://www.graphint.org/ver2

Human Genome Diversity Project of Iran (HGDPI): http://www.nrcgeb.ac.ir/Human%20Genom%20Diver.html

International Conference on Harmonization (ICH): http://www.ich.org/cache/compo/276-254-1.html

McLaughlin-Rotman Center for Global Health: http:// www.mrcglobal.org

National Institute for Genomic Medicine (INMEGEN), Mexico: http://www.inmegen.gob.mx

Pharmacogenetics for Every Nation Initiative (PGENI): http://pgeni.unc.edu

South African National Bioinformatics institute (SANBI): http://www.sanbi.ac.za

The Public Population Project in Genomics (P3G): http://www.p3gconsortium.org

UK Biobank: http://www.ukbiobank.ac.uk

US National Office of Public Health Genomics: http://www.cdc.gov/genomics/default.htm

7.

CONSUMER PERSPECTIVES ON GENOMICS AND HEALTHCARE

Alastair Kent

INTRODUCTION

In the half-century or so since Watson and Crick discovered the structure of DNA, and hence the mechanism for elucidating our genetic code, our understanding of the links between our genes and our health and/or disease state has progressed rapidly. Not only has our knowledge of our genome (and that of an increasing number of other species, too) increased at a rate faster than would have been imaginable only a relatively short time ago, but the rate of change is accelerating to the point where our ability to generate genomic data easily outstrips our ability to interpret it.

Paralleling this increase in our understanding of genetics has been the rise of new forms of communication that have, at least in terms of their own rhetoric, made scientific advances accessible to a wider public than has been possible previously, breaking down barriers created by professionals to keep their arcane and specialist knowledge safe from the misguided interpretations of the untutored populace. Now, with Wikipedia (www.wikipedia.org) and Google (www.google.com), everyone who has access to the Internet can be his or her own expert, seemingly able to challenge restrictive practices that would control access to the knowledge that would help them to gain and retain a degree of control over their own genetic future—to manage their own wellbeing, and take charge of their (and their family's) health. While this vision of the future may seem to be within our grasp, it is not yet a possibility, both because we do not have the science to pursue it yet, and even if we did, we lack the resources to make it a reality for all but a tiny fraction of the population of the planet—namely, the wealthy inhabitants of countries with developed healthcare systems (or those who have the means to access them).

For the overwhelming majority of the world's population, the ideal of a genetic future where complex diseases can be effectively diagnosed and then treated, prevented, or cured is still a pipedream (March of Dimes, 2008). Notwithstanding this fact, it remains the case that advances in our understanding of genetics and, more recently, of genomics do hold out considerable promise of being able to address currently unmet health needs. But for this to happen, it is important to be realistic about where we have come from, and where we need to go if we are to have the opportunity to address this in a rational manner.

It is also important not to allow ourselves to be carried away by the hype, by the exuberance of some protagonists, and by the desire of the media to find a good story that will engage listeners, viewers, or readers (Smith et al., 2005). Although we have made unimaginable progress in the six decades since Watson and Crick's discovery, it remains the case that the vast majority of diseases that afflict humanity remain, at best, intractable. Few are treatable, let alone curable even now, but options for prevention (where these exist) are either poorly understood, culturally inappropriate, or technically unfeasible in most of the world. Although we are the first generation since the emergence of Homo Sapiens as a distinct species to have the tools to understand the fundamental flaws in our biology that cause or predispose us to diseases with a genetic component, we must not allow the clarity of the vision of scientific possibilities to distract us from the length or the difficulties of the journey we face if these are to be realized.

GENOMICS AND HEALTHCARE FROM A PATIENT PERSPECTIVE

Chronic diseases are, by definition, not susceptible to quick fixes. There are no "magic bullets" that will restore those affected to a disease-free state, and for most patients, living with the disease requires a series of tradeoffs between interventional possibilities in order to establish a *modus vivendi* that, too often, represents the least worst rather than the best possible outcome. Most patients are realistic in their approach to their disease. They know that drugs do not work equally well in all who take them. They know that there may be adverse consequences (and that these may be serious) for some people who take them and they know that, for the most part, "miracle cures" tend, on closer examination, to bear out the adage that if something seems too good to be true, it probably is.

This is a counsel of realism, not one of despair. The support from patients and the public for biomedical research (e.g., Mori, 2009) is a palpable demonstration of the viewpoint that the problems of complex diseases are not beyond the wit of mankind to solve, given the time, resources, and opportunity to address them in a framework that allows for sustainable development and the application of new knowledge. So how do patients see genomics contributing to the resolution of these difficulties? Is it a matter of the application of science; are there other issues for patients, and, in the context of the present volume, what does this mean for patients and families in the developing world?

PERSONALIZED MEDICINE

As genomics progresses, so we see increasing claims for a dawning era of so called "personalized medicine," with its mantra of "the right medicine for the right patient at the right time and in the right dose." This creates the expectation of ever more targeted interventions, with greater health gains and reduced risks of side effects. Even in developed health economies this ideal is still the exception rather than the rule. Whatever the progress in biomedical research, the possibility of these interventions being available to patients and families in middle- and low-income developing countries in the foreseeable future seems remote for all but a privileged elite (Office of Science and Technology, 2009).

To outward appearances, rising opportunity in the application of genomic medicine to the resolution of serious, unmet health needs seems always to be accompanied by an increase in costs that is likely to place these developments even further outside the reach of patients and families in developing countries, opening up rather than diminishing the gap between the "haves" and "have-nots" in respect of access to healthcare. We do not have to accept that this as an inevitable consequence of progress in biomedical research, that new knowledge will widen gaps, increase inequality, and hold out prospects of change that are inevitably going to be dashed by a lack of infrastructure, an absence of skilled professionals, and inappropriate concepts regarding the technological imperatives in the application of genomic knowledge to the resolution of health problems.

Given that genetics and genomics operate at the outer reaches of knowledge and at the cutting edge of research, is it possible to envisage a healthcare system that is capable of delivering the benefits of genetic diagnosis and targeted intervention to individuals, to couples and to families living under the threat of life-limiting genetic disease?

THE IMPACT OF GENETIC DISEASE

For parents experiencing the birth of a wanted child, the first question asked of birth attendants is often "Is my baby alright?" The birth of a child with a genetic disease is rarely something to be aspired to—irrespective of how much the baby in question is valued, respected, and loved.

Similarly, the discovery that one is at risk of developing a significant disease as a consequence of a genetic predisposition is a life-changing event, not only for the individual directly at risk, but for those around him or her, altering aspirations and forcing a rethink of plans and ambitions for the future and, possibly, for further generations (McAllister et al, 2007).

The increasing availability of genetic testing provided direct to consumers, without the intervention of a doctor or otherwise qualified medical practitioner and by means of Internet-based sales from private companies, means that for some relatively wealthy individuals, getting a genetic diagnosis may at best become an increasingly curiosity-driven activity. For the vast majority of people (in resource-rich as well as resource-poor nations) this is not the case, and is unlikely to become so in the foreseeable future.

Apart from those wealthy enough to access genetics by private means, the route to a diagnosis of a genetic risk and a subsequent intervention is by means of the public healthcare system in their country of residence. For individuals and families in this position, the need for a genetic diagnosis may be triggered by some form of untoward event. This may arise from a known family history, through the detection of a possible "problem" in an examination for another reason (such as routine antenatal care), or one of a number of other reasons (for example, a series of unexplained miscarriages). Whatever the precipitating factor, the individual and the family close to him or her are likely to be anxious, uncertain, and under considerable emotional stress. Depending on their educational and cultural background, the notion of familial diseases may be unfamiliar, or their concept of inheritance may be based on ideas that are at a considerable remove from the laws of Mendel.

For individuals, couples, and families in this situation, the first hope is often that there has been some mistake. That what appeared to be a potentially serious problem is either not a problem at all, or something symptomatic of a minor issue that can be readily resolved, allowing them to proceed with their lives as previously planned. Failing this, then the hope is that this is a safe, effective, available, and affordable intervention that will cure the problem, or if this is not available then something can be done either to prevent it or prevent it from recurring.

Although great strides have been made in our understanding of genetic diseases in recent years, it remains the case that, for most of those genetic disorders identified and described in the Mendelian Inheritance in Man (OMIM) catalogue (www.ncbi.nlm.nih.gov/omim), there remains little that can be done to alter the natural progress of the conditions, and prevention may require serious interventions such as antenatal testing and termination of pregnancies identified as being affected by a major disorder. Such a course of action requires access to a level of service provision not often available outside the developed health economies of the richer nations, and where it does exist in developing countries it is usually concentrated in major centers of population and more likely to be accessed by the wealthier sections of society. This is quite apart from any legal, ethical, or religious constraints that may restrict or prevent access to such services in many parts of the world.

In the absence of the opportunity to cure or prevent a problem arising from a genetic cause, individuals, couples, and families at risk still need access to information, advice, counseling, and support, in order to enable them to understand the situation in which they find themselves—and, in so doing, to gain and retain a degree of control over their lives to the extent possible, given the limitations imposed by the emergence of the genetic condition that puts them at risk.

One thing is certain, and that is that genetic risk does not go away if it is ignored or if we are in ignorance of the potential threat it poses. While some people may decide not to act on the knowledge that they or their children may be at risk, preferring to accept what happens as and when it does, the absence of knowledge removes the possibility of choice, and deprives people of the opportunity to take action should they wish, and be able, to do so.

Despite opinions to the contrary, and notwithstanding the preceding paragraphs, it is not necessarily the case that genetic diseases require an expensive and elaborate scientific infrastructure to detect, or that any interventions to treat them are only available at a very high cost that is beyond the reach of most healthcare systems in the developing world. While it is certainly true that sophisticated and DNA-based diagnostics do require adequate investment in laboratory services, and therapies for orphan diseases can be expensive, there is much that can be done to improve the health and wellbeing of individuals and families at risk without spending disproportionate sums of money. Primary prevention measures focusing on maternal and child health, such as fortifying flour with folic acid to reduce the risk of neural tube defects, or salt with iodine to address avoidable mental retardation. Giving women control over their fertility, targeting maternal nutrition during pregnancy, and other measures can help reduce the risk of some genetic diseases in ways that produce real results at costs affordable in many parts of the world without the need for expensive laboratory or therapeutic expenditure. The impossibility of achieving a gold standard of sophistication in service provision should not blind us to the very real opportunities that may exist to devise and provide services that will increase the reproductive confidence of couples at risk, and provide interventions that will improve outcomes and reduce risks for individuals vulnerable to some later-onset genetic disorders.

GENETICS AND THE "CULTURE OF BLAME"

Given the (theoretical) possibility of detecting an ever greater number of genetic diseases antenatally, as the causative genes are identified and characterized—and given the prospect of moving the earliest date of detection back to around the eighth week of pregnancy, through the development of methods of noninvasive prenatal diagnosis using free fetal DNA in the maternal circulation—it might be feared that an over enthusiastic "seek and destroy" approach to antenatal testing might emerge, with those who subsequently went on to produce a child affected by a genetic disorder subjected to explicit or implicit criticism (or even sanctions) in some parts of the world. This would be a fundamental contradiction of the ethical principles of autonomy and consent, and would seriously undermine the legitimacy of the benefits that genetic information and support can bring.

Biology, social and cultural norms can cross over in other ways that may lead to a fear of being blamed, with the consequence that couples potentially at risk may avoid services potentially on offer. Consanguinity and attitudes to termination of pregnancy are cases in point. Many parts of the world have adopted the practice of consanguineous marriage. It is a valued tradition for many communities, bringing with it familial as well as societal benefits. In biological terms, however, consanguineous marriage may increase the risk of having a child affected by a recessive genetic disorder, if there is a relevant mutation in the family. This highlights the need for access to genetic testing and counseling if the couple are to have the opportunity to understand the risk they may face, and make informed choices. Unfortunately, in some quarters, genetic risks associated with consanguineous marriage has been overemphasized, leaving couples feeling that they are being blamed for conceiving a child affected by a genetic disorder.

The expectation of unjustified blame does not make the uptake of genetic services more likely, leaving the potential risk undiagnosed and denying the couple the option of choice. Too great an emphasis on the risks of consanguinity also flies in the face of couples' experiences and observations that most cousin marriages do not result in the birth of children with genetic disorders!

Beliefs about people's attitudes toward abortion can create problems where this procedure is a legal option. While Roman Catholic doctrine insists that life is sacred from the moment of conception, Islam and Judaism have the notion of "ensoulment," which happens around the 12th week of pregnancy. Despite this, it is commonly held that there is little point in offering Muslims a termination, as they will not accept. Awareness of "ensoulment" raises the importance of early access to genetic diagnosis in pregnancy for women from communities where this is the case, if choice is to be maintained. And in practice, given a supportive environment, couples will often make choices they feel to be right notwithstanding religion or cultural prohibitions (Darr, 2005).

GENETIC SERVICES IN DEVELOPING COUNTRIES

Clinical genetics is sometimes seen as a luxury in the developing world that can be set aside by publicly funded health care systems in these countries, on the grounds that it requires an expensive, high-tech environment to operate, and there is little that can be done for those affected. This view ignores the importance of primary preventative measures (mentioned earlier) when affordable interventions can make a substantial difference to maternal and child health. It is also based on a false premise about the cost of genetic services, as the recently concluded "Capability" project has demonstrated (www.capability.net.eu).

Capability (criteria for assessing clinical validity in national and international frameworks) was funded by the European Union, working with geneticists in Egypt, Argentina, and South Africa. Demonstration projects were established that offered targeted interventions to defined communities. In Egypt, materials explaining genetic risk were developed for a nonliterate population in a village community. These were used by specially trained community outreach workers (who were supporting pregnant women as part of their normal work, backed up by nurses and doctors in the community's primary health care center, who had also been given appropriate training and support). The intervention was so valued by the pilot community that they demanded its extension to other family members, friends and neighbors.

In the northern part of Argentina, in Chaco province, a syllabus for training primary healthcare workers in aspects of genetics—including a manual showing how to recognize commonly occurring congenital abnormalities in that part of the world—has resulted in significant service improvements and the establishment of a cytogenetic laboratory for that province and its neighbors, opening up access to diagnosis and subsequent advice and counseling about reproductive risk and the care and management of affected children to families who otherwise would have been unlikely to be able to access a genetic diagnosis due to poverty and geographical remoteness from expert centers.

In South Africa, community midwives were provided with mobile phones with which they could send pictures of newborns with suspected abnormalities to the genetic clinic in Johannesburg. They could receive immediate advice, a tentative diagnosis, and instruction as to what further information was needed (measurements to be taken, samples collected, etc.) in order to confirm the diagnosis and to advise the parents. Given the remoteness of the Greater Sekukhune area of Limpopo province (where this pilot was based), this helped avoid parents making long journeys to a hospital, and made effective use of scarce expertise in the provision of diagnosis and care.

Capability has attracted interest from China and the Philippines, and there are plans to transfer the relevant lessons learned for the development of services in these countries.

What the projects demonstrate is that small, targeted investments in locally appropriate genetic services can make a significant difference, supporting women and couples to have the healthy children they hope for. For such developments to be sustainable, there will need to be commitment from the health service, plus possibly the development of innovative and creative strategies to attract funding from a range of sources to recruit and retain appropriately trained professionals and other healthcare workers to work in these communities.

CONCLUSION

From the point of view of patients and families, the promise of genomic medicine and the fulfillment of the personalized healthcare agenda is still some way off. Even for those who live in the developed economies of the Western world, examples of the successful application of genomics to unmet health needs are thin on the ground. While there may be many interventions and novel therapies deriving from genomic research and development in the pipeline, patient and family experience to date has for the most part been characterized by hype rather than hope. If this is the case in the developed world, how much more is it so in the developing world, where resources and expertise are much more limited?

Where genetics can and does make a difference is in the opportunity that it provides for accurate, timely diagnosis of genetic conditions with, as a consequence, the opportunity to provide information, advice, and support to those at risk so that they can make more informed decisions regarding their future health and wellbeing. As the Capability project has demonstrated, small interventions can have a disproportionately large effect. Clinical genetics does not have to be an expensive branch of tertiary medical services. It can be a low-cost intervention, able to bring about dramatic improvements in maternal and child health, if available resources are used to best advantage, and focused on where they can do the most good. The challenge for healthcare planners and providers in the healthcare systems of the developing world is to resist the lure of inappropriate, expensive technology that demands a disproportionate share of scarce resources and expertise. Rather they should use their ingenuity and creativity to generate locally sustainable services, and to build professional competence in ways that will meet the identified needs of their people and which will build critical skills capable of utilizing technological and clinical advances as and when resources become more available. In so doing, a sound basis for the future exploitation of genetic and genomic advances will be created.

REFERENCES

Darr A (2005). "Cousin marriage is a social choice: it needn't be a problem." *Available at:* www.guardian.co.uk/news/2005/dec/02/mainsection.leadersandreply2.

Ipsos MORI. (2010). *Views on Animal Experimentation.* Available at: http://www.ipsos-mori.com/DownloadPublication/1343_sri-views-on-animal-experimentation-2010.pdf.

March of Dimes (2008). March of Dimes Foundation Global Programs: Building Global Networks of Experts in Lower Income Countries to improve the Health of Women, Mothers, Newborns and Babies. http://www.capabilitynet.eu/Documents/GPPHN%20Meeting.pdf.

McAllister M, Davies L, Payne K, Nicholls S, Donnai D, MacLeod R (2007). The emotional effects of genetic diseases: Implications for clinical genetics. *Am J Med Genet Part A 143A*, 2651–2661.

Nippert I, Christianson A. (2010). Introduction and overview: the CAPABILITY Project: Building for the translation of genetic knowledge into practice and prevention. In: *The CAPABILITY Report*, Capability Consortium. Available at: http://www.capabilitynet.eu.

Parliamentary Office of Science and Technology (2009). Personalised Medicine: Parliamentary Office of Science and Technology *(London 2009)* (www.parliament.uk/documents/post/postpn.329.pdf).

Smith GD, Ebrahim S, Lewis S, Hansell AL, Palmer LJ, Burton PR. (2005). Genetic epidemiology and public health: hope, hype and future prospects. *Lancet, 366* (9495), 1484–1498.

8.

TRANSCULTURAL PERSPECTIVES ON GENETICS AND GENOMICS

Dhavendra Kumar

Various chapters in this book have discussed key subjects relevant to genetic and genomic issues applicable to different subcontinents with emerging economies of the developing world. The key issues in an individual country center around ethnic and cultural diversity, genetic or genomic variation, perception and attitudes toward birth defects and inherited diseases, and the practice of inbreeding. This chapter primarily focuses on the sociocultural practices and beliefs common to many individuals. Most of the examples and references are made to the Indian subcontinent, by far culturally the most diverse and densely populated, with a huge burden of birth defects due to genetic and genomic factors. The information provided here is simply an overview, and should not be considered a detailed account of traditional cultural and spiritual beliefs. It is merely a glimpse of various practices of a vastly heterogeneous group of people. It is intended to be useful to various health professionals engaged in delivery of the clinical genetic services, both at the primary and secondary level. The term *transcultural genetic services* is used to emphasize understanding and appreciation of specific sociocultural values relevant to the delivery of the specialist medical service of clinical genetics (Naveed et al., 1992; Patil and Fisher, 1996).

The practice of transcultural medicine emphasizes three major areas—communication, awareness of social practices, and influence of the spiritual background (Qureshi, 1989). This is probably most relevant to genetic professionals in the provision of transcultural genetic services (Johnson and Nadirhaw, 1993).

LANGUAGE

Language forms the integral part of effective communication. The indigenous people would prefer to communicate in one of several languages spoken in the Indian subcontinent. There are over 20 different languages, each with its own linguistic basis and literature. In addition, there are several hundred dialects spread across the whole subcontinent. In broad terms, all languages of the Indian subcontinent belong to one of the language families, which include Indo-Aryan in the northwest and central India, Sino-Tibetan in the Himalayan states, and Dravidian in south India. With recurrent Muslim invasions, the Arabic and Persian languages were introduced, but were largely confined to the northwest region. However, subsequent Muslim rulers of the central and southern Indian states used it for official purposes. Gradually, its script and vocabulary were amalgamated with the commonly spoken dialect, often referred to as Hindustani, and led to the birth of Urdu. It has its own script and grammar, which is similar to Arabic/Persian, and enjoys the status of the national language of Pakistan. Although Hindi, a modern version of the ancient language Sanskrit, is the main spoken and written language of India, it is not recognized to be the national language of India. Most of the states within the Indian Republic have their own language. Hindi, however, is the official national language of the Himalayan Kingdom of Nepal. The official language of Bangladesh is Bangla. It is essentially similar to that spoken in the Indian state of West Bengal, well known for its famous city of Calcutta. Apart from Bangla, use of Urdu is also encouraged in Bangladesh, since it happens to be a predominantly Muslim country. There are two main languages in Sri Lanka—Sinhalese and Tamil. However, the majority of the people speak and write English, and it is effectively the official language.

Use of English is not uncommon in almost all countries of the Indian subcontinent. However, its use is confined to urban settings and limited to being a second language, often as a convenient adjuvant to the local language—or, in the corporate sector, to facilitate effective communication. Despite persistent opposition by certain religious, social, and political groups, English is accepted by most of the Indian subcontinent populations. The migrant South Asian people living in Western countries or other places with a Western lifestyle and environment are usually well conversant and adequately educated in the English medium. However, exceptions might include elderly parents, women, and children. Not surprisingly, this variation is fast disappearing due to environmental pressures and rapid acclimatization to the new adopted place and lifestyle. Although use of English in genetic consultation and counseling would remain unchanged, it is important to have some basic knowledge about various spoken and written languages from the Indian subcontinent. With increasing availability of language interpreters, the process of professional communication has become relatively smooth. However, use of such service requires some understanding about interpreters' skills and professional background. Some of them might even hold nursing and medical qualifications, and in this situation it is possible that the message could be either modified or phrased in a way to alter its meaning. Nevertheless, this kind of service should be welcomed. On the other hand, the interpreter might not be adequately skillful to translate the information and pass it on without changing its intended expression. It is not uncommon to note in such situations that the interpreter's use of phrase and manner of verbal expression could be seen as too directive, and hence influences the process of decision making. In order to ensure a "nondirective" and "nonjudgmental" approach, it is absolutely necessary that the person acting as an interpreter has received such instructions in fair detail from the clinical geneticist or genetic counselor.

SOCIAL AND CULTURAL PRACTICES

Every community or group of people forming the core group of a population are unique in terms of certain social and cultural beliefs and practices. These are often independent of language they speak or write, or their belief in a faith. This reflects their lifestyle to a large extent, and becomes symbolic for their recognition and acceptance in the society. When they migrate, this lifestyle follows and helps in establishing another mini-social group reflecting number of recognizable cultural practices.

The key areas to consider would include economy, family, and health.

ECONOMIC ASPECTS

Almost all countries in the Indian subcontinent are regarded as developing. Economically, they have done reasonably well over the last 20 years. Yet, a lot remains to be done. A large proportion of the population lives well below the poverty line. The net per capita income and the gross national income are far below that enjoyed by Western nations and other developed countries. It is not surprising to find a lot of people living in extreme poverty, and a significant proportion struggling to make ends meet. In strict economic terms, the lower-middle and the medium-middle class comprise about 50% of the population. Only 10% enjoy being classified as the upper middle class or the true upper class. Access to specialized health services such as a clinical genetics service may be restricted to middle and upper classes. However, a state-funded health service, such as the National Health Service in the United Kingdom, can ensure the availability of such services to all, irrespective of ethnic origin and social class.

Another quasi-economic method of separating people is dependent on the occupation of the individual or that of his or her family. It is often vertical from one generation to the next, and is similar to inheritance of property or a title. It is also known as the infamous "caste" system. Although it is often regarded as unique to the Indian society, among the predominantly Hindu section of the population, the caste system is practiced all over the Indian subcontinent, albeit in a covert or disguised manner. It comprises of four caste groups—Brahmin, Chatriya, Vashnawa, and Shudra. *Brahmin* refers to being "noble." These people belong to a religious and political establishment, and are assigned to rule. *Chatriya* refers to the warrior or protector group. Their main role is in administration and defense of the state. Members of the civil service, government employees, and other public service workers would belong to this group. *Vashnawa* refers to commerce. Anyone directly or indirectly associated with commerce or trading would belong to this group. The last group of *Shudra* is perhaps the largest, and includes people providing all kinds of services ranging from toilet cleaners, road sweepers, housekeepers, carpet weavers, textile workers, pottery makers, and even various kinds of peasants; the list is endless. Unfortunately, successive rulers and politicians have exploited these people for generations, as have those belonging to the so-called higher caste groups. The Shudras are regarded "untouchables," but Mahatma Gandhi called them *harijans*, the "people of God." Not surprisingly, these people do not enjoy the privileges of better education and health services. It is very probable that these people might not have access to specialized tertiary medical services, such as a clinical genetics service. This is true even if health services are divided between the state-run and the private sector. The state-run service has set itself different priorities, to focus on prevention of common illnesses related to malnutrition, infection, and cancer. The private sector is fast growing and getting technically strong and promising; several clinics across India and Pakistan are engaged in offering number of genetic services, including antenatal screening and diagnosis, prenatal sex selection, genetic testing and screening, and genetic counseling. The Indian parliament has recently passed a bill to ensure ethical and moral provision of feto-maternal services. The practice of prenatal sex selection for nonmedical reasons is forbidden.

Unwanted effects due to variations in economic background and the influence of the caste system have, to a large extent, not been seen in the migrant populations of the West and other developed nations. This is largely due to improvements in financial position and better access to healthcare facilities. However, this may not be true in places where certain services, including genetic services, are predominantly provided by the private sector; for example, in the United States. However, the majority of the states have statutory public health programs that ensure diagnosis and management of a large number of genetic diseases. Whether these services are equally available to all sections of the society remains unanswered. In the United Kingdom, the National Health Service (NHS) is based on the simple principle of "provision of healthcare dependent on clinical need and free at the point of delivery." The nationwide network of NHS-funded genetic centers and laboratories ensures a quality clinical genetic service shared by other specialists from pediatrics, oncology, obstetrics, radiology, and general practice. Despite an effective and efficient system of healthcare through the NHS, there is still some concern about racial discrimination among people from the Indian subcontinent (Bhopal, 2002). This is probably confined to only certain specialties, excluding clinical genetics.

FAMILY STRUCTURE AND DYNAMICS

The family in any Indian subcontinent region is the vital unit. The whole society is family-oriented, and loves to work endlessly and tirelessly to ensure the family's welfare. This is irrespective of faith, caste, or economic class. For several generations, emphasis was on the style of a "joint family." It was based on the concept of one person being head of the family, with all incomes collectively pooled in one pot. In the majority of cases, the leading member would be a male and the decision-making process would be executed through this person. All members of the family would share living accommodations and all facilities. All adults would be expected to be jointly responsible for the welfare of the women and children. This system worked well, especially when the income was derived from one source. It worked like a "mini welfare" system. The system is rewarding in crisis situations, including the need for expensive healthcare. Apart from the several benefits of this system, pitfalls include unequal distribution of wealth and resources, persistent stress on those actively pursuing earning more for the family, constant pressure on the womenfolk to ensure a regular supply of cooked meals, and carry out the daily laundry, housekeeping, and looking after children. Often, some of these tasks are delegated to the young and relatively healthy women of the family. Gradually, the joint family system is breaking up and is being replaced by the nuclear family set-up, comprising parents and two or three children. This is more commonplace in the fast-developing urban areas and established large cities and towns.

In the context of the family, a brief discussion about the practice of inbreeding is justified. The practice of inbreeding or consanguinity is not unheard of in most of the world's social structures. However, it is common in the Middle East and the Indian subcontinent, and it is more common in the northwest and the south, where the majority of the people happen to be Muslims. Not surprisingly, consanguinity is associated with being Muslim. This is not true, since it is not uncommon in certain Hindu families from the south, and some other ethnic minority groups. There are several explanations in favor of this custom. First, it is probably linked to uncertainty and fear for one's daughter's welfare should she marry an unknown person (not known to the family or community). Second, it might ensure keeping the family wealth, inheritance, or economic resources within easy reach of the daughter and son-in-law, or vice versa. Third, the elders and nobles of the family and community might be of the opinion that such a union would be best for the future progeny. It is likely that the community leaders, if they happen to be clerics or motivated by faith, would support this approach if it would safeguard preservation and continuity of the faith. In this regard, it does not matter what particular faith it might be. However, if this were the case, then this practice would be more popular among minority people due to constant fear of being outnumbered by the majority. On the other hand, the majority of people might not be too concerned, since they would not have much trouble finding someone of their choice. In either case, this kind of approach is a good example of assortative mating. Whether the practice of consanguinity is related to an excess of genetic diseases, increased incidence of birth anomalies, or extra burden of late-onset complex medical problems, it needs to be approached in a clearly nonjudgmental and unbiased manner. Making careless remarks or even expressing one's disapproval nonverbally is highly likely to hurt the person's feelings, initiate rage, and raise anxiety. This is particularly true in any clinical genetics activity, whether through telephone contact or during a clinical appointment.

RELIGION

The diversity among South Asians is reflected in its number of religious groups. These religions have a significant impact on the society and family life. The attitude toward health and perception of illness are largely influenced by the type of faith. This is probably relevant to understanding and appreciating genetics as a health or disease denominator. Among the various religious groups, Hindus, Sikhs, and Muslims outnumber several other small faith groups including Christianity, Buddhism, Jainism, and Zoroastrianism.

Hinduism

Hinduism is not a single faith, but rather an amalgamation of several beliefs that include diverse religious practices, attitudes toward life, and patterns of social behavior. Its origin dates back to about 5,000 years B.C. It is believed to have spread from the northwest region of the Indian subcontinent to the plains of north and central India, and deep into southern regions. Despite regular foreign invasions and constant migration to and from the subcontinent, the diversity and wide distribution of Hinduism is well maintained.

Hindus are identified with idol worship. There are various images of God with specific names or interpretations. This has resulted in various gods and goddesses being assigned with mythical roles. It is often mistaken as worshipping several different gods. Hindus believe in one God called *Brahma*, or the "supreme almighty." Lord Brahma is regarded as the *creator* of the universe along with Lord Shiva as the *destroyer* and Lord Vishnu as the *preserver*—the revered *Trinity* of Hinduism. Beliefs are taken from scriptures described in detail in such texts as the Vedas and Upanishads. The *Bhagavad Gita* contains the gist of these scriptures, and is the most sacred and revered Hindu text. Hindus believe that life is a continuous circle, without beginning or end. Good fortune in the present life depends on one's behavior or actions (Karma) in a former life. This is also known as *Law of Karma*. Each person is born in a Hindu family, and is expected to follow its rules and regulations. This is reflected in the caste system, which is an essential ingredient of the Hindu society. Hindus have a strong belief in astrology. They attach importance to luck or omens. Generally, Hindus have a fatalistic attitude. The left side of the body, such as the left hand or left foot, is considered inferior. The use of the left hand for greeting may not be seen positively.

The majority of Hindus are vegetarian. The cow is considered to be sacred, probably originating from its central role in agricultural life, and is worshipped like a mother. It is primarily linked to the valuable utilization of its several products. Milk is a valued food for babies and youngsters. It is also used in making butter and yogurt, which play a major role in the vegetarian diet and, as well, in religious ceremonies. Cow dung is used as fuel for daily cooking and for fertilizing crops. Cows produce bulls and oxen, which are used to plow the fields. Beef eating is strongly prohibited, probably due to a number of economic factors. If beef could be eaten, there would not be enough cows to fulfill these other essential functions.

Attitudes toward contraception and procreation may differ among Hindus, but are largely similar. In general, contraception is not preferred and procreation is desired. Having a male child first is generally preferred, and family size may depend on having a male child. The family decisions usually rest with the elder male member of the family, or the husband. Abortion could be a mild taboo, relating to the belief in sanctity of life. Sex education is done secretly in the home, from female to female or male to male. During menstruation, sexual intercourse and prayers are avoided. The family usually arranges marriages. Child marriages do happen, although illegally. The sexes are segregated, and common-law marriages, courtship, and homosexuality are forbidden. Childbirth is a major family and social occasion, and traditionally, female relatives usually support the mother for about forty days. A traditional birth ceremony usually takes place soon after delivery, and approximately after ten days there is a formal naming ceremony.

Death is accepted as a normal part of the life cycle. It is seen as the ultimate point of life when the soul leaves the body and travels far beyond to space in search of another physical place or body of any shape or species. Bereavement is conducted within the family. The funeral is arranged as soon as possible, and is followed by cremation. Burial is only done for infants or young children. Traditionally, for the next three days the family gathers at home and discusses the deceased and prayers are offered. The family offers meals. After the third day, members of the extended family and friends congregate as an expression of sympathy. The women and men do not work for the next ten days. On the tenth day the bereavement is closed, and after that day, the death is celebrated annually, usually with a prayer and communal meal.

Every year Hindus celebrate a number of festivals. Hindu women celebrate "brother's day" and fast for one day to pray for long life for their husbands. The Hindu festival of lights, or Diwali, is celebrated in October or November, according to the Hindu lunar calendar. This festival has different connotations in the various regions of India. In some areas Diwali is a renewal of life, and in some it is the start of the new year. The spring festival of Holi is celebrated in a variety of ways throughout India in February or March. Its highlights include throwing colored water or powder on one another, along with dancing in streets.

Islam

Islam is a major, fast-growing global religion, originating from the Arab subcontinent. Its followers are Muslims who submit themselves to the will of Allah. The word *Allah* is Arabic for "God." Muslims reject all other gods or faiths, and follow the teachings of the Prophet Mohammed. These teachings are contained in the Islamic scripture, Quran.

Islam is based on principles of generosity, fairness, respect, and honesty. Each Muslim individual is expected to practice the following way of life:

1. The belief in Allah and his messenger Mohammed.
2. Following the Holy Quran (Koran).
3. Offering five prayers a day in the direction of the Holy Mosque at Mecca. Men and women are separated.
4. Observe fasting (Ramadan) for one lunar month of the Islamic calendar.

5. Regular contributions to the benefit of the needy.
6. A pilgrimage to Mecca at least once in a lifetime (Haj).

The local mosque is the center of spiritual and social contact for the community. It is a place for prayer. It usually has facilities for free evening meal and free overnight accommodation. It is also the official residence for the designated priest (imam). He acts as community leader and offers spiritual support and counseling. He is also expected to be responsible for the educational, economic, and health welfare of the community.

The Islamic calendar is based in lunar months consisting of 29 or 30 days. The year is 355 days long. It is likely that the ages based on this calendar may be different then the Christian calendar. The holy fasting month (Ramadan) lasts for one full lunar month. At the end of this month, the holy festival of Eid is celebrated. Children are only expected to fast from age 12 onwards, beginning with a few days and gradually increasing to one full month as they grow older.

The family usually arranges a marriage, and marriage between cousins is common. Muslim women must wear a veil (purdah or burkha) in the presence of males who are not members of the extended family. A Muslim woman is usually touched by only one man, her husband, and would prefer a female physician. Couples are expected to wash or shower after sexual intercourse. Adultery is strictly forbidden and is punishable under the Islamic law (Shariah). Polygamy is permitted under the Islamic tradition, and it is not unusual to come across four wives for one male. However, this is limited to rural areas and among the tribal minority groups. This practice is not commonly encountered in urban areas and is legally prohibited in the West. It would not be inappropriate to elicit this information while drawing the pedigree.

Cleanliness is highly valued in Islamic culture. The pig is a symbol of uncleanliness because it lives in a filthy environment. Dogs are also considered unclean. Meat and poultry must be prepared by draining the blood to make it clean (Halal). There are several taboos in Islamic culture: gambling, consumption of alcohol, tobacco, pork, and lard are forbidden. Other taboos include owning statues, images, pictures, photographs, and crosses. Muslims may even refuse passport photographs, since images are believed to imitate God's creations. Sex education, postmortem examinations, and cremation are also forbidden. The preservation of the deceased body is considered important because of the belief in life after death. Muslims respect graves and value visits to graves and shrines of the dead.

Muslims believe in five modes of therapy, which they may use singularly or intermixed. Apart from modern scientific remedies, indigenous medicines offered by Hakim are popular. Faith healing (stones and charms), religious therapy, and food therapy have their place. Muslims believe in fate and accept serious and incurable diseases as the "will of Allah." Contraception and termination of pregnancy is unacceptable. The option for prenatal diagnosis with opportunity to selective termination of pregnancy is usually refused.

Sikhism

This is a relatively new religion, which was founded in the fifteenth century by Guru Nanak in northwest India. He believed in single God and rejected the Hindu caste system. Sikhism combines beliefs of both Hindus and Muslims. The Sikh believe in karma and reincarnation, and idol worship and representations of gods in pictures is forbidden. Guru Nanak had nine masters (*gurus*), and followers whose writings are compiled in the holy book, *Guru Granth Sahib.*

A Sikh is not necessarily born in a Sikh family. A newborn or a young child can be taken to a Sikh temple to be named or professed to be a Sikh. There is no set age for baptism, which is an important ceremony. However, following the ceremony an individual is expected to bear the responsibility of being a Sikh.

A male baptized Sikh is expected to wear turbans and not to cut his hair or shave. A Sikh is identified by five symbols, all beginning with *K*:

- Kesh (uncut hair)
- Kanga (comb)
- Kara (steel bangle, usually worn on the right wrist by both men and women)
- Kacha (pair of white undershorts, symbolic of chastity)
- Kirpan (short sword, 8 inches long and sheathed).

Use of alcohol and tobacco smoking is forbidden. Many Sikhs are vegetarians but there are no strict restrictions. Beef is generally not consumed. Social and religious activities take place in temple. Daily prayer meetings are held in the Sikh temple (*gurudwara*) and end with a communal meal (*langar*).

Sikhs do not have particular beliefs about health. They are liberal in choosing appropriate therapy. Contraception is prohibited and procreation is highly valued. Parental consent is needed for marriage, and marriage is usually arranged. Abortion may be permitted for specific purposes including sex selection. A woman usually takes her husband's surname upon marriage. The traditional Sikh method of greeting is by joining palms of the hands pressed together in front of the chest. Women do not shake hands. Direct eye contact is considered rude and a sign of disrespect. Generally all Sikhs are courteous, hardworking and kind people.

DELIVERY OF TRANSCULTURAL GENETIC SERVICES

Essentially, delivery of genetic services to a couple or family from the Indian subcontinent need not be different. However, all genetic professionals would require some knowledge and understanding of various sociocultural practices and beliefs. The concept of transcultural genetic counseling is relatively new. It includes all essential components of dealing with a genetics referral. Conventionally, it comprises assessment and explanation for the genetics referral, pre-clinic contact by a genetic associate or genetics nurse specialist, attending a clinic appointment, help in understanding and perception of genetic risk, assistance in appreciating various options, discussion about prenatal diagnosis, breaking bad news, explanation and consent for genetic testing, implications for the extended family, and post-clinic contact and follow-up. It would be helpful to explain how the genetic counseling session differs from other medical consultations. Generally, these people respect doctors and expect diagnosis and treatment. It is important to establish a positive relationship at the beginning of the genetic counseling session. Indirect counseling should be avoided, and the accepted format of "nondirective and nonjudgmental approach" should be followed.

Decision making usually rests with the husband or a senior male member of the family. It is important to keep this in mind if there should be a need for second appointment for tests or for other reasons. A thorough explanation for all diagnostic procedures should be provided. Carrier testing is not usually welcome, and may mean stigmatization if an individual is found to carry an altered gene. There is now increasing awareness of carrier status for important genetic diseases such as thalassemia or a metabolic disease. This may have a special impact on arranged marriages. If a wife is perceived to have the "defective" gene then the husband may go to the extreme of divorce and remarriage. This is relevant while discussing an X-linked recessive or autosomal dominant disorder. Extra care should be taken while discussing an autosomal recessive disease where both husband or wife would be heterozygous for the altered gene. Efforts should be made to minimize blame, feelings of guilt, and low self-esteem.

Lack of reliable family information may be a challenge to genetic professionals. It is not uncommon to come across individuals without a clear date of birth. Birth records are usually not kept by the family in almost all countries of the Indian subcontinent. This might pose a problem for the first-generation immigrants. Age is often calculated in association with an important event or celebration, such as birth or death of a family member. They are often assigned a hypothetical date of birth for the purpose of medical registration and immigration. However, this is being rectified to great extent in the second-generation immigrant population from the Indian subcontinent. The family may sometimes encourage use of astrological charts as a proof of age.

Another issue may be lack of reliable family information. Husband and wife may disclose different information. The wife may tend to prefer discussion about her children's health and development, while the husband may have different priorities. Undoubtedly, privacy and confidentiality must be observed. Marriage between cousins is not uncommon, and the genetic counselor should exercise a nonjudgmental approach without manifesting his or her bias either through verbal or nonverbal communication. It is likely that some individuals or couples may consider in-depth family history-taking to be irrelevant. A simple explanation at beginning of the consultation might be helpful.

In recent years, attitudes toward prenatal diagnosis have changed. Hindus and Sikhs generally have no reservations about the option of prenatal diagnosis, and would accept termination of an affected fetus. However, Muslims may not accept this option—but there is a slight shift in this approach. Some young couples from the second-generation immigrant group are inclined to accept this option. Religious leaders, previously strict against this option, may now condone it, provided that it would avoid birth of an affected child and that termination of the pregnancy is done prior to "quickening," when the fetus is not considered to be viable. A majority of couples and families would accept the adverse outcome as "Allah's will." The genetic counselor should approach this situation with extreme caution, and offer assistance in the decision making process. Prenatal diagnosis by amniocentesis and ultrasound examination for sex selection is not uncommon, particularly in India. Despite its ban through an act of parliament, it is still endemic in India. This practice is illegal in the United Kingdom and United States. Nevertheless, new techniques are being developed for sex selection including preconception sperm sorting, and pre-implantation diagnosis and selective embryo transfer. Alternative approaches to reduce genetic risk should be explored with the couple or the family. This would include artificial insemination by donor, use of donated egg for in-vitro fertilization, and pregnancy in a surrogate mother with the husband's sperm. Adoption is also acceptable in some cases.

Care should be taken while discussing children with learning or behavioral difficulties. Institutional care is not preferred compared to care provided by the family. Description of dysmorphic features may be meaningless to most. Discussion about a syndrome diagnosis should be done with extreme caution. Muslims and Sikhs may refuse photographs of such children. Muslim parents may decline an autopsy; however, it may be permitted if required for legal purposes. Informed consent from a Western perspective may not be fully appreciated.

Nonverbal communication is important, but direct eye contact may be considered disrespectful. Physical examination of a woman by a male examiner may not be acceptable. Whenever possible, a female geneticist should be assigned such procedures. It is reasonable to expect reaction to bad news, or grief, particularly from a woman. Embarrassing situations may manifest with silence or blushing. Wherever possible, use of idioms and technical phrases should be avoided. Use of practical aids and illustrations, for example pictorial atlases, is important.

SUMMARY

Understanding and appreciation of population and cultural heterogeneity among people from the Indian subcontinent is important in delivery of any specialist medical service. This is even more relevant to genetic services. All genetic professionals should have some knowledge about key issues important to the process of genetic counseling. These issues encompass different languages, family dynamics, social and cultural practices and beliefs, religious practices and beliefs, self-perception of health and disease, and attitudes toward conception, prenatal diagnosis, termination of pregnancy, and living with and caring for an individual with a chronic incurable disease. It is important to keep in mind that the decision-making process rests with the husband or a senior male member of the family. There may be some difference in attitudes toward genetic counseling between the first-generation and second-generation immigrant families.

REFERENCES

Bhopal R. (2002). Heterogeneity among Indians, Pakistanis, and Bangladeshi is key to racial inequalities. *British Medical Journal 325*, 903.

Johnson AM, Nadirhaw Z. (1993). Good practice in transcultural counselling: An Asian perspective. *British Journal of Guidance Counselling 21*, 20–29.

Naveed M, Phadke SR, Sharma A, Agarwal SS. (1992). Sociocultural problems in genetic counselling. *Journal of Medical Genetics 29*, 140.

Patil SR, Fisher NL. (1996). Traditional Asian Indian culture. In *Cultural and ethnic diversity: A guide for genetic professionals*, pp. 142–155. Johns Hopkins University Press, Baltimore and London.

Qureshi B. (1989). *Transcultural Medicine: Principles and Practice.* London: Kluwer Academic Publishers.

9.

MEDICAL GENETICS AND GENOMICS RESEARCH IN THE MUSLIM WORLD

Aida I. Al-Aqeel

INTRODUCTION

Medical genetics involves the application of genetic knowledge and technology to address specific clinical and epidemiologic issues. Using genetics to benefit society requires the use of empirically verified knowledge within an ethical framework that combines appeal to written precedent with sensitivity to the options of individuals and families dealing with choices and necessities within the laws, norms, and traditions of their society (Al-Aqeel, 2007). The practice of medical genetics is directly influenced by rules of bioethics. Bioethics is the study of ethical, social, legal, philosophical, and other related issues arising in healthcare and biological sciences (Al-Aqeel, 2005). Medical ethics are based on the moral, religious, and philosophical ideals and principles of the society in which they are practiced. It is, therefore, not surprising to find that what is ethical in one society might not be ethical in another society. It is mandatory for the practicing physicians to be aware of such backgrounds before they make their judgments on different medical practice decisions (Serouri & Dickens, 2001).

THE ISLAMIC ETHICAL FRAMEWORK FOR MEDICAL GENETICS AND GENOMICS

Islam is the continuum and the culmination of, not an alternative to, the sister Abrahamic faiths of Judaism and Christianity. In the *Quran*, they are linked by the one chain of God's message to humanity. The *Quran*, which Muslims believe is God's very word, says: *The same religion He enjoined on you as the one He enjoined on Noah, and this We reveal unto you and that We enjoined upon Abraham, Moses and Jesus, that you should uphold the faith and break not your unity therein* (The Holy Quran 42:13). This commonality we Muslims respect, although we are aware that on further comparison it is natural to find differences and distinct schools of thought between the faiths, and even within one faith. Islam is the predominant religion in the developing world; however, it is not an "Arab" religion. In fact, out of the 1.6 billion Muslims in the world, <10% of Muslims are Arabs. Among the 57 Islamic countries that make up the Organization of the Islamic Conference (OIC), the specific religious and cultural values of each country's population have a significant impact on the health, education, and social policies that drive the healthcare or models of care that patients receive (Al-Aqeel, 2005; 2007)

The essential core of Islamic teachings is the perfection of the ethical conduct of a human being. The road to moral and spiritual perfection is described as the "quest for God" in Islam. The seekers after God must satisfy two conditions: their actions must be governed by the prescriptions or ordinances of the "divine law," and they must ensure God is constantly present in their hearts. On the other hand, God gave humans life, and with that also gave humans the freedom and the authority to do good or evil. God also gave man the basic knowledge of "good" and "evil" at the time of his inception. The Holy *Quran* says (91: 7–10): *The human soul—the way He molded it and inspired it with knowledge of its evil and its goo —bears witness to the fact that indeed he who cleanses it (of all impiety) shall be successful, while he who corrupts it shall face down* (Figure 9-1).

In Islam, bioethical decision making is carried out within a framework of values derived from revelations and tradition. It is intimately linked to the broad ethical teachings of the Holy Qur'an and the statements, *Hadiths*, and traditions of the Prophet Mohammed, *Sunna*, and thus to the interpretation of *Shariai'ha* (Islamic law). *Shariah'ha* classifies all human actions without exception into one of five categories: obligatory, as fasting and praying; recommended, as marriage and family formation; permitted, as breaking fasting when sick or traveling; disapproved but not forbidden, as divorce; and absolutely forbidden, as killing. Development of *Shariai'ha* law in Islam over the ages has also required *Ijtihad,* the law of deductive logic, which includes *Ijmaa* (consensus) and *Qiyas* (analogy), which is the intelligent reasoning used to rule on events not mentioned by the Qur'an and Sunna, by matching against similar or equivalent events ruled upon, resulting in seven schools of *Figh* (jurisprudence). These are the four international *Sunni* schools: (*Hanbali, Maleki, Hanefi, Shafe'i),* inaddition to *Zaidyaha* in Yemen, and *Abaa'diyaha* in Oman, and a *Shiaha'a* (*Ja'fari*) school (Al-Aqeel 2005). Others are not Islamic but ethnic, and may even violate Islamic norms. Where appropriate,

7. By *Nafs* (Adam or a person or a soul), and Him Who perfected him in proportion;

وَنَفْسٍ وَمَا سَوَّىٰهَا ﴿٧﴾

8. Then He showed him what is wrong for him and what is right for him.

فَأَلْهَمَهَا فُجُورَهَا وَتَقْوَىٰهَا ﴿٨﴾

9. Indeed he succeeds who purifies his ownself (i.e. obeys and performs all that Allâh ordered, by following the true Faith of Islâmic Monotheism and by doing righteous good deeds).

قَدْ أَفْلَحَ مَن زَكَّىٰهَا ﴿٩﴾

10. And indeed he fails who corrupts his ownself (i.e. disobeys what Allâh has ordered by rejecting the true Faith of Islâmic Monotheism or by following polytheism or by doing every kind of evil wicked deeds).

وَقَدْ خَابَ مَن دَسَّىٰهَا ﴿١٠﴾

(The Holy Qur'an 91:7-10)

Figure 9-1 The essential core of Islamic teachings is the perfection of ethical conduct of a human being. These verses from the Holy *Quran* 91:7–10 stress that God (*Allaha*) gave man life, and with that, also gave man the freedom and the authority to do good or evil.

consideration is also given to *maslaha* (public interest) and *urf* (local customary precedent; see Kamali, 1991). Therefore, Islamic *Shariai'ha* is detailed in the endeavor of *Figh* and governed by its principles, *Usul* (Zaki, 1987). The *Shiaha'a* (*Ja'fari*) branch of Islam has in some cases developed its own interpretations, methodology, and authority systems, but on the whole its bioethical rulings do not differ fundamentally from the *Sunni* positions (Daar & Al Khitamy, 2001; Al-Aqeel, 2007).

The Islamic Jurisprudence (Figh) Council of the Organization of Islamic Countries (OIC countries), regularly holds conferences in situations that require specialist knowledge (for example, decisions concerning medical practice or research). A "*fatwa*" (consensus edict)—an authoritative ruling on a point of Islamic law—is what usually emerges. For rulings pertaining to medicine, these consensus groups will typically include a broad and diverse representation of *Ulema* (Islamic jurists) and specialists, including clinicians and scientists from relevant disciplines, the latter responsible for providing the necessary background information. The decision-making process is typically transparent, with members of the wider community able to scrutinize the arguments employed and the textual material underpinning these edicts. Counterarguments may be presented, and it is not unusual for two or more seemingly contrasting opinions to coexist. In such cases individuals are, in principle, free to choose whichever judgment they find most agreeable, though in practice many will choose to remain loyal to their particular school of thought (Al-Aqeel, 2005; Kamali, 1991; Darr & Al Khitamy, 2001). However, these consensus edicts are not binding by law, and every country has its own legislation (Al-Aqeel, 2007).

The Islamic *Shariai'ha* law has established many principles that can be applied to therapy and treatment. These principles are considered the most prominent Islamic phenomena of medicine. In his book, *Rules of Jurisdiction for the Welfare of Human Beings*, Ezzuddin Bin Abdul Salam indicates, "As the Islamic *Shariai'ha* is interested in specifying aims and objectives, it also regulates means and procedures. As the attainment of pure goals is very rare, the selection of the proper methods is necessary to achieve supremacy of interests over harm. Therefore, it is sufficient (when validating a certain action) to achieve goals that excel in some way, though tempered by some harm. (Al-sayad, no year).

In *Shariai'ha* there are judicial rules that encourage achieving ends, warding off corruption, and avoiding harm and evil; e.g., when removing harm is followed by an after-effect harm, a balance has to be sought to reach the lesser degree of the two harms. These general judicial rules are only entrusted to the jurisdiction of the scholars of jurisprudence, yet the doctor is required to be, to some degree, knowledgeable of the *Shariai'ha* laws of which he is in need, in order to practice his profession in keeping with the Islamic characteristics of medicine as represented in a comprehensive viewpoint, fair judgment, and righteousness of work (Al-Aqeel 2007).

Some of the Islamic *Shariai'ha* principles governing medicine, which are authoritative and agreed precedents are outlined in Box 9-1. (Ajlouni 2004; Al Othiamian 2006; Al-Aqeel 2007).

BOX 9-1 ISLAMIC SHARIAI'HA PRINCIPLES GOVERNING MEDICINE

1. Physical wellbeing has precedence over religious wellbeing: *"Sahat Al Abdan Mukadamaha Ala Sahat Al Adian"*
2. Necessities override prohibitions: *"Al Drawrat Tabieh Al mahdawrat"*

3. Preventing harm is preferable to procuring benefits: *"Draa Al Mafasad Mukadam Ala Jalab Al Manafaa"*
4. The basic concept in useful matters is permissiveness: *"Al Asal Fii Al Manafaa AL Abaha'a"*
5. The basic concept in harmful matters is prohibition: *"Al Asal Fii Al Madar Al Tahreem"*
6. Human life protection is one of the main principles of *Shariai'ha*: *"Al Mahafada'a Ala Al Nafas Al Bashariaha Ahda Al Maqasad Al Shariai'ha'a Al Raisayaha."*

"If any one saved one life, it would be as if he saved the lives of all mankind" — The Holy *Quran* 5:32.

GENETICS COUNSELING IN THE DEVELOPING WORLD

It has been estimated that 80% of the world's populations live in countries with low gross national product, and have not reached the levels of economic development of industrialized nations. Some may argue that there are many other priorities ahead of genetic diseases in developing countries where health, social, and educational facilities are poorly developed (Donnai, 2002).

Genetic counseling is the process in which an individual or a family obtains information and advice about a genetic condition that may affect the individual, his progeny, his relatives, or the family as a whole. Based on this knowledge he or she can make the pertinent decision regarding marriage, reproduction, abortion, and health management. Islamic teaching encourages counseling. The Prophet Mohammed said: "Religion (Islam) is sincere counseling and good advice." The Prophet Mohammed also said: "The counselor should be trustworthy." Indeed, a counselor cannot be trustworthy unless he or she is proficient in the field in which they are giving advice (Albar, 1999).

Counseling families in the Muslim world must be done in accordance with cultural and religious guidelines, to initiate treatment and to prevent these disorders. Counseling will include other family members, since the impact of these disorders includes other family members with recurrence risk. However, treatment is either difficult or expensive, with mixed outcomes, or unavailable in some centers. Therefore, prevention is of utmost importance (Al-Aqeel 2005, 2007).

A consanguineous marriage is usually defined as marriage between people who are second cousins or closer, and leads to and increases birth prevalence of infants with severe recessive disorders (El-Hashemite, 1997; Modell and Darr, 2002). It is customary in Middle Eastern populations, Irish travelers, Zoroastrians, some Jewish communities, and many tribes in sub-Saharan Africa and Southeast Asia. Although the custom is often perceived to be associated with Islam, it is in fact independent of religion (Al-Aqeel, 2005).

It is estimated that globally, at least 20% of the human population live in communities with a preference for consanguineous marriage, and that at least 8.5% of children in the world have consanguineous parents. This has led to an increase of the birth prevalence of infants with severe recessive disorders, but has a relatively small effect on the prevalence of dominant and X-linked disorders, and its role in complex disorders is still uncertain (Modell & Darr, 2002). Consanguineous unions account for 20%–70% of all marriages in the Middle East, excluding Israel and Cyprus (Teebi & El Shanti, 2006). In spite of the fact that Islamic teachings discourage it, consanguineous marriage is integral to the structure of many Muslim societies, including the Arabian Peninsula, and is known to have many social and economic benefits. It is narrated that Omer Ibn Al-Khatab, the second *khalifa*, noticed that the progeny of the tribe of Bani Althaa'b had become weak and unhealthy because of intermarriage of cousins. He advised the tribe to avoid close-cousin intermarriage and to seek wives and husbands from tribes further afield. He said: "Marry from far away tribes; otherwise you will be weak and unhealthy" (AlBar, 1999; Al-Aqeel, 2007).

Consanguinity, in combination with large family size and advanced maternal age, raises unique challenges and opportunities. It results in pedigrees with multiple affected individuals. This helps in homozygosity gene mapping of a multitude of novel Mendelian diseases (Al-Aqeel, 2000, 2005; Martignetti, 2001). The tradition of consanguineous unions is likely to persist, requiring the successful management of these disorders, which then incurs a high financial cost; e.g., the available injectable enzymes for Gaucher disease cost ~$30,000 per year for an infant. Consequently, it is a great burden on the healthcare system, the family, and the society, and carries a certain amount of stigmatization (Box 9-2). A review of the files of the Riyadh Military Hospital, Saudi Arabia, over a 10-year period, documented more than 150 varieties of neurodegenerative disease among 2,000 children. Some autosomal recessive disorders are common; e.g., sickle cell anemia. Others are unique; e.g., Al-Aqeel Sewairi syndrome (Al-Aqeel, 2005). In these disorders, the exact molecular defect is known. Therefore, prevention is preferred by either preimplantation genetics diagnosis, or prenatal diagnosis in accordance with the recommendations of Islamic scholars (Al-Aqeel, 2005; 2007).

BOX 9-2 IMPACT OF GENETIC DISEASES ON THE DEVELOPING WORLD POPULATION:

1. Impact on the patient and the family
   a. General quality of life
      - Patient suffering

- Parents' enormous guilt/shame
- Burden of care
- Psychological distress

b. Adverse effects on the family dynamics
- Focus on sick child, less attention on normal children
- Marital stress

c. Genetic stigmatization

2. Impact on healthcare professionals
   - Medical care - limited to supportive care
   - Frustration - untreatable diseases
   - Genetic counseling (prenatal diagnosis/termination)
   - Desperate therapies (mixed outcome)
3. Impact on the community and society
   - High medical and supportive costs
   - Community need to provide prevention and/or treatment

MEDICAL GENETICS AND GENOMICS IN THE DEVELOPING COUNTRIES

Medical genetics and genomics are rapidly evolving fields with a huge influx of knowledge. Such knowledge, especially when it involves humans, raises complex ethical, legal, social, and religious issues; many countries are struggling to establish their own frameworks. This is due to three general characteristics of genetic information: the implications of genetic information are simultaneously individual and familial; genetic information is often relevant to future disease; and genetic testing often identifies disorders for which there are no effective treatments or preventive measures (Burgess et al., 1998). Although there is no single ethical issue that unifies the field of genetics, informed consent, confidentiality, and the potential for social harm and psychological distress are issues that physicians involved with testing should understand (Al-Aqeel, 2007).

To insure that benefits of genomics are shared by developing countries, attention must be paid to complex ethical, legal, social, religious, and economic issues, as well as public education and engagement. Many of the advances in genomics were made, and in part are owned, by the developed world, which will further widen the equity gap in health between rich and poor nations, which is one of the concerns of the World Health Organization (Luri & Wolfe, 1994).

Genomics provides powerful means of discovering hereditary factors in disease. But even in the genomic era, it is not genes alone but the interplay of genetic and environmental factors that determines phenotype (Collins et al., 2003).

However, as is true for so much of the application of genomics, ethical, legal, and social issues will complicate this era unless complex issues regarding the patenting and licensing of gene-based knowledge and techniques are dealt with more successfully than they are today. Another social issue, with particular relevance in the United States, is the understandable concern of many patients that obtaining genetic information important to their healthcare is not worth the risk of discrimination stemming from the use of such information by potential insurers or employers. Other social issues require our attention if genomic medicine is to benefit our patients. How should genetic tests be regulated? What, if any, are the appropriate uses of direct-to-consumer marketing of genetic tests? How will healthcare providers and the public distinguish between these and responsible testing services, whether they are available through the Internet or in the hospital? It would be easy to assume that for the foreseeable future, the benefits of genomic medicine will accrue only to people in developed countries (Collins et al., 2003).

The Arab Gulf states offer good support for science and technology. Saudi Arabia approved a new national science and technology development plan in 2002, with ~$2.3 billion budget for the years 2008–2012. There is a move toward a knowledge-based economy (technology production, investment in technology, establishment of scientific incubators, new industrial knowledge cities (King Abdulaziz City for Science &Technology, 2008). Saudi Arabia's priorities are education and medical services, and there is a commitment to devoting 1.6% of the nation's GDP to research and development by 2020. Both Saudi Arabia and Kuwait are each investing around $2 billion in higher education institutes that include research centers. The Kingdom of Saudi Arabia (KSA) already has established a number of biomedical research institutes [Box 9-3], with priority being given to genomics research and biotechnology, using advanced techniques (e.g., microarrays); Kuwait has set up a new genetic center, and the United Arab Emirates is in the process of building a genetic center. However, the number of experts in these countries is still limited (Al-Aqeel, 2007).

BOX 9-3 SAUDI ARABIAN RESEARCH INSTITUTES

- King Abdulaziz City for Science & Technology (supporting basic science and medical research, with strong nanotechnology capabilities)
- King Abullah University for Science and Technology (with $10 billion endowment).
- King Abullah Economic City (extending over 168 square kilometers).
- Saudi-Italian Biomedical Institute by Saudi Arabian General Investment Agency (SAGIA; for cancer research)
- King Saud University (with many research chairs, and established nanotechnology unit)

- King Faisal Hospital Research Centre (established stem cell therapy program [SCTP] and gene therapy program)
- Prince Sultan City for Humanitarian Services (supporting humanitarian research services in all fields of medicine)
- Prince Salman Centre for Handicapped Children Research (supporting research to prevent handicap in children)

APPLICATIONS OF SHARIAI'HA PRINCIPLES TO GENOMICS RESEARCH AND BIOTECHNOLOGY

Islamic teachings offer a great deal related to the prevention and care of genetic diseases and birth defects, and for genomics research and biotechnology. In my opinion, practical applications of these Islamic *Shariai'ha* principles are as follows:

ENSOULMENT (INSPIRATION OF THE SOUL) *NAFKH AL ROUH* IN ISLAM

From an Islamic perspective, man is the vicegerent of Allah (God) on earth: *Behold thy Lord said to angels: I will create a vicegerent on earth.* (The Holy *Quran*, 2:30). The creation of human individuality and fetal development are referred to in several dozen verses of the *Quran* in various contexts, and also in *hadiths* (Aksoy 2005). Human life begins at the time of ensoulment (inspiration of the soul), *Nafkh Al Rouh*, which is stated in the *Hadith*, to be at the 120th day from the moment of conception, which is equivalent to 134 days from the last menstrual period (LMP) used by obstetricians. Prior to this moment, the embryo has sanctity but does not reach the status of a full human being; therefore, its eradication is a sin but its punishment for the abortion/eradication. (blood money) will be less than that after ensoulment (Al-Aqeel, 2005, 2007; Albar, 2002; El-Hashemite, 1997; Larijani & Zahadi, 2004).

The Holy *Quran* describes stages of the human creation in Surat 23 (Al-Muminun Sur'ah), verses 12–14:

> And indeed We created man from a quintessence of clay. Then We placed him as a small quantity of liquid (*nutfa*) in a safe lodging firmly established. Then We have fashioned the *nutfa* into something which clings ('alaqa), then We made *'alaqa* into a chewed lump of flesh (*mudgha*) and We made out of that chewed lump of flesh into bones, and clothed the bones with flesh. And then We brought it forth as another creation. So blessed be God, the Best to creators.

The role of *hadith* in Islamic teaching is to help us better understand and interpret the verses of the *Quran*, since Prophet Mohammad is the ultimate interpreter of the *Quran*. The Prophet Mohammad was not only a religious and political leader for his nation, but also a guide who would teach Muslims how to understand the *qur'anic* verses. He was reported to have said two *hadiths*, which will help us greatly to better understand *qur'anic* verse above, especially as they relate to the physical development of the embryo and the time of ensoulment (Aksoy, 2005).

The Prophet Mohammad has said:

> Verily your creation is on this wise. The constituents of one of you are collected for forty days in his mother's womb; it becomes something that clings (*'alaqa*) in the same (period) (*mithla dhalik*), then it becomes a chewed lump of flesh (*mudgha*) in the same (period) (*mithla dhalik*). And the angel is sent to him with instructions concerning four things, so the angel writes down his provision (sustenance), his death, his deeds, and whether he will be wretched or fortunate. Then the soul is breathed into him
>
> *(Aksoy 2005; Gatrad & Sheikh, 1999; Sahih AlBukhari, 1958).*

> And,
> After the zygote (*nutfa*) has been established in the womb for forty or forty-five nights, the angel comes and says: "My Lord, will he be wretched or fortunate?" And both these things would be written. Then the angel says: "My Lord, would he be male or female?" And both these things are written. And his deeds and actions, his death, his livelihood; these are also recorded. Then his document of destiny is rolled and there is no addition to or subtraction from it
>
> *(Sahih Al Bukhari, 1958; Aksoy, 2005).*

Many *hadith* scholars, referring to the first *hadith* only, and understanding the expression *mithla dhalik* as "time equal to this period" rather than "in the same period," have suggested that the angel comes to the prenate and breathes in the soul 120 days after conception. Some ancient scholars, as well as some contemporary researchers have disagreed about this interpretation, and concluded from both *hadiths* that, if one understands the expression *mithla dhalik* to mean "in the same period," then the completion of certain physical forms and ensoulment take place after the 40th day of conception. However, to interpret the expression *mithla dhalik* to mean "in the same period" is more accurate in this context from both an embryological and a theological perspective (Aksoy, 2005).

If it is accepted that ensoulment takes place at 120 days of gestation, the embryo should look like a *nutfa* (a drop of liquid: the zygote) between days 0 and 40; it should be something like *'alaqa* (something that clings: the implantation stage) between days 40 and 80, and it should be similar to *mudgha* (a chewed lump of flesh: the occurrence of somites—the appearance of the primitive streak) between days 80 and 120. As we know from modern embryology, these stages occur well before these times (England, 1990). It can be understood from the above verse and *hadiths* that, in order to receive the soul—that means to be a full human individual person—a prenate must pass through the various stages of conception: zygote (*nutfa*), implantation (*'alaqa*), formation of somites—the appearance of the primitive streak—(*mudgha*), and the beginning of the formation of bone and muscles. According to the embryological information given above, ensoulment cannot take place before seven weeks after conception, since these embryological stages are not completed before this time (Aksoy, 2005; England, 1990).

As pointed out previously, some of the judgments and decrees of Islamic jurisprudence regarding ensoulment originate from interpretations of the above verse of the Holy *Quran* and other similar verses and *hadiths*. Muslim jurists have made a clear distinction between the stages of embryo development before and after ensoulment (when God says, "then, We brought it forth as another creation";(Figure 9-2). It is quite clear that Muslims consider an embryo to acquire human status at the time when the soul is breathed into it. Prior to that moment, the embryo has sanctity but has not reached the status of a full human being. (Al-Aqeel, 2005; Aksoy, 2005; Daar et al., 2004; Larijani & Zahadi, 2004).

POPULATION GENETIC SCREENING PROGRAMS AND ISLAMIC ETHICS

Population genetic screening programs are public health programs targeted at populations or subgroups identified by their risk category. The goals are the detection or prevention of genetic disorders and birth defects at the population level (Donnai, 2002).

These screening programs are governed by several ethical principles, including: the rights of an individual to access appropriate information and to make choices about participation in the program; avoidance of social stigmatization of persons found to be at increased risk, or of those declining screening; and avoidance of the misuse of information and of discrimination based on the test results. Economic criteria alone cannot be used to justify a screening program (WHO, 1997; 1999). The Islamic *Shariai'ha* principle, "The basic concept in useful matters is permissiveness," which indicates that everything is lawful as long as it is useful to people (unless it is otherwise stipulated in a religious provision, or could be judged by analogy [*qiyas*] with unlawful things) can be applied to all methods of treatment or prevention (Al-Sayyad, no year; Al-Aqeel, 2007).

Primary Prevention Strategies

1. Control of Teratogens

Applying the Islamic *Shariai'ha* principle, "The basic concept in harmful matters is prohibition," prohibits performing any medical procedure where harm is absolute or predominantly outweighing benefit (Al-Aqeel, 2007). Therefore, rubella is virtually eliminated in many countries by vaccinating girls of school age. Syphilis and other STDs (sexually transmitted diseases) will not appear if all sexual desires are channeled through marriage, as Islamic teachings implies. Fornication, adultery, and sodomy are all harshly punished in Islamic legal code, and religiously they are considered the greatest of sins that each Muslim should avoid. Any substance that is going to be harmful to the baby (namely teratogens), e.g. alcohol, smoking, should be avoided, as the Prophet Mohammed said, "Do no harm" (Albar, 2002).

2. Premarital or Pre-pregnancy Genetic Screening

Genetic blood disorders (thalassemia and sickle cell anemia) and Tay-Sachs disease are the groups of disorders where there is most experience, and where outcome data are available. The birth rate of children with thalassemia major has fallen by at least 75% in Cyprus, Italy, and Greece, where national programs promote premarital screening, and where couples most at risk are identified before their first pregnancy (WHO, 1999).

The following disorders are transmitted by autosomal recessive inheritance and are quite common in the Islamic population: sickle cell anemia, thalassemias, and

14. Then We made the *Nutfah* into a clot (a piece of thick coagulated blood), then We made the clot into a little lump of flesh, then We made out of that little lump of flesh bones, then We clothed the bones with flesh, and then We brought it forth as another creation. So Blessed is Allâh, the Best of creators.

ثُمَّ خَلَقْنَا النُّطْفَةَ عَلَقَةً فَخَلَقْنَا الْعَلَقَةَ مُضْغَةً
فَخَلَقْنَا الْمُضْغَةَ عِظَامًا فَكَسَوْنَا الْعِظَامَ
لَحْمًا ثُمَّ أَنشَأْنَاهُ خَلْقًا آخَرَ فَتَبَارَكَ اللَّهُ
أَحْسَنُ الْخَالِقِينَ ﴿١٤﴾

(The Holy Qur'an 23:14)

Figure 9-2 Stages of embryo development as stated in the Holy *Quran* 23:14 ("*Nutfah*": mixed drops of male and female sexual discharge, "*AlAqa'a*": clot, "*Mudgha'a*": lump, each for forty days). On the basis of this text, and other texts in the Holy *Quran and Sunna,* many Muslim scholars conclude that fetal ensoulment occurs at 120 days postconception, when God says, "then we brought it forth as another creation."

glucose-6-phosphate dehydrogenase (G6PD) deficiency. They affect 20%–25% of the whole population in Hofuf and Qatif (eastern province of Saudi Arabia) and Jizan (southwest province of Saudi Arabia). As the carriers of the trait comprise one in four to one in five in the whole community, any carrier will have a high risk of marrying another carrier of the trait (Weatherall, 1998).

Applying the Islamic *Shariai'ha* principle, "the basic concept in useful matters is permissiveness," the Islamic Jurisprudence Council of the Organization of Islamic countries, in its 17th session (19–23.10.1424H/13–17 December 2003G), having looked into the legitimacy of premarital medical screening for genetic blood disorders, has decided:

> **First**: The marriage wedlock contract is governed by conditions of the Islamic *Shariai'ha,* from which legal consequences follow. Thus, additional conditions, such as enforcing premarital medical screening, are not permissible under the Islamic *Shariai'ha.*
> **Second**: The Council recommends that governments and Islamic institutions spread understanding of the importance of premarital genetic tests and encourage their use. They should facilitate such tests for those who wish to use them, while ensuring confidentiality so the results are not revealed except to the persons concerned.

Therefore, informed *premarital screening* for genetic blood disorders, thalassemia, and sickle cell anemia (which is not linked to any type of enforced prevention of marriage) has been mandatory in KSA for the past four years, after it was approved by a consensus edict. In the years 2004 and 2005, screening of almost a quarter of a million people from 130 Ministry of Health primary healthcare centers was carried out, and 2,003 and 2,441 cases of incompatibility were identified, respectively. Of these, sickle cell traits comprised 4.19% and 4.21%; actual sickle cell cases comprised 0.27% and 0.25%; β thalassemia minor comprised 3.2% and 3.24%; and β thalassemia major comprised 0.08% and 0.06%. This screening has subsequently led to a decrease in consanguineous marriages in the screened population of 9.2% and 11.6% for the years 2004 and 2005, respectively (Alhamdan 2007). However, the system is not yet able to follow up on all incompatible cases. In Bahrain, where newborn screening, school age screening, and premarital screening regularly takes place, the country has been able to reduce the incidence of blood disorders in newborns by 50% (Al-Gazali et al., 2006; Al-Aqeel, 2007).

The Iranian experience provides an instructive model from which to learn. Iran started its primary-care thalassemia prevention program in 1996, a well-coordinated national program in which screening is integrated with services for patients. The program aimed at identifying carrier couples before marriage and offered counseling, thus providing them with the opportunity to separate. Prenatal diagnosis and the option of selective abortion were not available in the country at that time. By 1999, an audit showed that couples were still opting to marry rather than separate (which had been assumed would happen), and that the people affected wanted prenatal diagnosis and selective abortion. Widespread discussions ensued, and the law was amended in 2001 to allow the option of selective abortion up to 15 weeks' gestation. A national DNA laboratory network was thereafter developed to offer prenatal diagnosis on chorionic villous samples; this reduced the birth of affected babies by 70% within five years (Samavat & Modell, 2004). However, the life expectancy of living patients has increased considerably as, out of 13,879 living patients, 23% are over 20 years of age. This has led to a new challenge, for which the healthcare system must be prepared (Abolghasemi et al., 2007). Many lessons could be learned from this program in both the developed and developing world [Box 9-4].

BOX 9-4 LESSONS LEARNED FROM THE IRANIAN PROGRAM IN BOTH THE DEVELOPED AND DEVELOPING WORLD (AL-AQEEL, 2007)

1. Screening programs must comply with societal and religious values. The messages conveyed in community education need to be clear, so that the public can understand the program's aims and objectives.
2. Working through primary care in implementing genetic screening programs in lower- and middle-resource countries is ideal, with widespread public health workers, (especially nongenetic trained health workers), and targeting education.
3. Establishing professional networks to provide genetic diagnostic services, genetic counseling, and evaluation (surveillance), with services developing in response to needs and wishes of the affected population and society in general, is critical.
4. Offering carrier identification to male army personnel and male high school students, with concomitant genetic counseling, followed by subsequent testing of the female spouse, if the man is proved to be a carrier, may avoid stigmatization of women, especially in lower- and middle-resource countries.
5. Expansion of premarital screening to other common recessive genetic disorders, including hereditary deafness and inherited metabolic disorders, is dependent upon molecular genetic screening.
6. The Iranian program provides lessons for high-income countries, too; for example, countries with minority

ethnic communities that originate from areas of high prevalence, such as the United Kingdom and the United States, and countries with indigenous populations with high carrier rates, such as Italy. The program provided evidence and supported the introduction of screening for the thalassemia trait for Bury schoolchildren in Lancashire, UK (Elton, 2005).

7. Western countries have been slow to adopt population-based genetic screening. One reason for this may be that status and interest focus on "cutting edge" science, rather than applying existing knowledge. Another reason may be the emphasis on the libertarian and individualistic meaning of "informed choice" without linking the societal debate to issues of the public's health and resources (Christianson, 2004).

Applying the same *Shariai'ha* principle, pre-pregnancy genetic screening could be done if a genetic disorder is known in a family, and the mutation for such disorder is already known. Populationwide pre-pregnancy carrier screening is possible, if the carriers' mutations are known in the population for common genetic disorders; for example, Tay-Sachs disease in the West. There is some experience with carrier screening for some of the common disorders in certain families in Saudi Arabia, for which pre-pregnancy screening is possible (Al-Aqeel et al., 2000, 2005; Martignetti et al., 2001).

Secondary Prevention Strategies

The aims of such screening programs are the early diagnosis of genetic disorders, with a view to preventing or ameliorating their effects.

Inborn errors of metabolism (IEMs) and other inherited Mendelian disorders are common in Saudi Arabia and throughout the Middle East, presumably because of the relatively high rates of consanguinity (Ozand et al., 1992). Twenty percent to 25% of all marriages in Saudi Arabia are of first cousins, another 20%–25% are second cousin marriages, and 15%–20% are family related, with a total of 60%–65% of consanguineous marriages (El-Hazemi, 1999; Al-Aqeel, 2005).

Even in segregated communities, IEMs are estimated to account for as much as 20% of disease among full-term neonates not known to have been at risk, and may affect as many as 1 in 5,000 live births (Rashed et al., 1997). Many of the IEMs carry serious clinical consequences for the affected neonate or young infant, including mild to severe mental retardation, physical handicap, and even fatality. Applying the Islamic *Shariai'ha* principle, "the basic concept in useful matters is permissiveness," such an important measure as preventing mental handicap in children is highly recommended by Islamic jurists. Therefore, neonatal screening by tandem mass spectroscopy for 16 metabolic disorders and for endocrine disorders, based on dried blood spots (DBS), was started in KSA in Ministry of Health hospitals in August 2005, with 1.2:1,000 newborns found to be positive (Al-Aqeel, 2007).

Automated whole-genome amplification on DBS on "IsoCode" paper (storing blood on an absorbent cellulose-based paper that contains chemicals to protect DNA from nuclease degradation and from bacterial growth), is an excellent approach for mass screening of heterozygotes in the extended family, as well as throughout the population in order to map genes, establish the haplotype, phenotype/haplotype, and geographical distribution, and to estimate the gene frequencies of common and complex disorders in any given country. However, limited forms of neonatal screening are available in other developing countries. Applying the sixth *Shariai'ha* principle (human life protection), the physician must follow up with treatment and counseling if neonatal screening is mandated (Al-Aqeel, 2007).

Prevention based on Reproductive Options

Reproductive options that are ethically approved by Western standards vary according to the condition for which an individual is being screened, and include prenatal diagnosis, pre-implantation diagnosis and sperm or egg donation, the avoidance of further pregnancy, or adoption.

1. Contraception and Sterilization

Applying the Islamic *Shariai'ha* principle, "preventing harm is preferable to procuring benefits," which indicates that a balance has to be established between potential benefits and harms of any method of treatment or research in use, on the condition that such use does not cause further social damage (Al-Sayyad, no year; Al-Aqeel, 2007), it is acceptable to use temporary means of contraception if the couple is agreeable and if no harm is likely to result. However, sterilization is not acceptable unless the health of the mother would be endangered by pregnancy. Nevertheless, in the situation where a couple already had two or three congenitally affected children and a lesser number unaffected, they might choose sterilization. In such a case they would find support from at least some Islamic jurists (Albar, 1999; Husain, 2003).

2. Adoption

Adoption has a different interpretation in Islam, though caring for orphan or children of unknown parents is encouraged and considered as a charity and a great act of worship. However, the lineage of the child should be kept to his biological parents. The Holy *Quran* (33:4) says: *God did not make your adopted ones your sons.* (Albar, 1999, 2002; Husain, 2003).

3. Donation of a Sperm, Ovum or Pre-embryo, or Motherhood Surrogacy

Applying the Islamic *Shariai'ha* principle, "the basic concept in harmful matters is prohibition," artificial insemination by a donor sperm, or egg donation, are all out of bounds in Islamic law. Procreation in Islamic law is limited to husband and wife, during the existence of matrimonial bondage. If divorce or death of a spouse occurs, no procreation will be allowed, including surrogacy (Albar, 1999, 2002; Husain, 2003).

A November 2000 workshop organized by the International Islamic Center for Population Studies and Research, Al-Azhar University, Cairo, considered use of assisted reproduction technologies (ART) in the Islamic world. The same above conclusions were drawn, including no embryo transplantation after husband death (Seroura & Dickens, 2001).

4. Pre-implantation Genetics Diagnosis

Pre-implantation genetic diagnosis (PGD) was introduced at the beginning of the 1990s as an alternative to prenatal diagnosis, to prevent termination of pregnancy in couples with a high risk for offspring affected by a sex-linked genetic disease (Sermon et al., 2004).

Expert groups have proposed a "family oriented" approach for communities in which consanguineous marriage is customary. Pre-implantation genetic diagnosis is a novel technique that allows the determination of genotype of an embryo before implantation, and an early form of prenatal diagnosis where in-vitro fertilization is carried out. The zygotes are grown to the 8-cell stage (morula stage). Embryos created in-vitro are analyzed for well-defined genetic defects; only those free of the defects are placed into the womb. The technique is used mainly in two broad indication groups. The first group comprises individuals at high risk of having a child with a genetic disease; e.g., carriers of a monogenic disease or of chromosomal structural aberrations, such as translocations. The second group consists of those being treated with in-vitro fertilization (IVF), who might have a low genetic risk but whose embryos are screened for chromosome aneuploidies to enhance their chance of an ongoing pregnancy (Sermon et al., 2004, ESHRE Task Force, 2001

The workshop organized by the International Islamic Center for Population Studies and Research, Al-Azhar University, Cairo recognized the importance of PGD, but was guarded about its use on nonmedical grounds, such as sex selection or family balancing, considering that each case should be treated on its own merits. Sex selection technologies have been condemned on the ground that their application is to discriminate against female embryos and fetuses (Seroura & Dickens, 2001).

Ethically Difficult Indications

For some diseases, PGD but not prenatal diagnosis can be defended from an ethical point of view. In nondisclosure PGD, which has been described for Huntington's disease but could also be applied to other late-onset diseases, patients do not wish to know their carrier status but want to have disease-free offspring (Stern et al., 2002).

Another new indication for PGD involves the selection of embryos according to their HLA type, so that a child born out of a PGD cycle can be a stem-cell donor for a sick sibling (Pennings et al., 2002). The use of PGD to diagnose risk of late-onset diseases (such as Huntington's disease and Alzheimer's disease) and to search for genes that predispose for cancer (BRCA1, BRCA2, Li-Fraumeni, neurofibromatosis 1 and 2) is also ethically debatable (Robertson, 2003; Verlinsky et al., 2002). Finally, several reports have been published on the use of sexing for social reasons and have provoked mixed reactions (Ray et al., 2003).

From an Islamic perspective, the Islamic Jurisprudence Council of the Organization of Islamic Countries has approved sex selection by PGD for sex-linked genetic disorders. Applying the fourth *Shariai'ha* principle (the basic concept in useful matters is permissiveness), this technique is ideal, as it is carried out in the pre-ensoulment (prior to the inspiration of the soul) stage at 48–72 hours of gestation (Al-Aqeel 2007). Therefore, embryo research before this time is allowed, including destroying surplus embryos and embryo freezing. However, the laboratory and medical personnel must take the necessary precautions to ensure that these embryos are not mixed when frozen. An important indication for PGD involves the selection of embryos for a certain HLA type, so that a child born out of a PGD cycle can be a stem-cell donor for a sick sibling. This application of PGD is controversial in the West. However, applying this sixth *Shariai'ha* principle (human life protection), such an indication is encouraged by Islamic scholars, on the assumption that the newly born child will be loved and well cared for by his parents and family. We have encountered this in a family with a child with leukemia; the mother underwent PGD to give birth to an HLA-compatible sibling, from whom umbilical cord stem cells were transplanted into his sick sibling. In the KSA, more than 60 cases with single gene or chromosomal disorders have undergone PGD, resulting in over 30 normal pregnancies (Hellani et al., 2004; Al-Aqeel et al., 2006).

5. Prenatal Diagnosis

A number of ethical considerations arise with regard to screening for, detecting, and managing fetal anomalies. The ethical principle of beneficence gives rise to a duty of the obstetrician to provide emotional support when needed in relation to screening, confirmatory testing, giving bad news, making abortion decisions, making management decisions after viability, and dealing with the grieving process. Other issues involve ethical decision making, such as deciding what recommendations to make concerning management of fetal anomalies after viability. The ethical principle of autonomy creates a duty of the obstetrician to help the pregnant woman make informed management decisions based on her values, religion, and goals (Strong, 2003).

The *fatwa* #4 of the Islamic Jurisprudence Council of the Organization of Islamic Countries, at its 12th session (15–22 Rajab 1410H/10–17 February, 1990G), agreed by a majority vote to allow for the option of abortion under certain specific conditions. The *fatwa* determined that an abortion may take place only if a committee of specialized, competent physicians has decided the fetus is grossly malformed, and that its life would be a calamity for both the family and itself. The malformation must be untreatable, unmanageable, and very serious, and the abortion may only be carried out prior to the 120th day of conception (computed from the date of fertilization, not the last menstrual cycle), but not including Down's syndrome or other serious conditions like thalassemia. Therefore, abortions due to serious fatal congenital disorders are carried out legally in some Muslim countries, such as Tunisia and Iran (Al-Gazali et al., 2006). However, applying the two principles, "physical wellbeing has precedence over religious wellbeing" and "necessities override prohibitions," abortion is allowed after 120 days if there is a danger to maternal life, regardless of whether the fetus is normal or abnormal. In this case, termination of pregnancy is against religious wellbeing, but it is done for the mother's physical health (Al-Aqeel, 2007).

GENOMIC RESEARCH IN THE DEVELOPING WORLD AND ISLAM

Many examples in genomics are implemented by applying the Islamic rule, "preventing harm is preferable to procuring benefits." The use of genomics research in Islam is encouraged, with a balance between potential benefits and harms, on the condition that such use does not cause social damage (Al-Aqeel, 2007).

In our experience, finding novel genes, mutations, or chromosomal defects in families with devastating metabolic or chromosomal disorders such as neuronopathic Gaucher disease (type II), allowed us to prenatally diagnose and abort these fetuses. This gives the family the benefit of such findings by pre-implantation genetic diagnosis of the embryos and the birth of normal babies. Extended genetic counseling and testing for other family members has allowed carriers to make informed choices from the widest range of available options, in order to reduce risk and to have normal babies (Al-Aqeel, 2007).

> "We "Allah" created Man in the most perfect form"
>
> *(The Holy Quran, 95:4).*

This citation from the Holy *Quran* is often used to explain that each human life has its own inherent value and goodness. While genomic research and gene therapy may have positive uses in serving to restore health (and in the process, integrity), care must be taken to ensure that other Islamic principles are not violated (Al-Aqeel, 2007).

The benefits of molecular genetics and bioengineering to Muslims have been discussed by the Islamic Jurisprudence Council of the Organization of Islamic countries in its 15th session (11th Rajab H/31 October 1998G), and they decided (Al-Aqeel, 2005):

1. to use genetic engineering for disease prevention, treatment, or amelioration, on the condition that such use does not cause further damage;
2. to forbid the use of genetic engineering for evil and criminal uses or that which is forbidden religiously;
3. to forbid using genetic engineering and its tools to change human personality and responsibility, or interfering with genes to improve the human race;
4. to forbid doing any research or therapy of human genes except in extreme need, after critical evaluation of its benefits and dangers and after an official consent from the concerned, respecting the extreme confidentiality of the information and human rights and dignity as dictated by Islamic *Shariai'ha*;
5. to allow the use of bioengineering in the field of agriculture and animals, on the condition that precautions are taken not to inflict harm (even in the long term) on humans, animals, or vegetation.

For DNA fingerprinting, the Islamic Jurisprudence Council of the Organization of Islamic countries in its 16th session (21–26.10.1422H/5–10 January 2002) has decided (Al-Aqeel, 2005):

1. It is religiously allowed to use DNA fingerprinting on judges' orders and performed in the state laboratories; in forensic interrogations to prove crimes which have no definite penalty in Islamic law (*Shariai'ha*); avoid punishment if there is any doubt, as doubt should always be used for the sake of the accused; this will lead to justice and to safety of the community, as the criminal will be punished and the innocent will be freed from guilt, which is one of the most important goals of *Shariai'ha*.
2. DNA fingerprinting may be used in lineage (genealogy) only with great caution and

confidentiality, as the *Shariai'ha* rules take precedence over DNA fingerprinting.

3. It is forbidden to use DNA fingerprinting in paternity (lineage) disputes, which should not precede the oath of condemnation (the sworn allegation of adultery committed by one's spouse).
4. It is forbidden to use DNA fingerprinting to confirm or refute legally proven lineage; the state should forbid this and inflict punishment, in order to protect people's honor and to preserve their lineage.
5. It is allowed to use DNA fingerprinting in proving lineage on the following conditions:
    - in cases of dispute about unknown lineage, as mentioned by the Islamic scholars because the evidence is either absent or equivocal, and to overcome the vagueness (suspicion).
    - in cases of disputes over babies in hospitals and nurseries or test-tube babies;
    - in cases of children lost because of war, accidents, or natural disasters, where their family could not be found;
    - to identify bodies or prisoners of war.

The challenge of obtaining a DNA diagnosis and achieving carrier testing is to ensure that it is carried out in accordance with Islamic rules. The consent has to be obtained according to Islamic ethical and cultural background of the individual, with community-based genetic counseling in one's own language. Essentially, the principles and components of consent that are generally acceptable in Western countries are also applicable to Muslims, although Muslims (depending on their level of education, background, and culture) will often want to consult with family members, and their own religious scholars, before consenting to major procedures (Al-Aqeel, 2005). The physician has to obtain the consent in the presence of two witnesses. However, if the Human Genome Project delivers as promised, DNA-based carrier diagnosis should become possible for single gene and chromosomal genetic disorders in the foreseeable future. The proposed approach has been explored in Pakistan using hemoglobin disorders as a model; because they are common, carriers (5% of the Pakistan population) are easily diagnosed and prenatal diagnosis is available locally (Modell & Darr, 2002).

While storing biological samples and establishing DNA databases may offer the opportunity to assess the population-based prevalence of specific genes and variants, simplify the search for molecular markers, define phenotype/genotype, improve targeted drug discovery and development, refine strategies for disease prevention, and provide data necessary for evidence-based decision making, serious scientific and social questions remain. Whether samples are identified, coded, or anonymous, biological banking raises profound ethical and legal issues pertaining to access, informed consent, privacy and confidentiality of genomic information and associated phenotype, civil liberties, patenting, and proprietary rights. These issues may be compounded by social and cultural factors pertaining to different habits and religions (Blatt, 2000). In the KSA and other OIC countries, these questions are still to be answered in a locally appropriate manner. This means that in practice they will be raised at the Islamic Jurisprudence Council of OIC for definitive rulings (Al-Aqeel, 2007).

Given the wide variations in the human genome, it would be useful to have standard criteria for databases of variation, with a universally acceptable system. Therefore, the World Health Organization cosponsored a meeting in June 2006, which launched the Human Variome Project (HVP). The project encouraged ethnicity and country-specific databases of mutations, to assist in the delivery of genetic health care. It also encouraged collaboration and involvement of emerging (developing) countries in the generation of molecular data, and in seeking pharmacogenetic knowledge. The project encourages and funds training programs and HVP reference centers within the emerging country groupings. It stresses that ethical and legal issues should be identified and followed (Cotton et al., 2007). An "Arab Variome Project" would build upon the sense of common purpose inherent in these distinctive populations and provide an appreciation of their place in the history of the human population (Al-Aqeel, 2007).

CLONING AND STEM CELL RESEARCH

Cloning is the field of medical science that can both have an enormous potential impact on human health and lead to new basic research discoveries. Islam allows procreation between man and woman only within marriage (Al-Aqeel, 2005, 2007). Human cloning is the production of a human being whose genetic makeup is nearly identical to that of a currently or previously existent individual. There are two kinds of cloning: therapeutic, which may allow the generation of stem cells in vitro, with their potential ability to repair damaged tissues. Because of this potential the ESHRE (European Society for Human Reproduction and Embryology) Ethics Task Force considers this technique acceptable. The second is reproductive cloning, which can be achieved by embryo-splitting or by somatic cells nuclear transfer (SCNT). SCNT is achieved by injecting a nucleus from a somatic cell of an adult into an enucleated egg (one whose nucleus has been removed). The cell would then grow into a fetus that would be a true genetic copy (clone) of the adult from which the somatic cell nucleus was taken (ESHRE Task Force, 2001).

In the Name of Allâh,
the Most Gracious, the Most Merciful.

1. **O mankind! Be dutiful to your Lord, Who created you from a single person (Adam), and from him (Adam) He created his wife [Hawwâ (Eve)], and from them both He created many men and women; and fear Allâh through Whom you demand (your mutual rights), and (do not cut the relations of) the wombs (kinship) . Surely, Allâh is Ever an All-Watcher over you.**

بِسْمِ ٱللَّهِ ٱلرَّحْمَٰنِ ٱلرَّحِيمِ

يَٰٓأَيُّهَا ٱلنَّاسُ ٱتَّقُوا۟ رَبَّكُمُ ٱلَّذِى خَلَقَكُم مِّن نَّفْسٍ
وَٰحِدَةٍ وَخَلَقَ مِنْهَا زَوْجَهَا وَبَثَّ مِنْهُمَا رِجَالًا كَثِيرًا
وَنِسَآءً وَٱتَّقُوا۟ ٱللَّهَ ٱلَّذِى تَسَآءَلُونَ بِهِۦ وَٱلْأَرْحَامَ
إِنَّ ٱللَّهَ كَانَ عَلَيْكُمْ رَقِيبًا ﴿١﴾

(The Holy Qur'an 4:1)

Figure 9-3 The importance of the concept of reproduction and family and preservation of kinship in Islam. The *Holy Quran 4:1*.

Applying the rule, "preventing harm is preferable to procuring benefits" to cloning, the real challenge is to find a balance between the need to preserve human dignity and the need for continued improvement in the quality of human life through research and development. Therefore, in addition to the fact that a resulting embryo from reproductive cloning would be severely deformed with developmental abnormalities and early aging (Bowring, 2004; Solter, 2000; Janisch & Wilmut 2001; Daar 2004), some Muslim references argue that major problems of reproductive cloning are:

1. the loss of kinship and lineage due to the unnaturalness of reproduction (mixing of kinship or the loss of it, would be considered *harram* [unlawful] in Islam) and undermining the concept of reproduction and family (Figure 9-3; the Holy *Quran* 4:1);
2. the social harms, problems of personal identity and the psychological development of a clone, and the disregard for human dignity;
3. the unjust eugenics (selecting genetic qualities by selective breeding), and trends toward "designer babies" (superior or inferior, depending upon the motives of the creator);
4. the contradiction of the Islamic belief (God is the only creator), as God Almighty says, *He it is who originates the creation, then He will repeat it (after it has been perished) and this is easier for Him* (the Holy *Quran* 30:27);
5. the contradiction of the principle of creation of all beings (human, animal, plants) in (male and female) pairs (Figure 9-4; the Holy *Quran* 36:36;51;49;75:39). Because cloning produces an exact copy of the previously existent individual, whether male or female, this leads to a disturbance of the natural existence of pairs; and finally,
6. the contradiction of the principle of diversity of creation of all beings, including humans, animals, and plants in various colors and shapes (Figure 9-5; The Holy *Quran* 35: 27–28).

Issues in public policy on cloning overlap somewhat with general stem cell matters. As previously stated, the potential harms of reproductive cloning outweigh its potential benefits; therefore, applying the *Shariai'ha* principle, "the basic concept in harmful matters is prohibition," Muslim scholars are unanimous in their opposition to cloning of humans (Al-Aqeel, 2007).

Stem cell research is morally controversial because it involves the deliberate production, use, and ultimate destruction of human embryos. The moral status of human embryos is a major ethical question in conducting this research. To create new cell lines, it is necessary to destroy pre-implantation blastocysts. The question is whether the destruction of human embryos in stem cell research amounts to the killing of human beings. Considerable differences of opinion exist with regard to the ontological and moral status of the pre-implantation embryo. On one side of the spectrum is the "conceptionalist" view. According to this view, the embryo is a "person" and because of the potential of the embryo to develop into a person, it ought to be considered as a person. On the other side of the spectrum we find the view that the embryo (and even the fetus) as a "nonperson" ought not to be attributed any moral status at all (de Wert & Mummery, 2003; Reiss, 2002).

36. **Glory is to Him Who has created all the pairs of that which the earth produces, as well as of their own (human) kind (male and female), and of that which they know not.**

سُبْحَٰنَ ٱلَّذِى خَلَقَ ٱلْأَزْوَٰجَ كُلَّهَا مِمَّا تُنۢبِتُ
ٱلْأَرْضُ وَمِنْ أَنفُسِهِمْ وَمِمَّا لَا يَعْلَمُونَ ﴿٣٦﴾

(The Holy Qur'an 36:36)

Figure 9-4 Creation of all beings in pairs; including human, animals and plants. The Holy *Quran* 51:49.

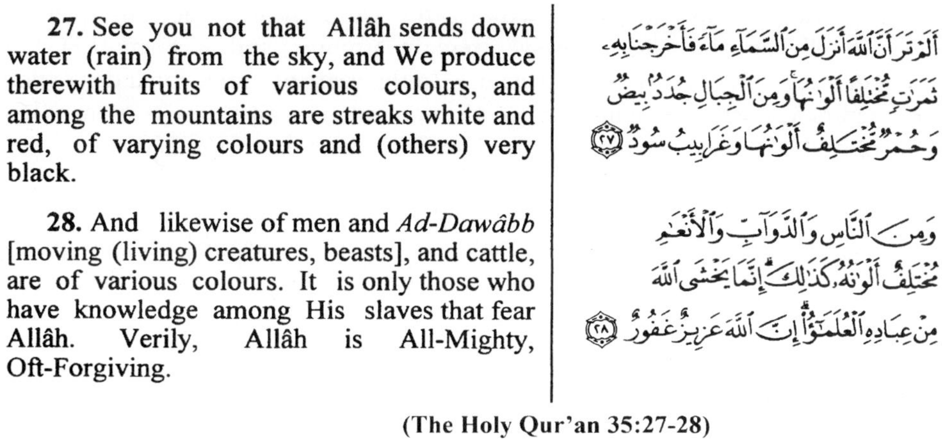

27. See you not that Allâh sends down water (rain) from the sky, and We produce therewith fruits of various colours, and among the mountains are streaks white and red, of varying colours and (others) very black.

28. And likewise of men and *Ad-Dawâbb* [moving (living) creatures, beasts], and cattle, are of various colours. It is only those who have knowledge among His slaves that fear Allâh. Verily, Allâh is All-Mighty, Oft-Forgiving.

ألم تر أن الله أنزل من السماء ماء فأخرجنا به
ثمرات مختلفا ألوانها ومن الجبال جدد بيض
وحمر مختلف ألوانها وغرابيب سود ﴿٢٧﴾

ومن الناس والدواب والأنعام
مختلف ألوانه كذلك إنما يخشى الله
من عباده العلماء إن الله عزيز غفور ﴿٢٨﴾

(The Holy Qur'an 35:27-28)

Figure 9-5 Diversity of creations of all beings; including human, animals and plants. The Holy *Quran* 35: 27–28.

Therefore, the central objective of therapeutic cloning is to cure disease, improve health, and hence strive for a better quality of life for humans. This calls for research and development in therapeutic cloning to improve the knowledge, skills, expertise, and techniques for achieving the stated objective. The problem, however, is that therapeutic cloning requires embryonic stem cells, and to acquire these, early-stage human embryos are needed. This raises serious ethical, moral, and religious concerns, for example, about the rights of the embryo, the timing of the onset of life, and the hazards to the woman (Bowring, 2004). With the widespread and unchecked use of stem cell technology, there is also the risk that it could fall into the wrong hands and be used for human cloning, with unimaginable consequences (United Nations Educational, Scientific and Culture Organization, 2004).

The Islamic Jurisprudence Council of the Organization of Islamic countries in Jeddaha in its 10th session (23–28.2.1418H/28 June-3 July 1997G) issued Decree #100/2/D10, on cloning (Al-Aqeel, 2005) containing the following points:

1. Human cloning is forbidden in any method that leads to human reproduction.
2. It is forbidden in all cases to introduce a third party into marriage, be it an egg donor, a surrogate womb, a sperm donor, or a cloned cell.
3. It is permissible to use genetic engineering and cloning in the fields of germs, microorganisms, plants, and animals, following legitimate rules which lead to benefits and prevent harm.
4. All Muslim countries are called upon to formulate the necessary legislation to prevent foreign research institutes, organizations, and experts from directly or indirectly using Muslim countries for experimentation on human cloning or promoting it.
5. Specialized committees should be set up to look into the ethics of biological research and adopt protocols for study and research in Muslim countries.
6. Biological and bioengineering research institutions (other than cloning research) should be supported and established, according to the Islamic rulings, so that the Muslim world will not be dependent on others in this field.
7. The communication media are called upon to deal with recent scientific advances from an Islamic perspective in a faithful way, and avoid employing their services against Islam, aiming to educate the public to be confident before any decision.

CORD BLOOD TRANSPLANTATION

Since the first successful use of cord blood as source of hematopoietic stem cells for transplantation in 1988, more than 2,000 patients with malignant or nonmalignant disorders have been treated with this procedure. Collection and storage of cord blood has prompted ethical considerations, mainly dealing with the issues of autonomy in making decisions about donation of cord blood, and of privacy and confidentiality in the tests required before use of placental cells for transplantation (Burgio et al., 2003).

Applying the first and last Islamic *Shariai'ha* (*figh*) principles ("necessities override prohibitions; human life protection is one of the main principles of *Shariai'ha*"), it can be argued that Islam does not totally prohibit early embryonic-stage research, especially if it is justified and deemed necessary (Al-Aqeel, 2005, 2007; Aksoy, 2005; Albar 2002; Larijani & Zahedi, 2004). However, the manner in which the early embryos may be obtained, and the inherent risks to women who would be the source of such embryos, pose serious ethical and social problems. Based on the majority of Islamic scholars, the use of an embryo for therapeutic or research purposes may be acceptable under necessity, if it takes place before the point at which the embryo is ensouled in its early stages of development (before 40–45 days), and the source is legitimate, since such cells could be used to save lives. Due to these clarifying decrees (*fatwa*), the Islamic Republic of Iran was one of the first countries to have produced human embryonic stem cells (Larijani & Zahedi, 2004; Al-Aqeel, 2005, 2007; Aksoy, 2005). At the same time, applying the Islamic *Shariai'ha* principle, "the basic concept in useful matters is permissiveness," the Islamic Jurisprudence Council of the Organization of Islamic Countries in Makkah Al-Mukarama in its 17th session (19–23.10.1424 H/13–17 December 2003G) have issued Decree # 3 on Stem Cell Therapy (Al-Aqeel, 2005):

- **First:** It is permissible to obtain stem cells, to be grown and used for therapy or for permissible scientific research, if its source is legitimate, as for example:
    1. Adults if they give permission, without inflicting harm on them.
    2. Children provided that their guardians allow it, for a legal benefit and without inflicting harm on the children.
    3. The placenta or the umbilical cord, with the parents' permission.
    4. A fetus if spontaneously aborted, or when aborted for a therapeutic reason permitted by Sharia, with the parents' permission. (Be reminded of decree # 7 of the Council in its 12th session on abortion.)
    5. Left over zygotes remaining from in-vitro fertilization, if donated by the parents, when it is ascertained that they will not be used in an illegal pregnancy.
- **Second**: It is forbidden to use stem cells, if their source is illegal. As for example:
    1. Intentionally aborted fetuses (that is, abortion without a legal medical reason).
    2. Intentional fertilization between a donated ovum and sperm.
    3. Therapeutic human cloning.

SOMATIC GENE THERAPY

Somatic gene therapy (SGT) involves introducing an exogenous gene sequence into an organism, to act as a substitute for an endogenous gene that produces inadequate or aberrant protein. SGT currently lies in the uncertain grey area between novel research topic and therapeutic reality. Clinical trials began in the early 1990s, and attempts to provide SGT for a number of conditions are underway, notably cancer, acquired immune deficiency syndrome, and inherited diseases. The clinical efficacy and safety of SGT, however, remain disputed, and no form of SGT is yet in routine use. In the years of professional discussion on human genetic manipulation, an ethical consensus has evolved. This views SGT as an extension of conventional medical interventions, and identifies the predominant ethical issues associated with SGT as: (i) the anticipated risk/benefit balance; (ii) the selection of appropriate patients; (iii) the provision of information to patients so that informed consent can be given; (iv) the preservation of patient confidentiality; and (v) the cost to the health-care system (Scully et al., 2001, 2004).

> We (God) created Man in the most perfect form
>
> *(The Holy Quran 95:4)*

This quotation is often used to explain that each human life has its own inherent value and goodness. While genetic research and gene therapy may have positive uses in serving to restore health (and in the process, integrity), care must be taken to ensure that other Islamic principles are not violated. An accurate and complete knowledge of one's pedigree is a fundamental human right; applying the *Shariai'ha* principle, "preventing harm is preferable to procuring benefits," only somatic cell lines should, therefore, be used in gene therapy, since parental integrity is then not compromised and there is no question of hereditary characteristics being influenced (Gatrad & Shiekh, 1999; Al-Aqeel, 2007).

CONCLUSIONS

Islam embraces scientific progress and research, and at no time during Islamic history has there ever been a conflict between Islam and science. The first verse from the Holy *Quran* which was revealed to Prophet Mohammed, says *Read (Akra)*. God Almighty also says, *He has taught man that which he knew not* (Figure 9-6; the Holy *Quran* 96:1–5). Islamic teachings offer a great deal to the Islamic community in the prevention and control of genetic diseases. It is important to educate people to the danger of consanguinity, which is very common in these countries. Premarital examination should be encouraged, which may detect the trait in those intending to get married. Proper counseling should be provided, the dangers explained, and the options discussed. Prenatal diagnosis and the option of abortion for serious, devastating diseases (prior to 120 days from conception) will reduce the incidence of such diseases. Pre-implantation genetics diagnosis is an excellent preventive measure for many genetic disorders. Neonatal screening can avert devastation by simple measures such as specific diets. Avoiding teratogens and provision of folate and iodine in the diet will help in reducing congenital diseases. Stem cell and gene therapy are very promising if used within the Islamic context.

In the Name of Allâh the Most Gracious, the Most Merciful.	بِسْمِ اللَّهِ الرَّحْمَٰنِ الرَّحِيمِ
1. Read! In the Name of your Lord Who has created (all that exists).	اقْرَأْ بِاسْمِ رَبِّكَ الَّذِي خَلَقَ ﴿١﴾
2. He has created man from a clot (a piece of thick coagulated blood).	خَلَقَ الْإِنْسَانَ مِنْ عَلَقٍ ﴿٢﴾
3. Read! And your Lord is the Most Generous.	اقْرَأْ وَرَبُّكَ الْأَكْرَمُ ﴿٣﴾
4. Who has taught (the writing) by the pen.	الَّذِي عَلَّمَ بِالْقَلَمِ ﴿٤﴾
5. He has taught man that which he knew not.	عَلَّمَ الْإِنْسَانَ مَا لَمْ يَعْلَمْ ﴿٥﴾

(The Holy Qur'an 96:1-5)

Figure 9-6 Islam embraces knowledge and science, as God Al Mighty says (*He has taught man that which he knew not*). The Holy *Quran* 95:1–5.

A minimum level of cultural awareness is a necessary prerequisite for the delivery of care that is culturally sensitive. Once equipped with such understanding, it is possible to move beyond the "recipe book" approach to dealing with minority traditions, offering the opportunity for experiential learning. Islamic principles offer an important framework and guide for addressing ethical, social, and legal issues surrounding the application of genetic services and genomics in Muslim countries (Al-Aqeel, 2005, 2007).

Marked health improvements derived from integrating genomics into individual and public healthcare depend on the effective education of health professionals and the public about the interplay of genetic and environmental factors in health and disease. The media are crucial sources of information about genomics and its societal implications (Collins et al., 2003).

Many ethical issues still need to be explored; e.g., who owns and controls genetic information, and does the physician have the right to release the patient's phenotype? While Islamic principles provide a moral and ethical basis for prevention and care in Muslim countries, there is much to learn from how other countries are struggling to do so within the context of their own moral and religious beliefs (Al-Aqeel, 2007).

Strategic plans need to be developed via establishment of strong international leadership by the scientific community, international organizations, governments, and industry, with critical analysis and systems development. The health system must balance two purposes. In the short term it must respond to the demands of the public for access to existing services. At the same time, it must try to improve the health of the whole population. In genetics, population screening programs for detection and prevention of genetic disorders and birth defects, through a genetics data registry and systematic follow-up of the probands, are of the utmost importance. This also includes automated testing, efficient recall systems, genetics counseling within the ethical context of religion and culture, availability of professionals, medications, diets, and a better-informed public, including high school and medical students. Therefore, a multidisciplinary agreement on a socially, religiously, and medically appropriate approach, with the development of population studies and national research priorities, would be needed. In addition, the development of population-based and family-oriented programs, together with the application of novel technologies, are central to preventive programs. This approach needs to be implemented and coordinated through public health authorities in order to ensure interdisciplinary, interregional, and international collaboration and networking, including national testing networks to share precise molecular and pedigree information that may need to be extended internationally (Al-Aqeel, 2007).

Developing science and technology for better health is a religious and moral obligation. Therefore, there is a need for a dialogue on healthcare ethics and human cloning at all levels, including interfaith dialogue as a means for enhancing dialogue among the world's peoples on challenges that are common to us all (Al-Aqeel, 2007).

REFERENCES

Abolghasemi H, Amid A, Zeinali S, et al. (2007). Thalassemia in Iran: epidemiology, prevention, and management. *J Pediat Hematol Oncol 29*(4), 233–238.

Aksoy S. (2005). Making regulations and drawing up legislation in Islamic countries under conditions of uncertainty, with special reference to embryonic stem cell research. *J Med Ethics 31,* 399–403

Al-Aqeel AI, Al Sewairi W, Edress B, et al. (2000). Inherited multicentric osteolysis with arthritis: a variant resembling Torg syndrome in a Saudi family. *Amer J Med Genet, 93,* 11–18.

Al-Aqeel AI (2005). Ethical guidelines in genetics and genomics: an Islamic perspective. *Saudi Med J 26*(12), 1862–1869.

Al-Aqeel AI. (2005). Al-Aqeel Sewairi syndrome: a new autosomal recessive disorder with multicentric osteolysis, nodulosis, arthropathy. The first genetic defect of matrix metalloproteinase 2 gene. *Saudi Med J 26*(1), 24–28.

Al-Aqeel AI. (2007). Islamic ethical framework for research into and prevention of genetic disease. Nat Genet; 39(11): 1293–1298.

Al-Aqeel AI, Serdar, C, Jaroudi, K, Ozand, PT, Hellani, A. (2006). Pre-implantation genetic diagnosis (PGD) for genetic disorders in Saudi Arabia. *J Inherited Metab Diseases 29*(1), 63.

Albar MA. (1999). Counseling about genetic disease: an Islamic perspective. *Easter Med Health J 5*(6), 1129–1133.

Albar MA. (2002). Ethical considerations in the prevention and management of genetic disorders with special emphasis on religious considerations. *Saudi Medical Journal 23*(6), 627–632.

Al-Gazali L, Hamamy H, Al-Arrayad S. (2006). Genetic disorders in the Arab world. *Brit Med J 333,* 831–834.

Alhamdan NA, Almazrou YY, Alswaidi FM, Choudhry AJ. (2007). Premarital screening for thalassemia and sickle cell disease in Saudi Arabia. *Genet Med 9*(6), 372–377.

Ajlouni K MS (2004). In: *Cloning between Science and Religion*, Ajial Press, Amman, Jordan; 31 (Available in Arabic) pp. 163–165.

Al Othiamian MS. (2006). In: *Fundamentals Of Islamic Jurisprudence "Isul AFagih".* Abn Al Jawzi Press, Damam, KSA (Available in Arabic).

Al-Sayyad IA. (no year of posting). *Legal directives for medical practice procedures: an Islamic approach to medicine.* Available at: http://www.islamicmedicine.org/SayadBook2.htm

Blatt RJR. (2000). Banking biological collections: data warehousing, data mining, and data dilemmas in genomics and global health policy. *Community Genet 3,* 204–211.

Burgess MM, Laberge CM, Knoppers BM. (1998). Bioethics for clinicians: 14. Ethics and genetics in medicine. *Canadian Med Assoc J 158,* 1309–1313.

Burgio GR, Gluckman E, Locatelli F. (2003). Ethical reappraisal of 15 years of cord-blood transplantation. *The Lancet 361,* 250–252.

Bowring F. (2004). Therapeutic and reproductive cloning: a critique. *Social Science in Medicine 58,* 401

Christianson A, Streetly A, Darr A. (2004). Lessons from thalassaemia screening in Iran. *Brit Med J 329,* 1115–1117.

Collins FS, Green ED, Guttmacher AE, Guyer MS. (2003). A vision for the future genomic research. *Nature 422*(6934), 835–847.

Cotton RGH, participants of the 2006. Human Variome Project meeting. (2007). Recommendations of the 2006 Human Variome Project Meeting. *Nature Genet 39*(4), 433–436.

Daar AS, Al Khitamy A. (2001). Bioethics for clinicians: 21. Islamic Bioethics. *Can Med Assoc J 164*(1), 60–63.

Daar AS, Cour BE, Singer PA. (2004). Stem cell research and transplantation: science leading ethics. *Transplantation Proceedings*; 36: 2504–2506.

England MA. (1990). *A colour atlas of life before birth.* London: Wolfe Medical Publications.

Janisch R, Wilmut J. (2001). Don't clone humans. *Science* 291, 2552.

De Wert G, Mummery CH. (2003). Human embryonic stem cells: research, ethics and policy. *Human Reproduction 18,* 672–682.

Donnai D. (2002). Genetic services. *Clinical Genetics* 61, 1–6.

El-Hashemite N. (1997). The Islamic view in genetic preventive procedures. *Lancet 350*(9072), 223.

El Hazmi M, Warsy A. (1996). Genetic disorders among Arab populations. *Saudi Med J 17,* 108–123.

Elton P. (2005). Leaning from low income countries: thalassaemia screening in Iran provides evidence for programme in Lancashire. *Brit Med J 330* (7489), 478–479.

ESHRE Task Force on Ethics and Law (2001). The moral status of the pre-implantation embryo. *Human Reproduction 16* (5), 1046–1048.

Gatrad AR, Sheikh A. (1999). Medical ethics and Islam: an Islamic perspective *Eastern Med Health J 5*(6), 1129–1133.

Hellani A, Al-Aqeel AI, Jaroudi K, Ozand P, Coskun S. (2004). Pregnancy after pre-implantation genetic diagnosis for Sanjad Sakati Syndrome *Prenat Diagn 24*(4), 302–306.

Husain FA. (2003). Reproductive issues from the Islamic perspective. *Hum Fertil (Camb.)3*(2), 124–128.

Kamali MH. (1991). Urf (custom). In Islamic Texts Society (eds.), *Principles of Islamic jurisprudence.* Cambridge: Islamic Texts Society, pp. 283–296.

King Abdulaziz City for Science &Technology (2008). King Abdulaziz City for Science &Technology and Ministry of Economy and Planning Strategic Priorities for Biotechnology Program in Saudi Arabia. Available at: www.KACST.edu.sa; http://www.mep.gov.sa.

Larijani B, Zahedi F. (2004). Islamic perspective on human cloning and stem cell research. *Transplantation Proceedings 36,* 3188–3189.

Luri P, Wolfe SM. (1994). Unethical trials of interventions to reduce perinatal transmission of human immunodeficiency virus in developing countries. *N Engl J Med 331,* 1173–1180.

Martignetti J, Al-Aqeel AI, Al Sewairi W et al (2001). Mutations of matrix metalloproteinase 2 gene (MMP_2) causes a multicentric osteolysis and arthritis syndrome. *Nature Genetics 28,* 261–265.

Modell B, Darr A. (2002). Genetic counseling and customary consanguineous marriage. *Nature Reviews/Genetics 3,* 225–229.

Ozand PT, Devol EB, Generoso GG. (1992). Neurometabolic diseases at a national referral center: five years experience at the King Faisal Specialist Hospital and Research Centre. *J Child Neurol 7*(Suppl), S4–S11.

Pennings G, Schots R, Liebaers I. (2002). Ethical considerations on pre-implantation genetic diagnosis for HLA typing to match a future child as a donor of haematopoietic stem cells to a sibling. *Hum Reprod 17,* 534–538.

Rashed MS, Bucknall MP, Little D, Awad A, Jacob M, Alamoudi M, Alwattar M, Ozand PT. (1997). Screening blood spots for inborn errors of metabolism by electrospray tandem mass spectrometry with a microplate batch process and a computer algorithm for automed flagging of abnormal profiles. *Clin Chem 43*(7), 1129–1141.

Ray P, Munnich A, Nisand I, Frydman R, Vekemans M, Viville S. (2003). Sex selection by pre-implantation genetic diagnosis: should it be carried out for social purposes?: Is pre-implantation genetic diagnosis for "social sexing" desirable in today's and tomorrow's society? *Hum Reprod 18,* 463–464.

Reiss MJ. (2002). Ethical dimensions of therapeutic human cloning. *Journal of Biotechnology 98,* 61–70.

Robertson J. (2003). Extending pre-implantation genetic diagnosis: the ethical debate—ethical issues in new uses of pre-implantation genetic diagnosis. *Hum Reprod 18,* 465–471.

Sahih Al-Bukhari (1958). *Matabi Asshab 4,* 135 (Available in Arabic).

Samavat A, Modell B. (2004). Iranian national thalassaemia screening programme. *Brit Med J 329* (7475), 1134–1137.

Scully JL, Rehmann-Sutter C. (2001). When norms normalize. The case of genetic enhancement. *Human Gene Therapy 12,* 87–95.

Scully JL, Rippberger C, Rehmann-Sutter C. (2004). Non-professionals' evaluations of gene therapy ethics. *Social Science & Medicine 58,* 1415–1425.

Sermon K, Van Steirteghem A, Liebaers I. (2004). Pre-implantation genetic diagnosis (Review). *The Lancet 363,* 1633–1641.

Seroura GI, Dickens BM. (2001). Ethical and legal issues in reproductive health. Assisted reproduction developments in the Islamic world. *International Journal of Gynecology & Obstetrics 74,* 187–193.

Solter D. (2000). Mammalian cloning: advances and limitations. *Nature Review Genetics 1,* 199.

Stern H, Harton G, Sisson M, et al. (2002). Non-disclosing pre-implantation genetic diagnosis for Huntington disease. *Prenat Diagn 22,* 303–307.

Strong C (2003). Fetal anomalies: ethical and legal considerations in screening, detection, and management. *Clin Perinatol. Mar 30(1),* 113–126.

Teebi AS, El-Shanti HI (2006). Consanguinity: implications for practice, research, and policy. *Lancet;* 367, 970–971.

United Nations Educational, Scientific and Cultural Organization (2004). Human cloning: ethical issues. France: UNESCO, SHS-2004. Available at: http://www.unesco.org/bioethics.

Verlinsky Y, Rechitsky S, Verlinsky O, Masciangelo C, Lederer K, Kuliev A. (2002). Pre-implantation genetic diagnosis for early-onset Alzheimer disease caused by V717L mutation. *JAMA 287,* 1018–1021.

Weatherall D. (1998). Some aspects of the hemoglobinopathies of particular relevance to Saudi Arabia and other parts of the Middle East. *Saudi Med J 9,* 107–115.

World Health Organization. (1997). Proposed international guidelines in ethical issues in medical genetics and genetic services. http://www.who.org/HGN/GL/ETH/98.1

World Health Organization. (1999). Services for the prevention and management of genetic disorders and birth defects in developing countries. http://www.who.org/HGN/GL/WAOPBD/99.1.

Zaki HA. (1987). Abu Hamid Al-Ghazali's juristic doctrine in *'Al-Mustasfa min 'ilm Al-Usul'* with a translation of volume one of *'Al-Mustasfa min 'ilm Al-Usul'.* Available at: http://www.ghazali.org/site/oeuvre-j.htm.

10.

THE PROGRESS OF GENOMICS IN THE DEVELOPING WORLD: RISING SOCIAL, ECONOMIC, AND ETHICAL CONCERNS

Minakshi Bhardwaj

INTRODUCTION

The developments in genomics have had an impact and provoked a cultural shift in the values and practice of medicine in developing countries as much as in the rest of the world. The range of the countries "developing" is difficult to comprehend, and includes some rapidly growing countries called *emerging economies*, such as India, China, and Mexico, where developments in science and technology have contributed significantly to economic growth. Most countries in Central and Eastern Europe are called *economies* or *nations in transition,* where the growth of science and technology is comparatively slower. Some extremely poor parts of Africa, Asia, and the Pacific are called *least developed countries* (LDCs), where basic healthcare provision is questionable. All these regions represent "the developing world," and huge potential for biomedical research. Unlike most developed countries, where there is relative homogeneity of ideology and policy over issues like consent, and regarding quality standards for research, there are huge geographical, economic, and sociocultural variations within the developing world. Therefore, it is difficult to comprehend the concerns that are raised by genomic technologies in all the developing countries. This chapter attempts to narrow down and take examples from some countries to highlight emerging ethical concerns arising in biomedicine and genomics, and the implications for healthcare that best reflect some of these issues in developing countries. A particular reference to the continent of Asia is made using some examples from other regions.

THE IMPACT OF GENOMICS: FROM AGRICULTURE TO HEALTH

There are different definitions and understanding of genetics, genomics, and related new technologies; however, in the context of this chapter, *genomics* is used as an umbrella term encompassing several technologies that use genetics and whole genome sequencing for gene identification, coding, annotation, and analysis in research, and for applied health purposes. The genomic revolution has made an impact in three areas globally: healthcare (biomedical research), agriculture (genetic engineering), and the economy (e.g., the pharmaceutical industry). Genomics came into the public limelight, with issues related to genetic engineering in agriculture in the developing countries, when debates about genetically modified (GM) crops (for example, BT cotton and Golden rice) raised huge controversies about biosafety, benefit sharing, and intellectual property rights for farmers, along with health and safety concerns for the ordinary public. Some countries—for example, India, Kenya, Brazil, and China—adopted national policies for GM crops, and in some respects these policies were actually more cautious than those adopted in Europe. The degree of caution is interesting, given the conspicuous unmet food production needs in some of these countries today, with basic food and nutrition being a fundamental priority in public policy.

Mapping of the human genome triggered enthusiasm over the possibilities of harnessing genomic knowledge in medical research among the populations of developing countries, which recognized the vast richness of genetic diversity in their populations. Although participation in the sequencing of the genome was limited to only Japan and China, it did considerably raise the profile of genetics in medicine, the media, and the public. Since then, the use of genomic technologies in biomedical research has been rapidly growing in most of the developing countries, in particular the Asian subcontinent (India, China, Hong Kong, South Korea, etc.) and some parts of Central and South America (Mexico, Brazil, etc.). Most large countries in the developing world recognize the capital value of the genome, and hence there are huge investments both at the public and private level to harness its potential and create a cutting-edge, competitive environment that is economically viable. The massive growth in the information technologies in India, and medical research in China and South Korea have made major contributions to the world health market today. Until recently, genetics was "hidden" in the clinic, laboratories, and among anthropologists—and so were the ethical concerns.

FROM TRADITIONAL MEDICINE TO MODERN MEDICINE

It is hard to predict the impact of genetics in medicine in developing countries and given that even today the health care priorities are not necessarily based on genomic technologies. Most regions of the developing world historically have followed traditional practices of medicine, such as Unani, Ayurveda and Chinese medicine that primarily focused on the holistic nature of a being, and were based on a relationship with nature and drawing upon medicinal plants from nature. Initially associated with a variety of negative terms during colonial times, more recently the "traditional medicine" has acquired largely positive connotations in the West. It needs to be acknowledged that a variety of practices and epistemologies fall under the rubric of "traditional" and most medical traditions are intrinsically plural in nature. The plurality lies both in the ways in which one tradition is interpreted and codified and also its practical applications, for example there are several versions of oriental/Chinese medicine seen in Buddhist countries ranging from Tibet to Thailand.

The distinction between traditional and modern medicine are seen in the ways in which treatment and therapy is viewed, and the ethical codes that guide them. Most traditional practices, including some African traditions, take a holistic approach and focus on balancing the body-mind with nature. The body and mind (spirit) are not separate, and exist in harmony with each other. Ancient Indian physicians asserted that both nature and man are made up of the same matter, the five gross elements: earth, fire, air, water and space; additionally, human beings have consciousness. The human body is a combination of body, mind, and self; physical and psychological processes are inseparable and interact, and both are expressions of the life force. Similar concepts are seen in African traditions focusing on harmony with nature, and most rural communities have relied on the spiritual and practical skills of traditional medical practitioners, whose botanical knowledge of plant species and their ecology are invaluable.

Recent progress in the field of environmental sciences and immunology has led researchers to appreciate the capacity and rationality of various traditional taxonomies, and the effectiveness of their treatments. The globalization of Chinese medicine and forms of Ayurveda has, in turn, led to the modernization of traditional medicine. A multitude of these traditional products and practices is increasingly in evidence in the European and American markets. Modernization is seen at the level of economics, as well as the ways in which these traditional approaches are subjected to clinical trials and then standardized, commodified, and mass distributed—such as the use curative use of artemesinin (*qing hao*) for malaria, and the controversial patent of the Neem tree (*Azadirachta indica*) for its medicinal properties (Ernst & Lo, 2005). Nevertheless, despite the modernization of traditional medicine, the traditional practices continue to thrive by default where provision of new biomedicine is not accessible. The rapid growth of traditional medicine also demonstrates that modern medicine/biomedicine is not always the treatment option preferred by patients.

TRADITIONAL MEDICINE AND GENOMICS

Medical advancements of the twentieth century—such as the discovery of DNA and its structure, renal and heart transplantation, oral contraceptives and safe abortions, in-vitro fertilization, and the ability to determine brain death—have pushed the boundaries of medicine and the ethical and social issues raised by advanced medical technologies. Traditional systems of medical teachings, though, have always focused on ethics, and the issues raised by developments in the new biomedicine were difficult to resolve within the traditional schools of ethical thought. One reason lies in the ways in which disease is conceptualized. In traditional medicine, illness is an imbalance between mind, body, and environment, whereas modern medicine focuses on the body. Medicine in the developing world has two arms, traditional practices and modern medicine, and both are practiced in parallel, although large populations within the developing world still depend on traditional medicine, given its accessibility in remote places and established networks and systems within the local communities.

Traditionally, the concept of " physical inheritance" didn't exist in medicine in the Asian medical system, and diseases were not thought to be passed from one generation to other. The type and the severity of the condition was considered based on one's Karma in the past. However, with the developments in genomics, attempts are made to find if there are any correlations between genetics and traditional medical systems. Ayurveda classifies human populations into three major constituents of *prakriti* (phenotypes); namely *vata* (motion), *pitta* (metabolism), and *kapha* (structure) as discreet phenotype groupings. This is generally independent of racial, geographic and ethnic considerations, and focuses only on the individual diseases, and physiological differences. However, researchers have now postulated that there are genetic connotations between HLA alleles and *tridosha* theory of individual *prakriti* types (Patwardhan & Bodekar, 2008). Furthermore, one of the arms of new biobanking project in India, called Indian Variome Project is trying to establish links between Ayurveda and genomics. A project under the name of "Ayurgenomics" is initiated that is focusing on the possibility to find correlations between genetic profile, Ayurvedic classification and biochemical data with an aim of identifying groups within normal populations that could be predisposed to certain kinds of diseases or might respond differently to drugs. The constitutional uniqueness of each individual is an important feature in Ayurvedic medicine; and prescription is individualized both in terms of composition of drugs and suitable diet. These features have similarities with modern day concept of pharmacogenomics and individualized medicine, though genetics is not clearly specified in traditional medicine. Also there are several examples in the religious and medical texts in India that reflect contemporary issues such as cloning and xenotransplantation; although never ethically challenged in the medical context.

Similarly, Chinese researchers have tried to link genomics with traditional Chinese medicine, where concepts related to choosing traits of potential children seemed to have emerged very early; for example, in the East Han Book written in 25AD–220AD during the Eastern Han Dynasty. The concepts of inheritance are also mentioned in some old literature; for example, the book *Zhu Shi Yi Shu*. In it, Jing Xue Ben says that children are similar to parents because their *jing* (essence) and *xue* (blood) are from parents, although genes are not specifically mentioned. Although Chinese medicine and modern medicine are based on totally different fundamentals, in recent years researchers have hypothesized using the concept of *chi* (electromagnetic waves) from Chinese medicine in the process of cellular treatment, through changing the abnormal electrocharged cell to normal, stimulating the cell function and activating gene functions. However, the challenge is how to detect, measure, and control the flow of *chi*. These are considered to be the basic directions in which genetics in Chinese medicine is likely to go forward (Chen, 2006).

GENETIC TESTING AND SCREENING IN MODERN MEDICINE

For a long time, genetic disorders remained largely unrecognized and underestimated in developing world. Haemoglobinopathies such as thalassemia and sickle cell anemia pose the largest health problems worldwide caused by single gene disorders especially in Africa, Middle East, the Mediterranean and South East Asia and that constitute majority of the developing world. However some conditions that are common in Europe and North America, such as cystic fibrosis and Phenylketonuria (PKU) are rare in developing countries, especially in Asia.

The basic underlying ethical issue in medicine, and in genetics, is equity. The disparities lie both between countries and within countries. Some nations, such as Switzerland, spend large sums per capita on public health ($2252) and others, such as India ($3), China ($10), and Columbia ($12), spend very little on their healthcare. The majority of those living in developing countries do not have health insurance, and only very few families can afford private genetic testing services. The common issues surrounding relative lack of genetic services include resource limitations and lack of expertise. In many developing countries, genetic services are not prioritized in healthcare because a majority of these services are perceived as expensive and applicable to rare disorders only. There is also a perception that such services are mostly diagnostic and not therapeutic. Therefore, a number of countries in Asia and Africa, including Bangladesh, Bhutan, Nepal, and Maldives, never prioritized genetic disorders. Hence, it is difficult to find epidemiological data on the prevalence of genetic diseases in these smaller developing countries.

At the time when human genome sequencing was underway, a number of programs focusing on genetics were initiated by international agencies, in particular the World Health Organization (WHO), to help poor countries develop screening programs and build capacity in research and infrastructure development. In Southeast Asia, screening for conditions such as thalassemia, sickle cell diseases, inborn errors of metabolism, hemoglobinopathies, and common genetic disorders such as neural tube defects and Down's syndrome, were initiated in the 1990s with the aid of international agencies (WHO, 2004). Similarly, WHO also initiated a special program for research and training in tropical disease (TDR) for US $1 million in Latin American countries, as support for training and scientific capacity building in disease-endemic countries.

In sub-Saharan Africa and India, infectious diseases and nutritional deficiencies still exceed the noncommunicable diseases such as heart disease and cancer as causes of mortality. Global patterns of genetic conditions are changing, and a number of noncommunicable diseases are considered to have a genetic component. By 2020, noncommunicable diseases will rise to about 73% of total deaths worldwide, with most of them occurring in the developing world (Murrey & Lopez, 1996). Developing countries have a great need for genetic services, to be merged into general healthcare. Nevertheless, the preferred approach in many developing countries will remain focused on single-gene disorders and strategies for prevention, prenatal diagnosis, and contraception, rather than treatment or creating expensive infrastructures that the healthcare budget couldn't support.

GENETICS: A FAMILY AFFAIR

CONSANGUINITY

The social and ethical perspectives in developing countries regarding genetic testing and screening are broad and wide, guided by sociocultural and religious beliefs. Within that framework, economics also plays a major role in decision-making. Epidemiological surveys demonstrated a direct link for mortality and morbidity to communicable diseases and malnutrition and provided data on the importance of congenital and hereditary genetic diseases as a major health burden with possible links to the practice of inbreeding in parts of the developing world. In the majority of the Arab world and parts of North Africa, the custom of inbreeding is very common, and even today, consanguineous marriage rates are in the range of approximately 40% to 50%, leading to a range of genetic disorders including congenital malformations, hemoglobinopathies, and glucose phosphate dehydrogenase deficiency (Teebi & Farag, 1997). There are also recessive disorders among certain groups and subpopulations. Consanguinity is also seen among certain populations in South India among certain Hindu groups, and also in some Christian communities; for example, Lebanon. Religious, cultural, and historical factors play a huge role in the practice of consanguineous marriages; however, Muslim scholars state that Islam does not advocate or encourage consanguinity. Consanguinity is also not mentioned in the Hindu and Buddhist religious texts, and one can say that it arose from social needs, the geographical remoteness of some communities, and as a way to keep family assets, protect the family members, and maintain a sense of reliability and trust in relations within extended families and communities.

TESTING AND SCREENING

Genetic testing and screening are not part of regular healthcare in most of the developing countries, and very few clinics in the public sector provide testing and screening for congenital disorders. Largely, screening and testing is privatized, and counseling by trained counselors is often not available for families. For the most part, clinical geneticists take personal responsibility for genetic counseling, and it is rarely given sufficient time during the process of testing and screening. The process of decision-making carries heavy cultural baggage in Asian traditions, and the ways in which information is communicated and risk presented affect choices and options for clients. Also, the relationship between mind–body, individual–community, and the concept of karma affects decision making to greater extent among patients.

The limited genetic testing and screening that is available is largely for single-gene disorders. Monogenic disorders are inherited and contained within families. The common knowledge about genetic disorders is gradually increasing; however, it has also raised issues about paternity testing, feticide for abnormal embryos, and predictive genetic testing. The decisions about taking genetic testing are very challenging, particularly where no cures and therapies are available and people don't have means to pay. The lack of trained counselors to help ethical decision-making on such issues hampers the development of services, and the ramifications are both personal and social. The positive outcome of a test is regarded as a burden, and even the counseling can be socially conditioned, especially in the ways it is provided, making it difficult for the bearer. Emotional, religious, and social pressures are brought into the process of decision making, and this makes both the authority and the choice of the final decision, for example, either to continue or discontinue with pregnancy, extremely difficult. Particularly in the two big countries of Asia, India and China, prenatal testing is not offered routinely to pregnant women. Fetal ultrasounds for gender tests are illegal in most of the Asian countries, but sometimes indirectly requested by the families and also illegally provided in some private practices. In many patriarchal societies, particularly in the rural areas, the views of men or the head of the family prevail in decision-making.

Although the one-child policy in China has helped to control the population a little, it has placed considerable pressure on women in a society where boys are traditionally preferred in a patriarchal family system. This is also seen in other countries in Asia. The notions of "abnormality" go beyond physical attributes, and become highly complex with the intertwining of socio-religious beliefs, such as "fate" and "the will of God." Positive results can be considered as a "shame" for the family and "burden" on society (Gupta, 2007). The social stigma attached to an inherited condition is huge, and affects all other aspects of life, such as college admissions and prospects of marriage in India. Hence, ordinary couples very rarely undergo carrier testing and counseling.

THE DOCTOR–PATIENT RELATIONSHIP

PATERNALISM

Paternalism has very negative connotations in the Western bioethics debates, particularly in reference to the emphasis on individual autonomy. However, paternalism is practiced commonly in most developing countries. In many developing countries doctors take paternalism as implicit within the goals and the role of a professional to provide advice, guidance, and direction. Generally, directive counseling is provided because of lack of knowledge among patients, and professionals take it as personal responsibility to "help" patients to make the "right" decision in cases where there is an abnormal fetus in pregnancy, which normally leads to abortion. Proponents of directive counseling argue that this sort of paternalism stems from the desire to prevent disease, to make the best use of scarce resources, and to protect the interests of patients. There are also concerns about the ways information is communicated within cultures where traditional medicine still plays an active role. The concept of *gene* doesn't exist in traditional medicine, and this, added to the lack of knowledge about genetics among ordinary public, makes it difficult to counsel patients. Hence, directive counseling is given.

In developing countries, even if physicians' personal moral values do not agree to perform a procedure, they can't make a referral, because of the financial constraints and legal difficulties that patients might face during the process. Although this form of individual eugenics is heavily criticized in the Western context, in the poorer parts of the world the focus is on "quality" of life rather than "sanctity" of life (Subramanian, 1986).

AUTONOMY, DUTY, AND RIGHTS

In the context of medical ethics, particularly genetics, *autonomy* relates to human rights. These rights include the right to know and not to know, access to healthcare, the right to privacy, etc. It implies that individuals are independent, separate entities who own their genetic information and may ask to share benefits and rights of the "intellectual properties" derived from research information. This concept of rights doesn't translate well in some Asian traditions. In cultures with Hindu, Buddhist, and Confucian ethics, individual identity is relational, within the context of family and community, and hence individuals do not exist as entities unto themselves. Encouraging individual rights has social implications as well; for example, the right to know, in societies where there is a preference for boys, may engender sex selection.

In cultures such as China, where a child becomes a person by virtue of belonging to a family and being accepted by the society, parents consider it a duty to bring a child who will be acceptable to the society as a whole. Hence, there is an indirect pressure from the larger communities. Confucian ethics prevails in China, and principle of filial piety is extremely important, which also includes the doctor–patient relationship. The second greatest Confucianist, Mencius, has said that there are three vices that violate the principle of filial piety, and the biggest is being without offspring. Also, many cultures in Asia consider the duty to reproduce integral to humanity, and a child with a disability is considered a sin and disgrace. Although Buddhism prohibits killing, pregnancy with an abnormal fetus is allowed to terminate with a view to avoiding suffering. Regardless of social status, a number of people in developing countries have a pessimistic view about disability, and families try to keep any handicap or genetic condition a secret. The human fetus has a value, but it is not treated as a person in several cultures. Although people believe in *karma* and disease is seen as a result of bad *karma* in previous life, the suffering of a person in this context is given a higher moral value. In some developing nations in Latin America with a strongly rooted Catholic tradition, a culture of preserving life and survival of the fetus is encouraged. It doesn't mean that abortion is not practiced; it is largely privatized, and sometimes done without any medical assistance, resulting in huge number of deaths.

PROGRESSION OF GENETIC SERVICES

Developments in genomics disproportionately originate in rich countries and are largely utilized for those countries' priorities; the benefits of this enormous progress are still out of reach for majority of the developing countries, where nearly 80% of the population lives. In the majority of the developing countries, pharmaceuticals are the second-biggest budget item in the health system after salaries; for example, lifesaving antiretroviral combinations cost between £38-£138 per year, whereas most poor countries have less than £1 per year per person available for all medicines (WHO, 2004). Hence, introducing genetic services for public health has enormous financial implications for developing countries. There is an epidemiological transition taking place within developing countries, where acute diseases are being controlled and chronic diseases are assuming an increasing role in the healthcare systems, as their prevention and management are recognized as a need rather than a luxury. Between 1960–1997, infant mortality rates fell from 138/1000 to 65/1000 in the developing world, but less progress was seen in Southeast Asia and sub-Saharan Africa. Consequently, genetic disorders and birth defects have started to increasingly influence general mortality and morbidity rates, and hence are gaining a public health significance (WHO, 1997).

THE MAGNITUDE OF GENETIC CONDITIONS

Genetic disorders are one of the leading causes of high mortality and morbidity in the developing world, and there is no cure or effective treatment for the majority of genetic conditions. Thanks to high birth rates, consanguinity, and procreation throughout later reproductive years, a large number of infants with genetic disorders are born every year to families in underserved populations. In India, for instance, approximately half a million children are born annually with congenital malformations, most of which are related to genetic or chromosomal aberrations. It is estimated that about 9000 babies with thalassemia major, 5200 babies with sickle cell disease, and 21,000 babies with Down syndrome are born in India each year (Verma, 2000). Such high frequencies have significant implications on national resources amidst competing health and social needs of the population.

The prevalence of genetic disorders, particularly single-gene disorders, also reflects geographical and racial characteristics. Common examples are sickle cell anemia in individuals of African descent, β-thalassemia in those from the Mediterranean, α-thalassemia in Southeast Asians, and Tay Sachs disease in Ashkenazi Jews. Of all these disorders, only sickle cell anemia has an incidence greater than 1 per 1000 in any ethnic group. The high frequency of carrier status of the disease among African populations, particularly those from West African countries, has ensured the persistence of the sickle cell gene. For instance, in Nigeria, where the prevalence of carriers of the gene is approximately 25%, about 1%–2% of children are born with sickle cell anemia (Akinyanju, 1989). The economic implications of providing care for affected children, coupled with the generally unfavorable outcome, have resulted in a rising demand for prenatal diagnosis of sickle cell anemia in Nigeria in recent years.

GRADUAL INTEGRATION OF GENETIC SERVICES

Genetic services deal with all medical aspects of genetic disorders, birth defects, and congenital malformations regardless of their causes. They also include care and management of those conditions. Due to lack of awareness at all levels, there is underreporting of genetic conditions and no recordkeeping, and it is difficult to comprehend the scale of genetic conditions and prevalence in populations in developing parts of the world. In 2001, the draft sequence of the human genome was completed and in 2002, WHO responded to this by publishing the Genomics and the World Health report. In this report, then Director-General of the WHO, Dr. Gro Harlem Bruntland, emphasized the need "to harness this knowledge and have it contribute to health equity, especially among developing nations"—acknowledging a concern that developments in genomics may add to greater disparity between industrialized nations and developing countries in access to healthcare and, in particular, medical genetics services.

Some lower-resource countries, including India, South Africa, and some Latin American countries, initiated medical genetics programs in the 1960s. The first medical genetic services initiated across the developing world included obstetrics, hematology, pediatrics, and neurology, but these were limited to academic institutions and research centers that provided very limited clinical services to the people—with some exceptions such as in Cuba, where genetic services have been integrated at all levels of healthcare since the beginning of the 1980s. However, soon it was realized that existing models in the rich countries were not suitable for the developing world (WHO, 1985). The challenge is how to apply general principles brought forth in basic medical genetics that focus on the individual and family units, to mass public health and primary healthcare systems of poor countries (Modell & Kuliev, 1998).

Nevertheless, genomically advanced developing countries such as India, China, Malaysia, Thailand, Brazil, Mexico, and some countries of the Arab peninsula have made significant commitments to the potential of genomics in healthcare (Pang, 2003). International agencies such as WHO, UNESCO, UNICEF, and the World Bank are important facilitators of this change, and several country-based regional programs have been initiated for research capacity building, improving awareness and infrastructure within primary healthcare, and establishing laws to protect patients. In India, tests are available for Down syndrome, hypothyroidism, congenital adrenal hyperplasia, galactosemia, and G-6-PD deficiency. These defects can be treated with medication and dietary support if detected early. At present, prenatal diagnosis is available in very few clinics, and mostly provided at a tertiary level and privately. Amniocentesis, ultrasonography, and chorionic villus sampling are commonly available in many developing countries, and expecting parents are encouraged to use them. In Africa, these techniques are popularly used for sickle cell trait, and about 90% of positive results end up in lead to the termination of pregnancies (Akinyanju et. al., 1999). Most developing countries are looking for cheap and reliable, simple techniques that could easily be used for large populations and be affordable for the ordinary public. However, as these new technologies are gradually becoming available in developing countries, termination of pregnancies is increasing in parallel, as prospective parents wish for a "perfect" child. Even for disorders for which antenatal surgery is possible, such as cleft palate, abortion is the preferred option for most couples.

CLONING AND RELIGIOUS PERSPECTIVES

Though cloning arguably may not come under the realm of genetics, the media frenzy coverage of the sheep Dolly, produced with cloning technique, had a global impact, including in the developing world. Cloning debates brought the issues raised by biomedical research into public knowledge, and were instrumental in raising awareness of the ethical issues in emerging technologies of biomedical research. The possibilities of reproductive cloning for human beings associated with gene therapy sparked ethical debates in the mid-1990s in Asia among lawyers, philosophers, ethicists, and scientists. It brought bioethics into the public domain through media coverage in several poor parts of the world. As stated earlier, Hindu literature mentions cloning in a religious context, so ethical concerns had never been raised, either about the possibility or the process. Nevertheless, the possibility of human cloning in modern medicine raised critical ethical issues concerning naturalness, the soul of the clone, therapeutic cloning, sex selection, and the right to choose, within different cultural and religious contexts. Cloning debates contributed immensely toward comparative studies of cross-cultural differences and similarities in bioethics.

In most of the countries with Hindu and Buddhist traditions, the laws on cloning have been liberal. In these traditions there is no single creator of the universe, no plan of creating the universe, and there may be several gods (Hindu) or no god (Shinto); the spirit is eternal, and individual virtue (*karma*) determines what happens to the spirit in reincarnation in the next life. Therefore, this view generally allows embryo research to support life and the creation of genetically modified crops. The four principles that generally guide decision making in the Buddhist tradition are: perpetual cycle of rebirth (*samsara*), the correlation of deed and its consequences (*kamma*), absolute emancipation (*nibbana*) and the value of compassion (*metta*; see Ratanakul, 2000). These are similar concepts to Hindu precepts; however, actions like selective abortion are still considered morally wrong in these traditions. In contrast, Judeo-Christian tradition has one God, the master creator of the universe, who gives souls to each individual; therefore, it is considered wrong to clone and create embryos for research. Similarly, different perspectives on cloning are also seen in Islam. It is argued that cloning doesn't harm any Islamic belief, as the creator of the universe has established a cause and effect relationship and all creations take place through his will; cloning is only manipulating his will, and therefore scientists will not become creators themselves. Human reproductive cloning is rejected in Islam on the basis of positive and negative aspects of this technique; for example, the number of eggs and attempts needed, and the destruction of those embryos, abuse of women, and birth of potentially disabled fetuses. In Islam, differences are also seen between different sects, just as there are different versions of Buddhist philosophy seen in different Buddhist countries. Islamic perspectives regard the right to life of the embryo from the day of its existence; however, distinction is made between different stages of human development in the uterus. The embryo, even in the pre-ensoulment stages, has sanctity; its eradication is a sin, but punishment for that sin will be less than for abortion after ensoulment (Inhorn, 2006). Given this view, abortion is permissible in Islam before the end of the fourth month of pregnancy when there is a physical or emotional necessity involved, like a life-threatening danger for the mother or a high possibility of severe fetal anomalies. In addition, using embryos for research may be acceptable if necessary, and prior to the ensoulment (Larijani, 2004).

REPRODUCTIVE TECHNOLOGIES AND COMMERCIALIZATION

IN-VITRO FERTILIZATION

Besides the abortion and termination of pregnancies due to the rise in prenatal diagnosis (PND) techniques, an inability to have children can similarly itself create enormous problems, particularly for the woman. The United Nations Population Division now estimates that 186 million women of reproductive age in developing countries (excluding China) are infertile, with more than 30% in many African countries unable to have children because of secondary infertility. Twenty-six countries in the world have primary infertility rates over 4%—all of them in sub-Saharan Africa, with the highest in Cameroon (7.3%) and the Central African Republic (10.5%). Prevalence studies in Asia and South America are very limited, given that infertility is not considered a life-threatening condition. In most developing countries, reproductive healthcare is synonymous with family planning and contraception, and infertility constitutes an important public health problem, though there is a general misconception that overpopulation occurs due to high fertility rates. The negative consequences of infertility are higher in developing countries in comparison to Western countries. In many cultures, the burden of infertility falls on women who can be stigmatized and ostracized, and subjected to domestic violence.

Conventional reproductive technologies such as In Vitro Fertilization (IVF) and Intra Cytoplasmic Sperm Injection (ICSI) are expensive, and therefore a very small proportion of people can afford such treatments. Subsequently, governments do not invest significant resources in introducing assisted reproductive technologies for public health, and focus on prevention and education instead. Furthermore, in the cultures where traditional medicine is practiced, local traditional practitioners and healers have greater advantages over using Western approaches in dealing with such sensitive issues, and there is an element of trust in local healers in remote places (Gerrits, 1997).

In the past two decades a number of low-cost infertility treatments were initiated in developing countries. The focus on simplifying diagnostic procedures, and simplifying assisted reproductive technologies and complication rates, were the most important need (Ombelet, 2008). Low-cost methodologies in infertility treatments have been developed in Africa with international aid; for example, less expensive medications for ovarian stimulation than the genetically engineered "recombined" follicle stimulating hormones, and replacing complex incubators with small plastic capsules (Mariani, 2009). Infertility treatments are increasingly becoming available in most developing countries, supported by governments as well as in private clinics. In Asian countries, particularly rapidly growing economies like India, China, Malaysia, Thailand, and Korea, infertility clinics are commonly available both in government hospitals and specialized private clinics. IVF treatment in Middle East countries is comparatively restricted for religious reasons. Islam is not against treating infertility; however, surrogacy and sperm or egg donation by other than husband or wife is prohibited. Research should always be limited to the time before ensoulment, and restricted to the use of extra embryos left behind by IVF techniques, according to Islamic scholars (Mousa et al., 2006).

In recent years, countries like India and China have come under criticism for misuse of IVF technologies for research on embryos, and commercialization of low-cost IVF techniques. Couples are encouraged to donate spare embryos for research without proper counseling and consent, and given paternalistic attitudes and low level of knowledge about its potential implications, several families may agree for gamete or embryo donation Infertility clinics have sprung up across India over the past decade, offering a range of modern assisted reproductive techniques, and some are engaged in unethical practices (Mudur, 2002). Given that the laws on embryo research are not very stringent in developing countries, and Islam, Hinduism, and Buddhism do not have strong religious objections based on the status of the embryo, it is easier to conduct research in developing countries. Restrictive embryo research regulation in the United States and Europe has forced scientists to look abroad for embryonic material, and hence the booming industry of infertility clinics in developing countries. Several big developing economies like India also have started to provide cryopreservation facilities for foreigners, where they can freeze embryos for long time at cheaper rates, and in some countries IVF is available for as low as US $200 (Ombelet, 2008), making it an attractive treatment for foreigners at very low cost, and free from religious objections. Similarly, South Korea opened one of the largest egg banks in the world for commercial purposes in 2005, designed for Japanese customers who might normally travel to United States for artificial insemination, the South Korean egg bank inexpensive treatment with "Asian" eggs, within East Asia, for under US $20,000 (Saito, 2003).

STEM CELLS

Building upon reproductive technologies, stem cell science is gaining momentum in rapidly developing economies such as India, China, and Korea. These countries provide world-leading centers for clinical translation of cell-based therapies. The government of India initiated three major programs on stem cell research aimed at treating blindness, some types of neurological disorders, and thalassemia, through the use of aborted fetuses and unused embryos from IVF clinics (Times of India, 2001). Given that there is a ban in India on export of embryos for research, many researchers are deprived of opportunities for potential foreign collaboration and undertaking clinical trials. To some extent this is being dealt with new national investments in biomedical research, for example China has encouraged a number of universities to invest heavily in stem cell research. Those universities have attracted public funds and substantial investment by private companies. The Chinese government also approved the establishment of state-run stem cell banks in Tianjin and Shanghai (Xu, 2008). As part of the Chinese National Stem Cell System Project, these banks will focus on harvesting stem cells from umbilical cord, and donors can store umbilical cord blood for potential future use.

In general China spends about $60 billion on research and development. About 60% of this comes from foreign and domestic firms, and the rest from the government (Du et al., 2004). A significant proportion is put aside for biomedical research including stem cell research. Some smaller countries have also invested in stem cell research and development; for example, Pakistan has created three stem cell centers within the country for bone marrow transplantation, there is a center for heart treatment in Thailand, and a center for beauty treatment in Vietnam. In Iran, which has come under strong sanctions for nuclear research, there is a proactive investment by the government in stem cell research, particularly in the field of dermatology and skin grafting (*Washington Times,* 2009). On the African continent, South Africa is the only country where some stem cell research is conducted. However, the majority of countries providing stem cell therapies claim to use autologous lines in their treatment.

Medical tourism is not new for routine treatments, but with new technologies emerging, desperate patients are also ready to take a chance on cheap, untested, and unproven treatments in the developing world. The emerging stem cell market in the developing countries has its roots in established commercial infertility treatments, and is growing as fast as fertility treatments. Stem cell treatments are increasingly provided in private hospitals, and there is growing criticism of the standards of safety and post-treatment care. Commercialization of stem cell techniques is also seen in smaller firms providing service facilities, and IT firms investing in software developments. Given that there are fewer moral objections on status of early embryos, the guidelines in stem cell research are more relaxed in several Asian countries than in the West.

POPULATION-BASED GENOMICS RESEARCH

HARNESSING GENOMIC DIVERSITY

The studies on human genetic diversity are not new, and go back to the twentieth century when they were based on blood groups. With developments in DNA-based research flourishing in the mid-1970s, the need was soon realized to understand human genetic diversity using DNA as the basis. Hence, in early 1990s, the Human Genome Diversity Project was initiated, focusing on understanding the genetic variations between different populations across the globe. The aim of the Human Genome Diversity Project was to understand how and when patterns of genetic diversity were formed, and create a resource for future biomedical research. It involved collecting storing "renewable samples" from populations from which any part of the genome could be examined, using lymphoblastoid cell lines (LCLs) from across the globe rather than DNA samples. These cell lines had been already collected worldwide for different purposes, and political and ethical difficulties rose with regard to exploitation of indigenous populations, lack of consent, and using samples for commercial purposes; controversies arose related to biopiracy and intellectual property rights (Cavelli-Sforza, 2005). The exploitation issue became a massive concern, and gave rise in some circles to calling the effort "the vampire project" (Crigger, 1995). The danger of racist interpretations of genetic differences worried some scientists, while others predicted that more research, accompanied by a public education campaign, would further undermine the outmoded concept of race. Among controversies related to the collection of renewable samples, and other ethical concerns, the HGDP funding was stopped and the project went out of the limelight.

Since the sequencing of the human genome, there has been an increasing understanding that most human diseases involve many genes interacting in very complex ways, with potentially important environmental factors. Moving away from the traditional Mendelian approach, population-based research needed new and faster methods to analyze different populations that might not necessarily require sequencing of the entire genome, but instead could focus on parts of the genome that contribute to disease. These efforts resulted in the HapMap project, based on the premise that full sequencing could be avoided by using single nucleotide polymorphisms (SNPs) as markers for disease variants, and using haplotypes to enhance the SNP-based mapping. Haplotypes are regions of chromosomes that have not been broken up by recombination during evolution, and they are separated by places where recombination has occurred. This could enable geneticists to search for genes involved in diseases and other medically important traits. It was hoped that by creating a catalogue of all the genetic variants in the form of a Hapmap, geneticists would be able to use them to locate genes involved in medically important traits, rather than fishing for genes in the entire genome. It is widely believed that there are only handful of haplotypes present in a human being. Hence, the project was initiated with the hope of providing a catalogue with disease-related SNPs in the human genome, enabling quicker, easier, more efficient and cost-effective analysis of the human genome.

The project aimed to compare the genetic sequences of different individuals in order to identify chromosomal regions where genetic variants are shared between different populations. The project involved collecting samples from populations of Asian, European, and African ancestry, with participation from six countries including Japan, China, the United Kingdom, United States, Canada, and Nigeria. Similar to the HGDP, the Hapmap project was also controversial, particularly regarding issues related to community engagement and consent from remote populations, and defining one's identity during consultation and sample collection. For example, participants were asked to contribute if their past four generations belonged to the same populations. Added to these concerns were delays in sample collection and lack of financial support for ethical, legal and social issues [ELSI] related work (Macer 2003). The Hapmap project is still active and claims to contribute towards gene-environment association studies as well as understanding human diversity.

PUBLIC HEALTH GENOMICS AND BIOBANKING

In parallel with HapMap developments, large-scale population-based studies were started within individual countries to create population-based biobanks, in an attempt to gain detailed understanding of the heritable variation in the human genome at the individual and population level. It is hoped that population-based genomics research will be crucial in understanding the differences in human susceptibility to diseases, drug responses, and the complex interaction of genetic and environmental factors in producing particular phenotypes (Department of Health UK, 2003). Iceland was the first country to create a national-level biobank of the entire Icelandic population. Since the development of the Icelandic Health Sector Database, the biobanking phenomenon is increasing its pace, and several countries of the world are starting several large-scale population-based genomics projects under the umbrella of biobanking and genetic databases.

In Asia, the heightened interest in and attention to genomics has led to investment in setting up biobanks and genetic databases for research and further investment opportunities. Some of the larger countries in Asia, including China, India, and Japan, and some smaller ones, such as Singapore and Taiwan, are in the process of setting up biobanks for research and development purposes, either independently or as collaborators in existing international projects. The well-publicized Guangzhou biobank project in China aims to create medical profiles for 50,000 Guangzhou individuals aged at least 50, and develop a database by examining genetic, lifestyle, occupational, and environmental factors emerging with economic development and causing illness. (Lam et al., 2006). In India, several projects have been initiated. IuVita BioBank, a private company, intends to develop a biospecimen repository of the Indian patient population by conducting genomics clinical trials in several therapeutic areas. Currently, a proposal for developing a biobank focusing on women's health in India is under debate, partly due to existing commitments to research into fertility control rather than developing biobanks for general research purposes. The National Chemical Laboratory (NCL) of India is developing a national database of defective proteins responsible for causing genetic disorders in humans (Kashyap, 2005). A new biotechnology company has launched another project to build a genetic, genealogical, and medical database of India's 69,000-strong Parsi community, similar to the Icelandic database (Jayaraman, 2007). The focus in India is on exploiting bioinformatics to develop linked databases of prevalent genetic disorders in the public domain. India is focusing on integrating genomics and information technology, and developing bioinformatics tools to use with existing public databases.

Rapidly growing economies in the developing countries are investing in publicly funded large-scale national genomic initiatives to explore genetic diversity and variation, in order to harness the benefits of genomic medicine, with the hope of cost-efficient drug development and potentially reducing healthcare costs. Such projects were also initiated in the remote parts of developing world, such as the Kingdom of Tonga, which later failed because the infrastructure required for such high-input technologies was difficult to create, and because of the lack of coordinated policy efforts on health and research. Similarly, several projects have been initiated in Latin America; some have included coordinating existing collections under one umbrella, such as in Mexico and Brazil, and new biobanks, as in Costa Rica. However, lack of appropriate regulation is common in most Latin American countries, and critical concerns have arisen about failing to protect the confidentiality of donors and using unauthorized samples in research (Godoy, 2008). Even in big economies like China and India, there is no committee with a specific responsibility to overview ethical and legal issues in population-based research, although the Chinese government has invested funds in developing a biobank, and has set up a national bioethics committee.

ECONOMICS, ETHICS AND POPULATION RESEARCH

CAPITALIZING FREE KNOWLEDGE

Developing countries are not only economically poor but also information poor. Genomics knowledge is expensive to create, and that is why much of it is created in the West. It is also a reason that applied research historically has been focused in developing countries, whereas basic research has been more in developed regions of the world. Sequencing genomes and deciphering different genes require massive investments to set up appropriate laboratory environments, and human skill, and it is also an ethical concern for many developing countries whether to invest in genomics research or to provide essential healthcare services to people. It has been argued that the bioinformatics approach to investing in genomics is far less expensive, requiring only initial training, access to computers, and access to the Internet. A large amount of information is available through the Internet, and it requires substantially less investment to develop computational tools for understanding the existing sequenced genomes than it is to sequence genomes (Pongor & Landsman, 1999).

SOCIETAL VALUES AND DECISION MAKING

In a survey conducted by Hitchcock in 1994, respondents among Asians and in the United States were asked about personal and societal values that were "critically important to the people in your country." The top six values chosen by Asians included: *(i)* orderly society, *(ii)* social harmony, *(iii)* accountability of public officials, *(iv)* openness to new ideas, *(v)* freedom of expression, and *(vi)* respect for authority. In contrast, the top six values chosen by respondents from the United States included: *(i)* freedom of expression, *(ii)* personal freedom, *(iii)* rights of the individual, *(iv)* open debate, *(v)* thinking for oneself, and *(vi)* accountability of public officials. Given the diversity of Asian cultures, although these findings cannot be conclusive regarding the differences in ethical values of the East and West, they do highlight some of the social and cultural values that play an important role in the decision making among people.

In cross-cultural medical ethics debates, the role of the family in decision-making is put forward as a considerable argument in Asia. The use of genetics brings informational risks and familial implications to matters of privacy and confidentiality that go beyond individual risk. New ethical principles, such as family-based individual informed consent, have been suggested (Zhang, 2005) in Asia, where the focus of consent is on the process of decision-making rather than on the individual participant in research and therapy. Given the complexities of genomics today, and with a focus on populations rather than individuals, there is a need for collective confidentiality. Collective confidentiality puts an obligation on researchers to collectively protect privacy and information of the participants in research involving multiple interlinked databases (Bhardwaj, 2004). Also, the distinction is made between confidentiality and privacy within different cultures. Confidentiality is considered as a duty of the doctor for his patients, and the boundaries of privacy are generally decided within that sphere in most Asian cultures.

One of the important concerns is about telling the truth regarding medical information. Although there is a rising trend and a constant demand to tell the truth, most of the information is filtered within the individual families. The family is one unit, and even in the present day in Asian societies, the ultimate decision of disclosure is negotiated between the experts and the family about whether or not to tell the affected individuals. The "relationship over time" is a major factor in the decision-making process in medicine in Asia, either among family members or between the family and physicians that constitute an element of "trust" installed and nurtured (Bhardwaj, 2010).

DEFICIENT GOVERNANCE STRUCTURES

One of the major criticisms of research & development projects is their rapidity, without the establishment of the appropriate regulatory mechanisms. The lack of stringent institutional review boards (IRBs) has resulted in the lack of trust in genomics projects, and particularly collection of biological samples. Given the lack of public knowledge about the nature and the potential of these projects, questions are raised regarding the autonomy of the participants, informed consent procedures, and quality of the data, as was evident in other debates on the HGDP and HapMap projects. At the same time, there are other ethical arguments made about promoting research on race and ancestry, which have had astounding implications for medicine and the ethics of healthcare. For example, healthcare in China has deteriorated since the 1980s, and where available it is not free. Similar to India, the government encourages liberalization and private investment. The Regulation for Biomedical Research (guidelines) was established on the basis of the Helsinki Declaration after China joined Human Genome Project in 2000. The Anhui case of Xu Xiping prompted the Chinese government to establish donation laws, and the Blood Donation Law was established in 1998 to protect patients and donors. The lack of stringent IRBs have not been able to instill trust large research projects, so that autonomy in informed consent procedures has been questioned (Lawler, 2002). Secondly, intellectual property rights (IPR) are not strongly protected in China. As late as March 2006, the government of China announced new guidelines for IPR in China. The lack of an adequate IPR and patent system is advantageous to foreign investors, and it is an important factor in the rapid growth of the private sector in China, including the pharmaceutical companies.

Similarly, the Indian Council of Medical Research (ICMR) drafted ethical guidelines for research on human subjects, as a code of conduct for research. These are neither legally binding nor stringent. The Research Safety Bill was submitted to parliament in January 2006 to make these guidelines legally binding. The guidelines provide a waiver of consent if there is a minimal risk, or in case of emergency. But it is debatable what constitutes a "minimal risk." The Medical Research Council of India's Code of Medical Ethics also permits the waiver of consent if the patient's identity is not disclosed. However, there is no legal mechanism to protect the patient's identity. Informed consent in India is legally located under the Indian Contract Act. There also have been cases of verbal consent in clinical trials in India (Mudur, 1997). But according to the Contract Act, verbal consent is approved if proven, but it is not clear what is considered to be evidence of consent, e.g., recording or a witness. However, such cases are rarely reported, and usually not by the participants, who have little understanding and no information about the tests and their legal rights.

In large genomics studies, the protection of privacy and maintaining of confidentiality of the participants is an immense concern. Secondly, comparative studies focusing on multiple populations and multiple agendas tend to also justify issues of privacy and confidentiality by double coding and prior anonymization. However, doubts can be raised if the anonymization is achieved during data transmission. Anonymization is also questionable when identifiers are removed, especially in studies that involve the entire population. These protection mechanisms should maximize societal benefits and reduce individual harm. However, it is debatable when the disclosure of personal data is justified for research and for providing the participants options for treatment in cases of unexpected findings. The level of zero risk in research is extremely difficult to achieve, and the more complex the research, the more difficulties arise in the protection of confidentiality.

NEW TECHNOLOGIES AND THE FUTURE OF GENOMICS

Besides genomics, several other technologies have been developed in the last decade in rich countries that take an "omics" approach; i.e., an integrated approach in the study of cell function. These "omics" approaches include the integration of genomics, proteomics, transcriptomics, and metabolomics. With these techniques emerging to understand the molecular functioning of the body, it is envisaged that they might be helpful in the future, through developments in pharmacogenomics and nutrigenomics, to enable giving personalized medical care and dietary advice to people, based on their individual genotypes. These technologies are building upon understandings of human genetic variation, and by measuring and understanding the influences of genetic, environmental, and immune factors on human variation, it will help medical science to classify patients and predict susceptibility to disease or clinical response. Therefore, medical treatments can be more precisely targeted. Similarly, it is projected that by understanding how individuals respond to specific nutritional exposures based on differences in their genotypes, it will be helpful in giving recommendations for specific nutritional regimens on an individual basis.

Developing countries suffer from a *double burden* of diseases—infectious diseases such as malaria, HIV, hepatitis C, etc., and fairly recent, diet-related chronic conditions such as such as obesity, cardiovascular diseases, diabetes, etc. Although genomic technologies have been successfully implemented in the case of infectious diseases, developing countries are struggling to develop disease-prevention strategies for genetic and diet related conditions. One reason is the lack of attention received prior to the epidemic, and underdeveloped facilities for research when compared to the industrialized countries. One can argue for the possibilities of screening for susceptibility biomarkers in populations with highly prevalent diseases such as hypertension, diabetes, coronary heart disease, and certain cancers. However, these new services do not have a place in developing countries as yet, because they do not help to contribute toward meeting the most urgent needs of preventing infectious disease and providing basic nutrition to the masses. The promissory nature of these technologies is fertile ground for its commercialization, but it is hard to imagine that many countries in the poor parts of the world would be proactive in their applications of these technologies today.

ACKNOWLEDGMENTS

The support of the Economic and Social Research Council (ESRC) is gratefully acknowledged. This work is part of the Research Programme of the ESRC Genomics Network at Cesagen (ESRC Centre for Economic and Social Aspects of Genomics).
I would also like to thank Prof. Ruth Chadwick for her critical comments on the earlier drafts of this paper.

REFERENCES

Akinyanju OO (1989). A profile of sickle cell disease in Nigeria. *Annals of NY Academy of Sciences.* 565: 126–136

Akinyanju OO et al. (1999). Initiation of prenatal diagnosis of sickle-cell disorders in Africa. Prenatal Diagnosis. April, 19 (4): 299–304

Bhardwaj M. (2004). Rich databases and poor people: opportunities for developing countries. *TRAMES, The Official Journal of Estonian Academy of Sciences* 8(1–2). 90–105.

Bhardwaj M. (2010). Bioethics in Asia: in transition. In: *Encyclopedia of Life Sciences.* John Wiley & Sons, Ltd., Chichester.

Cavalli-Sfroza L. The Human Genome Diversity Project: past, present and future. *Nature Reviews Genetics* 6, 333–340.

Chen E. (2006). Thinking process and principle of gene research in Chinese medicine. *Journal of Accord Integrative Medicine* 2(6), 148–150.

Crigger B-J. (1995). The Vampire Project. Hastings Center Report, 25

Department of Health UK (2003). *Our inheritance our future.* Government white paper.

Du J et al. (2004). *Stem cell mission to China, Singapore and South Korea.* Report of a DTI global watch mission. September 2004, Kings College London.

Ernst W, Lo V. (2005). Asian Medicine: tradition and modernity. Editorial. 1(1), 1–9.

Gerrits T. (1997). Social and cultural aspects of infertility in Mozambique. *Patient Education and Counselling 31,* 39–48.

Godoy E. (2008). Health–Latin America: Free rein for biobanks? Interpress service news agency. UNDP, UNEP and World Bank. Available at: http://ipsnews.net/print.asp?idnews=43570. (Accessed April 8, 2010).

Gupta J. (2007). Private and public eugenics: genetic testing and screening in India. *Bioethical Enquiry* 4(3), 217–228.

Hitchcock D. (1994). *Asian Values and the United States: How much conflict?* Washington DC: The Center for Strategic and International Studies.

Inhorn MC. (2006). Making Muslim babies: IVF and gamete donation in Sunni versus Shi'a Islam. *Culture Medicine and Psychiatry 30,* 427–450.

Jayaraman KS. (2007). Database targets Parsi genes. *Nature 446,* 475.

Kashyap P. (2005). "NCL develops database on genetic disorders." *Times News Network,* May 24, 2005, Delhi, India.

Lam TH et al. (2006). Cohort profile: The Guangzhou Biobank Cohort Study, a Guangzhou-Hong Kong–Birmingham collaboration. *International Journal of Epidemiology 35,* 844–852.

Larijani B, Zahedi F. (2004). Islamic perspective on human cloning and stem cell research. *Transplantation Proceedings,* 36(10), 3188–3189.

Lawler A. (2002). US questions Harvard Research in China. *Science 206,* 28.

Macer D. (2003). Ethical considerations in the HapMap project: An insider's personal view. *Eubios Journal of Asian and International Bioethics 13,* 125–127.

Mariani D. (2009). Africa obtains low-cost fertility treatment. http:www.swissinfo.ch.

Modell B, Kuliev A. (1998). The history of community genetics. The contribution of the haemoglobin disorders. *Community Genetics 1,* 3–11.

Mousa NA, Bedaiway MA, Casper RF. (2006). Biological and reproductive implications of stem cell research and therapeutics: prospects in the Middle East. *Middle East Fertility Society Journal 11*(3), 152–168.

Mudur G (1997). India to control foreign research involving Indian patients. *British Medical Journal.* 314: 165.

Mudur G. (2002). India considers government agency to license infertility clinics. *British Medical Journal 325,* 564.

Ombelet W. (2008). Infertility and provision of infertility medical services in developing countries. *Human Reproduction 14*(6), 605–621.

Pang T. (2003). Equal partnership to ensure that developing countries benefit from genomics. *Nature Genetics 33,* 18.

Patwardhan B, Bodekar G. (2008). Ayurveda genomics: establishing a genetic basis for mind-body typologies. *Journal of Alternative Complimentary Medicine 14*(5), 571–576.

Pongor S, Landsman D. (1999). Bioinformatics and the developing world. *Biotechnology and Development Monitor 40,* 10–13.

Ratanakul P. (2000). Buddhism, prenatal diagnosis and human cloning. In Norio Fujiki and Darryl Macer, eds. *Bioethics in Asia.* 405–407, Eubios Ethics Institute, Japan.

Saito M. (2003). Here comes the egg biz: ova operations open in Japan. June, 2003, *Japan Inc.* Available at: http://findarticles.com/p/articles/mi_m0NTN/is_44/ai_108881982/

Subramanian S. (1986). In India, Nepal and Sri Lanka quality of life weighs heavily. *Hastings Center Report,* August 1986, pp. 20–22.

Teebi A, Farag T. (1997). *Genetic disorders among Arab populations.* New York: Oxford University Press. Oxford Monographs on Medical Genetics, No 30.

Verma I. (2000). Burden of genetic disorders in India. *Indian Journal of Pediatrics 67*(12), 893–898.

Washington Times (2009). Iran at the forefront of stem cell research. *Washington Times*, April 15, 2009. Available at: http://www.washingtontimes.com/news/2009/apr/15/iran-at-forefront-of-stem-cell-research/. (Accessed April 7, 2010.)

World Health Organization (WHO). (1985). Community approaches to the control of hereditary diseases. WHO Advisory Group, Geneva. http://www.who.int/genomics/publications/WHOHGNWG85.10.pdf

WHO (2004). Identifying regional priorities in the area of human genetics in SEAR. WHO consultation document.

WHO (2004). WHO medicine strategy 2004–2007. Geneva.

WHO/EMRO (1997). Community control of genetic and congenital disorders. *Technical publication series, No. 24.* WHO office for the Eastern Mediterranean.

Xu G-T. (2008). Establishment of stem cell bank in China. *Cell Research 18,* 175.

Zhang X. (2005). Gene therapy in PR China: regulations and ethical concerns. *Journal of International Biotechnology Law 5,* 212–216.

11.

GENETIC OUTCOMES OF CONSANGUINEOUS MARRIAGES

Alan H. Bittles and Michael L. Black

INTRODUCTION

Marriage has long been the predominant institution within which procreation occurs and genes are transmitted. Therefore, a key initial step in the investigation of genetic differences at individual and population levels is to examine past and present marriage preferences in different societies. In terms of population genetics theory, infinite population size and random mating are important preconditions of the Hardy-Weinberg principle, which describes the intergenerational transmission of genes. It is, however, doubtful whether either of these assumptions has ever existed in human populations. For example, estimates of the effective population size of the Out-of-Africa migration some 60,000–70,000 years ago (Behar et al., 2008) range from approximately 10,000 (Harpending et al., 1998) to as few as 700 individuals (Zhivotovsky et al., 2003). Further, since the early human groups were itinerant hunter-gatherers, random mating would have been impossible and close kin mating almost inevitable.

Even in countries with large immigrant communities, such as the United States, Canada, and Australia, recent arrivals typically marry within their own ethnic and/or religious communities during the first and second post-migration generations. This practice, described as *positive assortative mating*, provides strong social advantages. However, it has important genetic implications, since it is probable that couples from the same national, ethnic, or religious sub-community will have a significant proportion of their genes in common. Therefore, as will be further discussed, their progeny are more likely to be homozygous (or, more correctly, autozygous) for a detrimental recessive disorder (Bittles, 2002).

The probability of genetic drift, i.e., chance fluctuations in gene frequencies in successive generations, is greatest in communities with restricted numbers of potential mating couples. In human populations this situation can arise in a variety of ways; e.g., through *founder effect*, when a subgroup of a population establishes a new breeding colony; via a demographic bottleneck following major disease- or disaster-related mortality (Auton et al., 2009); and in populations subdivided into multiple, strictly endogamous communities (Bittles, 2009a).

Virtually all traditional human societies are divided into long-established communities, with limited intercommunity marriage (Bittles, 2008), and genome-based association studies in industrialized Western societies have demonstrated the ongoing existence of similar, if less pronounced, subdivisions leading to significant population stratification (McEvoy et al., 2009). A higher probability of homozygosity results, with a consequent increased likelihood of recessive gene expression. Since a recessive founder or *de novo* mutation can rapidly increase in frequency within a small community by chance alone, the birth of an affected child may occur whether the parents are known to be consanguineous or consider themselves nonrelatives (Zlotogora et al., 2006).

CONSANGUINEOUS MARRIAGE

A couple are said to be consanguineous if they share one or more common ancestors, and pedigrees of the most common types of consanguineous unions are illustrated in Figure 11-1. As already noted, since most pairs of individuals living in the same location will have one or more common ancestors somewhere in their pedigrees, for practical purposes the search for a shared ancestor generally does not extend back more than three or four generations. In medical genetics, the definition of consanguinity is usually restricted to a preferential union between a couple related as second cousins or closer ($F \geq 0.0156$). But in countries such as Finland, where close kin marriage is avoided but marriage choice was limited in prior generations by geographical dispersal and severe population bottlenecks (Norio, 2003), multiple more remote consanguinity pathways exist. Under these circumstances, marriages between couples who are related as third cousins or beyond ($F \leq 0.0039$) may become clinically significant (Bittles, 2009a).

COEFFICIENT OF RELATIONSHIP AND COEFFICIENT OF INBREEDING

The coefficient of relationship (r) is the proportion of genes identical by descent (IBD) shared by two individuals, and is calculated from the formula:

$$R = (1/2)^n$$

where n is the number of steps apart on a pedigree for these two individuals via their common ancestor. Thus, for two persons related as first cousins:

$$R = (1/2)^4 + (1/2)^4 = 1/8$$

By comparison, the coefficient of inbreeding (F) is the proportion of gene loci at which an individual is homozygous by descent. The closest legally permissible consanguineous unions are between an uncle and niece, which mainly occur in South Indian Hindu communities, or between double first cousins, contracted in many Muslim populations in the Middle East and South Asia. In both of these forms of marriage the partners share a quarter of their genes ($r = 0.25$), and the coefficient of inbreeding in their progeny is $F = 0.125$. Double first cousins have both sets of grandparents in common, whereas in a first cousin marriage the couple shares two common grandparents (Figure 11-1).

Specific sub-communities and individual families often have long and unbroken traditions of close kin marriage, which can result in a cumulative coefficient of inbreeding that greatly exceeds the genetic influence of consanguinity in a single generation. To allow for the effect of consanguinity in preceding generations, a corrected coefficient of inbreeding can be calculated according to the formula:

$$F = \Sigma(½)^n(1+F_A)$$

where F_A is the ancestor's coefficient of inbreeding, n is the number of individuals in the path connecting the parents of the individual, and the summation (Σ) is taken over each path in the pedigree that goes through a common ancestor.

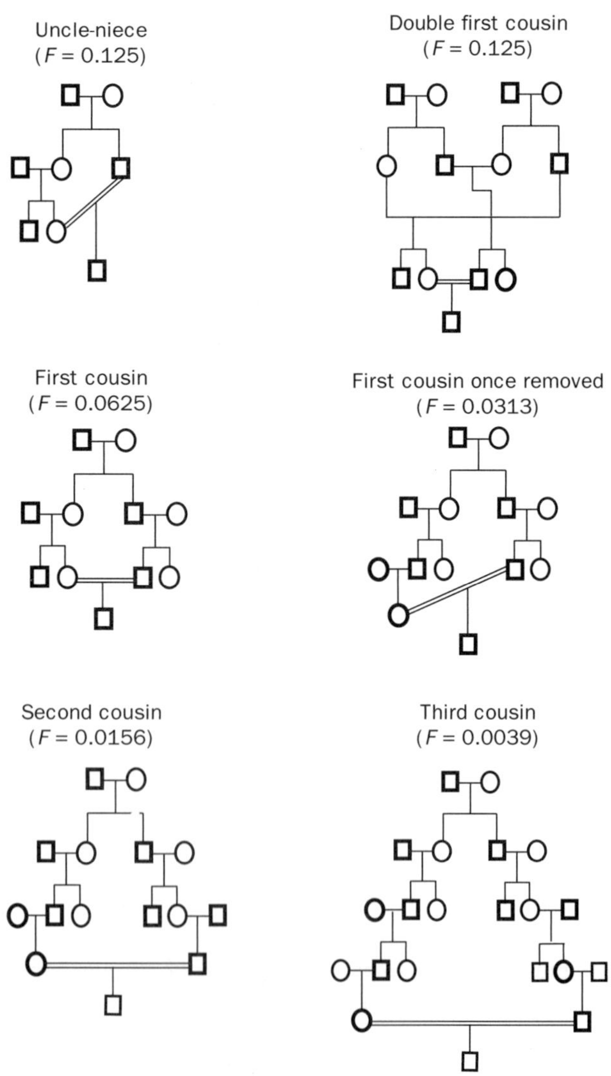

Figure 11-1 Pedigrees of Consanguineous Marriages

THE GLOBAL PREVALENCE OF CONSANGUINEOUS MARRIAGE

The preference for consanguineous marriage is primarily social and economic in nature, since it is believed that family ties will be strengthened, family honor will be preserved, health or financial uncertainties that may be encountered following marriage with a partner from another family or community are avoided, and family goods, monies, and landholdings are maintained (Bittles, 1994; Hamamy & Bittles, 2009). Throughout the twentieth century the lowest rates of consanguinity were reported in Western Europe, North America, and Oceania, where less than 1% of marriages were consanguineous (Figure 11-2). Because of recent large-scale immigration from developing countries in which consanguinity is favored, in countries such as France, Germany, and the United Kingdom it is highly probable that the overall prevalence of consanguinity now exceeds 1%. Approximately 1%–5% of current marriages are consanguineous in various regions of southern Europe, South America, and Japan, depending on local geography and social customs. The highest recorded rates of consanguinity are in North and sub-Saharan Africa, the Middle East, Turkey and Central Asia, and parts of South Asia, where unions between couples related as second cousins or closer account for 20% to over 50% of all marriages (www.consang.net).

It is difficult to derive a global prevalence for consanguineous marriage because of the lack of reliable data on consanguinity for many countries in Asia, including Bangladesh and Indonesia, and in sub-Saharan Africa. Worldwide, there seems to be a gradual decline in the prevalence of consanguineous marriage, although at the individual country level the picture is complex. For example, in the Middle East there has been a recent reduction in the prevalence of consanguineous marriages in Jordan (Hamamy et al., 2005), which may mirror growing urbanization, whereas in other neighboring Arab states, such as the United Arab Emirates and Qatar, increases have been

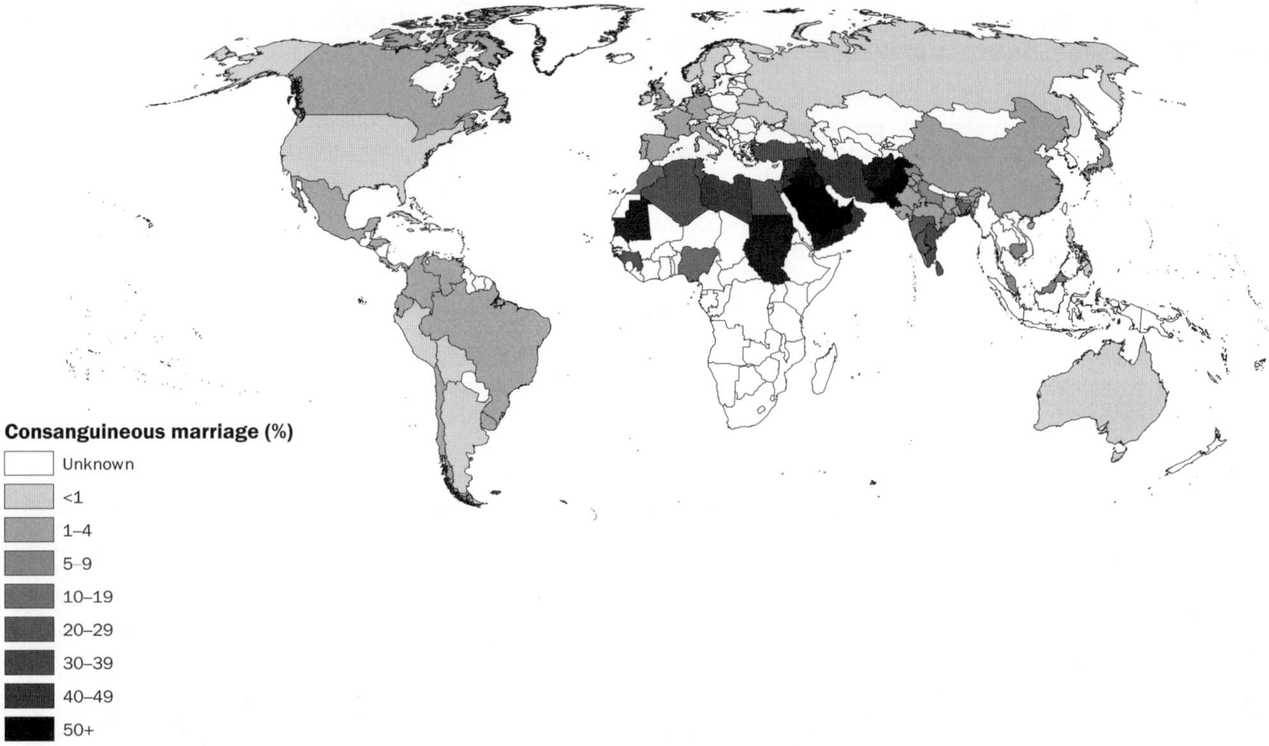

Figure 11-2 Global Distribution of Marriages between Couples Related as Second Cousins or Closer
SOURCES: *(F ≥ 0.0156). Reproduced from Bittles AH, Black ML. (2010a). Consanguinity, human evolution and complex diseases. Proceedings of the National Academy of Sciences USA. 107, 1779–1786.*

reported (Al-Gazali et al., 1997, Bener & Alali, 2006). During the latter decades of the twentieth century, little change in the prevalence of consanguinity has been reported in major South Asian countries, including India (Bittles et al., 1991), Pakistan (Ahmed et al., 1992), and Iran (Saadat et al., 2004). Based on the information available to date, it would appear that some 10.4% of all marriages worldwide have been contracted between couples who are related as second cousins or closer (Bittles & Black, 2010a).

If the specific types and frequencies of consanguineous marriage are known, the mean coefficient of inbreeding (α) can be calculated to provide a measure of the intensity of inbreeding in the population, according to the formula:

$$\alpha = \Sigma p_i F_i$$

where Σ is the summation of the proportion of individuals *pi* in each consanguinity category *Fi*. As indicated in Table 11-1, the values for α vary widely between populations and regions, from 0.0001–0.0006 in Western Europe and North America, to 0.0089 in Morocco, 0.0204 in Lebanon, 0.0332 in Pakistan, and 0.0333 in South India, where consanguineous marriage is widely preferential.

Some care is needed in interpreting these data, since different methods of assessment were used; e.g., Roman Catholic marriage dispensation records in the Republic of Ireland and Spain, as opposed to civil marriage registrations in Lebanon and Morocco. The actual types of consanguineous unions enumerated by investigators also varied, with double first cousin and first cousin unions only counted in the Palestinian Territories and Iraq, as opposed to double first cousin, first cousin, first cousin once removed, and second cousin unions in Egypt. Whatever the method of assessment, it is apparent that consanguineous marriage is not restricted to geographically remote communities or to religious, ethnic, or social isolates.

SPECIFIC TYPES OF CONSANGUINEOUS MARRIAGE CONTRACTED

The specific patterns of consanguineous marriage contracted in different populations largely reflect traditional customs and beliefs, with the highest levels of consanguineous marriage in a single generation reported from Pondicherry, South India, with 54.9% consanguinity (α = 0.0449; see Puri et al., 1978), and army recruits in Punjab, Pakistan, with 77.1% consanguinity (α = 0.0414; see Hashmi, 1997). In Pondicherry the most common form of consanguineous marriage was uncle–niece (F = 0.125), whereas in Punjab it was marriage between first cousins (F = 0.0625).

The influence of local customs and preferences also are apparent in the specific types of first cousin unions contracted (Table 11-2), so that in Arab Muslim communities a marriage between a man and his father's brother's daughter (FBD) is preferred, as opposed to the mother's

TABLE 11-1 PREVALENCE (%), MEAN COEFFICIENT OF INBREEDING (α) AND TYPES OF CONSANGUINEOUS MARRIAGE IN DIFFERENT REGIONS, COUNTRIES AND POPULATIONS

Region	Study Setting	Participants	Number	%	Relationships	α	Authors
WESTERN EUROPE							
Great Britain	Birmingham	Obstetric inpatients	2,431	0.2	1C, 2C	0.0001	Bundey *et al.* (1990)
Republic of Ireland	National	RC dispensation	149,029	0.5	UN, D1C, 1C, D2C, 2C	0.0002	Masterson (1970)
Norway	All-Norway	Civil registration	893,941	0.7	1C, 2C	0.0002	Magnus *et al.* (1985)
Spain	Alava	RC dispensation	11,700	2.5	UN, 1C, $1^{1/2}$C, 2C	0.0006	Calderón *et al.* (1998)
NORTH AMERICA							
Canada	Québec	RC dispensation	21,874	1.3	UN, 1C, $1^{1/2}$C, 2C	0.0003	De Braekeleer & Ross (1991)
U.S.A.	National	RC dispensation	133,228	0.2	1C, $1^{1/2}$C, 2C	0.0001	Freire-Maia (1968)
SOUTH AMERICA							
Argentina	National	Civil registration	212,320	0.4	1C	0.0002	Castilla *et al.* (1992)
Brazil	Rio Janeiro	RC dispensation	4,070	1.5	UN/AN, 1C, $1^{1/2}$C, 2C	0.0008	da Fonseca & Freire-Maia (1970)
	Paraiba	RC dispensation	9,521	12.8	UN/AN, 1C, $1^{1/2}$C, 2C	0.0058	da Fonseca & Freire-Maia (1970)
Chile	National	Hospital births	6,959	0.7	>1C, 1C, $1^{1/2}$C, 2C	0.0004	Liascovich *et al.* (2001)
NORTHERN AFRICA							
Morocco	National	Civil registration	4,773	19.9	1C, 2C	0.0089	Lamdouar Bouazzaoui (1994)
Egypt	National	Household/school/ school/workplace	26,554	29.0	D1C, 1C, $1^{1/2}$C, 2C	0.0101	Hafez *et al.* (1983)
Sudan	Khartoum	Blood donors	4,833	52.0	1C, 2C	0.0302	Saha & El Sheikh (1988)
WEST ASIA							
Lebanon	National	Household survey	1,556	35.5	IC, 2C	0.0204	Barbour & Salameh (2009)
Palestinian	National	Household survey	4,971	27.7	D1C, 1C	0.0186	Assaf &Khawaja (2009)
TERRITORIES							
Iraq	National	Household survey	23,937	33.0	D1C, 1C	0.0219	COSIT (2005)
Turkey	National	Household survey	7,435	20.1	1C, 2C	0.0110	Koç (2008)
SOUTH ASIA							
India-north	Lucknow (Hindu)	Obstetric inpatients	7,955	0.1	UN, 1C	0.0001	Agarwal *et al.* (1991)
India-south	Karnataka (Hindu)	Obstetric inpatients	86,448	33.5	UN, 1C, 2C	0.0333	Bittles *et al.* (1991)
Pakistan	National	Household survey	6,611	61.2	1C, 2C	0.0332	Ahmed *et al.* (1992)
OCEANIA							
Australia	Western Australia	Civil marriages	62,376	0.2	1C, $1^{1/2}$C, 2C	0.0001	Port & Bittles (2001)

α, the mean coefficient of inbreeding.

UN, uncle-niece ($F = 0.125$); AN (aunt-nephew ($F = 0.125$); D1C, double first cousin ($F = 0.125$); 1C, first cousin ($F = 0.0625$); $1^{1/2}$C, first cousin once removed ($F = 0.0313$); D2C, double second cousin ($F = 0.0313$); 2C, second cousin ($F = 0.0156$).

brother's daughter (MBD) pattern of first cousin marriage that predominates among Dravidian Hindus of South India, the Han Chinese, and the Tuareg of North Africa (Bittles, 2001). While the coefficient of inbreeding for FBD and MBD offspring is the same at autosomal loci ($F = 0.0625$), at X-chromosome loci $Fx = 0$ for FBD progeny but 0.125 for children born to MBD couples. The specific forms of first cousin union favored and contracted within particular populations may therefore influence the expression of X-linked disease genes, and so affect the incidence of disorders such as hemophilia A and glucose 6 phosphate dehydrogenase deficiency.

TABLE 11-2 PREFERRED PATTERNS OF FIRST COUSIN MARRIAGE IN DIFFERENT SOCIETIES

Country	Type I (%)	Type II (%)	Type III (%)	Type IV (%)	All First Cousin Unions (%)	Authors
MIDDLE EAST						
Lebanon	37.4	22.5	10.4	29.7	14.1	Khlat (1985)
Jordan	62.8	9.9	7.7	11.3	32.0	Khoury & Massad (1992)
United Arab Emirates	64.9	8.4	12.2	14.5	26.2	Al-Gazali et al. (1997)
Yemen	48.9	18.1	14.2	18.8	29.4	Gunaid et al. (2004)
Palestinian Territories	47.9	17.6	20.1	14.4	14.4	Assaf & Khawaja (2009)
SOUTH INDIA						
Tamil Nadu						
Hindu	1.0	0.9	35.1	63.0	25.0	Rao & Inbaraj (1977)
Muslim	11.1	8.9	31.1	48.9	21.2	
Christian	0	0	41.7	58.3	19.0	
Andhra Pradesh, tribal	0	0	39.4	60.3	35.8	Pingle (1983)
SOUTH AMERICA						
Argentina	19.6	38.3	18.7	23.8	0.4	Liascovich et al. (2001)

Parallel cousin marriage:

Type I: father's brother's daughter, $F = 0.0625$, $Fx = 0$

Type II: mother's sister's daughter, $F = 0.0625$, $Fx = 0.1875$

Cross-cousin marriage

Type III: father's sister's daughter, $F = 0.0625$, $Fx = 0$

Type IV: mother's brother's daughter, $F = 0.0625$, $Fx = 0.125$

RELIGION AND CONSANGUINEOUS MARRIAGE

As is apparent from Figure 11-2, the major world religions have a strong direct effect on the types of consanguineous marriage permitted or proscribed, and they can indirectly exert influence via the civil legislation enacted within individual countries (Bittles, 2009a, 2009b). Most of the major religions sanction consanguineous unions, although there are quite marked differences within each religious tradition. Judaism and Islam largely follow the guidelines provided in the Biblical Book of Leviticus 18:7–18, but with a Quranic prohibition on uncle–niece marriages, which are permitted within Judaism.

The Christian Orthodox and Roman Catholic churches restrict close-kin marriage, although the dispensation requirement for consanguineous unions in the Roman Catholic Church was relaxed in the early twentieth century and now applies only to couples related as first cousins or closer. By comparison, the Protestant denominations permit first cousin unions, following the Levitical marriage guidelines. Similar divergent attitudes towards consanguinity occur within Hinduism. While many Dravidian South Indians regard consanguineous marriage as preferential, in North India consanguinity is prohibited under the Indo-European Hindu tradition (Kapadia, 1958).

CIVIL LEGISLATION ON CONSANGUINEOUS MARRIAGE

First cousin marriage and lesser levels of consanguineous relationships are permissible under civil law in virtually all countries, and since 1987 marriage between half-sibs ($F = 0.125$) may be permitted in Sweden under specific circumstances (Bittles, 2001). The situation differs in the United States, where until 1861 first cousin marriage was universally legal (Ottenheimer, 1996), but through time, legislation to ban different types of consanguineous marriage was gradually introduced at the state level, the most recent example being the prohibition of first cousin marriage by the state of Texas in 2005.

First cousin unions are now a criminal offense in 10 states, and are illegal in a further 21 states (Paul & Spencer, 2008), despite a federal recommendation that all state laws on first cousin marriage should be rescinded (National Conference of Commissioners, 1970). As one of the very few countries to have enacted legislation banning first cousin marriage, alongside the People's Republic of China and the Democratic People's Republic of Korea, the widespread illegality of such marriages in the United States hinders assessment of their national prevalence and effects on health.

CONSANGUINITY AND REPRODUCTIVE BEHAVIOR

It has been proposed that an olfactory mate-choice system operates in humans and other mammals, which results in a lower than expected incidence of HLA haplotype matches between spouses, indicating the instinctive avoidance of partners with similar human leukocyte antigen (HLA) haplotypes (Ober et al., 1997). The phenomenon of mate choice was investigated in Swiss university students, with the claim that the level of attractiveness of male body odors to females was proportional to the degree of major histocompatibility complex (MHC) dissimilarity between the male subjects and the female testers (Wedekind et al., 1995).

Follow-up studies in other populations revealed a notable lack of uniformity in the results obtained (Bittles et al., 2002). A recent investigation using genome-wide analysis in conjunction with HLA typing to investigate these apparently contradictory results showed that African couples did not exhibit a significant pattern of similarity/dissimilarity in the MHC region, whereas across the genome they were more similar than random pairs of individuals. By comparison, European Americans were significantly more MHC-dissimilar than random pairs of individuals, to a greater extent than in other regions of the genome (Chaix et al., 2008).

These differences between Africans and European Americans indicate that, even if humans can discriminate between the MHC types of potential partners via odor cues, the system either is specific to certain ancestries, or in some cases it can be overridden by other, more potent signaling cues. In terms of actual marriage partner choice, the findings of the odor studies would seem to have little relevance in communities where consanguineous unions are strongly preferential, since in these more traditional societies marriage contracts are generally decided upon by parents (Bittles et al., 2002).

CONSANGUINITY AND FETAL LOSS RATES

Although it has been widely believed that consanguineous relationships are less fertile than unions between nonrelatives, reduced levels of pathologic sterility generally have been reported among consanguineous couples (Rao & Inbaraj, 1977; Edmond & De Braekeleer, 1993). Despite a positive association between parental HLA sharing and recurrent abortion, suggesting negative selection against individuals homozygous at HLA loci (Kostyu et al., 1993), there is no evidence of an increase in fetal loss rates among consanguineous couples (Jaber et al., 1997; Bittles et al., 2002).

Indirect indicators of fetal survival, such as multiple birth rates and the secondary sex ratio, also failed to show an adverse inbreeding effect (Bittles et al., 1988). Indeed, it has been suggested that intrauterine mortality may be reduced in the pregnancies of consanguineous couples, with lower rates of Rhesus (Rh) incompatibility (Stern & Charles, 1945) and preeclamptic toxemia (Stevenson et al., 1971), possibly due to the enhanced genetic compatibility that would be expected between mother and fetus in consanguineous unions.

COMPARATIVE FERTILITY IN CONSANGUINEOUS AND NONCONSANGUINEOUS COUPLES

As indicated in Table 11-3, in most populations where consanguineous marriage has been studied, maternal age at marriage and at first birth is lower in consanguineous pregnancies (Bittles et al., 1991; Bittles et al., 1993). There also is evidence that women in consanguineous unions continue to bear children at later ages (Tunçbilek & Koç, 1994), in part due to lower contraceptive usage (Hussain & Bittles, 1999). The net result is optimization of the female reproductive span with a higher mean number of children born to consanguineous couples.

Given their enhanced reproductive opportunities, most studies have reported higher mean levels of fertility among consanguineous couples, as demonstrated in a recent meta-analysis of 40 studies conducted in nine countries where first cousin couples had a higher mean number of live births in 33 of the 40 studies (Bittles & Black, 2010b). For the study sample as a whole, first cousin couples had 0.08 additional births when compared to co-resident nonconsanguineous couples ($r^2 = 0.67$, $P < 10^{-9}$). In Iceland, with access to good genealogical data and socioeconomic homogeneity, reproductive success

TABLE 11-3 BIOLOGICAL CHARACTERISTICS OF CONSANGUINEOUS COUPLES

Younger maternal and paternal ages at marriage
Lower spousal age differences
Younger maternal age at first birth
Shorter birth intervals
Extended maternal reproductive span
Larger completed family sizes

was shown to be greatest at a level of parental relatedness approximating to 3rd to 4th cousins ($F = 0.0039–0.00098$; see Helgason et al., 2008).

Reproductive compensation may be an additional explanation for the larger number of live births to consanguineous couples, with infants who die at an early age rapidly replaced (Schull et al., 1970; Ober et al., 1999). A reproductive strategy of this type could involve a conscious decision by parents to achieve their desired family size, although a pregnancy following the death of a breast-fed infant could also be due to the cessation of maternal lactational amenorrhea.

CONSANGUINITY, HEALTH AND INHERITED DISEASE

A significant positive association has been demonstrated between consanguinity and early mortality, with disorders involving the expression of detrimental recessive genes especially involved. However, as the highest rates of consanguineous marriage generally are found in the poorest and least educated sections of populations (Bittles, 1994), which are most disadvantaged in terms of health and healthcare provision, differentiating between the genetic and socioeconomic causes of morbidity and mortality can be difficult.

CONSANGUINITY AND HEALTH IN INFANCY AND CHILDHOOD

Data on the relationship between consanguinity and birth measurements have been mixed, with some studies suggesting that babies born to consanguineous parents are smaller and lighter, and therefore less likely to survive (Silbert et al., 1979; Ramankutty et al., 1983; Magnus et al., 1985; Kulkarni & Kurian, 1990), whereas others have failed to detect any significant difference (Rao & Inbaraj, 1980; Khlat, 1989; Saedi-Wong & Al-Frayh, 1989; Başaran et al., 1994). By comparison, there is a general consensus that postnatal mortality and morbidity are higher among the progeny of consanguineous unions, and the rarer the frequency of a deleterious recessive gene in a population the greater the proportional disadvantageous effect of inbreeding on its expression (Bittles, 2001). Estimates of the overall adverse effects of consanguinity have been highly variable and in many cases spuriously high, in large part due to inadequate control for important nongenetic variables that are known to influence childhood health, including maternal age and education, birth order, and birth intervals (Bittles, 2003).

Figure 11-3 is a regression plot based on a meta-analysis of 48 studies, conducted in 11 countries, on mortality during the first year of life, with deaths at first cousin level on average 1.2% higher than among the infants of unrelated parents ($r^2 = 0.71$, $P < 0.0001$). In the developing countries where consanguineous marriage is most prevalent, excess consanguinity-associated mortality often is concentrated during infancy, but in many cases no specific cause of death is determined because of inadequate diagnostic facilities and parental reluctance to sanction prenatal diagnosis or autopsy examinations (Nelson et al., 1997). Where a diagnosis has been possible, a clear link between consanguinity and autosomal recessive disorders is apparent, with multiple deaths reported in a proportion

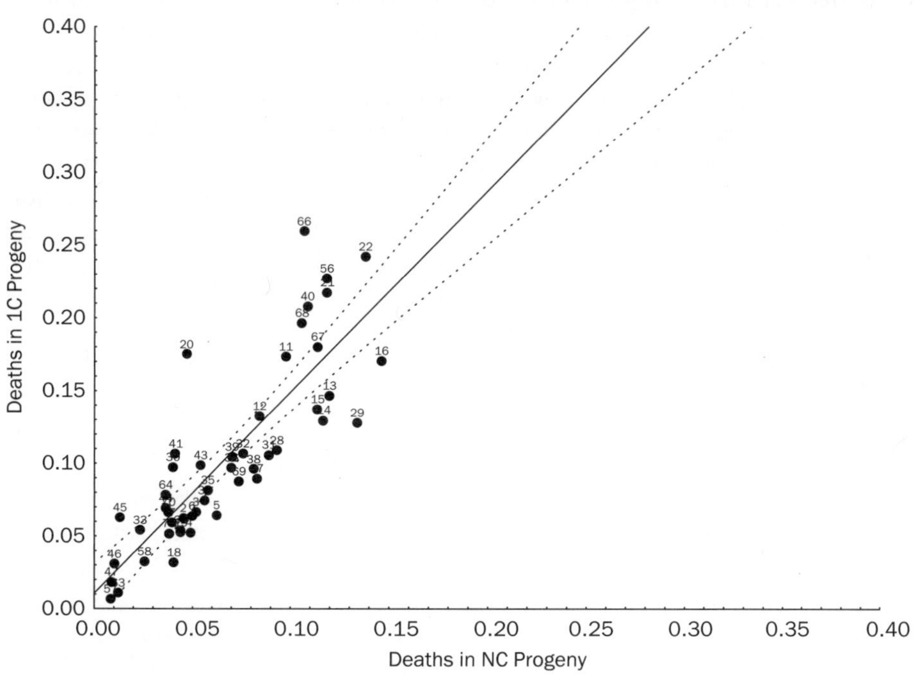

Figure 11-3 Comparative infant mortality in first cousin (1C; $F = 0.0625$: y-axis) versus nonconsanguineous marriages (NC; $F = 0$: x-axis) in 48 study populations
Reprinted from Early Human Development, vol. 86, A.H. Bittles and M.L. Black, The impact of consanguinity on neonatal and infant death, pp. 737-741, 2010 with permission from Elsevier.

of consanguineous families, the effect being proportional to the level of parental genetic relatedness (Stoltenberg et al., 1999; Bittles, 2001).

CONSANGUINITY AND CHILDHOOD MORBIDITY

Large, consanguineous pedigrees offer a cost- and time-effective strategy to locate recessive disease mutations via homozygosity mapping. Using this technique, the causative loci for a range of important childhood disorders have been identified in specific families and communities, including autosomal recessive nonsyndromal hearing loss (Al-Gazali, 1998; Sundstrom et al., 1999; Moynihan et al., 1999; Mustapha et al., 1999; Pulleyn et al., 2000; Dereköy et al., 2000; Shahin et al., 2002), and blindness caused by early onset retinal dystrophies (Rahi et al., 1995; Rogers et al., 1999; Elder & De Cock, 1993) and childhood glaucoma (Martin et al., 2000; Panicker et al., 2002).

The reported outcomes of studies into the prevalence of birth defects are dependent on the diagnostic criteria employed, and in less developed countries recognition of the symptoms of congenital disorders can often be dependent on neonatal survival rates. On average, 4%–5% of newborns in developed countries have some form of birth defect (Christianson et al., 2006). A significant excess of major congenital defects has been diagnosed in consanguineous offspring, especially disorders with a complex etiology and a higher rate of recurrence, but the reported rates of birth defects associated with consanguinity have varied quite widely. For example, among Israeli Arabs excess major malformations in first cousin progeny by comparison with nonconsanguineous controls ranged from 3.8% (Jaber et al., 1992) to 8.7% (Bromiker et al., 2004), while a 26-year study based on the Medical Birth Registry of Norway reported 1.9% excess birth defects in Norwegian first cousin couples and 2.4% among Pakistani migrant couples (Stoltenberg et al., 1997).

It is difficult to identify major categories of disease that are specifically overrepresented in consanguineous progeny, although cognitive impairment generally is more common, suggesting the expression of detrimental recessive genes. A study of Arab schoolchildren in Israel indicated a 0.8–1.3 point decrease in mean IQ scores among first cousin progeny by comparison with the children of unrelated parents, with a 2.6–5.9 point decline in the equivalent mean IQ scores of double first cousin progeny (Bashi, 1977). In Pakistan, mild and severe intellectual and developmental disability also have been associated with consanguinity (Durkin et al., 1998), although as with cognitive impairment, poor social conditions may play significant causative roles in such cases.

There is a consistent positive association between consanguinity and a number of congenital heart disorders, including ventricular septal defect and atrial septal defect, indicating the probable contribution of common genetic variants. But both positive and negative associations with patent ductus arteriosus, atrioventricular septal defect, pulmonary atresia, and tetralogy of Fallot have been reported in different populations (Gnanalingham et al., 1999; Becker et al., 2001; Nabulsi et al., 2003; Yunis et al., 2006). It is possible that these differences either simply reflect nonstandardized diagnostic protocols in different study centers or, alternatively, they may be indicative of community-specific founder mutations (Bittles & Black, 2010a).

According to the multinational Latin America Collaborative Study of Congenital Malformations (ECLAMC), which examined 34,102 newborn infants for major congenital anomalies, a significant association with consanguinity was found only for hydrocephalus, postaxial polydactyly, and bilateral cleft lip with or without cleft palate (Rittler et al., 2001). In the state of Tamil Nadu, South India, a community-based survey of visible congenital defects was conducted during 2004–2005, with 12.8 million children screened. A positive family history was identified in 9% of the 166,833 children with congenital defects, 32% of whom were born to consanguineous parents (Sridhar, 2009). While it was noted that the level of consanguinity determined in the group of affected children was similar to that of the state rural population, no information on the specific types of parental consanguineous unions was forthcoming.

CONSANGUINITY AND PRE-REPRODUCTIVE MORTALITY

To assess the overall effects of consanguinity on pre-reproductive deaths, from approximately six months gestation to a median 10 years of age, a meta-analysis was conducted directly comparing mortality in first cousin versus nonconsanguineous progeny in specific populations. The study sample comprised 69 populations resident in 15 countries located across four continents, with a total sample size of 2.14 million (Bittles & Black, 2010a). As with the data on infant deaths (Figure 11-3), an unweighted linear regression comparing mean mortality in first cousin versus nonconsanguineous progeny in each population was plotted.

The resultant scatter diagram (Figure 11-4) showed a mean excess mortality at first cousin level of 3.5% ($r^2 = 0.70$, $P < 0.00001$), which was consistent across the range of control mortalities; i.e., indicating that the level of excess consanguinity-associated mortality was independent of the basal (nonconsanguineous) death rate in each population. The estimate of 3.5% excess deaths among first cousin progeny compares with an earlier global estimate of 4.4% excess mortality (Bittles & Neel, 1994) calculated from 38 studies, each of which was included in

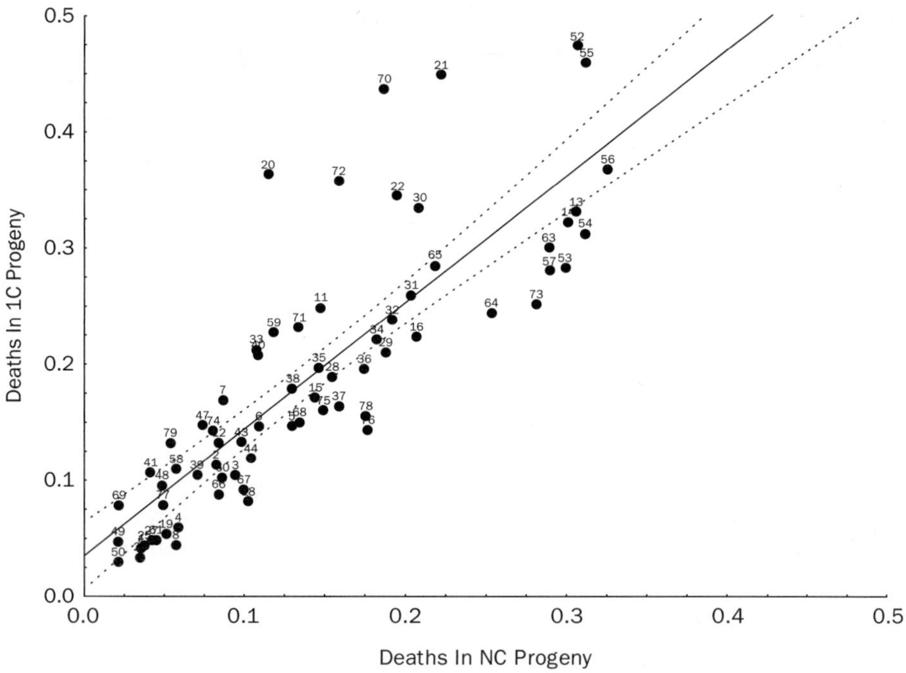

Figure 11-4 Comparative pre-reproductive mortality in first cousin (1C; *F* = 0.0625: *y*-axis) versus nonconsanguineous progeny (NC; *F* = 0: *x*-axis) in 69 study populations.
Reproduced from Bittles AH, Black ML. (2010a). Consanguinity, human evolution and complex diseases. *Proceedings of the National Academy of Sciences USA. 107,* 1779–1786.

the present analysis and matches the 3.5% excess mortality derived for Italian data of the early to mid-twentieth century (Cavalli-Sforza et al., 2004).

CONSANGUINITY AND ADULT MORBIDITY AND MORTALITY

Associations between consanguinity and complex diseases of adulthood have been under-investigated, and the few published studies mainly have relied on rudimentary sampling strategies, with simple consanguineous versus nonconsanguineous comparisons in disease prevalence. The lack of attention paid to possible genetic subdivisions or sociodemographic variables also has been a problem, and perhaps explains the often contradictory results obtained, with both positive and negative associations reported between consanguinity and breast cancer (Shami et al., 1991; Liede et al., 2002; Denic & Bener, 2001) and consanguinity and heart disease (Shami et al., 1991; Ismail et al., 2004; Jaber et al., 1997b).

As with childhood studies, the adult progeny of consanguineous unions have been reported to be over-represented in institutions caring for people with intellectual disability (Farag et al., 1993). A high prevalence of Alzheimer's disease was diagnosed in an Israeli Arab community, with over one-third of all cases being members of a single clan (*hamula*) (Farrer et al., 2003), which is strongly suggestive of a founder mutation. An earlier epidemiological survey undertaken in the demographically well-characterized Saguenay area of Québec, Canada had reported that cases of late-onset cases of Alzheimer's disease associated with the apolipoprotein (APOE) ε4 allele were significantly more inbred than controls (Vézina et al., 1999).

Long-term studies conducted on the Dalmatian Islands, Croatia have shown a positive association between inbreeding and a very wide range of common adulthood disorders, including hypertension, coronary heart disease, stroke, cancer, uni/bipolar depression, asthma, gout, and peptic ulcer (Rudan et al., 2003a, 2003b). Analysis of genealogical data covering 4–5 generations showed substantial levels of consanguinity in some of the study communities, with mean coefficients of inbreeding ranging from $\alpha = 0.002–0.049$ calculated at village level, thus supporting the causal role of rare autosomal recessive genes in adult-onset disease in this population (Rudan et al., 2003a). However, besides consanguinity, founder effect, genetic drift, and community endogamy also can result in substantial levels of random inbreeding in small population isolates of this type, and so may have contributed to the observed results.

GENOMIC MEASUREMENT OF INBREEDING AT INDIVIDUAL AND POPULATION LEVELS

Direct estimation of an individual's inbreeding coefficient by reference to genomic data offers many advantages over pedigree-based studies, since the effect of historical levels and patterns of inbreeding that are not identifiable in most pedigrees is included, and wrongly ascribed paternity also may be revealed. For this purpose, high-density genome scans have been developed to estimate individual autozygosity from uninterrupted runs of homozygosity (ROHs; see McQuillan et al., 2008). By testing individuals of differing ages, a steady decrease could be demonstrated through time in the size and frequency of ROHs in U.S. populations. These decreases were ascribed to expanded marriage pools and larger effective population sizes during the course of the twentieth century, and interpreted as indicating ongoing reductions in the frequency of rare recessive disorders (Nalls et al., 2009a).

Homozygosity mapping (see also Chapter 12 on autozygosity mapping) in consanguineous families has been used to identify autism loci and genes (Morrow et al., 2008), and a case-control study of bipolar disorder type 1 indicated an association with consanguinity (Mansour et al., 2009). ROHs have been shown to be significantly more common in patients with schizophrenia spectrum disorders, suggesting the involvement of recessive alleles in the etiology of the disorder (Lencz et al., 2007). Similarly, extended tracts of homozygosity have been reported in cases of late-onset Alzheimer's disease in an outbred U.S. population, allowing the identification of rare recessive gene variants contributing to the disease (Nalls et al., 2009b). In each of these examples, the ability to identify and utilize recent or historical consanguinity has greatly assisted the identification of causative recessive behavioral disease mutations.

CONCLUSIONS

It has been proposed that communities in which consanguineous marriage is preferential can be at an advantage when screening for deleterious mutations, as in the case of ß-thalassemia in Pakistan (Ahmed et al., 2002). Since there is a high probability that all family members will be homozygous for the same mutation, its identification in an affected individual could serve as a diagnostic marker for an extended family group at high genetic risk. Some caution is needed in this approach, since past intermarriage with other unrelated families and communities could have led to significant gene admixture, and hence increased the likelihood of compound heterozygosis. As earlier reported in Pakistan where 11.7% of first cousin progeny were compound heterozygotes for different ß-thalassemia mutations (Ahmed et al., 1996).

The future status and prevalence of consanguineous marriage is unclear. Currently, over 1,000 million people live in countries where from 20% to over 50% of marriages are consanguineous (Bittles, 2008), and it seems highly improbable that a form of marriage that remains so widely popular would rapidly decline in popularity, although in many developing countries, semi-official efforts are being made to lessen the appeal of consanguinity. Among migrant communities in Western societies there is the attraction of continuing with a form of marriage that has been a longstanding family and community tradition. At the same time, some younger members of migrant families may wish to adopt the social mores of their new homeland, including exogamous marriage.

Within migrant communities in Western countries a much greater awareness of genetic disease exists than would have been the case in their homelands, together with a more focused appreciation of the possible increased risks of affected children being born to consanguineous couples. These factors have led to increased demand for genetic counseling services (Bennett et al., 2002; Port et al., 2005), although this option is not necessarily the choice of all members of all migrant communities (Shaw & Hurst, 2008).

The fact that a reduction in consanguineous marriage should lead to a decrease in the incidence of inherited recessive disorders through increased genome-wide heterozygosity could act as a powerful incentive to opt for nonconsanguineous marriage (Campbell et al., 2009). Ultimately, however, declining family sizes and a reduction in the availability of potential marriage partners within the immediate family may prove to be the major factor in determining the future prevalence of consanguineous unions. Conversely, if several family members have a major disabling disorder, such as severe nonsyndromic intellectual disability, the marriage opportunities open to other family members may be limited, which increases the probability of further intrafamilial unions (Basel-Vanagaite et al., 2007).

Consanguineous marriage has been, and remains, a core organizational feature of many successful human societies. The advice offered by the World Health Organization (2006) that "Preference for consanguineous marriage is a feature of the sociocultural context within which medical genetic services must work" is appropriately nonjudgmental. Its adoption should ensure that the future health needs of consanguineous families and communities in both the developing and developed world can best be met.

REFERENCES

Agarwal SS, Singh U, Singh PS, Singh SS, Das V, Sharma A, et al. (1991). Prevalence and spectrum of congenital malformations in a prospective study at a teaching hospital. *Indian Journal of Medical Research 94*, 413–419.

Ahmed S, Petrou M, Saleem M. (1996). Molecular genetics of beta-thalassaemia in Pakistan: a basis for prenatal diagnosis. *British Journal of Haematology 94*, 476–482.

Ahmed S, Saleem M, Modell B, Petrou M. (2002). Screening extended families for genetic hemoglobin disorders in Pakistan. *New England Journal of Medicine 347*, 1162–1168.

Ahmed T, Ali SM, Aliaga A, Arnold F, Ayub M, Bhatti MH *et al.* (1992) *Pakistan Demographic and Health Survey 1990/91.* Pakistan National Institute of Population Studies and Macro International, Islamabad and Columbia, MD.

Al-Gazali LI. (1998). A genetic aetiological survey of severe childhood deafness in the United Arab Emirates. *Journal of Tropical Pediatrics 44*, 157–160.

Al-Gazali LI, Bener A, Abdulrazzaq YM, Micallef R, Al-Khayat AI, Gaber T. (1997). Consanguineous marriages in the United Arab Emirates. *Journal of Biosocial Science 29*, 491–497.

Assaf S, Khawaja M. (2009). Consanguinity trends and correlates in the Palestinian Territories. *Journal of Biosocial Science 41*, 107–124.

Auton A, Bryc K, Boyko AR, Lohmueller KE, Reynolds A. et al. (2009). Global distribution of genomic diversity underscores rich complex history of continental human populations. *Genome Research 19*, 795–803.

Barbour B. and Salameh, P. (2009). Consanguinity in Lebanon: prevalence, distribution and determinants. *Journal of Biosocial Science 41*, 505–517.

Başaran N. Artan S, Yazicioglu S, Şayli BS. (1994). Effects of consanguineous matings on anthropometric measurements of Saudi newborn infants. *Family Practice 6*, 217–220.

Bashi J. (1977). Effects of inbreeding on cognitive performance. *Nature 31*, 440–442.

Basel-Vanagaite L, Taub E, Halpern GJ, Drasinover V, Magal N. et al. (2007). Genetic screening for autosomal recessive nonsyndromic mental retardation in an isolated population in Israel. *European Journal of Human Genetics 15*, 250–253.

Becker SM, Al Halees Z, Molina C, Paterson RM. (2001). Consanguinity and congenital heart disease in Saudi Arabia. *American Journal of Medical Genetics 99*, 8–13.

Behar DM, Villems R, Soodyal HM, Blue-Smith J, Pereira L. et al. (2008). The dawn of human matrilineal diversity. *American Journal of Human Genetics 82*, 1130–1140.

Bener A, Alali KA. (2006). Consanguineous marriage in a newly developed country: the Qatari population. *Journal of Biosocial Science 38*, 239–246.

Bennett R, Motulsky AG, Bittles AH, Hudgins L, Uhlrich S. et al. (2002). Genetic counseling and screening of consanguineous couples and their offspring: Recommendations of the National Society of Genetic Counselors. *Journal of Genetic Counseling 11*, 97–119.

Bittles AH. (1994). The role and significance of consanguinity as a demographic variable. *Population and Development Review 20*, 561–584.

Bittles AH. (2001). Consanguinity and its relevance to clinical genetics. *Clinical Genetics 60*, 89–98.

Bittles AH. (2002). Endogamy, consanguinity and community genetics. *Journal of Genetics 81*, 91–98.

Bittles AH. (2003). Consanguineous marriage and childhood health. *Developmental Medicine and Childhood Neurology 45*, 571–576.

Bittles AH. (2008). A Community Genetics perspective on consanguineous marriage. *Community Genetics 11*, 324–330.

Bittles AH. (2009a). Consanguinity, genetic drift, and genetic diseases in populations with reduced numbers of founders. In: Vogel F, Motulsky AG, Antonarakis SE, Speicher M, eds. *Human Genetics - Principles and Approaches* 4th ed., pp. 507-528, Heidelberg, Springer.

Bittles AH. (2009b). The background and outcomes of the first cousin marriage controversy in Great Britain. *International Journal of Epidemiology 38*, 1429–1439.

Bittles AH, Black ML. (2010a). Consanguinity, human evolution and complex diseases. *Proceedings of the National Academy of Sciences USA. 107*, 1779–1786.

Bittles AH, Black ML. (2010b) Consanguinity and evolution. *Annual Review of Anthropology 39*, 193–207.

Bittles AH, Black ML. (2010c) The impact of consanguinity on neonatal and infant health. *Early Human Development 86*, 737–741.

Bittles AH, Grant JC, Shami SA. (1993). An evaluation of consanguinity as a determinant of reproductive behaviour and mortality in Pakistan. *International Journal of Epidemiology 22*, 463–467.

Bittles AH, Grant JC, Sullivan SG, Hussain R. (2002). Does inbreeding lead to decreased human fertility? *Annals of Human Biology 29*, 111–131.

Bittles AH, Mason WM, Greene J, Appaji Rao N. (1991). Reproductive behavior and health in consanguineous marriages. *Science 252*, 789–794.

Bittles AH, Neel JV. (1994). The costs of human inbreeding and their implications for variations at the DNA level. *Nature Genetics 8*, 117–121.

Bittles AH, Radha Rama Devi A, Appaji Rao N. (1988). Consanguinity, twinning and secondary sex ratio in Karnataka, South India. *Annals of Human Biology 15*, 455–460.

Bromiker R, Glam-Baruch M, Gofin R, Hammerman C, Amitai Y. (2004). Association of parental consanguinity with congenital malformations among Arab newborns in Jerusalem. *Clinical Genetics 66*, 63–66.

Bundey S, Alam H, Kaur A, Mir S, Lancashire RJ. (1990). Race, consanguinity and social features in Birmingham babies: a basis for a prospective study. *Journal of Epidemiology and Community Health* 44, 130–135.

Calderón R, Peña JA, Delgado J, Morales B. (1998). Multiple kinship in two Spanish regions: new model relating multiple and simple consanguinity. *Human Biology 70*, 535–561.

Campbell H, Rudan I, Bittles AH, Wright AF. (2009). Human population structure, genome autozygosity and human health. *Genome Medicine 1*, 91.

Castilla EE, Gomez MA, Lopez-Camelo JS, Paz JE. (1991). Frequency of first-cousin marriages from civil marriage certificates in Argentina. *Human Biology 63*, 203–210.

Cavalli-Sforza LL, Moroni A, Zei G. (2004). *Consanguinity, Inbreeding, and Genetic Drift in Italy.* Princeton University Press, Princeton, p. 292.

Chaix R, Cao C, Donnelly P. (2008). Is mate choice in humans MHC-dependent? *PLoS Genetics 4*, e10000184.

Christianson A, Howson CP, Modell M. (2006). *March of Dimes Global Report on Birth Defects.* March of Dimes, White Plains, NY.

COSIT (Central Organization for Statistics and Information Technology). (2005). *Iraq Living Conditions Survey 2004: Vol. II: Analytical Report.* Baghdad, Ministry of Planning and Development Cooperation.

da Fonseca LG, Freire-Maia N. (1970). Further data on inbreeding levels in Brazilian populations. *Social Biology 17*, 324–328.

De Braekeleer M, Ross M. (1991). Inbreeding in Saguenay-Lac-St-Jean (Québec, Canada): a study of Catholic Church dispensations. *Human Heredity 41*, 379–384.

Denic S, Bener A. (2001). Consanguinity decreases risk of breast cancer - cervical cancer unaffected. *British Journal of Cancer 85*, 1675–1679.

Dereköy FS. (2000). Etiology of deafness in Afyon school for the deaf in Turkey. *International Journal of Pediatric Otorhinolaryngology 55*, 125–131.

Durkin MS, Hasan ZM, Hasan KZ. (1998). Prevalence and correlates of mental retardation in Karachi, Pakistan. *American Journal of Epidemiology 147*, 281–288.

Edmond M, De Braekeleer M. (1993). Inbreeding effects on fertility and sterility: a case-control study in Saguenay-Lac-Saint-Jean (Québec, Canada) based on a population registry. *Annals of Human Biology 20*, 545–555.

Elder MJ, De Cock R. (1993). Childhood blindness in the West Bank and Gaza Strip: prevalence, aetiology, and hereditary factors. *Eye 7*, 580–583.

Farag TI, Al-Awadi SA, El-Badramary MH, Aref MA, Kasrawi B. et al. (1993). Disease profile of 400 institutionalized mentally retarded patients in Kuwait. *Clinical Genetics* 44, 329–334.

Farrer LA, Bowirrat A, Friedland RP, Waraska K, Kprczyn AD, Baldwin CT. (2003). Identification of multiple loci for Alzheimer disease in a consanguineous Israeli-Arab community. *Human Molecular Genetics 12*, 415–422.

Freire-Maia, N. (1968). Inbreeding levels in American and Canadian populations: a comparison with Latin America. *Eugenics Quarterly 15*, 22–33.

Gnanalingham MG, Gnanalingham KK, Singh A. (1999). Congenital heart disease and parental consanguinity in South India. *Acta Paediatrica 88*, 473–474.

Gunaid AA, Hummad NA, Tamim KA. (2004). Consanguineous marriage in the capital city Sana'a, Yemen. *Journal of Biosocial Science 36*, 111–121.

Hafez M, El-Tahan H, Awadalla M, El-Khayat H, Abdel-Gafar A, and Ghoneim M. (1983). Consanguineous matings in the Egyptian population. *Journal of Medical Genetics 20*, 58–60.

Hamamy H, Bittles AH. (2009). Genetic clinics in Arab communities: meeting individual, family and community needs. *Public Health Genomics 12*, 30–40.

Hamamy H, Jamhawi L, Al-Darawsheh J, Ajlouni K. (2005). Consanguineous marriages in Jordan: why is the rate changing with time? *Clinical Genetics 67*, 511–516.

Harpending HC, Batzer MA, Gurven M, Jorde LB, Rogers AR, Sherry ST. (1998). Genetic traces of ancient demography. *Proceedings of the National Academy of Sciences USA. 95*, 1961–1167.

Hashmi MA. (1997). Frequency of consanguinity and its effect on congenital malformation – a hospital based study. *Journal of the Pakistan Medical Association 47*, 75–78.

Helgason A, Pálsson S, Guðbjartsson DF, Kristjánsson P, Stefánsson K. (2008). An association between the kinship and fertility of human couples. *Science 319*, 813–816.

Hussain R, Bittles AH. (1999). Consanguinity and differentials in age at marriage, contraceptive use and fertility in Pakistan. *Journal of Biosocial Science 31*, 121–138.

Ismail J, Jafar TH, Jafary FH, White F, Faruqui AM, Chaturvedi N. (2004). Risk factors for non-fatal myocardial infarction in young South Asian males. *Heart 90*, 259–263.

Jaber L, Merlob P, Bu X, Rotter JI, Shohat M. (1992). Marked parental consanguinity as a cause for increased major malformations in an Israeli Arab community. *American Journal of Medical Genetics* 44, 1–6.

Jaber L, Merlob P, Gabriel R, Shohat M. (1997a). Effects of consanguineous marriage on reproductive outcome in an Arab community in Israel. *Journal of Medical Genetics 34*, 1000–1002.

Jaber L, Shohat T, Rotter JI, Shohat M. (1997b). Consanguinity and common adult diseases in Israeli Arab communities. *American Journal of Medical Genetics 70*, 346–348.

Kapadia KM. (1958). *Marriage and the Family in India*, 2nd ed. Oxford University Press, Calcutta, pp. 117–137.

Khlat M. (1985). Consanguineous marriages in Beirut: time trends, spatial distribution. *Social Biology 35*, 324–330.

Khlat M. (1989). Inbreeding effects on fetal growth in Beirut, Lebanon. *American Journal of Physical Anthropology 80*, 481–484.

Khoury SA, Massad D. (2000). Consanguinity, fertility, reproductive wastage, infant mortality and congenital malformations in Jordan. *Saudi Medical Journal 21*, 150–154.

Koç I. (2008). Prevalence and sociodemographic correlates of consanguineous marriages in Turkey. *Journal of Biosocial Science 40*, 137–148.

Kostyu DD, Dawson DV, Elias S, Ober C. (1993). Deficit of HLA homozygotes in a Caucasian isolate. *Human Immunology 37*, 135–142.

Kulkarni ML, Kurian M (1990). Consanguinity and its effect on fetal growth and development: a south Indian study. *Journal of Medical Genetics 27*, 348–352.

Lamdouar Bouazzaoui, N. (1994). Consanguinité et santé publique au Maroc. *Bulletin de l'Academie Nationale de Médicine 178*, 1013–1027.

Lencz T, Lambert C, DeRosse P, Burdick KE, Morgan TV. *et al.* (2007). Runs of homozygosity reveal highly penetrant recessive loci in schizophrenia. *Proceedings of the National Academy of Sciences USA 104*, 19942–19947.

Liascovich R, Rittler M, Castilla EE. (2001). Consanguinity in South America: demographic aspects. *Human Heredity 51*, 27–34.

Liede A, Malik IA, Aziz Z, Rios L, Kwan E, Narod SA. (2002). Contribution of BRAC1 and BRAC2 mutations of breast and overian cancer in Pakistan. *American Journal of Human Genetics 71*, 595–606.

Magnus P, Berg K, Bjerkedal T. (1985). Association of parental consanguinity with decreased birth weight and increased rate of early death and congenital malformations. *Clinical Genetics 28*, 335–342.

Mansour H, Klei L, Wood J, Talkowski M, Chowdari K. et al. (2009). Consanguinity associated with increased risk for bipolar I disorder in Egypt. *American Journal of Medical Genetics B: Neuropyschiatric Genetics 150B*, 879–885.

Martin NS, Sutherland J, Levin AV, Klose R. Proston M, Héon E. (2000). Molecular characterization of congenital glaucoma in a consanguineous Canadian community: a step towards preventing glaucoma related blindness. *Journal of Medical Genetics 37*, 422–427.

Masterson JG. (1970). Consanguinity in Ireland. *Human Heredity* 20, 371–382.

McEvoy BP, Montgomery GW, McRae AF, Ripatti S, Perola M. et al. (2009). Geographical structure and differential natural selection among North European populations. *Genome Research 19*, 804–814.

McQuillan R, Leutenegger A-L, Abdel-Rhaman R, Franklin CS, Pericic M. et al. (2008). Runs of homozygosity in European populations. *American Journal of Human Genetics 83*, 359–372.

Morrow EM *et al.* (2008) Identifying autism loci and genes by tracing recent shared ancestry. *Science 321*, 218–223.

Moynihan L, Houseman M, Newton V, Mueller R, Lench N. (1999). DFNB20: a novel locus for autosomal recessive, non-syndromal sensorineural hearing loss maps to chromosome 11q25-qter. *European Journal of Human Genetics 7*, 243–246.

Mustapha M, Weil D, Chardenoux S, Elais S, El-Zir E. et al. (1999). An α-tectorin gene defect causes a newly identified autosomal recessive from of sensorineural pre-lingual non-syndromic deafness. *Human Molecular Genetics* 8, 409–412.

Nabulsi MM, Tamim H, Sabbagh M, Obeid MY, Yunis KA, Bitar FF. (2003). Parental consanguinity and congenital heart malformations in a developing country. *American Journal of Medical Genetics 116A*, 342–347.

Nalls MA, Simon-Sanchez J, Gibbs JR, Paisan-Ruiz C, Bras JT. et al. (2009a). Measures of autozygosity in decline: globalization, urbanization, and its implications for medical genetics. *PLoS Genetics 5*, e1000415

Nalls MA, Guerreiro RJ, Simon-Sanchez J, Bras JT, Traynor BJ. et al. (2009b) Extended tracts of homozygosity identify novel candidate genes associated with late-onset Alzheimer disease. *Neurogenetics 10*, 183–190.

National Conference of Commissioners. (1970). *Handbook on Uniform State Laws and Proceedings of the Annual Conference Meeting in its Seventy-Ninth Year*, Port City Press, Baltimore.

Nelson J, Smith M, Bittles AH. (1997). Consanguineous marriage and its clinical consequences in migrants to Australia. *Clinical Genetics 52*, 142–146.

Norio R. (2003) Finnish Disease Heritage I: characteristics, causes, background. *Human Genetics 112*, 441–456.

Ober C, Hyslop T, Hauck WW. (1999). Inbreeding effects in humans: evidence for reproductive compensation. *American Journal of Human Genetics 64*, 225–231.

Ober C, Weitkamp LR, Cox N, Dytch H, Kostyu D, Elias S. (1997). HLA and mate choice in humans. *American Journal of Human Genetics 16*, 497–504

Ottenheimer M. (1996). *Forbidden Relatives: the American Myth of Cousin Marriage*, University of Illinois Press, Urbana, pp. 19–41.

Panicker SG, Reddy ABM, Mandal AK, Ahmed N, Nagarajaram HA. et al. (2002). Identification of novel mutations causing familial primary congenital glaucoma in Indian pedigrees. *Investigative Ophthalmology and Visual Science 43*, 1358–1366.

Paul DB, Spencer HG. (2008). "It's Ok, we're not cousin by blood"; the cousin marriage controversy in historical perspective. *PLoS Biology 6*, e320.

Pingle U. (1983). Comparative analysis of mating systems and marriage distance patterns in five tribal groups of Andhra Pradesh, India. *Social Biology 30*, 67–74.

Port KE, Bittles, AH. (2001). A population-based estimate of the prevalence of consanguineous marriage in Western Australia. *Community Genetics 4,* 97–101.

Port KE, Mountain H, Nelson J, Bittles AH. (2005). The changing profile of couples seeking genetic counseling for consanguinity in Australia. *American Journal of Medical Genetics 132A,* 159–163.

Pulleyn LJ, Jackson AP, Roberts E, Carridice A, Muxworthy C, et al. (2000). A new locus for autosomal recessive non-syndromal sensorineural hearing impairment (DFNB27) on chromosome 2q23-q31. *European Journal of Human Genetics 8,* 991–993.

Puri RK, Verma IC, Bhargava I. (1978) Effects of consanguinity in a community in Pondicherry. In: Verma IC (ed). *Medical Genetics in India,* Vol 2. Auroma, Pondicherry, pp. 129–139.

Rahi JS, Sripathi S, Gilbert CE, Foster A. (1995). Childhood blindness in India: causes in 1318 blind school students in nine states. *Eye* 9, 545–550.

Ramankutty P, Tikreeti RAS, Rasaam KW, Al-Thamery DM, Yacoub AAH, Mahmood DA. (1983). A study of birth weight in Iraqi children. *Journal of Tropical Pediatrics 29,* 5–10.

Rao PSS, Inbaraj SJ. (1977). Inbreeding effects on human reproduction in Tamil Nadu of South India. *Annals of Human Genetics 41,* 87–98.

Rao PSS, Inbaraj SJ. (1980). Inbreeding effects on fetal growth and development. *Journal of Medical Genetics 17,* 27–33.

Rittler M, Liascovich R, López-Camelo J, Castilla EE. (2001). Parental consanguinity in specific types of congenital anomalies. *American Journal of Medical Genetics 102,* 36–43.

Rogers NK, Gilbert CE, Foster A, Zakhidov BO, McCollum CJ. (1999). Childhood blindness in Uzbekistan. *Eye* 13, 65–70.

Rudan I, Smolej-Narancic N, Campbell H, Carothers A, Wright A, et al. (2003a). Inbreeding and the genetic complexity of human hypertension. *Genetics 163,* 1011–1021.

Rudan I, Rudan D, Campbell H, Carothers A, Wright A, et al. (2003b). Inbreeding and risk of late onset complex disease. *Journal of Medical Genetics 40,* 925–932.

Saadat M, Ansari-Lari M, Farhud DD. (2004). Consanguineous marriage in Iran. *Annals of Human Biology 31,* 263–269.

Saedi-Wong S, Al-Frayh AR. (1989). Effects of consanguineous matings on anthropometric measurements of newborn infants. *Family Practice 6,* 217–220.

Saha N, El Sheikh FS. (1988). Inbreeding levels in Khartoum. *Journal of Biosocial Science 20,* 333–336.

Schull WJ, Furusho T, Yamamoto M, Nagano H, Komatsu I. (1970). The effects of parental consanguinity and inbreeding in Hirado, Japan. IV. Fertility and reproductive compensation. *Humangenetik 9,* 294–315.

Shahin H, Walsh T, Sobe T, Lynch E, King E, et al. (2002). Genetics of congenital deafness in the Palestinian population: multiple connexin 26 alleles with shared origins in the Middle East. *Human Genetics 110,* 284–289.

Shami SA, Qaisar R, Bittles AH. (1991). Consanguinity and adult morbidity in Pakistan. *Lancet 338,* 954–955.

Shaw A, Hurst JA. (2008). "What is this genetics, anyway?" Understandings of genetics, illness causality and inheritance among British Pakistani users of genetic services. *Journal of Genetic Counseling 17,* 373–383.

Silbert JR, Jadhav M, Inbaraj SG. (1979). Fetal growth and parental consanguinity. *Archives of Diseases in Childhood 54,* 317–319.

Sridhar K. (2009). A community-based survey of visible congenital anomalies in rural Tamil Nadu. *Indian Journal of Plastic Surgery 42,* S184–S188.

Stern C, Charles DR. (1945). The Rhesus gene and the effect of consanguinity. *Science 101,* 305–307.

Stevenson AC, Davison BCC, Say B, Ustuoplu S, Liya D, et al. (1971). Contribution of feto-maternal incompatibility to aetiology of pre-eclamptic toxaemia. *Lancet ii,* 1286–1289.

Stoltenberg C, Magnus P, Lie RT, Dalveit AK, Irgens LM. (1997). Birth defects and parental consanguinity in Norway. *American Journal of Epidemiology 145,* 439–448.

Stoltenberg C, Magnus P, Skrondal A, Lie RT. (1999). Consanguinity and recurrence risk of stillbirth and infant death. *American Journal of Public Health 89,* 517–523.

Sundstrom RA, van Laer L, van Camo G, Smith RJH. (1999). Autosomal recessive nonsyndromic hearing loss. *American Journal of Medical Genetics 89,* 123–129.

Tunçbilek E, Koç I. (1994). Consanguineous marriage in Turkey and its impact on fertility and mortality. *Annals of Human Genetics 58,* 321–329.

Vézina H, Heyer É, Fortier I, Ouellette G, Robitaille Y, Gauvreau D. (1999). A genealogical study of Alzheimer disease in the Saguenay region of Québec. *Genetic Epidemiology 16,* 412–425.

Wedekind C, Seebeck T, Bettens F, Paepke AJ. (1995). MHC-dependent mate preference in humans. *Proceedings of the Royal Society of London Series B 260,* 245–249.

WHO. (2006). *Medical Genetics in Developing Countries: the Ethical, Legal and Social Implications of Genetic Testing and Screening.* World Health Organization, Geneva.

Yunis K, Mumtaz G, Bitar F, Chamseddine F, Kassar M, et al. (2006). Consanguineous marriage and congenital defects: a case-control study in the neonatal period. *American Journal of Medical Genetics 140A,* 1524–1530.

Zhivotovsky LA, Rosenberg NA, Feldman MW. (2003). Features of evolution and expansion of modern humans, inferred from genomewide microsatellite markers. *American Journal of Human Genetics 72,* 1171–1186.

Zlotogora J, Hujerat Y, Barges S, Shalev SA, Chakravarti A. (2006). The fate of 12 recessive mutations in a single village. *Annals of Human Genetics 71,* 202–208.

12.

AUTOSOMAL RECESSIVE CONDITIONS AND AUTOZYGOSITY MAPPING

Colin A. Johnson and Eamonn R. Maher

INTRODUCTION

Over a century ago, Sir Archibald Garrod (Garrod, 1902) provided the first proof of the validity of Mendel's laws in humans. Garrod showed that patients with recessive conditions such as alkaptonuria, cystinuria, and albinism (which he called "inborn errors of metabolism") have an unusually high frequency in the children of consanguineous parents. For alkaptonuria, he reported that the unaffected parents were first cousins in six of the ten families studied, whereas the incidence of first cousin marriages at the time was about 3%. Garrod had discussed this data with William Bateson, who reported to the Evolution Committee of the Royal Society in the same year that "the mating of first cousins gives exactly the conditions most likely to enable a rare, and usually recessive, character to show itself" (Bateson & Saunders, 1902). Bateson suggested that homozygosity-by-descent was the genetic explanation for Garrod's observations, since inbreeding would cause a relative increase of the frequency of homozygous individuals, and would therefore reveal rare recessive traits.

Half a century later Cedric Smith made the prescient suggestion that offspring of a consanguineous mating, and expressing a recessive trait, would be homozygous for the genetic markers near to the disease gene, and that these children could be used to detect linkage (Smith, 1953). Although he recognized that a linkage screen for regions homozygous by descent in consanguineous families offered a powerful strategy for mapping autosomal recessive disorders, the technical resources to do this were not available for many years, and Smith concluded that "the method is impractical." However, with the subsequent development of accurate genetic maps in humans, this type of mapping became feasible and was reintroduced by Lander and Botstein (1987). In a landmark paper, they described the use of restriction fragment length polymorphisms (RFLP) as markers for "homozygosity mapping" (Lander & Botstein, 1987), as it was then called. The strategy relies on the assumption that an affected individual inherits two identical alleles of a disease gene from a common ancestor, and that mapping a disease locus depends on the identification of chromosomal regions that are identical by descent (IBD, or *autozygous*) from the common ancestor. Searching for regions of homozygosity in affected individuals from inbred families, a strategy now known as *autozygosity mapping*, therefore represents a highly efficient and effective approach for localizing recessive disease genes (Mueller & Bishop, 1993), and has become the method of choice for mapping recessive traits.

PRINCIPLE OF AUTOZYGOSITY MAPPING

The principle of autozygosity mapping is shown schematically in Figure 12-1, for an idealized first cousin consanguineous family with two individuals affected with a hypothetical recessive condition. *Autozygosity mapping* has become the preferred term in more recent mapping studies, and will be used henceforth in this review, although *homozygosity mapping* has also been used in older literature. The mathematical basis for autozygosity mapping was formalized by Lander and Botstein (1987). A child from a consanguineous marriage has an inbreeding coefficient of F, which is defined simply as the probability that any given locus is will be IBD (or, equally, the fraction of the child's genome expected to be homozygous by descent). If a recessive disease allele occurs with frequency q, assuming that alleles are in Hardy-Weinberg equilibrium within the population, then the probability of the disease locus being IBD for an affected individual is Fq. The probability that a child is *not* IBD (in other words, identical by state, or IBS) at the disease locus, but still affected due to the random reassortment of the disease alleles, is $(1 - F)q^2$. The probability, α, that the recessive disease in an affected individual is due to IBD at the disease locus, is therefore: $Fq/[Fq + (1 - F)q^2]$. From this equation, it can be seen that if q is small (and for many recessive conditions $q < 0.005$) then $\alpha \approx 1$, thus validating the assumption that, for a rare recessive disorder, the disease gene will lie in a homozygous region. The converse argument explains the observation that the parents of children with a recessive condition are likely to be related.

An LOD score can be calculated from α and F, since the region at the disease locus is IBD with probability

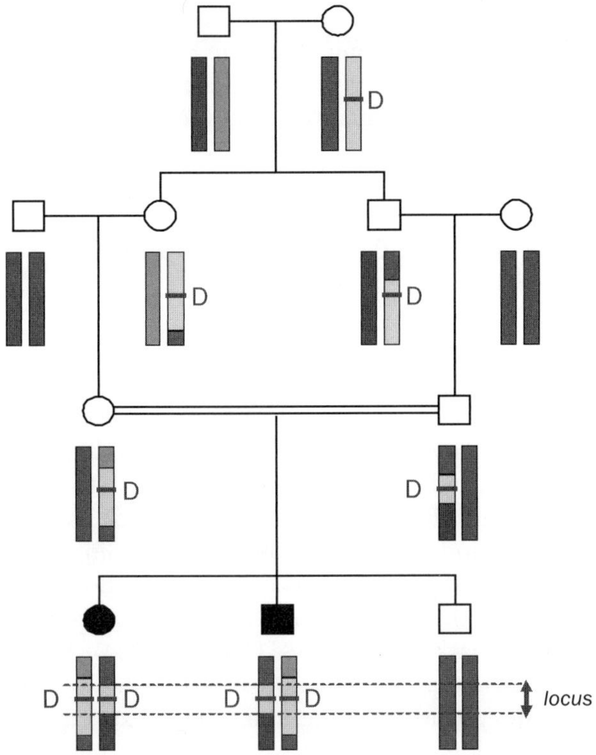

Figure 12-1 The principle of autozygosity mapping: a schematic representation of the inheritance of recessive disease gene in an idealized pedigree of a first cousin consanguineous family, leading to two individuals affected with a hypothetical recessive condition in the fourth (IV) generation. The great-grandmother in the first (I) generation is a carrier for one copy of the disease gene ("D"), indicated in red. Each of her children has a 50% chance of inheriting the disease allele, as well as the chromosomal segment or haplotype flanking the gene, shown in grey. Both children in generation II inherit the faulty gene, and then pass it on to their own children. As the disease gene is transmitted through the generations, the haplotype surrounding it becomes smaller due to meiotic recombination events. The parents in the third (III) generation are first cousins and carriers, each inheriting a copy of the disease gene from their grandmother, with a 1 in 4 risk that each of their children will inherit both copies of the disease gene. The two affected children (generation IV) in this pedigree are therefore *homozygous by descent* for this gene, but both are also homozygous for the chromosomal segment surrounding the disease gene. These shared homozygous regions can be identified by genotyping genetic markers spaced regularly throughout the genome in a genome-wide linkage scan. Refer color figure.

$\alpha \approx 1$, whereas unlinked regions are IBD with a probability of only F. The ratio α:F is therefore the odds ratio in favor of linkage (assuming the marker is informative and q is small, so that $\alpha \approx 1$). A single affected child of first cousin parents ($F = 1/16$) gives an odds ratio of 16:1 in favor of linkage, and therefore a LOD score of 1.204 (for a region that is clearly IBD). Overlapping regions of homozygosity that are IBD in just three separate first cousin marriages would give a statistically significant LOD score of 3.61. Although more distantly related families would increase the power to detect linkage (e.g., an affected child of a second cousin marriage could derive a maximum LOD score of 2.41), the region of IBD will decrease in size due to the increased number of meiotic recombinations, so more dense coverage of markers would be required to detect the autozygous segment.

In practice, the value of F, the inbreeding coefficient, is often an underestimate because of complex, hidden consanguinity in a pedigree (Woods et al., 2006), and the incorrect specification of the level of inbreeding can have a dramatic influence on the LOD scores (Miano et al., 2000). In some populations, specific subgroups may also have a background level of more distant, but ill-defined, distant consanguinity as a result of clan-based intermarriage or population isolates. This issue can be addressed by calculating the inbreeding coefficient to infer the pedigree structure and to allow an estimate of LOD scores (Leutenegger et al., 2003; Leutenegger et al., 2006).

AUTOZYGOSITY MAPPING TO GENE IDENTIFICATION

Although numerous recessive disorders are recognized, many are individually uncommon or cause early mortality, so that large cohorts of families suitable for a classical linkage-based study are often unavailable. Furthermore, in small families, only the diseased generation provides information for linkage, and the presence of locus heterogeneity greatly hinders mapping efforts using unrelated families. In contrast, the use of consanguineous families for autozygosity mapping greatly increases the power to find new disease loci and genes, even in the presence of locus heterogeneity (Lander & Botstein, 1987; Mueller & Bishop, 1993). Autosomal recessive diseases can be extremely heterogeneous, even within the same ethnic group, as has been shown for autosomal recessive primary microcephaly in the Pakistani community (Bond & Woods, 2006) and Bardet-Biedl syndrome (Beales, 2005). However, mapping in one large multiplex consanguineous family can often identify the locus for a disease gene, which can then be verified in other, smaller singleton families, thus overcoming the problem of locus heterogeneity.

The first disorders that had loci mapped by autozygosity mapping were Werner syndrome (Goto et al., 1992), ataxia with selective vitamin E deficiency (Ben Hamida et al., 1993), and alkaptonuria (Pollak et al., 1993). Other notable early advances were the identification of loci for Bardet-Biedl syndrome, (*BBS2* and *BBS3*; Kwitek-Black et al., 1993; Sheffield et al., 1994) and nonsyndromic sensorineural deafness, (*DFNB3*, *DFNB4*, *DFNB1* and *HJCD*; Friedman et al., 1995; Fukushima et al., 1995; Brown et al., 1996; Moynihan et al., 1998). These studies demonstrated the power of autozygosity mapping for gene localization even in the presence of genetic heterogeneity. The first disease genes to be identified through positional-candidate cloning, based on autozygosity mapping, were laminin α2-chain (*LAMA2*), with mutations causative for merosin-deficient congenital muscular dystrophy (Helbling-Leclerc et al., 1995), and human amiloride-sensitive epithelial sodium channel (hENaC) γ-subunit (*SCNN1G*) mutations as a cause of pseudohypoaldosteronism type 1 (Strautnieks et al., 1996). Connexin 26 mutations were identified as the most common cause of hereditary nonsyndromic sensorineural deafness (Kelsell et al., 1997).

The Sheffield laboratory in Iowa was amongst the first to use microsatellite markers in genome-wide linkage screens for autozygosity mapping. A genome scan used 300 to 400 evenly spaced, highly polymorphic microsatellite markers, to give an average coverage of one marker per 10cM. This was thought to be sufficient to detect most homozygous regions in first and second cousin consanguineous pedigrees (Lander & Botstein, 1987). Strategies that were developed to simplify autozygosity mapping included the initial genotyping of only affected individuals, with later verification of marker segregation in the rest of the family for homozygous regions of possible interest, and DNA pooling (Sheffield et al., 1994; Nystuen et al., 1996; Arbour et al., 1997), in which samples from affected and unaffected individuals were pooled separately and compared. DNA pooling reduced the volume and cost of initial genotyping, conserved often precious and rare DNA samples, and reduced the number of homozygous regions that need to be verified in subsequent studies. However, it came at the cost of reduced informativeness of individual genotypes, since a heterozygous genotype could easily mask a homozygous genotype. However, even modern autozygosity mapping studies with SNP chip genotyping, have used DNA pooling as a successful strategy (Parry et al., 2009; El-Sayed et al., 2009).

With the availability of human genome sequence data, a candidate-positional approach has superseded a purely positional cloning strategy for the identification of human disease genes. However, the essential prerequisite for candidate-positional gene identification is unequivocal mapping of the disease locus. Autozygosity mapping, in the past two decades, has mapped loci and led to the identification of a plethora of disease genes for recessive conditions. The *On-line Mendelian Inheritance in Man* database, as of October 2009, listed about 2700 autosomal Mendelian phenotypes with a known molecular basis, of which over 1100 were recessive genes identified by autozygosity mapping. However, two decades following the first positional identification of a human disease-causing gene, it is notable that an estimated 2300 autosomal recessive conditions, or 37% out of a total of about 6200 annotated Mendelian phenotypes, still do not have any identified molecular defect. Autozygosity mapping, therefore, remains a very active area of medical research, with a succession of major insights into disease pathogenesis resulting from the greater power of modern genetic technologies. Furthermore, the insights obtained from rare recessive disorders often elucidate the pathogenesis of more common complex diseases, including cancers, hypertension, neurodegenerative conditions, and diabetes (Peltonen et al., 2006).

PRACTICAL ASPECTS OF AUTOZYGOSITY MAPPING

Microsatellite markers are multiallelic and more informative than biallelic SNPs, but PCR reactions can be difficult to optimize and are prone to artefactual errors. In contrast, informative SNPs are abundant throughout the human genome, and can be easily genotyped with reduced time and cost in high-throughput formats such as SNP microarrays (Gunderson et al., 2004; Gibbs & Singleton, 2006). Previous studies have shown that high-density SNPs can extract more inheritance information than can microsatellite markers (Evans & Cardon, 2004). High-density SNP genotyping is therefore particularly advantageous for autozygosity mapping, since the markers can be analyzed as haplotypes, and extensive regions of homozygosity in affected individuals are readily detected (Carr et al., 2006; Carr et al., 2009a). Modern microarrays, such as the popular Affymetrix Genome-Wide Human SNP Array 6.0 and Illumina Infinium HD BeadChips, are both high-resolution arrays that allow SNP genotyping and the detection of copy number variation (CNV) by additional probes.

Meiotic recombinations reduce the size of the homozygous region, and the average length of homozygosity surrounding the disease gene has been estimated to be 28 cM for affected individuals from a first cousin mating, and 22 cM from a second cousin mating (Génin et al., 1998). However, this estimate is simplistic since it assumes a normal distribution for meiotic events during recombination, and neglects the effect of interference (the phenomenon of suppression of a second recombination event near to the first, original one). To overcome the problems of simulation studies, Woods et al. (2006) analyzed real-life SNP genotyping data for a number of successful autozygosity mapping studies that led to disease gene identification. In such data for any one affected individual, it was suggested that the largest contiguous homozygous segment would contain the disease gene in 17% of cases (Woods et al., 2006), and that the average size of the homozygous segment at the disease locus was 26 cM (n = 100; range 5–70 cM). However, these conclusions are likely to be biased toward successful mapping studies in which the segments were large and obvious, and would neglect those studies that were unsuccessful. A number of other studies have shown that the critical homozygous regions in one affected individual can be very small (about 0.5 cM), in otherwise large (>20 cM) loci (Gissen et al., 2004, Uppal et al., 2008). In practice, regions of homozygosity can be quite variable in size, even within the same consanguineous family, despite the relatively few recombination events that separate different affected individuals (Carr et al., 2009a).

LINKAGE ANALYSIS METHODS

Genome-wide linkage analyses were developed for microsatellite genotype data, but with the increasing use of SNP markers with low informativity (Gibbs & Singleton, 2006), family studies are usually limited by the small number of informative meioses. For most autozygosity mapping projects, multipoint linkage analyses under a recessive disease model have therefore been performed, with software such as GENEHUNTER (Kruglyak et al., 1996), SIMWALK2 (Sobel et al., 2002), MERLIN (Abecasis et al., 2002) or ALLEGRO (Gudbjartsson et al., 2000; Gudbjartsson et al., 2005). In general, haplotypes are then inspected manually for homozygous regions that are shared by all affected individuals, and can be inferred to be IBD if genotypes from parents or other close relatives are available. However, in practical terms, conventional parametric methods of multipoint linkage analysis for large datasets in complex consanguineous families are often difficult because of the excessive time required and computational limits, since consanguineous loops necessitate the inference of genotyping data from too many founder individuals. A nonparametric method that detects long, homozygous segments, but avoids the generation of LOD scores, is the program PLINK (Purcell et al., 2007).

Segments of homozygosity-by-descent (called "runs of homozygosity, ROH") of up to 4 Mb are common in outbred individuals that did not share maternal and paternal ancestors in five, and probably at least ten, generations (McQuillan et al., 2008). This observation prompted the suggestion that it may even be possible to find recessive disease genes by autozygosity mapping in single affected individuals from outbred populations (Hildebrandt et al., 2009). This study showed that homozygous mutations are frequent in recessive conditions (60% of a large, unselected cohort of patients had homozygous mutations in 13 different autosomal recessive disease genes), and that these mutations are often localized in segments of contiguous homozygous SNP markers. Furthermore, the detection of regions of homozygosity >2.1 Mb, by nonparametric linkage analysis of SNP genotyping data by ALLEGRO (Gudbjartsson et al., 2000), was able to detect the correct locus of the disease-causing mutation in 93% of cases (Hildebrandt et al., 2009). The detection of small or larger homozygous regions in both outbred and more inbred families may, therefore, be useful for the molecular genetic diagnosis of recessive conditions with extreme genetic heterogeneity, such as Bardet-Biedl syndrome (Harville et al., 2009).

ANALYSIS AND VISUALIZATION OF GENOTYPING DATA

Genotyping data from microsatellite markers for autozygosity mapping projects and for other genome-wide linkage screens is reasonably straight-forward, and can often be implemented in a simple spreadsheet (Forshew & Johnson, 2004). In contrast, the much higher densities of SNP markers need software approaches that will visualize large datasets in a clear and interactive fashion. Several programs have therefore been developed to either count SNPs or to visualize the raw genotyping data. These approaches generate either a numerical score or a graphical display that changes as the genotype data is scanned in linear map order. SNP-counting approaches include "AutoSNPa" and "IBDfinder" (Carr et al., 2006, 2009a) that generate a visual representation of the genotyping data in either consanguineous family or singleton affected individuals (Figure 12-2). The data does not undergo any statistical or linkage analyses that may be founded on assumptions that are invalid for the population under study. These nonparametric "model-free" methods therefore offer an empirical analysis that simply highlights the movement of haplotype segments from parent to child. Although they have undergone relatively few recombination events, these segments are quite variable in size (see above). Unlike other methods that use a likelihood approach to predict potential regions of IBD, both "AutoSNPa" and "IBDFinder" do not predict the presence of the disease gene from the length of an IBD region. Instead, these software tools have allowed the rapid prioritization of potential regions of interest in several recent autozygosity mapping studies (Uppal et al., 2008; Rice et al., 2009). Seelow et al. (2009) have used a website-based application called "HomozygosityMapper" (Figure 12-2) to allow researchers to upload unannotated SNP chip genotyping data, which implements a SNP-counting function and a visual representation of the data.

Sequential scoring has also been used in hidden Markov chain approaches that utilize allele frequencies to determine the likelihood of IBD from a long sequence of markers (Leutenegger et al., 2003). However, these likelihood approaches are dependent on supplementary information, such as accurate genetic maps and allele frequencies for the populations under study. These data are often unknown for the small and reproductively isolated populations that are most amenable to autozygosity mapping. In particular, likelihood predictions of IBD are prone to skewing by homozygous genotypes that are believed to be rare, but are actually of unknown frequency in the population under study.

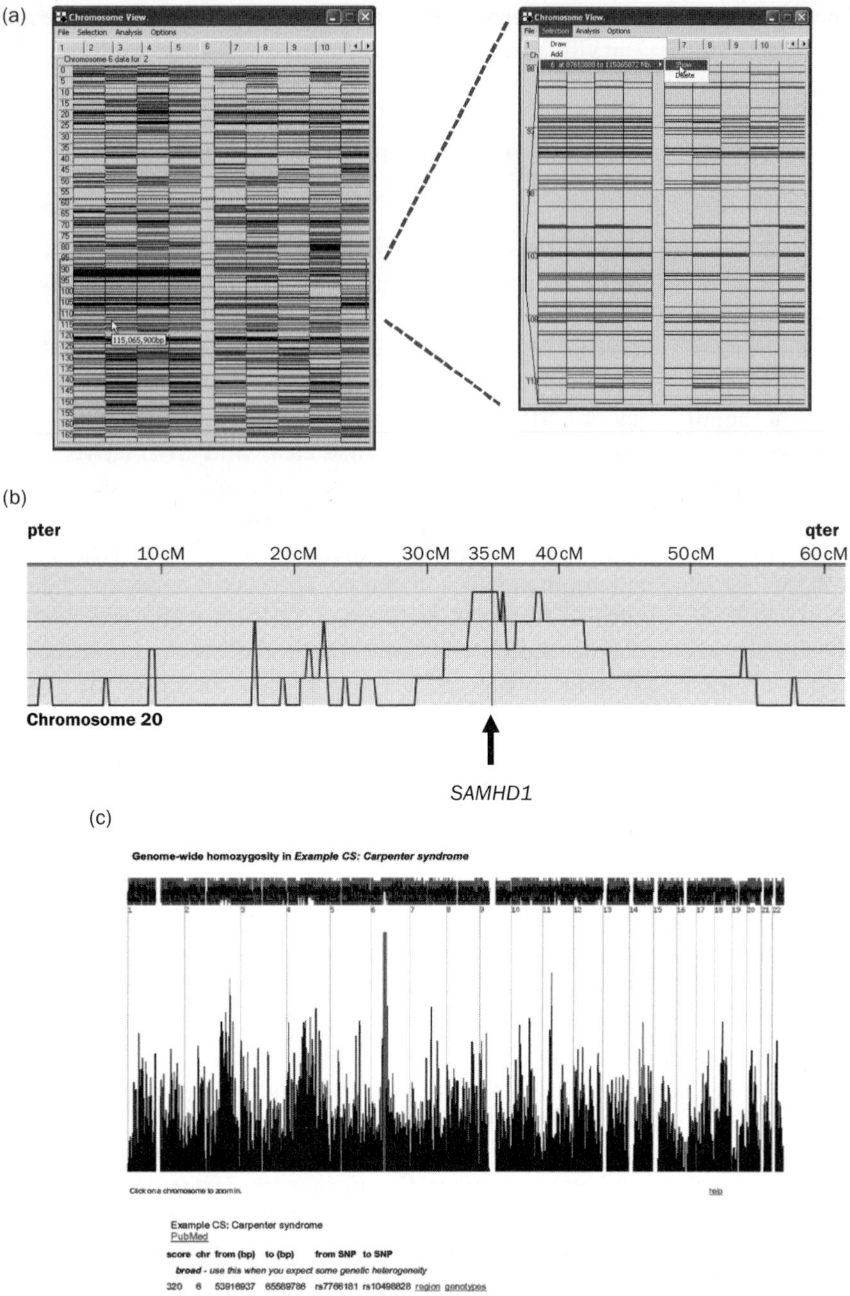

Figure 12-2 Programs to visualize chromosome and genome-wide SNP genotyping data for autozygosity mapping projects. (A) Regions of homozygosity on chromosome 6 displayed by "AutoSNPa" (Carr et al., 2006). The data is shown for one consanguineous family with affected and unaffected individuals on the left and right, respectively, in each "chromosome view" display panel. The extent of the chromosome is shown in the vertical axis, with physical location shown on the left. Homozygous genotypes are displayed in black, whereas heterozygous ones are in yellow. A clear shared region of homozygosity is seen in affected individuals, but is absent in unaffected individuals (red box), which can be expanded to show the region of interest in more detail, in this case chromosome 6, 876838888 to 115065872 bp. (B) "IBDFinder," a sequential SNP-counting program (Carr et al., 2009a), displays scores of homozygous genotypes along a chromosome to enable regions of homozygosity to be easily visualized in unrelated singleton cases. In the screenshot, the length of chromosome 20 is shown, marked with genetic distances, with a peak score identifying the causative gene, SAMHD1, for the AGS5 Aicardi-Goutieres syndrome locus (Rice et al., 2009). (C) Genome-wide homozygosity displayed by "Homozygosity Mapper" (Seelow et al., 2009). The screenshot shows the genome-wide homozygosity scores produced from SNP genotyping data in a Carpenter syndrome study (Jenkins et al., 2007). The scores are plotted as a bar chart with red bars indicating regions of interest, with chromosomes represented as bins along the horizontal axis. Above the bar chart, the excess or deficit of homozygous genotypes in cases versus controls is depicted. The program also automatically generates direct links to the most interesting regions, in this case on chromosome 6 between SNP markers rs7766181 and rs10498828 (53916937 to 65589786 bp). Refer color figure.

AUTOZYGOSITY MAPPING OF LETHAL RECESSIVE CONDITIONS

Autozygosity mapping has also been applied with great success to the study of the genetic basis of lethal autosomal recessive disorders (Morgan et al., 2006, Smith et al., 2006). Lethal recessive conditions provide a challenge for gene mapping, because DNA from the affected individuals may often be either unavailable or of insufficient quality for extensive molecular genetic investigations (e.g., if extracted from FFPE pathological samples). In such circumstances, genotyping of unaffected siblings can be utilized to progressively exclude genomic regions that cannot contain the disease gene. This approach, termed shadow autozygosity mapping by linkage exclusion (or SAMPLE) can be adopted successfully with data from high-density SNP arrays (Carr et al., 2009b).

FUTURE PROSPECTS

Over the past two decades, autozygosity mapping has proved to be a remarkably successful strategy for the identification of recessive disease genes. However, protein coding genes comprise only 1.6% of the human genome, but have an estimated 85% of the pathogenic mutations for monogenic conditions. Hence, with the advent of whole-exome sequencing (Choi et al., 2009, Ng et al., 2009), autozygosity mapping as a strategy for gene identification may decline in importance. Indeed, individuals with recessive disorders who are likely to be compound heterozygotes may prove to be more appropriate subjects for exome resequencing studies than consanguineous families with homozygous mutations. Nevertheless, as the vast majority of disease genes associated with coding sequence mutations are characterized, autozygosity mapping strategies may assume key roles in the identification of the molecular basis of recessive disorders caused by mutations, CNVs, or rearrangements in noncoding segments. After identification of a candidate interval by autozygosity mapping, the complete sequence of the candidate region could be determined by massively parallel sequencing analysis. In addition, after characterization of a disease gene, attention is likely to switch to the possible genetic basis of intra- and interfamilial phenotypic variability, and consanguineous families may be utilized to map genetic modifiers.

REFERENCES

Abecasis GR, Cherny SS, Cookson WO, Cardon LR. (2002). Merlin-rapid analysis of dense genetic maps using sparse gene flow trees. *Nat Genet 30*, 97–101.

Arbour NC, Zlotogora J, Knowlton RG, Merin S, Rosenmann A, Kanis AB, Rokhlina T, Stone EM, Sheffield VC. (1997). Homozygosity mapping of achromatopsia to chromosome 2 using DNA pooling. *Hum Mol Genet 6*, 689–694.

Bateson W, Saunders ER. (1902). Report to the Evolution Committee of the Royal Society. (1), 133n.

Beales PL. (2005). Lifting the lid on Pandora's box: the Bardet-Biedl syndrome. *Curr Opin Genet Dev 15*(3), 315–323.

Ben Hamida C, Doerflinger N, Belal S, et al. (1993). Localization of Friedreich ataxia phenotype with selective vitamin E deficiency to chromosome 8q by homozygosity mapping. *Nat Genet 5*(2), 195–200.

Bond J, Woods CG. (2006). Cytoskeletal genes regulating brain size. *Curr Opin Cell Biol 18*(1), 95–101.

Brown KA, Janjua AH, Karbani G, et al. (1996). Linkage studies of non-syndromic recessive deafness (NSRD) in a family originating from the Mirpur region of Pakistan maps DFNB1 centromeric to D13S175. *Hum Mol Genet 5*(1), 169–173.

Carr IM, Flintoff KJ, Taylor GR, Markham AF, Bonthron DT. (2006). Interactive visual analysis of SNP data for rapid autozygosity mapping in consanguineous families. *Hum Mutat 27*, 1041–1046.

Carr IM, Sheridan E, Hayward BE, Markham AF, Bonthron DT. (2009a). IBDfinder and SNPsetter: tools for pedigree-independent identification of autozygous regions in individuals with recessive inherited disease. *Hum Mutat 30*(6), 960–967.

Carr IM, Szymanska K, Sheridan E, Markham AF, Bonthron DT, Johnson CA. (2009b). Shadow autozygosity mapping by linkage exclusion (SAMPLE): A simple strategy to identify the genetic basis of lethal autosomal recessive disorders. *Hum Mutat 30*(12), 1642–1649.

Choi M, Scholl UI, Ji W, et al. (2009). Genetic diagnosis by whole exome capture and massively parallel DNA sequencing. *Proc Natl Acad Sci USA 106*, 19096–19101.

El-Sayed W, Parry DA, Shore RC, Ahmed M, Jafri H, Rashid Y, Al-Bahlani S, Al Harasi S, Kirkham J, Inglehearn CF, Mighell AJ. (2009). Mutations in the beta propeller WDR72 cause autosomal-recessive hypomaturation amelogenesis imperfecta. *Am J Hum Genet 85*, 699–705.

Evans DM, Cardon LR. (2004). Guidelines for genotyping in genomewide linkage studies: single-nucleotide-polymorphism maps versus microsatellite maps. *Am J Hum Genet 75*, 687–692.

Forshew T, Johnson CA. (2004). SCAMP: a spreadsheet to collate autozygosity mapping projects. *J Med Genet 41*, e125.

Friedman TB, Liang Y, Weber JL, Hinnant JT, Barber TD, Winata S, Arhya IN, Asher JH Jr. (1995). A gene for congenital, recessive deafness DFNB3 maps to the pericentromeric region of chromosome 17. *Nat Genet 9*(1), 86–91.

Fukushima K, Ramesh A, Srisailapathy CR, et al. (1995). Consanguineous nuclear families used to identify a new locus for recessive non-syndromic hearing loss on 14q. *Hum Mol Genet 4*(9), 1643–1648.

Garrod, AE. (1902). The incidence of alkaptonuria: a study in chemical individuality. *Lancet, Vol. ii*, 1616–1620.

Génin E, Todorov AA, Clerget-Darpoux F. (1998). Optimization of genome search strategies for homozygosity mapping: influence of marker spacing on power and threshold criteria for identification of candidate regions. *Ann Hum Genet 62*, 419–429.

Gibbs JR, Singleton A. (2006). Application of genome-wide single nucleotide polymorphism typing: simple association and beyond. *PLoS Genet 2*, e150.

Gissen P, Johnson CA, Morgan NV, et al. (2004). Mutations in VPS33B, encoding a regulator of SNARE-dependent membrane fusion, cause arthrogryposis-renal dysfunction-cholestasis (ARC) syndrome. *Nat Genet 36*, 400–404.

Goto M, Rubenstein M, Weber J, Woods K, Drayna D. (1992). Genetic linkage of Werner's syndrome to five markers on chromosome 8. *Nature 355*(6362), 735–738.

Gudbjartsson DF, Jonasson K, Frigge ML, Kong A. (2000). Allegro, a new computer program for multipoint linkage analysis. *Nat Genet 25*, 12–13.

Gudbjartsson DF, Thorvaldsson T, Kong A, Gunnarsson G, Ingolfsdottir A. (2005). Allegro version 2. *Nat Genet 37*, 1015–1016.

Gunderson KL, Kruglyak S, Graige MS, et al. (2004). Decoding randomly ordered DNA arrays. *Genome Res 14*, 870–877.

Harville HM, Held S, Diaz-Font A, et al. (2010). Identification of 11 novel mutations in 8 BBS genes by high-resolution homozygosity mapping. *J Med Genet 47*(4), 262–267.

Helbling-Leclerc A, Zhang X, Topaloglu H, et al. (1995). Mutations in the laminin alpha 2-chain gene (LAMA2) cause merosin-deficient congenital muscular dystrophy. *Nat Genet 11*(2), 216–218.

Hildebrandt F, Heeringa SF, Rüschendorf F, Attanasio M, et al. (2009). A systematic approach to mapping recessive disease genes in individuals from outbred populations. *PLoS Genet 5*, e1000353.

Jenkins D, Seelow D, Jehee FS, et al. (2007). RAB23 mutations in Carpenter syndrome imply an unexpected role for hedgehog signaling in cranial-suture development and obesity. *Am J Hum Genet 80*, 1162–1170.

Kelsell DP, Dunlop J, Stevens HP, Lench NJ, Liang JN, Parry G, Mueller RF, Leigh IM. (1997). Connexin 26 mutations in hereditary non-syndromic sensorineural deafness. *Nature 387*(6628), 80–83.

Kruglyak L, Daly MJ, Reeve-Daly MP, Lander ES. (1996). Parametric and nonparametric linkage analysis: a unified multipoint approach. *Am J Hum Genet 58*, 1347–1363

Kwitek-Black AE, Carmi R, Duyk GM, Buetow KH, Elbedour K, Parvari R, Yandava CN, Stone EM, Sheffield VC. (1993). Linkage of Bardet-Biedl syndrome to chromosome 16q and evidence for non-allelic genetic heterogeneity. *Nat Genet 5*(4), 392–396.

Lander ES and Botstein D. (1987). Homozygosity mapping: a way to map human recessive traits with the DNA of inbred children, Science *236*, 1567–1570.

Leutenegger AL, Prum B, Genin E, Verny C, Lemainque A, Clerget-Darpoux F, Thompson EA. (2003). Estimation of the inbreeding coefficient through use of genomic data. *Am J Hum Genet 73*, 516–523.

Leutenegger AL, Labalme A, Genin E, Toutain A, Steichen E, Clerget-Darpoux F, Edery P. (2006). Using genomic inbreeding coefficient estimates for homozygosity mapping of rare recessive traits: application to Taybi-Linder syndrome. *Am J Hum Genet 79*, 62–66.

McQuillan R, Leutenegger AL, Abdel-Rahman R, et al. (2008). Runs of homozygosity in European populations. *Am J Hum Genet 83*, 359–372.

Miano MG, Jacobson SG, Carothers A, et al. (2000). Pitfalls in homozygosity mapping. *Am J Hum Genet 67*, 1348–1351.

Morgan NV, Brueton LA, Cox P, et al. (2006). Mutations in the embryonal subunit of the acetylcholine receptor (CHRNG) cause lethal and Escobar variants of multiple pterygium syndrome. *Am J Hum Genet 79*, 390–395.

Moynihan LM, Bundey SE, Heath D, Jones EL, McHale DP, Mueller RF, Markham AF, Lench NJ. (1998). Autozygosity mapping, to chromosome 11q25, of a rare autosomal recessive syndrome causing histiocytosis, joint contractures, and sensorineural deafness. *Am J Hum Genet 62*(5), 1123–1128.

Mueller RF, Bishop DT. (1993). Autozygosity mapping, complex consanguinity, and autosomal recessive disorders. *J Med Genet 30*(9), 798–799.

Ng SB, Buckingham KJ, Lee C, et al. (2010). Exome sequencing identifies the cause of a Mendelian disorder. *Nat Genet 42*(1), 30–35.

Nystuen A, Benke PJ, Merren J, Stone EM, Sheffield VC. (1996). A cerebellar ataxia locus identified by DNA pooling to search for linkage disequilibrium in an isolated population from the Cayman Islands. *Hum Mol Genet 5*, 525–531.

Parry DA, Mighell AJ, El-Sayed W, et al. (2009). Mutations in CNNM4 cause Jalili syndrome, consisting of autosomal-recessive cone-rod dystrophy and amelogenesis imperfecta. *Am J Hum Genet 84*(2), 266–273.

Peltonen L, Perola M, Naukkarinen J, Palotie A. (2006). Lessons from studying monogenic disease for common disease. *Hum Mol Genet 15*, Spec No 1, R67–74.

Pollak MR, Chou YH, Cerda JJ, Steinmann B, La Du BN, Seidman JG, Seidman CE. (1993), Homozygosity mapping of the gene for alkaptonuria to chromosome 3q2. *Nat Genet 5*, 201–204.

Purcell S, Neale B, Todd-Brown K, et al. (2007). PLINK: a tool set for whole-genome association and population-based linkage analyses. *Am J Hum Genet 81*, 559–575).

Rice GI, Bond J, Asipu A, Brunette RL, et al. (2009). Mutations involved in Aicardi-Goutières syndrome implicate SAMHD1 as regulator of the innate immune response. *Nat Genet 41*, 829–832.

Seelow D, Schuelke M, Hildebrandt F, Nürnberg P. (2009). HomozygosityMapper—an interactive approach to homozygosity mapping. *Nucleic Acids Res*, Jul 1; 37(Web Server issue), W593–W599. Epub 2009 May 21.

Sheffield VC, Carmi R, Kwitek-Black A, Rokhlina T, Nishimura D, Duyk GM, Elbedour K, Sunden SL, Stone EM. (1994). Identification of a Bardet-Biedl syndrome locus on chromosome 3 and evaluation of an efficient approach to homozygosity mapping. *Hum Mol Genet 3*(8), 1331–1335.

Smith CAB. (1953). Detection of linkage in human genetics. *J R Stat Soc B 15*, 153–192.

Smith UM, Consugar M, Tee LJ, McKee BM, et al. (2006). The transmembrane protein meckelin (*MKS3*) is mutated in Meckel-Gruber syndrome and the wpk rat. *Nat Genet 38*, 191–196.

Sobel E, Papp JC, Lange K. (2002). Detection and integration of genotyping errors in statistical genetics. *Am J Hum Genet 70*, 496–508.

Strautnieks SS, Thompson RJ, Gardiner RM, Chung E. (1996). A novel splice-site mutation in the gamma subunit of the epithelial sodium channel gene in three pseudohypoaldosteronism type 1 families. *Nat Genet 13*(2), 248–250.

Uppal S, Diggle CP, Carr IM, et al. (2008). Mutations in 15-hydroxyprostaglandin dehydrogenase cause primary hypertrophic osteoarthropathy. *Nat Genet 40*, 789–793.

Woods CG, Cox J, Springell K, et al. (2006). Quantification of homozygosity in consanguineous individuals with autosomal recessive disease. *Am J Hum Genet 78*(5), 889–896.

13.

GENETIC CONTRIBUTION TO MALARIA SUSCEPTIBILITY AND RESISTANCE

Susana G. Campino and Taane G. Clark

INTRODUCTION

Despite the major efforts to try to control malaria, this infectious disease still persists as a major health burden, being responsible for over a million deaths each year, mainly children in sub-Saharan Africa (Snow et al., 2005). Malaria is caused by parasites of the genus *Plasmodium* (*P*) and is transmitted from person to person through the bite of a female mosquito of the genus *Anopheles*. In the human host the parasites first infect the liver cells and subsequently invade the erythrocytes, leading to malaria symptoms. The majority of disease and death is caused by *P. falciparum*, but the less deadly *P. malariae*, *P. ovale*, and mainly *P. vivax* are responsible for hundreds of millions of cases every year. The simian parasite *P. knowlesi* has also been reported in humans and is widely distributed in Southeast Asia (Singh et al., 2004).

The control of malaria has proven to be a great challenge: the complex biological organization of the pathogen, intricate host–pathogen interactions, two distinct hosts, and elaborate life cycle stages are some of the aspects that complicate the development of control strategies. Although several antimalaria drugs have been developed and have been successful, the parasite has shown its ability to acquire mechanisms to resist treatments (Hyde, 2005; White, 1998). Also, its impressive capacity to evade the host immune system holds back the development of an efficacious vaccine. Strategies to control the vector are also problematic, as the mosquito has evolved to resist the insecticides that are used to control malaria (Hemingway et al., 2002). To devise novel successful approaches, it is important to identify critical molecular pathways, both in the parasite and host, and the interactions between them. These can be achieved by uncovering the natural genetic factors that determine resistance or susceptibility to malaria, which may lead to an understanding of some important biological mechanisms of infection and disease progression.

Genetic epidemiology has demonstrated that genetic factors can influence significantly the response to malaria. The classical illustration of this genetic influence is the protective effect of certain erythrocyte polymorphisms. It was Haldane in the late 1940s who first pointed out that the high frequency of the genetic disorder thalassemia in the Mediterranean was the result of selective pressure by malaria (Haldane, 1949). Haldane's hypothesis stated that the disadvantage of the homozygous state, which results in severe anemia and can be fatal, would have been compensated by an advantage conferred by the heterozygous individuals who are healthy apart from a mild anemia. As a result, the mutant allele was maintained in the populations due to its selective advantage against malaria. This theory is now accepted for several hemoglobinopathies—the most prevalent monogenic diseases—as well as for other common red cell disorders, discussed further below (Feng et al., 2004; Hedrick, 2004; Min-Oo & Gros, 2005; Weatherall & Clegg, 2001). These examples point to the strong evolutionary force that malaria has exerted in the human genome, and the retention of some erythrocyte alleles as a consequence of their protective effect against death by malaria. Indeed, it is believed that malaria has had the strongest recent selective pressure on the human genome (Kwiatkowski, 2005; Sabeti et al., 2006), and that genetic studies will reveal further important signatures left by the disease. In this review, we will describe many of the polymorphisms and candidate genes that have been found to be associated with malaria phenotypes, the genetic/genomic epidemiological studies by which further variants are being discovered, and how these findings may be translated into effective control measurements, such as a vaccine. We will also discuss how new sequencing technologies are leading to insights into important features of the human, parasite, and mosquito genomes, and their potential impact on malaria control.

VARIABILITY IN RESPONSE TO MALARIA INFECTION AND HOST GENETIC INFLUENCE

Host response to malaria infection is variable, where a wide spectrum of phenotypes can be observed, varying from asymptomatic infections, fever, and mild symptoms to severe manifestation (severe malaria) and death. Severe malaria can comprise several complications such as hypoglycemia, metabolic acidosis, respiratory distress syndrome, cerebral malaria, and severe anemia (Marsh et al., 1996, 1995). It is thought that many of these complications result from the sequestration of infected red blood cells in small vessels and the formation of aggregates between noninfected and infected red cells (rosetting phenomena). Effective long-term immunity (clinical immunity) can be acquired in older children and adults after repeated infections, allowing infection to be tolerated without or with only mild symptoms. Therefore, in endemic regions where most people are repeatedly exposed to infection, probably only 1%–2% of individuals will develop severe malaria (Greenwood et al., 1991). Several factors are important for the different phenotypes observed, such as environmental conditions, parasite genetic makeup, and host age, state of immunity, and genetic background (Mackinnon et al., 2005). Host genetic factors might operate at any stage of the infection, between receiving a bite from a malaria-carrying mosquito to the development of severe malaria and death. The most remarkable evidence of the importance of host genetics on response to malaria is the Duffy blood group antigen, which is essential for *P. vivax* to infect red blood cells (Miller et al., 1976). Most of the sub-Sahara African populations are completely resistant to *P. vivax* infection due to a single nucleotide polymorphism (SNP) in the FY gene that results in the Duffy blood group–negative phenotype (Tournamille et al., 1995). This is a clear example of a single genetic factor providing resistance to infection, and leading to almost complete absence of *P. vivax* infections in this part of the world.

Striking differences in response to malaria infection have also been observed among ethnic groups who live in the same geographical region. For example, it has been observed that the Fulani of Burkina Faso (Modiano et al., 1995) and of Mali (Dolo et al., 2005) have a significantly lower prevalence of malaria parasitemia (i.e., parasite density in the blood) and fewer malaria clinical attacks, when compared to other ethnic groups living in neighboring villages. Although the Fulani have a distinctive culture and life style, different from the other groups, detailed epidemiological investigations indicate that their resistance to malaria arises primarily from genetic factors. It has also been reported that the Fulani have higher levels of antimalarial antibodies (Modiano et al., 1998; Modiano et al., 1999) and a low frequency of classic malaria resistance factors (Modiano et al., 2001a), indicating that unknown genetic factors could underlie the observed interethnic differences.

Despite the clear differences in response to malaria infection and disease progression at the population level, it is much more difficult to establish the extent of variability within and between families. This difficulty is mainly due to practical obstacles, such as some communities having incomplete or no medical records, and other infectious diseases being mistaken as malaria, making difficult to determine the affected status of the relatives. To perform a highly reliable family study, it is necessary to use a design that is longitudinal, where a set of individuals is monitored actively over a period of time. These longitudinal studies can potentially collect much data on uncomplicated malaria and infection intensity, which are affected by factors such as age, usage of bednets, and other environmental variables. Despite the difficulties, several studies have been successful in indicating the importance of host genetic factors to the response to malaria infection in families. A longitudinal study on parasite densities in Cameroon suggested the involvement of a major gene responsible for controlling blood infection levels (Abel et al., 1992), manifesting itself as a bimodal distribution of the parasite densities. This result was supported by a study of malaria infection during pregnancy in Burkina Faso (Cot et al., 1993). In both studies, data was adjusted for confounding factors such as age and parity. Subsequent longitudinal family studies on infection intensity, also in Cameroon and Burkina Faso, suggested a complex genetic model and linkage to the major histocompatibility complex (MHC) region on chromosome 6, and to the 5q31-33 region (Flori et al., 2003; Garcia et al., 1998a, 1998b; Rihet et al., 1998a, 1998b). Family studies have also revealed the contribution of host genetic factors to mild and severe malaria. A longitudinal study with Gambian twins showed that susceptibility to malaria fever episodes is determined partly by genetic factors, with monozygotic twins being more likely to both experience a malaria fever attack than dizygotic twins (Jepson et al., 1995). In Sri Lanka, a longitudinal family analysis estimated the heritability to be around 15% for both asymptomatic and symptomatic infections, and about 10% for the clinical malaria episodes (Mackinnon et al., 2000). Similar studies in two cohorts of Kenyan children estimated that 25% of the total variation in mild and severe malaria was explained by host genes, and that hemoglobin S, the strongest known resistance genetic factor (discussed below), explained only 2% of the total variation (Mackinnon et al., 2005). A family study conducted in Mali showed that the probability of a child developing severe disease (cerebral malaria or severe anemia) in a lifetime is greatly increased when a family member had a history of disease (Ranque et al., 2005). The immunological response to malaria antigens has also been analyzed in family studies, showing that it is at least partially genetically regulated (reviewed in Verra et al., 2009). Overall, there is substantial population genetic and epidemiological data indicating that genes influence malaria susceptibility and resistance.

GENES AND GENOMIC REGIONS INFLUENCING SUSCEPTIBILITY/ RESISTANCE TO MALARIA

LINKAGE AND ASSOCIATION STUDIES

Both family-based and population-based studies have highlighted the potential significance of a number of genes and genomic regions to malaria resistance/susceptibility. Genetic linkage is the phenomenon whereby alleles at genetic markers close together on the same chromosome will tend to be inherited together (Sham, 1998); it is rare for a crossover to occur between such proximal markers at meiosis. Although large stretches of each parental chromosome are preserved intact when new chromosomes are formed during meiosis, this phenomenon cannot be observed directly. Instead, one may observe cosegregation of a genetic marker with the disease in multiply affected pedigrees (Sham, 1998). Alternatively, one may test for increased marker allele-sharing between pairs of affected relatives, particularly affected sib pairs. Linkage is a very long-range phenomenon, so that only thousands of markers may be sufficient to screen the whole genome. However, this coarse resolution leads to large regions in which to discover the responsible gene and the functional loci.

Association between a disease and a polymorphism can occur if the disease gene is very tightly linked to the polymorphism, or if the genetic marker itself influences susceptibility to the disease. In the case-control setting, association arises as a difference in the frequency of alleles or genotypes between cases and unaffected controls at a genetic marker. Association is a very short-range phenomenon, and will only be observable for loci that are very tightly linked, and even then only if the disease and genetic marker happens to be in linkage disequilibrium (LD) or correlated (Sham, 1998). A direct association occurs when a test is performed on the genetic marker directly responsible for causing the disease. In contrast, an indirect association occurs where the genetic marker tested is not directly responsible for the disease, but is in LD with (probably near) the disease-causing marker. It is more common to detect an indirect association, and then perform follow-up (e.g., fine-mapping) experiments to detect or locate the causal polymorphism. Spurious or false positive associations can occur if the disease and genetic marker have different frequencies in different subpopulations, an effect known as *population stratification* or *structure* (Cardon & Palmer, 2003). An affected offspring–parental trio design is robust to population structure. The transmission disequilibrium (TDT) attempts to detect markers in LD with a disease, and by using parental genotypes avoids producing the spurious associations which can occur if controls are not well matched in case-control association studies (Spielman et al., 1993). Instead, the parental alleles that are not transmitted to an affected offspring act as controls. In general, in association studies, many hundreds of thousands or millions of genetic markers may be required to capture the LD patterns in the complete human sequence, and thus screen the whole genome. Before considering genome-wide studies, we consider genomic regions that have arisen because of their plausible disease etiology.

CANDIDATE GENES

Hemoglobins

The most compelling epidemiological evidence for protection from *P. falciparum* malaria is for the hemoglobinopathies, abnormalities of hemoglobin structure and synthesis. The adult hemoglobin, HbA, is a tetramer consisting of two α -globin chains and two β-globin chains (encoded by a cluster of genes localized on both chromosomes 16 and 11, respectively). Malaria has exerted a huge selection pressure on the structure and regulation of these genes, being responsible for the high frequency of a remarkable variety of genetic variants present in these genes (Min-Oo & Gros, 2005; Weatherall & Clegg, 2001; and reviewed in Kwiatkowski, 2005). Three different nonsynonymous coding SNPs in the HBB gene (which encodes β-globin) can confer resistance against clinical malaria. These SNPs result in different proteins known as the *Hemoglobin S* or sickle cell hemoglobin (HbS allele, glutamic acid to valine substitution on codon 6), the Hemoglobin C (HbC allele, glutamine to lysine substitution on codon 6), and Hemoglobin E (HbE allele, glutamic acid to lysine on codon 26). Individuals homozygous for the HbS allele have sickle cell disease, a debilitating and often fatal disease caused by red blood cell deformities, while heterozygous (designated HbAS) do not generally present any clinical abnormalities and have ~tenfold increase in protection against severe and complicated malaria in a number of populations (Aidoo et al., 2002; Allison, 1954a; Hill et al., 1991). Despite the lethal consequences of the homozygous state, the HbS allele is widespread all over malaria-endemic regions, and in sub-Saharan Africa, HbS carriers can reach 15%–20% (Allison, 1954b). This phenomenon makes the A/S genotype the best example of a balanced polymorphism in human biology, where despite the serious consequences of the homozygous state, the protected effect against malaria acts to maintain this polymorphism in the populations exposed to malaria. The precise biological mechanism has not been confirmed, but it may involve suppression of parasite growth in the red blood cell, an enhanced splenic clearance of the infected red blood cells, or reduced cytoadherence of infected erythrocytes containing sickle cell hemoglobin (Ayi et al., 2004; Cholera et al., 2008; Pasvol et al., 1978; Shear et al., 1993).

Although the hemoglobin C (HbC) allele is less common than HbS, it is found in several parts of West Africa. HbC homozygous have relative mild hemolytic anemia (clinically much less severe than sickle cell disease), while heterozygous do not experience a significant reduction in their hemoglobin levels. Several studies have shown the protective effect of both homozygous and heterozygous states against malaria clinical episodes (Agarwal et al., 2000; Mockenhaupt et al., 2004a; Rihet et al., 2004). An extensive study in Burkina Faso demonstrated that protection against clinical malaria is about 90% in homozygotes and 30% in heterozygotes (Modiano et al., 2001b). Protection may be the result of an increased immune clearance of infected erythrocytes, an abnormal parasite replication capacity, and/or impaired rosetting and cytoadherence of the infected red cells (Fairhurst et al., 2005; Fairhurst et al., 2003). Haplotype analysis (i.e., analysis of a set of SNPs that are transmitted together on the same chromosome) for both HbS and HbC suggested that these variants might have multiple origins, or have been redistributed on different genetic backgrounds (Chebloune et al., 1988; Flint et al., 1998; Sanchaisuriya et al., 2001).

The hemoglobin E (HbE) allele is extremely common in Southeast Asia, reaching rates of 50% in some regions (Flatz, 1967). HbE homozygous individuals generally have symptomless anemia. Epidemiological data obtained in a case-control study in Thailand indicates that HbE confers protection against severe malaria (Hutagalung et al., 1999). To support this evidence, erythrocyte invasion assays have shown that in heterozygous individuals, a subpopulation of HbE erythrocytes is relatively resistant to *P. falciparum* invasion (Chotivanich et al., 2002), which could confer protection against the development of high parasitemia and therefore severe disease.

The thalassemias are the most common hemoglobinopathies, and result from the deficient production of α- or β-globin chains, determining two major varieties α- and β-thalassemia. More than 200 disease-causing mutations have been identified, reflecting a vast spectrum of clinical phenotypes (reviewed in Weatherall, 2001; Williams et al., 1996). Additional complexity arises because α-globin is produced by two identical genes, HBA1 and HBA2. As a general rule, homozygous thalassemia causes severe disease or is fatal, and heterozygous individuals are generally healthy with possibly mild anemia. An exception to this rule is when either the HBA1 or the HBA2 gene, but not both, is disrupted, so some α-globin production is possible. This condition is referred as α^+ thalassemia and homozygous individuals are only mildly anemic.

The distribution of α- and β-thalassemias occurs throughout the Mediterranean region, sub-Saharan Africa, the Middle East, the Indian subcontinent, and Southeast Asia. Several population-based studies have provided strong evidence of the protective effect of both homozygous and heterozygous thalassemia disorders against severe malaria. As an example, the Tharu people have both a higher allele frequency of α-thalassemia and a lower incidence of malarial illness, compared to other ethnic groups that inhabit the same region of Nepal (Modiano et al., 1991). A case-control study in Papua New Guinea found that children homozygous for α^+ thalassemia had a reduced risk of severe malaria by 60% and a lesser degree in heterozygous children. However, this result has also been observed for other childhood infections (Allen et al., 1997). A study in Kenya found that both heterozygous and homozygous α^+ thalassemia children were protected against severe malaria (Williams et al. 2005b), while a study in Ghana found that heterozygous children were protected (Mockenhaupt et al., 2004b). Possible explanations for the nature of thalassemia protection against malaria include reduced parasite invasion and growth, better immune recognition of the infected red blood cells, and increased phagocytosis (Pattanapanyasat et al., 1999; Williams et al., 2002), and an alteration of cytoadherence and rosetting of *P. falciparum* infected red blood cells carrying β-thalassemia (Udomsangpetch et al., 1993). It has been proposed that individuals homozygous for α^+ thalassemia will have a biological advantage against the significant reduction in red cell counts that occurs during acute infection with the malaria parasite *P. falciparum*; their hematological profile may be the cause of a reduced risk of severe anemia (Fowkes et al., 2008).

Epistatic effects or interactions between genes are likely due to the co-occurrence at high frequency of several hemoglobinopathies that can protect against malaria and can be inherited together. A study in Kenya has shown that the protection afforded by HbS and α-thalassemia (both protect against malaria when inherited alone) was lost when the two conditions were inherited together (Williams et al., 2005a). Other combinations deserve attention in other populations, such as the co-occurrence of HbE and β-thalassemia in Southeast Asia.

RED BLOOD CELL SURFACE-RELATED GENES

Duffy Antigen

The FY gene encodes the glycoprotein Duffy antigen, a chemokine receptor expressed in various cell types. As mentioned above, this antigen is expressed in the red blood cells of most people, except in those populations from sub-Saharan Africa countries, which have a promoter SNP that disrupts a binding site for the erythroid transcription factor GATA-1, resulting in the Duffy blood group–negative phenotype (Tournamille et al., 1995). It was discovered over 30 years ago that *P. vivax* needs the Duffy antigen to invade red blood cells, hence the low prevalence of *P. vivax* in areas where individuals are Duffy negative (Miller et al., 1976). This discovery has now led to the development of a candidate vaccine against *P. vivax* (Yazdani et al., 2004).

Glycophorins

The invasion mechanisms used by *P. falciparum* are more complex than *P. vivax*, with available data suggesting the existence of multiple invasion pathways (Hadley et al.,

1987). Attention has been focused on variants at the genes encoding glycophorins A, B, and C (GYPA, GYPB and GYPC genes). GYPA and GYPB are major erythrocyte membrane sialoglycoproteins that carry the antigenic determinants for various blood groups. Erythrocytes deficient in any of these glycophorins have reduced invasion by *P. falciparum* (Facer, 1983). It is known that the *P. falciparum* erythrocyte-binding antigen (EBA) 175 recognizes specific sialic acid residues on the glycophorin A molecule (Mayor et al., 2005; Orlandi et al., 1992). The GYPC is a minor component of the red cell membrane that is recognized by the EBA-140 of *P. falciparum*. A deletion of the GYPC exon 3 is responsible for the Gerbich blood group–negative that reaches high frequency in coastal areas of Papua New Guinea, where malaria is hyperendemic (Maier et al., 2003). *P. falciparum* EBA-140 does not bind to Gerbish negative red blood cells, and invasion using this pathway is blocked (Maier et al., 2003). Although epidemiological studies indicate that this GYPC deletion does not alter the prevalence of malaria infection (Patel et al., 2004; Patel et al., 2001), further studies are needed to evaluate if it affects the progression to severe malaria.

ABO Blood Group

The ABO blood group antigens are glycoproteins attached to glycolipids on the erythrocyte surface. Association studies have consistently suggested that blood group O individuals are relatively protected from severe malaria (Migot-Nabias et al., 2000; Pathirana et al., 2005; Rowe et al., 2007). Protection seems to be due to reduced rosetting in blood group O individuals compared with non-O blood groups (Rowe et al., 2007). Evidence suggests that blood group A is associated with higher incidence of severe malaria (Barragan et al., 2000; Fischer & Boone, 1998; Lell et al., 1999). Also, studies using wild and laboratory parasite strains demonstrate blood group preferences (Heddini et al., 2001). Most epidemiological studies have used the phenotype of serology to determine host ABO genotype, but this phenotype is unable to discriminate all genotypes. A recent study, employing for the first time key molecular genetic variation underlying the ABO system, strongly supported the hypothesis that blood group O individuals are relatively protected from severe malaria in comparison to other blood groups (Fry et al., 2008b). The variable and complex global distribution of ABO blood groups in the human population might be another example of the selection pressure imposed by malaria (and maybe other pathogens) in the human genome.

Erythrocyte Band 3

The SLC4A1 gene encodes an anion exchanger in the red blood cell membrane known as the erythrocyte band 3 protein. A twenty-seven base pair deletion in this gene results in a form of ovalocytosis, a condition conferring an oval shape and increased rigidity to the membrane of red blood cells, and quite common in parts of Southeast Asia. Homozygosity is lethal, but heterozygosity seems to protect against malaria infection (Cattani et al., 1987; Foo et al., 1992) and against cerebral malaria (Allen et al., 1999; Genton et al., 1995). The mechanism that leads to protection is unknown; experimental work indicates that it could be due to reduced parasite invasion or growth (Cortes et al., 2004).

OTHER RED BLOOD CELL RELATED GENES

Glucose-6-phosphate-dehydrogenase (G6PD)

The G6PD gene on chromosome X encodes an enzyme that catalyzes the initial step of the pentose phosphate pathway and is involved in the control of oxidative stress. Several polymorphisms in this gene have been detected, and can cause variation in the enzyme activity and lead to hemolytic anemia. The geographic distribution of G6PD deficiency overlaps with malaria distribution, and analysis of haplotypes at the G6PD locus supports the hypothesis of recent positive selection, consistent with evolutionary selection pressure imposed by malaria on this gene (Ganczakowski et al., 1995; Tishkoff et al., 2001). Several lines of evidence link G6PD deficiency to protection from severe malaria, but the exact mechanism for the protection offered by G6PD deficiency is still uncertain (Nkhoma et al., 2009). Some, but not all, in vitro studies have suggested impaired parasite growth in deficient cells (reviewed in Scriver, 1995). More recent work has suggested that other mechanisms may be involved, such as early phagocytosis of infected erythrocytes (Cappadoro et al., 1998). It is widely considered that most G6PD deficiency in sub-Saharan Africa is caused by the G6PD A$^-$ variant (characterized mainly by two nonsynonymous polymorphisms in the coding sequence of the gene: A376G Asn→Asp and G202A Val→Met). Recent large association studies of G6PD deficiency and malaria have employed genotyping assays consisting of these two nonsynonymous polymorphisms as a convenient way to determine G6PD status (Clark et al., 2009b; Guindo et al., 2007; Ruwende et al., 1995). Selected large studies of severe disease have shown a decrease in the risk of severe malaria in hemizygous males, and either no association (Guindo et al., 2007; Ruwende et al., 1995) in heterozygous females. There has been greater inconsistency in studies of uncomplicated malaria, showing G6PD deficiency to be protective in heterozygous females (Ruwende et al., 1995; Bienzle et al., 1972; Bienzle et al., 1979), or to have no effect on the incidence of uncomplicated malaria in either hemizygous males or heterozygous females (Enevold et al., 2005).

Conversely, other studies showed an increase in the incidence of uncomplicated malaria in females heterozygous for G6PD deficiency (Lell et al., 1999; Parikh et al., 2004). Definitive conclusions based on these results are difficult because studies differed in design, clinical phenotype assessed (severe or uncomplicated malaria), or methods to identify G6PD deficiency (Johnson et al., 2009). Further insights have been shown in recent studies. A case-control study of severe malaria in a Gambian study using additional polymorphisms suggested that both G6PD-deficient male hemizygotes and female heterozygotes are protected from severe malaria (Clark et al., 2009b). Further, it was hypothesized that either non-A$^-$ variants are more common across African than previously thought, and/or the A376G and G202A alleles do not capture all genetic heterogeneity in the A$^-$ variant. Another study has suggested that G6PD-deficient females were significantly protected against uncomplicated malaria, but this protection was only seen when G6PD deficiency was described using enzyme activity (Johnson et al., 2009). These observations may help to explain the discrepancy in some published association studies involving G6PD deficiency and uncomplicated and severe malaria. Further studies correlating sequence data, enzymatic assays, and malaria phenotypes are ongoing.

A recent study investigated the effect of a common Southeast Asian G6PD deficiency variant on survival to *vivax* and *falciparum* malaria. The results showed that this genetic variant reduces *P. vivax* and not *P. falciparum* parasite density, an indication that *P. vivax* has also been a driving selection force in the G6PD gene (Louicharoen et al., 2009).

Haptoglobin

Haptoglobin, encoded by the gene HP on chromosome 16, is a hemoglobin-binding protein that is present in the plasma. It is an important factor regulating hemoglobin–iron metabolism after hemolysis, thus preventing hemoglobin-induced oxidative tissue damage.

HP is polymorphic with two codominant alleles, Hp1 and Hp2, resulting in three phenotypes, Hp11, Hp12, and Hp22, which are related with Hp plasma concentrations (Hp22 having the lowest; see Langlois & Delanghe, 1996). A fourth phenotype, Hp0, with no detectable circulating haptoglobin, has been reported in various African populations and has been associated with malaria endemicity and with parasite density (Boreham et al., 1981; Fowkes et al., 2006; Rougemont et al., 1988). Genetic association studies with HP and malaria have been controversial. Although two case-control studies in Sudan and Southern Ghana found that the Hp11 phenotype was associated with susceptibility to severe *P. falciparum* malaria (Elagib et al., 1998; Quaye et al., 2000), a study using haptoglobin genetic variants in a Gambian population failed to detect an association (Yano et al., 1998). A recent report showed that a promoter SNP, which reduces HP protein levels, was associated with protection from clinical malaria in older children (Cox et al., 2007). Experimental studies also support the importance of HP in malaria. It has been shown that HP can inhibit the development of malaria parasites in vitro and HP-knockout mice have higher parasite density than wild type mice (Hunt et al., 2001; Imrie et al., 2004).

CYTHOADHERENCE AND ROSETTING

P. falciparum infected red blood cells can adhere to the capillary endothelium (sequestration) and to other erythrocytes (rosetting phenomena). These mechanisms are proposed to be crucial for the specific virulence of *P. falciparum*. Some host candidate genes have been proposed as having an important role in these phenomena.

Intercellular Adhesion Molecule 1

The intercellular adhesion molecule type 1 (ICAM-1), encoded by the ICAM1 gene, is an endothelial- and immune-cell adhesion receptor for integrin-expressing leukocytes.

In vitro and autopsy studies suggest that the adhesion of erythrocytes infected with *P. falciparum* to ICAM-1 receptor is an important step in the pathogenesis of severe malaria. African populations have high frequencies of a coding polymorphism of ICAM-1 that reduces parasite binding to endothelium (Fernandez-Reyes et al., 1997). Conflicting results for this ICAM1 genetic variant have emerged from three different case-control studies. A study in Gabonese children found that this allele protects against severe malaria (Kun et al., 1999). However, a study in Kenya found that the allele increases susceptibility (Fernandez-Reyes et al., 1997), while in the Gambia no significant allelic effect has been found (Bellamy et al., 1998a). A recent large-scale study using trios from the Gambia, Malawi, and Kenya found no association with severe malaria in any of the 15 ICAM1 polymorphisms tested, including the allele previous found to be protective (Fry et al., 2008a).

CD36 Antigen

The CD36 antigen is another major receptor for infected erythrocytes (Barnwell et al., 1989). This antigen is a receptor for a range of different molecules, and is expressed by platelets and dendritic cells, as well as the endothelium. It has been reported that the CD36 antigen is necessary for the sequestration of the *P. falciparum* infected red blood cells and their binding to dendritic cells, which suppresses the antigen presenting capacity of these cells (Urban et al., 1999; Urban et al., 2001). Results of disease association between CD36 and malaria are also inconsistent. In the Gambia and Kenya three polymorphisms encoding truncated CD36 proteins were associated with susceptibility to cerebral malaria (Aitman et al., 2000). However, one of these polymorphisms was associated with protection against severe malaria in Kenya (Pain et al., 2001). Similar work conducted in patients with severe malaria from Thailand found that another variant, implicated in alternative splicing, was strongly associated with cerebral malaria protection (Omi et al., 2002; Omi et al., 2003). A recent study screened 70 variants across CD36 in a Gambian case-control study of 1350 subjects and a family study of 1288 parent–offspring trios, and found no significant association with malaria severity (Fry et al., 2009).

Complement Receptor

Complement receptor 1, encoded by the CR1 gene, is an immunoregulatory membrane glycoprotein present on erythrocytes and a variety of leucocytes. The function of CR1 includes opsonization, control of complement activation, and clearance of immune complexes. One of the mechanisms proposed for rosetting of infected erythrocytes is through the binding of the *P. falciparum* erythrocyte membrane protein (PfEMP-1) to the erythrocyte CR1 (Rowe et al., 1997; Rowe et al., 2000). CR1 deficiency is common in Melanesian populations and has been associated with protection against severe malaria in a study conducted in Papua New Guinea (Cockburn et al., 2004). In Thailand, a polymorphism associated with reduced expression of CR1 on the red blood cells has been associated with susceptibility to severe malaria (Nagayasu et al., 2001). CR1 deficiency seems not to be as common in Africa, but other polymorphisms are (Moulds et al., 2000). However, studies of *CR1* polymorphisms in the Gambia found no evidence of association with disease severity (Zimmerman et al., 2003).

GENES IN THE IMMUNE SYSTEM

Human Leukocyte Antigen Genes

The Human Leukocyte Antigen (HLA) genes, located on chromosome 6 within the MHC region, are the most polymorphic genes in the human genome. These genes are responsible for presenting foreign antigens to the host immune system. The extraordinary diversity and complexity of polymorphism in the HLA genes might have evolved through natural selection by infectious agents, such as malaria. This notion is supported by the data from a large case-control study performed in Gambian children indicating that the HLA-B53 class I antigen and the HLA class II haplotype DRQB1*1302-DQB1*0501 independently protect against severe malaria (Hill et al., 1991). These findings have boosted efforts to identify targets for the development of a liver-stage malaria vaccine, where, in contrast to liver cells, erythrocytes do not express MHC molecules. Subsequent molecular studies have shown that HLA-B53-restricted T-cells recognize conserved peptides from the parasite liver stage, suggesting that the HLA-B53 antigen may facilitate T-cells to kill infected hepatocytes (Hill et al., 1992). A longitudinal study in Gambian children did not find any association between the incidence of mild malaria episodes and HLA class I alleles; however, the incidence of malaria fever episodes was reported to show an overall association with the distribution of MHC class II haplotypes (Bennett et al., 1993). HLA associations might vary between populations according to the local prevalence of parasite–antigen polymorphisms, and therefore may have consequences for vaccine efficacy. Further association studies with larger sample sizes and finer genetic resolution in the HLA region are necessary.

Interleukin-4

The Interleukin-4, encoded by the IL4 gene, localized at the 5q31 region previously associated with *P. falciparum* infection (Flori et al., 2003; Rihet et al., 1998b), is a cytokine that participates in several B (and other)-cell activation processes, and promotes antibody development. A polymorphism at the promoter of IL4 (IL4-589T) was found to be associated with elevated antibody levels against malaria antigens in the Fulani people of Burkina Faso, who have an increased resistance to malaria (Luoni et al., 2001). Surprisingly, the same allele was found to be associated with higher infection prevalence in the Fulani from Mali (Vafa et al., 2007). The relationship of the same promoter polymorphism with malaria severity was analyzed in a case-control cohort in Burkina Faso. Although no association was found with severe malaria and the IL4-589T allele, higher total IgE levels were observed in the individuals that carry that allele (Verra et al., 2004). A case-control study in Ghana also revealed higher total IgE levels in cerebral malaria cases carrying the IL4-589T allelic variant (Gyan et al., 2004).

Tumor Necrosis Factor

The human MHC region contains many genes that are relevant to immunity and inflammation, such as the tumor necrosis factor (TNF). The TNF is a proinflammatory cytokine encoded by the TNF gene localized in the MHC class III region. Clinical and experimental evidence have shown that excessive TNF production is a major factor in the pathogenesis of cerebral malaria, while other studies have reported that this cytokine is critical for innate immunity against malaria parasites (reviewed by Clark et al., 2004; Kwiatkowski et al., 1990). Several TNF genetic variants localized at the promoter region have been found to be independently associated with severe malaria. Gambian children homozygous for the variant TNF-308A had a sevenfold increased risk of dying with cerebral malaria (McGuire et al., 1994). In Gabon, it was found that those carrying the same allele were more likely to have symptomatic reinfections with *P. falciparum* (Meyer et al., 2002), while in Kenya it is associated with increased infant mortality and morbidity (Aidoo et al., 2001). Also, the same allele was found in a study in Sri Lanka to increase the risk of severe infectious diseases (Wattavidanage et al., 1999). Another variant, TNF-376A, that confers allele-specific binding of the transcription factor OCT-1, has been associated with susceptibility to cerebral malaria in East and West Africans (Knight et al., 1999). In the Gambia, severely anemic children with *P. falciparum* infection had low plasma TNF levels—a condition associated with the presence of another TNF promoter allele, TNF-238A (McGuire et al., 1999). A recent large study involving more than 10,000 individuals and encompassing 3 African populations found evidence to support associations between the TNF -238A polymorphism and severe malaria in the Gambia (Clark et al., 2009a). However, no association was found in the Kenyan and Malawian cohorts. The MHC class III region contains a high density of immunological genes and a complex pattern of linkage disequilibrium or correlation. Therefore, it is possible that these polymorphisms are simply genetic markers for other variants in the neighboring region (Diakite et al., 2009).

Toll Receptors

The innate immune system has evolved to provide the first-line defense against infection by sensing conserved products or pathogen-associated molecular patterns, which are often essential for pathogen survival and are distinct from the host molecules. Toll-like receptors (TLR) recognize a diverse array of pathogens products, and initiate intracellular signaling that triggers a complex cascade of events, leading to the generation of an inflammatory response. Several reports have found that malaria parasites express molecules that are recognized by Toll-like receptors. For instance, the malaria pigment hemozoin has been shown to function as carrier for plasmodium DNA, which stimulates TLR9 and induces inflammation (Coban et al., 2005). The clinical relevance of TLR genes and malaria has been assessed in association studies. A case-control study conducted in northern Ghana revealed frequent polymorphisms at TLR4 that conferred an increased risk of severe malaria (Mockenhaupt et al., 2006a). Another study in southern Ghana found polymorphisms in both TLR4 and TLR9 that were associated with the clinical manifestation of malaria during pregnancy (Mockenhaupt et al., 2006b). Recently, a large study using both family- and population-based analysis in Malawian and Gambian populations found no strong evidence for TLR9 polymorphisms to be associated with severe malaria across all cohorts (Campino et al., 2009).

INTERFERON-γ AND RELATED GENES

The Interferon- γ cytokine, encoded by the interferon gamma gene (IFNG), is a key immunological mediator believed to play both a protective and a pathological role in malaria (reviewed in Stevenson & Riley, 2004). Analysis of SNPs in the IFNG gene region found several weak associations with severe malaria in a large case control of Gambian children (Koch et al., 2005). A study of affected-child parental trios from Mali also found that the IFNG-183T allele (which increases gene transcription (Bream et al., 2002; Chevillard et al., 2002) leads to a reduced risk of developing cerebral malaria (Cabantous et al., 2005).

The α subunit of the IFNG receptor, encoded by the IFNGR1 gene, plays an important role in receptor trafficking and signal transduction. A case-control study using Gambian children with severe malaria found that heterozygotes for the IFNGR1-56T/C polymorphism in the Mandinka ethnic group were protected against cerebral malaria and fatal outcome (Koch et al., 2002). In the same study, individuals carrying the IFNGR1-470 (insertion/deletion) polymorphism were protected against severe malaria.

Polymorphisms in the transcription factor interferon regulatory factor 1 (IRF-1) gene, a critical mediator of IFNG activity, have been recently reported to be associated with control of *P. falciparum* infection both in healthy adults and children with severe and uncomplicated malaria (Mangano et al., 2008). This gene lies in the 5q31 region, which, as mentioned earlier, has been shown to be in linkage with *P. falciparum* infection levels.

Nitric Oxide Synthase

The nitric oxide synthase type 2 (NOS2A) gene generates nitric oxide, a free radical with antiparasitic properties, but can also potentially contribute to cerebral malaria due to its role in neurotransmission (reviewed in Clark & Rockett, 1996). Some polymorphisms have been analyzed for possible implication with malaria, all have yielded inconsistent results between the different populations analyzed. A SNP in the promoter of the gene (NOS2A-1173C/T) was found to be associated with protection against severe malaria and resistance to reinfection in Tanzanian children, as well as protection from severe malarial anemia in Kenya children (Hobbs et al., 2002). The same study reported association of the same allele with increased fasting urine and plasma NO metabolite concentrations in Tanzanian children, suggesting that the polymorphism was functional in vivo (Hobbs et al., 2002). However, no protective effect against severe malaria was detected in the Gambia (Burgner et al., 2003). Shorter forms of a microsatellite polymorphism (CCTTTn) located upstream from the NOS2 transcription start site have been associated with susceptibility to fatal malaria in the Gambia (Burgner et al., 1998). Longer repeats of the same microsatellite polymorphism have been associated with severe malaria in Thai adults (Ohashi et al., 2002). Another polymorphism, the NOS2-954C, is associated with elevated NO synthase activity in cells from Gabonese individuals, as well as with protection from severe malaria and resistance to reinfection (Kun et al., 2001). Again, this SNP also failed to be associated in other studies (Burgner et al., 2003; Levesque et al., 1999).

Mannose-Binding Lectin

The mannose-binding lectin 2 (MBL2) gene encodes a serum protein, which initiates innate immune response against a wide range of microbial pathogens; by binding to nonself surface oligosaccharides it can activate the complement system or act directly as an opsonin. A study in Gabon reported that children with severe malaria had lower levels of MBL2 than children with mild malaria (Luty et al., 1998). In this study, two SNPs associated with lower levels of the protein were found at higher frequency in the severe malaria cases. These results were not confirmed in a study performed in the Gambia (Bellamy et al., 1998b). A recent study in a cohort of Ghanaian children, where six polymorphisms were analyzed, suggested that the MBL pathway of the complement system is a critical determinant of both susceptibility to *P. falciparum* infection and manifestation of severe malaria (Holmberg et al., 2008).

JUDGING GENETIC FINDINGS

Although several significant findings in the genetic response to malaria infection have been reported, it is important to highlight the need for caution when interpreting these results. Indeed, as suggested above, the results of a number of studies have not been replicated, nor supported by in vitro and gene expression studies that assess functional aspects. A number of methodological issues can be used to explain the nonreplication of results, including small sample size, variation in phenotype definition, the choice of controls, village surveys versus hospital-based studies, age or immune status of subjects, study location, and DNA quality issues with the need to whole-genome amplify samples. False positives may arise due to population structure and/or admixture, or by chance when performing many statistical tests. The success of family-based studies depends on the accurate ascertainment of the pedigree structure, and minimization of the number of family relationship misspecifications. In case-control studies, accounting for population structure and the absence of consanguineous relationships are important. These issues are not unique to malaria: in recent years, the importance of larger sample sizes, accounting for population structure, and careful phenotypic definitions have become appreciated increasingly, particularly for genome-wide studies of complex diseases (Burton et al., 2005; Cordell & Clayton, 2005). Until recently, research in the field of malaria genetics has not enjoyed the sort of funding afforded to similar work investigating diseases of importance to the developed world. Coupled with advances in genetic diagnostics that have led to massive automation and falling costs per gene explored, momentum has grown toward more generous funding that brings with it the opportunity for much larger, multisite cohesive studies.

GENOME-WIDE ASSOCIATION AND LINKAGE STUDIES OF MALARIA

A number of genome-wide studies, which attempt to survey the entire human genome, have recently been performed to identify genetic factors that confer protection to malaria in African populations (Jallow et al., 2009; Timmann et al., 2007). An autosome-wide linkage analysis based on 10,000 single nucleotide polymorphisms was conducted in 68 Ghanaian families (Timmann et al., 2007). In this study, malaria susceptibility was considered by closely monitoring parasite counts, malaria fever episodes, and anemia over an 8-month period. Several genomic regions were identified that showed evidence for linkage to the parasitological and clinical phenotypes studied, among them a prominent signal on the chromosome region 10p15.3–10p14 obtained with malaria fever episodes. Further regions providing evidence for linkage included chromosomes 13q (parasite density) and 1p36 (parasite prevalence, pack cell volume). These are large genomic regions and highlight a limitation of such studies that have employed a coarse map of the genome.

A genome-wide association study (GWAS) of severe malaria in the Gambia (~1000 mostly cerebral malaria cases, ~1500 umbilical cord controls) used an assay consisting of ~500,000 SNPs chosen pseudo-randomly across the genome (Jallow et al., 2009). They identified 19 regions of the genome in which there were at least two SNPs (P-value $<10^{-4}$) or a single SNP (P-value $<10^{-6}$). Three regions other than the HBB gene had signals of $P<10^{-6}$, and included chromosome regions 2q37 (closest genes: SPATA3, LOC257407, PSMD1, GPR55), 5p12 (genes encoding proteins of unknown function), and 14p1 (a gene desert). SNP effects found by combining a main and replication study were located in the HBB, SCO1, DDC and DDAH genes. SCO1 plays a role in cytochrome oxidase function and has been implicated in neurological disorders, while DDC is involved in dopamine and serotonin synthesis, linked to malaria refractory in mosquitoes, and DDAH1 regulates cellular methylarginine concentrations, which in turn inhibit nitric oxide synthase activity (Jallow et al., 2009). A number of candidate SNPs (described earlier) were not on the Affymetrix chip, but independent genotyping on the same Gambian cohorts highlighted associations of HbS (reduced risk of 90% for a single copy of the S allele, P-value $=10^{-28}$) and the ABO blood group (P-value = 10^{-4}), with severe malaria at the estimated genome-wide significance threshold (P-value < 0.0005). The SNP on the Affymetrix chip in greatest LD with the ABO polymorphism (r^2=0.15) was not statistically significant. By comparison, the strongest signal detected by SNPs in the vicinity of the HbS yielded a diminished protective effect of 39% at a statistical significance of 10^{-7}. These results demonstrate two necessary requirements for GWAS in African populations: large sample sizes and denser genotyping platforms. In general, the extent of LD is shorter in African populations, and denser genotyping platforms are needed for African populations in order to achieve the same level of genome coverage as European or Asian populations.

Several other GWAS are ongoing in Ghanaian, Kenyan, and Malawian populations, with larger samples sizes and using both case-control and affected-offspring parental trio designs (www.malariagen.net). The SNP arrays applied contain between 650,000 and 1.2 million SNPs, with a proportion being based on known common variants from catalogued genetic variations between Yoruba subjects from the Ibadan region of Nigeria and subjects with Asian and European ancestry (International HapMap Project et al., 2007). Highly variable regions in the human genome, such as the MHC and HLA, have very little coverage. However, arrays with many millions of SNPs are now available, including tagging polymorphisms for SNPs and copy-number variants found in expanded HapMap and 1000 genomes project populations.

Genomic variations can exist between individuals with different geographical or ethnic background, and in genetic epidemiological studies those differences may manifest themselves as differences in allele frequencies. These population structure and/or admixture effects lead to adverse effects on power and greater false positives (Balding & Nichols, 1995; Teo et al., 2009). Several analytical procedures have been devised to handle the presence of population structure with a large number of trait-independent SNPs (Devlin & Roeder, 1999; Price et al., 2006; Pritchard et al., 2000). In studies with fewer genetic markers, other approaches are adopted including sampling populations that are known a priori to be genetically homogeneous, using ancestral informative markers, or stratifying for self-reported sociodemographic variables (e.g., ethnicity, language).

FINE-MAPPING AND WHOLE-GENOME ASSOCIATION STUDIES

In the near future it will be possible at reasonable cost to sequence the complete genomes of large numbers of individuals, leading to whole-genome association studies (WGAS); such studies will not be reliant on LD to discover non-genotyped functional polymorphisms. At present it is still prohibitively expensive to sequence large numbers of individuals in sizable candidate regions. Until then, fine-mapping may be approached using *in silico* statistical approaches (Marchini et al., 2007; Li et al., 2009; Servin & Stephens, 2007). Here, a dense set of reference data (e.g., HapMap and 1000 genomes project) is used to infer the genetic composition of the samples at SNPs that are not assayed in the genome-wide scan. The accuracy of the inferred genotypes relies on the assumption of a similar LD structure in the reference data and in the samples in the study, and therefore the data sources should be related to the same populations.

Although the presence of longer LD translates into higher statistical power to detect disease association, it makes the identification of the causal variants more challenging, as there may be a need to systematically investigate a larger portion of the genome using sequencing techniques. In this respect, the shorter LD in African populations confers a useful advantage over genomic studies in non-African populations. In particular, fine-mapping a disease-associated region becomes easier in the presence of shorter LD, as the potential candidate region is smaller. This observation means that genetic epidemiological studies using African populations have a major role to play in identifying the functional variants for common diseases. A related approach, *transethnic mapping,* has been advocated as a means to fine-map in the setting of high LD, where differences in LD patterns between populations can be used to calibrate the resolution of association (McKenzie et al., 2001).

TOWARD THE GLOBAL PLASMODIUM AND ANOPHELES GENOME VARIATION MAPS

A barrier to successful malaria vaccine development and drug development has been our limited understanding of the genetic variability among *P. falciparum* parasites.

Genetic variability is thought to be substantial, with more than 100,000 SNPs identified (Jeffares et al., 2007; Mu et al., 2007; Volkman et al., 2007), but only a handful of *P. falciparum* parasite samples have been sequenced extensively. Among the barriers to routine sequencing have been the prohibitive cost and the AT-rich nature of the genome. However, next-generation sequencing technologies are facilitating the mapping of a number of genomes and cataloging of genomic variants (www.sanger.ac.uk). This map will include SNPs and structural variants (e.g., copy-number variants and deletions), and will contribute to the understanding of parasite diversity and population structure, assist future GWAS of parasite-specific phenotypes, as well as barcoding of parasites to follow them during intervention strategies and detect and prevent the spread of drug resistance. Already, some SNPs (e.g., the chloroquine resistance transporter [*pfcrt*] gene) and copy-number variants (e.g., multidrug resistance [pfmdrl] gene) have been implicated as being associated with drug resistance. A global surveillance system could detect entirely new forms of resistance that might arise after the deployment of a new antimalarial drug or vaccine, thus providing public health agencies with the information needed to deliver effective intervention strategies. It will soon be possible to sample thousands of parasites worldwide per annum, and use whole-genome sequencing data to detect genomic signatures of recent positive selection, such as that caused by antimalarial drugs. Statistical methods looking for these signatures, including a loss of heterozygosity or long-range haplotypes (Sabeti et al., 2002; Voight et al., 2006), have been applied primarily to studies of the host (Sabeti et al., 2006). There is evidence that antimalarial drug resistance in *P. falciparum* is amenable to this approach, with polymorphisms that confer resistance to chloroquine and sulfadoxine-pyrimethamine (pfdhfr, pfdhps genes) being in regions of the genome with a clear "trough of heterozygosity" when comparing drug-resistant and drug-sensitive parasites (Wootton et al., 2002).

There are ongoing efforts to whole-genome sequence the vector, *Anopheles gambiae*—the major malaria vector in sub-Saharan Africa—that may eventually lead to WGAS. At present, *gambiae* is the only Anopheles species for which there is a high-quality assembled and annotated genome (Holt et al. 2002; Wilding et al., 2009; Cohuet et al., 2008). Also, only a limited number of genomes have been sequenced to date, representing only a fraction of the true genetic diversity. The availability of sequencing data is making GWAS of Anopheles' natural populations a feasible option for medically relevant phenotypes such as insecticide resistance. However, it is first crucial to increase the understanding of the true levels of genetic diversity across the Anopheles genome, leading to knowledge of the impact of chemical control of vectors and the prospective use of transgenic technologies.

MALARIA PREVENTION AND CONTROL

NON-VACCINE INTERVENTIONS

Antimalarial drugs and mosquito control programs have historically been very successful in controlling malaria, and resulted in a significant reduction of malaria worldwide. However, the emergence and spread of drug-resistant parasites, and mosquito resistance to insecticides, have contributed to a substantial increase of malaria cases. In many parts of the world, *P. falciparum* has become resistant to chloroquine and to Fansidar (sulfadoxine-pyrimethamine), the two most commonly used and affordable antimalarial drugs. Although other antimalarial drugs have been introduced, such as mefloquine, artemisinin derivatives (e.g. artesunate), and atovaquone-proguanil (commercially available as Malarone), resistance in parasite strains (especially in Southeast Asia) has been reported (for a review of antimalarial drug resistance see Hyde, 2005). Artemisinin, obtained from the plant *Artemisia annua*, is currently the most effective antimalarial; it can rapidly reduce parasite numbers in a single 48-hour erythrocytic cycle. However, the (predictable) risk from using this type of drug in monotherapy was emphasized by a study where decreased sensitivity was observed in recrudescent parasites after seven days of treatment with artesunate alone (Menard et al., 2005). Artemisinin derivatives are now being used in pairwise combinations with several other drugs, such as mefloquine and lumefantrine, categorized as artemisinin combination therapies (ACTs). These drug combinations operate on longer time scales (Olliaro & Taylor, 2003), and are able to eradicate the small fraction of parasites that escape the rapidly metabolized artemisinin, as well as reduce the likelihood of drug-resistant parasites emerging. However, the urgency of discovering alternatives to artemisinins is highlighted by reports of the emergence of partial resistance (tolerance) in western Cambodia (Dondorp et al., 2009; Wongsrichanalai & Meshnick, 2008). Many factors contribute to the development and spread of resistance including the overuse of antimalarial drugs, inadequate or incomplete therapeutic treatments, and the great parasite versatility, adaptability, and high proliferation rate that permits selected populations to emerge relatively rapidly (White et al., 2009; Wongsrichanalai & Meshnick, 2008). In the absence of clinically proven vaccines, it is therefore critical to find new and effective drugs, and to design sustainable interventions and monitoring strategies (as suggested above) that will help prevent the emergence and spread of resistant parasites.

The World Health Organization (WHO, http://www.who.int/en/) has developed and implemented a protocol for assessing the therapeutic efficacy of antimalarial drugs in low to intense transmission areas. The main objectives are to evaluate the efficacy of first- and second-line malarial drugs and to provide information to ministries of health in order to revise their drug policies. Together with WHO, hundreds of other partners, like those collaborating in the Roll Back Malaria Partnership (http://www.rollbackmalaria.org/) and the Worldwide Antimalarial Resistance Network (WWARN; http://www.wwarn.org/), are developing tools and approaches to monitor resistance and to support global control and elimination. Existing preventive measures (reviewed in Greenwood, 2008; Greenwood et al., 2008), such as regular administration of antimalarials to high-risk groups (e.g., young children, pregnant women) either through chemoprophylaxis or intermittent preventive treatment (IPT), have been shown to reduce markedly mortality and morbidity from malaria in highly endemic regions. The use of mass drug administration (MDA) is potentially problematic because of its inappropriate delivery, and its possible selection for resistant parasites. The main objective is to administer drugs in the population that will kill both sexual and asexual stages of *P. falciparum* and interrupt malaria transmission. This approach needs careful assessment before being integrated into a malaria control and elimination program. Vector control measurements, such as insecticide-treated bed nets (ITNs) and indoor residual insecticide spraying (IRS), are also important; their efficacy is well established (reviewed in Greenwood, 2008). The results of the widespread implementation of simple control measures, made possible by a substantial increase in funding in recent years, have had a significant impact on the burden of malaria in several endemic countries. For example, the introduction of ACTs, ITNs, and IRS in Zanzibar, which previously had a relatively high level of malaria transmission, has almost eliminated malaria (Bhattarai et al., 2007). The coordinated combination of different intervention strategies will be crucial to malaria control, and possibly its elimination. It is necessary to evaluate how combinations can be used most effectively in different epidemiological situations and health system settings.

VACCINE DEVELOPMENT

In parallel to these important control measurements, the development of a safe and effective malaria vaccine is a key requirement long anticipated. Although this task has proved to be more difficult than expected, steady progress is being made. The majority of research into malarial vaccines has focused on *P. falciparum,* due to the high mortality it causes and the ease of performing in vitro/in vivo studies. The first report of human protection from malaria by vaccination consisted of exposing individuals to bites of a thousand mosquitoes infected with malaria parasites (sporozoite stage) that have been weakened by irradiation (Hoffman et al., 2002). These parasites can invade the liver but lose their capacity to complete liver-stage maturation and develop into blood stage (merozoite) parasites (Mellouk et al., 1990). Results showed that this strategy provided protection against subsequent malaria infection. However, the safety and efficacy of this whole-parasite-based vaccine relied on a precise irradiation dose. In addition, trial results have been variable, making it unlikely to be a practical means of mass vaccination. The development of subunit vaccines, based on partial or complex antigens of the malaria parasite, seems more feasible. However, a radiation attenuated sporozoite vaccine is still being pursued, as certain roadblocks related to safety and batch-to-batch variation have been met or are under investigation (reviewed in Matuschewski & Mueller, 2007). One possibility tested in rodent malaria models is to use genetically attenuated parasites that lack genes necessary to the invasion of red blood cells, making it extremely unlikely for the attenuated parasite vaccine to restore its capacity to multiply and lead to disease (Good, 2005; Mueller et al., 2005).

For the development of subunit vaccines, several antigens present throughout the parasite life cycle (both during the pre-erythrocytic or hepatic stage and asexual blood stage) are being researched as potential targets. Two parasite surface proteins of the invasive stages, circumsporozoite protein (CSP) expressed at the initial pre-erythrocytic stage, and the merozoite protein 1 (MSP1) expressed at the blood stage, remain the major candidate antigens that are being developed and tested in various formulations. The first malaria vaccine to be extensively evaluated in clinical trials was the SPf66, which used a combination of antigens from both stages (including the MSP1 and CSP; see Graves & Gelband, 2006). An initial trial reported a 30% efficacy against clinical episodes of malaria, but subsequent trials were unable to confirm the initial encouraging results and the development of this vaccine has been suspended (Graves et al., 1998). Many CSP based candidates (pre-erythrocyte stage vaccines) have meanwhile been evaluated in clinical trials, but resulted in only limited protection or were not suitable to further evaluation (reviewed in Matuschewski & Mueller, 2007). The most successful candidate of any life stage developed to date is the RTS,S recombinant vaccine (Heppner et al., 2005). It is based on the CSP antigen fused to part of the hepatitis B virus surface antigen, to form a hybrid DNA sequence. The RTS,S/AS02A, one of the key vaccines produced using this technique, afforded 30%–60% short-term protection in phase II trials (Bojang et al., 2001). Development of RTS,S has been accelerated by the Malaria Vaccine Initiative global program (http://www.malariavaccine.org), being the only vaccine that passed to phase III trials, tested now in several African countries. Several other pre-erythrocytic candidates have reached the clinical evaluation stage, such as the ICC-1132, which is being tested in different formulations in the United States, Germany, and in the United Kingdom (reviewed in Greenwood & Targett, 2009; Matuschewski & Mueller, 2007; Todryk & Hill, 2007).

Vaccines that target the blood stage are also being investigated and are mostly based on merozoite surface proteins, such as the MSP1. Recent progress has been made to advance a vaccine that uses a formulation similar to the RTS/s/AS02—the FMP1/AS02A, a recombinant protein based on a component of the MSP1 protein (Withers et al., 2006). Phase I trials have demonstrated encouraging safety and immunogenicity results in participating U.S. adults, and pediatric clinical trials are underway in field studies in Africa (Thera et al., 2006). Two additional targets, merozoite surface protein 3 (MSP3) and the glutamate rich protein (GLURP), stand out among the numerous potential blood stage antigens, and vaccine trials are ongoing.

Transmission-blocking vaccines, which will block parasite function in the gamete stage, have not received as much attention. However, they would add a critical and valuable target to reduce malaria transmission (reviewed in Sutherland, 2009). Moreover, the increasing interest in the prospect of malaria elimination will require more attention to be paid to vaccines with transmission-blocking potential.

The search for vaccines against *P. vivax* has been very much neglected compared with *P. falciparum*. Only two candidates are being tested in clinical trials, and a modest number of other candidates are being assessed in preclinical studies (Grimberg et al., 2007, and reviewed in Beeson & Crabb, 2007). More investment toward the development of a *P. vivax* vaccine is required mainly because of the increasing drug resistance and recent observations of severe and lethal *P. vivax* cases. Also, vaccine developers should recognize the overlapping distribution of *P. falciparum* and *P. vivax* malaria, and consider strategies for vaccination against both of these species.

The scientific advances and increased funding seen in the past years have laid the foundation for the development of an effective malaria vaccine. Although malaria vaccines offer the greatest hope for sustained protection and possible elimination of malaria, they should be considered as part of a multicomponent malaria prevention and control strategy.

CONCLUSIONS

International funding for malaria has increased substantially in recent years, and global initiatives have brought improved and formalized efforts to coordinate and monitor control activities. Nevertheless, new malaria control tools and intervention strategies need to be introduced, as many established methods appear to be failing, with effective chemotherapy being perhaps the most problematic. With the substantial contributions from funding agencies and donors (e.g., Bill and Melinda Gates Foundation), malaria research has been reinvigorated and is embracing new sequencing technologies and whole-genome studies. The study of the three genomes of malaria (parasite, vector, and host) will lead to an improved understanding of the interaction and coevolution of these organisms, and provide insights into malaria biology. It will assist with answering important questions, such as why only some individuals develop severe malaria, and how does the parasite overcome the antimalarial effect of artemisinins. Genomics has an important role in answering these and many other important questions, and also exposing novel targets that will allow the development of new approaches to control malaria.

ACKNOWLEDGMENTS

We wish to thank Elisa Anastasia and Sarah Auburn for useful comments.

REFERENCES

International HapMap Consortium, Frazer KA, Ballinger DG, Cox DR, et al. (2007). A second generation human haplotype map of over 3.1 million SNPs. *Nature 449*, 851–861.

Abel L, Cot M, Mulder L, Carnevale P, Feingold J. (1992). Segregation analysis detects a major gene controlling blood infection levels in human malaria. *Am J Hum Genet 50*, 1308–1317.

Agarwal A, Guindo A, Cissoko Y, et al. (2000). Hemoglobin C associated with protection from severe malaria in the Dogon of Mali, a West African population with a low prevalence of hemoglobin S. *Blood 96*, 2358–2363.

Aidoo M, McElroy PD, Kolczak MS, Terlouw DJ, et al. (2001). Tumor necrosis factor-alpha promoter variant 2 (TNF2) is associated with pre-term delivery, infant mortality, and malaria morbidity in western Kenya: Asembo Bay Cohort Project IX. *Genet Epidemiol 21*, 201–211.

Aidoo M, Terlouw DJ, Kolczak MS, et al. (2002). Protective effects of the sickle cell gene against malaria morbidity and mortality. *Lancet 359*, 1311–1312.

Aitman TJ, Cooper LD, Norsworthy PJ, et al. (2000) Malaria susceptibility and CD36 mutation. *Nature 405*, 1015–1016.

Allen SJ, O'Donnell A, Alexander ND, et al. (1997). alpha+-Thalassemia protects children against disease caused by other infections as well as malaria. *Proc Natl Acad Sci U S A 94*, 14736–14741.

Allen SJ, O'Donnell A, Alexander ND, et al. (1999). Prevention of cerebral malaria in children in Papua New Guinea by southeast Asian ovalocytosis band 3. *Am J Trop Med Hyg 60*, 1056–1060.

Allison AC. (1954a). Protection afforded by sickle-cell trait against subtertian malarial infection. *Br Med J 1*, 290–294.

Allison AC. (1954b). The distribution of the sickle-cell trait in East Africa and elsewhere, and its apparent relationship to the incidence of subtertian malaria. *Trans R Soc Trop Med Hyg 48*, 312–318.

Ayi K, Turrini F, Piga A, Arese P. (2004). Enhanced phagocytosis of ring-parasitized mutant erythrocytes: a common mechanism that may explain protection against falciparum malaria in sickle trait and beta-thalassemia trait. *Blood 104*, 3364–3371.

Balding DJ, Nichols RA. (1995). A method for quantifying differentiation between populations at multi-allelic loci and its implications for investigating identity and paternity. *Genetica 96*, 3–12.

Barnwell JW, Asch AS, Nachman RL, Yamaya M, Aikawa M, Ingravallo P. (1989). A human 88-kD membrane glycoprotein (CD36) functions in vitro as a receptor for a cytoadherence ligand on Plasmodium falciparum-infected erythrocytes. *J Clin Invest 84*, 765–772.

Barragan A, Kremsner PG, Wahlgren M, Carlson J. (2000). Blood group A antigen is a coreceptor in Plasmodium falciparum rosetting. *Infect Immun 68*, 2971–2975.

Beeson JG, Crabb BS. (2007). Towards a vaccine against Plasmodium vivax malaria. *PLoS Med 4*, e350.

Bellamy R, Kwiatkowski D, Hill AV. (1998a). Absence of an association between intercellular adhesion molecule 1, complement receptor 1 and interleukin 1 receptor antagonist gene polymorphisms and severe malaria in a West African population. *Trans R Soc Trop Med Hyg 92*, 312–316.

Bellamy R, Ruwende C, McAdam KP, et al. (1998b). Mannose binding protein deficiency is not associated with malaria, hepatitis B carriage nor tuberculosis in Africans. *QJM 91*, 13–18.

Bennett S, Allen SJ, Olerup O, et al. (1993) Human leucocyte antigen (HLA) and malaria morbidity in a Gambian community. *Trans R Soc Trop Med Hyg 87*, 286–287

Bhattarai A, Ali AS, Kachur SP, et al. (2007). Impact of artemisinin-based combination therapy and insecticide-treated nets on malaria burden in Zanzibar. *PLoS Med 4*, e309.

Bienzle U, Ayeni O, Lucas AO, Luzzatto L. (1972). Glucose-6-phosphate dehydrogenase and malaria. Greater resistance of females heterozygous for enzyme deficiency and of males with non-deficient variant. *Lancet 229 (7742)*, 107–110.

Bienzle U, Guggenmoos-Holzmann I, Luzzatto L. (1979). Malaria and erythrocyte glucose-6-phosphate dehydrogenase variants in West Africa. *Am J Trop Med Hyg 28*, 619–621.

Bojang KA, Milligan PJ, Pinder M, et al. (2001). Efficacy of RTS,S/AS02 malaria vaccine against Plasmodium falciparum infection in semi-immune adult men in The Gambia: a randomised trial. *Lancet 358*, 1927–1934.

Boreham PF, Lenahan JK, Port GR, McGregor IA. (1981). Haptoglobin polymorphism and its relationship to malaria infections in The Gambia. *Trans R Soc Trop Med Hyg 75*, 193–200.

Bream JH, Ping A, Zhang X, Winkler C, Young HA. (2002). A single nucleotide polymorphism in the proximal IFN-gamma promoter alters control of gene transcription. *Genes Immun 3*, 165–169.

Burgner D, Usen S, Rockett K, et al. (2003). Nucleotide and haplotypic diversity of the NOS2A promoter region and its relationship to cerebral malaria. *Hum Genet 112*, 379–386.

Burgner D, Xu W, Rockett K, et al. (1998). Inducible nitric oxide synthase polymorphism and fatal cerebral malaria. *Lancet 352*, 1193–1194.

Burton PR, Tobin MD, Hopper JL. (2005). Key concepts in genetic epidemiology. *Lancet 366*, 941–951.

Cabantous S, Poudiougou B, Traore A, et al. (2005). Evidence that interferon-gamma plays a protective role during cerebral malaria. *J Infect Dis 192*, 854–860.

Campino S, Forton J, Auburn S, et al. (2009). TLR9 polymorphisms in African populations: no association with severe malaria, but evidence of cis-variants acting on gene expression. *Malar J 8*, 44.

Cappadoro M, Giribaldi G, O'Brien E, et al. (1998). Early phagocytosis of glucose-6-phosphate dehydrogenase (G6PD)-deficient erythrocytes parasitized by Plasmodium falciparum may explain malaria protection in G6PD deficiency. *Blood 92*, 2527–2534.

Cardon LR, Palmer LJ. (2003). Population stratification and spurious allelic association. *Lancet 361*, 598–604.

Cattani JA, Gibson FD, Alpers MP, Crane GG. (1987). Hereditary ovalocytosis and reduced susceptibility to malaria in Papua New Guinea. *Trans R Soc Trop Med Hyg 81*, 705–709.

Chebloune Y, Pagnier J, Trabuchet G, et al. (1988). Structural analysis of the 5' flanking region of the beta-globin gene in African sickle cell anemia patients: further evidence for three origins of the sickle cell mutation in Africa. *Proc Natl Acad Sci U S A 85*, 4431–4435.

Chevillard C, Henri S, Stefani F, Parzy D, Dessein A. (2002). Two new polymorphisms in the human interferon gamma (IFN-gamma) promoter. *Eur J Immunogenet 29*, 53–56.

Cholera R, Brittain NJ, Gillrie MR, et al. (2008). Impaired cytoadherence of Plasmodium falciparum-infected erythrocytes containing sickle hemoglobin. *Proc Natl Acad Sci U S A 105*, 991–996.

Chotivanich K, Udomsangpetch R, Pattanapanyasat K, et al. (2002). Hemoglobin E: a balanced polymorphism protective against high parasitemias and thus severe P falciparum malaria. *Blood 100*, 1172–1176.

Clark IA, Alleva LM, Mills AC, Cowden WB. (2004). Pathogenesis of malaria and clinically similar conditions. *Clin Microbiol Rev 17*, 509–539, table of contents.

Clark IA, Rockett KA. (1996). Nitric oxide and parasitic disease. *Adv Parasitol 37*, 1–56.

Clark TG, Diakite M, Auburn S, et al.(2009a). Tumor necrosis factor and lymphotoxin-alpha polymorphisms and severe malaria in African populations. *J Infect Dis 199*, 569–575.

Clark TG, Fry AE, Auburn S, et al. (2009b). Allelic heterogeneity of G6PD deficiency in West Africa and severe malaria susceptibility. *Eur J Hum Genet 17*, 1080–1085.

Coban C, Ishii KJ, Kawai T, et al. (2005). Toll-like receptor 9 mediates innate immune activation by the malaria pigment hemozoin. *J Exp Med 201*, 19–25.

Cockburn IA, Mackinnon MJ, O'Donnell A, et al. (2004). A human complement receptor 1 polymorphism that reduces Plasmodium falciparum rosetting confers protection against severe malaria. *Proc Natl Acad Sci U S A 101*, 272–277.

Cohuet A, Krishnakumar S, Simard F, Morlais I, Koutsos A, Fontenille D, Mindrinos M, Kafatos FC (2008). SNP discovery and molecular evolution in Anopheles gambiae, with special emphasis on innate immune system. *BMC Genomics 9*, 227.

Cordell HJ, Clayton DG (2005) Genetic association studies. *Lancet 366*, 1121–1131

Cortes A, Benet A, Cooke BM, Barnwell JW, Reeder JC. (2004). Ability of *Plasmodium falciparum* to invade Southeast Asian ovalocytes varies between parasite lines. *Blood 104*, 2961–2966.

Cot M, Abel L, Roisin A, Barro D, et al. (1993). Risk factors of malaria infection during pregnancy in Burkina Faso: suggestion of a genetic influence. *Am J Trop Med Hyg 48*, 358–364.

Cox SE, Doherty C, Atkinson SH, et al. (2007). Haplotype association between haptoglobin (Hp2) and Hp promoter SNP (A-61C) may explain previous controversy of haptoglobin and malaria protection. *PLoS One 2*, e362.

Devlin B, Roeder K. (1999). Genomic control for association studies. *Biometrics 55*, 997–1004.

Diakite M, Clark TG, Auburn S, et al. (2009). A genetic association study in the Gambia using tagging polymorphisms in the major histocompatibility complex class III region implicates a HLA-B associated transcript 2 polymorphism in severe malaria susceptibility. *Hum Genet 125*, 105–109.

Dolo A, Modiano D, Maiga B, et al. (2005). Difference in susceptibility to malaria between two sympatric ethnic groups in Mali. *Am J Trop Med Hyg 72*, 243–248.

Dondorp AM, Nosten F, Yi P, et al. (2009). Artemisinin resistance in Plasmodium falciparum malaria. *N Engl J Med 361*, 455–467.

Elagib AA, Kider AO, Akerstrom B, Elbashir MI. (1998). Association of the haptoglobin phenotype (1-1) with falciparum malaria in Sudan. *Trans R Soc Trop Med Hyg 92*, 309–311.

Enevold A, Vestergaard LS, Lusingu J, et al. (2005). Rapid screening for glucose-6-phosphate dehydrogenase deficiency and haemoglobin polymorphisms in Africa by a simple high-throughput SSOP-ELISA method. *Malar J 4*, 61.

Facer CA. (1983). Merozoites of P. falciparum require glycophorin for invasion into red cells. *Bull Soc Pathol Exot Filiales 76*, 463–469.

Fairhurst RM, Baruch DI, Brittain NJ, et al. (2005). Abnormal display of PfEMP-1 on erythrocytes carrying haemoglobin C may protect against malaria. *Nature 435*, 1117–1121.

Fairhurst RM, Fujioka H, Hayton K, Collins KF, Wellems TE. (2003). Aberrant development of Plasmodium falciparum in hemoglobin CC red cells: implications for the malaria protective effect of the homozygous state. *Blood 101*, 3309–3315.

Feng Z, Smith DL, McKenzie FE, Levin SA. (2004). Coupling ecology and evolution: malaria and the S-gene across time scales. *Math Biosci 189*, 1–19.

Fernandez-Reyes D, Craig AG, Kyes SA, et al. (1997). A high frequency African coding polymorphism in the N-terminal domain of ICAM-1 predisposing to cerebral malaria in Kenya. *Hum Mol Genet 6*, 1357–1360.

Fischer PR, Boone P. (1998). Short report: severe malaria associated with blood group. *Am J Trop Med Hyg 58*, 122–123.

Flatz G. (1967). Hemoglobin E: distribution and population dynamics. *Humangenetik 3*, 189–234.

Flint J, Harding RM, Boyce AJ, Clegg JB. (1998). The population genetics of the haemoglobinopathies. *Baillieres Clin Haematol 11*, 1–51.

Flori L, Kumulungui B, Aucan C, et al. (2003). Linkage and association between Plasmodium falciparum blood infection levels and chromosome 5q31-q33. *Genes Immun 4*, 265–268.

Foo LC, Rekhraj V, Chiang GL, Mak JW. (1992). Ovalocytosis protects against severe malaria parasitemia in the Malayan aborigines. *Am J Trop Med Hyg 47*, 271–275.

Fowkes FJ, Allen SJ, Allen A, Alpers MP, Weatherall DJ, Day KP. (2008). Increased microerythrocyte count in homozygous alpha(+)-thalassaemia contributes to protection against severe malarial anaemia. *PLoS Med 5*, e56.

Fowkes FJ, Imrie H, Migot-Nabias F, et al. (2006). Association of haptoglobin levels with age, parasite density, and haptoglobin genotype in a malaria-endemic area of Gabon. *Am J Trop Med Hyg 74*, 26–30.

Fry AE, Auburn S, Diakite M, et al. (2008a). Variation in the ICAM1 gene is not associated with severe malaria phenotypes. *Genes Immun 9*, 462–469.

Fry AE, Ghansa A, Small KS, et al. (2009). Positive selection of a CD36 nonsense variant in sub-Saharan Africa, but no association with severe malaria phenotypes. *Hum Mol Genet 18*, 2683–2692.

Fry AE, Griffiths MJ, Auburn S, et al. (2008b). Common variation in the ABO glycosyltransferase is associated with susceptibility to severe Plasmodium falciparum malaria. *Hum Mol Genet 17*, 567–576.

Ganczakowski M, Town M, Bowden DK, et al. (1995). Multiple glucose 6-phosphate dehydrogenase-deficient variants correlate with malaria endemicity in the Vanuatu archipelago (southwestern Pacific). *Am J Hum Genet 56*, 294–301.

Garcia A, Cot M, Chippaux JP, Ranque S, Feingold J, Demenais F, Abel L. (1998a). Genetic control of blood infection levels in human malaria: evidence for a complex genetic model. *Am J Trop Med Hyg 58*, 480–488

Garcia A, Marquet S, Bucheton B, et al. (1998b). Linkage analysis of blood Plasmodium falciparum levels: interest of the 5q31-q33 chromosome region. *Am J Trop Med Hyg 58*, 705–709.

Genton B, al-Yaman F, Mgone CS, et al. (1995). Ovalocytosis and cerebral malaria. *Nature 378*, 564–565.

Good MF. (2005). Genetically modified Plasmodium highlights the potential of whole parasite vaccine strategies. *Trends Immunol 26*, 295–297.

Graves P, Gelband H. (2006). Vaccines for preventing malaria (SPf66). *Cochrane Database Syst Rev*, CD005966.

Graves P, Gelband H, Garner P. (1998). The SPf66 malaria vaccine: what is the evidence for efficacy? *Parasitol Today 14*, 218–220.

Greenwood B, Marsh K, Snow R. (1991). Why do some African children develop severe malaria? Parasitol Today 7, 277–281.

Greenwood B, Targett G. (2009). Do we still need a malaria vaccine? *Parasite Immunol 31*, 582–586.

Greenwood BM. (2008). Control to elimination: implications for malaria research. *Trends Parasitol 24*, 449–454.

Greenwood BM, Fidock DA, Kyle DE, et al. (2008). Malaria: progress, perils, and prospects for eradication. *J Clin Invest 118*, 1266–1276.

Grimberg BT, Udomsangpetch R, Xainli J, et al. (2007). Plasmodium vivax invasion of human erythrocytes inhibited by antibodies directed against the Duffy binding protein. *PLoS Med 4*, e337.

Guindo A, Fairhurst RM, Doumbo OK, Wellems TE, Diallo DA. (2007). X-linked G6PD deficiency protects hemizygous males but not heterozygous females against severe malaria. *PLoS Med 4*, e66.

Gyan BA, Goka B, Cvetkovic JT, et al. (2004). Allelic polymorphisms in the repeat and promoter regions of the interleukin-4 gene and malaria severity in Ghanaian children. *Clin Exp Immunol 138*, 145–150.

Hadley TJ, Klotz FW, Pasvol G, et al. (1987). *Falciparum* malaria parasites invade erythrocytes that lack glycophorin A and B (MkMk). Strain differences indicate receptor heterogeneity and two pathways for invasion. *J Clin Invest 80*, 1190–1193.

Haldane J. (1949). Disease and evolution. Ricerca Sci 19, 3–10.

Heddini A, Pettersson F, Kai O, et al. (2001). Fresh isolates from children with severe Plasmodium falciparum malaria bind to multiple receptors. *Infect Immun 69*, 5849–5856.

Hedrick P. (2004). Estimation of relative fitnesses from relative risk data and the predicted future of haemoglobin alleles S and C. *J Evol Biol 17*, 221–224.

Hemingway J, Field L, Vontas J. (2002). An overview of insecticide resistance. *Science 298*, 96–97.

Heppner DG, Jr., Kester KE, Ockenhouse CF, et al.(2005). Towards an RTS,S-based, multi-stage, multi-antigen vaccine against falciparum malaria: progress at the Walter Reed Army Institute of Research. *Vaccine 23*, 2243–2250.

Hill AV, Allsopp CE, Kwiatkowski D, et al. (1991). Common west African HLA antigens are associated with protection from severe malaria. *Nature 352*, 595–600.

Hill AV, Elvin J, Willis AC, et al. (1992). Molecular analysis of the association of HLA-B53 and resistance to severe malaria. *Nature 360*, 434–439.

Hobbs MR, Udhayakumar V, Levesque MC, et al. (2002). A new NOS2 promoter polymorphism associated with increased nitric oxide production and protection from severe malaria in Tanzanian and Kenyan children. *Lancet 360*, 1468–1475.

Hoffman SL, Goh LM, Luke TC, et al. (2002). Protection of humans against malaria by immunization with radiation-attenuated Plasmodium falciparum sporozoites. *J Infect Dis 185*, 1155–1164.

Holmberg V, Schuster F, Dietz E, et al. (2008). Mannose-binding lectin variant associated with severe malaria in young African children. *Microbes Infect 10*, 342–348.

Holt RA, Subramanian GM, Halpern A, et al. (2002). The genome sequence of the malaria mosquito *Anopheles gambiae. Science 298*, 129–149.

Hunt NH, Driussi C, Sai-Kiang L. (2001). Haptoglobin and malaria. *Redox Rep 6*, 389–392.

Hutagalung R, Wilairatana P, Looareesuwan S, Brittenham GM, Aikawa M, Gordeuk VR. (1999). Influence of hemoglobin E trait on the severity of Falciparum malaria. *J Infect Dis 179*, 283–286.

Hyde JE. (2005). Drug-resistant malaria. *Trends Parasitol 21*, 494–498.

Imrie H, Ferguson DJ, Day KP. (2004). Human serum haptoglobin is toxic to Plasmodium falciparum in vitro. *Mol Biochem Parasitol 133*, 93–98.

Jallow M, Teo YY, Small KS, et al. (2009). Genome-wide and fine-resolution association analysis of malaria in West Africa. *Nat Genet 41(6)*, 657-65.

Jeffares DC, Pain A, Berry A, et al. (2007). Genome variation and evolution of the malaria parasite Plasmodium falciparum. *Nat Genet 39*, 120–125.

Jepson AP, Banya WA, Sisay-Joof F, Hassan-King M, Bennett S, Whittle HC. (1995). Genetic regulation of fever in Plasmodium falciparum malaria in Gambian twin children. *J Infect Dis 172*, 316–319.

Johnson MK, Clark TD, Njama-Meya D, Rosenthal PJ, Parikh S. (2009). Impact of the method of G6PD deficiency assessment on genetic association studies of malaria susceptibility. *PLoS One 4*, e7246.

Knight JC, Udalova I, Hill AV, et al. (1999). A polymorphism that affects OCT-1 binding to the TNF promoter region is associated with severe malaria. *Nat Genet 22*, 145–150.

Koch O, Awomoyi A, Usen S, et al. (2002). IFNGR1 gene promoter polymorphisms and susceptibility to cerebral malaria. *J Infect Dis 185*, 1684–1687.

Koch O, Rockett K, Jallow M, Pinder M, Sisay-Joof F, Kwiatkowski D. (2005). Investigation of malaria susceptibility determinants in the IFNG/IL26/IL22 genomic region. *Genes Immun 6*, 312–318.

Kun JF, Klabunde J, Lell B, et al. (1999). Association of the ICAM-1Kilifi mutation with protection against severe malaria in Lambarene, Gabon. *Am J Trop Med Hyg 61*, 776–779.

Kun JF, Mordmuller B, Perkins DJ, et al. (2001) Nitric oxide synthase 2(Lambarene) (G-954C), increased nitric oxide production, and protection against malaria. *J Infect Dis 184*, 330–336

Kwiatkowski D, Hill AV, Sambou I, et al. (1990). TNF concentration in fatal cerebral, non-fatal cerebral, and uncomplicated Plasmodium falciparum malaria. *Lancet 336*, 1201–1204.

Kwiatkowski DP. (2005). How malaria has affected the human genome and what human genetics can teach us about malaria. *Am J Hum Genet 77*, 171–192.

Langlois MR, Delanghe JR. (1996). Biological and clinical significance of haptoglobin polymorphism in humans. *Clin Chem 42*, 1589–1600.

Lell B, May J, Schmidt-Ott RJ, et al. (1999). The role of red blood cell polymorphisms in resistance and susceptibility to malaria. *Clin Infect Dis 28*, 794–799.

Levesque MC, Hobbs MR, Anstey NM, et al. (1999). Nitric oxide synthase type 2 promoter polymorphisms, nitric oxide production, and disease severity in Tanzanian children with malaria. *J Infect Dis 180*, 1994–2002.

Li Y, Willer C, Sanna S, Abecasis G. (2009). Genotype imputation. *Annu Rev Genomics Hum Genet 10*, 387–406.

Louicharoen C, Patin E, Paul R, et al. (2009). Positively selected G6PD-Mahidol mutation reduces Plasmodium vivax density in Southeast Asians. *Science 326*, 1546–1549.

Luoni G, Verra F, Arca B, et al. (2001). Antimalarial antibody levels and IL4 polymorphism in the Fulani of West Africa. *Genes Immun 2*, 411–414.

Luty AJ, Kun JF, Kremsner PG. (1998). Mannose-binding lectin plasma levels and gene polymorphisms in Plasmodium falciparum malaria. *J Infect Dis 178*, 1221–1224.

Mackinnon MJ, Gunawardena DM, Rajakaruna J, Weerasingha S, Mendis KN, Carter R. (2000). Quantifying genetic and nongenetic contributions to malarial infection in a Sri Lankan population. *Proc Natl Acad Sci U S A 97*, 12661–12666.

Mackinnon MJ, Mwangi TW, Snow RW, Marsh K, Williams TN. (2005). Heritability of malaria in Africa. *PLoS Med 2*, e340.

Maier AG, Duraisingh MT, Reeder JC, et al. (2003). Plasmodium falciparum erythrocyte invasion through glycophorin C and selection for Gerbich negativity in human populations. *Nat Med 9*, 87–92.

Mangano VD, Luoni G, Rockett KA, et al. (2008). Interferon regulatory factor-1 polymorphisms are associated with the control of Plasmodium falciparum infection. *Genes Immun 9*, 122–129.

Marchini J, Howie B, Myers S, McVean G, Donnelly P. (2007). A new multipoint method for genome-wide association studies by imputation of genotypes. *Nat Genet 39*, 906–913.

Marsh K, English M, Crawley J, Peshu N. (1996). The pathogenesis of severe malaria in African children. *Ann Trop Med Parasitol 90*, 395–402.

Marsh K, Forster D, Waruiru C, et al. (1995). Indicators of life-threatening malaria in African children. *N Engl J Med 332*, 1399–1404.

Matuschewski K, Mueller AK. (2007). Vaccines against malaria – an update. *Febs J 274*, 4680–4687.

Mayor A, Bir N, Sawhney R, et al. (2005). Receptor-binding residues lie in central regions of Duffy-binding-like domains involved in red cell invasion and cytoadherence by malaria parasites. *Blood 105*, 2557–2563.

McGuire W, Hill AV, Allsopp CE, Greenwood BM, Kwiatkowski D. (1994). Variation in the TNF-alpha promoter region associated with susceptibility to cerebral malaria. *Nature 371*, 508–510.

McGuire W, Knight JC, Hill AV, Allsopp CE, Greenwood BM, Kwiatkowski D. (1999). Severe malarial anemia and cerebral malaria are associated with different tumor necrosis factor promoter alleles. *J Infect Dis 179*, 287–290.

McKenzie CA, Abecasis GR, Keavney B, et al. (2001). Trans-ethnic fine mapping of a quantitative trait locus for circulating angiotensin I-converting enzyme (ACE). *Hum Mol Genet 10*, 1077–1084.

Mellouk S, Lunel F, Sedegah M, Beaudoin RL, Druilhe P. (1990). Protection against malaria induced by irradiated sporozoites. *Lancet 335*, 721.

Menard D, Matsika-Claquin MD, Djalle D, et al. (2005). Association of failures of seven-day courses of artesunate in a non-immune population in Bangui, Central African Republic with decreased sensitivity of Plasmodium falciparum. *Am J Trop Med Hyg 73*, 616–621.

Meyer CG, May J, Luty AJ, Lell B, Kremsner PG. (2002). TNFalpha-308A associated with shorter intervals of Plasmodium falciparum reinfections. *Tissue Antigens 59*, 287–292.

Migot-Nabias F, Mombo LE, Luty AJ, et al. (2000). Human genetic factors related to susceptibility to mild malaria in Gabon. *Genes Immun 1*, 435–441.

Miller LH, Mason SJ, Clyde DF, McGinniss MH. (1976). The resistance factor to Plasmodium vivax in blacks. The Duffy-blood-group genotype, FyFy. *N Engl J Med 295*, 302–304.

Min-Oo G, Gros P. (2005). Erythrocyte variants and the nature of their malaria protective effect. *Cell Microbiol 7*, 753–763.

Mockenhaupt FP, Ehrhardt S, Cramer JP, et al. (2004a). Hemoglobin C and resistance to severe malaria in Ghanaian children. *J Infect Dis 190*, 1006–1009.

Mockenhaupt FP, Ehrhardt S, Gellert S, et al. (2004b). Alpha(+)-thalassemia protects African children from severe malaria. *Blood 104*, 2003–2006.

Mockenhaupt FP, Cramer JP, Hamann L, et al. (2006a). Toll-like receptor (TLR) polymorphisms in African children: common TLR-4 variants predispose to severe malaria. *J Commun Dis 38*, 230–245.

Mockenhaupt FP, Hamann L, von Gaertner C, et al. (2006b). Common polymorphisms of toll-like receptors 4 and 9 are associated with the clinical manifestation of malaria during pregnancy. *J Infect Dis 194*, 184–188.

Modiano D, Chiucchiuini A, Petrarca V, et al. (1998). Humoral response to Plasmodium falciparum Pf155/ring-infected erythrocyte surface antigen and Pf332 in three sympatric ethnic groups of Burkina Faso. *Am J Trop Med Hyg 58*, 220–224.

Modiano D, Chiucchiuini A, Petrarca V, et al. (1999). Interethnic differences in the humoral response to non-repetitive regions of the Plasmodium falciparum circumsporozoite protein. *Am J Trop Med Hyg 61*, 663–667.

Modiano D, Luoni G, Sirima BS, et al. (2001a). The lower susceptibility to Plasmodium falciparum malaria of Fulani of Burkina Faso (west Africa) is associated with low frequencies of classic malaria resistance genes. *Trans R Soc Trop Med Hyg 95*, 149–152.

Modiano D, Luoni G, Sirima BS, et al. (2001b). Haemoglobin C protects against clinical Plasmodium falciparum malaria. *Nature 414*, 305–308.

Modiano D, Petrarca V, Sirima BS, et al. (1995). Plasmodium falciparum malaria in sympatric ethnic groups of Burkina Faso, west Africa. *Parassitologia 37*, 255–259.

Modiano G, Morpurgo G, Terrenato L, et al. (1991). Protection against malaria morbidity: near-fixation of the alpha-thalassemia gene in a Nepalese population. *Am J Hum Genet 48*, 390–397.

Moulds JM, Kassambara L, Middleton JJ, et al. (2000). Identification of complement receptor one (CR1) polymorphisms in west Africa. *Genes Immun 1*, 325–329.

Mu J, Awadalla P, Duan J, et al. (2007). Genome-wide variation and identification of vaccine targets in the Plasmodium falciparum genome. *Nat Genet 39,* 126–130.

Mueller AK, Labaied M, Kappe SH, Matuschewski K. (2005). Genetically modified Plasmodium parasites as a protective experimental malaria vaccine. *Nature 433,* 164–167.

Nagayasu E, Ito M, Akaki M, et al. (2001). CR1 density polymorphism on erythrocytes of falciparum malaria patients in Thailand. *Am J Trop Med Hyg 64,* 1–5.

Nkhoma ET, Poole C, Vannappagari V, Hall SA, Beutler E. (2009). The global prevalence of glucose-6-phosphate dehydrogenase deficiency: a systematic review and meta-analysis. *Blood Cells Mol Dis 42,* 267–278.

Ohashi J, Naka I, Patarapotikul J, Hananantachai H, Looareesuwan S, Tokunaga K. (2002). Significant association of longer forms of CCTTT Microsatellite repeat in the inducible nitric oxide synthase promoter with severe malaria in Thailand. *J Infect Dis 186,* 578–581.

Olliaro PL, Taylor WR. (2003). Antimalarial compounds: from bench to bedside. *J Exp Biol 206,* 3753–3759.

Omi K, Ohashi J, Naka I, et al. (2002). Polymorphisms of CD36 in Thai malaria patients. *Southeast Asian J Trop Med Public Health 33* Suppl 3, 1–4.

Omi K, Ohashi J, Patarapotikul J, et al. (2003). CD36 polymorphism is associated with protection from cerebral malaria. *Am J Hum Genet 72,* 364–374.

Orlandi PA, Klotz FW, Haynes JD. (1992). A malaria invasion receptor, the 175-kilodalton erythrocyte binding antigen of Plasmodium falciparum recognizes the terminal Neu5Ac(alpha 2-3)Gal-sequences of glycophorin A. *J Cell Biol 116,* 901–909.

Pain A, Urban BC, Kai O, et al. (2001). A non-sense mutation in Cd36 gene is associated with protection from severe malaria. *Lancet 357,* 1502–1503.

Parikh S, Dorsey G, Rosenthal PJ. (2004). Host polymorphisms and the incidence of malaria in Ugandan children. *Am J Trop Med Hyg 71,* 750–753.

Pasvol G, Weatherall DJ, Wilson RJ. (1978). Cellular mechanism for the protective effect of haemoglobin S against P. falciparum malaria. *Nature 274,* 701–703.

Patel SS, King CL, Mgone CS, Kazura JW, Zimmerman PA. (2004). Glycophorin C (Gerbich antigen blood group) and band 3 polymorphisms in two malaria holoendemic regions of Papua New Guinea. *Am J Hematol 75,* 1–5.

Patel SS, Mehlotra RK, Kastens W, Mgone CS, Kazura JW, Zimmerman PA. (2001). The association of the glycophorin C exon 3 deletion with ovalocytosis and malaria susceptibility in the Wosera, Papua New Guinea. *Blood 98,* 3489–3491.

Pathirana SL, Alles HK, Bandara S, et al. (2005). ABO-blood-group types and protection against severe, Plasmodium falciparum malaria. *Ann Trop Med Parasitol 99,* 119–124.

Pattanapanyasat K, Yongvanitchit K, Tongtawe P, et al. (1999). Impairment of Plasmodium falciparum growth in thalassemic red blood cells: further evidence by using biotin labeling and flow cytometry. *Blood 93,* 3116–3119.

Price AL, Patterson NJ, Plenge RM, Weinblatt ME, Shadick NA, Reich D. (2006). Principal components analysis corrects for stratification in genome-wide association studies. *Nat Genet 38,* 904–909.

Pritchard JK, Stephens M, Rosenberg NA, Donnelly P. (2000). Association mapping in structured populations. *Am J Hum Genet 67,* 170–181.

Quaye IK, Ekuban FA, Goka BQ, et al. (2000). Haptoglobin 1-1 is associated with susceptibility to severe Plasmodium falciparum malaria. *Trans R Soc Trop Med Hyg 94,* 216–219.

Ranque S, Safeukui I, Poudiougou B, et al. (2005). Familial aggregation of cerebral malaria and severe malarial anemia. *J Infect Dis 191,* 799–804.

Rihet P, Abel L, Traore Y, Traore-Leroux T, Aucan C, Fumoux F. (1998a). Human malaria: segregation analysis of blood infection levels in a suburban area and a rural area in Burkina Faso. *Genet Epidemiol 15,* 435–450.

Rihet P, Flori L, Tall F, Traore AS, Fumoux F. (2004). Hemoglobin C is associated with reduced Plasmodium falciparum parasitemia and low risk of mild malaria attack. *Hum Mol Genet 13,* 1–6.

Rihet P, Traore Y, Abel L, Aucan C, Traore-Leroux T, Fumoux F. (1998b). Malaria in humans: Plasmodium falciparum blood infection levels are linked to chromosome 5q31-q33. *Am J Hum Genet 63,* 498–505.

RougemontA,DumboO,BouvierM,etal.(1988).Hypohaptoglobinaemia as an epidemiological and clinical indicator for malaria. Results of two studies in a hyperendemic region in West Africa. *Lancet 2(8613),* 709–712.

Rowe JA, Handel IG, Thera MA, et al. (2007). Blood group O protects against severe *Plasmodium falciparum* malaria through the mechanism of reduced rosetting. *Proc Natl Acad Sci U S A 104,* 17471–17476.

Rowe JA, Moulds JM, Newbold CI, Miller LH. (1997). *P. falciparum* rosetting mediated by a parasite-variant erythrocyte membrane protein and complement-receptor 1. *Nature 388,* 292–295.

Rowe JA, Rogerson SJ, Raza A, et al. (2000). Mapping of the region of complement receptor (CR) 1 required for Plasmodium falciparum rosetting and demonstration of the importance of CR1 in rosetting in field isolates. *J Immunol 165,* 6341–6346.

Ruwende C, Khoo SC, Snow RW, et al. (1995). Natural selection of hemi- and heterozygotes for G6PD deficiency in Africa by resistance to severe malaria. *Nature 376,* 246–249.

Sabeti PC, Reich DE, Higgins JM, et al (2002). Detecting recent positive selection in the human genome from haplotype structure. *Nature 419,* 832–837.

Sabeti PC, Schaffner SF, Fry B, et al. (2006). Positive natural selection in the human lineage. *Science 312,* 1614–1620.

Sanchaisuriya K, Fucharoen G, Sae-ung N, Siriratmanawong N, Surapot S, Fucharoen S. (2001). Molecular characterization of hemoglobin C in Thailand. *Am J Hematol 67,* 189–193.

Scriver C. (1995). Glucose-6-Phosphate Dehydrogenase Deficiency, In: The metabolic and molecular bases of inherited disease. *7th ed. McGraw- Hill, Inc.,* 3367–3398.

Servin B, Stephens M. (2007). Imputation-based analysis of association studies: candidate regions and quantitative traits. *PLoS Genet 3,* e114.

Sham P. (1998). *Statistics in Human Genetics.* 1st ed., Arnold, London.

Shear HL, Roth EF, Jr., Fabry ME, et al. (1993). Transgenic mice expressing human sickle hemoglobin are partially resistant to rodent malaria. *Blood 81,* 222–226.

Singh B, Kim Sung L, Matusop A, et al. (2004). A large focus of naturally acquired Plasmodium knowlesi infections in human beings. *Lancet 363,* 1017–1024.

Snow RW, Guerra CA, Noor AM, Myint HY, Hay SI. (2005). The global distribution of clinical episodes of Plasmodium falciparum malaria. *Nature 434,* 214–217.

Spielman RS, McGinnis RE, Ewens WJ. (1993). Transmission test for linkage disequilibrium: the insulin gene region and insulin-dependent diabetes mellitus (IDDM). *Am J Hum Genet 52,* 506–516.

Stevenson MM, Riley EM. (2004). Innate immunity to malaria. Nat Rev Immunol 4, 169–180.

Sutherland CJ. (2009). Surface antigens of Plasmodium falciparum gametocytes—a new class of transmission-blocking vaccine targets? *Mol Biochem Parasitol 166,* 93–98.

Teo YY, Small KS, Fry AE, Wu Y, Kwiatkowski DP, Clark TG. (2009). Power consequences of linkage disequilibrium variation between populations. *Genet Epidemiol 33,* 128–135.

Thera MA, Doumbo OK, Coulibaly D, et al. (2006). Safety and allele-specific immunogenicity of a malaria vaccine in Malian adults: results of a phase I randomized trial. *PLoS Clin Trials 1,* e34.

Timmann C, Evans JA, Konig IR, et al. (2007). Genome-wide linkage analysis of malaria infection intensity and mild disease. *PLoS Genet 3,* e48.

Tishkoff SA, Varkonyi R, Cahinhinan N, et al. (2001). Haplotype diversity and linkage disequilibrium at human G6PD: recent origin of alleles that confer malarial resistance. *Science 293,* 455–462.

Todryk SM, Hill AV. (2007). Malaria vaccines: the stage we are at. *Nat Rev Microbiol 5*, 487–489.

Tournamille C, Colin Y, Cartron JP, Le Van Kim C. (1995). Disruption of a GATA motif in the Duffy gene promoter abolishes erythroid gene expression in Duffy-negative individuals. *Nat Genet 10*, 224–228.

Udomsangpetch R, Sueblinvong T, Pattanapanyasat K, Dharmkrong-at A, Kittikalayawong A, Webster HK. (1993). Alteration in cytoadherence and rosetting of Plasmodium falciparum-infected thalassemic red blood cells. *Blood 82*, 3752–3759.

Urban BC, Ferguson DJ, Pain A, et al. (1999). Plasmodium falciparum-infected erythrocytes modulate the maturation of dendritic cells. Nature 400, 73–77.

Urban BC, Willcox N, Roberts DJ. (2001). A role for CD36 in the regulation of dendritic cell function. Proc Natl Acad Sci U S A 98, 8750–8755.

Vafa M, Maiga B, Berzins K, et al. (2007). Associations between the IL-4 -590 T allele and Plasmodium falciparum infection prevalence in asymptomatic Fulani of Mali. *Microbes Infect 9*, 1043–1048.

Verra F, Luoni G, Calissano C, et al. (2004). IL4-589C/T polymorphism and IgE levels in severe malaria. *Acta Trop 90*, 205–209.

Verra F, Mangano VD, Modiano D. (2009). Genetics of susceptibility to Plasmodium falciparum: from classical malaria resistance genes towards genome-wide association studies. *Parasite Immunol 31*, 234–253.

Voight BF, Kudaravalli S, Wen X, Pritchard JK. (2006). A map of recent positive selection in the human genome. *PLoS Biol 4*, e72.

Volkman SK, Sabeti PC, DeCaprio D, et al. (2007). A genome-wide map of diversity in Plasmodium falciparum. *Nat Genet 39*, 113–119.

Wattavidanage J, Carter R, Perera KL, et al. (1999). TNFalpha*2 marks high risk of severe disease during Plasmodium falciparum malaria and other infections in Sri Lankans. *Clin Exp Immunol 115*, 350–355.

Weatherall DJ. (2001). Phenotype-genotype relationships in monogenic disease: lessons from the thalassaemias. *Nat Rev Genet 2*, 245–255.

Weatherall DJ, Clegg JB. (2001). Inherited haemoglobin disorders: an increasing global health problem. *Bull World Health Organ 79*, 704–712.

White NJ. (1998). Drug resistance in malaria. *Br Med Bull 54*, 703–715.

White NJ, Pongtavornpinyo W, Maude RJ, et al. (2009). Hyperparasitaemia and low dosing are an important source of anti-malarial drug resistance. *Malar J 8*, 253.

Wilding CS, Weetman D, Steen K, Donnelly MJ. (2009). High, clustered, nucleotide diversity in the genome of Anopheles gambiae revealed through pooled-template sequencing: implications for high-throughput genotyping protocols. *BMC Genomics 10*, 320.

Williams TN, Maitland K, Ganczakowski M, et al. (1996). Red blood cell phenotypes in the alpha + thalassaemias from early childhood to maturity. *Br J Haematol 95*, 266–272.

Williams TN, Mwangi TW, Wambua S, et al. (2005a). Negative epistasis between the malaria-protective effects of alpha+-thalassemia and the sickle cell trait. *Nat Genet 37*, 1253–1257.

Williams TN, Wambua S, Uyoga S, et al. (2005b). Both heterozygous and homozygous alpha+ thalassemias protect against severe and fatal Plasmodium falciparum malaria on the coast of Kenya. *Blood 106*, 368–371.

Williams TN, Weatherall DJ, Newbold CI. (2002). The membrane characteristics of Plasmodium falciparum-infected and -uninfected heterozygous alpha(0)thalassaemic erythrocytes. *Br J Haematol 118*, 663–670.

Withers MR, McKinney D, Ogutu BR, et al. (2006). Safety and reactogenicity of an MSP-1 malaria vaccine candidate: a randomized phase Ib dose-escalation trial in Kenyan children. *PLoS Clin Trials 1*, e32.

Wongsrichanalai C, Meshnick SR. (2008). Declining artesunate-mefloquine efficacy against falciparum malaria on the Cambodia-Thailand border. *Emerg Infect Dis 14*, 716–729.

Wootton JC, Feng X, Ferdig MT, et al. (2002). Genetic diversity and chloroquine selective sweeps in Plasmodium falciparum. *Nature 418*, 320–323.

Yano A, Yamamoto Y, Miyaishi S, Ishizu H. (1998). Haptoglobin genotyping by allele-specific polymerase chain reaction amplification. *Acta Med Okayama 52*, 173–181.

Yazdani SS, Shakri AR, Mukherjee P, Baniwal SK, Chitnis CE. (2004). Evaluation of immune responses elicited in mice against a recombinant malaria vaccine based on Plasmodium vivax Duffy binding protein. *Vaccine 22*, 3727–3737.

Zimmerman PA, Fitness J, Moulds JM, et al. (2003). CR1 Knops blood group alleles are not associated with severe malaria in the Gambia. *Genes Immun 4*, 368–373.

14.

GLUCOSE 6-PHOSPHATE DEHYDROGENASE DEFICIENCY: GENETICS, PHARMACOGENETICS, AND MALARIA SELECTION

Lucio Luzzatto

INTRODUCTION

Glucose 6-phosphate dehydrogenase (G6PD) has been known in biochemistry since it was discovered in red cells by Otto Warburg in 1931 (Warburg & Christian, 1931); and it has been known in hematology since it was discovered by Paul Carson in 1956 that its genetically determined deficiency was responsible for the acute hemolytic anemia that follows administration of primaquine in some, but not in all, people (Carson et al., 1956). Since that time it has become clear that G6PD deficiency is one of the commonest genetic abnormalities worldwide (estimated over 500 million people affected), but it is generally benign, because for most of their lifetimes people with this trait remain asymptomatic. At the same time, G6PD deficiency accounts for well-characterized clinical syndromes. The biochemical genetics of G6PD and the clinical implications of G6PD deficiency have been extensively reviewed (Beutler, 1978, 1991; Cappellini & Fiorelli, 2008; Dacie, 1985; Luzzatto et al., 2001; Luzzatto & Poggi (2009) Here we will only summarize the essentials, and we will then focus on some features that are especially relevant to developing countries.

THE PHYSIOLOGICAL ROLE OF G6PD

Glucose 6-phosphate dehydrogenase (G6PD), the first enzyme of the pentose phosphate pathway, catalyzes the conversion of glucose 6-phosphate (G6P) and NADP to 6-phosphogluconolactone and NADPH (Lehninger et al., 2004). The latter is the more important reaction product: in all cells it is the electron donor for many biosynthetic reactions (in purine and pyrimidine synthesis, in fatty acid synthesis, and in several other pathways; Berg et al., 2002). In red cells, NADPH is required (see Figure 14-1) for the operation of glutathione (GSH) peroxidase (via GSH reductase), and at the same time it is essential for stabilizing catalase (Kirkman & Gaetani, 1984): these are the two enzymes able to detoxify hydrogen peroxide (Gaetani et al., 1989), the ultimate product of most types of oxidative processes. In red cells – where these biosynthetic pathways have been lost – G6PD is particularly important for several reasons. First, in these cells oxygen radicals are generated continuously with each cycle of oxygenation and deoxygenation of hemoglobin; second, red cells are directly exposed to a variety of exogenous oxidizing agents; thereby, oxygen radicals produced by such compounds are converted by superoxide dismutase to hydrogen peroxide, which is itself highly toxic; third, because mature red cells have no mitochondria, G6PD is the only source of NADPH. Normally, G6PD activity in red cells is such that NADPH is maintained at a high level and there is practically no NADP: the NADPH/NADP ratio plays a large part in the intracellular regulation of G6PD activity (Kirkman et al., 1975; Luzzatto, 1967).

The enzymatically active form of G6PD is either a dimer or a tetramer (see Figure 14-5) of a single protein subunit of 514 amino acids, which has a molecular mass of 59,096 Da. In addition to sites for the reversible binding of its two substrates (G6P and NADP), human G6PD has one molecule of NADP so tightly bound that it is regarded as "structural" (Au et al., 2000; De Flora et al., 1974).

GENETICS

Given the function of G6PD outlined above, it is not surprising that a decrease in G6PD activity would have potentially serious consequences, particularly in red cells. In spite of this, inherited G6PD deficiency is one of the commonest genetic abnormalities in the human species. The *G6PD* gene maps to the subtelomeric region of the long arm of the X-chromosome (band Xq28; see Martin-DeLeon et al., 1985), and therefore G6PD deficiency is inherited as a Mendelian X-linked trait (Adam, 1961). At the genomic level, the G6PD gene consists of 13 exons and spans some 18.5 kb (Martini et al., 1986). Structural and functional studies have revealed features of a "housekeeping gene" (Battistuzzi et al., 1985; Franze et al., 1998): this is in accord with the fact that G6PD is found in all cells.

As a result of X-linkage of *G6PD,* males have only one *G6PD* gene (i.e., they are hemizygous), and they must be either normal or G6PD-deficient. By contrast, females, having two G6PD genes, can be either normal or deficient (homozygous), or intermediate (heterozygous). Moreover, as a result of the phenomenon of X-chromosome

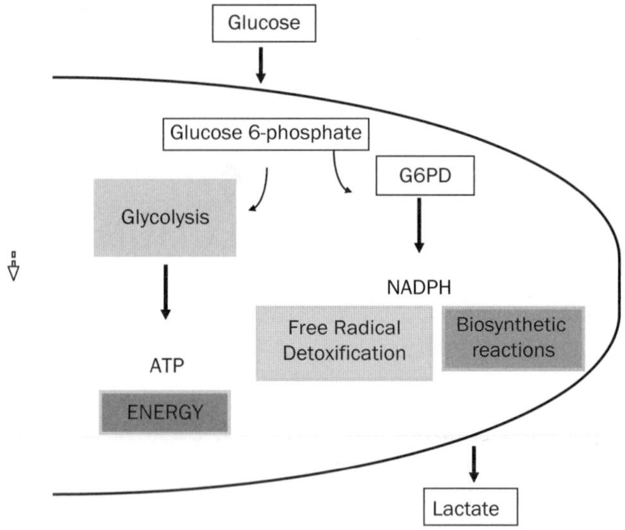

Figure 14-1 *Physiological role of G6PD.* Although G6PD is still referred to as the first enzyme of the pentose phosphate pathway, it is not essential for pentose synthesis: instead, it is a major provider of reductive potential in the form of NADPH. In most cells, NADPH is the coenzyme of several reactions in multiple biosynthetic metabolic pathways; in red cells, where these biosynthetic reactions are not operating, NADPH serves as the main agent of defense against oxidative injury.

inactivation, heterozygous females are genetic mosaics, and this in turn has clinical implications. Indeed, with most other (autosomal) enzyme deficiencies, heterozygotes are asymptomatic because their somatic cells have an enzyme level of about 50% of normal, which is usually sufficient for their metabolic functions. But in the case of G6PD, as a result of X-inactivation, the abnormal cells of a woman heterozygous for G6PD deficiency are just as deficient as those of a hemizygous deficient man, and therefore just as susceptible to pathology. Thus, although G6PD deficiency is still often referred to as an X-linked recessive trait, this is a misnomer because a recessive trait is, by definition, not expressed in a heterozygote: instead, G6PD deficiency is expressed in heterozygotes both biochemically and clinically. On average, the clinical manifestations will be less severe than in homozygotes (or in hemizygous males): however, there are exceptions (Roos et al., 1999), related to the broad variability of the X-chromosome inactivation ratio (Normah et al., 1991; Rinaldi et al., 1976).

G6PD is present in virtually all known organisms except strictly anaerobic bacteria; an analysis of sequence homology has helped to identify regions that are completely conserved, regions that are significantly conserved, and regions that are not conserved in evolution (Notaro et al., 2000).

EPIDEMIOLOGY OF G6PD DEFICIENCY

A remarkable feature of the population genetics of G6PD is that nearly 30 different alleles are present at polymorphic frequencies over the world (see Figure 14-2) that entail G6PD deficiency in red cells (Luzzatto et al., 2001; Mason et al., 2007): thus, the distribution of these alleles, taken together, represent the epidemiology of G6PD deficiency. Areas of high prevalence are found in Africa, Southern Europe, the Middle East, Southeast Asia, and Oceania. In the Americas and in parts of Northern Europe, G6PD deficiency was not common originally, but it has become quite prevalent as a result of migrations that have taken place in relatively recent historical times. A very rough estimate is that at least 500 million people have some form of G6PD deficiency.

CLINICAL MANIFESTATIONS

In view of the above, it is fortunate that the vast majority of G6PD-deficient people suffer no disease throughout their lifetime: indeed, they are not aware of this trait unless they are tested for it. However, all G6PD-deficient subjects can potentially become ill in two distinct situations.

a. In the newborn period, there is a much higher than average risk to develop neonatal jaundice (NNJ; see Bienzle et al., 1976; Doxiadis et al., 1961). The extent of the association between G6PD deficiency and NNJ appears to vary greatly in different populations and, within a certain population, the severity of this condition also varies widely, from subclinical, overlapping with "physiological jaundice," to imposing the threat of kernicterus if not treated. The reasons for this are not clear, but prematurity, infection, and environmental factors (for instance, naphthalene-camphor balls, used in babies' bedding and clothing) certainly play a part in making NNJ more severe and more dangerous. Coexistence with G6PD deficiency of the uridyl transferase mutation (*UDPG1A1*) responsible for Gilbert's syndrome increases at least twofold the risk of NNJ (Kaplan & Hammerman, 2010). From the point of view of public health this problem is important, because in some parts of the world G6PD deficiency is the commonest cause of severe NNJ which, if not adequately managed, can produce permanent neurological damage.
b. At any age a G6PD-deficient subject can develop acute hemolytic anemia in response to three types of triggers: (i) drugs (see Table 14-1), (ii) infections, and (iii) fava beans. The hemolytic attack can be from mild to very severe, and although in most cases the patient recovers fully, there are two situations when the attack can be life-threatening: in children, and in children or adults when there is a preexisting or concomitant other disease; for instance, anemia due to any other cause. Life is threatened mainly when the destruction of red cells is massive and abrupt:

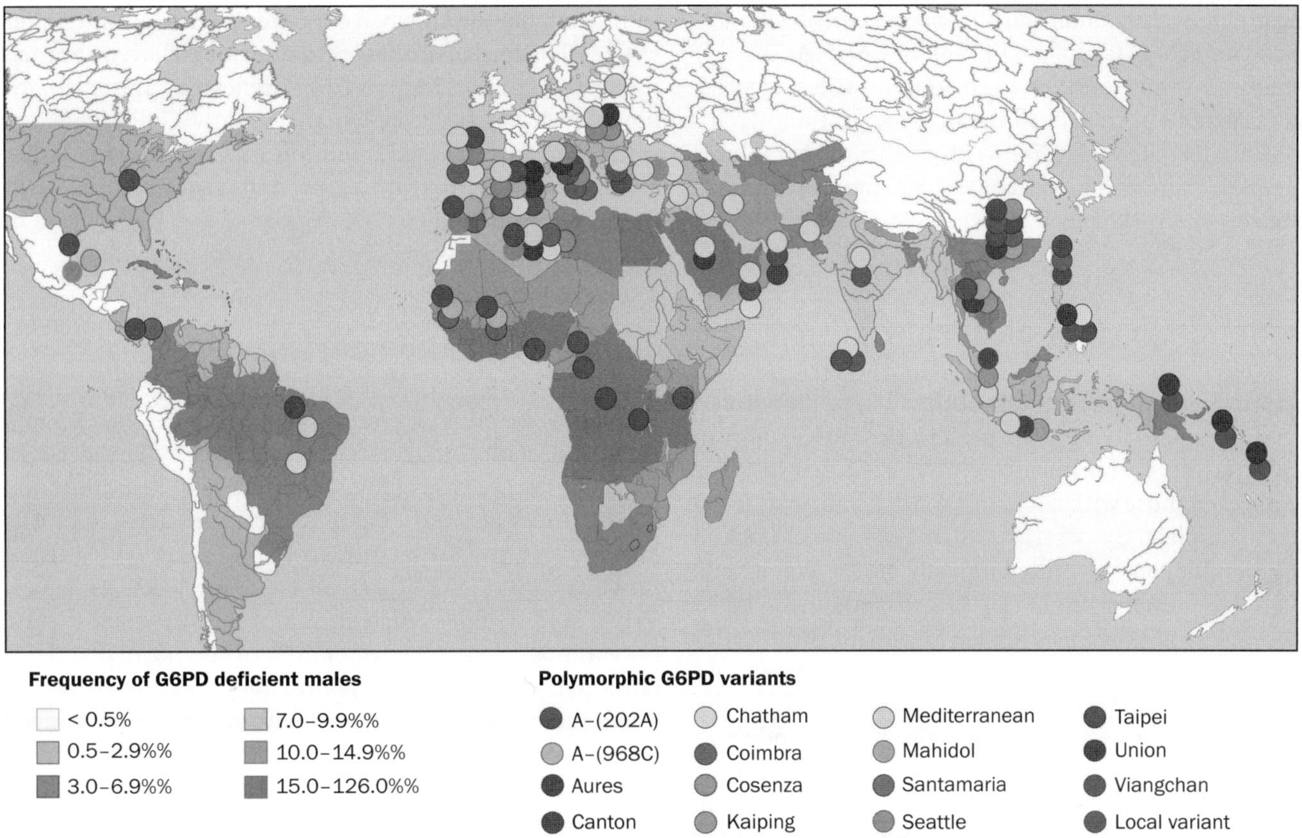

Figure 14-2 *World distribution of G6PD deficiency.* Shadings on the map indicate approximate frequencies of male G6PD deficiency frequencies in individual countries. Differently colored circles indicate where individual mutant G6PD alleles mostly contribute to the overall frequency of G6PD deficiency. (From Luzzatto & Notaro, 2001). Refer color figure.

in some cases the hemoglobin level has been known to drop from normal (14 G/dl) to 4 G/dl within 48 hours. A special example of AHA in G6PD-deficient subjects is known as *favism* (Luisada, 1941; Meloni et al, 1983), because it follows – by definition – a meal of fava beans; again, it can occur at any age, but mostly in children. The cause of favism is the presence in fava beans (or broad beans: *Vicia faba*) of vicine and convicine, two β-glycosides having as aglycones the substituted pyrimidines divicine and isouramil, which produce free radicals in the course of their autooxidation (Baker, et al., 1984; Chevion, et al., 1982). Eating fava beans is not followed by an attack of favism every time, probably on account of the quantity of fava beans consumed and of the amount of glycosides they contain, which vary in different cultivars. Bacterial infection is a significant trigger of hemolysis in G6PD-deficient subjects, and this has been documented particularly with pneumonia (Tugwell, 1973). It is also pertinent to note that when hemolysis takes place following administration of a drug for treating the infection, there has been a tendency to blame the drug rather than the infection (Youngster et al., 2010), when instead the reverse may be true.

TABLE 14-1 DRUGS THAT CARRY RISK OF ACUTE HEMOLYTIC ANEMIA IN SUBJECTS WHO ARE G6PD-DEFICIENT*

High Risk	Lower Risk
Dapsone (Lapdap)	Aspirin at high doses
Primachina	Chloroquine
Nitrofurantoin	Quinine
Ciprofloxacin	Quinidine
Nalidixic acid	Probenecid
Cotrimoxazole	Rasburicase

* In a recent review (Youngster et al., 2010) it was observed that the hemolytic potential of these drugs has not been always documented in accord with the principles of evidence-based medicine: in some cases only individual case reports have been published. However, I think it would not be prudent to ignore such reports, because many other similar episodes may have occurred but may have not have been published.

c. Whereas the above are relatively common events, a third type of clinical expression of G6PD deficiency, chronic nonspherocytic hemolytic anemia (CNSHA), is extremely rare, because it results exclusively from rare mutations of *G6PD*, which are distinct from the polymorphic mutations that are responsible for (a) and (b) (Mason, et al 1995). In CNSHA, hemolysis is not confined to episodic attacks triggered by exogenous agents: it is instead, by definition, continuous. Therefore, like in other chronic hemolytic anemias, there will be reticulocytosis, jaundice, and often an enlarged spleen.

GENETIC ANALYSIS OF G6PD FUNCTION

The clinical manifestations just outlined have pinpointed the crucial role of G6PD in preserving red cells from oxidative stress; and it must be emphasized that in these patients (even those with CNSHA), the deficiency of G6PD is never complete. By contrast, when *G6PD* was knocked out by homologous recombination in mouse embryonic stem cells (Pandolfi et al., 1995), the cells remained viable, but male embryos died (see Figure 14-3) with the onset of circulation, proving that G6PD is not essential for pentose synthesis and suggesting that it is instead essential in the presence of oxygen. Analysis of female embryos indicates that, probably for the same reason, the development of the placenta also requires G6PD activity (Longo et al., 2002). In contrast to the lethal knockout, a mouse with a *G6PD* point mutation obtained by random mutagenesis has mild G6PD deficiency in the red cells and does not have significant hemolysis when exposed to primaquine (Neifer et al., 1991); thus, in this respect we do not yet have an animal model that mimics human G6PD deficiency. Interestingly, however, it has been shown that mildly G6PD-deficient mice have an increased mutation rate in the brain (Felix et al., 2002), with features of mutagenesis consequent on oxidative damage to DNA, and also a higher rate of phenytoin-induced embryopathy (Nicol et al., 2000).

Generation of G6PD-*null* Mice

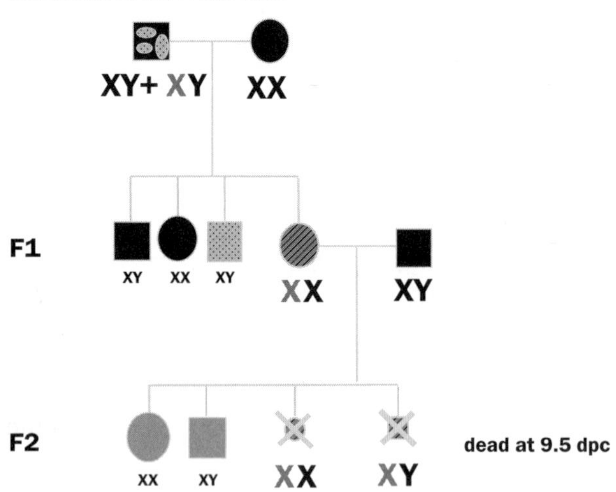

Figure 14-3 A G6PD-null mutation is an embryonic lethal. Chimeric mice were obtained by injecting into blastocysts G6PD knockout embryonic stem cells. Mating of a male chimeric mouse with a normal female (top line: the X-chromosome with the G6PD null mutation is in red) produced (F1), apart from normal offspring (in black), females heterozygous for the G6PD null mutation, but no hemizygous males, proving that G6PD null is an embryonic lethal. When a heterozygous female was mated with a hemizygous male, in the F2 both hemizygous male embryos and heterozygous female embryos died: the latter because of placental failure, since in the extraembryonic tissues, the normal paternal X is known to be selectively inactivated. (From Longo et al., 2002.) Refer color figure.

DIAGNOSIS AND SCREENING

Although the clinical picture of acute hemolytic anemia associated with G6PD deficiency is rather characteristic, especially in the case of favism, a final diagnosis must rely on the direct demonstration of decreased activity of this enzyme in red cells by a quantitative assay. The normal range is 7–10 IU/G Hb (Beutler, 1978; Dacie & Lewis, 1995), and G6PD-deficient males in the steady state have almost always less than 20% of the normal activity (although when reticulocytes are high the activity may be much increased, sometimes to within the normal range). Individual G6PD mutations can be tested for by appropriate PCR analysis: however, a G6PD test should not be regarded as superior simply because it is DNA-based, since in many populations more than one G6PD mutation may be prevalent. Several screening tests are available and have been tested in the field in various populations (Betke et al., 1967; Jalloh et al., 2004, Meissner et al., 2005).

PHARMACOGENETICS

Phenylthiocarbamide tasting had been known as a genetic trait since the 1930s, but pharmacogenetics came to be recognized as a branch of pharmacology in the 1960s, since succinylcholine sensitivity was shown to depend on genetic variation of pseudo-cholinesterase (Vogel & Motulsky, 1996). As mentioned in the introduction, G6PD deficiency was identified (Carson et al., 1956) after a prolonged investigation of former patients affected by what was known until then as "primaquine sensitivity" (Beutler, 1959), a term aptly describing an acute hemolytic anemia that developed in some but not all those who received primaquine as antimalarial prophylaxis. Thus, G6PD deficiency is probably the first example and remains a prototype of a condition that has been identified through pharmacogenetics[1]; only subsequently, and very promptly, did it become clear that the same biochemical defect was responsible for favism (Sansone & Segni, 1958).

1. Today, a more trendy term is *pharmacogenomics*, because genome-wide studies can be carried out in order to pinpoint genes that may influence the efficacy and/or the toxicity of drugs.

Today we have a well-established list of drugs that can cause hemolysis in G6PD-deficient subjects (Table 14-1). We must presume that these drugs have the capacity to inflict an oxidative stress to red cells, despite the fact that they do not share a common, or even an obviously related, chemical structure: therefore, at the moment, there is no way to predict whether any new drug is potentially hemolytic for G6PD-deficient subjects. Since we don't have an animal model (see above), there is no effective screening method either, and in practice the list of drugs has to be updated by empirical observation of hemolytic side effects, as has been the case not long ago, for instance, for rasburicase (Browning & Kruse, 2005; Youngster et al., 2010).

Paradoxically, however, serious complications have arisen quite recently from the use of a drug that was already on the list, dimethyl disulfone, or dapsone, a highly effective agent for the treatment of leprosy known to be potentially hemolytic, and much more so in G6PD-deficient subjects (Eppes et al., 1966). When in the 1990s the spread of chloroquine-resistant *Plasmodium falciparum* in Africa created a pressing need for new antimalarials, after preliminary tests, two major randomized clinical trials of Lapdap, a combination of chlorproguanil (lapudrine) and dapsone, were organized in Africa in children with malaria. In the first trial (Sulo et al., 2002) no G6PD testing was carried out, but the authors noted that anemia was more common in the Lapdap group than in the control group; in the next study (Alloueche et al., 2004) some but not all of the patients were tested for G6PD deficiency, and it was noted that G6PD-deficient children versus normal children were more likely to have a severe drop in hemoglobin levels: many required emergency blood transfusion. Lapdap was licensed in 2003, and it was marketed in many African countries until 2008, when it was withdrawn by the manufacturers in view of "the reduction in hemoglobin concentration observed in patients with G6PD deficiency taking Dacart" (a combination of Lapdap and artemisinin). As outlined recently (Luzzatto, 2010), several lessons that can be learned from this failure of heeding the principles of pharmacogenetics. *(1)* Drug-induced hemolysis in G6PD deficiency is markedly dose-dependent. The daily dosage of dapsone in a course of Lapdap was considerably higher than that normally used in the treatment of leprosy. *(2)* It is difficult to understand why, in the clinical trials, G6PD-deficient children were not excluded. *(3)* In general, serious adverse events (SAE) to a new drug occur unpredictably in a small fraction of patients. In the case of Lapdap, a potent hemolytic drug was administered to children who, predictably, had severe hemolysis: to call each case of this complication an SAE is a bit euphemistic. *(4)* It is part of pharmacogenetics teaching that, because of genetic differences, side effects of drugs may vary from one population to another. The frequency of G6PD deficiency was well known to be in the range from 10%–25% in the populations of sub-Saharan Africa (Beutler et al., 1985), at least one order of magnitude higher than in the United Kingdom. *(5)* It would have seem logical to formally list G6PD deficiency as a contraindication to Lapdap, but it was feared by some that, because of difficulties in providing the test in Africa, such a requirement would "kill the drug." In the end, withdrawal of Lapdap made the agent unavailable even for the large majority (more than 80% in most countries) of people who are not G6PD-deficient. *(6)* It would seem desirable, in the future, that a small fraction of the large cost of developing a new drug were invested in an ultra-simple, reliable test for G6PD deficiency.

G6PD DEFICIENCY AND MALARIA SELECTION

Since Haldane first suggested that thalassemia genes may have been subjected to malaria selection (Haldane, 1949), it became tempting to apply this notion to any polymorphic gene that is prevalent in areas of past or present malaria endemicity, especially if the gene is expressed in red cells. Half a century ago Allison (1960) and Motulsky (1960) independently suggested that G6PD deficiency might offer protection against *P. falciparum* malaria, and this became known as the "malaria hypothesis." Since then, the evidence that has accumulated in support of this hypothesis (see Table 14-2) is so overwhelming that it is somewhat surprising that the phrase "malaria hypothesis" is still used today: "malaria selection" would be probably a more appropriate terminology (see Figure 14-4).

Briefly, the evidence is threefold.

1. *Population studies.* As mentioned above, the worldwide correlation between malaria endemicity and geographic distribution of G6PD deficiency (Figure 14-2) was the first hint; and two further findings have further corroborated the significance of this correlation. First, when the frequency of G6PD deficiency was mapped carefully within areas where the transmission of malaria varied from high to low (*micromapping*), it was found that the frequency of G6PD deficiency was in direct correlation with malaria transmission, even when the population could be presumed to be genetically relatively homogeneous. This was found to be true, for instance, in areas as far apart as Sardinia (Siniscalco et al., 1966), Togo (Bienzle et al., 1972b) and Vanuatu (Ganczakowski et al., 1995). Second, once individual polymorphic mutations causing G6PD deficiency became known, it emerged that they mapped characteristically to specific parts of the world (Luzzatto & Mehta, 1995; Luzzatto & Notaro, 2001), suggesting that wherever a mutation entailing enzyme deficiency happened to take place, its population frequency increased if malaria was present—one of the best examples in the

TABLE 14-2 CLINICAL STUDIES OF *P. FALCIPARUM* MALARIA IN CHILDREN WITH G6PD DEFICIENCY

	Bienzle et al., 1972	Ruwende et al., 1995	Guindo et al., 2007
Population	Nigeria	Gambia + Kenya	Mali
Numbers	699	2000+	3197
Type of G6PD deficiency*	A-	A-	A-
Genotyping methods	Enzyme assay + electrophoresis + cytochemical test	Mutation analysis	Mutation analysis
Main criteria	Parasitemia	Severe malaria	Severe malaria
Main conclusion	Protection significant only in heterozygous females	Protection significant in both heterozygous females and hemizygous deficient males	Protection significant only in hemizygous deficient males

*It should be noted that G6PD A- has been originally defined by enzyme assay and electrophoresis. However, molecular analysis has subsequently revealed that there are three different mutations with the A- phenotype. Thus, correct classification of all genotypes can be only obtained by phenotypic analysis or by testing for all three mutations.

human species of convergent evolution resulting from one and the same selective agent. Extensive analysis of haplotypes surrounding the *G6PD* locus confirmed a selection signature there (Tishkoff et al., 2001).

2. *Clinical field studies.* If G6PD deficiency is selected by *P. falciparum* malaria, it should be possible to show that G6PD-deficient subjects have less malaria, or at least less severe malaria. It was immediately clear from preliminary reports that protection against malaria was not complete (Gilles et al., 1967). Subsequently, three major studies have been carried out in children in Africa (Bienzle et al., 1972a; Ruwende et al., 1995a; see Table 14-2). There were differences in the parameters used to measure the severity of malaria (parasitemia versus clinical criteria) and in how G6PD deficiency was assessed (quantitation and cytochemistry versus DNA analysis). The three studies were concordant in providing statistically significant evidence that G6PD-deficient subjects had less severe malaria, thus validating the "malaria hypothesis"; however, there were some surprising differences. There was evidence for protection, in the first study, only in heterozygotes; in the second, in both heterozygotes and hemizygotes; in the third, only in hemizygotes. These discrepancies are not trivial with respect to their clinical implications in areas where malaria is still rampant, and they are not all equally consistent with the population dynamics of genes that cause G6PD deficiency (Cavalli-Sforza & Bodmer, 1971). Indeed, with heterozygote advantage, the theory of balanced polymorphism is well worked out, and gene frequencies at equilibrium can be calculated for any given set of fitness values of male and female genotypes (or vice versa); on the other hand, if G6PD-deficient hemizygotes are at an advantage, it is not clear why the underlying allele has not reached fixation. Therefore, these discrepancies still need to be reconciled, and in this respect two issues deserve attention. (a) *Correct genotype classification.* This

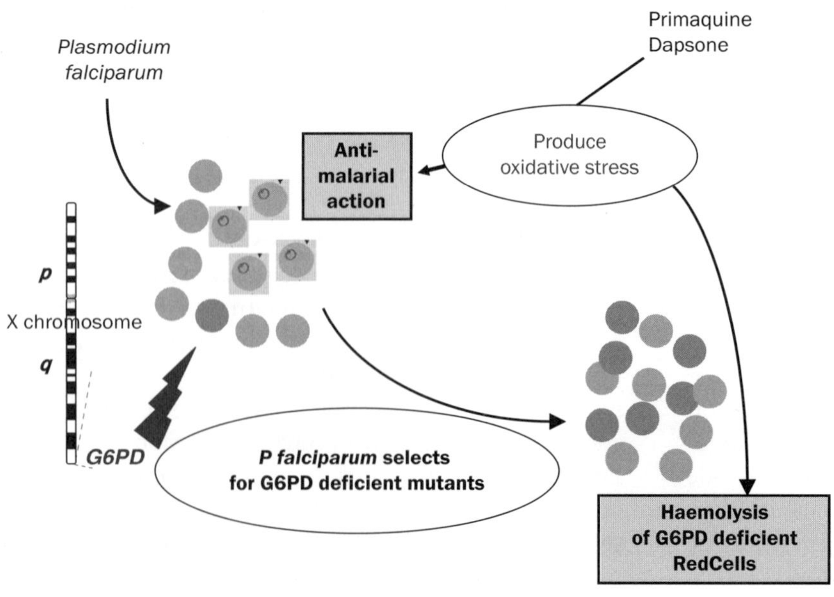

Figure 14-4 *Genetics and pharmacogenetics of G6PD deficiency.* Mutations of the *G6PD* gene arise spontaneously, like in any other gene. Some mutations entail deficiency of G6PD, and G6PD-deficient red cells confer protection against *P. falciparum*. As a result, genetically determined G6PD deficiency tends to become prevalent in areas where *P. falciparum* is endemic. Primaquine and dapsone are potent antimalarials, but unfortunately they are also potently hemolytic for G6PD-deficient red cells. Refer color figure.

has been highlighted in work recently carried out in Uganda, where different inferences on malaria protection were obtained, depending on whether G6PD deficiency was assessed from enzyme assays or from mutation analysis (Johnson et al., 2009). In principle, the latter approach ought to be superior, especially for heterozygotes—except that if more than one G6PD deficiency mutations are present in the population, each one of the mutations must be analyzed. In retrospect, this applies to one of the most widely quoted papers (Ruwende et al., 1995b), since we now know that in the Gambia, the commonest G6PD A-allele is not the one that was tested for (Clark et al., 2009), and therefore many G6PD-deficient subjects must have been misclassified as normal. (b) *Relevance of clinical criteria to Darwinian evolution.* What matters in terms of Darwinian selection is mortality from malaria. Most clinical studies currently tend to determine the relative risk, for a certain *G6PD* genotype, of "uncomplicated malaria" and/or "severe malaria:" the latter has been shown to correlate with malaria mortality in children as a whole, but we don't know whether this is true for a subset with a particular genotype. It is not impossible that some *G6PD* heterozygotes, for instance, might overcome "severe malaria" if they get it, and that the number of malaria attacks or the levels of parasitemia measured over a length of time might be more relevant to malaria mortality.

3. *Studies on the mechanism of protection.* The first evidence was obtained in vivo, when it was shown that in the blood of heterozygous girls who are genetic mosaics as a result of X-chromosome inactivation, there was an excess of parasites in the G6PD-normal red cells compared to the G6PD-deficient red cells (Luzzatto et al., 1969). This suggested that either the latter were able to "fend off" infection or, once they were infected, they were removed from the circulation. It became possible to test these alternatives only after William Trager had introduced in-vitro cultures of *P. falciparum* (Trager & Jensen, 1976), the greatest advance in malaria research in that generation, which suddenly made it possible to investigate the biology of this parasite anywhere in the world. Several studies suggested there might be a difference in the rate of infection and/or in parasite growth in G6PD-normal versus G6PD-deficient red cells (Friedman, 1979; Luzzatto et al., 1983; Roth et al., 1983a), and that this may depend on whether the parasite emerged from a G6PD-normal or a G6PD-deficient red cell (Nagel & Roth, 1989; Roth et al., 1983b; Usanga & Luzzatto, 1985). It seemed possible that in certain circumstances the parasite might depend on the host cell G6PD, especially since it was not known at the time whether *P. falciparum* had its own G6PD, as was in due course confirmed (O'Brien et al., 1994; Shahabuddin et al., 1994). However, in vivo studies have not thus far confirmed that the host red cell G6PD status influences the parasite G6PD expression (Sodeinde et al., 2003). By contrast, evidence has accumulated that G6PD-deficient *P. falciparum* infected red cells are recognized by autologous macrophages at a much earlier stage in parasite development than applies to G6PD-normal red cells (Cappadoro et al., 1998). Interesting details have been worked of this process (Arese et al., 2005; Ayi et al., 2004), which corresponds to the previously envisaged model of "suicidal infection" (Luzzatto et al., 1969): a phrase which has been recently resuscitated (Foller et al., 2009), along with the term *eryptosis*, which designates erythrocyte apoptosis, or accelerated destruction of red cells.

G6PD DEFICIENCY, ANTIMALARIALS, AND MALARIA: A TRIANGULAR RELATIONSHIP

Primaquine and other agents that are effective against *Plasmodia* are also toxic to G6PD-deficient red cells, and it is tempting to imagine that formation of free radicals may be the common mechanism that mediates damage to both[2]. If so, it was not a coincidence that G6PD deficiency was discovered via AHA triggered by primaquine: the red cells were "primaquine-sensitive" for the same reason *Plasmodia* are sensitive to primaquine. In addition, genetically determined G6PD deficiency has been selected by malaria, and one can ask whether this too may not be a coincidence (see Figure 14-4). It is now clear (see above) that G6PD-deficient red cells are relatively poor hosts to *P. falciparum* not because the parasite cannot enter them, but because infection itself produces some changes in the host cells that makes them liable to being selectively phagocytosed. It is not farfetched to surmise that such changes may result from oxidative damage imposed by the parasite onto the red cell. In brief, we can speculate that primaquine can cause oxidative damage to G6PD-deficient red cells as it does to parasites, whereas in a G6PD-deficient red cell, oxidative damage suffered by the parasite eventually brings about its demise—thus completing the triangle consisting of the parasite, the G6PD-deficient red cell, and primaquine.

2. In support of this notion, several 8-aminoquinoline antimalarials related to primaquine are still hemolytic, as though within this class of compounds the two activities cannot be separated. Of course there are other antimalarials that are not hemolytic for G6PD-deficient red cells: but in such cases we know, or we can presume, that the mechanism of the antimalarial action is different.

GENOTYPE-PHENOTYPE CORRELATIONS

There are some 150 known *G6PD* mutations causing G6PD deficiency (see Table 14-3): with a few exceptions they are single-point mutations in the coding region of the gene, entailing single amino acid replacements in the G6PD protein (Beutler et al., 2002; Vulliamy et al., 1993). The exceptions are small deletions of one to eight amino acids; one mutation causing abnormal splicing, and a few instances in which two point mutations rather than one are present (for instance, in G6PD A-, the variant most commonly encountered in Africa). Regulatory mutations have not yet been discovered.

Amino acid replacements can cause G6PD deficiency either by affecting its catalytic function [e.g., in G6PD Orissa (Kaeda et al., 1995), G6PD Union, G6PD Andalus (Wang et al., 2005)] or by decreasing the in-vivo stability of the protein, or by both of these mechanisms. Enzyme instability is the most common mechanism: this is consistent with the fact that G6PD deficiency is mostly expressed in red cells, where stability matters most, on

TABLE 14-3 CLASSIFICATION OF 154 MUTATIONS OF THE G6PD GENE

Type of mutation		Number of known mutants	Comments
Missense point mutations	Single	131	At least 35 of these are polymorphic
	Two	11	6 of these share the 126Asn->Asp mutation found on its own in the polymorphic non-deficient G6PD A
	Three	2	
In frame deletions	1 amino acid	6	None of these are polymorphic; all are associated with CNSHA
	2 amino acids	1	
	8 amino acids	1	
Affecting splicing		2	

Original references in Mason et al. (2007).

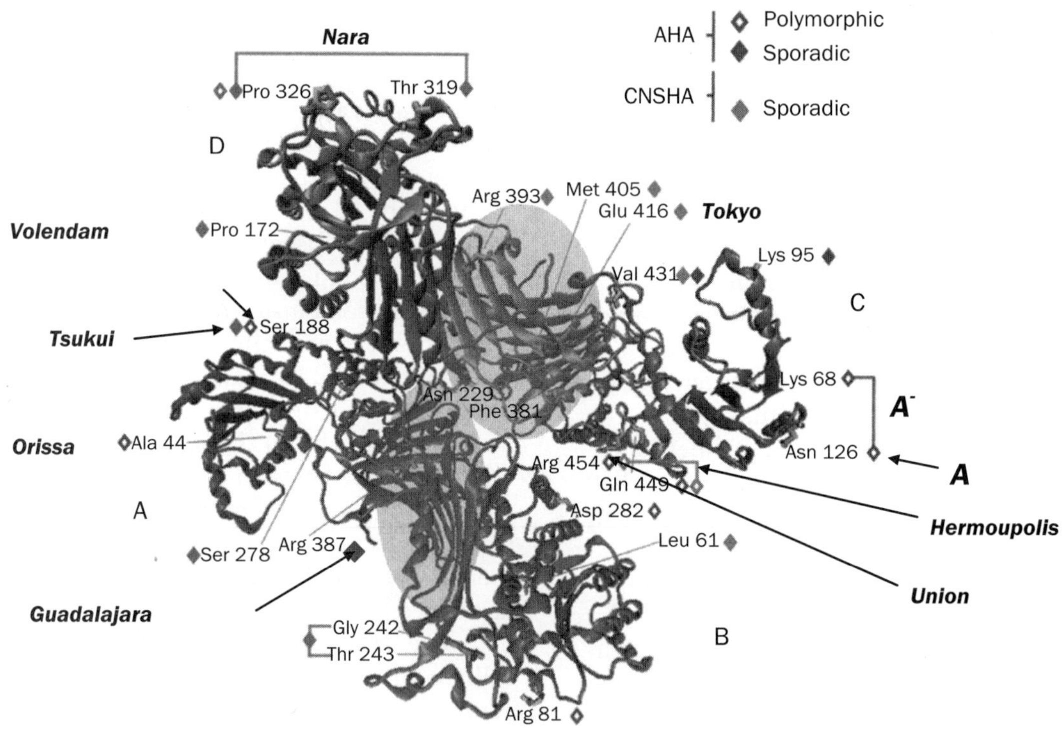

Figure 14-5 A three-dimensional view of G6PD. The G6PD tetramer is seen to be a dimer of a dimer of identical subunits (indicated by capital letters A–D). The two subunits forming a dimer are tightly held together by non-covalent bonding (note the numerous beta sheets at the interface between the two monomers, highlighted by gray background), and the dimer structure is stabilized by two molecules of "structural" NADP buried in the structure (not shown). The binding between the two dimers is more loose: indeed the tetramer can reversibly dissociate into two dimers without loss of activity (it is not known whether the dimer or the tetramer is the predominant form within the cell). A few of the 154 mutations are shown, with some of their names (in italics). Note numerous mutations causing the severe phenotype of CNSHA in the monomer–monomer interface within each dimer. The largest deletion (8 amino acids) of G6PD Nara is in an unstructured loop. The subtlety of genotype–phenotype correlations is illustrated, as an example, by the mutation 126Asn-> Asp in G6PD A, which has normal activity, versus G6PD A-, causing enzyme deficiency, when the same mutation is combined, for instance, with the 68Met-> Val mutation. (Modified from Mason, Bautista & Gilsanz, 2007.) Refer color figure.

account of the long life span of these cells after they have lost protein synthesis (Mason, 2007). By contrast, mutations affecting catalytic function would compromise G6PD function in all cells, potentially causing not just drug-induced hemolytic anemia, but more serious consequences that malaria selection may not have been sufficient to balance.

The molecular basis of CNSHA associated with G6PD deficiency is highly specific (see Figure 14-5), in the sense that the underlying mutations, while still within the coding region of the G6PD gene, are not the same as those underlying asymptomatic G6PD deficiency. The more severe clinical phenotype can be ascribed in some cases to adverse qualitative changes (for instance, a decreased affinity for the substrate, glucose 6-phosphate), or simply to the fact that the enzyme deficit is more extreme because of severe instability of the enzyme. For instance, a cluster of mutations map to the dimer interface, and clearly they compromise severely the formation of the dimer (Naylor et al., 1996). Tightly bound structural NADP has emerged as an important contributor to the long-term stability of G6PD (Wang et al., 2008), which in turn is crucially important in red cells, since they must survive on the G6PD endowment they have when they lose their ribosomes.

MANAGEMENT

The management of clinical manifestations of G6PD deficiency is covered in detail elsewhere (Luzzatto, 2009). Here, it seems important to mention that drug-induced AHA is preventable in most cases by choosing alternative drugs. Favism is in principle entirely preventable by G6PD-deficient people not eating fava beans: of course this implies population screening, which would be useful also in newborns in relation to NNJ.

ACKNOWLEDGMENTS

I wish to thank all colleagues with whom we worked on G6PD and on malaria, and particularly: Adeyinka Afolayan, Olaniyi Babalola, Giorgio Battistuzzi, Ulrich Bienzle, Michele D'Urso, Fola Esan, Felix Fasuan, Jaspal Kaeda, Buran Kurdi-Haidar, Cristina Mareni, Giuseppe Martini, PhilipMason, Guido Modiano, Estella O'Brien, Rosario Notaro, Gabriel Ogunmola, Lucia Perroni, Graziella Persico, Enzo Poggi, Roberta Rivi, Olugbemiro Sodeinde, Daniela Toniolo, Margaret Town, Essien Usanga, Tom Vulliamy. I am especially grateful to Herbert Gilles and to Adetokunbo Lucas, who inspired me and encouraged me in malaria research. Financial support was received from the World Health Organization, the National Institutes of Health (USA) and the Consiglio Nazionale delle Ricerche (Italy).

REFERENCES

Adam A. (1961). Linkage between deficiency of glucose 6-phosphate dehydrogenase and colour-blindness. *Nature 189,* 686–688.

Allison AC. (1960). Glucose 6-phosphate dehydrogenase deficiency in red blood cells of East Africans. *Nature 186,* 531–532.

Alloueche A, Bailey W, Barton S, et al. (2004). Comparison of chlorproguanil-dapsone with sulfadoxine-pyrimethamine for the treatment of uncomplicated falciparum malaria in young African children: double-blind randomised controlled trial. *Lancet 363,* 1843–1848.

Arese P, Turrini F, Schwarzer E. (2005). Band 3/complement-mediated recognition and removal of normally senescent and pathological human erythrocytes. *Cell Physiol Biochem 16,* 133–146.

Au SW, Gover S, Lam VM, Adams MJ. (2000). Human glucose-6-phosphate dehydrogenase: the crystal structure reveals a structural NADP(+) molecule and provides insights into enzyme deficiency [In Process Citation]. *Structure Fold Des 8,* 293–303.

Ayi K, Turrini F, Piga A, Arese P. (2004). Enhanced phagocytosis of ring-parasitized mutant erythrocytes: a common mechanism that may explain protection against falciparum malaria in sickle trait and beta-thalassemia trait. *Blood 104,* 3364–3371.

Baker MA, Bosia A, Pescarmona G, Turrini F, Arese P. (1984). Mechanism of action of divicine in a cell-free system and in glucose-6-phosphate dehydrogenase-deficient red cells. *Toxicol Pathol 12,* 331–336.

Battistuzzi G, D'Urso M, Toniolo D, Persico GM, Luzzatto L. (1985). Tissue-specific levels of human glucose-6-phosphate dehydrogenase correlate with methylation of specific sites at the 3' end of the gene. *Proc Natl Acad Sci USA 82,* 1465–1469.

Berg J, Tymoczko JL, Stryer L. (2002). *Biochemistry.* New York: Freeman.

Betke K, Brewer GJ, Kirkman HN, et al. (1967). Standardization of procedures for the study of glucose-6-phosphate dehydrogenase. *WHO Tech Rep Ser 366,* 53p.

Beutler E. (1959) The hemolytic effect of primaquine and related compounds. *Blood 14,* 103–139.

Beutler E. (1978). Glucose 6-phosphate dehydrogenase deficiency. In: *Hemolytic anemia in disorders of red cell metabolism* (ed. by Beutler, E.), pp. 23–167. New York: Plenum Medical Book Company.

Beutler E, Kuhl W, Ramirez E, Lisker R. (1991). Some Mexican glucose-6-phosphate dehydrogenase (G-6-PD) variants revisited. *Hum Genet 86,* 371–374.

Beutler E, Gaetani G, Der Kaloustian V, et al. (1985). Glucose-6-phosphate dehydrogenase deficiency (Report of a WHO working group, Geneva 3–4 September). *WHO.*

Beutler E, Vulliamy TJ, Williams TN, et al. (2002). Hematologically important mutations: glucose-6-phosphate dehydrogenase. *Blood Cells Mol Dis 28,* 93–103.

Bienzle U, Ayeni O, Lucas AO, Luzzatto L. (1972a). Glucose-6-phosphate dehydrogenase deficiency and malaria. Greater resistance of females heterozygous for enzyme deficiency and of males with non-deficient variant. *Lancet i,* 107–110.

Bienzle U, Effiong C, Luzzatto L. (1976). Erythrocyte glucose 6-phosphate dehydrogenase deficiency (G6PD type A-) and neonatal jaundice. *Acta Paediatr Scand 65,* 701–703.

Bienzle U, Okoye VC, Gogler H. (1972b). Haemoglobin and glucose-6-phosphate dehydrogenase variants: distribution in relation to malaria endemicity in a Togolese population. *Z Tropenmed Parasitol 23,* 56–62.

Browning LA, Kruse JA. (2005). Hemolysis and methemoglobinemia secondary to rasburicase administration. *Ann Pharmacother 39,* 1932–1935.

Cappadoro M, Giribaldi G, O'Brien E, et al. (1998) Early phagocytosis of glucose-6phosphate dehydrogenase (G6PD)-deficient erythrocytes parasitized by *Plasmodium falciparum* may explain malaria protection in G6PD deficiency. *Blood 92,* 2527–2534.

Cappellini MD, Fiorelli G. (2008). Glucose-6-phosphate dehydrogenase deficiency. *Lancet 371,* 64–74.

Carson PE, Flanagan CL, Ickes CE, Alving A. (1956). Enzymatic deficiency in primaquine-sensitive erythrocytes. *Science 124,* 484–485.

Cavalli-Sforza LL, Bodmer WF. (1971). *The genetics of human populations.* Freeman, San Francisco.

Chevion M, Navok T, Glaser G, Mager J. (1982). The chemistry of favism-inducing compounds. The properties of isouramil and divicine and their reaction with glutathione. *Eur J Biochem 127,* 405–409.

Clark TG, Fry AE, Auburn S, et al. (2009). Allelic heterogeneity of G6PD deficiency in West Africa and severe malaria susceptibility. *Eur J Hum Genet 17,* 1080–1085.

Dacie JV. (1985). Hereditary enzyme deficiency haemolytic anaemias. III: Deficiency of glucose-6-phosphate dehydrogenase. In: *Haemolytic Anaemias. The Hereditary Haemolytic Anaemias.* (ed. by Dacie, J.V.), Vol. 3, pp. 364–418. Churchill Livingstone, London.

Dacie JV, Lewis SM. (1995). *Practical Haematology.* Churchill Livingstone, London.

De Flora A, Morelli A, Giuliano F. (1974). Human erythrocyte G6PD. Content of bound coenzyme. *BBRC 59,* 406–413.

Doxiadis SA, Fessas P, Valaes T. (1961). Glucose-6-phosphate dehydrogenase deficiency. A new aetiological factor of severe neonatal jaundice. *Lancet i,* 297–301.

Eppes RB, Brewer GJ, De Gowin RL, et al. (1966). Oral glucose tolerance in Negro men deficient in G6PD. *New Engl J Med 275,* 855–850.

Felix K, Rockwood LD, Pretsch W, et al. (2002). Moderate G6PD deficiency increases mutation rates in the brain of mice. *Free Radic Biol Med 32,* 663–673.

Foller M, Bobbala D, Koka S, Huber SM, Gulbins E, Lang F. (2009). Suicide for survival - death of infected erythrocytes as a host mechanism to survive malaria. *Cell Physiol Biochem 24,* 133–140.

Franze A, Ferrante MI, Fusco F, et al. (1998). Molecular anatomy of the human glucose 6-phosphate dehydrogenase core promoter. *FEBS Lett 437,* 313–318.

Friedman MJ. (1979). Oxidant damage mediates variant red cell resistance to malaria. *Nature 280,* 245.

Gaetani GF, Galiano S, Canepa L, Ferraris A-M, Kirkman HN. (1989). Catalase and glutathione peroxidase are equally active in detoxification of hydrogen peroxide in human erythrocytes. *Blood 73,* 334–339.

Ganczakowski M, Town M, Bowden DK, et al. (1995). Multiple glucose 6-phosphate dehydrogenase-deficient variants correlate with malaria endemicity in the Vanuatu archipelago (southwestern Pacific). *Am J Hum Genet 56,* 294–301.

Gilles HM, Fletcher KA, Hendrickse RG, Lindner R, Reddu S, Allan N. (1967). Glucose-6-phosphate dehydrogenase deficiency, sickling, and malaria in African children in South Western Nigeria. *Lancet 1,* 138–130.

Haldane JBS. (1949). Disease and evolution. *Ricerca Sci 19*(Suppl. I), 68–76.

Jalloh A, Tantular IS, Pusarawati S, et al. (2004). Rapid epidemiologic assessment of glucose-6-phosphate dehydrogenase deficiency in malaria-endemic areas in Southeast Asia using a novel diagnostic kit. *Trop Med Int Health 9,* 615–623.

Johnson M, Clark TD, Njama-Meya D, Rosenthal PJ, Parikh S. (2009). Impact of the method of G6PD deficiency assessment on genetic association studies of malaria susceptibility. *PLOS One 4,* e7246.

Kaeda JS, Chhotray GP, Ranjit MR, et al. (1995). A new glucose-6-phosphate dehydrogenase variant, G6PD Orissa (44 Ala-->Gly), is the major polymorphic variant in tribal populations in India. *Am J Hum Genet 57,* 1335–1341.

Kaplan M, Hammerman C. (2010) Glucose 6-phosphate dehydrogenase deficiency and severe neonatal hyperbilirubinemia: a complexity of interactions between genes and environment. *Semin Fetal Neonatal Med 15,* 148–156.

Kirkman HN, Gaetani GD, Clemons EH, Mareni C. (1975). Red cell NADP and NADPH in glucose-6-phosphate dehydrogenase deficiency. *J Clin Invest 55* 875–880.

Kirkman HN, Gaetani GF. (1984). Catalase: a tetrameric enzyme with four tightly bound molecules of NADPH. *Proc Natl Acad Sci USA 81,* 4343–4347.

Lehninger A, Nelson DL, Cox MM. (2004) *Principles of Biochemistry.* Garland, New York.

Longo L, Vanegas OC, Patel M, et al. (2002). Maternally transmitted severe glucose 6-phosphate dehydrogenase deficiency is an embryonic lethal. *EMBO J 21,* 4229–4239.

Luisada L. (1941). Favism: a singular disease affecting chiefly red blood cells. *Medicine 20,* 229–250.

Luzzatto L. (1967). Regulation of the activity of glucose-6-phosphate dehydrogenase by $NADP^+$ and NADPH. *Biochim Biophys Acta 146,* 18–25.

Luzzatto L. (2010). The rise and fall of the antimalarial Lapdap: a lesson in pharmacogenetics. *Lancet 376,* 739–741.

Luzzatto L, Mehta A. (1995). Glucose 6-phosphate dehydrogenase deficiency. In: Scriver CR, Beaudet AL, Sly WS, Valle D, eds. *The metabolic and molecular bases of inherited disease* Vol. 7, pp. 3367–3398. McGraw-Hill, New York.

Luzzatto L, Mehta A, Vulliamy T. (2001). Glucose-6-phosphate dehydrogenase deficiency. In: Scriver C, Beaudet A, Sly W, Valle D, eds. *The Metabolic & Molecular Bases of Inherited Disease,* Vol. 3, pp. 4517–4553. New York: McGraw Hill.

Luzzatto L, Notaro R. (2001). Malaria. Protecting against bad air. *Science 293,* 442–443.

Luzzatto L, Poggi V. (2009) Glucose 6-phosphate dehydrogenase deficiency. In: Orkin S, Fisher DE, Look TA, Lux S, Ginsburg D, Nathan DG, eds. *Nathan & Oski's Hematology of Infancy and Childhood,* Elsevier, New York.

Luzzatto L, Sodeinde O, Martini G. (1983). Genetic variation in the host and adaptive phenomena in *Plasmodium falciparum* infection. *Ciba Found Symp 94,* 159–173.

Luzzatto L, Usanga EA, Reddy S. (1969). Glucose 6-phosphate dehydrogenase deficient red cells: resistance to infection by malarial parasites. *Science 164,* 839–842.

Martin-DeLeon PA, Wolf SF, Persico G, Toniolo D, Martini G, Migeon BR. (1985). Localization of glucose 6-phosphate dehydrogenase in mouse and man by in situ hybridization: evidence for a single locus and transposition of homologous X-linked genes. *Cytogenet Cell Genet 39,* 87–92.

Martini G, Toniolo D, Vulliamy T, et al. (1986). Structural analysis of the X-linked gene encoding human glucose 6-phosphate dehydrogenase. *EMBO J 5,* 1849–1855.

Mason PJ, Bautista JM, Gilsanz F. (2007) G6PD deficiency: the genotype-phenotype association. *Blood Rev 21,* 267–283.

Mason PJ, Sonati MF, MacDonald D, et al. (1995). New glucose 6-phosphate dehydrogenase mutations associated with chronic anemia. *Blood 85,* 1377–1380.

Meissner PE, Coulibaly B., Mandi G, et al. (2005). Diagnosis of red cell G6PD deficiency in rural Burkina Faso: comparison of a rapid fluorescent enzyme test on filter paper with polymerase chain reaction based genotyping. *Br J Haematol 131,* 395–399.

Meloni T, Forteleoni G, Dore A, Cutillo S. (1983). Favism and hemolytic anemia in glucose-6-phosphate dehydrogeanse deficiency subjects in North Sardinia. *Acta Haematol 70,* 83–80.

Motulsky AG. (1960). Metabolic polymorphisms and the role of infectious diseases in human evolution. *Hum Biol 32,* 28–62.

Nagel RL, Roth EF Jr. (1989). Malaria and red cell genetic defects. *Blood 74,* 1213–1221.

Naylor CE, Rowland P, Basak AK, et al. (1996). Glucose 6-phosphate dehydrogenase mutations causing enzyme deficiency in a model of the tertiary structure of the human enzyme. *Blood 87,* 2974–2982.

Neifer S, Jung A, Bienzle U. (1991). Characterization of erythrocytic glucose-6-phosphate dehydrogenase in a mouse strain with reduced G6PD activity. *Biomed Biochim Acta 3,* 233–238.

Nicol CJ, Zielenski J, Tsui LC, Wells PG. (2000). An embryoprotective role for glucose-6-phosphate dehydrogenase in developmental oxidative stress and chemical teratogenesis. *FASEB J 14,* 111–127.

Normah J, Choo KE, Oppenheimer SJ, Selamah G. (1991). Glucose-6-phosphate dehydrogenase enzyme activity in normal, hemizygote and heterozygote Kelantanese Malays. *J Paediat Child Health 27,* 376–379.

Notaro R, Afolayan A, Luzzatto L. (2000). Human mutations in glucose 6-phosphate dehydrogenase reflect evolutionary history. *FASEB J 14,* 485–494.

O'Brien E, Kurdi-Haidar B, Wanachiwanawin W, et al. (1994). Cloning of the glucose 6-phosphate dehydrogenase gene from *Plasmodium falciparum. Mol Biochem Parasitol 64,* 313–326.

Pandolfi PP, Sonati F, Rivi R, Mason P, Grosveld F, Luzzatto L. (1995). Targeted disruption of the housekeeping gene encoding glucose 6-phosphate dehydrogenase (G6PD): G6PD is dispensable for pentose synthesis but essential for defense against oxidative stress. *EMBO J 14,* 5209–5215.

Rinaldi A, Filippi G, Siniscalco M. (1976). Variability of red cell phenotypes between and within individuals in an unbiased sample of 77 certain heterozygotes for G6PD deficiency in Sardinians. *Am J Hum Genet 28,* 496–505.

Roos D, van Zwieten R, Wijnen JT, et al. (1999). Molecular basis and enzymatic properties of glucose 6-phosphate dehydrogenase volendam, leading to chronic nonspherocytic anemia, granulocyte dysfunction, and increased susceptibility to infections. *Blood 94,* 2955–2962.

Roth EF Jr, Raventos-Suarez C, Rinaldi A, Nagel RL. (1983a). Glucose-6-phosphate dehydrogenase deficiency inhibits in vitro growth of *Plasmodium falciparum. Proc Natl Acad Sci USA 80,* 298–299.

Roth EF Jr, Raventos Suarez, C, Rinaldi, A, Nagel RL. (1983b). The effect of X chromosome inactivation on the inhibition of *Plasmodium falciparum* malaria growth by glucose-6-phosphate-dehydrogenase-deficient red cells. *Blood 62,* 866–868.

Ruwende C, Khoo SC, Snow RW, et al. (1995a). Natural selection of hemi- and heterozygotes for G6PD deficiency in Africa by resistance to severe malaria. *Nature 376,* 246–249.

Ruwende C, Khoo SC, Snow RW, et al. (1995b). Natural selection of hemi- and heterozygotes for G6PD deficiency in Africa by resistance to severe malaria. *Nature 376,* 246–249.

Sansone G, Segni G. (1958). Nuovi aspetti dell'alterato biochimismo degli eritrociti dei favici: assenza pressoche` completa della glucos-6-P deidrogenasi. *Boll Soc Ital Biol Sper 34,* 327–329.

Shahabuddin M, Rawlings DJ, Kaslow DC. (1994). A novel glucose-6 -phosphate dehydrogenase in *Plasmodium falciparum*: cDNA and primary protein structure. *Biochim Biophys Acta 1219,* 191–194.

Siniscalco M, Bernini L, Filippi G, et al. (1966). Population genetics of haemoglobin variants, thalassaemia and glucose-6-phosphate dehydrogenase deficiency, with particular reference to the malaria hypothesis. *Bull WHO 34,* 379–393.

Sodeinde O, Clarke JL, Vulliamy TJ, Luzzatto L, Mason PJ. (2003). Expression of *Plasmodium falciparum* G6PD-6PGL in laboratory parasites and in patient isolates in G6PD-deficient and normal Nigerian children. *Br J Haematol 122,* 662–668.

Sulo J, Chimpeni P, Hatcher J, et al. (2002). Chlorproguanil-dapsone versus sulfadoxine-pyrimethamine for sequential episodes of uncomplicated falciparum malaria in Kenya and Malawi: a randomised clinical trial. *Lancet 360,* 1136–1143.

Tishkoff SA, Varkonyi R, Cahinhinan N, et al. (2001). Haplotype diversity and linkage disequilibrium at human G6PD: recent origin of alleles that confer malarial resistance. *Science 293,* 455–462.

Trager W, Jensen JB. (1976). Human malaria parasites in continuous culture. *Science 193,* 673–675.

Tugwell P. (1973). Glucose 6-phosphate dehydrogenase deficiency in Nigerians with jaundice associated with lobar pneumonia. *Lancet i,* 968–970.

Usanga EA, Luzzatto L. (1985). Adaptation of *Plasmodium falciparum* to glucose 6-phosphate dehydrogenase-deficient host red cells by production of parasite-encoded enzyme. *Nature 313,* 793–795.

Vogel F, Motulsky AG. (1996) *Human Genetics: Problems and Approaches.* Berlin: Springer.

Vulliamy T, Beutler E, Luzzatto L. (1993). Variants of glucose 6-phosphate dehydrogenase are due to missense mutations spread throughout the coding region of the gene. *Hum Mutat J1 - HM 2,* 159–167.

Wang X, Chan TF, Lam VMS, Engel PC. (2008). What is the role of the second "structural" NADP+-binding site in human glucose 6-phosphate dehydrogenase? *Protein Science 17,* 1403–1411.

Wang X, Lam VM, Engel PC. (2005). Marked decrease in specific activity contributes to disease phenotype in two human glucose 6-phosphate dehydrogenase mutants, G6PD(Union) and G6PD(Andalus). *Hum Mutat 26,* 284.

Warburg O, Christian W. (1931). šber Aktivierung der robinsonschen Hexosemono-phosphors "ure in roten Blutzellen und die Gewinnung aktivierender Fermentl" sung. *Biochem Z 242,* 206–227.

Youngster I, Arcavi L, Schechmaster R, et al. (2010). Medications and glucose-6-phosphate dehydrogenase deficiency: an evidence-based review. *Drug Saf 33,* 713–726.

15.

INFECTIONS, GENOMICS AND GLOBAL PUBLIC HEALTH

Tikki Pang

INTRODUCTION

Unprecedented advances have been made in the advanced genomic sciences with regard to new knowledge and powerful technologies related to infectious diseases. However, a major challenge is to ensure that these advances have a positive impact on public health and policies, especially in the developing world. The present chapter will identify the major infectious disease threats, the relevant technologies, the barriers and hurdles that need to be overcome, and the policy responses that need to be instituted to successfully meet the challenges.

INFECTIOUS DISEASE THREATS

Developing countries continue to struggle with emerging and reemerging infectious diseases. Recent outbreaks of cholera (in Zimbabwe), meningitis and Ebola (in Africa), chikungunya (in Southeast Asia, India, Sri Lanka, Kenya), foot-and-mouth disease (in China), and dengue (in Bolivia, Southeast Asia) underscore the continuing threat in many parts of the world. Avian influenza remains a major public health threat and, based on studies of the 1918 global flu pandemic, it has been estimated that a recurrence of such a pandemic will kill between 50 and 80 million people worldwide, 95% of them in the developing world (Murray et al., 2006), at an estimated economic cost of $3 trillion. In addition, the threat of new pathogens emerging was illustrated by the appearance of a new arenavirus in South Africa in 2008, which killed four out of five people it infected (Keeton, 2008). In addition, there are continued concerns about the possibility and risk of pathogens crossing the animal-to-human transmission barrier (Normile, 2009). Antibiotic resistance constitutes another major public health threat (Taubes, 2008) and includes MDR (multidrug resistant) and XDR (extensively drug-resistant) tuberculosis, MRSA (methicillin-resistant *Staphylococcus aureus*), vancomycin-resistant Enterococcus, and MDR *Salmonella typhi*. In addition, there are concerns with the recent appearance of resistance to artemisinin-combined therapy for malaria on the Thailand–Cambodia border, and emergence of resistance among influenza viruses to the antiviral agent oseltamivir (tamiflu). It has been estimated that by 2009, 98% of circulating influenza A/H1N1 virus strains in North America had become resistant to oseltamivir (Layne et al., 2009) and there are worrying reports of resistance developing among the much more virulent H5N1 strain of influenza. In a recent analysis of emerging and reemerging infectious disease threats globally, Jones et al. (2008) showed that approximately 60% of these threats came from zoonoses or wildlife origins (e.g., wild birds in the case of avian influenza, bats in the case of Nipah virus, etc.), 20% were vector-borne agents, and 20% were due to antibiotic-resistant bacteria. "Hot spots" of emergence of these threats were almost exclusively located in the developing world; i.e., in tropical Asia, central America, and tropical Africa. Importantly, and ironically, resources for surveillance and investigation of these threats are focused in areas where the next emerging pathogen is least likely to come from; i.e., in the developed world. Another important threat exists in the context of the current global financial crisis (Horton, 2009), and the reality that the health budget is often the first to be cut in developing countries in times of scarce resources. It is also likely that overseas development aid, which is relied upon by many low-income countries for a significant portion of their health budgets, may also be reduced in the coming years. This may further weaken many healthcare delivery systems in low- and middle-income countries that are already in a weak and fragile state, thus affecting their ability to effectively respond to infectious disease threats. xxx

TECHNOLOGIES DERIVED FROM GENOMICS

The genomics and post-genomics era has been accompanied by an impressive array of new and powerful technologies which have a direct impact on the diagnosis, control, and prevention of infectious diseases. An analysis of the top ten biotechnologies for improving health in developing countries identified diagnostics, vaccines, and drug and vaccine delivery as the top three technologies (Daar et al., 2002). In broader terms, the list of key technologies includes the following:

1. The rapid identification of pathogens—this is the cornerstone of surveillance and detection of new

or reemerging pathogens, and relies on the rapid sequencing of pathogen sequences followed by comprehensive, comparative analysis with genomic sequences present in openly accessible databases.

2. Rapid, cheap, easy to use point-of-care diagnostics—ideally, these tests should be able to detect multiple pathogens and be usable in least-developed settings. Better tests for TB and malaria, for example, can have a significant impact on the burden of disease in affected areas (Mabey et al., 2004).
3. Better, more effective, easily deployable vaccines—beyond the traditional vaccines used in the EPI (Expanded Programme on Immunization) there is also a need to better utilize existing vaccines (e.g., against typhoid, pneumonia) and develop more effective vaccines in the future (e.g., against TB, malaria, HIV/AIDS).
4. Genomics in drug discovery—pathogen genome sequences, combined with other approaches such as access to chemical libraries and molecular-target, high-throughput screening, are the starting points of many drug discovery and vaccine development initiatives directed at neglected diseases affecting developing countries; e.g., the African Network for Drugs and Diagnostics Innovation (ANDI) and Aeras, which aims to develop new, safe and effective vaccine regimens to prevent tuberculosis in children, adolescents, and adults.
5. Applications to vector control—the recent availability of complete genome sequences for some major disease vectors (e.g., *Aedes* spp., *Anopheles* spp.) will see some important future applications for the control of these disease-transmitting mosquitoes, including strategies to reduce or eliminate their competence as vectors of disease (Povelones et al., 2009).
6. Monitoring resistance (viral, bacterial, protozoal)—molecular markers for development of resistance to antimicrobial agents, rapidly deployed together with diagnostic tests, is critical for surveillance and appropriate therapeutic responses.
7. Translating new knowledge of host–pathogen interactions into better interventions—there is an increasingly important body of knowledge that links infections with their immune and inflammatory responses, including the genetic susceptibility to infectious diseases. Advances in this field have important implications for the development of therapeutics and diagnostics in the future. Pharmacogenomics and personalized medicine, for example, promises to improve drug safety and efficacy (McGuire et al., 2007; Katsanis et al., 2008). A recent study by Nijnik and Hancock (2009) described the antimicrobial and immunomodulatory activities of host defense peptides and their potential applications for tackling antibiotic-resistant infections.

With regard to future technologies, it is also useful to consider areas where technology development is needed for dealing with infectious disease threats. A "window" into what might constitute these priority research areas can be gleaned from the recently announced U.S. NIH Challenge Grants in Health and Science Research (National Institutes of Health, 2009) which highlighted the following as some of the priority areas:

1. Development of biomedical technologies and systems, including the provision of "immediate diagnostic information for multiple conditions at the point of care."
2. New computational and statistical methods for the analysis of large data sets from next-generation sequencing technologies.
3. *Theranostics*—the combined delivery of diagnostic and therapeutic agents.
4. Development of diagnostics and drugs for multiple- and extensively drug-resistant tuberculosis.
5. Development of drugs for neglected tropical diseases, with a special emphasis on malaria.
6. Novel approaches to improve immunogenicity of vaccines against small molecules.

It is also important to emphasize that genomics-derived technologies alone are not sufficient to deal with infectious disease threats. A "Foresight Report" from the United Kingdom has identified that, in addition to genomics and postgenomic approaches for rapid characterization of pathogens, it is also important to consider innovative IT (information technology) applications to capture, analyze and model data (e.g., the use of handheld devices); non-invasive mass screening of people, animals, and plants; and chip-based portable diagnostic devices (King et al., 2006).

BARRIERS AND CHALLENGES

The effective application of the technologies mentioned above faces some key barriers and challenges, which occur at three stages: from the research to the products, from the products to individual patients, and, finally, from patients to entire populations.

BARRIER 1—THE R&D BARRIER: FROM PUBLICATIONS TO PRODUCTS

The central question here is: Are we doing the right science to get the appropriate knowledge and interventions with regard to both the areas and the types of research? It has been estimated that approximately $160 billion was invested in global health research and development globally, compared to $85 billion in 1998 and $30 billion in 1986 (Burke & Matlin, 2008). Despite this influx of funds, gaps and imbalances continue to exist in the research process. The existence of the "10/90 gap", for example, is well known, where only 10% of global health research funding is spent on 90% of the global disease burden. There also exists a well-known market failure in producing drugs for neglected tropical diseases in the developing world—of 1,556 new drugs developed between 1975–2004, only 21 (1.3%) were for these diseases (Chirac & Torreele, 2006). In terms of support for various research areas, a recent report analyzing funding patterns among more than 100 donors found that 80% of the funding was allocated to HIV/AIDS, TB, and malaria (Moran et al., 2009). In another analysis of the types of research funded by the major U.K. agencies, it was found that an average of only 1% of funding was allocated to health services research (Rothwell, 2006) with the bulk of the funding going to biomedical, clinical, and epidemiological research.

More tellingly, most of the funding for health research is focused on the development of new technologies, rather than making use of existing interventions—in one analysis, 97% of grants were for the development of these new technologies, which was estimated to be able to reduce child mortality by 22% (Leroy et al., 2007). However, if instead, research were focused on making better use of existing technologies, the reduction in mortality can be much higher at 66%. In the future, a lot more attention needs to be paid to this field of implementation research.

BARRIER 2—THE TRANSLATION BARRIER: FROM PRODUCTS TO PATIENTS

In the context of this barrier, the key question is: Do we have the most efficient processes to ensure that effective products benefit the patients? In spite of the large amounts of resources going into global health R&D, the number of new drugs in the pipeline is actually decreasing, as exemplified by the lower number of new antibiotics coming on the market. Clinical trials, the cornerstone of getting drugs and other interventions to patients, are facing major problems with regard to high costs, difficulties in recruitment of trial participants, more regulations, and concerns about transparency and accountability. These concerns include unethical behavior on the part of pharmaceutical companies, publication bias, and the nonreporting of adverse events or negative results. Importantly, the developing countries will be affected, as it has been estimated that nearly 20% of clinical trials globally now take place in the developing world, up from in 9% in 2003 (Normile, 2008). Most of the increases are occurring in Russia, India, China, and Brazil. The main reasons for increasing interest in conducting trials in the developing world are related to lower costs, accelerated recruitment, and perceived ease of overcoming ethical and regulatory barriers.

There is also some concern about the overregulation of clinical trials, specifically the requirements to adhere to the ICH-GCP guidelines, which place too much emphasis on the process rather than on the benefits to the patients (White, 2006; Farrar, 2007). These bureaucratic requirements also place clinical trials beyond the reach of most developing country investigators, and disempower them from truly participating in the research and sharing in the benefits. Inefficiencies have also been noted in the diagnostics field, and an important question in this regard is: What are the human and economic costs of bad diagnostics? Many developing countries lack stringent evaluation of diagnostic tests and, in the case of dengue diagnostics, for example, the claims of manufacturers were found to be well short of acceptable standards (Blacksell et al., 2006). In a more comprehensive analysis of diagnostic evaluations reported in journals from 1978–1993, it was found that less than half of the studies fulfilled more than three of the seven methodological standards for diagnostics (Peeling et al., 2006).

BARRIER 3—THE IMPLEMENTATION BARRIER: FROM PATIENTS TO PEOPLE

In some ways this is the most challenging of the three barriers and focuses on the question: Do we have the needed delivery systems to get interventions to those in greatest need? Health systems failures and lack of access to the most basic interventions are at the heart of the problem. For example, Jones et al. (2003) found that less than 50% of children in the developing world were receiving basic, proven, effective and cheap interventions such as vitamin A, tetanus immunization, and insecticide-treated bednets. Tragically, the gap in coverage for four interventions (family planning, maternal and neonatal care, immunization, and treatment of sick children) was larger for the poorer segments of the population living in these countries (WHO, 2008). There are many reasons for these failures of delivery, but foremost among them are the shortage of health workers, lack of reliable health information, insufficient financing, and poor infrastructure. In relation to health workers, for example, Africa, which bears 25% of the global burden of disease, only has 2%–3% of the global health work force (WHO, 2006). In addition, in many low-income countries the recent deluge of external aid delivered through multiple global health initiatives has created additional problems and strains on already weak and fragile health systems. There is evidence of fragmentation and lack of coordination between these initiatives, and they tend to be "top down" and donor-driven, with a focus on short-term results around "big" diseases, thus creating vertical "silos" rather than overall health system strengthening. In Haiti, for example, a focus on HIV/AIDS testing and treatment resulted in the neglect of testing for syphilis, with the result that children born to HIV-positive mothers were avoiding HIV/AIDS but dying of syphilis (Peeling et al., 2004). With many of these initiatives there is also a lack of evaluation of the impact of their activities and limited accountability mechanisms, thus bringing into question their longer-term sustainability.

In the context of this third barrier, another important challenge is the weak linkages that exist between research and policy development. Policymakers and researchers have conflicting interests and often do not trust each other (WHO, 2004). Importantly, researchers need to present science in a way that is understandable to policymakers, and able to answer the three questions they are most concerned with: can it work, will it work, is it worth it? As stated previously, in a quote attributed to Sir Michael Marmot, the reality is that "scientific findings do not fall on blank minds that get made up as a result. Science engages with busy minds that have strong views about how things are and ought to be," and it is therefore important that researchers and policymakers work together, perhaps through "knowledge brokerage" mechanisms to more effectively use scientific evidence in health policy development. In the context of application of technological advances derived from genomics, this is particularly important, as the science is often complex and the benefits not immediately obvious to the policy- and decision makers.

CONCLUSIONS

Genomics has made great strides in the past decade and its importance for the diagnosis, treatment, prevention, and control of infectious diseases remains paramount. However, and in order to have an impact on global public health, the unprecedented scientific advances must be complemented with efforts to apply resulting knowledge and technologies to mitigate threats to health in the developing world. Barriers between research and public health application must be addressed, and the mantra of "sequencing, sequencing, sequencing" should be counterbalanced by a call for "analysis, analysis, analysis," to which one can arguably add a plea for "application, application, application." Effective application, in turn, depends on putting in place the appropriate policy responses, which should include better alignment, harmonization, and coordination of genomics research, so it addresses neglected areas. This should be accompanied by the development of appropriate capacity in developing countries through equitable partnerships, open and rapid sharing of information, developing intersectoral links, and facilitating translation of research to benefit patients. Finally, we need to develop a more holistic systems perspective to strengthen health systems, through better links of research to policy to achieve scale-up, impact, and, ultimately, better health outcomes and greater equity.

SUMMARY

Scientific and technological advances derived from the genomics revolution have a central role to play in dealing with continuing infectious disease threats in the developing world caused by emerging and reemerging pathogens. These techniques, coupled with increasing knowledge of host–pathogen interactions, can assist in the early identification and containment of outbreaks, as well as in the development of preventive vaccination and therapeutic interventions, including the urgent need for new antibiotics. However, the effective application of genomics technologies faces key barriers and challenges, which occur at three stages: from the research to the products, from the products to individual patients, and, finally, from patients to entire populations. There needs to be an emphasis on research in areas of greatest need, in facilitating the translation of research into interventions and, finally, the effective delivery of such interventions to those in greatest need. Ultimate success will depend on bringing together science, society, and policy to develop effective public health implementation strategies to provide health security and health equity for all peoples.

REFERENCES

Blacksell SD, Newton PN, Bell D, et al. (2006). The comparative accuracy of 8 commercial rapid immunochromatographic assays for the diagnosis of acute dengue virus infection. *Clin Infect Dis* *42*:1127–1134.

Burke MA, Matlin S (eds). (2008). *Monitoring Financial Flows for Health Research 2008—Prioritizing research for health equity.* Global Forum for Health Research, Geneva.

Chirac P, Torreele E. (2006). Global framework on essential health R&D. *Lancet 367*:1560–1561.

Daar AS, Thorsteinsdottir H, Martin DK, et al. (2002). Top ten biotechnologies for improving health in developing countries. *Nat Genet 32*:229–232.

Farrar J. (2007). Global health science: a threat and an opportunity for collaborative clinical science. *Nat Immunol 8*:1277–1279.

Horton R. (2009). The global financial crisis: an acute threat to health. *Lancet 373*:355–356.

Jones G, Steketee RW, Black RE, et al. (2003). How many child deaths can we prevent this year? *Lancet 362*:65–71.

Jones KE, Patel NG, Levy MA, et al. (2008). Global trends in emerging infectious diseases. *Nature 451*:990–993.

Katsanis SH, Javitt G, Hudson K, et al. (2008). A case study of personalized medicine. *Science 320*:53–54.

Keeton K. (2008). South African doctors move quickly to contain new virus. *Bull WHO 86*:912–913

King DA, Peckham C, Waage JK, et al. (2006). Infectious diseases: preparing for the future. *Science 313*:1392–1393.

Layne SP, Monto AS, Taubenberger JK. (2009). Pandemic influenza: an inconvenient mutation. *Science 323*:1560–1561.

Leroy JL, Habicht JP, Pelto G, et al. (2007). Current priorities in health research funding and lack of impact on the number of child deaths per year. *Am J Public Health 97*:219–223.

Mabey D, Peeling RW, Ustianowski A, et al. (2004). Diagnostics for the developing world. *Nat Revs Microbiol 2*:231–240.

McGuire AL, Cho MK, McGuire SE, et al. (2007). The future of personal genomics. *Science 317*:1687.

Moran M, Guzman J, Ropars AL, et al. (2009). Neglected disease research and development: how much are we really spending? *PLoS Med 6*:e1000030.

Murray CJL, Lopez AD, Chin B, et al. (2006). Estimation of potential global pandemic influenza mortality on the basis of vital registry data from the 1918–20 pandemic: a quantitative analysis. *Lancet 368*:2211–2218.

National Institutes of Health. NIH Challenge Grants in Health and Science Research. http://grants.nih.gov/grants/funding/challenge_award/ (Accessed on April 9, 2009).

Nijnik A, Hancock REW. (2009). Host defence peptides: antimicrobial and immunomodulatory activity and potential applications for tackling antibiotic-resistant infections. *Emerg Health Threats J 2*:e1.

Normile D. (2008). The promise and pitfalls of clinical trials overseas. *Science 322*:214–216.

Normile D. (2009). Scientists puzzle over Ebola-Reston virus in pigs. *Science 323*:451.

Peeling RW, Mabey D, Fitzgerald DW, et al. (2004). Avoiding HIV and dying of syphilis. *Lancet 364*:1561–1563.

Peeling RW, Smith PG, Bossuyt PMM. (2006). A guide for diagnostic evaluations. *Nat Revs Microbiol 4*:52–56.

Povelones M, Waterhouse RM, Kafatos F, et al. (2009). Leucine-rich repeat protein complex activates mosquito complement in defense against Plasmodium parasites. *Science 324*:258–261.

Rothwell PM. (2006). Funding for practice-oriented clinical research. *Lancet 368*:262–266.

Taubes G. (2008). The bacteria fight back. *Science 321*:356–361.

White NJ. (2006). Clinical trials in tropical diseases: a politically incorrect view. *Trop Med Intern Health 11*:1483–1484.

World Health Organization. (2004). World report on knowledge for better health. WHO, Geneva.

World Health Organization. (2006). World health statistics 2006. WHO, Geneva.

World Health Organization. (2008). World health statistics 2008. WHO, Geneva.

16.

MICROBIAL GENOMICS AND GLOBAL HEALTH

Rino Rappuoli

INTRODUCTION

Over the past two decades, microbiology and infectious disease research have undergone the most profound revolution since the times of Pasteur (McNeill, 1998). Genomic sequencing has revealed the much-awaited blueprint of most pathogens. Screening blood for the nucleic acids of infectious agents has blunted the spread of pathogens by transfusion, the field of antiviral therapeutics has exploded, and technologies for the development of novel and safer vaccines have become available (De Clercq, 2002). The quantum jump in our ability to detect, prevent, and treat infectious diseases resulting from improved technologies and genomics was moderated during this period by the greatest emergence of new infectious agents ever recorded, and a worrisome increase in resistance to existing therapies. Dozens of new infectious diseases are expected to emerge in the coming decades. Controlling these diseases will require a better understanding of the worldwide threat and economic burden of infectious diseases, and a global agenda. This chapter focuses on the major changes in infectious diseases and the scientific progress in the field during the last two decades, and how this may impact on global health.

THE MICROBIAL DISEASES

An overview of the principal infectious diseases of the last two decades shows that they can be divided into three major groups: those against which significant progress was made during this period, those that newly emerged, and those on which we had no impact (Figure 16-1 and Table 16-1). Considerable progress during the last 15 years was made against human immunodeficiency virus (HIV) and hepatitis C virus (HCV) in developed countries, poliomyelitis, and meningococcal and pneumococcal disease. The campaign for the eradication of poliomyelitis, involving vaccination of more than two billion children, reduced the global incidence of polio by 99.9% from 350,000 cases per year to less than 800–1600 during the period from 1998 to 2010 (Figure 16-1; Roberts, 2004), and today we believe that the disease will be eradicated from the world within the next few years. Similarly, in the United States, the introduction of conjugate vaccines against seven serotypes of pneumococcus decreased the incidence of invasive pneumococcal disease from 60 to less than 20 cases per 100,000 people (Flannery et al., 2004), while the proof-of-concept of conjugate vaccines against meningococcus in the early 1990s (Costantino et al., 1992) made possible the introduction of the conjugate vaccine against meningococcus C in the United Kingdom in 2000, virtually eliminating the disease (Ramsay et al., 2001; Trotter et al., 2004). Global elimination of bacterial meningitis may now be an achievable target, because conjugate vaccines against meningococcus ACYW have been licensed, and B is likely to become available within the next years. Significant progress during this period was also achieved for the control of HCV and HIV in developed countries. In 1987, the cloning of the hepatitis C genome (Choo et al., 1989) enabled the development of diagnostic methods to test blood and blood derivatives, which in the early 1990s resulted in the reduction of new HCV cases from >130,000 per year to 25,000 per year in the United States alone (Figure 16-1), while the introduction of a cocktail of drugs (highly active antiretroviral therapy, or HAART) as a standard of care for people with HIV8 transformed HIV in the Western world from a deadly disease into a manageable chronic disease. It is also worth mentioning that diphtheria (Finzi et al., 1997) and cholera (Sack et al., 2004), two diseases that had reemerged at the beginning of the 1990s, have been controlled during this period (Figure 16-1). Although control was achieved for some infections, an unprecedented number of infectious diseases emerged during this period. These include avian influenza, severe acute respiratory syndrome (SARS), West Nile, Ebola, and variant Creutzfeldt-Jakob disease (vCJD) (Glies, 2004). The last decade also saw an increase in the prevalence of antibiotic-resistant bacteria, and the reemergence of previously eradicated pathogens as agents of bioterror (Figure 16-1). Avian influenza was reported to infect humans seven times during the last several years—the last outbreak before this period was in 1968. The virus first reappeared in Hong Kong in 1997 as an H5N1 strain, infecting 18 humans and causing 6 deaths. Two years later, again in Hong Kong, a new avian virus, H9N2, infected two people: the same virus strain caused three infections and one death in 2003. In 2004, we witnessed an explosion of new H5N1 cases, with 43 cases and 31 deaths reported in a series of independent outbreaks in Thailand and Vietnam. The H5N1 outbreaks were controlled mainly by

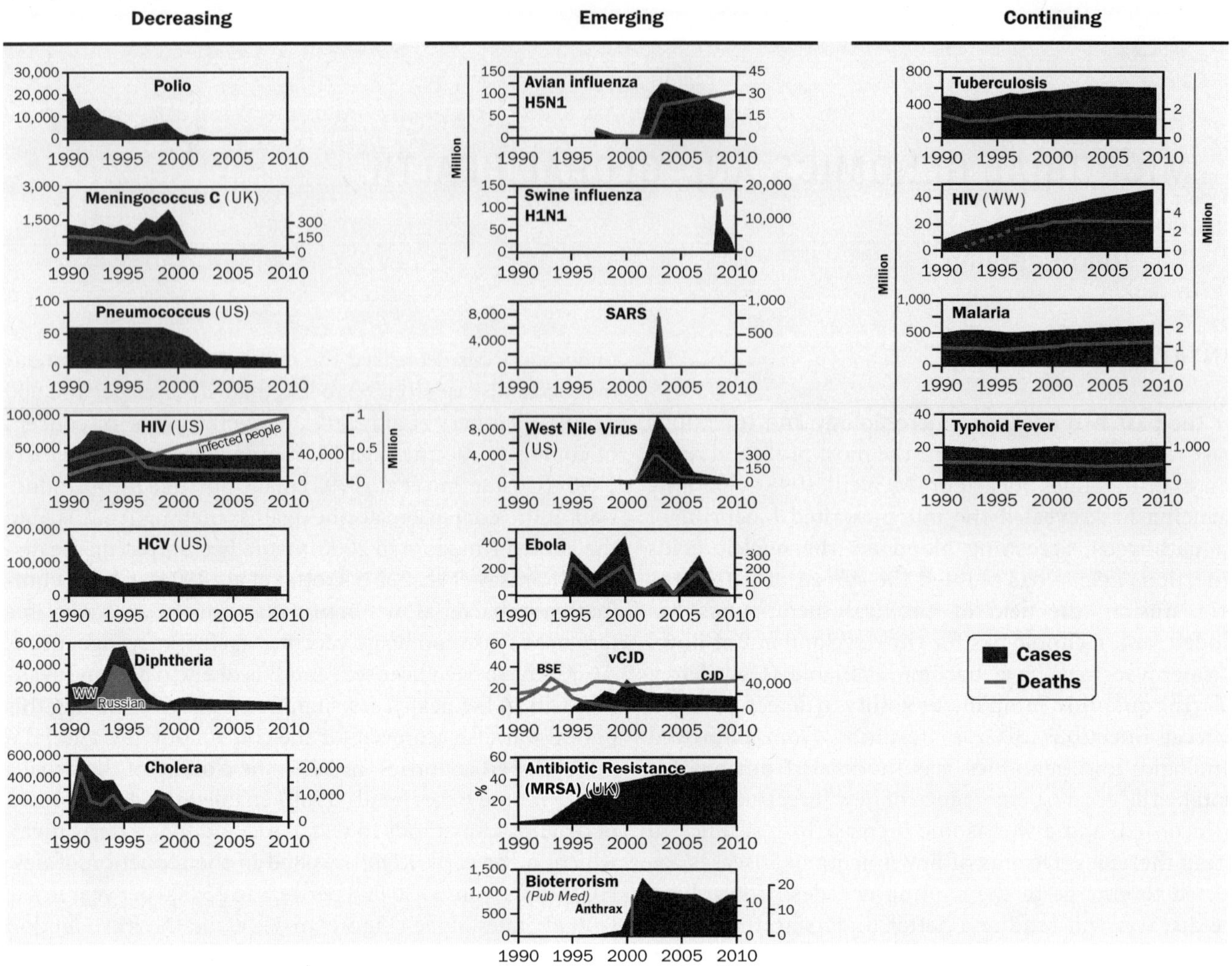

Figure 16-1 Major changes in infectious diseases during the last decade. Refer color figure.

culling more than 100 million fowl (Normile & Enserink, 2004), but the multiple, independent outbreaks during 2004–2007 suggested the proximity of a new flu pandemic (Guan et al., 2004; Li et al., 2004). Indeed, an independent pandemic came in 2009, but in a very unexpected way and took the world by surprise. In April 2009, a new H1N1 virus appeared in Mexico, and within a few months spread to all continents and infected hundreds of millions people (Layne et al 2009). The World Health Organization declared a pandemic on June 11, 2009, and hundreds of millions of doses of vaccine were prepared, licensed, and deployed. By December 2009 the peak of the pandemic was over. Fortunately, the new virus was not very virulent and most people, although they did not have virus-neutralizing antibodies, had immunological lining. This likely blunted the severity of the diseases. Overall, the pandemic caused >25,000 deaths, with high mortality in children and pregnant women. The last decade also saw the emergence of other new diseases caused by transmission of pathogens from animals to humans. The SARS coronavirus, which in the short period between February and July 2003 infected 8098 people and caused 774 deaths (Stadler et al., 2003; Figure 16-1), was likely to have been transmitted by increased contact between humans and virus-carrying wild animals that are sold for food in Asian markets (Stadler et al., 2003). Although the SARS outbreak was quickly resolved, it generated widespread panic, paralyzed travel, and threatened the global economy. Other diseases of animal origin that emerged over the past decade were Ebola hemorrhagic fever, which, after several independent instances of virus transmission from nonhuman primates to hunters (Leroy et al., 2004), caused a total of 264 deaths in Zaire, the Republic of Congo, and Gabon, and the variant form of Creutzfeldt-Jakob Disease (vCJD), a new type of lethal neurodegenerative disease affecting young people, which had been previously very rare and during this decade caused 143 cases (Figure 16-1). The disease was transmitted to humans through consumption of the meat of cows infected by the bovine spongiform encephalitis (BSE) prion (Harris, 2004). West Nile virus, a flavivirus described for the first time in 1937 in Africa and endemic in Europe and central Asia, was reported for the first time in New York

TABLE 16-1 ONLINE ADDRESSES OF MICROBIAL GENOMICS-RELATED WEBSITES

Antibiotic resistance	http://www.cdc.gov/drugresistance
Avian influenza	http://www.who.int/csr/disease/avian_influenza/en/
Bioterrorism	http://www.bt.cdc.gov/agent/agentlist-category.asp
Cholera	http://www.who.int/topics/cholera/en/
Diphtheria	http://www.who.int/topics/diphtheria/en/
Ebola	http://www.who.int/topics/haemorrhagic_fevers_viral/en/
HCV	http://www.cdc.gov/hepatitis/HCV/
HIV	http://www.cdc.gov/hiv/
Malaria	http://www.who.int/malaria/en/
Meningococcus C	http://www.who.int/topics/meningitis/en/
Pneumococcus	http://www.cdc.gov/vaccines/vpd-vac/pneumo/default.htm
Poliomyelitis	http://www.who.int/topics/poliomyelitis/en/
	http://www.polioeradication.org
SARS	http://www.who.int/topics/sars/en/
Tuberculosis	http://www.who.int/topics/tuberculosis/en/
vCJD and BSE	http://www.who.int/topics/creutzfeldtjakob_syndrome/en/
	http://www.who.int/topics/encephalopathy_bovine_spongiform/en/
West Nile	http://www.cdc.gov/ncidod/dvbid/westnile/index.htm

in 1999. Most likely air travel was responsible for the virus crossing continents. The virus infects birds and is transmitted from birds to mammals by the urban mosquito *Culex pipiens*—within four years the virus spread across most of the Unites States, infecting 9862 people and causing 264 deaths in 2003 (Granwehr et al. 2004; Figure 16-1). Another worrisome trend over the past 15 years has been the increased resistance of bacteria to multiple antibiotics (Lowy, 2003; Schmidt, 2004). In Figure 16-1, methicillin-resistant *Staphylococcus aureus* in the United Kingdom is used as an example for this global trend, which represents a serious threat to hospitalized patients. In addition to methicillin-resistant *S. aureus*, multiple antibiotic resistance is a problem also for *Streptococcus pneumoniae, Enterococcus faecalis, Pseudomonas aeruginosa, Mycobacterium tuberculosis*, and *Clostridium difficile* (Reacher et al., 2000; Cooper et al., 2004). Finally, humans have also propagated the use of microbial pathogens for bioterrorism. A few days after the September 11, 2001 terrorist attack in New York, an anthrax infection ignited by the deliberate release of bacterial spores was reported in Florida. Although the powdered spores, which were delivered by mail, caused only a total of 18 infections and 5 deaths (Morens, Folkers, & Fauci, 2004), they changed forever the public perception of microbial pathogens. Scientific publications on bioterrorism increased to more than 1000 per year (Figure 16-1). Smallpox vaccination was started again for the first time after it had been stopped following the eradication of the virus in 1977. Finally, the fear of biological weapons capable of mass destruction was a motivation for beginning the 2003–2010 war in Iraq.

Unfortunately, as shown in Figure 16-1, during the last decade, we had no impact on the three diseases which alone contribute to half of the global burden of infectious disease mortality: tuberculosis, malaria, and HIV (worldwide). Indeed, we have only seen a worsening of tuberculosis and malaria in the immunocompromised population infected by HIV. While the events summarized in Figure 16-1 were happening in the field, a parallel revolution was in progress in microbiology laboratories.

THE GENOMIC REVOLUTION AND CELLULAR MICROBIOLOGY

Since the times of Pasteur, the sequence of the first bacterial genome in 1995 (Fleischmann et al., 1995) and the subsequent genomic revolution represent the largest changes in microbiology.. In 15 years, genomics has provided the blueprints of microorganisms and the ability to explore, in a culture-independent manner, the global diversity of living organisms. Today, the availability in public databases of more than 190 bacterial genomes, 1600 viral genomes, and the malaria parasite sequence is incredible progress, considering that fewer than 15 years ago microbiologists spent most of their research efforts in cloning and sequencing one gene at a time. In addition to providing information about every gene of known pathogens, genome sequencing facilitated research on "old" infectious agents difficult or impossible to grow in vitro, such as HCV or the bacillus causing Whipple disease (Relman et al., 1992; Relman, 1999). For instance, HCV, which in the 1980s could not be grown in vitro and had never been seen by electron microscopy, was impossible to study using the basic technologies of conventional microbiology. But the cloning of the genome in 1987 (Choo et al., 1989) allowed the development of diagnostic methods to test blood and blood derivatives, which in the early 1990s virtually eliminated new transmissions of HCV by transfusion (Figure 16-1). In addition, the genomic sequence provided information on the proteins needed to make the virus, allowing the expression of recombinant viral proteins that have been used to develop prophylactic and therapeutic vaccines, or to test antiviral compounds that are now in clinical trials (Houghton & Abrignani, 2004). During the SARS outbreak, the genome sequence became available in less than a month from the identification of the virus, immediately enabling the development of nucleic acid tests for the detection of the virus, the optimization of measures to contain the spread, the understanding of the probable animal origin, and the design of therapeutics, monoclonal antibodies, and several effective vaccines that are moving toward clinical trials (Bisht et al., 2004; Traggiai et al., 2004). The SARS example shows how emerging pathogens can be identified, sequenced, and classified in real time, confirming that modern technologies can be very effective in handling unknown emerging infections. Another example of a pathogen that had escaped detection by conventional microbiology is human metapneumovirus. Isolated for the first time in 2001 from the nasopharyngeal aspirates of children in the Netherlands, today this negative RNA virus of the Paramyxoviridae family is a recognized cause of acute respiratory infections and considered to be responsible for morbidity and mortality worldwide. Together with respiratory syncytial virus (RSV) it is believed to cause a large fraction of severe acute respiratory tract infections in infant, elderly, and immunocompromised hosts (Hamelin, Abed, & Boivin, 2004). Although metapneumovirus can be handled by conventional technologies, the virus had probably been missed because it is rather difficult to grow in conventional cell cultures—molecular technologies, such as gene hybridization and sequencing, were instrumental in facilitating its identification.

In addition to better detection, the new genome-based technologies (complemented by great advances in confocal microscopy, fluorescent proteins, total body imaging, microarrays, and a number of animal models supported by genetic techniques such as signature-tagged mutagenesis and in vivo expression technology (McDaniel & Valdivia, 2000) have enabled the study of pathogens while they interact with their hosts. This is such a novel approach to microbial pathogenesis that a new discipline, "cellular microbiology" (Cossart et al., 1996), has been created to differentiate modern microbiology, which studies pathogens interacting with their hosts in their native environments, from classical microbiology, which studies pathogens grown in rich media under artificial conditions in the laboratory. These new approaches have contributed to the discovery of virulence factors not necessary to pathogens grown in the laboratory, but essential or preferentially expressed only in vivo during infection. Perhaps the most intriguing among these are the bacterial Type III and Type IV secretion systems, molecular syringes that inject proteins into host cells to facilitate infection and survival within the host (Cornelis & Van Gijsegem, 2000; Cascales & Christie, 2003). Similarly, the human genome and the clustering of the human populations using genetic tools, such as the single nucleotide polymorphisms (Clifford et al., 2004) are expected to provide the genetic basis for the susceptibility to infectious diseases and the pathologies deriving from infections. We expect to be able to predict who will be most likely to succumb to or resist infections, or respond to a therapy or vaccination. Preliminary examples of what may become routine in the future are the use of gene expression profiles to study the genes that are upregulated during clinical leprosy and define the clinical forms of the diseases (Bleharski et al., 2003) and the identification of a gene expression signature, which defines a successful vaccination and protection from infection from the *Helicobacter pylori* pathogen (Mueller et al., 2003).

MORE AND BETTER VACCINES

For over a century, vaccines were developed according to Pasteur's principles of isolating, inactivating, and injecting the causative microorganisms. These principles provided the killed, live attenuated, and subunit vaccines in use today (Figure 16-2) and are responsible for a large part of the control of infectious diseases achieved to date. But Pasteur's principles did not allow the development of vaccines against those microorganisms that cannot be cultivated in vitro, such as HCV, papillomavirus types 16 and 18, and *M. leprae*. Also, the principles of Pasteur did not provide vaccines for antigenically hypervariable microorganisms such as serogroup B meningococcus, gonococcus, malaria, and HIV, and did not teach how to induce cytotoxic T-cells, which kill infected host cells and therefore contribute to the control of viral replication. During the last decade we have overcome most of these technical limitations, even if we are still unable to use them effectively enough to solve challenges such as HIV. The availability

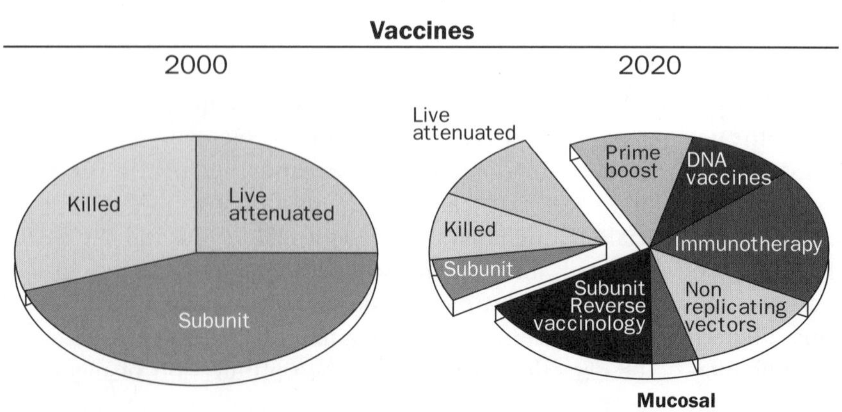

Figure 16-2 Projections of available vaccines from 2000 to 2020.

of genomes allowed the identification of novel vaccine candidates without the need to cultivate microorganisms, a process named "reverse vaccinology" (Rappuoli, 2000). In this approach, the genomic sequences of viral, bacterial, or parasitic pathogens are used to select by computer analysis, microarrays, proteomics, and other genome-based systematic approaches those antigens that are likely to confer protective immunity and eliminate potentially dangerous antigens, such as those showing homology to human proteins, which are therefore potentially able to induce autoimmunity. The predicted antigens are then expressed by recombinant DNA and tested in an animal model. This genome-based vaccine approach allowed the development of vaccines against HCV and human papillomavirus 16 and 18, which, being unable to grow in vitro, were not approachable following the principles of Pasteur. But once the genome sequence became available, it was used to predict the antigens coding for the surface proteins (E1 and E2 for HCV, and VP L1 for papillomavirus), express them, and develop vaccines that are presently in clinical trials, and licensed, respectively (Houghton & Abrignani, 2004; Koutsky et al., 2002). The most quoted example for reverse vaccinology is meningococcus B. At the end of the 1990s it was apparent that a universal meningococcal vaccine was beyond the reach of conventional vaccinology for two reasons: first, the polysaccharide antigen used successfully to make conjugate vaccines against other meningococci was not immunogenic (and, indeed, was a potential trigger of autoimmunity, being identical to a self-antigen), and second, the best protein-based vaccines induced immunity against antigenically variable proteins and had been shown to induce protective immunity mostly against the strain used to make the vaccine, without substantial cross-protection. In this case, the sequence of the bacterial genome allowed the computer prediction of approximately 600 novel vaccine candidates, 350 of which were expressed in *Escherichia coli* and tested for their ability to elicit protective immunity. Remarkably, the approach identified 29 novel protective antigens, some of which are conserved in all meningococcal strains and are now being tested in phase III clinical trials. The meningococcus B example has been followed for many other bacterial pathogens, including pneumococcus, group B streptococcus, chlamydia (Koutsky et al., 2002), *Staphylococcus aureus*, and group A streptococcus.

Although most vaccines available today work by inducing antibodies, it is believed that conquering the most difficult diseases, such as HIV, malaria, other chronic diseases and cancer, may require involvement of the T-cell arm of the immune system. During this period, several methods have been developed to induce cytotoxic T-cell responses following vaccination and, in the case of HIV, extensive proof of concept has been obtained in nonhuman primates that effective stimulation of cytotoxic T-cells is able to contain the viremia. The most effective methods to induce cytotoxic T-cells in nonhuman primates and in humans are engineered nonreplicating viral vectors, such as modified vaccinia Ankara/(MVA), replication-incompetent adenoviruses, and DNA vaccines (Shiver et al., 2002; Shiver & Emini, 2004). Mixed regimes, where DNA priming is followed by a viral vector boost, have been shown to further increase the cytotoxic response. Successful induction of malaria-specific CD8+ T-cells has been obtained in humans using a DNA-prime, MVA boost strategy (McConkey et al., 2003). An example of effective protection obtained in nonhuman primates by adenoviral vectors is the one-dose vaccine recently described against Ebola virus (Sullivan et al., 2004). While adenovirus-based vaccines inducing only CD8+ T-cells were found to be effective in preventing HIV infection or containing viral load, a live-boost strategy in Thailand provided for the first time proof of principle that HIV may be prevented by vaccination (Dolin, 2009).

One of the surprises of this decade has also been the understanding that living organisms have a conserved "innate" immune defense against pathogens, which is mediated by Toll-like receptors and Nod proteins (Sullivan et al., 2004), which sense molecules that have a signature of microbes, such as DNA containing unmethylated CpG sequences, double-stranded RNA, lipopolysaccharide, etc. The understanding of the molecular basis of the innate immune response provides further potential for manipulating the immune system. This represents one of the most intriguing developments that we expect in the next decade. Among the early results in this new area is the development of adjuvants based on unmethylated CpG motifs, which, being prevalent in bacterial but not in vertebrate genomic DNA, are seen as the signature of bacterial infection by the innate immune system. Synthetic CpG-containing oligonucleotides have been shown to stimulate Toll-like receptor 9 and to induce TH1-like proinflammatory cytokines, and to be effective adjuvants for many vaccines in animal models and in some preliminary clinical trials (Krieg, 2002). Figure 16-2 provides an example of what we predict will happen in the field of vaccines during the next decades. Reverse vaccinology, DNA vaccination, nonreplicating vectors, the understanding of the innate immune system, and the development of new, rationally designed adjuvants for mucosal and systemic delivery of vaccines are likely to be useful in driving the development of many novel vaccines to prevent and treat infectious diseases.

Finally, it should be mentioned that the safety of vaccines has improved enormously during the last decade. Very efficacious vaccines, such as the oral poliovirus and the whole-cell pertussis vaccines, have been removed from the vaccination schedules of Western countries because of very rare or perceived untoward effects, and replaced with less reactogenic vaccines. All new vaccines are based on technologies that deliver vaccines with a very high standard of safety.

THE ERA OF ANTIVIRALS

The large-scale screening for natural compounds able to kill bacteria in vitro, which was the basis for the boom of antibiotics in the 1950s, was not successful for antivirals. In fact, whereas bacteria are distinct free-living organisms that can be targeted by drugs, viruses are parasitic entities that invade healthy host cells and hijack their reproductive machinery, making it very difficult to kill viruses without also killing the cells they have infiltrated. The first antiviral, 5-iodo- 2'-deoxyuridine, was described in 1959; by 1990, only four were licensed (amantadine, ribavirin, acyclovir and derivatives, and AZT, against influenza, RSV, herpes, and HIV, respectively; see Field & De Clercq, 2004). Since then, more than 35 new antivirals have been approved, most of which combat HIV and others that combat hepatitis B virus, HCV, influenza, and herpes (Figure 16-3).

The driving force for the boom of antivirals in this period has been the pressure to contain the HIV pandemic, combined with the increased understanding of the molecular mechanisms of viral life cycles, which has allowed the identification of new targets for therapeutic intervention. At the beginning of the 1990s, HIV was growing in the United States and worldwide. In the United States, 49,000 new HIV cases and 31,000 deaths were recorded in 1990, growing to 72,000 and 50,000, respectively, in 1994 (Figure 16-1). Worldwide, there were eight million cases in 1990 and 18 million in 1994 (Figure 16-1). The first effective therapy for HIV came in 1992 with the approval of the first combination of reverse transcriptase inhibitors, 3′-azido-2′,3′-dideoxythymidine (AZT) and 2′,3′-dideoxycytidine (ddC), which prevent viral DNA synthesis by acting as chain terminators. The approval in 1993 of (–)-B-L-3′-thia-2′-3′-dideoxycytidine (3TC), a safer inhibitor of reverse transcriptase, and of the

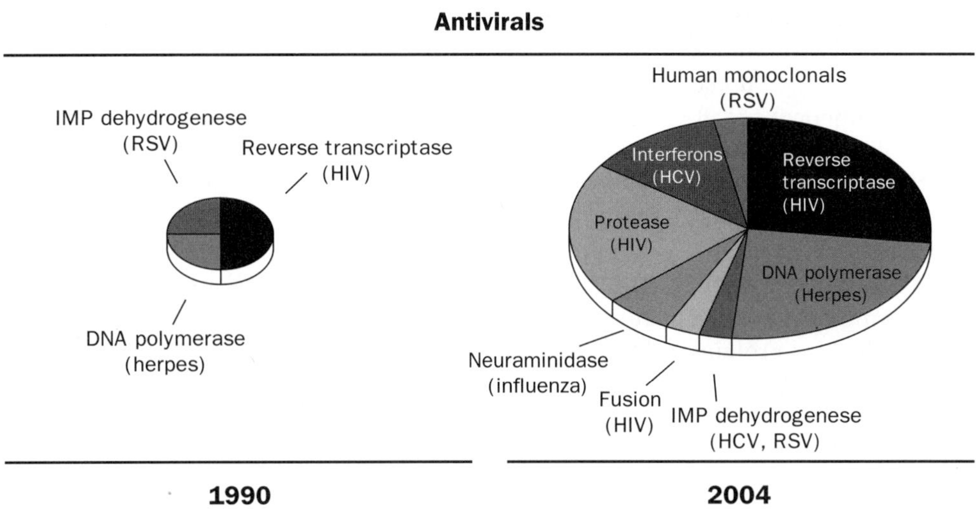

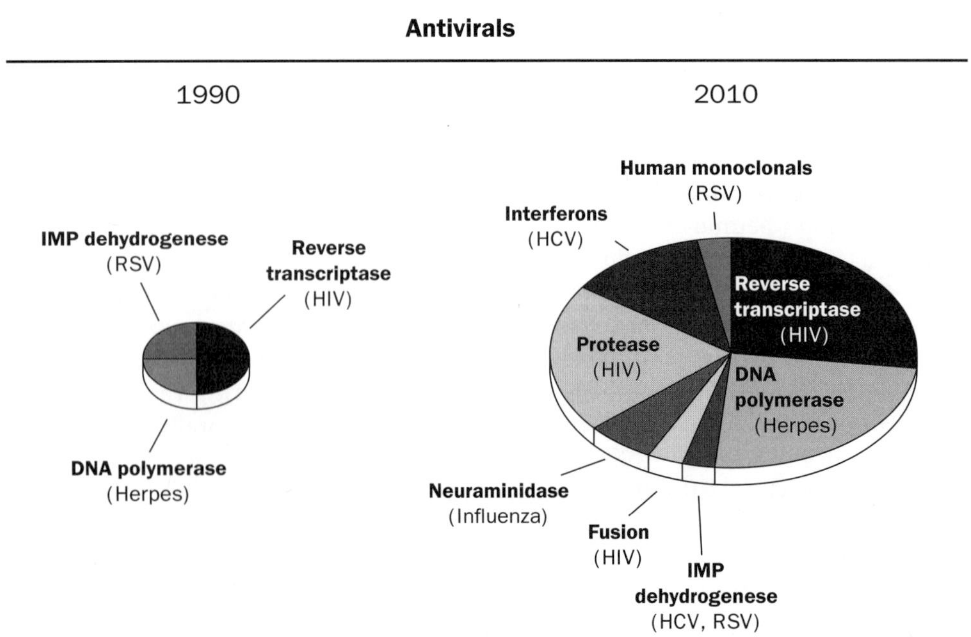

Figure 16-3 Antivirals approved by FDA in 1990 and 2004. Refer color figure.

first protease inhibitor in 1995, together with the ability to use the polymerase chain reaction to measure the amount of virus present in a given volume of blood (viral load), allowed the introduction of a cocktail of drugs as a standard of care for people with HIV (Finzi et al., 1997).

HAART increased life expectancy and transformed HIV in the Western world from a deadly disease into a more manageable chronic disease, which postpones but unfortunately does not solve the HIV problem. New infections and deaths from HIV started to plateau and decrease for the first time in the mid-1990s. Today in the United States, there are approximately 53,000 new cases and 18,000 deaths per year. But the number of people living with HIV has increased dramatically, from 30,000 in 1992 to more than one million (Figure 16-1), and currently we have no long-term solution for them. The management of the population now chronically infected by HIV provides new challenges, such as the continuous development of new antivirals active against the resistant HIV isolates selected by the present therapy and the control of side effects resulting from chronic treatment. Today the target enzymes for development of antiviral compounds are the viral enzymes necessary for the replication of the viral genome, such as the retroviral reverse transcriptases, DNA and RNA-dependent polymerases and helicases, the proteases necessary to cleave viral polyproteins, the influenza neuraminidase necessary for the release of the virus from the cell, and even cellular enzymes involved in the generation of the nucleotide pools, such as the inosine 5′-monophosphate deydrogenase45, that for unknown reasons are important for viral replication. The availability of validated target enzymes used for high-throughput screening of natural and combinatorial libraries is today complemented by the ability to co-crystallize target proteins and their inhibitors, allowing the use of structure-based drug design for the generation and optimization of more and better leads. Recently, the understanding of the molecular mechanisms of viral entry into host cells has allowed the development of a new class of antiviral compounds, known as *fusion inhibitors*, one of which (enfuvirtide) is already in use for the therapy of AIDS (Matthews et al., 2004). Enfuvirtide is a synthetic peptide that binds to a region of the HIV-1 envelope and prevents a conformational change necessary to drive the fusion of the viral membrane with that of the host cell. Very effective in vitro, although not yet clinically relevant, are the small interfering RNAs (siRNAs), which are double-stranded RNA oligonucleotides that use a cell-defense enzyme pathway to drive the selective degradation of viral RNA products (Howard, 2003). siRNAs have been shown to be effective in vitro against virtually all medically important viruses, including HIV and HCV; however, their development as drugs needs to meet the challenge of delivering effectively large and unstable molecules, such as oligonucleotides, to infected tissues. In conclusion, whereas antibiotics were the hot drugs of the past century, antivirals will be the hot drugs of the twenty-first century, during which time we expect to have effective drugs against most viral infections.

HUMAN MONOCLONAL ANTIBODIES

Treatment of infectious diseases made a quantum leap in 1890 when Behring and Kitasato discovered that immunization with low doses of sterile culture supernatants of tetanus or diphtheria induced sera that, after transfer to other animals, could cure them of symptoms of disease (Behring & Kitasato, 1890). During the following century, sera were one of the most important life-saving tools to treat bacterial infections such as diphtheria, tetanus, anthrax, botulism, pneumococcal disease, and viral infections such as rabies. But animal sera have largely disappeared from the anti-infection toolbox during the last 30 years, mostly because of the toxic reactions (serum sickness) caused by injecting humans with horse sera, while the decreased popularity of blood derivatives has reduced the use of human gammaglobulins to only a few applications such as cytomegalovirus and parvovirus B19. The development of the technology to commercialize human monoclonal antibodies by humanizing mouse antibodies allowed the licensure of the first human monoclonal antibody that neutralizes RSV in vitro and prevents RSV infection in newborns (Johnson et al., 1997). Today this represents an important tool for newborns, but it is still the only licensed humanized antibody for infectious diseases. The high doses required (10–15 mg/kg, that is >600 mg/adult person) are expensive and make administration impractical in adults. A few years ago, a pioneering new technology to efficiently clone human memory B cells from infected or immunized people transformed the field (Traggiai et al., 2004). This technology allows screening for very high-affinity human monoclonal antibodies against viruses, and the isolation of clones that produce antibodies may be able to be therapeutically effective at 0.1–1.0 mg/kg, making industrial development possible. Today several methods are available to efficiently make human monoclonal antibodies against infectious agents. Therefore, we expect that human monoclonal antibodies will become an increasingly important tool for the passive prevention of, and therapy for, infectious diseases (Casadevall, Dadachova, & Pirofski, 2004).

DIAGNOSIS AND SCREENING

The availability of the sequence of viral, bacterial, and parasite genomes provided the information necessary for the expression of recombinant microbial antigens that could be used to detect antibodies in patient sera, and the development of specific assays, based on nucleic acid amplification technologies that are able to directly detect a few genome copies of infectious agents. These new technologies allowed a timely response to the emergence of new pathogens, as in the case of SARS, or the development of assays to monitor and control the spread of infectious agents. The virtual elimination of serious infections transmitted by blood transfusion (Busch, Kleinman & Nemo, 2003), which is one of the milestones in the history of medicine that has been achieved during the last decade, was made possible by these technologies (Figure 16-4). In the early 1980s the risk of being infected by HIV or HCV by blood transfusion was in the range of 1.10–2. The introduction of screening for antibodies against HIV in 1984, and for HCV in 1990, tremendously increased the safety of blood and blood products, reducing the risk to less than 1.10–5. Finally, the introduction of nucleic acid testing in 1999 allowed the detection of those recently infected donors that had not had the time yet to develop antibodies, reducing the risk for both to one in two million (Stramer et al., 2004). Encouraged by the dramatic success of testing blood for HIV and HCV, new assays have been recently introduced to test blood for hepatitis B and West Nile viruses, virtually eliminating most common diseases that can be transmitted by blood transfusion. Similar technologies can be applied to the detection of other infectious agents such as parvovirus, malaria, leishmania, vCJD, pathogenic bacteria, or agents spread by bioterrorists.

MEASURES NECESSARY FOR THE CONTROL OF INFECTIOUS DISEASES

Increased global priority. The global threat and the economic burden of infectious diseases need to be visible to the public and to policymakers, both in developed and developing countries. Developing countries need to be aware that when they do not spend enough money on prevention, the cost of not doing so is very high (Bloom & Murray, 1992). Scientists and the media need to make an effort to improve effective communication so that infectious diseases will receive high priority in the political and economic agendas of every country in the world.

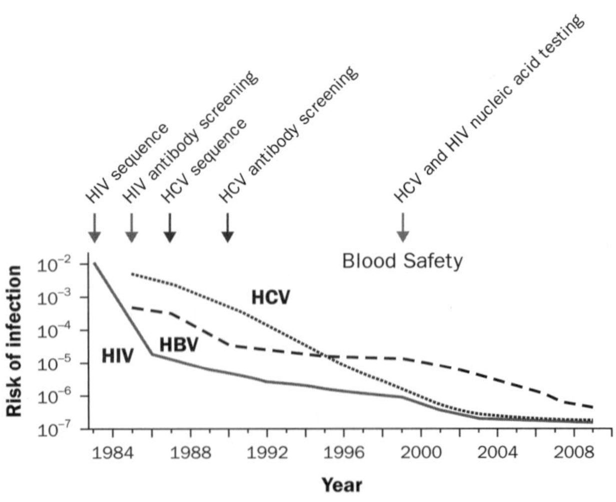

Figure 16-4 Risk of infection of human immunodeficiency virus (HIV), hepatitis C virus (HCV), and hepatitis B virus (HBV) per unit of blood transfusion during the last 20 years. The figure is revised from Busch et al. (2003).

Global legislation and collaboration. SARS, HIV, avian influenza, and H1N1 have clearly shown that whatever happens in one part of the world may very quickly become a global problem. An international law to regulate infectious diseases should be implemented (Gostin, 2004). This applies particularly to infectious diseases spread through various animals (Figure 16-5). International teams with power to oversee the global interest above local concerns, and promote effective international collaboration, should be in place.

Coordination of regulatory agencies. Agencies that regulate human health, animal health, trade, hunting, food, and travel have so far been acting independently. They need to talk to each other and coordinate their actions, because we have seen more than one example (BSE, SARS, H1N1) in which decisions by one agency resulted in strong implications for the spread of infectious diseases in other areas (Table 16-2).

Development of vaccines to prevent all diseases. Vaccines should be studied, developed, and licensed against all infectious agents, including those that are economically uninteresting. Smallpox and avian influenza have shown that orphaned infectious agents can become of global interest at any time, and that once they are a problem it is too late to develop measures to control them.

Sustainable economic system. We must develop a sustainable system for protecting the health of the world population. Public–private partnerships such as the

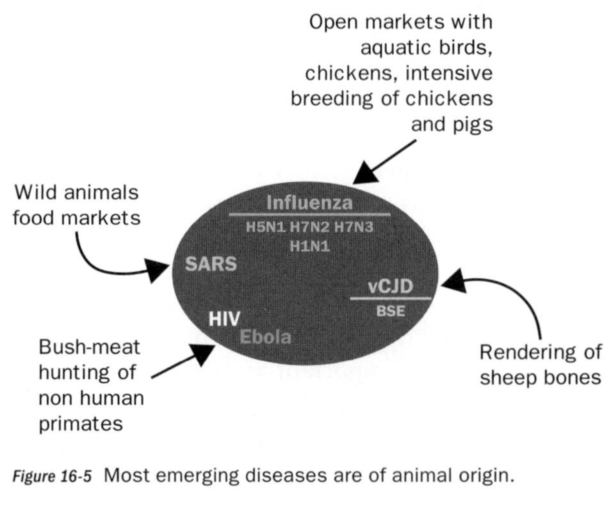

Figure 16-5 Most emerging diseases are of animal origin.

TABLE 16-2 HUMAN BEHAVIOR AND ECOSYSTEMS LINKED TO THE EMERGENCE OF NEW INFECTIOUS DISEASES

Disease	Human Behavior/Ecosystem	Year	Number Affected/Died	References
vCJD	Feeding of cows with infected cows & scrapie-infected sheep	1996	143/141	Harris, 2004; Giles, 2004
HCV	Unscreened blood products, medical practices involving unsterilized tools, including syringe reuse	1992	170 million	McHutchison (2004)
SARS	Live animals food markets, air travel	2003	8098/774 in 2002–2003	Stadler et al. (2003)
	Legionella pneumophila, Air-conditioning towers	1976	221/34 in 1976 outbreak	Roig et al. (2003)
Ebola	Bushmeat hunting	1976	313/264 during last 2 years	Leroy et al. (2004)
HIV	Bushmeat hunting, sex, travel	1981	40 million/3 million per year	Steinbrook (2004)
West Nile	Air travel, climatic change	1999	9858/262	Granwehr et al. (2004)
Avian Flu	Open markets with aquatic birds, chickens, intensive cultivation of chickens/pigs	1997	43/31 in 2004	Normile & Enserink (2004)
H1N1	Swine influenza	2009	>100 million/	Layne, Monto, & Taubenberger (2009)

Global Fund and the Gates Foundation. in the short term, seem to be the best way to make rapid progress. They triggered enormous progress never seen before. But in the long term, global health cannot rely on the good will of charities.

CONCLUSIONS

In the late 1960s the optimism deriving from the successes of vaccines and antibiotics had generated the belief that infectious diseases had been conquered and were no longer a problem for humanity. We now know that the optimism was premature, and that infectious diseases are here to stay. Indeed, human history has never seen so many new infectious diseases as in recent times. The shotgun sequencing of microorganisms captured by filtering hundreds of liters of water of the Sargasso sea, which identified 1.2 million new genes, 1800 species, and 148 novel bacterial phylotypes (Venter et al., 2004), showed that microbial diversity is orders of magnitude larger than had been previously estimated. This implies that in the environment and the animal world, there is a huge reservoir of microbes which, under the appropriate circumstances, may recombine, jump species, and generate new human pathogens. Modern society provides the most fertile environment for this to happen; the accelerated changes of human behavior and ecosystems continuously disturb the equilibrium between microbes and their hosts, exposing them to new environments where their behavior is sometimes unpredictable resulting in diseases. The factors involved are the increased global population, overcrowded cities, increased travel, intensive animal farming and food production, sexual practices, poverty, global warming, and the breakdown of public health measures. Some believe that humans are primarily responsible for this (Walters, 2003). Most infectious diseases are zoonoses (animal infections transferred to man) promoted by human behavior (Figure 16-5 and Table 16-2), such as bushmeat hunting (Ebola and probably HIV), the sale of live wild animals in open markets (SARS; see Guan et al., 2003), and intensive farming and sale of different species next to each other in open markets (avian influenza finds a favorable environment in the open markets in Asia, where migratory aquatic birds, in which the influenza virus grows without causing disease, are sold next to chickens and pigs, the cultivation of which has expanded enormously to feed the growing Asian population (Guan et al., 2004). It is therefore easy to predict that the next few decades will see dozens of new infectious agents emerging in humans. The question is whether they will be contained or cause new global pandemics such as HIV, influenza, or the Black Death in the Middle Ages, which have each caused several dozens of million deaths and destroyed economies and societies. Today, the most compelling new threat seems to be influenza. The H1N1 experience has somehow distracted us from H5N1; however, we should not forget that this deadly virus is still out there (Guan et al., 2004; Li et al., 2004). Fortunately, today's technologies can provide adequate tools to detect, manage, control, and prevent emerging infections. But, technologies are not sufficient on their own to contain the impact of infectious diseases, and we are likely to fail unless all governments increase the priority of infectious diseases in the political agenda and promote a coordinated, collaborative global action against them, both in developed and developing countries. A list of measures necessary for the global control of infectious diseases includes increased global priority, global legislation and collaboration, coordination of regulatory agencies, development of vaccines to prevent all diseases, and a sustainable economic system to support these initiatives.

An example showing how effective a global approach can be is represented by the campaign for eradication of poliomyelitis, which is the largest public health initiative ever tried. Launched in 1985 for South America, it was taken up by the World Assembly, the governing body of the World Health Organization, which in 1988 committed to the global eradication of poliovirus by the year 2000. During the period from 1998 to 2002, more than two billion children were immunized using the national immunization days and a budget of approximately US$ four billion (Minor, 2004). Progress toward eradication of the virus was fast. In 1988, 125 countries in the five continents reported endemic poliovirus. By 1994, 2000, and 2002, the American continent, the Western Pacific Region, and the European regions respectively were certified polio free (three years without a case of polio). In 2003 only six polio-endemic countries were reported (Niger, Nigeria, and Egypt in Africa and India, Afghanistan and Pakistan in Asia). During the same period, cases were reduced by 99.9% from 350,000/year to less than 800 (Figure 16-1). Unfortunately, in 2004 the virus was imported from these countries to eight African countries that were previously polio free (Flannery et al., 2004). Although the polio eradication missed the deadline of the year 2000 and has seen a setback in 2004, the initiative is one of the best examples of how a global effort can, within a short period of time, reduce the incidence of an infectious disease by more than 99.9%. The final effort in the remaining endemic regions will eliminate the disease from the world within the next few years, and leave us with the problem of whether and how long we should continue to vaccinate against a disease that no longer exists. Hopefully, the success against polio will be a catalyst to trigger similar global priority and collaboration against all infectious diseases.

REFERENCES

Athman R, Philpott D. (2004). Innate immunity via Toll-like receptors and Nod proteins. *Curr Opin Microbiol 7*, 25–32.

Behring EA, Kitasato S. (1890). Ueber das zustandekommen der diphtherie-immunitat und der tetanus-immunitat bei thieren. *Deutch Med Woch 49*, 1113–1114.

Bisht H, et al. (2004). Severe acute respiratory syndrome coronavirus spike protein expressed by attenuated vaccinia virus protectively immunizes mice. *Proc Natl Acad Sci USA 101*, 6641–6646.

Bleharski JR, et al. (2003). Use of genetic profiling in leprosy to discriminate clinical forms of the disease. *Science 301*, 1527–1530.

Bloom BR, Murray CJ. (1992). Tuberculosis: commentary on a reemergent killer. *Science 257*, 1055–1064.

Busch MP, Kleinman SH, Nemo GJ. (2003). Current and emerging infectious risks of blood transfusions. *JAMA 289*, 959–962.

Casadevall A, Dadachova E, Pirofski L. (2004). Passive antibody therapy for infectious diseases. *Nat Rev Microbiol 2*, 695–703.

Cascales E, Christie PJ. (2003). The versatile bacterial type IV secretion systems. *Nat Rev Microbiol 1*, 137–149.

Choo QL, et al. (1989). Isolation of a cDNA clone derived from a blood-borne non-A, non-B viral hepatitis genome. *Science 244*, 359–362.

Clifford RJ, et al. (2004). Bioinformatics tools for single nucleotide polymorphism discovery and analysis. *Ann NY Acad Sci 1020*, 101–109.

Cooper BS, et al. (2004). Methicillin-resistant Staphylococcus aureus in hospitals and the community: stealth dynamics and control catastrophes. *Proc Natl Acad Sci USA 101*, 10223–10228.

Cornelis GR, Van Gijsegem F. (2000). Assembly and function of type III secretory systems. *Annu Rev Microbiol 54*, 735–774.

Costantino P, et al. (1992). Development and phase 1 clinical testing of a conjugate vaccine against meningococcus A and C. *Vaccine 10*, 691–698.

Cossart P, Boquet P, Normark S, Rappuoli R. (1996). Cellular microbiology emerging. *Science 271*, 315–316.

De Clercq E. (2002). Strategies in the design of antiviral drugs. *Nat Rev Drug Discov 1*, 13–25.

Dolin R. (2009). HIV vaccine trial results—an opening for further research. *N Engl J Med 361*, 2279–2280.

Field HJ, De Clercq E. (2004). Antiviral drugs–a short history of their discovery and development. *Microbiology Today 31*, 58–61.

Finzi D, et al. (1997). Identification of a reservoir for HIV-1 in patients on highly active antiretroviral therapy. *Science 278*, 1295–1300.

Flannery B, et al. (2004). Impact of childhood vaccination on racial disparities in invasive Streptococcus pneumoniae infections. *JAMA 291*, 2197–2203.

Fleischmann RD, et al. (1995). Whole-genome random sequencing and assembly of Rd. *Science 269*, 496–512.

Giles J. (2004). Tissue survey raises spectre of 'second wave' of vCJD. *Nature 429*, 331.

Gostin LO. (2004). International infectious disease law: revision of the World Health Organization's International Health Regulations. *JAMA 291*, 2623–2627.

Granwehr BP, et al. (2004). West Nile virus: where are we now? *Lancet Infect Dis 4*, 547–556.

Guan, Y. et al. (2003). Isolation and characterization of viruses related to the SARS coronavirus from animals in southern China. *Science 302*, 276–278.

Guan Y, et al. (2004). H5N1 influenza: a protean pandemic threat. *Proc Natl Acad Sci USA 101*, 8156–3161.

Harris DA, ed. (2004). *Mad cow disease and related spongiform encephalopathies*, pp. 219. Berlin: Springer-Verlag.

Hamelin ME, Abed Y, Boivin G. (2004). Human metapneumovirus: a new player among respiratory viruses. *Clin Infect Dis 38*, 983–990.

Houghton M, Abrignani S. (2004). Vaccination against the hepatitis C viruses. In *New Generation Vaccines* (eds Levine MM, Kaper JB, Rappuoli R, Liu MA, Good MF), New York: Marcel Dekker, pp. 593–606.

Howard K. (2003). Unlocking the money-making potential of RNAi. *Nat Biotechnol 21*, 1441–1446.

Johnson S, et al. (1997). Development of a humanized monoclonal antibody (MEDI-493) with potent in vitro and in vivo activity against respiratory syncytial virus. *J Infect Dis 176*, 1215–1224.

Koutsky LA, et al. (2002). A controlled trial of a human papillomavirus type 16 vaccine. *N Engl J Med 347*, 1645–1651.

Krieg AM. (2002). CpG motifs in bacterial DNA and their immune effects. *Annu Rev Immunol 20*, 709–760.

Layne SP, Monto AS, Taubenberger JK. (2009). Pandemic influenza: an inconvenient mutation. *Science 323*, 1560–1561.

Leroy EM, et al. (2004). Multiple Ebola virus transmission events and rapid decline of central African wildlife. *Science 303*, 387–390.

Levy SB. (2001). Antibiotic resistance: consequences of inaction. *Clin Infect Dis 33*, S124–S129.

Li KS, et al. (2004). Genesis of a highly pathogenic and potentially pandemic H5N1 influenza virus in eastern Asia. *Nature 430*, 209–213.

Lowy FD. (2003). Antimicrobial resistance: the example of *Staphylococcus aureus*. *J Clin Invest 111*, 1265–1273.

McConkey SJ, et al. (2003). Enhanced T-cell immunogenicity of plasmid DNA vaccines boosted by recombinant modified vaccinia virus Ankara in humans. *Nat Med 9*, 729–735.

McDaniel TK, Valdivia RH. (2000). Promising new tools for virulence gene discovery. In: Cossart P, Boquet P, Normark S, Rappuoli R, eds. *Cellular Microbiology*, pp. 333–345. Washington DC: ASM Press.

McHutchison JG. (2004). Understanding hepatitis C. *Am J Manag Care 10*, S21–S9.

McNeill WH. (1998). *Plagues and Peoples.* p. 365 New York: Anchor Books, p. 365.

Markina SS, Maksimova NM, Vitek CR, Bogatyreva EY, Monisov AA. (2000). Diphtheria in the Russian Federation in the 1990s. *J Infect Dis 181*, S27–S34.

Matthews T, et al. (2004). Enfuvirtide: the first therapy to inhibit the entry of HIV-1 into host CD4 lymphocytes. *Nat Rev Drug Discov 3*, 215–225.

Minor PD. (2004). Polio eradication, cessation of vaccination and re-emergence of disease. *Nat Rev Microbiol 2*, 473–482.

Mora M, Veggi D, Santini L, Pizza M, Rappuoli R. (2003). Reverse vaccinology. *Drug Discov Today 8*, 459–464.

Morens DM, Folkers GK, Fauci AS. (2004). The challenge of emerging and reemerging infectious diseases. *Nature 430*, 242–249.

Mueller A, et al. (2003). Protective immunity against Helicobacter is characterized by a unique transcriptional signature. *Proc Natl Acad Sci USA 100*, 12289–12294.

Normile D, Enserink M. (2004). Infectious diseases. Avian influenza makes a comeback, reviving pandemic worries. *Science 305*, 321.

Ramsay ME, Andrews N, Kaczmarski EB, Miller E. (2001) Efficacy of meningococcal serogroup C conjugate vaccine in teenagers and toddlers in England. *Lancet 357*, 195–196.

Rappuoli R. (2000). Reverse vaccinology. *Curr Opin Microbiol 3*, 445–450 (2000).

Koutsky LA, et al. (2002). A controlled trial of a human papillomavirus type 16 vaccine. *N Engl J Med 347*, 1645–1651.

Reacher MH, et al. (2000). Bacteraemia and antibiotic resistance of its pathogens reported in England and Wales between 1990 and 1998: trend analysis. *BMJ 320*, 213–216.

Relman DA, Schmidt TM, MacDermott RP, Falkow S. (1992). Identification of the uncultured bacillus of Whipple's disease. *N Engl J Med 327*, 293–301.

Relman DA. (1999). The search for unrecognized pathogens. *Science 284*, 1308–1310.

Roberts L. (2004). Polio: Health workers scramble to contain African epidemic. *Science 305*, 24–25.

Sack DA, Sack RB, Nair GB, Siddique AK. (2004). Leading article: Cholera. *Lancet 363*, 223–233.

Schmidt FR. (2004). The challenge of multidrug resistance: actual strategies in the development of novel antibacterials. *Appl Microbiol Biotechnol 63*, 335–343.

Shiver JW, et al. (2002). Replication-incompetent adenoviral vaccine vector elicits effective anti-immunodeficiency-virus immunity. *Nature 415*, 331–335.

Shiver JW, Emini EA. (2004). Recent advances in the development of HIV-1 vaccines using replication-incompetent adenovirus vectors. *Annu Rev Med 55*, 355–372.

Stadler K, et al. (2003). SARS—beginning to understand a new virus. *Nat Rev Microbiol 1*, 209–218.

Steinbrook R. (2004). The AIDS epidemic in 2004. *N Engl J Med 351*, 115–117.

Stramer SL, et al. (2004). Detection of HIV-1 and HCV infections among antibody-negative blood donors by nucleic acid-amplification testing. *N Engl J Med 351*, 760–768.

Sullivan NJ, et al. (2003). Accelerated vaccination for Ebola virus hemorrhagic fever in non-human primates. *Nature 424*, 681–684.

Traggiai E, et al. (2004). An efficient method to make human monoclonal antibodies from memory B cells: potent neutralization of SARS coronavirus. *Nat Med 10*, 871–875.

Trotter CL, Andrews NJ, Kaczmarski EB, Miller E, Ramsay ME. (2004). Effectiveness of meningococcal serogroup C conjugate vaccine 4 years after introduction. *Lancet 364*, 365–367.

Venter JC, et al. (2004). Environmental genome shotgun sequencing of the Sargasso Sea. *Science 304*, 66–74.

Walters MJ. (2003). *Six Modern Plagues.* Washington DC: Island Press, pp. 19–46.

17.

PHARMACOGENETICS AND PHARMACOGENOMICS IN THE DEVELOPING WORLD

Warren E. Hochfeld, Sahle M. Asfaha, Marco Alessandrini, Tyren M. Dodgen, Renier Myburgh, Arinda Eloff, and Michael S. Pepper

INTRODUCTION

Half a century ago, the work of James D. Watson and Francis Crick divided world history in two eras: pre- and post-DNA double helix. The turn of the millennium saw the completion of the sequencing of the draft human genome. As a result of these and countless other landmarks in molecular cell biology, we know today that the haploid human genome contains 3 billion DNA base pairs tightly packaged into 2 sex chromosomes (X, Y) and 22 autosomal chromosomes. Although originally estimated at around 100,000, it is currently believed that the human genome contains 20,000–25,000 genes. Given the great complexity that characterizes the transcription and translation (including post-translational modifications) of genes into proteins, it is difficult to estimate the number of proteins synthesized in the human organism, but this is estimated to be well over 100,000.

We inherit hundreds of genetic mutations (germ-line) from our parents. Furthermore, the DNA in our own cells undergoes many new sporadic mutations during a lifetime. It is estimated that more than 10,000 human diseases are monogenic (single-gene disorders) and together these occur in about 1–2 out of every 200 births (0.5%–1.0% frequency). When a nucleotide sequence variant of a gene (allele) occurs in a given population with typically >1% allele frequency, which is greater than can be explained by recurrent mutation alone, this is called a *polymorphism.* Empowered by the completion of the sequence of the human genome, a haplotype map (HapMap) project is haplotyping the more than 1.4 million single nucleotide polymorphisms (SNPs), with over 60,000 of them occurring in gene coding regions. These polymorphisms are believed to be the source of interindividual variability revealed in several clinically important phenotypes, such as resistance or susceptibility to disease and variability in drug responses (Pearson & Manolio, 2008; Liu et al., 2004).

Interindividual variability in drug response and adverse drug reactions (ADRs) have been major challenges in the field of pharmacotherapy when attempting to design a standard drug dosage regimen that works across a wide patient spectrum. Therapeutic drug monitoring and application of kinetic-dynamic modeling are generally undertaken to improve precision and objectivity over the trial and error methods used to combat the problems of drug response variability. Two new disciplines have recently been redefined that have arisen out of the need to integrate polymorphism information from the human genome and HapMap into this classical scheme of optimizing drug therapy and rational drug design. They are *pharmacogenetics* and *pharmacogenomics.*

HISTORICAL BACKGROUND

The concept of chemical individuality developed by Sir Archibald Garrod in the early nineteenth century (Garrod, 1909) and the observation of racial differences in the incidence of the different taste phenotypes by Fox (1932) were probably the benchmark observations that foresaw the association of inherited components in the diversity of responses among individuals. About two decades later, in the 1950s, three independent studies clearly documented the emergence of the field of pharmacogenetics. These studies observed serious adverse drug reactions in some patients but not in others after the administration of primaquine and isoniazid, drugs used for the treatment of malaria and tuberculosis (TB) respectively. It was only later, in 1957, that Arno Motulsky clearly proposed a genetically determined individual variability in the incidence of these ADRs. Two years later, in 1959, Friedrich Vogel coined the term *pharmacogenetics.*

PHARMACOGENETICS AND PHARMACOGENOMICS

The aim of pharmacogenetic studies is to utilize the inherited genetic difference in drug metabolism among individuals to predict variations in drug efficacy and toxicity. Pharmacogenomics is a variation of pharmacogenetics that uses a genome-wide approach to study the entire spectrum of genes involved in a given drug response, and is used in the process of rational drug design. Two principal factors have motivated work in these fields: the presence of a large number of ADRs, and the lack of efficacy of certain drugs at conventional doses.

Subtherapeutic concentrations of drugs contribute to poor treatment outcome and help to drive the emergence or spread of drug resistance, whereas high concentrations are more likely to be associated with decreased safety and increased toxicity. This results in increased mortality, poor compliance, increased hospital stay and readmissions, which all lead to increased cost of treatment (Davies et al., 2007). To put this into perspective, in the United States alone (per annum), it has been estimated that 2 million hospital patients experience serious ADRs, leading to 141,000 deaths (Bond & Raehl, 2006). In the United Kingdom (per annum), 7% of all the patients are affected by ADRs and 1:10 National Health Service bed days are used by patients with ADRs, which costs an estimated £460 million (Wu et al., 2010). In many regions of the developing world, ADRs are underreported and statistics are largely lacking.

PHARMACOKINETICS AND PHARMACODYNAMICS

Harmonizing the dosage schedule with the time-frame concentration of the drug in the body, described by pharmacokinetics, and the association of the time-course concentration with efficacy/toxicity of the drug, described by pharmacodynamics, are the two most important parameters for optimizing drug treatment. Pharmacokinetics includes drug absorption, distribution, metabolism, and excretion, whereas pharmacodynamics addresses the physiological effects of drugs on the body and the relationship between drug concentration and effect (e.g., drug–receptor interactions). Pharmacodynamics is often seen as the study of what a drug does to the body, whereas pharmacokinetics is the study of what the body does to a drug.

DRUG ABSORPTION, DISTRIBUTION, AND EXCRETION

Except for drugs that act on the site of administration, the majority of drugs are passively absorbed through an epithelial barrier into the plasma. The drug in the plasma is then distributed to its site of action, where it must passively cross a cellular barrier into the inner compartment of the target tissue. Lastly, the drug is passively excreted from the body via body fluids, predominantly urine and bile and in some cases saliva, sweat, tears, milk, and exhaled air. Hence, the extent of drug absorption, distribution, and excretion from the body is a function of cellular membrane permeability to the drug, which in turn depends on its chemical and physical nature. There is also increasing evidence for the participation of many plasma-membrane embedded transport proteins in the active absorption, distribution, and excretion of drugs (Ayrton & Morgan, 2001). These proteins are involved in various phases, including the uptake of drugs from blood to liver, in hepatobiliary distribution and excretion, in active renal secretion, and in the influx/efflux of drugs from the intestine and brain.

One of the most studied drug transporter proteins is the P-glycoprotein encoded by the multidrug-resistant (MDR) 1 gene. The P-glycoprotein belongs to the superfamily of ATP-binding cassette transporters and actively effluxes a range of structurally diverse drugs including several anticancer drugs, cardiac glycosides, immunosuppressants, and others (Taubert et al., 2006). It limits accumulation of drugs in the intestine and brain by attenuating drug absorption and brain penetration. Other transporter proteins include multidrug-resistant protein-2 (MRP-2), serotonin transporter (SERT) and vesicular monoamine transporter (VMAT; see Dal, Sylte, & Ravna, 2004). Drug receptors such as β_2-adrenoreceptors, dopamine, mu opioid, and 5-hydroxy tryptamine 2C receptors have also been shown to influence drug absorption, distribution, and excretion (Schlag et al., 2004). Although very few are known, polymorphisms in these transporter and receptor proteins are likely to affect drug response, and therefore represent good candidates for pharmacogenetic studies.

DRUG METABOLISM

Most drugs are metabolized prior to being excreted from the body. Metabolism is the enzymatic biochemical transformation of lipophilic drugs into polar and readily excretable metabolites. Although its major role is detoxification and elimination of drugs and other xenobiotics, metabolism can also transform pro-drugs and procarcinogens into active therapeutics and potent carcinogenic compounds respectively (Zuber, Anzenbacherova, & Anzenbacher, 2002).

Metabolism is often divided into phase I and phase II reactions. Phase I reactions produce relatively more soluble metabolite through oxidation, hydroxylation, reduction, or hydrolysis. Phase II, which usually changes insufficiently polar metabolite products of phase I, consists of conjugation reactions including acetylation, glucuronidation, sulfation, and methylation (Crettol, Petrovic, & Murray, 2010). There are very high concentrations of phase I and II drug-metabolizing enzymes in the liver, which is the principal organ of drug metabolism. Other tissue types with the ability to metabolize drugs include the gastrointestinal tract, lungs, kidney, brain, and skin (Thomson et al., 2004).

FACTORS THAT AFFECT METABOLIC STATUS

Several factors have been reported to affect the pharmacokinetics of a drug and consequently the patient's response to drug therapy. The source of variability can be grouped into nongenetic and genetic factors.

NONGENETIC FACTORS

Several clinical parameters affect drug disposition. Generally, neonates and the elderly show slow metabolism compared to adults, and individuals with deteriorated organs such as the intestine, liver, or kidneys usually have a skewed drug response. Other clinical or environmental factors such as body weight, sex, pregnancy, concomitant disease, nutrition, and smoking contribute to interindividual variability in drug response. Drug disposition and response are also affected by interactions between co-administered drugs and the interaction of drugs with various environmental factors, such as dietary components (Katz et al., 2008).

GENETIC FACTORS

Genetic factors account for up to 95% of the interindividual difference in drug response. These differences are due to inherited DNA sequence variations (polymorphisms) in genes for drug-metabolizing enzymes, drug receptors, drug transporters, and molecules involved in drug-related signal transduction pathways. The polymorphisms in any of these genes are considered as the allelic variant of the wild-type allele. Many of these polymorphisms affect metabolizing enzymes by altering their catalytic activity, leading to one of four potential phenotypes: poor metabolizers (PM), intermediate metabolizers (IM), extensive metabolizers (EM) and ultra rapid metabolizers (UM).

PMs lack the functional enzyme (two defective allelic variants) needed to metabolize the drug. The outcome is a significantly elevated drug plasma concentration, making these patients more susceptible to adverse drug reactions at usual/recommended doses. UMs, who have duplications of one of the functional alleles, usually ranging from 3 to 13, have high enzyme activity and therefore an increased risk of therapeutic insufficiency due to diminished drug plasma levels at the standard dose. IMs are affected to a lesser degree by drug effects as they express an enzyme with a reduced activity. These individuals are either homozygous for low-activity alleles, or heterozygous for one reduced-activity allele and one null allele. EMs have a genotype that is homozygous for the normal alleles, or heterozygous for a normal allele and a defective/low activity allele. IMs and EMs show gene-dose effects, which is genotype-based variation in the mean half-life of drugs that, in turn, is a function of the extent of enzymatic activity.

POLYMORPHISMS IN DRUG-METABOLIZING ENZYMES

More than 30 families of drug-metabolizing enzymes have been identified, including cytochrome P450 (CYP450), N-acetyltransferase 2 (NAT2), thiopurine methyltransferase (TPMT), alcohol dehydrogenase (ADH3), and others. Compared to other drug-metabolizing enzyme systems, CYP450 is the most important, comprising drug-metabolizing enzymes that are involved in the metabolism of a significant proportion of commonly administered drugs (Eichelbaum, Ingelman-Sundberg, & Evans, 2006). Moreover, the CYP450s exhibit a relatively high degree of polymorphic variation, which is usually inherited as an autosomal recessive trait (Ingelman-Sundberg, 2005).

DRUG TRANSPORTERS AND TARGETS

Identification of polymorphisms with downstream functional effects in important drug-metabolizing enzymes, drug transporters, drug targets, and other drug-related genes allows for the potential prediction of therapeutic responses (efficacy and ADRs). As previously mentioned, drug transporter proteins play a significant role in the absorption, distribution, and excretion of parental drugs, as well as their metabolites (Ayrton & Morgan, 2001). Inhibition of transporter proteins has also been shown to play a role in drug–drug interactions, especially for poorly metabolized drugs. Moreover, for some drugs, such as fexofenadine, digoxin, talinolol, and pravastatin, transporter protein affinity can be the key rate-determining factor for their detoxification and elimination, where drug metabolism is negligible (Ayrton & Morgan, 2001).

The most commonly studied of these proteins is P-glycoprotein (see above). Two SNPs, 3435C>T and 2677G>T, which are in linkage disequilibrium, have been associated with enhanced P-glycoprotein *in vitro*, and lower plasma fexofenadine concentrations in humans (Ayrton & Morgan, 2001). Polymorphisms in drug targets such as β_2-adrenoreceptors have also been shown to alter therapeutic responses (Yuan et al., 2008). This makes it worth considering transporter proteins and drug targets for effective prediction of individualized drug therapeutic outcome.

PHARMACOGENETICS OF THE CYTOCHROME P450 ENZYME

Cytochrome P450 (CYP450) is a phase I enzyme primarily involved in Fe3+-dependent oxidative metabolism. Its expression is highly regulated and occurs principally in the liver, but is also found in the small intestine, lungs, and other organs. It is cytoplasmic and is associated with the endoplasmic reticulum (Ingelman-Sundberg, 2004a).

The human CYP450 superfamily consists of 57 active genes. The ones that appear to be commonly involved in metabolism occur in three families, namely CYP1, CYP2, and CYP3. Members of these families collectively account for approximately 80% of oxidative drug metabolism (Ingelman-Sundberg, 2005, 2004b).

CYP1 FAMILY

The CYP1 gene family consists of two subfamilies; namely CYP1A, consisting of CYP1A1 and CYP1A2 genes, and the subfamily CYP1B, with the CYP1B1 gene as the only known member. Even though these enzymes are involved in the metabolism of certain drugs, they are primarily involved in the detoxification or bioactivation of a number of environmental and dietary procarcinogens and some endogenous chemicals, such as estrogen (Elder et al., 2009; DeVito et al., 2000). These genes exhibit several polymorphisms with clear interethnic differences in the distribution of their allelic variants (Tang et al., 2000). However, the effect of these polymorphisms on the activity of the enzyme and their clinical importance is not yet clear.

CYP2 FAMILY

Members of the CYP2 family are the most highly polymorphic CYP450 enzymes. The CYP2 family includes the subfamilies CYP2A, CYP2B, CYP2C, CYP2D, and CYP2E. Some of the genes in these subfamilies include CYP2A6, CYP2B6, CYP2C8, CYP2C9, CYP2C18, CYP2C19, CYP2D6, and CYP2E1. Among these, CYP2C9, CYP2C19, and CYP2D6 are the most commonly studied, with far-reaching clinically significant roles.

CYP2C9 Gene Polymorphisms

The CYP2C9 gene is located at chromosomal region 10q24. The gene contains nine exons and encodes a protein of 490 amino acid residues. CYP2C9 is homologous to CYP2C19, differing only by 43 amino acids. The CYP2C9 enzyme is abundantly expressed in the human liver, constituting approximately 20% of the total hepatic CYP content. The enzyme catalyzes the metabolism of approximately 10% of therapeutically important drugs, including phenytoin, S-warfarin, tolbutamide, losartan, and torasemide, as well as diclofenac (Ahlstrom, Ridderstrom, & Zamora, 2007). It has also been shown to play a role in the metabolism of environmental procarcinogens such as benzo[a]pyrene, a major lung carcinogen (Yasar, Eliasson, & Dahl, 2002).

The two common CYP2C9 allelic variants are CYP2C9*2 and *3. These alleles are defined by nucleotide changes 430C>T (CYP2C*2) and 1075A>C (CYP2C9*3; see Ahlstrom, Ridderstrom, & Zamora, 2007). Interethnic variation in the distribution of the CYP2C9 variants has been documented. The wild-type variant, CYP2C9*1, is the most common allele in most populations studied so far. CYP2C9*2 and *3 are more frequent (6–11% allele frequency) in Caucasians, whereas they are rare (CYP2C9*3, 2%–3%) or absent (CYP2C9*2, 0%) in Orientals. In an African population, these alleles are more frequent in Egyptians, whereas they are relatively rare in Ethiopians and absent in Beninese.

Homozygous carriers of these variants with a genotype predictive of a PM phenotype have profoundly impaired metabolism of CYP2C9 substrate drugs. Homozygous carriers constitute 0.2%–1.0% of Caucasians and 2.8% of Egyptians, while they are absent from Ethiopians, Bennese, and Orientals (Xie et al., 2001; Scordo et al., 2001). These CYP2C9 polymorphisms play a dominant role in the dosage requirements and adverse effects of warfarin anticoagulant therapy (see "Diseases of Lifestyle" below). Like other enzymes, CYP2C9 is induced or inhibited by co-administered drugs.

CYP2C19 Gene Polymorphisms

Genetic polymorphisms in CYP2C19, which was first known as the *S-mephenytoin 4'-hydroxylation enzyme*, were reported in 1984. The gene is located on chromosome 10 and encodes an enzyme that affects the metabolism of more than 25 drugs, including S-mephenytoin, hexobarbital, diazepam, propranolol, imipramine, clomipramine, amitriptyline, omeprazole, proguanil, and chloroguanide (Foti & Wahlstrom, 2008). It is mainly expressed in the liver and the duodenum.

Up to 28 polymorphic variants of the CYP2C19 gene have been identified (see www.cypalleles.ki.se/cyp2c19.htm). The principal defective alleles in the PM phenotype are CYP2C19*2 and *3, while CYP2C19*1 is the normal allele. S-mephenytoin and omeprazole are usually the probe drugs used in CYP2C19 phenotyping studies to stratify populations into PM and EM phenotypes. The number of individuals classified as PMs of the CYP2C19 substrates is significantly higher in Orientals (14%–25%) compared to Caucasians (2%–7%), African Americans (1.05%), and in some of the African countries (4%–7.5%; see Edeki, 1996; Masimirembwa et al., 1995; Zhou & Liu, 2000; Daniel & Edeki, 1996).

There is a good correlation between phenotype and genotype in Orientals, where the defective alleles CYP2C19*2 and *3 have been shown to explain 100% of the PMs, with CYP2C19*2 responsible for 75% of the PMs and CYP2C19*3 for the rest (25%; see Edeki, 1996. The CYP2C19*3 allele is rare and/or absent in Caucasians, African Americans, and African populations; however, CYP2C19*2 seems to account for more than 75% of the

PMs in these populations. The less common defective alleles, CYP2C19*4, *5 and *6, were reported to contribute to the PM phenotype in Caucasians (Zhou & Liu, 2000). More recently, Sim et al. (2006) identified a novel CYP2C19 allele, CYP2C19*17, that causes increased CYP2C19 activity toward omeprazole and S-mephenytoin, leading to a UM phenotype (Sim et al., 2006). It is present in an allele frequency of 17.9% among Ethiopian populations (Sim et al., 2006).

CYP2D6 Gene Polymorphisms

The CYP2D6 gene is located on chromosomal band 22q13.1. It has nine exons and encodes a polypeptide of 497 amino acids. Although this mainly hepatic drug-metabolizing enzyme constitutes only 1%–2% of the total CYP protein in the liver, it is associated with the metabolism of almost one-fifth of all drugs currently on the market (Ingelman-Sundberg, 2005). These drugs include some antiarrhythmics, antidepressants, beta-blockers, neuroleptics, analgesics, antiemetics, and others (http://www.edhayes.com/startp450.html). The expression of this enzyme, which is not inducible and is extremely polymorphic, demonstrates clinically significant interindividual and interethnic variants. For these reasons, CYP2D6 has received much attention, and is at present an ideal candidate for diagnostic purposes (Ingelman-Sundberg, 2005).

Over 80 variants of CYP2D6 have been assigned to date (http://www.cypalleles.ki.se/cyp2d6.htm), comprising of point mutations, insertions or deletions as well as gene duplications. The effect of these alterations in the *in vivo* and/or *in vitro* enzyme activity of more than 52 of these alleles has been characterized. The majority of these alleles—including the most common, CYP2D6*3, *4, *5 and *40—have no enzyme activity. Some other common alleles that include CYP2D6*9, *10, *17, *29, and *41 have decreased activity, while the CYP2D6*$2_{n\geq2}$ allele is the genetic base for the UM phenotype. The prevalence of these CYP2D6 alleles tends to have a specific trend across different ethnic groups (Xie et al., 2001; Zhou & Liu, 2000; Aklillu et al., 2002; Masimirembwa et al., 1993; Sistonen et al., 2009).

The wild-type CYP2D6*1, as well as the CYP2D6*2 variant with normal enzyme activity, form the genetic basis for EM phenotype and are generally the predominant alleles in most of the populations genotyped so far.

The molecular basis for PM phenotype has been described as the occurrence of combinations of any of the detrimental alleles. The CYP2D6*5 allele, which has no enzyme activity due to gene deletion, is present with nearly the same allelic distribution in Caucasians (5.0%), Orientals (5.6%), African Americans (6.0%), and African populations (3.3%–10%). It contributes to some of the PM phenotypes in these populations. The CYP2D6*4 allele, which results from defective splicing, leads to an inactive enzyme that is highly prevalent in Caucasians (23%) while it is present in a lower frequency in African populations (1%–11.4%) and least so in Orientals (0.8%). The CYP2D6*3 allele, which is rare in Caucasians, is generally absent in Oriental and African populations. This is in parallel to the observation of the low frequency of PM in Oriental (0%–1%) and African populations compared to the high frequency of the phenotype in Caucasians (5%–10%; see Zhou & Liu, 2000).

Combinations of the alleles that code for low enzyme activity, as well as their heterozygous combination with the null alleles, are the genetic basis for the IM phenotype. The CYP2D6*10 allele with decreased enzyme activity is by far the most common allele in Orientals (50%), while it has a very low prevalence in Africans (3.1%–8.6%) and Caucasians (5%). The CYP2D6*17 and CYP2D6*29 alleles, which are absent or extremely rare in Caucasians and Orientals, are the two more common African-specific low-activity CYP2D6 alleles. CYP2D6*17, which has an allele frequency of 34% in Zimbabweans, has not been found in Oriental populations but occurs in Caucasians at an allele frequency of 0.3% (Masimirembwa et al., 1996). CYP2D6*29, which is highly prevalent among Tanzanians (20%), was identified in only 0.1% of individuals of European descent (Marez et al., 1997). Hence, CYP2D6*10 in Orientals and CYP2D6*17 and *29 in Africans explain the right shift in metabolic ratios of phenotyping studies in these populations (Wennerholm et al., 2001).

The UM phenotype, which occurs due to duplicated and multiduplicated CYP2D6 genes, is more common in Ethiopia (29%) and Saudi Arabia (20%). The phenotype is also relatively common in the Southern European populations (10%) compared to Northern European, Asian, and most Black populations, where the frequency is generally less than 2%.

CYP3 FAMILY

CYP3 consists of only one subfamily, CYP3A, which comprises four genes: CYP3A4, A5, A7, and A43. Members of this family are involved in the metabolism of several therapeutic drugs, carcinogens, fatty acids, and steroids (Baldwin, Marko, & Nelson, 2009). CYP3A4 is the most abundant, representing the largest CYP content of liver (30%) and intestine (70%), and has also recently been identified in the brain. Moreover, it oxidizes approximately half of all the drugs used today, including antiarrhythmics, antiretroviral, and antimalarial agents (Zuber, Anzenbacherova, & Anzenbacher, 2002). Numerous genetic variants have been identified especially in the CYP3A4 and CYP3A5 genes.

These polymorphisms are, however, not translated into significant interindividual variability in CYP3A-mediated drug metabolism, as the enzymes, particularly

CYP3A4, are highly inducible or easily inhibited by several substrates (Gellner et al., 2001). Moreover, the effect of polymorphisms in one of these genes may not be readily evident, as the CYP3A members show overlapping substrate specificity.

CLINICAL PHARMACOGENETICS

INFECTIOUS DISEASES

Infectious diseases have for centuries ranked with wars and famine as major challenges to human progress and survival, and they remain among the leading causes of death and disability worldwide (Morens, Folkers, & Fauci, 2004). Acquired immunodeficiency syndrome (AIDS), TB, and malaria remain the top three causes of mortality by infectious disease (Morens, Folkers, & Fauci, 2004).

Pharmacogenetics of Antiretrovirals

At present, more than 33 million people are infected with the human immunodeficiency virus (HIV), with 2.5 million new infections diagnosed in 2008 (Este & Cihlar, 2010). Sub-Saharan Africa remains most heavily affected by the pandemic, accounting for 67% of all people living with HIV and >70% of deaths from AIDS in recent years (Este & Cihlar, 2010; Asamoah-Odei, Garcia Calleja, & Boerma, 2004).

Since the approval of the first antiretroviral drug azidothymidine (AZT) in 1987, 31 approved single antiretrovirals have been developed in 6 mechanistic classes; these include nucleoside/nucleotide reverse transcriptase inhibitors (NRTIs), non-nucleoside reverse transcriptase inhibitors (NNRTIs), protease inhibitors (PIs), entry/fusion inhibitors (FIs) and, more recently, CCR5 antagonists and integrase inhibitors (Naeger et al., 2010; Tozzi, 2010).

International antiretroviral treatment guidelines currently recommend a daily combination of three or more antiretroviral drugs for the treatment of immune compromised HIV infected patients, a strategy referred to as *highly active antiretroviral therapy* (HAART). Although such regimens have considerable antiretroviral activity and have significantly decreased the mortality rates over the past few years, ARVs have multiple physiological effects other than the suppression of HIV, which can lead to adverse effects that include hypersensitivity reaction syndromes, hepatotoxicity, central nervous system toxicity, hyperbilirubinemia, peripheral neuropathy, lipodystrophy, hyperlipidemia, pancreatitis, and renal toxicity (Este & Cihlar, 2010; Tozzi, 2010; Tozzi et al., 1999, 2005; De Luca et al., 2000; Murri et al., 2003).

A relevant number of associations between human genetic variants and predisposition to these adverse events have been described, and for some antiretroviral drugs a clear genotype has been established. These genetic variants can play a causative role in treatment failure, either directly through subtherapeutic drug levels, which increase the risk of a poor virologic response, or indirectly, when toxic drug levels produce significant intolerability which leads to poor adherence.

Perhaps the most well-known genetic determinants of antiretroviral drug pharmacokinetics is the association of the hypersensitivity reaction to abacavir, related to the presence of the major histocompatibility complex class I allele HLA-B*5701 (Mallal et al., 2008; Phillips & Mallal, 2008; Saag et al., 2008). Abacavir is a nucleoside reverse transcriptase inhibitor available for once-daily use in combination with other antiretroviral agents. Abacavir has proven efficacy in HAART regimens, a favorable long-term toxicity profile, and results in few drug interactions (Tozzi, 2010; Phillips & Mallal, 2008). However, in 5%–9% of patients an immunologically mediated hypersensitivity reaction occurs that can lead to hospitalization and death (Tozzi, 2010). It is believed that this reaction is triggered when the HLA-encoded molecule presents the NRTI drug for T-cell activation, resulting in the release of inflammatory cytokines that initiate the potentially fatal hypersensitivity reaction. Studies on HIV infection confirm that avoiding abacavir in HLA-B*5701–positive patients can significantly reduce the incidence of the suspected hypersensitivity reaction, and as such, abacavir has become the preferred component of initial ART only for patients who test negative for HLA-B*5701 (Zucman et al., 2007). Considering that symptoms of the hypersensitivity reaction to abacavir are nonspecific and can be difficult to distinguish from concomitant infection and reaction to other drugs, this test could be extremely valuable in the clinical setting.

Another recently established pharmacogenetic association is that of UGT1A1 and MDR1 polymorphisms in atazanavir-associated hyperbilirubinemia (Park et al., 2010). Atazanavir is one of the latest protease inhibitors approved for the treatment of HIV infection. The drug has a relatively long half-life, a large inhibitory quotient, and is generally well tolerated. It was the first, and to date the only, protease inhibitor designed to be applied once daily, and was expected to overcome the adverse effects of earlier agents of this class of drugs, such as hyperlipidemia, diarrhea, and lipodystrophy (Rodriguez-Novoa et al., 2007). However, about 6% of patients who receive atazanavir present with hyperbilirubinemia, which can enter the range of clinical jaundice. This is due to a well-described association to a UDP-glucuronosyltransferase (UGT) promoter variant that contains 7 thymine adenine (TA) nucleotide repeats, A(TA)7TAA (UGT1A1*28), compared with the common promoter that contains only 6 TA repeats; the UGT1A1*28 promoter is less transcriptionally active than the common promoter (Park et al., 2010;

Rodriguez-Novoa et al., 2007). The underlying mechanism of this association is an inhibition of the UGT1A1 enzyme by atazanavir, which results in excess accumulation of bilirubin (hyperbilirubinemia). Studies across the world involving patients of different ethnicities treated with atazanavir have revealed that the frequency and severity of the hyperbilirubinemia correlates with these polymorphisms (Park et al., 2010; Rodriguez-Novoa et al., 2007). This risk is further increased by polymorphisms in the multidrug-resistant protein MDR1 gene, which is associated with increased atazanavir plasma levels.

Other antiretroviral drugs for which a clear genotype has been established include tenofovir-related kidney tubular dysfunction (KTD) mediated by MRP1 (Madeddu et al., 2008), NRTI associated peripheral neuropathy associated with point mutations in mitochondrial haplotype T (Keswani et al., 2002), and CYP2D6 polymorphisms associated with increased efavirenz plasma exposure and central nervous system side effects (Ribaudo et al., 2006; Haas et al., 2004). At this stage of treatment, patients could increasingly benefit from genotype-guided drug choice, which may become essential for choosing the optimal regimen, or avoiding potentially life-threatening adverse reactions.

Pharmacogenetics of Antimycobacterials

Tuberculosis (TB) is the fourth leading cause of death associated with infectious diseases globally, with 9 million new cases and 2 million deaths annually (Zhang, 2007; Waldman & Terzic, 2007a, 2007b). The World Health Organization (WHO) estimates that one-third of the world's population is infected by the bacterium that causes TB, and that the majority of new cases occur in developing countries, especially Southeast Asia and in sub-Saharan Africa. This has been amplified by the consequences of poverty and by its deadly synergy with HIV/AIDS.

Therapy of active TB includes a series of drugs, such as isoniazid (INH), ethambutol (EMB), rifampin (RMP), and pyrazinamide (PZA), which are administered in various combinations. However, because of the long-term nature of therapy, prolonged diagnostic delay, and failure to ensure optimal treatment, multi- and extensively drug-resistant strains of TB (MDR-TB and XDR-TB) have emerged (LoBue, 2009; Warren et al., 2009; Ingelman-Sundberg & Rodriguez-Antona, 2005). Treatment of these strains is severely compromised by the limited repertoire of drugs available, the toxicity of second-line compounds, and their relatively low bactericidal and sterilizing activity when compared to first-line anti-TB drugs.

In developing countries, most TB patients are tested for MDR-TB only after they fail to respond to standard treatments. However routine phenotypic drug susceptibility testing methods are often culture-based and can take between 3 and 6 weeks to develop. During this time, patients may be inappropriately treated, drug-resistant strains may continue to spread, and amplification of resistance may occur. With the aim of addressing these shortfalls, molecular techniques are now being applied widely to detect mutations associated with specific drug resistance.

The most recent example is the WHO-approved Hain GenoType® MTBDRplus assay (Hain Lifescience GmbH, Nehren, Germany) for the rapid diagnosis of INH and RMP resistance, which also allows for the simultaneous identification of *M. tuberculosis* complex strains in clinical isolates (Hillemann, Rusch-Gerdes, & Richter, 2007). A meta-analysis conducted in South Africa demonstrated that the test could reduce diagnostic delay to less than 48 hours, with reported sensitivities for RMP and INH resistance of ≥97% and ≥90% respectively (Bernard et al., 2008). Similar studies are being planned for Turkey, India, Vietnam, China, Thailand, and Uganda. As concluded in a manuscript recently published in the *American Journal of Respiratory and Critical Care Medicine*, the overall performance of this assay was superior to conventional testing (Bernard et al., 2008). Considering that the platform has the potential for high throughput at a lower cost than conventional testing (Bernard et al., 2008), this assay could also be applied for screening purposes to develop country-specific TB action plans.

The value of molecular techniques lies not only in the ability to rapidly confirm the diagnosis and refine drug-resistance patterns, but also to guide optimal therapy. For example, INH is the best current essential anti-TB drug to prevent the resistance in companion drugs, due to its high bactericidal activity and large therapeutic margin. More recently, studies have explored the benefits of high-dose INH, explicitly for TB-resistant cases (LoBue, 2009; Ingelman-Sundberg & Rodriguez-Antona, 2005; Katiyar et al., 2008). However, INH treatment can precipitate neurological and hepatic toxicity in a small percentage of patients. This is because INH is metabolized mainly by acetylation and dehydrazination in the liver, and this is genetically determined (Warren et al., 2009). Therefore, slow acetylators may incur higher blood concentrations and accompanying toxicity, which would be amplified in the case of the high-dose INH treatment groups. Rapid phenotypic or genotypic testing of slow or fast acetylator status could add an element of safety to reduce the potential for adverse effects and identify those who could stand to benefit from high-dose INH treatment in resistant TB cases.

Pharmacogenetic considerations could also provide information on the appropriateness of second-line therapies. For example, it has been suggested that mutations in the gyrA gene could determine whether moxifloxacin would be appropriate for the treatment of oxyfloxacin-resistant TB cases (Shi et al., 2006; Kam et al., 2006). A genotype microarray platform using common mutation markers to detect resistance to ethambutol, fluoroquinolones, and injectable antibiotics such as viomycin, kanamycin, amikacin, and capreomycin, is currently under development.

Antimalarial Pharmacogenetics

With approximately 243 million cases and 863,000 attributed deaths reported globally in 2009 (Gamo et al., 2010), malaria is one of the most severe infectious diseases, primarily affecting the world's most disadvantaged populations. Malaria is caused by protozoan parasites of the genus *Plasmodium* that infect and destroy red blood cells, leading to fever, severe anaemia, cerebral malaria and, if untreated, death. Malaria is both the cause and the consequence of poverty, and in countries with intense malaria transmission the economic impact of the disease slows growth considerably (Chuma, Okungu, & Molyneux, 2010).

One of the most important control measures in malaria is prompt diagnosis and treatment with an effective and safe combination of antimalarial drugs, as both drug efficacy and safety are strongly dependent on the achievement of appropriate circulating concentrations during treatment. Currently used antimalarials stem from seven main drug classes, although many more drugs are in clinical development. Over the past decade, treatment of acute uncomplicated malaria caused by all four species of *Plasmodium* involved the use of chloroquine for first-line treatment, and in combination with other drugs for second-line treatment (Sutanto et al., 2010). However, because of the increasing levels of resistance to conventional antimalarial drugs such as chloroquine, sulfadoxine-pyrimethamine, and amodiaquine, the WHO now recommends artemisinin-based combination therapies (ACTs) for the treatment of *falciparum* malaria (Mutabingwa, 2005). ACTs are extremely potent against both chloroquine- and sulfadoxine-pyrimethamine-resistant *P. falciparum in vitro* and *in vivo*, and can produce faster parasite clearance and fever resolution times than any other licensed antimalarial, including quinine. Combined with a favorable safety profile, even in children and pregnant women, ACTs are now used as first-line treatment throughout the world.

The clinical success of antimalarial treatment depends on a complex interplay of individual pharmacogenetic vulnerability and *Plasmodium* sensitivity in a multidrug environment (Kerb et al., 2009).

Most antimalarial drugs show large pharmacokinetic variability between individuals and undergo extensive metabolism, often involving various heritable genes. Some of the currently used antimalarials, and the metabolizing enzymes for which phenotypic and/or genotypic polymorphisms have been described, include CYP2A6 and CYP2B6 for the artemisinins (Svensson & Ashton, 1999), CYP2C8 for amodiaquine, chloroquine and dapsone (Gil, 2008), CYP2C9 and NAT2 for dapsone and sulfamethoxazole (Hodel et al., 2009), CYP2C19 for dapsone and proguanil (Hotel et al., 2009), CYP2D6 for chloroquine and halofantrine (Projean et al., 2003), CYP3A4 for the artemisinins, chloroquine, dapsone, halofantrine, lumefantrine, mefloquine, primaquine, and quinine (Kerb et al., 2009; Hodel et al., 2009), and CYP3A5 for chloroquine, mefloquine, quinine, and sulfadoxine (Grace et al., 1998).

The 8-aminoquinoline derivative primaquine is used in the treatment of *Plasmodium vivax* and *Plasmodium ovale* hypnozoites, and for malaria prophylaxis in endemic regions (Kerb et al., 2009). Single-dose primaquine has been widely recommended by malaria control programs for many years for these purposes. However, 8-aminoquinolines have oxidative properties that cause intravascular hemolysis in individuals with glucose-6-phosphate dehydrogenase (G6pD) deficiency (Leslie et al., 2010). At least 140 mutations have been described, which differ extensively in severity of the corresponding reduction in G6pD activity. G6pD deficiency differs greatly by region and ethnic group, and while G6pD deficiency is generally low (<2%), higher prevalence rates (with gene frequencies from 5%–25%) are found in tropical Africa, the Middle East, tropical and subtropical Asia, some areas of the Mediterranean, and Papua New Guinea (Mehta, Mason, & Vulliamy, 2000; Mehta, 1994). These findings are significant considering that over 90% of the annual malaria deaths occurring globally occur in 30 countries in sub-Saharan Africa and 5 countries in Asia (Jamison, Saxenian, & Bergevin, 1995).

Recently, attention has turned toward potential drug interactions between the combination partners, and between drugs simultaneously used to treat other diseases, such as infection with HIV or tuberculosis. In general, pharmacokinetic interactions involve mostly HIV PI and NNRTI classes, but more recently artemisinins have been demonstrated. PI (especially ritonavir) are among the most potent inhibitors of cytochrome P450 enzymes (CYP 3A4, CYP 2B6, CYP 2D6 and others; see He et al., 2009, 2010). Since P450 enzymes (in particular CYP 3A4) are central to the metabolism of a broad array of antimalarial drugs, these drug–drug interactions are an important and widely under recognized source of therapeutic insufficiency and adverse effects.

An essential component of malaria control will be the surveillance of therapeutic efficacy over time, in order to revise national drug policies and to ensure effective and safe treatment of malaria. A deeper insight into the interpopulation distribution of polymorphisms of genes encoding enzymes responsible for antimalarial drug metabolism could facilitate the selection of appropriate first-line treatments for uncomplicated malaria in a given population.

DISEASES OF LIFESTYLE

The diseases of lifestyle are becoming more prevalent as a result of industrialization and urbanization in the developing world. As the name states, lifestyle diseases manifest from unhealthy adaptations in lifestyle behavior. By far the most common in this category are the cardiovascular diseases (CVD) and the associated disorders of the metabolic syndrome, including diabetes and obesity.

Pharmacogenetics of Heart Disease and Stroke

Cardiovascular disease (CVD) is one of the leading causes of death worldwide, accounting for approximately one-third of the global mortality rate. An even more astounding statistic is that more than two-thirds of these deaths take place in the low- to middle-income populations. It is predicted that 23 million cardiovascular deaths will occur by the year 2030, with approximately 85% of these occurring in the developing world (Mathers & Loncar, 2006).

Although several treatment options are available to clinicians for the management of CVD, there is a great deal of interindividual variation in therapeutic responses. Although this variation can be attributed to several factors including age, gender, diet, drug interactions, and various nonspecific environmental factors, numerous reports demonstrate that variable response can be explained, at least to some extent, by genetic variation in the enzymes responsible for drug metabolism, as well as targets of commonly used CVD therapies.

Hypertension and the Pharmacogenetics of Antihypertensive Therapy

Hypertension is the most common risk factor for CVD globally. Approximately 50% of patients with coronary heart disease and 75% of those with stroke can be accounted for by the effects of hypertension. In the United States alone, approximately 30% of the adult population suffers from high blood pressure (Hajjar & Kotchen, 2003), while almost one billion people worldwide are hypertensive (Kearney et al., 2005). The role of pharmacogenetics in the response to antihypertensive treatment has been explored in several candidate genes.

Renin-Angiotensin-Aldosterone System Targeting Therapies The renin-angiotensin-aldosterone system (RAAS) plays a major role in the development and progression of cardiovascular diseases by promoting vasoconstriction, sodium reabsorption, cardiac remodeling, norepinephrine release, and other potentially detrimental effects (Atlas, Niarchos, & Case, 1983; Atlas et al., 1986). Drugs that inhibit or antagonize components of the RAAS system can effectively target blood pressure; however, there is often substantial variability in individual responses (Atlas, 2007). For example, fewer than 50% of hypertensive patients achieve adequate blood pressure control with ACE inhibitor monotherapy, and recent data suggest that therapeutic insufficiency runs within certain ethnic groups (Cusi, 1997; Yu, Zhang, & Liu, 2003).

Pharmacogenetic studies of drugs targeting the RAAS have investigated the genes encoding renin, angiotensinogen, ACE, the angiotensin II type I receptor, and CYP11B2 (aldosterone synthase). The most frequently investigated ACE variants with pharmacogenetic implications are the insertion/deletion (I/D) polymorphisms, as defined by an insertion or deletion of a 287 bp DNA sequence, in intron 16 of the gene (Rigat et al., 1990). There have been numerous studies reporting effect on blood pressure responses in individuals or subjects harboring either of the ACE insertion or deletion alleles (Yu, Zhang, & Liu, 2003; Hingorani et al., 1995). However, a homozygous genotype for insertion (II) has also been reported to be associated with decreased blood pressure (Ohmichi et al., 1997; Tamaki et al., 1997; Suwelack et al., 2002; Weekers et al., 2005). In stark contrast, there are also reports of a decrease in blood pressure in patients homozygous for the 287 bp deletion (DD) while receiving an ACE inhibitor (Hernandez et al., 2000; Li, Du, & Huang, 2003; Stavroulakis et al., 2000).

The gene that encodes angiotensinogen (AGT) was initially investigated by Hingorani et al. (1995) in patients receiving ACE inhibitors.. The M235T polymorphism in AGT was observed to be significantly associated with improved blood pressure responses in individuals or subjects harboring the T-allele. However, in a more recent study of almost 1,500 individuals, this associated result could not be replicated (Su et al., 2007). The angiotensin II type I receptor gene (AGTR1) was initially investigated by Benetos et al. (1996a, 1996b) in patients receiving an ACE inhibitor. Those that harbored the C1166 allele proved to respond better to the ACE inhibitors in terms of blood pressure. Unfortunately, this finding could not be repeated in a later study by Kurland et al. (2001).

The contradictory nature of these reports unfortunately does not allow for definitive and reliable use of ACE genotyping. However, these results are not surprising in view of the complex signaling pathway of the RAAS: numerous proteins are involved in the RAAS pathway, and many of these proteins exhibit functional polymorphisms. While studies of individual RAAS gene polymorphisms are simple to perform, it is unlikely that they will yield predictive accuracy sufficient for pre-prescription genotyping. Rather, future studies of drug responses should focus on multiple RAAS gene polymorphisms in the form of pharmacogenomic (as opposed to pharmacogenetic) approaches to evaluate these gene associations. There is hope that RAAS "polymorphism combinations" may provide more insight.

Beta-Blockers Although β-adrenergic receptor antagonists (β-blockers) are widely used in the treatment of cardiovascular disease, it is estimated that 25%–45% of patients with heart failure (HF) suffer adverse side effects from β-blocker therapy (Muszkat & Stein, 2005), 60% of patients fail to achieve adequate blood pressure control (Materson et al., 1993; Shin & Johnson, 2007), and 25% of patients require discontinuation of therapy due to intolerance (Tandon et al., 2004). Results of numerous studies now suggest that this variability may be accounted for, at least in part, by genetic polymorphisms that determine patient-to-patient variability in drug response (Azuma & Nonen, 2009).

Since the beta adrenergic receptors are G-protein coupled receptors, variations in the beta3 subunit of the G-protein gene (GNB3) have been investigated for pharmacogenetic potential. The C825T polymorphism is proposed to be associated with elevated blood pressure, acting through a mechanism involving stimulation of the Na+/K+-exchanger. Some studies have found a significantly lower use of antihypertensive medication among men with the CC genotype compared to men with CT and TT genotype, and a few studies have reported increased use of antihypertensive medication among hypertensive patients with the TT genotype. However, several follow-up studies examining the C825T polymorphism and susceptibility to hypertension have been inconsistent.

There are at least 6 papers in the literature reporting an association between blood pressure lowering with β-blockers and 1 of 2 genetic polymorphisms in the β_1-adrenergic receptor gene (ADRB1), Ser49Gly and Arg389Gly. The variant at codon 49 results in increased agonist-promoted desensitization, whereas the codon 389 polymorphism alters G-protein coupling and adenylyl cyclase activity. Various studies have implicated ethnicity in the variability of response to β-blocker therapy. For example, beta-blockers used as monotherapy appear to work least well in black hypertensive patients (Etuk et al., 2008). In addition, in the BEST (Beta-Blockers Evaluation Survival Trial) study, the use of bucindolol was associated with a nonsignificant increase in the risk of serious clinical events in black patients; however, Caucasian patients presented with reduced hospitalization and mortality (Sosin et al., 2004). This makes sense, considering that the less responsive Gly389 and Gly49 variants are approximately twice as frequent in Africans than in Caucasians. The allelic frequency of Arg389Gly polymorphism in Chinese, Japanese, and Indian populations closely match those of the Caucasian population. These differences in beta-blocker response appear to reflect ethnic differences in allelic frequencies in the polymorphisms that are critical to drug response, highlighting in this case that pre-prescription genotyping for cardiovascular treatment may be extremely beneficial in African populations.

Diuretics Diuretics act at various sites in the kidneys and result in initial plasma volume contraction and a decrease in cardiac output. This class of drug is also thought to block the sensitivity of blood vessels to catecholamines and reduce peripheral vascular resistance.

The α-adducin (ADD1) gene G460W polymorphism was one of the first polymorphisms to be investigated in hypertensive patients receiving diuretics. The polymorphism results in a glycine (G) to tryptophan (W) amino acid change at position 460, which in turn results in an increased affinity for sodium-potassium adenosine triphosphatase (ATPase). This activation results in increased renal tubular reabsorption of sodium (Manunta et al., 1999), lower plasma renin activity, and a more pronounced decrease in blood pressure after diuretic treatment (Cusi, 1997; Cusi et al., 1997).

In hypertensive patients receiving diuretic treatment, a mean blood pressure decrease of 14.7 mm Hg was observed in those heterozygous for the ADD1 W-allele, as opposed to a mean decrease of 6.8 mm Hg in patients that were homozygous for the G-allele (Glorioso et al., 1999). The significance of the ADD1 460W allele was further confirmed by Sciarrone et al. (2003), who reported a mean decrease of 12 mm Hg in arterial blood pressure in patients harboring at least one W-allele. Patients homozygous for the G-allele presented with a decrease of only 6 mm Hg. Although the previously mentioned studies were supportive of the role of the ADD1 W-allele, these results could not be confirmed in several later studies of larger patient populations (Schelleman et al., 2006a,b,c; Turner et al., 2003). The same variant was also investigated in other population groups where the risk of stroke and myocardial infarction were evaluated. In a single study (Psaty et al., 2002), it was demonstrated that the ADD1 W-allele was associated with a decreased risk for stroke and myocardial infarction for subjects undergoing diuretic treatment. This association could not be replicated in later studies Davis et al., 2007; Schelleman et al., 2007).

Frazier et al. (2004) also investigated the A1166C polymorphism in the AGTR1 gene in their cohort of African Americans and non-Hispanic whites receiving diuretic treatment. Upon stratification of the cohort for analysis, it was revealed that a significant association of the A1166 allele exists between African-American women and improved response to diuretics. This was, however, not the case for African-American males, nor for either of the non-Hispanic genders. There have been no follow-up studies that have been able to replicate these results, irrespective of the population group (Schelleman et al., 2006b; Jiang et al., 2007).

The C825T polymorphism of the GNB3 gene, as investigated in patients receiving beta-blockers, was also evaluated by Turner et al. (2001) in patients receiving diuretic treatment. The findings indicated that individuals homozygous for the T-allele (TT) responded more favorably in terms of blood pressure than individuals that were homozygous for the C-allele (CC). More recently, Schelleman et al. (2006a) confirmed this association and the benefits associated with the GNB3 TT genotype. Considering that the T-allele frequency is highest in black African (79%) and Asian populations (46%), and lowest in northeastern European populations (<25%), this polymorphism may be more relevant for the treatment of specific ethnic groups within the developing world.

Statins

Statin therapy aims primarily to reduce the risk of CVD by inhibiting 3-hydroxy-3-methylglutaryl coenzyme A (HMG CoA) reductase, an enzyme involved in cholesterol synthesis. Inhibition of HMG CoA reductase lowers LDL-C levels by slowing down the production of cholesterol in the liver and increasing the liver's ability to remove the LDL-C already in the blood. Although large clinical trials have established the beneficial effects of statin treatment in primary and secondary prevention of cardiovascular disease, considerable interindividual variation exists in response to therapy. The search for genetic determinants of treatment-effect heterogeneity has included more than 40 different genes involved in the pharmacokinetic and pharmacodynamic pathways of statin metabolism (Zintzaras et al., 2001).

Candidate genes for evaluating varied response profiles to statins include those involved in the cholesterol pathway. Among the most extensively studied is apolipoprotein E (ApoE), which plays a role in controlling cholesterol levels in the liver. Three major apolipoprotein E isoforms are coded by three alleles at the *APOE* gene, designated as E2, E3 and E4. The E2 variant of ApoE has been reported to be associated with improved response to statin therapy when compared to individuals harboring the E3 or E4 variants (Ordovas et al., 1995). However, the available evidence published to date is unreliable owing to inconsistent and inconclusive results of individual studies.

Being the target of statin therapy, the HMG CoA reductase gene has been investigated for variants influencing response to therapy. In a large study of 1,536 individuals on lipid lowering therapy (pravastatin), two key variants (SNP12 and SNP29) were found to be associated with the lowering of the levels of both total cholesterol and low density lipoprotein (LDL) cholesterol (Chasman et al., 2004). In contrast, though, Thompson et al. (2005) reported a lack of pharmacogenetic association of SNP29 and its lipid lowering effects in response to statin therapy.

The cholesteryl ester transfer protein (CETP) performs the function of reverse cholesterol transport by transporting of cholesterol from peripheral tissues back to the liver. A common variant of CETP is the Taq1B polymorphism. As originally reported by Kuivenhoven et al. (1998), individuals on pravastatin therapy and homozygous for the B1 variant progressed at a slower rate to coronary atherosclerosis. In addition, those that were homozygous for the B2 allele did not respond to the same therapy. These associations could not be confirmed in later studies (Klerkx et al., 2004; de Grooth et al., 2004; Marschang et al., 2006), hence placing doubt in the validity of the CETP polymorphism as a pharmacogenetic marker.

Other genes that have been investigated for variants that contribute to varied response profiles in patients taking statins include the LDL receptor, apolipoprotein A1, the paraoxonase enzyme, and the sterol regulatory elements binding factor 1.

Coagulation and the Pharmacogenetics of Anticoagulant Therapy

The worldwide standard of care for the major coagulation-related disorders, such as thromboembolic diseases and acute coronary syndromes, are warfarin oral anticoagulant and clopidogrel antiplatelet therapy (Manolopoulos, Ragia, & Tavridou, 2010; Sabatine et al., 2005a, 2005b).

Warfarin The anticoagulant warfarin is a coumarin vitamin K antagonist that has been widely used for over 60 years for the treatment and prevention of thrombotic diseases, including myocardial infarction, ischemic stroke, venous thrombosis, and following heart valve replacement and atrial fibrillation Schwarz et al., 2008). However, because of its narrow therapeutic index, it is one of the most common causes of emergency room visits for severe adverse drug reactions (Budnitz et al., 2007).

The two key enzymes responsible for the metabolism of warfarin are CYP4502C9 (CYP2C9 gene), which is involved in its metabolic clearance, and the C1 subunit of the vitamin K 2,3-epoxide reductase complex (VKORC1 gene) which recycles reduced vitamin K and is essential for the post-translational gamma-carboxylation of vitamin K–dependent clotting factors II (prothrombin), VII, IX, and X (Perez-Andreu et al., 2009).

A large genome-wide association study (GWAS) by Takeuchi et al. (2009) involving >1000 subjects demonstrated that polymorphisms occurring in VKORC1 and CYP2C9 had a significant effect on warfarin therapy, with a strong correlation between certain polymorphisms and anticoagulant properties/bleeding complications. Patients with CYP2C9*2 and CYP2C9*3 allelic variants and the VKORC1 haplotype A/A have been shown to require lower doses of warfarin to achieve an appropriate state of anticoagulation with a minimal risk of bleeding (Rieder et al., 2005; Higashi et al., 2002). However, the SNPs in

CYP2C9 account for only 6%–18% of the variance in final warfarin dose, the latter being more strongly influenced by VKORC1 haplotypes, which independently influences up to 30% of the variance in final dose requirements (Schwarz et al., 2008; Pereira & Weinshilboum, 2009).

Recently, Limdi and colleagues undertook a comprehensive assessment of the influence of 6 common VKORC1 SNPS and haplotypes on warfarin dose among various ethnic groups in the largest racially diverse cohort studied to date (Limdi & Veenstra, 2008; Limdi et al., 2010). They found that VKORC1 –1639G>A and 1173C>T individually explained the greatest variance in dose among the ethnic groups, but it was –1639A (or 1173T) that best explained the differences in the percentage of variance in dose across ethnic groups. For example, the low frequency of the derived T-allele observed among African populations (<10% with the exception of the San at 33%) is consistent with previous research indicating that individuals of African ancestry require a higher average warfarin dose than those of European or Asian descent. In contrast, the East Asian populations show extremely high frequencies for the derived T-allele, ranging from 75% (among the She) to 100% (Han and Oroqen).

In 2009, the international Warfarin Pharmacogenetics Consortium published a report showing the utility and success of using a genotype-based approach for predicting warfarin dosing. It was suggested that a pharmacogenetic approach was better at predicting the required warfarin dose than a clinical algorithm alone, or a fixed-dose approach (Pereira & Weinshilboum, 2009; Klein et al., 2009). Genetics-based dosing cannot eliminate clinical management-associated problems such as fragmented care and socioeconomic factors, which hinder frequent and effective INR measurement (Garcia & Hylek, 2009). However , by genotyping both VKORC1 and CYP2C9 in conjunction with physiological factors such as weight, height, and concomitant medications, clinicians will be able to use both pharmacokinetic and pharmacodynamic information to improve warfarin dose prediction and the efficiency of the dosage titration process, especially in susceptible patients (Klein et al., 2009).

Clopidogrel Clopidogrel is the standard of care in many patients undergoing percutaneous coronary intervention and those experiencing acute coronary syndromes. However, it has been suggested that response to clopidogrel varies widely, with nonresponse rates ranging from 4% to 30% at 24 hours (Mega et al., 2009). Following absorption through the ABCB1 p-glycoprotein protein transporter, clopidogrel (prodrug) requires several bioactivation steps largely mediated by CYP450 (CYP2B6, CYP2C9, CYP2C19, CYP3A4 and CYP3A5) to form an active metabolite (Taubert et al., 2006; Kurihara, Kazui, Ishizuka, Farid, & Ikeda, 2005; Kazui et al., 2010). This active metabolite binds irreversibly to the $P2RY_{12}$ platelet surface receptor, inhibiting adenosine diphosphate (ADP) induced platelet aggregation (Hollopeter et al., 2001; Savi et al., 2000). Suggested mechanisms for this variability include intrinsic interindividual differences resulting from genetic polymorphisms in the pathways of clopidogrel pharmacokinetics and pharmacodynamics.

Three SNPs identified in ABCB1 (C136T, G2677T and C3435T) have been investigated for association with reduced clopidogrel response; however, the function of these polymorphisms is unclear. In addition, several articles have seriously questioned the proposed correlation between genetic variations in CYP2B6, CYP3A4 and CYP3A5 and variations in clopidogrel response (Brandt et al., 2007; Hulot et al., 2006; Simon et al., 2009). However, CYP2C19 loss-of-function alleles (*2, *3, *4, or *5) appear to be strongly associated with variation in clopidogrel response. These loss-of-function alleles result in lower levels of active clopidogrel metabolite and reduced platelet inhibition, which are strongly associated with increased risk of death or major adverse cardiovascular events, including stent thrombosis and myocardial infarction or stroke (Mega et al., 2009; Simon et al., 2009). The link between decreased patient response to clopidogrel and common variations of CYP2C19 was particularly important with CYPC19*2, where patients were twice as likely as those without the variant to sustain an ischemic event. As CYP2C19*2 has been found at a frequency of 11%–22% in Africa, 11%–16% in Europe, 13%–34% in Asia, 8%-16% in the Americas, and high frequencies in Melanesia (60.2%) and Australia (35.5%), the clinical relevance is important for clopidogrel treatment around the world. For the reasons listed above, the U.S. Food and Drug Administration now recommends pre-prescription genotyping prior to clopidogrel treatment (Thumm et al., 1994). Patients with known clopidogrel susceptibility could receive a dosage adjustment or receive a drug alternative, such as prasugrel.

Diabetes, Obesity, and the Metabolic Syndrome

The metabolic syndrome is characterized by a group of conditions that include excess body fat, high blood pressure and cholesterol levels, and insulin resistance. Collectively, these conditions result in an increased risk of heart disease, stroke, and type 2 diabetes (T2D). Propelling the upsurge in cases of diabetes and hypertension is the growing prevalence of obesity, which has, during the past decade, joined malnutrition and infectious diseases as major health problems threatening the developing world (Hossain, Kawar, & El Nahas, 2007). The complexity of these disorders stems from the fact that they arise from an intricate interaction between the patient's genetic makeup and environment/lifestyle.

The cornerstone of treatment for the lifestyle disorders includes dietary modifications, exercise, and weight loss. However, several pharmacological options also exist. The biguanide metformin is an antihyperglycemic agent used in the treatment of T2D, particularly in patients not responding to dietary modifications. The drug is also used to promote weight loss in obese patients. Metformin plays its role in controlling hyperglycemia by decreasing hepatic glucose production and intestinal absorption of glucose, thereby improving insulin sensitivity by increasing peripheral glucose uptake and utilization. The sulfonylureas are used exclusively for the treatment of T2D. However, there is considerable interindividual response in patients receiving this treatment. The sulfonylureas bind to receptors on the pancreatic β-cells and result in the closure of potassium channels. Consequently, there is a depolarization of the cell, followed by a calcium influx and insulin release (unrelated to glucose concentration). Other treatment options for patients with T2D or the metabolic syndrome that are subject to pharmacogenetic investigations include the meglitinides, thiazolidinediones, alpha-glucosidase inhibitors, glucagon-like peptide-1 receptor agonists, and dipeptidyl peptidase-4 inhibitors.

The organic cation transporter 1 (OCT1), which is expressed in hepatocytes, has been demonstrated to play a significant role in mediating response to metformin. OCT1 is the major entry point for metformin into hepatocytes and enterocytes. A great deal of variation in the response to metformin is observed, and it is suggested that genetic variation in OCT1 may contribute to this (Shu et al., 2007). OCT1 is highly polymorphic, with the 420del and R61C variants being most prevalent in Europeans at frequencies of 19% and 7.2%, respectively (Bis et al., 2003; Shu et al., 2003). Furthermore, this variation appears to differ across ethnic groups, with the 420del polymorphism being present at a lower frequency (5%) in African Americans (Shu et al., 2003).

The potassium inwardly rectifying channel, subfamily J, member 11 (KCNJ11) is one of many genes of the sulfonylurea binding complex. It has been reported that the common Glu23Lys polymorphism in KCNJ11 is associated with increased risk for secondary failure to sulfonylurea treatment in patients with T2D (Sesti et al., 2006). Individuals harboring at least one copy of the Lys23 variant were reported to have a statistically significantly increased risk of secondary failure to treatment, when compared to individuals that were homozygous for the Glu allele. Furthermore, in a second study of a larger cohort, this finding was supported by the fact that individuals (although on metformin treatment) with the Lys23 variant progressed to diabetes at a higher rate than those that were homozygous for Glu23 (Sesti et al., 2006). These findings corroborated those previously reported by Gloyn et al. (2003) where it was indicated that homozygosity of Lys23 was associated with individuals with T2D.

The transcription factor 7-like 2 (TCF7L2) and variants thereof have been strongly associated with T2D risk. This was investigated by Pearson et al. (2007) in a large cohort of patients receiving either sulfonylurea or metformin treatment. The authors reported an almost twofold likelihood of sulfonylurea treatment failure in patients harboring a TT genotype at genomic position 14798892 (rs12255372).

The Sulfonylurea receptor 1 (SUR1), or the ABCC8 (ATP-binding cassette, subfamily c, member 8), gene has also been correlated with altered response to sulfonylurea treatment. The receptor is a subunit of the pancreatic β-cell potassium channel and plays a key role in the regulation of glucose-induced insulin secretion (Ueda et al., 1999). The common variant in intron 16, -3T/C, of SUR1 was investigated in T2D patients on sulfonylurea treatment, but could not be associated with an influence on plasma insulin, glucose, and cholesterol levels (Zychma et al., 2002). However, when it occurred in combination with a silent Thr755Thr change in exon 18, a significant reduction in insulin response was reported in diabetic patients on the sulfonylurea tolbutamide (Hansen et al., 1998).

Other genes that have been implicated in response to sulfonylurea treatment include calpain 10 (CAPN10), peroxisome proliferator-activated receptor γ (PPARγ), insulin receptor substrate 1 (IRS-1) and adiponectin.

NEUROPSYCHIATRIC CONDITIONS

Affective Disorders

Affective disorders are psychiatric diseases with multiple components, including biological, behavioral, social, and psychological. Although the affective disorders are not a clearly delineated group of diseases, they include conditions such as unipolar and bipolar depression, generalized anxiety disorder, and more specific anxiety disorders such as agoraphobia, panic disorder and social phobia, obsessive-compulsive disorder, and posttraumatic stress disorder (PTSD; see Strakowski, Delbello, & Adler, 2005; Keller, Schatzberg, & Maj, 2007). Major depressive disorder, psychosis, and bipolar disorders, which are the most common affective disorders, are highly prevalent and chronic, and therefore represent a substantial burden on public health.

Despite the availability of a wide range of different drug classes to treat these conditions, about 30%–50% of patients do not respond to acute treatment regardless of the initial choice of psychiatric medication (Steimer, 2010). Of those that do respond, 10%–20% suffer from a number of clinically relevant adverse effects including sedation, extrapyramidal side effects, insomnia, vertigo, and weight gain (Arranz & de Leon, 2007; Kirchheiner et al., 2004). The need to maximize clinical outcomes in major psychiatric conditions makes pharmacogenetics an attractive treatment strategy.

Antidepressant Pharmacogenetics

About one in six individuals in the United States will succumb to clinical depression during their lifetime (Krishnan & Nestler, 2008). Core symptoms include depressed mood, anhedonia (reduced ability to experience pleasure from natural rewards), irritability, difficulties in concentrating, and abnormalities in appetite and sleep.

The development of the selective serotonin-reuptake inhibitors (SSRIs), including fluoxetine, sertraline, fluvoxamine, paroxetine, and citalopram, represents an important advance in the pharmacotherapy of major depressive disorders. However, one of the major drawbacks of these treatments remains the high percentage of nonresponders (30%–40%) and the time that it can take for drugs to take effect (up to 6 weeks; see Drago, De Ronchi, & Serretti, 2009; Gorlyn et al., 2008). Genetic variation is responsible for a large percentage of variation in antidepressant response, both in terms of symptomatic improvement and tolerability.

CYP2D6 and CY2C19 are important enzymes for neuropsychiatric pharmacogenetics, and numerous studies have shown that their metabolic rates play an important role in the required therapeutic doses of many antidepressant and antipsychotic substrates (Kirchheiner et al., 2004). This could represent an area of pharmacogenetic potential, especially considering that variants in both of these genes vary geographically and within ethnic groups.

An important pharmacogenetic target for antidepressant therapy is the serotonergic system. An area in the promoter of the serotonin transporter (SERT) gene, the 5HTT gene-linked polymorphic region (5-HTTLPR), is comprised of short (s) and long (l) alleles. In terms of transcriptional activity, the l variant of 5-HTTLPR is more than twice as active as the s analogue, and therefore key in therapeutic responses (Kato & Serretti, 2010). Smeraldi et al. (2006) were the first to show the therapeutic association of the 5-HTTLPR polymorphism by demonstrating improved response to fluvoxamine in l allele carriers (l/l and l/s) when compared to s-variant (s/s) homozygotes. The majority of subsequent studies have produced similar findings. The s allele is more common in Asian (70%–80%) than Caucasian populations (38%–50%) whereas the l allele is more common in populations of African origin (>70%) compared to Japanese (19%) and Koreans (25%; see Kato & Serretti, 2010; Goldman et al., 2010; Durham et al., 2004).

Other genes that contribute to response determination in antidepressant therapy include brain-derived neurotrophic factor (BDNF) and its receptor, the neurotrophic tyrosine kinase receptor type 2 (NTRK2). Several recent reports have linked the loss of BDNF from the hippocampus to depression vulnerability, whereas upregulation has been associated with improved antidepressant efficacy and decreased drug-induced adverse effects (Lin et al., 2009). It has also been strongly suggested that acute reduction in central BDNF expression and dysfunction in the NTRK2 pathway are involved in treatment-related suicidal ideation (Perroud et al., 2009).

FKBP5, a glucocorticoid receptor–regulating co-chaperone of hsp-90, has also been associated with response to antidepressant drug treatment. The first study investigating FKBP5 variation and response to antidepressant treatment was performed in a cohort of 294 individuals suffering from depressive disorders (Binder et al., 2004). The strongest association was to rs1360780, an SNP located in the second intron of FKBP5, where patients presenting with a rs1360780 TT homozygous genotype demonstrated superior response over the 5-week treatment course (Binder et al., 2004). Subsequent attempts to replicate these findings have resulted in positive and negative associations between FKBP5 polymorphisms and antidepressant response.

Pharmacogenetics of Mood Stabilizers

Lithium, valproate, and carbamazepine are first-line agents for both acute and long-term treatments of bipolar disorders. Lithium is characterized by a narrow therapeutic index, and its clinical use is greatly impacted by important side effects, the most significant of which are an alteration in thyroid function and nephrogenic diabetes insipidus (Aurell, 2004). Given the paucity of clinical predictors of treatment response, genetic predictors would significantly assist clinicians. However, selecting candidate genes for pharmacogenetic investigation in bipolar and related conditions is difficult because the exact mechanism of action of established mood stabilizers remains uncertain.

The phosphoinositide pathway has been one of the major focal points for the investigation of lithium's mode of action. It has been postulated that the effects of lithium are transduced via G-protein-coupled phospholipase C (PLC) formation of inositol-1,4,5-triphosphate (IP3) and diacylglycerol (DAG) from phosphatidylinositol-4,5-bisphosphate (PIP2; see Meltzer, 1990). Several studies that have focused on the phosphoinositide pathway have identified two important genes that undergo lithium-induced inhibition, one coding for myo-inositol monophosphatase (IMAP1), and the other coding for inositol polyphosphate 1-phosphatase (INPP1; see Craddock & Jones, 2001). Although these genes have been thoroughly tested for possible association with lithium response, there has been a lack of conclusive data. Another protein that has been linked to lithium action is the myristoylated alanine-rich C-kinase strate (MARCKS). MARCKS is phosphorylated following the activation of PKC and is involved in the neuroplasticity of neurons. Lithium has been shown to decrease MARCKS protein expression in the brain, and this is hypothesized to occur in relation to the concentration of myo-inositol and the activation of the phosphoinositide pathway (Lenox & Wang, 2003).

A number of papers have also dealt with genes encoding elements of neurotransmitter systems. The weight of evidence suggests that the primary actions of lithium are on the serotonergic pathway, although the dopaminergic system has also been implicated. The 5-HTTLPR polymorphism has been investigated in lithium response by a number of authors, who have reported an association between the short allele (s) and/or genotypes carrying the s allele and a diminished response to lithium (Ferreira et al., 2009; Contreras et al., 2010). Another study investigating the interactive effect of 5-HTTLPR and BDNF Val66Met showed that patients with 5-HTTLPR s/s and l/s genotypes having the BDNF Val/Val genotype were the most frequent nonresponders (Bocchio-Chiavetto et al., 2008). However, other authors have failed to report positive findings for this polymorphism. Another candidate gene study examined the role of the Val158Met polymorphism in the catechol-O-methyl transferase (COMT) gene in response to mood stabilizers (Lee & Kim, 2010). The results showed that the Met/Met genotype was more frequent in nonresponders than in responders, while no differences were detected between bipolar patients and controls.

Pharmacogenetics of Antipsychotics

Psychosis is defined as a severe mental disorder characterized by gross impairment in reality testing, typically manifested by delusions, hallucinations, disorganized speech, or disorganized or catatonic behavior, and encompasses schizophrenia and several mood disorders. Drugs used in the treatment of these conditions are referred to as *antipsychotics* (Gaebel & Zielasek, 2009).

Antipsychotics target those neurotransmitter systems considered to be implicated in the pathophysiology of schizophrenia and, as such, the dopaminergic and serotonergic systems have been extensively studied. Several polymorphisms in dopaminergic receptors have been linked to clinical outcome in atypical antipsychotics. For example, polymorphisms in the dopamine D2 receptor gene (DRD2), such as -141C Ins/Del, affect DRD2 gene expression, while Ser311Cys causes diminished D2R receptor signal transduction. Therapeutic responses in patients treated for schizophrenia with haloperidol, clozapine, or risperidone, improves in patients with the Taq1D1D1, TaqIA2A2, or H313HCC DRD2 genotypes, in contrast to patients with the H313HTT or H313HTC DRD2 genotypes, which were associated with a poor response (Vijayan et al., 2007; Shen et al., 2009).

Second-generation antipsychotics display high affinities for serotonin (5-HT) receptors and are hypothesized to mediate, at least partially, their antipsychotic activity. For example, the 1354C>T polymorphism of the serotonin-2a receptor gene (5HTR2A) may affect treatment response to certain antipsychotics: a 5-HT2A 102-T/C silent polymorphism and a -1438-G/A reporter variant have been associated with increased response to the antipsychotics clozapine and risperidone, and with the antidepressant mirtazapine (Zandi & Judy, 2010; Catapano & Manji, 2007).

One particular area in antipsychotic treatment that could be of pharmacogenetic interest is the risk of developing tardive dyskinesia (TD), a serious complication afflicting a significant minority of subjects treated with antipsychotics on a long-term basis. Although the D2 receptor gene (DRD2) would appear to be the obvious candidate because it is the main site of action of typical antipsychotic medications, studies investigating the DRD2 have mostly yielded negative results. Interestingly, however, a D3 9Gly variant that confers higher binding affinity has been reported to increase the risk of TD (Wilffert, Zaal, & Brouwers, 2005). Polymorphisms (102T>C and 1354C>T) of the serotonin-2a receptor gene (5HTR2A), as well as -697G>C SNP in the serotonin-2c receptor gene (5HTR2C), were also found to be significantly associated with the risk of acquiring TD (Thelma, Srivastava, & Tiwari, 2008). Genetic investigations suggest that the risk of TD is also associated with variants in cytochrome P450 1A2 and SERT, although not all the studies show consistent results.

Additional research is required to dismiss current ambiguities concerning the associations between genes and drug response in the treatment of neuropsychiatric conditions. This will help to reveal the extent to which neuropsychiatric pharmacogenetics could be clinically important.

Neuropathic Pain

Neuropathic pain is triggered by lesions to the somatosensory nervous system that alter its structure and function so that pain occurs spontaneously, and responses to noxious and innocuous stimuli are pathologically amplified. The pain is often triggered by an injury, but this injury may or may not involve actual damage; pain may persist for a long time beyond the apparent healing of any damaged tissues. Some of the common types of peripheral neuropathic pain include painful diabetic neuropathy, HIV sensory neuropathy, and trigeminal neuralgia, whereas multiple sclerosis pain, spinal cord injury pain, central post-stroke pain, and syringomyelia are all common types of central neuropathic pain.

Unlike nociceptive pain, neuropathic pain is frequency chronic and tends to have a less robust response to traditional first-line analgesics. Favored treatments are certain antidepressants; e.g., tricyclic antidepressants and selective serotonin-norepinephrine reuptake inhibitors (SNRIs), anticonvulsants, especially pregabalin and gabapentin, and topical lidocaine. Opioid analgesics and tramadol are recognized as useful agents but are not recommended as first-line treatments. Other possible mechanisms of action include alpha–adrenergic blockade, sodium channel effects, and NMDA receptor antagonism (Sang, 2000). The existence of individual differences in the response to painful stimuli suggests that genetic factors are involved in its modulation.

Patients displaying inactive alleles of CYPP450s are characterized by deficient hydroxylation of several important analgesic drugs. For example, tramadol, a synthetic weak opioid, has been used therapeutically in paresthesias, allodynia, and touch-evoked pain. CYP450 metabolizes tramadol to 11-desmethylated compounds, of which M1 (O-desmethyltramadol) predominates and possesses analgesic properties Stamer & Stuber, 2005). As O-desmethylation to M1 requires CYP2D6 for its formation, poor metabolizers have a diminished ability to form the analgesic active M1 metabolite. Proof of principle can be seen in the percentage of nonresponders to postoperative tramadol administration, which was 46.7% higher in poor metabolizers than that in patients with functional CYP2D6 (Stamer et al., 2003). Similarly, enhanced adverse effects such as dizziness have been reported in patients with the UM phenotype. CYP2D6 is also involved in the metabolism of many other important analgesic drugs, including tricyclic antidepressants, codeine, and the NMDA antagonist dextromethorphan.

Several studies have been dedicated to variability in the clinical drug response to opioid medications. The mu-opioid receptor encoded by genetic locus OPRM1 is a prime site of action for endogenous opioid peptides, and thus is of interest. Clinical studies have shown increased opioid requirements in patients homozygous for the OPRM1 118G polymorphism (Kroslak et al., 2007). For example, in healthy patients administered alfentanil, those with one or two copies of the G variant allele showed a threefold reduced effect to experimentally induced pain (Somogyi, Barratt, & Coller, 2007). The frequency of patients homozygous for the G-allele ranges from 10% in Caucasians to almost 50% in Asians.

Catechol-O-methyltransferase (COMT), an enzyme involved in the catabolism of dopamine, epinephrine and norepinephrine, has wide variations in enzymatic activity that are underlined by common functional polymorphisms such as Val158Met. Individuals with a homozygous methionine-158 genotype showed diminished regional mu-opioid system response to pain when compared with heterozygotes, and also demonstrated higher sensory and affective ratings of pain (Zubieta et al., 2003). Cancer patients with the Met/Met genotype have also shown a heightened need for morphine.

Evidence emerging from a variety of experimental strategies, including genetic disruption studies and the development of selective antagonists, has indicated that the activation of purinergic P2X receptor subtypes is involved in chronic neuropathic pain (Donnelly-Roberts et al., 2008). Recently, it was reported that in mice lacking $P2X_7$, inflammatory and neuropathic hypersensitivity is completely absent to both mechanical and thermal stimuli, while normal nociceptive processing is preserved (Inoue, Tsuda, & Tozaki-Saitoh, 2007). Furthermore, $P2X_7$ receptors are upregulated in human dorsal root ganglia and injured nerves obtained from chronic neuropathic pain patients (Inoue, Tsuda, & Tozaki-Saitoh, 2007). The recent discovery of potent and highly selective antagonists for $P2X_7$ receptors should help to further clarify P2X receptor pharmacology and expand our understanding of chronic pain states.

FROM GENOTYPE TO PHENOTYPE

In line with the rapid expansion of detailed genomic information, cost effective genotyping technologies have been developed that can assay for hundreds of thousands of SNPs simultaneously. However, thus far many questionable genotype–phenotype associations have emerged. In order to fully understand the connection between genotype and phenotype, it is important to understand what each concept represents. Genotype is the genetic makeup of an individual organism. More specifically, genotype can be related to a specific gene or multiple genes that explain various hereditary traits in phenotype (Benfey & Mitchell-Olds, 2008). Phenotype, on the other hand, is the observable outward appearance of an organism, which is controlled by the genotype and its interaction with the environment. It is fundamentally important to realize, particularly in the genomic era, that phenotype is determined not by genes alone, but by the interplay of genetic and environmental factors. Compliance, bioavailability, drug–drug interactions, and drug–food interactions all affect variability in drug response (Ingelman-Sunderg, 2005, 2001).

Concomitant drug ingestion or "polypharmacy" is well documented to interfere with the direct correlation between genotype and phenotype, as several drugs have been documented to induce and inhibit the CYP450 enzymes (Ingelman-Sunderg, 2004a, 2004b). Enzyme capacity may also be overwhelmed by increased consumption of excessive amounts of a single substrate, or simultaneous consumption of multiple substrates. A case in point is the use of traditional medicine by many cultural groups in the developing world. *Hypoxis hemerocallidea* (African potato) is used in African countries as a natural remedy to treat an array of conditions, including AIDS. However, the use of African potato has been shown to significantly inhibit CYP3A4 activity, which is responsible for the metabolism of many drugs, including antiretroviral drugs (Owira & Ojewole, 2009). This underscores the need for appropriately designed pharmacokinetic studies to unveil the true interaction potential of herbal supplements with pharmacological agents.

Also relevant to many parts of Africa and the developing world is the role that malnutrition may play in the genotype/phenotype association. The pathological changes seen in malnutrition can impact on the drug pharmacokinetics in all stages of disposition in the body (Krishnaswamy, 1979). Therapeutic drug levels may be altered as a result of malnutrition-associated tissue receptor alterations, and in these patients drug therapy monitoring and modification of dosage is imperative (Krishnaswami 1979, 1989).

The fact that pharmacogenetic markers are currently more frequently identified, investigated, and validated in developed countries means that these markers will need to be reevaluated in developing countries, as these countries are likely to be subjected to different diseases, environmental influences, and diet. Determination of valid genotype–phenotype associations in the developing world will require logical, regional and ethnic group–specific strategies for conducting well-designed clinical studies, based on excellent quality control practices interwoven with sound analytical methods and sensible interpretation.

GENETIC PROFILING USING DNA ARRAY DIAGNOSTIC METHODS

In order to demonstrate that a gene is a "sound" pharmacogenetic marker for variability in efficacy and toxicity, diagnostic/prognostic tests must show clinical relevance, demonstrate high sensitivity and specificity, and genotype and phenotype should be closely associated. However, once these genetic mutations have been identified and characterized, they then need to be utilized. DNA microarray technology lends itself to many different applications, including expression studies for diagnostic use, pathogen detection (viral in particular), genetic biomarkers of disease, and can be used to assay for many pharmacogenetic polymorphisms simultaneously.

Gene expression and function have traditionally been studied on a "one gene at a time" basis using, for example, polymerase chain reaction (PCR), in situ hybridization, and Northern blot for mRNA analysis. However there is need for high-throughput methods to simultaneously screen many different expression products or genetic mutations from many different genes in a given individual or cohort. Since the inception of the microarray technology, it has been applied to almost every area of biomedical research, offering promising insights into diagnosis of disease, streamlining of personalized medicine, and drug discovery.

Capabilities

With the development of the Human Genome Project and the assistance of genome-wide association studies (GWAS), screening for the genetic biomarker will become more efficient and comprehensive. Using this tool, a variety of known mutations in the human genome can be used to ask a variety of different disease-based questions. Grant and Hakonarson (2008) reviewed microarray GWAS studies and highlighted the existence of reproducible association loci in T2D, obesity, ischemic heart disease, atrial fibrillation, lipid levels, prostate cancer, colorectal cancer, breast cancer, neuroblastoma, and host control of HIV-1 (Grant & Hakonarson, 2008; Hakonarson et al., 2008). DNA microarrays can be used for the discovery of new genetic markers of disease, the classification of pathogens, the implementation of correct therapy, and the tailoring of personalized medication, as well as the discovery and screening of new pharmacological agents. The integration of new human genetic variation, found in projects such as the 1000 genomes project, and advances in next-generation sequencing, has the potential to further open the GWAS door for applications in pharmacogenetics. Data mining from microarray data not fully utilized will also provide an important source of information. Thus, databases such as Gene Expression Omnibus or GEO (http://www.ncbi.nlm.nih.gov/geo/), ArrayExpress or AEX (http://www.ebi.ac.uk/microarray-as/ae), and Celsius (http://celsius.genomics.ctrl.ucla.edu/) offer a warehouse where microarray data from both healthy and pathological samples can be parked by some and evaluated by others.

Limitations

Initially, reproducibility of microarray data was a fundamental limitation, but this has been overcome by interlaboratory validation (Paakkonen & Tjaderhane, 2010). The major hurdle in microarray technology appears to no longer be in the lab but in bioinformatics. Researchers face serious issues due to the large amount of generated data. As a result different variables are ultimately built into the analysis, making an informed choice very difficult. Fortunately, guidelines for reporting microarray data, array design, and experimental design have been formulated by the Microarray Gene Expression Data Society (MGED), called *Minimum Information About Microarray Experiments* (MAIME; see http://www.mged.org/Workgroups/MIAME/miame.html. Used as a benchmark in many journals, this should assist in reducing publication of chance statistical findings (Paakkonen & Tjaderhane, 2010). It is also important to mention that with the development of DNA microarrays for diagnostic purposes and personalized medicine, one can only assess what is already known, which implies that novel or unknown variables cannot be accounted for. In addition, interethnic variability needs to be taken into account, as in many cases, DNA microarray technologies have been developed and validated on selected populations. Therefore, an all-inclusive approach will be beneficial for the relevance of pharmacogenetic assays. The 1000 Genome project hopes to increase our knowledge of human genetic variation using samples from all over the world (http://www.genome.gov/27528684) and will go a long way to bridge the chasm.

Additional limitations to implementing the technology in countries in the developing world include (but are not limited to): a lack of skilled human resources; ensuring sustainable funding and political will; improving the commercialization infrastructure in both the public and private sectors; developing a healthcare infrastructure that can address access and delivery issues of genomic medicine; training of healthcare workers; and engaging with the public to improve awareness and participation. Given the limitation of healthcare resources and the limited capacity of governments to finance all desired health interventions, the evaluation of these potentially expensive emerging health technologies is crucial and fundamental prior to their implementation.

EMERGING APPLICATIONS OF PHARMACOGENETIC INTEREST

Originally designed to measure gene expression, the use of microarrays has been adapted for transcriptomics (e.g., tumor profiling in cancer), proteomics, and mutation detection (e.g., pharmacogenetics – "gene on a chip"). With advancing knowledge of the human genome (Human Genome Project, http://www.ornl.gov/sci/techresources/Human_Genome/home.shtml research-based microarrays can be formulated to ask questions related to genetic variation with the objective of progressing toward prevention, diagnosis, and treatment. As a result, various diagnostic tests have been commercialized. The FDA approved the AmpliChip CYP450 test for CYP2D6 and CYP2C19 genotyping (Roche Molecular Systems, Inc. and Affymetrix, Inc. – first pharmacogenetic test using a DNA microarray) in 2005, and this has been applied to patients treated with risperidone (Sjoqvist & Eliasson, 2007; Gervasini, Benitez, & Carrillo, 2010). The MammaPrint, used to assess the risk of breast cancer metastasis (Agendia), was approved by the FDA in 2007. More recently, Pathwork Diagnostics was given FDA clearance for their gene expression microarray-based diagnostic test to aid in determining tumors of uncertain origin (Monzon & Dumur, 2010).

GENOMIC MEDICINE IN DEVELOPING COUNTRIES: CREATING A ROOM OF THEIR OWN

While it is true that many developing countries are burdened by poverty, a lack of infrastructure, diseases that are difficult to control, and poor governance, it can be argued that these are the countries most in need of emerging scientific and technological advances to ameliorate their situations (Seguin et al., 2008a; Singer et al., 2007). Science and technology have become increasingly recognized as vital components for national development, and several developing countries are building infrastructure to promote local innovation to retain the value of their human, plant, and microbial genomic diversity and research (Coloma & Harris, 2009a, 2009b).

The fact that certain genetic variants are more likely to be found within particular groups has implications for the design of clinical trials and the development of medicines. In the case of research and development, this means that it is necessary to compare trials conducted in different countries. Statements about efficacy based on evidence in one particular population may not be valid if there is a difference in the prevalence of relevant genetic variants in another. The best known example comes from the FDA approval of BiDil, a combination of 2 generic medications—hydralazine hydrochloride and isosorbide dinitrate—into a single tablet, for the treatment of heart failure specifically in black patients. Studies began as early as 2001, when NitroMed began the African-American Heart Failure Trial (A-HeFT), the first heart failure trial conducted exclusively in African-American patients. On claims that observed racial disparities in mortality and therapeutic response rates in black heart failure patients were predominantly due to ethnic differences, the drug has been promoted by some as a step toward the promised land of individualized pharmacogenetic therapies. In similar ways, genotyping projects can provide developing countries with the necessary tools to better understand drug response, disease mechanisms, and disease susceptibility in their own populations.

A special four-piece supplement to *Nature Reviews Genetics* (Seguin et al., 2008a, 2008b; Hardy et al., 2008a, 2008b; Singer et al., 2007; details how four countries with emerging economies—Mexico, India, Thailand and South Africa—are taking steps to build capacity for studies of human genomic variation. The study, which details how these four countries are actively establishing domestic capacity in genomic medicine, identified six major cross-cutting themes underlying initiatives: political will, institutional leadership, the goal of producing local health benefits, protecting genomic sovereignty, and promoting economic benefit. To break the cycle of dependence, these and other countries are beginning to build infrastructure for local innovation to capture the value of their own scientific research.

For instance, a new genetic database in India contains information about 15,000 people from across the country (Hardy et al., 2008a). According to local scientists, more than 10% of Northern India's population don't respond to up to 30 important drugs. With a probable genetic component, understanding more about this matter could lead to better, more tailored treatments in the future. Pharmacogenetic approaches could reduce costs in developing countries that can least afford to waste money on drugs that have no therapeutic effect. The same is true for adverse drug reactions and the emergence of resistant microorganisms.

Sub-Saharan African countries are also investing in this field. In the year 2000, a national DNA databank was initiated in the Gambia, containing samples from ~57,000 West Africans (Seguin et al., 2008a). More recently, a collaborative effort between African researchers resulted in the creation of a biobank and pharmacogenetics database representing several ethnic groups (Yoruba, Hausa, Ibo, Luo, Kikuyu, Maasai, Shona, San, and Venda) located at the African Institute of Biomedical Science and Technology (AibST) in Harare, Zimbabwe (Matimba et al., 2008).

Another admirable example is the Malaria Genomic Epidemiology Network (MalariaGEN), an international consortium funded by the National Institutes of Health (NIH), Wellcome Trust, the British Medical Research Council (MRC), and the Bill and Melinda Gates Foundation. This consortial approach brings together researchers from 21 countries to overcome some of the scientific, ethical, and practical challenges to conducting large-scale studies of genomic variation that could assist efforts in the fight against malaria (Coloma & Harris, 2009a; Achidi et al., 2008).

Although pharmacogenetic approaches may not replace the well-tried methods of culture and serology, they are turning out to be of considerable value in the developing world for the identification of organisms that are difficult or impossible to culture, and for assessing levels of activity of chronic viral infections. The increasing ability to rapidly monitor large samples of organisms for drug resistance, and to maintain regular surveillance of the emerging drug-resistant strains, is likely to be an important addition to public health measures. By adapting DNA technologies to the existing infrastructure, using homegrown solutions to reduce their cost and applying them to solve local health problems, on-site molecular approaches aimed at detecting and typing infectious agents (point-of-care), offer real value (Coloma & Harris, 2009a).

To share the benefits of this technology equitably worldwide, some have advocated that developed and developing countries alike should cooperatively participate in genomics research, to prevent a widening of the already large gap in global health resources (Acharya et al., 2004). In recognition of this consideration, in 2002 the World Health Organization (WHO) suggested that countries should "strengthen existing, or establish new centers and institutions engaged in genomics research with a view to strengthening national capacity and accelerating application of the advances in genomics relevant to developing countries' health problems." Developing countries might benefit from the experience of developed countries that are familiar with unique emerging issues associated with the regulation of genomic medicine products. One way to address these issues is the establishment of innovative industrial partnerships between developed and developing countries, which could provide both the training and technical expertise required by industry in the latter.

CONCLUSION

Today, with the help of rapidly evolving molecular technology and information from the human genome, pharmacogenetics promises to have an important impact on the major health problems of the developing world, especially with regard to infectious diseases such as malaria, tuberculosis, and HIV/AIDS. Africa is an important region to study human genetic diversity because of its complex population history, and the dramatic variation in climate, diet, and exposure to infectious diseases, which result in high levels of genetic and phenotypic variation in its resident populations. The ability to identify new targets for drug development, together with the potential application in identifying populations who will respond favorably to a particular drug, gives pharmacogenetics/pharmacogenomics a unique position to bridge the genomics divide between rich and poor nations. Despite the existing skepticism around their adoption in developing countries, the above examples are indicative of an effort to increase the likelihood that the knowledge they will generate will provide a competitive edge in emerging knowledge-based economies in developing countries.

REFERENCES

Acharya T, et al. (2004). Strengthening the role of genomics in global health. *PLoS Med 1*(3), e40.

Ahlstrom MM, M. Ridderstrom M, Zamora I. (2007). CYP2C9 structure-metabolism relationships: substrates, inhibitors, and metabolites. *J Med Chem 50*(22), 5382–5391.

Aklillu E, et al. (2002). Evidence for environmental influence on CYP2D6-catalysed debrisoquine hydroxylation as demonstrated by phenotyping and genotyping of Ethiopians living in Ethiopia or in Sweden. *Pharmacogenetics 12*(5), 375–383.

Arranz MJ, de Leon J. (2007). Pharmacogenetics and pharmacogenomics of schizophrenia: a review of last decade of research. *Mol Psychiatry 12*(8), 707–747.

Asamoah-Odei E, Garcia Calleja JM, Boerma JT. (2004). HIV prevalence and trends in sub-Saharan Africa: no decline and large subregional differences. *Lancet 364*(9428), 35–40.

Atlas SA. (2007). The renin-angiotensin aldosterone system: pathophysiological role and pharmacologic inhibition. *J Manag Care Pharm 13*(8 Suppl B), 9–20.

Atlas SA, Niarchos AP, Case DB. (1983). Inhibitors of the renin-angiotensin system. Effects on blood pressure, aldosterone secretion and renal function. *Am J Nephrol 3*(2–3), 118–127.

Atlas SA, et al. (1986). Effects of atrial natriuretic factor on blood pressure and the renin-angiotensin-aldosterone system. *Fed Proc 45*(7), 2115–2121.

Aurell M. (2004). Lithium nephropathy a severe adverse effect of lithium therapy. *Lakartidningen 101*(10), 934; author reply 934.

Achidi EA, et al. (2008). A global network for investigating the genomic epidemiology of malaria. *Nature 456*(7223), 732–737.

Ayrton A, Morgan P. (2001). Role of transport proteins in drug absorption, distribution and excretion. *Xenobiotica 31*(8–9): 469–497.

Azuma J, Nonen S. (2009). Chronic heart failure: beta-blockers and pharmacogenetics. *Eur J Clin Pharmacol 65*(1), 3–17.

Baldwin WS, Marko PB, Nelson DR. (2009). The cytochrome P450 (CYP) gene superfamily in Daphnia pulex. *BMC Genomics 10*, 169.

Barnard M, et al. (2008). Rapid molecular screening for multidrug-resistant tuberculosis in a high-volume public health laboratory in South Africa. *Am J Respir Crit Care Med 177*(7), 787–792.

Benetos A, et al. (1996a). Influence of the angiotensin II type 1 receptor gene polymorphism on the effects of perindopril and nitrendipine on arterial stiffness in hypertensive individuals. *Hypertension 28*(6), 1081–1084.

Benetos A, et al. (1996b). Influence of angiotensin-converting enzyme and angiotensin II type 1 receptor gene polymorphisms on aortic stiffness in normotensive and hypertensive patients. *Circulation 94*(4), 698–703.

Benfey PN, Mitchell-Olds T. (2008). From genotype to phenotype: systems biology meets natural variation. *Science 320*(5875), 495–497.

Binder EB, et al. (2004). Polymorphisms in FKBP5 are associated with increased recurrence of depressive episodes and rapid response to antidepressant treatment. *Nat Genet 36*(12), 1319–1325.

Bis JC, et al. (2003). Angiotensinogen Met235Thr polymorphism, angiotensin-converting enzyme inhibitor therapy, and the risk of nonfatal stroke or myocardial infarction in hypertensive patients. *Am J Hypertens 16*(12), 1011–1017.

Bocchio-Chiavetto L, et al. (2008). 5-HTTLPR and BDNF Val66Met polymorphisms and response to rTMS treatment in drug resistant depression. *Neurosci Lett 437*(2), 130–134.

Bond CA, Raehl CL. (2006). Adverse drug reactions in United States hospitals. *Pharmacotherapy 26*(5), 601–608.

Brandt JT, et al. (2007). Common polymorphisms of CYP2C19 and CYP2C9 affect the pharmacokinetic and pharmacodynamic response to clopidogrel but not prasugrel. *J Thromb Haemost 5*(12), 2429–2436.

Budnitz DS, et al. (2007). Medication use leading to emergency department visits for adverse drug events in older adults. *Ann Intern Med 147*(11), 755–765.

Catapano LA, Manji HK. (2007). G protein-coupled receptors in major psychiatric disorders. *Biochim Biophys Acta 1768*(4), 976–993.

Chasman DI, et al. (2004). Pharmacogenetic study of statin therapy and cholesterol reduction. *JAMA 291*(23), 2821–2827.

Chuma J, Okungu V, Molyneux C. (2010). The economic costs of malaria in four Kenyan districts: do household costs differ by disease endemicity? *Malar J 9*, 149.

Coloma J, Harris E. (2009a). Molecular genomic approaches to infectious diseases in resource-limited settings. *PLoS Med 6*(10), e1000142.

Coloma J, Harris E. (2009b). From construction workers to architects: developing scientific research capacity in low-income countries. *PLoS Biol 7*(7), e1000156.

Contreras J, et al. (2010). Association of serotonin transporter promoter gene polymorphism (5-HTTLPR) with depression in Costa Rican schizophrenic patients. *J Neurogenet 24*(2), 83–89.

Craddock N, Jones I. (2001). Molecular genetics of bipolar disorder. *Br J Psychiatry 41*(Suppl), s128–s133.

Crettol S, Petrovic N, Murray M. (2010). Pharmacogenetics of phase I and phase II drug metabolism. *Curr Pharm Des 16*(2), 204–219.

Cusi D. (1997). Genetic renal mechanisms of hypertension. *Curr Opin Nephrol Hypertens 6*(2), 192–198.

Cusi D, et al. (1997). Polymorphisms of alpha-adducin and salt sensitivity in patients with essential hypertension. Lancet *349*(9062), 1353–1357.

Dahl SG, Sylte I, Ravna AW. (2004). Structures and models of transporter proteins. *J Pharmacol Exp Ther 309*(3), 853–860.

Daniel HI, Edeki TI. (1996). Genetic polymorphism of S-mephenytoin 4'-hydroxylation. *Psychopharmacol Bull 32*(2), 219–230.

Davies EC, et al. (2007). Adverse drug reactions in hospitals: a narrative review. *Curr Drug Saf 2*(1), 79–87.

Davis BR, et al. (2007). Antihypertensive therapy, the alpha-adducin polymorphism, and cardiovascular disease in high-risk hypertensive persons: the Genetics of Hypertension-Associated Treatment Study. *Pharmacogenomics J 7*(2), 112–122.

de Grooth, GJ, et al. (2004). The cholesteryl ester transfer protein (CETP) TaqIB polymorphism in the cholesterol and recurrent events study: no interaction with the response to pravastatin therapy and no effects on cardiovascular outcome: a prospective analysis of the CETP TaqIB polymorphism on cardiovascular outcome and interaction with cholesterol-lowering therapy. *J Am Coll Cardiol 43*(5), 854–857.

De Luca A, et al. (2000). Benefits and risks of switching from protease inhibitors to nevirapine with stable background therapy in patients with low or undetectable viral load: a multicentre study. *AIDS 14*(11), 1655–1656.

DeVito MJ, et al. (2000). Dose-response relationships for induction of CYP1A1 and CYP1A2 enzyme activity in liver, lung, and skin in female mice following subchronic exposure to polychlorinated biphenyls. *Toxicol Appl Pharmacol 167*(3), 157–172.

Donnelly-Roberts D, et al. (2008). Painful purinergic receptors. *J Pharmacol Exp Ther 324*(2), 409–415.

Drago A, De Ronchi D, Serretti A. (2009). Pharmacogenetics of antidepressant response: an update. *Hum Genomics 3*(3), 257–274.

Durham LK, et al. (2004). The serotonin transporter polymorphism, 5-HTTLPR, is associated with a faster response time to sertraline in an elderly population with major depressive disorder. *Psychopharmacology (Berl) 174*(4), 525–529.

Edeki T. (1996). Clinical importance of genetic polymorphism of drug oxidation. *Mt Sinai J Med 63*(5–6), 291–300.

Eder C, et al. (2009). Ultrafine carbon particles down-regulate CYP1B1 expression in human monocytes. *Part Fibre Toxicol 6*, 27.

Eichelbaum M, Ingelman-Sundberg M, Evans WE. (2006). Pharmacogenomics and individualized drug therapy. *Annu Rev Med 57*, 119–137.

Este JA, Cihlar T. (2010). Current status and challenges of antiretroviral research and therapy. *Antiviral Res 85*(1), 25–33.

Etuk E, et al. (2008). Prescription pattern of anti-hypertensive drugs in a tertiary health institution in Nigeria. *Ann Afr Med 7*(3), 128–132.

Ferreira Ade A, et al. (2009). The role of 5-HTTLPR polymorphism in antidepressant-associated mania in bipolar disorder. *J Affect Disord 112*(1–3), 267–272.

Foti RS, Wahlstrom JL. (2008). CYP2C19 inhibition: the impact of substrate probe selection on in vitro inhibition profiles. *Drug Metab Dispos 36*(3), 523–528.

Fox AL. (1932). The relation between chemical constitution and taste. *Proc Natl acad Sci USA 18*, 115–120.

Frazier L, et al. (2004). Multilocus effects of the renin-angiotensin-aldosterone system genes on blood pressure response to a thiazide diuretic. *Pharmacogenomics J 4*(1), 17–23.

Garrod AE. (1909). *Inborn errors of metabolism*. London: Hoddler and Stoughton.

Gaebel W, Zielasek J. (2009). Future classification of psychotic disorders. *Eur Arch Psychiatry Clin Neurosci 259 Suppl 2*, S213–S218.

Gamo FJ, et al. (2010). Thousands of chemical starting points for antimalarial lead identification. *Nature 465*(7296), 305–310.

Garcia DA, Hylek E. (2009). Warfarin pharmacogenetics. *N Engl J Med 360*(23), 2474; author reply 2475.

Gellner K, et al. (2001). Genomic organization of the human CYP3A locus: identification of a new, inducible CYP3A gene. *Pharmacogenetics 11*(2), 111–121.

Gervasini G, Benitez J, Carrillo JA. (2010). Pharmacogenetic testing and therapeutic drug monitoring are complementary tools for optimal individualization of drug therapy. *Eur J Clin Pharmacol 66*(8), 755–774.

Gil JP. (2008). Amodiaquine pharmacogenetics. *Pharmacogenomics* 9(10), 1385–1390.

Glorioso N, et al. (1999). The role of alpha-adducin polymorphism in blood pressure and sodium handling regulation may not be excluded by a negative association study. *Hypertension 34*(4 Pt 1), 649–654.

Gloyn AL, et al. (2003). Large-scale association studies of variants in genes encoding the pancreatic beta-cell KATP channel subunits Kir6.2 (KCNJ11) and SUR1 (ABCC8) confirm that the KCNJ11 E23K variant is associated with type 2 diabetes. *Diabetes 52*(2), 568–572.

Goldman N, et al. (2010). The serotonin transporter polymorphism (5-HTTLPR): allelic variation and links with depressive symptoms. *Depress Anxiety 27*(3), 260–269.

Gorlyn M, et al. (2008). Neuropsychological characteristics as predictors of SSRI treatment response in depressed subjects. *J Neural Transm 115*(8), 1213–1219.

Grace JM, et al. (1998). Metabolism of beta-arteether to dihydroqinghaosu by human liver microsomes and recombinant cytochrome P450. *Drug Metab Dispos 26*(4), 313–317.

Grant SF, Hakonarson H. (2008). Microarray technology and applications in the arena of genome-wide association. *Clin Chem 54*(7), 1116–1124.

Haas DW, et al. (2004). Pharmacogenetics of efavirenz and central nervous system side effects: an Adult AIDS Clinical Trials Group study. *AIDS 18*(18), 2391–2400.

Hajjar I, Kotchen T. (2003). Regional variations of blood pressure in the United States are associated with regional variations in dietary intakes: the NHANES-III data. *J Nutr 133*(1), 211–214.

Hakonarson H, et al. (2008). A novel susceptibility locus for type 1 diabetes on Chr12q13 identified by a genome-wide association study. *Diabetes 57*(4), 1143–1146.

Hansen T, et al. (1998). Decreased tolbutamide-stimulated insulin secretion in healthy subjects with sequence variants in the high-affinity sulfonylurea receptor gene. *Diabetes 47*(4), 598–605.

Hardy BJ, et al. (2008a). From diversity to delivery: the case of the Indian Genome Variation initiative. *Nat Rev Genet 9*(Suppl 1), S9–S14.

Hardy BJ, et al. (2008b). South Africa: from species cradle to genomic applications. *Nat Rev Genet 9*(Suppl 1), S19–S23.

He Z, et al. (2009). Antiretroviral protease inhibitors potentiate chloroquine antimalarial activity in malaria parasites by regulating intracellular glutathione metabolism. *Exp Parasitol 123*(2), 122–127.

He Z, et al. (2010. In vitro interactions between antiretroviral protease inhibitors and artemisinin endoperoxides against Plasmodium falciparum. *Int J Antimicrob Agents 35*(2), 191–193.

Hernandez D, et al. (2000). Regression of left ventricular hypertrophy by lisinopril after renal transplantation: role of ACE gene polymorphism. *Kidney Int 58*(2), 889–897.

Higashi MK, et al. (2002). Association between CYP2C9 genetic variants and anticoagulation-related outcomes during warfarin therapy. *JAMA 287*(13), 1690–1698.

Hillemann D, Rusch-Gerdes S, Richter E. (2007). Evaluation of the GenoType MTBDRplus assay for rifampin and isoniazid susceptibility testing of Mycobacterium tuberculosis strains and clinical specimens. *J Clin Microbiol 45*(8), 2635–2640.

Hingorani AD, et al. (1995). Renin-angiotensin system gene polymorphisms influence blood pressure and the response to angiotensin converting enzyme inhibition. *J Hypertens 13*(12 Pt 2), 1602–1609.

Hodel EM, et al. (2009). A microarray-based system for the simultaneous analysis of single nucleotide polymorphisms in human genes involved in the metabolism of anti-malarial drugs. *Malar J 8*, 285.

Hollopeter G, et al. (2001). Identification of the platelet ADP receptor targeted by antithrombotic drugs. *Nature 409*(6817), 202–207.

Hossain P, Kawar B, El Nahas M. (2007). Obesity and diabetes in the developing world—a growing challenge. *N Engl J Med 356*(3), 213–215.

Hulot, JS, et al. (2006). Cytochrome P450 2C19 loss-of-function polymorphism is a major determinant of clopidogrel responsiveness in healthy subjects. *Blood 108*(7), 2244–2247.

Ingelman-Sundberg M. (2001). Pharmacogenetics: an opportunity for a safer and more efficient pharmacotherapy. *J Intern Med 250*(3), 186–200.

Ingelman-Sundberg M. (2004a). Pharmacogenetics of cytochrome P450 and its applications in drug therapy: the past, present and future. *Trends Pharmacol Sci 25*(4), 193–200.

Ingelman-Sundberg M. (2004b). Human drug metabolising cytochrome P450 enzymes: properties and polymorphisms. *Naunyn Schmiedebergs Arch Pharmacol 369*(1), 89–104.

Ingelman-Sundberg M. (2005). Genetic polymorphisms of cytochrome P450 2D6 (CYP2D6): clinical consequences, evolutionary aspects and functional diversity. *Pharmacogenomics J 5*(1), 6–13.

Ingelman-Sundberg M, Rodriguez-Antona C. (2005). Pharmacogenetics of drug-metabolizing enzymes: implications for a safer and more effective drug therapy. *Philos Trans R Soc Lond B Biol Sci 360*(1460), 1563–1570.

Inoue K, Tsuda M, Tozaki-Saitoh H. (2007). Modification of neuropathic pain sensation through microglial ATP receptors. *Purinergic Signal 3*(4), 311–316.

Jamison DT, Saxenian H, Bergevin Y. (1995). Investing in health wisely. The role of needs-based technology assessment. *Int J Technol Assess Health Care 11*(4), 673–684.

Jiang X, et al. (2007). Effect of renin-angiotensin-aldosterone system gene polymorphisms on blood pressure response to antihypertensive treatment. *Chin Med J (Engl) 120*(9), 782–786.

Kam KM, et al. (2006). Stepwise decrease in moxifloxacin susceptibility amongst clinical isolates of multidrug-resistant *Mycobacterium tuberculosis*: correlation with ofloxacin susceptibility. *Microb Drug Resist 12*(1), 7–11.

Katiyar SK, et al. (2008). A randomised controlled trial of high-dose isoniazid adjuvant therapy for multidrug-resistant tuberculosis. *Int J Tuberc Lung Dis 12*(2), 139–145.

Kato M, Serretti A. (2010). Review and meta-analysis of antidepressant pharmacogenetic findings in major depressive disorder. *Mol Psychiatry 15*(5), 473–500.

Katz DA, et al. (2008). Defining drug disposition determinants: a pharmacogenetic-pharmacokinetic strategy. *Nat Rev Drug Discov 7*(4), 293–305.

Kazui M, et al. (2010). Identification of the human cytochrome P450 enzymes involved in the two oxidative steps in the bioactivation of clopidogrel to its pharmacologically active metabolite. *Drug Metab Dispos 38*(1), 92–99.

Kearney PM, et al. (2005). Global burden of hypertension: analysis of worldwide data. *Lancet 365*(9455), 217–223.

Keller J, Schatzberg AF, Maj M. (2007). Current issues in the classification of psychotic major depression. *Schizophr Bull 33*(4), 877–885.

Kerb R, et al. (2009). Pharmacogenetics of antimalarial drugs: effect on metabolism and transport. *Lancet Infect Dis 9*(12), 760–774.

Keswani SC, et al. (2002). HIV-associated sensory neuropathies. *AIDS 16*(16), 2105–2117.

Kirchheiner J., et al. (2004). Pharmacogenetics of antidepressants and antipsychotics: the contribution of allelic variations to the phenotype of drug response. *Mol Psychiatry 9*(5), 442–473.

Klein TE, et al. (2009). Estimation of the warfarin dose with clinical and pharmacogenetic data. *N Engl J Med 360*(8), 753–764.

Klerkx AH, et al. (2004). Cholesteryl ester transfer protein concentration is associated with progression of atherosclerosis and response to pravastatin in men with coronary artery disease (REGRESS). *Eur J Clin Invest 34*(1), 21–28.

Krishnan V, Nestler EJ. (2008). The molecular neurobiology of depression. *Nature 455*(7215), 894–902.

Krishnaswamy K. (1979). Drug metabolism and pharmacokinetics in malnutrition. *Clin Pharmacokinet 3*(3), 216–240.

Krishnaswamy K. (1989). Drug metabolism and pharmacokinetics in malnourished children. *Clin Pharmacokinet 17 Suppl 1*, 68–88.

Kroslak T, et al. (2007). The single nucleotide polymorphism A118G alters functional properties of the human mu opioid receptor. *J Neurochem 103*(1), 77–87.

Kuivenhoven JA, et al. (1998). The role of a common variant of the cholesteryl ester transfer protein gene in the progression of coronary atherosclerosis. The Regression Growth Evaluation Statin Study Group. *N Engl J Med 338*(2), 86–93.

Kurihara AHK, Kazui M, Ishizuka T, Farid NA, Ikeda T. (2005). In vitro metabolism of antiplatelet agent clopidogrel: cytochrome P450 isoforms responsible for two oxidation steps involved in active metabolite formation. *Drug Metabolism Reviews 37*(2), 99.

Kurland L, et al. (2001). Angiotensin converting enzyme gene polymorphism predicts blood pressure response to angiotensin II receptor type 1 antagonist treatment in hypertensive patients. *J Hypertens 19*(10), 1783–1787.

Lee HY, Kim YK. (2010). Catechol-O-methyltransferase Val158Met polymorphism affects therapeutic response to mood stabilizer in symptomatic manic patients. *Psychiatry Res 175*(1–2), 63–66.

Lenox RH, Wang L. (2003). Molecular basis of lithium action: integration of lithium-responsive signaling and gene expression networks. *Mol Psychiatry 8*(2), 135–144.

Leslie T, et al. (2010). The impact of phenotypic and genotypic G6PD deficiency on risk of plasmodium vivax infection: a case-control study amongst Afghan refugees in Pakistan. *PLoS Med 7*(5), e1000283.

Li X, Du Y, Huang X. (2003). Correlation of Angiotensin-converting enzyme gene polymorphism with effect of antihypertensive therapy by Angiotensin-converting enzyme inhibitor. *J Cardiovasc Pharmacol Ther 8*(1), 25–30.

Limdi NA, et al. (2010). Warfarin pharmacogenetics: a single VKORC1 polymorphism is predictive of dose across 3 racial groups. *Blood 115*(18), 3827–3834.

Limdi NA, Veenstra DL. (2008). Warfarin pharmacogenetics. *Pharmacotherapy 8*(9), 1084–1097.

Lin E, et al. (2009). Gene-gene interactions of the brain-derived neurotrophic-factor and neurotrophic tyrosine kinase receptor 2 genes in geriatric depression. *Rejuvenation Res 12*(6), 387–393.

Liu T, et al. (2004). Sequencing complex diseases with HapMap. *Genetics 168*(1), 503–511.

LoBue P. (2009). Extensively drug-resistant tuberculosis. *Curr Opin Infect Dis 22*(2), 167–173.

Madeddu G, et al. (2008). Tenofovir renal safety in HIV-infected patients: results from the SCOLTA Project. *Biomed Pharmacother 62*(1), 6–11.

Mallal S, et al. (2008). HLA-B*5701 screening for hypersensitivity to abacavir. *N Engl J Med 358*(6), 568–579.

Manolopoulos VG, Ragia G, Tavridou A. (2010). Pharmacogenetics of coumarinic oral anticoagulants. *Pharmacogenomics 11*(4), 493–496.

Manunta P, et al. (1999). Adducin polymorphism affects renal proximal tubule reabsorption in hypertension. *Hypertension 33*(2), 694–697.

Marez D, et al. (1997). Polymorphism of the cytochrome P450 CYP2D6 gene in a European population: characterization of 48 mutations and 53 alleles, their frequencies and evolution. *Pharmacogenetics 7*(3), 193–202.

Marschang P, et al. (2006). Plasma cholesteryl ester transfer protein concentrations predict cardiovascular events in patients with coronary artery disease treated with pravastatin. *J Intern Med 260*(2), 151–159.

Masimirembwa C, et al. (1993). Genetic polymorphism of cytochrome P450 CYP2D6 in Zimbabwean population. *Pharmacogenetics 3*(6), 275–280.

Masimirembwa C, et al. (1995). Phenotyping and genotyping of S-mephenytoin hydroxylase (cytochrome P450 2C19) in a Shona population of Zimbabwe. *Clin Pharmacol Ther 57*(6), 656–661.

Masimirembwa C, et al. (1996). A novel mutant variant of the CYP2D6 gene (CYP2D6*17) common in a black African population: association with diminished debrisoquine hydroxylase activity. *Br J Clin Pharmacol 42*(6), 713–719.

Materson BJ, et al. (1993). Single-drug therapy for hypertension in men. A comparison of six antihypertensive agents with placebo. The Department of Veterans Affairs Cooperative Study Group on Antihypertensive Agents. *N Engl J Med 328*(13), 914–921.

Mathers CD, Loncar D. (2006). Projections of global mortality and burden of disease from 2002 to 2030. *PLoS Med 3*(11), e442.

Matimba A, et al. (2008). Establishment of a biobank and pharmacogenetics database of African populations. *Eur J Hum Genet 16*(7), 780–783.

Mega JL, et al. (2009). Cytochrome p-450 polymorphisms and response to clopidogrel. *N Engl J Med 360*(4), 354–362.

Mehta A, Mason PJ, Vulliamy TJ. (2000). Glucose-6-phosphate dehydrogenase deficiency. *Baillieres Best Pract Res Clin Haematol 13*(1), 21–38.

Mehta AB. (1994). Glucose-6-phosphate dehydrogenase deficiency. *Postgrad Med J 70*(830), 871–877.

Meltzer HL. (1990). Mode of action of lithium in affective disorders. An influence on intracellular calcium functions. *Pharmacol Toxicol 66*(Suppl 3), 84–99.

Monzon FA, Dumur CI. (2010). Diagnosis of uncertain primary tumors with the Pathwork tissue-of-origin test. *Expert Rev Mol Diagn 10*(1), 17–25.

Morens DM, Folkers GK, Fauci AS. (2004). The challenge of emerging and re-emerging infectious diseases. *Nature 430*(6996), 242–249.

Motulsky AG. (1957). Drug reactions enzymes and biochemical genetics. *J Am Med Assoc 165*, 835–837.

Murri R, et al. (2003). Determinants of health-related quality of life in HIV-infected patients. *AIDS Care 15*(4), 581–590.

Muszkat M, Stein CM. (2005). Pharmacogenetics and response to beta-adrenergic receptor antagonists in heart failure. *Clin Pharmacol Ther 77*(3), 123–126.

Mutabingwa TK. (2005). Artemisinin-based combination therapies (ACTs): best hope for malaria treatment but inaccessible to the needy! *Acta Trop 95*(3), 305–315.

Naeger LK, et al. (2010). Running a tightrope: regulatory challenges in the development of antiretrovirals. *Antiviral Res 85*(1), 232–240.

Ohmichi N, et al. (1997). Relationship between the response to the angiotensin converting enzyme inhibitor imidapril and the angiotensin converting enzyme genotype. *Am J Hypertens 10*(8), 951–955.

Ordovas JM, et al. (1995). Effect of apolipoprotein E and A-IV phenotypes on the low density lipoprotein response to HMG CoA reductase inhibitor therapy. *Atherosclerosis 113*(2), 157–166.

Owira PM, Ojewole JA. (2009). "African potato" (*Hypoxis hemerocallidea* corm): a plant-medicine for modern and 21st century diseases of mankind? A review. *Phytother Res 23*(2), 147–152.

Paakkonen V, Tjaderhane L. (2010). High-throughput gene and protein expression analysis in pulp biologic research: review. *J Endod 36*(2), 179–189.

Park WB, et al. (2010). Genetic factors influencing severe atazanavir-associated hyperbilirubinemia in a population with low UDP-glucuronosyltransferase 1A1*28 allele frequency. *Clin Infect Dis 51*(1), 101–106.

Pearson ER, et al. (2007). Variation in TCF7L2 influences therapeutic response to sulfonylureas: a GoDARTs study. *Diabetes 56*(8), 2178–2182.

Pearson TA, Manolio TA. (2008). How to interpret a genome-wide association study. *JAMA 299*(11), 1335–1344.

Pereira NL, Weinshilboum RM. (2009). Cardiovascular pharmacogenomics and individualized drug therapy. *Nat Rev Cardiol 6*(10), 632–638.

Perez-Andreu V, et al. (2009). Implications of pharmacogenetics for oral anticoagulants metabolism. *Curr Drug Metab 10*(6), 632–642.

Perroud N, et al. (2009). Genetic predictors of increase in suicidal ideation during antidepressant treatment in the GENDEP project. *Neuropsychopharmacology 34*(12), 2517–2528.

Phillips EJ, Mallal SA. (2008). Pharmacogenetics and the potential for the individualization of antiretroviral therapy. *Curr Opin Infect Dis 21*(1), 16–24.

Projean D, et al. (2003). In vitro metabolism of chloroquine: identification of CYP2C8, CYP3A4, and CYP2D6 as the main isoforms catalyzing N-desethylchloroquine formation. *Drug Metab Dispos 31*(6), 748–754.

Psaty BM, et al. (2002). Diuretic therapy, the alpha-adducin gene variant, and the risk of myocardial infarction or stroke in persons with treated hypertension. *JAMA 287*(13), 1680–1689.

Ribaudo HJ, et al. (2006). Pharmacogenetics of plasma efavirenz exposure after treatment discontinuation: an Adult AIDS Clinical Trials Group Study. *Clin Infect Dis 42*(3), 401–407.

Rieder MJ, et al. (2005). Effect of VKORC1 haplotypes on transcriptional regulation and warfarin dose. *N Engl J Med 352*(22), 2285–2293.

Rigat B, et al. (1990). An insertion/deletion polymorphism in the angiotensin I-converting enzyme gene accounting for half the variance of serum enzyme levels. *J Clin Invest 86*(4), 1343–1346.

Rodriguez-Novoa S, et al. (2007). Genetic factors influencing atazanavir plasma concentrations and the risk of severe hyperbilirubinemia. *AIDS 21*(1), 41–46.

Saag M, et al. (2008). High sensitivity of human leukocyte antigen-b*5701 as a marker for immunologically confirmed abacavir hypersensitivity in white and black patients. *Clin Infect Dis 46*(7), 1111–1118.

Sabatine MS, et al. (2005a). Addition of clopidogrel to aspirin and fibrinolytic therapy for myocardial infarction with ST-segment elevation. *N Engl J Med 352*(12), 1179–1189.

Sabatine MS, et al. (2005b). Effect of clopidogrel pretreatment before percutaneous coronary intervention in patients with ST-elevation myocardial infarction treated with fibrinolytics: the PCI-CLARITY study. *JAMA 294*(10), 1224–1232.

Sang CN. (2000). NMDA-receptor antagonists in neuropathic pain: experimental methods to clinical trials. *J Pain Symptom Manage 19*(1 Suppl), S21–S215.

Savi P, et al. (2000). Identification and biological activity of the active metabolite of clopidogrel. *Thromb Haemost 84*(5), 891–896.

Schelleman H, et al. (2006a). Drug-gene interaction between the insertion/deletion polymorphism of the angiotensin-converting enzyme gene and antihypertensive therapy. *Ann Pharmacother 40*(2), 212–218.

Schelleman H, et al. (2006b). The influence of the alpha-adducin G460W polymorphism and angiotensinogen M235T polymorphism on antihypertensive medication and blood pressure. *Eur J Hum Genet 14*(7), 860–866.

Schelleman H, et al. (2006c). Interactions between five candidate genes and antihypertensive drug therapy on blood pressure. *Pharmacogenomics* J *6*(1), 22–26.

Schelleman H, et al. (2007). Diuretic-gene interaction and the risk of myocardial infarction and stroke. *Pharmacogenomics J 7*(5), 346–352.

Schlag BD, et al. (2004). Ligand dependency of 5-hydroxytryptamine 2C receptor internalization. *J Pharmacol Exp Ther 310*(3), 865–870.

Schwarz UI, et al. (2008). Genetic determinants of response to warfarin during initial anticoagulation. *N Engl J Med 358*(10), 999–1008.

Sciarrone MT, et al. (2003). ACE and alpha-adducin polymorphism as markers of individual response to diuretic therapy. *Hypertension 41*(3), 398–403.

Scordo MG, et al. (2001). Genetic polymorphism of cytochrome P450 2C9 in a Caucasian and a black African population. *Br J Clin Pharmacol 52*(4), 447–450.

Seguin B, et al. (2008a). Genomic medicine and developing countries: creating a room of their own. *Nat Rev Genet 9*(6), 487–493.

Seguin B, et al. (2008b). Genomics, public health and developing countries: the case of the Mexican National Institute of Genomic Medicine (INMEGEN). *Nat Rev Genet 9*(Suppl 1), S5–S9.

Sesti G, et al. (2006). The E23K variant of KCNJ11 encoding the pancreatic beta-cell adenosine 5'-triphosphate-sensitive potassium channel subunit Kir6.2 is associated with an increased risk of secondary failure to sulfonylurea in patients with type 2 diabetes. *J Clin Endocrinol Metab 91*(6), 2334–2339.

Shen YC, et al. (2009). Effects of DRD2/ANKK1 gene variations and clinical factors on aripiprazole efficacy in schizophrenic patients. *J Psychiatr Res 43*(6), 600–606.

Shi R, et al. (2006). Emergence of ofloxacin resistance in Mycobacterium tuberculosis clinical isolates from China as determined by gyrA mutation analysis using denaturing high-pressure liquid chromatography and DNA sequencing. *J Clin Microbiol 44*(12), 4566–4568.

Shin J, Johnson JA. (2007). Pharmacogenetics of beta-blockers. *Pharmacotherapy 27*(6), 874–887.

Shu Y, et al. (2003). Evolutionary conservation predicts function of variants of the human organic cation transporter, OCT1. *Proc Natl Acad Sci U S A 100*(10), 5902–5907.

Shu Y, et al. (2007). Effect of genetic variation in the organic cation transporter 1 (OCT1) on metformin action. *J Clin Invest 117*(5), 1422–1431.

Sim SC, et al. (2006). A common novel CYP2C19 gene variant causes ultrarapid drug metabolism relevant for the drug response to proton pump inhibitors and antidepressants. *Clin Pharmacol Ther 79*(1), 103–113.

Simon T, et al. (2009). Genetic determinants of response to clopidogrel and cardiovascular events. *N Engl J Med 360*(4), 363–375.

Singer PA, et al. (2007). Applying genomics-related technologies for Africa's health needs. *Afr J Med Med Sci 36*(Suppl), 7–14.

Sistonen J, et al. (2009). Pharmacogenetic variation at CYP2C9, CYP2C19, and CYP2D6 at global and microgeographic scales. *Pharmacogenet Genomics 19*(2), 170–179.

Sjoqvist F, Eliasson E. (2007). The convergence of conventional therapeutic drug monitoring and pharmacogenetic testing in personalized medicine: focus on antidepressants. *Clin Pharmacol Ther 81*(6), 899–902.

Smeraldi E, et al. (2006). Serotonin transporter gene-linked polymorphic region: possible pharmacogenetic implications of rare variants. *Psychiatr Genet 16*(4), 153–158.

Somogyi AA, Barratt DT, Coller JK. (2007). Pharmacogenetics of opioids. *Clin Pharmacol Ther 81*(3), 429–444.

Sosin MD, et al. (2004). Heart failure – the importance of ethnicity. *Eur J Heart Fail 6*(7), 831–843.

Stamer UM, et al. (2003). Impact of CYP2D6 genotype on postoperative tramadol analgesia. *Pain 105*(1–2), 231–238.

Stamer UM, Stuber F. (2005). Pharmacogenetics of anesthetic and analgesic agents: CYP2D6 genetic variations. *Anesthesiology 103*(5), 1099; author reply 1101.

Stavroulakis GA, et al. (2000). Predicting response to chronic antihypertensive treatment with fosinopril: the role of angiotensin-converting enzyme gene polymorphism. *Cardiovasc Drugs Ther 14*(4), 427–432.

Steimer W. (2010). Pharmacogenetics and psychoactive drug therapy: ready for the patient? *Ther Drug Monit 32*(4), 381–386.

Strakowski SM, Delbello MP, Adler CM. (2005). The functional neuroanatomy of bipolar disorder: a review of neuroimaging findings. *Mol Psychiatry 10*(1), 105–116.

Su X, et al. (2007). Association between angiotensinogen, angiotensin II receptor genes, and blood pressure response to an angiotensin-converting enzyme inhibitor. *Circulation 115*(6), 725–732.

Sutanto I, et al. (2010). Evaluation of chloroquine therapy for vivax and falciparum malaria in southern Sumatra, western Indonesia. *Malar J 9*, 52.

Suwelack B, et al. (2002). Impact of ACE polymorphism on renal allograft function, blood pressure, and proteinuria under ACE inhibition. *Transplant Proc 34*(5), 1763–1766.

Svensson US, Ashton M. (1999). Identification of the human cytochrome P450 enzymes involved in the in vitro metabolism of artemisinin. *Br J Clin Pharmacol 48*(4), 528–535.

Takeuchi F, et al. (2009). A genome-wide association study confirms VKORC1, CYP2C9, and CYP4F2 as principal genetic determinants of warfarin dose. *PLoS Genet 5*(3), e1000433.

Tamaki S, et al. (1997). Effect of genotype on the angiotensin-converting enzyme mRNA level in human atria. *Clin Exp Pharmacol Physiol 24*(5), 305–308.

Tandon P, et al. (2004). The use of beta-blockers in a tertiary care heart failure clinic: dosing, tolerance, and outcomes. *Arch Intern Med 164*(7), 769–774.
Tang YM, et al. (2000). Human CYP1B1 Leu432Val gene polymorphism: ethnic distribution in African-Americans, Caucasians and Chinese; oestradiol hydroxylase activity; and distribution in prostate cancer cases and controls. *Pharmacogenetics 10*(9), 761–766.
Taubert D, et al. (2006). Impact of P-glycoprotein on clopidogrel absorption. *Clin Pharmacol Ther 80*(5), 486–501.
Thelma B, Srivastava V, Tiwari AK. (2008). Genetic underpinnings of tardive dyskinesia: passing the baton to pharmacogenetics. *Pharmacogenomics 9*(9), 1285–1306.
Thompson JF, et al. (2005). An association study of 43 SNPs in 16 candidate genes with atorvastatin response. *Pharmacogenomics J 5*(6), 352–358.
Thomson RE, et al. (2004). Tissue-specific expression and subcellular distribution of murine glutathione S-transferase class kappa. *J Histochem Cytochem 52*(5), 653–662.
Thumm M, et al. (1994). Isolation of autophagocytosis mutants of Saccharomyces cerevisiae. *FEBS Lett 349*(2), 275–280.
Tozzi V. (2010). Pharmacogenetics of antiretrovirals. *Antiviral Res 85*(1), 190–200.
Tozzi V, et al. (1999). Positive and sustained effects of highly active antiretroviral therapy on HIV-1-associated neurocognitive impairment. *AIDS 13*(14), 1889–1897.
Tozzi V, et al. (2005). Prevalence and risk factors for human immunodeficiency virus-associated neurocognitive impairment, 1996 to 2002: results from an urban observational cohort. *J Neurovirol 11*(3), 265–273.
Turner ST, et al. (2001). C825T polymorphism of the G protein beta(3)-subunit and antihypertensive response to a thiazide diuretic. *Hypertension 37*(2 Part 2), 739–743.
Turner ST, et al. (2003). Effects of endothelial nitric oxide synthase, alpha-adducin, and other candidate gene polymorphisms on blood pressure response to hydrochlorothiazide. *Am J Hypertens 16*(10), 834–839.
Ueda K, et al. (1999). Cooperative binding of ATP and MgADP in the sulfonylurea receptor is modulated by glibenclamide. *Proc Natl Acad Sci U S A 96*(4), 1268–1272.
Vijayan NN, et al. (2007). Association of dopamine receptor polymorphisms with schizophrenia and antipsychotic response in a South Indian population. *Behav Brain Funct 3*, 34.
Vogel F. (1959). Moderne problem der humangenetik. *Ergeb Inn Med U Kinderheilk 12*, 52–125.
Waldman S, Terzic A. (2007a). Targeted diagnostics and therapeutics for individualized patient management. *Biomark Med 1*(1), 3–8.
Waldman S, Terzic A. (2007b). Individualized medicine and the imperative of global health. *Clin Pharmacol Ther 82*(5), 479–483.
Warren RM, et al. (2009). The clinical relevance of Mycobacterial pharmacogenetics. *Tuberculosis (Edinb) 89*(3), 199–202.
Weekers L, et al. (2005). Modulation of the renal response to ACE inhibition by ACE insertion/deletion polymorphism during hyperglycemia in normotensive, normoalbuminuric type 1 diabetic patients. *Diabetes 54*(10), 2961–2697.
Wennerholm A, et al. (2001). Characterization of the CYP2D6*29 allele commonly present in a black Tanzanian population causing reduced catalytic activity. *Pharmacogenetics 11*(5), 417–427.
Wilffert B, Zaal R, Brouwers JR. (2005). Pharmacogenetics as a tool in the therapy of schizophrenia. *Pharm World Sci 27*(1), 20–30.
Wu TY, et al. (2010). Ten-year trends in hospital admissions for adverse drug reactions in England 1999–2009. *J R Soc Med 103*(6): 239–250.
Xie HG, et al. (2001). Molecular basis of ethnic differences in drug disposition and response. *Annu Rev Pharmacol Toxicol 41*, 815–850.
Yasar U, Eliasson E, Dahl ML. (2002). Association of CYP2C9 genotypes leading to high enzyme activity and colorectal cancer risk. *Carcinogenesis 23*(4), 665; author reply 667–668.
Yu H, Zhang Y, Liu G. (2003). Relationship between polymorphism of the angiotensin-converting enzyme gene and the response to angiotensin-converting enzyme inhibition in hypertensive patients. *Hypertens Res 26*(11), 881–886.
Yuan H, et al. (2008). Effects of polymorphism of the beta(1) adrenoreceptor and CYP2D6 on the therapeutic effects of metoprolol. *J Int Med Res 36*(6), 1354–1362.
Zandi PP, Judy JT. (2010). The promise and reality of pharmacogenetics in psychiatry. *Psychiatr Clin North Am 33*(1), 181–224.
Zhang Y. (2007). Advances in the treatment of tuberculosis. *Clin Pharmacol Ther 82*(5), 595–600.
Zhou HH, Liu ZQ. (2000). Ethnic differences in drug metabolism. *Clin Chem Lab Med 38*(9), 899–903.
Zintzaras E, et al. (2009). APOE gene polymorphisms and response to statin therapy. *Pharmacogenomics J 9*(4), 248–257.
Zuber RE, Anzenbacherova E, Anzenbacher P. (2002). Cytochromes P450 and experimental models of drug metabolism. *J Cell Mol Med 6*(2), 189–198.
Zubieta JK, et al. (2003). COMT val158met genotype affects mu-opioid neurotransmitter responses to a pain stressor. *Science 299*(5610), 1240–1243.
Zucman D, et al. (2007). Prospective screening for human leukocyte antigen-B*5701 avoids abacavir hypersensitivity reaction in the ethnically mixed French HIV population. *J Acquir Immune Defic Syndr 45*(1), 1–3.
Zychma MJ, et al. (2002). Sulfonylurea receptor gene 16–3 polymorphism - association with sulfonylurea or insulin treatment in type 2 diabetic subjects. *Med Sci Monit 8*(7), CR512–CR515.

18.

NUTRIGENOMICS RESEARCH IN WORLD POPULATIONS

Jim Kaput

INTRODUCTION

At the end of the first decade of the new millennium, over 1 billion of the world's population are still undernourished (FAO, 2009). Another one billion (http://www.who.int/dietphysicalactivity/publications/facts/obesity/en/) are overweight or obese from overconsumption of nutrient-poor, excess calories. A common feature of both forms of malnutrition is "hidden hunger," the lack of essential nutrients and micronutrients, which are required for maintaining health and preventing disease. Alleviating insufficiencies or frank deficiencies may seem straightforward: provide populations and individuals with recommended dietary allowances (RDA) or the more inclusive reference dietary intakes (RDIs). However, RDAs are calculated as intake of a nutrient that is two standard deviations from the estimated average requirement (EAR) such that 95%–97% of the population would meet their nutrient needs. EAR, in turn, is the intake at which 50% of the population meets some predetermined endpoint criterion (Suitor & Meyers, 2006). RDAs, EARs, RDIs, adequate intakes (AI), and the acceptable micronutrient distribution range (AMDR) are all calculated from the response of groups of individuals, so that some individuals in a population will need more and some may ingest too much for optimal health (Figure 18-1; Suitor & Meyers, 2006). Many of the studies used to determine these values were based on analyses of nutrient levels in European or European-descended populations. More recent research has oversampled other population groups in the United States (http://www.cdc.gov/nchs/nhanes/history.htm) for analyzing nutrient intakes.

Many populations and cultures, particularly in emerging economies, have not been included or studied, limiting the applicability of the recommendations. The genome research conducted to date has demonstrated that allele of genes regulating, interacting, or metabolizing nutrients varies in different ancestral populations. As an example of our published strategy for micronutrients (Wise & Kaput, 2009) and carbohydrate-metabolizing genes (Varma, Wise & Kaput 2010), pathway and interaction analyses programs (GeneGo MetaCore - http;//www.genego.com or WikiPathways - http://www.wikipathways.org) were used to identify genes in metabolic pathways or networks for 14 micronutrients (ascorbic acid, pyridoxal, pyridoxine, biotin, folate, niacin, pathothenic acid, riboflavin, thiamine, and vitamins A, B12, D, E and K). A network describes interactions between genes, metabolites, and proteins that have been reported in the literature, and typically extends beyond metabolic pathways described in textbooks. Of the 464 metabolic and network genes found for these 14 micronutrients, 106 were found in the interaction network of several micronutrients. Three were selected as examples and are shown in Table 18-1. Hepatocyte nuclear factor 4 alpha (*HNF4A*), a transcription factor that regulates the expression of a large number of genes, is in networks of pyridoxal, pyridoxine, biotin, folate, riboflavin, thiamine, retinol, and vitamin K2. Although dbSNP reports only one nonsynonymous SNP (p.Thr354Ile), this allele shows differences in population allele frequencies. Further characterization of other genomes, particularly by NextGen sequencing strategies, will likely identify other common and rare variants of these genes. A subset of these polymorphisms may change protein structure and function or alter the regulation of gene expression, generating differential responses to, or affecting, nutrient metabolism and, ultimately, health maintenance or disease initiation and progression. *HNF4A,* for example, has been associated with phenotypes (e.g., dyslipidemia and glucose metabolism) associated with type 2 diabetes (Andrulionyte et al., 2006; Barroso et al., 2003; Kathiresan et al., 2009). One of its roles in micronutrient biology may be in regulating the expression of genes involved in nutrient metabolism (e.g., Kang et al., 2009).

This chapter describes the current knowledge of genetic variation and current strategies for associating genes with phenotypes, including disease. A more significant task is to develop novel research strategies that generate personal risk factors, as opposed to population-attributable risks. Developing such strategies is necessary for identifying gene–nutrient interactions that underlie health and disease processes, and for optimizing nutrition for different populations.

GENETIC VARIATION

SINGLE NUCLEOTIDE POLYMORPHISM (SNP)

The initial sequencing of the human genome (Venter et al., 2001), subsequent analyses of human genetic

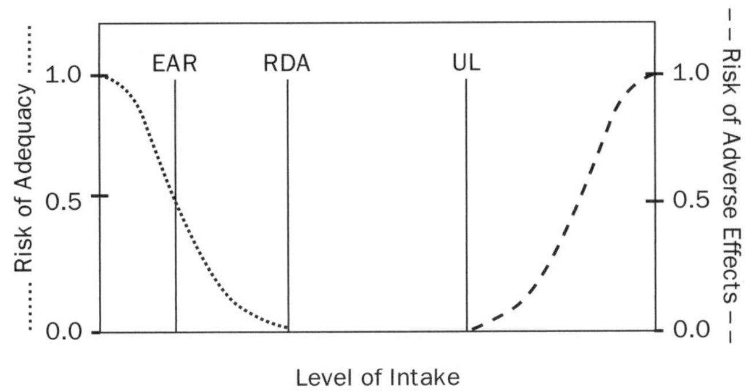

Figure 18-1 Definitions of Nutrient Intakes and Risks at Different Levels. RDAs are calculated as intake of a nutrient that is two standard deviations from the estimated average requirement (EAR), such that 95%–97% of the population would meet their nutrient needs. EAR, in turn, is the intake at which 50% of the population meets some predetermined endpoint criterion.

variation (Frazer et al., 2007; Silva-Zolezzi et al., 2009; Teo et al., 2009; The International HapMap, 2005), and ongoing resequencing of individual genomes has identified ~18 million single nucleotide polymorphisms (Baye et al., 2009; Johnson, 2009; Phillips, 2009), although estimates of uncharacterized and known range as high as 25 million. Several initiatives are underway to characterize more sequence variations in the human population. The 1000 Genomes Project (http://www.1000genomes.org) is an international consortium that is using genome-wide resequencing and targeted coding region sequencing in ~2500 individuals from 27 populations (Lang, 2008; Siva, 2008). The goal of this effort is to characterize alleles with frequencies of approximately 1% genome-wide, and less than 1% in coding regions. However, phenotypes of individuals sequenced in the project will not be analyzed. Hence, these data will be used for improving selection of reagents and designs for genetic and metabolomic studies and will not be directly focused on identification of disease-causing genes.

In contrast, the Human Variome Project is developing and fostering an international effort to systematically identify genes, their mutations, and their variants associated with phenotypic variability and indications of human disease (Kaput et al., 2009b; Ring et al., 2006). The HVP project connects clinical, medical, and research laboratories for developing knowledge housed within individual but "linkable" databases.

COPY-NUMBER VARIATION (CNV)

The in depth analyses of individual genomes led to (at the time) a surprising discovery: structural changes in DNA called *copy-number variations* (CNVs) or *copy-number polymorphisms* (CNPs) also produce variation in the genome. Current estimates indicate that about 13% of the variation in the human genome is CNVs (Barnes, 2010; Barnes & Breen, 2010). About 11,700 CNVs larger than ~400 bases have been identified which overlap more than 1000 genes. Deletions, insertions (INDELS), duplications, triplications, and translocations can all result in CNVs. These copy-number variations encompass more total nucleotides and arise more frequently than SNPs. CNVs have been shown to contribute to human evolution, genetic diversity between individuals, and a rapidly increasing number of traits or susceptibility to traits; that is, genomic disorders (Dhawan & Padh, 2009; Fanciulli et al., 2010; Ku et al., 2010). CNVs have also been found in genes involved in nutrient metabolism. Individuals from Europe, Japan, and Africa had between 2 and 16 copies of *AMY1*, the salivary amylase gene. The average number of *AMY1* genes differed between populations whose ancestors consumed high-starch diets compared to populations that historically ate low-starch diets (Perry et al., 2007). Although *AMY1* copy-number correlated with gene and protein expression, compensatory mechanisms may modulate expression of multiple copies of genes—that is, one cannot assume a one-to-one correlation between copy-number and the expression of a particular gene (Springer et al., 2010). Nonetheless, CNVs may alter nutrient metabolism and needs by altering gene dosage.

Epigenetic Mechanisms

Genetic variation provides the foundation for expression of phenotype. However, the expression of an individual's phenotype can be dramatically altered by epigenetic mechanisms, which include DNA and histone methylations, other covalent modifications of chromatin proteins, and noncoding RNAs (ncRNAs - review in Choudhuri, 2010). Changing or fixing chromosome structure occurs through DNA methylation and chromatin protein modifications (Delaval & Feil, 2004; Dolinoy et al., 2006; Fowler & Alarid, 2004; Jirtle & Skinner, 2007; Morgan et al., 2005), which alters gene expression without alterations in DNA sequence. Chromatin remodeling is an ongoing and active process, regulated in part by certain chemicals or metabolites derived from the diet (Cooney, 2006; Dolinoy et al., 2006; Fenech, 2005). For example, the substrate for DNA methylation is S-adenosylmethionine, a product of the one carbon pathway that requires folate, vitamins B2, B6, B12, and dietary precursors methionine, choline, and betaine (Mason, 2003). Chromatin remodeling responds to short- and long-term nutrient availability (Gallou-Kabani and Junien, 2005), including energy availability (Picard et al., 2004).

Many thousands of regulatory non-protein-coding RNAs, including microRNAs (miRNAs), small interfering RNAs (siRNA), PIWI-interacting RNAs (piwiRNA), and various classes of long ncRNA have recently been identified, all of which alter transcriptional or posttranscriptional

TABLE 18-1 EXAMPLES OF CODING SNPS AND POPULATION FREQUENCIES IN GENES INVOLVED IN NETWORKS WITH MICRONUTRIENTS*

NUTRIENT	GENE	CHROMOSOME	RS#	CHR POSITION	VARIANT SNP	REF SNP	AA	CODON	POPULATION	X/X	X/Y	Y/Y
Riboflavin	***ENPP2***	**8q24.1**	**rs1058913**	**120575184**	**A**	**C**	**p. F778L**	**3**		C/C	C/T	T/T
Panthotenate					**G**	**C**	**p. F778L**	**3**	European	0.733	0.250	0.017
					T	**C**	**p. F778L**	**3**	Asian-HCB	0.844	0.156	0.000
									Asian-JPT	0.977	0.023	0.000
									African	0.983	0.017	0.000
			rs4871364	**120594802**	**A**	**T**	**p. H528Q**	**3**		**A/A**	**A/T**	**T/T**
					G	**T**	**p. H528Q**	**3**	European	0.707	0.276	0.017
					C	**T**	**p. H528Q**	**3**	Asian-HCB	0.800	0.200	0.000
									Asian-JPT	0.841	0.136	0.023
									African	0.136	0.559	0.305
Nicotinamide	***CYP2D6***	**22q13.1**	**rs28371706**	**42525772**	**A**	**C**	**p. T410N**	**2**		**C/C**	**A/G**	**G/G**
VitA Retinol					**G**	**C**	**p.T410S**	**2**	CAUC3	1	0.000	0.000
					T	**C**	**p. T410I**	**2**	PAC3	1	0.000	0.000
									AFR3	0.757	0.203	0.041
									HISP3	0.980	0.020	0.000
Pyridoxal	***HNF4A***	**20q12-q13.1**	**rs1800961**	**43042364**	**C**	**T**	**p. T354I**	**2**		**C/C**	**C/T**	**T/T**
Pyridoxine									European	0.900	0.100	0.000
Biotin									Asian-HCB	0.978	0.022	0.000
Folate									Asian-JPT	1	0.000	0.000
Riboflavin									African	1	0.000	0.000
Thiamine												
VitA retinol												
Vit K2												

*Chromosome position and SNP related information were obtained from public-domain databases. RS# - Number of the reference SNP, AA - Amino Acid change and position, Codon - position of the SNP in the codon, Sample count - number of individuals of the indicated population analyzed. HapMap populations—European, Asian-HCB (Han Chinese in Beijing), Asian-JPT (Japanese in Tokyo), African (Sub-Saharan African), Other populations are from 1000Genome Project. X/X and Y/Y - representative homozygous alleles, X/Y - Representative heterozygous alleles. See text for more details.

regulation of gene expression. The effect of diet on the expression of ncRNAs has not been analyzed, although one study found that expression of certain microRNAs in adult offspring was correlated to control or high-fat diets fed to lactating dams (Zhang et al., 2009).

Since epigenetic mechanisms are affected by environment, the expression of sequence and copy-number variations that produce an individual's phenotype are regulated in a highly dynamic manner. When environment acts in utero or in early childhood, lifelong effects may be established. This process is called developmental plasticity (Bateson et al., 2004), nutritional epigenomics (Gallou-Kabani and Junien, 2005; Jirtle and Skinner, 2007), polypheny (Gluckman et al., 2007), and metabolic (DeRisi et al., 1997; Jackson, 2000; Lucas, 1998; McMillen and Robinson, 2005) or fetal (Burdge et al., 2007) programming. Plasticity is based on the response of a fetus to maternal environments, or the individual's early nutritional environment through epigenetic mechanisms. Programming may be advantageous because the fetus/child would be prepared for the existing environment (Bateson et al., 2004). When the environment no longer matches the programming, health outcomes are affected. Health outcomes are late onset, demonstrating the influence of long-term exposure to imbalanced (for one's genotype and epigenotype) nutrient intake. Such mechanisms could explain the projected increase in diabetes incidence in developing countries (~170%) versus that in developed countries (~45%) by 2025 (King et al., 1998)—that is, undernutrition establishes a disconnect with high-caloric-density foods available in many parts of the world.

The effects of environment on the fetus necessarily depend upon the genotype of the mother; specifically, her metabolic capacity to maintain (or not) the appropriate sensing of the environment (Kaput et al., 2009a). The response of the mother depends on the genes she inherited and their epigenetic modifications. Hence, genetic and physiological complexities, in addition to nutritional factors, will influence fetal programming. These concepts lead to an inescapable conclusion: improving diets of individuals throughout the world will require analyses of gene–nutrient interactions in more diverse populations than has previously been done. Hence, the need for an international effort in nutrigenomics research (Kaput, 2004; Kaput, 2007).

CURRENT RESEARCH STRATEGIES FOR GENE-PHENOTYPE ASSOCIATIONS

The predominant experimental design to associate genotype with phenotype associations has been genome-wide association studies (GWAS). These studies have generated hundreds of candidate loci for many phenotypes, including chronic diseases (see dbGaP - http://www.ncbi.nlm.nih.gov/projects/gap/cgi-bin/about.html). The majority of participants for these studies have been Europeans and their descendents. Analyses of allele frequencies of subsets of the candidate loci in non-European populations have demonstrated discordance with a role in the phenotype: some of these presumably causative alleles occurred in 100% or 0% in these populations (Adeyemo and Rotimi, 2009; Mattei et al., 2009; Myles et al., 2008).

Compelling arguments have been made that the GWAS data has generated novel candidate loci for chronic disease (Hardy and Singleton, 2009; Hirschhorn, 2009; Kraft and Hunter, 2009). However, the experimental designs of many GWA studies are based on false assumptions and have important limitations:

- Assume all control individuals are physiologically and culturally alike with either the same environmental exposures or that nongenetic variables cancel each other. The most important environmental factors are nutrients, which are not typically assessed. Exceptions have been published and demonstrate that diet alters the influence of an SNP (e.g., Corella et al., 2009; Ordovas et al., 2002; Warodomwichit et al., 2009).
- Assume all cases are produced by the same genetic pathways. Molecular and genetic cancer studies of the late 1980s demonstrated that cancer is a collection of diseases initiated by mutations in one or more genes in a tissue (e.g., breast, prostate, or colon). This lesson applies to chronic diseases: they are also caused by multiple genes in many pathways influenced by many environmental factors. An example for a nutritionally related chronic disease with complex phenotypes is diabetes, which can be demonstrated by the drugs used to control the condition. Diabetes drugs target insulin production in the pancreas, gluconeogenesis in the liver, glucose uptake in the intestines, and transcriptional regulation of PPAR-gamma in adipose and muscle, as an example of the various pathways (and therefore genes) that contribute to or are misregulated to produce T2DM (Kaput et al., 2007). All chronic diseases have similar physiological heterogeneity. Many GWAS designs included all subtypes of a disease in the case group, which likely increased the noise since unbalances in the multiple subtypes may have different collections of genes contributing to the same phenotype. Including individuals with any of the many diabetic complications (e.g., Calcutt et al., 2009) would further increase diversity of physiologies in the case group.
- Assume that all complex diseases result only from genetic contributions (Hardy & Singleton, 2009). In addition to the overwhelming published evidence that environmental factors, and particularly diet, alters the expression of genetic information, this assumption

is biologically naïve. All species are adapted to and respond to their environment, a lesson taught by Darwin. While some genes in humans (or any species) must be genetically regulated for the animal to survive, an indisputable body of literature exists that dietary chemicals alter the expression of genetic information at the transcriptional, protein translational, and protein activity levels. Data from fields of molecular biology, genetics, laboratory animal models of chronic disease and nutritional science, and nutritional sciences in humans have demonstrated that chemicals in the environment, and specifically chemicals in food, alter expression of genetic information (see extensive reviews in Bateson et al., 2004; Brigelius-Flohe' & Joost, 2006; Corella & Ordovas, 2005; Kaput & Rodriguez, 2004; Kaput, 2006; Ordovas & Corella, 2005 and 2006). As one example, genistein, a chemical found in a variety of foods, has been co-crystallized with estrogen receptor β (Gustafsson, 1999; Pike et al., 1999) and show to alter gene expression in laboratory animals (reviewed in Szkudelska & Nogowski, 2007). Epidemiology itself gives unequivocal data: the increase in obesity and diabetes during the past 20 years is too short a time period to exert selective pressure on allele frequencies in human populations.

- Although there may be genes regulated differently between individuals, and these may contribute to a chronic disease regardless of diet or other environmental factors, many genes involved in chronic disease and health are regulated by the interaction between environment and genotype (Bureau et al., 2008; Kaput et al., 1994; Park et al., 1997; Swartz et al., 1996). Hence, the statistical main effect is the genotype X environment term (Ottman, 1996), a value that can not be calculated in the absence of nutrient intake (calories and nutrients in) and physical activity (calories out). Failure to assess environmental and, in particular, dietary factors means that the GWA studies are incomplete at best, and inaccurate at worst.
- Assume that gene–gene interactions (Gao et al., 2010)—epistasis—are identical in all cases and all controls, best summarized by Scriver: the gene is not the disease and vice versa (Scriver, 1988). A gene variant does not exist in a vacuum or, more precisely, the gene product does not act alone. Proteins and enzymes are involved in interconnected networks that alter the influence of a gene variant. This has been shown from yeast through laboratory animals through humans.
- Assume that epigenetic factors such as DNA methylation or chromatin structure do not influence development of chronic diseases (Slatkin, 2009, and see above).

These limitations and assumptions may explain the apparent "missing heritability" (Bogardus, 2009; Clarke & Cooper, 2010; Eichler et al., 2010; Maher, 2008)—the very low percentage of disease (~2–10%) explained by the loci identified by GWAS. As one example, a genotype core calculated from the contributions of 18 SNPs found to be associated with type 2 diabetes mellitus (T2DM) in GWA studies increased the predictive value for diagnosis of the disease by ~4% (Meigs et al., 2008).

DEVELOPING NEW NUTRIGENOMIC EXPERIMENTAL DESIGNS

Improved research strategies can address some of the limitations of the genetic and nutritional experimental designs (e.g., Corella et al., 2009; Ordovas et al., 2002; Warodomwichit et al., 2009). However, the most significant challenge for developing knowledge about gene–nutrient interactions in individuals and groups remains: association studies, whether genetic, nutritional, or nutrigenomic, are based on population studies that yield the attributable fraction (AF)—"the proportional reduction in average disease risk over a specified time interval that would be achieved by eliminating the exposure of interest from the population" while other factors remain unchanged (Rockhill et al., 1998). For genetic association studies, the population-attributable factor is that proportion of cases in the population that would be avoided if nobody carried the risk allele (Hemminki & Bermejo, 2007). AF is often misinterpreted as a risk factor (Rockhill et al., 1998) rather than the fractional change in number of cases within the population. Perhaps most importantly, the attributable fraction is calculated from population models and data, and is not directly applicable to individuals (Vineis & Kriebel, 2006) because individuals may differ genetically, physiologically, and nutritionally from the population averages (Kaput, 2008). Novel research strategies and experimental designs are necessary for creating individual nutritional recommendations or personalized healthcare.

The experimental design that may have the most promise for understanding gene–nutrient interactions does not rely on intervention-control designs, since no one population or group within a population can be considered the reference (see Adeyemo & Rotimi, 2009; Cooper, 2009; Mattei et al., 2009; Myles et al., 2008). This conclusion was directly tested in an attempt to obtain the best overall (genetic) match for a group of over 2400 Europeans, each analyzed by genotyping ~500,000 SNPs (Lu et al., 2009): each individual differed from all others in the population tested. Adding individuals from the same or other ancestral groups would only increase the genetic (and other) heterogeneity of the population. In order to assess individual risk factors as opposed to the average response of the group studied, experimental designs must be based on measuring the quantitative response of individuals, a considerable challenge because individuals within a group show great variation in homeostasis (Williams, 1956). To circumvent this limitation, additional omics measurements may be obtained prior to and following a perturbation of homeostasis using well-characterized challenges (van Ommen et al., 2009). The most obvious example of a homeostatic challenge is the oral glucose tolerance test (OGT), a stressor that measures an individual's ability to respond to a glucose (e.g., Wopereis et al., 2009). A key aspect of the challenge homeostasis model is that multiple analytes are measured, rather than just the expected response to the challenge. In the case of OGT, one might measure amino acid or fatty acid metabolites in the serum or immune function markers (e.g., cytokines). Metabolomic, transcriptomic, proteomic, and genomic analyses would generate more complete analyses than the traditional method of preselecting specific biomarkers of some trait or condition.

Analyzing individual responses may appear to be statistically naïve and biologically difficult. However, as discussed, individual risk factors cannot be found by analyzing a group average, because of genetic and environmental diversity among humans. The high-dimensional data obtained by omic analyses can be analyzed by dimensionality reduction and classification algorithms (Kaput & Dawson, 2007), which can sort individuals into like genotype–environment interaction groups (Kaput, 2008). That is, each individual is analyzed by deep genotyping and deep phenotyping methodologies, particularly in response to homeostatic challenges, and all data are used to sort individuals into groups using classifiers. This approach eliminates one of the major limitations of the classic case (or intervention) control design, which often results in selection bias and misclassification (Ioannidis et al., 2009; Yesupriya et al., 2008).

The ideal experimental design would have participants from diverse genetic and cultural backgrounds without regard to nationalities or affiliations. Although a large amount of physiological data are available (Williams, 1956), for example, for reference values of serum metabolite concentrations (http://pathcuric1.swmed.edu/pathdemo/nrrt.htm), the full range of human metabolic responses has not been analyzed, since not all genotypes have been studied. Finding individuals of maximum gene–environment differences may allow for such a determination and for creating empirical groups between the extremes (Kaput, 2008). While many if not all biological traits are continuous, with no discrete breaks in the phenotypic or genetic continuum, "binning" is a standard for medical practice, using clinical measurements to group individuals into treatment options and for statistics that rely on tertiles, quartiles, quintiles, etc., to determine structure within experimental data.

Several reports support this overall design, although each dataset is limited. The aforementioned study (Perry et al., 2007) of *AMY1* showed that individuals within one population had different copy-number variants and the averages differed between high-starch and low-starch groups. One can use genotype or phenotype to select individuals to analyze. Metabolomic studies have shown that individuals can be clustered based on 52 urinary metabolites (Bernini et al., 2009). The closest match was twin sisters, and the next closest match was a father–son pair. Individuals from different ethnic populations could also be grouped based on cluster analyses of urinary metabolite levels and blood pressure measurements (Holmes et al., 2008). These studies demonstrate the proof-of-principle concept that individuals rather than groups can be analyzed. Each of these studies could be improved by more complete phenotyping [i.e., deep phenotyping (Tracy, 2008) or phenomics (Bilder et al., 2009)] and genotyping (whole-genome genotyping or whole-exome sequencing), which may provide more reliable classifications. Although such clusters or metabolic groups will exist within one ancestral population, including individuals from other ancestral groups will improve the reliability of classification schemes.

COMMUNITY-BASED AND PRIMARY CARE RESEARCH DESIGNS

While the challenge homeostasis research strategy provides a controlled method for analyzing gene–nutrient interactions, community-based participatory research (CBPR) and primary care research also provide a framework for analyzing individuals over time, in response to specific interventions or self-motivated changes in lifestyle in "real" time. Not all environmental variables can be controlled in a laboratory setting. Conducting research in an individual's or group's living environment therefore becomes a necessary component of translating research to the public. Recruiting, enrolling, and characterizing individuals of genetically and culturally diverse backgrounds by deep phenotyping and genotyping will remain a significant challenge if limited to laboratory or clinical settings.

Community-based research has historical roots in the 1940s but has been emerging from medical practice and from sociological/nutritional research efforts within the last two decades. A variety of different names describe these approaches, such as community-based participatory research (e.g., Boyer et al., 2005; Horowitz et al., 2008; Israel et al., 2005; O'Fallon et al., 2000), participatory action research, academic–community based participatory research, or primary care research (e.g., Beasley et al., 2007; Hueston et al., 2006; Mold and Peterson, 2005). The differences largely reflect the degree of control and involvement of community residents in all phases of the research process (Cornwall & Jewkes, 1995). The lowest level of community involvement is termed *contractual*, in which the researcher brings the proposal to the community and asks them to participate with no or little input or decision-making authority, while the researcher is in full control (Cornwall & Jewkes, 1995). At the next level (consultative), the researcher asks for the community's input and adopts some of the input, but the researcher retains full control. The third level is termed *collaborative,* wherein the community and researchers work together to design and implement the study, but the overall process is managed by researchers. This is a shared-control model and is the most frequently found model in today's community-based participatory research (CBPR). The fourth model is termed *collegiate*, wherein all parties work together, drawing upon different skills, while mutual learning takes place. In this desirable but seldom achieved model, the community is in full control (Cornwall & Jewkes, 1995). The collegiate model is found most often when community residents are well trained in research methods and have had previous experience in research studies.

While CBPR has been gaining much interest in the social and nutritional sciences fields (Chen et al., 2006; McCabe-Sellers et al., 2008; Plumb et al., 2008), relatively few studies have used this method for biomedical research (Boyer et al., 2007; Wells et al., 2006). Our team has collaborated with the USDA – Delta Obesity Prevention Research Unit and communities in the Mississippi Delta region of Arkansas on a CBPR nutritional intervention project that includes nutritional, genomic, metabolomic, and activity measures (McCabe-Sellers et al., 2008). CBPR may allow for the research results to be "personalized," since one individual at a time is assessed and informed. The applications are therefore more immediate than population-based methods, and are targeted to the community and individual. Since genetic and omic data developed from population studies cannot yet be reliably associated with health outcomes in individuals, the initial information flows between researcher and community collaborator focused on nutritional assessments and dietary advice. As more gene–nutrient or omic–nutrient associations are proven, the information flow will include biomedical data and results.

The development of relationships among researcher and community is challenging, but of critical importance for those who have typically been excluded from research studies or those who suffer from culturally based health disparities, whether nationally (Boyer et al., 2007; Chen et al., 2006; Plumb et al., 2008; Wells et al., 2006) or internationally. Individuals in these socioeconomically disadvantaged populations will not benefit from the advances in health research unless their genotypes and cultural environments are included in biomedical research studies. The U.S. NIH National Center for Minority Health and Health Disparities (NCMHD) recently echoed this paradigm shift by calling for the inclusion of minority, lower socioeconomic, and rural populations to be included in biomedical research studies (Dankwa-Mullan et al. 2010a,b). The University of Toronto McLaughlin-Rotman Centre for Global Health (http://www.mrcglobal.org) uses the term *community engagement* to include individuals and communities in developing countries in biomedical research (Tindana et al., 2007).

Although international agencies such as the World Bank, the United Nations, and its subsidiary organizations (World Health Organization [WHO], Food and Agricultural Organization [FAO], Children's Fund [UNICEF], Development Program [UNDP]), various bilateral agencies (e.g., United States Agency for International Development, USAID), and nonprofit organizations (Gates Foundation, Heifer Foundation, Clinton Foundation, and about 60 other major groups; see http://www.123world.com/organizations/) focus on health-related issues internationally, much of the science data used for policy and implementation decisions has been based on the "case-control" design that produces population-attributable risk in the populations studied. Since many world populations were not included in such studies, the results and implementation policies, regardless of how well-intentioned, may not produce the intended health outcomes when applied to unique cultures and genetic populations.

The research gap in international studies, and its consequences for health of the world's population, has been widely reported (Cohen et al., 2008; Daar et al., 2007; Daar et al., 2009; Melon et al., 2009; Seguin et al., 2008; Singer et al., 2007; Singh & Daar, 2008; Smith et al., 2005). Specific programs and initiatives have been undertaken to address these limitations—one of the more notable is the newly created Global Alliance for Chronic Disease (Daar et al., 2009), which will coordinate research activities. These efforts, however, have yet to address the complexity of characterizing the response of individual humans based on nutrigenomics. Only through the analyses of gene–environment interactions in diverse populations will the necessary data and understanding be developed for the world's groups and individuals. Such studies have not been undertaken in a systematic fashion.

Several initiatives have been started to include a wider representation in human studies of medical genetics: the Human Variome Project – HVP (Cotton et al., 2007; Kaput et al., 2009b; Ring et al., 2006), nutrition (Zehner, 2009), and nutrigenomics (Kaput, 2004; Kaput et al., 2005; Kaput & Rodriguez, 2004). The HVP and Nutrigenomics Organization (NuGO) have begun interactions to collaborate and coordinate the activities of assessing genetic variation and its response to the nutrient environment (Kaput et al., 2009b). International collaborations have a rich tradition in scientific research, and many arose through serendipity and personal contacts, although international foundations are now addressing the need for worldwide collaborations. The goal of the international nutrigenomic research community (Kaput et al., 2005) is to provide the means for making these collaborative efforts easier to initiate, maintain, and sustain. The development of best practices and sharing of data will produce more reliable results than can be obtained through individual efforts. Developing these strategies for individuals in developing countries will be a challenge, but surmountable through collaborations between basic researchers, institutions, nongovernment agencies, and governments.

THE NEXT STEPS

While many perceive that nutrigenomics has its "home" exclusively in the discipline of nutrition, it encompasses the concepts and technologies of numerous research and application fields, including genetics, molecular biology, physiology, food science, agriculture, behavioral science, anthropology, ethics, the food industry, and healthcare. The integration of these disciplines necessary for analyzing and understanding nutrient (and environment)–gene interactions for health and disease processes—that is, into a biological whole—requires intensive and extensive collaborations. The international research network will aid in focusing the talents and resources of individuals with diverse expertise in concepts and technologies.

The initial efforts at forming productive collaborations among researchers are being done through an international network initiated by the Nutrigenomics Organization (NuGO) in collaboration with the Human Variome Project. NuGO was initially developed and operated to foster communication among institutes, laboratories, and individuals in Europe; its extension to the international community builds on existing strengths and practices. Members of the international group are developing a nutrigenomics information portal called the *nutritional phenotype database* (van Ommen et al., 2010a) for scientists, healthcare workers, and the public (Kaput et al., 2006), and are developing a first international database and research effort entitled the *micronutrient genomics project* (van Ommen et al., 2010b). The Human Variome proposes a system of locus-specific databases that can be linked in a federated manner (Cotton et al., 2009). Analyzing genetic variation of genes involved in nutrient metabolism is a prerequisite for understanding gene–nutrient interactions. The goal of these efforts is to increase participation of international scientists in the development of best practices, and to collaborate on the research and applications of the science.

THE FUTURE OF NUTRIGENOMIC SCIENCE AND APPLICATIONS

Nutrigenomics researchers are advocating that an organism cannot be completely, or even adequately, analyzed or understood in isolation from its environment. That is, the reductionist view that genes or their variants alone will inform us about health or disease processes is incomplete at best, and perhaps provides erroneous results and conclusions. This strong statement is justified: all organisms respond at the molecular level to nutrients in their environment—if they did not, they would not survive that environment. Hence, nutrigenomic research is leading the development of a more complete scientific method, one which includes analyses of genes, epigenetic variables, and the environmental factors with which they interact.

The practical applications of nutrigenomics are immense: from delivering the right micronutrients in the optimum amount to the food-insecure, to developing novel foods that have taste, long shelf life, but, more importantly, will maintain or perhaps improve the health of the world's citizens.

SUMMARY

Current human nutritional and genetic epidemiological methods yield "risk factors" called *population attributable risks* (PAR). These risk factors are statistical estimates of the percentage reduction in disease in the population if the risk were to be avoided, or the gene variant were not present. These measures are often assumed to apply to individuals who are likely to differ in genetic make-up, lifestyle, and dietary patterns from those individuals in the study population. Developing individual risk factors in light of the genetic diversity of human populations, the complexity of foods, culture, and lifestyle, and the variety of metabolic processes that lead to health or disease, are significant challenges for personalizing dietary advice for the healthy, or medical treatments for individuals with chronic disease. New research and application strategies are needed for creating knowledge for personalizing nutrition advice and healthcare.

DECLARATION

The views presented here are those of the author and do not necessarily reflect official positions of the FDA.

REFERENCES

Adeyemo A, Rotimi C. (2009). Genetic Variants Associated with Complex Human Diseases Show Wide Variation across Multiple Populations. *Public Health Genomics. 13*(2), 72–79.

Andrulionyte L, Laukkanen O, Chiasson JL, Laakso M. (2006). Single nucleotide polymorphisms of the HNF4alpha gene are associated with the conversion to type 2 diabetes mellitus: the STOP-NIDDM trial. *J Mol Med 84*(8), 701–708.

Barnes MR. (2010). Genetic variation analysis for biomedical researchers: a primer. *Methods Mol Biol 628*, 1–20.

Barnes MR, Breen G. (2010). A short primer on the functional analysis of copy number variation for biomedical scientists. *Methods Mol Biol 628*, 119–135.

Barroso I, Luan J, Middelberg RP, et al. (2003). Candidate gene association study in type 2 diabetes indicates a role for genes involved in beta-cell function as well as insulin action. *PLoS Biol 1*(1), E20.

Bateson P, Barker D, Clutton-Brock T, et al. (2004). Developmental plasticity and human health. *Nature* 430(6998), 419–421.

Baye TM, Tiwari HK, Allison DB, Go RC. (2009). Database mining for selection of SNP markers useful in admixture mapping. *BioData Min 2*(1), 1.

Beasley JW, Starfield B, van Weel C, Rosser WW, Haq CL. (2007). Global health and primary care research. *J Am Board Fam Med 20*(6), 518–526.

Bernini P, Bertini I, Luchinat C, et al. (2009). Individual Human Phenotypes in Metabolic Space and Time. *J Proteome Res 8*(9), 4264–4271.

Bilder RM, Sabb FW, Cannon TD, et al. (2009). Phenomics: The systematic study of phenotypes on a genome-wide scale. *Neuroscience 164*(1), 30–42.

Bogardus C. (2009). Missing heritability and GWAS utility. *Obesity (Silver Spring) 17*(2), 209–210.

Boyer B, Mohatt GV, Pasker RL, Drew EM, McGlone KK. (2007). Sharing Results from Complex Disease Genetic Studies: A Community Based Participatory Research Approach. *International Jounal of Circumpolar Health 66*, 11–22.

Boyer BB, Mohatt GV, Lardon C, et al. (2005). Building a community-based participatory research center to investigate obesity and diabetes in Alaska Natives. *Int J Circumpolar Health 64*(3), 281–290.

Brigelius-Flohe' R, Joost H-G, eds.(2006). *Nutritional Genomics: Impact on Health and Disease.* Federal Republic of Germany: Wiley VCH Verlag GmbH & Co. 442 p.

Burdge GC, Hanson MA, Slater-Jefferies JL, Lillycrop KA. (2007). Epigenetic regulation of transcription: a mechanism for inducing variations in phenotype (fetal programming) by differences in nutrition during early life? *Br J Nutr 97*(6), 1036–1046.

Bureau A, Diallo MS, Ordovas JM, Cupples LA. (2008). Estimating interaction between genetic and environmental risk factors: efficiency of sampling designs within a cohort. *Epidemiology 19*(1), 83–93.

Calcutt NA, Cooper ME, Kern TS, Schmidt AM. (2009). Therapies for hyperglycaemia-induced diabetic complications: from animal models to clinical trials. *Nat Rev Drug Discov 8*(5), 417–430.

Chen DT, Jones L, Gelberg L. (2006). Ethics of clinical research within a community-academic partnered participatory framework. *Ethn Dis 16*(1 Suppl 1), S118–135.

Choudhuri S. (2010) Small noncoding RNAs: Biogenesis, function, and emerging significance in toxicology. *J Biochem Mol Toxicol. 24*(3), 1195–1216.

Clarke AJ, Cooper DN. (2010) GWAS: heritability missing in action? *Eur J Hum Genet 18*, 859–861.

Cohen ER, Masum H, Berndtson K, et al. (2008). Public engagement on global health challenges. *BMC Public Health 8*, 168.

Cooney CA. 2006. Maternal Nutrition: Nutrients and Control of Expression. In: Kaput, J, Rodriguez, RL, editor. Nutritional Genomics. Discovering the Path to Personalized Nutrition. Hoboken, NJ: John Wiley and Sons. p 219–254.

Cooper CB. (2009). Airflow obstruction and exercise. *Respir Med 103*(3), 325–334.

Corella D, Ordovas JM. (2005). Integration of environment and disease into 'omics' analysis. *Curr Opin Mol Ther 7*(6), 569–576.

Corella D, Peloso G, Arnett DK, et al. (2009). APOA2, dietary fat, and body mass index: replication of a gene-diet interaction in 3 independent populations. *Arch Intern Med 169*(20), 1897–1906.

Cornwall A, Jewkes R. (1995). What is participatory research? *Soc Sci Med 41*(12), 1667–1676.

Cotton RG, Al Aqeel AI, Al-Mulla F, et al. (2009). Capturing all disease-causing mutations for clinical and research use: toward an effortless system for the Human Variome Project. *Genet Med 11*(12), 843–849.

Cotton RG, Appelbe W, Auerbach AD, et al. (2007). Recommendations of the 2006 Human Variome Project meeting. *Nat Genet 39*(4), 433–436.

Daar AS, Berndtson K, Persad DL, Singer PA. (2007). How can developing countries harness biotechnology to improve health? *BMC Public Health 7*, 346.

Daar AS, Nabel EG, Pramming SK, et al. (2009). The global alliance for chronic diseases. *Science 324*(5935), 1642.

Dankwa-Mullan I, Rhee KB, Stoff DM, et al. (2010a). Moving toward paradigm-shifting research in health disparities through translational, transformational, and transdisciplinary approaches. *Am J Public Health 100*(Suppl 1), S19–24.

Dankwa-Mullan I, Rhee KB, Williams K, et al. (2010b). The science of eliminating health disparities: summary and analysis of the NIH summit recommendations. *Am J Public Health 100* Suppl 1, S12–18.

Delaval K, Feil R. (2004). Epigenetic regulation of mammalian genomic imprinting. *Curr Opin Genet Dev 14*(2), 188–195.

DeRisi JL, Iyer VR, Brown PO. (1997). Exploring the metabolic and genetic control of gene expression on a genomic scale. *Science 278*(5338), 680–686.

Dhawan D, Padh H. (2009). Pharmacogenetics: technologies to detect copy number variations. *Curr Opin Mol Ther 11*(6), 670–680.

Dolinoy DC, Weidman JR, Waterland RA, Jirtle RL. (2006). Maternal genistein alters coat color and protects Avy mouse offspring from obesity by modifying the fetal epigenome. *Environ Health Perspect 114*(4), 567–572.

Eichler EE, Flint J, Gibson G, et al. (2010). Missing heritability and strategies for finding the underlying causes of complex disease. *Nat Rev Genet 11*(6), 446–450.

Fanciulli M, Petretto E, Aitman TJ. (2010). Gene copy number variation and common human disease. *Clin Genet 77*(3), 201–213.

FAO. 2009. The State of Food Insecurity in the World 2009: Economic crises – impacts and lessons learned. Rome, Italy: Food and Agricultural Organization of the United Nations. 61 p.

Fenech M. (2005). The Genome Health Clinic and Genome Health Nutrigenomics concepts: diagnosis and nutritional treatment of genome and epigenome damage on an individual basis. *Mutagenesis 20*(4), 255–269.

Fowler AM, Alarid ET. (2004). Dynamic control of nuclear receptor transcription. *Sci STKE 2004*(256), pe51.

Frazer KA, Ballinger DG, Cox DR, et al. (2007). A second generation human haplotype map of over 3.1 million SNPs. *Nature 449*(7164), 851–861.

Gallou-Kabani C, Junien C. (2005). Nutritional epigenomics of metabolic syndrome: new perspective against the epidemic. *Diabetes 54*(7), 1899–1906.

Gao H, Granka JM, Feldman MW. (2010). On the classification of epistatic interactions. *Genetics 184*(3), 827–837.

Gluckman PD, Lillycrop KA, Vickers MH, et al. (2007). Metabolic plasticity during mammalian development is directionally dependent on early nutritional status. *Proc Natl Acad Sci U S A 104*(31), 12796–12800.

Gustafsson JA. (1999). Estrogen receptor beta—a new dimension in estrogen mechanism of action. *J Endocrinol 163*(3), 379–383.

Hardy J, Singleton A. (2009). Genomewide association studies and human disease. *N Engl J Med 360*(17), 1759–1768.

Hemminki K, Bermejo JL. (2007). Constraints for genetic association studies imposed by attributable fraction and familial risk. *Carcinogenesis 28*(3), 648–656.

Hirschhorn JN. (2009). Genomewide association studies—illuminating biologic pathways. *N Engl J Med 360*(17), 1699–1701.

Holmes E, Loo RL, Stamler J, et al. (2008). Human metabolic phenotype diversity and its association with diet and blood pressure. *Nature 453*(7193), 396–400.

Horowitz CR, Goldfinger JZ, Muller SE, et al. (2008). A model for using community-based participatory research to address the diabetes epidemic in east harlem. *Mt Sinai J Med 75*(1), 13–21.

Hueston WJ, Mainous AG 3rd, Weiss BD, Macaulay AC, Hickner J, Sherwood RA. (2006). Protecting participants in family medicine research: a consensus statement on improving research integrity and participants' safety in educational research, community-based participatory research, and practice network research. *Fam Med 38*(2), 116–120.

Ioannidis JP, Thomas G, Daly MJ. (2009). Validating, augmenting and refining genome-wide association signals. *Nat Rev Genet 10*(5), 318–329.

Israel B, Eng E, Schulz A, Parker E, eds. (2005) *Methods in Community-Based Participatory Research for Health.* San Francisco: Jossey-Bass.

Jackson AA. (2000). Nutrients, growth, and the development of programmed metabolic function. *Adv Exp Med Biol 478*, 41–55.

Jirtle RL, Skinner MK. (2007). Environmental epigenomics and disease susceptibility. *Nat Rev Genet 8*(4), 253–262.

Johnson AD. (2009). Single-nucleotide polymorphism bioinformatics: a comprehensive review of resources. *Circ Cardiovasc Genet 2*(5), 530–536.

Kang X, Zhong W, Liu J, et al. (2009). Zinc supplementation reverses alcohol-induced steatosis in mice through reactivating hepatocyte nuclear factor-4alpha and peroxisome proliferator-activated receptor-alpha. *Hepatology 50*(4), 1241–1250.

Kaput J. (2004). Diet-disease gene interactions. *Nutrition 20*(1), 26–31.

Kaput J. (2007). Developing the promise of nutrigenomics through complete science and international collaborations. *Forum Nutr 60*, 209–223.

Kaput J. (2008). Nutrigenomics research for personalized nutrition and medicine. *Curr Opin Biotechnol 19*(2), 110–120.

Kaput J, Astley, S, Renkema, M, Ordovas, J, van Ommen, B. (2006). Harnessing Nutrigenomics: Development of web-based communication, databases, resources, and tools. *Genes Nutr 1*(1), 5–11.

Kaput J, Chen J, Slikker W Jr. (2009a). Personalized Nutrition and Medicine in Perinatal Development. In: Hansen D, Abbott B, eds. *Developmental Toxicology*, 3rd ed., pp. 123–144. New York: Informa Healthcare USA.

Kaput J, Cotton RG, Hardman L, et al. (2009b). Planning the Human Variome Project: The Spain report. *Hum Mutat 30*(4), 496–510.

Kaput J, Dawson K. (2007). Complexity of type 2 diabetes mellitus data sets emerging from nutrigenomic research: a case for dimensionality reduction? *Mutat Res 622*(1–2), 19–32.

Kaput J, Noble J, Hatipoglu B, Kohrs K, Dawson K, Bartholomew A. (2007). Application of nutrigenomic concepts to Type 2 diabetes mellitus. *Nutr Metab Cardiovasc Dis 17*(2), 89–103.

Kaput J, Ordovas JM, Ferguson L, et al. (2005). The case for strategic international alliances to harness nutritional genomics for public and personal health. *Br J Nutr 94*(5), 623–632.

Kaput J, Rodriguez RL. (2004). Nutritional genomics: the next frontier in the postgenomic era. *Physiol Genomics 16*(2), 166–177.

Kaput J, Swartz D, Paisley E, Mangian H, Daniel WL, Visek WJ. (1994). Diet-disease interactions at the molecular level: an experimental paradigm. *J Nutr 124*(8 Suppl), 1296S-1305S.

Kaput, J and Rodriguez RL. (2006). *Nutritional Genomics. Discovering the Path to Personalized Nutrition.* Hoboken, NJ: John Wiley and Sons.

Kathiresan S, Willer CJ, Peloso GM, et al. (2009). Common variants at 30 loci contribute to polygenic dyslipidemia. *Nat Genet 41*(1), 56–65.

King H, Aubert RE, Herman WH. (1998). Global burden of diabetes, 1995–2025: prevalence, numerical estimates, and projections. *Diabetes Care 21*(9), 1414–431.

Kraft P, Hunter DJ. (2009). Genetic risk prediction—are we there yet? *N Engl J Med 360*(17), 1701–1703.

Ku CS, Loy EY, Salim A, Pawitan Y, Chia KS. (2010). The discovery of human genetic variations and their use as disease markers: past, present and future. *J Hum Genet. 55*, 403–415.

Lang L. (2008). Three sequencing companies join the 1000 genomes project. *Gastroenterology 135*(2), 336–337.

Lu TT, Lao O, Nothnagel M, et al. (2009). An evaluation of the genetic-matched pair study design using genome-wide SNP data from the European population. *Eur J Hum Genet 17*(7), 967–975.

Lucas A. (1998). Programming by early nutrition: an experimental approach. *J Nutr 128*(2 Suppl), 401S-406S.

Maher B. (2008). Personal genomes: The case of the missing heritability. *Nature 456*(7218), 18–21.

Mason JB. (2003). Biomarkers of nutrient exposure and status in one-carbon (methyl) metabolism. *J Nutr 133* (Suppl 3), 941S-947S.

Mattei J, Parnell LD, Lai CQ, et al. (2009). Disparities in allele frequencies and population differentiation for 101 disease-associated single nucleotide polymorphisms between Puerto Ricans and non-Hispanic whites. *BMC Genet 10*, 45.

McCabe-Sellers B, Lovera D, Nuss H, et al. (2008). Personalizing nutrigenomics research through community based participatory research and omics technologies. *OMICS 12*(4), 263–272.

McMillen IC, Robinson JS. (2005). Developmental origins of the metabolic syndrome: prediction, plasticity, and programming. *Physiol Rev 85*(2), 571–633.

Meigs JB, Shrader P, Sullivan LM, et al. (2008). Genotype score in addition to common risk factors for prediction of type 2 diabetes. *N Engl J Med 359*(21), 2208–2219.

Melon CC, Ray M, Chakkalackal S, et al. (2009). A survey of South-North health biotech collaboration. *Nat Biotechnol 27*(3), 229–232.

Mold JW, Peterson KA. (2005). Primary care practice-based research networks: working at the interface between research and quality improvement. *Ann Fam Med 3*(Suppl 1), S12–20.

Morgan HD, Santos F, Green K, Dean W, Reik W. (2005). Epigenetic reprogramming in mammals. *Hum Mol Genet 14*(Spec No 1), R47–58.

Myles S, Davison D, Barrett J, Stoneking M, Timpson N. (2008). Worldwide population differentiation at disease-associated SNPs. *BMC Med Genomics 1*(1), 22.

O'Fallon L, Tyson FL, Dearry A. (2000). Successful Models of Community Based Participatory Research. *Env. Health Persp. 110*(Suppl 2), 155–159.

Ordovas JM, Corella D. (2005). Genetic variation and lipid metabolism: modulation by dietary factors. *Curr Cardiol Rep 7*(6), 480–486.

Ordovas JM, Corella D. 2006. Gene-Environment Interactions: Defining the Playfield. In: Kaput, J & Rodriguez, RL, eds. *Nutritional Genomics. Discovering the Path to Personalized Nutrition*, p. 57–76. Hoboken, NJ: John Wiley and Sons.

Ordovas JM, Corella D, Cupples LA, et al. (2002). Polyunsaturated fatty acids modulate the effects of the APOA1 G-A polymorphism on HDL-cholesterol concentrations in a sex-specific manner: the Framingham Study. *Am J Clin Nutr 75*(1), 38–46.

Ottman R. (1996). Gene-environment interaction: definitions and study designs. *Prev Med 25*(6), 764–770.

Park EI, Paisley EA, Mangian HJ, et al. (1997). Lipid level and type alter stearoyl CoA desaturase mRNA abundance differently in mice with distinct susceptibilities to diet-influenced diseases. *J Nutr 127*(4), 566–573.

Perry GH, Dominy NJ, Claw KG, et al. (2007). Diet and the evolution of human amylase gene copy number variation. *Nat Genet 39*(10), 1256–1260.

Phillips C. (2009). SNP databases. *Methods Mol Biol 578*, 43–71.

Picard F, Kurtev M, Chung N, et al. (2004). Sirt1 promotes fat mobilization in white adipocytes by repressing PPAR-gamma. *Nature 429*(6993), 771–776.

Pike AC, Brzozowski AM, Hubbard RE, et al. (1999). Structure of the ligand-binding domain of oestrogen receptor beta in the presence of a partial agonist and a full antagonist. *Embo J 18*(17), 4608–4618.

Plumb M, Collins N, Cordeiro JN, Kavanaugh-Lynch M. (2008) Assessing process and outcomes: Evaluating community based participatory research. *Progress in Community Health Partnerships: Research, Education, and Action 2*(2), 87–97.

Ring HZ, Kwok PY, Cotton RG. (2006). Human Variome Project: an international collaboration to catalogue human genetic variation. *Pharmacogenomics 7*(7), 969–972.

Rockhill B, Newman, B, Weinberg C. (1998). Use and misuse of population attributable fractions. *Am J Public Health 88*(1), 15–19.

Scriver CR. (1988). Nutrient-gene interactions: the gene is not the disease and vice versa. *Am J Clin Nutr 48*(6), 1505–1509.

Seguin B, Hardy BJ, Singer PA, Daar AS. (2008). Genomic medicine and developing countries: creating a room of their own. *Nat Rev Genet 9*(6), 487–493.

Silva-Zolezzi I, Hidalgo-Miranda A, Estrada-Gil J, et al. (2009). Analysis of genomic diversity in Mexican Mestizo populations to develop genomic medicine in Mexico. *Proc Natl Acad Sci U S A 106*(21), 8911–8916.

Singer PA, Court EB, Bhatt A, et al. (2007). Applying genomics-related technologies for Africa's health needs. *Afr J Med Med Sci 36*(Suppl), 7–14.

Singh JA, Daar AS. (2008). The 20-year African biotech plan. *Nat Biotechnol 26*(3), 272–274.

Siva N. (2008). 1000 Genomes project. *Nat Biotechnol 26*(3), 256.

Slatkin M. (2009). Epigenetic inheritance and the missing heritability problem. *Genetics 182*(3), 845–850.

Smith AC, Mugabe J, Singer PA, Daar AS. (2005). "Harnessing genomics to improve health in Africa" - an executive course to support genomics policy. *Health Res Policy Syst 3*(1), 2.

Springer M, Weissman JS, Kirschner MW. (2010). A general lack of compensation for gene dosage in yeast. *Mol Syst Biol 6*, 368.

Suitor C, Meyers L. (2006). Dietary Reference Intakes Research Synthesis: Workshop Summary. National Academies Press.

Swartz DA, Park EI, Visek WJ, Kaput J. (1996). The e subunit gene of murine F1F0-ATP synthase. Genomic sequence, chromosomal mapping, and diet regulation. *J Biol Chem 271*(34), 20942–20948.

Szkudelska K, Nogowski L. (2007). Genistein—a dietary compound inducing hormonal and metabolic changes. *J Steroid Biochem Mol Biol 105*(1–5), 37–45.

Teo YY, Sim, X, Ong, RT, Tan, AK, Chen, J, Tantoso, E, Small, KS, Ku, CS, Lee, EJ, Seielstad, M, Chia, KS. (2009). Singapore Genome Variation Project: a haplotype map of three Southeast Asian populations. *Genome Res 19*(11), 2154–2162.

The International HapMap Consortium. (2005). A haplotype map of the human genome. *Nature 437*(7063), 1299–1320.

Tindana, PO, Singh, JA, Tracy, CS, Upshur, RE, Daar, AS, Singer, PA, Frohlich, J, Lavery, JV. (2007). Grand challenges in global health: community engagement in research in developing countries. *PLoS Med 4*(9), e273.

Tracy, RP. (2008). 'Deep phenotyping': characterizing populations in the era of genomics and systems biology. *Curr Opin Lipidol 19*(2), 151–157.

van Ommen B, Bouwman J, Dragsted L, et al. (2010a). Challenges of molecular nutrition research 6: The Nutritional Phenotype database to store, share and evaluate nutritional systems biology studies. *Genes & Nutrition 5*(3), 189–203.

van Ommen B, El-Sohemy A, Hesketh J, et al. (2010b). The Micronutrient Genomics Project: Creating a community driven knowledge base for micronutrient research. *British Journal of Nutrition*.

van Ommen B, Keijer J, Heil SG, Kaput J. (2009). Challenging homeostasis to define biomarkers for nutrition related health. *Mol Nutr Food Res 53*(7), 795–804.

Varma V, Wise C, Kaput J. (2010) Carbohydrate metabolic pathway genes associated with quantitative trait loci (QTLs) for obesity and type 2 diabetes. *Biotechnology Journal 5*(9), 942–949.

Venter JC, Adams MD, Myers EW, et al. (2001). The sequence of the human genome. *Science 291*(5507), 1304–1351.

Vineis P, Kriebel D. (2006). Causal models in epidemiology: past inheritance and genetic future. *Environ Health 5*, 21.

Warodomwichit D, Arnett DK, Kabagambe EK, et al. (2009). Polyunsaturated fatty acids modulate the effect of TCF7L2 gene variants on postprandial lipemia. *J Nutr 139*(3), 439–446.

Wells KB, Staunton A, Norris KC, et al. (2006). Building an academic-community partnered network for clinical services research: the Community Health Improvement Collaborative (CHIC). *Ethn Dis 16*(1 Suppl 1), S3–17.

Williams RP. (1956). *Biochemical Individuality: The Basis for the Genetotrophic Concept*. New Canaan, CT: Keats Publishing.

Wise C, Kaput J. (2009). A strategy for analyzing gene-nutrient interactions in type 2 diabetes. *J Diabetes Sci Technol 3*(4), 710–721.

Wopereis S, Rubingh CM, van Erk MJ, et al. (2009). Metabolic profiling of the response to an oral glucose tolerance test detects subtle metabolic changes. *PLoS ONE 4*(2), e4525.

Yesupriya A, Evangelou E, Kavvoura FK, et al. (2008). Reporting of human genome epidemiology (HuGE) association studies: an empirical assessment. *BMC Med Res Methodol 8*, 31.

Zehner ER. (2009). Integrating maternal, infant, and young child nutrition: report on the ten year strategy Infant and Young Child Nutrition (IYCN) Working Group October 2008 workshop. *Food Nutr Bull 30*(2 Suppl), S190–196.

Zhang J, Zhang F, Didelot X, et al. (2009). Maternal high fat diet during pregnancy and lactation alters hepatic expression of insulin like growth factor-2 and key microRNAs in the adult offspring. *BMC Genomics 10*, 478.

19.

TRANSGENIC FOODS AND NUTRITIONAL HEALTH IN DEVELOPING COUNTRIES

Dhavendra Kumar and Joel Cohen

INTRODUCTION

The concerns of food availability and nutritional intake are much greater in the developing world (least and lesser developed countries, LDCs) compared to the United States, Europe, and Japan. This is reflected in widely prevalent undernutrition and malnutrition across LDCs. Several developing countries have invested heavily to achieve increased crop yields and enhance the nutritional value of home-grown crops and plant foods. Despite successes in some regions, the food productivity and quality remain significantly low compared to the developed world.

Transgenic foods, or genetically modified foods (GM foods) are an excellent example of biotechnological achievement that has revolutionized the food industry. It offers opportunity for enhanced and improved agricultural productivity in low and middle economies of the developing world as duly acknowledged by the Organisation for Economic Co-operation and Development (OECD, 2003; www.OECD.org/). Recombinant DNA techniques provide plant breeders with abilities to introduce traits into plants not accomplished through traditional plant breeding. These new traits include enhanced resistance to insect pests or diseases responsible for significant yield loss, potential sources of tolerance to drought and soils with high concentration of salt or heavy metals, and improved productivity potential. However, there are several concerns and regulatory issues surrounding the potential adverse and/or human health consequences arising from the introduction of GM foods. Several developing countries have accepted the potential benefits of transgenic food biotechnology and have introduced biosafety regulatory measures for the growth of domestic biotechnology infrastructure and systems to assess GM foods currently available, and including all new products.

This chapter provides an overview of the current status of research and development in transgenic crops and plants, perceived benefits and risks, and regulatory systems in different developing countries. The chapter does not include information and discussion on genetically modified animal food products, as most developing countries do not have any major program for biotechnology development and production of GM animal products. The interested reader may refer to other resources for information on transgenic animal products for human consumption.

DIVERSITY OF TRANSGENIC CROPS AND PHENOTYPIC GROUPS

Several countries in the developed world are actively pursuing and harnessing transgenic biotechnology in producing a number of GM agriculture foods. In technical terms, the whole process of producing a specific transgenic phenotype, leading to commercialization, goes through four stages (Cohen, 2005):

1. Transformation events that produce stable transgenic plants derived from multiple generations at the laboratory/greenhouse/glasshouse scale
2. Transformation events expressing stable traits in small-scale, single, or multi-location confined trials
3. Transgenic plants advancing into larger, precommercial trials
4. Commercial release—products marketed to farmers through privately or publicly owned seed companies or other institutional mechanisms

In the above context, an event is defined as the stable transformation—the incorporation of foreign DNA into a living plant cell—undertaken by a single institute among the participating countries, thereby providing a unique crop-and-trait combination. So far, in fifteen 15 developing countries, 201 genetic transformation events have produced 45 different crops (Cohen, 2005; Table 19-1). The data is based on peer-reviewed information from 62 research institutions in 15 countries and assessed for 13 criteria: (i) country, (ii) food and fiber crops, (iii) source of germ plasm, (iv)

TABLE 19-1 TRANSFORMATION EVENTS GROUPED BY COUNTRY, CROPS AND PHENOTYPIC CATEGORY (COHEN, 2005).

Continent	Countries	No. Events[a]	Crops	Phenotypic Category[b]
Africa	Egypt	17	Cotton, cucumber, maize, melons, potatoes, squash and marrow, tomatoes, watermelons, wheat	AP, FR, FR/HT, HT, HT/IR, IR, OO, PQ, VR
	Kenya	4	Cotton, maize, sweet potatoes	HT, HT/IR, OO, PQ, VR
	South Africa	20	Apples, grapes, lupin, maize, melons, pearl millet, potatoes, sorghum, soybeans, strawberry, sugar cane, tomatoes, indigenous vegetables	AP, BR, FR, HT, HT/AP, IR, PQ, VR
	Zimbabwe	5	Cotton, cowpeas, maize, sweet potatoes, tomatoes	FR, HT/VR, VR
Asia	China	30	Cabbage, chili, cotton, maize, melons, papayas, potatoes, rice, soybeans, tomatoes	AP, FR, IR, VR
	India	21	Cabbage, cauliflower, chickpeas, citrus, eggplant, mung beans, muskmelon, mustard/ rapeseed, potatoes, rice, tomatoes	AP, FR, HT/AP, IR, IR/BR, OO, PQ, VR
	Indonesia	14	Cacao, cassava, chili pepper, coffee, groundnuts, maize, mung beans, papayas, potatoes, rice, shallot, soybeans, sugar cane, sweet potatoes	AP, FR, IR, PQ, VR
	Malaysia	5	Oil, palms, papayas, rice	HT, IR, VR
	Pakistan	5	Cotton, rice	HT, IR, PQ, VR
	Philippines	17	Bananas and plantains, maize, mangoes, papayas, rice, tomatoes	AP, OO, VR
	Thailand	7	Cotton, papayas, pepper, rice	AP, BR, IR, VR
Latin America	Argentina	21	Alfalfa, citrus, potatoes, soybeans, strawberry, sunflowers, wheat	AP, BR, FR, IR, IR/BR, OO, PQ, VR
	Brazil	9	Beans, maize, papayas, potatoes, soybeans	AP, BR, FR, HT, IR, PQ, VR
	Costa Rica	5	Bananas and plantains, maize, rice	AP, IR, VR
	Mexico	3	Bananas and plantains, maize, potatoes	IR, VR
Total		201		

[a] An event is defined as the stable transformation—incorporation of foreign DNA into a living plant cell—undertaken by a single institute among the participating countries, thereby providing a unique crop and trait combination.

[b] Phenotypes are defined as follows: AP, agronomic properties; BR, bacterial resistance; FR, fungal resistance; HT, herbicide tolerance; IR, insect resistance; OO, other; PQ, product quality; VR, virus resistance.

gene group, (v) gene, (vi) phenotype category, (vii) function, (viii) regulatory status, (ix) regulatory status by year, (x) lead research institutes (xi) collaborating institutes, (xii) institutional arrangement, and (xiii) dissemination.

The percentage of different phenotypic groups among the 201 transformation events identified is presented in Figure 19-1. Over half of the 201 transformation events involve single genes that confer biotic resistance to either viral or insect stresses to the host plant. In 11 events, stacked genes (those that simultaneously confer more than one trait) are being tested for phenotypic combinations. Some countries are working on five or fewer crops, whereas others, such as China and South Africa, are working on 15 or more.

The ten crops with the largest number of transformation events (Figure 19-2) have focused on cereals; significant numbers of a diverse range of transgenic vegetables, fruits, roots, and tubers have also been created. Significant progress has also been achieved in transforming orphan (noncereal) food staples and indigenous crops, including mung beans, beans, chickpeas, cowpeas, lupin, cacao, and coffee. The greatest numbers of transformation events to date are for rice, potatoes, maize and papaya, compared with cotton, which is used as an oil and fiber crop.

The largest number of transformation events were generated by the seven Asian countries surveyed (109), followed by the four African countries (54), and the four Latin American countries (38). However, Brazil also reported 37 events contracted by the private sector working with Embrapa (Brasilia), a public research institute associated with Brazil's Ministry of Agriculture (Brasilia), to address their market needs. Asian countries have products in all stages of the research pipeline, having made significant commitments to GM crops (Asian

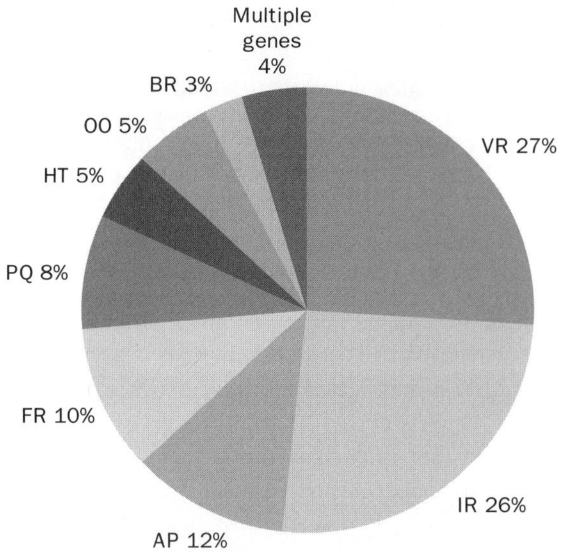

Figure 19-1 Total events distributed by phenotype. AP, agronomic properties; BR, bacterial resistance; FR, fungal resistance; HT, herbicide tolerance; IR, insect resistance; OO, Other; PQ, product quality; VR, virus resistance.

Development Bank, 2001; Cohen, 2001), and are already achieving significant success with insect-resistant GM cotton approvals (in China and to a lesser degree in India, and lastly, Indonesia). Despite the large number of transformation events in development in Asia, only the Philippines has approved a commercial feed crop for production, and China allows cultivation and use of publicly developed transgenic vegetables. Indonesia had approved commercial GM cotton, but it has now been taken off the market. Sub-Saharan Africa, with the exception of South Africa, lacks many capabilities and resources to advance such research (UNEC, 2002). Many countries are just considering whether to conduct research on, or to allow import of, GM crops or products. Research capacity and potential markets are evolving (e.g., for insect-resistant cotton), albeit subject to uncertainties regarding use and trade. Kenya and Egypt have demonstrated competence in regulatory and import approvals, but have still not approved any crop for open testing or commercial use.

Table 19-2 presents five of the eight phenotypic groups having the highest number of clearly identified genes or gene groups. Where the specific genes were not provided, the country's description of the trait being developed is retained. On the basis of the available data, three groups of genes appear to be important in conferring sufficiently

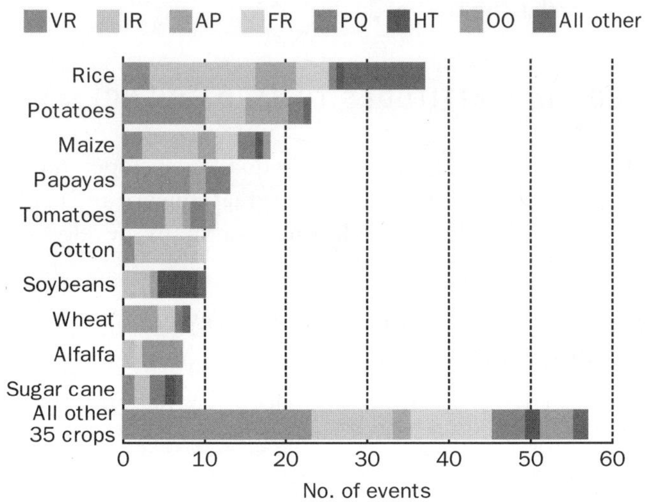

Figure 19-2 Phenotype characteristics sorted by number of transformation events among the top 10 crops in the study data set. AP, agronomic properties; BR, bacterial resistance; FR, fungal resistance; HT, herbicide tolerance; IR, insect resistance; OO, Other; PQ, product quality; VR, virus resistance.

TABLE 19-2 GENES AND GENE GROUPS IN FIVE PHENOTYPIC CATEGORIES

Phenotypic category Gene/gene group	Number of events[a]
Insect resistance	51
Bt	35
Galanthus niivalis agglutinin (Snowdrop lectin)	5
Pin	4
Trypsin inhibitor	2
Bt and trypsin inhibitor	2
Gall midge resistance gene (*Gm2*)	1
Alpha amylase inhibitor	1
Not disclosed	1
Viral resistance	53
Coat protein	47
Replicase	3
Coat protein and reporter genes	1
Coat protein and replicase	1
Antisense to tomato yellow leaf curl virus	1
Fungal resistance	21
Glucanase, chitinase	6
Glucanase, PGIP2	2
Chitinase, and *ap24* antifungal protein	2
Chitinase	2
Blast resistance	2
Not disclosed	2
PGIP1 and PGIP2 isolated in South Africa	1
Grape resveratrol	1
Glucanase (PGIP3)	1
*b*32, *PGIP*2 and other selected antifungal genes	1
AP24, CH5b, GLN3	1
Herbicide tolerance	11
5-enolpyruvylshikimate-3-phospate syntheses (EPSPS)	2
BAR encoding phosphinothricin acetyltransferase	6
Acetohydroxyacid synthase (*AHAS*)	2
PsbA encoding D1 polypeptide of photosystem II	1
Bacterial resistance	7
Xa21-resistance (*R*) gene	5
Unspecified antibacterial	1
Unspecified antimicrobial peptides	1

[a] An event is defined as the stable transformation—incorporating of foreign DNA into a living plant cell—undertaken by a single institute among the participating countries, thereby providing a unique crop and trait combination, PGIP, polygalacturnase-inhibiting protein.

robust utility and suitability for wide use. The first gene group consists of *Cry* genes from *Bacillus thuringiensis* (Bt) that confer resistance to lepidopteran insects. The second group consists of coat proteins of plant viruses used for inducing virus resistance; the third consists of genes conferring herbicide tolerance. Most other gene groups and their associated phenotypic traits have not yet demonstrated robust applicability in the field. For example, no gene group has yet to confer effective fungal resistance, although much experimental activity has been spent on investigating the glucanases and chitinases. Similarly, no group of genes has been shown to reliably confer bacterial resistance in the field, even though many investigators are studying the effects of antimicrobial peptides. Thus, success has been limited in developing crops with traits other than insect resistance, virus resistance, and herbicide tolerance. Among the genes and gene groups being tested, the *Cry* genes, coat protein genes, and herbicide tolerance genes are most likely to move through regulation with fewer requirements, assuming already packaged data are accepted by the developing country in which tests would occur. This is because numerous safety reviews have been conducted on these genes in several countries. However, this does not rule out tests to address specific environmental or biodiversity concerns, as such results may not be transferable from one country to another. The more unusual genes shown in Table 19-2 include different types of insect-resistance genes, replicase genes, antisense genes, and genes encoding antimicrobial peptides. Most countries are focusing on genes that are already available and have already been characterized, but a few are also investing in their own gene discovery and development, such as South Africa, Malaysia, Brazil, India, and China.

PERCEIVED RISKS AND BENEFITS OF TRANSGENIC (GM) FOODS

It is inevitable that any new consumer product introduced to the market will face uncertainties. The rational consumer makes decisions under uncertainty by assigning probabilities of occurrence to uncertain outcomes. Analysis of the benefits and potential costs of an action, such as consuming genetically modified foods, results in an expected utility payoff for each action.

FACTORS AFFECTING RISK PERCEPTIONS

Traditionally, the consumer weighs the expected benefits and expected costs depending on his or her risk tolerance (Figure 19-3). The scientific consensus is that GM foods do not pose any risk to consumers. However, there is a distinction between scientifically assessed risk and perceived risk. The public's beliefs about risk are often very different from the beliefs of experts. These perceived risks are seen as potential future costs by the consumer, and carry probabilities of occurrence assigned by the consumer, and are thus subjective (Curtis et al., 2004). The probability that the consumer assigns to each potential cost or risk primarily stems from three sources (Table 19-3):

a. the level of trust in government regulators regarding food supply safety
b. attitudes toward scientific discovery
c. the influence of media coverage.

		Opportunity	
		Yes	No
Danger	Yes	1. Fungability (costs and benefits)	2. "Better safe than sorry" (only costs)
	No	3. "Waste not, want not" (only benefits)	4. Indifference (neither costs nor benefits)

Figure 19-3 Risk Matrix (Margolis, 1996)

TRUST IN GOVERNMENT

Conclusions based on 599 Chinese consumers support the view that people largely put their trust in the government (Li et al., 2002; Curtis, 2003). The respondents were found to be trusting of government regulators concerning the safety of the food supply, and very positive toward science, including the use of biotechnology in agriculture. Most of the respondents were willing to pay a premium for genetically modified foods; many responded that they felt positively about science, were willing to try new products, or the price change was not enough to keep them from purchasing the products.

POSITIVE ATTITUDES TOWARD SCIENCE

European countries and Japan gradually developed modern capitalist societies while taking great concern and pride in preserving cultural traditions. For many developing

TABLE 19-3 INFLUENCES ON RISK PERCEPTION ASSOCIATED WITH GM FO ODS

Representative countries	Government regulation	Media coverage	Attitudes toward science
China/Colombia	+	+	+
"Developing nations"			
USA/Canada	+	+/−	+
Europe/Japan	−	−	−

Note. + positive influence; − negative influence; +/− ambiguous countries, history took another turn. For example, in China, a decade of Cultural Revolution from 1966 to 1976 systematically tore down historical and traditional structures in the society. The past was condemned as "feudal and superstitious" (Beech, 2002). The vacuum remaining was to some extent replaced by the communist state. Now, with a highly desired and incredibly rapid transition to capitalism, and with much of the old Chinese tradition crushed by the Cultural Revolution, the Chinese are forward-looking. Technological novelties from the rest of the world are often considered much-needed improvements, rather than reasons for concern.

In Colombia, Pachico and Wolf (2002) found that a positive predisposition toward scientific innovation was demonstrated by a strong agreement among 68% of the respondents that science improves the quality of life. Additionally, 75% of respondents agreed or strongly agreed that their government provides an adequate level of safety in their food supply.

MEDIA COVERAGE

Information from media is provided either by private firms with their own profit-maximizing and ideological objectives, or by public sources that may have the formal goal of providing objective information, but administrators and governments may have their own (private) incentives to bias the information (Swinnen, McCluskey, & Francken, 2003). Government-controlled media coverage in China concerning genetically modified crops has been very positive. China is the fourth largest producer of genetically modified crops in the world, and continues to support biotechnology research in an effort to sustain food self-sufficiency policies. In China, only around 9% of the population surveyed had a somewhat negative or very negative opinion concerning the use of biotechnology in foods. Fifty-four percent claimed to have no knowledge of genetically modified products at all, and only around 8% associated high risk with genetically modified foods. However, about two-thirds of the people relied upon advertising in making their food choices. In Colombia, the survey also indicated television as the major information source concerning genetically modified foods (Pachico & Wolf, 2002). However, nearly 75% of the consumers surveyed agreed that there may be some risk associated with genetically modified foods; nevertheless, most were willing to try genetically modified foods in any case.

FACTORS AFFECTING PERCEIVED BENEFITS FOR DEVELOPING COUNTRIES

The benefits of genetically modified foods are mainly cost-reducing and/or yield-enhancing attributes. The benefits of genetically modified foods to the developing world concern food availability, nutrition, and economic advantage. Food availability is a large problem in the developing world. Forty percent of the survey participants in Columbia (Pachico & Wolf, 2002) responded that they did not have adequate food for their families. China currently has almost 1.3 billion people and is likely to exceed 1.4 billion by 2050 (Population Reference Bureau, 2002). China recognizes that if it is going to continue to feed its people, it must find more efficient agricultural production methods. Additionally, more than a dozen genetically modified crops have been approved for development in China (Huang et al., 2001). Increased yields provided by many GM crops may provide answers to food availability issues in the developing world. James and Krattiger (1999) estimate that transgenic technology may increase rice production in Asia alone by 10%–20% in the next decade.

A second major problem in the developing world is malnutrition, especially vitamin A deficiency (VAD). It is estimated that a quarter to half a million Vitamin A–deficient children go blind each year (Zimmerman & Qaim, 2002). Because rice is widely consumed in developing countries, golden rice was genetically engineered to provide vitamin A, and thus reduce VAD. Zimmerman and Qaim (2002) estimate that golden rice could reduce related healthcare costs in the Philippines by up to 32%, and avert from 2200 to 10,200 cases of blindness each year. Consumers in developing countries are concerned with their nutritional intake. Li et al. (2003) found that consumers in China were willing to pay more for genetically modified rice than for genetically modified soybean oil, due to the existence of additional vitamin content in the rice product. In this context, recent developments in provitamin A biofortification of the maize germplasm, through incorporating the gene encoding β carotene hydroxylase, raise the prospects of growing a tropical maize crop enriched with vitamin A and thus significantly reduce the vitamin A deficiency–related blindness (Yan, et al., 2010).

A third problem facing developing nations is economic advantage. In order to be competitive in world markets, growers must find cost-efficient production

TABLE 19-4 ADVANTAGE OF IR OVER CONVENTIONAL COTTON (%)

	Argentina	China	India	Mexico	South Africa
Yield	33	19	34	11	65
Revenue	34	23	33	9	65
Pesticide costs	−47	−67	−41	−77	−58
Seed costs	530	95	17	165	89
Profit	31	340	69	12	299

Source: Author's calculations based on Argentina [23]; China [18]; India [36]; Mexico [30]; and South Africa [25].

methods. Genetically modified soybeans, also known as Roundup Ready (RR) soybeans, have provided Argentina an increase in total factor productivity in soybean production of 10% due to cost savings (Qaim & Traxler, 2002). A study by Kirsten, Gouse, and Jenkins (2002) found that both large-scale and small-scale Bt (*Bacillus thuringiensis*) cotton farmers in South Africa realized net income gains due to higher yields and savings on pesticides. These gains prevailed even with higher seed costs and technology fees not found with traditional seed varieties. The use of insect-resistant Bt cotton in China reduced production costs by 14%–33% (Pray, 2000).

WEIGHING THE PERCEIVED COSTS AND BENEFITS IN DEVELOPING COUNTRIES

As is evident from the above discussion, the developing world may benefit greatly from transgenic technology. When such benefits are compared with the relatively small perceived costs associated with GM foods due to consumer risk perceptions balanced by a trust in government regulation, positive media attention, and a positive predisposition for scientific discovery, it makes sense that consumers in developing countries would accept GM food products through cost-benefit analysis. Although genetically modified food acceptance is thus far relatively high in developing nations, recent surveys show that consumers wish to know which foods contain GM ingredients. In Columbia, (Pachico & Wolf, 2002) 90% of the survey respondents considered mandatory labeling of GM foods very important. However, only 64% of the respondents said they read food labels very often or somewhat often. Similarly, around 90% of the survey respondents in China considered labeling foods with GM ingredients very important (Curtis et al., 2004). In response to these concerns, China has required, since June 2001, that all GM products entering China for research, production, or processing have safety certificates from the agricultural ministry to ensure that they are safe for human consumption, animals, and the environment. China also requires that all listed transgenic biological products be labeled.

ECONOMIC IMPACT OF TRANSGENIC CROPS IN DEVELOPING COUNTRIES

Transgenic crops are spreading more rapidly than any other agricultural technology in history, suggesting that farmers perceive important economic benefits in growing them (Raney, 2006). Developing countries now account for 38% of global transgenic crop area (James, 2005), despite continuing controversy surrounding them. This section of this chapter attempts to identify the factors that influence the level and distribution of the economic value created by transgenic crops in developing countries. The availability of suitable transgenic crop cultivars in a country is the most basic requirement for successful adoption by farmers (Naylor et al., 2004; Huesing & English, 2004). Institutional factors such as national research capacity, intellectual property rights, environmental and food safety regulatory capacity, trade regulations, and the existence of functioning input markets are crucially important determinants of the level and distribution of gains (Raney, 2006).

The most extensive ex post studies of transgenic crop adoption in developing countries have been conducted for insect-resistant (IR) cotton in Argentina, China, India, Mexico, and South Africa. Transgenic herbicide-tolerant (HT) soybeans are being grown in Argentina, Brazil, Paraguay, and elsewhere, but Argentina is the only developing country for which peer-reviewed studies have been published. Some developing countries also produce HT and/or IR maize, but the only peer-reviewed ex post analyses of their impacts published so far are for Argentina and South Africa. Table 19-5 summarizes the results from the most comprehensive economic studies of the farm-level impacts of IR cotton in developing countries. Each of the studies was based on data from two or three seasons of commercial farm production. The figures reflect (Table 19-5) the average percentage difference between IR and conventional cotton for all farmers over all seasons covered in the study.

Although the averages conceal a high degree of temporal and spatial variation, they clearly indicate positive overall results. Farmers who adopted the transgenic varieties experienced higher effective yields (owing to less pest damage), higher revenue, and lower pesticide costs. These factors more than compensated for higher prices paid for IR seeds so that net profits increased for adopters.

CHINA

China, where some 7.5 million small farmers are growing IR cotton, represents the most successful case so far in terms

TABLE 19-5 PERFORMANCE ADVANTAGE OF IR OVER NON-IR COTTON IN INDIA; SHOWN AS % IN DIFFERENT SATES OF INDIA

	Yield	Revenue	Chemical costs	Total costs	Profits
Maharashtra	32***	29***	−44***	15**	56***
Karnataka	73***	67***	−49***	19**	172***
Tamil Nadu	43***	44***	−73***	5	229***
Andhra Pradesh	−3	−3	−19	13*	−40
National average	34***	33***	−41***	17***	69***

Values are calculated for the year 2002/03.

Source: [36**]. Statistically different from zero at the 10% (*), 5% (**), 1% (***) significance levels.

of productivity, farmer incomes, equity and sustainability (Pray, Huang, & Rozelle, 2002). Much of China's success rests on its highly developed public agricultural research system, which has independently produced two transgenic constructs that confer insect resistance. These have been incorporated into a large number of locally adapted cotton varieties, and compete directly with Monsanto's IR cotton varieties. As a result, transgenic seed prices are much lower in China than elsewhere and farmers reap substantially higher returns. The role of the public sector in developing and distributing IR cotton varieties has been instrumental in reducing the price premium. Lower costs and marginally higher yields translate into large net profit gains in China. Chinese farmers experience lower yield gains than in many other countries, because pest damage on conventional cotton is controlled by heavy pesticide use. China has been able to significantly reduce its use of chemical pesticides on cotton, with important environmental and farmer health benefits (Huang et al., 2003). Field trials for IR rice in China suggest a similar pattern of significantly lower pesticide use and small yield gains (Huang et al, 2003).

Pray and Huang (2003) analyzed the distribution of IR cotton benefits in China by farm size and found the innovation to be decidedly pro-poor. The smallest farms (less than 0.47 hectares [ha]) experienced the largest yield gains, and mid-size farmers (0.47–1.0 ha) had the largest reductions in total costs owing to less pesticide use. In terms of net income, the gains for the two smaller farm-size categories were more than twice those for the largest farms (over 1.0 ha).

ARGENTINA

The Argentine experience with IR cotton provides an interesting comparison with the Chinese case in terms of the effect of IPRs. Monsanto has strictly enforced its IPRs on IR cotton in Argentina, and charges significantly higher prices than for conventional cotton seed (Qaim & de Janvry, 2003). As a result, IR cotton offers relatively small returns and thus has not been widely adopted. This is in contrast with HT soybeans, which have been enthusiastically embraced by Argentine farmers. The authors conclude that the crucial difference is that Monsanto failed to patent its soybean innovation in Argentina, and thus has been unable to strictly enforce its IPRs.

HT soybeans are estimated to have increased total factor productivity in Argentina by 10% on average, with the cost savings being slightly greater for smaller farms (less than 100 ha) than for larger farms, owing mainly to lower seed prices among small farmers who are more likely than larger farms to use uncertified seed (Qaim & Traxler, 2004). Aggregate global welfare benefits from HT soybeans are estimated at more than US $1.2 billion, with the largest share going to consumers (53%), followed by seed and biotechnology firms (34%) and farmers (13%). Owing to comparatively weak intellectual property protection, Argentine soybean growers receive 90% of the benefits in that country. The only published study to look comprehensively at transgenic crop adoption in Argentina credits them with major responsibility for transforming the farming sector, although transgenic maize and cotton have been of less value to farmers than soybeans (Trigo & Cap, 2003).

SOUTH AFRICA

South Africa provides another important lesson about the role of institutions. South Africa has a dualistic farm sector, with large commercial farms served by a modern input supply system operating alongside smallholder, semi-subsistence farms. IR cotton and yellow IR maize were introduced in 1998/99. White IR maize varieties (preferred for food) were introduced in 2001–2002, and free seed was distributed by Monsanto to smallholders on a trial basis. For IR cotton, several studies (Bennett et al., 2004a,b; Thirtle et al., 2003) have found positive economic impacts for smallholder farmers in the Makhathini Flats of KwaZulu Natal province, where a local cooperative provided seed on credit along with technical advice. Within the smallholder category, in the first year IR adopters were more likely to be older males with larger farms who were specially targeted by the cooperative; by the second year both genders and all age groups and farm sizes were equally likely to adopt. The benefits were widely shared by all farm types, and both studies found significant pro-poor benefits. Pesticide use also declined significantly, having both environmental and health benefits. In a study comparing results for smallholders versus large farms, the smallholders were found to benefit more than large dryland farms, but less so than large irrigated farms (Gouse & Pray, 2004).

A comparison of IR maize varieties and their non-IR counterparts in South Africa in 2001–2002 found that large commercial farmers experienced yield, pesticide, and income advantages, while smallholders experienced higher yields (Gouse et al., 2005). The authors argue that the main constraint to smallholder IR maize adoption will be the ability of seed companies to provide seed at affordable prices.

MEXICO

The Mexican case provides a clear example of the importance of introducing effective innovations (Traxler & Godoy-Avila, 2004). The IR cotton varieties available in Mexico (as with most varieties available elsewhere) incorporate a gene from the soil bacterium *Bacillus thuringiensis* (Bt), which confers resistance against a narrow range of pests. Comarca Lagunera is the only state in Mexico where these pests constitute the major threat to cotton production; hence, it is the only state where IR cotton has been widely adopted. The major pests in other states require continued chemical pesticide use, making the IR varieties of less value. The authors estimate that farmers gained 83% of the total economic value created by the crop on average for the two years in the study.

INDIA

Despite the rapid expansion of IR cotton in India, its economic impact continues to be hotly contested, with some critics charging that IR adopters are worse off than conventional growers (Qayum & Sakkhari, 2005). The first economic studies were based on farm-level field trial data, and as such did not reflect the actual farm experience with commercial cultivation (Qaim & Zilberman, 2003; Qaim, 2003). These studies estimated potential yield benefits of 80%. Later farm-level research found smaller, but significant, yield advantages (Bennett et al., 2004a,b) even for unofficial varieties (Morse, Bennett, & Ismael, 2005). A subsequent study, summarized in Table 19-1 and shown in detail in Table 19-5, was based on farm-level data from four different states in India (Qaim et al., 2006). This study found large net gains from IR cotton adoption at the national level, although significant variation was observed across states, and one state, Andhra Pradesh, experienced negative results. The authors speculated that the lack of locally adapted cultivars was the main reason for poor performance in this state. At the time of the study, the Indian biosafety authorities had approved only four IR cotton varieties for use throughout the country. By 2005, that number had increased to 20, and the area of the country planted with IR cotton almost tripled from the previous year (James, 2005).

The evidence reviewed in this section suggests that farmers in developing countries can benefit from transgenic crops, but a fairly high level of national institutional capacity is required to ensure that farmers have access to suitable innovations on competitive terms. A certain level of national research and regulatory capacity are prerequisites, along with effective IPR management and input supply systems, especially for seeds. The economic results so far suggest that farmers in developing countries can benefit from transgenic crops, but for the poorest farmers in the poorest countries, where institutional conditions are weak, ensuring access will remain a formidable challenge.

BIOSAFETY AND REGULATORY SYSTEMS

Most developing countries harnessing the GM food biotechnology have put several biosafety measures in place, including strengthening regulatory systems and capacity, stimulating scientific risk assessment, and advancing efforts in the areas of public acceptance, technology transfer, and regulatory harmonization. Cohen and Paarlberg (2004) summarize the key actions that might ensure improving biosafety system efficiencies, capacity, and transparency, which are:

1. Emphasizing the central role of capacity: provide resources to stimulate the development of qualified individuals to enhance the knowledge and skills base available for regulatory review and decision making.
2. Revising national biosafety guidelines: provide greater clarity by stating purpose and objectives of biosafety reviews; outline procedural and facility requirements; provide a clear "road map" of approval processes, and examine the relationship between approvals and legal authority with sanctions to ensure compliance. Such work is supported through the UNEP/GEF project for developing biosafety framework and guidelines for signatories to the Cartagena Protocol on Biosafety (Briggs, 2001).
3. Enhancing the effectiveness of national biosafety committees: stimulate open and effective dissemination of information; promote cooperation with national and international bodies; and provide database listings of applications.
4. Improving biosafety procedures and decisions: increase the scientific base for decision making; identify research needs and collect data to support risk assessment; and adhere to realistic time frames for application decisions.
5. Building public awareness: information campaigns and outreach activities.
6. Strengthening institutional roles: define and clarify responsibilities among entities sharing responsibility for environmental, food safety and marketing reviews.
7. Enhancing transparency and efficiency: clarify review procedures to stakeholders; distribute responsibility for review of confidential business information between at least two biosafety officials; and organize "customer service" meetings.

INTERNAL MANAGEMENT OF BIOSAFETY MEASURES

CONCEPTUAL FRAMEWORK FOR IMPLEMENTING BIOSAFETY

Over the past two decades, developing countries have attempted to develop national biosafety frameworks and guidelines. These have often been implemented in a fragmented or ad hoc manner, due to particular needs and urgent pressures. Our studies, findings, and actions illustrate how difficult a task this is, as the design and implementation of a national biosafety system attempts to balance or align public policy goals, perceptions of safety and risk, and economic, political, and technical realities. Ideally, developing countries would benefit by working with a comprehensive plan to help with policy development and regulatory implementation. There is no single "best approach," however, that can reflect each country's environmental, political, financial, and scientific differences. Consequently, a comprehensive conceptual framework for biosafety implementation has often been lacking. For this reason, an international expert consultation was convened to explore how such a framework could be constructed.

The consultation reviewed prior studies, and concepts and lessons derived from other national, regional, and international experiences. The *Conceptual Framework* produced was developed with developing country partners, regulatory specialists from developed and developing countries, international agricultural research scientists, donor agencies, and capacity-building providers. It can systematically guide regulatory implementation and capacity building, and addresses concerns arising from the country studies. The Conceptual Framework identifies and discusses five elements fundamental to the development and implementation of national biosafety systems, as well as three crosscutting topics: transparency, public participation, and resources (McLean et al., 2002). The first two elements describe national policies, strategies, and research agendas regarding biotechnology and biosafety, and suggestions as to the use of national inventories and evaluation. These two elements provide the foundation for implementation, described in the next three elements. Requisite knowledge, skills, and capacity base is the third element, and describes the resource base within which the final two elements occur:

a. Development of regulations and implementation of regulations. Taken together, these elements build an integrated system, one which can avoid the difficulties encountered in the earlier ad hoc approaches, where elements and pieces were assembled on an as-needed basis, only then to be revisited each time that systems were advanced.

b. Implications of the conceptual framework – this broadly includes two elements of the framework: policies and capacity.

National Policies and Strategies

Biosafety policy should articulate a national approach for implementing regulatory guidelines or frameworks. This policy element serves to integrate political, social, ethical, health, economic, and environmental considerations into decisions regarding the safe use of biotechnology methods and products. These discussions are taken up in a national strategy that provides direction on many of the fundamental issues and public policy decisions considered during regulatory development. Paying explicit attention to this policy element helps ensure that links between biosafety and national food, sustainability, and environmental objectives are explored. This process and consultation helps define a biosafety system's goals and objectives, and provides a place for this effort and dialog to occur. In so doing, the Conceptual Framework anticipates a policy forum for discussing major issues affecting the goal and policies for biosafety, as these issues remain outside the sphere of implementing regulation, decision making, and determinations of biological safety.

Scientific Knowledge, Skills and Capacity Base

Building a strong base of scientific knowledge, and developing core competencies in biotechnology product evaluation, are fundamental to any national biosafety system. These activities allow an improved scientific basis for assessments of potential risks and/or benefits, and they strengthen the scientific capabilities for risk management, inspection, and monitoring. A deep and broad knowledge, skills, and capacity base tends to foster more latitude in regulatory development and more flexibility in regulatory implementation. Special attention to this matter is necessary, despite the work done in training and human development thus far. Many skills are involved, and require competency in the disciplines of biological science; expertise in information acquisition, communications, and management; and, experience in critical thinking, analysis, and decision making. Where these skills are found, they reside in very few people, and thus scientific safety review committees are strained and competent staff is lacking to staff the official regulatory office. The expert consultation identified key decision points and subsequent policy options for building scientific knowledge, skills, and capacity. The two key decision points include a) providing a coordinated approach to incorporating scientific advice into biosafety decision making, and b) locating the science evaluation function within the regulatory system.

Coordinating Scientific Expertise

As the science involved in the creation of GM crops advances, and the products themselves become more complex, there is an increasing need to strengthen the science base supporting risk assessment and regulation. Developing skills required for biotechnology product evaluation, and maintaining parity between risk assessors and their counterparts involved in developing new products, is of fundamental importance. It requires ongoing training about new scientific advances, without which a regulator's knowledge base has a limited life expectancy. The first policy option regarding coordination determines whether development of national capacity for scientific risk assessment should be given exclusive priority, or whether it is possible to coordinate risk assessment at a regional or subregional level. The second policy option is to determine if a country will rely on international experts versus domestic self-sufficiency and capability. Each of these policy options is being explored in various ways by developing countries, and in relation to expectations for adequate risk assessment of GM crops in relation to the Cartagena Protocol on Biosafety.

Locating the Science Evaluation Function

Maintaining access to scientific expertise is an issue for developed as well as developing countries. Structurally, different approaches can be taken to locating and securing scientific advice within the regulatory structure. In considering the risk assessment of biotechnology products, some countries have implemented a system of expert advisory committees, while others have relied primarily on scientists and professionals working within government agencies. In the latter approach, the mandate for risk assessment may be vested within a single agency exclusively tasked with regulating products of biotechnology (e.g., a gene technology regulator) or it may be distributed between agencies in accordance with their existing responsibilities (e.g., departments of health, agriculture, and/or environment).

In summary, implementing a comprehensive biosafety system, responsive to national regulatory needs and to various articles of the Cartagena Biosafety Protocol, is a complex, resource-intensive undertaking (Lichtenberg, 2000). The Conceptual Framework clarifies five system elements and critical decision points reached during its design and implementation. It examines choices among various policy options; and delineates scientific and social dimensions of these options (McLean et al., 2002). It is a tool to guide capacity-building as biosafety systems are developed or reevaluated.

EXTERNAL BIOSAFETY MEASURES

The case studies highlight some of the difficulties, challenges, and policies that affect biosafety systems in the developing world. The coming into force of the Cartagena Protocol on Biosafety (CBD, 2004), together with the UNEP/ GEF program, has given renewed attention to frameworks, legislation, policies, and trade-related measures regarding GM crops, or LMOs (living modified organisms). It is clear that decisions and actions outside the formal regulatory system, or those taken by individuals responding to external pressures, environmental advocacy, or judicial appeals, have caused serious delays and shifts in policy for approving GM crops.

Three actions are discussed below to help countries understand and evaluate factors external to regulatory decision making (Cohen & Paarlberg, 2002).

GETTING THE PROBLEM RIGHT

The regulatory decision making, including regulatory procedures, policies, and their implementation, should be the key focus of attention to unlock further testing and deployment of GM crops. This is important, while keeping in mind that much has been written regarding intellectual property right (IPR) systems and their need for revision (Victor & Ford Runge, 2002), and studies as to how IPR poses restrictions on the movement, testing, and commercial use of GM crops in developing countries (Glover & Yamin, 2003). There are, however, a growing number of cases where IPR agreements have been reached, thus creating platforms for technology and skill transfer. Placing emphasis on biosafety decision making calls attention to the responsibilities developing countries face for the Cartagena Protocol on Biosafety, preparations for guideline and framework development with UNEP/ GEF, and broader considerations as to how regulation takes into account development, trade, and environmental objectives. Indeed, biosafety now commands a multiplicity of actors and ministries. Consequently, diverse agendas find their way into regulatory and safety decision making and can cause regulators to vacillate, or safety decisions to stand still. This indicates how difficult it is to manage regulatory systems in developing countries. Focusing attention on regulation is one way to help policymakers and politicians understand this issue. Obviously, regulation is not the only consideration in setting a biotechnology strategy. But, it is the national biosafety committee that will evaluate GM crops, such as banana for Uganda (Smale & De Groote, 2003) and cassava for Kenya, to determine whether or not they are safe.

POLICY, TRADE, AND ADVOCACY POSITIONS COMPLICATING SAFETY APPROVALS

Difficulties in making regulatory decisions occur primarily at three points: approvals for confined trials; approval for scale-up, or multiple and larger location trials; and finally, approval for commercial use or release. For each step beyond the laboratory, there are specific criteria, which can vary from country to country, required for biosafety approval. As GM crops approach commercialization, broader decisions come into play. These often go beyond decisions regarding biological safety, and reside external to the official purview of regulators. Final decisions are then made by senior officials or ministers and then signed personally.

The ability to affect this individual's judgment, or suspend decision making, can result in serious delays, unofficial moratoriums, and in some cases overruling of regulatory council and scientific review committees. Commercial release decisions can also be affected by broader trade considerations—especially as regards differences between Europe and United States, given that current European markets are closed to GM products (Vogel, 2001)—whether or not a country wishes to remain or become GM free, or if GM crops will be acceptable for international export or trade (Victor & Ford Runge, 2002). These broader questions of policy and agricultural strategies go far beyond the responsibility of those performing regulatory, safety, and environmental reviews. Countries can use the interrelation of elements in the Conceptual Framework to establish clear relations between policy goals and objectives, and the functions of a national biosafety regulatory system. When specific attention is needed, as per trade, food aid shipments, acceptance, or advocacy positions affecting regulatory approval, then specific GM crops, by trait combinations and their potential impact, should be examined. Given the global debate regarding GM crops, while discussion and resolution can be sought on specific issues, reaching consensus among all parties may not be achievable.

ENSURING THE FARMER'S VOICE IS HEARD

The voice and perspective of farmers is often missing during regulatory review and national debates on GM crop approvals and their utility. This is a major concern to most farmers, because it is the farmers in many countries who are at the vanguard of using GM crops in smallholder settings and now covering thousands of hectares (Huang et al., 2002). While regular communication channels exist for advocacy groups, trade negotiators, and civil society, regulatory systems lack a parallel process to ensure that farmer's perceptions or needs are part of the regulatory review equation.

As shown in the studies of Egypt and Argentina, biosafety regulatory review focuses on risk analysis of perceived or hypothetical risks. There is, however, little opportunity for such assessments to consider the benefits from a new GM crop that could counterbalance any severe concerns identified in the risk analysis. If this voice is lacking, then regulators may not fully understand the demand or need for such technologies. Obviously, one would turn to the Ministry of Agriculture as the logical conduit for bringing the voice of farmers to regulatory decision makers. Such structures would enhance the accountability, stakeholder relations, and transparency of the regulatory system, as found to be needed for Argentina and Egypt.

THE FUTURE OF TRANSGENIC FOOD BIOTECHNOLOGY

The future of GM crop production in developing countries will largely depend on the public sector. Whether national policies in these countries stimulate or deter research and technology for publicly developed GM crops is unclear. As scientific capabilities and the number of research institutes increase, so will the diversity of crops and phenotypes. Greater attention is needed, however, for specific events where resources and knowledge are lacking to complete efficacy and safety testing. Otherwise, GM crops will remain in preliminary testing. It is estimated that approximately 22% of the 201 transformation events discussed in this chapter have remained in confined testing stage.

In contrast to achievements in R&D, most developing countries have only limited experience in compiling regulatory data, manifesting in lack of clearly defined all regulatory requirements. Despite positive attitudes in developing countries, few transformed crops have been released from confined stage to precommercial testing or into commercial use. This can be attributed to several factors (Cohen, 2005):

1. The overall isolation of public research institutes
2. The inability of public research to meet food safety and environmental regulatory requirements, and confusion regarding regulatory standards between confined versus open trials
3. Lack of regional abilities to exchange and evaluate regulatory data on specific transgenes and crops
4. Expertise with public genetic resources but few opportunities to use or evaluate proprietary germplasm
5. Difficulties in planning for advancement of specific products
6. Limited progress in determining authorities and frameworks for science-based decision making
7. Implementing processes arising from the international level, as well as at the regional level
8. External political barriers that halt regulatory review; for example, moratoriums resources will determine if local resources or adapted genes need intellectual property right protection. In this context, benefit distribution, accounting for the success in

transforming local genetic resources, can form the basis for agreements between public institutes, farmer organizations, and commercial producers (Mahony, Henson-Apolionio, & Odame, 2004).

Agreements can establish ownership among providers of transgenes (and the cost of their research) by equalizing investments with time and innovation provided by developing countries, creating combinations of genes in localized crops or genetic resources. Such decisions on ownership are made carefully to ensure an equitable arrangement between poor-country institutions and those supplying new technologies. Local and multinational companies could play a key role for specific local GM crops, given their experience in commercial development and regulatory information, including environmental and food safety studies. However, examples of successful public–private partnerships in plant biotech are still rare, even at international research centers (Spielman & Grebmer, 2004).

A number of factors may determine the future of transgenic (GM) food biotechnology in the developing world (Cohen, 2005):

FULLY EXPLOIT GENETIC RESOURCES

It is extremely important that locally available, agronomically productive genetic resources for transformation are exploited in public research. However, access to proprietary genetic resources in developing countries is extremely limited, amounting to only 6% of all transformation events. The ability to transform local, widely used public or indigenous genetic resources provides the potential for greater public and farmer acceptance. Using high-performance GM public germplasm means that farmers will not be prevented from saving seeds, nor will they potentially be under monopoly pricing of seeds. However, some private companies have promised free rights to their genes in specific crops, such as sweet potato and the rice genome for public research.

ENSURE RESEARCH SERVES THE PUBLIC GOOD

Examination of potential benefits and genetic resources will determine if local resources or adapted genes need IP protection. Benefit distribution, accounting for the success in transforming local genetic resources, can form the basis for agreements between public institutes, farmer organizations, and commercial producers. Agreements can establish ownership among providers of transgenes (and the cost of their research) by equalizing investments, with time and innovation provided by developing countries creating combinations of genes in localized crops or genetic resources.

CREATING EFFICIENCIES AND COMPETENCIES

Although limited collaboration does occur between developing countries and Western companies (Table 19-6), the study reveals that developing countries did not forge a single ("South-to-South") collaboration among themselves. Contacts with other countries of economic parity would create efficiencies by sharing knowledge on specific crops, traits, and regulatory dossiers. For example, by using data on genes and phenotypes under study (Table 19-5), countries could meet and assemble data and experience on specific genes and their constructs, making collected and relevant information available to their respective regulators. Scientists and regulators from developing countries can also meet to discuss specific crops, where common transformation events are occurring. Working from either specific crops or traits, joint studies can also highlight partnership models (or the lack of them) and address needs best suited for such collaboration. The same type of consultation can occur by examining crops at a particular stage in their regulation (Figure 19-4) that requires safety information and results from efficacy and safety trials. Such knowledge is valuable when selecting transgenes, considering regulatory requirements and determining which genetic resources are available or needed.

TABLE 19-6 DISTRIBUTION OF TRANSFORMATION EVENTS IN INDIVIDUAL DEVELOPING COUNTRIES IN MAJOR CONTINENTS

		Institutions		Events[a]	
Continent	Country	Number	With partners	Total	With partners
Africa	Egypt	1	1	17	12
	Kenya	1	1	4	4
	Zimbabwe	4	3	5	3
	South Africa	5	2	28	7
Asia	Malaysia	2	1	5	3
	Pakistan	3	3	5	5
	Philippines	3	3	17	17
	Thailand	3	2	7	6
	Indonesia	6	2	24	5
	China	9	1	30	1
	India	14	1	21	1
Latin America	Mexico	1	0	3	0
	Brazil	2	2	9	7
	Costa Rice	3	3	5	5
	Argentina	4	3	21	4

[a]An event is defined as the stable transformation—incorporating of foreign DNA into a living plant cell—undertaken by a single institute among the participating countries, thereby providing a unique crop and trait combination

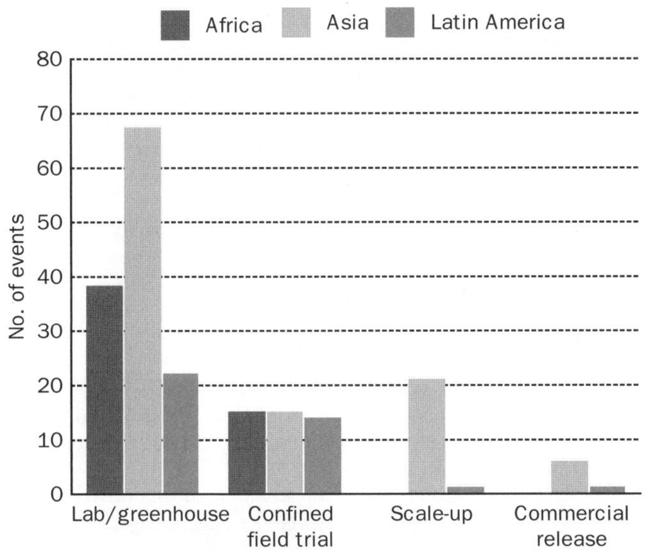

Figure 19-4 Number of publicly derived transformation events (in GM crops) by regulatory stage and region

THE BOTTOM LINE

Although some commercially developed GM products have a role to play, GM crops developed by public research institutes should be most relevant to local needs in poor countries. Paradoxically, because they are novel, locally developed products pose unique challenges for institutes seeking regulatory approval, and gaining approval can be one of the biggest obstacles facing public GM crops in developing nations. In contrast, commercial GM crops preapproved in Western markets are more successful in gaining approvals in developing countries. Demand for GM products by local farmers, combined with the established regulatory and production track record of Western products, sets the stage for interest in using GM crops in developing nations. This implies farmers may take advantage of options to grow locally unapproved Western products, thus avoiding licensing costs and IP issues. At the same time, locally produced GM crops remain in development and do not reach the same farmers, meaning their impact is yet to be seen. The sooner such evaluations occur, the faster GM crops unsuitable for field application can be discarded and successful GM crops moved forward, thus saving public funds and minimizing opportunity costs.

CONCLUSIONS

Worldwide consumer response toward food products made from genetically modified (GM) ingredients has been largely negative. However, the majority of the previous studies on consumer attitudes toward food products were conducted in developed countries in Europe as well as Japan. The small number of studies conducted in developing nations obtains results different from those of the developed world. This chapter considers the opportunities and motivations for consumer attitudes toward GM foods in developing nations. The generally positive perception towards genetically modified foods in developing nations stems from more urgent needs in terms of increased food availability and enhanced nutritional content. Additionally, perceived levels of risk may be smaller due to trust in government, positive perceptions of science, and positive media influences. This is contrary to the smaller benefits and higher perceived risks found in many developed countries that can be exploited in raising awareness, public–private partnership for infrastructure development, locally targeted research and field trials, setting biosafety standards, statutory regulatory measures, and commercial use including export.

ACKNOWLEDGMENT

The authors are grateful to Professor Robert Paarlberg of Wellesley Institute for allowing material from his publications to be used in this chapter. His critical review of the manuscript is acknowledged.

REFERENCES

Asian Development Bank. (2001). *Agricultural Biotechnology, Poverty Reduction, and Food Security. A Working Paper.* Manila, the Philippines: Asian Development Bank. http://www.adb.org/Documents/Books/Agri_Biotech/agribiotech.pdf.

Beech H. (2002). China's next cultural revolution. *TIME* (Asia Edition), *160*(18).

Bennett R, Ismael Y, Morse S, Shankar B. (2004a). Reductions in insecticide use from adoption of BT cotton in South Africa: impacts on economic performance and toxic load to the environment. *J Agric Sci* 142, 665–674.

Bennett R, Ismael Y, Kambhampati U, Morse S. (2004b). Economic impact of genetically modified cotton in India. *AgBioForum* 7, 96–100.

Briggs C. (2001). Planning for the development of nearly 100 national biosafety frameworks. Paper presented at the international consultation A framework for biosafety implementation—a tool for capacity building, organized by the International Service for National Agricultural Research (ISNAR) and Virginia Polytechnic and State University (Virginia Tech), Washington, DC (July 23–26).

CBD (Cartagena Protocol on Biosafety to the Convention on Biological Diversity). (2004). Status of ratification and entry into force. Available: http:// www.biodiv.org/biosafety/signinglist.aspx?sts¼rtf& ord¼dt.

Cohen JI. (2001). Harnessing biotechnology for the poor: challenges ahead for capacity, safety, and public investment. *J. Human Dev.* 2, 239–264.

Cohen JI & Paarlberg RL. (2002). Explaining restricted approval and availability of GM crops in developing countries. *AgBiotechNet 4*, 1–6.

Cohen JI & Paarlberg RL. (2004). Unlocking crop biotechnology in developing countries--a report from the field. *World Development 32*(9), 1563–1577.

Cohen JI. (2005). Poorer nations turn to publicly developed GM crops. *Nature Biotech 23*(1), 27–33.

Curtis KR. (2003). *The industrialized global market for processed potatoes: Contracting, westernization in Asia, and biotechnology.* Doctoral dissertation, Washington State University.

Curtis KR, McCluskey JJ & Wahl TI. (2004). Consumer acceptance of genetically modified food products in the developing world. *AgBioForum 7*(1&2), 72.

Glover D, Yamin F. (2003). Intellectual property rights, biotechnology and development. Democratizing Biotechnology: Genetically Modified Crops in Developing Countries Briefing Series. Briefing 4. Brighton, UK.

Gouse M, Pray C, Schimmelpfennig D. (2004). The distribution of benefits from Bt cotton adoption in South Africa. *AgBioForum 7*, 187–194.

Gouse M, Pray C, Kirsten J, Schimmelpfennig D. (2005). A GM subsistence crop in Africa: the case of Bt white maize in South Africa. *Int J Biotechnol* 7, 84–94.

Huang J, Rozelle S, Pray C, & Wang Q. (2001). *Plant biotechnology in the developing world: The case of China* (REAP working paper). Davis, CA: Center on Rural Economies of the Americas and Pacific Rim, University of California-Davis. Available on the World Wide Web: http://www.reap.ucdavis.edu/working_papers/plant_biotechnology.pdf.

Huang J, Rozell S, Pray C, & Wang Q. (2002). Plant biotechnology in China. *Science 295*, 674–677.

Huang J, Hu R, Pray C, Qiao F, Rozelle S. (2003). Biotechnology as an alternative to chemical pesticides: a case study of Bt cotton in China. *Agric Econ* 29, 55–67.

Huang J, Hu R, Rozelle S, Pray C. (2005). Insect-resistant GM rice in farmers' fields: assessing productivity and health effects. *Science* 308, 688–690.

Huesing J, English L. (2004). The impact of Bt crops on the developing world. *AgBioForum* 7, 84–95.

James C, & Krattiger A. (1999). The role of the private sector (brief 4 of 10). In Persley GJ. (ed.), *Biotechnology for Developing Country Agriculture: Problems and Opportunities: Focus 2: A 2020 Vision for Food, Agriculture, and the Environment.* Washington, DC: International Food Policy Research Institute.

James C. (2005). Preview: global status of commercialized biotech/GM cropsISAAA 2005. ISAAA Briefs No. 34.

Kirsten J, Gouse M, & Jenkins L. (2002). *Bt Cotton in South Africa: Adoption and the Impact on Farm Incomes Amongst Small-Scale and Large-Scale Farmers.* Paper presented at the 6th International ICABR Conference, Ravello, Italy.

Li Q, Curtis KR, McCluskey JJ, & Wahl TI. (2002). Consumer attitudes toward genetically modified foods in Beijing, China. *AgBioForum*, 5(4), 145–152. Available on the World Wide Web: http://www.agbioforum.org.

Lichtenberg E. (2000). Costs of regulating transgenic pest-protected plants. In *Genetically Modified Pestprotected Plants--Science and Regulation* (Appendix I) (pp. 217–243). Washington, DC: National Academy Press.

Mahony MM, Henson-Apolionio V, & Odame HH. (2004). *Strategies for Management of Intellectual Property Rights in Developing Countries and The Role of Farmers' Associations.* ISNAR.

McLean MA, Frederick R, Traynor PL, et al. (2002). *A Conceptual Framework for Implementing Biosafety: Policy, Capacity and Regulation.* ISNAR Briefing Paper 47, The Hague, The Netherlands.

Morse S, Bennett R, Ismael Y. (2005). Comparing the performance of official and unofficial genetically modified cotton in India. *AgBioForum 8*, 1–6.

Naylor R, Falcon W, Goodman R, Jahn M, Sengooba T, Tefera H, Nelson R. (2004). Biotechnology in the developing world: a case for increased investments in orphan crops. *Food Policy 29*, 15–44.

Organisation for Economic Co-operation and Development (OECD). (2003). *Accessing Agricultural Biotechnology in Emerging Economies.* Paris, France: OECD.

Pray C, Huang J, Hu R, Rozelle S. (2002). Five years of Bt cotton in China – the benefits continue. *Plant J 31*, 423–430.

Pachico D, & Wolf M. (2002). *Attitudes Toward Genetically Modified Food in Colombia.* Paper presented at the 6th International ICABR Conference, Ravello, Italy.

Population Reference Bureau. (2002). Available on the World Wide Web: http://www.prb.org.

Pray CE. (2001). China. In Pray CE & Fuglie KO (eds.), *Private investment in agricultural research and international technology transfer in Asia* (ERS agriculture economics report no. 805; chapter 7). Washington, DC: United States Department of Agriculture Economic Research Service. Available on the World Wide Web: http://www.ers.usda.gov/publications/aer805/aer805.pdf.

Pray C, Huang J. (2003). The Impact of Bt-cotton in China. In Kalaitzandonakes N. (ed.), *The Economic and Environmental Impacts of Agbiotech: A Global Perspective* (pp. 223–242). New York: Kluwer-Plenum Academic Publishers.

Qaim M, & Traxler G. (2002). *Roundup Ready Soybeans in Argentina: Farm Level, Environmental, and Welfare Effects.* Paper presented at the 6th International ICABR Conference, Ravello, Italy.

Qaim M, de Janvry A. (2003). Genetically modified crops, corporate pricing strategies, and farmers' adoption: the case of Bt cotton in Argentina. *Am J Agric Econ* 85, 814–828.

Qaim M, Traxler G. (2004). Roundup Ready soybeans in Argentina: farm level, environmental and welfare effects. *Agric Econ* 32, 73–86.

Qaim M, Zilberman D. (2003). Yield effects of genetically modified crops in developing countries. *Science* 299, 900–902.

Qaim M, Subramanian A, Naik G, Zilberman D. (2006). Adoption of Bt cotton and impact variability: insights from India. *Rev Agric Econ* 28, 48–58.

Qaim M. (2003). Bt cotton in India: field trial results and economic projections. *World Dev* 31, 2115–2127.

Qayum A, Sakkhari K. (2005). *Bt Cotton in Andhra Pradesh: A Three-Year Assessment.* Hyderabad: Deccan Development Society.

Raney T. (2006). Economic impact od transgenic crops in developing countries. *Curr Opin Biotechnol 17*, 1–5.

Smale M, & De Groote H. (2003). Diagnostic research to enable adoption of transgenic crop varieties by smallholder farmers in sub-Saharan Africa. *Afr J Biotechnol 2*(12), 586–595.

Spielman DJ, von Grebmer K. (2004). *Public-Private Partnerships in agricultural Research: An Analysis of Challenging Facing Industry and the Consultative Group on International Agricultural Research.* Washington, D.C., EPTD Discussion Paper No. 113, International Food Policy Research Institute (IFPRI).

Swinnen JFM, McCluskey JJ, & Francken N. (2003). *Food Safety, The Media, and The Information Market.* Plenary paper presented at the 25th International Conference of Agricultural Economists, Durban, South Africa.

Thirtle C, Beyers L, Ismael Y, Piesse J (2003). Can GM technologies help the poor? The impact of Bt cotton in Makhathini Flats, KwaZulu-Natal. *World Dev, 31*, 717–732.

Traxler G, Godoy-Avila S. (2004). Transgenic cotton in Mexico. *AgBioForum 7*, 57–62.

Trigo E, Cap E (2003). The impact of the introduction of transgenic crops in Argentinean agriculture. *AgBioForum 6*, 87–94.

United Nations Economic Commission for Africa. (2002). *Harnessing Technologies for Sustainable Development.* Addis Ababa: United Nations Economic Commission for Africa.

Victor DG, & Ford Runge C. (2002). *Sustaining A Revolution—A Policy Strategy for Crop Engineering.* A Council on Foreign Relations Paper, New York.

Vogel D. (2001). Ships passing in the night: the changing politics of risk regulation in Europe and the United States. EUI Working Papers RSC No. 2001/16, June 2001. Dan Domenico, Italy: European University Institute. Available: http://www.iue.it/RSC/WPTexts/01_16.pdf.

Yan J, Kandaianis B, Harjes CE, et al. (2010). Rare genetic variation at *Zea Mays crtRB1* increases β-carotene in maize grain. *Nature Genetics 42*(4), 322–327.

Zimmerman R, & Qaim M. (2002). *Projecting The Benefits of Golden Rice in the Philippines.* Paper presented at the 6th International ICABR Conference, Ravello, Italy.

20.

THE BURDEN OF GENETIC DISEASE IN THE DEVELOPING WORLD: THE EXAMPLE OF INHERITED BLOOD DISEASES

David J. Weatherall

It has been estimated that more than 7 million babies are born each year with either a congenital abnormality or a genetic disease, and that approximately 90% of these births occur in low- or middle-income countries (Christianson, Howson, & Modell, 2006). Approximately 25% of these births consist of only five disorders, two of which, the inherited disorders of hemoglobin and glucose-6-phosphate dehydrogenase (G6PD) deficiency, are monogenic diseases. Current estimates suggest that over 300,000 children are born each year with either sickle-cell anemia or one of its variants, or one or other forms of thalassemia (Modell & Darlison, 2008; Weatherall, 2010a). It is also estimated that more than 5 million infants are born with G6PD deficiency, although only 177,000 of these infants appear to be at risk for the severe adverse side effects of this condition, notably neonatal jaundice (Christianson, Howson, & Modell, 2006).

As well as these extremely common recessive hematological disorders, there is a wide range of single-gene disorders that result in hematological abnormalities, and that occur unevenly throughout the world population. Although they pose a less serious public health problem than the common recessive conditions, because of management difficulties and the fact that affected children may require treatment for the whole of their lives, these conditions cannot be disregarded in any assessment of the global burden of genetic hematological disease.

INHERITED DISEASES OF HEMOGLOBIN

The estimated numbers of annual births of the important hemoglobin disorders is summarized in Table 20-1. The extraordinary high frequency of these conditions reflects heterozygote advantage against malaria (Kwiatkowski, 2005; Weatherall, 2008), a high frequency in the numbers of consanguineous marriages in many countries (Bittles, 1990), population migration, and the effects of the epidemiological transition, whereby as public health measures and levels of nutrition improve in poor countries, babies with hemoglobin disorders who would have previously died are now surviving for diagnosis and management.

Sickle-cell anemia has arisen by independent mutation at least twice, and possibly on more occasions (Serjeant & Serjeant, 2001). Interestingly, and for reasons that are only partly understood, the form that arose in India, the Asian/Indian variety, has a milder phenotype than that which arose in Africa. Both forms of the disease are seen in some Middle Eastern populations. This condition occurs most frequently in sub-Saharan Africa and in countries with large African immigrant populations (Serjeant & Serjeant, 2001). It also occurs throughout the Middle East and in many parts of the Indian subcontinent. There are limited survival data, although the advent of neonatal screening and the use of prophylactic antibiotics have undoubtedly improved the survival in countries where these facilities are available, but the disease is still characterized by a wide variety of vascular complications. One of its major clinical problems is its remarkable phenotypic diversity, which is still largely unexplained.

The β thalassemias occur at the highest frequency across the tropical belt from Africa through the Mediterranean, Middle East to the Indian subcontinent, and throughout Southeast and East Asia (Weatherall & Clegg, 2001). There are two common forms. There are the β thalassemias, which in the homozygous state produce profound anemia from early in life, and which, without regular transfusion and the use of chelating agents, carry a poor prognosis. About half of the severe forms of β thalassemia globally are made up by the compound heterozygous state for β thalassemia and hemoglobin E, HbE β thalassemia. This is the commonest severe form of thalassemia throughout the eastern side of the Indian subcontinent and Southeast and East Asia (Weatherall & Clegg, 2001). Its management is made particularly difficult by its remarkably diverse clinical phenotype, ranging from a transfusion-dependent disorder to a condition compatible with reasonable growth and development

TABLE 20-1 APPROXIMATE NUMBERS OF BIRTHS OF COMMON HEMOGLOBIN DISORDERS

Hemoglobin Disorder	Number of Births
β thalassemia major	22,989
HbE β thalassemia	19,128
Hb H disease (α^{+}/α^{0})	9,568
Hb Bart's hydrops (α^{0}/α^{0})	5,183
Sickle cell disease	217,331
Hb S β thalassemia	11,070
Hb SC disease	54,736

From Modell and Darlison (2008) and Weatherall (2010a)

without treatment (Fucharoen & Winichagoon, 1997; Premawardhena et al., 2005). The homozygous states for the severe forms of α thalassemia cause stillbirth, while the interaction of the mild and severe forms result in a condition called *hemoglobin H disease,* which, again, shows remarkable phenotypic variability.

There are numerous challenges for the future for the better control and management of the hemoglobin disorders (Weatherall, 2010a). First, more information is required about their frequency. Most of the population surveys were carried out over twenty years ago and are based on limited population samples. It is now clear that these diseases have an extremely uneven distribution in high-frequency countries, and that to assess their true frequency, micromapping is required. This involves taking numerous samples from different parts of high-frequency countries. In the few reported studies of this type, the heterogeneity of distribution, possibly reflecting the high-transmission regions for malaria in the past, has become increasingly clear (Weatherall, 2010b). Only when data of this type are available will it be possible to provide a reasonably accurate picture of the burden of the disease for the purposes of public health approaches to its control.

While all the severe hemoglobin disorders are amenable to screening and prenatal diagnosis, and this approach has resulted in a remarkable fall in the birth of new cases of babies with severe β thalassemia (Alter, 1990; Cao, Galanello, & Rosatelli, 1998), so far it has been much less effective in the case of sickle-cell disease and hemoglobin E thalassemia, reflecting the difficulty in providing an accurate prognosis for the future of affected babies with these conditions. Clearly, therefore, much more needs to be learned about the mechanisms for such phenotypic diversity and whether it will be possible to identify those babies with a poor prognosis in the future.

Another major challenge is how to develop adequate diagnostic management programs for the thousands of patients with sickle-cell disease and thalassemia who are living in low-income countries. Over many years, the development of partnerships between centers in the rich countries and those in the developing countries, North/South partnerships, has been an extremely successful approach toward improving facilities for children in the poor countries (Weatherall, 2010a; World Health Organization, 2002). More recently thoughts have turned to the development of South/South partnerships; that is, partnerships between some of the poorer countries that have obtained expertise in the control and management of the hemoglobin disorders, with adjacent countries where no such expertise exists (Weatherall, 2010a). One of the major challenges is how to convince international health agencies and governments that these are extremely cost-effective approaches to the control and management of these conditions in the developing world.

Without progress along these lines the hemoglobin disorders will produce an increasingly severe global health problem. Recent work suggests that the high mortality in early life in Africa of babies with sickle-cell anemia results from infection due to organisms similar to those that affect children in the developed countries, and which can be controlled by prophylactic antibiotics given after neonatal screening has confirmed the diagnosis (Williams et al., 2009). It has been estimated that if the survival for children with sickle-cell anemia in Africa were to reach only 50% of that of normal children in the population there would be approximately 6 million children with sickle-cell anemia in sub-Saharan Africa in the near future (Modell & Darlison, 2008). It seems likely that there will be a similar burden on health services due to the different thalassemias in many Asian populations at the same time.

GLUCOSE-6-PHOSPHATE DEHYDROGENASE DEFICIENCY

While it is estimated that 5 million children with this condition are born each year in tropical countries and elsewhere, the global health burden caused by this condition is much less than that due to the thalassemias and sickle-cell disease (Luzzatoo, Mehta, & Vulliamy, 2001). However, since these children are sensitive to a variety of oxidant drugs, notably the antimalarial agent primaquine, this condition should not be underestimated. Recent studies suggest that malaria due to *P. vivax* is becoming an increasingly serious global health problem, and primaquine is still a vital agent for its management (Mendis, Sina, Marchesini, & Carter, 2001). Thus, it will become increasingly important to screen large populations for G6PD deficiency using simple and robust methods. Furthermore, in some, although not all populations, babies with this condition may develop severe jaundice after birth with the added complication of kernicterus (Luzzatoo, Mehta, & Vulliamy, 2001). In these populations neonatal screening for G6PD deficiency is vital.

RARE GENETIC HEMATOLOGICAL DISORDERS

There is a wide variety of genetic hematological disorders that occur at a low frequency, or at a frequency which varies and is higher in some population groups than others (Christianson, Howson, & Modell, 2006). While these conditions are of less potential global health importance, it is vital that facilities for their diagnosis and management are established in the poorer countries. Management of disorders like hemophilia, which require lifelong medical care, even if they occur at a lower frequency still have important public health implications. And although conditions like rhesus hemolytic disease of the newborn are much commoner in North European populations, where they can now be controlled, they also occur at lower frequencies in other parts of the world. There is growing evidence that genetic disorders of iron metabolism, some of which occur at relatively high frequencies in North European populations, are also being observed in many Asian countries (Lok et al., 2009). Clearly, future planning for the control of genetic blood disease must allow for these relatively rare but clinically important conditions.

THE FUTURE

Clearly, the common recessive diseases of the blood, notably the hemoglobin disorders and G6PD deficiency, have the potential to produce a major global health problem in the future. It is vital that more is learned about their frequency and that, through appropriate partnerships, the expertise required for their prevention and management is established in the developing countries. These countries also require at least one center where the rarer genetic blood diseases can be diagnosed and managed. While the current attitudes of the international health agencies toward genetic disease are understandable, given the problems of communicable disease and poverty, it is vital that they appreciate both the nature and extent of the global health problem that is waiting to explode due to the increasing numbers of children born with genetic blood diseases.

REFERENCES

Alter BP. (1990). Antenatal diagnosis. Summary of results. *Annals of the New York Academy of Sciences 612*, 237.

Bittles AH. (1990). Consanguineous marriage: current global incidence and its relevance to demographic research. Research Report No. 90–186, Population Studies Center: Detroit: University of Michigan.

Cao A, Galanello R, Rosatelli MC. (1998). Prenatal diagnosis and screening of the haemoglobinopathies. *Clinical Haematology 11*, 215–238.

Christianson A, Howson CP, Modell B. (2006). March of Dimes Global Report on Birth Defects. New York: March of Dimes Birth Defects Foundation.

Fucharoen S, Winichagoon P. (1997). Hemoglobinopathies in Southeast Asia: molecular biology and clinical medicine. *Hemoglobin 21*, 299–319.

Kwiatkowski DP. (2005). How malaria has affected the human genome and what human genetics can teach us about malaria. *Am J Hum Genet 77*(2), 171–192.

Lok CY, Merryweather-Clarke AT, Viprakasit V, et al. (2009). Iron overload in the Asian community. *Blood 114*(1), 20–25.

Luzzatto L, Mehta A, Vulliamy T. (2001). Glucose 6-phosphate dehydrogenase. In: Scriver CR, Beaudet AL, Sly WS, Valle D, Childs B, Vogelstein B, eds. *The Metabolic and Molecular Basis of Inherited Disease* (8th ed.). New York: McGraw Hill, pp. 4517–4554.

Mendis K, Sina BJ, Marchesini P, Carter R. (2001). The neglected burden of *Plasmodium vivax* malaria. *Am J Trop Med Hyg 64*(1–2 Suppl), 97–106.

Modell B, Darlison M. (2008). Global epidemiology of haemoglobin disorders and derived service indicators. *Bull World Health Organ 86*(6), 480–487.

Premawardhena A, Fisher CA, Olivieri NF, et al. (2005). Haemoglobin E ⊠ thalassaemia in Sri Lanka. *Lancet 366*, 1467–1470.

Serjeant GR, Serjeant BE. (2001). *Sickle Cell Disease* (3rd ed.). Oxford: Oxford University Press.

Weatherall DJ, Clegg JB. (2001). The *Thalassaemia Syndromes* (4th ed.). Oxford: Blackwell Science.

Weatherall DJ. (2008). Genetic variation and susceptibility to infection: the red cell and malaria. *Br J Haematol 141*(3), 276–286.

Weatherall DJ. (2010a). The inherited diseases of hemoglobin are an emerging global health burden. *Blood 115*(22), 4331–4336.

Weatherall DJ. (2010b). The importance of micromapping the gene frequencies for the common inherited disorders of haemoglobin. *Br J Haematol 149*, 635–637.

Williams TN, Uyoga S, Macharia A, et al. (2009). Bacteraemia in Kenyan children with sickle-cell anaemia: a retrospective cohort and case-control study. *Lancet 374*(9698), 1364–1370.

World Health Organization. (2002). *Genomics and World Health*. Geneva: WHO.

21.

HEALTH IMPLICATIONS OF INHERITED DISORDERS OF IRON OVERLOAD

Chun Yu Lok, Alison T. Merryweather-Clarke, and Kathryn J.H. Robson

INTRODUCTION

IMPORTANCE OF IRON IN HEALTH AND DISEASE

In an oxygen-rich environment, iron is a vital element for many biological processes, despite being required only in trace amounts in living organisms. Iron is a key component of many proteins. It is essential for the synthesis of heme, a protoporphyrin IX molecule complexed with Fe(II) that is a required component of cytochromes in redox reactions of the electron transport chain, and the oxygen-carrying proteins hemoglobin and myoglobin. It is also commonly found in the iron-sulfur clusters of many enzymes, and in the 4Fe-4S cluster of iron regulatory proteins (IRPs), which play a crucial role in regulating iron metabolism (Rouault et al., 1991; Haile et al., 1992). The total iron content of a healthy human adult is approximately 3–4 g, over two-thirds of which is found in hemoglobin in erythrocytes. Each of these erythrocytes contains over one billion atoms of iron, complexed in over 300 million hemoglobin molecules. It is therefore imperative that the mechanisms of iron sensing and uptake are effective in response to the body's demands, and that the utilization of iron is tightly controlled.

Iron is recognized as a transition metal with the characteristic property of variable oxidation states. It has the ability to accept and donate electrons readily, and is most commonly found in the ferrous (Fe^{2+}) and ferric (Fe^{3+}) forms. It is this redox activity that gives iron its biological importance. However, it is also due to this same property that molecular mechanisms are necessary to restrict iron uptake when levels are sufficient, as iron overload can lead to toxicity. Ferrous iron catalyzes the formation of reactive oxygen species (ROS) including the hydroxyl radical from hydrogen peroxide in the Fenton reaction, which can initiate further formation of free radicals in chain reactions. Although free radicals have roles essential in normal physiological processes, such as inflammation and phagocytosis, long-term accumulation may contribute to tissue damage. Free radical attack can lead to irreversible chemical modifications and result in adverse changes to macromolecular structure and function. These processes also contribute to the molecular changes that can be observed in aging, summarized in the "Free Radical Theory of Aging" (Harman, 1956). The loss of iron homeostasis resulting in iron deficiency anemia, or iron overload in hereditary hemochromatosis, can lead to many clinical disorders ranging from cognitive defects in one extreme, to liver and heart disease in the other.

To overcome the problems arising from the potential toxicity of iron and its insolubility in the ferric form, living systems have evolved complex mechanisms and pathways for handling this vital element. Body iron content is normally maintained at stable levels with little fluctuation. The absorption of dietary iron through the small intestines is about 1–2 mg per day and is distributed around the body to various tissues (Figure 21-1). This absorption is balanced by the loss of a similar amount. Perhaps one of the most surprising features of iron homeostasis is the absence of an active regulatory pathway for excreting iron in response to situations of iron overload (McCance & Widdowson, 1937). The mode of iron loss from the body is mainly through blood loss and sloughing of epithelial cells of the skin and gastrointestinal tract. Iron balance is therefore only achieved by the careful regulation of iron absorption.

IRON METABOLISM

In order to understand the diseases associated with misregulation of iron, it is important to grasp the process by which the body regulates iron. In the following paragraphs we outline some of the functions of the proteins involved in handling iron. Iron overload or anemia often results when these proteins are mutated or have altered expression levels.

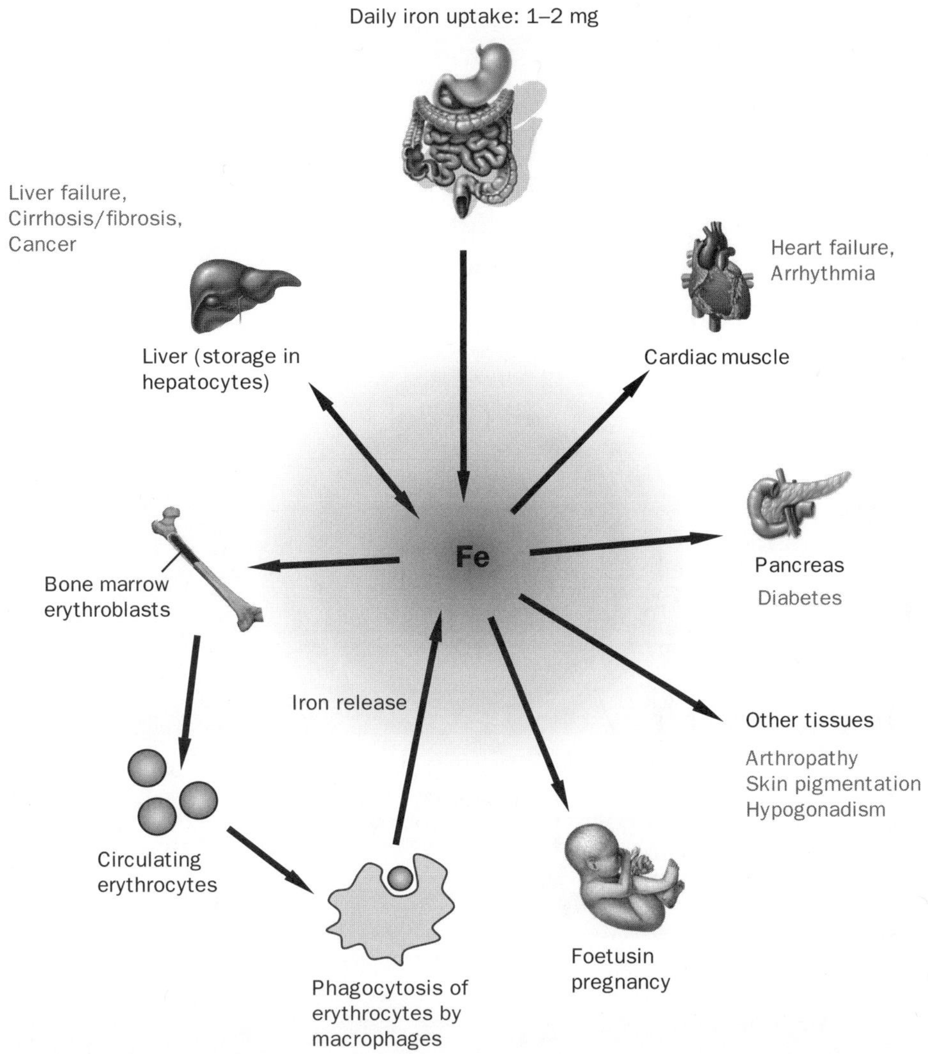

Figure 21-1 The Distribution of Iron

Transferrin and TFR1

Although the clinical complications associated with iron deficiency and overload have been well documented over several decades, the molecular mechanisms involved in iron metabolism have only recently been revealed to any extent. Iron exists as a large, insoluble ion, and its uptake and transport is a complex process involving several different proteins. One of the first components of the iron transport pathway to be identified was transferrin (Schade & Caroline, 1946). Transferrin (Tf) is a plasma glycoprotein with homologous N-terminal and C-terminal domains, each of which binds one ion of ferric iron with very high affinity. It plays a key role in transporting iron around the body to sites where iron is needed. It acts to solubilize the iron, sequester its potential free radical toxicity, and facilitate transport into different cells within the body (Figure 21-2). Tf-bound iron enters the cells by binding to transferrin receptor 1 (TFR1) on the cell membrane, and the ligand-receptor complex is internalized by receptor-mediated endocytosis.

DMT1 and Ferric Reductases

DMT1, a divalent metal transporter also known as NRAMP2/DCT1/SLC11A2, is the transmembrane transporter for intestinal absorption of iron and the release of iron into the cytoplasm from the Tf cycle (Fleming et al., 1997, Gunshin et al., 1997; Figure 21-2). DMT1 is localized to the intestinal brush-border and co-localizes with Tf in the endosomes of erythroid cells (Gruenheid et al., 1999, Canonne-Hergaux et al., 1999). As dietary non-heme iron predominantly exists in the insoluble ferric form, uptake into enterocytes through DMT1 first requires the reduction of iron to the ferrous form by a ferric reductase. This is carried out by duodenal cytochrome B (DCYTB), an ascorbate-dependent reductase (Figure 21-2; McKie et al., 2001). DCYTB is highly expressed in the enterocytes of the duodenal brush-border region. In other studies in mice, Steap3 (six-transmembrane epithelial antigen of the prostate 3) has been identified as the ferric reductase required for the Tf cycle in erythroid cells (Figure 21-2; (Ohgami et al., 2005). The human homolog is also known as *PHYDE* (Steiner et al., 2000). Knockout of *Steap3* in mice

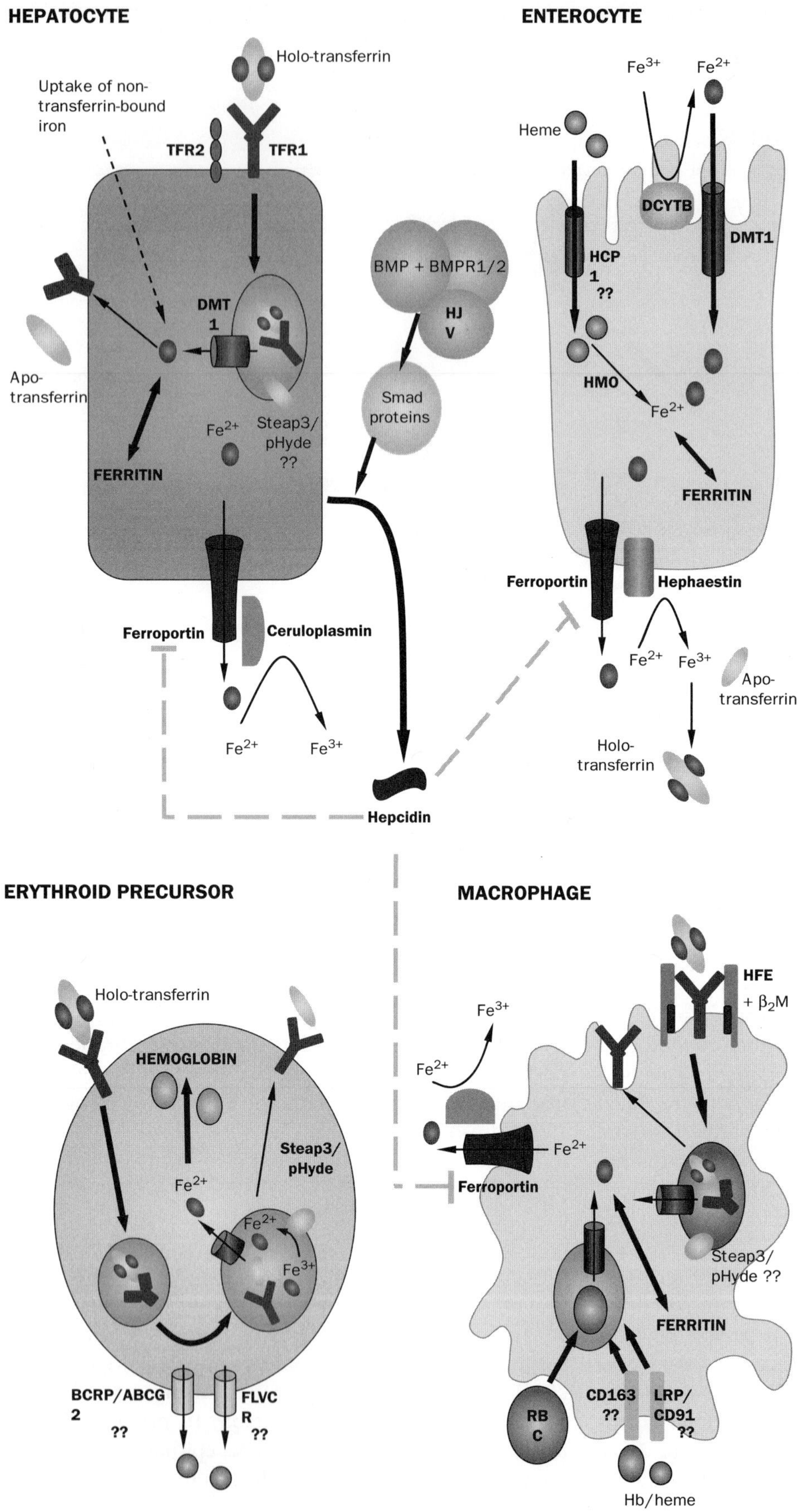

Figure 21-2 Schematic representation of iron transport and utilization by hepatocytes, enterocytes, erythroid cells, and macrophages. Hemojuvelin can upregulate hepcidin synthesis in the liver through the BMP-SMAD signaling pathway. When iron levels are high, increased hepcidin reduces iron uptake and availability by inhibitory binding to ferroportin. Different cell types are known to have different mechanisms of iron uptake. TFR1 levels in hepatocytes are low. It is likely that a non-transferrin-bound iron uptake pathway exists in the hepatocytes. In rat liver, it appears that *Hfe* is expressed predominantly in hepatocytes (Zhang et al., 2004). Arrows show the movement of iron, or the direction/activation of reactions/proteins. BCRP/ABCG2, breast cancer resistance protein/ABC transporter; β_2M, β_2-microglobulin; FLVCR, feline leukemia virus C receptor; HCP1, heme carrier protein 1; HMOX, heme oxygenase; LRP, low-density lipoprotein receptor-related protein; Steap3, six-transmembrane epithelial antigen of the prostate 3.

causes hypochromic, microcytic anemia. This is consistent with the fact that ferric iron released from Tf requires reduction to ferrous iron before DMT1 can transport it out of the endosomes. In humans, an E399D mutation in DMT1 causes hypochromic, microcytic anemia but also iron overload, presumably due to increased heme uptake. Iron levels in the reported case are high, and heavy iron staining can be seen in the liver (Mims et al., 2005).

Ferroportin

While the liver takes most of the burden for storage iron, the oxygen-carrying heme iron in red blood cells (RBCs) constitutes most of the total body iron. Red cells only have a lifespan of approximately 120 days, so the recycled heme iron from aging red cells is one of the most important sources of iron for erythropoiesis in the bone marrow. By acting as scavengers for RBCs and free heme molecules, macrophages therefore play a crucial role in this recycling. The heme iron taken up by macrophages is eventually exported as free iron into the body's circulation. Similarly, the pool of heme and non-heme iron absorbed by the intestinal enterocytes also requires transport across the basolateral membrane into the circulation. Absorbed heme is first degraded by heme oxygenase to release ferrous iron, which then joins the same pool of absorbed non-heme iron to be exported. One known mediator of this iron export is the transmembrane protein, ferroportin (FPN; Figure 21-2), encoded by the *SLC40A1* gene (also known as *IREG1/MTP1/SLC11A3*). Ferrous iron exported by ferroportin also requires oxidation to ferric iron before it is bound to Tf and transported to different tissues for utilization or storage. Identified simultaneously by several groups, ferroportin is found to be expressed in the liver, macrophages, placental syncytiotrophoblasts, and at the basolateral membrane of enterocytes (Abboud & Haile, 2000; Donovan et al., 2000; McKie et al., 2000). Human mutations to ferroportin result in an autosomal dominant form of iron overload known as *type 4 hemochromatosis* (Montosi et al., 2001; Njajou et al., 2001).

Hephaestin and Ceruloplasmin

At the basolateral surface of enterocytes, the oxidation of the ferrous iron exported by ferroportin is carried out by the ferroxidase hephaestin (Figure 21-2), which was identified to be defective in sex-linked anemia (*sla*) mice (Vulpe et al., 1999). Ceruloplasmin, an hephaestin homolog, serves the same purpose in iron export from other cells, including macrophages and hepatocytes (Figure 21-2), with defects to the protein causing aceruloplasminemia (Harris et al., 1995, Yoshida et al., 1995). This is a distinct form of iron loading characterized by anemia and iron accumulation in the central nervous system.

Ferritin

Iron surplus to the body's requirements is stored by intracellular ferritin, a 24-subunit protein shell capable of binding over 4000 atoms of iron, and consisting of two types of subunits, H and L (Arosio et al., 1978; Banyard et al., 1978). When iron is in excess, the binding of iron regulatory proteins (IRPs) to the iron-responsive element (IRE) in the 5'-untranslated region of ferritin mRNA is inactivated, leading to increased ferritin synthesis (Aziz & Munro 1987; Rouault et al., 1988). Mutations to the 5' IRE, although rare, can also cause iron metabolic disorders (Beaumont et al., 1995; Kato et al., 2001).

Hepcidin

In order to prevent excessive iron accumulation, it is now clear that iron absorption and release is modulated by the binding of the peptide hormone, hepcidin, to ferroportin. This binding inhibits ferroportin activity by inducing its internalization and degradation (Nemeth et al., 2004b). This consequently prevents the export of iron from enterocytes, and the release of iron by macrophages and the liver (Figure 21-2). First recognized as an antimicrobial peptide (Krause et al., 2000; Park et al., 2001), it has been demonstrated that hepcidin is regulated by anemia, hypoxia, and inflammation, and is an acute-phase protein (Nicolas et al., 2002b; Nemeth et al., 2003). Hepcidin has now been proven as a key regulator of iron homeostasis, and mutations to its gene (*HAMP*, hepatic antimicrobial peptide) cause a severe form of iron overload (Roetto et al., 2003). Hepcidin is stringently regulated by various molecules. With its involvement in the anemia of inflammation, it has been shown that *HAMP* expression can be induced by the inflammatory cytokine, interleukin-6 (IL-6; see Nemeth et al., 2003, 2004a). On the other hand, studies have suggested the role of other molecules, such as HIFs (hypoxia-inducible factors), in suppressing *HAMP* (Peyssonnaux et al., 2007).

HEMOJUVELIN

In recent years much research has shown that hepcidin can also be regulated by hemojuvelin (HJV). Mutations in hemojuvelin result in severe iron overload very similar to that observed in patients with hepcidin mutations (Papanikolaou et al., 2004). HJV is a member of the repulsive guidance molecule (RGM) family, and also recognized as RGMc. Studies have demonstrated that HJV can function as a BMP (bone morphogenetic protein) co-receptor, and upregulate *HAMP* expression via the BMP signaling pathway (Figure 21-2; see Babitt et al., 2006). The BMPs are a large subfamily of the transforming growth factor β (TGF-β) superfamily of ligands, and act by binding to complexes of BMP type 1 and type 2 kinase receptors

(BMPR1 and 2). This triggers a phosphorylation cascade involving SMAD proteins (homologs of the *Drosophila* gene, *mothers against decapentaplegic*), which leads to the upregulation of target gene transcription.

In contrast, it appears that a soluble form of HJV can instead suppress the expression of *HAMP*, and iron increase can reduce the amount of soluble HJV (Lin et al., 2005). Further studies suggest that this *HAMP* suppression may be due to the inhibition of BMP-SMAD activity by soluble HJV (Babitt et al., 2007). This soluble form appears to be derived from intracellular protease cleavage (Silvestri et al., 2008a). HJV is further regulated by transmembrane serine protease 6 (TMPRSS6; or matriptase-2), which cleaves and inactivates membrane-associated HJV, and therefore inhibits hepcidin activation (Silvestri et al., 2008b). Thus, the balance between membrane-associated HJV and soluble HJV provides one mechanism for modulating hepcidin levels according to systemic iron status.

HFE and TFR2

Although the first mutation causing the iron overload disorder hemochromatosis was identified in the *HFE* gene (Feder et al., 1996)—believed to be an abbreviation for *H*igh *Fe*—the normal function of the HFE protein is still not entirely clear. Previous reports have indicated that HFE interacts with TFR1, and competitively inhibits TFR1 binding to holo-transferrin, thereby reducing iron uptake (Feder et al., 1998; Lebrón et al., 1999). However, considering the high abundance of transferrin in vivo, the significance of HFE competition in inhibiting transferrin uptake by TFR1 is not entirely clear. HFE is also thought to interact with the cation-independent mannose-6-phosphate receptor (CI-MPR; also known as the insulin-like growth factor 2 receptor, IGF2R), a large transmembrane protein with multiple functions (Schimanski et al., 2009), but the purpose of this interaction is as yet unknown. The primary role of HFE may not be to prevent iron uptake, but rather to act as an iron sensor (Townsend & Drakesmith, 2002), and ultimately regulate hepcidin levels. This regulation also appears to involve transferrin receptor 2 (TFR2), in which mutations can also lead to iron overload. TFR2, like TFR1, is a transmembrane glycoprotein. It can facilitate iron uptake but has a lower affinity for Tf than TFR1 (Kawabata et al., 1999; West et al., 2000). It has been shown that HFE and TFR2 co-localize in the intestinal crypt cells (Griffiths & Cox, 2003), suggested to play a role in regulating iron absorption (Townsend & Drakesmith, 2002; Griffiths & Cox, 2003). Various studies indicate that TFR2 may be involved in transferrin-iron sensing, and subsequently modulates hepcidin expression rather than facilitating iron uptake (Johnson & Enns, 2004; Robb & Wessling-Resnick, 2004; Goswami & Andrews, 2006; Gao et al., 2009). The exact mechanisms involved in the regulation of hepcidin in response to body iron status are still not entirely understood. It appears likely that the different proteins involved will have multiple functions that are yet to be defined. What is clear, however, is that defects to these, as seen in animal models and patients, can lead to severe disorders of iron metabolism.

IRON OVERLOAD AND IRON DEFICIENCY ANEMIA

Although iron is one of the most abundant metals on Earth, it is estimated that the number of anemic people worldwide is over two billion, of which up to half is due to iron deficiency (WHO, 2001). On the other hand, approximately 1 in 300 individuals in northwestern Europe are at risk of developing hereditary hemochromatosis (HH; see Dadone et al. 1982), a disorder characterized by excessive iron absorption.

Iron deficiency anemia, as well as being due to social and environmental factors, can be due to rare inherited defects as seen in humans and animals. Mutations in TMPRSS6 cause iron-refractory iron deficiency anemia (IRIDA), which is a result of inappropriately high levels of hepcidin (Finberg et al., 2008). Hepcidin-induced anemia is not limited to IRIDA, however. Commonly described, hepcidin is also involved in the anemia of inflammation triggered by infection and inflammation (Roy & Andrews, 2005), and poses an everyday problem for many patients with chronic diseases in different populations.

Genetic variations appear to have a greater impact on iron overload than on iron deficiency, so far. The iron overload condition was first described by Trousseau in 1865, who observed liver cirrhosis and hyperpigmentation of the skin. In 1889, excessive iron deposition was recognized as the cause, and the term *hemochromatosis* was first used by von Recklinghausen to describe the disease (von Recklinghausen, 1889). However, it was not until 1935 that hemochromatosis was recognized by Sheldon to be a genetic disorder (Sheldon, 1935), and it was subsequently deduced that the common form of hemochromatosis was an autosomal recessive disorder (Saddi & Feingold, 1974). Biochemically, hemochromatosis is characterized by abnormally high transferrin saturation and serum ferritin concentration (Table 21-1). This excessive iron absorption and deposition can lead to the impairment of multiple organs with various clinical complications (Figure 21-1). Several distinct types of hemochromatosis are now recognized due to their different phenotypes, and mutations to several different genes (Table 21-2).

HFE HEMOCHROMATOSIS (TYPE 1)

HFE hemochromatosis is the most common form of hereditary hemochromatosis, and is almost entirely

TABLE 21-1 NORMAL HEMATOLOGY PARAMETERS

parameter	Normal Range
Hemoglobin (Hb; g/dl)	11.5–15.5 (females); 13.5–17.5 (males)
Transferrin saturation (Tsat; %)	15–50 (females); 15–55 (males)
Serum ferritin (μg/l)	15–200 (females); 40–300 (males)
Serum iron (μmol/l)	10–30
Total iron-binding capacity (TIBC; μmol/l)	40–75
Mean corpuscular volume (MCV; fl)	80–95

Normal ranges may vary with different laboratories and test kits.

restricted to people of northwestern European ancestry. The clinical symptoms of type 1 hemochromatosis include liver cirrhosis, arthritis, and diabetes, which are all symptoms associated with aging, confounding diagnosis. Type 1 hemochromatosis is an adult-onset disorder with a later age of onset in females than in males, due to menstrual iron loss. If diagnosis can be made before irreversible liver cirrhosis occurs, treatment by venesection returns life expectancy to normal (Niederau et al., 1996).

In 1996, the hemochromatosis gene *HFE* was identified, and homozygosity for a missense mutation causing an amino acid substitution of cysteine by tyrosine at position 282 of the protein (C282Y) was shown to be present in over 80% of hemochromatosis patients (Feder et al., 1996). Other patients have been found to be compound heterozygotes for C282Y and a milder *HFE* mutation, H63D. The C282Y frequency is up to 10% in some north European populations, while the H63D allele is as high as 20% (Merryweather-Clarke et al., 1997). *HFE* encodes an MHC class I-like protein (Feder et al., 1996), and migration of the protein to the cell surface requires the correct binding to β_2-microglobulin (β_2M). The C282Y mutation disrupts a disulfide bridge within the protein and prevents the necessary interaction of HFE with β_2M. Consequently, C282Y mutants are retained in the endoplasmic reticulum and middle Golgi, and eventually degraded (Feder et al., 1997; Waheed et al., 1997). H63D proteins, on the other hand, are able to associate with β_2M and localize to the cell surface, which may explain the rarity of H63D hemochromatosis patients. It is probable that homozygosity for H63D is necessary but not sufficient to develop some forms of iron overload. It is also widely recognized that the disease penetrance of the C282Y mutation is incomplete. This subject of penetrance has been at the center of much disagreement (reviewed in Rochette et al., 2010). Variations in disease penetrance have been reported between different populations. This suggests that, as well as environmental factors, other unknown genetic factors may be involved in modifying the phenotype.

TABLE 21-2 HEMOCHROMATOSIS TYPES

Type	Gene mutated	Onset	Inheritance
1	*HFE*	Adult	Autosomal recessive with incomplete penetrance
2A 2B	*HJV* *HAMP*	Juvenile	Autosomal recessive
3	*TFR2*	Adult	Autosomal recessive
4	*SLC40A1*	Adult	Autosomal dominant

JUVENILE HEMOCHROMATOSIS (TYPE 2)

Juvenile hemochromatosis (JH) is a more severe form of iron overload, with clinical symptoms including cardiomyopathy and hypogonadotrophic hypogonadism. Unlike type 1 hemochromatosis, the penetrance of JH is equal among males and females and is normally due to mutations in hemojuvelin (type 2A), or, more rarely, in hepcidin (type 2B; see Roetto et al., 2003; Papanikolaou et al., 2004). Homozygosity for mutations in both proteins leads to severe iron overload that arises before the age of 30. If untreated, premature death can result due to cardiac complications.

The *HAMP* gene mutated in JH consists of three exons, and encodes a prepropeptide of 84 amino acids that yields a propeptide of 60 amino acids after cleavage of the signal sequence. Further cleavage gives the mature peptide of 25 amino acids, encoded entirely by exon three (Park et al., 2001). Hepcidin was first identified as a cysteine-rich peptide with antimicrobial activity, synthesized in the liver (Krause et al., 2000; Park et al., 2001). Its function to suppress iron uptake in response to iron excess was later discovered in mouse models, which also demonstrated that the lack of hepcidin expression results in iron overload, and that upregulation leads to anemia (Pigeon et al., 2001; Nicolas et al., 2001, 2002a). Several *HAMP* mutations were later identified in JH patients. These included a nonsense mutation R56X, a frameshift mutation 93delG, and a missense mutation C70R, which disrupts a disulfide bond (Roetto et al., 2003).

Shortly after the identification of *HAMP*, the *HJV* gene (also known as *HFE2*) was also shown to be responsible for JH (Papanikolaou et al., 2004). A glycine to valine amino acid change at position 320 (G320V) of HJV is the most commonly observed mutation in JH patients originating from Europe, including Greece, Italy, and the United Kingdom. HJV upregulates *HAMP* expression as previously described. In patients with HJV mutations, the levels of hepcidin are low (Papanikolaou et al., 2004). Low hepcidin levels would result in continuous transport of iron by ferroportin into the body's circulation, resulting in iron overload. Recent reports have now demonstrated that BMP6 is an important regulator of hepcidin, with *Bmp6*-deficient mice exhibiting reduced *HAMP* expression and heavy iron accumulation, resembling severe hemochromatosis (Andriopoulos et al., 2009; Meynard et al., 2009). It is worthy of note that the inappropriately low hepcidin levels relative to body iron stores observed in JH are also seen in type 1 hemochromatosis (Bridle et al., 2003; Kulaksiz et al., 2004).

TFR2 HEMOCHROMATOSIS (TYPE 3)

Although relatively rare, hemochromatosis due to mutations in the transferrin receptor-2 gene displays clinical symptoms similar to that of type 1 hemochromatosis in most cases. Type 3 hemochromatosis often presents in young adults. The *TFR2* gene consists of 18 exons and is the largest of the hemochromatosis genes (Kawabata et al., 1999). The first mutation to be identified in *TFR2* was a nonsense mutation, Y250X, resulting in a truncated protein (Camaschella et al., 2000). Over the past few years, this has been followed by numerous other reports of mutations in this gene causing hemochromatosis. Interestingly, there are also reports of TFR2 mutations causing an early-onset form of hemochromatosis (Le Gac et al., 2004a; Piperno et al., 2004; Pietrangelo et al., 2005). This highlights the possibility that TFR2 may also be involved in the metabolic processes found to be defective in JH. Furthermore, patients with TFR2 mutations also display a decrease in hepcidin levels (Kawabata et al., 2005; Nemeth et al., 2005). Although suggested to be involved in various processes as discussed above, the exact mechanism of TFR2 function remains to be clearly defined.

TYPE 4 HEMOCHROMATOSIS (FERROPORTIN DISEASE)

Type 4 hemochromatosis is distinct from the other types of hemochromatosis, due to its autosomal dominant inheritance pattern. This may partly be explained by its iron export function, not only from duodenal enterocytes but also from the liver and iron-recycling macrophages. The exact mechanism underlying this inheritance pattern, however, remains unresolved and continues to be a subject of much debate. The disorder is due to mutations in the ferroportin gene, *SLC40A1*, and has a more global distribution than type 1 hemochromatosis. Its symptoms are similar to type 1, although it can sometimes lead to a more severe disease with cardiac complications, depending on the mutation. Many mutations in *SLC40A1* have now been identified.

There are two distinct groups of type 4 patients. In some type 4 cases, a high serum ferritin concentration but normal to low transferrin saturation is observed, and treatment by venesection often results in anemia (Lim et al., 2008). This group of patients is often described as having "ferroportin disease." Other type 4 patients, however, appear to have the same clinical symptoms as type 1 patients, with both high ferritin concentrations and transferrin saturation. This can be explained by the function of ferroportin and type of mutation. It is now recognized that there is a spectrum of ferroportin mutations that broadly fall into two classes: loss-of-function, and hepcidin-resistant mutations, at the opposite ends of the spectrum (Drakesmith et al., 2005; Schimanski et al., 2005). The loss-of-function mutations causing "ferroportin disease," such as A77D and V162del (Montosi et al., 2001; Cazzola et al., 2002; Devalia et al., 2002; Roetto et al., 2002; Wallace et al., 2002), lead to iron accumulation in the liver Kupffer cells and macrophages (reticuloendothelial system). This decrease in iron availability subsequently results in low transferrin saturation and limited erythropoiesis. The hepcidin-resistant mutations include N144D/H/T and C326S/Y (Njajou et al., 2001; Arden at al., 2003; Wallace et al., 2004; Drakesmith et al., 2005; Sham et al., 2005). These mutants are able to retain their export activity, and are instead associated with increased transferrin saturation and iron loading in hepatocytes. It appears that these mutants are unresponsive to negative feedback, and are able to completely or partially resist the inhibitory binding of hepcidin (Drakesmith et al., 2005). It has also been noted in cases of ferroportin disease due to the V162del mutation that hepcidin levels in the patients are relatively high before venesection therapy, which is in contrast to the other types of hemochromatosis (Papanikolaou et al., 2005; Zoller et al., 2005). Equally, patients with the C326S hepcidin-resistant mutation who have been treated by venesection also have high levels of hepcidin (Sham et al., 2009), highlighting the key importance of negative feedback by hepcidin.

DIGENIC INHERITANCE

The frequency of the HFE C282Y and H63D mutations in European and Asian populations, respectively, together with the increase in interracial marriages, means that there is the possibility of hemochromatosis due to combinations of mutations in HFE with those in other genes essential to iron metabolism. Digenic inheritance of hemochromatosis has previously been reported in European families who were heterozygous for mutations in both HFE and hepcidin (Merryweather-Clarke et al., 2003; Jacolot et al., 2004). Other combinations include double heterozygosity for mutations in *HFE* with *SLC40A1*, *HJV*, or *TFR2* mutations (Roetto et al., 2002; Le Gac et al., 2004b; Pietrangelo et al., 2005). Hence it is important to appreciate that if a patient has unexplained clinical iron overload, and only a heterozygous mutation is identified in one gene, disease may be due to additional mutations in other genes. There also remains the strong possibility that there are more genes with hemochromatosis-causing mutations yet to be identified.

IRON OVERLOAD IN ASIA

In contrast to European populations, reports of hereditary hemochromatosis in regions of Asia have been relatively rare and less well defined. There may be several explanations for this apparent rarity in Asia. Firstly, previous studies may well have been based on screening for known mutations common to Europe populations rather than a comprehensive analysis of each candidate gene for new mutations. Secondly, the presence of thalassemia and the associated secondary iron overload due to transfusions (Weatherall & Clegg, 2001) can make diagnosis of hereditary hemochromatosis difficult; and, thirdly, disease presentation may also be masked by the prevalence of iron deficiency in these regions (Stoltzfus et al., 2004). This, coupled with the lack of widespread screening, has largely led to the conclusion that hemochromatosis is not a significant clinical problem in Asia. However, the variability in iron loading in Asian patients with intermediate forms of thalassemia that are non-transfusion-dependent (Weatherall & Clegg, 2001) suggests that hemochromatosis may be more widespread than previously thought. Understanding the frequency of hemochromatosis in these populations would, therefore, be of benefit to the management of this disorder in these different environmental settings. In the following sections, the different types of hemochromatosis in Asian populations reported to date will be described.

HFE HEMOCHROMATOSIS IN ASIANS

HFE gene mutations in the Asian population have proved to be extremely rare. Where the C282Y allele has been found, it has been associated with the European haplotype, with one exception (Rochette et al., 1999). The C282Y allele is extremely scarce in Asia but H63D has a more universal distribution. Although found at a lower frequency in Asia, the H63D allele is as high as 10% in regions such as Pakistan and Sri Lanka, suggesting the presence of some selection pressures (Merryweather-Clarke et al., 1997, 2000). To date, the association of HFE mutations with iron overload outside Europe is very minimal and variable (Mortimore et al., 1999; Sohda et al., 1999; Tsui et al., 2000; Shiono et al., 2001; Pointon et al., 2003; Thakur et al., 2004; Viprakasit et al., 2004; Shukla et al.,, 2006). One case of hemochromatosis in Japan found to be homozygous for C282Y has been reported (Sohda et al., 2001), while another case was heterozygous for an A176V variant, although it is unclear whether this alone contributes to hemochromatosis (Imanishi et al., 2001). A novel *HFE* splice-site mutation has also been described in a Vietnamese family in Germany, and found in other Vietnamese populations (Steiner et al., 2002; Pointon et al., 2003). This will have implications on the expression of the disease where there is the likelihood of genetic admixture.

JUVENILE HEMOCHROMATOSIS IN ASIANS

Recent studies have identified a number of Asian families with juvenile hemochromatosis. Three JH cases have been described in Japan, and found to be due to two novel HJV mutations, D249H and Q312X (Koyama et al., 2005a). These patients had the typical clinical manifestations of JH, including cardiomyopathy. However, they displayed a surprisingly later onset of disease at around 50 years of age, atypical of JH. This is in fact similar to the age of onset of type 1 hemochromatosis. The Q312X mutation has since been found associated with cardiomyopathy and impotence in another Japanese family (Nagayoshi et al., 2008). In this case, the JH patients were also homozygous for the mutation and, in contrast to the first family, presented with disease in their early 20s, typical of JH. Consanguinity in the family was also reported.

Patients originating from the Indian subcontinent, a proportion of whom live in the United Kingdom, have also been identified to have novel HJV and HAMP mutations (Lok et al., 2009). These mutations are summarized in Tables 21-3 and 21-4. It should be noted that the proband in family A was a HJV C80Y/G320V compound heterozygote, inheriting the common European G320V mutation from her mother (Lok et al., 2009). This highlights the possibility of JH in families of mixed European and Asian ancestry. There has been a previous report of a HJV C80R mutation in Caucasians (Lee et al., 2004), where the cysteine is substituted by arginine rather than the tyrosine mutation of Bangladeshi origin. One HJV mutation, G99R, previously reported in an Albanian family (Lanzara et al., 2004), was found in two unrelated families of Pakistani origin, suggesting that it may be a recurrent mutation not restricted to Europe (Lok et al., 2009). The L194P mutation appears to result in a relatively later onset of disease. For all the HJV residues affected by mutations, an equivalent identical residue can be found across different species, with the exception of leucine 194 (Figure 21-3). L194 is only found in mammalian HJV. This may be one explanation for the later disease onset in case E

TABLE 21-3 IDENTIFIED JH MUTATIONS AND THEIR ORIGIN

Family	Protein	Exon	Nucleotide change	Amino acid change	Origin
A	Hemojuvelin	3	239G > A	C80Y	Bangladesh
B, C	Hemojuvelin	3	295G > A	G99R	Pakistan
D	Hemojuvelin	3	575C > T	P192L	Pakistan
E	Hemojuvelin	3	581T > C	L194P	Pakistan
F	Hemojuvelin	4	1026 del T	A343PfsX23	Sri Lanka
G	Hepcidin	2	126–127delAG	R42Sfs	Pakistan

TABLE 21-4 GENOTYPE AND PHENOTYPE OF JH PROBANDS FROM THE INDIAN SUBCONTINENT

Family/ patient	Gender	Age at Diagnosis (yrs.)	Tsat (%)	Ferritin (μg/l)	Hypogonadism	Diabetes	HJV Mutation	HAMP Mutation
A	F	19	100	5293	+	+	C80Y/G320V	WT
B	M	26	Raised	4485	+	+	G99R	WT
C	F	11	ND	3772	ND	ND	G99R	WT
D/ II-3	M	23	Raised	>6000	+	ND	P192L	WT
E	M	32	100	3756	+	-	L194P	WT
F	M	17	94	5999	+	+	A343PfsX23	WT
G/ V-2	M	21	>97	5686	+	-	WT	R42Sfs

All patients shown are homozygous for their mutations except the proband in family A, who is a compound heterozygote. Normal values for transferrin saturation (Tsat) 15%–45%; ferritin 15–200 μg/l for females, 40–300 μg/l for males. + indicates presence of symptom; –, absence of symptom; ND, not determined.

```
                        C80Y ↓              G99R ↓
Human HJV         GGVGSGGLCRALRSYALCTRRTARTCRGDL 101
Mouse HJV         -GLASGGLCRALRSYALCTRRTARTCRGDL 94
Zebrafish HJV     -------YCSALRSYALCTQQTARACRGDL 92
Human RGMA        -------FCAALRSYALCTRRTARTCRGDL 97
Mouse RGMA        -------FCAALRTYALCTRRTARTCRGDL 98
Zebrafish RGMA    -------FCTALRAYNSCVRRTARTCRGDL 74
Zebrafish RGMB    -------FCKALRAYSACTQRTAKSCRGNL 94
                         * ***:*  *.::**::***:*
```

```
                     P192L ↓ ↓L194P
Human HJV         TCRVQGAWPLLDNDFLFVQATSSPMALGAN 213
Mouse HJV         TCRVQGAWPLLDNDFLFVQATSSPVSSGAN 206
Zebrafish HJV     TCAVQGAWPLIDNQYLYIQATSSPTRESSD 199
Human RGMA        TCKVQGAWPLIDNNYLNVQVTNTPVLPGSA 209
Mouse RGMA        TCKVQGAWPLIDNNYLNVQVTNTPVLPGSA 210
Zebrafish RGMA    TCKVEGAWPLIHNKYLSVQVTNTPVVVGSS 192
Zebrafish RGMB    TCKVEGAWPLIDNNYLSVQVTNVPVVYGSS 204
                  ** *:*****:.*.:* :*.*. *    .:
```

```
                                          ↓G320V
Human HJV         DVAMAFS--AEQDLQLCVGGCPPSQRLSRS- 330
Mouse HJV         DVARAFS--AEQDLQLCVGGCPPSQRLSRS- 323
Zebrafish HJV     AIVESYT--PEQDLQLCVWGCPISQRLEML- 316
Human RGMA        EVVNAVEDWDSQGLYLCLRGCPLNQQIDFQ- 328
Mouse RGMA        EVVNAVEDRDSQGLYLCLRGCPLNQQIDFQ- 329
Zebrafish RGMA    EVVNSVEDQDNQDLYLCLHGCPANQRIDFR- 311
Zebrafish RGMB    ELAMAFD--ETQDLQLCMNGCPTSERIDQEG 321
                   :. :        *.* **: ***  .:::.
```

```
                          ↓A343PfsX23
Human HJV         IDTARRLCKEGLPVEDAYFHSCVFDVLISG 369
Mouse HJV         IDTARRLCKEGLPVEDAYFQSCVFDVSVSG 362
Zebrafish HJV     FDPAYTHCSSLFPGRDVYFQACLFDVQVTG 350
Human RGMA        YETAVAKCKEKLPVEDLYYQACVFDLLTTG 386
Mouse RGMA        YETAVAKCKEKLPVEDLYYQACVFDLLTTG 385
Zebrafish RGMA    YQSAMAKCKERLPVEDLYFQSCVFDLLSSG 366
Zebrafish RGMB    LESASRRCRDQLEVKDIYFHSCVFDLLTTG 381
                   :.*     *  .  :   .* *:::*:**:  :*
```

Figure 21-3 Multiple sequence alignment of hemojuvelin protein sequence. Alignment of selected regions of human hemojuvelin with the mutations, against the mouse and zebrafish hemojuvelin, and human, mouse, and zebrafish RGMs. Identified mutations are indicated by arrows. Red indicates small, hydrophobic, and aromatic residues. Blue indicates acidic residues. Magenta indicates basic residues. Green indicates hydroxyl, amine, and basic residues. * indicates identical residues in the alignment; colon (:), conserved substitutions; period (.), semi-conserved substitutions of residues. (Alignment produced using ClustalW version 1.83.)

Family A

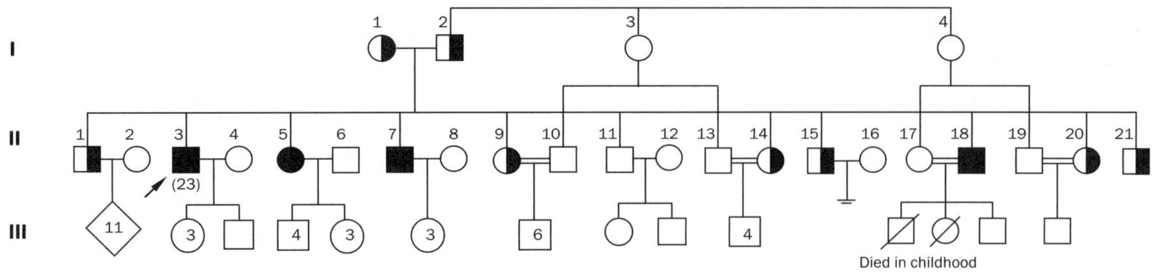

Figure 21-4 Pedigree of family D with the HJV P192L mutation. The proband (II-3) has 10 siblings, of whom three are also homozygous for the mutation. Six siblings are heterozygotes, and sibling II-11 is the only wild-type. The proband's parents are unaffected. Arrow indicates proband; age at diagnosis is shown in brackets; diamond symbol indicates gender unspecified; numbers inside symbols indicate the number of offspring.

(Table 21-4). The JH patients from the Indian subcontinent have a high frequency of hypogonadism associated with increased serum iron parameters (Table 1-4). Surprisingly, none of the cases presented with the classical cardiomyopathy as observed in European patients (Lok et al., 2009).

The variation in the age of onset and the lack of cardiac complications in some of these patients may reflect the influence of further environmental or genetic factors on disease expression. Furthermore, the tradition of first and second cousin marriages, as observed in some of the affected families (Family A & B) in these studies, may increase the frequency of this disorder in these populations (illustrated in Figures 21-4 and 21-5).

TFR2 HEMOCHROMATOSIS IN ASIANS

Type 3 hemochromatosis in Asian populations is rare. Mutations in this gene have mainly been reported in a few cases in Japan. The first report was a deletion of four amino acids, AVAQ at position 594-597 of TFR2 (Hattori et al., 2003). A second report described two novel mutations, L490R and V561X, causing hemochromatosis with the typical clinical features described in European patients (Koyama et al., 2005b).

Type 3 hemochromatosis has also been described in a Taiwanese patient. In this case, an R481H amino acid substitution was reported, but in a heterozygous state (Hsiao et al., 2007). Whether any other mutations were present is unknown. A TFR2 polymorphism, I238M, has also been found in different Asian populations, but of no pathogenic association (Lee et al., 2001; Chan et al., 2003; Ma et al., 2003; Koyama et al. 2005b).

Family B

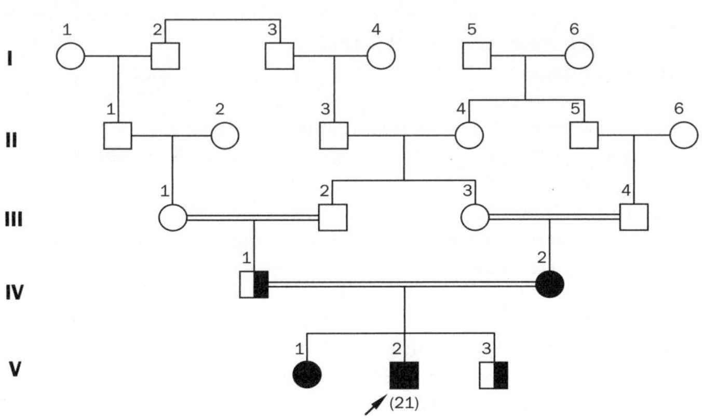

Figure 21-5 Pedigree of family G with the *HAMP* 126-127delAG mutation. The proband (V-2) is the progeny of multiply consanguineous parents who are unaffected heterozygotes. The proband's older sister (V-1), who is a homozygote, is also affected. The proband, siblings, and parents were the only ones available for testing.

TYPE 4 HEMOCHROMATOSIS IN ASIANS

A number of mutations in ferroportin causing the autosomal dominant form of iron overload have been described in Asia. The earliest report of iron overload with this mode of inheritance was described 20 years ago in a large family of 81 members from the Solomon Islands (Eason et al., 1990). It is a possibility that this was due to the ferroportin N144T mutation, as a patient of Solomon Islands ancestry living in Australia was identified with this mutation (Arden et al., 2003). Some of the other ferroportin mutations that have been described in Asians include the V162 deletion that now turns out to be a mutation found in a number of different geographical regions, including Sri Lanka (Devalia et al., 2002; Roetto et al., 2002; Wallace et al., 2005). V162 is one of three contiguous valine residues in the protein, each encoded by the same codon (GTT), such that slippage during replication is the likely mutation mechanism and probably explains the high frequency of V162del. The A77D mutation is the only other ferroportin mutation found in the Indian subcontinent so far (Agarwal et al., 2006).

Type 4 hemochromatosis has also been reported in Eastern Asia. The G490D mutation causing "ferroportin disease" was reported of possible Chinese origin (Jouanolle et al., 2003). A further report identified a G267D mutation in another family of Chinese descent with hyperferritinemia (Cremonesi et al., 2005). In Japan, a novel R489S mutation was found in a family causing mild disease and predominant iron overload in liver Kupffer cells, also suggesting "ferroportin disease" (Koyama et al., 2005c). Another report identified a patient with high ferritin and transferrin saturation, who inherited a novel point mutation in the 5'-UTR close to the IRE of ferroportin (Liu et al., 2005). Type 4 hemochromatosis is not the only autosomal dominant form of the disorder found in Asia. A mutation in the IRE of H-ferritin, causing autosomal dominant iron overload, has previously been found in a Japanese family (Kato et al., 2001).

More recently, the hepcidin-resistant mutation, C326Y (Drakesmith et al., 2005), has been found in a Thai family, adding to the increasing number of cases of non-HFE hemochromatosis. The phenotype of the subjects was indistinguishable from classical type 1 hemochromatosis, and iron loading was associated with macrocytic red cells (Lok et al., 2009). To date, type 4 hemochromatosis appears to be the most widespread form of the disorder.

CONCLUSIONS

The steadily increasing number of cases of hemochromatosis identified in different Asian countries suggests that this disorder may be more common than previously thought. Furthermore, the high frequency of consanguineous marriages in many Asian populations might increase the frequency of this disorder. Regions such as Azad Kashmir of Pakistan are remote and mountainous areas. Historically, such geographical conditions, along with cultural practices, are contributing factors favoring consanguinity.

While the prevalence of iron deficiency in some of these populations may have prevented the clinical expression of hemochromatosis, as these countries undergo the demographic transition associated with better public health measures and diet, iron overload may begin to have an impact on healthcare resources in those regions. It is also important to determine the frequency of hemochromatosis, because of its potential in influencing iron loading in thalassemia patients.

The occurrence of hemochromatosis mutations also has consequences for the Asian diaspora in regions such as the United Kingdom and North America. Currently, around 10% of the population in the United Kingdom is of South Asian origin, a significant proportion having come from the Kashmir region of Pakistan, where JH mutations have been found. There are even larger numbers of South Asians living in the United States and Canada. It is therefore important to keep a vigilant watch for the early signs of hemochromatosis in developed countries with Asian populations, as well as in regions where resources are limited, to allow for early therapeutic intervention and avoid prolonged, complicated treatment of patients. It should be remembered that unlike a number of other genetic diseases, hemochromatosis is very easily treated by venesection and, with early diagnosis and treatment, life-threatening complications can be avoided. One of the problems is that some of the presenting features of hemochromatosis may not trigger the request for simple diagnostic measurements. At present, there are an increasing number of cases of unexplained iron overload of Asian origin, and the current prevalence of liver disease in South Asian communities due to underlying hemochromatosis is unclear.

While much remains to be learned about hemochromatosis in Asian populations, the knowledge to date highlights the need for greater awareness of this condition, and the need to initiate appropriate screening for individuals that are likely to be at risk.

ACKNOWLEDGMENTS

The work in the laboratory of KJHR is supported by the Hemochromatosis Society.

REFERENCES

Abboud S, Haile DJ. (2000). A novel mammalian iron-regulated protein involved in intracellular iron metabolism. *J Biol Chem 275*, 19906–19912.

Agarwal S, Sankar VH, Tewari D, Pradhan M. (2006). Ferroportin (SLC40A1). gene in thalassemic patients of Indian descent. Clin Genet *70*, 86–87.

Andriopoulos B Jr, Corradini E, Xia Y, et al. (2009). BMP6 is a key endogenous regulator of hepcidin expression and iron metabolism. Nat Genet *41*, 482–487.

Arden KE, Wallace DF, Dixon JL, et al. (2003). A novel mutation in ferroportin1 is associated with hemochromatosis in a Solomon Islands patient. Gut *52*, 1215–1217.

Arosio P, Adelman TG, Drysdale JW. (1978). On ferritin heterogeneity. Further evidence for heteropolymers. *J Biol Chem 253*, 4451–4458.

Aziz N, Munro HN. (1987). Iron regulates ferritin mRNA translation through a segment of its 5' untranslated region. *Proc Natl Acad Sci U S A 84*, 8478–8482.

Babitt JL, Huang FW, Wrighting DM, et al. (2006). Bone morphogenetic protein signaling by hemojuvelin regulates hepcidin expression. *Nat Genet 38*, 531–539.

Babitt JL, Huang FW, Xia Y, Sidis Y, Andrews NC, Lin HY. (2007). Modulation of bone morphogenetic protein signaling in vivo regulates systemic iron balance. *J Clin Invest 117*, 1933–1939.

Banyard SH, Stammers DK, Harrison PM. (1978). Electron density map of apoferritin at 2.8-A resolution. *Nature 271*, 282–284.

Beaumont C, Leneuve P, Devaux I, et al. (1995). Mutation in the iron responsive element of the L ferritin mRNA in a family with dominant hyperferritinaemia and cataract. *Nat Genet 11*, 444–446.

Bridle KR, Frazer DM, Wilkins SJ, et al. (2003). Disrupted hepcidin regulation in HFE-associated hemochromatosis and the liver as a regulator of body iron homeostasis. *Lancet 361*, 669–673.

Camaschella C, Roetto A, Cali A, et al. (2000). The gene *TFR2* is mutated in a new type of hemochromatosis mapping to 7q22. *Nature Genetics 25*, 14–15.
Canonne-Hergaux F, Gruenheid S, Ponka P, Gros P. (1999). Cellular and subcellular localization of the Nramp2 iron transporter in the intestinal brush border and regulation by dietary iron. *Blood 93*(12), 4406–4417.
Cazzola M, Cremonesi L, Papaioannou M, et al. (2002). Genetic hyperferritinaemia and reticuloendothelial iron overload associated with a three base pair deletion in the coding region of the ferroportin gene (SLC11A3). *Br J Haematol 119*, 539–546.
Cremonesi L, Forni GL, Soriani N, et al. (2005). Genetic and clinical heterogeneity of ferroportin disease. *Br J Haematol 131*, 663–670.
Chan V, Wong MS, Ooi C, et al. (2003). Can defects in transferrin receptor 2 and hereditary hemochromatosis genes account for iron overload in HbH disease? *Blood Cells Mol Dis 30*, 107–111.
Dadone MM, Kushner JP, Edwards CQ, Bishop DT, Skolnick MH. (1982). Hereditary hemochromatosis. Analysis of laboratory expression of the disease by genotype in 18 pedigrees. *Am J Clin Pathol 78*, 196–207.
Devalia V, Carter K, Walker AP, et al. (2002). Autosomal dominant reticulo-endothelial iron overload associated with a three base pair deletion in the ferroportin 1 gene (*SLC11A3*). *Blood 100*, 695–697.
Donovan A, Brownlie A, Zhou Y, et al. (2000). Positional cloning of zebrafish *ferroportin1* identifies a conserved vertebrate iron exporter. *Nature 403*, 776–781.
Drakesmith H, Schimanski LM, Ormerod E, et al. (2005). Resistance to hepcidin is conferred by hemochromatosis-associated mutations of ferroportin. *Blood 106*(3), 1092–1097.
Eason R, Adams P, Aston C, Searle J. (1990). Familial iron overload with possible autosomal dominant inheritance. *Australian and New Zealand Journal of Medicine 20*, 226–230.
Feder JN, Gnirke A, Thomas W, et al. (1996). A novel MHC class I-like gene is mutated in patients with hereditary hemochromatosis. *Nature Genetics 13*, 399–408.
Feder JN, Penny DN, Irrinki A, et al. (1998). The hemochromatosis gene product complexes with the transferrin receptor and lowers its affinity for ligand binding. *Proc Natl Acad Sci U S A 95*, 1472–1477.
Feder JN, Tsuchihashi Z, Irrinki A, et al. (1997). The hemochromatosis founder mutation in HLA-H disrupts b_2-microglobulin interaction and cell surface expression. *J Biol Chem 272*, 14025–14028.
Finberg KE, Heeney MM, Campagna DR, et al. (2008). Mutations in TMPRSS6 cause iron-refractory iron deficiency anemia (IRIDA). *Nat Genet 40*, 569–571.
Fleming MD, Trenor CC 3rd, Su MA, et al. (1997). Microcytic anemia mice have a mutation in Nramp2, a candidate iron transporter gene. *Nat Genet 16*, 383–386.
Gao J, Chen J, Kramer M, Tsukamoto H, Zhang AS, Enns CA. (2009). Interaction of the hereditary hemochromatosis protein HFE with transferrin receptor 2 is required for transferrin-induced hepcidin expression. *Cell Metab 9*, 217–227.
Goswami T, Andrews NC. (2006). Hereditary hemochromatosis protein, HFE, interaction with transferrin receptor 2 suggests a molecular mechanism for mammalian iron sensing. *J Biol Chem 281*, 28494–28498.
Griffiths WJ, Cox TM. (2003). Co-localization of the mammalian hemochromatosis gene product (HFE). and a newly identified transferrin receptor (TfR2). in intestinal tissue and cells. *J Histochem Cytochem 51*, 613–624.
Gruenheid S, Canonne-Hergaux F, Gauthier S, Hackam DJ, Grinstein S, Gros P. (1999). The iron transport protein Nramp2 is an integral membrane glycoprotein that colocalizes with transferrin in recycling endosomes. *J Exp Med 189*(5), 831–841.
Gunshin H, Mackenzie B, Berger UV, et al. (1997). Cloning and characterization of a mammalian proton-coupled metal-ion transporter. *Nature 388*, 482–488.
Haile DJ, Rouault TA, Tang CK, Chin J, Harford JB, Klausner RD. (1992). Reciprocal control of RNA-binding and aconitase activity in the regulation of the iron-responsive element binding protein: role of the iron-sulfur cluster. *Proc Natl Acad Sci U S A 89*, 7536–7540.
Harman D. (1956). Aging: a theory based on free radical and radiation chemistry. *J Gerontol 11*, 298–300.
Harris ZL, Takahashi Y, Miyajima H, Serizawa M, MacGillivray RT, Gitlin JD. (1995). Aceruloplasminemia: molecular characterization of this disorder of iron metabolism. *Proc Natl Acad Sci U S A 92*, 2539–2543.
Hattori A, Wakusawa S, Hayashi H, et al. (2003). AVAQ 594-597 deletion of the TfR2 gene in a Japanese family with hemochromatosis. *Hepatol Res 26*, 154–156.
Hsiao PJ, Tsai KB, Shin SJ, et al. (2007). A novel mutation of transferrin receptor 2 in a Taiwanese woman with type 3 hemochromatosis. J *Hepatol 47*(2), 303–306.
Imanishi H, Liu W, Cheng J, Ikeda N, Amuro Y, Hada T. (2001). Idiopathic hemochromatosis with the mutation of Ala176Val heterozygous for HFE gene. *Intern Med 40*(6), 479–483.
Jacolot S, Le Gac G, Scotet V, Quere I, Mura C, Ferec C. (2004). HAMP as a modifier gene that increase the phenytypic expression of the HFE p.C282Y homozygous genotype. *Blood 103*, 2835–2840.
Johnson MB, Enns CA. (2004). Diferric transferrin regulates transferrin receptor 2 protein stability. *Blood 104*, 4287–4293
Jouanolle AM, Douabin-Gicquel V, Halimi C, et al. (2003). Novel mutation in ferroportin 1 gene is associated with autosomal dominant iron overload. *J Hepatol 39*, 286–289.
Kato J, Fujikawa K, Kanda M, et al. (2001). A mutation, in the iron-responsive element of H ferritin mRNA, causing autosomal dominant iron overload. *Am J Hum Genet 69*, 191–197.
Kawabata H, Fleming RE, Gui D, et al. (2005). Expression of hepcidin is down-regulated in TfR2 mutant mice manifesting a phenotype of hereditary hemochromatosis. *Blood 105*, 376–381.
Kawabata H, Yang R, Hirama T, et al. (1999). Molecular cloning of transferrin receptor 2. *J Biol Chem 274*, 20826–20832.
Koyama C, Hayashi H, Wakusawa S, et al. (2005a). Three patients with middle-age-onset hemochromatosis caused by novel mutations in the hemojuvelin gene. *J Hepatol 43*, 740–742.
Koyama C, Wakusawa S, Hayashi H, et al. (2005b). Two novel mutations, L490R and V561X, of the transferrin receptor 2 gene in Japanese patients with hemochromatosis. *Haematologica 90*, 302–307.
Koyama C, Wakusawa S, Hayashi H, et al. (2005c). A Japanese family with ferroportin disease caused by a novel mutation of SLC40A1 gene: hyperferritinemia associated with a relatively low transferrin saturation of iron. *Intern Med 44*, 990–993.
Krause A, Neitz S, Magert HJ, et al. (2000). LEAP-1, a novel highly disulfide-bonded human peptide, exhibits antimicrobial activity. *FEBS Letters 480*, 147–150.
Kulaksiz H, Gehrke SG, Janetzko A, et al. (2004). Pro-hepcidin: expression and cell specific localization in the liver and its regulation in hereditary hemochromatosis, chronic renal insufficiency, and renal anemia. *Gut 53*, 735–743.
Lanzara C, Roetto A, Daraio F, et al. (2004). Spectrum of hemojuvelin gene mutations in 1q-linked juvenile hemochromatosis. *Blood 103*(11), 4317–4321.
Le Gac G, Mons F, Jacolot S, Scotet V, Ferec C, Frebourg T. (2004a). Early onset hereditary hemochromatosis resulting from a novel TFR2 gene nonsense mutation (R105X). in two siblings of north French descent. *Br J Haematol 125*, 674–678.
Le Gac G, Scotet V, Ka C, et al. (2004b). The recently identified type 2A juvenile hemochromatosis gene (HJV), a second candidate modifier of the C282Y homozygous phenotype. *Hum Mol Genet 13*, 1913–1918.
Lebrón JA, West AP, Jr., Bjorkman PJ. (1999). The hemochromatosis protein HFE competes with transferrin for binding to the transferrin receptor. *J Mol Biol 294*, 239–245.
Lee PL, Beutler E, Rao SV, Barton JC. (2004). Genetic abnormalities and juvenile hemochromatosis: mutations of the HJV gene encoding hemojuvelin. *Blood 103*, 4669–4671.

Lee PL, Halloran C, West C, Beutler E. (2001). Mutation analysis of the transferrin receptor-2 gene in patients with iron overload. *Blood Cells Mol Dis 27*(1), 285–289.

Lim FL, Dooley JS, Roques AW, Grellier L, Dhillon AP, Walker AP. (2008). Hepatic iron concentration, fibrosis and response to venesection associated with the A77D and V162del "loss of function" mutations in ferroportin disease. *Blood Cells Mol Dis 40*, 328–333.

Lin L, Goldberg YP, Ganz T. (2005). Competitive regulation of hepcidin mRNA by soluble and cell-associated hemojuvelin. *Blood 106*, 2884–2889.

Liu W, Shimomura S, Imanishi H, et al. (2005). Hemochromatosis with mutation of the ferroportin 1 (IREG1) gene. *Intern Med 44*(4), 285–289.

Lok CY, Merryweather-Clarke AT, Viprakasit V, et al. (2009). Iron overload in the Asian community. *Blood 114*, 20–25.

Ma ES, Lam KK, Chan AY, Ha SY, Au WY, Chan LC. (2003). Transferrin receptor-2 polymorphisms and iron overload in transfusion independent b-thalassemia intermedia. *Haematologica 88*(3), 345–346.

McCance RA, Widdowson EM. (1937). Absorption and excretion of iron. *Lancet 230*(5951), 680–684.

McKie AT, Barrow D, Latunde-Dada GO, et al. (2001). An iron-regulated ferric reductase associated with the absorption of dietary iron. *Science 291*, 1755–1759.

McKie AT, Marciani P, Rolfs A, et al. (2000). A novel duodenal iron-regulated transporter, IREG1, implicated in the basolateral transfer of iron to the circulation. *Molecular Cell 5*, 299–309.

Merryweather-Clarke AT, Cadet E, Bomford A, et al. (2003). Digenic inheritance of mutations in *HAMP* and *HFE* results in different types of hemochromatosis. *Human Molecular Genetics 12*, 2241–2247.

Merryweather-Clarke AT, Pointon JJ, Jouanolle AM, Rochette J, Robson KJH. (2000). Geography of HFE C282Y and H63D mutations. *Genetic Testing 4*, 183–198.

Merryweather-Clarke AT, Pointon JJ, Shearman JD, Robson KJ. (1997). Global prevalence of putative hemochromatosis mutations. *J Med Genet 34*, 275–278.

Meynard D, Kautz L, Darnaud V, Canonne-Hergaux F, Coppin H, Roth MP. (2009). Lack of the bone morphogenetic protein BMP6 induces massive iron overload. *Nat Genet 41*, 478–481.

Mims MP, Guan Y, Pospisilova D, et al. (2005). Identification of a human mutation of DMT1 in a patient with microcytic anemia and iron overload. *Blood 105*, 1337–1342.

Montosi G, Donovan A, Totaro A, et al. (2001). Autosomal-dominant hemochromatosis is associated with a mutation in the ferroportin (SLC11A3) gene. *Journal of Clinical Investigation 108*, 619–623.

Mortimore M, Merryweather-Clarke AT, Robson KJ, Powell LW. (1999). The hemochromatosis gene: a global perspective and implications for the Asia-Pacific region. *J Gastroenterol Hepatol 14*, 838–843.

Nagayoshi Y, Nakayama M, Suzuki S, et al. (2008). A Q312X mutation in the hemojuvelin gene is associated with cardiomyopathy due to juvenile haemochromatosis. *Eur J Heart Fail 10*(10), 1001–1006.

Nemeth E, Rivera S, Gabayan V, et al. (2004a). IL-6 mediates hypoferremia of inflammation by inducing the synthesis of the iron regulatory hormone hepcidin. *J Clin Invest 113*, 1271–1276.

Nemeth E, Tuttle MS, Powelson J, et al. (2004b). Hepcidin regulates cellular iron efflux by binding to ferroportin and inducing its internalization. *Science 306*, 2090–2093.

Nemeth E, Roetto A, Garozzo G, Ganz T, Camaschella C. (2005). Hepcidin is decreased in TFR2 hemochromatosis. *Blood 105*, 1803–1806.

Nemeth E, Valore EV, Territo M, Schiller G, Lichtenstein A, Ganz T. (2003). Hepcidin, a putative mediator of anemia of inflammation, is a type II acute-phase protein. *Blood 101*, 2461–2463.

Nicolas G, Bennoun M, Devaux I, et al. (2001). Lack of hepcidin gene expression and severe tissue iron overload in upstream stimulatory factor 2 (*USF2*). knockout mice. *Proc Natl Acad Sci U S A 998*, 8780–8785.

Nicolas G, Bennoun M, Porteu A, et al. (2002a). Severe iron deficiency anemia in transgenic mice expressing liver hepcidin. *Proc Natl Acad Sci U S A 99*, 4596–4601.

Nicolas G, Chauvet C, Viatte L, et al. (2002b). The gene encoding the iron regulatory peptide hepcidin is regulated by anemia. *Journal of Clinical Investigation 110*, 1037–044.

Niederau C, Fischer R, Purschel A, Stremmel W, Haussinger D, Strohmeyer G. (1996). Long-term survival in patients with hereditary hemochromatosis. *Gastroenterology 110*, 1107–1119.

Njajou OT, Vaessen N, Joosse M, et al. (2001). A mutation in *SLC11A3* is associated with autosomal dominant hemochromatosis. *Nature Genetics 28*, 213–214.

Ohgami RS, Campagna DR, Greer EL, et al. (2005). Identification of a ferrireductase required for efficient transferrin-dependent iron uptake in erythroid cells. *Nat Genet 37*, 1264–1269.

Papanikolaou G, Samuels ME, Ludwig EH, et al. (2004). Mutations in *HFE2* cause iron overload in chromosome 1q-linked juvenile hemochromatosis. *Nature Genetics 36*, 77–82.

Papanikolaou G, Tzilianos M, Christakis JI, et al. (2005). Hepcidin in iron overload disorders. *Blood 105*, 4103–4105

Park CH, Valore EV, Waring AJ, Ganz T. (2001). Hepcidin, a urinary antimicrobial peptide synthesized in the liver. *J Biol Chem 276*, 7896–7910

Peyssonnaux C, Zinkernagel AS, Schuepbach RA, et al. (2007). Regulation of iron homeostasis by the hypoxia-inducible transcription factors (HIFs). *J Clin Invest 117*, 1926–1932

Pietrangelo A, Caleffi A, Henrion J, et al. (2005). Juvenile hemochromatosis associated with pathogenic mutations of adult hemochromatosis genes. *Gastroenterology 128*, 470–479.

Pigeon C, Ilyin G, Courselaud B, et al. (2001). A new mouse liver-specific gene, encoding a protein homologous to human antimicrobial peptide hepcidin, is overexpressed during iron overload. *J Biol Chem 276*, 7811–7819.

Piperno A, Roetto A, Mariani R, et al. (2004). Homozygosity for transferrin receptor-2 Y250X mutation induces early iron overload. *Haematologica 89*, 359–360.

Pointon JJ, Viprakasit V, Miles KL, et al. (2003). Hemochromatosis gene (HFE). mutations in South East Asia: a potential for iron overload. *Blood Cells Molecules and Diseases 30*, 302–306

Robb A, Wessling-Resnick M. (2004). Regulation of transferrin receptor 2 protein levels by transferrin. *Blood 104*, 4294–4299

Rochette J, Le Gac G, Lassoued K, Ferec C, Robson KJ. (2010). Factors influencing disease phenotype and penetrance in HFE haemochromatosis. *Hum Genet 128*(3), 233–248.

Rochette J, Pointon JJ, Fisher CA, et al. (1999). Multicentric origin of hemochromatosis gene (*HFE*). mutations. *American Journal of Human Genetics 64*, 1056–1062.

Roetto A, Merryweather-Clarke AT, Daraio R, et al. (2002). A valine deletion in ferroportin 1: a common mutation in hemochromatosis type 4? *Blood 100*, 733–734.

Roetto A, Papanikoloau G, Politou M, et al. (2003). Mutant antimicrobial peptide hepcidin is associated with severe juvenile hemochromatosis. *Nature Genetics 33*, 21–22.

Rouault TA, Hentze MW, Caughman SW, Harford JB, Klausner RD. (1988). Binding of a cytosolic protein to the iron-responsive element of human ferritin messenger RNA. *Science 241*, 1207–1210.

Rouault TA, Stout CD, Kaptain S, Harford JB, Klausner RD. (1991). Structural relationship between an iron-regulated RNA-binding protein (IRE-BP). and aconitase: functional implications. *Cell 64*, 881–883.

Roy CN, Andrews NC. (2005). Anemia of inflammation: the hepcidin link. *Curr Opin Hematol 12*, 107–111.

Saddi R, Feingold J. (1974). Idiopathic hemochromatosis: an autosomal recessive disease. *Clinical Genetics 5*, 234–241.

Schade AL, Caroline L. (1946). An Iron-binding Component in Human Blood Plasma. *Science 104*, 340–341.

Schimanski LM, Drakesmith H, Merryweather-Clarke AT, et al. (2005). In vitro functional analysis of human ferroportin (FPN).

and hemochromatosis-associated FPN mutations. *Blood 105*, 4096–4102.

Schimanski LM, Drakesmith H, Sweetland E, et al. (2009). In vitro binding of HFE to the cation-independent mannose-6 phosphate receptor. *Blood Cells Mol Dis 43*, 180–193.

Sham RL, Phatak PD, Nemeth E, Ganz T. (2009). Hereditary hemochromatosis due to resistance to hepcidin: high hepcidin concentrations in a family with C326S ferroportin mutation. *Blood 114*(2), 493–494.

Sham RL, Phatak PD, West C, Lee P, Andrews C, Beutler E. (2005). Autosomal dominant hereditary hemochromatosis associated with a novel ferroportin mutation and unique clinical features. *Blood Cells Mol Dis 34*, 157–161.

Sheldon JH. (1935). *Hemochromatosis.* Oxford University Press, Oxford, England.

Shiono Y, Ikeda R, Hayashi H, et al. (2001). C282Y and H63D mutations in the HFE gene have no effect on iron overload disorders in Japan. *Intern Med 40*, 852–856.

Shukla P, Julka S, Bhatia E, Shah S, Nagral A, Aggarwal R. (2006). HFE, hepcidin and ferroportin gene mutations are not present in Indian patients with primary hemochromatosis. *Natl Med J India 19*, 20–23.

Silvestri L, Pagani A, Camaschella C. (2008a). Furin-mediated release of soluble hemojuvelin: a new link between hypoxia and iron homeostasis. *Blood 111*, 924–931.

Silvestri L, Pagani A, Nai A, De Domenico I, Kaplan J, Camaschella C. (2008b). The serine protease matriptase-2 (TMPRSS6). inhibits hepcidin activation by cleaving membrane hemojuvelin. *Cell Metab 8*, 502–511.

Sohda T, Okubo R, Kanimura S, Ohkawara T. (2001) Hemochromatosis with HFE gene in a Japanese patient. *Am J Gastroenterol 96*, 2487–2488.

Sohda T, Yanai J, Soejima H, Tamura K. (1999). Frequencies in the Japanese population of HFE gene mutations. *Biochem Genet 37*, 63–68.

Steiner M, Ocran K, Genschel J, et al. (2002). A homozygous *HFE* gene splice mutation (IVS5+1G/A). in a hereditary hemochromatosis patient of Vietnamese origin. *Gastroenterology 122*, 789–795.

Steiner MS, Zhang X, Wang Y, Lu Y. (2000). Growth inhibition of prostate cancer by an adenovirus expressing a novel tumor suppressor gene, pHyde. *Cancer Res 60*, 4419–4425.

Stoltzfus RJ, Mullany L, Black RE. (2004). Iron deficiency anaemia. In: Ezzati M, Lopez AD, Rodgers A, Murray CJL (eds.), *Comparative Quantification of Health Risks: Global and Regional Burden of Disease Attributable to Selected Major Risk Factors.* (vol. 1, pp. 163–209). World Health Organization, Geneva.

Thakur V, Guptan RC, Hashmi AZ, Sakhuja P, Malhotra V, Sarin SK. (2004). Absence of hemochromatosis associated Cys282Tyr HFE gene mutation and low frequency of hemochromatosis phenotype in nonalcoholic chronic liver disease patients in India. *J Gastroenterol Hepatol 19*, 86–90.

Townsend ARM, Drakesmith H. (2002). Role of HFE in iron metabolism, hereditary hemochromatosis, anemia of chronic disease, and secondary iron overload. *Lancet 359*, 786–790.

Tsui WM, Lam PW, Lee KC, et al. (2000). The C282Y mutation of the HFE gene is not found in Chinese hemeochromatotic patients: multicentre retrospective study. *Hong Kong Med J 6*, 153–158.

Viprakasit V, Vathesathokit P, Chinchang W, et al. (2004). Prevalence of HFE mutations among the Thai population and correlation with iron loading in hemoglobin E disorder. *Eur J Haematol 73*, 43–49.

von Recklinghausen FD. (1889). *Über Hämochromatose.* Tagebl Versamml Natur Ärtze Heidelberg.

Vulpe CD, Kuo YM, Murphy TL, et al. (1999). Hephaestin, a ceruloplasmin homologue implicated in intestinal iron transport, is defective in the sla mouse. *Nat Genet 21*, 195–199.

Waheed A, Parkkila S, Zhou XY, et al. (1997). Hereditary hemochromatosis: Effects of C282Y and H63D mutations on association with b_2-microglobulin, intracellular processing, and cell surface expression of the HFE protein in COS-7 cells. *Proc Natl Acad Sci U S A 94*, 12384–12389.

Wallace DF, Browett P, Wong P, Kua H, Ameratunga R, Subramaniam VN. (2005). Identification of ferroportin disease in the Indian subcontinent. *Gut 54*, 567–568.

Wallace DF, Clark RM, Harley HAJ, Subramaniam VN. (2004). Autosomal dominant iron overload due to a novel mutation of ferroportin1 associated with parenchymal iron loading and cirrhosis. *J Hepatol 40*, 710–713.

Wallace DF, Pedersen P, Dixon JL, et al. (2002). Novel mutation in ferroportin1 is associated with autosomal dominant hemochromatosis. *Blood 100*, 692–694.

Weatherall DJ, Clegg JB. (2001). *The Thalassaemia Syndromes.* 4th ed. Oxford Blackwell Science, Oxford, England.

West AP, Jr., Bennett MJ, Sellers VM, Andrews NC, Enns CA, Bjorkman PJ. (2000). Comparison of the interactions of transferrin receptor and transferrin receptor 2 with transferrin and the hereditary hemochromatosis protein HFE. *J Biol Chem 275*, 38135–38138.

WHO, UNICEF, UNU. (2001). Iron deficiency anaemia: assessment, prevention, and control (WHO/NHD/01.3). World Health Organization, Geneva.

Yoshida K, Furihata K, Takeda S, et al. (1995). A mutation in the ceruloplasmin gene is associated with systemic hemosiderosis in humans. *Nat Genet 9*, 267–272.

Zhang AS, Xiong S, Tsukamoto H, Enns CA. (2004). Localization of iron metabolism-related mRNAs in rat liver indicate that HFE is expressed predominantly in hepatocytes. *Blood 103*(4), 1509–1514.

Zoller H, McFarlane I, Theurl I, et al. (2005). Primary iron overload with inappropriate hepcidin expression in V162del ferroportin disease. *Hepatology 42*, 466–472.

22.

INHERITED FACTORS IN DIABETES MELLITUS AND RELATED METABOLIC DISORDERS IN THE DEVELOPING WORLD

Dwaipayan Bharadwaj, Nikhil Tandon, and Anubha Mahajan

INTRODUCTION

The syndrome of diabetes mellitus (DM) is clinically and genetically heterogeneous, characterized by hyperglycemia and glucose intolerance due to insulin deficiency, impaired effectiveness of insulin action (resistance), or both. On the basis of etiology and clinical characteristics of the disorder, diabetes mellitus is classified into type 1 diabetes (T1D), type 2 diabetes (T2D), gestational diabetes mellitus, and other specific types like maturity onset diabetes of the young (MODY). This chapter discusses the inherited factors in diabetes mellitus from the developing world perspective, with emphasis on the Indian subcontinent.

The prevalence and incidence of diabetes have reached epidemic proportions globally. An estimated 285 million people have diabetes worldwide, accounting for 7% of the world's population. In developed regions the prevalence is very high and increasing exponentially, but in developing regions, like India, the situation is much more extreme and alarming. It has been recognized that the low- and middle-income countries face the greatest burden of diabetes. India leads the world in the looming epidemic of diabetes and is referred to as the "diabetes capital of the world." According to the 2009 Diabetes Atlas, published by the International Diabetic Federation (IDF), India currently has the highest number, 50.8 million, of adults suffering from diabetes, followed by China with 43.2 million and the United States with 26.8 million. The report projected that 58.7 million people would have diabetes in India by 2010, approximately 7% of the adult population. By 2030, an alarmingly high 8.4% of the Indian adult population is anticipated to suffer from diabetes. These rapid epidemiological shifts are the result of increasing life expectancy, urbanization, and changes in lifestyle associated with physical inactivity. Most prominently, in urban Indians residing in South India, there has been a very steep increase in the prevalence of T2D over the last two decades, from 8.3% in 1992, to 11.6% in 1997, to 15.5% in 2005 (Ramachandran et al., 1992, 1997; Mohan et al., 2005; Gupta & Kumar, 2008). The 2009 IDF report also states that of the estimated 480,000 children with T1D worldwide, South Asians accounts for 114,000 cases (24%).

Despite such a high prevalence of diabetes, large-scale genetic studies have not been carried out in most developing countries. Few studies have been carried out in India, and the true spectrum of genetic factors predisposing the Indian subcontinent populations to DM remains unclear. The Indian population is heterogeneous and is conventionally divided into four different linguistic groups—Austro-Asiatic, Dravidian, Tibeto-Burman, and Indo-European (see Chapters 73 and 74). Individuals residing in the northern and western regions of India generally belong to the Indo-European ethnicity, and those from South India belong to the Dravidian ethnicity. In recent research done on the genetic diversity of Southeast and East Asian populations, it was noted that the genetic subgroups in Asia, including India, had a geographical and linguistic basis, therefore warranting caution when conducting genetic studies in this region due to inherently high population substructure (HUGO Pan-Asian SNP Consortium, 2009; Reich, 2009). A relatively low admixture and high prevalence of endogamy present in this population distinguishes it from most other populations presently used in genetic studies. Furthermore, Asian Indian diabetics exhibit unique phenotypic characteristics that distinguish them from the others, suggesting that unique causative factors underlie this and possibly other, related diseases. This chapter focuses on the current status of the genetics of diabetes in Asian Indians.

Both T1D and T2D are known to be multifactorial diseases, strongly influenced by a combination of multiple genetic loci (inheritance) and environmental factors (diet and lifestyle). The remarkable interethnic differences in T1D and T2D incidence point toward an inheritance component. In contrast, MODY is a class of monogenic form of diabetes, is predominantly due to single gene defects, and accounts for less than 5% of all cases of diabetes. They manifest early, have complete penetrance, and are less influenced by environmental factors. Until a few years back, two major approaches have commonly been used in the search for diabetes genes: the candidate gene approach, and genome-wide linkage studies. Noteworthy progress has been made in the characterization of monogenic forms of diabetes using these techniques, but the study of complex polygenic diabetes has, until recently, been much slower. Significant advances in recent years, following improved genotyping techniques and the completion of the Human Genome Project, have allowed the examination of genetic variations across the genome, commonly known as genome-wide association studies (GWAS). This has led to the identification of several susceptibility loci, offering new insights into the genetic basis of these complex conditions.

GENETIC EPIDEMIOLOGY OF T1D

The type 1 diabetes mellitus (T1D) is comparatively more common in the developed world than in developing countries. The incidence of T1D is variable in different populations, ranging from approximately 20 cases yearly per 100,000 individuals among European populations to only 0.4–1.1 per 100,000 among Asians (Park & Eisenbarth, 2001). Although definitive figures are not available, the incidence of T1D in South Indian regions has been reported to be 10.1–10.6/100,000 per year and no data on T1D prevalence is available for North Indian regions (Ramachandran et al., 1996).

GENETICS OF T1D

T1D is highly heritable, with first-degree relatives of affected cases having a 6% lifetime risk, as compared with a 0.4% risk in the general population. However, the inheritance pattern is complex. Data from twin studies indicate that approximately one-third of the disease susceptibility is genetic. The estimated concordance for T1D in monozygotic twins (30%–50%) is significantly higher than in dizygotic twins (4.8%–27%) and siblings (4.4%–12.5%), suggesting that susceptibility is determined partly by genetic factors. However, the relatively low concordance rate among identical twins suggests that the susceptibility genes have low penetrance (Degnbol & Green, 1978; Tillil & Köbberling, 1987; Kumar et al., 1993; Nerup et al., 1994; Kyvik et al., 1995).

Since the early 1990s, employing various approaches, different studies have identified T1D susceptibility loci scattered throughout the genome. These studies have consistently reported evidence of linkage to the human leukocyte antigens (HLA) region on chromosome 6p21, accounting for almost 50% of the T1D genetic risk (Risch, 1987). Additionally, over 20 genomic intervals (IDDM1, IDDM2, etc.) with variable degrees of linkage evidence have been revealed. Although several of these reported associations remain controversial, a few have been established beyond doubt. These non-HLA loci confer more modest effects: the insulin gene (*INS*) VNTR on 11p15 (Bell et al., 1984; Awata et al., 1997; Kawaguchi et al., 1997), the cytotoxic T-lymphocytes-associated antigen 4 (*CTLA-4*) locus on 2q31 (Marron et al., 1997; Takara et al., 2000), the protein tyrosine phosphatase non-receptor type 22 (*PTPN22*) gene on 1p13 (Bottini et al., 2004; Chelala et al., 2007) and the regions around the interleukin-2 receptor α (*IL2RA/CD25*) on 10p15 (Chistiakov et al., 2008; Qu et al., 2009). With advent of GWAS, more loci have now been uncovered as being robustly associated with T1D. In fact, T1D is one of the few success stories of the GWAS approach, and by early 2009 sixteen loci, including the abovementioned five, had been found to play a major role in T1D pathogenesis (Grant & Hakonarson, 2009). Results of a number of case-control studies conducted on the Indian population have been published and are discussed below.

THE HLA REGION

As mentioned earlier, the major contribution to familial clustering of T1D comes from genes located in the major histocompatibility region (MHC) on chromosome 6p21 (IDDM1), although the specific haplotypes and the strength of genetic associations differ between various ethnic groups.

In several populations, HLA-DR3 and DR4 alleles individually or together have been reported to be strongly associated with T1D. On the other hand, HLA-DR2 and DR15 alleles have been shown to be protective (Rani et al., 1999; Mehra et al., 2002; Witt et al., 2002; Kanga et al., 2004; Rani et al., 2004). Furthermore, the HLA-DQ heterodimer DQ2 (DQB1*0201-DQA1*0501) and DQ8 (DQB1*0302-DQA1*0301) occur in strong linkage disequilibrium with DR3 and DR4 respectively, and have been shown to be strongly associated with T1D, especially when present in the DR3/DR4 heterozygous state (Field, 2002). HLA-DR3-DQ2 and DR4-DQ8 are referred to as "high-risk class II haplotypes" in T1D. In addition to these high-risk haplotypes, other, less common HLA haplotypes have also been associated with diabetes risk in different populations (Erlich et al., 1996; Noble et al., 2000; Cucca et al., 2001). Studies on extended DR3 haplotypes, especially ancestral haplotype (AH) 8.1 (HLA-A1-B8-DR3-DQ2), which is observed at a very high frequency in Caucasian populations, have reported association with T1D.

The incidence of T1D in Asia is very low; hence, it is not easy to find multiplex cases of T1D, or to get a large dataset of incidence cases. Therefore, it is not easy to find a study dissecting genes determining T1D susceptibility applying whole-genome approaches in India. The first HLA study in the Indian cohort with T1D was carried out over two decades ago by Serjeantson et al. (1987), wherein they showed that DR, DQ linkage arrangements in South Indian subjects for DR2, DR4, and DRw6 were different from those commonly seen in Europeans, and implicated at least one DQ β allele in the T1D pathogenesis. Further studies on T1D patients have revealed a positive association of DR3 with the disease, but not of DR4, in eastern and northern Indian subjects (Sanjeevi et al., 1999; Mehra et al., 2002). However, patients from the South India show positive association with the DR4 family of alleles (Kanga et al., 2004). Also, in eastern Indians, DQ2 was seen to associate with T1D (Sanjeevi et al., 1999).

In the North Indian population, multiple DR3-positive haplotypes have been revealed. The widely implicated classical common Caucasian ancestral haplotype-AH8.1 (HLA-B8-DR3) is rare in the Indian population. It has been replaced by other DR3 haplotypes: AH8.2 (HLA-A26-B8-DR3), AH8.3 (HLA-A24-B8-DR3), AH8.4 (A3-B8-DR3), AH8.5 (A31-B8-DR3), AH8.6 (A2-B8-DR3), AH8.7 (A11-B8-DR3), and AH8.8 (A33-B8-DR3). The exact distribution of AH8.2 and its association with T1D in major population groups of India has not been fully elucidated. However, it has been seen that in North Indians, the AH8.2 haplotype is the most common and accounts for 43% of the total B8-DR3 haplotypes, and a significant association of AH8.2 with TID has been observed, along with AH8.3 and AH8.6. Also, B58-DR3 (AH58.1) and B50-DR3 (50.2) have been shown to be positively associated with T1D in the Indian population (Mehra et al., 2007). The mechanism of generation and preference of AH8.2 over AH8.1 in the Indian population is not yet clear. It is postulated that due to a survival advantage of B8-DR3 haplotypes in North Indians, imposed by vigorous immune response to various pathogens, might have led to an independent evolution of AH8.2 and other related haplotypes (Mehra & Kaur, 2006).

Kirk et al. (1985) studied South Indian T1D patients from three different centers and showed that HLA-B8 and BF*F were significantly increased, and C4*A6 decreased, in those patients. Another study revealed an increased frequency of T2ω/T6, Taq 1 DQ β RFLPs and the homozygous 4.6 kb allele of Taq 1 DQ α in the South Indian T1D patients (Hitman et al. 1988).

NON-HLA MARKERS

The second largest contribution in causing T1D has come from the polymorphic repeat in the 5′ regulatory region of the *INS* gene (*IDDM2*). Two discrete classes of *INS* VNTRs have been observed: class I alleles are recessive and predispose to T1D, while class III alleles confer a dominant protective effect (Vafiadis et al., 1997). In North Indian T1D patients, class I homozygosity was significantly increased, and the simultaneous co-occurrence of DRB1*0301 predisposed and conferred a relative risk (RR) of 70.81 (Rani et al., 2004).

CTLA-4 is another T1D candidate, and is also known as *IDDM12*. In North Indian T1D subjects, a promoter region SNP (–1661A) of the *CTLA-4* gene was detected to confer susceptibility (Baniasadi et al., 2006). Another study involving North Indian subjects showed that the variant allele of exonic SNP (49A/G) of the *CTLA-4* gene conferred susceptibility to T1D. In the same study, when patients were categorized on the basis of the age of onset into three groups (age of onset <10 years, 10–20 years, and >20 years), and the genotype frequencies compared, it was observed that the homozygous variant genotype was associated with younger age of onset (unpublished data).

A novel exonic SNP (M55V) located in evolutionally conserved domain of the *SUMO4* gene, present in the IDDM5 interval, has been found to show significant association with susceptibility to T1D (Guo et al., 2004). However, the association results have been inconsistent. Interestingly, while a positive association has been detected in studies with subjects from Asian populations, subjects of European descent show lack of association and, in contrast, an opposite association has been observed in British subjects (Park et al., 2005; Smyth et al 2005; Qu et al., 2005; Wang et al., 2006; Noso et al., 2006). No significant association of this variant was detected in North Indian patients (Sedimbi et al., 2007).

Studies in South Indian T1D subjects have explored the association of T1D with other putative candidates. Investigations have suggested that a polymorphism within or close to the Vitamin D receptor (*VDR*) gene may modify susceptibility in South Indian T1D subjects (McDermott et al., 1997). Further, suggestive evidence for association with T1D was demonstrated for the region around the translation initiation factor 2-α kinase 3 gene (*EIF2AK3*) in South Indian subjects (Allotey et al., 2004). Also, IL1 receptor type 1 gene (*IL1R1*) variants were examined for association with T1D in this ethnic group; however, no association was detected (Metcalfe et al., 1996). Interestingly, suggestive evidence for association with T1D was also detected for the 2q12–21 region telomeric to that harboring *IL1* and *IL1R* gene clusters in the South Indian T1D patients (Ogunkolade et al., 2000).

Although several candidate loci have been examined and implicated in T1D susceptibility in the Indian population, there is still no clear picture of the genetic determinants of the disease. Many of the confirmed candidate loci (from studies on Caucasian subjects) still remain largely unexplored in the context of Indian population.

TYPE 2 DIABETES

The complex pattern of inheritance in T2D, with both gene–gene and gene–environment interactions impacting on disease risk, has made it difficult to elucidate the genetic influences of genes involved, and the progress of discovery has been very slow.

GENETIC PREDISPOSITION TO T2D IN INDIANS

The strong influence of genetics in T2D manifestation is evident from familial aggregation and a high concordance in identical twins. Individuals with T2D have a relatively greater proportion of family members with the disease (10%–30%) than do nondiabetic individuals (1%–6%). Studies have shown 45%–96% concordance for T2D among monozygotic twins and 3%–37% for dizygotic twins (Scheuner et al., 1997).

Indians, an ethnically distinct population, lead the world in the number of people with T2D (Wild et al., 2004; International Diabetes Federation, 2009). Studies have indicated that genetic causes may account for predisposition to insulin resistance and T2D in South Asians (Mather & Keen, 1985; Abate & Chandalia, 2003). A greater familial aggregation and higher prevalence of diabetes in offspring of two diabetic parents is also observed in Indians (Viswanathan et al., 1985; Mohan et al., 1986). Employing nontwin sib concordance rates, Shanker et al. (2007) demonstrated a highly probable genetic component in the causation of T2D in Indians.

Panikar et al. (2008) showed that the age of onset of diabetes is much earlier in the present generation of Indians, being 38.4 years (SD ±9.2) as compared to 55.95 years (SD ±9.98) in the previous generation. A very high proportion, 62.63%, of diabetic siblings had metabolic syndrome (MS) as compared to 28.45% in nondiabetic siblings. In summary, this study found that T2D and MS are both more prevalent, and T2D occurred at an earlier age, in the current generation of Indians.

Another family study of large multiplex families from Chennai, South India, reported on the familial aggregation of quantitative traits associated with T2D in these pedigrees (Mathias et al., 2009). Five hundred twenty-four individuals older than 19 years, from 26 large multiplex pedigrees, were ascertained. The heritability estimates for traits most directly related to T2D were 0.24 ± 0.08 for fasting blood glucose and 0.41 ± 0.09 for fasting insulin. In addition, there was evidence for common genetic control for many pairs of these traits. Bivariate analyses suggested common genes for fasting insulin and central obesity measures (BMI, waist, and hip), with complete genetic correlation between fasting insulin and waist. The study concluded that quantitative traits associated with T2D have heritabilities suggestive of some familial or genetic effect.

In addition to this, various studies on migrant Indians have consistently shown a higher degree of insulin resistance, central body obesity, and diabetes in Indians compared to the indigenous population (Zimmet, 1983; Mohan et al., 1986; Samanta et al., 1987; Sharp et al., 1987; McKeigue et al., 1992; Simmons & Powell, 1993; Misra & Vikram, 2002; Abate & Chandalia, 2003; Barnett et al., 2006). While environmental factors are known to greatly influence disease susceptibility, they alone do not fully explain the differences between ethnic groups. These observations suggest that Indians have a stronger predilection to diabetes, probably due to genetic predisposition.

GENETIC STUDIES FOR T2D IN INDIANS

Until recently, very few well-designed T2D genetic association studies performed on the Indian population were available. Overall, the proportion of studies involving Indian populations compared to Western populations is very small, and the majority of these have been replications, although some demonstrate interesting differences.

GENOME-WIDE LINKAGE APPROACH

Genome-wide linkage scans have led to identification of very few chromosomal regions with significant or consistent correlation to T2D in different populations. Loci that have shown significant linkage to T2D with LOD scores >3.6, and have been supported by evidence for linkage by more than one study, are 1q21-q24, 1q25.3, 2q37.3, 3p24.1, 3q28, 10q26.13, 12q24.31, and 18p11.22 (Florez et al., 2003). However, so far there is only one genome-wide linkage analysis for T2D involving Indian populations (Francke et al., 2001). A 10cM density genome-wide scan with 403 microsatellite markers was performed in a primary set of 99 independent complex families comprising 535 individuals of North Indian origin (whose ancestors had migrated from the port of Calcutta), ascertained from Mauritius. A second cohort comprising 35 multiplex South Indian (Tamil) families ascertained in Pondicherry (India) through at least two T2D-affected individuals was also used. A suggestive linkage with T2D was observed for chromosomal regions 1q44 (MLS-LOD = 2.14) at marker D1S2836, 3q22 (MLS-LOD = 2.06) at marker D3S1292, and 8q23 (MLS-LOD = 1.73) at marker D8S1784 in the Mauritian families. Dense marker analysis led to the improvement in linkage signal at 8q23 (MLS-LOD = 2.55: D8S1779). In the replication cohort of families from Pondicherry, nominal evidence of linkage was detected only at chromosomal region 3q22 (MLS-LOD = 1.36: D3S1292). However, in addition, weak evidence for linkage was detected at the chromosomal region 16p13-pter (MLS-LOD = 1.15) at marker D16S407. In ordered-subset analysis based on family BMI, suggestive evidence for linkage with T2D (MLS = 3.03) was detected at position 273 cM on chromosome 2q37 in the 24 T2D families with the lowest BMI. Other interesting results were also observed on chromosome 8q23, 16q12 and 19p13.3.

CANDIDATE GENE APPROACH

In candidate gene analysis, genes are selected based on their known biological function. Over several decades, extensive attempts have been made to identify T2D-associated variants in genes that encode proteins in the pathways of β-cell development, glucose-induced insulin secretion, insulin signaling, and different mechanisms of peripheral insulin resistance.

Initial genetic studies have been predominantly performed on T2D patients of South Indian origin (Dravidian). These studies analyzed the association of variants of islet amyloid polypeptide (*IAPP*), glucokinase (*GCK*), apolipoprotein D (*ApoD*), *GLUT1*, *GLUT2*, *GLUT4*, tumor necrosis factor-β (*TNF-β*), and insulin receptor substrate-1 (*IRS-1*) for association with T2D, and were hampered by relatively small sample sizes (<250), rendering their findings largely inconclusive (McCarthy et al., 1992, 1993; Baker et al., 1994; Hitman et al., 1995; Pontiroli et al., 1996; Hawrami et al., 1996).

β-CELL DEVELOPMENT AND INSULIN SECRETION

Studies involving South Indian Subjects

Variants from three candidate genes involved in insulin secretion—glucagon receptor (*GCG-R*), mitochondrial tRNALeu (*UUR*), and sulfonylurea receptor (*SUR1*)—were investigated for T2D association in 49 families (567 individuals) with at least two affected siblings, in South Indians hailing from Pondicherry (Lepretre et al., 1998). No evidence of association between any of these commonly found gene variants and T2D or related quantitative traits (including BMI, waist/hip ratio, insulinemia, glycemia, triglycerides, and total cholesterol) was detected.

Uncoupling protein 2 (UCP2) is a negative modulator of glucose-stimulated insulin secretion (Zhang et al., 2001). The exon 8 insertion-deletion polymorphism (45bpI/D) of UCP2 was showed to be associated with BMI in women, in a study conducted on 453 South Indian subjects (Cassell et al., 1999). These findings were replicated in a separate group of 143 South Indian subjects irrespective of sex. Nonetheless, no T2D association was found using either a family-based (85 families) or case-control design (normal glucose tolerance $n = 335$, impaired glucose tolerance $n = 42$, T2D $n = 76$). The polymorphism was again not found to be associated with T2D in another study done recently on a larger sample size in South Indian subjects (487 with T2D, and 919 subjects with normal glucose tolerance ([NGT]; see Vimaleswaran et al., 2009). Two more variants were investigated in this study (−866G/A and Ala55Val) and the Ala55Val variant showed a significant protective effect against T2D with an odds ratio (OR) of 0.72 (adjusted for age, sex, and BMI) for individuals carrying Ala/Val, and 0.37 for individuals carrying Val/Val.

The transcriptional coactivator, peroxisome proliferator activated receptor-γ coactivator (PGC-1α), has been suggested to be involved in the regulation of UCP2 (Ek et al., 2001; Oberkofler et al., 2006). In two separate studies, PGC-1α variants were investigated for association with T2D and body fat in South Indian subjects. In the first study, three variants—Thr394Thr, Gly482Ser, and +A2962G—were investigated for association with T2D in the study group composed of 515 T2D and 882 NGT subjects (Vimaleswaran et al., 2005). The Thr394Thr polymorphism was associated with T2D with OR = 1.68 for individuals with the GA genotype compared with GG genotype taken as the reference. The authors further extended the study and investigated the relation of these polymorphisms with body fat in 82 T2D and 82 NGT subjects (Vimaleswaran et al., 2006). The genotype (0.59 vs 0.32) and allele (0.30 vs 0.17) frequencies of Thr394Thr polymorphism were significantly higher in T2D subjects compared to those in NGT subjects. The OR for diabetes for the susceptible genotype, XA, was 2.53. Visceral and subcutaneous fat were significantly higher in NGT subjects with XA genotype compared to those with GG genotype. Abdominal, central abdominal, and non-abdominal fat were also significantly higher in the NGT subjects with XA genotype compared to those with GG genotype. These studies showed that among South Indian subjects the Thr394Thr polymorphism is associated with T2D, and also with increased total, visceral, and subcutaneous body fat.

The individual and additive role of polymorphisms of three transcription factors involved in β-cell development—neurogenic differentiation-1 (*NEUROD1*-Ala45Thr), neurogenin-3 (*NEUROG3*-Ser199Phe), and hepatic nuclear factor-1α (*TCF1*-Ala98Val)—was investigated in 454 subjects from South India (Jackson et al., 2004). Additionally, 97 South Indian parent–offspring trios were studied. Only the *TCF1* polymorphism was found to be associated with T2D in case-control analysis. Family-based analysis was not performed due to insufficient power resulting from low variant allele frequency. All the three polymorphisms were associated with either fasting blood glucose and/or 2-hr blood glucose in either the total data set or when restricted to NGT subjects. The study demonstrated the additive effect of the three polymorphisms on both fasting blood glucose and/or 2-hr blood glucose, indicating the possible role of these polymorphisms in overall glucose intolerance in the South Indian population.

The role of another regulator of insulin secretion, lipoprotein lipase (*LPL*), was investigated in an urban

South Indian population (619 T2D and 731 NGT subjects) and the variant T93G was shown to be associated with obesity but not T2D, whereas the G53C variant was shown to be protective against both obesity and T2D (Radha et al., 2007).

Studies involving North Indian Subjects

Lately, studies to find T2D susceptibility loci have also been performed on North Indian T2D subjects. In a replication study carried out by Bhat et al. (2007) on subjects from two North Indian regions (Kashmir and Punjab & Jammu: 351 T2D patients and 471 controls), variants Thr394Thr and Gly482Ser of the *PGC-1α* gene were checked for association with T2D. The Thr394Thr variant showed significant association with T2D in both groups, while the Gly482Ser variant was associated with T2D only in the subjects from Kashmir. Comparing their results with the observations made by Vimaleswaran (see above), the authors suggested that perhaps the Thr394Thr variant predisposes most of the Indian population to T2D, while the Gly482Ser variant has population-specific influence.

Mitochondria are known to play an important role in insulin secretion, and genetic variants in mitochondrial genes have been investigated by various groups for their association with T2D (Lowell & Shulman, 2005). Bhat et al. (2007) examined mtDNA variants mt-ND3: G10398A and T16189C for their role in T2D susceptibility in North Indian subjects. This replication study involved two North Indian population cohorts of 312 T2D subjects and 466 controls. A statistically significant association was observed for the 10398A allele in both cohorts (cohort1 [OR = 2.67]; cohort2 [OR = 1.76]). The analysis of G10398A/T16189C haplotypic combinations revealed that the 10398A/16189C haplotype provided a risk in both North Indian cohorts. In the follow-up work done by Rai et al. (2007) checked for an additive interaction between the polymorphisms of the candidate genes—UCP2-866G>A, mt-ND3: 10398G>A and PGC-1α Thr394Thr (rs2970847) and Gly482Ser (rs8192673)—in a total of 1,686 individuals (762 T2D subjects and 924 controls) from North India. A comparison of risk genotype combinations of UCP2-866GG, mtDNA 10398A, and PGC-1α p.Thr394Thr or p.Gly482Ser, against the protective genotypes UCP2-866XA, mtDNA 10398G, and PGC1α p.Thr394Thr (OR = 5.29) or PGC1α p.Gly482Ser (OR = 8.59), showed a highly significant difference and increased ORs.

In another study involving North Indian subjects, for the first time, variants of *FOXA2* were investigated for T2D and related phenotype associations. *FOXA2* is an upstream activator of the β-cell transcription factor network. The study involved 1,031 T2D patients and 625 controls (Tabassum et al., 2008). A strong association of $(TCC)_n$ common allele A5 with T2D (OR = 1.66 for A5 homozygotes) was detected. Obese individuals with A5A5 genotype had enhanced risk when segregated from normal-weight subjects (OR = 1.92). A5 was also nominally associated with higher fasting glucose and lower fasting insulin and C-peptide levels among controls. Also, the common allele (G) of variant rs1055080 provided reduced risk among normal-weight subjects (OR = 0.59). Combination of protective GG and non-risk genotypes of $(TCC)_n$ showed reduced risk of T2D both among normal-weight (OR = 0.43) and obese individuals (OR = 0.47). This was the first report demonstrating that *FOXA2* variants may affect T2D susceptibility and metabolic traits. Therefore, replication studies in other cohorts are required to confirm the findings.

INSULIN SIGNALING

Studies involving South Indian Subjects

In the study conducted by Lepretre et al. in 1998, ten candidate genes were investigated for their association with T2D. Along with a few genes presumably involved in insulin secretion, genes with a possible role in insulin signaling were also examined; viz, variants of *IRS-1*, insulin receptor (*INSR*), human β 3 adrenergic receptor (*Hb3AR*), fatty acid binding protein 2 (*FABP2*), *UCP1* and the glycogen-associated regulatory subunit of protein phosphatase-1 (*PPP1R3*). No evidence of association between any of these gene variants and T2D or related traits was found. Negative association with *FABP2* variant Ala54Thr was also reported in another study on South Indian subjects (773 T2D and 899 NGT; Vimaleswaran et al., 2006). However, this study revealed that the NGT subjects carrying the Thr54 allele had significantly higher 2-hour plasma glucose, glycated hemoglobin, 2-hour insulin, and fasting low-density lipoprotein cholesterol levels compared with those with the Ala54 allele. Carriers of the Thr54 allele in this South Indian population showed associations with MS and hypertriglyceridemia.

Calpain-10 gene (*CAPN10*) was the first important gene associated with T2D to be identified using the genome-wide scanning technique, and calpain-10 has been suggested to influence both insulin secretion and resistance (Horikawa et al., 2000; Sreenan et al., 2001). A haplotype of three important polymorphisms (UCSNP 43, −19, and −63) of this gene was originally shown to be associated with an increased risk of T2D in Mexican-American populations. An evaluation of the possible contribution of this haplotype combination and its constituent haplotypes and variants to increased susceptibility to IFG/IGT and T2D was carried out in a South Indian population (Cassell et al., 2002). Two study groups were used: 95 families ascertained through a proband with T2D, and 468 subjects recruited as part of an urban survey

(69.1% with NGT, 12.8% with IFG/IGT, and 18.2% with T2D). The four-locus haplotype combination 1112/1121 (UCSNP 44, −43, −19, and −63) in South Indians conferred both a 10.7-fold increased risk for IFG/IGT and a 5.78- to 6.52-fold increased risk for T2D in the two study groups. A combination of the 1112 haplotype with the 1221 haplotype also increased the risk for both IFG/IGT and T2D. However, although the original haplotype increases the risk of T2D in South Indians, the frequency of this "at-risk" combination is very low in this population and therefore the study suggests that its contribution to the risk of T2D is likely to be small here.

In another study, the association of insulin receptor substrate-2 (*IRS-2*) G1057D polymorphism with T2D and obesity was investigated in South Indian subjects (Bodhini et al., 2007). The study comprised 1193 NGT subjects and 1018 T2D subjects. The genotype frequency of the G1057D polymorphism was significantly different between the NGT and T2D groups in the total study subjects, as well as among the obese subjects. The DD genotype showed an increased susceptibility to diabetes with an OR (adjusted for age and sex) of 2.19 when compared to the GG + GD genotype among the obese subjects, but not in non-obese subjects. The coefficient corresponding to the interaction parameter (genotype x obesity) was found to be significant. The authors thus concluded that the DD genotype increased susceptibility to T2D by interacting with obesity.

Several studies are available reporting positive association between polymorphisms of the insulin signaling pathways genes, plasma cell membrane glycoprotein (*ENPP1*) and *IRS-1*. A study was carried out to determine their contribution to ethnic susceptibility to insulin resistance among Asian Indians. Variants of *K121Q* and G972A of *ENPP1* and *IRS-1* respectively were analyzed in 638 Asian Indians (originating from India, Pakistan, and Bangladesh) living in Texas in the United States, and 738 Caucasian subjects. Asian Indians harboring the variant *ENPP1* allele had a significantly higher insulin area under the curve, and lower insulin sensitivity, indicating that it was associated with a higher degree of insulin resistance in Asian Indians (Abate et al., 2003). In a later study, the contribution of *ENPP1* variant *K121Q* was evaluated for its role in prediction of T2D in three populations, differing in susceptibility to diabetes and environmental exposure (Abate et al., 2005). These populations included 679 non-migrant South Indians living in Chennai, India (223 with T2D), 1,083 migrant South Asians living in Dallas, Texas, and 858 Caucasians living in Dallas (141 with T2D). The prevalence of subjects carrying the variant allele was 25% vs 34%, 33% vs 45%, and 26% vs 39% in control and diabetic groups for South Asians living in Chennai, in Dallas, and Caucasians respectively. This study thus supported the hypothesis that *ENPP1* 121Q predicts genetic susceptibility to T2D in both South Asians and Caucasians, reinforcing the possibility of it being a functional variant that mediates its effect on the phenotype by modulating insulin signaling.

The contribution to T2D susceptibility of an intronic variant (+10211T>G) in the adiponectin (*ADIPOQ*) gene was evaluated in South Indians (Vimaleswaran et al., 2008). The study comprised 2,000 NGT and 2,000 unrelated T2D subjects, and revealed that among the South Indian subjects this polymorphism was associated with T2D, obesity, and hypoadiponectinemia.

The common Pro12Ala polymorphism in the peroxisome proliferator activator γ (*PPARG*) gene was also studied for its possible role in modulating susceptibility to diabetes in South Asians residing in Dallas, Texas (n = 697), South Indians living in Chennai (n = 1,619), and Caucasians (n = 457) living in Dallas. This polymorphism was present at the same frequency in both subjects with and without diabetes among the South Asians living in Dallas (20% vs. 23%) and in India (19% vs. 19.3%) and was therefore concluded not to associate with either improved insulin sensitivity or decreased risk of T2D (Radha et al., 2006). Recently, the group further analyzed three variants of *PPARG* inclusive of 1279G/A, Pro12Ala, and His478His for association with T2D and its related traits, in 1000 T2D and 1000 NGT subjects from South India, and again found that none of these associated with T2D (Vimaleswaran et al., 2010). However, in the presence of Pro/Pro genotype of the Pro12Ala variant, the −1279G/A promoter variant showed increased susceptibility to T2D (OR = 2.09), whereas in the presence of 12Ala allele, the −1279G/A showed a protective effect against T2D (OR = 0.27). The A-Ala-T (−1279G/A-Pro12Ala-His478His) haplotype was also shown to be associated with a reduced risk of T2D. Although the data again indicated that the *PPARG* gene variants, independently, did not associate with T2D, the 2-loci genotype and the 3-loci haplotype analyses showed a significant association with T2D in South Indian population.

Recently, along with the *UCP2* variants, −55C/T variant of *UCP3* gene was also examined for association with T2D in 1406 South Indian subjects, and was found to have a significant protective effect against T2D (Vimaleswaran et al., 2009). The OR for diabetes for individuals carrying CT and TT was 0.79 and 0.61, respectively. The association of Ala55Val with diabetes was further confirmed through the haplotype analyses.

Vimaleswaran et al. (2007) also evaluated the association of *PGC-1*, *PPARG*, and *UCP1* gene polymorphisms with metabolic syndrome (MS) in South Indian subjects. Among the nine common polymorphisms genotyped in 950 NGT and 550 T2D subjects, only the Thr394Thr variant of the *PGC-1* gene was significantly associated with diabetes and obesity. The heterozygote genotype frequency of Thr394Thr variant was lower in the non-MS group compared to the MS group (16% vs 22%) with an OR of 1.41. A similar trend was observed both in males and females. However, in logistic regression analysis, no independent association of this variant with MS was detected. Hence, the authors concluded that the polymorphisms in the *PGC-1, PPARG* and *UCP1* genes are not associated with MS in South Indian subjects.

Studies involving North Indian Subjects

Very few studies have evaluated the role of genetic variants from the genes involved in insulin signaling in T2D and related traits in North Indian subjects of Indo-European ancestry. In two such studies involving North Indian subjects, the association of variants in *ADIPOQ*, *PPARG 1* and *PPARG 2*, and tumor necrosis factor receptor superfamily 1B (*TNFRSF1B*) genes with T2D has been investigated.

The association of *TNFRSF1B* polymorphisms—rs1061622 (M196R) and rs3397 and $(CA)_n$ repeat—was evaluated with T2D and associated peripheral neuropathy and hypertension in 1,852 North Indian subjects (1,040 T2D subjects and 812 controls; Tabassum et al., 2008). The $(CA)_n$ polymorphism has been shown to be associated with diabetic neuropathy in Caucasians; however, the association could not be replicated in this study. None of the polymorphisms were found to be associated with T2D or associated peripheral neuropathy and hypertension, or any of the T2D-related traits.

Five tag SNPs in the *ADIPOQ* gene and 14 tag SNPs in the *PPARG1* and *PPARG2* genes were examined for their effect on T2D and related subphenotypes in Asian Indian Sikhs of Indo-European ethnicity (Sanghera et al., 2010). A total of 554 T2D subjects and 527 NGT controls were examined, and with the exception of a strong association of PPARG2/Pro12Ala with T2D (OR = 0.13), no other significant association with T2D was revealed in this population. Haplotype analysis did provide strong evidence of protective association of the *ADIPOQ* GA haplotype (rs182052 and rs7649121) with T2D. An association of *ADIPOQ* rs12495941 with total body weight, waist, and hip was also detected, although these associations were not significant after adjusting for multiple testing. These findings strongly suggest that the genetic variations at the *ADIPOQ* locus could contribute to the risk of T2D in Indian Sikhs. Haplotype analysis of the *PPARG* locus provided evidence of association with T2D of two haplotypes harboring rs11715073, rs3892175, and rs1801282. However, conditional analysis indicated that the major effect was driven by Pro12Ala.

GWA LOCI

The advent of high-throughput, cost-effective genotyping technologies has fostered spectacular progress in the identification of loci influencing predisposition to various diseases. A series of publications has convincingly identified and replicated genetic associations for T2D. Around 20 such signals have been identified, including several unsuspected loci with previously unknown functions. Unfortunately, until now, Indian subjects have not been included in any of these GWA studies. Nonetheless, a number of investigations have been undertaken by various groups to replicate the findings of these studies in Indians, a high-risk population.

In separate studies, replication of various GWA loci were carried out in Khatri Sikh subjects from North India. In the first such study, Sanghera et al. (2008a) sought to confirm the already reported association of six *TCF7L2* variants in 556 T2D subjects and 537 controls from North India. They reported replication of association of four of the six SNPs with T2D (rs7903146, rs11196205, rs10885409, and rs4918789) under a dominant model, conferring ORs of 1.39, 1.44, 1.57, and 1.36, respectively. Further evidence of association was provided by showing a significant difference between haplotype frequencies in cases and controls. The study also provided new information about the significant impact of *TCF7L2* gene variants on plasma cholesterol levels that appear to be independent of BMI.

Another study investigated the role of the nine most significant SNPs reported in various GWA studies—*PPARG2*: rs1801282; *IGF2BP2*: rs4402960; *CDK5*: rs7754840; *SLC30A8*: rs13266634; *CDKN2A*: rs10811661; *HHEX*: rs1111875; *TCF7L2*: rs10885409; *KCNJ11*: rs5219; and *FTO*: rs9939609—for T2D association in 918 individuals consisting of 532 T2D and 386 NGT subjects (Sanghera et al., 2008b). Four of the nine SNPs revealed a significant association with T2D—*PPARG2* (OR = 0.12), *IGF2BP2* (OR = 1.37), *TCF7L2* (OR = 1.64) and *FTO* (OR = 1.46)—after adjusting for age, sex, and BMI. Interestingly, in North Indian subjects the association of the *FTO* variant did not seem to be mediated entirely through BMI, as was the case in Caucasians (Frayling et al., 2007). The risk allele of *CDK5*: rs7754840 was significantly associated with decreased HDL cholesterol levels in both NGT and combined (NGT and T2D) groups. The risk allele of *TCF7L2*: rs10885409 was associated with increased LDL-cholesterol in the NGT group, and total and LDL-cholesterol levels in the combined cohort.

This group later also attempted to confirm the association of the six loci (*NOTCH2, THADA, ADAMTS9, JAZF1, CDC123/CAMKID* and *TSPAN8/LGRS*) identified to be associated with T2D through a meta-analysis on three GWAs (Sanghera et al., 2009). The highly significant variants from each locus were investigated in 680 T2D subjects and 637 NGT controls. Only *CDC123/CAMKID* (rs12779790) replicated earlier evidence of association with T2D under a dominant model (observed ratio, OR = 1.27) during initial testing. However, this association did not remain significant after multiple testing corrections. The same variant in the *CDC123/CAMKID* independently revealed a marked decrease in fasting insulin levels among the risk allele carriers in NGT controls and in T2D subjects, as well as in the combined sample after adjusting for covariates. Evidence of impaired β-cell function was also observed among risk allele carriers in T2D subjects and in a combined cohort. The authors could not confirm the role of the remaining variants with risk either for T2D or quantitative phenotypes measuring insulin secretion or insulin resistance.

They also evaluated the association of the two variants (rs12970134 and rs4450508) near the melanocortin 4 receptor (*MC4R*) and the nonsynonymous (Gln241His) variant (rs3812316) in MLX interacting protein-like (*MLXIPL*) gene with T2D and obesity-related traits in 745 T2D subjects and 783 controls (Been et al., 2010). The association of *MC4R* (rs12970134) with BMI, total weight, and waist circumference was successfully replicated. However, the association of *MLXIPL* variant rs3812316 with triglyceride levels could not be replicated. These observations demonstrated that the genetic variation in *MC4R* locus could have a moderate contribution in the regional fat deposition and development of central obesity in Asian Indians.

In the recent past, a few studies investigating genetic predisposition toward T2D and related traits have also been taken up in Asian Indian subjects living abroad. However, these studies have ignored the inherent heterogeneity that exists among these subpopulations, and have taken a mix of Indian subjects from different subpopulations. Following are the few reports of that kind.

In a large cohort of Asian Indian subjects presently residing in the United Kingdom, a study was conducted examining the role in various metabolic traits of a large number of variants. The research was carried out as part of the London Life Sciences Population (LOLIPOP) study of environmental and genetic causes of cardiovascular disease, diabetes, and obesity in approximately 30,000 U.K. citizens of Indian Asian and European ancestry. Indian Asians were selected if all four grandparents originated from the Indian subcontinent. A GWA was carried out (318,237 SNPs) for insulin resistance and related phenotypes in 2,684 Indian Asian men from the LOLIPOP study, with further testing in individuals of Indian Asian (n = 7,394) or European ancestry (n = 4,561; Chambers et al 2008). Association was detected for rs12970134 near *MC4R* with waist circumference and, independently, with insulin resistance. Homozygotes for the risk allele of rs12970134 had ~2 cm increased waist circumference ~10% higher HOMA-IR compared with wild type. Risk allele frequencies of rs12970134 were higher among individuals of Indian Asian ancestry than those of European ancestry. The increased frequency of the risk allele in Indian Asians has been proposed as an explanation for the increased burden of central adiposity and insulin resistance in this group.

In another study, 267,000 SNPs were tested in 1,005 Northern Europeans and 248,000 in 1,006 Indian Asians, with replication in 10,536 subjects to identify the genetic polymorphisms associated with component phenotypes of the MS viz. triglyceride and HDL cholesterol level (Kooner et al., 2008). From the 180,410 successfully genotyped variants, rs1558861 and rs17120139, both flanking the *APOA1-APOC3-APOA4-APOA5* gene cluster on chromosome 11q23, were associated with triglycerides in Asian Indian men. Also, variants in the *MLXIPL* region (rs3812316) and *LPL* cluster (rs326) were associated with raised triglyceride levels. Variants of the *CETP* (rs7205804), *LPL* (rs326), and *LIPC* gene (rs11858164) were found to associate with HDL cholesterol levels. The authors proposed that since MLXIPL coordinates transcriptional regulation of enzymes that channel glycolytic end-products into lipogenesis and energy storage, *MLXIPL* is a plausible "thrifty gene."

Another GWA was carried out for fasting glucose in 6,776 nondiabetic and 698 T2D Indian Asians (Chambers et al., 2009). Three SNPs (rs2166706, rs1387153, and rs10830963) associated with glucose among Indian Asians, all near melatonin receptor *MTNR1B*. The variant rs2166706 was in moderate LD with rs1387153 (r^2 = 0.60) and rs10830963 (r^2 = 0.45). Risk allele frequency and effect sizes for rs2166706 were similar among Indian Asians and European Caucasians. SNP rs2166706 was also associated with T2D in Indian Asians (OR = 1.21). SNPs at the rs4607517 (*GCK*), rs1260326 (*GCKR*) and rs560887 (*G6PC2*) loci were also associated with glucose among Indian Asians. Genetic variation at the *MTNR1B*, *GCK*, *GCKR*, and *G6PC2* loci may contribute to abnormal glucose metabolism and related metabolic disturbances among Indian Asians.

Newton-Cheh et al. (2009) tested 2.5 million genotyped and imputed SNPs for association with systolic and diastolic blood pressure in a large and diverse sample set. Twelve SNPs selected from the first stage were analyzed for association in the 12,899 individuals from Indian Asian ancestry (from the LOLIPOP study). The study replicated association of rs16998073 at 4q21 near fibroblast growth factor 5 (FGF5), and rs11191548 at 10q24 near *CYP17A1*. However, association of rs1378942 at *CYP1A2* could not be replicated, probably due to limited power to detect the modest effect size, differences in LD patterns in Indian Asians compared to Europeans, or simply lack of association in subjects of Indian Asian ancestry. The authors concluded that these associations offer mechanistic insights into the regulation of blood pressure, and may point to novel targets for interventions to prevent cardiovascular diseases.

Tan et al. (2010) examined the effect of eight loci identified through GWA studies (*CDKAL1, CDKN2A/B,*

IGF2BP2, HHEX, SLC30A8, PKN2, LOC387761, and *KCNQ1*) on risk of T2D in three major ethnic groups in Asia, inclusive of Asian Indians living in Singapore (364 controls and 246 T2D subjects). The combined analysis of the three ethnic groups revealed significant associations between SNPs in *CDKAL1* (OR = 1.13), *CDKN2A/B* (OR = 1.16), *HHEX* (OR = 1.14), and *KCNQ1* (OR = 1.16) with T2D. *SLC30A8* (OR = 1.06) showed association only after adjustment for gender and BMI. Failure to detect these effects across different populations may be due to issues of power owing to limited sample size, lower minor allele frequency, or differences in genetic effect sizes. In another study, this group examined the association between four previously reported *KCNQ1* polymorphisms with insulin resistance, β-cell function, and other T2D-related traits in 3,734 subjects inclusive of 521 Asian Indians (Tan et al., 2009). However, no association was detected for any of the variant in Asian Indians, maybe due to limited sample size.

MODY GENETICS

Much of the understanding of MODY, the most common form of monogenic diabetes, comes from studies in the Caucasian population. To date, it is known that MODY can result from mutations of at least six different genes. One of them encodes the glycolytic enzyme glucokinase, while the other five are transcription factors: hepatocyte nuclear factor 4α (*HNF4A*), *HNF1A*, insulin promoter factor-1 (*IPF1*), *HNF1B*, and neurogenic differentiation 1/B-cell E-box transactivator 2 (*NEUROD1*). Protein products of all these genes are involved in pancreatic β-cell metabolism, causing decreased insulin secretion, and are responsible for MODY forms 2, 1, 3, 4, 5, and 6, respectively. A few MODY families have still not been associated with any of the MODY genes identified so far. Therefore, it can be assumed that new genes are still to be identified.

Studies involving the Indian population are very rare, and there is a considerable clinical overlap between T2D with MODY, making it difficult to estimate its true prevalence. In a study on South Indian subjects, the prevalence of MODY was estimated to be around 4.8% (Mohan et al., 1985).

MODY3, resulting from mutations in the *HNF1A* gene, is the commonest form of MODY worldwide (up to 60%). To date, more than 200 different mutations in *HNF1A* have been identified in patients of different ethnic backgrounds. However, not many MODY3 mutations have been described in Indian patients. Anuradha et al. (2005) examined the prevalence of Ala98Val polymorphism of the *HNF1A* gene in different T2D in South Indians. The analysis was performed in the following groups: *(1)* MODY, defined as non-insulin-dependent diabetes (age at onset <25 years) and vertical transmission of diabetes through at least three generations (n = 122); *(2)* very-early-onset T2D (age at onset <25 years) without family history (n = 23); *(3)* early-onset T2D (age at onset between 26 and 40 years, n = 171); *(4)* late-onset T2D (age at onset >40 years, n = 133); *(5)* T1D (n = 150); and *(6)* normal glucose tolerance (n = 130). The frequency of the Val allele was significantly higher in MODY patients compared with control groups. In a recent study of South Indian subjects (96 clinically diagnosed MODY patients and 57 controls), mutation screening in *HNF1A* revealed the presence of nine mutations and twelve polymorphisms, of which seven mutations and two polymorphisms were novel (Radha et al., 2009). Functional studies revealed reduced transcriptional activity of the *HNF1A* promoter for two promoter variants. Cosegregation of the Arg263His coding region mutation was observed in eight members of one MODY family, whereas it was absent in nondiabetic subjects of this family. This study demonstrated that mutations in *HNF1A* comprised about 9% of clinically diagnosed MODY patients, suggesting the need to explore other MODY genes to understand the spectrum of the different types of MODY seen in the Asian Indian population. Due to its high prevalence in Caucasian subjects, the mutation P291fsinsC is known as a mutational "hotspot" (Kaisaki et al., 1997). However, its prevalence has not been reported in Indian subjects. A study of 32 clinically diagnosed MODY patients from North India also failed to detect its presence. In the same study, another missense mutation (R200W) that was previously described in European and Japanese studies was investigated, and its prevalence was found to be 3% among the study population (Sahu et al., 2007).

As in the case of abovementioned studies, it is seen that most of the studies on MODY have been replication studies. The variants shown to be associated in other populations are replicated here, and the results obtained are predominantly negative. However, when new variants in the same gene are analyzed, association signals are detected. These studies have not been replicated in any other subgroup of South Asians. Considering that many of the genes involved in MODY are also candidate genes for polygenic T2D, there is a need for more studies in Indian Asians.

GENES ASSOCIATED WITH DIABETES COMPLICATIONS

Diabetes is associated with a series of complications, such as nephropathy, retinopathy, and cardiovascular disease. In addition to a genetic predisposition to developing diabetes, accumulating evidence suggests a genetic component for developing these complications as well. Although a wide range of genes have been assessed for their association with increased risk of such complications, only a few studies have revealed significant and consistent results. Furthermore, the predisposition to diabetic complications also varies significantly between ethnic groups. In general, the prevalence of diabetic nephropathy and retinopathy is higher in South Asians than in Caucasians (Mather et al., 1998; Raymond et al., 2009). There are reports on familial clustering of diabetic retinopathy and nephropathy in Indian subjects (Rema et al., 2002; Vijay et al., 1999). Furthermore, there is evidence indicating that Indian T2D patients, as well as U.K.-migrated Indians, have a higher prevalence of nephropathy than Caucasians (Mather et al., 1998; Chandie Shaw et al., 2006). Various groups have explored genetic variants of different candidate genes for association with diabetic complications; their findings, however, have not been conclusive and need to be verified in larger cohorts.

RETINOPATHY

Various studies have been undertaken to investigate the genetic variants located in different candidate genes predisposing to diabetic retinopathy in South Indian and North Indian subjects. Similar to the genetic studies examining association of variants with T2D, the initial studies done in this direction have been performed on limited sample sizes and have been underpowered.

In two different studies performed on South Indian T2D subjects (n = 138, 154 respectively), Hawrami et al. (1991, 1996) investigated association of diabetic retinopathy with six candidate loci, namely, HLA-DQB1, HLA-DQA1, HLA-DRA, INS-VNTR (the switch region of the immunoglobulin IgM heavy chain gene), and microsatellite dinucleotide $(GT)_n$ upstream of *TNF-β*. The polymorphism of the heavy chain immunoglobulin genes located on chromosome 14 and the *TNF-β* repeat were detected to be associated with retinopathy.

Polymorphisms of the receptor for AGE (*RAGE*) gene (–429T/C, –374T/A, and 63 bp deletion) were evaluated for association with diabetic retinopathy in South Indian subjects (NGT n = 149, diabetes with retinopathy [DR] n = 189, and diabetes without retinopathy [DNR] n = 190; Ramprasad et al., 2007). The –374 polymorphism was found to be associated with nonproliferative retinopathy when this subgroup was compared to the DNR group (OR = 1.81). However, this association did not remain significant when both the subphenotypes of DR (the nonproliferative and proliferative DR groups) were studied jointly.

The TNF-β $(GT)_n$ repeat was also studied in Asian Indian subjects (mixed South and North Indian) with 15-year duration of T2D, including subjects with DR (n = 100) and DNR (n = 107), and the length of the repeat was correlated with susceptibility for the development of retinopathy (Kumaramanickavel et al., 2001). Allele 4 (103 bp) had a higher prevalence (9.81%) in the DNR group compared to that (2.5%) in the DR group. Even after separating the allele 4 data into those of North and South Indian subjects, results were still significant. Patients with retinopathy and allele 8 (111 bp) had a tendency to develop proliferative diabetic retinopathy (PDR). It was therefore suggested that in Indian subjects, allele 4 is a low-risk allele for developing retinopathy and allele 8 (111 bp) is associated with PDR.

This group further investigated the association of a pentanucleotide microsatellite repeat, located 2.5 kb upstream of the transcription start site of the inducible nitric oxide synthase (*iNOS*) gene and Gly82Ser polymorphism in exon 3 of *RAGE* gene, with retinopathy in Asian Indians (Kumaramanickavel, 2002a, 2002b). In the 199 individuals genotyped, eleven alleles, 175–225 bp, were identified for the *iNOS* repeat, and the individuals carrying allele 210 bp had twice the risk of developing retinopathy compared with those who did not carry this allele (OR = 2.03). Also, alleles of 200 and 220 bp were found to protect an individual from developing retinopathy or its complications. The frequency of the Ser82 allele was observed to be significantly higher—18% in the DNR group (n = 100) compared to 7% in the DR group (n = 100). The same genotype was 2% in the control group (n = 50), suggesting that *RAGE* Ser82 allele is a low-risk allele for developing DR in Asian Indian T2D subjects.

Kumaramanickavel et al. (2003) also studied the association between the $(CA)_n$ polymorphism located 2.1kb upstream of the promoter of the aldose reductase (*ALR2*) gene in T2D Asian Indian subjects, with (n = 105) and without (n = 109) retinopathy. In these subjects, 13 alleles were identified and the Z-2 allele (136 bp) showed an association with the DR group (13.81%) when compared with the DNR group (7.34%). The Z-2 allele also showed a significant association with those DR subjects who had PDR and maculopathy. The Z-2 allele was, therefore, concluded to be a high-risk allele for diabetic retinopathy in the Asian Indian patients.

NEPHROPATHY

Among the candidate genes examined for susceptibility to diabetic nephropathy in Indian population, angiotensin-converting enzyme (*ACE*) gene is one. The ins/del polymorphism of this gene was studied in South Indian T2D subjects (nephropathic DNP n = 86, and normoalbuminuric patients n = 23; Viswanathan et al., 2001). In the nephropathic patients, ins/del and del/del genotypes were present in 52.3% and 27.9% of the patients, respectively, as compared to 34.8% and 21.7%, respectively, in those with normoalbuminuria. The del allele was present in 80.2% of the DNP group and 56.5% of the normoalbuminuric patients (OR = 3.12), showing a positive association between the del allele and diabetic proteinuria in

South Indian T2D subjects. Another study in the South Indian population showed an association between *ACE* del/del genotype and hypertensive T2D (Bhavani et al., 2005). The polymorphism was studied for association with T2D and/or hypertension in 200 hypertension patients, 100 T2D subjects, and 200 age- and sex-matched controls from South India. The del/del genotype was strongly associated with hypertension (OR = 2.02), and remained so even when patients with T2D were excluded from the analysis (OR = 2.07).

In North Indian subjects, different candidate genes have been evaluated for their role in diabetic nephropathy. Ahluwalia et al. (2008) examined the association of –786T>C, 894G>T, and a 27 bp repeat polymorphism in Intron 4 (27VNTR) of the constitutive endothelial nitric oxide synthase gene (*eNOS*) with T2D nephropathy in T2D subjects with nephropathy (n = 195) and in T2D subjects without nephropathy (n = 255). The homozygous minor genotype of –786T>C, 894G>T and aa of 27VNTR were found to be associated with higher risk of nephropathy (OR = 5.5, OR = 1.8, and OR = 6.2, respectively). Haplotype with all minor alleles was associated with higher risk of diabetic nephropathy (OR = 2.6). The serum nitric oxide levels were observed to be significantly lower in minor allele carriers of T-786C and G894T within each of the subject groups (with and without nephropathy). The authors concluded that the *eNOS* gene locus is associated with diabetic nephropathy, and the functional polymorphisms (–786T>C, 894G>T) might lead to a decreased expression of the *eNOS* gene.

In another study, the combined effect of eight polymorphisms of five pro-inflammation cytokine genes (*TGFB1*: T869C [Leu10Pro] and Tyr81His; *CCL2*: A-2518G and Insertion/Deletion [I/D]; *CCR5*: Insertion/Deletion [I/D] and G59029A; *IL8*: T-251A; *MMP9*: Arg279Gln [G>A]) were investigated for association with diabetic nephropathy in two independently recruited groups of T2D subjects and controls (Ahluwalia et al., 2009a,b). The first group consisted of 255 unrelated individuals with T2D and no nephropathy (DM), and 240 consecutive diabetic nephropathy patients (DN) from North India. Replication studies were performed on 92 DM and 96 DN patients recruited from South India. The *IL8*, *CCL2*, *CCR5*, and *MMP9* polymorphisms were found to be associated with increased risk of diabetic nephropathy in both North and South Indian T2D subjects. Furthermore, the co-occurrence of risk-associated genotypes (*CCL2*, *CCR5*, and *MMP9*) conferred a tenfold increased risk of nephropathy among T2D subjects.

In a recent study on North Indian subjects, the association of *ACE* haplotypes and the interactions of *ACE*, angiotensinogen (*AGT*), and angiotensin II receptor type 1 (*AGTR1*) gene polymorphisms with DNP was evaluated (Ahluwalia et al., 2009a,b). Interaction of the seven variants with each other and with ACE ins/del polymorphism was assessed in 440 individuals (240 DNP and 200 DM patients). The del/del genotype of *ACE* ins/del polymorphism was found to be an independent risk factor for DNP, as evidenced by a twofold increase in DNP risk. The risk of events was significantly associated with the *ACE* ins/del polymorphism even after adjustments for potential confounders, such as gender, age, and duration of diabetes. The findings were consistent with the observations made earlier by Viswanathan et al. (2001), where a positive association had been reported between the del allele of the *ACE* polymorphism and diabetic nephropathy in South Indian T2D patients.

In North Indian subjects, association of genetic variants of proteins involved in lipid metabolism—namely, cholesteryl ester transfer protein (CETP: TaqI B and D442G) and apolipoprotein (APOE: Hha I)—were investigated for their role in determining the susceptibility for complications associated with T2D in 136 T2D subjects and 264 healthy controls (Dixit, 2005). Although *CETP* TaqI B polymorphism was not associated with the T2D, B1B2 genotype was significantly associated with high risk of hypertension in diabetic patients (OR = 3.07). In North Indians, the D442G variation in the *CETP* gene was found to be absent. Frequency of *APOE* HhaI polymorphism was also not different between patients and controls. Lipid levels did not show any correlation with the *CETP* TaqI B and *APOE* Hha I polymorphisms. The results suggest that *CETP* TaqI B and *APOE* HhaI polymorphisms may not be associated with T2D in North Indians; however, *CETP* TaqI B polymorphism may be associated with hypertension accompanying T2D.

There are significant regional variations in prevalence of diabetic chronic renal insufficiency (CRI) in India. The contribution of renin-angiotensin-aldosterone system (RAAS) polymorphisms to CRI among T2D Asian Indian subjects was investigated by Prasad et al. (2006). Twelve polymorphisms from six genes—namely, renin (*REN*), angiotensinogen (*ATG*), angiotensin-converting enzyme I (*ACE*), angiotensin II type 1 receptor (*AT1*), the aldosterone synthase (*CYP11B2*) gene from the RAAS pathway and one from the chymase pathway—were tested for their association with diabetic CRI in 196 T2D subjects with CRI, and 225 T2D subjects with no evidence of renal disease. Glu53Stop in *AGT* and A>T (–777) in *AT1* genes were found to be monomorphic. Met235Thr:*ATG* (OR = 2.68), –344:*CYP11B2* (OR = 1.57), and –1903:chymase (OR = 2.06) were significantly associated with diabetic chronic renal insufficiency in Indian patients, although replication in larger sample sets is warranted. The authors suggested that use of such markers for prediction of susceptibility to diabetes-specific renal disease in the ethnic Indian population appears promising.

In a similar study design, Prasad et al. (2007) tested association of nine polymorphisms transforming growth factor β1 (*TGFβ1*), tumor necrosis factor α (*TNFα*), and chemokine receptor 2 and 5 (*CCR2* and *CCR5*) with CRI. Polymorphisms Tyr81His and Thr263Ile in the *TGFβ1* gene were monomorphic, and Arg25Pro in the *TGFβ1* gene and Δ32 polymorphism in the *CCR5* gene were minor variants (MAF<0.05). A minor allele of 59029G>A SNP of *CCR5* was observed to confer predisposition to development of diabetic CRI (OR = 1.39). In CRI subjects, a compound group of genotypes "GA and AA" of SNP G>A −800 was found to confer predisposition for proliferative retinopathy (OR = 3.03) among Asian Indians.

Twelve polymorphisms from five more genes—dopamine receptor 1 (*DRD1*), *DRD2*, *DRD3*, *DRD4*, and catechol-*O*-methyltransferase (*COMT*)—were then examined for association with diabetic CRI using the same sample set (Prasad et al., 2008). Polymorphisms of the *DRD2* gene (−141 ins/del C and intronic G>A) and *COMT* gene (900 ins/del C) were found to predispose the Asian Indian population to CRI. The authors stated that, if confirmed, this observation would be consistent with a suggested role of dopamine metabolism in disease occurrence.

Twenty-six polymorphisms from 13 genes from the oxidative stress pathway were analyzed for association with CRI in North Indian ($n = 90$) and South Indian ($n = 194$) T2D subjects with CRI, and diabetic subjects with no evidence of renal disease (North Indian $n = 75$ and South Indian $n = 149$; Tiwari et al., 2009). Genes included were superoxide dismutases (*SOD1, 2, 3*), *UCP1, 2*, endothelial nitric oxide synthase (*NOS3*), glutathione-S-transferases (GST) (*M1, T1, P1*), vascular endothelial growth factor (*VEGF*), paraoxonase (*PON*) 1 and 2, and nicotinamide adenine dinucleotide phosphate reduced, oxidase p22(*phox*). In South Indian subjects there was significant association of the wild-type allele in *SOD2* (Ala9Val), *UCP1* (−112 T>G; Ala64Thr), *NOS3* (Glu298Asp) and *GSTP1* (Ile105Val) genes with development of CRI. However, none of these observations were replicated in the North Indian subjects. Association was observed for two other markers, *VEGF* (−460 T>C) and *PON1* (Arg192Gly) among North Indian diabetic CRI subjects. The nonreplication of association suggests differential genetic susceptibility of the two populations to diabetic chronic renal insufficiency.

A few studies have also been conducted to decipher the role of genetic variants in susceptibility to peripheral neuropathy, dyslipidemia, and MS. The Ala54Thr polymorphism in the *FABP2* gene, as well as the T-455C and C-482T polymorphisms in the apolipoprotein C-III (*APOC3*) gene promoter, were evaluated in a small number (70 controls and 110 subjects with T2D) of South Indian subjects for association with MS and dyslipidemia (Guettier et al., 2005). Controls carrying FABP2 Thr54 were more likely to have MS than noncarriers (OR = 6.9). Those carrying at least one polymorphic allele in both genes had a higher likelihood of having MS than wild type (OR = 12.1). Dyslipidemia was associated with the polymorphism as well. The association of the polymorphisms with MS and dyslipidemia shows that they could contribute to the high cardiovascular disease prevalence in the high-risk South Indian population. In a larger sample sized study (T2D subjects $n = 773$, IGT $n = 899$) conducted later, Vimaleswaran et al. (2006) also observed that carriers of the Thr54 allele showed associations with MS and hypertriglyceridemia in this South Indian population.

Three common functional polymorphisms of tumor necrosis α (*TNFα*) −308G/A, interferon γ (*IFNγ*) +874A/T and interleukin (*IL*) *10* −1082G/A were examined for their association with peripheral neuropathy in diabetic peripheral neuropathy (DPN) in South Indian subjects (Kolla et al., 2009). The polymorphisms in *IFNγ* (OR = 2.0) and *IL-10* (OR = 2.9) genes, but not the *TNFα* gene, were significantly associated with peripheral neuropathy in South Indian T2D subjects. The study shows that the "high-producer" *IL*-10 −1082 G/G genotype and the "low-producer" *IFNγ* +874 A/A genotype may be responsible for the downregulation of immune response leading to inflammation in this setting.

The prevalence of gestational diabetes mellitus has increased considerably during the last decade. However, genetic studies exploring the role of genetic variants in its manifestation have not been performed yet in the Indian population.

CONCLUSIONS

India harbors the greatest number of diabetic subjects as compared to any other country, but the genetic basis of diabetes in the Indian population is poorly understood. It is known that the genetic makeup of Asian Indians is different from that of other populations. It is known that in a polygenic complex disease, the contribution of individual variations may be small, and therefore gene–gene and gene–environment interactions may play a more pronounced role. However, comprehensive studies, where multiple alleles and environmental effects have been examined simultaneously for determining the risk of diabetes mellitus and related traits, are still minimal in the Indian population. Several studies done on relatively small sample sizes have concentrated mainly on replicating the findings of the other populations, and have met with small success.

Although differences in the incidence and subtypes of T1D and T2D are observed between Caucasian and Asian Indian populations, the association of candidate genes with both diseases is generally similar in both populations. HLA are consistently associated with T1D in both populations, but apparent differences in disease-associated alleles and haplotypes are observed because of differences in the presence and absence of disease-associated alleles in the general population. In North Indians, DR3-DQ2 is the dominant HLA locus, unlike Caucasians, wherein DR4-DQ8 is the dominant association. Also, the widely implicated classical common Caucasian haplotype AH8.1 is rare in Indians and has been replaced by AH8.1v, which has been shown to be associated with T1D in North Indians.

With regards to T2D, the replications of signals identified by GWA studies in Caucasian populations are providing "similar" though not identical hits. *TCF7L2* variants have been consistently seen to be associated with T2D in all populations worldwide, inclusive of Indians. The observation made in the case of the *FTO* gene, where the T2D susceptibility effect is not mediated by BMI/obesity in the Indian population, in contrast to that observed in Caucasians, is also very important. Also, the association seen for the first time with a β-cell pathway gene, *FOXA2*, in the Indian population, highlights the fact that while there are broad areas of similarity, there are equally important differences—which could, in part, explain differences in the presentation of T2D in Indians (younger onset, at lower BMI, etc.) compared with Caucasians.

There is a need to carry out more large-scale, hypothesis-free studies to actually pinpoint the susceptibility markers in this population. A large number of samples in each ethnic group, and comparative studies in different populations, including GWA studies, will contribute to the identification of susceptibility genes for diabetes mellitus, enable better understanding of the disease mechanisms, and help to establish effective methods for disease prediction, prevention, and intervention.

REFERENCES

Abate N, Carulli L, Cabo-Chan A Jr, Chandalia M, Snell GP, Grundy SM. (2003). Genetic polymorphism PC-1 K121Q and ethnic susceptibility to insulin resistance. *J Clin Endocrinol Metab 88*, 5927–5934.

Abate N, Chandalia M. (2003). The impact of ethnicity on type 2 diabetes. *J Diabetes Complications 17*, 39–58.

Abate N, Chandalia M, Satija P, Adams-Huet B, Grundy SM, Sandeep S, Radha V, Deepa R, Mohan V. (2005). ENPP1/PC-1 K121Q polymorphism and genetic susceptibility to type 2 diabetes. *Diabetes 54*(4), 1207–1213.

Ahluwalia TS, Ahuja M, Rai TS, Kohli HS, Sud K, Bhansali A, Khullar M. (2008). Endothelial nitric oxide synthase gene haplotypes and diabetic nephropathy among Asian Indians. *Mol Cell Biochem 314*(1–2), 9-17.

Ahluwalia TS, Ahuja M, Rai TS, et al. (2009a). ACE Variants Interact with the RAS Pathway to Confer Risk and Protection against Type 2 Diabetic Nephropathy. *DNA Cell Biol 28*(3), 141–150.

Ahluwalia TS, Khullar M, Ahuja M, Kohli HS, Bhansali A, Mohan V, Venkatesan R, Rai TS, Sud K, Singal PK. (2009b). Common variants of inflammatory cytokine genes are associated with risk of nephropathy in type 2 diabetes among Asian Indians. *PLoS One 4* (4), e5168.

Allotey RA, Mohan V, McDermott MF, et al. (2004) The EIF2AK3 gene region and type I diabetes in subjects from South India. *Genes Immun 5*(8), 648–652.

Anuradha S, Radha V, Deepa R, et al. (2005). A prevalent amino acid polymorphism at codon 98 (Ala98Val) of the hepatocyte nuclear factor-1alpha is associated with maturity-onset diabetes of the young and younger age at onset of type 2 diabetes in Asian Indians. *Diabetes Care 28*(10), 2430–2435.

Awata T, Kurihara S, Kikuchi C, et al. (1997). Evidence for association between the class I subset of the insulin gene minisatellite (IDDM2 locus) and IDDM in the Japanese population. *Diabetes 46*(10), 1637–1642.

Baker WA, Hitman GA, Hawrami K, et al. (1994). Apolipoprotein D gene polymorphism: a new genetic marker for type 2 diabetic subjects in Nauru and south India. *Diabet Med 11*(10), 947–952.

Baniasadi V, Narain N, Goswami N, Das SN. (2006). Promoter region -318 C/T and -1661 A/G CTLA-4 single nucleotide polymorphisms and type 1 diabetes in North Indians. *Tissue Antigen 67*(5), 383–389.

Barnett AH, Dixon AN, Bellary S, Hanif MW, et al. (2006). Type 2 diabetes and cardiovascular risk in the UK South Asian community. *Diabetologia 49*(10), 2234–2246.

Been LF, Nath SK, Ralhan SK, et al. (2010). Replication of association between a common variant near melanocortin-4 receptor gene and obesity-related traits in Asian Sikhs. *Obesity 18*(2), 425–429.

Bell GI, Horita S, Karam JH. (1984). A polymorphic locus near the human insulin gene is associated with insulin-dependent diabetes mellitus. *Diabetes 33*(2), 176–183.

Bhat A, Koul A, Rai E, Sharma S, Dhar MK, Bamezai RN. (2007). PGC-1alpha Thr394Thr and Gly482Ser variants are significantly associated with T2DM in two North Indian populations: a replicate case-control study. *Hum Genet 121*(5), 609–614.

Bhat A, Koul A, Sharma S, et al. (2007). The possible role of 10398A and 16189C mtDNA variants in providing susceptibility to T2DM in two North Indian populations: a replicative study. *Hum Genet 120*(6), 821–826.

Bhavani BA, Padma T, Sastry BKS, Krishna Reddy N, Nausheen K. (2005). The Insertion I/Deletion D polymorphism of Angiotensin-Converting Enzyme (ACE) Gene Increase the Susceptibility to Hypertension and/or Diabetes. *Int J Hum Genet 5*(4), 247–252.

Bodhini D, Radha V, Deepa R, et al. (2007) The G1057D polymorphism of IRS-2 gene and its relationship with obesity in conferring susceptibility to type 2 diabetes in Asian Indians. *Int J Obes (Lond) 31*(1), 97–102.

Bottini N, Musumeci L, Alonso A, et al. (2004). A functional variant of lymphoid tyrosine phosphatase is associated with type I diabetes. *Nat Genet 36*(4), 337–338.

Cassell PG, Jackson AE, North BV, et al. (2002). Haplotype combinations of calpain-10 gene polymorphisms associate with increased risk of impaired glucose tolerance and type 2 diabetes in South Indians. *Diabetes 51*(5), 1622–1628.

Cassell PG, Neverova M, Janmohamed S, et al. (1999). An uncoupling protein 2 gene variant is associated with a raised body mass index but not Type II diabetes. *Diabetologia 42*(6), 688–692.

Chambers JC, Elliott P, Zabaneh D, et al. (2008). Common genetic variation near MC4R is associated with waist circumference and insulin resistance. *Nat Genet 40*(6), 716–718.

Chambers JC, Zhang W, Zabaneh D, et al. (2009). Common genetic variation near melatonin receptor MTNR1B contributes to raised plasma glucose and increased risk of type 2 diabetes among Indian Asians and European Caucasians. *Diabetes 58*(11), 2703–2708.

Chandie Shaw PK, Baboe F, van Es LA, et al. (2006). South-Asian type 2 diabetic patients have higher incidence and faster progression of renal disease compared with Dutch-European diabetic patients. *Diabetes Care 29*, 1383–1385.

Chelala C, Duchatelet S, Joffret ML, et al. (2007). PTPN22 R620W functional variant in type 1 diabetes and autoimmunity related traits. *Diabetes 56*(2), 522–526.

Chistiakov DA, Voronova NV, Chistiakov PA. (2008). The crucial role of IL-2/IL-2RA-mediated immune regulation in the pathogenesis of type 1 diabetes, an evidence coming from genetic and animal model studies. *Immunol Lett 118*(1), 1–5.

Cucca F, Dudbridge F, Loddo M, et al. (2001). The HLA-DPB1 associated component of the IDDM1 and its relationship to the major loci HLA-DQB1, -DQA1, and -DRB1. *Diabetes 50*(5), 1200–1205.

Degnbol B, Green A. (1978). Diabetes mellitus among first- and second-degree relatives of early onset diabetics. *Ann Hum Genet 42*(1), 25–47.

Dixit M, Bhattacharya S, Mittal B. (2005). Association of CETP TaqI and APOE polymorphisms with type II diabetes mellitus in North Indians: a case-control study. *BMC Endocr Disord 5*, 7.

Ek J, Andersen G, Urhammer SA, et al. (2001). Mutation analysis of peroxisome proliferator activated receptor-γ coactivator-1 (PGC-1) and relationships of identified amino acid polymorphisms to Type II diabetes mellitus. *Diabetologia 44*, 2220–2225.

Erlich HA, Rotter JI, Chang JD, et al. (1996). Association of HLA-DPB1*0301 with IDDM in Mexican-Americans. *Diabetes 45*(5), 610–614.

Field LL. (2002). Genetic linkage and association studies of type 1 diabetes, challenges and rewards. *Diabetologia* 45, 21–35.

Florez JC, Hirschhorn J, Altshuler D. (2003). The inherited basis of diabetes mellitus, implications for the genetic analysis of complex traits. *Annu Rev Genomics Hum Genet 4*, 257–291.

Francke S, Manraj M, Lacquemant C, et al. (2001). A genome-wide scan for coronary heart disease suggests in Indo-Mauritians a susceptibility locus on chromosome 16p13 and replicates linkage with the metabolic syndrome on 3q27. *Hum Mol Genet 10*(24), 2751–2765.

Frayling TM, McCarthy MI. (2007). Genetic studies of diabetes following the advent of the genome-wide association study: where do we go from here? *Diabetologia 50*(11), 2229–2233.

Frayling TM, Timpson NJ, Weedon MN, et al. (2007). A common variant in the FTO gene is associated with body mass index and predisposes to childhood and adult obesity. *Science 316*(5826), 889–894.

Grant SF, Hakonarson H. (2009). Genome-wide association studies in type 1 diabetes. *Curr Diab Rep 9*(2), 157–163.

Guettier JM, Georgopoulos A, Tsai MY, et al. (2005). Polymorphisms in the fatty acid-binding protein 2 and apolipoprotein C-III genes are associated with the metabolic syndrome and dyslipidemia in a South Indian population. *J Clin Endocrinol Metab 90*(3), 1705–1711.

Guo D, Li M, Zhang Y, et al. (2004). A functional variant of SUMO4, a new I kappa B alpha modifier, is associated with type 1 diabetes. *Nat Genet 36*, 837–841.

Gupta R, Kumar P. (2008). Global diabetes landscape-type 2 diabetes mellitus in South Asia, Epidemiology, risk factors, and control. *Insulin 3*(2), 78–94.

Hawrami K, Hitman GA, Rema M, et al. (1996). An association in non-insulin-dependent diabetes mellitus subjects between susceptibility to retinopathy and tumor necrosis factor polymorphism. *Hum Immunol 46*(1), 49–54.

Hawrami K, Mohan R, Mohan V, Hitman GA. (1991). A genetic study of retinopathy in south Indian type 2 (non-insulin-dependent) diabetic patients. *Diabetologia 34*(6), 441–444.

Hitman GA, Hawrami K, McCarthy MI, et al. (1995). Insulin receptor substrate-1 gene mutations in NIDDM; implications for the study of polygenic disease. *Diabetologia 38*(4), 481–486.

Hitman GA, Karir PK, Sachs JA, et al. (1988). HLA-D region RFLPs indicate that susceptibility to insulin-dependent diabetes in South India is located in the HLA-DQ region. *Diabet Med 5*(1), 57–60.

Horikawa Y, Oda N, Cox NJ, et al. (2000). Genetic variation in the gene encoding calpain-10 is associated with Type 2 diabetes mellitus. *Nature Genetics 26*(2), 163–175.

HUGO Pan-Asian SNP Consortium, Indian Genome Variation Consortium, et al. (2009). Mapping human genetic diversity in Asia. *Science 326*(5959), 1541–1545.

Jackson AE, Cassell PG, North BV, et al. (2004). Polymorphic variations in the neurogenic differentiation-1, neurogenin-3, and hepatocyte nuclear factor-1alpha genes contribute to glucose intolerance in a South Indian population. *Diabetes 53*(8), 2122–2125.

Kaisaki PJ, Menzel S, Lindner T, et al. (1997). Mutations in the hepatocyte nuclear factor-1α gene in MODY and early-onset NIDDM: Evidence for a mutational hotspot in exon 4. *Diabetes 46*, 528–535.

Kanga U, Vaidyanathan B, Jaini R, Menon PS, Mehra NK. (2004). HLA haplotypes associated with type 1 diabetes mellitus in North Indian children. *Hum Immunol* 65, 47–53.

Kawaguchi Y, Ikegami H, Shen GQ, et al. (1997). Insulin gene region contributes to genetic susceptibility to, but may not to low incidence of, insulin-dependent diabetes mellitus in Japanese. *Biochem Biophys Res Commun 233*(1), 283–287.

Kirk RL, Ranford PR, Serjeantson SW, Thompson AR, et al. (1985). HLA, complement C2, C4, properdin factor B and glyoxalase types in South Indian diabetics. *Diabetes Res Clin Pract 1*(1), 41–47.

Kolla VK, Madhavi G, Pulla Reddy B, et al. (2009). Association of tumor necrosis factor alpha, interferon gamma and interleukin 10 gene polymorphisms with peripheral neuropathy in South Indian patients with type 2 diabetes. *Cytokine 47*(3), 173–177.

Kooner JS, Chambers JC, Aguilar-Salinas CA, et al. (2008). Genome-wide scan identifies variation in MLXIPL associated with plasma triglycerides. *Nat Genet 40*(2), 149–151.

Kumar D, Gemayel NS, Deapen D, Kapadia D, Yamashita PH, Mack TM. (1993). North-American twins with IDDM. Genetic, etiological, and clinical significance of disease concordance according to age, zygosity, and the interval after diagnosis in first twin. *Diabetes 42*, 1351–1363.

Kumar R, Thompson EB. (2003). Transactivation functions of the N-terminal domains of nuclear hormone receptors, protein folding and coactivator interactions. *Mol Endocrinol 17*, 1–10.

Kumaramanickavel G, Ramprasad VL, Sripriya S, Upadyay NK, Paul PG, Sharma T. (2002a). Association of Gly82Ser polymorphism in the RAGE gene with diabetic retinopathy in type II diabetic Asian Indian patients. *J Diabetes Complications 16*(6), 391–394.

Kumaramanickavel G, Sripriya S, Ramprasad VL, Upadyay NK, Paul PG, Sharma T. (2003). Z-2 aldose reductase allele and diabetic retinopathy in India. *Ophthalmic Genet 24*(1), 41–48.

Kumaramanickavel G, Sripriya S, Vellanki RN, et al. (2001). Tumor necrosis factor allelic polymorphism with diabetic retinopathy in India. *Diabetes Res Clin Pract 54*(2), 89–94.

Kumaramanickavel G, Sripriya S, Vellanki RN, et al. (2002b). Inducible nitric oxide synthase gene and diabetic retinopathy in Asian Indian patients. *Clin Genet 61*(5), 344–348.

Kyvik KO, Green A, Beck-Nielsen H. (1995). Concordance rates of insulin dependent diabetes mellitus: A population based study of young Danish twins. *BMJ 311*, 913–917.

Lacquemant C, Chikri M, Boutin P, Samson C, Froguel P. (2002). No association between the G482S polymorphism of the proliferatoractivated receptor-γ coactivator-1 (PGC-1) gene and Type II diabetes in French Caucasians. *Diabetologia 45*, 602–603.

Leprêtre F, Vionnet N, Budhan S, et al. (1998). Genetic studies of polymorphisms in ten non-insulin-dependent diabetes mellitus candidate genes in Tamil Indians from Pondichery. *Diabetes Metab 24*(3), 244–250.

Lowell BB, Shulman GI. (2005). Mitochondrial dysfunction and type 2 diabetes. *Science* 307, 384–387.

Marron MP, Raffel LJ, Garchon HJ, et al. (1997). Insulin-dependent diabetes mellitus (IDDM) is associated with *CTLA-4* polymorphisms in multiple ethnic groups. *Hum Mol Genet 6*(8), 1275–1282.

Mather HM, Chaturvedi N, Kehely AM, et al. (1998). Comparison of prevalence and risk factors for microalbuminuria in South Asians and Europeans with Type 2 diabetes mellitus. *Diabet Med 15*(8), 672–677.

Mather HM, Keen H. (1985). The Southall Diabetes Survey: prevalence of known diabetes in Asians and Europeans. *Br Med J 291*, 1081–1084.

Mathias RA, Deepa M, Deepa R, Wilson AF, Mohan V. (2009). Heritability of quantitative traits associated with type 2 diabetes mellitus in large multiplex families from South India. *Metabolism 58*(10), 1439–1445.

McCarthy MI, Hitchins M, Hitman GA, et al. (1993). Positive association in the absence of linkage suggests a minor role for the glucokinase gene in the pathogenesis of type 2 (non-insulin-dependent) diabetes mellitus amongst south Indians. *Diabetologia 36*(7), 633–641.

McCarthy MI, Hitman GA, Mohan V, Ramachandran A, Snehalatha C, Viswanathan M. (1992). The islet amyloid polypeptide gene and non-insulin-dependent diabetes mellitus insouth Indians. *Diabetes Res Clin Pract 18*(1), 31–34.

McDermott MF, Ramachandran A, Ogunkolade BW, et al. (1997). Allelic variation in the vitamin D receptor influences susceptibility to IDDM in Indian Asians. *Diabetologia 40*, 971–975.

McKeigue PM, Pierpoint T, Ferrie JE, Marmot MG. (1992). Relationship of glucose intolerance and hyperinsulinaemia to body fat pattern in south Asians and Europeans. *Diabetologia 35*, 785–791.

Mckeigue PM. (1992). Coronary heart disease in Indians, Pakistanis and Bangladeshis, aetiology and possibilities for prevention. *Br Heart J 67*, 341–342.

Mehra N, Kaur G. (2006). HLA genetics and disease with particular reference to Type 1 diabetes and HIV infection in Asian Indians. *Expert Rev Clin Immunol 2*(6), 901–913.

Mehra NK, Kaur G, Kanga U, Tandon N. (2002). Immunogenetics of autoimmune diseases in Asian Indians. *Ann NY Acad Sci 958*, 333–336.

Mehra NK, Kumar N, Kaur G, Kanga U, Tandon N. (2007). Biomarkers of susceptibility to type 1 diabetes with special reference to the Indian population. *Indian J Med Res 125*(3), 321–344.

Metcalfe KA, Hitman GA, Pociot F, et al. (1996). An association between type 1 diabetes and the interleukin-1 receptor type 1 gene. The DiMe Study Group. Childhood Diabetes in Finland. *Hum Immunol 51*(1), 41–48.

Misra A, Vikram NK. (2002). Insulin resistance syndrome (metabolic syndrome) and Asian Indians. *Curr Sci 83*, 1483–1496.

Mohan V, Deepa R, Deepa M, Somannavar S, Datta M. (2005). A simplified Indian Diabetes Risk Score for screening for undiagnosed diabetic subjects. *J Assoc Physicians India 53*, 759–763.

Mohan V, Ramachandran A, Snehalatha C, Mohan R, Bharani G, Viswanathan M. (1985). High prevalence of maturity-onset diabetes of the young (MODY) among Indians. *Diabetes Care 8*, 371–374.

Mohan V, Sharp PS, Aber VR, Mather HM, Kohner EM. (1986). Family histories of Asian Indian and Europeans non-insulin-dependent diabetic patients. *Pract Diabetes 3*, 254–256.

Mohan V, Sharp PS, Cloke HR, Burrin JM, Schumer B, Kohner EM. (1986). Serum immunoreactive insulin responses to a glucose load in Asian Indian and European Type 2 (non-insulin-dependent) diabetic patients and control subjects. *Diabetologia 29*, 235–237.

Nerup J, Mandrup-Poulsen T, Helqvist S, Andersen HU, Pociot F, Lorenzen T, et al. (1994) On the pathogenesis of IDDM. *Diabetologia 37*, S82–S89.

Newton-Cheh C, Johnson T, Gateva V, Wellcome Trust Case Control Consortium, et al. (2009). Genome-wide association study identifies eight loci associated with blood pressure. *Nat Genet 41*, 666–676.

Noble JA, Valdes AM, Thomson G, Erlich HA. (2000). The HLA class II locus DPB1 can influence susceptibility to type 1 diabetes. *Diabetes 49*(1), 121–125.

Noso S, Ikegami H, Fujisawa T, et al. (2006). Association of SUMO4, as a candidate gene for IDDM5, with susceptibility to type 1 diabetes in Asian populations. *Ann N Y Acad Sci 1079*, 41–46.

Oberkofler H, Klein K, Felder TK, Krempler F, Patsch W. (2006). Role of peroxisome proliferator-activated receptor-gamma coactivator-1alpha in the transcriptional regulation of the human uncoupling protein 2 gene in INS-1E cells. *Endocrinology 147*(2), 966–976.

Ogunkolade WB, Ramachandran A, McDermott MF, et al. (2000). Family association studies of markers on chromosome 2q and Type 1 diabetes in subjects from South India. *Diabetes Metab Res Rev 16*(4), 276–280.

Panikar VK, Joshi SR, Kakraniya P, Nasikkar N, Santavana C. (2008). Inter-generation comparison of type-2 diabetes in 73 Indian families. *J Assoc Physicians India 56*, 601–604.

Park Y, Eisenbarth GS. (2001). Genetic susceptibility factors of Type 1 diabetes in Asians. *Diabetes Metab Res Rev 17*(1), 2–11.

Park Y, Park S, Kang J, et al. (2005). Assessing the validity of the association between the SUMO4 M55V variant and risk of type 1 diabetes. *Nat Genet 37*, 112.

Pontiroli AE, Capra F, Veglia F, et al. (1996). Genetic contribution of polymorphism of the GLUT1 and GLUT4 genes to the susceptibility to type 2 (non-insulin-dependent) diabetes mellitus in different populations. *Acta Diabetol 33*(3), 193–197.

Prasad P, Tiwari AK, Kumar KM, et al. (2006). Chronic renal insufficiency among Asian Indians with type 2 diabetes: I. Role of RAAS gene polymorphisms. *BMC Med Genet 7*, 42.

Prasad P, Tiwari AK, Kumar KM, Ammini AC, Gupta A, Gupta R, Thelma BK. (2007). Association of TGFbeta1, TNFalpha, CCR2 and CCR5 gene polymorphisms in type-2 diabetes and renal insufficiency among Asian Indians. *BMC Med Genet 8*, 20.

Prasad P, Kumar KM, Ammini AC, Gupta A, Gupta R, Thelma BK. (2008). Association of dopaminergic pathway gene polymorphisms with chronic renal insufficiency among Asian Indians with type-2 diabetes. *BMC Genet 9*, 26.

Qu H, Bharaj B, Liu XQ, et al. (2005). Assessing the validity of the association between the SUMO4 M55V variant and risk of type 1 diabetes. *Nat Genet 37*, 111–112.

Qu HQ, Bradfield JP, Bélisle A, Grant SF, Hakonarson H, Polychronakos C; Type I Diabetes Genetics Consortium. (2009). The type I diabetes association of the IL2RA locus. *Genes Immun 10* Suppl 1, S42–S48.

Radha V, Ek J, Anuradha S, Hansen T, Pedersen O, Mohan V. (2009). Identification of novel variants in the hepatocyte nuclear factor-1alpha gene in South Indian patients with maturity onset diabetes of young. *J Clin Endocrinol Metab* 94(6), 1959–1965.

Radha V, Mohan V. (2007). Genetic predisposition to Type 2 diabetes among Asian Indians. *Indian J Med Res 125*(3), 259–274.

Radha V, Vimaleswaran KS, Babu HN, et al. (2006a). Role of genetic polymorphism peroxisome proliferator-activated receptor-gamma2 Pro12Ala on ethnic susceptibility to diabetes in South Asian and Caucasian subjects: evidence for heterogeneity. *Diabetes Care 29*(5), 1046–1051.

Rai E, Sharma S, Koul A, Bhat AK, Bhanwer AJ, Bamezai RN. (2007). Interaction between the UCP2-866G/A, mtDNA 10398G/A and PGC1alpha p.Thr394Thr and p.Gly482Ser polymorphisms in type 2 diabetes susceptibility in North Indian population. *Hum Genet 122*(5), 535–540.

Ramachandran A, Snehalatha C, Dharmaraj D, Viswanathan M. (1992). Prevalence of glucose intolerance in Asian Indians. Urban-rural difference and significance of upper body adiposity. *Diabetes Care 15*(10), 1348–1355.

Ramachandran A, Snehalatha C, Krishnaswamy CV. (1996). Incidence of IDDM in children in urban population in southern India. Madras IDDM Registry Group Madras, South India. *Diabetes Res Clin Pract 34*(2), 79–82.

Ramachandran A, Snehalatha C, Latha E, Vijay V, Viswanathan M. (1997). Rising prevalence of NIDDM in an urban population in India. *Diabetologia 40*(2), 232–237.

Ramachandran A, Snehalatha C, Viswanathan V, Viswanathan M, Haffner SM. (1997). Risk of non-insulin-dependent diabetes mellitus conferred by obesity and central obesity in different ethnic groups. A comparative analysis between Asian Indians, Mexican Americans and whites. *Diabetes Res Clin Pract 36*, 121–125.

Ramachandran A, Snehlatha C, Abdul Khadeer OMS, Annie Joseph T, Viswanathan M. (1992). Prevalence of childhood diabetes in urban population in South India. *Diab Res Clin Pract 17*, 227–231.

Ramachandran A, Snehlatha C, Krishnaswamy CV. (1996). Incidence of IDDM in children in urban poulation in Southern India. *Diab Res Clin Pract 34*, 79–82.

Ramprasad S, Radha V, Mathias RA, Majumder PP, Rao MR, Rema M. (2007). Rage gene promoter polymorphisms and diabetic retinopathy in a clinic-based population from South India. *Eye (Lond) 21*(3), 395–401.

Rani R, Sood A, Goswami R. (2004). Molecular basis of predisposition to develop type 1diabetes mellitus in North Indians. *Tissue Antigens 64*, 145–155.

Rani R, Sood A, Lazaro AM, Stastny P. (1999). Associations of MHC class II alleles with insulin-dependent diabetes mellitus (IDDM) in patients from North India. *Hum Immunol 60*(6), 524–531.

Raymond NT, Varadhan L, Reynold DR, et al. (2009). Higher prevalence of retinopathy in diabetic patients of south Asian ethnicity compared to white Europeans in the community: a cross sectional study. *Diabetes Care 32*(3), 410–415.

Reich D, Thangaraj K, Patterson N, Price AL, Singh L. (2009). Reconstructing Indian population history. *Nature 24*, 489–494.

Rema M, Saravanan G, Deepa R, Mohan V. (2002). Familial clustering of diabetic retinopathy in south Indian Type 2 diabetic patients. *Diabet Med 19*, 910–916.

Risch N. (1987). Assessing the role of HLA-linked and unlinked determinants of disease. *Am J Hum Genet 40*(1), 1–14.

Sahu RP, Aggarwal A, Zaidi G, et al. (2007). Etiology of early-onset type 2 diabetes in Indians, islet autoimmunity and mutations in hepatocyte nuclear factor 1alpha and mitochondrial gene. *J Clin Endocrinol Metab 92*(7), 2462–2467.

Samanta A, Burden AC, Fent B. (1987). Comparative prevalence of non-insulin-dependent diabetes mellitus in Asian and white Caucasian adults. *Diabetes Res Clin Pract 4*(1), 1–6.

Sanghera DK, Been L, Ortega L, et al. (2009). Testing the association of novel meta-analysis-derived diabetes risk genes with type II diabetes and related metabolic traits in Asian Indian Sikhs. *J Hum Genet 54*(3), 162–168.

Sanghera DK, Demirci FY, Been L, et al. (2010). PPARG and ADIPOQ gene polymorphisms increase type 2 diabetes mellitus risk in Asian Indian Sikhs: Pro12Ala still remains as the strongest predictor. *Metabolism 59*, 492–501..

Sanghera DK, Nath SK, Ortega L, et al. (2008a). TCF7L2 polymorphisms are associated with type 2 diabetes in Khatri Sikhs from North India: genetic variation affects lipid levels. *Ann Hum Genet 72*, 499–509.

Sanghera DK, Ortega L, Han S, et al. (2008b). Impact of nine common type 2 diabetes risk polymorphisms in Asian Indian Sikhs: PPARG2 (Pro12Ala), IGF2BP2, TCF7L2 and FTO variants confer a significant risk. *BMC Med Genet 9*, 59–67.

Sanjeevi CB, Kanungo A, Shtauvere A, Samal KC, Tripathi BB. (1999). Association of HLA class II alleles with different subgroups of diabetes mellitus in Eastern India identify different associations with IDDM and malnutrition-related diabetes. *Tissue Antigens 54*, 83–87.

Scheuner MT, Wang SJ, Raffel LJ, Larabell SK, Rotter JI. (1997). Family history, a comprehensive genetic risk assessment method for the chronic conditions of adulthood. *Am J Med Genet 71*(3), 315–324.

Sedimbi SK, Kanungo A, Shastry A, Park Y, Sanjeevi CB. (2007). No association of SUMO4 M55V with autoimmune diabetes in Asian-Indian patients. *Int J Immunogenet 34*(2), 137–142.

Serjeantson SW, Ranford PR, Kirk RI, et al. (1987). HLA-DR and DQ DNA genotyping in insulin-dependent diabetes patients in South India. *Dis Markers 5*, 101–108.

Shanker J, Kanjilal S, Rao VS, et al. (2007). Adult nontwin sib concordance rates for type 2 diabetes, hypertension and metabolic syndrome among Asian Indians: the Indian Atherosclerosis Research Study. *Vasc Health Risk Manag 3*(6), 1063–1068.

Sharp PS, Mohan V, Levy JC, Mather HM, Kohner EM. (1987). Insulin resistance in patients of Asian Indian and European origin with non-insulin dependent diabetes. *Horm Metab Res 19*, 84–85.

Simmons D, Powell MJ. (1993). Metabolic and clinical characteristics of south Asians and Europeans in Coventry. *Diabet Med 10*, 751–758.

Smyth DJ, Howson JM, Lowe CE, et al. (2005). Assessing the validity of the association between the SUMO4 M55V variant and risk of type 1 diabetes. *Nat Genet 37*, 110–111.

Sreenan SK, Zhou YP, Otani K, et al. (2001). Calpains play a role in insulin secretion and action. *Diabetes 50*(9), 2013–2020.

Tabassum R, Chavali S, Dwivedi OP, Tandon N, Bharadwaj D. (2008). Genetic variants of FOXA2: risk of type 2 diabetes and effect on metabolic traits in North Indians. *J Hum Genet 53*(11–12), 957–965.

Tabassum R, Chavali S, Mahajan A, et al. (2008). Association analysis of TNFRSF1B polymorphisms with type 2 diabetes and its related traits in North India. *Genomic Med 2*(3–4), 93–100.

Takara M, Komiya I, Kinjo Y, et al. (2000). Association of CTLA-4 gene A/G polymorphism in Japanese type 1 diabetic patients with younger age of onset and autoimmune thyroid disease. *Diabetes Care 23*(7), 975–978.

Tan JT, Ng DP, Nurbaya S, et al. (2010). Polymorphisms Identified through Genome-Wide Association Studies and Their Associations with Type 2 Diabetes in Chinese, Malays, and Asian-Indians in Singapore. *J Clin Endocrinol Metab 95*, 390–397.

Tan JT, Nurbaya S, Gardner D, Ye S, Tai ES, Ng DP. (2009). Genetic variation in KCNQ1 associates with fasting glucose and beta-cell function: a study of 3,734 subjects comprising three ethnicities living in Singapore. *Diabetes 58*(6), 1445–1449.

Tillil H, Köbberling J. (1987). Age-corrected empirical genetic risk estimates for first-degree relatives of IDDM patients. *Diabetes 36*(1), 93–99.

Tiwari AK, Prasad P, Thelma BK, et al. (2009). Oxidative stress pathway genes and chronic renal insufficiency in Asian Indians with Type 2 diabetes. *J Diabetes Complications 23*(2), 102–111.

Vafiadis P, Bennett ST, Todd JA, et al. (1997). Insulin expression in human thymus is modulated by INS VNTR alleles at the IDDM2 locus. *Nat Genet 15*(3), 289–292.

Vijay V, Snehalatha C, Sinha K, Lalitha S, Ramachandran A. (1999). Familial aggregation of diabetic kidney disease in type 2 diabetes in south India. *Diabetes Res Clin Pract 43*, 167–171.

Vimaleswaran KS, Radha V, Anjana M, et al. (2006). Effect of polymorphisms in the PPARGC1A gene on body fat in Asian Indians. *Int J Obes (Lond) 30*(6), 884–891.

Vimaleswaran KS, Radha V, Deepa R, Mohan V. (2007). Absence of Association of Metabolic Syndrome with PPARGC1A, PPARG and UCP1 Gene Polymorphisms in Asian Indians. *Metab Syndr Relat Disord 5*(2), 153–162.

Vimaleswaran KS, Radha V, Ghosh S, et al. (2005). Peroxisome proliferator-activated receptor-gamma co-activator-1alpha (PGC-1alpha) gene polymorphisms and their relationship to Type 2 diabetes in Asian Indians. *Diabet Med 22*(11), 1516–1521.

Vimaleswaran KS, Radha V, Ghosh S, Majumder PP, Rao MR, Mohan V. (2009). Association of uncoupling protein 2 and 3 gene polymorphisms with type 2 diabetes in Asian Indians. *J Endocrinol Invest* (in press).

Vimaleswaran KS, Radha V, Jayapriya MG, et al. (2010). Evidence for an association with type 2 diabetes mellitus at the PPARG locus in a South Indian population. *Metabolism 59*, 457–462.

Vimaleswaran KS, Radha V, Mohan V. (2006). Thr54 allele carriers of the Ala54Thr variant of FABP2 gene have associations with metabolic syndrome and hypertriglyceridemia in urban South Indians. *Metabolism 55*(9), 1222–1226.

Vimaleswaran KS, Radha V, Ramya K, et al. (2008). A novel association of a polymorphism in the first intron of adiponectin gene with type 2 diabetes, obesity and hypoadiponectinemia in Asian Indians. Hum Genet *123*(6), 599–605

Viswanathan M, Mohan V, Snehalatha C, Ramachandran A. (1985). High prevalence of type 2 (non-insulin-dependent) diabetes among the offspring of conjugal type 2 diabetic parents in India. *Diabetologia 28*, 907–910.

Viswanathan V, Zhu Y, Bala K, et al. (2001). Association between ACE gene polymorphism and diabetic nephropathy in South Indian patients. *JOP 2*(2), 83–87.

Wang CY, Podolsky R, She JX. (2006). Genetics and functional evidence supporting SUMO4 as a type 1 diabetes susceptibility gene. *Ann N Y Acad Sci 1079*, 257–267.

Wild S, Roglic G, Green A, Sicree R, King H. (2004). Global prevalence of diabetes, Estimates for the year 2000 and projections for 2030. *Diabetes Care 27*, 1047–1053.

Witt CS, Price P, Kaur G, et al. (2002). The common HLA-B8-DR3 haplotype in northern India is different from that found in Europe. *Tissue Antigens 60*, 474–480.

Zhang CY, Baffy G, Perret P, et al. (2001). Uncoupling protein-2 negatively regulates insulin secretion and is a major link between obesity, ß cell dysfunction, and type 2 diabetes. *Cell 105*, 745–755.

Zimmet P, Taylor R, Ram P, et al. (1983). Prevalence of diabetes and impaired glucose tolerance in the biracial (Melanesian and Indian) population of Fiji, a rural-urban comparison. *Am J Epidemiol 118*, 673–688.

23.

GLOBAL HEALTH IMPLICATIONS OF GENETIC STUDIES OF OBESITY AND TYPE 2 DIABETES

Jessica Buxton, Julia El-Sayed Moustafa, Haya Al-Saud, and Alexandra Blakemore

INTRODUCTION

Type 2 diabetes (T2D) and obesity are two of the most significant health problems currently facing high-income and middle-income countries and, increasingly, developing nations. An estimated 285 million adults worldwide are affected by T2D, and this figure has been predicted to rise to 439 million by 2030 (Shaw et al., 2010). The resulting increase in mortality and morbidity rates due to common diabetes-related complications, such as renal failure, cardiovascular disease, neuropathy, and retinopathy, will be substantial. In 2005, an estimated 1.1 million people died from diabetes, but this figure is predicted to double by 2030 (WHO, 2009). Much of the recent rise in prevalence of T2D can be attributed to the concurrent increase in obesity rates worldwide, together often termed the "diabesity" epidemic. Over 400 million adults worldwide are obese (BMI >30kg/m^2), a figure that is expected to reach 700 million by 2015, with obesity levels also rising rapidly in children (WHO, 2006). In addition to its role in T2D, obesity is associated with an increased risk of cardiovascular disease, hypertension, stroke, and certain cancers.

While the prevalence of both T2D and obesity is rising worldwide, the recent increase in obesity observed in developing nations, and predicted sharp rise in T2D in regions such as the Middle East, sub-Saharan Africa, Southeast Asia, and India, are a particular cause for concern (Hossain et al., 2007, Shaw et al., 2010). Until recently, the leading causes of morbidity and mortality in the developing world have been infectious diseases and malnutrition, with much of the worldwide aid effort being aimed at the eradication of these conditions. However, the impact of long-term chronic conditions in these regions, some of which are associated with dietary and lifestyle changes, is becoming increasingly evident (Miranda et al., 2008). Developing nations now make up a significant proportion of the top ten countries with the highest proportions of individuals with BMI >30kg/m^2 (Ono et al., 2005; Figure 23-1). Larger population sizes in developing nations means that an increasingly large proportion of obese individuals reside in the developing world, placing a significant burden on the healthcare systems and economies of affected countries (Kelly et al., 2008).

The rise in levels of obesity has been attributed to the increasingly sedentary lifestyle and high-calorie diet adopted by many populations around the world. However, in addition to these "obesogenic" environmental influences, there is a strong genetic component to both obesity and T2D, which contributes to differences in individual risk *within* populations; some individuals carry genetic variants that make them more sensitive to an obesogenic environment, and it is these people who are at increased risk of obesity. Furthermore, the relationship between obesity and T2D is not always a straightforward one. In general, obesity increases the risk of insulin resistance, which causes an increase in fasting plasma glucose levels. It is thought that this, in turn, places greater demands on the pancreatic beta cells, increasing the likelihood of impaired glucose tolerance or overt T2D developing in susceptible individuals. Although obesity is the most important risk factor for T2D, approximately a third of obese individuals remain metabolically healthy (Wildman et al., 2008), and only a minority of morbidly obese individuals (BMI >40 kg/m^2) are diabetic. Conversely, it is estimated that around 20% of European individuals with T2D have a normal weight (Vaag & Lund, 2007). The "non-obese" form of T2D is characterized by reduced insulin secretion and less insulin resistance, compared to the typical phenotype seen in obese individuals with T2D.

A growing number of rare forms of obesity and T2D inherited in a simple Mendelian manner are being identified. Such highly penetrant genetic causes are currently estimated to account for around 5% of cases of morbid obesity (Farooqi & O'Rahilly, 2006; Walters et al., 2010) and an estimated 2%–5% of T2D (Doria et al., 2008). In most cases, however, T2D and obesity are multifactorial conditions, caused by multiple inherited susceptibility variants

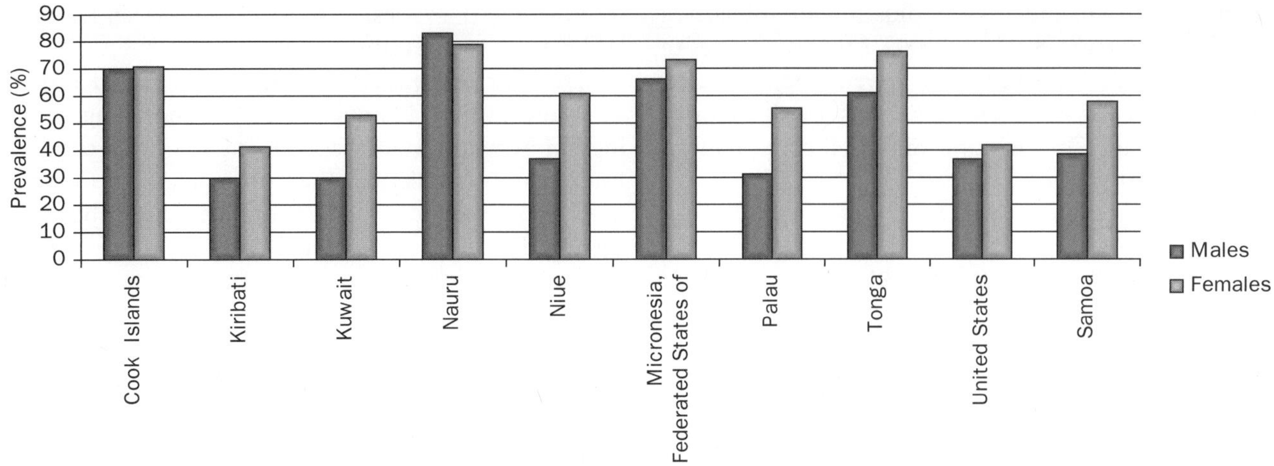

Figure 23-1 Top Ten Countries with the Highest Percentage Prevalence of Obesity. Proportion of males and females in each country with a BMI >30kg/m^2.
SOURCE: Data compiled from Ono et al., 2005).

interacting with environmental triggers in a way that is complex and currently poorly understood. This chapter will review established risk factors for T2D and obesity, with particular emphasis on those most relevant to developing nations; summarize recent advances in the genetics of T2D and obesity; and, finally, consider how studying the genetic architecture of these conditions in a wider range of human populations will extend understanding of their pathology. Such an approach will permit the elucidation of novel disease mechanisms and the development of optimal therapies and management strategies, tailored to geographically and genetically distinct patient populations. In particular, understanding the contribution of both rare, highly penetrant mutations and common genetic variants, with more subtle effects, to susceptibility to diabesity, and their interplay with environmental factors, may be important in framing specific services and informing public health strategies in developing nations.

ENVIRONMENTAL RISK FACTORS FOR T2D AND OBESITY

A well-established association exists between obesity and T2D, with obesity acting as a risk factor for T2D, and several risk factors being common to both disorders. As stated above, risk factors for obesity and T2D shared by developing and developed nations include increasingly sedentary lifestyles and the increased consumption of readily available foods high in fat and refined sugar. The resulting imbalance between energy intake and expenditure places individuals at higher risk of obesity and insulin resistance, with a subsequently increased risk of T2D (Hossain et al., 2007). In many developing nations there has been a recent increase in urbanization. This has resulted in a transition in forms of employment from occupations involving a significant amount of physical labor, such as farming, to more sedentary occupations within urban environments involving less energy expenditure (Misra & Khurana, 2008).

Undernutrition is a serious problem affecting many developing countries, ranging from mild caloric insufficiency to starvation conditions. In an effort to eradicate malnutrition, feeding programs have been introduced in many developing countries. However, a subsequent rise in levels of obesity has been noted in some countries implementing such feeding programs, such as Chile and India (Misra & Khurana, 2008, Uauy et al., 2001). This highlights the importance of achieving a balance between combating caloric deficit and overcompensation, resulting in increased risk of obesity. Cultural attitudes toward obesity have also been shown to modulate disease prevalence. This is reflected in the fact that a higher prevalence of obesity has been noted in countries where it is perceived as a sign of wealth, happiness, and success (Misra & Khurana, 2008). Related to this strongly held cultural belief, there is anecdotal evidence for the force-feeding of women and young girls in some countries. While this practice is aimed at increasing marriage prospects, it can also lead to obesity and subsequent ill-health.

Smoking has been identified as a risk factor for T2D in numerous studies (Foy et al., 2005, Houston et al., 2006). A meta-analysis carried out in 2007 examining data from a pooled dataset that included 1.2 million individuals, found a combined relative risk of 1.44 (95% confidence interval 1.31–1.58), with risk increasing in a dose-dependent manner (Willi et al., 2007). As 80% of the world's 1.1 billion smokers live in developing countries, smoking is a particularly important risk factor for T2D in these regions (Jha et al., 2002).

An association between hepatitis C infection (HCV) and T2D has also been demonstrated in numerous studies (Imazeki et al., 2008, Lecube et al., 2004). The relationship appears to be bidirectional, with HCV infection increasing risk of the condition, and T2D modulating progression of HCV infection (Negro & Alaei, 2009). Prevalence rates of HCV infection are high in several developing countries, particularly parts of Africa, so this too is a risk factor of particular relevance to low- and middle-income nations (Lavanchy, 2009).

GENETICS OF T2D AND OBESITY

While environmental risk factors play a significant role in modulating disease severity and onset, numerous twin and family studies have demonstrated that both obesity and T2D are highly heritable traits (Maes et al., 1997; Moore & Florez, 2008; Wardle et al., 2008). An estimated two-thirds of the variance in BMI is attributed to genetic factors, while siblings of a person with T2D have an approximate threefold increased risk of developing the condition, compared to the general population. Rare syndromic and monogenic forms of both conditions exist, for which many of the chromosomal regions and genes involved have already been identified (Farooqi & O'Rahilly, 2006; McCarthy & Hattersley, 2008; Walters et al., 2010). Such work has proved valuable for elucidating underlying molecular mechanisms, and in some cases has led to dramatic improvements in treatment for monogenic forms of T2D (Murphy et al., 2008) and obesity (Farooqi et al., 2002). However, these account for a minority of clinical cases and, until recently, the genetic variants that influence the common, complex forms of both conditions have remained elusive.

The advent of the genome-wide association study (GWAS) has transformed the field of complex disease genetics. This hypothesis-free approach to identifying common single nucleotide polymorphism (SNP) variants that contribute to the risk of multifactorial conditions such as T2D—and disease-relevant phenotypic traits such as BMI and fasting blood glucose—was made possible by the advent of high-throughput, relatively low-cost SNP genotyping technologies, combined with the completion of the "HapMap" project (International HapMap Consortium, 2007). This international collaborative effort examined linkage disequilibrium (LD) relationships between SNPs to provide genome-wide haplotype information in populations with African, Asian, and European ancestry. The identification of common haplotype blocks that exist in populations worldwide allows researchers to select "tag" SNPs for GWAS, which can be used to predict the likely alleles at the other SNP loci within that particular block, facilitating large-scale, genome-wide searches for risk variants. Typically, several hundred thousand to over a million SNPs have been analyzed in thousands of individuals in the search for variants that influence the risk of disease. So far, over 150 common susceptibility loci for more than 60 different complex diseases and traits have been identified in this way, most with modest effect sizes of between 1.1 and 1.5-fold (Manolio & Collins, 2009).

Our current understanding of the genetics of both T2D and obesity is reviewed below, including the findings of recent GWAS carried out for both conditions. Interestingly, studies carried out to date have found little overlap between loci for T2D and those for obesity, suggesting different physiological bases for these conditions (Walley et al., 2009).

MONOGENIC AND SYNDROMIC OBESITY

Studies of monogenic obesity have contributed to our understanding of some of the pathways involved in body weight regulation, and the mechanisms by which obesity develops and is maintained (Blakemore & Froguel, 2008; O'Rahilly, 2009). Mutations in the gene encoding the adipocyte-expressed hormone leptin were the first variants to be identified as a cause of a nonsyndromic, monogenic form of obesity (Montague et al., 1997). Leptin acts within the hypothalamic leptin-melanocortin pathway, which is involved in appetite and satiety regulation, and leptin gene mutations have been shown to result in obesity through hyperphagia (Montague et al., 1997). Subsequent identification of mutations in the leptin receptor gene that also cause monogenic obesity confirms the central role of this pathway in appetite regulation (Clement et al., 1998).

Mutations in the melanocortin-4 receptor gene (*MC4R*) – which encodes a neuroendocrine factor in the leptin-melanocortin pathway implicated in appetite control—cause autosomal dominant forms of monogenic obesity in up to 6% of childhood and 1.8% of adult obesity cases (Farooqi et al., 2003; Stutzmann et al., 2008; reviewed in Blakemore & Froguel, 2008). *MC4R* mutations have also recently been shown to contribute to common forms of obesity (Loos et al., 2008). Pre-pro-opiomelanocortin (*POMC*) encodes a prohormone processed to produce several hormones involved in a number of pathways, one of which is alpha melanocyte-stimulating hormone (α-MSH), a hormone known to interact with the melanocortin-4 receptor (Lu et al., 1994). Mutations in *POMC* have also been shown to cause monogenic obesity (Krude et al., 1998).

In addition to monogenic forms of obesity, a number of syndromes have been identified in which obesity and developmental delay are presenting clinical features. Prader-Willi syndrome (PWS), whose primary clinical features include developmental delay, obesity, hyperphagia, and hypogonadism, has been associated with the imprinted 15q region (Butler et al., 1986; Goldstone, 2004). Both deletions and imprinting defects in this region have been shown to result in PWS (Ledbetter et al., 1981; Mascari et al., 1992). The recent identification of a patient with some of the features of PWS—including hyperphagia and obesity—with a smaller deletion within the disease critical region has provided evidence for the role of the HBII-85 snoRNA family in human feeding behavior (de Smith et al., 2009).

Syndromes with subphenotypic groups distinguishable by the presence of obesity have identified further novel genes and genomic regions involved in the pathogenesis of obesity. One such syndrome is Wilms tumor, Aniridia, Genitourinary anomalies and Mental Retardation (WAGR) syndrome, whose classic clinical features include Wilms tumors, cognitive impairment, and genitourinary abnormalities (reviewed in Fischbach et al., 2005). Reduced expression of brain-derived neurotrophic factor (BDNF) resulting from point mutations and deletions in the 11p12p14 region have been identified in a subset of individuals suffering from WAGR syndrome, who, in addition to classic clinical features of WAGR syndrome, also develop severe childhood obesity (Gray et al., 2006; Han et al., 2008). Support for the role of the BDNF pathway in syndromes encompassing both obesity and developmental delay is provided by the fact that mutations in the BDNF receptor (encoded by the *NTRK2* gene) result in the development of a phenotype characterized by severe obesity and cognitive impairment (Yeo et al., 2004).

Other syndromes that have implicated genomic regions that may be involved in the development of obesity include Bardet-Biedl syndrome and pseudo-hypoparathyroidism (Badano et al., 2006; Spiegel, 1990). However, although they provide useful biological insights and have led to improved management of affected individuals in some cases, monogenic and syndromic forms of obesity account for only a small proportion of obesity cases worldwide. The overwhelming majority of these are, instead, thought to arise from a more complex interplay of multiple genetic susceptibility factors and environmental pressures.

COMMON OBESITY

While the mutations identified through the investigation of monogenic and syndromic forms of obesity are mainly rare variants with large effect sizes, the genetic contribution to common obesity is somewhat more complex, with multiple susceptibility loci, each individually accounting for a smaller proportion of overall risk of disease (Walley et al., 2009). Recent GWAS have uncovered several genomic loci showing association with BMI and obesity. The locus showing the strongest and most widely replicated association is a region on chromosome 16q12, encompassing the fat mass and obesity associated (*FTO*) gene. Association of this locus with obesity and BMI has been replicated in numerous GWAS (Cotsapas et al., 2009; Dina et al., 2007; Hinney et al., 2007; Loos et al., 2008; Meyre et al., 2009; Scuteri et al., 2007; Thorleifsson et al., 2009; Willer et al., 2009). Functional analyses of *FTO* have shown it plays a role in DNA demethylation, suggesting a potential involvement in regulation of gene expression (Gerken et al., 2007). *FTO* is highly expressed in multiple tissues, showing particularly high expression in the hypothalamus and pancreatic islet cells (Dina et al., 2007, Frayling et al., 2007).

As previously discussed, rare mutations in the melanocortin-4 receptor gene (*MC4R*) are known to result in autosomal dominant forms of monogenic obesity. Providing support for the suggestion that regions known to harbor rare variants of large effect size may also contain common variants associated with less severe phenotypes, SNPs within the *MC4R* region are associated with common forms of obesity (Loos et al., 2008; Meyre et al., 2009; Thorleifsson et al., 2009; Willer et al., 2009).

Neuronal growth regulator 1 (*NEGR1*) is a member of the IgLON family of cell adhesion molecules (Funatsu et al., 1999), homologues of which have been shown to be involved in neuronal development (Hashimoto et al., 2008; Marg et al., 1999; Schafer et al., 2005). Associations at genome-wide levels of significance were detected at SNPs within the *NEGR1* region in two recent studies (Thorleifsson et al., 2009; Willer et al., 2009).

In addition to these three major regions, numerous additional susceptibility loci have been identified through GWAS, including variants at the *TMEM18, GNPDA2, NPC1, PCSK1, PTER* and *MAF* gene loci (Benzinou et al., 2008; Meyre et al., 2009; Thorleifsson et al., 2009; Willer et al., 2009). The fact that a large proportion of these susceptibility genes encode products that act within neurological pathways, as opposed to pathways involved in fat storage, has altered our perception of the pathogenesis of obesity, suggesting it is largely a disorder of appetite regulation rather than metabolic disruption (reviewed in O'Rahilly, 2009).

MONOGENIC AND SYNDROMIC T2D

The first attempts to study the genetics of T2D focused on rare monogenic disorders in which insulin production is disrupted, the largest group of which is collectively known as *Maturity Onset Diabetes of the Young* (MODY). All forms of MODY have an autosomal dominant inheritance pattern, and efforts to identify the causative genes through linkage analysis have proved extremely successful (reviewed in McCarthy & Hattersley, 2008). The majority of MODY cases are accounted for by mutations in one of six genes: *GCK,* which encodes the glucose "sensor" glucokinase, and the transcription factor genes *TCF1, TCF2, HNF4A, PDX1* and *NEUROD1*. Other, very rare forms are caused by mutations in *KLF11*, another transcription factor gene, and *CEL*, which encodes the lipolytic enzyme carboxyl ester lipase.

Nearly all MODY genes affect pancreatic beta cell development or function, and thus cause a reduction or disruption of insulin production, rather than insulin resistance. However, mutations have been identified in other genes that result in severe monogenic forms of insulin resistance (reviewed in O'Rahilly, 2007); namely, the insulin receptor gene (*INSR*); *AKT2*, which encodes a protein kinase involved in insulin signaling; and the peroxisome proliferator-activated receptor gene *PPARG*. Finally, mutations in the *ABCC8* and *KCNJ11* genes, which encode the two subunits of the ATP-sensitive K+ channel, can result in both transient and permanent forms of neonatal diabetes. The identification of these last two monogenic forms of T2D has important clinical applications, since both can be successfully treated with sulfonylurea tablets rather than insulin injections (Gloyn et al., 2004; Pearson et al., 2006). Permanent neonatal diabetes can also be caused by mutations in the insulin gene (*INS*) itself (Stoy et al., 2007), while another, transient form of neonatal diabetes is associated with a global disruption of imprinting caused by mutations in the *ZFP57* gene (Mackay et al., 2008).

Studies of syndromes that include T2D as part of their phenotype have also yielded novel mechanistic insights, in particular those caused by mitochondrial genome mutations, the most common of which is maternally inherited diabetes and deafness (MIDD). This results from a heteroplasmic mitochondrial gene mutation at position 3243, first identified in a large Dutch family (van den Ouweland et al., 1994). Another syndrome that includes T2D as part of is phenotype is Wolfram syndrome, a recessive condition also known as DIDMOAD (Diabetes Insipidus, Diabetes Mellitus, Optic Atrophy, and Deafness). It is caused by mutations in the *WFS1* gene, variants in which have also recently been associated with the common form of T2D (Sandhu et al., 2007). However, in general, the gene mutations that underlie syndromic and monogenic forms of T2D account for a minority of cases of T2D, necessitating wider searches for novel T2D genes.

COMMON T2D

Following the identification of the majority of the genes involved in monogenic forms of T2D and insulin resistance, efforts focused on exploring their possible role in the common form of the disease. This candidate gene approach lead to the identification of both *PPARG*, mutations in which cause severe insulin resistance, and the neonatal diabetes gene *KCNJ11* as risk variants for T2D (Beamer et al., 1998, Barroso et al., 2003). However, a large case-control association study of SNPs within the causative genes for the six major types of MODY revealed only a modest effect on T2D for variants within *TCF2* (Winckler et al., 2007)—an association confirmed in a later study, which found that different variants in the same gene are associated with prostate cancer risk (Gudmundsson et al., 2009). In the pre-GWAS era, linkage studies implicated multiple genomic candidate regions for common T2D, but most of these have proved inconclusive. One notable exception is the *TCF7L2* gene, initially discovered through linkage analysis (Grant et al., 2006), but since replicated in numerous GWA studies, and which remains the major locus for T2D. Individuals homozygous for the risk allele have a twofold increased risk of developing T2D, particularly among the "lean" diabetic patient population (Cauchi et al., 2008; Watanabe et al., 2007). The gene encodes a transcription factor involved in the WNT signaling pathway, which is critical for cell proliferation in many organs, including the pancreatic beta cells (McLin et al., 2007). Several other loci have been associated with T2D, such as those in the *CAPN10* (Horikawa et al., 2000) and *ENPP1* (Pizzuti et al., 1999) genes, but these have not been confirmed in later GWAS.

The first GWAS for T2D was reported by Sladek et al. in 2007, and was rapidly followed by four others (Saxena et al., 2007; Scott et al., 2007; Steinthorsdottir et al., 2007; Zeggini et al., 2007) and several smaller studies (reviewed by Doria et al., 2008). These hypothesis-free studies confirmed that *TCF7L2*, *KCNJ11*, and *PPARG* are risk loci for T2D, and identified an additional 11 genes. Some T2D GWAS also identified the *FTO* gene, though this has since been shown to mediate its effect through BMI, and is thus a risk locus for obesity rather than T2D. A meta-analysis of GWAS data led to the identification of a further six loci (Zeggini et al., 2008), and a GWAS of fasting glucose identified the melatonin receptor 1B (*MTNR1B*) as a key locus for this quantitative trait, and also as a risk locus for T2D (Bouatia-Naji et al., 2009; Lyssenko et al., 2009; Prokopenko et al., 2009). A further meta-analysis of GWA studies yielded more association signals, bringing the current number of confirmed T2D risk loci to 38 (Voight et al., 2010). Thus, the GWAS approach has already led to significant advances in the genetics of the common form of T2D. Interestingly, the majority of genetic risk variants identified so far are associated with insulin secretion pathways, rather than insulin sensitivity of peripheral tissues and, overall, there is evidence for enrichment of genes involved in cell cycle regulation (Voight et al., 2010).

THE "MISSING HERITABILITY" ISSUE

While the numbers of novel genes implicated in obesity and T2D identified through GWAS are impressive, the common variants identified to date confer only modest effects on risk, and so the majority of the observed heritability for both T2D and obesity remains unexplained (Bogardus, 2009; Stolerman & Florez, 2009). Thus, although such studies have proved invaluable in the identification of novel pathways involved in these conditions, the genetic information yielded so far is not useful for assessing disease risk in particular individuals. For example, even when considering 18 common variants associated with altered risk of T2D, there is no improved prediction value over a model utilizing solely clinical measurements, such as systolic blood pressure, levels of triglycerides, HDL cholesterol and fasting glucose, BMI, and family history (Lango et al., 2008). Similarly, a combined analysis of 20 susceptibility variants for obesity found that although they have some predictive value when used in combination with other obesity risk factors, this new genetic information did not have sufficient discriminative power to be of clinical use (Sandholt et al., 2010).

The issue of "missing heritability" has been encountered for almost all of the common, complex diseases for which GWAS have so far been carried out, and debate continues over alternative explanations for this observation (Manolio & Collins, 2009). One possibility is that current heritability estimates are inflated, and so will never be completely explained, even if all variants contributing to the genetic risk are identified. Alternatively, it is likely that in many cases, the SNP identified as being associated with a disease is merely "tagging" the causal variant responsible, and so the effect size of the associated SNP does not accurately reflect that of the risk locus. Thirdly, it could be that gene–gene and gene–environment interactions could explain more of the heritability than a simple additive model of individual risk variants; i.e., the overall effect of all common SNP variants identified so far may be much greater than their summed individual effects. Fourthly, it may be that for some traits, much of the heritability *can* be explained, if all loci implicated in GWA studies are considered, rather than just those that demonstrate highly significant associations—for example, a recent study found that 45% of the variance for human height can be explained by considering all genotyped GWAS SNPs simultaneously (Yang et al., 2010).

Another strong possibility is that the genetic architecture of many complex conditions, including T2D and obesity, may be partly refractory to dissection using the GWAS approach. GWAS are predicated on the "common disease, common variant" model of complex disease, which assumes that the effects of several variants present in >1% of the population will account for most of the genetic risk for a condition. However, alternative models are now increasingly being considered, such as the possibility that several rare variants of much stronger effect may explain a large proportion of the genetic risk for a disease such as T2D. The first example of such a "high penetrance" variant associated with adult obesity has recently been reported (Walters et al., 2010). This study revealed that rare deletions of at least 593 kilobases at chromosome 16p11.2 are associated with a highly penetrant form of morbid obesity; it was present in 0.7% of the affected individuals studied, but not in any normal weight controls. In contrast to the small effect sizes reported for common SNP variants, the deletion is associated with a 43-fold increase in risk of morbid obesity. It is likely that several more rare variants of similar strong effect remain to be identified, both for obesity and other conditions. The identification of a rare deletion associated with obesity also highlights another highly active area of research in complex disease genetics: the potential contribution of genomic structural variants (GSVs). GSVs involve deletion, duplication, or inversion of genetic material, varying widely in size from a few base pairs to many megabases of DNA, and are currently estimated to encompass up to 34% of the human genome (Database of Genomic Variants, http://projects.tcag.ca/variation/).

Although the role of GSVs in rare disorders is well established, their contribution to common, complex disease is only now beginning to be elucidated. As is the case for simple DNA sequence changes, GSVs contributing to phenotype may include both rare, highly penetrant variants of strong influence, and common variants with more subtle effects. It is thus likely that at least some of the missing heritability reported for common diseases will be accounted for by structural variants, as exemplified by reported associations of common GSVs with conditions such as psoriasis (de Cid et al., 2009; Hollox et al., 2008) and autoimmune glomerulonephritis (Aitman et al., 2006; Fanciulli et al., 2007) and rare GSVs resulting in autism (Sebat et al., 2007) and schizophrenia (McCarthy et al., 2009; Walsh et al., 2008).

In addition to the direct benefit in diagnosis of subgroups of carrier individuals, the identification of rare, highly penetrant structural variants is also useful in that it can highlight regions of the genome that are likely to contain variants that exert a more subtle effect on disease risk. The identification of the role of the 16p11.2 deletion in obesity illustrates the power of identifying rare variants in patients with extreme phenotypes, information that can then be applied in targeted case-control and population study cohorts (Froguel & Blakemore, 2008).

STUDYING DIFFERENT HUMAN POPULATIONS TO ELUCIDATE DIABESITY GENETICS

The vast majority of studies on the genetics of obesity and diabetes described above, although fruitful, have been performed in populations of European origin and thus have been limited in their ability to interrogate the full spectrum of human genetic diversity. It is only by studying other ethnic groups that a true picture of genetic variation and its role in health and disease may be gained. Each different human population group has its own genetic profile in terms of diversity. In particular, African populations tend to have a higher degree of polymorphism than other human groups, and so are a rich source of novel genetic diversity for study. Thus, both African and African-admixed populations are particularly useful for "transethnic mapping," an approach discussed in more detail below.

Other population groups, for example the Pima Indians, Nauru Islanders, and South Asians have a particularly high incidence of diabesity and thus have provided important information on risk factors for this trait—particularly in comparative studies of individuals who have emigrated to an obesogenic environment (Dowse et al., 1991; Knowler et al., 1978; Ramachandran et al., 2001). The specific history and social structure of certain populations may also have particular advantages for genetic study. Populations with a high degree of consanguinity, or descended from a small founder group, may be enriched for genetic variants that are rare in other human groups—these can be of particular interest in identification of recessive conditions. For example, the population isolates of Iceland and Finland, with their well-developed health and public-records infrastructure, have yielded a number of important findings, including the discovery of the contribution of *TCF7L2* variants to T2D, and additional novel T2D loci identified through GWAS in these populations (Grant et al., 2006; Scott et al., 2007; Steinthorsdottir et al., 2007). Additionally, populations with a high frequency of consanguineous unions offer opportunities for identification of rare, recessive mutations by "homozygosity mapping," another approach discussed in more detail below.

GAINING ADDITIONAL GENETIC INFORMATION BY TRANSETHNIC MAPPING IN GWA STUDIES

Transethnic mapping in GWAS analyses will be of increasing importance, both to maximize the scientific information gained from such work and, more importantly, as novel findings are translated into clinical practice, to avoid health disparities arising from the bias toward European study populations and the consequent underidentification of genetic factors important in other human population groups.

Due to the nature of GWAS, it has become commonplace for these studies to be carried out in a two-stage process, with a first stage genome-wide analysis carried out in a discovery cohort, followed by a targeted replication stage in one or more independent and often larger sample sets. A threshold for genome-wide significance, often in the range of 5×10^{-8} (based on correction for multiple testing on a genome-wide platform) is employed to select SNPs for targeted replication in the second stage (Hirschhorn & Daly, 2005; McCarthy et al., 2008). While this approach has proven successful in selecting SNPs with association signals most likely to represent genuine associations with disease—and, thus, most likely to replicate in other independent study cohorts—it also carries the risk of missing potentially genuine associations for which the discovery set is insufficiently powered to detect at the significance threshold. The diversity in genetic architecture of different populations, particularly differences in minor allele frequencies and linkage disequilibrium structures, has already proven instrumental in the identification of additional novel SNP associations (Adeyemo & Rotimi, 2010; Teo et al., 2009). The identification of significant association of the *KCNQ1* locus with T2D, a finding that was subsequently replicated in populations of European origin, highlights the valuable contribution that can be made by carrying out studies in non-European cohorts (Unoki et al., 2008; Yasuda et al., 2008).

In addition, a deeper exploration of regions already associated with conditions such as T2D and obesity in a wider range of populations permits fine-mapping of confirmed risk loci. The heterogeneity of linkage disequilibrium (LD) blocks across diverse populations may enable the identification of some elusive causal variants; i.e., SNPs found to be associated with disease that are located within particular LD blocks in European populations may be distinguishable from the true causal variants in other populations with different patterns of LD. The greatest levels of human genetic diversity are found in Africa—the ancestral home of all humans (Teo et al., 2010; Tishkoff et al., 2009)—due to its complex population history and the variation in climate, diet, and exposure to infectious disease (Campbell & Tishkoff, 2008). Genetic association studies in Africa will thus be of particular importance in identifying novel risk variants for disease (Tishkoff & Williams, 2002; Teo et al., 2010). The genetic structure of the African population is different from that of any other, with higher levels of haplotype diversity and lower levels of LD (Bryc et al., 2010). However, this diversity presents challenges as well as opportunities for genetic research.

A survey of 927 samples representing 52 different worldwide populations found that there is a steady trend of reduction in haplotype diversity with geographical distance from Africa. The authors also investigated what fraction of worldwide haplotype variation is captured by HapMap samples, and found greater representation of European and East Asian populations than of the African population (Conrad et al., 2006). A more recent review of GWAS in diverse populations also identifies potential issues with SNP selection, and additionally with genotype imputation (the prediction of SNP genotypes that have not been experimentally determined), which is more accurate in high-LD populations (Rosenberg et al., 2010). Despite the challenges, however, transethnic mapping is now being widely applied and has already proved useful in fine-mapping SNP association signals from GWAS studies originally performed in European subjects—for example, in refining our understanding of the *FTO* locus (Grant et al., 2008). New approaches to interpretation of transethnic mapping data are under active development (Hao et al., 2010) and, as sample collections from different ethnic groups become available, may offer significant scientific advantages in mapping of causative variants in GWAS-identified loci.

IDENTIFYING RARE, HIGHLY PENETRANT VARIANTS IN CONSANGUINEOUS POPULATIONS

A consanguineous marriage is most often one between cousins (Campbell et al., 2009), but may also be defined as "unions of individuals with at least one common ancestor such as occurring within population isolates, small towns, and tribes; intra-community or endogamous marriages" (Tadmouri et al., 2009). These unions are often favored in some populations, due to sociocultural factors such as maintaining family structure, financial advantages for the families involved, ease of marital arrangements, and the belief that they are more stable than marriages between nonrelatives (Khlat & Khoury, 1991; Campbell et al., 2009).

The overall frequency of consanguineous marriages is estimated to be around 20% worldwide, but this figure varies widely between different populations, depending on religion, culture, and geography (Tadmouri et al., 2009). A decline in consanguineous marriages has been seen since the beginning of this century in Western countries, but this is not the case in developing countries, such as parts of Asia, Africa, and the Middle East, with endogamous populations (Bittles et al., 1993; Campbell et al., 2009). Many of the Arab regions still have the highest consanguineous marriages rates in the world; e.g., Kuwait (54%), Morocco (30%), and Palestine (44.3%; see Teebi, 1994; Zlotogora, 1997a, 1997b; Jaouad et al., 2009).

Consanguinity increases the frequency of recessive disorders (often with an unusual pattern of prevalence specific to the particular population), since there is a higher risk that both parents will carry the same deleterious gene mutation, and therefore a 25% chance that any child they have will inherit two copies of the disease allele—thus, the health risks to the offspring of first cousin marriage are increased compared to nonconsanguineous marriages (Tadmouri et al., 2009).

The first two causes of human monogenic obesity—mutations in the genes encoding leptin and the leptin receptor—were originally identified by analysis of consanguineous pedigrees, although these discoveries were made by direct sequencing of candidate genes (Clement et al., 1998; Montague et al., 1997). Where the disease etiology is unclear, however, no candidates may suggest themselves and hypothesis-free homozygosity mapping may be carried out. High-density SNP analysis has been used to investigate individual and community levels of homozygosity in highly consanguineous populations, and to locate disease gene loci in consanguineous families affected by rare recessive conditions (Campbell et al., 2009). This is done by performing genetic linkage studies, and mapping blocks of homozygosity using genome-wide SNP genotyping tools (White et al., 2007; Khlat and Khoury, 1991; Kurian et al., 2009). These regions of homozygosity can enable the rapid identification of disease loci (White et al., 2007; Khaliq et al., 2007; Kurian et al., 2009). Fine-mapping can then be carried out using highly polymorphic microsatellite markers to exclude or fine-map the regions of interest (Kurian et al., 2009). Once the critical region has been tightly defined, candidate genes within it can be sequenced in order to identify the disease-causing mutation.

Homozygosity mapping has thus proved to be a powerful tool for identifying the causative mutations for many different diseases in highly consanguineous families and populations. Studies carried out on Saudi, Pakistani, and Omani families have helped in the rapid and efficient identification of mutations that cause retinitis pigmentosa, Knobloch syndrome, and Bardet-Biedl syndrome (White et al., 2007; Khaliq et al., 2007; Aldahmesh et al., 2009). In many parts of developing countries, the population is still tribal, leading to an unusual burden of genetic diseases in these populations. Thus the structure of consanguineous families in developing countries has led to unique distribution patterns for genetic diseases that are not found in any other population (Tadmouri et al., 2009). Homozygosity mapping studies in such families may identify multiple rare, highly penetrant variants for a range of conditions, including novel T2D and obesity loci.

THE IMPACT OF NEXT GENERATION SEQUENCING

In order to identify novel causal variants for obesity and T2D, it will be necessary to sequence candidate genomic regions identified through GWAS or homozygosity mapping. Over the last few years there has been rapid progress both in the technology available for massively parallel DNA sequencing, also known as "next generation sequencing," and in the bioinformatics capacity to store and interpret the resulting data. Costs have fallen rapidly, so that at the time of writing, it cost approximately $3,000 to carry out full exome sequencing (including exons and known control regions) and around $10,000 per sample for whole-genome sequencing. As costs continue to fall, genome sequencing is likely to become accessible as a routine research tool for study of human disease (Mardis, 2008). Initial results have already demonstrated that, at least in the case of recessive disorders, exome sequencing can be an effective strategy for revealing novel mutations; for example, identification of de novo mutations in the *SETBP1* gene as the cause of Schinzel-Giedion syndrome (Hoischen et al., 2010).

To date, large-scale sequencing initiatives, such as the 1,000 Genomes project (www.1000genomes.org), have concentrated on mapping diversity in itself, without regard to phenotype. More recent ventures include plans to sequence individuals with disease, including

T2D (Henderson, 2010). While commendable in aims and gargantuan in scale, these projects may suffer from some disadvantages: not all population groups are well represented, phenotyping quality may be variable and, critically, much of the sequencing will be carried out on DNA from cell lines. This is problematic, since cell lines accumulate mutations during culture and a nontrivial proportion of the variants thus detected may be artefactual, in that they do not represent the genome of the original human subject. Nonetheless, such studies are expected to reveal a wealth of new genetic variants for study, although each must be evaluated carefully for the reasons explained. However, a more targeted approach is also possible—by sequencing individuals with "extreme" phenotypes, for example "obesity-plus" (high BMI coupled with developmental delay or dysmorphic features) or neonatal diabetes, new highly penetrant rare variants may be revealed, highlighting new genes or genomic regions with a causative role in diabesity.

Once issues of cost, throughput, and data handling are addressed, the main challenge of next generation sequencing projects will be to identify which of the many thousands of genetic variants identified in each human is responsible for the phenotype of interest. Elucidation of causative variants will require a combination of approaches: comparison with genomes of subjects lacking the trait of interest; family studies to determine which variants are co-inherited with the trait of interest; bioinformatic prediction of sequence conservation and predicted functional implications of the variant; and, finally, a return to basic biological study of altered function. In this way, genetics is coming full circle: after huge technological advances in data generation, there will be a return to examination of familial inheritance patterns and laboratory investigation of biochemistry, cell biology, and physiology.

FROM DEEPER GENETIC KNOWLEDGE TO IMPROVED HEALTHCARE

The genetic analysis of obesity and diabetes has two major goals. Firstly, scientists aim to understand the causes and progression of these conditions, in order to design new approaches to prevention, therapy, and management of diabesity. Secondly, such knowledge holds out the prospect of personalized medicine. While the current generation of GWAS-identified common variants have very modest predictive power, the picture is very different for the significant minority of individuals whose diabesity results from carriage of a rarer variant of strong effect. For these subjects, and their affected relatives, predictive power is very high and the potential for personalized medicine is very real.

A crucial area of future human genetic research into the genetics of obesity and T2D will be an exploration of the variation that exists in non-European populations, and the contribution of both common and rare variants to disease risk. The search for additional common variants will be carried out by extending GWAS studies into non-European populations, while additional rare variants may be identified through examination of extreme cases and by homozygosity mapping in consanguineous families. Additionally, as the cost of high-throughput sequencing continues to fall, it will become increasingly feasible to carry out whole-exome or whole-genome sequencing studies—both of "normal" subjects from a range of populations or, in a more targeted strategy, of individuals with "extreme" phenotypes of relevance to obesity and T2D (for example BMI, insulin resistance, and other measures of glycemic control). Such samples are likely to be enriched for rare, highly penetrant variants, which in turn will reveal novel genes and pathways involved in these conditions.

For the study of T2D and obesity, such research will help to paint a clearer picture of the genetic landscape of these complex conditions, and the way in which population-specific genetic and environmental risk factors interact to cause adverse metabolic phenotypes. The genetic risk variants in non-European populations may be distinct from those already identified in European countries, and some may point to entirely new underlying biological pathways. Such differences may reflect different evolutionary pressures or migration patterns, or may simply have arisen by chance in human populations separated by culture or geography. A detailed analysis of this variation will permit a deeper understanding of the development of T2D and obesity, paving the way for more targeted prevention, lifestyle interventions, and treatments for these increasingly common conditions.

In conclusion, it is clear that the rich diversity (social, environmental, and genetic) of populations worldwide represents an important resource for genetic studies of diabesity. Increasingly, subjects from the developing world will be asked to contribute their "genetic wealth" to the global research effort. In return, it is hoped that with suitable economic and political conditions, improved infrastructure may also allow them a full share of the health benefits of that research.

REFERENCES

Adeyemo A, Rotimi C. (2010). Genetic variants associated with complex human diseases show wide variation across multiple populations. *Public Health Genomics 13*, 72–79.

Aitman TJ, Dong R, Vyse TJ, et al. (2006). Copy number polymorphism in Fcgr3 predisposes to glomerulonephritis in rats and humans. *Nature, 439*, 851–855.

Aldahmesh MA, Safieh LA, Alkuraya H, et al. (2009). Molecular characterization of retinitis pigmentosa in Saudi Arabia. *Mol Vis 15*, 2464–2469.

Badano JL, Mitsuma N, Beales PL, Katsanis N. (2006). The ciliopathies: an emerging class of human genetic disorders. *Annu Rev Genomics Hum Genet 7*, 125–148.

Barroso IS, Luan JA, Middelberg RPS. et al. (2003). Candidate gene association study in type 2 diabetes indicates a role for genes involved in beta-cell function as well as insulin action. *PLoS Biol 1(1)*, E20.

Beamer BA, Yen CJ, Andersen RE, et al. (1998). Association of the Pro12Ala variant in the peroxisome proliferator-activated receptor-gamma2 gene with obesity in two Caucasian populations. *Diabetes 47*, 1806–1808.

Benzinou M, Creemers JW, Choquet H. et al. (2008). Common nonsynonymous variants in PCSK1 confer risk of obesity. *Nat Genet, 40*, 943–945.

Bittles AH, Grant JC, Shami SA. (1993). Consanguinity as a determinant of reproductive behaviour and mortality in Pakistan. *Int J Epidemiol 22*, 463–467.

Blakemore AI, Froguel P. (2008). Is obesity our genetic legacy? *J Clin Endocrinol Metab 93*, S51–56.

Bogardus C. (2009). Missing heritability and GWAS utility. *Obesity 17*, 209–210.

Bouatia-Naji N, Bonnefond A, Cavalcanti-Proenca C. et al. (2009). A variant near MTNR1B is associated with increased fasting plasma glucose levels and type 2 diabetes risk. *Nat Genet 41*, 89–94.

Bryc K, Auton A, Nelson MR, et al. (2010). Genome-wide patterns of population structure and admixture in West Africans and African Americans. *Proc Natl Acad Sci USA 107*, 786–791.

Butler MG, Meaney FJ, Palmer CG. (1986). Clinical and cytogenetic survey of 39 individuals with Prader-Labhart-Willi syndrome. *Am J Med Genet 23*, 793–809.

Campbell H, Rudan I, Bittles AH, Wright AF. (2009). Human population structure, genome autozygosity and human health. *Genome Med 1*, 91.

Campbell MC, Tishkoff SA. (2008). African genetic diversity: implications for human demographic history, modern human origins, and complex disease mapping. *Annu Rev Genomics Hum Genet 9*, 403–433.

Cauchi S, Choquet H, Gutierrez-Aguilar R, et al. (2008). Effects of TCF7L2 polymorphisms on obesity in European populations. *Obesity 16*, 476–482.

Clement K, Vaisse C, Lahlou N, et al. (1998). A mutation in the human leptin receptor gene causes obesity and pituitary dysfunction. *Nature 392*, 398–401.

Conrad DF, Jakobsson M, Coop G, et al. (2006). A worldwide survey of haplotype variation and linkage disequilibrium in the human genome. *Nat Genet 38*, 1251–1260.

Cotsapas C, Speliotes EK, Hatoum IJ, et al. (2009). Common body mass index-associated variants confer risk of extreme obesity. *Hum Mol Genet 18*, 3502–3507.

de Cid R, Riveira-Munoz E, Zeeuwen PLJM, et al. (2009). Deletion of the late cornified envelope LCE3B and LCE3C genes as a susceptibility factor for psoriasis. *Nat Genet 41*, 211–215.

de Smith AJ, Purmann C, Walters RG, et al. (2009). A deletion of the HBII-85 class of small nucleolar RNAs (snoRNAs) is associated with hyperphagia, obesity and hypogonadism. *Hum Mol Genet 18*, 3257–3265.

Dina C, Meyre D, Gallina S, et al. (2007). Variation in FTO contributes to childhood obesity and severe adult obesity. *Nat Genet 39*, 724–726.

Doria A, Patti ME, Kahn CR. (2008). The emerging genetic architecture of type 2 diabetes. *Cell Metab 8*, 186–200.

Dowse GK, Zimmet PZ, Finch CF, Collins VR. (1991). Decline in incidence of epidemic glucose intolerance in Nauruans: implications for the "thrifty genotype." *Am J Epidemiol 133*, 1093–1104.

Fanciulli M, Norsworthy PJ, Petretto E, et al. (2007). FCGR3B copy number variation is associated with susceptibility to systemic, but not organ-specific, autoimmunity. *Nat Genet 39*, 721–723.

Farooqi IS, Keogh JM, Yeo GS, Lank EJ, Cheetham T, O'Rahilly S. (2003). Clinical spectrum of obesity and mutations in the melanocortin 4 receptor gene. *N Engl J Med 348*, 1085–1095.

Farooqi IS, Matarese G, Lord GM, et al. (2002). Beneficial effects of leptin on obesity, T cell hyporesponsiveness, and neuroendocrine/metabolic dysfunction of human congenital leptin deficiency. *J Clin Invest 110*, 1093–1103.

Farooqi IS, O'Rahilly S. (2006). Genetics of obesity in humans. *Endocr Rev 27*, 710–718.

Fischbach BV, Trout KL, Lewis J, Luis CA, Sika M. (2005). WAGR syndrome: a clinical review of 54 cases. *Pediatrics 116*, 984–988.

Foy CG, Bell RA, Farmer DF, Goff DC Jr., Wagenknecht LE. (2005). Smoking and incidence of diabetes among U.S. adults: findings from the Insulin Resistance Atherosclerosis Study. *Diabetes Care 28*, 2501–2507.

Frayling TM, Timpson NJ, Weedon MN, et al. (2007). A common variant in the FTO gene is associated with body mass index and predisposes to childhood and adult obesity. *Science 316*, 889–894.

Froguel P, Blakemore AI. (2008). The power of the extreme in elucidating obesity. *N Engl J Med 359*, 891–893.

Funatsu N, Miyata S, Kumanogoh H, et al. (1999). Characterization of a novel rat brain glycosylphosphatidylinositol-anchored protein (Kilon), a member of the IgLON cell adhesion molecule family. *J Biol Chem 274*, 8224–8230.

Gerken T, Girard CA, Tung YC, et al. (2007). The obesity-associated FTO gene encodes a 2-oxoglutarate-dependent nucleic acid demethylase. *Science 318*, 1469–1472.

Gloyn AL, Pearson ER, Antcliff JF, et al. (2004). Activating mutations in the gene encoding the ATP-sensitive potassium-channel subunit Kir6.2 and permanent neonatal diabetes. *N Engl J Med 350*, 1838–1849.

Goldstone AP. (2004). Prader-Willi syndrome: advances in genetics, pathophysiology and treatment. *Trends Endocrinol Metab 15*, 12–20.

Grant SFA, Li M, Bradfield JP, et al. (2008). Association analysis of the *FTO* gene with obesity in children of Caucasian and African ancestry reveals a common tagging SNP. *PLoS ONE 3*, e1746.

Grant SFA, Thorleifsson G, Reynisdottir I, et al. (2006). Variant of transcription factor 7-like 2 (TCF7L2) gene confers risk of type 2 diabetes. *Nat Genet 38*, 320–323.

Gray J, Yeo GS, Cox JJ, et al. (2006). Hyperphagia, severe obesity, impaired cognitive function, and hyperactivity associated with functional loss of one copy of the brain-derived neurotrophic factor (BDNF) gene. *Diabetes 55*, 3366–3371.

Gudmundsson J, Sulem P, Gudbjartsson DF, et al. (2009). Genome-wide association and replication studies identify four variants associated with prostate cancer susceptibility. *Nat Genet 41*, 1122–1126.

Han JC, Liu QR, Jones M, et al. (2008). Brain-derived neurotrophic factor and obesity in the WAGR syndrome. *N Engl J Med 359*, 918–927.

Hao K, Chudin E, Greenawalt D, Schadt EE. (2010). Magnitude of stratification in human populations and impacts on genome wide association studies. *PLoS ONE 5*, e8695.

Hashimoto T, Yamada M, Maekawa S, Nakashima T, Miyata S. (2008). IgLON cell adhesion molecule Kilon is a crucial modulator for synapse number in hippocampal neurons. *Brain Res 1224*, 1–11.

Henderson M. (2010) Landmark DNA study of 3,000 people to unlock mystery of type 2 diabetes. *The Times, 5/1/10.*

Hinney A, Nguyen TT, Scherag A, et al. (2007). Genome wide association (GWA) study for early onset extreme obesity supports the role of fat mass and obesity associated gene (FTO) variants. *PLoS ONE 2*, e1361.

Hirschhorn JN, Daly MJ. (2005). Genome-wide association studies for common diseases and complex traits. *Nat Rev Genet 6*, 95–108.

Hoischen A, van Bon BWM, Gilissen C, et al. (2010). De novo mutations of SETBP1 cause Schinzel-Giedion syndrome. *Nat Genet 42*, 483–485.

Hollox EJ, Huffmeier U, Zeeuwen PLJM, et al. (2008). Psoriasis is associated with increased [beta]-defensin genomic copy number. *Nat Genet 40*, 23–25.

Horikawa Y, Oda N, Cox NJ, et al. (2000). Genetic variation in the gene encoding calpain-10 is associated with type 2 diabetes mellitus. *Nat Genet 26*, 163–175.

Hossain P, Kawar B, EL Nahas M. (2007). Obesity and diabetes in the developing world—a growing challenge. *N Engl J Med 356*, 213–215.

Houston TK, Person SD, Pletcher MJ, Liu K, Iribarren C, Kiefe CI. (2006). Active and passive smoking and development of glucose intolerance among young adults in a prospective cohort: CARDIA study. *BMJ 332*, 1064–1069.

Imazeki F, Yokosuka O, Fukai K, Kanda T, Kojima H, Saisho H. (2008). Prevalence of diabetes mellitus and insulin resistance in patients with chronic hepatitis C: comparison with hepatitis B virus-infected and hepatitis C virus-cleared patients. *Liver Int 28*, 355–362.

International HapMap Consortium. (2005). A haplotype map of the human genome. *Nature 437*, 1299–1320.

Jaouad IC, Elalaoui SC, Sbiti A, Elkerh F, Belmahi L, Sefiani A. (2009). Consanguineous marriages in Morocco and the consequence for the incidence of autosomal recessive disorders. *J Biosoc Sci 41*, 575–581.

Jha, P., Ranson, M. K., Nguyen, S. N. & Yach, D. (2002) Estimates of global and regional smoking prevalence in 1995, by age and sex. *Am J Public Health 92*, 1002–1006.

Kelly T, Yang W, Chen CS, Reynolds K, He J. (2008). Global burden of obesity in 2005 and projections to 2030. *Int J Obes (Lond) 32*, 1431–1437.

Khaliq S, Abid A, White DR, et al. (2007). Mapping of a novel type III variant of Knobloch syndrome (KNO3) to chromosome 17q11.2. *Am J Med Genet A 143A*, 2768–2774.

Khlat M, Khoury M. (1991). Inbreeding and diseases: demographic, genetic, and epidemiologic perspectives. *Epidemiol Rev 13*, 28–41.

Knowler WC, Bennett PH, Hamman RF, Miller M. (1978). Diabetes incidence and prevalence in Pima Indians: a 19-fold greater incidence than in Rochester, Minnesota. *Am J Epidemiol 108*, 497–505.

Krude H, Biebermann H, Luck W, Horn R, Brabant G, Gruters A. (1998). Severe early-onset obesity, adrenal insufficiency and red hair pigmentation caused by POMC mutations in humans. *Nat Genet 19*, 155–157.

Kurian MA, Zhen J, Cheng SY, et al. (2009). Homozygous loss-of-function mutations in the gene encoding the dopamine transporter are associated with infantile parkinsonism-dystonia. *J Clin Invest 119*, 1595–1603.

Lango H, Palmer CNA, Morris AD, et al. (2008). Assessing the combined impact of 18 common genetic variants of modest effect sizes on type 2 diabetes risk. *Diabetes 57*, 3129–3135.

Lavanchy D. (2009). The global burden of hepatitis C. *Liver Int 29*(Suppl 1), 74–81.

Lecube A, Hernandez C, Genesca J, Esteban JI, Jardi R, Simo R. (2004). High prevalence of glucose abnormalities in patients with hepatitis C virus infection: a multivariate analysis considering the liver injury. *Diabetes Care 27*, 1171–1175.

Ledbetter DH, Riccardi VM, Airhart SD, Strobel RJ, Keenan BS, Crawford JD. (1981). Deletions of chromosome 15 as a cause of the Prader-Willi syndrome. *N Engl J Med 304*, 325–329.

Loos RJ, Lindgren CM, Li S, et al. (2008). Common variants near MC4R are associated with fat mass, weight and risk of obesity. *Nat Genet 40*, 768–775.

Lu D, Willard D, Patel IR, et al. (1994). Agouti protein is an antagonist of the melanocyte-stimulating-hormone receptor. *Nature 371*, 799–802.

Lyssenko V, Nagorny CLF, Erdos MR, et al. (2009). Common variant in MTNR1B associated with increased risk of type 2 diabetes and impaired early insulin secretion. *Nat Genet 41*, 82–88.

Mackay DJG, Callaway JLA, Marks SM, et al. (2008). Hypomethylation of multiple imprinted loci in individuals with transient neonatal diabetes is associated with mutations in ZFP57. *Nat Genet 40*, 949–951.

Maes HH, Neale MC, Eaves LJ. (1997). Genetic and environmental factors in relative body weight and human adiposity. *Behav Genet 27*, 325–351.

Manolio TA, Collins FS. (2009). The HapMap and genome-wide association studies in diagnosis and therapy. *Annu Rev Med 60*, 443–456.

Mardis, E. R. (2008) The impact of next-generation sequencing technology on genetics. *Trend Genet*, 24, 133–141.

Marg A, Sirim P, Spaltmann F, et al. (1999). Neurotractin, a novel neurite outgrowth-promoting Ig-like protein that interacts with CEPU-1 and LAMP. *J Cell Biol 145*, 865–876.

Mascari MJ, Gottlieb W, Rogan PK, et al. (1992). The frequency of uniparental disomy in Prader-Willi syndrome. Implications for molecular diagnosis. *N Engl J Med 326*, 1599–1607.

McCarthy MI, Abecasis GR, Cardon LR, et al. (2008). Genome-wide association studies for complex traits: consensus, uncertainty and challenges. *Nat Rev Genet 9*, 356–369.

McCarthy MI, Hattersley AT. (2008). Learning from molecular genetics: novel insights arising from the definition of genes for monogenic and type 2 diabetes. *Diabetes 57*, 2889–2898.

McCarthy SE, Makarov V, Kirov G, et al. (2009). Microduplications of 16p11.2 are associated with schizophrenia. *Nat Genet 41*, 1223–1227.

McLin VRA, Rankin SA, Zorn AM. (2007). Repression of Wnt/beta-catenin signaling in the anterior endoderm is essential for liver and pancreas development. *Development 134*, 2207–2217.

Meyre D, Delplanque J, Chevre JC, et al. (2009). Genome-wide association study for early-onset and morbid adult obesity identifies three new risk loci in European populations. *Nat Genet 41*, 157–159.

Miranda JJ, Kinra S, Casas, JP, Davey Smith G, Ebrahim S. (2008). Non-communicable diseases in low- and middle-income countries: context, determinants and health policy. *Trop Med Int Health 13*, 1225–1234.

Misra A, Khurana L. (2008). Obesity and the metabolic syndrome in developing countries. *J Clin Endocrinol Metab 93*, S9–30.

Montague CT, Farooqi IS, Whitehead JP, et al. (1997). Congenital leptin deficiency is associated with severe early-onset obesity in humans. *Nature 387*, 903–908.

Moore AF, Florez JC. (2008). Genetic susceptibility to type 2 diabetes and implications for antidiabetic therapy. *Annu Rev Med 59*, 95–111.

Murphy R, Ellard S, Hattersley AT. (2008). Clinical implications of a molecular genetic classification of monogenic [beta]-cell diabetes. *Nat Clin Pract End Met 4*, 200–213.

Negro F, Alaei M. (2009). Hepatitis C virus and type 2 diabetes. *World J Gastroenterol 15*, 1537–1547.

O'Rahilly S. (2007). Human obesity and insulin resistance: lessons from experiments of nature. *Biochem Soc Trans 35*, 33–36.

O'Rahilly S. (2009). Human genetics illuminates the paths to metabolic disease. *Nature 462*, 307–314.

Ono T, Guthold R, Strong K. (2005). *WHO Global Comparable Estimates*. WHO Global Infobase.

Pearson ER, Flechtner I, Njolstad PR, et al., The Neonatal Diabetes International Collaborative Group. (2006). Switching from insulin to oral sulfonylureas in patients with diabetes due to Kir6.2 mutations. *N Engl J Med 355*, 467–477.

Pizzuti A, Frittitta L, Argiolas A, et al. (1999). A polymorphism (K121Q) of the human glycoprotein PC-1 gene coding region is strongly associated with insulin resistance. *Diabetes 48*, 1881–1884.

Prokopenko I, Langenberg C, Florez JC, et al. (2009). Variants in MTNR1B influence fasting glucose levels. *Nat Genet 41*, 77–81.

Ramachandran A, Snehalatha C, Kapur A, et al. (2001). High prevalence of diabetes and impaired glucose tolerance in India: National Urban Diabetes Survey. *Diabetologia 44*, 1094–1101.

Rosenberg NA, Huang L, Jewett EM, Szpiech ZA, Jankovic I, Boehnke M. (2010). Genome-wide association studies in diverse populations. *Nat Rev Genet 11*, 356–366.

Sandholt CH, Sparsã T, Grarup N, et al. (2010). Combined analyses of 20 common obesity susceptibility variants. *Diabetes 59*, 1667–1673.

Sandhu MS, Weedon MN, Fawcett KA, et al. (2007). Common variants in WFS1 confer risk of type 2 diabetes. *Nat Genet 39*, 951–953.

Saxena R, Voight BF, Lyssenko V, et al. (2007). Genome-wide association analysis identifies loci for type 2 diabetes and triglyceride levels. *Science 316*, 1331–1336.

Schafer M, Brauer AU, Savaskan NE, Rathjen FG, Brummendorf T. (2005). Neurotractin/kilon promotes neurite outgrowth and is expressed on reactive astrocytes after entorhinal cortex lesion. *Mol Cell Neurosci 29*, 580–590.

Scott LJ, Mohlke KL, Bonnycastle LL, et al. (2007). A genome-wide association study of type 2 diabetes in Finns detects multiple susceptibility variants. *Science 316*, 1341–1345.

Scuteri A, Sanna S, Chen WM, et al. (2007). Genome-wide association scan shows genetic variants in the FTO gene are associated with obesity-related traits. *PLoS Genet 3*, e115.

Sebat J, Lakshmi B, Malhotra D, et al. (2007). Strong association of de novo copy number mutations with autism. *Science 113*, 8659.

Shaw JE, Sicree RA, Zimmet PZ. (2010). Global estimates of the prevalence of diabetes for 2010 and 2030. *Diabetes Res Clin Pract 87*, 4–14.

Sladek R, Rocheleau G, Rung J, et al. (2007). A genome-wide association study identifies novel risk loci for type 2 diabetes. *Nature 445*, 881–885.

Spiegel AM. (1990). Albright's hereditary osteodystrophy and defective G proteins. *N Engl J Med 322*, 1461–1462.

Steinthorsdottir V, Thorleifsson G, Reynisdottir I, et al. (2007). A variant in CDKAL1 influences insulin response and risk of type 2 diabetes. *Nat Genet 39*, 770–775.

Stolerman ES, Florez JC. (2009). Genomics of type 2 diabetes mellitus: implications for the clinician. *Nat Rev Endocrinol 5*(8), 429–436.

Stoy J, Edghill EL, Flanagan SE, et al. (2007). Insulin gene mutations as a cause of permanent neonatal diabetes. *Proceedings of the National Academy of Sciences USA 104*, 15040–15044.

Stutzmann F, Tan K, Vatin V, et al. (2008). Prevalence of melanocortin-4 receptor deficiency in Europeans and their age-dependent penetrance in multigenerational pedigrees. *Diabetes 57*, 2511–2518.

Tadmouri GO, Nair P, Obeid T, al Ali MT, al Khaja N, Hamamy HA. (2009). Consanguinity and reproductive health among Arabs. *Reprod Health 6*, 17.

Teebi AS. (1994). Autosomal recessive disorders among Arabs: an overview from Kuwait. *J Med Genet 31*, 224–233.

Teo YY, Fry AE, Bhattacharya K., Small, KS, Kwiatkowski, DP, Clark TG. (2009). Genome-wide comparisons of variation in linkage disequilibrium. *Genome Res 19*, 1849–1860.

Teo YY, Small KS, Kwiatkowski DP. (2010). Methodological challenges of genome-wide association analysis in Africa. *Nat Rev Genet 11*, 149–160.

Thorleifsson G, Walters GB, Gudbjartsson DF, et al. (2009). Genome-wide association yields new sequence variants at seven loci that associate with measures of obesity. *Nat Genet 41*, 18–24.

Tishkoff SA, Reed FA, Friedlaender FR, et al. (2009). The genetic structure and history of Africans and African Americans. *Science 324*, 1035–1044.

Tishkoff SA, Williams SM. (2002). Genetic analysis of African populations: human evolution and complex disease. *Nat Rev Genet 3*, 611–621.

Uauy R, Albala C, Kain J. (2001). Obesity trends in Latin America: transiting from under- to overweight. *J Nutr 131*, 893S–899S.

Unoki H, Takahashi A, Kawaguchi T, et al. (2008). SNPs in KCNQ1 are associated with susceptibility to type 2 diabetes in East Asian and European populations. *Nat Genet 40*, 1098–1102.

Vaag A, Lund SS. (2007). Non-obese patients with type 2 diabetes and prediabetic subjects: distinct phenotypes requiring special diabetes treatment and (or) prevention? *Appl Physiol Nutr Metab 32*, 912–920.

Van Den Ouweland JM, Lemkes HH, Trembath RC, et al. (1994). Maternally inherited diabetes and deafness is a distinct subtype of diabetes and associates with a single point mutation in the mitochondrial tRNA(Leu(UUR)) gene. *Diabetes 43*, 746–751.

Voight BF, Scott LJ, Steinthorsdottir V, et al. (2010). Twelve type 2 diabetes susceptibility loci identified through large-scale association analysis. *Nat Genet 42*, 579–589.

Walley AJ, Asher JE, Froguel P. (2009). The genetic contribution to non-syndromic human obesity. *Nat Rev Genet 10*, 431–442.

Walsh T, McClellan JM, McCarthy SE, et al. (2008). Rare structural variants disrupt multiple genes in neurodevelopmental pathways in schizophrenia. *Science 320*, 539–543.

Walters RG, Jacquemont S, Valsesia A, et al. (2010) A new highly penetrant form of obesity due to deletions on chromosome 16p11.2. *Nature 463*, 671–675.

Wardle J, Carnell S, Haworth CM, Plomin R. (2008). Evidence for a strong genetic influence on childhood adiposity despite the force of the obesogenic environment. *Am J Clin Nutr 87*, 398–404.

Watanabe RM, Allayee H, Xiang AH, et al. (2007). Transcription factor 7-like 2 (TCF7L2) is associated with gestational diabetes mellitus and interacts with adiposity to alter insulin secretion in Mexican Americans. *Diabetes 56*, 1481–1485.

White DR, Ganesh A, Nishimura D, et al. (2007). Autozygosity mapping of Bardet-Biedl syndrome to 12q21.2 and confirmation of FLJ23560 as BBS10. *Eur J Hum Genet 15*, 173–178.

Wildman RP, Muntner P, Reynolds K, et al. (2008). The obese without cardiometabolic risk factor clustering and the normal weight with cardiometabolic risk factor clustering: prevalence and correlates of 2 phenotypes among the US population (NHANES 1999–2004). *Arch Intern Med 168*, 1617–1624.

Willer CJ, Speliotes EK, Loos RJ, et al. (2009). Six new loci associated with body mass index highlight a neuronal influence on body weight regulation. *Nat Genet 41*, 25–34.

Willi C, Bodenmann P, Ghali WA, Faris PD, Cornuz J. (2007). Active smoking and the risk of type 2 diabetes: a systematic review and meta-analysis. *JAMA 298*, 2654–2664.

Winckler W, Weedon MN, Graham RR, et al. (2007). Evaluation of common variants in the six known maturity-onset diabetes of the young (MODY) genes for association with type 2 diabetes. *Diabetes 56*, 685–693.

World Health Organization [WHO]. (2006). Obesity and overweight. *Factsheet No. 311.*

World Health Organization [WHO]. (2009). Diabetes. *Factsheet No. 312.*

Yang J, Benyamin B, Mcevoy BP, et al. (2010). Common SNPs explain a large proportion of the heritability for human height. *Nat Genet 42*, 565–569.

Yasuda K, Miyake K, Horikawa Y, et al. (2008). Variants in KCNQ1 are associated with susceptibility to type 2 diabetes mellitus. *Nat Genet 40*, 1092–1097.

Yeo GS, Connie Hung CC, Rochford J, et al. (2004). A de novo mutation affecting human TrkB associated with severe obesity and developmental delay. *Nat Neurosci 7*, 1187–1189.

Zeggini E, Scott LJ, Saxena R, et al. (2008). Meta-analysis of genome-wide association data and large-scale replication identifies additional susceptibility loci for type 2 diabetes. *Nat Genet 40*, 638–645.

Zeggini E, Weedon MN, Lindgren CM, et al. (2007). Replication of genome-wide association signals in UK samples reveals risk loci for type 2 diabetes. *Science 316*, 1336–1341.

Zlotogora J. (1997a). Autosomal recessive diseases among Palestinian Arabs. *J Med Genet 34*, 765–766.

Zlotogora J. (1997b). Genetic disorders among Palestinian Arabs: 1. Effects of consanguinity. *Am J Med Genet 68*, 472–475.

24.

HEALTH IMPLICATIONS OF FAMILIAL HYPERLIPIDEMIA

Deepak Bhatnagar

INTRODUCTION

LIPID AND LIPOPROTEIN METABOLISM

Lipids play a pivotal role in energy metabolism, the structural integrity of the cell, in the formation of steroid hormones, intracellular signaling, and nerve function. In addition, they have a crucial part in regulation of inflammation and protein modification. There are several classes of lipids, but this chapter is limited to the clinically conventional lipids, cholesterol and triglycerides. In order to maintain the above functions, the metabolism of lipids is finely regulated and integrated with many nonlipid pathways (Durrington, 2007). Essentially there is a "delivery system" that ensures the tissues renew lipid stores to maintain functionality. Broadly speaking, it involves the movement of cholesterol and triglycerides from the intestine to the liver, and then to the tissues. For every delivery system there needs to be a "removal system" that prevents excess lipid accumulation, which can be harmful. This is termed *reverse cholesterol transport,* with much of the excess cholesterol recycled back to the liver.

Lipids are not water soluble, and the need to transport them across a predominantly water-based milieu in the body means that they need to form macromolecular complexes with proteins. This combination of lipids and proteins is called *lipoproteins.* The proteins in lipoproteins are called *apolipoproteins,* and with a diverse range of structures they play an important role in regulation of lipid metabolism. Lipoproteins can be classified according to their hydrated density or composition, or molecular mass or molecular diameter. They are taken up by the tissues by receptor-mediated uptake, with the apolipoproteins playing a key role in recognition of lipoproteins. A brief outline of lipid and lipoprotein metabolism is outlined in Figure 24-1 (Bhatnagar et al., 2008).

FACTORS THAT INFLUENCE SERUM LIPIDS

The concept of "normality" is challenging in clinical medicine, and this is particularly so with serum lipid concentrations (Durrington, 2007). The main influence on serum lipid concentrations is nutritional, with a partial contribution of genetics, perhaps, at a general population level. The many observational studies and treatment trials indicate that in the prevention of CHD, the lower the serum cholesterol or LDL cholesterol the less the chances of CHD (Bhatnagar, 1998). This often translates into debate around diet, affluence, and increasing urbanization, "Westernization," or globalization (Narayan et al., 2010; Reddy, 2002; Mendis, 2005). Should the normal serum cholesterol in an adult be the same as that of a newborn baby, or people living subsistence or aboriginal lives in primordial conditions? Should we be aiming at serum cholesterol concentrations seen in rural Japanese farmers or Chinese peasants, where CHD is much lower than that seen in more industrialized societies? These arguments have largely been overtaken by the inexorable increase in prosperity of the emerging economies, and the accompanying increase in energy intake and overnutrition.

Serum triglycerides, the other serum lipid seen on laboratory reports, are also influenced by lifestyle. One example of this is seen in subjects who are obese or who drink excess alcohol. Both are lifestyle disorders, and are associated with mild hypertriglyceridemia (up to 3 to 4mmol/L). In most emerging economies, rates of obesity are increasing and, where culturally permitted, alcohol consumption is increasing. As a result, in many of these countries both serum cholesterol and triglyceride levels are rising (Bhatnagar et al., 1995; Patel et al., 2006; Jeemon et al., 2009. Similarly, HDL cholesterol, which is inversely related to CHD in population studies and also to serum triglycerides, is lower in developing nations. However, HDL cholesterol is often low in societies consuming a high-carbohydrate diet.

Ideally, the results of serum lipids should be interpreted in the fasting state, as serum triglycerides increase postprandially, but serum cholesterol and HDL cholesterol are unaffected by fasting. There is an inherent biological variation in serum lipids, which in women is also affected by the menstrual cycle. In most societies, serum lipids increase with age, and there appears to be a seasonal variation in temperate climates. By far the greatest variability

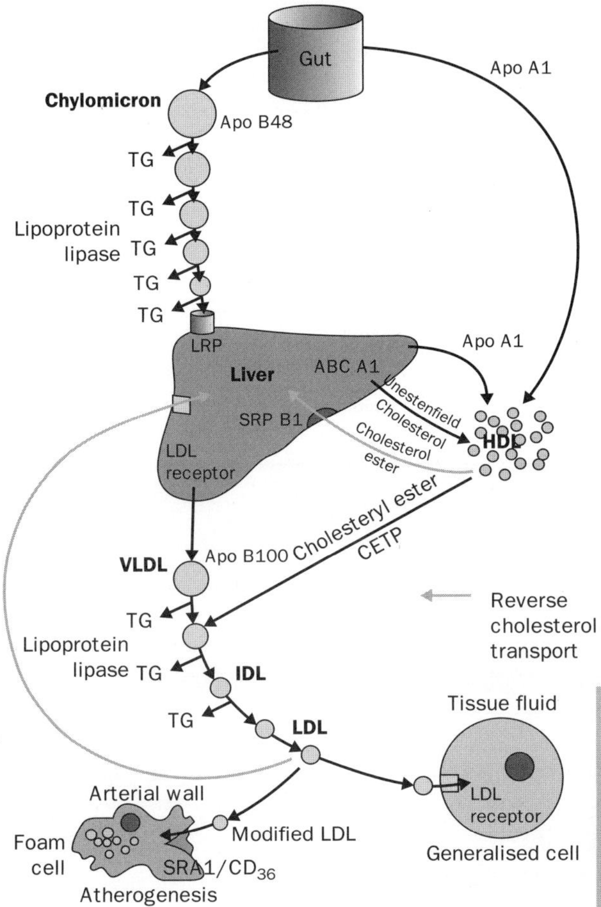

Dietary cholesterol absorbed in the gut is transported to the liver in chylomicrons, which are rich in triglyceride. Cholesterol is also synthesised de novo in the liver, gut, and central nervous system. Cholesterol in the liver is packaged along with triglycerides and secreted as VLDL, which is then converted into LDL in the circulation. LDL delivers cholesterol to most tissues, and the uptake of cholesterol is facilitated by LDL receptors. Excess cholesterol is removed from cells by HDL and is then either transferred to LDL and VLDL, whence it can return to the liver via hepatic LDL receptors or be taken up directly from HDL via the hepatic class B scavenger receptors. cholesterol is also excreted in bile both as free cholesterol and as bile acids, which are both subject to enterohepatic circulation. Most of the plasma cholesterol is present in LDL, with smaller amounts in VLDL and HDL. Total plasma cholesterol mainly reflects LDL cholesterol. In the fasting state, plasma triglyceride generally indicates the VLDL concentration.

LDL – low density lipoprotein
HDL – High density lipoprotein
VLDL – Very low density lipoprotein
IDL – Intermediate density lipoprotein
CETP – cholesteryl ester transfer protein
ABCAI – ATP binding cassette transporter 1
SRB 1 – class B scavenger receptors
LRP – lipoprotein related receptor
TG – triglyceride
Apo – apolipoprotein
SRA1 – steroid receptor RNA activator
$CD_{3/4}$ – thrombospondin receptor

Figure 24-1 The Role of Cholesterol in the Metabolism of Lipoprotein.
SOURCE: Adapted, with permission, from Bhatnagar et al., 2008

in serum lipids comes not just from physiological factors, but also the variability in laboratory measurements.

INFLUENCE OF GENOMIC VARIATION ON DISTRIBUTION OF LIPID LEVELS AND EFFECT ON DISEASE

The many small case-control studies of the past decade, describing linkages using candidate genes or markers, have made a limited contribution to the understanding of the genetic influences on lipoproteins. Of course, monogenic disorders of lipoprotein metabolism have helped greatly in delineating lipoprotein metabolism. This initial approach, to characterize the mutations in those with monogenic disorders or study those at the tail-end of normal distribution, has now been overtaken by advances in technology and the rapidity with which genome-wide association studies (GWAS) can be done (Hegele, 2009; Kullo & Cooper, 2010; Holmes et al., 2010)

GWAS of serum lipids indicate that the effect of common variants is small (Pirruccello & Kathiresan, 2010). Increasingly, studies are being carried out in large numbers, and this will allow the identification of large gene loci with common variants. A worldwide collaborative approach will yield more information (Rotimi & Jorde, 2010), and although at present some of the data highlight heritability, they are not yet robust enough to allow for genetic risk stratification (Hofman et al., 2010; Shuldiner & Pollin, 2010). However, GWAS has helped to identify many proteins which are closely related to lipid metabolism, but whose functions are not yet clear. These may well help in finding new drug targets. While modeling environmental factors in GWAS may help find new loci, whole-genome sequencing in the future may be more useful (Igl et al., 2010).

GENETIC DISORDERS OF LIPID METABOLISM

The majority of hyperlipidemia seen in routine clinical practice is diet and lifestyle-related. Guidelines vary between countries on when to intervene with lipid-lowering drugs in such people. Of course, factors other than serum lipids are also responsible for CHD, and most risk estimate equations take this into account. However, risk equations should not be applied to those with secondary hyperlipidemias (Table 24-1) or primary hyperlipidemias (Table 24-2). This chapter will touch on two genetic hyperlipidemias that present as mixed hyperlipidemia, and focus in particular on familial hypercholesterolemia (FH) which,

TABLE 24-1 CAUSES OF SECONDARY HYPERLIPIDEMIA

Disorders/Drugs causing increase in serum cholesterol	Disorders/Drugs causing increase in serum triglycerides	Disorders/Drugs causing mixed hyperlipidemia
Nephrotic syndrome	Uncontrolled diabetes mellitus	Type2 diabetes mellitus
Obstructive jaundice	Chronic renal failure	Ethanol excess
Hypothyroidism	Peritoneal dialysis	Renal transplantation
Anorexia nervosa	Systemic lupus erythematosus	Hemodialysis
	Acromegaly	Parenteral nutrition
	Partial lipodystrophy	
Thiazide diuretics	Glycogen storage disorders	Glucocorticoids
Cyclosporine	Beta blockers	Phenytoin
	Anabolic steroids	Protease inhibitors
	Oral contraceptives	
	Estrogens	
	Isotretinoin	
	Protease inhibitors	

TABLE 24-2 CAUSES OF PRIMARY HYPERLIPIDEMIA

Disorder	Remarks
DISORDERS DUE TO DEFECTS IN LIPOPROTEIN RECEPTORS	
Familial Hypercholesterolemia	Serum cholesterol elevated since birth Premature CAD Defect in LDL receptor
DISORDERS DUE TO OVERPRODUCTION OF LIPOPROTEINS	
Familial combined hyperlipidemia	Mixed hyperlipidemia, usually apparent around 30–40 years of age No characteristic clinical signs Premature CAD Overproduction of VLDL
Familial hypertriglyceridemia	Moderate to severe hypertriglyceridemia Risk of pancreatitis Overproduction of VLDL
DISORDERS DUE TO DEFECTS IN APOLIPOPROTEINS	
Remnant hyperlipidaemia	Mixed hyperlipidemia Characteristic clinical signs Premature CHD and PVD Apolipoprotein E2 homozygosity
Familial defective apolipoprotein B	Biochemical and clinical features similar to heterozygous FH Defective ligand for LDL receptor
Apolipoprotein CII deficiency	Severe hypertriglyceridemia due to lack of lipoprotein lipase activity
Familial hypoalphalipoproteinemia	Low HDL cholesterol values
Abetalipoproteinemia	Very low serum triglycerides and cholesterol Progressive neurological disorder Steatorrhoea
DISORDERS DUE TO DEFECTS OF ENZYMES AND PROTEINS INVOLVED IN LIPOPROTEIN	
metabolism	
Familial hyperchylomicronemia	Severe hypertriglyceridemia due to lack of lipoprotein lipase
Hepatic lipase deficiency	Predominant hypertriglyceridemia Clinical signs similar to remnant hyperlipidemia
α Lecithin cholesterol acyltransferase deficiency	Hypertriglyceridemia Proteinuria Corneal opacities
Cholesteryl ester hydrolase deficiency	Hypercholesterolemia Hepatosplenomegaly
Cholesteryl ester transfer protein deficiency	High HDL cholesterol

(Continued)

TABLE 24-2 (CONTINUED)

Disorder	Remarks
MISCELLANEOUS DISORDERS	
Polygenic hypercholesterolemia	Hypercholesterolemia associated with a family history
Hyperalphalipoproteinemia	High HDL-cholesterol levels; may be related to CETP deficiency
Hyperapobetalipoproteinemia	Elevated apolipoprotein B in the presence of normal LDL cholesterol values; associated with increased CHD risk
Elevated Lp(a)	Partial homology with plasminogen; associated with increased CHD risk
LDL oxidation	Oxidative modification of LDL leads to increased uptake by scavenger pathway, thereby leading to atherogenesis

(Adapted with permission from Bhatnagar, 1998).

with a general prevalence of 1in 500, is a relatively common genetic disorder. Most primary disorders of lipid metabolism generally have signs associated with them, and either increase the risk of premature CHD or pancreatitis (Table 24-3).

FAMILIAL COMBINED HYPERLIPIDEMIA (FCH)

FCH appears to be a relatively common disorder that appears around the age of 35 or 40 years. The genetic basis of FCH is not clear, but there is a distinct inheritance pattern (Shoulders et al., 2004). Patients present with a mixed hyperlipidemia and have premature CHD. There are no particular distinguishing clinical features, although many will have corneal arcus and some will have xanthelasmata. Most patients are obese, and the condition clusters with hypertension and diabetes mellitus. There is much overlap with a newly recognized syndrome of the metabolic syndrome, which unfortunately has multiple definitions and no firm genetic basis apart from the atherogenic lipoprotein phenotype, seen typically in what is known as FCH (Durrington, 2007).

TYPE III OR REMNANT HYPERLIPIDEMIA

Type III or remnant hyperlipidemia has a distinct genetic basis with homozygosity for apolipoprotein E2. It is thought to result from accumulation of intermediate-density lipoproteins. It is has an estimated frequency of 1 in 10,000, and patients can present with mixed hyperlipidemia soon after puberty. Untreated, they have striate palmar xanthomata and eruptive xanthomata on the pressure points, such as elbows, shoulders and buttocks. Over time, most develop CHD and have a predilection for peripheral vascular disease (Durrington, 2007).

TABLE 24-3 SOME PHYSICAL SIGNS ENCOUNTERED IN FAMILIAL HYPERLIPIDEMIAS

Eyes	Skin	Systemic
Corneal arcus	Xanthelasmata	Aortic stenosis
Corneal opacities	Tendon xanthomata	Hepatosplenomegaly
Lipaemia retinalis	Eruptive xanthomata	Peripheral neuropathy
	Tubero-eruptive xanthomata	Ataxia
	Planar xanthomata	

(Adapted with permission from Bhatnagar, 1998).

FAMILIAL HYPERCHOLESTEROLEMIA

Familial hypercholesterolemia (FH) is an autosomal dominant inherited disorder of lipoprotein metabolism, characterized by increased LDL cholesterol that is present since birth (Brown & Goldstein, 1986). If untreated it leads to premature CHD (Austin et al., 2004a). It is distinct from simple heritable conditions that lead to increased serum cholesterol. It results from a defect in the LDL receptor gene (Lindgren et al., 1985), but other single-gene mutations causing hypercholesterolemia can also produce a clinical phenotype similar to FH (Bhatnagar, 2006; Table 24-1). The LDL receptor gene is situated on the short arm of chromosome 19, has 18 exons, and stretches over 45 kb (Yamamoto et al., 1984; Schneider, 1982), and its product, the LDL receptor, is responsible for removing LDL from plasma. More than 800 mutations have been described worldwide (Fokkema et al., 2005) and are recorded on an online database (http://www.ucl.ac.uk/fh – accessed October 31, 2010). A review of genetic causes of FH is also available on the Human Genome Epidemiology network (Austin et al., 2004b). Most mutations seem to occur in exon 4, but variants occur along all 18 exons and in the promoter region. A small number of mutations are deletions or duplications, and are found in exons 1–18 and intron 12. It is possible to classify the mutations on the basis of the loss of function they produce, but this classification can be misleading, as some mutations may fall into two or more functional classes. Generally, there is geographical variation of the mutations, except in populations who have been genetically isolated. For example, the Afrikaans, French Canadians, Tunisians, Lebanese,

Danish, Japanese, and North Koreans have a common gene pool, and FH is more common (Austin et al., 2004b). A variety of other genes that are linked with the clinical phenotype of FH have recently been identified. Notable among these is *familial defective apolipoprotein B* (FDB; see Myant et al., 1991). This condition results from a mutation in apolipoprotein B, the ligand for the LDL receptor, which itself is normal.

CHARACTERISTIC CLINICAL FEATURES OF FH

The hypercholesterolemia of FH results from an increase in LDL particle numbers, resulting from decreased catabolism due to an LDL receptor defect (Langer et al., 1972), but there may also be some overproduction of LDL (Soutar et al., 1977). HDL levels are generally low in FH patients, and lipoprotein (a) levels are increased (Utermann et al., 1989; Mbewu et al., 1991). The FH clinical phenotype (Table 24-4) is variable, and as is the case with serum lipids in the general population, is affected by environmental, metabolic, and other genetic factors (Thompson et al., 1989; Hill et al., 1991).

Patients with heterozygous FH characteristically have tendon xanthomata present in the extensor tendons. They are commonly found in Achilles tendons or the dorsum of the hands, and rarely in the triceps tendons and over the tibial tuberosity. They can be unilateral, and generally appear after the age of 30, although there is some evidence that they appear at an earlier age in FH due to certain mutations. In some patients, Achilles tendinitis can be a presenting symptom, especially so in patients with ill-fitting shoes, or those who are physically active. The presence of tendon xanthomata in Achilles tendons weakens them and can lead to tendon rupture. It is unclear, but perhaps there is a relationship between tendon xanthomata, and the onset or severity of CHD. Most patients will also have corneal arcus, but this can be seen in other genetic lipid disorders, too. Cutaneous xanthelasmata, especially around the eyes, are also nonspecific but are frequently encountered in FH. They are often bilateral and are frequently removed for cosmetic reasons. Aortic stenosis is also seen in patients with heterozygous FH, and is nearly always present in patients with homozygous FH (Beppu et al., 1983).

Patients with heterozygous FH, especially untreated, have premature CHD and peripheral vascular disease (Goldstein, 1972; Beaumont et al., 1976; Mabuchi et al., 1986; Hutter et al., 2004; Koeijvoets et al., 2005), but the relationship between ischemic stroke and FH is unclear (Huxley et al., 2003). In patients with homozygous FH, sudden death from CHD is not uncommon and often, despite aggressive management with LDL apheresis and prophylactic coronary artery bypass grafting, the prognosis remains poor (Goldstein, 1972).

The nature and onset of CHD in heterozygotes is variable. The natural history of CHD in FH is drawn from previous studies carried out in the United States, United Kingdom, Scandinavia, Canada, France, Holland, and Japan (Stone et al., 1974; Slack, 1969; Jensen et al., 1967; Heiberg, 1975). Not all these studies were well designed, and in some cases patients were on non-statin lipid-lowering drugs. The association between the severity of CHD and specific FH mutations has been sought (Gudnason et al., 1994; Vuorio et al., 2001; Sun et al., 2006) but are difficult to establish, as the series of patients examined have been small, with large allelic variation and a variation in the definition of CHD.

The more severe form of homozygous FH results when a double dose of the defective gene is inherited, or the subject is a compound heterozygote. These patients invariably develop CHD before the age of 25 years with homozygous FH, and in addition to tendon xanthomata they can develop cutaneous and tuberous xanthomata on the knees, elbows, buttocks, and the dorsum of the hands. In many cases CHD appears in childhood, but is difficult

TABLE 24-4 MONOGENIC DISORDERS THAT CAN PRESENT WITH A CLINICAL PHENOTYPE OF FAMILIAL HYPERCHOLESTEROLEMIA

Disorder	Gene involved	Inheritance
Classical familial hypercholesterolemia	LDL receptor	Codominant
Familial defective apolipoprotein B	Apolipoprotein B	Codominant
Autosomal dominant familial hypercholesterolemia	PCSK9	Dominant
FH3 (NARC 1)	convertase subtilisin/kexin 9) (unknown function)	
Autosomal recessive hypercholesterolemia (ARH)	ARH (involved in internalization of LDL receptor)	Recessive
Cholesterol 7 alpha hydroxylase deficiency	CYP7A1 (involved in bile acid synthesis) Recessive	Recessive
Sitosterolemia	ABC G5 and ABC G8	Recessive

to diagnose unless there is a high degree of suspicion (Goldstein, 1972; Sprecher et al., 1984).

Hypertension, hypertriglyceridemia, or hyperuricemia, often an association with other forms of hyperlipidemia, are uncommon in patients with FH (Durrington, 2007). The reason for this is not clear. Hypertension and hypertriglyceridemia are also recognized features of familial combined hyperlipidemia (FCH) and, clinically, this difference may help distinguish FH from FCH.

It is now recognized that the FH clinical phenotype is variable. People with the same LDL receptor mutation can have varying degrees of hypercholesterolemia, although generally patients from the same family have similar serum cholesterol levels (Kotze et al., 1993). Not all the classical features of FH originally described are present in all patients. Genetic, metabolic, and environmental factors can all influence the phenotypic variability, but the studies to establish these data have small numbers and are derived mainly from case-control studies. Attempts have been made to determine the effects of mutations on phenotypic variability by pooling details from different, mainly European populations. The results are variable, but appear to suggest that receptor-negative mutations are associated with a more severe phenotype, compared with receptor-defective mutations (Vohl et al., 1997; Gaudet et al., 1999; Bertolini et al., 2000).

The FH phenotype can be affected by genes that influence the assembly or secretion or remodeling of lipoproteins. The apolipoprotein E genotype, which has some effect on serum cholesterol concentrations, also has some effect on the FH phenotype, but this varies with the population studied (Eto et al., 1988; Duly et al., 1997; Friedlander & Leitersdorf, 1996). The effect of genetic factors such as lipoprotein lipase, cholesteryl ester transfer protein (CETP), microsomal triglyceride transfer protein (MTP), hepatic lipase, fatty acid-binding protein, methylene tetrahydrofolate reductase (MTHFR), ATP-binding cassette A1 (ABCA1), renin-angiotensin, and paraoxonase have been described in FH, but it is unusual for them to exert a major influence on the FH phenotype (Bhatnagar, 2006). Gender does seem to have a major influence on the FH phenotype, with women demonstrating clinical features of CHD later than men, but earlier than women without FH.

The expression of LDL receptors can be influenced by variation in physiological or pathological in factors in thyroid hormone or estrogen levels. Other metabolic factors, such as prolonged fasting or dramatic weight loss, can also influence LDL cholesterol levels in FH. Lipoprotein (a), which has been an established as a risk factor for CHD, is increased in FH (Seed et al., 1990). However, its role in increasing CHD risk in FH is unclear. The presence of small, dense LDL particles is associated with increased CHD risk in FH. Remnant-like proteins are increased in FH and may lead to increased CHD in patients. The role of plasma fibrinogen, C-reactive protein, and homocysteine has been explored in FH, but there appears to be no certain link in contributing to increasing risk of CHD in FH.

Increased LDL cholesterol and premature CHD are significantly influenced by environmental factors such as diet, behavior, and cultural factors, and also affect the FH phenotype. For example, LDL cholesterol levels and CHD are reported to be lower in FH heterozygotes in China (Pimstone et al., 1998) and Cuba (Pereira et al., 1995). With increasing recognition of elevated serum cholesterol as a risk factor, physicians are increasingly treating hypercholesterolemia at an earlier stage, often without establishing its origin or nature. Therefore, statin treatment often modifies the FH phenotype, delaying the appearance of signs such as tendon xanthomata. Of course, the added advantage of early cholesterol-lowering treatment is to decrease the chance of premature CHD. This is, of course, beneficial, but would affect screening of relatives of the proband.

SCREENING STRATEGIES FOR FH

With its prevalence of about 1 in 500, FH is a relatively common genetic disorder, but the population at risk is much smaller compared with other lipid-related coronary risk factors. However, the relative odds of a patient with FH developing CHD are the highest compared with common hypercholesterolemia, as the exposure to elevated serum cholesterol is present from birth. FH fulfills the Wilson and Jungner (1968) criteria for screening, but as yet there are very few programs for the condition, as a lack of resources has made it less of a priority for public health bodies. Moreover, a wide variety of mutations and phenotypic variability have added to the difficultly in establishing universal diagnostic criteria. Worldwide, currently three diagnostic criteria for FH are used. These include the Simon Broome Register criteria put forward in 1991 in the United Kingdom (Scientific Steering Committee, 1991), the US MEDPED criteria, published in 1993 (Williams et al., 1993), and the Dutch Lipid Clinic Networks criteria for FH published in 1999 (WHO, 1999).

The various options for screening include: (a) population screening; (b) screening in the clinical setting; (c) screening patients with premature CHD; (d) screening relatives of FH probands with diagnostic criteria based on laboratory and clinical findings; and (e) screening relatives of FH probands with genetic testing. Ideally, screening programs should be based on high-quality randomized controlled trials, with an agreed set of quality assurance standards for managing and monitoring the screening program (McQueen, 2002). In practice, screening is often driven by patients, their relatives and patient organizations, or interested clinicians or the availability of a new test. Psychological and economic issues are important, as are issues regarding ethics and logistics (Marteau & Croyle, 1998; Andersen et al., 1997; Senior et al., 1999). In addition, the sensitivity and specificity of a test are also important, as is an understanding as to how clinicians will use a particular test for making clinical decisions.

Universal screening is feasible, but is unlikely to be cost-effective even in founder populations where the prevalence of FH is high. Screening in the clinical setting can be carried out in an outpatient setting using clinical and biochemical features of FH. For example, most patients with FH will have pure hypercholesterolemia and will generally have serum cholesterol values greater than 7 or 7.5 mmol/L in a Western setting. Once secondary causes of hyperlipidemia, such as thyroid, liver, or renal function are ruled out, a personal, social, and occupational history will help establish any lifestyle factors that may lead to a poor diet. However, it is unusual to see serum cholesterol concentrations greater than 7mmol/L simply as a result of a diet high in saturated fat. Tendon xanthomata are, of course, diagnostic, but it is also worthwhile listening for the murmur of aortic stenosis. Not all patients with FH have clinical signs (especially those on treatment), which are, in any case, rarely seen in patients below the age of 30 years. The presence of corneal arcus and/or xanthelasmata, particularly in association with hypercholesterolemia, is suggestive of a genetic disorder of lipoprotein metabolism.

It may be possible to distinguish other lipid disorders, such as familial combined hyperlipidemia (FCH), type III or remnant hyperlipidemia, and secondary hyperlipidemia due an obstructive liver disorder from FH, in a clinical setting. Patients with FCH often have a strong family history of hyperlipidemia and premature CHD, but no tendon xanthomata. Untreated patients with type III hyperlipidemia may have palmar crease xanthomata and eruptive xanthomata on pressure points. The characteristic mixed hyperlipidemia seen in FCH and type III hyperlipidemia can help distinguish the disorders from the pure hypercholesterolemia seen in FH. However, with increasing obesity and excess alcohol consumption, hypertriglyceridemia can also be seen in patients with FH. Patients with biliary obstruction have abnormal liver function tests, jaundice, pure hypercholesterolemia, and often florid xanthelasmata. It is possible to arrive at a clinical diagnosis of FH either by exclusion of other disorders, or by a combination of clinical signs and personal and family history, but the yield of new patients is likely to be low with opportunistic screening in the clinic. Screening patients with premature-CHD FH may be better than opportunistic screening, but there will be disorders other than FH, and certain polymorphisms will cause premature CHD.

Screening relatives of FH probands is also termed *cascade testing* (Bhatnagar et al., 2000; Leren, 2004) or family tracing, and can be achieved by testing serum cholesterol concentrations or by genetic testing (Hadfield & Humphries, 2005). This method of identifying FH has been tried in several settings in the United Kingdom, Holland, Norway, Iceland, and the United States. The Manchester group from the United Kingdom was able to find 121 new patients from a register of 262 probands, using cascade testing and determining serum cholesterol in 200 relatives of the 262 probands (Bhatnagar et al., 2000). In contrast, if the prevalence of FH is assumed to be one in 500, universal screening of 60,000 tests would have to be carried out to detect a similar number of patients. Most cascade testing is ideally carried out in specialist centers, such as lipid or cardiovascular risk management or metabolic clinics. In the United Kingdom, detailed guidance on screening for FH was published in 2008 (DeMott et al., 2008). Recommendations include clinics run by nurses trained in genetic counseling and cardiovascular risk management, and increased genetic testing.

It is important to establish a formal structured program for cascade testing with a recognized definition of FH, rather than sporadic testing (Newson & Humphries, 2005; Hadfield et al., 2009). Genetic testing for diagnosis has cost implications, but it is hoped that with high-throughput techniques, the cost of finding mutations will drop in due course. Before asking for a genetic screening test, it is best to try and characterize the diagnosis of FH as definite FH, or probable or possible FH, rather than using the term "familial" in a nonspecific or colloquial way.

Probands and their relatives should be approached with sensitivity in order to minimize the anxieties that can be associated with screening. Some young relatives delay having a test FH, and others prefer not to be tested at all. It is reassuring that most studies show that the psychological effects of screening among relatives of probands are no greater than that seen in the general population.

There are no data on the sensitivity and specificity of the different clinical criteria used for screening for FH. A Danish study (Damgaard et al., 2005) compared the Simon Broome, Dutch, and MEDPED criteria in a group of 408 probands with a known FH mutation. The specificity for the presence of a mutation reached 90% in the definite group, but was lower for the possible and probable categories defined by both the Simon Broome and Dutch criteria. All three sets of FH criteria had a low sensitivity (proportion of mutation-positive diagnosis) with figures varying between 34% and 42% in definite FH and about 90% in possible or probable FH.

It is unclear if genetic testing of relatives of probands with FH will be a better option for cascade testing. Allelic heterogeneity, and the other genetic and environmental influences, may limit the utility of genetic testing. Moreover, any approach to screening will probably need to be country- or community-specific. New techniques will not necessarily overcome the issue of the lack of detection of a mutation in someone in whom the clinical type is not suggestive of FH. The current techniques are relatively insensitive in screening the whole LDL or apo B gene, and there are also likely to be genes responsible for FH which are yet to be discovered. The Dutch experience of genetic screening of relatives with FH found that 80% of patients of Dutch origin were known to have different LDL receptor mutations. The group tested the relatives with those known to have a particular mutation. The Dutch program has identified nearly 6000 new patients by this method (Fouchier et al., 2005; Defesche et al., 2004).

Programs for genetic screening are being developed in Denmark, Spain, Northern Ireland, Scotland, and New Zealand. The Icelandic approach suggests that screening all individuals genealogically traced to a common ancestor from a proband may be better than cascade testing (Thorsson et al., 2003). This model is unlikely to be adopted outside Iceland, as most populations are much more heterogeneous. Data indicate that genetic cascade testing would identify between 60% and 80% of patients with definite FH, with much poorer yields in probable and possible FH. Alongside developing genetic services, it is also important to develop proficiency testing of clinical genetic testing laboratories.

Some data on cost-effectiveness of cascade testing for FH are available from the United States (Goldman et al., 1993) and Holland (Marang-van de Mheen et al., 2002; Umans-Eckenhausen et al., 2001; Wonderling et al., 2004). The studies show that identifying cholesterol results in patients leads to benefit from increased and appropriate use of statins, and increased awareness of treatment. Marks et al. (2000) have published a detailed analysis of cost-effectiveness of screening for FH. They conclude that cascade testing using clinical criteria was the most cost-effective, followed closely by genetic cascade testing.

REFERENCES

Andersen LK, Jensen HK, Juul S, Faergeman O. (1997). Patients'attitudes towards detection of heterozygous familial hypercholesterolaemia. *Arch Intern Med 157*, 553–560.

Austin MA, Hutter CM, Zimmern RL, Humphries SE. (2004a). Familial hypercholesterolemia and coronary heart disease: a HuGE association review. *Am J Epidemiol 160*, 421–429.

Austin MA, Hutter CM, Zimmern RL, Humphries SE. (2004b). Genetic causes of monogenic heterozygous familial hypercholesterolemia: a HuGE prevalence review. *Am J Epidemiol 160*, 407–420.

Beaumont V, Jacotot B, Beaumont JL. (1976). Ischaemic disease in men and women with familial hypercholesterolaemia and xanthomatosis. A comparative study of genetic and environmental factors in 274 heterozygous cases. *Atherosclerosis 24*, 441–450.

Beppu S, Minura Y, Sakakibara H, et al. (1983). Supravalvular aortic stenosis and coronary ostial stenosis in familial hypercholesterolemia: two-dimensional echocardiographic assessment. *Circulation 67*, 878–884.

Bertolini S, Cantafora A, Averna M, et al. (2000). Clinical expression of familial hypercholesterolemia in clusters of mutations of the LDL receptor gene that cause a receptor-defective or receptor negative phenotype. *Arterioscler Thromb Vasc Biol 20*, E41–52.

Bhatnagar D, Morgan J, Siddiq S, Mackness MI, Miller JP, Durrington PN. (2000). Outcome of case finding among relatives of patients with known heterozygous familial hypercholesterolaemia. *BMJ 321*, 1497–1500.

Bhatnagar D, Soran H, Durrington PN. (2008). Hypercholesterolaemia and its management. *BMJ 21*, 337.

Bhatnagar D. (2006) Diagnosis and screening for familial hypercholesterolaemia *Ann Clin Biochem 43*, 441–456

Bhatnagar D. (1998). Lipid-lowering drugs in the management of hyperlipidaemia. *Pharmacol Ther 79*, 205–230.

Brown MS, Goldstein JL. (1986). A receptor-mediated pathway for cholesterol homeostasis. *Science 232*, 34–47.

Damgaard D, Larsen ML, Nissen PH, et al. (2005). The relationship of molecular genetic to clinical diagnosis of familial hypercholesterolemia in a Danish population. *Atherosclerosis 180*, 155–160.

Defesche JC, Lansberg PJ, Umans-Eckenhausen MA, Kastelein JJ. (2004). Advanced method for the identification of patients with inherited hypercholesterolemia. *Semin Vasc Med 4*, 59–65.

DeMott K, Nherera L, Shaw EJ, et al. (2008). Clinical Guidelines and Evidence Review for familial hypercholesterolaemia: the identification and management of adults and children with familial hypercholesterolaemia. London: National Collaborating Centre for Primary Care and Royal College of General Practitioners.

Duly EB, Ward AJ, Kirk CW, et al. (1997). APO E genotype and familial hypercholesterolaemia. *Ann Clin Biochem 34*, 534–536.

Durrington PN. (2007). *Hyperlipidaemia: Diagnosis and Management.* Oxford: Butterworth Heinemannn.

Eto M, Watanabe K, Chonan N, Ishii K. (1988). Familial hypercholesterolemia and apolipoprotein E4. *Atherosclerosis 72*, 123–128.

Fokkema IFAC, Den Dunnen JT and Taschner PEM. (2005). LOVD: easy creation of a locus-specific sequence variation database using an "LSDB-in-a-Box" approach. *Hum Mutat 26*(2), 63–68.

Fouchier SW, Kastelein JJ, Defesche JC. (2005). Update of the molecular basis of familial hypercholesterolemia in The Netherlands. *Hum Mutat 26*, 550–556

Friedlander Y, Leitersdorf E. (1996). Influence of apolipoprotein E genotypes on plasma lipid and lipoprotein concentrations: results from a segregation analysis in pedigrees with molecularly defined familial hypercholesterolemia. *Genet Epidemiol 13*, 159–177.

Gaudet D, Vohl MC, Couture P, et al. (1999). Contribution of receptor negative versus receptor defective mutations in the LDL-receptor gene to angiographically assessed coronary artery disease among young (25–49 years) versus middle-aged (50–64 years) men. *Atherosclerosis 143*, 153–161.

Goldman L, Goldman PA, Williams LW, et al. (1993). Cost-effectiveness considerations in the treatment of heterozygous familial hypercholesterolemia with medications. *Am J Cardiol 72*, 75–79D.

Goldstein JL. (1972). The cardiac manifestations of the homozygous and heterozygous forms of familial type II hypercholesterolaemia. *Birth Defects 8*, 202.

Gudnason V, Day IN, Humphries SE. (1994). Effect on plasma lipid levels of different classes of mutations in the low-density lipoprotein receptor gene in patients with familial hypercholesterolemia. *Arterioscler Thromb 14*, 1717–1722.

Hadfield SG, Horara S, Starr BJ, et al.; Steering Group for the Department of Health Familial Hypercholesterolaemia Cascade Testing Audit Project. (2009). Family tracing to identify patients with familial hypercholesterolaemia: the second audit of the Department of Health Familial Hypercholesterolaemia Cascade Testing Project. *Ann Clin Biochem 46*, 24–32.

Hadfield SG, Humphries SE. (2005). Implementation of cascade testing for the detection of familial hypercholesterolaemia. *Curr Opin Lipidol 16*, 428–433.

Hegele RA. (2009). Plasma lipoproteins: genetic influences and clinical implications. *Nature Reviews Genetics 10*, 109–121.

Heiberg A. (1975). The risk of atherosclerotic vascular disease in subjects with xanthomatosis. *Acta Med Scand 198*, 249–261.

Hill JS, Hayden MR, Frohlich J, Pritchard PH. (1991). Genetic and environmental factors affecting the incidence of coronary artery disease in heterozygous familial hypercholesterolemia. *Arterioscler Thromb 11*, 290–297.

Hofman N, van Langena I, Wilde AAM. (2010). Genetic testing in cardiovascular diseases. *Current Opinion in Cardiology 25*, 243–248.

Holmes MV, Shah SH, Angelakopoulou A, et al. (2010). A report on the Genetics of Complex Diseases meeting of the British Atherosclerosis Society, Cambridge, UK, 17–18 September 2009. *Atherosclerosis 208*(2), 599–602.

Hutter CM, Austin MA, Humphries SE. (2004). Familial hypercholesterolemia, peripheral arterial disease, and stroke: a HuGE minireview. *Am J Epidemiol 160*, 430–435

Huxley RR, Hawkins MH, Humphries SE, Karpe F, Neil HA, Simon Broome Familial Hyperlipidaemia Register Group and Scientific Steering Committee. (2003). Risk of fatal stroke in patients with treated familial hypercholesterolemia: a prospective registry study. *Stroke 34*, 22–25. Erratum in: *Stroke 34*, 826.

Igl W, Johansson Å, Wilson JF, Wild SH, Polašek O, et al. (2010). Modeling of environmental effects in genome-wide association studies identifies *SLC2A2* and *HP* as novel loci influencing serum cholesterol levels. *PLoS Genet 6*(1), e1000798.

Jeemon P, Neogi S, Bhatnagar D, Cruickshank JK, Prabhakaran D. (2009). The impact of migration on cardiovascular disease and its risk factors among people of Indian origin. *Current Science 97*, 378–384.

Jensen J, Blankenhorn DH, Kornerup V. (1967). Coronary disease in familial hypercholesterolaemia. *Circulation 36*, 77–82.

Koeijvoets KC, Wiegman A, Rodenburg J, Defesche JC, Kastelein JJ, Sijbrands EJ. (2005). Effect of low-density lipoprotein receptor mutation on lipoproteins and cardiovascular disease risk: a parent–offspring study. *Atherosclerosis 180*, 93–99.

Kotze MJ, De Villiers WJ, Steyn K, et al. (1993). Phenotypic variation among familial hypercholesterolemics heterozygous for either one of two Afrikaner founder LDL receptor mutations. *Arterioscler Thromb 13*, 1460–1468.

Kullo IJ, Cooper LT. (2010). Early identification of cardiovascular risk using genomics and proteomics. *Nat Rev Cardiol 7*, 309–317.

Langer T, Strober W, Levy RI. (1972). The metabolism of low density lipoprotein in familial type II hyperlipoproteinemia. *J Clin Invest 51*, 1528–1536.

Leren TP. (2004). Cascade genetic screening for familial hypercholesterolemia. *Clin Genet 66*, 483–487.

Lindgren V, Luskey KL, Russell DW, Francke U. (1985). Human genes involved in cholesterol metabolism: chromosomal mapping of the loci for the low density lipoprotein receptor and 3-hydroxy-3 -methylglutaryl-coenzyme A reductase with cDNA probes. *Proc Natl Acad Sci USA 82*, 8567–8571.

Mabuchi H, Miyamoto S, Ueda K, et al. (1986). Causes of death in patients with familial hypercholesterolemia. *Atherosclerosis 61*, 1–6.

Marang-van de Mheen PJ, ten Asbroek AH, Bonneux L, et al. (2002). Cost-effectiveness of a family and DNA based screening programme on familial hypercholesterolaemia in the Netherlands. *Eur Heart J 23*, 1922–1930.

Marks D, Wonderling D, Thorogood M, et al. (2000). Screening for hypercholesterolaemia versus case finding for familial hypercholesterolaemia: a systematic review and cost-effectiveness analysis. *Health Technol Assess 4*, 1–123.

Marteau TM, Croyle RT. (1998). The new genetics. Psychological responses to genetic testing. *BMJ 316*, 693–696.

Mbewu AD, Bhatnagar D, Durrington PN, et al. (1991). Serum lipoprotein (a) in patients heterozygous for familial hypercholesterolemia, their relatives, and unrelated control populations. *Arterioscler Thromb 11*, 940–946.

McQueen MJ. (2002). Screening for the early detection of disease, the need for evidence. *Clin Chim Acta 315*, 5–15.

Mendis S. (2005). Cardiovascular risk assessment and management in developing countries. *Vascular Health and Risk Management 1*, 15–18.

Myant NB, Gallagher JJ, Knight BL, et al. (1991). Clinical signs of familial hypercholesterolemia in patients with familial defective apolipoproteins B-100 and normal low density lipoprotein receptor function. *Arterioscler Thromb 11*, 691–703.

Narayan KM, Ali MK, Koplan JP. (2010). Global noncommunicable diseases—where worlds meet. *N Engl J Med 363*, 1196–1198.

Newson AJ, Humphries SE. (2005). Cascade testing in familial hypercholesterolaemia: how should family members be contacted? *Eur J Hum Genet 13*, 401–408.

Patel JV, Vyas A, Cruickshank JK, et al. (2006). Impact of migration on coronary heart disease risk factors: comparison of Gujaratis in Britain and their contemporaries in villages of origin in India. *Atherosclerosis 185*, 297–306.

Pereira E, Ferreira R, Hermelin B, et al. (1995). Recurrent and novel LDL receptor gene mutations causing heterozygous familial hypercholesterolemia in La Habana. *Hum Genet 96*, 319–322.

Pimstone SN, Sun XM, du Souich C, Frohlich JJ, Hayden MR, Soutar AK. (1998). Phenotypic variation in heterozygous familial hypercholesterolemia: a comparison of Chinese patients with the same or similar mutations in the LDL receptor gene in China or Canada. *Arterioscler Thromb Vasc Biol 18*, 309–315.

Pirruccello J, Kathiresan S. (2010). Genetics of lipid disorders. *Current Opinion in Cardiology 25*, 238–242.

Reddy KS. (2002). Cardiovascular diseases in the developing countries: dimensions, determinants, dynamics and directions for public health action. *Public Health Nutr 5*, 231–237.

Rotimi CN, Jorde LB. (2010). Ancestry and disease in the age of genomic medicine. *N Engl J Med 363*(16), 1551–1558.

Schneider WJ, Beisiegel U, Goldstein JL, Brown MS. (1982). Purification of the low density lipoprotein receptor, an acidic glycoprotein of 164,000 molecular weight. *J Biol Chem 257*, 2664–2673.

Scientific Steering Committee on behalf of the Simon Broome Register Group. (1991). Risk of fatal coronary heart disease in familial hypercholesterolaemia. *BMJ 303*, 893–896.

Seed M, Hoppichler F, Reaveley D, et al. (1990). Relation of serum lipoprotein (a) concentration and apolipoprotein(a) phenotype to coronary heart disease in patients with familial hypercholesterolemia. *N Engl J Med 322*, 1494–1499.

Senior V, Marteau TM, Peters TJ. (1999). Will genetic testing for predisposition for disease result in fatalism? A qualitative study of parents responses to neonatal screening for familial hypercholesterolaemia. *Soc Sci Med 48*, 1857–1860.

Shoulders CC, Jones EL, Naoumova RP. (2004). Genetics of familial combined hyperlipidemia and risk of coronary heart disease. *Hum Mol Genet 13*(Spec No 1), R149–160.

Shuldiner AR, Pollin TI. (2010). Genomics: Variations in blood lipids. *Nature 466*, 703–704.

Slack J. (1969). Risks of ischaemic heart-disease in familial hyperlipoproteinaemic states. *Lancet ii*, 1380–1382.

Soutar AK, Myant NB, Thompson GR. (1977). Simultaneous measurement of apolipoprotein B turnover in very-low and low-density lipoproteins in familial hypercholesterolaemia. *Atherosclerosis 28*, 247–256.

Sprecher DL, Schaefer EJ, Kent KM, et al. (1984). Cardiovascular features of homozygous familial hypercholesterolemia: analysis of 16 patients. *Am J Cardiol 54*, 20–30.

Stone NJ, Levy RI, Fredrickson DS, Verter J. (1974). Coronary artery disease in 116 kindred with familial type II hyperlipoproteinemia. *Circulation 49*, 476–488.

Sun XM, Patel DD, Bhatnagar D, et al. (1995). Characterization of a splice-site mutation in the gene for the LDL receptor associated with an unpredictably severe clinical phenotype in English patients with heterozygous FH. *Arterioscler Thromb Vasc Biol 15*, 219–227.

Thompson GR, Seed M, Niththyananthan S, McCarthy S, Thorogood M. (1989). Genotypic and phenotypic variation in familial hypercholesterolemia. *Arteriosclerosis 9*(1 Suppl), I75–180.

Thorsson B, Sigurdsson G, Gudnason V. (2003). Systematic family screening for familial hypercholesterolemia in Iceland. *Arterioscler Thromb Vasc Biol 23*, 335–338

Umans-Eckenhausen MA, Defesche JC, Sijbrands EJ, Scheerder RL, Kastelein JJ. (2001). Review of first 5 years of screening for familial hypercholesterolaemia in the Netherlands. *Lancet 357*, 165–168.

Utermann G, Hoppichler F, Dieplinger H, Seed M, Thompson G, Boerwinkle E. (1989). Defects in the low density lipoprotein receptor gene affect lipoprotein (a) levels: multiplicative interaction of two gene loci associated with premature atherosclerosis. *Proc Natl Acad Sci USA 86*, 4171–4174.

Vohl MC, Gaudet D, Moorjani S, et al. (1997). Comparison of the effect of two low-density lipoprotein receptor class mutations on coronary heart disease among French–Canadian patients heterozygous for familial hypercholesterolaemia. *Eur J Clin Invest 27*, 366–373.

Vuorio AF, Aalto-Setala K, Koivisto UM, et al. (2001). Familial hypercholesterolaemia in Finland: common, rare and mild mutations of the LDL receptor and their clinical consequences. Finnish FH-group. *Ann Med 33*, 410–421.

Williams RR, Hunt SC, Schumacher MC, et al. (1993). Diagnosing heterozygous familial hypercholesterolemia using new practical criteria validated by molecular genetics. *Am J Cardiol 72*, 171–176.

Wilson JM, Jungner G. (1968). Principles and Practice of Screening for Disease. (Public Health Paper Number 34). Geneva: World Health Organization.

Wonderling D, Umans-Eckenhausen MA, Marks D, Defesche JC, Kastelein JJ, Thorogood M. (2004). Cost-effectiveness analysis of the genetic screening program for familial hypercholesterolemia in The Netherlands. *Semin Vasc Med 4*, 97–104.

World Health Organization. (1999). Familial hypercholesterolaemia; Report of a Second WHO Consultation, WHO/HGN/FH/CONS/99.2. Geneva: World Health Organization.

Yamamoto T, Davis CG, Brown MS, et al. (1984). The human LDL receptor: a cysteine-rich protein with multiple Alu sequences in its mRNA. *Cell 39*, 27–38.

25.

GLOBAL HEALTH IMPLICATIONS OF INHERITED RENAL DISEASES

Wendy Jones and Anand Saggar

INTRODUCTION

Chronic kidney disease (CKD) is a global health problem, yet a treatable and preventable condition. The worldwide incidence of end-stage renal failure (ESRF) is growing by approximately 8% annually (Schieppati, 2005) and the number of patients requiring renal replacement therapy is threatening to reach epidemic proportions (El Nahas & Bello, 2005). More than 100 developing countries, with a combined population in excess of 600 million, do not have the ability to provide renal replacement therapy and, as a result, in excess of 1 million people die every year worldwide from end-stage renal disease (Lacson Jr et al., 2005). Although world data on inherited renal disease (IRD) are lacking, IRDs account for approximately 9% of adult patients with end-stage renal failure, and larger numbers of pediatric patients. Renal failure is often asymptomatic until function is severely impaired, so by identifying the at-risk population of patients with IRD we can not only screen for, and hopefully delay, if not prevent, progression of renal disease. Furthermore, given that inherited renal diseases by definition will identify other family members at risk, they too can be identified before they become symptomatic. Similarly, in the long term, early identification may prevent or delay the need for financially burdensome renal replacement therapy with the associated morbidity and mortality.

This chapter discusses global health implications of inherited renal diseases with particular reference to ethnic minorities and in low- and middle-income countries (LMIC) as defined by the World Bank, based mainly on gross national income (GNI) per capita (World Bank Monetary Fund, 2009). Countries classed as LMIC are often referred to as *developing countries.*

Inherited renal disorders account for about 7% of all renal disease (Table 25-1), and autosomal dominant polycystic kidney disease (ADPKD) accounts for 5%–10% of patients on dialysis in most Western renal register populations (Byrne, 2010a; Byrne, 2010b). It is the commonest inherited renal disease, with an affected person having a 50% risk of passing the gene to offspring, regardless of sex, in each pregnancy. ADPKD forces consideration of the practical significance of genetic variation in disease liability. Certainly, society will have to come to terms with complex genetic variation in the context of adult-onset variability as genomic knowledge increases.

Whatever the renal disease, it is necessary to be encouraging and to set a treatment plan that emphasizes health maintenance through proper diet, exercise, and attention to potential complications, e.g., urine tract infections and hypertension. Patients can now also, in many countries, access the Internet or alternative literature to get more detailed information from websites, and find patient groups for more detailed information and support.

INCIDENCE AND PREVALENCE OF RENAL DISEASE WORLDWIDE

In the developed countries, registers for patients on renal replacement therapy (RRT) are well established (Table 25-2). Little is known of prevalence and incidence of chronic kidney disease and of end-stage renal disease (ESRD) in middle-income and low-income countries, where the use of RRT is scarce or nonexistent.

The renal replacement data on the incidence and prevalence of renal disease and its underlying etiology is still limited globally (Table 25-2). Measures to improve this are being made through pilot projects and screening initiatives (Eastwood et al., 2010; Plange-Rhule et al., 2002; Park et al., 2005). More recently, a joint consensus led by the World Health Organization, the International Society of Nephrology, and the Transplantation Society–Global Alliance for Transplantation has called for action in emerging countries to start chronic kidney disease prevention and screening programs, develop end-stage renal disease registries, and start or further elaborate transplantation programs (Persy et al., 2008).

It is not surprising, therefore, that epidemiological data regarding the prevalence of IRD in LMIC is severely limited, and in some countries no information is known. This reflects lack of national registries and representative

TABLE 25-1 PRIMARY RENAL DISEASE IN PREVALENT RRT ADULT PATIENTS, U.K. RENAL REGISTRY (BYRNE, 2010)[a]

Primary diagnosis	% all patients
Etiology uncertain/Glomerulonephritis (not biopsy proven)	20.5
Glomerulonephritis (biopsy proven)	16.0
Pyelonephritis	12.0
Diabetes	14.1
Polycystic kidney	9.6
Hypertension	5.6
Renal vascular disease	3.5
Other	14.5
Not sent	4.4

Excluded centers with ≥40% primary renal diagnosis etiology uncertain/glomerulonephritis (not biopsy proven) as well as centers with ≥50% primary diagnosis not sent.

surveys, inaccurate data, and investigators using a variety of different collection methods. Some information is, however, available from the USRDS about ethnic group–specific renal diseases (Table 25-3).

In fact, the lack of knowledge of renal diseases in the developing world is a major challenge. The paucity of data almost certainly reflects underdiagnosis and documentation, rather than being a true reflection of underrepresentation. Given the lack of investigative tools and funding for medical services, many patients with renal diseases die undiagnosed, particularly pediatric patients, and likewise metabolic conditions are very likely to be underrepresented in any data collection. International as well as regional collaborations need to be stimulated and strengthened, in order to allow emerging countries to acquire the much-needed demographic and epidemiological data (Eastwood et al., 2001).

TABLE 25-2 INCIDENCE AND PREVALENCE OF END-STAGE RENAL FAILURE IN THE ADULT POPULATION BY COUNTRY IN 2008 (U.S. RENAL DATA SYSTEM, 2010)

Country	Incidence of ESRD (per million per year)	Prevalence of ESRF (per million)	% of incident patients with ESRF secondary to diabetes mellitus
UK	108	774	20.6
USA	362	1752	43.8
Australia	116	803	34
Russia	35	157	17.2
Argentina	144	756	35.4
Bangladesh	13	112	
Bosnia/ Herzegovina	149	637	16.3
Brazil	148	415	34.5
Chile	153	1065	
Malaysia	139	743	55.8
Philipines	87	110	40.7
Russia	35	157	17.2
Thailand	100	497	40.6

TABLE 25-3 PREVALENCE OF INHERITED RENAL DISEASE BY ETHNIC GROUP BY PERCENT OF TOTAL (U.S. RENAL DATA SYSTEM, 2010)

	White	African American	Native American	Asian
Total cystic/hereditary/ congenital diseases	9.5	2.7	3.6	4.8
ADPKD	6.3	1.8	1.8	3.2
ARPKD	0.1	–	0.1	0.1
Medullary cystic disease, including nephronophthisis	0.2	–	0.1	0.1
Alport syndrome	0.6	0.1	0.3	0.2
Cystinosis	0.1	–	–	–
Congenital nephritic syndrome	0.1	0.1	0.1	0.1
Renal hypoplasia, dysplasia, oligonephronia	0.8	0.3	0.5	0.4
Prune belly syndrome	0.1			
Other (congenital malformation syndromes)	0.2		0.1	0.1

The large Western registers, such as that in United States, have managed to collect data from LMIC of the incidence and prevalence of end-stage renal failure but do not provide published data on the diagnosis (Table 25-2). By examining the U.S. and the U.K. renal registers it is possible to get some idea of the proportion of ethnic minorities affected by renal disease.

The U.K. renal register similarly contains data on ethnic groups. In the United Kingdom the incidence of ESRF is currently 108 new patients per million population (Byrne, 2010b). In urban areas, large ethnic populations have high prevalence of ESRF, standardized for age and gender (Byrne, 2010a). The South Asian populations, for example, have a prevalence of end-stage renal failure 2.5 times that of the white population. The study of ethnic subpopulations in large urban conurbations may therefore provide clues to a more global prevalence.

Given the universality of the U.K.-based National Health Service healthcare access, it might be assumed that the diseases seen in the ethnic groups in the United Kingdom are more representative (Figure 25-1).

Given that data from LMIC as defined by the World Bank (World Bank Monetary Fund, 2009) regarding the incidence and prevalence of inherited adult renal disease are severely limited, we set out to obtain a more directly informed view of the major renal diseases prevalent in

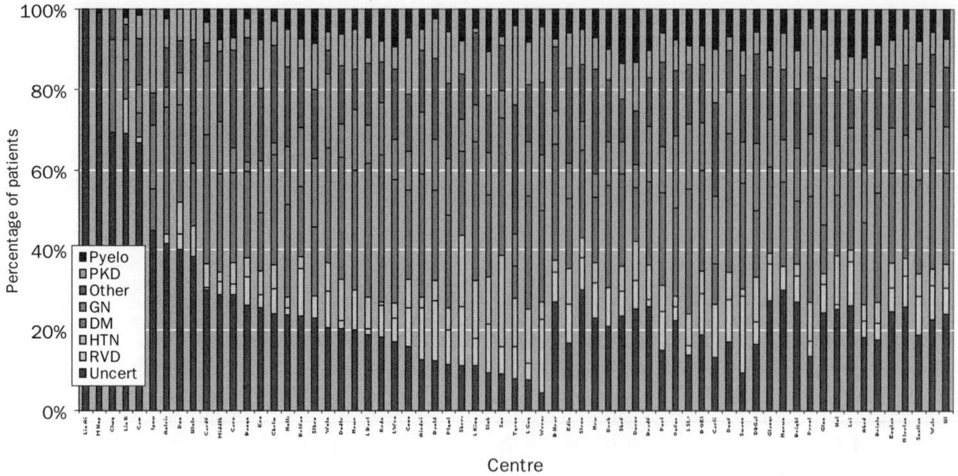

Figure 25-1 Proportion of patients with primary renal disease based on UK Renal Registry 2009 (Byrne, 2010).

LMICs. We contacted senior adult nephrologists or senior hospital physicians in 80 LMICs by email, and requested information as to the most commonly reported inherited causes of renal disease in their country according to their experience. (Jones & Saggar, 2012, unpublished). Table 25-4 lists both the countries that were approached, and from whom we received a response. Figure 25-2 shows overwhelmingly how polycystic kidney disease is the most common inherited renal disease, with Alports syndrome, vesicoureteric reflux, and renal tubular acidosis making up the other main inherited disorders seen (Figures 25-2a and 25-2b).

The small amount of data derived from our data collection is in line with findings from the registers of developed nations that the most common inherited renal disease in adult patients is autosomal dominant polycystic kidney disease (ADPKD).

These personal communications from renal physicians in LMICs serve to highlight the common conditions, but we recognize that this most probably reflects

TABLE 25-4 LIST OF COUNTRIES WHO WERE APPROACHED FOR DATA ON INHERITED RENAL DISEASES

Countries contacted from whom data was received	Countries contacted from whom no reply was received			
Russia	Albania	Ecuador	Malaysia	Uruguay
Bangladesh	Algeria	El Salvador	Moldova	Venezuela
Argentina	Benin	Estonia	Nepal	Vietnam
Croatia	Bhutan	Egypt	Nicaragua	
Guatemala	Bolivia	Gabon	Niger	
Hungary	Bosnia-Herzegovina	Ghana	Nigeria	
India	Brazil	Guinea	Pakistan	
Morocco	Bulgaria	Honduras	Panama	
Puerto Rico	Cambodia	Hong Kong	Peru	
República Dominicana	Cape Verde	Jamaica	Poland	
Thailand	Chad	Kenya	Qatar	
Congo	Cameroon	Kiribati	Romania	
Indonesia	Chile	Korea	Senegal	
South Africa	Colombia	Lao	Togo	
Burkina Faso	Cook Islands	Latvia	Tonga	
Serbia	Costa Rica	Lebanon	Tunisia	
Mexico	Cote d'Ivoire	Lithuania	Turkey	
Sudan	Cuba	Macedonia	Ukraine	

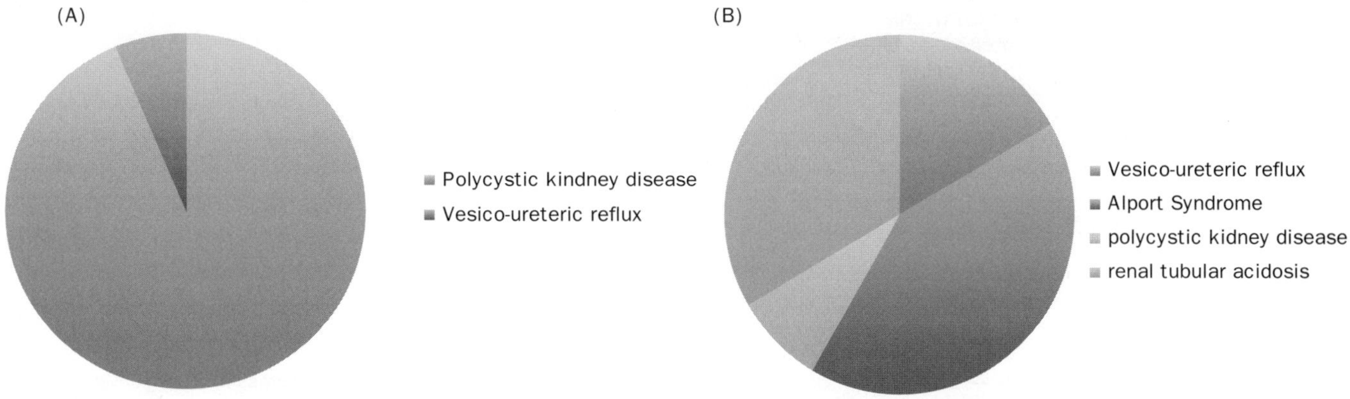

Figure 25-2 Pie chart showing the most common inherited causes of renal failure reported in adults in 17 worldwide LMIC responding to survey (Saggar & Jones, 2012, unpublished). Chart A shows the most commonly reported inherited causes. Chart B shows the second most commonly reported inherited causes. Data from: Russia, Bangladesh, Argentina, Croatia, Guatemala, Hungary, India, Mexico, Morocco, Puerto Rico, Republican Dominicana, Thailand, Congo, Indonesia, South Africa, Burkina Faso and Sudan.

a predominantly adult population, and it is undoubtedly biased in its ascertainment based upon local centers, large centers in big cities, and doesn't reflect the vast number of pediatric patients who die young with other inherited renal diseases.

END-STAGE RENAL FAILURE IN PEDIATRIC POPULATIONS

In the U.K. pediatric population, the most common cause of end-stage renal failure in 2008 was renal dysplasia, accounting for 33% of total cases, followed by glomerular disorders (19%) and obstructive uropathy (12%; see Lewis et al., 2010 and Table 25-5). The incidence of ESRF in the first four years of life is almost half that seen in the second four years of life. The peak in the first four years is felt to represent the presentation of many children with obstructive uropathy and renal dysplasia at this age.

In all age groups there were significantly more males affected with ESRF, with an overall male to female ratio of pediatric patients in 2008 of 1.5:1 (Lewis et al., 2010). This gender discrepancy is also seen in the U.K. adult population. In the pediatric population, this gender discrepancy is thought to be due to the preponderance of renal dysplasia and obstructive uropathy causing end-stage renal failure in males (Lewis et al., 2007).

This high prevalence of disease in the U.K. South Asian population is thought to be due to the high incidence of autosomal recessive conditions in this community (Lewis et al., 2007). In 2008, 20.5% of prevalent patients overall had end-stage renal failure secondary to a condition with autosomal recessive inheritance. There were comparatively few numbers of patients affected with autosomal dominant, sex-linked, or mitochondrial disorders (Lewis et al., 2010; Table 25-5).

KIDNEY MALFORMATIONS AND TERATOGENS

The human kidney is derived from the ureteric bud and the renal mesenchyme, both components of the embryonic metanephros. Kidney formation is a complex process, and one that is commonly perturbed. Congenital kidney and urinary tract abnormalities are common, and make up one-third of all anomalies diagnosed on fetal ultrasound and are likely due to a multitude of different chromosomal abnormalities, single gene disorders and polygenic and or

TABLE 25-5 ETHNIC DISTRIBUTION OF END-STAGE RENAL FAILURE GROUPS (LEWIS ET AL., 2007) IN 2006 FOR PATIENTS PRESENTING AFTER 1996 AND BEFORE THE AGE OF 16 YEARS

Diagnostic group	White No (%)	South Asian No (%)	Black No (%)	Other No (%)
Dysplasia	193 (26.2)	23 (16.4)	5 (25)	0 (0.00)
Glomerular diseases	161 (21.8)	29 (20.7)	11 (55.0)	4 (26.7)
Obstructive uropathy	117 (15.9)	17 (12.1)	0 (0.00)	2 (13.3)
Tubulointerstitial diseases	52 (7.1)	17 (12.1)	0 (0.00)	4 (26.7)
Reflux nephropathy	60 (8.1)	6 (4.3)	1 (5.0)	2 (13.3)
Congenital nephrotic syndrome	30 (4.1)	16 (11.4)	0 (0.0)	0 (0.0)
Metabolic diseases	31 (4.2)	13 (9.3)	0 (0.0)	0 (0.0)
Renovascular	31 (4.2)	2 (1.4)	1 (5.0)	0 (0.0)
Unknown	17 (2.3)	9 (6.4)	1 (5.0)	2 (13.3)
ADPKD	20 (2.7)	5 (3.6)	1 (5.0)	1 (6.7)
ERF from drug nephrotoxicity	17 (2.3)	2 (1.4)	0 (0.0)	0 (0.0)
Malignant disease	9 (1.2)	1 (0.7)	0 (0.0)	0 (0.0)

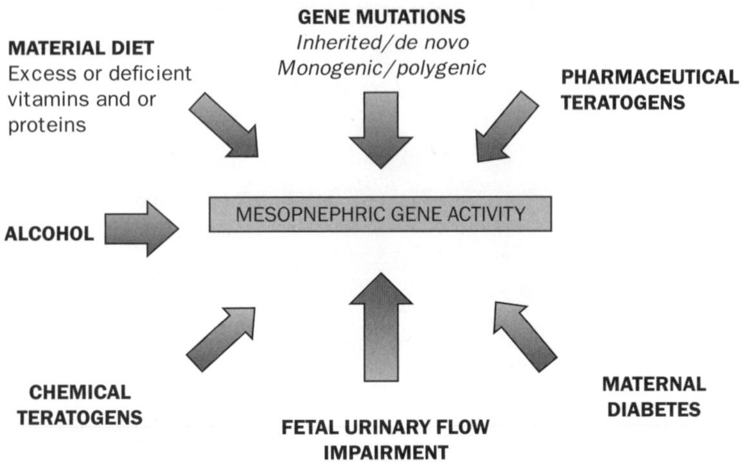

Figure 25-3 Influences on gene expression during development of the urinary tract: mutations, chemical and pharmaceutical teratogens, physical obstruction of urine flow in the fetus and alterations in maternal diet.

environmental factors (Figure 25-3; Woolf, 2000), these are not discussed in this chapter.

Many teratogenic agents may affect renal development. These, together with dietary deficiencies, may adversely influence renal development in the background of several genetic factors. However, gene-environment interactions largely remain unknown for many nephrotoxic substances. Variable renal effects may depend on the nature of the substance, precise timing of the insult and the level of concentration (Figure 25-3). There is increasing knowledge about genetic susceptibility factors including genetic imprinting with some evidence for genotype-phenotype correlations.

HOW WILL DEVELOPMENT OF GENETICS AND GENOMICS INFLUENCE MANAGEMENT OF INHERITED RENAL DISEASES?

The management of end-stage renal failure is expensive. With development of fetal medicine, nephrology, and genetic services in LMICs, it is important that focus is placed not only on the management of ESRF but also on the identification and prevention of end-stage renal failure. Early identification of genetic conditions that may predispose to renal disease resulting in ESRF must form the mainstay of preventing progression to end stage, given that there is a lack of funds, services, and infrastructure for chronic dialysis.

At present there are no genetic molecular treatments or interventions for inherited renal diseases, but it is feasible they may play a role in the future.

Certainly there is increasing knowledge about genetic susceptibility factors and underlying dietary influences, which may involve the emergence of imprinting and clear genotype/phenotype correlations.

The value of identifying genetic diseases early, and the development of clinical genetics as a diagnostic specialty, will help in the earlier diagnosis of potentially life-threatening conditions. Examples of this are renal coloboma syndrome, where 50% of patients will develop renal disease. Another example is nail patella syndrome, where proteinuria may represent underlying glomerulonephritis (GN), and early treatment with ACE inhibitors may slow renal disease progression and reduce proteinuria.

Secondary prevention involves the tight control of blood pressure and the use of angiotensin-converting enzyme inhibitors and angiotensin II blockers concomitantly, to reduce proteinuria (to less than 1g/24 hours). Given that almost all patients with established renal diseases will develop hypertension, regular monitoring of blood pressure in patients at risk may help to reduce cardiovascular risk factors and slow renal disease progression, and is a simple, reliable, reproducible measurement, the treatment of which is likely affordable in developing countries.

It is also often assumed that in order to make a genetic diagnosis, be it renal disease or some other disease, requires a genetic test; however, this is not the case. A good example of this is polycystic kidney disease, where the finding of large kidneys, often with the cysts palpable abdominally, and hypertension with or without ultrasound, can establish a genetic diagnosis without the need for any molecular analysis. In those countries where ultrasound is available, it can be more reliable than a genetic test for diagnosing or excluding an individual affected with ADPKD, after the age of 30.

It is true that in some cases a genetic test may be useful to establish a diagnosis or to exclude another condition, but very often the clinical phenotype is sufficiently clear to give a diagnosis. This, as in the case of renal disorders, may be very helpful, as phakomata (superficial external markers of an internal disease) can serve as indicators of affected individuals without the need to resort to expensive or unavailable genetic testing.

The identification of gene mutations has nonetheless allowed the diagnosis of patients with milder or unusual phenotypes into one group; e.g., it was always assumed that recessive polycystic kidney disease (ARPKD) resulted in end-stage renal failure in the first decade of life with liver abnormalities. However, it is now recognized that some patients may present even as adults primarily with liver fibrosis, and portal hypertension with bleeding varies. In such circumstances a genetic test can be very useful.

One further area is predictive testing, and genetics again has a value here. However, in renal diseases there are also *forme fruste,* or subclinical markers of an individual being affected. For example, in Alport syndrome over 90% of carrier females have microscopic hematuria, branchial cysts, or fistulae in branchiootorenal syndrome (BOR).

Our greater understanding of genetics and genomics continues to help us to more fully understand renal disease phenotypes. For example, *renal cysts and diabetes syndrome* (RCAD) is the name generally applied to patients with HNF1ß mutations. However, renal cysts and diabetes is only one phenotype seen in this variable condition, and those affected may have renal disease or diabetes, or neither disease. This therefore makes diagnosis challenging. It also provides a management challenge, since the phenotype in the next generation, being a dominant condition with a 50% risk of being passed on, can be highly variable.

The advancement of knowledge in the field of genomics and genetics has helped us both understand the correlations between different genetic conditions, and elucidate and differentiate conditions that present with similar features. For example, patients with tuberous sclerosis complex (TSC) may in some cases also present with renal cysts, hypertension, and in some cases, progress to renal failure. A tuberous sclerosis gene, TSC2, lies immediately adjacent to PKD1, which is also located on chromosome 16. Genomic progress has determined that such patients have a contiguous gene deletion spanning from the TSC2 gene on chromosome 16 into the adjacent PKD1 gene (Figure 25-4a; Sampson, 1997). In this instance, therefore, the patient has two genetic conditions, and this knowledge is important in the management of their care. For example, patients with TSC and ADPKD are more prone to develop hypertension and ESRD, in addition to

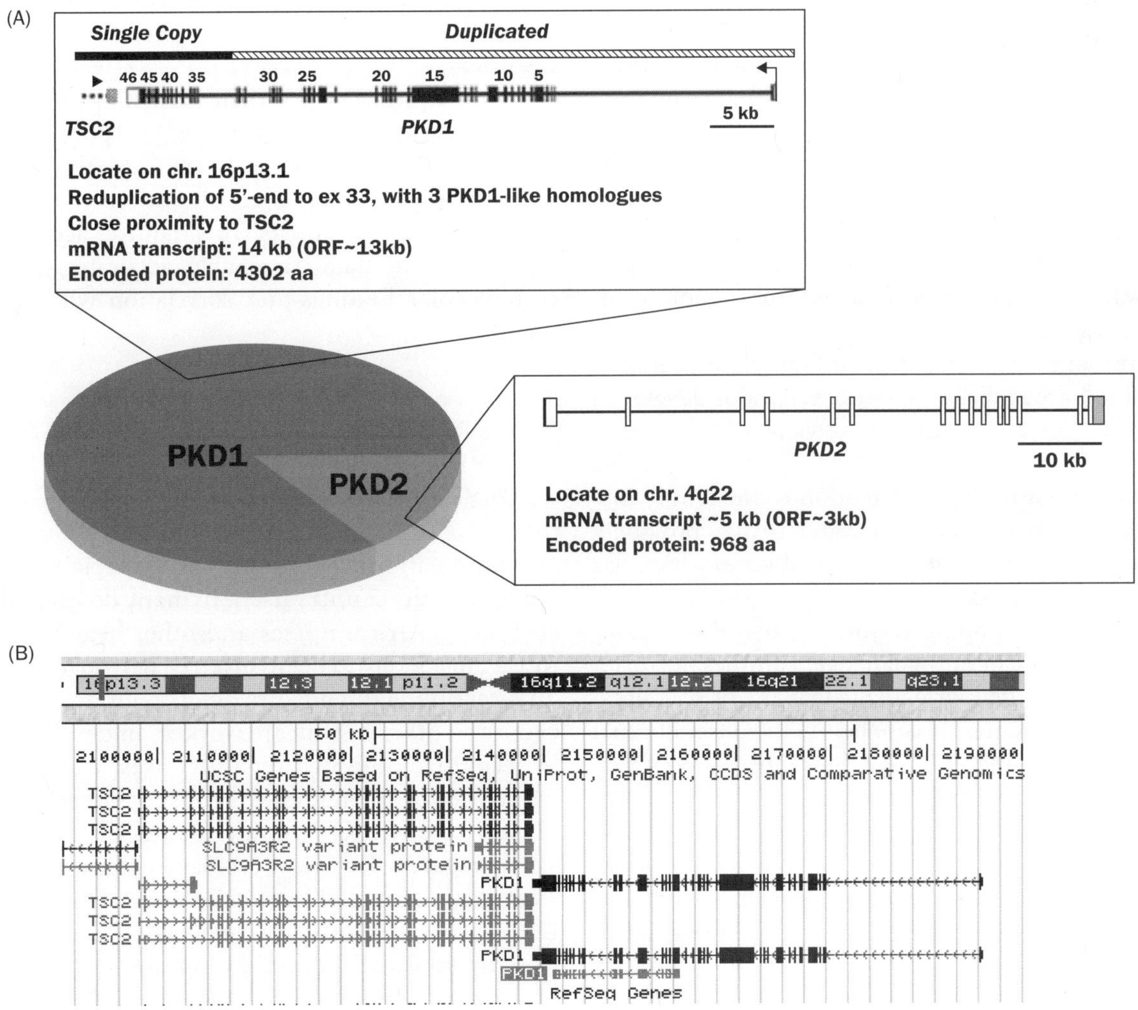

Figure 25-4 a) The two genes (PKD1 & PKD2) causing ADPKD and the contiguous gene TSC2 in close proximity to PKD1. b) Detailed molecular genetics of TSC2 and PKD1 on chromosome 16p.

the regular complications of TS, such as AMLs (Brook-Carter, 1994).

More recently it has been determined that isolated autosomal dominant polycystic liver disease (ADPLD) is a disease entity separate from ADPKD, with massive polycystic liver disease (Qian et al., 2003) with the underlying mutation being mutations in PRKCSH and SEC63(Davila et al., 2004; Li et al., 2003). In the past, ADPLD may have been misdiagnosed as polycystic kidney disease, as both conditions are dominantly inherited and present with massive liver cysts.

Genomics continues to lead to better understanding of the diverse phenotype that can occur in genetic conditions, and greater understanding of the modifier genes involved. More recently there has been a greater recognition of hypomorphic alleles, e.g., in ADPKD, where the presence of such an allele in unison with a more usual mutation leads to a more severe neonatal phenotype (Pei, 2001).

Different mutations, such as in APKD, have a different prognosis. It is recognized that patients with PKD2 gene mutations have less cysts, later onset of hypertension and ESRD as a group; however, for the individual patient it is more difficult to give any degree of certainty. Using genomics and clinical observation, it is possible then to establish clinical markers of assumed genotype. Using ADPKD again as an example, it is possible to identify which patients that have PKD2 are likely to have the milder PKD2 phenotype, by examining the family history.

Barua et al (2009) found that the presence of at least one affected family member who developed ESRD at age < or = 55 was highly predictive of a PKD1 mutation (positive predictive value 100%; sensitivity 72%). In contrast, the presence of at least one affected family member who continued to have sufficient renal function, or developed ESRD at age >70, was highly predictive of a PKD2 mutation (positive predictive value 100%; sensitivity 74%). These data suggest that close attention to the family history of renal disease severity in ADPKD may provide a simple means of predicting the mutated gene, which has prognostic implications.

However, in developing countries obtaining a family history in itself can be a challenge. Families' histories may be not be recorded, or patients may die of accidental causes or cardiovascular disease before the onset of renal failure is known or manifest. Furthermore, patients may die undiagnosed with renal disease.

Having established a diagnosis, it allows other individuals to be informed and allows the identification of at-risk individuals so that they too can be followed up. Having established a diagnosis in one individual, it makes it easier to understand and reliably advise how at-risk individuals in the family should be followed up. In the case of renal diseases, very often such follow-up is not an expensive form of healthcare (e.g., urine dipsticks for glycosuria in RCAD, or proteinuria in Nail-Patella syndrome or renal coloboma syndrome, and blood pressure measurement for ADPKD and most other renal diseases). Genomics will undoubtedly continue to be important in renal diseases to establish treatments and cures. An increased understanding of the genetic, molecular, and cellular mechanisms responsible for the development of polycystic kidney disease has laid out the foundation for the development of rational therapies (Figure 25-5). Many animal models also now exist, so that these therapies can be tested. As an example, understanding the biochemical pathway of action of the MAP kinase and mTor pathways has led to the development of novel treatments for ADPKD (Gattone et al., 2009).

Genomics has also led to a greater understanding of the commonality of different diseases. The finding that ADPKD like many other kidney-related disorders—such as Bardet-Beidel, Joubert syndrome, Alstroms syndrome, and nephronophthisis, to name a few—are due to mutations in cilia-regulating genes has led to developments and discoveries from one cilial disorder that have relevance and impact upon the whole family of ciliopathy disorders. This may also be true for potential treatments that can then be applied across the group of cilial disorders.

Until countries have their own molecular and cytogenetic services up and running, one possibility may be biobanking of samples until genetic analysis is possible, either locally or as part of collaborations with other countries or research scientists. This requires accurate recording of patient data, to enable accurate interpretation of molecular findings and correlation with the clinical picture.

A PRACTICAL GUIDE FOR RECOGNIZING INHERITED RENAL DISEASES IN THE DEVELOPING WORLD

In many nations, nephrology is a specialty developing along with the country itself. In many developed nations, such as in Africa, nurses and other health workers are those most likely to be involved in delivering the service. This section is intended to act as a guide to help physicians and other healthcare workers in developed nations to recognize inherited renal diseases, and to serve as a guideline to differential diagnosis. Plange-Rhule et al. (2002) have written a guide for the primary care physician regarding clinical assessment, investigation, and treatment of renal disease in Africa—although for many LMICs, the aim of delivering effective services would initially be the detection, prevention, and management of the most common problems within that community.

However, it is important to recognize how inherited renal conditions affect prognosis and management, and

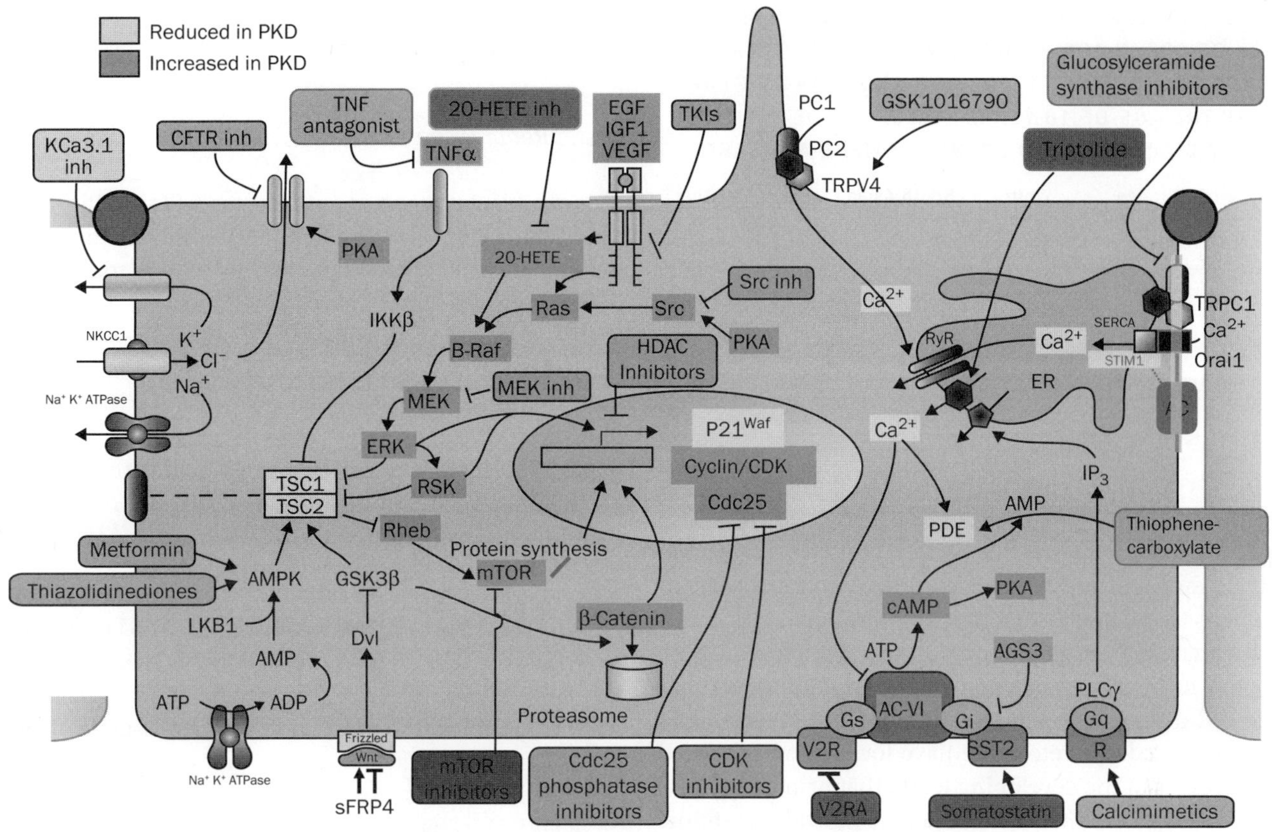

Figure 25-5 Possible sites for drug development shown in colored boxes (Adapted from Torres et al., 2007).

the patient may have other manifestations of the condition. There may be other family members at risk who would benefit from the detection and management of renal disease and/or management of other manifestations. As countries develop nephrology services and imaging, and histological services become available, inherited renal disease detection will become more important. The development of genetics and prenatal imaging, and knowledge of inherited renal diseases, will become more important still, and significant as countries develop.

DIAGNOSING INHERITED RENAL DISEASES IN LMICS

Since ADPKD is the most common inherited renal disease worldwide, we have focused examples using this disease.

Family History

When taking a history from a renal patient, it is important to ask whether there is anyone else in the family affected with renal disease, and if so, the age of onset and severity of their renal disease and whether their diagnosis is known. The other family members affected should be recorded, and this information can be used to consider the possible mode of inheritance. However, there may be no family history in the case of de novo gene mutations, or in some families with autosomal recessive or X-linked conditions. It is important also to bear in mind the possibility of nonpaternity.

By taking a family history, it may be possible to identify the pattern of inheritance, and so not only narrow down the differential but also identify more easily other individuals at risk. Table 25-7 lists the most common inherited renal diseases and the pattern of inheritance. Clinical information may then help to further narrow down the possible diagnosis. An excellent review by Joosten et al (2010) has outlined how the use of urine dipstick testing can be a valuable aid to diagnosis (Table 25-8).

URINE ANALYSIS

See Table 25.6

DISCUSSION OF INHERITED RENAL DISEASES

In discussing inherited renal diseases in developing countries, we have limited the number of diseases for several reasons. Exceptions are where diseases affect only certain populations, e.g., Finnish nephropathy, where these

TABLE 25-6 PATTERN OF INHERITANCE OF THE PRIMARY CAUSE OF ERF IN 190 PATIENTS WITH INHERITED RENAL DISEASE PRESENTING AFTER 1996 AND STARTING ERF BEFORE THE AGE OF 16 YEARS FOR WHOM DETAILS OF PRIMARY DIAGNOSIS AND ETHNICITY WERE AVAILABLE

Disease inheritance	White	South Asian	Black	Other
Autosomal recessive	120	47	1	4
Autosomal dominant	5	0	0	0
Sex linked	6	2	1	0
Mitochondrial disease	3	1	0	0

conditions are not knowingly increased in developing countries. Similarly, diabetes (Table 25-2), which affects all populations (see also Chapter 22), and HIV-associated renal disease will not be discussed here, but are covered elsewhere (see Chapters 15 and 16). Similarly, sickle cell disease may lead to papillary necrosis, which would be expected to be more common in Black populations, but sickle cell disease is not primarily a renal disease and so is not discussed. Therefore, we have focused on inherited renal diseases in the developing world that primarily cause kidney disease, even though there may be wider systemic problems.

CHROMOSOMAL DISORDERS

A wide variety of chromosomal disorders have an association with renal abnormalities, but as they are not primarily renal disorders, these are not discussed here, since the mainstay of problems and management is focused on the other extrarenal problems, most often developmental delay, feeding, and growth.

TABLE 25-7 INHERITED RENAL DISEASES GROUPED BY MODE OF INHERITANCE. (ADAPTED FROM JOOSTEN ET AL., 2010)

Autosomal Dominant	Autosomal recessive	X-Linked
Autosomal dominant Polycystic kidney disease (ADPKD)	Autosomal recessive Polycystic kidney disease	Fabry disease
Alport syndrome	Alport syndrome	Alport syndrome
Nail-patella syndrome	APRT deficiency	
Glomerulopathy with fibronectin deposits (GFND)	Cystinosis	
Renal cysts and diabetes syndrome (RCAD)	Nephronophthisis	
Nephropathies associated with uromodulin (UMOD)	L-CAT deficiency	
	Primary oxalosis	

Cystic Renal Diseases

This is a diverse group of renal diseases (Table 25-9). Cystic diseases can be congenital, inherited, or acquired. They may be an isolated abnormality, or be one manifestation of a systemic disease. There is great genetic heterogeneity within the group of inherited renal cystic diseases, and the presence of modifier genes adds to the complexity of presentations. There is great potential for misdiagnosis upon noting the presence of cysts on a renal ultrasound scan. Therefore, it is important to take a thorough personal and family history, and carry out a thorough assessment of patients with renal cysts.

AUTOSOMAL DOMINANT POLYCYSTIC KIDNEY DISEASE (ADPKD)

ADPKD is one of the most common single-gene disorders of humans, and is also the most frequent inherited form of cystic kidney disease

ADPKD typically presents in the third or fourth decade with hematuria, loin (flank) pain, or hypertension. It is characterized by the progressive development and enlargement of bilateral renal cysts, and often leads to end-stage renal disease (ESRD) by late middle-age. It is well recognized now that ADPKD is a multisystem disorder with many extrarenal changes and complications. These extrarenal manifestations include hepatic cysts, cerebral aneurysms, and cardiac valve abnormalities. With increasing awareness of potential complications in the West/developed countries, there has been a greater drive to diagnose patients presymptomatically, such that many patients may now be asymptomatic for many years.

PATHOGENESIS OF CYSTIC DEVELOPMENT

Cysts in ADPKD vary considerably in size and appearance (Figure 25-6), from a few millimeters to many centimeters, and can produce an enlarged kidney as big as 40 cm in length and weighing as much as 8 kg (Gabow, 1993). Rarely, cysts may be unilateral (Bear, 1974; Fick-Brosnahan, 1999) or may be seen in a fetus following biallelic inheritance (Pei, 2001). Cysts increase in size and number with time; however, some patients remain asymptomatic and may only be diagnosed at autopsy (Iglesias, 1983). Detectable renal cysts, by ultrasound scanning, are often only first seen in the second decade of life. The relation between size and degree of renal functional impairment has been studied in the CRISP study, showing that total kidney volume and total cyst volume

TABLE 25-8 INHERITED RENAL DISORDERS GROUPED BY URINE ANALYSIS

Normal urine	Macroscopic hematuria	Microscopic hematuria	Proteinuria >1.5g	Fanconi	Concentration defect	Kidney stones or crystaluria
ADPKD	Alport syndrome	Nail-patella syndrome	Nail-patella syndrome	Cystinosis	Nephronophthisis	Oxalosis
UMOD associated nephropathies	ADPKD	LCAT	LCAT		UMOD associated nephropathies	APRT-deficiency
Nephronophthisis	APRT-deficiency	Fabry disease	Fabry disease			
RCAD		Alport syndrome	Alport syndrome			
		GFND	GFND			

(Adapted from Joosten et al.,2010)

increase exponentially, a result consistent with an expansion process dependent on growth. A baseline total kidney volume above 1500 ml in 51 patients was associated with a declining glomerular filtration rate (by 4.33+/−8.07ml per minute per year, P<0.001). Continuing expansion of cysts in ADPKD is quantifiable and associated with the decline of renal function, and higher rates of kidney enlargement are associated with a more rapid decrease in renal function (Grantham et al., 2006).

The cystic cavity fluid distinguishes cysts from benign epithelial tumors or adenomas. The cysts themselves are fluid-filled structures, which vary in diameter. Most cysts (70%) exist as noncommunicating structures (Gardner, 1992). It is believed that overall, only 5% of tubules are affected by early cystic expansion (Grantham, 1987). Typically, only a few renal cysts are detectable in an affected individual during the first two decades of life. By the fifth decade, however, hundreds of renal cysts can be found in most patients. Additionally, significant intrafamilial variability of ADPKD has been well documented. These findings suggest that factors in addition to the germline mutation of a polycystic kidney disease gene might be required for individual cyst formation. There is now compelling evidence in support of a "two-hit" model of cystogenesis in ADPKD. In this model, inactivation of both copies of a polycystic kidney disease gene by germline and somatic mutations within one renal epithelial cell provides growth advantages for it to proliferate clonally into a cyst (Pei, 2001).

TABLE 25-9 RENAL CYSTIC DISEASES

Congenital Renal Cystic Diseases
Chromosomal abnormalities
Unknown etiology
Hereditary Renal Cystic Diseases
Autosomal Recessive Cystic Diseases
Autosomal recessive polycystic kidney disease
Meckel's syndrome
Fryn's syndrome
Oro-facial-digital syndrome
Zellweger syndrome
Glutaric academia type II
Autosomal Dominant Cystic Diseases
Autosomal dominant polycystic kidney disease
Tuberous sclerosis
Von-Hippel-Lindau syndrome
Acquired Renal Cystic Diseases
Simple cysts
Hypokalemia-induced cysts (primary aldosteronism)
Acquired cystic disease

EPIDEMIOLOGY

In Europe and North America, the prevalence of end-stage renal disease (ESRD) due to ADPKD is approximately 1 in 2500, but the prevalence of ADPKD is estimated at 1 in a 1000 to 1 in 400 (Bear et al., 1992; Chapman, 1991; Davies, 1991; Iglesias, 1983). However, at least half of the individuals with ADPKD are asymptomatic at death (Iglesias, 1983). This approximates to 4 to 6 million affected individuals worldwide. ADPKD had a prevalence of almost 9% in the United Kingdom (Byrne, 2010 p. 1540) with similar figures in North America, Europe, and Australasia. However, in Japan there appears

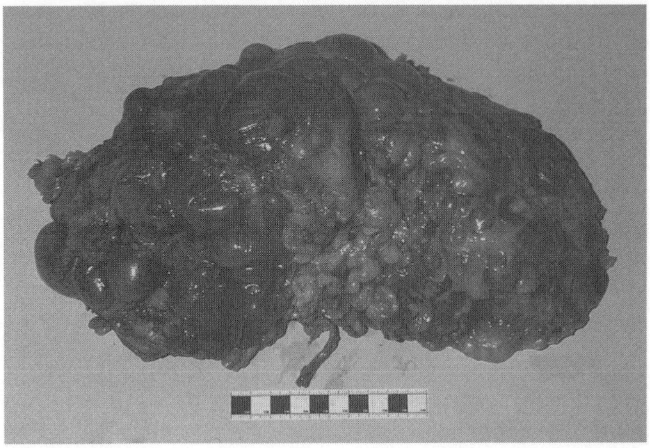

Figure 25-6 Typical appearances of an enlarged cystic kidney removed from a patient with ADPKD.

to be a low rate of prevalence, approximating 3% at most (Higashihara, 1998). In Europe it ranks as one of the four most common causes of ESRF. Our own research has identified that ADPKD is the single most common genetic diagnosis in LMICs (Jones & Saggar, 2012, personal communication)

Ethnic Distribution

Historically, it has been suggested that ADPKD is less common in the nonwhite Caucasian population. This, in our opinion, is not true and cannot be justified. Certainly there have been differences in numbers of reported cases, but these almost certainly reflect underdiagnosis, underinvestigation, or misdiagnosis. ADPKD has been described to occur in Africans and Asians and many other races (Ding, 2002; Iglesias, 1997; Lee, 2001; Mizoguchi, 2002; Phakdeekitcharoen, 2000; Yersin, 1997), but again, underlying the lack of study in different racial groups, the prevalence of ESRD in non-European subjects has not been extensively determined. Although the disease is said to be less common in some parts of Africa, little information exists (Seedat, 1984). In U.S. Blacks, ESRD from ADPKD is about as frequent as in Whites, but age at ESRD is more variable and lower in Blacks than in Whites (Freedman, 2000). Furthermore, ESRD in ADPKD Blacks with sickle cell trait is earlier than in those without sickle trait (Yium, 1994).

CLINICAL ASPECTS

Diagnosis and Diagnostic Criteria

Autosomal dominant polycystic kidney disease presents mainly in adult life. The clinical expression is variable, ranging from incidental findings at autopsy in old age to end-stage renal failure by the age of 30. Developing renal cysts can usually be detected well before the onset of symptoms. In one study (Bear, 1974), cysts were visualized by ultrasound in 56% of gene carriers by 10 years of age, and in 90% of gene carriers by the age of 20.

Neonatal development of renal cysts and/or early clinical manifestations of ADPKD are known to occur with an estimated frequency of 2% (Zerres, 1993). In these families there is a 40% recurrence risk for similar early presentation in subsequent affected children. The biological basis of this marked change in clinical expression of the disease from parent to offspring is now thought to be due to the presence of biallelic inheritance, with hypomorphic alleles in one parent (Harris, 2010).

ADPKD is most readily diagnosed using ultrasonography. A positive ultrasound diagnosis is often unambiguous, but a negative scan is complicated by the fact that cysts develop gradually over time. (Figure 25-7).

In the past, Bear et al.'s (1984) classification of three cysts between two kidneys, in a person at 50% risk, was used. Certainly, using this classical definition, a false positive diagnosis before the age of 30 by this criterion is virtually impossible, but leaves many false negatives. The recent recognition of the milder phenotype, and later presentation of cysts in patients with the PKD2 mutation, where probability of false negative ultrasound diagnosis is higher at all ages, has led to new revised, combined criteria for ultrasound-based diagnosis (Pei et al., 2009). These criteria have now replaced those of Ravine et al (1994; see Tables 25-10 and 25-11, and Figure 25-8).

An isolated kidney cyst in a child is very rare (Ravine, 1993) and raises a strong suspicion of ADPKD in a child even if there is no positive family history (Fick, 1994; Gabow, 1997).

Genetic diagnosis is possible, but since ultrasound is so good, genetic testing presently is reserved for special circumstances. Linkage testing is feasible if the family structure permits; more recently, sequence-based direct gene mutation detection has become commercially available. The sensitivity and specificity of this test is complicated by the difficulties in mutation detection in large genes without any predominant mutations in the population, and experience is just now beginning to accrue on how to best utilize gene-based diagnosis (Harris, 2010; Barua, & Pei, 2010). Important prognostic information, from knowing if one has PKD1 or PKD2, is now available (Hateboer, 1999), but current understanding, even in the year 2008, does not permit reliable application of this

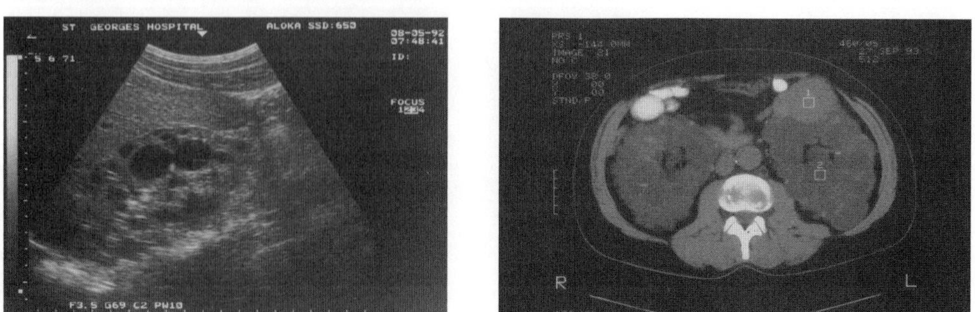

Figure 25-7 Typical ultrasound (left) and CT (right) appearances of an ADPKD kidney.

TABLE 25-10. ULTRASOUND CRITERIA FOR DIAGNOSIS OF ADPKD (GIVES THE POSITIVE AND NEGATIVE PREDICTIVE VALUES OF VARIOUS ULTRASOUND DIAGNOSTIC CRITERIA; PEI, 2009).

Age, y	PKD1	PKD2	Unknownn
15–29	≥3 cysts PPV 100% SEN 94%	PPV 100% SEN 69%	PPV 100% SEN 82%
30–39	≥3 cysts PPV 100% SEN 97%	PPV 100% SEN 95%	PPV 100% SEN 95%
40–59	≥2 cysts bilat PPV 100% SEN 93%	PPV 100% SEN 89%	PPV 100% SEN 90%

*≥60y ≥4 cysts in each kidney PPV and SEN 100%

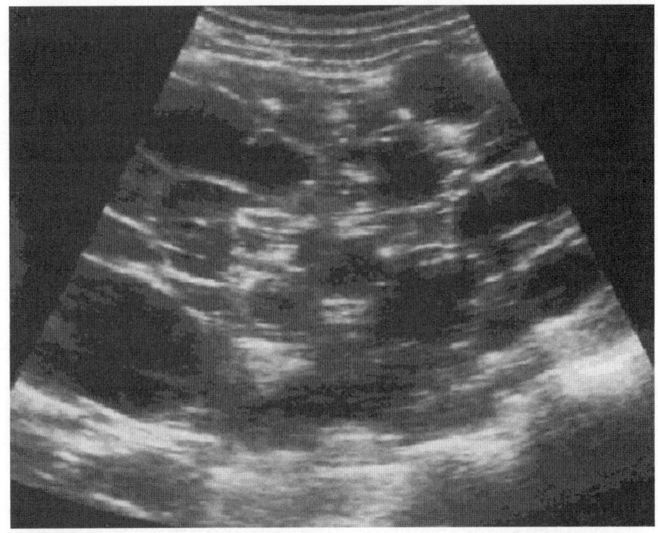

Ultrasound appearance of ADPKD

Age	cysts
15-30	>3
30-39	≥ 3
40-59	≥ 2 in each kidney

(Modified Ravine criteria: individuals at 50% risk)

Figure 25-8 Summary of numbers of cysts required to establish a diagnosis by Ultrasound (Pei, 2009)

knowledge to the prognosis for *individual* patients; only generalizations can be made.

In summary, using modern-day ultrasound scanners that have a much higher resolution, general figures for counseling the risk of a false positive scan in a first-degree relative approximate to <10% at the age of 20, and much less than 1% at the age of 30. These figures can be reduced further by the use of fine-cut renal CT scanning or MRI (Figure 25-9).

THE SPECTRUM OF PHENOTYPE OF ADPKD

Clinical Aspects

Common clinical problems in ADPKD described in the literature are summarized in Table 25-12 and Table 25-13. These can be divided into renal and extrarenal. There is some degree of correlation between renal size and symptoms or signs (Grantham et al., 2006). The larger kidneys are more often associated with pain, and are associated with hypertension (Gabow, 1990). Pain can be difficult to control, and in some, can require narcotic analgesia. An excellent review of pain management in ADPKD is provided by Bajwa (2001). In a few patients where there is an easily identifiable and accessible renal or liver cyst, percutaneous drainage and alcohol sclerosis may provide relief. Open surgical drainage has also been undertaken where cysts are smaller and more numerous. Neither of these procedures have been reported to enhance or prevent renal deterioration.

Hematuria occurs in 30%–50% of ADPKD patients and may be the first presenting sign. The mean age at first episode is usually about 30 years, and repeated episodes are associated with a worse renal outcome (Gabow, 1992).

TABLE 25-11 ULTRASOUND CRITERIA FOR EXCLUSION OF ADPKD

Age, y	PKD1	PKD2	Unknownn
15–29	≤1 cyst NPV 99% SPEC 98%	NPV 83% SPEC 97%	NPV 91% SPEC 97%
30–39	≤1 cyst NPV 100% SPEC 96%	PV 97% SPEC 94%	NPV 98% SPEC 95%
40–59	≤1 cyst NPV 100% SPEC 94%	NPV 100% SPEC 94%	NPV 100% SPEC 94%

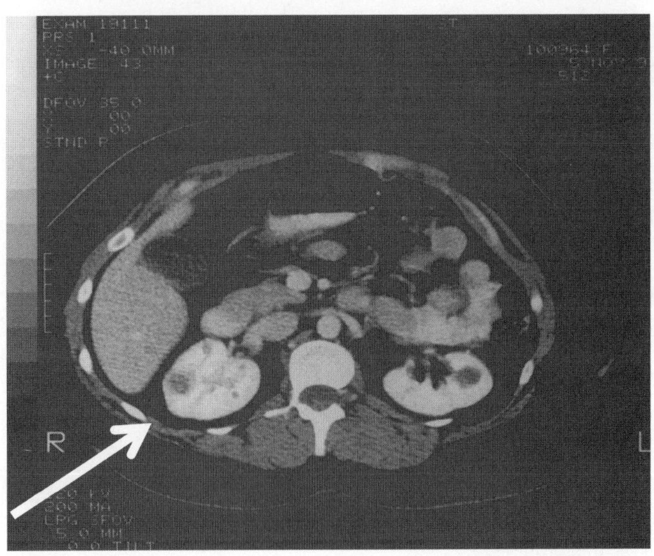

Figure 25-9 CT scan of early bilateral polycystic kidney disease. Multiple darker areas, indicated by white arrow, are the cystic lesions.

TABLE 25-12 COMMON CLINICAL FEATURES ASSOCIATED WITH ADPKD

Kidney cysts	Hematuria
Liver cysts	Loin pain
Hypertension	Nephrolithiasis
Intracranial aneurysms	Mitral valve prolapse
Increased left ventricular mass	Progressive renal impairment

Hematuria is self-limiting and usually lasts no more than a few days. Management involves bed rest, analgesia, fluids, and the exclusion of urine infection. Proteinuria is a poor prognostic indicator. More rarely, bleeding may also occur into a renal cyst.

Urinary tract infection also commonly occurs in the bladder, the kidney tissue, or cyst fluid. Infection within cystic fluid often fails to respond to initial antibiotic treatment because many of the commonly used first-line antibiotics, e.g., penicillin or cephalosporin, fail to concentrate sufficiently within cysts (Schwab, 1987). Ideally, non-ionized lipid-soluble antibiotics should be used.

Nephrolithiasis occurs in about 25% of ADPKD patients and is one of the differential diagnoses for patients presenting with pain or hematuria. Renal scanning may be required to identify stones, given the distortion and acoustic shadowing that can obscure a clear image on ultrasound. ADPKD patients are prone to develop calculi because of the relative intrarenal stasis and distortion produced by the cysts. Furthermore, a common metabolic abnormality is hypocitraturia (Torres, 1988), a known risk factor for nephrolithiasis.

TABLE 25-13 THE FREQUENCY OF RENAL AND EXTRARENAL MANIFESTATION OF ADPKD

RENAL	
Cysts	100%
Cyst calcification	Common
Hypertension (adult)	60% (pre ESRF) 80% (post ESRF)
Hypertension (children)	30%
Hematuria	50%
Renal pain	60%
Urinary tract infection	Common
Nephrolithiasis	30%
ESRF	45% by age 60
EXTRARENAL	
Hepatic cysts	50%
Pancreatic cysts	7%
Colonic diverticulae	80% (post ESRF)
Cardiac valve abnormalities	30%
Intracranial aneurysm	8%
Thoracic aortic aneurysm	8%
Arachnoid cysts	8%

(adapted from Gabow, 1993)

Hypertension

Hypertension is a well-recognized complication of ADPKD, precedes renal failure in 50% to 80% of patients with ADPKD, and is an early event. The well-established influence of blood pressure in deteriorating renal function, and the increasing use of ultrasound in screening at-risk family members for the presence of the disease, have led to great interest in the early detection of hypertension, its mechanisms, and the initiation of early effective treatment as a means of delaying progression of renal disease.

Despite this, hypertension remains inadequately managed in ADPKD (Ecder, 2001). One important justification for the increased use of early ultrasound screening of at-risk family members has been that early initiation of effective antihypertensive treatment may delay progression of renal disease, although frequent monitoring of blood pressure in at-risk individuals can be used in place of early presymptomatic ultrasound diagnosis.

Hypertension in children presents a particular problem. In children with ADPKD, approximately 33% are reported to be hypertensive (Cadnapaphornchai et al., 2008). Milder forms of hypertension are more difficult to assess in children. In Sedman's study (1987), 13% of children had abnormal renal function at the time of presentation.

Hypertension, Renal Function, and Renal Disease Progression

In common with other renal diseases resulting in chronic renal failure, the deleterious effect of systemic hypertension on renal function is well recognized. ADPKD appears to be no exception, in that patients with hypertension appear to have a worse renal outcome. Treatment of hypertension in ADPKD may be effective in deferring end-stage renal failure, as it has been shown in the essential hypertension. However, recent studies examining patients with quite severe established renal failure show no slowing of progression (Klahr, 1995; Locatelli, 1996; Maschio, 1996). This may well reflect the need to target patients much earlier. Nonetheless, adequate treatment of hypertension in ADPKD subjects is important. Detection and early treatment of hypertension, on the assumption that it will delay early progression of renal disease and other types of target organ damage, represents one of the best justifications for early diagnosis of ADPKD, and is the basis for the recent commencement of the HALT Progression of Polycystic Kidney Disease (HALT PKD) trial, in which the efficacy of interruption of the renin-angiotensin-aldosterone system on the progression of cystic disease, and on the decline in renal function in autosomal dominant kidney disease (ADPKD), will be assessed in two multicenter randomized clinical trials targeting different levels of kidney function—GFR >60 mL/min/1.73 m2 (Study A), and moderately advanced disease defined by GFR 25–60 mL/min/1.73 m2 (Study B). Patients who have blood pressures that are exceptionally well controlled will be in one arm, at maximum 110 mmHg. Participants will be followed four to six years.

Rate of Renal Disease Progression

The most important functional abnormality is progressive renal failure. The rate of loss of glomerular filtration rate (GFR) is therefore commonly used to assess progression. Approximately 50% of patients will have ESRD by the age of 60 years (Hateboer, 1999; Figure 25-10). Age at onset of renal failure is variable (in the range of 2 to 80 years), even within families. This variation is presumed due to the stochastic nature of pathogenic "second hits," and to genetic modifiers within and between families. Table 25-14 lists some of the factors associated with a worse renal prognosis. There seems to be much hope that renal prognosis will improve further, since the quoted figures above are based on a previous generation, not all of whom were treated adequately or sufficiently early.

In a review by Grunfeld (1998), the mean rate of decline of creatinine clearance in 2 groups of ADPKD

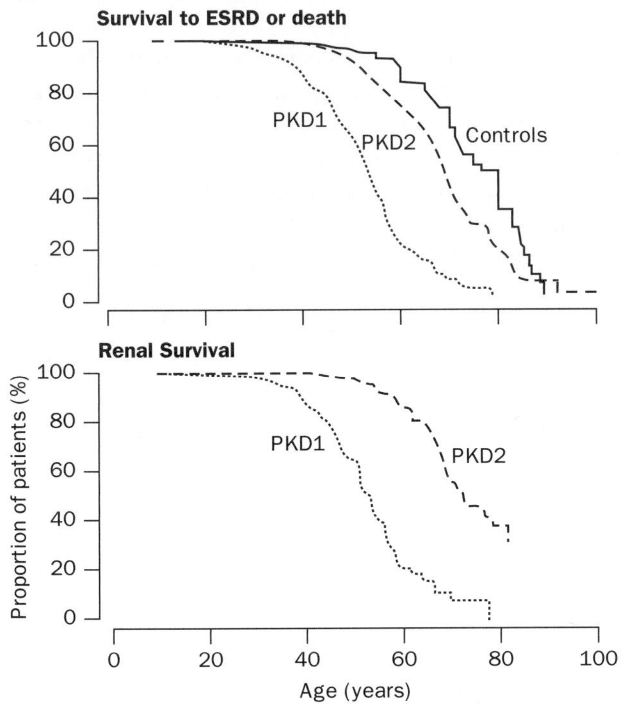

Figure 25-10 Cumulative survival to ESRD or (death from kidney failure) in ADPKD compared to unaffected relatives (Hateboer, 1999).

TABLE 25-14 FACTORS ASSOCIATED WITH A WORSE RENAL PROGNOSIS (ADAPTED FROM GABOW, 1992, P.1117).

History of hypertension	Male gender
Repeated haematuria	Black race
Urine infection (in males)	Concurrence of sickle-cell trait
More than 2 children (in women)	Massive polycystic liver disease

patients was about 4mls/min/year. One group, in whom moderate renal impairment was documented between 21 and 40 years of age (n = 23 males), the mean monthly decline in creatinine clearance 0.53, while in patients whose renal failure started between 51 and 65 years of age (n = 38), the monthly decline was 0.40. These data by Grunfeld also agree with those collected prospectively in the U.S. Modification of Diet in Renal Disease (MDRD) study (Klahr, 1995). In these and other studies, no correlation was found between rate of progression and protein intake. Age distribution of end-stage renal disease in the study showed that approximately 15% of the PKD population progressed to end stage before the age of 40, 10% after the age of 65, and 75% between 41 and 65.

A consistent problem with these studies, from the ADPKD point of view, has been the low level of renal function at the inception of the trial and the short duration of follow-up. A recent study by Ecder (2000) found that while enalapril, compared to amlodipine, reduced proteinuria more significantly, there was no difference between the groups in the level of blood pressure reduction, and both drugs equally slowed renal functional loss over a 5-year period. In this study, all patients began with a creatinine clearance greater than 50 mls/min/1.73m^2. If the rate of progression is to be modified, early identification of predictive factors is vital for early intervention, since therapeutic modifications are a very real possibility now. The results of the HALT trial will, therefore, be of great interest, as it will also determine whether a low blood pressure target (95–110/65–75 mmHg) is superior to a standard blood pressure target (120–130/70–80 mmHg) in patients with preserved renal function (Chapman, 2008).

CLINICAL FACTORS AFFECTING PROGRESSION

ESRD is used, since it is a hard endpoint in ADPKD. This obviously can only reflect in part the rate of progression in a given family, since in affected members of the same generation, some progress to ESRD while others will not progress within their lifetime.

CARDIOVASCULAR RISK FACTORS

The most common cause of death in ADPKD patients is cardiovascular disease and, as such, patients with ADPKD are at increased risk of sudden death and premature cardiovascular disease. One recent paper from Toronto, Canada, comparing causes of death in ADPKD patients, described a cardiac cause for death in 34% of patients and a 12% cause of death for vascular/cerebrovascular disease (Roscoe, 1993). Rupture of intracranial aneurysms (ICAs) is an important cause of sudden death in polycystic patients, but it only accounts for 14% of all such deaths (Iglesias, 1983). This leaves many patients with no apparent cause for their sudden demise.

The mechanisms underlying the increased cardiovascular risk are not fully understood, but recent work has clearly demonstrated expression of PKD genes in vascular tissue (Torres, 2001). Hypertension and the suggested defect of extracellular matrix seen in ADPKD may result in extrarenal manifestations associated with sudden death, including cardiac valve defects, thoracic, iliac, and abdominal aneurysms, arterial dolichoectasia, and intracranial arterial dissection.

There is a higher incidence of atypical chest pain, palpitations, and cardiac valvular defects described in ADPKD (Gabow, 1984; Harrap, 1991; Timio, 1992), and an almost 30% incidence of valvular heart abnormalities. It is possible that some of the cardiovascular deaths seen in ADPKD result from arrhythmias secondary to this ventricular enlargement, or valvular changes.

Left Ventricular Hypertrophy

The previous reports described above, of cardiac and vascular lesions and early abnormalities in young, normotensive, asymptomatic polycystic patients, reinforces the need for early presymptomatic diagnosis and assessment of possible risk factors in individuals at risk within affected families. Left ventricular hypertrophy (LVH) has been clearly established as a complication of ADPKD, with or without hypertension, and LVH is a well-known risk factor for premature cardiovascular disease and may be a consequence of early elevation in blood pressure load.

EXTRARENAL MANIFESTATIONS

Hepatobiliary Cystic Disease

Simple hepatic cysts occur in 2.5% to 4.6% of patients referred for ultrasound examination of the abdomen. Such cysts are solitary in two-thirds of patients, but no more than three cysts are usually found in those with simple multiple lesions (Figure 25-11). Hepatic cysts (Table 25-15) are seen in a number of other conditions, and are very common in ADPKD (Table 25-15). Hepatic cysts are seen in approximately 50% of ADPKD patients overall (Torres, 1996). They increase in size and number, as in the kidney, but rarely result in any problems or compromise hepatic function. The incidence of cysts increases with age from approximately 20% in the third, to 75% in the seventh decade of life (Torres, 1996). They are more common in women than in men, and the cysts are often more numerous and larger than in men. Women with ADPKD who have had no estrogens or previous pregnancies are less likely to have hepatic cysts than those who have used hormones, been pregnant, or both (Torres, 1996).

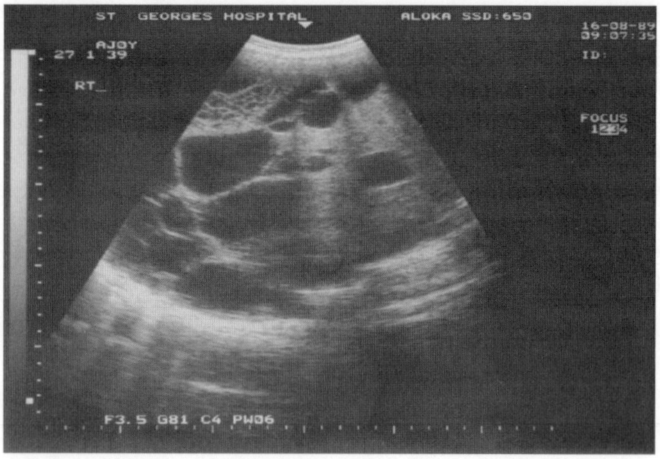

Figure 25-11 Ultrasound of a polycystic liver.

The natural history of PLD has been described in patients with ADPKD (Torres, 1996). Hepatic cysts are exceptionally rare in children.

Most patients with liver cysts are asymptomatic and have normal liver function tests. Symptoms are usually due to mass effect, or a massively enlarged polycystic liver with or without rare complications, e.g., hepatic vein thrombosis (Budd-Chiari syndrome). Furthermore, hemorrhage into a cyst can cause severe acute abdominal pain, fever, elevation of liver enzymes, and imitate acute cholecystitis or hepatic abscess. Ultrasound imaging or CT scanning is a sensitive technique to differentiate the complicated from an uncomplicated hepatic cyst.

Congenital hepatic fibrosis, a hallmark of autosomal recessive polycystic kidney disease, has been reported in 22 patients from 13 families with ADPKD (Hadimeri, 1997; Tamura, 1994; Torres, 1996).

In general, where treatment is required it is directed at reducing the volume of cystic tissue. The approach can

TABLE 25-15 NON-ADPKD CONDITIONS ASSOCIATED WITH CYSTIC LIVER DISEASE

Dilatation of the extra and/or intrahepatic bile ducts
Segmental dilatation of intrahepatic bile ducts (Caroli's disease)
Congenital hepatic fibrosis
Autosomal recessive polycystic kidney disease
Autosomal dominant polycystic liver disease

be preventative or surgical, which in turn can be open or closed.

Larger superficial hepatic cysts can be treated by laparoscopic fenestration, while in some extreme cases patients have had liver transplantation (Torres, 1996).

INTRACRANIAL ANEURYSMS

The association of intracranial aneurysms (ICAs) and ADPKD has been established for many years. Aneurysm rupture is life-threatening and entails a 30% to 50% mortality rate in the general population. Although an uncommon manifestation, ICA rupture is part of the spectrum of extrarenal features that may occur at any time in ADPKD patients. Subarachnoid hemorrhage as a cause of death is found in about 0.7% of the general population, and in about 6% in ADPKD patients.

Pathogenesis

The mechanisms leading to the formation of ICA in ADPKD are largely unknown. As in non-ADPKD patients, ICAs mostly develop at bifurcation sites of the circle of Willis. These areas are exposed to maximum blood flow–induced sheer stress.

The prevalence of hypertension in polycystic kidney disease may be responsible for this increased frequency of vascular associations. More recently, evidence suggests that both polycystin-1 and polycystin-2 are strongly expressed in vascular smooth muscle cells of large elastic distributive arteries (Torres, 2001 p. 306). Therefore, vascular manifestations may be a direct consequence of the basic genetic defect in ADPKD. A mutational basis for ADPKD-associated ICA is supported by both the young age of some affected patients, and the familial aggregation of the association (Chauveau, 1994 p. 986). ICAs are associated with the both the PKD1 and PKD2 forms of ADPKD.

Besides the PKD mutation, other genetic and nongenetic modifiers could play a role in the development of ICA, and so explain the frequent discordance for ICA within ADPKD families, even in monozygotic twins (Chauveau, 1994).

Clinical Features

Symptoms arising from an ICA are virtually always due to its rupture; they include nausea and vomiting, photophobia, focal neurological deficit, seizures, lethargy, and loss of consciousness. Neck stiffness may take up to six hours to develop. ICA rupture may also mimic an acute psychiatric illness. However, as in the general population, ICA rupture in polycystic patients can be preceded by herald bleeds in almost 40% of cases (Leblanc, 1987 p. 1489); i.e., headaches due to minor leaks of blood, a few hours to two weeks before rupture. The onset of headaches should be investigated urgently when they are of sudden onset, unusual character or severity in an ADPKD patient. Recognizing such warning headaches is important, since delay in the diagnosis of ICA rupture increases morbidity and mortality. Rupture has a 35% to 55% risk of combined mortality and morbidity (Chauveau, 1994 p. 986; Schievink, 1992 p. 1107). These high rates of mortality and morbidity are attributed mainly to the consequences of damage from initial bleeding, rebleeding, and cerebral ischemia. Intracerebral hematomas are also not uncommon with ruptured ICA. Early diagnosis and management will hopefully improve prognosis.

Rarely, ICA becomes symptomatic through local expansion. Focal findings such as cranial nerve palsy (particularly oculomotor and optic nerves) or seizures may result from compression by a large ICA. Transient ischemic attacks may result either from an embolus from within an ICA, or from direct compression of adjacent vessels.

Investigation and Management

The principles of management in ADPKD patients are fundamentally similar to those of ruptured ICA in the general population. Management includes a medical and surgical approach, and patients ideally require management in a neurosurgical unit. Full details of the investigation, management, and follow-up of patients with ruptured ICA are beyond the scope of this work and the reader is referred to more standard neurosurgical texts on this subject.

The mainstay of surgical treatment is ICA occlusion. Surgically clipping the ruptured ICA at its neck is the standard method (see Figure 25-12). Morbidity and mortality rates remain at about 10% because of occasional intraoperative ICA rupture. Asymptomatic ICAs accessible during the surgical exposure should also be clipped.

New techniques of selective endovascular occlusion now represent valuable alternatives to surgery. Detachable balloons were first used, but this technique has now been largely abandoned. A detachable platinum coil system has been reported to show, at 6 months follow-up evaluation, that 84% of patients had a good recovery; 4% had residual disability, and 11% died. In 84% of cases, the ICAs were completely occluded (Casasco, 1993 p. 1503). Undoubtedly, endovascular therapy will continue to play a major role in the treatment of ICA.

Management of Unruptured ICA

The key question is whether ICAs in ADPKD are more likely to rupture than aneurysms in the general

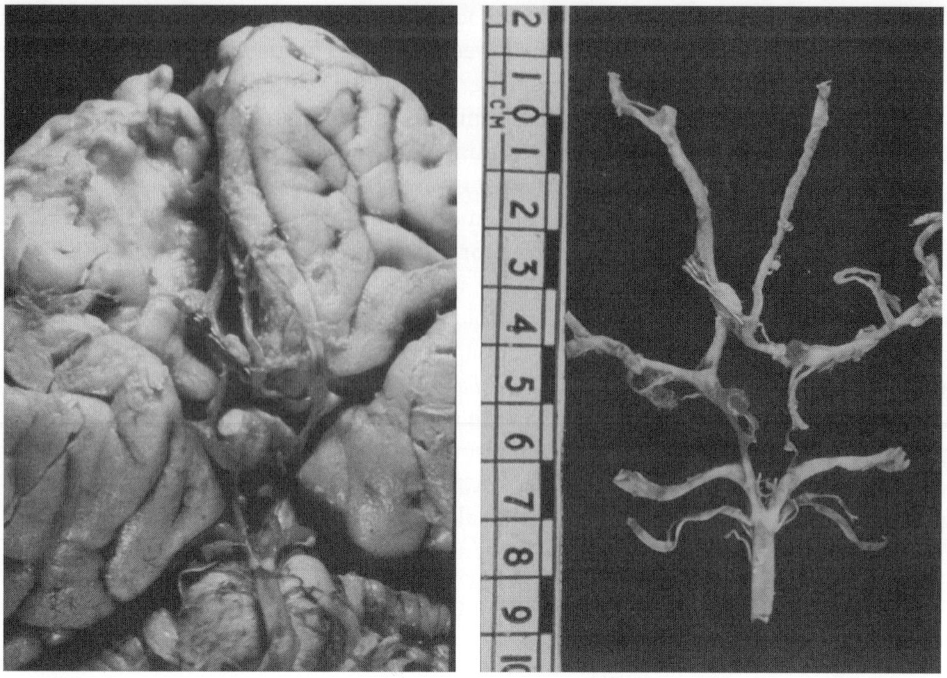

Figure 25-12 Aneurysm rupture with clipping showing (left panel).

population. Management is based on this risk of rupture. Available evidence suggests that the majority of asymptomatic ICAs do not rupture, although the natural history remains largely unknown. However, where there is a previous history of rupture, risk of a further bleed is inversely correlated with age. In addition to size, location is also important and aneurysms in the posterior circulation are more likely to rupture than aneurysms in the anterior circulation. Other possible factors include family history of subarachnoid hemorrhage, smoking, binge drinking, sudden blood pressure elevations, the presence of daughter sacs, and possibly the distensibility of the aneurysm.

Screening for Unruptured ICA

The multiplicity and the new development of an ICA, together with the severe outlook following ICA rupture, has prompted a suggestion that in at least a subset of ADPKD patients, regular screening for another ICA is justified (Figure 25-13).

Presymptomatic screening is mostly indicated in patients who are at a high risk of having ICA; i.e., patients who have a strong family history of ICA or personal history of previous bleed. Screening is also indicated in special cases where the patient has a high-risk occupation, or for patients planning major elective surgery.

MRA has emerged as the best noninvasive screening test, with a sensitivity of 100% for ICA >6 mm, which is appropriate since this is also the size threshold for surgical repair. Specificity is at least 92% (Horikoshi, 1994 p. 1484).

It remains to be determined how frequently at-risk patients with negative MRA should be reevaluated. At St. George's Hospital, London, and also the Mayo Clinic,

Figure 25-13 Angiokeratomas of the skin seen in a patient with Fabry disease caused by deposition of globotriaosylceramide (GL-3) in skin capillaries. Lesions are dark red to blue-black and punctate, they appear in clusters in the superficial layers of the skin. The lesions are often symmetrical, and most dense between the umbilicus and the knees. Mucosal areas are commonly involved.

Rochester, an MRA is recommended every five years. Patients surviving a first ICA rupture warrant MRA rescreening every 2–3 years, with early intervention at low ICA size. The age threshold is unknown, but it seems reasonable today to screen ADPKD patients with a family history of ICA from the ages 18 to 35 years. Guidelines of surgical intervention should be as for the general population, where prophylactic surgery is undertaken for asymptomatic ICA greater than 6–7 mm.

ADPKD GENETICS

The table 25-16 summarizes the two ADPKD genes (PKD1 & PKD2) encoding polycystin 1 and polycystin 2 respectively. The *PKD1* gene had been cloned by the European Polycystic Kidney Disease consortium (EPKDC), led by Peter Harris and coworkers in Oxford (EPKDC, 1994 p.1396). Subsequently, news broke of the cloned gene sequence for *PKD2* on chromosome 4 (Mochizuki, 1996 p. 773).

Present estimates indicate that ~85% of ADPKD is due to *PKD1* in a white European population. Mutations at the *PKD2* locus on chromosome 4q3-q23 probably account for ~15% of ADPKD. Recent reports have suggested that this may be even higher, at 20%–25%, in nonwhite European populations.

However, the prevalence of *PKD2* disease among elderly patients with ADPKD (~40%) is almost three times the prevalence of the disease in the general ADPKD population (Torra, 2000 p. 329). It is also possible that the proportion of PKD2 families is greater outside of white European populations. Mutations at the PKD1 locus are suggested to produce a clinically more severe form of the disease (mean age of ESRD, ~53 yrs.) than mutations in the PKD2 gene (mean age of ESRD, ~69 yrs.; see Hateboer, 1999 p. 510). The population frequency of ADPKD is relatively high compared to that of other autosomal dominant conditions, at least in part due to the fact that the clinical manifestations rarely occur before those affected have passed down mutations to their offspring.

An essential insight to explain both the significant intrafamilial variability, and the focal nature of the disease in the kidney and liver, can possibly be the somatic acquisition of second hits (i.e., mutations) in the normal allele at the respective PKD gene (Pei, 2001 p. 1550). This model first proposed by Knudson (1971 p. 1464) to explain the variation in retinoblastoma has been applied to ADPKD. Wild-type inactivation of polycystin 1 and polycystin 2 has been reported by several laboratories, and it has been shown that almost a quarter of cysts examined show loss of heterozygosity (LOH). Similar findings have been reported for liver cyst cells. The loss of both alleles provides a loss-of-function model for ADPKD. Table 25-16 summarizes the salient features of *PKD1* and *PKD2*.

Incidence of PKD1 and PKD2

The assessment of the true incidence of PKD1 and PKD2 disease has provoked considerable debate recently.

Racial differences exist. ESRD has been shown to develop almost ten years earlier in Blacks with ADPKD than Whites (Freedman, 2000 p. 411; Yium, 1994). Among Blacks with sickle cell trait, there has been an even more rapid progression of disease (Yium, 1994).

Phenotype and Genotype

PKD1 and PKD2 forms of ADPKD differ in clinical course. In PKD1 disease, mean age of onset of ESRD is 55–60 years, while for PKD2 it is at least 70 years. Cyst formation occurs later in persons with PKD2; disease and hypertension is less frequent (Hateboer, 1999).

TABLE 25-16 COMPARISON OF POLYCYSTIN 1 AND POLYCYSTIN 2

	PKD1	PKD2
Genomic size	~50 kb	~68 kb
Exons	46	15
Transcript	14.8 kb	5.1 kb
Localization	Integral membrane protein: basolateral plasma membrane and cilium(?)	Integral membrane protein: endoplasmic reticulum and cilium
Transmembrane domains	11	6
COOH terminus	intracellular (~210 aa)	intracellular (~280 aa)
NH2 terminus	extracellular (~3000 aa)	intracellular (~225 aa)
Homology	PKD2; sea urchin receptor for egg jelly (suREJ3); nematode LOV-1 protein	TRP family of Ca^{2+} channels; voltage activated Ca^{2+} channels
Function	Cell surface receptor	Cation channel (Ca^{2+})

PATIENT INFORMATION/GENETIC COUNSELING

Since 5%–10% of all patients on dialysis suffer from cystic kidney disease, and because in pediatric autopsy material, cystic changes of the kidneys are among the most common malformations, there has been much debate about early and often presymptomatic diagnosis in ADPKD.

Evidence is now emerging about factors and new treatments that will modulate renal disease progression, but these have to be formally tested in a prospective and randomized study.

With the localization of the responsible gene, predictive testing as well as prenatal diagnosis has become possible, in principle. It had been thought that prenatal testing would become important with respect to "prevention" of polycystic kidney disease in the population. However, there is very little demand in the West for prenatal as well as predictive testing, and this is undoubtedly in part due to subtle attitudes to adult-onset diseases, particularly those such as ADPKD, which have a variable onset and prognosis and are not characterized by obvious external stigmata. It is hoped that future success in molecular biology, in discovering the basic defect in ADPKD, could make real prevention a possibility. For now, adequate knowledge by the patients, their families, and the physicians, of the natural history and factors affecting progression, is needed.

This is an argument against early diagnosis; however, against this must be weighed the fact that although ADPKD has no cure, there are important treatable components that do make a strong argument for early diagnosis. Early treatable complications appear quite common in children (Cadnapaphornchai et al., 2008; Ravine, 1991). Ravine (1991) and others (Cadnapaphornchai et al., 2008) demonstrated that undiagnosed cases had one or more treatable complications at the time of diagnosis. These complications included hypertension, impaired renal function, and bacterial urinary tract infection.

Thus, there appears to be a clear and strong need for genetic counseling and clinical information in ADPKD families.

PATIENT REGISTERS

In its 1972 report, the World Health Organization (WHO, 1972) recommended setting up family-oriented genetic registries as part of a system to provide counseling and diagnostic services, treatment, and long-term follow-up for patients with genetic disorders. In 1978, the definition of genetic register functions was divided into five main roles, which are not mutually exclusive (Emery, 1978). These were the clinical or therapeutic role (follow-up and recall); the reference list; to monitor outcomes of service provision; to act as a research tool; and to assist in the prevention of genetic disease through complete ascertainment and family follow-up. Since that time, the use of genetic registers for family follow-up at predetermined times has been advocated to inform younger family members of their genetic risks when they reach maturity, or to carry out interval screening for complications of genetic disorders. The concept of a register may prove of value to LMICs; although it is appreciated that there may be population movement, in time a dataset will emerge.

Recognition that adequate treatment of blood pressure in essential hypertension and in those with renal disease needed better control and management, coupled with early reports that there were treatable complications, provided strong support and encouragement to develop a cascade screening program that would better manage the common disease complications and sequelae of ADPKD—in particular, hypertension, progressive renal impairment, loin pain, and urinary tract infection.

The high incidence of hypertension with early renal impairment among ADPKD patients suggests that greater awareness and earlier informed diagnosis, in those adults most at risk, will reduce cardiovascular morbidity mortality and possibly delay renal disease progression, although further work needs to be done in this respect.

The value of creating a genetic patient register has been well demonstrated for centers in the West and similar, albeit less elaborate, registers are possible in LMICs. The population drift and the self-selection of patients attending the clinic—based on distance, time, and financial cost—does mean that such registers are limited, but even documentation of the cases and the etiology will go a long way in collecting demographic data that in itself can be used for future public health planning.

In summary, ADPKD advances have moved forward in a dramatic and exciting way. First, there has been substantial progress in understanding the molecular consequences of the two major mutations that account for this disease, PKD1 and PKD2. Second, following studies in Caenorhabditis elegans, it has been recognized that a malfunction of cilia in the tubular lumen underlies ADPKD, which is presumably a defect of cell cytoskeleton and receptor signaling/placement and, accordingly, cell growth.

Furthermore, it is recognized that there are even milder forms of the disease, where individuals carry the ADPKD mutation but never develop cystic kidneys at all. Environmental influences or exposure, and their influence on progression, remain poorly understood. What is clear is that the polycystins play an important role in organizing and integrating signals that regulate and maintain the highly specialized structure of the kidney. Knowledge about the gene mutation, combined with clinical studies examining specific risk factors (e.g., left ventricular hypertrophy, hypertension), will allow a better understanding of the pathogenesis, pathophysiology, and progression of the disease and its associations.

The future requires there to be some process or treatment whereby progression of renal disease in ADPKD can be modified. With careful control of blood pressure, microalbuminuria, and early treatment of complications, the rate of renal disease progression can be reduced.

The use of novel agents to specifically target mediators of cyst formation, e.g., use of tyrosine kinase inhibitors (Sweeney, 2000), and potential novel therapeutic approaches currently under investigation as national trials may prove useful. These approaches include the use of antagonists of the vasopressin receptor type 2, and inhibitors of the mTOR kinase (e.g., rapamycin).

The finding of early abnormalities in young, normotensive, asymptomatic polycystic patients reinforces the need for early presymptomatic diagnosis and assessment of possible risk factors in individuals at risk within affected families. Earlier diagnosis and treatment of complications will undoubtedly help to reduce mortality and morbidity, and delay progression in this common condition.

ACQUIRED CYSTIC KIDNEY DISEASE

Renal cysts may result from a variety of causes. Table 25-17 shows a simple classification of renal cystic kidney disease. Acquired cysts can result from a variety of insults, and appear to be a common final end result of injury to the kidney. A diagnosis of multiple renal cysts is often confused with autosomal dominant polycystic kidney disease (ADPKD). Acquired cysts are particularly common in elderly patients and those on chronic hemodialysis, with the occurrence dependent upon the duration of dialysis, previous transplantation, and gender. There is no known genetic susceptibility.

TABLE 25-17 COMMON CAUSES OF RENAL CYSTS

SIMPLE RENAL CYSTS
Common in aging population
Unilocular
Cortical (most commonly)
Increase in number with age
Increase in size with age
Acquired renal cystic diseases
Chronic uremia
Hypokalemia-induced cysts
MULTICYSTIC KIDNEYS
No known disease associations
RENAL SYNDROMIC DISORDERS
Acromandibular syndrome
ADPKD
Autosomal trisomy syndromes 13 and 18
Chromosomal translocation syndromes
Congenital cutis laxa syndrome
Di George syndrome
Ehlers-Danlos syndrome
Fryns syndrome
Glutaricaciduria Type II
Goldenhar syndrome
Lissencephaly syndrome
Meckels syndrome
Medullary cystic kidney disease
Noonan syndrome
Oral-facial-digital syndrome
Recessive polycystic kidney disease
Rubenstein Taby syndrome
Short rib polydactyly syndrome
Tuberose sclerosis
Turner syndrome
Von Hippel Lindau disease
Zellweger-cerebrohepatorenal syndrome
RENAL NONSYNDROMIC DISORDERS
Medullary cystic disease
Medullary sponge kidneys
Obstructive uropathy
Renal dysplasia

CYSTIC RENAL DYSPLASIA

Renal cystic dysplasia most probably represents a nonspecific reaction to an early abnormality of development, and so is seen in genetic and nongenetic syndromes. The genetic basis of isolated renal dysplasia remains unknown and the recurrence risk is low.

However, in some families there have been described patterns of autosomal dominant dysplasia (termed *adysplasia*), autosomal recessive dysplasia, and also X-linked inherited dysplasia. These conditions are characterized by a highly variable degree of kidney involvement ranging from virtually normal kidneys to aplasia.

GLUTARIC ACIDURIA TYPE IIA

Glutaric aciduria Type IIA is a rare recessive condition that is characterized by neonatal acidosis, hypoglycemia, polycystic kidneys, and a strong "sweaty feet" odor. Hepatomegaly, respiratory distress, and muscle hypotonia are also typical clinical features. The odor is due to high concentrations of glutaric acid in blood and urine. The fundamental defect in this disorder is due to abnormal metabolism of the acyl-CoA compounds. There is, therefore, often co-excretion of excess lactic, ethylmalonic, butyric, isobutyric, 2-metyl-butyric, and isovaleric acids, which differentiates it from glutaric aciduria type I. Inhibition of acyl-CoA pathways is also seen in cases of Jamaican vomiting, caused by eating unripe *ackee*, which contains the acyl-CoA dehydrogenase inhibitor hypoglycin.

For those cases that do not die of neonatal acidosis, symptoms later include muscle weakness, nausea, vomiting, and hypoglycemia; occasionally cataracts, cerebral pachygyria, pulmonary hypoplasia, genital abnormalities and dysmorphism may also occur. Facial dysmorphic features include macrocephaly, large anterior fontanelle, high forehead, flat nasal bridge, and telecanthus. However, the dysmorphic features appear to be confined to the group of patients with the severe type, IIC. Fatty infiltration of the liver, heart, and kidney invariably occurs. Selective tubular damage leads to glycosuria and aminoaciduria. Unique ultrasound findings suggestive of early renal cyst formation have been reported, which the authors claim may be useful for establishing a diagnosis when enzyme studies are not available (Wilson, 1989).

Improvement in muscle power using a low-fat diet has been observed, as has successful treatment with riboflavin in a male child with progressive spastic ataxia and leukodystrophy, but without acidosis.

The glutaric acidurias are subdivided into type I and type II. The latter is further subdivided on the basis of which electron transfer flavoprotein subunit (alpha, IIA; beta, IIB) or flavoprotein dehydrogenase (IIC) is defective. The phenotype of these three subtypes is largely indistinguishable. All can lead to mild or severe cases, dependent upon the intragenic mutation. The gene locus for GA IIA has been cloned and sequenced at chromosome 15q23-25.

NEPHROPATHIES ASSOCIATED WITH UROMODULIN

These were previously referred to as medullary cystic disease (MCD-II) and familial juvenile hyperuricemic nephropathy (FJHN).

There are a group of inherited cystic nephropathies characterized by juvenile-onset recessive inheritance, previously termed *familial juvenile nephronophthisis* (FJN); or by adult-onset dominant inheritance, termed *medullary cystic disease*, previously termed MCD. They share clinical and pathological features and were traditionally grouped together under the term *FJN/MCD complex*. The main clinical features are renal cyst formation in the medulla or the corticomedullary junction, although this finding is inconsistent, as is salt wasting. Earlier reports had suggested that one single gene may be responsible for these diseases, but increasingly it has become apparent that the recessive and dominant forms are not allelic, and even within both the FJN and the ADMCKD complexes there is genetic heterogeneity.

Autosomal dominant medullary cystic kidney disease (ADMCKD) is an adult-onset inherited nephropathy which leads to end-stage renal failure (ESRF) in adulthood.

ADMCKD is associated with polyuria, polydipsia, and anemia, and the renal histology shows interstitial fibrosis and disintegration of the tubular basement membrane, together with bilateral corticomedullary cyst formation. All of these features are in common with FJN.

However, it is now apparent that ADMCKD is a genetically distinct condition from FGN; it has an autosomal dominant mode of inheritance affecting mainly adults, with a mean age of 30 to 47 years. Many patients are hypertensive at an early stage, but some later develop hypotension due to excessive salt wasting. Medullary cysts are present in over 75% of patients, but may not always be seen on ultrasound or CT scan due to their size. ADMCKD does have its clinical variants, and has been associated with such disparate renal manifestations as hyperuricemia and gouty arthritis, peripheral dysostosis, mental retardation, postaxial polydactyly, cerebellar abnormalities, hypogonadism, obesity, renal tubular acidosis, parathyroid insufficiency, and congenital amaurosis.

ADMCKD has now been shown to have three different loci, the first (MCKD1) localized on chromosome 1q21 and a second locus (MCKD2) linked to chromosome 16p12 in an Italian pedigree (Scolari et al., 1999).

Naturally, one of the obvious differential diagnoses for adult-onset cystic kidney disease is ADMCKD. However, important differences include a typical absence of flank pain, hypertension, or hematuria. Furthermore, the kidneys are very often non-enlarged. The average of onset of symptoms is 23 years, with renal failure after the third decade.

RENAL CYSTS AND DIABETES SYNDROME (RCAD)

This has historically been a general term used to describe a heterogeneous group of cystic kidney diseases, characterized by cystic dilatation of Bowman's spaces. Such changes are often seen within other systemic renal disorders, e.g., tuberous sclerosis, trisomy 13, orofacial-digital type1, and brachymesomelia-renal syndrome, to name but a few. Although usually sporadic, there is described a rare distinct dominant form with hypoplastic kidneys. Renal function is usually impaired from infancy but stable. The gene is distinct from PKD1 and PKD2.

More recently, hypoplastic forms have been mapped to the HNF1β gene and associated with a history of diabetes mellitus. In this form it is a variable autosomal dominant condition, with a variety of renal abnormalities with extrarenal manifestations and diabetes. It should be considered in all cases of unexplained cystic kidney disease and developmental renal abnormalities, especially in the presence of diabetes (Edghill, 2006, 2008). The underlying genetic defect is a mutation or deletion of TCF2, a gene encoding hepatocyte nuclear factor 1β (HNF1β). HNF1β is a transcription factor of vital importance in pancreatic and renal development (Coffinier et al., 1999). There is a high incidence of de novo mutations and deletions of around 30%–60% (Edghill, 2006, 2008).

Patients usually present with renal disease. The most common phenotype found to be associated with HNF1β gene mutations and deletions is renal cysts (in around 85% of patients) (Edghill, 2006, 2008). Renal malformations are the second most common abnormality seen (Edghill, 2006). Cysts or echobright kidneys are commonly seen on antenatal scanning (Edghill, 2008). However, the condition is extremely variable, even within the same family. Severity of renal disease varies from a report of grossly enlarged, nonfunctioning cystic kidneys detected prenatally (Bingham, 2000 p. 76) to normal renal function.

Renal disease is frequently associated with early-onset diabetes. In the majority of cases, diabetes develops after the onset of renal disease (Edghill, 2006, 2008), with a mean age of onset of diabetes of age 17, and requirement for treatment with insulin is common (Edghill, 2008).

Other features seen in this condition include uterine abnormalities, deranged liver function tests, and hypomagnesemia. One study of pediatric patients with renal malformations and chronic kidney disease found 44% of mutation carriers had hypomagnesaemia, versus 2% of mutation-negative patients (Adalat, 2009). The authors subsequently found HNF1β to be a regulator of FXYD2, which participates in the tubular handling of Mg^{2+}.

Management involves screening for and managing renal failure and diabetes, and monitoring for liver dysfunction and hypomagnesaemia.

NAIL PATELLA SYNDROME

Nail patella syndrome (NPS) is an autosomal dominant condition characterized by nail changes (98%), knee abnormalities, elbow abnormalities, and bony pelvic abnormalities detectable on pelvic X-ray. Thirty to fifty percent of patients develop renal failure, with 5% of patients developing end-stage renal failure. The first detectable sign of renal involvement is proteinuria, which may be associated with hematuria. Other associations of NPS are primary open-angle glaucoma and ocular hypertension; these occur at a younger age and more frequently in patients with NPS than in the general population.

NEPHRONOPHTHISIS (NPHP)

This is the name given to a group of rare autosomal recessive conditions that present with polydipsia, polyuria, and decreased urinary concentrating ability, and progress to end-stage renal failure, usually in adolescence. It has a recessive mode of inheritance with a neonatal juvenile or adolescent onset. The most reliable way to differentiate between NPHP and ADMCKD is by family history and mode of inheritance.

It was described as the original disease in 10%–32% of children with uremia and 7% of children with ESRF in the EDTA registry. There is equal sex distribution, and it has been reported in all races. Children initially develop progressive symptoms in the first decade of life if not in the first few months of life, with ESRF occurring on average at the age of 12. However, the age range is considerable, with one study reporting an age range from 4 to 23 years.

Nephronophthisis, previously called *familial juvenile nephronophthisis*, was recently mapped and localized to chromosome 2q13, (NPH1; see Antignac, 1993). Within NPH, genetic heterogeneity has been identified (Antignac, 1993; Haider, 1998; Schuermann, 2002) on 1p36. Two reports have also identified different loci for Senior-Loken syndrome, also known as "NPH with extrarenal manifestations," mapping to NPH3 (Omran, 2002) and NPH4 (Schuermann, 2002).

FABRY DISEASE

Fabry disease is an X-linked semidominant lysosomal storage disorder, caused by deficiency of the enzyme α-galactosidase (α-Gal A). Absence of this lysosomal hydrolase leads to the accumulation of glycosphingolipids, mainly globotriaosylceramide (GL-3), in the lysosomes of cells throughout the body. Affected organs include the kidneys, the vascular endothelium, the heart, the brain, and the peripheral and central nervous system. The disease is more severe in men, and is seen in varying severity in affected women depending on the pattern of X-inactivation. In untreated males, median life expectancy is 48–49 years. Death is most commonly due to renal failure or stroke.

The disease is characterized by raised, red, vascular cutaneous lesions of the skin (angiokeratomas), episodes of severe neuropathic pain in the extremities (acroparesthesias), and cardiac involvement. Cardiac involvement is seen frequently and involves cardiac hypertrophy (especially left ventricular hypertrophy), valvular insufficiency, diastolic dysfunction, and conduction disorders. Ocular abnormalities include characteristic ocular opacities (cornea verticillata), lenticular opacities, and tortuosity of conjunctival and retinal vessels.

Renal failure occurs secondary to toxic glycosphingolipid deposition in the renal tubular cells, podocytes, and renal vessel walls. Microalbuminuria or proteinuria is the first manifestation of renal involvement that can be detected in the clinic. As well as proteinuria, casts, red cells, and birefringent lipid globules with characteristic "Maltese crosses" can be observed in the urinary sediment in childhood.

In untreated males, elevated serum creatinine and a decline in GFR generally develop during the third decade. Without renal replacement therapy, the average life expectancy in a male patient with Fabry disease is 41 years (Colombi, 1967).

Renal tubular acidosis and nephrogenic diabetes insipidus can occasionally occur.

In males, diagnosis can be made by demonstrations of deficient α-galactosidase enzyme activity in the plasma, leukocytes, or isolated cells. This is unreliable in females. The most reliable way of testing carrier females is by molecular genetic testing of the GLA gene, the only gene known to be associated with Fabry disease.

Renal biopsy findings include a foamy appearance of the glomeruli, with diffuse swelling and vacuolization of visceral podocytes on light microscopy (Breunig, 2008). Electron microscopy shows podocytes and mesangial cells filled with lysosomal electron-dense granules arranged in a lamellar myelin pattern (Breunig, 2008).

Although further long-term studies are needed in patients with severe organ dysfunction, enzyme replacement therapy has been shown to be efficacious and safe in treating Fabry disease (Breunig, 2008). Additional treatment to manage risk factors is advised, including ACE inhibitors/angiotensin receptor blockers to reduce protein and albumin excretion, and treatment to manage cardiovascular risk factors.

ALPORT SYNDROME AND THIN BASEMENT MEMBRANE DISEASE

Alport syndrome is a collagen IV–related nephropathy encompassing a spectrum from Alport syndrome to thin basement membrane nephropathy (TBMN).

Subtypes of collagen IV–related nephropathies are X-linked Alport syndrome (XLAS), autosomal recessive Alport syndrome (ARAS), autosomal dominant Alport syndrome (ADAS), and TBMN.

Alport syndrome is characterized by renal, cochlear, and ocular involvement. Renal disease progresses from microscopic hematuria to proteinuria, progressive renal insufficiency, and end-stage renal disease (ESRD) in all males with X-linked disease, and in all males and females with autosomal recessive disease. Progressive sensorineural hearing loss (SNHL) is usually present by late childhood or early adolescence. Ocular findings include anterior lenticonus, maculopathy (whitish or yellowish flecks or granulations in the perimacular region), or recurrent corneal erosions.

Thin basement membrane disease (TBMN) is characterized by persistent microscopic hematuria often first observed in childhood; progressive renal involvement and extrarenal abnormalities are rare.

The prevalence of Alport syndrome has been estimated at approximately 1:50,000 live births (Levy & Feingold, 2000). Data from several series suggest that approximately one-fifth of children evaluated by pediatric nephrologists for isolated microhematuria have a diagnosis of Alport syndrome. According to the United States Renal Data System (USRDS), approximately 0.2% of adults and 3% of children in the United States with ESRD carry a diagnosis of Alport syndrome. Accurate data on ethnic-group or world prevalence is unknown, but from our research (Jones & Saggar, 2012, personal communication) it ranks as the second commonest disease in LMIC. TBMN disease is heterogeneous with an estimated population prevalence of around 1%. A small number of TBMN is caused by *COL4A3* and *COL4A4* gene mutations.

The diagnosis of collagen IV–related nephropathies rests on history and physical examination, which may include hearing and ophthalmic evaluation, family history, and urinalyses. Immunohistochemical analysis of basement membrane type IV collagen expression using skin and/or renal biopsy specimens, and examination of renal biopsy specimens by electron microscopy, will allow confirmation of the diagnosis because of the characteristic changes seen. Using these methods, the diagnosis can be confirmed in most cases. Molecular genetic testing of the type IV collagen genes, *COL4A3*, *COL4A4*, and *COL4A5*, is also available.

The mainstay of management is angiotensin-converting enzyme inhibitor and/or angiotensin receptor blocker in proteinuric individuals; routine treatment of hypertension; dialysis and renal transplantation for ESRD; routine treatment of SNHL and cataracts; and surgical intervention for symptomatic leiomyomas and protection of corneas in those with recurrent corneal erosions.

Follow-up of all individuals with a collagen IV–related nephropathy is required, and carrier females with XLAS need regular measurement of blood pressure and renal function. At-risk children need audiology evaluation every one to two years beginning at age 6 to 7 years.

At-risk family members can be screened by urinalysis, or if the disease-causing mutation in the family is known, by molecular genetic testing. Potential living related donors must be evaluated carefully to avoid nephrectomy in an affected individual.

Clinical Diagnosis

Clinical and pedigree features that support a diagnosis of Alport syndrome include (Kashtan, 2004):

Renal: Hematuria; in XLAS, 100% of affected males and more than 90% of affected females have microhematuria. One hundred percent of males and females with ARAS have hematuria. Episodic gross hematuria is not unusual.

Proteinuria, hypertension, and renal insufficiency develop with advancing age in all males with XLAS, and in all males and females with ARAS.

Cochlear: Bilateral high-frequency sensorineural hearing loss (SNHL) typically becomes apparent in late childhood or early adolescence in males with XLAS, and in males and females with ARAS. In XLAS, SNHL eventually develops in 80%–90% of affected males, as well as in some affected females; the incidence of SNHL in males and females with ARAS is probably similar to the incidence in males with XLAS. In some families with XLAS, and in ADAS, SNHL may not be detectable until well into adulthood.

Ocular: Anterior lenticonus is pathognomonic of Alport syndrome where there is a family history. It occurs in 15%-20% of those with XLAS or ARAS, and typically becomes apparent in late adolescence or early adulthood. Perimacular flecks occur in approximately 30% of individuals with Alport syndrome.

Family history: May be negative for renal failure or a history of hematuria because 10%-15% of males with XLAS represent de novo mutations, and approximately 15% of affected individuals have ARAS.

Urinalysis: In males with XLAS, and males and females with ARAS, urinalysis typically shows hematuria. For female heterozygotes for XLAS, approximately 95% of carriers will have persistent or intermittent microhematuria. Heterozygotes for ARAS: Approximately 50% of carriers of ARAS exhibit persistent or intermittent microhematuria. Individuals with TBMD typically also have microscopic hematuria.

Males with XLAS typically show complete absence of immunostaining for the collagen α3(IV) chain, α4(IV) chain, and α5(IV) chain on renal biopsy. Approximately 20% of males with XLAS show normal staining of renal basement membranes for the collagen α3(IV) chain, α4(IV) chain, and α5(IV) chain. Females typically exhibit patchy loss of staining for the collagen α3, α4, and α5(IV) chains in GBMs and tubular basement membranes (Kashtan, 2004). Individuals with TBMN exhibit normal GBM staining for the collagen α3(IV) chain, α4(IV) chain, and α5(IV) chain.

Light and immunofluorescence microscopy are unremarkable in individuals with typical TBMN, but with electron microscopy (EM), when diffusely present, the following three alterations are pathognomonic of Alport syndrome.

- The lamina densa appears to be split into multiple interlacing strands of electron-dense material, resembling basket-weaving.
- The lacunae between these strands are frequently occupied by round, electron-dense bodies (possibly entrapped cytoplasm).
- The glomerular capillary wall is diffusely thickened and its epithelial aspect is scalloped.

Molecular Genetic Testing

Three genes, *COL4A3*, *COL4A4*, and *COL4A5*, have been associated with collagen IV–related nephropathies.

Approximately 80% of Alport syndrome is caused by mutations in *COL4A5*, inherited in an X-linked manner.

Approximately 15% of Alport syndrome is caused by mutations in *COL4A3* and *COL4A4*, inherited in an autosomal recessive manner.

Approximately 5% of Alport syndrome is caused by mutations in *COL4A3* and *COL4A4*, inherited in an autosomal dominant manner.

Many individuals with autosomal dominant TBMN (precise percentage unknown) have heterozygous mutations in *COL4A3* or *COL4A4*.

In collagen IV–related nephropathies, a spectrum of phenotypes from progressive renal disease with extrarenal abnormalities (Alport syndrome) to isolated hematuria with a typically benign course (TBMN) is observed.

All males with XLAS develop proteinuria and, eventually, progressive renal insufficiency, which leads to end-stage renal disease (ESRD). Overall, an estimated 60% reach ESRD by age 30 years, and 90% by age 40 years (Jais et al., 2000). The rate of progression to ESRD is also influenced by the nature of the *COL4A5* mutation (Jais et al., 2000).

Approximately 12% of females with XLAS develop ESRD before age 40 years, increasing to 30% by age 60 years and 40% by age 80 years (Jais et al., 2003). Most individuals with ARAS develop significant proteinuria in late childhood or early adolescence, and ESRD before age 30 years. Progression to ESRD occurs at a slower pace in individuals with ADAS than in those with XLAS or ARAS.

Hearing loss in Alport syndrome is never congenital. Diminished hearing is usually detectable by late childhood or early adolescence in boys with XLAS. In affected males, the hearing loss is progressive. Hearing loss is frequently identifiable by formal assessment of hearing in late childhood, but in some families is not detectable until relatively late in life. SNHL develops in 80%–90% of males with XLAS by age 40 years (Jais, 2003). The course of the hearing loss depends on the causative mutation. Hearing impairment in members of families with Alport syndrome is always accompanied by evidence of renal involvement. There is no convincing evidence that deaf males lacking renal disease can transmit Alport syndrome to their offspring. In females with XLAS, hearing loss is less frequent and tends to occur later in life.

There do not appear to be gender differences in the incidence or course of hearing loss in ARAS. Individuals with ARAS typically exhibit juvenile onset of hearing loss. Hearing loss may also be a very late development in individuals with ADAS.

Ocular lesions are common in Alport syndrome, occurring in 30%–40% of individuals with XLAS. The spectrum of ocular lesions appears to be similar in XLAS and ARAS. Ocular lesions seem to be relatively uncommon in ADAS.

Aneurysms of the thoracic and abdominal aorta have been described in a small number of males with Alport syndrome (Kashtan et al., 2010; Lyons et al., 2007).

Like Alport syndrome, TBMN is characterized clinically by persistent microscopic hematuria, often first observed in childhood. In some individuals, microhematuria is intermittent and may not be detected until adulthood. Episodic gross hematuria, frequently in association with upper respiratory infections, is not unusual. The hematuria of TBMN appears to be lifelong.

TBMN differs clinically from Alport syndrome in the following important respects. It is only rarely associated with extrarenal abnormalities. Proteinuria and hypertension and progression to ESRD are unusual. Gender differences in expression of TBMN are not apparent.

ORAL-FACIAL-DIGITAL TYPE I SYNDROME (OFDI)

This is an X-linked dominant condition, which is lethal in males. Polycystic kidneys are typical in this condition. Additionally, like other types of oral-facial-digital syndrome, the condition almost invariably causes abnormalities of the oral cavity, face, and digits, with marked phenotypical variability that correlates with genotype (Thauvin-Robinet et al., 2006). Oral and craniofacial abnormalities include cleft palate (30%–80%), bifid or lobulated tongue (30%–45%) and lingual hamartomas (70%). Limb abnormalities include brachydactyly, syndactyly, and clinodactyly. Polydactyly is seen rarely. Renal cysts are seen in 50% of patients. Central nervous system malformations are also common in OFDI (Gurrieri, 2007).

The underlying genetic defect is a mutation in the OFD1 gene (Ferrante, 2001 p. 80). The OFD1 protein was thought to be important in the formation of primary cilia and in left–right asymmetry (Ferrante, 2004 p. 83; Giorgio, 2007 p. 82). However, there is recent evidence to suggest that the absence of primary cilia seen in this condition is a consequence of cystic renal disease, rather than the primary cause. (Zullo et al., 2010)

IGA NEPHROPATHY

Since immunoglobulin-A nephropathy (IgAN) was first described by Jean Berger in 1968, It has been identified as the commonest cause of glomerulonephritis in both adult and pediatric populations in many countries worldwide. This is highlighted in data from Korea and Japan, where there are annual urine screening programs for all children (Kitagawa, 2008).

PRENATALLY DIAGNOSED RENAL DISEASE

With the introduction of fetal ultrasound as countries develop, there will be an increase in the number of renal and urinary tract disorders diagnosed antenatally, with these disorders making up 30% of all fetal anomalies diagnosed antenatally on ultrasound (Noia, 1996).

Of the total number of patients with end-stage renal failure in the United Kingdom in 2008, 20.5% had an underlying autosomal recessive condition causing renal failure; 10.6% of patients had other family members affected by the same disorder; however, an antenatal diagnosis was only recorded as having been made in 4% (Lewis et al., 2010). Thus, there is a strong potential role for antenatal diagnosis and intervention in inherited renal diseases. This would not only enable identification of conditions early, which would allow intrauterine intervention or early treatment from birth, but it would enable termination of affected pregnancies if this was something the families in question would want to consider.

Abnormalities Detected Antenatally

1. Renal tract dilatation
2. Cystic or bright kidneys
3. Oligohydramnios
4. Renal agenesis

HOW TO APPROACH A PATIENT WITH POSSIBLE INHERITED RENAL DISEASE

The following algorithm (Figure 25-14) gives a good working plan to follow, especially when no molecular testing is available.

CONCLUSIONS

Medicine and society faces a flood, or possibly an epidemic, of genetic information. Particularly problematic is information on genetic variation in adult-onset disorders, chiefly cardiovascular disease, cancer, and renal disease. This chapter highlights the following major aspects of inherited renal diseases with significant health implications:

- ADPKD illustrates dramatically that genes are not disease, and that patients, doctors, and the insurance companies must evaluate how any one person will respond to his or her own genetic susceptibility.

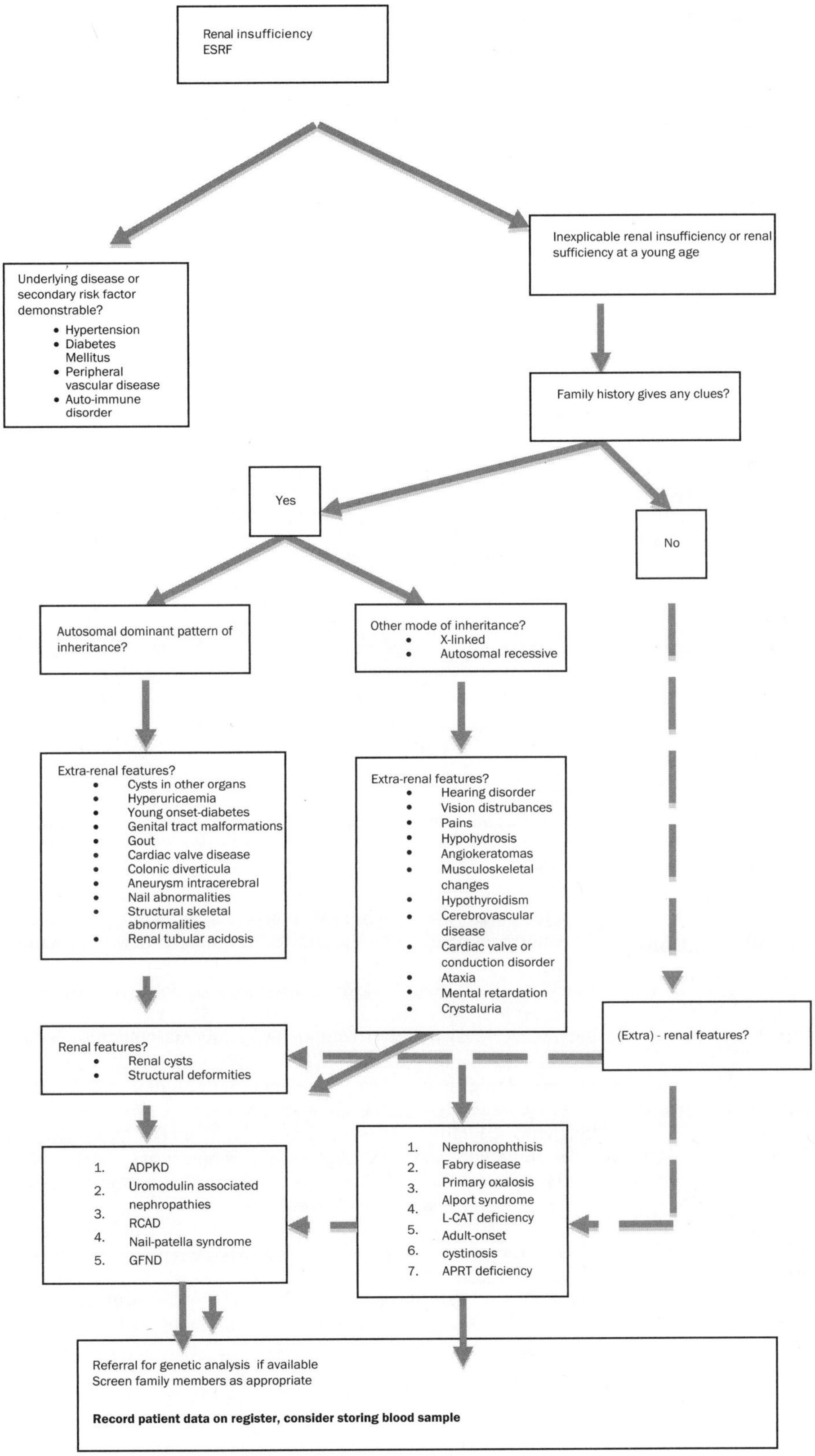

Figure 25-14 Management plan for end-stage renal disease secondary to renal cystic disease.

Certainly there are strong arguments for and against testing, and certain benefits to affected siblings if there are unaffected siblings diagnosed. In particular, unaffected siblings are acceptable kidney donors for those who have the disease. There are also several treatable complications of renal diseases that might benefit from early awareness, including hypertension, urinary tract infection, the impact of repeated pregnancies, and impaired renal function. Early treatable complications appear quite common in children. Knowledge of the correct diagnosis may also be of value in marital and family planning.

- Emphasis on identification of patients at risk of inherited renal disease, and preventing progression of renal failure, is required, in addition to the much-needed focus on how to manage the increasing number of patients with ESRF globally. Knowledge of inherited genetic diseases gives us additional power in fighting the global burden of chronic kidney disease. Renal failure is often asymptomatic until function is severely impaired, so the early identification of at-risk family members may, in the long term, prevent or delay the need for financially burdensome renal replacement therapy and the associated morbidity and mortality.
- The importance of renal registries, including more detailed information on U.K. patients, is much needed.
- Taking a family history may yet be the most important weapon in combatting the rise of CKD due to IRD.
- Chronic kidney disease is a global health problem and there is a need for all countries to have a public health policy for CKD (Levey, 2007).
- It is imperative that complete and good-quality data is stored in renal registries, which cost money to set up and store patient data (£17 per patient per year in the United Kingdom in 2009). It is also important that adult and pediatric data are stored together to prevent data analysis being affected by the constraints of dividing a population by age.
- Multicenter biobanks and international collaboration projects are important for the collection of clinical data and, where possible, biological samples for the identification of genes conferring susceptibility to renal disease.
- Identifying renal disease antenatally and in childhood will benefit children with IRD.
- It is also important to recognize the disparity in ethnic minorities living in developed nations, and to encourage living related donation in these communities, as there has been shown to be a paucity of living related organ donation in ethnic minorities such as the U.K. southeast Asian population (Lewis et al., 2007 p. 174).

In the future, more will be known about the survival of people of different ethnicity with established renal failure—but first we need to collect the information!

ACKNOWLEDGMENTS

Prof Visith Sitprija, Thailand. Dra. Sandra Rodriguez and Dr. Emilio Mena Castro, Dominican Republic. Dr. José L. Cangiano, Puerto Rico. Dr. Jorge Luna, Guatemala. Dr Walter Douthat, Argentina. Dr Prodjosudjadi, Indonesia. Professor A Lengani, Burkina Faso. Professor Kirpal S Chugh, India. Professor Djordjevic, Serbia. Prof. Petar Kes, Croatia. Prof. Mohammed Benghanem Gharbi, Morrocco. Dr. Harun ur Rashid, Bangladesh. Dr. Guillermo García García, Mexico. Dr Boris Bikbov, Professor Natalia Tomilina, Russia, Professor Sarala Naicker, South Africa, Dr Túri Sándor, Hungary, Dr Omar Abboud and Professor Nazaire Nseka, Congo.

REFERENCES

Adalat S, Woolf AS, Johnstone KA, Wirsing A, Harries LW, Long DA, et al. (2009). HNF1B mutations associate with hypomagnesemia and renal magnesium wasting. *J Am Soc Nephrol 20*(5), 1123–1131.

Antignac C, Arduy CH, Beckmann JS, Benessy F, Gros F, Medhioub M, et al. (1993). A gene for familial juvenile nephronophthisis (recessive medullary cystic kidney disease) maps to chromosome 2p. *Nat Genet 3*(4), 342–345.

Bajwa ZH, Gupta S, Warfield CA, & Steinman TI. (2001). Pain management in polycystic kidney disease. *Kidney Int 60*(5), 1631–1644.

Barua M, Cil O, Paterson AD, Wang K, He N, Dicks E, Parfrey P, Pei Y. (2009). Family history of renal disease severity predicts the mutated gene in ADPKD. *J Am Soc Nephrol. 20*(8), 1833–1838.

Barua M, Pei Y. (2010). Diagnosis of autosomal-dominant polycystic kidney disease: an integrated approach. *Semin Nephrol. 30*(4), 356–365.

Bear JC, McManamon P, Morgan J, Payne RH, Lewis H, Gault MH, et al. (1984). Age at clinical onset and at ultrasonographic detection of adult polycystic kidney disease: data for genetic counselling. *Am J Med Genet 18*(1), 45–53.

Bear JC, Parfrey PS, Morgan JM, Martin CJ, & Cramer BC. (1992). Autosomal dominant polycystic kidney disease: new information for genetic counselling. *Am J Med Genet 43*(3), 548–553.

Bear RA. (1974). Solitary kidney affected with polycystic disease: a report of 2 cases. *J Urol 111*(5), 566–567.

Beerman I, Novak J, Wyatt RJ, Julian BA, Garavi AG. (2007). The genetics of IgA nephropathy. *Nat clin Pract Nephrol 3*(6), 325–338.

Berger J, Hinglais N. (1968). Les dépôts intercapillaries d'IgA-IgG. *J Urol Nephrol* (Paris) *74*, 694–695.

Bingham C, Ellard S, Allen L, Bulman M, Shepherd M, Frayling T, et al. (2000). Abnormal nephron development associated with a frameshift mutation in the transcription factor hepatocyte nuclear factor-1 beta. *Kidney Int 57*(3), 898–907.

Breunig F, & Wanner C. (2008). Update on Fabry disease: kidney involvement, renal progression and enzyme replacement therapy. *J Nephrol 21*(1), 32–37.

Byrne C, Ford D, Gilg J, Ansell D, & Feehally J. (2010). UK Renal Registry 12th Annual Report (December 2009): Chapter 3: UK ESRD incident rates in 2008: national and centre-specific analyses. *Nephron Clin Pract 115*(Suppl 1), c9–c39.

Byrne C, Steenkamp R, Castledine C, Ansell D, & Feehally J. (2010). UK Renal Registry 12th Annual Report (December 2009): Chapter 4: UK ESRD prevalent rates in 2008: national and centre-specific analyses. *Nephron Clin Pract 115*(Suppl 1), c41–c67.

Cadnapaphornchai MA, Kim McFann, Strain JD, et al. (2008). Increased left ventricular mass in children with autosomal dominant polycystic kidney disease and borderline hypertension. *Kidney Int. 74*(9), 1192–1196.

Casasco AE, Aymard A, Gobin YP, Houdart E, Rogopoulos A, George B, et al. (1993). Selective endovascular treatment of 71 intracranial aneurysms with platinum coils. *J Neurosurg 79*(1), 3–10.

Chapman AB, Gabow PA, & Schrier RW. (1991). Reversible renal failure associated with angiotensin-converting enzyme inhibitors in polycystic kidney disease. *Ann Intern Med 115*(10), 769–773.

Chapman AB, Johnson A, Gabow PA, & Schrier RW. (1990). The renin-angiotensin-aldosterone system and autosomal dominant polycystic kidney disease. *N Engl J Med 323*(16), 1091–1096.

Chapman AB, & Schrier RW. (1991). Pathogenesis of hypertension in autosomal dominant polycystic kidney disease. *Semin Nephrol 11*(6), 653–660.

Chapman AB. (2008). Approaches to testing new treatments in autosomal dominant polycystic kidney disease: insights from the CRISP and HALT-PKD studies. *CJASN 3*(4):1197–1204.

Chauveau D, Pirson Y, Verellen-Dumoulin C, Macnicol A, Gonzalo A, & Grunfeld JP. (1994). Intracranial aneurysms in autosomal dominant polycystic kidney disease. *Kidney Int 45*(4), 1140–1146.

Coffinier C, Barra J, Babinet C, & Yaniv M. (1999). Expression of the vHNF1/HNF1beta homeoprotein gene during mouse organogenesis. *Mech Dev 89*(1–2), 211–213.

Coffinier C, Thepot D, Babinet C, Yaniv M, & Barra J. (1999). Essential role for the homeoprotein vHNF1/HNF1beta in visceral endoderm differentiation. *Development 126*(21), 4785–4794.

Colombi A, Kostyal A, Bracher R, Gloor F, Mazzi R, & Tholen H. (1967). Angiokeratoma corporis diffusum–Fabry's disease. *Helv Med Acta 34*(1), 67–83.

Davies F, Coles GA, Harper PS, Williams AJ, Evans C, & Cochlin D. (1991). Polycystic kidney disease re-evaluated: a population-based study. *Q J Med 79*(290), 477–485.

Davila S, Furu L, Gharavi AG, Tian X, Onoe T, Qian Q, et al. (2004). Mutations in SEC63 cause autosomal dominant polycystic liver disease. *Nat Genet 36*(6), 575–577.

Ding L, Zhang S, Qiu W, Xiao C, Wu S, Zhang G, et al. (2002). Novel mutations of PKD1 gene in Chinese patients with autosomal dominant polycystic kidney disease. *Nephrol Dial Transplant 17*(1), 75–80.

Eastwood JB, Kerry SM, Plange-Rhule J, Micah FB, Antwi S, Boa FG, Banerjee D, Cappuccio FP. (2010). Assessment of GFR by four methods in adults in Ashanti, Ghana: the need for an eGFR equation for lean African populations. *Nephrol Dial Transplant 25*(7), 2178–2187.

Eastwood JB, Plange-Rhule J, Parry V, Tomlinson S. (2001). Medical collaborations between developed and developing countries. *QJM Int J Med 94*(11), 637–641.

Ecder T, Chapman AB, Brosnahan GM, Edelstein CL, Johnson AM, & Schrier RW. (2000). Effect of antihypertensive therapy on renal function and urinary albumin excretion in hypertensive patients with autosomal dominant polycystic kidney disease. *Am J Kidney Dis 35*(3), 427–432.

Ecder T, Edelstein CL, Fick-Brosnahan GM, Johnson AM, Chapman AB, Gabow PA, et al. (2001). Diuretics versus angiotensin-converting enzyme inhibitors in autosomal dominant polycystic kidney disease. *Am J Nephrol 21*(2), 98–103.

Ecder T, & Schrier RW. (2001). Hypertension in autosomal-dominant polycystic kidney disease: early occurrence and unique aspects. *J Am Soc Nephrol 12*(1), 194–200.

Edghill EL, Bingham C, Ellard S, & Hattersley AT. (2006). Mutations in hepatocyte nuclear factor-1beta and their related phenotypes. *J Med Genet 43*(1), 84–90.

Edghill EL, Oram RA, Owens M, Stals KL, Harries LW, Hattersley AT, et al. (2008). Hepatocyte nuclear factor-1beta gene deletions—a common cause of renal disease. *Nephrol Dial Transplant 23*(2), 627–635.

El Nahas M, Bello AK. (2005). Chronic kidney disease: the global challenge. *Lancet 9456*(365), 331–340.

Emery AE, Brough C, Crawfurd M, Harper P, Harris R, & Oakshott G. (1978). A report on genetic registers. Based on the report of the Clinical Genetics Society Working Party. *J Med Genet 15*(6), 435–442.

EPKDC. (1994). The polycystic kidney disease 1 gene encodes a 14 kb transcript and lies within a duplicated region on chromosome 16. The European Polycystic Kidney Disease Consortium. *Cell 77*(6), 881–894.

Ferrante MI, Giorgio G, Feather SA, Bulfone A, Wright V, Ghiani M, et al. (2001). Identification of the gene for oral-facial-digital type I syndrome. *Am J Hum Genet 68*(3), 569–576.

Ferrante P, Messali S, Ballabio A, & Meroni G. (2004). Identification and biochemical characterization of an avian sulfatase homologous to the human ARSE, the gene for X-linked chondrodysplasia punctata. *Gene 336*(2), 155–161.

Fick-Brosnahan G, Johnson AM, Strain JD, & Gabow PA. (1999). Renal asymmetry in children with autosomal dominant polycystic kidney disease. *Am J Kidney Dis 34*(4), 639–645.

Fick GM, Duley IT, Johnson AM, Strain JD, Manco-Johnson ML, & Gabow PA. (1994). The spectrum of autosomal dominant polycystic kidney disease in children. *J Am Soc Nephrol 4*(9), 1654–1660.

Flinter F. (Ed.) (2004). *The Genetics of Renal Disease*. Oxford: Oxford University Press.

Freedman BI, Soucie JM, Chapman A, Krisher J, & McClellan WM. (2000). Racial variation in autosomal dominant polycystic kidney disease. *Am J Kidney Dis 35*(1), 35–39.

Gabow PA. (1990). Autosomal dominant polycystic kidney disease—more than a renal disease. *Am J Kidney Dis 16*(5), 403–413.

Gabow PA. (1993). Autosomal dominant polycystic kidney disease. *N Engl J Med 329*(5), 332–342.

Gabow PA, Chapman AB, Johnson AM, Tangel DJ, Duley IT, Kaehny WD, et al. (1990). Renal structure and hypertension in autosomal dominant polycystic kidney disease. *Kidney Int 38*(6), 1177–1180.

Gabow PA, Duley I, & Johnson AM. (1992). Clinical profiles of gross hematuria in autosomal dominant polycystic kidney disease. *Am J Kidney Dis 20*(2), 140–143.

Gabow PA, Ikle DW, & Holmes JH. (1984). Polycystic kidney disease: prospective analysis of nonazotemic patients and family members. *Ann Intern Med 101*(2), 238–247.

Gabow PA, Johnson AM, Kaehny WD, Kimberling WJ, Lezotte DC, Duley IT, et al. (1992). Factors affecting the progression of renal disease in autosomal-dominant polycystic kidney disease. *Kidney Int 41*(5), 1311–1319.

Gabow PA, Johnson AM, Kaehny WD, Manco-Johnson ML, Duley IT, & Everson GT. (1990). Risk factors for the development of hepatic cysts in autosomal dominant polycystic kidney disease. *Hepatology 11*(6), 1033–1037.

Gabow PA, Kimberling WJ, Strain JD, Manco-Johnson ML, & Johnson AM. (1997). Utility of ultrasonography in the diagnosis of autosomal dominant polycystic kidney disease in children. *J Am Soc Nephrol 8*(1), 105–110.

Gahl WA, Thoene JG, Schneider JA. (2002). Cystinosis. *N Engl J Med 347*, 111–121.

Gardner KD, Jr. (1992). Phenotype recognition. Clinicians' contributions to molecular genetics. *West J Med 156*(5), 491–494.

Gattone VH, Sinders RM, Hornberger TA, & Robling AG. (2009). Late progression of renal pathology and cyst enlargement is reduced by rapamycin in a mouse model of nephronophthisis Rapamycin ameliorates murine nephronophthisis. *Kidney Int 76*, 178–182.

Giorgio G, Alfieri M, Prattichizzo C, Zullo A, Cairo S, & Franco B. (2007). Functional characterization of the OFD1 protein reveals a nuclear localization and physical interaction with subunits of a chromatin remodeling complex. *Mol Biol Cell 18*(11), 4397–4404.

Grantham JJ. (2000). Time to treat polycystic kidney diseases like the neoplastic disorders that they are. *Kidney Int 57*(1), 339–340.

Grantham JJ, Geiser JL, & Evan AP. (1987). Cyst formation and growth in autosomal dominant polycystic kidney disease. *Kidney Int 31*(5), 1145–1152.

Grantham JJ, Torres VE, Chapman AB, Guay-Woodford LM, Bae KT, et al. (2006). Volume progression in polycystic kidney disease. *N Engl J Med. 354*(20), 2122–2130.

Grunfeld JP. (1998). Factors influencing progression of renal failure in autosomal dominant polycystic kidney disease. *Am J Kidney Dis 32*(5), xlvi–xlviii.

Gurrieri F, Franco B, Toriello H, & Neri G. (2007). Oral-facial-digital syndromes: review and diagnostic guidelines. *Am J Med Genet A 143A*(24), 3314–3323.

Hadimeri H, Norden G, Friman S, & Nyberg G. (1997). Autosomal dominant polycystic kidney disease in a kidney transplant population. *Nephrol Dial Transplant 12*(7), 1431–1436.

Haider NB, Carmi R, Shalev H, Sheffield VC, & Landau D. (1998). A Bedouin kindred with infantile nephronophthisis demonstrates linkage to chromosome 9 by homozygosity mapping. *Am J Hum Genet 63*(5), 1404–1410.

Hall YN, Fuentes EF, Chertow GM, Olson JL. (2004). Race/ethnicity and disease severity in IgA nephropathy. *BMC Nephrol 5*, 10.

Harrap SB, Davies DL, Macnicol AM, Dominiczak AF, Fraser R, Wright AF, et al. (1991). Renal, cardiovascular and hormonal characteristics of young adults with autosomal dominant polycystic kidney disease. *Kidney Int 40*(3), 501–508.

Hateboer N, v Dijk MA, Bogdanova N, Coto E, Saggar-Malik AK, San Millan JL, et al. (1999). Comparison of phenotypes of polycystic kidney disease types 1 and 2. European PKD1-PKD2 Study Group. *Lancet 353*(9147), 103–107.

Higashihara E, Nutahara K, Kojima M, Tamakoshi A, Yoshiyuki O, Sakai H, et al. (1998). Prevalence and renal prognosis of diagnosed autosomal dominant polycystic kidney disease in Japan. *Nephron 80*(4), 421–427.

Iglesias CG, Torres VE, Offord KP, Holley KE, Beard CM, & Kurland LT. (1983). Epidemiology of adult polycystic kidney disease, Olmsted County, Minnesota: 1935–1980. *Am J Kidney Dis 2*(6), 630–639.

Iglesias DM, Martin RS, Fraga A, Virginillo M, Kornblihtt AR, Arrizurieta E, et al. (1997). Genetic heterogeneity of autosomal dominant polycystic kidney disease in Argentina. *J Med Genet 34*(10), 827–830.

Jais JP, Knebelmann B, Giatras I, De Marchi M, Rizzoni G, et al. (2000). X-linked Alport syndrome: natural history in 195 families and genotype- phenotype correlations in males. *J Am Soc Nephrol 11*, 649–657.

Jais JP, Knebelmann B, Giatras I, De Marchi M, Rizzoni G, (2003). X-linked Alport syndrome: natural history and genotype-phenotype correlations in girls and women belonging to 195 families: a "European Community Alport Syndrome Concerted Action" study. *J Am Soc Nephrol 14*, 2603–2610.

Jones W, Saggar A (2012). Personal communication.

Joosten H, Strunk AL, Meijer S, Boers JE, Aries MJ, Abbes AP, et al. (2010). An aid to the diagnosis of genetic disorders underlying adult-onset renal failure: a literature review. *Clin Nephrol 73*(6), 454–472.

Kashtan CE. (2004). Diagnosis of Alport syndrome. *Kidney Int 66*, 1290–1291.

Kashtan CE, Segal Y, Flinter F, Makanjuola D, Gan JS, Watnick T. (2010). Aortic abnormalities in males with Alport syndrome. *Nephrol Dial Transplant 25*(11), 3554–3560.

Kitagawa T. (1988). Lessons learned from the Japanese nephritis screening study. *Pediatr Nephrol 2*(2), 256–263.

Klahr S, Breyer JA, Beck GJ, Dennis VW, Hartman JA, Roth D, et al. (1995). Dietary protein restriction, blood pressure control, and the progression of polycystic kidney disease. Modification of Diet in Renal Disease Study Group. *J Am Soc Nephrol 5*(12), 2037–2047.

Knudson AG, Jr. (1971). Mutation and cancer: statistical study of retinoblastoma. *Proc Natl Acad Sci USA 68*(4), 820–823.

Kopitinik TA, & Sanmson DS. (1993). Management of subarachnoid haemorrage. *J Neurol 56*, 947–957.

Lacson Jr E, Kuhlmann MK, Shah K, Yoshino M, Levin NW. (2005). Outcomes and economics of ESRF. In El Nahas AM (ed.), *Kidney Diseases in the Developing World and Ethnic Minorities* (pp. 15–37). New York: Taylor and Francis.

Leblanc R. (1987). The minor leak preceding subarachnoid hemorrage. *J Neurosurg 66*, 986–994.

Lee JG, Lee KB, Kim UK, Ahn C, Hwang DY, Hwang YH, et al. (2001). Genetic heterogeneity in Korean families with autosomal-dominant polycystic kidney disease (ADPKD): the first Asian report. *Clin Genet 60*(2), 138–144.

Levey AS, Andreoli SP, DuBose T, Provenzano R, & Collins AJ. (2007). Chronic kidney disease: common, harmful, and treatable—World Kidney Day 2007. *J Am Soc Nephrol 18*(2), 374–378.

Levy M, & Feingold J. (2000). Estimating prevalence in single-gene kidney diseases progressing to renal failure. *Kidney Int 58*, 925–943.

Lewis M, Shaw J, Reid C, Evans J, Webb N, & Verrier-Jones K. (2007). Demography and management of childhood established renal failure in the UK (chapter 13). *Nephrol Dial Transplant 22*(Suppl 7), 165–175.

Lewis MA, Shaw J, Sinha MD, Adalat S, Hussain F, Castledine C, et al. (2010). UK Renal Registry 12th Annual Report (December 2009): Chapter 14: Demography of the UK paediatric renal replacement therapy population in 2008. *Nephron Clin Pract 115*(Suppl 1), c279–c288.

Li Kam Wa TC, Macnicol AM, & Watson ML. (1997). Ambulatory blood pressure in hypertensive patients with autosomal dominant polycystic kidney disease. *Nephrol Dial Transplant 12*(10), 2075–2080.

Li A, Davila S, Furu L, Qian Q, Tian X, Kamath PS, King BF, Torres VE, Somlo S. (2003). Mutations in PRKCSH cause isolated autosomal dominant polycystic liver disease. *Am J Hum Genet 72*(3), 691–703.

Locatelli F, Marcelli D, Comelli M, Alberti D, Graziani G, Buccianti G, et al. (1996). Proteinuria and blood pressure as causal components of progression to end-stage renal failure. Northern Italian Cooperative Study Group. *Nephrol Dial Transplant 11*(3), 461–467.

Lyons OT, St. John ERC, Morales JP, Chan YC, Taylor PR. (2007). Ruptured thoracoabdominal aortic aneurysm in a renal transplant patient with Alport's syndrome. *Ann Vasc Surg* 21, 816–818.

Maschio G, Alberti D, Janin G, Locatelli F, Mann JF, Motolese M, et al. (1996). Effect of the angiotensin-converting-enzyme inhibitor benazepril on the progression of chronic renal insufficiency. The Angiotensin-Converting-Enzyme Inhibition in Progressive Renal Insufficiency Study Group. *N Engl J Med 334*(15), 939–945.

Mizoguchi M, Tamura T, Yamaki A, Higashihara E, & Shimizu Y. (2002). Genotypes of autosomal dominant polycystic kidney disease in Japanese. *J Hum Genet 47*(1), 51–54.

Mochizuki T, Wu G, Hayashi T, Xenophontos SL, Veldhuisen B, Saris JJ, et al. (1996). PKD2, a gene for polycystic kidney disease that encodes an integral membrane protein. *Science 272*(5266), 1339–1342.

Noia G, Masini L, De Santis M, Caruso A. (1996). The impact of invasive procedures on prognostic, diagnostic and therapeutic aspects of urinary tract anomalies. In Cataldi L, Fanos V, Simeoni U, (eds.), *Neonatal Nephrology in Progress* (pp. 67–84). Agora, Lecce.

Ogborn MR, Bankovic-Calic N, Shoesmith C, Buist R, & Peeling J. (1998). Soy protein modification of rat polycystic kidney disease. *Am J Physiol 274*(3 Pt 2), F541–F549.

Omran H, Sasmaz G, Haffner K, Volz A, Olbrich H, Melkaoui R, et al. (2002). Identification of a gene locus for Senior-Loken syndrome in the region of the nephronophthisis type 3 gene. *J Am Soc Nephrol 13*(1), 75–79.

Park YH, Choi JY, Chung HS, Koo JW, et al. (2005). Hematuria and proteinuria in a mass school urine screening test. *Pediatr Nephrol 20*(8), 1126–1130.

Pei Y. (2001). A "two-hit" model of cystogenesis in autosomal dominant polycystic kidney disease? *Trends Mol Med 7*(4), 151–156.

Pei Y, Obaji J, Dupuis A, Paterson AD, Magistroni R, Dicks E, et al. (2009). Unified criteria for ultrasonographic diagnosis of ADPKD. *J Am Soc Nephrol. 20*(1), 205–212.

Persy VP, Remuzzi G, Perico N, Benghanem Gharbi M, Adu D, Jha V, Rizvi A, Ben Ammar M, Fongoro S, De Broe ME. (2008). *Prevention and Transplantation in Chronic Kidney Disease: What is*

Achievable in Emerging Countries? Meeting report: Bamako meeting December 4–6, 2008.

Plange-Rhule J, et al. (2002). In El Nahas AM. (ed.), *Kidney Diseases in the Developing World and Ethnic Minorities* (pp. 15–37). New York: Taylor and Francis.

Phakdeekitcharoen B, Watnick TJ, Ahn C, Whang DY, Burkhart B, & Germino GG. (2000). Thirteen novel mutations of the replicated region of PKD1 in an Asian population. *Kidney Int 58*(4), 1400–1412.

Qian Q, Li A, King BF, Kamath PS, Lager DJ, Huston J III, Shub C, et al. (2003). Clinical profile of autosomal dominant polycystic liver disease. *Hepatology 37*(1), 164–171.

Ravine D, Gibson RN, Donlan J, & Sheffield LJ. (1993). An ultrasound renal cyst prevalence survey: specificity data for inherited renal cystic diseases. *Am J Kidney Dis 22*(6), 803–807.

Ravine D, Gibson RN, Walker RG, Sheffield LJ, Kincaid-Smith P, & Danks DM. (1994). Evaluation of ultrasonographic diagnostic criteria for autosomal dominant polycystic kidney disease 1. *Lancet 343*(8901), 824–827.

Ravine D, Walker RG, Gibson RN, Sheffield LJ, Kincaid-Smith P, & Danks DM. (1991). Treatable complications in undiagnosed cases of autosomal dominant polycystic kidney disease. *Lancet 337*(8734), 127–129.

Renigunta A, Renigunta V, Saritas T, Decher N, Mutig K, Waldegger S. (2011). Tamm-Horsfall Glycoprotein interacts with renal outer medullary potassium channel ROMK2 and regulates its function. *J Biol Chem. 286*(3), 2224–2235.

Roscoe JM, Brissenden JE, Williams EA, Chery AL, & Silverman M. (1993). Autosomal dominant polycystic kidney disease in Toronto. *Kidney Int 44*(5), 1101–1108.

Saveland H, et al. (1986). Outcome evaluation following subarachnoid hemorrage. *J Neurosurg 64*, 191–196.

Schena FP, Cerullo G, Torres DD, Scolari F, Foramitti M. et al. (2005). The IgA nephropathy Biobank. An important starting point for the genetic dissection of a complex trait. *BMC Nephrol 5*, 6–14.

Schieppati A, & Remuzzi G. (2005). Chronic renal diseases as a public health problem: epidemiology, social, and economic implications. *Kidney Int Suppl 98*, S7–S10.

Schievink WI, Torres VE, Piepgras DG, & Wiebers DO. (1992). Saccular intracranial aneurysms in autosomal dominant polycystic kidney disease. *J Am Soc Nephrol 3*(1), 88–95.

Schmidt UD, Steiger HJ, & Huber P. (1987). Accuracy of high resolution CT in direct diagnosis of cerebral aneurysms. *Neuroradiology 29*, 152–159.

Schuermann MJ, Otto E, Becker A, Saar K, Ruschendorf F, Polak BC, et al. (2002). Mapping of gene loci for nephronophthisis type 4 and Senior-Loken syndrome, to chromosome 1p36. *Am J Hum Genet 70*(5), 1240–1246.

Schwab SJ, Bander SJ, & Klahr S. (1987). Renal infection in autosomal dominant polycystic kidney disease. *Am J Med 82*(4), 714–718.

Schmid M, Prajczer S, Gruber LN, Bertocchi C, Gandini R, Pfaller W, Jennings P, Joannidis M. (2010). Uromodulin facilitates neutrophil migration across renal epithelial monolayers. *Cell Physiol Biochem.* 26(3), 311–318.

Scolari F, Puzzer D, Amoroso A, Caridi G, Ghiggeri GM, Maiorca R, et al. (1999). Identification of a new locus for medullary cystic disease, on chromosome 16p12. *Am J Hum Genet 64*(6), 1655–1660.

Sedman A, Bell P, Manco-Johnson M, Schrier R, Warady BA, Heard EO, et al. (1987). Autosomal dominant polycystic kidney disease in childhood: a longitudinal study. *Kidney Int 31*(4), 1000–1005.

Seedat YK, Naiker S, Rawat R, & Parsoo I. (1984). Racial Differences in the causes of end-stage renal failure in Natal. *S Afr Med J 65*, 956–958.

Sweeney WE, Chen Y, Nakanishi K, Frost P, & Avner ED. (2000). Treatment of polycystic kidney disease with a novel tyrosine kinase inhibitor. *Kidney Int 57*(1), 33–40.

Tamura H, Kato H, Hirose S, Itoyama S, Matsumura O, Nagasawa R, et al. (1994). An adult case of polycystic kidney disease associated with congenital hepatic fibrosis. *Nippon Jinzo Gakkai Shi 36*(8), 962–967.

Thauvin-Robinet C, Cossee M, Cormier-Daire V, Van Maldergem L, (2006). Clinical, molecular, and genotype-phenotype correlation studies from 25 cases of oral-facial-digital syndrome type 1: a French and Belgian collaborative study. *J Med Genet 43*, 54–61.

Timio M, Monarca C, Pede S, Gentili S, Verdura C, & Lolli S. (1992). The spectrum of cardiovascular abnormalities in autosomal dominant polycystic kidney disease: a 10-year follow-up in a five-generation kindred. *Clin Nephrol 37*(5), 245–251.

Torra R, Badenas C, San Millan JL, Perez-Oller L, Estivill X, & Darnell A. (1999). A loss-of-function model for cystogenesis in human autosomal dominant polycystic kidney disease type 2. *Am J Hum Genet 65*(2), 345–352.

Torres VE. (1996). Polycystic liver disease. In Watson ML, & Torres VE (eds.), *Polycystic Kidney Disease* (pp. 500–529). Oxford: Oxford University Press.

Torres VE, Cai Y, Chen X, Wu GQ, Geng L, Cleghorn KA, et al. (2001). Vascular expression of polycystin-2. *J Am Soc Nephrol, 12*(1), 1–9.

Torres VE, Erickson SB, Smith LH, Wilson DM, Hattery RR, & Segura JW. (1988). The association of nephrolithiasis and autosomal dominant polycystic kidney disease. *Am J Kidney Dis 11*(4), 318–325.

Torres VE, Harris PC, Pirson Y. (2007). Autosomal dominant polycystic kidney disease. *Lancet 369*(9569), 1287–1301.

United States Renal Data System. (2010). Available at http://www.usrds.org/ Accessed September 2010.

World Health Organisation. (1972). *Genetic Disorders: Prevention, Treatment and Rehabilitation*. Report of a WHO scientific group. Geneva: WHO.

Wilson GN, de Chadarevian JP, Kaplan P, Loehr JP, Frerman FE, & Goodman SI. (1989). Glutaric aciduria type II: review of the phenotype and report of an unusual glomerulopathy. *Am J Med Genet 32*(3), 395–401.

Woolf AS. (2000). A molecular and genetic view of human renal and urinary tract malformation *Kidney Int 58*, 500–512.

World Bank Monetary Fund. (2009). Data Report. http://data.worldbank.org/about/country-classifications.

Yersin C, Bovet P, Wauters JP, Schorderet DF, Pescia G, & Paccaud F. (1997). Frequency and impact of autosomal dominant polycystic kidney disease in the Seychelles (Indian Ocean). *Nephrol Dial Transplant, 12*(10), 2069–2074.

Yium J, Gabow P, Johnson A, Kimberling W, & Martinez-Maldonado M. (1994). Autosomal dominant polycystic kidney disease in blacks: clinical course and effects of sickle-cell hemoglobin. *J Am Soc Nephrol, 4*(9), 1670–1674.

Zerres K, Rudnik-Schoneborn S, & Deget F. (1993). Childhood onset autosomal dominant polycystic kidney disease in sibs: clinical picture and recurrence risk. German Working Group on Paediatric Nephrology (Arbeitsgemeinschaft fur Padiatrische Nephrologie). *J Med Genet 30*(7), 583–588.

Zullo A, Iaconis D, Barra A, Cantone A, Messaddeq N, Capasso G, et al. (2010). Kidney-specific inactivation of Ofd1 leads to renal cystic disease associated with upregulation of the mTOR pathway. *Hum Mol Genet 19*(14), 2792–2803.

26.

THE BURDEN OF CONGENITAL ANOMALIES ON GLOBAL HEALTH

Dhavendra Kumar

INTRODUCTION

Developmental abnormalities presenting at birth as single or multiple anatomic alterations are not restricted to any geographic region or specific human population. Irrespective of definition, birth defects can cause spontaneous abortions and stillbirths and are a significant but under-recognized cause of mortality and disability among infants and children under five years of age. They can be life-threatening, result in long-term disability, and negatively affect individuals, families, healthcare systems, and societies. Every year about 3% of all children born in any given hospital in any country will have a significant congenital abnormality. The World Health Organization (WHO) estimates that some 260,000 deaths worldwide (about 7% of all neonatal deaths) are caused by congenital anomalies (WHO, 2008). This is more apparent as a cause of death in regions with overall lower mortality rates; for example, in the European region 25% of all neonatal deaths are due to congenital anomalies. Thus it is clear that congenital anomalies may be of more than cosmetic concern and, if uncorrected, will interfere with normal functioning (Aase, 1990). Such anomalies occur in only a small fraction of all newborns. However, these collectively attribute to about 30% of all neonatal and infant deaths, and children born with birth defects make up about 30% of all admissions in pediatric hospitals. In addition, these children present with a range of problems requiring medical support from various specialists. These problems usually start early in life, and may require chronic care for decades. The burdens imposed on these children and their families, and on society at large, may be enormous. The great majority of birth defects are neither detectable by prenatal diagnosis nor preventable. Thus, the impact of these problems has not decreased despite advances in medicine.

The term *malformation* or *anomaly* is interchangeably used in conjunction with a birth defect. The prefix "congenital" simply denotes detection of an anomaly at birth. The study of the scientific basis of congenital anomalies and related clinical applications are now included within the discipline of *dysmorphology*. The emphasis is generally on delineation and etiologic description of multiple malformation syndromes, or *dysmorphic* syndrome. The *International Statistical Classification of Diseases and Related Health Problems, 10th revision* (ICD-10), includes birth defects in "Chapter XVII: Congenital malformations, deformations and chromosomal abnormalities." Birth defects can be defined as structural or functional abnormalities, including metabolic disorders, which are present from birth. The term *congenital disorder* is considered to have the same definition; the two terms are used interchangeably (WHO, 2006). The eleventh ICD revision of the classification provides an opportunity for a review of the current entry.

The word *syndrome* is derived from the Greek, "running together." The term has a wide and ill-defined usage. In clinical medicine, a syndrome refers to a set of symptoms and signs (clinical, radiological, laboratory etc.) that occur in a consistent pattern. In clinical dysmorphology, it simply refers to a particular set of anomalies occurring together more than would be expected by chance and with a common etiological relationship. A lot of researchers and epidemiologists have expressed concerns about the confusion surrounding the concept and lack of an acceptable definition of a "syndrome." Jablonski (1991) and other epidemiologists working in the field of birth defects epidemiology have reviewed this problem in some detail (Khoury, 1989).

Clinical geneticists and epidemiologists are expected to have good understanding in using the word *syndrome* to describe an infant with multiple congenital anomalies (MCA), recognized to have a single gene disease or a chromosome abnormality (Aase, 1990). Recognizable patterns of MCA related to fetal teratogenic exposures are often described as a "syndrome"; for example, fetal warfarin syndrome, fetal hydantoin syndrome, fetal valproate syndrome, etc. It is likely that demarcation between a "dysmorphic syndrome" or an "association" may be very subtle, and there may be some common etiologic relationship (Stevenson and Hall, 1993).

Nonsyndromal patterns of MCA (Jones, 1995) include *associations* (for example VACTERL: Vertebral, Anal, Cardiac, Tracheo-Esophageal, Renal and Limb; CHARGE:

Coloboma, congenital Heart defect, choanal Atresia, Retarded Growth and abnormal Ears), *developmental field defects or sequences* (holoprosencephaly, caudal dysgenesis, etc.) and *complexes* or *spectra* (hypoglossia-hypodactyly; facio-auriculo-vertebral spectrum, etc.). These terms are often inappropriately used as convenient labels and do not help to clarify underlying cause and pathogenesis. There are MCA patterns known to occur that are neither syndromal nor nonsyndromal. These are often erroneously described under any of the above categories because of lack of more precise information, such as karyotype, autopsy, imaging studies, etc. (Khoury et al., 1994). A good example in this category would be a *sequence*. A "sequence" refers to the occurrence of a number of anomalies in succession to a single, early developmental abnormality; for example, the frontonasal dysplasia sequence, early urethral obstruction sequence (prune belly), and athyrotic hypothyroidism sequence (Jones, 1995).

The present chapter provides some insight into the global burden of congenital anomalies, with particular reference to developing countries.

EPIDEMIOLOGY

Congenital anomalies or "birth defects" occur in all populations. More than seven thousand different birth defects have been identified to date. Every year an estimated 7.9 million children—6 percent of total births worldwide—are born with a serious birth defect of genetic or partially genetic origin (Bittles, 2002). Additional hundreds of thousands more are born with serious birth defects of postconception origin, including maternal exposure to environmental agents (teratogens), such as alcohol, rubella, syphilis, and iodine deficiency, that can harm a developing fetus. Some birth defects are clinically obvious at birth; others may only be diagnosed later in life. Spina bifida is one example of a structural defect that is obvious at birth. The bleeding disorder hemophilia is a functional defect, usually not clinically obvious until infancy or childhood. The author accepts that the term *birth defect* is not considered appropriate by some, but it has been used extensively in medical literature over time, and is widely understood by the broad audience in developing countries. The March of Dimes survey lists five major types of birth defects that pose a huge burden on global health (MOD, 2006). These include: *(1)* congenital heart defects (1,040,835 births); *(2)* neural tube defects (323,904 births); *(3)* the hemoglobin disorders, thalassemia, and sickle cell disease (307,897 births); *(4)* Down syndrome (trisomy 21; 217,293 births); and *(5)* glucose-6-phosphate dehydrogenase (G6PD) deficiency (177,032 births). Combined, these five conditions account for about 25% of all of birth defects of genetic or partially genetic origin (Figure 26-1).

This chapter focuses on congenital anomalies within the broad category of birth defects. The true incidence and prevalence of anomalies in a given population is not possible to determine, due to inconsistent ascertainment and reporting. This may be different in various countries, and among various population subgroups. Most congenital anomalies occur at a constant rate and degree of prevalence. Nonlethal anomalies would be expected to have higher prevalence compared to lethal or potentially lethal congenital anomalies (Table 26-1).

The general rule that prevalence would be lower for lethal or potentially lethal congenital anomalies applies to all ethnic

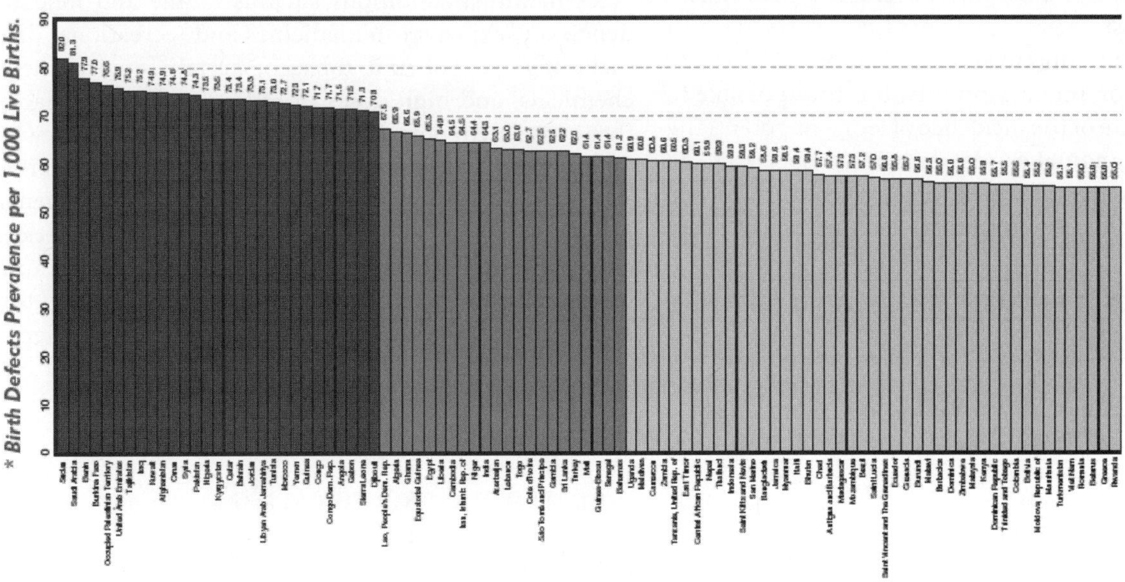

Figure 26-1 The Prevalence of "Birth Defects" in Developing Countries
SOURCE: adapted from Executive Summary, March of Dimes Global Report on Birth Defects, 2006

TABLE 26-1 INCIDENCE OF COMMON LETHAL OR SEVERE MALFORMATIONS

Prevalence per 1000 births

TYPE OF MALFORMATION		UNITED STATES (SINGLE BIRTHS)		U.K.
	Japan	Black	White	White
Anencephaly	0.6	0.2	1.0	1.6
Spina bifida aperta	0.2	0.7	0.7	2.0
Heart anomalies	7.0	7.5	8.2	6.9
Cleft lip+/-Palate	2.2	0.7	1.4	1.2
Infantile hypertrophic pyloric stenosis	Not included	0.8	3.2	3.4
Clubfoot	1.4	4.2	3.6	6.2
Dislocation of hip	7.1	0.8	4.0	3.2
Down syndrome	0.9	1.0	1.2	1.6

Adapted from Stevenson and Hall, 1993 (Table 4-2).

groups. Traditionally, all anomalies are classified into major or minor categories. Various reports estimate an incidence of major congenital anomalies ranging from 2%–7%, and minor anomalies at birth from as low as 7% to 40% (Table 26-2). This variation probably reflects differences in sample size and observation bias. It is possibly related to variation in the examiners' clinical perception of an anomaly, which might be reflected in wide-ranging incidence figures.

Lack of reliable incidence and prevalence data is also due to the lack of a coherent public health policy. In the very few surveys carried out, the data was based on hospital births, and thus reflected a gross underestimate of true incidence. Further, regional, statewide, or national surveys have not been carried out. Some data may be derived from perinatal or infant mortality figures, which are reasonably accurate.

The Southeast Asia office of the World Health Organization has maintained a good resource of information and data on infant mortality. It is thus possible to have some estimate of the incidence of lethal or potentially lethal congenital anomalies. However, it is suspected to be a gross underestimate of the true incidence and prevalence of isolated or multiple congenital anomalies. Another significant problem is related to detection and proper delineation of an anomaly. Paramedical workers, who are often assigned the duty of reporting, are not adequately trained in most of the Indian subcontinent countries, for example. Further internal anomalies may be missed due to lack of diagnostic resources such as ultrasound.

Routine detection and reporting of congenital anomalies is being carried out in the Western and developed countries. A congenital anomalies registry is maintained as part of statutory public health policy or, in some cases, in the form of longitudinal surveys supervised and funded by state-run medical research councils. The Trent Congenital Anomaly Register (CAR) and Wales Congenital Anomaly Register and Information Service (CARIS) in the United Kingdom are good examples. The data is collected and classified into various categories, including ethnic origin. Similar kinds of birth registries are also maintained in some states of North America and Australia. Although the data is generally hospital-based, the network is extended to include primary healthcare providers. It is thus possible that the data in these registries may reflect nearly true incidence and prevalence figures.

TABLE 26-2 INCIDENCE OF MAJOR AND MINOR ANOMALIES AT BIRTH

	Leppig et al. (1987)	Marden et al. (1964)	Mehes (1983)	Myrianthopoulas and Chung (1974)
Major malformations (%)	3.8	2.1	2.2	7.1
Minor malformations (%)	40.7	14.7	17.2	7.26
Sample size	4305	4412	4589	53,257

Adapted from Stevenson and Hall, 1993 (Table 2-1)

COMMON CAUSES OF BIRTH DEFECTS

Birth defects are a diverse group of disorders of prenatal origin, which can be caused by single gene defects, chromosomal disorders, multifactorial inheritance, environmental teratogens, and micronutrient deficiencies. Maternal infectious diseases such as syphilis and rubella are a significant cause of birth defects in low-income and middle-income countries. Maternal illnesses like diabetes mellitus, conditions such as iodine and folic acid deficiency, exposure to medicines and recreational drugs including alcohol and tobacco, certain environmental chemicals, and high doses of radiation are other factors that cause birth defects.

Congenital anomalies are etiologically heterogeneous. Apart from environmental factors, various genetic mechanisms are involved (Table 26-3). Nongenetic or environmental factors are varied and include intrauterine infections, maternal drugs, maternal smoking, toxic effects of alcohol, and mechanical constraints. Genetic mechanisms may include one of the Mendelian patterns or a chromosomal aberration.

ISOLATED MAJOR ANOMALIES

An anomaly is usually considered to be "major" if it is associated with or resulting in significant anatomic or functional abnormality. A major is anomaly may be associated with a minor anomaly. It is essential to look for

TABLE 26-3 THE ETIOLOGY OF HUMAN MALFORMATIONS (JONES, 1995)

Etiology	Number of Cases	Percent
Autosomal recessive	83	30
Autosomal dominant	74	27
Unknown	34	12
Chromosomal	31	11
Sporadic	24	8
X-linked	19	7
Teratogenic	13	5
Total	278	100

*Adapted from Stevenson and Hall, 1993 (Table 10-2)

minor anomalies if a major anomaly is present. Table 26-4 provides some figures on the number of minor anomalies observed in relation to a major anomaly. This is based on European and North American data. There are no such figures published from developing countries. The major anomalies are not unique to any one ethnic group. The majority is either detected immediately after birth or during infancy, except for certain internal anomalies involving the heart, gastrointestinal tract, and genitourinary system. A few major congenital anomalies are discussed here, in view of certain characteristics relevant to South Asians.

NEURAL TUBE DEFECTS

Neural tube defects (NTDs)) include spina bifida (occulta or cystica), meningocele (without neural tissue), meningomyelocele (with neural tissue), anencephaly (absence of major part of the brain), and some types of hydrocephalus. Heterogeneous developmental anomalies of the vertebral column, collectively called *spinal dysraphism*, are also included as NTDs. They occur in about 0.002% of all live births (about two per 1000 live births). NTDs develop early in gestation as a result of defective embryonic mesoderm development, resulting in either failure of closure of the neural tube (primary neurulation) or incomplete closure and/or separation of the neural tube (abnormal canalization).

The general population incidence may range from 1–4 per 1000 live births. Racial and regional variation in incidence of NTDs has been recognized for many decades. For example, a higher incidence of NTDs in the United Kingdom was recorded in South Wales and South Yorkshire. Some parts of Ireland also had higher live birth rates of NTD. Incidence has gradually declined in these regions, probably as a result of improved social and economic conditions. Incidence in North America has been consistently lower among people of European descent. However, higher birth prevalence has been recorded among Hispanics living in the Mexican border regions in the United States and among Sikhs of British Columbia, Canada (Baird, 1983; Hall et al., 1988). Baird (1983) reported relatively higher incidence of NTD in the Sikh population of British Columbia (2.86 per 1,000 births) compared to the total births in the state (1.26 per 1,000). This difference was found to be statistically significant. Further analysis indicated that a greater proportion of Sikh cases were stillborn, and comprised much of the excess number of cases. Background information on the incidence of NTD in Sikhs in India and elsewhere was also included. Hall et al. (1988) found higher frequency of cervico-thoracic spina bifida and multiple NTDs among the Sikhs compared to the native peoples of the state of British Columbia (P less than 0.02). The same study also observed higher frequency of non-NTD congenital anomalies among siblings of the Sikh probands (8.8%) than among siblings of other probands (2.4%; p less than 0.05).

TABLE 26-4 CONCURRENCE OF MAJOR AND MINOR ANOMALIES AT BIRTH

	Percent with major malformations		
No. of minor anomalies	Leppig et al. (1987)	Marden et al. (1964)	Mehes (1983)
0	2.3	1.4	1.2
1	3.7	2.9	3.8
2	6.7	10.8	12.5
>3	19.6	90	26

Adapted from Stevenson and Hall, 1993 (Table 2-3)

CLEFT LIP AND PALATE

Cleft lip with or without cleft palate (CL±P) is a major public health problem worldwide, with an incidence at birth of about 1 in 500–1,000 that varies by race and nationality (Melnick, 1992). Asians (Chinese, Koreans, Filipinos, and Japanese) are at higher risk than Caucasians or Blacks (Chung et al., 1974; Hu et al., 1982; Melnick et al., 1986). It is increasingly clear that the same cause is probably not applicable in all races, or even across national groupings within races. It may be argued that an important genetic component to the etiology is applicable in all populations, if not in all individual cases (Melnick, 1992). However, despite intense study spanning over half a century, the etiology of CL±P remains an enigma. The precise roles played by genes, environment, and chance are poorly understood. This is probably due to lack of phenotypic clarity in CL±P cases, which appears to be a late manifestation of "inborn errors of morphogenesis." Apart from Mendelian inheritance in a small subset of CL±P cases, the multifactorial threshold model (MF/T) has been tested and discussed to explain the etiology of CL±P (Carter, 1976). All cases are conventionally grouped as nonsyndromic or syndromic CL±P. There are over several multiple anomaly syndromes with CL±P listed in the dysmorphology databases [London Dysmorphology Data Base (LDDB), 2000; www.lmdatabases.com] and birth defects text, such as Gorlin's *Syndromes of the Head and Neck* (Gorlin et al., 2001). Interested readers may refer to these resources for further information.

True incidence of nonsyndromic CL±P cases in the Indian subcontinent is unknown. Based on two reports from South India (Nemana et al., 1992) and the eastern state of Bengal (Ray et al., 1993), birth incidence is estimated to be 1.5 to 2 per 1,000 live births. This figure might be an underestimate, since both reports are based on a very selective hospital-based cohort. The purpose of these studies was to establish the most likely etiology. The South Indian study (Nemana et al., 1992) examined the MF/T model, while the Bengal study analyzed their data searching for another major autosomal locus (Ray et al., 1993).

The South Indian study (Nemana et al., 1992) examined birth records of all children born over a three-year period in the Madras (now Chennai) region. There were 54 nonsyndromic CL±P cases recorded among 34,267 newborn infants, an incidence of 1.6/1,000 births; the incidence was 2.14/1,000 in males and 1.06/1,000 in females. For family studies, nonsyndromic CL±P probands were ascertained from surgical records. There were 331 surgical probands identified over a period of 6 years, 127 with cleft lip alone and 204 with cleft lip and cleft palate. The data was analyzed for predictions of an MF/T model (Carter, 1976, Melnick et al., 1980). The MF/T model was then investigated using the goodness-of-fit test (PGOODFIT) described by Gladstein et al. (1978). The M:F proband sex ratio of 1.24 was not significantly greater than expected. However, the M:F ratio of the affected sibs was significantly greater than expected; the estimated proband ascertainment probability was only 2.35, and the population estimate of the affected M:F ratio from birth records was greater than 2.0. This difference was considered to be due to unexplained ascertainment bias in favor of surgical female cases. The authors concluded that the data were inconclusive for a sex-influenced, inheritance or sex-biased natural prenatal selection as seen in other populations (Melnick et al., 1980, 1986). Although females were only marginally less affected in this sample than males, the sex of the proband was independent of the risk to sibs. Further risk estimates based on the cleft type were not significantly different using various combinations. The authors acknowledged that it was difficult, due to the small sample size. However, when the data were analyzed using a complex segregation analysis method, it favored the possibility of the model best fitting a major locus with a reduced transmission. The non-Mendelian transmission probability associated with this conclusion could be due to phenocopies or incomplete penetrance. Further, the manifestation of CL±P also depends on in-utero exposure to harmful environmental agents during the critical period of embryonic development (Melnick et al., 1981). Further, the results of the Madras study are quite similar to those in other populations of Europe and Asia (Chung et al., 1986; Marazita et al., 1984, 1989; Melnick et al., 1980, 1986), albeit with some idiosyncratic differences in each population.

The West Bengal study (Ray et al., 1993) examined 90 extended families with nonsyndromic CL±P. These families were ascertained from rural areas and included one or more affected individuals with nonsyndromic cleft lip, with or without cleft palate. These families included 138 affected individuals; 66% were males and 64% had cleft lip (CL) alone. The "multiplex" families (multiple affected members) were less common than "simplex" (single affected member) pedigrees, comprising only 34% of all extended pedigrees. No significant difference was observed for frequency of CL between multiplex and simplex pedigrees. However, multiplex pedigrees had fewer affected males (58%) compared to simplex pedigrees (76%; $p \leq 0.02$). The authors examined the data using the POINTER computer program, and rejected both hypotheses, of no familial transmission ($p<0.0001$) and that the familial occurrence could be solely explained by the MF/T model ($p<0.05$). Although not statistically significant, the data could not reject the hypothesis of a major-locus model. Among the major-locus models, strictly recessive inheritance was rejected ($P<0.0001$), but codominant and dominant models were not. In conclusion, the results of complex segregation analysis were consistent with a dominant or codominant major-locus mode of inheritance of CL±P in the West Bengal families. The authors recommended looking for a major autosomal locus controlling this common birth anomaly. It is likely to locate the major CL±P locus, in this part of the Indian subcontinent, using genetic linkage between the CL/P trait and markers at candidate or random (genome screen) loci.

MULTIPLE ANOMALY SYNDROMES

A number of dysmorphic syndromes and associations of multiple congenital anomalies have been described in children born to either indigenous or migrant parents from different parts of the world. Some of these are unique, and are assigned as "new" syndromes. A large number of reports describe syndromes or associations that are not restricted to any particular ethnic group. Major databases, such as McKusick's catalog of Mendelian disorders (www.OMIM.org), the London dysmorphology database (www.lmdatabases.com), and the Australian database (www.possum.net.au), list these syndromes with adequate references. The interested reader may search these databases for detailed information.

PREVENTION AND MANAGEMENT OF BIRTH DEFECTS

The wide range of causes of birth defects means that a portfolio of prevention approaches is needed. Most birth defects of environmental origin can be prevented by public health approaches, including prevention of sexually transmitted infections, legislation controlling management of toxic chemicals (e.g., certain agricultural chemicals), vaccination against rubella, and fortification of basic foods with micronutrients (iodine and folic acid). Prevention may be considered in terms of life stage (Table 26-5). WHO recommends several country-level actions that can support the development of services for the prevention and care of birth defects. Prevention requires basic public health approaches to be integrated into health systems, including maternal and child health services. Many of the services and interventions proposed are already within the reach of low-income and middle-income countries, while others can be added as needs and resources determine.

TABLE 26-5 WHO RECOMMENDATIONS FOR PREVENTION AND TREATMENT OF CONGENITAL ANOMALIES: PRECONCEPTION CARE, PREGNANCY CARE AND NEWBORN INFANT AND CHILD CARE

FAMILY PLANNING
• Introducing women to the concept of reproductive choice
• Reducing the total number of children born with a birth defect
• Reducing the proportion of mothers of advanced maternal age, which reduces the birth prevalence of autosomal trisomies, particularly Down syndrome
• Allowing women with affected children the option of not having further children
PRECONCEPTION SCREENING AND COUNSELING
• Using family history taken in primary healthcare facilities to identify individuals at risk of having affected children
• Undertaking carrier screening for common recessive disorders (e.g., thalassemia and sickle cell disorders)
OPTIMIZING WOMEN'S DIETS BEFORE AND THROUGHOUT PREGNANCY
• Promoting the use of salt fortified with iodine to prevent iodine deficiency disorder
• Promoting the use of a staple food fortified with folic acid and of supplementary multivitamins with folic acid to prevent neural tube defects and other malformations
• Promoting avoidance of alcohol, tobacco, and cocaine
• Promoting an adequate general diet (e.g., with sufficient protein, calories, iron)
• Ensure adequate general diet (protein, calories, iron)
PREVENTING AND TREATING TERATOGEN-INDUCED INFECTIONS BEFORE AND THROUGHOUT PREGNANCY
Such infections include:
• syphilis
• rubella (67 countries do not have national rubella immunization programs)
OPTIMIZING PRECONCEPTION MATERNAL HEALTH AND TREATMENT FOR:
• Insulin-dependent diabetics
• Women on treatment for epilepsy
• Women on treatment with warfarin
ANTENATAL SCREENING FOR:
• Rhesus status
• Syphilis
• Individuals at risk of having children with birth defects using a family history
• Down syndrome (advanced maternal age; maternal serum screening; early ultrasound scanning)
• Neural tube defects (maternal serum screening)
• Major malformations (ultrasound fetal anomaly scanning @ 18+ weeks gestation)
• Carriers of common recessive disorders (e.g., thalassemia and sickle cell disorders)
PRENATAL DIAGNOSIS
• Ultrasound
• Amniocentesis
• Chorionic villus biopsy
FETAL TREATMENT FOR:
• Syphilis
• Fetal anemia with intrauterine transfusion
NEWBORN INFANT EXAMINATION
• Trained examiner clinically examining all newborn infants for birth defects
NEWBORN INFANT SCREENING
• Congenital hypothyroidism
• Phenylketonuria
• Cystic fibrosis
• Others, as dictated by each country's needs and circumstances
MEDICAL TREATMENT
Examples:
• Neonatal jaundice in glucose-6-phosphate dehydrogenase deficiency and Rhesus incompatibility
• Treatment and care for children with blood disorders such as sickle cell disorder, thalassemia, etc.
• Treatment of some inborn errors of metabolism
• Care of children with cystic fibrosis

(Continued)

TABLE 26-5 (CONTINUED)

SURGERY
Examples include the correction of:
• Simple congenital heart defects
• Cleft lip and palate
• Club foot
• Congenital cataracts
REHABILITATION AND PALLIATIVE CARE
As appropriate

NATIONAL PROGRAM FOR THE PREVENTION AND CARE OF BIRTH DEFECTS

A national program for the prevention and care of birth defects should include:

a. Commitment of policymakers and provision of adequate managerial support
b. Core network of appropriate specialist clinical and laboratory services that can be expanded in response to demand
c. Integration of approaches to the prevention and care of birth defects into primary healthcare, with an emphasis on maternal and child health
d. Education and training for healthcare providers, particularly those in primary healthcare
e. Organization of health education programs for the general population and recognized high-risk groups
f. Establishment of effective mechanisms to foster development of patient/parent support organizations, and collaboration with them in caring for people with birth defects and their families
g. Definition of the ethical, legal, religious, and cultural issues relevant to formulating services appropriate for the local population
h. Initiation and monitoring of population screening programs, such as screening of newborn infants, premarital/pre-pregnancy screening, and screening during pregnancy
i. Establishment of appropriate surveillance systems for birth defects; such support may be obtained by collaborating with existing birth defect surveillance systems, including the International Clearinghouse for Birth Defects Surveillance and Research, which includes: the Latin American Collaborative Study of Congenital Malformations, the WHO-supported International Database of Craniofacial Anomalies, and the European Registration of Congenital Anomalies.

NATIONAL PROGRAMS FOR THE CONTROL OF BIRTH DEFECTS

These programs depend on regularly updated technical guidance; in this regard, there are several priority actions for the international community:

a. Resolve currently divergent opinions on the health burden of both environmental and constitutional birth defects, using the revision of ICD-10 to draw on expert review of available data and to consider broadening the groups of conditions beyond those currently included in the classification of congenital anomalies.
b. Promote legislation and public health activities to minimize exposure of the population, and particularly of pregnant women, to potentially teratogenic infections, chemicals, and other environmental risk factors.
c. Define effective community services, and support the integration of the prevention and care of birth defects into maternal and child health programs. Support the provision to health ministries of an organized assessment of requirements and costs, and support them in choosing priorities.
d. Identify successful models that can be applied in low-income and middle-income countries.
e. Facilitate and support international networking on birth defect prevention and care programs, with an emphasis on developing common approaches and optimizing instruments for information, education, cost analysis, and surveillance, among others.
f. Promote translational genomic research assisted by bioinformatics tools in developing cost-effective, high-throughput diagnostic methods, and the discovery and development of new pharmacological agents.

DEVELOPMENT OF SERVICES FOR BIRTH DEFECTS

Multidisciplinary services and interventions for the prevention and care of birth defects should be part of existing healthcare services in all low-income to middle-income developing countries. In particular, this should be integrated with maternal and child health services. They should combine the best possible patient care with a preventive strategy encompassing education, preconception care, population screening, genetic counseling, and the availability of diagnostic services. That strategy must deliver services for the prevention and care of birth defects as part of a continuum of interventions for maternal and child health. Depending on the individual nation's healthcare capacities, these services should go beyond primary healthcare to include obstetric, pediatric, surgical, laboratory, radiological, and, if available, clinical genetic services in secondary and tertiary healthcare.

Effective delivery of services for the prevention and care of birth defects depends on the availability of a range of specialist clinical and diagnostic services, and a primary healthcare system that is able to use them. A nucleus of expertise in medical genetics, pediatric surgery, imaging, and fetal medicine is required, with the potential to expand to meet needs. Conventional laboratory services (hematological, microbiological, and biochemical) need to be supplemented with cytogenetic and DNA-based diagnostic services. Introducing such services may need to be a gradual process. Over time, the new technologies (for example, array-comparative genomic hybridization, array-CGH) will support more efficient and cost-effective service delivery.

The diversity of priority conditions, social structures, cultural conventions, and healthcare capabilities means that countries need to be able to consider a range of possible services, assessing costs and relative effectiveness, in order to make a selection and decide the sequence of implementation. However, no organized guidance is yet available on this. The leading health organizations, such as WHO, have an important potential role to play in identifying successful models, and providing coherent information on community genetics that is accessible to public health policymakers.

REFERENCES

Aase JM. (1990). *Diagnostic Dysmorphology*. New York: Plenum Medical Book Co., pp. 7–8.

Baird PA. (1983). Neural tube defects in the Sikhs *Am J Med Genet 16*(1), 49–56.

Bittles AH. (2002). Endogamy, consanguinity and community genetics. *J Genet 81*, 91–98.

Carter CO. (1976). Genetics of common single malformations. *Br Med Bull 32*, 21–26.

Chung CS, Ching GHS, Morton NE. (1974). A genetic study of cleft lip and palate in Hawaii. II Complex segregation analysis and genetic risk. *Am J Hum Genet 16*(2), 177–188.

Gladstein K, Lange K, Spence MA. (1978). A goodness-of-fit test for the polygenic threshold model: application to pyloric stenosis. *Am J Med Genet* 2, 7–13.

Gorlin RJ, Cohen MM Jr., Hennekam RCM. (2001). *Syndromes of the Head and Neck* (fourth edition). Oxford: Oxford University Press.

Hall JG, Friedman JM, Kenna BA, Popkin J, Jawanda M, Arnold W. (1988). Clinical, genetic, and epidemiological factors in neural tube defects. *Am J Hum Genet 43*(6), 827–837.

Hu DN, Li JH, Chen HY, Chang HS, Wu BX, Wang DZ, Liu XG. (1982). Genetics of cleft palate in China. *Am J Hum Genet* 34, 999–1002.

Chung CS, Bixler D, Watanabe T et al. (1986). Segregation analysis of cleft lip with or without cleft palate: a comparison of Danish and Japanese data. *Am J Hum Genetics* 39(5), 603–611.

Jablonski S. (1991). Syndrome: Le Mot de Jour. *Am J Med Genet 39*, 342–346.

Jones KL. (1995). *Smith's Recognizable Patterns of Human Malformation*. Philadelphia: WB Saunders Co., pp. 1–5.

Khoury MJ. (1989). Epidemiology of birth defects. *Epidemiol Rev 11*, 244–248.

Khoury MJ, Moore CA, Evans JA. (1994). On the use of the term "syndrome" in clinical genetics and birth defects epidemiology. *Am J Med Genet 49*, 26–28.

Leppig KA, Werler MM, Cann CI. (1987). Predictive value of minor anomalies. I. Association with major malformations. *J Pediatr* 110, 531.

March of Dimes. (2006). The executive summary: Global report on birth defects—The hidden tool of dying and the disabled (www.marchofdimes.org–accessed July 21, 2010).

Marazita ML, Spence MA, Melnick M. (1984). Genetic analysis of cleft lip with or without cleft palate in Danish kindreds. *Am J Med Genet* 19, 9–18.

Marazita ML, Melnick M, Spence MA, Hu Din. (1989). Family study of cleft lip and cleft palate in Shanghai, China. *Am J Hum Genet* 45, A243.

Marden PM, Smith DW, McDonald MJ. (1964). Congenital anomalies in the newborn infant. *J Pediatr* 64, 357.

Mehes K. (1983). Minor malformations in the neonate. Budapest: Akademiao Kiado.

Melnick ML, Bixler D, Gogh-Anderson P, Conneally PM. (1980). Cleft lip-cleft palate: an overview of the literature and an analysis of Danish cases born between 1941 and 1968. *Am J Med Genet 6*, 83–97.

Melnick M, Jaskoll T, Slavkin HC. (1981). Corticosteroid-induced cleft lip with in mice: a teratogenic, topographic, and histologic investigation. *Am J Med Genet 10*, 333–350.

Melnick M. (1992). Cleft lip (+/-cleft palate) etiology: a search for solutions. *Am J Med Genet 42*, 10–14.

Melnick M, Marazita ML, Hu DN. (1986). Genetic analysis of cleft lip with or without cleft palate in Chinese kindreds. *Am J Med Genet* (Suppl 2), 183–190.

Myrianthopoulas NC, Chung CS. (1974). Congenital malformations in singletons: epidemiologic survey. *BDOAS X*(11), 1.

Nemana LJ, Marazita ML, Melnick M. (1992). A genetic analysis of cleft lip and with or without cleft palate in Madras, India. *Am J Med Genet 42*, 5–10.

Ray AK, Leigh Field L, Marazita ML. (1993). Nonsyndromic cleft lip with or without palate in West Bengal, India: evidence for an autosomal major locus. *Am J Hum Genet 52*, 1006–1011.

Stevenson, RE, Hall, JG (eds.). (1993). *Human Malformations and Related Anomalies*. Oxford & New York: Oxford University Press.

World Health Organization (WHO, 2006). *Management of Birth Defects and Haemoglobin Disorders: Report of a Joint WHO-March of Dimes Meeting (May 17-19, 2006)*. Geneva, Switzerland.

World Health Organization (WHO, 2008). *The Global Burden of Disease: 2004 Update*. WHO, Geneva, Switzerland.

27.

INHERITED EYE DISEASES IN DEVELOPING COUNTRIES

Tiziana Cogliati, James Friedman, Norimoto Gotoh, Neeraj Agarwal, and Anand Swaroop

INTRODUCTION

Visual perception of the surrounding world is a complex and important physiological function that, in most instances, mediates our primary interaction with the environment. Furthermore, with approximately 30% sensory input to the brain, visual stimuli affect several higher cognitive functions, including learning, memory, and behavior (Rodieck, 1998). In 2004, World Health Organization (WHO) estimated that over 314 million individuals worldwide were visually impaired; of these, 45 million were legally blind (Resnikoff et al., 2004). Strikingly, the number of people visually impaired due to uncorrected refractive errors was over 153 million (145 million with low vision, and 8 million blind), accounting for almost half of the affected population. The causes of blindness and vision loss are in part dependent on socioeconomic conditions and access to quality healthcare (Dandona & Dandona, 2001). Unlike other human disorders affecting the central nervous system, visual dysfunction is relatively easy to diagnose and quantify, making it amenable to phenotyping in genetic studies. A testament to the importance of vision and ease of diagnosis is that almost 30% of the entries in the Online Mendelian Inheritance in Man (OMIM) website (www.ncbi.nlm.nih.gov/omim/) refer to the eyes and/or vision.

THE EYE, RETINA, AND VISION

The eye is comprised of three tunics or layers. The outermost layer is the fibrous tunic, which gives the shape to the eye. It consists of the sclera that extends into the cornea to the front, at the corneoscleral junction (Figure 27-1A). The middle layer is the vascular tunic, which gives rise to the iris in the anterior eye, the ciliary body, and the choroid. The ciliary body is responsible for the production of aqueous humor, and the vasculature in the choroid supplies nourishment to the neural retina. The innermost layer is the photoreceptive tunic that comprises the retina. The space between the cornea and the iris, called the *anterior chamber*, is filled with aqueous humor, while the space between the lens and the retina contains vitreous humor. Light comes into the eye through the cornea, passes through the pupil aperture in the center of the iris, and is more finely shifted by the lens before being focused on the retina at the back of the eye. The retina is the sensory element of the eye and is part of the central nervous system.

The retina is a layered structure that includes distinct neuronal and non-neuronal cell types. The sensory portion of the retina is composed of six types of neurons (rod and cone photoreceptors, bipolar, horizontal, amacrine, and retinal ganglion cells (RGCs)), and one type of (Müller) glia (Figure 27-1B). These cells are organized in three nuclear layers: the outer nuclear layer (ONL) containing rod and cone cell nuclei; the inner nuclear layer (INL) containing the cell bodies of bipolar, amacrine, horizontal, and Müller glia cells; and the ganglion cell layer (GCL) containing RGC bodies. Synapses are localized in two layers: the outer plexiform layer (OPL) that connects photoreceptors to bipolar and horizontal cells, and the inner plexiform layer (IPL) that connects bipolar, RGCs, and amacrine cells. RGC axons extend along the inner surface of the retina to form the nerve fiber layer (NFL) and converge toward the optic disc to form the optic nerve (Figure 27-1A). Rod and cone photoreceptors exist alongside each other, project their outer segments (OS) toward the outer retina, and form the photoreceptor layer (Figure 27-1B). Rods account for 95% of photoreceptors and are responsible for vision in low light conditions, whereas cones (though only 5% of photoreceptors) mediate daylight and high resolution color vision. The retinal pigment epithelium (RPE) has an intimate relationship with photoreceptor OS, and is essential for functional maintenance of photoreceptors, the light-capturing neurons in the retina. The RPE has several functions, including phagocytosis of shed rod and cone outer segments and regeneration of 11-cis retinal (a vitamin A derivative), important for the visual cycle. Disruption of the interaction between RPE and retina is detrimental to photoreceptor function. Rod and cone photoreceptors are differentially distributed within the human retina. At

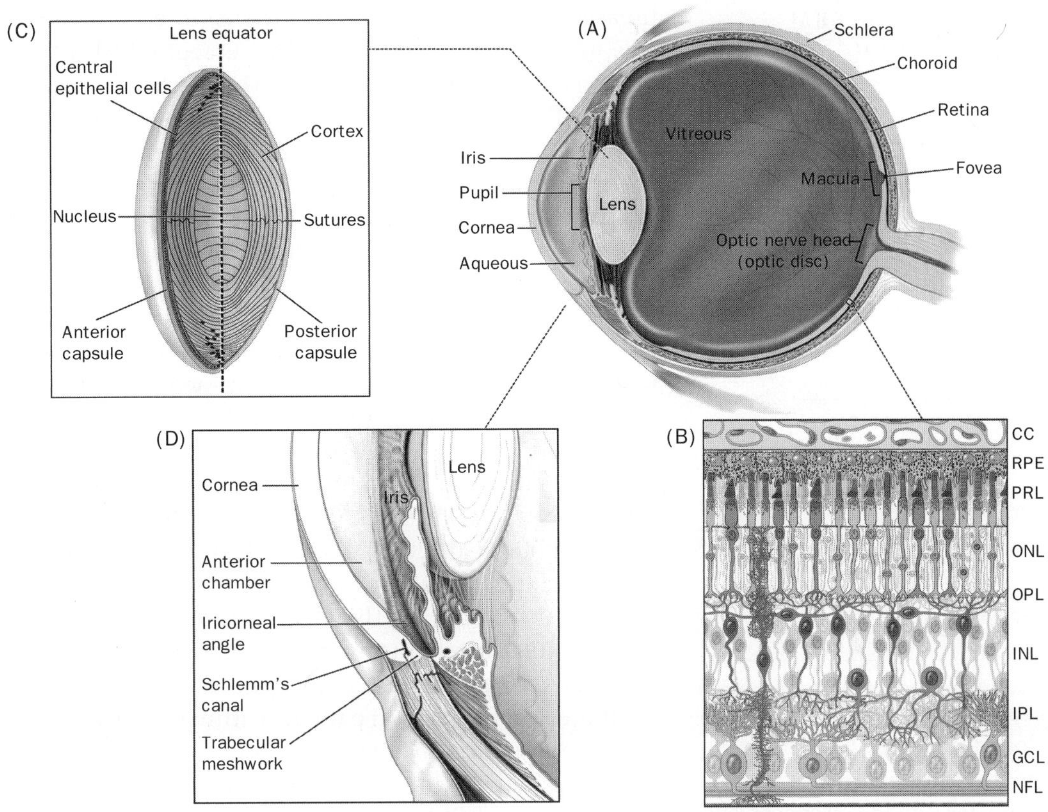

Figure 27-1 Schematic representation of the eye and anatomical structures affected in disease. A. Sagittal view of the eye structure; B. Layers and cell types in the retina; C. Histological organization of the lens; D. Magnification of the irido-corneal angle and trabecular meshwork. CC = choriocapillaris; RPE = retinal pigment epithelium; PRL = photoreceptor layer; ONL = outer nuclear layer; OPL = outer plexiform layer; INL = inner nuclear layer; IPL = inner plexiform layer; GCL=ganglion cell layer; NFL=neural fiber layer. Refer color figure.

the center of the retina (on the same axis as the pupil and lens and eccentric to the optic disc), a specialized region called the *macula* contains a higher density of cones compared with the rest of the retina (Figure 27-1A). This area is responsible for central vision. In the middle of the macula, the retina thins to form the fovea, which contains only cones and is avascular in its center (foveola). Visual acuity is highest at the fovea, partly because of the 1:1 ratio of ganglion cells to cone photoreceptors.

Light, and with it visual stimulus, travels through all layers of the neural retina until it reaches the photoreceptor OS, where it is captured by a photosensitive pigment (rhodopsin in rods and cone opsins in cones). This event initiates the phototransduction cascade, leading to hyperpolarization of the photoreceptor membrane and a decrease in glutamate neurotransmitter release at the photoreceptor–bipolar cell synapse. From bipolar cells, the visual stimulus is conveyed to RGCs directly or via amacrine cells. RGC axons converge into the optic nerve and extend outside of the orbit to relay information to several processing areas in the brain. Horizontal and amacrine cells integrate and modulate visual stimuli as they are transmitted from photoreceptors to RGCs.

PREVALENCE OF VISUAL IMPAIRMENT AND BLINDNESS

A global survey of the World Health Organization, based on data from 55 countries, estimated that over 2% of individuals suffered from visual impairment (excluding that caused by uncorrected refractive errors) in 2002, with a 3.7 median ratio of people with low vision to those with blindness (Resnikoff et al., 2004). When estimated data from 31 countries on visual impairment consequent to uncorrected refractive errors was added to the above figures, the worldwide percentage of visually impaired people in 2004 reached 4.7% and of blind individuals reached 0.7% (Resnikoff et al., 2008b). In 2002, the highest prevalence of low vision and blindness was reported for the African region (3% and 1%, respectively). The eastern Mediterranean and South Asia had 0.9% and 0.8% prevalence of blindness, respectively (Table 27-1). Epidemiological data from 2004 showed that the leading cause of blindness worldwide was cataract, with an average 39.1% prevalence (Figure 27-2). The second cause of blindness was uncorrected refractive error, with 18.2% prevalence; a major cause of visual impairment in the world population. Glaucoma and age-related macular degeneration (AMD) followed, with 10.1% and 7.1% prevalence, respectively. While cataract was most prevalent in developing countries, glaucoma was somewhat more prevalent (up to 18%) and AMD reached 50% prevalence in developed countries (Table 27-2). The latter figures should be interpreted taking into account the proportion of aging (over 60 years old) individuals, who represented more than 30% of the population in developed countries and only 15% in developing countries at the time of the report in 2002. Similarly, with increase in the incidence of diabetes in developed countries, diabetic retinopathy

TABLE 27-1 WORLD HEALTH ORGANIZATION GLOBAL ESTIMATE OF VISUAL IMPAIRMENT IN 2002[1]

Region[2]	Prevalence of blindness (%)	Prevalence of low vision (%)
Africa	1.0	3.0
Americas[3]	0.4	1.9
Eastern Mediterranean	0.9	2.7
Europe[4]	0.4	1.3
South-East Asia	0.8	2.2
Western Pacific[5]	0.6	1.4
World	0.6	2.0

[1] Excluding from uncorrected refractive errors.

[2] Classification based on WHO Global Burden of Disease 2000 Project (Murray et al., 2001).

[3] USA excluded (blindness, 0.2%; low vision, 1.2%).

[4] Western Europe excluded (blindness, 0.2%; low vision, 1.3%).

[5] Australia excluded (blindness, 0.3%; low vision, 1.2%).

Adapted from Resnikoff et al. (2004).

scored sixth among the causes of blindness (3.9% worldwide and over 10% in developed countries). In contrast, corneal opacities and childhood blindness (fifth and seventh leading causes of blindness worldwide, respectively) are more prevalent in developing than developed countries (Resnikoff et al., 2008b, and Figure 27-2).

GENETICS OF EYE DISEASES

The genetic pattern of diseases of the eye can range from simple (i.e., disease caused by a single gene defect) to complex (i.e., multiple genes and/or environmental factors contributing to the disease phenotype). Monogenic diseases typically follow a Mendelian inheritance pattern (autosomal dominant, autosomal recessive, or X-linked) and can be traced within a single family. In addition to one gene–one disease model, digenic effects have been observed in retinal degenerative diseases. In complex diseases, sequence changes (mutations) in each involved gene may influence the likelihood (susceptibility or protection) of developing the disease. Multifactorial diseases (e.g., AMD, refractive error, and open-angle glaucoma) are influenced by both genetics and environment.

The identification of disease-associated genes was accelerated with the advances in and the completion of the Human Genome Project, and with new high-throughput technologies for sequencing and data analysis. The latter has further expedited the identification of Mendelian inherited diseases, as families that are too small for standard linkage analysis can now be examined for mutations on a whole exome level (Ng et al., 2010; Raychaudhuri et al., 2010). Genome-wide association studies (GWAS) are facilitating the identification of common variants implicated in complex diseases (Chen et al., 2010; Manolio, 2009; Manolio et al., 2009). Finally, the study of numerous mouse mutants, including those with retinal degenerative phenotypes, is allowing the identification of novel genes and molecular mechanisms that can be further analyzed in humans. Genetic investigations of ocular diseases present unique problems as (i) patients within a family having the same mutation may exhibit distinct phenotypes, (ii) different mutations in the same gene give different clinical phenotypes, and (iii) similar clinical manifestations may be caused by mutations in different genes.

This chapter focuses on myopia and refractive error, as they are the most common ocular disorders, and on cataract and glaucoma, as they are the most prevalent causes of blindness in the world, including in developing countries. Although the prevalence of AMD is currently highest in industrialized countries in the population over the age of 60, it has been included in this review since its prevalence is also rapidly increasing in developing countries. Other ocular diseases that are most prevalent in developing countries, such as trachoma, onchocerciasis (river blindness), and corneal dystrophies (reviewed in Klintworth, 2009, and Poulaki & Colby, 2008), will not be discussed as they are infectious diseases primarily arising from environmental risk factors and can be avoided by increased hygiene and

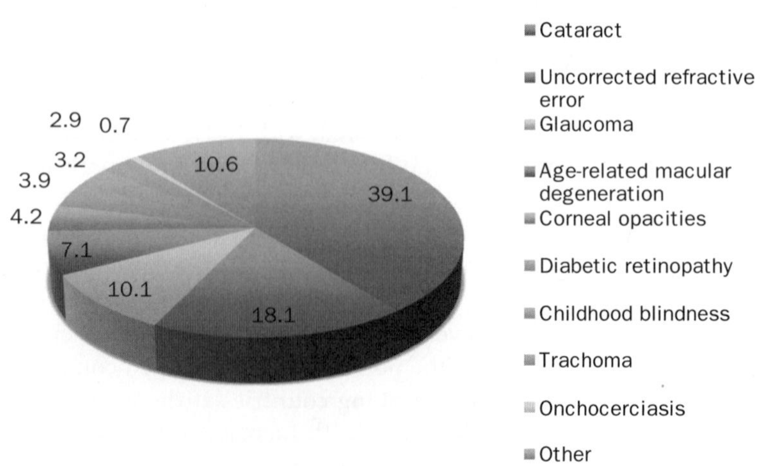

Figure 27-2 Global causes of blindness as a percentage of total blindness in 2004.
SOURCE: Reproduced from ref. (Resnikoff et al, 2008a).

TABLE 27-2 WORLD HEALTH ORGANIZATION GLOBAL ESTIMATE OF THE CAUSES OF BLINDNESS IN 2002

Region[1]	Cataract (%)	Uncorrected refractive error	Glaucoma (%)	AMD (%)
Africa	53	n.c.[2]	15	-
Americas[3]	49	n.c.	12	5
Eastern Mediterranean	49	n.c.	11	3
Europe[4]	32	n.c.	16	15
South-East Asia	55	n.c.	12	4
Western Pacific[5]	60	n.c.	8	8
World	47.8	n.c.	12.3	8.7

[1] Classification based on WHO Global Burden of Disease 2000 Project (Murray et al., 2001).

[2] Not calculated. The other percentages were calculated without considering the contribution made by uncorrected refractive error.

[3] USA excluded (cataract, 5%; glaucoma, 18%; AMD, 50%).

[4] Western Europe excluded (cataract, 5%; glaucoma, 18%; AMD, 50%).

[5] Australia excluded (cataract, 5%; glaucoma, 18%; AMD, 50%).

AMD, Age-related Macular Degeneration.

Adapted from Resnikoff et al. (2004).

antibiotic administration. Similarly, childhood blindness (ranked the sixth cause of blindness worldwide) exhibits a higher prevalence in developing compared to industrialized countries (Resnikoff et al., 2004); however, it will be mentioned in the paragraphs on cataract and glaucoma since it is largely associated with these diseases and primarily determined by socioeconomic conditions and lack of access to comprehensive healthcare. A review of ocular genetics can be found in (MacDonald et al., 2004).

REFRACTIVE ERRORS AND MYOPIA

Images of the surrounding world are focused on the retina after being refracted by the cornea, the lens, and the aqueous and vitreous humor. Sharp images are produced when they are converged precisely on the surface of the fovea (emmetropia). The axial length (the distance between the anterior and posterior poles of the eye) determines the refractive power of the eye; refraction is generated by the cornea and lens, and is contributed to a lesser extent by the aqueous and vitreous humor. Changes in any of these parameters can result in refractive errors classified as myopia, hyperopia, and astigmatism. In eyes of individuals with myopia or hyperopia, the images are focused in front of or behind the surface of the retina, respectively, resulting in a blurred vision. In these cases, abnormal axial length contributes for the most part to the refractive error. In contrast, in individuals with astigmatism, the cornea displays non-spherical shapes causing different degrees of refraction of the vertical and horizontal planes of view and consequently poor detailed vision. Refractive errors are easily diagnosed and quantified, and can be rectified with corrective lenses. However, a large proportion of refractive errors in developing countries remains uncorrected and can lead to visual impairment and blindness (Resnikoff et al., 2004, 2008b).

EPIDEMIOLOGY

Uncorrected refractive errors (mostly myopia) are a leading global cause of visual impairment, accounting for approximately 50% of cases, and are the second major cause of blindness (Resnikoff et al., 2008b) because of secondary ocular morbidity, such as choroidal neovascularization, retinal detachment, glaucoma, and cataract (Montero & Ruiz-Moreno, 2010; Praveen et al., 2008; Wang et al., 2009; Xu et al., 2007). The prevalence of visual impairment consequent to uncorrected (or inadequately corrected) refractive error in 5 to 15-year-old children is 0.96%, and is highest in urban areas in Southeast Asia and China (Resnikoff et al., 2008b). In older cohorts, the prevalence of visual impairment from uncorrected refractive error increases to 1.1% and 2.45% in people aged 16–39 and 40–49 years, respectively. In 2004, approximately 95 million people over 50 years of age were estimated to be visually impaired because of uncorrected refractive error; 7.3% of these were blind. The prevalence was highest in India and Southeast Asia (20%) and in China (10%), but ranged from 2% to 5% in other regions (Resnikoff et al., 2008b).

GENETICS OF MYOPIA

Myopia is the most common and most studied of the refractive errors. It can be classified as low (< −3.00 diopters [D]), medium (−3.00 to −6.00 D), and high or pathologic (> −6.00 D). Low and medium myopia are also known as "juvenile-onset myopia" because of their manifestation at an early age of 10–15 years, contrary to pathologic myopia that develops at even younger ages (Grosvenor, 1987; Gwiazda, 2009). The latter is a leading cause of blindness in several developed countries as well, especially in the Asian region (Young et al., 2007). Phenotypically, the disease is quantified by measuring axial length, anterior chamber depth, corneal curvature, and lens thickness. Each variable can be treated as a quantitative trait and correlated with the clinical phenotype. For the most part, axial length is the highest contributing variable to the refractive error (Hornbeak & Young, 2009). Family and twin studies have conclusively shown that inheritance is a major determinant of axial length, and of size and shape of the cornea and lens (reviewed in Baird et al., 2010; Hornbeak & Young, 2009; Young et al., 2007). Linkage studies and recently genome-wide association studies (GWAS) have identified several disease-causing genetic loci (Table 27-3). Some of these (MYP1,

MYP3, MYP6, MYP9, MYP11, MYP12, and MYP14) have been reproduced in more than one study. Other novel loci have been identified but await confirmation in similar or independent populations (Table 27-4).

Based on pathophysiology, several candidate genes have been screened and found to be associated with myopia in patient cohorts (reviewed in Baird et al., 2010). Among these are genes involved in eye development (*PAX6*, *SOX2*) and those encoding enzymes that modulate scleral extensibility (matrix metalloproteases, *MMP-1*, *MMP-3*, and *MMP-9*). Genes for collagen type I alpha 1 (*COL1A1*) and type II alpha 1 (*COL2A1*) have been evaluated in myopia patient populations. While initial studies failed to reach a consensus, defects in type II collagen are thought to be causative of myopia. Susceptibility gene loci have been tested as a follow-up to GWAS and found to be associated with myopia; these include the mitochondria-related genes mitofusin-1 (*MFN1*) and mitochondrial rhomboid protease (*PSARL*), and *SOX20T* in the MYP8 linkage region (Andrew et al., 2008). Associations in the MYP1 locus are shown with myocillin (*MYOC*) in European and Chinese populations (Tang et al., 2007; Vatavuk et al., 2009; Zayats et al., 2009), uromodulin-like 1 (*UMODL1*) in Japanese patients (Nishizaki et al., 2009), and with *TEX28* copy-number variations in a North American cohort (Metlapally et al., 2009). The following candidate genes have not been validated in other studies: transforming growth factor beta 1 (*TGFB1*), TGFB-induced factor homeobox 1 (*TGIF*), *Lumican*, hepatocyte growth factor (*HGF*), and membrane-type frizzled-related protein (*MFRP*), the latter involved in axial length regulation (Sundin et al., 2005).

More recently, two novel susceptibility loci (at 15q14 and 15q25) have been identified through GWAS in European populations (Hysi et al., 2010; Solouki et al., 2010). Regulation of expression of three candidate genes (*GJD2* and *ACTC1*, and *RASGRF1*, respectively) appears to be affected by SNPs in those loci (Table 27-4). *RASGRF1* is a promising candidate gene for refractive error based on expression, predicted function, and phenotype in a mouse null mutant. These studies also suggest that refractive error is determined by multiple genetic variants with low to moderate penetrance. The emerging paradigm of myopia is that of a complex genetic disease, which does not follow a single-gene Mendelian inheritance pattern, and therefore several genes—potentially under the influence of environmental factors—contribute to its phenotype, allowing its classification as a multifactorial disease (Rose et al., 2002).

TABLE 27-3 IDENTIFIED MYOPIA LOCI LISTED BY THE HUGO GENE NOMENCLATURE COMMITTEE (HGNC)

Trait[1] (age of onset)	Locus	Chromosome	OMIM
High myopia (1.5–15 yrs.)	MYP1	Xq28	310460
High myopia (avg. 6.8 yrs.)	MYP2	18p11.31	160700
High myopia (avg. 5.9 yrs.)	MYP3	12q21-q23	603221
High myopia	MYP4	7q36	608367
High myopia (avg. 8.9 yrs.)	MYP5	17q21-q22	608474
Refraction (QTL)	MYP6	22q12	608908
Refraction (QTL)	MYP7	11p13	609256
Refraction (QTL)	MYP8	3q26	609257
Refraction (QTL)	MYP9	4q12	609258
Refraction (QTL)	MYP10	8p23	609259
High myopia (<5–6 yrs.)	MYP11	4q22-q27	609994
High myopia (<12 yrs.)	MYP12	2q37.1	609995
High myopia (<5–6 yrs.)	MYP13	Xq23-q25	300613
Refraction (QTL)	MYP14	1p36	610320
High myopia	MYP15	10q21.1	612717
High myopia	MYP16	5p15.33-p15.2	612554
Refraction (QTL)	MYP17	7p15	608367
High myopia	MYP18	14q22.1-q24.2	(Yang et al., 2009)

[1] Severity: high, −5.00 D and below; mild–moderate, −1.00 D and below; variable, −12.12 to +7.25 D.
avg.=average; D=diopters; QTL=quantitative trait locus; OMIM=Online Mendelian Inheritance in Man.
Adapted from Hornbeak & Young (2009).

CURRENT TREATMENTS

Myopia is an important public health problem. It is common and, if not corrected, increases the risk of other vision-threatening abnormalities and blindness. The easiest and relatively inexpensive solution is the use of corrective lenses (i.e., spectacles and contact lenses). However, this measure has not yet been widely implemented in several developing countries. The reasons are geographical, economical, and cultural. Access to eye care may not be readily available in remote rural regions. Furthermore, the costs of corrective lenses may not be affordable for a part of the population. Finally, cultural isolation may limit information on the disease, its consequences if untreated, and interventions available.

In developed countries, pharmacological treatment and optical devices have been tested to slow the progression of myopia. The use of cycloplegic eye drops (e.g., atropine), although effective to a certain degree, is limited by its short-term (photophobia) and potentially long-term (light-induced retinal damage and cataract) side effects. Other molecules, more selective for muscarinic receptors, have also been evaluated (Saw et al., 2002).

CATARACT

Cataract is a disease of the lens, a transparent and flexible disc-shaped structure that refracts light and focuses it on the retinal surface (Figure 27-1A). Lens transparency and

TABLE 27-4 NEWLY IDENTIFIED MYOPIA LOCI

Trait[1]	Type of study	Geographical area or ethnic origin	Chromosome (*Gene*)	Reference
Refraction (QTL)	Linkage	African -American	2p	Ciner et al. (2008)
Refraction (QTL)	Linkage	African -American	3p	Ciner et al. (2008)
Refraction (QTL)	Genome-wide linkage scan	U.S.A. Caucasian	4q21	Wojciechowski et al. (2009)
Refraction (QTL)	Meta-analysis	U.S.A.	4q21–22	Wojciechowski et al. (2009)
Refraction (QTL)	Genome-wide linkage scan	U.S.A. Amish	5qter	Wojciechowski et al. (2009); Zhu et al. (2008)
Refraction (QTL)	Linkage	African -American	10p	Ciner et al. (2008)
Refraction (QTL)	Genome-wide linkage scan	U.S.A. Caucasian	12q24	Wojciechowski et al. (2009)
Refraction (QTL)	Genome-wide association	European German	15q14 (*GJD, AcTC1*)	Solouki et al. (2010)
Refraction (QTL)	Genome-wide association	European German	15q25 (*Rasgrf1*)	Hysi et al. (2010)
Mild-moderate myopia	Linkage	U.S.A.	20	Ciner et al. (2009)
High myopia	Genome-wide linkage scan	Several	9q34.11	Li et al. (2009)
High myopia	Linkage	Chinese	15q12–13	Yu et al. (2007)
High myopia	Genome-wide association	Japanese	21q22	Nishizaki et al. (2009)
Corneal curvature (QTL)	Genome-wide linkage scan	Caucasian	2p25.3	Biino et al. (2005)
Corneal curvature (QTL)	Genome-wide linkage scan	Caucasian	3p26	Biino et al. (2005)
Corneal curvature (QTL)	Genome-wide linkage scan	Caucasian	7q22	Biino et al. (2005)
Anterior chamber depth (QTL)	Genome-wide linkage scan	Caucasian	1p32.2	Biino et al. (2005)
Axial length (QTL)	Genome-wide linkage scan	Caucasian	2p24	Biino et al. (2005)

[1] Severity: high, −5.00 D and below; mild–moderate, −1.00 D and below; variable, −12.12 to +7.25 D. QTL=quantitative trait locus.

Adapted from Baird et al. (2010); Hornbeak & Young (2009).

function are determined by its fiber cell architecture and by the structured arrangement of its high protein content (Delaye & Tardieu, 1983). The mature lens (Figure 27-1C) comprises a central area of organelle-free nuclear fiber cells, surrounded by a layer of nucleated cortical fiber cells. Once lens development is complete, new fibers are constantly added from a pool of epithelial cells in the germinative zone, anterior to the lens equator, in a process called *secondary fiber formation* that maintains the cytoarchitecture. Lens fiber cells contain the highest proportion (30%–35%) of soluble proteins; 80%–90% of which are crystallins, a family of proteins critical for transparency (Graw, 2009; Morner, 1893). Pathologic changes in the cytoarchitecture and arrangement of the lens fibers, or in the stability, solubility, and molecular structure of its protein content, result in altered lens transparency and focusing capacity (Kuszak et al., 2004).

The generic term *cataract* defines the pathological condition in which the lens of the eye is affected by opacity, resulting in reduced vision. Moderate opacity can cause light sensitivity defects. However, severe opacity causes reduced visual perception and can lead to loss of vision. Cataracts can be categorized in four major groups: (1) congenital or infantile cataract, which normally manifests within the first year of age; (2) cataracts associated with systemic or genetic diseases, with onset before the fourth decade of life; (3) traumatic cataracts consequent to serious eye injury, which mostly occur in working-age males; and (4) age-related cataracts affecting the over-60 population (Shiels & Hejtmancik, 2007).

EPIDEMIOLOGY

An assessment by WHO of the prevalence of visual impairment in 2002 ranked cataract as the leading cause of blindness, accounting for 47.8% of all cases worldwide (Table 27-2). A great disparity is observed between developed and developing countries. While cataract is responsible for only 5% of blindness in North America, Western Europe, and Australia, it accounts for over 50% of cases in Africa, south-east Asia, and the western Pacific.

Age-related cataract accounts for almost half of blindness worldwide and is one of the major public health problems in developing countries, where the disease remains untreated because of the lack of information and clinical resources. Contributing risk factors include smoking, obesity, diabetes, high myopia, exposure to UV light, alcohol consumption, and malnutrition. In contrast, inherited cataract has a different clinical impact since it manifests in newborn children. The visual stimulus perceived by the retina is transmitted and modified by complex neural networks and sent to the brain. Visual cognitive functions develop upon stimulation of these neural networks during the early postnatal years (newborn to infant). Without the appropriate stimuli, visual functions do not develop correctly and subsequently only allow discrimination of light and dark, or blurred figures at most (called *amblyopia*). For the most part, inherited cataract develops during uterine growth and can progress into school age. It needs to be treated with specialized surgery, as early as possible depending on the severity, and preferably before 3 years of age. Current surgical techniques are remarkably successful in restoring vision in young children affected by hereditary cataract. In managing visual impairment among children, hereditary cataract is one of the few clinical conditions in which risk assessment based on family history, early diagnosis, and early intervention can have significant impact in reducing the number of visually impaired or blind individuals.

GENETICS

Both congenital and age-related cataracts have a strong genetic component (Shiels & Hejtmancik, 2007). Studies of congenital cataracts have greatly benefited from genetic linkage and candidate gene screening, whereas the genetics of age-related cataracts has lagged behind until recently, when model-free linkage analysis and association studies on candidate genes have been applied (Shiels & Hejtmancik, 2007).

Approximately 30% of congenital cataracts and 50% of age-related cortical cataracts have a genetic component (Haargaard et al., 2004; Hammond et al., 2001; Hejtmancik, 2008). Over 30 disease-causing loci have been linked to congenital cataract, and mutations in more than 20 genes have been reported, with the numbers constantly growing. Mutations in genes coding for lens proteins that alter the protein structure and cause aggregation usually result in congenital cataracts (Francois, 1982) and are transmitted in a Mendelian fashion with high penetrance. Congenital cataract can also be present in association with other ocular abnormalities (e.g., microphthalmia, aniridia, retinal degeneration) or can be part of syndromic genetic disorders (e.g., Lowe syndrome). Mutations (often in more than one gene) causing milder structural changes in a protein may confer increased susceptibility to damaging environmental stimuli (e.g., light and/or oxidative damage) and cause age-related cataracts (Shiels & Hejtmancik, 2007). Although aging is the major contributing factor to age-related cataract, it can also arise secondary to diabetes, trauma, lightning strike, or ocular inflammation.

Lens transparency depends on the ordered arrangement of fiber cells and of soluble proteins they contain (Delaye & Tardieu, 1983). Lens soluble proteins in humans belong to three crystallin families, α-, β-, and γ-. Mutations in crystallin genes have been identified in almost half of the families with cataract (Table 27-5). Approximately 30% of lens soluble proteins are αA-crystallin and αB-crystallin. They form large complexes that have chaperone-like functions, e.g., to maintain in solution β- and γ-crystallins (Horwitz, 1992), but also contribute to the integrity of the fiber cytoskeleton, inhibition of apoptosis, response to stress insults, and direct the migration of lens epithelial cells (reviewed in Andley, 2007). These functions are distinct and make crystallins particularly interesting from a biological perspective. αA-crystallin and αB-crystallin are coded for by the *CRYAA* and *CRYAB* genes (Quax-Jeuken

TABLE 27-5 CATARACT CRYSTALLIN LOCI LISTED BY THE HUGO GENE NOMENCLATURE COMMITTEE (HGNC)

Trait	Inheritance pattern	Locus	Chromosome	OMIM
Congenital/Presenile	AD/AR/Sporadic	CRYAA	21q22.3	123580
Congenital	AD	CRYAB	11q22.3-q23.1	123590
Congenital	AD	CRYBA1	17q11.2-q12	123610
Congenital	AD	CRYBA4	22q12.1	123631
Congenital	AD/AR	CRYBB1	22q12.1	600929
Congenital	AD	CRYBB2	22q11.23	123620
Congenital	AR	CRYBB3	22q11.23	123630
Congenital	AD	CRYGB	2q33-q35	123670
Congenital	AD	CRYGC	2q33-q35	123680
Congenital	AD	CRYGD	2q33q-35	123690
Congenital	AD	CRYGS	3q26.3-qter	123730

AD = autosomal dominant; AR = autosomal recessive; OMIM = Online Mendelian Inheritance in Man.

Adapted from Graw (2009); Hejtmancik (2008).

et al., 1985; Quax-Jeuken, 1985), located on chromosome 21q22 and 11q22, respectively. A total of nine mutations, seven dominant (p.R116C, p.R49C, p.R116C, p.G98R, p.R12C, p.R21W, p.R116H), one recessive (p.W9X), and one sporadic (p.R21L), have been identified in *CRYAA*, mostly causing only cataract. Single nucleotide polymorphisms (SNPs) have also been described that, although silent, may affect the amount of synthesized protein (Graw, 2009; Hejtmancik, 2008). Of the nine mutations identified in the *CRYAB* gene, six result in amino acid substitutions (p.P20S, p.R120G, p.D140N, p.G154S, p.R157H, p.A171T), two in frameshifts (p.K150fs, and p.P155fs), and one generates a shorter protein (p.Q151stop). Contrary to mutations in *CRYAA*, mutations in *CRYAB* appear to be causative not only of dominant cataracts but also of desmin-related myopathy (Vicart et al., 1998) or dilated cardiomyopathy (Inagaki et al., 2006a). Furthermore, four of the SNPs identified in the *CRYAB* gene are associated with changes in the coding sequence, although their association with the disease remains to be confirmed.

β-crystallins are a heterogeneous group of proteins that undergo extensive posttranslational modifications and form different oligomers. They are encoded by six genes, *CRYBA1*, -2, -4, *CRYBB1*, -2, -3, located on chromosome 17q11 (*CRYBA1*), 2q34 (*CRYBA2*), and 22q11 (*CRYBA4*, *CRYBB1*, -2, -3). Mutations have been identified in all but the *CRYBA2* gene (Table 27-5). Noteworthy is the p.Q155X mutation that converts *CRYBB2* into a pseudogene and leads to copy-number variations. Several SNPs are reported to cause amino acid substitutions in the CRYB family members.

γ-crystallins are low molecular weight proteins that are present in the lens fibers as monomers. Based on sequence homology, these proteins can be divided in two groups: γA/B/C-crystallins and γD/E/F-crystallins. All γ-crystallin genes (*CRYG*) are located on the long arm of chromosome 2 (q), where *CRYGA* through *CRYGE* are clustered and separated from *CRYGF* (Shiloh et al., 1986). Mutations have been identified in *CRYGC*, *CRYGD*, and *CRYGS*, and display autosomal dominant inheritance (Table 27-5). Variants have been reported for *CRYGB*. However, variable phenotypes caused by each individual mutation do not allow genotype–phenotype correlations. The presence of additional mutations and/or the contribution of environmental risk factors may affect the phenotype. Five reported variants, some of which have been found in families in Asia and Africa, change the amino acid sequence of γC- and D-crystallins. Mutations in β- and γ-crystallins tend to alter the protein structure, facilitating the formation of precipitates that accumulate in the lens fiber and cause opacity.

A group of proteins known as *connexins* form gap junctions that play important roles in mediating cell communication and transport of nutrients in the lens fibers, and in maintaining the lens cytoarchitecture and its transparency (Goodenough, 1992). One-quarter of cataract families are reported to contain mutations in connexin genes. Several mutations in *GJA3* on chromosome 13q11–13 and *GJA8* on chromosome 1q21–25, coding for connexin 46 and 50, respectively, have been linked to autosomal dominant cataracts with similar clinical manifestations.

Other families of proteins encoded by genes mutated in families with cataracts include: membrane proteins such as the water channel aquaporin (*AQP0/MIP*), lens intrinsic membrane protein 2 (*LIM2*), and transmembrane protein (*TMEM114*); beaded filament proteins such as *BFSP1/CP115/filensin* and *BFSP2/CP49/phakinin*, that interact with α-crystallin; the heat-shock transcription factor family member *HSF4* that controls the expression of αB-crystallin, associated with dominant and recessive cataracts; and the *MAF* transcription factor, associated with congenital and juvenile-onset cataract. Mutations in transcription factors (e.g., *FOXE3*, *PITX3*, *PAX6*, and *CHX10*) have been described for isolated congenital cataracts, mostly in association with ocular developmental defects. Other genes mutated in patients with syndromic cataract include *GCNT2* (a β-1 6-N-acetylglucosaminyltransferase), chromatin modifying protein-4B (*CHMP4B*), and the Nance-Horan syndrome (*NHS*) gene, that is mutated in the X-linked NHS. Notably, a mutation in ferritin light chain (*FTL*) gene causes overexpression and crystallization of ferritin L in the lens (Cazzola et al., 1997).

CURRENT TREATMENTS

Cataracts are routinely treated by removing a patient's opaque lens and implanting an artificial intraocular lens through a relatively simple surgical procedure. WHO lists cataract surgery among the first five most cost-effective health interventions. However, in some parts of the world, lack of appropriate information and limited geographical and financial accessibility to quality care are responsible for the inadequacy of the surgical treatment and the high prevalence of cataract-related blindness.

GLAUCOMA

Glaucoma is a heterogeneous group of optic neuropathies characterized by slow, progressive loss of RGCs and optic nerve degeneration with concurrent loss of supporting glia and vasculature. The resulting pathology comprises variable degrees of visual field deficits and, if untreated, loss of vision. Elevated intraocular pressure (IOP) is one of the most common risk factors in glaucoma. Optic nerve pathology is associated with elevated IOP, but this phenotype can also occur without elevation of IOP.

Patients with glaucoma initially lose their peripheral vision, while central vision loss occurs at later stages of the disease. Total loss of vision can be prevented with early intervention. However, as early symptoms often go unnoticed, diagnosis may be delayed to a point when optic nerve damage is already severe. Clinical diagnosis of glaucoma is based on the characteristic appearance, called cupping, of the optic nerve head upon fundus examination. Each human retina has more than one million ganglion cells. Their axons converge at the back of the eye to form the optic nerve and project toward the brain (Figure 27-1A). The light depression in the retina where all axons converge to exit the eye is called the *optic nerve head* (or optic disc). As ganglion cells die by apoptosis, their axons degenerate and the optic nerve thins, the depression of the optic nerve head becomes larger (cupping) and can be readily detected with an ophthalmoscope. The degree of visual impairment is generally proportional to the number of degenerated ganglion cells.

Elevation of the IOP, when present, is mostly caused by inefficient drainage of aqueous humor through the trabecular outflow pathway (Figure 27-1D). The aqueous humor is produced by the ciliary body and secreted into the anterior chamber to provide nutrients to the avascular lens and cornea. It is then drained through the trabecular meshwork (TM) and Schlemm's canal (part of the trabecular outflow pathway) in the iridocorneal angle, where cornea and iris intersect.

Glaucomas are classified, based on the configuration of the iridocorneal angle, as open-angle, angle-closure, and congenital. Each group can be further distinguished in primary and secondary subtypes. Open-angle glaucoma presents with or without elevation of the IOP; the latter is often classified as *normal tension glaucoma* (NTG). The most common form has adult onset (> 40 years of age) and is known as *primary open-angle glaucoma* (POAG). The rare juvenile form (JOAG) has variable onset between 3 and 40 years of age. Current diagnostic criteria for open-angle glaucoma include the presence of visual field deficiency and optic disc cupping (Leske, 2007). The most common form of secondary open-angle glaucoma is associated with (pseudo) exfoliation syndrome (XFS). Primary angle-closure glaucoma (PACG) can develop from aberrations of the iris that ultimately obstruct access to the TM and physically disrupt drainage of the aqueous humor. Secondary angle-closure glaucoma may be of inflammatory or neovascular origin. Primary congenital glaucoma (PCG), or buphthalmos, is a developmental form of glaucoma. Children usually have the unmistakable sign of large, bulging eyes (hence the term *buphthalmos*) that derive from excessively high IOP. Secondary developmental glaucomas are associated with anterior segment dysgenesis syndromes, such as aniridia and Axenfeld-Rieger syndrome, in which patients may present with other systemic deficiencies. The molecular events leading to increased IOP and those responsible for glaucoma in the absence of elevated IOP are still unknown.

EPIDEMIOLOGY

Glaucoma is the third leading cause of blindness in the world after cataract and uncorrected refractive error (Figure 27-2). Prevalence is around 2% in individuals over 40 years of age (Rudnicka et al., 2006). Notably, the rate of undiagnosed glaucoma ranges between 33% (USA) to over 90% (Brazil, South Africa, and India; Cedrone et al., 2008). POAG is the most common subtype, with over 20% of cases presenting without the elevation of the IOP (NTG). Prevalence of POAG increases exponentially with age, irrespective of ethnic group (Rudnicka et al., 2006). Although variable, populations of African descent above the age of 40 show, on average, a much higher prevalence, (Resnikoff et al., 2004) ranging from 1% in Nigeria, to 4% in African Americans, to 8.5% in Ghana, to 7%–9% in African-Caribbeans; Leske, 2007; Leske et al., 1994; Mason et al., 1989; Murdoch et al., 2001; Ntim-Amponsah et al., 2004; Tielsch et al., 1991), making it the leading cause of blindness in this ethnic group. PACG is most common in Asia, excluding Japan (Cedrone et al., 2008; Yamamoto et al., 2005), but it is extremely rare in Australian Aborigines. Epidemiologic data on gender-based prevalence are inconsistent, with studies showing higher frequency in men, in women, or none (Leske, 2007). However, a projected estimate of the prevalence of glaucoma for 2010 indicated that women tend to suffer more from glaucoma than men because of their relatively higher longevity compared to men and their higher susceptibility to PACG (Quigley & Broman, 2006). In recent years, the prevalence of POAG has undergone an apparent increase, in part because of improved diagnostic methods based on visual field testing. On the other hand, the prevalence of PACG has shown an apparent decrease, due to more stringent criteria for its diagnosis (Cedrone et al., 2008) that include not only the presence of a narrow drainage angle and elevated IOP, but also the presence of visual field deficiencies and optic disc cupping.

GENETICS

Glaucoma is a heterogeneous group of diseases, genetically characterized by Mendelian (early-onset types) and multifactorial (polygenic and environmental) traits (adult-onset types), as suggested by diverse phenotypes that can be detected even within families carrying the same mutation. Family history has been widely recognized as a major risk factor for POAG (Weih et al., 2001). Fifty percent of patients with POAG in cross-sectional studies have reported a family history (Tielsch et al., 1994).

Furthermore, first-degree relatives of affected individuals are estimated to have a three to nine fold higher risk of disease compared to the general population (Tielsch et al., 1994; Wolfs et al., 1998). Twin studies and racial differences in the prevalence of POAG further support its genetic basis. For the most part, POAG is inherited as a complex trait (Sieving & Collins, 2007) and multiple genes and/or environmental factors are likely to contribute to the pathological phenotype. By convention, POAG loci are designated as GLC1, followed by a capital letter assigned in alphabetical order at the time of identification. The current list includes GLC1A through GLC1O (Table 27-6). GLC1J, GLC1K, GLC1M, and GLC1N have been associated with juvenile-onset glaucoma. The genes for three of the GLC1 loci have been identified: myocilin (*MYOC*) for GLC1A, optineurin (*OPTN*) for GLC1E, and WD repeat domain 36 (*WDR36*) for GLC1G. One additional gene (*NTF4*) for GLC1O has been recently suggested.

MYOC, initially named *TM inducible glucocorticoid response* (TIGR) *protein*, was first identified by studies of TM cells in culture; its gene was mapped to the GLC1A locus and subsequently identified with *MYOC*

TABLE 27-6 GLAUCOMA LOCI AND GENES

Trait	Inheritance Pattern	Locus (Gene)	Chromosome	OMIM
JOAG, POAG	JOAG, AD; POAG, Complex	GLC1A (*MYOC*)	1q23-q24	137750
POAG	AD	GLC1B	2cen-q13	606689
POAG	AD	GLC1C	3q21-q24	601682
POAG	AD	GLC1D	8q23	602429
POAG, NTG	AD	GLC1E (*OPTN*)	10p14	606657
POAG	AD	GLC1F	7q35-q36	603383
POAG	AD/Complex	GLC1G (*WDR36*)	5q21.3-q22.1	609887
JOAG, POAG	AD	GLC1H	2p16.3-p15	611276
POAG	Complex	GLC1I	15q11-q13	609745
JOAG	AD	GLC1J	9q22	608695
JOAG	AD	GLC1K	20p12	608696
POAG	AD	GLC1L	3p21-p22	(Sherwin et al., 2009)
JOAG	AD	GLC1M	5q22.1-q32	610535
JOAG	AD	GLC1N	15q22-q24	611274
POAG/NTG	Sporadic/Complex	GLC1O (*NTF4*)	19q13.3	613100
Congenital	AD	GLC3A (*CYP1B1*)	2p22-p21	231300
Congenital	AR	GLC3B	1p36.2-p36.1	600975
Congenital	AR	GLC3C	14q24.3	613085
Congenital	AR	GLC3D (*LTBP2*)	14q24	613086
(Pseudo)exfoliation Syndrome (XFS)	AD	LOXL1	15q24	177650
Nanophthalmos	AD	NNO1	11p13	600165
Nanophthalmos	AR	NNO2 (*MFRP*)	11q23	609549
Nanophthalmos	AD	NNO3	2q11-q14	611897
Axenfeld-Rieger syndrome	AD	RIEG1 (*PITX2*)	4q25-q26	180500
Axenfeld-Rieger syndrome	AD	RIEG2	13q14	601499
Axenfeld-Rieger syndrome/ Iridodysgenesis	AD	RIEG3/IRID1 (*FOXC1*)	6p25	602482/601631
Aniridia	AD	AN (*PAX6*)	11p13	106210
Nail-patella syndrome	AD	NPS (*LMX1B*)	9q34.1	161200
Pigment dispersion syndrome	AD	GPDS1	7q35-q36	600510

JOAG = juvenile open-angle glaucoma; POAG=primary open-angle glaucoma (adult onset); NTG=normal tension glaucoma; AD=autosomal dominant; AR=autosomal recessive; OMIM = Online Mendelian Inheritance in Man.

Adapted from Wiggs (2007).

on chromosome 1q (Stone et al., 1997). MYOC is an evolutionarily conserved glycoprotein of 504 amino acids comprised of an N-terminal myosin-like domain and a C-terminal olfactomedin-like domain. Over 70 disease-causing mutations, clustered in the olfactomedin-like domain, have been identified. Together, they cause approximately 3%–5% of POAG and up to 20% of JOAG cases. Ninety percent of the individuals carrying *MYOC* mutations develop the disease during their lifespan. The most common mutation worldwide is p.Q368stop, which is suggested to descend from a single ancestor. p.T377M is the second most common mutation, which is associated with increased IOP and an autosomal dominant pattern of inheritance (Allingham et al., 2009). MYOC is expressed in the eye and extraocular tissues. Intact MYOC function is not necessary for physiological aqueous humor outflow (Kim et al., 2001; Lam et al., 2000; Wiggs & Vollrath, 2001). In the cells of the TM, it interacts with optimedin (Torrado et al., 2002), an olfactomedin-related protein, and with flotillin-1, a lipid raft protein (Joe et al., 2005). The secreted form of MYOC is a multimer. Although *MYOC* function remains unknown, several lines of evidence suggest that the disease is not due to loss-of-function or haploinsufficiency. Instead, gain-of-function or a dominant-negative effect may underlie the disease-causing mechanism (Wiggs, 2007). First, individuals who do not synthetize the MYOC protein appear normal. Similarly, IOP is not increased either in mice null for *MYOC* or in mice overexpressing the wild type TM gene. In addition, mutant forms of *MYOC* are misfolded and tend to aggregate in the endoplasmic reticulum. They could be toxic for the TM cells, or may interfere with the normal secretion of wild type MYOC (or other protein) causing congestion of the TM secretory pathway (Aroca-Aguilar et al., 2005; Gobeil et al., 2006; Wiggs, 2007). Interaction of mutated MYOC with the wild type protein appears to be important for the pathogenesis of the disease, as suggested by the variable and more severe phenotypes observed in individuals carrying heterozygous mutations compared to individuals carrying homozygous mutations. Notably, individuals homozygous for both p.Q368stop and p.K423E mutations do not seem to develop the phenotype. The severity of the phenotype also reflects the type of mutation and the age of onset (e.g., p.P370L causes severe JOAG, while p.Q368stop has a later onset), as well as the presence and degree of elevated IOP. Finally, the variability of phenotypes within families may also reflect a multigenic effect, such as the earlier onset of POAG in individuals carrying a combination of p.G388V MYOC and p.R368H CYP1B1 mutations.

Mutations in the gene encoding OPTN, previously known as FIP-2 (Li et al., 1998), on chromosome 10p14 (GlC1E locus) were first identified in families with autosomal dominant adult-onset POAG without elevation of IOP (Rezaie et al., 2002). Mutations in *OPTN* account for approximately 17% of normal tension POAG, but only for 0.1% of the total POAG (Aung et al., 2003; Rezaie et al., 2002). The gene encodes a 577 amino acid protein (Rezaie et al., 2002) that is expressed in ocular and non-ocular tissues, including the TM, and appears to be involved in several cellular functions: exocytosis, Golgi ribbon formation, FAS-ligand, and tumor necrosis factor alpha (TNF-α) apoptotic pathways (Rezaie et al., 2002; Sahlender et al., 2005; Sarfarazi et al., 1995; Zhu et al., 2007). OPTN is suggested to have a protective role in TNF-α-mediated optic nerve degeneration (Wiggs, 2007). Its detection in the aqueous humor indicates that it is a secreted protein (Rezaie et al., 2002). Induction of expression in experimental models, either by elevated IOP, or dexamethasone, or TNF-α, suggests that OPTN is produced and secreted in response to stress (Vittitow & Borras, 2002) and may play a protective role in reducing RGC susceptibility to apoptosis (Chalasani et al., 2007; De Marco et al., 2006). The most common mutation in OPTN is p.E50K that causes early-onset severe NTG (Aung et al., 2003, 2005; Rezaie et al., 2002). The p.M98K mutation appears to increase susceptibility to NTG and may act as a modifier for MYOC (Melki et al., 2003; Rezaie et al., 2002). Three OPTN variants, p.T34T, p.E163E, and p.555-5C, are shown to be associated with increased risk of glaucoma (Forsman et al., 2003). Finally, the association has been reported between *OPTN* variants combined with *TNF-α* variants and POAG, suggesting a multigenic component (Funayama et al., 2004).

The third gene that has been associated with POAG is *WDR36*, mapped to the *GLC1G* locus (Kramer et al., 2006; Monemi et al., 2005). It is a large, 23-exon gene, expressed in the eye and other organs, encoding a protein of 951 amino acids (Kramer et al., 2006). WDR36 protein is among the regulators of T-cell activation and proliferation (Mao et al., 2004), is involved in ribosome biogenesis, and interacts with the p53 stress-response pathway (Skarie & Link, 2008). The prevalence of *WDR36* sequence variants in POAG patients is variable, with several studies finding little or no association (reviewed in Allingham et al., 2009). The observation of a more severe disease phenotype in patients carrying sequence variants in *WDR36* has raised the possibility that, although not disease-causing, *WDR36* variants increase susceptibility or have a modifier effect on the disease phenotype (Hauser et al., 2006).

Mutations in the neurotrophin 4 (*NTF4*) gene at the *GLC1O* locus have recently been reported in a case-control study of German patients (Pasutto et al., 2009), but not in a similar study of U.S. patients of European ancestry (Liu et al., 2010). Variations in other genes are reportedly associated with POAG; these include apolipoprotein E (APOE), heat shock protein (HSPA1A), insulin-like growth factor 2 (IGF2), nitric oxide synthase 3 (NOS3), toll-like receptor 4 (TLR4), TNF-α, and tumor suppressor protein p53 (Table 27-7). However, further investigations are required in independent studies.

TABLE 27-7 CANDIDATE GLAUCOMA GENES

Trait	Geographical area or ethnic origin	Gene	Chromosome	Reference
NTG	Japanese	*AGTR2*	Xq22-q23	Hashizume et al. (2005)
POAG	Caucasian	*ANP*	1p36.2	Tunny et al. (1996)
NTG, POAG	Japanese, Chinese, Tasmanian, French	*APOE*	19q13.2	Copin et al., (2002); Lam et al. (2006); Mabuchi et al. (2005); Vickers et al. (2002)
POG	Chinese	*CDH-1*	16q22.1	Lin et al. (2006)
NTG	Korean, Japanese	*EDNRA*	4q31.2	Ishikawa et al. (2005)
POAG	Arab, Turkish, Estonian	*GSTM1*	1p13.3	Abu-Amero et al. (2008); Juronen et al. (2000); Unal et al. (2007)
POAG	Chinese	*IGF2*	11p15.5	Tsai et al. (2003)
POAG	Chinese	*IL1α*	2q14	Wang et al. (2006)
POAG	Chinese	*IL1β*	2q14	Lin et al. (2003b)
NTG, POAG	Korean, German	*MTHFR*	1p36.3	Junemann et al. (2005); Woo et al. (2009)
POAG	Caucasian	*NOS3*	7q36	Logan et al. (2005); Tunny et al. (1998)
POAG	Japanese	*OCLM*	1q31.1	Fujiwara et al. (2003)
POAG	Japanese	*OLFM2*	19p13.2	Funayama et al. (2006)
NTG	Japanese, Caucasian	*OPA1*	3q28-q29	Aung et al., (2002); Mabuchi et al. (2007)
POAG	Chinese	*p21*	6p21.2	Tsai et al. (2004)
POAG	Chinese, Caucasian	*p53*	17p13.1	Lin et al. (2002)
NTG	Japanese	*PON1*	7q21.3	Inagaki et al. (2006b)
POAG	Chinese	*TAP1*	6p21.3	Lin et al. (2004)
NTG	Japanese	*TLR4*	9q32-q33	Shibuya et al. (2008)
POAG	Japanese, Chinese	*TNF-α*	6p21.3	Lin et al. (2003a)

NTG = normal tension glaucoma; POAG=primary open angle glaucoma.

Adapted from Allingham et al. (2009).

Secondary open-angle glaucoma in XFS is inherited as a multifactorial, late-onset disease. It is more prevalent in women and older individuals. Depending on the geographical region, XFS accounts for 20%–60% of open-angle glaucoma cases and can be a major cause of vision loss (Forsius, 1988; Tarkkanen, 1986). XFS is characterized by systemic abnormalities, including elevation of homocysteine (Altintas et al., 2005; Bleich et al., 2004; Leibovitch et al., 2003) and deposition of fibrillary materials in connective tissues, including the TM and around the Schlemm's canal (Richardson & Epstein, 1981; Schlotzer-Schrehardt & Naumann, 1995). Recently, variants in the lysyl oxidase-like 1 (*LOXL1*) gene at the 15q24 locus were found to associate with XFS (Thorleifsson et al., 2007). The p.G153D variant appears to be the major risk allele and has been extensively replicated in several populations (Challa, 2008). The p.R141L variant is associated with the disease in some populations but not others (Ramprasad et al., 2008; Hayashi et al., 2008).

PCG is the most severe and common form of glaucoma in infants, affecting males more frequently than females before their third birthday. PCG can be inherited as an autosomal recessive trait and is prevalent among consanguineous groups. In 80% of cases, it manifests within the first year of life (Chandler & Grant, 1965) and rapidly leads to blindness if the elevated IOP is not treated (deLuise & Anderson, 1983). The etiology probably involves a developmental defect of the anterior chamber angle that critically impairs aqueous outflow and causes very high IOP (Anderson, 1981). PCG has been linked to four genetic loci (Table 27-6): GLC3A (2p22-p21) (Sarfarazi et al., 1995), GLC3B (1p36.2-p36.1) (Akarsu et al., 1996), GLC3C (14q24.3) (Stoilov et al., 1997), and GLC3D (14q24.2-24.3) (Firasat et al., 2008).

The *CYP1B1* gene encodes a component of the human cytochrome p450 pathway and is linked to PCG at the GLC3A locus (Stoilov et al., 1997). It was identified by studies of consanguineous pedigrees from Saudi Arabia and Turkey. Mutations in *CYP1B1* (most of which are insertions, deletions, or missense mutations in highly conserved protein regions) have been observed in 20%–30% of ethnically mixed populations, in 50%–80% of more consanguineous groups, and in almost 100% of highly consanguineous groups (Challa, 2008). *CYP1B1* mutations may affect normal development of the anterior chamber during ocular differentiation (Nebert, 1991; Stoilov et al., 1997).

Variants in a second gene, latent transforming growth factor-beta-binding protein 2 (*LTBP2*), have been linked to glaucoma in Pakistani, European Gypsy, and Iranian patients (Ali et al., 2009; Narooie-Nejad et al., 2009b). *LTBP2* maps to the GLC3D locus and partially overlaps with the GLC3C locus. LTBP2 is an extracellular matrix protein, involved in the development of the trabecular meshwork and anterior chamber angle (Ali et al., 2009).

The existence of an additional locus was suggested by an Iranian study, based on the absence of linkage to GLC3B and GLC3C in at least nine families without *CYP1B1* gene mutations (Narooie-Nejad et al., 2009a). Loci/genes associated with other primary and secondary glaucomas are summarized in Tables 27-6 and 27-7.

Standard genetic approaches to glaucoma have not yet revealed the causative mutations in a majority of patients. Certain anatomical and physiological features of the eye (e.g., IOP, retinal NFL thickness, optic disk parameters, and thickness of central cornea) can constitute risk factors for glaucoma and have been analyzed as quantitative traits for linkage studies, yielding novel candidate genomic regions on chromosome 2 and 10, and candidate genes, *COL1A1*, *COL1A2*, and *ATOH7* (Dimasi et al., 2010; Duggal et al., 2007; Macgregor et al., 2010; Ramdas et al., 2010; van Koolwijk et al., 2007). Large-scale projects, such as the NEIGHBOR and GLAUGEN studies sponsored by the National Eye Institute (NEI) in the United States, have been initiated to examine a large set of glaucoma patients and controls for GWAS to identify genetic variants with more modest effects on glaucoma phenotype. (http://www.nei.nih.gov/news/statements/glaucoma_initiatives.asp)

CURRENT TREATMENTS

Blindness from glaucoma can be prevented with early diagnosis and surgical/pharmacological intervention. Therapies aimed at relieving the elevated IOP can reduce ganglion cell damage and prevent progression to blindness, if performed before severe loss of ganglion cells has occurred. Early diagnosis is possible if reliable biomarkers are available. Genetic testing is currently useful in early-onset glaucoma, where 30% of cases can be identified (Fan & Wiggs, 2010). For adult-onset POAG, it is foreseeable that genetic determinants will be used for genetic screening. However, loss of ganglion cells and optic nerve degeneration are irreversible. No therapy is currently available for optic nerve atrophy without elevation of IOP. To address ganglion cell loss, current research is focused on delivering neuroprotective molecules (e.g., neurotrophic factors), gene replacement, and cell-based therapies that are being evaluated in animal models (Fan & Wiggs, 2010).

AGE-RELATED MACULAR DEGENERATION (AMD)

AMD is a late-onset, progressive retinal degenerative disease that affects individuals over 50 years of age. The Age-Related Eye Disease Study (AREDS) describes the progression of AMD through three stages: early, intermediate, and advanced (AREDS, 2001). Each stage is defined by characteristic lesions (called *drusen*) that are detected upon fundoscopy as yellow spots in the macular region of the retina. Drusen are a hallmark of early and intermediate macular degeneration and consist of lipids and proteins (e.g., glycoconjugates, vitronectin, apolipoproteins, α-crystallin, and complement) that are deposited between the basal lamina of the RPE and the outer collagenous layer of Bruch's membrane, just beneath the choroid (Crabb et al., 2002; Malek et al., 2003; Mullins et al., 2000). The mechanism(s) underlying initiation and progression of drusen formation are still not delineated. In early AMD, several small drusen (<63 μm diameter) or at least one intermediate size drusen (63–125 μm) are observed, accompanied by pigment abnormalities. Intermediate AMD is characterized by widespread intermediate size drusen or at least one large drusen (≥125 μm), and often by hyper- or hypopigmentation of the RPE. As AMD progresses, focal loss of photoreceptors and RPE occurs, and in some instances small blood vessels may invade the space between the RPE and the photoreceptor layer. Two separate clinical features can characterize advanced AMD: geographic atrophy (GA) and neovascular (NV) AMD. GA is defined by the presence of areas (> 175 μm in diameter) of retinal depigmentation exposing the choroidal vasculature. At the cellular level, RPE is abnormal, Bruch's membrane is thinner, and degeneration of photoreceptors and choriocapillaries is evident (Green, 1999; Kim et al., 2002; Lutty et al., 1999). NV AMD presents with exudative/hemorrhagic detachment of the RPE or retina and the presence of subretinal fibrous tissue, and eventually progression to RPE atrophy. New blood vessels that have sprouted from the choroid expand through the RPE (Grossniklaus & Green, 2004; Sarks et al., 1997) resulting in three possible clinical phenotypes: (i) NV AMD, with accumulation of lipid-rich fluid; (ii) hemorrhagic AMD, with blood extravasation to the subretinal space; and (iii) disciform AMD, with replacement of neural retina with RPE, new vessels, and fibrous tissue.

From a clinical perspective, patients initially manifest minimal visual impairment, up to the stage of intermediate AMD. Neovascularization plays an important role in causing vision loss and is observed in 10% of AMD patients with advanced disease. As the disease progresses, loss of central vision accrues due to the loss of macular photoreceptors.

EPIDEMIOLOGY

Although AMD was the fourth most common cause of blindness and accounted (Li et al., 2006) for only 7% of cases in 2004 worldwide (Figure 27-2), AMD is a leading cause of blindness accounting for 54% of cases in the United States and other developed countries (Congdon et al., 2004) because of the aging population and access to treatments for other potentially blinding conditions. Differences in prevalence are based on ethnic origins, with 5.4% prevalence in white, 4.6% in Chinese, 4.2% in Hispanic, and 2.4% in black people. The highest prevalence for all groups is observed at the age of 75–84 years, with 15.8% prevalence in whites and Chinese and 7.4% in black individuals (Klein et al., 2006). Since the primary risk factor for AMD is aging, these numbers rise as the aging population increases. NV is reported to have an incidence twice the incidence of GA (Smith et al., 2001) and is more common in whites than Hispanic or blacks, yet less common compared to Chinese (Klein et al., 2006). Studies in China suggest variations in prevalence of AMD according to ethnicity and location. Prevalence is highest among Tibetans (15.6%) living at high altitude, compared to Uighur (11.3%) and Han (6.4%), living at lower altitudes (Klein et al., 1999). Low prevalence of AMD has also been reported for black populations in Rhodesia and of Bantu ethnicity in southern Africa (Klein et al., 1999).

GENETICS

AMD is a complex multifactorial disease that emerges from three fundamental factors: environment, advanced age, and genetics (Swaroop et al., 2009). Among environmental factors, smoking has been consistently associated with AMD (AREDS, 2000; McCarty et al., 2001; Tomany et al., 2004; Vingerling et al., 1996). The risk of developing advanced AMD is doubled in smokers, with suggested dose-response effect (Khan et al., 2006). Other environmental factors that have been associated with AMD include cardiovascular disease, sunlight exposure, alcohol use, and bacterial infections (Swaroop et al., 2009). The protective effect of diets that contain greens high in antioxidants (e.g., lutein and zeaxanthin), and fish high in omega-3 fatty acids has also been suggested (Flood et al., 2002; Seddon et al., 2003). The contribution of aging as a risk factor for AMD is substantial. The progressive decline in organ and tissue functionality that accompanies aging is manifested in the eye especially in rod photoreceptors, ganglion cells, and RPE. It is common to observe the accumulation of cholesterol in Bruch's membrane and the presence of small drusen in older individuals. Similarly, lipofuscin accumulates in the RPE causing ultimately mitochondrial dysfunction, deficient phagocytic activity, and activation of the complement system (Bonnel et al., 2003; Hageman et al., 2001; Jackson et al., 2002; Zhou et al., 2006). In recent years, significant progress has been made in understanding the genetics of AMD. Genetic linkage analysis in affected relative pairs has led to the identification of strong susceptibility loci at chromosomes 1q25–31 and 10q26 (Fisher et al., 2005) and of potential other loci on several other chromosomes (Figure 27-3). Subsequently, allelic variants at complement factor H (*CFH*) and age-related maculopathy susceptibility 2 (*ARMS2*)/HtrA serine peptidase 1 (*HTRA1*) genes were discovered and validated by association studies in AMD patients worldwide (Dewan et al., 2006; Edwards et al., 2005; Hageman et al., 2005; Haines et al., 2005; Klein et al., 2005; Li et al., 2006; Zareparsi et al., 2005; Maller et al., 2006). CFH is involved in the regulation of the alternative complement pathway. In the eye, it may protect RPE cells and choriocapillaries from attacks by the immune system by inhibiting complement activation. The initially described p.Y402H predisposing allele is thought to impair CFH normal biding to and inhibition of C-reactive protein, an activator of the complement cascade, leading to dysregulated complement activation and damage to the RPE and choriocapillaries. However, numerous genetic association studies have highlighted the role of multiple variants in noncoding and intronic regions around the *CFH* locus, and suggested that the contribution to AMD susceptibility of differential expression of *CFH* and other adjacent genes might be more important than changes in *CFH* coding sequence alone (Chen et al., 2010; Li et al., 2006; Swaroop et al., 2009). Furthermore, deletion of *CFHR1* and *CFHR3* genes in the environs of *CFH* is suggested to confer reduced risk of AMD (Hughes et al., 2006; Spencer et al., 2008) though it may not be independent of other reported variants (Raychaudhuri et al., 2010). Complement 2 (C2) and/or complement factor B (CFB), an activator of the complement pathway (Gold et al., 2006), complement 3 (C3; Yates et al., 2007), and complement factor I (CFI; Fagerness et al., 2009) have been associated with AMD and validated in genome-wide studies (Chen et al., 2010; Figure 27-3). The ARMS2/HTRA1 region in chromosome 10 carries the other strongest genetic risk allele for AMD. Though further investigation is required to clearly establish the contribution of ARMS2 and/or HTRA1, a combined effect of the two genes cannot be excluded (Kanda et al., 2007; Swaroop et al., 2009). Candidate gene association studies have also revealed other potential AMD susceptibility genes (Figure 27-3); these include *APOE* and *ABCA4* among others (see Swaroop et al., 2009 for a review).

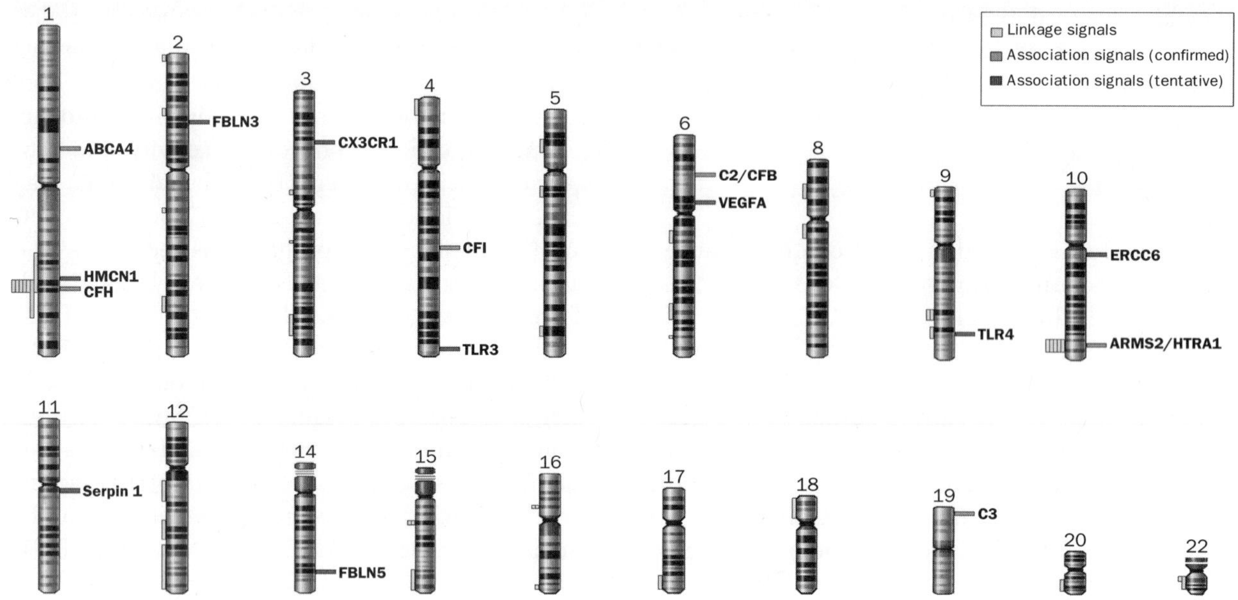

Figure 27-3 Summary of regions of linkage (yellow) and association signals (confirmed in green and tentative in red) for AMD susceptibility loci/genes. Refer color figure.
SOURCE: Reproduced from ref. (Swaroop et al, 2009).

CURRENT TREATMENTS

AMD remains mostly an untreatable disease. In the past, palliative therapies focused on delaying vision loss included laser photocoagulation, photodynamic therapy (PDT) with verteporfin (Visudyne®), transpupillary thermotherapy (TTT), and specialized surgery. Recently, inhibition of vascular endothelial growth factor (VEGF) has proven successful for the treatment of NV AMD. Intravitreal injection of VEGF inhibitors prevents angiogenesis and leakage from abnormal blood vessels. Three molecules, pegaptanib (Macugen®), ranibizumab (Lucentis®), and bevacizumab (Avastin®) have been tested in clinical trials and are being used widely for NV AMD. Pegaptanib is an oligonucleotide aptamer that binds VEGF with high specificity and affinity. Ranibizumab and bevacizumab are humanized monoclonal antibodies raised against VEGF. They are administered as injections into the eye every 4–6 weeks for variable periods of time, and have been shown to be superior to previously applied therapies in slowing the progression of vision loss in AMD patients. Pegaptanib and ranibizumab are approved by the U.S. Food and Drug Administration (FDA) for use in NV AMD, whereas bevacizumab is approved for the treatment of solid tumors, and its use in AMD patients is off-label. Nonetheless, bevacizumab has been preferentially used in many countries (including US) over ranibizumab because of its considerably cheaper cost. The safety and efficacy of bevacizumab in the treatment of AMD has been demonstrated in a recently published small multicenter randomized double blind study (Tufail et al., 2010). In May 2011, the Comparison of Age-Related Macular Degeneration Treatments Trials (CATT) research group published the results of their multicenter randomized clinical trial conducted in the United States on the use of bevacizumab versus ranibizumab for AMD. (The CAAT Research Group, 2011). They conclude that bevacizumab and ranibizumab have equivalent effect on visual acuity at one year follow-up when administered according to the same schedule. Following from the success of the above-mentioned anti-VEGF treatments and the questions still open on their safety, cost and dose frequency, a novel anti-VEGF molecule (VEGF trap) has been recently engineered and tested in phase 1/2 clinical trials. Since VEGF trap is designed to bind VEGF with higher affinity than currently used VEGF inhibitors, it would offer the advantage of longer intervals between injections (reviewed in Schmidt-Erfurth et al., 2010, and Campa & Harding, 2010).

The progressive and age-related nature of AMD suggests that preventive strategies should also be adopted and might yield beneficial effects. The first level of intervention is the reduction of potential risk factors such as smoking, hypertension, and elevated body mass index (BMI). Data on the positive effects of certain types of diets (Flood et al., 2002; Seddon et al., 2003) suggest that a second level of prevention could be the introduction of specific dietary supplements. Some of these were tested in two clinical trials sponsored by the NEI in the United States. The first multicenter randomized and controlled clinical trial (AREDS) has shown that dietary supplementation with combined high doses of vitamin C, vitamin E, β-carotene, zinc, and copper can reduce the chance of developing advanced AMD by 25% for people at high risk (AREDS, 2001). A second study (AREDS2) is underway to evaluate the efficacy of lutein, zeaxanthin, and/or omega-3 fatty acids in the prevention of advanced AMD (http://www.areds2.org/). In the future, genetic testing should help in identifying patients at risk and may suggest personalized preventive strategies.

FUTURE PROSPECTS

Among 6.8 billion people in the world, 314 million individuals are visually impaired; of these 45 million are blind (http://www.who.int/mediacentre/factsheets/fs282/en/index.html, http://www.prb.org/pdf09/09wpds_eng.pdf). A majority (>85%) of visually impaired individuals live in developing countries, with the highest prevalence in India, China, and Africa, where the predominant cause of blindness is cataract. It is estimated that 85% of all visual impairment and 75% of blindness worldwide is avoidable through cost-effective prevention or cure. Refractive errors in children and childhood blindness remain a significant global problem, even though vision can be improved in as many as 12 million children (5–15 years old) with the use of corrective lenses or surgery. During the last 20 years, the number of individuals losing their sight as a consequence of infectious diseases has decreased due to effective eye health strategies and the Alliance for Global Elimination of Blinding Trachoma by 2020 (GET 2020), an international program under the sponsorship of the WHO (http://www.who.int/mediacentre/factsheets/fs282/en/index.html). In contrast, the number of aging individuals that are either visually impaired or blind is increasing because of the higher life expectancy in some developing countries.

The largest initiative and global partnership addressing the need for higher awareness and stronger intervention in the prevention and cure of visual impairment and blindness is "Vision 2020: the Right to Sight" (http://www.vision2020.org/main.cfm), an international program that aims at the elimination of avoidable blindness by the year 2020. The initiative was launched in 1999, jointly by WHO and International Agency for the Prevention of Blindness (IAPB) in partnership with nongovernmental organizations, professional associations, eye care institutions, and corporations. All 193 WHO member states pledged support and commitment to this initiative. In countries such as India, several charitable private foundations and hospitals (e.g., Aravind Eye System, Shankar Nethralaya, and L.V. Prasad Eye Institute) have made a major difference in the lives of rural and poor populations. Eleven years into the program, 15 million fewer people are blind compared to projections made when the initiative was launched (http://www.vision2020.org/main.cfm?type=ACHIEVEMENTS). In 2010, the priority remains to expand cataract surgery and correction of refractive errors, especially to the poor and underserved populations.

Analyses of the geographic and socioeconomic distribution of blindness have shown variations in prevalence of blindness not only between developed and developing countries, but also between urban and poorer rural areas within a country (Dandona & Dandona, 2001; Limburg et al., 2008). The relationship between socio-economic status and blindness clearly emerges from these studies. The socio-economic status is determined to different degrees not only by the geographic location but also by income, education, ethnicity, and gender. Factors that affect the epidemiology of visual impairment and blindness include awareness (of the disease condition as well as of treatment options), access to (i.e., availability and affordability) and quality of available health services.

Although clinical evaluation remains the first line of intervention, research in genetics could contribute to efforts for prevention and treatment of visual impairment and blindness. Because early detection is fundamental to efficacious treatment of ocular diseases, identification of at risk populations based on genetic background and presymptomatic (genetic) testing, would guide targeted clinical screens and timely plans for corrective interventions (Fan et al., 2006). Common variants might be tested initially; however, disease-causing mutations or sequence variations may vary in separate geographic and ethnic groups. Therefore, standardized protocols and data analyses will have to be adopted to map populations and determine common mutations/variations, evaluate prevalence, and study patterns of inheritance. Once the predictive value of genetic testing is established, quick, low-cost, and robust assays to detect mutations/variations should be made available, together with access to appropriate preventive and corrective therapies. Because of the technology and costs involved, genetic evaluation of populations in developing countries will rely on the commitment of governments and communities, as well as on international collaborations. The medical and scientific enterprise will need to go in parallel with the promotion of awareness globally, and development of eye healthcare services for wider access to screening, prevention and treatment.

ACKNOWLEDGMENTS

The authors are grateful to Tiansen Li and Yi-Sheng Chang for critical review of the manuscript, and to Lydia Kibiuk, NIH Medical Arts and Printing Services, for medical illustrations.

The authors are supported by the Intramural and Extramural Research Program of the National Eye Institute, National Institutes of Health, Bethesda, MD, USA.

REFERENCES

Abu-Amero KK, Morales J, Mohamed GH, Osman MN, Bosley TM. (2008). Glutathione S-transferase M1 and T1 polymorphisms in Arab glaucoma patients. *Mol Vis* 14:425–430.

Akarsu AN, Turaci ME, Aktan SG, et al. (1996). A second locus (GLC3B) for primary congenital glaucoma (Buphthalmos) maps to the 1p36 region. *Hum Mol Genet* 5(8):1199–1203.

Ali M, McKibbin M, Booth A, et al. (2009). Null mutations in LTBP2 cause primary congenital glaucoma. *Am J Hum Genet* 84(5):664–671.

Allingham RR, Liu Y, Rhee DJ. (2009). The genetics of primary open-angle glaucoma: a review. *Exp Eye Res* 88(4):837–844.

Altintas O, Maral H, Yuksel N, Karabas VL, Dillioglugil MO, Caglar Y. (2005). Homocysteine and nitric oxide levels in plasma of patients with pseudoexfoliation syndrome, pseudoexfoliation glaucoma, and primary open-angle glaucoma. *Graefes Arch Clin Exp Ophthalmol* 243(7):677–683.

Anderson DR. (1981). The development of the trabecular meshwork and its abnormality in primary infantile glaucoma. *Trans Am Ophthalmol Soc* 79:458–485.

Andley UP. (2007). Crystallins in the eye: Function and pathology. *Prog Retin Eye Res* 26(1):78–98.

Andrew T, Maniatis N, Carbonaro F, et al. (2008). Identification and replication of three novel myopia common susceptibility gene loci on chromosome 3q26 using linkage and linkage disequilibrium mapping. *PLoS Genet* 4(10):e1000220.

AREDS. (2000). Risk factors associated with age-related macular degeneration. A case-control study in the age-related eye disease study: Age-Related Eye Disease Study Report Number 3. *Ophthalmology* 107(12):2224–2232.

AREDS. (2001). A randomized, placebo-controlled, clinical trial of high-dose supplementation with vitamins C and E, beta carotene, and zinc for age-related macular degeneration and vision loss: AREDS report no. 8. *Arch Ophthalmol* 119(10):1417–1436.

Aroca-Aguilar JD, Sanchez-Sanchez F, Ghosh S, Coca-Prados M, Escribano J. (2005). Myocilin mutations causing glaucoma inhibit the intracellular endoproteolytic cleavage of myocilin between amino acids Arg226 and Ile227. *J Biol Chem* 280(22):21043–21051.

Aung T, Ebenezer ND, Brice G, et al. (2003). Prevalence of optineurin sequence variants in adult primary open angle glaucoma: implications for diagnostic testing. *J Med Genet* 40(8):e101.

Aung T, Ocaka L, Ebenezer ND, et al. (2002). A major marker for normal tension glaucoma: association with polymorphisms in the OPA1 gene. *Hum Genet* 110(1):52–56.

Aung T, Rezaie T, Okada K, et al. (2005). Clinical features and course of patients with glaucoma with the E50K mutation in the optineurin gene. *Invest Ophthalmol Vis Sci* 46(8):2816–2822.

Baird PN, Schache M, Dirani M. (2010). The Genes in Myopia (GEM) study in understanding the aetiology of refractive errors. *Prog Retin Eye Res* 29(6):520–542.

Biino G, Palmas MA, Corona C, et al. (2005). Ocular refraction: heritability and genome-wide search for eye morphometry traits in an isolated Sardinian population. *Hum Genet* 116(3):152–159.

Bleich S, Roedl J, Von Ahsen N, Schlotzer-Schrehardt U, et al. (2004). Elevated homocysteine levels in aqueous humor of patients with pseudoexfoliation glaucoma. *Am J Ophthalmol* 138(1):162–164.

Bonnel S, Mohand-Said S, Sahel JA. (2003). The aging of the retina. *Exp Gerontol* 38(8):825–831.

Campa C, Harding SP. (2010). Anti-VEGF Compounds in the Treatment of Neovascular Age Related Macular Degeneration. *Curr Drug Targets* 12(2):173–181.

Cazzola M, Bergamaschi G, Tonon L, et al. (1997). Hereditary hyperferritinemia-cataract syndrome: relationship between phenotypes and specific mutations in the iron-responsive element of ferritin light-chain mRNA. *Blood* 90(2):814–821.

Cedrone C, Mancino R, Cerulli A, Cesareo M, Nucci C. (2008). Epidemiology of primary glaucoma: prevalence, incidence, and blinding effects. *Prog Brain Res* 173:3–14.

Chalasani ML, Radha V, Gupta V, Agarwal N, Balasubramanian D, Swarup G. (2007). A glaucoma-associated mutant of optineurin selectively induces death of retinal ganglion cells which is inhibited by antioxidants. *Invest Ophthalmol Vis Sci* 48(4):1607–1614.

Challa P. (2008). Glaucoma genetics. *Int Ophthalmol Clin* 48(4):73–94.

Chandler PA, Grant WM. (1965). *Lectures on glaucoma [by] Paul A. Chandler [and] W. Morton Grant*, Philadelphia: Lea & Febiger.

Chen W, Stambolian D, Edwards AO, et al. (2010). Genetic variants near TIMP3 and high-density lipoprotein-associated loci influence susceptibility to age-related macular degeneration. *Proc Natl Acad Sci U S A* 107(16):7401–7406.

Ciner E, Ibay G, Wojciechowski R, et al. (2009). Genome-wide scan of African-American and white families for linkage to myopia. *Am J Ophthalmol* 147(3):512–517, e512.

Ciner E, Wojciechowski R, Ibay G, Bailey-Wilson JE, Stambolian D. (2008). Genomewide scan of ocular refraction in African-American families shows significant linkage to chromosome 7p15. *Genet Epidemiol* 32(5):454–463.

Congdon N, O'Colmain B, Klaver CC, et al. (2004). Causes and prevalence of visual impairment among adults in the United States. *Arch Ophthalmol* 122(4):477–485.

Copin B, Brezin AP, Valtot F, Dascotte JC, Bechetoille A, Garchon HJ. (2002). Apolipoprotein E-promoter single-nucleotide polymorphisms affect the phenotype of primary open-angle glaucoma and demonstrate interaction with the myocilin gene. *Am J Hum Genet* 70(6):1575–1581.

Crabb JW, Miyagi M, Gu X, et al. (2002). Drusen proteome analysis: an approach to the etiology of age-related macular degeneration. *Proc Natl Acad Sci U S A* 99(23):14682–14687.

Dandona R, Dandona L. (2001). Socioeconomic status and blindness. *Br J Ophthalmol* 85(12):1484–1488.

De Marco N, Buono M, Troise F, Diez-Roux G. (2006). Optineurin increases cell survival and translocates to the nucleus in a Rab8-dependent manner upon an apoptotic stimulus. *J Biol Chem* 281(23):16147–16156.

Delaye M, Tardieu A. (1983). Short-range order of crystallin proteins accounts for eye lens transparency. *Nature* 302(5907):415–417.

deLuise VP, Anderson DR. (1983). Primary infantile glaucoma (congenital glaucoma). *Surv Ophthalmol* 28(1):1–19.

Dewan A, Liu M, Hartman S, et al. (2006). HTRA1 promoter polymorphism in wet age-related macular degeneration. *Science* 314(5801):989–992.

Dimasi DP, Chen JY, Hewitt AW, et al. (2010). Novel quantitative trait loci for central corneal thickness identified by candidate gene analysis of osteogenesis imperfecta genes. *Hum Genet* 127(1):33–44.

Duggal P, Klein AP, Lee KE, Klein R, Klein BE, Bailey-Wilson JE. (2007). Identification of novel genetic loci for intraocular pressure: a genomewide scan of the Beaver Dam Eye Study. *Arch Ophthalmol* 125(1):74–79.

Edwards AO, Ritter R, 3rd, Abel KJ, Manning A, Panhuysen C, Farrer LA. (2005). Complement factor H polymorphism and age-related macular degeneration. *Science* 308(5720):421–424.

Fagerness JA, Maller JB, Neale BM, Reynolds RC, Daly MJ, Seddon JM. (2009). Variation near complement factor I is associated with risk of advanced AMD. *Eur J Hum Genet* 17(1):100–104.

Fan BJ, Tam PO, Choy KW, Wang DY, Lam DS, Pang CP. (2006). Molecular diagnostics of genetic eye diseases. *Clin Biochem* 39(3):231–239.

Fan BJ, Wiggs JL. (2010). Glaucoma: genes, phenotypes, and new directions for therapy. *J Clin Invest* 120(9):3064–3072.

Firasat S, Riazuddin SA, Hejtmancik JF, Riazuddin S. (2008). Primary congenital glaucoma localizes to chromosome 14q24.2-24.3 in two consanguineous Pakistani families. *Mol Vis* 14:1659–1665.

Fisher SA, Abecasis GR, Yashar BM, et al. (2005). Meta-analysis of genome scans of age-related macular degeneration. *Hum Mol Genet* 14(15):2257–2264.

Flood V, Smith W, Wang JJ, Manzi F, Webb K, Mitchell P. (2002). Dietary antioxidant intake and incidence of early age-related maculopathy: the Blue Mountains Eye Study. *Ophthalmology* 109(12):2272–2278.

Forsius H. (1988.) Exfoliation syndrome in various ethnic populations. *Acta Ophthalmol Suppl* 184:71–85.

Forsman E, Lemmela S, Varilo T, et al. (2003). The role of TIGR and OPTN in Finnish glaucoma families: a clinical and molecular genetic study. *Mol Vis* 9:217–222.

Francois J. (1982). Genetics of cataract. *Ophthalmologica* 184(2):61–71.

Fujiwara N, Matsuo T, Ohtsuki H. (2003). Protein expression, genomic structure, and polymorphisms of oculomedin. *Ophthalmic Genet* 24(3):141–151.

Funayama T, Ishikawa K, Ohtake Y, et al. (2004). Variants in optineurin gene and their association with tumor necrosis factor-alpha polymorphisms in Japanese patients with glaucoma. *Invest Ophthalmol Vis Sci* 45(12):4359–4367.

Funayama T, Mashima Y, Ohtake Y, et al. (2006). SNPs and interaction analyses of noelin 2, myocilin, and optineurin genes in Japanese patients with open-angle glaucoma. *Invest Ophthalmol Vis Sci* 47(12):5368–5375.

Gobeil S, Letartre L, Raymond V. (2006). Functional analysis of the glaucoma-causing TIGR/myocilin protein: integrity of amino-terminal coiled-coil regions and olfactomedin homology domain is essential for extracellular adhesion and secretion. *Exp Eye Res* 82(6):1017–1029.

Gold B, Merriam JE, Zernant J, et al. (2006). Variation in factor B (BF) and complement component 2 (C2) genes is associated with age-related macular degeneration. *Nat Genet* 38(4):458–462.

Goodenough DA. (1992). The crystalline lens. A system networked by gap junctional intercellular communication. *Semin Cell Biol* 3(1):49–58.

Graw J. (2009). Genetics of crystallins: cataract and beyond. *Exp Eye Res* 88(2):173–189.

Green WR. (1999). Histopathology of age-related macular degeneration. *Mol Vis* 5:27.

Grossniklaus HE, Green WR. (2004). Choroidal neovascularization. *Am J Ophthalmol* 137(3):496–503.

Grosvenor T. (1987). A review and a suggested classification system for myopia on the basis of age-related prevalence and age of onset. *Am J Optom Physiol Opt* 64(7):545–554.

Gwiazda J. (2009). Treatment options for myopia. *Optom Vis Sci* 86(6):624–628.

Haargaard B, Wohlfahrt J, Fledelius HC, Rosenberg T, Melbye M. (2004). Incidence and cumulative risk of childhood cataract in a cohort of 2.6 million Danish children. *Invest Ophthalmol Vis Sci* 45(5):1316–1320.

Hageman GS, Anderson DH, Johnson LV, et al. (2005). A common haplotype in the complement regulatory gene factor H (HF1/CFH) predisposes individuals to age-related macular degeneration. *Proc Natl Acad Sci U S A* 102(20):7227–7232.

Hageman GS, Luthert PJ, Victor Chong NH, Johnson LV, Anderson DH, Mullins RF. (2001). An integrated hypothesis that considers drusen as biomarkers of immune-mediated processes at the RPE-Bruch's membrane interface in aging and age-related macular degeneration. *Prog Retin Eye Res* 20(6):705–732.

Haines JL, Hauser MA, Schmidt S, et al. (2005). Complement factor H variant increases the risk of age-related macular degeneration. *Science* 308(5720):419–421.

Hammond CJ, Duncan DD, Snieder H, et al. (2001). The heritability of age-related cortical cataract: the twin eye study. *Invest Ophthalmol Vis Sci* 42(3):601–605.

Hashizume K, Mashima Y, Fumayama T, et al. (2005). Genetic polymorphisms in the angiotensin II receptor gene and their association with open-angle glaucoma in a Japanese population. *Invest Ophthalmol Vis Sci* 46(6):1993–2001.

Hauser MA, Allingham RR, Linkroum K, et al. (2006). Distribution of WDR36 DNA sequence variants in patients with primary open-angle glaucoma. *Invest Ophthalmol Vis Sci* 47(6):2542–2546.

Hayashi H, Gotoh N, Ueda Y, Nakanishi H, Yoshimura N. (2008). Lysyl oxidase-like 1 polymorphisms and exfoliation syndrome in the Japanese population. *Am J Ophthalmol* 145(3):582–585.

Hejtmancik JF. (2008). Congenital cataracts and their molecular genetics. *Semin Cell Dev Biol* 19(2):134–149.

Hornbeak DM, Young TL. (2009). Myopia genetics: a review of current research and emerging trends. *Curr Opin Ophthalmol* 20(5):356–362.

Horwitz J. (1992). Alpha-crystallin can function as a molecular chaperone. *Proc Natl Acad Sci U S A* 89(21):10449–10453.

Hughes AE, Orr N, Esfandiary H, Diaz-Torres M, Goodship T, Chakravarthy U. (2006). A common CFH haplotype, with deletion of CFHR1 and CFHR3, is associated with lower risk of age-related macular degeneration. *Nat Genet* 38(10):1173–1177.

Hysi PG, Young TL, Mackey DA, et al. (2010). A genome-wide association study for myopia and refractive error identifies a susceptibility locus at 15q25. *Nat Genet* 42(10):902–905.

Inagaki N, Hayashi T, Arimura T, et al. (2006a). Alpha B-crystallin mutation in dilated cardiomyopathy. *Biochem Biophys Res Commun* 342(2):379–386.

Inagaki Y, Mashima Y, Funayama T, et al. (2006b). Paraoxonase 1 gene polymorphisms influence clinical features of open-angle glaucoma. *Graefes Arch Clin Exp Ophthalmol* 244(8):984–990.

Ishikawa K, Funayama T, Ohtake Y, et al. (2005). Association between glaucoma and gene polymorphism of endothelin type A receptor. *Mol Vis* 11:431–437.

Jackson GR, Owsley C, Curcio CA. (2002). Photoreceptor degeneration and dysfunction in aging and age-related maculopathy. *Ageing Res Rev* 1(3):381–396.

Joe MK, Sohn S, Choi YR, Park H, Kee C. (2005). Identification of flotillin-1 as a protein interacting with myocilin: implications for the pathogenesis of primary open-angle glaucoma. *Biochem Biophys Res Commun* 336(4):1201–1206.

Junemann AG, von Ahsen N, Reulbach U, et al. (2005). C677T variant in the methylentetrahydrofolate reductase gene is a genetic risk factor for primary open-angle glaucoma. *Am J Ophthalmol* 139(4):721–723.

Juronen E, Tasa G, Veromann S, et al. (2000). Polymorphic glutathione S-transferase M1 is a risk factor of primary open-angle glaucoma among Estonians. *Exp Eye Res* 71(5):447–452.

Kanda A, Chen W, Othman M, Branham KE, Brooks M, Khanna R, He S, Lyons R, Abecasis GR, Swaroop A (2007). A variant of mitochondrial protein LOC387715/ARMS2, not HTRA1, is strongly associated with age-related macular degeneration. *Proc Natl Acad Sci U S A* 104:16227–16232.

Khan JC, Thurlby DA, Shahid H, et al. (2006). Smoking and age related macular degeneration: the number of pack years of cigarette smoking is a major determinant of risk for both geographic atrophy and choroidal neovascularisation. *Br J Ophthalmol* 90(1):75–80.

Kim BS, Savinova OV, Reedy MV, et al. (2001) Targeted disruption of the myocilin gene (Myoc) suggests that human glaucoma-causing mutations are gain of function. *Mol Cell Biol* 21(22):7707–7713.

Kim SY, Sadda S, Pearlman J, et al. (2002). Morphometric analysis of the macula in eyes with disciform age-related macular degeneration. *Retina* 22(4):471–477.

Klein R, Klein BE, Cruickshanks KJ. (1999). The prevalence of age-related maculopathy by geographic region and ethnicity. *Prog Retin Eye Res* 18(3):371–389.

Klein R, Klein BE, Knudtson MD, et al. (2006). Prevalence of age-related macular degeneration in 4 racial/ethnic groups in the multi-ethnic study of atherosclerosis. *Ophthalmology* 113(3):373–380.

Klein RJ, Zeiss C, Chew EY, et al. (2005). Complement factor H polymorphism in age-related macular degeneration. *Science* 308(5720):385–389.

Klintworth GK. (2009) Corneal dystrophies. *Orphanet J Rare Dis* 4:7.

Kramer PL, Samples JR, Monemi S, Sykes R, Sarfarazi M, Wirtz MK. (2006). The role of the WDR36 gene on chromosome 5q22.1 in a large family with primary open-angle glaucoma mapped to this region. *Arch Ophthalmol* 124(9):1328–1331.

Kuszak JR, Zoltoski RK, Sivertson C. (2004). Fibre cell organization in crystalline lenses. *Exp Eye Res* 78(3):673–687.

Lam CY, Fan BJ, Wang DY, et al. (2006). Association of apolipoprotein E polymorphisms with normal tension glaucoma in a Chinese population. *J Glaucoma* 15(3):218–222.

Lam DS, Leung YF, Chua JK, Baum L, Fan DS, Choy KW, Pang CP. (2000) Truncations in the TIGR gene in individuals with and without primary open-angle glaucoma. *Invest Ophthalmol Vis Sci* 41(6):1386–1391.

Leibovitch I, Kurtz S, Shemesh G, et al. (2003). Hyperhomocystinemia in pseudoexfoliation glaucoma. *J Glaucoma* 12(1):36–39.

Leske MC. (2007). Open-angle glaucoma—an epidemiologic overview. *Ophthalmic Epidemiol* 14(4):166–172.

Leske MC, Connell AM, Schachat AP, Hyman L. (1994). The Barbados Eye Study. Prevalence of open angle glaucoma. *Arch Ophthalmol* 112(6):821–829.

Li M, Atmaca-Sonmez P, Othman M, et al. (2006). CFH haplotypes without the Y402H coding variant show strong association with susceptibility to age-related macular degeneration. *Nat Genet* 38(9):1049–1054.

Li Y, Kang J, Horwitz MS. (1998). Interaction of an adenovirus E3 14.7-kilodalton protein with a novel tumor necrosis factor alpha-inducible cellular protein containing leucine zipper domains. *Mol Cell Biol* 18(3):1601–1610.

Li YJ, Guggenheim JA, Bulusu A, et al. (2009). An international collaborative family-based whole-genome linkage scan for high-grade myopia. *Invest Ophthalmol Vis Sci* 50(7):3116–3127.

Limburg H, Barria von-Bischhoffshausen F, Gomez P, Silva JC, Foster A. (2008) Review of recent surveys on blindness and visual impairment in Latin America. *Br J Ophthalmol* 92(3):315–319.

Lin HJ, Chen WC, Tsai FJ, Tsai SW. (2002). Distributions of p53 codon 72 polymorphism in primary open angle glaucoma. *Br J Ophthalmol* 86(7):767–770.

Lin HJ, Tsai CH, Tsai FJ, Chen WC, Chen HY, Fan SS. (2004). Transporter associated with antigen processing gene 1 codon 333 and codon 637 polymorphisms are associated with primary open-angle glaucoma. *Mol Diagn* 8(4):245–252.

Lin HJ, Tsai FJ, Chen WC, Shi YR, Hsu Y, Tsai SW. (2003a). Association of tumour necrosis factor alpha-308 gene polymorphism with primary open-angle glaucoma in Chinese. *Eye (Lond)* 17(1):31–34.

Lin HJ, Tsai FJ, Hung P, et al. (2006). Association of E-cadherin gene 3'-UTR C/T polymorphism with primary open angle glaucoma. *Ophthalmic Res* 38(1):44–48.

Lin HJ, Tsai SC, Tsai FJ, Chen WC, Tsai JJ, Hsu CD. (2003b). Association of interleukin 1beta and receptor antagonist gene polymorphisms with primary open-angle glaucoma. *Ophthalmologica* 217(5):358–364.

Liu Y, Liu W, Crooks K, Schmidt S, Allingham RR, Hauser MA. (2010). No evidence of association of heterozygous NTF4 mutations in patients with primary open-angle glaucoma. *Am J Hum Genet* 86(3):498–499; author reply 500.

Logan JF, Chakravarthy U, Hughes AE, Patterson CC, Jackson JA, Rankin SJ. (2005). Evidence for association of endothelial nitric oxide synthase gene in subjects with glaucoma and a history of migraine. *Invest Ophthalmol Vis Sci* 46(9):3221–3226.

Lutty G, Grunwald J, Majji AB, Uyama M, Yoneya S. (1999). Changes in choriocapillaris and retinal pigment epithelium in age-related macular degeneration. *Mol Vis* 5:35.

Mabuchi F, Tang S, Ando D, et al. (2005). The apolipoprotein E gene polymorphism is associated with open angle glaucoma in the Japanese population. *Mol Vis* 11:609–612.

Mabuchi F, Tang S, Kashiwagi K, Yamagata Z, Iijima H, Tsukahara S. (2007). The OPA1 gene polymorphism is associated with normal tension and high tension glaucoma. *Am J Ophthalmol* 143(1):125–130.

MacDonald IM, Tran M, Musarella MA. (2004). Ocular genetics: current understanding. *Surv Ophthalmol* 49(2):159–196.

Macgregor S, Hewitt AW, Hysi PG, et al. (2010). Genome-wide association identifies ATOH7 as a major gene determining human optic disc size. *Hum Mol Genet* 19(13):2716–2724.

Malek G, Li CM, Guidry C, Medeiros NE, Curcio CA. (2003). Apolipoprotein B in cholesterol-containing drusen and basal deposits of human eyes with age-related maculopathy. *Am J Pathol* 162(2):413–425.

Maller J, George S, Purcell S, et al. (2006). Common variation in three genes, including a noncoding variant in CFH, strongly influences risk of age-related macular degeneration. *Nat Genet* 38(9):1055–1059.

Manolio TA. (2009). Cohort studies and the genetics of complex disease. *Nat Genet* 41(1):5–6.

Manolio TA, Collins FS, Cox NJ, et al. (2009). Finding the missing heritability of complex diseases. *Nature* 461(7265):747–753.

Mao M, Biery MC, Kobayashi SV, et al. (2004). T lymphocyte activation gene identification by coregulated expression on DNA microarrays. *Genomics* 83(6):989–999.

Mason RP, Kosoko O, Wilson MR, et al. (1989). National survey of the prevalence and risk factors of glaucoma in St. Lucia, West Indies. Part I. Prevalence findings. *Ophthalmology* 96(9):1363–1368.

McCarty CA, Mukesh BN, Fu CL, Mitchell P, Wang JJ, Taylor HR. (2001). Risk factors for age-related maculopathy: the Visual Impairment Project. *Arch Ophthalmol* 119(10):1455–1462.

Melki R, Belmouden A, Akhayat O, Brezin A, Garchon HJ. (2003). The M98K variant of the OPTINEURIN (OPTN) gene modifies initial intraocular pressure in patients with primary open angle glaucoma. *J Med Genet* 40(11):842–844.

Metlapally R, Michaelides M, Bulusu A, et al. (2009). Evaluation of the X-linked high-grade myopia locus (MYP1) with cone dysfunction and color vision deficiencies. *Invest Ophthalmol Vis Sci* 50(4):1552–1558.

Monemi S, Spaeth G, DaSilva A, et al. (2005). Identification of a novel adult-onset primary open-angle glaucoma (POAG) gene on 5q22.1. *Hum Mol Genet* 14(6):725–733.

Montero JA, Ruiz-Moreno JM. (2010). Treatment of choroidal neovascularization in high myopia. *Curr Drug Targets* 11(5):630–644.

Morner CT. (1893). Untersuchungen der Proteinsubstanzen in den lichtbrechenden Medien des Auges. *Z Physiol Chem* 18:61–106.

Mullins RF, Russell SR, Anderson DH, Hageman GS. (2000). Drusen associated with aging and age-related macular degeneration contain proteins common to extracellular deposits associated with atherosclerosis, elastosis, amyloidosis, and dense deposit disease. *FASEB J* 14(7):835–846.

Murdoch IE, Cousens SN, Babalola OE, Yang YF, Abiose A, Jones BR. (2001). Glaucoma prevalence may not be uniformly high in all 'black' populations. *Afr J Med Med Sci* 30(4):337–339.

Murray CJL, Lopez AD, Mathers CD, Stein C. (2001). The Global Burden of Disease 2000 Project: aims, methods and data sources. *Global Programme on Evidence for Health Policy Discussion.* Geneva: World Health Organization.

Narooie-Nejad M, Chitsazian F, Khoramian Tusi B, et al. (2009a). Genotyping results of Iranian PCG families suggests one or more PCG locus other than GCL3A, GCL3B, and GCL3C exist. *Mol Vis* 15:2155–2161.

Narooie-Nejad M, Paylakhi SH, Shojaee S, et al. (2009b). Loss of function mutations in the gene encoding latent transforming growth factor beta binding protein 2, LTBP2, cause primary congenital glaucoma. *Hum Mol Genet* 18(20):3969–3977.

Nebert DW. (1991). Proposed role of drug-metabolizing enzymes: regulation of steady state levels of the ligands that effect growth, homeostasis, differentiation, and neuroendocrine functions. *Mol Endocrinol* 5(9):1203–1214.

Ng SB, Nickerson DA, Bamshad MJ, Shendure J. (2010). Massively parallel sequencing and rare disease. *Hum Mol Genet* 19(R2):R119–R124.

Nishizaki R, Ota M, Inoko H, et al. (2009). New susceptibility locus for high myopia is linked to the uromodulin-like 1 (UMODL1) gene region on chromosome 21q22.3. *Eye (Lond)* 23(1):222–229.

Ntim-Amponsah CT, Amoaku WM, Ofosu-Amaah S, et al. (2004). Prevalence of glaucoma in an African population. *Eye (Lond)* 18(5):491–497.

Pasutto F, Matsumoto T, Mardin CY, et al. (2009). Heterozygous NTF4 mutations impairing neurotrophin-4 signaling in patients with primary open-angle glaucoma. *Am J Hum Genet* 85(4):447–456.

Poulaki V, Colby K. (2008). Genetics of anterior and stromal corneal dystrophies. *Semin Ophthalmol* 23(1):9–17.

Praveen MR, Vasavada AR, Jani UD, Trivedi RH, Choudhary PK. (2008). Prevalence of cataract type in relation to axial length in subjects with high myopia and emmetropia in an Indian population. *Am J Ophthalmol* 145(1):176–181.

Quax-Jeuken Y, Quax W, van Rens G, Khan PM, Bloemendal H. (1985). Complete structure of the alpha B-crystallin gene: conservation of the exon-intron distribution in the two nonlinked alpha-crystallin genes. *Proc Natl Acad Sci U S A* 82(17):5819–5823.

Quax-Jeuken Y, Quax, W, van Rens G, Meera Khan P, Bloemendal H. (1985). Assignment of the human alpha-A-crystallin gene (CRYA1) to chromosome 21. *Cytogenet Cell Genet* 40:727–728.

Quigley HA, Broman AT. (2006). The number of people with glaucoma worldwide in 2010 and 2020. *Br J Ophthalmol* 90(3):262–267.

Ramdas WD, van Koolwijk LM, Ikram MK, et al. (2010). A genome-wide association study of optic disc parameters. *PLoS Genet* 6(6):e1000978.

Ramprasad VL, George R, Soumittra N, Sharmila F, Vijaya L, Kumaramanickavel G. (2008). Association of non-synonymous single nucleotide polymorphisms in the LOXL1 gene with pseudo-exfoliation syndrome in India. *Mol Vis* 14:318–322.

Raychaudhuri S, Ripke S, Li M, et al. (2010). Associations of CFHR1-CFHR3 deletion and a CFH SNP to age-related macular degeneration are not independent. *Nat Genet* 42(7):553–555; author reply 555–556.

Resnikoff S, Kocur I, Etya'ale DE, Ukety TO. (2008a). Vision 2020—the right to sight. *Ann Trop Med Parasitol* 102 Suppl 1:3–5.

Resnikoff S, Pascolini D, Etya'ale D, et al. (2004). Global data on visual impairment in the year 2002. *Bull World Health Organ* 82(11):844–851.

Resnikoff S, Pascolini D, Mariotti SP, Pokharel GP. (2008b). Global magnitude of visual impairment caused by uncorrected refractive errors in 2004. *Bull World Health Organ* 86(1):63–70.

Rezaie T, Child A, Hitchings R, et al. (2002). Adult-onset primary open-angle glaucoma caused by mutations in optineurin. *Science* 295(5557):1077–1079.

Richardson TM, Epstein DL. (1981). Exfoliation glaucoma: a quantitative perfusion and ultrastructural study. *Ophthalmology* 88(9):968–980.

Rodieck RW. (1998). *The first steps in seeing*, Sunderland, MA: Sinauer Associates.

Rose KA, Morgan IG, Smith W, Mitchell P. (2002). High heritability of myopia does not preclude rapid changes in prevalence. *Clin Experiment Ophthalmol* 30(3):168–172.

Rudnicka AR, Mt-Isa S, Owen CG, Cook DG, Ashby D. (2006). Variations in primary open-angle glaucoma prevalence by age, gender, and race: a Bayesian meta-analysis. *Invest Ophthalmol Vis Sci* 47(10):4254–4261.

Sahlender DA, Roberts RC, Arden SD, et al. (2005). Optineurin links myosin VI to the Golgi complex and is involved in Golgi organization and exocytosis. *J Cell Biol* 169(2):285–295.

Sarfarazi M, Akarsu AN, Hossain A, et al. (1995). Assignment of a locus (GLC3A) for primary congenital glaucoma (Buphthalmos) to 2p21 and evidence for genetic heterogeneity. *Genomics* 30(2):171–177.

Sarks JP, Sarks SH, Killingsworth MC. (1997). Morphology of early choroidal neovascularisation in age-related macular degeneration: correlation with activity. *Eye (Lond)* 11 (Pt 4):515–522.

Saw SM, Gazzard G, Au Eong KG, Tan DT. (2002). Myopia: attempts to arrest progression. *Br J Ophthalmol* 86(11):1306–1311.

Schlotzer-Schrehardt U, Naumann GO. (1995). Trabecular meshwork in pseudoexfoliation syndrome with and without open-angle glaucoma. A morphometric, ultrastructural study. *Invest Ophthalmol Vis Sci* 36(9):1750–1764.

Schmidt-Erfurth U, Pollreisz A, Mitsch C, Bolz M. (2010). Antivascular endothelial growth factors in age-related macular degeneration. *Dev Ophthalmol* 46:21–38.

Seddon JM, Cote J, Rosner B. (2003). Progression of age-related macular degeneration: association with dietary fat, transunsaturated fat, nuts, and fish intake. *Arch Ophthalmol* 121(12):1728–1737.

Sherwin JC, Hewitt AW, Bennett SL, Baird PN, Craig JE, Mackey DA. (2009). Primary open angle glaucoma in subjects harbouring the predicted GLC1L haplotype reveals a normotensive phenotype. *Clin Experiment Ophthalmol* 37(2):201–207.

Shibuya E, Meguro A, Ota M, et al. (2008). Association of toll-like receptor 4 gene polymorphisms with normal tension glaucoma. *Invest Ophthalmol Vis Sci* 49(10):4453–4457.

Shiels A, Hejtmancik JF. (2007). Genetic origins of cataract. *Arch Ophthalmol* 125(2):165–173.

Shiloh Y, Donlon T, Bruns G, Breitman ML, Tsui LC. (1986). Assignment of the human gamma-crystallin gene cluster (CRYG) to the long arm of chromosome 2, region q33–36. *Hum Genet* 73(1):17–19.

Sieving PA, Collins FS. (2007). Genetic ophthalmology and the era of clinical care. *JAMA* 297(7):733–736.

Skarie JM, Link BA. (2008). The primary open-angle glaucoma gene WDR36 functions in ribosomal RNA processing and interacts with the p53 stress-response pathway. *Hum Mol Genet* 17(16):2474–2485.

Smith W, Assink J, Klein R, et al. (2001). Risk factors for age-related macular degeneration: Pooled findings from three continents. *Ophthalmology* 108(4):697–704.

Solouki AM, Verhoeven VJ, van Duijn CM, et al. (2010). A genome-wide association study identifies a susceptibility locus for refractive errors and myopia at 15q14. *Nat Genet* 42(10):897–901.

Spencer KL, Hauser MA, Olson LM, et al. (2008). Deletion of CFHR3 and CFHR1 genes in age-related macular degeneration. *Hum Mol Genet* 17(7):971–977.

Stoilov I, Akarsu AN, Sarfarazi M. (1997). Identification of three different truncating mutations in cytochrome P4501B1 (CYP1B1) as the principal cause of primary congenital glaucoma (Buphthalmos) in families linked to the GLC3A locus on chromosome 2p21. *Hum Mol Genet* 6(4):641–647.

Stone EM, Fingert JH, Alward WL, et al. (1997). Identification of a gene that causes primary open angle glaucoma. *Science* 275(5300):668–670.

Sundin OH, Leppert GS, Silva ED, et al. (2005). Extreme hyperopia is the result of null mutations in MFRP, which encodes a Frizzled-related protein. *Proc Natl Acad Sci U S A* 102(27):9553–9558.

Swaroop A, Chew EY, Rickman CB, Abecasis GR. (2009). Unraveling a multifactorial late-onset disease: from genetic susceptibility to disease mechanisms for age-related macular degeneration. *Annu Rev Genomics Hum Genet* 10:19–43.

Tang WC, Yip SP, Lo KK, et al. (2007) Linkage and association of myocilin (MYOC) polymorphisms with high myopia in a Chinese population. *Mol Vis* 13:534–544.

Tarkkanen AH. (1986). Exfoliation syndrome. *Trans Ophthalmol Soc U K* 105 (Pt 2):233–236.

The CAAT Research Group (2011). Ranibizumab and bevacizumab for neovascular age-related macular degeneration. *N Engl J Med* 364:1897–1908.

Thorleifsson G, Magnusson KP, Sulem P, et al. (2007). Common sequence variants in the LOXL1 gene confer susceptibility to exfoliation glaucoma. *Science* 317(5843):1397–1400.

Tielsch JM, Katz J, Singh K, et al. (1991). A population-based evaluation of glaucoma screening: the Baltimore Eye Survey. *Am J Epidemiol* 134(10):1102–1110.

Tielsch JM, Katz J, Sommer A, Quigley HA, Javitt JC. (1994). Family history and risk of primary open angle glaucoma. The Baltimore Eye Survey. *Arch Ophthalmol* 112(1):69–73.

Tomany SC, Wang JJ, Van Leeuwen R, et al. (2004). Risk factors for incident age-related macular degeneration: pooled findings from 3 continents. *Ophthalmology* 111(7):1280–1287.

Torrado M, Trivedi R, Zinovieva R, Karavanova I, Tomarev SI. (2002). Optimedin: a novel olfactomedin-related protein that interacts with myocilin. *Hum Mol Genet* 11(11):1291–1301.

Tsai FJ, Lin HJ, Chen WC, Chen HY, Fan SS. (2003). Insulin-like growth factor-II gene polymorphism is associated with primary open angle glaucoma. *J Clin Lab Anal* 17(6):259–263.

Tsai FJ, Lin HJ, Chen WC, Tsai CH, Tsai SW. (2004). A codon 31ser-arg polymorphism of the WAF-1/CIP-1/p21/tumour suppressor gene in Chinese primary open-angle glaucoma. *Acta Ophthalmol Scand* 82(1):76–80.

Tufail A, Patel PJ, Egan C, et al. (2010). Bevacizumab for neovascular age related macular degeneration (ABC Trial): multicentre randomised double masked study. *BMJ* 340:c2459.

Tunny TJ, Richardson KA, Clark CV. (1998) Association study of the 5' flanking regions of endothelial-nitric oxide synthase and endothelin-1 genes in familial primary open-angle glaucoma. *Clin Exp Pharmacol Physiol* 25(1):26–29.

Tunny TJ, Richardson KA, Clark CV, Gordon RD. (1996). The atrial natriuretic peptide gene in patients with familial primary open-angle glaucoma. *Biochem Biophys Res Commun* 223(2):221–225.

Unal M, Guven M, Devranoglu K, et al. (2007). Glutathione S transferase M1 and T1 genetic polymorphisms are related to the risk of primary open-angle glaucoma: a study in a Turkish population. *Br J Ophthalmol* 91(4):527–530.

van Koolwijk LM, Despriet DD, van Duijn CM, et al. (2007). Genetic contributions to glaucoma: heritability of intraocular pressure, retinal nerve fiber layer thickness, and optic disc morphology. *Invest Ophthalmol Vis Sci* 48(8):3669–3676.

Vatavuk Z, Skunca Herman J, Bencic G, et al. (2009). Common variant in myocilin gene is associated with high myopia in isolated population of Korcula Island, Croatia. *Croat Med J* 50(1):17–22.

Vicart P, Caron A, Guicheney P, et al. (1998). A missense mutation in the alphaB-crystallin chaperone gene causes a desmin-related myopathy. *Nat Genet* 20(1):92–95.

Vickers JC, Craig JE, Stankovich J, et al. (2002). The apolipoprotein epsilon4 gene is associated with elevated risk of normal tension glaucoma. *Mol Vis* 8:389–393.

Vingerling JR, Hofman A, Grobbee DE, de Jong PT. (1996). Age-related macular degeneration and smoking. The Rotterdam Study. *Arch Ophthalmol* 114(10):1193–1196.

Vittitow J, Borras T. (2002). Expression of optineurin, a glaucoma-linked gene, is influenced by elevated intraocular pressure. *Biochem Biophys Res Commun* 298(1):67–74.

Wang CY, Shen YC, Lo FY, et al. (2006). Polymorphism in the IL-1alpha (-889) locus associated with elevated risk of primary open angle glaucoma. *Mol Vis* 12:1380–1385.

Wang NK, Chen YP, Lai CC, et al. (2009). Paediatric retinal detachment: comparison of high myopia and extreme myopia. *Br J Ophthalmol* 93(5):650–655.

Weih LM, Nanjan M, McCarty CA, Taylor HR. (2001). Prevalence and predictors of open-angle glaucoma: results from the visual impairment project. *Ophthalmology* 108(11):1966–1972.

Wiggs JL. (2007). Genetic etiologies of glaucoma. *Arch Ophthalmol* 125(1):30–37.

Wiggs JL, Vollrath D. (2001). Molecular and clinical evaluation of a patient hemizygous for TIGR/MYOC. *Arch Ophthalmol* 119(11):1674–1678.

Wojciechowski R, Stambolian D, Ciner E, Ibay G, Holmes TN, Bailey-Wilson JE. (2009). Genomewide linkage scans for ocular refraction and meta-analysis of four populations in the Myopia Family Study. *Invest Ophthalmol Vis Sci* 50(5):2024–2032.

Wolfs RC, Klaver CC, Ramrattan RS, van Duijn CM, Hofman A, de Jong PT. (1998). Genetic risk of primary open-angle glaucoma. Population-based familial aggregation study. *Arch Ophthalmol* 116(12):1640–1645.

Woo SJ, Kim JY, Kim DM, Park SS, Ko HS, Yoo T. (2009). Investigation of the association between 677C>T and 1298A>C 5,10-methylenetetra- hydrofolate reductase gene polymorphisms and normal-tension glaucoma. *Eye (Lond)* 23(1):17–24.

Xu L, Wang Y, Wang S, Jonas JB. (2007). High myopia and glaucoma susceptibility the Beijing Eye Study. *Ophthalmology* 114(2):216–220.

Yamamoto T, Iwase A, Araie M, et al. (2005). The Tajimi Study report 2: prevalence of primary angle closure and secondary glaucoma in a Japanese population. *Ophthalmology* 112(10):1661–1669.

Yang Z, Xiao X, Li S, Zhang Q. (2009). Clinical and linkage study on a consanguineous Chinese family with autosomal recessive high myopia. *Mol Vis* 15:312–318.

Yates JR, Sepp T, Matharu BK, et al. (2007). Complement C3 variant and the risk of age-related macular degeneration. *N Engl J Med* 357(6):553–561.

Young TL, Metlapally R, Shay AE. (2007). Complex trait genetics of refractive error. *Arch Ophthalmol* 125(1):38–48.

Yu ZQ, Li YB, Huang CX, et al. (2007). [A genome-wide screening for pathological myopia suggests a novel locus on chromosome 15q12—13]. *Zhonghua Yan Ke Za Zhi* 43(3):233–238.

Zareparsi S, Branham KE, Li M, et al. (2005). Strong association of the Y402H variant in complement factor H at 1q32 with susceptibility to age-related macular degeneration. *Am J Hum Genet* 77:149–153.

Zayats T, Yanovitch T, Creer RC, et al. (2009). Myocilin polymorphisms and high myopia in subjects of European origin. *Mol Vis* 15:213–222.

Zhou J, Jang YP, Kim SR, Sparrow JR. (2006). Complement activation by photooxidation products of A2E, a lipofuscin constituent of the retinal pigment epithelium. *Proc Natl Acad Sci U S A* 103(44):16182–16187.

Zhu G, Hewitt AW, Ruddle JB, et al. (2008). Genetic dissection of myopia: evidence for linkage of ocular axial length to chromosome 5q. *Ophthalmology* 115(6):1053–1057 e1052.

Zhu G, Wu CJ, Zhao Y, Ashwell JD. (2007). Optineurin negatively regulates TNFalpha-induced NF-kappaB activation by competing with NEMO for ubiquitinated RIP. *Curr Biol* 17(16):1438–1443.

28.

PRINCIPLES AND PRACTICE OF NEWBORN SCREENING: THE LESSONS FROM THE DEVELOPED WORLD

Rodney J. Pollitt

INTRODUCTION

Newborn screening for inherited metabolic disease is founded on the pioneering work of Robert Guthrie in the early 1960s. Driven partly by tragic experiences in his own family, Guthrie developed a practicable method of screening newborn babies for phenylketonuria. If left untreated the condition leads to severe mental retardation, and to be fully effective treatment must begin in the first weeks of life. Guthrie's approach was to use dried blood samples collected on a filter paper card in the newborn nursery, and a bacterial growth assay for measuring phenylalanine. Though the assay itself has largely been abandoned, blood collection using the "Guthrie card" remains the mainstay of screening programs worldwide.

THE EXPANDING RANGE OF NEWBORN SCREENING

In the decade following Guthrie's early demonstration project (Guthrie & Susi, 1963) many countries in the developed world instituted screening programs for phenylketonuria. This was driven partly by humanitarian concern, but economic considerations also came into play. Though relatively rare, incidence between 1 in 10,000 and 1 in 20,000 births in most European-derived populations, untreated phenylketonuria poses a significant economic burden on affected families and, because patients live for many years, on society at large. There have been several economic analyses, all showing a cost/benefit ratio of less than 1 but with absolute costs and cost/benefit ratios showing very striking overall variations by factors of >10 (Lord et al., 1999). The outcome of this type of analysis is greatly influenced by the incidence of the disease, and the medical and social background within which the screening program is set.

Once the sample collection and screening laboratory infrastructures had been set up, it was relatively easy to extend the range of tests performed. By 1980 Guthrie was able to list twenty conditions for which blood-spot screening tests had been developed in his laboratory (Guthrie, 1980; listed in Pollitt, 2009). Very few of the conditions on Guthrie's list have been widely screened for. Most are generally considered to be too rare to justify standalone screening, galactosemia and the sickle cell anemias proving the main exceptions. For other conditions early detection offers little clear benefit. One such, histidinemia, was originally thought to cause mental retardation but turned out to be benign (Lam et al., 1996), although some centers continued to screen for it for many years.

Developments in immunoassay technology during the 1970s led to the introduction of screening for congenital hypothyroidism. Though some affected babies show clear clinical signs at birth, the majority remain undetected for some time and this usually results in significant intellectual impairment. The disorder is common worldwide (screening incidence about 1 in 3000 births), survival into adulthood is usual and, as with phenylketonuria, the condition imposes a significant economic burden. Initially screening was based on radio-immunoassay of thyroxine, but once a sufficiently sensitive assay for thyrotropin (thyroid stimulating hormone) became available this was adopted as the primary screen in most European countries. When the sample is taken very early in life, as is usual in the United States, for example, thyroxine may be more suitable as the primary screen. Congenital hypothyroidism may be diagnosed using cord blood, which allows screening to be based entirely in local hospitals. However, many inherited metabolic disorders are not detectable in cord blood so that expansion to cover additional conditions would require major restructuring of the screening program (Autti-Rämö et al., 2005).

Two other relatively common disorders were added to the blood-spot screening repertoire following the development of reliable immunoassays. Congenital adrenal hyperplasia, which may be due to any one of five enzymes involved in cortisol synthesis, can be detected by increased levels of 17-hydroxyprogesterone (van der Kamp and Wit 2004). Screening for cystic fibrosis relies on increased immunoreactive trypsinogen in the newborn period (Wilcken, 2007). In both disorders the initial screening test lacks specificity, and a second-tier test is required to provide an acceptable positive predictive value. Both disorders have a wide range of clinical presentations and severity; the advantages of early detection are diverse and less clear-cut than for congenital hypothyroidism.

Over the past two decades, developments in tandem mass spectrometry (MS-MS) have greatly increased the scope of blood-spot screening (Carpenter & Wiley, 2002). The technique offers improved sensitivity and specificity when screening for phenylketonuria, particularly when using samples taken very early in life, and requires minimal preliminary sample processing. At little additional cost it can be extended to include rare but well-characterized and treatable amino acid disorders such as Maple Syrup Urine Disease and pyridoxine-nonresponsive homocystinuria. Adding acylcarnitines to the profile detects a wide range of organic acid and fatty acid oxidation disorders that were not previously amenable to screening. In theory, combined analysis of amino acids and acylcarnitines can detect 54 individual disorders (American College of Medical Genetics, 2006). Within limits it is possible to select which compounds are to be measured and thus which diseases will be detected. Deciding which disorders should be included is one of the major challenges of screening policy.

More recently, multiplex assays using novel substrates and MS-MS for product detection have developed for a number of lysosomal enzymes and are being evaluated for newborn screening. MS-MS may also be used to assay specific proteins by determining characteristic peptides after tryptic digestion; a potential application is screening for Wilson disease by assaying ceruloplasmin.

GEOGRAPHICAL COVERAGE

Very few of the available newborn screening tests are in widespread use. Data for the calendar year 2004 collected from 44 European countries showed that all but two were screening their entire populations for congenital hypothyroidism and 41 were screening for phenylketonuria (Loeber, 2007). Otherwise, there was little common ground and even countries with broadly similar population genetics and levels of economic development showed great variation in their disease panels. Twelve of the counties surveyed had fully developed screening programs for congenital adrenal hyperplasia, seven for galactosemia, six for cystic fibrosis, and six for biotinidase deficiency. A number of pilot studies for these disorders were also in progress. MS-MS screening for medium-chain acyl-CoA dehydrogenase deficiency is becoming more widespread but there is no general agreement on which, if any, of the other disorders detectable by MS-MS should be included (Pollitt, 2007). Until very recently a similar degree of variation existed between individual states of the United States, and there are now moves toward a more uniform approach (American College of Medical Genetics, 2006). Reviews of the status of newborn screening in Latin America (Borrajo, 2007), the Asia Pacific region (Padilla & Therrell, 2007), and the Middle East and North Africa (Saadallah & Rashed, 2007) reveal the complexity of the situation and the size of the challenge.

BOX 28-1 PRINCIPLES AND PRACTICE OF SCREENING FOR DISEASE (WILSON & JUNGNER 1968)

1. The condition sought should be an important health problem.
2. There should be an accepted treatment for patients with recognized disease.
3. Facilities for diagnosis and treatment should be available.
4. There should be a recognized latent or early symptomatic stage.
5. There should be a suitable test or examination.
6. The test should be acceptable to the population.
7. The natural history of the condition, including development from latent to declared disease, should be adequately understood.
8. There should be an agreed policy on whom to treat as patients.
9. The costs of case-finding (including diagnosis and treatment of patients diagnosed) should be economically balanced in relation to possible expenditure on medical care as a whole.
10. Case-finding should be a continuous process and not "a once and for all" project.

FORMULATING SCREENING POLICY

Attempts go back over forty years to provide a general framework for assessing whether a screening program that is technically possible is also desirable and cost-effective. The classic WHO report by Wilson and Jungner (1968) enunciated ten "principles" of screening that have been the starting point of almost all discussion since (Box 28-1). The principles themselves met with almost universal approval but they were couched mainly in qualitative terms with no clear endpoints. There have been numerous attempts to develop them into more objective decision tools. Some policymaking bodies now require detailed information on all aspects of a disease: its incidence, the performance of the screening test, and the effects of screening on treatment costs and outcomes (Downing & Pollitt, 2008). However, as most of the disorders in question are rare and clinically heterogeneous, it is impossible to obtain statistically significant data within a realistic time scale and, ultimately subjective judgments still need to be made. Extrapolation from experience in other countries is of limited value as, despite increasing population mobility, there remain large geographical variations in the incidence of individual disorders. The cost-effectiveness of screening also depends crucially on the socioeconomic climate in the country concerned.

A more recent set of criteria (Andermann et al., 2008), also produced under the banner of the WHO, is more explicit and at first sight appears much easier to evaluate (Box 28-2). Seven of the criteria could be rephrased as direct instructions on how to design and implement a screening program. Only three (numbers 1, 4, and 10) require the evaluation of external data. However, the requirement that the benefits of screening should outweigh the harm again requires subjective judgments and reminds us that screening policy falls into the class of "wicked problems," in that it raises complex issues to which most people expect science to provide objective answers, but which are actually questions that require deliberations about ethical values as well as scientific and technological facts. In reality, most screening programs require some form of state or local government involvement, and pressure from parent groups and wider political considerations also come into play.

BOX 28-2 SYNTHESIS OF EMERGING SCREENING CRITERIA PROPOSED OVER THE PAST 40 YEARS

1. The screening program should respond to a recognized need.
2. The objectives of screening should be defined at the outset.
3. There should be a defined target population.
4. There should be scientific evidence of screening program effectiveness.
5. The program should integrate education, testing, clinical services, and program management.
6. There should be quality assurance, with mechanisms to minimize potential risks of screening.
7. The program should ensure informed choice, confidentiality, and respect for autonomy.
8. The program should promote equity and access to screening for the entire target population.
9. Program evaluation should be planned from the outset.
10. The overall benefits of screening should outweigh the harm.

(Andermann et al., 2008)

IMPLEMENTING THE PROGRAM

No matter how good the intentions when planning a screening program, and no matter how effective it turns out to be at detecting the target conditions, inevitably there are adverse effects. Early experience with screening for phenylketonuria in the United States highlighted the level of parental stress that a poorly coordinated program can generate (Rothenberg & Sills, 1968). Despite much improvement in practice since then, a request for a repeat blood sample inevitably causes the parents anxiety, although this can be minimized by providing clear information both prior to the start of screening and when communicating the need for any additional samples (Hewlett & Waisbren, 2006). Newborn screening in the United States has tended to generate a particularly high proportion of "false positive" results, partly reflecting the early age at which the blood sample is taken, usually in the newborn nursery prior to the mother's discharge from the hospital. A survey revealed that in 1993–1994, an initial-positive result for phenylketonuria typically had a positive predictive value of about 3%, for congenital hypothyroidism below 2%, and for galactosemia, approximately 0.6% (Kwon & Farrell, 2000). In the United Kingdom the sample is taken at 5–8 days of age, and in many other European countries 3–5 days of age is the norm. However, European countries differ greatly both in their cutoff points and "recall rates" (for a second blood sample) for any given screen, with little apparent relationship between the two, an "obvious and important area for improvement" (Loeber, 2007). Differences in analytical methodology will affect screening performance, and the need for objective metrics, particularly with the greatly expanded disease range now available through MS-MS, is being recognized (Rinaldo et al., 2006).

The social and financial aspects of treatment must also be taken into account. It is pointless, indeed cruel, to screen for a disorder unless all cases so detected have ready access to appropriate medical services. This was clearly recognized by the pioneers of newborn screening in the United States (Guthrie, 1980; Koch, 1980) but unfortunately their vision has still not been fully realized in their home country (Buist & Huntingdon, 2007; Howell & Engelson, 2007). A comprehensive, integrated program is easier to achieve when there is national (or state) provision for health services, though striking the optimum balance between local flexibility and central direction can be difficult (Downing & Pollitt, 2008).

CLINICAL MANAGEMENT

The Wilson and Jungner principles require that for a condition to merit screening there should be an accepted treatment and an agreed policy on whom to treat. Many policymaking bodies go further and demand evidence that the "accepted" treatment is effective. There is convincing evidence of the short-term effectiveness of treatment for most of the conditions currently being considered for screening, though for some of the disorders newly detectable through MS-MS screening, early success has been tempered by the emergence of previously unsuspected late complications.

Formulating an agreed policy on whom to treat may be more problematic. In the context of a diagnostic service, there is often a clear causal link between the abnormal biochemical finding and the presenting symptoms. A similar biochemical abnormality discovered by screening does not necessarily predict disease. Medium-chain acyl-CoA dehydrogenase (MCAD) deficiency is a relatively common condition in European populations, and under conditions of severe metabolic stress, such as a childhood infection, manifests as sudden death or a Reye-like syndrome that leaves a proportion of those who recover with severe neurological sequelae. However, MCAD deficiency may be regarded more as a risk factor than a disease in itself, as sometimes an affected older sibling, or occasionally a parent of a clinically presenting case has remained asymptomatic, presumably never having encountered sufficient metabolic stress to trigger an episode of decompensation. Newborn screening for MCAD deficiency has generally revealed more cases than expected, and a genotype distribution notably different from that in clinically diagnosed cases. Some genotypes found to be relatively common in screening-detected cases have never been recorded in the clinically presenting disease. Such patients tend to show less marked biochemical abnormalities than recovered clinically presenting cases, and presumably have a much lower risk of experiencing a metabolic crisis. How much lower, and should the management regime be moderated correspondingly?

An analogous problem occurs with 3-methylcrotonyl-CoA carboxylase deficiency, which only occasionally presents with Reye-like symptoms, the majority of cases remaining asymptomatic. Wilcken (2008) discusses how this and similar disorders should be managed, and whether they should be included in the panel for MS-MS screening at all. Screening for phenylketonuria will also detect milder forms, often described as non-PKU hyperphenylalaninemia, mild hyperphenylalaninemia, or benign hyperphenylalaninemia. Deciding whether such cases require any form of treatment should present little difficulty, as the relationship between exposure to increased blood phenylalanine concentration and the degree of intellectual impairment is well established.

IMPLICATIONS FOR DEVELOPING COUNTRIES

The case for screening for congenital hypothyroidism seems almost universal and the International Atomic Agency, through its Technical Cooperation Program, has been very active in promoting it. Their handbook (International Atomic Energy Agency, 2005) contains useful guidance on the general requirements of a successful blood-based newborn screening program. Support has also come from other sources, such as the Japan International Cooperation Agency, which has provided somewhat broader training and technical assistance programs to a wide range of developing countries (Fukushi, 2007). The Newborn Screening Branch of the U.S. Centers for Disease Control and Prevention (http://www.cdc.gov/labstandards/nsqap.html) has long opened its quality assurance and proficiency testing programs to newborn screening laboratories worldwide. Nevertheless, starting a newborn screening program remains a formidable undertaking. Prerequisites include effective systems for collecting samples at an appropriate age, transporting them to the screening laboratory, and the ability to contact babies' families again should further action be required. Medical services and scientific and technical expertise in the relevant fields may initially be in short supply, but much can be achieved by collaborative ventures with more established services (Lindner, 2007)

As in the developed world, the choice of which disorders to include in the program is difficult. The incidence of the disorders concerned and the effectiveness of treatment are major factors. Screening for phenylketonuria will almost certainly prove cost-effective in large parts of China, for example (Zhan et al., 2009). Wide-range MS-MS screening is usually most productive in areas with high rates of consanguineous marriage and, paradoxically, its economic benefits become stronger as counties expand their medical infrastructures and diagnostic services and treatment facilities improve.

The many challenges of managing patients with inherited metabolic disorders in a developing country are reviewed by Acosta and Sabo (2003). Cost is likely to be a major factor and ultimately, much may depend on the development of local resources as opposed to imported commercially available diets. Modification of the routine local diet may have limited success, but for many conditions a manufactured product will be essential for complete metabolic control (Kabra, 2002).

Public expectation is likely to be a powerful factor in driving the spread of newborn screening. As economies develop social patterns based on large family size, high childhood morbidity and mortality tend to evolve into patterns with a lower birth rate and improved survival. Genetic disorders will supplant infectious disease as a major cause of childhood morbidity and mortality and newborn screening will assume increasing importance.

However, experience in the developed world shows that plotting a well-balanced and appropriate course remains a considerable challenge.

REFERENCES

Acosta PB, Sabo R. (2003). Challenges of managing patients with inherited metabolic disorders in a developing country. *Southeast Asian Journal of Tropical Medicine and Public Health 34* (Suppl 3), 202–207.

American College of Medical Genetics. (2006). Newborn screening: towards a uniform screening panel and system. *Genetics in Medicine* 8(Suppl 1), 1S-252S.

Andermann A, et al. (2008). Revisiting Wilson and Jungner in the genomic age: a review of screening criteria over the past 40 years. *Bulletin of the World Health Organisation* 86, 317–319.

Autti-Rämö I, Mäkelä M, Sintonen H, et al. (2005). Expanding screening for rare metabolic disease in the newborn: an analysis of costs, effect and ethical consequences of decision-making in Finland. *Acta Paediatr 94*, 1126–1136.

Carpenter K, Wiley V. (2002). Application of tandem mass spectrometry to biochemical genetics and newborn screening. *Clinica Chimica Acta 322*, 1–10.

Borrajo GJC. (2007). Newborn screening in Latin America at the beginning of the 21st century. *Journal of Inherited Metabolic Disease 30*, 466–481.

Buist RM, Huntingdon K. (2007). Scene from the USA: The illogic of mandating screening without also providing for treatment. *Journal of Inherited Metabolic Disease 30*, 445–446.

Downing M, Pollitt RJ. (2008). Newborn blood-spot screening in the UK – past, present and future. *Annals of Clinical Biochemstry 45*, 11–17.

Fukushi M. (2007). An international training and support programme for the establishment of neonatal screening in developing countries. *Journal of Inherited Metabolic Disease 30*, 593–595.

Guthrie R, Susi A. (1963). A simple phenylalanine method for detecting phenylketonuria in large populations of newborn infants. *Pediatrics 32*, 338–343.

Guthrie R. (1980). Organisation of a regional newborn screening laboratory. In: Bickel H, Guthrie R, Hammersen G, (eds.) *Neonatal Screening for Inborn Errors of Metabolism*. Berlin: Springer-Verlag, pp. 259–270.

Hewlett J, Waisbren SE. (2006). A review of the psychological effects of false-positive results on parents and current communication practices in newborn screening. *Journal of Inherited Metabolic Disease 29*, 677–682.

Howell RR, Engelson G. (2007). Structures for clinical follow-up: Newborn screening. *Journal of Inherited Metabolic Disease 30*, 600–605.

International Atomic Energy Agency. (2005). *Screening of Newborns for Congenital Hypothyroidism. Guidance for Developing Programmes.* Vienna. (http://www-pub.iaea.org/MTCD/publications/PDF/Pub1234_web.pdf; accessed 17/08/2011).

Kabra M. (2002). Dietary management of inborn errors of metabolism. *Indian Journal of Pediatrics 69*, 421–426.

Koch R. (1980). Medical backup needed for newborn screening programmes. In: Bickel H, Guthrie R, Hammersen G, (eds.). *Neonatal screening for inborn errors of metabolism*. Berlin: Springer-Verlag, pp. 271–273.

Kwon C, Farrell PM. (2000). The magnitude and challenge of false-positive newborn screening test results. *Archives of Pediatrics and Adolescent Medicine 154*, 714–718.

Lam WK, et al. (1996). Histidinaemia: a benign metabolic disorder. *Archives of Disease in Childhood 74*, 343–346.

Lindner M, et al. (2007). Implementation of extended neonatal screening and a metabolic unit in the state of Qatar: Developing and optimising strategies in cooperation with the neonatal screening centre in Heidelberg. *Journal of Inherited Metabolic Disease 30*, 522–529.

Loeber JG. (2007). Neonatal screening in Europe in 2004. *Journal of Inherited Metabolic Disease 30*, 430–438.

Lord J, et al. (1999). Secondary analysis of economic data: a review of cost-benefit studies of neonatal screening for phenylketonuria. *Journal of Epidemiology and Community Health 53*, 179–186.

Padilla CD, Therrell BL. (2007). Newborn screening in the Asia Pacific Region. *Journal of Inherited Metabolic Disease 32*, 490–506.

Pollitt RJ. (2007). Introducing new screens: Why are we all doing different things? *Journal of Inherited Metabolic Disease 30*, 423–429.

Pollitt RJ. (2009). Newborn blood spot screening: new opportunities, old problems. *Journal of Inherited Metabolic Disease 32*, 395–399.

Rinaldo P, et al. (2006). Making the case for objective performance metrics in newborn screening by tandem mass spectrometry. *Mental Retardation and Developmental Disability Research Review12*, 255–261.

Rothenberg MB, Sills EM. (1968). Iatrogenesis: the PKU anxiety syndrome. *Journal of the American Academy of Child Psychiatry 7*, 689–692.

Saadallah AA, Rashed MS. (2007). Newborn screening: experiences in the Middle-East and North America. *Journal of Inherited Metabolic Disease 30*, 482–489.

van der Kamp HJ, Wit JM. (2004). Neonatal screening for congenital adrenal hyperplasia. *European Journal of Endocrinology 151*(Suppl 3), U71–75.

Wilcken B. (2007). Newborn screening for cystic fibrosis: techniques and strategies. *Journal of Inherited Metabolic Disease 30*, 537–543.

Wilcken B. (2008). The consequences of extended newborn screening programmes: Do we know who needs treatment? *Journal of Inherited Metabolic Disease 31*, 173–177.

Wilson JMG, Jungner G. (1968). Principles and practice of screening for disease. Geneva: WHO. Available from: http://www.who.int/bulletin/volumes/86/4/07-050112BP.pdf.

Zhan JY, et al. (2009). Neonatal screening for congenital hypothyroidism and phenylketonuria in China. *World Journal of Pediatrics 5*, 136–139.

29.

DIAGNOSTIC GENETIC LABORATORY SERVICES: MODELS FOR THE DEVELOPING WORLD

Peter Lunt

INTRODUCTION

Diagnostic genetic laboratory service is an integral part of any clinical genetic service. In most developed nations in the West, this is in majority of cases either part of a comprehensive genetic service or an independent laboratory genetic service provider working parallel to one or more clinical genetic service providers. This chapter provides the structure and function of a typical diagnostic genetic laboratory setting that could be adopted, with appropriate local modifications, as a practical model in developing countries.

Genetic laboratory testing can or will be required in a clinically presenting individual for making the primary diagnosis, informing treatment options, prognosis prediction, establishing the inheritance pattern and for determining the family-specific mutation to enable family risk evaluation and testing. It is also applicable for monitoring extent and control of a disease in populations, individual presymptomatic risk assessment (single gene and multifactorial) with or without a specific family history, presymptomatic disease screening in a person or a fetus, genetic risk screening (e.g., carrier testing for hemoglobinopathy) in family members, presymptomatic prediction for targeted clinical surveillance, and genetic risk assessment including carrier testing and prenatal diagnosis.

TYPES OF DIAGNOSTIC GENETIC TESTS

Genetic tests can be divided into those based on:

i. DNA analysis of known single genes

ii. DNA analysis of polymorphic variation

iii. Chromosomes

iv. Molecular cytogenetics (i.e., detecting genome imbalance)

v. Biochemistry

vi. Diagnostic Imaging

vii. Other clinical test

These broad differences between genetic tests and other laboratory tests necessitate a careful, considered approach to their development and introduction (Table 29-1). What tests are introduced will also depend on the needs of the particular population, in some cases influenced by ethnic genetic differences. Their introduction will ideally be accompanied by availability of trained counselor support, as well as the trained laboratory scientists. An approach that combines local, regional, national, and international components is likely to be required, with decisions needed as

Table 29-1 Differences in Perspective between Genetic Tests and other Clinical or Laboratory Tests

	GENETIC TESTS	OTHER LAB TESTS
Result implications:	Permanent	Often transient
Implications for others:	Often exposing genetic risk to close family members and relatives in the extended family	May be marked with infectious disease, especially sexually transmitted disease.
Social effects	Risk of discrimination-career prospects and life insurance	Possible segregation for fear of infection to others
Cost:	Most are high cost, low volume	Most are low cost, high volume
Interpretation:	Often required, usually involving lab scientist, particularly to distinguish pathogenicity from variation.	Mostly straightforward for referrer
Delivery to patient/family:	Often requires counseling support	Mostly to person already aware of symptoms
Wait for result:	Usually several weeks	Most within 1–7 days
Durability:	Tests change as technology and efficiency evolve rapidly	Many are long established

to which tests to provide or access, alone or in partnership, at which organizational level. The increasing development of commercially available testing kits, but also the cost of analytic hardware, will help inform decisions of the flexibility or standardization of tests offered, but also may give opportunity for partnership discussion with commercial firms to develop kits tailored to individual ethnic population groups.

One of the main questions is in relation to the degree of oversight that should be considered for services introducing diagnostic genetic laboratory testing. This could range from no regulation—the only limiting factor being the cost to whoever is paying—to tightly managed testing with restriction of availability of tests to those with independently endorsed validity and clinical utility (i.e., relevance), restriction of referrers to those from certain specialties only, and restriction to patients who meet established clinical testing criteria. In the United Kingdom, the escalating cost of tests to the NHS (National Health Service) has resulted in the establishment of a nationally funded body, UK Genetic Testing Network (UKGTN) to promote quality and equity for genetic testing. In the United States, a similar regulatory framework is under discussion. In the United Kingdom, through the UKGTN, any laboratory wishing to introduce a new genetic test to be made available for purchase by the NHS for the wider UK population must first demonstrate that it fits broadly within the "ACCAE framework" of test evaluation. This framework is based on:

A: analytical validity

C: clinical validity

C: clinically utility

E: ethical soundness

Laboratory applicants for new test availability are asked to define an appropriate target population (i.e., which patients would be offered the test), the sensitivity and specificity of the test in that population, the clinical utility as the effect of a positive or negative result and the likelihood of each, and the predicted consequence to NHS patients of not providing the test. Clinical testing criteria are defined (i.e., what features a patient must have in order for the test to be appropriate, and who can request the test), and are often rediscussed between UKGTN and the applicant to try to ensure that the test is focused most appropriately. The application is based on completion of a descriptive "gene dossier" which documents these aspects.

Following specific clinical examples may help to illustrate the types of genetic tests that are available and for consideration in developing a service:

i. Child with learning difficulty, dysmorphism, or multiple congenital malformations
ii. Child with clinically diagnosed known genetic condition
iii. Adult at risk for a late-onset degenerative condition
iv. Adult with cancer in context of strong family history of the same
v. Adult with a neuromuscular condition concerned for his or her present or future children
vi. Adult with or without family history of heart disease, concerned to know if he/she is at increased risk.
vii. Routine pregnancy
viii. Pregnancy with fetal malformation on scan
ix. Pregnancy at risk for an X-linked recessive condition

The above clinical situations could be considered in detail for a specific requirement as outlined at the beginning of this chapter.

Making a Primary Diagnosis in a Child (or Adult) with Combinations Of Learning Difficulty/Delay, Dysmorphism, Congenital Malformation

A specific and accurate diagnosis is usually the key for providing information to parents on etiology; (explanation as to why their child has problems), long-term prospects (prognosis), management/treatment, inheritance pattern and recurrence risk, availability of genetic testing as prenatal test to avoid recurrence and genetic risk assessment more widely applicable to the family. In many cases the clinician may suspect a specific diagnosis (e.g., Williams syndrome, or CHARGE syndrome) for which specifically targeted testing is appropriate. Alternatively, it may be a *type* of diagnosis which is suspected, such as a chromosome anomaly, or a probable developmental pathway anomaly (e.g., RAS-MAPK pathway in Noonan, CFC, Costello syndromes), for which testing may be more generalized. In most cases, if no specific diagnosis is clear, a geneticist or other clinician will wish to know whether there is evidence of obvious genome imbalance, or, alternatively, in cases with learning difficulty alone, whether there may be a prevalent, readily testable single gene disorder (e.g., Fragile-X syndrome).

Genome Imbalance

For genome imbalance (i.e., where there are too few or too many copies of some sections of the genome), the standard analysis to date has been a routine karyotype, involving counting and studying under the light microscope a chromosome preparation obtained from cultured lymphocytes from peripheral blood. This is labor-intensive, and needs skilled cytogeneticists for analysis and for interpretation of minor variants. In antenatal work, where the main question (arising from increased maternal age, or from raised indices on a maternal serum screening program, or following abnormality detected on ultrasound scan), is usually whether the fetus may have a trisomy of chromosomes 21, 18, or 13 (Downs, Edwards, or Patau's syndromes, respectively), full karyotyping from amniocentesis is in many centers now preceded or even replaced by direct qualitative Fluorescent Polymerase Chain Reaction (qFPCR) analysis on DNA of amniocytes. Similarly, qFPCR may now be the first test in newborns suspected of having trisomy 21, 18, or 13. However, given the variety of different structural chromosomal anomalies that can present in the newborn, qFPCR cannot be a substitute for full chromosome analysis in neonates, and also any positive qFPCR trisomy result must be followed by full karyotype to identify any translocation case, in view of the potential implications for recurrence and need then to test the parents.

For some conditions presenting with combinations of learning difficulty/delay, dysmorphism, and/or congenital malformation, there may be suspected genome imbalance, but at a specific site and at submicroscopic level. For such cases (e.g., Williams, Miller-Dieker, or 22q11 syndromes), specific FISH testing (fluorescent in-situ hybridization) using the appropriate FISH probe on a chromosomal preparation is appropriate. Alternatively, MLPA techniques using site-specific PCR probes on a DNA sample can similarly detect site-specific imbalance (e.g., duplication of PLP gene in X-linked Pelizeus-Merzbacher disease). There are now many different recognized microdeletion syndromes for which these approaches can be appropriate. Microdeletion is also recognized on the Y-chromosome (in the AZFa, b or c region) as one of the more common genetic causes of infertility in men, and is detected by specific FISH, or by DNA testing.

In other cases, the clinical presentation may be due to chromosome abnormality that has arisen post-zygotically (i.e., after conception), and is present only in some rather than all cells. Recognition of such mosaic genome imbalance may require analysis of cell lines other than those in peripheral blood—e.g., by direct analysis and culture of skin fibroblasts (obtained by skin biopsy), or by FISH testing on buccal cells obtained simply from a buccal brush or mouth-spit sample. Cells in urine may also sometimes be used. The author was able to diagnose a boy aged 14 years as having trisomy 8 mosaic through showing uniform 47,XY+8 karyotype in a skin fibroblast culture, when peripheral blood lymphocytes had shown a uniform normal 46,XY karyotype.

MICROARRAY COMPARATIVE GENOME HYBRIDIZATION (ARRAY CGH)

In any discussion now for development of laboratory genetic services, it is the technique of microarray CGH which must be considered. This is emerging as a technique to replace routine standard karyotyping for genome imbalance, as well as multiple individual FISH tests or tests for lymphocyte-involved mosaicism. The technique involves hybridization of fluorescently tagged whole-genome DNA, usually from peripheral lymphocytes, in equal quantity with different-color fluorescently tagged standard genomic DNA arranged on a pre-prepared microarray plate. Hybridization in a normal 1:1 ratio gives a neutral mix color, but imbalance is seen as a color shade nearer to one of the 2 original fluorescent tag colors, according to whether the test DNA is deleted or duplicated at that genomic site. Significance of imbalance is usually taken as requiring 3 adjacent probes to appear consistently reduced or increased from the 1:1 ratio. As well as the microarray plates, the main equipment required is the analyzer for reading the fluorescence intensity by wavelength, and the software for displaying this as dosage along each chromosome length.

Current use of the microarray CGH for patients selected by clinical geneticists following a normal standard karyotype is finding significant imbalance in around 13% of those tested. Since the technique will also detect microscopically visible imbalance (around 8% of all standard chromosomal referrals), an overall detection rate up to 21% can be anticipated if CGH were to be used instead of karyotyping as the standard first-line investigatory test. The main difficulty at present with CGH is the interpretation of genome imbalance identified. Within the normal population this is considerable, and it is only from accumulated experience that normal copy number variants (CNVs) of no apparent consequence are being characterized. Typically, one or more novel CNVs will be found in around 10%–20% of samples analyzed by array CGH. The number and frequency of CNVs complicates the interpretation of array CGH results, and certainly requires considerable input from experienced trained molecular-cytogenetic scientists. The main indications that a detected imbalance is pathogenic are if it is either *de novo* (i.e., not present in either parent), or alternatively co-segregates in a family with a potential phenotype (e.g., parent also having learning difficulty), or involves more than 3 adjacent probes, or contains or is in close proximity to genes which, by their nature, might be contributing to the phenotype if in altered dosage, and finally has been observed with a similar clinical presentation or dysmorphic features. Information on this is becoming available through the establishment of worldwide available databases (such as DECIPHER (**D**atabas**e** of **C**hromosomal **I**mbalance and **P**henotype in **H**umans Using **E**nsemble **R**esources; http://decipher.sanger.ac.uk/) to document phenotype–genotype correlations. Conversely, one of the following might indicate that an apparent imbalance might be a coincidental CNV:

i. presence also in one of the parents (who are unaffected), +/- in wider family

ii. nvolves no more than 3 adjacent probes

iii. does not seem to contain any known functional gene, or any that seem relevant to the presentation

iv. has previously been observed in a normal individual, or with a non-overlapping (very different) presentation.

Again, the establishment and free availability of databases of recognized CNVs will be invaluable for any service introducing CGH. However, these may well need to be independently established for each different genetic ethnic group.

Hence, in many cases, interpretation of apparent imbalance requires also obtaining and testing DNA samples from both parents. In the United Kingdom, our experience is that this is only possible in around 80%–90% of cases. The typical cost for microarray CGH analysis, where this is run on a high-volume basis, is now around £300, compared with £150-£200 for standard cytogenetic analysis. The array cost is, of course, increased if parents' samples are also required, and the clinical input required for liaison with parents and other clinicians may be much greater.

Modifications to the array CGH technique, or the way that it is offered, can reduce cost by running one patient against another rather than each against control DNA, although care must be taken to pair patient samples as those having very different clinical presentations, and detected imbalance must be further investigated on both samples to see whether one is deleted or the other one duplicated. The array platform can also be designed to intensify only areas of the genome that have well-characterized microdeletion or duplication syndromes, and have a relatively blunter coverage of the rest of the genome to detect chromosome imbalance that would have been detectable cytogenetically. This reduces the likelihood of finding a CNV of uncertain significance, and is a promising tool for prenatal testing.

It should be remembered also that microarray CGH will only detect chromosomal imbalance, and so cannot replace standard karyotyping in testing for true balanced chromosome translocation (e.g., in investigation of recurrent miscarriage), or where a balanced translocation may be the cause of a condition by cutting through a specific gene. Also, standard karyotyping is essential for investigation of parents or siblings of a recognized case of genome imbalance if one of them might have a balanced translocation from which the imbalance has arisen. On the other hand, many de novo, apparently balanced chromosome translocations, particularly if ascertained in a child with delay/malformation/dysmorphism may have more complicated breakpoints involving cryptic deletion or duplication, discovered only by performing array CGH.

FUTURE DEVELOPMENT IN DETECTING GENOME IMBALANCE

Microarray CGH should not be seen as the final successor to standard cytogenetics. At this time, other DNA-based techniques for whole-genome exon resequencing (whole-exome array) are being researched, and may well have application for genome imbalance as well as for single gene mutation analysis. However, the relative role of this, and of microarray CGH, of single gene DNA analysis, and of symptom-related multigene DNA analysis (e.g., for hereditary peripheral neuropathy) will emerge over the next two decades or so.

DIAGNOSIS OF OTHER "CYTOGENETIC" DISORDERS WITH COMBINATIONS OF LEARNING DIFFICULTY/DELAY, DYSMORPHISM, CONGENITAL MALFORMATION

The cytogenetic laboratory is the one primarily involved for diagnosis of chromosome breakage syndromes. This group of disorders—Fanconi anemia (FA), Ataxia telangiectasia (AT), Bloom's syndrome (BS), Nijmegen breakage syndrome (NBS)—are recessive disorders, which are recognized by specific in-vitro tests on peripheral lymphocyte chromosome preparations. FA, BS, and NBS usually present clinically with microcephaly and growth delay, while AT usually presents with ataxia. In FA and AT there are increased chromosome breaks or rearrangements on exposure of cell cultures to cytotoxic agents (alkylating agents such as BrDU or X-irradiation, respectively), while BS is recognized through demonstrating increased sister-chromatid exchange on radioactivity incorporation. Investigation for these disorders should be accessible to populations where they may be encountered, although in time it is conceivable that DNA resequencing techniques may lead to kits for simultaneous DNA analysis of the various different genes involved in these and other similar microcephaly disorders.

A further "cytogenetic"-based type of disorder is uniparental disomy (UPD), whereby a child or adult has an apparently normal karyotype but with both copies of one chromosome, or even part of a chromosome, originating from the same parent. The phenotypic effect will usually differ according to the parent of origin of the two copies of the chromosome. Examples are UPD 15mat. (in some cases of Prader-Willi syndrome), UPD 15pat. (in some cases of Angelman syndrome), UPD7 (in some cases of Russell-Silver syndrome), UPD 14mat, and UPD14pat.

Primary Diagnostic Testing by DNA Mutation Searching

Until DNA analysis to establish a primary diagnosis becomes high-throughput, efficient, and inexpensive, its role in this setting must be considered circumspectly in each case before making a test request. Where a test request is appropriately focused to have a high chance of establishing diagnosis, or a good chance of excluding a particular diagnosis very easily (and cheaply), the advantages over further clinical testing can be:

i. Speed (no need to await organizing further clinical investigation)

ii. Cost (the DNA test may be cheaper than involving clinical diagnostics)

iii. Accuracy (a mutation may be more definitive than equivocal clinical results)

iv. Providing a tool to enable ready testing of other family members

v. Comfort for the patient (a blood sample is less traumatic than EMG testing for a diagnosis of myotonic dystrophy)

The disadvantages of the above approach may be:

i. High cost (especially where a mutation is likely to be family-specific, and one of multiple genes could be involved)

ii. Uncertainty of pathogenicity and therefore relevance (especially for novel sequence variants)

iii. Unrecognized genetic heterogeneity (falsely assuming exclusion of diagnosis from a negative result)

Depending on the ease and cost of gene sequencing techniques for finding mutation, the availability of primary diagnosis by DNA has depended until recently on the nature of the mutation in the condition. These could be:

i. Uniform mutation mechanism (e.g., trinucleotide repeats in Fragile-X, myotonic dystrophy or Huntington's disease)

ii. Prevalent mutation (or mutation-type) in the specific population (e.g., dF508 in cystic fibrosis; exon7 deletion of SMN1 in SMA)

iii. Family-specific mutation (e.g., in FBN1 gene in Marfan syndrome)

UNIFORM MUTATION MECHANISM

Where all patients with a condition have the same mutation, diagnostic testing is definitive, efficient, and usually technically very simple and cheap. The test can then be used both to confirm diagnosis in typical or probable clinical cases, and to exclude diagnosis in people who may have only partial overlap with the typical presentation (e.g., testing to exclude myotonic dystrophy in patients whose presentation is more suggestive for congenital myotonia)—although test requests should still fall within agreed clinical test criteria, so as not to overload the laboratory with irrelevant requests. The test also does not rely on first confirming the mutation in an affected family member, although this is preferable.

PREVALENT MUTATION (OR MUTATION TYPE)

This may often be the case for recessive conditions, but the most prevalent mutations may also be individual for the particular ethnic population concerned. Testing for these, in the absence of full gene sequencing, can be helpful. The finding of one mutation in a known recessive condition in a patient with typical symptoms for that condition, is strong evidence in favor of that diagnosis, particularly if the ethnically matched population carrier frequency is low. In families where it is not possible to test an affected person or obligate carriers directly to identify the particular mutations involved, carrier risk estimates can also be modified considerably if someone does not carry one of the normally prevalent mutations. In a dominant condition such as familial hypercholesterolemia (FH), diagnosis is confirmed in most cases by finding one of several prevalent mutations in one of three possible genes, and is followed by full sequencing of one of these genes (LDLR).

FAMILY-SPECIFIC MUTATION

In many genetic conditions, even the most common mutations account for only a small percentage of cases. It is in these cases where a DNA mutation-finding technique, traditionally involving a first-step by SSCP, DGGE, or mismatch splicing to identify the sections of the gene to be sequenced is required. Increasingly, the initial steps are becoming redundant as analysis moves to primary direct sequencing of the whole gene, which in turn will become more efficient through high-throughput resequencing on "next-generation" equipment, and in genetically heterogeneous conditions applicable to multiple genes simultaneously.

Key to the efficient use of DNA techniques in primary diagnosis is the setting of clinical criteria that should be met in order to be able to request the specific DNA analysis. Ideally, these could be accessible online, and where possible linked to online ordering of tests on a DNA sample. The UKGTN (UK Genetic Testing Network) has adopted this approach for requests from member laboratories, to provide tests that can be available for commissioning for

NHS use. An example of clinical criteria is given below for Marfan syndrome.

Even with such an approach, the clinician should not request DNA analysis primarily to make a diagnosis if there is an alternative, cheaper, and reliable method. Thus, for example, a patient with multiple cafe au lait patches, who clinically has multiple neurofibromas, can be diagnosed clinically as having neurofibromatosis (NF1) without resorting to a molecular test.

GENETIC TESTING TO INFORM PROGNOSIS AND TREATMENT

In many cases where treatment/clinical management follows from diagnosis, the genetic test is only the means to establishing the diagnosis (e.g., FBN1 mutation confirming Marfan syndrome, or LDLR mutation confirming FH). However, increasingly there are examples where the purpose of the genetic test is to direct treatment or management according to the specific test result. For example, in familial adenomatous polyposis coli (FAP), the nature and site of mutation in the APC gene may determine whether the patient will present with typical FAP or an attenuated form, and clinical management will vary accordingly. Similarly, where there is genetic locus heterogeneity, the prognosis, and hence management, may differ according to the specific gene involved (e.g., in autosomal dominant polycystic kidney disease). Perhaps the most familiar examples are in trinucleotide repeat disorders, where the number of repeats correlates inversely with age at onset, in Xp21 muscular dystrophy where the effect of deletion mutation on the reading frame of dystrophin largely determines whether this gives a Duchenne or a Becker phenotype, and in FSH muscular dystrophy, where the residual number of copies of a 3.3 kb repeat(D4Z4) is a strong influence (inversely) on severity of manifestation. Prediction of prognosis in these conditions can be made from the DNA result combined with the actual clinical history and presentation.

It is, however, where a specific DNA result directly affects treatment that provides the strongest argument for DNA testing to determine and influence prognosis. In focal segmental glomerulosclerosis (FSGS), cases caused by or associated with a *podocin* gene mutation will recur following renal transplant, and should not be considered as candidates for this. Those cases with no mutation respond much better to renal transplantation. For Duchenne muscular dystrophy, or for some recessive conditions, preliminaries to treatment trials have begun, where the treatment will aim to block a specific exon in the gene according to the specific deletion present, or will be a specific drug to promote read-through of any aberrant STOP-codon.

ESTABLISHING THE INHERITANCE PATTERN

GENETIC LOCUS AND MUTATION HETEROGENEITY

There are many types of genetic conditions where there is more than one gene that can cause the condition, and/or where different mutations can behave as recessive or dominant, either in the same gene or in the different genes. Common examples are the hereditary spastic paraplegias, hereditary motor and sensory neuropathies (HMSN), hereditary hearing loss, or retinitis pigmentosa. DNA testing may not only identify the specific gene, but establish the inheritance pattern according to whether the gene is X-linked or autosomal, and whether one or both copies have mutation. However, there are examples such as Myotonia congenita, where CLCN1 gene mutation can present as a dominant (Thomsen's disease) or a recessive (Becker type) condition, where the finding of a single mutation in an isolated clinical case, and in one of the person's (unaffected) parents, leaves doubt as to whether the mutation is sufficient on its own to cause the disease. Indeed, the same mutation may appear dominant in one family and recessive in another.

The more widely encountered situation of establishing inheritance pattern is in the identification of high-risk cancer families from those where multiple cases of cancer, or an early-onset cancer, have occurred by chance. For setting up DNA mutation testing for cancer, clinical and family history criteria to determine the entry point are essential, in order that the genetic mutation searching is appropriately targeted. This can be aided by genetic testing of the cancer tissue itself. Thus, in colorectal cancer, it is the presence of positive MSI (microsatellite instability), with or without specific gene product immunohistochemistry (IHS), which, together with family history and age at onset, determine the likely chance for there to be an underlying constitutional mutation in a mismatch repair gene (MMR gene) such as MLH1, MSH2 or MSH6. Algorithms and computer programs for calculation of cancer risk according to age-onset and family history parameters are available, but may be limited by genetic ethnic population specificity.

Family-Specific Mutation in Known Diagnosis

Where diagnosis is certain, apart from any prognostic prediction or aspects influencing treatment, the main role of specific mutation testing in an index case in a family is for risk evaluation in other family members, including, where appropriate, prenatal testing. This therefore encompasses presymptomatic testing or clarification of equivocal clinical status (mostly in dominant conditions), carrier testing in recessive or X-linked conditions, and prenatal or preimplantation testing. It is therefore usually entirely appropriate to request DNA analysis to identify the underlying specific mutation in the index case where these subsequent testing requests from family members can be anticipated, and where the risk is best assessed by a DNA test rather than an alternative clinical or biochemical test. In many cases, it is perhaps the responsibility of the doctor involved with the index case to request such tests, so that the information is then available for the other family members. However, not all genetic conditions will necessarily require DNA mutation analysis, and in some recessive inborn errors of metabolism, prenatal diagnosis may be efficiently performed biochemically on chorionic villus biopsy or amniocentesis samples, and carrier testing (by DNA) only be relevant if there is likely to be consanguinity, since otherwise the chance that an unrelated partner of a family member is also a carrier may be very low.

Monitoring Extent and Control of a Disease

Although the emphasis for most genetic laboratories for molecular genetic and cytogenetic analyses is on constitutional genomic disorders, the same laboratory techniques are used in characterizing hematological disorders (particularly malignancies) and some solid tumors, while related techniques are used in diagnosing and characterizing infection, or in assessing endocrine gland dysfunction. Not only can the techniques be applied to diagnosis, but also in clarifying the spread of disease and monitoring response to treatment.

GENETIC TESTING IN FAMILY MEMBERS

In general, testing family members of an index case requires less consideration for laboratory resource than does the original primary diagnostic or mutation-finding test on the index case. For DNA mutation, the test is usually for that specific mutation only, and may involve a single PCR reaction, or a specific enzyme digestion. Chromosomal testing (e.g., for a balanced translocation) may still require a full karyotype, but the analysis can largely be focused to the chromosomal region involved. The main resource implication is usually for the sample collection and the clinical counseling input required, particularly where the test may have a direct predictive consequence.

Presymptomatic Prediction +/– Targeted Clinical Surveillance

In most cases, DNA testing will require knowledge of the specific mutation in the index case, although testing can be carried out without that knowledge where there is a known uniform mutation mechanism underlying the condition, and the index case diagnosis is certain. Tests for prevalent mutation can also be offered in the absence of index case information, but will only give a modified rather than absolute predictive risk estimate. This is, for example, often the approach taken in testing couples (one of whom starts at population risk) for carrier risk in recessive disorders.

One of the main reasons for testing a family member for presymptomatic prediction is where clinical surveillance for early detection of that condition can improve clinical outcome. In particular, this can apply to inherited cancer predispositions, such as familial polyposis coli (FAP) or dominant colorectal cancer, or to familial cardiac disorders such as HOCM (hypertrophic cardiomyopathy) or long QT syndromes. By offering specific genetic testing to those at risk, the clinical surveillance can be targeted to those for whom it is needed, while those who have not inherited the condition can be reassured. The importance of first knowing the family-specific mutation in the index case cannot be overemphasized.

Genetic Risk Assessment (Carrier Testing and Prenatal Diagnosis)

Carrier testing similarly requires prior knowledge of the specific mutation(s) in the index case to enable definitive testing, or at least knowledge of the most prevalent mutations in the population to be able to give a modified risk estimate. Testing of unrelated partners for recessive carrier risk relies on knowledge of the most prevalent mutations in that particular local or genetic/ethnic population.

Prenatal diagnosis may involve DNA testing, chromosomal testing, or biochemical (metabolic) testing, or all three. In establishing a diagnostic genetic laboratory service, the urgency to obtain a result, the need for accuracy, and checks to exclude maternal cell contamination are important considerations. Samples may come as a chorion villus biopsy, taken usually at 11 weeks gestation, or as an amniocentesis, usually at 16–20 weeks, and may require both direct analysis and analysis of cell culture.

Other prenatal techniques to consider in planning services are fetal blood sampling (usually after 20 weeks), free fetal DNA (see above), and preimplantation genetic diagnosis (PGD). An important application of FFDNA, apart from Y-sequence determination in pregnancies at risk for an X-linked recessive condition, is in identifying Rhesus genotype by a noninvasive method. Rhesus-negative mothers who have previously experienced Rhesus isoimmunization can know from a FFDNA test on a blood sample whether FFDNA shows evidence of Rhesus positivity or not, and therefore whether an amniocentesis to assess bilirubin levels, etc., is required.

For PGD, as an alternative to invasive prenatal testing, one cell is taken from an 8-cell blastocyst following in-vitro fertilization or ICSI (in-vitro cytoplasmic sperm injection). Analysis is usually now performed by PCR of closely linked marker haplotypes, requiring a prior linkage study of the immediate family members. For dominant conditions, this requires DNA samples across two generations in a family (e.g., parents and the relevant grandparents), which adds to the complexity of prior work-up. A laboratory performing PGD must have isolation-work facilities, so as to avoid contamination when working with single cells. However, single cells are transportable between laboratories, and participating labs may wish to consider how best to collaborate to provide service.

CELL FREE FETAL DNA

A further current area of development is the study of minute amounts of fetal/placental-derived DNA, which is present in maternal serum from the first trimester onwards (free fetal DNA/FFDNA). Analysis of FFDNA from 9 weeks gestation onwards (or possibly from earlier in gestation) is already possible for Y-chromosome DNA. This enables the sex of a fetus to be determined in pregnancies at risk for an X-linked recessive condition, and therefore enables mothers to avoid undergoing invasive prenatal testing in the 50% of cases where the fetus is predicted to be female. Expansion of this technique to other conditions (e.g., where the genetic risk is for a dominant mutation from the father) can be anticipated over the next decade, quite possibly including ways of screening for common trisomies from FFDNA. Any current proposal to set up traditional antenatal screening techniques may wish to bear in mind this area of potential future scientific development.

GENETIC TESTING IN POPULATIONS

Presymptomatic Risk Assessment (single gene and multifactorial)

Individuals may be concerned to know their estimated risk for developing a disorder, either as an absolute risk figure, or to know whether the risk is increased or not compared with the general population. For people with a clear family history of a single gene (Mendelian) disorder this is discussed in Section 10 below, and the approach may depend on whether the disorder would have a family-specific mutation, a prevalent mutation, or a uniform mutation mechanism. For people with a more nonspecific family history (such as for heart disease or diverse cancers), or none, the value of genetic testing is at best not well established, unless it is incorporated in a population screening program. It would, for example, be possible to test someone with no known family history for any dominant single gene disorder that has a uniform mutation mechanism (e.g., Huntington's disease or spinocerebellar ataxias), but it would not seem genetically or ethically appropriate in the absence of specific indication and accompanying counseling. Similarly, it would be possible to test on request for certain dominant mutations conferring cancer predisposition in a particular ethnic population (e.g., 185delAG common mutation in BRCA1 gene in the Ashkenazi Jewish population). Carrier testing on demand for prevalent recessive mutations, or in females for prevalent X-linked recessive ones (e.g., Fragile-X), would perhaps raise less ethical concern, although unless there is a current screening program in that population for these mutations, testing may still be treated with caution as an inappropriately targeted use of resources. However, the possibility to screen someone for carrier status simultaneously for multiple prevalent recessive mutations is becoming feasible, and likely to be available as a marketed test within the next few years.

However, where the question is of testing for susceptibility to multifactorial disease, it is the interpretation of results that needs particular caution, and that questions current test value.

MULTIFACTORIAL DISEASE SUSCEPTIBILITY

The Human Genome Project enabled the identification of many polymorphic sites across the genome, particularly for SNPs (single nucleotide polymorphisms), which have been employed in a nonspecific "trawling" approach in studying disease susceptibility. In most cases these are not directly related to specific genes, and the mechanism of conferring susceptibility is unknown. Also, the incremental steps for conferring risk, even in combination, are small in comparison with risk conferred by environmental and lifestyle factors. Given the complexity of likely interaction in the human genome, and the likelihood that a "risk" allele for one disease may well be a "protector" for another disease, it would seem naïve to assume that a risk allele would confer the same strength of effect, or possibly the same direction of effect, in diverse genetic backgrounds. Hence, interpretation of susceptibility profiles according to combinations of polymorphisms is likely to be genetically ethnically specific. The marketing of direct-to-consumer tests, which report profiles of apparent susceptibility to several different multifactorial diseases, will be at best premature for the populations of many developing world countries, and at worst mere commercial opportunism.

Evidence for different degrees of susceptibility in different populations can be found in conditions where a variable mutation size correlates with severity of presentation. For example, in facioscapulohumeral (FSH) muscular dystrophy (FSHD), the extent of deletion required to give symptoms seems to be greater in Southeast Asian populations than in Caucasians.

Polymorphic SNPs as component susceptibility factors for disease represent one end of a spectrum of susceptibility. At the other end would be fully penetrant, single-gene Mendelian disorders, while in between would be various levels of reduced penetrance for a recognized mutation. Testing for reduced penetrance mutations is a risk predictor, the accuracy of which will become better defined as the additional factors controlling gene expression become understood.

Presymptomatic Disease Screening

In developing diagnostic laboratory services, consideration needs to be given to the nature of appropriate population screening programs, and whether they would be managed and run with the support of the laboratory service. In general, for a screening program to be appropriate a condition should:

i. be serious (i.e., high disease burden to individual person or family)
ii. be of sufficiently high incidence or prevalence for screening to be cost-effective
iii. be one where presymptomatic or early detection provides a treatment or management option that improves outcome
iv. have high sensitivity and specificity

In the United Kingdom, screening is now offered at birth from a blood spot card, for phenylketonuria (Guthrie test), hypothyroidism, cystic fibrosis (CF), and sickle cell anemia, an MCADD. The CF testing is based on immunoreactive trypsin (IRT) as an initial screen, and DNA testing for one of the four most common mutations in the CFTR gene in those with >99th percentile level for IRT on the first screen. The program detects equal numbers of heterozygous carriers and affected individuals, and may require sweat testing of those where only a single mutation was found, in order then to give diagnostic information to the parents and to target those cases where full CFTR gene sequencing may be required. Other programs trialed in the United Kingdom have included newborn screening for serum CK as a screen for Duchenne muscular dystrophy, in order to be able to alert parents to the potential risk of further born children, and to avoid long-delayed and unrecognized diagnosis, which can be distressing for families.

Screening programs need to be tailored to the specific genetic ethnic population concerned. One can anticipate that the diversity of programs will expand, including perhaps voluntary ones becoming available for carrier testing for prevalent recessive conditions within individual populations.

Participation in antenatal (fetal) screening programs will, by its nature, need to remain voluntary. Currently in the United Kingdom, programs are available mainly aimed at detection of Down syndrome (trisomy 21), and are based on combinations of maternal serum screening and fetal ultrasound scan. Trisomy 13 and 18 may be identified by the same scheme. Amniocentesis or chorion villus biopsy sample analysis, if taken as a result of maternal serum parameters +/scan findings, may be initially analyzed by qFPCR for chromosomes 21, 18, and 13; the role of full karyotyping, if qFPCR is normal, is under debate. However, the advent of array CGH based on a DNA sample may lead to adoption of a prenatal array platform covering each chromosome region for partial or complete aneuploidy (gross copy-number variation), and with specific targeting for well-established microdeletion or microduplication syndromes.

Genetic Risk Screening

Genetic screening for carriers of recessive or X-linked conditions, rather than affected persons themselves, is likely to be applicable only to the reproductive age-group. However, programs for hemoglobinopathies in particular, and for recessive conditions prevalent within the specific genetic ethnic population may be appropriate, particularly also where there may be a significant level of consanguinity. These tests are likely to be DNA-based, but any screening program needs to be subject to appropriate regulation to ensure that sensitivity and specificity are at a sufficiently high level, and that results can be conveyed appropriately to those participating.

SAMPLE TRANSPORT AS A CONSIDERATION IN LABORATORY SERVICE DEVELOPMENT

In developing laboratory genetic services, one of the overall decisions is what to develop in-house, and what tests to outsource to other labs. The transportability of specimens may be a significant factor in this discussion. Generally, EDTA samples or DNA can be held at ambient temperature for 24 hours (although 4°C is preferable), and so can be satisfactorily posted for next-day delivery. Samples for pulse-field gel electrophoresis (where intact longer lengths of DNA are required) may need more direct processing. Lithium heparin blood samples for cytogenetic preparation can survive at ambient temperature for 2–3 days. Skin cells for biopsy (e.g., from a fetus) are usually still viable for up to 3 days following death. DNA samples, once extracted, can be readily transported between labs, making it sensible to concentrate laboratory (and clinical) expertise for particular rare conditions, and to concentrate expensive equipment, in a few laboratories in a country rather than having many smaller diagnostic labs.

REGULATION OF DIAGNOSTIC GENETIC LABORATORY SERVICES

In the West, all diagnostic genetic laboratory services are regulated by one of the laboratory accreditation agencies, for example the Clinical Pathology Accreditation (CPA) in the UK and the European Genetic Testing network (EuroGeneTest) in Europe. Similar arrangements are in force in Australasia, Canada and the USA. In the UK, specific genetic tests that a genetic laboratory can offer is determined by UK Genetic Testing Network (UKGTN), an independent professional organization administered by the Department of Health and with professional input from a broad range of professionals including clinical geneticists, clinical cytogeneticists, clinical molecular geneticists and clinical biochemical geneticists. It is possible that the same genetic test is offered by one or more diagnostic genetic laboratories. The process by which a particular genetic test is approved and assigned to a genetic laboratory requires following the stringent guidelines and application procedure. The key to UKGTN approval is the 'Gene Dossier' that includes all scientific, technical and administrative (including costs) aspects of the specific genetic test in consideration.

SUMMARY

This chapter provides practical approaches in carrying out diagnostic genetic testing in some of the key situations encountered by a clinical geneticist and other team workers. The main areas of diagnostic genetic testing are outlined with reference to most commonly used laboratory methods. There are other important aspects that are not discussed in detail due to space constraints. It is important that any diagnostic genetic laboratory service in a developing nation with limited resources should focus on the most common genetic tests. Only selected genetic laboratories should be assigned more complicated methodology dependent genetic tests including those for rare diseases. It is essential that there should be clear distinction made between the research and diagnostic genetic laboratories. Any result generated from the research genetic laboratory should be replicated and verified in an independent diagnostic genetic laboratory. The interested reader is recommended to visit one of the following resources for more information.

ON-LINE RESOURCES FOR FURTHER INFORMATION

Several professional organizations and agencies have produced a number of guidelines for setting up genetic laboratories, accreditation requirements, types and list of genetic tests available and administrative information including costs (Table 29-2). A current or prospective genetic laboratory service provider in one of the developing countries may find useful information through browsing and accessing relevant information in these on-line portals.

TABLE 29-2 ON-LINE RESOURCES FOR FURTHER INFORMATION ON LABORATORY GENETIC TESTING

www.acmg.net	American College of Medical Genetics
www.bshg.org.uk	British Society of Human Genetics
www.ccmg-ccgm.org	Canadian College of Medical Genetics
www.cmgs.org	Clinical Molecular Genetic Society- UK
www.cytogenetics.org.uk	Association of Clinical Cytogeneticists in UK
www.Eurogentest.org	Harmonizing genetic testing across Europe
www.embl.org	European Molecular Biology Laboratory
www.esp-pathology.org	European Society of Pathology
www.geneticalliance.org.uk	Genetic Alliance-UK
www.rcpath.org	Royal College of Pathologists- UK
www.ukgtn.org	United Kingdom Genetic Testing Network

30.

ETHICAL ISSUES ARISING IN MEDICAL GENETICS IN DEVELOPING COUNTRIES

Angus Clarke

As a clinical geneticist based in the United Kingdom with little experience of medicine in developing countries, my focus in this chapter is on two topics. First, the ethical issues likely to arise in clinical genetics and genetic counseling practice—in the consultation—for a professional trained in the Western tradition but practicing in a very different social context. This entails a discussion of (i) relationships, (ii) decisions, and (iii) the approach to disease and disability in the context of a developing country. Second, the organization and purpose of genetics services in the developing world: How do the different contexts shape what is "appropriate," when the standard formulations of the ethical issues in genetics practice have been drawn up in the developed world? In what way do these different contexts alter what is ethically permissible?

AUTONOMY AND THE PROFESSIONAL

Before looking at the conduct of genetic counseling in more detail, however, we should first consider the Western notion of "autonomy." Conventional Western views often assume the centrality of autonomy and downplay the weight and scope of obligations to others. This view of autonomy, which is widely promulgated in both popular discourse and much of the literature on the ethics of genetics, can be seen as rather shallow. Individual choice is valued for its own sake, and is often taken to trump other principles as well as the practical consequences of any decision being made. This "thin" notion of autonomy has been roundly criticized in a number of contexts but is still widely employed. Individuals are taken to be rootless, as if they could be located outside—abstracted from—any social context of mutual obligations and concerns; i.e., "in vacuo." The choices that are to be accorded such weight may seem to be little more than whim or caprice, and need not reflect a person's commitments and sense of values. Even an ill-conceived or thoughtless preference is given much greater respect than it may warrant, simply because it has been uttered; resistance to a person's stated wish is the ultimate provocation to the True Believer in autonomy. This tends to devalue the role of the professional, reducing professionalism to the provision of services to a customer.

An exaggerated respect for autonomy can blind all parties to the burdensome aspects of choice in health care, especially when decisions have to be made at difficult times by those who remain competent but are distressed and unwell. Patients may then be given the relevant technical information on which to base their decision, but may feel emotionally abandoned and without human or professional support in making their decisions. This having been stated, one must acknowledge that personal autonomy may be undervalued in other societies where obligations to close family or a wider kinship network may be identified as the ruling principle, especially if a senior family figure expects to make the decisions about how others should honor and enact those obligations.

One factor that may powerfully influence the professional–patient relationship is the matter of payment for health care. The cash nexus of purchaser–provider may be seen as strengthening patient autonomy or, alternatively, as weakening the role of professional judgment and ethics, with implications for the conduct of medicine in all areas. This may make it more difficult for professionals to resist the demands of those purchasing healthcare. In settings where the professional structures are weak, the power gained by those purchasing healthcare may also lead to practices that would be professionally unacceptable, or even frankly illegal in other contexts. Examples would include the systematic breaches of privacy and consent processes that occur when decisions are made about a competent adult (often younger or female) by one or more others (often older or male), and in particular fetal sexing by ultrasound (with a view to the termination of female pregnancies) or the mutilation of female "circumcision."

While the rhetoric of Western autonomy usually acknowledges only two parties, the professional and the (paying) patient, there will often in fact be three parties:

the patient, the professional, and the purchaser. One must therefore ask whose autonomy is being respected by the professional—the patient's, or the purchaser's? And who is the purchaser? Is it the family patriarch? In patriarchal communities, the distribution of power in medical practice will be highly complex. Both the professional and the patient may be subject to strong social pressures, over and above any question of monetary influence.

RELATIONSHIP

In contemporary Western countries, the relationship between the two parents of a child—the relationship of the couple, the partners or spouses—may be regarded as essentially optional and potentially transient. Of course, many Western couples succeed in establishing enduring relationships that provide a stable home for the couple's children, whether they are married or not, but this cannot be assumed. The children in a household may be full siblings, half-siblings or step-siblings. It may be difficult to maintain cordial communication between parents and their children when the parents' social relationship has fractured. This in turn may make it difficult for genetic information to be transmitted even to close (biological) relatives, if they are not also close "socially." Even when family relationships are not socially awkward, of course, the transmission of "unwelcome" genetic information to relatives may be compromised by factors such as distance and opportunity, or concern that the information may cause distress or may trigger anger directed at the informant ("shooting the messenger"; Forrest et al., 2003; Featherstone et al., 2006).

While we have no choice about our parents or siblings, social relationships in the West are distinguished from biological ones, and are regarded as being at least as important. In addition, genetic testing has given us some control over the type (nature, quality) of our children; not only over whether to conceive, or whether to terminate a pregnancy, but also over our children's disease-related characteristics. This already means that some genetic disorders can be understood as being "optional," in that antenatal screening (to detect chromosome anomalies and structural malformations) and prenatal diagnosis (in the context of a family history of a specific disorder) can be used by families to prevent the birth of children with a wide range of conditions. This technologically enabled parental choice can be morally troubling when applied to the parent–child relationship, because that is often valued most highly when seen as a relationship of unconditional love and nurturing (McDougall, 2007).

In contrast to Western social relationships, those in less developed countries may feel very different. It would be all too easy to stereotype, making false assumptions of cultural homogeneity within any one "culture" or "community," and to exaggerate the differences between cultures and communities. Nonetheless some patterns can be seen in the differences between Western and other societies. One difference is the question of the marital relationship, often represented as being much more stable in less economically developed societies. This is not the place to debate that question, but it must be remembered that marriages that are difficult to annul may lead to great unhappiness and even domestic violence, while domestic violence remains a persistent problem in Western societies in which divorce and cohabitation without marriage both frequently occur. There are two particular patterns of relationship that must be mentioned in this context: (i) arranged marriages, and (ii) customary consanguinity, that of course overlap in some communities.

In a society where marriages are arranged for young people, those encountered in the social world of the parents (those thinking about arranging marriages for their children) will fall into two categories: those who (or whose children) might be eligible as a match for one's child—and "the rest." The fact of "having a child who will require a spouse" must color the dealings of parents with many others, at least until a betrothal has been arranged.

Where consanguinity is not a frequent arrangement, so that a marriage will usually be a marriage "out" into the wider society (exogamy), it will be of great importance for one's children that their prospects of a good match are not spoiled by association with genetic disease, or simply by the rumor of such problems in the family. One would anticipate that families in such a culture or community would often strive hard to suppress any suggestion that their "blood line" might be tainted. Of course, some families, especially educated or professional families, will prefer to investigate the possibility of a genetic disorder and confront it—perhaps taking advantage of genetic testing to identify carriers, and perhaps choosing to use prenatal diagnosis. Other families, however, might feel the need to choose the path of secrecy and tight information control as the only way to deal with the "courtesy stigma" that could otherwise prove highly disadvantageous for their children.

This response to malformation, genetic disorders, and also psychiatric disease is well known in all societies and was, indeed, a frequent theme in the English novel of the nineteenth century. Secrecy about diseases in the family—"skeletons in the cupboard"—are substantially less common now in Western societies than they used to be, even just 30 or 40 years ago. This change may be the result of a trust that medical science will help to minimize the risk of such problems occurring. It remains true, however, that information about some adult-onset, autosomal dominant disorders—e.g. Huntington's disease, polycystic kidney disease—is still not transmitted effectively through many families in the United Kingdom, with parents often anxious not to complicate the lives of their children before this becomes "necessary." The fact that such complications may be preferred by their children to the later feeling of having been deceived—that important information had been withheld from them—is often not well recognized until bitterness and resentment have developed. In Western countries, at least, anger at such concealment can be profound and may lead to family estrangement once it becomes impossible to sustain the secrecy.

Arranged marriages within a community practicing customary consanguinity face some different challenges. It will usually be much more difficult for brothers and sisters—who will often be the parents of the couple to be wed—to conceal from one another any problems that have occurred in their children or other family members, and of course the condition in question may well be viewed as a common (i.e., joint) concern by the parents of the young couple. If the prospective couple are more distantly related, then there may be the possibility of concealment; in that case, the implicated branch of the family will face a dilemma between deceiving the prospective in-laws or risking their loss of interest in arranging a marriage at all.

The extent to which these "cultural" issues are specific to a person's culture, community, or population of origin is not clear. If one adopts a historical perspective from which to view Western society, described in fiction as the social dynamics of blood and money, many of the same social arrangements can be seen to have played out here, too, in "the West"—if, indeed, one can regard both Jane Austen and Lev Tolstoy as "Westerners." This suggests that the different patterns of family relationship and communication found in different population groups today may be at least as much the result of differences in their levels of economic and technological development and education as anything intrinsic to their cultural identity. Clearly, these different factors will often be impossible to disentangle, but it would be a mistake to think of "cultures" as more fundamentally distinct than they are.

DIFFICULT DECISIONS

There are at least five categories of difficult decisions to be made by patients or clients about genetic testing. These are decisions about predictive genetic testing (except where the test results are clearly helpful for medical management), carrier screening to identify those in a population who carry an autosomal recessive disease, the genetic testing of children for disease prediction or carrier state (when this does not bring any clear health benefits in childhood), prenatal and preimplantation diagnostic testing and, finally, participation in "routine" antenatal screening programs, which may be regarded as a distinct category of test. Each has an extensive literature describing it, but we do not have the space to look at these in detail: for further insight into these, the reader may look elsewhere (e.g., Parker 2004; Clarke 2007, 2008, 2009).

There are generally four types of outcome to the process of deciding about a genetic test: the result may be favorable (also termed "negative"), unfavorable ("positive"), complicated (e.g., results of uncertain, indeterminate significance, or results that were completely unanticipated), and finally, the test may be deferred if the person decides not to go ahead with testing, at least for the moment. The clinician's approach in Western-style genetic counseling will be to help the client think through for themselves how they—and others—may respond to the different possible outcomes of the testing process. Not all clients wish to engage in such reflective work—it can be emotionally challenging—but it does help those who are looking for "reassurance" (a good result), and could not accept the opposite, to recognize that testing may not be the right step for them to take at that time.

In at least some developing countries, and indeed in some developed countries, the myth of the autonomous individual making decisions in a social vacuum is replaced by that of "the family" making "family decisions" for the good of all. Such an idyllic system would, of course, be wonderful if it lived up to its rhetoric of making decisions "for the good of all." While this will often be the intention of those involved, decisions will in practice usually be made by a family boss, often a patriarch but sometimes a matriarch, who takes a particular view of what would be best for the family. The decision-making family boss may be present in the consultation, which gives an opportunity for the counselor to engage them in reflection and see if their decision can at least be informed by an appreciation of the biological facts, rather than being based on a misunderstanding. If they are not present but operate remotely—perhaps after the consultation, or even by mobile telephone during the consultation—it will be more difficult to counter their influence when it appears to work against, perhaps to sacrifice, the interests of those who have come to the consultation and who would be the ones making the decision in a Western context.

Even in a series of consultations, it may be neither possible nor desirable for a genetics professional to resist the influence of a dominant family figure, when the client has lived within such a system of control for his or her entire life. Such occasions, however, will be difficult for a Western or Western-trained professional; it will feel unnatural and "wrong" for anyone other than the at-risk person to make a decision about predictive testing, or anyone but the pregnant woman, in discussion with her partner, to make a decision about prenatal testing or screening, and especially about the question of a termination of pregnancy.

In a Western setting, as elsewhere, decisions will be taken within, and constrained by, a social context. Let us think of the questions that might be raised in discussion with a woman confronting the possibility of a predictive or prenatal test: How would others in the family react to the different possible outcomes of this process? What care and support would be available—from within the family or from society more generally—if you became unwell (if you developed an adult-onset condition)? Or if a baby was born with serious developmental or health problems? How would you cope with the distress you are likely to feel? How would your partner feel? Would he support you (or you and the baby)? If you already have children, what help would be available for looking after them, if you have a child likely to be unwell and to need additional support?

If we transpose this clinical situation to the social context of a developing country, what will be different? If the family has resources, then more care may be available through the family but there will probably be less support from "society" or the state. The degree of support available from the family might also depend upon the social ranking of the woman; if she is living among her husband's kin, and has few allies within the household, then she may be vulnerable. Whether a baby was given much support might depend upon how weak the baby was thought to be (how likely to survive in that setting)...and upon the baby's gender...and perhaps upon the state's policy on permitted family size. As in developed countries, the practical implications of having a child with malformation or trisomy in a developing country depend greatly upon the prospects of their long-term survival, so that realism about the prospects for survival beyond infancy of a child with (for example) trisomy 21 is important.

One particular clinical issue that is likely to play out differently in different social contexts is that of the genetic testing of children. In many Western countries, predictive testing to identify those likely to develop adult-onset disorders (e.g., neurodegenerative conditions, familial cancers) would usually be deferred until the child had become adult and could make the decision in person. Families that are accustomed to being managed by a patriarch or other elder may seek testing at an early age so that decisions could be made about the child's life (e.g., education, employment, marriage) in the light of "the facts." Is it an unacceptable contemporary manifestation of a colonialist mentality to suggest that the Western defense of the individual child against family-imposed genetic testing is the "correct" approach? I would argue for the "defense" of the rights of the child, but it might be very difficult for a practitioner to implement the same policy if they came from the same community as the client family, and would themselves be expected to comply with the traditional expectations of doctor–patient–family relationships and behavior; to do anything else would require great courage.

In the context of the carrier testing of young children, the information might be wanted to permit the early planning of arranged marriages, especially within the family, taking into account the child's carrier status for recessive disease. Is this use of genetic testing sufficiently benign that we could countenance providing the service in our clinics in the United Kingdom? Refusal to do this could be seen as a dogmatic rejection of the community's preferred approach to marriage, and would deny them the benefits of modern genetics in making such arrangements. It would be seen as unacceptably rigid for a genetics professional in the UK to implement an invariant policy that could not be modified in the light of a specific set of circumstances; our professional guidance discourages the adoption of a general approach for or against such testing. Certainly, the concerns on behalf of the child may play out differently in different societies. We might like to protect children's autonomy, so that they can make their own decisions about genetic testing in the future and their privacy can be protected, but our approach might simply antagonize or estrange another community without any compensating benefits.

In addition to the two issues identified earlier, concerning arranged marriages and customary consanguinity, we have two additional "problems" that need to be addressed here: (i) who is making the decisions, and (ii) what is the status of women in the society? These issues will be core for any genetics health professional to consider. Within a patriarchal system, of course, the answers to these questions are often closely connected.

It will never be possible for a health professional, or even a health profession, to change an entire social system. As a minimum, however, the professionals must find ways of not making matters worse for those who are already vulnerable; a family's engagement with genetics services must not lead to the vulnerable experiencing, on balance, a worse life.

When a patriarch or matriarch usually makes the important decisions in a family, others will be unaccustomed to making their own decisions and may be confused when presented with choices. It has to be a matter of professional judgment when to promote the patient's independence from their family and when, in contrast, that would be too disruptive and damaging, so that the professional should support them within their traditional family structure. This calls for a ruthless emotional honesty on the part of the professional. This ruthless honesty is likely to require supervision for the clinical geneticist or genetic counselor, perhaps by a counselor or a psychologist, if such honesty is to be sustained throughout a professional career.

One especially difficult area may be that of fetal sex selection. In parts of India, the birth ratio has been seriously distorted by ultrasound scans assigning fetal sex and then leading to terminations of pregnancy when the woman is thought to carry a girl. For health professionals to perpetuate and to profit from this system is not only illegal (in India) but also a great abuse of professional privilege and status. For women to be obliged to terminate a female fetus, on the grounds that females are of such little value, must cause great distress, in addition to the distress felt by the women experiencing the termination of a (very likely wanted) pregnancy; this must "add insult to injury." In part, this may be driven by the financial burden of dowries and weddings (although dowries are also illegal in India), but the preference for sons is more deeply rooted than these monetary considerations. With the sex ratio in India having fallen below 900 females for 1,000 males overall, and being much lower in some areas and for subsequent births after the birth of one or two daughters (Jha et al., 2006), it seems that there may be a troubled future for the relationship between men and women in coming decades (Moazam, 2004). A more rigorous enforcement of the ban on fetal sexing may limit such problems, but they are still likely to be very substantial. Those working in genetic services need to be absolutely clear that such practices are degrading to women and are inflicting great damage on the future wellbeing of the population; it will require sustained insight and vigilance if unintended acts of collusion are to be avoided.

Sons are valued in China, too, and the preference for males may be especially strong when a couple are usually permitted to have only one child (Hesketh & Xing, 2006), although the One Child policy is said to be less strict in rural areas where agricultural labor is required. However, we do not know to what extent female fetuses are aborted, or whether the girls are born but simply not registered. While the latter would seem to be less grim, it may lead to its own problems with girls being denied education and other rights, as their existence would not be formally recognized.

DISEASE AND DISABILITY

In any given social setting, the genetics professional needs to be aware of the natural history of a range of diseases. If the chance of an infant with Down syndrome surviving to 2 years of age is remote, because cardiac surgery may not be available and because of the continuous exposure to a wide range of infectious diseases, then the nature and purpose of antenatal screening in that community—its impact and relevance—will be very different from that in the United Kingdom. By the same token, other conditions that weaken the body's immune defenses, or its resistance to UV light, will have a different time course in developing countries when compared to the developed world.

In a remote, rural community, the stigma associated with mild cognitive impairment may be less than in London, Paris, or New York. On the other hand, the social stigma of some disfiguring conditions may be just as intense. Such stigma may be a burden for females more than for males, if the primary role of a woman is understood to be as a wife and mother, so that a "flawed" physical appearance may be especially difficult (Rozario, 2007; Gaff & Clarke, 2007). While the Western cult of fame and beauty receives much publicity, there is no reason to think that the expectations for physical attractiveness in a spouse will be any less intense in developing countries. Particularly where marriages are arranged, there may indeed be less opportunity for a young adult suffering from disfigurement to find a partner who can see past this to value her for who she is; the "marriage-arrangers" may feel more constrained to find a woman without obvious "blemish" than a man might feel if seeking a partner for himself. There may also be a wider range of socially acceptable occupations for women in developed countries, leaving a woman with more career options than wife/mother.

Levels of physical disability are higher in the developing world because of the high rate of road traffic accidents and work-related injuries coupled with poor trauma care. In some countries there is also an enormous legacy from decades of conflict, and even fresh post-conflict injuries from land mines. Physical disability is therefore less associated with genetic disorders or malformation than in the West. One might therefore anticipate rather less prejudice and stigma but, if this is the case, it will often be accompanied by a less developed system of healthcare and social services.

The contribution of genetic factors to the causation of "ordinary" (not especially genetic) disease is important, as is the question of how this genetic contribution to the common, complex disorders is understood. It may appear strange to mention such conditions—the "ordinary" diseases such as coronary artery disease, type 2 diabetes (T2D), stroke, and the common cancers—but some of these occur at very high frequency in the greatly disadvantaged indigenous populations, such as the First Nations peoples (in North America) and aboriginal communities in Australia. There are several competing explanations for high frequencies of such "Western lifestyle" diseases in these often remote and always disadvantaged and stigmatized communities, including genetic selection for the ability to survive on low-calorie intakes, a mismatch between nutritional deprivation in utero and a subsequent excess of calories, the chronic stress of social deprivation and powerlessness, and simply an unhealthy lifestyle. The causal model adopted has major implications for the public policy response to poor health in marginalized communities (McDermott, 1998; Räisänen et al., 2006; Gluckman et al., 2008, 2010) and may be more generally relevant to the epidemic of T2D in South Asians in both South Asia and in the United Kingdom.

One other aspect of the social response to disease and disability needs to be mentioned as a contrast between developing countries and the West; this is fatalism, which may be thought of as the willingness to accept suffering or, perhaps, as resignation and passivity in the face of suffering. Physicians from developing countries often refer to the fatalism of their fellow nationals, especially in poor rural districts, and it may be attributed to religion. While this attitude may accompany a simple religious faith, it is not clear that it can be said to be caused by it. Poverty and a low level of education are associated with low expectations of life in general, and with "fatalism." While the great world religions may inspire the holy to lead a life of poverty and to neglect their physical wellbeing, they often also inspire lives of sacrifice and action on behalf of others that is the absolute converse of fatalism. There are, therefore, good reasons to be highly suspicious of any claim that fatalism is a freely chosen path through life and that, consequently, sophisticated medical services would be irrelevant if not intrusive. Experience with minority ethnic groups in the United Kingdom indicates that access to all health services is low when a group has recently arrived and has low expectations, and only a limited ability to speak English, but that demand for healthcare (including genetic services) increases once the group is more settled and the ability to communicate with health professionals is enhanced. Experience with African and South Asian families in England, for example, has shown that the uptake of genetic services for the hemoglobin disorders and for antenatal screening are affected by many complex social factors (Green & France-Dawson, 1997). Once specific barriers to access are addressed, however, interest in genetic services increases to become comparable to that in other communities.

PROFESSIONAL PRACTICE, THE ORGANIZATION OF GENETIC SERVICES, AND CONSANGUINITY

In many developing countries, health service provision is through a complex mix of providers that may include the state, various charities and international aid organizations, a few large, commercial providers and, often predominantly, a large number of small-scale private practices. This applies to pathology laboratory services as much as to clinical services. The need for most clinical and laboratory services to be financially viable must therefore be acknowledged and respected, but this can constrain service provision and prevent it from meeting the community's needs in the most effective fashion. Private practitioners will establish themselves in large cities and compete for the custom of the wealthy and the middle classes, often leaving rural areas and the poor with little access to good healthcare.

In such an environment, it may be difficult for genetic services to be shaped "rationally" unless government has the resources to invest in a planned system. The use of genomic approaches in guiding the choice of therapies within oncology will doubtless be developed for the wealthy few in some developing countries, but wider access to such opportunities seems unlikely in the short to medium term. A number of wealthy developing countries have established genetic screening programs—such as the carrier screening programs for thalassemia (and other autosomal recessive disorders) in Iran and Saudi Arabia—but poorer countries are often unable to implement such schemes.

We should take care here to explain that this chapter is addressing the questions of clinical genetic testing and counseling and screening; genetic technologies may well be introduced into developing countries for other tasks (such as microbial diagnostics or tumor characterization) but we are not concerned with those developments here. There may also be good reasons—scientific, educational, industrial—for conducting genomic research among the populations of developing countries (Alwan & Modell, 2003; Hardy et al., 2008; Séguin et al., 2008) but those factors are beyond the scope of this chapter.

The genetic services likely to be offered through private providers in developing countries will, in general, be those that appeal to the wealthy, such as predictive and prenatal testing within defined high-risk families, and may include Internet access to the clinically unwarranted—at best clinically marginal—SNP-based disease risk profiles. Genetic counseling will often be regarded as an expensive luxury, perhaps even positively unhelpful if it might lead some potential clients to decide against proceeding with the fee-earning genetic test. Only in wealthy countries (including those in the Middle East) are genomic technologies likely to be applied to the health of the general population.

A question of central importance to the future application of genetics to human health in the developing world is whether it is an appropriate goal for health services to seek to reduce the birth incidence of genetic disorders. Many Western countries have, in effect, answered this in the affirmative through the introduction of antenatal screening for malformation and for chromosomal aneuploidy (especially Down syndrome) over the past several decades. These programs were often introduced with the explicit goal of sparing resources, through the termination of pregnancies in which the fetus would be affected by a serious (costly) disorder, although it is now politically unacceptable to use such arguments. There are questions in these developed countries about the routinization of screening that undermines the autonomy of women, and the lack of respect implicit in many programs toward those affected by such conditions. How do these issues play out in the developing world?

Where childhood mortality remains very high, so that life expectancy is short—as in parts of Africa—the morbidity and mortality of genetic disorders is effectively invisible, as it is obscured by the mortality in early childhood from infectious disease and malnutrition. As countries pass through the demographic transition, with a fall in childhood mortality and a rise in life expectancy and in total population, the burden of genetic disorders becomes more apparent. The hemoglobinopathies, in particular, present a major challenge, as they are so common. Once it is usual to diagnose β-thalassemia in childhood, and if a country can afford to treat it with blood transfusion and iron chelation (to avoid the iron toxicity from regular transfusion), then a cohort of thalassemia survivors is formed who should now face a long and healthy life. This cohort will swell each year, however, with children who would previously have died without a diagnosis. The cost of treating this steadily expanding cohort for life then becomes apparent; projections make it clear that, as it enlarges, it could swallow the whole healthcare budget of low-income or even middle-income countries (WHO, 2002). One possible solution is to make population carrier screening and prenatal diagnosis available, as this may lead to fewer affected births, so that a policy of treating those affected becomes financially sustainable. The question then arises about the balance between "making carrier screening available" and "promoting" it or making it obligatory: what is an acceptable balance? Can developing countries afford the subtleties of the debates we have in wealthy Western countries about nondirectiveness and respect for those with disease and disability?

A number of countries now have experience of preconception carrier population screening (Zlotogora, 2009). China has introduced a system of premarital health checks under its Eugenics Law (Hesketh, 2003), with reproduction being effectively banned for couples deemed to be inappropriate parents. In Cyprus, there has been a policy of strongly encouraging premarital carrier testing for β-thalassemia, promoted by the Orthodox Church through its insistence upon carrier screening before marriage. This has led to an almost complete halt to the births of affected children through prenatal diagnosis in pregnancies at high risk, with the selective termination of affected pregnancies—not through the use of test results to alter the choice of a marital partner. In other communities—most famously in the case of the Dor Yeshorim movement in some Orthodox Jewish communities—decisions about arranged marriages take premarital genetic carrier testing into account, to avoid the birth of infants with Tay Sachs disease and other disorders (Raz & Vizner, 2008). Iran and several Arab countries have effectively mandatory screening programs.

If programs of genetic screening and prenatal diagnosis are to be promoted within an ethically sound framework, many would agree that they must be voluntary and must operate alongside, not in place of, good facilities for the treatment and care of those affected by the relevant disorders. This demonstrates in action—more reliably than in rhetoric—the respect in which affected individuals are held. It also allows families to learn about their disease, as they see it managed, and to find out about prenatal diagnosis from trusted professionals whom they know rather than impersonal health promotion leaflets or publicity. The degree of coercion that appears to operate in China (and some other countries) makes many uncomfortable (Harper, 1997). Even the church-supported initiative in Cyprus may seem harsh to those who live in societies where treatment for an affected child would be readily available. On the other hand—suggesting that a concern for the experiences and feelings of individuals is of significance even in much poorer countries—Akinyanju gives an account of the importance of making good medical treatment facilities available before introducing genetic screening programs, from his Nigerian perspective (Akinyanju, 1997). Respect for autonomy is not solely the concern of wealthy Western liberals.

An interesting development that appears to be imminent in a few oil-rich Middle Eastern countries is likely to be the use of genomic technologies (including exomic and whole-genome sequencing) to identify the rare, often private, recessive mutations common within particular kindreds, and that result in disease because of the high frequency of consanguinity in some of those communities. This could represent an application of truly personalized medicine in the sphere of reproduction. In most developing countries, however, this incorporation of the latest genetic science into standard clinical practice appears a more remote prospect.

Given that such technological interventions will remain unavailable in many countries as solutions to the increased reproductive loss, and the increased early childhood morbidity and mortality that result from customary consanguinity, what responses to this burden of disease might be appropriate and helpful within the relevant communities in developing countries? Recent estimates of the increased rates of stillbirth, and of neonatal and post-neonatal infant mortality, are each of the order of 1%, and estimates of the increase in birth defects are in the range 2%–4%; all these figures vary between communities, but the overall effect contributes substantially to disease in these communities (Bittles & Black, 2010; Zlotogora & Shalev, 2010). How should health professionals approach this topic of consanguinity and health outcomes?

The approach of health services to the question of consanguinity is littered with ill feeling and intercommunity tensions that should, at least with hindsight, have been avoided. In England, over the last 30 years there have been several cycles of publicly expressed concern about the consequences of consanguinity, with politicians calling for public health services to tackle the problem in apparent ignorance of the social functions of consanguinity (Darr, 1997). This has sometimes been exacerbated by insensitive media reports (Shaw, 2009) and has had predictably unhelpful consequences, with families of South Asian origin (as the most affected in this way within United

Kingdom) understandably feeling mistrust towards health professionals requesting information about family history. Within Israel, there have been efforts to make genetic services available to communities in which consanguinity is widespread; uptake of genetic services was shown to be influenced by many social factors, rather than the actual degree of genetic risk (Lewando-Hundt et al., 2001; Falik-Zaccai et al., 2008). Customary consanguinity is globally widespread, does not usually result in genetic disease, and is a practice reinforced by many social benefits (Modell & Darr, 2002). For example, a man's wife may move to her bridegroom's family's home, but will not be isolated: she will already know her parents-in-law as her uncle and aunt. If a child has a serious developmental problem or becomes ill, she can rely on the support of a family she knows well. Confrontational calls for a community to change its longstanding traditions will simply cause animosity, whereas offering education and explanations, and making modern genetic services available that can identify the cause of serious genetic disorders and provide reproductive options, will allow families to make informed decisions for the future. Perhaps the wealthy Middle Eastern countries, with longstanding traditions of consanguinity, will show the way for developed countries, and others, to follow.

LANGUAGE AND CULTURE

From the experience of genetic service delivery within developed countries, and the more limited experience within developing countries, it is clear that there are many barriers to the introduction of genetic services and their operation in practice. The most obvious cultural barrier is that of language; many developing countries have dozens or scores of languages used within their territory, and some have hundreds (India, Indonesia, China). This raises more problems in discussing the sensitive topics of genetic counseling than when treating malaria or a bone fracture. The question of who can act as an interpreter raises awkward issues of privacy, especially if a close relative (perhaps a young child) or a passing acquaintance is expected to serve as interpreter for a very personal discussion.

The scale of this problem of language—of being able to communicate with words at all—varies enormously between countries. In those with hundreds of languages, many of which may have no written form—as in Brazil or India—the practical difficulties may be immense. In other developing countries, the problems may be more tractable (as in South Africa, perhaps).

Another problem is the difficulty of explaining genetic concepts—to the extent that is necessary, which is sometimes overestimated—to a client or patient with little education or scientific background (Shaw & Hurst, 2008). Other, nonscientific, explanations may be favored in accounting for diseases or misfortunes in the past or the present (Shaw, 2009); for example, birth defects may be caused by "The Evil Eye," or by the mother having seen a person with a similar deformity during the pregnancy (Rozario & Gilliatt-Ray, 2007). In the face of nonrational but conventional, accepted explanations, it may be difficult to persuade a family that genetic services have anything to offer. In presenting such social science research findings, one must recall that Western societies may also resist the clinical, biomedical models of disease causation (Davison et al., 1992) and may retain superstitious beliefs while acknowledging the applicability of a scientific causal account (Weil, 1991).

One response to the diversity of cultures is to teach health professionals a degree of competence in facing those of specific different cultures. If the aim is to teach openness and to encourage the wish to communicate, then such teaching may be productive. Where the object is to convey specific items of cultural information—to instill "cultural competence"—then one must be cautious, because of the danger of stereotyping clients or patients, and of treating different cultural categories and traditions as if they were monolithic and homogeneous.

It is far more important for a health professional to have an open and respectful approach to all their clients and patients, of whatever age, gender, social status, or population group, than it is for them to "know" some specific "facts" supposed to apply to a traditional community's belief system.

CONCLUSION

The challenge of making genetic services available to the people of developing countries is prodigious, but of enormous importance. Genetic services have the potential to help the people of these countries to emerge from poverty and disease, while there are also real dangers from the too forceful application of genetics. Both the imposition of a centrally planned "public health genetics" approach to reproduction, and the inequity of an unregulated free market in healthcare, including genetic services, will be unacceptable to any health professional concerned with the welfare—in the broad sense—of the peoples of the developing world.

The challenge is to move forward on multiple fronts simultaneously, including community outreach of specialist services, the education of professionals and the public, developing an expectation of patient choice, respect for women, and delivering effective treatments for genetic disease, as well as ensuring the availability of prevention.

REFERENCES

Akinyanju O. (1997). Coping with the sickle cell gene in Africa. Chapter 9, pp. 133–146. In Clarke A, Parsons EP, eds. Culture, Kinship and Genes. Basingstoke: Macmillan.

Alwan A, Modell B. (2003). Recommendations for introducing genetics services in developing countries. Nature Reviews Genetics 4, 61–68

Bittles AH, Black ML. (2010). The impact of consanguinity on neonatal and infant health. Early Human Development 86(11), 737–41

Clarke A. (2007). Ethical and social issues in clinical genetics. Chapter 38, pp 799–824. In Rimoin DL, Connor JM, Pyeritz RE, Korf BR, eds. Emery & Rimoin's Principles and Practice of Medical Genetics, Vol. 1, 5th edition. London, New York, and Edinburgh: Churchill Livingstone.

Clarke A. 2008. Genetic testing and genomic screening, Chapter 39, pp. 595–612. In Kumar D, Weatherall D, eds. Genomics and Clinical Medicine. Oxford: Oxford University Press.

Clarke A. 2009. Genetic counseling, testing and screening. Chapter 22, pp. 245–259. In Kuhse H, Singer P, eds. A Companion to Bioethics, 2nd edition. Chichester: Wiley-Blackwell.

Darr, A. 1997. Consanguineous marriage and genetics: a positive relationship. Chapter 5, pp 83–96. In Clarke A, Parsons EP, eds. Culture, Kinship and Genes. Basingstoke: Macmillan.

Davison C, Frankel S, Smith GD. 1992. The limits of lifestyle: re-assessing 'fatalism' in the popular culture of illness prevention. Social Science & Medicine 34(6), 675–685.

Falik-Zaccai TC, Kfir N, Frenkel P, et al. (2008). Population screening in a Druze community: the challenge and the reward. Genetics in Medicine 10(12), 903–909.

Featherstone K, Bharadwaj A, Clarke A, Atkinson P. (2006). Risky Relations. Family and Kinship in the Era of New Genetics. Oxford: Berg Publishers.

Forrest K, et al. (2003). To tell or not to tell: Barriers and facilitators in family communication about genetic risk. Clinical Genetics 64, 317–326.

Gaff CL, Clarke A. (2007). Stigmatization, culture and counseling a commentary on growing up and living with NF1: a UK-Bangladeshi case study by Santi Rozario. Journal of Genetic Counseling 16, 561–565.

Gluckman PD, Hanson MA, Cooper C, Thornburg KL. (2008). Effect of in utero and early-life conditions on adult health and disease. New England Journal of Medicine 359, 61–73.

Gluckman PD, Hanson MA, Mitchell MD. (2010). Developmental origins of health and disease: reducing the burden of chronic disease in the next generation. Genome Medicine 2, 14.

Green J, France-Dawson M. (1997). Women's experiences of screening in pregnancy: ethnic differences in the West Midlands. Chapter 8, pp. 120–132. In Clarke A, Parsons EP, eds. Culture, Kinship and Genes. Basingstoke: Macmillan.

Hardy BJ, Séguin B, Goodsaid F, Jimenez-Sanchez G, Singer PA, Daar AS. (2008). The next steps for genomic medicine: challenges and opportunities for the developing world. Nature Reviews Genetics 9 (Suppl 1), S23–S27.

Harper PS. (1997). China's genetic law. Chapter 18, pp. 237–246. In Harper PS, Clarke A. Genetics, Society and Clinical Practice. Oxford: Bios Scientific Publishers.

Hesketh T. (2003). Getting married in China: pass the medical first. British Medical Journal 326, 277–279

Hesketh T, Xing ZW. (2006). Abnormal sex ratios in human populations: Causes and consequences. Proceedings of the National Academy of Sciences of the USA 103(36), 13271–13275.

Jha P, Kumar R, Vasa P, Dhingra N, Thiruchelvam D, Moineddin R. (2006). Low male-to-female sex ratio of children born in India: national survey of 1.1 million households. Lancet 367(9506), 211–218.

Lewando-Hundt G, Shoham-Vardi I, Beckerleg S, Belmaker I, Kassem F, Jaafar AA. (2001). Knowledge, action and resistance: the selective use of pre-natal screening among Bedouin women of the Negev, Israel. Social Science & Medicine 52(4), 561–569.

McDermott R. (1998). Ethics, epidemiology and the thrifty gene: biological determinism as a health hazard. Social Science & Medicine 47, 1189–1195.

McDougall R. (2007). Parental virtue: a new way of thinking about the morality of reproductive actions. Bioethics 21, 181–189.

Moazam F. (2004). Feminist discourse on sex screening and selective abortion of female fetuses. Bioethics 18, 205–220.

Modell B, Darr A. (2002). Genetic counselling and customary consanguineous marriage. Nature Reviews Genetics 3, 225–229.

Parker M. (2004). Ethical issues in clinical genetics and the Indian subcontinent. Chapter 28, pp. 555–572. In Kumar D, ed. Genetic Disorders of the Indian Subcontinent. Dordrecht: Kluwer Academic Publishers.

Räisänen U, Bekkers M-J, Boddington P, Sarangi S, Clarke A. (2006). The causation of disease: the practical and ethical consequences of competing explanations. Medicine, Health Care and Philosophy 9, 293–306.

Raz AE, Vizner Y. (2008). Carrier matching and collective socialization in community genetics: Dor Yeshorim and the reinforcement of stigma. Social Science & Medicine 67(9), 1361–1369.

Rozario S. (2007). Growing up and living with neurofibromatosis1 (NF1): A British Bangladeshi case-study. Journal of Genetic Counseling 16, 551–560.

Rozario S, Gilliatt-Ray S. (2007). Working Paper 93. Genetics, religion and identity: a study of Bangladeshi Muslims in Britain. Cardiff University School of Social Sciences. Available at: http://www.cardiff.ac.uk/socsi/resources/wrkgpaper-93.pdf.

Séguin N, Hardy B-J, Singer PA, Daar AS. (2008). Genomics, public health and developing countries: the case of the Mexican National Institute of Genomic Medicine (INMEGEN). Nature Reviews Genetics 9 (Suppl. 1), S5–S9.

Shaw A. (2009). Negotiating Risk. British Pakistani experiences of genetics. New York and London: Bergbahn Books.

Shaw A, Hurst JA. (2008). What is this genetics, anyway? Understandings of genetics, illness causality and inheritance among British Pakistani users of genetic services. Journal of Genetic Counseling 17(4), 373–383.

Weil J. 1991. Mothers' postcounseling beliefs about the causes of their children's genetic disorders. American Journal of Human Genetics 48(1), 145–153.

WHO Advisory Committee on Health Research. (2002). Genomics and World Health. The Potential of Genomics for the Health of the Developing Countries. Chapter 5, pp. 79–105. Geneva: World Health Organization.

Zlotogora J. (2009). Population programs for the detection of couples at risk for severe monogenic genetic diseases. Human Genetics 126(2), 247–253.

Zlotogora J, Shalev SA. (2010). The consequences of consanguinity on the rates of malformations and major medical conditions at birth and in early childhood in inbred populations. American Journal of Medical Genetics A. 152A(8), 2023–2028.

31.

GENETIC AND GENOMIC SERVICES IN EMERGING ECONOMIES OF THE DEVELOPING WORLD

Robert George Elles

This chapter will examine some of the issues and trends that need to be taken into account in evolving policy and developing opportunities in genetic and genomic (testing) services in emerging economies. This is based on a case study selected from among the leading group of fast-emerging economies of the developing world recognized by international intergovernmental agencies, such as the G20 and Organization for Economic Cooperation and Development (OECD). This cluster of countries includes Brazil, Russia, India, China, and South Africa (BRICS). Each economy is relatively large and is characterized by rapid and sustained growth, but typically with large disparities between rich and poor classes. Each has a different health system, but only in Russia is the state the main provider of healthcare. In each of the remaining countries, healthcare is provided by a mixture of public and commercial institutions.

India is the example used in this chapter. In India, healthcare policy is set by the Ministry of Health and Family Welfare, although responsibility for public health is largely devolved to individual States. Around 6% of GDP is spent on healthcare, of which only 21% is provided by the central government and states. In contrast to almost all countries in the world, in India the proportion of healthcare spending provided by the state has fallen in the last decade, even though GDP has risen in excess of 6% per year. Most spending and the majority of growth in healthcare is activity in the private sector.

In the public sector, the emphasis is on preventive public health measures and basic, e.g., primary, health provision through local clinics in rural and poorer urban areas. The main issues arise from poor nutrition, especially in childhood, and waterborne and other infectious disease including HIV/AIDS. The emphasis is on public health and preventive measures, including vaccination programs. Some of these programs, for example to counter leprosy, polio, and smallpox have been very successful. Other efforts, for example providing access to antiretroviral drugs, have been less effective than programs in other developing countries.

In this context, genetic healthcare in the public sector is a low priority. The few genetic services that exist are generally aligned to medical research. For example, the All India Institute for Medical Sciences (AIIMS) has a genetic program that organizes genetic clinics, tertiary genetic advisory services, and laboratory genetic services, including tissue typing for transplantation surgery. AIIMS is a one of the major tertiary public health institutions exclusively administered and funded by the central government of India. The Sir Ganga Ram Hospital in Delhi is a charitable semiprivate institution that provides clinical and laboratory genetic services to the public (on a payment basis) and to public and private health providers across the Indian subcontinent, including the Middle East. Across India there are 15 genetic counseling centers; for example, five institutions provide prenatal diagnosis for Duchenne muscular dystrophy.

DEMOGRAPHIC PROFILE

India has a population of 1.1 billion, often divided into socioeconomic groups that include the poor (23%), aspirants and consuming classes (47%), the very rich (1%), and middle class Indians (9%) with disposable income, a proportion of which is spent on healthcare. In this chapter we concentrate on the latter group as a potential economic driver for the development of genetic healthcare services in India. The size of the middle class is approximately 90 million (more than the population of the United Kingdom at around 60 million, which contributes to and is served by its socialized National Health Service). The middle class population is concentrated in west and south India. On average, families spend 11% of their household income on healthcare, with a middle class family of four spending £80–£140 per year. Around 2m of this group constitute the population with health insurance, which is growing at 13% per year. Overall the Indian economy has escaped the global recession, and is predicted to expand by 8.5% in 2010 and to outgrow China for the next decade. India's working age population will grow to 44% of the total population and will increase by 136m by 2020, while China's will grow by only 23m.

The population of India is young, and its working age population proportionate to children and aged dependents is significantly greater than in China. In China the impact of the one child policy has given it a legacy of an aging population profile resembling that of Japan, Korea, and much of western Europe. This population profile has been called India's *demographic dividend* and is predicted to produce a wave of economic dynamism, allowing India to emulate and perhaps even overtake China in its rate of growth. However, India as a decentralized capitalist economy and a democracy has taken a different path to China, and has not yet developed its economy as a major locus of manufactured products for export. India is inhibited by poor physical infrastructure, for example roads. However, India has a strong and outward looking service industry. This is based on cheap and reliable telecommunications allied to a strong information technology sector, a respect for copyright and intellectual property laws, an adequate educational base (adult literacy 66%, and over 80% among 15–24-year-olds), and its adoption of the English language as a business and educational medium.

The independent healthcare sector has played a significant part in this service-dominated economy. For example, the commercial hospital sector provides 60%–80% of health services in India. Currently it is highly fragmented, and major cities are served by many individual or small hospital groups with perhaps 200–300 inpatient beds. However, within this sector a small number of regional or all-India hospital chains have developed, including Apollo, Fortis, and Wockhardt.

Even smaller individual hospitals see incoming health tourism as a significant income stream, and some have developed satellite hospitals outside India. For example, Apollo has hospitals in the Middle East and Fiji. In 2005, 150,000 patients came from outside India for healthcare treatments and medical tourism is estimated to generate 2 billion USD by 2012. Healthcare employed 4m people in 2002, and by 2012 the private healthcare industry is predicted to contribute 8.5% to GDP. This emphasis has driven the adoption of high standards of care and access to advanced technologies in these hospitals, often recognized by external accreditation, for example through the US Joint Commission International agency.

In Europe, genetic medicine developed first as a specialty based on pediatrics and developmental conditions, and on the diagnosis and characterization of a large number of relatively rare diseases. National and European rare disease strategies continue to play a major role in the direction of genetic health services, as it is recognized that rare diseases remain collectively a significant health burden and are individually poorly served in terms of care management strategies and drug development. More latterly, adult-onset conditions, especially Mendelian subsets of common conditions beginning with breast and colorectal cancers, began to constitute a major part of the workload of clinical and laboratory geneticists. A strong ethos in this practice is the idea of the family rather than the individual as the unit of care. A genetic service may maintain formal family registries, but almost always need to keep genetic registers at the center of their practice mechanisms to maintain contact with families extended horizontally from the proband, and over the long term, as new generations become at risk of adverse outcomes from inherited conditions. This continuity of care model does not fit well in the Indian commercial healthcare model, which is based on a fixed menu of medical or surgical procedures (or diagnostic screens), with the healthcare episode being closely defined and ending with a bill being issued. This may drive a different model for the adoption of genetic technologies in healthcare in India.

HEALTH PROFILE IN INDIA

Infectious disease remains a major health burden in India. However, improvements in sanitation and availability of clean water, and the success of public health preventative programs, along with access to reproductive health options and vaccination programs (for example, for polio and smallpox) are changing the health burden even for the rural and urban poor. The result is that congenital, including genetic, disease is becoming more evident. In Sri Lanka it was estimated that treating thalassemia would consume 5%–10% of the total health budget. In practice, few patients access the most effective treatments; in India, with an estimated 100,000 beta thalassemia patients, only 100 bone marrow transplants for this condition are completed each year. Cardiac conditions have replaced infectious disease as the single most common cause of death among adults in urban and rural India. Worldwide, cardiovascular conditions are the most frequent cause of death, and the Indian subcontinent has the highest rate of cardiovascular deaths of any world region, at between 3.6 m–4.2 m per year in 2008 and predicted to rise to over 4.2 m by 2015. For the middle classes, the health burden has already shifted from infectious hazards to lifestyle-related illnesses including obesity and diabetes, and the health burden in other classes is expected to follow this pattern. Superimposed on this picture is a similar background rate of congenital or adult-onset genetic disease, as is found in Western populations. In addition, those communities that commonly practice consanguineous and caste-based marriages experience an enhanced rate of rare recessive conditions over that found in communities where this practice is rare. WHO surveys show that more than 50% of all recessive disease worldwide is associated with consanguineous relationships.

MODELS FOR THE DEVELOPMENT OF GENETIC SERVICES IN THE INDIAN COMMERCIAL HEALTHCARE SECTOR

In India, the shape of the health economy and the trajectory of demographic and economic growth point to the commercial sector as having an important potential role in the delivery of genetic medicine.

Clinical genetic services are relevant to all fields of adult and childhood disease. The birth frequency of single-gene and chromosomal disorders is around 2%, while at least 3% of couples have a high risk of a recurrent condition in their children. Genetic factors are prominent in about 30% of serious chronic conditions, and specific genetic variants have been associated with more than 1000 single-gene disorders.

A referral to a clinical genetic service is appropriate for individuals concerned about a disorder with a significant genetic component (this includes inherited and sporadic genetic disorders). A service typically includes the following:

- Accurate clinical and genetic laboratory diagnosis, risk estimation, genetic counseling for individuals and families to support their decision making concerning family planning, prenatal and predictive testing, prevention of a disorder or complications, referral for surveillance for tumors, or other symptoms of disease progression. In addition, referral to other health professionals for expert advice, specific treatment, or trials and treatment.
- Clinical genetic services are underdeveloped in India. In the short term, an in-house comprehensive clinical and diagnostic genetic service could provide benefits to patients under the care of clinical specialties within a commercial hospital group and be a unique selling point to attract new patients from India and overseas, and generate income for the company.
- In the longer term, genetics will be central to the provision of healthcare. Establishing an integrated clinical and core technology genetic services platform will position Indian commercial hospitals to maximize benefits and income from the provision of genetic healthcare.

TWO PROJECT MODELS FOR A COMMERCIAL HOSPITAL GROUP

In this model, a commercial hospital group develops a comprehensive clinical, counseling, and laboratory genetic medicine capability, accepting secondary referrals mostly from other medical specialties within the group. It provides consultations from a fixed base through satellite clinics in neighboring cities, and to more remote hospitals by telemedicine.

The project would initially require clinical and laboratory experts from a developed economy to work with the commercial hospital group and establish a readymade clinical genetic service. It would also require the recruitment and training of Indian staff, commitment of space, and purchase of equipment and consumable materials. The two major components of the project are the establishment of two or three fixed-site and satellite or telemedicine-linked clinical genetic and counseling clinics within a multicenter hospital group in India, and secondly, development of a centralized and standalone genetic diagnostic laboratory facility to serve multiple hospitals.

CLINICAL SERVICE

As an example, a clinical service would require sessions in one or more base hospitals, of a minimum of one consultant clinical geneticist, two trainee nonmedical genetic counselors, and two support staff (clerical/secretarial). In addition, an outpatient facility would be required comprising offices, consulting room, examination/clinic room, medical records, and full access to diagnostic and telemedicine facilities. In each satellite hospital, access to an outpatient facility should be available for a number of days each month according to demand. Each base clinic could serve two or more satellite clinics, in other group hospitals in neighboring cities, per month.

LABORATORY SERVICE

A new laboratory service would require a minimum of a fully equipped laboratory space and support facilities, standing alone and dedicated to diagnostics, with funding of short-term visits to expert laboratories in Europe or the United States, or elsewhere. A budget for initial startup support from abroad would include purchasing specialized genetic tests from over 1000 services available worldwide.

The second model is more directly aligned to individual medical specialties within the group. In this model, surgical or medical procedures are identified where a genetic test is key to the success or effectiveness of the procedure. This test is then offered either from a source in India, or initially from foreign laboratories with a view to a technology transfer to bring the test within India. Examples may include sets of "genetic test and procedure" that illustrate economic and ethical use of medical and economic resources. For example, genetic tests for familial mutations in genes specific for inherited cardiac conditions, such as arrhythmogenic right ventricular cardiomyopathy (ARVC) and catecholaminergic polymorphic ventricular tachycardia (CPVT), may help in the best case selection for the requirement or contraindication for expensive intracardiac devices, for example of implantable defibrillators that cost the UK health service around £20,000 plus additional costs incurred in long-term clinical surveillance. Another category is companion diagnostics; for example, testing for mutations in the epidermal growth factor receptor (EGFR) gene, indicative of resistance to anti-EGFR therapies in non–small cell lung cancer patients.

Although genetic testing and clinical genetic services are not widespread in India, the rapid growth and high standards and availability of technology make the commercial sector a promising vehicle for introducing genetic services, training personnel, and transferring technologies. The public–private partnership is the key for effective genetic services of high clinical utility for less wealthy citizens in the developing world, through philanthropic efforts and affordable insurance provision.

FURTHER READING

1. The Economist October 2010.
2. Trustech Market intelligence report 2005
3. WHO Genetics programme; Genetics, Genomics and Public Health: The Implications of testing and screening for health in developing countries 2005.

PART II.

GENOMICS IN MEDICINE AND HEALTH—REGIONAL

AFRICA

32.

ANTHROPOLOGY AND POPULATION HETEROGENEITY IN NORTH AFRICA

Rym Kefi

INTRODUCTION

North Africa has represented a coveted area for many civilizations since ancient times because of its strategic location. Indeed, it is located at a crossroad between Europe, Africa, and Asia (Figure 32-1). It is delimited by the Mediterranean and the Atlantic Ocean in the north and the east, and by the Sahara and the Red Sea in the south and the west. Only 25 km separate it from the Iberian Peninsula, and a hundred kilometers separate it from Italy. Moreover, the Sinai in the west ensures the connection of North Africa to Asia. Recently, there was a resurgence of interest from anthropologists and geneticists to better understand and trace the origins of the different settlements of this Mediterranean area.

PREHISTORIC AND HISTORIC DATA

The six North African countries (Egypt, Libya, Tunisia, Algeria, Morocco, and Mauritania) did not know the same prehistoric civilizations. Thus in all the studies describing the prehistory of North Africa, there is a distinction between the Maghreb (Libya, Tunisia, Algeria, Morocco, Mauritania) and the Valley of the Nile represented by Egypt. The most ancient human presence in the Maghrebis dated to 700,000 years before present (BP), and corresponds to bone remains of *Atlanthropus maurétanicus* discovered in Ternifine in Algeria (Arambourg, 1963). Specimens excavated from archaeological sites of Sidi Abderrahman (200,000 YBP), Djebel Irhoud (100,000 YBP), Ain Hanech, Salé, and Rabat (160,000 YBP) in Morocco (Figure 32-1) highlight the transition of *Homo erectus* to *Homo sapiens sapiens* (Ferembach, 1986; Aouraghe, 2006).

Around 20,000 YBP to 10,000 YBP, western North Africa has been occupied by Mechta-Afalou Men, also called Iberomaurusian (in reference to late Paleolithic industry- who replaced the Aterian aborigines (in reference to middle Paleolithic industry).

Regarding the Nile Valley, the most ancient Human remains in this region were discovered in the archeological site of Nazlet Khater (33,000 years BP) (Figure 32-1). The anatomic features of the man of Nazlet Khater are incontestably more modern than the Aterian Man (Dutours 1995, 2001).

In the eighth millennium BP, a new type of *Homo sapiens* appears in the oriental part of the Maghreb (Tunisia and Eastern Algeria). They already presented the morphological features of certain current Mediterranean populations. They were referred to as "proto-Mediterranean," and were the authors of the Capsian culture. Proto-Mediterraneans were the ancestors of the Berbers.

During the historic period, North Africa has been colonized by Phoenicians, Greeks, Romans, Vandals, Byzantine, Arabs, Turkish, Spanish, English, and French.

From 1950 to 1970, all of the North African states gained independence from their colonial European rulers, except for a few small Spanish colonies on the far northern tip of Morocco.

GENETIC DATA

The richness of the cultural inheritance of North African populations has made them the subject of many molecular investigations based on the use of a large variety of genetic markers. All these studies highlighted the richness and the complexity of the genetic pool of the North African populations (Plaza et al., 2003; Kefi et al., 2005; Gonzalez et al., 2006).

Statistical analyses calculating the specific parameters to each population; i.e., haplotypic diversity, and the mean number of pairwise differences, showed that at the regional scale, the North African populations have a high haplotypic diversity for mitochondrial DNA polymorphism (Table 32-1). With the exception of Algeria, this diversity is similar to that observed in the Arabic peninsula (Iraq, Syria, and Palestine), and higher than that observed in Europe (except Andalusia).

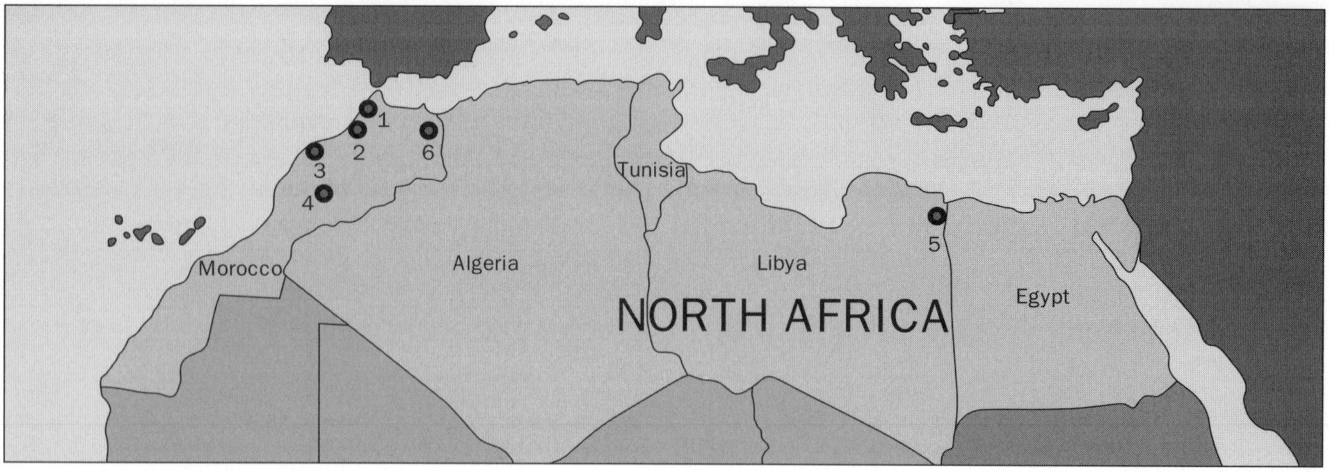

Figure 32-1 Map of North Africa.

TABLE 32-1 GENETIC DIVERSITY PARAMETERS IN SOME NORTH AFRICAN, EUROPEAN AND ASIAN POPULATIONS

Countries	Populations	N	N Haplotype	N of Polymorphic Sites	Gene Diversity	MNP	References
Tunisia	Makthar	49	37	57	0.9855 ± 0.007	7,349	Kefi (2005)
Tunisia	Tunis	47	41	58	0.9889 ± 0.009	6,041	Plaza et al. (2003)
Tunisia	Kesra	47	20	43	0.932 ±0.021	6,265	Cherni et al. (2009)
Tunisia	Zriba	50	16	31	0.904±0.022	3,949	Cherni et al. (2005)
Algeria	Mzab	85	30	36	0.9431 ± 0.010	4,855	Côrte-Real et al. (1996)
Algeria	Algeria	47	24	44	0.9399 ± 0.018	5,219	Plaza et al. (2003)
Morocco	Souss	50	34	35	0.9608 ± 0.018	4,312	Brakez et al. (2001)
Morocco	North Morocco	60	38	50	0.9633 ± 0.014	4,514	Plaza et al. (2003)
Morocco	Arab speakers	50	46	60	0.9943 ± 0.006	6,747	Plaza et al. (2003)
Mauritania	Mauritania	30	23	30	0.9747 ± 0.017	6,192	Rando et al. (1998)
Libya	Tuareg	23	21	40	0.992 ±0.015	6,838	Ottoni et al. (2009)
Morocco	Sahara	56	41	45	0.9760 ± 0.012	5,499	Plaza et al. (2003)
Egypt	Egypt	102	87	80	0.9950 ± 0.002	7,816	Stevanovitch et al. (2004)
Egypt	Gurna	34	29	57	0.9893 ± 0.010	8,670	Stevanovitch et al. (2004)
Spain	Catalonia	61	37	40	0.9290 ± 0.027	3,870	Plaza et al. (2003)
France	France	109	62	57	0.963 ± 0.010	3,921 ± 1,980	Plaza et al. (2003)
Italia	South Italia	86	68	71	0.962 ± 0.016	4,714 ± 2,330	Francalacci (1996); Richards et al. (2000)
Italia	Sardinia	69	44	48	0.9369 ± 0.023	4,043 ± 2,043	Di Rienzo et al. (1991)
Italia	Sicilia	169	94	83	0.9333 ± 0.016	3,842 ± 1,941	Cali et al. (2001); Richards et al. (2000)
Greece	Greece	186	88	73	0.9677± 0.007	4,344 ± 2,158	HVR database (www.hvrbase.org)
Turkey	Turkey	213	152	94	0.9877 ± 0.003	4,972 ± 2,428	Comas et al. (1997); Richards et al. (2000)
Saudi Arabia	Saudi Arabia	43	37	56	0.9922 ± 0.007	6,851 ± 3,288	Di Rienzo et al. (1991)
Palestine	Palestine	117	100	90	0.9944 ± 0.003	5,799 ± 2,793	Richards et al. (2000)
Syria	Syria	69	56	70	0.9881 ± 0.007	5,432 ± 2,648	Richards et al. (2000)
Iraq	Iraq	113	91	85	0.9904 ± 0.004	5,659 ± 2,733	Richards et al. (2000)

In North African countries, the haplotypic diversity of the mitochondrial DNA is quite variable as a function of the geographical localization and the ethnic group (Table 32-2). Indeed, a difference in genetic diversity was observed between the population of the Berber city of Sejnane and the population of the Berber village of Takrouna, in spite of their geographic proximity in the north of Tunisia (Frigi et al., 2006).

For ethnic origins, as an example: the Berber population of Kesra in Tunisia very different from the population (not Berber) of Zriba, separated only by 40 kilometers (Cherni et al., 2005a). In the same vein, a differentiation was found between Arab populations and Berber populations from the island of Djerba in Tunisia (Yacoubi-Loueslati et al., 2006).

The variability of the haplotypic diversity in North Africa has been also signaled in molecular studies using other genetic markers, such as Alu polymorphisms (Ennafaa et al., 2006) and Y-chromosome polymorphisms (Khodjet El Khil et al., 2001; Cherni et al., 2005b).

The genetic structure of North African populations is a mosaic composed of North African, Eurasian, and sub-Saharan components.

NORTH AFRICAN COMPONENT

The North African component represents the Paleolithic indigenous substrate (up to 50,000 years). This ancestral origin is revealed by the presence of the mitochondrial haplogroup U6.

The haplogroup U6 was observed for the first time in Mozabits, among the Berber population from the south of Algeria (Côrte-Real et al., 1996). Mozabites show a high rate of haplotypes sharing the following polymorphisms: 16172C, 16189C, 16219G, and 16278T compared to the mitochondrial DNA reference sequence (Anderson et al., 1980; Andrews et al., 1999).

Other studies analyzing Berber and non-Berber populations from North Africa defined the specific motif of U6 as 16172C–16219G. Phylogenetic analysis (Macaulay et al., 1999; Maca-Meyer et al., 2001), showed that U6 is the "brother" of U5, the oldest haplogroup in Europe (up to 50,000 years BP; Richards et al., 1998). U5 and U6 would have a common ancestor in the Near East, from which two branches diverged—one populating the Northern Mediterranean coast, giving rise to the haplogroup U5 in Europe, the second colonizing the southern Mediterranean coast, giving rise to the haplogroup U6 in North Africa.

Thus, the haplogroup U6 marks the signature of the return of the modern man toward Africa, around 39,000–52,000 years ago (Rando et al., 1998; Macaulay et al., 1999; Maca-Meyer et al., 2001, Maca-Meyer et al., 2003).

In Egypt the haplogroup U6 is only slightly represented, and even sometimes absent (Krings et al., 1999; Stévanovitch et al., 2004) however, an ancestral component is highlighted in this region represented by the haplogroup M1 (Table 32-2).

M1 originated in east Africa. Its high frequencies and diversities are found in Ethiopia (20.3%) and in Gurna in Egypt (17.6%). The haplogroup M1 marks an exit from Africa to the Near East (Quintana-Murci et al., 1999; Stevanovitch et al., 2003).

U6 and M1 show a differentiation, in the Paleolithic period, between the east and the west of North Africa, which is in agreement with the anthropological data (Dutour, 1995).

The presence of an indigenous ancestral component in North Africa was also highlighted by the analysis of the Alu polymorphism and the microsatellites of gene *CD4* (*CD4* STR). It was shown that the haplotype 110(–), specific to North Africa, represents the ancestral component in the genetic pool of the current populations (Flores et al., 2000). Indeed, the haplotype 110 (-) was reported with frequencies of 2% to 4% in North Africa, whereas it is missing in sub-Saharan Africa and Europe (except the Iberian peninsula, 2%).

Studies analyzing the polymorphisms of Y-chromosome also detected an ancestral indigenous component in the genetic pool of the current North African populations, represented by the polymorphism M81. This polymorphism is shared by 43% to 75% of Northern African populations (Bosch et al., 2001; Underhill et al., 2000). However, the age of the ancestral component given by the paternal lineage data is about 4200–5600 years (Arredi et al., 2004, Cruciani et al., 2004), more recent than that given by the maternal lineage data of around 50,000 years (Rando et al., 1998; Macaulay et al., 1999; Maca-Meyer et al., 2001).

EURASIATIC COMPONENTS

Genetic flow from Eurasia is highlighted by the presence of the haplotypes $Gm^{5*;3;23}$ - $Gm^{5*;1;17;...}$ - $Gm^{5*;3;...}$ for the marker IgG (Yacoubi-Loueslati et al., 2001), by the alleles 186 and 190 of the locus DYS19, allele 254 of the locus DYS392 for the Y-chromosome STR (Khodjet El Khil et al., 2001), and by the haplogroups H, V, T, J, U, K, X for the mitochondrial DNA polymorphism (Brakez et al., 2001; Stevanovitch et al., 2004; Plaza et al., 2003; Fadhlaoui-Zid et al., 2004, Cherni et al., 2009).

The contribution of the Eurasiatic component in the genetic pool of North African populations decreases from east to west and from north to south (Krings et al., 1999; Bosch et al. 2000). This gradient cannot be applied for Berber populations. In fact, Berbers are characterized by a major Eurasiatic component, which brings them closer

TABLE 32-2 MITOCHONDRIAL DNA HAPLOGROUP DISTRIBUTIONS IN NORTHERN AFRICANS POPULATIONS

Countries	Population	N	L1	L2	L3	M1	N	R	I	W	X	J/T	J	T	U	U5	U6	K	H	preHV	HV	V	Others	Ref
Tunisia	Jerba (Berbers)	30	–	3.3	10	–	–	–	–	–	–	–	16.7	13.3	20	3.3	–	3.3	23.3	–	–	6.7	–	1
Tunisia	Jerba (Arab speakers)	29	3.4	7	10.3	–	–	–	–	–	–	–	7	20.7	13.8	–	10.4	3.4	17.2	–	–	3.4	3.4	1
Tunisia	Makthar	49	14.3	8.2	20.5	–	2	–	–	–	2	–	2	10.2	2	–	6.2	6.2	18.4	4.1	–	2	2	2
Tunisia	Tunis	47	–	12.8	14.9	4.2	–	–	2.1	–	2.1	–	4.2	6.4	6.4	4.2	–	6.4	23.4	6.4	–	–	6.4	3
Tunisia	Zriba (non Berbers)	35	–	11	11	3	–	–	–	–	–	–	3	3	15	6	9	6	34	–	–	–	–	4
Tunisia	Kesra (Berbers)	43	7	12	5	–	–	–	–	–	–		2	5	11	2	9	14	19		–	7	7	5
Algeria	Algeria	47	6.3	6.3	14.9	12.8	–	–	–	–	2.1	–	13	4.2	2.1	–	–	4.2	34	–	–	–	–	2
Algeria	Mzab	85	–	5.9	7	4.7	–	–	–	–	–	–	3.5	4.7	12.9	–	28.2	–	24.7	–	–	8.2	–	6
Morocco	Asni (Berbers)	53	7.5	5.7	7.5	3.75	1.9	3.75	–	–	1.85	–	–	–	1	1	11.3	3.75	37.7	–	7.5	3.75	1.9	7
Morocco	Bouhria (Berbers)	70	7.2	2.8	2.9	4.3	2.8	–	–	1.4	2.8	–	–	9.3	5.7	–	1.4	5.7	37.1	–	15	1.4	–	7
Morocco	Figuig (Berbers)	78	6.4	4.3	29.7	2.1	–	–	–	–	1.1	–	11.7	3.2	1.1	–	3.2	–	24.4	–	5.3	3.2	4.3	7
Morocco	Marrakech	52	9.6	1.9	1.9	3.8	–	–	1.9	–	1.9	–	5.8	–	–	5.8	3.8	3.8	23.1	1.9	–	7.7	26.9	8
Morocco	Moroccans (Arab speakers)	50	10	6	16	2	–	–	–	–	4	–	4	4	12	–	8	4	26	–	–	4	–	3
Morocco	Moroccans Berber (North)	60	1.6	–	1.6	–	–	–	–	–	–	1.6	9.4	15.6	6.2	–	7.8	7.8	42.2	–	–	6.2	–	3
Morocco	Souss	50	6	10	10	–	–	–	–	–	–	–	10	4	8	–	6	2	32	2	–	10	–	9
Egypt	Siwa (Berbers)	78	1.3	5.1	12.8	16.7	–	2.6	–	–	3.8	–	5.1	1.3	1.3	16.7	–	11.5	2.6	–	14	–	5.2	7
Egypt	Egyptians	102	4.5	6	7.5	7.7	7.5	–	4.5	–	–	–	6	22.7	4.5	–	1.5	3	18.1	3	–	–	1.5	10
Egypt	Gurna	34	17.6	2.9	8.8	17.6	8.8	–	5.9	–	–	–	5.9	5.9	8.8	–	–	–	14.7	–	–	–	2.9	10
Lybia	Touareg	129	4	12	12	2	–	–	–	–	–	–	–	–	–	–	–	–	–	61	–	4	6	11
Mauritania	Mauritania	30	–	23.4	13.4	6.7	–	–	–	–	–	–	3.3	–	3.3	–	20	6.6	20	–	–	3.3	–	12

N: Number; MNP: Mean Number of Pairwise.

References: (1) Loueslati et al. (2006). (2) Kefi (2005). (3) Plaza et al. (2003) (4) Cherni et al. (2005) (5) Cherni et al. (2009). (6) Côrte-Real et al. (1996). (7) Coudray et al. (2009). (8) Falchi et al. (2006). (9) Brakez et al. (2001). (10) Stevanovitch et al. (2004). (11) Ottoni et al. (2009). (12) Rando et al. (1998).

to European populations than to sub-Saharan populations (Fadhlaoui-Zid et al., 2004; Cherni et al., 2005; Coudray et al., 2009; Ottoni et al., 2009). This observation highlighted the interpopulation heterogeneity in North Africa.

SUB-SAHARAN COMPONENT

The existence of the sub-Saharan type was confirmed by the presence of the haplotypes 115(+) and 85(–) of CD4 STR/Alu polymorphism in high frequency, in particular in the populations of western North Africa (Flores et al., 2000; Bosch et al., 2000; Esteban et al., 2004).

Regarding the mitochondrial data, the proportion of the sub-Saharan component in the genetic pool of the current North African populations varies from 4% in the Berber from the north of Morocco to 43% in the Mauritanians (Plaza et al., 2003; Turchi et al., 2009). Previous studies showed the presence of a south/north gradient in the sub-Saharan gene flow, observed as well in northwestern populations (Rando et al., 1999; Brakez et al., 2001; Plaza et al., 2003) and in Nile Valley populations (Krings et al., 1999; Stévanovitch et al., 2004). However, while studying the population of Makthar, in the center of Tunisia, we found a high proportion of the sub-Saharan haplotypes (43%), as significant as that observed in the Mauritanian population (43%; Kefi et al., data in submission). Thus, the south/north gradient of the sub-Saharan component, suggested by previous studies, cannot be generalized for all the populations of North Africa. This reflects the high genetic heterogeneity in North Africa.

The chronology of the sub-Saharan genetic flow was defined by the analysis of mitochondrial DNA variability in the prehistoric population of Taforalt from Morocco. The genetic structure of this population, dated from 23,000 to 13,000 years ago, is composed of North African and the Eurasiatic components. The absence of sub-Saharan haplotypes suggests that the genetic flow from sub-Sahara to North Africa took place less than13,000 years ago (Kefi et al., 2005), probably in the prehistoric period during trans-Saharan trade of gold, salt, and slaves.

Molecular investigation was also focused on the impact on the genetic pool of North African populations of the Arab Moslem invasion during the seventh century. The results of these works are controversial depending on the type of the molecular markers used. Studies based on autosomal markers, such as the ABO/Rhesus system, IgG polymorphism, and Alu polymorphism, show no significant difference between the Berber (excluding Touaregs) and Arab populations (Sanchez-Mazas, 2000; Yacoubi-Loueslati et al., 2001; Ennafaa et al., 2006), whereas mitochondrial DNA and the STR of Y-chromosome data underline a clear differentiation between these two groups (Loueslati et al., 2006; Cherni et al., 2005a, 2005b; Khodjet Elkhil et al., 2005).

It is interesting to note that genetic differentiation was also observed among Berbers from Maghreb and Egyptian Berbers. This can be explained by the prehistorical and historical events that are different between Maghrebian and Egyptian populations. Influences from the Middle East and east Africa are marked in the Siwa region, while southeastern European influences are observed in the Maghreb (Coudray et al., 2009).

Situated at the crossroads of Africa, Europe, and Asia, North Africa has received many migratory waves since the Paleolithic period, which caused a complexity in the settlement of this region. Except some ethnic groups such as Mozabite and Touaregs, molecular investigations carried out on Berbers and non-Berber groups show a high genetic diversity in North African populations.

Despite these several molecular studies on North African populations, many issues are yet to be resolved, such as the impact of genetic diversity on monogenetic and multifactorial disorders and the influence of genetic polymorphisms on drug responses.

REFERENCES

Anderson S, Bankier AT, Barrell BG, et al. (1980). Sequence and organization of the human mitochondrial genome. *Nature 290,* 457–465

Andrews RM, Kubacka I, Chinnery PF, Lightowlers RN, Turnbull DM, Howell N. (1999). Reanalysis and revision of the Cambridge reference sequence for human mitochondrial DNA. *Nat Genet 23,* 147.

Aouraghe H. (2006). Histoire du peuplement paléolithique de l'Afrique du Nord et dynamique des interactions entre l'homme et son environnement. *C.R.Pale* vol 5, 237–242.

Arambourg, C. (1963). Le gisement de Ternifine II-l'Atlanthrope mauretanicus. *Archives de l'I.P.H 32,* 37–190.

Arredi B, Poloni ES, Paracchini S, et al. (2004). A predominantly neolithic origin for Y-chromosomal DNA variation in North Africa. *Am J Hum Genet 75*(2), 338–345

Brakez Z, Bosch E, Izaabel H, et al. (2001). Human mitochondrial DNA sequence variation in the Moroccan population of the Souss area. *Ann Hum Biol 28,* 295–307.

Bosch E, Calafell F, Perez-Lezaun A, et al. (2000). Genetic structure of north-west Africa revealed by STR analysis. *Eur J Hum Genet* 8(5), 360–366.

Bosch E, Calafell F, Comas D, et al. (2001). High-resolution analysis of human Y-chromosome variation shows a sharp discontinuity and limited gene flow between north western Africa and the Iberian Peninsula. *Am J Hum Genet 68,* 1019–1029.

Cali F, Le Roux Mg, Anna R, et al. (2001). MtDNA control region and RFLP data for Sicily and France. *Int J Legal Med 114,* 229–231.

Cherni L, Loueslati BY, Pereira L, Ennafaâ H, Amorim A, El Gaaied AB (2005a). Female gene pools of Berber and Arab neighboring communities in central Tunisia: microstructure of mtDNA variation in North Africa. *Hum Biol 77*(1), 61–70.

Cherni L, Pereira L, Goios A, et al. (2005b). Y-chromosomal STR haplotypes in three ethnic groups and one cosmopolitan population from Tunisia. *Forensic Sci Int 152*(1), 95–99.

Cherni L, Fernandes V, Pereira JB,et al. (2009) Post-last glacial maximum expansion from Iberia to North Africa revealed by fine characterization of mtDNA H haplogroup in Tunisia. *Am J Phys Anthropol 139*(2), 253–260.

Comas D, Calafell F, Mateu E, et al. (1997). Mitochondrial DNA variation and the origin of the Europeans. *Hum Genet 99*(4), 443–449.

Côrte-Real H, Macaulay V, Richards M., et al. (1996). Genetic diversity in the Iberian Peninsula determined from mitochondrial sequence analysis. *Ann Hum Genet 60*, 331–350.

Coudray C, Olivieri A, Achilli A, et al. (2009). The complex and diversified mitochondrial gene pool of Berber populations. *Ann Hum Genet 73*(2), 196–214.

Cruciani F, La Fratta R, Santolamazza P, et al. (2004). Phylogeographic analysis of haplogroup E3b (E-M215) Y chromosomes reveals multiple migratory events within and out of Africa. *Am J Hum Genet 74*(5), 1014–1022.

Di Rienzo A, Wilson AC. (1991). Branching pattern in the evolutionary tree for human mitochondrial DNA. *Proc Natl Acad Sci USA 88*(5), 1597–1601.

Dutour O. (1995). Le peuplement moderne de l'Afrique septentrionale et ses relations avec celui du Proche Orient. *Paleorient 21*(2), 97–109.

Ennafaa H, Amor MB, Yacoubi-Loueslati B, et al. (2006). Alu polymorphisms in Jerba Island population (Tunisia): comparative study in Arab and Berber groups. *Ann Hum Biol 33*(5–6), 634–640.

Esteban E, Gonzalez-Perez E, Harich N, et al. (2004). Genetic relationships among Berbers and South Spaniards based on CD4 microsatellite/Alu haplotypes. *Ann Hum Biol 31*(2), 202–212

Fadhlaoui-Zid K, Plaza S, Calafell F, Ben Amor M, Comas D, Bennamar El gaaied A. (2004) Mitochondrial DNA heterogeneity in Tunisian Berbers. *Ann Hum Genet 68*(Pt3), 222–233

Falchi A, Giovannoni L, Calo CM, et al. (2006) Genetic history of some western Mediterranean human isolates through mtDNA HVR1 polymorphisms. *J Hum Genet 51*(1), 9–14.

Ferembach D. (1986). Les hommes du Paléolithique supérieur autour du bassin méditerranéen. *L'Anthropologie, Paris 90*, 579–587.

Flores C, Maca-Meyer N, Gonzalez AM, Cabrera VM. (2000). Northwest African distribution of the CD4/Alu microsatellite haplotypes. *Ann Hum Genet 64*, 321–327

Francalacci P, Bertranpetit J, Calafell F, Underhill PA. (1996). Sequence diversity of the control region of mitochondrial DNA in Tuscany and its implications for the peopling of Europe. *Am J Phys Anthropol 100*, 443–460.

Frigi S, Yacoubi B, Pereira F, Pereira L, et al.(2006). MtDNA lineages in two Tunisian Berber communities: Comparing diversities between villages and towns. *International Congress Series 1288*, 121–123.

González AM, Cabrera VM, Larruga JM, et al.(2006). Mitochondrial DNA variation in Mauritania and Mali and their genetic relationship to other Western Africa populations. *Ann Hum Genet 70*, 631–657.

Kefi R, Stevanovitch A, Bouzaid E, Béraud-Colomb E. (2005). Diversité mitochondriale de la population de Taforalt (12.000 ans, Maroc): une approche génétique à l'étude du peuplement de l'Afrique du Nord. *Anthropologie 1*, 55–64.

Krings M, Salem A, Bauer K, et al. (1999). mtDNA analysis of Nile River Valley populations: A genetic corridor or a barrier to migration? *Am J Hum Genet 64*, 1166–1176

Khodjet El Khil H, Triki Marrakchi R, Yacoubi Loueslati B, et al. (2001). Y chromosome microsatellite variation in three populations of Jerba Island (Tunisia). *Ann Hum Genet 65*(Pt 3), 263–270.

Khodjet El Khil H, Marrakchi RT, Loueslati BY, Langaney A, Fellous M, BenAmmar Elgaaied A. (2005). Distribution of Y chromosome lineages in Jerba island population. *Forensic Sci Int 148*(2–3), 211–218.

Macaulay V, Richards M, Hickey E, et al. (1999). The emerging tree of West Eurasian mtDNAs: a synthesis of control-region sequences and RFLPs. *Am J Hum Genet 64*, 232–249.

Maca-Meyer N, Gonzalez AM, Larruga JM, et al. (2001). Major genomic mitochondrial lineages delineate early human expansions. *BMC Genet* 2, 13.

Maca-Meyer N, Gonzalez AM, Pestano J, Flores C, Larruga JM. (2003). Cabrera VM Mitochondrial DNA transit between West Asia and North Africa inferred from U6 phylogeography. *BMC Genet 4*(1), 15.

Ottoni C, Martinez-Labarga C, Vitelli L, et al.(2009). Human mitochondrial DNA variation in Southern Italy. *Ann Hum Biol 73*, 438–448.

Plaza S, Calafell F, Helal A, et al. (2003). Joining the pillars of Hercules: mtDNA sequences show multidirectional gene flow in the western Mediterranean. *Ann Hum Gene 67*(Pt 4), 312–328

Quintana-Murci L, Semino O, Bandelt HJ, Passarino G, McElreavey K, Santachiara-Benerecetti AS. (1999). Genetic evidence of an early exit of Homo sapiens sapiens from Africa through eastern Africa. *Nat Genet 23*, 437–441.

Rando JC, Pinto F, Gonzalez AM, et al. (1998). Mitochondrial DNA analysis of northwest African populations reveals genetic exchanges with European, near-eastern, and sub-Saharan populations. *Ann Hum Genet 62*, 531–550.

Rando JC, Cabrera VM, Larruga JM, Hernandez M, Gonzalez AM, Pinto F, Bandelt HJ. (1999). Phylogeographic patterns of mtDNA reflecting the colonization of the Canary Islands. *Ann Hum Genet 63*(Pt 5), 413–428.

Richards MB, Macaulay VA, Bandelt HJ, Sykes BC. (1998). Phylogeography of mitochondrial DNA in western Europe. *Ann Hum Genet 62*, 241–260.

Richards M, Macaulay V, Hickey E, et al. (2000). Tracing European founder lineages in the near eastern mtDNA pool. *Am J Hum Genet 67*, 1251–1276.

Sanchez-Mazas A. (2000). The Berbers of North Africa: genetic relationships to HLA and other polymorphisms. In: Arnaiz-Villena A, ed. *Prehistoric Iberia: Proceedings of "Genetics, Anthropology and Linguistics."* p. 65. New York: Kluwer Academic/Plenum Publishers.

Stévanovitch A, Gilles A, Bouzaid E, et al. (2004). Mitochondrial DNA sequence diversity in a sedentary population from Egypt. *Ann Hum Genet 68*(1), 23–39.

Turchi C, Buscemi L, Giacchino E, et al. (2009). Polymorphisms of mtDNA control region in Tunisian and Moroccan populations: an enrichment of forensic mtDNA databases with Northern Africa data. *Forensic Sci Int Genet 3*(3), 166–172.

Underhill PA, Shen P, Lin AA, et al. (2000). Y chromosome sequence variation and the history of human populations: *Nat Genet 26*, 358–361.

Yacoubi-Loueslati B, Ennefaa H, Marrakchi R, et al. (2001). A study of Gm allotypes and immunoglobulin heavy gamma IGHG genes in Berbers, Arabs and Sub-Saharan Africans from Jerba Island, Tunisia. *Eur J Imm Genet 28*(5), 531–538.

Yacoubi-Loueslati, Cherni L, Khodjet-Elkhil H, et al. (2006). Islands inside an island: reproductive isolates on Jerba island. *Am J Hum Biol 18*, 149–153.

33.

THE GENETIC LANDSCAPE OF SUB-SAHARAN AFRICAN POPULATIONS

Himla Soodyall and Carina M. Schlebusch

INTRODUCTION

The phrase "every disease has two histories" was coined by Mayr (1961) and used by Scriver (1993), who argued that these two histories are linked to (1) the *demographic* history—the history of a population; and (2) the *genetic* history—the origin and distribution of genes in a population. These two histories together begin to provide the history of the genome, where "demography" explains the distribution of disease, and "genetic" explains its cause and the effects of mutant gene(s) on health (Mayr, 1961; Scriver, 1993).

Most studies on disease, however, focus on the latter component of the two-history scenario mentioned above. Human population genetic studies have, over the years, provided a wealth of information on the frequencies and distribution of genes in global populations, and have contributed to the management of disease by assisting genetic counselors with risk prediction, and clinicians and geneticists in the efficient detection of mutations associated with disease in a rapid and cost-effective way.

In trying to understand population demographic history, scholars have adopted several approaches, with varying degrees of input with respect to depth of resolution on a timescale. Recorded history, in the form of the written record, is useful up to about 4,000 years before present (BP). Linguistic data, based on the relatedness of languages spoken by different groups of people, retain evidence of their origin and are related to other languages in a measurable fashion; this data also has a relatively shallow time history, and linguists have suggested that languages do not retain evidence of their origin beyond ~10,000 years (Jobling et al., 2004). Archaeological data harbors information about the past with greater antiquity. These data are gleaned from evidence captured in physical remains, such as bones, stone tools, pottery, waste deposits, and dwellings left over by past human groups. Paleontological data, based on investigations carried out on fossilized remains, have the deepest and most direct account of our ancestral past. However, this data is fragmentary and incomplete.

The molecular genetics approach related to human history and origins grew since the beginning of the twentieth century, following the demonstration by Landsteiner of variation in populations in the ABO blood groups (Landsteiner, 1901). The principle of this approach is to reconstruct the history of mutations found in the DNA of contemporary individuals, and to trace its origin to a common ancestor, who would have lived at some point in the past. Several demographic events, such as population migrations, dramatic reduction in numbers constituting a population, a so-called bottleneck, and increase in population numbers—so-called population expansions—leave imprints, in the form of altered allele frequencies on the collective human genome. Since these imprints are transmitted to succeeding generations, the genomes of living peoples are packaged with "stories" depicting events in our evolutionary past. Thus, by studying human variation at the molecular or gene level, not only can we learn more about our evolutionary history, but we can also better understand the genetic contribution to health and disease—in keeping with Mayr's two histories of disease scenario.

Various studies have shown that there is marked genetic substructure in human populations, and that individuals within certain groups are genetically more similar to each other than to individuals from other groups (Cavalli-Sforza et al., 1994; Jakobsson et al., 2008; Li et al., 2008; Rosenberg et al., 2002; Tishkoff et al., 2009). Some of these differences account for differences in the type of mutations and frequencies of disease mutations between populations. Several genetic studies have shown that most of the variation found in human population occurs among African populations, and only a subset of this diversity is found in non-African populations (Garrigan & Hammer, 2006; Jobling & Tyler-Smith, 2003; Torroni et al., 2006; Underhill & Kivisild, 2007).

Until recently, the origin and dispersal of human populations focused on the two haploid genetic markers;

namely, mitochondrial DNA (mtDNA) and the non-recombining region of the Y-chromosome (Forster, 2004; Jobling & Tyler-Smith, 2000; Jobling & Tyler-Smith, 2003; Torroni et al., 2006; Underhill & Kivisild, 2007). Recent advances following the completion of the first human genome in 2003 (Consortium, 2004; Lander et al., 2001), and the recent complete sequencing of two southern African genomes (Schuster et al., 2010) have improved on the resolution of genomic data in assessing populations structure, and in studies examining gene function, genetic susceptibility to disease and human evolutionary studies.

MITOCHONDRIAL DNA (MTDNA) VARIATION IN AFRICA

Since the demonstration by Cann et al. (1987) of the value of mtDNA in reconstructing human origins, mtDNA studies have continued to advance our understanding on historical human migration routes and assessing population affinities (Cann et al., 1987; Kayser et al., 2008; Kivisild et al., 2006; Kivisild et al., 2002; Metspalu et al., 2004; Salas et al., 2002; Salas et al., 2004; Torroni et al., 2006; Torroni et al., 2000). More specifically, mtDNA harbors information on the transmission of the contemporary divergent mtDNA lineages exclusively along maternal lines in the population over time, and since this DNA does not recombine, changes in mtDNA result from mutations only. The relationship of the divergent lineages can be deduced by phylogenetic analysis and linked to a most recent common ancestor (MRCA); such studies have shown that the MRCA for all contemporary mtDNA lineages has a geographic region of origin in sub-Saharan Africa, dating to about 150,000–200,000 years ago (Behar et al., 2008; Forster, 2004; Reed and Tishkoff, 2006; Torroni et al., 2006).

The mtDNA diversity in Africa is broadly assigned into seven macro-haplogroups (L0“1“2“3“4“5“6), while the rest of the worlds' lineages are classified as subgroups of macro-haplogroups M, N and R (Figure 33-1; Behar et al., 2008). The first split in the human mtDNA phylogeny is between the two daughter branches, L0 (formerly L1a, L1d, L1f and L1k) and L1“2“3“4“5“6 (L1-6), located on opposite sides of the root (Figure 33-1). This divergence has been dated to 133,000–155,000 years BP (Behar et al., 2008). The distribution of the haplogroups within the L1-6 branch is far more widespread among contemporary African populations, whereas the L0 haplogroup and some of its sub-haplogroups has a limited distribution among other sub-Saharan African populations. Haplogroup L1 is found at moderately high frequencies in western and central sub-Saharan Africa, including in the pygmy populations of the central equatorial forest region (Salas et al., 2002). Haplogroup L2 is common in western and southeastern sub-Saharan Africa and is the most numerous and widespread of the four major haplogroups, accounting for as much as 25% of indigenous haplotype variation (Salas et al., 2002). The youngest of the major haplogroups, L3, is most common in western and eastern/southeastern sub-Saharan Africa, particularly among speakers of the Bantu language family, and is thought to have originated in eastern Africa, where it accounts for half of all types (Salas et al., 2002). Two other haplogroups descended from L3 (M and N) are the only mtDNA lineages observed outside Africa (Atkinson et al., 2009).

Previous studies reported high frequencies of haplogroups L0d and L0k among Khoe-San groups from southern Africa (Chen et al., 2000; Tishkoff et al., 2007; Vigilant et al., 1991) with the TMRCA dating to 124,000–172,000 years ago (Atkinson et al., 2009). Haplogroup L0d was found in San groups currently living in the Northern Cape Province in South Africa at frequencies of 51%–61% (!Xun) and 16–22% (Khwe) while L0k was found at frequencies of 26% in the !Xun, and 23% in the Khwe (Chen et al., 2000; Tishkoff et al., 2007). In addition, L0d has been found at low frequencies in the Sandawe (5%) but not in the Hadzabe, the two click-speaking groups from Tanzania (Tishkoff et al., 2007). In the Ju\'hoansi, a San group from Botswana, L0d was found to be the most prevalent haplogroup (96%), while the remaining mtDNA lineages (4%) were resolved into haplogroup L0k (Vigilant et al., 1991). More recently, Schlebusch et al. (2009) reported frequencies of 75.9% of L0d and 10.7% of L0k among Khoe-San groups from southern Africa. Haplogroups L0d and L0k are absent, or found at low frequencies, in other sub-Saharan African populations, usually the result of gene flow from Khoe-San-speaking populations into Bantu-speaking groups in southern Africa (Pereira et al., 2001; Salas et al., 2002; Soodyall et al., 2008). Frequencies of up to 28.5% of L0d were found among southeastern Bantu-speakers and 70.5% among non-white peoples in South Africa (Schlebusch et al., 2009).

More recent mtDNA studies in Africa based on whole mitochondrial genome analysis (Behar et al., 2008; Gonder et al., 2007; Kivisild et al., 2006; Mishmar et al., 2003; Torroni et al., 2006) resulted in a revised phylogeny and nomenclature of the mtDNA haplogroups and sub-haplogroups found among African populations (Behar et al., 2008). Atkinson and colleagues took advantage of the different geographical distributions of the four common haplogroups and used them as markers to examine broad population demographic processes within Africa over the last 150,000 years (Atkinson et al., 2009). These data have been examined with statistical tools like the Bayesian skyline plot (BSP) which is a useful coalescence procedure that uses Bayesian theory to infer relative population sizes through time (Drummond & Rambaut, 2007; Drummond et al., 2005).

Using this approach, the authors found that the African mtDNA data were consistent with the expansion of four

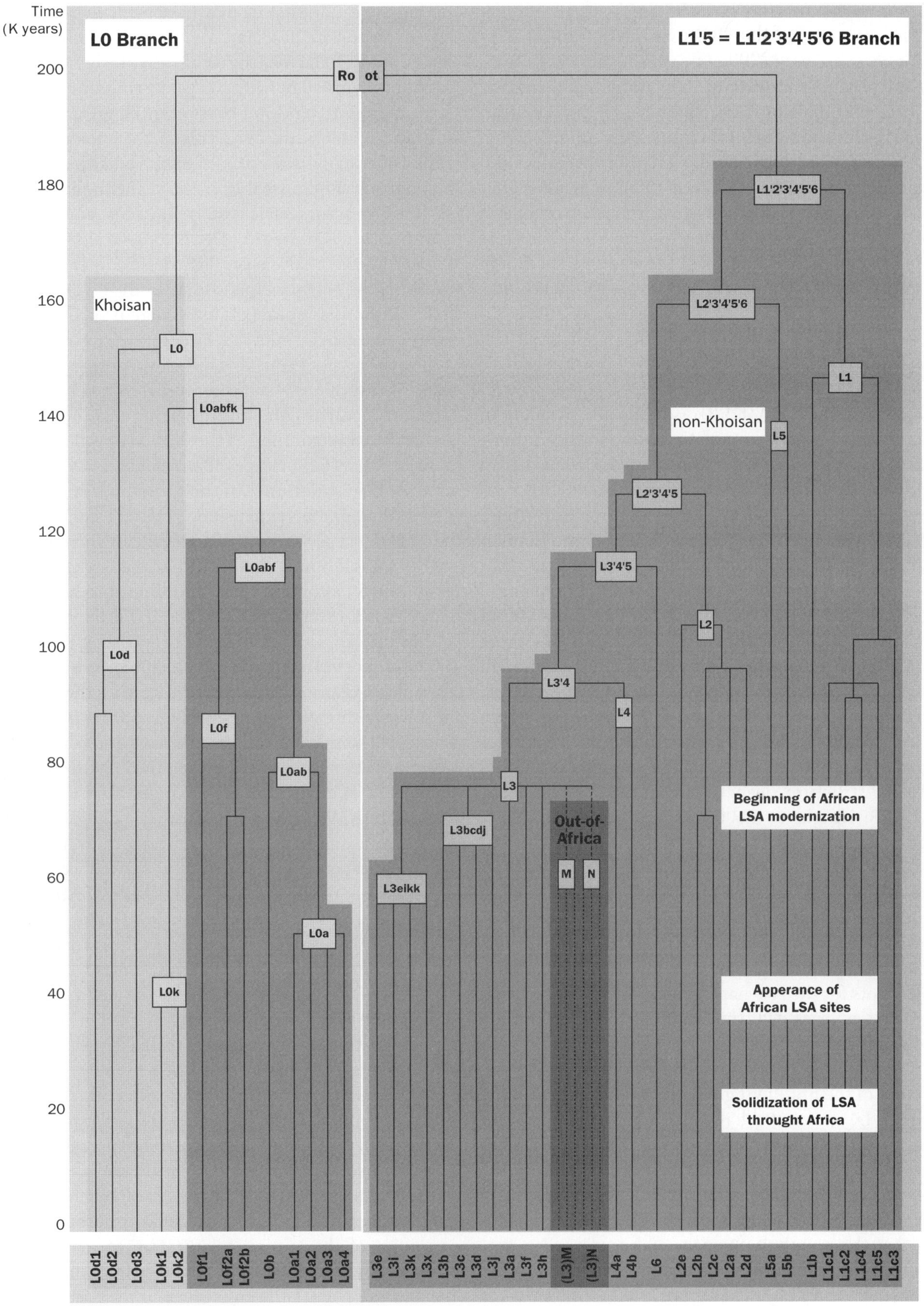

Figure 33-1 Tree showing global mtDNA macro-haplogroups. (Behar et al., 2008). Note: non-African mtDNA haplogroups trace ancestry to haplogroups M and N on this tree. Refer color figure.

separate mitochondrial lineages among African populations. However, the separate BSPs for the four major haplogroups revealed very different patterns of growth from the overall picture, particularly for haplogroups L2 and L3. Examination of the plots indicated that growth signals in haplogroups L0 and L1 are not significantly different from each other, showing substantial overlap in their distributions through time. However, distributions for haplogroups L2 and L3 were significantly different from each other and from haplogroups L0 and L1. It was suggested that these differences may reflect changes in the size or structure of the populations carrying these haplotypes, or could indicate the influence of selection pressures causing advantageous new variants to sweep through the population.

Haplogroups L0 and L1 revealed slow constant growth over the last 100,000–200,000 years, and estimates of effective population size did not differ significantly throughout the history of the two lineages. According to the authors, this pattern would result if both lineages formed part of an early panmictic African population and contributed relatively equally to the haplotype diversity present in Africa today. However, recent evidence indicates some population structure within Africa from an early stage. Analysis of genetic data from four African populations using a coalescent-based model of population divergence found support for population structure in Africa from more than 50,000 years ago when applied to mtDNA, although not when applied to Y-chromosome data (Garrigan et al., 2007). Others have argued that the distribution of ancient mtDNA L0d and L0k lineages among the hunter–gatherer speakers of the African "click" languages also supported population structure deep in the mtDNA tree (Behar et al., 2008; Knight et al., 2003; Tishkoff et al., 2007). Repeating the BSP analysis on only the L0d and L0k lineages, Atkinson et al. did not find any significantly different growth profiles from the L0 BSP. The authors suggest that it is unlikely that such putative deep divergence events were associated with substantial changes in available territory or mode of living. However, the effective population size plots for the two haplogroups do appear to be on different trajectories over the last 15,000 years, although they do not differ significantly; L0 was consistent with an increasing trend, while L1 showed a decreasing trend. Inspection of the underlying phylogeny indicates that the increase in L0 is probably a result of a rapid emergence of lineages within the L0a haplogroup at this time. L0a has been proposed as a marker of the Bantu expansion 3,000–4,000 years ago (Bandelt et al., 1995; Chen et al., 1995). The time scale estimated by Atkinson and colleagues suggested that the population expansion began earlier, perhaps linked to the expansion of L2a.

From the BSP for L2, a TMRCA of 73,000–127,000 years ago was estimated (Atkinson et al., 2009). This range is consistent with previous age estimates for the haplogroup based on D-loop and complete sequence data and has led to the suggestion that L2 may have been involved in population expansions associated with the later African exodus (Salas et al., 2002; Watson et al., 1997). However, the BSP profile is consistent with L2 occurring at relatively low frequency until a period of substantial growth beginning 12,000–20,000 years ago. This is the most pronounced expansion signal the authors observed among the major haplogroups and probably explains the high frequency of L2 lineages across Africa today. The most common haplogroup within L2 is L2a (TMRCA 25,000–31,000 years ago). Repeating the analysis on just the L2a lineages confirms that they are the principal source of the L2 expansion signal. While this rapid expansion signal in L2 could have been caused by a selective sweep, the substitutions that define the L2 (and L2a) lineages are synonymous changes, making it unlikely that these were new and advantageous variants capable of quickly displacing existing haplotypes. As with L0a mentioned above, L2a has been linked to the spread of Bantu languages (Bandelt et al., 2001). However, as with L0a, the timing of the growth we infer here predates the Bantu expansion. This discrepancy could be explained as a result of the shift in inferred rates of mtDNA evolution at shallow time depths identified by Ho et al. (2005), although rates would need to be four to five times faster to bring the start of the growth phase seen in L2a into agreement with the proposed spread of Bantu 3,000–4,000 years ago (Ho et al., 2005). While the Bantu expansion undoubtedly played a role in the spread of a number of mtDNA haplotypes, L2 appears to have begun to expand somewhat earlier. Atkinson and colleagues suggest that L2 lineages (and perhaps the L0a lineages) spread as a result of environmental changes associated with the Last Glacial Maximum (Salas et al., 2002). Particularly arid conditions at this time are thought to have resulted in the enlargement of the Sahara, as well as the conversion of much of the forested area of central Africa to open savannah and woodlands (Adams & Faure, 1997), which may have allowed an expansion of human populations into new territories, particularly in central Africa.

The BSP for haplogroup L3 was interpreted by Atkinson et al. (2009) to have resulted from a marked increase in effective population size from an estimated TMRCA of 61,000–86,000 years ago. L3 is the only lineage to show such marked growth at this time. Again, the substitutions that define L3 are synonymous changes; hence, the inferred increase in L3 frequency is unlikely to be the result of a selective sweep. Instead, it seems likely that L3 spread due to demographic expansion within Africa. L3 haplotype frequencies are highest in eastern Africa (Salas et al., 2002), the proposed launching point of the human colonization of the globe. More importantly, the authors contend that the timing of the L3 expansion predates the emergence of the first non-African lineages (haplogroups M and N) by 8,000–12,000 years; they estimated an age of 53,000–69,000 years ago for haplogroup M, and 50,000–64,000 years ago for haplogroup N (Atkinson et al., 2009). The fact that L3 is the only haplogroup with descendants outside Africa, and shows a clear growth signal 8,000–12,000 years prior to the emergence of its non-African descendants, strongly suggests that L3 did not simply spill over into Eurasia, but was driven as part of an expansion that had begun in sub-Saharan Africa thousands of years earlier.

There are a number of possible explanations for such an expansion—climatic (Scholz et al., 2007), or cultural advantage that allowed its population to outcompete rival groups (Bowler and Thorne, 1976; Marean et al., 2007; Mellars, 2006). Determining which, if any, of these factors were involved in the L3 expansion will require the synthesis of multiple lines of evidence from archaeology, palaeoclimatology, historical linguistics, and population genetics (Atkinson et al., 2009).

Y-CHROMOSOME DNA STUDIES

The Y-chromosome contains the largest nonrecombining block in the human genome, and is therefore extremely important for evolutionary genetic studies. While studies on mtDNA describe the maternal history of a population, the paternal history can be described through using the Y-chromosome. The first Y-chromosome polymorphism was reported in 1985 (Casanova et al., 1985), but more than a decade elapsed before a well-resolved Y-chromosome tree was available (Hammer et al., 2001; Underhill et al., 2000; YCC, 2002).

The Y-chromosome tree structure is primarily based on binary polymorphisms, and specific branches are assigned to haplogroups following a hierarchical pattern. Currently the Y-chromosome tree consists of 20 major clades containing 311 distinct haplogroups, defined by 599 mutational events (Karafet et al., 2008). Furthermore, by typing Y-chromosome short tandem repeats (Y-STRs) haplotypes are generated, which are then used for finer resolution within the haplogroups (Underhill & Kivisild, 2007).

The two primary splits in the Y-chromosome tree leads to two branches, Haplogroup A and B, which have a distribution restricted to Africa (Figures 33-2a and 33-2b). These two clades are genetically diverse, and its haplogroups have different geographical distribution patterns. This suggests population fragmentation, isolation and re-expansions in prehistoric Africa. Haplogroups A and B are associated with the distribution of ancient hunter-gatherer tribes before the expansions of pastoralists (Underhill & Kivisild, 2007; Underhill et al., 2001). The rest of the Y-chromosome tree is defined by the M168 mutation, which represents the most common African lineages (Haplogroup E) as well as all the non-African clades (see Figure 33-2b).

Haplogroup A is defined by the M91 and P97 mutations and contains 12 branches determined by 45 (internal) mutations. A strict regional distribution is particularly pronounced for haplogroup A. Within Haplogroup A, A1 is found in Mali and Morocco (Scozzari et al., 2001; Underhill et al., 2000), A3b2 is found in east Africa (Sudan, Ethiopia, Tanzania, Kenya) and in lower frequencies in north Cameroon (Cruciani et al., 2002; Knight et al., 2003; Scozzari et al., 1999; Semino et al., 2002; Underhill et al., 2000), while A3b1 and A2 are found exclusively among the Khoe-San (Scozzari et al., 1999; Underhill et al., 2000). Haplogroup B is defined by four mutations (M60, M181, P85, and P90) and contains 17 branches with 28 internal markers (Karafet et al., 2008). Haplogroup B occurs throughout Africa but with high frequencies among Pygmies, Khoe-San, and Hadza, with some lineages being restricted to them (Cruciani et al., 2002; Knight et al., 2003; Semino et al., 2002; Underhill et al., 2000; YCC, 2002). There is a clear-cut difference between the B haplogroups associated with the Pygmies, Khoe-San, and Hadza vs. all the other African populations. Pygmies, Khoe-San, and Hadza populations have mainly Haplogroup B haplotypes defined by the M112 mutation, while other populations have the M150 mutation. Within haplogroup B-M112, haplogroups B2b2, B2b3, and B2b4b are restricted to the Pygmy populations, while B2b1 (P6) and B2b4a (P8) are restricted to Khoe-San groups (Hadza groups were not typed for B-M112 subgroups). The B2b* ancestral haplotype occurs in both Pygmy and Khoe-San groups (Cruciani et al., 2002; Knight et al., 2003; Semino et al., 2002; Underhill et al., 2000; YCC, 2002).

Eighteen mutations currently define haplogroup E. Haplogroup E is the most diverse of all the major Y-chromosome clades, and contains 83 polymorphisms that define 56 distinct haplogroups (Karafet et al., 2008). The E haplogroups are found at high frequencies in Africa, at moderate frequencies in the Middle East and southern Europe, and have sporadic occurrences in Central and South Asia. Although Haplogroup E groups are widespread all over Africa, the distributions of the numerous distinctive haplogroups are not homogeneous across the continent (see Figure 33-2b; Bosch et al., 2001; Cruciani et al., 2004; Cruciani et al., 2002; Hammer and Horai, 1995; Hammer et al., 2001; Hammer et al., 1997; Qamar et al., 1999; Underhill et al., 2001).

Haplotypes carrying the mutations M75 (E2) and M33 (E1a) are present at low frequencies across Africa but with different individual distributions. Haplogroups E1b1a and E1b1b are the most frequent and widespread of the E haplogroups (Cruciani et al., 2004; Cruciani et al., 2002; Hammer et al., 2001; Semino et al., 2004; Underhill et al., 2001). E1b1a, defined by M2 and seven other mutations, is mainly limited to sub-Saharan populations and is associated with the expansion of Bantu-speaking populations (Hammer et al., 1998; Passarino et al., 1998; Scozzari et al., 1999). The E1b1a subgroups have differential distributions and frequencies. The M191 mutation defines the most frequent E-M2 subgroup and is evident of a founder

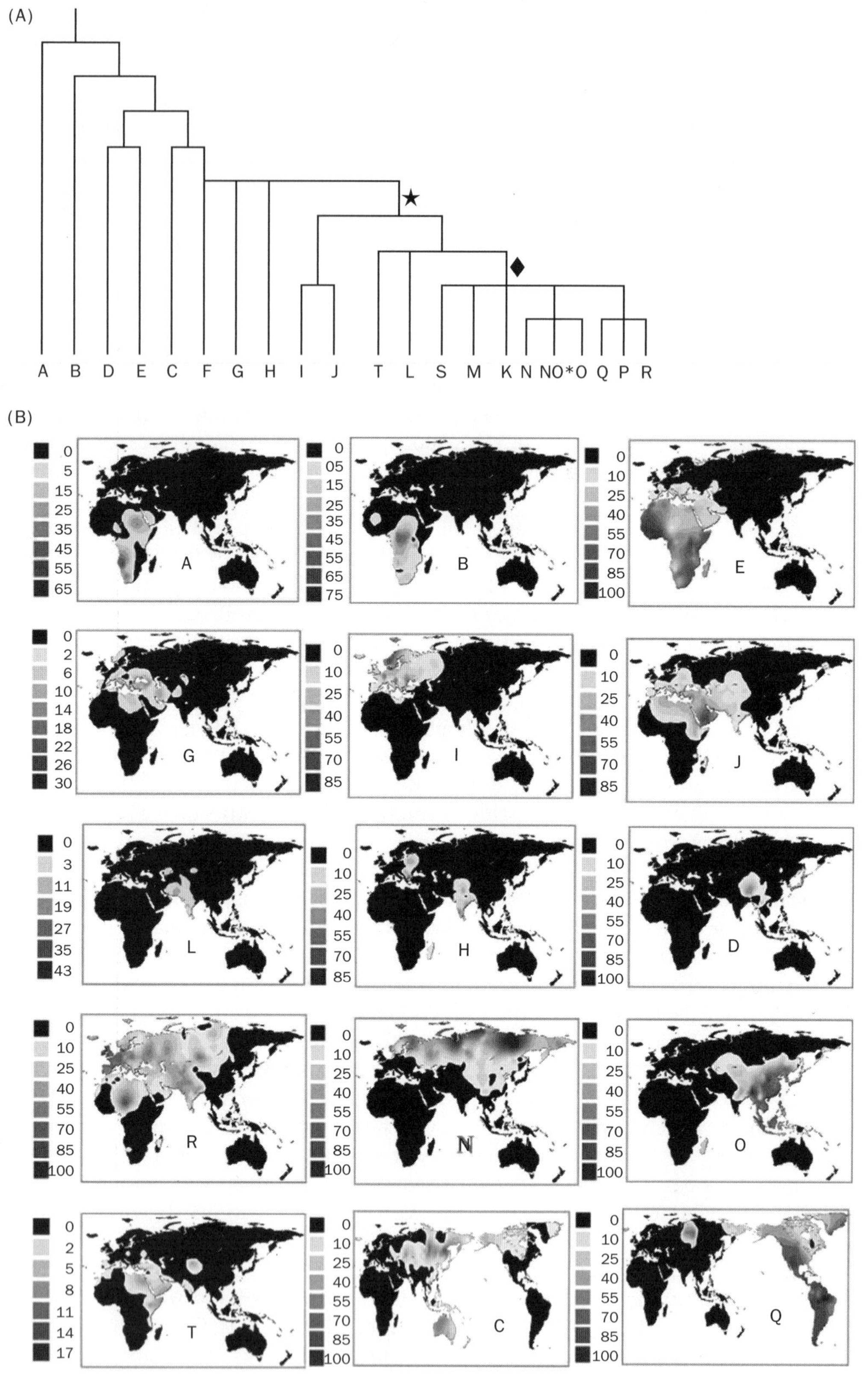

Figure 33-2 Y chromosome haplogroup tree showing the relationship of 20 major Y haplogroups (a) and the geographic frequency maps of some haplogroups (b) (Chiaroni et al., 2009). Haplogroups A, B and E are the common ones found in Africa.

effect that resulted from the Bantu expansions (Cruciani et al., 2004; Cruciani et al., 2002; Hammer et al., 2001; Semino et al., 2004; Underhill et al., 2001).

The non-African distribution of haplogroup E is associated with haplogroup E1b1b, characterized by the M35 and M215 mutations (Hammer et al., 1998; Semino et al., 2004; Semino et al., 2000; Underhill et al., 2001). This haplogroup, however, also has a widespread African representation (Cruciani et al., 2004, 2002; Hammer et al., 2001; Semino et al., 2004; Underhill et al., 2001). Compared to other E haplogroups, M35 occurs at very low frequencies within Bantu speakers but is widely though not uniformly dispersed throughout Africa. Among the different lineages carrying the M35 mutation, haplotypes defined by M78 occur in east Africa, North Africa, the Middle East, and Europe. It is the E-M35 subgroup that has the highest frequency and the widest distribution outside Africa. This marker has a northeastern African origin, and multiple exodus routes out of Africa have been demonstrated (Cruciani et al., 2007). M123 haplotypes are present in eastern Africa, northeast Africa, the Middle East, and southeast Europe, but do not reach western Europe. M81 is found at high frequencies only in northern Africa and is almost absent in Europe, with the exception of Sicily and Iberia (Bosch et al., 2001; Cruciani et al., 2004; Semino et al., 2004). In addition to these differentiated E1b1b lineages, there were many haplotypes that were classified as E-M35*, which occurred in high frequencies particularly in Ethiopian, Kenyan, Tanzanian, and Khoe-San groups (Cruciani et al., 2004; Semino et al., 2004).

Recently a new Y-chromosome polymorphism (M293) was discovered, which grouped these previously paraphyletic E-M35 groups into a monophyletic group (Henn et al., 2008). This E-M293 haplogroup has a spread concentrated in eastern and southern Africa with maximum frequencies in Tanzania and southern Africa. In eastern Africa, high frequencies of M293 are observed in the Datog (43%), Burunge (28%), Sandawe (24%), and Hadza (11%). The Datog are pastoralists who speak a Southern Nilotic language and the Burunge are Afro-Asiatic agropastoralists. In southern Africa it was observed in the Khwe (31%) and !Xun (11%). The Khwe and the !Xun were the only Khoe-San groups included in the study. Network analysis revealed haplotype sharing and close similarities between Khwe/!Xun haplotypes and Hadza/Sandawe haplotypes. M293 also occurs at low to moderate frequencies in Bantu-speaking populations of eastern and southern Africa, which likely reflects recent admixture with local populations after the Bantu expansions (Henn et al., 2008).

E-M293 data from the study supported a demic diffusion model correlated with the spread of sheep, cows, and pottery along a Tsetse fly-free corridor between eastern and southern Africa, 2,000 years BP (Gifford-Gonzalez, 2000; Henn et al., 2008; Sadr, 1998; Smith, 2005). A previous model, where pastoralism was transmitted from eastern Africa to southern central Africa with little to no population movement, was thus rejected (Sadr, 1998; Smith, 2005). The new model suggested that a small pastoralist population carrying M293 migrated from east Africa into southern central Africa with their livestock (Henn et al., 2008). After arriving in southern Africa, these pastoralists could have mixed with local populations, or expanded without substantial genetic exchange with local groups. Without representation of more Khoe-San groups the study, however, researchers could not address the question of how pastoralism spread after it reached south central Africa. The scale of the migration from east Africa may have been small, minimally four E-M293 male individuals. It is possible that other male individuals who did not carry M293 were also involved. For instance, E-M2 individuals could have been involved but it would not be possible to distinguish these from the E-M2 introduced later during the Bantu expansions. The Henn et al. study thus supports a migration, independent of the Bantu expansion, of east Africans harboring the E-M293 marker, which initially brought pastoralism to southern Africa (Henn et al., 2008).

AUTOSOMAL DNA VARIATION

When compared to the haploid inherited Y-chromosome DNA and mtDNA markers, the diploid inherited autosomal DNA markers carry more biological information from both parents, but the interpretation of the data is complicated because of recombination. To overcome this problem, some studies have made use of short stretches of linked polymorphisms to infer haplotypes that have been analyzed using computational tools using homozygous group frequencies to infer the phase of heterozygous loci (Excoffier & Slatkin, 1995; Niu et al., 2002; Scheet & Stephens, 2006; Stephens et al., 2001). Consequently, these short stretches of inferred haplotypes can be treated as lineages in the same way that the nonrecombining mtDNA and Y-chromosome DNA are treated. An early example of an autosomal haplotype study is the 2.7 kb region on chromosome 11 that encompasses the β-globin gene (Harding et al., 1997). The phylogeny obtained from the 326 haplotypes reflected results from Y-chromosome and mtDNA studies. The root of the tree was in Africa, with many lineages that were exclusive to Africa. Since then, several other loci have been studied, all supporting an African root (Clark et al., 1998; Harding et al., 2000; Harris & Hey, 1999). Similar to mtDNA and Y-chromosome studies, however, the history and dynamics of the lineage under investigation is the history coupled to a certain locus, and is reflective of only a small part of the genome. Some of these loci might be heavily influenced by selection, which would violate the assumptions of population genetic models and in the end would not give a true picture of the population history. Ultimately, to get the true history of a population or the human species, one should take into account all of the separate loci.

Another way to utilize information contained in the autosomes is to use genotypes of unlinked markers spread over the whole genome, instead of inferred haplotypes. Through using AMOVA analysis on such multilocus genotypes (microsatellites, single nucleotide polymorphisms (SNPs), and insertion/deletions) it was found that 79%–94% (depending on the marker type) of variation represents variation between individuals within the same population (Barbujani et al., 1997; Jorde et al., 2000; Romualdi et al., 2002; Rosenberg et al., 2002). This thus means that genotypic variation is not homogenous across the human species but 6%–21% of the variation is due to differences between populations and continental groups. This leads to the question of whether a genotype from an individual can be correctly assigned to the correct population or continent of origin.

The earliest method used to explore this question was to calculate pairwise individual distances based on allele sharing (Bowcock et al., 1994). These distances were then used to construct a tree of genotypes from individuals without taking into account any prior population origin. The aim was to see if the tree showed clusters according to populations or continents. The tree that resulted correctly assigned 88% of genotypes to continent specific clusters. The population specificity was less precise, but 64% of populations formed clusters that included more than half of their individuals (Bowcock et al., 1994). Since then, more powerful genotype assigning methods have been developed (Corander et al., 2003; Falush et al., 2003; Falush et al., 2007; Francois et al., 2006; Pritchard et al., 2000).

A widely used technique, implemented in the program STRUCTURE, is based on the Bayesian clustering of individuals into K number of clusters (Falush et al., 2003; Falush et al., 2007; Pritchard et al., 2000). The user specifies the K number of clusters, and the program assigns a genotype or a proportion of a genotype to a certain cluster with a certain probability. A signature of population structure will then emerge (if there is structure) through the unequal assignment of individuals or partial genotypes to certain clusters. For instance: if K=2, the program will divide the total variation of the whole study group optimally into two clusters, and then assign each individual with a certain probability to each of the two clusters. When there is structure in the sample group, individuals from population x will be preferentially assigned to a certain cluster, for instance cluster 1, while individuals of population y will be preferentially assigned to cluster 2. If, for instance, individuals from population z resulted from an admixture event between population x and y, these individuals will be assigned with certain probabilities to both clusters 1 and 2, depending on the marker contribution from each population to the individual. When an admixture model is assumed, individuals are not assigned to a cluster with a certain probability; rather, a part of their genome (made up by the markers included in the study) is assigned to a certain cluster. The procedure usually followed when running STRUCTURE is to assign K clusters from K=2 to K=10, and then test which K number of clusters has the highest likelihood by looking at the posterior likelihood scores, or by using the deltaK method that takes into account the rate of change between successive K clusters (Evanno et al., 2005).

The first genotypic studies were based on a limited number of markers and individuals (Bowcock et al., 1987., 1991a, 1991b; Nei & Livshits, 1989). RFLPs were individually typed from isolated DNA, which were cloned or transformed to increase the quantity. These laborious processes limited the experimental size. During the past 20 years, however, techniques rapidly developed that enabled high-throughput marker typing. The newest techniques are able to type thousands of markers (Jakobsson et al., 2008; Li et al., 2008; Tishkoff et al., 2009).

Despite the small size of the first studies, it was immediately apparent that African and non-African genetic variations represent the earliest diversion in human history (Bowcock et al., 1987, 1991a, 1991b, 1994; Nei & Livshits, 1989). Africans had higher levels of nucleotide diversity compared to non-Africans. Furthermore, the genetic diversity in non-African populations represents a subset of the genetic diversity in sub-Saharan Africa. Also, more private alleles and haplotypes are observed in Africa than in other regions. All of this strongly supported the "Out of Africa" model that was suggested by mtDNA studies. Additionally, these low-resolution studies were already able to distinguish individuals on a continental basis.

Increasing the number of loci increased the accuracy of the continental assignment of genotypes, and facilitated the emergence of subclusters which correspond to populations within continents (Jakobsson et al., 2008; Li et al., 2008; Rosenberg et al., 2005, 2002; Tishkoff et al., 2009). Most of these studies utilized the HGDP-CEPH panel. The panel consists of cell lines of 1064 individuals from 51 populations from sub-Saharan Africa, North Africa, Europe, the Middle East, South/Central Asia, East Asia, Oceania, and the Americas (Cann et al., 2002). This sample set is freely available and allows a detailed characterization of worldwide genetic variation.

The study by Rosenberg et al. (2002) used 377 autosomal microsatellite loci on the HGDP-CEPH panel. They found that worldwide variation could be clustered into six clusters, of which five correspond to major geographic locations. Furthermore, they could infer subclusters within these major regions. The sub-Saharan African cluster is optimally divided into four subclusters, which represent Bantu-speaking + pre-Bantu-speakers, San, Mbuti Pygmy, and Biaka Pygmy clusters (Rosenberg et al., 2005, 2002).

Li et al. (2008) used 650 000 SNP markers on the HGDP-CEPH panel and also found clustering into the five continental groups at K=5. At K=6 South/Central Asia separates from Europe and the Middle East; and at K=7 the Middle East separates from Europe. Many populations, however, have representation from more than one cluster. This can be an indication of recent admixture or shared ancestry before divergence. Additionally, PCA showed that the largest part of variation (56%) can be summarized as variation between African and non-African populations. In the population distance tree, African populations lie closest to the root. The San group forms the earliest branch, followed by the Mbuti Pygmy group, the Biaka Pygmy group, and thereafter the Bantu-speakers (Li et al., 2008).

Jakobsson et al. (2008) typed SNPs (525 910) and copy number variants (396) on the HGDP-CEPH panel. This study found that on the global scale, similar to Rosenberg et al. (2002), populations optimally grouped into six clusters, of which five correspond to major geographic regions. The clustering within Africa, however, yielded interesting results. While Rosenberg et al. optimally identified four clusters corresponding to the two Pygmy groups, the San and the Bantu + pre-Bantu speakers, Jakobsson et al. (2008) only identified three clusters. One of the three clusters represented Bantu + pre-Bantu-speakers (the Bantu-speakers from South Africa showed the largest contributions from the San/Pygmy clusters additional to the Bantu-speaking cluster, followed by the Kenyans, the Yoruba, and the Mandenka). The remaining two clusters were present at highest frequency in the Pygmy and San populations. Aside from small amounts of admixture from the Bantu/pre-Bantu-speaking cluster, the Mbuti belonged almost exclusively to one of these clusters. The Biaka predominantly belonged to a third cluster, but also had large contributions from the Mbuti cluster. The San contained both the Mbuti and Biaka cluster, but with a larger contribution from the Mbuti cluster. It thus appears that the San and Mbuti group are more closely related (Jakobsson et al., 2008).

A recent study by Tishkoff et al. (2009) included 2432 African individuals from 113 geographically diverse populations. For evaluation against non-African groups, the HGDP-CEPH panel was also included. In these samples, 1327 polymorphic markers (microsatellites and insertion/deletions) were typed. Similar to previous studies, African populations contained the highest levels of genetic diversity. Globally, diversity declines with distance from Africa. Within Africa, the Pygmy and San populations had the highest genetic diversities, while the San groups had the most private alleles. In the tree analysis, the two Khoe-San populations cluster together and are most distant from the other populations. Using PCA, 72 significant global principal components (PCs) were identified. The first PC (19.5%) separates African from non-African populations. The Hadza is separated from other populations at PC3 (3.5%).

Using STRUCTURE analysis, the populations showed clustering according to major geographic region, both on a global scale and within Africa (Figure 33-3; Tishkoff et al., 2009). Globally, 14 ancestral population clusters were identified, while nine of these were found in Africa (Tishkoff et al., 2009). A cluster emerged (at K=5) that is present in the Hadza, and to a lesser extent the Pygmy, San, and Sandawe hunter-gatherers. Subsequently (at K=6) the cluster split into a Hadza/Sandawe and Pygmy/Khoe-San cluster. The Mbuti Pygmy and San groups split from the other Pygmy groups at K=11, indicating common ancestry between these groups. Results from this study showed that the San, Hadza, Sandawe, and Pygmy populations contain shared genetic variation that distinguishes them

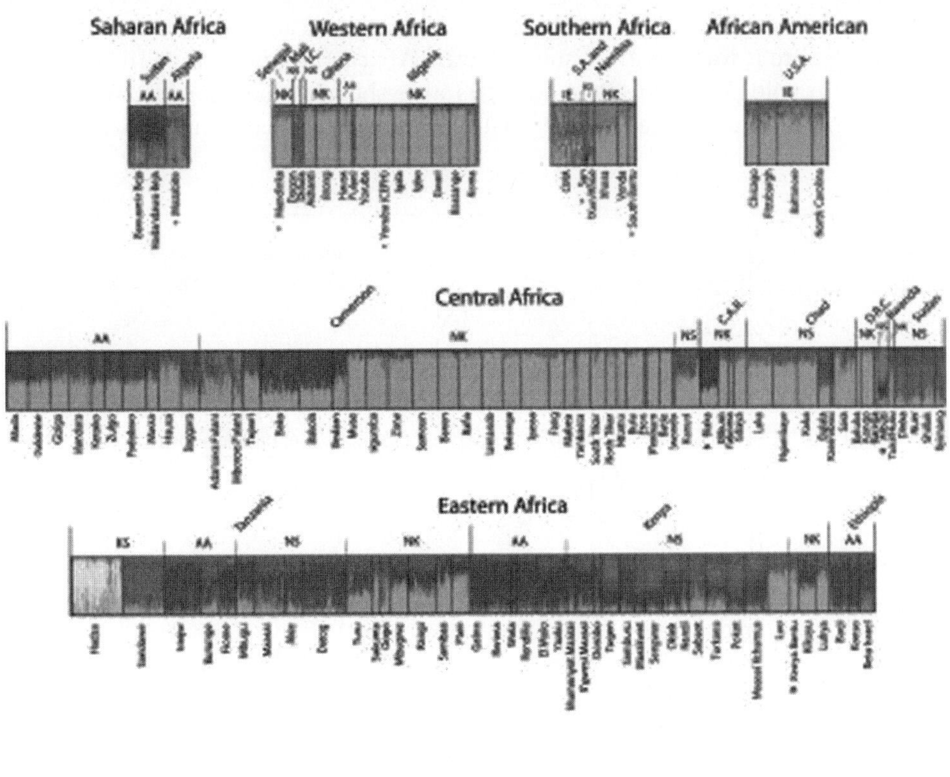

Figure 33-3 Geographic and genetic structure of populations within Africa. (Tishkoff et al., 2009).

from other African populations (Tishkoff et al., 2009). This led to the suggestion that these groups were the remnants of a proto-Khoe-San/Pygmy/Hadza/Sandawe population of hunter-gatherers. MtDNA and Y-chromosome analysis suggest a divergence of >35 000 years BP (Behar et al., 2008; Gonder et al., 2007; Semino et al., 2002; Tishkoff et al., 2007, 2009).

The Hadza are genetically the most distinct from the other African groups (Tishkoff et al., 2009), which is consistent with linguistic evidence that the Hadza language is unrelated to other Khoisan languages (Sands, 1998). The Hadza is an isolated population that has had little interaction with surrounding groups, and has maintained their hunter-gatherer lifestyle up until recent times. They show only very low levels of asymmetric gene flow from surrounding groups. The Sandawe, on the other hand, adopted mixed farming practices and show evidence of bidirectional gene flow with neighboring groups (Newman, 1995). Populations from northern Tanzania, Southern Ethiopia, and northern Kenya show evidence of the Sandawe-associated genetic cluster (Tishkoff et al., 2009). Aside from the association proven by autosomal DNA results, other commonalities between these two east African groups and the Khoe-San groups are: the language connection between Sandawe and Khoisan, similarities between Tanzanian and San rock art, the Sandawe formerly performed a trance dance similar to San trance dances, and there is evidence of pan San belief systems across all of southern Africa to as far north as Zimbabwe (Huffman, 1983, p. 2116; Lewis-Williams, 1986, p. 2115).

The clustering of the Khoe-San groups with the Pygmies (Tishkoff et al., 2009) suggests that they may have a common genetic history. Pygmy populations might have had a Khoisan-related language before it was replaced by Bantu-speaking language. Anthropological support for this theory comes from the shared music styles between the Khoe-San and Pygmy groups (Lomax, 1968; Tishkoff et al., 2009). The San populations show a closer shared genetic ancestry to the Mbuti Pygmy than the Biaka Pygmy groups (Jakobsson et al., 2008; Tishkoff et al., 2009). The Mbuti lives in the Ituri rainforest of the eastern DRC, while the Biaka (also called Baka, part of the Mbenga group) live to the west of the Mbuti in Cameroon, Gabon, and the Republic of Congo. Another main group of Pygmies, the Twa or Ba-Twa and Cwa, live in dispersed groups south-central to the Mbuti and Mbenga (Cavalli-Sforza, 1986). These groups live in swamps and deserts far from the forest, there are no genetic data available for them, and it is not known if they are indigenous to the area or more recent migrants from the forest. It may be that before the Bantu expansions, these Pygmy groups formed a continuous network of related groups that also had contact and gene flow with their Khoe-San neighbors to the south, and Hadza and Sandawe neighbors to the east.

WHOLE-GENOME ANALYSIS STUDIES IN AFRICA

A recent study by Schuster et al. (2010) discussed the value of whole-genome sequence analysis on two southern Africans—Archbishop Desmond Tutu, a Bantu-speaking individual from South Africa, and !Gubi, a San individual from Namibia. At the time of publication, 14 whole-genome sequences were published, among them a Yoruban individual from Nigeria. The three sub-Saharan African genomes, when analyzed and compared with non-Africans, clearly demonstrated the partitioning in the data between Africans and non-Africans and also population structure within sub-Saharan Africa (Figure 33-4a). In addition, the PCA analysis also showed that there is substantial genomic variation among sub-Saharan Africans that

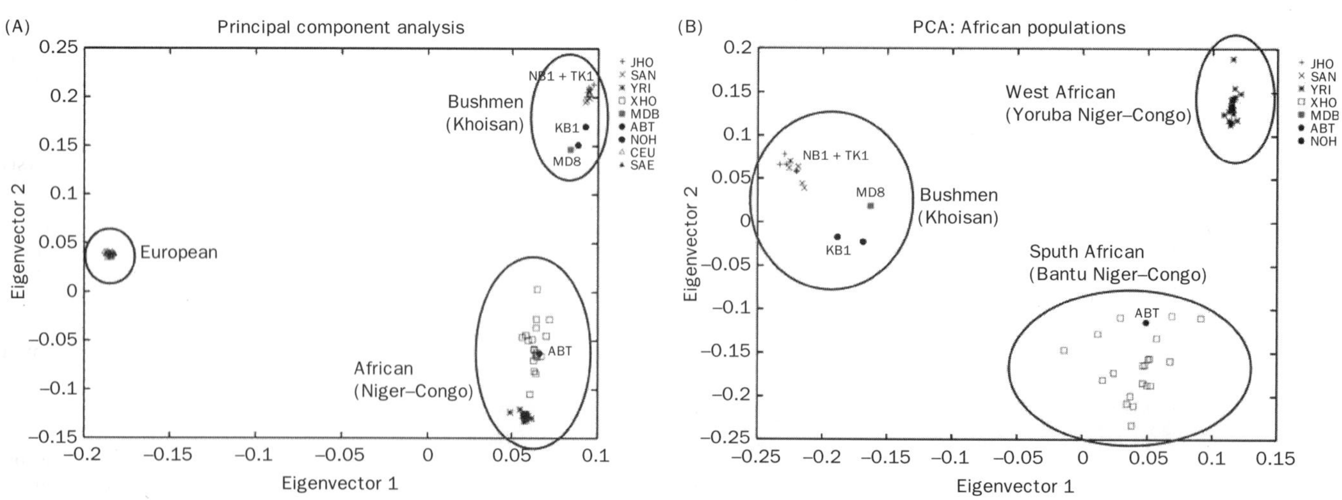

Figure 33-4 Three-way population structure based on 174,272 autosomal SNPs using PCA. (Schuster et al., 2010).

clearly separated Khoe-San from West African (Yoruba Niger-Congo) from Southern African (Bantu Niger-Congo) groups (Figure 33-4b; Schuster et al., 2010). Whole-genome sequence analysis reaffirms the findings from SNP array studies outlined in the previous section, of the levels of population structure within Africa.

CONCLUSION

Irrespective of the choice of genetic marker(s) used in population genetic studies, this brief review of some of the recent DNA studies conducted among African populations highlights the value of the inclusion of population demographic histories in understanding the genetic landscape of African populations. These studies will underpin genetic studies trying to understand population susceptibility to disease, and to explain why some diseases occur only in some populations and in certain parts of the world. Use of data based on an understanding of the factors that contribute to population structure; i.e., the demographic component of the two-history scenario of disease alluded to earlier, would greatly enhance global management of disease.

In addition, molecular anthropological studies, like the ongoing international Genographic Project being conducted by the National Geographic Society in partnership with IBM and the Waitt Family Foundation, would vastly improve our understanding of the migrations and dispersal of human populations throughout the world. After all, where we came from and how to understand our ancestral past are questions that are the focus of many societies.

ACKNOWLEDGEMENTS

The authors are most grateful to the South African Medical Research Council, the National Health Laboratory Service, and the University of the Witwatersrand for the financial support making this research possible.

REFERENCES

Adams JM, Faure H. (1997). *Review and atlas of palaeovegetation: preliminary land ecosystem maps of the world since the Last Glacial Maximum.* (published online only: http://www.esd.ornl.gov/ern/qen/adams1.html) Oak Ridge National Laboratory, TN.

Atkinson QD, Gray RD, Drummond AJ. (2009). Bayesian coalescent inference of major human mitochondrial DNA haplogroup expansions in Africa. *Proc Biol Sci 276*, 367–373.

Bandelt HJ, Alves-Silva J, Guimaraes PE, et al., (2001). Phylogeography of the human mitochondrial haplogroup L3e: a snapshot of African prehistory and Atlantic slave trade. *Ann Hum Genet 65*, 549–563.

Bandelt HJ, Forster P, Sykes BC, Richards MB. (1995). Mitochondrial portraits of human populations using median networks. *Genetics 141*, 743–753.

Barbujani G, Magagni A, Minch E. Cavalli-Sforza LL. (1997). An apportionment of human DNA diversity. *Proc Natl Acad Sci U S A 94*, 4516–4519.

Behar DM, Villems R, Soodyall H, et al. (2008). The dawn of human matrilineal diversity. *Am J Hum Genet 82*, 1130–1140.

Bosch E, Calafell F, Comas D, Oefner PJ, Underhill PA, Bertranpetit J. (2001). High-resolution analysis of human Y-chromosome variation shows a sharp discontinuity and limited gene flow between northwestern Africa and the Iberian Peninsula. *Am J Hum Genet 68*, 1019–1029.

Bowcock AM, Bucci C, Hebert JM, et al. (1987). Study of 47 DNA markers in five populations from four continents. *Gene Geogr 1*, 47–64.

Bowcock AM, Hebert JM, Mountain JL, et al. (1991a). Study of an additional 58 DNA markers in five human populations from four continents. *Gene Geogr 5*, 151–173.

Bowcock AM, Kidd JR, Mountain JL, et al.. (1991b). Drift, admixture, and selection in human evolution: a study with DNA polymorphisms. *Proc Natl Acad Sci U S A 88*, 83–9843.

Bowcock AM, Ruiz-Linares A, Tomfohrde J, Minch E, Kidd JR, Cavalli-Sforza LL. (1994). High resolution of human evolutionary trees with polymorphic microsatellites. *Nature 368*, 455–457.

Bowler JM, Thorne AG. (1976). Human remains from Lake Mungo: discovery and excavation of Lake Mungo III. *In* Kirk RL, Thorne AG, eds. *The Origin of the Australians*. Australian Institute of Aboriginal Studies, Canberra, Australia.

Cann HM, de Toma C, Cazes L, et al. (2002). A human genome diversity cell line panel. *Science 296*, 261–262.

Cann RL, Stoneking M, Wilson AC. (1987). Mitochondrial DNA and human evolution. *Nature 325*, 31–36.

Casanova M, Leroy P, Boucekkine C, et al. (1985). A human Y-linked DNA polymorphism and its potential for estimating genetic and evolutionary distance. *Science 230*, 1403–1406.

Cavalli-Sforza LL. (1986). *African Pygmies*. Orlando, FL: Academic Press.

Cavalli-Sforza LL, Menozzi P, Piazza A. (1994). *The History and Geography of Human Genes*. Princeton, NJ: Princeton University Press.

Chen YS, Olckers A, Schurr TG, Kogelnik AM, Huoponen K, Wallace DC. (2000). mtDNA variation in the South African Kung and Khwe-and their genetic relationships to other African populations. *Am J Hum Genet 66*, 1362–1383.

Chen YS, Torroni A, Excoffier L, Santachiara-Benerecetti AS, Wallace DC. (1995). Analysis of mtDNA variation in African populations reveals the most ancient of all human continent-specific haplogroups. *Am J Hum Genet 57*, 133–149.

Chiaroni J, Underhill PA, Cavalli-Sforza LL. (2009). Y chromosome diversity, human expansion, drift and cultural evolution. *Proc Natl Acad Sci, USA. 106*, 20174–20179.

Clark AG, Weiss KM, Nickerson DA, et al. (1998). Haplotype structure and population genetic inferences from nucleotide-sequence variation in human lipoprotein lipase. *Am J Hum Genet 63*, 595–612.

Consortium IHGS. (2004). Finishing the euchromatic sequence of the human genome. *Nature 431*, 931–945.

Corander J, Waldmann P, Sillanpaa MJ. (2003). Bayesian analysis of genetic differentiation between populations. *Genetics 163*, 367–374.

Cruciani F, La Fratta R, Santolamazza P, et al. (2004). Phylogeographic analysis of haplogroup E3b (E-M215). Y chromosomes reveals multiple migratory events within and out of Africa. *Am J Hum Genet 74*, 1014–1022.

Cruciani F, La Fratta R, Trombetta B, et al. (2007). Tracing past human male movements in northern/eastern Africa and western Eurasia: new clues from Y-chromosomal haplogroups E-M78 and J-M12. *Mol Biol Evol 24*, 1300–1311.

Cruciani F, Santolamazza P, Shen P, et al. (2002). A back migration from Asia to sub-Saharan Africa is supported by high-resolution analysis of human Y-chromosome haplotypes. *Am J Hum Genet 70*, 1197–1214.

Drummond AJ, Rambaut A. (2007). BEAST: Bayesian evolutionary analysis by sampling trees. *BMC Evol Biol 7*, 214.

Drummond AJ, Rambaut A, Shapiro B, Pybus OG. (2005). Bayesian coalescent inference of past population dynamics from molecular sequences. *Mol Biol Evol 22*, 1185–1192.

Evanno G, Regnaut S, Goudet J. (2005). Detecting the number of clusters of individuals using the software STRUCTURE: a simulation study. *Mol Ecol 14*, 2611–2620.

Excoffier L, Slatkin M. (1995). Maximum-likelihood estimation of molecular haplotype frequencies in a diploid population. *Mol Biol Evol 12*, 921–927.

Falush D, Stephens M, Pritchard JK. (2003). Inference of population structure using multilocus genotype data: linked loci and correlated allele frequencies. *Genetics 164*, 1567–1587.

Falush D, Stephens M, Pritchard JK. (2007). Inference of population structure using multilocus genotype data: Dominant markers and null alleles. *Mol Ecol Notes* In Press

Forster P. (2004). Ice Ages and the mitochondrial DNA chronology of human dispersals: a review. *Philos Trans R Soc Lond B Biol Sci 359*, 255–264; discussion 264.

Francois O, Ancelet S, Guillot G. (2006). Bayesian clustering using hidden Markov random fields in spatial population genetics. *Genetics 174*, 805–816.

Garrigan D, Hammer MF. (2006). Reconstructing human origins in the genomic era. *Nat Rev Genet 7*, 669–680.

Garrigan D, Kingan SB, Pilkington MM, et al. (2007). Inferring human population sizes, divergence times and rates of gene flow from mitochondrial, X and Y chromosome resequencing data. *Genetics 177*, 2195–2207.

Gifford-Gonzalez D. (2000). Animal disease challenges to the emergence of pastoralism in sub-Saharan Africa. *Afr Archaeol Rev 17*, 95–139.

Gonder MK, Mortensen HM, Reed FA, de Sousa A, Tishkoff SA. (2007). Whole-mtDNA genome sequence analysis of ancient African lineages. *Mol Biol Evol 24*, 757–768.

Hammer MF, Horai S. (1995). Y chromosomal DNA variation and the peopling of Japan. *Am J Hum Genet 56*, 951–962.

Hammer MF, Karafet T, Rasanayagam A, et al. (1998). Out of Africa and back again: nested cladistic analysis of human Y chromosome variation. *Mol Biol Evol 15*, 427–441.

Hammer MF, Karafet TM, Redd AJ, et al. (2001). Hierarchical patterns of global human Y-chromosome diversity. *Mol Biol Evol 18*, 1189–1203.

Hammer MF, Spurdle AB, Karafet T, et al. (1997). The geographic distribution of human Y chromosome variation. *Genetics 145*, 787–805.

Harding RM, Fullerton SM, Griffiths RC, et al. (1997). Archaic African and Asian lineages in the genetic ancestry of modern humans. *Am J Hum Genet 60*, 772–789.

Harding RM, Healy E, Ray AJ, et al. (2000). Evidence for variable selective pressures at MC1R. *Am J Hum Genet* 66, 1351–1361.

Harris EE, Hey J. (1999). X chromosome evidence for ancient human histories. *Proc Natl Acad Sci U S A 96*, 3320–3324.

Henn BM, Gignoux C, Lin AA, et al. (2008). Y-chromosomal evidence of a pastoralist migration through Tanzania to southern Africa. *Proc Natl Acad Sci U S A 105*, 10693–10698.

Ho SY, Phillips MJ, Cooper A, Drummond AJ. (2005). Time dependency of molecular rate estimates and systematic overestimation of recent divergence times. *Mol Biol Evol 22*, 1561–1568.

Jakobsson M, Scholz SW, Scheet P, et al. (2008). Genotype, haplotype and copy-number variation in worldwide human populations. *Nature 451*, 998–1003.

Jobling MA, Hurles ME, Tyler-Smith C. (2004). *Human Evolutionary Genetics. Origins, Peoples & Disease.* New York: Garland Publishing.

Jobling MA, Tyler-Smith C. (2000). New uses for new haplotypes the human Y chromosome, disease and selection. *Trends Genet 16*, 356–362.

Jobling MA, Tyler-Smith C. (2003). The human Y chromosome: an evolutionary marker comes of age. *Nat Rev Genet 4*, 598–612.

Jorde LB, Watkins WS, Bamshad MJ, et al. (2000). The distribution of human genetic diversity: a comparison of mitochondrial, autosomal, and Y-chromosome data. *Am J Hum Genet 66*, 979–988.

Karafet TM, Mendez FL, Meilerman MB, Underhill PA, Zegura SL, Hammer MF. (2008). New binary polymorphisms reshape and increase resolution of the human Y chromosomal haplogroup tree. *Genome Res 18*, 830–838.

Kayser M, Choi Y, van Oven M, et al. (2008). The impact of the Austronesian expansion: evidence from mtDNA and Y chromosome diversity in the Admiralty Islands of Melanesia. *Mol Biol Evol 25*, 1362–1374.

Kivisild T, Shen P, Wall DP, et al. (2006). The role of selection in the evolution of human mitochondrial genomes. *Genetics 172*, 373–387.

Kivisild T, Tolk HV, Parik J, et al. (2002). The emerging limbs and twigs of the East Asian mtDNA tree. *Mol Biol Evol 19*, 1737–1751.

Knight A, Underhill PA, Mortensen HM, et al. (2003). African Y chromosome and mtDNA divergence provides insight into the history of click languages. *Curr Biol 13*, 464–473.

Lander ES, Linton LM, Birren B, et al. (2001). Initial sequencing and analysis of the human genome. *Nature* 409, 860–921.

Landsteiner K. (1901). Uber Agglutinationserscheinungen normalen menschlichen. *Wiener Klin. Wochenschr. 14*, 1132–1134.

Li JZ, Absher DM, Tang H, et al. (2008). Worldwide human relationships inferred from genome-wide patterns of variation. *Science 319*, 1100–1104.

Lomax A. (1968). *In Folk Song Style and Culture*. National Association for the Advancement of Science, Washington, DC.

Marean CW, Bar-Matthews M, Bernatchez J, et al. (2007). Early human use of marine resources and pigment in South Africa during the Middle Pleistocene. *Nature 449*, 905–908.

Mayr E. (1961). Cause and effect in biology. *Science 134*, 1501–1506.

Mellars P. (2006). Why did modern human populations disperse from Africa ca. 60,000 years ago? A new model. *Proc Natl Acad Sci U S A 103*, 9381–9386.

Metspalu M, Kivisild T, Metspalu E, et al. (2004). Most of the extant mtDNA boundaries in south and southwest Asia were likely shaped during the initial settlement of Eurasia by anatomically modern humans. *BMC Genet 5*, 26.

Mishmar D, Ruiz-Pesini E, Golik P, et al. (2003). Natural selection shaped regional mtDNA variation in humans. *Proc Natl Acad Sci U S A 100*, 171–176.

Nei M, Livshits G. (1989). Genetic relationships of Europeans, Asians and Africans and the origin of modern Homo sapiens. *Hum Hered 39*, 276–281.

Newman JL. (1995). *The peopling of Africa* Yale University Press, New Haven, CT.

Niu T, Qin ZS, Xu X, Liu JS. (2002). Bayesian haplotype inference for multiple linked single-nucleotide polymorphisms. *Am J Hum Genet 70*, 157–169.

Passarino G, Semino O, Quintana-Murci L, Excoffier L, Hammer M, Santachiara-Benerecetti AS. (1998). Different genetic components in the Ethiopian population, identified by mtDNA and Y-chromosome polymorphisms. *Am J Hum Genet 62*, 420–434.

Pereira L, Macaulay V, Torroni A, Scozzari R, Prata MJ, Amorim A. (2001). Prehistoric and historic traces in the mtDNA of Mozambique: insights into the Bantu expansions and the slave trade. *Ann Hum Genet 65*, 439–458.

Pritchard JK, Stephens M, Donnelly P. (2000). Inference of population structure using multilocus genotype data. *Genetics 155*, 945–959.

Qamar R, Ayub Q, Khaliq S, et al. (1999). African and Levantine origins of Pakistani YAP+ Y chromosomes. *Hum Biol 71*, 745–755.

Reed FA, Tishkoff SA. (2006). African human diversity, origins and migrations. *Curr Opin Genet Dev 16*, 597–605.

Romualdi C, Balding D, Nasidze IS, et al. (2002). Patterns of human diversity, within and among continents, inferred from biallelic DNA polymorphisms. *Genome Res 12*, 602–612.

Rosenberg NA, Mahajan S, Ramachandran S., Zhao C, Pritchard JK, Feldman MW. (2005). Clines, Clusters, and the Effect of Study Design on the Inference of Human Population Structure. *PLoS Genet 1*, e70.

Rosenberg NA, Pritchard JK, Weber JL, et al. (2002). Genetic structure of human populations. *Science 298*, 2381–2385.

Sadr K. (1998). The first herders at the Cape of Good Hope. *Afr Archaeol Rev 15*, 101–132.

Salas A, Richards M, De la Fe T, et al. (2002). The making of the African mtDNA landscape. *Am J Hum Genet 71*, 1082–1111.

Salas A, Richards M, Lareu MV, et al. (2004). The African diaspora: mitochondrial DNA and the Atlantic slave trade. *Am J Hum Genet 74*, 454–465.

Sands B. (1998). *In Language, Identity and Conceptualization Among the Khoisan*. Köln: Rudiger Kupper.

Scheet P, Stephens M. (2006). A fast and flexible statistical model for large-scale population genotype data: applications to inferring missing genotypes and haplotypic phase. *Am J Hum Genet 78*, 629–644.

Schlebusch CM, Naidoo T, Soodyall H. (2009). SNaPshot minisequencing to resolve mitochondrial macro-haplogroups found in Africa. *Electrophoresis 30*, 3657–3664.

Scholz CA, Johnson TC, Cohen AS, et al. (2007). East African megadroughts between 135 and 75 thousand years ago and bearing on early-modern human origins. *Proc Natl Acad Sci U S A 104*, 16416–16421.

Schuster SC, Miller W, Ratan A, et al. (2010). Complete Khoisan and Bantu genomes from southern Africa. *Nature 463*, 943–947.

Scozzari R, Cruciani F, Pangrazio A, et al. (2001). Human Y-chromosome variation in the western Mediterranean area: implications for the peopling of the region. *Hum Immunol 62*, 871–884.

Scozzari R, Cruciani F, Santolamazza P, et al. (1999). Combined use of biallelic and microsatellite Y-chromosome polymorphisms to infer affinities among African populations. *Am J Hum Genet 65*, 829–846.

Scriver CR. (1993). Every disease has two histories. *Transactions of the Royal Society of Canada, Sixth Series* Volume IV, 25–38.

Semino O, Magri C, Benuzzi G, et al. (2004). Origin, diffusion, and differentiation of Y-chromosome haplogroups E and J: inferences on the neolithization of Europe and later migratory events in the Mediterranean area. *Am J Hum Genet 74*, 1023–1034.

Semino O, Passarino G, Oefner PJ, G., et al. (2000). The genetic legacy of Paleolithic Homo sapiens sapiens in extant Europeans: a Y chromosome perspective. *Science 290*, 1155–1159.

Semino O, Santachiara-Benerecetti AS, Falaschi F, Cavalli-Sforza LL, Underhill PA. (2002). Ethiopians and Khoisan share the deepest clades of the human Y-chromosome phylogeny. *Am J Hum Genet 70*, 265–268.

Smith A. (2005). The concepts of "Neolithic" and "Neolithisation" for Africa? *Before Farming 1*, 1–6.

Soodyall H, Makkan H, Haycock P, Naidoo T. (2008). The prehistory of the Khoe and San. *Southern African Humanities 20*, 37–48.

Stephens M, Smith NJ, Donnelly P. (2001). A new statistical method for haplotype reconstruction from population data. *Am J Hum Genet 68*, 978–989.

Tishkoff SA, Gonder MK, Henn BM, et al. (2007). History of click-speaking populations of Africa inferred from mtDNA and Y chromosome genetic variation. *Mol Biol Evol 24*, 2180–2195.

Tishkoff SA, Reed FA., Friedlaender FR, et al. (2009). The genetic structure and history of Africans and African Americans. *Science 324*, 1035–1044.

Torroni A, Achilli A, Macaulay V, Richards M, Bandelt HJ. (2006). Harvesting the fruit of the human mtDNA tree. *Trends Genet 22*, 339–345.

Torroni A, Richards M, Macaulay V, et al., (2000). mtDNA haplogroups and frequency patterns in Europe. *Am J Hum Genet 66*, 1173–1177.

Underhill PA, Kivisild T. (2007). Use of y chromosome and mitochondrial DNA population structure in tracing human migrations. *Annu Rev Genet 41*, 539–564.

Underhill PA, Passarino G, Lin AA, et al., (2001). The phylogeography of Y chromosome binary haplotypes and the origins of modern human populations. *Ann Hum Genet 65*, 43–62.

Underhill PA, Shen P, Lin AA, et al., (2000). Y chromosome sequence variation and the history of human populations. *Nat Genet 26*, 358–361.

Watson E, Forster P, Richards M, Bandelt HJ. (1997). Mitochondrial footprints of human expansions in Africa. *Am J Hum Genet* 61, 691–704.

Vigilant L, Stoneking M, Harpending H, Hawkes K, Wilson AC. (1991). African populations and the evolution of human mitochondrial DNA. *Science 253*, 1503–1507.

YCC (2002). A nomenclature system for the tree of human Y-chromosomal binary haplogroups. *Genome Res 12*, 339–348.

34.

GENETIC DISORDERS IN NORTH AFRICAN POPULATIONS

Lilia Romdhane and Sonia Abdelhak

INTRODUCTION

North African populations—encompassing, from the east to the west, Egypt, Libya, Tunisia, Algeria, and Morocco—constitute a large and heterogeneous group generated as a consequence of the admixture with many other populations throughout history. Most of them are Muslim of Arab and Berber descents, with a minority of Jewish and Christian communities. The practice of consanguineous marriages in the North African populations appears to be a deep-rooted norm of mating and dates back to ancient civilizations. Inbreeding in pedigrees of pharaohs are well recorded in Egypt, and the most specific types of marriages were incestuous. Queen Hatshepsut of the Eighteenth Dynasty (1580–1350 BC) was the daughter of the half-brother/sister marriage of Thothmus and Aahmes. Her mother, Aahmes, was the result of two successive generations of marriages of brothers and sisters (Schulz & Seidel, 1999). Recent studies performed on old mummies, statues, and reliefs suggest that genetic diseases were present in those times. Osteogenesis imperfecta was suspected in a 3000-year-old mummy (Lowenstein, 2009), and the distinctive physical features of Akhenaten (1385–1359 BC) are explained by either of two different familial disorders: the aromatase excess syndrome and the sagittal craniosynostosis syndrome, or a variant of the Antley-Bixler syndrome (Braverman et al., 2009). This shows that inbreeding and, consequently, autosomal recessive genetic disorders are ancient in North Africa. In this chapter we report on genetic diseases affecting the North African population on the basis of literature mining, and discuss their spectrum, phenotypic peculiarities, and consequences for public health.

DEMOGRAPHIC FEATURES AND THE FAMILY STRUCTURE

As do the other Arab communities, the North African populations present some demographic, cultural, social, and geographic features that markedly affect the prevalence of genetic diseases in this area (Table 34-1). Some of these characteristics include: high population growth rate (0.98%–2.17%); high birth rates (15.42%–25.15%); high infant mortality rates (21.05%–36.88%); presence of isolates like Jews in Djerba and Nubians in Egypt; high rate of consanguinity and inbreeding within the same village, isolates, or social units; marriage at low age in rural areas; childbearing at older maternal ages; and, most often, lack of public health measures for both genetic and congenital disorders prevention and control. In the middle of the twentieth century, North African families were as large as the Arab families, whose sizes reach more than 6 children per woman. The total fertility per woman was averaged to 7.59 children born between 1970 and 1975 (United Nations, Department of Economic and Social Affairs, Population Division, 2010). This extended family structure, most often associated with inbreeding, displays a unique distribution for genetic diseases compared to Occidental societies. Mutation carriers in the Western populations tend to be scattered in the general population after many generations, and their relationship is dissipated, in contrast to mutation carriers in the North African societies who remain clustered in the extended family. Nowadays, North African families show declined rates of fertility, reaching 2.04 children born per woman by an assessment performed between 2005 and 2010 (United Nations, Department of Economic and Social Affairs, Population Division, 2010). The lowest rate is attributed to Tunisia, which has undergone a considerable reduction in fertility averaged by 1.06 children per decade (United Nations, Department of Economic and Social Affairs, Population Division, 2010). Many factors have contributed to the fertility decline; notably, familial planning, contraception, and nationwide education, especially for women.

INHERITANCE PATTERNS OF GENETIC DISORDERS

In order to assess the burden of genetic diseases affecting the North African populations, a review of the

TABLE 34-1 DEMOGRAPHIC FEATURES OF NORTH AFRICAN POPULATIONS (2008)

	Algeria	Egypt	Libya	Morocco	Tunisia
Population growth rate (%)	1.5	1.8	2.0	1.2	1.0
Birth rates (births/1000 population)	21	25	23	20	18
Infant mortality rates (deaths/1000 live births)	36	20	15	32	18
Life expectancy at birth (years)	72	70	74	71	74
Total fertility rate (children born/ woman)	2.36	2.86	2.70	2.35	2.06

(Adapted from http://da ta.worldbank.org/indicator)

scientific literature was performed. This allowed the identification of a nonexhaustive list of 532 inherited conditions. Classification of genetic diseases reported in the region according to the mode of inheritance revealed a high proportion of autosomal recessive disorders (AR) that reaches 60%, followed by autosomal dominant (AD) conditions (27.2%), while the X-linked (X) diseases represent 5.8%. Other modes of transmission are also reported at lower frequencies as sporadic (3%), mitochondrial (0.3%), and Y-linked (0.3%). For genetic diseases that could be transmitted according to more than one mode of inheritance, i.e., as AD/AR, AD/sporadic, AD/X, AR/sporadic, and AR/X, these entities are rare (3.4%). In addition, some others (0.35%) remain without a specific mode of transmission (Figure 34-1, Figure 34-2).

The high frequency of recessive genetic disorders confirms the association between inbreeding and high prevalence of these kind of inherited diseases as previously reported in consanguineous populations (Teebi, 1994; Hamamy et al., 2007). Indeed, the main biologic consequence of inbred mating in the population is the incrementation of the probability that two gametes carrying the same genetic information meet each other, and thus of the frequency of alleles at homozygous state compared to the general population (Jakobi et al., 1971). As first cousin unions are the most frequent among consanguineous mating in North Africa, it has been demonstrated that first cousins inherit one quarter of their genes from each of their common grandparents, and hence, one-eighth of their genes are identical by inheritance. Therefore, the children of first cousin parents inherit identical gene copies from each parent at one-sixteenth (6.25%) of all gene loci. This degree of homozygosity by descent, also referred to as *autozygosity*, is expressed as a coefficient of inbreeding, noted "F," of 0.0625. In double first cousin marriages, where the spouses have both sets of grandparents in common, 12.5% of children's gene pairs are identical by descent;

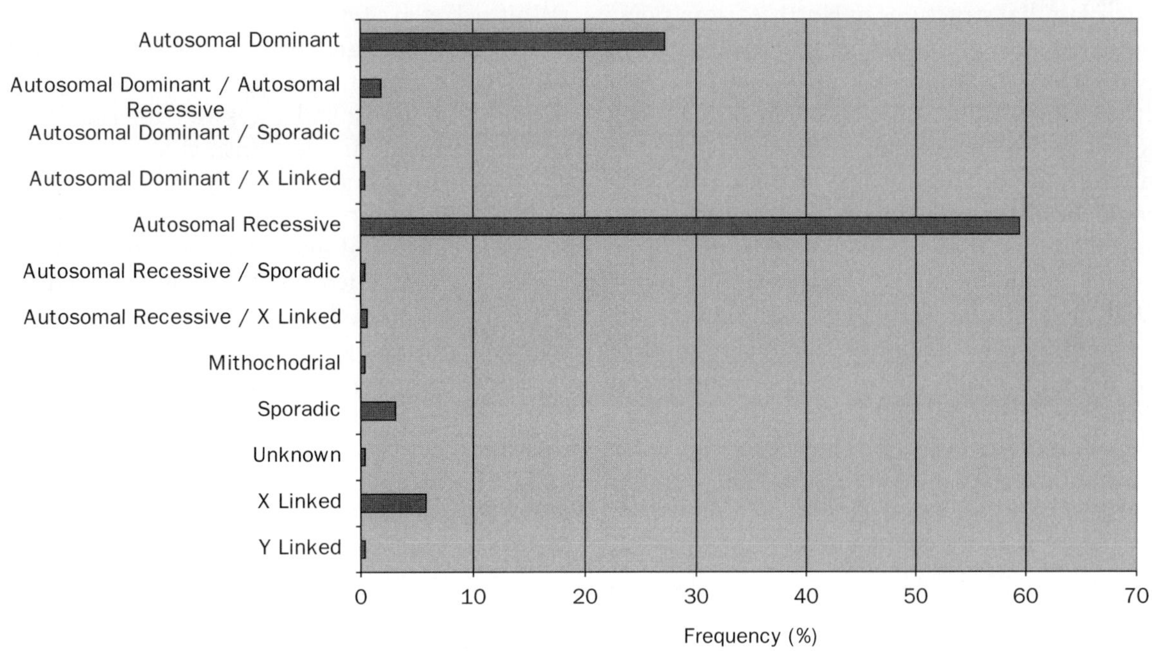

Figure 34-1 Inheritance Patterns of Genetic Disorders in North Africa

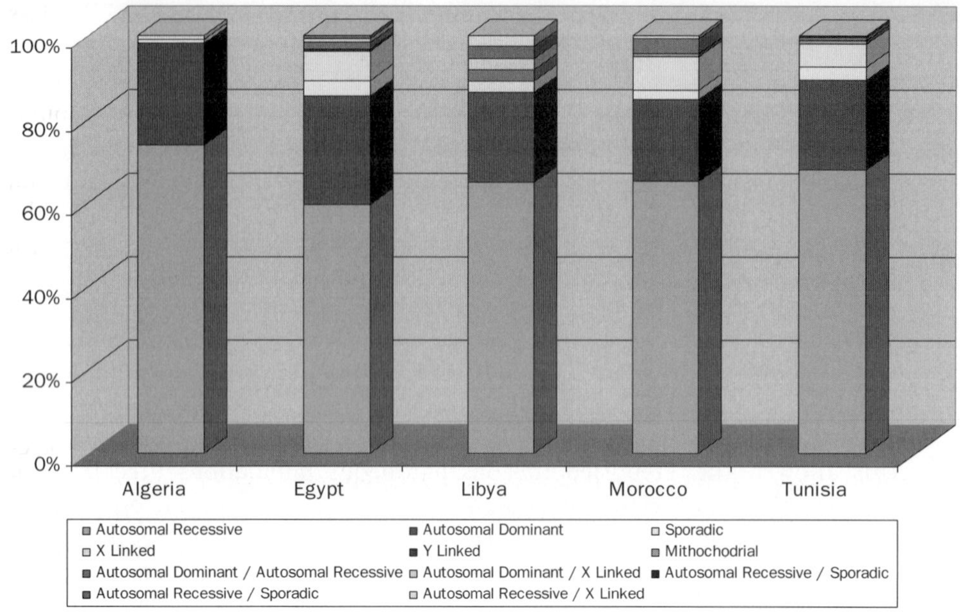

Figure 34-2 Mode of Inheritance of Genetic Disorders in North African Countries Refer color figure.

therefore, F is equal to 0.125; this is equivalent to F value for children of uncle–niece unions among Sephardic Jews (Hafez et al., 1983). However, this correlation has not been proved for dominant and X-linked conditions (Hamamy et al., 2007). As one copy of a deleterious mutation leads to the expression of the morbid phenotype of an autosomal dominant condition, two related parents—each one of whom has an autosomal dominant mutation—do not have more probability of having an affected child than a nonconsanguineous couple. Similarly, the X-linked diseases require that the hemizygote has one abnormal allele for their expressions in males, explaining their rarity in females.

Particular familial structures in North African populations allowed identification of inheritance patterns, and allowed better characterization of some genetic diseases. For Parkinson's disease, initially considered as exclusively multifactorial, genetic epidemiological investigation of large Tunisian families allowed demonstration that this disease could be inherited with at least two transmission patterns: autosomal dominant and autosomal recessive (Gouider-Khjoua et al., 2000). Similarly, a report of Marden-Walker syndrome in two Egyptian first cousins born from consanguineous unions, allowed researchers to attribute the autosomal recessive mode of transmission to this phenotype (Temtamy et al., 1975). More recently, a rare constellation of malformations has been reported as following an autosomal dominant mode of inheritance in a Tunisian family (Al Kaissi et al., 2007a).

In some cases, even for well-known genetic diseases, peculiar inheritance patterns have been reported among North African families. For Mal de Meleda, an autosomal recessive palmoplantar keratoderma, females heterozygous for different *ARS* gene mutations expressed minor clinical signs of the skin disease (Mokni et al., 2004). Moderate clinical features in heterozygous carriers have been reported in very rare cases of AR diseases (Beighton, 1997; Croft et al., 1995; Schoser et al., 2008), but specific gender expression for heterozygous carriers has never been reported.

In a large family showing a wide range of clinical manifestations of dystrophic epidermolysis bullosa, another rare genetic skin disorder (Ouragini et al., 2009), the causative mutation was identified and showed an autosomal semidominant pattern of inheritance with incomplete penetrance and variable expression. The same mutation was previously reported as recessive in a European patient (Whittock et al., 1999).

On the other hand, when studying a genetic disease in consanguineous affected individuals born from healthy parents, the autosomal recessive inheritance pattern is usually assumed as the transmission mode. That is not always true, as demonstrated by the following example. A linkage study to known genes associated with autosomal recessive congenital muscular dystrophy (AR CMD) and a genome-wide microsatellite analysis failed to identify a linking chromosomal region, by assuming autosomal recessive inheritance and homozygosity by descent, in two Algerian consanguineous sisters with AR CMD. Sequencing of the *LMNA* gene, previously excluded in this study, revealed the recurrent heterozygous missense change (c.1580G>C) only in the two affected siblings. This was explained by a paternal germinal mosaicism, and illustrates that de novo dominant mutations in lamin-related congenital muscular dystrophy should be kept in mind as the most frequent causes responsible for congenital muscular dystrophy, followed by autosomal recessive inheritance (Makri et al., 2009).

SPECTRUM AND PHENOTYPIC PECULIARITIES OF GENETIC DISEASES

Birth defects are defined as abnormalities of structure or function, including metabolism, which are present at birth. They are of genetic or partially genetic origin. Some of them are clinically obvious at birth; others are diagnosed later in life. Serious birth defects threaten life or result in physical, sensorial, or mental disabilities. Thus, they are considered as a global health problem in middle-income and low-income countries. More than 94% of children born with birth defects, and 95% of the deaths of these affected children, occur in these regions (Christianson et al., 2006). The highest birth defect prevalences are encountered among the poorest countries of the world (Christianson et al., 2006). In North Africa, these statistics range from 65.3 to 73.1 birth defects per 1000 live births (Christianson et al., 2006; Figure 34-3).

Except for a few birth defects due to genetic factors, like thalassemias, known as the commonest genetic disease of hemoglobin (Weatherall, 2008), precise epidemiological data are lacking in North Africa, although many reports indicate that they are relatively frequent.

The absence of epidemiological data could be explained mainly because of underdiagnosis, lack of genetic studies, difficulties in establishing nationwide registries, and insufficient dedicated funding, as priorities are in favor of infectious diseases and for common noncommunicable diseases. Few studies have attempted to evaluate the spectrum of well-known common inherited metabolic diseases (IMD), such as cystic fibrosis and phenylketonuria. Cystic fibrosis (CF), frequent in European populations, seems not exceptional in North Africa. On the basis of patient records, 390 affected subjects from 383 families were recently reported in Tunisia (Messaoud et al., 2005). In Algeria, the disease appears to be less frequent, as only 81 affected individuals were described (Loumi et al., 2008). On the other hand, and in order to evaluate the prevalence of CF in the Moroccan population in the absence of patient data, 150 healthy individuals were screened for *CFTR* gene mutations (Ratbi et al., 2008). The study concluded that CF prevalence could be in the range of that found in European population (Ratbi et al., 2008).

With regard to phenylketonuria, two strategies were adopted to evaluate its incidence, mainly by a limited local screening program. In Egypt, a newborn screening program of some IMD was initiated in Alexandria city (Ismail et al., 1996) followed, recently, by molecular investigations of a large number of patients both in Egypt and Morocco (Effat et al., 2008; Dahri et al., 2010). In Tunisia, an approximate evaluation of PKU frequency of 1.32% was estimated through screening mentally retarded persons from different centers for mentally disabled in the country (Khemir et al., 2009). Unfortunately, no data on PKU were recorded from Algeria and Libya.

In order to estimate the most frequent types of genetic and congenital disorders in the Egyptian population, specifically from the Giza governorate, more than 73,000 individuals attending different healthcare centers in the region were analyzed. Their global prevalence was 1.16%. Neurologic disorders and chromosomal disorders were the most prevalent (9.5% and 9.3%, respectively). Genetic syndromes, growth disorders, and mental retardation and behavioral disorders were also frequent, with an average rate of 8.2%. Nevertheless, neuromuscular disorders, metabolic and endocrinal disorders, and skeletal disorders represent a group with a mean of frequency of 5.3%. Dermatological and renal disorders represented the least common cases (1.1% and 0.5%, respectively; Afifi et al., 2010).

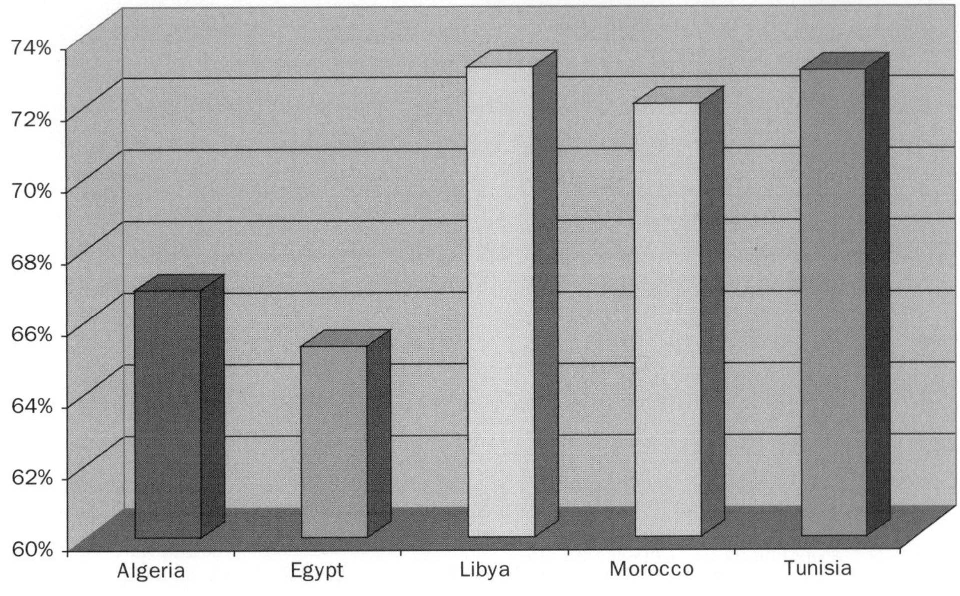

Figure 34-3 Birth Defects Prevalence per 1000 Live Births in North Africa

In the absence of quantitative data for the whole region, an estimation of the spectrum of genetic disorders affecting the North African population could be assessed by classification according to the affected organ or tissue, or affected biological process of published data on these disorders. Using the World Health Organization International Classification of Disease (WHO ICD), it was possible to arrange the inherited diseases according to different categories. Congenital malformations, deformations, and chromosomal abnormalities account for the majority of the inherited disorders (29.4%), followed by the endocrine, nutritional, and metabolic disorders (22.4%). Diseases of the nervous system (14.3%) represent the third most prevalent class of conditions (Figure 34-4).

Some genetic diseases appear to be frequent and, in several cases with founder mutations, confined to North African populations, as they are reported in more than 3 of the 5 countries of the region (Table 34-2). This could be explained as a consequence of combinations of many factors, including a common genetic background and historical events. On the other hand, some others are only reported in one population. The most illustrative examples are the following:

- "Maghrebian myopathy" or "Tunisian muscular dystrophy" was the first term used for a new type of recessively inherited muscular dystrophy described by Ben Hamida and Fardeau in 1980. It was later called *limb-girdle muscular dystrophy type 2C* (LGMD2C) after the identification of the morbid locus on chromosome 13q in Tunisian families (Ben Othman et al., 1992; Teebi & Farag, 1997).
- The Algerian type of spondylometaphyseal dysplasia is another example of North African–specific genetic disorder. This rare and severe, dominantly inherited malformation is characterized by a unique clinical and radiological set of features that were reported for the first time in five members of an Algerian family (Kozlowski et al., 1988).
- Bare lymphocyte syndrome type II is an autosomal recessive defect of the expression of the major histocompatibility complex class II genes, caused by mutations in the *RFXANK* gene. The majority of the patients described with the disease were of North African origin (Wiszniewski et al., 2000).
- Creutzfeldt-Jacob disease (CJD), the most prevalent of the human spongiform encephalopathies, is a progressive neurodegenerative disease. The largest cluster of CJD occurs among Libyan Jews, where the incidence (1/10,000) is 100 times higher than in other populations elsewhere (Colombo, 2000).
- Xeroderma pigmentosum (XP) is photodermatosis recessively inherited. It is characterized by hypersensitivity, DNA repair defects, and a high

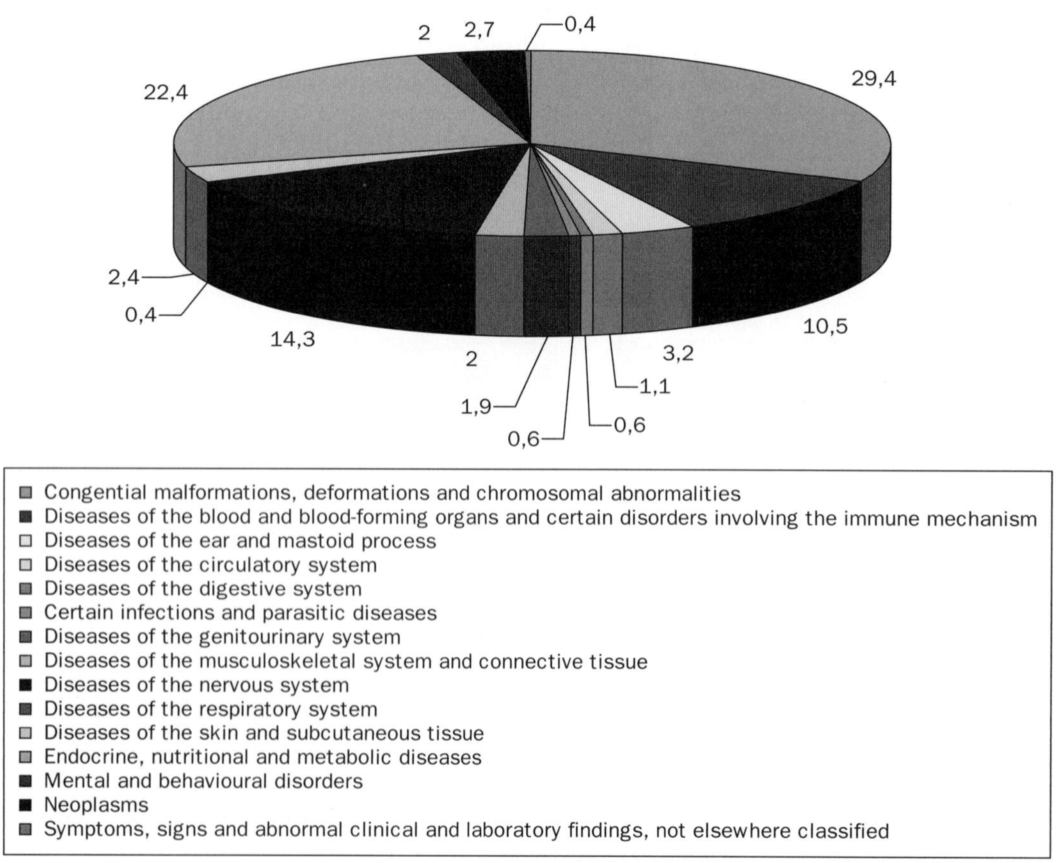

Figure 34-4 WHO-ICD10 Classification of Genetic Disorders in North Africa. Refer color figure.

TABLE 34-2 GENETIC DISORDER REPARTITION AMONG NORTH AFRICAN POPULATIONS

Disease	OMIM	Transmission	Country
3-M Syndrome	273,750	AR	Algeria, Egypt, Morocco
Achalasia-Addisonianism-Alacrima Syndrome	231,550	AR	Algeria, Morocco, Tunisia
Adrenal Hyperplasia, Congenital, due to 21-Hydroxylase Deficiency	201,910	AR	Egypt, Morocco, Tunisia
Autosomal Recessive Parkinson Disease 2	600,116	AR	Algeria, Morocco, Tunisia
Bare Lymphocyte Syndrome, Type II	209,920	AD	Algeria, Egypt, Morocco, Tunisia
Beta-Thalassemia	141,900	AR	Algeria, Egypt, Mauritania, Morocco, Tunisia
Chanarin-Dorfman Syndrome	275,630	AR	Algeria, Egypt, Morocco, Tunisia
Charcot-Marie Tooth Disease, Type 4A	214,400	AR	Algeria, Morocco, Tunisia
Colorectal Cancer Hereditary Nonpolyposis	120,435	AD	Egypt, Morocco, Tunisia
Congenital Myasthenia	608,931	AR	Algeria, Libya, Morocco, Tunisia
Creutzfeld-Jakob Disease	123,400	AD	Egypt, Morocco, Libya, Tunisia
Cystic Fibrosis	219,700	AR	Algeria, Egypt, Libya, Morocco, Tunisia
Darier Disease	124,200	AD	Egypt, Morocco, Tunisia
Down Syndrome	190,685	Sporadic	Egypt, Tunisia, Libya
Duchenne/Becker Muscular Dystrophy	310,200	X-Linked	Egypt, Morocco, Tunisia
Dyggve-Melchior-Clausen Syndrome	223,800	AR	Egypt, Morocco, Tunisia
Familial Atypical Mycobacteriosis	209,950	AR	Algeria, Morocco, Tunisia
Familial Mediterranean Fever	249,100	AR	Algeria, Egypt, Libya, Morocco, Tunisia
Farber Disease	228,000	AR	Egypt, Morocco, Tunisia
Fibrodysplasia Ossificans Progressiva	135,100	AD	Egypt, Morocco, Tunisia
Fragile X Mental Retardation Syndrome	300,624	X-Linked	Egypt, Morocco, Tunisia
GAPO Syndrome	230,740	AR	Algeria, Egypt, Tunisia
Glucose-6-Phosphate Dehydrogenase	305,900	X-Linked	Algeria, Egypt, Libya, Morocco, Mauritania, Tunisia
Hemochromatosis	235,200	AR	Algeria, Egypt, Morocco, Tunisia
Hemoglobin – Alpha Locus 1	141,750	AD	Algeria, Egypt, Mauritania, Tunisia
Hurler syndrome	607,014	AR	Libya, Morocco, Tunisia
Ichthyosiform Erythroderma, Congenital, Nonbullous, 1	242,100	AR	Algeria, Egypt, Tunisia
Infertility Associated with Multi-Tailed Spermatozoa and Excessive DNA	243,060	AR	Libya, Morocco, Tunisia
Kindler Syndrome	173,650	AR	Algeria, Egypt, Tunisia
Leber Congenital Amaurosis	204,000	AR	Algeria, Morocco, Tunisia
Leigh Syndrome	256,000	Mitochondrial	Mauritania, Morocco, Tunisia
Lipodystrophy, Congenital Generalized, Type 1	608,594	AR	Algeria, Egypt, Morocco, Tunisia
Morquio Syndrome A, Mucopolysaccharidosis Type IVA	253,000	AR	Egypt, Morocco, Tunisia
Muscular Dystrophy, Congenital, 1C	606,612	AR	Algeria, Libya, Tunisia

(Continued)

TABLE 34-2 (CONTINUED)

Disease	OMIM	Transmission	Country
Muscular Dystrophy, Limb-Girdle, Type 2C	253,700	AR	Algeria, Egypt, Libya, Morocco, Tunisia
Niemann-Pick Disease Type B	607,616	AR	Algeria, Morocco, Tunisia
Parkinson Disease 8	607,060	AD	Algeria, Morocco, Tunisia
Phenylketonuria	261,600	AR	Algeria, Egypt, Tunisia
Prader-Willi Syndrome	176,270	Sporadic	Algeria, Egypt, Morocco
Schwartz-Jampel Syndrome Type 1	255,800	AR	Algeria, Egypt, Tunisia
Sickle Cell Anemia	603,903	AR	Algeria, Egypt, Mauritania, Tunisia
Spinal Muscular Atrophy 1	253,300	AR	Egypt, Morocco, Libya, Tunisia
Spinal Muscular Atrophy 3	253,400	AR	Egypt, Morocco, Tunisia
Spinal Muscular Atrophy type II	253,550	AR	Egypt, Morocco, Tunisia
Spinocerebellar ataxia 2	183,090	AD	Algeria, Egypt, Tunisia
Tay-Sachs Disease	272,800	AR	Algeria, Egypt, Tunisia
Thrombophilia due to activated protein C	188,055	AD	Algeria, Egypt, Tunisia
Usher Syndrome Type 1B	276,900	AR	Algeria, Morocco, Tunisia
Xeroderma Pigmentosum Complementation Group A	278,700	AR	Egypt, Libya, Tunisia
Xeroderma Pigmentosum Complementation Group C	278,720	AR	Egypt, Morocco, Tunisia
Xeroderma Pigmentosum Complementation variant type	278,750	AR	Algeria, Egypt, Tunisia

incidence of skin cancers (Lehmann, 2003). Its prevalence in the United States and Europe is estimated to 1/300,000, but it seems to be more frequent in the Maghreb, and especially in Tunisia, since its prevalence is estimated to 1/10,000 (Zghal et al., 2006).

- Nonsyndromic hearing loss accounts for a major proportion of sensorial handicaps in the region. Its prevalence could be especially high in isolates, in which it ranges from 2% to 8% (Ben Arab et al., 2004). Genetic heterogeneity for this kind of impairment has been demonstrated, and many loci were also identified by investigating North African families. The autosomal recessive forms, named DFNB, are the most frequent of the prelingual genetic forms of deafness, as they represent 80% among the studied cases (Ben Arab et al., 2004).

Underlying founder mutations for all these diseases are reported further in this chapter.

In addition to the categorization of the new form of Duchenne-like muscular dystrophy in North Africa, new syndromes or variants of very rare syndromes were also described in the last decades. They encompass bone malformations, such as SEMDJL and Desbuquois syndromes (Al Kaissi et al., 2008, 2007a, 2005), and a constellation of new clinical features associated to already-known syndromes (Al Kaissi et al., 2007b, 2009). Extremely rare syndromes have also been highlighted in the region. A third case of acro-fronto-facio-nasal dysostosis, combined with genitourinary anomalies, was recently described (Chaabouni et al., 2008). Recently, unreported neurological manifestations have been described in XP Group A, in compound heterozygous individuals harboring one recurrent founder mutation (p.R228X), and a new likely private one (p.V241GfsX5) in a consanguineous family of Tunisian origin (Messaoud et al., unpublished data).

MAPPING GENETIC DISEASE LOCI IN NORTH AFRICAN FAMILIES

The North African populations, probably because they were isolated by the Mediterranean Sea to the north and the Sahara to the south, were genetically isolated and, in addition, they exhibit high frequency of inbreeding as a common social feature. Their families raised geneticists' interests because they are large and span many generations. As a consequence, recessive disorders have emerged at increased prevalence, and informative families are common. From a historical point of view, the study of such inbred populations, especially in the region, has been successful in the fine-mapping of recessive traits using a strategy known as *homozygosity* or *autozygosity mapping* (Lander and Botstein, 1987). Its rationale assumes that an affected offspring co-inherits two copies of a chromosomal segment related to the disease from a common ancestor. In fact, the effect of consanguinity tends to reduce allelic and non-allelic heterogeneity and increases linkage disequilibrium, allowing the detection of associations between genetic markers and diseases (Sheffield et al. 1998). Consequently, positional cloning of recessive genes in families bearing rare genetic variants, most often at the homozygous state, has provided valuable insight into the localization and identification of genetic factors responsible for diseases with similar clinical phenotypes. As eminent examples of the strategy's success, the AVED locus was identified in one Tunisian family (Ben Hamida et al., 1993), followed by other morbid loci involved in neurogenetic diseases, mainly from Tunisia and Morocco (Table 34-3). In addition, it was possible to identify the two first AR deafness loci DFNB1 and DFNB2 loci (Guilford et al., 1994; Boulila-Elgaied et al., 1997). New loci and causative mutations are continuously revealed, thus enlarging the mutation spectra and knowledge of the genetics of sensorial impairment (Table 34-4).

COMORBIDITY IN NORTH AFRICAN FAMILIES

Comorbidity, literally, in medicine refers to either the presence of one or more disorders in addition to a primary disease, or to the effect of such additional disorders (Valderas et al., 2009). This striking phenomenon was noticed in inherited disorders in North African families as a consequence of the high rate of inbreeding. Two or more phenotypes could occur with an autosomal recessive mode of inheritance within the same large family. Two consanguineous Tunisian female siblings presented clinical manifestations of Mal de Meleda and congenital cataract—a rare genodermatosis, and a major visual defect that leads to blindness, respectively (Bchetnia et al., 2010 a). The coexistence of these two uncommon conditions was first reported in North Africa, and could be a random phenomenon due to consanguinity or to the fact that lens and skin share the same embryological origin (Bchetnia et al., 2010a). Autosomal dominant and autosomal recessive genodermatoses were associated in one Tunisian patient (Bchetnia et al., 2010b) suffering from Darier disease, which is a rare dominantly inherited disorder characterized by warty papules and plaques in seborrheic areas, palmoplantar pits, and distinctive nail dystrophy. The second genetic defect of the skin was a form of an autosomal recessive ichthyosis, which affects the keratinization process characterized by desquamation over the whole body (Bchetnia et al., 2010b). In addition, the same patient presented other clinical features such as diffuse alopecia, ovarian cancer, corneal opacity, and mental retardation.

In other, more complex cases and in highly inbred families, the comorbidity may lead to erroneous diagnosis, genetic mapping, and renders genetic counseling challenging, in particular for genetic heterogeneous phenotypes (Hmani-Aifa et al., 2009). As an example, a large consanguineous family with both Usher syndrome, an autosomal recessive disorder featured by sensorineural hearing loss and retinal degeneration, and autosomal recessive retinitis pigmentosa was previously used to define USH2B, a new locus involved in Usher syndrome (Hmani et al., 1997). Further clinical and genetic reanalysis of the extended family showed phenotypic heterogeneity within USH2 patient and excluded linkage to USH2B locus. Molecular investigations unraveled the coexistence of mutations affecting two different genes, *GPR98* and *PDE6B*, thus excluding the involvement of the

TABLE 34-3 NOVEL LOCI OF NEUROGENETIC DISEASES IN TUNISIA

Disorder	OMIM	Novel loci	Reference
Autosomal recessive cerebellar ataxia, Charlevoix-Saguenay type; SACS	270,550	13q11-12	Mrissa et al., 2000
Autosomal recessive Charcot-Marie-Tooth disease type 4B2; CMT4B2	604,563	11p15	Ben Othmane et al., 1999
Autosomal recessive limb-girdle muscular dystrophy 2I; LGMD2I	607,155	19q13.3	Driss et al., 2000
Autosomal recessive limb-girdle muscular dystrophy; LGMD2C	253,700	13q	Ben Othmane et al., 1992
Charcot-Marie-Tooth disease type 4A; CMT4A	214,400	8q21.3	Barhoumi et al., 2000
Friedreich's ataxia with vitamin E deficiency; AVED	277,460	8q	Ben Hamida et al., 1993
Giant axonal neuropathy 1; GAN1	256,850	16q24.1	Ben Hamida et al., 1997
Recessive familial amyotrophic lateral sclerosis; ALS2	205,100	2q33-q35	Hentati et al., 1994

TABLE 34-4 NEW LOCI AND MUTATIONS IDENTIFIED IN SENSORIAL GENETIC CONDITIONS

Disease	OMIM	Locus/Gene	Mutation	Population	References
Autosomal dominant deafness, DFNA36	606,705	*TMC1*	p.W588X	Tunisia	Tlili et al., 2008
Autosomal dominant deafness, DFNA36	606,705	*TMC1*	p.W588X	Tunisia	Tlili et al., 2008
Autosomal recessive hearing loss; DFNB59	610,220	*DFNB59*	c. 2189+1G>T	Morocco	Ebermann et al., 2007
Autosomal recessive nonsyndromic hearing loss, DFNB32	608,653	1q13.3-22.1		Tunisia	Masmoudi et al., 2003
Autosomal recessive nonsyndromic hearing loss; DFNB79	613,307	*DFNB79*	c.42_52del	Morocco	Li et al., 2010
Autosomal recessive nonsyndromic deafness, DFNB66	610,212	6q21.2-22.3		Tunisia	Tlili et al., 2005
Autosomal recessive nonsyndromic sensorineural hearing loss, DFNB1A	220,290	*GJB2*	p.A88G	Tunisia	Alemanno et al., 2009
Autosomal recessive retinitis pigmentosa	180,072	*PDE6B*	p.W807R	Tunisia	Hmani-Aifa et al., 2009
Brittle cornea syndrome	229,200	Locus mapped on 16q24		Tunisia	Abu et al., 2006
Nonsyndromic autosomal recessive deafness, DFNB10	605,316	*TMPRSS3*	W251C P404L	Tunisia	Masmoudi et al., 2001
Nonsyndromic deafness DFNB13	603,098	Refined the DFNB13 to an area of 2.2 Mb		Tunisia	Masmoudi et al., 2004
Nonsyndromic hearing loss, DFNB3	600,316	*MYO15A*	p.C1666X c.9229 + 1G>A c.7395 +3G>C	Tunisia	Belguith et al., 2009
Nonsyndromic hearing loss, DFNB63	611,451	11q13.3-q13.4		Tunisia	Tlili et al., 2007
Nonsyndromic recessive hearing loss; DFNB31	607,084	*WHRN*	c.2423delG	Tunisia	Tlili et al., 2005
Thiamine-responsive megaloblastic anemia syndrome with diabetes and deafness	249,270	*SLC19A2*	287delG	Tunisia	Gritli et al., 2001
Usher syndrome type 1 B	600,060	*MYO7A*	c. 1687G>A	Morocco	Boulouiz et al., 2007
Usher syndrome type 2 A	605,472	*GPR98*	p.Y6044C	Tunisia	Hmani-Aifa et al., 2009

same gene for both forms of hereditary retinal degeneration (Hmani-Aifa et al., 2009). Furthermore, the severe and early onset of the retinal attack in one individual was attributed to the overlapping of these two impairments (Hmani-Aifa et al., 2009). Another example is illustrated by Fendri et al. (2006): three siblings from the same consanguineous Tunisian families were diagnosed as having autosomal recessive limb girdle muscular dystrophy (AR-LGMD2). Immunohistochemical study revealed heterogeneous patterns, thus suggesting genetic heterogeneity. Linkage and mutational screening showed that two siblings shared the 157G>A homozygous mutation in the *SGCA* gene leading to the LGMD2D, while the other sibling had the most widely distributed del521T mutation in the *SGCG* gene, causing the LGMD2C phenotype (Fendri et al., 2006). Another example of the complexity of such phenomena is illustrated by the genetic heterogeneity involved in autosomal recessive ataxia in a single large consanguineous Tunisian family (Bouhlal et al., 2008). Autosomal recessive cerebellar ataxias are a group of clinically neurodegenerative disorders. Mutation analysis in eleven patients revealed the presence of three distinct gene defects in the *FXN*, *TTPA* and *SACS* genes (Bouhlal et al., 2008).

In addition to the comorbidity of two or more monogenic disorders, co-occurrence of two numerical chromosomal abnormalities in the same individual can exist. This chromosomal event, known as a *double aneuploidy*, is very rare and could involve both autosomal and sex chromosomes. Down syndrome, which is known to be the major result of chromosomal disorder associated with mental retardation and is reported in North Africa and worldwide, is caused by three genetic mechanisms: free trisomy, mosaic trisomy, and translocation (Stoll et al.,

1998). The most represented sex chromosome disorders are Turner and Klinefelter syndromes. They are mainly due, respectively, to the absence or the addition of an X-chromosome. For the first time in a North African country, a double aneuploidy was highlighted in three Egyptian patients. They presented both combinations of Down–Turner and Down–Klinefelter syndromes (Zaki et al., 2005). The phenomenon of either the simple or double aneuploidy in the same family was also noticed (Zaki et al., 2005). Factors leading to either aneuploidy or familial aggregation of such chromosomal abnormalities remain unknown. Nevertheless, many efforts have been deployed to give explanations to these chromosomal disorders. Advanced maternal age is reported to be among the most likely factors (Antonarakis, 1998), but this hypothesis is controversial, since young women of less than 30 years also have babies with Down syndrome (James et al., 1999). Besides the father's age, the grandmother's age at which she gave birth to her daughter (the Down syndrome baby's mother) could be additional factors to Down syndrome occurrence (Malini & Ramachandra, 2006). Consanguinity was suggested as an additional factor in the emergence of such anomalies (Amudha et al., 2005; Malini & Ramachandra 2006). It was also proposed that an autosomal recessive gene may exist that results in meiotic nondisjunction in the homozygous parents and, therefore, parents in highly inbred families have a probability to be homozygous for that gene (Alfi et al., 1980). A more plausible explanation to the occurrence of Down syndrome or another type of aneuploidy and their familial recurrence is cytoplasmic inheritance. In fact, the number of mitochondrial DNA mutations increases with age in different cells, particularly in oocytes, and animal studies have shown the effect of mitochondrial DNA mutations on meiotic apparatus, leading to a high incidence of ovulated diploid oocytes (Keefe et al., 1995, Beerman et al., 1988).

MOLECULAR ETIOLOGY AND FOUNDER EFFECT OF GENETIC DISORDERS

A great proportion of the genetic disorders reported in North Africa remain reduced to clinical observations and are poorly investigated on the molecular level. However, mutation studies are well performed in the Sephardic Jewish community (Colombo, 2000; White et al., 1991; Gilad et al., 1996; Benayoun et al., 2007; Gershoni-Baruch et al., 1994). Nearly half of the inherited diseases in Tunisia have a known molecular etiology. In the remaining fraction, the gene or the causative mutation has not yet been identified (Romdhane & Abdelhak, 2011). Two initiatives to gather molecular data on genetic disorders have been realized: The Tunisian National Mutation Database (www.goldenhelix.org/tunisian/) and the Moroccan Human Mutation Database (http://www.sante.gov.ma/Departements/INH/MoHuMuDa/index.htm). Because they report only on diseases for which mutations were identified, they do not reflect the real situation of genetic diseases affecting the North African population. In addition, they are rarely updated.

As one or more genes could lead to the expression of the same morbid phenotype (genetic heterogeneity) in the region, allelic heterogeneity was also observed. Eight mutations could be responsible for familial Mediterranean fever in the Tunisian population, and more than 45 mutations can lead to β-thalassemia in Tunisia, Algeria, and Morocco, of whom 29 were identified in the Tunisian population (Haj Khelil et al., 2010; Romdhane & Abdelhak, 2011). The accumulation and the recurrence of many mutations in an isolated population may be the consequence of many factors: this could be due to chance (a random phenomenon; Zlotogora, 2007), a selective advantage of heterozygous carriers (Allison, 1954; Ruwende et al., 1995), or it may be due to high mutation rates because the variants occur in a region of particular risk for nucleotide changes called a "hot spot" (Crow, 1997). The high prevalence of β -thalassemia, sickle cell anemia, hereditary elliptocytosis, and G6PD deficiency among North Africans was explained by the exposure of the region to environmental factors that led to natural selection of carriers such as malaria, in the case of sickle cell anemia, and dietary tradition in the case of glucose-6-dehydrogenase deficiency (Roux et al., 1989). In the first part of the twentieth century, the low tuberculosis mortality among Jews of Tunis City, compared with the other communities (Moslem and Christians) of the same town, was attributed to a selective advantage of heterozygote carriers of the deleterious alleles of the *MEFV* gene that leads to the familial Mediterranean fever and additional, other factors (Cattan et al., 2007). Indeed, on the one hand, these alleles are noticeably more frequent in the North African Jewish populations than in the others (Belmahi et al., 2006; Stoffman et al., 2000), and on the other hand, it was suggested that the heterozygous state increases the biological indicators of inflammation and is thought to protect against infectious diseases (Lachmann et al., 2006).

Another cause that contributes to high frequencies for a deleterious allele in a population is a phenomenon known as *genetic drift*; the fluctuation in mutation frequency operates by chance on the small gene pool within a small population. One form of genetic drift is the *founder effect*. This happens when one of the founders of a new group carries a relatively rare allele; this allele will have a higher frequency than it would have in the larger group from which the new group was originated (Nussbaum et al., 2004). Within families where consanguinity is culturally favored, individuals resulting from common ancestors maintain genotypes and haplotypes bearing the mutated gene (Teebi & El-Shanti, 2006). Haplotypic analysis is used to confirm a founder effect. In this case, all affected patients share the same chromosomal segments for markers around the deleterious mutations (Marrakchi et al., 2003).

It has been established that the number of different mutations in a given population, encompassing private and recurrent allelic variants, is variable, and there are substantial differences in the mutational spectra among populations (Mollas et al., 1999). In fact, compilation of founder mutations reported in North African populations revealed that some ethnic mutations could be specific to a country. A recent collection of founder variants in the Tunisian population unraveled 27 genetic disorders due to at least one founder mutation not reported elsewhere (Romdhane et al., unpublished data). This fact could be explained by a sole event of creation of that mutation in a founder ancestor belonging to that population (local origin of the founder mutation), or simply by the fact that these variants have not yet been identified elsewhere. Other founder variants could be shared with other populations, mainly from North Africa and the Middle East, and other populations from the Mediterranean area (Romdhane et al., unpublished data; Table 34-5), thus showing the considerable homogeneity of the genetic background of these populations.

Different categories of genetic diseases due to a founder mutation were reported. They include genodermatosis, cancers, neurogenetic diseases, and metabolic disorders. The most frequent genetic disorders in the Mediterranean area, known to be the result of an advantage selection of heterozygous carriers, have also been proved to be the result of a combination with a founder effect. In Tunisia and Algeria, the most frequent mutations of the *HBB* gene leading to β-thalassemia are codon 39 (C>T) and IVS-I-110 (G>A), which account for more than 50% of all mutations (Haj Khelil, 2010). However, in Morocco, the predominant mutations are codon 39 (C>T) and frameshift codon 8 (-AA) (Haj Khelil, 2010). Furthermore, only the Glu6Val mutation affecting the Beta-globin results in sickle cell anemia in these populations (Mears et al., 1981). Haplotype analysis revealed different common chromosomal segments bearing the sickle cell variant, codon 39 (C>T) and IVS-I-110 (G>A), and allowed the hypothesis of a Benin origin for sickle cell anemia, a local North African origin for codon 39 (C>T), and an Eastern Mediterranean origin for IVS-I-110 (G>A) (Haj Khelil, 2010). Familial Mediterranean fever (FMF) is the main cause of periodic fever syndrome due to mutations in the *MEFV* gene at a homozygous state. Five founder mutations—M694V, V726A, M680I, M694I, and E148Q—account for 74% of FMF chromosomes in patients from the Mediterranean basin (Touitou, 2001). In a cohort of 209 unrelated Arab patients from Maghreb (85 Algerians, 87 Moroccans, and 37 Tunisians), M694V and M694I were the most frequent mutations, with some differences (Belmahi et al., 2006). Their frequencies were estimated at 5% and 80% in Algeria, 49% and 37% in Morocco, and 50% and 25% in Tunisia, respectively (Belmahi et al., 2006). M694I mutation appears to be specific to the Arab population from North Africa (Belmahi et al., 2006).

Limited heterogeneity was shown for founder mutations. In the case of xeroderma pigmentosum group A, only the recurrent mutation p.R228X in the *XPA* gene was identified in six unrelated Tunisian families (Nishigori et al., 1993; Messaoud et al, 2010). The XP group C is due to mutations in the *XPC* gene. The p.V548Afs572X mutation was previously described in three unrelated families from Italy, Algeria, and Morocco (Chavanne et al., 2000; Khan et al., 2006). The same deleterious variant was prioritized in a molecular diagnosis of 20 XPC individuals belonging to 14 Tunisian families (Ben Rekaya et al., 2009). All patients harbored this mutation (Ben Rekaya et al., 2009). The founder origin for these two mutations was confirmed by haplotype analysis in this population (Ben Rekaya et al., 2009; Messaoud et al., 2010), and further validated in a larger XP cohort, mainly from Algeria, Morocco, and Tunisia (Soufir et al., 2010). The prevalence of p.V548Afs572X was 96% in Maghrebi XP-C patients, and the same *XPA* p.R228X mutation was found in 16 North African XP-A patients of whom 9 were from Tunisia (Messaoud et al., 2010; Soufir et al., 2010).

Taking into account these observations, founder mutations are of particular interest because they can be used to quantify the degree of homogeneity within a population and help to design suitable molecular diagnosis tools (Bellayou et al., 2009). Indeed, testing for one or a few prevalent founder mutations may be more efficient and at lower cost than testing for many rare mutations (Pastinen et al., 2001; Bellayou et al., 2009; Romdhane et al., unpublished).

Structural mutations, when combined with founder effect, could be bound with particular clinical manifestations. In this case, the patient's ethnic origin could either orientate the diagnosis or predict the clinical course of the disease. Carbonic anhydrase II deficiency syndrome is usually characterized by osteopetrosis, renal tubular acidosis, and cerebral calcification. Patients from North African and of Arab descent with carbonic anhydrase II deficiency bear the "Arabic mutation," a splice-site mutation of the *CA II* gene (Fathallah et al., 1997). In these patients, mental retardation and infrequency of skeletal features distinguish their clinical course from those of American and Belgian origins harboring the p.His107Tyr mutation (Hu et al., 1992).

TABLE 34-5 EXAMPLES OF FOUNDER MUTATIONS IN NORTH AFRICAN COUNTRIES

Disease	Gene	Mutation	Population/Ethnic groups	Reference
5 alpha-reductase type 2 deficiency	*SRD5A2*	p.G34R	Founder effect in Egypt	Kou et al., 2008
Adenomatous Polyposis of the colon	*MYH*	c.1186_1187insGG	Maghreb: Tunisia and Morocco	Baert-Desurmont et al., 2007
Ataxia-telangiectasia	*ATM*	c.103C>T	Jews from North Africa: Morocco and Tunisia	Gilad S et al., 1996
Autosomal recessive nonsyndromal hearing loss	*GJB2*	c.35delG	Tunisia, Morocco, Lebanon	Masmoudi et al., 2000 Belguith et al., 2005
			Founder mutation in Morocco	Abidi et al. 2008
Autosomal recessive axonal Charcot-Marie-Tooth disease CMT2K/4A	*GDAP1*	p.S194X	Founder effect in Morocco and mutation reported in Tunisia and Spain	Bouhouche et al., 2007 Baxter et al., 2002 Cuesta et al., 2002
Autosomal recessive axonal Charcot-Marie-Tooth disease CMT2B1	*LMNA*	p.R298C	Founder effect in Morocco (private mutation) and Algeria	Bouhouche et al., 2007 Tazir et al., 2004 De Sandre-Giovannoli et al., 2002 Hamadouche et al., 2008
Autosomal recessive syndromic congenital Ichthyosis with sclerosing cholangitis, hypotrichosis of the frontal scalp, and vacuolated leukocytes	Mapped to chromosome 3	Common haplotype: 1(D3S1580)-1(D3S2747)-1(D3S1601)	Founder effect in Morocco	Baala et al., 2002
Bare Lymphocyte syndrome type II	*RFXANK*	c.752delG-25	Founder effect in North Africa	Wiszniewski et al., 2000
Carbonic anhydrase II deficiency	*CA II*	c.IVS II, G>A	Founder effect in Tunisia, Arabic Peninsula	Fathallah et al., 1997
Chylomicronemia	*LPL*	p.Ser259Arg	Founder effect in Morocco	Foubert et al., 1997
Combined factor V and factor VIII deficiency	*LMAN1*	c.IVS9 +2 T>G	Founder effect in Tunisia	Segal et al., 2004
Congenital muscular dystrophy	*FKRP*	c.1364C>A	Founder effect in Tunisia	Louhichi et al., 2004
Congenital myasthenic syndrome	*CHRNE*	c.1293insG	Founder effect in North Africa: Tunisia, Algeria, Morocco, Libya	Richard et al., 2008
Complete hydatidiform mole	*NLRP7*	intragenic tandem duplication including exons 2 through 5	Founder effect in Egypt	Akhavan et al. 2007
Creutzfeldt-Jacob disease	*PRNP*	p.E200K	Libya, Tunisia, Spain, Italy, Chile	Colombo, 2000 Hee Suk et al., 1999
Crigler–Najjar type I syndrome	*UGT1A1*	c.1070A>G	Founder effect in Tunisia, Kuwait	Petit et al., 2008
Dubin-Johnson syndrome	*MRP2*	p.R1150H	Founder effect in Morocco Jews	Mor-Cohen et al., 2007
Factor VII deficiency	*F7*	p.A244V	Founder effect in Morocco Jews	Mor-Cohen et al., 2007
Factor X deficiency	*FX*	p.F31S	Founder effect in Algeria	Akhavan et al., 2007
Factor XIII deficiency	*F13A*	c.869insC	Founder effect in Tunisia	Louhichi et al., 2010
Familial hypercholesterolemia	*LDLR*	p.S493RfsX44	Founder effect in Tunisia	Slimane et al., 2001
Familial Mediterranean fever	*MEFV*	p.M680I	Tunisian, Arab, Moroccan, Algerian, Jordanian, Palestinian, Armenians, Turkish	Gershoni-Baruch et al., 2001 Majjed et al., 2005 Chaabouni et al., 2007 Touitou, 2001 Al-Alami et al., 2003 Mattit et al., 2006

(*Continued*)

TABLE 34-5 (CONTINUED)

Disease	Gene	Mutation	Population/Ethnic groups	Reference
		p.M694V	Arabs, North, Tunisian, Jordanian, Egyptian, Armenian, Lebanese, Algerian, Moroccan African Jews	
		p.V726A	Tunisian, Arabs, Lebanese, Egyptian, Armenian, Turkish, Saudi Arabian	
		p.M694I	Tunisian, Arab, Syrian, Jordanian, Egyptian, Maghrebian, Lebanese, Palestinian, Algerian, Moroccan	
		p.E148Q	Tunisian, Syrian, Jordanian, Arab, Lebanese, Egyptian, North African, Saudi Arabian, Armenian, Turkish, Palestinian	
Fanconi anemia group A	*FANCA*	c.890-893del	Founder effect in Tunisia	Tamary et al., 2000
Fragile X syndrome	*FMR1*	Common haplotype DXS548-FRAXAC1-FRAXAC2 (7+ -4- 6+)	Tunisian Jews, Sephardic Jews, Arabs	Falik-Zaccai et al., 1997
Hereditary breast and ovarian cancer	*BRCA1*	c.798_799delTT	Non-Jewish from North Africa: Algeria, Tunisia	Uhrhammer et al., 2008
Hepatocerebral mitochondrial DNA depletion syndrome	*DGUOK*	c.444-62C>A	Founder mutation in Tunisia and Morocco	Brahimi et al., 2009
Hirschsprung's disease	*RET*	p.C618R	Founder mutation in Moroccan Jews	Peretz et al., 1997
Hurler Syndrome	*IDUA*	p.P533R	Founder effect in Tunisia, mutation reported in Morocco and Maghrebian population living in France	Chkioua et al., 2007 Alif et al., 1999
Idiopathic BCGitis	*IL12p40*	c.297del8	Founder effect in Tunisia	Elloumi-Zghal et al., 2002
Lamellar ichthyosis type 3	*FLJ39501*	c.1303C>T	Founder effect in Algeria	Lefèvre et al. 2006
Leucocyte adhesion deficiency	*CD18*	c.119del10	Founder effect in Tunisia	Ben Mustapha et al., 2008
Mal de Meleda	*SLURP-1*	c.82delT	Tunisia, Algeria	Charfeddine et al., 2003 Marrakchi et al., 2003 Fischer et al., 2001
			Algeria, Croatia	
		c.178+1G>A	Founder effect in Algeria	
		p.C99Y	Founder effect in Tunisia	
Male infertility	*AURKC*	c.144delC	Tunisia, Algeria, Morocco	Dieterich et al. 2009
Meckel syndrome type 2	*MKS2*	Common haplotype encompassing D11S916, D11S911 and D11S906	Founder effect in Tunisia and Algeria	Roume et al., 1998
Megaloblastic anemia 1	*AMN*	c.208-2A>G	Non-Tunisian Jews, Tunisian Jews, Mediterranean basin	Bouchlaka et al., 2007
Muscular dystrophy, limb-girdle, type 2C	*SGCG*	c.521delT	Founder effect in Tunisia	Kefi et al., 2003
Triple-A syndrome	*ALADIN*	c.IVS14 +1 G>A	Founder effect in Tunisia, Algeria and Spain	Tullio-Pelet et al., 2000
Parkinson's disease	*LRRK2*	p.G2019S	Tunisia, Morocco, Algeria	Warren et al., 2008
Primary congenital glaucoma	*CYP1B1*	c.4339delG	Founder effect in Morocco	Belmouden et al., 2002
Spastic paraplegia	*SPG15*	p.Q493X	Founder effect in Tunisia	Hanein et al. 2008

(Continued)

TABLE 34-5 (CONTINUED)

Disease	Gene	Mutation	Population/Ethnic groups	Reference
Unverricht-Lundborg disease	*CSTB*	Expansion of a dodecamer repeat (CCCCGCCCCGCG)n in the cystatin B gene Common haplotype: 1(D21S1885)-1(D12S2040)-A(D12S1259)-1(D12S59)	Founder effect in Tunisia, Morocco, Algeria and Scandinavia	Moulard et al., 2002
Xeroderma pigmentosum	*XPA*	p.R228X	Founder effect in North Africa	Messaoud et al., 2010; Soufir et al., 2010
Xeroderma pigmentosum	*XPC*	p.V548AfsX572	Founder effect in Tunisia, mutation reported in Algeria, Morocco, Italy	Ben Rekaya et al., 2009

CONCLUSION

Genetic disorders in North Africa constitute a real public health problem and expose the affected persons to economic and social disadvantages. Autosomal recessive disorders are prevalent probably due to a combination of high rates of consanguinity and endogamy, selective pressure, and founder effect. Comorbidity renders efficient genetic counseling and diagnosis challenging in deeply inbred communities. Further investigations are necessary to help elucidate their molecular pathogenesis in order to design suitable diagnosis tools. Taking into account the common genetic background shared by the populations in North Africa and Middle East, the founder mutations could be screened directly and would provide an appropriate diagnosis tool since it is precise, noninvasive when applied to newborns, and low cost. Unfortunately, no national screening programs for heterozygous carriers of deleterious mutations exist in any of the five North African countries, and premarital counseling and prenatal diagnosis are the only way to avoid having an affected child with a genetic disease in at-risk families. Educational preventive programs do exist but are insufficient; interactions between researchers from biomedical as well as from social sciences domains should help find means and ways for adequate measures to reduce the burden of genetic diseases in the region.

REFERENCES

Abidi O, Boulouiz R, Nahili H, et al. (2008). The analysis of three markers flanking GJB2 gene suggests a single origin of the most common 35delG mutation in the Moroccan population. *Biochemical and Biophysical Research Communications 377*(3), 971–974.

Abu A, Frydman M, Marek D, et al. (2006). Mapping of a gene causing brittle cornea syndrome in Tunisian jews to 16q24. *Invest Ophthalmol Vis Sci 47*(12), 5283-5287.

Afifi HH, El-Ruby MO, El-Bassyouni HT, et al. (2010). The most encountered groups of genetic disorders in Giza Governorate, Egypt. *Bratisl Lek Listy. 111*(2), 62–69.

Akhavan S, Chafa O, Obame FN, et al. (2007). Recurrence of a Phe31Ser mutation in the Gla domain of blood coagulation factor X, in unrelated Algerian families: a founder effect? *Eur J Haematol 78*(5), 405–409.

Al-Alami JR, Tayeh MK, Najib DA, et al. (2003). Familial mediterranean fever mutation frequencies and carrier rates among a mixed Arabic population. *Saudi Medical Journal* 24(10), 1055–1059

Al Kaissi A, Nessib N, Ghachem MB, Hammou A, Guiddana N, Kozlowski K. (2005). A novel syndrome resembling Desbuquois dysplasia. *Am J Med Genet A 132A*(1), 68–75.

Al Kaissi A, Ben Chehida F, Gharbi H, et al. (2007a). Persistent torticollis, facial asymmetry, grooved tongue, and dolicho-odontoid process in connection with atlas malformation complex in three family subjects. *Eur Spine J 16* (Suppl 3), 265–270.

Al Kaissi A, Grill F, Safi H, Ben Ghachem M, Ben Chehida F, Klaushofer K. (2007b). Craniocervical junction malformation in a child with Oromandibular-limb hypogenesis-Möbius syndrome. *Orphanet J Rare Dis 8*, 2, 2.Al.

Al Kaissi A, Chehida FB, Ghachem MB, Klaushofer K, Grill F. (2008). A hypoplastic atlas and long odontoid process in a girl manifesting phenotypic features resembling spondyloepimetaphyseal dysplasia joint laxity syndrome. *Skeletal Radiol 37*(5), 469–473.

Al Kaissi A, Chehida FB, Ghachem MB, Grill F, Klaushofer K. (2009). Atlanto-axial segmentation defects and os odontoideum in two male siblings with opsismodysplasia. *Skeletal Radiol 38*(3), 293–296.

Alemanno MS, Cama E, Santarelli R, et al. (2009). A novel missense mutation in the Connexin 26 gene associated with autosomal recessive nonsyndromic sensorineural hearing loss in a consanguineous Tunisian family. *Int J Pediatr Otorhinolaryngol 73*(1), 127–131.

Alfi OS, Chang R, Azen SP. (1980). Evidence for Genetic Control of Nondisjunction in Man. *Am J Hum Genet 32*, 477–483.

Alif N, Hess K, Straczek J, et al. (1999). Mucopolysaccharidosis type I: characterization of a common mutation that causes Hurler syndrome in Moroccan subjects. *Ann Hum Genet 63*(Pt1), 9–16.

Allison AC. (1954). The distribution of the sickle-cell trait in East Africa and elsewhere, and its apparent relationship to the incidence of subtertian malaria. *Trans R Soc Trop Med Hyg 48*(4), 312–318.

Amudha S, Aruna N, Rajangam S. (2005). Consanguinity and chromosomal abnormality. *Indian Journal of Human Genetics 11*(2) 108–110.

Antonarakis SE. (1998). 10 years of Genomics, chromosome 21, and Down syndrome. *Genomics 51*(1), 1–16.

Baala L, Hadj-Rabia S, Hamel-Teillac D, et al. (2002). Homozygosity Mapping of a Locus for a Novel Syndromic Ichthyosis to Chromosome 3q27±q28. *J Invest Dermatol 119*(1), 70–76.

Baert-Desurmont S, Rouquette A, Mauillon J, et al. (2007). Une mutation à effet fondateur du gène MYH associée à la polypose adénomateuse et au cancer colorectal dans les pays du Maghreb. SNFGE (www.snfge.asso.fr)

Barhoumi C, Amouri R, Ben Hamida C, et al. (2000). Linkage of a new locus for autosomal recessive axonal form of Charcot-Marie-Tooth disease to chromosome 8q21.3. *Neuromuscul Disord 11*(1), 27–34.

Baxter RV, Ben Othmane K, Rochelle JM, et al. (2002). The gene encoding ganglioside-induced differentiation-associated protein 1 in mutated in Charcot-Marie-Tooth type 4A/8q21. *Nat Genet 30*(1), 21–22.

Bchetnia M, Merdassi A, Charfeddine C, et al. (2010a). Coexistence of mal de Meleda and congenital cataract in a consanguineous Tunisian family: two case reports. *J Med Case Reports 4*, 108.

Bchetnia M, Charfeddine C, Kassar S, et al. (2010b). Clinical and mutational heterogeneity of Darier disease in Tunisian families. *Arch Dermatol 145*(6), 654–656.

Beerman F, Hummler E, Franke U, Hansmann I. (1988). Maternal modulation of the inheritable meiosis I error Dipl I in mouse oocytes is associated with the type of mitochondrial DNA. *Hum Genet 79*(4), 338–340.

Beighton P. (1997). Heterozygous manifestations in the heritable disorders of the skeleton. *Pediatr Radiol* 27(5), 397–401.

Belguith H, Aifa-Hmani M, Dhouib H, et al. (2009). Screening of the DFNB3 locus: identification of three novel mutations of MYO15A associated with hearing loss and further suggestion for two distinctive genes on this locus. *Genet Test Mol Biomarkers 13*(1), 147–151.

Belguith H, Hajji S, Salem N, et al. (2005). Analysis of GJB2 mutation: evidence for a Mediterranean ancestor for the 35delG mutation. *Clin Genet 68*(2), 188–189.

Bellayou H, Dehbi H, Bourezgui M, Slassi I, Nadifi S. (2009). Ataxia with vitamin E deficiency (AVED); an example of the contribution of research in molecular genetic to counselling in Morocco. *Pathol Biol (Paris) 57*(5), 425–426.

Belmahi L, Sefiani A, Fouveau C, et al. (2006). Prevalence and distribution of MEFV mutations among Muslims from the Maghreb patients suffering from familial Mediterranean fever. *Comptes Rendus Biologies 329*(2), 71–74.

Belmouden A, Melki R, Hamdani M, et al. (2002). A novel frameshift founder mutation in the cytochrome P450 1B1 (CYP1B1) gene is associated with primary congenital glaucoma in Morocco. *Clin Genet 62*(4), 334–339.

Ben Arab S, Masmoudi S, Beltaief N, et al. (2004). Consanguinity and endogamy in Northern Tunisia and its impact on nonsyndromic deafness. *Genet Epidemiol* 27(1), 74–79

Ben Hamida M, Fardeau M. (1980). Severe autosomal recessive limb-girdle muscular dystrophies frequent in Tunisia. In: Angelini C, Danieli GA, Fontanari D, eds. *Muscular Dystrophy Research: Advances and New Trends.* Amsterdam: Excerpta Medica, pp. 143–146.

Ben Hamida C, Cavalier L, Belal S, et al. (1997). Homozygosity mapping of giant axonal neuropathy gene to chromosome 16q24.1. *Neurogenetics 1*(2), 129–133.

Ben Hamida C, Doerflinger N, Belal S, et al. (1993). Localization of Friedreich ataxia phenotype with selective vitamin E deficiency to chromosome 8q by homozygosity mapping. *Nat Genet 5*(2), 195–200.

Ben Mustapha I, Kammoun A, Mellouli F, et al. (2008). Effet fondateur d'une délétion de 10pb au niveau du gène CD18 chez les malades Nord-Africains atteints de déficit en molécules d'adhésion leucocytaires. African Society for Immunodeficiencies, ASID.

Ben Othmane K, Ben Hamida M, Pericak-Vance MA, et al. (1992). Linkage of Tunisian autosomal recessive Duchenne-like muscular dystrophy to the pericentromeric region of chromosome 13q. *Nat Genet 2*(4), 315–317.

Ben Rekaya M, Messaoud O, Talmoudi F, et al. (2009). High frequency of the V548A fs X572 XPC mutation in Tunisia: implication for molecular diagnosis. *J Hum Gene 54*(7), 426–429.

Benayoun L, Granot E, Rizel L, Allon-Shalev S, Behar DM, Ben-Yosef T. (2007). Abetalipoproteinemia in Israel: evidence for a founder mutation in the Ashkenazi Jewish population and a contiguous gene deletion in an Arab patient. *Molec Genet Metab 90*(4), 453–457.

Bouchlaka C, Maktouf C, Mahjoub B, et al. (2007). Genetic heterogeneity of megaloblastic anaemia type 1 in Tunisian patients. *J Hum Genet 52*(3), 262–270.

Bouhlal Y, Zouari M, Kefi M, et al. (2008). Autosomal recessive ataxia caused by three distinct gene defects in a single consanguineous family. *J Neurogenet 22*(2), 139-148.

Bouhouche A, Birouk N, Azzedine H, et al. (2007). Autosomal recessive axonal Charcot-Marie-Tooth disease (ARCMT2): phenotype-genotype correlations in 13 Moroccan families. *Brain 130*(Pt 4), 1062–1075.

Boulila-Elgaied A, Masmoudi S, Drira M, et al., (1997). Contribution of DFN1 and DFNB2 loci to neurosensory deafness in affected Tunisian families. *Arch Inst Pasteur Tunis 74*(1-2), 5–8.

Boulouiz R, Li Y, Abidi O, et al. (2007). Analysis of MYO7A in a Moroccan family with Usher syndrome type 1B: novel loss-of-function mutation and non-pathogenicity of p.Y1719C. *Mol Vis 2*(13), 1862–1865.

Brahimi Na, Jambou M, Sarzi E, et al. (2009). The first founder DGUOK mutation associated with hepatocerebral mitochondrial DNA depletion syndrome. *Mol Genet Metab 97*(3), 221–226.

Braverman IM, Redford DB, Mackowiak PA. (2009). Akhenaten and the strange physiques of Egypt's 18th dynasty. *Ann Intern Med. 150*(8), 556–560.

Cattan D, Mallet A, Feingold J. (2007). Tuberculosis Mortality among the Jews of Tunis (Tunisia) in the First Half of the Twentieth Century. *Population 62*(3), 523–543.

Chaabouni HB, Ksantini M, M'rad R, et al. (2007). MEFV mutations in Tunisian patients suffering from familial Mediterranean fever. *Semin Arthritis Rheum 36*(6), 397–401.

Chaabouni M, Maazoul F, Ben Hamida A, et al. (2008). Autosomal recessive acro-fronto-facio-nasal dysostosis associated with genitourinary anomalies: A third case report. *AmJ Med Genet Part A* 146A(14), 1825–1827.

Charfeddine C, Mokni M, Ben Mousli R, et al. (2003). A novel missense mutation in the gene encoding SLURP-1 in patients with Mal de Meleda from northern Tunisia. *British Journal of Dermatology 149*(6), 1108–1115.

Chavanne F, Broughton BC, Pietra D, et al. (2000). Mutations in the XPC gene in families with xeroderma pigmentosum and consequences at the cell, protein, and transcript levels. *Cancer Res 60*(7), 1974–1982.

Chkioua L, Khedhiri S, Jaidane Z, et al. (2007). Mucopolysaccharidosis type I: identification of alpha-L-iduronidase mutations in Tunisian families. *Archives de pédiatrie 14*(10), 1183–1189.

Colombo R. (2000). Age and Origin of the PRNP E200K Mutation Causing Familial Creutzfeldt-Jacob Disease in Libyan Jews. *Am J Hum Genet 67*(2), 528–531.

Christianson A, Howson CP, Modell B. (2006). Global report on birth defects. The hidden toll of dying and disabled children. March of Dimes Birth Defects Foundation, White Plains, New York. (http://www.marchofdimes.com/mission/globalprograms_birthdefectsreport.html)

Croft JB, Morrell D, Chase CL, Swift M. (1995). Obesity in heterozygous carriers of the gene for the Bardet-Biedl syndrome. *Am J Med Genet 55*(1), 12–15.

Crow JF. (1997). The high spontaneous mutation rate: Is it a health risk? *Proc Natl Acad Sci USA 94*(16), 8380–8386.

Cuesta A, Pedrola L, Sevilla T, et al. (2002). The gene encoding ganglioside-induced differentiation-associated protein 1 in mutated in axonal Charcot-Marie-Tooth type 4A disease. *Nat Genet 30*(1), 22–25.

Dahri S, Desviat LR, Pérez B, et al. (2010). Mutation analysis of phenylketonuria patients from Morocco: high prevalence of mutation G352fsdelG and detection of a novel mutation p.K85X. *Clin Biochem* 43(1-2), 76-81.

De Sandre-Giovannoli A, Chaouch M, Kozlov S, et al. (2002). Homozygous defects in LMNA, encoding lamin A/C nuclear-envelope proteins, cause autosomal recessive axonal neuropathy in human (Charcot-Marie-Tooth Disorder Type 2) and mouse. *Am J Hum Genet 70*(3), 726–736.

Dieterich K, Zouari R, Harbuz R, et al. (2009). The Aurora Kinase C c.144delC mutation causes meiosis I arrest in men and is frequent in the North African population. *Human Molecular Genetics 18*(7), 1301–1309.

Driss A, Amouri R, Ben Hamida C, et al. (2000). A new locus for autosomal recessive limb-girdle muscular dystrophy in a large consanguineous Tunisian family maps to chromosome 19q13.3. *Neuromuscul Disord 10*(4–5), 240–246.

Ebermann I, Walger M, Scholl HP, et al. (2007). Truncating mutation of the DFNB59 gene causes cochlear hearing impairment and central vestibular dysfunction. *Human Mutation 28*(6), 571–577.

Effat LK, Essawi ML, Abd El Hamid MS, Hawari N, Gad YZ. (2008). Screening for six Mediterranean mutations in 90 Egyptian patients with phenylketonuria. *Bratisl Lek Listy 109*(1), 17–19.

Elloumi-Zghal H, Ridha Barbouche M, Chemli J, et al. (2002). Clinical and genetic heterogeneity of inherited autosomal recessive susceptibility to disseminated *Mycobacterium bovis* Bacille Calmette-Guérin infection. *J Infect Dis 185*(10), 1468–1475.

Falik-Zaccai TC, Shachak E, Yalon M, et al. (1997). Predisposition to the fragile X syndrome in Jews of Tunisian descent is due to the absence of AGG interruptions on a rare Mediterranean haplotype. *Am J Hum Genet 60*(1), 103–112.

Fathallah DM, Bejaoui M, Lepaslier D, Chater K, Sly WS, Dellagi K. (1997). Carbonic anhydrase II (CA II) deficiency in Maghrebian patients: evidence for founder effect and genomic recombination at the CA II locus. *Hum Genet. 99*(5), 634–637.

Fendri K, Kefi M, Hentati F, Amouri R. Genetic heterogeneity within a consanguineous family involving the LGMD 2D and the LGMD 2C genes (2006). *Neuromuscul Disord 16*(5), 316–320.

Fischer J, Bouadjar B, Heilig R, et al. (2001). Mutations in the gene encoding SLURP-1 in Mal de Meleda. *Human Molecular Genetics 10*(8), 875–880.

Foubert L, Bruin T, De Gennes JL, et al. (1997). A single Ser259Arg mutation in the gene for lipoprotein lipase causes chylomicronemia in Moroccans of Berber ancestry. *Hum Mutat 10*(3), 179–185.

Gershoni-Baruch R, Rosenmann A, Droetto S, Holmes S, Tripathi RK, Spritz RA. (1994). Mutations of the tyrosinase gene in patients with oculocutaneous albinism from various ethnic groups in Israel. *Am J Hum Genet 54*(4), 586–594.

Gershoni-Baruch R, Shinawi M, Leah K, Badarnah K, Brik R. (2001). Familial Mediterranean fever: prevalence, penetrance and genetic drift. *European Journal of Human Genetics* 9(8), 634–637.

Gilad S, Bar-Shira A, Harnik R, et al. (1996). Ataxia-telangiectasia: founder effect among North African Jews. *Hum Mol Genet 5*(12), 2033–2037.

Gouider-Khouja N, Belal S, Hamida MB, Hentati F. (2000). Clinical and genetic study of familial Parkinson's disease in Tunisia. *Neurology 54*(8), 1603–1609.

Gritli S, Omar S, Tartaglini E, et al. (2001). A novel mutation in the SLC19A2 gene in a Tunisian family with thiamine-responsive megaloblastic anaemia, diabetes and deafness syndrome. *Br J Haematol 113*(2), 508–513.

Guilford P, Ben Arab S, Blanchard S, et al. (1994). A non-syndrome from of neurosensory, recessive deafness maps to the pericentromic region of chromosome 13q. *Nat Genet 6* (1), 24–28.

Hafez M, El-Tahan E, Awadalla M, El-Khayat H, Abdel-Gafar A, Ghoneim M. (1983). Consanguineous matings in the Egyptian population. *J Med Genet 20*(1), 58–60.

Haj Khelil A, Denden S, Leban N, et al. (2010). Hemoglobinopathies in North Africa: a review. *Hemoglobin 34*(1), 1–23.

Hamadouche T, Poitelon Y, Genin E, et al. (2008). Founder Effect and Estimation of the Age of the c.892C>T (p.Arg298Cys) Mutation in LMNA Associated to Charcot-Marie-Tooth Subtype CMT2B1 in Families from North Western Africa. *Annals of Human Genetics 72*(Pt 5), 590–597.

Hamamy HA, Masri AT, Al-Hadidy AM, Ajlouni KM. (2007). Consanguinity and genetic disorders. Profile from Jordan. *Saudi Med J 28*(7), 1015–1017.

Hanein S, Martin E, Boukhris A, et al. (2008). Identification of the SPG15 gene, encoding spastizin, as a frequent cause of complicated autosomal-recessive spastic paraplegia, including Kjellin syndrome. *Am J Hum Genet 82*(4), 992–1002.

Hentati A, Bejaoui K, Pericak-Vance MA, et al. (1994). Linkage of recessive familial amyotrophic lateral sclerosis to chromosome 2q33-q35. *Nat Genet 7*(3), 425–428.

Hmani M, Ghorbel A, Boulila-Elgaied A et al. (1997). A novel locus for Usher syndrome type II, USH2B, maps to chromosome 3 at p23–24.2. *Eur J Hum Genet 7*(3), 363–367.

Hmani-Aifa M, Benzina Z, Zulfiqar F, et al. (2009). Identification of two new mutations in the GPR98 and the PDE6B genes segregating in a Tunisian family. *Eur J Hum Genet* 17(4), 474–482.

Hu PY, Roth DE, Skaggs LA, et al. (1992). A splice junction mutation in intron 2 of the carbonic anhydrase II gene of osteopetrosis patients from Arabic countries. *Hum Mutat 1*(4), 288–292.

Ismail SR, Abdel-Rahim N, Hashishe MM, Abdallah EM. (1996). Newborn screening for certain treatable inborn errors of metabolism in Alexandria. *J Egypt Public Health Assoc 71*(5–6), 495–520.

Jakobi L, Jacquard A. (1971). Consanguinité proche, consanguinité éloignée. Essai de mesure dans un village breton. Cahier n°60: 263–268. INED, Paris.

James SJ, Pogribna M, Pofribny IP, et al. (1999). Abnormal folate metabolism and mutation in the methylenetetrahydrofolate reductase gene may be maternal risk factors for Down syndrome. *American J Cli Nut 70*(4), 495–501.

Keefe DL, Niven-Fairchild T, Powell S, Buradagunta S. (1995). Mitochondrial deoxyribonucleic acid deletions in oocytes and reproductive aging in women. *Fertil Steril 64* (3), 577–583.

Kefi M, Amouri R, Driss A, et al. (2003). Phenotype and sarcoglycan expression in Tunisian LGMD 2C patients sharing the same del521-T mutation. *Neuromuscular Disorders 13*(10), 779–787.

Khan SG, Oh KS, Shahlavi T, et al. (2006). Reduced XPC DNA repair gene mRNA levels in clinically normal parents of xeroderma pigmentosum patients. *Carcinogenesis 27*(1), 84–94.

Khemir S, Tebib N, Nasrallah F, et al. (2009). Phenylketonuria in Tunisian institutions for the mentally handicapped. *Arch Dis Child 94*(8), 647–648.

Kou YC, Shao L, Peng HH, et al. (2008). A recurrent intragenic genomic duplication, other novel mutations in NLRP7 and imprinting defects in recurrent biparental hydatidiform moles. *Mol Hum Reprod 14*(1), 33–40.

Kozlowski, K., Bacha, L., Massen, R., Ayati, M., Sator, S., Brahimi, L. (1988). A new type of spondylo-metaphyseal dysplasia—Algerian type: report of five cases. *Pediat. Radiol. 18*(3), 221–226.

Lachmann HJ, Sengül B, Yavuzşen TU, et al. (2006). Clinical and subclinical inflammation in patients with familial Mediterranean fever and in heterozygous carriers of MEFV mutations. *Rheumatology* 45(6), 746–750.

Lander ES, Botstein D. (1987). Homozygosity mapping: a way to map human recessive traits with the DNA of inbred children. *Science 236*(4808), 1567–1570.

Lefèvre C, Bouadjar B, Ferrand V, et al. (2006). Mutations in a new cytochrome P450 gene in lamellar ichthyosis type 3. *Human Molecular Genetics 15*(5), 767–776.

Lehmann AR. (2003). DNA repair-deficient diseases, xeroderma pigmentosum, Cockayne syndrome and trichothiodystrophy. *Biochimie 85*(11), 1101–1111.

Li Y, Pohl E, Boulouiz R, et al. (2010). Mutations in TPRN cause a progressive form of autosomal-recessive nonsyndromic hearing loss. *Am J Hum Genet 86*(3), 479–484.

Louhichi N, Medhaffar M, Hadjsalem I, et al. (2010). Congenital factor XIII deficiency caused by two mutations in eight Tunisian families: molecular confirmation of a founder effect. *Ann Hematol 89*(5), 499–504.

Louhichi N, Triki C, Quijano-Roy S, et al. (2004). New FKRP mutations causing congenital muscular dystrophy associated with mental retardation and central nervous system abnormalities. Identification of a founder mutation in Tunisian families. *Neurogenetics 5*(1), 27–34.

Loumi O, Ferec C, Mercier B, et al. (2008). CFTR mutations in the Algerian population. *Journal of Cystic Fibrosis* 7(1), 54–59.

Lowenstein EJ. (2009). Osteogenesis imperfecta in a 3,000-year-old mummy. *Childs Nerv Syst. 25*(5), 515–516.

Majjed HA, El-Khateeb M, El-Shanti H, Abu Rabaiha Z, Tayeh M, Najib D. (2005). The spectrum of familial Mediterranean fever gene mutations in Arabs: report of a large series. *Semin Arthritis Rheum 34*(6), 813–818.

Makri S, Clarke NF, Richard P, et al. (2009). Germinal mosaicism for LMNA mimics autosomal recessive congenital muscular dystrophy. *Neuromuscul Disord 19*(1), 26–28.

Malini SS, Ramachandra BN. (2006). Influence of advanced age of maternal grandmothers on Down syndrome. *BMC Medical Genetics 7*, 4 doi:10.1186/1471-2350-7-4.

Marrakchi S, Audebert S, Bouadjar B, et al. (2003). Novel mutations in the gene encoding secreted lymphocyte antigen-6/urokinase-type plasminogen activator receptor-related protein-1 (SLURP-1) and description of five ancestral haplotypes in patients with Mal de Meleda. *J Invest Dermatol 120*(3), 351–355.

Masmoudi S, Antonarakis SE, Schwede T, et al. (2001). Novel missense mutations of TMPRSS3 in two consanguineous Tunisian families with non-syndromic autosomal recessive deafness. *Hum Mutat 18*(2), 101–108.

Masmoudi S, Charfedine I, Rebeh IB, et al. (2004). Refined mapping of the autosomal recessive non-syndromic deafness locus DFNB13 using eight novel microsatellite markers. *Clin Genet 66*(4), 358-364.

Masmoudi S, Elgaied-Boulilia A, Kassab I, et al. (2000). Determination of the frequency of connexin26 mutations in inherited sensorineural deafness and carrier rates in the Tunisian population using DGGE. *J Med Genet 37*(11), E39.

Masmoudi S, Tlili A, Majava M, et al. (2003). Mapping of a new autosomal recessive nonsyndromic hearing loss locus (DFNB32) to chromosome 1p13.3–22.1. *Eur J Hum Genet 11*(2), 185–188.

Mattit H, Joma M, Al-Cheikh S, et al. (2006). Familial Mediterranean fever in the Syrian population: gene mutation frequencies, carrier rates and phenotype-genotype correlation. *Eur J Med Genet 49*(6), 481–486.

Mears JG, Beljord C, Benabadji M. (1981). The sickle gene polymorphism in North Africa. *Blood 58*(3), 599–601.

Messaoud O, Ben Rekaya M, Cherif W, et al. (2010). Genetic homogeneity of mutational spectrum of group-A xeroderma pigmentosum in Tunisian patients. *Int J Dermatol 49*(5), 544–548.

Messaoud T, Bel Haj Fredj S, Bibi A, Elion J, Férec C, Fattoum S. (2005). Molecular epidemiology of cystic fibrosis in Tunisia. *Ann Biol Clin 63*(6), 627–630.

Mokni M, Charfeddine C, Ben Mously R, et al. (2004). Heterozygous manifestations in female carriers of Mal de Meleda. *Clin Genet 65*(3), 244–246.

Mor-Cohen R, Zivelin A, Fromovich-Amit Y, Kovalski V, Rosenberg N, Seligsohn U. (2007). Age estimates of ancestral mutations causing factor VII deficiency and Dubin-Johnson syndrome in Iranian and Moroccan Jews are consistent with ancient Jewish migrations. *Blood Coagul Fibrinolysis 18*(2), 139–144.

Moulard B, Genton P, Grid D, et al. (2002). Haplotype study of West European and North African Unverricht-Lundborg chromosomes: evidence for a few founder mutations. *Hum Genet 111*(3), 255–262.

Mrissa N, Belal S, Hamida CB, et al. (2000). Linkage to chromosome 13q11-12 of an autosomal recessive cerebellar ataxia in a Tunisian family. *Neurology 54*(7), 1408–1414.

Nishigori C, Zghal M, Yagi T, et al., (1993). High prevalence of the point mutation in exon 6 of the xeroderma pigmentosum group A-complementing (XPAC) gene in xeroderma pigmentosum group A patients in Tunisia. *Am J Hum Genet 53*(5), 1001-1006.

Othmane KB, Johnson E, Menold M, et al. (1999). Identification of a new locus for autosomal recessive Charcot-Marie-Tooth disease with focally folded myelin on chromosome 11p15. *Genomics 62*(3), 344–349.

Ouragini H, Cherif F, Kassar S, et al. (2009). Dystrophic epidermolysis bullosa phenotypes in a large consanguineous Tunisian family. *J Dermatol Sci 54*(2), 114–120.

Pastinen T, Perola M, Ignatius J, et al. (2001). Dissecting a population genome for targeted screening of disease mutations. *Hum Mol Genet 10*(26), 2961–2972.

Peretz H, Luboshitsky R, Baron E, et al. (1997). Cys 618 Arg mutation in the RET proto-oncogene associated with familial medullary thyroid carcinoma and maternally transmitted Hirschsprung's disease suggesting a role for imprinting. *Human Mutation 10*(2), 155–159.

Petit FM, Bézieau S, Gajdos V, et al. (2008). The Tunisian population history through the Crigler-Najjar type I syndrome. *Eur J Hum Genet 16*(7), 848–853.

Ratbi I, Génin E, Legendre M, et al. (2008). Cystic fibrosis carrier frequency and estimated prevalence of the disease in Morocco. *J Cyst Fibros 7*(5), 440–443.

Richard P, Gaudon K, Haddad H, et al. (2008). The CHRNE 1293insG founder mutation is a frequent cause of congenital myasthenia in North Africa. *Neurology 71*(24), 1967–1972.

Nussbaum RL, McInnes RR, Willard HF, Thompson MW. (2004). *Genetics in Medicine*, 6th ed., W.B. Saunders Company, Philadelphia, p. 106.

Romdhane L, Abdelhak S (2011). Genetic diseases in the Tunisian population. *Am J Med Genet A 155A*(1), 238-267.

Romdhane L, Kefi-Ben Atig R, Azaiez H, et al. (2011). Founder mutations in the Tunisian Population: Implications for diagnosis in North Africa and Middle East. Unpublished.

Roume J, Genin E, Cormier-Daire V, et al. (1998). A gene for Meckel syndrome maps to chromosome 11q13. *Am J Hum Genet 63*(4), 1095–1101.

Roux AF, Morlé F, Guetarni D, et al. (1989). Molecular basis of Sp alpha I/65 hereditary elliptocytosis in North Africa: insertion of a TTG triplet between codons 147 and 149 in the alpha-spectrin gene from five unrelated families. *Blood 73*(8), 2196–2220.

Ruwende C, Khoo SC, Snow RW, et al. (1995). Natural selection of hemi- and heterozygotes for G6PD deficiency in Africa by resistance to severe malaria. *Nature. 376*(6537), 246–249.

Schoser B, Gläser D, Müller-Höcker J. (2008). Clinicopathological analysis of the homozygous p.W1327X AGL mutation in glycogen storage disease type 3. *Am J Med Genet A 146A*(22), 2911–2915.

Schulz R, Seidel M. (1999). L'Egypte Sur Les Traces De La Civilisation Pharaonique. Konemann edition, Germany.

Segal A, Zivelin A, Rosenberg N, Ginsburg D, Shpilberg O, Seligsohn U. (2004). A mutation in LMAN1 (ERGIC-53) causing combined factor V and factor VIII deficiency is prevalent in Jews originating from the island of Djerba in Tunisia. *Blood Coagul Fibrinolysis 15*(1), 99–102.

Sheffield VC, Stone EM, Carmi R. (1998). Use of isolated inbred human populations for identification of disease genes. *Trends Genet 14*(10), 391–396.

Slimane MN, Lestavel S, Sun X, et al. (2001). FH-Souassi: a founder framshift mutation in exon 10 of the LDL-receptor gene, associated with a mild phenotype in Tunisian families. *Atherosclerosis 154*(3), 557–565.

Soufir N, Ged C, Bourillon A, et al. (2010). A prevalent mutation with founder effect in xeroderma pigmentosum group C from north Africa. *J Invest Dermatol 130*(6), 1537–1542

Stoffman N, Magal N, Shohat T, et al. (2000). Higher than expected carrier rates for familial Mediterranean fever in various Jewish ethnic groups. *Eur J Hum Genet 8*(4), 307–310.

Stoll C, Alembik Y, Dott B, Roth MP. (1998). Study of DS in 238942 consecutive births. *Ann Genet 41*(1), 44–51.

Tamary H, Bar-Yam R, Shalmon L, et al. (2000). Fanconi anaemia group A (FANCA) mutations in Israeli non-Ashkenazi Jewish patients. *Br J Haematol 111*(1), 338–343.

Tazir M, Azzedine H, Assami S, et al. (2004). Phenotypic variability in autosomal recessive Charcot-Marie-Tooth disease due to the R298C mutation in lamin A/C. *Brain 127*(Pt 1), 154–163.

Teebi AS. (1994). Autosomal recessive disorders among Arabs: an overview from Kuwait. *31*(3), 224–233.

Teebi A, Farag T. (1997). *Genetic Disorders among Arab Populations*. New York: Oxford University Press.

Teebi AS, El-Shanti HI. (2006). Consanguinity: implications for practice, research, and policy. *The Lancet 367*(9515), 970–971.

Temtamy SA, Shoukry AS, Raafat M, Mihareb S. (1975). Probable Marden-Walker syndrome: evidence for autosomal recessive inheritance. *Birth Defects Orig Artic Ser 11*(2), 104–108.

Tlili A, Charfedine I, Lahmar I, et al. (2005). Identification of a novel frameshift mutation in the DFNB31/WHRN gene in a Tunisian consanguineous family with hereditary non-syndromic recessive hearing loss. *Hum Mutat* 25(5), 503.

Tlili A, Männikkö M, Charfedine I, et al. (2005). A novel autosomal recessive non-syndromic deafness locus, DFNB66, maps to chromosome 6p21.2-22.3 in a large Tunisian consanguineous family. *Hum Hered 60*(3), 123–128.

Tlili A, Masmoudi S, Dhouib H, et al. (2007). Localization of a novel autosomal recessive non-syndromic hearing impairment locus DFNB63 to chromosome 11q13.3-q13.4. *Ann Hum Gene 71*(Pt 2), 271–275.

Tlili A, Rebeh IB, Aifa-Hmani M, et al. (2008). TMC1 but not TMC2 is responsible for autosomal recessive nonsyndromic hearing impairment in Tunisian families. *Audiol Neurootol 13*(4), 213–218.

Touitou I. (2001). The spectrum of Familial Mediterranean Fever (FMF) mutations. *Eur J Hum Genet 9*(7), 473–83.

Tullio-Pelet A, Salomon R, Hadj-Rabia S, et al. (2000). Mutant WD-repeat protein in triple-A syndrome. *Nat Gene 263*, 332–335.

Uhrhammer N, Abdelouahab A, Lafarge L, Feillel V, Ben Dib A, Bignon YJ. (2008). BRCA1 mutations in Algerian breast cancer patients: high frequency in young, sporadic cases. *Int J Med Sci 5*(4), 197–202.

United Nations, Department of Economic and Social Affairs, Population Division (2011). World Population Prospects: The 2010 Revision, CD-ROM Edition. (http://esa.un.org/unpd/wpp/Excel-Data/fertility.htm).

Valderas JM, Starfield B, Sibbald B, Salisbury C, Roland M. (2009). Defining comorbidity: implications for understanding health and health services. *Ann Fam Med 7*(4), 357–363.

Warren L, Gibson R, Ishihara L, et al. (2008). A founding LRRK2 haplotype shared by Tunisian, US, European and Middle Eastern families with Parkinson's disease. *Parkinsonism Relat Disord 14*(1), 77–80.

Weatherall DJ. (2008). Hemoglobinopathies worldwide: present and future. *Curr Mol Med 8*(7), 592–599.

White PC, Dupont J, New MI, Leiberman E, Hochberg Z, Rösler A. (1991). A mutation in CYP11B1 arg448-to-his associated with steroid 11-beta-hydroxylase deficiency in Jews of Moroccan origin. *J. Clin. Invest. 87*(5), 1664–1667.

Whittock NV, Ashton GH, Mohammedi R, et al. (1999). Comparative mutation detection screening of the type VII collagen gene (COL7A1) using the protein truncation test, fluorescent chemical cleavage of mismatch, and conformation sensitive gel electrophoresis. *J Invest Dermatol 113*(4), 673–686.

Wiszniewski W, Fondaneche MC, Lambert N, et al. (2000). Founder effect for a 26-bp deletion in the RFXANK gene in North African major histocompatibility complex class II-deficient patients belonging to complementation group B. *Immunogenetics 51*(4–5), 261–267.

Zaki MS, Kamel AA, El-Ruby M. (2005). Double aneuploidy in three Egyptian patients: Down-Turner and Down-Klinefelter syndromes. *Genet Couns 16*(4), 393–402.

Zghal M, Fazaa B, Kamoun M. (2006). Xeroderma pigmentosum. *EMC (Elsevier SAS, Paris), Dermatologie, 10*(A), 98–660.

Zlotogora J. (2007). Multiple mutations responsible for frequent genetic diseases in isolate populations. *Eur J Hum Genet 15*(3), 272–278.

35.

GENETIC NEUROLOGICAL DISEASES IN TUNISIA

Neziha Gouider-Khouja and Ibtihel Rebai

INTRODUCTION

This chapter is based on data published on genetic neurological disorders in Tunisian patients. It is now well known from papers published on genetic neurological diseases in Tunisian patients of Arab-Berber origin that wide variation exists compared to patients of Caucasian origin. These differences are probably based on genetic heterogeneity, an elevated consanguinity rate in Tunisia, and variable drug responses indicating pharmacogenomic variation. Studies performed by Tunisian neurology teams have described at least two previously unknown neurological disorders and several novel genes and/or mutations among Tunisian patients. Furthermore, many mutations found worldwide in common neurological disorders have been linked to founding mutations going back several centuries in Tunisia and North Africa. A large web search using "neurological diseases in Tunisia" key words revealed 625 published papers from 1980 to date. This included 87 publications related to genetic disorders of nonmetabolic origin (degenerative).

GENETIC DISEASES OF THE CENTRAL NERVOUS SYSTEM

The systematic search for genetic central nervous system disorders in Tunisia began with the foundation of the National Institute of Neurology in Tunis in 1975 by the late Professor Mongi Ben Hamida, the founder of neurosciences in Tunisia. This institution, unique in Africa and in the Arab world, includes several subspecialist fields including neurology, neurophysiology, neurosurgery, neuroresuscitation, neurorehabilitation, neuropathology, neurogenetics, and molecular biology. This has allowed a "bench to bedside" approach to neurosciences, and explains the scientific fertility of Mongi Ben Hamida's group and that of other groups. Main contributions of Tunisian neurosciences to worldwide neurology include insights into genetics of Parkinson's disease, hereditary ataxias, hereditary spastic paraplegia, and many other uncommon neurogenetic disorders.

PARKINSON'S DISEASE

Parkinson's disease (PD) is the second most common neurodegenerative disorder, causing rest tremor, bradykinesia, rigidity, and postural instability. It affects 1% of the population over the age of 65 years and increases to 4% to 5% of the population by the age of 85 years (Guttmacher & Collins, 2003). Since its description in 1817 by Sir James Parkinson, it was considered as a sporadic disorder. However, early in the twentieth century, familial forms with Mendelian patterns of inheritance were reported, leading to the discovery of the first gene involved in PD in 1995 (Polymeropoulos et al., 1996).

To date, 15 loci have been described in PD (PARK 1 through PARK 15) and six genes with several causative mutations have been identified confirming heterogeneity across different ethnic population groups. In autosomal dominant (AD) forms of PD, 3 main genes are involved: the alpha-synuclein gene (SNCA) on 4q22.1, responsible for PARK1 and PARK4, due respectively to mutation in or triplication of SCNA (Polymeropoulos et al., 1997; Farrer et al., 1999); PARK5 due to mutation in the UCHL1 gene on 4p14 (Leroy et al., 1998) and PARK8, due to mutation in the LRRK2 gene on 12q12 (Funayama et al., 2002).

Three genes are associated with the autosomal recessive (AR) early-onset PD, including PARK2, caused by mutation in the gene encoding for a protein called parkin on 6q25.2–q27 (Kitada et al., 1998); PARK6 caused by mutations in the PINK1 gene (Valente et al., 2001), and PARK7, caused by mutation in the DJ1 gene on 1p36, 27 cM centromeric to PARK 6 (Van Duijn et al., 2001).

Studies on Tunisian patients were initiated by Gouider-Khouja and Ben Hamida in 1992, and the results of the first study of familial PD in Tunisia was acknowledged by the International American Academy Award on Neuroepidemiology in 1998. In 2000, Gouider-Khouja et al. published the core paper reporting on these results in

21 families and showed that in Tunisian patients, clinical features in familial forms were similar to those in sporadic forms and to the literature on Caucasian patients, except for the age of onset, which was much lower in Tunisian patients. Intra- and interfamilial variability of all clinical features was emphasized. (Gouider-Khouja et al., 2000). Subsequently, all published papers on familial PD in Tunisia were based on international collaborations. In the first collaborative study with a Japanese group, Gouider-Khouja et al. described three siblings with autosomal recessive juvenile Parkinsonism linked to a microdeletion of the parkin gene, in whom clinical and pathological characteristics were similar to those previously reported in Japanese and European families, except for more important involvement of locus coeruleus in the Tunisian patients (Gouider-Khouja et al., 2003). The subsequent publications on the Tunisian familial PD patients derived from the cohort collected over 10 years (1992–2002), focused particularly on parkin, PINK1, and LRRK2 mutations.

A comparative study including 92 Tunisian kindreds with 208 affected individuals suggested that only four PINK1 mutations were pathogenic (Q129X, Q129fsX157, G440e and Q456X), that other sequence variants of PINK1 gene are potentially benign, and heterozygous individuals were not more susceptible to developing an idiopathic form of the disease (Ishihara et al, 2008).

Similar studies were carried out on ADPD linked to LRRK2 gene mutations. These studies focused on analyzing causative mutations, genotype–phenotype correlations, and comparison of homozygous and heterozygous carriers. Studies in different ethnic population groups enabled selection of the founder haplotype or mutation with an estimate of its age. Two studies showed frequency of the LRRK2 gene in Tunisian PD patients with a founding mutation (G2019S) of this gene going back 1300 to 2600 years (Lesage et al., 2005; Warren et al., 2008). In a collaborative descriptive report including 23 Tunisian individuals presenting with familial PD linked to LRRK2 G2019S mutation, Ishihara et al. (2006) demonstrated clinical similarities between patients with homozygous and heterozygous mutations with no evidence of gene dosage effect. In another comparative study between Tunisian and North American families, the same author reported an epidemiological analysis of the LRRK2 G2019S substitution in Tunisian families, estimated at 42%, with few differences compared to families negative for the substitution, and insisted that no phenotypic correlation was due to homozygosity although estimated penetrance was greater in homozygous carriers (Ishihara et al., 2007). A large study over 18 countries including Tunisia showed Japanese patients with the G2019S mutation unlikely to have a single founder effect (Tomiyama et al., 2006). A more recent case-control study demonstrated that penetrance was variable among young and old patients, and that susceptibility to neuronal degeneration seemed to be increased in carriers of the G2019S mutation, although many triggers could be implicated (Hulihan et al., 2008).

A recent study on 231 Tunisian PD patients compared LRRK2 patients, PINK1, and genetically undefined familial PD from the Tunisian patient collection, and found LRRK2 patients had more severe motor symptoms and a higher rate of dyskinesia than PINK1 and genetically undefined patients, and that PINK1 patients had younger age of onset (Nishioka et al., 2010). The genetically undefined group worths a comment related to the history of PD in Tunisia. Indeed, since the very first clinical report of Ben Hamida and Gouider-Khouja (Gouider-Khouja, 1991), it was clear that in North Africa a very frequent form of PD existed, characterized by the association of rest with postural tremor. This phenotype was much more frequent in Tunisia and North Africa than in Europe. It was termed "Maghrebian form of PD." The recent comparative study of Nishioka et al. demonstrates that the genetic background of this phenotype is yet to be found.

HEREDITARY ATAXIAS

Molecular genetics improvement in the last decades has been a helpful tool in identifying causal genes implicated in previously called "spinocerebellar heredodegeneration," and thus to classify different related disorders depending on phenotypic presentation and genetic findings. Three inheritance patterns are described in hereditary ataxia (HA): autosomal recessive, autosomal dominant, and X-linked.

The study of HA in Tunisia was started by Ben Hamida himself in the 1980s. In 1986, he conducted a descriptive study including 99 Tunisian families presenting with spinocerebellar heredodegeneration that enabled him to specify transmission inheritance for each clinical group. Autosomal recessive pattern was reported in almost all cases of Friedreich's ataxia (FA) and all cases of Pierre Marie's heredo-ataxia (Ben Hamida et al., 1986). Five years later, he reported an association of both subgroups to a peroneal atrophy of Charcot-Marie-Tooth type, incomplete or complex forms, and isolated Charcot-Marie-Tooth or spastic paraplegia in other kindred individuals, which was the most striking finding, suggesting clinical and genetic heterogeneity (Ben Hamida M et al., 1991, 1993). Focusing on FA, Belal, a member of Ben Hamida's group, performed a linkage study of three Tunisian families, allowing to narrow the gene locus to the D9S5 region on chromosome 9q13–q21 (Belal et al., 1992). Based on these findings, Rodius et al. refined the gene locus to a 450kb interval, a smaller region for direct search of candidate gene (Rodius et al., 1994).

The story of the discovery of ataxia with vitamin deficiency (AVED) is unique and amazing. Rarely in the history of scientific medical research has the same group, on the same collection of patients, performed entirely or participated in all the research steps that can be done on a disease: clinical delineation, biochemical defect, molecular background, pathogenic mechanisms, and curative treatment. Actually, this was the case for AVED (Ben Hamida et al., 1993Doerflinger et al., 1995; Ouahchi et al., 1995; Gabsi et al., 2001). In 1993, Ben Hamida himself published a paper on Tunisian families with ataxia strictly similar to FA phenotype but not linked to chromosome 9, and associated to selective autosomal recessive vitamin E deficiency (Ben Hamida et al., 1993). International collaborative studies allowed a description of the genetic background of this newly described entity, and C. Ben Hamida mapped it to chromosome 8q on a 2.4cM interval (Ben Hamida et al., 1993; Doerflinger et al., 1995). Two years later, Ouahchi, a Ben Hamida fellow, published a paper in *Nature Genetics* on the pathogenic mechanism of the disease: an impaired ability of α-tocopherol incorporation into lipoproteins secreted by the liver (Ouahchi et al., 1995). Three frameshift mutations were implicated—one of which, the 744delAmutation, was correlated to a severe phenotype. Subsequently Gabsi, from Ben Hamida's group, started an open-label trial of vitamin E supplementation in Tunisian patients with AVED, and results showed improvement of all clinical features in symptomatic cases; furthermore, vitamin E supplementation allowed presymptomatic cases to remain totally free of symptoms (Gabsi et al., 2001).

Some other papers relate to rare forms of AR ataxias such as sacsin-related AR ataxia (El Euch-Fayache et al., 2003; Bouhlal et al., 2008) with a novel mutation of SACS gene in a Tunisian family (Bouhlal et al., 2009).

As for autosomal dominant cerebellar ataxia, Tunisian patients were involved in a collaborative study on AD cerebellar ataxia (SCA2), and with progressive macular dystrophy (SCA7; see Cancel et al., 1997; Belal et al., 1994; David et al., 1998; Stevanin et al., 1998).

SPASTIC PARAPLEGIA

Hereditary spastic paraplegia (HSP) is a heterogeneous group of neurodegenerative disorders, including pure and complex forms. Pure forms are characterized by a pyramidal motor neuron dysfunction, whereas complex forms are associated to additional neurological features such as ataxia, mental retardation, dementia, extrapyramidal signs, visual impairment, and epilepsy. Extraneurological signs, including cataract, retinal degeneration, ichthyosis, and skeletal abnormalities may also be seen. At least 35 different genes/loci are associated with HSP (Salinas et al., 2008)

Starting with pure forms, A. Hentati (1994) from Ben Hamida's group conducted an international collaborative linkage analysis including five Tunisian families with pure autosomal recessive spastic paraplegia. Results showed genetic heterogeneity: 4 families were linked to chromosome 8 and one family showed no linkage (Hentati et al., 1994a). In the same year, the same authors suggested a new classification of spastic paraplegia based on genetic identification, emerging from the genetic analysis of five families who presented with the pure autosomal dominant form; four of them showed locus linkage to chromosome 2p, and one of them to chromosome 14q (Hentati et al., 1994b).

Genetic heterogeneity was also observed in complex forms. In a form associating mild ataxia and sensory neuropathy, Klebe et al. (2006) reported on 11 families from Tunisia, Algeria, France, and Czech Republic, with mapping of one Algerian family to a new locus on chromosome 2q37.3, corresponding to SPG30, and that was not found in other families including the Tunisian family (Klebe et al., 2006). Association to thin corpus callosum (TCC) was one of the complex forms reported in the Tunisian population. The subsequent international studies involved mainly the group of C Mhiri, another Ben Hamida's fellows (Boukhris et al., 2008). In 2007, Stevanin et al. described 12 families (including one Tunisian) who had AR spastic paraplegia with TCC. They refined the SPG11 candidate interval on chromosome 15, and identified ten mutations of the KIAA1840 gene encoding spatacsin, which was helpful to determine pathogenic mechanisms (Stevanin et al., 2007). The same authors performed another study one year later dealing with SPG11 mutations. They reported 22 mutations; 19 of them were novel, two recurrent mutations were shared by Portuguese and North Africans, suggesting a founder effect in both populations, and they concluded that the association to mental retardation, to later cognitive deterioration, or to white matter lesions increases the chance of identifying the SPG11 mutations (Stevanin et al., 2008). Goizet et al. reported another complex form with TCC and variable phenotypes (cognitive impairment, ataxia, axonal neuropathy, white matter abnormalities, cerebellar atrophy, and macular pigmentation) in eight families classified in the subgroup of SPG15 and carrying mutations in the recently identified gene, ZFYVE26. They demonstrated complexity of gene analysis, since mutations were distributed along the entire coding sequence, and considered this form as the second most common cause of spastic paraplegia with TCC, after SPG11 (Goizet et al., 2009a). Among trials to determine pathogenic mechanisms of neuronal loss, Tsaousidou et al. (2008) focused on the first pure autosomal recessive HSP locus defined, SPG5 linked to chromosome 8q12.3. They performed a sequence analysis of genes in this region and discovered sequence alterations in the cytochrome P450-7B1 (CYP7B1) which is implicated in alternative pathway for cholesterol degradation in the liver and offers, in the brain, the primary metabolic route for the modification of dehydroepiandrosterone neurosteroids. These results were the first steps in understanding the pivotal role of altered cholesterol metabolism and thus, to therapeutic proposals. Recently, Goizet et al. (2009b) reported eight mutations in this coding region, in 82 individuals with autosomal recessive HSP (pure and complex forms); six of them were novel mutations. These findings would highlight the molecular mechanism involved in the pathogenesis of the disease.

A recent study performed by Boukhris et al., from C. Mhiri's group, exposed the clinical variability of HSP in a southern Tunisian population. The minimal prevalence of hereditary spastic paraplegia in the district of Sfax was 5.75/100.000, 69% of whom had a complicated form (Boukhris et al., 2009).

HEREDITARY NEUROPATHIES

CHARCOT-MARIE-TOOTH DISEASE

Inherited neuropathies, and particularly the Charcot-Marie-Tooth disease, consist of a heterogeneous group of diseases with a high prevalence about 1/2500 (Scherer, 2006), in which more than 30 genes have been implicated currently (Vallat et al., 2009). Clinical, electrophysiological, and histological features show great variability between and within families and from different ethnic populations, which goes along with genetic heterogeneity. Many Tunisian papers contributed to figure out phenotypic and genetic presentations.

It is classically grouped into two categories according to electrophysiological and nerve biopsy findings: CMT1 with demyelinating features on EMG and histological analysis, and CMT2 with almost normal conduction velocities and pathological signs of axonal degeneration and regeneration. A dominant X-linked category is also included in the CMT1 group. CMT3 (or Dejerine-Sottas syndrome) applied to early and severe autosomal recessive or sporadic demyelinating neuropathy is another subtype. The CMT4 group, with autosomal recessive inheritance, similar to CMT1, is characterized by early age of onset and progressive distal limb weakness and atrophy. It is subdivided into eight categories according to causal genes: CMT4A, CMT4B, CMT4C CMT4D, CMT4E, CMT4F, CMT4H, CMT4J (Payerson et al., 2006).

An attempt to determine the CMT4A locus through an international collaborative genetic analysis of four inbred families was published in *Human Molecular Genetics* by Ben Othmane et al. from Ben Hamida's group, in 1993, which led to linkage evidence to a 5-cM region of chromosome 8q13–21 (Ben Othmane et al., 1993). Two years later, the same authors conducted a collaborative international study which reduced the flanking interval to less than 1 cM region, and proved the exclusion of the myelin protein PMP2 gene, localized to chromosome 8, as the causal defect in CMT4A, since it lies distal to the reduced gene region (Ben Othmane et al., 1995). In 1998, using a PAC contig and haplotype analysis, which is a useful tool in gene mapping among isolated populations, they succeeded in identifying a conserved haplotype in affected chromosomes from three families, reducing the CMT4A minimal candidate region and allowing the start of transcript identification and mutation search (Ben Othmane et al., 1998). Baxter et al. revealed, in 2002, three different mutations (two nonsense, G92A and C581, and one missense, G482A) related to the candidate gene previously mapped, and which encode the ganglioside-induced, differentiation-associated protein-1 (GDPA1) (Baxter et al., 2002; Ammar et al., 2003).

In a study on CMT4B, another form of AR CMT disease, Azzedine et al. described two families (one Tunisian and one Morrocan) presenting with a demyelinating autosomal recessive CMT associated to early-onset glaucoma. Two groups from Tunisia participated in the study, both being headed by Ben Hamida's fellows (Azzedine et al., 2003). The entity was mapped to chromosome 11p15; haplotype reconstruction narrowed the implicated locus to a 4.6 cM region, which overlaps the locus of CMT4B2 previously identified by Ben Othmane et al. in 1999 (glaucoma was not reported previously). Myotubularin-related 13 gene (MTMR13) was identified as the causal gene, with two nonsense mutations. MTMR13 was one of the MTMR protein family including proteins with phosphoinositide phosphatase activity implicated in X-linked myotubular myopathy and CMT4B1, and other proteins in which key catalytic residues are missing, called *pseudophosphatases* (Azzedine et al., 2003). The MTMR13 identification was to be the first reported disease related to a pseudophosphatase subgroup.

GIANT AXONAL NEUROPATHY

Giant axonal neuropathy (GAN) is a severe recessive neurodegenerative disorder which affects both peripheral and central nervous system, characterized by variable combinations of progressive sensory motor neuropathy and curly or kinky hair (Gordon., 2004). Cytopathologically, the disorder is characterized by giant axons due to neurofilament accumulation (Yang et al., 2007). To date, many distinct mutations scattered throughout the coding region of the locus, localized by homozygosity mapping to chromosome 16q24.1, have been reported.

C. Ben Hamida et al., from M. Ben Hamida's group, demonstrated genetic homogeneity of the disease among Tunisian population, referred to as GAN1, and suggested a possible genetic heterogeneity between Tunisian and non-Tunisian patients regarding the slow course of the disease shared by three nonrelated Tunisian families

(Ben Hamida et al., 1997). In 2000, Bomont et al. identified the GAN1 gene, which encodes a ubiquitously expressed protein they called *gigaxonin*. Nine missense, four nonsense, and one frameshift mutations were found. Today, despite many suggested possibilities to explain axonal loss, underlying mechanisms remain unclear (Bomont et al., 2000).

INHERITED MUSCULAR DISORDERS

Inherited muscular disorders are clinically and genetically heterogenous. Muscular dystrophies, particularly limb girdle dystrophy (LGMD), was the main topic on which Tunisian papers focused, contributing to the identification of new nosologic entities and consequently to their recent classification.

Studies about other muscular disorders, such as congenital muscular dystrophy, inclusion body myositis, Schwartz Jampel syndrome, and congenital myasthenic syndromes were also reported.

LIMB GIRDLE MUSCULAR DYSTROPHIES

The story of the limb girdle muscular dystrophies is as amazing as that of AVED: a new entity emerging from the careful study of a very old and well-known one. In the 1980s and even well before, it was known that girls could be affected by Duchenne muscular dystrophy, an X-linked disorder. The explanation was an abnormal inactivation of the normal chromosome X at embryonic stages of development, leading to the expression of the mutated chromosome X in these girls. In 1980, M. Ben Hamida and D. Marrakchi published an intriguing paper reporting on 31 Tunisian patients with a myopathy strictly similar to Duchenne muscular dystrophy, but of autosomal recessive inheritance (Ben Hamida & Marrakchi, 1980).

In the same year, they conducted a study including 73 patients diagnosed as having progressive muscular dystrophy with X-linked, autosomal dominant and autosomal recessive inheritance, in which they insisted on classification difficulties and their genetic determination, since a new phenotype of "Duchenne fille" was described (Ben Hamida & Marrakchi, 1980). This new entity was emphasized in collaboration with Fardeau in 1983, and was termed "autosomal recessive childhood severe proximal myopathy" (Ben Hamida et al., 1983a; Ben Hamida M et al., 1983b). They proved the presence of normal dystrophin in this clinical variety by immunohistochemical testing (Ben Jalloun-Dellagi et al., 1990). Other terms used in the literature for this newly defined entity reflected its geographical origin (Tunisian myopathy, Maghrebian myopathy) and in some instances, it was termed "Ben Hamida's myopathy." This denomination remained, unfortunately. almost confidential (mostly used by Ben Hamida's fellows) even though the late Mongi Ben Hamida deserves this term to be more widely used due to his enormous contribution to its description.

In 1992, Ben Othmane et al. revealed its linkage to a pericentromeric region of the chromosome 13q (Ben Othmane et al., 1992) and the entity was termed "Duchenne-like limb girdle muscular dystrophy" (LGMD2C) and evidence of a linkage disequilibrium was shown (Ben Othmane et al., 1995). Through an histoenzymological, immunocytochemical, and biochemical analysis of muscle biopsies belonging to Tunisian patients with LGMD2 compared to those of patients with Duchenne muscular dystrophy, C. Ben Hamida et al. (1994a) demonstrated that protein expression of myosin isoforms, desmin, vimentin, and titin were less affected than in Duchenne muscular dystrophy.

In 1995, 35-kilodalton dystrophin-associated glycoprotein, called the γ sarcoglycan, was identified. Causal gene mutations were linked to chromosome 13q12; they were incriminated in the integrity disruption of the entire sarcoglycan complex (Noguchi et al., 1995), and clinical phenotype was referred to as LGMD 2C. These findings encouraged an emphasis on a clinical approach as a differential diagnosis of Duchenne muscular dystrophy (Ben Hamida et al., 1996). Kefi et al. (2003) led a study including 132 patients carrying LGMD 2C and sharing the same del521-T mutation in order to compare clinical and immunocytochemical findings. Heterogeneity of clinical phenotypes, severity of the disease, sarcoglycan expression abnormalities (γ, α and β) and their phenotypic correlations supported the hypothesis of a modifying gene that controls the course of the disease.

Other autosomal recessive LGMD and their causal mutations were recognized, proving genetic heterogeneity of this group even in the same population. Indeed, Fendri et al. (2006) reported three Tunisian patients belonging to the same consanguineous family and sharing LGMD 2 phenotype. Both siblings carried the homozygous del521-T mutation corresponding to LGMD 2C form (γ sarcoglycanopathy), and their inbred carried a 157G>A homozygous mutation referring to LGMD 2D (α sarcoglycanopathy), which was a proof of genetic counseling complexity. The first family presenting with LGMD 2E due to β sarcoglycanopathy was reported in 1998. Direct gene sequencing led to the identification of a homozygous mutation (G272-->TArg91Leu) in exon 3 on chromosome 4q (Bönnemann et al., 1998). Using an inversed process to determine other subgroups, Driss et al. (2000). came to identify in a large consanguineous Tunisian family a new locus for another LGMD2, mapped to chromosome 19q13.3, and referred to as LGMD 2I. The candidate region was further narrowed down to a 1.1 Mb, including the fukutin-related protein (FKRP) gene implicated in congenital muscular dystrophy, on which a new homozygous mutation was identified, suggesting its involvement

TABLE 35-1 SUMMARY OF OTHER PUBLICATIONS ON RARE DISORDERS IN TUNISIAN PATIENTS

Disease	Main Findings	References
Congenital muscular dystrophy	3 Tunisian families unlinked to LAMA2, FCMD, MEB and CMD1B loci	Triki et al., 2003
	2 new FKRP mutations with identification of a founder mutation (A455D) in 6 Tunisian families	Louihichi et al., 2004
	2 novel mutation (C8005delT, 8244+1G<A) in LAMA2 in 2 Tunisian families	Louihichi et al., 2006
	1 new homozygous splicing mutation of LAMA2 leading to decrease of mRNA level in a Tunisian family	Siala et al., 2007, 2008
Congenital myasthenic syndromes	First Tunisian family described with the AD CMS Slow Channel Syndrome CHRNE 1293insG found as a founder mutation frequent in North Africa 11 (5 novel) mutations in DOK7 gene	Ben Youssef-Turki et al., 2008 Richard et al., 2008 Ben Ammar et al., 2010
Myotonic dystrophies	Evaluation of frequency of cardiac abnormalities in Steinert myotonic dystrophy type 1	Miladi et al., 2009
Schwartz Jampel syndrome	Localization of the SJS locus to chromosome 1P34-36 in 3 Tunisian families Confirmation of linkage of SJS to chromosome 1 and narrowing of the interval to 3cM	Nicole et al., 1995 Fontaine et al., 1996
Inclusion body myositis	Allelic heterogeneity of GNE mutations in 2 Tunisian families	Amouri et al., 2005
Spinal muscular atrophy	Clinical, electrophysiological and genetic data on 101 Tunisian patients with SMA Immunohistochemichal and histochemical analysis and classification of SMA	Ben Hamida et al., 1988 Ben Hamida et al., 1994
Amyotrophic lateral sclerosis	Clinical characterization of Juvenile ALS in 43 patients and 17 families Linkage of AR ALS to chromosome 2q33–35 Linkage of a commoner form of AR ALS to chromosome 15q15–22	Ben Hamida et al., 1990 Hentati et al., 1994 Hentati et al., 1998

in the interaction between extracellular matrix components (Driss et al. 2003),. Clinically, patients from this original Tunisian family sharing a homozygous 1486T > A mutation did not display any cardiac involvement. Four siblings, from a second family carrying LGMD 2I mutations, shared clinical cardiac involvement (mild in two, and severe leading to death in the two others), and genetically, a compound heterozygous 1486T > A AND 1364C > A mutation in the FKRP gene. These findings supported the hypothesis that cardiac involvement might be related to the association of a compound heterozygous state of mutations responsible for LGMD 2I and congenital muscular dystrophy (Kefi et al., 2008).

Among the muscular dystrophy group, an autosomal recessive distal myopathy, the Miyoshi myopathy, was reported in 12 Tunisian families by Bejaoui et al. (1995), who contributed to the identification of the implicated locus linked to chromosome 2p12–14.

OTHER RARE DISORDERS

Other studies on neurological diseases including Tunisian patients are related to hereditary inclusion body myositis (HIBM; see Hentati et al., 1998, congenital muscular dystrophies (Triki et al, 2003; Louhichi et al., 2004, 2006; Siala et al., 2007, 2008a and b), Congenital Myasthenic Syndromes (CMS; Ben Youssef-Turki et al., 2008; Richard et al., 2008; Ben Ammar et al., 2010), myotonic dystrophy (Miladi et al., 2009), spinal muscular atrophy (M. Ben Hamida et al., 1998, C. Ben Hamida et al., 1994b) and amyotrophic lateral sclerosis (M Ben Hamida et al., 1990, 1995; Hentati et al., 1994, 1998) and Schwartz Jampell syndrome (Nicole et al., 1995; Fontaine et al., 1996). Data related to these rare disorders are summarized in Table 35-1.

CONCLUSION

Tunisia is a small country of 10 million inhabitants located in North Africa. It has its roots in North Africa but its population is of Arab-Berber origin, making its population characteristics somehow different from that of sub-Saharan Africa. Its geographical position at the crossroads of Africa, Asia, and Europe, the multiple invasions it has witnessed (since the Phoenicians founded Carthage until the Arab invasion) made it a unique place where the genetic background of the population reflects the heritage of many ethnicities. A strong tradition of consanguineous marriages gave it some characteristics of an isolate, where the pioneering studies of the late Mongi Ben Hamida and the subsequent studies of his fellows' groups allowed participating in advances in neurosciences and elucidation of several diseases. Indeed, 90 international studies, related to more than 20 neurological disease entities, have involved Tunisian patients and Tunisian authors. The results of these studies gave and continue to give insights into the pathogenetic mechanisms of several prevalent diseases; they also probably contribute to the advance of protein-targeted therapies and gene therapy. The future of Tunisian neurosciences appears to be promising, with the new advances providing insights into past observations.

REFERENCES

Ammar N, Nelis E, Merlini L, et al. (2003). Identification of novel GDAP1 mutations causing autosomal recessive Charcot-Marie-Tooth disease. *Neuromuscul Disord 13* (9): 720–728.

Azzedine H, Bolino A, Taieb T, et al. (2003). Mutations in MTMR13, a new pseudophosphatase homologue of MTMR2 and Sbf1, in two families with an autosomal recessive demyelinating form of Charcot-Marie-Tooth disease associated with early-onset glaucoma. *Am J Hum Genet 72*: 1141–53.

Baxter RV, Ben Othmane K, Rochelle JM, et al. (2001). Ganglioside-induced differentiation associated protein-1 is mutant in Charcot-Marie-Tooth disease type 4A/8q21. *Nat Genet 30* (1): 21–22.

Bejaoui K., Hirabayashi K, Hentati F, et al. (1995). Linkage of Miyoshi myopathy (distal autosomal recessive muscular dystrophy) locus to chromosome 2p12–14. *Neurology 45* (4): 768–72.

Belal S, Cancel G, Stevanin G, et al. (1994). Clinical and genetic analysis of a Tunisian family with autosomal dominant cerebellar ataxia type 1 linked to the SCA2 locus. *Neurology 44* (8): 1423–26.

Belal S, Panayides K, Sirugo G, et al. (1992). Study of Large Inbred Friedreich Ataxia Families Reveals a Recombination between D9S15 and the disease Locus. *Am J Hum Genet 51* (6): 1372–76.

Ben Ammar A, Petit F, Alexandri N, et al. (2010). Phenotype genotype analysis in 15 patients presenting a congenital myasthenic syndrome due to mutations in DOK7. *J Neurol 257* (5): 754–66.

Ben Hamida C, Cavalier L, Belal S, et al. (1997). Homozygosity mapping of giant axonal neuropathy gene to chromosome 16q24.1. *Neurogenetics 1* (2): 129–33.

Ben Hamida C, Doerflinger N, Belal S, et al. (1993). Localization of Friedreich ataxia phenotype with selective vitamin E deficiency to chromosome 8q by homozygosity mapping. *Nat Genet 5* (2): 195–200.

Ben Hamida C, Soussi-Yanicostas N, Bejaoui K, Butler-Browne GS, Hentati F, Ben Hamida M. (1994a). Expression of myosin isoforms and of desmin, vimentin, and titin in Tunisian Duchenne-like autosomal recessive muscular dystrophy. *J Neurol Sci 123* (1–2): 114–21.

Ben Hamida C, Soussi-Yanicostas N, Butler-Browne GS, Bejaoui K, Hentati F, Ben Hmida M. (1994b). Biochemical and immunocytochemical analysis in chronic proximal spinal muscular atrophy. *Muscle Nerve 17* (4): 400–10.

Ben Hamida M, Attia-Romdhane N, Triki CH, Oueslati S, Hentati F. (1991). Analyse clinique et génétique de 188 familles d'hérédo-dégénérescence spino-cérébelleuse. Maladies de Friedrich et hérédo-ataxies de P. Marie. *Rev Neurol 147* (12): 798–808.

Ben Hamida M, Attia N, Chaabouni H, Fardeau M. (1983). Autosomal recessive severe, proximal myopathy in children, common in Tunisia. *Rev Neurol 139* (4): 289–97.

Ben Hamida M, Attia-Romdhane N, Hentati F, Triki C. (1993). Early onset ataxias in Tunisia. Intrafamilial heterogeneity. *Adv Neurol 61*: 105–14.

Ben Hamida M, Belal S, Sirugo G, et al. (1993). Friedreich's ataxia phenotype not linked to chromosome 9 and associated with selective autosomal recessive vitamin E deficiency in two inbred Tunisian families. *Neurology 43* (11): 2179–83.

Ben Hamida M, Ben Hamida C, Zouari M, Belal S, Hentati F. (1996). Limb-girdle muscular dystrophy 2C: Clinical aspects. *Neuromuscul Disord 6* (6): 493–94.

Ben Hamida M, Chaabouni H, Madani S, et al. (1986). Genetic study of Spinocerebellar hereditary degeneration in Tunisia. The role of consanguinityEtude génétique des hérédo-dégénérescences spino-cérébelleuses en Tunisie. Le rôle de la consanguinité dans leur survenue). *J Genet Hum 34* (3–4): 267–74.

Ben Hamida M, Fardeau M, Attia N. (1993). Severe childhood muscular dystrophy affecting both sexes and frequent in Tunisia. *Muscle Nerve 6* (7): 469–80.

Ben Hamida M, Hentati F, Ben Hamida C. (1990). Hereditary motor system diseases (chronic juvenile amyotrophic lateral sclerosis). Conditions combining a bilateral pyramidal syndrome with limb and bulbar amyotrophy. *Brain 113*: 347–63.

Ben Hamida M, Hentati F, Chebbi N, Mhiri C, Ayed K, Hamzaoui K. (1998). Chronic proximal spinal amyotrophy in Tunisia (Amyotrophies spinals proximales et chroniques en Tunisie). *Rev Neurol 144* (11): 737–47.

Ben Hamida M, Hentati F. (1995). Atypical amyotrophic lateral sclerosis. *Adv Neurol 68*: 153–55.

Ben Hamida M, Marrakchi D. (1980). Duchenne muscular dystrophy in Tunisia: 31 cases in 13 families with autosomal recessive inheritance. *J Genet Hum 28* (1): 1–9.

Ben Hamida M, Marrakchi D. (1980). Etude génétique des myopathies en Tunisie. *Union Medicale Canada 109* (11): 1609–28.

Ben Jelloun-Dellagi S, Chaffey P, Hentati F, et al. (1990). Presence of normal dystrophin in Tunisian severe childhood autosomal recessive muscular dystrophy. *Neurology 40* (12): 1903.

Ben Othmane K, Ben Hamida M, Pericak-Vance MA, et al. (1992). Linkage of Tunisian autosomal recessive Duchenne-like muscular dystrophy to the pericentromeric region of chromosome 13q. *Nat Genet 2* (4): 315–7.

Ben Othmane K, Hentati F, Lennon F, et al. (1993). Linkage of a locus (CMT4A) for autosomal recessive Charcot-Marie-Tooth disease to chromosome 8q. *Hum Mol Genet 2* (10): 1625–8.

Ben Othmane K, Loeb D, Hayworth-Hodgte R, et al. (1995). Physical and Genetic mapping of the CMT4A locus and exclusion of PMP-2 as the defect in CMT4A. *Genomics 28* (2): 286–90.

Ben Othmane K, Rochelle JM, Ben Hamida M, et al. (1998). Fine localization of the CMT4A locus using a PAC contig and haplotype analysis. *Neurogenetics 2* (1): 18–23.

Ben Othmane K, Speer MC, Stauffer J, et al. (1995). Evidence for linkage Disequilibrium in chromosome13-Linked Duchenne-like Muscular Dystrophy (LGMD2C). *Am J Hum Genet 57* (3): 732–4.

Ben Youssef-Turki I, Kraoua I, Gargouri A, Akopova-Larbi R, Gouider R, Gouider-Khouja N. (2008). Aspects cliniques et neurophysiologiques de syndrome de canal lent Slow Channel Syndrome). *Tunisie Médicale 86* (2): 202–4.

Bomont P, Cavalier L, Blondeau F, et al. (2000). The gene encoding gigaxonin, a new member of the cytoskeletal BTB/Kelch repeat family, is mutated in giant axonal neuropathy. *Nat Genet 26* (3): 370–4.

Bönnemann CG, Wong J, Ben Hamida C, Ben Hamida M, Hentati F, Kunkel LM. (1998). LGMD 2E in Tunisia is caused by a homozygous missense mutation in beta-sarcoglycan exon 3. *Neuromuscul Disord 8* (3–4): 193–7.

Bouhlal Y, El Euch-Fayeche G, Hentati F, Amouri R. (2009). A novel SACS gene mutations in a Tunisian family. *J Mol Neurosci 3* (3): 333–6.

Bouhlal Y, Zouari M, Kefi M, Ben Hamida C, Hentati F, Amouri R. (2008). Autosomal recessive ataxia caused by three distinct gene defects in a single consanguineous family. *J Neurogenet 22* (2): 139–48.

Bouhouche A, Birouk N, Azzedine H, et al. (2007). Autosomal recessive axonal Charcot-Marie-Tooth disease (ARCMT2), phenotype - genotype correlations in 13 Moroccan families. *Brain 130*: 1062–1075.

Boukhris A, Stevanin G, Feki I, et al. (2008). Hereditary spastic paraplegia with mental impairment and thin corpus callosum in Tunisia: SPG11, SPG15, and further genetic heterogeneity. *Arch Neurol 65* (3): 393–402.

Boukhris A, Stevanin G, Feki I, et al. (2009). Tunisian hereditary spastic paraplegias: clinical variability supported by genetic heterogeneity. *Clin Genet 75* (6): 527–36.

Cancel G, Durr A, Didierjean O, et al. (1997). Molecular and clinical correlations in spinocerebellar ataxia 2: A study of 32 families. *Hum Mol Genet 6* (5): 709–15.

David G, Durr A, Stevanin G, et al. (1998). Molecular and clinical correlations in autosomal dominant cerebellar ataxia with progressive macular dystrophy (SCA7). *Hum Mol Genet 7* (2): 165–70.

Doerflinger N, Linder C, Ouahchi K, et al. (1995). Ataxia with Vitamin E deficiency: refinement of genetic localization and analysis of linkage disequilibrium by using new marker in 14 families. *Am J Hum Genet 56* (5): 1116–24.

Driss A, Amouri R, Ben Hamida C, et al. (2000). A new locus for autosomal recessive limb-girdle muscular dystrophy in a large consanguineous Tunisian family maps to chromosome 19q13.3. *Neuromuscul Disord 10* (4–5): 240–6.

Driss A, Noguchi S, Amouri R, et al. (2003). Fukutin-related protein gene mutated in the original kindred limb-girdle MD 2I. *Neurology 60* (8): 1341–4.

El Euch-Fayache G, Lalani I, Amouri R, et al. (2003). Phenotypic features and genetic findings in sacsin-related autosomal recessive ataxia in Tunisia. *Arch Neurol 60* (7): 982–8.

Farrer M, Gwinn-Hardy K, Muenter M, et al. (1999). A chromosome 4p haplotype segregating with Parkinson's disease and postural tremor. *Hum Mol Genet 8* (1);81–5.

Fendri K, Kefi M, Hentati F, Amouri R. (2006). Genetic heterogeneity within a consanguineous family involving the LGMD 2D and the LGMD 2C genes. *Neuromuscul Disord 16* (5): 316–20.

Fontaine B, Nicole S, Topaloglu H, et al. (1996). Recessive Schwartz-Jampel syndrome (SJS), confirmation of linkage to chromosome 1p, evidence of genetic homogeneity and reduction of the SJS locus to a 3-cM interval. *Hum Genet 98* (3): 380–5.

Funayama M, Hasegawa K, Kowa H, Saito M, Tsuji S, Obata F. (2002). A new locus for Parkinson's disease (PARK8) maps to chromosome 12p11.2-q13,1. *Ann Neurol 51* (3): 296–301.

Gabsi S, Gouider-Khouja N, Belal S, et al. (2001). Effect of Vitamin E supplementation in patients with ataxia with vitamin E deficiency. Eur *J Neurol 8* (5): 477–81.

Goizet C, Boukhris A, Maltete D, et al. (2009a). SPG15 is the second most common cause of hereditary spastic paraplegia with thin corpus callosum. *Neurology 73* (14): 1111–9.

Goizet C, Boukhris A, Durr A, et al. (2009b). CYP7B1 mutations in pure and complex forms of hereditary spastic paraplegia type 5. *Brain 133*: 1589–600.

Gordon N. (2004). Giant axonal neuropathy. *Dev Med Child Neurol 46* (10): 717–19.

Gouider-Khouja N, Belal S, Ben Hamida M, Hentati F. (2000). Clinical and genetic study of familial Parkinson's disease en Tunisia. *Neurology 54* (8): 1603–09.

Gouider-Khouja N, Larnaout A, Amouri R, et al. (2003). Autosomal recessive parkinsonism linked to parkin gene in a Tunisian family. Clinical, genetic and pathological study. *Parkinsonism Relat Disord 9* (5): 247–51.

Gouider-Khouja N. (1991). [Parkinson disease in Tunisia: Epidemiological, clinical, follow-up and therapeutic aspects.] La maladie de Parkinson en Tunisie. Aspects épidémiologiques, cliniques, évolutifs et thérapeutiques. Faculty of Medicine of Tunis.

Nussbaum RL, Ellis CE. (2003). Alzheimer's Disease and Parkinson's Disease. *NEJM 348* (14): 1356–64.

Hentati A, Bejaoui K, Pericak-Vance MA, et al. (1994). Linkage of recessive familial amyotrophic lateral sclerosis to chromosome 2q33-q35. *Nat Genet 7* (3): 4258.

Hentati A, Ouahchi K, Pericak-Vance MA, et al. (1998). Linkage of a commoner form of recessive amyotrophic lateral sclerosis to chromosome 15q-q22 markers. *Neurogenetics 2* (1):55–60.

Hentati A, Pericak-Vance MA, Hung WY, et al. (1994a). Linkage of "pure" autosomal recessive familial spastic paraplegia to chromosome 8 markers and evidence of genetic locus heterogeneity. *Hum Mol Genet 3* (8): 1263–1267.

Hentati A, Pericak-Vance MA, Lennon F, et al. (1994b). Linkage of a locus for autosomal dominant familial spastic paraplegia to chromosome 2p markers. *Hum Mol Genet 3* (10): 1867–71.

Hentati F, Ben Hamida C, Belal S, Tomé F, Fardeau M, Ben Hamida M. (1998). Familial autosomal-recessive inclusion-body myositis with asymptomatic leukoencephalopathy. In: Askanas V, Serratrice G, King Engel W eds. Inclusion body myositis and myopathies. Cambridge University Press, 211–220.

Hulihan MM, Ishihara-Paul L, Kachergus J, et al. (2008). LRRK2 Gly2019Ser penetrance in Arab-Berber patients from Tunisia: A case-control genetic study. *Lancet Neurol 7* (7): 591–4.

Ishihara L, Gibson RA, Warren L, et al. (2007). Screening for Lrrk2 G2019S and clinical comparison of Tunisian and North American Caucasian Parkinson's Disease families. *Mov Disord 22* (1): 55–61.

Ishihara L, Warren L, Gibson R, et al. (2006). Clinical features of Parkinson Disease patients with homozygous leucine-rich repeat kinase 2 G2019S mutations. *Arch Neurol 63* (9): 1250–4.

Ishihara-Paul L, Hulihan MM, Kachergus J, et al. (2008). PINK1 mutations and parkinsonism. *Neurology 71* (12): 896–902.

Kefi M, Amouri R, Chabrak S, Mechmeche R, Hentati F. (2008). Variable cardiac involvement in Tunisian siblings harboring FKRP gene mutations. *Neuropediatrics 39* (2): 113–5.

Kefi M, Amouri R, Driss A, et al. (2003). Phenotype and sarcoglycan expression in Tunisian LGMD 2C patients sharing the same del521-T mutation. *Neuromuscul Disord 13* (1): 779–87.

Kitada T, Asakawa S, Hattori N, et al. (1998). Mutations in the parkin gene cause autosomal recessive juvenile parkinsonism. *Nature 392* (6676): 605–8.

Klebe S, Azzedine H, Durr A, et al. (2006). Autosomal recessive spastic paraplegia (SPG30) with mild ataxia and sensory neuropathy maps to chromosome 2q37.3. *Brain 129*: 1456–62.

Leroy E, Boyer R, Auburger G, et al (1998). The ubiquitin pathway in Parkinson's disease. *Nature 395* (6701): 451–452.

Lesage S, Leutenegger AL, Ibanez P, et al. (2005). LRRK2 haplotype analyses in European and North African families with Parkinson disease: a common founder for the G2019S mutation dating from the 13th century. *Am J Hum Genets 77* (2): 330–2.

Louhichi N, Richard P, Triki C, et al. (2006). Novel mutations in LAMA2 gene responsible for a severe phenotype of congenital muscular dystrophy in two Tunisian families. *Arch Inst Pasteur Tunis 83* (1–4): 19–23.

Louhichi N, Triki C, Quijano-Roy S, et al. (2004). New FKRP mutations causing congenital muscular dystrophy associated with mental retardation and central nervous system abnormalities. Identification of a founder mutation in Tunisian families. *Neurogenetics 5* (1): 27–34.

Miladi MI, Charfeddine H, Feki I, et al. (2009). [Cardiac involvement in Steinert myotonic dystrophy.] *Rev Méde Int 30* (7): 573–7.

Nicole S, Ben Hamida C, Beighton P, et al. (1995). Localization of the Schwartz-Jampel Syndrome (SJS) locus to chromosome 1p34-p36.1 by homozygosity mapping. *Hum Mol Genet 4* (9): 1633–6.

Nishioka K, Kefi M, Jasinska-Myga B, et al. (2010). A comparative study of LRRK2, PINK1 and geneticallay undefined familial Parkinson's disease. *J Neurol Neurosurg Psy 81* (4): 391–5.

Noguchi S, McNally EM, Ben Othmane K, et al. (1995). Mutations in the dystrophin-associated protein gamma-sarcoglycan in chromosome 13 muscular dystrophy. *Science 270* (5237): 781–4.

Ouahchi K, Arita M, Kayden H, et al. (1995). Ataxia with isolated vitamin E deficiency is caused by mutations in the α-tocopherol transfer protein. *Nat Genet 9* (2): 141–5.

Payerson D, Scaioli V, Laura M. (2006). Clinical and electrophysiological aspects of Charcot Marie Tooth Disease. *Neuromol Med 8* (1–2): 3–23.

Polymeropoulos MH, Higgins JJ, Golbe LI, et al. (1996). Mapping of a gene for Parkinson's disease to chromosome 4q21-q23. *Science 274*: 1197–8.

Polymeropoulos MH, Lavedan C, Leroy E, et al. (1997). Mutation in the alpha-synuclein gene identified in families with Parkinson's disease. *Science 276* (5321): 2045–7.

Richard P, Gaudon K, Haddad H, et al. (2008). The CHRNE 1293insG founder mutation is a frequent cause of congenital myasthenia in North Africa. *Neurology 71* (24): 1967–72.

Rodius F, Duclos F, Wrogemann K, et al. (1994). Recombinations in individuals homozygous by descent localize the Friedreich ataxia locus in a cloned 450-kb interval. *Am J Hum Genet 54* (6): 1050–9.

Salinas S, Proukakis C, Crosby A, Warner TT. (2008). Hereditary spastic paraplegia: clinical features and pathogenetic mechanisms. *Lancet Neurol 7* (12): 1127–38.

Scherer SS. (2006). Finding the causes of inherited neuropathies. *Arch Neurol 63* 812–6.

Siala O, Louhichi N, Triki C, et al. (2007). Severe MDC1A congenital muscular dystrophy due to a splicing mutation in the LAMA2 gene resulting in exon skipping and significant decrease of mRNA level. *Genet Test 11* (3): 199–207.

Siala O, Louhichi N, Triki C, et al. (2008a). LAMA2 mRNA processing alterations generate a complete deficiency of laminin-alpha2 protein and a severe congenital muscular dystrophy. *Neuromuscl Disord 18* (2): 137–45.

Siala O, Kammoun-Feki F, Louhichi N. et al. (2008b). Molecular prenatal diagnosis of muscular dystrophies in Tunisia and postnatal follow-up role. *Genet Test 12* (4): 581–6.

Stevanin G, Azzedine H, Denora P, et al. (2008). Mutations in SPG11 are frequent in autosomal recessive spastic paraplegia with thin corpus callosum, cognitive decline and lower motor neuron degeneration. *Brain 131* (3): 772–84.

Stevanin G, Giunti P, David GD, et al. (1998). De novo expansion of intermediate alleles in spinocerebellar ataxia 7. *Hum Mol Genet 7* (11): 1809–13.

Stevanin G, Santorelli FM, Azzedine H, et al. (2007). Mutations in SPG11, encoding spatacsin, are a major cause of spastic paraplegia with thin corpus callosum. *Nat Genet 39* (3): 366–72.

Tomiyama H, Li Y, Funayama M, et al. (2006). Clinicogenetic study of mutations in LRRK2 exon 41 in Parkinson's disease patients from 18 countries. *Mov Disord 21* (8): 1102–8.

Triki C, Louhichi N, Méziou M, et al. (2003). Merosin-deficient congenital muscular dystrophy with mental retardation and cerebellar cysts, unlinked to the LAMA2, FCMD, MEB and CMD1B loci, in three Tunisian patients. *Neuromusul Disord 13* (1): 4–12.

Tsaousidou MK, Ouahchi K, Warner TT, et al. (2008). Sequence alterations within CYP7B1 implicate defective cholesterol homeostasis in Motor-neuron degeneration. *Am J Hum Genet 82*: 510–5.

Valente EM, Bentivoglio AR, Dixon PH, et al. (2001). Localization of a novel locus for autosomal recessive early-onset parkinsonism, PARK6, on chromosome 1p35-p36. *Am J Hum Genet 68*: 895–900.

Vallat JM, Tazir M, Calvo J, Funalot B. (2009). Hereditary peripheral neuropathy. *Presse Méd 38* (9): 1325–4.

Van Duijn CM, Dekker MCJ, Bonifati V, et al. (2001). PARK7, a novel locus for autosomal recessive early-onset parkinsonism, on chromosome 1p36. *Am J Hum Genet 69*: 629–34.

Warren L, Gibson R, Ishihara L, et al. (2008). A founding LRRK2 haplotype shared by Tunisian, US, European and Middle Eastern families with Parkinson's disease. *Parkinsonism Relat Disord 1*: 77–80.

Yang Y, Allen E, Ding J, et al. (2007). Giant axonal neuropathy. *Cell Mol Life Sci 64* (5): 601–9.

36.

HEMOGLOBINOPATHIES IN AFRICA

Ambroise Wonkam and Amanda Krause

INTRODUCTION

Inherited hemoglobin disorders fall into two main groups: the structural and qualitative hemoglobin variants (e.g., sickle cell anemia), and the quantitative changes in hemoglobin production, the thalassemias. The structural hemoglobin variants mostly result from single amino acid substitutions in the α or β chains that may alter the stability or functional properties of the hemoglobin and lead to a clinical disorder (Weatherall & Clegg, 2001). The thalassemias are classified according to the particular globin chains that are ineffectively synthesized. From a public health viewpoint, the α and β thalassemias are sufficiently common to be of importance.

As a result of heterozygote advantage against malaria, the inherited hemoglobin disorders are the commonest monogenic group of disorders (WHO, 2008). The World Health Organization (WHO) has estimated that at least 7% of the world's population carries a hemoglobinopathy, alpha- or beta-thalassemias, or one of the structural variant hemoglobins S, C, and E (Figure 36-1; Clegg & Weatherall, 1999). Up to 300,000–400,000 babies with severe forms of these diseases are born each year (Weatherall, 2008). Approximately 80% of these births occur in low-income or middle-income countries (Weatherall, 2010). Many of these inherited hemoglobin disorders, if left untreated, result in death in the first few years of life. Their effect on the burden of healthcare is only now being appreciated in many parts of the world, as they require expensive treatment, including transfusion, chelation therapy, and bone marrow transplantation.

INHERITED HEMOGLOBIN DISORDERS

NORMAL HUMAN HEMOGLOBIN

The structure of human hemoglobin (Hb) changes from embryonic to fetal to adult life (Weatherall & Clegg, 2001). All the normal hemoglobins are tetramers of two pairs of α-like and β-like globin chains. Adult and fetal hemoglobin have α β (Hb A, α2β2), δ (Hb A2, α2δ2), or γ chains (Hb F, α2γ2), whereas in the embryo, α-like chains—termed ζ γ (Hb Portland, ζ 2γ2) or ε ζ2ε2— and α and ε chains form Hb Gower 2 (α2ε2) (see Figure 36-2; Weatherall, 2001).

Embryonic hemoglobin production is confined to the yolk sac stage of development and thereafter is replaced by fetal hemoglobin. HbF is the predominant type of hemoglobin in fetal life, but around birth there is a switch from fetal to adult globin gene expression, when HbF is gradually replaced by adult hemoglobin, such that by 6 months of age the major Hb is HbA (α2β2) (Stamatoyannopoulos et al., 1994). Residual amounts of HbF, however, continue to be synthesized throughout adult life, and the amounts vary considerably, with the majority of adults having less than 1% HbF.

SICKLE CELL DISEASES

Although over 700 structural hemoglobin variants have been identified, only two (Hb S, Hb C) reach high frequencies in Africa. Sickle hemoglobin (Hb S) causes a group of related blood disorders called *sickle cell disease* (SCD). SCD is the most common "monogenic" disease in man. Sickle cell hemoglobin (Hb S) is caused by the amino acid substitution of valine for glutamic acid at the sixth position of the β-globin chain, and is due to a single nucleotide substitution, GAG->GTG in codon 6 of the β-globin gene on chromosome 11p15.5.

The homozygous state for the sickle cell gene results in sickle cell anemia, which is the most common and severe form. SCD also includes compound heterozygous states for Hb S and other hemoglobinopathies, such as Hb C and β-thalassemia, which, although milder than sickle cell anemia, also has important public health implications (Weatherall, 2010). It has been estimated that SCD results in the annual loss of several millions of disability-adjusted life years, particularly in the developing world (Weatherall, 2008).

The first description of SCD, including the clinical signs and the sickle shape of erythrocytes, was published in 1910 and concerned a black American patient (Serjeant, 2001). However, SCD was known in some parts of Africa before the twentieth century: inhabitants of western Africa gave

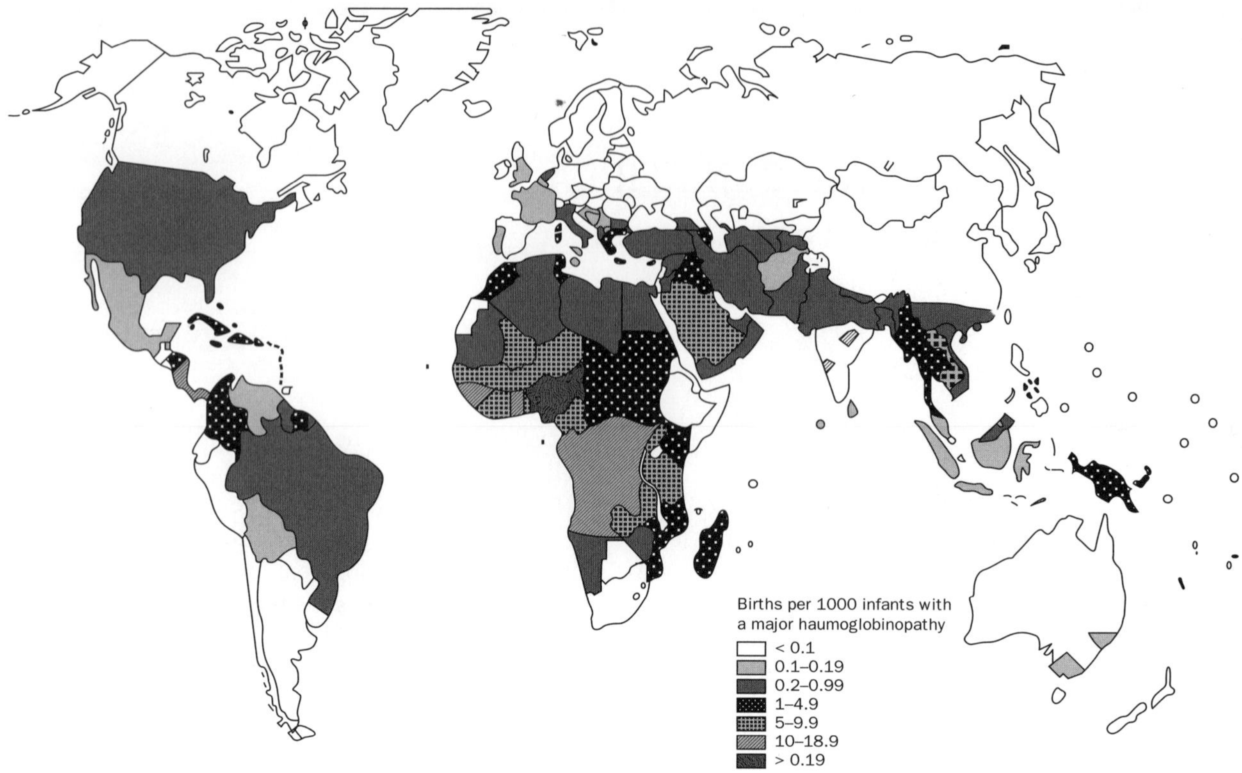

Figure 36-1 Global distribution of hemoglobin disorder, in term of births of affected infants per 1000 births. (WHO, 1996)

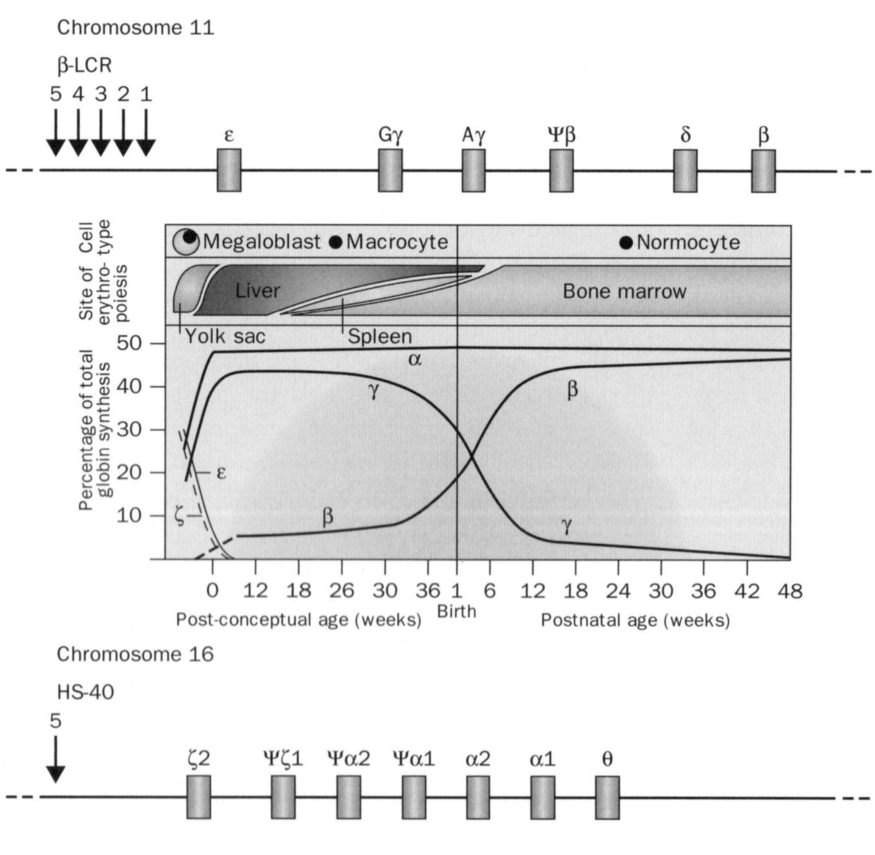

Figure 36-2 The structure of human hemoglobin (Hb) changes from embryonic to fetal and adult life. (Weatherall, 2001)

the disease specific names that evoke acute, painful episodes or death, or refer to children destined to die and to be reborn as their own siblings (Onwubalili, 1983; Nzewi, 2001).

Africa is the main birthplace of sickle mutations (Figure 36-3; Antonarakis et al., 1984). In sub-Saharan Africa, SCD occurs at its highest frequency; the prevalence of sickle cell trait is estimated to be 5% to 38% in Tanzania (Weatherall & Clegg, 2001; Aliyu et al., 2008). Despite the fact that >70% of sufferers live in Africa (about 10 million people), expenditure on the related care and research in the continent is negligible, and most advances in the understanding and management of this condition have been based on research conducted in the North (Makani et al., 2007). Sickle cell anemia is also observed in Africa in patients originating from the Indian subcontinent, with carrier rates of 0.9%–1.3% (Krause, 1994), although the disease tends to be milder in these patients.

CLINICAL FEATURES OF SICKLE CELL DISEASE

The clinical features of sickle cell disorders reflect the propensity of the red cells to assume a sickled configuration when blood is oxygenated, leading to a shortened red-cell survival and a tendency to vasoocclusion (Bunn, 1997). Although patients with sickle cell anemia may adapt to their anemia quite well, their illness is interspersed with acute episodes, or crises, which include episodes of sequestration of blood into the lungs, liver, or spleen, or the occlusion of cerebral vessels with resulting stroke. Furthermore, they are particularly prone to infection in early childhood and, indeed, at all ages. In Cameroon, SCD is the main etiological factor (31.4%) of stroke in children (Obama et al., 1994). We have recently found a high stroke prevalence of 6.7%, with an annual recurrence rate of 25% in SCD patients (Njamnshi et al., 2006a,b). In addition, 16.5% of Cameroonian SCD patients feature abnormal transcranial Doppler velocity (MCA >200; see Njamnshi et al., 2009). Abnormal transcranial Doppler velocity (TCD) and pulmonary hypertension, which are diagnosed by Doppler echocardiography, are proxy for stroke episodes risk and cause of death, respectively. TCD and pulmonary hypertension are reliable noninvasive clues for the identification of these complications before end-stage progression occurs (Nichols et al., 2001; Gladwin & Vichinsky, 2008). These instruments need to be used more widely in clinical practice in Africa.

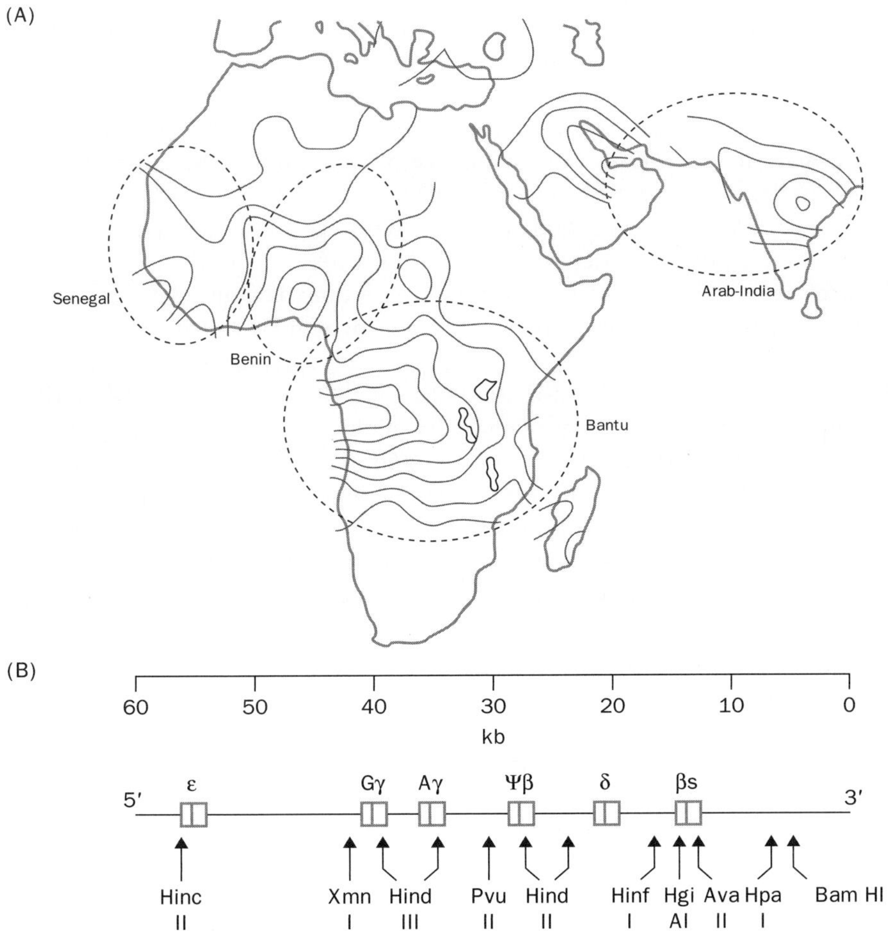

Figure 36-3 ORIGIN OF THE BETA S-GLOBIN GENE. (A). The S Hb gene variant appeared independently in three sites on the African continent, as illustrated by the 3 Hb gene cluster haplotypes that have been described on the continent (Senegal, Benin Bantu). (B). B-globin haplotypes are generated by linkage patterns of the restriction fragment length polymorphisms (RFLPs).

Although little is known about the mortality from SCD in Africa, it is clear that many patients die in early childhood. Indeed, only 7% of a random sample of 210 Cameroonian sickle cell patients were more than 30 years old (Wonkam, 2009). An earlier study reported that 50% of Nigerian SCD patients died before the age of one year (Fleming, 1989), and this is unlikely to have altered dramatically. Previously, median survival in the US for those with SCD was estimated to be age 42 years for men and age 48 years for women (Platt et al., 1994); however, considerable improvements in longevity have been suggested (Wierenga et al., 2001; Serjeant et al., 2007). Nevertheless, significant morbidity remains; the main causes of death are infection, acute chest syndrome, pulmonary artery hypertension, and cerebrovascular events (Steinberg et al., 2003; Bakanay et al., 2005).

Although Hb SC disease is milder than sickle cell anemia, it is associated with similar important complications, including a high frequency of aseptic necrosis of the femoral or humoral heads, hematuria, proliferative retinopathy, and a thrombotic tendency which, particularly in pregnancy and the puerperium, may lead to massive pulmonary thromboembolic disease and death (Serjeant, 1992).

Because of the high morbidity and mortality of sickle cell disease, all 50 U.S. states provide universal newborn screening for sickle cell disease. The vast majority of new cases are diagnosed at birth. International screening is limited: the United Kingdom and Bahrain screen regularly, while other countries such as France do limited screening. Screening is minimal in Africa, where the majority of affected individuals reside, although some countries, such as Ghana, are starting programs (Bender & Hobbs, 2009). Early intervention is associated with decreased morbidity and mortality. New treatments, such as hydroxyurea and stem cell transplantation, are available to SCD patients in a number of Western countries, but few African patients receive anything but the most basic services.

However, clinical severity of SCD is extremely variable, from death in childhood to being relatively well even until old age. The reasons for this heterogeneity are not fully understood (Steinberg, 1998). Interindividual variation in fetal hemoglobin (HbF) levels is one of the main modifiers that contribute to the clinical heterogeneity observed in SCD patients. Higher expression of HbF in adulthood ameliorates morbidity and mortality in SCD (Platt et al., 1991, 1994; Bunn, 1997; Castro et al., 1994). The median of level of HbF in Cameroonian SCD patient is 14.9% (ranging from 0% to 37.2%); up to 32.4% of Cameroonian SCD patients had HbF below 5%, and only 10.6% above 25% (Wonkam, 2009). Experimental and clinical work in this area has led to the use of hydroxyurea, an agent that was found empirically to increase the production of HbF (Platt, 2008).

It is now clear that common HbF variation is a quantitative genetic trait shaped by common polymorphisms. Multiple genes, together with an environmental component, determine the measured value of HbF in any given individual. Genetic variation at three major loci accounts for a relatively large proportion (20%–50%) of the phenotypic variation in HbF levels: (1) a single-base substitution (T/C) at position –158 of the $^{G}\gamma$ globin gene, termed *Xmn*I $^{G}\gamma$ site (Labie et al., 1985); (2) the *HMIP* locus (*HBS1L-MYB Intergenic Polymorphism*) on chromosome 6q (Creary et al., 2009); and (3) the oncogene *BCL11A* on chromosome 2 (Uda et al., 2008). These variants have been well reported in non-anemic Northern Europeans and Sardinians, a β-thalassemia cohort, in SCD patients from Brazil, and in the African-American Cooperative Study of Sickle Cell Disease (CSSCD) (Lettre et al., 2008; Sedgewick et al., 2008, Menzel & Thein, 2009). There is very little description of the three main genetic polymorphisms explaining phenotypic variation in HbF levels and clinical phenotype in native African SCD patients (Creary et al., 2009).

β-THALASSEMIAS

β-thalassemia is a group of genetic diseases caused by mutations in the β-globin gene resulting in reduced (β^{+}) or no synthesis (β^{0}) of β-globin. The pathophysiology of the disease is, therefore, the lack of β-globin and the precipitation of excess α-globin chains, which precipitate out and form inclusion bodies known as *Heinz bodies* (Schrier et al., 1989). The free α-hemoglobin chains oxidize and create reactive oxygen species (ROS), which causes oxidative damage to red blood cells and leads to apoptosis (Schrier et al., 1989). Untreated, this leads to anemia and a range of other symptoms due to compensatory attempts of the body. Treatment is by regular transfusion and iron chelation, or stem cell transplantation. β-thalassemia is an autosomal recessive disease with manifestations occurring when individuals are either homozygous for β^{0}/β^{+} mutations, or are compound heterozygotes (Kazazian & Boehm, 1988). Carriers of the trait do not usually manifest with the disease, but coinheritance of other genetic traits can result in a β-thalassemia phenotype (Advani et al., 1992). β-thalassemia patients can be divided into two groups: β-thalassemia major, which has a clinically severe phenotype, and β-thalassemia intermedia, which results in a milder clinical phenotype (Rund & Rachmilewitz, 2005).

The thalassemias are extremely heterogeneous at the molecular level, with over 200 different mutations of the β globin genes have been found in patients' with β thalassemia, most of which are point mutations or small insertions and deletions (Weatherall, 2001). Few large deletions have been reported (Weatherall, 2001), Importantly, every

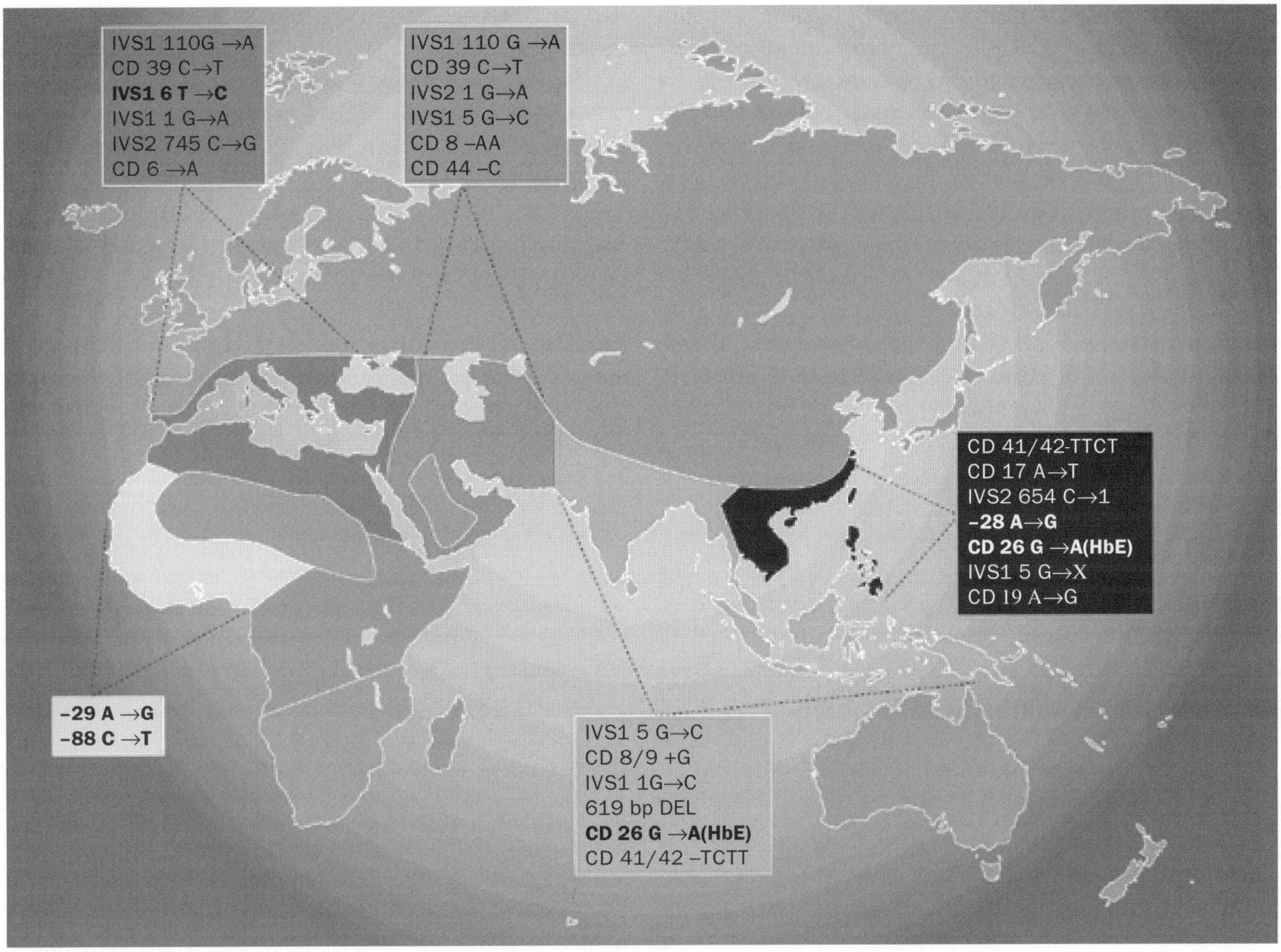

Figure 36-4 Global distribution of β-thalassemia mutations (Weatherall, 2001). Refer color figure.
source: While β-thalassemia is not endemic in southern Africa, patients with β-thalassemia are managed in the hematology clinic. They originate predominantly from two areas, the Mediterranean (Greece and Cyprus) and the Indian subcontinent. The local mutation profiles differ somewhat from those of the parent population to local founder effects. For example, in the local Asian Indian population, the codon 8–9 mutation has never been observed, while the IVS1nt5 mutation predominates (Krause, 1994; see Table 36-1). Rare β-thalassemia patients from other population groups are also seen.

high-frequency population in the world seems to carry a few common mutations that are unique to a particular region, together with varying numbers of rare ones (Figure 36-4).

The carrier frequency for β-thalassemia in Africa ranges from 1% to 10%, rarely greater (Weatherall & Clegg, 2001; Brabin et al., 2004). In the South African Asian Indian population, carrier rates range from 1% to 2%, but with high rates of consanguinity in some groups, thus increasing the number of homozygotes (Krause, 1994).

The homozygous or compound heterozygous states for β thalassemia also run a variable course, although in a majority of cases, without transfusion, death occurs in the first few years of life (Weatherall & Clegg, 2001). With adequate transfusion and the administration of the chelating agent desferrioxamine, children may grow and develop well and survive into adult life (Brittenham et al., 1994; Olivieri et al., 1994; Cao & Galanello, 2002). The situation is further complicated by the fact the common β thalassemias of intermediate severity, the most prevalent in North Africa, exhibit a very wide clinical spectrum ranging from a transfusion-dependent disease to a condition that is compatible with normal survival and growth into adult life without treatment (Haj Khelil et al., 2010). Thus, even though β-thalassemia is a monogenic disease there is a large amount of phenotypic heterogeneity. This heterogeneity cannot be explained entirely by the different mutations which result in β-thalassemia (Chen et al., 2010; Sharma et al., 2010). It has been shown that much of the variation can be attributed to other genetic loci, such as those which cause an increase in fetal hemoglobin (HbF) levels in adults (Uda et al., 2008; Sedgewick et al., 2008) and those which play a role in α-hemoglobin precipitation in red blood cells.

Because there is so much phenotypic heterogeneity between β-thalassemia patients, there has been a great deal of investigation to try and understand the molecular and the genetic basis of the heterogeneity. This is required in order to understand the relationship between genotype and phenotype, and find new ways of treating the disease. As with SCD, fetal hemoglobin is one of the major modifiers of β-thalassemia. This is due to the fact that γ-globin chains can substitute for β-globin chains and dimerize with α-globin to form fetal hemoglobin (HbF). The same 3 loci that have been shown to affect hemoglobin F production in SCD are implicated in HbF production in β-thalassemia. It was shown that not only individuals with β-thalassemia (Thein & Craig, 1998) had elevated levels of HbF, but other family members as well, indicating that that there are other genetic factors accounting for increased HbF (Thein & Craig, 1998). The coinheritance of α-thalassemia, in which there is a mutation that results in decreased α-hemoglobin synthesis, results in a milder phenotype (Galanello et al., 2009). This is because there is a balance in the synthesis of hemoglobin chains and there is a decrease in the excess of α-hemoglobin chains. Alternatively, when there is inheritance of an extra copy of the α-hemoglobin (ααα/αα) gene, this results in a more severe phenotype (Ho and Luo, 1998).

α-THALASSEMIAS

The milder forms of thalassemia (single α-gene deletions) are the most prevalent in Africa, (termed α + thalassemias), since two functional α-genes remain. -α(3.7)/αα and -α(3.7)/-α(3.7), though extremely common, only produce mild hypochromic anemia in their homozygous states (Weatherall, 2001). Carrier rate of α-thalassemia is much higher, ranging from 10% to 20% in parts of Africa (Table 36-1).

Chromosomes with double delation (--/αα) are rare in Africa, and thus HbH disease and Hb Barts hydrops fetalis are rarely seen in endemic African populations. They are, however, seen in immigrant populations, such as individuals with Southeast Asian or Mediterranean ancestry.

Most low-income and middle-income countries have undergone gradual improvements in key national

TABLE 36-1 SOME REPRESENTATIVE CARRIER FREQUENCIES THE MAIN HEMOGLOBIN DISORDERS IN AFRICA*

Hemoglobin S		β Thalassemia		α + Thalassemia	
Country	Carrier Frequency (%)	Country	Carrier Frequency (%)	Country	Carrier Frequency (%)
Nigeria	6–24	Algeria	3	Nigeria	10–50
The Gambia	6–24	Libya	7	Kenya	40–50
Senegal	1–23	Egypt	2–7	Zambia	40
Guinea	2–28	Sudan	5	Senegal	20
Liberia	2–15	Burkina Faso	2.6	Benin	40
Côte d'Ivoire	3–22	Mali	3	Algeria	10
Mali	7–29	Liberia	1–12		
Ghana	3–22	Nigeria	1		
Benin	7–29				
Niger	5–23				
Cameroon	8–33				
Central Africa Republic	1–21				
Gabon	8				
Congo (RDC)	2–20				
Angola	4–24				
Zambia	6–27				
Uganda	4–30				
Tanzania	10–38				
Kenya	2–32				
Sierra Leone	16–30				

* Sources: Weatherall & Clegg (2001).

sectors over 40 years, including economic and health sectors. These improvements have been most notable in countries of Latin America, East Asia, Middle Asia, and North Africa, and underpin a gradual, consistent shift in the major causes of population mortality and morbidity. This health transition is marked by the decreasing occurrence of infectious diseases and malnutrition and increasing population life expectancy. Unfortunately, rates of birth defects and their associated developmental disabilities have not decreased in the same populations within the same time period. In Africa, noncommunicable diseases including hemoglobinopathies are emerging in both rural and urban areas, and are resulting in increased pressure on acute and chronic healthcare services. Major factors include demographic change, leading to a rise in the proportion of people older than 60 years, despite the negative effect of HIV/AIDS on life expectancy (Mayosi et al., 2009). The scale of the challenge posed by the combined and growing burden of HIV/AIDS and noncommunicable diseases including hemoglobinopathies demands an extraordinary response that few African countries are able to provide. Concerted action is needed to strengthen the district-based primary healthcare system, to integrate the care of birth defects and management of risk factors, to develop national surveillance systems, and to apply interventions of proven cost-effectiveness in the primary and secondary prevention of hemoglobinopathies within populations and health services.

POPULATION GENETICS AND DYNAMICS: MALARIA, MIGRATION, CONSANGUINITY

Compared to noncarriers, healthy carriers of recessive genes for SCD and thalassemias have a well-documented survival advantage against the lethal effects of malaria. In cross-sectional malariological surveys of 3,739 human subjects, and transmission experiments involving 60 children and 6,446 mosquitoes in Burkina Faso, West Africa, protective hemoglobins C (HbC, beta6Glu—>Lys) and S (beta6Glu—>Val) were associated with a twofold in vivo (odds ratio 2.17, 95% CI 1.57–3.01, $P = 1.0 \times 10^{-6}$) and a fourfold ex vivo (odds ratio 4.12, 95% CI 1.90–9.29, $P = 7.0 \times 10^{-5}$) increase of parasite transmission from the human host to the Anopheles vector (Gouagna et al., 2010). As a result, carriers are more likely to reach reproductive age. Consequently, the birth prevalence of SCD and thalassemia, is high in Africa (Enevold et al., 2008; Vento et al., 2006; Williams et al., 2005; Aidoo et al., 2002). The resurgence of malaria in many parts of the world will serve to maintain these polymorphisms, but even if this selective force were removed it would take many generations for the gene frequencies of these conditions to fall significantly (Weatherall & Clegg, 2001). Any changes resulting from variation in selection or population dynamics will, however, be very small compared with the effect of the demographic transition that many countries have undergone over recent years (Weatherall & Clegg, 2001).

Specifically, there is a high prevalence of hemoglobin S (HbS) in Africa and hemoglobin C (HbC) in parts of West Africa (Modiano et al., 2008). Since subjects that are homozygous for HbC do not feature any severe disease like those that are homozygous for HbS, it is anticipated that the frequency of HbC will progressively increase over that of HbS in the population, if malaria is not controlled. Indeed, in Ouagadougou (Burkina Faso) the gene frequency in the city schools was 0.111 for the betaC gene and 0.051 for the betaS gene; in five villages near Ouagadougou, it was 0.122 for the betaC gene and 0.047 for the betaS gene (Simpore et al., 2002).

β-thalassemia is spread throughout the Mediterranean basin, including North Africa (Tunisia, Algeria, and Morocco) where the population prevalence ranges between 3%–10% (Table 36-1). In Tunisia, Algeria, and Morocco, more than 45 mutations have been identified on the β-globin gene (Godet et al., 1977; Chouk et al., 2004; Haj Khelil et al., 2010). The most common in Tunisia and in Algeria are codon 39 (C>T) and IVS-I-110 (G>A), which together account for more than 50% of all mutations (Figure 36-4). In Morocco, the predominant mutations are codon 39 and frameshift codon (FSC) 8 (-AA) (Haj Khelil et al., 2010). Various population migrations also contributed to β-thalassemia epidemiology in North Africa, resulting in a great amount of heterogeneity in the mutations that cause β-thalassemia—an example being in Tunisia, where the Romans introduced the codon 39(C→T) mutation. This mutation is found throughout the Mediterranean basin with the highest frequency found in Sardinia (Chouk et al., 2004). The haplotypes of the β^S, codon 39 and IVS-I-110 origins allowed the hypothesis of a Benin origin for β^S, a local North African origin for codon 39, and an eastern Mediterranean origin for IVS-I-110 (Haj Khelil et al., 2010).

Internal migration in Africa has led to SCD, which was previously rare, being introduced in South Africa through an influx of migrants from west and central Africa (Beighton & Botha, 1986). β-thalassemia is also present within three South African populations as a result of migrations. It has been documented in the Greek (Bonafede et al., 1979), mixed ancestry (Bird et al., 1987, 1982), and Indian populations in South Africa (Poole et al., 1989). Thus, although this might be a disease that was originally thought to affect the east, it has become a world health problem due to migration and the fact that developing countries are now realizing the cost of the disease and its treatment (Ho et al., 2006). Therefore, it is important to understand the disease and factors that contribute to its heterogeneity in order to offer better disease treatment options.

The highest rates of consanguineous marriage occur in North Africa, where 25%–80% of unions involve related family members (Abdel-Meguid et al., 2000; Ben Arab et al., 2004). The practice is also accepted in parts of sub-Saharan Africa, and even couples who regard themselves as unrelated may exhibit high levels of homozygosity because marriage within clan, tribe, caste, or *biradari* boundaries has been a long-established tradition (Bittles & Black, 2010; Hampshire & Smith, 2001). Parental consanguinity increases the birth prevalence of autosomal recessive birth defects, but little data have been reported on the effects of consanguinity on the frequency of hemoglobinopathies in Africa.

FREQUENCY AND HEALTH BURDEN OF HEMOGLOBIN DISORDERS IN AFRICA

The sickle cell gene is distributed widely throughout sub-Saharan Africa, where carrier frequencies range from 5% to 38% or more of the population (Table 36-1). Thalassemia has a high incidence in a broad band extending from the Mediterranean basin and parts of Africa (Table 36-1). The carrier frequency for β-thalassemia in these areas ranges from 1% to 7%, rarely greater, while that for the milder forms of α-thalassemia is much higher, ranging from 10% to 40% in parts of sub-Saharan Africa (Table 36-1). The α0 thalassemias are more restricted in their distribution, occurring at high frequencies only in parts of South-East Asia and the Mediterranean basin and therefore pose less of a global health problem that the β-thalassemias (Weatherall, 2001).

Globally, WHO's Southeast Asia Region, where the thalassemias and Hb E predominate, is the most severely affected; sub-Saharan Africa has the second highest burden, reflecting the very high frequency of the sickle cell gene. Further information about the different thalassemia mutations has been reported (Figure 36-4). These data give only a very approximate estimate of the problems for healthcare services that will be posed by the hemoglobin disorders in the future. Unfortunately, few of them are based on micromapping of their frequency in different populations.

Given these limitations, the data shown in Table 36-1 represent only the basis for a minimal estimate of the likely health burden that will be posed by the hemoglobin disorders in the present and future health system in Africa. Furthermore, they need adaptation related to projected increases in birth rates for particular countries. Because of these uncertainties about the frequencies of the hemoglobin disorders, and the time it will take for many countries to pass through the demographic transition such that babies with these conditions will present for treatment, it is even more difficult to assess the burden that such disorders will impose on health provision in Africa.

It should also be remembered that, as more of these babies survive and present for treatment, the population of patients on long-term therapy will steadily expand; the more effective the treatment, the greater the burden on healthcare services.

Since these patients require red cells, blood has to be processed and also screened for hepatitis B and C viruses, human immunodeficiency virus (HIV), and, in many countries, malarial parasites. Because some children involved die of iron overload if iron derived from transfusion is not removed, and because current iron chelation regimens are extremely expensive, basic treatment of these diseases is very costly. The cost of diagnostic technology, medical staff, hospital admissions for complications, and other items that are difficult to define, all add to this burden.

Although the management of SCD is less expensive, over the lifetime of a patient it is also very costly because of the propensity to infection and high frequency of vascular occlusive complications. There have been very few attempts to quantify the economics of managing the hemoglobin disorders. WHO has made a very approximate estimate (WHO, 2006), although because of variation in costs of healthcare between countries the data offer only the broadest guidelines. It has been estimated that SCD results in the annual loss of several millions of disability-adjusted life years, particularly in the developing world (Weatherall, 2006). Hemoglobinopathies alone represent a health burden comparable to that of communicable and other major diseases (Weatherall, 2008).

In a study in Sri Lanka, where the frequency of β-thalassemia and HbE is relatively low compared with many other populations of the Indian subcontinent and Southeast Asia, it was estimated that the future management of thalassemia would consume 5%–10% of the island's current expenditure on health (De Silva et al., 2000).

CONTROL AND MANAGEMENT OF HEMOGLOBIN DISORDERS

Countries can be divided into three general categories in terms of the patterns of control and management of the hemoglobin disorders (Angastiniotis & Modell, 1998). First, in some Mediterranean countries, control programs consisting of screening and antenatal diagnosis have succeeded in reducing the frequency of new births of β-thalassemia by between 80% and close to 100% (Cao & Galanello, 2002). Specialized clinics also provide optimum management of established cases. Second, in some of the richer industrialized countries, where the prevalence is increasing because of immigration of patients from high-frequency regions, facilities for control and management have been developed but there are still major problems in reaching immigrant groups with different cultural backgrounds.

Finally, in many of the countries of the developing world including Africa that have undergone a demographic transition, provision of service is still hampered by major economic and organizational difficulties, and many affected infants and children are dying with undiagnosed, untreated, or undertreated disease. The mortality rate associated with SCD has remained high despite the use of appropriate interventions to manage the various forms of crisis. The organisms causing bacteremia in African children with sickle cell anemia are the same as those in developed countries: Streptococcus pneumoniae (44/108 isolates; 41%), non-typhi Salmonella species (19/108; 18%), Haemophilus influenzae type b (13/108; 12%), Acinetobacter species (7/108; 7%), and *Escherichia coli* (7/108; 7%) (Williams et al., 2009; Serjeant, 2005). Introduction of conjugate vaccines against *S pneumoniae* and *H influenzae* into the childhood immunization schedules of African countries could substantially affect survival of children with sickle cell anemia. Methods for screening SCD are now well established and inexpensive, and it is vital that they are applied to define the magnitude of the problem that will be posed by these conditions in the future. Effective enrollment of affected neonates in a comprehensive follow-up program adapted to the socioeconomic conditions, health structures, and cultural background of African setting is essential (De Montalembert & Tsilolo, 2007). In developed countries, newborn screening has resulted in a reduction in the SCD mortality rate from 16% to <1% (Streetly et al., 2009, Komba et al., 2009). Similarly, in the Republic of Benin, following neonatal screening, counseling, and follow-up, the under-five mortality rate was 15.5 per 10,000, a figure that is 10 times lower than the general rate recorded (Rahimy et al., 2009). It is clear, however, that many patients with sickle cell anemia require the expertise of specialized centers and lifelong medical care and surveillance.

The life expectancy of patients with SCD has improved in the United States and Europe thanks to the use of penicillin prophylaxis, appropriate immunizations, neonatal screening, implementation of a quality transfusion policy, hydroxyurea therapy, detection and treatment of cerebral vasculopathy, recognition of situations that can benefit from allogenic marrow transplantation, and improvements in bone marrow transplantation techniques. The cost of almost all these techniques is far beyond the means of healthcare systems in Africa, where they can be hardly used (De Montalembert & Tsilolo, 2007). Hematopoietic stem cells transplantation (SCT) is now an established treatment modality with definitive indications for many hematological disorders. However, this line of treatment requires tremendous resources, and it becomes increasingly difficult for transplanters practicing in the developing world to reconcile the difference between what is possible and what is available. On the basis of 18 years of experience and more than 1300 transplants performed in Egypt, this could inspire approaches to addressing special issues that are important for hematopoietic SCT practices in developing countries (Mahmoud et al., 2008). A great deal can be done to help the situation by developing further North/South and South/South partnerships for disseminating better practice (Weatherall, 2008).

Interestingly, up to 95% of Cameroonian parents with a SCD-affected child indicated they would request prenatal genetic diagnosis (PND) for SCD, and 65% would opt for abortion of an affected SCD pregnancy. The effective initiation of PND for SCD (amniocentesis and fetal DNA analysis) has laid the foundation for the first medical genetic service in Cameroon, and demonstrates that the practice of medical genetics is possible in a sub-Saharan African country (Wonkam et al., 2010). The use of SCD prenatal diagnosis as point of entry of genetic medicine was also reported in Nigeria (Akinyanju et al., 1999). There is a need for widespread screening and counseling programs directed at developing PND expertise where this is not available (Weatherall, 2005).

FUTURE CHALLENGES

Because of their uneven distribution in high-frequency populations, reflecting their complex population genetics, the true magnitude of burden of hemoglobinopathy alone is still unknown. In many African countries there are few or virtually no facilities for the diagnosis and management of genetic and partially genetic birth defects, and there are limited data about their frequency, clinical course, or mortality. Without this information it will be impossible to persuade governments of the increasing importance of these diseases.

In addition, increased HbF is a major determinant of the phenotype and can contribute significantly to SCD and β-thalassemia. It is expected that the identification of genetic polymorphisms that modulate HbF levels will shed light on the molecular mechanisms that control HbF expression, and on the etiology of the clinical heterogeneity observed in these hemoglobinopathies. Data from the African continent is urgently needed. Such insights might ultimately identify novel drug targets for new treatments of SCD and β-thalassemia. An immediate application would be an improved prediction of one's ability to produce HbF, which, in turn, may improve prediction of disease severity.

There are several critical pieces of information that are urgently required as a background to developing control and management programs. First, so that governments can be given more accurate and realistic data about the likely burden of the disease, micromapping studies need to be carried out to obtain better data about its frequency. Screening for the important forms of thalassemia and sickle cell disease is accurate and cheap. Second, more needs to be known about the natural history, mechanisms for clinical variability, and most appropriate ways to control SCD and thalassemias in Africa.

Screening and antenatal diagnosis programs have been applied widely in Mediterranean populations. Their success has been based on the excellence of public education programs about the disease, followed by the development of effective screening regimens and facilities for antenatal detection. Programs of this type need to be established in Africa. Fortunately, DNA diagnostics are becoming cheaper and more efficient and it should be possible to provide services of this type on a centralized basis, to form the background for both screening and diagnostic services for large communities. Similarly, efforts to provide more accurate methods of cost control and treatment of these diseases are required.

The situation will only be improved by concerted action on the part of the international community of the richer countries, together with input from the major international health organizations and funding agencies (Weatherall, 2010).

REFERENCES

Abdel-Meguid N, Zaki MS, Hammad SA. (2000). Premarital genetic investigations: effect of genetic counselling. *East Mediterr Health J. 6*(4), 652–60.

Advani R, Sorenson S, Shinar E, Lande W, Rachmilewitz E, Schrier S. (1992). Characterization and comparison of the red blood cell membrane damage in severe human alpha-and beta-thalassemia. *Blood 79*, 1058.

Aidoo M, Terlouw DJ, Kolczak MS, et al. (2002). Protective effects of the sickle cell gene against malaria morbidity and mortality. *Lancet 359*(9314), 1311–1312.

Akinyanju OO, Disu RF, Akinde JA, et al. (1999). Initiation of prenatal diagnosis of sickle cell anaemia in Africa. *Prenat Diagn 19*(4), 299–304.

Aliyu ZY, Kato GJ, Taylor J, et al. (2008). Sickle cell disease and pulmonary hypertension in Africa: a global perspective and review of epidemiology, pathophysiology, and management. *Am J Hematol* 83, 63–70

Angastiniotis M, Modell B. (1998). Global epidemiology of hemoglobin disorders. *Ann N Y Acad Sci 850*, 251–269.

Antonarakis SE, Boehm CD, Serjeant GR, Theisen CE, Dover GJ, Kazazian HH Jr. (1984). Origin of the beta S-globin gene in blacks: the contribution of recurrent mutation or gene conversion or both. *Proc Natl Acad Sci U S A* 81(3), 853–856.

Bakanay SM, Dainer E, Clair B, et al. (2005). Mortality in sickle cell patients on hydroxyurea therapy. *Blood* 105, 545–547.

Beighton P, Botha MC. (1986). Inherited disorders in the black population of southern Africa. Part I. Historical and demographic background; genetic haematological conditions. *S Afr Med J 69*(4), 247–249.

Ben Arab S, Masmoudi S, Beltaief N, Hachicha S, Ayadi H. (2004). Consanguinity and endogamy in Northern Tunisia and its impact on non-syndromic deafness. *Genet Epidemiol 27*(1), 74–79.

Bender MA, Hobbs W (2009). Sickle Cell Disease, *GeneReviews*, http://www.ncbi.nlm.nih.gov/books/NBK1377/

Bird AR, Ellis P, Wood K, Mathew C, Karabus C. (1987). Inherited hemoglobin variants in a South African population. *J Med Genet 24*, 215.

Bird AR, Karabus CD, Hartley PS. (1982). Mycrocytic anaemia and hemoglobinopathy in Cape Town Children. *S Afr Med J 62*, 429.

Bittles AH, Black ML. (2010). Evolution in health and medicine Sackler colloquium: Consanguinity, human evolution, and complex diseases. *Proc Natl Acad Sci U S A 107*(Suppl 1), 1779–86.

Bonafede RP, Botha MC, Beighton P. (1979). Inherited anaemias in the Greek community of Cape Town. *Br Med J 16*, 197.

Brittenham GM, et al. (1994). Efficacy of deferoxamine in preventing complications of iron overload in patients with thalassemia major. *N Eng J Med 331*, 567–573.

Bunn HF. (1997). Pathogenesis and treatment of sickle cell disease. *N Eng J Med 337*, 762–769.

Cao A, Galanello R. (2002). Effect of consanguinity on screening for thalassemia. *N Engl J Med 347*(15), 1200–1202.

Castro O, et al. (1994). The acute chest syndrome in sickle cell disease: Incidence and risk factors. The Cooperative Study of Sickle Cell Disease. *Blood 84*, 643–649.

Chen W, Zhang X, Shang X, et al. (2010). The molecular basis of beta-thalassemia intermedia in southern China: genotypic heterogeneity and phenotypic diversity. *BMC Medical Genetics11*, 31.

Chouk I, Daoud BB, Mellouli F, et al. (2004). Contribution to the description of the β-thalassemia spectrum in Tunisia and the origin of mutation diversity. *Hemoglobin 28*, 189–195.

Clegg JB, Weatherall DJ. (1999). Thalassemia and malaria: new insights into an old problem. *Proc Assoc Am Physicians 111*(4), 278–282.

Creary LE, Ulug P, Menzel S, et al. (2009). Genetic variation on chromosome 6 influences F cell levels in healthy individuals of African descent and HbF levels in sickle cell patients. *PLoS ONE 4*, e4218.

De Montalembert M, Tshilolo L. (2007). Is therapeutic progress in the management of sickle cell disease applicable in sub-Saharan Africa? *Med Trop (Mars) 67*(6), 612–616.

De Silva S, et al. (2000). Thalassaemia in Sri Lanka: implications for the future health burden of Asian populations. *Lancet 355*, 786–791.

Enevold A, Lusingu JP, Mmbando B, et al. (2008). Reduced risk of uncomplicated malaria episodes in children with alpha+-thalassemia in northeastern Tanzania. *Am J Trop Med Hyg 78*(5), 714–720.

Fleming AF. (1989). The presentation, management and prevention of crisis in sickle cell disease in Africa. *Blood Rev 13*(1), 18–28.

Galanello R, Sanna S, Perseu L, et al. (2009). Amelioration of Sardinian {beta} 0 thalassemia by genetic modifiers. *Blood 114*, 3935.

Garner CP, Tatu T, Best S, Creary L, Thein SL. (2002). Evidence of Genetic Interaction between the [beta]-Globin Complex and Chromosome 8q in the Expression of Fetal Hemoglobin. *Am J Hum Genet 70*, 793–799.

Gilman J and Huisman T. (1985). DNA sequence variation associated with elevated fetal G gamma globin production. *Blood 66*, 783.

Gladwin MT, Vichinsky E. (2008). Pulmonary complications of sickle cell disease. *N Engl J Med 359*(21), 2254–2265.

Godet J, Verdier G, Nigon V, et al. (1977). beta-O-thalassemia from Algeria: genetic and molecular characterization. *Blood 50*, 463.

Gouagna LC, Bancone G, Yao F, et al. (2010). Genetic variation in human HBB is associated with Plasmodium falciparum transmission. *Nat Genet 42*(4), 328–331.

Haj Khelil A, Denden S, Leban N, et al. (2010). Hemoglobinopathies in North Africa: a review. *Hemoglobin 34*(1), 1–23.

Hampshire KR, Smith MT. (2001). Consanguineous marriage among the Fulani. *Hum Biol 73*(4), 597–603.

Ho H and Luo W. (1998). Beta-thalassaemia intermedia: is it possible consistently to predict phenotype from genotype? *Br J Haematol 100*, 70–78.

Ho WL, Lin KH, Wang JD, et al. (2006). Financial burden of national health insurance for treating patients with transfusion-dependent thalassemia in Taiwan. *Bone Marrow Transplant 37*, 569–574.

Kazazian Jr H, Boehm C. (1988). Molecular basis and prenatal diagnosis of beta-thalassemia. *Blood 72*, 1107.

Komba AN, Makani J, Sadarangani M, et al. (2009). Malaria as a cause of morbidity and mortality in children with homozygous sickle cell disease on the coast of Kenya. *Clin Infect Dis 49*(2), 216–222.

Labie D, Pagnier J, Lapoumeroulie C, et al. (1985). Common haplotype dependency of high G gamma-globin gene expression and high Hb F levels in beta-thalassemia and sickle cell anemia patients. *PNAS 82*, 2111–2114.

Lettre G, Sankaran VG, et al. (2008). DNA polymorphisms at the BCL11A, HBS1L-MYB, and beta-globin loci associate with fetal hemoglobin levels and pain crises in sickle cell disease. *PNAS* 105(33), 11869–11874.

Mahmoud H, El-Haddad A, Fahmy O, et al. (2008). Hematopoietic stem cell transplantation in Egypt. *Bone Marrow Transplant 42*(Suppl 1), S76-S80.

Makani J, Williams TN, Marsh K. (2007). Sickle cell disease in Africa: burden and research priorities. *Ann Trop Med Parasitol 101*(1), 3–14.

Mayosi BM, Flisher AJ, Lalloo UG, Sitas F, Tollman SM, Bradshaw D. (2009). The burden of non-communicable diseases in South Africa. *Lancet 374*(9693), 934–947. Review.

Menzel S, Thein SL. (2009). Genetic architecture of hemoglobin F control. *Curr Opin Hematol. 16*, 179–186.

Modiano D, Bancone G, Ciminelli BM, et al. (2008). Hemoglobin S and hemoglobin C: 'quick but costly' versus 'slow but gratis' genetic adaptations to Plasmodium falciparum malaria. *Hum Mol Genet 17*(6), 789–799.

Nichols FT, Jones AM, Adams RJ. (2001). Stroke prevention in sickle cell disease (STOP) study guidelines for transcranial Doppler testing. *J Neuroimaging 11*(4), 354–362.

Njamnshi AK, Wonkam A, Djientcheu V. de P, et al. (2006a). Stroke may appear to be rare in Saudi-Arabian and Nigerian children with sickle cell disease, but not in Cameroonian sickle cell patients. *Br J Haematol 133*(2), 210.

Njamnshi K, Mbong EN, Wonkam A, et al. (2006b). The epidemiology of stroke in sickle cell anemia patients in Yaoundé-Cameroon. *J Neurosci 250*, 79–84.

Njamnshi AK, Wonkam A, Ongolo-Zogo P, et al. (2009). The risk of stroke in sickle cell subjects in Cameroon: the first report *J Neurol Sci 285*, S161-S161.

Nzewi E. (2001). Malevolent ogbanje: recurrent reincarnation or sickle cell disease? *Soc Sci Med 52*(9), 1403–1416.

Obama MT, Dongmo L, Nkemayim C, Mbede J, Hagbe P. (1994). Stroke in children in Yaounde, Cameroon. *Indian Pediatr 31*(7), 791–795.

Olivieri NF, et al. (1994). Survival of medically treated patients with homozygous ß thalassemia. *N Engl J Med 331*, 574–578.

Onwubalili JK. (1983). Sickle cell anaemia: an explanation for the ancient myth of reincarnation in Nigeria. *Lancet 2*(8348), 503–505.

Platt OS, et al. (1994). Mortality in sickle cell disease. Life expectancy and risk factors for early death. *N Engl J Med 330*, 1639–1644.

Platt OS, et al. (1991). Pain in sickle cell disease. Rates and risk factors. *N Engl J Med 325*, 11–16.

Platt OS. (2008). Hydroxyurea for the treatment of sickle cell anemia. *N Engl J Med 358*, 1362–1369.

Poole JE, Chon RJ, Roode H, Spector I. (1989). Beta-thalasssemia—the Johannesburg experience. *S Afr Med J 75*, 367–370.

Rahimy MC, Gangbo A, Ahouignan G, Alihonou E. (2009). Newborn screening for sickle cell disease in the Republic of Benin. *J Clin Pathol 62*(1), 46–48.

Rund D, Rachmilewitz E. (2005). Beta-thalassemia. *N Engl J Med 353*, 1135.

Schrier SL, Rachmilewitz E, Mohandas N. (1989). Cellular and membrane properties of alpha and beta thalassemic erythrocytes are different: implication for differences in clinical manifestations. *Blood 74*, 2194.

Sedgewick AE, Timofeev N, Sebastiani P, et al. (2008). BCL11A is a major HbF quantitative trait locus in three different populations with β-hemoglobinopathies. *Blood Cells Mol Dis 41*, 255–258.

Serjeant GR. (2005). Mortality from sickle cell disease in Africa. *BMJ 330*(7489), 432–433.

Serjeant GR. (1992). *Sickle Cell Disease*. New York: Oxford University Press.

Serjeant GR. (2001). The emerging understanding of sickle cell disease. *Br J Haematol 112*(1), 3–18.

Serjeant GR, Higgs DR, Hambleton IR. (2007). Elderly survivors with homozygous sickle cell disease. *N Engl J Med* 356, 642–643.

Sharma N, Das R, Kaur J, et al. (2010). Evaluation of the genetic basis of phenotypic heterogeneity in north Indian patients with Thalassemia major. *Eur J Haematol 84*, 531–537.

Simpore J, Pignatelli S, Barlati S, Musumeci S. (2002). Modification in the frequency of Hb C and Hb S in Burkina Faso: an influence of migratory fluxes and improvement of patient health care. *Hemoglobin 26*(2), 113–120.

Stamatoyannopoulos G, Nienhuis AW. (1994). Hemoglobin switching. In: Stamatoyannopoulos G, Nienhuis AW, Majerus PW, Varmus H, eds. *The Molecular Basis of Blood Diseases*, 2nd ed., pp. 107–155. Philadelphia: W.B. Saunders.

Steinberg MH. (1998). Pathophysiology of sickle cell disease. *Clin Haematol 11*, 163–184.

Steinberg MH, Barton F, Castro O, et al (2003). Effect of hydroxyurea on mortality and morbidity in adult sickle cell anemia: risks and benefits up to 9 years of treatment. *JAMA 289*, 1645–1651.

Streetly A, Latinovic R, Hall K, Henthorn J. (2009). Implementation of universal newborn bloodspot screening for sickle cell disease and other clinically significant hemoglobinopathies in England: screening results for 2005–7. *J Clin Pathol 62*(1), 26–30.

Thein S, Craig I. (1998). Genetics of Hb F/F cell variance in adults and heterocellular hereditary persistence of fetal hemoglobin. *Hemoglobin 22*, 401–414.

Uda M, et al. (2008). Genome-wide association study shows BCL11A associated with persistent fetal hemoglobin and amelioration of the phenotype of beta-thalassemia. *PNAS 105*, 1620–1625.

Vento S, Cainelli F, Cesario F. (2006). Infections and thalassaemia. *Lancet Infect Dis 6*(4), 226–233. Review.

Weatherall DJ, Clegg JB. (2001). Inherited hemoglobin disorders: an increasing global health problem. *Bull World Health Organ 79*(8), 704–712.

Weatherall DJ, Clegg JB. (2001). *The Thalassaemia Syndromes*. Oxford: Blackwell Science.

Weatherall DJ. (2001). Phenotype-genotype relationships in monogenic disease: lessons from the thalassaemias. *Nat Rev Genet 2*(4), 245–255.

Weatherall DJ. (2006). Inherited Disorders of Hemoglobin. In: Jamison DT et al. (eds.) *Disease control priorities in developing countries*, 2nd ed., pp. 663–680. Oxford University Press: New York.

Weatherall DJ. (2008). Hemoglobinopathies worldwide: present and future. *Curr Mol Med 8*(7), 592–599.

Weatherall DJ. (2010). The inherited diseases of hemoglobin are an emerging global health burden. *Blood 115*(22), 4331–4336.

Williams TN, Mwangi TW, Wambua S, et al. (2005). Negative epistasis between the malaria-protective effects of alpha+-thalassemia and the sickle cell trait. *Nat Genet 37*(11), 1253–1257.

Williams TN, Uyoga S, Macharia A, et al. (2009). Bacteraemia in Kenyan children with sickle cell anaemia: a retrospective cohort and case-control study. *Lancet 374*(9698), 1364–1370.

Wierenga KJ, Hambleton IR, Lewis NA. (2001). Survival estimates for patients with homozygous sickle-cell disease in Jamaica: a clinic-based population study. *Lancet* 357, 680–683.

Wonkam A, Muna W, Ramesar R, Rotimi CN, Newport MJ. (2010). Capacity-building in human genetics for developing countries: initiatives and perspectives in sub-Saharan Africa. *Public Health Genomics. 13*(7–8), 492–494.

Wonkam A. (2009). Phenotype, burden and perspectives of sickle cell anemia studies in sub-Saharan Africa: the case of Cameroon. *Proceedings, ICBD*, New Delhi, India, Ocober 4, 2009.

World Health Organization (WHO). (1996). Control of Hereditary Diseases. WHO Technical Report Series 865. WHO, Geneva, Switzerland.

World Health Organization (WHO). (2006). *Management of Birth Defects and Hemoglobin Disorders*. Report of a Joint WHO–March of Dimes Meeting, 17–19 May, 2006.

World Health Organization (WHO). (2008). Management of Hemoglobin Disorders: Report of Joint WHO-TIF Meeting Nicosia, Cyprus, 16–18 November 2007 Geneva, World Health Organization, 2008. *www.who.int/genomics/WHO-TIF_genetics_final.pdf.*

37.

THE BURDEN OF BIRTH DEFECTS AND GENETIC DISEASE IN SUB-SAHARAN AFRICA

Ambroise Wonkam

INTRODUCTION

A birth defect is defined as any abnormality affecting body structures or function that is present from birth, or may be diagnosed only later in life. Causes can be divided broadly into two groups: genetic and partially genetic causes originating mostly before conception, and causes developing after conception but before birth. Most birth defects originate before conception. In an industrialized country genetic abnormalities account for around 7% of birth defects including numerical or structural chromosomal abnormalities, e.g. Down syndrome or single genes defects, e.g., sickle cell anemia. However, in most cases birth defects are due to a combination of genes that puts the fetus at risk in the presence of specific environmental factors, and account for an estimated 20%–30% of all birth defects, a number of which are often lethal, e.g., congenital heart disease (Turnpenny & Ellard, 2007). Causes of birth defects originating after conception are primarily nongenetic. These include teratogens that interfere with normal growth and development of the embryo or fetus (maternal illness, maternal infections, drugs, radiation, and environmental pollutants), mechanical forces that deform the fetus, and vascular accidents that disrupt the normal growth of organs. This category accounts for an estimated 5%–10% of all birth defects (Turnpenny & Ellard, 2007).

Some of the birth defects with unknown cause (50%) may be due to new mutations, submicroscopic chromosome deletions, or epigenetic modification. Advances of genomic technology, especially affordable individual whole-genome sequencing, analysis of copy-number variants, methylation, and genome-wide association studies will allow causes of birth defects to be continue identified (Lupski et al., 2010, Vissers et al., 2010; Stefansson et al., 2008; Mangold et al., 2009, Butler, 2009)

THE TOLL AND MISUNDERSTANDING

According to the World Health Organization (WHO), 7.6 million children are born yearly with severe genetic conditions or malformations that are predominantly genetic in origin, and 90% of these children are born in low- or middle-income countries (WHO, 2005). When disability is considered in addition to mortality, the global toll of birth defects reflects a particularly harsh reality. Many infants burdened with serious birth defects die early in life, particularly in middle- and low-income countries, and those who survive almost always have the potential to be disabled, either as a direct or a indirect result of their condition (Christianson & Modell, 2004). However, genetic conditions outside the industrialized world receive little attention due to the focus on communicable diseases such as HIV and malaria, where most resources are diverted.

Several misunderstandings underlie the reasons why care and prevention of birth defects have been accorded relatively low priority by international donors, health agencies, and national governments (Mach of Dimes Report, 2006). The first is that health policymakers are generally unaware of the global toll of birth defects and associated disabilities, because data documenting the extent of the problem have been lacking. A second misunderstanding is the widespread belief that effective care and prevention of birth defects require costly, high-technology interventions that are beyond the national health budgets of low- and middle-income countries. A third misunderstanding, which follows from the last, is that programmatic attention to care and prevention of birth defects will necessarily draw funding away from other high-priority interventions in maternal and child health.

THE CHALLENGES

There is an increasing trend in developing countries, where the demographic and socioeconomic transition imposes more constraints on dealing with the double burden of infectious and noninfectious diseases. It is predicted that, by 2020, noncommunicable diseases including genetic conditions will cause seven out of every ten deaths in developing countries (Bongani et al., 2009). Up to 85% of the world's six billion people live middle- and low-income countries (UNICEF, 2003), many of which are characterized by poor environments and inadequate healthcare systems. This is a tragedy, because 70% of birth defects could either be prevented or the children affected offered care that would be life-saving or significantly reduce disability (Christianson & Modell, 2004). To achieve the Millennium Development Goal for child survival (MDG-4), neonatal deaths need to be prevented (Martines et al., 2005). The data in the March of Dimes Global Report on Birth Defects show that such efforts must also recognize the global contribution of birth defects to infant and childhood mortality if they are to succeed (March of Dimes Report, 2006). In sub-Saharan Africa (SSA), where sickle cell disease (SCD) occurs at its highest frequency, with up to 224,200 new affected babies yearly, hemoglobinopathies alone represent a health burden comparable to that of communicable and other major diseases (Weatherall, 2008).

Care and preventative strategies must take into account the growing trend of risk factors correlated to birth defects in developing countries. This chapter is a brief review of recent literature dealing with birth defects in SSA. It gives a global view of the main diseases and their impact on populations living in SSA.

GLOBAL BURDEN OF BIRTH DEFECTS IN AFRICA

HEALTH TRANSITION IN AFRICA

Most low- and middle-income countries have undergone gradual improvements in key national sectors over 40 years, including economic and health sectors. These improvements have been most notable in countries of Latin America, East Asia, Middle Asia, and North Africa, and underpin a gradual, consistent shift in the major causes of population mortality and morbidity. This positive health transition is marked by the decreasing occurrence of infectious diseases and malnutrition, mirrored by declining infant and under-5 mortality rates, and increasing population life expectancy. Unfortunately, rates of birth defects and their associated developmental disabilities have not decreased in the same populations within the same time period. In fact, a growing body of data suggests that birth defects are a far more significant contributor to infant childhood mortality and disability in low- and middle-income countries than previously estimated (IOM, 2003).

In SSA, noncommunicable diseases are emerging in both rural and urban areas, most prominently among poor people living in urban settings, and are resulting in increasing pressure on acute and chronic healthcare services. Major factors include demographic change, leading to a rise in the proportion of people older than 60 years, despite the negative effect of HIV/AIDS on life expectancy (Mayosi et al., 2009). The scale of the challenge posed by the combined and growing burden of HIV/AIDS and noncommunicable diseases including genetic conditions demands an extraordinary response that few SSA countries are able to provide. Concerted action is needed to strengthen the district-based primary healthcare system, to integrate the care of birth defects and management of risk factors, to develop national surveillance systems, and to apply interventions of proven cost-effectiveness in the primary and secondary prevention of specific genetic conditions like sickle cell disease (SCD) within populations and health services.

BIRTH PREVALENCE, POPULATION PREVALENCE, DEATH, AND DISABILITY

Birth prevalence is the number of infants affected in one or a defined collective group of birth defects per 1000 live births. A few specific conditions, however, including the hemoglobin disorders, G6PD deficiency, oculocutaneous albinism, and neural tube defects have higher birth prevalence in SSA. The birth prevalence of birth defects is about 20% higher SSA than in high-income countries (March of Dimes Report, 2006).

The birth prevalence of postconception birth defects due to teratogens differs widely, but is generally higher in middle- and low-income countries. In SSA, there tend to be few, if any, regulations governing the use of some of these substances and health services are not directed toward identifying and controlling exposure (Penchaszadeh, 2002). An undetermined number, but undoubtedly in the hundreds of thousands, of infants are born annually with serious birth defects of teratogenic origin.

Population prevalence refers to the number of affected individuals per 1000 in a defined population. For birth defects, this number is usually far lower than birth prevalence, especially in middle- and low-income countries, because serious birth defects shorten life. In SSA, the difference between birth prevalence and population prevalence is so marked that studies may identify few or no living individuals with serious birth defects. In rural South Africa in the early and mid-1990s, for example, the birth prevalence of Down syndrome was reported to be 2.09 per 1000 live births and the population prevalence to be only 0.75 per 1000 children aged two to nine years (Christianson et al., 2002; Venter et al., 1995).

Similarly, after the first description of SCD in a Grenadian dental student in Chicago in 1910, it was not until the late 1940s that reports began to appear from Africa itself (Serjeant, 2005). The prevalence of SCD in the newborn population was 3% and 2% in Nigeria and Cameroon, respectively (Odunvbun et al., 2008, Weatherall, 2001) but the population prevalence is likely less. While median survival for patients with sickle cell anemia (HbSS) in North America, Europe, and the Caribbean is now estimated to be 45–55 years (Platt et al., 1994; Wierenga et al., 2001), Only 7% of Nigerians and 2% of Cameroonian compared with 46% of US adults with SCD were >35 years (Aliyu et al., 2008, Wonkam, 2009). Approximately two-thirds of the children (60.9%) were under 5 years of age in Gabon (Koko et al., 1998). Moreover 90% of children with SCD die before the diagnosis can be made (Williams et al., 2009).

As effective care becomes available, population prevalence rises gradually to approximate birth prevalence. In 1970s, 50% of sickle cell anemia patients died by the age of 5 years in Nigeria (Tattah & Ekere, 1975). The recognition of SCD, and near-total extinction of the people's belief in reincarnation as cause of SCD (Onwubalili, 1983), the provision of well-organized holistic care, including counseling, preventive health, and nutritional education, prompt treatment of illness and free supplies of vitamin supplements, malarial prophylactics and other necessary medication, significantly reduced illness and deaths and improved the quality of lives of people living with HbSS in Nigeria. The mortality rate fell from 20.6% in 1988 to 0.6% in 1995 ($P < 0.0001$); the number of hospital admissions fell from 350 (119%) in 1988 to 30 (4%) in 1995 ($P < 0.0001$); the number of patients transfused with blood fell from 260 (90%) in 1988 to 25 (2%) in 1995 ($P < 0.00001$) (Akinyanju et al., 2005).

By conservative estimates, some 3.3 million deaths each year are associated with birth defects. This estimate includes approximately 50% who die in low-income countries, including SSA. It is also estimated that without appropriate care another 3.2 million children born with a serious birth defect are disabled each year (Christianson & Modell, 2004).

DETERMINANTS INFLUENCING BIRTH DEFECTS IN AFRICA

Malaria

Compared to noncarriers, healthy carriers of recessive genes for sickle cell anemia, thalassemia, and G6PD deficiency have a well-documented survival advantage against the lethal effects of malaria. As a result, carriers are more likely to reach reproductive age. Consequently the birth prevalence of thalassemia, SCD, and G6PD deficiency is high in sub-Saharan Africa (Enevold et al., 2008; Föller et al., 2009; Vento et al., 2006; Williams et al., 2005; Aidoo et al., 2002).

Consanguineous Marriage

Currently, couples related as second cousins or closer ($F >$ or $= 0.0156$) and their progeny account for an estimated 10.4% of the global population (Bittles & Black, 2010). The highest rates of consanguineous marriage occur in North Africa, where 25%–80% of unions involve related family members (Abdel-Meguid et al., 2000; Ben Arab et al., 2004; Bittles et al., 1991). The practice is also accepted in parts of SSA, and even couples who regard themselves as unrelated may exhibit high levels of homozygosity because marriage within clan, tribe, caste, or biraderi boundaries has been a long-established tradition (Bittles & Black, 2010; Saka et al., 2009; Hampshire & Smith, 2001). Parental consanguinity increases the birth prevalence of autosomal recessive birth defects. The risk for neonatal and childhood death, intellectual disability, and serious birth defects is almost doubled for first-cousin unions (Bittles & Black, 2010; Mansour et al., 2009; Nouira et al., 2008)

Advanced Maternal Age and Chromosomal Anomalies

The increasing risk of chromosomal abnormalities, particularly Down syndrome, with advancing maternal age is well recognized. The percentage of women of advanced maternal age (greater than 35 years) delivering infants is high in middle- and low-income countries and sub-Saharan Africa, which do not have universal screening, prenatal diagnosis, and associated services (average 11%–18%; see Oloyede, 2008; Watcham et al., 2007; Tebeu et al., 2007; Orji & Ndububa, 2004; Mokhtar & Abdel-Fattah, 2001; Obed et al., 1995).

Migration

The sickle cell disease (SCD), which was previously rare in South Africa, is being introduced through an influx of migrants from west and central Africa (Beighton & Botha, 1986). Other well-known examples of the introduction of single-gene disorders into the African continent or country by migration include porphyria and Fanconi anemia, which appears in high frequency in Afrikaner population in South Africa (Parker et al., 2008; Tipping et al., 2001; Jenkins, 1990).

Migration also occurs as a result of increased urbanization, as people move from rural settings and traditional lifestyles to town and cities. The rapid growth of cities led to an increase in urban populations of common disorders with a genetic predisposition, such as insulin-dependent diabetes mellitus (Bongani et al., 2009). Etiologic factors involved include not only genetic predisposition, but also lifestyle and dietary changes (e.g., high-fat, high-caloric diets and lack of exercise leading to obesity, increased risk of diabetes and coronary artery disease) and exposure to occupational, recreational, and environmental toxins and pollutants, particularly alcohol and smoke (WHO, 1999).

National Level of Healthcare and Poverty

The proportion of gross domestic product spent on research and development is extremely low in SSA (Hardy et al.,

2008). Most SSA countries lack clinical and diagnostic facilities for genetic medicine and have limited capacity for research to address African genetic problems (Wonkam et al., 2006). Furthermore, human genetics research in SSA has often failed to address the ethical obligation to build capacity. For example, we identified 50 studies on PubMed published since 1988 that used Cameroonian DNA samples. Only seven publications (14%) included a Cameroonian collaborative center, and only 14 (28%) included a Cameroonian author (Wonkam et al., 2010).

As services become available, though, the condition can be alleviated for an increasing proportion of children at birth. For example, structural birth defects including congenital heart defects, cleft lip and palate, and clubfoot, can be corrected with pediatric surgery. Children with functional problems, including thalassemia, sickle cell disorders, and congenital hypothyroidism, also can survive with appropriate treatment. In addition, many children with chronic disabilities due to birth defects, such as Down syndrome or neural tube defects, can be treated so they will live longer, more productive lives (Christianson & Modell, 2004).

DISTRIBUTION OF SPECIFIC BIRTH DEFECTS IN SUB-SAHARAN AFRICA

The categories include birth defects of genetic or partially genetic origin, including single-gene defects, chromosomal disorders, multifactorial disorders, and birth defects caused by teratogens.

BIRTH DEFECTS OF GENETIC OR PARTIALLY GENETIC ORIGIN

Single-Gene Defects

Common Recessive Disorders

These include three major disorders in SSA: the hemoglobin disorders, sickle cell anemia and thalassemia, and oculocutaneous albinism.

Sickle Cell Anemia The WHO has estimated that at least 5% of the population are carriers for one or other of the most serious forms, the alpha- and beta-thalassemias and the structural variant hemoglobins S, C, and E, which are found at polymorphic frequencies in many countries. All these hemoglobinopathies are believed to provide protection against malaria (Clegg & Weatherall, 1999). It is estimated that in excess of 300,000 children are born each year with a severe inherited disorder of hemoglobin, and that approximately 80% of these births occur in low- or middle-income countries (Weatherall, 2010)

Sickle cell disease (SCD) is caused by a point mutation leading to a single amino acid substitution in the beta subunit of hemoglobin (Hb), the principal oxygen transporter in red blood cells. In SSA, SCD occurs at its highest frequency; the prevalence of sickle cell trait is estimated to be 5% to 40% (Weatherall, 2001). It has been estimated that SCD results in the annual loss of several millions of disability-adjusted life years, particularly in the developing world (Weatherall, 2006). A notable feature of SCD is the frequent occurrence of pain and acute clinical episodes, which are generally attributable to vasoocclusive crises (Castro et al., 1994, Platt et al., 1991). As a consequence of these complications, SCD patients have increased mortality as compared with control populations (Platt et al., 1994).

SCD has recently been recognized as a problem of major public health significance by the WHO. Despite the fact that >70% of sufferers live in Africa, expenditure on the related care and research in the continent is negligible, and most advances in the understanding and management of this condition have been based on research conducted in the north. African countries need to focus research and interventions on areas that will lead to the maximum impact (Makani et al., 2007). Typically, clinical severity of SCD is extremely variable, from death in childhood to being relatively well even until old age. The reasons for this heterogeneity are not fully understood. Abnormal transcranial Doppler velocity (TCD) and pulmonary hypertension, which are diagnosed by Doppler echocardiography, are proxy of stroke episodes risk and cause of death, respectively. TCD and pulmonary hypertension are reliable clues, using a noninvasive tool, for the identification of these complications before end-stage progression occurs (Nichols et al., 2001; Gladwin & Vichinsky, 2008). Interindividual variation in fetal hemoglobin (HbF) levels is one of the main modifiers that contribute to the clinical heterogeneity observed in SCD patients. Higher expression of HbF in adulthood ameliorates morbidity and mortality in SCD. It is now clear that common HbF variation is a quantitative genetic trait shaped by common polymorphisms. Genetic variation in three major loci account for a relatively large proportion (20%–50%) of the phenotypic variation in HbF levels: (1) a single-base substitution (T/C) at position −158 of the $^{G}\gamma$ globin gene, termed *Xmn*I $^{G}\gamma$ site (Nagel & Platt, 2001); (2) the *HMIP* locus (*HBS1L-MYB Intergenic Polymorphism*) on chromosome 6q (Labie et al., 1985); (3) the third major locus is the oncogene *BCL11A* on chromosome; see Menzel & Thein, 2009; Uda et al., 2008). It is expected that the identification of genetic polymorphisms that modulate HbF levels will shed light on the molecular mechanisms that control HbF expression, and on the etiology of the clinical heterogeneity observed in general in SCD patients. In addition, it could open perspectives into new therapeutic approaches for HbF reactivation, but an immediate application would be an improved prediction of one's ability to produce HbF, which, in turn, may improve prediction of disease severity.

The mortality rate associated with SCD has remained high despite the use of appropriate interventions to manage the various forms of crisis. The organisms causing bacteremia in African children with sickle cell anemia are the same as those in developed countries (Streptococcus pneumoniae (44/108 isolates; 41%), non-typhi Salmonella species (19/108; 18%), Haemophilus influenzae type b (13/108; 12%), Acinetobacter species (seven of 108; 7%), and Escherichia coli (seven of 108; 7%) (Williams et al., 2009; Serjeant, 2005). Introduction of conjugate vaccines against S pneumoniae and H influenzae into the childhood immunization schedules of African countries could substantially affect survival of children with sickle cell anemia. Methods for screening SCD are now well established and inexpensive, and it is vital that they are applied to define the magnitude of the problem that will be posed by these conditions in the future, and effective enrollment of affected neonates in a comprehensive follow-up program adapted to the socioeconomic conditions, health structures and cultural background of the African setting (De Montalembert & Tshilolo, 2007). In developed countries, newborn screening has resulted in a reduction in the SCD mortality rate from 16% to <1% (Streetly et al., 2009, Komba et al., 2009). Similarly, in the Republic of Benin, following neonatal screening, counseling, and follow up, the under-5 mortality rate was 15.5 per 10,000, a figure that is 10 times lower than the general rate recorded (Rahimy et al., 2009).

The life expectancy of patients with SCD has improved in the United States and Europe thanks to the use of penicillin prophylaxis, appropriate immunizations, neonatal screening, implementation of a quality transfusional policy, hydroxyurea therapy, detection and treatment of cerebral vasculopathy, recognition of situations that can benefit from allogenic marrow transplantation, and improvements in bone marrow transplantation techniques. The cost of almost all these techniques is far beyond the means of healthcare systems in Africa, where they can hardly be used (De Montalembert & Tsilolo, 2007). Hematopoietic stem cell transplantations (SCT) is now an established treatment modality with definitive indications for many hematological disorders. However, this line of treatment requires tremendous resources, and it becomes increasingly difficult for transplanters practicing in the developing world to reconcile the difference between what is possible and what is available. On the basis of 18 years of experience and more than 1300 transplants that were performed in Egypt, this could inspire how to address special issues that are important for hematopoietic SCT practices in developing countries (Mahmoud et al., 2008). A great deal can be done to help the situation by developing further North/South and South/South partnerships for disseminating better practice (Weatherall, 2008).

Interestingly, up to 95% of Cameroonian parents with a SCD-affected child indicated they would request prenatal genetic diagnosis (PND) for SCD, and 65% would opt for abortion of a sickle cell anemia affected pregnancy. The effective initiation of PND for SCD (amniocentesis and fetal DNA analysis) has laid the foundation for the first medical genetic service in Cameroon, and demonstrates that the practice of medical genetics is possible in an SSA country (Wonkam et al. 2010). The use of SCD prenatal diagnosis as point of entry of genetic medicine was also reported in Nigeria (Akinyanju et al., 1999). There is a need for widespread screening and counseling programs directed at developing PND expertise where this is not available (Weatherall, 2005).

Thalassemias The carrier frequency for β-thalassemia in SSA ranges from 1% to 10%, rarely greater, while that for the milder forms of α-thalassemia is much higher, ranging from 10%–20% in parts of SSA (Weatherall, 2001; Brabin et al., 2004). The α-thalassemia genotypes described in SSA (-alpha (3.7)/alpha alpha and –alpha (3.7)/-alpha (3.7)) could not lead to hydrops fetalis as observed in Southeast Asia. Similarly, the nature of mild beta mutations in SSA is not usually associated with children with a β-thalassemia major that will require transfusion. Nevertheless, a varying proportion of HbS-β thalassemia and SC forms of SCD do exist, and studies are needed to determine their frequency in SSA (Weatherall, 2001).

Oculocutaneous Albinism Oculocutaneous albinism (OCA) is a genetically inherited autosomal recessive condition and OCA2, tyrosine-positive albinism, is the most prevalent type found throughout SSA. Due to the lack of melanin, people with albinism are more susceptible to the harmful effects of ultraviolet radiation exposure. This population must deal with issues such as photophobia, nystagmus, decreased visual acuity, extreme sun sensitivity, and skin cancer (Hong et al., 2006, Asuquo et al., 2009). As a result, oculocutaneous albinism is associated with early death (George et al., 2005; Adegbidi et al., 2007). In Tanzania and Nigeria, for example, only 10% of those affected survive beyond 30 years of age (Kromberg, 1992; Kromberg et al., 1989; Onuigbo, 2006). An overall estimate of albinism prevalence ranges from 1/5000–1/15,000 (Hong et al., 2006). From epidemiologic data on OCA available for South Africa, Zimbabwe, Tanzania, Cameroon, and Nigeria, population prevalences as high as 1 in 1000 were reported for selected populations in Zimbabwe and other specific ethnic groups in southern Africa. One clan, the Vhatavhatsindi in South Africa, had a significantly higher frequency of 1 in 832 (Lund et al., 2007).

OCA is a highly visible birth defect in dark-skinned people. As consequence, people with albinism also face social discrimination as a result of their difference in appearance. Several myths account for this event. Albinos are believed to possess good and evil magical powers (Imperato & Imperato, 2006; McBride & Leppard, 2002;

Lund, 2005). On the white magic side, some organs are believed to confer luck, health, and prosperity. For this reason albinos in Tanzania and Burundi are still, in 2008, prey to ritual murders and mutilations to obtain various body parts such as arms, legs, and genitals for preparation of amulets (Aquaron et al., 2009). The WHO is currently investigating the issues concerning this vulnerable population. Disseminating accurate medical information explaining the genetic basis of albinism will be necessary to eliminate ignorance and superstition.

Children with albinism living away from home in rural areas appear to have inadequate sun protection strategies (Lund & Taylor, 2008). Changes in health policy could address these deficiencies by providing more detailed healthcare information, giving advice on appropriate styles of hat to wear, and how to assess commercial sun-protection products. The existence of tens of thousands of people living with albinism in Africa reiterates the need for increased awareness of, and public health interventions for, albinism in order to better address the medical, psychological, and social needs of this vulnerable population.

Cystic Fibrosis Mutations of the cystic fibrosis transmembrane conductance regulator (CFTR) result in defective chloride secretion and excessive sodium reabsorption, and a depletion of the airway surface liquid causing impaired mucociliary clearance. The ongoing infection and inflammation leads to persistent symptoms and irreversible lung damage (Mogayzel & Flume, 2010). Cystic fibrosis (CF) is the most common single-gene birth defect in Caucasian populations, with a birth prevalence of approximately one in 2000 live births. The defect had been considered rare in other populations, but reports of birth prevalence of one in 2560 and one in 2608 live births in Jordan and Egypt, respectively, challenge that assumption. In addition, studies in migrants from the Middle East and Africa to Europe is revealing different CFTR mutations in patients (Spinelli et al., 2009, Ratbi et al., 2008).

The assumption that CF is considered as a rare disease in black Africans seems to be also incorrect. Previous studies indicate that African Americans have their own set of "common" CF mutations that originate from the native African population. One novel mutation, 3120+1G->A, occurred with a frequency of 12.3% and 46% in the black South African population, indicating a common ethnic origin of the mutation in different African populations (Macek et al., 1997; Dörk et al., 1998). Since this mutation accounts for between 15% and 65% of CF chromosomes in South African blacks, a corrected CF carrier frequency would be between 1 in 14 and 1 in 59. Hence, the incidence of CF would be predicted to be between 1 in 784 and 1 in 13,924 live births in this population (Padoa et al., 1999). Four other mutations have been detected, resulting in the identification of a total of 62.5% of mutations in black South Africans (Goldman et al., 2001). Screening for 16 "common Caucasian" mutations identified 52% of CF alleles in African Americans, while screening for 8 "common African" mutations accounted for an additional 23%. The combined detection rate of 75% was comparable to the sensitivity of mutation analysis in Caucasian CF patients (Macek et al., 1997). There are several possible reasons why CF in African populations is not being detected. Some of these are misdiagnosis as chronic pulmonary infection, malnutrition, tuberculosis, infantile diarrhea, failure to thrive, or a high infant mortality rate (Feuillet-Fieux et al., 2004).

Rare Single-Gene Defects

The birth prevalence of rare single-gene defects is generally one in 10,000, or more. However, in some clusters of people in Africa, gene frequencies of rare single-gene defects are greater than expected due to the founder effect, genetic drift, and geographic or cultural isolation (WHO, 1999). Typical examples include porphyria in South African Africa. In spite of their infrequency, certain rare single-gene disorders for which treatment is available, for example hemophilia, are important in Africa because of the high costs associated with the care of affected individuals. Collectively, these disorders also add significantly to the infant and child health burden.

Genetic Predisposition to Hemolysis and Neonatal Jaundice: G6PD Deficiency

G6PD deficiency, an X-linked recessive disorder, is the most common enzyme deficiency disease in humans. An estimated 7.5% of the world's population carries a G6PD-deficiency gene. Hundreds of variants of the disorder with differing clinical severity have been described, with approximately 90% of those affected being male. Individuals may be asymptomatic or may experience varying degrees of anemia caused by rapid breakdown of red blood cells. Acute crises can be induced by infection or the ingestion of drugs, or certain foods, such as fava beans. In newborns with severe G6PD deficiency, neonatal jaundice due the rapid breakdown of red blood cells may result in damage to the brain and auditory nerves, a condition called *kernicterus*. Kernicterus can lead to intellectual disability, cerebral palsy, deafness and death.

More than 5 million infants are born annually with G6PD deficiency of varying severity, mainly in tropical SSA, the eastern Mediterranean, and North Africa (Luzzatto et al., 1995). G6PD deficiency in Nigerian males from the western region of the country is between 20% and 26% of the population (Egesie et al., 2008). A similar prevalence was found in Sierra Leone, with 11.3% in male subjects (Jalloh et al., 2008). In a Nigerian study,

30.9% of the jaundiced neonates were G6PD-deficient, compared with 13.3% of controls (Sodeinde et al., 1995). G6PD deficiency (40%) was the major etiological factor of kernicterus (Ahmed H et al 1995, Owa, 1989). More data are needed on the adverse effects of G6PD deficiency in African populations.

Rhesus Negativity

When a Rhesus negative woman has a Rhesus positive partner, the fetus is often Rhesus positive. In such cases, fetal red cells may pass into the mother's circulation and stimulate the production of antibodies that cross the placenta and destroy fetal red cells, causing severe anemia. An affected fetus may be stillborn or become severely jaundiced soon after birth, with the risk of severe mental and neurological disability. Rhesus negativity values in blood donors range from 4% in Guinée (Loua et al., 2007), 6% in Kenya (Mwangi, 1999) and 7.26% In Nigeria (Nwauche & Ejele, 2004.). Avoidable fetal and neonatal death from Rhesus hemolytic disease will continue in any country that does not have a systematic prevention program. This problem seems to have received relatively little attention in SSA.

CHROMOSOMAL DISORDERS: DOWN SYNDROME

Birth prevalence of trisomy 21 may be as high as 2 per 1000 live births in Africa (WHO, 1996). Early infant or childhood death from congenital heart disease and infection is common among infants and children with Down syndrome (DS) in SSA (Ali, 2009; Lawrenson et al., 2006; Faye et al., 2004). As a result, the population prevalence of chromosomal disorders is low (Christianson & Modell, 2004; Adeyokunnu, 1982). Still, there is reason suspect an underestimation, inasmuch as in previous studies it was suggested that there may be clinical difficulties in making the diagnosis in African neonates (Christianson, 1996). These findings suggest that if mothers themselves cannot see the differences between their DS children and normal children, clinical diagnosis based on physical stigmata may be difficult. Furthermore, acceptance of the diagnosis may be retarded until delayed milestones can be observed in the affected infants (Christianson & Kromberg, 1996). The difficulties experienced by medical and nursing staff in diagnosing DS in Black neonates extends into infancy and childhood, despite the fact that the clinical features of Black DS infants and children do not differ from those seen in DS patients in other ethnic groups (Christianson, 1997)

Prenatal screening for Down syndrome is not routine in many antenatal clinics in SSA (Oloyede, 2007). It has been demonstrated that professionals and women in SSA are favorably disposed to first trimester prenatal diagnosis services where available (Wonkam et al., 2006; Salako et al 2006). In North Africa, many countries and Egypt and Tunisia have established experience in biochemical or echographic screening followed by invasive procedure (amniocentesis) for DS (Chelli et al., 2008; Baghagho et al., 2004; Mokhtar & Abdel-Fattah, 2001). In SSA, however, only a few centers in South Africa have a long experience in screening diagnosis and treatment of children with DS (Dangor and Ross, 2006; Venter et al., 1995). We report a single experience from Cameroon of available reports on prenatal genetic diagnosis options for DS in SSA (Wonkam et al., 2010).

It is important to improve professional and public awareness of DS prenatal screening options. Many more personnel and centers should be encouraged to provide first trimester screening, and the NT scan may be the most feasible. Collaboration with African countries with long experience of PND may help to cost-effectively build capacity where it's lacking and offer this service to the public. QF-PCR appears as an interesting option in the context, since no culture of amniocytes will be required. When initiating prenatal genetic diagnosis in Cameroon, we successfully established such collaboration with Geneva University hospitals.

MUTIFACTORIAL DISORDERS

Key multifactorial disorders include congenital heart defects, neural tube defects, and cleft lip with or without cleft palate.

Congenital Heart Defects

Congenital heart defects (CHD) are the most common form of birth defects, occurring in 4–8 per 1000 live births. The majority of congenital heart defects, about 90%, have a multifactorial cause. An estimated 1,040,800 infants are born each year with a multifactorial congenital heart defect. (Turnpenny & Ellard, 2007). The remaining congenital heart defects are associated with chromosome abnormalities (5%–8%), single-gene defects (3%–5%), and teratogens (2%–3%; see Seashore & Wappner, 1996). With epidemiologic transition, congenital heart disease rather than rheumatoid heart defects are dominant in some African settings, and their relative frequencies are similar to reports elsewhere—ventricular septal defects, oval fossa type atrial septal defects, Fallot's tetralogy, and patent arterial duct together accounted for 77.8% of the structural heart defects in Africa (Edwin et al., 2010; Ali, 2009; Ejim et al., 2009; Okoromah et al., 2008).

Christiaan Barnard (1922–2002) introduced heart transplantation in humans for the first time, on the African continent (Toledo-Pereyra, 2010). Despite this historical fact, the wide range of children with diverse native CHD is a reflection of non-availability of definitive surgical facilities in some African countries. Humanitarian organizations are very active on the African continent as part of an outreach project that sends teams of specialist doctors and health professionals to perform open heart surgery on children and adults with congenital heart abnormalities (Lindblom, 2009; Tippett et al., 2008). Regional and international collaboration could be mutually beneficial, inasmuch as studies demonstrate feasibility of surgery for adolescents and adults with congenitally malformed hearts in Soudan, Kenya, Nigeria, and Cameroon, with a good outcome (Edwin et al. 2010; Tantchou Tchoumi & Butera, 2009; Sulafa & Karani, 2007). Nevertheless, due to poverty and illiteracy of parents, the postsurgical follow-up of patients is challenging (Sulafa & Karani, 2007; Tantchou Tchoumi & Bureta, 2009). In addition, the mean age at referral to a pediatric cardiologist is estimated to be 16.9 months in Kenya (Awori et al., 2007). This suggests that a significant number of patients may miss the opportunity to have optimal surgical intervention, indicating a need of policies for early diagnosis, close monitoring, and timely intervention (Sani et al., 2007).

With the establishment of tertiary healthcare institutions in many SSA countries, availability of echocardiographic facilities, as well as increasing numbers of pediatric cardiologists, more of these cases are likely to be seen in the future. There is an urgent need for the government to establish a well-equipped cardiothoracic surgical center to cater to these patients, either free or at highly subsidized rates.

Single-gene congenital heart defects are starting to be recognized and characterized, especially in South Africa, with molecular characterization of Lond QT-syndome (Brink et al., 2005), arrhythmogenic right ventricular cardiomyopathy (Matolweni et al., 2006), or Marfan syndrome (Child et al., 2007).

Neural Tube Defects

Neural tube defects (NTD), including spina bifida, anencephaly, and encephalocele, occur when part of the neural tube, which forms the spine, spinal cord, skull, and brain, fails to close between 21 and 28 days after conception—before women realize they are pregnant. A paleopathological study reports a meningocele in an ancient Egyptian infant mummy from the XI dynasty (2100–1955 B.C.; see Boano et al., 2009). Many children affected by neural tube defects have multiple lifelong disabilities, including varying degrees of lower limb paralysis, bowel and bladder incontinence, hydrocephalus, and intellectual and learning disabilities. Each year spina bifida and anencephaly, the two most common forms of neural tube defects, occur in an estimated 300,000 newborns worldwide (Shin et al., 2010). Although there is scarcity of information, NTDs are not rare and the burden of disease in SAA and has been estimated to 1.99 cases per 1000 births in Cameroon (Njamnshi et al, 2008). The cranial lesions are mainly congenital primary hydrocephalus in 44% of cases, encephalocele in 28%, and microcephaly in 17%. The spinal lesions were spinal dysraphism in more than 90%, all in the lumbosacral region (Adeleye & Olowookere, 2009; Ba et al., 2007). Cervical spina bifida is rarely reported in SSA (Andronikou et al., 2006).

Some African countries are developing considerable experience in surgical treatment, like endoscopic third ventriculostomy or choroid plexus cauterization (ETV/CPC) in Uganda or South Africa (Warf, 2005; Warf & Campbell, 2008; Warf et al., 2009). Still, delay in presentation even with associated malnutrition and local sepsis seems not to significantly affect the outcome of repair. Intracranial sepsis is the main cause of death in neural tube defect patients who are not operated on time in Nigeria and Cameroon (Alatise et al., 2006, Idowu & Apemiye, 2008; de Paul Djientcheu et al., 2008).

Prenatal management and outcome at birth are limited by poverty and cultural beliefs. Prevention is possible and may be better than palliative care in developing countries. A call is made for the establishment of nationwide efforts to fully clarify the epidemiology of this silent epidemic and develop appropriate preventive strategies such as folic acid awareness among women of reproductive age, and pre-conception folic acid supplements (Sayed et al. 2006; Van't Veer et al, 2008).

Cleft Lip With or Without Cleft Palate

Cleft lip, with or without cleft palate, is a congenital malformation etiologically different from isolated cleft palate. It comprises a partial or complete cleft of the upper lip, with or without a cleft of the alveolar ridge (gum) or the palate. Multifactorial inheritance accounts for the majority of cleft lip, with or without cleft palate, with birth prevalence ranging from 0.3 per 1000 live births in African-American populations, to one per 1000 in Caucasians. A number of potentially causative genes are under investigation, though certain drugs and maternal smoking are considered to play a role (WHO, 2002). Cleft lip with and without cleft palate is also associated with other syndromes caused by chromosomal abnormalities, single-gene defects, and teratogens.

Cleft lip and palate is a significant congenital malformation in SSA, with a variable birth prevalence based on hospital records on the continent from 0.3, 0.6, and 0.9 per 1000 live births in Zambia, Zimbabwe and Sudan, respectively (Elliott et al., 2008; Pham & Tollefson, 2007; Suleiman et al., 2005). There seems to be a higher incidence of familial tendency and associated congenital malformations than that reported elsewhere. In a large Kenyan study there was a male predominance, and most clefts were on the left side. A retrospective and prospective study featured a positive family history in 3.5% and 30.9%, respectively. Associated congenital malformations were 8.2% for the retrospective study and 25% for the prospective study (Spritz et al., 2007; Wanjeri & Wachira, 2009). A seasonal variation was observed in Zambia that could be explained, at least in part, by environmental factors affecting the development of CL+/-P. But the genetic background also affects the distribution, since there was a considerable variation between the groups of orofacial clefts occurring in the Caucasian versus South African Black patients with cleft deformities (Bütow et al., 2007). Bantu tribes have lower risk of orofacial clefts than Nilotic-Paranilotic tribes in Kenya (Spritz et al., 2007)

The awareness of oral and maxillofacial surgery by both the public and medical specialties has led to rapid development and expansion of the oral and maxillofacial surgery specialty, with management of diverse and complex problems within a well-defined anatomical region. There is an indication of low awareness of the specialty by the public and the professionals in Nigeria (Adewole & Akinwande, 2007). Access to treatment is likely the major determinant of regional disparity in clefts in Zambia and Zimbabwe (Elliott et al., 2008, Pham & Tollefson, 2007). Humanitarian plastic surgery training missions have been essential for the development of plastic surgery in some SSA countries (Saboye, 1999). Financial, logistical, and training support to cleft teams has led to more surgeries for children with cleft in Ghana and Uganda (Hodges & Hodges, 2000; Donkor et al., 2007).

There is an associated substantial clinical and psychosocial impact on patients with cleft lip/palate. Compliance with primary surgical treatment in people with cleft lip and palate is a well-known problem, especially in developing countries fraught with poverty and ignorance. The West African subregion has a high dropout rate (20%) after lip repair (Onah et al., 2008). Cultural and religious factors seem to have an important role in the explanations, labels, and treatment that follows the birth of a child with CLP in Nigeria (Olasoji et al., 2007). Studies emphasized the need for greater collaboration and sharing of information between modern medical practitioners and traditional healers in Nigeria and South Africa (Dagher & Ross, 2004; Ross, 2007, Olasoji et al., 2007). By viewing cultural differences on a continuum, following the asset-based approach, applying knowledge based on contextually relevant research, and recognizing family uniqueness, families may be empowered to participate fully in the team approach to support their children with cleft lip and palate and craniofacial anomalies in attaining their full potential in the SSA context (Louw et al., 2006).

A larger population-based study is warranted to further clarify these various epidemiological findings, to provide a basis for further prevention and therapeutic approaches of orofacial clefts in SSA.

INDIVIDUAL TERATOGEN-ASSOCIATED BIRTH DEFECTS

Birth defects due to teratogens are among the more readily preventable. Pregnancies in middle- and low-income countries, by comparison to high-income countries, are more likely to be at risk from potential teratogens for several reasons, including increased frequency of intrauterine infection, poor maternal nutrition, low socioeconomic and educational levels, lack of environmental protection policies, and poorly regulated access to medication (Penchaszadeh, 2002). Broadly, three of the five causal groups of teratogens have the most significance: (1) congenital infections, (2) maternal illness and altered maternal metabolism, and (3) recreational and therapeutic drugs.

CONGENITAL INFECTION

The TORCH organisms—toxoplasmosis, other (syphilis, varicella-zoster, human parvovirus B19), rubella, cytomegalovirus (CMV), and herpes—account for the most common infections associated with birth defects (Stegmann & Carrey, 2002).

Toxoplasmosis

Toxoplasma gondii's importance for humans refers mainly to primary infection during pregnancy, resulting in abortion/stillbirth or congenital toxoplasmosis. Foci of high prevalence exist in Latin America, parts of Eastern/Central Europe, the Middle East, parts of southeast Asia, and Africa (Pappas et al., 2009). The results indicated that the prevalence of T. gondii among pregnant women was relatively high in the rural (57.6%) compared to urban (46.5%) areas. The IgG positivity was correlated, in general, with age, as it was higher in the older age groups in Egypt (El-Gozamy et al., 2009). Nevertheless, there are significant variations in IgG seroprevalence in pregnant women in SSA. It was reported to be 25.3% in Burkina Faso (Simpore et al. 2006), 35% in Tanzania (Doehring et al. 1995), 40% in Senegal (Ndir et al, 2004), 50.6% in Morroco (El Mansouri et al., 2007), 53.6% in Benin (Rodier et al., 1995), 60% in Ethiopia and Côte d'Ivoire (Negash et al., 2008; Adou-Bryn et al., 2004), 71.2% in Gabon (Nabias et al, 1998), 75.4% in Nigeria (Onadeko et al., 1996), and 77.1% in Cameroon (Ndumbe et al., 1992). Despite the high IgG seroprevalence, country-specific awareness and screening is needed, since cases of congenital toxoplasmosis do occur in Africa (Ben Abdallah et al., 2009).

Congenital Syphilis

Syphilis is a common sexually transmitted infection. Over 11 million new infections occur each year worldwide.

The continued occurrence of congenital syphilis is an indictment of the inadequate antenatal care services and poor quality of programs to control sexually transmitted infections. More than 1 million infants are born with congenital syphilis each year. In SSA alone, previous study showed 6%–16% of pregnant women have active syphilis (Romoren &Rahman, 2006; Walker & Walker, 2007; Gichangi et al., 2004; Kamanga, 1995; Jenniskens et al., 1995; Ndumbe et al., 1992). The burden of congenital syphilis is also increasing along with AIDS pandemics (Walker & Walker, 2002; Mullick et al., 2004; Saloojee et al., 2004).

Unless the mother is treated with penicillin, as many as 40% of embryos or fetuses with syphilis die before or shortly after birth. Those who survive are at risk for brain damage, blindness, hearing loss, and bone and tooth problems if they are not treated with an antibiotic shortly after birth (Fitzgerald et al., 1998). Although screening and appropriate treatment of infected pregnant women can prevent fetal and maternal complications, traditional screening algorithms requiring multiple tests have proven to be difficult to implement in resource-poor settings because of the lack of investigative ability and the pressure on health facilities to discharge infants early (Lech, 2003; Walker & Walker, 2002; Saloojee et al., 2004). Diagnosis of congenital syphilis is problematic, since more than half of all infants are asymptomatic, and signs in symptomatic infants may be subtle and nonspecific. Newer diagnostic tests such as enzyme immunoassays, polymerase chain reaction, and immunoblotting have made diagnosis more sensitive and specific, but are largely unavailable in the settings where they are most needed (Saloojee et al., 2004). Universal prenatal syphilis screening using rapid point-of-care tests will improve both maternal and infant outcomes and is cost-effective (Rydzak & Goldie, 2008).

In SSA, with high maternal syphilis prevalence, important goals for patients, healthcare workers, and health authorities are earlier antenatal care attendance, rapid on-site testing, improved partner treatment, and a repeat test late in pregnancy to manage incident cases. Using the decision analysis to estimate the incremental cost-effectiveness of two on-site antenatal syphilis screening strategies to avert congenital infections—qualitative RPR (on-site RPR) and treponemal immunochromatographic strip assay (on-site ICS)—compared to the current practice (off-site RPR/TPHA), on-site antenatal screening with ICS was a cost-effective approach to reduce the incidence of congenital syphilis in a South African setting (Blandford et al., 2007; Bronzan et al., 2007). It was shown to be feasible in Bolivian, Kenya, South Africa, and Mozambique to introduce rapid syphilis testing in settings without laboratory services at a small incremental cost per woman screened (Deperthes et al., 2004, Levin et al., 2007). In settings with laboratories, the cost of ICS is similar to that of RPR. Nevertheless, in a recent study in Nigeria, the seroprevalence rate of syphilis was extremely low (12/18101). The peak age-specific incidence of 0.02% was in the 20–24-year age group. This calls for a review the policy of routine antenatal serology screening for syphilis according to the setting (Bukar et al., 2009).

Varicellla-Zoster Virus and Herpes Simplex Virus

Primary infection with varicella-zoster virus (VZV; chickenpox) can pose risks to the fetus if a mother is infected during pregnancy. A susceptible pregnant woman who is exposed to an infected household member has a 90% risk of contracting the illness. Maternal infection can lead to congenital varicella syndrome, a group of birth defects that can include defects of muscle and bone, malformed and paralyzed limbs, a smaller-than-normal head, blindness, seizures, and intellectual disability. This syndrome affects about 2% of babies whose mothers are infected with varicella during the first 20 weeks of pregnancy, but is rare if infection occurs after 20 weeks (MOD, 2006). The birth prevalence of VZV in SSA countries is unknown.

Herpes simplex virus type 2 (HSV-2) can cause genital infection and can be passed to the fetus by the mother in utero, during birth, and postnatally. HSV infection in the newborn can present as isolated skin, eye, or mouth infection, or as disseminated disease with CNS involvement that can result in microcephaly, intellectual disability, and seizures. Mortality rates as high as 25% have been reported for disseminated disease. Morbidity is also high. While 95% of newborns with isolated infection are normal at 2 years of age, only 60% of newborns who survive disseminated disease are normal at that age. (Dunkle et al., 1991). Unfortunately, many women are asymptomatic with no evidence of genital lesions at the time of primary infection, complicating control efforts. Maternal HSV seroprevalences of greater than 50% have been documented in specific populations in Cental America, Africa, and Asia (Ghebrekidan et al., 1999; Kaur et al 1999; Nahmias et al., 1990; Oberle et al., 1989).

In recent decades the alarming spread of HIV infection in Africa, and associated frequent reactivation of herpes infections, is leaving less room for neglect (Ghebrekidan et al., 1999). In Côte d'Ivoire, the herpes viral isolation and the direct immunofluorescence (IF) assay of the genital secretions were found to have a prevalence of 12.7% among the pregnant women who tested positive (the majority of whom were from modest socioeconomic origins and young), 10% had antecedents of genital herpes, 2.7% showed an asymptomatic excretion, and 36.8% had an acute episode during their pregnancy including a primary infection in the 20th week. The scarcity of data that exists in regard to herpes virus infections in many African regions could partly be due to the mild nature of their primary infections and the lack of means for their proper diagnosis (Duff et al., 2009).

Human Parvovirus

Human parvovirus (HPV) B19 causes Fifth Disease (erythema infectosum), a mild viral illness usually seen in children. When the infection occurs in pregnant women, it acts as a teratogen and can cause fetal death. The virus disrupts the fetus's ability to produce red blood cells, leading to anemia, heart failure, and up to a 9% risk of miscarriage or stillbirth. Fetal deaths are more likely when a pregnant woman contacts the infection in the first 20 weeks of pregnancy. An estimated one per 3333 newborns are infected in high-income countries (Markenson & Yancey, 1998). In South Africa, in a prospective study of 1967 pregnant women who were routinely screened for recent human B19 parvovirus infection, 64 (3.3%) were identified as being IgM positive. No adverse effects were documented by ultrasound in any of the fetuses. The outcome of pregnancy was favorable in 95.1% of these women, with no evidence of hydrops fetalis or any congenital abnormalities. Two neonates (3.4%) were small for gestational age, and there was one abortion. Samples of blood obtained from 20 neonates born to women with evidence of recent infection were B19 parvovirus IgM negative. Recent infection with human B19 parvovirus in pregnancy constitutes a low risk for the development of adverse fetal effects; hence, routine antenatal screening is not warranted (Guidozzi et al., 1994).

Congenital Rubella Syndrome

Rubella poses a serious threat to the fetus if the mother contracts it during the first 16 weeks of pregnancy. The annual incidence of rubella worldwide is difficult to ascertain because the disease is often mild and overlooked, or, when it is more severe is often misdiagnosed as measles. About 25% of babies whose mothers contract rubella during the first trimester of pregnancy are born with one or more birth defects, which, together, are referred to as *congenital rubella syndrome*. These birth defects include blindness, hearing impairment, heart defects, intellectual disability and, less frequently, movement disorders. The infection occasionally causes miscarriage and stillbirth. Children with CRS also are at increased risk of diabetes, which may develop during childhood or adulthood (MOD, 2002). Although the situation is improving, more than 100,000 children are born annually with congenital rubella syndrome (WHO, 2000b). In high-income countries, successful rubella immunization programs have virtually eliminated CRS.

In the African region only 2% of countries had rubella vaccine programs; nevertheless, the few available data indicate a high immunization rate. Within the Western Cape a total of 95.3% of women in the 15–24-year age group, 97.5% in the 25–34-year group, and 98% in the 35–45-year age group were immune to rubella (Corcoran & Hardie, 2005); an 83.9% immunization rate was reported in Cameroon (Ndumbe et al., 1992). Any vaccination program has to take in account the natural immunization rate in a given country.

Cytomegalovirus

Cytomegalovirus (CMV) infection is the most common congenital infection in high-income countries. Intrauterine infections occur in mothers who have developed a first-time CMV infection during pregnancy. These women have a 30%–40% risk of passing the virus to their fetus (MOD, 2001). Most of the time, the infected infant shows no symptoms at birth, but 14% of asymptomatic newborns with infection develop serious disabilities over the next several years. These include neurological, growth, and developmental problems; sight or hearing problems; and dental abnormalities. About 10% of infected newborns have symptomatic CMV disease. Of these, most who survive suffer from progressive deafness and/or intellectual disability. An estimated four per 10,000 newborns are affected in middle- and low-income countries (Christianson & Modell, 2004). The overall prevalence of congenital CMV was 5.71% among neonate in intensive care units in Egypt, with higher frequency in premature and low-birth-weight neonates (Morgan et al., 2003). Up to 7.3% of 68 infants with congenital anomalies were tested positive for CMV IgM antibodies by microassay ELISA technique in an Egyptian cohort (Shaaban et al., 1995). CMV was found to be the most prevalent infection in both groups of women (19.2%) and in the babies (24.2%), indicating the importance of this virus in intrauterine infection, in a community study in South Africa (Bos et al., 1995). A 15-month longitudinal study of CMV infection in 178 Gambian mothers and their babies was undertaken; 25 (14%) of the babies were congenitally infected despite the fact that 87% of their mothers were antibody positive to the virus (Bello & Whittle, 1991). The few data available seem to indicate that CMV could play a considerable role in congenital anomalies in SSA.

It is anticipated that with HIV, the prevalence of the TORCH organisms related to birth defects will not decrease unless serious preventive actions are taken. This will require more research for a better understanding of the problem with a country-specific approach, and public health action for screening and preventions.

MATERNAL ILLNESS AND ALTERED MATERNAL METABOLISM

Maternal Diabetes Mellitus

Pregnancy in type 1 and type 2 diabetes mellitus (DM) is associated with an increased rate of adverse outcomes for both mother and fetus. IDDM (type 1) affects 0.5% of pregnancies in high-income countries (Khoury et al., 1989), and increases the risk of serious birth defects up to threefold. The risk of fetal abnormalities due to maternal IDDM can be reduced by good periconception diabetic control (Kinsley, 2007).

The number of cases of diabetes worldwide has increased significantly in the last decade. Characteristically, the incidence of gestational diabetes (GDM) reflects the incidence of type 2 diabetes mellitus (T2DM) in the background population, which is a warning that a rapid increase in the incidence is to be expected, concomitant with the already-observed increase in the incidence of T2DM. Although the majority of all deliveries worldwide take place in the so-called developing world, little is known about the prevalence of diabetes in pregnancy in rural areas of East Africa (Zeck & McIntyre, 2008).

Gestational diabetes mellitus complicates about 7.7% of pregnancies in Morocco (Bouhsain et al., 2009); 5.4% in Nigeria (Adegbola & Ajayi, 2008); 8.8% in Limpopo, South Africa (Mamabolo et al., 2007); and 11% in Mozambique (Challis et al., 2002). The prevalence rate of gestational diabetes mellitus was found to be 3.7% in rural Ethiopia (Seyoum et al., 1999).

The perinatal mortality was 6.1%, 10%, and 12.5% in South Africa, Tanzania, and Nigeria, respectively (Huddle et al., 1993; Ozumba et al., 2004). Major congenital anomalies in pregnancy in diabetic women in 1.7% and 2% in South Africa and Tanzania, respectively (Huddle et al., 1993; Lutale et al., 1991). Maternal diabetes is a risk factor for congenital heart diseases in Alexandria, Egypt (Bassili et al., 2000). A study in Libya showed that 13.8% of infants of mothers with insulin-dependent diabetes mellitus had easily recognizable birth defects, compared to 3% in the nondiabetes maternal population (Malik & Mir, 1992). Maternal hyperglycemia is an important factor affecting maternal wellbeing and neonatal morbidity and mortality in Sudan (Abdelgadir et al., 2003).

The institution of a specialized service for pregnant diabetic women from Soweto for a sustained period of 20 years proved feasible and was of significant benefit. The interventions, centered around a diabetes nurse educator, were relatively simple and easy to implement, even though late presentation/referral of patients remains a problem (Huddle, 2005). Programs and policies have to be established, including organization of the health system to provide care, medicines, and other tools necessary for management of diabetes in pregnancy, consideration of accessibility and affordability of care, education for healthcare workers, and education of pregnant and nonpregnant women of reproductive age.

Maternal Epilepsy

Children of epileptic mothers have a greater risk for birth defects than is seen in the general population. Risks of major malformations, growth retardation, and hypoplasia of the midface and fingers, known as *anticonvulsant embryopathy*, increases in infants exposed to anticonvulsant drugs in utero, including sodium valproate and phenytoin.

One-quarter of the 40 million individuals estimated to have epilepsy in the developing world live in Africa (Onwuekwe et al., 2009). Data on the incidence of and prognosis for epilepsy in SSA are scarce, but prevalence data show that epilepsy is two or three times more common than in industrialized countries in nontropical areas. Door-to-door surveys and neurological assessment showed a prevalence of 11.2/1000 in Tanzania (Winkler at al., 2009), 12.5/1000 in rural Zambia (Birbeck & Kalichi, 2004); 14.2/1000 in Senegal (Ndoye et al., 2005) 15.7/1000 in Togo (Balogou et al., 2007); 15.9/1000 in Benin (Debrock et al, 2000), 29.5% in Ethiopia (Almu et al., 2006), and 41/1000 in rural Kenya (Edwards et al., 2008); 64/1000 in rural Ethiopia (Tekle-Haimanot et al., 1997); 85 /1000 in Tanzania (Moser et al., 2007); and up to 105/1000 in a rural area in Cameroon (Prischich et al., 2008). Active convulsive epilepsy was estimated to be 4.5/1000 in rural Kenya (Edwards et al., 2008) and 7.4/1000 in Tanzania (Dent et al., 2005). The high prevalence of epilepsy and low life expectancy indicate that incidence is high. Reasons for this type of frequency could be ascribed to malnutrition, trauma, genetic conditions, antenatal and prenatal care, diseases resulting in febrile convulsions, and/or cerebral infections, and may even result in death in specifically diseased individuals. Relative contributions of each cause of epilepsy are difficult to determine. Only a few case-control studies have been done in SSA. Infections, particularly cysticercosis in its endemic areas, cause most cases of epilepsy. The implementation of studies to accurately determine the causes of epilepsy in SSA is urgently required. Such studies will help to lower the incidence of epilepsy in this region and better understand the etiology of epilepsy in other areas (Preux & Druet-Cabanac, 2005)

Despite the high frequency of epilepsy in SSA, There is only anecdotal data concerning birth defects due to antiepileptic drugs (Lekwuwa et al., 1995). In developing countries, the fetus of an epileptic mother is at even greater risk because anti-epileptic therapy is less likely to be well controlled, multiple drug therapy is more common, and cheaper drugs, more likely to cause fetal damage, are used in place of more expensive, less teratogenic anticonvulsant drugs (Shorvon & Farmer, 1988). The general fertility rate for reproductive-aged epileptic women for the 3-year period 2003–2006 preceding the study was 46 live births per 1000 women-years, decreased by two-thirds in Kenya (Kariuki et al., 2008). The social burden of the disease largely outweighed the medical burden in women living with epilepsy in SSA (Birbeck et al., 2008). Phenobarbitone was the most frequently prescribed drug

(Preux et al., 2000), and the use of phenytoin therapy is not unusual (Valodia et al., 1999). Fetal valproate syndromes have been reported in sibling pairs in South Africa (Christianson et al., 1994).

Maternal Malnutrition

Recent estimates indicate that globally over two billion people are at risk for vitamin A, iodine, and/or iron deficiency, in spite of recent efforts in the prevention and control of these deficiencies. The prevalence is especially high in Southeast Asia and SSA, and pregnant women and young children are at greatest risk. Other micronutrient deficiencies of public health concern include zinc, folate, and the B vitamins. However, there is limited data on the actual prevalence of these deficiencies. Finally, in many settings, more than one micronutrient deficiency exists, suggesting the need for simple approaches that evaluate and address multiple micronutrient malnutrition (Ramakrishnan, 2002).

Folic Acid

Overwhelming evidence links an increased intake of synthetic folic acid (the bioavailable form of folate) to a reduced risk of neural tube defects. Because neural tube defects result from early failure of neural tube closure, which occurs before pregnancy can be confirmed, research indicates that at least half the cases of neural tube defects can be prevented if women consume folic acid before conception and during early pregnancy (CDC, 2005). These and other findings lent support to the 1992 recommendation by the U.S. Public Health Service, and the 1998 recommendation of the Institute of Medicine of the U.S. National Academy of Sciences that all women of childbearing age should consume 0.4 mg (400 µg) of synthetic folic acid daily, in addition to folate provided by a varied diet, in order to reduce their risk of a neural tube defects–affected pregnancy (IOM, 1998).

A 1997 demographic and health survey in Mozambique showed that 47% of girls 15 to 19 years old living in Manica province (west-central Mozambique) were pregnant or had already had a child. Supplementation periods can have an important impact on girls' hematological status and birth defect prevention (Horjus et al., 2005). There was a high prevalence of anemia, and folate deficiency varies among different population in SSA. Up to 57.7% of pregnant women were reported to have anemia and folate deficiency in Sudan (Abdelrahim et al., 2009), Studies also evidenced folate deficiency in Nigerian pregnant women (Vanderjagt et al., 2007, Kamau-Mbuthia & Elmadfa, 2007); Folate deficiency, defined by red blood cell folate of less than 100 micrograms/l, was observed in 6% of pregnant women in Congo (Mashako et al., 1991). However, only 2% had low plasma folate (< 6.8 nmol/L) in Ethiopia (Gibson et al., 2008), and in a cross-sectional study of 150 pregnant women in Blantyre, Malawi, 23% were iron-deficient with no evidence of folate deficiencies (van den Broek & Letsky, 2000).

In SSA, effective use of iron/folic acid during pregnancy was estimated without intervention to be 40% and 86% with improving women's access to supplements in Senegal (Seck & Jackson, 2009). A lack of awareness and a lack of supplies and materials were the main barriers to the use of iron/folic acid during pregnancy in Cameroon (Tita et al, 2005). Maternal international multiple micronutrient preparations (UNIMMAP), rather than iron and folic acid alone, modestly but significantly increased fetal growth in Senegal, Niger, Guinea-Bissau, Zimbabwe, and Tanzania (Roberfroid et al., 2008; Zagré et al., 2007; Kaestel et al., 2005; Friis et al., 2004; Makola et al., 2003).

If a woman has already had a baby with a neural tube defects, studies have shown that a larger daily dose of folic acid (4000 µg or 4 mg), beginning at least one month before pregnancy and throughout the first trimester, reduces the risk of having another affected pregnancy by about 70%. Women with diabetes, epilepsy being treated with sodium valproate and, possibly, obesity also are at increased risk of having a baby with a neural tube defect and should discuss the possibility of taking a higher level of folic acid with their primary care provider (MOD, 2004). An increasing body of evidence suggests that folic acid may help prevent other major malformations including congenital heart defects, as well as coronary heart disease, certain forms of cancer, and possibly dementia (Canfield et al., 2005; McIlroy et al., 2002).

A decrease in NTD rates post-fortification was reported in Black populations in South Africa (Sayed et al., 2008). Nutrient education for dietary diversification, and nutrient supplementation should be explored as interventions for improving nutrient intake among pregnant women. The economic benefit flowing from the prevention of NTDs greatly exceeds the costs of implementing folic acid fortification.

Iodine

Iodine is part of thyroid hormones, which contribute to brain development in the fetus and regulates human metabolism; iodine deficiency is the most frequent cause of preventable mental retardation. Related disorders include lethargy, physical disabilities, goiter, stillbirth, and neonatal death. UNICEF considers iodine deficiency the single most important cause during pregnancies of brain damage and intellectual disability, with most cases caused before birth (UNICEF, 1998). Despite these efforts, though, approximately 1.6 billion people in more than 100 countries live in areas where iodine is not available in sufficient amounts; those most at risk include about one-third of China's population. It is also a severe problem in some areas in SSA (Boulet, 1997; Cherinet & Kelbessa, 2000; Abuye et al., 2008). In 1998, 60,000 babies were born worldwide with severe congenital hypothyroidism and an estimated 28 million pregnancies were still at risk from maternal iodine deficiency (UNICEF, 1998). About 50 million people worldwide live with intellectual, motor, and hearing disabilities due to iodine deficiency disorders (UNICEF, 2000).

The gross prevalence of iodine deficiency disorders (IDD) among schoolchildren was 53.3% in Ethiopia (Cherinet & Kelbessa, 2000). The virtual absence of iodized salt in Ethiopian households shows that currently there is no salt iodization program in the country (Abuye et al., 2008). A study in Sudan suggests that, in areas with mild iodine deficiency, neonates may be at the limit of decompensation as evidenced by their enhanced TSH and T3 levels, as well as increased T4 compared to their mothers (Eltom et al., 2001). Similarly, the incidence of neonatal chemical hypothyroidism (NCH) in Saki (Nigeria) was 14.7 per 1000 babies. The results suggest environmental iodine deficiency and relative chemical hypothyroidism of mothers and neonates in Saki when compared to mothers and noenates from non-iodine-deficient areas (Ojule & Osotimehin, 1998). In school-aged children in South Africa, surveys demonstrated goiter prevalence in the 20%–29.9% range, and a median urinary iodine level in the 2–4.9 micrograms/dl range, indicating iodine deficiency of moderate severity. Iodized salt was not available in any of the three community shops studied (Benedé et al., 1997). Study in a tertiary care infant welfare clinic in Ibadan, southwestern Nigeria, a geographical area recognized to be outside Nigeria's endemic goiter belt (goiter prevalence < 5.0%), suggests that the breast-fed child of an iodine-replete mother resident in a non-iodine-deficient area may be iodine-deficient and in need of iodine supplementation. Deficiency affects most of the Zairian population and requires public health measures on a larger scale than previously estimated. Combined iodine and selenium deficiency affects Equateur, Haut-Zaire, and Kivu, where endemic myxoedematous cretinism occurs, but also in Shaba, where it was not previously described. Besides combined iodine and selenium deficiency, another factor (thiocyanate) must be taken into account to explain the peculiarly elevated prevalence of endemic myxoedematous cretinism in Central Africa (Ngo et al., 1997). Besides low levels of iodine intake, cassava consumption and living in a high altitude were responsible for the observed variation and severity in goiter rates. These observations have important implications for planning IDD control programs in Africa (Akanji et al., 1996).

Progress has been made due to affordable, cost-effective technology and alliances between government, the private sector, and international agencies. Almost all but Ethiopia, of the 53 countries of Africa that were identified as having IDD problems, had implemented IDD control programs by 1995. The programs emphasized the long-term use of iodized salt. By February 1996, over 50% of the salt consumed in Africa was estimated to be iodized. It also appears that the proportion of households using iodized salt is increasing. Fortification of household salt appears to be an appropriate method of controlling iodine deficiency in Benin (Melse-Boonstra et al., 1998). Preliminary evidence suggests that the prevalence of goiter in three districts of Kenya declined rapidly after the introduction of legislation authorizing use of iodized salt. Success is not just limited to declines in IDD (Kavishe, 1997). In South Africa, mandatory iodization of salt virtually eradicated iodine deficiency within one year in South African schoolchildren, but the goiter rate in these children did not decline (Kalk, 1998; Jooste et al., 2000). Countrywide, around 30% of low socioeconomic households obtained their salt from unconventional sources compared to less than 5% in high socioeconomic households, emphasizing the vulnerability of low socioeconomic groups to the use of non-iodized salt. Similarly, a short prospective study was conducted to assess thyroid status in healthy full-term newborns (n=90) of a large maternity hospital in Marrakech (Morocco), as part of the validation of a national salt iodization program. High TSH (>5mU/l) was detected in 89% of infants tested; however, iodine deficiency seems to remain a widespread problem in this population and justifies a large-scale survey of iodine status (Zaida et al., 2006).

Intervention strategies should mobilize all players involved in unconventional marketing channels of household salt to provide only iodized salt to consumers, as required by law (Jooste, 2005). A follow-up of proper implementation of the program is required to reach the realistic goal to expect IDD to be eradicated in the next century.

RECREATIONAL AND THERAPEUTIC DRUGS

Fetal Alcohol Spectrum Disorder

Fetal alcohol spectrum disorder (FASD) encompasses a range of effects that can occur in an individual whose mother drank alcohol during pregnancy (SAMHSA, 2005). These effects, which can range from mild to severe, include physical, behavioral, mental, and learning disabilities with lifelong implications for affected individuals (Hoyme et al., 2005). The only cause of FASD is alcohol use during pregnancy, and no threshold for safety is known. Multidisciplinary approaches include a dysmorphology examination, developmental and neuropsychological testing, and a structured maternal interview to gather data about prenatal drinking practices and other demographic and family information, as recommended in published guidelines, are used for diagnosis (Hoyme et al., 2005, Peadon et al., 2008). The facial anomalies associated with fetal alcohol syndrome (FAS), some of which are also present in individuals with less severe forms of the broader category of fetal alcohol spectrum disorders (FASD), are typically identified with the aid of linear distance measurements taken between facial landmarks (smooth philtrum, small palpebral fissures, and thin upper lip; see Douglas & Mutsvangwa, 2010).

The prevalence of FAS appears to be much higher in certain other regions of the world. In the wine-growing Western Cape Province of South Africa, more than a quarter of pregnant women abuse alcohol during pregnancy and around 4% of six- to seven-year-old school children had physical features suggestive of FAS (the highest rate ever reported for a stable community; Viljoen et al., 2003). Comparable studies in urban Johannesburg found 2.7% of children had fetal alcohol syndrome (Croxford & Viljoen, 1999; May et al., 2000; Viljoen et al., 2003). This finding has raised concerns about the prevalence of FAS in other middle- and low-income countries where alcohol is available and used by women of reproductive age (Rosenthal et al., 2005).

THERAPEUTIC DRUGS

A number of therapeutic drugs have been implicated as teratogens, and their adverse effects linked to the timing of use during pregnancy, dose, genetic susceptibility, and other factors. Jolted into action by the thalidomide tragedy, developed Western countries began to establish pharmacovigilance systems for identifying and responding to adverse drug reactions and events about 40 years ago. These systems focus on side effects, adverse reactions, and drug interactions. In developing countries, especially in Africa, the scope for pharmacovigilance needs to be broader (despite the additional challenges this brings) because of growing problems with substandard and counterfeit drugs, and the need to have an early warning signal system for the development of antimicrobial resistance to the "new essential drugs" that are barely beyond the clinical trial stage in Africa; e.g. artemisinin-combination therapy (ACT) for malaria, and antiretrovirals (ARV) for HIV/AIDS. Zambia learned important lessons from its own initial experiences in attempting to use ACT as a pathfinder for pharmacovigilance, as well as its experience with other drug information systems (Huff-Rousselle et al., 2007).

Thalidomide causes severe limb-reduction birth defects and was formerly used to combat morning sickness, but is also used to treat leprosy, macular degeneration in AIDS, and psoriasis, and also inhibits monocyte tumor necrosis factor (TNF)-alpha production and co-stimulates T cell. It was tested for immune modulation in patients with human immunodeficiency virus (HIV) infection and tuberculosis (TB) in a placebo-controlled study. In the presence of thalidomide, antigen-specific immune responses, in vitro and in patients with HIV/TB, were enhanced (Bekker et al. 2000, Baccaglini et al., 2007). There is a need to monitor such limited anecdotal reports of benefit existing for thalidomide or other agents on HIV patients (Marais et al., 2009). Indeed, implications for birth defects specifically in pregnant women affected with HIV should not be underestimated.

Anticonvulsant drugs are associated with major malformations, including neural tube defects, microcephaly, intellectual disability, growth restriction, and malformations of the face and fingers. Anticoagulants have been linked to nasal hypoplasia, stippling of bones, optic atrophy, microcephaly, growth and intellectual disability, and fetal and neonatal hemorrhage (Adab et al., 2001; Holmes et al., 2001). These medications are cause for concern in SSA, where use is less regulated and obstetric or fetal outcomes are unmonitored; the data are needed.

Unlike the Western world, valvular disease ranks among the major cardiovascular afflictions in Africa. Acute rheumatic fever and chronic rheumatic valvular disease in their most virulent form are still commonly encountered and impose a huge burden on limited healthcare resources (Essop & Nkomo, 2005). Pregnant patients with mechanical valves require careful attention to ensure maternal survival and prevent fetal warfarin embryopathy. Prolonged subcutaneous heparin and frequent monitoring of the partial thromboplastin time are sometimes impractical in this setting. The pregnancy rate in open heart surgical patients at Kenyatta National Hospital was approximately one-quarter of that in the general Kenyan population. The risks of maternal and fetal complications with heparin seem to be greater than the likely drawbacks associated with warfarin use in these patients (Ogendo et al., 1999). The mean fetal birth weight was low (2.54 +/– 0.98 kg). In South Africa, a study on maternal and fetal sequelae of anticoagulation during pregnancy in patients with mechanical heart valve prostheses, revealed 9 abortions (18%), 7 stillbirths (14%), 2 neonatal deaths (4%) and 2 instances of warfarin embryopathy (4%) (Sareli et al., 1989).

CONCLUSION AND PERSPECTIVES

As revealed and emphasized by the present review, as African countries go through an epidemiological transition, with a reduction in childhood and infant mortality due to improved public health measures, babies who would have previously died of these diseases before they were recognized are now surviving to present for diagnosis and treatment. Hence, they are presenting an increasing global health burden. In addition some specific factors like malaria, migration, consanguineous marriages, maternal age, and the poor level of national health systems contribute to the bulk of the burden of birth defects in SSA.

In the context of sickle cell anemia, there is an urgent need to develop projects to gather more epidemiological data on birth defects, the stage in development at which each country needs to introduce medical genetic services, the relative roles of care and prevention in public health policies, the appropriate application of knowledge and technology to these purposes, and the financial implications. WHO, in partnership with a March of Dimes affiliate, the World Alliance of Organizations for the Prevention of Birth Defects, identified the need to integrate medical genetic services into primary healthcare and to further integrate these with secondary and tertiary medical genetic services and with reproductive, obstetric, pediatric, and other medical and social services, and also considered how this could be achieved within the limits of local circumstances, customs, available manpower, and material resources (WHO, 1999). The objectives are to care for those with a disorder in order to reduce suffering and to improve health by prevention.

When raised in the context of developing health policy for many middle- and low-income countries, the need for prevention is usually given priority over the requirement for patient care. In fact, the bulk of care and prevention of birth defects is most appropriately carried out routinely in primary and secondary care settings (Christianson & Modell, 2004; WHO, 1999). In low-income countries the most effective interventions for prevention include family planning, optimizing women's diets, managing maternal health problems, and avoiding maternal infections. These are feasible and affordable in financially constrained environments and have a cascade of beneficial effects in addition to preventing birth defects. Because risk factors for birth defects (including advanced maternal age, poverty, maternal medical complications, infection, poor nutrition, smoking, alcohol and drug use) are common to other adverse maternal and child health outcomes, interventions to reduce birth defects will contribute to overall women's, maternal, neonatal, and child health. Care and prevention of birth defects should be considered an integral and cost-effective arm of public health programs directed at saving the lives of and reducing disability among women, newborns, and children. In fact, some interventions like fortification of food with micronutrients benefit the entire population (Christianson & Modell, 2004).

Our review suggests that the goals of prenatal care in SSA should be revisited in order to implement a program based on real scientific evidence. Specific purposes of consultations would be: (1) to screen birth defect risk factors, which, when recognized, lead to specific action including: determination of advanced maternal age, nucal translucency screening, TORCH screening, micronutrient inquiry and supplementation when necessary (iodine, folate), history of therapeutic or recreational drugs (anticonulsivant and alcohol); (2) to prevent and/or detect (and treat) specific complications of pregnancy including infection (venereal disease, HIV), anemia and trace element deficiencies, and gestational diabetes mellitus; (3) to provide counseling, support, and information for pregnant women and their families (including the partner) concerning sickle cell disease screening (prenatal and neonatal) and, if possible, prenatal genetic diagnosis and reproductive options. These potentially effective actions aimed at preventable birth defects can only have a real public health impact if implemented within an organized maternal health system with a functional network of delivery units, if truly quality care is given, and if the relationships between healthcare providers and the population are based on mutual respect. SSA women use prenatal care extensively when it is accessible; this opportunity must be used to implement evidence-based actions with appropriate and realistic goals of birth defect prevention.

Because of their uneven distribution in high-frequency populations, reflecting their complex population genetics, the true magnitude of burden of hemoglobinopathies alone is still unknown. In many SSA countries there are few facilities for the diagnosis and management of genetic and partially genetic birth defects, and there are limited data about their frequency, clinical course, or mortality. Without this information it will be impossible to persuade governments of the increasing importance of these diseases. The situation will only be improved by concerted action on the part of the international community of the richer countries together with input from the major international health organizations and funding agencies (Weatherall, 2010).

REFERENCES

Abdelgadir M, Elbagir M, Eltom A, Eltom M, Berne C. (2003). Factors affecting perinatal morbidity and mortality in pregnancies complicated by diabetes mellitus in Sudan. *Diabetes Res Clin Pract 60*(1), 41–47.

Abdel-Meguid N, Zaki MS, Hammad SA. (2000). Premarital genetic investigations: Effect of genetic counselling. *East Mediterr Health J* 6(4), 652–660.

Abdelrahim II, Adam GK, Mohmmed AA, et al. (2009). Anaemia, folate and vitamin B12 deficiency among pregnant women in an area of unstable malaria transmission in eastern Sudan. *Trans R Soc Trop Med Hyg 103*(5), 493–496.

Abuye C, Berhane Y, Ersumo T. (2008). The role of changing diet and altitude on goitre prevalence in five regional states in Ethiopia. *East Afr J Public Health* 5(3), 163–168.

Adab N, Jacoby A, Smith D, Chadwick D. (2001). Additional educational needs in children born to mothers with epilepsy. *J Neurol Neurosurg Psychiatry 70*(1), 15–21.

Adegbidi H, Yedomon H, Atadokpede F, et al. (2007). Skin cancers at the National University Hospital of Cotonou from 1985 to 2004. *Int J Dermatol* 46 Suppl 1, 26–29.

Adegbola O, Ajayi GO. (2008). Screening for gestational diabetes mellitus in Nigerian pregnant women using fifty-gram oral glucose challenge test. *West Afr J Med 27*(3), 139–143.

Adeleye AO, Olowookere KG. (2009). Central nervous system congenital anomalies: A prospective neurosurgical observational study from Nigeria. *Congenit Anom (Kyoto) 49*(4), 258–261.

Adewole RA, Akinwande JA. (2007). Public and professional perception of oral and maxillofacial surgery (a pilot study). *Nig Q J Hosp Med 17*(1), 8–12.

Adeyokunnu AA. (1982). The incidence of Down's syndrome in Nigeria. *J Med Genet 19*(4), 277–279.

Adou-Bryn KD, Ouhon J, Nemer J, Yapo CG, Assoumou A. (2004). Serological survey of acquired toxoplasmosis in women of childbearing age in Yopougon (Abidjan, Côte d'Ivoire). *Bull Soc Pathol Exot 97*(5), 345–348.

Ahmed H, Yukubu AM, Hendrickse RG. (1995). Neonatal jaundice in Zaria, Nigeria—a second prospective study. *West Afr J Med 14*(1), 15–23.

Aidoo M, Terlouw DJ, Kolczak MS, et al. (2002). Protective effects of the sickle cell gene against malaria morbidity and mortality. *Lancet 359*(9314), 1311–1312.

Akanji AO, Mainasara AS, Akinlade KS. (1996). Urinary iodine excretion in mothers and their breast-fed children in relation to other childhood nutritional parameters. *Eur J Clin Nutr 50*(3), 187–191.

Akinyanju OO, Disu RF, Akinde JA, et al. (1999). Initiation of prenatal diagnosis of sickle-cell anaemia in Africa. *Prenat Diagn 19*(4), 299–304.

Akinyanju OO, Otaigbe AI, Ibidapo MO. (2005). Outcome of holistic care in Nigerian patients with sickle cell anaemia. *Clin Lab Haematol 27*(3), 195–199.

Akoua-Koffi GC, Bakayoko S, Tanoh G, et al. (1999). Prevalence of herpes simplex virus infection in pregnant women in Yopougon (Côte d'Ivoire). *Bull Soc Pathol Exot 92*(1), 3–5.

Alatise OI, Adeolu AA, Komolafe EO, Adejuyigbe O, Sowande OA. (2006). Pattern and factors affecting management outcome of spina bifida cystica in Ile-Ife, Nigeria. *Pediatr Neurosurg 42*(5), 277–283.

Ali SK. (2009). Cardiac abnormalities of Sudanese patients with Down's syndrome and their short-term outcome. *Cardiovasc J Afr 20*(2), 112–115.

Aliyu ZY, Gordeuk V, Sachdev V, et al. (2008). Prevalence and risk factors for pulmonary artery systolic hypertension among sickle cell disease patients in Nigeria. *Am J Hematol 83*(6), 485–490.

Almu S, Tadesse Z, Cooper P, Hackett R. (2006). The prevalence of epilepsy in the Zay Society, Ethiopia—an area of high prevalence. *Seizure 15*(3), 211–213.

Andronikou S, Wieselthaler N, Fieggen AG. (2006). Cervical spina bifida cystica: MRI differentiation of the subtypes in children. *Childs Nerv Syst 22*(4), 379–384.

Aquaron R, Djatou M, Kamdem L. (2009). Sociocultural aspects of albinism in sub-Saharan Africa: Mutilations and ritual murders committed in east Africa (Burundi and Tanzania) *Med Trop 69*(5), 449–453. [French].

Asuquo ME, Ngim O, Ebughe G, Bassey EE. (2009). Skin cancers amongst four Nigerian albinos. *Int J Dermatol 48*(6), 636–638.

Awori MN, Ogendo SW, Gitome SW, Ong'uti SK, Obonyo NG. (2007). Management pathway for congenital heart disease at Kenyatta National Hospital, Nairobi. *East Afr Med J 84*(7), 312–317.

Ba MC, Ly Ba A, Hossini A, et al. (2007). The occipital encephaloceles: Report of 16 cases. *Mali Med 22*(2), 54–57.

Baccaglini L, Atkinson JC, Patton LL, Glick M, Ficarra G, Peterson DE. (2007). Management of oral lesions in HIV-positive patients. *Oral Surg Oral Med Oral Pathol Oral Radiol Endod 103*(Suppl S50), 1–23.

Baghagho EE, Kharboush IF, El-Kaffash DM, KarKour TA, Ismail SR, Mortada MM. (2004). Maternal serum alpha fetoprotein among pregnant females in Alexandria. *J Egypt Public Health Assoc 79*(1–2), 59–81.

Balogou AA, Grunitzky EK, Belo M, et al. (2007). Management of epilepsy patients in Batamariba district, Togo. *Acta Neurol Scand 116*(4), 211–216.

Bassili A, Mokhtar SA, Dabous NI, Zaher SR, Mokhtar MM, Zaki A. (2000). Risk factors for congenital heart diseases in Alexandria, Egypt. *Eur J Epidemiol 16*(9), 805–814.

Beighton P, Botha MC. (1986). Inherited disorders in the black population of southern Africa. Part I. Historical and demographic background; genetic haematological conditions. *S Afr Med J 69*(4), 247–249.

Bekker LG, Haslett P, Maartens G, Steyn L, Kaplan G. (2000). Thalidomide-induced antigen-specific immune stimulation in patients with human immunodeficiency virus type 1 and tuberculosis. *J Infect Dis 181*(3), 954–965.

Bello C, Whittle H. (1991). Cytomegalovirus infection in Gambian mothers and their babies. *J Clin Pathol 44*(5), 366–369.

Ben Abdallah R, Aoun K, Siala E, et al. (2009). Congenital toxoplasmosis: Clinical and biological analysis of 11 cases in Tunisia. *Arch Pediatr 16*(2), 118–121.

Ben Arab S, Masmoudi S, Beltaief N, Hachicha S, Ayadi H. (2004). Consanguinity and endogamy in Northern Tunisia and its impact on non-syndromic deafness. *Genet Epidemiol 27*(1), 74–79.

Benedé JG, Oelofse A, van Stuijvenberg ME, Jooste PL, Weight MJ, Benadé AJ. (1997). Endemic goitre in a rural community of KwaZulu-Natal. *S Afr Med J 87*(3), 310–313.

Birbeck GL, Chomba E, Atadzhanov M, Mbewe E, Haworth A. (2008). Women's experiences living with epilepsy in Zambia. *Am J Trop Med Hyg 79*(2), 168–172.

Birbeck GL, Kalichi EM. (2004). Epilepsy prevalence in rural Zambia: A door-to-door survey. *Trop Med Int Health 9*(1), 92–95.

Bittles AH, Black ML. (2010). Evolution in health and medicine Sackler colloquium: Consanguinity, human evolution, and complex diseases. *Proc Natl Acad Sci U S A 107*(Suppl 1), 1779–1786.

Bittles AH, Mason WM, Greene J, Rao NA (1991). Reproductive behavior and health in consanguineous marriages. *Science 252*(5007), 789–794.

Blandford JM, Gift TL, Vasaikar S, Mwesigwa-Kayongo D, Dlali P, Bronzan RN. (2007). Cost-effectiveness of on-site antenatal screening to prevent congenital syphilis in rural eastern Cape Province, Republic of South Africa. *Sex Transm Dis 34*(Suppl 7), S61–S66.

Boano R, Fulcheri E, Martina MC, et al. (2009). Neural tube defect in a 4000-year-old Egyptian infant mummy: A case of meningocele from the museum of anthropology and ethnography of Turin (Italy). *Eur J Paediatr Neurol 13*(6), 481–487.

Mayosi BM, Flisher AJ, Lalloo UG, et al. (2009). The burden of non-communicable diseases in South Africa. *Lancet 374*(9693), 934–947.

Bos P, Steele D, Alexander J. (1995). Prevalence of antibodies to rubella, herpes simplex 2 and cytomegalovirus in pregnant women and in neonates at Ga-Rankuwa. *Cent Afr J Med 41*(1), 14–17.

Bouhsain S, Dami A, Elannaz H, et al. (2009). A critical study of the screening practices of gestational diabetes of a service of gynecology and obstetrics. *Ann Biol Clin (Paris) 67*(2), 159–162.

Boulet M. (1997). Micronutrient deficiencies. Reports from the field—Africa. *Glob Impacts 13*.

Brabin BJ, Prinsen-Geerligs PD, et al. (2004). Haematological profiles of the people of rural southern Malawi: An overview. *Ann Trop Med Parasitol 98*(1), 71–83.

Brink PA, Crotti L, Corfield V, et al. (2005). Phenotypic variability and unusual clinical severity of congenital long-QT syndrome in a founder population. *Circulation 112*(17), 2602–2610.

Bronzan RN, Mwesigwa-Kayongo DC, Narkunas D, et al. (2007). On-site rapid antenatal syphilis screening with an immunochromatographic strip improves case detection and treatment in rural South African clinics. *Sex Transm Dis 34*(Suppl 7), S55–S60.

Bütow KW, van Wyk PJ, Zwahlen RA. (2007). Differences in the clinical appearances of white versus black patients with facial cleft deformities: A retrospective study of a South African clinic. *SADJ 62*(7), 298, 300–304.

Bukar M, Audu BM, Takai UI, Ajayi BB, Kullima AA. (2009). Is routine antenatal screening for syphilis in Nigeria still justified clinically and economically? *Saudi Med J 30*(10), 1311–1315.

Butler MG. (2009). Genomic imprinting disorders in humans: A mini-review. *J Assist Reprod Genet 26*(9–10), 477–486. Review.

Canfield MA, Collins JS, Botto LD, et al. (2005). Changes in the birth prevalence of selected birth defects after grain fortification with folic acid in the United States: Findings from a multi-state population-based study. *Birth Defects Research, Part A* 73, 679–689.

Castro O, Brambilla DJ, Thorington B, et al. (1994). The acute chest syndrome in sickle cell disease: Incidence and risk factors. The Cooperative Study of Sickle Cell Disease. *Blood 84*(2), 643–649.

Challis K, Melo A, Bugalho A, Jeppsson JO, Bergström S. (2002). Gestational diabetes mellitus and fetal death in Mozambique: an incident case-referent study. *Acta Obstet Gynecol Scand. 81*(6), 560–563.

Chelli D, Dimassi K, Chaabouni M, et al. (2008). Prenatal diagnosis of trisomy 21: The Tunisian experience. *Santé 18*(4), 199–203.

Cherinet A, Kelbessa U. (2000). Determinants of iodine deficiency in school children in different regions of Ethiopia. *East Afr Med J 77*(3), 133–137.

Child A, Comeglio P, Arno G, Beighton P. (2007). Marfan syndrome in South Africa: A molecular genetic approach to diagnosis. *S Afr Med J 97*(9), 845–847.

Christianson A, Modell B. (2004). Medical genetics in developing countries. *Annu Rev Genomics Hum Genet 5*, 219–265.

Christianson AL, Chesler N, Kromberg JG. (1994). Fetal valproate syndrome: Clinical and neuro-developmental features in two sibling pairs. *Dev Med Child Neurol 36*(4), 361–369.

Christianson AL, Kromberg JG. (1996). Maternal non-recognition of Down syndrome in black South African infants. *Clin Genet 49*(3), 141–144.

Christianson AL, Zwane ME, Manga P, et al. (2002). Children with intellectual disability in rural South Africa – prevalence and associated disabilities. *Journal of International Disability Research 46*: 179–186.

Christianson AL. (1996). Down syndrome in sub-Saharan Africa. *J Med Genet. 33*(2), 89–92.

Christianson AL. (1997). Down syndrome in black South African infants and children—clinical features and delayed diagnosis. *S Afr Med J 87*(8), 992–995.

Clegg JB, Weatherall DJ. (1999). Thalassemia and malaria: New insights into an old problem. *Proc Assoc Am Physicians 111*(4), 278–282.

Corcoran C, Hardie DR. (2005). Seroprevalence of rubella antibodies among antenatal patients in the Western Cape. *S Afr Med J 95*(9), 688–690.

Croxford J, Viljoen D. (1999). Alcohol consumption by pregnant women in the Western Cape. *S Afr Med J 89*, 962–965.

Dagher D, Ross E. (2004). Approaches of South African traditional healers regarding the treatment of cleft lip and palate. *Cleft Palate Craniofac J 41*(5), 461–469.

Dangor T, Ross E. (2006). Western and traditional medicine: Cultural beliefs and practices of South African Muslims with regard to Down syndrome. *S Afr J Commun Disord 53*, 27–38.

De Montalembert M, Tshilolo L. (2007). Is therapeutic progress in the management of sickle cell disease applicable in sub-Saharan Africa?. *Med Trop (Mars) 67*(6), 612–616.

de Paul Djientcheu V, Njamnshi AK, Wonkam A, et al. (2008). Management of neural tube defects in a Sub-Saharan African country: The situation in Yaounde, Cameroon. *J Neurol Sci 275*(1–2), 29–32.

Debrock C, Preux PM, Houinato D, et al. (2000). Estimation of the prevalence of epilepsy in the Benin region of Zinvié using the capture-recapture method. *Int J Epidemiol 29*(2), 330–335.

Dent W, Helbok R, Matuja WB, Scheunemann S, Schmutzhard E. (2005). Prevalence of active epilepsy in a rural area in South Tanzania: A door-to-door survey. *Epilepsia 46*(12), 1963–1969.

Deperthes BD, Meheus A, O'Reilly K, Broutet N. (2004). Maternal and congenital syphilis programmes: Case studies in Bolivia, Kenya and South Africa. *Bull World Health Organ 82*(6), 410–416.

Doehring E, Reiter-Owona I, Bauer O, et al. (1995). Toxoplasma gondii antibodies in pregnant women and their newborns in Dar es Salaam, Tanzania. *Am J Trop Med Hyg 52*(6), 546–548.

Donkor P, Bankas DO, Agbenorku P, Plange-Rhule G, Ansah SK. (2007). Cleft lip and palate surgery in Kumasi, Ghana: 2001–2005. *J Craniofac Surg 18*(6), 1376–1379.

Dörk T, El-Harith EH, Stuhrmann M, et al. (1998). Evidence for a common ethnic origin of cystic fibrosis mutation 3120+1G->A in diverse populations. *Am J Hum Genet 63*(2), 656–662.

Douglas TS, Mutsvangwa TE. (2010). A review of facial image analysis for delineation of the facial phenotype associated with fetal alcohol syndrome. *Am J Med Genet A 152A*(2), 528–536.

Duff P, Barth WH Jr, Post MD. (2009). Case records of the Massachusetts General Hospital. Case 4-2009. A 39-year-old pregnant woman with fever after a trip to Africa. *N Engl J Med 360*(5), 508–516.

Dunkle LM, Arvin AM, Whitley RJ, et al. (1991). A controlled trial of acyclovir for chickenpox in normal children. *N Engl J Med. 325*(22), 1539–1544.

Edwards T, Scott AG, Munyoki G, et al. (2008). Active convulsive epilepsy in a rural district of Kenya: A study of prevalence and possible risk factors. *Lancet Neurol 7*(1), 50–56.

Edwin F, Sereboe LA, Tettey MM, et al. (2010). Experience from a single centre concerning the surgical spectrum and outcome of adolescents and adults with congenitally malformed hearts in West Africa. *Cardiol Young 4*, 1–6.

Egesie OJ, Joseph DE, Isiguzoro I, Egesie UG. (2008). Glucose-6-phosphate dehydrogenase (G6PD) activity and deficiency in a population of Nigerian males resident in Jos. *Niger J Physiol Sci 23*(1–2), 9–11.

Ejim EC, Ike SO, Anisiuba BC, Onwubere BJ, Ikeh VO. (2009). Ventricular septal defects at the University of Nigeria Teaching Hospital, Enugu: A review of echocardiogram records. *Trans R Soc Trop Med Hyg 103*(2), 159–161.

El Mansouri B, Rhajaoui M, Sebti F, et al. (2007). Seroprevalence of toxoplasmosis in pregnant women in Rabat, Morocco. *Bull Soc Pathol Exot 100*(4), 289–290.

El-Gozamy BR, Mohamed SA, Mansour HA. (2009). Toxoplasmosis among pregnant women in Qualyobia Governorate, Egypt. *J Egypt Soc Parasitol 39*(2), 389–401.

Elliott RF, Jovic G, Beveridge M. (2008). Seasonal variation and regional distribution of cleft lip and palate in Zambia. *Cleft Palate Craniofac J 45*(5), 533–538.

Eltom A, Eltom M, Idris M, Gebre-Medhin M. (2001). Thyroid function in the newborn in relation to maternal thyroid status during labour in a mild iodine deficiency endemic area in Sudan. *Clin Endocrinol (Oxf) 55*(4), 485–490.

Enevold A, Lusingu JP, Mmbando B, et al. (2008). Reduced risk of uncomplicated malaria episodes in children with alpha+-thalassemia in northeastern Tanzania. *Am J Trop Med Hyg 78*(5), 714–720.

Essop MR, Nkomo VT. (2005). Rheumatic and nonrheumatic valvular heart disease: Epidemiology, management, and prevention in Africa. *Circulation 112*(23), 3584–3591.

Faye M, Hennequin M, Yam AA, Ba I. (2004). Evaluation of oral health and access to care in Senegalese children with Down syndrome: Preliminary study. *Dakar Med 49*(1), 64–69.

Feuillet-Fieux MN, Ferrec M, Gigarel N et al. (2004). Novel CFTR mutations in black cystic fibrosis patients. *Clin Genet. 65*(4), 284–287.

Fitzgerald DW, Behets FM, Lucet C, Roberfroid D. (1998). Prevalence, burden, and control of syphilis in Haiti's rural Artibonite region. *Int J Infect Dis 2*(3), 127–131.

Föller M, Bobbala D, Koka S, Huber SM, Gulbins E, Lang F. (2009). Suicide for survival—death of infected erythrocytes as a host mechanism to survive malaria. *Cell Physiol Biochem 24*(3–4), 133–140. Review.

Friis H, Gomo E, Nyazema N, et al. (2004). Effect of multimicronutrient supplementation on gestational length and birth size: A randomized, placebo-controlled, double-blind effectiveness trial in Zimbabwe. *Am J Clin Nutr 80*(1), 178–184.

George AO, Ogunbiyi AO, Daramola OO, Campbell OB. (2005). Albinism among Nigerians with malignant melanoma. *Trop Doct 35*(1), 55–56.

Ghebrekidan H, Rudén U, Cox S, Wahren B, Grandien M. (1999). Prevalence of herpes simplex virus types 1 and 2, cytomegalovirus, and varicella-zoster virus infections in Eritrea. *J Clin Virol 12*(1), 53–64.

Gibson RS, Abebe Y, Stabler S, et al. (2008). Zinc, gravida, infection, and iron, but not vitamin B-12 or folate status, predict hemoglobin during pregnancy in Southern Ethiopia. *J Nutr 138*(3), 581–586.

Gichangi P, Renterghem LV, Karanja J, Bwayo J, Kiragu D, Temmerman M. (2004). Congenital syphilis in a Nairobi maternity hospital. *East Afr Med J 81*(11), 589–593.

Gladwin MT, Vichinsky E. (2008). Pulmonary complications of sickle cell disease. *N Engl J Med 359*(21), 2254–2265.

Goldman A, Labrum R, Claustres M, et al. (2001). The molecular basis of cystic fibrosis in South Africa. *Clin Genet 59*(1), 37–41.

Guidozzi F, Ballot D, Rothberg AD. (1994). Human B19 parvovirus infection in an obstetric population. A prospective study determining fetal outcome. *J Reprod Med 39*(1), 36–38.

Hampshire KR, Smith MT. (2001). Consanguineous marriage among the Fulani. *Hum Biol 73*(4), 597–603.

Hardy BJ, Séguin B, Ramesar R, Singer PA, Daar AS. (2008). South Africa: From species cradle to genomic applications. *Nat Rev Genet 9*(Suppl 1), S19–S23.

Hodges AM, Hodges SC. (2000). A rural cleft project in Uganda. *Br J Plast Surg 53*(1), 7–11.

McIlroy SP, Dynan KB, Lawson JT, Patterson CC, Passmore AP. (2002). Moderately elevated plasma homocysteine, methylenetetrahydrofolate reductase genotype, and risk for stroke, vascular dementia, and Alzheimer disease in Northern Ireland. *Stroke 33*(10), 2351–2356.

Holmes LB, Harvey EA, Coull BA, et al. (2001). The teratogenicity of anticonvulsant drugs. *N Engl J Med. 344*(15), 1132–1138.

Hong ES, Zeeb H, Repacholi MH. (2006). Albinism in Africa as a public health issue. *BMC Public Health 6*, 212.

Horjus P, Aguayo VM, Roley JA, Pene MC, Meershoek SP. (2005). School-based iron and folic acid supplementation for adolescent girls: Findings from Manica Province, Mozambique. *Food Nutr Bull 26*(3), 281–286.

Hoyme HE, May PA, Kalberg WO, et al. (2005). Pediatrics. A practical clinical approach to diagnosis of fetal alcohol spectrum disorders: Clarification of the 1996 institute of medicine criteria. *Pediatrics 115*(1), 39–47.

Huddle K, England M, Nagar A. (1993). Outcome of pregnancy in diabetic women in Soweto, South Africa 1983–1992. *Diabet Med 10*(3), 290–294.

Huddle KR. (2005). Audit of the outcome of pregnancy in diabetic women in Soweto, South Africa, 1992–2002. *S Afr Med J 95*(10), 789–794.

Huff-Rousselle M, Simooya O, Kabwe V, et al. (2007). Pharmacovigilance and new essential drugs in Africa: Zambia draws lessons from its own experiences and beyond. *Glob Public Health 2*(2), 184–203.

Idowu OE, Apemiye RA. (2008). Outcome of myelomeningocoele repair in sub-Saharan Africa: The Nigerian experience. *Acta Neurochir (Wien) 150*(9), 911–913; discussion 913.

Imperato GH, Imperato PJ. (2006). Beliefs and practices concerning twins, hermaphrodites, and albinos among the Bamana and Maninka of Mali. *J Community Health 31*(3), 198–224.

IOM (Institute of Medicine). (2003). *Reducing Birth Defects. Meeting the Challenge in the Developing World.* Board on International Health, Institute of Medicine, National Academy of Sciences. Washington, DC: National Academy Press.

IOM. Howson CP, Kennedy ET, Horwitz, A (eds.) (1998). *Prevention of Micronutrient Deficiencies: Tools for Policymakers and Public Health Workers.* Board on International Health, Institute of Medicine, National Academy of Sciences. Washington, DC: National Academy Press.

Jalloh A, Jalloh M, Gamanga I, et al. (2008). G6PD deficiency assessment in Freetown, Sierra Leone, reveals further insight into the molecular heterogeneity of G6PD A-. *J Hum Genet 53*(7), 675–679.

Jenkins T. (1990). Medical genetics in South Africa. *J Medl Genet 27*, 760–779.

Jenniskens F, Obwaka E, Kirisuah S, et al. (1995). Syphilis control in pregnancy: Decentralization of screening facilities to primary care level, a demonstration project in Nairobi, Kenya. *Int J Gynaecol Obstet 48*(Suppl), S121–S128.

Jooste PL, Weight MJ, Lombard CJ. (2000). Short-term effectiveness of mandatory iodization of table salt, at an elevated iodine concentration, on the iodine and goiter status of schoolchildren with endemic goiter. *Am J Clin Nutr 71*(1), 75–80.

Jooste PL. (2005). Sources of household salt in South Africa. *Asia Pac J Clin Nutr 14*(2), 159–162.

Kaestel P, Michaelsen KF, Aaby P, Friis H. (2005). Effects of prenatal multimicronutrient supplements on birth weight and perinatal mortality: A randomised, controlled trial in Guinea-Bissau. *Eur J Clin Nutr 59*(9), 1081–1089.

Kalk WJ. (1998). Iodine deficiency disorders in South Africa. *S Afr Med J 88*(3 Endocrinology), 352–354.

Kamanga J. (1995). Maternal and congenital syphilis in Zambia. *Afr Health 18*(1), 15–17.

Kamau-Mbuthia E, Elmadfa I. (2007). Diet quality of pregnant women attending an antenatal clinic in Nakuru, Kenya. *Ann Nutr Metab 51*(4), 324–330.

Kariuki JG, Joshi MD, Adam AM, Kwasa TO, Machoki M. Fertility rate of epileptic women at Kenyatta National Hospital. East Afr Med J (2008). *85*(7), 341–346.

Kaur R, Gupta N, Nair D, Kakkar M, Mathur MD. (1999). Screening for TORCH infections in pregnant women: A report from Delhi. *Southeast Asian J Trop Med Public Health 30*(2), 284–286.

Kavishe F. (1997). Salt iodization programmes in Africa. *SCN News.* 15, 10.

Khoury MJ, Becerra JE, Coredo JF, Erickson JD. (1989). Clinical-epidemiological assessment of pattern of birth defects associated with human teratogen: Application to diabetic embryopathy. *Pediatrics 84*, 658–665.

Kinsley B. (2007). Achieving better outcomes in pregnancies complicated by type 1 and type 2 diabetes mellitus. *Clin Ther 29*(Suppl D), S153–S160.

Koko J, Dufillot D, M'Ba-Meyo J, Gahouma D, Kani F. (1998). Mortality of children with sickle cell disease in a pediatric department in Central Africa. *Arch Pediatr 5*(9), 965–969.

Komba AN, Makani J, Sadarangani M, et al. (2009). Malaria as a cause of morbidity and mortality in children with homozygous sickle cell disease on the coast of Kenya. *Clin Infect Dis 49*(2), 216–222.

Krimmel M, Opitz C, Lenz JH, et al. (2009). Genome-wide linkage scan of nonsyndromic orofacial clefting in 91 families of central European origin. *Am J Med Genet A 149A*(12), 2680–2694.

Kromberg J. (1992). Albinism in the South African Negro: IV: Attitudes and the death myth. *Birth Defects 28*, 159–166.

Kromberg JGR, Castle D, Zwane EM, Jenkins T. (1989). Albinism and skin cancer in southern Africa. *Clinical Genetics 36*, 43–52.

Labie D, Pagnier J, Lapoumeroulie C, et al. (1985). Common haplotype dependency of high G gamma-globin gene expression and high Hb F levels in beta-thalassemia and sickle cell anemia patients. *PNAS 82*, 2111–2114.

Lawrenson JB, Kalis NN, Pribut H, et al. (2006). Why are some South African children with Down syndrome not being offered cardiac surgery? *S Afr Med J 96*(9 Pt 2), 914–919.

Lech MM. (2003). Non-effective partner notification system: A missed opportunity for the reduction of sexually transmitted infections in sub-Saharan Africa. *Med Wieku Rozwoj 7*(4 Pt 1), 503–509.

Lekwuwa GU, Adewole IF, Thompson MO. (1995). Antiepileptic drugs and teratogenicity in Nigerians. *Trans R Soc Trop Med Hyg 89*(2), 227.

Levin CE, Steele M, Atherly D, et al. (2007). Analysis of the operational costs of using rapid syphilis tests for the detection of maternal syphilis in Bolivia and Mozambique. *Sex Transm Dis 34*(Suppl 7), S47–S54.

Lindblom D. (2009). Heart surgery in Sudan. Experiences from a highly specialized cardiac surgicalcentre managed by an Italian humanitarian organization. *Lakartidningen 106*(38), 2370–2373.

Loua A, Lamah MR, Haba NY, Camara M. (2007). Frequency of blood groups ABO and rhesus D in the Guinean population. *Transfus Clin Biol 14*(5), 435–439.

Louw B, Shibambu M, Roemer K. (2006). Facilitating cleft palate team participation of culturally diverse families in South Africa. *Cleft Palate Craniofac J 43*(1), 47–54.

Lund PM, Maluleke TG, Gaigher I, Gaigher MJ. (2007). Oculocutaneous albinism in a rural community of South Africa: A population genetic study. *Ann Hum Biol 34*(4), 493–497.

Lund PM. (2005). Oculocutaneous albinism in southern Africa: Population structure, health and genetic care. *Ann Hum Biol 32*(2), 168–173.

Lund PM, Taylor JS. (2008). Lack of adequate sun protection for children with oculocutaneous albinism in South Africa. *BMC Public Health* 8, 225.

Lupski JR, Reid JG, Gonzaga-Jauregui C, et al. (2010). Whole-genome sequencing in a patient with Charcot-Marie-Tooth neuropathy. *N Engl J Med 362*(13), 1181–1191.

Lutale JK, Justesen A, Lema RS, Swai AB, McLarty DG. (1991). Outcome of pregnancy in diabetic patients in Dar es Salaam, Tanzania. *Diabet Med 8*(9), 881–884.

Luzzatto L, Mehta A, Meloni T. (1995). Haemoglobinuria and haptoglobin in G6PD deficiency. *Br J Haematol 91*(2), 511–512.

Macek M Jr, Mackova A, Hamosh A, et al. (1997). Identification of common cystic fibrosis mutations in African-Americans with cystic fibrosis increases the detection rate to 75%. Am J Hum Genet *60*(5), 1122–1127.

Mahmoud H, El-Haddad A, Fahmy O, et al. (2008). Hematopoietic stem cell transplantation in Egypt. *Bone Marrow Transplant 42*(Suppl 1), S76-S80.

Makani J, Williams TN, Marsh K. (2007). Sickle cell disease in Africa: Burden and research priorities. *Ann Trop Med Parasitol 101*(1), 3–14.

Makola D, Ash DM, Tatala SR, Latham MC, Ndossi G, Mehansho H. (2003). A micronutrient-fortified beverage prevents iron deficiency, reduces anemia and improves the hemoglobin concentration of pregnant Tanzanian women. *J Nutr 133*(5), 1339–1346.

Malik SJ, Mir NA. (1992). Perinatal mortality in high risk pregnancy: a prospective study of preventable factors. *Asia Oceania J Obstet Gynaecol. 18*(1), 45–48.

Mamabolo RL, Alberts M, Levitt NS, Delemarre-van de Waal HA, Steyn NP. (2007). Prevalence of gestational diabetes mellitus and the effect of weight on measures of insulin secretion and insulin resistance in third-trimester pregnant rural women residing in the Central Region of Limpopo Province, South Africa. *Diabet Med 24*(3), 233–239.

Mangold E, Reutter H, Birnbaum S, Walier M, et al. (2009). Consanguinity associated with increased risk for bipolar I disorder in Egypt. *Am J Med Genet B Neuropsychiatr Genet 150B*(6), 879–885.

Mansour H, Klei L, Wood J, et al. (2009). Consanguinity associated with increased risk for bipolar I disorder in Egypt. *Am J Med Genet B Neuropsychiatr Genet. 150B*(6), 879–885.

Marais S, Wilkinson RJ, Pepper DJ, Meintjes G. (2009). Management of patients with the immune reconstitution inflammatory syndrome. *Curr HIV/AIDS Rep 6*(3), 162–171.

Markenson G, Yancey M. (1998). Parvovirus B19 infections in pregnancy. *Obstetrics & Gynecology 91*, 125–128.

Martines J, Paul VK, Bhutta ZA, et al. (2005). Lancet Neonatal Survival Steering Team. Neonatal survival: A call for action. *Lancet 365* (9465), 1189–1197.

Mashako L, Preziosi P, Nsibu C, et al. (1991). Iron and folate status in Zairian mothers and their newborns. *Ann Nutr Metab 35*(6), 309–314.

Matolweni LO, Bardien S, Rebello G, et al. (2006). Arrhythmogenic right ventricular cardiomyopathy type 6 (ARVC6), support for the locus assignment, narrowing of the critical region and mutation screening of three candidate genes. *BMC Med Genet 7*, 29.

May PA, Brooke L, Gossage JP, Croxford J, et al. (2000). Epidemiology of fetal alcohol syndrome in a South African community in the Western Cape Province. *Am J Public Health 90*(12), 1905–1912.

Mayosi BM, Flisher AJ, Lalloo UG, Sitas F, Tollman SM, Bradshaw D. (2009). The burden of non-communicable diseases in South Africa. *Lancet 374*(9693), 934–947. Review.

McBride SR, Leppard BJ. (2002). Attitudes and beliefs of an albino population toward sun avoidance: Advice and services provided by an outreach albino clinic in Tanzania. *Arch Dermatol 138*(5), 629–632.

McIlroy SP, Dynan KB, Lawson JT, Patterson CC, Passmore AP. (2002). Moderately elevated plasma homocysteine, methylenetetrahydrofolate reductase genotype, and risk for stroke, vascular dementia, and Alzheimer disease in Northern Ireland. *Stroke 33*(10), 2351–2356.

Melse-Boonstra A, Rozendaal M, Rexwinkel H, et al. (1998). Determination of discretionary salt intake in rural Guatemala and Benin to determine the iodine fortification of salt required to control iodine deficiency disorders: Studies using lithium-labeled salt. *Am J Clin Nutr 68*(3), 636–641.

Menzel S. and Thein SL. (2009). Genetic architecture of hemoglobin F control. *Current Opinion in Hematology 16*, 179–186.

MOD (March of Dimes Global). (2006). *The March of Dimes Global Report on Birth Defects, The Hidden Toll of Dying and Disabled Children*. White Plains, New York: March of Dimes Birth Defects Foundation.

MOD. (2001). *Cytomegalovirus in Pregnancy: Fact Sheet.* Order Number: 09-1473-00. White Plains, New York.

MOD. (2004). *Folic Acid: Fact Sheet.* Order Number: 09-1144-98. White Plains, New York.

MOD. (2002). *Sexually Transmitted Infections in Pregnancy: Fact Sheet.* Order Number: 09-1651-02. White Plains, New York.

Mogayzel PJ Jr, Flume PA. (2010). Update in cystic fibrosis 2009. *Am J Respir Crit Care Med 181*(6), 539–544.

Mokhtar MM, Abd el-Aziz AM, Nazmy NA, Mahrous HS. (2003). Cytogenetic profile of Down syndrome in Alexandria, Egypt. *East Mediterr Health J 9*(1–2), 37–44.

Mokhtar MM, Abdel-Fattah MM. (2001). Consanguinity and advanced maternal age as risk factors for reproductive losses in Alexandria, Egypt. *Eur J Epidemiol 17*(6), 559–565.

Morgan MA, el-Ghany el-SM, Khalifa NA, Sherif A, Rasslan LR. (2003). Prevalence of cytomegalovirus (CMV) infection among neonatal intensive care unit (NICU) and healthcare workers. *Egypt J Immunol 10*(2), 1–8.

Mullick S, Broutet N, Htun Y, Temmerman M, Ndowa F. (2004). Controlling congenital syphilis in the era of HIV/AIDS. *Bull World Health Organ 82*(6), 431–432.

Mwangi J. (1999). Blood group distribution in an urban population of patient targeted blood donors. *East Afr Med J 76*(11), 615–618.

Nabias R, Ngouamizokou A, Migot-Nabias F, Mbou-Moutsimbi RA, Lansoud-Soukate J. (1998). Serological investigation of toxoplasmosis in patients of the M.I.P. center of Franceville (Gabon). *Bull Soc Pathol Exot 91*(4), 318–320.

Nagel RL, Platt OS. (2001). General pathophysiology of sickle cell anemia. In: Steinberg MH, Forget BG, Higgs DR, Nagel RL, eds. *Disorders of Hemoglobin: Genetics, Pathophysiology, and Clinical Management*, pp. 494–526. Cambridge, UK: Cambridge University Press.

Nahmias AJ, Lee FK, Beckman-Nahmias S. (1990). Sero-epidemiological and -sociological patterns of herpes simplex virus infection in the world. *Scand J Infect Dis Suppl 69*, 19–36.

Ndir I, Gaye A, Faye B, Gaye O, Ndir O. (2004). Seroprevalence of toxoplasmosis among women having spontaneous abortion and pregnant women following in a center of health up-town in Dakar. *Dakar Med 49*(1), 5–9.

Ndoye NF, Sow AD, Diop AG, et al. (2005). Prevalence of epilepsy its treatment gap and knowledge, attitude and practice of its population in sub-urban Senegal an ILAE/IBE/WHO study. *Seizure 14*(2), 106–111.

Ndumbe PM, Andela A, Nkemnkeng-Asong J, Watonsi E, Nyambi P. (1992). Prevalence of infections affecting the child among pregnant women in Yaounde, Cameroon. *Med Microbiol Immunol 181*(3), 127–130.

Negash T, Tilahun G, Medhin G. (2008). Seroprevalence of Toxoplasma gondii in Nazaret town, Ethiopia. *East Afr J Public Health 5*(3), 211–214.

Ngo DB, Dikassa L, Okitolonda W, et al. (1997). Selenium status in pregnant women of a rural population (Zaire) in relationship to iodine deficiency. *Trop Med Int Health 2*(6), 572–581.

Nichols FT, Jones AM, & Adams RJ. (2001). Stroke prevention in sickle cell disease (STOP) study guidelines for transcranial Doppler testing. *J Neuroimaging 11*(4), 354–362.

Niger J. (2006). *Clin Pract 9*(1), 26–36.

Njamnshi AK, Djientcheu Vde P, Lekoubou A, et al. (2008). Neural tube defects are rare among black Americans but not in sub-Saharan black Africans: The case of Yaounde–Cameroon. *J Neurol Sci 270*(1–2), 13–17.

Nouira S, Kamoun I, Ouragini H, et al. (2008). Clinical and genetic investigation of atrial septal defect with atrioventricular conduction defect in a large consanguineous Tunisian family. *Arch Med Res 39*(4), 429–433.

Nwauche CA, Ejele OA. (2004). ABO and rhesus antigens in a cosmopolitan Nigeria population. *Niger J Med 13*(3), 263–266.

Obed SA, Armah JO, Wilson JB. (1995). Advanced maternal age and pregnancy. *West Afr J Med 14*(4), 198–201.

Oberle MW, Rosero-Bixby L, Lee FK, Sanchez-Braverman M, Nahmias AJ, Guinan ME. (1989). Herpes simplex virus type 2 antibodies: High prevalence in monogamous women in Costa Rica. *Am J Trop Med Hyg 41*(2), 224–229.

Odunvbun ME, Okolo AA, Rahimy CM. (2008). Newborn screening for sickle cell disease in a Nigerian hospital. *Public Health 122*(10), 1111–1116.

Ogendo SW. (1999). Pregnancy in open heart surgical patients at Kenyatta National Hospital. *East Afr Med J 76*(1), 19–22.

Ojule AC, Osotimehin BO. (1998). Maternal and neonatal thyroid status in Saki, Nigeria. *Afr J Med Med Sci 27*(1–2), 57–61.

Okoromah CA, Ekure EN, Ojo OO, Animasahun BA, Bastos MI. (2008). Structural heart disease in children in Lagos: profile, problems and prospects. *Niger Postgrad Med J. 15*(2), 82–88.

Olasoji HO, Ugboko VI, Arotiba GT. (2007). Cultural and religious components in Nigerian parents' perceptions of the aetiology of cleft lip and palate: Implications for treatment and rehabilitation. *Br J Oral Maxillofac Surg 45*(4), 302–305.

Oloyede OA. (2008). Down syndrome screening in Nigeria. *Int J Gynaecol Obstet 100*(1), 88–89.

Onadeko MO, Joynson DH, Payne RA, Francis J. (1996). The prevalence of toxoplasma antibodies in pregnant Nigerian women and the occurrence of stillbirth and congenital malformation. *Afr J Med Med Sci. 25*(4), 331–334.

Onah II, Opara KO, Olaitan PB, Ogbonnaya IS. (2008). Cleft lip and palate repair: The experience from two West African sub-regional centres. *J Plast Reconstr Aesthet Surg 61*(8), 879–882.

Onuigbo WI. (2006). Epidemiology of skin cancer arisen from the burn scars in Nigerian Ibos. *Burns 32*(5), 602–604.

Onwubalili JK. (1983). Sickle-cell anaemia: An explanation for the ancient myth of reincarnation in Nigeria. *Lancet 2*(8348), 503–505.

Onwuekwe IO, Onodugo OD, Ezeala-Adikaibe B, et al. (2009). Pattern and presentation of epilepsy in Nigerian Africans: A study of trends in the southeast. *Trans R Soc Trop Med Hyg 103*(8), 785–789.

Orji EO, Ndububa VI. (2004). Obstetric performance of women aged over forty years. *East Afr Med J 81*(3), 139–141.

Owa JA. (1989). Relationship between exposure to icterogenic agents, glucose-6-phosphate dehydrogenase deficiency and neonatal jaundice in Nigeria. *Acta Paediatr Scand 78*(6), 848–852.

Ozumba BC, Obi SN, Oli JM. (2004). Diabetes mellitus in pregnancy in an African population. *Int J Gynaecol Obstet 84*(2), 114–119.

Padoa C, Goldman A, Jenkins T, Ramsay M. (1999). Cystic fibrosis carrier frequencies in populations of African origin. *J Med Genet 36*(1), 41–44.

Pappas G, Roussos N, Falagas ME. (2009). Toxoplasmosis snapshots: Global status of Toxoplasma gondii seroprevalence and implications for pregnancy and congenital toxoplasmosis. *Int J Parasitol 39*(12), 1385–1394.

Parker M, Corrigall AV, Hift RJ, Meissner PN. (2008). Molecular characterization of erythropoietic protoporphyria in South Africa. *Br J Dermatol 159*(1), 182–191.

Peadon E, Fremantle E, Bower C, Elliott EJ. (2008). International survey of diagnostic services for children with fetal alcohol spectrum disorders. *BMC Pediatr 8*, 12.

Penchaszadeh VB. (2002). Preventing congenital anomalies in developing countries. *Community Genetics 5*, 61–69.

Pham AM, Tollefson TT. (2007). Cleft deformities in Zimbabwe, Africa: Socioeconomic factors, epidemiology, and surgical reconstruction. *Arch Facial Plast Surg 9*(6), 385–391.

Platt OS, Brambilla BJ, Rosse WF, et al. (1994). Mortality in sickle cell disease. Life expectancy and risk factors for early death. *N Engl J Med 330*, 1639–1644.

Platt OS, Thorington BD, Brambilla DJ, et al. (1991). Pain in sickle cell disease. Rates and risk factors. *N Engl J Med 325*(1), 11–16.

Preux PM, Druet-Cabanac M. (2005). Epidemiology and aetiology of epilepsy in sub-Saharan Africa. *Lancet Neurol 4*(1), 21–31.

Preux PM, Tiemagni F, Fodzo L, et al. (2000). Antiepileptic therapies in the Mifi Province in Cameroon. *Epilepsia 41*(4), 432–439.

Prischich F, De Rinaldis M, Bruno F, et al. (2008). High prevalence of epilepsy in a village in the Littoral Province of Cameroon. *Epilepsy Res 82*(2–3), 200–210.

Prual A, De Bernis L, El Joud DO. (2002). Potential role of prenatal care in reducing maternal and perinatal mortality in sub-Saharan Africa. *J Gynecol Obstet Biol Reprod (Paris) 31*(1), 90–99.

Rahimy MC, Gangbo A, Ahouignan G, Alihonou E. (2009). Newborn screening for sickle cell disease in the Republic of Benin. *J Clin Pathol 62*(1), 46–48.

Ramakrishnan U. (2002). Prevalence of micronutrient malnutrition worldwide. *Nutr Rev 60*(5 Pt 2), S46–S52.

Ratbi I, Génin E, Legendre M, et al. (2008). Cystic fibrosis carrier frequency and estimated prevalence of the disease in Morocco. *J Cyst Fibros 7*(5), 440–443.

Roberfroid D, Huybregts L, Lanou H, et al. (2008). MISAME Study Group. Effects of maternal multiple micronutrient supplementation on fetal growth: A double-blind randomized controlled trial in rural Burkina Faso. *Am J Clin Nutr 88*(5), 1330–1340.

Rodier MH, Berthonneau J, Bourgoin A, et al. (1995). Seroprevalences of Toxoplasma, malaria, rubella, cytomegalovirus, HIV and treponemal infections among pregnant women in Cotonou, Republic of Benin. *Acta Trop 59*(4), 271–277.

Romoren M, Rahman M. (2006). Syphilis screening in the antenatal care: A cross-sectional study from Botswana. *BMC Int Health Hum Rights 6*, 8.

Rosenthal J, Christianson A, Cordero J. (2005). Fetal alcohol syndrome prevention in South Africa and other low-resource countries. *Am J Public Health 95*(7), 1099–1101.

Ross E. (2007). A tale of two systems: Beliefs and practices of South African Muslim and Hindu traditional healers regarding cleft lip and palate. *Cleft Palate Craniofac J 44*(6), 642–648.

Rydzak CE, Goldie SJ. (2008). Cost-effectiveness of rapid point-of-care prenatal syphilis screening in sub-Saharan Africa. *Sex Transm Dis 35*(9), 775–784.

Saboye J. (1999). Plastic surgery training missions in developing countries. A 10-year experience at missions in Mali. *Ann Chir Plast Esthet. 44*(1), 35–40.

Saka B, Mouhari-Touré A, Kombaté K, et al. (2009). Scattered papules in three Togolese children from a consanguineous marriage: epidermodysplasia verruciformis. *Med Trop 69*(3), 293–294.

Salako AA, Iyaniwura CA, Jeminusi OA, Sofowora R. (2006). Sexual behaviour, contraception and fertility among in-school adolescents in Ikenne Local Government, south-western Nigeria. *Niger J Clin Pract. 9*(1), 26–36.

Saloojee H, Velaphi S, Goga Y, Afadapa N, Steen R, Lincetto O. (2004). The prevention and management of congenital syphilis: An overview and recommendations. *Bull World Health Organ 82*(6), 424–430.

SAMHSA (Substance Abuse and Mental Health Services Administration). (2005). *The language of Fetal Alcohol Spectrum Disorders.* Center for Substance Abuse Prevention. http://www.samhsa.gov.

Sani MU, Mukhtar-Yola M, Karaye KM. (2007). Spectrum of congenital heart disease in a tropical environment: An echocardiography study. *J Natl Med Assoc 99*(6), 665–669.

Sareli P, England MJ, Berk MR, et al. (1989). Maternal and fetal sequelae of anticoagulation during pregnancy in patients with mechanical heart valve prostheses. *Am J Cardiol 63*(20), 1462–1465.

Sayed AR, Bourne D, Pattinson R, Nixon J, Henderson B. (2008). Decline in the prevalence of neural tube defects following folic acid fortification and its cost-benefit in South Africa. *Birth Defects Res A Clin Mol Teratol 82*(4), 211–216.

Sayed AR, Nixon J, Bourne DE. (2006). Folic acid awareness among women of reproductive age in Cape Town. *S Afr Med J 96*(1), 12.

Seashore MR, Wappner RS. (1996). *Genetics in Primary Care & Clinical Medicine.* New York: Prentice-Hall International Inc.

Seck BC, Jackson RT. (2009). Providing iron/folic acid tablets free of charge improves compliance in pregnant women in Senegal. *Trans R Soc Trop Med Hyg 103*(5), 485–492.

Serjeant GR. (2005). Mortality from sickle cell disease in Africa. *BMJ 330*(7489), 432–433.

Seyoum B, Kiros K, Haileselase T, Leole A. (1999). Prevalence of gestational diabetes mellitus in rural pregnant mothers in northern Ethiopia. *Diabetes Res Clin Pract 46*(3), 247–251.

Shaaban AA, Abdel-Raoof TA, Mohamed MA. (1995). Role of cytomegalovirus in inducing congenital anomalies. *J Egypt Public Health Assoc 70*(3–4), 343–355.

Shin M, Besser LM, Siffel C, et al. and the Congenital Anomaly Multistate Prevalence and Survival Collaborative (2010). Prevalence of spina bifida among children and adolescents in 10 regions in the United States. *Pediatrics 126*(2), 274–279.

Shorvon SD, Farmer PJ. (1988). Epilepsy in developing countries: A review of epidemiological, sociocultural, and treatment aspects. *Epilepsia 29*(Suppl 1), S36–S54.

Simpore J, Savadogo A, Ilboudo D, et al. (2006). Toxoplasma gondii, HCV, and HBV seroprevalence and co-infection among HIV-positive and -negative pregnant women in Burkina Faso. *J Med Virol 78*(6), 730–733.

Sodeinde O, Chan MC, Maxwell SM, Familusi JB, Hendrickse RG. (1995). Neonatal jaundice, aflatoxins and naphthols: Report of a study in Ibadan, Nigeria. *Ann Trop Paediatr 15*(2), 107–113.

Spinelli E, Seia M, Melotti P, et al. (2009). CFTR mutation in an Arab patient: clinical and functional features of 875+1G->A/875+1G->A genotype. *J Cyst Fibros. 8*(4), 282–284.

Spritz RA, Arnold TD, Buonocore S, et al. (2007). Distribution of orofacial clefts and frequent occurrence of an unusual cleft variant in the Rift Valley of Kenya. *Cleft Palate Craniofac J 44*(4), 374–377.

Stefansson H, Rujescu D, Cichon S, et al. (2008). Large recurrent microdeletions associated with schizophrenia. *Nature 455*(7210), 232–236.

Stegmann BJ, Carey JC. (2002). TORCH infections. *Current Women's Health Reports 2*, 253–258.

Streetly A, Latinovic R, Hall K, Henthorn J. (2009). Implementation of universal newborn bloodspot screening for sickle cell disease and other clinically significant haemoglobinopathies in England: Screening results for 2005–7. *J Clin Pathol 62*(1), 26–30.

Sulafa KM, Karani Z. (2007). Diagnosis, management and outcome of heart disease in Sudanese patients. *East Afr Med J 84*(9), 434–440.

Suleiman AM, Hamzah ST, Abusalab MA, Samaan KT. (2005). Prevalence of cleft lip and palate in a hospital-based population in the Sudan. *Int J Paediatr Dent. 15*(3), 185–189.

Tantchou Tchoumi JC, Butera G. (2009). Rheumatic valvulopathies occurrence, pattern and follow-up in rural area: The experience of the Shisong Hospital, Cameroon. *Bull Soc Pathol Exot 102*(3), 155–158.

Tattah EB, Ekere MC. (1975). Death patterns in sickle cell anemia. *JAMA 233*(8), 889–890.

Tebeu PM, Kouam L, Nghonguia MF, Obama MT, Major AL, Ludicke F, Ngassa P, Wamba TM, Moyo JK, Doh AS. (2007). Historical cohort study of delivery in women of forty years of age and older. *Rev Med Liege 62*(7–8), 509–514.

Tekle-Haimanot R, Forsgren L, Ekstedt J. (1997). Incidence of epilepsy in rural central Ethiopia. Epilepsia. *38*(5), 541–546.

Tellal S, Dhayni M, Moussaoui D, Derouiche M. (2009). A critical study of the screening practices of gestational diabetes of a service of gynecology and obstetrics. *Ann Biol Clin (Paris) 67*(2), 159–162.

Tippett B, Wilson J, Shepherd J, Cutler R, Webster P. (2008). Operation Open Heart—Rwanda. *Heart Lung Circ 17*(Suppl 4), S82–S83.

Tipping AJ, Pearson T, Morgan NV, et al. (2001). Molecular and genealogical evidence for a founder effect in Fanconi anemia families of the Afrikaner population of South Africa. *Proc Natl Acad Sci USA 98*(10), 5734–5739.

Tita AT, Selwyn BJ, Waller DK, Kapadia AS, Dongmo S. (2005). Evidence-based reproductive health care in Cameroon: Population-based study of awareness, use and barriers. *Bull World Health Organ 83*(12), 895–903.

Toledo-Pereyra LH. (2010). Heart transplantation. *J Invest Surg 23*(1), 1–5.

Turnpenny P, Ellard S. (2007). *Emery's Elements of Medical Genetics,* 13th edition. Edinburgh, United Kingdom: Churchill Livingstone.

Uda M, Galanello R, Sanna S, et al. (2008). Genome-wide association study shows BCL11A associated with persistent fetal hemoglobin and amelioration of the phenotype of beta-thalassemia. *Proc Natl Acad Sci U S A 105*(5), 1620–1625.

Ugwu RO, Eneh AU, Oruamabo RS. (2007). Neural tube defects in a university teaching hospital in southern Nigeria: Trends and outcome. *Niger J Med 16*(4), 368–371.

UNICEF (United Nations International Children's Fund). (1998). *The State of the World's Children.* UNICEF.

UNICEF. (2000). *The State of the World's Children.* New York: UNICEF.

UNICEF. (2003). *The State of the World's Children.* New York: UNICEF.

Valodia PN, Seymour MA, Kies BM, Folb PI. (1999). Optimization of phenytoin therapy in adults with epilepsy in the Western Cape, South Africa. *J Clin Pharm Ther 24*(5), 381–385.

van den Broek NR, Letsky EA. (2000). Etiology of anemia in pregnancy in south Malawi. *Am J Clin Nutr 72*(Suppl 1), S247- S256.

Vanderjagt DJ, Brock HS, Melah GS, El-Nafaty AU, Crossey MJ, Glew RH. (2007). Nutritional factors associated with anaemia in pregnant women in northern Nigeria. *J Health Popul Nutr 25*(1), 75–81.

van't Veer T, Meester H, Poenaru D, Kogei A, Augenstein K, Bransford R. (2008). Quality of life for families with spina bifida in Kenya. *Trop Doct 38*(3), 160–162.

Venter PA, Christianson AL, Hutamo CM, Makhura MP, Gericke GS. (1995). Congenital anomalies in rural black South African neonates – a silent epidemic? *S Afr Med J 85*(1), 15–20.

Vento S, Cainelli F, Cesario F. (2006). Infections and thalassaemia. *Lancet Infect Dis 6*(4), 226–233. Review.

Viljoen D, Craig P, Hymbaugh K, Boyle C, Blount S. (2003). Fetal alcohol syndrome – South Africa. *MMWR 52,* 660–662.

Vissers LE, de Vries BB, Veltman JA. (2010). Genomic microarrays in mental retardation: From CNV to gene, from research to diagnosis. *J Med Genet. 47*(5), 289–297.

Walker DG, Walker GJ. (2002). Forgotten but not gone: The continuing scourge of congenital syphilis. *Lancet Infect Dis 2*(7), 432–436.

Walker GJ, Walker DG. (2007). Congenital syphilis: A continuing but neglected problem. *Semin Fetal Neonatal Med 12*(3), 198–206.

Wanjeri JK, Wachira JM. (2009). Cleft lip and palate: A descriptive comparative, retrospective, and prospective study of patients with cleft deformities managed at 2 hospitals in Kenya. *J Craniofac Surg 20*(5), 1352–1355.

Warf B, Ondoma S, Kulkarni A, et al. (2009). Neurocognitive outcome and ventricular volume in children with myelomeningocele treated for hydrocephalus in Uganda. *J Neurosurg Pediatr 4*(6), 564–570.

Warf BC, Campbell JW. (2008). Combined endoscopic third ventriculostomy and choroid plexus cauterization as primary treatment of hydrocephalus for infants with myelomeningocele: Long-term results of a prospective intent-to-treat study in 115 East African infants. *J Neurosurg Pediatr 2*(5), 310–316.

Warf BC. (2005). Comparison of endoscopic third ventriculostomy alone and combined with choroid plexus cauterization in infants younger than 1 year of age: A prospective study in 550 African children. *J Neurosurg 103*(Suppl 6), 475–481.

Watcham SJ, Schön S, Christianson AL. (2007). Neglect in the care of pregnant South African women of advanced maternal age. *S Afr Med J 97*(11), 1064, 1068–1069.

Weatherall DJ. (2008). Hemoglobinopathies worldwide: Present and future. *Curr Mol Med 8*(7), 592–599.

Weatherall DJ. (2006). *Inherited Disorders of Hemoglobin. Disease Control Priorities in Developing Countries,* 2nd ed., pp. 663–680. New York, NY: Oxford University Press.

Weatherall DJ. (2001). Phenotype-genotype relationships in monogenic disease: Lessons from the thalassaemias. *Nat Rev Genet 2*(4), 245–255.

Weatherall DJ. (2005). The global problem of genetic disease. *Ann Hum Biol 32*(2), 117–122.

Weatherall DJ. (2010). The inherited diseases of hemoglobin are an emerging global health burden. *Blood 115*(22), 4331–4336.

WHO. (2005). *Control of Genetic Diseases.* Report 2005. Available at http://apps.who.int/gb/ebwha/pdf_files/ EB116/B116_3-en.pdf.

WHO. (1996). *Control of Hereditary Diseases.* WHO Technical Report Series 865. Geneva, Switzerland: WHO.

WHO. (2002). *Global Strategies to Reduce the Health Care Burden of Craniofacial Anomalies.* Geneva, Switzerland: WHO.

WHO. (2002b). *Report of a Meeting on Preventing Congenital Rubella Syndrome: Immunization Strategies, Surveillance Needs.* Geneva, Switzerland: WHO.

WHO/ World Alliance for the Prevention of Birth Defects. (1999) *Services for the Prevention and Management of Genetic Disorders and Birth Defects in Developing Countries.* Geneva, Switzerland: WHO.

Wierenga KJJ, Hambleton IR, Lewis NA. (2001). Survival estimates for patients with homozygous sickle-cell disease in Jamaica: A clinic-based population study. *Lancet 357,* 680–683.

Williams TN, Mwangi TW, Wambua S, et al. (2005). Negative epistasis between the malaria-protective effects of alpha+-thalassemia and the sickle cell trait. *Nat Genet 37*(11), 1253–1257.

Williams TN, Uyoga S, Macharia A, et al. (2009). Bacteraemia in Kenyan children with sickle-cell anaemia: A retrospective cohort and case-control study. *Lancet 374*(9698), 1364–1370.

Winkler AS, Kerschbaumsteiner K, Stelzhammer B, Meindl M, Kaaya J, Schmutzhard E. (2009). Prevalence, incidence, and clinical characteristics of epilepsy – a community-based door-to-door study in northern Tanzania. *Epilepsia 50*(10), 2310–2313.

Wonkam A. (2009). *Phenotype, burden and perspectives of sickle cell anaemia studies in sub-Saharan Africa: The case of Cameroon.* New Delhi, India: ICBD Proceedings, October 4, 2009.

Wonkam A, Muna, Ramesar R, Rotimi C, Newport M. (2010). Capacity-building in Human Genetics for Developing Countries: Initiatives and perspectives in sub-Saharan Africa. *Public Health Genomics 13*(7–8), 492–494.

Wonkam A, Njamnshi AK, Angwafo III FF. (2006). Knowledge and attitudes concerning medical genetics amongst physicians and medical students in Cameroon (sub-Saharan Africa). *Genet Med 8*, 331–338.

Zagré NM, Desplats G, Adou P, Mamadoultaibou A, Aguayo VM. (2007). Prenatal multiple micronutrient supplementation has greater impact on birthweight than supplementation with iron and folic acid: a cluster-randomized, double-blind, controlled programmatic study in rural Niger. *Food Nutr Bull. 28*(3), 317–327.

Zaida F, Sedki A, Laroche D, Lekouch N, Bouglé D. (2006). High cord blood TSH in Morocco: Iatrogenic hypothyroidism? *Ann Endocrinol (Paris) 67*(1), 39–41.

Zeck W, McIntyre HD. (2008). Gestational diabetes in rural East Africa: A call to action. *J Womens Health (Larchmt) 17*(3), 403–411.

38.

GENETIC EPIDEMIOLOGY OF SOUTH ATLANTIC OCEANIC ISLANDS

Peter Beighton

INTRODUCTION

St. Helena and Tristan da Cunha are isolated mid-oceanic islands in the South Atlantic. St. Helena is a British Crown colony and Tristan da Cunha, which lies 3200 km to the south, is a Dependency of St. Helena. A third island, Ascension, situated approximately 1600 km to the northwest of St. Helena, is also a Dependency; there is no settled population on Ascension, but since the conflict in the Falkland Islands in 1985 it has been an important military base. The population of St. Helena is approximately 5000, including 300 belonging to the Tristan community. The islanders have ancestral and family links with each other and with South Africa.

Due to their special circumstances, oceanic island communities provide unique opportunities for research, especially in the fields of population and medical genetics (Roberts & Beighton, 1986). In this context, a variety of Mendelian disorders have been investigated on St. Helena, while the small but well-documented community of Tristan has been the focus of extensive studies of genealogy and population genetics.

A brief account of these islands and their communities is provided in this chapter, together with details of their medical and genetic circumstances.

TRISTAN DA CUNHA

GEOGRAPHY AND HISTORY

The island of Tristan da Cunha is situated in the South Atlantic Ocean, on latitude 37° south, longitude 12° west. The nearest major mainland habitation is Cape Town in South Africa, 2800 km to the east. A fishing company based in Cape Town deep-freezes crayfish in a factory on Tristan. Company vessels and a U.K. government supply ship, the *St. Helena,* make intermittent visits to the island.

Tristan da Cunha is a volcanic cone, 24 km in diameter and 2000 meters in height, which arises abruptly out of the ocean. The inhabitants, numbering approximately 300, live together in a village, Edinburgh Settlement, which is situated on a small area of flat land close to a permanent spring. Many of the houses are of stone, thatched with flax, and lined with planking obtained from wrecked ships. The design is similar to that of the houses of crofters in the western highlands of Scotland. The climate is also similar, with temperate summers and cold, wet, windy winters. Every home is now supplied with running water, electric light, and a drainage system.

The island was discovered in 1506 by a Portuguese admiral, Tristoa da Cunha. Permanent settlement commenced in 1816, when a small British military unit was landed. The purpose of this garrison was to prevent the use of Tristan as a base in any French attempt at the rescue of Napoleon Bonaparte, who had been exiled on St. Helena after his defeat at the battle of Waterloo. When the garrison was withdrawn in 1817, Corporal William Glass and his family stayed on Tristan. Other castaways and wanderers arrived in the years that followed, and womenfolk of mixed ancestry from Cape Town and St. Helena came as wives for these lonely single men. At the present time, 7 family names are represented on Tristan (Glass, Lavarello, Repetto, Hagan, Swain, Rogers, and Green).

During the nineteenth century, sailing vessels frequently called for fresh water and supplies. After the opening of the Suez Canal in 1869 and the decline of sail, Tristan's isolation increased, and many months passed between visits from the outside world. The community survived, but in 1961 a volcanic eruption at the edge of Edinburgh Settlement necessitated their evacuation to England, where extensive medical investigations were undertaken (Black et al., 1963). Disenchanted with the outside world, the majority returned to Tristan in 1963 after the volcano became quiescent.

Prior to the 1950s, the islanders' economy was based upon fishing and the cultivation of potatoes, supplemented by barter with passing ships. In this period, and in the years that followed, the islanders' diet consisted almost entirely of fish and potatoes, together with protein and fat derived from seabirds and their eggs (Chambers & Lewis, 1969; Richardson et al., 1975). A wide variety of foodstuffs is now available in the government store, the islanders are relatively affluent, and the daily diet is probably very similar to that of an average fishing community in Western Europe.

POPULATION GENETICS

The genealogy and history of the Tristan population has been extensively documented from oral and written sources by several researchers, notably Munch (1945), Roberts (1968) and Crawford (1982). Details of the early settlers provided by these authors, and obtained anecdotally by the author while on Tristan in 1971, are summarized below.

Of the men who formed the garrison of Tristan in 1815, only Corporal William Glass remained permanently on the island with his wife, Maria Leenders of Cape Town. Thomas Swain arrived in 1826, and in the following year married Sarah Jacobs, a widow from St. Helena who arrived with her daughter from a previous marriage, Mary Jacobs. Swain had served in the Royal Navy and was said by the family to have fought at the battle of Trafalgar in *HMS Victory,* and been present at the death of Lord Nelson. Pieter William Groen, born in Katwijk, Holland, in 1808, was shipwrecked on Tristan in 1836 in the schooner *Emily.* He subsequently Anglicized his name to Peter Green and married the aforementioned Mary Jacobs. In the same year, Thomas Rogers, a seal hunter from the United States, married Jane Glass, the daughter of William and Maria Glass. In 1849 the captain of an American whaling vessel, Andrew Hagan, retired from the sea, settled in Tristan, and married Selenia Glass. (Another account of his arrival, which could be a confabulation, holds that Hagan was marooned on Tristan by his dissatisfied crew after a disastrous and unproductive whaling season in the Antarctic). The final input to the present-day family names was made when Gaetano Lavarello and Andre Repetto, both from Camogli in Italy, reached Tristan in 1892 when their ship, the barque *Italia,* caught fire and was purposely beached on the island.

An additional female contribution to the Tristan gene pool was made by Maria Williams, who arrived from St. Helena in 1827 and married Andrew Cotton; this surname is now extinct on Tristan. In 1862 Susanna Philips also arrived from St. Helena and married a Tristanian. It is likely that these ladies and those that preceded them (i.e., Mary Leenders, Sarah and Mary Jacobs) were all of mixed ancestry (Munch 1945). The final female input came in 1908 from Elizabeth and Agnes Smith, who were of Irish stock. Although it is recorded that these women were sisters, molecular evidence suggests that they were unrelated (Soodyall et al., 1997).

Population genetics has played a major role in the scientific approach to the Tristan community. In particular, in the 1960s, the British Medical Research Council (MRC) established a program of investigations under Dr. H.E. Lewis, a member of the medical team that examined the islanders when they were evaluated in the United Kingdom. During this investigation gene markers in the form of red cell enzymes and protein polymorphisms were studied by Harris and Robson (1963). Further studies on material from the same source were undertaken by Ikin and Mourant (1976). The MRC project was subsequently coordinated by Professor D.F. Roberts of the University of Newcastle, who published detailed accounts of the history and genealogy of the islanders (Roberts, 1968, 1971, 1980). In the 1970s additional gene markers became available and fresh biological specimens were obtained. In this way, the historical origins of the women from St. Helena who married into the Tristan community were confirmed (Jenkins et al., 1985).

Advances in molecular biology have made a major impact on population genetic research. In particular, investigation of mitochondrial DNA facilitates the determination of relationships in the female lineage. These techniques enabled Soodyall and her colleagues to link 161 persons in the Tristan community to five female founders (Soodyall et al., 1997).

GENETIC DISORDERS ON TRISTAN DA CUNHA

As genetic disorders are individually uncommon, and in view of the small size of the Tristan population, it is unlikely that heritable conditions would be present in any numbers. This has proved to be the case, and research investigations have been mainly focused on genealogy and population genetics.

Following the volcanic eruption in 1961, the islanders were evacuated to England via Cape Town and initially accommodated at Pendell Camp in Surrey. The health status of all 259 islanders was then comprehensively investigated by a medical team established by the British Medical Research Council, and the detailed findings were subsequently published (Black et al., 1963). This article is a major source of information concerning the Tristan population.

The islanders were found to have good general health, although they were susceptible to respiratory infections. Indeed, almost 50% of the adults experienced intermittent asthma, with wheezing, tightness of the chest and dyspnea. This condition, which was usually precipitated by a minor respiratory infection, often developed in mid-childhood. In the island dialect, it was known as the "hashmeer." The high population prevalence of asthma may be indicative of a genetic component in the predisposition to development of this condition.

Retinitis Pigmentosa

The most important genetic observation made during the 1963 MRC survey was the recognition at fundoscopy of retinitis pigmentosa (RP) in a young woman with night blindness and tunnel vision. Night blindness was also found in eight adult males and three females, of whom three had some restriction of the visual fields. The retinal changes were designated "less evident" and the diagnosis of a genetic form of RP in these latter persons remained unsubstantiated. It seems likely, however, that the condition is the autosomal recessive form of RP and that these individuals could be either homozygotes in the early stages of the disorder or manifesting heterozygotes.

The presence of night blindness in the Tristan community had previously been recorded by Barrow (1910), and it is of considerable significance that RP has also been documented in the population of St. Helena. Shine (1970) made specific mention of an elderly St. Helenian lady with RP who was ancestrally related to a woman who married into the Tristan community in 1827. It is possible that the latter lady introduced the determinant RP gene to Tristan in the heterozygous state.

Congenital Malformations

In their examination of the Tristan population, Black et al. (1963) listed a total of 36 persons with varying combinations of perceptive deafness, mental dullness, abnormal shape of the external ears, and high palates. Other minor malformations included clinodactyly in 37 persons, short terminal phalanges in 3, and hypertelorism in 3. Accessory nipples were present in 18, and isolated individuals had micrognathia, a supernumerary toe, and hypospadias. Congenital cardiac defects were present in two persons. Crawford (1982) mentioned that three individuals in three generations of a Tristan family had polydactyly, and that ten persons in three generations had "slight external malformation of the external ear." The syndromic status of these abnormalities is uncertain.

Bone and Joint Disorders

In the early 1970s there was considerable academic interest in the epidemiology of rheumatological disorders and the possible contribution of genetic factors in their etiology. In this context, the author visited Tristan in order to assess the musculoskeletal status of the adult population (Beighton & Valkenburg, 1974). A total of 111 persons were examined, comprising 87% of the adult population. Limited radiographic studies were undertaken and blood specimens were obtained for serological studies of rheumatoid factors by Dr. H.A. Valkenburg of the University of Rotterdam. Sera for investigation of genetic polymorphisms were also obtained.

Various osteoarthropathies were identified in a few persons, but there was no evidence of any familial preponderance. Three men and three women had osteoarthritis of the knee associated with marked genu valgum. In view of the ancestral links between the Tristan and St. Helena populations, it is possible, with hindsight, that these persons might have had the genetic disorder that was subsequently designated "St. Helena Genu Valgum" (Beighton et al., 1986). Clinodactyly of the 5th fingers was present as an isolated anomaly in 12 members of the Glass family or in the descendants of Granny Jane Lavarello, the granddaughter of the community's founder, William Glass.

MEDICAL RESOURCES ON TRISTAN DA CUNHA

Prior to WWII, the island community was dependent upon passing ships for medical assistance. A small garrison spent the war on the island, and a hutted hospital with basic facilities was established. Thereafter, a medical officer was maintained on the island by the British government, usually on a two-year contract. A new purpose-built hospital is now in place, and the contract physician has been given the status of "Director of Medical Services." Four locally recruited nurses, trained at the hospital on St. Helena, complete the medical team.

The problems of geographical isolation, the paucity of regular contact by sea, and the absence of suitable terrain for an airfield make medical self-sufficiency a necessity. In this context, radio and telephone connections with a designated surgeon in Cape Town are maintained and specialist advice can readily be obtained.

ST. HELENA

GEOGRAPHY AND HISTORY

St. Helena is an oceanic island approximately 10 km in width and 16 km in length, situated in the South Atlantic 16° south of the equator and 1600 km from the shores of Africa. The island is the remnant of an extinct volcano and a narrow central ridge, average height 800 meters, which divides the fertile north side from the barren south. This stark distinction is the result of climatic influences, notably the lack of rainfall. Deep valleys radiate from the central ridge and there are very few flat regions. The coastline is rocky and there are no natural harbors.

St. Helena was discovered by the Portuguese in 1502 and was at first used as a source of fruit and water for ships sailing on voyages of exploration and commerce. In 1659 the island came under British rule, and was colonized with settlers from England and slaves from Africa. The opening of the Suez Canal in 1869, and the replacement of sail by steam, resulted in a dramatic decline in the number of ships calling at St. Helena. Thereafter, the island economy and the size of the population remained static. In recent years many members of the St. Helena population have been employed on contract on Ascension at the military base and airfield. As a Crown colony, the administration and public services are provided by the British government.

The origins of the population of St. Helena are very diverse. The first settlers were Portuguese, Dutch, and English, and these groups brought slaves from the Malabar Coast of India, Bombay, Bengal, Central Africa, West Africa, and Madagascar. The Great Fire of London in 1666 precipitated the immigration of English settlers, and they were followed by Huguenot refugees. Chinese indentured laborers arrived in the middle of the eighteenth century, and in the same period African slaves, freed by the British Navy from slave ships, made a further contribution to the island's gene pool. The current population of St. Helena numbers about 5000, and over the years several hundred St. Helenians have settled in Britain and Cape Town. Shine (1970) stated that in 1800 or thereabouts, approximately 50% of the islanders were of African stock, 25% were European and 25% were Chinese. He went on to conclude that by 1960, there were very few persons of unmixed descent. As far as the author is aware, there have been no formal investigations of population genetics using serum or molecular markers.

The majority of the community are domiciled in Jamestown, in a long deep valley on the north side of the island. Ships calling at St. Helena anchor at the seaward end of the valley, and cargo and visitors are brought ashore in small boats. A castle and a cathedral face the landing place, and early colonial-style houses line the route to the interior. A hospital is situated at the apex of the valley, about a mile from the shore. The road then climbs to the central ridge through lush vegetation and reaches the historical estate "Longwood," where Napoleon lived in exile following his defeat at the battle of Waterloo in 1815.

The island is linked to the outside world by the mail ship *St. Helena*, which calls at regular intervals on voyages from Cape Town and England. Visits by other ships are sporadic, a determining factor being the lack of any harbor and the necessity to anchor offshore. Proposals have been made for the construction of an airfield, but to date these have not materialized.

GENETIC DISORDERS ON ST. HELENA

Interest in genetic conditions on St. Helena was sparked by an English physician, Dr. Ian Shine, who served as island medical officer in 1960–1962. He examined 4259 members of the population, undertook family studies, tabulated his findings, and wrote a doctoral thesis for the University of Oxford after his return to the United Kingdom. His contributions culminated in 1970 in the publication of his classical monograph, "Serendipity in St. Helena; a genetical and medical study of an isolated community" (Shine, 1970). This excellent book, written in the early days of medical genetics, is commended to colleagues with an interest in oceanic islands and the medical aspects of population genetics. Important Mendelian genetic conditions documented in detail by Shine include:

Christmas disease (Factor IX deficiency)	XL
Albinism	AR
Retinitis Pigmentosa	AR?
"Brachydactylous Dwarfism" (Acromesomelic dysplasia)	AR
"Familial knock knees" (St. Helena genu valgum)	AD
Dwarfism with infantilism (Pituitary dwarfism)	AR

Genetic links between St. Helena and South Africa became apparent when family studies of three Cape Town sisters with gross shortening of stature due to a skeletal dysplasia revealed that they had progenitors on the island (Beighton, 1974). Thereafter, a male with severe genu valgum was referred from St. Helena for corrective orthopedic surgery in Cape Town. He was a member of the kindred described by Shine as having "Familial Knock Knees." Radiographic skeletal survey revealed hypoplasia of the lateral femoral condyles and upper tibial regions, with similar but lesser changes in the lateral humeral condyles. These observations were subsequently documented as "St. Helena Familial Genu Valgum" (Beighton et al 1986).

In 1984, Dr. Sibylle Eickhoff of the Department of Human Genetics, University of Cape Town, was seconded for a 3-month stay on St. Helena as an assistant medical officer, in order to study the families documented by Shine, update pedigree data, and investigate other genetic disorders and malformations. Eickhoff undertook a detailed study of the two families in which 15 persons had AD hereditary angioneurotic edema (HANO), identified a common progenitor, and commented that management of this life-threatening disorder was a serious problem on St. Helena. She also documented a mother and three children with an AR form of Charcot-Marie-Tooth disease and mental retardation. Two other sets of siblings, both with consanguineous parents, had apparent AR spastic quadriplegia and intellectual disability. Her identification of two unrelated adults with retinitis pigmentosa provided additional evidence for the presence of this AR form of RP in the St. Helena community (Eickhoff & Beighton, 1985).

With the continuing evolution of concepts concerning the heritable skeletal disorders, and the eventual prospects for genetic management, the issue of diagnostic precision in the St. Helena form of brachydactylous dwarfism became relevant. Accordingly, in 1997, the author sailed to St. Helena in order to address this issue and to identify other heritable conditions on the island. The hospital in Jamestown provided consulting, radiological, and laboratory facilities, and with the assistance of a local nursing sister the affected family was restudied. Pedigree data were entirely consistent with AR inheritance and the clinical and radiographic phenotype was confirmed as acromesomelic dysplasia. With informed consent, blood specimens were obtained and subsequent homozygosity mapping by Dr. Petros Tsipouras and his group in the United States located the determinant gene in the pericentric region of chromosome 9 (Ianakiev et al., 2000). Acromesomelic dysplasia, Maroteaux type, maps to the same region, but molecular characterization has not been undertaken and the possibility of intramolecular heterogeneity remains unresolved. The issue of possible homology with the Cape Town family with acromesomelic dysplasia is also unsettled.

A family with AD symphalangism was identified during the author's stay in St. Helena. The progenitors of this family were said to have immigrated to St. Helena after the Great Fire of London in 1866, and they claimed links with King Richard II and the Earls of Shrewsbury. As this rare but relatively minor Mendelian disorder has been suspected in these aristocratic English families, the possibility of a founder effect arises. Another autosomal disorder was the Waardenburg syndrome, which was encountered in siblings with heterochromia, a white forelock, and deafness. It is possible that this condition was represented among the persons with nonspecific deafness listed by Shine. The fetal alcohol syndrome is also present on the island. The relative affluence of the islanders employed on Ascension, the low cost of alcohol on St. Helena, and the abundance of leisure time have all contributed to the emergence of this condition.

MEDICAL RESOURCES ON ST. HELENA

A 58-bed hospital provided by the British government was opened in 1905, and radiographic and laboratory facilities are available. Medical care is provided by four doctors on contract from the U.K. government, a resident dentist, and a well-developed district nursing service. Patients requiring highly specialized investigations or treatment are transported by sea to Cape Town, where they are seen at the Red Cross Children's Hospital or Groote Schuur Hospital.

At present, the general state of health of the islanders can be compared to that of any developed society, and there is a significant prevalence of hypertension, carcinoma, and atherosclerotic disease, while diabetes and asthma occur in high frequency. The issues concerning provision of psychiatric services have been raised (Aldridge, 1994). Medical reports concerning the island children published during the past two decades include reviews of behavioral problems (Lovemore et al., 1995) and dental caries in preschool children (Entwistle, 2004).

IMPLICATIONS OF GENETIC DISORDERS ON TRISTAN DA CUNHA AND ST. HELENA

Retinitis Pigmentosa

Retinitis Pigmentosa has been recognized in the past on several occasions in the populations of both Tristan and St. Helena. In 2009 the senior medical officer of St. Helena, Dr. Sarel Bloem, provided information that currently there were 7 adults on St. Helena in whom RP had been diagnosed. Pedigree data are consistent with autosomal recessive inheritance and there is evidence for ancestral links between affected persons on both islands. If this mode of genetic transmission is substantiated, it follows that for every affected homozygote, there will probably be several asymptomatic heterozygotes. As both populations are endogamous, the likelihood of union between heterozygotes is comparatively high. The risks are further increased on Tristan by the geographical isolation, with consequent lack of choice of marriage partners and the fact that only 7 interrelated families make up the small population of three hundred persons.

Progressive loss of vision is a serious complication of AR RP. Molecular diagnostic techniques are available for the detection of the heterozygous "carriers" of the determinant gene. The confirmation of the diagnosis in an affected individual, and the identification of the precise mutation by means of molecular analysis, would be the starting point for a genetic approach to the condition. Thereafter the issue of population screening on Tristan, or cascade family screening on St. Helena, would warrant consideration.

By 2010, more than 15 different genes and 500 mutations had been recognized worldwide in the AR form of RP. On this basis it is possible that affected persons on Tristan and St. Helena could be genetic compounds, having heterozygosity for two separate mutations rather than homozygosity for the same mutation. This issue could be resolved by the use of microarray techniques for microarray mutation screening of affected persons on both islands.

Bronchial Asthma

Bronchial asthma reaches a high population prevalence on both Tristan and St. Helena. It is likely that predisposition to asthma has a strong genetic component, and in view of the ancestral links between the two communities, it is possible that the same genes are involved. The fact that the islands are very different in terms of their climates and environments increases the likelihood that genetic factors are operative.

In view of the endogamy on Tristan and the small size of the population, a molecular approach to asthma using SNP markers and microarrays for disease association studies could prove to be fruitful. Success in this respect would have immense implications for the pharmaceutical industry, and in this respect it is vital that the interests of the Tristan community are fully supported by the relevant governmental authorities.

OPPORTUNITIES

Short-term medical locum posts on Tristan da Cunha and St. Helena offer an opportunity for wide medical experience and the advancement of medical genetics (www.nico.org.uk). In this context, the following quotation may be relevant:

> The Medical Officer, Dr Wignall was always drunk and nearly killed the Governor by prescribing unsuitable medicines, his excuse being that he had nothing else. For drunken and disorderly conduct, Dr Wignall was placed in the stocks for an hour; he sung, shouted and swore for the whole time.
>
> *4th January 1725. St. Helena consultations. Cited by Shine (1970)*

ACKNOWLEDGEMENTS

I am grateful to my wife, Greta for her assiduous preparation of the manuscript and for her contributions on St. Helena during our research project.

I thank the Tristan da Cunha and St. Helena communities for their kindness and goodwill during my earlier investigations on these islands.

My archival research for this chapter was supported by the National Research Foundation and the Medical Research Council of South Africa.

REFERENCES

Aldridge SJ. (1994). Where there is no psychiatrist: the development of a psychiatric service on St. Helena. *J R Soc Med 87*(12), 790–793.

Barrow KM. (1963) (1910). Three years on Tristan da Cunha. 1910. London. Cited by Black et al

Beighton P. (1974). Autosomal recessive inheritance in the mesomelic dwarfism of Campailla and Martinelli. *Clin Genet 5*, 363–367.

Beighton P, Valkenburg HA. (1974). Bone and joint disorders on Tristan da Cunha. *S Afr Med J 48*, 743–747.

Beighton P, Myers HS, Aldridge SJ, Sedgewick J, Eickhoff S. (1986). St. Helena familial genu valgum. *Clin Genet 30*, 309–314.

Black JA, Lewis HE, Thacker CKM, Thould AK (1963). Tristan da Cunha: general medical investigations. *Brit Med J ii*, 1018–1024.

Chambers MA, Lewis HE. (1969). Nutrition study of the islanders on Tristan da Cunha. *Br J Nut 23*, 237–239.

Crawford A. (1982). *Tristan da Cunha and the roaring forties.* Skilton Ltd, Edinburgh & London.

Eickhoff S, Beighton P. (1985). Genetic disorders on the island of St. Helena. *S Afr Med J 68*, 475–478.

Entwistle NB. (2004). The dental caries experience of five-year-old children in St. Helena. *J R Soc Promot Health 124*(4), 177–179.

Harris H, Robson EB. (1974) (1963). Haptoglobins in Tristan da Cunha. *Vox Sanguinis 8*, 226–230.

Ianakiev P, Kilpatrick MW, Daly MJ, et al. (2000). Localization of an acromesomelic dysplasia on chromosome 9 by homozygosity mapping. *Clin Genet 57*, 278–283.

Ikin EW, Mourant AE. (1976). Unpublished observations. Quoted in Mourant AE, Kopéc AC, Domaniewska-Sobczak K. *The Distribution of the Human Blood Groups.* London: Oxford University Press.

Jenkins T, Beighton P, Steinberg AG (1985). Serogenetic studies on the inhabitants of Tristan da Cunha. *Ann Hum Biol 12*(4), 363–371.

Lovemore T, Flores H, Charlton T. (1995). Rates of problem behaviour among preschoolers attending nursery classes in St. Helena, South Atlantic. *Percept Mot Skills 81*(3 Pt 1), 817–818.

Munch P. (1945). *Sociology of Tristan da Cunha.* Oslo: Det Norske Videnskaps-Akademi.

Richardson H, Watermeyer G, Soskolne CL, Beighton P. (1975). Serum lipid patterns in the islanders of Tristan da Cunha. *S A J Sci 71*, 243–245.

Roberts DF. (1968). Genetic effects of population size reduction. *Nature* 220, 1084–1088.

Roberts DF. (1971). The demography of Tristan da Cunha. *Population Studies 25*, 465–479.

Roberts DF, Beighton P. (1986). Oceanic islands - a unique opportunity for medical research. *S Afr Med J 70*, 608–610.

Roberts DF. (1980). Genetic structure and the pathology of an isolated population. In: *Population Structure and Genetic Disorders,* Eds. AW Erikson, H Forsius, HR Nevenlinna, PL Workman, RK Norio. Academic Press, New York.

Shine I. (1970). *Serendipity in St. Helena: a genetical and medical study of an isolated community.* Pergamon Press, Oxford.

Soodyall H, Jenkins T, Mukherjee A, Du Toit E, Roberts DF, Stoneking M. (1997). The founding mitochondrial DNA lineages of Tristan da Cunha islanders. *Am J Phys Anthropol 104*, 157–166.

39.

GENETIC DISORDERS IN MINORITY GROUPS IN SOUTH AFRICA

Peter Beighton

INTRODUCTION

The evolution of medical genetic services and research in South Africa has been influenced by the presence or absence of specific genetic disorders in different population groups. This issue was initially addressed more than 30 years ago, when available information concerning the population distribution of heritable conditions was documented (Beighton, 1976).

In the decade that followed, delineation and elucidation of unique genetic conditions in different communities was the main focus of the Department of Human Genetics, University of Cape Town, while highly productive investigations in the field of population genetics were undertaken by Professor Trefor Jenkins and his team at the University of the Witwatersrand, Johannesburg. The inception of molecular technology in the 1980s facilitated further understanding of genetic disorders in this country (Beighton et al., 1987).

The total population of South Africa numbers about 50 million and comprises several distinct communities, each with its own pattern of genetic disorders. The indigenous African community is by far the largest, totaling 40 million, and the heritable disorders and birth defects in this group have been reviewed in a series of articles (Kromberg & Jenkins, 1982; Beighton & Botha, 1986a,b,c). In view of the numerical predominance of this population, it is the subject of a separate chapter in this book.

The terms *San* and *Khoi* pertain to the early populations of present-day South Africa. The San (formerly Bushman) population of South Africa were hunter-gatherers, and by far the earliest inhabitants of this country (Jenkins, 1979). They were joined by the Khoi (formerly Hottentot) cattle herders, who migrated southwards down the east coast of Africa about 10,000 years ago; collectively, this group are termed the *Khoisan*. In turn, the Khoisan were assimilated into the present-day Cape Mixed Ancestry population, in which their genes are enshrined. The Khoisan are introduced at this point, as their historical status underlies fundamental aspects of current genetic epidemiology in the country. Further details of the Khoisan are provided in the last section of this chapter.

The South African minority groups are listed below:

1. Afrikaner	3 million
2. Mixed Ancestry	4.5 million
3. British	2 million
4. Indian	1 million
5. Jewish	60,000
6. Greek	60,000
7. Others	10,000

The "Mixed Ancestry" category is a source of possible confusion and misunderstanding and for these reasons, an explanation is necessary. This term has had political connotations, and although regarded as pejorative by some, it is still in general use, especially in the field of genetic epidemiology. It is in this context that the heritable conditions in the community are addressed in this chapter.

The Mixed Ancestry group as a whole is regarded as having two major subdivisions, the Cape Mixed Ancestry community, numbering about 3 million, and the Mixed Ancestry population of Johannesburg and other major centers in the north and east of South Africa, totaling about 1.5 million. It is clearly recognized that the concept of "mixed ancestry" is an artificial construct, with social and political nuances that are dependent upon an individual's own view of her or his identity. Nevertheless, the genetic endowments of these two groups differ, as do the genetic disorders that occur in these communities.

The Cape Mixed Ancestry community had its origins in the 1670s, when indigenous Khoisan persons in the Cape began to be assimilated into the recently arrived Dutch community. Thereafter, over the following 300 years, further genetic input was made by individuals from numerous sources including Java, India, Angola, and Europe. During this period, the community was domiciled in the relative isolation of the Cape; hence the designation, "Cape Mixed Ancestry" community. The heritable disorders that are present in this population are derived from their disparate origins and are discussed in detail in Section "Cape Mixed Ancestry Community" of this chapter.

The Mixed Ancestry population of the major cities in the north and east of South Africa is predominantly of indigenous African and European admixture, with small contributions from Indian and other sources. This population has evolved during the past 150 years, largely following the discovery of precious minerals, influx of immigrants, urbanization, and industrialization of the country. The genetic conditions in this group reflect the heritable conditions in their parent population.

With the exception of the Khoisan progenitors of the Mixed Ancestry community (vide infra), the populations listed above have all immigrated to South Africa at one time or another during the past 350 years. The process of immigration can be a powerful factor in changing the frequency of a particular gene in an immigrant community (i.e., the founder effect), and this mechanism is apparent in several of these communities. Other processes such as endogamy are also operative and, in essence, the presence of specific genetic conditions in each of these populations is the result of historical accidents, biological pressures, and demographic factors. The genetic disorders in each of these communities form the subject of this chapter.

As would be expected, common Mendelian disorders that have a worldwide distribution are present in the minority groups in South Africa. Genetic conditions of this type, including retinal disorders, familial colonic carcinoma, Huntington disease, and hypercholesterolemia have been the subject of extensive investigations in South Africa. These studies have progressed through delineation, genealogy, genetic epidemiology, biochemistry, and molecular biology. In order to avoid repetition, and where relevant, these conditions are mentioned and cross-referenced in this chapter in the context of the population in which they predominate.

Several doctoral theses in the Department of Human Genetics, UCT, have been based upon investigations of genetic disorders in different communities in South Africa. Those which pertain to material presented in this chapter are listed below:

M Hayden (1979) PhD. *Huntington Chorea in South Africa.*

RP Bonafede (1979) MD. *Hereditary haematological disorders in the Greek population of Cape Town.*

J Goldblatt (1980) MD. *Gaucher disease in the Ashkenazi Jewish community of South Africa.*

CE Wallis (1988) MD. *Genetic disorders on the island of Mauritius.*

LJ Greenberg (1990) PhD. *A molecular approach to Huntington disease in South Africa.*

A Bryer (1994) PhD. *Spinocerebellar ataxias.*

R Ballo (1998) PhD, *An investigation into the molecular basis of familial forms of osteoarthropathy in South Africa.*

AFRIKANER COMMUNITY

The present-day Afrikaans-speaking community originated in 1652 with the establishment of a permanent settlement in Cape Town. The initial purpose was the provision of fresh food and water for the ships of the Dutch East India Company on their voyages between Holland and their outpost of Batavia in Java. Farming became the predominant occupation, and during the years that followed, the population expanded and rural communities were established. Dutch immigration continued, with input from France and Germany. Smaller contributions to the gene pool were also made by Southeast Asians and indigenous Khoisan people. The Afrikaans language, derived from Dutch, together with religion and social customs, define the Afrikaner community.

The original settlers were few in number but in successive generations, Afrikaner families were large, and for this reason the founder effect for important AD genetic disorders was operative. Equally, the relative homogeneity of this community is reflected in the presence of certain rare AR conditions.

The history and demography of the Afrikaners, together with accounts of the genetic disorders in this community, has been documented by Botha and Beighton (1983a,b).

Important or unique heritable conditions that are present in the Afrikaner community are listed below.

1.1 Familial hypercholesterolemia

1.2 Progressive familial heart block

1.3 Huntington disease

1.4 Porphyria variegata

1.5 Gaucher disease

1.6 Cystic fibrosis

1.7 Familial adenomatous polyposis (FAP)

1.8 Sclerosteosis

1.9 Spondyloepimetaphyseal dysplasia with joint laxity (SEMDJL)

1.10 Keratolytic winter erythema

1.11 Hemochromatosis

FAMILIAL HYPERCHOLESTEROLEMIA

Familial hypercholesterolemia (FH) predisposes to cardiovascular disorders and represents a major problem in the Afrikaner community (Jenkins et al., 1980). This AD disorder is heterogeneous, several distinct mutations have been documented, and the founder effect has been recognized (Seftel et al., 1980, 1981). The molecular basis and diagnosis of FH in the Afrikaners has been reviewed (Kotze et al., 1991), and the role of DNA screening in this population has been addressed (Kotze et al., 1992). A molecular diagnostic service in which multiple mutations can be detected is available to persons at high risk of cardiovascular disease (Kotze et al., 2003).

PROGRESSIVE FAMILIAL HEART BLOCK

Progressive familial heart block is a heterogeneous AD disorder in which abnormalities in the cardiac conducting mechanism predispose to syncope and sudden death. Different forms of the condition, designated types I and II, can be recognized by electrocardiography (ECG).

The condition was initially recognized in an Afrikaner family in which 22 sudden deaths had occurred in three generations. The type I ECG changes were subsequently detected in 31 asymptomatic family members and the determinant gene was traced back to a couple who married in SA in 1735. The husband was an immigrant from Portugal, while his wife had French origins. A second family with type II ECG changes has been linked to an immigrant from Amerfoort in Holland, who married an Afrikaner woman in 1720 (Brink & Torrington, 1977). The determinant gene in type I has been mapped to the locus 19q13 (Brink et al., 1995).

HUNTINGTON DISEASE

Huntington disease (HD) is a well-known AD disorder in which neurological disturbance and dementia commence in adulthood. The condition is progressive and eventually lethal. As persons with the determinant gene are usually asymptomatic before procreation is contemplated, there are few barriers to transmission from generation to generation, and long dynasties of affected family members have been recognized.

Major contributions to the understanding of Huntington disease have been made by Professor Michael Hayden. In Cape Town in the early phase of his career, he wrote a doctoral thesis on the condition (Hayden, 1979), followed by a classical monograph (Hayden, 1981).

The findings in a nationwide survey of HD in South Africa were reported by Hayden et al. (1980a). In the Afrikaner community, 250 affected persons were identified and traced to common ancestors through 14 generations. A major input was made by a Dutch immigrant from Rotterdam and a woman of German stock, who married in South Africa in 1668. Huntington disease was also introduced into an Afrikaner kindred of Huguenot stock via the island of Mauritius, from an aristocratic French family who had left France at the time of the Revolution (Hayden et al., 1980b).

In Cape Town, attention became focused on the molecular basis of HD (Hayden et al., 1987) and initial proposals were made for the implementation of a molecular service (Greenberg, 1989). The molecular aspects of HD were further examined by Professor J. Greenberg in her doctoral thesis (Greenberg, 1990). Thereafter, the role of the base pair repeat mechanism was addressed (Squitieri et al., 1994), and a predictive testing service was established (Futter et al., 2009).

PORPHYRIA VARIEGATA

The South African form of porphyria variegata is an AD disorder of heme metabolism characterized by skin photosensitivity and urine that darkens on standing, due to the presence of porphyrin metabolites. Although apparently innocuous, the condition is potentially lethal due to a propensity to acute neuropsychiatric and gastrointestinal crises on exposure to certain medicinal or anesthetic agents.

Porphyria was investigated in South Africa more than 40 years ago by an Irish physician, Dr. Geoffrey Dean in Port Elizabeth. In a classical genealogical investigation, he traced the condition to a marriage in 1668 between an immigrant from Deventer, Holland, and an orphan girl from Rotterdam (Dean, 1972).

Porphyria variegata has been the focus of a research unit at the Groote Schuur Hospital, Cape Town, for four decades (Meissner et al., 1987). The determinant gene is widespread, predominantly in the Afrikaner community, and more than 30,000 persons are known to be affected (Kirsch et al., 1998). The biochemical pathogenesis was elucidated by the late professor Lennox Eales (1975), and the molecular determinants have been identified (Meissner et al., 1996; Warnich et al., 1996; Jenkins, 1997). These developments in the understanding of the disorder have been reviewed (Hift et al., 1997). At the public health level, family tracing and testing has been ongoing, genetic counseling is provided, and "medic alert" bracelets are worn by persons known to have the mutation.

Acute intermittent porphyria is another AD disorder in which a founder effect has been recognized in South Africa. The acute attacks and their precipitants differ in this form of porphyria from those of the more common variegata type (Hift & Meissner, 2005). Erythropoietic protoporphyria is yet another condition in which molecular characterization has been achieved by the Groote Schuur group (Parker et al., 2008).

GAUCHER DISEASE

Although the chronic non-neuropathic form of Gaucher disease is predominant in the Ashkenazi Jewish community, this AR disorder is also present in the Afrikaner population. The Afrikaner form is precocious and progressive, skeletal and hematological complications are frequent, and the condition is potentially lethal in early adulthood.

This form of Gaucher disease was initially recognized in 10 affected Afrikaners, the majority of whom are now deceased. It was estimated that the minimum prevalence was 1 in 200,000 in this community, and that about 1 in every 220 persons of Afrikaner stock were asymptomatic heterozygous gene carriers (Goldblatt & Beighton, 1979b).

CYSTIC FIBROSIS

Cystic fibrosis (CF) is a well-known AR condition that typically manifests in infancy with alimentary disturbance and pulmonary dysfunction. The condition is progressive, and demise before adulthood is frequent. Cystic fibrosis is comparatively common in Europe, where the birth incidence is 1 in 1600, with a heterozygous carrier rate of 1 in 20.

In the Afrikaans population of South Africa, CF is generally less frequent, the incidence at birth being about 1 in 5000. Paradoxically, the Afrikaner community in the Mariental region of Namibia, who migrated to that country early in the previous century, has a CF birth incidence of 1 in 635, with a carrier rate of 1 in 12. This perturbation of the usual situation was initially ascribed to the positive influence of malaria on survival of heterozygotes, as this disease was responsible for the deaths of many of the Afrikaner migrants during their journey to Namibia (Super, 1975). It subsequently emerged that this community had their origins in the Calvinia district of the northern Cape, and it has been postulated that the gene frequency in Mariental is the consequence of the founder effect, together with endogamy in this relatively isolated community (Super, 1978a,b).

FAMILIAL ADENOMATOUS POLYPOSIS

Familial adenomatous polyposis (FAP) is an AD trait in which there is a serious risk of the development of carcinoma in adulthood. This condition is comparatively common in the Afrikaner community, a determinant gene has been recognized, and screening in affected families is feasible (Grobbelaar et al., 1995). The molecular basis and diagnosis of FAP has been reviewed by Grobbelaar et al. (2000) and it has been the subject of an editorial (Goldberg et al., 2000).

Hereditary nonpolypotic colorectal carcinoma (HNPCC) is a second AD disorder that is an important cause of familial cancer in adulthood. Although present in the Afrikaner population, HNPCC predominates in the Mixed Ancestry Community, and it has been addressed in this context (see Section "Cape Mixed Ancestry Community").

SCLEROSTEOSIS

Sclerosteosis is an AR bone dysplasia in which progressive skeletal overgrowth leads to tall stature, mandibular enlargement, and facial asymmetry. Entrapment of the 7th and 8th cranial nerves causes deafness and facial palsy, while progressive hyperostosis of the calvarium leads to potentially lethal elevation of intracranial pressure. Syndactyly of the 2nd and 3rd fingers is an additional syndromic manifestation.

Since 1970, more than 70 affected persons have been identified in the Afrikaner population, and a heterozygous gene carrier rate of 1 in 120,000 individuals has been estimated (Hamersma et al., 2003).

Further details of Sclerosteosis are given in Chapter 41, "Heritable disorders of the skeleton in South Africa."

SPONDYLOEPIMETAPHYSEAL DYSPLASIA WITH JOINT LAXITY (SEMDJL)

SEMDJL is a rare AR dwarfism skeletal dysplasia, which has been identified in 30 children in the Afrikaner community. Gross articular hypermobility is a notable feature, which leads to significant complications, including club feet, dislocation of the radial heads, and potentially lethal spinal malalignment (Beighton, 1994). The determinant gene was probably introduced into South Africa by German immigrants who arrived in the eastern Cape region in 1856. These individuals had fought with the British during the Crimean war of 1853–1855, and for political reasons they were unable to return to their homes in Germany.

Further details of SEMDJL are given in Chapter 41.

KERATOLYTIC WINTER ERYTHEMA

Keratolytic winter erythema is an AD disorder, which manifests in the skin as attacks of a red papular eruption followed by peeling. The palms and soles are most often affected, but lesions occur elsewhere on the body. The condition usually has its onset in childhood and tends to subside after the age of 30 years. Acute episodes are related to climate and are precipitated by cold weather.

A significant proportion of families with the disorder have ancestral links with Oudtshoorn and the southwestern districts of South Africa (Findlay et al., 1977). The determinant gene has been localized to 8p22-23, and a founder effect has been postulated (Starfield et al., 1997).

HEMOCHROMATOSIS

Idiopathic hemochromatosis is an AR disorder in which clinically significant iron overload can develop in homozygotes. The clinical manifestations are very variable, and often trivial, but it is possible that the iron overload can play a role in the pathogenesis of a wide range of multifactorial disorders (Milani & Kotze, 1999). Hemochromatosis occurs in significant frequency in the Afrikaner community, and HLA linkage to an iron overload gene has been identified (Meyer et al., 1987). Further investigations have been undertaken by Meyer et al. (1988) and a molecular approach to diagnosis has been developed (de Villiers et al., 1999).

CAPE MIXED ANCESTRY COMMUNITY

As set out in the introduction to this chapter, the Cape Mixed Ancestry (CMA) community numbers approximately 3 million. This population evolved following the early Dutch settlement, and their ancestral origins include Europeans, indigenous Khoisan, persons of Indian ancestry and Javanese (Indonesian) stock brought by the Dutch East India Company, and African slaves from Mozambique and Angola. The majority of the community is domiciled in greater Cape Town and in small centers in adjacent country districts. Although the CMA population is recognized as a distinct social and cultural entity, it is not necessarily homogenous and estimations of the different contributions to the gene pool are dependent upon the geographical sites that have been sampled and investigated. For instance, there is an endogamous "Cape Malay" Muslim community in Cape Town, whose gene pool has a significant Indonesian component. The HLA status of the CMA population has been extensively reviewed by Oudtshoorn et al. (1986) and Du Toit et al. (1988).

Approximately one million persons of mixed ancestry reside in Johannesburg and other major centers. These individuals are predominantly of European and indigenous African stock, and differ in their genetic constitution from the CMA community. In addition, a small group of persons of mainly Khoisan, Cape Mixed Ancestry, and Afrikaner admixture broke away from the Cape in the early 1800s. One group migrated with their cattle to present-day Namibia, where they settled in Rehoboth, south of Windhoek. Another community, who became known as *Griqua*, moved to Campbell, west of Kimberly; this area was subsequently named *Griqualand West* (Nurse & Jenkins, 1975). A section of this group continued their travels to the southern border of Natal, (Griqualand East). No specific genetic disorders have been documented in any of these mixed-ancestry communities.

A few rare AD disorders in the CMA community have arisen by mutation or represent a founder effect. Equally, endogamy has influenced the occurrence of some AR conditions. Genetic disorders that have been recognized in this community are listed below, with relevant references and brief comment in the text that follows.

- Huntington disease
- Spinocerebellar ataxia
- Hereditary nonpolypotic colorectal cancer (HNPCC)
- Lipoid proteinosis
- Cleidocranial dysplasia
- Fetal alcohol syndrome

HUNTINGTON DISEASE

Huntington disease (HD) is a relatively common disorder in the CMA community of Cape Town (Hayden & Beighton, 1977). The determinant gene probably entered this group from Dutch immigrants during the early years of settlement.

Ongoing investigations into HD in Cape Town have included all communities, and specific aspects of the condition in the CMA group have been documented (Greenberg, 1989). These studies gave impetus to the establishment of a predictive testing and counseling service for persons at risk of developing the disorder (Futter, Heckman, & Greenberg, 2009). *(see Section Afrikaner community).*

SPINOCEREBELLAR ATAXIA

The spinocerebellar ataxias (SCA) are a heterogeneous group of genetic neurological disorders in which progressive instability and disturbance of gait can lead to considerable physical handicap. The different forms of SCA are categorized in terms of age of onset, rate of progression, severity, and presence of additional manifestations. The SCAs are comparatively common in the CMA community. Investigations in the Department of Human Genetics and the Department of Neurology in University of Cape Town culminated in a doctoral thesis by Professor A. Bryer (1994).

The initial investigation into the SCAs concerned vestibular function in nine affected adults in three generations of a family (Philcox et al., 1975). Thereafter, linkage with the HLA system was demonstrated (Bryer et al., 1992) and a predictive testing service for late-onset SCA was established (Greenberg et al., 1996). Expanded CAG repeats segregating with distinct haplotypes were subsequently demonstrated in 16 families with SCA (Ramesar et al., 1997). In a collaborative analysis between the Cape Town and Johannesburg investigators, members of 54 South African families with AD SCA were evaluated clinically, and molecular analysis for SCA expansions was undertaken (Bryer et al., 2003). Forty per cent of these affected kindreds had expansions in SCA1.

HEREDITARY NONPOLYPOTIC COLORECTAL CANCER (HNPCC)

Carcinoma of the colon and rectum is frequent, universal, and usually nongenetic. Nevertheless, a proportion of these cancers are heritable, occurring as Familial Adenomatous Polyposis (FAP) and Hereditary Nonpolypotic Colorectal Cancer (HNPCC).

These two AD genetic disorders occur in most populations, and HNPCC is well recognized in the CMA group, while FAP is particularly common in the Afrikaner community of South Africa. These distributions are probably related to founder effects and demographical factors. It is relevant that the onset of potentially lethal carcinoma in these disorders occurs in early adulthood, usually after the family has been completed. For this reason there is no barrier to transmission of the determinant gene to the next generation.

Hereditary nonpolypotic colorectal cancer was identified in a large CMA kindred in Namaqualand, in the north of Cape Province, and 16 affected persons in three generations were investigated (Goldblatt et al., 1990). The genetic basis of HNPCC was reviewed (Madden & Ramesar, 1991) and program of colonoscopy and molecular testing was implemented. Thereafter, the role of mutation identification in the management of the condition was addressed in a cohort of 27 affected members of the kindred (Goldberg et al., 1998).

The diagnosis and management of HNPCC were reviewed by Goldberg et al. (2000) and Ramesar et al. (2000). A diagnostic surgical service for families with the HNPCC has been established and molecular testing, colonoscopy, genetic counseling, and surgical resection are now offered (Anderson et al., 2006; Stupart et al., 2008).

LIPOID PROTEINOSIS

Lipoid proteinosis is a rare AR disorder characterized by the deposition of hyaline material in the skin and mucous membranes. Thickening and induration of these structures, particularly in the head and neck, leads to hoarseness, alopecia, scarring, and dermal pigmentation. Salivary gland obstruction may occur, and ocular complications have been reported. Laryngeal involvement may cause airway obstruction (Sellars, 1979). The condition is progressive but does not usually shorten the life span.

More than 50 affected persons have been documented in South Africa, and lipoid proteinosis is probably more common in this country than elsewhere in the world (Heyl, 1970). The condition is present in significant frequency in the Mixed Ancestry community of Namaqualand, where the determinant gene was introduced in the heterozygous state by an asymptomatic German progenitor (Gordon et al., 1971). This individual had several partners, each of whom produced offspring. In this way the gene spread through the isolated community, and homozygous affected children began to emerge in consanguineous unions in subsequent generations. The genetic defect has been elucidated (Ramsey & Jenkins, 2003) and molecular characterization has been achieved (Van Hougenhouck-Tulleken et al., 2004).

CLEIDOCRANIAL DYSPLASIA

Cleidocranial dysplasia (CCD) is an AD disorder in which variable hypoplasia of the clavicles is associated with defective mineralization of the cranial sutures. Stunted stature, supernumerary teeth, short digits, and a variety of minor skeletal changes are additional features. The intellect is normal, general health is good, and affected persons usually have little disability.

The condition is of special importance in the Cape, where several hundred affected persons are descended from the same ancestor. This individual was a sailor from China whose ship called at Cape Town in 1896. He decided to remain onshore, Anglicized his name to Arnold, became a Muslim, and married several wives (Jackson 1951).

These unions produced four affected sons and two affected daughters, all of whom went on to create dynasties of persons with CCD. In his study of the extended Cape Town kindred, Jackson was able to trace 70 of their affected descendants. By 1988, during a survey of dental abnormalities in CCD, 851 descendants were identified in genealogical studies, and the dental status of 68 affected persons was documented in detail (Staz, 1988).

The determinant gene in the Cape Town family has been mapped to a locus at 6p21.1-6p21 (Ramesar, 1996). It can be estimated that several hundred affected persons are currently domiciled in Cape Town, but as the condition is usually innocuous, they do not often come to medical attention.

FETAL ALCOHOL SYNDROME

The fetal alcohol syndrome, now defined more broadly as the fetal alcohol spectrum disorder (FASD) results from damage to the fetus by maternal alcohol consumption during pregnancy. The main features of this very variable disorder are low birth weight, small stature, delayed developmental milestones, and intellectual disability. Microcephaly, short palpebral fissures, a smooth philtrum and a thin upper lip are recognizable craniofacial features. It is possible that a genetic component in maternal and fetal metabolism of alcohol plays a role in the defective embryogenesis.

Fetal alcohol syndrome was first diagnosed in the Cape in the late 1970s, following an academic visit from the late Professor D.W. Smith of Vancouver. It became apparent that the syndrome was very frequent among children in the CMA community, especially in small country towns near Cape Town. Professor Denis Viljoen instituted a series of population surveys and it emerged that the frequency of FAS was of the order of 6% in schoolchildren in these communities. Many of the affected children had serious educational and behavioral difficulties, with potential social handicap in later life, and it was evident that the condition was a problem at the public health level. These alarming findings gave impetus to the formation of the Foundation for Alcohol Related Research. This organization has undertaken many aspects of the disorder, including research into diagnostic, educational, and social implications of FAS.

Research into FAS has continued, and the epidemiology of the condition in the western Cape has been documented (May et al., 2000), maternal factors have been investigated (Viljoen et al., 2002) and a biomarker for fetal exposure to alcohol has been identified (Bearer et al., 2003). It is evident that FAS is a widespread problem in South Africa and in a large-scale survey of a railway town in the center of the country, a prevalence in schoolchildren of 6%–7% has been established (Viljoen et al., 2005).

BRITISH COMMUNITY

The British occupied the Cape in the 1790s, following warfare with France and Holland. Sporadic immigration followed, with significant input in 1820, when settlers from England arrived in the eastern Cape. The discovery of diamonds in the 1870s in the Kimberley region, followed by gold in the 1890s on the Witwatersrand, prompted further British settlement. The benign climate and the availability of employment were other positive factors that attracted British immigrants, notably after the Anglo-Boer War of 1899–1902 and following World Wars I and II. At the present time, the English-speaking community is predominantly urbanized, while the Afrikaner community is both rural and urban. Bilingualism is frequent, and with increasing social interaction and marriage, the distinctions between these two communities are becoming blurred.

The English-speaking South Africans are derived from a very wide social and geographical background in Britain. The presence and frequency of genetic disorders in this community is very similar to that of the United Kingdom. Cystic fibrosis and other relatively common AR and XL disorders exemplify this situation. As immigration has been relatively recent, and as families are generally small, genetically disparate, and nonconsanguineous, no unique genetic disorders have emerged. A few uncommon nonlethal AD disorders can be traced to antecedents in the United Kingdom, as summarized below.

ADAMS-OLIVER SYNDROME

The Adams-Oliver syndrome is a form of ectrodactyly, characterized by hypoplastic digits, scalp defects, and lacrimal duct aplasia. This AD condition, which is very variable in clinical expression, is present in four generations of a large South African family of European stock (Bonafede & Beighton, 1979).

CRANIOMETAPHYSEAL DYSPLASIA

Craniometaphyseal dysplasia is a sclerosing bone dysplasia characterized by hyperostosis of the cranial vault and base, with expansion of the metaphyses of the tubular bones of the limbs. Paranasal bossing and compression of the 7th and 8th cranial nerves are variable features (Beighton, 1995). This AD disorder has been documented in 15 affected persons in five generations of a family in South Africa, with antecedents in the United Kingdom (Beighton et al., 1979). Collaborative molecular investigations involving this large kindred resulted in the identification of the determinant gene (Reichenberger et al., 2001).

DIGITOTALAR DYSMORPHISM

Digitotalar dysmorphism is an AD disorder in which variable flexion contractures of the digits are associated with vertical orientation of the talus bone of the ankle. Due to this positioning of the talus, the soles of the feet are convex, with a "rocker-bottom" configuration. The condition has been traced through five generations of a family of European stock (Sallis & Beighton, 1972).

HUNTINGTON DISEASE

As in other populations in South Africa, the AD Huntington disease can be linked to an immigrant group. In this instance, the determinant gene was brought to South Africa among Cornish miners who came to the country in the 1880s after the tin mines in Cornwall failed and gold mining commenced on the Witwatersrand (Hayden et al., 1980b).

INDIAN COMMUNITY OF SOUTH AFRICA

The late Professor WS Winship provided pediatric and medical genetic services to the Indian community of Natal for more than 50 years. His experience of heritable disorders in this group was encapsulated in an article published in recognition of the 150th anniversary of their arrival in Natal (Winship and Beighton 2011).

The Asian Indian community of South Africa had its origins in 1860, with the arrival of batches of indentured laborers and their families for employment in the sugar fields of Natal. The majority of these individuals were Tamils of the Hindu faith from Southern India, who departed from Calcutta and Madras. A few Moslems from Gujarat in the Punjab also arrived in South Africa from 1870 onwards, as did Hindus from the Surath coast in present-day Pakistan. Many of these individuals were entrepreneurs, and became shopkeepers in the urban environment. There are currently about 1 million persons of Indian ancestry in South Africa, most of whom are domiciled in Durban, with smaller communities in Johannesburg and other centers.

Unusual genetic conditions present in the Indian community are summarized below, and others which have been encountered are listed:

FAMILIAL DEAFNESS

In 1987 a diagnostic survey was undertaken in the V.A. Naik special school, Natal, for profoundly deaf children from the Indian community. Of the 212 deaf scholars, 48% were of Tamil stock, 29% were Hindus from Surath, and 11% were Muslims.

Familial profound hearing loss without any additional manifestations (undifferentiated deafness), which was suggestive of AD, AR, and XL modes of genetic transmission, was recognized. An excess of AR deafness that was present in Muslim kindreds might reflect the prevalence of parental consanguinity (Beighton et al., 1987b). Molecular investigations could be informative in these forms of familial deafness but, thus far, no studies of this type have been undertaken in the affected families.

Other well-known genetic deafness syndromes which were diagnosed in the deaf scholars include the Waardenburg, Pendred, Treacher-Collins, and Alport syndromes.

NEUROFIBROMATOSIS

Neurofibromatosis (NF), is a common AD disorder characterized by multiple papules and pigmented macules on the skin. This condition has been investigated in large families in all three subgroups of the Indian community of South Africa. In a study of 45 affected persons in three kindreds, a surprising absence of Lisch nodules on the iris was documented following slit lamp examination. Similarly, histological study of skin specimens biopsied from café-au-lait macules failed to reveal the presence of the characteristic macromelanosomes (Wallis & Slater, 1987).

CONGENITAL CONTRACTUAL ARACHNODACTYLY

Congenital contractual arachnodactyly, also known as Beals syndrome, is an AD disorder which resembles Marfan syndrome, with the additional features of crumpled external ears, flexion deformities of the digits, and limitation of extension of the elbows and knees. The condition is present in several generations of a large Hindu family with ancestral origins in Surath (Viljoen et al., 1991). Linkage investigations with serum and protein markers, and with RFLP probes for type I and II collagen, did not yield any positive results.

FRAGILE X SYNDROME

The Fragile X mental disability syndrome was confirmed cytogenetically in a large Cape Town kindred of predominantly Indian stock. Seven male cousins were severely affected and four female cousins were intellectually compromised to a lesser extent (Gardner et al., 1983). By 2010 another generation had been born, and diagnostic confirmation by molecular techniques had become a routine procedure.

INVERTED Y-CHROMOSOME

An innocuous cytogenetic condition in which the Y-chromosome is inverted is present in the Gujarati Muslim community of South Africa (Bernstein et al.,

1986). A single origin for this anomaly has been demonstrated (Spurdle & Jenkins, 1992).

OSTEOPOROSIS-PSEUDOGLIOMA SYNDROME

The osteoporosis-pseudoglioma syndrome is an AR disorder in which skeletal fragility is associated with ocular lesions that resemble gliomata. Frequent fractures and visual disturbance are the main manifestations. This condition is present in an Indian family in Durban, and in a collaborative investigation, DNA from this kindred facilitated identification of the determinant gene (Gong et al., 1996).

DIABETES MELLITUS

Although the condition is multifactorial, the presence of diabetes mellitus in the Indian community of Cape Town warrants mention. Almost 50 years ago, the pedigree of a very extensive Tamil-speaking Hindu family with diabetes was documented by the late Professor W.P.U. Jackson (Jackson et al., 1974). There is a high frequency of diabetes mellitus in the Indian community of South Africa as a whole, and in this context, the interaction of genetic and environmental factors has been explored (Conn et al., 1987).

MISCELLANEOUS GENETIC CONDITIONS IN THE INDIAN COMMUNITY

Genetic hematological disorders, notably thalassemia and, to a lesser extent, G6PD deficiency (Cayanis et al., 1977) and sickle cell anemia, are present in the Indian community. Likewise, hypercholesterolemia is frequent (Langenhoven et al., 1996; Kotze et al., 1997). The HLA status of the Indian community of Johannesburg has been investigated by Wadee & Du Toit, (1989).

The author has encountered Wilson's disease (AR hepatolenticular degeneration) and a severe undelineated AR form of osteogenesis imperfecta in consanguineous families. Sialidosis is another AR condition which has been documented (Bhigjee et al., 1991).

COMPLEX CHROMOSOMAL REARRANGEMENT

Three generations of an extensive Indian family in Natal have a balanced reciprocal translocation involving chromosomes 5, 8, 12 and 13. In the first generation studied there were 12 pregnancies, of which three were aborted spontaneously and four offspring died in the neonatal period. Of the survivors, only one male had a normal karyotype while two males and two females had the balanced reciprocal translocation. Thereafter, another nine members of the family were identified with the reciprocal translocation. Three children with an unbalanced translocation resulting in partial Trisomy 8 and partial Monosomy 12 survived in this family, and all had dysmorphic features and severe intellectual handicap.

Genetic Disorders in the Indian Community of Natal

Together with Namitha Chabilal, a genetic nursing sister, Professor WS Winship reviewed case records of patients whom they had encountered. The diagnoses that have been established are listed below:

AUTOSOMAL DOMINANT	
Achondroplasia	Brachman de Lange syndrome
Hypochondroplasia	Apert syndrome
Spondyloepiphyseal dysplasia	Crouzon syndrome
Spondylometaphyseal dysplasia	Treacher-Collins syndrome
Marfan syndrome	Waardenberg syndrome types 1 and 2
Ehlers-Danlos syndrome types 1 and 2	Neurofibromatosis type 1
Larsen syndrome	Osteogenesis imperfecta type 1
Familial adenomatous polyposis	Beals syndrome
Robinow syndrome	
AUTOSOMAL RECESSIVE	
Thalassemias (several types)	Sickle cell disease
Osteogenesis imperfecta type III	Seckel syndrome
Hurler syndrome (MPS I)	Sanfilippo syndrome (MPS III)
Morquio syndrome (MPS IV)	TAR syndrome
Ellis-van Creveld syndrome	Escobar syndrome
Dyggve-Melchior-Clausen syndrome	Spinal muscular atrophy
Epidermolysis bullosa dystrophica	Xeroderma pigmentosa
Femoral abnormal facies syndrome	Albinism
Cystic fibrosis (not Delta 508)	Bardet-Biedl syndrome
X-LINKED RECESSIVE	
Hemophilia A	Hunter syndrome (MPS II)
Androgen insensitivity syndrome (XY female)	Fragile X syndrome
Duchenne muscular dystrophy	Becker muscular dystrophy
X-LINKED DOMINANT	
Incontinentia pigmenti	
Hypophosphatemic rickets	
Goltz syndrome	

MULTIFACTORIAL
Cleft lip and palate, including Robin Sequence.
Neural Tube defects—hydrocephalus, encephaloceles, and meningomyeloceles.
CHROMOSOMAL DISORDERS
Trisomy 21, 18, and 13 occur as frequently in Indian families as in other populations.
Velo-cardio-facial syndrome 22 q 11
Williams syndrome 7 q 11
Prader-Willi syndrome 15 q 11
Y-chromosome inversion in males from Gujarat, of no clinical significance.

Although based upon experience in Natal, these genetic disorders are probably broadly representative of the gene pool of the Indian community of South Africa as a whole. Nevertheless, this community is derived from three distinct populations in different geographical regions of India. For this reason, it is to be expected that detailed analysis of the data listed above would reveal discrepancies in the relative frequencies of some of the genetic conditions in these population subgroups.

JEWISH COMMUNITY OF SOUTH AFRICA

The Ashkenazi Jewish community of South Africa had their origins in immigrants from Lithuania, who began to arrive in the 1880s. Discrimination and persecution precipitated their departure from their homeland, and economic opportunities and access prompted the choice of South Africa as a destination. The early Jewish immigrants were involved in entrepreneurial activities in the diamond and gold fields of Kimberley and the Witwatersrand, and also in the boom in ostrich feathers in the Oudtshoorn region in the 1890s. Their descendants in South Africa are involved in commerce, the arts, and the professions. The majority of the Jewish community is currently living in Johannesburg and Cape Town. In 1970 they totaled 120,000, but diminution in family size and emigration have reduced this number to about 70,000. A detailed genetic profile of the South African Ashkenazi Jewish community has been published (Lane et al., 1985).

A small Sephardic Jewish community of about 1000 is present in South Africa, mainly in Cape Town. A few families arrived from Iraq and Egypt following World War II. The majority of the community had their origins on the Mediterranean island of Rhodes, and arrived in the Cape from the former Belgian Congo, following disruption in 1960. No specific genetic disorders have been identified in this group.

The Lemba are an indigenous African group living in the far northeastern region of South Africa. They claim Jewish heritage and follow cultural and religious practices which are in keeping with this contention. Investigations of Y-chromosomal markers have provided supportive evidence, and it is likely that early trading links with merchants from the present-day Arabian peninsula were responsible for the input of genes and cultural concepts into this African community (Spurdle & Jenkins, 1996).

The most important genetic disorders in the Ashkenazi Jewish community of South Africa are the non-neuropathic form of Gaucher disease and Tay-Sachs disease. These conditions are reviewed below and others are briefly mentioned.

GAUCHER DISEASE

Gaucher disease, non-neuropathic type, is an AR trait characterized by splenomegaly, episodes of pseudo-osteomyelitis, hip pain due to collapse of the femoral capital epiphyses, and hematological problems.

Gaucher disease is well known in South Africa, and extensive investigations commenced in the 1970s (Beighton & Sachs, 1974). More than 40 affected persons (homozygotes) were identified, mainly in Johannesburg and to a lesser extent in Cape Town. The radiological manifestations were documented (Myers et al., 1975), as were the orthopedic aspects (Goldblatt et al., 1978). Thereafter, Gaucher disease in the South African Jewish community was the subject of a doctoral thesis (Goldblatt, 1980).

Reviews of the condition have included cutaneous features (Goldblatt & Beighton, 1984) and obstetric complications (Goldblatt & Beighton, 1985). The pathogenesis of thrombocytopenia in Gaucher disease was addressed by Lester et al. (1984), and tomographic scanning and radionucleotide assessment of the bone marrow was undertaken (Goldblatt et al., 1986). The longstanding involvement of the orthopedic surgeons culminated in a review of successful total hip arthroplasty in affected persons (Goldblatt et al., 1988).

Gaucher disease is an AR trait and approximately 1 in 20 persons in the Ashkenazi Jewish community of South Africa is an asymptomatic heterozygous carrier of the determinant gene (Goldblatt & Beighton, 1979a). The condition is also present in appreciable but lower frequency in the Afrikaner community (Goldblatt & Beighton, 1979b), and a few affected individuals have been recognized in other populations in South Africa (Jacobs et al., 1983). There is considerable intramolecular heterogeneity in the Gaucher disease gene, and the major determinant mutations in this country have been documented (Morar & Lane, 1996).

In view of the high frequency of the Gaucher disease gene in the Ashkenazi Jewish community, a service for carrier detection by estimation of β-glucosidase in white blood cells was established in the 1970s. Genetic counseling was an inherent component of this process. Molecular technology has now replaced biochemical investigation for diagnostic confirmation, heterozygote detection, and prenatal diagnosis.

Therapy by infusion of glucocerebrosides has proved to be very effective in Gaucher disease, and a service for treatment of affected persons from all populations in South Africa is available in Johannesburg, funded from governmental resources.

TAY-SACHS DISEASE

Tay-Sachs disease is a progressive AR metabolic disorder, which is lethal in infancy. The enzyme hexosaminidase-A is defective, and accumulation of the substrate gives rise to the disease manifestations. As with Gaucher disease, the heterozygote gene carrier rate in the Ashkenazi Jewish community is about 1 in 20; programs for family screening for heterozygosity, antenatal diagnosis, and genetic counseling were established in South Africa in the mid-1970s (Lane et al., 1976; Jenkins et al., 1977).

OTHER GENETIC DISORDERS IN THE ASHKENAZIM OF SOUTH AFRICA

The BRCA1 and BRCA2 genes, which predispose to familial carcinoma of the breast, show great intramolecular heterogeneity. Three specific mutations predominate in women in Israel, and the same mutations have been identified in the South African Ashkenazi population (*BRCA1 185delAG; BRCA1 5382insC; BRCA2 6174delT*).

Other genetic conditions that occur in this community include familial hypercholesterolemia (Durst et al., 2001; Seftel et al., 1989) and cystic fibrosis (Goldman et al., 1994). Multifactorial disorders that are well recognized in the Jewish community include Crohn's disease and diabetes mellitus.

Several genetic conditions that are present in the Ashkenazi populations of the United States and Europe are absent or extremely rare in the South African community. These include familial dysautonomia, Canavan disease, Fanconi anemia, mucolipidosis IV, Niemann-Pick disease, glycogen storage disease (type 1a) and Bloom's syndrome. This discrepancy reflects the different geographical origins in Eastern Europe of these groups of Jewish immigrants.

GREEK COMMUNITY OF SOUTH AFRICA

The Greek community of South Africa had their origins in the wave of general immigration which took place in the 1880s and 1890s, following the discovery of diamonds and gold. The Greek immigrants mainly became entrepreneurs and shopkeepers, and several thousands eventually settled in Johannesburg and neighboring towns. The Cape Town community currently numbers about 5000, and smaller communities are domiciled in urban centers throughout South Africa. Individual immigrants encouraged members of their extended families to join them, and the different Greek communities in South Africa tend to have links with particular geographical regions in Greece. For instance, the majority of the Greek community of Cape Town have their antecedents on the island of Lesbos. Other Grecian origins include Cyprus, Lemnos, Cephalonia, Athens, and the Peloponnese.

Hereditary anemias constitute a major public health problem in Greece, and it is to be expected that these disorders would be represented in immigrants from that country. This is indeed the case, and thalassemia and glucose-6-phosphate dehydrogenase deficiency (G6PD deficiency) occur in the Greek population of South Africa.

THALASSEMIA

Homozygous β-thalassemia, or thalassemia major, manifests with anemia, a typical facies due to hyperplasia of the bone marrow of the maxilla, growth retardation, hepatosplenomegaly, and dermal pigmentation. Heterozygous carriers of the determinant gene (thalassemia minor) may have mild anemia, but are otherwise usually asymptomatic.

Homozygous alpha-thalassemia results in hydrops fetalis and is lethal, while the heterozygous state is innocuous.

A formal survey of the hereditary anemias in the Greek community of Cape Town was undertaken as a doctoral project by Dr. Peter Bonafede (Bonafede, 1979). In 250 volunteers, the carrier rates of alpha and β-thalassemia were 9.2% and 1.2%, respectively (Bonafede et al., 1979). Two symptomatic persons with homozygous β-thalassemia who had been born in Cape Town were subsequently traced through the local blood transfusion service (Bonafede et al., 1983).

GLUCOSE-6-PHOSPHATE DEHYDROGENASE DEFICIENCY

Persons with this X-linked disorder have a propensity to acute hemolytic anemia following inappropriate medicinal therapy. In the Bonafede project, 10 males had G6PD deficiency and two of these individuals had previously needed hospitalization for this complication.

The prevalence of the thalassemia and G6PD gene in different parts of South Africa probably reflects the situation in the corresponding community of origin in Greece.

In turn, the distribution of these genes in Greece has been influenced by the protection that they provide against falciparum malaria. This determinant factor is not present in most of South Africa, and the biological advantage is therefore no longer operative or relevant in this country. Indeed, as this evolutionary process takes place over many generations, and as malaria has now largely been prevented in Greece, an individual's current genetic status is a legacy of the past.

In addition to the above conditions, hereditary spherocytosis was diagnosed during laboratory screening in three asymptomatic persons. This disorder can cause anemia and lead to cholelithiasis due to chronic hemolysis. In view of the AD mode of genetic transmission, a founder effect may have been operative.

OTHER COMMUNITIES IN SOUTH AFRICA

There are a few small but distinct communities in South Africa. In some, genetic conditions that have been identified can be directly related to their countries of origin.

San Community

There are no San currently living in South Africa apart from a small group displaced from Angola in the recent past, although they are still extant in the Kalahari Desert of neighboring Botswana (formerly Bechuanaland). The San have been extensively investigated in terms of population genetics, principally by Professor Trefor Jenkins and his group at the University of the Witwatersrand, Johannesburg (Nurse et al., 1977; Jenkins et al., 1987; Ramsey & Jenkins, 1988). The HLA antigenic status of the San has been addressed (Du Toit, 1990; Arendse et al., 1992).

In the molecular era, Professor Himla Soodyall of the same university has made notable research contributions (Soodyall & Jenkins, 1992). The San have also featured in a large-scale investigation of the evolutionary history of African populations using multiple nucleus microsatellite markers and insertion/deletion markers (Tishkoff et al., 2009).

In a major project of considerable genetic and anthropological importance, the complete genome of a San individual from Namibia has been documented (Schuster et al., 2010). In addition, the exomes of three elderly San were compared with the reference human genome and all published human genetic variation. No less than 1.3 million novel genome variants were detected.

As far as the author is aware, few if any specific genetic disorders have been documented in the San population. It has been suggested, however, that there is an increased frequency of color blindness in this community, and lactose intolerance may also be present (Nurse & Jenkins, 1974). In this context, it is of interest that all four persons of San stock mentioned in the preceding paragraph were homozygous for the gene that is associated with lactose intolerance. This genetic situation presumably reflects evolutionary adaptation to the diet inherent in a hunter-gatherer lifestyle.

Khoi Community

The Khoi, who remain unassimilated into the Mixed Ancestry population, numbering about 25,000, are resident in Keetmanshoop in southern Namibia. As with the San, population genetic studies have been undertaken using conventional gene markers (Balinsky & Jenkins, 1966). An epidemiological investigation of bone and joint disorders in several hundred persons was also completed in the early 1970s, but no specific genetic disorders or malformations were identified (Beighton et al., 1974). The status of the Khoi in respect of HLA haplotypes and C4 polymorphisms has been documented (Creemers & Du Toit, 1996).

Lebanese Community

A small community have their origins in Lebanon, at the eastern end of the Mediterranean Sea. In this group a rare AR disorder, the Dyggve-Melchior-Clausen syndrome, has been diagnosed in three adult siblings born in a consanguineous marriage (Bonafede & Beighton, 1978). This unusual condition manifests with stunted stature, skeletal dysplasia, and intellectual disability; it is of interest that the majority of other reported cases have also concerned individuals of Lebanese stock.

Portuguese Community

The Portuguese community of South Africa have long-standing links with this country. Indeed, the first Europeans to set foot on the coast of present-day Southern Africa were fifteenth-century Portuguese navigators. In the early 1900s, immigrants from the Portuguese island of Madeira commenced settlement in Cape Town and other regions of South Africa. More recently, many persons of Portuguese ancestry have settled in this country, following disturbances in the former colonies of Angola and Mozambique.

Machado-Joseph disease, also known as spinocerebellar ataxia type 3, is an AD disorder in which progressive disturbance of balance and neurological dysfunction can cause severe handicap (Bryer et al., 2003). The disorder is well documented on the Azores islands in the Atlantic, to the southwest of Portugal. The condition is present in the

Portuguese community in South Africa, and it is likely that the determinant gene arrived from that source, possibly via Madeira.

An autosomal recessive form of idiopathic dilated cardiomyopathy has been documented in the section of the South African Portuguese community with Madeira connections (Goldblatt et al., 1987). The condition was recognized in a brother and sister born to consanguineous parents, and in a female cousin. All three developed congestive cardiac failure at the respective ages of 20 years, 21 years, and 11 years. The diagnosis was established after extensive investigations during the course of their illnesses.

ACKNOWLEDGEMENTS

I am most appreciative of the contribution made by my wife Greta in the preparation of this chapter.

The investigations upon which the manuscript was based were supported by the South African Medical Research Council and the National Research Foundation.

The material in this chapter forms part of the Cape Genetic Heritage project, in the Division of Human Genetics, University of Cape Town.

REFERENCES

Anderson DW, Goldberg PA, Algar U, Felix R, Ramesar RS. (2006). Mobile colonoscopic surveillance provides quality care for hereditary nonpolyposis colorectal carcinoma families in South Africa. *Colorectal Disease 9*, 509–514.

Arendse B, Martell RW, Jacobs M, Oosthuizen O, Du Toit ED. (1992). HLA-DP polymorphism in the present day San (Bushmen) and Khoi (Hottentots) using polymerase chain reaction and sequence-specific oligonucleotide typing. *Tissue Antigen 40*, 38–40.

Balinsky D, Jenkins T. (1966). Glucose-6-phosphate dehydrogenase: occurrence of the "A" electrophoretic variant in Hottentots, Bushmen and Bantu. *S Afr J Med Sci 31*, 124.

Bearer CF, Jacobson JL, Jacobson SW, et al. (2003). Validation of a new biomarker of fetal exposure to alcohol. *J Pediatr 143*(4), 463–469.

Beighton P, Sacks S. (1974). Gaucher's disease in Southern Africa. *S Afr Med J 48*, 1295–1299.

Beighton P, Solomon L, Soskolne CL, Sweet B. (1974). Serum uric acid levels in a Nama (Hottentot) community in South West Africa. *S A J Sci 70*, 281–283.

Beighton P (1976). Genetic disorders in Southern Africa. *S Afr Med J 50*, 1125–1128.

Beighton P, Hamersma H, Horan F. (1979). Craniometaphyseal dysplasia - variability of expression within a large family. *Clin Genet 15*, 252–258.

Beighton P, Botha MC. (1986a). Inherited disorders in the black population of Southern Africa: Part I. Historical and demographic background; genetic haematological conditions. *S Afr Med J 69*, 247–249.

Beighton P, Botha MC. (1986b). Inherited disorders in the black population of Southern Africa: Part II. Gene disorders. *S Afr Med J 69*, 293–296.

Beighton P, Botha MC. (1986c). Inherited disorders in the black population of Southern Africa: Part III. Multifactorial, chromosomal and congenital conditions. *S Afr Med J 69*, 375–378.

Beighton P, Goldblatt J, Wallis G (1987a). Genetic disease in South Africa. A molecular approach. *S Afr Med J 72*, 766–769.

Beighton P, Sellars SL, Goldblatt J, Viljoen DL, Beighton G. (1987b). Childhood deafness in the Indian population of Natal. *S Afr Med J 72*, 209–211.

Beighton P. (1994). Spondyloepimetaphyseal dysplasia with joint laxity (SEMDJL). *J Med Genet 31*, 136–140.

Beighton P. (1995). Craniometaphyseal dysplasia (CMD), autosomal dominant form. *J Med Genet 32*, 370–374.

Bernstein R, Wadee A, Rosendorff J, Wessels A, Jenkins T. (1986). Inverted Y chromosome polymorphism in the Gujerati Muslim Indian population of South Africa. *Hum Genet 74*, 223–229.

Bhigjee AI, Seebaran AR, Petersen EM, Bill PLA. (1991). Sialidosis type I: first report in the Indian population. *Clin Neurol Neurosurg 93*(2), 115–118.

Bonafede RP, Beighton P. (1978). The Dyggve-Melchior-Clausen syndrome in adult siblings. *Clin Genet 14*, 24–30.

Bonafede RP, Beighton P. (1979). Autosomal dominant inheritance of scalp defects with ectrodactyly. *Am J Med Genet 3*, 35–41.

Bonafede RP. (1979). *Hereditary haematological disorders in the Greek population of Cape Town.* Doctoral Thesis (MD), University of Cape Town.

Bonafede RP, Botha MC, Beighton P. (1979). Inherited anaemias in the Greek community of Cape Town. *J Med Genet 16*(3), 197–200.

Bonafede RP, Botha MC, Beighton P. (1983). Thalassaemia in the Greek community of Cape Town. *S Afr Med J 64*, 860–863.

Botha MC, Beighton P. (1983a). Inherited disorders in the Afrikaner population of Southern Africa. Part I. *S Afr Med J 64*, 609–612.

Botha MC, Beighton P. (1983b). Inherited disorders in the Afrikaner population of Southern Africa. Part II. *S Afr Med J 64*, 664–667.

Brink AJ, Torrington M. (1977). Progressive familial heart block – two types. *S Afr Med J 7*, 52–53.

Brink PA, Ferreira A, Moolman JC, Weymar HW, van der Merwe P-L, Corfield VA. (1995). Gene for progressive familial heart block type I maps to chromosome 19q13. *Circulation 91*, 1633–1640.

Bryer A, Martell RW, du Toit ED, Beighton P. (1992). Adult onset of spinocerebellar ataxia linked to HLA in a South African kindred of mixed ancestry. *Tissue Antigens 40*, 111–115.

Bryer A. (1994). *Spinocerebellar Ataxias.* Doctoral Thesis (PhD), University of Cape Town.

Bryer A, Krause A, Bill P, et al. (2003). The hereditary adult onset ataxias in South Africa. *J Neurol Sci 216*, 47–54.

Bhigjee AI, Seebaran AR, Petersen EM, Bill PLA. (1991). Sialidosis type I: first report in the Indian population. *Clin Neurol Surg 93*(2), 115–118.

Cayanis E, Lane AB, Jenkins T, Nurse GT, Balinsky D. (1977). Glucose-6-phosphate dehydrogenase Porbandar: A new slow variant with slightly reduced activity in a South African family of Indian descent. *Biological Genetics 15*, 7665–773.

Conn V, Ramsey M, Jenkins T. (1987). Non-insulin-dependent diabetes mellitus and the 5' hypervariable region of the insulin gene in two South African Indian families. *S Afr Med J 71*, 174–178.

Creemers PC, Du Toit ED. (1996). C4 polymorphism and extended HLA haplotypes in Namibian San and Khoi and in South African Xhosa. *Tissue Antigens 47*, 111–116.

Dean G. (1972). *The Porphyrias.* 2nd ed London, Pitman.

de Villiers JNP, Hillermann R, de Jong G, et al. (1999). High prevalence of the Cys282Tyr HFE mutation facilitates an improved diagnostic service for hereditary haemochromatosis in South Africa. *S Afr Med J 89*, 279–282.

Durst R, Columbo R, Shpitzen S, et al. (2001). Recent origin and spread of a common Lithuanian mutation (G197delLDLR) causing familial hypercholesterolaemia: Positive selection is not always necessary to account for disease incidence among Ashkenazi Jews. *Am J Hum Genet 68*, 1172–1188.

Du Toit ED, MacGregor KJ, Taljaard DG, Oudtshoorn M. (1988). HLA-A, B, C, DR and DQ polymorphisms in three South African population groups: South African Negroes, Cape Coloureds and South African Caucasoids. *Tissue Antigens 31*, 109–125.

Du Toit ED, Taljaard DG, Marshall J, Ritchie C, Oudtshoorn M. (1990). The HLA genetic constitution of the Bushmen (San). *Human Immunol 28*, 406–415.

Eales L, Grosser Y, Sears WG. (1975). The clinical biochemistry of the human hepatocutaneous porphyrias in the light of recent studies of newly identified intermediates and porphyrin derivatives. *Ann NY Acad Sci 244*, 441–471.

Findlay GH, Nurse GT, Heyl T, et al. (1977). Keratolytic winter erythema or 'Oudtshoorn Skin'. A newly recognized inherited dermatosis prevalent in South Africa. *S Afr Med J 52*, 871–874.

Futter MJ, Heckman JM, Greenberg LJ. (2009). Predictive testing for Huntington disease in a developing country. *Clin Genet 75*(1), 549–554.

Gardner RJM, Smart RD, Cornell JM, Merckel LM, Beighton P. (1983). The fragile X chromosome in a large Indian kindred. *Clin Genet 23*, 311–317.

Goldberg PA, Madden MV, Harocopos, C, Felix R, Westbrook C, Ramesar RS. (1998). In a resource-poor country, mutation identification has the potential to reduce the cost of family management for hereditary nonpolyposis colorectal cancer. *Dis Colon Rectum 41*(10), 1250–1255.

Goldberg PA, Madden MV, Harocopos C, et al. (2000*)*. Inherited Colon Cancers. (Editorial) *S Afr Med J, 90*(7), 703–705.

Goldblatt J, Sacks S, Beighton P. (1978). The orthopedic aspects of Gaucher disease. *Clin Orthop 137*, 208–214.

Goldblatt J, Beighton P. (1979a). Gaucher's disease in South Africa. *J Med Genet 16*(4), 302–305.

Goldblatt J, Beighton P. (1979b). Gaucher disease in the Afrikaner population of South Africa. *S Afr Med J 55*, 209–210.

Goldblatt J. (1980). *Gaucher disease in the Ashkenazi Jewish community of South Africa.* Doctoral thesis (MD), University of Cape Town.

Goldblatt J, Beighton P. (1984). Cutaneous manifestations of Gaucher disease. *Brit J Derm III*, 331–334.

Goldblatt J, Beighton P. (1985). Obstetric aspects of Gaucher disease. *Brit J Obstet & Gynaec 92*, 145–149.

Goldblatt J, Hermann G, Levy RN, Goldsmith S, Grabowski GA, Desnick RJ. (1986). Computed tomographic scan and radionuclide assessment of marrow involvement in type I Gaucher disease. *AJR 147*(5), 943–948.

Goldblatt J, Melmed J, Rose AG. (1987). Autosomal recessive inheritance of idiopathic dilated cardiomyopathy in a Madeira Portuguese kindred. *Clin Genet 32*, 249–254.

Goldblatt J, Sacks S, Dall D, Beighton P. (1988). Total hip arthroplasty in Gaucher's disease. *Clin Orthop 228*, 94–98.

Goldblatt J, Madden MV, Boshoff PJD, Wallis C, Price SK. (1990). Hereditary non-polyposis colorectal cancer in a Namaqualand kindred. *S Afr Med J 77*, 42–44.

Goldman A, Jenkins T, Ramsey M. (1994). Analysis of 40 known cystic fibrosis mutations in South African patients. *Clinical Genetics 46*, 398–400.

Gong, Y, Vikkula M, Boon L, et al. (1996). Osteoporosis-pseudoglioma syndrome, a disorder affecting skeletal strength, is assigned to chromosome region 11q12–13. *Am J Hum Genet 59*, 146–151.

Gordon H, Gordon W, Botha V. (1971). Lipoid proteinosis. In *Birth Defects: Original Article Series 8*(8), 164–169, New York: March of Dimes.

Greenberg J. (*1989*). A molecular service for Huntington disease in Southern Africa.(editorial) *S Afr Med J 76*, 135–137.

Greenberg LJ. (1990). *A molecular approach to Huntington disease in South Africa.* Doctoral Thesis (PhD), University of Cape Town.

Greenberg J, Beatty S, Soltau H, Bryer A. (1996). Huntington disease and late-onset spinocerebellar ataxia: A predictive testing service in Cape Town. *CME 14*(9), 1364–1367.

Grobbelaar JJ, Oosthuizen CJJ, Maddin M, Retief AE, Kotze MJ. (1995). The use of DNA markers in the pre-clinical diagnosis of familial adenomatous polyposis in families in South Africa. *S Afr Med J* 85, 269–271.

Grobbelaar JJ, Fortuin R, Scholtz CL, et al. (2000). Familial adenomatous polyposis coli (FAP) in South Africa - Molecular basis and diagnosis. *S Afr Med J 90*, 715–719.

Hamersma H, Gardner J, Beighton P. (2003). The Natural History of Sclerosteosis. *Clin Genet 63*, 191–196.

Hayden MR, Beighton P. (1977). Huntington's Chorea in the Cape Coloured community of South Africa. *S Afr Med J 52*, 886–888.

Hayden MR. (1979). *Huntington Chorea in South Africa.* Doctoral Thesis (PhD), University of Cape Town.

Hayden MR, MacGregor JM, Beighton P. (1980a). The prevalence of Huntington's Chorea in South Africa. *S Afr Med J 58*, 193–196.

Hayden MR, Hopkins HC, Macrae M, Beighton PH. (1980b). The origin of Huntington's Chorea in the Afrikaner population of South Africa. *S Afr Med J 58*, 197–200.

Hayden MR. (1981). *Huntington's Chorea.* Springer-Verlag, Berlin, Heidelberg.

Hayden MR, Goldblatt J, Wallis G, Winship IM, Beighton P. (1987). Molecular genetics and Huntington's disease. The South African situation. *S Afr Med J 71*, 683–686.

Heyl T. (1970). Genealogic study of lipoid proteinosis I South Africa. *Br J Dermatol 83*, 338–341.

Hift RJ, Meissner PN, Corrigall AV, et al (1997). Variegate porphyria in South Africa, 1688–1996 –new developments in an old disease. *S Afr Med J 87*(6), 722–731. Review.

Hift RJ, Meissner PN. (2005). An analysis of 112 acute prophyric attacks in Cape Town, South Africa: evidence that acute intermittent porphyria and variegate porphyria differ in susceptibility and severity. *Medicine (Baltimore) 84*(1), 48–60.

Jackson WPU. (1951). Osteo-dental dysplasia (Cleido-cranial dysostosis). The Arnold Head. *Acta Med Scand 139*, 292–307.

Jackson WPU, Van Mieghem W, Marine N, Keller P, Edelstein I. (1974). Diabetes among the Tamilian Indian community in Cape Town: Re-examination after five years. *S Afr Med J 48*, 1839–1843.

Jacobs P, Tribe R, Petersen EM, Beighton P. (1983). Gaucher's disease in the Black population of South Africa. *S Afr Med J 64*(13), 490–492.

Jenkins T, Lane AB, Kromberg JGR. (1977). Tay-Sachs Disease screening and prevention in South Africa. *S Afr Med J 51*, 95–98.

Jenkins T. (1979). The Bushmen, San hunters and herders of Southern Africa. *S Afr J Sci 75*, 280–282.

Jenkins T, Nicholls E, Gordon E, Mendelsohn D, Seftel HC, Andrew MJA. (1980). Familial hypercholesterolaemia - a common genetic disorder in the Afrikaans population. *S Afr Med J 57*, 943–947.

Jenkins T, Joffe BI, Panz VR, Ramsey M, Seftel HC. (1987). Transition from a hunter-gatherer to a settled life-style among the !Kung San (Bushmen): effect on glucose tolerance and insulin secretion. *S Afr J Sci 83*, 410–412.

Jenkins T. (1997). The molecular basis of South African genetic porphyria established at last! *SAMJ 87*, 733–735.

Kirsch RE, Meissner PN, Hift RJ. (1998). Variegate porphyria. *Semin Liver Dis 18*(1), 33–41. Review.

Kotze MJ, Langenhoven E, Warnich L, Du Plessis L, Retief AE. (1991). The molecular basis and diagnosis of familial hypercholesterolaemia in South African Afrikaners. *Ann Hum Genet 55*, 115–121.

Kotze MJ, Langenhoven E, Kriek JA, Oosthuizen CJJ, Retief AE. (1992). DNA screening of hyperlipidemic Afrikaners for familial hypercholesterolemia. *Clin Genet 42*, 43–46.

Kotze MJ, Loubser O, Thiart R, et al. (1997). CpG hotspot mutations at the LDL receptor locus are a frequent cause of familial hypercholesterolemia among South African Indians. *Clin Genet 51*, 394–398.

Kotze MJ, Kriegshäuser G, Thiart R, et al. (2003). Simultaneous detection of multiple familial hypercholesterolaemia mutations facilitates an improved diagnostic service in South African patients at high risk of cardiovascular disease. *Mol Diagn 7*, 169–174.

Kromberg JGR, Jenkins T. (1982). Common birth defects in South African Blacks. *S Afr Med J 62*, 599–602.

Lane AB, Skikne MI, Jenkins T. (1976). The prenatal diagnosis of Tay-Sachs Disease. *S Afr Med J 50*, 1553–1555.

Lane AB, Dunn DS, Jenkins T. (1985). A genetic profile of the South African Ashkenazi Jewish population. *S Afr Med J 68*, 935–939.

Langenhoven E, Warnich L, Thiart R, et al. (1996). Two novel point mutations causing receptor-negative familial hypercholesterolemia in a South African Indian homozygote. *Atherosclerosis 125*, 111–119.

Lester TJ, Grabowski GA, Goldblatt J, Leidermann IZ, Zaroulis CG. (1984). Immune thrombocytopenia and Gaucher's disease. *Am J Med 77*, 569–571.

Madden M, Ramesar R. (1991). How genes cause inherited large-bowel cancer (editorial). *S Afr Med J 80*, 69–70.

May PH, Brook L, Gossage JP, et al. (2000). Epidemiology of fetal alcohol syndrome in a South African community in the Western Cape. *Am J Public Health 90*(12), 1905–1912.

Meissner PN, Meissner DM, Sturrock ED, Davidson B, Kirsch RE. (1987). Porphyria – the UCT experience. *S Afr Med J 72*(11), 755–61. Review.

Meissner PN, Dailey TA, Hift RJ, et al. (1996). A R59W mutation in human protoporphyrinogen oxidase results in decreased enzyme activity and is prevalent in South Africans with variegate porphyria. *Nat Genet 13*(1), 95–7.

Meyer TE, Ballot D, Bothwell TH, et al. (1987). The HLA linked iron loading gene in an Afrikaner population. *J Med Genet 24*, 348–356.

Meyer TE, Baynes RD, Bothwell TH, et al. (1988). Idiopathic Haemochromatosis in an Afrikaner population. *Ann NY Acad Sci 526*, 328–330.

Milani MY, Kotze MJ. (1999). Molecular diagnosis of hereditary haemochromatosis: identify an affected person and save a family. *S Afr Med J 89*, 263–264.

Morar B, Lane AB. (1996). The molecular characterization of Gaucher disease in South Africa. *Clinical Genetics 50*, 78–84.

Myers HS, Cremin B, Beighton P, Sacks S. (1975). Chronic Gaucher's disease: Radiological findings in 17 South African cases. *Brit J Radiol 48*, 465–469.

Nurse GT, Jenkins T (1974). Lactose intolerance in San populations. *Brit Med J 3*, 809.

Nurse GT, Jenkins T. (1975). The Griqua of Campbell, Cape Province. *Am J Phys Anthropol 43*, 71–78.

Nurse GT, Botha MC, Jenkins T. (1977). Sero-genetic studies on the San of South West Africa. *Hum Hered 27*, 81–89.

Oudtshoorn M, Du Toit ED, Taljaard DG, MacGregor KJ. (1986). A Study of the HLA-Dw determinants and their relationship to DR and DQ antigens in three South African Population groups: South African Caucasoids, South African Negroes and Cape Coloureds. *Human Immunol 17*, 273–287.

Parker M, Corrigall AV, Hift RJ, Meissner PN. (2008). Molecular characterization of erythropoietic protoporphyria in South Africa. *Br J Dermatol 159*(1), 182–191.

Philcox DV, Sellars S, Pamplett R, Beighton P. (1975). Vestibular dysfunction in hereditary ataxia. *Brain 98*(II), 309–316.

Ramesar RS, Greenberg J, Martin R, et al. (1996). Mapping of the gene for cleidocranial dysplasia in the historical Cape Town (Arnold) kindred and evidence for locus homogeneity. *J Med Genet 33*, 511–514.

Ramesar RS, Bardien S, Beighton P, Bryer A. (1997). Expanded CAG repeats in spinocerebellar ataxia (SCA1) segregate with distinct haplotypes in South African families. *Hum Genet 100*, 131–137.

Ramesar RS, Madden MV, Felix R, et al. (2000). Molecular genetics improves the management of hereditary non-polyosis colorectal cancer. *S Afr Med J 90*(7), 709–714.

Ramsey M, Jenkins T. (1988). Alpha-Globin Gene Cluster Haplotypes in the Kalahari San and Southern African Bantu-speaking Blacks. *Am J Hum Genet 43*(4), 527–533.

Ramsey M, Jenkins T. (2003). The genetic defect identified and a founder effect confirmed in Namaqualand families with lipoid proteinosis. *S Afr J Sci 98*, 9–10

Reichenberger E, Tiziani V, Watanabe S, et al. (2001). Autosomal dominant craniometaphyseal dysplasia is caused by mutations in the transmembrane protein ANK. *Am J Hum Genet 68*(6), 1321–1326.

Sallis JG, Beighton P. (1972). Dominantly inherited digito-talar dysmorphism. *J Bone Jt Surg 54B*, 509–515.

Schuster SC, Miller W, Ratan A et al. (2010). Complete Khoisan and Bantu genomes from southern Africa. *Nature 463*(7283), 943–947.

Seftel HC, Baker SG, Sandler M, et al. (1980). A host of hypercholesterolaemic homozygotes in South Africa. *Brit Med J 281*, 633–636.

Seftel HC, Baker SG, Sandler MP, et al. (1981). Familial hypercholesterolaemia (FH) in South Africa with special reference to the homozygous form. *The Leech 50*, 47–52.

Seftel HC, Baker SG, Jenkins T, Mendelsohn D. (1989). Prevalence of familial hypercholesterolemia in Johannesburg Jews. *Am J Med Genet 34*, 545–547.

Sellars SL. (1979). Benign tumours of the larynx. *S Afr Med J*56, 943–946.

Soodyall H, Jenkins T. (1992). Mitochondrial DNA polymorphisms in Khoisan populations from southern Africa. *Annals of Human Genetics 56*, 315–324.

Spurdle A, Jenkins T. (1992). The inverted Y chromosome polymorphism in the Gujerati Muslim Indian Population of South African has a single origin. *Human Heredity 42*, 330–332.

Spurdle AB, Jenkins T. (1996). The origins of the Lemba "Black Jews" of Southern Africa: evidence from P12F2 and other Y chromosome markers. *Am J Hum Genet 59*, 1126–1133.

Squitieri F, Andrew AE, Goldberg YP, et al. (1994). DNA haplotype analysis of Huntington disease reveals clues to the origins and mechanisms of CAG expansion and reasons for geographic variations of prevalence. *Hum Mol Genet 3*(12), 2103–2114.

Starfield M, Hennies HC, Jung M, et al. (1997). Localization of the gene causing keratolytic winter erythema to chromosome 8p22-p23, and evidence for a founder effect in South African Afrikaans-speakers. *Am J Hum Genet 61*, 370–378.

Staz J. (1988). Cleidocranial dysplasia in the South Western Cape: preliminary report. *J Dent Assoc SA 43*, 547–551.

Stupart DA, Goldberg PA, Algar U, Ramesar R. (2008). Surveillance colonoscopy improves survival in a cohort of subjects with a single mismatch repair gene mutation. *Colorectal Disease 1192*, 126–130.

Super M. (1975). Cystic fibrosis in the South West African Afrikaner: an example of population drift possibly with heterozygote advantage. *S Afr Med J 49*, 818–820.

Super M. (1978a). *Cystic fibrosis in South West Africa.* Thesis for the Doctorate in Medicine, University of Cape Town.

Super M. (1978b). Cystic fibrosis in southern Africa, including the preparation of a register of carriers and potential carriers. *S Afr Med J 54*, 18–21.

Tishkoff SA, Reed FA, Friedlander FR, et al. (2009). The genetic structure and history of Africans and African Americans. *Science 324*, 1035–1044.

van Hougenhouck-Tulleken W, Chan I, Hamada T, et al. (2004). Clinical and molecular characterization of lipoid proteinosis in Namaqualand, South Africa. *Brit J Dermatology 151*(2), 413–423.

Viljoen D, Ramesar R, Behari D. (1991). Beals syndrome: clinical and molecular investigations in a kindred of Indian descent. *Clin Genet 39*, 181–188.

Viljoen D, Croxford J, Gossage JP. (2002). Characteristics of mothers of children with fetal alcohol syndrome in the Western Cape Province of South Africa: a case control study. *J Stud Alcohol 63*(1), 6–16.

Viljoen DL, Gossage JP, Brooke L, et al. (2005). Fetal alcohol syndrome epidemiology in a South African community: a second study of a very high prevalence area. *J Stud Alcohol 66*(5), 593–604.

Wadee AA, and Du Toit ED. (1989). HLA frequencies in the Asian Indian population resident in Johannesburg. *S Afr Med J 76,* 331–334.
Wallis CE, Slater CP. (1987). Neurofibromatosis in the South African Indian community – further evidence for heterogeneity? *S Afr Med J 72,* 478–480.
Warnich L, Kotze MJ, Groenewald IM, et al. (1996). Identification of three mutations and associated haplotypes in the protoporphyrinigen oxidase gene in South African families with variegate porphyria. *Hum Mol Genet 5,* 981–984.
Winship WS, Beighton P. (2011). Genetic disorders in the Indian Community of South Africa. *S Afr Med J 101,* 481–484.

40.

HEREDITARY DEAFNESS IN SOUTHERN AFRICA

Peter Beighton and Sean Sellars

INTRODUCTION

Profound childhood deafness is a major disability that results in defective speech and impaired social interaction. Damage to the hearing mechanism can result from numerous different genetic and acquired factors, and accurate diagnosis of the underlying cause is crucial for preventative measures (Beighton, 1990).

In southern Africa, approximately 5000 profoundly deaf children attend 25 special educational facilities, most of which are residential. Between 1975 and 1999, regular clinical assessment and diagnostic surveys were undertaken; the findings are reviewed in the first section of this chapter. This chapter focuses on prevalence of genetic deafness and introduces genetic conditions that are of special importance in South Africa.

EPIDEMIOLOGY

DIAGNOSTIC SURVEYS IN SPECIAL SCHOOLS FOR DEAF CHILDREN

The children attending special schools in Southern Africa are profoundly deaf, but otherwise educable. The schools were established on a linguistic basis, catering for the English, Afrikaans, and a variety of indigenous African languages. These establishments are mainly situated in urban environments and in rural administrative centers. Prior to 1996 they were ethnically segregated, but they are now fully integrated. Sisters of the Dominican Order founded several of the schools, while others were established by the government; all are now fully supported by the South African Department of Education.

The diagnostic surveys were carried out by the authors, with the assistance of nursing sisters from the Department of Human Genetics, University of Cape Town. Each affected child had a clinical and otological examination, and available case notes were scrutinized for relevant diagnostic and family data. Members of the school staff participated in the organization of the survey and assisted with communication with the deaf scholars. If concurrent illness or abnormality was recognized, treatment was arranged. Where necessary, the affected child was transported to Cape Town or other major center for specialized medical or surgical care. In the early surveys, referral for previously unrecognized cardiac or orthopedic malformations was not uncommon, but as the overall standard of healthcare improved, referral of this type became less frequent.

At the time of the surveys, the majority of the children had been categorized by the school authorities in terms of the severity of their hearing deficit and very few had etiological diagnoses. In view of the implications for amplification of sound, however, the conductive or sensorineural (perceptive) nature of child hearing loss had usually been determined. A minority of children were regarded as having "familial" deafness on a basis of deaf relatives who had previously attended the same school. In others, the hearing deficit was attributed to unspecified illness, such as encephalitis or meningitis during infancy and early childhood. The presence or absence of residual speech was relevant in the identification of a prelingual or postlingual cause of deafness.

A notable success of the surveys was the recognition of children with the Treacher-Collins syndrome and other forms of conductive deafness, in which the deafness could be ameliorated by surgical correction.

At the conclusion of each survey, a full report was sent to the school principal for inclusion in the case notes. Relevant information was also provided for medical colleagues who were involved with individual children. In the later surveys, genetic counseling was provided for the school leavers.

During the period of the surveys (1975–1999), some schools in the Cape were visited annually, while others in distant parts of South Africa were only visited on single occasions. By 1983, surveys had been undertaken in 19 schools and 3064 deaf children had been examined. During this period the schools were segregated on an ethnic and linguistic basis, and the findings were published in this context (Sellars, Napier, & Beighton, 1975; Sellars, Groeneveldt, & Beighton, 1976; Sellars, Beighton G, Horan, & Beighton P, 1977; Sellars & Beighton, 1978;

Beighton et al., 1987). The authors' observations also formed the subject of a monograph entitled "Genetics and Otology" (Beighton & Sellars, 1982).

SURVEY FINDINGS

The diagnoses that were established in 3064 deaf scholars were summarized by Sellars and Beighton (1983). The numbers and proportions of children in each of the diagnostic categories in the surveys are shown below:

Syndromic genetic deafness	7%
Nonsyndromic familial deafness	11%
Acquired deafness	25%
Undifferentiated deafness with additional abnormalities	11%
Undifferentiated deafness without additional abnormalities	46%

SYNDROMIC GENETIC DEAFNESS

Of the 203 children with genetic or multifactorial syndromes, no less that 89 (44%) had the Waardenburg syndrome. The Treacher-Collins syndrome (16%), Branchial Arch syndrome (11%) and the Pendred syndrome (8%) were well represented, but other genetic disorders in this category were individually rare.

SYNDROME	MODE OF INHERITANCE	TOTAL
Waardenburg	AD	89
Treacher-Collins	AD	28
Branchial arch	AD	22
Pendred	AR	17
Usher	AR	9
Sclerosteosis	AR	8
Marfan	AD	4
Onychodystrophy	AR	2
Vitilogo	AD	2
Nager	AD/AR	2
Osteopetrosis	AD	2
Crouzon	AD	1
Holt-Oram	AR	1
Trichorhinophalangeal	AD	1
Tuberous sclerosis	AD	1
Lentigenes (LEOPARD)	AD	1
Hunter	XL	1
Peripheral dysostosis	?	1
Klippel-Feil	-	3
Goldenhar	-	2
Wildervanck	-	1
Sturge-Weber	-	1
		202

NONSYNDROMIC FAMILIAL DEAFNESS

Three hundred and forty-seven children (187 boys and 160 girls) had a familial distribution of nonsyndromic deafness, which permitted determination of the presumptive mode of inheritance. These children lacked nonotological manifestations and no syndromic diagnosis was possible. In 65 children (i.e., approximately 20% of those with nonsyndromic familial deafness) transmission was autosomal dominant, and in 282 (80%) it was recessive. This latter category includes both X-linked and autosomal forms but, as these could not always be distinguished with certainty, they were grouped together in this analysis.

ACQUIRED DEAFNESS

A well-documented history or objective evidence of a presumptive acquired cause of deafness was obtained in 765 children (442 boys and 323 girls). In certain of the special schools, especially in rural areas, case notes were scanty. A nonspecific diagnosis of "severe illness" was accepted for the purposes of analysis, if the authors were reasonably confident that the deafness was indeed "acquired," and in the absence of evidence to the contrary. The types of acquired deafness and numbers of affected children are listed in Table 40-1.

UNDIFFERENTIATED DEAFNESS WITH ADDITIONAL ANOMALIES

Undifferentiated deafness of unknown cause, together with a wide variety of nonspecific minor anomalies in other systems, was present in 331 children (190 boys and 141 girls). The etiological interrelationships were uncertain and it is not known to what extent these additional abnormalities were chance concomitants.

UNDIFFERENTIATED DEAFNESS WITHOUT OTHER ANOMALIES

In 1,418 deaf children (781 boys and 637 girls) there were no phenotypical stigmata other than deafness. In many of these affected children, a genetic etiology was possible in view of the absence of any history or clinical manifestations of an acquired causation such as intrauterine damage

TABLE 40-1 THE AETIOLOGY OF "ACQUIRED" DEAFNESS IN SOUTHERN AFRICA

	M	F	TOTAL
PRENATAL			
Maternal rubella	68	75	143
Maternal toxemia	1	1	2
Maternal Alcohol	0	2	2
Maternal therapy with			
Quinine	3	0	2
Streptomycin	0	1	1
Thalidomide	0	1	1
Conjoined twin	0	1	1
	71	81	152
PERINATAL			
Jaundice	49	17	66
Birth trauma	29	19	48
Prematurity	18	6	24
Anoxia	10	4	14
	106	46	152
POSTNATAL			
Meningoencephalitis	138	92	230
Streptomycin therapy	11	16	27
Burns therapy	16	26	
Middle-ear disease	15	7	22
Tuberculous meningitis	12	7	19
Head injury	3	3	6
Congenital syphilis	1	0	1
	190	141	331
POSTNATAL			
"Severe illness"	62	47	109
Measles	8	4	12
Pneumonia	2	2	4
Poliomyelitis	1	0	1
Gastroenteritis	1	1	2
Malaria	1	0	1
Typhoid	0	1	1
	75	55	130
Grand total	442	323	765

(e.g., maternal rubella), birth injury, or infectious disease in infancy such as meningoencephalitis. Further evidence was provided by audiometrical studies, as the hearing loss is usually of the "perceptive" or "neurosensory" type in the genetic forms of profound deafness. It is generally assumed that many of these children have AR deafness and it follows that accurate diagnosis at the molecular level will play a crucial role in genetic counseling and the formation of appropriate regimes of medical management.

During the past decade, mutations at more than 30 different gene loci have been proposed as determinants of AR nonsyndromic deafness. Of these, the connexin 26 gene (*CX26*) has proved to be the most common etiological factor, with important implications for diagnostic screening by means of molecular techniques. This gene, which is located on chromosome 13q11, is 5.5 kilobases in length, and contains two exons. A number of different mutations in this gene can cause perceptive deafness in homozygotes.

Mutations in *CX26* have been demonstrated in Europe in 50% of a series of families with profound perceptive deafness. In the families of persons with this genotype, 86% had 35 del G mutations. In another study in England, the *CX26* genotype was implicated in 27.8% of persons with AR familial undifferentiated deafness and 7.9% of sporadic persons with this form of hearing loss. In both instances, 70% had 35 del G mutations. In other populations, different mutations predominate. In particular, 176 del T is frequent in Ashkenazi Jews, although the status of this particular mutation in the Ashkenazi community of South Africa has not been determined. In addition to these two mutations, approximately 60 additional mutations have been identified in *CX26*. These are all individually rare and genotype–phenotype correlations are incomplete. As mutations in *CX26* are responsible for a significant proportion of undifferentiated AR deafness, it follows that diagnostic screening by molecular techniques has a place in routine genetic management.

ADDITIONAL DIAGNOSTIC SURVEYS

By 1991, a total of 4452 deaf scholars had been examined, including 885 children of Shona (634) and Ndebele (222) stock attending five special schools in Zimbabwe (Viljoen et al., 1988) The findings were further analyzed and it was evident that the broad categorization of the causes of hearing loss has remained substantially unchanged (Beighton et al., 1991). Anomalous population distribution of certain genetic disorders was noted:

1. High frequency of Waardenburg syndrome in scholars of Southern Sotho stock (7.8%).
2. High frequency of AR undifferentiated deafness in Muslim scholars in Natal.
3. High frequency of AR undifferentiated deafness in scholars of Shona stock in Zimbabwe.
4. Paucity of Usher syndrome in all South African populations (0.3% of deaf scholars).
5. High frequency of sclerosteosis (AR) in Afrikaner scholars.

6. Absence of structural malformations among deaf Zulu scholars.

The genetic founder effect allied to social constraints and demographic factors might account for these observations.

Details of the individual school surveys from which the above data were obtained are given in the articles referenced in the paragraph preceding the list of genetic syndromes.

A further survey of an additional 1060 scholars attending 5 schools for the deaf, undertaken by Dr. Jessica Gardner in the late 1990s, was the subject of a doctoral thesis (Gardner, 1999). Again, the general proportions of the etiological categories of deafness remained unchanged.

DIAGNOSTIC SURVEYS IN OTHER FORMS OF DEAFNESS

Partial Deafness

It was noteworthy that investigation of 223 children attending three special schools for the partially deaf revealed that the proportions of the etiological categories were very similar to those of profoundly deaf children. In particular, 11% had a specific genetic syndrome and of these, 18 had middle-ear abnormalities that were potentially correctable by surgery (Sellars & Beighton, 1978).

Deafness and Blindness

In the 12 deaf and blind children attending the Worcester School for the Blind, the firm diagnosis of rubella syndrome was made in 4 boys and in 2 girls. In one boy the combination of retinitis pigmentosa and perceptive deafness permitted diagnosis of the Usher syndrome. A tentative diagnosis of the Refsum syndrome (retinitis pigmentosa, polyneuritis, and deafness) was made in the second, and of the Kniest syndrome (deafness, cataracts, skeletal dysplasia, and cleft palate) in a third. The fourth boy had the progressive, potentially lethal rod-cone dystrophy-deafness-chronic renal dysfunction syndrome (OMIM 268315). This autosomal recessive disorder is well known in the Cape, and affected siblings from 3 other kindreds have at some time attended the Worcester School for the Blind (Sellars & Beighton, 1976).

SYNDROMIC DEAFNESS IN SOUTH AFRICA: MANIFESTATIONS AND IMPLICATIONS

The recognition of specific syndromic forms of genetic deafness in children in the special schools gave impetus to family studies. In addition, other persons with deafness syndromes were frequently seen at the genetic counseling clinics in Cape Town, and in the outreach clinics throughout the country. With increasing interest in medical genetics in the 1970s, and the inception of relevant molecular technology in the 1980s, these affected individuals provided the basis for wide-ranging research. In this context, genetic deafness disorders of special importance in South Africa, in terms of practical implications or research directions, are reviewed below.

Waardenburg Syndrome (OMIM 193500)

The Waardenburg syndrome (WS) is an AD disorder characterized by heterochromia irides, a white forelock, perceptive deafness, and dystopia canthorum. Hypopigmentation of the skin, facial and body hair are additional inconsistent features. Clinical expression is very variable, and it is estimated that only about 15% of affected persons have a significant hearing deficit. Waardenburg syndrome is conventionally subcategorized into the common type I, in which dystopia canthorum is present, and the uncommon type II, in which this feature is absent.

Waardenburg syndrome was by far the most common genetic deafness disorder in South Africa, and in the period 1974–1983, 89 affected children were documented among 3006 deaf scholars (Sellars & Beighton, 1983). The overall frequency of WS in the schools was 3% but there were considerable discrepancies, especially in the different linguistic groups of indigenous Africans. For instance, in a school predominantly catering for children of southern Sotho stock, the frequency was 7.8%, while in a school in Zululand, the comparable frequency was 1.4%.

Although complete or segmental differences in iris color are regarded as a major feature of WS, in our experience, eyes of a striking sapphire blue color was a frequent finding. In several families with the classical manifestations of WS in several generations, this impressive eye color was the only feature in some individuals, with or without deafness. Equally, a number of deaf children without any additional clinical manifestations were members of families with classical WS (Beighton et al., 1991). It is apparent that the clinical expression of the WS gene can be highly variable, and for every family member with deafness there may be several relatives with the determinant gene in the absence of significant clinical manifestations (Winship & Beighton 1992).

In an international collaborative linkage investigation, which included 7 South African families, the determinant gene for WS I was assigned to a locus on 2q (Farrer et al., 1992). A mutation in the PAX 3 gene at 2q35 was subsequently identified in one of the SA families (Butt et al., 1994). Thereafter, in an investigation that included a South African family with WS II, molecular evidence for the autonomous identity of these two forms of WS was obtained by the international consortium. The consortium subsequently demonstrated a high level of intramolecular heterogeneity in WS I by mutation analysis. Some genotype correlations were established with eye, skin, and hair phenotypes. Nevertheless, there was no obvious correlation between genotype and the presence or absence of deafness in the affected persons (DeStefano et al., 1998).

Piebaldism with Deafness (OMIM 172800)

Piebaldism is a rare AD disorder in which extensive patchy skin pigmentation is associated with a white forelock. Involvement of the irides has been reported and there are a few reports of sporadic piebald individuals with deafness. These manifestations represent considerable phenotypic overlap with WS, and the question of syndromic identity arises.

A South African family with gross piebaldism in 7 persons in three generations, but without deafness, has been documented (Winship et al., 1991). Syndromic identity was confirmed by demonstration of a splice junction mutation of IVS12 in the KIT gene (Spritz et al., 1992). In another sporadic South African child of Xhosa stock, with piebald skin and profound perceptive deafness but normal eye color, initially diagnosed as having WS, heterozygosity for a novel missense mutation in the KIT gene was identified (Spritz & Beighton, 1998). This finding provided evidence that deafness can be a genuine but infrequent component of piebaldism.

X-linked Deafness with Perilymphatic Gusher (OMIM 304400)

Bilateral mixed conductive and perceptive deafness associated with fixation of the stapedial footplate and profuse perilymphatic otorrhoea on attempted stapedectomy is inherited as an X-linked trait. Audiometry reveals characteristic conductive and perceptive hearing loss. This condition was recognized in Cape Town when a deaf medical student expressed an interest in genetics. It emerged that a perilymphatic gusher had occurred during diagnostic tympanotomy a few years previously, and that his brother and maternal uncle were similarly affected. He then undertook a genealogical study in his own family and determined that 7 male relatives on his mother's side of the family had been deaf. The pedigree data were entirely consistent with X-linked transmission (Thorpe et al., 1974). A few years later, he went on to document his own experiences under the title, "The Deaf Doctor" (Thorpe, 1978).

The Department of Human Genetics at the University of Cape Town had a special interest in genetic disorders on oceanic islands (see Chapter 39), and as part of this initiative a survey of heritable conditions was undertaken on the island of Mauritius, in the Indian Ocean, by Dr. Colin Wallis. His findings, which were the subject of a doctoral thesis (Wallis, 1985) included documentation of a large family of Hindu stock, in which 12 males had conductive deafness with stapes fixation, inherited as an X-linked trait. Molecular investigations undertaken under the guidance of his sister, Dr. Gillian Wallis, mapped the determinant gene to the locus Xq13-21.1 (Wallis et al., 1988). Microdeletion screening further refined the localization of the gene in the Xq region (Bach et al., 1992).

In view of the implications for surgical management and for genetic transmission, diagnostic accuracy is crucial, and in the future, diagnostic screening by molecular characterization might be helpful in the special schools. At present, however, molecular testing is only available on a limited individual basis.

X-linked Ocular Albinism with Late-Onset Sensorineural Deafness (OMIM30065)

During an outreach clinic in the Eastern Cape province, 7 males in a large family of Afrikaner stock were found to have ocular albinism and late-onset perceptive deafness. The affected males all had pale blue eyes, horizontal nystagmus, and significant impairment of visual acuity, with patchy hypopigmentation of the retina. Family studies revealed that the 7 males were members of three generations, with typical X-linked transmission, and that 12 female relatives were obligate gene carriers (Winship et al., 1984). To the best of the authors' knowledge, this condition had not previously been documented.

Molecular studies in the affected family revealed tight linkage with the DX545 2 locus at Xp22 (Winship et al., 1993). Further investigations indicated that the determinant gene was situated on the telomeric side of the conventional ocular albinism (OAI) gene in the Xp23 chromosomal region (Bassi et al., 1999).

Usher Syndrome (OMIM 276900)

Usher syndrome is a genetically heterogeneous autosomal recessive disorder, in which perceptive deafness is associated with progressive visual loss due to retinitis pigmentosa. Vestibular dysfunction and disturbed balance may also be present. The severity and progression of the syndromic components varies in different forms of the disorder, but deafness and blindness may be the eventual outcomes.

The 1993 diagnostic review of 3064 deaf scholars revealed that Usher syndrome was present in 7 boys and 2 girls. Several of these children had not previously been diagnosed, and during the surveys diagnostic suspicion was raised when bespectacled scholars were observed sitting at the front of the class so that they could see the blackboard. In view of the ultimate prognosis, it is crucial that appropriate education is provided while hearing and vision are still functional. Equally, the AR mode of genetic transmission has important implications for recurrence risks in the family.

The investigation of retinal disorders became a major initiative of the molecular laboratory in the Department of Human Genetics and other, additional persons with Usher syndrome were ascertained in special schools or facilities for the blind. In collaborative international studies, a gene for Usher syndrome type I in South African and overseas families was localized to chromosome 11q (Kimberling et al., 1992). Thereafter, the gene for Usher syndrome Type IIa was localized to chromosome 1q41 (Kimberling et al 1995). By the millennium, 6 genetic subtypes of Usher syndrome type I had been recognized (Weston et al., 2000; Astuto et al., 2000).

Rod-Cone Dystrophy, Sensorineural Deafness, and Renal Dysfunction (OMIM 268315)

In the context of retinal disorders with deafness, an unusual syndrome of rod-cone dystrophy with perceptive deafness and Fanconi type renal dysfunction has been documented in 14 children in 9 families of Afrikaner stock (Beighton et al., 1993). These children had all initially been diagnosed as having "retinitis pigmentosa" or "Usher syndrome." This hitherto unrecognized disorder manifests in early childhood with visual impairment, progressive deafness, and rickets-like skeletal changes due to renal failure. Management is difficult due to progressive visual and auditory dysfunction, together with stunted stature, limb malalignment, and physical handicap. Eight affected persons had died from renal failure between the ages of 3 and 20 years. Pedigree data are consistent with AR inheritance.

Sclerosteosis (OMIM 269500)

Sclerosteosis is an autosomal recessive sclerosing bone dysplasia, in which skeletal hyperostosis is associated with variable syndactyly, usually of the second and third fingers. The condition predominates in the Afrikaner community of South Africa, where more than 70 affected homozygotes have been studied during the last 40 years (Hamersma, Gardner, & Beighton, 2003).

Progressive bone overgrowth leads to gigantism, distortion of the facies, and entrapment of the 7th and 8th cranial nerves. Hearing loss commences in childhood and it is variable in severity. For these reasons, children with sclerosteosis are encountered in the schools for the partially hearing, rather than schools for the deaf.

Familial Microtia, Facial Palsy, and Conductive Deafness (Sellars Syndrome)

Three affected siblings (a boy and three girls) in a South African family of Indian stock had profound bilateral conductive deafness with facial paralysis and malformations of the external ears. The deafness resulted from variable abnormalities of the middle ear ossicles, and it was ameliorated to some extent by surgical intervention. The children's mother was similarly affected, while two younger sisters were normal, and it seemed likely that the condition was transmitted as an AD trait (Sellars & Beighton, 1983). Two decades later, three siblings in the third generation were found to be affected and AD transmission was substantiated (Gardner, 1999). No similar reports could be found in the literature, and it was concluded that the condition was an autonomous or "private" syndrome, resulting from a unique new mutation. In view of the lengthy title of the condition, the eponymous designation "Sellars syndrome" is appropriate and expedient.

Molecular studies involving candidate loci, undertaken during the doctoral project of Gardner (1999), were uninformative. Further investigations using current molecular technology are now underway.

Charcot-Marie-Tooth Syndrome with Sensorineural Deafness

Two brothers of Indian stock were encountered during a routine diagnostic survey in a special school for the deaf. Both had severe bilateral sensorineural deafness which had been present since birth, and normal speech had not developed. They also had bilateral weakness and wasting of the calf, foot, and intrinsic hand muscles. Nerve conduction studies were compatible with the Charcot-Marie-Tooth group of disorders. A third elder brother was similarly affected, while two brothers and two sisters were normal. The unaffected parents were consanguineous. No similar reports could be identified in the literature, and it seemed likely that the condition was a private syndrome. Pedigree data were compatible with either AR or XL inheritance, but in view of the parental consanguinity, the former was probably more likely (Cornell, Sellars, & Beighton, 1984).

Pendred Syndrome (OMIM 274600)

The Pendred syndrome is an AR disorder characterized by the combination of perceptive deafness with high tone loss and enlargement of the thyroid gland. The goiter usually develops in late childhood, and mild hypothyroidism may occur.

In the absence of an affected sibling, diagnosis is difficult prior to the development of the goiter. This problem is compounded by the fact that normal physiological enlargement of the thyroid gland commonly occurs at puberty. Lack of iodine in the diet, as in regions remote from the sea, can also cause thyroid enlargement; in the authors' experience the above factors have led to an incorrect diagnosis of Pendred syndrome in deaf scholars on several occasions.

Familial Streptomycin Ototoxicity

Streptomycin is frequently used in the treatment of tuberculosis, and severe permanent perceptive hearing loss is a well-recognized complication of this form of therapy. During an investigation of familial deafness in a small coastal town about 150 miles from Cape Town, a large kindred was encountered in which eight persons in two generations had lost their hearing during treatment with streptomycin for pulmonary tuberculosis. By contrast, other members of the family with tuberculosis had retained their hearing despite full courses of streptomycin therapy. The eight affected persons were admitted to the Groote Schuur Hospital, Cape Town, for full diagnostic assessment and appropriate management. It was concluded that their perceptive deafness had resulted from damage to the cochlea in the inner ear by streptomycin sulfate. As this therapy had been administered in standard doses, it appeared that the affected persons had a familial propensity to cochlear hair cell hypersensitivity to aminoglycosides.

Males and females in two generations were affected, and it was postulated that this propensity to ototoxicity was an autosomal dominant trait (Viljoen, Sellars, & Beighton, 1983). Thereafter, Hu et al. (1991) pointed out that aminoglycoside ototoxicity could be transmitted by mitochondria inheritance, and a causative mitochondrial mutation was then recognized (Hutchin et al., 1993). On reassessment, pedigree data of the South African family was seen to be consistent with this mode of genetic transmission, and subsequent molecular investigations revealed that affected persons had the characteristic mitochondrial DNA (mtDNA) 1555 A to G point mutation in the 12S ribosomal RNA gene (Gardner et al., 1997). Further investigation of 97 persons in the extended family demonstrated that 76 individuals had the mutation and were therefore at risk if treated for tuberculosis using streptomycin or other aminoglycosides (Human et al., 2009).

The frequency of tuberculosis in disadvantaged communities in South Africa is very high. Treatment protocols with multiple drug therapy include aminoglycosides such as streptomycin, and the presence of familial propensity to ototoxicity could pose a serious threat at the community level. The frequency of the specific mutation beyond the affected family is unknown, but in view of the serious sequelae of damage to the hearing mechanism, screening of all persons with tuberculosis in South Africa has been advocated (Bardien, de Jong, et al., 2009a). To this end, a cost-effective multiple screening method has been developed that includes the five sequence variants thus far recognized worldwide, and a pilot survey is underway (Bardien, Human, et al., 2009b).

Aminoglycoside Ototoxicity in Burn Victims

In the recent past, the management of severe and extensive burns has included the liberal application of neomycin and therapy with gentamycin to prevent or control infection in the damaged skin. In disadvantaged social circumstances in South Africa, heating and cooking are often undertaken on open fires or using paraffin heaters. Household accidents occur and, regrettably, extensive burns in young children are a relatively frequent medical emergency.

In the context of childhood deafness, heavily scarred deaf children were often encountered in early surveys in the special schools. With the recognition of the ototoxic potential of the aminoglycosides, and the modification of management protocols for children with extensive burns, this problem is now rarely encountered.

Rubella Embryopathy

Rubella is a viral infection that causes a mild transient illness characterized by a red macular rash and enlarged lymph glands in the neck. Maternal infection during pregnancy can be catastrophic for the fetus, with a 90% risk of damage during the first ten weeks. The manifestations of the congenital rubella syndrome (CRS) in the neonate are protean, and include malformation of the external, middle, and inner ears, eyes, heart, and brain.

The clinical impact of the CRS is entirely dependent upon the severity and anatomical distribution of the malformations. The condition is frequently encountered in special schools for the deaf, but for educational requirements the deaf scholars must be educable and therefore do not have severe intellectual disability or neurological and visual dysfunction. Nevertheless, deaf scholars with CRS not infrequently have unilateral microphthalmia, and this feature represents a significant diagnostic indicator. The importance of full clinical examination is emphasized by the fact that the affected children may also have unsuspected structural cardiac abnormalities, which can be recognized on auscultation. The situation was sometimes encountered in the early school surveys in the 1970s, and referral for full cardiological assessment was a routine matter. In several instances, cardiac surgery was subsequently undertaken. Due to greater diagnostic awareness, this problem is no longer commonplace.

Rubella embryopathy can produce profound sensorineural deafness without any additional manifestations, and in these circumstances the diagnosis is established on a basis of a firm history of maternal rubella during the first 16 weeks of pregnancy. Case notes containing serological confirmation of maternal infection during pregnancy provide valuable objective support for this diagnosis. In the authors' experience, a major diagnostic dilemma can arise if serological tests are not undertaken, as a maternal history of a "rash" during early pregnancy is often misinterpreted as rubella. Indeed, parents of a deaf child may be impelled to seek a cause of the disability, and an unintentional but spurious diagnosis of maternal rubella is a frequent problem. This issue is relevant in genetic counseling, as an individual with rubella embryopathy is not at risk of transmitting the condition to the next generation.

The prevalence of CRS in the special schools varied geographically and with the passage of time. This situation is directly linked to the uptake of maternal vaccination against rubella, and the occurrence of rubella epidemics when immunity in the community is low. On epidemiologic data it has been estimated that 654 neonates with CRS would have been born in South Africa during 2005 (Schoub et al., 2009). The proportion of these children subsequently entering special schools for the deaf is uncertain.

MEDICAL AND GENETIC SERVICES FOR THE DEAF IN SOUTH AFRICA

The care and education of deaf children in special schools in South Africa continues, generally with high levels of care and expertise. Sisters of the Dominican Order still play an important role, although funding and administrative control are derived from the government.

Diagnostic surveys are occasionally undertaken in the larger centers, but the extent and scope have been greatly reduced. The genetic counseling services that were formerly provided for school leavers by the state genetic services have also dwindled, partly due to the higher priority now accorded to communicable disorders such as AIDS and tuberculosis.

Advances in molecular technology have implications for diagnosis and genetic management in some hereditary deafness disorders (Gardner et al., 1997). Investigations of this type are available in major academic centers on an individual basis, but they have not been implemented for diagnostic screening in special schools. Nevertheless, this approach would be appropriate for any person with familial AR nonsyndromic deafness, and for any sporadic individual with perceptive deafness of unknown cause. If positive results emerged, the option of screening parents and siblings for carrier status would arise. Equally, a positive diagnosis in a person with hearing defect would facilitate appropriate medical care. At the present time, however, financial constraints at the government level preclude wide implementation of this technology.

ACKNOWLEDGEMENTS

The surveys of the special schools upon which this manuscript is based were supported by the South African Medical Research Council. The preparation of the manuscript was undertaken with support from the National Research Foundation of South Africa.

Our analysis of accumulated data is part of the Cape Genetic Heritage initiative in the Division of Human Genetics, Faculty of Health Sciences, University of Cape Town.

REFERENCES

Astuto LM, Weston MD, Carney CA, et al. (2000). Genetic heterogeneity of Usher Syndrome Type I: mutation and linkage analysis of 155 families. *Am J Hum Genet 67*(6), 1569–1574.

Bach I, Brunner HG, Beighton P, et al. (1992). Microdeletions in Patients with Gusher-Associated, X-linked Mixed Deafness (DFN3). *Am J Hum Genet 50*, 38–44.

Bardien S, de Jong G, Schaaf HS, Harris T, Fagan J, Petersen L. (2009). Aminoglycoside-induced hearing loss: South Africans at risk. *S Afr Med J 99*(6), 440–441.

Bardien S, Human H, Harris T, et al. (2009). A rapid method for detection of five unknown mutations associated with aminoglycoside-induced deafness. *BMC Med Genet 10*, 2.

Bassi MT, Ramesar RS, Caciotti B, et al. (1999). X-Linked Late-Onset Sensorineural Deafness Caused by a Deletion Involving OA1 and a Novel Gene Containing WD-40 Repeats. *Am J Hum Genet 64*, 1604–1616.

Beighton P, Sellars S. (1982). *Genetics and Otology.* Edinburgh: Churchill Livingstone.

Beighton P, Sellars SL, Goldblatt J, Viljoen DL, Beighton G. (1987). Childhood deafness in the Indian population of Natal. *S Afr Med J 72*, 209–211.

Beighton P. (1990). In Emery EH, Rimoin DL, eds. *Principles and Practice of medical Genetics* (2nd ed., pp. 733–748). Edinburgh: Churchill Livingstone.

Beighton P, Ramesar R, Winship I, et al. (1991). Hearing impairment and pigmentary disturbance. *Ann NY Acad Sci 630*, 152–166.

Beighton P, Viljoen D, Winship I, Beighton G, Sellars S. (1991). Profound childhood deafness in Southern Africa. *Ann NY Acad Sci 630*, 290–291.

Beighton P, Bartman L, Bingham G, Sellars S. (1993). Rod-cone dystrophy, sensorineural deafness and renal dysfunction; an autosomal recessive syndrome? *Am J Med Genet 47*, 832–836.

Butt J, Greenberg J, Winship I, Sellars S, Beighton P, Ramesar R. (1994). A splice junction mutation in PAX3 causes Waardenburg syndrome in a South African family. *Hum Mol Genet 3*(1), 197–198.

Cornell J, Sellars S, Beighton P. (1984). Autosomal recessive inheritance of Charcot-Marie-Tooth disease associated with sensorineural deafness. *Clin Genet 25*, 163–165.

DeStefano AL, Cupples LA, Arnos KS, et al. (1998). Correlation between Waardenburg syndrome phenotype and genotype in a population of individuals with identified PAX3 mutations. *Hum Genet 102*, 499–506.

Farrer LA, Grundfast KM, Amos J, et al. (1992). Waardenburg syndrome (WS) Type I is caused by defects at multiple loci, one of which is near ALPP on chromosome 2: first report of the WS Consortium. *Am J Hum Genet 50*, 902–913.

Gardner J, Beighton P, Sellars SL. (1997). The genetics of childhood deafness. *S Afr Med J 87*(12), 1661–1662.

Gardner JC, Goliath R, Viljoen D, et al. (1997). Familial streptomycin ototoxicity in a South African family: a mitochondrial disorder. *J Med Genet 34*, 904–906.

Gardner JC. (1999). Profound Childhood Deafness in South Africa; a clinical and molecular approach. *Thesis for the Doctorate of Medicine*. University of Cape Town.

Hamersma H, Gardner J, Beighton P. (2003). The Natural History of Sclerosteosis. *Clin Genet 63*, 191–196.

Hu DN, Qiu WQ, Wu BT, et al. (1991). Genetic aspects of antibiotic induced deafness: mitochondrial inheritance. *J Med Genet 28*, 79–83.

Human H, Lombard D, de Jong G, Bardien S. (2009). A South African family with the mitochondrial A1555G mutation L0d. *Biochem Biophys Res Commun 382*(2), 390–394.

Hutchin T, Haworth I, Higashi K, et al. (1993). A molecular basis for human hypersensitivity to aminoglycoside antibiotics. *Nucleic Acids Res 21*, 4174–4179.

Kimberling WJ, Möller CG, Davenport S, et al. (1992). Linkage of the Usher syndrome type I gene (USH1B) to the long arm of chromosome 11. *Genomics 14*, 988–994.

Kimberling WJ, Weston MD, Moller C, et al. (1995). Gene Mapping of Usher Syndrome Type 11a: Localization of the Gene to a 2.1cM Segment on Chromosome 1q41. *Am J Med Genet 56*, 216–223.

Schoub BD, Harris BN, McAnerny J, Blumberg L. (2009). Rubella in South Africa: an impending Greek tragedy? *S Afr Med J 99*, 515–519.

Sellars S, Napier E, Beighton P. (1975). Childhood deafness in Cape Town. *S Afr Med J 49*, 1135–1138.

Sellars S, Groeneveldt L, Beighton P. (1976). Aetiology of deafness in white children in the Cape. *S Afr Med J 50*, 1193–1197.

Sellars S, Beighton G, Horan F, Beighton P. (1977). Deafness in Black children in Southern Africa. *S Afr Med J 51*, 309–312.

Sellars S, Beighton P. (1978). The aetiology of partial deafness in childhood. *S Afr Med J 54*, 811–813.

Sellars S, Beighton P. (1983). Autosomal dominant inheritance of conductive deafness due to stapedial anomalies, external ear malformations and congenital facial palsy. *Clin Genet 23*, 376–379.

Sellars S, Beighton P. (1983). The Waardenburg syndrome in deaf children in Southern Africa. *S Afr Med J 63*, 725–728.

Sellars S, Beighton P. (1983). Childhood deafness in Southern Africa. *J Laryngol & Otol 97*, 885–889.

Spritz RA, Holmes SA, Ramesar R, Greenberg J Curtis D, Beighton P. (1992). Mutations of the*KIT* proto-oncogene account for a continuous range of phenotypes in human Piebaldism. *Am J Hum Genet 51*, 1058–1065.

Spritz RA, Beighton P. (1998). Piebaldism with deafness: molecular evidence for an expanded syndrome. *Am J Med Genet* 75, 101–103.

Thorpe P, Sellars S, Beighton P (1974). X-linked deafness in a South African kindred. *S Afr Med J 48*, 587–590.

Thorpe PM (1978). The deaf doctor. *S Afr Med J 54*, 326–328.

Viljoen DL, Sellars SL, Beighton P. (1983). Familial aggregation of streptomycin ototoxicity: autosomal dominant inheritance. *J Med Genet 20*(5), 357–360.

Viljoen DL, Dent GM, Sibanda AG, et al. (1988). Childhood deafness in Zimbabwe. *S Afr Med J 73*, 286–288.

Wallis CE. (1985). Genetic Disorders on the Island of Mauritius. *Thesis for the Doctorate of Medicine*, University of Cape Town.

Wallis C, Ballo R, Wallis G, Beighton P, Goldblatt J. (1988). X-linked mixed deafness with stapes fixation in a Mauritian kindred: linkage to Xq probe pDP34. *Genomics 3*, 299–301.

Weston MD, Eudy JD, Fujita S, et al. (2000). Genomic Structure and Identification of Novel Mutations in Usherin, the Gene Responsible for Usher Syndrome Type IIa (USH2A). *Am J Hum Genet 66*(4), 1199–1210.

Winship I, Gericke G, Beighton P. (1984). X-linked inheritance of ocular albinism with late-onset sensorineural deafness. *Am J Med Genet 19*, 797–803.

Winship I, Young K, Martell R, Ramesar R, Curtis D, Beighton P. (1991). Piebaldism: an autonomous autosomal dominant entity. *Clin Genet 39*, 330–337.

Winship I, Beighton P. (1992). Phenotypic discriminants in the Waardenburg syndrome. *Clin Genet 41*, 181–188.

Winship I, Babaya M, Ramesar R. (1993). X-linked ocular albinism and sensorineural deafness: linkage to Xp22.3. *Genomics 18*, 444–445.

41.

HERITABLE DISORDERS OF THE SKELETON IN SOUTH AFRICA

Peter Beighton

During the past 40 years in South Africa (SA), genetic skeletal conditions have been a major interest of the author. In this period more than 3000 affected persons and families have been examined and documented in genetic clinics, special institutions for the physically handicapped, and at outreach clinics throughout the country. In the period 1982–1998, the project was undertaken under the auspices of a unit funded by the Medical Research Council of South Africa and established within the Department of Human Genetics, University of Cape Town (UCT).

A decade ago, archived case notes concerning the osteochondrodysplasias were reviewed, and the numbers of affected persons with different conditions in this broad category who had been examined in South Africa were documented (Beighton, 1996). These figures were subsequently updated in the Cape Genetic Heritage compilation, which reposes in the Division of Human Genetics, University of Cape Town, and numbers quoted in this chapter pertain to this latter document.

The genetic bone disorders encountered in South Africa have formed the basis of the monographs listed below. The conditions that are reviewed in this chapter are depicted in these books.

Bone Dysplasias of Infancy. B Cremin and P Beighton. Springer-Verlag, Heidelberg, 1978.

Sclerosing Bone Dysplasias. P Beighton and B Cremin. Springer-Verlag, Heidelberg, 1980.

Orthopaedic Problems in Inherited Skeletal Disorders. F Horan and P Beighton. Springer-Verlag, Heidelberg, 1982.

Inherited Disorders of the Skeleton (2nd ed.). P Beighton. Churchill Livingstone, Edinburgh, 1988.

Gamut Index of Skeletal Dysplasias (3rd ed.). K Kozlowski and P Beighton. Springer, Heidelberg, 2001.

Skeletal disorders that are of special interest or importance in South Africa are reviewed in this chapter.

1. Achondroplasia
2. Spondyloepimetaphyseal dysplasia, SA type (SEMDJL)
3. Dense bone dysplasias
4. Sclerosteosis
5. Bone dysplasias of infancy
6. Namaqualand hip dysplasia
7. Beukes hip dysplasia
8. Mseleni joint disease
9. Schwartz syndrome
10. Stippled epiphyses
11. Ectrodactyly
12. Miscellaneous skeletal disorders

1. ACHONDROPLASIA

Achondroplasia is the most common form of short-limbed dwarfism, and more than 110 affected persons in all population groups in South Africa have been documented. The lumbar spine in Achondroplasia is lordotic, but paradoxically, affected persons of indigenous African stock often have a lumbar gibbus, which poses a threat to the functional integrity of the spinal cord (Beighton & Bathfield, 1981). This anomalous situation may be the consequence of the traditional way in which a mother or helpful female relative carries an infant on her back, supported by a blanket, with the child's legs around her waist. In this position the hips are flexed and abducted, while the vulnerable lumbar spine is flexed. With the ongoing abandonment of traditional practices, it can be foreseen that this problem of a gibbal angulation of the spine will diminish and eventually disappear.

2. SPONDYLOEPIMETAPHYSEAL DYSPLASIA WITH JOINT LAXITY (SEMDJL)

The South African form of SEMDJL is a severe AR dwarfing skeletal dysplasia, which was delineated three decades ago (Beighton & Kozlowski, 1980). Affected neonates have generalized joint laxity and a characteristic facies (Beighton et al., 1983). Club foot, dislocation of the hips, and kyphoscoliosis are often present. The recognition of specific radiological changes permits diagnosis in the neonate (Kozlowski & Beighton, 1984). Progressive spinal malalignment leads to cord compression, and affected children are often chairbound; survival into adulthood is unusual (Beighton, 1994). This condition has now been documented in about 30 affected children in South Africa.

The determinant gene was probably brought to this country by immigrants from the German states, who began to arrive in the eastern region of South Africa in 1856 (Torrington & Beighton, 1991). Despite extensive investigations, the molecular basis of SEMDJL has not yet been elucidated.

3. SCLEROSTEOSIS

Sclerosteosis is a severe craniotubular hyperostosis in which overgrowth of the skeleton leads to tall stature, mandibular asymmetry, and distortion of the facies. Syndactyly, usually of the 2nd and 3rd fingers, is a syndromic component. Compression of the 7th and 8th cranial nerves causes deafness and facial palsy. Involvement of the calvarium and base of the skull leads to diminution of the volume of the cranial cavity and a potentially lethal elevation of intracranial pressure (Beighton et al., 1976). Prior to current management with prophylactic craniectomy in childhood, this complication often caused sudden death in adulthood (Beighton, 1988).

Sclerosteosis is an AR disorder that predominates in the Afrikaner community of South Africa, where more than 70 affected persons have been documented (Hamersma et al., 2003). Obligate heterozygotes are clinically normal, but densitometry has revealed increased bone density in these individuals (Gardner et al., 2005). A phenotypically similar but much milder AR disorder in which syndactyly is absent, van Buchem disease (VBD), has been diagnosed in about 20 persons in Holland. As the Afrikaners have Dutch antecedents, it could be speculated that the two conditions might share syndromic identity (Beighton et al., 1984). Nevertheless, they differ at the molecular level, as VBD results from a mutation in the promoter region of the determinant gene. The gene *SOST* and the gene product, sclerostin, are the focus of considerable scientific interest (Beighton et al., 2007).

4. SCLEROSING BONE DYSPLASIAS

Genetic skeletal dysplasias in which bone density is increased have been extensively investigated over the past 40 years, and the contributions of Professor Herman Hamersma of Pretoria are notable in this respect. Conditions in this category that have been documented in South Africa are listed below, together with pertinent references.

i. Craniometaphyseal dysplasia. (Spiro et al., 1975; Beighton et al., 1979; Beighton, 1995; Reichenberger et al., 2001)
ii. Osteopetroses. (Beighton et al., 1977)
iii. Osteopathia striata with cranial sclerosis. (Horan & Beighton, 1978)
iv. Metaphyseal dysplasia. (Raad & Beighton, 1978; Heselson et al., 1979; Beighton, 1987)
v. Oculodento-osseous dysplasia. (Barnard et al., 1981)
vi. Frontometaphyseal dysplasia. (Beighton & Hamersma, 1980)
vii. Distal osteosclerosis. (Beighton et al., 1980)
viii. Dysplasia epiphysealis hemimelica. (Connor et al., 1983)
ix. Osteoectasia. (Saxe & Beighton, 1982)

5. BONE DYSPLASIAS OF INFANCY

Bone dysplasias of infancy have been documented in Cape Town since 1972. The majority of affected infants were referred for diagnostic assessment with a view to the provision of genetic counseling for the parents (Beighton & Cremin, 1974). Radiographic investigations played a key role in this process, and early studies culminated in an atlas, *Bone Dysplasias of Infancy* (Cremin & Beighton, 1978). A review of skeletons of infants with genetic conditions in the Museum of Pathology, Vienna, undertaken in collaboration with Professor Eva Sujanski of the University of Denver, proved to be academically productive (Beighton et al., 1993). In recent years, autopsy material has been evaluated in collaboration with Professor Helen Wainwright of the Division of Pathological Anatomy, University of Cape Town.

In order to provide a perspective of the genetic infantile bone dysplasias that have been investigated in Cape Town, these are listed below with relevant references.

i. Lethal chondrodysplasia punctata. (Heselson et al., 1978)
ii. Lethal non-rhizomelic dysplasia epiphysealis punctata. (Kozlowski et al., 2002)

iii. Chondrodysplasia punctata. (Kozlowski et al., 2006)

iv. Infantile Metaphyseal dysplasia. (Horan & Beighton., 1980)

v. Atelosteogenesis. (Maroteaux et al., 1982)

vi. Boomerang dysplasia. (Winship et al., 1990)

vii. Platyspondylic osteochondrodysplasias. (Kozlowski et al., 1999)

viii. Spondylometaphyseal dysplasia Sutcliffe type. (Beighton & Kozlowski, 2007)

ix. Osteogenesis imperfecta with holoprosencephaly. (Wainwright & Beighton, 2007)

x. Dyssegmental dysplasia. (Winship & Beighton, 2008; Wainwright & Beighton, 2010a)

xi. Hypochondrogenesis. (Wainwright & Beighton, 2008a)

xii. Achondrogenesis type II with cutaneous hemangiomata. (Wainwright & Beighton, 2008b)

xiii. Osteogenesis imperfecta with pseudo-anencephaly. (Wainwright & Beighton, 2008c)

xiv. Osteogenesis imperfecta type IIA. (Spranger et al., 1982)

xv. Tetraphocomelia with Waardenburg syndrome. (Wu et al., 2009)

xvi. Sonic hedgehog phenotype. (Wainwright & Beighton, 2009)

xvii. SEMDJL. (Christianson & Beighton, 1996)

6. NAMAQUALAND HIP DYSPLASIA

Namaqualand hip dysplasia (NHD) is a unique AD familial disorder in which progressive collapse of the femoral capital epiphyses leads to an osteoarthropathy of the hip joints in early adulthood (Beighton et al., 1984; Learmonth et al., 1987). Minor involvement of the lumbar vertebrae is sometimes present, and NHD is categorized as a mild form of spondyloepiphyseal dysplasia. As with other conditions in this category, the COL2A1 gene has been implicated (Sher et al., 1991). The condition has been documented in 45 persons in 5 generations of a Cape Mixed Ancestry family; their early origins in Namaqualand in the northern region of the Cape has prompted the geographical appellation.

7. BEUKES HIP DYSPLASIA

Beukes hip dysplasia is an AD disorder that has been documented in 47 persons in 6 generations of an Afrikaner family domiciled in the Orange Free State in South Africa (Cilliers & Beighton, 1990). The condition presents with hip pain in early adulthood and progressive deformation of the femoral neck and capital epiphyses, producing a characteristic radiographical appearance. Professor H.J. Cilliers of Bloemfontein has operated upon many family members, and prosthetic hip joint replacement produced good functional results. The condition has formed the subject of his doctoral thesis.

The disorder, which was named by the affected family, appears to be a private syndrome. It has not been recognized in Europe and it seems likely that it has arisen by new mutation in a progenitor of the kindred after his or her arrival in SA. The determinant gene has been shown to be unlinked to COL2A1 (Beighton et al., 1994).

8. MSELENI JOINT DISEASE

Mseleni joint disease (MJD) is a precocious, progressive, generalized degenerative osteoarthropathy, which has been documented in several hundred persons of indigenous African stock in a remote coastal region in northern Zululand. Affected persons are severely handicapped by early adulthood, and the disorder, which is geographically circumscribed, poses a considerable burden on the community. Mseleni joint disease clusters in families but does not Mendelise, and there is no obvious mode of genetic transmission. The genetic situation is further complicated by the presence of 10–20 members of the community who have the MJD clinical and radiological phenotype plus severe brachydactylous dwarfism (Viljoen et al., 1993). Family data do not support the possibility that these persons are homozygous for a gene that causes MJD in the heterozygote. Molecular studies have been uninformative (Ballo et al., 1996), while extensive environmental and dietary investigations over the past 30 years have failed to yield any positive information. Interestingly, there is now good anecdotal evidence that the frequency of MJD in the community is diminishing.

The issue of syndromic identity has been complicated by the emergence of a very similar disorder in Central India. This condition, which has the geographical appellation "Handigodu joint disease" (HJD), is present in two villages in a tropical environment. The characteristic course and prognosis closely resembles those of MJD, but AD transmission has been postulated, and HJD is currently regarded as a form of spondyloepiphyseal dysplasia (Agarwal et al., 1997).

9. SCHWARTZ SYNDROME

The Schwartz syndrome is an AR disorder characterized by the combination of myotonia and a skeletal dysplasia.

Affected persons have an immobile "crying" facies and stunted stature with spinal and limb malalignment (Beighton, 1973). The condition predominates in the indigenous African community, although it has also been documented in brothers of predominantly Khoisan stock (Horan & Beighton, 1975; Viljoen & Beighton, 1992).

Homozygosity mapping revealed that the determinant gene in an affected boy from Cape Town was situated at 1p34-1p36.1 (Nicole et al., 1995). The same research group subsequently provided evidence for genetic homogeneity (Fontaine et al., 1996) and determined that the proteoglycan component of the basement membrane, perlecan, is altered in the Schwartz syndrome (Nicole et al., 2000).

The impact of the facial myotonia and anesthetic problems upon dental management has been reviewed in the faculty of dentistry, University of the Western Cape (Stephen & Beighton, 2002).

10. STIPPLED EPIPHYSES

Punctate stippling of the epiphyses is an important radiographic manifestation of a number of genetic skeletal dysplasias and related disorders, some of which are lethal or potentially lethal. These epiphyseal changes, which, are evident in the fetus and usually disappear by the age of two years are an important diagnostic sign. In later life the sequelae of stippling are variable disturbances in skeletal growth in the affected regions. Depression of the bridge of the nose is frequent, due to involvement of the nasal cartilage.

The Conradi-Hunermann group of disorders are a well-known cause of stippling; indeed, the generic term "stippled epiphyses" is often loosely used as a designation for this heterogeneous condition. More than 20 persons with different forms of this disorder have been documented in South Africa. In later childhood, the diagnosis has been suspected in individuals with stunted stature and limb asymmetry and confirmed retrospectively by the recognition of stippling in archived radiographs which had been obtained during infancy (Kozlowski et al., 2006).

It is of interest that stippling of the epiphyses and phenotype suggestive of the Conradi-Hunermann syndrome can occur in the offspring of mothers with lupus erythematosus, and affected brothers with this disorder have been recognized in Cape Town (Kozlowski et al., 2004).

Warfarin embryopathy is an important and preventable cause of stippling, which is most often encountered in the indigenous African population. Cardiac damage, notably mitral stenosis or incompetence is a frequent complication of beta hemolytic streptococcal infection in this community. Maternal anticoagulation is a component of routine medical management, and if Warfarin is used during pregnancy, embryopathy may result. This condition enters into the differential diagnosis of stippling in any neonate (Wainwright & Beighton 2010b).

A perspective on the relative frequencies of the various causes of stippling in South Africa has been obtained from an analysis of thirty fetuses and neonates in whom stippling was detected radiologically, prior to autopsy (Wainwright & Beighton, 2010c).

11. ECTRODACTYLY

Ectrodactyly is a generic term which embraces "blighted digits," lobster claw deformity and the split hand-foot syndrome. These conditions are genetically heterogeneous, with considerable inter and intra familial variability, both anatomically and in clinical severity. Over the years, more than 100 affected persons have been documented in South Africa.

The combination of scalp defects and shortened (blighted) digits, termed the Adams-Oliver syndrome is present in four generations of a South African family of British origin (Bonafede & Beighton, 1979).

The classical split hand-foot syndrome was first photographed more than 60 years ago in the former Rhodesia in an indigenous African community, the Wadoma, in a remote region of the Zambesi valley. Following upheaval in that country a section of the Wadoma emigrated to Botswana, where they are now domiciled near the western border of South Africa. The manifestations of this AD condition, which are very variable, have been documented in 30 persons in 4 generations of the kindred (Viljoen & Beighton, 1984).

A large Mixed Ancestry kindred with AD split hand-foot syndrome is domiciled in Blikkiesdorp, about 30 miles east of Cape Town. The pedigree of this family is remarkable for the presence of the condition in three out of eight offspring of unaffected non-consanguineous parents. Each of the affected individuals went on to produce a dynasty of descendants with the disorder (Spranger & Schapera 1988).

12. MISCELLANEOUS SKELETAL DISORDERS

The genetic skeletal disorders which are listed below have all been investigated in South Africa and documented in this context. In order to facilitate access to additional information, relevant references are provided.

i. Morquio syndrome (Beighton & Craig, 1973)

ii. Acromesomelic dysplasia (Beighton, 1974)

iii. Parastremmatic dwarfism (Horan & Beighton, 1976)

iv. Pseudoachondroplasia (Heselson et al., 1977)

v. Dyggve-Melchior-Clausen syndrome (Bonafede & Beighton, 1978; Beighton, 1990)

vi. Hypochondroplasia (Heselson et al., 1979)

vii. Osteoglophonic dwarfism (Beighton et al., 1980; Beighton, 1989; Farrow et al., 2006; Roberts et al., 2006)

viii. Spondylometaphyseal dysplasia (Kozlowski et al., 1980)

ix. Spondylocostal dysostosis (Beighton & Horan, 1981)

x. Femoral hypoplasia-unusual facies syndrome (Lord & Beighton, 1981)

xi. Langer mesomelic dysplasia (Goldblatt et al., 1987)

xii. Rhizomelic dysplasia (Viljoen et al., 1987)

xiii. Beta-galactosidase deficiency (Beck et al., 1987)

xiv. Kyphomelic dysplasia (Viljoen & Beighton, 1988)

xv. Enchondromatosis (Wallis et al., 1989)

xvi. Spondyloenchondromatosis (Honey et al., 2003)

xvii. Brachyolmia (Gardner & Beighton, 1994)

xviii. Multiple epiphyseal dysplasia (Ballo et al., 1997)

xix. Spondyloepiphyseal dysplasia (Anderson et al., 1990; Eyre et al., 2002)

xx. Cleidocranial dysplasia (Bertie et al., 2011)

ACKNOWLEDGEMENTS

I am most appreciative of the contribution made by my wife Greta in the preparation of this chapter.

The investigations upon which the manuscript was based were supported by the South African Medical Research Council and the National Research Foundation.

The material in this chapter forms part of the Cape Genetic Heritage project, which is archived in the Division of Human Genetics, University of Cape Town.

REFERENCES

Agarwal SS, Phadke SR, Fredlund V, Viljoen DL, Beighton P. (1997). Mseleni and Handigodu familial osteoarthropathies: syndromic identity? *Am J Med Genet 72*(4): 435–439.

Anderson IJ, Tsipouras P, Scher C, Ramesar RS, Martell RW, Beighton P. (1990). Spondyloepiphyseal dysplasia: mild autosomal dominant type is not due to primary defects of Type II collagen. *Am J Med Genet 37*, 272–276.

Ballo R, Viljoen D, Machado M, et al. (1996). Mseleni joint disease – a molecular genetic approach to defining the aetiology. *S Afr Med J 86*(8): 956–958.

Ballo, R, Briggs MD, Cohn DH, Knowlton RG, Beighton PH, & Ramesar RS. (1997). Multiple epiphyseal dysplasia, Ribbing type: A novel point mutation in the COMP gene in a South African family. *Am J Genet 68*, 396–400.

Barnard A, Hamersma H, De Villiers JC, Beighton P. (1981). Intracranial calcification in oculodento-osseous dysplasia. *S Afr Med J 59*, 758–762.

Beck M, Petersen EM, Spranger J, Beighton P. (1987). Morquio's disease type B (beta galactosidase deficiency) in three siblings. *S Afr Med J 72*, 704–707.

Beighton P. (1973). The Schwartz syndrome in Southern Africa. *Clin Genet 4*, 548–555.

Beighton P, Craig J. (1973). Atlanto-axial subluxation in the Morquio syndrome. *J Bone Jt Surg 55B*(3), 478–481.

Beighton P. (1974). Autosomal recessive inheritance in the mesomelic dwarfism of Campailla and Martinelli. *Clin Genet 5*, 363–367.

Beighton P, Cremin BJ. (1974). Dwarfism in the newborn. *Brit J Radiol 47*, 77–93.

Beighton P, Durr L, Hamersma H. (1976). The clinical features of sclerosteosis. *Ann Int Med 84*(4), 393–397.

Beighton P, Horan F, Hamersma H. (1977). A review of the osteopetroses. *Postgrad Med J 53*, 507–515.

Beighton P, Hamersma H, Horan F. (1979). Craniometaphyseal dysplasia – variability of expression within a large family. *Clin Genet 15*, 252–258.

Beighton P, Kozlowski K. (1980). Spondylo-epi-metaphyseal dysplasia with joint laxity and severe progressive kyphoscoliosis. *Skel Radiol 5*, 205–212.

Beighton P, Hamersma H. (1980). Frontometaphyseal dysplasia: autosomal dominant or X-linked? *J Med Genet 17*(1) 53–56.

Beighton P, Macrae M, Kozlowski K. (1980). Distal osteosclerosis. *Clin Genet 18*, 298–304.

Beighton P, Cremin BJ, Kozlowski K. (1980). Osteoglophonic dwarfism. *Pediatr Radiol 10*, 46–50.

Beighton P, Horan F. (1981). Spondylocostal dysostosis in South African sisters. *Clin Genet 19*, 23–25.

Beighton P, Bathfield CA. (1981). Gibbal achondroplasia. *J Bone Jt Surg 63B*(3), 328–329.

Beighton P, Kozlowski K, Gericke G, Wallis G, Grobler L. (1983). Spondylo-epimetaphyseal dysplasia with joint laxity and severe, progressive kyphoscoliosis. *S Afr Med J 64*, 772–775.

Beighton P, Christy G, Learmonth ID. (1984). Namaqualand Hip Dysplasia: An autosomal dominant entity. *Am J Med Genet* 19(1), 161–169.

Beighton P, Barnard A, Hamersma H, van der Wouden A. (1984). The syndromic status of sclerosteosis and van Buchem disease. *Clin Genet 25*, 175–181.

Beighton P. (1987). Pyle disease (metaphyseal dysplasia). *J Med Genet 24*, 321–324.

Beighton P. (1988). *Inherited Disorders of the Skeleton, 2nd ed.* Churchill Livingstone, Edinburgh.

Beighton P. (1988). Sclerosteosis. *J Med Genet 25*, 200–203.

Beighton P. (1989). Osteoglophonic dysplasia. *J Med Genet 26*, 572–576.

Beighton P. (1990). Dyggve-Melchior-Clausen syndrome. *J Med Genet 27*, 512–515.

Beighton P, Sujansky E, Patzak B, Portele KA. (1993). Genetic skeletal dysplasias in the Museum of Pathological Anatomy, Vienna. *Am J Med Genet 47*, 843–847.

Beighton P. (1994). Spondyloepimetaphyseal dysplasia with joint laxity (SEMDJL). *J Med Genet 31*, 136–140.

Beighton P, Cilliers HJ, Ramesar R. (1994). Autosomal dominant (Beukes) premature degenerative osteoarthropathy of the hip joint unlinked to COL2A1. *Am J Med Genet 53*, 348–351.

Beighton P. (1995). Craniometaphyseal dysplasia (CMD), autosomal dominant form. *J Med Genet 32*, 370–374.

Beighton, P. (1996). Osteochondrodysplasia in South Africa. *Am J Med Genet* 63, 7–11.

Beighton P, Hamersma H, Brunkow M. (2007). SOST-Related sclerosing bone dysplasias. In: *GeneReviews at GeneTests: Medical Genetics Information Resource [database online]*. Available at http://www.ncbi.nlm.nih.gov/books/NBK1228.

Beighton P, Kozlowski K. (2007). Spondylometaphyseal dysplasia Sutcliffe type – case report. *Pol J Radiol 72*(1), 76–78.

Bertie JD, Thompson D, Beighton P. (2011). Cleidocranial dysplasia presenting as familial coxa vara in a South African family. *SA Orthopaed J 10*(2), 34–38.

Bonafede RP, Beighton P. (1978). The Dyggve-Melchior-Clausen syndrome in adult siblings. *Clin Genet 14*, 24–30.

Bonafede RP, Beighton P (1979). Autosomal dominant inheritance of scalp defects with ectrodactyly. *Am J Med Genet 3*, 35–41.

Christianson, A, Beighton P. (1996). Spondyloepimetaphseal dysplasia with joint laxity (SEMDJL) in three neonates. *Genetic Counselling 7*(3), 219–225.

Cilliers HJ, Beighton P. (1990). Beukes familial hip dysplasia: an autosomal dominant entity. *Am J Med Genet 36*, 386–390.

Connor JM, Horan FT, Beighton P. (1983). Dysplasia epiphysialis hemimelica. *J Bone Jt Surg 65B*(3), 350–354.

Cremin B, Beighton P. (1978). *Bone Dysplasias of Infancy.* Springer-Verlag, Heidelberg.

Eyre S, Roby P, Wolstencroft K, et al. (2002). Identification of a locus for a form of spondyloepiphyseal dysplasia on chromosome 15q26.1: exclusion of aggrecan as a candidate gene. *J Med Genet 39*(9), 634–638.

Farrow EG, Davis SI, Mooney SD, et al. (2006). Extended mutational analysis of FGFR1 in osteoglophonic dysplasia (OD). *Am J Med Genet 140A*(5), 537–539.

Fontaine B, Nicole S, Topaloglu H, et al. (1996). Recessive Schwartz-Jampel syndrome (SJS): confirmation of linkage to chromosome 1p, evidence of genetic homogeneity and reduction of the SJS locus to a 3-cM interval. *Hum Genet 98*(3), 380–385.

Gardner J, Beighton P. (1994). Brachyolmia: an autosomal dominant form. *Am J Med Genet 49*, 308–312.

Gardner JC, vanBezooijen RL, Mervis B, et al. (2005). Bone mineral density in sclerosteosis; affected individuals and gene carriers. *J Clin Endocrinol Metab 90*(12), 6392–6395.

Goldblatt J, Wallis C, Viljoen D, Beighton P. (1987). Heterozygous manifestations of Langer mesomelic dysplasia. *Clin Genet 31*, 19–24.

Hamersma H, Gardner J, Beighton P. (2003). The Natural History of Sclerosteosis. *Clin Genet 63*, 191–196.

Heselson NG, Cremin BJ, Beighton P. (1977). Pseudochondroplasia. A report of 13 cases. *Brit J Radiol 50*, 473–482.

Heselson NG, Cremin BJ, Beighton P. (1978). Lethal chondrodysplasia punctata. *Clin Radiol 29*, 679–684.

Heselson NG, Cremin BJ, Beighton P. (1979). The radiographic manifestations of hypochondroplasia. *Clin Radiol 30*, 79–85.

Heselson NG, Raad M, Hamersma H, Cremin B, Beighton P. (1979). The radiological manifestations of metaphyseal dysplasia (Pyle disease). *Brit J Radiol 52*, 431–440.

Honey EM, van Rensburg M, Knoll DP, Mienie LJ, van der Werke L, Beighton P. (2003). Spondyloenchondromatosis with D-2-hydroglutaric aciduria: a report of a second patient with this unusual combination. *Clin Dysmorphol 12*(2), 95–99.

Horan F, Beighton P. (1975). Orthopaedic aspects of the Schwartz syndrome. *J Bone Jt Surg 57A*(4), 542–544.

Horan F, Beighton P. (1976). Parastremmatic dwarfism. *J Bone Jt Surg 58B*(3), 343–346.

Horan F, Beighton P. (1978). Osteopathia striata with cranial sclerosis. An autosomal dominant entity. *Clin Genet 13*, 201–206.

Horan FT, Beighton P (1980). Infantile metaphyseal dysplasia or "Battered Babies?" *J Bone Jt Surg 62B*(2), 243–247.

Kozlowski K, Cremin B, Beighton P. (1980). Variability of spondylometaphyseal dysplasia, common type. *Radiol Diagn 21*, 682–686.

Kozlowski K, Beighton P. (1984). Radiographic features of spondylo-epimetaphyseal dysplasia with joint laxity and progressive kyphoscoliosis. Review of 19 cases. *Fortschr Röntgenstr 141*(3), 337–341.

Kozlowski K, Bieganski T, Gardner J, Beighton P. (1999). Osteochondrodystrophies with marked platyspondyly and distinctive peripheral anomalies. *Pediatr Radiol 29*(1), 1–5.

Kozlowski K, Godlonton J, Gardner J, Beighton P. (2002). Lethal non-rhizomelic dysplasia epiphysealis punctata. *Clin Dysmorphol 11*(3), 202–208.

Kozlowski K, Basel D, Beighton P. (2004). Chondrodysplasia punctata and maternal lupus erythematosus. *Clin Genet 66*(6), 545–549.

Kozlowski K, Basel D, Beighton P. (2006). Retrospective diagnosis of chondrodysplasia punctata. *Australas Radiol 50*, 55–58.

Learmonth ID, Christy G, Beighton P. (1987). Namaqualand Hip Dysplasia. Orthopaedic implications. *Clin Orthop 218*, 142–147.

Lord J, Beighton P (1981). The femoral hypoplasia-unusual facies syndrome: A genetic entity? *Clin Genet 20*, 267–275.

Maroteaux P, Spranger J, Stanescu V, et al. (1982). Atelosteogenesis. *Am J Med Genet 13*, 15–25.

Nicole S, Ben Hamida C, Beighton P, et al. (1995). Localization of the Schwartz-Jampel syndrome (SJS) locus to chromosome 1p34-p36.1 by homozygosity mapping. *Hum Mol Genet 4*(9), 1633–1636.

Nicole S, Davoine C-S, Topaloglu H, et al. (2000). Perlecan, the major proteoglycan of basement membranes, is altered in Schwartz-Jampel syndrome (chondrodystrophic myotonia). *Nature Genetics 26*(4), 480–483.

Raad M, Beighton P. (1978). Autosomal recessive inheritance of metaphyseal dysplasia (Pyle disease). *Clin Genet 14*, 251–256.

Reichenberger E, Tiziani V, Watanabe S, et al. (2001). Autosomal dominant craniometaphyseal dysplasia is caused by mutations in the transmembrane protein ANK. *Am J Hum Genet 68*(6), 1321–1326.

Roberts TS, Stephen L, Beighton P. (2006). Osteoglophonic dysplasia: dental and orthodontic implications. *Orthod Craniofacial Res* 9,153–156.

Saxe N, Beighton P (1982). Cutaneous manifestations of osteoectasia. *Clin Exp Derm 7*, 605–609.

Sher C, Ramesar R, Martell R, Learmonth I, Tsipouras P, Beighton P. (1991). Mild spondyloepiphyseal dysplasia (Namaqualand type): genetic linkage to the type II collagen gene (COL2A1). *Am J Hum Genet 48*, 518–524.

Spiro PC, Hamersma H, Beighton P. (1975). Radiology of the autosomal dominant form of craniometaphyseal dysplasia. *S Afr Med J 49*, 839–842.

Spranger J, Cremin B, Beighton P. (1982). Osteogenesis imperfecta congenita. *Pediatr Radiol 12*, 21–27.

Spranger M, Schapera J. (1988). Anomalous inheritance in a kindred with split hand, split foot malformation. *Eur J Pediatr 147*, 202–205.

Stephen LXG, Beighton PH. (2002). Oro-dental manifestations of the Schwartz-Jampel syndrome. *J Clin Pediatr Dent 27*(1), 67–70.

Torrington M, Beighton P. (1991). The ancestry of spondyloepimetaphyseal dysplasia with joint laxity (SEMDJL) in South Africa. *Clin Genet 39*(3), 210–213.

Viljoen D, Beighton P. (1984). The split-hand and split-foot anomaly in a Central African Negro population. *Am J Med Genet 19*, 545–552.

Viljoen D, Goldblatt J, Wallis C, Beighton P. (1987). Familial rhizomelic dysplasia: Phenotypic variation or heterogeneity? *Am J Med Genet 26*, 941–947.

Viljoen D, Beighton P. (1988). Kyphomelic dysplasia: further delineation of the phenotype. *Dysmorph Clin Genet 1*, 136–141.

Viljoen D, Beighton P. (1992). Schwartz-Jampel syndrome (chondrodystrophic myotonia). *J Med Genet 29*, 58–62.

Viljoen D, Fredlund V, Ramesar R, Beighton P. (1993). Brachydactylous dwarfs of Mseleni. *Am J Med Genet 46*, 636–640.

Wainwright H, Beighton P. (2007). Osteogenesis imperfecta and holoprosencephaly. *Clin Dysmorphol 16*(3), 189–191.

Wainwright H, Beighton P (2008b). Achondrogenesis type II with cutaneous hamartomata. *Clin Dysmorphol 17*(3), 207–209.

Wainwright H, Beighton P. (2008c). Osteogenesis imperfecta type IIa with pseudoanencaphaly. *Clin Dysmorphol 17*(4), 271–272.

Wainwright H, Beighton P. (2008a). Visceral manifestations of hypochondrogenesis. *Virchows Arch 453*, 203–207.

Wainwright H, Beighton P. (2009). Multiple malformations; a possible Sonic hedgehog phenotype? *Virchows Archiv 454*(3), 345–347.

Wainwright H, Beighton P. (2010a). Dyssegmental dysplasia with bone entrapment and visceral manifestations. *Pediatr Dev Pathol 13*(1), 46–49.

Wainwright H, Beighton P. (2010b). Warfarin embryopathy; foetal manifestations. *Virchows Archiv 457*(6), 735–739.

Wainwright H, Beighton P. (2010c). Lethal epiphyseal stippling in the foetus and neonate; pathological implications. *Virchows Archiv 456*(3), 301–308.

Wallis C, Cremin BJ, Beighton P. (1989). Enchondromatosis with dwarfism and deafness. *Dysmorph Clin Genet 3*(4), 108–111.

Winship I, Cremin B, Beighton P. (1990). Boomerang dysplasia. *Am J Med Genet 36*, 440–443.

Winship W, Beighton P. (2008). Dyssegmental Dysplasia. *Clin Dysmorphol 17*(2), 95–98.

Wu H, Wainwright HH, Beighton P. (2009). Tetraphocomelia with the Waardenburg syndrome. *Clin Dysmorphol 18*(2), 112–115.

42.

HERITABLE DISORDERS OF CONNECTIVE TISSUE IN SOUTH AFRICA

Peter Beighton

INTRODUCTION

The heritable disorders of connective tissue have been the focus of ongoing interest and, since 1972, numerous affected persons have been investigated in Cape Town and outreach clinics. Several of the conditions reviewed in this chapter were the subject of doctoral theses in the Department of Human Genetics, University of Cape Town, and these are listed below:

GS Gericke (1984) MD. Clinical and genetic implications of congenital contractural disorders.

IM Winship (1986) MD. Epidermolysis bullosa in South Africa.

GA Wallis (1987) PhD. Type I collagen involvement in South African patients with osteogenesis imperfecta.

DL Viljoen (1991) MD. Pseudoxanthoma elasticum in South Africa.

LXJ Stephen (2000) PhD. Dental manifestations of osteogenesis imperfecta; pathogenesis, management and nosological significance.

Experience gained in these conditions is encapsulated in the 5th edition of *McKusick's Heritable Disorders of Connective Tissue* (Beighton, 1993). Articular laxity is a major feature of some conditions in this category; the pathogenesis and clinical implications of this unusual attribute have formed the basis for successive editions of the monograph "*Hypermobility of Joints*" (Beighton, Grahame & Bird, 2010).

Genetic connective tissue disorders that are of special importance or interest in South Africa form the subject of this chapter.

1. Osteogenesis imperfecta
2. Other thin bone and skeletal fragility syndromes
3. Marfan syndrome
4. Ehlers-Danlos syndrome
5. Pseudoxanthoma elasticum
6. Epidermolysis bullosa
7. Fibrodysplasia ossificans progressiva
8. Congenital contractural syndromes

OSTEOGENESIS IMPERFECTA

The major manifestation of osteogenesis imperfecta (OI) is skeletal fragility, which leads to frequent fractures. Blueness of the sclerae, opalescent teeth (dentinogenesis imperfecta), and occasional deafness in early adulthood are other very variable features. The condition is clinically and genetically heterogeneous, and the different forms of OI are categorized numerically.

Large numbers of persons and families with all forms of OI were seen at clinics and institutions throughout South Africa, and by 1983 it was possible to review the skeletal complications in 153 affected persons (Beighton et al., 1983). Biological material from these persons subsequently formed the basis for a doctoral thesis on the subject of the involvement of type I collagen in OI (Wallis, 1987).

Osteogenesis imperfecta type I, formerly designated *OI tarda*, in which fracturing is usually mild to moderate in severity, is transmitted as an AD trait and it is by far the most common form of OI. Although OI-I is frequent in persons of European ancestry, it is distinctly uncommon in the indigenous African population; indeed, only a few affected persons have been identified among the 40 million members of this community. The absence of a founder effect and the negative influence of sociobiological factors in previous generations are possible explanations of this discrepancy.

The issue of presence or absence of dentinogenesis imperfecta (DI) as a marker for subcategorization of OI has been the subject of considerable debate. This issue was addressed in a doctoral project by Professor Lawrence Stephen of the University of the Western Cape (Stephen, 2000), and it is now generally accepted that DI is not indicative of heterogeneity in OI.

In contrast to OI-I, the severe, progressive potentially lethal OI type III predominates in the indigenous African population (Beighton & Versfeld, 1985). This AR disorder is characterized by numerous fractures in childhood, with disturbed growth and limb deformity (Versfeld et al., 1985). The sclerae are usually white, and DI is infrequent. Affected children often become chairbound, cardiothoracic insufficiency supervenes, and survival into adulthood is unusual. In recent years, infusion therapy with bisphosphonates has proved to be effective and in this respect, the prognosis has greatly improved.

This form of OI has been encountered in all linguistic groups of the indigenous African population (Viljoen & Beighton, 1987) and more than 100 affected children have been identified. There are no clinical or radiographic manifestations in obligate heterozygotes, and nothing to suggest any biological advantage of this status. In laboratory investigations, the molecular and protein determinants of other forms of OI have been excluded (Wallis et al., 1993). The FKBP10 gene has recently been implicated in the molecular pathogenesis of the South African form of OI-III (Kelley et al., 2011).

OTHER THIN BONE AND SKELETAL FRAGILITY SYNDROMES

Skeletal lucency and/or a fracturing tendency are a feature of several rare genetic conditions that are present in South Africa.

OSTEOPOROSIS-PSEUDOGLIOMA SYNDROME

In the osteoporosis-pseudoglioma syndrome, thin bones and frequent fractures are associated with changes in the globe of the eye, which resemble those of a malignant glioma. This condition is present in a consanguineous family of Indian stock in Natal, and it was initially documented as the ocular form of OI (Beighton et al., 1985). After the correct diagnosis had been established, collaborative molecular investigations facilitated the localization of the determinant gene to 11q 12-13 (Gong et al., 1996). Thereafter, the role of the protein product LRP5 in the pathogenesis of the bone and eye changes was recognized (Gong et al., 2001).

FAMILIAL OSTEOLYSES

The familial osteolyses are a group of very rare genetic disorders in which progressive resorption of bone leads to severe physical handicap. The carpus and tarsus are most often affected, but the condition may spread to other regions of the skeleton and even become generalized. Renal involvement leading to kidney failure is a late manifestation. In a nationwide collaborative project, the manifestations in three affected boys of indigenous African stock were documented (Beighton et al., 2007). The only other instances of conditions of this type in South Africa known to the author were two adolescents of European ancestry, with the enigmatic and transient idiopathic juvenile osteoporosis syndrome, encountered more than 30 years ago.

ARENDSE SYNDROME

The Arendse syndrome is an AD disorder in which marked skeletal fragility, reminiscent of OI, is associated with severe nodal osteoarthropathy and a tendency to develop bony ankylosis. This disorder has been documented in a family living in a rural area near Cape Town (Arendse et al., 2009). The molecular pathogenesis of this condition has not yet been addressed.

BRUCK SYNDROME

Several neonates of indigenous African stock have been encountered in whom widespread joint rigidity and skeletal lucency prompted an initial diagnosis of arthrogryposis multiplex congenita. A fracturing tendency developed, and an independent diagnosis of osteogenesis imperfecta was subsequently made in these infants (Viljoen et al., 1989; Mokete et al., 2005). By virtue of historical precedent, the condition was given the eponymous title, "Bruck syndrome."

MARFAN SYNDROME

Marfan syndrome is characterized by tall stature, an asthenic habitus, thoracic asymmetry, arachnodactyly, a high palate, joint laxity and occasional dislocation of ocular lenses. Cardiac valvular incompetence may develop, and aortic dissection is potentially lethal in middle age. This AD disorder is encountered in all population groups in South Africa, with an estimated overall prevalence of about 1 in 5000 persons.

Between 1972 and 1990, more than 160 persons with a provisional diagnosis of Marfan syndrome were referred to the adult genetic clinic in Cape Town. The establishment of a firm diagnosis upon which prognostication and management could be based in these individuals was a major problem. Indeed, for every person in whom a definite diagnosis could be reached, several others had "partial" forms of the syndrome (Viljoen & Beighton, 1990). The availability of formal diagnostic criteria alleviated this situation to some extent, but the overall problem remains unresolved. In a pilot survey, in collaboration with colleagues at St George's Hospital, London, molecular investigations were undertaken on DNA from persons from different population groups in Cape Town with a putative diagnosis of Marfan syndrome (Child et al., 2007). Diagnostic confirmation and mutation detection was achieved in the majority; to date, however, financial pressures at the national level have precluded the establishment of a dedicated molecular diagnostic service for Marfan syndrome in South Africa.

EHLERS-DANLOS SYNDROME

Ehlers-Danlos syndrome (EDS) is characterized by articular hypermobility, skin hyperextensibility, and tissue fragility, which leads to scarring over the bony prominences. EDS is very heterogeneous and the manifestations are variable and inconsistent. Elucidation of the underlying molecular and biochemical determinants has facilitated ongoing recategorization, and at present, six major forms of EDS are recognized (Beighton et al., 1998). By 2005, a total of 64 affected individuals with different forms of EDS had been documented in Cape Town.

The archetype classical form (EDS-I), in which the diagnostic triad is transmitted as an AD trait, is encountered in most genetic clinics in South Africa. The potentially lethal arterial or vascular form (EDS-IV) is also seen occasionally. These conditions predominate in persons of European stock and they are rare in the indigenous African population of South Africa (Winship, 1985). Although EDS-IV is regarded as having a recognizable phenotype and facies, the arterial and alimentary tract catastrophes that typify this disorder have also been encountered in South Africa by the author in persons with ostensible EDS-I. Molecular characterization has not yet been undertaken, and the question arises as to whether this anomalous situation indicates that further genetic heterogeneity is present in this country.

Articular hypermobility in the absence of other stigmata of EDS is a frequent problem, which arises in all population groups in South Africa. Whether some or all of these persons have EDS-III is debatable. The basic defect has not been determined and this issue remains unresolved.

An objective clinical assessment of the extent of hypermobility in an individual is conventionally based upon a nine-point scale that is in wide general use. It is of historical interest that this scale was developed for rapid assessment of the range of joint movements in large numbers of persons in an epidemiological survey of the arthritides in South Africa; incorporation into routine clinical practice was not foreseen (Beighton et al., 1973).

PSEUDOXANTHOMA ELASTICUM

Pseudoxanthoma elasticum (PXE) is a heterogeneous disorder, in which yellowish induration of the skin occurs in flexural creases, notably the neck and groin. In the later stages, the skin of affected areas becomes lax, hanging in loose folds. The cutaneous changes are associated with progressive visual dysfunction due to breaks in Bruch's membrane in the retina of the eye. Less frequently, involvement of the media of the arteries may lead to thrombotic and hemorrhagic episodes. The condition is very variable and severity in one system does not necessarily correlate with the degree of involvement in another.

The condition has been encountered worldwide, and frequency of more than 1 in 160,000 is generally quoted. In South Africa, the condition is most often encountered in persons of European stock (Viljoen et al., 1984).

In the endogamous Afrikaner population of South Africa, 58 affected persons with an AR form of PXE were identified. In this disorder the eye manifestations were severe, while skin involvement was comparatively mild (Viljoen et al., 1987). The Afrikaner community has origins in Holland and neighboring regions of Europe, and Professor Ann De Paepe of the University of Ghent was able to establish molecular identity between the condition in South Africa and PXE in Belgium (De Paepe et al., 1991). This form of PXE subsequently formed the subject of Denis Viljoen's doctoral thesis (Viljoen, 1991).

EPIDERMOLYSIS BULLOSA

Epidermolysis bullosa (EB) is a clinically and genetically heterogeneous disorder in which blistering of the skin is the presenting feature. The different forms of EB vary greatly in severity and, in some, involvement of the mucous membranes can be potentially lethal (Winship, 1988).

The South African experience of EB was reviewed after a nationwide survey that formed the subject of a doctoral thesis (Winship, 1986). Thereafter, Dr. Ingrid Winship and her pediatrician father, Professor W. (Bill) Winship documented three children with EB in whom a misdiagnosis of child abuse had been made (Winship & Winship, 1988). Biological material from South Africa was used in a collaborative investigation in which the type VII collagen gene was shown to be linked to a severe AR form of EB (Dunnill et al., 1994).

FIBRODYSPLASIA OSSIFICANS PROGRESSIVA

Fibrodysplasia ossificans progressiva (FOP) is a rare genetic disorder in which progressive ossification of fibrous tissue leads to generalized restriction of movements. In the later stages of the condition, affected persons may be able to move only their fingers, eyes and tongue. Despite these limitations, survival into middle age is not unusual but death eventually supervenes from respiratory insufficiency.

In view of the severity of the handicap, affected persons need comprehensive support or institutional care. In a nationwide survey, six affected persons of different population and linguistic groups were identified in South Africa (Connor & Beighton, 1982), and radiographs pertaining to three more were found in the hospital archives (Cremin et al., 1982). In the same decade two additional adult SA males of European stock were documented by other researchers under the title "myositis ossificans" (Shipley et al., 1979; Shipton et al., 1985).

The identification of a specific mutation (R206H) in the ACVR1 gene at 2q23-q24 has given impetus to the understanding of FOP, and it has emerged that this mutation has a worldwide distribution. Interest in FOP in Cape Town has been revived with a new diagnosis in a young girl of Xhosa stock, and adults who were first examined almost thirty years ago have been reassessed (Scott et al., 2011; Roberts et al., 2011). Mutation analysis revealed that all three had the conventional ACVR1 mutation, which is apparently ubiquitous.

CONGENITAL CONTRACTURAL SYNDROMES

Congenital contractures of joints are a component of several autonomous genetic syndromes. Involvement of the connective tissues, notably the tendons and ligaments, is the primary defect in these disorders. Conditions of this type that have been encountered in South Africa are outlined below.

CONGENITAL CONTRACTURAL ARACHNODACTYLY

Congenital contractural arachnodactyly (Beals syndrome) is an AD disorder which resembles the Marfan syndrome, but differs by virtue of contractures of the digits and crumpling of the pinnae of the ears. This disorder was documented in a large Indian kindred in Natal (Viljoen, 1994). Genetic material from this family was used in early linkage investigations, in which the fibrillin genes were implicated (Tsipouras et al., 1992). It is relevant that the congenital contractions are a component of a wide range of rare genetic disorders and malformation complexes.

These disorders were reviewed and discussed in detail in a doctoral thesis (Gericke, 1984).

DIGITOTALAR DYSMORPHISM

Digital contractures and "rocker bottom" feet, due to a vertical alignment of the talus bone, are the main features of digitotalar dysmorphism. This AD condition has been documented in 5 generations of a South African family of British stock (Sallis & Beighton, 1972).

GORDON SYNDROME

The Gordon syndrome is an AD syndrome comprising variable cleft lip, club feet, and digital contractures (Basel et al., 2000). This condition, which is present in four generations of a Mixed Ancestry family in Cape Town, is named after the late Dr. Hymie Gordon, who initially documented the disorder four decades ago.

FREEMAN-SHELDON SYNDROME

Freeman-Sheldon syndrome is an uncommon AD disorder in which flexion contractures and ulnar deviation of the fingers and talipes equinovarus are associated with microstomia and myotonia of the facial muscles. The alternative designation, "Whistling Face syndrome," pertains to the characteristic facial appearance. The condition has been documented and depicted in a father and daughter of Tswana stock (Beighton, 1988), and orodental implications in a Xhosa kindred have been recorded (Roberts et al., 2005).

MEDICAL AND GENETIC SERVICES

Persons and families with the heritable disorders of connective tissue in South Africa are generally under the care of their general practitioners, and specialists whose skills are appropriate for their clinical problems. Increasingly, affected individuals are referred for genetic counseling, which is available in major university centers. The actual number of medical geneticists in South Africa is well below the norm for Europe and the United States, and few are being trained. Nevertheless, the establishment of formal training programs for genetic counselors has alleviated the problem to some extent. (see Chapter 44).

A major difficulty is posed by the absence of laboratory facilities for diagnostic appraisal of various forms of the Ehlers-Danlos syndrome, osteogenesis imperfecta, and Marfan syndrome. The state health laboratories are largely focused upon acquired disorders, such as AIDS and tuberculosis, while laboratories in the private sector have little concern for rare genetic disorders. For these reasons it is necessary to send specimens to overseas laboratories, at considerable expense. If physical handicap is severe, as in OI-III and FOP, institutional care may be needed. There are several facilities of this type in South Africa for affected children but, at present, provision for affected adults is inadequate.

ACKNOWLEDGMENTS

The investigations upon which this chapter was based were supported by the South African Medical Research Council and the National Research Foundation. This study forms part of the Cape Genetic Heritage project, in the Division of Human Genetics, University of Cape Town.

I am most appreciative of the contributions made by my wife Greta in the preparation of the manuscript.

REFERENCES

Arendse R, Brink P, Beighton P. (2009). Hereditary bone dysplasia with pathological fractures and nodal osteoarthropathy. *Skel Radiol 38*, 1197–1203.

Basel D, Sobey G, Gardner J, Beighton P. (2000). The Gordon Syndrome Revisited. *S Afr Med J 90*(9), 864–867.

Beighton P, Solomon L, Soskolne CL. (1973). Articular mobility in an African population. *Ann Rheum Dis 32*(5), 413–418.

Beighton P, Spranger J, Versveld G. (1983). Skeletal complication in osteogenesis imperfecta; A review of 153 South African patients. *S Afr Med J 64*, 565–568.

Beighton P, Versfeld GA. (1985). On the paradoxically high relative prevalence of osteogenesis imperfecta type III in the Black population of South Africa. *Clin Genet 27*, 398–401.

Beighton P, Winship I, Behari D. (1985). The ocular form of osteogenesis imperfecta: a new autosomal recessive syndrome. *Clin Genet 28*, 69–75.

Beighton P. (1988). *Inherited Disorders of the Skeleton* (2nd ed.). Edinburgh: Churchill Livingstone, pp. 240–241.

Beighton P, De Paepe A, Steinmann B, Tsipouras P, Wenstrup RJ. (1998). Ehlers-Danlos Syndromes: Revised Nosology, Villefranche. *Am J Med Genet 77*, 31–37.

Beighton P, ed. (1993). *McKusick's Heritable Disorders of Connective Tissue (5th Ed).* St. Louis: C.V. Mosby.

Beighton P, Mennen U, Golele SS, Urban M. (2007). Orthopaedic implications of heritable osteolysis in South Africa. *SA Orthopaedic Journal* 6(2), 26–32.

Beighton P, Grahame R, Bird H. (2010) *Hypermobility of Joints (4th Ed).* Springer-Verlag, Heidelberg.

Child A, Comeglio P, Arno G, Beighton P. (2007). Marfan syndrome in South Africa – a molecular genetic approach to diagnosis. *S Afr Med J 97*(9), 845–847.

Connor JM, Beighton P (1982). Fibrodysplasia ossificans progressiva in South Africa. *S Afr Med J 61*, 404–406.

Cremin B, Connor JM, Beighton P. (1982). The radiological spectrum of fibrodysplasia ossificans progressiva. *Clin Radiol 33*, 499–508.

De Paepe A, Viljoen D, Matton M, et al. (1991). Pseudoxanthoma elasticum: similar autosomal recessive subtype in Belgian and Afrikaner families. *Am J Med Genet 38*(1), 16–20.

Dunnill MGS, Richards AJ, Milana G, et al. (1994). Genetic linkage to the type VII collagen gene (COL7A1) in 26 families with generalised recessive dystrophic Epidermolysis Bullosa and anchoring fibril abnormalities. *J Med Genet 31*, 745–748.

Gericke GS. (1984). *Clinical and genetic implications of congenital contractural disorders.* Doctoral Thesis (MD), University of Cape Town.

Gong, Y, Vikkula M, Boon L, et al. (1996). Osteoporosis-pseudoglioma syndrome, a disorder affecting skeletal strength, is assigned to chromosome region 11q12-13. *Am J Hum Genet 59*, 146–151.

Gong Y, Slee RB, Fukai N, et al; Osteoporosis-Pseudoglioma Syndrome Collaborative Group (2001). LDL receptor-related protein 5 (LRP5) affects bone accrual and eye development. *Cell 107*(4), 513–523.

Kelley BP, Symoens S, Bonafe L et al. (2011). Mutations in the FKBP10 cause recessive osteogenesis imperfecta and Bruck syndrome. *J Bone Miner Res 26*(3), 666–672.

Mokete L, Robertson A, Viljoen D, Beighton P. (2005). Bruck syndrome: congenital joint contractures with bone fragility. *J Orthop Sci 10*(6), 641–646.

Roberts T, Stephen LXG, Naidoo T, Fieggen K, Beighton P. (2005). Freeman-Sheldon Syndrome; orthodontic implications. *J Clin Pediatr Dent 29*(2), 267–273.

Roberts T, Stephen LXG, Scott C et al. (2011). Fibrodysplasia ossificans progressive (FOP) in South Africa: dental implications in 5 cases. *Oral Surg Oral Med Oral Pathol Oral Radiol Endod 112*, 11–18.

Sallis JG, Beighton P. (1972). Dominantly inherited digito-talar dysmorphism. *J Bone Jt Surg 54B*, 509–515.

Scott C, Arendse R, Dandara C et al. (2011). Fibrodysplasia ossificans progressive in South Africa. *J Clin Rheumatol 17*(1), 37–41.

Shipley JA, Pompe Van Meerdervoort HF. (1979). Myositis Ossificans Progressiva. *S Afr Med J* 55, 947–950.

Shipton EA, Retief LW, Theron HD, de Bruin FA. (1985). Anaesthesis in myositis ossificans progressiva: a case report and clinical review. *S Afr Med J 67*, 26–28.

Stephen LXJ. (2000) *Dental manifestations of osteogenesis imperfecta; pathogenesis, management and nosological significance.* Doctoral thesis (PhD), University of Cape Town.

Tsipouras P, Mastro RD, Sarfarazi M, et al. (1992). Genetic linkage of the Marfan syndrome, ectopia lentis, and congenital contractural arachnodactyly to the fibrillin genes on chromosomes 15 and 5. *New Eng J Med 326*, 905–909.

Versfeld GA, Beighton PH, Katz K, Solomon A. (1985). Costovertebral anomalies in osteogenesis imperfecta. *J Bone Jt Surg 67B*(4), 602–604.

Viljoen D, Beighton P, Mabin T, Woods K, Saxe N, Bonafede P. (1984). Pseudoxanthoma elasticum in South Africa – genetic and clinical implications. *S Afr Med J 66*, 813–816.

Viljoen DL, Pope FM, Beighton P. (1987). Heterogeneity of pseudoxanthoma elasticum: delineation of a new form? *Clin Genet 32*, 100–105.

Viljoen D, Beighton P. (1987). Osteogenesis imperfecta type III: An ancient mutation in Africa? Am J Med Genet *27*(4), 907–912.

Viljoen D, Versfeld G, Beighton P. (1989). Osteogenesis imperfecta with congenital joint contractures (Bruck syndrome). *Clin Genet 36*(2), 122–126.

Viljoen D, Beighton P. (1990). Marfan syndrome: a diagnostic dilemma. *Clin Genet 37*, 417–422.

Viljoen DL. (1991). *Pseudoxanthoma elasticum in South Africa.* Doctoral Thesis (MD), University of Cape Town.

Viljoen D. (1994). Congenital contractural Arachnodactyly (Beals syndrome). *J Med Genet 31*, 640–643.

Wallis GA. (1987). *Type I collagen involvement in South African patients with osteogenesis imperfecta.* Doctoral Thesis (PhD), University of Cape Town.

Wallis GA, Sykes B, Byers PH, Mathew CG, Viljoen D, Beighton P. (1993). Osteogenesis imperfecta type III: mutations in the type I collagen structural genes, COL1A1 and COL1A2, are not necessarily responsible. *J Med Genet 30*, 492–496.

Winship IM. (1985). Ehlers-Danlos syndrome in the Western Cape. *S Afr Med J 67*, 509–511.

Winship IM. (1986). *Epidermolysis bullosa in South Africa.* Doctoral Thesis (MD), University of Cape Town.

Winship I. (1988). Epidermolysis bullosa in South Africa. *S Afr Med J 69*, 743–746.

Winship IM, Winship WS. (1988). Epidermolysis Bullosa misdiagnosed as child abuse. *S Afr Med J 73*, 369–370.

43.

GENETIC PERSPECTIVES OF TUBERCULOSIS IN SOUTHERN AFRICA

Erika de Wit, Marlo Möller, and Eileen Hoal

INTRODUCTION

Tuberculosis (TB) is a common infectious disease in the developing world and can be fatal even when treated. It is estimated that the TB-causing bacterium *Mycobacterium tuberculosis* infected more than 9 million individuals during 2007, and that 44% of these new infections progressed to active disease (World Health Organization, 2009). Africa accounts for 31% of global TB cases, the majority of which are from sub-Saharan Africa (Figure 43-1, World Health Organization, 2009). However, individuals respond differently to infection, and the presence of the bacterium is not enough to cause disease. In fact, 90% of infected individuals who are immunocompetent will never develop clinical TB. This observed difference in disease susceptibility can be explained by both environmental and host factors, and the complex interactions between them.

A plethora of studies (reviewed by Möller et al., 2010a) have proven that genetic features contribute to the outcome of TB, which makes deciphering the human genes involved in TB susceptibility a vital element in the ongoing fight against the bacterium. One of the observations substantiating the involvement of host genes in TB susceptibility is the seemingly increased resistance of European individuals to TB. It has been speculated that this is the result of centuries of exposure to TB in Europe, which may have resulted in a degree of selection for a more resistant population, whereas sub-Saharan Africa was only relatively recently exposed to the disease (Dubos & Dubos, 1952). However, the natural selection process may be hampered in Africa due to the use of TB drugs, which save lives but consequently facilitate the inheritance of susceptibility variants. A study done in a U.S. nursing home determined that the TB disparity observed is not only due to socioeconomic factors, since individuals with African ancestry were twice as likely to be infected with *M. tuberculosis* than individuals with European ancestry, even though they shared the same environment (Stead et al., 1990). A more recent study found that white U.S.-born TB cases were more likely to be homeless than black U.S.-born TB cases, which substantiated the finding by Stead et al. (Serpa et al., 2009). In the United States, TB rates for U.S.-born black individuals are eight times higher than for U.S.-born whites (Centers for Disease Control and Prevention [CDC], 2008), suggesting ongoing transmission of TB among this group instead of reactivation. Although the environmental and social factors are difficult to control for, it does appear that the underlying susceptibility is different and that individuals with African ancestry could be more susceptible to TB. For this reason, populations with African genetic contributions are ideal for the investigation of TB susceptibility genes.

More than 2000 distinct ethnic groups exist on the African continent, and these are largely responsible for the genetic diversity among the different African populations (Tishkoff & Williams, 2002). Extreme population heterogeneity can also be found in southern Africa, which consists of the countries of Botswana, Lesotho, Namibia, South Africa, and Swaziland (United Nations Statistics Division, 2009); e.g., in South Africa alone there are 11 official languages, which reflect the main ethnic groups. The genetic diversity between various South African populations has been shown by studies that determined the human leukocyte antigen (HLA) class I and II alleles frequencies in different population groups (Cao et al., 2004; Lombard et al., 2006b). The majority of published TB genetic studies in southern Africa have been done in two populations from South Africa, namely the South African Coloured (SAC) population and the Venda.

The SAC population is a highly admixed group, with genetic contributions from ancestral populations such as San, Khoe, Indian, African blacks, and Europeans (Barreiro et al., 2006; Tishkoff et al., 2009). Official segregation served to maintain the population makeup of this group. According to the 2001 census, the SAC population currently comprises approximately 9% of South Africa, and constituted 53.9% of the population of the Western

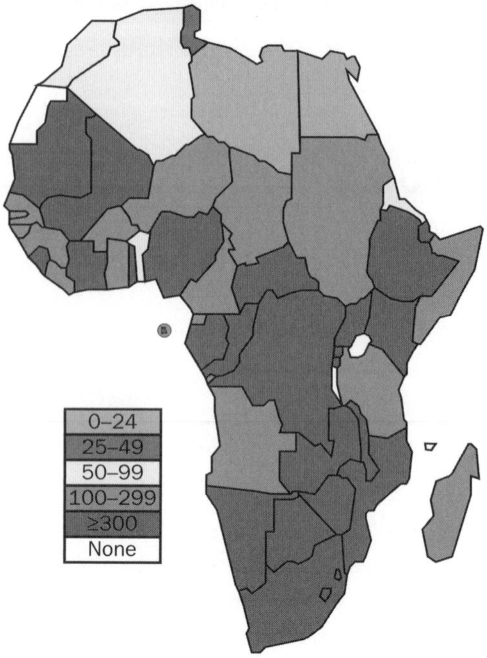

Figure 43-1 Estimated new TB cases (all forms) in African countries per 100,000 population (World Health Organization, 2009). Refer color figure.

Cape province of South Africa, which was also home to the majority of this population (61.1%). The Venda population consists of approximately 1 million individuals and may have been one of the last groups to migrate from either the Congo or from the Great Rift Valley to South Africa during the Bantu expansion (Lombard et al., 2006b; Kruger National Park, 2009; Tishkoff et al., 2009). This idea is supported by the substantial levels of southern African Khoesan and western African Bantu ancestry found in this population (Tishkoff et al., 2009), but a low level of east African Bantu genetic contribution. The majority of the population resides in a region located in the Limpopo Province of South Africa.

Several study designs, such as heritability analyses, linkage studies, and candidate gene association studies, have been used to investigate genetic susceptibility to TB in southern Africa. These studies will be considered below.

HERITABILITY ANALYSES

Heritability analysis is a population genetics technique which measures the strength of a genetic influence on a particular phenotype. This type of study has shown that an individual's immune response to infection with *M. tuberculosis* will be regulated by the genetic background, with an estimated heritability ranging from 36% to 80% (Comstock, 1978; Jepson et al., 2001; Kallmann & Reisner, 1942; Newport et al., 2004; Kimman et al., 2006).

A recent heritability analysis (Cobat et al., 2010) used 155 families from the SAC population and measured in vitro secretion of tumor necrosis factor-α (TNF-α) and interferon gamma (IFN-γ). This study also determined the frequency of IFN-γ^{+}CD4^{+} and IFN-γ^{+}CD8^{+} cells after whole blood was stimulated with bacillus Calmette-Guérin (BCG), purified protein derivate (PPD), and early secreted antigenic target 6 kDa protein (ESAT6) antigens. Heritability estimates from this study were high, and support the idea that host genetic factors regulate antimycobacterial immune responses to a large extent. For TNF-α production elicited by co-incubation with BCG alone, or with BCG and IFN-γ, heritability was found to be 30% and 39% respectively. This supports the study done by Stein et al. (2005), which estimated the heritability of TNF-α production at 68%, using segregation analysis in a Ugandan population, and predicted the existence of a gene that determines one-third of TNF-α production (Stein et al., 2005). Estimates of IFN-γ production heritability in the SAC population varied depending on the nature of the stimulating antigen, and ranged from 43% to 58%, with ESAT6-triggered IFN-γ production showing the highest heritability level (Cobat et al., 2010). The study done by Cobat et al. (2010) was the first to report on the heritability of the frequency of IFN-γ^{+}CD4^{+} $^{+}$ and IFN-γ^{+}CD8^{+} cells, with estimates ranging from 53% to 74% and the frequency of BCG-specific IFN-γ^{+}CD8^{+} cells having the highest heritability.

LINKAGE STUDIES

Genome-wide genetic linkage studies can identify genomic regions containing genes that predispose to disease, based on the assumption that susceptibility regions will segregate nonrandomly with the disease of interest in large affected families (Alcaïs & Abel, 2004). This approach has been very successfully implemented for monogenic diseases, but can only identify large regions (sometimes tens of cM) containing many plausible candidate genes for complex diseases (Teare & Barrett, 2005). Linkage is typically reported as a logarithm of odds (LOD) score, where higher LOD scores gives greater evidence of linkage. As a rule, an LOD score of 3 is considered to give evidence of significant linkage (Teare & Barrett, 2005). To date, seven genome-wide TB linkage scans have been done worldwide, using populations from Brazil (Miller et al., 2004), Malawi (Cooke et al., 2008), Morocco (Baghdadi et al., 2006), South Africa (Bellamy et al., 2000; Cooke et al., 2008), Thailand (Mahasirimongkol et al., 2009), the Gambia (Bellamy et al., 2000), and Uganda (Stein et al., 2008).

Two TB genome-wide linkage studies have been done using samples from the SAC population (Table 43-1) with TB disease as the phenotype of interest. The first of these studies used sib pairs from the Gambia and South Africa, and chromosomes Xq and 15q were identified as

TABLE 43-1 CHROMOSOMAL REGIONS SHOWING SUGGESTIVE EVIDENCE OF LINKAGE TO TB IN POPULATIONS FROM SOUTHERN AFRICA

POPULATION	TB PHENOTYPE	CHROMOSOMAL REGION	LOD[a]	REFERENCE
SAC[b] and The Gambia	TB	15q11-13	2	(Bellamy et al., 2000)
		Xq	1.77	
SAC and Malawi	TB	6p21-q23	1.9	(Cooke et al., 2008)
		20q13.31-33	3.1	
SAC	TST[c] reactivity	Submitted	–	(Cobat et al., 2009)

[a] LOD, logarithm of odds score

[b] SAC, South African Coloured

[c] TST, tuberculin skin test

possible TB susceptibility regions (Bellamy et al., 2000). A subsequent study (Cervino et al., 2002) did fine-mapping of chromosome 15q11-13 and found evidence that the ubiquitin protein ligase E3A (human papilloma virus E6-associated protein, Angelman syndrome) gene (*UBE3A*) or a flanking gene was involved in TB susceptibility. Maternally inherited deletion of the gene causes Angelman syndrome (Van Buggenhout & Fryns, 2009) and the enzyme targets proteins destined for degradation inside cells (Jiang & Beaudet, 2004). Chromosomes 6p21-q23 and 20q13.31-33 showed linkage to TB susceptibility in the second genome-wide linkage study, in the SAC and Malawian populations (Cooke et al., 2008). A linkage study done in Uganda validated the chromosome 20q13 susceptibility region identified (Stein et al., 2008). Two genes located in the chromosome 20q13.31-33 region, namely cathepsin Z (*CTSZ*) and melanocortin 3 receptor (*MC3R*), were implicated in TB susceptibility in association studies that followed the original linkage study (Cooke et al., 2008). MC3R is involved in the regulation of energy homeostasis in humans (Wang et al., 2008) and was previously associated with obesity (Lee et al., 2002), while CTSZ may be involved in tumorigenesis (Nagler et al., 2004). The genome-wide linkage studies in the SAC population have therefore identified genes that, based on their known functions at the time, would not normally have been considered as TB susceptibility factors.

A third genome-wide linkage study in the SAC population investigated the genetic basis of tuberculin skin test (TST) reactivity, which is used as a diagnostic test for TB (Cobat et al., 2009). This study was based on the observation that approximately 20% of individuals living in regions severely burdened with TB will never show TST reactivity, and appear to be naturally resistant to infection with *M. tuberculosis*. A major locus was identified which controlled T-cell-independent resistance to *M. tuberculosis* infection. Varying intensities of TST responses are also observed in individuals who do become infected with *M. tuberculosis*. A second major locus found in the study done by Cobat et al. controlled the intensity of T-cell mediated, delayed type hypersensitivity to tuberculin.

CANDIDATE GENE ASSOCIATION STUDIES

The genetic susceptibility component of TB seems to be scattered across many genes (Figure 43-2) as well as different genes being implicated in different populations (Maartens & Wilkinson, 2007). Genes may not only play a role in the risk of infection upon exposure, they might also influence the risk of disease development and the course of disease (Daniel, 1997). Many studies indicated that *M. tuberculosis* applies a powerful selective pressure on the human genome (Bloom & Small, 1998). The genetic component of susceptibility to TB has been studied extensively in the past few years, resulting in a rapid increase in available information. Since TB is a complex disease, it can be assumed that susceptibility to *M. tuberculosis* has multiple genetic causative factors (as well as environmental causes); therefore, a single approach would not be able to identify all the genes involved (Bellamy, 2006).

The most popular study design for investigating host susceptibility to TB is association studies, since it was shown that this design has much greater power than linkage studies for detecting genes of small effect (Risch & Merikangas, 1996). Linkage analysis and case-control association studies are based on two completely different approaches (Table 43-2). Population-based case-control association studies are the classic association studies, where the allele frequency of a specific marker is compared between unrelated cases (affected individuals) and controls (unaffected individuals), when Hardy-Weinberg equilibrium (HWE) holds (Lander & Schork, 1994). It is very important to address population stratification and to replicate (in the same population) and validate (in a different population) case-control association studies to confirm the results.

Several genes, including *HLA*, *NRAMP1*, *IFNG*, *NOS2A*, *MBL*, *CD209* and *VDR* have been associated with TB in the SAC and/or the Venda populations in southern

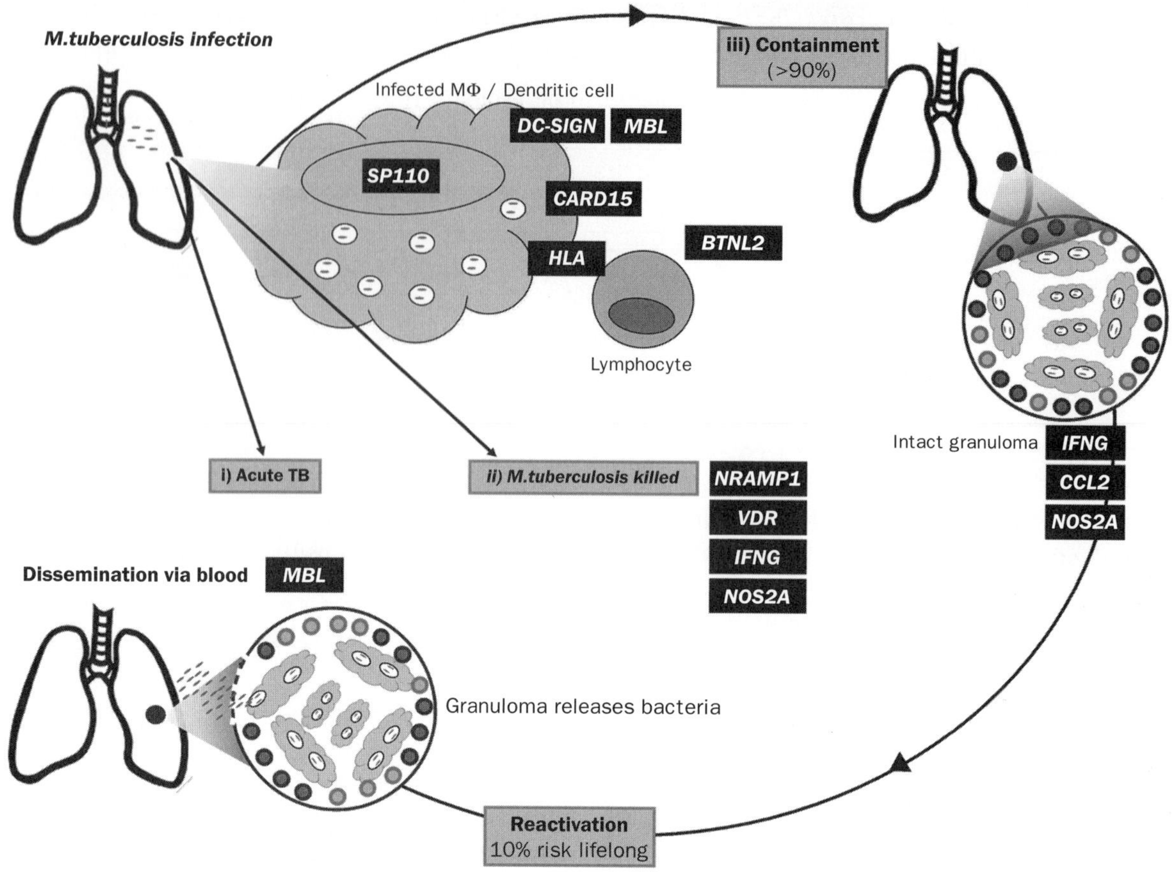

Figure 43-2 Genetic involvement in the TB disease process. A simplified representation of different outcomes after *M. tuberculosis* infection, and some of the genes that may be involved at various stages. The bacteria enter the respiratory system of the host via inhaled droplets and are engulfed by macrophages (MΦ) and dendritic cells. There are three potential outcomes after inhalation of *M. tuberculosis*: (i) infection advances to active TB, (ii) *M. tuberculosis* is instantly killed by the pulmonary immune system, or (iii) infection does not develop into active disease, because the bacteria are controlled in granulomas. This containment can last for decades, or the lifetime of the individual. *M. tuberculosis* can disseminate from granulomas, causing active TB (reactivation disease) (Adapted from Kaufmann & McMichael, 2005). Refer color figure.

Africa. Some of these associations were replicated, but none were validated. This emphasizes the complexities of characterizing host susceptibility in different ethnic populations living in diverse environments (Ardlie et al., 2002). A summary of the studies discussed below is given in Table 43-3. Polymorphisms in several biologically plausible candidate genes, and previously reported susceptibility genes, have also been investigated in southern Africa using genetic association studies, but were not associated with TB. It is possible that certain TB genes are unique susceptibility factors in non-African populations, which would explain the lack of association found. Negative association studies are seldom, if ever, published, due to publication bias, and there may therefore be many unpublished studies that were done in southern Africa.

TABLE 43-2 A COMPARISON BETWEEN LINKAGE STUDIES AND ASSOCIATION STUDIES

LINKAGE STUDIES	ASSOCIATION STUDIES
Test whether they show correlated transmission within a pedigree	Test whether a disease and an allele show correlated occurrence in a population
Focus on concordant inheritance	Focus on population frequencies
Can detect linkage without association	Can detect association without linkage

Lander & Schork, 1994.

HUMAN LEUKOCYTE ANTIGEN REGION

The human leukocyte antigen (HLA) region contains the genes of the major histocompatibility complex (MHC) in humans. The majority of these genes have a function in the immune system, such as the class I and class II genes, which are involved in antigen presentation to T-cells, and were therefore logical first candidates to test as susceptibility factors for TB. HLA-DR subtypes, in particular, were consistently associated with TB and other mycobacterial diseases (Amirzargar

TABLE 43-3 ASSOCIATION STUDIES ON TB SUSCEPTIBILITY CANDIDATE GENES IN POPULATIONS FROM SOUTHERN AFRICA

GENE	POLYMORPHISM	POPULATION	PHENOTYPE	CASES	CONTROLS	P VALUE (PC)[A]	OR[B]/RR[C]	REFERENCE
HLA	DQB1*0301–0304	Venda	TB	95	117	0.001	2.58	(Lombard et al., 2006b)
	DRB1*1302	Venda	TB	92	117	< 0.001	5.05	(Lombard et al., 2006b)
	DRB1*03	SAC[d]	TB	106	107	Not given		(Lombard et al., 2006a)
BTNL2	rs2076530	SAC	TB	432	482	ns[e]		(Möller et al., 2007a)
CARD15	R702W	SAC	TB	431	482	Ns		(Möller et al., 2007b)
	G908R	SAC	TB	431	482	Ns		(Möller et al., 2007b)
	1007fs	SAC	TB	431	482	Ns		(Möller et al., 2007b)
CCL2	–2518	SAC	TB	431	482	Ns		(Möller et al., 2009)
CD209	–336A	SAC	TB	351	360	0.01	1.48	(Barreiro et al., 2006)
	–871G	SAC	TB	351	360	8.2×10^{-4}	1.85	(Barreiro et al., 2006)
IFNG	+874T	SAC	TB	313	235	0.0055	1.64	(Rossouw et al., 2003)
	+874T	SAC	TB	131 families		0.005		(Rossouw et al., 2003)
MBL	B	SAC	TBM[f]	91	79	0.017		(Hoal-van Helden et al., 1999)
NOS2A	rs9282799 & rs8078340	SAC	TB	431	482	0.011 (0.029)	1.4	(Möller et al., 2009)
SLC11A1	GT9	SAC	TB	265	224	0.002	0.6	(Hoal et al., 2004)
	TGTG+/del	SAC	TB	265	224	0.013	5.19	(Hoal et al., 2004)
SP110	rs2114592	SAC	TB	381	417	ns		(Babb et al., 2007a)
	rs3948464	SAC	TB	381	417	ns		(Babb et al., 2007a)
VDR ApaI	Venda	TB	~85	~88		ns		(Lombard et al., 2006b)
	ApaI	SAC	TB	249	352	ns		(Babb et al., 2007b)
	BsmI	Venda	TB	~85	~88	ns		(Lombard et al., 2006b)
	FokI	Venda	TB	~85	~88	ns		(Lombard et al., 2006b)
	FokI	SAC	TB	249	352	ns		(Babb et al., 2007b)
	TaqI	Venda	TB	~85	~88	ns		(Lombard et al., 2006b)
	TaqI	SAC	TB	249	352	ns		(Babb et al., 2007b)

[a] (pc), corrected p value

[b] OR, odds ratio

[c] RR, relative risk

[d] SAC, South African Coloured

[e] ns, not significant

[f] TBM, TB meningitis

et al., 2004; Bothamley et al., 1989; Chandanayingyong et al., 1988; Kim et al., 2005; Singh et al., 1983).

Despite the seemingly important role of HLA genes in disease susceptibility, there are few published studies investigating the genetic association with TB in southern Africa. Lombard et al. (2006b) studied the association between HLA and TB susceptibility in 104 patients with pulmonary or meningeal disease and 117 controls from the Venda population of South Africa. The DQB1*03010304 and DRB1*1302 phenotypes were associated with TB, with the DRB1*1301 and DRB1*1302 alleles found at significantly higher frequencies in cases and controls respectively. Interestingly, the DR2 (DRB1*15 and DRB1*16) association, frequently observed in other populations, was not identified in their study, but DRB1*1101-1121 and DRB1*1302 associations, during single-point or haplotype analyses, were detected. A second study found a marginal association of DRB1*03 with TB in 106 TB patients and 107 controls from the SAC population (Lombard et al., 2006a). The South African results are similar to those in

Thailand, which also found the DQB1*0301 to be associated with TB (Vejbaesya et al., 2002).

Heterozygosity for the HLA-DRB1 allele may protect human immunodeficiency virus (HIV-1) infected individuals coinfected with *M. tuberculosis* from quickly developing AIDS, possibly due to a more efficient control of TB infection during HIV-1 disease. A study in Zimbabwe found that HIV+ patients with TB who were homozygous for the HLA-DRB1 allele had a higher virus load, which conferred a risk for rapid progression to AIDS (Zijenah et al., 2002). The authors speculated that their findings lend support to the heterozygote advantage hypothesis.

NATURAL RESISTANCE ASSOCIATED MACROPHAGE PROTEIN 1

The natural resistance associated macrophage protein 1 (NRAMP1, renamed *solute carrier family 11A member 1* [SLC11A1]) was the first susceptibility gene to be identified from a mouse model of mycobacterial disease, and is the most extensively studied non-HLA TB candidate gene. The mouse gene, Nramp1, was identified by positional cloning and it was determined that a nonsynonymous nucleotide substitution in codon 169 led to an amino acid change (glycine to aspartic acid), which caused susceptibility to mycobacteria (BCG) in mice (Malo et al., 1994; Vidal et al., 1993). Additional proof of the effect of Nramp1 was acquired when an Nramp1 knockout mouse strain, and a mouse strain with the nonsynonymous nucleotide substitution, had identical phenotypes (Vidal et al., 1995). However, Nramp1 does not seem to affect susceptibility to *M. tuberculosis* in mice (North et al., 1999). The human homologue of this candidate gene, derived from the mouse model, was named NRAMP1 and mapped to chromosome 2q35 (Cellier et al., 1994). The protein product of NRAMP1 is a divalent cation transporter, which is recruited to the phagolysosomal membrane when the macrophage is activated (Gruenheid et al., 1997). Several TB association studies in different populations have considered NRAMP1 as a candidate gene (Hoal et al., 2004). Combined, these studies suggest that NRAMP1 is involved in *M. tuberculosis* susceptibility in humans. However, the effect of this gene is not strong enough to suggest that it is the major gene involved in TB.

Our study investigating *NRAMP1* and *NRAMP2* in the SAC population determined that two polymorphisms, one in the 5' and the other in the 3' untranslated region (UTR) of *NRAMP1*, independently contributed to TB in this population (Hoal et al., 2004). The 5'(GT)9 allele, which drives the highest rate of transcription of the *NRAMP1* gene, was associated with protection against the disease in the SAC population, and was also less frequent in TB cases from other published studies, making it one of the few consistent genetic associations with TB throughout a number of distinct populations (Awomoyi et al., 2002; Bellamy et al., 1998; Gao et al., 2000; Ma et al., 2002).

Two other African studies, although not from southern Africa, also found statistical evidence that *NRAMP1* is involved in TB susceptibility. A study in West Africa genotyped more than 800 cases and controls, and two polymorphisms (intronic and 3'UTR) were associated with TB (Bellamy et al., 1998). A large case-control association study in northern Malawi found that a CAAA insertion/deletion polymorphism in the 3'UTR of *NRAMP1* was associated with TB in both HIV+ and HIV− individuals. This study provided evidence that the NRAMP1 protein shaped innate immune responses that influence TB susceptibility in both immunocompetent and compromised patients (Fitness et al., 2004).

INTERFERON-GAMMA AND GENES MODULATING INTERFERON-GAMMA

When exposed to *M. tuberculosis*, most individuals initially respond by secreting IFN-γ. This cytokine is essential to control and contain *M. tuberculosis* during infection, as proven by numerous observations, which include the increased susceptibility seen in individuals suffering from Mendelian susceptibility to mycobacterial disease (MSMD), caused by rare inherited mutations in the genes of the interleukin (IL)-12/IL-23/IFN-γ axis (reviewed by Al-Muhsen & Casanova, 2008). However, production of IFN-γ on its own is not sufficient to protect against TB (Leal et al., 2001)—various other cytokines, such as IL-4, IL-10 and IL-12, and their receptors may modulate the effects of IFN-γ.

The functional +874A→T polymorphism, a well-studied common variant in intron 1 of the IFN-γ gene (*IFNG*), was associated with TB in the SAC population (Rossouw et al., 2003). This association was observed in both a case-control ($p = 0.0055$) and family-based study ($p = 0.005$). The T allele of this SNP, which was over represented in controls in the SAC, has NFκB binding activity and is associated with an increased IFN-γ expression in vitro (Pravica et al., 2000). A meta-analysis, which considered 11 association studies including that of Rossouw et al., also found evidence of a protective effect for the T allele (Pacheco et al., 2008). More recently it was shown that individuals in Japan with the AA genotype were less likely to become smear-negative after treatment (Shibasaki et al., 2009), which strengthens the case for the involvement of this polymorphism in TB susceptibility. A recent association study in the SAC population, which genotyped 54 polymorphisms from 8 IFN-γ modulating genes (IL-4, IL-10, IL-12B, IL-12 receptor beta 1, IL-12 receptor beta 2, IL-18, wingless-type MMTV integration site family member 5A, and frizzled homolog 5) in more than 800 individuals, reached a different conclusion as no significant associations were detected (Möller et al., 2010b).

NITRIC OXIDE SYNTHASE 2

Nitric oxide (NO) is an essential component in the host defense against TB and other intracellular bacteria, and an increase of this gaseous signaling molecule may protect against infectious disease (Jamieson et al., 2004). In humans, NO is produced by three different nitric oxide synthases (Levesque et al., 1999). Neuronal NOS (nNOS) is constitutively available in the central and peripheral nervous systems, where NO functions as a neurotransmitter. Endothelial NOS (eNOS) is constitutively produced by endothelium and other cell types, and is important in cardiovascular homeostasis (Miller et al., 2004). Inducible NOS (iNOS), encoded by the nitric oxide synthase 2 gene (*NOS2A*), is a cytoplasmic protein and absent in resting cells, but is rapidly produced in response to stimuli such as infections and cytokines (Burgner et al., 1999). The proximal *NOS2A* promoter is functionally important, because of its involvement in the regulation of NO production, and contains several polymorphisms that have been associated with numerous diseases including TB. Functional effects for some of the SNP alleles have been defined by in vitro data.

A 2-SNP *NOS2A* haplotype consisting of the functional promoter polymorphisms rs9282799 and rs8078340 was associated with TB in the SAC population (Möller et al., 2009). These haplotypes contained different alleles of the rs8078340 SNP (C and T), with the T allele present in the haplotype more commonly found in TB cases. Previous studies have shown that the T allele of rs8078340 reduced the quantity of DNA-protein complex bound, and reduced the binding time to this DNA region, which contains possible transcription factor binding sites (Burgner et al., 2003). *NOS2A* polymorphisms were also associated with TB in studies done in Brazil (rs2779249 and rs2301369; Jamieson et al., 2004), Colombia (CCTTT microsatellite; Gomez et al., 2007) and in African Americans (rs2274894 and rs7215373; Velez et al., 2009).

MANNOSE BINDING LECTIN

The gene encoding the collectin mannose-binding lectin (MBL), also known in the past as mannose-binding protein (MBP), is mapped to chromosome 10. MBL is a C-type lectin that plays a major role in the pulmonary innate immune system (Ji et al., 2005). Microbial carbohydrate structures are recognized by pathogen-associated molecular pattern (PAMP) receptors, such as MBL (Buzas et al., 2006). MBL binds to the surface carbohydrates of several microorganisms including *M. tuberculosis* to initiate the lectin pathway, which acts as the first line of defense against microbes and therefore plays an important role in the modulation of inflammation (Holmskov et al., 2003; Turner et al., 2003). It has been suggested that MBL is one of the molecules that have been linked to human susceptibility or protection against TB (Garred et al., 1992). Low levels of MBL have previously been associated with protection against tuberculosis in Turkish children (Cosar et al., 2008).

The three mutations (rs5030737, rs1800450 and rs1800451) in codons 52, 54, and 57 respectively in exon 1, have been associated with increased susceptibility to infectious diseases (Summerfield et al., 1997). However, inconsistent results have been found in the case of TB, where the variant alleles have been associated with protection against TB. The B allele was associated particularly with tuberculosis meningitis (TBM) in the SAC population (Hoal-van Helden et al., 1999).

DENDRITIC CELL-SPECIFIC INTRACELLULAR ADHESION MOLECULE 3-GRABBING NON-INTEGRIN

The transmembrane C-type lectin, the dendritic cell-specific intracellular adhesion molecule (ICAM) – grabbing non-integrin (DC-SIGN), also known as *CD209*, is located on chromosome 19. It has been shown that DC-SIGN allows monocyte-derived dendritic cells (DCs) to identify a large range of microbes including viruses and bacteria (Cambi et al., 2005). DC-SIGN is an important *M. tuberculosis* receptor on the surface of human monocyte-derived DCs (Neyrolles et al., 2006). DC-SIGN interactions with mycobacteria or its products can either be advantageous to the pathogen, by downmodulating the functions of DCs, or to the host by restricting tissue inflammation (Geijtenbeek and van Kooyk, 2003; Tailleux et al., 2003; Tailleux et al., 2005). Three studies, investigating promoter polymorphisms in DC-SIGN, found positive associations between DC-SIGN and infectious diseases (Barreiro et al., 2006; Martin et al., 2004; Sakuntabhai et al., 2005). Therefore, it was suggested that DC-SIGN promoter polymorphisms may be involved in determining susceptibility to infectious diseases.

CD209 is highly polymorphic, and numerous SNPs have been reported (Martin et al., 2004; Sakuntabhai et al., 2005). Two promoter polymorphisms—871A/G and 336A/G—were investigated in the SAC population with respect to TB susceptibility. The -871G / -336A haplotype was significantly more frequent among healthy controls (Barreiro et al., 2006), and it is of interest to note that this haplotype is also more frequent in the European population.

VITAMIN D RECEPTOR

The vitamin D receptor gene (*VDR*), located on chromosome 12, spans 79kb and regulates a range of metabolic pathways involved in the immune response. Vitamin D is metabolized to the active steroid hormone 1,25-dihydroxyvitamin D3, also known as calcitriol. Calcitriol is produced within granulomas by activated monocytes and macrophages, and previous in vivo studies showed that calcitriol suppresses the growth of *M. tuberculosis* (Denis, 1991; Rockett et al., 1998). Conflicting results have been found in association studies of *VDR* polymorphisms in TB (Bellamy et al., 1999; Selvaraj et al., 2004; Bornman et al., 2004). A meta-analysis of *VDR* polymorphisms indicated that most studies were underpowered and therefore inconclusive (Lewis et al., 2005).

The four most extensively studied *VDR* polymorphisms in a variety of diseases include *FokI* (rs10735810), *ApaI* (rs7975232), *TaqI* (rs731236) and *BsmI* (rs1544410). Although no association was found with TB in a case-control analysis in the SAC population (Babb et al., 2007b), the *ApaI* "AA" genotype and "T"-containing *TaqI* genotypes predicted a faster response to TB treatment. Another association study investigating TB susceptibility in southern Africa found no significant association with the *ApaI, BsmI, FokI* and *TaqI* polymorphisms in the Venda population (Lombard et al., 2006b).

BUTYROPHILIN-LIKE 2 AND CASPASE RECRUITMENT DOMAIN-CONTAINING PROTEIN 15

Granulomas are present in a number of diseases and can develop due to infectious diseases, such as TB, or because of idiopathic disorders such as sarcoidosis and Crohn's disease (CD; see Table 43-4). Given this similarity, susceptibility genes in other granulomatous diseases provide worthwhile candidate genes for testing in TB. A sarcoidosis susceptibility gene, butyrophilin-like 2 (*BTNL2*) (Valentonyte et al., 2005) was tested as a TB candidate gene in the SAC population, but no evidence of association was found (Möller et al., 2007a). This result was validated by a study using 92 TB families from Brazil (Johnson et al., 2007). Three variants in the caspase recruitment domain-containing protein 15 (*CARD15*) gene, which encodes the nucleotide-binding oligomerization domain 2 (NOD2) protein, and that were previously associated with CD (Hampe et al., 2001; Ogura et al., 2001), were tested as TB susceptibility factors in the SAC population. These CD-associated mutations occurred at extremely low frequencies in the SAC population and were not associated with TB (Möller et al., 2007b). A study in the Gambia did not detect any association between *CARD15* promoter polymorphisms and TB (Stockton et al., 2004), but it was recently reported that common nonsynonymous SNPs of the gene, not tested in the two previous studies, were associated with TB in African Americans (Austin et al., 2008). *CARD15* therefore remains a possible TB candidate gene, as the NOD2 protein is an essential recognition system of *M. tuberculosis* (Ferwerda et al., 2005).

C-C CHEMOKINE LIGAND 2

The functional rs1024611 polymorphism in the C-C chemokine ligand 2 (*CCL2*) gene was previously associated with TB in Mexico and Korea (Flores-Villanueva et al., 2005), but not in Brazil (Jamieson et al., 2004). The protein encoded by *CCL2*, namely monocyte chemoattractant protein-1, is essential for the recruitment of T lymphocytes and monocytes to the site of *M. tuberculosis* infection (Hasan et al., 2005), but a study in the SAC population did not validate the previously reported genetic association (Möller et al., 2009).

SP110 NUCLEAR BODY PROTEIN

The SP110 nuclear body protein gene (*SP110*) is the closest human homologue of the intracellular pathogen resistance 1 (*Ipr1*) gene, which determines resistance to pulmonary TB in mice by controlling the intracellular growth of the bacterium and by affecting the mechanism of macrophage death (Pan et al., 2005; Kramnik, 2008). Three

TABLE 43-4 COMPARISON OF TB, SARCOIDOSIS AND CROHN'S DISEASE.

	TB	SARCOIDOSIS	CROHN'S DISEASE
Incidence	↑ in developing countries	~ in developing and developed countries	↑ in developed countries
Granuloma	Caseating—Mostly in the lungs	Non-caseating—In two or more organs	Non-caseating—Mostly in the digestive tract
Aetiology	Infectious pathogen	Unknown	Unknown
Susceptibility factors	Genetic Environmental Bacterial	Genetic Environmental	Genetic Environmental
References	(World Health Organization, 2009; Quast & Browning, 2006)	(American Thoracic Society, 1999)	(Marsh & McLeod, 2003; Matsumoto et al., 2001)

polymorphisms in *SP110* were associated with TB in West Africa (Tosh et al., 2006), but no association was detected in a large study done in the SAC population (Babb et al., 2007a). This negative association with *SP110* was validated in studies using samples from Ghana (Thye et al., 2006) and Russia (Szeszko et al., 2006).

FUTURE STUDIES

GENE-GENE INTERACTION STUDIES

Due to the complex nature of the immune system and the polygenic nature of complex diseases, it has become increasingly evident that gene–gene interactions play a far more important part in an individual's susceptibility to a complex disease than single polymorphisms would on their own (Ritchie et al., 2001; Tsai et al., 2003; Williams et al., 2000). Gene–gene interaction, also known as *epistasis*, can be defined as the interaction between genes that are not on the same locus, and which can be very distant from each other in the genome. In a gene–gene interaction, one gene can enhance (synergistic epistasis) or hinder (antagonistic epistasis) the expression of another gene (Nagel, 2005).

Gene–gene interaction analysis is still a fairly new approach to elucidate susceptibility to complex diseases; therefore, little has been published regarding gene–gene interactions in infectious diseases such as TB. One study (De Wit et al., 2011) has been published in southern Africa regarding gene–gene interactions in TB. Since this is a multilocus and multigene approach, it may provide the paradigm for future genetic studies of TB.

GENE-TRAIT INTERACTION STUDIES

In addition to the host genotype conferring susceptibility, it has been hypothesized that the interaction between the genotype of the human host, and the bacterial strain genotype, could influence the susceptibility of the host in terms of both the progression to disease and perhaps the type of disease seen. Several *M. tuberculosis* strains exist all over the world, of which the W-Beijing family of strains has been studied most extensively (Malik & Godfrey-Faussett, 2005). The Beijing family of strains is probably the most dominant or distinctive strain worldwide, and has been reported in many countries including South Africa (Warren et al., 2004). It has been suggested that Beijing strains are hypermutable and have greater virulence compared to other strains, and they have also been associated with a higher frequency of drug resistance (van der Spuy et al., 2009).

An investigation into 875 strains from 80 different countries led to the hypothesis that ancestral *M. tuberculosis* originated and drifted together with humans "out of Africa" (Gagneux et al., 2006). Hanekom et al. (2007) showed that strains from a defined sublineage may have been selected by a human population in a defined South African geographical setting. This argument is supported by the fact that the HLA allele frequencies vary widely between human populations with different historical backgrounds, with some alleles completely absent in certain populations (Lombard et al., 2006a), and that the HLA genotype has been associated with susceptibility to *M. tuberculosis*. No studies have been published in southern Africa regarding gene–trait interactions in TB.

ADMIXTURE MAPPING

Admixture mapping is a novel approach for disease gene discovery. This technique requires a population that has developed from two or more genetically diverse parent populations, where the incidence of the disease, and therefore presumably of the underlying risk polymorphisms, is different in the original populations. The aim is then to find the regions of the genome inherited by the cases from a specific ancestral population, to identify the locus responsible for the disease. The resolution of admixture mapping is higher than that of linkage analysis, but lower than that of association studies (Chakraborty & Weiss, 1988; Stephens et al., 1994). The study design has become feasible as a genome-wide approach in admixed populations, and has been used in a variety of complex diseases; e.g., hypertension (Zhu et al., 2005), multiple sclerosis (Reich et al., 2005), prostate cancer (Freedman et al., 2006), diabetes (Elbein et al., 2009), and obesity (Cheng et al., 2009) in African Americans, as well as for asthma in a population from the Caribbean coast of Colombia (Vergara et al., 2009). Preparation for studies in admixed populations from Mexico (Bonilla et al., 2005; Martinez-Marignac et al., 2007), Hispanic/Latino populations (Mao et al., 2007; Price et al., 2007), and Uyghur populations from China (Xu & Jin, 2008) are also underway, but none of these studies have involved TB. This approach could yield novel susceptibility genes for TB. An admixture mapping study with respect to TB is underway for an admixed population in South Africa, by our group.

KEY POINTS

- Tuberculosis (TB) is a common infectious disease in the developing world.
- Sub-Saharan Africa accounts for the majority of TB cases in Africa.
- Only 10% of immunocompetent individuals infected with *Mycobacterium tuberculosis* will develop active

TB, an observed difference determined by host factors as well as the environment.

- The majority of published TB genetic studies in southern Africa have been done in two populations from South Africa, namely the South African Coloured (SAC) population and the Venda.
- Numerous genetic study designs, such as heritability analyses, linkage studies, and candidate gene association studies, have been used to investigate genetic susceptibility to TB in southern Africa.

CONCLUSIONS

The origins of modern humans lie in sub-Saharan Africa, a region not only characterized by linguistic and cultural diversity, but also by extreme genetic variation (Hardy et al., 2008; Sirugo et al., 2008). TB is a relatively new infectious disease in Africa (Dubos & Dubos, 1952) and, together with the HIV pandemic, has a devastating effect on the health of Africans (World Health Organization, 2009). As proven by heritability analyses, TB carries a significant genetic risk, and studies of the genetic factors involved in susceptibility could, in the future, help to improve health in southern Africa. However, as is evident from the studies described here, very few southern African populations have been investigated in TB genetic research, even though this region is heavily burdened by the disease. These studies have nevertheless made a significant contribution to the field of TB host genetic susceptibility, yielding new and frequently unexpected conclusions.

ACKNOWLEDGMENTS

The authors thank Paul van Helden for critical reading of the manuscript and helpful suggestions.

REFERENCES

Al-Muhsen S, Casanova JL. (2008). The genetic heterogeneity of mendelian susceptibility to mycobacterial diseases. *J Allergy Clin Immunol 122*, 1043–1051.

Alcaïs A, Abel L. (2004). Application of genetic epidemiology to dissecting host susceptibility/resistance to infection illustrated with the study of common mycobacterial infections. *In* Richard Bellamy (ed.), *Susceptibility to infectious diseases – The importance of host genetics* (pp. 7–43). Cambridge: Cambridge University Press.

American Thoracic Society (1999). Statement on sarcoidosis. Joint Statement of the American Thoracic Society (ATS), the European Respiratory Society (ERS) and the World Association of Sarcoidosis and Other Granulomatous Disorders (WASOG) adopted by the ATS Board of Directors and by the ERS Executive Committee, February 1999. *Am J Respir Crit Care Med 160*, 736–755.

Amirzargar AA, Yalda A, Hajabolbaghi M, et al. (2004). The association of HLA-DRB, DQA1, DQB1 alleles and haplotype frequency in Iranian patients with pulmonary tuberculosis. *Int J Tuberc Lung Dis 8*, 1017–1021.

Ardlie KG, Kruglyak L, Seielstad M. (2002). Patterns of linkage disequilibrium in the human genome. *Nat Rev Genet 3*, 299–309.

Austin CM, Ma X, Graviss EA. (2008). Common nonsynonymous polymorphisms in the *NOD2* gene are associated with resistance or susceptibility to tuberculosis disease in African Americans. *J Infect Dis 197*, 1713–1716.

Awomoyi AA, Marchant A, Howson JM, McAdam KP, Blackwell JM, Newport MJ. (2002). Interleukin-10, polymorphism in *SLC11A1* (formerly *NRAMP1*), and susceptibility to tuberculosis. *J Infect Dis 186*, 1808–1814.

Babb C, Keet EH, van Helden PD, Hoal EG. (2007a). *SP110* polymorphisms are not associated with pulmonary tuberculosis in a South African population. *Hum Genet 121*, 521–522.

Babb C, van der Merwe L, Beyers N, et al. (2007b). Vitamin D receptor gene polymorphisms and sputum conversion time in pulmonary tuberculosis patients. *Tuberculosis 87*, 295–302.

Baghdadi JE, Orlova M, Alter A, et al. (2006). An autosomal dominant major gene confers predisposition to pulmonary tuberculosis in adults. *J Exp Med 203*, 1679–1684.

Barreiro LB, Neyrolles O, Babb CL, et al. (2006). Promoter Variation in the DC-SIGN Encoding Gene CD209 Is Associated with Tuberculosis. *PLoS Medicine 3*, e20.

Bellamy R. (2006). Genome-wide approaches to identifying genetic factors in host susceptibility to tuberculosis. *Microbes Infect 8*, 1119–1123.

Bellamy R, Beyers N, McAdam KP, et al. (2000). Genetic susceptibility to tuberculosis in Africans: a genome-wide scan. *Proc Natl Acad Sci U S A 97*, 8005–8009.

Bellamy R, Ruwende C, Corrah T, et al. (1999). Tuberculosis and chronic hepatitis B virus infection in Africans and variation in the vitamin D receptor gene. *J Infect Dis 179*, 721–724.

Bellamy R, Ruwende C, Corrah T, McAdam KP, Whittle HC, Hill AV. (1998). Variations in the *NRAMP1* gene and susceptibility to tuberculosis in West Africans. *N Engl J Med 338*, 640–644.

Bloom BR, Small PM. (1998). The evolving relation between humans and *Mycobacterium tuberculosis*. *N Engl J Med 338*, 677–678.

Bonilla C, Gutierrez G, Parra EJ, Kline C, Shriver MD. (2005). Admixture analysis of a rural population of the state of Guerrero, Mexico. *Am J Phys Anthropol 128*, 861–869.

Bornman L, Campbell SJ, Fielding K, et al. (2004). Vitamin D receptor polymorphisms and susceptibility to tuberculosis in West Africa: a case-control and family study. *J Infect Dis 190*, 1631–1641.

Bothamley GH, Beck JS, Schreuder GM, et al. (1989). Association of tuberculosis and *M tuberculosis*-specific antibody levels with HLA. *J Infect Dis 159*, 549–555.

Burgner D, Rockett K, Kwiatkowski D. (1999). Nitric oxide and infectious diseases. *Archives of Disease in Childhood 81*, 185–188.

Burgner D, Usen S, Rockett K, et al. (2003). Nucleotide and haplotypic diversity of the *NOS2A* promoter region and its relationship to cerebral malaria. *Human Genetics 112*, 379–386.

Buzas EI, Gyorgy B, Pasztoi M, Jelinek I, Falus A, Gabius HJ. (2006). Carbohydrate recognition systems in autoimmunity. *Autoimmunity 39*, 691–704.

Cambi A, Koopman M, Figdor CG. (2005). How C-type lectins detect pathogens. *Cell Microbiol 7*, 481–488.

Cao K, Moormann AM, Lyke KE, et al. (2004). Differentiation between African populations is evidenced by the diversity of alleles and haplotypes of HLA class I loci. *Tissue Antigens 63*, 293–325.

Cellier M, Govoni G, Vidal S, et al. (1994). Human natural resistance-associated macrophage protein: cDNA cloning, chromosomal mapping, genomic organization, and tissue-specific expression. *J Exp Med 180*, 1741–1752.

Centers for Disease Control and Prevention (CDC) (2008). Trends in tuberculosis-United States, 2007. *MMWR Morb Mortal Wkly Rep 57*, 281–285.

Cervino AC, Lakiss S, Sow O, et al. (2002). Fine mapping of a putative tuberculosis-susceptibility locus on chromosome 15q11-13 in African families. *Hum Mol Genet 11*, 1599–1603.

Chakraborty R, Weiss KM. (1988). Admixture as a tool for finding linked genes and detecting that difference from allelic association between loci. *Proc Natl Acad Sci U S A 85*, 9119–9123.

Chandanayingyong D, Maranetra N, Bovornkitti S. (1988). HLA antigen profiles in Thai tuberculosis patients. *Asian Pac J Allergy Immunol 6*, 77–80.

Cheng CY, Kao WH, Patterson N, et al. (2009). Admixture mapping of 15,280 African Americans identifies obesity susceptibility loci on chromosomes 5 and X. *PLoS Genet 5*, e1000490.

Cobat A, Gallant CJ, Simkin L, et al. (2009) Two loci control tuberculin skin test reactivity in an area hyperendemic for tuberculosis. *J Exp Med 206*, 2583–2591.

Cobat A, Gallant CJ, Simkin L, et al. (2010). High heritability of antimycobacterial immunity in a hyper-endemic area for tuberculosis disease. *J Infect Dis 201*, 15–19.

Comstock GW. (1978). Tuberculosis in twins: a re-analysis of the Prophit survey. *Am Rev Respir Dis 117*, 621–624.

Cooke GS, Campbell SJ, Bennett S, et al. (2008). Mapping of a novel susceptibility locus suggests a role for *MC3R* and *CTSZ* in human tuberculosis. *Am J Respir Crit Care Med 178*, 203–207.

Cosar H, Ozkinay F, Onay H, et al. (2008). Low levels of mannose-binding lectin confers protection against tuberculosis in Turkish children. *Eur J Clin Microbiol Infect Dis 27*, 1165–1169.

Daniel TM. (1997). *Captain of Death: the story of tuberculosis.* University of Rochester Press, New York.

De Wit E, van der Merwe L, van Helden PD, Hoal EG. (2011). Gene-gene interaction between tuberculosis candidate genes in a South African population. *Mamm Genome* 22, 100–110.

Denis M. (1991). Killing of *Mycobacterium tuberculosis* within human monocytes: activation by cytokines and calcitriol. *Clin Exp Immunol 84*, 200–206.

Dubos R, Dubos J. (1952). *The White Plague: Tuberculosis, Man and Society.* Boston: Little, Brown & Co.

Elbein SC, Das SK, Hallman DM, Hanis CL, Hasstedt SJ. (2009). Genome-wide linkage and admixture mapping of type 2 diabetes in African American families from the American Diabetes Association GENNID (Genetics of NIDDM) Study Cohort. *Diabetes 58*, 268–274.

Ferwerda G, Girardin SE, Kullberg BJ, et al. (2005). NOD2 and toll-like receptors are nonredundant recognition systems of *Mycobacterium tuberculosis. PLoS Pathog 1*, 279–285.

Fitness J, Floyd S, Warndorff DK, et al. (2004). Large-scale candidate gene study of tuberculosis susceptibility in the Karonga district of northern Malawi. *Am J Trop Med Hyg 71*, 341–349.

Flores-Villanueva PO, Ruiz-Morales JA, Song CH, et al. (2005). A functional promoter polymorphism in monocyte chemoattractant protein-1 is associated with increased susceptibility to pulmonary tuberculosis. *J Exp Med 202*, 1649–1658.

Freedman ML, Haiman CA, Patterson N, et al. (2006). Admixture mapping identifies 8q24 as a prostate cancer risk locus in African-American men. *Proc Natl Acad Sci U S A 103*, 14068–14073.

Gagneux S, Deriemer K, Van T, et al. (2006). Variable host-pathogen compatibility in *Mycobacterium tuberculosis. Proc Natl Acad Sci U S A 103*, 2869–2873.

Gao PS, Fujishima S, Mao XQ, et al. (2000). Genetic variants of *NRAMP1* and active tuberculosis in Japanese populations. International Tuberculosis Genetics Team. *Clin Genet 58*, 74–76.

Garred P, Madsen HO, Kurtzhals JA, et al. (1992). Diallelic polymorphism may explain variations of the blood concentration of mannan-binding protein in Eskimos, but not in black Africans. *Eur J Immunogenet 19*, 403–412.

Geijtenbeek TB, van Kooyk. (2003). Pathogens target DC-SIGN to influence their fate DC-SIGN functions as a pathogen receptor with broad specificity. *APMIS 111*, 698–714.

Gomez LM, Anaya JM, Vilchez JR, et al. (2007). A polymorphism in the inducible nitric oxide synthase gene is associated with tuberculosis. *Tuberculosis (Edinb) 87*, 288–294.

Gruenheid S, Pinner E, Desjardins M, Gros P. (1997). Natural resistance to infection with intracellular pathogens: the Nramp1 protein is recruited to the membrane of the phagosome. *J Exp Med 185*, 717–730.

Hampe J, Cuthbert A, Croucher PJ, et al. (2001). Association between insertion mutation in *NOD2* gene and Crohn's disease in German and British populations. *Lancet 357*, 1925–1928.

Hanekom M, van der Spuy GD, Streicher E, et al. (2007). A recently evolved sublineage of the *Mycobacterium tuberculosis* Beijing strain family is associated with an increased ability to spread and cause disease. *J Clin Microbiol 45*, 1483–1490.

Hardy BJ, Seguin B, Ramesar R, Singer PA, Daar AS. (2008). South Africa: from species cradle to genomic applications. *Nat Rev Genet 9*(Suppl 1), S19–S23.

Hasan Z, Zaidi I, Jamil B, Khan MA, Kanji A, Hussain R. (2005). Elevated ex vivo monocyte chemotactic protein-1 (CCL2) in pulmonary as compared with extra-pulmonary tuberculosis. *BMC Immunol 6*, 14.

Hoal EG, Lewis L-A, Jamieson SE, et al. (2004). *SLC11A1 (NRAMP1)* but not *SLC11A2 (NRAMP2)* polymorphisms are associated with susceptibility to tuberculosis in a high-incidence community in South Africa. *Int J Tuberc Lung Dis 8*, 1464–1471.

Hoal-van Helden EG, Epstein J, et al. (1999). Mannose-binding protein B allele confers protection against tuberculous meningitis. *Pediatr Res 45*, 459–464.

Holmskov U, Thiel S, Jensenius JC. (2003). Collections and ficolins: humoral lectins of the innate immune defense. *Annu Rev Immunol 21*, 547–578.

Jamieson SE, Miller EN, Black GF, et al. (2004). Evidence for a cluster of genes on chromosome 17q11-q21 controlling susceptibility to tuberculosis and leprosy in Brazilians. *Genes Immun 5*, 46–57.

Jepson A, Fowler A, Banya W, et al. (2001). Genetic regulation of acquired immune responses to antigens of *Mycobacterium tuberculosis*: a study of twins in West Africa. *Infect Immun 69*, 3989–3994.

Ji X, Gewurz H, Spear GT. (2005). Mannose binding lectin (MBL) and HIV. *Mol Immunol 42*, 145–152.

Jiang YH, Beaudet AL. (2004). Human disorders of ubiquitination and proteasomal degradation. *Curr Opin Pediatr 16*, 419–426.

Johnson CM, Traherne JA, Jamieson SE, et al. (2007). Analysis of the *BTNL2* truncating splice site mutation in tuberculosis, leprosy and Crohn's disease. *Tissue Antigens 69*, 236–241.

Kallmann FJ, Reisner D. (1942). Twin studies on the significance of genetic factors in tuberculosis. *Am Rev Tuberc 47*, 549–547.

Kaufmann SH, McMichael AJ. (2005). Annulling a dangerous liaison: vaccination strategies against AIDS and tuberculosis. *Nat Med 11*, S33–S44.

Kim HS, Park MH, Song EY, et al. (2005). Association of HLA-DR and HLA-DQ genes with susceptibility to pulmonary tuberculosis in Koreans: Preliminary evidence of associations with drug resistance, disease severity, and disease recurrence. *Hum Immunol 66*, 1074–1081.

Kimman TG, Janssen R, Hoebee B. (2006). Future prospects in respiratory syncytial virus genetics. *Future Virology 1*, 483–492.

Kramnik I. (2008). Genetic dissection of host resistance to *Mycobacterium tuberculosis*: the sst1 locus and the Ipr1 gene. *Curr Top Microbiol Immunol 321*, 123–148.

Kruger National Park. (2009). Venda | African Tribe. http://www.krugerpark.co.za/africa_venda.html.

Lander ES, Schork NJ. (1994). Genetic dissection of complex traits. *Science 265*, 2037–2048.

Leal IS, Smedegard B, Andersen P, Appelberg R. (2001). Failure to induce enhanced protection against tuberculosis by increasing T-cell-dependent interferon-gamma generation. *Immunology 104*, 157–161.

Lee YS, Poh LK, Loke KY. (2002). A novel melanocortin 3 receptor gene (*MC3R*) mutation associated with severe obesity. *J Clin Endocrinol Metab 87*, 1423–1426.

Levesque MC, Hobbs MR, Anstey NM, et al. (1999). Nitric oxide synthase type 2 promoter polymorphisms, nitric oxide production, and disease severity in Tanzanian children with malaria. *J Infect Dis 180*, 1994–2002.

Lewis SJ, Baker I, Davey SG. (2005). Meta-analysis of vitamin D receptor polymorphisms and pulmonary tuberculosis risk. *Int J Tuberc Lung Dis 9*, 1174–1177.

Lombard Z, Brune AE, Hoal EG, et al. (2006a). HLA class II disease associations in southern Africa. *Tissue Antigens 67*, 97–110.

Lombard Z, Dalton DL, Venter PA, Williams RC, Bornman L. (2006b). Association of HLA-DR, -DQ, and vitamin D receptor alleles and haplotypes with tuberculosis in the Venda of South Africa. *Hum Immunol 67*, 643–654.

Ma X, Dou S, Wright JA, et al. (2002). 5'-dinucleotide repeat polymorphism of *NRAMP1* and susceptibility to tuberculosis among Caucasian patients in Houston, Texas. *Int J Tuberc Lung Dis 6*, 818–823.

Maartens G, Wilkinson RJ. (2007). Tuberculosis. *Lancet 370*, 2030–2043.

Mahasirimongkol S, Yanai H, Nishida N, et al. (2009). Genome-wide SNP-based linkage analysis of tuberculosis in Thais. *Genes Immun 10*, 77–83.

Malik AN, Godfrey-Faussett P. (2005). Effects of genetic variability of *Mycobacterium tuberculosis* strains on the presentation of disease. *Lancet Infect Dis 5*, 174–183.

Malo D, Vogan K, Vidal S, et al. (1994). Haplotype mapping and sequence analysis of the mouse *Nramp* gene predict susceptibility to infection with intracellular parasites. *Genomics 23*, 51–61.

Mao X, Bigham AW, Mei R, et al. (2007). A genomewide admixture mapping panel for Hispanic/Latino populations. *Am J Hum Genet 80*, 1171–1178.

Marsh S, McLeod HL. (2003). Crohn's disease: ethnic variation in *CARD15* genotypes. *Gut 52*, 770.

Martin MP, Lederman MM, Hutcheson HB, et al. (2004). Association of *DC-SIGN* promoter polymorphism with increased risk for parenteral, but not mucosal, acquisition of human immunodeficiency virus type 1 infection. *J Virol 78*, 14053–14056.

Martinez-Marignac VL, Valladares A, et al. (2007). Admixture in Mexico City: implications for admixture mapping of type 2 diabetes genetic risk factors. *Hum Genet 120*, 807–819.

Matsumoto T, Nakamura S, Jin-No Y, et al. (2001). Role of granuloma in the immunopathogenesis of Crohn's disease. *Digestion 63*(Suppl 1), 43–47.

Miller EN, Jamieson SE, Joberty C, et al. (2004). Genome-wide scans for leprosy and tuberculosis susceptibility genes in Brazilians. *Genes Immun 5*, 63–67.

Möller M, Kwiatkowski R, Nebel A, van Helden PD, Hoal EG, Schreiber S. (2007a). Allelic variation in *BTNL2* and susceptibility to tuberculosis in a South African population. *Microbes Infect 9*, 522–528.

Möller M, Nebel A, Kwiatkowski R, van Helden PD, Hoal EG, Schreiber S. (2007b). Host susceptibility to tuberculosis: *CARD15* polymorphisms in a South African population. *Mol Cell Probes 21*, 148–151.

Möller M, Nebel A, Valentonyte R, van Helden PD, Schreiber S, Hoal EG. (2009). Investigation of chromosome 17 candidate genes in susceptibility to TB in a South African population. *Tuberculosis (Edinb) 89*, 189–194.

Möller M, de Wit E, Hoal EG. (2010a). Past, present and future directions in human genetic susceptibility to tuberculosis. *FEMS Immunol Med Microbiol 58*, 3–26.

Möller M, Nebel A, Valentonyte R, van Helden PD, Schreiber S, Hoal EG. (2010b). Analysis of eight genes modulating interferon gamma and human genetic susceptibility to tuberculosis: a case-control association study. *BMC Infect Dis 10*, 154.

Nagel R.L. (2005). Epistasis and the genetics of human diseases. *C R Biol 328*, 606–615.

Nagler DK, Kruger S, Kellner A, et al. (2004). Up-regulation of cathepsin X in prostate cancer and prostatic intraepithelial neoplasia. *Prostate 60*, 109–119.

Newport MJ, Goetghebuer T, Weiss HA, Whittle H, Siegrist CA, Marchant A. (2004). Genetic regulation of immune responses to vaccines in early life. *Genes Immun 5*, 122–129.

Neyrolles O, Gicquel B, Quintana-Murci L. (2006). Towards a crucial role for DC-SIGN in tuberculosis and beyond. *Trends Microbiol 14*, 383–387.

North RJ, LaCourse R, Ryan L, Gros P. (1999). Consequence of Nramp1 deletion to *Mycobacterium tuberculosis* infection in mice. *Infect Immun 67*, 5811–5814.

Ogura Y, Bonen DK, Inohara N, et al. (2001). A frameshift mutation in *NOD2* associated with susceptibility to Crohn's disease. *Nature 411*, 603–606.

Pacheco AG, Cardoso CC, Moraes MO. (2008). IFNG +874T/A, IL10 −1082G/A and TNF −308G/A polymorphisms in association with tuberculosis susceptibility: a meta-analysis study. *Hum Genet 123*, 477–484.

Pan H, Yan BS, Rojas M, et al. (2005). *Ipr1* gene mediates innate immunity to tuberculosis. *Nature 434*, 767–772.

Pravica V, Perrey C, Stevens A, Lee JH, Hutchinson IV. (2000). A single nucleotide polymorphism in the first intron of the human IFN-gamma gene: absolute correlation with a polymorphic CA microsatellite marker of high IFN-gamma production. *Hum Immunol 61*, 863–866.

Price AL, Patterson N, Yu F, et al. (2007). A genomewide admixture map for Latino populations. *Am J Hum Genet 80*, 1024–1036.

Quast TM, Browning RF. (2006). Pathogenesis and clinical manifestations of pulmonary tuberculosis. *Dis Mon 52*, 413–419.

Reich D, Patterson N, De Jager PL, et al. (2005). A whole-genome admixture scan finds a candidate locus for multiple sclerosis susceptibility. *Nat Genet 37*, 1113–1118.

Risch N, Merikangas K. (1996). The future of genetic studies of complex human diseases. *Science 273*, 1516–1517.

Ritchie MD, Hahn LW, Roodi N, et al. (2001). Multifactor-dimensionality reduction reveals high-order interactions among estrogen-metabolism genes in sporadic breast cancer. *Am J Hum Genet 69*, 138–147.

Rockett KA, Brookes R, Udalova I, Vidal V, Hill AV, Kwiatkowski D. (1998). 1,25-Dihydroxyvitamin D_3 induces nitric oxide synthase and suppresses growth of *Mycobacterium tuberculosis* in a human macrophage-like cell line. *Infect Immun 66*, 5314–5321.

Rossouw M, Nel HJ, Cooke GS, van Helden PD, Hoal EG. (2003). Association between tuberculosis and a polymorphic NFkappaB binding site in the interferon gamma gene. *Lancet 361*, 1871–1872.

Sakuntabhai A, Turbpaiboon C, Casademont I, et al. (2005). A variant in the *CD209* promoter is associated with severity of dengue disease. *Nat Genet 37*, 507–513.

Selvaraj P, Kurian SM, Chandra G, Reetha AM, Charles N, Narayanan PR. (2004). Vitamin D receptor gene variants of *BsmI, ApaI, TaqI, and FokI* polymorphisms in spinal tuberculosis. *Clin Genet 65*, 73–76.

Serpa JA, Teeter LD, Musser JM, Graviss EA. (2009). Tuberculosis disparity between US-born blacks and whites, Houston, Texas, USA. *Emerg Infect Dis 15*, 899–904.

Shibasaki M, Yagi T, Yatsuya H, et al. (2009). An influence of Interferon-gamma gene polymorphisms on treatment response to tuberculosis in Japanese population. *J Infect 58*, 467–469.

Singh SP, Mehra NK, Dingley HB, Pande JN, Vaidya MC. (1983). Human leukocyte antigen (HLA)-linked control of susceptibility to pulmonary tuberculosis and association with HLA-DR types. *J Infect Dis 148*, 676–681.

Sirugo G, Hennig BJ, Adeyemo AA, et al. (2008). Genetic studies of African populations: an overview on disease susceptibility and response to vaccines and therapeutics. *Hum Genet 123*, 557–598.

Stead WW, Senner JW, Reddick WT, Lofgren JP. (1990). Racial differences in susceptibility to infection by *Mycobacterium tuberculosis*. *N Engl J Med 322*, 422–427.

Stein CM, Nshuti L, Chiunda AB, et al. (2005). Evidence for a major gene influence on tumor necrosis factor-alpha expression in tuberculosis: path and segregation analysis. *Hum Hered 60*, 109–118.

Stein CM, Zalwango S, Malone LL, et al. (2008). Genome scan of *M tuberculosis* infection and disease in Ugandans. *PLoS ONE 3*, e4094.

Stephens JC, Briscoe D, O'Brien SJ. (1994). Mapping by admixture linkage disequilibrium in human populations: limits and guidelines. *Am J Hum Genet 55*, 809–824.

Stockton JC, Howson JM, Awomoyi AA, McAdam KP, Blackwell JM, Newport MJ. (2004). Polymorphism in *NOD2*, Crohn's disease, and susceptibility to pulmonary tuberculosis. *FEMS Immunol Med Microbiol 41*, 157–160.

Summerfield JA, Sumiya M, Levin M, Turner MW. (1997). Association of mutations in mannose binding protein gene with childhood infection in consecutive hospital series. *BMJ 314*, 1229–1232.

Szeszko JS, Healy B, Stevens H, et al. (2006). Resequencing and association analysis of the *SP110* gene in adult pulmonary tuberculosis. *Hum Genet 121*, 155–160.

Tailleux L, Pham-Thi N, Bergeron-Lafaurie A, et al. (2005). DC-SIGN induction in alveolar macrophages defines privileged target host cells for mycobacteria in patients with tuberculosis. *PLoS Med 2*, e381.

Tailleux L, Schwartz O, Herrmann JL, et al. (2003). DC-SIGN is the major *Mycobacterium tuberculosis* receptor on human dendritic cells. *J Exp Med 197*, 121–127.

Teare MD, Barrett JH. (2005). Genetic linkage studies. *Lancet 366*, 1036–1044.

Thye T, Browne EN, Chinbuah MA, et al. (2006). No associations of human pulmonary tuberculosis with *Sp110* variants. *J Med Genet 43*, e32.

Tishkoff SA, Williams SM. (2002). Genetic analysis of African populations: human evolution and complex disease. *Nat Rev Genet 3*, 611–621.

Tishkoff SA, Reed FA, Friedlaender FR, et al. (2009). The genetic structure and history of Africans and African Americans. *Science 324*, 1035–1044.

Tosh K, Campbell SJ, Fielding K, et al. (2006). Variants in the *SP110* gene are associated with genetic susceptibility to tuberculosis in West Africa. *Proc Natl Acad Sci U S A 103*, 10364–10368.

Tsai CT, Fallin D, Chiang FT, et al. (2003). Angiotensinogen gene haplotype and hypertension: interaction with *ACE* gene I allele. *Hypertension 41*, 9–15.

Turner MW, Johnson M, Booth C, Klein N, Rolland J, Davies J. (2003). Assays for human mannose-binding lectin. *J Immunol Methods 276*, 147–149.

United Nations Statistics Division. (2009.) Standard country and area codes classifications. http://milleniumindicators.un.org/unsd/methods/m49/m49regin.htm.

Valentonyte R, Hampe J, Huse K, et al. (2005). Sarcoidosis is associated with a truncating splice site mutation in *BTNL2*. *Nat Genet 37*, 357–364.

Van Buggenhout G, Fryns JP. (2009). Angelman syndrome (AS, MIM 105830). *Eur J Hum Genet* doi:10.1038/ejhg.2009.67.

van der Spuy GD, Kremer K, Ndabambi SL, et al. (2009). Changing *Mycobacterium tuberculosis* population highlights clade-specific pathogenic characteristics. *Tuberculosis (Edinburgh, Scotland) 89*, 120–125.

Vejbaesya S, Chierakul N, Luangtrakool K, Srinak D, Stephens HA. (2002). Associations of HLA class II alleles with pulmonary tuberculosis in Thais. *Eur J Immunogenet 29*, 431–434.

Velez DR, Hulme WF, Myers JL, et al. (2009). *NOS2A*, *TLR4*, and *IFNGR1* interactions influence pulmonary tuberculosis susceptibility in African-Americans. *Hum Genet 126*, 643–653.

Vergara C, Caraballo L, Mercado D, et al. (2009). African ancestry is associated with risk of asthma and high total serum IgE in a population from the Caribbean Coast of Colombia. *Hum Genet 125*, 565–579.

Vidal S, Tremblay ML, Govoni G, et al. (1995). The *Ity/Lsh/Bcg* locus: natural resistance to infection with intracellular parasites is abrogated by disruption of the *Nramp1* gene. *J Exp Med 182*, 655–666.

Vidal SM, Malo D, Vogan K, Skamene E, Gros P. (1993). Natural resistance to infection with intracellular parasites: isolation of a candidate for Bcg. *Cell 73*, 469–485.

Wang SX, Fan ZC, Tao YX. (2008). Functions of acidic transmembrane residues in human melanocortin-3 receptor binding and activation. *Biochem Pharmacol 76*, 520–530.

Warren RM, Victor TC, Streicher EM, et al. (2004). Patients with active tuberculosis often have different strains in the same sputum specimen. *Am J Respir Crit Care Med 169*, 610–614.

Williams SM, Addy JH, Phillips JA, et al. (2000). Combinations of variations in multiple genes are associated with hypertension. *Hypertension 36*, 2–6.

World Health Organization. WHO report 2009: Global Tuberculosis Control 2009 – epidemiology, strategy, financing. WHO/HTM/TB/2009.411. 2009. Geneva.

Xu S, Jin L. (2008). A genome-wide analysis of admixture in Uyghurs and a high-density admixture map for disease-gene discovery. *Am J Hum Genet 83*, 322–336.

Zhu X, Luke A, Cooper RS, et al. (2005). Admixture mapping for hypertension loci with genome-scan markers. *Nat Genet 37*, 177–181.

Zijenah LS, Hartogensis WE, Katzenstein DA, et al. (2002). Association of high HIV-1 RNA levels and homozygosity at HLA class II DRB1 in adults coinfected with *Mycobacterium tuberculosis* in Harare, Zimbabwe. *Hum Immunol 63*, 1026–1032.

44.

FAMILIAL CANCER SYNDROMES IN AFRICA

Yonglan Zheng, Toshio F. Yoshimatsu, and Olufunmilayo I. Olopade

INTRODUCTION

CANCER AND AFRICA

Data of the World Health Organization (WHO) indicates that cancer accounted for approximately 13% of all deaths (7.4 million) in 2004. Lung, stomach, colorectal, liver, and breast cancer are the top five leading causes of cancer death (WHO, http://www.who.int). The majority of all cancer deaths (>70%) occur in low- and middle-income countries, and worldwide cancer death rates are expected to continue to rise.

Cancer has diverse and complex causes, both environmentally and genetically. Mutations (changes in genetic material, DNA) caused by carcinogens lead to cellular abnormalities and may cause errors in DNA replication. Subsequent accumulation of such genetic alterations may lead to the development and progression of cancer. Unlike sporadic cancers, which represent 90%–95% of all cancers, familial cancers are those that seem to be caused by inherited mutations, suggesting that genetic factors play a significant role in cancer etiology. Based on findings from the Human Genome Project, and novel sequencing technologies that have high throughput and are relatively inexpensive, many cancer susceptibility genes have been and will continue to be discovered. Inherited predispositions deemed to be responsible for familial/hereditary cancer syndromes are indicated in Table 44-1 (Kufe, 2003; Hecht, 2007; Forbes, Bhamra et al., 2008; Lindor, McMaster et al., 2008). There is a paucity of data on cancer predisposition syndromes from Africa. This review represents a summary of publications written in English currently available in the public electronic database, PubMed.

The African continent is the second largest in the world and second most populous after Asia. It is home to over 800 million people in more than fifty countries, with extensive cultural and genetic diversity. The breadth of diversity is partially represented with over 2000 ethnolinguistic groups that exist in Africa (http://www.ethnologue.com). Some studies have suggested that the highest levels of human genetic diversity in Africa, both within and between populations, may be best explained by the multiregional origin model for modern humans within Africa, followed by massive migration and admixture (McBrearty & Brooks, 2000; Reed & Tishkoff, 2006; Yotova, Lefebvre et al., 2007) that led to the formation of morphological traits, and the "out of Africa" model of modern human origins. In fact, migration of Niger-Congo Bantu language–speaking agricultural tribes from southern Cameroon to subequatorial Africa began about 4000 years ago (Holden, 2002; Lane, Soodyall et al., 2002; Destro-Bisol, Coia et al., 2004; Wood, Stover et al., 2005). The subsequent admixture with indigenous hunter-gatherer populations played a fundamental role in the spread of farming, as well as the weaving of genetic diversity that constitutes the modern black African population (Tishkoff, Gonder et al., 2007; Pilkington, Wilder et al., 2008; Quintana-Murci, Quach et al., 2008). Furthermore, full genome sequencing of different indigenous hunter-gatherers from Southern Africa has shown the genetic manifold between neighboring tribes to be more diverse than that between Europeans and Asians, demonstrating an added complexity of African population heterogeneity (Schuster, Miller et al., 2010). Such diversity provides an opportunity to observe rare elements that may play a critical role in phenotypic expression, but otherwise would exist in obscurity if studied in less diverse populations. In addition, high consanguineous cohorts in Northern Africa, for example in Tunisia, could benefit from mutation discovery of recessive disease genes. Thus, studying African populations at a molecular level does not only reveal information about the origin and evolution of humans, but also lends itself to the discovery of key factors of human health that nature has inlaid in genetic content throughout history. Future efforts in cancer genetics should aim at characterizing the genetic basis of inherited cancer predisposition in African cancer populations, as these will likely yield important results concerning the role of gene–environment interactions in cancer progression, and penetrance of high-risk alleles.

TABLE 44-1 INHERITED PREDISPOSITIONS RESPONSIBLE FOR FAMILIAL/HEREDITARY CANCER SYNDROMES. OMIM, ONLINE MENDELIAN INHERITANCE IN MAN (HTTP://WWW.NCBI.NLM.NIH.GOV/OMIM/).

FAMILIAL CANCER SYNDROME - SUBTYPE	OMIM #	CAUSATIVE GENE SYMBOL	CHROMOSOMAL POSITION	MODE OF INHERITANCE
Aicardi Syndrome	304050	*AIC*	Xp22	X-linked dominant
Alpha-1-Antitrypsin Deficiency	107400	*SERPINA1*	14q32.13	autosomal recessive
Ataxia Telangiectasia	208900	*ATM*	11q22.3	autosomal recessive
Barrett Esophagus/Barrett's Syndrome	109350	*GER*	13q14	autosomal dominant
Beckwith-Wiedemann Syndrome/Wiedemann-Beckwith Syndrome	130650	-	11p15.5	autosomal dominant (10%–15%)
Birt-Hogg-Dubé Syndrome	135150	*FLCN*	17p11.2	autosomal dominant
Bloom Syndrome/Bloom-Torre-Machacek Syndrome	210900	*BLM*	15q26.1	autosomal recessive
Carney Complex/LAMB Syndrome/NAME Syndrome - CNC1	160980	*PRKAR1A*	17q24.2	autosomal dominant
Carney Complex/LAMB Syndrome/NAME Syndrome - CNC2	605244	*CNC2*	2p16	
Congenital Central Hypoventilation Syndrome/Ondine's Curse/Undine's Curse	209880	*PHOX2B*	4p13	autosomal dominant
		RET	10q11.21	autosomal dominant
		GDNF	5p13.2	
		EDN3	20q13.32	
		BDNF	11p14.1	
		ASCL1	12q23.2	
Costello Syndrome/Faciocutaneoskeletal Syndrome	218040	*HRAS*	11p15.5	autosomal dominant
Cowden Syndrome/Multiple Hamartoma Syndrome	158350	*PTEN*	10q23.31	autosomal dominant
Dyskeratosis Congenita	305000	*DKC1*	Xq28	X-linked
	127550	*TERC*	3q26.2	autosomal dominant
		TINF2	14q12	autosomal dominant
		TERT	5p15.33	autosomal recessive
	224230	*NHP2 (NOLA2)*	5q35.3	autosomal recessive
		NOP10 (NOLA3)	15q14	autosomal recessive
Familial Adenomatous Polyposis	175100	*APC*	5q22.2	autosomal dominant
Familial Chordoma	215400	*CHDM*	7q33	autosomal dominant
Familial Cylindromatosis	132700	*CYLD*	16q12.1	autosomal dominant
Familial Essential Thrombocythemia	187950	*MPL*	1p34.2	autosomal dominant
Familial Hepatic Adenoma	142330	*HNF1A (TCF1)*	12q24.31	autosomal recessive
Familial Hyperparathyroidism – Hyperparathyroidism - Jaw Tumor Syndrome	145001	*CDC73 (HRPT2)*	1q31.2	autosomal dominant
Familial Hyperparathyroidism - Familial Isolated Hyperparathyroidism	145000	*MEN1*	11q13.1	
Familial Intestinal Carcinoid	114900	*SDHD*	11q23.1	autosomal dominant
Familial Acute Myeloid Leukemia - Familial Acute Myeloid Leukemia	601626	*CEBPA*	19q13.11	uncertain
		NPM1	5q35.1	
Familial Acute Myeloid Leukemia - Familial Acute Myelogenous Leukemia	602439	–	16q22	uncertain
Familial Acute Myeloid Leukemia - Familial Platelet Disorder with associated Myeloid Malignancy	601399	*RUNX1 (CBFA2)*	21q22.12	autosomal dominant

(Continued)

TABLE 44-1 (CONTINUED)

FAMILIAL CANCER SYNDROME - SUBTYPE	OMIM #	CAUSATIVE GENE SYMBOL	CHROMOSOMAL POSITION	MODE OF INHERITANCE
Familial Chronic Lymphocytic Leukemia	151400	–	2q21.2, 6p22.1, 18q21.1, etc.	heterogeneous
Familial Pleuropulmonary Blastoma	601200	*DICER1*	14q32.13	autosomal dominant
Fanconi Anemia	607139	*FANCA*	16q24.3	autosomal recessive, except X-linked for FANCB
	300515	*FANCB*	Xp22.2	
	227645	*FANCC*	9q22.32	
	605724	*BRCA2 (FANCD1)*	13q13.1	
	227646	*FANCD2*	3p25.3	
	600901	*FANCE*	6p21.31	
	603467	*FANCF*	11p14.3	
	602956	*XRCC9 (FANCG)*	9p13.3	
	611360	*FANCI*	15q26.1	
	605882	*BRIP1 (FANCJ/ BACH1)*	17q23.2	
	608111	*FANCL*	2p16.1	
	609644	*FANCM*	14q21.3	
	610832	*PALB2 (FANCN)*	16p12.1	
Gastrointestinal Stromal Tumor	606764	*KIT*	4q12	autosomal dominant
		PDGFRA	4q12	
Hereditary Breast Cancer	114480	*BRCA1*	17q21.31	autosomal dominant
		BRCA2	13q13.1	
		TP53	17p13.1	
		PTEN	10q23.31	
		CHEK2	22q12.1	
		ATM	11q22.3	
		BRIP1 (BACH1)	17q23.2	
		PALB2	16p12.1	
Hereditary Multiple Exostoses/Hereditary Multiple Osteochondromas - Type I	133700	*EXT1*	8q24.11	autosomal dominant
Hereditary Multiple Exostoses/Hereditary Multiple Osteochondromas - Type II	133701	*EXT2*	11p11.2	
Hereditary Multiple Exostoses/Hereditary Multiple Osteochondromas - Type III	600209	*EXT3*	19p	
Hereditary Diffuse Gastric Cancer	192090	*CDH1*	16q22.1	autososmal dominant
Hereditary Leiomyomatosis and Renal Cell Cancer	605839	*FH*	1q43	autosomal dominant
Hereditary Multiple Melanoma - CMM1	155755		1p36	autosomal dominant
Hereditary Multiple Melanoma - CMM2	155601	*CDKN2A (MTS1)*	9p21.3	
Hereditary Multiple Melanoma - CMM3	609048	*CDK4*	12q14.1	
Hereditary Multiple Melanoma - CMM4	608035		1p22	
Hereditary Neuroblastoma	256700	*KIF1B*	1p36.22	autososmal dominant
	613013	*PHOX2B*	4p13	autososmal dominant
	613014	*ALK*	2p23.2–p23.1	autososmal dominant

TABLE 44-1 (CONTINUED)

FAMILIAL CANCER SYNDROME - SUBTYPE	OMIM #	CAUSATIVE GENE SYMBOL	CHROMOSOMAL POSITION	MODE OF INHERITANCE
Hereditary Nonpolyposis Colon Cancer–HNPCC1	120435	*MSH2*	2p21	autososmal dominant
Hereditary Nonpolyposis Colon Cancer–HNPCC2	120436	*MLH1*	3p22.2	
Hereditary Nonpolyposis Colon Cancer–HNPCC3	600258	*PMS1*	2q32.2	
Hereditary Nonpolyposis Colon Cancer–HNPCC4	600259	*PMS2*	7p22.1	
Hereditary Nonpolyposis Colon Cancer–HNPCC5	600678	*MSH6*	2p16.3	
Hereditary Nonpolyposis Colon Cancer–HNPCC6	190182	*TGFBR2*	3p24.1	
Hereditary Nonpolyposis Colon Cancer–HNPCC7	604395	*MLH3*	14q24.3	
Hereditary Non-Polyposis Colon Cancer–HNPCC8	613244	*EPCAM (TACSTD1)*	2p21	
Hereditary Ovarian Cancer	167000	*BRCA1*	17q21.31	autosomal dominant
		BRCA2	13q13.1	
		MLH1	3p22.2	
		MSH2	2p21	
		MSH6	2p16.3	
		PMS2	7p22.1	
Hereditary Pancreatic Cancer	260350	*PALLD*	4q32.3	autososmal dominant
Hereditary Paraganglioma-Pheochromocytoma	168000	*SDHD (PGL1)*	11q23.1	autosomal dominant
	613019	*SDHAF2 (PGL2)*	11q12.2	
	605373	*SDHC (PGL3)*	1q23.3	
	115310	*SDHB (PGL4)*	1p36.13	
Hereditary Prostate Cancer	176807	*RNASEL (HPC1)*	1q25.3	heterogeneous
		ELAC2 (HPC2)	17p12	
		–	1p36, Xq27-q28, etc.	
Juvenile Polyposis Syndrome	174900	*SMAD4*	18q21.2	autosomal dominant
		BMPR1A	10q23.2	
Li-Fraumeni Syndrome	151623	*TP53*	17p13.1	autosomal dominant
		CHEK2	22q12.1	
Mosaic Variegated Aneuploidy	257300	*BUB1B*	15q15.1	autosomal recessive
Multiple Endocrine Neoplasia–Type I	131100	*MEN1*	11q13.1	autosomal dominant
Multiple Endocrine Neoplasia–Type IIA, IIB	171400, 162300	*RET*	10q11.21	
Multiple Familial Trichoepithelioma–Type 1	601606	*CYLD*	16q12.1	autosomal dominant
Multiple Familial Trichoepithelioma–Type 2	612099		9p21	
MYH-Associated Polyposis	608456	*MUTYH (MYH)*	1p34.1	autosomal recessive
Neurofibromatosis 1/Von Recklinghausen Disease	162200	*NF1*	17q11.2	autosomal dominant
Neurofibromatosis 2	101000	*NF2*	22q12.2	autosomal dominant
Nevoid Basal Cell Carcinoma Syndrome/Basal Cell Nevus Syndrome	109400	*PTCH*	9q22.32	autosomal dominant
Nijmegen Breakage Syndrome/Ataxia Telangiectasia Variant 1	251260	*NBN*	8q21.3	autosomal recessive

(Continued)

TABLE 44-1 (CONTINUED)

FAMILIAL CANCER SYNDROME - SUBTYPE	OMIM #	CAUSATIVE GENE SYMBOL	CHROMOSOMAL POSITION	MODE OF INHERITANCE
Oculocutaneous Albinism–Type 1	203100, 606952	*TYR*	11q14.3	autosomal recessive
Oculocutaneous Albinism–Type 2	203200	*OCA2 (P)*	15q12-13.1	
Oculocutaneous Albinism–Type 3	203290	*TYRP1*	9p23	
Oculocutaneous Albinism–Type 4	606574	*SLC45A2 (MATP)*	5p13.3	
Peutz-Jeghers Syndrome/Hereditary Intestinal Polyposis Syndrome	175200	*STK11*	19p13.3	autosomal dominant
Retinoblastoma	180200	*RB1*	13q14.2	autosomal dominant
Rhabdoid Predisposition Syndrome	609322	*SMARCB1 (SNF5/ INI1)*	22q11.23	autosomal dominant
Rothmund-Thomson Syndrome	268400	*RECQL4*	8q24.3	autosomal recessive
Severe Congenital Neutropenia	202700	*ELANE (ELA2)*	19p13.3	autosomal dominant
Shwachman-Diamond Syndrome	260400	*SBDS*	7q11.21	autosomal recessive
Simpson-Golabi-Behmel Syndrome–Type I	312870	*GPC3*	Xq26.2	X-linked
Simpson-Golabi-Behmel Syndrome–Type II	300209	*OFD1 (CXorf5)*	Xp22.2	
Testicular Germ-Cell Tumor	300228	-	Xq27	X-linked
Tuberous Sclerosis Complex	191100	*TSC1*	9q34.13	autosomal dominant
		TSC2	16p13.3	autosomal dominant
Tylosis with Esophageal Cancer	148500	*TOC*	17q25	autosomal dominant
Tyrosinemia Type 1	276700	*FAH*	15q25.1	autosomal recessive
Von Hippel-Lindau Syndrome	193300	*VHL*	3p25.3	autosomal dominant
Werner Syndrome	277700	*WRN*	8p12	autosomal recessive
Wilms Tumor 1	194070	*WT1*	11p13	autosomal dominant
Wiskott-Aldrich Syndrome	301000	*WAS*	Xp11.23	X-linked recessive
X-Linked Lymphoproliferative Disease/Duncan Disease - XLP1	308240	*SH2D1A*	Xq25	X-linked
X-Linked Lymphoproliferative Disease/Duncan Disease - XLP2	300635	*XIAP (BIRC4)*	Xq25	
Xeroderma Pigmentosum–XPA	278700	*XPA*	9q22.33	autosomal recessive
Xeroderma Pigmentosum–XPB	610651	*ERCC3 (XPB)*	2q14.3	
Xeroderma Pigmentosum–XPC	278720	*XPC*	3p25.1	
Xeroderma Pigmentosum–XPD	278730	*ERCC2 (XPD)*	19q13.32	
Xeroderma Pigmentosum–XPE	278740	*DDB2 (XPE)*	11p11.2	
Xeroderma Pigmentosum–XPF	278760	*ERCC4 (XPF)*	16p13.12	
Xeroderma Pigmentosum–XPG	278780	*ERCC5 (XPG)*	13q33.1	
Xeroderma Pigmentosum–XPV	278750	*POLH (XP-V)*	6p21.1	

HEREDITARY PREDISPOSITION OF FAMILIAL CANCER SYNDROMES IN AFRICA

In humans, a familial disease can be passed down through families in several ways. Autosomal diseases can be classified into two groups: autosomal dominant disease (one copy of abnormal gene) and autosomal recessive disease (two copies of abnormal genes). Sex-linked diseases are inherited through one of the sex chromosomes (chromosome X or Y). The following are some examples of familial cancer syndromes and their corresponding mutation spectra that occur within African populations.

HEREDITARY BREAST-OVARIAN CANCER (HBOC)

Globally, cancer is one of the top ten leading causes of death, and breast cancer comprises around 10% of all cancer incidences. Among women, breast cancer is the leading cause of cancer mortality, accounting for almost 1% of all deaths and 16% of cancer deaths in adult women (World Health Statistics 2008, http://www.who.int/). Around the world, the incidence of breast cancer varies, being lower in developing countries and greater in industrialized countries.

HBOC is a syndrome characterized by the aggregation of a substantial number of family members diagnosed with breast cancer, or both breast and ovarian cancers. Familial breast cancer accounts for 5%–10% of all breast cancer patients (Fackenthal & Olopade, 2007), whereas familial cases of ovarian cancer comprise approximately 10% of invasive ovarian cancer (Sowter & Ashworth, 2005). Early-age onset of disease, multiple affected individuals within a family (especially multiple first-degree relatives), existence of bilateral malignant neoplasms, and males with breast cancer have been classified as indicators for HBOC (Fackenthal & Olopade, 2007). The past two decades have witnessed the discovery of two highly penetrant tumor suppressor genes, *BRCA1* (17q21.31) and *BRCA2* (13q13.1) that contribute to breast and ovarian cancer risk. In general, mutations in either *BRCA1* or *BRCA2* account for about 10%–20% of inherited susceptibility of HBOC (Ponder, Antoniou et al., 2005; Fackenthal & Olopade, 2007; Lindor, McMaster et al., 2008). *BRCA1*-associated breast cancer has been reported to be associated with high grade, lymph node positive, estrogen receptor negative (ER-), progesterone receptor negative (PR-), HER2/neu negative (HER2-), and basal-like phenotype (expression of basal or myoepithelial markers as determined by immunohistochemistry; see Foulkes, Stefansson et al., 2003; Foulkes, Metcalfe et al., 2004; Honrado, Benitez et al., 2005; Lakhani, Reis-Filho et al., 2005). Compared to breast cancer, ovarian cancer is observed less frequently than in *BRCA1* carriers. Clinicopathological features such as high histological grade, no special histological subtype, luminal phenotype, ER+, and HER2/neu overexpression are associated with *BRCA2*-related breast cancer, compared with sporadic breast cancer (Bane, Beck et al., 2007). Males with breast cancer or pancreatic cancer were frequently found to carry mutations in *BRCA2*.

The transmission of *BRCA1* and *BRCA2* follow an autosomal dominant pattern. As large genes, *BRCA1* (24 exons) and *BRCA2* (27 exons) encodes a 220-kd protein of 1863 amino acids and a 380-kd protein of 3418 amino acids, respectively. BRCA1 interacts with several proteins involved in cell cycle control, transcription regulation, DNA damage response, and ubiquitination (Deng 2006; Rosen, Fan et al., 2006). At sites of DNA damage, BRCA1 co-localizes with BRCA2 and RAD51 and subsequently activates RAD51-mediated homologous recombination repair of DNA double-strand breaks (Cousineau, Abaji et al., 2005).

An open-access online breast cancer mutation database, Breast Cancer Information Core (BIC, http://research.nhgri.nih.gov/bic/), categorizes the DNA sequence variants of *BRCA1* and *BRCA2*. According to the last modified version (February 13, 2007), the mutation spectrums in *BRCA1/2* are diverse, consisting of deleterious mutations of frameshift deletions, insertions, or nonsense mutations leading to premature truncation of protein transcription, and a small number of missense mutations, as well as other variants with unknown impact. In total, 1650 and 1846 distinct mutations, polymorphisms, and variants were identified throughout the *BRCA1* and *BRCA2* genes, respectively. The distribution of all DNA sequence variants in *BRCA1* and *BRCA2* varies, in Africa and worldwide (Figure 44-1). It appears that missense mutations in both *BRCA1* and *BRCA2* are frequently identified in Africans, 56% and 82%, respectively. Tremendous differences exist for theoretically deleterious mutations (frameshift and nonsense mutations) in *BRCA1* (25% vs. 45%) and *BRCA2* (6% vs. 39%) in Africans, compared to those discovered in other populations worldwide. Furthermore, it has been suggested that families with mutations in the ovarian cancer cluster region (OCCR) of exon 11 (3059–6629 bp) of the *BRCA2* gene have a higher ratio of ovarian to breast cancer than families with mutations elsewhere in the *BRCA2* gene. In Ashkenazi Jews with HBOC, three founder mutations (*BRCA1* 185delAG, *BRCA1* 5382insC, and *BRCA2* 6174delT) account for 80%–90% of *BRCA1/2* mutations. However, it is difficult to identify founder mutations in African-American and African populations, partly due to the high diversity of genetic background or the paucity of attention (Ferla, Calo et al., 2007).

In spite of the fact that many publications about breast and ovarian cancer from Africa are case reports and retrospective studies based on clinical records, there is a paucity of epidemiological factor exploration and population studies that investigate cancer risk for mutations in genes predisposing to inherited disease susceptibilities. The majority of studies on epidemiology, tumor biology, and genetics in black Africans are based on Nigerian breast and ovarian cancer patients. An approximate 20% of Black individuals around the world originated from Nigeria, the most populous country in Africa, which has a steeply increasing prevalence of breast cancer (Adebamowo & Ajayi, 2000). Thus, the behavior pattern of breast cancer in African American women differs from that of white non-Hispanic women, but is similar to that of African women with whom they share a similar genetic background. The incidence rate of breast cancer in Africa is about 25% of that in the United States; however, the mortality rates in these two geographic areas seem to be nearly the same (WHO, http://www.who.int). It has been

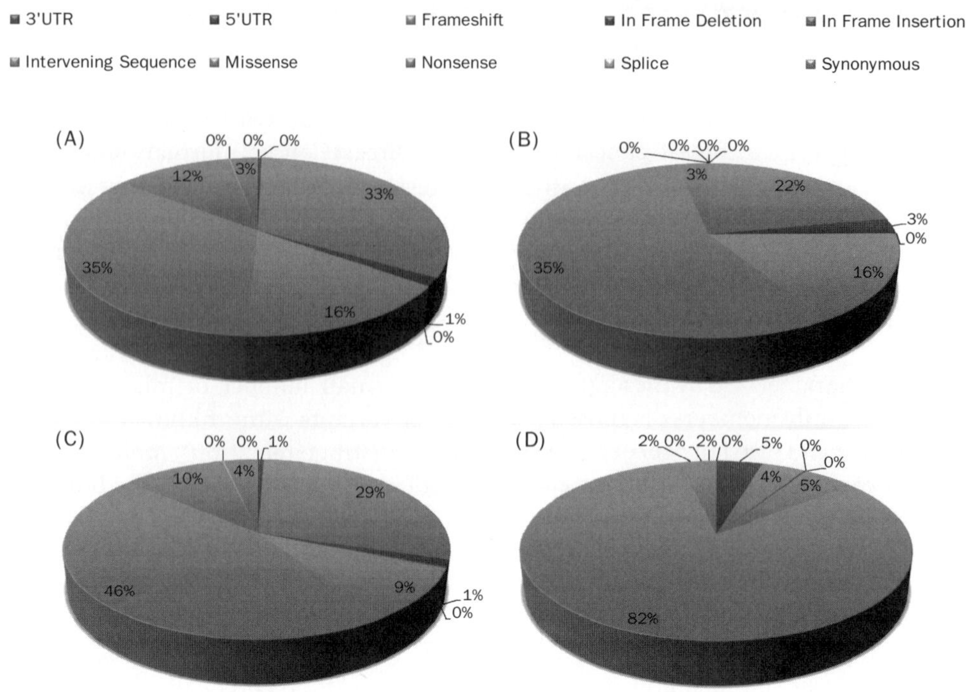

Figure 44-1 Pie charts of distinct mutations, polymorphisms and variants in *BRCA1* and *BRCA2*. Data were retrieved from Breast Cancer Information Core (BIC): http://research.nhgri.nih.gov/bic/). (A) *BRCA1*, worldwide; (B) *BRCA1*, Africa only; (C) *BRCA2*, worldwide; and (D) *BRCA2*, Africa only. Refer color figure.

revealed that women with breast cancer in Africa have had a higher risk for early onset (premenopausal, incidence peaks between 35 and 45 years old, 10 to 15 years younger than white patients), large tumors, node-positive, high grade, advance stage, higher proliferative activity, poor prognosis and overrepresentation of triple negative tumors (ER-, PR- and HER2-; see Ijaduola & Smith, 1998; Ikpatt, Kuopio et al., 2002; Fregene & Newman, 2005; Gukas, Jennings et al., 2005; Adesunkanmi, Lawal et al., 2006; Huo, Ikpatt et al., 2009).

A small number of genetic studies have been published for inherited predisposition to HBOC in Africa. By investigating *BRCA1* mutations in 103 women with familial breast cancer and 57 women with HBOC from France, Italy, Poland, Spain, Northern Africa, and Western Africa, Stoppa-Lyonnet et al. (1997) found one frameshift mutation (943ins10) resulting in a truncated protein in one family originating from Western Africa (Côte d'Ivoire, or Ivory Coast). This mutation was also identified in families from the African diaspora (Bahamas and the United States), and it has been suggested to be a potential founder mutation of African origin with a mutation history of >200 years (Mefford, Baumbach et al., 1999; Panguluri, Brody et al., 1999). Wagner et al. (1999) performed a global sequence diversity analysis for *BRCA2* using 71 Austrian HBOC families and 95 controls of various ethnicities including 21 Africans, and detected 33 of 45 (73%) sequence variants (not disease-associated) as being unique to the African population. Yawitch et al. (2000) screened 206 black South African women with breast cancer for recurring *BRCA1* mutations but could not identify any. One study investigated *BRCA1* and *BRCA2* mutations in 70 Nigerian women with breast cancer (64 of them were diagnosed at or under the age of 40 years; see Gao, Adebamowo et al., 2000). In total, two protein truncating mutations (Q1090X and 17422insG) in *BRCA1* and one truncating *BRCA2* mutation (3034del4 or 3036del4) were identified. An allelic ascertainment of *BRCA1* and *BRCA2* variants in 39 breast cancer patients from Nigeria was conducted by Fackenthal et al. (2005). They found 29 (74%) patients had at least one *BRCA1* or *BRCA2* sequence variant, with 69% carrying a genetic variant in *BRCA2*. Only one truncating mutation (3034del4), which was previously reported, along with 24 exonic single-base substitutions were detected. Another *BRCA1* mutation assessment of Nigerian women with breast cancer was published recently (Zhang, Fackenthal et al., 2009). Zhang et al. screened the *BRCA1* Y101X mutation in 365 breast cancer cases and 177 controls of Yoruba ancestry from Nigeria, unselected for age of onset or family history. Four patients were found to carry a recurrent *BRCA1* Y101X mutation, and this ancient mutation occurred on a single, rare haplotype. They also characterized the same mutation on the same haplotype in a subject of European ancestry with a strong family history of HBOC.

Besides the abovementioned *BRCA1/2* mutation analyses in black Africans, other studies from Southern and Northern Africa have also been published. Mestiri et al. (2000) performed a mutational analysis to detect *BRCA1* germline mutations in 12 Tunisian families with breast cancer, as well as the 185delAG and 1294del40 mutations (previously reported to be overrepresented in Ashkenazi

Jewish and Irish breast cancer families, respectively) in 150 Tunisian women with sporadic breast cancer. A nonsense mutation (in exon 11) and 1294del40 were found in the *BRCA1* gene in one case of familial breast cancer and in a sporadic patient, respectively, whereas the 185delAG mutation was absent in all cases of breast cancer. Charef-Hamza et al. reported loss of heterozygosity (LOH) occurred at the *BRCA1* gene in 21 tumors from Tunisian women diagnosed with sporadic breast cancer (Charef-Hamza, Trimeche et al., 2005). Troudi et al. (2007) screened 36 patients who had at least one first-degree relative with HBOC and detected six deleterious mutations (330dupA, 2789delG, 4160delAG, and 5385insC in *BRCA1*; 1537del4 and 5909insA in *BRCA2*) in 7 familial cases. Their findings also indicated an increased risk of identifying a *BRCA1/2* mutation with the presence of at least one ovarian cancer case in the family. Later, they continued ascertainment of characterizing *BRCA1* mutations and haplotypes in 32 unrelated HBOC Tunisian cases of Arabic/Berber ethnicity (Troudi, Uhrhammer et al., 2008). A total of four frameshift mutations, along with six unclassified variants and twelve polymorphisms were found. Additionally, 5 of 32 familial cases were found to carry deleterious *BRCA1* mutations, several of which were assigned to the most frequent ancestral *BRCA1* haplotype. Their results were also consistent with the previous finding that the Tunisian population has a greater genetic diversity and ethnic background as compared to Caucasian populations. By studying young familial and sporadic Algerian breast cancer cases, Uhrhammer et al. (2008) found a much higher frequency of *BRCA1* mutations than that in Europeans. Also, mutation screening and haplotype analysis suggested that one deleterious mutation, c.798_799delTT, might be a non-Jewish founder mutation in Northern Africa. Ninety HBOC families from South Africa (60 Afrikaners, 11 Ashkenazi Jews, 8 British, 3 Scottish, 2 Dutch, and 6 from multiple ethnic groups) were screened for *BRCA1* mutations by Reeves et al. (2004). Six distinct frameshift mutations (185delAG, 448insA, 1127insA, 1493delC, 4957insC and 5382insC) and two nonsense mutations (S451X and E881X) were found in 18 (20%) families, of which 12 were HBOC families, 4 were breast cancer families, and 2 were ovarian cancer families. Haplotype analysis with five microsatellite markers revealed that the five Afrikaner families with the E881X mutation shared a common haplotype, and further genealogic studies identified three potential French founders. Masri et al. (2002) performed mutation detection for germline and somatic mutations in *BRCA2* exon 11 and the main conserved region of *TP53*, in 20 Sudanese breast cancer patients. However, their failure to detect mutations in *TP53* but instead, one somatic mutation and one polymorphism in *BRCA2* exon 11, was possibly because of the small sample set. Awadelkarim et al. (2007) conducted a combined study of breast cancer epidemiology and *BRCA1/2* mutation discovery in a constitution of Central Sudanese tribes. The epidemiological data consisted of 456 breast cancer cases from a total of 2233 cancer patients. One male and 34 female breast cancer patients from 18 Central Sudanese tribes were screened for *BRCA1/2* mutations. Overall, 32 *BRCA1* and 28 *BRCA2* variants were detectable with high diversity in 33 of the 35 cases, and the observation of *BRCA1/2* variants were clustered independently of tribe origins. Five (14%) truncating mutations (two in *BRCA1* and three in *BRCA2*) were found in five patients.

HEREDITARY NONPOLYPOSIS COLON CANCER (HNPCC)

Hereditary nonpolyposis colorectal cancer (HNPCC), also known as Lynch Syndrome, is an autosomal dominant, genetically heterogeneous disease that accounts for about 4%–8% (Lynch, Lanspa et al., 1991) of colorectal cancer (CRC). Three percent of incidences are mutation-positive, caused by germline mutations in the DNA mismatch repair genes (i.e., *MSH2*, *MLH1*, *PMS1*, *PMS2*, *MSH6*, *TGFBR2*, and *MLH3*), but most remaining cases are still molecularly unexplained. The OMIM (Online Mendelian Inheritance in Man, http://www.ncbi.nlm.nih.gov/omim/) categorizes HNPCC to eight subtypes corresponding to the genes mutated. Even though HNPCC is a type of CRC, the phenotypic features of HNPCC differ from that of CRC. Most cases of HNPCC (>70%) occur proximally to the splenic flexure, and show: proclivity to early onset (mean age at diagnosis is 40–45 years); excess of multiple primary (both synchronous and metachronous) cancers; gene penetrance of >85%–90%; decreased number of rectal locations; reduced metastatic tendency and better prognosis than sporadic CRC; better survival. Pathological features (including poor differentiation, increased signet cells, medullary features, peritumoural lymphocytic infiltration, Crohn's-like reaction, and tumor infiltrating lymphocytes mixed with tumor cells) are often distinguishable (Myrhoj, Bisgaard et al., 1997; Lynch & de la Chapelle, 1999; Rijcken, Hollema et al., 2002). Clinically, HNPCC can be subdivided into Lynch syndrome I and Lynch syndrome II, where Lynch syndrome II is classified by increased risk for malignancy at several extracolonic sites, whereas Lynch syndrome I is site-specific. An international standard diagnostic criteria, called the Amsterdam criteria, indicates *(1)* families should exhibit three histologically verified cases of colorectal cancer, of which at least one should be diagnosed before the age of 50 years; *(2)* there should be affected individuals in two generations, one individual being a first-degree relative to the other two; and *(3)* familial adenomatous polyposis should be excluded (Vasen, Mecklin et al., 1991). It is important to screen for germline mutations for risk allele also, because risk of developing CRC increases to about 75% for mutation carriers (Barrow, Alduaij et al., 2008). Carrier frequencies in men are generally higher than women (Barnetson, Tenesa et al., 2006), but CRC risks were comparable in males and females (Parc, Boisson et al., 2003). One indicative marker of DNA mismatch repair genes has been the microsatellite instability (MSI). MSI is often divided into two forms: MSI-high (MSI-H) and MSI-low (MSI-L), based quantitatively on the observed frequency of genomic mutations (Boland, Thibodeau et al., 1998). Germline mutations in mismatch repair genes are common with HNPCC families showing MSI; thus, MSI-H phenotype is often used as an inclusion criterion for mutation testing of mismatch repair genes.

HNPCC type 1 refers to the disorder caused by mutations in the *MSH2* gene. Defects in the *MSH2* gene account for approximately 60% of HNPCC cases, where patients with heterozygous mutations have disease onset at an average age of 45 years (Pensotti, Radice et al., 1997; Muller, Schackert et al., 2006). The *MSH2* gene is located on chromosome 2p22-p21 and contains 16 exons. It is involved in the repair of defective DNA replication errors that create insertion-deletion loop-type mismatched nucleotides (Fishel, Ewel et al., 1994; Pearson, Ewel et al., 1997; Ni, Marsischky et al., 1999), as well as safeguarding the genome from promiscuous recombination (de Wind, Dekker et al., 1995). In conjunction with other proteins, *MSH2* also plays a role in critical cell regulatory mechanisms, such as S-phase checkpoint activation (Brown, Rathi et al., 2003), estrogen-dependent gene expression (Wada-Hiraike, Yano et al., 2005), and sensor of abnormal DNA structures (Gradia, Subramanian et al., 1999; Lamers, Perrakis et al., 2000; Wang, Cortez et al., 2000). It is known that the genomic mutation rate in *MSH2*-deficient tumor cells is at least 100-fold higher than the wild types (Parsons, Li et al., 1993), and *MSH2*-deficient mice develop tumors containing microsatellite instability two months after birth (Reitmair, Schmits et al., 1995), indicating a direct link between *MSH2* deficiency and the pathogenesis of cancer. Thus, even though mutations in the mismatch repair pathway leading to defective protein are only found in about 20% of patients presenting CRC (Cunningham, Kim et al., 2001), deficiencies of *MSH2,* along with other mismatch repair enzymes, are considered to increase cancer risk of HNPCC (DeWeese, Shipman et al., 1998). Among the variety of mutations found, genomic deletions in *MSH2* or *MLH1* are a frequent cause of HNPCC (Taylor, Charlton et al., 2003). Small deletions in poly(dA/dT) sequences are frequent (Ionov, Peinado et al., 1993; Aquilina, Hess et al., 1994; Umar, Boyer et al., 1994; Kraus, Kastl et al., 1999; van der Klift, Wijnen et al., 2005), though a large deletion that removes exons 1–6 have been reported to occur in multiple populations (Wang, Friedl et al., 2002; Lynch, Coronel et al., 2004; Stella, Surdo et al., 2007). A large proportion of mutations detected in *MSH2* are point mutations (Fishel, Lescoe et al., 1993; Leach, Nicolaides et al., 1993; Peltomaki, Aaltonen et al., 1993; Wijnen, Vasen et al., 1995; Ellison, Lofing et al., 2001), where CpG dinucleotides within the coding domain were found to be hotspots for single base pair substitutions (Maliaka, Chudina et al., 1996). Some of the missense mutations were also found to affect the splicing site, altering the splicing pattern and producing truncated products (Maliaka, Chudina et al., 1996; Pagenstecher, Wehner et al., 2006; Tournier, Vezain et al., 2008). However, no definitive evidence has been found that indicate causal association of missense mutations or splice site mutations with an increased risk of developing colorectal cancer (Cravo, Afonso et al., 2002). On the other hand, profound mismatch repair gene deficiencies were found to affect mutation frequency of *PTEN*, which is thought to relate to the increased occurrence of *HNPCC*-related endometrial cancer (Zhou, Kuismanen et al., 2002). Germline allele-specific and mosaic hypermethylation of the *MSH2* promoter are also considered to play a role in the incidence of *HNPCC,* as hypermethylation can silence gene expression, and the methylation patterns can be inherited (Chan, Yuen et al., 2006). Studying the mutation spectrum of the *MSH2*, however, several groups have come to a conclusion that the Amsterdam criteria for *HNPCC* is neither sufficiently sensitive nor specific enough for use as the sole criteria for determining which families should undergo for *MSH2* screening (Wijnen, Vasen et al., 1998; Bapat, Madlensky et al., 1999; Syngal, Fox et al., 2000). One group in South Africa has reported a disease-causing mutation in exon 13 within a family meeting the Amsterdam criteria (Goldberg, Madden et al., 1998), and one case report of HNPCC with germline *MSH2* mutation in Nigeria has been published to date (Adebamowo, Adeyi et al., 2000).

HNPCC type 2 comprises up to 30% of total HNPCC cases. Loss of heterozygosity (LOH) of the *MLH1* gene results in HNPCC2 (Papadopoulos, Nicolaides et al., 1994), since LOH of markers within or adjacent to the *MLH1* gene occurs nonrandomly in tumors, suggesting the 2-hit Knudson model (Hemminki, Peltomaki et al., 1994). HNPCC patients with *MLH1* mutations are known to have a high level of microsatellite instability (MSI-H), be affected at younger ages, and develop multiple other cancers, including breast cancer (Bapat, Madlensky et al., 1999; Scott, McPhillips et al., 2001; Clyne, Offman et al., 2009). The overall risk of *MLH1* mutation carriers developing colorectal cancer by age 70 years is estimated to be around 75% (Quehenberger, Vasen et al., 2005; Barrow, Alduaij et al., 2008). The *MLH1* gene is located at 3p21.3 and consists of 19 coding exons spanning approximately 100 kb. MLH1 is known to form a heterodimer with PMS2 to function as part of mismatch repair mechanism (Li & Modrich, 1995; Kadyrov, Dzantiev et al., 2006), and to interact with BRCA1 and other proteins to serve as a sensor of abnormal DNA structure (Wang, Cortez et al., 2000). A wide range in mutation spectrum has been found (Han, Maruyama et al., 1995), and 50% of mutations described to date have been found clustered within the region encompassing exon 15 and 16, which alter an evolutionarily conserved domain (Wijnen, Khan et al., 1996). Large proportions of missense mutations seemed to reduce the functional efficiency of the protein, but do not silence it (Bronner, Baker et al., 1994; Sasaki, Horii et al., 1996; Wang, Friedl et al., 1997; Liu, Tannergard et al., 1999; Ellison, Lofing et al., 2001; Cravo, Afonso et al., 2002; Lipkin, Rozek et al., 2004). Large deletions, which were found to be caused by Alu-mediated recombination, and truncating mutations were found to be associated with increased risk of colorectal cancer (Papadopoulos, Nicolaides et al., 1994; Yuan, Kasprzak et al., 1998; Chan, Yuen et al., 2001). Since *MLH1* is known to have multiple splice isoforms, mutations on splicing factors lead to modified splicing patterns, which were sometimes found to increase the risk of extracolonic tumor formation (Nystrom-Lahti, Kristo et al., 1995; Moisio, Sistonen et al., 1996; Jager, Bisgaard et al., 1997; Gorlov, Gorlova et al., 2003; McVety, Li et al., 2006; Pagenstecher, Wehner et al., 2006; Tournier, Vezain et al., 2008). One unique aspect of *MLH1* in colorectal cancer is that some MSI-positive patients show malignant phenotype while retaining wild-type allele (Katabuchi, van Rees et al., 1995; Liu, Nicolaides et al., 1995). In Turkey, researchers have investigated an HNPCC family and confirmed the occurrence of missense mutation (676C>T) that causes altered splicing, leading to truncated peptide lacking PMS2 binding domain (Ricciardone, Ozcelik et al., 1999; Huang, Lavine et al., 2001). On the other hand, frequency of *MLH1* silencing due to hypermethylation in a Turkish population was found to be higher than the Egyptian and Jordanian populations, and the molecular characteristics of Turkish colorectal carcinoma were found to be most similar to those reported from Western countries (Chan, Soliman et al., 2005). The C1528T mutation in exon 13 is known to be common among South African patients, whose risk of developing colorectal cancer by age 65 is increased to 92% and leads to development of extracolonic malignancies (breast and endometrial being the most common; see Blokhuis, Goldberg et al., 2008; Stupart, Goldberg et al., 2009). Similarly, approximately 40% of colorectal and gastric carcinomas in South Africa were correlated with loss of expression of MLH1, but no mutation analysis was further performed in these cases (Chetty, Naidoo et al., 2002; Hameed, Goldberg et al., 2006). In Tunisia, large deletions due to Alu-mediated recombination that removes exon 6 of *MLH1* (Aissi-Ben Moussa, Moussa et al., 2009), and silencing of gene expression by hypermethylation leading to development of breast cancer in 24% of cases, were observed (Karray-Chouayekh, Trifa et al., 2009).

HNPCC type 3 is caused by mutations in the *PMS1* gene. First identified by screening for genes homologous to yeast and bacterial *mutL* genes, the human *PMS1* gene is a homolog of the yeast *PMS* gene (Papadopoulos, Nicolaides et al., 1994) and is thought to be involved in DNA mismatch repair by forming heterodimer with the MLH1 protein (Li and Modrich 1995; Kadyrov, Dzantiev et al., 2006). It is known that there are at least 11 genes in the human genome that share highly similar sequences to *PMS1*, constituting a multiple gene family (Horii, Han et al., 1994). The *PMS1* gene comprises 13 exons and is located on chromosome 2q31-q33 (Nicolaides, Papadopoulos et al., 1994). Thus far, only one *PMS1* mutation, a nonsense mutation (C-to-T at codon 233) that results in skipping of that exon, has been reported (Nicolaides, Papadopoulos et al., 1994). This mutation has not been identified in an African cohort; however, studies performed in Egypt reported that increased risk of colorectal cancer is associated with reduced expression of PMS1 and PMS2 (Soliman, Bondy et al., 1998). In particular, low PMS2 expression in colon cancer seems to be a unique characteristic within the Egyptian population, even though there seems to be no predilection to a specific age group associated with colorectal cancer (Abou-Zeid, Khafagy et al., 2002).

HNPCC type 4 is caused by somatic and germline mutations in the *PMS2* gene (Ma, Xia et al., 2000; Niessen, Kleibeuker et al., 2009). The *PMS2* gene, at 7p22, contains 15 exons (Nicolaides, Carter et al., 1995). Small insertions and deletions of 1 to 3 bp, as well as nucleotide transition in the 3' end exons (mostly exon 10 and 11), which cause frameshifts leading to premature termination or aberrant splicing, are commonly observed among HNPCC4 patients (Thompson, Meldrum et al., 2004; Worthley, Walsh et al., 2005; Hendriks, Jagmohan-Changur et al., 2006; Clendenning, Senter et al., 2008). It was found that the single functional *PMS2* allele is sufficient to generate normal levels of somatic mosaicism; thus, germline mutations often lead to loss of heterozygosity before malignant phenotypes appear (Nicolaides, Papadopoulos et al., 1994; Gomes-Pereira, Fortune et al., 2004; Nakagawa, Lockman et al., 2004). No genetic studies have been done in an African cohort, but low PMS2 expression has been reported as a unique characteristic of Egyptian colorectal cancer patients, as mentioned above.

Compared to other subtypes, HNPCC type 5 patients display a very high frequency of atypical hyperplastic lesions and carcinomas of the endometrium, delayed age of cancer onset, as well as incomplete penetrance (Wijnen, de Leeuw et al., 1999; Wagner, Hendriks et al., 2001; Suchy, Kurzawski et al., 2002; Goodfellow, Buttin et al., 2003; Buttin, Powell et al., 2004). The causative gene, *MSH6* (also known as *GTBP*) at 2p16, is a member of the MutS homolog family, which is required for mismatch-specific binding. MSH6 itself primarily functions as base–base mismatch correction; thus, MSH6 deficiency largely causes alterations in mononucleotide tracts (Drummond, Li et al., 1995;

Papadopoulos, Nicolaides et al., 1995). *MSH6* mutations seem to cluster on the 5' end of the gene as several small (usually 1 to 5 bp) polymorphisms, insertions, and deletions that cause frameshift and truncation have been observed (Wijnen, de Leeuw et al., 1999; Huang, Kuismanen et al., 2001; Wagner, Hendriks et al., 2001; Wu, Berends et al., 2001; Suchy, Kurzawski et al., 2002; Ostergaard, Sunde et al., 2005). To date, no genetic studies for *MSH6* mutation and HNPCC5 have been done in an African cohort.

The primary cause of HNPCC type 6 is germline and somatic mutations in the 7-exon *TGFBR2* gene at 3p22-p21. This gene encodes a transmembrane receptor belonging to the serine-threonine kinase family. TGFBR2 inactivation is often found in colon cancer exhibiting microsatellite instability within the coding regions of *TGFBR2* (Eshleman and Markowitz 1995; Markowitz, Wang et al., 1995; Souza, Appel et al., 1996). Frameshift mutations are among the most common mutation types, where small insertions or deletions within small adenine mononucleotide repeats result in synthesis of the truncated product (Markowitz, Wang et al., 1995; Parsons, Myeroff et al., 1995; Tannergard, Liu et al., 1997). Other mutations reported include germline mutation of nucleotide transition that result in amino acid substitution of T315M (Lu, Kawabata et al., 1998) and R528H (Grady, Myeroff et al., 1999; Loeys, Chen et al., 2005). Currently, there are no genetic studies investigating *TGFBR2* mutations in African populations.

The *MLH3* gene is another mismatch repair gene found at 14q24.3, and it is highly expressed in gastrointestinal epithelium and testis. The gene plays a critical role in mammalian meiosis, since MLH3 is required for MLH1 when binding to meiotic chromosome (Lipkin, Moens et al., 2002), and interacts with MSH4 in mammalian meiotic recombination (Santucci-Darmanin, Neyton et al., 2002). Colorectal cancers with a *MLH3* mutation show no involvement of microsatellite instability, indicating that *MLH3* itself is not contributing to carcinogenesis through impaired DNA mismatch repair function (Liu, Zhou et al., 2003). Several nonsense, missense, and frameshift mutations have been reported to occur primarily on Exon1 and 7 of the *MLH3* gene (Lipkin, Wang et al., 2001; Wu, Berends et al., 2001; Liu, Zhou et al., 2003; Kim, Roh et al., 2007), yet it is thought that germline mutation is uncommon (Lipkin, Wang et al., 2001). Most of the mutations studied showed reduced penetrance, suggesting that *MLH3* is a low risk factor and may work together with other factors in an additive manner (Liu, Zhou et al., 2003). Thus, even though it is not confirmed yet (Ou, Rasmussen et al., 2009), mutations in the *MLH3* gene are thought to play a role in the progression of HNPCC type 7 tumorigenesis. Unfortunately, no mutation studies have been conducted using an African cohort.

On the other hand, HNPCC type 8 is a unique variant of HNPCC because it is not caused directly by mutations in DNA mismatch repair gene. Rather, it is caused by a combination of inherited germline allele-specific, mosaic hypermethylation of the *MSH2* gene, and heterozygous deletion of intergenic region directly upstream of *MSH2,* as well as 3' exons of the *EPCAM* (*TACTD1*) gene. The *EPCAM* gene is located at 2p21, and its molecular function pertaining to carcinogenesis is based more on its role in signaling and transcription regulation (Osta, Chen et al., 2004). Patients diagnosed with HNPCC8 often show microsatellite instability and loss of MSH2 protein expression, but no evidence of DNA mismatch repair gene mutation has been shown. This perplexing phenomena was explained when deletions of the 3' end of the *EPCAM* gene was found to correlate with epigenetic inactivation of MSH2 (Ligtenberg, Kuiper et al., 2009). Deletion of 3' *EPCAM* exons causes loss of normal polyadenylation signal, and subsequently, leads to transcriptional read-through and induction of promoter methylation of the neighboring gene (Ligtenberg, Kuiper et al., 2009). Several deletions encompassing the 3' end of *EPCAM*, ranging from 5 to 22.8 kb, have been reported (Ligtenberg, Kuiper et al., 2009). Again, there is a lack of genetic study on these mutations in Africa.

FAMILIAL ADENOMATOUS POLYPOSIS (FAP)

FAP is an autosomal dominant disorder with high penetrance, variable expressivity (Naylor & Gardner, 1977), and high susceptibility to cancer. Patients with FAP usually develop numerous adenomatous polyps in the colon and rectum, a small proportion of which may progress to colorectal cancer, without surgical intervention, by the sixth decade of life in nearly all affected individuals (Szodoray 1975; Nishisho, Nakamura et al., 1991). Affected individuals are at risk of developing colonic and extracolonic features such as multiple intestinal polyps, osteomas, fibromas, lipomas, fibrosarcomas of the skin and mesentery, sebaceous cyst, desmoid tumors, dental abnormalities, carcinoma of ampulla of Vater, and thyroid carcinoma (Gardner 1951; Gardner & Plenk, 1952; Gorlin & Chaudhary, 1960; Gardner 1962; Jones & Nance, 1977; Denzler, Harned et al., 1979; Harned & Williams, 1982; Sugihara, Muto et al., 1982; Walsh, Qizilbash et al., 1987; Iida, Yao et al., 1988).

The *adenomatous polyposis coli* (*APC*) gene (at 5q22) is thought to play a central role in the occurrence of FAP, as several groups reported heterozygous inactivating mutations (Groden, Thliveris et al., 1991), point mutations (Nishisho, Nakamura et al., 1991), and germline mutations (primarily nonsense, frameshift; see Fodde, van der Luijt et al., 1992; Miyoshi, Ando et al., 1992) of the *APC* gene that highly correlated with FAP. The *APC* gene is classified as a tumor suppressor, where the functional protein product plays a role in controlling phenotypes such as cell division, adherence to other cells within tissue, and motility. The *APC* gene also takes part in ensuring the maintenance of correct chromosome number during cell division (Peifer 1993; Kaplan, Burds et al., 2001; Etienne-Manneville & Hall, 2003; Hanson & Miller, 2005). The *APC* gene is known to have 15 exons, 3 transcript initiation sites, and 2 *APC*-specific promoter elements spanning over 110 kb. Over 700 mutation variants have been reported (Beroud & Soussi, 1996; Beroud, Collod-Beroud et al., 2000), where the vast majority comprises small deletions and point mutations resulting in stop codons, but insertions and missense mutations are rare. Nearly all mutations (>90%) result in premature polypeptide chain terminations, due to either frameshift (68%) or nonsense (30%) mutations (Miyoshi, Ando et al., 1992; Beroud & Soussi, 1996), suggesting that the C terminal of the protein is vital for proper function. Two germline mutation hotspots exist, one at codon 1061 and the second at 1309, while two somatic mutation hotspots occur at codons 1309 and 1450. The great majority of the mutations were found in the central region of the *APC* gene (between codons 1240 and 1513), known as a mutation cluster region (Polakis 1995; Dihlmann, Gebert et al., 1999) and thought to be associated with a severe phenotype of thousands of polyps at a young age (Nagase, Miyoshi et al., 1992; Gayther, Wells et al., 1994; Davies, Armstrong et al., 1995). On the other hand, mutations in the first or last third of the gene are generally thought to be associated with an attenuated polyposis with late onset and small number of polyps (Spirio, Olschwang et al., 1993; Friedl, Meuschel et al., 1996; van der Luijt, Meera Khan et al., 1996).

Colorectal cancer (CRC) is considered the leading digestive cancer in the world, and FAP is one of the prime hereditary causative syndromes. However, due to genetic heterogeneity, the rate of progression to CRC in Africans is smaller than less diverse populations such as Ashkenazi Jews in Israel (Shtoyerman-Chen, Friedman et al., 2001). The disease onset from adenomatous polyposis in Africans has a predilection to younger adults (mean age of disease onset is 10 years younger than reported for occidental series, and 20 years younger than European counterparts), tendency to express less pathological biomarkers, and proliferation of malignant tumor is generally more aggressive in Africans than in Europeans (Mourali, Tabbane et al., 1986; Jaskiewicz, Lancaster et al., 1998; Hsairi, Fakhfakh et al., 2002). Some rare mutations were found in different African populations (Mengesha, Johnson et al., 1994; Grobbelaar, Ziskind et al., 1996; Kotze, Grobbelaar et al., 1996; Grobbelaar, Fortuin et al., 2000; Grobbelaar, Wilken et al., 2002; Bougatef, Marrakchi et al., 2008; Bougatef, Ouerhani et al., 2008), most of them occurring on the mutation-prone 5' half of the *APC* gene and causing non-functional truncated proteins. Within the African continent, it seems to be a general trend that separates Arabic populations in Northern Africa from sub-Saharan black Africans on a genetic scale, since common mutations in the mutation cluster region (namely codons 1061, 1309, and 1450) were found in the black African population but missing in Arabic population (Grobbelaar, Oosthuizen et al., 1995; Grobbelaar, Wilken et al., 2002; Bougatef, Ouerhani et al., 2008). Those common variants were reported to be responsible for FAP in 18%–35% of affected individuals in sub-Saharan Africans, a frequency similar to that described in several other non-African populations (Grobbelaar, Oosthuizen et al., 1995; Grobbelaar, Wilken et al., 2002). The Arabic population, on the other hand, seems to possess a higher frequency of other sets of mutations, which may suggest a population-specific mutation hotspot for the *APC* gene. Furthermore, it was once reported that a missing predominant haplotype was associated with FAP within the South African population, suggesting the absence of founder mutations (Hayes, Kotze et al., 1996). However, another study has shown the presence of five haplotypes within the South African population, two of which were predominantly associated with FAP and may be causing disease in about 70% of families in this population (Grobbelaar, Oosthuizen et al., 1995). Since loss of heterozygosity (LOH) at *APC* locus was only observed in 52% of tumor cases (Bougatef, Ouerhani et al., 2008), the apparent low incidence of CRC in the African population may be due to the rare occurrence of LOH, or low mutation frequency in the *APC* gene.

MULTIPLE ENDOCRINE NEOPLASIA (MEN)

Multiple endocrine neoplasia (MEN) consists of two clinical components, multiple endocrine neoplasia type 1 (MEN1) and multiple endocrine neoplasia type 2 (MEN2), each characterized by different and specific phenotypic features. MEN1 is an autosomal dominant inherited cancer syndrome associated with several tumors, such as parathyroid, endocrine pancreas, anterior pituitary, and gastrointestinal endocrine tumors. A tumor suppressor gene, *MEN1* (at 11q13) has been found to be responsible for this syndrome.

All hereditary MEN2 forms are transmitted in an autosomal dominant manner, with an incidence rate of one in 30,000 births. Based on clinical and biological presentations, MEN2 has been classified as having three distinct subtypes: MEN2A (>90% of MEN2 cases), MEN2B (5%), and familial medullary thyroid carcinoma (FMTC, <5% see Eng, 1996). Medullary thyroid carcinoma (MTC) occurs in >95% of MEN2 patients, and it presents in all three MEN2 subclasses. Besides MTC, families with MEN2A develop either pheochromocytoma (PC, in about 50% cases), HPT (20%–30%), or both; however, MEN2A is typically not associated with other abnormalities (Eng, 1996; Margraf, Crockett et al., 2009). Onset of MTC is often younger (10 years earlier) in MEN2B, which also includes symptoms of PC (50%) and additional developmental abnormalities; for example, mucosal ganglioneuromas (70%–100%) and gastrointestinal ganglioneuromas (40%–90%), whereas HPT seldom occurs (Eng, 1996; Margraf, Crockett et al., 2009). Familial medullary thyroid carcinoma (about 25%–40% of MTC cases) is the third form of MEN2 (Komminoth, Kunz et al., 1995; Benazzouz, Chraibi et al., 2006), presenting only one clinical feature of MTC with late onset (often after the age of 50 years).

Subsequent to a linkage study (Simpson, Kidd et al., 1987), the *RET* protooncogene has been initially identified as a susceptibility gene that predisposes patients to MEN2 (Donis-Keller, Dou et al., 1993; Mulligan, Kwok et al., 1993). This 21-exon gene has been mapped to the long arm of chromosome 10 (10q11.21) and encodes a transmembrane receptor tyrosine kinase that functions crucially in the normal development, differentiation, and neoplastic growth of neural crest lineages (Takahashi 2001). Deleterious germline *RET* mutations are identifiable in more than 98% of cases of MEN2A and 85% of FMTC (Frank-Raue, Rondot et al., 2007; Margraf, Crockett et al., 2009). To date, 139 DNA sequence changes in *RET*, including mutations (54%), variants (40%), and polymorphisms (6%), have been deposited in the MEN2 *RET* database (last update January 2010; also see Margraf, Crockett et al., 2009). The majority of MEN2 causative mutations are missense mutations that lie in *RET* exon 10, 11, 13 and 14. There appear to be fairly consistent genotype–phenotype correlations in three MEN2 subtypes: MEN2A (exon 5, 8, 10, 11, 13-15), MEN2B (exon 15-16), and FMTC (exon 8, 10, 11, 13-15). C634 mutation and M918T mutation contribute to the high risk of early aggressive penetrance in MEN2A and MEN2B, respectively (Machens, Gimm et al., 2001; Sanso, Domene et al., 2002). It is worth noting that the pivotal role of *RET* in MEN2 and Hirschsprung's disease (HSCR) has been documented (van Heyningen, 1994). *RET* mutations account for 51% of familial HSCR in females and 72% familial HSCR in males, and 7%–35% of sporadic HSCR cases (Eng, 1996; Moore & Zaahl, 2008).

In a *RET* mutation detection study of a Moroccan family with MEN2A, Benazzouz et al. (2006) discovered that C634Y mutation was present in the heterozygous state, while six asymptomatic members of this family were not C364Y carriers. Benazzouz et al. later (2008) investigated the relationship between mutations of the protooncogene *RET* and MEN2 in a large Moroccan family with MEN2A. Mutation screening in *RET* exon 8, 10, 13, 14, 15, and 16 was performed, and a C634R mutation was discovered in the index and three additional individuals, as well as in three young carriers who were apparently asymptomatic. An MTC family from northwest Morocco, consisting of a 18-year-old female index patient and her five relatives from two generations, was screened for germline *RET* mutations in exon 10-11 and 13-16 by sequencing (Ainahi, Kebbou et al., 2006). No mutations were detectable in these exons except a heterozygous missense mutation (C634R) in exon 11. To evaluate the risk to children in families with MEN2, Moore et al. (2007) screened *RET* mutations (in all 21 exons) in 25 patients from 10 South African families of three ethnic groups: whites, blacks and mixed ancestry. A total of 11 potential disease-causing *RET* sequence changes were found, including nine mutations and two splicing variants, all of which were observed in a heterozygous state. Three cases from different families were detected with two mutations each. A missense mutation (C620W in exon 10) was detected in a patient with HSCR and a family member (mother diagnosed with MEN2A). In this study, a fairly high incidence of *RET* gene variations (>40%) was observed to occur throughout the gene in MEN2/MTC carriers. Moore and Zaahl (2009) also carried out a *RET* mutation detection study in 22 total colonic aganglionosis (TCA) patients from 114 HSCR cases (including 16 kindreds), in a diverse South African population. *RET* variations were detected in 82% (18/22) of TCA, with 11 (50%) having multiple genetic *RET* variations. A cysteine radical mutation (C620R) was found in a patient with TCA and a MEN2 case in the family.

HEREDITARY MULTIPLE OSTEOCHONDROMAS (HMO)

HMO, previously called hereditary multiple exostoses (HME), is characterized by the formation of multiple cartilage-capped protuberances, or osteochondromas, developing from the juxtaepiphyseal regions of the long bones and causing orthopedic deformities and occasionally sarcomatous degeneration (Stieber & Dormans, 2005; Bovee, 2008). It is an autosomal dominant condition, with a penetrance of 96% (Schmale, Conrad et al., 1994). It is genetically heterogeneous, and it is associated with mutations in *EXT1* (8q24.11), *EXT2* (11p11.2), and *EXT3* (19p) for subtypes I, II, and III, respectively (Bovee, 2008; Jennes, Pedrini et al., 2009). *EXT1* and *EXT2* encode exostosins involved in the chain elongation step of heparan sulfate biosynthesis. As the most common benign bone neoplasmata in humans, approximate 15% of osteochondromas are HMO (Hennekam, 1991). This skeletal disorder has an estimated prevalence of 1/50,000 in the Western population and seems to be higher in males (Wicklund, Pauli et al., 1995). Disproportionately short stature and chondrosarcoma are observed in HMO patients. Size, number, and location of osteochondromas, and the degree of deformities vary significantly (Hennekam, 1991). Previous studies have suggested a more severe phenotype to be associated with *EXT1* mutations than is associated with *EXT2* (Francannet, Cohen-Tanugi et al., 2001; Porter,

Lonie et al., 2004; Alvarez, Tredwell et al., 2006). About 10% of affected individuals have a de novo mutation, and 56%–78% and 21%–44% of HMO families have mutations harbored in tumor suppressor genes, *EXT1* and *EXT2*, respectively. To date, an *EXT1* or *EXT2* mutation is detected in 70%–95% of affected individuals with HMO. As of July 28, 2009, the online Multiple Osteochondromas Mutation Database (MOdb) contains a total of 895 entries for multiple osteochondromas, including 351 and 166 unique mutations for *EXT1* and *EXT2*, respectively (Jennes, Pedrini et al., 2009).

In 26 HMO families originating from nine countries (two Moroccan families), Wuyts et al. (1998) conducted a mutation identification study for *EXT1* and *EXT2*. They found that 10 families had an *EXT1* mutation and 10 had an *EXT2* mutation. A single HMO patient from one Moroccan family had a splicing variant at the first nucleotide of intron 7. Sfar et al. (2009) investigated two HMO families (36 members including 21 affected individuals) originating from two neighboring regions in the middle coast of Tunisia. Linkage analysis and mutation detection in *EXT1* were conducted. In the larger multigenerational family, evidence for linkage to *EXT1* was observed, and a heterozygous missense mutation in exon 2 was subsequently identified. A heterozygous truncating mutation in exon 1 was also detected in the second family. Genetic testing of *EXT1* and *EXT2* mutations in black Africans is lacking, and most of the publications about osteochondromas and exostoses are case reports (Govender & Parbhoo, 1999; Odetayo 2001; Ntsiba & Bazebissa, 2002; Mohammed & Isa, 2007; Obalum, Eyesan et al., 2008).

MULTIPLE FAMILIAL TRICHOEPITHELIOMA (MFT)

Multiple familial trichoepithelioma (MFT), also known as epithelioma adenoides cysticum (EAC), is a familial syndrome characterized by skin appendage neoplasms occurring on the face. This disorder, inherited in an autosomal dominant model, has been thought to represent a benign hamartoma of the pilosebaceous apparatus (Lee, Grossman et al., 2005). While MFT2 was mapped to chromosome 9p21, and LOH was found at 9q22.3 in around 50% of sporadic trichoepitheliomas, MFT1 was reported to be caused by pathogenic mutations spanning the *CYLD* gene on chromosome 16q12.1 (Blake & Toro, 2009). Since cylindromas, trichoepitheliomas, and/or spiradenomas have been documented to coexist within a single family, allelic diseases may represent a phenotypic spectrum of a single disease entity, Brooke-Spiegler syndrome (BSS), and also share the same causative gene, *CYLD* (Bowen, Gill et al., 2005; Lee, Grossman et al., 2005; Young, Kellermayer et al., 2006; Saggar, Chernoff et al., 2008; Blake & Toro, 2009). The *CYLD* gene encodes the CYLD protein, which is a deubiquitinating (DUB) enzyme and plays a role in inflammation and cell proliferation. To date, 51 unique *CYLD* germline mutations have been identified in 73 pedigrees from the Unite States, the United Kingdom, Germany, China, Algeria, among others (Blake and Toro 2009). Of the 51 mutations, 21 (41%) are frameshift, 18 (35%) are nonsense, 7 (14%) are missense, and 5 (10%) are splicing variants. The majority (44/51, 86%) of the mutations resulted in truncated proteins. In addition, it seems that there lacks a correlation between *CYLD* mutations and a specific phenotype (Bowen, Gill et al., 2005; Blake & Toro, 2009).

Because of the rare incidences of these diseases, molecular and genetic investigations were seldom carried out in Africa. Using a large family of Algerian origin, with 21 individuals affected with MFT in four consecutive generations, Salhi et al. (2004) performed linkage, LOH analysis, and mutation screening, and ruled out the possibility that disease-causing mutations lie on 9p, as well as LOH in tumors on 9q. They identified a 1 bp germline frameshift deletion in *CYLD*, along with LOH or a somatic point mutation at 16q detected in individual tumors.

AUTOSOMAL RECESSIVE DISEASE

FANCONI ANEMIA (FA)

Fanconi anemia (FA) is an inherited genomic instability disorder. FA is characterized by diverse clinical features including physical abnormalities (congenital abnormalities and endocrine abnormalities such as skin pigmentation, cardiac/renal/limb malformations, skeletal deformities, neurological abnormalities, ear anomalies, genitalia abnormalities, short stature), bone marrow failure (progressive pancytopenia), and increased risk of cancers (such as leukemia and breast cancer; see Kalb, Neveling et al., 2006; Lindor, McMaster et al., 2008; Auerbach 2009). It is estimated that the prevalence of FA is 1 in 100,000 live births, and the carrier (heterozygote) frequency is 1 in 300 within the general populations of the United States, Europe, and Japan (Swift, 1971), while higher prevalence is reported within Ashkenazi Jew, Spanish gypsy, as well as white Afrikaans-speaking and black Bantu-speaking South African populations (Rosendorff, Bernstein et al., 1987; Kutler & Auerbach, 2004; Callen, Casado et al., 2005; Morgan, Essop et al., 2005).

Genetically, FA is a heterogeneous syndrome; thirteen complementation groups and their corresponding genes have been identified to date (see Table 44-1). All the FA genes are inherited in an autosomal recessive manner except *FANCB*, which is X-linked recessive. The common complementation groups are A, C, and G; group A accounts for approximately 66%, and groups C and G account for 9.5% and 8.7%, respectively. The molecular basis of FA genes has been elucidated in recent years.

The proteins encoded by these genes participate in the FA pathway, which regulates cellular resistance to DNA damage and is involved in the maintenance of DNA stability during the S phase of the cell cycle (Joenje & Patel, 2001; Mathew, 2006; Jacquemont & Taniguchi, 2007; Patel & Joenje, 2007; Wang, 2007; Moldovan & D'Andrea, 2009). Most FA genes have a wide spectrum of mutations in FA patients that include nonsense mutations, deletions, frameshift mutations, splicing variants, and missense mutations (Joenje & Patel, 2001). The Fanconi Anemia Mutation Database (http://www.rockefeller.edu/fanconi/mutate/) was established to collect mutations in FA genes. To date, there are 631 unique mutations deposited in the database, with mutations in *FANCA* predominating (56%) among all FA genes. Founder mutations have been identified in different populations, such as Ashkenazi Jewish, Afrikaners, Spanish gypsies, and sub-Saharan Africans (Whitney, Saito et al., 1993; Tipping, Pearson et al., 2001; Callen, Casado et al., 2005; Morgan, Essop et al., 2005). Correlation between mutations in FA genes and the severity of the clinical phenotypes has been pointed out. In addition, associations of FA heterozygosity with breast and other cancers have been reported. Recent discoveries have demonstrated that the similar clinical phenotypes caused by biallelic mutations in *BRCA2/FANCD1* and *PALB2/FANCN* (PALB2 protein is a molecular partner of BRCA2) may strengthen the association between FA and hereditary breast and/or ovarian cancer (Reid, Schindler et al., 2007; Xia, Dorsman et al., 2007). However, researchers also found that there was no increased risk for cancer among FA heterozygote relatives overall, although *FANCC* mutations might contribute to breast cancer (Berwick, Satagopan et al., 2007).

Some FA studies in Africa have been published, including chromosomal instability/aberration studies in Egyptian FA patients (Al-Haggar, Al-Morsy et al., 2008) and families from South Africa (Pearson, Jansen et al., 2001), as well as other genetic investigations for mutations in FA genes. Macdougall et al. suggested that FA in blacks appears to be clinically indistinguishable from that in other groups, except the Afrikaners (Macdougall, Greeff et al., 1990; Alter 1992). However, Macdougall et al. later pointed out that there was no notable difference between black (40 Bantu-speaking children) and white (31 Afrikaans-speaking children) FA patients (Macdougall, Rosendorff et al., 1994). Exceptionally high incidence of FA in white, Afrikaans-speaking South Africans was reported to be at least 1 in 22,000 and it could be due to founder mutations (Rosendorff, Bernstein et al., 1987). This was strongly solidified by a linkage analysis of FA complementation group A gene on chromosome 16q242.3 in a total of 21 Afrikaner FA families (both parents and at least three grandparents were Afrikaans-speaking) (Pronk, Gibson et al., 1995). Gibson et al. (1994) conducted a genetic mapping study of the *FANCC* gene, using 36 multi-ethnic FA families composed of 13 Afrikaner and 2 black South African families, as well as 21 families from the Middle East, Europe, and Brazil. They identified the candidate locus on 9q22.3, and also found there was no linkage signal at the *FANCC* locus in those 13 Afrikaner FA families, possibly because of its nature of predisposing to the *FANCA* gene instead. Later, Gibson et al. (1996) screened the *FANCC* gene looking for mutations in 76 unrelated FA patients of diverse ethnic origins, including 24 South African Afrikaners and 9 South African blacks, followed by a population investigation of polymorphism frequency in 33 Tswana or 15 Sotho chiefdoms of South Africa. In total, five mutations were detected in 4 out of 76 patients (5.3%), including three previously identified truncating mutations (Q13X, R185X, and R548X), one novel missense mutation L496R, and one novel truncating mutation W22X, while no mutations were found in the 24 Afrikaner FA families. They also identified nine polymorphisms, six of them detectable in patients or controls from the Tswana or Sotho chiefdoms, but absent in 50 unrelated European controls. In addition, their finding of FA complementation group C subtype accounted for less than 10% for all the cases. Twenty-six FA families from the Afrikaner population of South Africa were enlisted for a molecular and genealogical study for a founder effect by Tipping et al. (2001). Four microsatellite markers flanking the *FANCA* gene and six single nucleotide polymorphisms (SNPs) in intron 31 were genotyped. Five haplotypes were found to be associated with four mutations (Del E12-31, Del E11-17, 3398delA and 795-808del), while the first deletion accounted for 60% of all FA alleles and the second and third mutations for 20% in 46 unrelated Afrikaner FA patients. Additional FA patients, including 26 French, 18 German, and 8 Dutch, were then screened for these four mutations. The result showed that one patient from the western Ruhr region of Germany harbored a heterozygous major deletion (Del E12-31) in Afrikaners. Another mutation investigation was performed in the *FANCG* gene by Morgan et al. (2005) in a total of 40 unrelated black FA patients from South Africa, Swaziland, Mozambique, and Malawi. All patients were found to carry a homozygous deletion (c.637_643delTACCGCC), which was present in 1% of control populations and found to be in strong linkage disequilibrium with a SNP in intron 1 (c.84 + 77A > C). The comparison of clinical data from 20 black South Africans and 20 Europeans indicated a remarkable frequency of distal radial ray abnormalities, growth retardation, and eye abnormalities in black FA-G patients. Evidence of biallelic mutations in *PALB2* causing Fanconi anemia subtype FA-N in seven families was described by Reid et al. in 2006 (Reid, Schindler et al., 2007). Parents of one affected individual were found to carry 395delT and 3113+5G>C mutations related to parental origins of Albanian and Moroccan, respectively.

ALPHA-1 ANTITRYPSIN DEFICIENCY (α1ATD, AATD)

AATD is a heritable autosomal recessive metabolic disease that is regarded as one of the most common serious hereditary disorders in the world. Patients with AATD often present early onset of panlobular pulmonary emphysema in adults, especially in habitual tobacco smokers. The disease also presents development of neonatal cholestasis that may progress to infant and juvenile cirrhosis. This degeneration of liver cells may slowly progress to liver disease, such as hepatocellular cancer (Fargion, Klasen et al., 1981), in up to 40%–50% of the adult patients due to a pathologic aggregation of abnormal AAT polymers in hepatocytes. The risk of cirrhosis and liver cancer was shown to increase with AATD homozygous males but not with females, suggesting additive effects of exogenous factors (Eriksson, Carlson et al., 1986). Other common clinical symptoms that AATD is suspected to promote are asthma, bronchiectasis, systemic vasculitis, relapsing panniculitis, rheumatoid arthritis, intracranial arterial dissections, multiple sclerosis, as well as lung and liver cancer.

The *AAT* (*SERPINA1*) gene resides within a cluster of 11 different *SERPIN* genes (SERine Proteinase INhibitor) at 14q32 (Brantly, Nukiwa et al., 1988), with three noncoding and four coding exons spanning over 12 kb (Blank & Brantly, 1994). The *AAT* gene is highly polymorphic, as more than 75 different alleles were identified (Axelsson & Laurell, 1965; Hug, Chuck et al., 1981; Nukiwa, Brantly et al., 1988). The normal gene product is designated PIM. The two most frequent deficient alleles are PIS (50%–60%) and PIZ (10%–20%). In clinical practice, about 95% of AAT deficiency–related diseases are linked with the PIZZ phenotype. It was shown that non-M phenotypes are common among AAT-deficient hepatocellular carcinoma patients (Fargion, Klasen et al., 1981). Moreover, patients with liver cancer and non-M phenotype are of a lower average age than those with M phenotype. Although both carriers and deficiency allele combinations for the PIS and PIZ defective alleles (namely PIMS, PIMZ, PISS, PISZ, and PIZZ phenotypes) are at risk for all the aforementioned adverse health effects and others (Carrell & Lomas, 2002), clear scientific evidence of the relationship among AATD and some of these diseases remain to be established. At the moment, the only definitive evidence for lung and liver disease and cancer is with the PIMZ, PISZ, and PIZZ phenotypes, with the level of risk being much higher in the latter phenotypic class (Aagenaes, Matlary et al., 1972; Lomas, Evans et al., 1992; Sigsgaard, Brandslund et al., 1994; Wu, Whitman et al., 1994; Rodriguez-Cintron, Guntupalli et al., 1995; Dahl, Tybjaerg-Hansen et al., 2002).

AATD was once believed to be a disease primarily affecting Northern European whites, and was considered virtually nonexistent in Africa. This distorted view was a consequence of early studies in which small samples of patients who showed symptoms of AATD were most common in Northern European populations and absent elsewhere. However, global meta-analysis and population genetics studies have shown that AATD is ubiquitous (Blanco, Bustillo et al., 2001; de Serres 2002), and has been clinically misinterpreted most of the time as episodes of lung/liver disease (de Serres, 2003). In fact, deficiency alleles were found to vary widely in different populations. For example, the Z allele was reported in 1%–2% of U.S. whites, 0.48% in Africans, 0.4% in central Asians, and 1.51% of Australians and New Zealanders. The S allele, on the other hand, was reported in 2%–4% of U.S. Whites, 3.1% of Africans, 0.43% in central Asians, and 3.95% in Australians and New Zealanders. Focusing on sub-Sahara African populations, Angola and Namibia have the highest deficiency genotypes (54 per 1000 and 22 per 1000, respectively) followed by Cape Verde (15 per 1000), despite the prevalence of AATD being lower compared to Europeans (Spinola, Brehm et al., 2010; Ankrah & Ekuban, 1994; Tete-Benissan, Duriez et al., 2000). Similarly, the frequency of the S allele was found to be most common in Angola (18.8%), Namibia (14.7%), Nigeria (6.4%), Botswana (4.5%), South Africa (3.3%), and Cape Verde (3.3%). On the other hand, the Z allele was found in Somalia (1.15%), Mali (0.98%), Nigeria (0.36%), and Cape Verde (0.2%), but missing in other sub-Saharan populations including those in Botswana, Congo, Cameroon, Angola, Gambia, South Africa, Mozambique, and Namibia. In general, the PiM1 allele is highly frequent in the Black population, but the PiM2 allele is less common compared to other populations in the world (Spinola, Brehm et al., 2010; Massi & Vecchio, 1977; Frants & Eriksson, 1978; Houvet & Clerc, 1978; Ojutiku & Ladipo, 1978; Constans, Viau et al., 1981; Bowry, Pade et al., 1982; Olusi, Ladipo et al., 1982; Dunn, Madhoo et al., 1986; Pascali, Ranalletta et al., 1986; Willcox, Beckman et al., 1986; Hitzeroth, Bissbort et al., 1987; Onwuameze 1989; DeCroo, Kamboh et al., 1991). Very little has been done in regard to Northern African populations, and thus, it is only speculated that both S and Z allele are extremely rare or missing in Tunisian, Algerian, and Moroccan populations (Ward, Pickering et al., 1977; Chaabani, Martin et al., 1984; Nevo & Cleve, 1991; Sebetan 1992; de Serres 2002; Denden, Haj Khelil et al., 2008; Denden, Braham et al., 2009). One characteristic outcome from African population studies is that some rare alleles not found in other continents were observed several times, some of which implied inefficient transcript products but not defective alleles (Vandeville, Martin et al., 1974; Khitri, Benlatrache et al., 1977; Welch, McGregor et al., 1980; Chaabani, Martin et al., 1984; Dunn, Madhoo et al., 1986; Sebetan 1992; Hayes 2003).

ATAXIA TELANGIECTASIA (AT)

As an autosomal recessive disorder with an estimated incidence of 1 in 30,000–100,000 live births in Caucasian populations, ataxia telangiectasia (AT) has a variable clinical manifestation. In most countries, progressive cerebellar ataxia (in 100% of cases) occurs in early childhood, while oculocutaneous telangiectasia begins in late childhood. Furthermore, oculomotor apraxia, sinopulmonary infections, endocrine dysfunction (glucose intolerance, hypogonadism, and impaired fertility), immunodeficiency (IgE and IgA deficiency in 85% and 75% of patients, respectively), and choreoathetosis are frequently involved in AT (Lindor, McMaster et al., 2008). There is also a high risk of the development of malignancy and increased sensitivity to ionizing radiation. Patients with AT have a 100-fold increased risk for developing cancer, and 15% of all AT patients will die of cancer (Morrell, Cromartie et al., 1986; Lindor, McMaster et al., 2008). AT has been reported to be associated with an increased risk of malignancy, such as non-Hodgkin's lymphoma, leukemia, ovarian cancer, and gastric cancer. In addition, heterozygous mutation carriers with AT reveal high risk of breast cancer (Swift, Reitnauer et al., 1987; Swift, Morrell et al., 1991; Athma, Rappaport et al., 1996).

The gene predisposed to AT, *ATM* (at 11q22.3) was initially mapped by linkage and positional cloning (Gatti, Berkel et al., 1988; Savitsky, Bar-Shira et al., 1995). Four genetically heterogeneous complementation groups of AT were found to be due to mutations in the same gene. The *ATM* gene has 66 exons and encodes a protein of 3056 amino acids. The protein belongs to the PI3K-related protein kinases (PIKK) family and is involved in cellular processes ranging from the repair of DNA double-strand breaks to regulation of telomere length (Ahmed & Rahman, 2006). *ATM* mutations responsible for AT in biallelic carriers are breast cancer susceptibility alleles in monoallelic carriers, with an estimated relative risk of 2.37 (Renwick, Thompson et al., 2006). Tavtigian et al. (2009) found stronger evidence of a subset of rare, evolutionarily unlikely missense substitutions contributing to breast cancer risk, besides marginal evidence of *ATM* protein-truncating and splice-junction variants. Accumulative mutations in *ATM* were identified and deposited in the online *ATM* mutation database (http://www.LOVD.nl/ATM). On November 23, 2007, a total of 432 unique DNA variants had been reported in various ethnic groups, with a wide spectrum and truncating mutations predominating. Global analysis of *ATM* polymorphism using 93 individuals from 7 major human populations (18 Africans) revealed a total of 7 haplotypes of *ATM* (Thorstenson, Shen et al., 2001). Haplotype diversity was greatest in Africans, as well as nucleotide diversity, compared to other populations. Two haplotypes sharing the same D126E missense mutation accounted for 33% of chromosomes in Africa but absent elsewhere, while an A259S missense mutation was found in heterozygous status in nearly half of non-Africans but was never identified in Africans. Jain et al. (2003) also reported three haplotypes were concentrated in 39 samples from Northern Africa, 27 from Eastern Africa, 54 from Western Africa, 20 from Central Africa, and 13 from Southern Africa.

AT is a rare disease in Africa. Of 2.1 million patients seen in 25 years at the University College Hospital, Ibadan, Nigeria, Aiyesimoju et al. (1984) reported only two AT patients. One study in Libya gave a crude prevalence of 4.8/100,000 in Benghazi for spinocerebellar degenerations. Their samples consisted of 13 patients of early onset cerebellar ataxia with retained tendon reflexes (EOCA), 5 patients of late onset cerebellar ataxia (LOCA) with pigmentary retinal degeneration, 6 single cases of LOCA, and 1 AT patient (Sridharan, Radhakrishnan et al., 1985). Bejaoui et al. (1997) performed the immunological investigations in 153 cases and discovered a cellular or combined immunodeficiency with a majority of the 53 AT cases. They suggested that the relative frequency of primary immunodeficiencies in Tunisia was probably due to the high rate of consanguinity among the general population. One study in South Africa indicated that the studied nonconsanguineous family of mixed ancestry (four siblings) might present a rare variant of AT (de Graaf, de Jong et al., 1995). By investigating 33 Jewish AT patients originating from various regions of Morocco and Tunisia, Gilad et al. (1996) found a single truncating mutation (103C>T) which was proposed to be a founder mutation among Northern African Jews. In 24 cases of AT seen between 1992 and 1997 in 17 families in southern Tunisia, Triki et al. (2000) found that low levels of IgA were present in 18 patients (low IgE in 7 and low IgG in all 24 patients). Among 12 patients, chromosomal instability was observed in five patients with balanced rearrangements involving chromosomes 2, 7, 14, 22, 1, 3 and 11.

OCULOCUTANEOUS ALBINISM (OCA)

Oculocutaneous albinism (OCA) is a genetically heterogeneous, autosomal recessive congenital disorder characterized by a decrease in or absence of pigmentation in hair, skin, and eyes. There are four types of oculocutaneous albinism (namely, OCA1, OCA2, OCA3, and OCA4), each having different severity of affliction and phenotypic traits, as they are caused by mutations in different genes. In general, clinical features of OCA depend on the level of pigmentation of the affected individual, which makes OCA a quantitative trait disorder with phenotypic variation in individuals of different ethnic background (Chiang, Drautz et al., 2008). No sex bias has been reported, thus this disorder equally affects both males and females (Scialfa, 1972; O'Donnell, King et al., 1978). Among the array of clinical traits observed, one feature OCA patients commonly experience is ophthalmologic anomalies; iris transillumination, nystagmus, strabismus, high refractive errors, foveal dysgenesis, chorioretinal hypopigmentation, and albinotic optic disks are some of the commonly observed complications in OCA patients (Lee, Nicholls et al., 1994). Studies have shown that optic irregularities are mainly caused by deformities in the visual system of the central nervous system (e.g., optic axon and chiasm) (Creel, Garber et al., 1980; Summers, Creel et al., 1991; Schmitz, Schaefer et al., 2003; Brodsky & Fray, 2004) and also by abnormal development of the eye (e.g., foveal pit; see Witkop, Jay et al., 1982; Meyer, Lapolice et al., 2002; Recchia, Carvalho-Recchia et al., 2002). Since the prevalence of OCA is significantly higher in isolated populations or tribes, consanguinity has been suggested to be an important factor that causes OCA (Okoro, 1975; Aquaron, 1980; Kromberg & Jenkins, 1982; Aquaron, 1990; De Jong & Fryns, 1991; Durham-Pierre, Gardner et al., 1994; Kagore & Lund, 1995; Lund, Puri et al., 1997; Lund, Maluleke et al., 2007). The evident difference in physical appearance of those with OCA in African populations can lead to discrimination of these individuals and possibly the entire tribe to which they belong (Gaigher, Lund et al., 2002; McBride & Leppard, 2002; Hong, Zeeb et al., 2006). Furthermore, since OCA patients are photosensitive, chronic sun exposure, as well as lack of education and poor political intervention to support affected individuals, have led to high incidence of skin cancer among affected individuals in Africa (Luande, Henschke et al., 1985; Kromberg, Castle et al., 1989; Lund 2005; Adegbidi, Yedomon et al., 2007; Lund & Taylor, 2008).

The human tyrosinase (*TYR*) gene is the causative gene of OCA1; based on activity level of the *TYR* gene, OCA1 is further divided clinically into OCA1A and OCA1B. OCA1A is characterized by a complete lack of TYR activity due to production of inactive enzymes, whereas OCA1B has reduced activity of TYR (King & Witkop, 1977). TYR is involved in the conversion of tyrosine to melanin using copper molecules (Spritz 1994; Petris, Strausak et al., 2000; Setty, Tenza et al., 2008), and OCA1B patients are known to retain residual amounts of eumelanin in their body, unlike patients with OCA1A (Nance, Jackson et al., 1970; Hu, Hanifin et al., 1980). Patients with OCA1B are considered dead white at birth, with serious ocular abnormalities like OCA1A patients, but since they retain some function of the *TYR* gene, develop minimal amounts of pigment as they develop (King, Wirtschafter et al., 1986). Consequently, they also acquire some color in their hair, iris, and skin as they mature (Nance, Jackson et al., 1970). The tyrosinase gene is located on chromosome 11q14-q21, and has 4 introns and 5 exons (Barton, Kwon et al., 1988; Giebel, Strunk et al., 1991; Ponnazhagan, Hou et al., 1994). Over 60 independent *TYR* gene alleles known to cause albinism have been reported, and about 90% of OCA in Caucasians were accounted for by 29 mutations (Oetting & King, 1992; Tripathi, Strunk et al., 1992; Oetting & King. 1993, 1999; Passmore, Kaesmann-Kellner et al., 1999). Some mutations have been suggested to be specific to OCA1B (Giebel, Strunk et al., 1990; Giebel, Musarella et al., 1991; King, Townsend et al., 1991), and two SNPs were indicated to increase susceptibility to melanoma (Gudbjartsson, Sulem et al., 2008; Bishop, Demenais et al., 2009). Eighty percent of the mutations cluster within two relatively small regions of the *TYR* gene, one involved in the copper binding site (King, Mentink et al., 1991) and the other suggested to play a role in protein folding domain. Since OCA1 is rare in Africa, only two mutations, Gly47Asp (G47D) and c.781_784delAACT, have been reported among Moroccan Jews and a Cameroonian black patient, respectively (Gershoni-Baruch, Rosenmann et al., 1994; Badens, Courrier et al., 2006).

Worldwide, OCA type 2 is the predominant type, with highly variable phenotypes (Lee, Nicholls et al., 1994). A large fraction of OCA morbidity throughout sub-Saharan Africa is attributed to OCA2, commonly causing complications such as skin cancer and gross visual impairment (Lee, Nicholls et al., 1994). The symptoms, however, are milder than those associated with OCA1, and some cases have similar characteristics of OCA1B patients who acquire pigmentation as they mature (O'Donnell, King et al., 1978; Lee, Nicholls et al., 1994). In Africans and African Americans, individuals with OCA2 typically have yellow hair, blue-gray or hazel irides, and develop many pigmented spots in the skin. The brown OCA (BOCA), a phenotypic variant, has been described in African populations, with an estimated gene frequency of 0.025 +/– 0.007 (King & Rich, 1986). Affected individuals are characterized by light brown hair and skin color, and gray to tan irides (Kromberg, Castle et al., 1989). OCA2 is caused by mutation in the *OCA2* gene, a homolog to the pink-eyed dilution mouse gene (*P* gene; see Ramsay, Colman et al., 1992; Kedda, Stevens et al., 1994). The human iris color is linked to *OCA2,* which is located at 15q11.2-q12 (Zhu, Evans et al., 2004; Duffy, Montgomery et al., 2007). The gene contains 25 exons, and its product plays a role in regulating the pH of melanosomes (Donlon, Lalande et al., 1986; Knoll, Nicholls et al., 1990; Robinson, Bottani et al., 1991; Gardner, Nakatsu et al., 1992; Rosemblat, Durham-Pierre et al., 1994; Lee, Nicholls et al., 1995; Yuasa, Umetsu et al., 2007). Over 50 different mutations have been reported (Lee, Nicholls et al., 1994; Kerr, Stevens et al., 2000; Kawai, Suzuki et al., 2005), and certain mutations in *OCA2* have been suggested to be associated with skin cancer risk in Caucasian populations (Jannot, Meziani et al., 2005). A 2.7 kb deletion, which causes premature truncation of the protein product by removing exon 7 (Durham-Pierre, Gardner et al., 1994; Chiang, Spector et al., 2008), is the most common mutation found in black Africans, but rarely found in Caucasians (Stevens, van Beukering et al., 1995). A large proportion of BOCA and OCA patients in black African populations are compound heterozygotes for the 2.7 kb deletion and represent a high proportion of mutated *P* alleles in central, eastern and southern African countries (Spritz, Fukai et al., 1995; Stevens, van Beukering et al., 1995; Puri, Durbam-Pierre et al., 1997; Stevens, Ramsay et al., 1997), but are rarely found in western Africa (Spritz, Fukai et al., 1995; Aquaron, Soufir et al., 2007). The predominant occurrence of mutated *P* alleles in Africa suggests that this mutation originated in Africa about 3000 to 5600 years ago, and its association with common haplotype indicates a founder effect (Stevens, van Beukering et al., 1995; Stevens, Ramsay et al., 1997). As OCA2 is common in Africa, the incidence frequency among Black populations is significantly higher compared to Caucasians (Barnicot, 1952; Okoro, 1975; Aquaron, 1980; Kromberg & Jenkins, 1982; Aquaron, Kamdem et al., 1984; Luande, Henschke et al., 1985; Aquaron 1990; Rinchik, Bultman et al., 1993; Kagore & Lund, 1995; Durham-Pierre, King et al., 1996; Lund 1996; Hong, Zeeb et al., 2006; Aquaron, Soufir et al., 2007; Lund, Maluleke et al., 2007).

OCA type 3, also known as *red* or *rufous* oculocutaneous albinism (ROCA), is characterized by bright copper-red coloration of skin and hair, nystagmus, and dilution of the iris color (Manga, Kromberg et al., 1997). First found in southern African blacks, the characteristic freckled skin, reddish hair, and red reflex on transillumination of the iris, led OCA3 to be classified as a new variant of tyrosine-positive OCA (e.g., OCA1B or OCA2; see King, Lewis et al., 1985). Further research made evident that OCA3 is caused by mutation in tyrosinase-related protein-1 (*TYRP1*) on chromosome 9p23, a human homolog of the mouse *brown (b)* locus gene (Cohen, Muller et al., 1990; Abbott, Jackson et al., 1991; Chintamaneni, Ramsay et al., 1991; Murty, Bouchard et al., 1992). *TYRP1* contains 7 exons spread over 24 kb of genomic DNA, has a LINE-1 repeat element immediately adjacent to its 5-prime promoter region influencing the gene expression, and is speculated to have derived from duplication of the *TYR* gene (Sarangarajan & Boissy, 2001). The *TYRP1* gene encodes a glycoprotein that functions as a catalase involved in melanogenesis, which stabilizes tyrosinase and increases the yield of melanin production (Halaban & Moellmann, 1990; Sarangarajan & Boissy, 2001). There are a few well-characterized mutations in the *TYRP1* gene, including SNPs associated with increased susceptibility to cutaneous malignant melanoma (Forshew, Khaliq et al., 2005; Rooryck, Roudaut et al., 2006; Gudbjartsson, Sulem et al., 2008; Sulem, Gudbjartsson et al., 2008). Of all the known mutations, however, 1 bp deletion (368delA) and nonsense mutation (S166X) are the most commonly recurring mutations that cause premature truncation of products that lead to reduced stability of the protein (Boissy, Zhao et al., 1996). In southern African blacks, OCA accounts for three types, of which OCA2 is the most common and BOCA and ROCA account for the majority of the remaining cases. The incidence rate of ROCA was estimated to be approximately 1 in 8580 with a gene frequency of 0.025 +/− 0.007, and 368delA and S166X were the two most common mutations in *TYRP1* (King & Rich, 1986; Kromberg, Castle et al., 1990; Manga, Kromberg et al., 1997). In addition, the characteristic reddish skin color of OCA3 patients in black Africans is the result of increased pheomelanin synthesis rather than reduction of pigments; thus, OCA3 patients have less sun sensitivity compared to other OCA types (Kidson, Richards et al., 1993).

OCA type 4 is a rare form of OCA, most common among Japanese patients (Inagaki, Suzuki et al., 2004). It is caused by mutations in the *MATP* gene on chromosome 5p13.3, and contains 7 exons spanning approximately 40 kb (Newton, Cohen-Barak et al., 2001; Rundshagen, Zuhlke et al., 2004). The *MATP* is transcriptionally modulated by melanocyte-specific transcription factor, MITF, and encodes a melanocyte differentiation antigen that is highly expressed in melanoma cells (Du & Fisher, 2002). The product is thought to span the lipid bilayer several times, functioning as a transporter that mediates melanin synthesis (Fukamachi, Shimada et al., 2001). Several population-specific studies have identified polymorphisms in the *MATP* gene causing normal skin color variation, of which promoter haplotypes are known to decrease MATP transcription activity (Graf, Hodgson et al., 2005; Graf, Voisey et al., 2007; Stokowski, Pant et al., 2007). Since OCA4 is very rare in Africa, it is only documented in Turkish (Newton, Cohen-Barak et al., 2001) and Moroccan reports (Rooryck, Morice-Picard et al., 2008; Konno, Abe et al., 2009), with two known population-specific mutations, p.K247X and p.H38R.

XERODERMA PIGMENTOSUM (XP)

Xeroderma pigmentosum (XP) is inherited in an autosomal recessive manner; namely, heterozygotes are unaffected but homozygotes have severe clinical phenotype, like extraordinary hypersensitivity to sun exposure. Other clinical features include photophobia, early onset of freckling, and subsequent cutaneous malignancy on skin (basal cell and squamous cell carcinomas and melanomas) and eyes (epitheliomas, squamous cell carcinomas and melanomas) (Kraemer, Herlyn et al., 1989; Kraemer, Lee et al., 1994; Gratchev, Strein et al., 2003). Xeroderma pigmentosum is a rare disease, with the incidence of approximately 1 in 250,000 in the United States (Robbins, Kraemer et al., 1974) but with higher incidence in Japan (1 in 40,000; see Takebe, Nishigori et al., 1987) and in Mediterranean areas (Cleaver, Thompson et al., 1999). Prevalence is about 1 in 22,000 in Japan (Hirai, Kodama et al., 2006) but is more common in Northern Africa (Tunisia, Algeria, Morocco,

Libya, and Egypt) and the Middle East (Turkey, Israel, and Syria), especially in communities in which consanguinity is common (GeneReviews, http://www.ncbi.nlm.nih.gov/bookshelf/br.fcgi?book=gene&part=xp). XP is genetically heterogeneous, and several causative genes have been identified among XP patients: *XPA* (9q22.33), *ERCC3* (*XPB*, 2q14.3), *XPC* (3p25.1), *ERCC2* (*XPD*, 19q13.32), *DDB2* (*XPE*, 11p11.2), *ERCC4* (*XPF*, 16p13.12), *ERCC5* (*XPG*, 13q33.1), and *POLH* (*XP-V*, 6p21.1). The first seven genes are involved in nucleotide excision repair, and the last gene is involved in replication of damaged DNA on the leading strand (Cleaver, Thompson et al., 1999; Gratchev, Strein et al., 2003). An online database (Allelic Variations of the XP Genes, http://xpmutations.org/) hosts the mutations that have been detected (Cleaver, Thompson et al., 1999).

Besides some epidemiological studies or case reports in Tunisia (Fazaa, Zghal et al., 2001; Zghal, Fazaa et al., 2003; Maalej, Hentati et al., 2007), Morocco (Moussaid, Benchikhi et al., 2004), Egypt (Hashem, Bootsma et al., 1980; Cleaver, Zelle et al., 1981; German, Hashem et al., 1984), Cameroon (Moussala, Behar-Cohen et al., 2000), Nigeria (Ahmed, Hassan et al., 2001), Libya (Khatri, Shafi et al., 1992; Visweswara, Osman et al., 1997; Khatri, Bemghazil et al., 1999), Zimbabwe (Chidzonga, Mahomva et al., 2009), and South Africa (Jacyk 1999), few genetic studies have been performed to examine mutations of XP genes in Africans. Nishigori et al. (1993) found 6 out of 7 (86%) Tunisian XPA patients carried a homozygous nonsense mutation at codon 228 in exon 6 of the *XPA* gene, with an absence of a mutation at the splicing junction of intron 3 and exon 4 predominating in Japanese XPA patients. Given the milder clinical symptoms of those R228X mutation carriers, their results were consistent with the findings that more severe cases tend to have homozygous truncating mutations within the DNA binding region of the protein (States, McDuffie et al., 1998). Khan et al. (2006) conducted mutation detection and real-time quantitative reverse transcriptase PCR to study the *XPC* mRNA level changes. They detected 15 culprit truncating mutations (5 frameshift, 6 nonsense, and 4 splicing) in the *XPC* gene in cells from 16 XP probands in which salient evidence of reduced *XPC* mRNA levels was observed, compared to heterozygous parents and normal controls. Nine males and 11 females with XPC from 14 families, mainly originating from coastal regions of Tunisia, were screened for the same V548AfsX572 mutation in exon 9 of *XPC* (Ben Rekaya, Messaoud et al., 2009). All the patients were found to bear this truncating mutation, and founder effect of this mutation was also investigated and confirmed.

SUMMARY

To date, more than 50 human familial cancer syndromes have been identified, while most of them are rare and follow the role of Mendelian inheritance. In spite of the fact that many publications are case reports and retrospective studies based on clinical records, few genetic ascertainments focusing on familial cancers have been carried out in Africa, in view of the rarity of diseases, as well as the consciousness and capability of African countries to perform academic and clinical research. Only a handful of familial cancer syndromes have been investigated, including some examples described in this chapter. Several familial cancers were found to be common or to have higher prevalence in the continent of Africa. For instance, breast cancer is the leading cause of cancer deaths among women in Nigeria, and this disease is characterized with younger age onset, severity of tumor development, and poor clinical outcomes. Fanconi anemia has been reported to have a higher prevalence in white Afrikaans-speaking and black Bantu-speaking South African populations, compared to developed geographic regions like the United States and Europe. Many mutational screenings of cancer-causative genes have been conducted in developing countries like Nigeria (*BRCA1/2*), Morocco (*RET*), Tunisia (*EXT1*), Algeria (*CYLD*), and South Africa (*FANCC*), as the first three are lower–middle income countries and the last two are upper–middle income countries. As the founder population of Blacks around the world during the African diaspora, the Nigerian population has suffered a marked burden of breast cancer and thus has attracted active international collaborations for breast cancer research. Similarly, Moroccan and Tunisian populations were involved in several international studies because Moroccan Jews historically connect to other Jewish populations in the United States or Israel, and the Tunisian population is a highly consanguineous cohort that is beneficial for mutation detection and risk assessment of recessive inherited disease-causing genes. However, it is difficult to investigate genotype–phenotype correlations in cancer susceptibility genes, simply because there is a lack of available mutation profiles and clinical details. Remarkably, the paucity of cancer investigations presents in central Africa. Africa is at a time of unique opportunities to significantly reform healthcare, notably through ensuring adequate investments in health delivery systems. The past decade marked improvements in public sectors with the help of foreign aid. Accelerated involvement of private sectors has been heavily promoted also. The wellness of the continent of Africa is in dire condition, and there is much to be done. An improved socioeconomic situation, awareness of cancer perniciousness, and enhanced medical care systems are necessary to accelerate the implementation of cancer studies in Africa.

- In general, familial cancer syndromes are rare, accounting for approximately 5%–10% of all cancers. Both oncogenes and tumor suppressor genes are involved in familial cancer etiology; they can be transmitted in an autosomal dominant, autosomal recessive, or X-linked manner.

- Other than rudimental studies like case reports, a limited number of mutation analyses of cancer susceptibility genes have been performed in Africa, especially in northern (e.g., Morocco, Tunisia), southern (e.g., South Africa), and western Africa (e.g., Nigeria). The struggling African economy limits funding and effective implementation of healthcare policy, which impedes the delivery of adequate healthcare to the population. Moreover, the weak healthcare system negatively impacts societal involvement and investment in biomedical research.
- Spectra of mutations found in some causative genes appear to be wide, and the mutational distributions can significantly vary between distinct ethnic groups, which is the case with *BRCA1* and *BRCA2* mutations. Some research findings have revealed obvious differences between clinicopathological features in African breast cancer patients as compared to Caucasians in developed countries. Genotype–phenotype correlations of cancer syndromes have been seldom analyzed, considering the insufficient number of deleterious mutations detected in corresponding cohorts.
- African populations have strikingly higher levels of genetic diversity and ethnic constitutes. However, unique genetic manifolds of African tribes and familial clustering within a small area may help genetic mapping of familial trait/disease genes. Also, consanguineous cohorts in northern Africa will benefit from increased understanding of causative genes of human recessive diseases.
- Several founder mutations in some hereditary cancers have been identified in African populations. The findings will subsequently benefit clinical screening, guide cancer treatment, and expand translational research in the continent of Africa.

REFERENCES

Aagenaes O, Matlary A, et al. (1972). Neonatal cholestasis in alpha-1-antitrypsin deficient children. Clinical, genetic, histological and immunohistochemical findings. *Acta Paediatr Scand 61*(6), 632–642.

Abbott C, Jackson IJ, et al. (1991). The human homolog of the mouse brown gene maps to the short arm of chromosome 9 and extends the known region of homology with mouse chromosome 4. *Genomics 11*(2), 471–473.

Abou-Zeid AA, Khafagy W, et al. (2002). Colorectal cancer in Egypt. *Dis Colon Rectum 45*(9), 1255–1260.

Adebamowo CA, Adeyi O, et al. (2000). Case report on hereditary non-polyposis colon cancer (HNPCC) in Nigeria. *Afr J Med Med Sci 29*(1), 71–73.

Adebamowo CA, Ajayi OO. (2000). Breast cancer in Nigeria. *West Afr J Med 19*(3), 179–191.

Adegbidi, H., H. Yedomon, et al. (2007). Skin cancers at the National University Hospital of Cotonou from 1985 to 2004. *Int J Dermatol 46 Suppl 1*: 26–29.

Adesunkanmi AR, Lawal OO, et al. (2006). The severity, outcome and challenges of breast cancer in Nigeria. *Breast 15*(3), 399–409.

Ahmed H, Hassan RY, et al. (2001). Xeroderma pigmentosum in three consecutive siblings of a Nigerian family: observations on oculocutaneous manifestations in black African children. *Br J Ophthalmol 85*(1), 110–111.

Ahmed M, Rahman N. (2006). ATM and breast cancer susceptibility. *Oncogene 25*(43), 5906–5911.

Ainahi A, Kebbou M, et al. (2006). Study of the RET gene and his implication in thyroid cancer: Morocco case family. *Indian J Cancer 43*(3), 122–126.

Aissi-Ben Moussa S, Moussa A, et al. (2009). Identification and characterization of a novel MLH1 genomic rearrangement as the cause of HNPCC in a Tunisian family: evidence for a homologous Alu-mediated recombination. *Fam Cancer 8*(2), 119–126.

Aiyesimoju AB, Osuntokun BO, et al. (1984). Hereditary neurodegenerative disorders in Nigerian Africans. *Neurology 34*(3), 361–362.

Al-Haggar M, Al-Morsy Z, et al. (2008). Correlation of thyroid and growth hormones to chromosomal instability in Egyptian Fanconi anemia patients. *Indian J Pediatr 75*(7), 679–684.

Alter BP. (1992). Fanconi anemia in blacks. *Am J Med Genet 42*(3), 393–394.

Alvarez C, Tredwell S, et al. (2006). The genotype-phenotype correlation of hereditary multiple exostoses. *Clin Genet 70*(2), 122–130.

Ankrah NA, Ekuban FA. (1994). Evaluation of alpha-1-antitrypsin, alpha-1-acid glycoprotein and pre albumin in serum of Ghanaian subjects. *East Afr Med J 71*(7), 429–431.

Aquaron R. (1980). [Oculocutaneous albinism in Cameroun. About 216 cases (author's transl)]. *Rev Epidemiol Sante Publique 28*(1), 81–88.

Aquaron R. (1990). Oculocutaneous albinism in Cameroon. A 15-year follow-up study. *Ophthalmic Paediatr Genet 11*(4), 255–263.

Aquaron R, Kamdem L, et al. (1984). [Seroanthropological studies of albino and melanoderm Bamileke populations (Cameroon). ABO and Rhesus erythrocyte groups, hemoglobin S and taste sensitivity to phenylthiocarbamide]. *Med Trop (Mars) 44*(4), 311–318.

Aquaron R, Soufir N, et al. (2007). Oculocutaneous albinism type 2 (OCA2) with homozygous 2.7-kb deletion of the P gene and sickle cell disease in a Cameroonian family. Identification of a common TAG haplotype in the mutated P gene. *J Hum Genet 52*(9), 771–780.

Aquilina G, Hess P, et al. (1994). A mismatch recognition defect in colon carcinoma confers DNA microsatellite instability and a mutator phenotype. *Proc Natl Acad Sci U S A 91*(19), 8905–8909.

Athma P, Rappaport R, et al. (1996). Molecular genotyping shows that ataxia-telangiectasia heterozygotes are predisposed to breast cancer. *Cancer Genet Cytogenet 92*(2), 130–134.

Auerbach AD. (2009). Fanconi anemia and its diagnosis. *Mutat Res 668*(1–2), 4–10.

Awadelkarim KD, Aceto G, et al. (2007). BRCA1 and BRCA2 status in a Central Sudanese series of breast cancer patients: interactions with genetic, ethnic and reproductive factors. *Breast Cancer Res Treat 102*(2), 189–199.

Axelsson U, Laurell CB. (1965). Hereditary variants of serum alpha-1-antitrypsin. *Am J Hum Genet 17*(6), 466–472.

Badens C, Courrier S, et al. (2006). A novel mutation (delAACT) in the tyrosinase gene in a Cameroonian black with type 1A oculocutaneous albinism. *J Dermatol Sci 42*(2), 121–124.

Bane AL, Beck JC, et al. (2007). BRCA2 mutation-associated breast cancers exhibit a distinguishing phenotype based on morphology and molecular profiles from tissue microarrays. *Am J Surg Pathol 31*(1), 121–128.

Bapat BV, Madlensky L, et al. (1999). Family history characteristics, tumor microsatellite instability and germline MSH2 and MLH1 mutations in hereditary colorectal cancer. *Hum Genet 104*(2), 167–176.

Barnetson RA, Tenesa A, et al. (2006). Identification and survival of carriers of mutations in DNA mismatch-repair genes in colon cancer. *N Engl J Med 354*(26), 2751–2763.

Barnicot NA. (1952). Albinism in south-western Nigeria. *Ann Eugen 17*(Part 1), 38–73.

Barrow E, Alduaij W, et al. (2008). Colorectal cancer in HNPCC: cumulative lifetime incidence, survival and tumour distribution. A report of 121 families with proven mutations. *Clin Genet 74*(3), 233–242.

Barton DE, Kwon BS, et al. (1988). Human tyrosinase gene, mapped to chromosome 11 (q14—q21), defines second region of homology with mouse chromosome 7. *Genomics 3*(1), 17–24.

Bejaoui M, Barbouche MR, et al. (1997). [Primary immunodeficiency in Tunisia: study of 152 cases]. *Arch Pediatr 4*(9), 827–831.

Ben Rekaya M, Messaoud O, et al. (2009). High frequency of the V548A fs X572 XPC mutation in Tunisia: implication for molecular diagnosis. *J Hum Genet 54*(7), 426–429.

Benazzouz B, Chraibi A, et al. (2006). [Characterization of RET proto-oncogene C634Y mutation in a Moroccan family with multiple endocrine neoplasia type 2A]. *Ann Endocrinol (Paris) 67*(1), 21–26.

Benazzouz B, Hafidi A, et al. (2008). [C634R mutation of the proto-oncogene RET and molecular diagnosis in multiple endocrine neoplasia type 2 in a large Moroccan family]. *Bull Cancer 95*(4), 457–463.

Beroud C, Collod-Beroud G, et al. (2000). UMD (Universal mutation database), a generic software to build and analyze locus-specific databases. *Hum Mutat 15*(1), 86–94.

Beroud C, Soussi T. (1996). APC gene: database of germline and somatic mutations in human tumors and cell lines. *Nucleic Acids Res 24*(1), 121–124.

Berwick M, Satagopan JM, et al. (2007). Genetic heterogeneity among Fanconi anemia heterozygotes and risk of cancer. *Cancer Res 67*(19), 9591–9596.

Bishop DT, Demenais F, et al. (2009). Genome-wide association study identifies three loci associated with melanoma risk. *Nat Genet 41*(8), 920–925.

Blake PW, Toro JR. (2009). Update of cylindromatosis gene (CYLD) mutations in Brooke-Spiegler syndrome: novel insights into the role of deubiquitination in cell signaling. *Hum Mutat 30*(7), 1025–1036.

Blanco I, Bustillo EF, et al. (2001). Distribution of alpha1-antitrypsin PI S and PI Z frequencies in countries outside Europe: a meta-analysis. *Clin Genet 60*(6), 431–441.

Blank CA, Brantly M. (1994). Clinical features and molecular characteristics of alpha 1-antitrypsin deficiency. *Ann Allergy 72*(2), 105–120; quiz 120–122.

Blokhuis MM, Goldberg PA, et al. (2008). The extracolonic cancer spectrum in females with the common 'South African' hMLH1 c.C1528T mutation. *Fam Cancer 7*(3), 191–198.

Boissy RE, Zhao H, et al. (1996). Mutation in and lack of expression of tyrosinase-related protein-1 (TRP-1) in melanocytes from an individual with brown oculocutaneous albinism: a new subtype of albinism classified as OCA3. *Am J Hum Genet 58*(6), 1145–1156.

Boland CR, Thibodeau SN, et al. (1998). A National Cancer Institute Workshop on Microsatellite Instability for cancer detection and familial predisposition: development of international criteria for the determination of microsatellite instability in colorectal cancer. *Cancer Res 58*(22), 5248–5257.

Bougatef K, Marrakchi R, et al. (2008). First genetic analysis in Tunisian familial adenomatous polyposis probands. *Oncol Rep 19*(5), 1213–1218.

Bougatef K, Ouerhani S, et al. (2008). Prevalence of mutations in APC, CTNNB1, and BRAF in Tunisian patients with sporadic colorectal cancer. *Cancer Genet Cytogenet 187*(1), 12–18.

Bovee JV. (2008). Multiple osteochondromas. *Orphanet J Rare Dis 3*: 3.

Bowen S, Gill M, et al. (2005). Mutations in the CYLD gene in Brooke-Spiegler syndrome, familial cylindromatosis, and multiple familial trichoepithelioma: lack of genotype-phenotype correlation. *J Invest Dermatol 124*(5), 919–920.

Bowry TR, Pade J, et al. (1982). African specific phenotypes of alpha-1-antitrypsin. *East Afr Med J 59*(3), 185–188.

Brantly M, Nukiwa T, et al. (1988). Molecular basis of alpha-1-antitrypsin deficiency. *Am J Med 84*(6A), 13–31.

Brodsky MC, Fray KJ. (2004). Positive angle kappa: a sign of albinism in patients with congenital nystagmus. *Am J Ophthalmol 137*(4), 625–629.

Bronner CE, Baker SM, et al. (1994). Mutation in the DNA mismatch repair gene homologue hMLH1 is associated with hereditary non-polyposis colon cancer. *Nature 368*(6468), 258–261.

Brown KD, Rathi A, et al. (2003). The mismatch repair system is required for S-phase checkpoint activation. *Nat Genet 33*(1), 80–84.

Buttin BM, Powell MA, et al. (2004). Penetrance and expressivity of MSH6 germline mutations in seven kindreds not ascertained by family history. *Am J Hum Genet 74*(6), 1262–1269.

Callen E, Casado JA, et al. (2005). A common founder mutation in FANCA underlies the world's highest prevalence of Fanconi anemia in Gypsy families from Spain. *Blood 105*(5), 1946–1949.

Carrell RW, Lomas DA. (2002). Alpha1-antitrypsin deficiency—a model for conformational diseases. *N Engl J Med 346*(1), 45–53.

Chaabani H, Martin JP, et al. (1984). Genetic study of Tunisian Berbers. II. Alpha 1-antitrypsin (Pi) polymorphism: report of a new allele (Pi S Berber). *Exp Clin Immunogenet 1*(1), 19–24.

Chan AO, Soliman AS, et al. (2005). Differing DNA methylation patterns and gene mutation frequencies in colorectal carcinomas from Middle Eastern countries. *Clin Cancer Res 11*(23), 8281–8287.

Chan TL, Yuen ST, et al. (2001). A novel germline 1.8-kb deletion of hMLH1 mimicking alternative splicing: a founder mutation in the Chinese population. *Oncogene 20*(23), 2976–2981.

Chan TL, Yuen ST, et al. (2006). Heritable germline epimutation of MSH2 in a family with hereditary nonpolyposis colorectal cancer. *Nat Genet 38*(10), 1178–1183.

Charef-Hamza S, Trimeche M, et al. (2005). Loss of heterozygosity at the BRCA1 locus in Tunisian women with sporadic breast cancer. *Cancer Lett 224*(2), 185–191.

Chetty R, Naidoo R, et al. (2002). Chromosome 2p, 3p, 5q and 18q status in sporadic gastric cancer. *Pathology 34*(3), 275–281.

Chiang PW, Drautz JM, et al. (2008). A new hypothesis of OCA1B. *Am J Med Genet A 146A*(22), 2968–2970.

Chiang PW, Spector E, et al. (2008). Evidence suggesting the inheritance mode of the human P gene in skin complexion is not strictly recessive. *Am J Med Genet A 146A*(11), 1493–1496.

Chidzonga MM, Mahomva L, et al. (2009). Xeroderma pigmentosum: a retrospective case series in Zimbabwe. *J Oral Maxillofac Surg 67*(1), 22–31.

Chintamaneni CD, Ramsay M, et al. (1991). Mapping the human CAS2 gene, the homologue of the mouse brown (b) locus, to human chromosome 9p22-pter. *Biochem Biophys Res Commun 178*(1), 227–235.

Cleaver JE, Thompson LH, et al. (1999). A summary of mutations in the UV-sensitive disorders: xeroderma pigmentosum, Cockayne syndrome, and trichothiodystrophy. *Hum Mutat 14*(1), 9–22.

Cleaver JE, Zelle B, et al. (1981). Xeroderma pigmentosum patients from Egypt: II. Preliminary correlations of epidemiology, clinical symptoms and molecular biology. *J Invest Dermatol 77*(1), 96–101.

Clendenning M, Senter L, et al. (2008). A frame-shift mutation of PMS2 is a widespread cause of Lynch syndrome. *J Med Genet 45*(6), 340–345.

Clyne M, Offman J, et al. (2009). The G67E mutation in hMLH1 is associated with an unusual presentation of Lynch syndrome. *Br J Cancer 100*(2), 376–380.

Cohen T, Muller RM, et al. (1990). Nucleotide sequence of the cDNA encoding human tyrosinase-related protein. *Nucleic Acids Res 18*(9), 2807–2808.

Constans J, Viau M, et al. (1981). Gc, Tf, Hp subtype and alpha 1-antitrypsin polymorphisms in a Pygmy Bi-Aka sample. *Hum Hered 31*(3), 129–137.

Cousineau I, Abaji C, et al. (2005). BRCA1 regulates RAD51 function in response to DNA damage and suppresses spontaneous sister chromatid replication slippage: implications for sister chromatid cohesion, genome stability, and carcinogenesis. *Cancer Res 65*(24), 11384–11391.

Cravo M, Afonso AJ, et al. (2002). Pathogenicity of missense and splice site mutations in hMSH2 and hMLH1 mismatch repair genes: implications for genetic testing. *Gut 50*(3), 405–412.

Creel D, Garber SR, et al. (1980). Auditory brainstem anomalies in human albinos. *Science 209*(4462), 1253–1255.

Cunningham JM, Kim CY, et al. (2001). The frequency of hereditary defective mismatch repair in a prospective series of unselected colorectal carcinomas. *Am J Hum Genet 69*(4), 780–790.

Dahl M, Tybjaerg-Hansen A, et al. (2002). Change in lung function and morbidity from chronic obstructive pulmonary disease in alpha1-antitrypsin MZ heterozygotes: A longitudinal study of the general population. *Ann Intern Med 136*(4), 270–279.

Davies DR, Armstrong JG, et al. (1995). Severe Gardner syndrome in families with mutations restricted to a specific region of the APC gene. *Am J Hum Genet 57*(5), 1151–1158.

de Graaf AS, de Jong G, et al. (1995). An early-onset recessive cerebellar disorder with distal amyotrophy and, in two patients, gross myoclonia: a probable ataxia telangiectasia variant. *Clin Neurol Neurosurg 97*(1), 1–7.

de Jong G, Fryns JP. (1991). Oculocerebral syndrome with hypopigmentation (Cross syndrome), the mixed pattern of hair pigmentation as an important diagnostic sign. *Genet Couns 2*(3), 151–155.

de Serres FJ. (2002). Worldwide racial and ethnic distribution of alpha1-antitrypsin deficiency: summary of an analysis of published genetic epidemiologic surveys. *Chest 122*(5), 1818–1829.

de Serres FJ. (2003). Alpha-1 antitrypsin deficiency is not a rare disease but a disease that is rarely diagnosed. *Environ Health Perspect 111*(16), 1851–1854.

de Wind N, Dekker M, et al. (1995). Inactivation of the mouse Msh2 gene results in mismatch repair deficiency, methylation tolerance, hyperrecombination, and predisposition to cancer. *Cell 82*(2), 321–330.

DeCroo S, Kamboh MI, et al. (1991). Population genetics of alpha-1 -antitrypsin polymorphism in US whites, US blacks and African blacks. *Hum Hered 41*(4), 215–221.

Denden S, Braham W, et al. (2009). [Biochemical and molecular diagnosis of alpha 1 antitrypsin deficiency in a Tunisian family]. *Ann Biol Clin (Paris) 67*(4), 441–445.

Denden S, Haj Khelil A, et al. (2008). Alpha 1 antitrypsin polymorphism in the Tunisian population with special reference to pulmonary disease. *Pathol Biol (Paris) 56*(3), 106–110.

Deng CX. (2006). BRCA1: cell cycle checkpoint, genetic instability, DNA damage response and cancer evolution. *Nucleic Acids Res 34*(5), 1416–1426.

Denzler TB, Harned RK, et al. (1979). Gastric polyps in familial polyposis coli. *Radiology 130*(1), 63–66.

Destro-Bisol G, Coia V, et al. (2004). The analysis of variation of mtDNA hypervariable region 1 suggests that Eastern and Western Pygmies diverged before the Bantu expansion. *Am Nat 163*(2), 212–226.

DeWeese TL, Shipman JM, et al. (1998). Mouse embryonic stem cells carrying one or two defective Msh2 alleles respond abnormally to oxidative stress inflicted by low-level radiation. *Proc Natl Acad Sci U S A 95*(20), 11915–11920.

Dihlmann S, Gebert J, et al. (1999). Dominant negative effect of the APC1309 mutation: a possible explanation for genotype-phenotype correlations in familial adenomatous polyposis. *Cancer Res 59*(8), 1857–1860.

Donis-Keller H, Dou S, et al. (1993). Mutations in the RET proto-oncogene are associated with MEN 2A and FMTC. *Hum Mol Genet 2*(7), 851–856.

Donlon TA, Lalande M, et al. (1986). Isolation of molecular probes associated with the chromosome 15 instability in the Prader-Willi syndrome. *Proc Natl Acad Sci U S A 83*(12), 4408–4412.

Drummond JT, Li GM, et al. (1995). Isolation of an hMSH2-p160 heterodimer that restores DNA mismatch repair to tumor cells. *Science 268*(5219), 1909–1912.

Du J, Fisher DE. (2002). Identification of Aim-1 as the underwhite mouse mutant and its transcriptional regulation by MITF. *J Biol Chem 277*(1), 402–406.

Duffy DL, Montgomery GW, et al. (2007). A three-single-nucleotide polymorphism haplotype in intron 1 of OCA2 explains most human eye-color variation. *Am J Hum Genet 80*(2), 241–252.

Dunn DS, Madhoo B, et al. (1986). Alpha-1-antitrypsin variation in Southern Africa. *Hum Hered 36*(4), 238–242.

Durham-Pierre D, Gardner JM, et al. (1994). African origin of an intragenic deletion of the human P gene in tyrosinase positive oculocutaneous albinism. *Nat Genet 7*(2), 176–179.

Durham-Pierre D, King RA, et al. (1996). Estimation of carrier frequency of a 2.7 kb deletion allele of the P gene associated with OCA2 in African-Americans. *Hum Mutat 7*(4), 370–373.

Ellison AR, Lofing J, et al. (2001). Functional analysis of human MLH1 and MSH2 missense variants and hybrid human-yeast MLH1 proteins in Saccharomyces cerevisiae. *Hum Mol Genet 10*(18), 1889–900.

Eng C. (1996). Seminars in medicine of the Beth Israel Hospital, Boston. The RET proto-oncogene in multiple endocrine neoplasia type 2 and Hirschsprung's disease. *N Engl J Med 335*(13), 943–951.

Eriksson S, Carlson J, et al. (1986). Risk of cirrhosis and primary liver cancer in alpha 1-antitrypsin deficiency. *N Engl J Med 314*(12), 736–739.

Eshleman JR, Markowitz V. (1995). Microsatellite instability in inherited and sporadic neoplasms. *Curr Opin Oncol 7*(1), 83–89.

Etienne-Manneville S, Hall A. (2003). Cdc42 regulates GSK-3beta and adenomatous polyposis coli to control cell polarity. *Nature 421*(6924), 753–756.

Fackenthal JD, Olopade OI. (2007). Breast cancer risk associated with BRCA1 and BRCA2 in diverse populations. *Nat Rev Cancer 7*(12), 937–948.

Fackenthal JD, Sveen L, et al. (2005). Complete allelic analysis of BRCA1 and BRCA2 variants in young Nigerian breast cancer patients. *J Med Genet 42*(3), 276–281.

Fargion S, Klasen EC, et al. (1981). Alpha 1-antitrypsin in patients with hepatocellular carcinoma and chronic active hepatitis. *Clin Genet 19*(2), 134–139.

Fazaa B, Zghal M, et al. (2001). [Melanoma in xeroderma pigmentosum: 12 cases]. *Ann Dermatol Venereol 128*(4), 503–506.

Ferla R, Calo V, et al. (2007). Founder mutations in BRCA1 and BRCA2 genes. *Ann Oncol 18 Suppl 6*: vi93–98.

Fishel R, Ewel A, et al. (1994). Binding of mismatched microsatellite DNA sequences by the human MSH2 protein. *Science 266*(5189), 1403–1405.

Fishel R, Lescoe MK, et al. (1993). The human mutator gene homolog MSH2 and its association with hereditary nonpolyposis colon cancer. *Cell 75*(5), 1027–1038.

Fodde R, van der Luijt R, et al. (1992). Eight novel inactivating germ line mutations at the APC gene identified by denaturing gradient gel electrophoresis. *Genomics 13*(4), 1162–1168.

Forbes SA, Bhamra G, et al. (2008). The Catalogue of Somatic Mutations in Cancer (COSMIC). *Curr Protoc Hum Genet Chapter 10*: Unit 10 11.

Forshew T, Khaliq S, et al. (2005). Identification of novel TYR and TYRP1 mutations in oculocutaneous albinism. *Clin Genet 68*(2), 182–184.

Foulkes WD, Metcalfe K, et al. (2004). Estrogen receptor status in BRCA1- and BRCA2-related breast cancer: the influence of age, grade, and histological type. *Clin Cancer Res 10*(6), 2029–2034.

Foulkes WD, Stefansson IM, et al. (2003). Germline BRCA1 mutations and a basal epithelial phenotype in breast cancer. *J Natl Cancer Inst 95*(19), 1482–1485.

Francannet C, Cohen-Tanugi A, et al. (2001). Genotype-phenotype correlation in hereditary multiple exostoses. *J Med Genet 38*(7), 430–434.

Frank-Raue K, Rondot S, et al. (2007). Change in the spectrum of RET mutations diagnosed between 1994 and 2006. *Clin Lab 53*(5–6), 273–282.

Frants RR, Eriksson AW. (1978). Reliable classification of six Pi M subtypes by separator isoelectric focusing. *Hum Hered 28*(3), 201–209.

Fregene A, Newman LA. (2005). Breast cancer in sub-Saharan Africa: how does it relate to breast cancer in African-American women? *Cancer 103*(8), 1540–1550.

Friedl W, Meuschel S, et al. (1996). Attenuated familial adenomatous polyposis due to a mutation in the 3' part of the APC gene. A clue for understanding the function of the APC protein. *Hum Genet 97*(5), 579–584.

Fukamachi S, Shimada A, et al. (2001). Mutations in the gene encoding B, a novel transporter protein, reduce melanin content in medaka. *Nat Genet 28*(4), 381–385.

Gaigher RJ, Lund PM, et al. (2002). A sociological study of children with albinism at a special school in the Limpopo province. *Curationis 25*(4), 4–11.

Gao Q, Adebamowo CA, et al. (2000). Protein truncating BRCA1 and BRCA2 mutations in African women with pre-menopausal breast cancer. *Hum Genet 107*(2), 192–194.

Gardner EJ. (1951). A genetic and clinical study of intestinal polyposis, a predisposing factor for carcinoma of the colon and rectum. *Am J Hum Genet 3*(2), 167–176.

Gardner EJ. (1962). Follow-up study of a family group exhibiting dominant inheritance for a syndrome including intestinal polyps, osteomas, fibromas and epidermal cysts. *Am J Hum Genet 14*: 376–390.

Gardner EJ, Plenk HP. (1952). Hereditary pattern for multiple osteomas in a family group. *Am J Hum Genet 4*(1), 31–36.

Gardner JM, Nakatsu Y, et al. (1992). The mouse pink-eyed dilution gene: association with human Prader-Willi and Angelman syndromes. *Science 257*(5073), 1121–1124.

Gatti RA, Berkel I, et al. (1988). Localization of an ataxia-telangiectasia gene to chromosome 11q22-23. *Nature 336*(6199), 577–580.

Gayther SA, Wells D, et al. (1994). Regionally clustered APC mutations are associated with a severe phenotype and occur at a high frequency in new mutation cases of adenomatous polyposis coli. *Hum Mol Genet 3*(1), 53–56.

German J, Hashem N, et al. (1984). Xeroderma pigmentosum in Egypt. III. ABO blood grouping in 22 affected families. *Ann Hum Genet 48*(Pt 1), 61–64.

Gershoni-Baruch R, Rosenmann A, et al. (1994). Mutations of the tyrosinase gene in patients with oculocutaneous albinism from various ethnic groups in Israel. *Am J Hum Genet 54*(4), 586–594.

Gibson RA, Ford D, et al. (1994). Genetic mapping of the FACC gene and linkage analysis in Fanconi anaemia families. *J Med Genet 31*(11), 868–871.

Gibson RA, Morgan NV, et al. (1996). Novel mutations and polymorphisms in the Fanconi anemia group C gene. *Hum Mutat 8*(2), 140–148.

Giebel LB, Musarella MA, et al. (1991). A nonsense mutation in the tyrosinase gene of Afghan patients with tyrosinase negative (type IA) oculocutaneous albinism. *J Med Genet 28*(7), 464–467.

Giebel LB, Strunk KM, et al. (1990). A frequent tyrosinase gene mutation in classic, tyrosinase-negative (type IA) oculocutaneous albinism. *Proc Natl Acad Sci U S A 87*(9), 3255–3258.

Giebel LB, Strunk KM, et al. (1991). Organization and nucleotide sequences of the human tyrosinase gene and a truncated tyrosinase-related segment. *Genomics 9*(3), 435–445.

Gilad S, Bar-Shira A, et al. (1996). Ataxia-telangiectasia: founder effect among north African Jews. *Hum Mol Genet 5*(12), 2033–2037.

Goldberg PA, Madden MV, et al. (1998). In a resource-poor country, mutation identification has the potential to reduce the cost of family management for hereditary nonpolyposis colorectal cancer. *Dis Colon Rectum 41*(10), 1250–1253; discussion 1253–1255.

Gomes-Pereira M, Fortune MT, et al. (2004). Pms2 is a genetic enhancer of trinucleotide CAG.CTG repeat somatic mosaicism: implications for the mechanism of triplet repeat expansion. *Hum Mol Genet 13*(16), 1815–1825.

Goodfellow PJ, Buttin BM, et al. (2003). Prevalence of defective DNA mismatch repair and MSH6 mutation in an unselected series of endometrial cancers. *Proc Natl Acad Sci U S A 100*(10), 5908–5913.

Gorlin RJ, Chaudhary AP. (1960). Multiple osteomatosis, fibromas, lipomas and fibrosarcomas of the skin and mesentery, epidermoid inclusion cysts of the skin, leiomyomas and multiple intestinal polyposis: a heritable disorder of connective tissue. *N Engl J Med 263*: 1151–1158.

Gorlov IP, Gorlova OY, et al. (2003). Missense mutations in hMLH1 and hMSH2 are associated with exonic splicing enhancers. *Am J Hum Genet 73*(5), 1157–1161.

Govender S, Parbhoo AH. (1999). Osteochondroma with compression of the spinal cord. A report of two cases. *J Bone Joint Surg Br 81*(4), 667–669.

Gradia S, Subramanian D, et al. (1999). hMSH2-hMSH6 forms a hydrolysis-independent sliding clamp on mismatched DNA. *Mol Cell 3*(2), 255–261.

Grady WM, Myeroff LL, et al. (1999). Mutational inactivation of transforming growth factor beta receptor type II in microsatellite stable colon cancers. *Cancer Res 59*(2), 320–324.

Graf J, Hodgson R, et al. (2005). Single nucleotide polymorphisms in the MATP gene are associated with normal human pigmentation variation. *Hum Mutat 25*(3), 278–284.

Graf J, Voisey J, et al. (2007). Promoter polymorphisms in the MATP (SLC45A2) gene are associated with normal human skin color variation. *Hum Mutat 28*(7), 710–717.

Gratchev A, Strein P, et al. (2003). Molecular genetics of Xeroderma pigmentosum variant. *Exp Dermatol 12*(5), 529–536.

Grobbelaar JJ, Fortuin R, et al. (2000). Familial adenomatous polyposis coli in South Africa—molecular basis and diagnosis. *S Afr Med J 90*(7), 715–719.

Grobbelaar JJ, Oosthuizen CJ, et al. (1995). Screening South African familial adenomatous polyposis families for the five-nucleotide deletion at codon 1309 of the APC gene. *Mol Cell Probes 9*(1), 49–51.

Grobbelaar JJ, Oosthuizen CJ, et al. (1995). The use of DNA markers in the pre-clinical diagnosis of familial adenomatous polyposis in families in South Africa. *S Afr Med J 85*(4), 269–271.

Grobbelaar JJ, Wilken E, et al. (2002). Familial adenomatous polyposis in two Black South African families. *Clin Genet 61*(3), 214–217.

Grobbelaar JJ, Ziskind A, et al. (1996). A novel deletion at codon 441 of the APC gene associated with ophthalmic lesions (CHRPE) in a South African family. *J Med Genet 33*(5), 384–386.

Groden J, Thliveris A, et al. (1991). Identification and characterization of the familial adenomatous polyposis coli gene. *Cell 66*(3), 589–600.

Gudbjartsson DF, Sulem P, et al. (2008). ASIP and TYR pigmentation variants associate with cutaneous melanoma and basal cell carcinoma. *Nat Genet 40*(7), 886–891.

Gukas ID, Jennings BA, et al. (2005). Clinicopathological features and molecular markers of breast cancer in Jos, Nigeria. *West Afr J Med 24*(3), 209–213.

Halaban R, Moellmann G. (1990). Murine and human b locus pigmentation genes encode a glycoprotein (gp75) with catalase activity. *Proc Natl Acad Sci U S A 87*(12), 4809–4813.

Hameed F, Goldberg PA, et al. (2006). Immunohistochemistry detects mismatch repair gene defects in colorectal cancer. *Colorectal Dis 8*(5), 411–417.

Han HJ, Maruyama M, et al. (1995). Genomic structure of human mismatch repair gene, hMLH1, and its mutation analysis in patients with hereditary non-polyposis colorectal cancer (HNPCC). *Hum Mol Genet 4*(2), 237–242.

Hanson CA, Miller JR. (2005). Non-traditional roles for the Adenomatous Polyposis Coli (APC) tumor suppressor protein. *Gene 361*: 1–12.
Harned RK, Williams SM. (1982). Familial polyposis coli and periampullary malignancy. *Dis Colon Rectum 25*(3), 227–229.
Hashem N, Bootsma D, et al. (1980). Clinical characteristics, DNA repair, and complementation groups in xeroderma pigmentosum patients from Egypt. *Cancer Res 40*(1), 13–18.
Hayes VM. (2003). Genetic diversity of the alpha-1-antitrypsin gene in Africans identified using a novel genotyping assay. *Hum Mutat 22*(1), 59–66.
Hayes VM, Kotze MJ, et al. (1996). Presymptomatic diagnosis of familial adenomatous polyposis using intragenic polymorphisms and CA repeats flanking the APC gene. *Genet Couns 7*(1), 1–7.
Hecht F. (2007). Familial cancer syndromes: catalog with comments. *Cytogenet Genome Res 118*(2–4), 222–228.
Hemminki A, Peltomaki P, et al. (1994). Loss of the wild type MLH1 gene is a feature of hereditary nonpolyposis colorectal cancer. *Nat Genet 8*(4), 405–410.
Hendriks YM, Jagmohan-Changur S, et al. (2006). Heterozygous mutations in PMS2 cause hereditary nonpolyposis colorectal carcinoma (Lynch syndrome). *Gastroenterology 130*(2), 312–322.
Hennekam RC. (1991). Hereditary multiple exostoses. *J Med Genet 28*(4), 262–266.
Hirai Y, Kodama Y, et al. (2006). Heterozygous individuals bearing a founder mutation in the XPA DNA repair gene comprise nearly 1% of the Japanese population. *Mutat Res 601*(1–2), 171–178.
Hitzeroth HW, Bissbort S, et al. (1987). Alpha-1-antitrypsin genetic polymorphism in South Africa. *S Afr Med J 71*(12), 755–759.
Holden CJ. (2002). Bantu language trees reflect the spread of farming across sub-Saharan Africa: a maximum-parsimony analysis. *Proc Biol Sci 269*(1493), 793–799.
Hong ES, Zeeb H, et al. (2006). Albinism in Africa as a public health issue. *BMC Public Health 6*: 212.
Honrado E, Benitez J, et al. (2005). The molecular pathology of hereditary breast cancer: genetic testing and therapeutic implications. *Mod Pathol 18*(10), 1305–1320.
Horii A, Han HJ, et al. (1994). Cloning, characterization and chromosomal assignment of the human genes homologous to yeast PMS1, a member of mismatch repair genes. *Biochem Biophys Res Commun 204*(3), 1257–1264.
Houvet D, Clerc M. (1978). [Variation of the profile phenotype M of alpha-1-chymotrypsine: trial of interpretation (author's transl)]. *Ann Biol Clin (Paris) 36*(1), 1–9.
Hsairi M, Fakhfakh R, et al. (2002). [Assessment of cancer incidence in Tunisia 1993–1997]. *Tunis Med 80*(2), 57–64.
Hu F, Hanifin JM, et al. (1980). Yellow mutant albinism: cytochemical, ultrastructural, and genetic characterization suggesting multiple allelism. *Am J Hum Genet 32*(3), 387–395.
Huang J, Kuismanen SA, et al. (2001). MSH6 and MSH3 are rarely involved in genetic predisposition to nonpolypotic colon cancer. *Cancer Res 61*(4), 1619–1623.
Huang SC, Lavine JE, et al. (2001). Germline characterization of early-aged onset of hereditary non-polyposis colorectal cancer. *J Pediatr 138*(5), 629–635.
Hug G, Chuck G, et al. (1981). PiPclifton: a new alpha 1-antitrypsin allele in an American Negro family. *J Med Genet 18*(1), 43–45.
Huo D, Ikpatt F, et al. (2009). Population differences in breast cancer: survey in indigenous African women reveals over-representation of triple-negative breast cancer. *J Clin Oncol 27*(27), 4515–4521.
Iida M, Yao T, et al. (1988). Natural history of gastric adenomas in patients with familial adenomatosis coli/Gardner's syndrome. *Cancer 61*(3), 605–611.
Ijaduola TG, Smith EB. (1998). Pattern of breast cancer among white-American, African-American, and nonimmigrant west-African women. *J Natl Med Assoc 90*(9), 547–551.
Ikpatt OF, Kuopio T, et al. (2002). Proliferation in African breast cancer: biology and prognostication in Nigerian breast cancer material. *Mod Pathol 15*(8), 783–789.
Inagaki K, Suzuki T, et al. (2004). Oculocutaneous albinism type 4 is one of the most common types of albinism in Japan. *Am J Hum Genet 74*(3), 466–471.
Ionov Y, Peinado MA, et al. (1993). Ubiquitous somatic mutations in simple repeated sequences reveal a new mechanism for colonic carcinogenesis. *Nature 363*(6429), 558–561.
Jacquemont C, Taniguchi T. (2007). The Fanconi anemia pathway and ubiquitin. *BMC Biochem 8 Suppl 1*: S10.
Jacyk WK. (1999). Xeroderma pigmentosum in black South Africans. *Int J Dermatol 38*(7), 511–514.
Jager AC, Bisgaard ML, et al. (1997). Reduced frequency of extracolonic cancers in hereditary nonpolyposis colorectal cancer families with monoallelic hMLH1 expression. *Am J Hum Genet 61*(1), 129–138.
Jain M, Thorstenson YR, et al. (2003). Genotyping African haplotypes in ATM using a co-spotted single-base extension assay. *Hum Mutat 22*(3), 214–221.
Jannot AS, Meziani R, et al. (2005). Allele variations in the OCA2 gene (pink-eyed-dilution locus) are associated with genetic susceptibility to melanoma. *Eur J Hum Genet 13*(8), 913–920.
Jaskiewicz K, Lancaster E, et al. (1998). Proliferative activity of normal and neoplastic colonic mucosa in population groups with high and low risk for colorectal carcinoma. *Anticancer Res 18*(6B), 4641–4644.
Jennes I, Pedrini E, et al. (2009). Multiple osteochondromas: mutation update and description of the multiple osteochondromas mutation database (MOdb). *Hum Mutat 30*(12), 1620–1627.
Joenje H, Patel KJ. (2001). The emerging genetic and molecular basis of Fanconi anaemia. *Nat Rev Genet 2*(6), 446–457.
Jones TR, Nance FC. (1977). Periampullary malignancy in Gardner's syndrome. *Ann Surg 185*(5), 565–573.
Kadyrov FA, Dzantiev L, et al. (2006). Endonucleolytic function of MutLalpha in human mismatch repair. *Cell 126*(2), 297–308.
Kagore F, Lund PM. (1995). Oculocutaneous albinism among schoolchildren in Harare, Zimbabwe. *J Med Genet 32*(11), 859–861.
Kalb R, Neveling K, et al. (2006). Fanconi anemia: causes and consequences of genetic instability. *Genome Dyn 1*: 218–242.
Kaplan KB, Burds AA, et al. (2001). A role for the Adenomatous Polyposis Coli protein in chromosome segregation. *Nat Cell Biol 3*(4), 429–432.
Karray-Chouayekh S, Trifa F, et al. (2009). Clinical significance of epigenetic inactivation of hMLH1 and BRCA1 in Tunisian patients with invasive breast carcinoma. *J Biomed Biotechnol 2009*: 369129.
Katabuchi H, van Rees B, et al. (1995). Mutations in DNA mismatch repair genes are not responsible for microsatellite instability in most sporadic endometrial carcinomas. *Cancer Res 55*(23), 5556–5560.
Kawai M, Suzuki T, et al. (2005). A patient with subclinical oculocutaneous albinism type 2 diagnosed on getting severely sunburned. *Dermatology 210*(4), 322–323.
Kedda MA, Stevens G, et al. (1994). The tyrosinase-positive oculocutaneous albinism gene shows locus homogeneity on chromosome 15q11-q13 and evidence of multiple mutations in southern African negroids. *Am J Hum Genet 54*(6), 1078–1084.
Kerr R, Stevens G, et al. (2000). Identification of P gene mutations in individuals with oculocutaneous albinism in sub-Saharan Africa. *Hum Mutat 15*(2), 166–172.
Khan SG, Oh KS, et al. (2006). Reduced XPC DNA repair gene mRNA levels in clinically normal parents of xeroderma pigmentosum patients. *Carcinogenesis 27*(1), 84–94.
Khatri ML, Bemghazil M, et al. (1999). Xeroderma pigmentosum in Libya. *Int J Dermatol 38*(7), 520–524.
Khatri ML, Shafi M, et al. (1992). Xeroderma pigmentosum. A clinical study of 24 Libyan cases. *J Am Acad Dermatol 26*(1), 75–78.
Khitri A, Benlatrache K, et al. (1977). [Pi W3 Constantine, new allele of Pi system]. *Sem Hop 53*(16), 909–910.
Kidson SH, Richards PD, et al. (1993). An ultrastructural study of melanocytes and melanosomes in the skin and hair bulbs of rufous albinos. *Pigment Cell Res 6*(4 Pt 1), 209–214.

Kim JC, Roh SA, et al. (2007). MLH3 and EXO1 alterations in familial colorectal cancer patients not fulfilling Amsterdam criteria. *Cancer Genet Cytogenet 176*(2), 172–174.

King RA, Lewis RA, et al. (1985). Brown oculocutaneous albinism. Clinical, ophthalmological, and biochemical characterization. *Ophthalmology 92*(11), 1496–505.

King RA, Mentink MM, et al. (1991). Non-random distribution of missense mutations within the human tyrosinase gene in type I (tyrosinase-related) oculocutaneous albinism. *Mol Biol Med 8*(1), 19–29.

King RA, Rich SS. (1986). Segregation analysis of brown oculocutaneous albinism. *Clin Genet 29*(6), 496–501.

King RA, Townsend D, et al. (1991). Temperature-sensitive tyrosinase associated with peripheral pigmentation in oculocutaneous albinism. *J Clin Invest 87*(3), 1046–1053.

King RA, Wirtschafter JD, et al. (1986). Minimal pigment: a new type of oculocutaneous albinism. *Clin Genet 29*(1), 42–50.

King RA, Witkop CJ. (1977). Detection of heterozygotes for tyrosinase-negative oculocutaneous albinism by hairbulb tyrosinase assay. *Am J Hum Genet 29*(2), 164–168.

Knoll JH, Nicholls RD, et al. (1990). Angelman syndrome: three molecular classes identified with chromosome 15q11q13-specific DNA markers. *Am J Hum Genet 47*(1), 149–154.

Komminoth P, Kunz EK, et al. (1995). Analysis of RET protooncogene point mutations distinguishes heritable from nonheritable medullary thyroid carcinomas. *Cancer 76*(3), 479–489.

Konno T, Abe Y, et al. (2009). Oculocutaneous albinism type IV: A boy of Moroccan descent with a novel mutation in SLC45A2. *Am J Med Genet A 149A*(8), 1773–1776.

Kotze MJ, Grobbelaar JJ, et al. (1996). Identification of a novel single base insertion in the adenomatous polyposis coli gene. *Hum Mutat 7*(1), 68–69.

Kraemer KH, Herlyn M, et al. (1989). Reduced DNA repair in cultured melanocytes and nevus cells from a patient with xeroderma pigmentosum. *Arch Dermatol 125*(2), 263–268.

Kraemer KH, Lee MM, et al. (1994). The role of sunlight and DNA repair in melanoma and nonmelanoma skin cancer. The xeroderma pigmentosum paradigm. *Arch Dermatol 130*(8), 1018–1021.

Kraus C, Kastl S, et al. (1999). A proven de novo germline mutation in HNPCC. *J Med Genet 36*(12), 919–921.

Kromberg JG, Castle D, et al. (1989). Albinism and skin cancer in Southern Africa. *Clin Genet 36*(1), 43–52.

Kromberg JG, Castle DJ, et al. (1990). Red or rufous albinism in southern Africa. *Ophthalmic Paediatr Genet 11*(3), 229–235.

Kromberg JG, Jenkins T. (1982). Prevalence of albinism in the South African negro. *S Afr Med J 61*(11), 383–386.

Kufe DW, Raphael P, et al. (2003). *Holland-Frei Cancer Medicine, 6th edition*, B.C. Decker, ISBN-10: 1-55009-213-8 (http://www.ncbi.nlm.nih.gov/books/NBK12354/).

Kutler DI, Auerbach AD. (2004). Fanconi anemia in Ashkenazi Jews. *Fam Cancer 3*(3–4), 241–248.

Lakhani SR, Reis-Filho JS, et al. (2005). Prediction of BRCA1 status in patients with breast cancer using estrogen receptor and basal phenotype. *Clin Cancer Res 11*(14), 5175–5180.

Lamers MH, Perrakis A, et al. (2000). The crystal structure of DNA mismatch repair protein MutS binding to a G x T mismatch. *Nature 407*(6805), 711–717.

Lane AB, Soodyall H, et al. (2002). Genetic substructure in South African Bantu-speakers: evidence from autosomal DNA and Y-chromosome studies. *Am J Phys Anthropol 119*(2), 175–185.

Leach FS, Nicolaides NC, et al. (1993). Mutations of a mutS homolog in hereditary nonpolyposis colorectal cancer. *Cell 75*(6), 1215–1225.

Lee DA, Grossman ME, et al. (2005). Genetics of skin appendage neoplasms and related syndromes. *J Med Genet 42*(11), 811–819.

Lee ST, Nicholls RD, et al. (1994). Mutations of the P gene in oculocutaneous albinism, ocular albinism, and Prader-Willi syndrome plus albinism. *N Engl J Med 330*(8), 529–534.

Lee ST, Nicholls RD, et al. (1995). Organization and sequence of the human P gene and identification of a new family of transport proteins. *Genomics 26*(2), 354–363.

Lee ST, Nicholls RD, et al. (1994). Diverse mutations of the P gene among African-Americans with type II (tyrosinase-positive) oculocutaneous albinism (OCA2). *Hum Mol Genet 3*(11), 2047–2051.

Li GM, Modrich P. (1995). Restoration of mismatch repair to nuclear extracts of H6 colorectal tumor cells by a heterodimer of human MutL homologs. *Proc Natl Acad Sci U S A 92*(6), 1950–1954.

Ligtenberg MJ, Kuiper RP, et al. (2009). Heritable somatic methylation and inactivation of MSH2 in families with Lynch syndrome due to deletion of the 3' exons of TACSTD1. *Nat Genet 41*(1), 112–117.

Lindor NM, McMaster ML, et al. (2008). Concise handbook of familial cancer susceptibility syndromes - second edition. *J Natl Cancer Inst Monogr 38*, 1–93.

Lipkin SM, Moens PB, et al. (2002). Meiotic arrest and aneuploidy in MLH3-deficient mice. *Nat Genet 31*(4), 385–390.

Lipkin SM, Rozek LS, et al. (2004). The MLH1 D132H variant is associated with susceptibility to sporadic colorectal cancer. *Nat Genet 36*(7), 694–699.

Lipkin SM, Wang V, et al. (2001). Germline and somatic mutation analyses in the DNA mismatch repair gene MLH3: Evidence for somatic mutation in colorectal cancers. *Hum Mutat 17*(5), 389–396.

Liu B, Nicolaides NC, et al. (1995). Mismatch repair gene defects in sporadic colorectal cancers with microsatellite instability. *Nat Genet 9*(1), 48–55.

Liu HX, Zhou XL, et al. (2003). The role of hMLH3 in familial colorectal cancer. *Cancer Res 63*(8), 1894–1899.

Liu T, Tannergard P, et al. (1999). Missense mutations in hMLH1 associated with colorectal cancer. *Hum Genet 105*(5), 437–441.

Loeys BL, Chen J, et al. (2005). A syndrome of altered cardiovascular, craniofacial, neurocognitive and skeletal development caused by mutations in TGFBR1 or TGFBR2. *Nat Genet 37*(3), 275–281.

Lomas DA, Evans DL, et al. (1992). The mechanism of Z alpha 1-antitrypsin accumulation in the liver. *Nature 357*(6379), 605–607.

Lu SL, Kawabata M, et al. (1998). HNPCC associated with germline mutation in the TGF-beta type II receptor gene. *Nat Genet 19*(1), 17–18.

Luande J, Henschke CI, et al. (1985). The Tanzanian human albino skin. Natural history. *Cancer 55*(8), 1823–1828.

Lund PM. (1996). Distribution of oculocutaneous albinism in Zimbabwe. *J Med Genet 33*(8), 641–644.

Lund PM. (2005). Oculocutaneous albinism in southern Africa: population structure, health and genetic care. *Ann Hum Biol 32*(2), 168–173.

Lund PM, Maluleke TG, et al. (2007). Oculocutaneous albinism in a rural community of South Africa: a population genetic study. *Ann Hum Biol 34*(4), 493–497.

Lund PM, Puri N, et al. (1997). Oculocutaneous albinism in an isolated Tonga community in Zimbabwe. *J Med Genet 34*(9), 733–735.

Lund PM, Taylor JS. (2008). Lack of adequate sun protection for children with oculocutaneous albinism in South Africa. *BMC Public Health 8*: 225.

Lynch HT, Coronel SM, et al. (2004). A founder mutation of the MSH2 gene and hereditary nonpolyposis colorectal cancer in the United States. *JAMA 291*(6), 718–724.

Lynch HT, de la Chapelle A. (1999). Genetic susceptibility to non-polyposis colorectal cancer. *J Med Genet 36*(11), 801–818.

Lynch HT, Lanspa S, et al. (1991). Hereditary nonpolyposis colorectal cancer (Lynch syndromes I & II). Genetics, pathology, natural history, and cancer control, Part I. *Cancer Genet Cytogenet 53*(2), 143–160.

Ma AH, Xia L, et al. (2000). Somatic mutation of hPMS2 as a possible cause of sporadic human colon cancer with microsatellite instability. *Oncogene 19*(18), 2249–2256.

Maalej M, Hentati D, et al. (2007). [Skin cancer in Tunisia: a retrospective study: 1379 cases and risk factors]. *Tunis Med 85*(9), 728–733.

Macdougall LG, Greeff MC, et al. (1990). Fanconi anemia in black African children. *Am J Med Genet 36*(4), 408–413.
Macdougall LG, Rosendorff J, et al. (1994). Comparative study of Fanconi anemia in children of different ethnic origin in South Africa. *Am J Med Genet 52*(3), 279–284.
Machens A, Gimm O, et al. (2001). Genotype-phenotype correlations in hereditary medullary thyroid carcinoma: oncological features and biochemical properties. *J Clin Endocrinol Metab 86*(3), 1104–1109.
Maliaka YK, Chudina AP, et al. (1996). CpG dinucleotides in the hMSH2 and hMLH1 genes are hotspots for HNPCC mutations. *Hum Genet 97*(2), 251–255.
Manga P, Kromberg JG, et al. (1997). Rufous oculocutaneous albinism in southern African Blacks is caused by mutations in the TYRP1 gene. *Am J Hum Genet 61*(5), 1095–1101.
Margraf RL, Crockett DK, et al. (2009). Multiple endocrine neoplasia type 2 RET protooncogene database: repository of MEN2-associated RET sequence variation and reference for genotype/phenotype correlations. *Hum Mutat 30*(4), 548–556.
Markowitz S, Wang J, et al. (1995). Inactivation of the type II TGF-beta receptor in colon cancer cells with microsatellite instability. *Science 268*(5215), 1336–1338.
Masri MA, Abdel Seed NM, et al. (2002). Minor role for BRCA2 (exon11) and p53 (exon 5–9) among Sudanese breast cancer patients. *Breast Cancer Res Treat 71*(2), 145–147.
Massi G, Vecchio FM. (1977). Alpha-1-antitrypsin phenotypes in a group of newborn infants in Somalia. *Hum Genet 38*(3), 265–269.
Mathew CG. (2006). Fanconi anaemia genes and susceptibility to cancer. *Oncogene 25*(43), 5875–5884.
McBrearty S, Brooks AS. (2000). The revolution that wasn't: a new interpretation of the origin of modern human behavior. *J Hum Evol 39*(5), 453–563.
McBride SR, Leppard BJ. (2002). Attitudes and beliefs of an albino population toward sun avoidance: advice and services provided by an outreach albino clinic in Tanzania. *Arch Dermatol 138*(5), 629–632.
McVety S, Li L, et al. (2006). Disruption of an exon splicing enhancer in exon 3 of MLH1 is the cause of HNPCC in a Quebec family. *J Med Genet 43*(2), 153–156.
Mefford HC, Baumbach L, et al. (1999). Evidence for a BRCA1 founder mutation in families of West African ancestry. *Am J Hum Genet 65*(2), 575–578.
Mengesha B, Johnson O, et al. (1994). Familial polyposis in two Ethiopians. *Ethiop Med J 32*(1), 49–55.
Mestiri S, Monastiri K, et al. (2000). [Mutational analysis of breast/ovarian cancer hereditary predisposition gene BRCA1 in Tunisian women]. *Arch Inst Pasteur Tunis 77*(1–4), 11–15.
Meyer CH, Lapolice DJ, et al. (2002). Foveal hypoplasia in oculocutaneous albinism demonstrated by optical coherence tomography. *Am J Ophthalmol 133*(3), 409–410.
Miyoshi Y, Ando H, et al. (1992). Germ-line mutations of the APC gene in 53 familial adenomatous polyposis patients. *Proc Natl Acad Sci U S A 89*(10), 4452–4456.
Mohammed A, Isa HA. (2007). Pattern of primary tumours and tumour-like lesions of bone in Zaria, northern Nigeria: a review of 127 cases. *West Afr J Med 26*(1), 37–41.
Moisio AL, Sistonen P, et al. (1996). Age and origin of two common MLH1 mutations predisposing to hereditary colon cancer. *Am J Hum Genet 59*(6), 1243–1251.
Moldovan GL, D'Andrea AD. (2009). How the fanconi anemia pathway guards the genome. *Annu Rev Genet 43*: 223–249.
Moore SW, Appfelstaedt J, et al. (2007). Familial medullary carcinoma prevention, risk evaluation, and RET in children of families with MEN2. *J Pediatr Surg 42*(2), 326–332.
Moore SW, Zaahl M. (2009). Clinical and genetic differences in total colonic aganglionosis in Hirschsprung's disease. *J Pediatr Surg 44*(10), 1899–903.
Moore SW, Zaahl MG. (2008). Multiple endocrine neoplasia syndromes, children, Hirschsprung's disease and RET. *Pediatr Surg Int 24*(5), 521–530.
Morgan NV, Essop F, et al. (2005). A common Fanconi anemia mutation in black populations of sub-Saharan Africa. *Blood 105*(9), 3542–3544.
Morrell D, Cromartie E, et al. (1986). Mortality and cancer incidence in 263 patients with ataxia-telangiectasia. *J Natl Cancer Inst 77*(1), 89–92.
Mourali N, Tabbane S, et al. (1986). [Cancer of the rectum in Tunisia]. *Tunis Med 64*(4), 343–349.
Moussaid L, Benchikhi H, et al. (2004). [Cutaneous tumors during xeroderma pigmentosum in Morocco: study of 120 patients]. *Ann Dermatol Venereol 131*(1 Pt 1), 29–33.
Moussala M, Behar-Cohen F, et al. (2000). [Xeroderma pigmentosum and its ocular manifestations. The first case in Cameroon]. *J Fr Ophthalmol 23*(4), 369–374.
Muller A, Schackert HK, et al. (2006). A novel MSH2 germline mutation in homozygous state in two brothers with colorectal cancers diagnosed at the age of 11 and 12 years. *Am J Med Genet A 140*(3), 195–199.
Mulligan LM, Kwok JB, et al. (1993). Germ-line mutations of the RET proto-oncogene in multiple endocrine neoplasia type 2A. *Nature 363*(6428), 458–460.
Murty VV, Bouchard B, et al. (1992). Assignment of the human TYRP (brown) locus to chromosome region 9p23 by nonradioactive in situ hybridization. *Genomics 13*(1), 227–229.
Myrhoj T, Bisgaard ML, et al. (1997). Hereditary non-polyposis colorectal cancer: clinical features and survival. Results from the Danish HNPCC register. *Scand J Gastroenterol 32*(6), 572–576.
Nagase H, Miyoshi Y, et al. (1992). Correlation between the location of germ-line mutations in the APC gene and the number of colorectal polyps in familial adenomatous polyposis patients. *Cancer Res 52*(14), 4055–4057.
Nakagawa H, Lockman JC, et al. (2004). Mismatch repair gene PMS2: disease-causing germline mutations are frequent in patients whose tumors stain negative for PMS2 protein, but paralogous genes obscure mutation detection and interpretation. *Cancer Res 64*(14), 4721–4727.
Nance WE, Jackson CE, et al. (1970). Amish albinism: a distinctive autosomal recessive phenotype. *Am J Hum Genet 22*(5), 579–586.
Naylor EW, Gardner EJ. (1977). Penetrance and expressivity of the gene responsible for the Gardner syndrome. *Clin Genet 11*(6), 381–393.
Nevo S, Cleve H. (1991). PI polymorphism in Israel: report on six Jewish population groups. *Am J Med Genet 39*(4), 399–403.
Newton JM, Cohen-Barak O, et al. (2001). Mutations in the human orthologue of the mouse underwhite gene (uw) underlie a new form of oculocutaneous albinism, OCA4. *Am J Hum Genet 69*(5), 981–988.
Ni TT, Marsischky GT, et al. (1999). MSH2 and MSH6 are required for removal of adenine misincorporated opposite 8-oxo-guanine in S. cerevisiae. *Mol Cell 4*(3), 439–444.
Nicolaides NC, Carter KC, et al. (1995). Genomic organization of the human PMS2 gene family. *Genomics 30*(2), 195–206.
Nicolaides NC, Papadopoulos N, et al. (1994). Mutations of two PMS homologues in hereditary nonpolyposis colon cancer. *Nature 371*(6492), 75–80.
Niessen RC, Kleibeuker JH, et al. (2009). PMS2 involvement in patients suspected of Lynch syndrome. *Genes Chromosomes Cancer 48*(4), 322–329.
Nishigori C, Zghal M, et al. (1993). High prevalence of the point mutation in exon 6 of the xeroderma pigmentosum group A-complementing (XPAC) gene in xeroderma pigmentosum group A patients in Tunisia. *Am J Hum Genet 53*(5), 1001–1006.
Nishisho I, Nakamura Y, et al. (1991). Mutations of chromosome 5q21 genes in FAP and colorectal cancer patients. *Science 253*(5020), 665–669.

Ntsiba H, Bazebissa R. (2002). [Multiple exostoses. The first 4 Congolese cases]. *Bull Soc Pathol Exot 95*(1), 20–22.

Nukiwa T, Brantly ML, et al. (1988). Characterization of the gene and protein of the common alpha 1-antitrypsin normal M2 allele. *Am J Hum Genet 43*(3), 322–330.

Nystrom-Lahti M, Kristo P, et al. (1995). Founding mutations and Alu-mediated recombination in hereditary colon cancer. *Nat Med 1*(11), 1203–1206.

O'Donnell FE Jr, King RA, et al. (1978). Autosomal recessively inherited ocular albinism. A new form of ocular albinism affecting females as severely as males. *Arch Ophthalmol 96*(9), 1621–1625.

Obalum DC, Eyesan RA, et al. (2008). Pattern of osteochondromas in Lagos, Nigeria. *Nig Q J Hosp Med 18*(2), 69–71.

Odetayo OO. (2001). Pattern of bone tumours at the National Orthopaedic Hospital, Lagos, Nigeria. *West Afr J Med 20*(2), 161–164.

Oetting WS, King RA. (1992). Molecular analysis of type I-A (tyrosinase negative) oculocutaneous albinism. *Hum Genet 90*(3), 258–262.

Oetting WS, King RA. (1993). Molecular basis of type I (tyrosinase-related) oculocutaneous albinism: mutations and polymorphisms of the human tyrosinase gene. *Hum Mutat 2*(1), 1–6.

Oetting WS, King RA. (1999). Molecular basis of albinism: mutations and polymorphisms of pigmentation genes associated with albinism. *Hum Mutat 13*(2), 99–115.

Ojutiku OO, Ladipo GO. (1978). Alpha-1-antitrypsin levels and phenotypes in some healthy Nigerians—A preliminary report. *Niger Med J 8*(6), 511–513.

Okoro AN. (1975). Albinism in Nigeria. A clinical and social study. *Br J Dermatol 92*(5), 485–492.

Olusi SO, Ladipo GO, et al. (1982). Alpha 1-antitrypsin phenotypes in a Nigerian population. *Afr J Med Med Sci 10*(3–4), 113–117.

Onwuameze IC. (1989). Specific protein pattern in adult healthy Nigerians. *Afr J Med Med Sci 18*(1), 49–53.

Osta WA, Chen Y, et al. (2004). EpCAM is overexpressed in breast cancer and is a potential target for breast cancer gene therapy. *Cancer Res 64*(16), 5818–5824.

Ostergaard JR, Sunde L, et al. (2005). Neurofibromatosis von Recklinghausen type I phenotype and early onset of cancers in siblings compound heterozygous for mutations in MSH6. *Am J Med Genet A 139A*(2), 96–105; discussion 96.

Ou J, Rasmussen M, et al. (2009). Biochemical characterization of MLH3 missense mutations does not reveal an apparent role of MLH3 in Lynch syndrome. *Genes Chromosomes Cancer 48*(4), 340–350.

Pagenstecher C, Wehner M, et al. (2006). Aberrant splicing in MLH1 and MSH2 due to exonic and intronic variants. *Hum Genet 119*(1–2), 9–22.

Panguluri RC, Brody LC, et al. (1999). BRCA1 mutations in African Americans. *Hum Genet 105*(1–2), 28–31.

Papadopoulos N, Nicolaides NC, et al. (1995). Mutations of GTBP in genetically unstable cells. *Science 268*(5219), 1915–1917.

Papadopoulos N, Nicolaides NC, et al. (1994). Mutation of a mutL homolog in hereditary colon cancer. *Science 263*(5153), 1625–1629.

Parc Y, Boisson C, et al. (2003). Cancer risk in 348 French MSH2 or MLH1 gene carriers. *J Med Genet 40*(3), 208–213.

Parsons R, Li GM, et al. (1993). Hypermutability and mismatch repair deficiency in RER+ tumor cells. *Cell 75*(6), 1227–1236.

Parsons R, Myeroff LL, et al. (1995). Microsatellite instability and mutations of the transforming growth factor beta type II receptor gene in colorectal cancer. *Cancer Res 55*(23), 5548–5550.

Pascali VL, Ranalletta D, et al. (1986). Antitrypsin and Gc polymorphisms in some populations of Congo: an unusual, highly frequent mutant, PIS, in Bateke and Babenga. *Ann Hum Biol 13*(3), 267–271.

Passmore LA, Kaesmann-Kellner B, et al. (1999). Novel and recurrent mutations in the tyrosinase gene and the P gene in the German albino population. *Hum Genet 105*(3), 200–210.

Patel KJ, Joenje H. (2007). Fanconi anemia and DNA replication repair. *DNA Repair (Amst) 6*(7), 885–890.

Pearson CE, Ewel A, et al. (1997). Human MSH2 binds to trinucleotide repeat DNA structures associated with neurodegenerative diseases. *Hum Mol Genet 6*(7), 1117–1123.

Pearson T, Jansen S, et al. (2001). Fanconi anemia. a statistical evaluation of cytogenetic results obtained from South African families. *Cancer Genet Cytogenet 126*(1), 52–55.

Peifer M. (1993). Cancer, catenins, and cuticle pattern: a complex connection. *Science 262*(5140), 1667–1668.

Peltomaki P, Aaltonen LA, et al. (1993). Genetic mapping of a locus predisposing to human colorectal cancer. *Science 260*(5109), 810–812.

Pensotti V, Radice P, et al. (1997). Mean age of tumor onset in hereditary nonpolyposis colorectal cancer (HNPCC) families correlates with the presence of mutations in DNA mismatch repair genes. *Genes Chromosomes Cancer 19*(3), 135–142.

Petris MJ, Strausak D, et al. (2000). The Menkes copper transporter is required for the activation of tyrosinase. *Hum Mol Genet 9*(19), 2845–2851.

Pilkington MM, Wilder JA, et al. (2008). Contrasting signatures of population growth for mitochondrial DNA and Y chromosomes among human populations in Africa. *Mol Biol Evol 25*(3), 517–525.

Polakis P. (1995). Mutations in the APC gene and their implications for protein structure and function. *Curr Opin Genet Dev 5*(1), 66–71.

Ponder BA, Antoniou A, et al. (2005). Polygenic inherited predisposition to breast cancer. *Cold Spring Harb Symp Quant Biol 70*: 35–41.

Ponnazhagan S, Hou L, et al. (1994). Structural organization of the human tyrosinase gene and sequence analysis and characterization of its promoter region. *J Invest Dermatol 102*(5), 744–748.

Porter DE, Lonie L, et al. (2004). Severity of disease and risk of malignant change in hereditary multiple exostoses. A genotype-phenotype study. *J Bone Joint Surg Br 86*(7), 1041–1046.

Pronk JC, Gibson RA, et al. (1995). Localisation of the Fanconi anaemia complementation group A gene to chromosome 16q24.3. *Nat Genet 11*(3), 338–340.

Puri N, Durbam-Pierre D, et al. (1997). Type 2 oculocutaneous albinism (OCA2) in Zimbabwe and Cameroon: distribution of the 2.7-kb deletion allele of the P gene. *Hum Genet 100*(5–6), 651–656.

Quehenberger F, Vasen HF, et al. (2005). Risk of colorectal and endometrial cancer for carriers of mutations of the hMLH1 and hMSH2 gene: correction for ascertainment. *J Med Genet 42*(6), 491–496.

Quintana-Murci L, Quach H, et al. (2008). Maternal traces of deep common ancestry and asymmetric gene flow between Pygmy hunter-gatherers and Bantu-speaking farmers. *Proc Natl Acad Sci U S A 105*(5), 1596–601.

Ramsay M, Colman MA, et al. (1992). The tyrosinase-positive oculocutaneous albinism locus maps to chromosome 15q11.2-q12. *Am J Hum Genet 51*(4), 879–884.

Recchia FM, Carvalho-Recchia CA, et al. (2002). Optical coherence tomography in the diagnosis of foveal hypoplasia. *Arch Ophthalmol 120*(11), 1587–1588.

Reed FA, Tishkoff SA. (2006). African human diversity, origins and migrations. *Curr Opin Genet Dev 16*(6), 597–605.

Reeves MD, Yawitch TM, et al. (2004). BRCA1 mutations in South African breast and/or ovarian cancer families: evidence of a novel founder mutation in Afrikaner families. *Int J Cancer 110*(5), 677–682.

Reid S, Schindler D, et al. (2007). Biallelic mutations in PALB2 cause Fanconi anemia subtype FA-N and predispose to childhood cancer. *Nat Genet 39*(2), 162–164.

Reitmair, A. H., R. Schmits, et al. (1995). MSH2 deficient mice are viable and susceptible to lymphoid tumours. *Nat Genet 11*(1), 64–70.

Renwick A, Thompson D, et al. (2006). ATM mutations that cause ataxia-telangiectasia are breast cancer susceptibility alleles. *Nat Genet 38*(8), 873–875.

Ricciardone MD, Ozcelik T, et al. (1999). Human MLH1 deficiency predisposes to hematological malignancy and neurofibromatosis type 1. *Cancer Res 59*(2), 290–293.

Rijcken FE, Hollema H, et al. (2002). Proximal adenomas in hereditary non-polyposis colorectal cancer are prone to rapid malignant transformation. *Gut 50*(3), 382–386.

Rinchik EM, Bultman SJ, et al. (1993). A gene for the mouse pink-eyed dilution locus and for human type II oculocutaneous albinism. *Nature 361*(6407), 72–76.

Robbins JH, Kraemer KH, et al. (1974). Xeroderma pigmentosum. An inherited diseases with sun sensitivity, multiple cutaneous neoplasms, and abnormal DNA repair. *Ann Intern Med 80*(2), 221–248.

Robinson WP, Bottani A, et al. (1991). Molecular, cytogenetic, and clinical investigations of Prader-Willi syndrome patients. *Am J Hum Genet 49*(6), 1219–1234.

Rodriguez-Cintron W, Guntupalli K, et al. (1995). Bronchiectasis and homozygous (P1ZZ) alpha 1-antitrypsin deficiency in a young man. *Thorax 50*(4), 424–425.

Rooryck C, Morice-Picard F, et al. (2008). Molecular diagnosis of oculocutaneous albinism: new mutations in the OCA1-4 genes and practical aspects. *Pigment Cell Melanoma Res 21*(5), 583–587.

Rooryck C, Roudaut C, et al. (2006). Oculocutaneous albinism with TYRP1 gene mutations in a Caucasian patient. *Pigment Cell Res 19*(3), 239–242.

Rosemblat S, Durham-Pierre D, et al. (1994). Identification of a melanosomal membrane protein encoded by the pink-eyed dilution (type II oculocutaneous albinism) gene. *Proc Natl Acad Sci U S A 91*(25), 12071–12075.

Rosen EM, Fan S, et al. (2006). BRCA1 regulation of transcription. *Cancer Lett 236*(2), 175–185.

Rosendorff J, Bernstein R, et al. (1987). Fanconi anemia: another disease of unusually high prevalence in the Afrikaans population of South Africa. *Am J Med Genet 27*(4), 793–797.

Rundshagen U, Zuhlke C, et al. (2004). Mutations in the MATP gene in five German patients affected by oculocutaneous albinism type 4. *Hum Mutat 23*(2), 106–110.

Saggar S, Chernoff KA, et al. (2008). CYLD mutations in familial skin appendage tumours. *J Med Genet 45*(5), 298–302.

Salhi A, Bornholdt D, et al. (2004). Multiple familial trichoepithelioma caused by mutations in the cylindromatosis tumor suppressor gene. *Cancer Res 64*(15), 5113–5117.

Sanso GE, Domene HM, et al. (2002). Very early detection of RET protooncogene mutation is crucial for preventive thyroidectomy in multiple endocrine neoplasia type 2 children: presence of C-cell malignant disease in asymptomatic carriers. *Cancer 94*(2), 323–330.

Santucci-Darmanin S, Neyton S, et al. (2002). The DNA mismatch-repair MLH3 protein interacts with MSH4 in meiotic cells, supporting a role for this MutL homolog in mammalian meiotic recombination. *Hum Mol Genet 11*(15), 1697–1706.

Sarangarajan R, Boissy RE. (2001). Tyrp1 and oculocutaneous albinism type 3. *Pigment Cell Res 14*(6), 437–444.

Sasaki S, Horii A, et al. (1996). Somatic mutations of a human mismatch repair gene, hMLH1, in tumors from patients with multiple primary cancers. *Hum Mutat 7*(3), 275–278.

Savitsky K, Bar-Shira A, et al. (1995). A single ataxia telangiectasia gene with a product similar to PI-3 kinase. *Science 268*(5218), 1749–1753.

Schmale GA, Conrad EU 3rd, et al. (1994). The natural history of hereditary multiple exostoses. *J Bone Joint Surg Am 76*(7), 986–992.

Schmitz B, Schaefer T, et al. (2003). Configuration of the optic chiasm in humans with albinism as revealed by magnetic resonance imaging. *Invest Ophthalmol Vis Sci 44*(1), 16–21.

Schuster SC, Miller W, et al. (2010), Complete Khoisan and Bantu genomes from southern Africa. *Nature 463*(7283), 943–947.

Scialfa AC. (1972). Ocular albinism in a female. *Am J Ophthalmol 73*(6), 943–948.

Scott RJ, McPhillips M, et al. (2001). Hereditary nonpolyposis colorectal cancer in 95 families: differences and similarities between mutation-positive and mutation-negative kindreds. *Am J Hum Genet 68*(1), 118–127.

Sebetan IM. (1992). PI*E Tripoli: a new allele in the alpha-1-antitrypsin system. *Hum Hered 42*(3), 206–208.

Setty SR, Tenza D, et al. (2008). Cell-specific ATP7A transport sustains copper-dependent tyrosinase activity in melanosomes. *Nature 454*(7208), 1142–1146.

Sfar S, Abid A, et al. (2009). Genetic analysis of hereditary multiple exostoses in Tunisian families: a novel frame-shift mutation in the EXT1 gene. *Mol Biol Rep 36*(4), 661–667.

Shtoyerman-Chen R, Friedman E, et al. (2001). The I1307K APC polymorphism: prevalence in non-Ashkenazi Jews and evidence for a founder effect. *Genet Test 5*(2), 141–146.

Sigsgaard T, Brandslund I, et al. (1994). Low normal alpha-1-antitrypsin serum concentrations and MZ-phenotype are associated with byssinosis and familial allergy in cotton mill workers. *Pharmacogenetics 4*(3), 135–141.

Simpson NE, Kidd KK, et al. (1987). Assignment of multiple endocrine neoplasia type 2A to chromosome 10 by linkage. *Nature 328*(6130), 528–530.

Soliman AS, Bondy ML, et al. (1998). Reduced expression of mismatch repair genes in colorectal cancer patients in Egypt. *Int J Oncol 12*(6), 1315–1319.

Souza RF, Appel R, et al. (1996). Microsatellite instability in the insulin-like growth factor II receptor gene in gastrointestinal tumours. *Nat Genet 14*(3), 255–257.

Sowter HM, Ashworth A. (2005). BRCA1 and BRCA2 as ovarian cancer susceptibility genes. *Carcinogenesis 26*(10), 1651–1656.

Spinola C, Brehm A, et al. (2010). Alpha-1-antitrypsin deficiency in the Cape Verde islands (Northwest Africa), High prevalence in a sub-Saharan population. *Respir Med 104*(7), 1069–1072.

Spirio L, Olschwang S, et al. (1993). Alleles of the APC gene: an attenuated form of familial polyposis. *Cell 75*(5), 951–957.

Spritz RA. (1994). Molecular genetics of oculocutaneous albinism. *Hum Mol Genet 3 Spec No*: 1469–1475.

Spritz RA, Fukai K, et al. (1995). Frequent intragenic deletion of the P gene in Tanzanian patients with type II oculocutaneous albinism (OCA2). *Am J Hum Genet 56*(6), 1320–1323.

Sridharan R, Radhakrishnan K, et al. (1985). Prevalence and pattern of spinocerebellar degenerations in northeastern Libya. *Brain 108 (Pt 4)*, 831–843.

States JC, McDuffie ER, et al. (1998). Distribution of mutations in the human xeroderma pigmentosum group A gene and their relationships to the functional regions of the DNA damage recognition protein. *Hum Mutat 12*(2), 103–113.

Stella A, Surdo NC, et al. (2007). Germline novel MSH2 deletions and a founder MSH2 deletion associated with anticipation effects in HNPCC. *Clin Genet 71*(2), 130–139.

Stevens G, Ramsay M, et al. (1997). Oculocutaneous albinism (OCA2) in sub-Saharan Africa: distribution of the common 2.7-kb P gene deletion mutation. *Hum Genet 99*(4), 523–527.

Stevens G, van Beukering J, et al. (1995). An intragenic deletion of the P gene is the common mutation causing tyrosinase-positive oculocutaneous albinism in southern African Negroids. *Am J Hum Genet 56*(3), 586–591.

Stieber JR, Dormans JP. (2005). Manifestations of hereditary multiple exostoses. *J Am Acad Orthop Surg 13*(2), 110–120.

Stokowski RP, Pant PV, et al. (2007). A genomewide association study of skin pigmentation in a South Asian population. *Am J Hum Genet 81*(6), 1119–1132.

Stoppa-Lyonnet D, Laurent-Puig P, et al. (1997). BRCA1 sequence variations in 160 individuals referred to a breast/ovarian family cancer clinic. Institut Curie Breast Cancer Group. *Am J Hum Genet 60*(5), 1021–1030.

Stupart DA, Goldberg PA, et al. (2009). Cancer risk in a cohort of subjects carrying a single mismatch repair gene mutation. *Fam Cancer 8*(4), 519–523.

Suchy J, Kurzawski G, et al. (2002). Ovarian cancer of endometrioid type as part of the MSH6gene mutation phenotype. *J Hum Genet 47*(10), 529–531.

Sugihara K, Muto T, et al. (1982). Gardner's syndrome associated with periampullary carcinoma, duodenal and gastric adenomatosis. Report of a case. *Dis Colon Rectum 25*(8), 766–771.

Sulem P, Gudbjartsson DF, et al. (2008). Two newly identified genetic determinants of pigmentation in Europeans. *Nat Genet 40*(7), 835–837.

Summers CG, Creel D, et al. (1991). Variable expression of vision in sibs with albinism. *Am J Med Genet 40*(3), 327–331.

Swift M. (1971). Fanconi's anaemia in the genetics of neoplasia. *Nature 230*(5293), 370–373.

Swift M, Morrell D, et al. (1991). Incidence of cancer in 161 families affected by ataxia-telangiectasia. *N Engl J Med 325*(26), 1831–1836.

Swift M, Reitnauer PJ, et al. (1987). Breast and other cancers in families with ataxia-telangiectasia. *N Engl J Med 316*(21), 1289–1294.

Syngal S, Fox EA, et al. (2000). Sensitivity and specificity of clinical criteria for hereditary non-polyposis colorectal cancer associated mutations in MSH2 and MLH1. *J Med Genet 37*(9), 641–645.

Szodoray L. (1975). [Some observations and considerations on reticuloses]. *Dermatol Monatsschr 161*(10), 842–850.

Takahashi M. (2001). The GDNF/RET signaling pathway and human diseases. *Cytokine Growth Factor Rev 12*(4), 361–373.

Takebe H, Nishigori C, et al. (1987). Genetics and skin cancer of xeroderma pigmentosum in Japan. *Jpn J Cancer Res 78*(11), 1135–1143.

Tannergard P, Liu T, et al. (1997). Tumorigenesis in colorectal tumors from patients with hereditary non-polyposis colorectal cancer. *Hum Genet 101*(1), 51–55.

Tavtigian SV, Oefner PJ, et al. (2009). Rare, evolutionarily unlikely missense substitutions in ATM confer increased risk of breast cancer. *Am J Hum Genet 85*(4), 427–446.

Taylor CF, Charlton RS, et al. (2003). Genomic deletions in MSH2 or MLH1 are a frequent cause of hereditary non-polyposis colorectal cancer: identification of novel and recurrent deletions by MLPA. *Hum Mutat 22*(6), 428–433.

Tete-Benissan AC, Duriez P, et al. (2000). [Study of the protein profile of the Adele tribe of Togo]. *Sante 10*(4), 261–266.

Thompson E, Meldrum CJ, et al. (2004). Hereditary non-polyposis colorectal cancer and the role of hPMS2 and hEXO1 mutations. *Clin Genet 65*(3), 215–225.

Thorstenson YR, Shen P, et al. (2001). Global analysis of ATM polymorphism reveals significant functional constraint. *Am J Hum Genet 69*(2), 396–412.

Tipping AJ, Pearson T, et al. (2001). Molecular and genealogical evidence for a founder effect in Fanconi anemia families of the Afrikaner population of South Africa. *Proc Natl Acad Sci U S A 98*(10), 5734–5739.

Tishkoff SA, Gonder MK, et al. (2007). History of click-speaking populations of Africa inferred from mtDNA and Y chromosome genetic variation. *Mol Biol Evol 24*(10), 2180–2195.

Tournier I, Vezain M, et al. (2008). A large fraction of unclassified variants of the mismatch repair genes MLH1 and MSH2 is associated with splicing defects. *Hum Mutat 29*(12), 1412–1424.

Triki C, Feki I, et al. (2000). [Clinical, biological and genetic study of 24 patients with ataxia telangiectasia from southern Tunisia]. *Rev Neurol (Paris) 156*(6–7), 634–637.

Tripathi RK, Strunk KM, et al. (1992). Tyrosinase gene mutations in type I (tyrosinase-deficient) oculocutaneous albinism define two clusters of missense substitutions. *Am J Med Genet 43*(5), 865–871.

Troudi W, Uhrhammer N, et al. (2008). Complete mutation screening and haplotype characterization of BRCA1 gene in Tunisian patients with familial breast cancer. *Cancer Biomark 4*(1), 11–18.

Troudi W, Uhrhammer N, et al. (2007). Contribution of the BRCA1 and BRCA2 mutations to breast cancer in Tunisia. *J Hum Genet 52*(11), 915–920.

Uhrhammer N, Abdelouahab A, et al. (2008). BRCA1 mutations in Algerian breast cancer patients: high frequency in young, sporadic cases. *Int J Med Sci 5*(4), 197–202.

Umar A, Boyer JC, et al. (1994). Defective mismatch repair in extracts of colorectal and endometrial cancer cell lines exhibiting microsatellite instability. *J Biol Chem 269*(20), 14367–14370.

van der Klift H, Wijnen J, et al. (2005). Molecular characterization of the spectrum of genomic deletions in the mismatch repair genes MSH2, MLH1, MSH6, and PMS2 responsible for hereditary nonpolyposis colorectal cancer (HNPCC). *Genes Chromosomes Cancer 44*(2), 123–138.

van der Luijt RB, Meera Khan P, et al. (1996). Germline mutations in the 3' part of APC exon 15 do not result in truncated proteins and are associated with attenuated adenomatous polyposis coli. *Hum Genet 98*(6), 727–734.

van Heyningen V. (1994). Genetics. One gene—four syndromes. *Nature 367*(6461), 319–320.

Vandeville D, Martin JP, et al. (1974). Alpha 1-antitrypsin polymorphism of a Bantu population: description of a new allele PiL. *Humangenetik 21*(1), 33–38.

Vasen HF, Mecklin JP, et al. (1991). The International Collaborative Group on Hereditary Non-Polyposis Colorectal Cancer (ICG-HNPCC). *Dis Colon Rectum 34*(5), 424–425.

Visweswara RN, Osman AR, et al. (1997). Xeroderma pigmentosum with adult Wilms' tumour. A rare association. *Scand J Urol Nephrol 31*(6), 569–571.

Wada-Hiraike O, Yano T, et al. (2005). The DNA mismatch repair gene hMSH2 is a potent coactivator of oestrogen receptor alpha. *Br J Cancer 92*(12), 2286–2291.

Wagner A, Hendriks Y, et al. (2001). Atypical HNPCC owing to MSH6 germline mutations: analysis of a large Dutch pedigree. *J Med Genet 38*(5), 318–322.

Wagner TM, Hirtenlehner K, et al. (1999). Global sequence diversity of BRCA2: analysis of 71 breast cancer families and 95 control individuals of worldwide populations. *Hum Mol Genet 8*(3), 413–423.

Walsh N, Qizilbash A, et al. (1987). Biliary neoplasia in Gardner's syndrome. *Arch Pathol Lab Med 111*(1), 76–77.

Wang W. (2007). Emergence of a DNA-damage response network consisting of Fanconi anaemia and BRCA proteins. *Nat Rev Genet 8*(10), 735–748.

Wang Y, Cortez D, et al. (2000). BASC, a super complex of BRCA1-associated proteins involved in the recognition and repair of aberrant DNA structures. *Genes Dev 14*(8), 927–939.

Wang Y, Friedl W, et al. (1997). Hereditary nonpolyposis colorectal cancer: causative role of a germline missense mutation in the hMLH1 gene confirmed by the independent occurrence of the same somatic mutation in tumour tissue. *Hum Genet 100*(3–4), 362–364.

Wang Y, Friedl W, et al. (2002). A modified multiplex PCR assay for detection of large deletions in MSH2 and MLH1. *Hum Mutat 19*(3), 279–286.

Ward AM, Pickering JD, et al. (1977). Pi (antitrypsin) typing methods. A comparison of acid starch-gel electrophoresis and isoelectric focusing. *Hum Hered 27*(4), 292–297.

Welch SG, McGregor IA, et al. (1980). Alpha 1-antitrypsin (pi) phenotypes in a village population from The Gambia, West Africa. Evidence of a new variant occurring at a polymorphic frequency. *Hum Genet 53*(2), 233–235.

Whitney MA, Saito H, et al. (1993). A common mutation in the FACC gene causes Fanconi anaemia in Ashkenazi Jews. *Nat Genet 4*(2), 202–205.

Wicklund CL, Pauli RM, et al. (1995). Natural history study of hereditary multiple exostoses. *Am J Med Genet 55*(1), 43–46.

Wijnen J, de Leeuw W, et al. (1999). Familial endometrial cancer in female carriers of MSH6 germline mutations. *Nat Genet 23*(2), 142–144.

Wijnen J, Khan PM, et al. (1996). Majority of hMLH1 mutations responsible for hereditary nonpolyposis colorectal cancer cluster at the exonic region 15-16. *Am J Hum Genet 58*(2), 300–307.

Wijnen J, Vasen H, et al. (1995). Seven new mutations in hMSH2, an HNPCC gene, identified by denaturing gradient-gel electrophoresis. *Am J Hum Genet 56*(5), 1060–1066.

Wijnen JT, Vasen HF, et al. (1998). Clinical findings with implications for genetic testing in families with clustering of colorectal cancer. *N Engl J Med 339*(8), 511–518.

Willcox M, Beckman G, et al. (1986). Serum protein polymorphisms in a Liberian population. *Hum Hered 36*(1), 54–57.

Witkop CJ Jr, Jay B, et al. (1982). Optic and otic neurologic abnormalities in oculocutaneous and ocular albinism. *Birth Defects Orig Artic Ser 18*(6), 299–318.

Wood ET, Stover DA, et al. (2005). Contrasting patterns of Y chromosome and mtDNA variation in Africa: evidence for sex-biased demographic processes. *Eur J Hum Genet 13*(7), 867–876.

Worthley DL, Walsh MD, et al. (2005). Familial mutations in PMS2 can cause autosomal dominant hereditary nonpolyposis colorectal cancer. *Gastroenterology 128*(5), 1431–1436.

Wu Y, Berends MJ, et al. (2001). A role for MLH3 in hereditary nonpolyposis colorectal cancer. *Nat Genet 29*(2), 137–138.

Wu Y, Whitman I, et al. (1994). A lag in intracellular degradation of mutant alpha 1-antitrypsin correlates with the liver disease phenotype in homozygous PiZZ alpha 1-antitrypsin deficiency. *Proc Natl Acad Sci USA 91*(19), 9014–9018.

Wuyts W, Van Hul W, et al. (1998). Mutations in the EXT1 and EXT2 genes in hereditary multiple exostoses. *Am J Hum Genet 62*(2), 346–354.

Xia B, Dorsman JC, et al. (2007). Fanconi anemia is associated with a defect in the BRCA2 partner PALB2. *Nat Genet 39*(2), 159–161.

Yawitch TM, van Rensburg EJ, et al. (2000). Absence of commonly recurring BRCA1 mutations in black South African women with breast cancer. *S Afr Med J 90*(8), 788.

Yotova V, Lefebvre JF, et al. (2007). Tracing genetic history of modern humans using X-chromosome lineages. *Hum Genet 122*(5), 431–443.

Young AL, Kellermayer R, et al. (2006). CYLD mutations underlie Brooke-Spiegler, familial cylindromatosis, and multiple familial trichoepithelioma syndromes. *Clin Genet 70*(3), 246–249.

Yuan Z Q, Kasprzak L, et al. (1998). I1307K APC and hMLH1 mutations in a non-Jewish family with hereditary non-polyposis colorectal cancer. *Clin Genet 54*(4), 368–370.

Yuasa I, Umetsu K, et al. (2007). OCA2 481Thr, a hypofunctional allele in pigmentation, is characteristic of northeastern Asian populations. *J Hum Genet 52*(8), 690–693.

Zghal M, Fazaa B, et al. (2003). [A whole family affected by xeroderma pigmentosum: clinical and genetic particularities]. *Ann Dermatol Venereol 130*(1 Pt 1), 31–36.

Zhang B, Fackenthal JD, et al. (2009). Evidence for an ancient BRCA1 mutation in breast cancer patients of Yoruban ancestry. *Fam Cancer 8*(1), 15–22.

Zhou XP, Kuismanen S, et al. (2002). Distinct PTEN mutational spectra in hereditary non-polyposis colon cancer syndrome-related endometrial carcinomas compared to sporadic microsatellite unstable tumors. *Hum Mol Genet 11*(4), 445–450.

Zhu G, Evans DM, et al. (2004). A genome scan for eye color in 502 twin families: most variation is due to a QTL on chromosome 15q. *Twin Res 7*(2), 197–210.

45.

GENETIC COUNSELING IN SOUTH AFRICA

Jacquie Greenberg, Jennifer Kromberg, Kelly Loggenberg, and Tina-Marié Wessels

INTRODUCTION—BACKGROUND

South Africa, although classified as a middle-income country, is the richest country in Africa. It is a country with many peoples of different origins, some ancient and some modern, many cultures, and a multitude of different natural resources. It has a democratically elected government, several large, thriving cities, with extensive urbanization, a large rural population, and much migration between rural and urban communities. It is a country in transition at all levels, including the healthcare systems and services: "South African society is fluid and undergoing rapid social changes unparalleled in the rest of Africa" (Binka, 2007). There are gaps in health itself, in health research funding, and in access to information concerning health conditions (Wall, 2007), which are typical of developing countries. Fifteen years after the political liberation of South Africa, the country is struggling to live up to the ideals articulated by those with a vision, including Nobel laureates Nelson Mandela and Desmond Tutu (Tollman & Kahn, 2007). There are many difficulties in building a stable system of government and a social environment responsive to the needs of the whole community.

There are 49.32 million people living in South Africa (estimate as of mid-2009: www.statssa.gov.za). The majority (79%) of the population is Black, and the major ethnic groups include the Nguni (e.g., Zulu and Xhosa) and Sotho-Tswana. A further 9.6% of the population is White, mainly of Dutch or English origin, and the remainder (11.4%) is of mixed ancestry, and/or Indian or Asian. There are 11 different official languages and many different religions. About 14% of the people have no education, but 32% have completed their schooling and have a Grade 12 certificate. There are many disparities in access to healthcare, especially between the private and a state health sectors, and staff shortages are exacerbating the problem. South Africa also has one of the highest HIV-positive rates in the world, and the HIV/AIDS epidemic places a significant burden on the country's health budget and the provision of healthcare services (Doherty & Colvin, 2004). According to official estimates, as of mid-2009 (www.statssa.gov.za), there are 5.2 million people living with HIV/AIDS in South Africa and the overall HIV positive prevalence rate is 10.6% of the population. Life expectancy is 53.5years for males and 57.2 for females, while infant mortality is 45.7 per 1000 live births.

There is still much poverty in the country and, in 1994, 6.3% of the population were living on less than US$1 per day, while in 2005 this number had risen to 11% (Editorial, *Lancet*, 2008). In addition, rates of under-5 mortality have increased since 1990. Many of these childhood deaths are due to the high prevalence of HIV-positive individuals in South Africa. Almost one-third of all pregnant women are HIV positive and, in 2003, HIV-related infection caused more than half the deaths in children aged below 5 years. Another major problem is that one in seven nurses and nursing students are HIV+ and there are more nurses with full-blown AIDS than are being trained (Bateman, 2008), which could shortly cripple the health system. However, some progress is being made and in 2001 almost 21% of 15–49-year-old people were HIV+, whereas in 2005 this figure was about 19%. Life expectancy is dropping and morbidity increasing, due to the epidemic. Nevertheless, access to safe water and sanitation has improved, school enrollment is high, and the difference in literacy rates for boys and girls is negligible (Editorial, *Lancet*, 2008). Also, the number of Internet users increased from 0.83% of the population in 1996 to 10.75% in 2005 and the number of mobile phone users has risen astronomically. These are encouraging steps in South Africa's development.

South Africa is bordered on the northwest, north, and northeast by several countries. These include Namibia, Botswana, Zimbabwe and Mozambique, while Swaziland and Lesotho are landlocked and surrounded by South Africa. These countries and their citizens often seek help and expertise in healthcare (particularly in the field of human genetics) from South African experts, and migration to the country, particularly from Zimbabwe due to its present crisis situation, is extensive.

EPIDEMIOLOGY OF COMMON GENETIC CONDITIONS

It is widely recognized that in the world population, about 6% of all infants born have some degree of congenital or birth defect, and that these defects are responsible for many neonatal deaths (Christianson & Modell, 2004). Furthermore, it is estimated that one person in ten will experience a multifactorial disease in her/his lifetime. The types of genetic disorders in a community depend on the genetic profile of that community, some being common to all the ethnic groups in the community and some occurring at much higher rates in specific ethnic groups (Beighton & Botha, 1986a,b,c).

In South Africa the more serious common monogenic or single-gene disorders that occur in all groups include Huntington disease (Hayden, 1981; Krause & Greenberg, 2008), Marfan syndrome, myotonic dystrophy, neurofibromatosis (Beighton et al., 1991), polycystic kidneys, tuberous sclerosis, and osteogenesis imperfecta, among others (Beighton, 1976). In addition, those that occur at higher rates than usual in specific groups include: oculocutaneous albinism in the Black population (Kromberg & Jenkins, 1982a); cystic fibrosis in the White groups (Denter et al., 1992); Tay Sachs and Gaucher disease in the Jewish population (Jenkins et al., 1977; Goldblatt & Beighton, 1979); familial hypercholesterolemia and porphyria in the Afrikaans (of Dutch origin) population (Jenkins, 1990); beta thalassemias in the Greek, and various thalassemias in the Indian populations (Krause, 1994). The X-linked disorders, such as hemophilia, Duchenne muscular dystrophy, fragile X syndrome, and some inherited forms of retinal degenerative disease (Greenberg et al., 1994) appear to occur in all populations at more or less the same rates as elsewhere.

The multifactorial or polygenic disorders have also been found in all the population groups at varying rates. These disorders include the neural tube disorders, talipes equinovarus, isolated hydrocephalus, facial clefts (Kromberg & Jenkins, 1982b), epilepsy (Christianson et al., 2000), and intellectual disability (Kromberg et al., 2008). Fetal alcohol syndrome, however, occurs at unusually high rates in specific areas of the country where the predisposing environmental factors predominate. For example, it is found in 6%–7% of children in the Western Cape, and at similar levels in the Northern Cape province (Viljoen et al., 2005).

Chromosome disorders are found at varying rates in different ethnic groups. Down syndrome (Smart, 1981; Kromberg et al., 1992) and trisomies 18 and 13 occur more frequently, as expected, in the infants of older women in all groups, as has been observed worldwide. Sex chromosome abnormalities, such as Klinefelter (XXY) and Turner (XO) syndrome, as well as triple-X syndrome also occur.

While knowledge of the inherited cancer syndromes is increasing in South Africa, few epidemiological data are yet available. However, cases of inherited colorectal cancer —such as hereditary nonpolyposis colon cancer (HNPCC) and familial adenomatous polyposis (FAP); Stupart et al., 2008; Anderson et al., 2007; Goldberg et al., 2000; Ramesar et al., 2000) and breast cancer—have been observed in all groups. Data on a variety of cancers are collected by the National Cancer Registry of South Africa, and in the case of breast cancer (the commonest cancer in women) in general, the incidence is highest in White women (lifetime risk 1/12) followed by Asian women (risk 1/18), those of mixed ancestry (risk 1/18), and Black (risk 1/49) women (Mqoqi et al., 2004). The incidence of colorectal cancer in the Black population appears to be reduced, but at least two cases (one Zulu and one Xhosa family) have been reported (Grobbelaar et al., 2002).

LOCAL BELIEFS ABOUT GENETIC DISORDERS

Cultural transformation has been taking place in the country for centuries, particularly since the influx of foreign traders and explorers in the 1400s. However, when modern health services are offered to developing communities, who, despite adopting some of the ways of the foreign culture still retain many of their own cultural ways, many problems may ensue. It is therefore essential for the providers of the new services to have a good understanding of the cultural influences on the perception of such services—in this case genetic services. Some of these influences include: ideas about disease causation; common taboos and practices; myths and superstitions; use of traditional healers; systems of thought; and language issues (Kromberg & Jenkins, 1997).

Causes of disease are not easily understood in any developing population with limited scientific insight. Ideas about causation are, therefore, often associated with traditions and myths and old wives tales. Ancestor worship is common in various South African ethnic groups, and transgression of the rules is believed to lead to ill-health in the transgressors and/or their child. Such ill-health can include congenital abnormalities in the offspring. For example, 51% off the rural traditional healers interviewed by Kromberg et al. (2008) considered the cause of childhood disabilities to be the breaking of taboos.

The different local ethnic groups have differing taboos about abortion, but in general, especially previously, they were totally opposed to it. One proverb stated that "to force out the womb is grievous, and the knot of the cradle skin is a flower"—i.e., women derive great pleasure from carrying a baby slung on the back (Schapera, 1940). This taboo, however, did not apply to newborn infants, and infanticide and euthanasia were practiced, historically, with regard to defective children and sometimes one or both twins (Jeffreys, 1953). Livingstone (1857) described the case of a mother who was excommunicated from her village with her albino child until she agreed to put him to death; she was then allowed to return. Generally, the midwife smothered or did away with the abnormal child and told the father it was stillborn, and if the chief became aware of the case it was never prosecuted (Schapera, 1940). However, these attitudes and taboos are changing and moving toward acceptance; abortion in certain circumstances (such as a severe genetic disorder in the fetus) is becoming acceptable, and if infanticide is still occurring, it is carried out in secret.

Another common practice with relevance to genetic services is that associated with mate selection. In several groups, the preferred mating was with a first cousin, such as a mother's brother's daughter (Krige, 1937). Such marriages might be motivated by the keeping of the bridewealth (often cattle) or dowry in the family, keeping the power in the family, the gaining of political benefits, or greater family cohesion or solidarity (Bittles, 2008). However, locally these cultural practices affect the rates of the common recessive conditions such as oculocutaneous albinism, and this condition is found at higher rates in groups that favor consanguineous marriages than in those for whom this practice is taboo (Kromberg & Jenkins, 1982b). Polygamy and polyandry are also practiced locally, with both advantages and disadvantages. Polygamy might alter the sex ratio, with more females being born (Miller, 1993) and sequential polyandry might mean that a woman would seldom have two children with the same recessive disorder.

Myths and superstitions have been encountered with regard to genetic disorders, and one of the most prevalent is the widely held belief that albinos do not die (Kromberg, 1992). This myth affects the acceptance of people with albinism in the community, and contributes to the anxieties of mothers who give birth to infants with albinism (Kromberg et al., 1987). The genetic counselor needs to be aware of such myths if the genetic counseling is to be culturally relevant, sensitive, and appropriate for the client (Kromberg & Jenkins, 1984)

Traditional healers are still powerful in the local community, and many people use the services of both Western medical doctors and traditional healers (Freeman et al., 1994). These healers believe that childhood disabilities are caused by breaking taboos (51%) or fate (12%), using contraception (4%), inheritance (4%), or consanguinity (3%) (Kromberg et al., 2008). However, they do influence the way in which people see the causes of disease and seek treatment. Their influence needs to be taken into consideration by those offering services, since, for example, facts given in genetic counseling can be disregarded by the patient if the traditional healer gives a different opinion.

Systems of thought in the Black population appear to favor collective rather than individual thinking (Hammond-Tooke, 1989). This system may be accompanied by an external locus of control, and the social schema of the body may predominate over the individual schema. A balance between the individual and sociological schemata of the body is essential for competent and self-steering behavior (Manganyi, 1973). If there is imbalance, slow decision-making processes may occur and individuals may be reluctant to make any decisions or to act without consulting another person.

Another philosophy related to this worldview is that of fatalism. One of the major components of culture involves the complex of beliefs which surround the relation of man to nature, man to his fellow man, and man to the supernatural powers that are believed to control the universe (Read, 1966). If this control is believed to be absolute, then individuals cannot change what happens to them. Kuppermann et al. (2006) reported that African-American women have higher levels of fatalism than women from other ethnic groups. Furthermore, in another U.S. study, higher scores on a fatalism scale were found to be associated with lower willingness to request termination of pregnancy when the fetus had a severe abnormality (Learman et al., 2005). This philosophy has many implications for genetic counseling, since, if patients do not make decisions or act on the genetic information given, for fatalistic reasons, they may not benefit fully from the knowledge and choices available to them.

INTERNATIONAL DEVELOPMENT OF GENETIC COUNSELING

Genetic counseling is a young and growing healthcare profession available mostly in first-world countries. It has emerged as the Human Genome Project gained publicity, and as the genetic causes of many disorders became understood and recognized such that their prevention and, in some cases, even treatment, becomes a possibility.

The term *genetic counseling* is broadly defined as a dynamic psychoeducational process centered on genetic information (Biesecker & Peters, 2001). The goal of this process is to facilitate the client's ability to use genetic information in a personally meaningful way, so that choices become available, psychological distress is minimized, and personal control is maximized. Genetic counseling aims to preserve client autonomy, self-directedness, competence, and feelings of worth, and to be client-centered and nondirective (Kessler, 1979).

The profession emerged partly in response to the fact that people were living in the community with a genetic disorder in the family, but a specific informed health professional was not available to counsel them. Also, it was recognized that about 1 in 10 people will have a genetic disorder in their lifetime, and many of these people will request and make use of accessible genetic services.

Genetic advice about inherited traits was offered as early as 1906, soon after Bateson suggested that the study of human heredity be called "genetics" (Walker, 1998). The first people who provided genetic advice were generally physicians and PhD geneticists, and the first genetic clinic was set up by Dr. Dight in Michigan in 1940. However, in 1947, Sheldon Reed coined the phrase "genetic counseling" and explained that it was a kind of genetic social work, without eugenic connotations and with concern for the nonmedical impact of genetic diseases (Reed, 1955). The necessity to educate the family through the decision-making process was recognized. The family history was assessed, and the risk of recurrence of their family condition was estimated. Reed (1980) correctly predicted that the demand for genetic counseling would soon exceed the supply of genetic counselors.

The first genetic clinic in the United Kingdom was set up in 1946 at the Hospital for Sick Children, Great Ormond Street, London, by Professor Cedric Carter, assisted by his social worker, Kathleen Evans (J. Kromberg, personal communication). By 1955 there were dozens of clinics in the United States. In the 1960s and 1970s, the prime motivation of medical practitioners was the elimination of genetic disorders and achieving a significant reduction in the healthcare burden incurred by such disorders. In 1972 the World Health Organization (WHO) stated that the role of the genetic counselor (usually, at that time, understood to be medically trained) should be to assist the physician in diagnosis, in estimating recurrence risks, in interpreting the information for the clients in meaningful terms, and in helping them to reach and act upon an appropriate decision.

The training of nonmedical genetic counselors began in 1969 at the Sarah Lawrence College in New York (McCarthy Veach et al., 2003). The program was organized over two years, and by 1973 four more programs had developed. In 1989 the first Asilomar Conference was held in California, and heads of genetic counseling training programs were invited to discuss the syllabus, selection of students, practical work, ethical issues, roles of counselors, definitions of terms, and other components of the program. Professor Jennifer Kromberg, from Johannesburg, South Africa, was invited to participate. She had set up the first local training program early in 1989 at the University of the Witwatersrand, Johannesburg.

RECENT INTERNATIONAL DEVELOPMENTS AND THE TRANSNATIONAL ALLIANCE FOR GENETIC COUNSELING (TAGC)

Professor Janice Edwards, director of the genetic counseling program at the University of South Carolina, USA, was the Jane Engelberg Fellowship awardee of the National Society of Genetic Counselors (NSGC) for 2004/5. In 2004, she invited five international leaders in the genetic counseling profession from Australia, Japan, South Africa, the United States, and the United Kingdom, who shared an interest in genetic counseling education worldwide, to join together and organize an international meeting of genetic counselors. Professor Jacquie Greenberg from Cape Town, South Africa, was invited to participate, as she had just established the training program at the University of Cape Town in 2004. (http://tagc.med.sc.edu/).

The first meeting of the group was held in Manchester, UK, in May 2006, and was attended by 72 delegates from 15 countries and the Transnational Alliance for Genetic Counseling (TAGC) was formed. The aim of the meeting was to create an interactive opportunity for international genetic counseling program directors, professional organizations, as well as credentialing bodies that represent genetic counselors, to consider the profession in its rapidly evolving international context and to create sharing that would enhance all the educational programs and foster the transnational development of the genetic counseling profession.

The group met again in Barcelona in 2008 to develop the TAGC mission and vision. The TAGC Standing Committee was created, and one appointee from each of the fifteen countries represented at the Manchester meeting in 2006 was nominated and elected. The Alliance is now an independent and autonomous organization, and the principal office is located in the United States, in the County of Richland, state of South Carolina. Issues such as international student/faculty exchange, teaching collaborations, research collaborations, and core competencies/credentialing are among the organization's priorities.

TAGC strives to foster communication and collaboration among the international genetic counseling community and enhance genetic counseling education transnationally. The general consensus from the Barcelona meeting was that, with regard to core competencies/credentialing, there is agreement that reciprocal registration arrangements would be of great value and could be on a country-by-country basis. In addition, the establishing of core competencies between countries would be helpful. At the Barcelona meeting, a working document from the United Kingdom entitled "Registration of Overseas Genetic Counsellors Working Group, on behalf of the Genetic Counsellor Registration Board UK and Eire and the Association of Genetic Nurses and Counsellors UK and Eire Committee" was presented and discussed. The UK working group had been in contact with members of each registration/certification board from the United States, Canada, Australia (which includes New Zealand and is jointly called Australasia), and South Africa, and they reported that they had identified that currently, no reciprocal agreements exist for registration across countries.

The website at http://tagc.med.sc.edu/ serves as the hub for the TAGC, continuing and building upon the global connections and is a resource site for the greater genetic counseling community. There is an open invitation to explore the site and provide input; international perspectives are most welcome and strongly encouraged.

After the first meeting of this TAGC group in 2006, Janice Edwards, Jacquie Greenberg, and Margaret Sahhar compiled the following data, based on personal communication with members of TAGC, which summarizes the evolution of genetic counseling in developing and developed countries. Since the creation of Master's level education for genetic counseling in the United States in 1969, thirty-two U.S. programs have emerged. Canada has been training genetic counselors for over twenty years, the first program having been established in 1985; three others are currently training, and another two are proposed. South Africa was the third country to initiate postgraduate training for genetic counselors, and its first program was introduced in 1989 with a second program being added in 2004. Europe's first program was established in Manchester, UK, in 1992, and an additional training site for the United Kingdom was founded in Cardiff in 2000. Cuba has trained over 500 genetic counselors since 1995 in a unique model, providing a Master's degree in genetic counseling to family physicians, who then serve in community health centers throughout the country. The Dutch also have had a national training program since 1996. Four educational programs emerged in Australia between 1996 and 2000 as one-year post-graduate certificate programs requiring on-the-job training to achieve full recognition as a genetic counselor. Two of these programs are currently shifting toward a two-year Master's degree curriculum. Israel has trained over 40 genetic counselors since 1997. Japan saw the creation of seven programs between 2002 and 2005. Taiwan created their first program in 2003, and Norway soon thereafter. The French government decreed genetic counseling as a new health profession with the initiation of their first training program in 2004; twenty-three genetic counselors have formed the French association of genetic counselors. Saudi Arabia formed a program in 2005 and has educated four Saudi nationals at home, and sent three abroad for education, bringing the trained cohort of counselors to seven in a very short time. Spain joined the ranks of established genetic counseling programs when their first students graduated in January 2008. Other programs are envisioned for Finland, India, and China, as well as in other countries who hope to develop the capacity to bring the genetic counseling profession into their genetic services delivery. However, at present, apart from South Africa, no other genetic counselor training programs are being developed in Africa, and very few in other developing countries. (http://precedings.nature.com/documents/1574/version/1)

DEVELOPMENT OF THE GENETIC COUNSELING PROFESSION IN SOUTH AFRICA

In the 1950s Professor Phillip Tobias (anatomy), Professor Lewis Hurst (psychiatry), and Dr. I. Anderson started providing heredity counseling at the University of the Witwatersrand Medical School in Johannesburg (Jenkins, 1990). Around the same time, a specialist physician in Cape Town, Dr. Hymie Gordon, also started to see patients requiring genetic advice. In 1972 the University of Cape Town established a Chair in Human Genetics, and Professor Peter Beighton took the position and set up genetic counseling clinics at Groote Schuur Hospital, the Red Cross Children's Hospital, and Princess Alice Hospital. Also in 1972, Professor Trefor Jenkins (together with Professor J. Hansen, Pediatrics Department) set up a formal genetic counseling clinic in Johannesburg (Jenkins et al., 1973) and in 1975 Professor Jenkins was offered a Chair in Human Genetics at the University of the Witwatersrand. These two universities are presently the only two with such Chairs. Both professors established genetic counseling clinics and began counseling people with many different conditions—for example, 241 patients were seen in a six-month period in 1982 in Johannesburg (Kromberg & Berkowitz, 1986)—and conducting research projects on a variety of genetic disorders.

In the early 1980s it was decided that it was unnecessary for the medically trained staff at the clinic to see all the patients and that graduates, initially graduate social workers, could be trained to take selected cases at the University of the Witwatersrand. One such experienced staff member with a PhD, Professor Kromberg, had been involved in counseling since the service was first offered formally in Johannesburg in 1972. She was asked not only to provide genetic counseling services to selected clients, but also to develop a training course for genetic counselors at the Master's level, in 1988. She consulted widely and visited the courses at Sarah Lawrence College in New York and at the University of California, Los Angeles. The course was then set up in the Department of Human Genetics at the University of the Witwatersrand, Johannesburg. The first two students were accepted in 1989, and both were graduates with BSc (Hons) degrees in genetics. They completed the coursework in principles and practice of genetic counseling and medical genetics over a two-year period, and also wrote up a research project, before being awarded the MSc degree. The University of Cape Town started offering a similar MSc program, by course work and dissertation, in 2004.

In 1975 the South African Medical and Dental Council registered two cytogeneticists (one working at UCT and one at Stellenbosch University) under the category of genetic counselors. These were the first genetic counselors in South Africa who were not registered medical doctors. Their role mostly concerned chromosome disorders, and patients were mostly seen either in the laboratory or at a pregnancy counseling clinic. In 1992, the Health Professions Council of South Africa (HPCSA) was approached to set up a register of genetic counselors. This was completed in due course, and in 1996 a dedicated category was introduced by the HPCSA and a total of 10 individuals were placed on this new register, initially through a grandfather clause (as they had PhD or MSc degrees, many years of experience working in various fields of human genetics, and offered some genetic counseling, but they had little or no relevant training) has recently been excluded, as most of these counselors are no longer active and the majority of the counselors now on the register are fully qualified, with the relevant Master's degree. At present there are 23 people on the register, and graduates must be officially registered with the HPCSA and must maintain that registration throughout their careers, if they are to be employed as genetic counselors in South Africa.

In the early days, graduates in South Africa had to take jobs in genetics laboratories and/or in research programs, or they emigrated to countries where jobs were available (and several graduates are now working in the United States, United Kingdom, and Australia). Job creation was, and still is, a serious problem, and it was only in 2000 that the new professor of human genetics in Johannesburg, Professor Denis Viljoen, was successful in motivating the creation of full-time permanent jobs for genetic counselors in the National Health Department. These posts now fall under the National Health Laboratory Service (NHLS) in Johannesburg. There are four junior posts and one senior post available at present in Johannesburg, and these are all filled with fully qualified graduates. There are no official permanent posts in the other provinces, although there is one intern post and one 6/8ths genetic counseling post at UCT/NHLS in Cape Town. Human genetics staff members at the University of Cape Town (UCT) are strongly advocating the creation of permanent full-time posts in the city, and at present, two qualified HPCSA registered genetic counselors are employed in part-time research posts.

Genetic counselors are currently all required to be registered with the HPCSA in order to practice (HPCSA Regulations, 2009) and a genetic counselor representative sits on the Committee of Medical Scientists at the HPCSA. This committee recently embarked on a process to revise the scope of the profession of medical scientists (including genetic counselors) to ensure that these professionals perform acts for which they are appropriately trained. The legislation has been passed (Government Gazette, May 22, 2009) and since 2010, genetic counselors are required to pass a national assessment of competency to be registered with the HPCSA. In the past this registration was automatic. As genetic counselors are required to do a two-year internship training (one year concurrent with the second academic year, and one year post academic qualification), drawing up of national intern training guidelines is part of the standardization process and it is currently underway. At the same time, the South African Qualifications Authority (SAQA) embarked on a process of establishing generic academic qualifications. This, together with the HPCSA initiatives, will provide appropriate training that is in line with that of other countries such as the United States, Canada, Britain, and Australia.

The South African Society of Human Genetics (SASHG) was established in 1983, when a group of human geneticists broke away from the South African Genetics Society (set up in 1956) to form their own group. Genetic counselors have been active members of the SASHG from the beginning. The main function of the Society is to hold biennial congresses for everyone working in the field. The recent 13th congress was held in Stellenbosch in 2009, and genetic counselors played an active part. They invited an expert international speaker from Australia, and had a full afternoon session dedicated to genetic counseling issues. They also had a national genetic counselors' meeting to discuss current issues, future plans, and a proposal to form a subcommittee of genetic counselors under the auspices of the SASHG. Several counselors have also attended international congresses of the Psychosocial Issues in Genetics group, the National Society of Genetic Counselors, USA, the British, American, Australian, European, and international societies of human genetics, international birth defects congresses, where they have presented some of their research work, and both meetings of the recently established Transnational Alliance for Genetic Counseling (TAGC). In this way genetic counselors in South Africa keep up-to-date with what is happening in the field, even though they are somewhat isolated geographically, being at the tip of the African continent.

ORGANIZATION OF GENETIC COUNSELING SERVICES IN SOUTH AFRICA

The governing of South Africa is organized through a national and provincial system. There are nine provinces (Mpumalanga, North West, Limpopo, Kwazulu/Natal, Gauteng, Free State, Western Cape, Northern Cape and Eastern Cape). Health services are governed by the Minister of Health, with a Director General at the national level. The provinces are autonomous and, although the role of national government is to advise and guide, implementation of programs is undertaken at the discretion of the provincial heads, taking the needs of each specific province into account. This has been a major challenge for the country. The key legislation governing the health sector in South Africa is the National Health Act. Implementation of this has been slow, but progress has been made in some areas (Gray & Jack, 2008).

There are many disparities in health and access to healthcare in South Africa (Benatar, 2004). The government has, since 1994, attempted to address these inequalities by establishing a district-based system of primary healthcare, nationalization of health laboratory services, regulation of healthcare professionals, and reforming health legislation. However, staff shortages and allocation of resources remain a challenge. Social inequalities, such as poverty, negative attitudes toward women and girls, racial and other discrimination, disempowerment, and the HIV/AIDS pandemic are some of the issues that the country still faces. Furthermore, health services are funded through state organizations and the private sector. State health services are provided to the majority of the population by hospitals (quaternary, tertiary, and secondary) and clinics. The private health sector, which services a minority of the population, is privately funded by medical aid societies to which members belong by paying monthly fees. Professionals offering services in private practice are paid by the patient's medical aid society. As a result, there are many disparities between services offered in the private and state sectors.

The "Human Genetics Policy Guidelines for the Management and Prevention of Genetic Disorders, Birth Defects and Disabilities" of 2001 provides a framework for the planning of genetic service delivery nationally. It constitutes a "hub and spoke model" whereby the main centers are the hubs, providing outreach services to more distant parts of the country. Genetic counseling services are offered, at present, in the capital cities of four of the nine provinces: Gauteng, Western Cape, Free State, and Kwazulu/Natal. Only two of these provinces (Gauteng at the University of the Witwatersrand (Wits) and Western Cape at the University of Cape Town (UCT) have trained genetic counselors with Master's degrees and offer a fully integrated service including molecular, cytogenetics (and serogenetics at Wits), laboratory, and clinical support. All four of the provinces, however, provide outreach clinics to other regions within their province and occasionally to other provinces.

The Division of Human Genetics of the National Health Laboratory Service (NHLS, previously the South Africa Institute for Medical Research – SAIMR) and the University of the Witwatersrand's first genetic counseling clinic was established at the Transvaal Memorial Hospital (TMI) in 1972 (Jenkins et al., 1973). This clinic, due to its siting, mainly provided a service to the White population of Johannesburg. Later, in 1987, a genetic counseling clinic was set up at Chris Hani Baragwanath Hospital (CHB) (Kromberg & Jenkins, 1988) to provide an accessible service to the Black population from Soweto and other peri-urban areas. The Division of Human Genetics at Wits now offers a genetic counseling service at all three state-funded tertiary teaching hospitals and at a private hospital in Johannesburg (Gauteng), as well as regular national outreach clinics. The service is provided to patients from all the population groups. On average, about 80% of the patients seen are from the state healthcare sector. The services are provided to antenatal, pediatric, and adult patients with many different diagnoses, from common chromosomal disorders to rare single-gene conditions, as well as multifactorial and mitochondrial conditions. Altogether, a total of 1837 patients received genetic counseling in the year 2008, and 1975 in 2007. Diagnostic testing is performed in the Division for a wide range of conditions, by DNA, chromosome and enzyme based investigations, and/or for prenatal genetic diagnosis, diagnostic confirmations, and predictive testing. A cancer genetics service was introduced in 2006, and the number of patients seen is increasing annually. The genetic counseling service is provided by HPCSA registered professionals. The division has ten clinicians (three in training), nine genetic counselors (two in training), and one genetic nurse counselor.

In the Western Cape (WC), medical services to state patients are provided by the University of Cape Town (UCT) and the University of Stellenbosch (US). Medical genetic services in the WC work on the recommended "hub and spoke" model with the two academic hospitals: Groote Schuur Hospital (GSH)/UCT and Tygerberg Hospital/US (funded from the provincial budget), constituting the "hub" of the model, with spokes extending to regional and district services via outreach and support to regional specialists and general practitioners. The different responsibilities of the clinic team relate to service provision, treatment protocols, skills development, service planning for specific conditions, along with teaching, training, and research. Screening clinics at LSEN schools (for learners with special educational needs) and institutions (approximately 20 clinics per year), as well as genetic outreach programs in the WC (two outreach weeks per year), are undertaken by the various teams. These teams can comprise medical geneticists, genetic counselors, nurses, and scientists. The primary role of the team is to provide genetic services within the WC provincial health system, and a small private-sector genetic service is also offered.

The Division of Human Genetics at UCT has five clinicians (one part-time and one in training) and five HPCSA-registered genetic counselors, two of whom are primarily involved with research programs. There are also two genetic counselors-in-training and two full-time genetic nursing sisters. The genetic service includes antenatal, pediatric, and adult services, as well as outreach programs. Within the genetic service provided by UCT in the western Cape (WC), genetic counseling is offered by medical geneticists, genetic counselors, and genetic nurses. The five medical geneticists are medical doctors with a special interest in the diagnosis, care, and prevention of genetic conditions and birth defects.

For diagnosed conditions, genetic counseling is frequently provided by the genetic counselors or genetic nurses. Three of the genetic counselors have a Master's level qualification and relevant experience in genetic counseling. The remaining two are PhD graduates with many years of experience in human genetics. The two genetic nurses have a long track record and extensive counseling experience within the public service. At the University of Cape Town's various genetics clinics, approximately 5000 individuals/families were seen and counseled from 1972–1988, and then from 1989–2008 approximately 2000 patients/family members received genetic counseling every year (e.g., 2024 in 2006, 2141 in 2007, and 2249 in 2008, as reflected in the annual reports of the relevant years).

The two academic hospitals (provincially funded) in the WC therefore provide outreach services ("spokes") to a number of sites within their drainage areas, as well as to schools. For these genetic outreach programs, as well as the school visits, the "clinical" group works as a team. Therefore, a good number of new patient children and adults, and school and outreach cases seen by clinicians, have their initial family/clinical history taken by a genetic nurse or counselor. Similarly, a number of nurse and counselor cases will receive input from clinicians during these outreach programs.

CURRENT TRAINING OF GENETIC COUNSELORS IN SOUTH AFRICA

TRAINING OF GENETIC COUNSELORS AT THE UNIVERSITY OF THE WITWATERSRAND (WITS)

http://www.wits.ac.za/pathology/humangenetics/9225/home.html

The Division of Human Genetics of the National Health Laboratory Service (NHLS) first offered a Master's degree in genetic counseling in 1989. Postgraduate students from a variety of backgrounds, including genetics (the majority of students), social work, psychology, and medicine have been enrolled. There is much competition for places and only about 10% of the applicants succeed in entering the program. A total of 16 students have successfully completed the coursework and research report, and obtained the degree. The course extends over two years, during which the students learn the principles and applications of medical genetics and genetic counseling. There is a major coursework component comprising tutorials, essays, case reports, and experiential practical exercises. The students learn to apply the information and skills through providing a genetic counseling service to patients and through clinical work under supervision by a clinical psychologist, as well as the more experienced genetic counselors. They also learn to work in multidisciplinary teams and to interact with many other academics, health professionals, social science professionals, and international exchange students who seek genetic counseling work experience in Johannesburg.

A supervised research project is carried out and written up in the second year. This requires the students to decide on a feasible project, study the relevant literature, write a protocol and see it through the university higher degrees committee, apply for ethics clearance, assess the funding requirements, collect the data, computerize the data, use statistical tests where necessary, analyze the findings, discuss results, draw conclusions, and write up an acceptable research report (assessed by external examiners). Furthermore, students should produce a paper for publication on their project. Projects that have been completed in this way in the division have contributed to the local understanding of people with genetic disorders and their problems and needs. Topics have included expectations of genetic counseling, perceptions of chorionic villus sampling, psychosocial issues and Fragile X syndrome, trisomies 13,18, and 21 in an African population, breast cancer in the Jewish community, spina bifida in Gauteng, Huntington disease and adolescents, fetal abnormalities, and termination of pregnancy, among others.

In addition, experience is gained by attendance at departmental meetings, journal club presentations and meetings, seminars and conferences, educational meetings, and community visits. Students are also encouraged to present their research work at the biennial human genetics conference of the SASHG, to get involved with local genetic support groups, and to give talks on genetic conditions to the lay public when requested.

The students who complete the course successfully have to undertake an internship year before they can be registered as a genetic counselor. On completion of this three-year program, the students are well honed and experienced, higher-degree graduates with unique research credentials, as well as both the psychosocial and genetic skills to provide a comprehensive genetic counseling service within a variety of health settings.

TRAINING OF GENETIC COUNSELORS AT THE UNIVERSITY OF CAPE TOWN (UCT)

http://web.uct.ac.za/depts/genetics/counselling/

The Human Genome Project, by facilitating the identification of actual disease-causing genes, as well as predisposing genes for many human disorders, has led to a dramatic transformation in the practice of medicine and primary healthcare. The impetus for the establishment of a Master's program in genetic counseling at UCT was the urgent need in the Cape Province and other parts of southern Africa for genetic counselors that are adequately educated to provide genetic-oriented and supportive counseling to families and individuals with genetic conditions. In addition, it was recognized that there is a need for genetic counseling researchers who will broaden the scope of this new profession in southern Africa by generating research literature from and for the continent of Africa. The overall benefit of such a postgraduate program will be to fulfill the need for genetic health education and service development for the people of Africa who suffer from a genetic condition.

Students were first admitted onto the UCT program in 2004 and five have graduated, completed their internship, and are registered with the Health Professionals Council of South Africa (HPCSA) to practice professionally in South Africa (although one is presently working as a genetic counselor in the United Kingdom).

The purpose of the MSc in Genetic Counseling program at UCT is to provide students with a wide range of knowledge and skills unique to genetic counseling, which will enable them to register with the HPCSA and to work in the public or private sectors in communities, hospitals, clinics, special schools, and laboratories. There is an explicit commitment to a primary healthcare approach of client management throughout the program.

The second purpose of the qualification is to equip the students with management, leadership, and research skills that will enable them to participate in national healthcare development and policymaking, and to contribute to the international body of knowledge in genetic counseling. The third purpose of the UCT qualification is to endow South Africa with a nucleus of genetic counselors to provide effective, efficient, and accessible genetic counseling services while continuing to improve their knowledge and skills. In addition, it also affords the UCT MSc graduates in genetic counseling the opportunity of proceeding to a PhD degree.

The program provides students with genetic counseling knowledge, attitudes, and skills to manage genetic diseases in southern Africa and to contribute to the international body of research in genetic conditions. Interdisciplinary collaboration occurs by means of lectures and workshops, in the Medical Genetics and Principles of Genetic Counseling modules, where students from the MMed program in palliative medicine and MMed in medical genetics participate as well. Students are exposed to and work with other health disciplines in various clinical areas during their clinical practice. In order to progress into the second year of study, all courses completed in the first year have to be passed and a research proposal has to be completed and approved by the Clinical Laboratory Sciences Research Committee and the Faculty of Health Sciences Ethics Committee. The UCT program requires students to be competent in analytical and critical thinking, problem-solving, numeracy, computer literacy, appropriate writing and verbal language proficiency, and to demonstrate an enthusiasm for learning. They need to be able to search for new knowledge and research and, given the prevalence of HIV/AIDS in the country, they are also involved with HIV/AIDS-related issues during the course of the program.

The Division of Human Genetics at UCT also accommodates elective students from other countries, who work with the UCT students in the clinics, thereby increasing their awareness of genetic counseling programs and practice in other countries. Inter-institutional collaboration also occurs with the Department of Genetics and Obstetrics and Gynecology at Tygerberg Hospital (U.S.) so that students from UCT gain part of their counseling experience at other hospitals, as well as in various clinics.

In summary, the UCT program aims to produce healthcare professionals equipped with psychosocial and communication skills to relay medical, genetic, and scientific information to patients and their families, as well as to other healthcare professionals and the wider community. They are trained to provide support and management for all patients and their families with a history of genetic disease, as part of a primary healthcare approach of client management in South Africa (http://web.uct.ac.za/depts/genetics/counselling/).

ROLES AND FUNCTIONS OF GENETIC COUNSELORS IN SOUTH AFRICA

The roles and functions of genetic counselors in South Africa are developing and emerging out of insights gained from both appropriate international contacts and a deeper relevant knowledge and understanding of the local situation. Kessler (1980) described the roles, in general, as including: healthcare professional providing supportive care; education; resources and referrals; advocacy; grief counseling; and research. However, roles in specific situations are determined by professional responsibilities, employers' requirements, service site, needs of the surrounding communities, genetic disorders found in those communities, and the experience and vision of the counselors themselves. Roles also vary and change as counselors adjust to the insights they receive from patients in genetic counseling interactions, from supervision discussions, and from other learning experiences.

A study on roles of genetic counselors in South Africa showed that roles fall into five basic fields: counseling, teaching, research, marketing, and administration (Kromberg et al., 2009). Regarding the counseling role, genetic counselors are being trained to counsel people with a diverse range of genetic conditions, and these increase every year as the counselors gain experience. In the two-year period from 2007 to 2008, for example, counselors counseled 2554 cases in South Africa, and they presented with 44 different diagnoses (Kromberg, personal communication). In comparison, in Queensland, Australia, a more experienced team of counselors counseled for 79 different diagnoses (Kromberg et al., 2006). Advanced maternal age and Down syndrome were still the commonest indications for genetic counseling in the country, but abnormal ultrasound, and/or fetal abnormalities, was the third most common indication. Although South Africa has had a sophisticated prenatal diagnosis service for many years (Kromberg et al., 1989) the new advances in ultrasound technology mean that fetal abnormalities can be detected more reliably and more specifically in utero, and skilled genetic counseling is required. Other disorders, such as retinal disorders (http://web.uct.ac.za/depts/genetics/retina) and cancer (particularly colorectal cancers) (http://web.uct.ac.za/depts/genetics/cancer), were seen more often by Cape Town counselors, due to the research interests of scientists in their university departments, (http://web.uct.ac.za/depts/genetics/neuro) while hemophilia and cystic fibrosis were seen more commonly in Johannesburg due to the attachment of counselors to the large local clinics treating those conditions (Macaulay, 2008). The patients being counseled were mostly from the Black population (65%) in Johannesburg, while those from the Mixed Ancestry group (63%) were counseled more frequently in the Cape. This ethnic distribution of patients was appropriate for the population structure found in the two different areas. The vast majority of patients were seen in the state hospital system (80%), which is the primary responsibility of the counselors; the remainder were seen either in private practice (12%) or in rural outreach clinics (8%).

The teaching role of counselors is increasing, and all are involved in teaching trainees (both medical geneticist and genetic counselor trainees), medical and other students, health professionals, learners in schools, and the lay public. The counselors' teaching role includes a supervision role, in which counselors supervise the learning experiences of students and new counselors, who are developing their interviewing skills and expertise in practice. In South Africa there are no genetic counselors who have specialized, and there are no specialization courses; also, there are no counselors in full-time education roles (such as are found in Australia), and all are generalists. Some counselors, however, have become particularly interested in a specific field, often due to having carried out their Master's research project in that field—for example, cancer genetic counseling—and may do more counseling for and teaching on the condition for which they have developed expertise (Greenberg, 1989, 2008; Greenberg et al., 2009; Loubser et al., 2009).

Research has always been an integral part of the local MSc in genetic counseling program, unlike the equivalent degree gained in Australia and the United States. Every qualified counselor has completed a research project and research report, to the satisfaction of external examiners, and a few have proceeded to a PhD (one is presently involved in a PhD degree, partly due to the fact that there is no counseling job available for her at present). Altogether, 69% of counselors are presently involved in research (mostly as a member of a research team) and 37% are published authors. Recent projects have included: general practitioners' (GP) knowledge of cancer genetics, cystic fibrosis testing in relatives of an affected child, an investigation into the level of genetic knowledge and family communication about genetic risk in parents of children with cystic fibrosis, decisions made after the diagnosis of a fetal abnormality, utilization of the Huntington disease (HD) genetic service, the use of genetic tests by the life insurance industry, and Duchenne and Becker muscular dystrophy and the implications for at-risk individuals, as well as another study regarding the level of genetic knowledge of parents of sons with Duchenne muscular dystrophy (Loggenberg et al., 2007). In addition, the perceptions of patients with inherited forms of retinal disorders while participating in a genetic research program was examined (Basson et al., 2007), as well as an investigation for a PhD on the HD predictive testing service in the WC over the past 11 years (Futter et al., 2009).

Counselors also have a responsibility to market the genetic counseling service, and they all receive training in community genetics and the public health aspects of human genetics. Generally, marketing is carried out by distributing pamphlets and articles on, for example, genetic counseling, prenatal diagnosis, and specific genetic disorders; giving talks to the public or on radio or TV; writing articles for newspapers, popular magazines, or local professional or medical journals (such as Greenberg et al., 1996; Krause & Craig, 2001); and holding meetings for specific purposes (such as providing feedback to GPs from a research project on their knowledge of cancer genetics). All the patient information documents on the UCT web page have been compiled by genetic counselors during the course of their training (http://www.uct.ac.za/depts/genetics).

Administration duties are involved in every genetic counseling job. These duties include client-related activities (mainly the compiling, keeping, and organizing of client records), clinic coordination duties, being on call and/or on receiving office, and organizing meetings and/or in-service seminars.

Counselors therefore contribute a great deal to the provision of genetic services in South Africa. They counsel approximately one-third of all the cases seen for genetic counseling, and the numbers are increasing, and they play many other roles that are essential to the smooth and efficient running of the service.

Essentially, genetic counselors in genetic counseling sessions elicit a relevant family and medical history, construct an accurate pedigree, and then calculate risk status of the individual and family and offer choices for dealing with this risk. In the process they need to be able to obtain an appropriate social and psychosocial history of the individual and family. They are also expected to be able to interpret special investigations associated with the genetic condition, and apply theoretical scientific knowledge, sensitively. As part of this process, they also need to be able to adapt to change and tolerate uncertainty. They need to plan and execute a range of interviewing and counseling techniques, and then also manage the clients, patients, and families with empathy, caring, compassion, sensitivity, acceptance of diversity, and respect for their dignity, privacy, and confidentiality. Genetic counselors are trained to observe and interpret interactions between family members and counselors and inform patients of local and international resources and services. They also critically evaluate genetic counseling techniques used in practice, identify their own personal limitations, and recognize the need to seek help when necessary. It is essential that they are able to identify the limitations of the scope of genetic counseling and refer to other health practitioners appropriately and in a timely way. In addition, they need to know how to identify relevant ethical and legal aspects associated with clients and their families, and then to refer patients/families to specific genetic support groups and nongovernmental organizations (NGOs) where relevant. In South Africa, they have to be in a position to apply knowledge of human rights as referred to in the constitution, and recognize the cultural, social, religious, and historical contexts of their clients and families.

In summary, genetic counseling has several broad goals in the South African setting that are generally similar to those in other countries:

- Firstly, once a clinical diagnosis has been made, it is essential to assist the family in addressing their concerns relating to the development or transmission of their genetic disorder, to assess their risks and give choices, and to facilitate informed decision making that promotes patient autonomy as well as quality of life for the family.
- Secondly, genetic counselors should help individuals and their families adjust to difficult situations in a way that involves a realistic review of the positive and negative aspects of possible outcomes, promotes competence and mobilizes support, and is consistent with the family's beliefs, values, and culture. Where necessary, genetic counselors should ensure that informed consent, based on an adequate understanding of the technical information and its implications for the affected individual and family members, is promoted.
- Lastly, genetic counselors should aim to develop and promote a relationship of trust, understanding, and support that encourages the family to continue utilizing the counseling service, as well as the services of other appropriate healthcare professionals.

CHALLENGES FOR THE GENETIC COUNSELING PROFESSION IN SOUTH AFRICA

One of the biggest challenges genetic counselors face in South Africa today is to find ways to make genetic counseling more understandable and accessible to the local people. Many languages spoken in the country have no words to describe genetic concepts, and there is as yet no African language-speaking registered genetic counselor. In many communities there are myths and stigmas surrounding birth defects that are being explored, but their relevance to genetic concepts is not well understood by genetics health professionals. In the Third World and developing countries, like South Africa, underdevelopment causes marked social differences, and among people of lower socioeconomic strata there is less dissemination and understanding of scientific concepts. Partly due to the persistence of traditional medicine in many developing countries, and often due to a lack of written records, a large number of people with a genetic disorder in the family do not know how to explain what has happened in their family (De Pina-Neto & Petean, 1999). Individuals often draw on other belief systems which may be held in addition to scientific explanations. The genetic explanation may be accepted on one level, but the personal meaning of the situation comes in large part from other beliefs (Weil, 2000). Research in the South African context is currently being undertaken to investigate individual beliefs regarding genetic disorders, as well as aspects relating to communication in multicultural genetic counseling settings (research currently being undertaken by Wessels et al., 2009).

When a genetic condition is diagnosed it has a profound impact on the family both emotionally and practically. South Africans who are affected by a genetic condition are often confronted with unique circumstances and barriers, including poverty and minimal access to resources. Affected individuals and families may not be afforded information and interventions to address basic health issues, let alone to maximize their quality of life. Members of rural communities may have to travel hundreds of kilometers to tertiary hospital settings to obtain comprehensive genetic services. Genetic counseling often has to take place through a translator with no training in genetics or counseling, leading to possible misunderstanding between the patient and the counselor. In a developing country like South Africa, genetic counseling will not only be influenced by psychosocial issues within a family but by problems related to poverty, low level of education, unemployment, and inaccessibile services.

Disability creates different burdens for each family. Maslow's (1970) complete hierarchy of needs should be considered when examining family strengths and weaknesses. A family cannot be expected to follow up on recommendations for therapeutic treatment for their child, or to attend clinic regularly, if, for example, they are not able to provide food or heat for the family. The health-care professionals involved with these families should be familiar with local resources or services that are available to assist families, where necessary, with food, clothing, shelter, and financial assistance (Ziolko, 1991)

Participants with little formal education may benefit from additional counseling. Although it is possible that some may still not understand genetic concepts after several sessions, it is nevertheless the counselor's obligation to try and find ways of communicating the relevant complex information (Somer et al., 1988). In order for family members to incorporate risk information, they need an opportunity to discuss how new information about inheritance fits into, or supersedes, previously held beliefs. This type of discussion has been identified as a useful method for helping individuals internalize information about inheritance and reproductive risk (James et al., 2006).

In a recent research project, local genetic counselors, themselves, suggested areas for role expansion, in an attempt to meet these challenges. These areas include raising the visibility of the profession, increasing the numbers and variety of suitable referrals, and making the service more widely known and accessible (Kromberg et al., 2009).

Clearly, more education of the medical and lay public is needed. If the health professionals were better informed, they would refer a wider range of patients, more people would benefit from the counseling services, and counselors could expand their expertise to a broader range of genetic conditions. Furthermore, if the general public becomes more knowledgeable about genetic conditions they will understand their relevant family history better and will demand an appropriate and more accessible service. It is recognized, too, that genetic testing is being done, often in private practices, without informed genetic counseling pre- and post-testing—regardless of the international injunction that such testing may be debatable, the results may be complex, and skilled genetic counseling is essential if the service is to be used appropriately (Kromberg, 1993). Patients are still having genetic testing, even prenatal genetic diagnosis testing, without understanding what the results might show and what the implications of the results could be, while genetic counselors are trained to provide the service associated with this scenario.

Employment for genetic counselors remains a huge challenge; currently the NHLS provides some employment, and there no genetic counselors employed by the state-funded hospitals. As in many other countries, such as Australia (Kromberg et al., 2006), most of the genetic services are based in the big cities and the rural populations have little access to them. The challenge for South African counselors is to increase their outreach services, and/or to train and place counselors in outlying towns (as happens in Queensland, Australia; see Kromberg et al., 2006), so that more at-risk people can have their needs met. Similarly, many of the local population who have private medical aids do not attend the state hospitals, and therefore do not come into contact with the genetic clinics; they may not be referred, and may not make use of genetic services. For this reason, private practice may need to increase and become more widely offered, in settings such as private laboratories where genetic testing is carried out, or infertility clinics, or cancer clinics (which are all settings for genetic counselors in other countries) so that this section of the population has more access to services. Another model would be to have genetic counselors attached to groups of general practitioners, so that patient management improves for those with a history of genetic disorders. Genetic counselors should also become more involved with genetic support groups in educational and supportive roles, as well as to provide information for members regarding the available genetic services.

Cancer genetic counseling is a growing field in South Africa, as it is in Queensland, Australia (Kromberg et al., 2006) and other countries. The number of referrals has increased in Johannesburg since the completion of two cancer related research projects. However, cancer testing facilities are not yet fully developed in the country, and these are required if a functional testing and counseling service is to be developed. In the meantime, genetic counselors are counseling more cases every year, and this part of their service is likely to be much in demand when the laboratory service is up and running.

Offering genetic counseling services within the local health system has many challenges, not the least of which is the selection, training, and integrating into the health system of the staff required to run this new and developing service. The report from the United Kingdom on "Clinical Genetic Services into the 21st Century" recommended that two full-time clinicians and eight genetic counselors are required per 1 million people of a population (Royal College of Physicians, 1996). By this standard (which was supported by the World Health Organization) approximately 98 clinicians and 392 genetic counselors are required to provide an adequate service in South Africa. With the number of practicing medical geneticists and genetic counselors (some in training) currently being about the same and, importantly, less than 20 for each profession, the country has a very serious shortage of fully trained staff. Both professions are therefore attempting to attract students, train them, find employment for them, and retain them in South Africa once they are qualified and experienced. This situation requires the urgent attention of the National Health Department (in terms of job creation), the provincial health departments, and academic institutions (who can train more counselors if jobs become available).

In general, these are the challenges of any new profession in a young and developing country, and especially in one overwhelmed by an AIDS epidemic that demands a large slice of the health budget. However, the numbers of genetic counselors are increasing in other countries, and if the need is to be met they need to increase much more quickly in South Africa. The recent Law Reform Commission in Australia (ALRC, 2003) inquired into genetic privacy, a problem that will have to be faced in the future locally, and recommended that increased resources for genetic counseling services were a priority if problems such as invasion of privacy were to be prevented.

THE FUTURE OF THE PROFESSION IN SOUTH AFRICA

The future of the genetic counseling profession in South Africa will be determined by the way the challenges currently presented are faced and managed. One of the most important tasks is ensuring that more genetic counselors are trained, and that the WHO-recommended ratio of counselors to population size becomes a more attainable goal. The genetic counseling profession in South Africa has modeled itself partly on genetic counseling services in First World countries such as the United States, United Kingdom, and Australia. However, South Africa is a developing country with a very diverse population, and the challenges faced by local genetic counselors will differ somewhat from those of other countries. It is therefore acknowledged that specific attention needs to be paid to local diversity and cultural issues during training and when seeking areas of potential employment for genetic counselors. Due to some of the day-to-day challenges genetic counselors in South Africa face—for example, dealing with and being dependent on interpreters and/or counseling individuals with little or no education—the workload may be somewhat greater than in developed countries. Local clients may need more sessions to ensure their understanding of the necessary information. For these and many other reasons it is essential that the number of genetic counselors in South Africa is increased. Ensuring employment opportunities for such counselors is a vital element that will influence both the future of the profession and the availability of genetic services in South Africa.

Apart from the required extra positions which need to be made available, primarily by the Department of Health, a secondary practical model that has been proposed to increase employment opportunities is that trained genetic counselors are attached to groups of general practitioners (GPs) or specialist clinics (e.g., in-vitro fertilization [IVF] clinics) and are then jointly paid by public and private funding. This could maximize the access of both the GPs/specialists and the clients to current information and genetic services, while at the same time providing ongoing genetic management and follow-up. The added advantage of this model would be that it could bring current information on the new genetic tests and technologies to the GPs in the community and rural areas, thereby facilitating and improving the outcome of the primary healthcare model in South Africa.

CONCLUSION

One of the other developing countries in the process of establishing genetic services in Africa is Tunisia. Chaabouni-Bouhamed (2008) states that there is no national program in Tunisia dedicated to hereditary disorders, and even the national program on maternal and child healthcare does not include these conditions. However, genetic counseling and prenatal diagnosis seem to be the methods of choice for the prevention of genetic diseases, and therefore such services should be prioritized and developed, despite the financial and emotional costs of such programs. He adds that genetic services need support, and community genetic services need initiating. In particular, the following is required:

> a comprehensive national program to develop basic genetic services; the strengthening of human resources in the practice sector (clinical geneticists, social workers and trained laboratory personnel); training courses for primary care worker so that they can recognize at-risk families; population screening programs and a prenatal screening program based on ultrasonology; and a national registry for congenital malformations. Support for such programs should come mainly from public funds, but also from the private sector.

Chaabouni-Bouhamed concludes that as genetic disorders are prevalent in the region, and as they have literature on these conditions in Tunisia, the country should benefit from modern scientific advances in the field in the same way that industrialized countries do. Basic medical genetic services should be implemented and supported to capitalize on these benefits. South Africa, at the other end of the continent from Tunisia, is some way along this path. However, it is imperative that further support comes from government, if a comprehensive program as described by Chaabouni-Bouhamed)—but including the new professionals, genetic counselors, who play a great part in providing services in developed countries—is to be strengthened and offered more equitably and with accessibility, locally.

Genetics and genomics are integral components of medical practice today. No longer confined to dysmorphology and Mendelian genetic disorders, it is anticipated that genetic understanding will play a crucial role in the management of common illnesses such as cancer, diabetes, psychiatric disorders, hypertension, and infectious diseases in the future. Genetic specialists have a central role to play in education and research in order to relate these advances in developing countries in a way that will maximize impact on healthcare. Professional genetic counselors are an integral part of any holistic genetic service and it is their duty to provide the communication interface between the patient and the clinical and laboratory teams. This new complex science should be made more accessible to everyone, and it is essential that the understanding and ethical practice of genetic and genomic medicine be promoted in all developing countries today. In addition, academic research and teaching in genetics needs to be promoted in a manner that will benefit all nations. Education needs to be provided to professionals and laypersons to support the integration of genetics into the primary healthcare services. Today's research should provide a sound scientific foundation on which to develop a practical and realistic genetic service for tomorrow, for and in Africa and the developing world.

ACKNOWLEDGMENTS

Dr. Merle Futter, who reviewed the UCT MSc in Genetic Counseling program in 2008/9.

Dr. Mike Urban and Dr. Karen Fieggen, who compiled the Medical Genetics UCT Vision Document, from which some of the background information related to genetic services in South Africa was taken.

Emeritus Professor Peter Beighton for the historical background related to human genetics at UCT.

Mr. Ronnie Smart for the historical background related to genetic counseling at UCT.

The Mellon Foundation is acknowledged for funding the post for JGRK at the University of the Witwatersrand.

REFERENCES

Anderson DW, Goldberg PA, Algar U, Felix R, Ramesar RS. (2007) Mobile colonoscopic surveillance provides quality care for hereditary nonpolyposis colorectal carcinoma families in South Africa. *Colorectal Dis. 9*, 509–514.

Australian Law Reform Commission, and Australian Health Ethics committee of the national Health and Medical Research Council Report (ALRC/AHEC) (2003). *Essentially yours: The Protection of Human Genetic Information in Australia*. Sidney: Commonwealth of Australia.

Basson F, Futter M, Greenberg J. (2007). Qualitative research methodology in the explorations of patients' perceptions of participation in a genetic research programme. *Ophthalmic Genet 28*, 143–149.

Bateman C. (2008). Lack of 'warm hands' will render health care system 'impotent'. *S Afr Med J 98*, 510–511.

Beighton P. (1976). Genetic disorders in Southern Africa. *S Afr Med J 50*(29), 1125–1128.

Beighton P, Botha MC. (1986a). Inherited disorders in the Black population of southern Africa. Part I. Historical and demographic background; genetic haematological conditions. *S Afr Med J 69*(4), 247–249.

Beighton P, Botha MC. (1986b). Inherited disorders in the black population of southern Africa. Part II. Gene disorders. *S Afr Med J 69*(5), 293–296. Review.

Beighton P, Botha MC. (1986c). Inherited disorders in the black population of southern Africa. Part III. Multifactorial, chromosomal and congenital conditions. *S Afr Med J 69*(6), 375–378.

Beighton P, Ramesar R, Winship I, et al. (1991). Hearing impairment and pigmentary disturbance. *Ann N Y Acad Sci 630*, 152–166. Review.

Benatar SR. (2004). Health reform and the crisis of HIV and AIDS in South Africa. *N Engl J Med 351*(1), 81–92.

Biesecker B, Peters K. (2001). Process studies in genetic counseling: peering into the black box. *Am J Med Genet 106*, 191–198.

Binka F. (2007). Foreword. *Scand J Public Health 35*(Suppl 69), 3.

Bittles AH. (2008). A community genetics perspective on consanguineous marriage. *Community Genet 11*, 324–330.

Chaabouni-Bouhamed H. (2008). Tunisia: Communities and Community Genetics. *Community Genet 11*, 313–323.

Christianson AL, Zwane ME, Manga P, Rosen E, Venter A, Kromberg JGR. (2000). Epilepsy in rural South Africa – prevalence associate disability and management. *S Afr Med J 90*, 262–266.

Christianson AL, Modell B. (2004). Medical genetics in developing countries. Annual review. *Genomics and Human Genetics 5*, 219–265.

Denter M, Ramsay M, Jenkins T. (1992). Cystic fibrosis. Part 1. Frequency of the delta F 508 mutation in South African families with cystic fibrosis. *S Afr Med J 82*, 7–10.

Doherty T, Colvin M. (2004). *HIV/AIDS*. Chapter 14 in: Ijumba P, Day P, Ntuli A (eds). *South African Health Review 2003–2004* (pp. 191–211). Durban: Health Systems Trust.

De Pina-Neto JM, Petean EBL. (1999). Genetic counselling follow-up – A retrospective study with a quantitative approach. *Genet Mol Biol 22*(3), 295–307.

Freeman M, Lee T, Vivian W. (1994). Evaluation of the mental health services in the Orange Free State. Johannesburg: University of the Witwatersrand Centre for Health Policy.

Futter MJ, Heckmann JM, Greenberg LJ. (2009). Predictive testing for Huntington disease in a developing country. *Clin Genet 75*(1), 92–97

Goldberg PA, Madden MV, Harocopos C, et al.(2000). Inherited colon cancers. *S Afr Med J 90*(7), 703–704.

Goldblatt J, Beighton P. (1979). Gaucher's disease in South Africa. *J Med 16*(4), 302–305.

Gray A, Jack C. (2008) *Health Legislation and Policy*. In Barron P, Roma-Reardon J.(eds.), *South African Health Review* (pp. 31–49). Durban: Health Systems Trust.

Greenberg LJ. (1989). A molecular service for Huntington disease in Southern Africa. *S Afr Med J 76*, 135–136. Editorial.

Greenberg J, Ramesar R, Beighton P. (1994). Genetic mapping of retinitis pigmentosa – implications for South African patients. *S Afr Med J 84*, 410–412.

Greenberg J, Beatty S, Soltau H, Bryer A. (1996). A predictive testing service for Huntington disease (HD) and late onset spinocerebellar ataxia (SCA1) in Cape Town. *CME 14*, 1364–1367. Review.

Greenberg J. Genetic counselling challenges with a family with HDL2: From the bedside to the bench & back to the bedside.(2008). *European Human Genetics Conference*. Barcelona, Spain. Poster presentation. P09.37 P 425. Available at: http://www.eshg.org/eshg2008/.

Greenberg J, Roberts LJ, Schoeman M, Bruwer Z, Loggenberg K, Loubser F. (2009). Delivery of an ophthalmic genetic service including a telephone counselling model: a bench-to-bedside review. Abstract book, *13th Congress of the SA Society of Human Genetics*, 5–8 April 2009. Stellenbosch, Cape Town, South Africa.

Grobbelaar JJ, Wilken E, de Ravel TJ, Nicholson DL, Kotze MJ. (2002). Familial adenomatous polyposis in two Black South African families. *Clin Genet 61*, 214–217.

Hammond-Tooke D. (1989). *Rituals and Medicine*. Johannesburg: AD Donker.

Hayden MR. (1981) *Huntington's Chorea*. Berlin, Heidelberg: Springer-Verlag.

James CA, Hadley DW, Holtzman NA, Winkelstein JA. (2006). How does the mode of inheritance of a genetic condition influence families? A study of guilt, blame, stigma, and understanding of inheritance and reproductive risks in families with X-linked and autosomal recessive diseases. *Genet Med* 8(4), 234–242.

Jeffreys MDW. (1953). Twin births among Africans. *S Afr J Sci 50*, 89–93.

Jenkins T. (1990). Medical genetics in South Africa. *J Med Genet* 27, 760–779.

Jenkins T, Lane AB, Kromberg JGR. (1977). Tay Sachs disease screening and prevention in South Africa. *S Afr Med J 51*, 95–98

Jenkins T, Wilton E, Bernstein R, Nurse GT. (1973). The genetic counselling clinic at a children's hospital. *S Afr Med J 47*, 1834–1838.

Kessler S. (1979). *Genetic Counseling: Psychological Dimensions* (pp. 17–33). New York: Academic Press.

Kessler S. (1980). The psychological paradigm shift in genetic counseling. *Social Biology 27*, 167–185.

Krause A. (1994). *Globin genes of the South Africa Indian population group*. PhD (Med) thesis. University of the Witwatersrand, Johannesburg.

Krause A, Craig P. (2001). Genetic counseling – what the medical practitioner needs to know. *CME 19*, 490–492

Krause A, Greenberg J. (2008). Genetic testing for Huntington's disease in South Africa. Editorial. *S Afr Med J 98*(3), 193–194.

Krige EJ. (1937). Individual development. In: Schapera I (ed.), *The Bantu-speaking Tribes of South Africa* (pp. 95–118). Cape Town: Maskew Miller.

Kromberg JGR, Jenkins T. (1982a). Albinism in the South African Negro. II. Prevalence. *S Afr Med J 61*, 383–386.

Kromberg JGR, Jenkins T. (1982b). Common birth defects in South African Blacks. *S Afr Med J 62*, 599–602.

Kromberg JGR, Jenkins T. (1984). Albinism in the South African Negro. III. Genetic counselling issues. *J Biosoc Sci 16*, 99–108.

Kromberg JGR, Berkowitz DE. (1986). Managing genetic disorders. *SA Fam Pract* 28, 279–285.

Kromberg JGR, Zwane EM, Jenkins T. (1987). The response of black mothers to the birth of an albino infant. *Am J Dis Child 141*, 911–916.

Kromberg JGR, Jenkins T. (1988) Marketing and providing an outreach genetic counselling service in Southern Africa: Experience with 40 families in a new clinic. In: *Proceedings of the National Society of Genetic Counsellors 8th Annual Conference*, New Orleans.

Kromberg JGR, Bernstein R, Jacobson MJ, Rosendorff J, Jenkins T. (1989). A decade of mid-trimester amniocentesis in Johannesburg. *S Afr Med J 76*, 344–349.

Kromberg J. (1992). Albinism in the South African Negro: IV. Attitudes and the Death Myth. *Birth Defects: Original Article Series 28*, 159–166. New York: March of Dimes Birth Defects Foundation.

Kromberg JGR, Christianson AL, Duthie-Nurse G, Zwane E, Jenkins T. (1992). Down syndrome in the black population. *S Afr Med J 81*, 337.

Kromberg JGR. (1993). Prenatal genetic diagnosis. *CME 11*, 1293–1302.

Kromberg J, Jenkins T. (1997). Cultural Influences on the perception of genetic disorders in the black population of South Africa. In Clarke A, Parsons E, eds. *Culture Kinship and Genes. Towards Cross-Cultural Genetics*. London: Macmillan, pp. 147–157.

Kromberg JGR, Parkes J, Taylor S. (2006). Genetic counseling as a developing healthcare profession: A case study in the Queensland context. *Aust J Prim Health 12*, 5–11

Kromberg J, Zwane E, Manga P, Venter A, Rosen E, Christianson A. (2008). Intellectual disability in the context of a South African population. *J Pol Pract Intellect Disabil 5*, 89–95.

Kromberg JGR, Krause A, Wessels T. (2009). Roles of genetic counsellors in South Africa. Abstract book, *13th Congress of the SA Society of Human Genetics*, April 5–8, 2009. Stellenbosch, Cape, South Africa, p. 36.

Kuppermann M, Learman LA, Gates E, Gregorich S, Nease RF, Lewis J, Washington AE (2006). Beyond race or ethnicity and socioeconomic status: Predictors of prenatal testing for Down syndrome. *Obstet Gynecol 107*, 1087–1097.

Lancet (2008). South Africa in the spotlight. *The Lancet 371*, 1215. Editorial.

Learman L, Drey E, Gates E, Kang M-S, Washington A, Kuppermann M. (2005). Abortion attitudes of pregnant women in prenatal care. *Am J Obstet Gynecol 192*, 1939–1947.

Livingstone D. (1857). *Missionary Travels and Researchers in South Africa*. London: John Murray.

Loggenberg K, Futter M, Greenberg J. (2007). A qualitative investigation of the impact of DMD on South African families. Poster presentation at the *Annual Congress of the European Society of Human Genetics. Nice, France*. Available at https://www.eshg.org/eshg2007/downloads/ESHG2007Abstracts.pdf p. 332, Poster P1336.

Loubser F, Loggenberg K, Roberts L, Greenberg LJ. (2009). Genetic counselling experiences in delivery of genetic research results to patients affected with ABCA4 associated retinal degenerative disorders. (Poster) Abstract book, *13th Congress of the SA Society of Human Genetics, April 5–8, 2009*. Stellenbosch, Cape Town, South Africa.

Macaulay S. (2008). *Cystic Fibrosis Genetic Counselling: an audit of counselees and their at-risk relatives*. MSc (Med) Research Report. Johannesburg: University of the Witwatersrand.

Manganyi NC. (1973). *Being Black in the World*. Johannesburg: Ravan Press.

Maslow AH. (1970). *Motivation and Personality*. 2nd edition. New York: Harper and Row.

McCarthy Veach P, LeRoy BS, Bartels DM. (2003) *Facilitating the Genetic Counselling Process*. New York: Springer.

Miller SK. (1993). Many wives means many daughters. *New Scientist 1876*, 5.

Mqoqi N, Kellett P, Sitas F, Jula M. (2004). Incidence of histologically diagnosed cancer in South Africa, 1998–1999. *National Cancer Registry of South Africa,* National Health Laboratory Service, Johannesburg.

Ramesar RS, Madden MV, Felix R, et al. (2000). Molecular genetics improves the management of hereditary non-polyposis colorectal cancer. *S Afr Med J 90*(7), 709–714.

Read M. (1966). *Culture, Health and Disease*. London: Tavistock.

Reed S. (1955). *Counselling in Medical Genetics*. Philadelphia: WB Sanders.

Reed S. (1980). *Counselling in Medical Genetics*. 3rd edition. New York: Alan Liss.

Schapera I. (1940). *Married Life in an African Tribe*. London: Faber and Faber.

Smart RD. (1981). Down syndrome in the Cape Peninsula and the value of amniocentesis as a preventive measure. *S Afr Med J 59*(19), 670–672.

Somer M, Mustonen H, Norio R. (1988) Evaluation of genetic counselling: recall of information, post-counselling reproduction, and attitude of the counsellees. *Clin Genet 34*, 352–265.

Statistics South Africa (2009). Mid year population estimates. (www.statssa.gov.za).

Stupart DA, Goldberg PA, Algar U, Ramesar R. (2008) Surveillance colonoscopy improves survival in a cohort of subjects with a single mismatch repair gene mutation. *Colorectal Dis 11*(2), 126–130

Tollman SM, Kahn K. (2007) Health, population and social transition in rural South Africa. *Scand J Public Health 35*(Suppl 69), 4–7.

Viljoen DL, Gossage JP, Brooke L, et al. (2005). Fetal alcohol syndrome epidemiology in a South African community: a second study of a very high prevalence area. *J Stud Alcohol 66*(5), 593–604.

Walker AP. (1998). The practice of genetic counseling. In: Baker DL, Schuette JL, Uhlmann WR (eds.), *A Guide to Genetic Counseling* (pp. 1–20). New York: Wiley Liss.

Wall S. (2007). Foreword. *Scand J Public Health 35*(Suppl 69), 2.

Wessels TM, Penn C, Kromberg JGR. (2009). "Do you know why you are here today?" Genetic Counselling in an antenatal multicultural setting. Abstract book, *13th Congress of the SA Society of Human Genetics*, (p. 36). April 5–8, 2009. Stellenbosch, Cape Town, South Africa.

Weil J. (2000). *Psychosocial Genetic Counselling.* New York: Oxford University Press.

Ziolko ME. (1991). Counseling parents of children with disabilities: a review of the literature and implications for practice. *J Rehabil 57*, 29–34.

WEBSITES AND BIBLIOGRAPHY

http://www.doh.gov.za/docs/policy/humangenetics.pdf. Policy Guidelines for the management & prevention of genetic disorders, birth defects & disabilities. Department of Health, Pretoria, SA. 2001.

http://www.uct.ac.za/depts/genetics/

http://web.uct.ac.za/depts/genetics/counselling/

http://web.uct.ac.za/depts/genetics/retina/

http://web.uct.ac.za/depts/genetics/neuro/

http://web.uct.ac.za/depts/genetics/cancer/

http://www.genetichealthvic.net.au

http://www.nhs.uk/Pages/homepage.aspx

www.statssa.gov.za

http://tagc.med.sc.edu/

Edwards J, Greenberg J, Sahhar M. (2008). Global Awakening in Genetic Counseling. Available from Nature Proceedings http://hdl.handle.net/10101/npre.2008.1574.1.

46.

PRENATAL DIAGNOSTIC SERVICES AND PREVENTION OF BIRTH DEFECTS IN SOUTH AFRICA

Michael Urban and Lut Geerts

INTRODUCTION

In order to adequately describe prenatal diagnostic (PND) services in South Africa and how they have developed, we first outline the context of South Africa, its population, health problems and services, and also consider the priority genetic disorders and birth defects from a prenatal diagnosis viewpoint. We then summarize the literature regarding prenatal diagnosis in South Africa, and provide information regarding the current state of PND services throughout South Africa. To this end we have been helped by numerous colleagues throughout the country. Nevertheless, much of the data presented regarding the current status of PND services pertains to our respective services in the western and eastern halves of the Cape Town metropole. Further information regarding genetic disorders and services in South Africa is found in several other chapters of this book.

THE POPULATION, HEALTH, AND HEALTH SERVICES OF SOUTH AFRICA

The Republic of South Africa is a middle-income country with a population of 48 million people. As is well known, a constitutional democracy has been in place since 1994. Archbishop Desmond Tutu coined the epithet "the Rainbow Nation," which aptly captures the diversity of this population: the large majority comprises black African individuals from nine main ethnic groups, and there are distinct White, mixed ancestry and Indian minorities (these four categories are often termed "population groups"). In addition, several million migrants or refugees, mainly from other African countries, now reside in South Africa. Since 1994, the country has comprised nine provinces (see Figure 46-1), each with a different ethnic and language mix. A total of 11 official languages are recognized nationally.

South Africa remains a very economically inequitable society. This is strongly influenced by "population group" background: although the middle class has increasingly deracialized since 1994, the working class consists disproportionately of people with black African and mixed ancestry. Inequity also exists geographically, with considerable variability in economic and health indicators between provinces, and between urban and rural areas. For example, the estimated under-5 mortality varies from 46/1000 in the Western Cape, to over 100/1000 in the Eastern Cape and Kwazulu-Natal provinces (Bradshaw et al., 2004).

Urbanization has proceeded apace in the democratic era, and an increasing proportion of the population is thus concentrated in the large urban centers: census data from 2001 indicate that approximately 57% of the South African population was urbanized and 43% rural-based (Kok & Collinson, 2006), although black South Africans were significantly less urbanized (47% in urban areas) than the minority population groups. Gauteng province is largely metropolitan, with a population of over eight million distributed between Johannesburg, Soweto, and Tshwane (Pretoria). Cape Town and Durban, each with approximately 3.5 million people, are the other metropolitan centers, while other large cities include Port Elizabeth, East London, Bloemfontein, and Polokwane (referred to as Pietersburg on the map).

Since 1994, South Africa has made progress in many areas of service delivery, including initiatives that would be expected to improve child health, such as a substantial increase in provision of housing and basic services. Despite this, the country has relatively poor (and worsening) child health indicators for a country of its level of development; for example, it is estimated that infant mortality increased during the 1990s, from 45/1000 to 60/1000 by the year 2000 (Bradshaw et al., 2003). The baseline level of mortality reflects a legacy of poor social development, whereas the recent deterioration is ascribed largely to the impact of HIV/AIDS.

Health services in South Africa have evolved rapidly since the 1980s. The apartheid era was characterized by

Figure 46-1 Map of South Africa
From: www.africanvacation.co.za/africanvacations/ (accessed 17/09/2009).

wide disparities between population groups in health and access to healthcare. Services were extremely fragmented, with well-developed services largely limited to urban areas and geared to serving the needs of the White population (Jenkins, 1990). The democratic era has brought legislation aimed at achieving a more equitable health system. The Constitution of the Republic of South Africa (Act 108 of 1996) requires that "the State must take reasonable legislative and other measures within its available resources to achieve the progressive realization of the right of the people of South Africa to have access to health care services, including reproductive health care."

To achieve this, the National Department of Health has aimed to develop a district based system of primary healthcare, with redistribution of resources from tertiary services toward primary care. For reasons including the effects of the HIV/AIDS epidemic and limitations in management and professional capacity, the public health sector remains stressed at all levels of care, and equity of access to healthcare has been difficult to achieve.

The provision of health services is largely determined at the provincial level, with considerable variation in the quality of health services provided by different provinces. In addition, there has been a dramatic expansion of the private health sector in South Africa. Benatar (2004) argues that this has created "a two-tier health care system... with discrimination in access to care on economic grounds replacing the racial discrimination of the past."

The above discussion outlines the challenging context within which PND services are provided in South Africa, and highlights the need to address both the relevance and the equitable provision of PND services. We will discuss these issues intermittently, but will focus on how PND services have been provided through the past 40 years, their current status, and possible future approaches.

ANTENATAL HEALTH SERVICES IN SOUTH AFRICA

Services for PND in South Africa are located within broader antenatal healthcare services, which provide for the approximately 1.1 million pregnancies that take place annually. Since 1994, comprehensive antenatal and delivery care has been available free of charge to all South African women through the public health service, and is the responsibility of the national and provincial Women's Health and Genetics Directorates.

The Department of Health has adopted policy regarding nationally agreed minimum components of basic antenatal care (BANC) at the primary care level, and criteria for referral to other levels of care (Pattinson, 2005). These recommend that women should initiate antenatal care as soon as pregnancy is diagnosed. Following the initial "classifying visit," low-risk women should then receive four follow-up antenatal visits in primary care, whereas higher-risk women should be referred to secondary or tertiary levels of care. Approximately 94% of pregnant women receive some antenatal care, although only 27% initiate antenatal care before 20 weeks gestation (Pattinson et al., 2007).

One criterion for referral to a higher level of care is maternal age of above 40 years, since this is associated with increased risk of a variety of poor obstetric outcomes. In addition, it is national policy that women who initiate antenatal care before 20 weeks gestation and who are 35–40 years of age (the cutoff varies by province) should be offered referral regarding prenatal diagnosis of Down syndrome. There has been no national policy decision regarding obstetric ultrasound services, the delivery of which is left to the local or regional authorities.

Approximately 830,000 deliveries took place in the public sector in a one-year period in 2006–2007 (Pattinson, 2009), with 59% at primary care level (clinic and district hospital), 30% at secondary level (regional) hospitals and 11% at tertiary level (Sanders et al., 2007). Home births are unusual in South Africa and mainly practiced by "traditional midwives" in rural areas.

Approximately 20% of all deliveries occur in the private sector. In this sector, obstetric services are provided both by obstetricians and general practitioners, with a small contribution from registered midwives. Although most deliveries are carried out by obstetricians, many patients receive their antenatal care from a general practitioner until the third trimester. The service is generally characterized by early initiation of antenatal care, frequent antenatal visits and ultrasound assessments (often monthly), hospital deliveries, and a high caesarean section rate. Most private medical insurance schemes provide cover for this, with reimbursement for ultrasound examinations generally limited to two visits. Although a number of private sector fetal medicine centers exist in metropolitan areas, only a relatively small proportion of women attending private antenatal care access dedicated fetal medicine services.

BOX 46-1

- In the public sector in South Africa, approximately 94% of pregnant women receive some antenatal care, but only 27% initiate antenatal care before 20 weeks gestation.

THE CAUSES OF CHILD MORTALITY AND DISABILITY IN SOUTH AFRICA

For the purposes of our discussion we define birth defects as abnormalities of bodily structure or function that are present, although not necessarily obvious, at the time of birth. Fetal anomalies refer to structural birth defects that can be detected on prenatal ultrasound examination.

Information will be presented on the contribution of birth defects to perinatal and child mortality, but some qualifications need to be kept in mind. Firstly, accurate information on the frequency of birth defects in South Africa is limited, as it is in many other middle-income countries. In addition, it has been cogently argued that there is an international tendency for the frequency of birth defects to be underestimated (Christianson et al., 2006), both because of limited recordkeeping in many countries and because birth defects usually constitute an underlying rather than a final cause of death, and therefore do not necessarily feature on death certificates. Another limitation is that birth defects often cause physical or mental disability rather than death, and it is increasingly recognized that disability is a problem of great magnitude on which limited information is available (Grantham-McGregor et al., 2007). A national Birth Defects Surveillance notification system has been recently initiated in South Africa, and may eventually allow for quantification of some conditions which cause physical disability. However, the quantity and quality of surveillance data currently available precludes meaningful interpretation.

Bradshaw et al. (2003) have estimated the magnitude of the commonest causes of mortality in South African children under 5 years of age. The list is dominated by infectious and nutritional causes of mortality, and HIV/AIDS alone causes 40% of deaths. However, several birth defects, including congenital heart disease, neural tube defects, and chromosome disorders, feature among the top twenty causes of mortality. When combined, these four birth defects contribute 3% of deaths and rank sixth in the list of causes.

The Perinatal Problem Identification Programme (Pattinson, 2009) is a national sentinel surveillance program for perinatal mortality, which includes numerous health facilities at different levels of health care. According to this program, congenital anomalies were estimated to cause 4% of perinatal mortality in 2007 nationwide, and were the sixth leading cause after preterm birth, intrapartum asphyxia, maternal hypertension, antepartum hemorrhage, and congenital and perinatal infections. It should be noted that this is a minimum estimate, in part because of the limited extent of investigation of the causes of many neonatal deaths, and especially stillbirths.

Information regarding the prevalence of intellectual and physical disability in South Africa is limited to a few individual studies. A population-based study in a rural area of Limpopo province, conducted in the 1990s, found a relatively high rate of intellectual disability of 35/1000 (Christianson et al., 2002). Of note is that this geographical area is anticipated to have a low frequency of fetal alcohol syndrome, which is a major contributor to intellectual disability in some parts of South Africa.

The South African National Department of Health has published guidelines for human genetics (National Department of Health, 2001), which lists several individual birth defects as priority targets for prevention and care. The most important of these from a prevention point of view are Down syndrome, neural tube defects, and fetal alcohol syndrome. Although ultrasound-identifiable fetal anomalies are an important category from a prenatal diagnosis point of view, the policy guidelines do not list these as a group.

DOWN SYNDROME

Down syndrome (DS) is relatively common in South Africa. As is well known, the birth prevalence of DS is determined by the maternal age structure of the population, and the availability and utilization of prenatal diagnosis. A relatively high proportion of South African mothers are of advanced maternal age, with approximately 17.2% of pregnant women being over 35 years of age (United Nations demographic yearbook series, 2003), compared to 14% as an average for the European Union in the 1990s (Dolk et al., 2005). Live-birth prevalence of DS appears to be relatively high, with studies in the 1990s showing figures of 1.33 (Delport et al., 1995) and 1.8 per 1000 live births (Christianson & Kromberg, 1996) reported from two urban areas in Gauteng, and 2.1 per 1000 reported from a rural area in Limpopo province (Venter et al., 1995). These are higher live birth rates than reported from many developed countries in recent years: 1.365 per 1000 as a national estimate at the turn of the millennium in the United States (Canfield et al., 2006), 1.08 per 1000 over the last 20 years in North England (Irving et al., 2008), and 0.7 per 1000 in Paris (de Vigan et al., 2008). Conversely the population prevalence of DS in South Africa is likely to be relatively low, because life expectancy of children with DS is believed to be relatively low, especially in parts of the country with less access to health services (Christianson & Kromberg, 1996).

NEURAL TUBE DEFECTS

Historically, neural tube defects (NTDs) have been relatively common in South Africa, and particularly frequent in rural areas. Prior to the introduction of folate fortification of staple foods, several studies reported on NTDs. In urban areas, the birth prevalence of NTDs was reported as 0.99 per 1000 live births in Cape Town (Buccimazza et al., 1994), 1.18 per 1000 in Johannesburg (Kromberg & Jenkins, 1982) and 0.99 per 1000 in Pretoria (Delport et al., 1995). In rural areas, a rate of 3.5 per 1000 was reported from Limpopo province (Venter et al., 1995) and a rate of 6.1 per 1000 from the Eastern Cape (Ncayiyana, 1986).

A program of folic acid fortification of staple foods was introduced in October 2003, and the birth prevalence of NTDs was followed at a number of sentinel sites before and after this intervention. Folic acid fortification resulted in a 30.5% reduction in birth prevalence of NTDs from 1.41/1000 to 0.98/1000 (Sayed et al., 2008). There was a 41.6% reduction in the rate of spina bifida compared to a 10.9% reduction in the rate of anencephaly. The cost-benefit ratio in averting NTDs was 46:1, indicating that this intervention has excellent cost-benefit characteristics in the South African context.

FETAL ALCOHOL SYNDROME

Fetal alcohol syndrome (FAS) was first reported in South Africa by Palmer (1985) and it has now been recognized that South African communities have the world's highest documented prevalence rates of FAS. Fetal alcohol spectrum disorder has been surveyed in three of the nine South African Provinces – generally by active case ascertainment in school-entry children, with particularly high rates found in small towns and rural areas of the Western and Northern Cape. In the Western Cape, rates of FAS and partial FAS of between 40.5 and 89 per 1000 were described in serial studies among school-entry children in the town of Wellington (May et al., 2000, 2007). In two towns in the Northern Cape province, rates of FAS and partial FAS were found to be 74.7/1000 and 119.1/1000 respectively (Urban et al., 2008). In addition, it is anticipated that there are significant numbers of children with neurological deficits caused by prenatal alcohol exposure but without facial features of FAS. However, this category, designated fetal alcohol spectrum disorder, is more difficult to quantify.

The primary risk factor for FAS is heavy episodic (binge) drinking among reproductive-age women, which is common in some South African communities (Croxford & Viljoen, 1999), and is believed to relate historically to the "tot system" by which laborers on wine farms were paid part of their wage in the form of alcohol. Other known risk factors for FAS include: rural residence, low socioeconomic status and maternal education levels, maternal depression, concomitant cigarette smoking and other drug abuse, low body-mass indices suggestive of chronic maternal malnutrition, and disrupted family structure; e.g., single-parent families. Many of these risk factors are common among women in high-prevalence areas of South Africa (Urban et al., 2008). The impact of maternal alcohol abuse during pregnancy extends far beyond FASD and includes health aspects (concomitant HIV/AIDS and tuberculosis infections), educational and social developmental factors, economic consequences, poor housing, and a lack of work opportunities.

Although public and political awareness of the problem of FASD is growing, efforts at prevention are currently very limited in scope. These include advice from health workers during antenatal clinic visits, use of media to increase public awareness, and ongoing research studies of brief educational and motivational interventions to address risky drinking behavior.

FETAL ANOMALIES

The fact that fetal anomalies as a group are not included in the policy guidelines for human genetics possibly reflects the fact that fetal ultrasound services are relatively restricted to larger urban areas. This is in part because the value of routine fetal ultrasound at a public health level has been considered questionable—for example, a Cochrane review found that perinatal outcome is only improved when the prenatal detection (and termination) rate for severe abnormalities is high (Neilson, 2002). However, routine ultrasound services have additional benefits for obstetric services in general, mainly due to accurate pregnancy dating and early detection of unexpected problems, as illustrated in a few local studies (Geerts et al., 2004; van Dyk et al., 2007). Fetal ultrasound services are therefore increasingly used, and in metropolitan areas fetal medicine is emerging as an important discipline.

In addition to the abovementioned broad information regarding the contribution of birth defects to perinatal and child mortality, more detailed data is available from the Cape Town Metropole East region, which documents changes in the contribution of fetal anomalies to perinatal loss over time. A survey was performed in the same geographical area in three different decades (in 1986, 1993, and 2006–2007), and clearly illustrates an increase in the relative contribution of fetal anomalies: they were responsible for 7.9% of perinatal losses in 1986, 8.0% in 1993, and 11.4% in 2007 (Losper et al., 2008). Although the relative position of fetal anomalies as a cause of perinatal mortality has not changed (fifth position in 1986 and 2007), the actual numbers have increased by 50%. In the latest era, fetal anomalies were the second most common cause of neonatal deaths (17%) after severe prematurity. These changes may be due to other causes of perinatal loss being successfully addressed, or may reflect improvements in prenatal diagnosis, especially of those anomalies that have no obvious external signs. The first option would be related to progress in general obstetric services, while the latter would reflect improved access and quality of ultrasound services. The two do not necessarily go hand in hand in all regions.

BOX 46-2

- In descending order of frequency, the priority birth defects in South Africa are: fetal alcohol syndrome, Down syndrome, and neural tube defects.
- Fetal anomalies cause up to 11.4% of perinatal mortality in South African settings, but are underrecognized.

HISTORICAL DEVELOPMENT OF PRENATAL DIAGNOSTIC SERVICES

Modern PND services require an enabling legislative and policy framework, as well as a sophisticated mix of fetal medicine and biochemical screening laboratory services, clinical genetic and genetic counseling services, and genetic diagnostic laboratory services. We outline the evolution and current state of prenatal diagnostic services in South Africa since the late 1960s.

LEGISLATION PERTINENT TO PRENATAL DIAGNOSIS

A key issue facilitating prenatal diagnosis is legislation regarding termination of pregnancy. Until the 1990s, prenatal diagnosis in South Africa was regulated by the relatively restrictive abortion laws prescribed by the Abortion and Sterilization Act of 1975, which permitted medical termination of pregnancy in prescribed circumstances, but had the effect of inhibiting the development of fetal blood sampling in the 1970s and chorionic villous sampling in the 1980s (Jenkins, 1990). The Choice on Termination of Pregnancy Act of 1996 provided for a more liberal approach to termination of pregnancy. Some provisions of this legislation are open to wide interpretation, and it is noteworthy that no legislation or national policy exists regulating the use of feticide, which is left up to individual obstetric units.

BOX 46-3 THE CHOICE ON TERMINATION OF PREGNANCY ACT OF 1996

A pregnancy may be terminated

(a) upon request of a woman during the first 12 weeks of the gestation period in her pregnancy

(b) from the 13th up to and including the 20th week of the gestation period if a medical practitioner, after consultation with the pregnant woman, is of the opinion that –

(i) the continued pregnancy would pose a risk of injury to the woman's physical or mental health; or

(ii) there exists a substantial risk that the fetus would suffer from a severe physical or mental abnormality; or

(iii) the pregnancy resulted from rape or incest

(iv) the continued pregnancy would significantly affect the social or economic circumstances of the woman; or

(c) after the 20th week of the gestation period, if a medical practitioner, after consultation with and medical practitioner or a registered midwife, is of the opinion that the continued pregnancy -

(i) would endanger the life of the woman

(ii) would result in severe malformations to the fetus; or

(iii) would pose a risk of injury to the fetus

PRENATAL DIAGNOSTIC TECHNIQUES AND GENETIC TESTING

Prenatal diagnosis has been available in South Africa for several decades. The development of university-associated cytogenetics laboratories and the advent of the technique of (blind) amniocentesis, allowed the introduction of prenatal karyotyping, first in Johannesburg in 1969, and subsequently in Cape Town in 1973 and several other urban centers (Jenkins, 1990).

PND was well established in the public sector by the mid-1970s (and in private practice by the late 1980s). Early experiences with 434 amniocenteses in Cape Town were recorded by Nelson et al. (1978) when the main indication was advanced maternal age (AMA). Retief et al. (1983) reviewed data on cytogenetic services in South Africa, and obtained data from all cytogenetic laboratories over a three-year period in the late 1970s, when prenatal samples comprised nearly a quarter of all karyotypes performed (2150 of 9661 specimens). Kromberg et al. (1989) reported on 4554 amniocenteses performed over a 10-year period (1976–1985) in Johannesburg. The primary indications were risk of chromosome abnormalities (83.4%), and neural tube defects (NTDs) or other causes of increased amniotic fluid alpha-fetoprotein (AFP; 10.6%).

Overall, the numbers of amniocenteses in the 1970s were small—less than 0.5% of all pregnancies (Jenkins, 1990) but increasing, while the positive yield decreased over time from 6% abnormal results reported by Nelson et al. (1978) to 3.2% reported by Kromberg et al. (1989), as the age cutoff for AMA was relaxed from 40 years to 35–38 years of age (each service sets its own age cutoff). The annual demand for amniocentesis increased considerably during the 1980s, but this was not sustained: Kromberg et al. (1989) found that in Johannesburg during the 1980s there was a fivefold rise in the use of amniocentesis, followed by a plateau toward the end of the study period. Availability of amniocentesis was restricted to a few metropolitan centers and samples were not representative of the different ethnic groups of the country (Jenkins, 1990), probably reflecting greater accessibility to white South Africans. During this period, the service in Johannesburg was used mainly by white South Africans (Kromberg et al., 1989) and it was noted that the great majority of counselees accepted amniocentesis.

Although many factors have likely shaped the trends in utilization of public-sector PND services (and some are examined further in the sidebar), one significant factor has been the growth of private health care since the 1980s. During this period, there has been more frequent start-up and greater growth of private than public sector cytogenetic laboratories.

Through all the years, the primary prenatal genetic test has been chromosome analysis. In the last decade this has been supplemented in urgent cases by aneuploidy fluorescence in-situ hybridization (FISH) testing in some laboratories, and by aneuploidy quantitative fluorescence polymerase chain reaction (qF-PCR) in others. Some laboratories in the public sector have recently begun using aneuploidy testing (mainly qF-PCR) as a "standalone" method for cases in which Down syndrome is suspected, in order to alleviate cost and staffing constraints. The cytogenetic laboratory services for the public and private sectors remain largely separate, with only one laboratory providing services to both sectors (in this case, aneuploidy qF-PCR).

Although available for research purposes, microarray techniques are not available for routine cytogenetic diagnosis in South Africa. The field of molecular cytogenetics is still in its infancy, whereas molecular genetic testing for single gene disorders has expanded significantly since the 1980s and testing is offered for an increasing number of the commoner single-gene disorders. Access to genetic tests for rare disorders remains limited, and neither the National Department of Health nor the majority of private health insurance schemes will fund tests done abroad.

Relatively little epidemiological data regarding prenatal diagnosis in South Africa has been published since the 1980s, but trends in one service are outlined in the case study (see sidebar).

BIOCHEMICAL AND ULTRASOUND SCREENING FOR ANEUPLOIDY

When obstetric ultrasonography was introduced in developed countries, South Africa followed soon after and obstetrics was among the first disciplines in the country to employ ultrasound for diagnostic purposes in the early 1970s. Gradually, the field of fetal medicine developed and it became clear that many features of abnormal development in aneuploid fetuses were visible with ultrasound. With the advent of real-time ultrasound technology by the mid-1980s, the continuous advances in image quality, and the recognition of more and stronger signs of aneuploidy risk, the indications for prenatal karyotyping have considerably expanded. The introduction in the private healthcare sector of serum screening by means of alphafetoprotein (AFP), beta-human chorionic gonadotropin (b-HCG) and unconjugated estriol (uE3) in the mid-1990s caused a significant increase in the prenatal detection of trisomy 21. This was followed by the introduction of nuchal translucency screening in 1998, and first-trimester serum screening by means of free b-HCG and pregnancy-associated placental protein-A (PAPP-A) in 2002. With the advent of ultrasound-guided invasive procedures in the early 1980s, the safety of prenatal diagnosis improved considerably (Smart et al., 1985; Kromberg et al., 1989), and with the introduction of chorionic villus sampling (CVS) in the 1980s (Rosendorff et al., 1989) and cordocentesis, the gestational age window for diagnostic procedures could be extended as well.

South Africa does not have an official accreditation framework for obstetric scanning, and even in 2009, detailed fetal assessment is not a compulsory part of the postgraduate training curriculum for registrars in obstetrics and gynecology. To promote the discipline of fetal medicine and to establish and define the role of antenatal ultrasound, SASUOG (the South African Society of Ultrasound in Obstetrics and Gynecology) was founded in 1988. This society hosts a biennial educational congress with practical workshops, and has presented several international meetings in collaboration with ISUOG, the International Society of Ultrasound in Obstetrics and Gynecology. Further ultrasound courses are offered by individual academic departments (and recently also by private individuals), and a nuchal translucency theoretical course (Fetal Medicine Foundation, UK) is offered locally at least once a year and is now available online. Maternal and fetal medicine was officially recognized by the HPCSA (Health Professions Council of South Africa) as a subspecialty of obstetrics and gynecology in 2006, and to date all (six) major academic centers in the country have a well-established fetal medicine unit led by a subspecialist to provide services to the public sector. Many of the subspecialists have spent time overseas in recognized training centers, and several have obtained the Fetal Medicine

Diploma offered by the Fetal Medicine Foundation, UK (first graduate in 2000). A few similar referral centers exist within the private sector.

MEDICAL GENETIC AND GENETIC COUNSELING SERVICES

Prenatal genetic counseling in the 1970s and 1980s was provided by medical geneticists and genetic nurses. An M.Sc.(Med) in genetic counseling has been offered since 1989, and a relatively small cadre of genetic counselors has largely replaced the role of genetic nurses. Currently, genetic counselors are not available at all fetal medicine

BOX 46-4

CASE STUDY: 30 YEARS OF PRENATAL DIAGNOSIS IN THE WESTERN CAPE TOWN METROPOLE

This region comprises a large and diverse urban population, with a majority of "mixed ancestry" individuals and minorities of Whites and black Africans, mainly of Xhosa ethnicity. The service was always, at least theoretically, available to women of different population groups. It is currently used mainly by the mixed ancestry population of Cape Town. This community has its roots in the era of slavery, when slaves were brought to the Cape from the Dutch East Indies and various parts of Africa, and it includes strong Christian and Muslim religious traditions. Over the time period of the study, many demographic changes have occurred in the population of Cape Town, including a progressive shift of middle class individuals, including most Whites, into private healthcare. Over the years, disproportionately few Xhosa women have utilized the service, although the reasons for this have not been well explored.

The obstetric service is large, with approximately 34,000 deliveries in 2005 (Patel, unpublished data). In this time period, obstetric healthcare services were provided by primary care clinics to 45%, level 2 hospitals to 35% and level 3 hospitals to 20% of women respectively, whereas approximately 2.5% of women did not attend antenatal care and 3.3% did not deliver in a health facility.

A retrospective review of the University of Cape Town (UCT) cytogenetics laboratory database was conducted for a 30-year period, 1976–2005 inclusive. Most comparisons were restricted to samples referred from within the provincial health service in this region to allow for comparisons between prenatal and postnatal diagnoses (all prenatal, neonatal, and childhood services in this region refer to this laboratory). It is likely that the neonatal diagnoses are drawn from the same cohort of pregnancies as the prenatal diagnoses, whereas later postnatal diagnoses are drawn from a larger referral area. In this time period, 5943 amniotic fluid samples were tested in the laboratory, of which 5319 were from women within the Cape Town Metro West area. The vast majority of prenatal diagnostic tests done within the service have historically been by means of amniocentesis, although chorionic villous sampling and cordocentesis are done in small numbers.

Indications for amniocentesis have varied over the years:

- AMA has always been the commonest indication, but the cutoff has varied over the years. Initially the age threshold was 35 years, but this was increased to 40 years in 1983, and then lowered progressively to 39 years in 1984, 38 years in 1992, and 37 years in 2004.
- A previous child with DS has always been an indication for invasive testing.
- Biochemical screening for DS is not available in the public sector, and therefore is only an indication for amniocentesis in the occasional patient referred from the private sector.
- In recent years, increasing numbers of women have been identified as high risk based on first-trimester nuchal translucency screening, but the numbers remain small because no formal screening program exists.
- Fetal anomaly detected by ultrasound has been an indication for amniocentesis since the 1970s, with considerably increased use since the 1990s (Figure 46-3).
- Prior to the 1980s, a significant number of amniocenteses were performed for prenatal detection of NTDs by means of elevated amniotic AFP levels (Figure 46-3). This indication has been replaced by detailed ultrasound examination.
- Although amniocentesis for single-gene disorders has increased in frequency over the years due to increased availability of specific genetic tests, the numbers of tests remain small and are not included.

Genetic counseling for women at increased risk of DS requires referral to the fetal medicine unit at 16–20 weeks of pregnancy. Depending on the indication, genetic counseling is provided by a genetic counselor, genetic nurse, or medical geneticist, and aims to provide full relevant information in a nondirective manner, in order to facilitate informed and autonomous decision making. Consenting women receive an amniocentesis for karyotype (supplemented, in recent years, by an aneuploidy FISH test in urgent cases).

The results are presented as bar graphs showing trends over time, stratified into 5-year periods. Figure 46-3 shows the trend in number of amniocenteses performed. The total

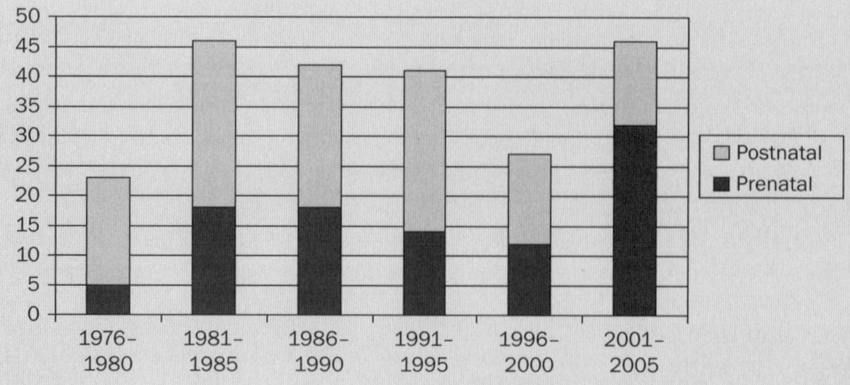

Figure 46-3 Trend in frequency of amniocentesis by indication.

number peaked in the early 1980s and then decreased, with a plateau beyond 1990. The reduction reflects a cessation of amniocentesis for family history of NTD, and a continuing reduction in amniocenteses for AMA (despite a concurrent reduction in the cutoff age). The plateau arises because increasing numbers of amniocenteses done for fetal anomalies detected on ultrasound balance out the reduction in amniocenteses for AMA.

Figure 46-4 demonstrates that the prenatal detection of trisomy 18 has increased over time, while the total number of cases has remained relatively stable. In the last 5-year period, 32 of 46 (70%) cases of trisomy 18 were diagnosed prenatally. Of these, 26 (81%) were diagnosed by ultrasound, while only 5 (16%) were detected on amniocentesis for AMA. These results suggest that fetal ultrasound is increasingly available at some stage in pregnancy, and that uptake of amniocentesis for trisomy 18 is relatively high. Despite the fact that DS is much more common than trisomy 18, only 13 cases of DS were diagnosed prenatally in the last 5-year period, with approximately equal numbers referred for fetal anomalies and for AMA.

Figure 46-5 demonstrates that the total number of DS cases detected in the neonatal period, and the number detected at any age postnatally, has remained relatively constant over time. The same is true of prenatal diagnoses, but only a small proportion of DS were diagnosed prenatally in any time period. In the last 5-year period, only 13 (13%) DS cases were diagnosed prenatally out of a total of 104 cases diagnosed by the end of the neonatal period.

In conclusion, although the numbers of amniocenteses done for AMA has reduced over time, the proportion of DS diagnosed prenatally has remained relatively constant, and the proportion of trisomy 18 diagnosed prenatally has increased, due to greater numbers being detected on fetal anomaly scan. Currently, the majority of cases of trisomy 18 are detected prenatally in our service, most likely reflecting the ease with which it can be detected by routine ultrasound, and the fact that trisomy 18 is often associated with clinically obvious pregnancy complications (like growth restriction or polyhydramnios) that would trigger a referral for ultrasound in women who had not yet received a routine ultrasound examination in early pregnancy. Only a small minority of DS cases are detected antenatal. The reasons for this are discussed further in the section on "Current status of prenatal screening and diagnosis in the public health sector."

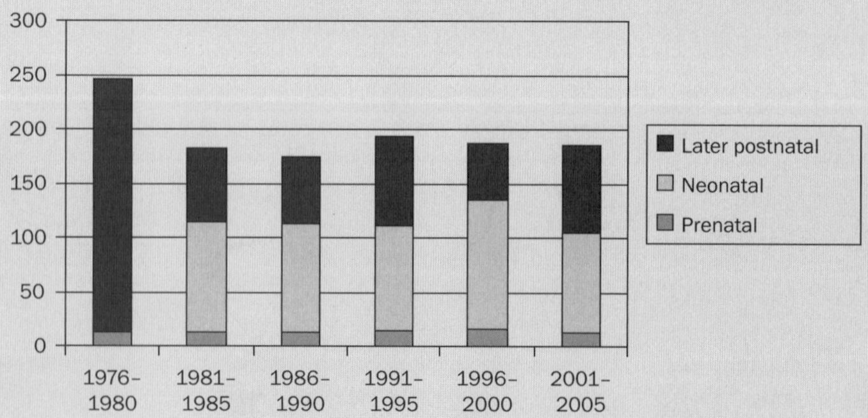

Figure 46-4 Frequency of prenatal versus postnatal trisomy 18 diagnosis per time period.

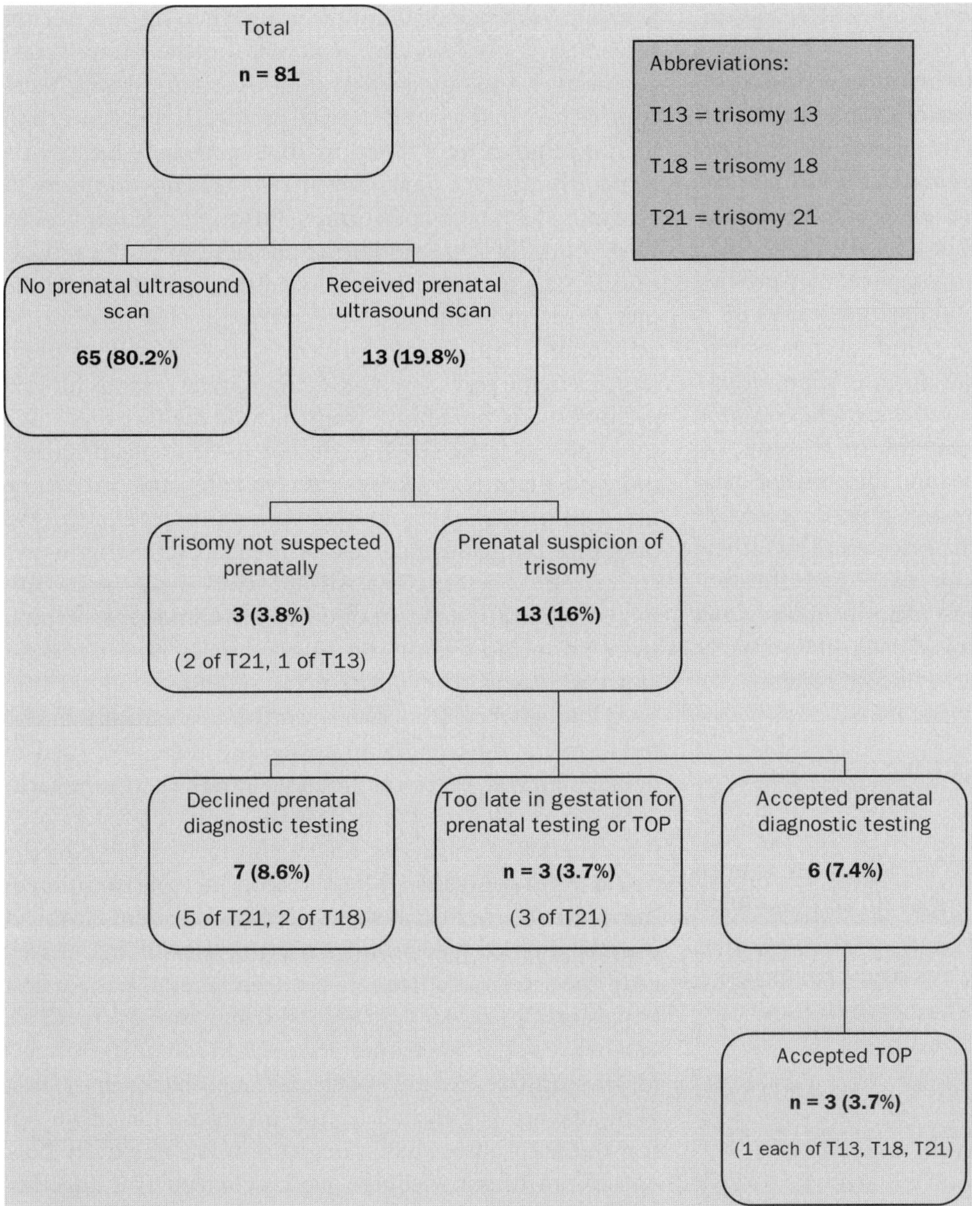

Figure 46-5 Prenatal versus postnatal Down syndrome diagnosis.

units. There is no legislative or policy requirement that genetic counseling be provided by a specifically trained professional and, particularly in the private sector, genetic counseling is often provided by obstetricians and fetal medicine practitioners themselves.

Medical geneticists play an important role in prenatal genetic counseling, particularly regarding fetal anomalies. However, medical geneticists are in short supply in South Africa, and there are few medical geneticists with a specialist interest in prenatal medical genetics. While medical geneticists and/or genetic counselors are available at most fetal medicine units, the level of integration between medical genetic and fetal medicine services is widely variable around the country. Some units conduct regular joint meetings, while in others there is limited contact between the disciplines.

CURRENT STATUS OF PRENATAL SCREENING AND DIAGNOSIS IN THE PRIVATE HEALTH SECTOR

The situation of prenatal screening and diagnosis for chromosomal abnormalities is currently very different in the private and public sectors, each of which has its own issues and limitations.

Approximately 20% of pregnant women in South Africa receive antenatal care in the private sector. Although it is the accepted norm that some form of biochemical and/or ultrasound screening for aneuploidy is offered to all women who present in early pregnancy, and although many medical insurance companies cover the cost thereof, there is no uniformity in screening strategy except that the maternal age cutoff for AMA-based screening is 35 years.

GENETIC LABORATORY TESTING

We conducted a nationwide questionnaire survey of cytogenetic/molecular cytogenetic laboratories in South Africa. There are five laboratories serving the private sector (three in Gauteng, one in Cape Town and one in Port Elizabeth), of which three responded. Results are presented as a comparison between private and public sector (Table 46-1). Although the data are somewhat incomplete, they provide useful insight into trends in prenatal diagnostic testing.

The survey showed that, in the private sector, amniocentesis is by far the most common diagnostic procedure. Approximately 1 in 20 samples showed one of the common trisomies (Down syndrome, trisomy 18, or trisomy 13). Maternal age alone is still the indication for prenatal diagnostic testing in almost half of invasive procedures, while fetal anomaly appears to be underrepresented as an indication for testing (the reason for this is unclear). While chromosome analysis is performed in the vast majority of cases, half received an additional rapid aneuploidy test—presumably, in most cases, to reduce the waiting period for results.

BIOCHEMICAL AND ULTRASOUND SCREENING FOR DOWN SYNDROME

Data are entirely lacking on how many younger women are not offered any form of screening at all, and how often women of any age decline screening when offered. Serum screening is used widely, however, and the three largest chemical pathology laboratories in South Africa were approached with a survey and all provided data. More than 52,700 samples were tested in 2008, and there has been a trend over recent years toward increasing numbers of screening tests being performed. Positive screening results were found in 2494 samples, and these were the indication for the majority of amniocenteses performed in the private sector (Table 46-1). However, it is clear that serum screening is still not performed in the majority of pregnancies in the private sector. Although the reasons for this are unknown, possible reasons include the fact that some medical insurance schemes do not cover prenatal screening, and that many women receive antenatal care in the first half of pregnancy by their general practitioner, who often has limited knowledge of prenatal screening issues compared to obstetricians. All laboratories have noticed an increasing relative contribution of first trimester screening since its introduction, and early samples now represent approximately 44% of all serum screening.

TABLE 46-1 PRENATAL CYTOGENETIC TESTING PERFORMED IN PUBLIC AND PRIVATE SECTOR LABORATORIES IN SOUTH AFRICA, 2008

	PUBLIC SECTOR	PRIVATE SECTOR
PRENATAL TECHNIQUE:	n = 1212 (100%)	n = 2677 (100%)
Amniocentesis	1063 (88%)	2418 (90%)
CVS	28 (2%)	232 (9%)
Cordocentesis	121 (10%)	27 (1%)
INDICATIONS:	n = 928 (100%)	n = 2170 (100%)
Advanced maternal age	526 (57%)	954 (44%)
Screen positive	266 (29%)	1164 (54%)
Fetal anomaly on ultrasound	136 (15%)	52 (2%)
LABORATORY TEST:	n= 1156 (100%)	n = 2399 (100%)
Karyotype	807 (70%)	2136 (89%)
Aneuploidy FISH	88 (8%)	90 (4%)
Aneuploidy qF-PCR	375 (32%)	1358 (57%)
Standalone aneuploidy screens**	348 (30%)	263 (11%)
PRENATAL DIAGNOSES:	n (%)*	n (%)*
Down syndrome	61 (8.4%)	86 (39%)
Trisomy 18	48 (36%)	32 (62%)
Trisomy 13	15 (19%)	4 (40%)
POSTNATAL DIAGNOSES:		
Down syndrome	660	134
Trisomy 18	87	20
Trisomy 13	65	6

* Expressed as percentage of total prenatal and postnatal diagnoses of each condition

** Subset of aneuploidy FISH and PCR Groups

Although there is consistency in the biochemical markers tested in the different laboratories and the cutoff used to define high risk, there are differences in the software package used to calculate the risk, as well as in the risk quoted (i.e., whether it is the risk for an affected live birth or the risk at the time of testing). There is considerable variation in the maternal age distribution of screened patients (women over 35 years of age varying from 10.5% to 19.7%), as well as in the screen positive rate between laboratories (in first trimester 1.9% to 4.9%; in second trimester 6.2% to 9.3%), and between the two trimesters (2.8% to 7.8%). The reasons for these differences are not clear. The problems in the first trimester are aggravated by the incorporation of nuchal translucency–based risk calculations for non-accredited operators without adequate quality control. This, together with inaccuracies in the clinical and gestational age data provided by the referring clinicians, is of great concern to the laboratories, since all of these factors are known to negatively influence screening performance.

Ultrasound and combined screening services are offered in several dedicated private fetal medicine units in the major cities that probably represent less than 15% of the total screening offered nationwide. These centers are all externally accredited for nuchal translucency screening (and other first trimester markers), and their performance indices are generally comparable with the world literature. The majority of women with private healthcare, however, have their routine ultrasound examinations performed by obstetricians, radiologists, radiographers, or general practitioners with highly variable diagnostic skills and varying case loads. There is no consensus on how ultrasound markers should be used to assess aneuploidy risk, and racial differences in the presence of markers like echogenic foci, nasal bone length, etc., are often not appreciated or accounted for. Registered genetic counselors are, as a rule, not used, with counseling provided by fetal medicine professionals themselves. Ultrasound findings (either gross anomalies or soft markers) are quoted to the laboratories as the reason for karyotyping in only a very small number of cases (Table 46-1). The reason for this is unclear, but may relate to the fact that private patients have limited access to dedicated fetal medicine services.

Due to the separation between chemical pathology laboratories and cytogenetic services, as well as between obstetricians and pediatricians, no accurate information is available on the uptake rate of invasive testing after a positive screening result, nor on the false-negative rates of screening. It is noted, however, that the number of karyotypes done in private laboratories for a positive screening test is considerably smaller than the number of positive serum screening results. Although a small number of these women may have had karyotyping done in the public sector, the overall uptake of invasive testing after a positive screening result is probably no higher than 50%, since a considerable number of karyotypes are still done when screening results indicate low risk (data not available). In addition, when comparing the number of prenatally and postnatally diagnosed trisomies, it is clear that no more than 43% of trisomies are detected prenatally. For trisomy 21 in particular, no more than 39% are diagnosed prenatally, despite widespread acceptance of screening.

BOX 46-5 (INSIDE SIDEBAR)

- Historical records from one service in South Africa show that AMA screening is functioning ineffectively in the prevention of Down syndrome, with no more than 13% of DS cases detected prenatally

BOX 46-6

In the South African private health sector:

- PND services are mainly based on AMA and serum screening approaches
- Uptake of invasive testing after positive screening is quite low
- Less than 40% of DS cases are detected prenatally, although epidemiological information is not easily accessible

CURRENT STATUS OF PRENATAL SCREENING AND DIAGNOSIS IN THE PUBLIC HEALTH SECTOR

GENETIC LABORATORY TESTING

Table 46-1 gives information regarding six of the seven public-sector cytogenetic laboratories services approached to participate in our questionnaire (3 in Gauteng, 2 in Cape Town, and 1 each in Durban and Port Elizabeth). All laboratories provide postnatal cytogenetic testing, and in theory are all able to process prenatal samples.

In the public sector, laboratories receive prenatal samples from a limited number of fetal medicine centers that are situated in major urban areas. Some laboratories receive almost no prenatal samples, indicating that no or minimal prenatal diagnostic services exist in the provinces using these laboratories. Without exception, these are provinces lacking fetal medicine centers (Limpopo, Mpumalanga, Northwest and Eastern Cape provinces).

As shown in Table 46-1, amniocentesis is the predominant technique in all centers. There is some variation between the different fetal medicine centers as to the diagnostic techniques used in specific situations. For example, in late-presenting patients, one center does cordocentesis for rapid karyotyping, whereas others use FISH or qF-PCR on amniotic fluid samples. In cases where the indication for testing is increased risk of DS, one laboratory performs standalone aneuploidy qF-PCR, whereas other centers offer aneuploidy FISH or qF-PCR mainly as an adjunct to karyotyping.

When comparing the data from the genetic laboratories serving the public sector with those for the private sector, it is clear that in the public sector, advanced maternal age and fetal anomaly on ultrasound are more common indications for prenatal diagnostic testing, and screening tests are less common than in the private sector.

Serum screening is not really feasible in the public sector of South Africa because many women initiate antenatal care later than the optimal gestation, and because it would need to be preceded by a routine ultrasound service for dating purposes (Roberts et al., 1979; Benn et al., 1997). The AFP-screening program in Kwazulu-Natal has shown that 50% of positive results were false due to inaccurate dating (Grace et al., 1984), and several other local studies have illustrated the inaccuracy of either historical or clinical dating methods in the indigenous population (Geerts et al., 1996, 2004; Van Dyk et al., 2007). Even in developed countries, where recall of the menstrual history may be better or where women book earlier for antenatal care, it is now accepted that dating without ultrasound confirmation is inaccurate (Gardosi, 1997; Gardosi & Geirsson, 1998), and that ultrasound dating in all women improves the efficiency of serum screening (Benn et al., 1999). An additional barrier to serum screening is the difficulty in tracing of patients with abnormal results in a timely manner (Grace et al., 1984), since intervals between routine antenatal visits in the public sector are wide. The major difficulties in tracing patients are well recognized in the country, and have led to a national move toward on-site testing, for example for common infections, to ensure immediate initiation of treatment (Bronzan et al.,

2007). However, on-site screening services for PND, such as available in some developed countries (Bindra et al., 2002), are not a realistic option for a developing country like South Africa.

ADVANCED MATERNAL AGE SCREENING

Advanced maternal age is a nationally accepted indication for direct referral to a tertiary center for genetic counseling, and is still the indication for the majority of amniocenteses (Table 46-1), although to a lesser extent than a few years ago. For example, comparing data for the Cape Metropole West from 1992 (Viljoen et al., 1996) to audit data from 2008 for the same service, the proportion of women in whom the indication for prenatal genetic counseling was AMA decreased from 55% to 44% (Table 46-2). During a recent survey of the six academic fetal medicine units in the country, it transpired that the latest gestational age for AMA referral varies between provinces (from 20 to 23 weeks) and the maternal age cutoff varies from 35 to 40 years. These differences may partly explain why the actual number of older women seen in these units varies more than tenfold. It is clear, however, that only a small fraction of older women in a region (probably less than 20%) receive genetic counseling, while a significant proportion of postnatal diagnoses of aneuploidy continue to occur in AMA women (Geerts, 2008).

The very small proportion of DS cases being diagnosed prenatally is in keeping with absence of serum screening and low use of AMA screening, but AMA is, in addition, an inefficient screening method. A local study showed that an age cutoff of 37 years had a maximal detection rate of 43.3%, and lowering this to age 35 would have only marginally improved this to 52% (while significantly increasing the number of procedures; Geerts, 2008). An age cutoff of 37 years had a positive predictive value of only 2.2% for all aneuploidies (1 in 34 samples) and 1.9% for autosomal trisomies, which is in line with the international literature. The high proportion of invasive tests being performed for AMA, together with a lowering of the age cutoff, has caused a steady decrease in the percentage of samples with abnormal karyotypes over the years—from 6% (Nelson et al., 1978) to 3.2% (Kromberg et al., 1989). Another deficiency of AMA as a screening strategy (by comparison with fetal anomaly ultrasound, for example), is that AMA is a risk factor for the autosomal trisomies only, and not for other chromosomal abnormalities of clinical importance.

While it is problematic that very few eligible women receive counseling, the underlying reasons are often unknown and it is too easily assumed that low utilization rates are mainly caused by patient-related factors. We will examine factors affecting availability, accessibility, affordability, and acceptability of prenatal diagnosis and termination of pregnancy (TOP) on the one hand, and the alternative of having a disabled child on the other.

TABLE 46-2 UPTAKE OF AMNIOCENTESIS PER INDICATION, IN CAPE TOWN METROPOLE WEST—COMPARISON OF TWO TIME PERIODS

	1992–1993 (Two year period)*	2008–2009 (18 month period)
TOTAL	466	872
AMA women (based on age cutoff in study period)	n = 258 (age >39 years)	n = 382 (age >37 years)
Uptake of amniocentesis (%)	191/258 (74%)	113/369 (31%)
AMA women >39 years	As above	n = 267 87/257 (34%)
Fetal anomaly on ultrasound	n = 66	n = 302
Uptake of amniocentesis (%)	17/66 (unknown how many offered amnio)	85/129 (66% of those offered)
Ultrasound screening (First trimester nuchal or second trimester soft markers)	–	n = 63
Uptake of amniocentesis (%)		34/63 (54%)
Other	n = 142	n = 105

*Data extracted from Viljoen et al., 1996

AVAILABILITY AND ACCESSIBILITY OF PND SERVICES

As described above, prenatal diagnosis is simply not available in large parts of South Africa. Reasonable physical access exists only in the urban centers with genetic laboratory services, as these typically have an integrated genetic counseling and fetal medicine service in addition. Even for women living in close proximity to a tertiary center, access may be limited by factors such as lack of knowledge about prenatal diagnosis, lack of information about availability of the service, or lack of efficient referral systems from primary care.

Lack of knowledge and information: In an interview survey of AMA women, Pelser (1998) found that 90% of Caucasian women but only 52% of African women were aware of the availability of prenatal diagnosis, and that they were poorly informed about it by healthcare professionals. Watcham et al. (2007) interviewed 70 AMA pregnant women who were attending antenatal care in the public sector and had not received genetic counseling. They found low prior knowledge of prenatal diagnosis (12%) and of DS (31%). Low awareness of DS among black South Africans may be due to a lower prevalence of the condition (disproportionately high early mortality rate of affected infants) and less frequent recognition of the condition by clinicians and parents alike (Christianson

& Kromberg, 1996). In addition, the relative contribution of congenital or genetic abnormalities is lower in communities with a high background infant mortality rate, because other causes for poor pregnancy outcome dominate (Jenkins, 1990).

Late initiation of antenatal care: Late initiation of antenatal care, especially after 20 weeks gestation, limits access to prenatal diagnosis. Unfortunately, late initiation of antenatal care remains the norm in South Africa (Pattinson et al., 2007). To explore the reasons for low rates of referral of AMA women from primary care to genetic counseling services, an audit was conducted at a single, well-functioning primary care facility in the Cape Metropole West in 2008. The audit showed that, of 61 AMA women delivering at the facility only 19 women (31%) initiated antenatal care within the first 20 weeks of gestation, 10 women (16%) attended the genetic counseling clinic and only 3 women (5%) opted for amniocentesis (Mentoor & Urban 2008, unpublished data). Staff members at this clinic were interviewed and were aware of the existence of genetic counseling services, and of the fact that this is a free service for AMA women. However, staff members had an incorrect understanding of the cutoff age, gave women little information about the reason for referral, and did not feel that referral for AMA was important. On the other hand, they had a positive attitude to referral for fetal anomaly ultrasound scan. Facilities in the same region that serve predominantly the black African population make almost no referrals for genetic counseling, and the reasons for this need further exploration.

There are many possible reasons for late initiation of antenatal care. A study in peri-urban Cape Town found that socioeconomic factors related to the cost of transport and lack of child care, lack of knowledge of the importance of antenatal care, as well as healthcare services that were very busy and perceived to be unwelcoming, all affected health-seeking behavior of pregnant women (Abrahams et al, 2001). A study conducted in rural Kwazulu-Natal found that most women did not perceive significant health threats during pregnancy, and therefore attended antenatal care only near the time of delivery (Myer & Harrison, 2003).

Inadequate referral from primary care: A study by Watcham et al. (2007), conducted in Gauteng, showed that more than half of AMA women were not referred to genetic counseling in spite of presenting to healthcare providers in the first trimester for pregnancy confirmation. In most of these cases the missed opportunities were not in antenatal clinics but in other primary care services (70%) and many of those were in the private sector (60%). This study clearly indicated that poor compliance of primary healthcare staff with existing referral policies considerably limits access to existing prenatal counseling services.

A qualitative study conducted in 2009 with twelve AMA women attending a genetic counseling clinic in Cape Town Metropole West (Perrott, unpublished data) found that women received very little information in primary care regarding either the reason for their referral, or regarding the fact that they would need to make an important decision fairly rapidly, and that it would be valuable to have their partner present. This slowed the process of decision making.

Language barriers: Language differences are another frequent barrier to prenatal diagnosis, since South Africa has 11 official languages, and health providers/genetic counselors are often not able to communicate with patients in their mother tongue. Women who have English as their second language often prefer to receive counseling in their home language even if they speak English quite fluently (Perrott, unpublished data). Unfortunately, despite recognition of the importance of language barriers by the National Department of Health, very few professional interpreters are employed in the public sector and translation is often an informal and inadequate process.

AFFORDABILITY OF PND SERVICES

Although prenatal counseling and diagnostic testing is offered free of charge to eligible women, affordability may be reduced by factors such as transport costs and loss of income on the day of the visit. Conversely, the affordability of having a child with a disability has improved considerably in recent years with the introduction of a means-tested Care Dependency Grant for children with severe disability, which includes DS. Whether the availability of this financial assistance has been important in altering uptake of prenatal diagnostic testing is unknown, but such grants not infrequently provide an important source of income for an entire family.

ACCEPTABILITY OF PND SERVICES

Among AMA women receiving genetic counseling, the uptake of amniocentesis has decreased over the past two decades, as shown in Table 46-2, which compares current data for the genetic counseling service in the Cape Town Metropole West to previously published data for 1992–1993 (Viljoen et al., 1996). The cutoff age for referral was reduced from 39 to 37 years in the intervening years, and there was an increase in absolute numbers of women counseled for AMA. However, the proportion of women counseled for AMA fell, and the uptake of amniocentesis for AMA more than halved—even if the analysis was restricted to women of over 39 years.

Similarly low levels of amniocentesis uptake for AMA are also found in Johannesburg, with recent data indicating that, of 696 consecutive women offered amniocentesis, 208 (30%) accepted (Wessels, personal communication). Data from Cape Town Metropole East were not strictly comparable, since ultrasound soft markers are used to modify age-related risk, but the overall uptake of amniocentesis in this service was 54% for 2008.

The reasons for these changes in uptake, and the low current levels of amniocentesis uptake for AMA in particular, are uncertain. There have been significant shifts in the demographic characteristics of women using public-sector antenatal services: the majority of higher-income women now attend antenatal care in the private sector, which may be important since women of higher socioeconomic status are known to be more likely to accept amniocentesis (Julian-Reynier et al., 1994). The introduction of a democratic culture since 1994 has also significantly increased public awareness of "patient rights" and it is plausible that this could affect reproductive choices as well.

It is expected that the acceptability of prenatal diagnosis is strongly influenced by personal and cultural values. Cultural values are diverse, and limited research has been done on how they impact on reproductive decisions in a South African context. Evidence from the World Values Survey indicates that many basic values are interrelated, such that over 70% of cross-national variation in values can be described by just two "dimensions:" whether the society is traditional/religious or secular/rational, and whether the focus is on survival (economic and physical security) or self-expression. Like much of Africa, South Africa remains predominantly traditional/religious and survival oriented (Ingelhart & Welzel, 2005).

The high levels of traditional and religious beliefs in South Africa are expected to reduce utilization of PND services. To some extent this is borne out by the reasons that women advance for declining invasive testing. Clinic data from 2008 in the Cape Town Metropole West region found that, of 145 who gave reasons for declining an amniocentesis, 94 (66%) would not consider TOP, 50 (34%) felt they would accept the condition and 34 (23%) said that this was on the basis of religious beliefs. A further 28 (19%) were undecided or wanted to consult other family members, but did not return for amniocentesis, and 14 (10%) feared the amniocentesis procedure itself or the possibility that it may result in a miscarriage.

One might anticipate that availability of disability services, and public attitudes regarding physical and intellectual disability, would impact on the use of prenatal diagnostic services. Jenkins (1990) reported on the great differences in unmet needs for services for the intellectually disabled between population groups, with very limited services available to the black African population. There is little doubt that this inequality persists. The issue of disability has certainly become more visible in the democratic era: the rights of people with disability are specifically safeguarded by the South African Constitution, and the profile of physically disabled athletes has increased through the Paralympic movement. On the other hand, a recent survey commissioned by the Special Olympics did not find a particularly knowledgeable or positive attitude to intellectual disability: as in many other developing countries, South Africans tend to underestimate the abilities of intellectually disabled individuals and to believe that their schooling, work, and living arrangements should not be included with the mainstream (Bardon et al., 2006). However, this does not appear to translate into high utilization of PND services.

In HIV-infected women, the risk of mother-to-child transmission of HIV associated with invasive prenatal testing may affect the parents' decision regarding amniocentesis. In a survey of AMA women in 2004 (Bee, 2005), significantly fewer HIV-infected women (4/29 = 14%) accepted amniocentesis compared to their HIV-negative peers (39/100 = 39%). Given the magnitude of the HIV epidemic in South Africa, this difference in uptake rate could have a significant impact on overall amniocentesis rates. It should be noted that this study was conducted at a time when no antiretroviral (ARV) drugs, or only single-dose Nevirapine, was offered as prophylaxis. Subsequently, dual ARV prophylaxis and triple therapy for women requiring treatment for HIV have become more widely available. This is anticipated to reduce the risk of HIV transmission, but its impact on uptake of amniocentesis has not been studied.

Women who already have a child with DS appear more likely to want prenatal diagnosis. Lampret (2006) conducted a study with 50 nonpregnant South African women from diverse backgrounds (but predominantly of black African ethnicity) who had a previous child with DS. Of these women, 38 (76%) felt they would want PND in any future pregnancy and 21 (42%) predicted that they would opt for termination of an affected pregnancy (TOP).

Acceptance of amniocentesis is not synonymous with acceptance of termination of pregnancy. A recent retrospective study in Johannesburg found that, of 116 women offered TOP for fetal abnormalities (almost half were more than 24 weeks pregnant) 73 (63%) accepted (Todd et al., 2010). A decision in favor of TOP was significantly associated with earlier gestation at diagnosis and at offer of TOP, as well as with White rather than Black ethnicity, while there was a trend (which did not reach statistical significance) toward more frequent TOP in younger women. A recent 2-year survey in Cape Town (Stewart, 2008) showed the acceptance rate of TOP for fetal anomalies to be fetal anomaly–specific, and to vary from 17% for congenital heart defects to 63% for holoprosencephaly—for DS with associated fetal anomalies, it was 43%.

BOX 46-7

- Factors limiting use of prenatal diagnostic services in the South African public health sector include:
  - Lack of fetal medicine services in 4 of the provinces

- Limited awareness of Down syndrome and prenatal diagnosis among reproductive-age women
- Lack of timely referral by primary care practitioners or clinics who confirm most pregnancies early
- Reduced uptake of amniocentesis in the context of maternal HIV infection

- Uptake of amniocentesis in women receiving genetic counseling for AMA appears to have reduced over the years and is currently ~30%

FETAL MEDICINE SERVICES

A major change has occurred in the public sector with the dramatic improvement of obstetric scanning services in recent years, due to the increasing availability of ultrasound equipment and the creation of posts for sonographers. A significant advantage of ultrasound is that most women are very keen to make use of existing ultrasound services, while they may not be aware of age-related screening for AMA women. Ultrasound services are, however, still fragmented and access is neither uniform nor equitable at present.

Data on access to these services is difficult to obtain, since many examinations are unrecorded and performed outside formal ultrasound units by self-trained but non-accredited health care workers. Even within formal ultrasound units, the total number of patients reached is unknown, since patients are sometimes scanned more than once, either in the same or at a different facility. Coverage for tertiary obstetric patients is generally very high, but when looking at a whole region, far less than half of all pregnant women in the public sector appear to have access to a routine fetal anatomical scan by a trained sonographer. Figures are much lower for patients at level 1 and 2 facilities and further away from metropolitan areas, with data almost entirely lacking in rural areas.

An exception is found in two large rural districts of the Western Cape: the rural area of the Cape Winelands East and the Overberg district. In these districts a major effort has been made to roll out a routine ultrasound service to remote rural areas by the use of a traveling sonographer. Meticulous statistics show that of the more than 13,200 women who initiated antenatal care in this area in 2008, 53% initiated before 20 weeks and of these, approximately 88.5% (47% of the total) received a detailed fetal scan before 24 weeks. Attempts are being made to expand a similar service in other areas of the Western Cape, and the provincial plan for obstetric services now makes provision for a routine detailed fetal assessment at 18–20 weeks for all pregnant women in the short-to-medium term. This plan involves strict criteria for ultrasound requests, the employment of more traveling sonographers to reach remote areas, defined criteria for ultrasound skills, and the provision of outreach courses to train non-accredited professionals outside the main centers. For optimal use of the service in the public sector, it is essential to establish clearly defined indications and to have measures in place to ensure adherence to referrals in order to limit overuse (Tautz et al., 2000).

Routine obstetric ultrasound examinations are well accepted by patients and staff alike. Male partners more often accompany pregnant women for the ultrasound scan than for AMA counseling. It is even speculated that wider access to ultrasound within a community may improve the rate of early initiation of antenatal care, although there is a need to explore in more detail how the local population perceives the value of ultrasound examinations. It has been recognized in other African settings that clients often overestimate the diagnostic power of ultrasound, and of therapeutic possibilities when a problem is detected (Tautz et al., 2000), and therefore patient education is essential to ensure more realistic expectations and a true informed decision regarding utilization. An evaluation of the expectations and experiences of women having the routine fetal anomaly scan (Patel & Stewart, 2008) found good background knowledge and showed that for 83% of women it was a positive experience, and there was a significant reduction in the mother's concern for the baby's health after the scan.

The value of ultrasound for aneuploidy screening should certainly not be underestimated, since scanning skills in the public-sector ultrasound services are usually quite high. Nuchal translucency screening is only offered in tertiary units, but numbers are very small (less than 2000 cases per year in total) and vary hugely between provinces (from 25 to more than 1200 per year). Karyotyping for ultrasonographically diagnosed structural fetal abnormalities in referral centers is quite uniform, but the use of soft markers for risk assessment is ill-defined and variable, and little work has been done on ethnic differences in their occurrence. For many units the sensitivity and specificity of ultrasound screening are unknown, but some encouraging local data have been published recently. A study on first-trimester screening with nuchal translucency in Gauteng showed a high detection rate (6/7) for a false positive rate of 14% (Naidoo et al., 2008). A recent 3-year survey on ultrasound screening in Cape Town Metropole East showed a prenatal aneuploidy detection rate in excess of 92% (Geerts, 2008) in women of all ages exposed to a formal ultrasound-based risk assessment at any opportunity throughout gestation. The overall prenatal identification rate for autosomal trisomies in the region was 40.4%, with an overall prenatal confirmation rate of 28.9%, which indicated a doubling compared to AMA alone (see Table 46-3).

Prenatal detection of DS was considerably higher using this approach than achieved in the Cape Town Metropole

TABLE 46-3 RESULTS OF ULTRASOUND-BASED RISK ASSESSMENT IN 114 FETUSES WITH AUTOSOMAL TRISOMY DIAGNOSED PRENATALLY AND POSTNATALLY

Trisomy diagnoses	Timing of karyotypic diagnosis	Ultrasound performed	Ultrasound high risk	Ultrasound not done
Trisomy 13 19 cases	Prenatal 5 (26.3%)*	5	5	-
	Postnatal 14	2	1	12
	All 19	**7**	**6 (31.6%)****	**12**
Trisomy 18 24 cases	Prenatal 10 (41.4%)*	10	10	
	Postnatal 14	3	3	11
	All 24	**13**	**13 (54.2%)****	**11**
Down syndrome 71 cases	Prenatal 18 (25.3%)*	18	18	
	Postnatal 53	11	9	42
	All 71	**29**	**27 (38.0%)****	**42**
Total diagnoses 114 cases	Prenatal 33 (28.9%)	33	33	
	Postnatal 81	16	13	65
	All 114	**49**	**46 (40.4%)**	**65**

* p>0.05; ** p=0.07 (adapted from Geerts, 2008)

West region, where genetic counseling was predominantly for AMA or structural defects detected at 20–22 week scan (see Sidebar). The aneuploidies detected after abnormal ultrasound findings were more frequent (1 in 9.4 versus 1 in 34.3 results abnormal) and of more clinical relevance than those found after normal scan results, and the uptake of invasive testing was also higher than when the ultrasound was reassuring.

The trend to increased uptake of amniocentesis after abnormal ultrasound findings was also noted in the audit in Cape Town Metropole West: uptake among women offered amniocentesis after detection of a fetal anomaly was 66%, while it was 54% for positive ultrasound screening only (either first-trimester nuchal or second-trimester soft markers) and 31% for AMA alone. The considerable variations between tertiary centers in uptake rates for invasive testing (after counseling) probably reflect not only differences in culture or counseling, but also in ultrasound assessment.

Additional advantages of routine ultrasound-based aneuploidy screening compared to serum screening are that communication of abnormal results can happen directly and immediately without the need to trace patients in between visits, that many other fetal anomalies will be detected, and that referrals for invasive testing will be at a more optimal time due to accurate determination of gestational age. Absence of any ultrasound markers can reduce the risk in many older women to such an extent that invasive testing may no longer be indicated, and this could significantly reduce laboratory-related costs while maintaining a very high detection rate (Geerts, 2008). As shown by Geerts (Figure 46-2), the main reason for lack of prenatal diagnosis of trisomies in this Cape region was lack of ultrasound assessment, and not missed ultrasound diagnosis.

The current reality is that less than 0.15% of all pregnant women in the public sector undergo an invasive procedure, and there are great discrepancies in numbers performed in the different fetal medicine centers (from 80 to 350 per year) as well as in choice of procedure (e.g., cordocentesis varies from 2% to more than 50% of all procedures performed). In part, this may reflect differences in selection criteria, which are probably responsible for differences in the proportion of abnormal results (varying from 1 in 6 to 1 in 20). Other factors causing variation in the numbers of prenatal diagnostic tests done by fetal medicine centers are the widely variable rate of women receiving antenatal care in different parts of the country, and of early initiation of antenatal care (before 20 weeks). The proportion of women attending at least some antenatal care varies from 64% to close to 100%, and of early initiation of antenatal care varies from less than 30% to more than 60% (South African Strategic Plan, 2008). In addition, varying proportions of antenatal clinic attendees are cared for by doctors rather than nurses (15% to 82% depending on population group and location in rural rather than urban areas), as well as varying incidences of HIV (varying from 15% to 40%) with great variation in measures for ARV prophylaxis.

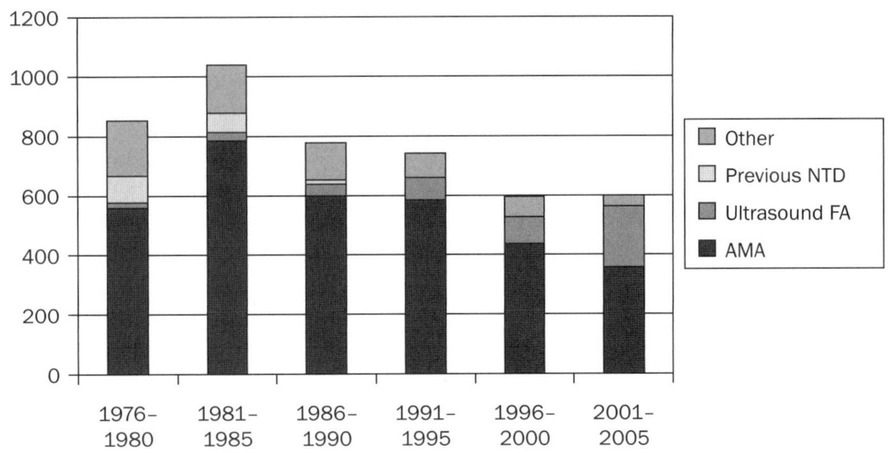

Figure 46-2 Reasons for lack of prenatal diagnosis in 81 fetuses with autosomal trisomy not detected prenatally. (Adapted from Geerts, 2008)

BOX 46-8

- Referral for prenatal ultrasound is well accepted by women and primary care professionals
- In a fetal medicine service that has prioritized the use of prenatal ultrasound (including the use of second trimester soft markers) on a communitywide basis, 38.0% of Down syndrome cases were suspected and 25.3% confirmed prenatally

CONCLUSIONS AND VISION FOR THE FUTURE

There is mounting evidence that for middle-income countries such as South Africa to meet the United Nations Millenium Development Goal of reducing under-5 mortality rate by two thirds, they need to address the issue of birth defects (A Christianson, personal communication). As a component of this, we believe that PND services should become a greater priority in both the public and the private sectors in South Africa, and certain observations need to be taken into account.

1. Although there is considerable variation in effectiveness of PND services by region within South Africa, at least one regional service is showing reasonable impact at a population level. This service emphasizes ultrasound screening and access to PND from primary care. To assess the potential for improving effectiveness in other regional services, further studies are urgently required to obtain broader national data on issues such as women's attitudes to prenatal screening and diagnosis, termination of pregnancy, and disability.
2. Even in ideal circumstances, the strategy of AMA screening is suboptimal, and expected to be associated with a low yield and many missed diagnoses. In addition, AMA screening is not functioning effectively within the public sector, in spite of it being offered as a free service. Current evidence suggests that ultrasound screening and diagnosis is better accepted than AMA screening, but further research is required.
3. Current PND services are not representative of the population's ethnic and socioeconomic diversity. In early studies of PND services, all nonwhite population groups were significantly underrepresented (Jenkins, 1990). Although these differences remain, they have been to some extent replaced by differences between the private and public sector.
4. The lower utilization of PND services by black South African women could certainly be by choice, but our evidence suggests that it is at least in part due to barriers to access. Researchers have highlighted the need for equitable prenatal screening services offered by the state (Gear, 1986), and demonstrated, in France at least, that widespread access to effective screening can reduce disparities in prenatal diagnosis across different ethnic and socioeconomic groups (Khoshnood et al., 2008). Locally, solutions are required to make access progressively more equitable, in line with the South African Constitution.
5. Several barriers to accessing PND services have been identified. Awareness of PND services among reproductive-age women needs to be addressed. Health system factors limiting utilization of prenatal counseling and diagnosis include a lack of availability of PND services in some areas, and a lack of implementation of referral policy within increasingly decentralized primary care services in others. Limited utilization of PND services is at least partly dependent on the knowledge and attitudes of staff serving these

communities, and illustrates how disease prevention relies heavily on education of health professionals.

6. As regards laboratory testing, traditional karyotyping is both costly and operator dependent. Cheaper or more efficient strategies have been initiated in certain localities. In Gauteng province, the use of standalone qF-PCR testing has been implemented. Further research is required.
7. An assessment of screening methods indicates that biochemical screening is still not used optimally in the private sector, and is not really feasible as an option in the public sector. Ultrasound is quite widely available in South Africa, in both the public and private sectors and, in experienced hands, is an effective and efficient screening method for fetal anomalies and for aneuploidy screening. Because it can be applied to women of all ages and over a wide range of gestational ages, it has the potential of greatly increasing the total number of diagnosed cases, as well as reducing the need (and laboratory cost) for invasive testing in selected AMA women.

Some broad points would need to be considered in the development of an effective nationwide screening program. For public health reasons, the most inclusive but cost-effective approach to prenatal screening and diagnosis should be provided, and this would need to be measured against other health priorities. Ideally, a single program should provide screening both for aneuploidies and for other fetal anomalies. The development of such a screening program should include plans to ensure that informed decision making is facilitated regarding prenatal screening, diagnosis, and reproductive choices. In addition, the development of PND services cannot ethically be separated from services for care of physical and mental disability, which would require concurrent attention.

Although there is substantial scope to improve the screening procedure chosen for South Africa, it is clear that only a wider access to screening, and better implementation of existing referral policies, will address the current inequalities and inadequacies of the PND services. Education of both professionals and patients is indicated to ensure optimal use of existing services.

For the private sector, informed decision making by patients deserves attention. The use of serum screening can be increased, and professionals should be educated on the importance of providing complete and accurate data to laboratories to allow optimal biochemistry risk assessment. Combined first-trimester screening (ultrasound + serum biochemistry) should be promoted as the preferred screening method for women of all ages. Nuchal translucency screening in accredited hands is highly effective and should continue, but the number of trained practitioners needs to be expanded and other early markers included. Qualified sonographers are well-suited to meet this demand, and ultrasound skills of general obstetricians should be increased. Collaboration between pathology and genetic laboratories toward a national database would allow determination of sensitivity, false positive rates, and optimal risk cutoffs. This would provide further insight into the causes for missed diagnoses and point to the way forward.

For the public sector, healthcare professionals within primary healthcare services need to be educated to ensure compliance with existing referral policies. A number of theoretical and practical deficiencies of an age-based approach to prenatal screening have been outlined, although it does have the advantage of simplicity. We propose that wide and equitable access to high quality obstetric ultrasound services should become a priority for local health authorities, moving away from "many scans for the lucky few and no scans for many" toward "one scan for all." Age should increasingly be seen as a criterion for referral and further screening, rather than for expensive invasive testing per se.

A highly cost-effective model for prenatal screening and diagnosis is a routine ultrasound service by trained sonographers within the community (Geerts et al, 2004), and this is achievable in the medium term for many urban and surrounding areas, provided the service is well planned to ensure maximally rational use. The Western Cape Provincial plan can also serve as a model to expand access to rural areas. Apart from the many advantages for general obstetric management, this service will allow the detection of a large proportion of all fetal anomalies and serve the purpose of aneuploidy screening. Local data should guide the rational selection of a limited number of the strongest and most relevant ultrasound signs of aneuploidy, and sonographer training should target these skills. Examples of such signs are an abnormal appearance of the crux of the heart, nuchal thickening, shortening of the nasal bone, ventriculomegaly, and abnormal position of the fingers (Geerts, 2008). Written policies should be drawn up locally for the choice and use of ultrasound soft markers, the method to calculate the risk, as well as the cutoff risk that requires referral by sonographers directly to fetal medicine units. This cutoff can be flexible and modified depending on (current and future) availability of local resources. By providing training in basic genetic counseling to sonographers, women at increased risk could even be given high quality information at their scanning visit, which may help them decide whether to accept referral or not (this often involves significant transport costs to the patient).

As fetal medicine services expand, the need for prenatal genetic counseling will continue to increase (this process is already under way). The currently variable links between fetal medicine services, and medical genetic and genetic counseling services, need to be stronger and more uniform. In the context of decentralizing services, it is likely that at some stage, decisions will need to be made as to where (at tertiary centers or more locally) and by which professionals (medical geneticists, genetic counselors, or midwives) this genetic counseling is provided.

Genetic diagnostic laboratory services currently provide full karyotyping for fetal anomalies detected by ultrasound, but for increased risk of DS the approach is more variable (either full karyotype or standalone aneuploidy FISH or qF-PCR). Given our cost-constrained environment, the appropriate match-up between the findings of the screening procedure and the genetic test with the best cost and benefit characteristics will need to be determined.

Causes for inequalities between different regions should be addressed, and an effort must be made to reach out to all levels of care and to include rural areas. Innovative approaches like, e.g., telemedicine to provide Internet-based training in basic obstetric ultrasound skills and interpretation, may be of assistance (Solano et al., 2009). Inequalities in postnatal care for affected infants also needs urgent attention, since mortality rates in certain areas are unacceptably high.

We have felt it beyond the scope of this chapter to discuss family planning services and their important role in the prevention of birth defects, including the fetal anomalies, aneuploidies, and teratogenic disorders discussed in this chapter. It is worth noting that many cases of aneuploidy and fetal alcohol syndrome, especially those diagnosed postnatally, occur in unintended pregnancies and commonly in women of advanced age or high parity. Appropriate family planning for these target groups alone would already reduce birth prevalence considerably. In addition, family planning services could also play a role in the development of preconceptional care and counseling. This is currently undeveloped but could play a significant role in education of reproductive-age women.

In short, although there are active PND services in the larger urban areas of South Africa, much remains to be done in order achieve an effective, equitable, and cost-effective service. We have outlined some directions for the future.

The elements of an appropriate PND service for South Africa should include:

- The aim to provide an equitable, cost-effective service
- An integrated approach for the prenatal diagnosis of both DS and fetal anomalies, based on fetal ultrasound
- Sufficient trained sonographers to make prenatal ultrasound services widely accessible
- Appropriate and accessible genetic counseling
- Cost-effective laboratory testing, possibly based on molecular cytogenetic approaches

ACKNOWLEDGMENTS

We would like to thank the following individuals who provided assistance and information:

Dr. Francois Cilliers, Prof. Greetje de Jong, Dr. Stefan Gebhardt, Dr. Logie Govender, Ms. Shelagh Kavonic, Ms. Adele Kirsten, Dr. Esme Hitchcock, Dr. Anthony Lane, Dr. Hennie Lombaard, Dr. Munro Marx, Prof. Ermos Nicolau, Dr. Bhavna Patel, Dr. Vanessa Perrott, Ms. Wilma Pfaffenzeller, Ms. Elsabe Poggenpoel, Dr. Kogie Reddi, Dr. Jennifer Rosendorff, Ms. Theresa Ruppelt, Dr. Chantal Stewart, Ms. Arina van Litsenborgh, Dr. Leentje van Niekerk, Prof. Denis Viljoen, Ms. Tina-Marie Wessels

REFERENCES

Abrahams N, Jewkes R, Mvo Z. (2001). Health care-seeking practices of pregnant women and the role of the midwife in Cape Town, South Africa. *J Midwifery Womens Health 46*, 240–247.

Bardon JN, Siperstein GN, Parker RC, Corbin SB. (2006). Attitudes toward people with intellectual disabilities. www.specialolympics.org, accessed 10 September 2009.

Bee J. (2005). The influence of HIV status on women of advanced maternal age presenting for genetic counseling [Research report]. Johannesburg: Division of Human Genetics, University of Witwatersrand.

Benatar SR. (2004). Health care reform and the crisis of HIV and AIDS in South Africa. *New Engl J Med 351*(1), 81–92.

Benn PA, Borgida A, Horne D, Briganti S, Collins R, Rodis JF. (1997). Down syndrome and neural tube defect screening: the value of using gestational age by ultrasonography. *Am J Obstet Gynecol 176*(5), 1056–1061.

Benn PA, Rodis JF, Beazoglou T. (1999). Cost-effectiveness of estimating gestational age by ultrasonography in Down syndrome screening. *Obstet Gynecol 94*(1), 29–33.

Bindra R, Heath V, Liao A, Spencer K, Nicolaides KH. (2002). One-stop clinic for assessment of risk for trisomy 21 at 11–14 weeks: a prospective study of 15 030 pregnancies. *Ultrasound Obstet Gynecol 20*(3), 219–225.

Bradshaw D, Bourne D, Nannan N. (2003). What are the leading causes of death among South African children. MRC Policy Brief, number 3.

Bradshaw D, Nannan N, Laubsher R, et al. (2004). South African National Burden of Disease study 2000 – estimates of provincial mortality. Cape Town: Medical Research Council of South Africa, Burden of Disease Unit. http://www.mrc.ac.za/bod/estimates.htm (accessed 2 October 2009)

Bronzan RN, Mwesigwa-Kayongo DC, Narkunas D, et al. (2007). On-site rapid antenatal syphilis screening with an immunochromatographic strip improves case detection and treatment in rural South African clinics. *Sex Transm Dis 34*(7 Suppl), S55–60.

Buccimazza SS, Molteno CD, Dunne TT, Viljoen DL. (1994). Prevalence of neural tube defects in Cape Town. *Teratology 50*, 194–199.

Canfield MA, Honein MA, Yuskiv N, et al. (2006). National estimates and race/ethnic-specific variation of selected birth defects in the United States, 1999–2001. *Birth Defects Res (Part A) 76*, 747–756.

Christianson AL, Kromberg JGR. (1996). Maternal non-recognition of Down syndrome in black South African infants. *Clin Genet 49*, 141–144.

Christianson AL, Zwane ME, Manga P, et al. (2002). Children with intellectual disability in rural South Africa: prevalence and associated disability. *J Intellect Disabil Res 46*(2), 179–186.

Christianson AL, Howson CP, Modell B. (2006). The Hidden Toll of Dying and Disabled Children. New York: March of Dimes Global Report on Birth Defects. Report No: 31 – 2008–2005.

Croxford J and Viljoen D. (1999). Alcohol consumption by pregnant women in the Western Cape. *S Afr Med J 89*, 962–965.

Delport SD, Christianson AL, van den Berg HJS, Wolmarans L, Gericke GS. (1995). Congenital anomalies in black South African liveborn neonates at an urban academic hospital. *S Afr Med J 85*, 11–15.

de Vigan C, Khoshnood B, Cadio E, Vodovar V, Goffinet F. (2008). Prenatal diagnosis and prevalence of Down syndrome in the Parisian population, 2001–2005. *Gynecol Obstet Fertil 36*(2), 146–150.

Dolk H, Loane M, Garne E, et al. (2005). Trends and geographic inequalities in prevalence of Down syndrome in Europe: 1980–1999. *Rev Epidemiol Sante Publique 53*, Spec 2, 2S87–2S95.

Gardosi J. (1997). Dating of pregnancy: time to forget the last menstrual period. *Ultrasound Obstet Gynecol 9*, 367–368.

Gardosi J, Geirsson RT. (1998). Routine ultrasound is the method of choice for dating pregnancy. *Br J Obstet Gynaecol 105*, 933–936.

Gear J. (1986). Screening for disease: is it worth it? *SA J Cont Med Educ 4*, 15–20.

Geerts LT, Brand EJ, Theron GB. (1996). Routine obstetric ultrasound examinations in South Africa: cost and effect on perinatal outcome--a prospective randomised controlled trial. *Br J Obstet Gynaecol 103*, 501–507.

Geerts L, Theron AM, Grove D, Theron GB, Odendaal HJ. (2004). A community-based obstetric ultrasound service. *Int J Gynaecol Obstet 84*, 23–31.

Geerts L. (2008). Prenatal diagnosis of chromosomal abnormalities in a resource-poor setting. *Int J Gynaecol Obstet 103*, 16–21.

Grace HJ, Hitzeroth HW, Op't Hof J. (1984). Maternal serum alphafetoprotein screening in Natal. Results of over 12 000 tests. *S Afr Med J 65*, 8–11.

Grantham-McGregor S, Cheung YB, Cueto S, Glewwe P, Richter L, Strupp B, and the International Child Development Steering Group (2007). Developmental potential in the first five years for children in developing countries. *Lancet 369*, 60–70.

Ingelhart R, Welzel C. (2005). Modernization, cultural change and democracy. New York, Cambridge University Press, p 64.

Irving C, Basu A, Richmond S, Burn J, Wren C. (2008). Twenty year trends in prevalence and survival of Down syndrome. *Eur J Hum Genet 16*(11), 1336–1340.

Jenkins T. (1990). Medical genetics in South Africa. *J Med Genet 27*, 760–779.

Julian-Reynier C, Macquart-Moulin G, Moatti JP, Aurran Y, Chabal F, Aymé S. (1994). Reasons for women's non-uptake of amniocentesis. *Prenat Diagn 14*(9), 859–864.

Khoshnood B, De Vigan C, Blondel B, Vodovar V, Cadio E, Goffinet F. (2008). Long-term trends for socio-economic differences in prenatal diagnosis of Down syndrome: diffusion of services or persistence of disparities? *Br J Obstet Gynaecol 115*, 1087–1095.

Kok P, Collinson M. (2006). Migration and urbanization in South Africa. Report 03-04-02, Pretoria: Statistics South Africa.

Kromberg JGR, Jenkins T. (1982). Common birth defects in South African blacks. *S Afr Med J 62*, 599–602.

Kromberg JG, Bernstein R, Jacobson MJ, Rosendorff J, Jenkins T. (1989). A decade of mid-trimester amniocentesis in Johannesburg. Prenatal diagnosis, problems and counseling. *S Afr Med J 76*, 344–349.

Lampret J. (2006). The reproductive choices of South African mothers who have children with Down syndrome [Research report]. Johannesburg: Division of Human Genetics, University of Witwatersrand.

Losper Q, Theron GB, De Jong G, Steyn DW, Hall DR. (2008). Total perinatal-related losses in the Tygerberg subdistrict, Cape Town. *S Afr J Obstet Gynaecol 14*, 132. (abstract)

May PA, Brooke LB, Gossage JP, et al. (2000). Epidemiology of fetal alcohol syndrome in a South African community in the Western Cape province. *Am J Pub Health 90*(12), 1905–1912.

May PA, Gossage JP, Marais AS, et al. (2007). The epidemiology of fetal alcohol syndrome and partial FAS in a South African community. *Drug Alcohol Depend 88*, 259–271.

Mentoor & Urban M (2008), unpublished data.

Myer L, Harrison AJ. (2003) Why do women seek antenatal care late? Perspectives from rural South Africa. *J Midwifery Womens Health 48*, 268–272.

Naidoo P, Erasmus I, Jeebodh J, Nicolaou E, van Gelderen CJ. (2008). Nuchal translucency as a method of first trimester screening for aneuploidy. *S Afr Med J 98*, 295–299.

National Department of Health, Subdirectorate Human Genetics. (2001). Human genetic policy guidelines for the management and prevention of genetic disorders, birth defects and disabilities. Department of Health, Pretoria, South Africa. URL: http://www.doh.gov.

Ncayiyana DJ. (1986). Neural tube defects among rural blacks in a Transkei District. *S Afr Med J 69*, 618–620.

Nelson M, Henderson HE, Petersen EM, Smart RD, Harris C, Coetzee EJ. (1978). Antenatal diagnosis in practice. *S Afr Med J 54*, 305–309.

Neilson JP. (2002). Ultrasound for fetal assessment in early pregnancy. *Cochrane Database Syst Rev 2000*(2), CD000182.

Palmer C. (1985). Fetal alcohol effects – incidence and understanding in the Cape. *S Afr Med J 68*(11), 779–780.

Patel M, Stewart CJM. (2008). An evaluation of the expectations and experiences of women having the routine ultrasound examination for fetal abnormalities at 22 weeks of pregnancy. *S Afr J Obstet Gynaecol 14*, 140–141.

Pattinson RC. (2005). *The Basic Antenatal Care Manual.* MRC, South Africa.

Pattinson RC, Etsane, Snyman JS, et al. (2007). Report to UNICEF on the scaling-up of the Basic Antenatal Care quality improvement programme in two sub-districts per province in South Africa (Draft). Pretoria: MRC Maternal and Infant Health Care Strategies Research Unit and the University of Pretoria; 2007.

Pattinson RC, ed. (2009). *Saving babies 2006–2007: Sixth perinatal care survey of South Africa.* Tshepesa Press, Pretoria.

Pelser J. (1998). *Knowledge and attitudes of South African women to prenatal diagnosis* [Research Report]. Pretoria: Division of Human Genetics, University of Pretoria.

Patel M, Stewart CJM. (2008). An evaluation of the expectations and experiences of women having the routine ultrasound examination for fetal abnormalities at 22 weeks of pregnancy. *S Afr J Obstet Gynaecol 14*, 140–141.

Retief AE, Bernstein R, Grace HJ, et al. (1983). A 3-year cytogenetic survey of 9 661 patients in South Africa. *S Afr Med J 63*, 48–53.

Roberts CJ, Hibbard BM, Evans DR, et al. (1979). Precision in estimating gestational age and its influence on sensitivity of alphafetoprotein screening. *Br Med J 1*(6169), 981–983.

Rosendorff J, Jacobson MJ, Morris D, Ramsay M, Lane AB, Bernstein R. (1989). First trimester prenatal diagnosis by chorionic villus sampling. The Johannesburg experience with 48 cases. *S Afr Med J 75*, 15–17.

Sanders D, Reynolds L, Eley B, et al. (2007). Western Cape Burden of Disease Reduction Project, Volume 7: *Decreasing the Burden of childhood disease.* Appendix 2: low birth weight. MRC of South Africa.

Sayed AR, Bourne D, Pattinson R, Nixon J, Henderson B. (2008). Decline in prevalence of neural tube defects following folic acid fortification and its cost benefit in South Africa. *Birth Defects Res (Part A) 82*, 211–216.

Smart RD, Jordaan H, Ross J, Nelson MM, Coetzee EJ. (1985). Ultrasound examination before amniocentesis and success of culture in genetic referrals. Six years' experience in Cape Town. *S Afr Med J 67*, 368–369.

Solano M, Kim E-H, Christiansen M, et al. (2009). *Asynchronous Telemedicine with Ultrasound: Improving Maternal Health in Developing Countries.* Institute of Electrical and Electronics Engineers (IEEE) Ultrasonics Symposium (abstract book).

Stewart C. (2008). Termination of pregnancy for fetal abnormality. *S Afr J Obstet Gynaecol 14*, 144. (abstract)

Tautz S, Jahn A, Molokomme I, Görgen R. (2000). Between fear and relief: how rural pregnant women experience foetal ultrasound in a Botswana district hospital. *Soc Sci Med 50*, 689–701.

Todd C, Haw T, Kromberg J, Christianson A. (2010). Genetic counseling for fetal abnormalities in a South African community. *J Genet Couns 19*, 247–254.

United Nations. (2003). *Demographic yearbook series.* New York: United Nations.

Urban M, Chersich M, Fourie LA, Chetty C, Olivier L, Viljoen D. (2008). Fetal alcohol syndrome among Grade 1 School children in Northern Cape Province: Prevalence and risk factors. *S Afr Med J 98*, 877–882.

van Dyk B, Motto JA, Buchmann EJ. (2007). Routine second-trimester ultrasound for low risk pregnancies in a South African community. *Int J Gynaecol Obstet 98*, 257–258.

Venter PA, Christianson AL, Hutamo CM, Mahura MP, Gericke GS. (1995). Congenital anomalies in rural black South African neonates – a silent epidemic? *S Afr Med J 85*, 15–20.

Viljoen D, Oosthuizen C, van der Westhuizen S. (1996). Patient attitudes to prenatal screening and termination of pregnancy at Groote Schuur Hospital: a two year prospective study. *East Afr Med J 73*, 327–329.

Watcham SJ, Schön S, Christianson AL. (2007). Neglect in the care of pregnant South African women of advanced maternal age. *S Afr Med J 97*, 1064–1069.

47.

SOCIOCULTURAL PERSPECTIVES OF INHERITED DISEASES IN SOUTHERN AFRICA

Claire Penn and Jennifer Watermeyer

INTRODUCTION

The South African context provides an interesting and particularly challenging backdrop for the provision of genetic counseling services. Given the legacies of apartheid, poverty, and lack of access, which have profoundly affected most healthcare interactions, as well as the numerous barriers to care experienced by patients, the provision of appropriate and effective services remains a pressing field of concern.

In response to a range of such challenges in the health sector, our research group (the Health Communication Project) has examined communication practices in various crosscultural healthcare interactions and, recently, our focus has included the field of genetic counseling. Some of our findings will form the basis of this chapter, and will demonstrate the powerful influence of context on the delivery of service.

As discussed elsewhere in this volume, genetic counseling is a relatively new profession in South Africa. It was introduced in the 1960s and was offered initially only to the urban educated population (Jenkins, 1990). During the 1990s, some attempts were made to address this imbalance by setting up clinical genetics centers in rural areas (Christianson et al., 2000). However, despite a recognized need for such services, the number of genetic counselors in South Africa remains extremely low and there is a serious lack of available posts for counselors, especially in the public sector. Genetic counseling services that have been established around the country are generally linked to academic institutions and are limited to major cities (Kromberg, Sizer & Christianson, 2010). Thus, access to genetic counseling services remains restricted.

Though genetic counseling interactions have been described as "resolutely medico-scientific," with a focus on the provision of complex biomedical information and the presentation of risk (Rapp, 1988, p.146), our research suggests that social and cultural variables may have a profound impact on the process of genetic counseling interactions and their outcomes. Historically, genetic counseling models from developed countries have been used to train genetic counselors in South Africa and to provide services, but these are not necessarily appropriate given our multicultural heritage and unique disease profile. Our goal in this chapter is, therefore, to demonstrate how context shapes the process and, through some research evidence, we will derive some suggestions for a model of culturally attuned genetic counseling practice.

CONTEXTUAL INFLUENCES

A consideration of genetic counseling in context requires an understanding of a range of influences that may affect practice. Some of the major factors are summarized in Figure 47-1.

This model demonstrates that a number of variables (e.g., historical, political, institutional, and personal) will have an influence on genetic counseling. While these domains are of course not separate but overlapping, they appear to be useful categories to frame our understanding of potential barriers and facilitators to effective service delivery. It is clear that, on a micro level, what happens within each genetic counseling session is influenced by these broader factors.

SOCIOPOLITICAL FACTORS AND HEALTHCARE IN SOUTH AFRICA

South Africa presents a juxtaposition of sorts: the country is advanced industrially and financially, yet disadvantaged and underdeveloped areas remain. The impact of apartheid processes, and specifically the discriminative allocation of resources and denial of access to healthcare, has meant that the health needs of many South Africans are often neglected. Many disparities continue to exist for individuals and groups, in terms of the availability of and access to healthcare services.

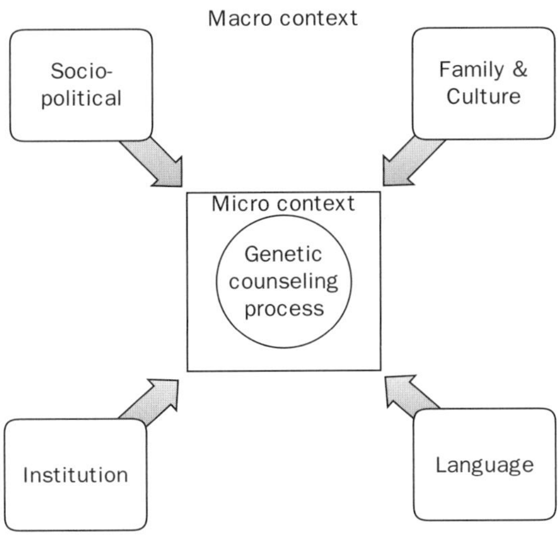

Figure. 47-1 An ecological model of genetic counseling

For example, approximately 85% of the country's population is dependent on public healthcare services (Statistics South Africa General Household Survey, 2007, as reported in Day & Gray, 2008). High rates of unemployment, poverty, and migrant labor practices have compounded these disparities (Coovadia et al., 2009).

The country also carries an exceptionally high disease burden. The epidemic of HIV/AIDS is highest in sub-Saharan Africa, and affects many individuals (UNAIDS/WHO, 2008). With this disease (and with the development and rollout of antiretrovirals) has come a huge necessity for outreach of medical services, community education and intervention, and regular interface between the community and the healthcare sector.

In an attempt to address past imbalances, the government has initiated various health reforms with the goal of achieving equitable access to healthcare. However, much work remains in addressing the "two-class" nature of the health system—"a weak public sector that caters 'second-class' services to that majority of the population dependent upon the state, and a strong private health sector providing 'first-class' services for the wealthy and insured minority" (van Rensburg & Ngwena, 2001, p. 378). Healthcare infrastructure remains poor in many areas, and numerous South African hospitals and clinics are experiencing a human resource crisis because of the health burden imposed by the HIV/AIDS epidemic (Schneider, 2006). Many South African patients thus continue to experience numerous potential barriers to accessing the healthcare system, interacting with health professionals, or even adhering to treatment regimens. These barriers are linked to factors such as HIV/AIDS, stigma and discrimination, poverty, unemployment, gender, pregnancy, language, education, literacy, culture, access to treatment and healthcare, financial resources, different world views, and trust of the healthcare system or health professional (Nachega et al., 2004).

Documented too are a general lack of agency in individuals, and a lack of assertiveness and knowledge about patient rights. As noted by Ellis (2004, p. 44), "in the past many patients' initiative was taken away from them by the apartheid regime which invaded communication in all aspects of South African life, including the health care worker–patient relationship." Cultural norms of politeness and deference to the authority of the health professional also mean that patients are often passive participants in healthcare interactions. For many individuals, the Western biomedical healthcare framework operates alongside an established, centuries-old system of traditional medicine (Herselman, 2007). Some patients may utilize traditional medicine or Western biomedicine only, while others may seek treatment from both systems (a practice known as *dual consultation*; Kale, 1995). Healers have great status, authority, and influence within communities, and patients may refrain from explicitly questioning advice or decisions regarding diagnosis and treatment of disease (Raubenheimer, 1987).

Of particular relevance to genetic counseling is the fact that there are highlighted differences between traditional medicine and Western biomedicine models in the conceptualization of health, disease, and illness. Traditional approaches believe that all diseases have a specific cause, and that it is important to discover who or what caused the disease. This cause may be natural (e.g., germs), through pollution (ritual impurities associated with death, the reproductive system, or violation of sexual prohibitions), from disharmony with the ancestors, the failure of human relations, or witches (van Dyk, 2001).

INSTITUTIONAL CONTEXT OF GENETIC COUNSELING

Genetic counseling usually takes place in the context of a hospital and, as such, is subject to the characteristics of institutional talk described by Drew and Heritage (1992) and others. Genetic counseling happens as part of a chain of events, and must thus be contextualized as an experience for the patient that occurs between other episodes of institutional talk. Such experiences may well influence the dynamics and expectations of the client. For example, institutional talk is typically highly goal-oriented, and does not therefore follow the normal give-and-take of conversational interaction; its vocabulary and syntax place special constraints on communication, and there are frameworks and procedures peculiar to this context. Further, there are recognized asymmetries in health professional–patient relationships, which are heightened by historical inequities in the health system (Lupton, 2003). Barriers to care also include systemic obstacles, such as time and space and the use of a third party in the interaction who acts as interpreter. For reasons described above, many South African patients are not accustomed to models of shared decision making.

As the genetic counseling service is, by definition, embedded into other disciplines (at the interface between clinical and laboratory services), it implies particularly the relevance of good communication at a systemic level, including patterns of referral and follow-up. As Jenkins (1990) notes, in a context of poverty, genetic disorders and genetic counseling services may not be considered important services. In some areas in South Africa, and especially among poorer, disadvantaged patients, there appears to be poor uptake of prenatal genetic testing—studies have found uptake rates of as low as 14% for invasive tests (Wessels & Zwane, 2006) and 29% for amniocentesis (Urban, 2009). Some exploratory work suggests that these poor figures are linked to factors such as late presentation of pregnant women at antenatal clinics (Watcham et al., 2007), poor understanding of genetic disorders in terms of a biomedical model (Lampret & Christianson, 2007), poor understanding on the part of clients and health professionals about the scope and availability of such services, and reluctance to terminate pregnancy (Jenkins, 1990). The general disease burden in South Africa, including the emphasis in prenatal clinics on HIV/AIDS management, may also detract from additional referrals. Further, access to such services may be affected by factors such as distance and lack of transport.

CULTURAL AND FAMILY ASPECTS

Although there is some excellent consideration of cultural issues in relation to genetic counseling in the broad literature (see Clarke & Parsons, 1997), limited research exists that specifically investigates cultural beliefs about genetic disorders in the South African context, nor how such contextual influences may affect the actual process of genetic counseling and on the communication features central to this process. Models of causation, beliefs about inheritance, contextual and family structures, all affect health-seeking behavior in general, and the genetic counseling process in particular (Modell, 1997; Richards, 1996). Perceptions of the features of a disorder, prognosis, recurrence risks, ideas about kinship and inheritance, possible alternatives to transmission of the condition to the next generation, as well as beliefs and practices, are some of the issues which are fundamentally influenced by culture (Meiser et al., 2008; Weil, 2001). Prior research also suggests that there are intergenerational differences regarding models of illness causation, which may influence treatment-seeking paths (e.g., Liddell et al., 2006).

In many cultures "the patient" is not just the individual but the whole family. In South Africa, the family, and particularly the older generation, are often closely involved in caring for children and in many cases may act as primary caregivers to a child (Barratt & Penn, 2009; King, 2008). Elders hold a position of great respect in the community and the family (Kasanga & Lwanga-Lumu, 2007) and they are often the first to be consulted when there is illness in the family. The birth of children is very important because of the need to sustain a clan and a family tree (Ingstad et al., 1997). Further, as Petersen points out (2006) there may be cultural differences regarding autonomy and protection of confidentiality in families (Forrest et al., 2003; Clarke & Parsons, 1997), and a tendency toward collective decision making and a social responsibility for health in some groups (Ware et al., 2009). A woman coming alone to the genetic counseling clinic after referral is being asked to make a decision that may well be subsequently queried or challenged by the family.

In a similar way, religious beliefs have been found to impinge significantly on clients' understanding of the cause of genetic disorders, their acceptance of genetic counseling and recommendations, and their acceptance of a child with a disability (see for example Greeson et al., 2001, who report on the religious beliefs of Somali immigrants, and Raz and Atar, 2003, who discuss religious beliefs among Jewish and Arab populations).

LANGUAGE AND COMMUNICATION ASPECTS

South Africa is a country rich in multilingual diversity, with eleven official languages. Although English and Afrikaans remain the two most prominent languages spoken in the public and healthcare sectors, the majority of the population is not fully proficient in these languages. For example, language audits at a number of hospitals revealed that few patients have English or Afrikaans as a first language, and few doctors are able to conduct the interviews in the home languages of their patients (Deumert, 2010). Even when healthcare professionals and patients are culturally and linguistically matched, interactions may take place in a language that is not the patient's home language (Watermeyer & Penn, 2008). Therefore, for the most part (at least in the public sector), there is a lack of concordance between the language and cultures of health professionals and their clients, and the situation is no different in the field of genetic counseling. Aside from the eleven official languages spoken in the country, a number of other African languages, as well as European languages such as Portuguese and French, are encountered in everyday healthcare practice. We have doctors from other parts of Africa, Cuba, and Eastern Europe, who cannot speak any of the local languages very well, and migration patterns have led to an increasing number of patients who come from other African states.

Despite constitutional recommendations for access to interpreters, a lack of institutional resources and infrastructure means that the provision of interpreting services is sorely lacking (Levin, 2006). Our research has demonstrated a variety of institutional language practices, and differences in the preparedness of clinicians and interpreters (Evans, 2000; Fisch, 2001; Penn, 2007). Common practice indicates that in most cases a trained interpreter is usually not available, and, at best, interpreting is performed in an *ad hoc* way using nurses, counselors, family members, cleaners, or fellow patients.

In the field of genetic counseling, language barriers can be particularly problematic, as sessions generally involve the provision of large amounts of complex information often based in modern scientific terminology, descriptions of complicated medical procedures, expression of uncertainty, and explanations of difficult concepts such as "risk." These terms are not necessarily amenable to translation. Further, the nature of the setting, power relations, the materials, and the task often lead to the use of English. Those seeking healthcare services often have low levels of literacy and education, limited exposure to the media, and a lack of knowledge of medical terminology. Several studies that we have conducted across diseases and healthcare contexts in South Africa (summarized in Penn, 2007; Penn et al., 2011) have highlighted the pivotal roles played by communication, language, and cultural barriers, in terms of access to care and the successful outcomes of the consultation and the care provided.

To summarize, there are a range of contextual features that provide a backdrop for understanding barriers and facilitators to health interactions in general, and genetic counseling in particular. Many prior approaches to the issue of culture in genetic counseling appear to have been predicated within a so-called "crosscultural" framework, in which beliefs and practices of different cultural and ethnic groups are contrasted. This terminology implies an "othering" of clinical and cultural groups, rather than an awareness both that cultural factors impinge on all clients, regardless of language or ethnicity (Daniels & Swartz, 2007), and that health professionals and clients bring their own values, beliefs, and experiences to an interaction. A move toward a more culturally embedded model is clearly required. We will argue that South Africa, with its very diverse cultural and linguistic heritage, provides an exceptionally fruitful testing ground for the study of such issues, and we will claim that the development of local rather than universal models of genetic counseling is a desirable goal.

SOME EVIDENCE

The main source from which we draw our data is the body of research emerging from the Health Communication Project. In a series of studies over the past decade, some of the challenges to effective communication have been highlighted by research at several sites. This research has used qualitative methods derived from anthropology and linguistics to explore the dynamics of same-language and cross-language intercultural health interactions, across a number of different professions and contexts countrywide.

Our methods of data collection include audio and video recording of healthcare interactions, measurement of participant perceptions (including patients, interpreters, caregivers, and health practitioners) through interviews, observation and description of content and process factors using ethnography, and development and measurement of training programs and materials. Analysis methods have been drawn from approaches such as Conversation Analysis, Discourse Analysis and Thematic Content Analysis. Further details regarding the methods and findings of this work may be found in Penn (2007) and Penn et al. (2011).

Two lines of evidence from the field of genetic counseling will be described here. The first is some data from a series of studies conducted on cultural beliefs about genetic disorders. The second illustrates some interactional dynamics in intercultural prenatal genetic counseling sessions.

EXPLORING TRADITIONAL BELIEFS ABOUT THE CAUSATION OF GENETIC DISORDERS

Several studies from the project have yielded information pertaining to traditional beliefs about genetic disorders. The first, described as our "Grandmother Project," involved focus-group research with a group of 15 grandmothers living in an urban community. This study explored traditional beliefs about the causes of various common childhood genetic disorders, as well as traditional methods of treatment, management, and prevention (Penn et al., 2010).

The participants represented a wide cultural and geographical mix: the study included grandmothers who originated from different cultural groups across five of the nine provinces in South Africa, including Sotho, Xhosa, Tswana, and so-called Coloured groups ("Coloured" is a term used to describe a group of people in South Africa with mixed ancestry, often including sub-Saharan, European, and Asian ancestry). The disorders we discussed included Down syndrome and albinism—two of the most prevalent congenital conditions in the South African population (Christianson, 2000)—as well as cleft lip and palate, club foot, congenital deafness, and cerebral palsy. (While the latter is not generally considered a genetic condition, some recent research does suggest the involvement of a genetic element in impaired cortical function; see Morris-Rosendahl, 2009). Another source of data on causation is derived from a study of cerebral palsy in a rural area where caregivers were interviewed about their experiences (Barratt & Penn, 2009). Finally, we draw from Solomon's study (2009),of beliefs about hemophilia among a group of Xhosa-speaking caregivers in the Western Cape province. The main findings of this research are summarized in Table 47-1.

MODELS OF CAUSATION

The grandmother study revealed a wide range of different beliefs held by the different cultural groups about the causes of genetic disorders. The participants were generally familiar with all of the genetic disorders and conditions that were discussed. However, as Table 47-1 shows, beliefs were not specific to individual cultural groups. In addition, the beliefs discussed by the participants were not necessarily specific to each condition, but were often related to disability in general.

The participants discussed a number of different causes for genetic conditions, which we have classified into themes.

1. Lifestyle and Behavior-Related Causes

The participants frequently cited eating culturally forbidden foods as a cause of genetic disorders or disability. In particular, young unmarried girls should not indulge in forbidden foods such as eggs, tripe, animal innards, duck, or duck feet, as these are thought to possess aphrodisiac properties and may "*cause your body to develop too quickly.*" Only married people are permitted to eat certain foods. Eating these foods might also cause a child to be born disabled – "*So if you eat as you please, it will be said that your child is disabled because you ate inappropriately.*" One participant related a story of a disabled child: "*The child passed away, the feet were like a duck's... they say the mother liked eating ducks... this is why the child's legs and fingers were like a duck's.*"

In addition, not eating nutritious foods during pregnancy ("*true Tswana food*"), as well as smoking, drinking, and taking too many pills, can cause disability – "*sometimes it's the things you eat or drink.*" The use of alcohol

TABLE 47-1 TRADITIONAL BELIEFS REGARDING CAUSES OF VARIOUS GENETIC DISORDERS

	Name given by participants	Causes discussed by participants
Albinism	*inkau, maswefe* (albinos) albinism	· not eating correct foods · alcoholism · smoking during pregnancy · advanced maternal age ("mother too old") · punishment from God or ancestors · gift from God · in the family · born with first layer of skin and not second layer
Club foot	*sebopiwa, segole* (God's creature) retard disabled child family children	· mother was mischievous and went to the zoo in secret; child was therefore born disabled · advanced maternal age · God's gift · laughing at others in the street or specifically at disabled people · mother did not eat well or ate culturally forbidden foods (e.g., eggs, tripe, duck's feet) · cultural rituals (especially rituals of the father's family) were not performed, therefore child is not whole/complete · incest
Cleft lip	family children	· consanguinity · "something wrong inside," a natural thing · punishment from God · something you eat or drink during pregnancy · being hit while pregnant
Cleft palate		· cancer · God's creation or a gift from God · incorrect eating or drinking during pregnancy · taking pills while pregnant (teratogens)
Down syndrome	*sebopiwa* (God's creature) *isidatwa* (cripple)	· combining different blood (across cultures, or incest) · heredity (both mental retardation and physical disabilities) · advanced maternal age · alcohol · wearing tight clothes during pregnancy
Cerebral palsy		· heredity (both mental retardation and physical disabilities) · incorrect eating or drinking during pregnancy · family children · being hit while pregnant · child not immunized, therefore became ill (linked with polio) · *umuthi* (medicine or poison) (Barratt & Penn, 2009) · birthing difficulties (Barratt & Penn, 2009) · gift from God (Barratt & Penn, 2009)
Deafness		· unsure about cause · "born like that," "it just happens"
Hemophilia (Solomon, 2009)		· unsure about cause · result of injury · result of medical treatment · a type of cancer
General discussion of causes of disability		· mother got a fright while pregnant (e.g., she saw an animal) · eating forbidden foods, e.g., ducks or duck feet · not eating nutritious foods during pregnancy · child is "born tired" · domestic violence during pregnancy · blood is dirty from sleeping around too much · punishment from God (esp. linked to promiscuity) · not following cultural practices, therefore infants are stillborn · not allowing a period of healing after birth/pregnancies in quick succession can lead to disability · child conceived in impurity can lead to disability

Penn et al., 2010; Solomon, 2009; Barratt & Penn, 2009.

during pregnancy was referred to fairly frequently as a cause of childhood disorders. One participant stated that "*alcohol maybe, or smoking, they say it is not right when you are pregnant*," but her use of the word "they" implies that this is not necessarily something that she herself believes, but rather something that she has been told, perhaps by a health professional. In Wessels' ongoing study of prenatal genetic counseling interactions (Wessels & Penn, 2008), one client discussed a previous miscarriage and attributed this to eating incorrectly and indulging in foods such as sugar. This client appeared to experience a high degree of guilt over her actions.

Other lifestyle-related causes were related to behaviors such as wearing clothes that are too tight during pregnancy. For example, one participant related that "*Today our children wear pants that are so tight. It's those pants that choke them and then the children are born with squint eyes.*"

Promiscuity was also suggested as a cause of disability, specifically that a child "*conceived in impurity*" might be born disabled as a punishment from God. In addition, getting pregnant rapidly after a birth, and having "*dirty blood*" as a result of promiscuous behavior, may also cause disability, as one participant related: "*Our children are not strong because they do not obey as we did. They rush into things; that's why they lose their babies. She has not completed her cleansing so her blood is still not clean. You must clean her first, then she may begin with that stuff (reference to sex). If you are not honest, you will never come right. She doesn't even come back home, she goes to spend the night. You understand how dishonest that is, looking after yourself. If it happens that the after birth is still intact, and she is with that boy, maybe she falls pregnant again, is that not already resulting in a disability? That child will be born deformed because it was conceived in impurity and that is dishonest! If you are honest, all will fall into place. I am sure even the doctors warn them against this Monday Tuesday business. You must give the first child a chance to grow before you make the next one. Our children don't heed this, they come from hospital with the baby and three days later she is sleeping around. She is breastfeeding, and nine months later another one. These are the kinds of things that lead to deformed babies. You have not healed or recovered and yet you're pregnant again.*"

These results confirm the findings of Ingstad et al. (1997) and others who reported on the widely held belief in traditional communities that at critical periods after confinement, delivery, and abortion, sexual abstinence should be observed and then purification rituals followed. Failure to do this will anger the ancestors, and this will lead to punishment by the ancestors in the form of illness including birth defects. The participants' reference to "*dirty blood*" also confirms previous research describing how some cultures believe that a mother can acquire "bad blood" or become polluted after engaging in sexual relations with a man who has slept with another woman, and this may cause something to be wrong with the mother's uterus and consequently with the infant (Haram, 1991).

Some interesting explanations emerged regarding the behavior of a pregnant woman that may lead to her child being born with a disorder or disability. For example, laughing at someone with albinism may cause a woman to give birth to a child who has albinism; similarly, laughing at a disabled person may lead to a disabled child being born. One participant mentioned that pregnant women should not be allowed to visit the zoo, as they might get a fright that would affect the baby, or their child might be born with the appearance of a monkey: "*There is a woman who gave birth to a boy who looks like a monkey; he actually really looks like one—a monkey-man. They say his mother used to secretly visit the zoo.*" In rural areas, there are reportedly restrictions placed on pregnant women because there is fear that being frightened by something (e.g., a wild animal) may affect the pregnancy.

Interestingly, the participants acknowledged differences in belief and lifestyle not only between themselves and the younger generation (e.g., "*we wore loose clothes,*" "*children no longer listen to their elders,*" "*our children are not strong because they do not obey as we did and that's why they lost their babies,*" "*we believed that babies came from airplanes*"), but also between themselves and their own grandparents ("*the old people*") who lived long lives and were "*not lazy like us.*" Although they considered themselves industrious, they did not feel that they worked as hard as their own grandmothers did.

2. *Social Causes*

Abuse and domestic violence were cited as causes of disabled children. Some participants told stories about women who had been assaulted during pregnancy, and had subsequently delivered disabled children: "*There's another woman, she's dead now, her sister hit her with the flower vase. And then when the child was born the child had a crooked head.*" "*There is this tendency, when they are pregnant their husbands beat them. When they beat them, the baby is affected and will arrive with a scare.*"

3. *Family Causes*

Consanguinity was discussed in several of the groups as a possible cause of genetic disorders, and this was distinguished from incest: *Interviewer: "So the main cause of this would be incest?" Grandmother: "No we just say it's a family child, same blood."* There was acknowledgment that some problems come from "*intercultural relations*" or "*sleeping with your brother*" and that the offspring have the "*same blood*" and are called "*family children.*"

4. Cultural Causes

A number of causes related to culture and tradition were discussed by participants. Not following cultural rituals, especially those of the father's family, were often cited by participants as causes of disability: *"Often it happens when the mother does not perform the rituals of the father's family." "If you have a child with someone in whose family these rituals are not performed, it will be said that your child is not whole."*

These rituals were spoken about by some participants, and they appear to include practices such as cleansing and abstinence as part of a mourning process after miscarriage or the death of a husband (confirmed in Barratt and Penn's 2009 study): *"Your family does not fix...not even a simple bath [to cleanse]...you will just take plain water in the morning and bath...there are some traditional herbs you must still take...and mourning, you must spend at least one month without lying with a man...so these things are what causes these strange things to happen...like for instance children born with abnormalities."*

Participants reported that promiscuity might lead to punishment from the ancestors in the form of a disabled child: *"What big offense must you have committed for the ancestors to punish you with a child like this?" "Ancestors get angry with you when you do wrong things; that is why they give you a disabled child." "Sometimes it is not that they are angry but that they see you are becoming a loose woman...maybe they are forcing you to grow up."*

Another explanation related to punishment was that if a mother had one disabled child but she did not love this child, then the ancestors might punish her and give her another disabled child: *"...there is another [woman]...the younger sister was telling me that her sister doesn't want the albino...she didn't love the child, didn't need her, when the child came she would keep quiet, she got another one, now she has two...then, this shows punishment...if you don't love her, feeling disgusted."*

Solomon's (2009) study revealed that an illness may be a warning from the ancestors that a specific ritual or ceremony needs to be performed or that a ritual may have been overlooked. The child's illness may then disappear or be cured when the ancestors are appeased. This belief, as reported in Solomon's (2009) study, appeared in some cases to be linked to anecdotal evidence in the community. Jewkes and Wood (1998) also discuss various rituals that may be performed in certain communities when a child is born, in order to introduce the child to the ancestors and gain protection from illness. Interestingly, and perhaps contrary to what is widely assumed about African cultural beliefs regarding illness and disability (Herselman, 2007; Masasa et al., 2005), the participants in the grandmother project did not discuss witchcraft or evil spirits as a cause of genetic disorders or disability. However, this was mentioned as a cause of childhood disability by some participants in Barratt and Penn's (2009) study, as was *umuthi* (a substance or poison used to achieve either healing, or illness and ailment).

5. Religious Explanations

Religion appeared to play an important role in discussions of causality, and disability was seen by some participants as a gift or punishment from God (depending on individual perspectives and cultural beliefs). One participant indicated that although having a disabled child is considered a punishment, it is more appropriate to refer to the child as a gift: *"It is thought to be a punishment from God, though that is not 'kosher' so then it is sometimes thought of as a gift from God."*

Participants also referred to the Bible when discussing topics such as infanticide and consanguinity. References were also infrequently and rather uncertainly made to disability being "God's will": *"Maybe, with other people it's God's will, we don't know." "If it is yours, you must accept that, you can't question God, why God gave you a baby like that, you accept the baby."*

There was some discussion and disagreement about whether it is the ancestors or God who plays a role in causing disability. In one group, there was also disagreement as to whether ancestors do in fact exist: *Interviewer: "What do you know about genes or do you not believe in them and believe it is the ancestors?" Grandmother 1: "I believe it is God, I don't believe in these ancestors because they are human like me. When they die, we call them ancestors. They play no role in my life, but God does." Interviewer: "Tell me, do you also believe that ancestors don't exist and play no significant role?" Grandmother 2: "I think it is all God's hand." Interviewer: "What about the ancestors?" Grandmother 2: "Hey, never, never! Ancestors are just dead people. When this kind of thing happens, it is God's doing." Interviewer: "Because he created all things. So tell me mama, what have you to say about this matter?" Grandmother 3: "As I understand it, ancestors do exist." Interviewer: "Can the ancestors give you a disabled child?" Grandmother 3: "No, I don't think so. I don't really know whether they are involved or not so I don't want to say too much about it. In my experience...people believe in ancestors for other things like callings. But in such a matter, I believe it can only be God. I agree with...I agree that it is only God who gives."*

The participants in both Barratt's and Solomon's studies also make references to God and religion. In Barratt's study, some participants referred to their disabled children as "gifts from God." Interestingly, Solomon's study also highlights the conflict between traditional ancestor beliefs and the church, particularly related to approaches to treatment.

6. Genetic Explanations

Participants recognized advanced maternal age as a cause for some of the disorders discussed (particularly in relation to Down syndrome) – *"Maybe the time has come not to have a baby because of age…so maybe when you are over age you are not healthy to have a baby."*

The participants appeared to have a good understanding of the concepts of "heredity" and "genetics," and they expressed this as a possible cause of some of the disorders. Explanations of "hereditary" included *"getting bad genes from your parents," "someone in the blood line was like that too,"* and *"because of my blood, my child can have any characteristic of mine."* One participant explained that she had a history of mental illness in her family, which she attributed to heredity: *"But another [cause] is heredity. I say this because I don't know about my grandmother's grandmother or who she was. She could have been disabled mentally. Perhaps there was a man in my blood line who was ill and it got passed on to my mother. It keeps going down. I am my mother's first-born and I am fine. There were six of us and the last-born was mentally ill. Our sister gave birth to a mentally ill child. I was born of a mentally ill person and I turned out to be normal. My sister after me is also normal. The last-born was mentally ill and her last-born is also mentally ill. It keeps coming down like that. And with my own children, to prove that it is indeed hereditary, there is some sign of mental illness…even physical disabilities are hereditary."*

When asked about their understanding of "genes," participants were able to link the concept of "genes" to "heredity" and "inheritance": *Interviewer: "Okay, do you know about genes? What are genes?" Participant: "Maybe if he inherited something, like the illness. Family illness. Maybe you inherited something from your granny."*

The gendering of cause was interesting, and there appeared to be some conflict among the participants as to whether genetic disorders were passed down through females or males. One participant reported that good things come from men and bad things come from women. This belief aligns with the work of Ingstad et al. (1997) and Solomon (2009), which report that the woman is always blamed, because according to cultural beliefs, pollution lies in the womb of the woman. However, several of the grandmothers, as well as participants in Barratt's study, reported the father as a source of disability or disorder: *"We are often told it is the father, that is why they say the father's family has given this gift. Perhaps there was a man in my blood line who was ill and it got passed on to my mother."*

Participants also referred to *"something [being] wrong inside"* or *"while inside the womb… maybe he/she was born like that"*—i.e., a congenital malformation in the infant or something being wrong with the mother. Another point was raised: that a child would be more likely to be disabled if the mother failed to go the antenatal clinic.

PREVENTION AND MANAGEMENT STRATEGIES

A number of potential strategies for prevention and management of genetic disorders and resultant disabilities were suggested by the grandmothers.

Infanticide, or killing disabled infants at birth, was mentioned by some of the grandmothers, although the topic appeared to be rather taboo and was sometimes spoken about indirectly using gestures or references to infanticide in the Bible: *"Something was wrong inside. The child was born on Sunday then they called me Monday, I must come look. We came there, my sister and I and my aunt. Then the child looks so* ((points to picture)) *and then my aunt said they should* ((gestures to indicate 'letting go')) *because he doesn't even have legs, how will he even walk?"*

One participant indicated that this was a common practice in previous generations, but because infants are now born in hospitals rather than at home in the rural villages, infanticide is no longer possible: *"We gave birth in the rural areas. Nowadays we give birth the Western way and the number of such [disabled] people has increased. The doctors give you the child as is. [In the old days] they (midwives/reltives) would say the child was born tired. They would blame you, the mother, saying 'we told you to open wide but you tightened, you suffocated it'* ((laughter))*… That was their way of distracting you."*

A number of traditional medicines were mentioned across the groups, although among the African participants there appeared to be a reluctance to discuss specifics of the remedies: *Interviewer: "I want to know about traditional things, those traditional herbs, that is what we want, you know, expose them." Participant: "No, I don't want to expose it." Interviewer: "Yes, we are supposed to expose it."*

Grandmothers appear to play an important role in selecting and preparing traditional medicine. Some of the remedies that were discussed include the following:

- "Dassie pis"—*"Dassie pis is a wood, then you crush it, and eat it. My granny used to cook it and then she gave it to the child, then she chewed it and spat the child wet with it."* This was described as "*serokolo*" in Sesotho. From Afrikaans, "*dassie urine.*" A dassie is a small rock hyrax, a common small African mammal. The herbal concoction presumably smells similar.
- *Harmsdruppels, lewensessens, wonderkroon*—Afrikaans words for old Dutch remedies used during menstruation.
- Various herbs ("*inhlonyani*") or roots, which are dug up, cooked, and mixed into traditional remedies—"*herbs they would give to children to drink and get well… They trusted the herbs… my grandmother used to help those people who couldn't conceive, when the*

person was unable to have a baby, like when you were going to menstruate you first get sick, she was able to cure that, they call it period pain."

Such remedies align with the range of traditional medicines reported by Jewkes et al.'s informants in their 1998 study of healthcare-seeking practices in the Cape.

Other suggested prevention measures included the following:

- delaying sexual maturity (by not eating culturally forbidden foods, which are thought to act as aphrodisiacs);
- cleansing rituals after birth, miscarriage, or widowhood;
- abstinence or "cooling off."

As with the discussion of causes, the management strategies suggested by the grandmothers were not mentioned in relation to any one specific genetic condition but were considered to be general ways of managing disease and reproductive responsibility. However, there were some interesting features unique to specific diseases, and our data conformed to some of the work in the area of albinism related to infanticide (e.g., Mesaki, 2008, as cited in Stensson, 2008). The findings on cerebral palsy linked to attitudes expressed by informants in Barratt's study, for example, resorting to infanticide or seeking traditional healing and herbal remedies for a disabled child.

Interestingly, the grandmothers did not directly mention taking a child to a traditional healer as a possible treatment strategy, although it is assumed that some of the traditional medicines they mentioned might well be obtained from a healer. In Barratt's and Solomon's studies, there was more explicit discussion of the role of the traditional healer. For example, in Barratt's study, two-thirds of the participants had taken their child with cerebral palsy to a traditional healer on the advice of elderly relatives, although no improvements in the child's condition were noted in any of these cases. In Solomon's study, participants indicated that the traditional healer's role may not necessarily be to cure the child's condition or provide treatment, but rather to give advice about the reason for the child's condition.

In the grandmother study, there was, however, an acknowledgment of parallel routes of healing—even in the participants' own discourses there did not appear to be a conflict or competition between parallel routes of explanation for the disorders studied, and in some of the groups it was acknowledged that both Western and traditional healing systems could and should be accessed: *"These doctors, you could swear they create these illnesses or they conjure them up…so if we obey tradition we can help the doctor." "Yes, the doctor is in line with the customs, just like God is in line with the sangoma"* (a traditional healer).

There appears to be acceptance of mixed beliefs in ancestors and God, natural causes and God, and following rituals and God for healing (dual consultation). There are various reasons for the existence of this practice: clients may desire to obtain different perspectives on their illness, or the traditional healer might be able to identify a cultural cause or conduct a healing ritual, while the doctor provides medication for the symptoms.

ATTITUDES TOWARD TERMINATION OF PREGNANCY

There was some discussion about abortion or termination of pregnancy, related to diagnostic technologies for detecting disability prenatally, and how the grandmothers' communities would respond to a woman who decides to terminate a pregnancy.

Attitudes toward termination appeared to be mixed. One grandmother reported that in her culture, a woman would far rather have a disabled child than have a termination, because of the stigma placed upon abortion and miscarriage and the value attached to having children and a family tree. Some religious references were also made; specifically, that aborting a child may be considered a sin: *Interviewer: "Now how do you see that, if there is something like this doing abortion to you, the child being like this, they will tell you, how they will take it out?" Grandmother: "I will rather say they leave the child until I give birth you see. It's your child, because it is a God-given thing. It's a Godly thing. Because if I allow them to do abortion, it is a sin before God."*

One grandmother, however, indicated that the option of termination might be better in the case of some causes of disability (e.g., when there are physical disabilities) because the problems of caring for a child and physically managing the child may be too great (*"we would have to lock it up"… "we would have to carry him"*).

FAMILY INVOLVEMENT IN DECISION MAKING

The role of grandmothers in keeping the family name was particularly interesting. In families where there are no sons born, the grandmothers would look around for a suitable family bride (a *makoti*) who would be expected to bear children (preferably sons). This *makoti* would then find someone to be her secret lover, a man not known or chosen by the family. The resulting children would then carry the family name. The grandmothers would then keep the secret and, according to the participants, *"this cover-up worked like a charm"* because the mothers and grandmothers were the ones who raised the children while the men were away working in the cities.

Participants also described how, in families where a son could not bear children, the grandmothers in the family will intervene by telling the son's wife to find another man (a secret lover), but to be discreet about the affair. When the wife then falls pregnant by this man, the family will automatically accept the child as the son's, and the child will carry the family name. This process was facilitated by the fact that African men reportedly are often vague about the timing of pregnancies, and it is not customary to indicate exact dates of delivery because of susceptibility to witchcraft (Mathole et al., 2004). Grandmothers also shielded extramarital affairs and, if questioned or in a suspect case, they would indicate that the resulting grandchildren looked like a long-dead relative.

Grandmothers are also involved in the birthing process, and participants described how grandmothers hold wisdom and authority in relation to the identification and management of disability. In previous times they could make the decision to terminate the life of a child born with a disability or obvious congenital abnormality. It was also said that the grandmother would then fabricate a reason to explain the child's death to the mother; e.g., "*the child was too weak.*" Grandmothers may thus play a potential role as gatekeepers who hold the power of life or death.

The grandmother study also revealed the involvement of the family and the community in termination-of-pregnancy decisions. For example, abortion and miscarriage require a cleansing ritual, including a gathering of the whole family, cleansing of the house, and the slaughtering of a goat. Thus, a woman's decision to terminate a pregnancy would not be a confidential or individual decision because of its implications for the family and the community.

CULTURE IN THE GENETIC COUNSELING INTERACTION

The above discussion suggests that within a particular context there may be a number of sociocultural influences that may impinge on the beliefs and perspectives of the client who is referred for genetic counseling. The question remains as to how such factors may influence the effectiveness and outcome of the actual process of genetic counseling in an intercultural context. How does the genetic counselor incorporate an understanding of these variables into the session, and how can a model of shared decision making be reached?

By way of exemplification, we will present extracts from a recorded single crosscultural genetic counseling interaction (Tovell, Wessels, & Penn, 2005) which highlights some of these challenges. We have used an analysis technique similar to that used by Pilnick (2004), and our transcription symbols are presented in the chapter Appendix. The extracts are from a session with a female client aged 23 years in the late stage of the first trimester of pregnancy. She was born in Swaziland, used Zulu as a home language, and was living with her partner, an immigrant from Nigeria. She had some tertiary education at a computer college, and at the time of the research was employed as a waitress. Her reason for attending the clinic was that she had given birth to a baby with a neural tube defect, specifically anencephaly, three years previously. Given the relatively high risk of having another baby with this deficit, this mother had been referred to the clinic to receive genetic counseling regarding the options of prenatal diagnosis. She arrived alone for the session and indicated that she was satisfied to receive this counseling in English, and she declined the offer of an interpreter. The counselor was an experienced genetic counselor whose home language is Afrikaans. The session proceeded in a standard way, moving from the opening and introduction to a collection of demographic and case history information, followed by patient counseling and education, an explanation of various procedures, and discussion of follow-up and recommendations. Interestingly, while cultural issues were for the most part not discussed explicitly in the interaction, such factors seem to frame the whole interaction, as the following illustrative themes from the transcript demonstrate. Notable is the asymmetry in the session and the high number of questions asked by the counselor, which is indicative of the contrasting communication styles of each participant.

Background and Family Structures

As the genetic counselor obtains the pedigree or family tree from the patient, it becomes apparent that the patient comes from a large extended family in which polygamy is practiced. In the following extract, for example, although this is not explicitly discussed or commented on by the genetic counselor, she appears to understand that this is the case in this family situation.

> 222 GC: But now your mother had children with another man.
> 223 Pt: mm.
> 224 GC: Okay. How many? Is your mother and your father divorced?
> 225 Pt: No.
> 226 GC: They still together.
>
> ...
>
> 239 GC: And then your *father* had children with another wife. How many?
> 240 Pt: Four.
>
> ...

247 GC: So your father must be happy that he has the two girls from your
248 mother? ((smiles))
249 Pt: ((laughs))

In the next interaction, the counselor initiates a question regarding language and cultural background, but does not use this as a potential opening to inquire about the client's attitudes or beliefs around relevant issues.

349 GC: Are you Zulu or (.) Tswana?
350 Pt: I'm a Swazi.
351 GC: Swazi. And he's from Nigeria hey? ((referring to Pt's boyfriend))352 Pt: ((nods))
353 GC: So that's all the information that I need from you.

DISCUSSION OF MEDICAL CONDITIONS

While asking about family history of illness or genetic conditions, the genetic counselor in the case above refers to "high blood" and "sugar sickness." These terms are frequently used by patients in the South African context. However, when patients use these terms they do not necessarily mean "high blood pressure" or "diabetes." In fact, our research (Legg & Penn, 2008) has revealed that patients' use of these terms may mean something completely different to a biomedical understanding of these conditions. This has been confirmed by studies that have revealed that in some communities, the term "high blood" actually refers to a condition characterized by fatigue, burning sensations, and difficulty breathing. The cause is linked to leading a stressful life, eating too much sugar or oil, or not exercising enough (Thorogood et al., 2007).

Although the genetic counselor chooses to use these lay terms in this interaction, there is no evidence from the transcript to indicate whether she understands them from a biomedical point of view, or as the patient would possibly understand them, and she does not actively explore the patient's understanding of these conditions.

Delicate Topics: Discussion of Termination of Pregnancy

The nature of a genetic counseling session is such that it necessitates discussion of delicate topics, which may be heavily embedded in cultural beliefs. In this interaction, that topic is whether or not the patient would terminate this pregnancy if it was found that this fetus also has a neural tube defect.

The discussion is preempted by an explanation from the genetic counselor of why anencephaly can occur. She explains that there is nothing that can be done to treat this condition. The patient raises the topic of abortion, as well as the alternative option of carrying the fetus to full term even though it will not survive. The genetic counselor's discussion of this topic is filled with euphemisms and hesitations. Such verbal evidence highlights the complexity of genetic counseling and demonstrates the use of mitigation (Delbene, 2004).

465 GC:... we couldn't have *stopped* the problem, we couldn't have fixed the problem.
466 Pt: So there is nothing they can do about it?
467 GC: ((shakes head)) Nothing we can do.
468 Pt: They uh (.) abortion.
469 GC: ((nods)) Take the pregnancy out.
470 Pt: To continue or or wait until delivery and after that he'll die.
471 GC: Die ja [Afrikaans word for 'yes'], if the baby's brain is not working, it won't, it can't, the babies can't breathe, they they can't function, so they can't live.

The topic of abortion is returned to at a later stage in the interaction and we see that the patient seems to be against the idea of possibly terminating her pregnancy. Earlier in the interaction, the patient indicated that this pregnancy was planned, and there is a sense that she really wants to have this baby.

568 GC: So if we do the sonar, you feel that that's good enough?
569 Pt: mm hm. ((nods))
570 GC: And if we *miss* the problem with the sonar?
571 Pt: Then you do again.
572 GC: Then?
573 Pt: Then we'll find out. ((laughs))
574 GC: And you'll have the baby, and if the baby has a problem?
575 Pt: Maybe there is no problem.
576 GC: Would you consider to terminate, make the abortion if there's any problem with the baby?
577 Pt: ((looks away)) If there is no way (.) yes. ((nods))

Although this discussion is not specifically linked to cultural issues, the results of the grandmother study highlight possible reasons why this client might be against a decision to terminate, for a number of culturally related reasons. For example, she has not yet had a healthy child—we know that her culture values children, and there would be an expectation from her family and community that she should produce children. Thus, she might rather have a child who is potentially disabled, than no child at all. In addition, we know that her family and community may not approve of a decision to terminate her pregnancy and she may be afraid of stigmatization should she make such a decision. However, at no point in the interview are such reasons raised or explored, and the topic remains unresolved.

The Issue of Blood

An interesting discussion arises about one of the genetic tests that the genetic counselor offers to the patient, which involves taking blood and testing it. The genetic counselor begins to explain the blood test and how it can be used to check for neural tube defects. What becomes evident, though, is that the introduction of this topic leads to a disconnection between the counselor and the patient. This disconnection culminates at the end of turn 517, where the patient yawns and looks sufficiently disinterested for the genetic counselor to stop her explanation and change topic completely. This appears to be an uncomfortable topic for the patient, and her discomfort is confirmed later in the interaction.

515 GC: You want to know. Because there are different ways for us to check it.
516 Pt: Ja.
517 GC: The one way is the sonar, like I said to you, we can see the problem and the other test that we do is we take the *blood* from you. And in your *blood*, we check (.) you see because everything is open. You know we have got water around our heads, so when we shake our brain doesn't (.) bash against the head, there is fluid that keeps it nicely. So when there's no, the the bone is over then the fluid comes out and if we take your blood, we can actually see if *that* fluid is leaking out of the baby's head into the water into your *blood*. So we can check whether that protein is *high* or whether it is *normal*. So that's the test we do, so we take your *blood*. And then in that *blood* we can see whether the baby's head is open and the water leaks that *protein*, we call it *AFP*. Don't know if you will remember Alpha Fetal Protein, if it leaks out. So that's the one test we can do. The other one which is more accurate, is when we-((Pt yawns)) you tired from all the working hey?
518 Pt: Ja. ((laughs))
519 GC: Did you work over the week, did you work yesterday?
520 Pt: Ja. ((521–526: discussion about why the Pt is tired))

After this disconnection occurs, the counselor attempts to return to her explanation of the blood test. She also offers an amniocentesis test to the patient and shows her a picture of the test, specifically of a needle inserted into a pregnant stomach.

527 GC: Ok, let's just talk about this thing, because now we take your blood (.)to measure this protein *but* we can also at sixteen weeks put the needle in. ((directs attention to picture)) This is the womb, the baby, you know there's water around the baby, if we put the needle in, we take that water out then we check if the brain is open. If the brain is open, that protein will leak into the water and if we measure the water again, we will see if it's high or if it's normal. So this is called a amniocentesis. This is a *more* accurate test than the blood test. The blood test has no risk, but this one (.) there is a chance that you can have a miscarriage.
528 Pt: ((nods))
529 GC: It's about less than one percent.
530 Pt: ((shakes head))
531 GC: You don't like to hear about miscarriages.
532 Pt: I would prefer sonar.
533GC: You prefer sonar. And the blood?
534 Pt: You make two?
535 GC: You can. But not really the blood? ((smiles))
536 Pt: ((directs gaze away from GC)) No, I don't want the blood. ((laughs))
537 GC: Ok. Why don't you want the blood?
538 Pt: Because I think that the blood (.) too much.

As our grandmother data suggests, the client's apparent aversion to this idea may well be rooted in cultural beliefs. Blood has particular cultural significance in relation to disease, health, and strength. Research conducted in other African countries has revealed that disturbances to the flow or quantity of blood in the body are believed to cause illness. Thus, if blood is taken out of the body, this may lead to fatigue, weakness, and illness (Leach & Fairhead, 2007). Other studies have reported the abundance of myths about blood stealing and blood selling in many African communities (Fairhead, Leach, & Small, 2006; Geissler & Pool, 2006).

Within the context of the HIV/AIDS epidemic, blood has taken on a particular significance for South African patients. The diagnosis of and the monitoring of the disease is dependent on a blood test, which obviously has life-changing implications and may mean a highly stigmatized and traumatic diagnosis. In the initial stages of the disease, there were also rumors in many communities about tainted blood (Stadler, 2003) and, particularly in a context where there is a high incidence of HIV/AIDS and a stigma attached to its diagnosis, many African patients may thus view with deep suspicion formal blood testing in the context of the genetics clinic.

Without asking this client, we cannot know what her perspective is on the issue. Nor can we ascertain why she is also not willing to have an amniocentesis test. We do not know whether the patient's aversion to having this test may be linked to a fear of needles or a fear of pain, both of which have been found to be important factors in patients' decision making about whether or not to have an amniocentesis test (Gordon et al., 2002; Peters, 1984). Research in the South African context has also indicated that some patients believe that needles (used for vaccinations, for example) may cause HIV, and there is a fear that health professionals may reuse needles (Lindegger, Quayle, & Ndlovu, 2007). Such information in the case would be useful to discuss in order to alleviate anxiety, but again, the nature of the power relations in this setting, and the cultural and linguistic barriers between client and counselor, probably make it impossible to explore.

While the examples above emanate from a single session, a similar pattern is emerging in an ongoing study by Wessels, who has recorded and analyzed 17 prenatal intercultural genetic counseling sessions with mothers of advanced maternal age. Very few opportunities emerge to probe possible patient perspectives about cultural variables that may impinge on the decision making process and, in the light of very limited uptake of services, this information seems critical.

TOWARD A CULTURAL FRAMEWORK OF GENETIC COUNSELING

From some of the emerging evidence described above, it is possible to see the powerful influence of contextual variables on the process of genetic counseling. Such evidence implies the need for a critical reevaluation of knowledge bases, and a rethinking of client centeredness, particularly in intercultural contexts. Based on the results of our research in the diverse South African context, we believe that a revised model of genetic counseling is needed—one that incorporates some aspects of culture, and acknowledges that each genetic counseling session has unique characteristics framed by the context, by the clinician's own background and experience, as well as the client's expectations, culture, and experience. Such *lifeworld issues* (to use Mishler's 1984 term), far from being irrelevant to the interview, are critical to explore and understand, as they appear to have a major impact on uptake and success of services. Without an understanding of what our clients believe the causation of inherited illness is, we cannot hope to counsel them effectively about inheritance patterns, molecular diagnostic testing, and predictive options. Sensitivity to cultural aspects, kinship terms, decision-making structures, acceptance and awareness of dual routes of consultation, and an awareness of religious factors, all become important in delivering effective genetic counseling services.

It is unwise to assume that because a client has come for genetic counseling, he or she automatically accepts the biomedical approach to healing. The referral structures in the particular context, and inefficient communication at an institutional level may be factors that may influence the client's journey. Understanding the story of a client's illness and their previous experiences of healthcare may have a great influence on their response to the process and their uptake of genetic testing.

The measurement of effectiveness in genetic counseling may thus link not only to standard variables such as outcomes, or in a quantitative measure such as uptake of services or return to clinic, but to process variables, and the extent to which we have developed cultural safety in the genetic counseling session. This implies a move beyond simply cultural awareness to a situation where "a client feels that their cultural, social and human values are respected, and that an organization providing services to that client re-orients its institutional practices, values, resource and governance arrangements accordingly" (Phillips, 2007).

Reconciling a cultural and biomedical view of genetic counseling thus requires a special set of skills and awareness, including (we suggest) an alignment to the following principles (described in Table 47-2).

PATIENTS ARE GOOD CONSUMERS OF HEALTHCARE SERVICES AND SHOULD BE CONSIDERED EXPERTS IN THEIR OWN CONDITIONS AND LIFEWORLD

Despite language and culture barriers in interactions, as well as various other barriers to accessing genetic counseling services, our research in this field and in other healthcare professions has taught us that patients are good consumers and good appraisers of communication efficiency. They are able to judge who is a "good" doctor or what constitutes a "bad" interaction, as well as the quality of communication.

Part of attending to the client's lifeworld also involves finding out about their cultural beliefs, particularly those related to genetic disorders and disability, and then to incorporate these beliefs and practices into the client's treatment. It is important to recognize that each patient has unique beliefs, values, attitudes, expectations, and

TABLE 47-2 PRINCIPLES OF CULTURAL SAFETY IN A GENETIC COUNSELING CONTEXT

1. Cultural beliefs regarding illness and disability have a pervasive impact on genetic counseling interactions and outcomes.
2. Patients are experts of their own lifeworld and this information should be tapped.
3. Facilitators and barriers to care are identifiable and can be addressed effectively through a focus on communication and process aspects, and the adoption of new methods for deriving information and checking understanding.
4. Cultural incompetence and reflexivity should be seen as a goal.
5. Stereotypes need to be avoided.
6. The use of cultural brokers should be promoted.

fears. The aim is not to change these beliefs or align them with a biomedical point of view, but to engage with and develop a shared understanding of the patient's viewpoint, so that mutually agreed-upon goals can be negotiated in the interaction. The critical feature seems to be to have relevant methods of asking.

Shared decision making and active participation may be a goal of genetic counseling, but not something to which clients are accustomed. Some clients may not want to be active participants in interactions or in decision-making processes. It is important to recognize that patient-centered care is not just about involving the client but about finding out about their preferences, family decision-making patterns, and adopting a communication style that is in accordance with these factors. Guiding the client toward collaboration will require a full understanding of barriers and facilitators to that goal.

FACILITATORS AND BARRIERS TO COMMUNICATION CAN BE IDENTIFIED IN INTERACTIONS

Our choice of research method has enabled us to identify when things go well in interactions, and also when disconnection happens between patients and health professionals. These variables, once identified, can be used in order to modify and enhance the process and patterns of mutual understanding. What is crucial, though, is a focus not just on content, but also on process factors—communication variables are critical in ensuring mutual understanding.

Some facilitators may include a more detailed negotiation of language rules, the conscious deployment of nonverbal strategies (e.g., the use of gesture, attention to seating, facial expression), recognition of family patterns of decision making, gaining insight into cultural models of disability, culturally relevant methods of assessing comprehension (rather than assuming that concepts are understood), or ways of incorporating a third party. Some of these have been documented in training programs we have developed for various health professionals (e.g., pharmacists; Watermeyer & Penn, 2009). Our research has shown that communication may be enhanced through training, through expansion and elaboration of methods that are culturally specific, and through careful attention to barriers and facilitators.

One area of critical relevance to genetic counseling is the development of an approach to mutually understood terminology. Solomon (2009) in her study found that her participants had a completely different conception of the notion of inheritance from that of the researcher, and there appeared to be no equivalent Xhosa word for the term "genes." The development of community advisory boards and suitable methods to explore communication issues and to select by consensus the appropriate vocabulary seems very important.

IT IS WISE TO FEEL "CULTURALLY INCOMPETENT"

Working in a context of cultural diversity can be challenging, overwhelming, and even frightening to health professionals. Authors such as Daniels and Swartz (2007) suggest that perhaps the key to working with these challenges lies in acknowledging such uncertainty and accepting that a feeling of discomfort and insecurity and disempowerment is natural. Such a stance requires an attitude of reflexivity, in which the counselor attempts to understand how their own cultural background affects communication, acknowledges their own cultural biases, and reflects on their cultural identity as a health professional and as an individual.

It also implies the adoption of new methods that are sensitive to cultural issues; for example, asking questions about culture and actively showing an interest in the cultures served, using cultural narratives to explore clients' cultural beliefs (Penn, 2002; Petersen, 2006), and using health-related and illness-related cultural metaphors (Henneman et al., 2008).

AVOID STEREOTYPES

When working in a multicultural context, it is easy to fall into the trap of applying stereotypes about culture. Although some cultural norms may apply to a specific group in general, members of the same cultural group may not necessarily hold the same beliefs about health, disease, and treatment (as our grandmother project has shown). It is important not to assume uniformity within a cultural group, but to seek to determine individual perceptions and beliefs.

INCLUDE CULTURAL BROKERS

Training of the mid-level worker becomes an important issue for consideration in intercultural interactions. We believe that in the South African context the selection, training, and involvement of such an individual in genetic counseling could well act as an important resource, and in our opinion should be formalized. Such an individual can serve as both a linguistic and cultural mediator—a "cultural broker" who acts as a mediator between the genetic counselor and client (Jezewski & Sotnik, 2001). We believe that valuable resources already exist in the community in this regard, and that in the genetic counseling context, such a mediator, who is familiar with local cultural beliefs about disability and illness, could as act an advocate for the client and assist the genetic counselor in exploring the client's beliefs about genetic disorders, preferences for decision making, and preferences for inclusion of family members.

The role of such mediating individuals in South Africa is poorly understood and (at least in our experience) informal, erratic, poorly paid, and often superimposed upon another job description (e.g., nursing). However, considerable opportunity exists for expanding their role, particularly in a specialist field such as genetic counseling. In a diverse context, even if the interview takes place in a common language, there are aspects beyond the words which need interpretation. In the case study presented above, for example, the client declined the offer of an interpreter. However, there were clear communication difficulties and asymmetries, which might have been overcome by the presence of a trained individual to mediate not only linguistic aspects, but also cultural aspects of the session.

CONCLUSIONS

A cultural model of genetic counseling suggests that what happens in the individual genetic counseling session is influenced significantly by a range of contextual factors linked to sociopolitical, linguistic, institutional, and cultural variables such as kinship patterns and religion.

While the illustrative evidence we have presented here is from a unique context, it does highlight the relevance of such cultural and contextual variables to genetic counseling in general and, in the context of increasing globalization and patterns of migration, will hopefully have broader application. A move toward a local rather than a universal model of genetic counseling will require further collaborations with communities. The development of trust relates not only to the individual relationship between healthcare practitioners and patients, but is linked to perceptions of the whole unit and the institution. This has a powerful resonance for the genetic counseling service, as well as the content of training programs, and implies particularly the relevance of good communication at the level of the individual client relationship, and at institutional and community levels. Importantly, in line with a philosophy of cultural safety, these changes should extend beyond a mere understanding of such barriers to care, to a level of fluent and committed implementation of suitable methods, as well as the development of relevant sustainable policies.

ACKNOWLEDGMENTS

Some of the data in this chapter was presented at the 13th South African Society for Human Genetics Conference in Stellenbosch, in April 2009. This research was funded by grants from the South African Netherlands Research Programme on Alternatives in Development (SANPAD) and the South African Medical Research Council. We gratefully acknowledge the research completed by Rae Tovell, some of which is included in this chapter.

APPENDI1X

CONVERSATION ANALYSIS TRANSCRIPTION CONVENTIONS (JEFFERSON, 2004)

GC:	genetic counselor
Pt:	patient
word	some form of stress, via pitch or amplitude
((laugh))	transcriber's descriptions, e.g., laughter or head nod
(.)	short pause
°word°	softly spoken, quieter than the surrounding talk

REFERENCES

Barratt J, Penn C. (2009). Listening to the voices of disability: Experiences of caring for children with cerebral palsy in a rural South African setting. In MacLachlan M, Swartz L (eds.), *Disability and International Development: Toward Inclusive Global Health* (pp. 191–212). New York: Springer.

Christianson A. (2000). Community Genetics in South Africa. *Community Genetics 3*(3), 128–130.

Christianson A, Venter P, Modiba J, Nelson M. (2000). Development of a primary health care clinical genetic service in rural South Africa – the Northern Province experience, 1990–1996. *Community Genetics 3*(2), 77–84.

Clarke A, Parsons E. (1997). *Culture, Kinship and Genes.* Basingstoke: Macmillan.

Coovadia H, Jewkes R, Barron P, Sanders D, McIntyre D. (2009). Health in South Africa 1: The health and health system of South Africa: historical roots of current public health challenges. *The Lancet 374*, 817–834.

Daniels K, Swartz L. (2007). Understanding health care workers' anxieties in a diversifying world. *PLoS Medicine 4*(11): e319.

Day C, Gray A. (2008). Health and related indicators. In: Barron Roma-Reardon (eds.), *South African Health Review 2008* (pp. 239–395). Durban: Health Systems Trust.

Delbene R. (2004). The function of mitigation in the context of a socially stigmatized disease: a case studying a public hospital in Montevideo, Uruguay. *Spanish in Context 1*, 241–266.

Deumert A. (2010). 'It would be nice if they could give us more language': serving South Africa's multilingual patient base. *Social Science & Medicine 71*, 53–61.

Drew P, Heritage J. (1992). *Talk at work: Interaction in institutional settings.* Cambridge: Cambridge University Press.

Ellis C. (2004). *Communicating with the African Patient.* Scottsville: University of KwaZulu-Natal Press.

Evans M. (2000). *The communicative, interpersonal and therapeutic dynamics of the audiologist-interpreter-caregiver triad in interpreted aural rehabilitation consultations.* Unpublished honors dissertation. University of Cape Town, South Africa.

Fairhead J, Leach M, Small M. (2006). Where techno-science meets poverty: Medical research and the economy of blood in The Gambia, West Africa. *Social Science & Medicine 63*, 1109–1120.

Fisch M. (2001). *Interpreting practices in health care: an investigation of differences across trained and untrained interpreters within the fields of speech-language and hearing therapy.* Unpublished Masters dissertation, University of Cape Town, South Africa.

Forrest K, Simpson S, Wilson B, et al. (2003). To tell or not to tell: barriers and facilitators in family communication about genetic risk. *Clinical Genetics 64*, 317–326.

Geissler PW, Pool R. (2006). Editorial: Popular concerns about medical research projects in sub-Saharan Africa – a critical voice in debates about medical research ethics. *Tropical Medicine and International Health 11*(7), 975–982.

Gordon C, Prince MB, Benkendorf JL, Hamilton HE. (2002). "People say it's a little uncomfortable": prenatal genetic counselors' use of constructed dialogue to reference procedural pain. *Journal of Genetic Counseling 11*(4), 245–263.

Greeson C, McCarthy Veach P, LeRoy BS. (2001). A qualitative investigation of Somali immigrant perceptions of disability: implications for genetic counseling. *Journal of Genetic Counseling 10*(5), 359–378.

Haram L. (1991). Tswana medicine in interaction with biomedicine. *Social Science & Medicine 33*, 167–175.

Henneman L, Marteau TM, Timmermans DRM. (2008). Clinical geneticists' and genetic counselors' views on the communication of genetic risks: A qualitative study. *Patient Education and Counseling 73*(1), 42–49.

Herselman S. (2007). 'Health care through a cultural lens': insights from medical anthropology. *Current Allergy & Clinical Immunology 20*(2), 62–65.

Ingstad B, Bruun F, Tlou S. (1997). AIDS and the elderly Tswana: The concept of pollution and consequences for AIDS prevention. *Journal of Cross Cultural Gerontology 12*, 357–372.

Jefferson, G. (2004). Glossary of transcript symbols with an Introduction. In Lerner, G. (Ed.), *Conversation analysis: Studies from the first generation* (pp. 13–23). Philadelphia: John Benjamins.

Jenkins T. (1990). Medical genetics in South Africa. *Journal of Medical Genetics 27*, 760–779.

Jewkes R, Abrahams N, Mvo Z. (1998). *Study of Health Care Seeking Practices of Pregnant Women in Cape Town.* Report Four: Summary of Findings, Conclusions and Policy Action. Tygerberg: Medical Research Council.

JewkesR, Wood K. (1998). Competing discourses of vital registration and personhood: perspectives from rural South Africa. *Social Science & Medicine 46*(8), 1043–1056.

Jezewski MA, Sotnik P. (2001). *Culture brokering: Providing culturally competent rehabilitation services to foreign-born persons.* Center for International Rehabilitation Research Information and Exchange. CIRRIE Monograph Series, John Stone, Ed. Buffalo, NY: CIRRIE.

Kale R. (1995). South Africa's health: Traditional healers in South Africa: a parallel health care system. *British Medical Journal 310*(6988), 1182–1185.

Kasanga L, Lwanga-Lumu J. (2007). Cross-cultural linguistic realization of politeness: a study of apologies in English and Setswana. *Journal of Politeness Research 3*, 65–92.

King SV. (2008). Introduction to the journal of cross-cultural gerontology, special issue on aging and social change in Africa. *Journal of Cross-Cultural Gerontology 23*(2), 107–110.

Kromberg J, Sizer E, Christianson A. (2010). *Genetic Testing in Emerging Economies: An International Pilot Study. Preliminary Report on South Africa.* Retrieved September 19, 2011 from the World Wide Web: https://www.eshg.org/fileadmin/www.eshg.org/documents/received/GenTEE_final_draft_3.pdf

Lampret J, Christianson A. (2007). Reproductive choices made by South African mothers who have a child with Down syndrome. *South African Medical Journal 97*(7), 515–516.

Leach M, Fairhead J. (2007). *Vaccine Anxieties: Global Science, Child Health and Society.* London: Earthscan.

Legg C, Penn C. (2008). *An ethnography of adults living with aphasia in Khayelitsha Cape Town.* Paper presented at Communication, Medicine and Ethics (COMET) Conference, Cape Town, July 2–4, 2008.

Levin M. (2006). Language as a barrier to care for Xhosa-speaking patients at a South African paediatric teaching hospital. *South African Medical Journal 96*(10), 1076–1079.

Liddell C, Barrett L, Bydawell M. (2006). Indigenous beliefs and attitudes to Aids precautions in a rural South African community: an empirical study. *Annals of Behavioral Medicine 32*(3), 218–225.

Lindegger G, Quayle M, Ndlovu M. (2007). Local knowledge and experiences of vaccination: implications for HIV-preventive vaccine trials in South Africa. *Health Education and Behavior 34*, 108–123.

Lupton D. (2003). *Medicine as Culture: Illness, Disease and the Body in Western Societies* (2nd ed.). London: Sage Publications.

Masasa T, Irwin-Carruthers S, Faure M. (2005). Knowledge of, beliefs about and attitudes to disability: implications for health professionals. *South African Family Practice 47*(7), 40–44.

Mathole T, Lindmark G, Majoko F, Maina Ahlberg B (2004). A qualitative study of women's perspectives of antenatal care in a rural area of Zimbabwe. *Midwifery 20*, 122–132.

Meiser B, Irle J, Lobb E, Barlow-Stewart K. (2008). Assessment of the content and process of genetic counseling: a critical review of empirical studies. *Journal of Genetic Counseling 17*(5), 434–451.

Mishler E. (1984). *The Discourse of Medicine: Dialectics of Medical Interviews.* Norwood: Ablex.

Modell B. (1997). Kinship and medical genetics: a clinician's perspective. In Angus Clarke and Evelyn Parsons (Eds.), *Culture, kinship and genes* (pp. 27–39). London: Macmillan.

Morris-Rosendahl D. (2009). *Molecular genetic analysis of patients with malformations of cortical development.* Paper presented at 13th South African Society for Human Genetics Congress, Stellenbosch, 5–8 April 2009.

Nachega J, Stein D, Lehman D, et al. (2004). Adherence to antiretroviral therapy in HIV-infected adults in Soweto, South Africa. *Aids Research and Human Retroviruses 20*(10), 1053–1056.

Penn C. (2002). Cultural narratives: bridging the gap. *Folia Phoniatrica et Logopaedica 54*(2), 95–99.

Penn C. (2007). Factors affecting the success of mediated medical interviews in South Africa. *Current Allergy & Clinical Immunology 20*(2), 66–72.

Penn C, Watermeyer J. Evans M. (2011). Why don't patients take their drugs? The role of communication, context and culture in patient adherence and the work of the pharmacist in HIV/Aids. *Patient Education and Counseling 83*, 310–318.

Penn C, Watermeyer J, MacDonald C, Moabelo C. (2010). Grandmothers as gems of genetic wisdom: exploring South African traditional beliefs about the causes of childhood genetic disorders. *Journal of Genetic Counseling, 19*, 9–21.

Peters JA. (1984). Amniocentesis and rhaphiphobia: A preliminary survey. Strategies in Genetic Counseling: Clinical Investigation Studies. *Birth Defects Original Articles Series 20*(6), 201–207.

Petersen A. (2006). The best experts: the narratives of those who have a genetic condition. *Social Science & Medicine 63*, 32-42.

Phillips G. (2007). *Healing, Cultural Safety and the Criminal Justice System.* Paper presented at the 3rd National Indigenous Justice CEO Forum, Brisbane, November 21, 2007.

Pilnick A. (2004). 'It's just one of the best tests that we've got at the moment': the presentation of screening for fetal abnormality in pregnancy. *Discourse and Society 15*(4), 451–465.

Rapp R. (1988). Chromosomes and Communication: The Discourse of Genetic Counseling. *Medical Anthropology Quarterly 2*(2), 143–157.

Raubenheimer J. (1987). Counselling across cultural borders in South Africa. *International Journal for the Advancement of Counselling 10*, 229–235.

Raz AE, Atar M. (2003). Nondirectiveness and its lay interpretations: the effect of counseling style, ethnicity and culture on attitudes towards genetic counseling among Jewish and Bedouin respondents in Israel. *Journal of Genetic Counseling 12*(4), 313–332.

Richards M. (1996). Families, kinship and genetics. In Marteau T, Richards M (eds.). *The troubled helix: social and psychological implications of the new human genetics* (pp. 249–73). Cambridge: Cambridge University Press.

Schneider H. (2006). Reflections on ART policy and its implementation: rebuilding the ship as we sail? *Acta Academica Supplementum* 1, 17–43.

Solomon G. (2009). *An Investigation into the Understanding of Basic Genetic Inheritance amongst AmaXhosa Caregivers of Haemophilia Patients.* Unpublished Masters dissertation, University of Cape Town, South Africa.

Stadler J. (2003). Rumor, gossip and blame: implications for HIV/Aids prevention in the South African Lowveld. *AIDS Education & Prevention 15*(4), 357–369.

Stensson E. (2008). The social stratification of albinos in Tanzania - a case study from Babati. Unpublished Bachelors Thesis, Södertörn University College, Sweden.

Thorogood M, Connor MD, Hundt GL, Tollman SM. (2007). Understanding and managing hypertension in an African sub¬district: A multidisciplinary approach. *Scandinavian Journal of Public Health 35*, 52–59.

Tovell R, Wessels T-M, Penn C. (2005). *Dispensing Words not Tablets: Using conversation analysis to identify the conversational characteristics of a genetic counseling session.* Paper presented at the first combined congress of the Southern African Society for Human Genetics and the Africa Society of Human Genetics, Johannesburg, March 2005.

UNAIDS/WHO (2008). *Epidemiological Fact Sheet on HIV and AIDS – Core data on epidemiology and response.* South Africa. Retrieved September 18, 2009 from http://apps.who.int/globalatlas/predefinedReports/EFS2008/full/EFS2008_ZA.pdf

Urban M. (2009). *Prenatal Screening for Down Syndrome in the Peninsula Maternity and Neonatal Service: Past, Present and Future.* Paper presented at the 13th South African Society for Human Genetics Conference, Stellenbosch, April 5–8, 2009.

van Dyk A. (2001). Traditional African beliefs and customs: Implications for AIDS education and prevention in Africa. *South African Journal of Psychology 31*, 60–66.

van Rensburg HCJ, Ngwena C. (2001). Health and health care in South Africa: Against an African background. In Cockerham W (ed.). *The Blackwell Companion to Medical Sociology* (pp. 365–391). Oxford: Blackwell Publishers.

Ware N, Idoko J, Kaaya S, et al. (2009). Explaining adherence success in sub-Saharan Africa: an ethnographic study. *PLoS Medicine 6*, 39–47.

Watcham S, Schön S, Christianson A. (2007). Neglect in the care of pregnant South African women of advanced maternal age. *South African Medical Journal 97*(11), 1064–1066.

Watermeyer J, Penn C. (2009). *Working across language and culture barriers: communication skills for pharmacists.* Cape Town: PATA.

Watermeyer J, Penn C. (2008). 'They take positive people': An investigation of communication in the informed consent process of an HIV/AIDS Vaccine Trial in South Africa. *Critical Inquiry in Language Studies 5*(2), 81–108.

Weil J. (2001). Multicultural education and genetic counseling. *Clinical Genetics 59*, 143–149.

Wessels T-M, Penn C. (2008). *"Do you know why the doctor sent you?" – Characteristics of the genetic counselling process in a multicultural antenatal context in Johannesburg, South Africa.* Poster presented at the European Human Genetics Conference, Barcelona, Spain, May 31, 2008 – June 3, 2008.

Wessels T-M, Zwane T. (2006). *Invasive procedure uptake rate among women of advanced maternal age in Johannesburg – cultural aspects.* Paper presented at the European Human Genetics Conference, Amsterdam, May 6–9.

48.

ETHICAL, LEGAL, AND SOCIOCULTURAL ISSUES AND GENETIC SERVICES IN SOUTHERN AFRICA

Jennifer G. R. Kromberg and Trefor Jenkins

INTRODUCTION

What is special about the sociocultural aspects of the practice of medical genetics in South Africa? Are they very different from those in other countries? We hope to show that they are, indeed, different when it comes to the care of members of the majority population, even though they may be similar for a small but significant number—those who belong to the previously privileged minority; i.e., those of European descent. When considering the ethical and legal aspects, we shall assume that there are no differences and that all "patients" presenting for genetic counseling or for invasive procedures are entitled to the uncompromised and highest standards of care, and are under the same dictates of the law.

Racially segregated healthcare has been the rule in South Africa ever since the "colonization" of the country in the seventeenth century. From 1652 to the end of the eighteenth century, Dutch settlers ruled the Cape of Good Hope and many of the indigenous people (Khoi or "Hottentot," and San or "Bushmen") were, in effect, slaves. In addition, there were many imported slaves from the Dutch East Indies, Madagascar, and East Africa, and there was, in fact, one shipload of slaves from West Africa. British rule, from 1795, continued the slavery until its abolition in 1837. By this time the British ruled the whole of the Cape Province, including the Transkei and, soon after, Natal also became part of the British colony. The short-lived independent Dutch or Afrikaner republics, Orange Free State, Transvaal, and other, smaller ones, became part of the Union of South Africa (in 1910). During the period of Nationalist Party rule from 1948 until 1994 (during which time Nelson Mandela and other political prisoners were released and other Black leaders returned from exile), there was strict segregation of the different race groups (apartheid), with Blacks effectively deprived of political power—except in the clearly defined "independent," so-called, homelands.

In the year 1990, one of the present authors (TJ) published an extensive review of "Medical Genetics in South Africa" as one of a series of articles solicited by Professor Peter Harper for the *Journal of Medical Genetics*, of which he was editor (Jenkins, 1990). This comprehensive article summarized the developments in the field of human genetics in South Africa up to that time.

HEALTHCARE IN SOUTH AFRICA: CHANGES SINCE THE ABOLITION OF APARTHEID

Benatar (2004), in a review of healthcare reform in South Africa, following the peaceful transition in 1994 from an oligarchy controlled by the White minority to a truly democratic republic, carefully assesses the changes that have taken place. It is acknowledged that the narrowing of racial disparities in healthcare is a formidable challenge for the new government and, very significantly, has had to be implemented in the context of an almost uncontrollable epidemic of the human immunodeficiency virus (HIV) and full-blown AIDS. There is the acknowledgment that the "new government can be proud of its many achievements that have improved health among the nation's approximately 45 million people" (Benatar, 2004, p. 81), including stabilization of the economy and significant economic growth, and the reversal of discriminatory legislation. Providing access to clean drinking water, the building of more than a million homes and installing electricity and telephone connections to most of them, are impressive achievements. Hundreds of new clinics, providing free, desegregated primary healthcare services to pregnant women and children under 5 years of age, have been built; new food programs reach over 5 million children.

Several shortcomings to the improved facilities and programs have been listed by Benatar (2004, p. 81), the most serious of which is the poor functionality in many of the new clinics due to "lack of personnel and finances, poor administration, and expanding demands." A particularly disturbing feature is that these changes have been accompanied by the eroding of public tertiary health services. The HIV and AIDS pandemic had a low profile for over 10 years, due partly to the denialism of the Minister of Health, apparently encouraged by the former president (Thabo Mbeki), who was peacefully "deposed" in September 2008.

Eleven percent of the government budget for healthcare ($5.3 billion) funds all the healthcare needs of approximately 80% of the population who do not have health insurance, as well as all healthcare training and education in the country. Private sector healthcare expenditure has grown over the past 35 years, and today approximately 9% of GDP is spent on healthcare with 60% of it used for the care of 18% of citizens who have private insurance. The same trend is evident in the deployment of doctors in the private sector, which has grown from 40% of all doctors in the 1970s to 66% today (Benatar, 2004). It can be assumed that a similar disproportionate distribution of nurses exists between the two healthcare sectors.

How are these trends in the distribution of healthcare personnel and resources reflected in the provision of genetic services in the country? Prior to approximately 1998, the national Department of Health had a genetic services department with two PhD geneticists and a small team (numbering around 17) of genetic nurses, who tried to increase the public awareness of inherited disorders, presenting courses of instruction in medical genetics for medical and nursing personnel throughout the country. The staff of the two academic departments of human genetics at Cape Town University (established in 1972) and the University of the Witwatersrand (established in 1975) were involved in teaching these courses. There was no recognized registrable status for medical genetics at this time. With the abolition of apartheid in 1994, the recognition of medical genetics as a medical specialty, and the registration of genetic counselors (in 1989), the university departments were able to employ specialists in the field: doctors, genetic counselors (with the financial support of the Department of Health), and nurses.

A two-tier health care system continues in South Africa, with the public system deteriorating due to shortages of staff (medical and nursing) and resources—evident in annual per capita spending on healthcare, at about $US 150 in the public sector and about $US 850 in the private sector. Thus, "nationally, there are 8.7 cardiac surgeons per one million population, who perform 774 operations per one million people annually in the private sector; in the public sector, there are 0.6 cardiac surgeons per one million people, who perform 69 operations per one million population" (Benatar, 2004, p. 82). This two-tier health care system continues with "discrimination in access to care on economic grounds replacing the racial discrimination of the past" (Benatar, 2004, p. 82).

Benatar (2004) has drawn attention to the impact that new government policy is having on tertiary care services in the public sector—adverse trends, in his opinion—in both the Western Cape and Gauteng. In the former, between 1995 and 2000 it meant a reduction of hospital beds by 3601 (24, 4%) and 9282 fewer health and support personnel (27.9%), while the local population increased by 8%; cardiac operations in adults were reduced from 700 to 250 per year. In orthopedics, joint replacement dropped from 350 in 1993 to 60 in 2003! Similar reductions were evident in the number of patients undergoing long-term dialysis, and in the number receiving renal transplantation.

While some are optimistic about these changes, interpreting them as temporary and the price to be paid to secure a more equitable system of public health, Benatar (2004) fears that these trends threaten postgraduate training and a drop in the standard of surgical skills, and in the research necessary to maintain standards. The very survival of the academic institutions is under threat.

But running through all the concerns for the continuing high standards of medical research and service delivery, there is the specter of HIV/AIDS, which is showing few signs of being brought under control. After many years of equivocating over a national policy to deal with the pandemic, the government seems, at last, to be acknowledging the magnitude of the epidemic and is deploying appropriate resources to coping with it. The desired results are yet to be seen—denialism is still common and the public remains ignorant about safe sex, or, fails to practice it. Poverty, which is still rife among the disadvantaged in society, is a likely motive for becoming pregnant—the meager state subsidies for the resulting children seem to encourage the unsafe sexual practices.

Complicating the situation, in the opinion of some observers, is the claim of researchers that male circumcision reduces the risk of acquiring HIV from the female sexual partner by nearly 60%. It is debatable how the males in question interpret such a claim. Does it give a feeling of false reassurance? A number of the tribal groups in South Africa already practice male circumcision, but it is carried out post-puberty and up to the age of 18 or 19 years. It is often performed by untrained, inexperienced men, who see it as a money-making exercise—as evidenced by the fact that dozens of the initiates die each year as a result of this procedure. It is argued that the "operation" used to be performed by well-trained chiefs or their designated assistants, but this is no longer the case. The circumcisions carried out in the trials, in South and East Africa, were performed by skilled medical practitioners under ideal conditions.

The number of mother-to-child transmissions of HIV is unacceptably high—for a number of reasons, ranging from the introduction of antiretrovirals to the HIV-positive mother too late in the pregnancy, and extending to the complete failure of administering the antiretrovirals. There is a serious shortage of trained nurses and medical practitioners in the rural areas—as well as in predominantly Black townships, even those near large cities like Johannesburg and Cape Town.

The expulsion of missionary doctors in the 1970s and 1980s, or, more importantly, the difficulties put in the way of foreign doctors wishing to practice in the rural areas of South Africa, meant a serious depletion of the numbers of doctors working in rural areas, where over 50% of the population resided. Attempts were made to staff some of the rural hospitals with young South African doctors when doing their compulsory 2 years of military service, but these inexperienced doctors found it difficult to function optimally in the absence of more experienced colleagues. The standard of medical practice in the rural areas of South Africa, provided by dedicated missionary doctors, was difficult to maintain after the state decided to "take over" these hospitals in the 1980s.

The migrant labor system had been introduced by the mining industry in the 1880s and grew to become an enormous disaster for involved individuals and families. Its effects ranged from the disruption of family life in the rural areas to the introduction of tuberculosis (TB) into communities, brought there by discharged TB-affected mine workers. In addition, the repatriation, to their rural homes, of the sick miners with active TB resulted in spread of the disease to the local population, where, in the main, the facilities for providing continuing healthcare to the discharged miner and his TB-infected family members were grossly inadequate. Professor John Hargrove (2007), in his inaugural address at the University of Stellenbosch, traces this history and its disastrous consequences, including the widespread dissemination of HIV/AIDS throughout South Africa and the neighboring countries.

Within this context various ethical, legal, and sociocultural issues have faced health professionals setting up and providing genetic services. Some of these issues will be discussed below.

ETHICAL ISSUES

DELIVERY OF GENETIC SERVICES

Genetic services have been provided in South Africa for nearly four decades, and the first formal genetic counseling clinics were set up both in Cape Town and Johannesburg in 1973. These services have developed comprehensive programs comparable with the best available internationally. Genetic services were later initiated, usually by staff of departments of pediatrics at the local medical school, in the smaller cities of Durban, Bloemfontein, Pretoria, and Stellenbosch. However, accessibility to all these services was always an issue, and to meet this need, those programs with sufficient staff established outreach clinics to underserved populations. The provincial health department in Gauteng (the smallest but wealthiest province, boasting two medical schools) has taken responsibility for developing and funding the services in the province, but the other provinces still need to recognize the demand for the services and start funding them. The universities play an essential role in the training and education of genetic professionals and medical scientists, as well as in auditing the quality of the service and providing expertise in the laboratory services. Since there are no genetic services available in the other countries of the subcontinent, these genetics professionals occasionally provide genetic services to patients from other Southern African countries, who are willing and able to travel to South Africa to consult with them.

South Africa is one of the few countries on the continent that has not been plagued by the problem of sickle cell anemia (SCA) or, indeed, the thalassemias. The sickle cell trait has virtually zero frequency in its indigenous population, even though increasing numbers of recent immigrants from central and, to a lesser extent, West Africa, are bringing the $Hb\beta^{s}$ gene into the country. Furthermore, parents of children with sickle cell anemia (or pregnant women at high risk of having a child with the condition) are traveling to South Africa for treatment and/or investigation. Families of Indian origin have, for many years, presented for management of their children with a thalassemia syndrome and/or for investigation of high-risk pregnancies. During recent years, daylong courses on SCA and the thalassemias have been held at the University of the Witwatersrand medical school, in order to update clinical hematologists on developments in the field—experts from London and other overseas centers have been guest lecturers at these courses. For the present, however, there is no ethical justification for the expansion of facilities for the laboratory investigations or treatment of the hemoglobin disorders. Experienced hematologists in academic institutions in both Johannesburg and Cape Town are adequately meeting the needs of the country, as well as those of other countries to the north.

The delivery of genetic services raises ethical issues, which include the requirement that they be accessible, voluntary, of high quality, and high ethical standards. They should offer nondirective counseling, options, and freedom of choice to patients, and should be provided by well-trained professionals, respectful of human diversity and individual differences. The European Alliance of Genetic Support Groups has called for specific standards to be met by genetic services, and these include: *(1)* equal access for all individuals of families with congenital or genetic disorders to full information; *(2)* early diagnosis in accredited centers; *(3)* the maintenance of confidentiality; and *(4)* the freedom of choice for all within the legal framework of each country (Kromberg & Jenkins, 1997a).

SURVEY ON ETHICS AND GENETICS IN SOUTH AFRICA

In 1994 South Africa was involved in the second worldwide survey, on ethics and genetics, in 37 nations. The country had been excluded from the first survey (1985–1986) on 17 nations, due to being ostracized as a result of the apartheid policies of the government of that time. The findings of the first survey were published in a book by Wertz and Fletcher (1989). The second survey was also organized by Wertz and Fletcher (1993) from the United States, and it was coordinated by the present authors in South Africa. Some of the South African findings, integrated and compared with those for other countries, have been published by Wertz (1998, 1999), and some appear in a paper by the present authors (Kromberg & Jenkins, 1997a), but the remaining results were never published due to Professor Wertz's untimely death, prior to the completion of the project. The aim of the survey was to encourage medical geneticists to reflect on their preferred approaches to the most frequent of the difficult moral choices encountered in the practice of medical genetics. A questionnaire (obtained from Professor Wertz and modified slightly by us, by the addition of items of local interest) was used to obtain the relevant data. Altogether, 16 of the 21 possible respondents completed the questionnaire—a 76% response rate compared with a 60% (16 participants) rate for Australia (Wertz, 1999),—and they were working in six different cities in the country. There were 10 males and 6 females; their average age was 48 years, 11 (69%) had medical degrees, 4 (25%) had PhDs, and one had an MSc. They had had, on average, 14 years of experience and they were consulting, on average, 6 cases per week. However, the subjects were generally inexperienced regarding ethical issues, and many stated that they had not counseled cases in which such issues arose. About 62% had some personal experience with disability in relatives or friends, and religion was very important to 50% of the participants (Kromberg & Jenkins, unpublished data).

The subjects showed consensus concerning issues on patient autonomy, confidentiality of genetic data, the right to choose, nondirective and unbiased counseling methods, the right to abortion for fetal abnormalities, and sex selection (prenatal diagnosis should not be used for this purpose). Differing from the views of other developing countries, South African respondents supported nondirective counseling (which was generally a practice supported only by the English-speaking countries, and an aberration from the more general practice of giving advice) and rejected the statement that an important goal of genetic counseling was "to reduce the number of deleterious genes in the population" (in China 100% of respondents agreed with this statement; Wertz, 1998). However, similar to the respondents from other English-speaking countries, South Africans had a generally pessimistic view of disability (Wertz, 1999). Most thought that society would never provide enough services (75%), that some disabilities would never be overcome, even with maximum support (81%), and that care usually falls on the mother (75%), while few thought that people with severe disabilities made society richer and more varied (14%).

All the participants stated that they would request abortion for themselves if faced with a diagnosis of trisomies 13 and 21, spina bifida, anencephaly, HIV, rubella, and Hurler's syndrome; however, only 25%, 25%, 12%, 56%, 25%, 31%, and 25%, respectively, would urge termination in a counseling situation (showing their nondirective approach; Kromberg & Jenkins, unpublished findings). None would urge termination for conditions such as Turner syndrome, 47XXY, clefts, or achondroplasia. Also, none approved of abortion for the wrong fetal sex, and some stated that abortion should be illegal in this case. Several admitted that they would not give completely unbiased counseling in cases where trisomy 21, cystic fibrosis, or Huntington's disease has been diagnosed prenatally, nor would they give it in cases of a diagnosis of clefting (suggesting that it was not a bad defect).

In the case of a divorcing couple, with regard to unexpected results, most participants would respect the privacy of the wife and would not disclose nonpaternity to the ex-husband (but 25% would tell him if he asked), and only a few (25%) would inform him of an XXY diagnosis in the fetus. However, in the case of a diagnosis of Huntington's disease (where the patient refuses to disclose the result to the relatives), only 13% of the participants would respect confidentiality, compared with 40% in the United Kingdom (Kromberg & Jenkins, unpublished data).

Again, in the case of a bus driver at high risk for familial hypercholesterolemia (and the issue of confidentiality versus public safety) only 31% of the participants would keep this information confidential, while 65% in the United Kingdom and 53% in Australia would do so (Kromberg & Jenkins, unpublished findings). Regarding prevention of genetic disease, 19% of participants agreed that in the case of high-risk pregnancies all should be tested (the figure was 3% in the United Kingdom and 86% in Cuba; Kromberg & Jenkins, unpublished findings).

In general, the South African participants were found to be somewhat paternalistic in their attitudes, and since not many had been exposed to much training in the field of medical ethics, such training was recommended. However, this survey was undertaken nearly 15 years ago, attitudes have probably changed somewhat over time, as has been shown in other countries, and a new survey should be conducted. Nevertheless, the survey played an important role in raising the awareness of local genetic counselors and alerting them to ethical issues at the time. Presently, the code of ethics of the National Society for Genetic Counselors of the USA (Eunpu, 1998), which provides a general framework for professional conduct, is used as a basis for training local genetic counselors on the ethical issues they may encounter in their genetic counseling practice.

USE OF THE GENETIC INFORMATION

The collection and dissemination of genetic information is surrounded by ethical issues. These include issues of confidentiality, informed consent, patient autonomy, allocation of scarce resources, and the imperative to do good (beneficence) and do no harm (non-maleficence). In South Africa, medical geneticists and genetic counselors are taught the basic ethical principles, and genetic counselors take the Hippocratic Oath upon qualifying. In general, both groups of professionals agree that the genetic information they collect from their patients is the property of the patient; it is kept confidential and not divulged to any third parties except with the written permission or informed consent of the patient.

Patients also have the right of access to all information on themselves obtained by the counselor/physician counselor. However, two ethical principles can be in conflict; for example, if, in the course of molecular studies for albinism on two "parents" and an affected child, nonpaternity is uncovered, this information may be withheld if there is no advantage to the family in knowing about it. In such a case, encountered by one of us (JGRK), the father believed the 10-year-old child to be his, and he had loved and supported him all his life. However, the right to full information was in conflict with the imperative to do no harm, and the counselor chose to take the latter path.

The ability of the life insurance industry to access genetic test results has raised public concern regarding loss of privacy and the possibility of discrimination, and this led to a research project recently being carried out in Johannesburg (Kinsley, 2009). The findings showed that the responses of the companies involved in the study, to the collection of genetic information, particularly genetic test results, were fair and aligned with the regulatory guidelines of the industry. However, some irregularities were found that could lead to discrimination. These were due to misunderstandings relating to genetic disease mechanisms, the assigning of genetic risk, and the interpretation of genetic information. It was concluded that such misunderstandings could be corrected by support for the industry from specialist genetic services.

PRENATAL DIAGNOSIS

The field of prenatal diagnosis is well developed in South Africa, and the service, initially provided by the academic departments of human genetics at the University of the Witwatersrand in Johannesburg (Kromberg et al., 1989) and the University of Cape Town, has been extended to other major cities, where private pathology practices generally provide the testing. One of the ethical issues arising in this field involves the diagnosis of fetal abnormalities. A recent study shows that even in the state hospital system, White patients are receiving their diagnostic results of fetal abnormalities earlier than are Black patients (Todd et al., 2010). The principles of equity and accessibility and quality of services are involved in such a situation. The patients all have access to the service, but the Black patients present for the service later; whether this is due to the problem of poorer assessment and detection of problem pregnancies—associated with staff shortages and/or staff morbidity due to the high rates of HIV in nurses (Bateman, 2008)—at the primary healthcare level, and/or patients delaying acting on advice, or misunderstanding the urgency of the situation, or consulting family members or traditional healers first, or lack of funding for transport, or other reasons, is not known.

According to the findings from the South African survey on Genetics and Ethics, medical geneticists and genetic counselors believe in giving options and the freedom of choice in the prenatal diagnosis setting (Kromberg & Jenkins, 1997a), as well as the provision of full information on the fetal anomaly, and obtaining written informed consent to termination of pregnancy, if that is requested, or support if the pregnancy continues. The attitudes of women to termination are generally positive in this situation and, in a recent study, 70% requested termination (for conditions such as trisomies 21, 18, 13; Todd et al., 2010).

However, the attitudes of South African geneticists and counselors to using prenatal diagnosis for the purposes of fetal sexing (with the aim of asking for a termination if the fetus is of the unwanted sex, usually female) are generally negative (Kromberg & Jenkins, 1997a). Most counselors are opposed to fetal sexing as an indication for prenatal diagnosis, unless it is performed for a high risk of severe X-linked conditions, which cannot be diagnosed in males and where the intention would be to request termination if the fetus were found to be a high-risk male. Although the authors know of a few cases, requests for sexing of the fetus are rare in South Africa (the fertility rate is 1.9 per woman in the Johannesburg area, and our observation is that most couples accept the two children they have, although in some cultures there is still the widely held belief that a "pigeon pair," i.e., a boy and a girl, is the ideal).

Prenatal diagnostic services are offered in some private practices in South Africa. This situation is not ideal, as such patients seldom receive a comprehensive service because they are generally not referred for genetic counseling. This means that they often do not fully understand what tests they are having, what the results might show, what the possible outcomes are, and the choices available to them. Most medical geneticists in South Africa believe that the highest standard of care includes at least one session with a genetic counselor prior to prenatal testing being performed, even in routine cases such as women of advanced maternal age (50% of such women have other, additional indications which may not be detected when genetic counseling is not included in the prenatal diagnosis service; Wessels T, personal communication, 2009).

NEWBORN SCREENING

Very little newborn screening is undertaken in South Africa, since the ethical issue of "allocation of scarce resources" comes into play in this setting and the epidemiology regarding the common diseases of infants has not been fully investigated. Without good evidence, the rates of the diseases were believed to be so low that testing might not be cost-effective. New information, however, shows that taken together, the rates of these diseases are as high as in other countries where screening is performed (for example, galactosemia alone occurs in 1 in 18,000 in South Africa; Manga et al., 1999). In a recent project, testing for 33 diseases in 35,000 newborns identified 13 affected infants, giving a rate of 1 in 3700 births, which is arguably high enough to make newborn screening worthwhile and cost-effective in South Africa, provided that an efficient, effective, well-organized, all-round service is available (J Cartwright, personal communication, 2009).

Ideally, such screening should only be performed if the results will be of benefit to the child; if the disease to be screened for is treatable; if there is benefit to diagnosing it at birth, and experts are available to treat it; if such treatment would prevent morbidity in the child; if the necessary laboratory tests (quality controlled, reproducible, and costing only about $1 each) and trained laboratory staff are available; if the community is educated regarding such testing and the possible results; if the mother and child can be traced once the results become available; and if genetic counseling services are available and accessible. For example, a small newborn screening project is taking place in a state hospital in the Cape (J Cartwright, personal communication, 2009), on a trial basis, but only mothers who have had caesarian sections are being included, since they are usually still in the hospital three days after the birth (whereas most mothers with normal deliveries only stay 24 hours), when the results become available. Tracing newborns and mothers with no fixed abode or address, or living in shacks, is often difficult in South Africa's cities, which have inadequate housing, many peri-urban informal settlements, and where rapid urbanization is occurring. This has been one of the several stumbling blocks in the introduction of newborn screening.

SCREENING OF CHILDREN

Screening of children for genetic disease, or for predictive testing for such a disease, is generally not approved of in South African genetic services. Such testing needs to be in the child's best interests, preserving their future choices, and undertaken preferably with informed consent (Harper & Clarke, 1990) and not at the request of a third party.

The age of informed consent is still debatable, but the age of 18 years is usually considered appropriate in the case of genetic testing. Exceptions to this rule include some of the inherited cancers in families with a strong family history of, for example, familial adenomatous polyposis and colorectal cancers, where regular screening and treatment could be required in children carrying the gene. Furthermore, girls at 50% risk of being carriers for hemophilia are sometimes tested for carrier status at the molecular level, in Johannesburg, since some are at risk of excessive bleeding at, for example, tonsillectomy.

There is some discrepancy between the decisions in this regard made in academic circles and in private practice, and there are scientists in private laboratories who undertake testing of children on request without considering the ethical implications. South Africa needs a national policy and ethical guidelines on these issues, so that the Health Professional Council of SA, one of whose main roles is to protect the public, can monitor the situation and, where necessary, censure unethical behavior.

ADULT SCREENING AND DIAGNOSIS

Ethical issues that have presented in this area include the testing of people of European extraction for the cystic fibrosis gene. It is accepted that the presently available DNA technology can only detect approximately 80% of the mutations for cystic fibrosis in these populations, and therefore about 1 in 5 carriers will be missed using the modern technology. Also, anxiety may be raised in carriers who might misunderstand the seriousness of having an abnormal gene (thinking it more serious than it is), and may not receive adequate and informed counseling. These issues, together with the problem of allocation of scarce resources and the prevalence being high in only one minority population group, have led to the rejection of such screening in South Africa. Cystic fibrosis was thought to be uncommon in the Black population. However, it is now known that it occurs at a rate of about 1 in 4600 births (compared with 1 in 2000 for infants of European extraction), and different mutations from those causing the disease in the White population are responsible (Padoa et al., 1999). However, although population screening is not available, cascade screening of the adult relatives of an affected child is encouraged (Macaulay,

2007). Similarly, the screening of adult female relatives of a boy with Duchenne muscular dystrophy, where the molecular abnormality has been defined, is recommended (Erasmus, 2009)

On the other hand, population screening for Tay-Sachs disease was promoted in the local Jewish population due to the high rate of carriers (1 in 26), the effective testing available, the requests of the local community, and seriousness of this lethal disease (Jenkins et al., 1977). As a result of this testing, the birth rate of affected children fell by about 80% within a few years of initiating the carrier testing (AB Lane, personal communication, 2009).

Another contentious issue is the testing of the majority Black population for the genes for albinism (the carrier rate is about 1 in 30). The majority (90%) of medical students, when asked (informally by one of us, JGRK, at a lecture on the subject) if they would choose to be tested, indicated that they would do so and if a child was diagnosed in utero many would request a termination of the pregnancy. The effect such screening would have on community attitudes toward albinism is of great concern. It is noted that just because a test can be done and is available does not mean it should be offered to the public. This situation presents particular ethical challenges, because of the various social and cultural motives that may be involved, and much research needs to be carried out before such screening is undertaken.

RESEARCH ISSUES

Research in developing countries should be focused on current practical problems and ways of solving them. In South Africa, most research on human subjects is carried out by the academic staff at universities, the Medical Research Council, Human Sciences Research Council, and other similar bodies. Most of these groups accept that all research protocols should go through ethics committees of the institutions where the researchers work. At the University of the Witwatersrand, the Human Research Ethics committee screens all applications for ethics clearance. If there are objections the applicant has to meet the demands of the committee, resubmit the application, and await approval before the project can begin. This procedure is mainly to protect the public from unethical research practices and to make sure that the project is realistic and feasible, that research subjects are voluntarily recruited, understand the purpose of the research (they must receive an information sheet), give informed consent (the consent form must be comprehensive and approved by the committee) to participation, are not subjected to any unnecessary or harmful (both physically and psychologically) procedures, and will be informed of the outcome of the study, if they wish to know about it. These processes ensure that research is carefully screened by experienced people, although there are still some problematic exceptions. This is a better picture than that found elsewhere in Africa, where concerns have been raised about inadequate research ethics regulations, the responsibilities of foreign research teams to vulnerable populations in these countries, and the fact that research was seen as cheaper, easier, and less likely to meet with legal opposition and stringent ethics reviews in developing countries, including South Africa, in many respects (Anya & Raine, 2008).

The International Research Ethics Network for Southern Africa has been set up to assess capacity, increase awareness, and train key people in the field. There is increased emphasis on the need to benefit the populations among whom research is taking place. In South Africa there has been a successful introduction of statutory comprehensive ethics teaching programs in medical and nursing courses; other African countries, such as Nigeria, are interested in following suit (Anya & Raine, 2008). Anya and Raine (2008, p. 1596) state that "South Africa has a longer history of clinical and research ethics than other African countries and therefore the potential for shared learning is great." The same would apply to the field of medical genetics. Since South Africa has probably the oldest and most sophisticated medical genetics service available in Africa, and consequently geneticists have faced many of the complex ethical issues occurring in the provision of such services, professionals from other countries could benefit by learning from this experience.

FUTURE ISSUES TO BE CONSIDERED

South Africa can be expected to face various contextual ethical issues in the field of medical genetics in the future. The allocation of research funds raises some contentious issues. However, allocation is already being redirected and tightened up, so that research relevant to the needs of a developing country, and which will lead to the improvement of circumstances for the majority, has priority for funding bodies. Furthermore, research projects leading to outcomes with practical applications, which are sustainable at the development level, are more likely to be prioritized. Nevertheless, the many international advances being made in medical genetics still need attention, as well as the focus of local researchers, so that the country does not fall behind in this regard.

Genetic services are just one, and a very small one at that, of the many health services required, and a balance with regard to other services, as well as the allocation of scarce resources and societal issues, needs to be considered in the delivery of these services. Presently genetic counseling is frequently crosscultural counseling, and as such, the counselors need to develop a sound insight into the cultural practices, especially those related to the utilization of healthcare services, of the patients being counseled. More research is required on the ways in which these cultural issues affect genetic services, as well as the ways in which acculturation is taking place. Such research should be carried out by local researchers in local settings. At present, genetic services are run mostly by academics, and it would be preferable for the country if the development of the service was taken over by the state, so that statewide programs could be developed. Those areas where services are not available could be prioritized, appropriate research projects could be set up, and genetic education of the community could continue. Coordinating bodies need to be established so that ethical practice in genetic services can be formalized, a code of practice drawn up, criteria for genetic testing can be agreed upon (and abuse of testing minimized), and outcomes can be monitored.

LEGAL ISSUES

South Africa has a fine new constitution and bill of rights, which were drawn up in 1993 prior to the first democratic election in 1994 (Republic of South Africa, 1996).

There are several legal issues that are relevant to the delivery of genetic services. The new government reviewed the Termination of Pregnancy Act, and the Choice on Termination of Pregnancy Act was passed in 1996. It states that termination of pregnancy may be carried out at any stage of pregnancy when two medical practitioners, or one medical practitioner and one registered midwife, believe that the continued pregnancy would result in a severe malformation in the fetus. "Severe malformation" is not defined in the Act, and the decision to offer termination for fetal abnormalities is left to the discretion of the doctors involved.

Other laws that reflect on genetic services include the Health Act, Sterilization Act of 1998, Mental Health Care Act of 2002, Children's Act of 2005 and the Children's Amendment Bill of 2006, the Employment Equity Act (M Christianson, personal communication, 2007), the Health Professions Act No 56 of 1974, and its amendments. In addition, the National Committee for Confidential Enquiry into Maternal Deaths, National Contraceptive Policy, and National Maternity Care Guidelines, have relevance for genetic service delivery.

The revised Mental Health Care Act, No 17, was passed in 2002. This bill covers the rights of people with mental health problems, as well as those with intellectual disabilities (families with these problems, often undiagnosed, present frequently at local genetic counseling clinics). The constitution states that "Everyone is equal before the law and has the right to equal protection and benefit of the law," and "The state may not unfairly discriminate against... anyone on one or more grounds, including... disability," and "Everyone has inherent dignity and the right to have their dignity respected and protected."

Regarding children, the Constitution (1996, Section 28) states that: "Every child has the right... to family care or parental care, or to appropriate alternative care when removed from the family environment... to basic nutrition, basic health care services and social services... to be protected from maltreatment, neglect, abuse... to be treated in a manner and kept in conditions that take account of the child's age..."(<18 years). The Children's Act No 38 of 2005 (Section 11) states that, concerning a child with a disability, due consideration must be given to providing special care when appropriate; making it possible for the child to participate in social, cultural, religious, and educational activities, recognizing the special needs the child may have; providing conditions that ensure dignity, promote self-reliance and facilitate active participation in the community; and providing the child and child's caregiver with the necessary support services. Children with genetic disorders resulting in disabilities are, therefore, well covered by these acts.

The preamble of the Sterilization Act of 1998 states: "To provide for the right to sterilization... and in particular the circumstances under which sterilization may be performed on persons incapable of consenting or incompetent to consent due to mental disability...." and that "the inability to give consent should not automatically entail the loss of constitutional rights... it is necessary to ensure that mentally disabled persons are able to exercise these rights as far as possible." Regarding persons not capable of consenting, the Act states that no one is prohibited from having sterilization if over 18 years and capable of consenting; if not capable of consenting, due to severe mental disability, then upon request of a parent, spouse, guardian, or curator to the person in charge of a hospital; a panel (consisting of a psychiatrist, or doctor, psychologist or social worker, and a nurse) must be set up to consider all information and possible alternatives; the termination must be performed at a designated facility. The Act also states that if a severely mentally disabled woman is incapable of understanding and appreciating the nature of the consequences of a termination, her pregnancy may be terminated during the first 12 weeks or from 13–20 weeks upon request of and with consent of her natural guardian.

The professions of medical geneticist and genetic counselor are at present being regulated by the National Department of Health. Registration is available under the Health Act, and the Health Professionals Council of South Africa has outlined the scope of practice and monitors the training and registration of these professionals, in line with the performance expected from other health professionals. The genetic professionals receive some teaching in their courses on the laws that apply in genetic service delivery, and the legal implications of providing genetic counseling.

SOCIOCULTURAL ISSUES

The concept of *culture* indicates difference and the recognition of difference (Strathern, 1997). But what work does culture do? Firstly, the concept is associated with the opening up of assumptions, or an openness to cultural context, not taking things for granted, and remembering that when opinions are formed they are made up from personal opinions as well as values and assumptions shared with others. Secondly, it is associated with working through the ramifications of issues; we need to think about the way opinions mobilize values, as well as be open to their wider social implications. Thirdly, it draws attention to the work of ideas, and the way things are conceptualized. Values are based on ideas about the world, but ideas also determine how people think, act, and react. Fourthly, culture is associated with the invitation to work with specifics, since a culture is specific to time and place. Views of the world tend to be associated with the time and place of people's lives, and operate relative to particular circumstances. Fifthly, the concept of culture may be an aid to reflection. Studying cultures outside those that are Eurocentric means becoming attuned to the assumptions with which we (Europeans) start, and to what makes us cultural beings. Lastly, there is no substitute for empirical inquiry, and investigation is essential. Important to genetic counseling is that culture consists of the techniques and practices that mediate the way people interact and communicate with one another. It offers both a resource and a limitation, enabling communication and constraining it. Therefore, empirical inquiry is essential if understanding about what different people mean, in a variety of circumstances, is to be advanced.

Cultural and ethnic backgrounds of individuals and their families, as well as the background differences between these individuals and the health professionals offering genetic services, may influence the way problems are approached, the receptiveness of the counselees to assistance, guidance, and interventions, communication, the impact the genetic disorder has, and the way in which the family accepts it (Murray, 1984**)**. Some of the sociocultural issues and the way in which they impact on genetic counseling locally will be discussed below.

SYSTEMS OF THOUGHT

Providing genetic counseling to African patients whose worldview might involve an amount of fatalism (Kromberg & Jenkins, 1997b) and/or strong beliefs in destiny (Jegede, 2009) can be problematic. One of the major components of culture, according to Read (1966), involves the complex beliefs that surround the relation of people to nature, people to their fellows, and people to the supernatural powers that are believed to control the universe. If this control is seen as absolute, then individuals cannot alter what happens to them. Such strong beliefs can lead to patients rejecting prenatal diagnosis and selective abortion. Higher levels of fatalism have been found in a group of African-American women, compared with women of other ethnic groups (Kuppermann et al., 2006). Also, higher scores on a fatalism scale were shown to be associated with lower willingness to request termination of pregnancy for fetal abnormalities in another U.S. sample (Learman et al., 2005). Individuals who are fatalistic believe that all events are predetermined and inevitable.

Another system of thought that may affect genetic counseling in the South African setting is associated with communal decision making. Such decision making has been pursued in the past because what affects the individual also affects the other members of the group. As has been reported from Nigeria (Jegede, 2009), the well-being of the individual affects the well-being of the community, and vice versa. The traditional worldview tends to favor collective rather than individual thinking, according to South African anthropologists (Hammond-Tooke, 1989), and this view may also be associated with an external locus of control. The social schema of the body might predominate over the individual schema, and a balance between these two schemas is essential to promote self-steering behavior (Manganyi, 1973). In line with group decision-making practices and fatalistic beliefs, is a diffuse body-image boundary differentiation. For example, people with albinism were found to have a significantly less diffuse body-image boundary differentiation than did the normally pigmented controls, although both groups had diffuse boundaries associated with difficulties in differentiating themselves from the group (Manganyi et al., 1974). Sociocultural issues tend to nurture the development of body boundary characteristics, and these are related to many different areas of personality functioning.

Integrated with these systems of thought is the undefined and/or indistinct line between life and death, the strong belief that there is life after death (Hammond-Tooke, 1989). Furthermore, there was the belief that some people became ancestral spirits after death and continued to take an unpredictable interest, and interfere, in the lives of their descendants. These beliefs can affect both the perception of genetic disorders and ideas about their causes.

Although these systems of thought used to be widespread, they might now be found more commonly among people living in the rural areas than among those in the urban areas of the country, where acculturation is occurring more rapidly. However, no studies have been carried out to establish that this is the case.

DISEASE CAUSATION

Initially, all illness was attributed to the ancestors, particularly among one local ethnic group, the Southern Sotho, who believed that the ancestors would try to cause the death of people in order to secure their companionship (Hammond-Tooke, 1993, p. 153). Traditions and customary rituals, which have to be followed in the correct way, surrounded ancestor worship, and ancestors could complain of neglect. Transgressing the rules could anger them and they might then cause illness, reproductive difficulties, congenital abnormalities, and other problems. We are aware of parents of children with albinism and with Down syndrome blaming the ancestors (who were accused of punishing them) for the child's problem. Some illnesses, however, could be caused by witchcraft or ritual pollution. Witchcraft was associated with some individuals wishing to harm others, while pollution came from being involved, usually inadvertently, in a contaminating situation (such as with a corpse).

Childhood diseases, however, were also sometimes attributed to maternal impression; for example, if one laughed at an albino when pregnant, one could have a child with albinism (Kromberg & Jenkins, 1984). The breaking of strict taboos (e.g., having sexual intercourse too soon after the birth of a baby) could also lead to congenital disorders or intellectual disability in the next child (Schapera, 1940). In a study on 100 rural traditional healers, it was found that 51% ascribed the cause of childhood disabilities to the breaking of taboos, such as those associated with the rights of first wives and the rules of polygamy (Kromberg et al., 2008).

A preventable cause of childhood disorders, however, that needs to be mentioned, is fetal alcohol syndrome. High rates of this syndrome have been recognized in South Africa, particularly in the Cape Province among the children of mixed race parents (May et al., 2000). Medical geneticists have recognized and researched this problem and the sociocultural circumstances that surround it. Consequently, intervention and education programs for women of childbearing age have been initiated. These programs are teaching them about the dangers of consuming alcoholic drinks in pregnancy, and the risks for the fetus, and it is hoped that prevalence rates will drop in future.

COMMON PRACTICES AND TABOOS

Children were much desired and marriages were believed to be incomplete without them, according to reports from South Africa (Hammond-Tooke, 1989), as well as Nigeria (Jegede, 2009). Sons were essential for inheritance purposes, for continuity of the family name and descent group, to care for the parents in old age and to sacrifice to their spirits after death. Infertility was therefore greatly feared and potent medicine was required, as well as fertility symbols, to cure the condition (Hammond-Tooke, 1989). A pregnant woman was subject to many taboos to protect the fetus and ease the delivery. She was required to avoid certain foods (such as guinea fowl, to prevent a long flat head in the offspring, or rabbit to prevent big ears and/or teeth) and some outside activities, so that she would not come into contact with pollution or the medicines of sorcerers. Food taboos for pregnant women have also been reported among the Yoruba in Nigeria (Jegede, 2009). Recent research shows that at least 80% of urban women in the Johannesburg area still use traditional drugs at some time during pregnancy to protect, particularly from evil spirits, and strengthen the fetus (M Glass, personal communication, 2009).

Previously, twins were considered unnatural (and therefore polluting) and one or both were killed in some groups, particularly among the Sotho and Venda, who believed they caused droughts. Infanticide was also reported in connection with congenital abnormalities, and those born with a tooth or feet first (Schapera, 1940). No recent data on these issues exist.

Cultural practices also used to provide rules for mate selection, and the ideal partner was often a first cousin, with a different surname (i.e., a mother's brother's child) in some ethnic groups. Two Tswana proverbs illustrate this situation: "Child of my paternal uncle marry me, so that the cattle should return to our kraal" (referring to the payment of the bride-price; Schapera, 1940, p. 42), and, "Side by side with his cousin a man is always happy" (Schapera, 1940, p. 41). Cousin marriages were also promoted among the Swazi ruling parties to keep the political power in the family. These practices led to a high rate of recessive conditions, the commonest of which was oculocutaneous albinism, in groups practicing consanguineous marriages (Kromberg & Jenkins, 1982). However, in some groups, such as the Zulu, such marriages were generally taboo (Schapera, 1937).

Polygamy is widely practiced in South Africa (as is the case in Nigeria and elsewhere in Africa; Jegede, 2009), and many men may have one wife in the town where they work and another in the rural area from which they come. This custom is practiced by the South African president, Jacob Zuma, who has five wives, but in other cases serial polygamy occurs. Some polyandry exists too, but it is generally sequential and a woman may have several husbands or partners, one after another. In the genetic counseling situation it is important to establish the practices in any one family, as they may affect the recurrence risk for recessive conditions, which is 1in 4 only if the next child is from the same two parents as the affected child. The frequency of migrant labor adds another dimension to these issues. Also, the collecting of pedigree information can be complicated by these practices and by the naming of various relatives (e.g., a mother's younger sister may be called a younger mother).

USE OF TRADITIONAL HEALERS

Traditional healers still play an important part in the health services of South Africa. There is one available in almost every village in the country, and about half the people in the urban areas use their services, often in combination with Western medical services (Freeman et al., 1994). In a study on epilepsy in a rural province, the findings showed that 42% of parents consulted traditional healers alone, and a further 22% consulted both traditional healers and Western medical services (Christianson et al., 2000).

Traditional healers involved in the study by Kromberg et al. (2008) were interested in knowing more about genetic diseases, and at the conclusion of the project they requested and participated in tutorials on the topic given by the research team, and a workshop on epilepsy and its treatment given by community workers from the SA National Epilepsy league. It is hoped that as a result of this action-oriented research they were willing to refer their cases of epilepsy to the nearest hospital, where treatment, including anti-epileptic medication, was available.

MYTHS AND SUPERSTITIONS

There are many myths and superstitions which, if pervasive in the patient's system of thought, can interfere in the process of genetic counseling and the development of necessary insights and good understanding. Some of the relevant myths include those surrounding reproduction and embryology. Since menstruation stops at the beginning of a pregnancy, some local mothers used to believe that the child received all its blood from the mother (Schapera, 1940). This myth needs to be understood because, if the patient has this belief, it will interfere with her developing an understanding of the genetic contribution of both parents. Alternatively, as other mothers believe, they may think that the father gives the life-force to the infant and the mother is only the incubator for the developing child. More research in this field is necessary to gain insight into how common these beliefs are in the country, and how they affect the acceptance and effectiveness of genetic counseling services.

The only inherited condition that has been very thoroughly investigated on the psychosocial and cultural level in South Africa is albinism. Several of the myths that surround this condition have been uncovered (Kromberg & Jenkins, 1984; Kromberg et al., 1987; Kromberg, 1992). Maternal impression is believed to be a possible cause for the condition (although if there is a family history, which occurs quite frequently, due to the fact that about 1 in 30 black people carries the gene, then patients may recognize the part played by inheritance), as is punishment for bad behavior toward people with albinism. The possibility that albinism is infectious is also considered by many people and, furthermore, most affected people are not accepted as marriage partners by the community. Also, it is widely believed—by the community, by people with albinism themselves and their relatives—that albinos do not die natural deaths (Kromberg, 1992). This myth causes them and their families to feel insecure and unsure of what the future holds, and it needs to be discussed in genetic counseling sessions with affected individuals and/or their families (Kromberg & Jenkins, 1984).

ATTITUDES

An attitude is an idea charged with emotion, which predisposes a class of actions to a particular class of situations (Triandis, 1971). Three components are involved: *(1)* a cognitive component; *(2)* an affective component; and *(3)* a behavioral component. These three components interact, the cognitive being the minimum condition for having the attitude and becoming associated with pleasant or unpleasant events and/or desirable or undesirable goals. The category then becomes charged with affect. The more pleasant the events (or the more desirable the goals) and the more frequently they occur in the presence of the cognitive component, the greater is the amount of affect that becomes attached and the more likely that a behavioral component will result. In other words: attitudes involve what people think about, feel about, and how they would like to behave toward an attitude object. Behavior is not only determined by what people would like to do, but also by what they think they should do (that is, social norms), by what they have usually done (that is, habits), and by the expected consequences of the behavior.

We have studied, with various colleagues, attitudes to two common inherited disorders in the local Black population; namely, oculocutaneous albinism and Down syndrome. Albinism is the commonest recessive condition in Southern Africa, and it is obvious to all, especially in the Black population. Attitudes have varied from complete rejection (with the behavioral component of infanticide, as described by Livingstone, 1857) to adoration (one mother, known to one of us, JGRK, was a traditional healer and after the birth of her child with albinism, whom she named Gobela, or spirit, her practice thrived and she desperately wanted the next child to be affected, too). Attitudes are partly determined by beliefs, and in some African communities, as well as in Papua New Guinea, albinos were said to be very strong and were therefore admired and respected. Although attitudes have been found to be quite positive (Kromberg, 1992), as social distance became closer they became more negative. Albinos themselves did not feel fully accepted and tried to make themselves more acceptable by, for example, dyeing their hair black. Furthermore, Livingstone (1857) was asked by an albino woman for medicine to turn her skin black. Our study on adolescents in Soweto, Johannesburg, showed that those with albinism were generally accepted and that attitudes were positive, but affected girls were significantly better accepted than affected boys (Kromberg,

1992). Acceptance, however, stopped short of marriage, although this did not preclude childbearing. Our study on bonding in a normally pigmented mother with a child with albinism showed that significantly more mothers rejected their children emotionally for the first three months, in comparison with controls with normally pigmented infants (Kromberg et al., 1987). Possibly this behavior was due partly to community attitudes and the insecurity caused by the death myth, but also to the mother's difficulty in identifying with an infant who looks so different from herself. However, they then slowly started to bond with the affected infant, and bonding behavior was no different from the controls by the time the infants were 9 months of age. Nevertheless, our experience shows that in some situations people with albinism are so well integrated into their community that they have had the terminal phalanx of their fifth finger amputated like all the others in their group.

In the case of Down syndrome, on the other hand, attitudes are determined by other factors. Down syndrome is common in Southern Africa, occurring in about 1 in 600 infants. It is not always easy to recognize, and therefore attitudes are not so clearly formed when the child is young and they do not interfere with bonding initially. Later, however, when the mothers observe the slow development of the child, they realize the nature of the condition and find it difficult to cope, attitudes become more negative and mothers show signs of stress. One study on stress in these mothers showed that there was a significant difference between mothers with affected children and the controls, on psychosomatic factors related to stress (Kromberg & Zwane, 1993). Mothers complained significantly more often of poor general health, depression, tiredness, sleeplessness, and bad dreams. In terms of attitude theory, the mothers' attitudes became affected and emotionally charged by their experiences of regular difficulties with their child, and this influenced their attitudes in a negative direction. Mothers stated that their children were always sick and they could not communicate with them.

Attitudes, however, can change in both directions. Understanding the condition and its cause, and sharing the problems with people in the same situation, can help to improve the mother's attitude to her child and help her cope with her situation. For this reason we initiated facilitated mothers' groups, in the 1970s, where mothers with children with Down syndrome could benefit from obtaining new knowledge and supporting each other. There is now a very active nationwide Down Syndrome Association. Similarly, an active Albinism Society of South Africa was formed some years ago. However, more work is required, from professionals and involved lay people, to change community attitudes and dispel the prevailing myths; such an awareness program should help to turn attitudes in a more positive direction and improve the lives of affected children in the future.

LANGUAGE ISSUES, COMMUNICATION AND TRANSLATORS

Due to the fact that most of the professionals working in the field of human genetics in Southern Africa are English or Afrikaans first-language speakers, there are often language barriers in the provision of genetic services to patients who have other languages as their first language. Trained translators are provided in situations such as in the courts of justice, where it is essential for the people involved to understand the proceedings. However, no translators are available in hospitals or outpatient clinics, where patients need to understand their condition and its importance for their health (sometimes in life-and-death circumstances). This is a great problem for genetic services, especially as the concepts are complicated and the languages spoken by the patients often do not have words for genetic terms such as *chromosome*, *recessive inheritance*, etc., that have to be explained. The only resource available is usually a hospital nurse who is willing to translate and can speak the patient's language fairly well (there are 11 official local languages, but most Black people can speak some Zulu and/or some Sotho, the languages of the largest groups in the country). Post-counseling sessions should be offered to all those patients where either simplified English or Afrikaans had to be used, or where an untrained translator (a nurse, nursing assistant, or relative) was used, so that the understanding of the patient can be checked, corrected, and reinforced where necessary.

CONCLUSION

In this chapter we have attempted to cover some of the many ethical and sociocultural issues and situations that have arisen or may arise in the delivery of genetic services in South Africa. We have placed these in the context of the present health and service delivery issues facing the country, as it struggles to meet the many demands of a developing nation. Many of these genetic issues we have encountered personally, and some we have researched at some depth. We acknowledge that there is still a dearth of research in this field, and that much still needs to be done. On the ethical front, South Africa requires ethical guidelines for genetic services, and these guidelines need to be developed by the Department of Health, in consultation with the active professionals providing services in the field. The Health Professionals Council of South Africa has its own Code of Ethics for health professionals, and this code at present is the only code officially available for guiding their professional behavior. A code of ethics (or a Code of Ethical Behavior, such as that developed by the Ethics Committee of the Council of Regional Networks in the United States; Baumiller et al., 1996a, 1996b) specifically for practitioners in the field of human genetics may be useful, as a guide for ethical behavior in this context. Further training of genetic professionals, both during their qualifying courses and in service, is also needed. However, another survey is required to assess the attitudes to ethical issues among professionals presently working in genetic service delivery, since many have entered the field since the last survey was conducted.

Regarding sociocultural aspects of genetic services, one needs to note that research in Africa is in its infancy, and the little that there is mostly emanates from South Africa. We have given a brief overview of some of the cultural issues, and brief summaries of the findings of selected relevant projects and, in some cases, described their effects on the communities where they were undertaken. Such research needs to show benefits for the people who are investigated, and we have illustrated where this has occurred. A more in-depth understanding of sociocultural issues is required if genetic counseling is to become more culturally relevant and effective (and this effectiveness itself, and the interfering stumbling blocks, need to be researched, too) in this pluralistic society, with its multitude of cultures and varying levels of cultural adaptation and development.

South African genetic professionals are developing closer ties with professionals working in the field in other African countries, and this has led to, for example, one of the authors (JGRK) being invited to discuss our research on the psychosocial, health, and genetic issues of albinism with dermatologists from all over Africa at an annual dermatology conference in 2008 in Tanzania. Also, our research findings on intellectual disability and stress in rehabilitation workers have been presented at the annual congress for occupational therapists, from various central and southern African countries, in 2007, in Zanzibar. Such opportunities for interaction with other African countries should lead to a greater accumulation of in-depth knowledge in the future, and possibly interdisciplinary cooperative research projects, to the benefit of all concerned.

We have briefly mentioned South Africa's new constitution, and some of the laws that have relevance to the practice of genetic counseling in South Africa. These laws are generally acceptable to the genetics community and appear to cover the necessary issues, to the satisfaction of the professionals working in the field.

South Africa is a developing country, and the genetic health needs of the population have been quite overwhelmed by the other desperate health needs experienced by many of its people. We hope that as the AIDS epidemic and infectious diseases are brought under control, and as the socioeconomic status of the average person improves, genetic services will receive the attention, funding, and staffing required, so that the country has a service that meets the genetic needs of its diverse ethnic groups. Such staffing and funding will mean that the ethical, sociocultural, and legal issues that surround genetic counseling service delivery can be researched more comprehensively, and the findings applied, so that the country can develop a service that is more directly relevant and appropriate for the people it seeks to serve.

ACKNOWLEDGMENTS

The authors acknowledge the support from the Mellon Foundation Mentorship scheme of the University of the Witwatersrand, Johannesburg, South Africa.

REFERENCES

Anya I, Raine R. (2008). Strengthening clinical and research ethics in Nigeria – an agenda for change. *Lancet 372*, 1594–1597.

Bateman C. (2008). Lack of warm hands will render health care system impotent. *S Afr Med J 98*, 510–511.

Baumiller RC, Comley S, Cunningham G, et al. (1996a). Code of ethical principles for genetics professionals. *Am J Med Genet 65*, 177–178.

Baumiller RC, Cunningham G, Fisher N, et al. (1996b). Code of ethical principles for genetics professionals: an explication. *Am J Med Genet 65*, 179–183.

Benatar S. (2004). Health care reform and the crisis of HIV and AIDS in South Africa. *New Engl J Med 351*, 81–92.

Christianson AL, Zwane ME, Manga P, Rosen E, Venter A, Kromberg JGR. (2000). Epilepsy in rural South African children – prevalence, associated disability and management. *S Afr Med J 90*, 262–266.

Eunpu DL. (1998). Relationally based professional conduct for genetic counselors. In: Baker DL, Schuette JL, Uhlmann WR (eds.). *A Guide to Genetic Counseling* (pp. 277–293). New York: Wiley-Liss.

Erasmus S. (2009). *Duchenne and Becker muscular dystropyhy: Implications for at risk individuals.* MSc(Med) research report. Johannesburg: University of the Witwatersrand.

Freeman M, Lee T, Vivian W. (1994). *Evaluation of mental health services in the Orange Free State of South Africa.* Centre for Health Policy. Johannesburg: University of the Witwatersrand.

Hammond-Tooke D. (1989) *Rituals and Medicines.* Johannesburg: A.D. Donker.

Hammond-Tooke D. (1993) *The Roots of Black South Africa.* Johannesburg: Jonathan Ball.

Hargrove J. (2007). *Migration, Mines, and Mores: The HIV Epidemic in Southern Africa.* Inaugural address, 14/11/2007. University of Stellenbosch, Stellenbosch.

Harper PS, Clarke A. (1990). Should we test children for adult genetic diseases? *Lancet 335*, 1205–1206.

Jegede AS. (2009). Culture and Genetic Screening in Africa. *Developing World Bioeth 9*, 128–137.

Jenkins T, Lane AB, Kromberg JGR. (1977). Tay-Sachs disease screening and prevention in South Africa. *S Afr Med J 51*, 95–98.

Jenkins T. (1990). Medical genetics in South Africa. *J Med Genet 27*, 760–779.

Kinsley N. (2009). The use of genetic tests by the individual life insurance industry in South Africa. MSc(Med) Research report, University of the Witwatersrand, Johannesburg.

Kromberg JGR, Jenkins T. (1982). Albinism in the South African negro. II. Prevalence. *S Afr Med J 61*, 383–386.

Kromberg JGR, Jenkins T. (1984). Albinism in the South African Negro. III. Genetic Counselling Issues. *J Biosoc Sci 16*, 99–108.

Kromberg JGR, Zwane E, Jenkins T. (1987). The response of black mothers to the birth of an albino infant. *Am J Dis Child 141*, 911–916.

Kromberg JGR, Bernstein R., Jacobson M, Rosendorff J, Jenkins T. (1989). A decade of mid-trimester amniocentesis in Johannesburg. *S Afr Med J 76*, 344–349.

Kromberg JGR. (1992). Albinism in the South African Negro. IV. Attitudes and the death myth. In: Evers-Kiebooms G, Fryns JP, Cassiman JJ, Van den Berghe H (eds.), *Psychosocial Aspects of Genetics* (pp. 159–166). March of Dimes, Birth Defects Original Article Series 28. New York: Wiley.

Kromberg JGR, Zwane E. (1993). Down syndrome in the black population of the Southern Transvaal. *SA Journal of Child and Adolescent Psychiatry 5*, 30–33.

Kromberg JGR, Jenkins T. (1997a). Genetic Services in South Africa and common ethical issues related to genetic information. *Specialist Medicine May*, XIX, 54–60.

Kromberg JGR., Jenkins T. (1997b). Cultural influences on the perception of genetic disorders in the black population of Southern Africa. In: Clarke A, Parsons E (eds.), *Culture, Kinship and Genes.* London: Macmillan. pp. 147–157.

Kromberg J, Zwane E, Manga P, Venter A, Rosen E, Christianson A. (2008). Intellectual disability in the context of a South African population. *J. Policy Pract Intellect Disabilities 5*, 89–95.

Kuppermann M, Learman LA, Gates E, et al. (2006). Beyond race or ethnicity and socioeconomic status: predictors of prenatal testing for Down syndrome. *Obstet Gynecol 107,* 1087–1097.

Learman LA, Drey E, Gates E, Kang M-S, Washington AE, Kuppermann M. (2005). Abortion attitudes of pregnant women in perinatal care. *Am J Obstet Gynecol 192*, 1937–1947.

Livingstone D. (1857). *Missionary Travels.* London: John Murray.

Macaulay S. (2007). *Cystic fibrosis genetic counselling: an audit of counselees and their at-risk relatives.* MSc(Med) research report. University of the Witwatersrand, Johannesburg.

Manga N, Jenkins T, Jackson H, Whittaker DA, Lane AB. (1999). The molecular basis of transferase galactosaemia in South African negroids. *J Inherit Metab Dis 22*, 37–42.

Manganyi NC. (1973). *Being Black in the World.* Johannesburg: Ravan Press.

Manganyi NC, Kromberg JGR, Jenkins T. (1974). Studies on albinism in the South African negro. I. Intellectual maturity and body image differentiation. *J Biosoc Sci 6*, 107–112.

May PA, Brooke LE, Gossage JP, et al. (2000). Epidemiology of fetal alcohol syndrome in a South Africa community in the Western Cape. *Am J Public Health 90*, 1905–1912.

Murray RF. (1984). Cultural and Ethnic Influences on the Genetic Counselling Process. In: Weiss JO, Bernhardt BA, Paul NW (eds.), *Genetic disorders and Birth defects in families and society: towards interdisciplinary understanding* (pp. 71–74). Birth Defects Original Article Series 20, March of Dimes, New York.

Padoa C, Goldman A, Jenkins T, Ramsay M. (1999). Cystic fibrosis carrier frequencies in populations of African origin. *J Med Genet 36*, 41–44.

Read M. (1966) *Culture, Health and Disease.* London: Tavistock.

Republic of South Africa (1996). *Constitution of the Republic of South Africa* Act 108 of 1996. Government Printer, Pretoria.

Schapera I. (1940) *Married life in an African tribe.* London: Faber and Faber.

Schapera I. (1937). Law and Justice. In Schapera I, ed. *The Bantu-speaking tribes of South Africa.* Cape Town: Maskew Miller. pp. 197–219.

Strathern M. (1997). The work of culture: an anthropological perspective. In Clarke A, Parsons E (eds.), *Culture, Kinship and Genes* (pp. 40–53). London: Macmillan.

Todd C, Haw T, Kromberg J, Christianson A. (2010). Genetic counselling for fetal abnormalities in a South African community. *J Genet Couns 19*(3), 247–254.

Triandis HC. (1971). *Attitude and Attitude Change.* New York: John Wiley.

Wertz DC, Fletcher JC (eds.) (1989). *Ethics and Human Genetics. A Cross-Cultural Perspective.* New York: Springer-Verlag.

Wertz DC, Fletcher JC. (1993). Geneticists approach ethics: an international survey. *Clin Genet 43*, 104–110.

Wertz DC. (1998). Eugenics is alive and well: a survey of genetic professionals around the world. *Science in Context 11*, 493–510.

Wertz DC. (1999). Patients' and professionals views on autonomy, disability, and "discrimination." Results of a 36-Nation Survey. In: Caulfield TA, Williams-Jones B (eds.), The *Commercialization of Genetic Research: Ethical, Legal and Policy Issues* (pp. 171–180). New York: Kluwer Academic/Plenum.

49.

THE GENOMIC RESEARCH AND HEALTH APPLICATIONS IN SOUTH AFRICA[1,2]

Billie-Jo Hardy, Béatrice Séguin, Raj Ramesar, Peter A. Singer, and Abdallah S. Daar

INTRODUCTION

The South African government is committed to science and technology innovation, to establishing a knowledge-based economy, and to harnessing life sciences research for health and economic development. Given the constraints and the early stage of development of the field as a whole in South Africa, an impressive amount of research on human genomic variation has been carried out in this country. Encouragingly, South Africa is beginning to apply genomics to address local health needs, including human immune deficiency virus (HIV) and tuberculosis (TB) infections. This chapter, originally published as part of a Supplement in Nature Reviews Genetics in 2008, documents a number of initiatives in South Africa that are beginning to study genetic variation within the various local indigenous populations. Other early initiatives focus on pharmacogenetic studies, mutation characterization in individual disease genes, and genome-wide association studies. Public engagement in genomic issues is spearheaded by the Africa Genome Education Institute (AfGEI).

Sub-Saharan Africa, where modern humans arose (Jakobsson et al., 2008; Li et al., 2008; Lohmueller et al., 2008), is home to a wealth of human genetic, linguistic, and cultural diversity (Tishkoff & Williams, 2002). There is thus much to be learned from characterizing human genomic variation in this part of Africa, especially regarding health applications. In recent years, the African Society of Human Genetics (AfSHG) has begun to raise awareness of genomics and genetics in Africa with a focus on networking and capacity building throughout the continent. South Africa in particular has generated new scientific capacity in the field of genomic sciences, including the recently launched high-throughput genomics platform at Lifelabs, located at the Nelson R. Mandela School of Medicine at the University of KwaZulu-Natal, and the Centre for Proteomic and Genomic research (CPGR) at the University of Cape Town (UCT). These initiatives are in line with the recent report by the African Union high-level panel on modern biotechnology, entitled *Freedom to Innovate*, which recommends that Southern Africa focuses on a "core mission" in health biotechnologies, while other regions direct biotechnology efforts toward animal health (East Africa), agriculture (West Africa), and forestry (Central Africa).

South Africa has developed a potential roadmap toward the development of large-scale genotyping projects similar to other emerging economies and the developing world, such as Mexico, India, and Thailand (Séguin et al., 2008a). The present chapter includes findings of a study carried out in South Africa. Using previously described qualitative methods (Séguin et al., 2008a,b) semistructured interviews with members of the African Genome Education Institute (AGEI), and the Division of Human Genetics at UCT were conducted as key informants from diverse backgrounds, including institutional research units, government officials, funding agencies, genomics/pharmacogenetics experts, and bioethicists from both developed and developing countries. This is similar to the analysis reported in other chapters on Mexico, India and Thailand: political will, genomic sovereignty, institutional leadership, local health benefits, and knowledge-based economy (Séguin et al., 2008a).

POLITICAL WILL

The value of political will is acknowledged in creating the necessary policy and legislation, public funding, and support to encourage the development of genomic medicine in emerging economies and developing nations. For instance, government support through funding and legislation was crucial to the establishment of large-scale

1 Reprinted from Hardy B-J, Séguin B, Ramesar R, Singer PA, Daar AS. (2008). South Africa: from species cradle to genomic applications. Copyright 2008 by Nature Publishing. Reproduced with permission of Nature Publishing.

2 Author's Note: The research presented in this chapter was conducted between 2006 and 2008, it has not been updated to reflect the many changes in the field.

genotyping initiatives for health in Mexico, India, and Thailand (Séguin et al., 2008a).

In 2002, the *National Biotechnology Strategy* report for South Africa (Parker et al., 2002), commissioned by the South African government, recommended that the country focus on documenting the genomic diversity contained within the local indigenous and immigrant populations, including: "the mapping and identification of genes underlying disease structure, elucidation of the molecular basis of diseases common to South Africa, and the development of technology to develop medicines and diagnostic tests." The Division of Human Genetics at UCT is currently involved in elucidating the genetic basis of diseases (retinal degeneration, hypertension, familial hereditary nonpolyposis colorectal cancer, and bipolar disorder) in Southern Africa, and exploring genomic diversity in indigenous Southern African populations. In collaboration with the Human Genomic Diversity and Disease research unit at the University of the Witwatersrand in Johannesburg, they are characterizing human genomic diversity using biological samples collected from several indigenous tribes in Southern Africa—Zulu, Xhosa, Herero, San, and Sotho–Tswana. Although this study does not look at disease genes per se, it will provide data on baseline variation across these populations. The Division of Human Genetics at UCT has also recently submitted a proposal to the South African government to initiate a national genomic medicine research program with three components: characterization of human genomic variation within the South African populations, identification of the genomic basis of susceptibility to common diseases (both chronic and infectious), and pharmacogenomics. The research program will be supported by a national biobank, a high-throughput technology and bioinformatics network, and an economic, education, ethical, legal, and social issues (EEELSI) unit. They propose to use this platform to generate research on human genomic variation within the South African population to explore the underlying mechanisms of disease and health, as well as for pharmacogenomic and genetics applications. The proposal is currently awaiting approval from government funding agencies. The 2002 National Biotechnology Strategy report has resulted in approximately 450 million South African rand (ZAr) in funding for the development of a biotechnology sector in South Africa. This investment has largely gone toward the establishment of biotechnology regional innovation centers and incubator centers (Cloete et al., 2006; Gastrow, 2008; Louet, 2006; Motari et al., 2004).

Key informants highlighted that, although successful, these centers are directed at late-stage research and commercialization. Human genomic variation studies, on the other hand, are considered basic research, and thus local support for these studies continues to come mainly from traditional government funding agencies, such as the Medical research Council (MrC), the Department of Science and Technology, and the National research Foundation (NrF), all of which are constrained by how much they can direct into any single research platform project. For instance, the MrC usually provides about 400,000 ZAr (~US$50,000–$60,000) per research unit. In some instances, government funding agencies will match the level of funding for researchers involved in international collaborations that contribute to local capacity building. Thus, key informants felt that despite the infrastructure put in place to serve innovation within the biotechnology sector (including genomics-related research) there remained a gap between policy and action on the part of the South African government. Specifically, there is a lack of available funds to follow up on the above-mentioned 2002 National Biotechnology Strategy recommendations. It is difficult to predict when the current situation will change, given the other pressing priorities faced by the South African government. As summarized by one key informant, "I'd like to think that the South African government will come to the party, but then when you look at the nature of the other challenges that the country is facing: lack of [electrical] power, need to improve education, need to improve general healthcare provision for the general population, need to improve housing provision for the general population... well, where does scientific research fit into all of that? And it is easy for educated scientists who've got a vested interest to say, well of course what we are doing is incredibly important and an integral part of all of those challenges for the country."

The lack of funding for genomics-related research and development (R&D) leads to additional limiting factors, such as the lack of sufficiently trained human resources. Interviewees emphasized their inability to retain locally trained research scientists and attract foreign postdoctoral candidates. They pointed out that they are unable to provide the adequate compensation and the state-of-the-art facilities desired by leaders in the field. In order to address these limitations, South African researchers rely upon international collaborations supported by philanthropic foundations and public institutes, such as the Gates Foundation or the Wellcome Trust, as well as multinational pharmaceutical companies, such as Pfizer and GlaxoSmithKline. International collaborations form the backbone of many research projects focusing on human genomic variation in South Africa, and all of our key informants considered these partnerships as crucial. However, as a result of these collaborations, key informants also expressed concern that the biological samples, and consequently information, might be leaving the country at a regular pace, potentially affecting the ability of South Africans, and more generally Africans, to capitalize on the results of the associated research.

GENOMIC SOVEREIGNTY

The concept of genomic sovereignty was previously identified by us as representing a nation's ability to capture the value of its investments in the field of genomic medicine. In Mexico and India, unique population resources, and the desire to prevent exploitation of these resources, in part drove the creation of their national gene banks (Séguin et al., 2008a). Genomic diversity within sub-Saharan Africa, and for that matter the entire African continent, is relatively under-studied, despite being home to significant portions of human genomic diversity (Tishkoff & Williams, 2002). South Africa in particular contains a wealth of different population groups. It is home to the indigenous Khoisian, Xhosa, Zulu, venda, and Sotho Pedi groups, the Afrikaners, and the Colored, the latter being a uniquely admixed population of immigrant Europeans, Asians, and the indigenous populations (Hodgson, 2001). Accordingly, all key informants indicated that the South African population, with its rich diversity and multiple isolated populations, is an excellent resource that could provide South African researchers with a unique research advantage. In this context, we were told by participants that South Africa, and Africa generally, also has a long history of "parachute research," whereby foreign scientists obtain human samples from the African continent with no intention of contributing to Africa's development. To prevent the parachute research phenomenon, the South African government now requires, through the Human Tissues Act, that individuals exporting biological samples from the country apply for a permit from the Ministry of Health. However, our informants pointed out that this law is difficult to enforce, and it is possible for samples to continue to leave the country undocumented and unaccounted for at a national level. As suggested by one key informant, "if the information to drive that new diagnostic test fundamentally is derived from clinical samples that arose in South Africa, then we, as a nation, have to do something about controlling that value. And controlling where that value is generated and who owns it. Who owns the information and any spin-offs that come from that downstream." Consequently, we learned of discussions in South Africa on how to best capture this value. For example, should South Africans consider extending the current Biodiversity Act to include human samples, or should they take a different approach? It is important, however, that the vital role of international collaborations for South African research is not unduly restricted to the detriment of South Africa. Thus, many key informants view this as a delicate balancing act, in which they appreciate that research is a global enterprise and that knowledge should remain public, but also emphasize the importance to South Africa of generating equitable international collaborations that can provide much-needed financing and potentially contribute to local scientific capacity building.

GENOMIC SOVEREIGNTY

The Public Understanding of Biotechnology program, initiated by the Department of Science and Technology in South Africa, is mandated to educate the public and stimulate debate on the broad issues of biotechnology. Nevertheless, it is the AGEI that has taken the lead on communicating issues of genomic sciences to the general public. The AGEI is devoted to engaging with and educating the public about the structure and function of the genome, and about the genetic basis of diseases relevant to the South African population. The AGEI has developed a number of programs toward this effort. For instance, the Living History Project provides the public with the opportunity to discover their genetic ancestry. And the Skin Color Education Project provides schools and museums with education materials, including published literature and interactive media, on human genomic variation and skin pigmentation. These types of projects are designed, in part, to address the public concerns associated with decades of apartheid rule, thereby establishing trust between the public and researchers. The AGEI also hosts regular seminars, such as the public Darwin Seminar Series, which give the general public the opportunity to engage with leaders in the field. In addition, and in recognition of the limitations of these types of initiatives, as part of the proposed national EEELSI unit, the Division of Human Genetics is developing a pilot proposal entitled *Heritage*. By proposing that human genomic research relates and incorporates the contextual realities of how individuals and communities already understand themselves, Heritage aims to transform human genomics research from parachute research to an interactive, ongoing process that involves both the humanities and the sciences. For example, through an interactive online system, Heritage could link genetic researchers, anthropologists, linguists, historians, clinicians, government policymakers, the community, and the public, with each group contributing and commenting on the description of population groups. If successful, Heritage could transform the traditional research process and result in innovative models, unique to South Africa, for raising public awareness and for obtaining public support. For instance, it could serve as a mechanism for the government to develop contextual public health policy that incorporates genomic, environmental, cultural, and health data.

LOCAL HEALTH BENEFITS

South Africa is faced with formidable health challenges, notably its current HIV crisis—HIV is its leading cause of death in both men and women (see the MrC's Burden of Disease Research Unit FAQs). It has one of the world's lowest life expectancies, 50 for men and 53 for women (see the world Health Organization's South Africa overview), and its long-term economic stability is threatened by its current health crisis (Dixon et al., 2002). Accordingly, it is not surprising that public investments directed toward innovation in the health sciences must align with South Africa's health priorities, such as the development of an HIV vaccine. There are only few relevant HIV-related human genomic variation projects that have been recently implemented or that are in the pipeline. For instance, a study at Stellenbosch University is investigating the genetic basis of the observed differences in response to antiretroviral therapies in HIV-positive patients of various South African ethnicities (Parathyras, 2008). Another major project is a public–private admixture mapping study recently initiated between researchers at the DST/NRF Centre of Excellence for Biomedical TB Research at Stellenbosch University and GlaxoSmithKline. The study involves up to 1,000 samples from the South African Colored population. Previous genome-wide scans and genetic variation studies in this population have contributed to increasing understanding of genetic susceptibility to TB (Babb et al., 2007; Barreiro et al., 2006; Cooke et al., 2008), including revealing novel target loci MC3R (melanocortin 3 receptor) and CTSZ (cathepsin Z; Cooke et al., 2008). It is hoped that the data generated by the admixture study will make similar contributions. Confirming the limited local capacity faced by South African researchers when conducting scaled-up human genomic variation studies, the admixture-mapping study samples were analyzed by SNP chip technology and the data generated at the Affymetrix laboratories in the United States. Raw data was provided to Stellenbosch University. Other, smaller-scale studies explore variation within single candidate genes and/or involve pharmacogenetics studies in several chronic conditions, including cardiovascular disease, obesity, hypertension, and fetal alcohol syndrome. For instance, the Institute for Cellular and Molecular Biology—recently established at the University of Pretoria and funded by the university, the NRF, the National Health Laboratory Trust, and a private pathology group in South Africa—is exploring pharmacogenetics and obesity in sub-Saharan populations. It is clear that the results of these and other future studies would be easier to interpret and utilize if there existed a single national database providing details of baseline variation within South Africa. Not surprisingly, a few key informants suggested that they were, in fact, investigating the possibility of independently developing databases (either national or regional), with some interviewees intimating that they might introduce fees for commercial use.

It is generally agreed that the lack of local collaborations between university-based academics limited R&D in the field of human genomic variation. We did learn of recent initiatives through the Department of Science and Technology and the MrC, to generate multidisciplinary collaborations between local researchers and institutions in an effort to address this challenge. An example of a successful collaboration is provided by the recent collaboration between the Division of Human Genetics at UCT and the Human Genomic Diversity and Disease research unit of the University of the Witwatersrand, which was set up to explore genetic variation within indigenous Southern African populations. With shared resources and Affymetrix's donation of GeneChip Human Mapping 6.0 Array Sets, and despite the lack of funding, the researchers were able to complete their study, which they are currently preparing for publication. In addition, some researchers are attempting to develop a National Functional Genomics Network, with one of the mandates being to generate shared access between existing South African genomic and genetic databases.

KNOWLEDGE-BASED ECONOMY

In Mexico, India, and Thailand, we found that investment in genomic medicine platforms is in part driven by the desire of these nations to participate in a global knowledge-based economy (Séguin et al., 2008a). The National Biotechnology Strategy for South Africa also highlights the importance of developing a national system of innovation in the biotechnology sector, which includes genomics, in order to enhance global competitiveness and local economic growth. Genomic medicine, specifically pharmacogenomics, is highlighted in this document as a potential industrial opportunity (AUTHOR, 2001). More recently, the Department of Science and Technology in South Africa produced a report entitled "Innovation toward a knowledge-based economy: ten-year plan (2008–2018) for South Africa" (Department of Science and Technology, 2007). In this report, the government highlights genomics and suggests that South Africa should position itself as a major producer in the pharmaceutical and nutraceutical industries through the development of appropriate technology platforms that would be based on local strengths. So far, South Africa has no national genomic medicine platform comparable to Mexico's National Institute of Genomic Medicine (INMEGEN) (Séguin et al., 2008c), with its focus on commercialization of locally generated knowledge. Nevertheless, there is evidence of entrepreneurial activity around genomic-based health sciences in South Africa, in both the public and private sectors. The CPGR is a not-for-profit service provider based in Cape Town that was created under the National Biotechnology Strategy. Currently it receives public funds from the Cape Biotech Trust (CBT) and Plantbio, two of the biotechnology regional innovation centers, to create local capacity and a knowledge base to increase its scientific output, and to support or drive existing or new commercial interests in the genomic and the proteomic sectors. At current capacity, the CPGR is servicing up to 80 clients, both local and international.

Its recent successes, together with the requirement to become self-sustainable over the next few years, have led the CPGR to consider adopting a commercial business model in addition to its not-for-profit core. Toward this aim, the CPGR is currently exploring R&D opportunities within genomic medicine, including the development of biomarkers and personalized medicine. Members of the CPGR cite tremendous scientific and business opportunities in the biotech sector in South Africa, and their ability to provide low-cost services on the international market, as a significant advantage for driving commercial ventures. Another example is Gknowmix, a private wellness service, which was established in collaboration with Kopano Life Sciences and the South African MRC, the Cape Biotech Trust, and Acorn Technologies (a regional business incubator). Although it is too early to determine Gknowmix's success, it does provide an example of how basic academic human genomic research can be translated into a product or service that uses the existing infrastructure designed to generate a biotechnology sector in South Africa. Gknowmix offers two genetic tests for multifactorial conditions: the Cardiovascular Genescreen, and the Wellness Genescreen. The interpretation of the results in the context of specific lifestyle and clinical data constitutes Gknowmix's intellectual property. By opting to provide healthcare professionals with the test results, along with their analysis, Gknowmix's approach addresses the challenge of limited knowledge of genomics on the part of healthcare professionals. Nevertheless, key informants raised a number of challenges associated with commercialization of genomics-based research. Some suggested that despite the existence of technology transfer units located in the universities, spinning out companies from basic genomics-based research is not well established in South Africa and, across the life sciences, "researchers struggle with it." We were also told that the Gknowmix model was not well received by the scientific community, with academics questioning the ethics and usefulness of introducing what they considered to be largely unproven, costly services that will only benefit private healthcare payers. As in other countries where such services are being introduced, the regulatory system is not in place to either support or regulate the products and services of genomic medicine as they become available. For instance, Gknowmix is registered with the board of Healthcare Funders, and operates according to the Human Tissue Act, although the type of services that can be offered is not regulated. The South African government will need to anticipate these issues as they arise, and develop the appropriate legislation and guidelines to address them. For example, interviewees cited the lack of a genetic privacy act as a challenge. Currently, insurance companies are self-regulated in this respect, and it is not known how insurance companies plan to manage genetic information.

CONCLUDING REMARKS

Given the constraints and the early stage of development of the field as a whole, we found an impressive amount of research on human genomic variation in South Africa. Because of the predominance of HIV and TB infections as public health threats, there is an appropriate emphasis on using human genetic variation research to improve healthcare for these scourges, focusing on drug metabolism and disease susceptibility at the individual and population level. At the basic population-profiling level, there has been at least one recently completed substantive collaborative project looking at human genetic diversity among indigenous population groups in South Africa. Moreover, a national large-scale genotyping project to profile the South African population is being considered, along the lines of those established in Mexico and India (Séguin et al., 2008a).

As compared with other case studies conducted by our group, South Africa provides a unique perspective of the feasibility of a privately funded population database; this is an important consideration in countries where public funding is limited. However, in these cases, stakeholders should develop guidelines on how to maximize access to private resources by local academic researchers in order to encourage basic research. The well-developed public engagement programs that are in place will go a long way toward facilitating the implementation of genomics-based products and services as they become a reality. However, in order to progress, South Africa will need to pay attention to some of the challenges we identified here—for instance, additional government support in terms of available public funding to help establish research platforms. In addition, although international collaborations contribute to capacity building in South Africa, we learned that they can occur at the expense of local collaborations. The recent report, discussed above, by the Department of Science and Technology on knowledge-based innovation in South Africa (Department of Science and Technology, 2007), affirms many of the challenges highlighted by our key informants, including the lack of domestic collaboration and the need for funding to close the gap between basic research and commercialization, and also suggests methods for addressing them. There is recent evidence of research activity in other sub-Saharan countries. A collaborative effort between African researchers has resulted in the creation of a biobank and pharmacogenetics database representing several ethnic groups (Yoruba, Hausa, Ibo, Luo, Kikuyu, Maasai, Shona, San, and Venda) located at the African Institute of Biomedical Science and Technology (AIBST) in Harare, Zimbabwe. However, the ongoing political turmoil in that country could limit the potential impact of this resource.

The South African government, on the other hand, is committed to science and technology innovation, to establishing a knowledge-based economy, and to harnessing life sciences research for health and economic development. Although the relevant policies are in place, what remains to be seen is to what extent they will be translated into concrete support for research, development, and delivery of products and services in the human genomics field. Identifying the needs, encouraging local collaborations, and helping to form R&D networks will go a long way toward establishing South Africa in this field.

ACKNOWLEDGMENTS

This chapter was previously published in 2008 as "South Africa: from species cradle to genomic applications" by Billie-Jo Hardy, Béatrice Séguin, Raj Ramesar, Peter A. Singer, and Abdallah S. Daar, published in *Nature Reviews Genetics*, special issue on "Genomics Medicine in Developing Countries," October 2008, pp. S19–S23. Authors and editors are grateful to the Nature Publishing Group for their permission to use part of the text in this chapter.

This project was funded by Genome Canada through the Ontario Genomics Institute. The McLaughlin–Rotman Centre for Global Health, Program on Life Sciences, Ethics and Policy is primarily supported by Genome Canada through the Ontario Genomics Institute, the Ontario Research Fund, and the Bill and Melinda Gates Foundation. Other matching partners are listed at The McLaughlin–Rotman Centre for Global Health web site. A.S.D. and P.A.S. are supported by the McLaughlin Centre for Molecular Medicine.

REFERENCES

Babb C, et al. (2007). Vitamin D receptor gene polymorphisms and sputum conversion time in pulmonary tuberculosis patients. *Tuberculosis (Edinb.) 87*, 295–302.

Barreiro LB, et al. (2006). Promoter variation in the DC-SIGN-encoding gene *CD209* is associated with tuberculosis. *PLoS Med 3*, e20.

Cloete TE, Nel LH, Theron J. (2006). Biotechnology in South Africa. *Trends Biotechnol 24*, 557–562.

Cooke GS, et al. (2008). Mapping of a novel susceptibility locus suggests a role for MC3R and CTSZ in human tuberculosis. *Am J Respir Crit Care Med 178*, 203–207.

Department of Science and Technology. (2007). *Innovation Towards a Knowledge-Based Economy Ten-Year Plan for South Africa (2008–2018)*. Available at: http://unpan1.un.org/intradoc/groups/public/documents/CPSI/UNPAN027810.pdf.

Dixon S, McDonald S, Roberts J. (2002). The impact of HIV and AIDS on Africa's economic development. *BMJ 324*, 232–234.

Gastrow M. (2008). Great expectations: the state of biotechnology research and development in South Africa. *African Journal of Biotechnology 7*, 342–348.

Hodgson J. (2001). South Africa seeks genomic partners. *Nature Biotechnol 19*, 601–602.

Jakobsson M, et al. (2008). Genotype, haplotype and copynumber variation in worldwide human populations. *Nature 451*, 998–1003.

Li JZ, et al. (2008). Worldwide human relationships inferred from genome-wide patterns of variation. *Science 319*, 1100–1104.

Lohmueller KE, et al. (2008). Proportionally more deleterious genetic variation in European than in African populations. *Nature 451*, 994–997.

Louet S. (2006). Rainbow biotech—South Africa's emerging sector. *Nature Biotechnol 24*, 1313–1316.

Motari M, et al. (2004). South Africa—blazing a trail for African biotechnology. *Nature Biotechnol 22*(Suppl), DC37–DC41.

Parathyras JB. (2008). *Molecular Genetic Analysis of Human Immunodeficiency Virus Antiretroviral Therapy Response in South Africa—a Pharmacogenetics Study.* Thesis, Stellenbosch University. Available at: http://scholar.sun.ac.za/handle/10019.1/2638.

Parker et al. (2001). Public Understanding of Biotechnology. National Biotechnology Strategy for South Africa. Available at: http://www.pub.ac.za/resources/docs/biotechstrategy_2002.pdf.

Séguin B, Hardy B, Singer PA, Daar AS. (2008a). Genomic medicine and developing countries: creating a room of their own. *Nature Rev Genet 9*, 487–493.

Séguin B, Hardy B, Singer PA, Daar AS. (2008b). Human genomic variation initiatives in emerging economies and developing countries. *Nature Rev. Genet. 9*(Suppl), S3–S4.

Séguin B, Hardy B, Singer PA, Daar AS. (2008c). Genomics, public health and developing countries: the case of the Mexican National Institute of Genomic Medicine (INMEGEN). *Nature Rev Genet 9*(Suppl), S5–S9.

Tishkoff SA, Williams SM. (2002). Genetic analysis of African populations: human evolution and complex disease. *Nature Rev Genet 3*, 611–621.

FURTHER INFORMATION

African Genome education institute (AGei): http://www.africagenome.com

African institute of Biomedical science and technology (AiBst): http://www.aibst.com

African society of Human Genetics (AfsHG): http://www.afshg.org

Biodiversity Act: http://www.info.gov.za/gazette/acts/2004/a10-04.pdf

Burden of Disease research Unit FAQs, Medical research council: http://www.mrc.ac.za/bod/faqdeath.htm

Centre for Proteomic and Genomic research (cPGr): http://www.cpgr.org.za

Freedom to innovate report by the African Union High Level Panel on Modern Biotechnology: http://www.nepadst.org/doclibrary/pdfs/biotech_africa_2007.pdf

Gknowmix: http://www.gknowmix.com

Lifelabs: http://www.lifelab.co.za

McLaughlin–Rotman centre for Global Health: http://www.mrcglobal.org

National institute of Genomic Medicine (INMEGEN), Mexico: http://www.inmegen.gob.mx

the Division of Human Genetics, University of cape town: http://web.uct.ac.za/depts/genetics

The Living History Project: http://www.africagenome.com/index.php?option=com_content&task=category&sectionid=4&id=13&Itemid=43

The Public Understanding of Biotechnology program: http://www.pub.ac.za

The skin color education Project: http://www.africagenome.com/index.php?option=com_content&task=view&id=17&Itemid=39

The World Health Organisation's south Africa overview: http://www.who.int/countries/zaf/en

ARAB AND MIDDLE EAST

50.

HEREDITARY DISORDERS IN ARABS

Ghazi Omar Tadmouri and Mahmoud Taleb Al Ali

INTRODUCTION

Genetic and inherited disorders have accompanied the human race since its earliest existence. Many prehistoric and historic sites have revealed archeological remains with pathologies suggestive of inherited disorders. One of the oldest of such records includes a 1.5 million-year-old fossil of a 2-year-old *Homo erectus* child with amelogenesis imperfecta (Zilberman et al., 2004). In Egypt, scientific investigation of mummies from the huge necropolis of Thebes-West in Upper Egypt revealed osseous manifestations suggestive of metabolic and chronic anemia in high frequencies in populations of the "Middle Kingdom" (2050–1750 BCE; Nerlich et al., 2002). Interestingly, ancient DNA analysis revealed a beta-thalassemia mutation in the skeletal remains of an Ottoman child with severe porotic hyperostosis (Filon et al., 1995).

Concomitantly, almost all prehistoric and ancient civilizations suffered from the severe consequences of many infectious diseases. In fact, many of these diseases played major roles in the development of human civilization, impacting and altering the course of wars, migrations, population growth, and urbanization. In many world countries, including Arab States, breakthroughs in managing such diseases only happened few decades ago with the clinical availability of specific drugs and vaccines. This progress decreased the impact of such disorders in favor of an increased understanding of the molecular basis of heredity, hence, a better recognition of genetically transmitted conditions as a major cause of morbidity and mortality (Tadmouri, 2008a).

- Genetic and inherited disorders have accompanied the human race since its earliest existence.
- Many prehistoric and historic sites have revealed archeological remains with pathologies suggestive of inherited disorders.

GENETIC DISORDERS IN WORLD POPULATIONS

To date, records for more than 6000 inherited disorders, known to afflict world populations, are indexed and maintained at the Online Mendelian Inheritance in Man (OMIM) database, which is a comprehensive compendium of human genes and genetic phenotypes (McKusick, 2007). Many of these disorders are rare entities that occur at higher frequencies in particular ethnic, racial, or demographic groups. More common disorders show a worldwide spread supported by natural selective forces or stimulated by socially oriented reproductive choices.

Systematic attempts to catalog genetic disorders in world populations are almost nonexistent. The only available examples include the Finnish Disease Database (www.findis.org) and Rare Diseases Sweden database (www.sallsyntadiagnoser.nu) that represent national efforts to highlight the disease heritage of their respective populations. Hence, to try to give an approximate visualization of the number of genetic disorders in various world populations, the only comprehensive source that could be utilized would be the OMIM database (www.omim.org). Using proper search syntaxes it is possible to deduce the populations in which most of the genetic disorders were described; as of September 2011, these include: the Japanese (~750), Italian (~550), German (~450), Turkish (~400), Chinese (~350), French (~300), Canadian (~300), Finnish (~300), Pakistani (~300), and many other populations. However, this strategy only gives a rough estimate on the number of genetic disorders in a population, with possibly very large deviations from actual figures (reviewed in Tadmouri, 2008b).

ARAB POPULATIONS: ORIGINS AND DIASPORAS IN THE WORLD

Defining the term *Arab* is a challenging task. Linguistically, Arab populations encompass a vast geographical region that extends from south of Iran in the east to Morocco in the west, including parts in the southeast of Asia Minor, East Africa, and West Africa. However, the political definition of Arab populations is more conservative, as it only includes those populations residing in 23 Arab states, namely: Algeria, Bahrain, Comoros, Djibouti, Egypt, Eritrea, Iraq, Jordan, Kuwait, Lebanon, Libya, Mauritania, Morocco, Oman, Palestine, Qatar, Saudi Arabia, Somalia,

Sudan, Syria, Tunisia, United Arab Emirates (UAE), and Yemen.

Many of the most important events during the early evolutionary history of modern humans took place in this vast geographical region. DNA evidence indicates that modern humans originated in East Africa about 100,000–200,000 years ago (Liu et al., 2006), then established regional populations throughout the continent. Archeological artifacts indicate that human inhabitation of modern day's Maghreb region (modern day Morocco, Algeria, Tunisia, and Libya) date back to some 82,000 years ago (Taforalt, Morocco). According to the "Recent Out-of-Africa model," members of one branch of anatomically modern humans left Africa to the Near East some 60,000–70,000 years ago. Phylogenies constructed on the basis of mitochondrial DNA comparisons are indicative for two possible migration routes in this episode of human history. One route laid across the Bab-el-Mandeb straits in the Red Sea, linking modern day Eritrea and Djibouti in Africa to Yemen in the Arabian Peninsula (Bailey et al., 2007). This route maintained its important role in influencing gene flow from Africa into the region until nearly 2500 years ago (Cadenas et al., 2008). Another route followed the Nile from East Africa, heading northwards and crossing through Sinai into the Levant, between 45,000–50,000 years ago (Cann et al., 1987; Ingman et al., 2000). Human populations in the Near East then branched in several directions, some moving into Europe and others heading east into Asia (Ke et al., 2001; Maca-Meyer et al., 2001; Underhill et al., 2001; Lahr & Field, 2005).

Around 11,500 years ago, Neolithic farmers settled in mosquito-infested soft and marshy soil, causing large malarial outbreaks (Grmek, 1994; de Zulueta, 1994) that imposed the greatest selective pressure on the human genome and amplified the frequencies of several genetic disorders including sickle cell disease, beta-thalassemia, and glucose-6-phosphate dehydrogenase (G6PD) deficiency (Angel, 1966; Carter & Mendis, 2002; Kwiatkowski, 2005). Undoubtedly, dependence on agriculture helped in the establishment of major historical settlements with sizable densities that could have contributed enormously to the genetic makeup of modern Arab populations. Some of these earliest civilizations in the Maghreb region include immigrant Berbers who originated from the Sahara 10,000 years ago, followed by the Mesolithic Capsians who settled in the area 2000 years later (Irish, 2000). In the Arabian Peninsula, Semitic-speaking peoples of Arabian origin migrated into the valley of the Tigris and Euphrates rivers in Mesopotamia some 7000–5500 years ago (Beech et al., 2005). Archeological evidence further indicates that another group of Semites left Arabia around 4500 years ago during the Early Bronze Age and settled along the Levant, mixing in with the local populations there. Some 3500 years ago, the Phoenician civilization of Lebanon became a developed, enterprising maritime trading culture. Subsequently, Phoenician traders spread across the Mediterranean and established major cities and colonies that harbored their pathologic or polymorphic gene variations (Tadmouri et al., 2001; Zalloua et al., 2008a). Results of the Genographic Consortium from Y-chromosome variations indicate that as many as 1 in 17 men living today on the coasts of North Africa and southern Europe may have a Phoenician direct male-line ancestor (Zalloua et al., 2008b). The genetic pool was further enriched in Mesopotamia through Persians, while Romans gained a 600-years-long period of settlements throughout most of the region and were subsequently replaced by the Byzantines (Zahed et al., 2002). Soon after the rise of Islam 1400 years ago, the Arab caliphates unified the entire region as a distinct territory and amalgamated the dominant ethnic identity that persists today as far as the Maghreb region in North Africa, as well as Andalusia in the Iberian Peninsula (Fattoum & Abbes, 1985; Ben Abdeladhim et al., 1987). Similarly, the Arabian Peninsula linked distant populations of China and India to communities of the Mediterranean and beyond. At the beginning of the sixteenth century, major parts of the eastern Arabian Peninsula fell under the influence of the Portuguese, and remained as such for the following 150 years. Simultaneously, Ottomans controlled much of the Mediterranean, then expanded their influence to cover all the Arabian Peninsula and further contributed to the enrichment of the genetic pool in the region (Haj Khelil et al., 2004). After the nineteenth century, areas of the Maghreb were colonized by France, Spain, and Italy, while the remaining areas where colonized by France and England.

Throughout modern history, Arab emigrants formed many diasporas in world continents (Box 50-1). At present, the main countries of immigration within the Arab world are countries of the Gulf Cooperation Council that host more than 2 million emigrant Arabs mostly from Egypt, Iraq, Lebanon, Palestine, and Syria. Outside the Arab world, the largest Arab communities live in Latin America, with an estimated presence of more than 16 million people of Arab (Lebanese, Syrian, and Palestinian) descent, according to statistics compiled by the Arab Federation of Latin America (Salloum, 2000; Luxner, 2001). Arabs in North America and Europe total some 14 million people. In North America, Arabs arrived in several waves of immigrations from countries of southwestern Asia (Lebanon, Syria, and Palestine) and North Africa (Morocco, Algeria, and Egypt) since the 1880s (The Canadian Arab Federation, 1999; Zogby International, 2000). In Europe, the history of Arab immigrations started as early as 1898 with the arrivals of Somali and Yemeni communities in the United Kingdom. However, the European country with the largest number of Arab immigrants is France, with nearly 7 million Arabs, mostly from North Africa and Lebanon. In Southeast Asia, most of the prominent Indonesians, Malaysians, and Singaporeans of Arab descent have their origins in the southern part of the Arabian Peninsula, especially the coastal Hadhramaut region of Yemen and Oman. Estimates indicate that as many as 4 million Hadramis live in Indonesia, and 10,000 are in Singapore (Talib, 1997). In Australia, Arabic is the fourth most widely spoken second language, due to the presence of a considerable Arab community with roughly 400,000 people from Arab countries (mostly Lebanese; Price, 1999). Arab diasporas are also noted in Africa, especially those from Lebanese descent residing in several countries in Central and West Africa (Handloff, 1988).

BOX 50-1

- The political definition of Arab populations only includes those populations residing in 23 Arab States, namely: Algeria, Bahrain, Comoros, Djibouti, Egypt, Eritrea, Iraq, Jordan, Kuwait, Lebanon, Libya, Mauritania, Morocco, Oman, Palestine, Qatar, Saudi Arabia, Somalia, Sudan, Syria, Tunisia, UAE, and Yemen.
- Many of the most important events during early evolutionary history of modern humans took place in this vast geographical region.
- Throughout modern history, Arab emigrants formed many diasporas in world continents.
- Outside the Arab world, the largest Arab communities live in Latin America, with an estimated presence of more than 16 million people of Arab descent.
- In North America, Arabs arrived in several waves of immigrations from countries of southwestern Asia and North Africa since the 1880s.

DEMOGRAPHIC AND MEDICO-ECONOMIC CHARACTERISTICS OF ARAB POPULATIONS

Current estimates indicate that the total population size in Arab countries is approximately 350 million (reviewed in Tadmouri, 2008b). Although Arabs may share a commonality of language, history, and religion, their societies are at variance when it comes to demographic and medico-economic characteristics. Some of the aspects that markedly affect the prevalence and natural history of genetic diseases in the Arab World include:

1. High fertility rates (1.7–6.6 children born/woman).
2. High birth rates (15.5–44.1 births/1,000 people).
3. High annual population growth rates of 1.0%–3.8%.
4. Extremely high rates of birth defects (63–82/1000 live births; Christianson et al., 2006; Figure 50-1).
5. High rates of inbreeding or consanguineous marriages (Table 50-1).
6. Large family sizes (2–7 children born/woman).
7. Childbearing at either very early or old maternal ages.
8. The presence of isolates (e.g., Armenians, Bedouins, Druzes, Jews, Kurds, and Nubians).

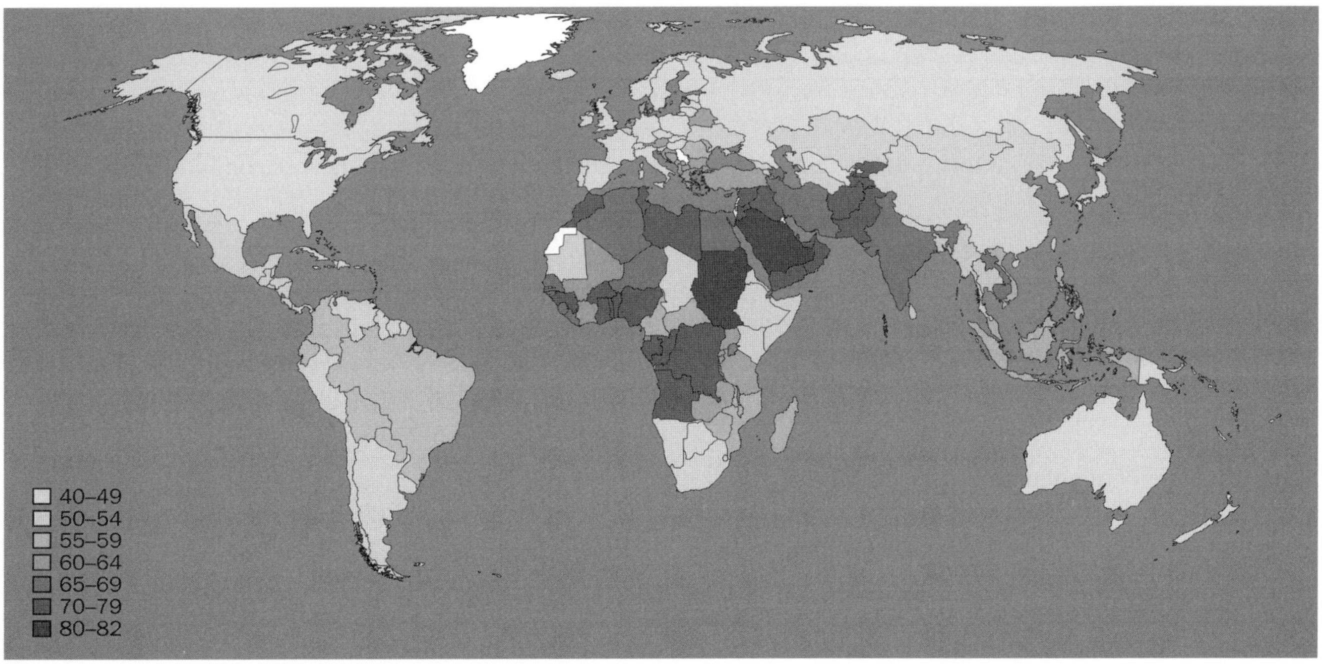

Figure 50-1 Global birth defects prevalence per 1000 live births (adapted from Christianson et al., 2006). Refer color figure.

TABLE 50-1 CONSANGUINITY RATES IN ARAB POPULATIONS. MINIMUM AND MAXIMUM REPORTED RATES ARE INDICATED WHEN AVAILABLE. FOR COMPREHENSIVE DETAILS AND ADDITIONAL DATA, SEE CHAPTER 51

Country	>1C,1C*	Overall consanguinity	References
Algeria	11.3	22.6–34	Benallegue & Kedji, 1984; Zaoui & Biemont, 2002
Bahrain	11.4–24.5	20–45.5	Al Arrayed, 1995, 1999, 2006
Egypt	14.3–23.2	20.9–32.8	Stevenson et al., 1966; Hafez et al., 1983; Habib & Book, 1983; Mohamed, 1995; Mokhtar & Abdel-Fattah, 2001
Iraq	29–33	47–60	Hamamy et al., 1986; Hamamy & Al-Hakkak, 1989
Jordan	19.5–39	28.5–63.7	Khoury & Massad, 1992; Al-Salem & Rawashdeh, 1993; Nabulsi, 1995; Hamamy et al., 2005
Kuwait	16.9–31.7	22.5–64.3	Al-Awadi et al., 1985; Radovanovic et al., 1999
Lebanon	6.7–31.6	12.8–42	Khlat et al., 1986; Khlat, 1988; Tamim et al., 2003; Barbour & Salameh, 2009
Libya		48.4	Broadhead & Sehgal, 1981
Mauritania		47.2	Hammami et al., 2005
Morocco	8.6–10	15.2–28	Lamdouar Bouazzaoui, 1994; Cherkaoui et al., 2005; Talbi et al., 2007; Jaouad et al., 2009
Oman	24.1	56.3	Rajab & Patton, 2000
Palestine	13.6- 34.2	17.5–66.3	Freundlich & Hino, 1984; Jaber et al., 1994; Jaber et al., 2000; Vardi-Saliternik et al., 2002; Zlotogora et al., 2002; Assaf and Khawaja, 2008; Sharkia et al., 2008
Qatar	34.8	22-54	Bener & Alali, 2006; Sandridge et al., 2010
Saudi Arabia	24.6- 42.3	42.1–66.7	El-Hazmi et al., 1995; Al Husain M, Al Bunyan, 1997; Al-Abdulkareem & Ballal, 1998; El-Mouzan et al., 2007
Sudan	44.2–49.5	44.2–63.3	Ahmed, 1979; Saha & El Sheikh, 1988; Saha et al., 1990
Syria	28.7	30.3–39.8	Othman & Saadat, 2009
Tunisia	17.4–23	20.1–39.3	Riou et al., 1989; Ben Arab et al., 2004; Kerkeni et al., 2006; Kerkeni et al., 2007
UAE	20.7–28.2	40–54.2	Fahmy et al., 1993; Al-Gazali et al., 1997
Yemen	32–34	40–44.7	Jurdi & Saxena, 2003; Gunaid et al., 2004

*Abbreviations: [>1C] = Double first-cousin marriage; [1C] = First-cousin marriage.

9. The lack of public health measures directed toward control and prevention of congenital and genetically determined disorders.

HEALTHCARE IN THE ARAB WORLD

Despite the alarming demographic characteristics, unprecedented achievements have been especially prominent in the healthcare sector during the last few decades in almost all Arab countries. Improvements have been reflected by the reducing rates of infant mortality below the global average of 42.6 per 1000 live births. Other positive indicators include the increasing rate of literacy, far beyond the global level of 82% of total populations, and the increase in life expectancies beyond the global level of 66.1 years (ranging from 70 to 78.7 years; reviewed in Tadmouri, 2008b).

However, these improvements will only solidify once attention is focused on basic health problems, as well as on underlying factors such as poverty and illiteracy. As matters currently stand, many factors combine to undermine commitment to such priorities including rapid population growth and urban expansion, spiraling health and medical costs, fascination with high-tech medicine and sophisticated high-profile hospitals, and inappropriate medical education and emigration of trained personnel (Stephen, 1992).

OUTLOOK OF GENETIC RESEARCH IN THE REGION

Science in the Arab world is generally suffering from serious lack of funds. For this reason, science groups working in the highly demanding field of human genetics conduct very limited studies that mostly result in clinical reports rather than molecular analyses (Tadmouri & Bissar-Tadmouri, 2003). In fact, the most recent survey conducted by the Centre for Arab Genomic Studies, on articles discussing genetic disorders in Arab patients, indicate that gene-pathology articles constitute only 28% of the total number of articles analyzed (Tadmouri, 2008b). Maghreb countries (Algeria, Tunisia, and Morocco) are leading in terms of molecular genetic studies (Figure 50-2). However, careful investigation of corresponding research articles indicate the significant role of French science groups in diagnosing disorders in Algerian, Tunisian, or Moroccan patients living in France (Tadmouri & Bissar-Tadmouri, 1999). While, at present, this remark could be preliminary, solid results could be pronounced for countries in which exhaustive surveys have been conducted (Qatar: 16%, Oman: 16%, UAE: 15%, and Bahrain: 14%). In general, the majority of molecular studies are the fruits of collaborations with international groups that usually cover the needed expenses and provide the sophisticated equipment (Selamnia & Tali-Maamar, 2003; Dakik et al., 2006). Accordingly, large-scale data production of DNA or protein sequences, mutations, and single nucleotide polymorphisms (SNPs) is seriously lacking in the region and cannot be foreseen in the near future.

On the bright side, however, noteworthy achievements do also exist in the region. Recently, a scientific team from Al Jawhara Center for Molecular Medicine, Genetics, and Inherited Disorders at the Arabian Gulf University isolated a new protein called ISRAA (immune system released activating agent), which represents the first molecule that acts as a mediator between the central nervous system and the immune system. This protein is produced in the spleen when the host is infected. It then activates the immune system and gives it the capability to fight against the infection. This unprecedented discovery may help to find a possible cure for killer diseases including AIDS and cancer (Bakhiet & Taha, 2008). Studies are now being developed to study expression analyses of related genes, as well as functional studies of possibly associated proteins (Bakhiet & Taha, 2008). In Oman, Sultan Qaboos University, in collaboration with several international institutions, established a research program called "Oman Family Study." The aim of this program is to establish an Omani model for the study of the genetics of complex diseases, such as diabetes, obesity, dyslipidemia, and hypertension. Individual studies in this program resulted in identifying several loci that may be implicated in the pathogenesis of these diseases (Bayoumi et al., 2008). Similarly, the Molecular Pathology Unit at the University of Kuwait specializes in studying early-stage colorectal cancer at the genomic level, using genome-wide scanning DNA methodologies (Al-Mulla et al., 2006). Likewise, in Bethlehem University, genomic analyses conducted over extended consanguineous families have resulted in the determination of multiple novel alleles for inherited hearing loss in the Palestinian population (Walsh et al., 2006).

CONSANGUINITY IN ARAB PEOPLE

Many Arab countries display some of the highest rates of consanguineous marriages in the world (Table 50-1, see also Chapter 51). Sociocultural factors, such as maintenance of family structure and property, ease of marital arrangements, better relations with in-laws, and financial advantages relating to dowry, play a crucial role in the preference of consanguinity in Arab populations (Khlat et al., 1986).

In Arab populations, consanguinity rates are changing in either way. In Bahrain, Egypt, Iraq, Jordan, Kuwait, Lebanon, Libya, Palestine, and Tunisia, decreasing trends of consanguineous marriages have been recorded among previous and present generations (Khlat, 1988; Jaber et al., 2000; Hamamy et al., 2005; Sharkia et al., 2008). On the other hand, increasing consanguinity rates have been recorded among generations in Algeria, Morocco, Oman, Qatar, Saudi Arabia, Sudan, Syria, the UAE, and Yemen (Bener et al., 2004; Bener & AlAli, 2006; Talbi et al., 2007).

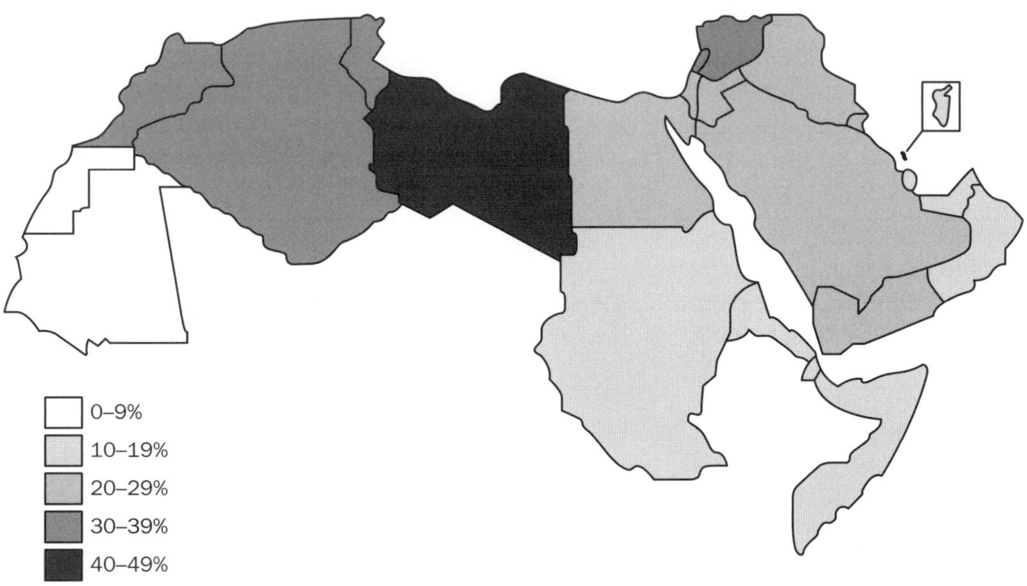

Figure 50-2 Schematic representation of the percent gene-related articles of the total number of biomedical records surveyed at the Centre for Arab Genomic Studies (December 2010).

Some of the plausible reasons behind the rising trend in consanguinity include urban–rural residence ratios of families, education levels, and the increase in the availability of cousins due to high fertility (Jurdi & Saxena, 2003).

In Arab populations and diasporas, the deep-rooted norm of consanguineous marriage has been widely accused of being an important factor contributing to the preponderance of autosomal recessive genetic disorders (El-Shafei et al., 1986; Hoodfar & Teebi, 1996; Abdulrazzaq et al., 1997; Mokhtar et al., 1998; Hamamy et al., 2007; Kerkeni et al., 2007; El-Mouzan et al., 2007). Actual data from countries for which surveys on the occurrence of genetic disorders have been completed by the Centre for Arab Genomic Studies (Bahrain, Oman, Qatar, and the UAE) indicate that recessive disorders are greater in number than the dominant ones (reviewed in Tadmouri, 2008b). Interestingly, these data demonstrate a direct correlation between the increase in consanguinity rates in a population accompanied with an increase in the share of autosomal recessive disorders, and a decrease in the number of autosomal dominant disorders in the respective population (Box 50-2; see Chapter 51).

BOX 50-2

- Many Arab countries display some of the highest rates of consanguineous marriages in the world.
- In Arab populations, consanguinity rates are changing in either way.
- Increasing consanguinity rates have been recorded among past and present generations in Algeria, Morocco, Oman, Qatar, Saudi Arabia, Sudan, Syria, the UAE, and Yemen.
- In Arab populations and diasporas, the deep-rooted norm of consanguineous marriage has been widely accused of being an important factor contributing to the preponderance of autosomal recessive genetic disorders.

THE CTGA DATABASE

Several attempts have been made to catalog genetic disorders in certain Arab countries (Der Kaloustian et al., 1980; Hamamy & Alwan, 1994; Al-Gazali et al., 1995; Teebi & Farag, 1997). Yet, similar publications are rapidly outdated as new disorders are continually described in Arabs. To overcome this problem, Tadmouri & Bissar-Tadmouri (1999) initiated an early attempt to maintain offline tabular lists of genetic disorders described in Arab individuals, with corresponding references, by monitoring international disease databases and scanning bibliographic indices. Using this strategy, 374 entries for genetic disorders in Arabs were recorded in 1999. This number increased to 752 entries early in 2004. In March 2004, the name "Catalogue of Transmission Genetics in Arabs" (CTGA) Database was coined for any future online database that may materialize out of this survey (Tadmouri, 2004).

To this end, the Centre for Arab Genomic Studies (CAGS) adopted the proposal to launch a pilot project to construct the CTGA Database with the aim of educating the medical community and raising public awareness in at-risk populations. Initially, major software components of the CTGA Database were constructed, and tested separately, early in year 2004. The database was then assembled and tested with limited amounts of data prior to its public release on November 30, 2004 (Tadmouri et al., 2006).

The CTGA Database (Box 50-3) is a continuously updated catalog of bibliographic material and observations on human gene variants, and inherited, or heritable, genetic diseases in Arab individuals (Tadmouri et al., 2006). Since the public release of the CTGA Database, knowledge on the presence of genetic disorders in Arab populations is continuously expanding. This process is largely driven by the diverse methods used at the Centre for Arab Genomic Studies to collect data and information on genetic conditions in Arab patients, including dependence on bibliographic indices, regional peer-reviewed medical publications, and personal submissions (Tadmouri, 2008b). In fact, the survey to collect data for the CTGA Database in the period between years 2004 and 2008 involved screening 61,000 international and national articles. From this, nearly 4700 articles were investigated in detail during the surveys, to collect data for the UAE, Bahrain, Oman, and Qatar. Of these, 2713 articles contained informative details about the occurrence of genetic disorders in Arab populations in the region.

BOX 50-3

- Several attempts have been made to catalog genetic disorders in certain Arab countries.
- The CTGA Database is a continuously updated catalog of bibliographic material and observations on human gene variants and inherited, or heritable, genetic diseases in Arab individuals.
- Since the public release of the CTGA Database, knowledge about the presence of genetic disorders in Arab populations is continuously expanding.

GENETIC DISORDERS IN ARAB POPULATIONS

As of September 2011, the CTGA Database has recorded the presence of 976 phenotype/disease entries in Arab

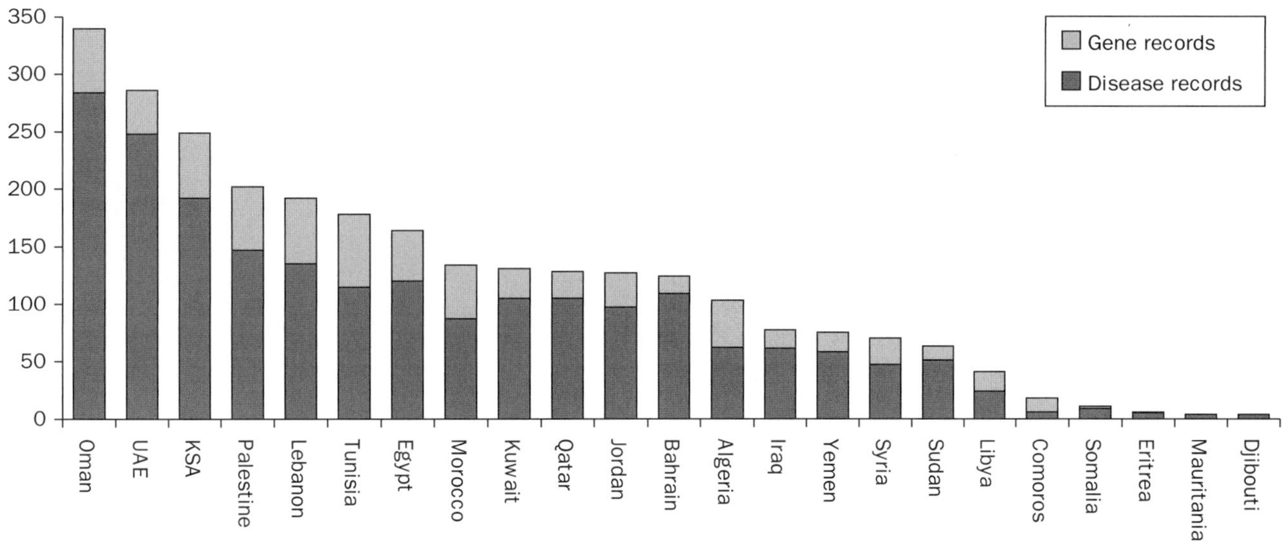

Figure 50-3 Disease and gene records in Arab countries according to the CTGA Database (September 2011).

individuals (Box 50-4). However, data on only about 387 related genes are available in the CTGA Database, reflecting the dominance of clinical observations over molecular analyses in most of the research conducted in the region (Figure 50-3).

By employing the World Health Organization International Classification of Disease version 10 (WHO ICD-10), it is possible to categorize the distribution of genetic disorders in Arab people according to disease taxonomies. Nearly one-third of genetic disorders in Arab individuals result from congenital malformations and chromosomal abnormalities (34%). They are then followed by endocrine and metabolic disorders (19%) and diseases of the nervous system (11%). Other types of disorders seem to occur at lower frequencies in the Arab population (Figure 50-4); this finding may be due to the lack of specific regional research or expertise in these entities (Tadmouri, 2006).

The overwhelming proportion of genetically transmitted diseases in Arab patients is inherited through autosomal recessive modes (approximately 61%). These are followed by autosomal dominant (28%) and X-linked traits (6%; Figure 50-5). High consanguinity rates and the

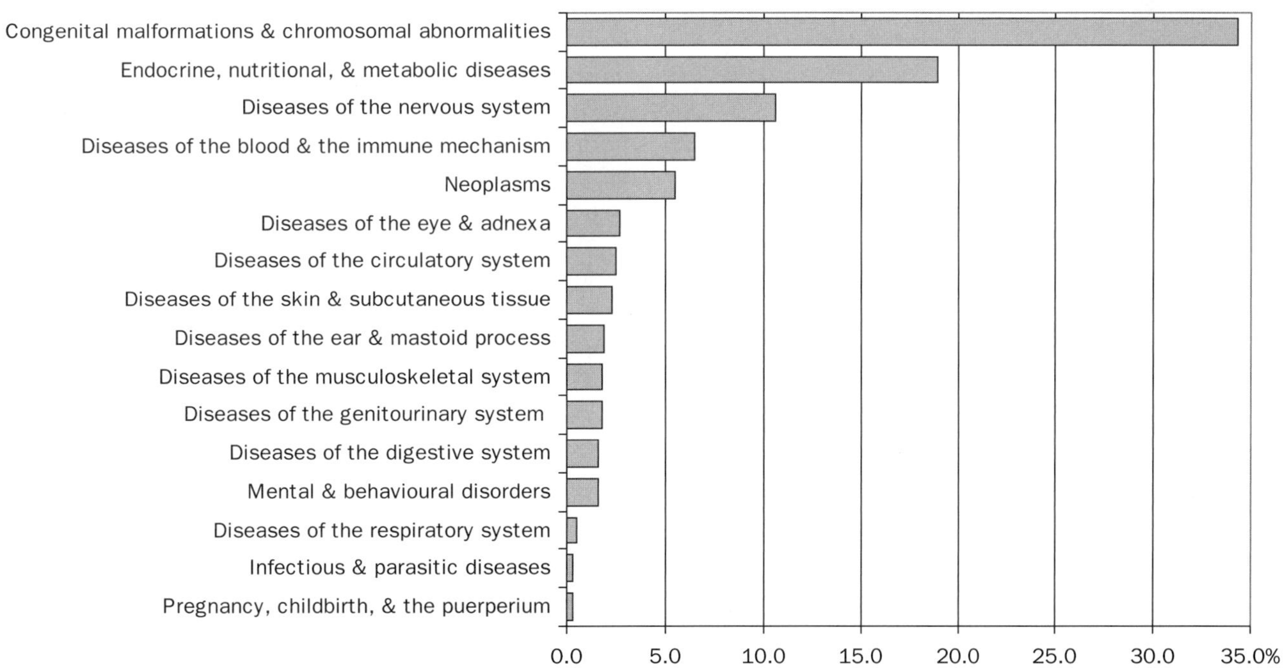

Figure 50-4 Classification of genetic disorders in Arabs according to the WHO ICD-10 (September 2011).

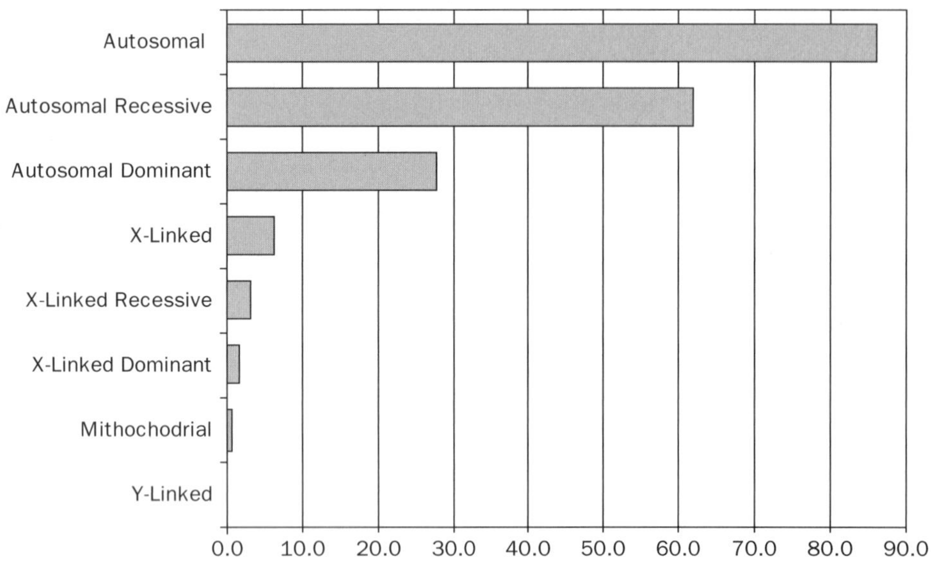

Figure 50-5 Classification of genetic disorders in Arabs according to mode of inheritance (September 2011).

extended family structure, commonly present in Arab societies, are likely explanations for these observations (Tadmouri, 2006; see also "Consanguinity as a Driving Force for Autosomal Recessive Disorders" in Chapter 51).

BOX 50-4

- The CTGA Database has recorded the presence of 976 phenotype/disease entries in Arab individuals.
- Nearly one-third of genetic disorders in Arab individuals result from congenital malformations and chromosomal abnormalities.
- The overwhelming proportion of genetically transmitted diseases in Arab patients is inherited through autosomal recessive modes.

THE SPECTRUM OF GENETIC DISORDERS IN ARAB POPULATIONS

The CTGA Database includes many records of genetic disorders that have reached epidemic values and occur at extremely high annual incidences (i.e., > 100 cases/100,000 live births). This group encompasses all hemoglobin disorders (thalassemias, sickle cell disease, and hemoglobin variants), G6PD deficiency, Down syndrome, breast cancer, diabetes, anencephaly, Graves disease, Caffey disease, Takayasu arteritis, polycystic kidneys, hypercholesterolemia, orofacial cleft, pyloric stenosis, and others. On the other hand, many other disorders do occur in the Arab world at higher incidence rates when compared to world data, these include: tetralogy of Fallot, familial Mediterranean fever, deafness, Noonan syndrome, Meckel syndrome, and spondyloarthropathy. The overwhelming distribution of these diseases in Arabs is best explained by the exposure of Arab countries to common environmental factors that encouraged natural selection for these disorders, such as malaria in the case of hemoglobinopathies and dietary traditions in the case of G6PD deficiency (reviewed in Tadmouri, 2008b).

In many instances, however, genetic disorders exhibit peculiar geographical distributions encompassing one or more neighboring regions, such as in the cases of hypoparathyroidism-retardation-dysmorphism (also known as Sanjad-Sakati) syndrome and Laurence-Moon syndrome in the Gulf region (Figure 50-6). Nevertheless,

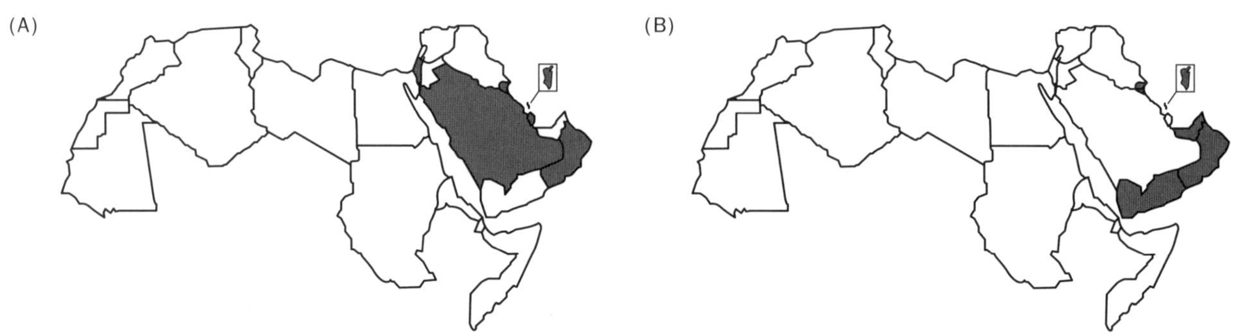

Figure 50-6 Computer generated maps indicating the relative geographic distribution of (A) hypoparathyroidism- retardation-dysmorphism syndrome and (B) Laurence-Moon syndrome in the Arab World according to the CTGA Database (September 2011).

some genetic disorders seem to be specific to certain Arab populations, such as: the Lebanese type of mannose 6-phosphate receptor recognition defect (Alexander et al., 1984), the Bedouin spastic ataxia syndrome (Mousa et al., 1986), the Algerian type of spondylometaphyseal dysplasia (Kozlowski et al., 1988), the Kuwaiti type facio-digitogenital syndrome (Teebi et al., 1988), the Kuwaiti type of cardioskeletal syndrome (Reardon et al., 1990), the Yemenite deaf-blind hypopigmentation syndrome (Warburg et al., 1990), the Nablus mask-like facial syndrome (Teebi, 2000), the Jerash type of distal hereditary motor neuropathy (Christodoulou et al., 2000), Karak syndrome (Mubaidin et al., 2003), and the Omani type of spondyloepiphyseal dysplasia (Rajab et al., 2004).

It is important to note that many new syndromes and variants have recently been described in Arab people. In many cases, Arab scholars and researchers were the first to describe some of these disorders, such as: Najjar syndrome (Najjar et al., 1973), Barakat syndrome (Barakat et al., 1977), Abdallat syndrome (Abdallat et al., 1980), Fadhil syndrome (Fadhil et al., 1983), Woodhouse-Sakati syndrome (Woodhouse & Sakati, 1983), Al-Awadi-Raas-Rothschild syndrome (Al-Awadi et al., 1985), Malouf syndrome (Malouf et al., 1985), the Teebi type of hypertelorism (Teebi, 1987), Jalili syndrome (Jalili & Smith, 1988), the Teebi-Shaltout syndrome (Teebi & Shaltout, 1989), Sanjad-Sakati syndrome (Sanjad et al., 1991), Majeed syndrome (Majeed et al., 1989), Teebi Al Saleh Hassoon syndrome (Teebi et al., 1989), Teebi-Shaltout syndrome (Teebi & Shaltout, 1989), Dudin-Thalji syndrome (Dudin & Thalji, 1991), Sanjad-Sakati syndrome (Sanjad et al., 1991), Al-Gazali syndrome (Al Gazali et al., 1994), Temtamy preaxial brachydactyly syndrome (Temtamy et al., 1998), Al Aqeel-Al Sewairi syndrome (Al Aqeel et al., 2000), Megarbane syndrome (Megarbane et al., 2001), El-Shanti syndrome (El-Shanti et al., 2003), Bosley-Salih-Alorainy syndrome (Tischfield et al., 2005), and Megarbane-Jalkh syndrome (Megarbane et al., 2008).

DISEASE AND GENE DIVERSITY IN THE REGION

Although extensive surveys to catalog genetic disorders in Arab populations are complete for Bahrain, Oman, Qatar, and the UAE, many important observations could be deduced with respect to the diversity of genetic disorders in Arab people. CTGA Database records indicate the presence of 506 genetic disorders in the combined Arab populations of Bahrain, Oman, Qatar, and the UAE. This number is by any standards extremely high, considering the relatively small sizes of the populations involved. Among these four countries, however, the number of genetic disorders surveyed seems to follow a recognizable trend. Oman and the UAE, with large population sizes, report the largest number of disorders, whereas Qatar and Bahrain report a smaller number of genetic disorders.

Despite the close proximity of these four Gulf States to each other and the long-shared history of their populations, each of the countries report many exclusive diseases not described in the other countries. Major explanations for this observation include: (1) the high heterogeneity of the populations due to the residence of many Arab expatriate populations in the region and (2) the diverse historical relations that took place between the national populations of Bahrain, Oman, Qatar, and the UAE on one side with Iran, East Africa, and the Indian subcontinent on the other side through trade routes and immigration flows (reviewed in Tadmouri, 2008b).

Numerical diversities of genetic disorders in the region are further exemplified at three interrelated levels that give altogether an unparalleled heterogeneity at the clinical level, and the level of molecular pathologies as well. The first level of complexity is derived from the fact that many genetic disorders are occurring at epidemic levels that make them occur in many individuals with a variety of other distinct disorders at a rate higher than expected by chance; thus the term *comorbidity*. In fact, many studies reviewed in the CTGA Database reveal the presence of comorbidity in patients from the region, especially with major disorders including thalassemias, sickle cell disease, cystic fibrosis, Down syndrome, G6PD deficiency, and many others (Table 50-2). In certain cases, the striking occurrences of such comorbidities motivated scientists to explore the clinical outcomes, such as in the case of Mohammad and colleagues (1998) who studied the coexistence of sickle cell disease with G6PD deficiency and found out that severe G6PD deficiency occurs in 47% of individuals with sickle hemoglobin in Bahrain. In an independent study of six patients with cystic fibrosis, Khan and Mohammad (1985) described a reduced G6PD enzyme activity in four of their patients. A decade later, Al Arrayed and Abdulla (1996) studied the incidence of cystic fibrosis in Bahrain retrospectively, by reviewing the records of patients diagnosed with the disorder during a 17-year period in a major hospital in Bahrain. The survey included a total of 27 patients, including 25 Bahrainis, with cystic fibrosis, among whom 98% also had G6PD deficiency. The common presenting clinical picture was failure to thrive (66%), pneumonia (62%), hypochloremic alkalosis (44%), and anemia (37%), with a mortality rate of 60%. Certainly, the systemic study of comorbidity would represent a main approach to study the clinical complexity in Arab patients with genetic disorders. Once epidemiologically established through population or community surveys, the study of the comorbidity direction and of the chronological patterns of associated clinical entities may then be translated into enhanced care of patients, selection of initial treatment, evaluation of treatment effectiveness, and improvement of prognosis.

TABLE 50-2 SELECTED EXAMPLES OF GENETIC DISEASE COMORBIDITIES RECORDED IN THE CTGA DATABASE (SEPTEMBER 2010).

Disease	Comorbidity
Alpha-thalassemia	Wolman disease
Beta-thalassemia	Alpha-thalassemia Cardiomyopathy Dyskeratosis Congenita, X-linked G6PD deficiency Pycnodysostosis
Cystic fibrosis	Beta-thalassemia Ehlers-Danlos syndrome type III G6PD deficiency Glioblastoma multiforme Infantile hypertrophic pyloric stenosis 1 Sickle cell disease Sickle/beta-thalassemia
Down syndrome	Absence of abdominal muscles with urinary tract abnormality and cryptorchidism Choanal Atresia, Posterior Moyamoya syndrome Tetralogy of Fallot
G6PD deficiency	Dyskeratosis Congenita, X-linked Pycnodysostosis
Sickle cell disease	Alpha-thalassemia Beta-thalassemia Blepharophimosis, Ptosis, and Epicanthus Inversus G6PD deficiency Hand-foot syndrome
Tetralogy of Fallot	Chromosome 18p deletion syndrome Down syndrome Frontonasal dysplasia Holt-Oram syndrome
Cardiomyopathy	Crohn's disease Ebstein disease Familial antiphospholipid syndrome HbH disease

At another clinical level, many of the broad groupings of genetic disorders are further classified into types and subtypes in Arab patients, indicating a heterogeneity in the features associated with these disorders (Box 50-5). For example, mucopolysaccharidosis occurs in at least five types in the south of the Arabian Peninsula (IIIA, IIIB, IVA, VI, and VII). In Oman, glycogen storage disease is reported into four distinct types (I, II, III, and IV) and spinal muscular atrophy is associated with three types (I, II, and III). This latter disease also occurs in three variants in the Bahraini, Saudi, and Kuwaiti populations. In the UAE population, three variants occur for epidermolysis bullosa (Junctional Herlitz type, junctional non-Herlitz type, and junctionalis with pyloric atresia) as well as osteogenesis imperfecta (I, IIA, and III). Some of the disorders that occur in other Arab regions with multiple variants include: Charcot-Marie-Tooth disease (Axonal 2B1, 2H, 2K, 4A, 4B2, 4C, and 4H), Ehlers-Danlos syndrome (III, VI, VIB, and the progeroid form), lamellar ichthyosis (1, 2, 3, and 5), limb-girdle muscular dystrophy (2A, 2B, 2C, 2D, and 2I), spinocerebellar ataxia (1, 2, 7, autosomal recessive 2 and 5, and autosomal recessive with axonal neuropathy), Usher syndrome (ID, IE, IG, IIA, IIB, and III), Parkinson's disease (2, 6, 7, and 9), and many others.

At the molecular level, this diversity is further expressed. In contrast to many world populations, 50 mutations are responsible for the molecular etiology of beta-thalassemia in the UAE population, making it one of the most heterogeneous populations with regard to mutations in the beta-globin gene (Baysal, 2005). Besides that, many novel mutations have been identified in common genetic disorders in the region. In cystic fibrosis, worldwide studies indicate that more than 70% of patients with cystic fibrosis show a single mutation that involves the deletion of three nucleotides of exon 10, within the first nucleotide binding domain, resulting in the deletion of phenylalanine at position 508 in the protein product: the delta F508. In countries situated east in the Arabian Peninsula, this picture seems to be different. In Bahrain, the disease is associated with eight mutations including 2043delG (30.8%) and, to a lesser extent, delta F508 (7.7%). This latter mutation occurs in Bahrainis of Persian origins (Eskandarani, 2002). In Oman and the UAE, the picture seems to be different, with the S549R (T-G) mutation being most common and the delta F508 occurring at relatively low frequencies, but exclusively in patients of Baluchi descent (Frossard et al., 2000; Dawson & Frossard, 2000). At the clinical level, patients homozygous for the mutations delta F508 and S549R (T-G) have a severe clinical presentation and illness and are indistinguishable on clinical grounds; hence, emphasizing the importance of gene testing to explain the processes converting distinct genotypes into highly severe and similar phenotypes (Dawson et al., 2001).

On the other hand, in a comprehensive molecular study in Omani patients with familial hemophagocytic lymphohistiocytosis, seven distinct mutations were identified in the coding region of the Perforin gene, of which five were novel (Muralitharan et al., 2007). This molecular heterogeneity is also expressed in linkage analyses that excluded classical gene loci known to be involved in the pathogenesis of certain genetic disorders, and indicated the possibility of other genes for modeling these disorders in Arab patients (Al-Yahyaee et al., 2006; Bayoumi et al., 2006; White et al., 2007).

BOX 50-5

- CTGA Database records indicate the presence of 506 genetic disorders in the combined Arab populations of Bahrain, Oman, Qatar, and the UAE.
- The systematic study of comorbidity would represent a main approach to study the clinical complexity in Arab patients with genetic disorders.

- Many of the broad groupings of genetic disorders are further classified into types and subtypes in Arab patients indicating a heterogeneity in the features associated with these disorders.
- The UAE population is one of the most heterogeneous world populations with regard to mutations in the beta-globin gene.
- Many novel mutations have been identified in common genetic disorders in the region.

THE ECONOMIC IMPACT OF GENETIC DISORDERS

Genetic disorders are chronic in nature and often require lifelong management with no definitive cure. In the Arab world, several disorders, including chromosomal (Down syndrome, Turner syndrome), single-gene (sickle cell disease, thalassemia, G6PD deficiency, hemophilia, inborn errors of metabolism) and multifactorial disorders (coronary artery disease, arteriosclerosis, diabetes mellitus, hypertension, obesity) are common. Some of these disorders have assumed epidemic proportions, as in the cases of sickle cell disease, alpha-thalassemia, hypertension, and diabetes mellitus. The economic impact of each of these disorders differs according to their severity, many of which involve medical, surgical, or cosmetic interventions. Generally, these conditions are a leading cause of spontaneous abortion, neonatal death, and increased morbidity and mortality in both children and adults. They are a significant healthcare and psychosocial burden for the patient, the family, the healthcare system, and the community as a whole (El-Hazmi, 1999).

In terms of economic burden, patients with genetic or partly genetic disorders have longer and more frequent hospital admissions with a higher number of surgeries than other patients (Carnevale et al., 1985; McCandless et al., 2004). Additionally, the total costs paid by patients with genetic conditions are slightly greater (Hall et al., 1978), and these patients often must travel significant distances to get specialized treatment (Carnevale et al., 1985).

In recent years, heath economists have made significant advances in calculating the costs of genetic disorders, as well as disabilities caused by various congenital abnormalities. There are now generally accepted annual "cost of illness" estimates per patient for all common genetic conditions including: Down syndrome (US$36,000; Boulet et al., 2008), cystic fibrosis (US$28000; Eidt et al., 2009), Niemann-Pick disease type C (US$27000; Imrie et al., 2009), inherited leukodystrophy (US$22579; Bonkowsky et al., 2010), sickle cell disease (US$16668; Kauf et al., 2009), beta-thalassemia (US$7000; Karnon et al., 1999), diabetes mellitus (US$1500; Wang et al., 2009), asthma (US$1250; Bahadori et al., 2009), retinal dystrophy (US$1000; Porz et al., 2010), hemophilia (US$1000; Meyers et al., 1972), and atopic dermatitis (US$1000; Fowler et al., 2007). Overall, the total annual cost of the most common 20 genetic ailments in the Arab world is estimated not to be less than $13 billion per year (reviewed in Tadmouri, 2008b).

FAMILY PERCEPTION OF GENETIC DISORDERS

Arab families are highly close-knit units, integrated within the mesh of the larger, culture-bound socioeconomic network. Families form the center of the Arabian society and religion is an integral part of the culture. The major involvement of the family and society on genetic disorders in the Arab world does not come, thus, as a surprise. The most important ways in which the society plays its role in the manifestation of genetic diseases includes, but is not necessarily restricted to, the perception of family history and compliance to medical and genetic advice.

Family History: Close to 140 diseases catalogued in the CTGA Database document a positive family history of corresponding conditions. Interestingly, within this group, a large number of records document consanguinity within the family. This could again be due to the high prevalence of consanguinity in the region, an assumption highly supported by the autosomal recessive mode of inheritance of most of these disorders. In such a scenario, traditional genetic counseling cannot be expected to be as successful as reported in other societies with lower rates of inbreeding and less emphasis on families. Instead, genetic counseling in the Arab society needs to be oriented toward families. Such family-oriented counseling can have an enormous impact on multiply consanguineous kindred where rare genetic disorders are clustered (Al-Gazali et al., 2006). Unfortunately, the medical community in the Arab region demonstrates a very high turnover rate. This is especially true of the Gulf States, where expatriates form the major mass of the medical community. As a result, there is a dearth of family physicians who have been involved with families for a fairly long enough time to know the genetic background of the entire family. In Western societies, family physicians routinely use their experience with the history of the family right from prenatal and newborn screening to the early recognition of individuals at risk of disease (Feder & Modell, 1998). Since such a system is lacking in the region, specialists may find it difficult to diagnose a genetic disorder from consultation with the patient in isolation.

Noncompliance: Therapeutic noncompliance is a feature frequently encountered by physicians. This phenomenon is characterized by a refusal of the patient to abide by the recommendations of the healthcare provider. Apart from the direct effect on the treatment outcome of the patient, noncompliance can also cause an indirect increase in the financial burden on the society by way of excess urgent care visits, hospitalizations, and higher treatment costs (Jin et al., 2008). It has been seen that chronic diseases tend to lead to lower rates of compliance than do diseases of a short duration (Farmer et al., 1994; Gascon et al., 2004). Genetic diseases tend to require lifelong treatment, and thus patients are highly susceptible to resorting to noncompliance with the treatment strategies. The CTGA Database documents 17 cases of noncompliance within the patient population of the four GCC countries studied in detail. Data on this subject may not be complete, since most authors do not deem it necessary to concentrate on this seemingly non-vital aspect. CTGA Database records clearly demonstrate that noncompliance can occur in any of the phases of management of the genetic screening. Thus, there are instances of parents at risk of having an affected child refusing to get themselves screened (Joshi et al., 2002), and of parents not complying with treatment options for their children (Al Ansari, 1984; Soliman et al., 1995). More dangerous is the practice of not only disregarding the medical advice, but also seeking the help of traditional "healers," often leading to disastrous consequences (Ahmed & Farooqui, 2000). In most cases of noncompliance by parents, however, parents tend to get back to the physician and restart treatment once the symptoms in the children begin to worsen and they see that there is no option left. Patient refusal, on the other hand, is more disturbing, since those patients who demonstrate noncompliance on their own do not generally restart treatment later, even with adequate counseling (Khandekar et al., 2005). Patients may show noncompliance by refusing invasive diagnostic methods (Bhat & Hamdi, 2005), discontinuing medication (Khandekar et al., 2005), refusing surgery (Venugopalan et al., 1997), or refusing a medically advised termination of pregnancy (Gowri & Jain, 2005).

A characteristic feature of the Arab society is the extremely low level of genetic literacy, which makes the role of the genetic counselor quite difficult. Instances of successful applications of genetic counseling in affected families, which have helped in the early diagnoses and management of disease conditions, can be extracted from the CTGA Database (Subramanyan & Venugopalan, 2002). However, generally, people in these populations are reluctant to refer to a genetic counselor under the erroneous assumption that counselors would advise them on personal issues, such as the choice of a marriage partner or on their reproductive options. Unfortunately, such misconceptions exist not only among the patients themselves, but also among the healthcare providers (Hamamy & Bittles, 2009).

A direct correlation has been demonstrated between the level of education and awareness, and early detection and better management of genetic disorders in Arab populations (Al-Mahroos, 1998; Al-Saweer et al., 2003; Al-Moosa et al., 2006). Of note is the observation that compulsory premarital screening for identifying thalassemia carriers increased public awareness on genetic diseases in general, with requests for counseling on other conditions increasing (Hamamy & Bittles, 2009). It may also be a useful strategy to integrate genetic counseling into the procedure of routine prenatal care, so that it may be better accepted by the population. The Box 5-6 summarizes family and social attitudes toward genetic disorders.

BOX 50-6

- Close to 140 diseases catalogued in the CTGA Database document a positive family history of corresponding conditions.
- Therapeutic noncompliance is a feature frequently encountered by physicians.
- Patients may show noncompliance by refusing invasive diagnostic methods, discontinuing medication, refusing surgery, or refusing a medically advised termination of pregnancy.
- A characteristic feature of the Arab society is the extremely low level of genetic literacy, which makes the role of the genetic counselor quite difficult.

PREVENTIVE ASPECTS OF GENETIC DISORDERS

A detailed analysis of the molecular basis of defined genetic diseases indicates that approximately half of the genetic disorders described in Arabs (59%) result from single-gene or gene loci alterations (Figure 50-7). Hence, in the presence of the necessary technical infrastructure, diagnostic services may well be available for people at risk, and preventive programs may be successfully applied in many Arab communities (Box 50-7).

The successful management of genetic disorders also incurs a high financial cost, which could be eased by the application of effective prevention programs in populations at risk of genetic disease (WHO, 1996). Prevention programs are effective in decreasing the impact of genetic disorders on families and societies, and also lead to early treatment and improvements in outcome and prognosis (Al-Odaib et al., 2003). A majority of Arab countries have the expertise and resources to apply most of these preventive measures, especially in the areas of newborn screening and carrier screening for prevalent genetic disorders. However, having the technology and resources alone are not enough. To start an effective program it has to be orchestrated by different strata of the society, including patient representatives, medical geneticists, physicians, public health physicians, sociologists, ethicists, pharmaceutical industries, policy commentators, and policymakers within governments. Moreover, distinct

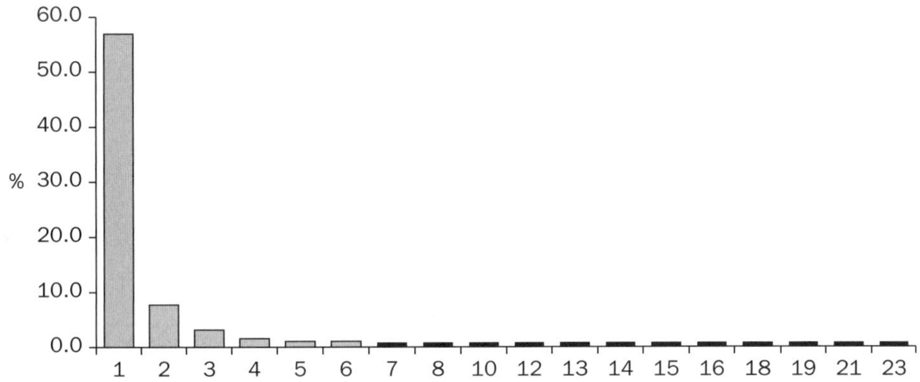

Figure 50-7 Distribution of genetic disorders in Arabs according to the number of causative gene loci (September 2011).

policy areas have to be defined when dealing with a prevention program for genetic disorders; these include:

1. **The science base**: Through which disease conditions could be defined according to their attributed burden, prevalence of the genetic trait, and natural history in the target population, from susceptibility to latent and to overt disease. Experts have to also evaluate the safety and effectiveness of possible tests involved. Most important, population screening should be performed only if the abnormal finding in question can change the clinical management, and this management will improve the prognosis.
2. **Educational strategies**: Should be built based on the concept of "genetic literacy." This could be best achieved by applying interactive processes of dialog and discussion in educational processes. For this, new biology has to be integrated as a necessary component of general education, as well as in the education and training of all health professionals. In this framework, medical geneticists, genetic counselors, and clinical psychologists have a crucial role in making the prevention program acceptable to the target population by clear explanation of associated risks, proper parental consenting, and counseling (Khalifa, 1999).
3. **Regulatory framework**: Has to be directed to resolve the widespread misunderstanding about genetics, especially the concept of genetic risk for a disease. In this area, scientists and the media have a great role. Politicians also have a responsibility in leading the public debate and making available adequate infrastructures, surveillance policies, and supportive financing mechanisms.
4. **Cultural sensitivity**: is a required safeguard to protect the public from premature and inappropriate use of genetic information, stigmatization, and discrimination, and to avoid coercion or manipulation (Meyer, 2005).
5. **Financial framework**: Is a vital factor in securing the continuity of a preventive program. In most cases, the annual cost of a nationwide disease prevention program would not exceed the cost of treating one annual birth cohort of patients with a genetic condition for one year (WHO, 1996).

Several Arab States have initiated systematic, cost-effective prevention programs for certain common genetic conditions, using a variety of approaches:

1. **Neonatal screening**: This involves the establishment of national or hospital-based registries for congenital abnormalities (e.g., Oman and the UAE) and the implementation of biochemical technologies to depict the incidence of many inherited disorders. Currently, Egypt, Oman, Palestine, Qatar, Saudi Arabia, and the UAE execute national newborn screening programs varying from one disease to 23. With an estimated 10 million newborns per year, a wide coverage of such programs in the Arab world becomes an important challenge (reviewed in Saadallah & Rashed, 2007).
2. **Family screening**: In Arab countries this proves to be more effective than a population screening program, due to the high level of consanguinity and the relative clustering of genetic diseases in specific population groups (Defesche et al., 2004).
3. **Premarital screening**: Several Arab countries have introduced premarital screening, especially for hemoglobin disorders, as in Bahrain (Al Arrayed, 2005), Jordan (Hamamy et al., 2007), Lebanon, Saudi Arabia (Al-Suliman, 2006; Alhamdan et al., 2007), Tunisia (Al-Gazali et al., 2006), and the UAE (Al-Gazali, 2005).
4. **Prenatal diagnosis**: Is an important component that requires an extensive social framework and without which a prevention program is practically
5. **Preimplantation genetic diagnosis**: This is a more welcomed approach, since it allows the avoidance of termination. Successful applications of this approach have been reported in Jordan (Kilani & Haj Hassan, 2002; Abdelhadi et al., 2003) and Saudi Arabia (Ozand

et al., 2005). A recent study from the UAE found that most people favor this mode of prevention (Al-Gazali, 2005).

BOX 50-7

- Prevention programs are effective in decreasing the impact of genetic disorders on families and societies, and also lead to early treatment and improvements in outcome and prognosis.
- Several Arab States have initiated systematic, cost-effective prevention programs for certain common genetic conditions using a variety of approaches.
- Egypt, Oman, Palestine, Qatar, Saudi Arabia, and the UAE execute national newborn screening programs varying from one disease to 23.
- In Arab countries, family screening proves to be more effective than a population screening program due to the high level of consanguinity and the relative clustering of genetic diseases in specific population groups.
- Several Arab countries have introduced premarital screening, especially for hemoglobin disorders.

CONCLUDING REMARKS

At present, congenital malformations are the second leading cause of infant mortality in countries of the Gulf Cooperation Council, including Bahrain, Kuwait, Oman, and Qatar. Reports from Saudi Arabia indicate that congenital malformations account for about 30% of perinatal deaths (Hamamy & Alwan, 1994). Additionally, in most Arab populations the birth prevalence of severe recessively inherited disorders may approach that of congenital malformations (Alwan & Modell, 1997).

Approximately 27% of reported genetic disorders in Arabs remain confined to clinical observations, with no significant attempts to depict their molecular pathologies. A large number of these disorders are confined to local families and communities and have not been described elsewhere. Mummifying these disorders at the clinical level represents a very serious loss for the global scientific community, since permanently burying information regarding hundreds or thousands of human gene variants might lead to the loss of important information that can be used for research, and potential cures for genetically transmitted conditions (Editorial, 2006). Unfortunately, no established system is yet available in many Arab medical research institutions to translate clinical observations into genetic data. The limited examples available in the Arab world are usually local efforts, mostly exerted by medical practitioners and clinical geneticists who have developed a particular interest or have specialized in molecular studies (Naveed et al., 2006, 2007). Increasing the emphasis on subjects such as molecular genetics in medical schools in the region will help to create future generations of physicians and other medical personnel capable of establishing the phenotype/genotype correlations that are key elements in the modern medical applications of genetics.

Databasing prevalence data as well as the molecular pathologies leading to genetic disorders in Arabs offers solid groundwork to promote proper education in the field and employ knowledge-driven development to address urgent regional health needs. The organization of such information also promotes Arab scientists to a position of strength and allows them to contribute to global research efforts in the field, and build sustainable research activities based upon education and the improvement of human health.

REFERENCES

Abdallat A, Davis SM, Farrage J, McDonald WI. (1980). Disordered pigmentation, spastic paraparesis and peripheral neuropathy in three siblings: a new neurocutaneous syndrome. *J Neurol Neurosurg Psychiatry 43*(11), 962–926.

Abdelhadi I, Colls P, Sandalinas M, Escudero T, Munné S. (2003). Preimplantation genetic diagnosis of numerical abnormalities for 13 chromosomes. *Reprod Biomed Online 6*(2), 226–231.

Abdulrazzaq YM, Bener A, Al-Gazali LI, Al-Khayat AI, Micallef R, Gaber T. (1997). A study of possible deleterious effects of consanguinity. *Clin Genet 51*(3), 167–173.

Ahmed A, Farooqui MA. (2000). Naphthalene-associated severe intravascular hemolysis in a G6PD deficient patient. *Oman Med J 17*(2), 33–35.

Ahmed AH. (1979). Consanguinity and schizophrenia in Sudan. *Br J Psychiatry 134*, 635–636.

Al Ansari AM. (1984). Gilles De La Tourette's syndrome. *Bahrain Med Bull 6*(3), 109–110.

Al Aqeel A, Al Sewairi W, Edress B, Gorlin RJ, Desnick RJ, Martignetti JA. (2000). Inherited multicentric osteolysis with arthritis: a variant resembling Torg syndrome in a Saudi family. *Am J Med Genet 93*(1), 11–18.

Al Arrayed S. (2005). Campaign to control genetic blood diseases in Bahrain. *Community Genet 8*(1), 52–55.

Al Arrayed SS. (2006). Genetic diseases in Bahrain. In: Tadmouri GO, Taleb Al Ali M, Al Khaja N. (eds.), *Genetic Disorders in the Arab World: Bahrain* (pp. 72–91). Centre for Arab Genomic Studies. Dubai, United Arab Emirates.

Al Arrayed SS, Abdulla F. (1996). Incidence of cystic fibrosis in Bahrain. *J Bahrain Med Soc 8*(3), 157–160.

Al Arrayed SS. (1995). The frequency of consanguineous marriages in the State of Bahrain. *Bahrain Medical Bull 17*(2), 63–66.

Al Gazali LI, Al Talabani J, Mosawi A, Lytle W. (1994). Anterior segment anomalies of the eye, clefting and skeletal abnormalities in two sibs of consanguineous parents: Michels syndrome or new syndrome? *Clin Dysmorphol 3*(3), 238–244.

Al Husain M, Al Bunyan M. (1997). Consanguineous marriages in a Saudi population and the effect of inbreeding on prenatal and postnatal mortality. *Ann Trop Paediatr 17*(2), 155–160.

Al-Abdulkareem AA, Ballal SG. (1998). Consanguineous marriage in an urban area of Saudi Arabia: rates and adverse health effects on the offspring. *J Community Health 23*(1), 75–83.

Al-Arrayed SS. (1999). Review of the spectrum of genetic diseases in Bahrain. *East Mediterr Health J 5*(6), 1114–1120.

Al-Awadi SA, Moussa MA, Naguib KK, et al. (1985). Consanguinity among the Kuwaiti population. *Clin Genet 27*(5), 483–486.

Al-Awadi SA, Teebi AS, Farag TI, Naguib KM, el-Khalifa MY. (1985). Profound limb deficiency, thoracic dystrophy, unusual facies, and normal intelligence: a new syndrome. *J Med Genet 22*(1), 36–38.

Alexander D, Dudin G, Talj F, et al. (1984). Five related Lebanese individuals with high plasma lysosomal hydrolases: a new defect in mannose-6-phosphate receptor recognition? *Am J Hum Genet 36*(5), 1001–1014.

Al-Gazali L, Hamamy H, Al-Arrayad S. (2006). Genetic disorders in the Arab World. *BMJ 333*, 831–834.

Al-Gazali LI, Bener A, Abdulrazzaq YM, Micallef R, Al-Khayat AI, Gaber T. (1997). Consanguineous marriages in the United Arab Emirates. *J Biosoc Sci 29*(4), 491–497.

Al-Gazali LI, Dawodu AH, Sabarinathan K, Varghese M. (1995). The profile of major congenital abnormalities in the United Arab Emirates (UAE) population. J *Med Genet 32*(1), 7–13.

Al-Gazali LI (2005). Attitudes toward genetic counseling in the United Arab Emirates. *Community Genet 8*(1), 48–51.

Alhamdan NA, Almazrou YY, Alswaidi FM, Choudhry AJ. (2007). Premarital screening for thalassemia and sickle cell disease in Saudi Arabia. *Genet Med 9*(6), 372–377.

Al-Mahroos F. (1998). Cystic fibrosis in Bahrain incidence, phenotype, and outcome. *J Trop Pediatr 44*(1), 35–39.

Al-Moosa S, Allin S, Jemiai N, Al-Lawati J, Mossialos E. (2006). Diabetes and urbanization in the Omani population: an analysis of national survey data. *Popul Health Metr 4*, 5.

Al-Mulla F, Hagan S, Behbehani AI, et al. (2006). Raf kinase inhibitor protein expression in a survival analysis of colorectal cancer patients. *J Clin Oncol 24*(36), 5672–5679.

Al-Odaib AN, Abu-Amero KK, Ozand PT, Al-Hellani AM. (2003). A new era for preventive genetic programs in the Arabian Peninsula. *Saudi Med J 24*(11), 1168–1175.

Al-Salem M, Rawashdeh N. (1993). Consanguinity in north Jordan: prevalence and pattern. *J Biosoc Sci 25*(4), 553–556.

Al-Saweer A, Yacoub F, Mohammed N. (2003). The prevalence of risk factors among women diagnosed with breast cancer. *Bahrain Med Bull 25*(4), 156–158.

Al-Suliman A. (2006). Prevalence of beta-thalassemia trait in premarital screening in Al-Hassa, Saudi Arabia. *Ann Saudi Med 26*(1), 14–16.

Alwan A, Modell B. (1997). Hereditary disorders in the Eastern Mediterranean Region: Role of customary consanguineous marriage. EMRO Technical Publications Series 24. Alexandria, Egypt: WHO.

Al-Yahyaee S, Al-Gazali LI, De Jonghe P, et al. (2006). A novel locus for hereditary spastic paraplegia with thin corpus callosum and epilepsy. *Neurology 66*(8), 1230–1234.

Angel JL. (1966). Porotic hyperostosis, anemias, malarias, and marshes in the prehistoric Eastern Mediterranean. *Science 153*, 760–763.

Assaf S, Khawaja M. (2008). Consanguinity trends and correlates in the Palestinian Territories. *J Biosoc Sci 13*, 1–18.

Bahadori K, Doyle-Waters MM, Marra C, et al. (2009). Economic burden of asthma: a systematic review. *BMC Pulm Med 9*, 24.

Bailey GN, Flemming NC, King GCP, et al. (2007). Coastlines, submerged landscapes, and human evolution: The Red Sea Basin and the Farasan Islands. *J Island Coastal Archaeol 2*, 127–160.

Bakhiet M, Taha S. (2008). A novel nervous system-induced factor inducing immune responses in the spleen. *Immunol Cell Biol 86*(8), 688–699.

Barakat AY, D'Albora JB, Martin MM, Jose PA. (1977). Familial nephrosis, nerve deafness, and hypoparathyroidism. *J Pediatr 91*(1), 61–64.

Barbour B, Salameh P. (2009). Consanguinity in Lebanon: prevalence, distribution and determinants. *J Biosoc Sci 41*(4), 505–517.

Bayoumi R. Al-Yahyaee S, Albarwani S, et al. (2008). Some genetic determinants of obesity, type 2 diabetes and dyslipidemias in "Oman Family Study." In: Tadmouri GO, Taleb Al Ali M, Al Khaja N, eds. *Genetic Disorders in the Arab World: Oman*. Centre for Arab Genomic Studies, Dubai.

Bayoumi RA, Eapen V, Al-Yahyaee S, Al Barwani HS, Hill RS, Al Gazali L. (2006). The genetic basis of inherited primary nocturnal enuresis: A UAE study. *J Psychosom Res 61*(3), 317–320.

Baysal E. (2005). Molecular heterogeneity of beta-thalassemia in the United Arab Emirates. *Community Genet 8*(1), 35–39.

Beech M, Cuttler R, Moscrop D, Kallweit H, Martin J. (2005). New evidence for the Neolithic settlement of Marawah Island, Abu Dhabi, United Arab Emirates. *PSAS 35*, 37–56.

Ben Abdeladhim A, Aïssaoui B, Boussen M. (1987). Homozygous O Arab hemoglobinopathy in a Tunisian family. Apropos of a case. Tunis Med 65, 571–574.

Ben Arab S, Masmoudi S, Beltaief N, Hachicha S, Ayadi H. (2004). Consanguinity and endogamy in Northern Tunisia and its impact on nonsyndromic deafness. *Genet Epidemiol 27*(1), 74–79.

Benallegue A, Kedji F. (1984). Consanguinite et sante publique: une etude Algerrienne. *Arch Fr Pediatr 41*(6), 435–440.

Bener A, Alali KA. (2006). Consanguineous marriage in a newly developed country: the Qatari population. *J Biosoc Sci 38*(2), 239–246.

Bener A, Al-Suwaidi J, Al-Jaber K, Al-Marri S, Elbagi IE. (2004). Epidemiology of hypertension and its associated risk factors in the Qatari population. *J Hum Hypertens 18*(7), 529–530.

Bhat SM, Hamdi IM. (2005). Primary amenorrhea. Varied etiology. Saudi Med J *26*(9), 1453–1455.

Bonkowsky JL, Nelson C, Kingston JL, et al. (2010). The burden of inherited leukodystrophies in children. *Neurology 75*(8), 718–725.

Boulet SL, Molinari NA, Grosse SD, Honein MA, Correa-Villaseñor A. (2008). Health care expenditures for infants and young children with Down syndrome in a privately insured population. *J Pediatr 153*(2), 241–246.

Broadhead RC, Sehgal KC. (1981). Consanguinity and congenital anomalies in East Libya. *Garyounis Med J 4*(1), 3–6.

Cadenas AM, Zhivotovsky LA, Cavalli-Sforza LL, Underhill PA, Herrera RJ. (2008). Y-chromosome diversity characterizes the Gulf of Oman. *Eur J Hum Genet 16*(3), 374–386.

Cann RL, Stoneking M, Wilson AC. (1987). Mitochondrial DNA and human evolution. *Nature 325*, 31–36.

Carnevale A, Hernandez M, Reyes R, Paz F, Sosa C. (1985). The frequency and economic burden of genetic disease in a pediatric hospital in Mexico City. *Am J Med Genet 20*(4), 665–675.

Carter R, Mendis KN. (2002). Evolutionary and historical aspects of the burden of malaria. *Clin Microbiol Rev 15*, 564–594.

Chaabouni-Bouhamed H. (2008). Tunisia: communities and community genetics. *Community Genet 11*(6), 313–323.

Cherkaoui M, Baali A, Larrouy G, Sevin A, Boetsch G. (2005). Consanguinity, fertility of couples and mortality of children in the high Atlas population (commons of Anougal and Azgour, Marrakesh, Morroco). *Intl J Anthropol 20*(3–4), 199–206.

Christianson A, Howson CP, Modell B. (2006). March of Dimes global report on birth defects: The hidden toll of dying and disabled children. March of Dimes Birth Defects Foundation, New York.

Christodoulou K, Zamba E, Tsingis M, et al. (2000). A novel form of distal hereditary motor neuronopathy maps to chromosome 9p21.1-p12. *Ann Neurol 48*(6), 877–884.

Dakik HA, Kaidbey H, Sabra R. (2006). Research productivity of the medical faculty at the American University of Beirut. *Postgrad Med J 82*(969), 462–464.

Dawson KP, Frossard PM, Al-Awar B. (2001). Disease severity associated with cystic fibrosis mutations deltaF508 and S549R(T-->G). *East Mediterr Health J 7*(6), 975–980.

Dawson KP, Frossard PM. (2000). The geographic distribution of cystic fibrosis mutations gives clues about population origins. *Eur J Pediatr 159*(7), 496–499.

de Zulueta J (1994). Malaria and ecosystems: from prehistory to posteradication. *Parassitologia 36*, 7–15.

Defesche JC, Lansberg PJ, Umans-Eckenhausen MA, Kastelein JJ. (2004). Advanced method for the identification of patients with inherited hypercholesterolemia. *Semin Vasc Med 4*(1), 59–65.

Der Kaloustian VM, Naffah J, Loiselet J. (1980). Genetic diseases in Lebanon. *Am J Med Genet 7*(2), 187–203.

Dudin AA, Thalji. (1991). Diaphragmatic hernia and epidermolysis bullosa in two sibs. *Am J Med Genet 39*(4), 498–499.

Editorial. (2006). The germinating seed of Arab genomics. *Nat Genet 38*(8), 851.

Eidt D, Mittendorf T, Wagner TO, Reimann A, Graf von der Schulenburg JM. (2009). [Cost analysis for ambulatory treatment of cystic fibrosis patients in Germany. Overview of the prospective study results]. *Med Klin (Munich) 104*(7), 529–535.

El-Hazmi MA, Al-Swailem AR, Warsy AS, Al-Swailem AM, Sulaimani R, Al-Meshari AA. (1995). Consanguinity among the Saudi Arabian population. *J Med Genet 32*(8), 623–626.

El-Hazmi MA. (1999). Spectrum of genetic disorders and the impact on health care delivery: an introduction. *East Mediterr Health J 5*(6), 1104–1113.

El-Mouzan MI, Al-Salloum AA, Al-Herbish AS, Qurachi MM, Al-Omar AA. (2007). Regional variations in the prevalence of consanguinity in Saudi Arabia. *Saudi Med J 28*(12), 1881–1884.

El-Shafei A, Rao PS, Sandhu AK. (1986). Congenital malformations and consanguinity. *Aust N Z J Obstet Gynaecol 26*(3), 168–172.

El-Shanti H, Ahmad M, Ajlouni K. (2003). Alopecia universalis congenita, XY gonadal dysgenesis and laryngomalacia: a novel malformation syndrome. *Eur J Pediatr 162*(1), 36–40.

Eskandarani HA. (2002). Cystic fibrosis transmembrane regulator gene mutations in Bahrain. *J Trop Pediatr 48*(6), 348–350.

Fadhil M, Ghabra TA, Deeb M, Der Kaloustian VM. (1983). Odontoonychodermal dysplasia: a previously apparently undescribed ectodermal dysplasia. *Am J Med Genet 14*(2), 335–346.

Fahmy NA, Benson PF, Al-Garrah DB. (1993). Consanguinity in UAE: Prevalence and analysis of some risk factors. *Emirates Med J 1*, 39–41.

Farmer KC, Jacobs EW, Phillips CR. (1994). Long-term patient compliance with prescribed regimens of calcium channel blockers. *Clin Ther 16*, 316–326.

Fattoum S, Abbes S. (1985). Some data on the epidemiology of hemoglobinopathies in Tunisia. *Hemoglobin 9*, 423–429.

Feder G, Modell M. (1998). Cancer genetics in primary care. In: Foulkes WD, Hodgson SV. (eds.), *Inherited Susceptibility to Cancer: Clinical, Predictive, and Ethical Perspectives* (pp. 96–107). Cambridge University Press, Cambridge, UK.

Filon D, Faerman M, Smith P, Oppenheim A. (1995). Sequence analysis reveals a beta-thalassaemia mutation in the DNA of skeletal remains from the archaeological site of Akhziv, Israel. *Nat Genet 9*(4), 365–368.

Fowler JF, Duh MS, Rovba L, et al. (2007). The direct and indirect cost burden of atopic dermatitis: an employer-payer perspective. *Manag Care Interface 20*(10), 26–32.

Freundlich E, Hino N. (1984). Consanguineous marriage among rural Arabs in Israel. *Isr J Med Sci 20*(11), 1035–1038.

Frossard PM, Dawson KP, Das SJ, Alexander PC, Girodon E, Goossens M. (2000). Identification of cystic fibrosis mutations in Oman. *Clin Genet 57*(3), 235–236.

Gascon JJ, Sanchez-Ortuno M, Llor B, et al (2004). Treatment Compliance in Hypertension Study Group. Why hypertensive patients do not comply with the treatment: results from a qualitative study. *Fam Pract 21*, 125–130.

Gowri V, Jain R. (2005). Ollier's disease complicating pregnancy. *BJOG 112*(3), 384–385.

Grmek MD. (1994). Malaria in the eastern Mediterranean in prehistory and antiquity. *Parassitologia 36*, 1–6.

Gunaid AA, Hummad NA, Tamim KA. (2004). Consanguineous marriage in the capital city Sana'a, Yemen. *J Biosoc Sci 36*(1), 111–121.

Habib Z, Böök JA. (1983). Consanguinity and incidence of thalassaemia in Egypt. *Hereditas 99*(2), 215–217.

Hafez M, El-Tahan H, Awadalla M, El-Khayat H, Abdel-Gafar A, Ghoneim M. (1983). Consanguineous matings in the Egyptian population. *J Med Genet 20*(1), 58–60.

Haj Khelil A, Laradi S, Miled A, Tadmouri GO, Ben Chibani J, Perrin P. (2004). Clinical and molecular aspects of haemoglobinopathies in Tunisia. *Clin Chim Acta 340*, 127–137.

Hall JG, Powers EK, McLlvaine RT, Ean VH. (1978). The frequency and financial burden of genetic disease in a pediatric hospital. *Am J Med Genet 1*(4), 417–436.

Hamamy H, Al-Bayati N, Al-Kubaisy W. (1986). Consanguineous marriages in Iraqi urban population and the effect on pregnancy outcome and infant mortality. *Iraqi Med J 34*, 76–80.

Hamamy H, Al-Hait S, Alwan A, Ajlouni K. (2007). Jordan: communities and community genetics. *Community Genet 10*(1), 52–60.

Hamamy H, Alwan A. (1994). Hereditary disorders in the Eastern Mediterranean Region. *Bull World Health Organ 72*(1), 145–154.

Hamamy H, Bittles AH. (2009). Genetic clinics in Arab communities: meeting individual, family and community needs. *Pub Health Genom 12*, 30–40.

Hamamy H, Jamhawi L, Al-Darawsheh J, Ajlouni K. (2005). Consanguineous marriages in Jordan: why is the rate changing with time? *Clin Genet 67*(6), 511–516.

Hamamy HA, Al-Hakkak ZS. (1989). Consanguinity and reproductive health in Iraq. *Hum Hered 39*(5), 271–275.

Hamamy HA, Masri AT, Al-Hadidy AM, Ajlouni KM. (2007). Consanguinity and genetic disorders. Profile from Jordan. *Saudi Med J 28*(7), 1015–1017.

Hammami A, Elgazzeh M, Chalbi N, Mansour BA. (2005). [Endogamy and consanguinity in Mauritania] *Tunis Med 83*(1), 38–42.

Hoodfar E, Teebi AS. (1996). Genetic referrals of Middle Eastern origin in a western city: inbreeding and disease profile. *J Med Genet 33*(3), 212–215.

Imrie J, Galani C, Gairy K, Lock K, Hunsche E. (2009). Cost of illness associated with Niemann-Pick disease type C in the UK. *J Med Econ 12*(3), 219–229.

Ingman M, Kaessmann H, Pääbo S, Gyllensten U. (2000). Mitochondrial genome variation and the origin of modern humans. *Nature 408*, 708–713.

Irish JD. (2000). The Iberomaurusian enigma: north African progenitor or dead end? *J Hum Evol 39*, 393–410.

Jaber L, Bailey-Wilson JE, Haj-Yehia M, Hernandez J, Shohat M. (1994). Consanguineous matings in an Israeli-Arab community. *Arch Pediatr Adolesc Med 148*(4), 412–415.

Jaber L, Halpern GJ, Shohat T. (2000). Trends in the frequencies of consanguineous marriages in the Israeli Arab community. *Clin Genet 58*(2), 106–110.

Jalili IK, Smith NJ. (1988). A progressive cone-rod dystrophy and amelogenesis imperfecta: a new syndrome. *J Med Genet 25*(11), 738–740.

Jaouad IC, Elalaoui SC, Sbiti A, et al. (2009). Consanguineous marriages in Morocco and the consequence for the incidence of autosomal recessive disorders. *J Biosoc Sci 41*(5), 575–581.

Jin J, Sklar GE, Min Sen Oh V, Chuen Li S. (2008). Factors affecting therapeutic compliance: A review from the patient's perspective. *Ther Clin Risk Manag 4*(1), 269–286.

Joshi SN, Hashim J, Venugopalan P. (2002). Pattern of inborn errors of metabolism in an Omani population of the Arabian Peninsula. *Ann Trop Paediatr 22*(1), 93–96.

Jurdi R, Saxena PC. (2003). The prevalence and correlates of consanguineous marriages in Yemen: similarities and contrasts with other Arab countries. *J Biosoc Sci 35*(1), 1–13.

Karnon J, Zeuner D, Brown J, Ades AE, Wonke B, Modell B. (1999). Lifetime treatment costs of beta-thalassaemia major. *Clin Lab Haematol 21*(6), 377–385.

Kauf TL, Coates TD, Huazhi L, Mody-Patel N, Hartzema AG. (2009). The cost of health care for children and adults with sickle cell disease. *Am J Hematol 84*(6), 323–327.

Ke Y, Su B, Song X, et al. (2001). African origin of modern humans in East Asia: a tale of 12,000 Y chromosomes. *Science 292*, 1151–1153.

Kerkeni E, Monastiri K, Saket B, Guediche MN, Ben Cheikh H. (2007). Interplay of socio-economic factors, consanguinity, fertility, and offspring mortality in Monastir, Tunisia. *Croat Med J 48*(5), 701–707.

Kerkeni E, Monastiri K, Saket B, Rudan D, Zgaga L, Ben Cheikh H. (2006). Association among education level, occupation status, and consanguinity in Tunisia and Croatia. *Croat Med J 47*(4), 656–661.

Khalifa MM. (1999). Preventive aspects of genetic morbidity: experiences of the Canadian model. *East Mediterr Health J 5*(6), 1121–1128.

Khan IM, Mohammad AM. (1985). Muscoviscidosis (cystic fibrosis of the pancreas) in Bahrain, Arabian Gulf. *Bahrain Med Bull 7*(1), 17–23.

Khandekar R, Shama Mel-S, Mohammed AJ. (2005). Noncompliance with medical treatment among glaucoma patients in Oman--a cross-sectional descriptive study. *Ophthalmic Epidemiol 12*(5), 303–309.

Khlat M, Halabi S, Khudr A, Der Kaloustian VM. (1986). Perception of consanguineous marriages and their genetic effects among a sample of couples from Beirut. *Am J Med Genet 25*(2), 299–306.

Khlat M. (1988). Consanguineous marriages in Beirut: time trends, spatial distribution. *Soc Biol 35*(3–4), 324–330.

Khoury SA, Massad D. (1992). Consanguineous marriage in Jordan. *Am J Med Genet 43*(5), 769–775.

Kilani Z, Haj Hassan L. (2002). Sex selection and preimplantation genetic diagnosis at The Farah Hospital. *Reprod Biomed Online 4*(1), 68–70.

Kozlowski K, Bacha L, Massen R, Ayati M, Sator S, Brahimi L. (1988). A new type of spondylometaphyseal dysplasia--Algerian type. Report of five cases. *Pediatr Radiol 18*(3), 221–226.

Kwiatkowski DP. (2005). How malaria has affected the human genome and what human genetics can teach us about malaria. *Am J Hum Genet 77*, 171–192.

Lahr MM, Field JS. (2005). Assessment of the southern dispersal, GIS-based analyses of potential routes at oxygen isotopic stage 4. *J World Prehistory 19*, 1–45.

Lamdouar Bouazzaoui N. (1994). [Consanguinity and public health in Morocco] Bull *Acad Natl Med 178*(6), 1013–1025.

Liu H, Prugnolle F, Manica A, Balloux F. (2006). A geographically explicit genetic model of worldwide human-settlement history. *Am J Hum Genet 79*, 230–237.

Luxner M. (2001). The Arabs of Honduras. *Saudi Aramco World*. July/August, pp. 34–37.

Maca-Meyer N, González AM, Larruga JM, Flores C, Cabrera VM. (2001). Major genomic mitochondrial lineages delineate early human expansions. *BMC Genet 2*, 13.

Majeed HA, Kalaawi M, Mohanty D, et al. (1989). Congenital dyserythropoietic anemia and chronic recurrent multifocal osteomyelitis in three related children and the association with Sweet syndrome in two siblings. J Pediatr *115*(5 Pt 1), 730–734.

Malouf J, Alam S, Kanj H, Mufarrij A, Der Kaloustian VM. (1985). Hypergonadotropic hypogonadism with congestive cardiomyopathy: an autosomal-recessive disorder? *Am J Med Genet 20*(3), 483–489.

McCandless SE, Brunger JW, Cassidy SB. (2004). The burden of genetic disease on inpatient care in a children's hospital. *Am J Hum Genet 74*(1), 121–127.

McKusick VA. (2007). Mendelian Inheritance in Man and its online version, OMIM. *Am J Hum Genet 80*(4), 588–604.

Mégarbané A, Ruchoux MM, Loeys B, Ayoub N, Nuytinck L. (2001). Short stature, abnormal face, joint laxity, dislocation, hernias, delayed bone age, and severe psychomotor retardation in two brothers: previously undescribed MCA/MR syndrome. *Am J Med Genet 104*(3), 221–224.

Mégarbané A, Samaras L, Chédid R, et al. (2008). Developmental delay, dysmorphic features, neonatal spontaneous fractures, wrinkled skin, and hepatic failure: a new metabolic syndrome? *Am J Med Genet A 146A*(24), 3198–3201.

Meyer BF. (2005). Strategies for the prevention of hereditary diseases in a highly consanguineous population. *Ann Hum Biol 32*(2), 174–179.

Meyers RD, Adams W, Dardick K, et al. (1972). The social and economic impact of hemophilia--a survey of 70 cases in Vermont and New Hampshire. *Am J Public Health 62*(4), 530–535.

Mohamed MS. (1995). An epidemiological study on consanguineous marriage among urban population in Alexandria. *J Egypt Public Health Assoc 70*(3–4), 293–305.

Mohammad AM, Ardatl KO, Bajakian KM. (1998). Sickle cell disease in Bahrain: coexistence and interaction with glucose-6-phosphate dehydrogenase (G6PD) deficiency. J *Trop Pediatr 44*, 70–72.

Mokhtar MM, Abdel-Fattah MM. (2001). Consanguinity and advanced maternal age as risk factors for reproductive losses in Alexandria, Egypt. *Eur J Epidemiol 17*(6), 559–565.

Mokhtar MM, Kotb SM, Ismail, SR. (1998). Autosomal recessive disorders among patients attending the genetic clinic in Alexandria. *East Mediterr Health J 4*(3), 470–479.

Mousa AR, Al-Din AS, Al-Nassar KE, et al. (1986). Autosomally inherited recessive spastic ataxia, macular corneal dystrophy, congenital cataracts, myopia and vertically oval temporally tilted discs. Report of a Bedouin family—a new syndrome. *J Neurol Sci 76*(1), 105–121.

Mubaidin A, Roberts E, Hampshire D, et al. (2003). Karak syndrome: a novel degenerative disorder of the basal ganglia and cerebellum. *J Med Genet 40*(7), 543–546.

Muralitharan S, Wali YA, Dennison D, et al. (2007). Novel spectrum of perforin gene mutations in familial hemophagocytic lymphohistiocytosis in ethnic Omani patients. *Am J Hematol 82*(12), 1099–1102.

Nabulsi A. (1995). Mating patterns of the Abbad tribe in Jordan. *Soc Biol 42*(3–4), 162–174.

Najjar SS, Kaloustian VM der, Nassif SI. (1973). Genital anomaly, mental retardation, and cardiomyopathy: a new syndrome? *J Pediatr 83*(2), 286–288.

Naveed M, Al-Ali MT, Murthy SK, et al. (2006). Ectrodactyly with aplasia of long bones (OMIM; 119100) in a large inbred Arab family with an apparent autosomal dominant inheritance and reduced penetrance: clinical and genetic analysis. *Am J Med Genet A 140*(13), 1440–1446.

Naveed M, Nath SK, Gaines M, et al. (2007). Genomewide linkage scan for split-hand/foot malformation with long-bone deficiency in a large Arab family identifies two novel susceptibility loci on chromosomes 1q42.2-q43 and 6q14.1. *Am J Hum Genet 80*(1), 105–111.

Nerlich AG, Rohrbach H, Zink A. (2002). [Paleopathology of ancient Egyptian mummies and skeletons. Investigations on the occurrence and frequency of specific diseases during various time periods in the necropolis of Thebes-West] *Pathologe 23*(5), 379–385.

Othman H, Saadat M. (2009). Prevalence of consanguineous marriages in Syria. *J Biosoc Sci 41*(5), 685–692.

Ozand PT, Odaib AA, Sakati N, Al-Hellani AM. (2005). Recently available techniques applicable to genetic problems in the Middle East. *Community Genet 8*(1), 44–47.

Porz G, Scholl HP, Holz FG, Finger RP. (2010). [Methods for estimating personal costs of disease using retinal diseases as an example]. *Ophthalmologe 107*(3), 216–220 and 222.

Price C. (1999). Australian population: Ethnic origins. *People and Place 7*(4), 12–16.

Radovanovic Z, Shah N, Behbehani J. (1999). Prevalence and social correlates to consanguinity in Kuwait. *Ann Saudi Med 19*(3), 206–210.

Rajab A, Kunze J, Mundlos S. (2004). Spondyloepiphyseal dysplasia Omani type: a new recessive type of SED with progressive spinal involvement. *Am J Med Genet A 126*(4), 413–419.

Rajab A, Patton MA. (2000). A study of consanguinity in the Sultanate of Oman. *Ann Hum Biol 27*(3), 321–326.

Reardon W, Hurst J, Farag TI, Hall C, Baraitser M. (1990). Two brothers with heart defects and limb shortening: case reports and review. *J Med Genet 27*(12), 746–751.

Riou S, El Younsi C, Chaabouni H. (1989). [Consanguinity in the population of northern Tunisia] *Tunis Med 67*(3), 167–172.

Handloff RE. (1988). Ivory Coast: A Country Study. GPO for the Library of Congress, Washington.

Saadallah AA, Rashed MS. (2007). Newborn screening: experiences in the Middle East and North Africa. *J Inherit Metab Dis 30*(4), 482–489.

Saha N, El Sheikh FS (1988). Inbreeding levels in Khartoum. *J Biosoc Sci 20*(3), 333–336.

Saha N, Hamad RE, Mohamed S. (1990). Inbreeding effects on reproductive outcome in a Sudanese population. *Hum Hered 40*(4), 208–212.

Salloum H. (2000). Arabs making their mark in Latin America: Generations of immigrants in Colombia, Venezuela and Mexico. *Al Jadid 6*(30), winter issue.

Sandridge AL, Takeddin J, Al-Kaabi E, Frances Y. (2010). Consanguinity in Qatar: knowledge, attitude and practice in a population born between 1946 and 1991. *J Biosoc Sci 42*(1), 59–82.

Sanjad SA, Sakati NA, Abu-Osba YK, Kaddoura R, Milner RD. (1991). A new syndrome of congenital hypoparathyroidism, severe growth failure, and dysmorphic features. *Arch Dis Child 66*(2), 193–196.

Selamnia M, Tali-Maamar H. (2003). Biomedical research in developing countries. I--The Algerian case (1993–1998). *Tunis Med 81*(7), 456–460.

Sharkia R, Zaid M, Athamna A, Cohen D, Azem A, Zalan A. (2008). The changing pattern of consanguinity in a selected region of the Israeli Arab community. *Am J Hum Biol 20*(1), 72–77.

Soliman AT, Bappal B, Darwish A, Rajab A, Asfour M. (1995). Growth hormone deficiency and empty sella in DIDMOAD syndrome: an endocrine study. Arch *Dis Child 73*(3), 251–253.

Stephen WJ. (1992). Primary health care in the Arab world. Wells, Somerset House.

Stevenson AC, Johnston HA, Stewart MI, Golding DR. (1966). Congenital malformations. A report of a study of series of consecutive births in 24 centres. *Bull World Health Organ 34*(Suppl), 9–127.

Subramanyan R, Venugopalan P. (2002). Malignant familial long QT syndrome. *Saudi Med J 23*(6), 738–742.

Tadmouri GO, Al Ali MT, Al-Haj Ali S, Al Khaja N. (2006). CTGA: the database for genetic disorders in Arab populations. *Nucleic Acids Res 34*(Database issue), D602–D606.

Tadmouri GO, Bissar-Tadmouri N. (1999). Genetic disorders among Arabs as for OMIM™. *Saudi Med J 20*, 4–18.

Tadmouri GO, Bissar-Tadmouri N. (2003). Biomedical publications in an unstable region: the Arab world, 1988–2002. *Lancet 362*(9397), 1766.

Tadmouri GO, Garguier N, Demont J, Perrin P, Basak AN. (2001). History and origin of beta-thalassemia in Turkey: sequence haplotype diversity of beta-globin genes. *Hum Biol 73*, 661–674.

Tadmouri GO. (2004). Genetic diseases in the Arab populations. Dubai International Genetics Conference 2004 (21–22 March 2004). Dubai, United Arab Emirates.

Tadmouri GO. (2006). Genetic disorders in Arab Populations: A 2006 Update. In: Tadmouri GO, Taleb Al Ali M, Al Khaja N (eds.), *Genetic Disorders in the Arab World: Bahrain*. Centre for Arab Genomic Studies, Dubai, United Arab Emirates.

Tadmouri GO. (2008a). Genetic disorders in Arabs. In: Nasir LS, Abdul-Haq AK (eds.), *Caring for Arab Patients: A Biopsychosocial Approach*. Radcliffe Publishing, Abingdon.

Tadmouri GO. (2008b). Genetic disorders in Arabs: A 2008 Update. In: Tadmouri GO, Taleb Al Ali M, Al Khaja N (eds.), *Genetic Disorders in the Arab World: Oman*. Centre for Arab Genomic Studies, Dubai, United Arab Emirates.

Talbi J, Khadmaoui AE, Soulaymani AEM, Chafik AEA. (2007). Etude de la consanguinité dans la population marocaine. Impact sur le profil de la santé. *Antropo 15*, 1–11.

Talib AA. (1997). Hadramis in Singapore. *J Muslim Minority Affairs 17*(1), 89–96.

Tamim H, Khogali M, Beydoun H, Melki I, Yunis K; National Collaborative Perinatal Neonatal Network. (2003). Consanguinity and apnea of prematurity. *Am J Epidemiol 158*(10), 942–946.

Teebi AS, Al-Saleh QA, Hassoon MM, Farag TI, Al-Awadi SA. (1989). Macrosomia, microphthalmia, +/- cleft palate and early infant death: a new autosomal recessive syndrome. *Clin Genet 36*(3), 174–177.

Teebi AS, Farag TI. (1997). Genetic Disorders among Arab Populations. Oxford University Press.

Teebi AS, Naguib KK, Al-Awadi S, Al-Saleh QA. (1988). New autosomal recessive faciodigitogenital syndrome. *J Med Genet 25*(6), 400–406.

Teebi AS, Shaltout AA. (1989). Craniofacial anomalies, abnormal hair, camptodactyly, and caudal appendage. *Am J Med Genet 33*(1), 58–60.

Teebi AS. (2000). Nablus mask-like facial syndrome. *Am J Med Genet 95*(4), 407–408.

Teebi AS. (1987). New autosomal dominant syndrome resembling craniofrontonasal dysplasia. *Am J Med Genet 28*(3), 581–591.

Temtamy SA, Meguid NA, Ismail SI, Ramzy MI. (1998). A new multiple congenital anomaly, mental retardation syndrome with preaxial brachydactyly, hyperphalangism, deafness and orodental anomalies. *Clin Dysmorphol 7*(4), 249–255.

The Canadian Arab Federation. (1999). A profile of Arabs in Canada. Arab Community Centre of Toronto.

Tischfield MA, Bosley TM, Salih MA, et al. (2005). Homozygous HOXA1 mutations disrupt human brainstem, inner ear, cardiovascular and cognitive development. *Nat Genet 37*(10), 1035–1037.

Underhill PA, Passarino G, Lin AA, et al. (2001). The phylogeography of Y chromosome binary haplotypes and the origins of modern human populations. *Ann Hum Genet 65*, 43–62.

Vardi-Saliternik R, Friedlander Y, Cohen T. (2002). Consanguinity in a population sample of Israeli Muslim Arabs, Christian Arabs and Druze. *Ann Hum Biol 29*(4), 422–431.

Venugopalan P, Akinbami FO, Elnour IB. (1997). Cardiovascular manifestations of Marfan's syndrome a case report with review of literature. *Oman Med J 14*(2), 46–49.

Walsh T, Abu Rayan A, Abu Sa'ed J, et al. (2006). Genomic analysis of a heterogeneous Mendelian phenotype: multiple novel alleles for inherited hearing loss in the Palestinian population. *Hum Genomics 2*(4), 203–211.

Wang W, McGreevey WP, Fu C, et al. (2009). Type 2 diabetes mellitus in China: a preventable economic burden. *Am J Manag Care 15*(9), 593–601.

Warburg M, Tommerup N, Vestermark S, et al. (1990). The Yemenite deaf-blind hypopigmentation syndrome. A new oculo-dermato-auditory syndrome. *Ophthalmic Paediatr Genet 11*(3), 201–207.

White DR, Ganesh A, Nishimura D, et al. (2007). Autozygosity mapping of Bardet-Biedl syndrome to 12q21.2 and confirmation of FLJ23560 as BBS10. *Eur J Hum Genet 15*(2), 173–178.

WHO (1996). Control of Hereditary Diseases: Report of a WHO Scientific Group. World Health Organization.

Woodhouse NJ, Sakati NA. (1983). A syndrome of hypogonadism, alopecia, diabetes mellitus, mental retardation, deafness, and ECG abnormalities. *J Med Genet 20*(3), 216–219.

Zahed L, Demont J, Bouhass R, et al. (2002). Origin and history of the IVS-I-110 and codon 39 beta-thalassemia mutations in the Lebanese population. *Hum Biol 74*, 837–847.

Zalloua PA, Platt DE, El Sibai M, et al; Genographic Consortium. (2008b). Identifying genetic traces of historical expansions: Phoenician footprints in the Mediterranean. *Am J Hum Genet 83*, 633–642.

Zalloua PA, Xue Y, Khalife J, et al; Genographic Consortium. (2008a). Y-chromosomal diversity in Lebanon is structured by recent historical events. *Am J Hum Genet 82*, 873–882.

Zaoui S, Biemont C. (2002). Frequency of consanguineous unions in the Tlemcen area (West Algeria). *Sante 12*(3), 289–295.

Zilberman U, Smith P, Piperno M, Condemi S. (2004). Evidence of amelogenesis imperfecta in an early African Homo erectus. *J Hum Evol 46*(6), 647–653.

Zlotogora J, Habiballa H, Odatalla A, Barges S. (2002). Changing family structure in a modernizing society: a study of marriage patterns in a single Muslim village in Israel. *Am J Hum Biol 14*(5), 680–682.

Zogby International. (2000). The global Arab demographics report. Zogby International.

51.

COMMUNITY HEALTH IMPLICATIONS OF CONSANGUINITY IN ARAB POPULATIONS

Ghazi Omar Tadmouri, Pratibha Nair, Tasneem Obeid, and Hanan Hamamy

INTRODUCTION

Linguistically, *consanguinity* is a term that is derived from two Latin words, *con* meaning common and *sanguis* meaning blood; hence, referring to a relationship between two people who share a common ancestor or blood. In other words, consanguineous marriage refers to unions contracted between biologically related individuals. In clinical genetics, a consanguineous marriage means union between couples who are related as second cousins or closer (Modell & Darr, 2002; Alwan & Modell, 1997). Among Arabs, this would include double first cousins, first cousins, first cousins once removed, and second cousins. Uncle–niece marriage is prohibited in Islam, so it is absent among Arabs. In population genetics, consanguinity may also refer to unions of individuals with at least one common ancestor, such as those occurring within population isolates, small towns, and tribes—also referred to as *intracommunity* or *endogamous* marriages.

CONSANGUINITY IN WORLD POPULATIONS

Consanguineous marriages have been practiced since the early existence of modern humans. At present, about one-fifth of world human populations live in communities with a preference for consanguineous marriage (Modell & Darr, 2002). Consanguinity rates vary from one population to another depending on religion, culture, and geography. The practice of consanguineous marriage declined substantially in industrialized countries. In Sweden, for example, a study of 14,639 marriages that took place during 1720–1899 indicated that 20.8% of marriages were between couples related as sixth cousins or closer (Bittles & Egerbladh, 2005). In the 1950s, the rate of consanguineous marriages in Sweden decreased to around 0.6% (Romanus, 1953). In Japan, a substantial decrease in consanguineous marriage occurred between 1947 and 1972, from 13% in urban and 21% in rural areas to 2.9% in urban and 4.3% in rural areas (Saito, 1988). Similar observations could also be seen in the United States and in many European countries (Stevenson et al., 1966). On the other hand, consanguinity rates are increasing in some populations in Asia and Africa (Al-Gazali et al., 1997; Bener & AlAli, 2006; Mukherjee et al., 2007). Noticeably, many Arab countries display some of the highest rates of consanguineous marriages in the world, and specifically first cousin marriages (Table 51-1; Figure 51-1).

CONSANGUINITY IN ARAB POPULATIONS

Sociocultural factors, such as maintenance of family structure and property, ease of marital arrangements, better relations with in-laws, and financial advantages relating to dowry seem to play a crucial role in the preference of consanguinity in Arab populations (Bittles, 2008). Among women in Beirut, the most cited argument in favor of cousin marriages was protection of women and strengthening of family ties. In such urban settings, the maintenance of family cohesions takes precedence over financial concerns (Khlat et al., 1986). Consanguineous marriages are generally thought to be more stable than marriages between nonrelatives, though there are no studies to compare divorce rates of consanguineous and nonconsanguineous marriages among Arabs. It is generally believed that the husband's family would side with the consanguineous wife in marital disputes, since she is considered part of the extended family. When there are children with disabilities, more family members share in caring for these children. Unlike what is widely thought, consanguinity in the Arab world is not only confined to Muslim communities. Several other communities, including the Lebanese, Jordanian, and Palestinian Christian populations, also practice consanguinity, but to a lesser extent than Muslims (Khlat, 1988a,b; Khoury & Massad, 1992; Vardi-Saliternik et al., 2002). For example, in Beirut, first cousin marriages among Muslims and Christians were 17.3% and 7.9%, respectively (Khlat, 1988a).

TABLE 51-1 CONSANGUINITY RATES IN ARAB POPULATIONS

Country	Location	Collection Period	Sample Size	Percent Consanguineous Marriage Rates [Degree of Consanguinity]			Mean inbreeding coefficient	Reference
				1C	>1C-2C	Overall		
Algeria		1979	120,491		22.6 [1C,2C]	22.6	0.0136–0.0169	Benallegue and Kedji, 1984
		1988	239			16.7–27 [?]		Grassivaro Gallo and Florio, 1993
		1992 [?]				40 [?]		ENAF, 1992
		1994 [?]				36.4 [?]		ANSO, 1994
	Tiemcen	2002 [?]	3,983	11.3		34 [?]		Zaoui and Biémont, 2002
Bahrain		1983	141			10.6 [?]		El-Shafei et al., 1986
		1989	10,070		45.5 [1C<1C]	45.5	0.0166	Al-Naser, 1944
		1989	500	21	31.8 [1C,1.5C,2C]	38.6	0.0152	Al Arrayed, 1994
		1995	26,723	24		32		Ministry of Health [Bahrain], 1995
		1995 (Past generation)	100			45.5 [?]		Al Arrayed, 1995
		1995 (Present generation)	500			39.4 [?]		Al Arrayed, 1995
		1999	1,000	24.5	36 [1C,1.5C,2C]	43.1		Al Arrayed, 1999
		2002	4,000	12.5	7	20		Al Arrayed, 2006
		2006	1,000	11.4	8.6	20		Al Arrayed, 2006
Egypt	Alexandria	1961–1964	9,475			32.8 [>1C,1C,2C.<2C]		Stevenson et al., 1966
	Nubia	1965–1967	281	39	60.5 [1C,<1C]	60.5	0.0245	Hussien, 1971
	All Nubia	1967–1968	1,782	41.5–47.2	63.6 [1C,2C]	72.9–80.4	0.0334	Badr, 1972
		1970s	505	23.2		23.2	0.0145	Habib and Böök, 1983
		1983 [?]	26,554	14.3	24.54 [>1C,1C,2C]	29 [>1C,1C,2C,<2C]	0.0101	Hafez et al., 1983
		1989				41 [?]		ENPC, 1989
	Alexandria	1995 [?	500	15.8		22.8 [?]	0.01172	Mohamed, 1995
		1996 [?]				39 [?]		ENPC,1996
	Alexandria	1998–2000	2,081	15.9		20.9 [?]	0.0099	Mokhtar and Abdel-fattah (2001)
Iraq	Kurds	1951	61			32.8 [?]		Barth, 1954
	Baghdad	1986	4,491	29.2	46.4	57.8 [1C,1.5C,2C,<2c]	0.0225	Hamamy et al., 1986
	Baghdad	1989	382	29–32.2	47.1–57.8 [1C,2C]	47.1–57.8	0.0208–0.0241	Hamamy and Al-Hakkak, 1989
	Central and South Iraq	1999	23,105			60 [?]		Ali and Shah, 2000
	North Kurdish Region	1999	14,035			47 [?]		Ali and Shah, 2000
	All Iraq	2001	23,937	33		33	0.0219	COSIT, 2005
Jordan	Irbid (Rural)	1963–1964	1,097	39	52.1 [1C,2C]	52.1	0.0230	Cook and Hanslip, 1966
		1969–1979	1,989	32.8	39.7 [1C,1.5C,2C]	50.3 [1C,1.5C,2C,<2C]	0.0225	Khoury and Massad, 1992
	Balqa	1986–1989	340		40.3 [1C,1.5C]	40.3	0.018	Nabulsi, 1995
	Irbid	1993 [?]	900	37.3	50.6 [1C,1.5C,2C]	63.7 [1C,1.5C,2C,<2C]	0.0284	Al-Salem and Rawashdeh, 1993
	South Ghor	2003 [?]	608			58.1 [?]		Sueyoshi and Ohtsuka, 2003
	Amman	2004 (3 generations)	2,594	21.8	25.6 [1C,1.5C,2C]	28.5 [1C,1.5C,2C,<2C]	0.0135–0.02	Hamamy et al., 2005
	North of Jordan	2007	3,269			49 [>1C,1C,2C,<2C]		Obeidat et al., 2008
Kuwait		1967–1968	2,220			53.9 [?]		El-Alfi et al., 1968
		1969 [?]	2,133		34.3 [1C,1.5C,2C]	38.4	0.0174	El-Alfi et al., 1969

Country	Location	Collection Period	Sample Size	Percent Consanguineous Marriage Rates [Degree of Consanguinity] 1C	>1C–2C	Overall	Mean inbreeding coefficient	Reference
		1983	5,007	30.2	34.3 [1C,1.5C,2C]	64.3 [1C,1.5C,2C]	0.0219	Al-Awadi et al., 1985
		1989 [?]				54 [?]		Al-Nasser et al., 1989
		1996 [?]				36 [?]		Ministry of Health [Kuwait], 1996
		1996–1997	555	31.7	42.2 [1C,2C]	42.2	0.0251	Radovanovic et al., 1999
		1996–1997	404	16.39	22.5 [1C,2C]	22.5	0.0139	Radovanovic et al., 1999
	Jahra	2002	7,315			54.8 [?]		Al-Kandari, 2007
	Kuwait City							
Lebanon	Beirut	1981–1982	750		26 [1C,1.5C,2C]	26	0.0116	Klat and Khudr, 1986
	Beirut	1983–1984	2,2854	14		25 [1C,1.5C,2C,<2C]	0.0088	Khlat, 1988a
	Beirut	1983–1984	3,033			25 [?]		Khlat, 1988b
		1998 [?]				21 [?]		Ministry of Public Health
	Beirut	1998–2001	21,723	6.7		12.8 [1C,2C,<2C]		[Lebanon], 1998
	Beirut	2007 [?]	552	31		42 [?]		Tamim et al., 2003
		2009 [?]	1,556 (women)	31.6	35.5 [1C,1.5C,2C]	37.8 [1C,1.5C,2C,<2C]	0.02	Kanaan et al., 2008
								Barbour and Salameh, 2009
Libya	Benghazi	1981 [?]	<500			48.4 [?]		Broadhead and Sehgal, 1981
		1996 [?]				46.5 [?]		GPCHSI, 19996
Mauritania		1990–1991				60.1 [?]		National Statistical Office
		2005 [?]	2,413			47.2 [?]		[Mauritania], 1992
								Hammami et al., 2005
Morocco	High-Atlas Valleys	1904–1985				23.1 [?]	0.0098	Baali, 1994
		1940	582	10	13.4 [>1C,1C,2C]	21.5	0.0091	Talbi et al., 2007
		1984	291	8.6	15.1 [>1C,1C,2C]	25.4	0.0094	Talbi et al., 2007
	All Morocco	1982–1992	4,773		19.9 [1C,2C]	19.9	0.0089	Lamdouar Bouazzaoui, 1994
		1987 [?]				29 [?]		Azelmat et al., 1987
		1992 [?]				33 [?]		Azelmat et al., 1992
	High-Atlas Valleys	2005 [?]	438			28 [?]		Cherkaoui et al., 2005
		2009 [?]				15.2 [?]	0.0065	Jaouad et al., 2009
Oman		1994–1997	60,895	24.1	35.9 [1C,1.5C,2C]	66.3 [1C,1.5C,2C,<2C]	0.0198	Rajab and Patton, 2000
		1996				54 [?]		Ministry of Health [Oman], 1996
Palestine		1977	3,203	34.2		34.2	0.0238	Bashi, 1977
	Western Galilee	1976–1983	1,546		32.2 [1C,1.5C,2C]	32.2	0.0163	Gev et al., 1986
	Western Galilee	1984	550		39 [1C,1.5C,2C]	39	0.0137–0.0252	Freundlich and Hino, 1984
		1980–1985 (Past		23.9		33.1 [?]		Sharkia et al., 2008
		generation)		13.6		25.9		Sharkia et al., 2008
		2000–2004 (Present	3,328	23.5	26.6 [1C,1.5C,2C]	40.6	0.0153	Jaber et al., 2000
	Muslim	generation)	1,025		32.1 [1C,1.5C,2C]	32.1	0.0177	Vardi-Salitemik et al., 2002
	Bedouin	1981–1985	278		45.3 [1C,1.5C,2C]	45.3	0.0253	Vardi-Salitemik et al., 2002

(Continued)

TABLE 51-1 (CONTINUED)

Country	Location	Collection Period	Sample Size	Percent Consanguineous Marriage Rates [Degree of Consanguinity] 1C	>1C–2C	Overall	Mean inbreeding coefficient	Reference
Palestine	Christian	1990–1992	107		20.6 [1C,1.5C,2C]	20.6	0.0119	Vardi-Salitemik et al., 2002
	Druze	1990–1992	115		40.9 [1C,1.5C,2C]	40.9	0.0233	Vardi-Salitemik et al., 2002
		1990–1992	8,521 (total)	25.3	29.4 [1C,1.5C]	44.3 [?]	0.0192	Jaber et al., 1994
		1990–1992	5,098 (rural)		30.6 [1C,1.5C]	30.6	0.0194	Jaber et al., 1994
		1992	2,267		28.2 [1C,1.5C]	28.2	0.0178	Jaber et al., 1994
			(suburban)		24.2 [1C,1.5C]	24.2	0.0153	Jaber et al., 1994
	West Bank		1,156 (urban)	27.2		66.3 [>1C,1C,2C,<2C]	0.017	Pedersen, 2002
	Gaza		10,409	31.6		64.5 [>1C,1C,2C,<2C]	0.0198	Pedersen, 2002
	Palestinians in Jordan	1995	5,778	26.1		47.1 [>1C,1C,2C,<2C]		Pedersen, 2002
	Palestinians in	1995	4,950	19.9		32.3 [>1C,1C,2C,<2C]		Pedersen, 2002
	Lebanon	1996	3,972	26		46.9 [>1C,1C,2C,<2C]		Pedersen, 2002
	Palestinians in Jordan	1998	2,888	18.3		30.5 [>1C,1C,2C,<2C]		Pedersen, 2002
	Palestinians in Syria	1999	4,195	17.5		17.5	0.0104	Zlotogora et al., 2002
	Lower Galilee	2001	483	29.2		49.8 [>1C,1C,2C,<2C]	0.0183	Assai and Khawaja, 2009
		1970–2000	16,171	27.7		45.4 [>1C,1C,2C,<2C]	0.0186	Assai and Khawaja, 2009
		1995	4,971					
		2004						
Qatar		1999 [?]				46 [?]		Ministry of Health [Qatar], 1999
		2004				41.8 [?]		Bener et al., 2004
	Doha	2004	1,515	34.8	41.4 [1C,1.5C,2C]	54 [>1C,1C,2C,<2C]	0.02706	Bener and Alali, 2006
		2004–2005	876			51 [?]		Bener et al., 2007
		2007–2008 (Past	337			22 [?]		Sandridge et al., 2010
		generation)	229			35 [?]		Sandridge et al., 2010
		2007–2008 (Present						
		generation)						
Saudi		1980s	143		18.9 [1C,2C]		0.0085	Chaleby and Tuma, 1987
Arabia	Riyadh	1979–1980	1,149		55 [1C,2C]		0.033	Seremius et al., 1988
	Riyadh	1983–1986	4,497	31		54.3 [?]	0.196	Saed-Wong and Al-Frayh, 1989
	Riyadh	1989–1990	6,421		45.1 [1C,2C]		0.174	Zakzouk et al., 1993
	Riyadh	1993	2,001	28.4	41.1 [1C,1.5C,2C]	51.1 [1C,1.5C,2C,<2C]	0.02265	Al Hussain and Al Bunyan, 1997
		1995 [?]	3,212	25.8	40.6 [1C,1.5C,2C]	56.8 [1C,1.5C,2C,<2C]	0.0241	El-Hazami et al., 1995
	Dammam	1998 [?]	1,307	39.3	43.4 [1C,1.5C,2C]	52[1C,1.5C,2C,<2C]	0.0312	Al-Abdulkareem and Ballal, 1998
	Al-Baha	2004–2005	487	29		42.1 [?]		El-Mouzan et al., 2007
	Al-Jowf	2004–2005	593	34.8		53.5 [?]		El-Mouzan et al., 2007
	Assir	2004–2005	833	24.6		44.5 [?]		El-Mouzan et al., 2007
	Eastern Province	2004–2005	1,032	33.3		57.8 [?]		El-Mouzan et al., 2007
	Gizan	2004–2005	565	33		53.5 [?]		El-Mouzan et al., 2007
	Hail	2004–2005	505	25.1		48.9 [?]		El-Mouzan et al., 2007
	Madinah	2004–2005	618	39.2		67.2 [?]		El-Mouzan et al., 2007
	Makkah	2004–2005	2,278	32.4		55.9 [?]		El-Mouzan et al., 2007

Country	Location	Collection Period	Sample Size	Percent Consanguineous Marriage Rates [Degree of Consanguinity]			Mean inbreeding coefficient	Reference
				1C	>1C–2C	Overall		
Saudi Arabia	Najram	2004–2005	472	28.4		66.7 [?]		El-Mouzan et al., 2007
	Northern Borders	2004–2005	504	31.4		63.9 [?]		El-Mouzan et al., 2007
	Qassim	2004–2005	713	29.6		46.7 [?]		El-Mouzan et al., 2007
	Riyadh	2004–2005	2,522	42.3		60 [?]		El-Mouzan et al., 2007
	Tabuk	2004–2005	432	28.3		60 [?]		El-Mouzan et al., 2007
	All Saudi Arabia	2004–2005	11,554	33.6		56 [?]		El-Mouzan et al., 2007
Sudan	Gezira	1969–1974	2,999	44.2		44.2	0.028	Ahmed, 1979
	Khartoum	1988 [?]	4,833		52 [1C,2C]	52	0.0302	Saha and El Sheikh, 1988
	Khartoum	1990 [?]	926	49.5	63.3 [1C,<1C]	63.3		Saha and Hamad, 1990
		1995 [?]				65 [?]		FMHNCHS, 1995
Syria		1974 [?]				33 [?]		Prothro and Diab, 1974
		1995 [?]				38 [?]		Central Bureau of Statistics [Syria], 1995
Tunisia		1988				49?		Alowi et al., 1988
	Northern Tunisia	1983–1985	5, 767	23	26.9 [1C,2C]	26.9	0.0213	Riou et al., 1989
	Monastir	1989–1990	1,741	17.4	63.3 [1C,2C]	24.8		Kerkeni et al., 2007
		1996 [?]				40 [?]		Ministry of Public Health [Tunisia], 1996
	Monastir	2003–2004	1,016		20.1 [1C,2C]	20.1		Kerkeni et al., 2006
	Bizerte	2000–2002	4,860	20.8		39.33 [1C,2C,<2C]	0.01707–0.03493	Ben Arab et al., 2004
UAE		1993 [?]				51 [?]		Fahmy et al., 1993
	Al Ain	1992–1994	16, 419 [?]			54 [?]		Al-Gazali et al., 1995
	Al Ain	1994–1995	1,502	28.2	37.4 [1C,1.5C,2C]	54.2 [1C,1.5C,2C,<2C]	0.0245	Al-Gazali et al., 1997
	Dubai	1994–1995	531	20.7	32 [1C,1.5C,2C]	40 [1C,1.5C,2C,<2C]	0.0159	Al-Gazali et al., 1997
	Al Ain & Dubai		2,033	26.2		50.5 [1C,2C,<2C]	0.022222	Al-Gazali et al., 1997
Yemen		1997	9,762	34		40 [?]	0.0212	Jurdi and Saxena, 2003
	Sanaa	2000	1,050	32	44.7 [>1C,1C,1.5C,2C]	44.7	0.02442	Gunaid et al., 2004

Abbreviations: [?] = Unknown year of sampling or unknown types of consanguineous marriages; [>1C] = Double first cousin marriages; [1C] = Marriage beyond first cousins; [1.5C] = First cousin once removed marriage; [2C] = Second cousin marriage; [<2C] = Marriage between distant relatives beyond second cousins.

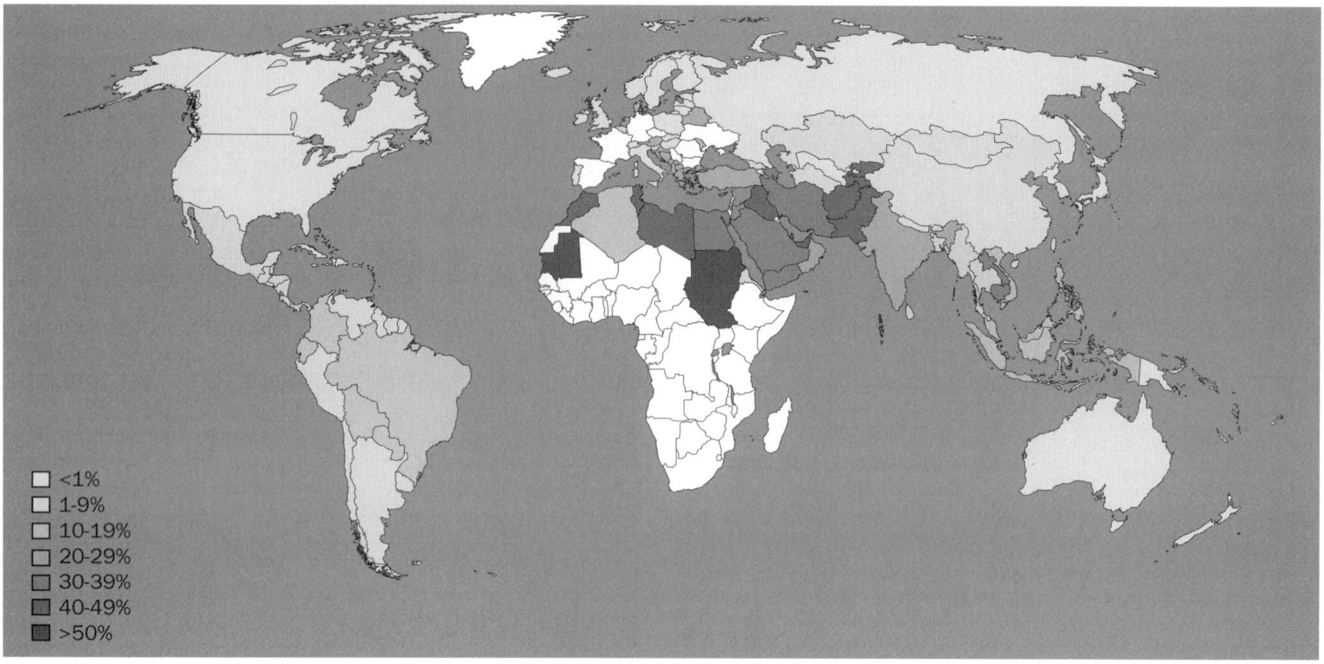

Figure 51-1 Schematic representation of consanguineous marriage rates worldwide (adapted from Table 51-1, Tadmouri, 2008, and www.consang.net). Only second cousin and closer marriages are represented.

In Islam, the Holy Quran contains no specific guidance that could be interpreted as encouraging consanguinity. To the contrary, the Quran includes indications that encourage human dispersal: "*O mankind! We created you from a single (pair) of a male and a female, and made you into nations and tribes, that ye may know each other*" (AlHujurat 049.013), and the appreciation of human variation: "*And among His Signs is the creation of the heavens and the earth, and the variations in your languages and your colours: verily in that are Signs for those who know.*" (AlRum 030.022).

Consanguinity rates show wide variations among Arab countries, as well as within the same country (Table 51-1). However, reports from Arab countries on consanguinity rates may sometimes include marriages between third cousins or far relatives within the consanguineous category. Although this discrepancy affects the total consanguinity rate, it does not markedly alter the average inbreeding coefficient. Therefore, for comparison of consanguinity rates among populations, two parameters are best used: the mean inbreeding coefficient (F), and marriages between first cousins. However, Arab societies have a long tradition of consanguinity, and the cumulative estimate of (F) may exceed the estimated value that is calculated for a single generation (Bittles et al., 1993). Several factors may have an influence on the consanguinity rates in Arab countries. These factors include urban/rural residence of families within the country, education levels, religion, and cultural variations. Generally, the highest rates of marriages to close relatives are consistently reported in the more traditional rural areas and among the poorest and least educated in society (Bittles et al., 1993).

Reports from some Arab countries have shown that consanguinity rates are lower in urban when compared to rural settings. Urban to rural first cousin rates in Algeria were 10% and 15% (Zaoui & Biémont, 2002), in Egypt 8.3% and 17.2% (Hafez et al., 1983), and in Jordan 29.8% and 37.9% (Khoury & Massad, 1992), respectively. In Jordan, it was evident that the higher the level of education of the female partner, the lower the consanguinity rate. Only 12% of university-educated females would marry their first cousins, whereas 25% of university-educated males tend to marry first cousins (Khoury & Massad, 1992). Similar trends of lower consanguinity rates among educated women, but not educated men, were also noticed in Yemen (Jurdi & Saxena, 2003) and Tunisia (Kerkeni et al., 2006). In the UAE, Bener et al. (1996) reported a higher rate of consanguinity among women with more educated husbands. In Arab populations, consanguinity rates are changing in either way (Table 51-1; Figure 51-2). In Jordan, first cousin marriage rates showed a significant decline among marriages contracted after 1980 (19.5%), compared to marriages contracted between 1950 and 1979 (28.5%; Hamamy et al., 2005). In Lebanon, consanguinity rates significantly decreased from 30% before 1950, to 25% in 1950–1969, to about 20% since 1970 (Khlat, 1988b). Among Palestinians, the frequency of consanguineous marriage decreased from a high of 50.6% in the period 1961–1965, to 32.8% in the period 1991–1998 (Jaber et al., 2000). In another study in four villages, the frequency of consanguineous marriages decreased from 33.1% in the period 1980–1985 to 25.9% in the period 2000–2004, with a parallel decrease in first cousin marriage rate from 22.3% to 16.25% (Sharkia et al., 2008). In Gaza and the West Bank,

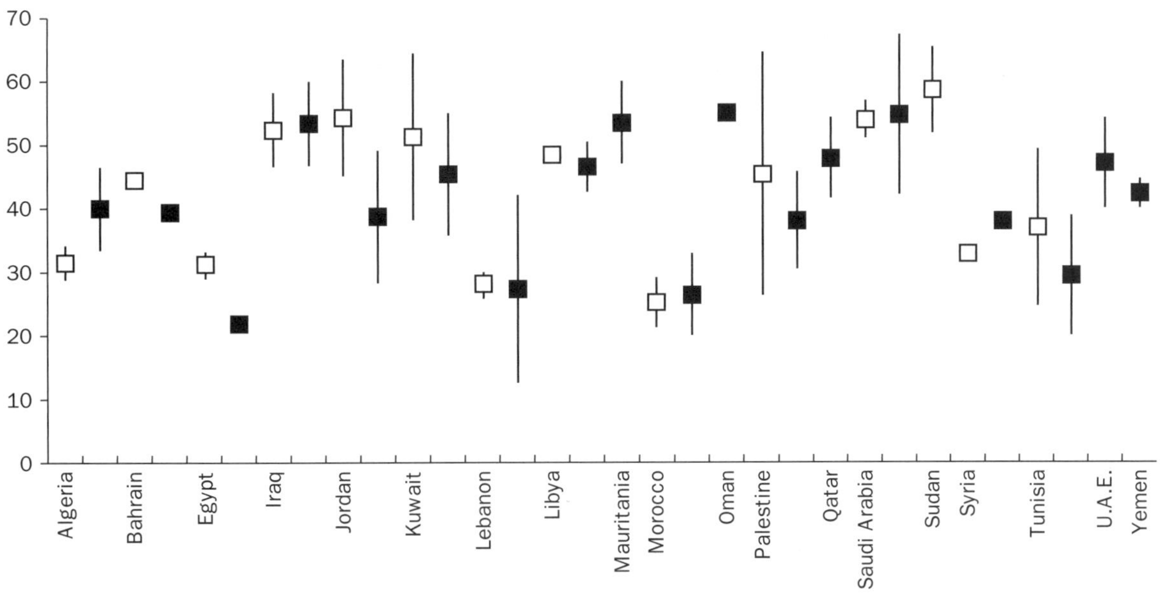

Figure 51-2 Average percent rates of past (white boxes) and present (black boxes) consanguineous marriages in Arab States (adapted from Table 51-1).

first cousin marriage rates decreased from 32.7% of marriages contracted before 1983, to 22.2% of marriages contracted during the period 2000–2004 (Assaf & Khawaja, 2009). Additionally, detailed analysis of data from Bahrain indicates a relative decrease in consanguinity rate between past (45.5%) and present (39.4%) generations (Al Arrayed, 1995). This rate decreased further to 20%, as indicated in subsequent studies (reviewed in Al Arrayed, 2006).

In Kuwait, consanguinity levels differ across the country, and they might be decreasing in Kuwait City but not in Jahra (Radovanovic et al., 1999). Several factors may be playing a role in decreasing the consanguinity rates in Arab countries. Among these factors are the increasing higher female education levels, the declining fertility resulting in lower numbers of suitable relatives to marry, more mobility from rural to urban settings, and the improving economic status of families. Moreover, genetic diseases may be feared more now that infectious diseases are on the decline as causes of severe morbidity and mortality. The decreasing family size means that couples cannot take many chances for having normal children. On the other hand, social, religious, cultural, political, and economic factors still play roles in favoring consanguineous marriages among the new generations, just as strongly as they did among the older generations, particularly in rural areas. Consanguinity rates seem to be increasing at a higher pace in Qatar, since first cousin marriage rate (including double first cousins) increased from 22.8%–27.9% in the previous generation to 37.9% in the current generation (Bener & AlAli, 2006; Figure 51-3), with a parallel increase in inbreeding coefficient from 0.016410 to 0.023724 (Bener et al., 2007). An increase in consanguineous marriages has also been reported for the current generation in Yemen (Jurdi & Saxena, 2003) and the United Arab Emirates (UAE), where the average inbreeding coefficient rose from 0.0158 in the previous generation to 0.023 in this generation (Al-Gazali et al., 1997). In Algeria, the consanguinity rate has increased from 34.1% in 1951–1960 to 46.4% in 1981–1990 in rural

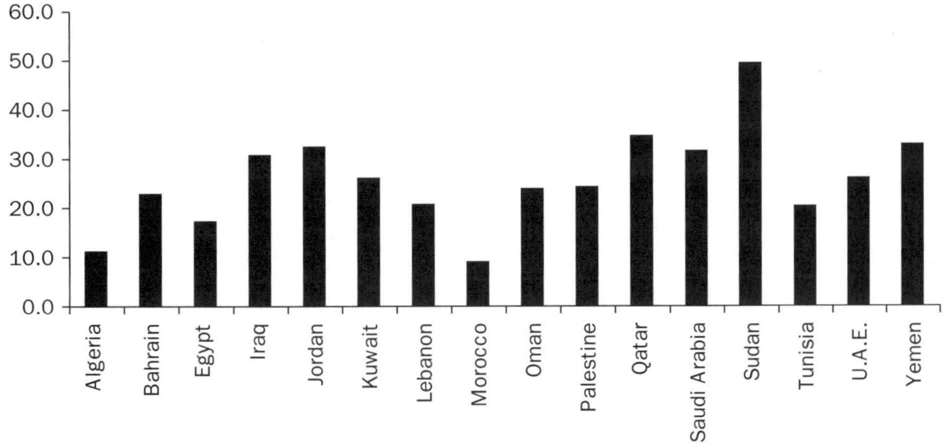

Figure 51-3 Average percent rates of first-cousin marriages in Arab States (adapted from Table 51-1).

Tlemcen (Zaoui & Biémont, 2002). In Morocco, a study indicated an increasing consanguinity rate from the previous (21.5%) to the present (25.4%) generations (Talbi et al., 2007), while another study indicated a decreasing consanguinity rate (Lamdouar Bouazzaoui, 1994). Consanguinity rates are not declining in some Arab countries, because it is generally accepted that the advantages of consanguinity outweigh the disadvantages (Jaber et al., 1998), and consanguinity is regarded as a deeply rooted cultural trend. It is believed that the practice of consanguinity promotes family stability and has significant social and economic advantages. Better health parameters in the population as a whole provides more relatives to intermarry, while the adverse genetic effects on health are seen not to affect the majority of consanguineous marriages. In all cases, reports on secular trends in consanguinity need to be treated with some caution. In countries where consanguinity is favored, major regional and ethnic differences in prevalence are commonly observed (Bittles, 2008). At the time of preparation of this manuscript, consanguinity data are not available for the Arab Eastern African countries of Djibouti, Eritrea, and Somalia. However, genetic studies from these countries do refer to cases affected by genetic disorders and born to consanguineous parents (Riepe et al., 2009).

NEGATIVE EFFECTS OF CONSANGUINITY

In Arab populations and diasporas, the deep-rooted norm of consanguineous marriage has been widely accused of being an important factor contributing to the preponderance of autosomal recessive genetic disorders (El-Shafei et al., 1986; Hoodfar & Teebi, 1996; Abdulrazzaq et al., 1997; Mokhtar et al., 1998; Hamamy et al., 2007; Kerkeni et al., 2007; El Mouzan et al., 2008). An analysis of data in the Catalogue for Transmission Genetics in Arabs (CTGA) database, maintained by the Center for Arab Genomic Studies, indicates that of the 451 genetic disorders reported in the United Arab Emirates, Bahrain, and Oman, 36.6% document the presence of patients resulting from consanguineous marriages, mostly among first cousins (Tadmouri, 2008). Undoubtedly, this is a high frequency, and researchers looking at this figure in isolation can easily be influenced in concluding that consanguinity is a direct cause of most of these disorders. Many studies from the region have drawn correlations between consanguinity and different types of morbid conditions.

CONSANGUINITY AND CONGENITAL MALFORMATIONS

Approximately 3%–5% of all live newborns have a medically significant birth defect. The recent report by March of Dimes gave higher estimated levels for birth defects in Arab countries (Christianson et al., 2006). Lower observed rates of 7.92/1000 births and 12.5/1000 births were registered in the UAE and Kuwait, respectively (Al Hosani et al., 2005; Madi et al., 2005). In Oman, among 21,988 births, 24.6 per 1000 births had major malformations (Sawardekar, 2005). Differences in birth-defect rates in different countries and studies could be attributed to true differences among different populations, or to different definitions of birth defects, different methods, and different time periods for ascertainment. The risk of birth defects in first cousin marriages may be estimated to be 2–2.5 times the population rate, mainly due to the expression of autosomal recessive disorders (Stoltenberg et al., 1997; Bromiker et al., 2004; Zlotogora, 2002; Jaber et al., 1998). Another estimate puts the offspring of first cousin unions at a 1.7%–2.8% increased risk for congenital defects above the population background risk (Bennett et al., 2002). However, these risk figures need validation for Arab countries through further, well controlled, evidence-based research. Frequency of consanguineous marriages was higher among parents of offspring with congenital malformations compared with the figures for the general population in all studies reported among Arabs, including in the UAE (Al-Gazali et al., 1995; Abdulrazzaq et al., 1997, Al Hosani et al., 2005; Dawodu et al., 2005), Kuwait (Madi et al., 2005), Oman (Sawardekar, 2005; Hasab & Jaffer, 1997; Patel, 2007), Iraq (Hamamy & Al-Hakkak, 1989; Mahdi, 1992), Jordan (Khoury & Massad, 2000; Obeidat et al., 2008), Egypt (Temtamy et al., 1998), Lebanon (Khlat, 1988a; Bittar, 1998), Tunisia (Khrouf et al., 1986), Arabs in Jerusalem (Bromiker et al., 2004), and Saudi Arabia (El Mouzan et al., 2008). Among two million births in nine South American countries, a significant association with parental consanguinity was observed for three congenital anomaly types: hydrocephalus, postaxial hand polydactyly, and bilateral cleft lip +/- cleft palate (Rittler et al., 2001). Among 238,942 consecutive births in France, birth defects showed higher rates among consanguineous versus nonconsanguineous couples (Stoll et al., 1999). After controlling for confounders, first cousin consanguinity remained significantly associated with an increased risk of congenital heart defects (CHD), where infants born to first cousin marriages had a higher risk of having a CHD diagnosed at birth compared to those born to unrelated parents in Lebanon (Yunis et al., 2006). Similar findings were reported from other studies in Lebanon (Nabulsi et al., 2003; Yunis et al., 2006), Saudi Arabia (Seliem et al., 2007; Becker & Al Halees, 1999; El Mouzan et al., 2008), Egypt (Bassili et al., 2000), and Arabs in the Occupied Territories (Gev et al., 1986). Conversely, the overall incidence of CHD among about 140,000 newborns in Oman, a country with high consanguinity rate, was similar to that reported from developed countries in Europe and America, insinuating that consanguinity is not a risk factor for CHD (Subramanyan et al., 2000). It could be argued, however, that although the overall incidence is not increased, the rates among consanguineous and nonconsanguineous marriages may be different, a point that was not investigated in the study. Consanguinity rates were noted to be higher among parents of newborns with congenital hydrocephalus (Murshid et al., 2000) and neural tube defects than in the general population in some studies (Rajab et al., 1998; Murshid, 2000,) but not in others (Asindi & Shehri, 2001). A positive association of consanguinity with cleft lip and/or palate was reported in the Palestinians (Zlotogora, 1997) and the Lebanese (Kanaan et al., 2008), but not from studies in Kuwait and Saudi Arabia (Aljohar et al., 2008; Al-Bustan et al., 2002).

CONSANGUINITY AND POSTNATAL MORTALITY

Countries with high rates of consanguineous marriages generally report smaller effects of consanguinity on mortality than populations with low rates of consanguineous marriages (Khlat & Khoury, 1991). The most recent mortality estimate, derived from a multinational study comprising 69 populations from 15 countries, with a total sample size of 2.14 million, is that first cousin progeny experience 3.5% more pre-reproductive deaths than the offspring of nonconsanguineous unions (Bittles and Black, 2010). Most studies among Arabs have indicated that postnatal mortality is higher among offspring of consanguineous parents than among nonrelated parents (Saha et al., 1990, Al Husain and Al Bunyan, 1997, Khoury & Massad, 2000; Kerkeni et al., 2007, Wong & Anokute, 1990; Pedersen, 2002; Khlat, 1988a; Mokhtar & Abdel-Fattah, 2001). Few studies have not detected this increase in postnatal mortality (Al-Awadi et al., 1986; Abdulrazzaq et al., 1997). The increased postnatal mortality among the offspring of consanguineous parents may be related to the action of deleterious recessive genes and multi-gene complexes inherited from a common ancestor. The higher parity rate among consanguineous couples counterbalances the higher infant mortality; as a result, there is equality in the number of living children among consanguineous and nonconsanguineous couples.

CONSANGUINITY AND CHILDHOOD BLINDNESS AND DEAFNESS

Environmental factors in the causation of childhood blindness and deafness are declining with the general improvement in healthcare facilities among Arabs. Concurrently, genetic factors are becoming more apparent as causative agents in these disorders (Kotb et al., 2006, Al-Alawi et al., 2004). In a study from Saudi Arabia on 913 blind school students, congenital (63%) and hereditary (25%) causes were more common than acquired (12%) causes (Al Rajhi et al., 2004). A large number of autosomal recessive gene loci have been identified to cause hearing and vision impairment and, as a consequence, these disorders are seen to be associated with consanguinity (Bener et al., 2004). For example, in Tunisia, parental first cousin rate was 61.25% among 160 deaf-mutes as compared to 20.75% in the general population (Ben Arab et al., 2004). Among 1400 Omani children suffering from severe to profound levels of hearing impairment, the rate of consanguineous marriage in their parents was 70% (Al Khabori, 2004).

THE OTHER SIDE OF THE COIN

Parallel to the huge body of literature detailing the negative effects of consanguinity on human health, there also exists a considerable amount of data that suggests that the practice of consanguinity is not the great evil that it is generally thought to be. In addition, several new studies have actually presented beneficial or neutral effects of the practice in some morbid conditions.

FETAL WASTAGE

Multiple studies in highly consanguineous world populations have noted that fetal loss has no significant association with consanguinity. In Sudan, among 4,471 pregnancies, no significant difference in the reproductive loss was observed between the inbred and outbred groups (Saha et al., 1990). In a study in Saudi Arabia, total prenatal losses were essentially the same among consanguineous and nonconsanguineous couples (Al Husain & Al Bunyan, 1997). Among 1867 married couples in Jordan, abortion rate was not affected by consanguinity (Khoury & Massad, 2000). Other studies have reported similar results (Jaber et al., 1997; Al-Awadi et al., 1986; Khlat, 1988a; Saad & Jauniaux, 2002; Al-Abdulkareem & Ballal, 1998; Kerkeni et al., 2007; Abdulrazzaq et al., 1997). Fewer studies noticed a higher rate of prenatal losses among consanguineous couples (Mokhtar & Abdel-Fattah, 2001; Assaf et al., 2009; Bener & Hussain, 2006).

FERTILITY

Consanguinity was generally not found to be associated with a significant positive or negative effect on fertility (Devi et al., 1981; Bittles et al., 2002; Hussain & Bittles, 2004; Saad & Jauniaux, 2002), although many studies report a higher fertility among consanguineous couples (Fuster, 2003; Helgason et al., 2008). Among Arabs, higher fertility rates and higher rates of live births were reported among first cousin couples than nonconsanguineous couples in Qatar (Bener & Hussain, 2006), Kuwait (Kandari, 2007), Saudi Arabia (Al-Abdulkareem & Ballal, 1998), and

Tunisia (Kerkeni et al., 2007). For all the Mauritanian ethno groups, the consanguineous couples had averages of fertility significantly higher than those of nonconsanguineous (Hammami et al., 2005). Researchers tend to think that this increase in fertility could be a biological means of compensating for the increased risk of postnatal loss expected in related marriages (Bittles et al., 2002), or possibly to the earlier age at marriage and first maternity among consanguineous as compared to nonconsanguineous couples (Fuster, 2003). Effects of consanguineous marriages on couples' fertility and on offspring mortality were investigated in Beirut through a population-based health survey of 2,752 households. Total pregnancies, live births, and living children were significantly higher among consanguineous couples than among nonconsanguineous ones, as was the proportion dead among children ever born.

However, no difference remained in either fertility or mortality, when allowance was made for socioeconomic status, religious affiliation, and marriage duration. The lack of significant pattern in the final analysis is interpreted as resulting from a long-term practice of consanguineous marriages (Khlat, 1988a).

CONSANGUINITY AND BIRTH ANTHROPOMETRIC MEASUREMENTS

Studies among Arabs related to the effect of consanguinity on anthropometric measurements such as birth weight gave conflicting results. Studies from Saudi Arabia (Al-Abdulkareem & Ballal, 1998; Al-Eissa et al., 1991; Al-Sekait, 1989; Saedi-Wong & Al-Frayh, 1989) and from Lebanon (Khlat, 1989; Mumtaz et al., 2007) presented conflicting findings. Studies from Jordan (Obeidat et al., 2008) and Arabs in the Occupied Territories (Jaber et al., 1997) detected a significant reduction in birth weight with consanguinity. An inbreeding depression has been observed for stature among children in studies conducted in Egypt (Abolfotouh et al., 1990).

It seems that there is no definite correlation between consanguinity and anthropometric measurements in populations with high consanguinity rates. More studies using standardized methodology are recommended to verify any such correlation, taking into consideration the changing socioeconomic and nutritional parameters.

THE UNKNOWN SIDE OF THE COIN

CONSANGUINITY AND COMMON ADULT DISEASES

Although the effect of consanguinity on multifactorial late-onset diseases gets little attention, and most studies to date are not conclusive as to whether consanguinity increases the risk for multifactorial disorders (Bennett et al., 2002), it is proposed that consanguinity may have a greater influence on late-onset traits than on traits that are subject to early natural selection (Hughes et al., 2002). A study on genetic isolate communities in Croatia found consanguinity as a risk factor for coronary heart disease, stroke, cancer, uni/bipolar depression, asthma, gout, peptic ulcer, and osteoporosis (Rudan et al., 2003, 2004), with the conclusion that inbreeding causes an increase in homozygosity at many genetic loci with small deleterious effects on homeostatic pathways, resulting in increased disease risk (Rudan et al., 2003). Studies reported among Arabs gave variable results on the association of consanguinity with noncommunicable disorders. For example, in Qatar, a study that included 876 married females showed that in a population with a high rate of consanguinity, there is a significant increase in the prevalence of common adult diseases like cancer, mental disorders, heart diseases, gastrointestinal disorders, and hypertension (Bener et al., 2005), and that conditions like asthma, epilepsy, and diabetes were more common in children of consanguineous than nonconsanguineous couples (Bener & Hussain, 2006). Another study among Arab communities of the Occupied Territories, showed no difference in the prevalence between the offspring of consanguineous versus nonconsanguineous matings for myocardial infarction, bronchial asthma, or duodenal ulcer, suggesting that in a population with a high rate of consanguinity, there is no significant increase in the prevalence of these common adult diseases (Jaber et al., 1997).

Studies among Arabs on the correlation of consanguinity with diabetes mellitus gave conflicting findings. Some studies reported a positive association (Bessaoud et al., 1990; Bener & Hussain, 2006; Bener et al., 2005), while other studies concluded that there is no association between consanguinity and the risk to develop diabetes mellitus (Al-Mahroos & McKeigue, 1998; Jaber et al., 1997; Zalloua et al., 2003).

Similarly, studies on the association of consanguinity with cancers, leukemias, and lymphomas (Bener et al., 2001) in populations with high consanguinity rates gave variable results. For example, a positive association between consanguinity and breast cancer was found in patients under 40 years in Pakistan (Liede et al., 2002). On the other hand, parental consanguinity in Arabs, even when a marriage is between first cousins or double first cousins, was not associated with an altered risk of breast cancer (Denic et al., 2005). Moreover, having consanguineous parents decreased the risk of breast cancer especially in younger women, while the risk of cervical cancer remained unaffected (Denic & Bener, 2001). More studies are needed to evaluate the association of consanguinity and risks to develop cancer among Arabs.

In communities with high consanguinity rates, such as the Lebanese and Jordanian communities, the proportions of hereditary conditions and consanguinity were increased among hemodialysis patients (Barbari et al., 2003; Hamed, 2002). A study from Kuwait reported that schoolchildren whose parents were consanguineous were much more likely to develop hypertension (Saleh et al., 2000), whereas a study from Saudi Arabia did not reveal any association between consanguinity and blood pressure levels (Soyannwo et al., 1998). A positive association with apnea of prematurity was documented in Lebanon (Tamim et al., 2003).

Scanty data has been published on the effects of consanguinity on adult and child psychiatric and neurological conditions, with conflicting results (Ahmed, 1979; Chaleby & Tuma, 1987; Al-Rajeh et al., 1990, 1991; Vézina et al., 1999; Eapen et al., 2004; Dobrusin et al., 2008; Mansour et al., 2009). In conclusion, the studies on the association of consanguinity with noncommunicable disorders among Arabs are not presently conclusive, and further standardized research is recommended.

CONSANGUINITY AND CHROMOSOME ABERRATIONS

Controversial results on the association of Down syndrome with consanguinity have previously been reported in populations with high and low consanguinity rates. It was suggested that consanguinity should be considered another risk factor for Down syndrome (DS), in support of the hypothesis that there exists an autosomal recessive gene that results in a predisposition toward nondisjunction of chromosome 21 (Alfi et al., 1980; Naguib et al., 1989; Pineda de Del Villar et al., 1991; de Braekeleer et al., 1994). On the other hand, it was concluded that available information does not support the presence of a "nondisjunction gene" in man (Basaran et al., 1992), since no increase in Down syndrome cases were seen in consanguineous parents or grandparents above the population levels in several studies (Devoto et al., 1985; Hamamy et al., 1990; Zlotogora, 1997; Sayee & Thomas, 1998; Chaabouni, 1999; El Mouzan et al., 2008). It is important to clarify this discrepancy because of the high rate of Down syndrome compared to other genetic conditions, and because the culture in many Arab countries perceives the birth of a Down syndrome baby as a considerable burden that imposes great psychological stress on the parents. Among some of the public, there is a belief that Down syndrome is inherited, runs in families, and is associated with consanguinity; this is an issue that needs to be scientifically clarified.

CONSANGUINITY, INTELLIGENCE, AND SPEECH DISORDERS

Studies on the association of consanguinity and intelligence quotient are scanty among Arabs. Results of studies on the association of consanguinity and mental retardation performed in Western countries cannot be applied directly to societies with high consanguinity rates, such as the Arab society.

Among the few studies on consanguinity and mental subnormality, a study from Bahrain indicated a positive association of mild mental retardation with consanguinity among 109 schoolchildren born between 1967–1982 (Al-Ansari, 1993). Another study from Cairo, Egypt showed that the consanguinity and first cousin marriage rates were higher among 420 mentally retarded patients than their rates among the general population (Salem & Lashin, 1994). Consanguinity rates among parents of mentally retarded children are expected to be higher due to the action of several autosomal recessive conditions associated with mental subnormality.

Among 814 pupils in the Arab community in the Occupied Territories, rate of reading disabilities in children of first cousin parents was higher than that of children of second cousin parents, distantly related parents, or unrelated parents (Abu-Rabia & Maroun, 2005). Speech disorders rate was 31% in offspring of consanguineous versus 22.4% in nonconsanguineous couples among 1282 respondent Arab parents in the Occupied Territories (Jaber et al., 1997).

CONSANGUINITY AS A PROTECTIVE EVOLUTIONARY MECHANISM

Several studies demonstrated that consanguinity could have a positive influence on weeding out deleterious alleles from the population. A computer-simulated stochastic modeling study has suggested that carrier rates for breast cancer genes BRCA1/BRCA2 decrease rapidly in consanguineous populations, and that increased rates of consanguinity could be the reason for lower incidences of breast cancer seen in some populations (Denic & Al-Gazali, 2002). The main argument in support of this observation is that inbreeding increases the number of homozygotes, therefore decreasing the frequency of lethal genes in consanguineous than in nonconsanguineous populations (reviewed in Denic & Nicholls, 2007). Similar studies on malaria offer another good example of how consanguinity has been beneficial in increasing the fitness of a population. In a simulated study of the evolution of alpha-thalassemia in populations exposed to malaria, consanguineous populations grew faster than randomly mating populations, and became significantly larger after a relatively small, but variable, number of generations (Denic et al., 2007, 2008). Although the deleterious effect of inbreeding in populations with a longstanding practice of consanguineous mating is known to be reduced through the progressive elimination of lethal equivalents (Rittler et al., 2001), it seems that recessive alleles in populations with high incidence of recessive disorders are maintained through reproductive compensation. If child loss is due to a genetic disorder, then the surviving replacement is more likely to be a carrier (67%) than a noncarrier, hence maintaining the existence of the recessive mutation in the population. Studies have demonstrated that reproductive compensation can effectively counteract the rate at which deleterious alleles are eliminated from the gene pool of consanguineous populations (Bittles, 2001).

OTHER CONSIDERATIONS

There are a number of situations where consanguinity could be erroneously accused for the high rates of genetic diseases and congenital anomalies. In the Arabian Peninsula, alpha-thalassemia alleles enjoy very high frequencies. Within the United Arab Emirates, one in every two citizens can be expected to be a carrier of the mutated allele (El-Kalla & Baysal, 1998). In a population with such high allele frequency, there is no difference left with respect to the risk of having an alpha-thalassemia allele in consanguineous and nonconsanguineous unions. At another extreme lies the concept that consanguinity could help in preserving desirable physical characteristics within families (Rushton, 1989; Morgan & Tooker, 1997). An interesting hypothesis for the extreme longevity in the island of Sardinia is that the high rate of inbreeding and low immigration rates have progressively decreased the variability of the genetic pool and facilitated the emergence of genetic characteristics that protect individuals from diseases that are major causes of mortality, particularly in older individuals (Poulain et al., 2004).

CONSANGUINITY AS A DRIVING FORCE FOR AUTOSOMAL RECESSIVE DISORDERS

The extended family structure, commonly present in Arab societies and mostly associated with consanguinity, tends to display unique distribution patterns for genetic diseases that are not present in many other societies. A major model that explains this concept is the difference in the vertical dissemination of a recessive genetic mutation between Arab and Western families. In the typical Western family, carriers of mutations usually become scattered through the general population due to marriages contracted outside the family. After a few generations, the genetic relationship of these family members becomes unrecognizable due to elevated gene flow mechanisms; therefore, a family history suggesting a genetic basis for their predisposition to recessive disease states is easily missed. On the other hand, in an Arab society, mutation carriers mostly remain concentrated within the extended family, and the genetic nature of their disease predisposition is often much more obvious due to decelerated human dispersal. This situation is further accentuated when consanguinity is practiced, which, in population genetic concepts, leads to further slowdown of gene flow and significant heterozygote deficiencies (Rajkumar & Kashyap, 2004; Cadenas et al., 2008). This condition is further perplexed with the fact that in a consanguineous marriage, autosomal recessive genes could be hidden within the family for generations and then surface in a new consanguineous marriage within the family, hence adding an extra element of surprise (Hamamy et al., 2007).

On individual bases, when a severe autosomal recessive condition segregates in a family, genetic counseling becomes of utmost importance in prevention of the disorder by minimizing further intermarriages in the family. Arab families are inclined to large sibship size; early genetic counseling after the birth of one affected child could prevent the birth of further affected children, and reduce the prevalence of inherited disease by 50% (Alwan & Modell, 1997). Unfortunately, diagnostic facilities for prenatal diagnosis of genetic disorders are not generally available, and therapeutic abortion of an affected fetus faces variable limitations in Arab countries, thus limiting the use of this prevention tool.

Efforts to minimize further intermarriages in families with segregating autosomal recessive disorders face many difficulties, due to the existence of deeply rooted social beliefs. Repeated long counseling sessions involving various family members are usually required. The term *hereditary disorder* may arouse a multitude of feelings of hostility and denial, often leading to disregard of the risk information given to the family. Families fear that their daughters will be stigmatized and will not get married.

Theoretically, recessive traits are expected to be less expressed than dominant traits, of equal frequencies, in any randomly mating population. This is attributed to the fact that while a recessive gene could be inherited by 75% of the population at risk, it will manifest only in 25% of the population, compared to 50% in dominant traits.

In mathematical terms, consanguinity does not alter the allele frequencies of common disorders, but increases the probability of a mating between two individual heterozygotes for the same recessive mutant allele. In this regard, the risk for birth defects in the offspring of first cousin marriage is expected to increase sharply compared to nonconsanguineous marriages. While spouses share one-eighth of their genes inherited from a common ancestor, their progeny become homozygous (i.e., autozygous) at one-sixteenth of all loci. Accordingly, the progeny are predicted to inherit identical gene copies from each parent at 6.25% of all gene loci, over and above the baseline level of homozygosity in the general population—hence leading to an increased exposure of rare recessive traits (Gelehrter et al., 1998). To investigate this subject, the CTGA Database was queried for the occurrence of autosomal recessive and dominant genetic disorders in Arab populations. The database indexes more than 976 genetic disorders in Arabs, for which data are derived from over 2100 independent studies (Tadmouri, 2008). The overwhelming proportion of these disorders follow a recessive mode of inheritance (62%), compared to a smaller proportion of dominantly inherited traits (28%). A detailed study of countries for which surveys on the occurrence of genetic disorders have been completed (United Arab Emirates, Bahrain, and Oman) indicates that recessive disorders are more in number than the dominant ones (Tadmouri, 2004, 2006, 2008). As explained above, given the high rates of consanguinity in the countries, this pattern is not entirely surprising. Data from Bahrain indicate that autosomal recessive and dominant disorders comprise 44% and 39%, respectively, of all genetic disorders in the country. The frequency gap between autosomal recessive and dominant disorders increases in the United Arab Emirates, and increases further in Oman (Figure 51-4). Despite the preliminary nature of disease records from Saudi Arabia, data from this country further support this observation, with the widest variance between frequencies of autosomal recessive (61%) and autosomal dominant (25%), accompanied with an elevated rate of consanguineous marriages (55%). These indicators clearly reveal the strength of the CTGA Database as a population-based information hub, with a capacity to draw direct relations between variables that could influence the type of genetic disorders common in the region.

Although preliminary, these data show for the first time a quantitative measure of the association between consanguinity and autosomal recessive traits. Certainly, the completion of other surveys on genetic disorders in Arab populations, accompanied with continuous surveillance of this pattern, will provide further insights on this subject.

CONCLUDING REMARKS

Consanguineous marriages are widely practiced in several global populations, with one of the highest rates observed in the Arab world. Reports abound on both the negative and positive biological effects of consanguinity. In net terms, the reproductive criteria related to consanguineous versus nonconsanguineous couples include earlier parental age at marriage, younger maternal age at first live birth, higher number of infants born to consanguineous parents, similar rates of primary sterility and abortions, and higher rates of postnatal mortality and birth defects in offspring of consanguineous parents. Furthermore, consanguineous unions lead to increased expression of autosomal recessive disorders. The CTGA Database offers clear evidence for a direct correlation between these two factors.

Further studies in the Arab world, with its presently high consanguinity rates, are needed to validate the role of consanguinity in multifactorial chronic noncommunicable diseases and chromosomal disorders. The effects of consanguinity on normal human variations, such as intelligence and height, are still unknown, and it would be interesting to investigate any correlation of these variables with consanguinity. With the achievements made in the control of communicable diseases of childhood and nutritional disorders, the infant mortality rates in most Arab countries have declined. When the infant mortality rate declines, the proportion of deaths due to genetic and congenital disorders becomes higher, thus increasing public and governmental concerns related to these disorders. In the minds of people, consanguinity has always been linked to hereditary disorders, but with vague and nonscientific misconceptions such as the misconception that all genetic diseases are due to consanguinity, that Down syndrome is more common among first cousin marriages, and that autosomal dominant conditions like achondroplasia in the father will not be transmitted to his progeny if he does not marry a blood relative.

Young Arabs contemplating marriage are nowadays seeking a scientifically sound answer to their questions: "Will our children be physically or mentally abnormal if I marry my cousin?" "How can we prevent having abnormal children?" Research on inbreeding is considered a priority in societies with high consanguinity rates, to help understand and prevent the deleterious impact of consanguinity on health, and to provide standardized and evidence-based

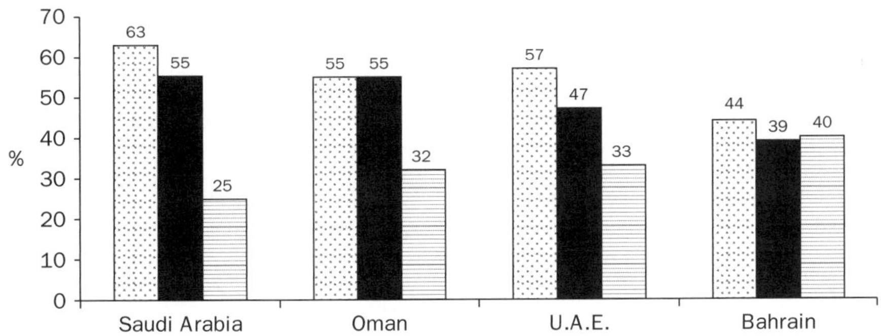

Figure 51-4 A comparison between percent consanguinity rates (black columns) and frequency of recessive (dotted columns) and dominant (striated columns) disorders in Bahrain, Oman, the UAE, and Saudi Arabia in the CTGA Database (September 2011).

guidelines for healthcare professionals in providing genetic counseling for consanguineous couples.

SUMMARY

Consanguineous marriages have been practiced since the early existence of modern humans. Until now, consanguinity is widely practiced in several global communities with variable rates depending on religion, culture, and geography. Arab populations have a long tradition of consanguinity, due to sociocultural factors such as maintenance of family structure and property, and ease of marital arrangements. Many Arab countries display some of the highest rates of consanguineous marriages in the world, specifically first cousin marriages.

Improvement in female education levels, more mobility from rural to urban settings, and other factors may be decreasing the consanguinity rates in Arab countries at the present time. Yet, in countries like Qatar, Yemen, and UAE, consanguinity rates are increasing in the current generation. There are huge arguments on both the negative and positive or neutral biological impact of consanguinity. Research among Arabs and worldwide has indicated that consanguinity could have an effect on some reproductive health parameters, such as postnatal mortality and rates of congenital malformations. The main impact of consanguinity, however, is an increase in the rate of homozygotes for autosomal recessive genetic disorders. The Catalogue of Transmission Genetics in Arabs (CTGA) database offers clear evidence for a direct correlation between these two factors.

REFERENCES

Abdulrazzaq YM, Bener A, Al-Gazali LI, Al-Khayat AI, Micallef R, Gaber T. (1997). A study of possible deleterious effects of consanguinity. *Clin Genet 51*(3), 167–173.

Abolfotouh MA, Nofal LM, Safwat H. (1990) Growth and nutritional status of preschool children attending the well-baby clinics. *J Egypt Public Health Assoc 65*(5–6), 485–507.

Abu-Rabia S, Maroun L. (2005). The effect of consanguineous marriage on reading disability in the Arab community. *Dyslexia 11*(1), 1–21.

Ahmed AH. (1979). Consanguinity and schizophrenia in Sudan. *Br J Psychiatry 134*, 635–636.

Al Arrayed SS. (1995). The frequency of consanguineous marriages in the State of Bahrain. *Bahrain Medical Bull 17*(2), 63–66.

Al Arrayed, S. (1994). Consanguinity in the state State of Bahrain. In *Medical Genetics in the Setting of Middle Eastern Populations*, El-Hazmi MAF (ed.), King Abdul Aziz City for Science and Technology, Riyadh, Saudi Arabia.

Al Arrayed SS. (2006). Genetic diseases in Bahrain. In: Tadmouri GO, Taleb Al Ali M, Al Khaja N. (eds.), *Genetic Disorders in the Arab World: Bahrain* (pp. 72–91). Centre for Arab Genomic Studies, Dubai, United Arab Emirates.

Al Hosani H, Salah M, Abu-Zeid H, Farag HM, Saade D. (2005). The National Congenital Anomalies Register in the United Arab Emirates. *East Mediterr Health J 11*(4), 690–699.

Al Husain M, Al Bunyan M. (1997). Consanguineous marriages in a Saudi population and the effect of inbreeding on prenatal and postnatal mortality. *Ann Trop Paediatr 17*(2), 155–160.

Al Khabori M. (2004). Causes of severe to profound deafness in Omani paediatric population. *Int J Pediatr Otorhinolaryngol 68*(10), 1307–1313.

Al Rajhi AAA, Awad AA, Badeeb O, et al. (2004). Causes of blindness in students attending schools for the blind in Saudi Arabia. *Saudi J Ophthalmol 18*(Special Issue), 73–77.

Al-Abdulkareem AA, Ballal SG. (1998). Consanguineous marriage in an urban area of Saudi Arabia: rates and adverse health effects on the offspring. *J Community Health 23*(1), 75–83.

Al-Ansari A. (1993). Etiology of mild mental retardation among Bahraini children: a community-based case control study. *Ment Retard 31*(3), 140–143.

Al Arrayed SS. (1999). Review of the spectrum of genetic diseases in Bahrain. *East Mediterr Health J 5*(6), 1114–1120.

Al-Alawi E, Ahmed A, Fataha B. (2004). Causes of visual disability at the Al-Noor institute for the visually impaired in Bahrain. *Saudi J. Ophthalmol 18*(Special Issue), 47–50.

Al-Awadi SA, Moussa MA, Naguib KK, et al. (1985). Consanguinity among the Kuwaiti population. *Clin Genet 27*(5), 483–486.

Al-Awadi SA, Naguib KK, Moussa MA, Farag TI, Teebi AS, El-Khalifa MY. (1986). The effect of consanguineous marriages on reproductive wastage. *Clin Genet 29*(5), 384–388.

Al-Bustan SA, El-Zawahri MM, Al-Adsani AM, et al. (2002). Epidemiological and genetic study of 121 cases of oral clefts in Kuwait. *Orthod Craniofac Res 5*(3), 154–160.

Al-Eissa YA, Ba'Aqeel HS, Haque KN. (1991). Low birthweight in Riyadh, Saudi Arabia: incidence and risk factors. *Ann Trop Paediatr 11*(1), 75–82.

Alfi O, Chang R, Azen S. (1980). Evidence for genetic control of nondisjunction in man. *Am J Hum Genet 32*, 477–483.

Al-Gazali LI, Bener A, Abdulrazzaq YM, Micallef R, Al-Khayat AI, Gaber T. (1997). Consanguineous marriages in the United Arab Emirates. *J Biosoc Sci 29*(4), 491–497.

Al-Gazali LI, Dawodu AH, Sabarinathan K, Varghese M. (1995). The profile of major congenital abnormalities in the United Arab Emirates (UAE) population. *J Med Genet 32*(1), 7–13.

Ali MM, Shah IH. (2000). Sanctions and childhood mortality in Iraq. *Lancet 355*(9218), 1851–1857.

Aljohar A, Ravichandran K, Subhani S. (2008). Pattern of cleft lip and palate in hospital-based population in Saudi Arabia: retrospective study. *Cleft Palate Craniofac J 45*(6), 592–596.

Al-Kandari YY. (2007). Fertility and its relationship with sociocultural factors in Kuwaiti society. *East Mediterr Health J 13*(6), 1364–1371.

Al-Mahroos F, McKeigue PM. (1998). High prevalence of diabetes in Bahrainis. Associations with ethnicity and raised plasma cholesterol. *Diabetes Care 21*(6), 936–942.

Al-Naser YE. (1994). The role of consanguineous marriage on infant and child mortality in Bahrain. In: El-Hazmi MAF, ed. *Medical Genetics in the Setting of Middle Eastern Populations*. Riyadh, Saudi Arabia: King Abdul Aziz City for Science and Technology.

Al-Nasser KE, Kelly CL, El-Kazimi A. (1989). Patterns of consanguinity in the population of Kuwait. *Am J Hum Genet 45*(Suppl 4), 0915A.

Aloui T, Ayad M, Habib F. (1988). Enquete demographique et de sante en Tunisie. Office National de la Famille et de la Population, Tunisia.

Al-Rajeh S, Abomelha A, Awada A, Bademosi O, Ismail H. (1990). Epilepsy and other convulsive disorders in Saudi Arabia: a prospective study of 1,000 consecutive cases. *Acta Neurol Scand 82*(5), 341–345.

Al-Rajeh S, Bademosi O, Awada A, Ismail H, Al-Shammasi S, Dawodu A. (1991). Cerebral palsy in Saudi Arabia: a case-control study of risk factors. *Dev Me a Child Neurol 33*(12), 1048–1052.

Al-Salem M, Rawashdeh N. (1993). Consanguinity in north Jordan: prevalence and pattern. *J Biosoc Sci 25*(4), 553–556.

Al-Sekait MA. (1989). Maternal influences on birth weight. *J R Soc Health 109*(2), 69–70.

Alwan A, Modell B. (1997). Community control of genetic and congenital disorders. *EMRO Technical Publication Series* 24, WHO Regional Office for the Eastern Mediterranean Region, Egypt.

ANSO (Algerian National Statistics Office). (1994). *The Algeria maternal and child health survey.* The Republic of Algeria and the League of Arab State. Pan Arab Project for Child Development, Algeria.

Asindi A, Shehri A. (2001). Neural tube defects in the Asir region of Saudi Arabia. *Ann Saudi Med 21*(1–2), 26–29.

Assaf S, Khawaja M. (2009). Consanguinity trends and correlates in the Palestinian Territories. *J Biosoc Sci 41*(1), 107–124.

Assaf S, Khawaja M, DeJong J, Mahfoud Z, Yunis K. (2009). Consanguinity and reproductive wastage in the Palestinian Territories. Paediatr Perinat Epidemiol 23(2), 107–115.

Azelmat M, Ayad M, Belhachmi H. (1987). Enquete nationale sur la planification familiale et de la sante de la population au Maroc. Ministere de la Sante Publique, Services des Etudes et de l'Information Sanitaire, Rabat, Morocco.

Azelmat M, Ayad M, Housni EA. (1992). Enquete nationale sur la population et la sante (ENPS-II). Ministere de la Sante Publique, Rabat, Morocco.

Baali A. (1994). *Etude anthropobiologique d'une population berbere semi-isolee du Hgaut-Atlas: Valle d'Azgour, cercle d'Amizmiz, Marrakech.* PhD. Thesis, Semlalia Faculty of Sciences, Marrakech, Morocco.

Badr FM. (1972). Genetic studies of Egyptian Nubian populations. I. Frequency and types of consanguineous marriages. *Hum Hered 22*(4), 387–398.

Barbari A, Stephan A, Masri M, et al. (2003). Consanguinity-associated kidney diseases in Lebanon: an epidemiological study. *Mol Immunol 39*(17–18), 1109–1114.

Barbour B, Salameh P. (2009). Consanguinity in Lebanon: prevalence, distribution and determinants. *J Biosoc Sci 28*, 1–13.

Barth F. (1954). Father's brother's daughter marriage in Kurdistan. *South Western J Anthropol 10*, 164–171.

Basaran N, Cenani A, Sayli BS, et al. (1992). Consanguineous marriages among parents of Down patients. *Clin Genet 42*(1), 13–15.

Bashi J. (1977). Effects of inbreeding on cognitive performance. *Nature 266*(5601), 440–442.

Bassili A, Mokhtar SA, Dabous NI, Zaher SR, Mokhtar MM, Zaki A. (2000). Risk factors for congenital heart diseases in Alexandria, Egypt. *Eur J Epidemiol 16*(9), 805–814.

Becker S, Al Halees Z. (1999). First-cousin matings and congenital heart disease in Saudi Arabia. *Community Genet 2*(2–3), 69–73.

Ben Arab S, Masmoudi S, Beltaief N, Hachicha S, Ayadi H. (2004). Consanguinity and endogamy in Northern Tunisia and its impact on non-syndromic deafness. *Genet Epidemiol 27*(1), 74–79.

Benallegue A, Kedji F. (1984). Consanguinite et sante publique: une etude Algerrienne. *Arch Fr Pediatr 41*(6), 435–440.

Bener A, Abdulrazzaq YM, Al-Gazali LI, Micallef R, Al-Khayat AI, Gaber T. (1996). Consanguinity and associated socio-demographic factors in the United Arab Emirates. *Hum Hered 46*(5), 256–264.

Bener A, Alali KA. (2006). Consanguineous marriage in a newly developed country: the Qatari population. *J Biosoc Sci 38*(2), 239–246.

Bener A, Al-Suwaidi J, Al-Jaber K, Al-Marri S, Elbagi IE. (2004). Epidemiology of hypertension and its associated risk factors in the Qatari population. *J Hum Hypertens 18*(7), 529–530.

Bener A, Denic S, Al-Mazrouei M. (2001). Consanguinity and family history of cancer in children with leukemia and lymphomas. *Cancer 92*(1), 1–6.

Bener A, Hussain R, Teebi AS. (2007). Consanguineous marriages and their effects on common adult diseases: studies from an endogamous population. *Med Prin Pract 16*(4), 262–267.

Bener A, Hussain R. (2006). Consanguineous unions and child health in the State of Qatar. *Paediatr Perinat Epidemiol 20*(5), 372–378.

Bener A, Zirie M, Al-Rikabi A. (2005). Genetics, obesity, and environmental risk factors associated with type 2 diabetes. *Croat Med J 46*(2), 302–307.

Bennett R, Motulsky A, Bittles A, et al. (2002). Genetic counseling and screening of consanguineous couples and their offspring: recommendations of the National Society of Genetic Counselors. *J Genet Counseling 11*, 97–119.

Bessaoud K, Boudraa G, Deschamps I, Hors J, Benbouabdallah M, Touhami M. (1990). [Epidemiology of juvenile insulin-dependent diabetes in Algeria (Wilaya of Oran)]. *Rev Epidemiol Sante Publique 38*(2), 91–99.

Bittar Z. (1998). Major congenital malformations presenting in the first 24 hours of life in 3865 consecutive births in south of Beirut. Incidence and pattern. *J Med Liban 46*(5), 256–260.

Bittles AH, Black ML. (2010). Evolution in health and medicine Sackler colloquium: Consanguinity, human evolution, and complex diseases. *Proc Natl Acad Sci USA 107*(Suppl 1), 1779–1786.

Bittles AH, Egerbladh I. (2005). The influence of past endogamy and consanguinity on genetic disorders in northern Sweden. *Ann Hum Genet 69*(Pt 5), 549–558.

Bittles AH, Grant JC, Shami SA. (1993). Consanguinity as a determinant of reproductive behaviour and mortality in Pakistan. *Int J Epidemiol 22*(3), 463–467.

Bittles AH, Grant JC, Sullivan SG, Hussain R. (2002). Does inbreeding lead to decreased human fertility? *Ann Hum Biol 29*(2), 111–130.

Bittles AH. (2008). A community genetics perspective on consanguineous marriage. *Community Genet 11*(6), 324–330.

Bittles AH. (2001). Consanguinity and its relevance to clinical genetics. *Clin Genet 60*, 89–98.

Broadhead RC, Sehgal KC. (1981). Consanguinity and congenital anomalies in East Libya. *Garyounis Med J 4*(1), 3–6.

Bromiker R, Glam-Baruch M, Gofin R, Hammerman C, Amitai Y. (2004). Association of parental consanguinity with congenital malformations among Arab newborns in Jerusalem. *Clin Genet 66*(1), 63–66.

Cadenas AM, Zhivotovsky LA, Cavalli-Sforza LL, Underhill PA, Herrera RJ. (2008). Y-chromosome diversity characterizes the Gulf of Oman. *Eur J HumGenet 16*(3), 374–386.

Central Bureau of Statistics [Syria]. (1995). *Maternal and child health survey in the Syrian Arab Republic.* Syrian Arab Republic and the League of Arab States, PanArab Project for Child Development, Syria.

Chaabouni H. (1999). Etude epidemiologique et genetique de la trisomie 21 en Tunisie. *Tunisie Med 77*(8–9), 407–414.

Chaleby K, Tuma TA. (1987). Cousin marriages and schizophrenia in Saudi Arabia. *Br J Psychiatry 150*, 547–549.

Cherkaoui M, Baali A, Larrouy G, Sevin A, Boetsch G. (2005). Consanguinity, fertility of couples and mortality of children in the High Atlas population (commons of Anougal and Azgour, Marrakesh, Morroco). *Intl J Anthropol 20*(3–4), 199–206.

Christianson A, Howson C, Modell B. (2006). *Global Report on Birth Defects. The Hidden Toll of Dying and Disabled Children.* White Plains, New York: March of Dimes Birth Defects Foundation.

Cook R, Hanslip A. (1966). Mortality among offspring of consanguineous marriage in a rural area of East Jordan. *J Trop Pediatr Afr Child Health 11*(4), 95–99.

COSIT (Central Organization for Statistics and Information Technology). (2005). Iraq Living Conditions Survey 2004: Vol. II: Analytical Report. Ministry of Planning and Development Cooperation, Baghdad, Iraq.

Dawodu A, Al-Gazali L, Varady E, Varghese M, Nath K, Rajan V. (2005). Genetic contribution to high neonatally lethal malformation rate in the United Arab Emirates. *Community Genet 8*(1), 31–34.

de Braekeleer M, Landry T, Cholette A. (1994). Consanguinity and kinship in Down syndrome in Saguenay Lac-Saint-Jean (Quebec). *Ann Genet 37*(2), 86–88.

Denic S, Al-Gazali L. (2002). Breast cancer, consanguinity, and lethal tumor genes, simulation of BRCA1/2 prevalence over 40 generations. *Int J Mol Med 10*(6), 713–719.

Denic S, Bener A, Sabri S, Khatib F, Milenkovic J. (2005). Parental consanguinity and risk of breast cancer: A population-based case-control study. *Med Sci Monit 11*(9), CR415–419.

Denic S, Bener A. (2001). Consanguinity decreases risk of breast cancer—cervical cancer unaffected. *Br J Cancer 85*(11), 1675–1679.

Denic S, Frampton C, Nagelkerke N, Nicholls MG. (2007). Consanguinity affects selection of alpha-thalassemia genotypes and the size of populations under selection pressure from malaria. *Ann Hum Biol 34*(6), 620–631.

Denic S, Nagelkerke N, Agarwal MM. (2008). Consanguineous marriages and endemic malaria: can inbreeding increase population fitness? *Malar J 7*, 150.

Denic S, Nicholls MG. (2007). Genetic benefits of consanguinity through selection of genotypes protective against malaria. *Hum Biol 79*(2), 145–158.

Devi AR, Rao NA, Bittles AH. (1981). Consanguinity, fecundity and post-natal mortality in Karnataka, South India. *Ann Hum Biol 8*(5), 469–472.

Devoto M, Prosperi L, Bricarelli FD, et al. (1985). Frequency of consanguineous marriages among parents and grandparents of Down patients. *Hum Genet 70*(3), 256–258.

Dobrusin M, Weitzman D, Levine J, et al. (2008). The rate of consanguineous marriages among parents of schizophrenic patients in the Arab Bedouin population in Southern Israel. *World J Biol Psychiatry 11*, 1–3.

Eapen V, Al-Sabosy M, Saeed M, Sabri S. (2004). Child psychiatric disorders in a primary care Arab population. *Int J Psychiatry Med 34*(1), 51–60.

El Mouzan MI, Al Salloum AA, Al Herbish AS, Qurachi MM, Al Omar AA. (2008). Consanguinity and major genetic disorders in Saudi children: a community-based cross-sectional study. *Ann Saudi Med 28*(3), 169–173.

El-Alfi OS, Bahig AH, Abdul Salam T, Shaath R. (1969). Birth weights in Kuwait, and their relation to consanguinity and to birth order. *J Kuwait Medical Assoc 3*, 227–231.

El-Alfi OS, Shaker Y, Shaath R, Abdul Salam T. (1968). Congenital malformations in Kuwait. *J Kuwait Medical Assoc 2*, 99–108.

El-Hazmi MA, Al-Swailem AR, Warsy AS, Al-Swailem AM, Sulaimani R, Al-Meshari AA. (1995). Consanguinity among the Saudi Arabian population. *J Med Genet 32*(8), 623–626.

El-Kalla S, Baysal E. (1998). Alpha-thalassemia in the United Arab Emirates. *Acta Haematol 100*(1), 49–53.

El-Mouzan MI, Al-Salloum AA, Al-Herbish AS, Qurachi MM, Al-Omar AA. (2007). Regional variations in the prevalence of consanguinity in Saudi Arabia. *Saudi Med J 28*(12), 1881–1884.

El-Shafei A, Rao PS, Sandhu AK. (1986). Congenital malformations and consanguinity. *Aust N Z J Obstet Gynaecol 26*(3), 168–172.

ENAF (Enquete Nationale Algerrienne sur la Fecondite). (1992). *In Familles, femmes et contraception* Ed. Kouaouci A. Contribution des Nations Unies pour la Population et Centre National d'Etudes et d'Analyses pour la Planification, Algeria.

ENPC (Egypt National Population Council). (1989). *Demographic and health survey 1988*. Cairo, Egypt: National Population Council.

ENPC (Egypt National Population Council). (1996). *Egypt demographic and health survey*. Calverton, Maryland: Arab Republic of Egypt and Macro International Inc.

Fahmy NA, Benson PF, Al-Garrah DB. (1993). Consanguinity in UAE: Prevalence and analysis of some risk factors. *Emirates Med J 1*, 39–41.

FMHNCHS (Federal Ministry of Health and the National Center of Health Statistics) [Sudan]. (1995). *Sudan maternal and child health survey*. The Republic of Sudan and the League of Arab State, Pan Arab Project for Child Development, Sudan.

Freundlich E, Hino N. (1984). Consanguineous marriage among rural Arabs in Israel. *Isr J Med Sci 20*(11), 1035–1038.

Fuster V. (2003). Inbreeding pattern and reproductive success in a rural community from Galicia (Spain). *J Biosoc Sci 35*(1), 83–93.

Gelehrter TD, Collins FS, Ginsburg D. (1998). *Principles of Medical Genetics*. Baltimore, MD: Lippincott Williams & Wilkins.

Gev D, Roguin N, Freundlich E. (1986). Consanguinity and congenital heart disease in the rural Arab population in northern Israel. *Hum Hered 36*(4), 213–217.

GPCHSI (General People's Committee for Health and Social Insurance) [Libya]. (1996). *Arab Libyan maternal and child health survey*. The Great Socialist People's Libyan Arab Republic and the League of Arab State, Pan Arab Project for Child Development, Libya.

Grassivaro Gallo P, Florio A. (1993). [Female fertility in Algeria: biodemographic and psychosocial aspects]. *Genus 49*(3–4), 115–134.

Gunaid AA, Hummad NA, Tamim KA. (2004). Consanguineous marriage in the capital city Sana'a, Yemen. *J Biosoc Sci 36*(1), 111–121.

Habib Z, Böök JA. (1983). Consanguinity and incidence of thalassaemia in Egypt. *Hereditas 99*(2), 215–217.

Hafez M, El-Tahan H, Awadalla M, El-Khayat H, Abdel-Gafar A, Ghoneim M. (1983). Consanguineous matings in the Egyptian population. *J Med Genet 20*(1), 58–60.

Hamamy H, Jamhawi L, Al-Darawsheh J, Ajlouni K. (2005). Consanguineous marriages in Jordan: why is the rate changing with time? *Clin Genet 67*(6), 511–116.

Hamamy HA, Al-Hakkak ZS, Al-Taha S. (1990). Consanguinity and the genetic control of Down syndrome. *Clin Genet 37*(1), 24–29.

Hamamy HA, Al-Hakkak ZS. (1989). Consanguinity and reproductive health in Iraq. *Hum Hered 39*(5), 271–275.

Hamamy H, Al-Bayati N, Al-Kubaisy W. (1986). Consanguineous marriages in Iraqi urban population and the effect on pregnancy outcome and infant mortality. *Iraqi Med J 34*, 76–80.

Hamamy HA, Masri AT, Al-Hadidy AM, Ajlouni KM. (2007). Consanguinity and genetic disorders. Profile from Jordan. *Saudi Med J 28*(7), 1015–1017.

Hamed RM. (2002). The spectrum of chronic renal failure among Jordanian children. *J Nephrol 15*(2), 130–135.

Hammami A, Elgazzeh M, Chalbi N, Mansour BA. (2005). [Endogamy and consanguinity in Mauritania] *Tunis Med 83*(1), 38–42.

Hasab A, Jaffer A. (1997). Congenital anomalies among Omani births: a case control approach. *Bull. High Inst. Public Health; 27*(Supp. 1), 93–99.

Helgason A, Pálsson S, Gudbjartsson DF, Kristjánsson T, Stefánsson K. (2008). An association between the kinship and fertility of human couples. *Science 319*(5864), 813–816.

Hoodfar E, Teebi AS. (1996). Genetic referrals of Middle Eastern origin in a western city: inbreeding and disease profile. *J Med Genet 33*(3), 212–215.

Hughes KA, Alipaz JA, Drnevich JM, Reynolds RM. (2002). A test of evolutionary theories of aging. *Proc Natl Acad Sci U S A 99*(22), 14286–14291.

Hussain R, Bittles AH. (2004). Assessment of association between consanguinity and fertility in Asian populations. *J Health Popul Nutr 22*(1), 1–12.

Hussien FH. (1971). Endogamy in Egyptian Nubia. *J Biosoc Sci 3*(3), 251–257.

Jaber L, Bailey-Wilson JE, Haj-Yehia M, Hernandez J, Shohat M. (1994). Consanguineous matings in an Israeli-Arab community. *Arch Pediatr Adolesc Med 148*(4), 412–415.

Jaber L, Halpern GJ, Shohat M. (1998). The impact of consanguinity worldwide. *Community Genet 1*(1), 12–17.

Jaber L, Halpern GJ, Shohat T. (2000). Trends in the frequencies of consanguineous marriages in the Israeli Arab community. *Clin Genet 58*(2), 106–110.

Jaber L, Merlob P, Gabriel R, Shohat M. (1997). Effects of consanguineous marriage on reproductive outcome in an Arab community in Israel. *J Med Genet 34*(12), 1000–1002.

Jaber L, Nahmani A, Shohat M. (1997). Speech disorders in Israeli Arab children. *Isr J Med Sci 33*(10), 663–665.

Jaber L, Shohat T, Rotter JI, Shohat M. (1997). Consanguinity and common adult diseases in Israeli Arab communities. *Am J Med Genet 70*(4), 346–348.

Jaouad IC, Elalaoui SC, Sbiti A, et al. (2009). Consanguineous marriages in Morocco and the consequence for the incidence of autosomal recessive disorders. *J Biosoc Sci 41*(5), 575–581.

Jurdi R, Saxena PC. (2003). The prevalence and correlates of consanguineous marriages in Yemen: similarities and contrasts with other Arab countries. *J Biosoc Sci 35*(1), 1–13.

Kanaan ZM, Mahfouz R, Tamim H. (2008). The prevalence of consanguineous marriages in an underserved area in Lebanon and its association with congenital anomalies. *Genet Test 12*(3), 367–372.

Kandari Y. (2007). Fertility and its relationship with sociocultural factors in Kuwaiti society. *Eastern Medit Health J 13*(6), 1364–1371.

Kerkeni E, Monastiri K, Saket B, Guediche MN, Ben Cheikh H. (2007) Interplay of socio-economic factors, consanguinity, fertility, and offspring mortality in Monastir, Tunisia. *Croat Med J 48*(5), 701–707.

Kerkeni E, Monastiri K, Saket B, Rudan D, Zgaga L, Ben Cheikh H. (2006). Association among education level, occupation status, and consanguinity in Tunisia and Croatia. *Croat Med J 47*(4), 656–661.

Khlat M, Halabi S, Khudr A, Der Kaloustian VM. (1986). Perception of consanguineous marriages and their genetic effects among a sample of couples from Beirut. *Am J Med Genet 25*(2), 299–306.

Khlat M, Khoury M. (1991). Inbreeding and diseases: demographic, genetic, and epidemiologic perspectives. *Epidemiol Rev 13*, 28–41.

Khlat M. (1988a). Consanguineous marriage and reproduction in Beirut, Lebanon. *Am J Hum Genet 43*(2), 188–196.

Khlat M. (1988b). Consanguineous marriages in Beirut: time trends, spatial distribution. *Soc Biol 35*(3–4), 324–330.

Khlat M. (1989). Inbreeding effects on fetal growth in Beirut, Lebanon. *Am J Phys Anthropol 80*(4), 481–484.

Khoury SA, Massad D. (1992). Consanguineous marriage in Jordan. *Am J Med Genet 43*(5), 769–775.

Khoury SA, Massad DF. (2000). Consanguinity, fertility, reproductive wastage, infant mortality and congenital malformations in Jordan. *Saudi Med J 21*(2), 150–154.

Khrouf N, Spang R, Podgorna T, Miled SB, Moussaoui M, Chibani M. (1986). Malformations in 10,000 consecutive births in Tunis. *Acta Pediatr Scand 75*(4), 534–539.

Klat M, Khudr A. (1986). Religious endogamy and consanguinity in marriage patterns in Beirut, Lebanon. *Soc Biol 33*(1–2), 138–145.

Kotb AA, Hammouda EF, Tabbara KF. (2006). Childhood blindness at a school for the blind in Riyadh, Saudi Arabia. *Ophthalmic Epidemiol 13*(1), 1–5.

Lamdouar Bouazzaoui N. (1994). [Consanguinity and public health in Morocco] *Bull Acad Natl Med 178*(6), 1013–1025.

Liede A, Malik IA, Aziz Z, Rios Pd Pde L, Kwan E, Narod SA. (2002). Contribution of BRCA1 and BRCA2 mutations to breast and ovarian cancer in Pakistan. *Am J Hum Genet 71*(3), 595–606.

Madi SA, Al-Naggar RL, Al-Awadi SA, Bastaki LA. (2005). Profile of major congenital malformations in neonates in Al-Jahra region of Kuwait. *East Mediterr Health J 11*(4), 700–706.

Mahdi A. (1992). Consanguinity and its effect on major congenital malformations. *Iraqi Med J* 40–42, 170–176.

Mansour H, Klei L, Wood J, et al. (2009). Consanguinity associated with increased risk for bipolar I disorder in Egypt. *Am J Med Genet B Neuropsychiatr Genet 150B*(6), 879–885.

Ministry of Health [Bahrain]. (1995). Bahrain family health survey. State of Bahrain and Gulf Family Health Survey, Bahrain.

Ministry of Health [Kuwait]. (1996). Kuwait family health survey. State of Kuwait and Gulf Family Health Survey, Kuwait.

Ministry of Health [Oman]. (1996). Oman family health survey. Sultanate of Oman and Gulf Family Health Survey, Oman.

Ministry of Health [Qatar]. (1999). Qatar family health survey. State of Qatar.

Ministry of Public Health [Lebanon]. (1998). Lebanon maternal and child health survey: Summary report. Republic of Lebanon and the League of Arab State, Pan Arab Project for Child Development, Lebanon.

Modell B, Darr A. (2002). Science and society: genetic counselling and customary consanguineous marriage. *Nat Rev Genet 3*(3), 225–229.

Mohamed MS. (1995). An epidemiological study on consanguineous marriage among urban population in Alexandria. *J Egypt Public Health Assoc 70*(3–4), 293–305.

Mokhtar MM, Abdel-Fattah MM. (2001). Consanguinity and advanced maternal age as risk factors for reproductive losses in Alexandria, Egypt. *Eur J Epidemiol 17*(6), 559–565.

Mokhtar MM, Kotb SM, Ismail, SR. (1998). Autosomal recessive disorders among patients attending the genetic clinic in Alexandria. *East Mediterr Health J 4*(3), 70–479.

Morgan LH, Tooker E. (1997). *Systems of Consanguinity and Affinity of the Human Family.* University of Nebraska Press, USA.

Mukherjee DP, Das S, Banik SD. (2007). Trends of consanguineous marriages in a Sunni Muslim population of West Bengal, India. *Anthropol Anz 65*(3), 253–262.

Mumtaz G, Tamim H, Kanaan M, et al; National Collaborative Perinatal Neonatal Network. (2007). Effect of consanguinity on birth weight for gestational age in a developing country. *Am J Epidemiol 165*(7), 742–752.

Murshid WR, Jarallah JS, Dad MI. (2000). Epidemiology of infantile hydrocephalus in Saudi Arabia: birth prevalence and associated factors. *Pediatr Neurosurg 32*(3), 119–123.

Murshid WR. (2000). Spina bifida in Saudi Arabia: is consanguinity among the parents a risk factor? *Pediatr Neurosurg 32*(1), 10–12.

Nabulsi A. (1995). Mating patterns of the Abbad tribe in Jordan. *Soc Biol 42*(3–4), 162–174.

Nabulsi MM, Tamim H, Sabbagh M, Obeid MY, Yunis KA, Bitar FF. (2003). Parental consanguinity and congenital heart malformations in a developing country. *Am J Med Genet A 116A*(4), 342–347.

Naguib K, Al-Awai S, Moussa M, Farag T, Teebi A. (1989). Effect of Parental age, birth order and consanguinity on nondisjunction in the population of Kuwait. *J Kuwaiti Medical Association 23*(1), 37–43.

National Statistical Office [Mauritania]. (1992). Mauritania maternal and child health survey (1990–1991). Islamic Republic of Mauritania and the League of Arab States, Pan Arab Project for Child Development, Mauritania.

Obeidat BR, Khader YS, Amarin ZO, Kassawneh M, Al Omari M. (2008). Consanguinity and Adverse Pregnancy Outcomes: The North of Jordan Experience. *Matern Child Health J 14*(2), 283–289.

Patel PK. (2007). Profile of major congenital anomalies in the Dhahira region, Oman. *Ann Saudi Med 27*(2), 106–111.

Pedersen J. (2002). The influence of consanguineous marriage on infant and child mortality among Palestinians in the West Bank and Gaza, Jordan, Lebanon and Syria. *Community Genet 5*(3), 178–181.

Pineda de Del Villar L, Chacon I, Villalobos de Michelena M. (1991). [Genealogic evidence of genetic control in Down syndrome] *Acta Cient Venez 42*(3), 134–137.

Poulain M, Pes GM, Grasland C, et al. (2004). Identification of a geographic area characterized by extreme longevity in the Sardinia island: the AKEA study. *Exp Gerontol 39*(9), 1423–1429.

Prothro ET, Diab LN. (1974). *Changing family patterns in the Arab East.* Beirut: American University of Beirut, Lebanon.

Radovanovic Z, Shah N, Behbehani J. (1999). Prevalence and social correlates to consanguinity in Kuwait. *Ann Saudi Med 19*(3), 206–210.

Rajab A, Patton MA. (2000). A study of consanguinity in the Sultanate of Oman. *Ann Hum Biol 27*(3), 321–326.

Rajab A, Vaishnav A, Freeman NV, Patton MA. (1998). Neural tube defects and congenital hydrocephalus in the Sultanate of Oman. *J Trop Pediatr 44*(5), 300–303.

Rajkumar R, Kashyap VK. (2004). Genetic structure of four socio-culturally diversified caste populations of southwest India and their affinity with related Indian and global groups. *BMC Genet 5*, 23.

Riepe FG, van Bemmelen MX, Cachat F, et al. (2009). Revealing a subclinical salt-losing phenotype in heterozygous carriers of the novel S562P mutation in the alpha subunit of the epithelial sodium channel. *Clin Endocrinol (Oxf) 70*(2), 252–258.

Riou S, El Younsi C, Chaabouni H. (1989). [Consanguinity in the population of northern Tunisia] *Tunis Med 67*(3), 167–172.

Rittler M, Liascovich R, Lopez-Camelo J, Castilla EE. (2001). Parental consanguinity in specific types of congenital anomalies. *Am J Med Genet 102*(1), 36–43.

Romanus T. (1953). Frequency of consanguineous relations among applicants for legal abortion and among their parents in Sweden. *Acta Genet Stat Med 4*(2–3), 266–273.

Rudan I, Rudan D, Campbell H, et al. (2003). Inbreeding and risk of late onset complex disease. *J Med Genet 40*(12), 925–932.

Rudan I, Skarić-Jurić T, Smolej-Narancić N, et al. (2004). Inbreeding and susceptibility to osteoporosis in Croatian island isolates. *Coll Antropol 28*(2), 585–601.

Rushton JP. (1989). Genetic similarity, human altruism, and group selection. *Behav Brain Sci 12*, 503–559.

Saad FA, Jauniaux E. (2002). Recurrent early pregnancy loss and consanguinity. *Reprod Biomed Online 5*(2), 167–170.

Saedi-Wong S, Al-Frayh AR. (1989). Effects of consanguineous matings on anthropometric measurements of Saudi newborn infants. *Fam Pract 6*(3), 217–220.

Saha N, El Sheikh FS. (1988). Inbreeding levels in Khartoum. *J Biosoc Sci 20*(3), 333–336.

Saha N, Hamad RE, Mohamed S. (1990). Inbreeding effects on reproductive outcome in a Sudanese population. *Hum Hered 40*(4), 208–212.

Saito T. (1988). An expected decrease in the incidence of autosomal recessive disease due to decreasing consanguineous marriages. *Genet Epidemiol 5*(6), 421–432.

Saleh EA, Mahfouz AA, Tayel KY, Naguib MK, Bin-al-Shaikh NM. (2000). Hypertension and its determinants among primary-school children in Kuwait: an epidemiological study. *East Mediterr Health J 6*(2–3), 333–337.

Salem K, Lashin S, El Dib A. (1994). Consanguinity and mental retardation. *Med J Cairo Univ 62*(Supp. 4), 177–186.

Sandridge AL, Takeddin J, Al-Kaabi E, Frances Y. (2010). Consanguinity in Qatar: knowledge, attitude and practice in a population born between 1946 and 1991. *J Biosoc Sci 42*(1), 59–82.

Sawardekar KP. (2005). Profile of major congenital malformations at Nizwa Hospital, Oman: 10-year review. *J Paediatr Child Health 41*(7), 323–330.

Sayee R, Thomas IM. (1998). Consanguinity, non-disjunction, parental age and Down's syndrome. *J Indian Med Assoc 96*(11), 335–337.

Seliem MA, Bou-Holaigah IH, Al-Sannaa N. (2007). Influence of consanguinity on the pattern of familial aggregation of congenital cardiovascular anomalies in an outpatient population: studies from the eastern province of Saudi Arabia. *Community Genet 10*(1), 27–31.

Serenius F, Edressee AW, Swailem AR. (1988). Characteristics of the obstetric population in a Saudi maternity hospital. *Acta Paediatr Scand Suppl 346*, 29–43.

Sharkia R, Zaid M, Athamna A, Cohen D, Azem A, Zalan A. (2008) The changing pattern of consanguinity in a selected region of the Israeli Arab community. *Am J Hum Biol 20*(1), 72–77.

Soyannwo MA, Kurashi NY, Gadallah M, et al. (1998). Blood pressure pattern in Saudi population of Gassim. *Afr J Med Med Sci 27*(1–2), 107–116.

Stevenson AC, Johnston HA, Stewart MI, Golding DR. (1966). Congenital malformations. A report of a study of series of consecutive births in 24 centres. *Bull World Health Organ 34*(Suppl), 9–127.

Stoll C, Alembik Y, Roth MP, Dott B. (1999). Parental consanguinity as a cause for increased incidence of births defects in a study of 238,942 consecutive births. *Ann Genet 42*(3), 133–139.

Stoltenberg C, Magnus P, Lie RT, Daltveit AK, Irgens LM. (1997). Birth defects and parental consanguinity in Norway. *Am J Epidemiol 145*, 439–448.

Subramanyan R, Joy J, Venugopalan P, Sapru A, Al Khusaiby SM. (2000). Incidence and spectrum of congenital heart disease in Oman. *Ann Trop Paediatr 20*(4), 337–341.

Sueyoshi S, Ohtsuka R. (2003). Effects of polygyny and consanguinity on high fertility in the rural Arab population in South Jordan. *J Biosoc Sci 35*(4), 513–526.

Tadmouri GO. (2004). Genetic disorders in Arab Populations. In: Tadmouri GO, Taleb Al Ali M, Al Khaja N. (eds.), *Genetic Disorders in the Arab World: United Arab Emirates* (pp. 7–33). Centre for Arab Genomic Studies, Dubai, United Arab Emirates.

Tadmouri GO. (2006) Genetic disorders in Arab populations: A 2006 update. In: Tadmouri GO, Taleb Al Ali M, Al Khaja N. (eds.), *Genetic Disorders in the Arab World: Bahrain* (pp. 10–40). Centre for Arab Genomic Studies, Dubai, United Arab Emirates.

Tadmouri GO. (2008) Genetic disorders in Arab Populations: A 2008 Update. In: Tadmouri GO, Taleb Al Ali M, Al Khaja N. (eds.), *Genetic Disorders in the Arab World: Oman* (pp. 8–43). Centre for Arab Genomic Studies, Dubai, United Arab Emirates.

Talbi J, Khadmaoui AE, Soulaymani AEM, Chafik AEA. (2007) Etude de la consanguinité dans la population marocaine. Impact sur le profil de la santé. *Antropo 15*, 1–11.

Tamim H, Khogali M, Beydoun H, Melki I, Yunis K, National Collaborative Perinatal Neonatal Network. (2003). Consanguinity and apnea of prematurity. *Am J Epidemiol 158*(10), 942–946.

Temtamy SA, Abdel Meguid N, Mazen I, Ismail SR, Kassem NS, Bassiouni R. (1998). A genetic epidemiological study of malformations at birth in Egypt. *East Mediterr Health J 4*(2), 252–259.

Vardi-Saliternik R, Friedlander Y, Cohen T. (2002). Consanguinity in a population sample of Israeli Muslim Arabs, Christian Arabs and Druze. *Ann Hum Biol 29*(4), 422–431.

Vézina H, Heyer E, Fortier I, Ouellette G, Robitaille Y, Gauvreau D. (1999). A genealogical study of Alzheimer disease in the Saguenay region of Quebec. *Genet Epidemiol 16*(4), 412–425.

Wong SS, Anokute CC. (1990). The effect of consanguinity on pregnancy outcome in Saudi Arabia. *J R Soc Health 110*(4), 146–147.

Yunis K, Mumtaz G, Bitar F, et al. (2006). Consanguineous marriage and congenital heart defects: a case-control study in the neonatal period. *Am J Med Genet A 140*(14), 1524–1530.

Zakzouk S, El-Sayed Y, Bafaqeeh SA. (1993). Consanguinity and hereditary hearing impairment among Saudi population. *Ann Saudi Med 13*(5), 447–450.

Zalloua PA, Terwedow H, Shbaklo H, Halaby G, Xu X, Azar ST. (2003). Host and environmental factors defining the epidemiology of type 1 diabetes mellitus in a group of Lebanese children and young adults. *J Pediatr Endocrinol Metab 16*(5), 759–769.

Zaoui S, Biémont C. (2002). [Frequency of consanguineous unions in the Tlemcen area (West Algeria)]. *Sante*; *12*(3), 289–295.

Zlotogora J, Habiballa H, Odatalla A, Barges S. (2002). Changing family structure in a modernizing society: a study of marriage patterns in a single Muslim village in Israel. *Am J Hum Biol 14*(5), 680–682.

Zlotogora J. (1997). Genetic disorders among Palestinian Arabs: 1. Effects of consanguinity. *Am J Med Genet 68*(4), 472–475.

Zlotogora J. (2002). What is the birth defect risk associated with consanguineous marriages? *Am J Med Genet 109*(1), 70–71.

52.

CLINICAL DYSMORPHOLOGY IN THE ARAB WORLD

Lihadh Al-Gazali and Hanan Hamamy

INTRODUCTION

The art of dysmorphology involves "a detailed history and clinical evaluation of a potentially dysmorphic individual with precise observation and collating of all physical features, followed by interpretation, categorization, and diagnosis" (Hall, 1993). Clinical dysmorphology is an active clinical and research discipline supported by important databases such as the London Dysmorphology and the POSSUM databases. (Donnai & Read, 2003), and updated publications of reference books such as *Smith's Recognizable Patterns of Human Malformation* (Jones, 1996).

Dysmorphology evolved from a small nucleus of clinicians in the 1950s into a recognized and widely practiced discipline, and more recently has incorporated translational research into developmental biology, molecular genetics, and metabolic medicine (Allanson et al., 2009). The Arab world has contributed to the identification of many rare abnormal genes that cause dysmorphic syndromes, through the active collaboration between Arab medical geneticists and scientists in Europe and North America.

This chapter will present available data on dysmorphology syndromes among Arabs with examples of common, rare, and new syndromes reported in this population. Available epidemiologic data on dysmorphism among Arabs is generally scanty; reports hitherto published on prevalence rates among selected groups of populations will be presented.

EPIDEMIOLOGY OF BIRTH DEFECTS AMONG ARABS

Available data suggest that genetic and congenital disorders are more common in Arab countries than in high-income countries (Al-Gazali et al., 2006). Studies on the birth prevalence of congenital and genetic disorders that are lethal, or cause lifelong impairment if untreated, indicated that, of the six World Health Organization (WHO) regions, the highest rate of 65 affected children/1000 live births was reported in the Eastern Mediterranean Region, which covers the majority of Arab countries (Alwan & Modell, 2003). These figures were supported by a recent March of Dimes report, which estimated birth defects to be 69.9/1000 live births in most Arab countries, as opposed to 52.1/1000 live births in Europe, North America, and Australia (Christianson et al., 2006).

The relatively high prevalence of genetic and congenital disorders among Arabs can be attributed to several factors (Al-Gazali et al., 2006, Hamamy & Alwan, 1994) including:

1. A high rate of traditional consanguineous marriages, which involve around 25%–60% of all marriages among most Arabs, mostly marriages among first cousins (Tadmouri et al., 2009; also Chapter 51, Consanguinity Among Arabs). High consanguinity rates are expected to increase the frequency of recessively inherited disorders, which account for a substantial proportion of dysmorphic syndromes, physical, and mental handicaps (Teebi & Farag, 1997; Alwan & Modell, 1997; Hamamy et al., 2007a).
2. A relatively high birth rate of infants with chromosomal disorders related to advanced maternal age, such as Down syndrome and other trisomies.
3. A relatively high birth rate of infants with malformations due to new dominant mutations related to advanced paternal age.
4. A high frequency of hemoglobinopathies and enzymopathies, such as glucose-6-phosphate dehydrogenase deficiency.
5. A large family size, which contributes to the increase in number of affected children in families with autosomal recessive conditions.
6. A general inadequacy of birth-defects care and prevention.

For example, among the few reports on the prevalence rates of dysmorphic syndromes among Arabs, a study on

3000 hospital live births and stillbirths in Giza, Egypt, showed that the prevalence rates were 1.7/1000 for polydactyly with or without syndactyly, 0.3/1000 for each of achondroplasia, chondrodysplasia punctata, thanatophoric dysplasia, Freeman-Sheldon, osteogenesis imperfecta, and oto-palato-digital, pterygium, and Larsen syndromes. (Temtamy et al., 1998a). In Oman, analysis of 3000 records in the years 1993–2002 using a hospital-based registry of genetic diseases was attempted, indicating the birth rates of a number of autosomal recessive diseases (Rajab et al., 2005a).

HIGH PREVALENCE OF AUTOSOMAL RECESSIVE DISORDERS AMONG ARABS

In addition to the high consanguinity rates among Arabs, isolated subpopulations with a high level of inbreeding exist, with many parts of the Arab world still exhibiting a tribal trend of society. The epidemiology of genetic disorders is thus not uniform among Arabs, since families and tribal groups that descend from a limited number of ancestors could harbor specific autosomal recessive syndromes, and some such syndromes are confined to specific villages, families, and tribal groups (Al-Gazali et al., 2005; Hamamy et al., 2007a). The importance of autosomal recessive syndromes in the etiology of dysmorphism among Arabs is highlighted in several reports. For example, a publication on genetic disorders among Arab populations lists 115 new genetic syndromes, among which 100 syndromes are autosomal recessive, 10 are autosomal dominant, and five are possibly X-linked recessive or autosomal recessive (Teebi & Farag, 1997). An analysis of data in the Catalogue for Transmission Genetics in Arabs (CTGA), a database on genetic disorders in Arab populations maintained by the Centre for Arab Genomic Studies (CAGS), indicates that in contrast to international databases, the overwhelming proportion of the disorders in the CTGA database follow a recessive mode of inheritance (63%) compared to the smaller proportion of dominantly inherited traits (27%).

As a consequence of the high consanguinity rate among the Palestinian Arabs, an autosomal recessive disease was diagnosed or strongly suspected in 601 of the 2000 different families seen at a genetic clinic. Disorders found in higher than expected frequencies included Bardet-Biedl syndrome, Meckel syndrome, autosomal recessive congenital hydrocephalus, and recessive osteopetrosis (Zlotogora, 1997a). Similarly, in a genetic clinic in Alexandria, Egypt, 33.6% of 660 families had an autosomal recessive condition (Mokhtar et al., 1998), and in a genetic clinic in Amman, Jordan, 32% of all presentations were attributed to autosomal recessive inheritance (Hamamy et al., 2007b). Several studies from the UAE showed that a high proportion of malformations, including brain malformations and skeletal dysplasias, are caused by autosomal recessive genes (Al-Gazali et al., 1995a, 1999a, 2003a).

EPIDEMIOLOGY OF DOWN SYNDROME (DS) AMONG ARABS

Chromosomal aberrations are among the most important causes of congenital malformation and mental handicap. The risk of having a child with Down syndrome increases with increased maternal age; available data for Arab countries suggest that in some countries, 50% of Down syndrome children are born to mothers over 40 years (Alwan & Modell, 1997). The overall incidence of Down syndrome in Dubai during a 5-year period (1999–2003) was 2.2 per 1000 live births; with a high rate of 3.13 per 1000 among 24,250 UAE nationals and a lower rate of 1.66 per 1000 among 39,148 non-UAE nationals (Murthy et al., 2007). Another study among 24,233 births in Abu Dhabi gave a prevalence rate for Down syndrome of 2.06/1000 (al Talabani et al., 1998)

In Nizwa, Oman, 44 cases of Down syndrome were diagnosed among 21,988 births (2/1000; Sawardekar, 2005). Among 30,076 Qatari babies born in the country, 65 cases of Down syndrome were diagnosed, giving a rate of 1.83/1000 live births. This high rate was attributed to the tendency toward increased maternal age, with a mean maternal age of 35.4 years and a mean paternal age of 40 years (Wahab et al., 2006).

The observed prevalence of Down syndrome among 3000 births in Egypt was 1.3/1000 total births (Temtamy et al., 1998a) in one study, and 1.8 per 1000 births in another study (Hafez et al., 1983).

In Libya, Down syndrome was identified in 1.7/1000 live births in a report where, over a period of two years, 32,332 live-born infants were screened for the presence of identifiable congenital malformations (Mir et al., 1992), while in Bahrain, Down syndrome prevalence rate was reported to be 1.14 per 1000 births (Al-Arrayet, 1987).

Most of the literature studying the association of DS with consanguinity concluded that DS is not related to consanguinity. A few studies gave controversial results, such as the study on 20 cases in Kuwait reporting that the incidence of DS among offspring of consanguineous parents was higher than among offspring of nonconsanguineous parents, putting forward a hypothesis that there is a gene that influences meiotic or mitotic nondisjunction (Alfi et al., 1980). This hypothesis was not verified in many subsequent studies, such as the study in Turkey showing that parental consanguinity and the inbreeding coefficient were lower among parents of 1598 patients with trisomy 21 than in control parents (Basaran et al., 1992). Similarly, studying parental consanguinity among 85 cases of DS in Palestinian Arabs (Zlotogora, 1997b) and 83 cases in Iraq (Hamamy et al., 1990) did not reveal an association between trisomy 21 and consanguinity. Furthermore, no increase in the frequency of consanguinity among maternal and/or paternal grandparents was detected to support the presence of an autosomal recessive gene causing meiotic nondisjunction (Zlotogora, 1997b).

SPECIFIC SYNDROMES

MALFORMATION SYNDROMES (INCLUDING BRAIN MALFORMATIONS)

Bardet-Biedel Syndrome (BBS)

This is a rare autosomal recessive disorder characterized by obesity, retinitis pigmentosa, polydactyly, renal failure, and delayed development (Figure 52-1). At the molecular level it is highly heterogeneous, with more than 11 genes identified so far. The prevalence of this syndrome among Arabs, particularly Arab Bedouins in Kuwait, is relatively high, affecting 1 in 13,500 individuals (Farag & Teebi, 1989) compared to 1 in 150,000 individuals in North America and Europe. A study of the prevalence of commonly diagnosed autosomal recessive disorders in Oman from a hospital-based register in the years 1993–2002 revealed 14 cases of BBS, giving a prevalence of 1 in 30,000 births (Rajab et al., 2005a).

Although common in almost all Arab countries, the molecular basis of this syndrome has not been established in the Arabs except in Lebanon, Tunisia, and Saudi Arabia. Homozygosity mapping in a large Lebanese family identified a new BBS locus (BBS10) with a mutation (S311A) in *FLJ23650* gene in affected individuals (Stoetzel et al., 2006). Molecular study on 19 Tunisian families with 34 individuals affected with BBS revealed 7 different homozygous mutations; 5 of them were new: C448T (R146X) in exon 5 of BBS1, C565T (R189X) in exon 5 of BBS2, 123delA (novel), C1436G (S479X), a large novel deletion of 20 Kb from exon 9 through 15 0f BBS7, insertion of 16-bp in exon 4 of BBS8 (novel), 459+1G>A (novel; see Smaoui et al., 2006). In Jordan, 20 families with multiple members affected with BBS were seen at the genetic counseling clinic in Amman over a period of 5 years. The molecular diagnosis was reached in only one patient that had a mutation in BBS 4).

Among 19 patients with BBS in seven Saudi families, seven novel mutations were identified; three in BBS1, three in BBS3 and one in BBS4. Only one BBS gene was sequenced per family, which is the one that falls within a block of homozygosity shared between affected individuals. The splicing mutations, which represented more than 70% of all mutations identified in this study, resulted in exon-skipping in BBS1 (c.136+1G>A and c.178-2A>G) or intron inclusion in BBS4 (c.963+58C>T). One family had a mutation in BBS1 that resulted in an aberrant transcript in which the entire intron 10 is retained. A mutation in BBS3 led to a 4kb deletion that starts in intron 7 and ends in intron 8 (c.732+1952_899-3806del4139). The two missense mutations identified in BBS3 (c.266C>T and c.431C>T) were not found in 192 control Saudi individuals, and they both replace highly conserved amino acids. This study also revealed interesting phenotypic aspects of BBS, including the first report of nonsyndromic retinitis pigmentosa as a novel BBS phenotype (AbuSafieh et al., 2009).

Cohen Syndrome

This autosomal recessive syndrome was first described in 1973. It is characterized by distinctive face with prominent central incisors, a prominent nasal bridge, and a short philtrum with maxillary and mandibular hypoplasia. There is usually microcephaly, short stature, and

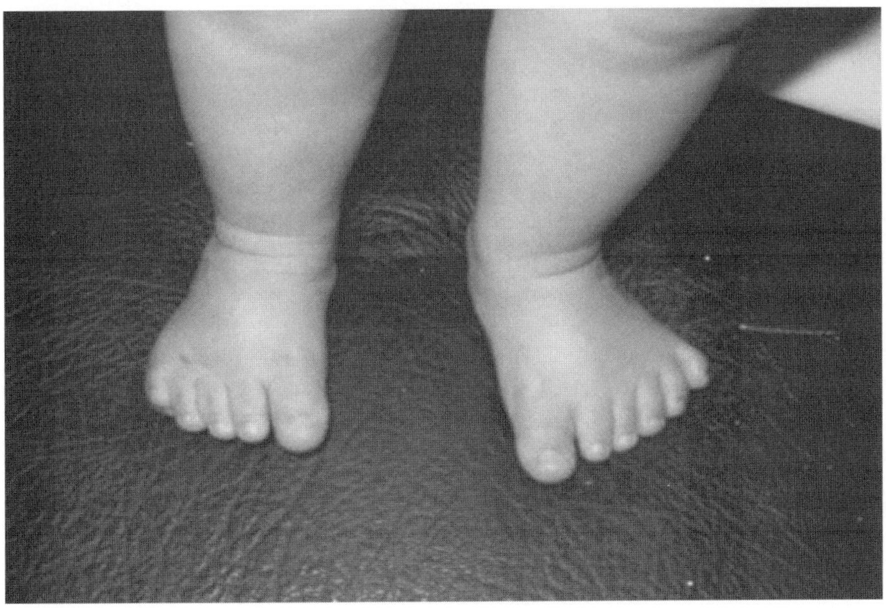

Figure 52-1 Postaxial polydactyly in Bardet-Biedel syndrome.

truncal obesity. The fingers are characteristically long and tapered. Ophthalmological features and neutropenia could also be part of the syndrome. It is caused by mutations in COH1 gene.

Cohen syndrome has been reported in different Arab countries, including Lebanon (Horn et al., 2000; Hennies et al., 2004), Oman, Saudi Arabia (Mochida et al., 2004; Hennies et al., 2004), and the United Arab Emirates (Chandler et al., 2003). The families from UAE originated from Lebanon and Syria. Molecular study revealed different mutations in COH1 gene in these families. These mutations included: homozygous c.6530_6732del (p.Val2245fsX16), homozygous c.1225G>T (p.Glu409X) (Taban et al., 2007), homozygous c.7934G>A (p.Gly2645Asp), homozygous c.9406-1G>T (p.Tyr3136fsX16) (Hennies et al., 2004), homozygous 1219C>T (Q407X) (Mochida et al., 2004).

Sanjad-Sakati Syndrome (Hypoparathyroidism-Retardation-Dysmorphism, HRD)

This autosomal recessive syndrome is characterized by prenatal and postnatal growth retardation, mental retardation, hypoparathyroidism, and dysmorphic features that include microcephaly, deep-set eyes, micrognathia, and large floppy ears. It is known to be common in Arabs, although the exact incidence is not known. It is particularly common in Bedouins of the Gulf region, but has been reported also in Palestine, Egypt, and Jordan (Draz, 2006; El-Sawy et al., 2004; Albaramki et al., manuscript accepted for publication). This syndrome is allelic to the AR form of Kenny-Caffey syndrome. Both syndromes are caused by mutations in TBCE gene (Parvari et al., 2002; Hershkovitz et al., 2004). Most Arab patients with this syndrome have a single 12bp deletion in the second coding exon of the TBCE gene (Hershkovitz et al., 2004). Preimplantation diagnosis was successful in a Saudi Arabian family (Hellani et al., 2004).

Hennekam Syndrome

This is an autosomal recessive syndrome characterized by lymphangiectasia, severe peripheral lymphedema, facial anomalies, seizures, mild growth retardation, and variable mental retardation (Hennekam et al., 1989; Figure 52-2). Six children from 5 different families (4 published) with this syndrome were evaluated. One was from UAE, 2 from Oman, 2 from Palestine, and 1 from Tunisia. All had flat midface with hypertelorism, but only 2 of them had dysplastic ear with atretic ear canal (Al-Gazali et al., 2003b and unpublished data). Intestinal lymphangiectasia was present in 5 of the 6 patients. Three of the six children had mild/borderline mental retardation, and 1 had moderate retardation (Al-Gazali et al., 2003b and unpublished data). Molecular studies identified a homozygous mutation in *CCB1* gene in 1 of these families (Alders et al., 2009).

Multiple Pterygium Syndrome (MPS)

This autosomal recessive syndrome is characterized by multiple joint contractures including camptodactyly, talipes equinovarus and rocker-bottom feet, short and webbed neck. The face is relatively immobile with downturned corners to the mouth, epicanthic folds and mild ptosis. Webs develop later on in childhood usually at the neck, axillae, antecubital, groin, and popliteal regions. This syndrome has been reported in patients from UAE, Oman (Rajab et al., 2005b), Saudi Arabia, Palestine, Kuwait, and Jordan.

While reviewing data of a community genetic survey at a district hospital which serves a mixed Arab population

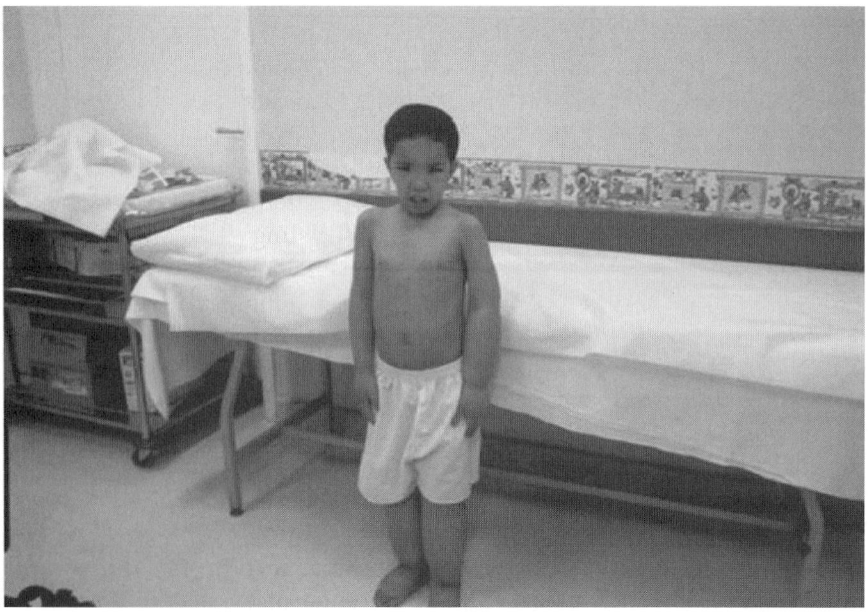

Figure 52-2 Generalized lymphedema and flat face in Hennekam syndrome.

of 400,000 in Kuwait, Teebi and Daoud (1990) ascertained the presence of 13 cases of MPS in 6 sibships in 4 Arab families. The authors estimated the minimum prevalence MPS in the general population of that district as approximately 1: 31,000, indicating the relatively high occurrence of this syndrome in the Arab population.

Morgan et al. (2006) reported a large consanguineous family from Saudi Arabia with both lethal and nonlethal variants of MPS. All affected individuals were found to be homozygous for missense mutation (Val1107Gly) in the embryonal acetylcholine receptor subunit gene (*CHRNG*).

Two children from a UAE family of Pakistani origin had the typical features of this syndrome. Genetic study on these children revealed a homozygous mutation in the *CHRNG* gene (IVS4-9T>C; Morgan et al., 2006).

Six children from 2 Omani families from the same tribe were reported to have MPS (Rajab et al., 2005b) with a homozygous mutation in the CHRNG gene detected in an affected individual in the second family (Hoffmann et al., 2006).

Weill-Marchesani Syndrome

This is a rare autosomal recessive syndrome characterized by face and eye abnormalities, short stature with brachydactyly, and joint stiffness. Eye abnormalities include ectopia of the lens, severe myopia, glaucoma, and microspherophakia. This syndrome has been reported from Lebanon, Saudi Arabia, and Oman. Three cases from 2 sibships of a consanguineous family with this syndrome were reported from Lebanon (Megarbane et al., 2000). Clinical features included limitation of joint movement, microspherophakia, dislocated lenses, glaucoma, and severe myopia. Two other families were reported from Saudi Arabia and from Oman (Dagoneau et al., 2004a; Riad et al., 2006). Molecular study on the 2 families from Lebanon and Saudi Arabia identified 2 different mutations in the ADAMTS gene in these families (Dagoneau et al., 2004a).

Donnai-Barrow Syndrome

Donnai -Barrow syndrome is a rare autosomal recessive disorder characterized by hypertelorism with down-slanting palpebral fissures, diaphragmatic hernia, exomphalos, absent corpus callosum, myopia, and sensorineural hearing loss. Other ophthalmological features include iris coloboma and retinal detachment. Intellectual impairment is variable. Proteinuria is a feature. There is overlap with the facio-oculo-acoustico-renal syndrome, which was found to be allelic. Both syndromes are caused by mutations in the low-density lipoprotein receptor 2 gene (LRP2; Kantarci et al., 2007). This syndrome has been reported from the UAE, Qatar, and Saudi Arabia (Kantarci et al., 2007; Gripp et al., 1997). Molecular study on the family from UAE showed a missense mutation in the LRP2 gene (c. 756T>C, p.Y2522H), and a homozygous frameshift mutation was found in the Qatari family (c.9358_9359delAG, p.S3120Wfs26) (Kantarci et al., 2007).

Robinow Syndrome

The face in this syndrome resembles that of a fetus. There is usually a prominent forehead, hypertelorism, wide mouth with gum hyperplasia, and short small nose with anteverted nostrils. Other features include mesomelic limb shortening, micropenis, and urinary tract abnormalities. Two modes of inheritance have been described in this syndrome, autosomal dominant and autosomal recessive. The phenotype in the recessive type is more severe, with more pronounced short stature and with severe brachymelia associated with multiple rib and vertebral anomalies (Figure 52-3). It is caused by homozygous loss of function mutations in the gene encoding receptor orphan receptor tyrosine kinase 2 (ROR2). This syndrome has been reported from Kuwait, Oman, Saudi Arabia, UAE, and Jordan, but there are no prevalence data available (Teebi, 1990; Soliman et al., 1998; Nazer et al., 1990, Al Talabani et al., 1998; Hamamy unpublished data). However, a study from Oman on the prevalence of commonly diagnosed AR disorders from 1993–2002 revealed a prevalence of 1/35,000 births (Rajab et al., 2005a). In 14 patients with recessive Robinow syndrome from 7 families originating from Oman, a gln502-to-ter (Q502X) nonsense mutation in exon 9 of the ROR2 gene was identified that removed the tyrosine kinase domain and all subsequent 3-prime regions (Afzal et al., 2000).

Joubert Syndrome

Joubert syndrome is an autosomal recessive developmental brain disorder characterized by congenital cerebellar ataxia, hypotonia, oculomotor apraxia, mental retardation, and variably associated retinal dystrophy, nephronophthisis, liver fibrosis, and polydactyly. The neuroradiological hallmark of the disorder is a malformation of the midbrain–hindbrain junction known as the "molar tooth sign," consisting of cerebellar vermis hypoplasia or dysplasia, thick horizontally-oriented superior cerebellar peduncles, and abnormally deep interpeduncular fossa. This syndrome is included in the newly emerging group of "ciliopathies." It is seen frequently in the Arabs, with an estimated birth prevalence in the UAE of 1 in 5000 (Al-Gazali et al., 1999a). Forty cases from 17 families with Joubert syndrome were evaluated at the clinical and molecular level. These families were either of UAE, Omani, Palestinian, or Jordanian origin. Using homozygosity mapping, several genes and mutations were identified in some of these families (AHI,;

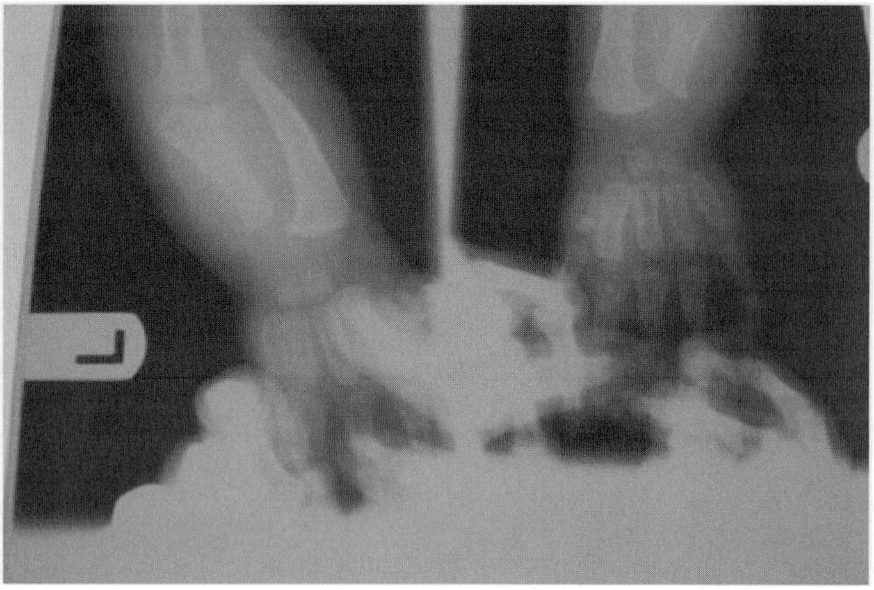

Figure 52-3 X-ray of the forearm in Robinow syndrome showing hypoplastic radius and ulna.

787insC (fsX270), CEP290; 5668G>T (G1890X), INPP5E, c.1537C>T (p.R512W), c.1546C>T (p.R515W), c.1691G>A (p.R563H) (Dixon-Salazar et al., 2004; Valente et al., 2006; Biels et al., 2009).

Meckel Syndrome

This autosomal recessive syndrome is characterized by occipital meningoencephalocele, infantile polycystic kidneys, with multicystic dysplasia and fibrotic changes in the portal area of the liver and postaxial polydactyly. It is commonly seen in Arabs. It has a birth prevalence of 1/3500 live births in Kuwait and 1/5000 live births in the UAE (Teebi & Teebi, 2005; Al-Gazali et al., 1999a). It is also quite common in Palestinian Arabs, with a prevalence of 1/2000 (Teebi & Teebi, 2005). The prevalence in Oman was estimated to be 1/50,000 births (Rajab et al., 2005a). It has also been reported from Saudi Arabia and Tunisia (Ramadani & Nasrat 1992; Boutheina et al., 2000). One UAE family originally from Oman was studied at the molecular level and was found to have a homozygous mutation in mecklin (*MKS3*) gene (383-384delAC, p.H128fsX140 (Smith et al., 2006).

Woodhouse–Sakati Syndrome

Two inbred Saudi Arabian families with 6 patients affected with hypogonadism, sparse hair, diabetes mellitus, mental retardation, and mild deafness were described (Woodhouse & Sakati 1983). Subsequently, several other families of different ethnic origin, including a Jordanian family from Kuwait, were described (Al-Awadi et al., 1985b; Oerter et al., 1992; Gül et al., 2000). The condition has been mapped to 2q33-q35, and mutations found in C2orf37 in several Saudi patients and in the original Woodhouse et al. family (Alazami et al., 2008).

Teebi Hypertelorism Syndrome

This autosomal dominant syndrome was described originally by Teebi in 1987 in 16 individuals from a 4-generation Arab family. Affected individuals had hypertelorism, a broad forehead with a widow's peak, a mild antimongoloid eye slant, ptosis, and a broad nasal bridge. The midface was hypoplastic with a hypoplastic maxilla, a long philtrum, and a thin upper lip. The hands were short and broad with 5th-finger clinodactyly. Males were reported to have a shawl scrotum. The intelligence and stature are normal. Several other families of different ethnic origin were subsequently reported (Tsai et al., 2002; Machado-Paula & Guion-Almeida, 2003; Han et al., 2006).

Faciodigital Syndrome, Teebi Type

Teebi et al. (1988) described three brothers and two sisters, the offspring of consanguineous Arab parents, with a condition resembling Aarskog syndrome. The main features of the condition were short stature, a short nose, low-set ears with an over-folded upper helix, a triangular face, and coarse, dry, hypopigmented hair. The males had a degree of shawl scrotum. Further cases were reported subsequently (Teebi & Al-Awadi, 1991)

Bifid Nose Associated with Renal Agenesis and Anorectal Malformations

Al-Gazali et al. (2002) reported a consanguineous Arab family from Egypt in which 4 sibs had bifid nose associated with renal and anorectal anomalies (Figure 52-4). All had median nose clefts, wide bulbous nasal tip, short philtrum, but no hypertelorism. All had renal agenesis, which was unilateral in 3 and bilateral in 1. Development was normal for all the affected children. Two more families were identified, one of Pakistani origin and one of Afghani origin, and molecular study identified a homozygous mutation in *FREM1* gene [c.2721delG (p.V908fsX17)] (Alazami et al., 2009).

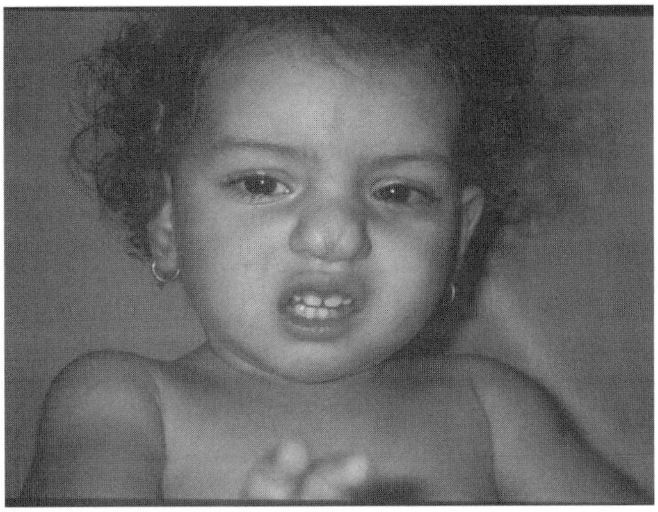

Figure 52-4 Bifid nose in the syndrome of Bifid nose, renal agenesis and anorectal malformation.

Anterior Segment Anomalies of the Eye Associated with Skeletal Abnormalities and Early Lethality (Al-Gazali Syndrome)

Al- Gazali et al. (1994) reported two sibs from a consanguineous Sudanese family with anterior chamber abnormalities, with clouding of the cornea and skeletal abnormalities. There was arachnodactyly with clinodactyly of the 5th finger, and dislocation of some of the interphalangeal joints. Radiographs revealed radio-ulnar synostosis, bowing of forearm bones, and long bowed fibulae (Figure 52-5a,b,c). Other features included talipes equinovarus, cleft lip and palate (in one case), intrauterine growth retardation, and congenital heart disease. Both children died in the first two months of life. Two cousins

(A)

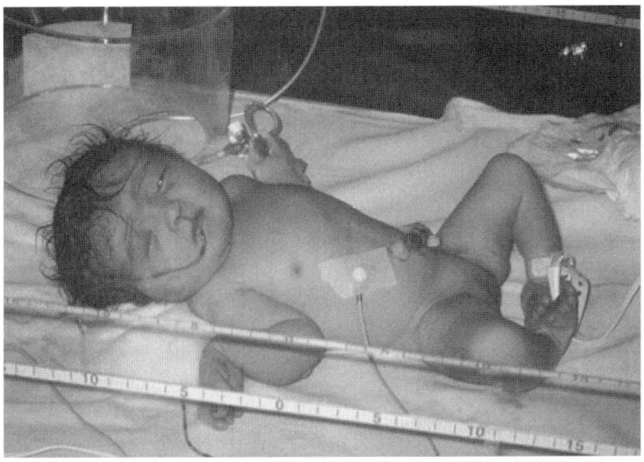

(B)

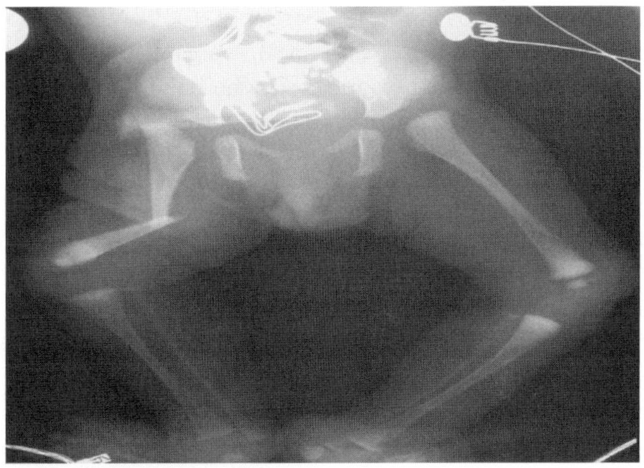

(C)

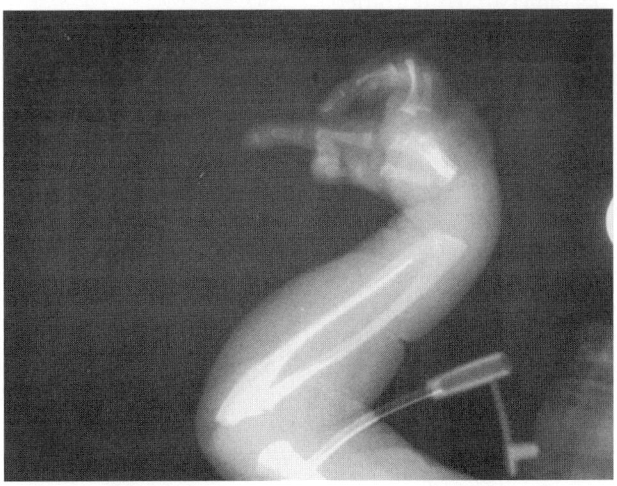

Figure 52-5 (A) Neonate with Al-Gazali syndrome showing widely spaced eyes, small mouth and bilateral wrist drop with joint. (B) X-ray of the lower limbs in Al-Gazali syndrome showing fracture of the femur and thin and long fibula. (C) Radio-ulnar synostosis in Al-Gazali syndrome.

of the sibs had a similar condition. Two further sibs from a consanguineous family were reported by Al-Gazali et al. (1999b). Multiple fractures secondary to thinning of the bones was part of the condition. One case had a hypoplastic inferior cerebellar vermis, and hyperechogenic areas in the basal ganglia.

Two sibs from a consanguineous Malaysian family with this syndrome were subsequently reported (Thong et al., 2005).

Agenesis of Corpus Callosum-Coloboma-Mental Retardation (Temtamy Syndrome)

Temtamy et al. (1996) reported 2 sibs from a consanguineous Egyptian family with macrodolichocephaly, a narrow face, micrognathia, low-set simple ears, hypertelorism, colobomata of the iris, retina, and choroid, a beaked nose, and crowding, malposition, and hypoplasia of the teeth. A brain CT scan in all cases showed agenesis of the corpus callosum with dilatation of the fourth ventricle. In two cases, cardiovascular examination revealed a dilated aorta and early aortic regurgitation. One sib died at 22 years from sudden heart failure. Chan et al. (2000) reported a similar case. Talisetti et al. (2003) reported a 5-year-old girl with features of the condition who had a (2p24;9q32), apparently balanced, de novo translocation. Other similar cases were reported (Ramocki et al., 2003; Li et al., 2007).

Temtamy Preaxial Brachydactyly Hyperphalangism Syndrome

Temtamy et al. (1998b) reported a child from a consanguineous Egyptian family with unusual facies, mental retardation, deafness, and unusual hand, mouth malformations and dental anomalies consisting of prominent talon cusps of the upper central incisors. The face was rounded with midface hypoplasia, apparent hypertelorism, thin arched eyebrows, synophrys, a pinched nose, a small mouth, micrognathia, and low-set ears. There was a shawl scrotum with retractile testes. There was short stature with mild microcephaly and learning difficulties. The hands were short with proximally placed abducted thumbs, radial deviation of the fingers, camptodactyly, and partial syndactyly and symphalangism of the radial digits. Radiographs showed a short oval-shaped first metacarpal with a delta epiphysis, two triangular ossification centers of the proximal phalanx of the thumb, short second metacarpals, an accessory phalanx of the second proximal phalanx, and short middle phalanges. Similar abnormalities were found in the feet. There was a similarly affected brother. Clarkson et al. (2004) reported a case from a consanguineous family with similar anomalies but thought the child had Catel-Manzke syndrome. Temtamy (2005) suggested that this case was similar to the syndrome she had described. In two Egyptian families, a homozygous 1-bp deletion c.14delG and a homozygous 30-bp deletion, c.44_73del30, were located in exon 1 of the TPBS1 gene (Li et al., 2010a). A different mutation in the same gene was identified in a Jordanian family with two affected sibs (Tian et al., 2010) (Figure 52-6).

Severe Hypertelorism, Midface Prominence, Prominent/Simple Ears, Severe Myopia, Borderline Intelligence, and Bone Fragility

Three siblings, born to double-first-cousin Jordanian Arab parents, presented with a new autosomal recessive syndrome of severe hypertelorism with up-slanted palpebral fissures, brachycephaly, abnormal ears, sloping shoulders, enamel hypoplasia, osteopenia with repeated fractures, severe myopia, mild to moderate sensorineural hearing loss, and borderline intelligence (Hamamy et al., 2007c; Figure 52-7). Similar cases were identified in Turkey (Kayserili et al., 2010.).

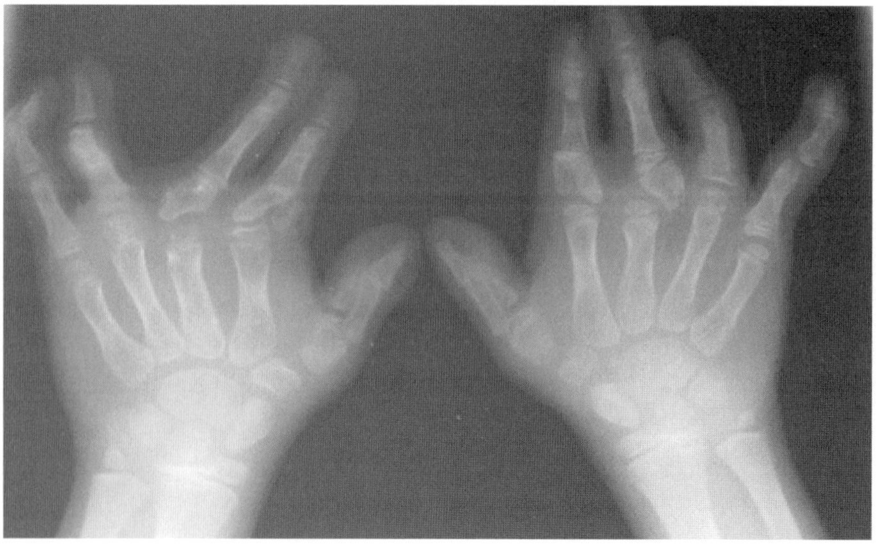

Figure 52-6 Temtamy preaxial brachydactyly hyperphalangism syndrome.

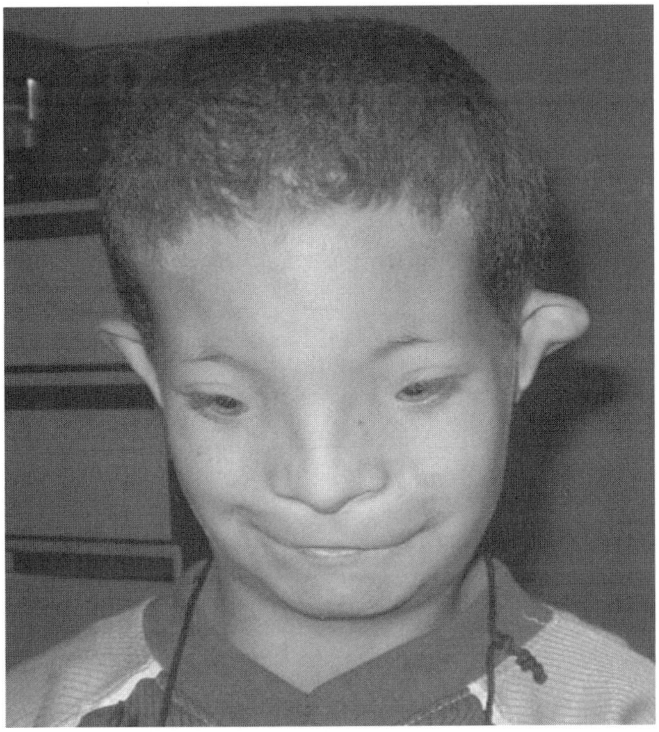

Figure 52-7 Severe hypertelorism, midface prominence, prominent/simple ears, severe myopia, borderline intelligence, and bone fragility.

Cenani-Lenz Syndrome

This syndrome was originally described by Cenani and Lenz in 1967, in 2 brothers who had a form of syndactyly similar to that seen in Apert syndrome, with less severe involvement of the feet. In addition, the brothers had severe shortening of the ulna and radius, fusion of the metacarpals, and disorganization of phalangeal development. Temtamy et al. (2003) reported 2 inbred Egyptian families with features of this syndrome who, in addition, had facial dysmorphic features that included a high, broad and prominent forehead, hypertelprism, depressed nasal bridge with short nose, and prominent short philtrum.

A female child from Jordan with Cenani–Lenz syndactyly showed bilateral malformations of the hands and feet, comprising a complex syndactyly and disorganization of the fingers and partial disorganization of the toes. The patient also had a prominent forehead, deep-set eyes, low-set ears, retrognathia, high arched narrow palate, a short beaked nose, and high nasal bridge (Jarbhou et al., 2008; Figure 52-8). By sequencing candidate genes, recessive LRP4 mutations in 12 families with CLS were identified (Li et al., 2010b).

SKELETAL DYSPLASIAS

Ellis-van-Creveld Syndrome

The cardinal features of this autosomal recessive syndrome are postaxial polydactyly, mostly involving the hands but also occasionally involving the feet, meso/acromelic shortening of the limbs, small deep-set nails, multiple oral frenula, and a congenital heart defect, predominantly an ASD. Mutations in EVC1 and EVC2 genes have been reported in the majority of patients with Ellis-van-Creveld syndrome.

This syndrome is seen frequently in the Arabs. It has been reported from Egypt, Jordan, Saudi Arabia (Mostafa et al., 2005; Reddy & Madenlioglu, 1967), Oman, and UAE. Rajab et al. (2005a) estimated the birth prevalence in Oman to be 1 in 25,000 births, while in the UAE it was 0.52/10,000 births (Al-Gazali et al., 2003a).

Five children from 4 families in the UAE with this syndrome were studied at the clinical and molecular level. Molecular study revealed a homozygous mutation in exon 13 of EVC1 Q605X in one family, a homozygous deletion (c.981delG) in exon (8) of EVC2 in the second, and splice-site homozygous change (IVS13-1G>T) in exon 14 of EVC2 gene in the third, with no mutation identified in the fourth family (Ali et al. submitted).

Stuve-Wiedemann Syndrome

The syndrome is characterized by camptomelia, camptodactyly, and pursing of the mouth on stimulation (Figure 52-9a). Radiological features include short, bowed

(A)

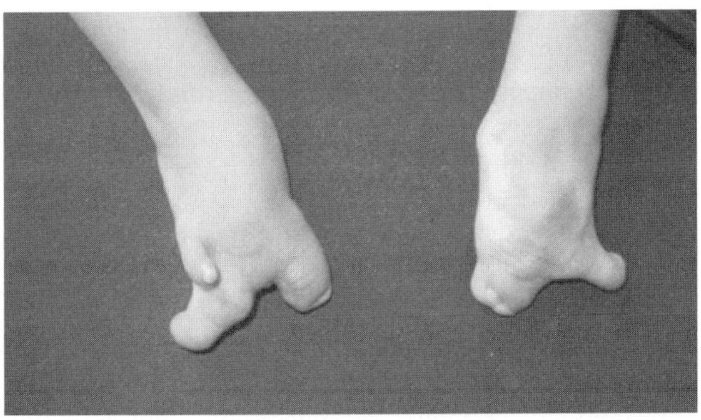

(B)

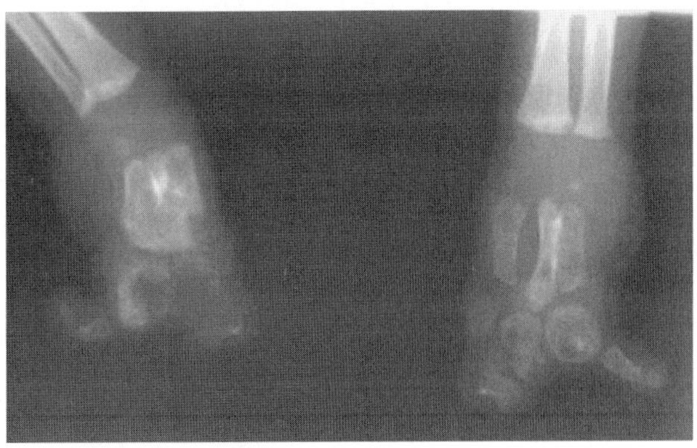

Figure 52-8 (A) Hands in Cenani-Lenz syndrome. (B) Hands X-ray in Cenani-Lenz syndrome.

(A) (B) (C)

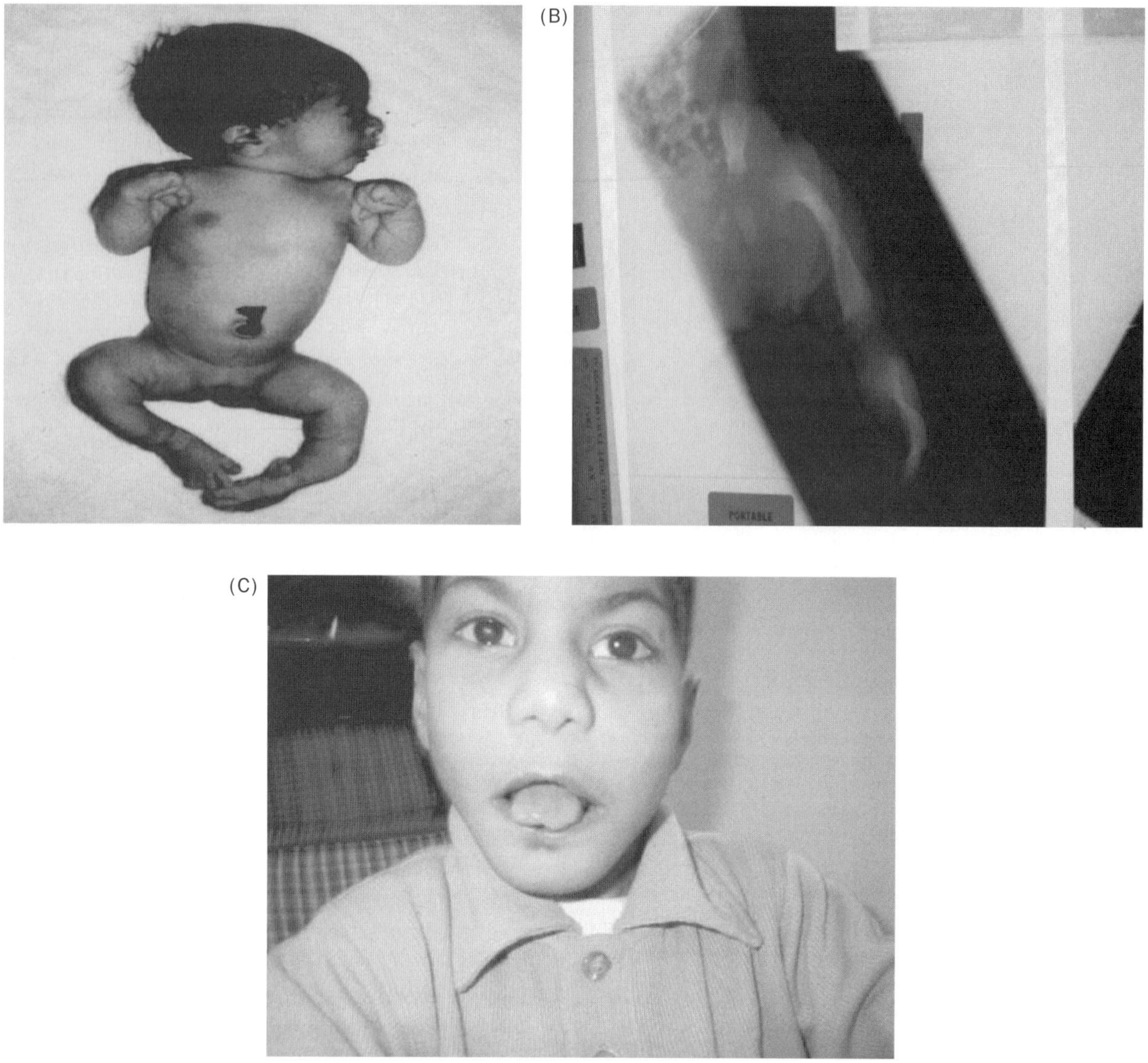

Figure 52-9 (A) Stuve-Wiedemann syndrome in the neonatal period showing camptomelia and camptodactyly. (B) X-ray of the lower limb in Stuve-Wiedemann syndrome showing bowing of the femur and tibia with increased thickness in the concavity of the curve. (C) Stuve-Wiedemann syndrome in childhood. Note smooth tongue and corneal clouding due to keratitis.

bones with increased thickness of the bone in the concavity of the curve, and wide osteoporotic metaphyses with striation (Figure 52-9b). The course of the disorder is complicated by hyperthermia, and feeding and swallowing difficulties leading to frequent aspirations and respiratory problems. Most children die in the first year of life. However, if they survive they develop progressive spinal deformity associated with insensitivity to pain, leading to self-mutilating behavior and corneal scarring (Al-Gazali et al., 2003c; Figure 52-9c). At the molecular level this syndrome is caused by mutations in the leukemia inhibitory factor receptor gene (LIFR; Dagoneau et al., 2004b).

This rare autosomal recessive syndrome is relatively common in the UAE (Al-Gazali et al., 1996a), but has not been reported from other Arab countries. Forty cases from 25 families have been evaluated. Most of these families originated from Oman and Yemen. Molecular study revealed a founder mutation in the leukemia inhibitory factor receptor gene (LIFR) [653_654 insT at exon 6] in these families (Dagoneau et al., 2004b; Al-Gazali unpublished data).

Microcephalic Osteodysplastic Primordial Dwarfism Type II

This syndrome is characterized by severe prenatal and postnatal growth retardation, microcephaly, a prominent nose, micrognathia, mental retardation, and mesomelic limb shortening (Figure 52-10). Radiographs show short first metacarpals, short middle phalanges, short, curved, forearm bones, a V-shaped flare of the distal femoral metaphyses, coxa vara with a fragmented and flattened proximal femoral epiphysis, triangular distal femoral epiphyses, a high and narrow pelvis, and pseudoepiphyses of the hands.

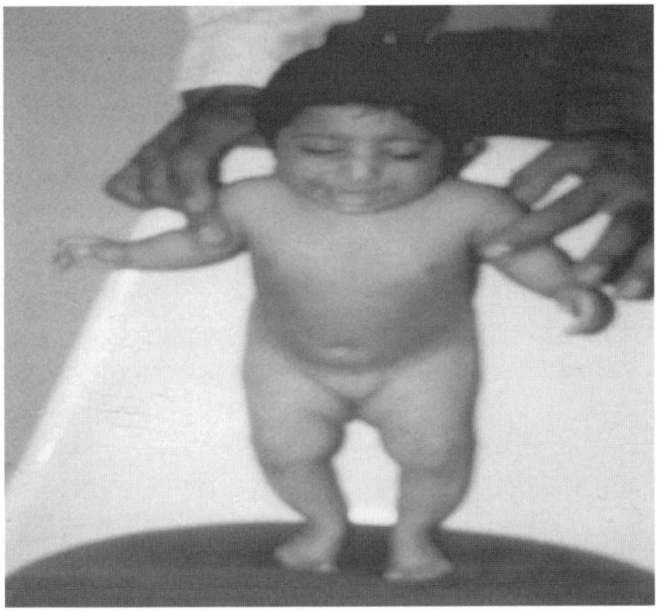

Figure 52-10 Microcephalic osteodysplastic primordial dwarfism type II.

This disorder is seen in the UAE, most commonly in families of Omani and Pakistani origin. Six children from 6 families have been evaluated. Using homozygosity mapping, the gene and mutations were identified in some of these families. Two mutations in the pericentrin gene were identified in some of these families c.5767C>T (p.R1923X), c.1336C>T (p.Q446X) (Rauch et al., 2008).

Desbuquois Syndrome

This bone dysplasia is characterized by joint laxity with multiple dislocations and flat face similar to Larsen syndrome (Figure 52-11a). Radiological changes are distinctive. There are supernumerary phalanges, characteristically situated between the metacarpal and proximal phalanx of the index finger, osteoporosis, a short narrow thorax, metaphyseal enlargement, platyspondyly, and prominence of the lesser trochanter, giving the appearance of Swedish key (Maroteaux et al., 1986; Figure 52-11b). Ossification in the carpal centers may be advanced, whereas the epiphyses of the long bones can have retarded development.

This syndrome has been reported from Morocco, Tunisia, and UAE (Gillessen-Kaesbach et al., 1995; Al Kaissi et al., 2005; Al-Gazali et al., 1996b). Affected members of the Arab Bedouin family reported by Al-Gazali et al. (1996b) had typical features of this syndrome including a midface hypoplasia and joint laxity. This was probably the first report on Desbuquois syndrome in Arab Bedouins. Using homozygosity mapping, the gene responsible for Desbuquois syndrome was mapped to chromosome 17q25.3 with a possible genetic homogeneity of the clinical subtype with hand anomalies (Faivre et al., 2003) and recently, a mutation in exon4 of *CANT1* gene was identified in this family (Huber et al., 2009). Another consanguineous family from the UAE, with one child affected with Desbuquois syndrome without hand anomalies, was also seen (Al-Gazali unpublished data).

Limb/Pelvis/Hypoplasia/Aplasia Syndrome

This syndrome was originally reported from Kuwait in a non-Bedouin Palestinian family (Al-awadi et al., 1985b), followed by a more complete report in a second Arab family by Raas-Rothschild et al. (1988). The malformations consist of hypoplasia of the proximal part, or marked angulation and hypoplasia, of the femur, absence of the fibula, and a malformed hypoplastic ulna (Figure 52-12). There is usually monodactyly. Both lower limbs are usually involved and may have the appearance of small, stick-like projections with very hypoplastic or absent feet. Molecular study in a family, originally Syrian Bedouins, who had 3

(A)

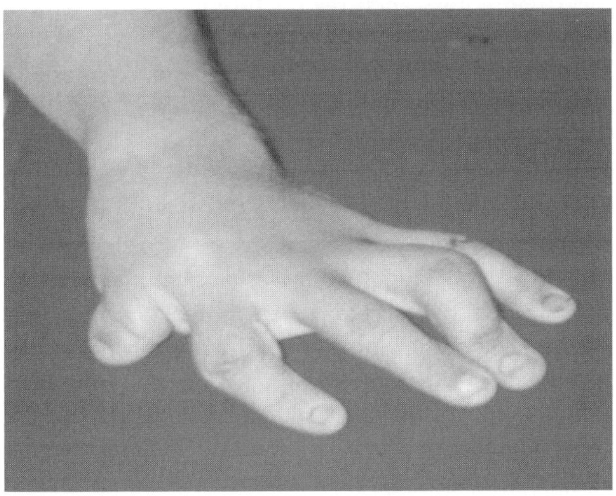

(B)

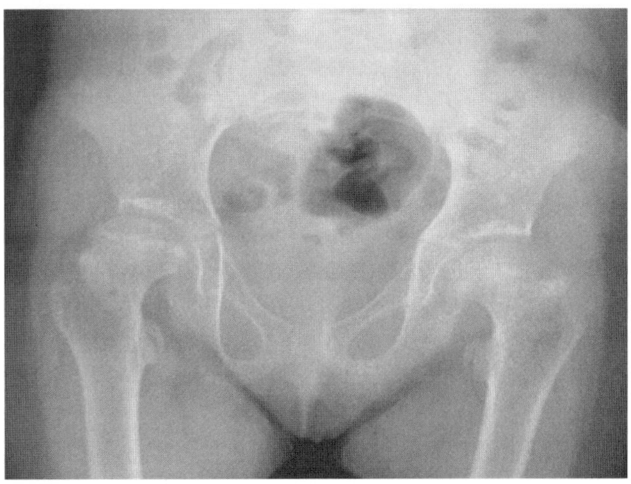

Figure 52-11 (A) The hand in Desbuquois syndrome. Note multiple joint dislocations, camptodactyly and abnormal thumb. (B) X-ray of the hip in Desbuquois syndrome showing prominent lesser trochanter (Swedish key appearance).

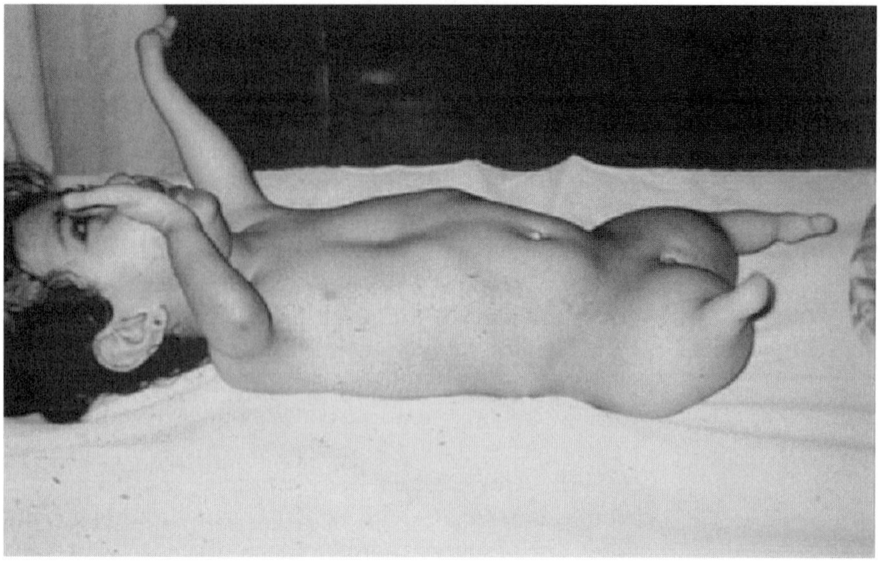

Figure 52-12 Limb/pelvis/hypoplasia/aplasia syndrome.

children in 2 branches with the typical phenotype of this syndrome, revealed a homozygous missense mutation in exon 4 of *WENT7A* gene (c.1179C>T) leading to substitution of arginine to cysteine (Woods et al., 2006).

Dyggve-Melchior-Clausen Syndrome

The main features of this autosomal recessive bone dysplasia are mental retardation, short stature, progressive sternal bulging, and delayed bone age. The skeletal changes include flattening of the vertebral bodies with radiolucent depression in their midportion. The vertebral bodies become progressively flat and there is a lacy appearance of the iliac crest on X-ray of the pelvis. The femoral heads are delayed in their ossification and they might be fragmented. Progressive hip dislocation might occur. Many cases have severe mental retardation; however, the degree of retardation can be variable. This syndrome is more common in Lebanon than elsewhere (Afifi et al., 1974; Naffah and Taleb, 1974). It has also been reported from Egypt, Palestine, Tunisia, and Morocco (Hosny & Fabry, 1998; Sbihi et al., 1980; Schorr et al., 1977). Molecular study showed several homozygous mutations in Dymeclin gene in families of Moroccan origin, Tunisian origin, and Lebanese origin [c.610C>T, c.1447C>T, c.1877delA, IVS10 1125+1G>T, c.580C>T, IVS 11 1252-1G>A]. One family of Moroccan origin was double-heterozygous for 2 mutations [656T>G/1877delA] (El Ghouzzi et al., 2003).

Wolcott-Rallison Syndrome

This rare autosomal recessive disorder is characterized by permanent diabetes developing in the neonatal period or early infancy, requiring insulin, and spondyloepiphyseal dysplasia developing at a later stage. The syndrome results from mutations in the gene encoding the eukaryotic translation initiation factor 2-kinase 3 (EIF2AK3; see Delepine et al., 2000).

It has been reported from Kuwait, UAE, Saudi Arabia, and Tunisia (Marafie et al., 2004; Al-Gazali et al., 1995b; Bonthron et al., 1998; Abdelrahman et al., 2000; Bin-Abbas et al., 2002; Nicolino et al., 1998). Several mutations in the EIF2AK3 gene were identified in Arab families of different origins (Brickwood et al., 2003; Delepine et al., 2000).

Spondylometaepiphyseal Dysplasia (SMED), Short Limb-Abnormal Calcification Type

SMED short limb-abnormal calcification type is a rare autosomal recessive type of SMED, initially reported by Borochowitz et al. (1993). It is characterized by short stature with short limbs and very short hands, narrow chest, and normal intelligence (Figure 52-13a,b). Radiological abnormalities include short tubular bones with very abnormal metaphyses, and epiphyses with premature calcification of cartilaginous structures including epiphyses and costal cartilage, short ribs, and platyspondyly. Abnormal calcifications are a diagnostic feature of this dysplasia. Many children with this disorder die in childhood due to respiratory and neurological complications (Al-Gazali et al., 1996c). It is caused by mutations in the DDR2 gene (Bargal et al., 2009). Most patients reported with this disorder were Arabs from Egypt and Palestine, and Middle Eastern Jews from Iraq and Egypt; there was one family from Algeria (Bargal et al., 2009). One mutation was identified in all the Palestinian families and one Egyptian family (c.2254C>T [R752]). The Algerian family had a different mutation (c.2177T>G [1726R] (Bargal et al., 2009).

Fibrochondrogenesis

This rare lethal chondrodysplasia is characterized by very short long bones, particularly the upper segment, with

(A) (B)

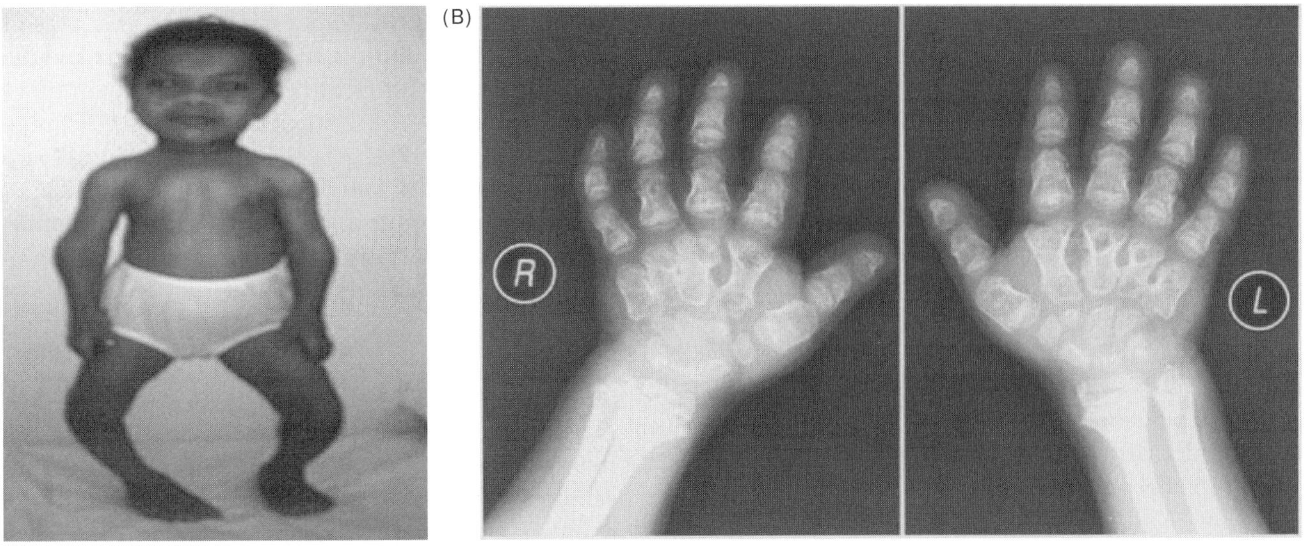

Figure 52-13 (A) SMED-abnormal calcification type. (B) X-ray of the hands in SMED-abnormal calcification type.

narrow chest and short neck. The face is distinctive with very small nose and protuberant eyes (Figure 52-14a). Radiologically, the long bones are short and very broad with slightly irregular metaphyses and peripheral spur formation. The iliac bones are small and inferiorly broad, and the lower margins have a caudally directed bump-like configuration bordered by spurs (Figure 52-14b). The vertebral bodies are flat with a sagittal midline cleft. The ribs are short and cupped at their anterior ends.

This bone dysplasia was reported from Lebanon and UAE (Megarbane et al., 1998; Al-Gazali et al., 1999c, 1997). One consanguineous UAE family, originally from Oman, had 6 affected children. One of the children was still surviving at the age of 4 years after prolonged respiratory support in the neonatal period. The child had developmental delay and severe skeletal deformities (Al-Gazali et al., 1997 Akawi et al., 2011)). Two other UAE families, one originally from Yemen and one originally from Oman, had one affected child each with this bone dysplasia (Akawi et al., 2011, Al-Gazali unpublished data). Mutations in COL11A1 was identified in two of the families (Akawi et al., 2011).

(A) (B)

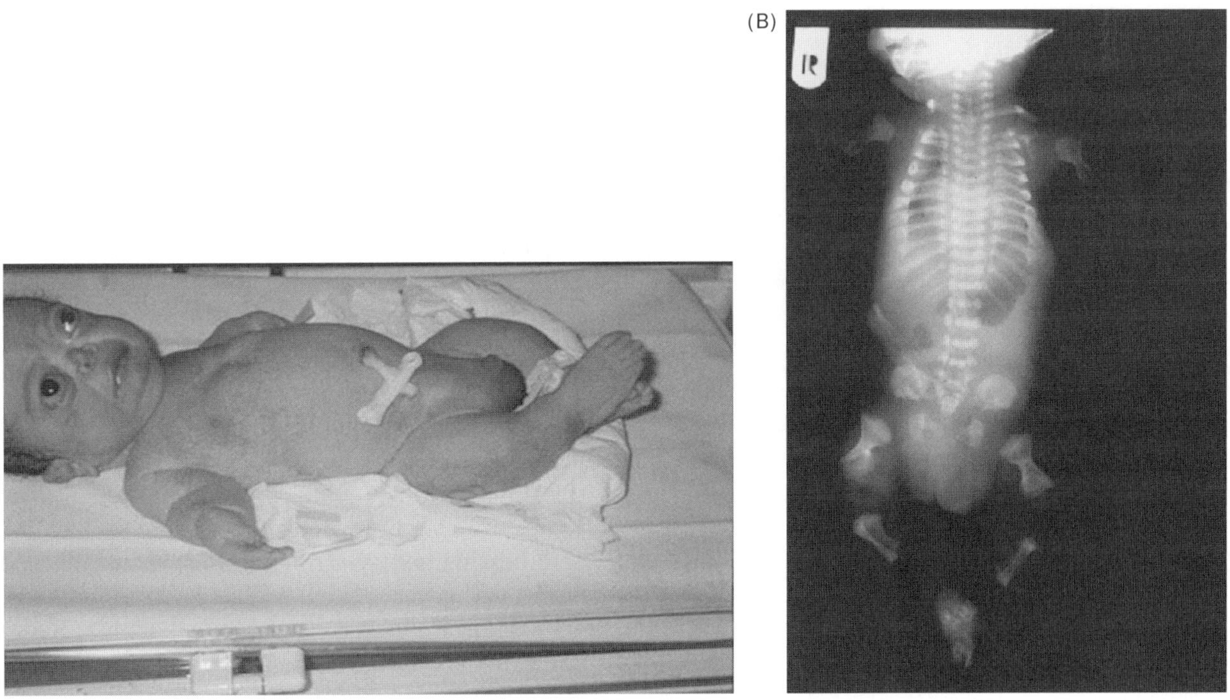

Figure 52-14 (A) Protuberant eyes, very small nose and short limbs in Fibrochondrogenesis. (B) Babygram in Fibrochondrogenesis. Note short bones with dumb-bell appearance and typical appearance of the ilium.

Raine Syndrome

Raine syndrome is a lethal bone dysplasia characterized by generalized osteosclerosis with periosteal new bone formation and distinctive facies (Al-Gazali et al., 2003d).

This bone dysplasia has been reported frequently in the Arabs. It has been reported in Saudi Arabian families, Palestinian, Egyptian, and UAE families (Al-Mane et al., 1998; Shalev et al., 1999; AlGazali et al., 2003d, and unpublished data).

Four consanguineous families with several affected children were seen in Al-Ain (Al-Gazali et al., 2003d; unpublished data) Molecular study in one of these families revealed a mutation in *FAM20C* gene (C915-3C>G) (Simpson et al., 2007).

Larsen Syndrome

Larsen syndrome is characterized by joint hypermobility, multiple joint dislocations, especially of the knees, and talipes equinovarus (Figure 52-15). The midface is hypoplastic with a depressed nasal bridge. Characteristically there is a bifid calcaneus and advanced bone age in the carpals, or extra carpal bones on radiological examination. There is evidence that both autosomal dominant and recessive types exist.

The recessive type has been reported from Egypt, UAE, and Jordan (Knoblauch et al., 1999; Topley et al., 1994; Hamamy unpublished data). A large Tunisian family with the dominant type was also reported (Al-Kaissi et al., 2003).

Acromesomelic Dysplasia

Two families from the UAE with acromesomelic dysplasia, Maroteaux type, and Langer mesomelic dysplasia were evaluated. One was originally from Oman and had 3 affected sibs with acromesomelic dysplasia. One of the affected had, in addition, cervical spina bifida and died in the first few months of life. Molecular study in this family showed c.2869C>T change leading to R957C missense mutation in the NPR2 gene (Bartels et al., 2004). Two families, one from UAE and one from Jordan, were diagnosed with Langer mesomelic dysplasia, which was confirmed by molecular study showing a homozygous deletion in SHOX gene (personal data).

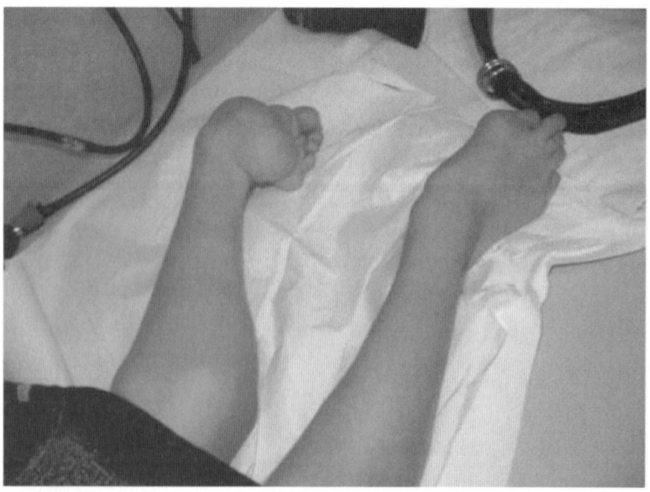

Figure 52-15 Joint dislocations in Larsen syndrome.

Spondyloepiphyseal Dysplasia, Omani Type

A new type of spondyloepiphyseal dysplasia was reported in 9 patients from 2 Omani families originating from the same tribe (Rajab et al., 2004). Short stature was apparent at birth, but body length dropped off further to below the 3rd centile by 6 months of age. In later childhood, length dropped off even further. There were limited movements of most joints, camptodactyly, and the knee joints were prominent. There was a severe, progressive osteoarthritis, and a progressive kyphoscoliosis. The short stature was particularly truncal. Thiele et al. (2004) identified the gene and mutation in these patients, a missense mutation (R304Q) in the chondroitin-6-O-sulfotransferase gene (CHST3). Mutations in the same gene were also identified in patients diagnosed as having AR Larsen syndrome and spondylo-humeral dysostosis, making these 3 conditions allelic (Hermanns et al., 2008).

Osteopetrosis Autosomal Recessive

Autosomal recessive osteopetrosis is a severe hereditary bone disease whose cellular basis is in the osteoclast (Figure 52-16), but with heterogeneous molecular defects. Among 4 families from Lebanon and two families from Syria with autosomal recessive osteopetrosis, mutations were identified in two genes: TCIRG1 and OSTM1. The mean age of diagnosis of the patients was 3 months. Failure to thrive, prominent forehead, exophthalmia, optic atrophy, hepatosplenomegaly, neurological manifestations, anemia, thrombocytopenia, and an early fatal outcome were common (Souraty et al., 2007).

Torg-Winchester Syndrome

Al-Mayouf et al. (2000) reported six females and four males from six unrelated consanguineous Saudi Arabian families, with a syndrome characterized by nodulosis, arthropathy, osteolysis. Affected individuals presented with painful, clawed, or deformed hands. Radiological examination showed osteopenia and undertubulation of the bones of the hands and feet together with osteolysis of the tarsals and carpals. Al Aqeel et al. (2000) reported a consanguineous Saudi Arabian family where a brother

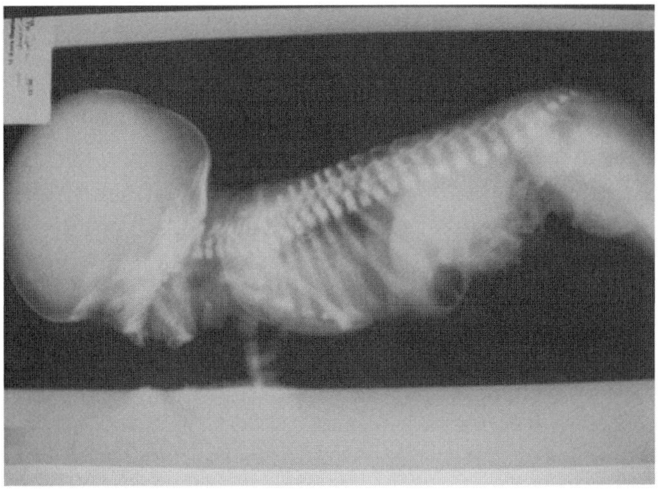

Figure 52-16 Osteopetrosis.

and sister had features overlapping with Torg syndrome and with the syndrome described in the family reported by Al-Mayouf et al. (2000). Martignetti et al. (2001) demonstrated mutations in the matrix metalloproteinase 2 gene (MMP2) in both families. Zankl et al. (2007, 2005) reported a family, with a diagnosis of Winchester syndrome, who had a homozygous mutation in the MMP2 gene.

It therefore seems that the 3 syndromes—Winchester, Torg, and the 2 families described by Al- Mayouf et al. (2000) and Al-Aqeel et al. (2000)—are the same disorder.

PROGEROID SYNDROMES

Gerodermia Osteodysplastica

This autosomal recessive disorder is characterized by droopy, jowled, and prematurely aged face. The bones are usually osteoporotic with susceptibility to fracture and biconcave vertebrae. It is caused by mutations in the SCYL1BP1 gene (Hennies et al., 2008).

Al-Torki et al. (1997) reported a Bedouin family from Kuwait in which 2 sisters had the disorder. Rajab et al. (2008) reported 22 Omani patients from 11 consanguineous families with the diagnosis of GO or "wrinkly skin syndrome" (WSS). However, only one patient from family B in that report was found to have mutation in SCL1BP1 gene (Hennies et al., 2008).

Autosomal Recessive Cutis laxa, Type IIA

This type of cutis laxa is evident from birth and is characterized by severe prenatal and postnatal growth failure, wide fontanelle, and congenital hip dislocation. This disorder is frequently seen in Saudi Arabia (Sakati et al., 1983; Allanson et al., 1986). It has also been reported from Oman and in an Iraqi Jewish family (Rajab et al., 2008; Reisner et al., 1971). Kornak et al. (2008) investigated glycosylation of serum proteins isolated from affected individuals and found a CDG type II pattern corresponding to a defect of N-glycosylation at the level of processing in the golgi apparatus. The gene was localized to chromosome 12q24, and mutations in the ATP6V0A2 gene were identified in patients of different ethnic origin (Kornak et al., 2008).

Autosomal Recessive Cutis Laxa, Type IIB (Cutis Laxa with Progeroid Features)

Al-Gazali et al. (2001) reported 2 families, Palestinian and Syrian, with 5 children affected with a disorder overlapping with gerodermia osteodysplastica and wrinkly skin syndrome. Features included broad and prominent forehead, hypotelorism with epicanthic folds, prominent bulbous nose, flat malar region, and large protruding ears. All had, in addition, wrinkling of the skin more marked on the dorsum of the hands, feet, and abdomen with hyperextensibility of joints of hands and an aged appearance. All had prenatal and postnatal growth retardation, developmental delay, and variable degrees of osteoporosis. Progressive prognathism developed with age. Hamamy et al. (2005) reported a Palestinian child with similar features (Figure 52-17a,b,c). Nanda et al. (2008) reported 3 children from 3 unrelated consanguineous Kuwaiti families with the same disorder. In addition, one of the families reported by Rajab et al. (2008) as having typical GO was found to have the same disorder. Two more families, one of Bahraini origin and one of Qatari origin, were identified (Reversade et al., 2009). Reversade et al. sequenced the PYCR1 gene and found different mutations in all the above families in addition to other families of different ethnic origin.

Ehler-Danlos Syndrome (EDS), Progeroid Form

Two members of an inbred Arab family from Qatar were reported with this form of EDS. Both had less severe manifestations, in that the skin was not remarkably loose, and their appearance was not progeroid. Both were floppy and developmentally delayed. The younger patient had a small face with frontal bossing, sparse scalp hair, down-slanting palpebral fissures, low-set ears, and yellow teeth with defective enamel. Her facial skin was loose and thin. There was joint laxity with limitation of movement at the elbow. A skeletal survey showed radioulnar synostosis, bowing of the radius and ulna, and an abnormal appearance of the distal metaphyses of the humeri and ulnae. Her paternal uncle was also affected. Both children had B4GALT7 mutations (Faiyaz-Ul-Haque et al., 2004).

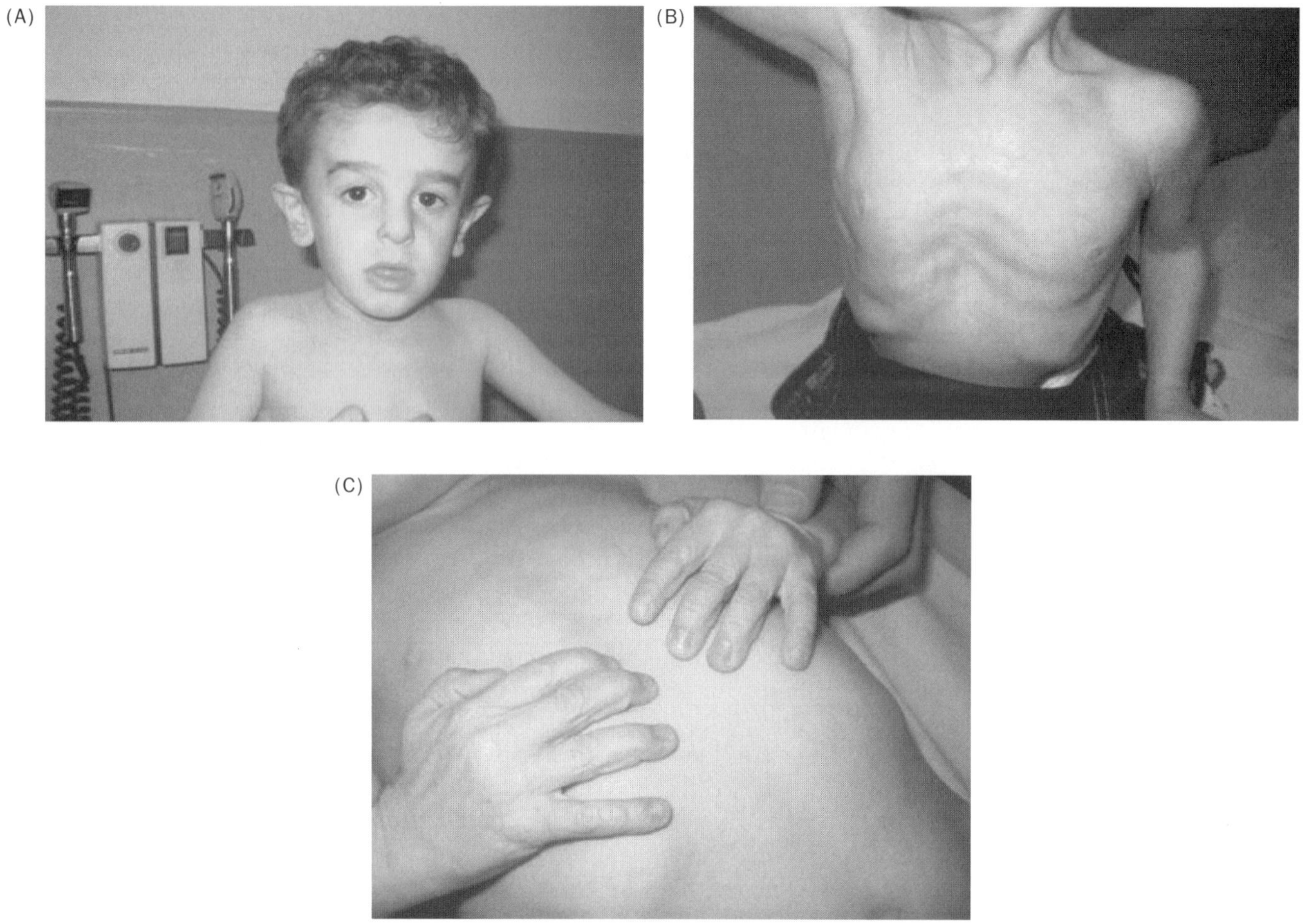

Figure 52-17 (A) Cutis laxa type IIB. (B,C) Wrinkled skin particularly over the dorsum of the hands in Cutis laxa type IIB.

Congenital Generalized Lipodystrophy of Seip

This autosomal recessive disorder is characterized by muscle hypertrophy which is, at least in part, due to a lack of subcutaneous adipose tissue. There may be increased height in childhood with organomegaly and a hypertrophic cardiomyopathy. There is usually hyperlipidemia and hyperinsulinemia with insulin-resistant diabetes. It is caused by mutations in AGPAT2 and BSCL2 genes.

This disorder has been reported from Palestine, Egypt, Tunisia, Algeria, Morocco, Oman, and UAE (Van Maldergem et al., 2002; Soliman et al., 1995; Heathcote et al., 2002; Agarwal et al., 2003). Rajab et al. (2005) estimated the prevalence of this disorder in Oman to be 1 in 25,000 births.

Four families from UAE, each with one affected child with this disorder, were evaluated at the molecular level: one was of Moroccan origin, one of Syrian origin, and 2 of UAE origin. Molecular study revealed mutations in *AGPAT2* in 2 families *(CG 8100, del 158G* homozygous [*H52sx59*]) and a mutation in *BSCL2* gene *(658 del GTATC [F105fsx111]* (Agarwal et al., 2003; personal data).

Leprechaunism and Leprechaunism-like Syndromes

Der Kaloustian et al. (1971) described two unrelated cases with leprechaunism born to consanguineous parents.

In a Yemeni family living in the UAE, 5 out of 8 children were affected with a syndrome very similar to leprechaunism but with a milder phenotype (Al-Gazali et al., 1993). All the affected individuals are still alive in their 20s. Molecular study revealed a new mutation in the insulin receptor gene (Ile^{119}-Met; see Hone et al., 1994).

In 2 other families of UAE origin, each had 2 affected children with classical leprechaunism.

Setleis Syndrome

This syndrome is characterized by distinctive bi-temporal, scar-like depression resembling forceps marks, lateral deficiency of the eyebrows, double-row eyelashes, and coarse, aged facial appearance. An inbred family from the UAE had 2 children and an uncle from the father's side affected with this syndrome (Al-Gazali et al., 1996d).

Molecular study in this family identified a nonsense mutation in the *TWIST1* gene (Q119X; see Desnick et al., 2004).

CONCLUSIONS AND FUTURE PROSPECTS

Dysmorphic syndromes, specifically those that follow an autosomal recessive mode of inheritance, constitute a considerable proportion of the birth defects pool among Arabs. The persistent commitment of Arab geneticists to their patients and to advancing research has contributed to the description of a number of rare and new autosomal recessive conditions, with mapping of novel loci. Molecular diagnosis of a large number of dysmorphic syndromes has been reached through the fruitful collaboration of Arab clinical geneticists with geneticists and molecular research laboratories in high-income countries. This collaboration has yielded important information on the responsible genes, underlying mutations, and consequently the functional genomics related to these syndromes.

Arab societies are characterized by a wide range of family and social structures, religious and legal conventions, and highly variable economic resources. As might be expected under these circumstances, genetic services cannot readily be devised, delivered, and assessed according to a single model. However, in general terms, the provision of genetic services in Arab communities at all population levels is still inadequate given the prevalence and burden of genetic diseases, including dysmorphic syndromes, due to the major impediments of paucity of resources and trained health professionals in the area of medical genetics. Furthermore, genetic diagnosis is often difficult given the potential diversity of the conditions involved, and the dearth of appropriately qualified clinical geneticists in the Arab world imposes a heavy responsibility on the few trained staff. Improving this situation calls for major educational efforts that include increasing the genetic literacy of the general public, comprehensive courses and campaigns to familiarize primary health care workers with counseling needs and skills, and with referral guidelines for high-risk families, updating medical, nursing, and paramedical curricula to incorporate information on genetics, and training clinical and laboratory genetic specialists to meet the short-term and long-term goals of genetic disease prevention and care. As in non-Arab societies, meeting the needs of individuals and communities in a culture-sensitive manner, and understanding and respecting the ethical, moral, cultural, social, and religious beliefs of clients, are essential prerequisites for the successful delivery of genetic services (Hamamy & Bittles, 2009). Since the majority of dysmorphic syndromes among Arabs tend to cluster in certain tribes and families, the most appropriate and perhaps the most effective approach to prevention of genetic disorders in these communities is to use the family-oriented approach. In this approach, carrier screening and counseling of the extended family members is initiated after the diagnosis is established in the presenting individual (Al-Gazali et al., 2005). The aim is to inform families of the implications of a carrier status, and where the carrier status cannot be diagnosed, to inform them of the possibility that future affected pregnancies could be circumvented by avoidance of marriage with close biological relatives. It is still unusual for Arab couples to learn of a major genetic risk before marriage or before starting a family, and most clinic sessions are retrospective, with couples learning of the risks to future pregnancies after the birth of an affected child. Couples who wish to avoid this possibility, most commonly due to an autosomal recessive condition, are faced with several alternative strategies: (1) divorce and remarriage to a noncarrier; (2) the husband considering a second polygamous marriage, which is accepted in Islam; (3) to have no more children; (4) to opt for prenatal diagnosis with selective termination of the pregnancy. As Arab culture favors large families with many offspring, if the affected infant is the first or second child, most couples would not accept the option of curtailed reproduction, whereas for couples with many children family planning does become a viable choice (Hamamy & Bittles, 2008).

Although still uncommon, the choice of prenatal diagnosis with selective termination of an affected fetus is a slowly evolving option in most Arab countries. Tunisia is at present the only clear example of an Islamic Arab country where selective abortion of an affected fetus is permissible under civil law and, for example, cytogenetic prenatal diagnosis followed by selective pregnancy termination has been permitted by the religious authorities and accepted by parents (Chaabouni-Bouhamed, 2008; Chaabouni et al., 2001). There is diversity of opinion among Islamic institutions on the issue of pregnancy termination, ranging from an absolute prohibition of abortion at any time to permission for pregnancy termination before the 120th day of gestation under specific circumstances (Al Aqeel, 2007; Alkuraya & Kilani, 2001). All Arab countries permit abortion to save the life of a pregnant woman, but otherwise they differ quite widely in their legal indications (Hamamy & Bittles, 2008). Preimplantation genetic diagnosis (PGD) is now emerging in some Arab countries as an additional alternative reproductive choice for high-risk couples who do not favor prenatal genetic diagnosis and selective termination of an affected fetus (see also Chapter 59). For example, in a family from Jordan where the husband was affected by Brachydactyly type B, and the specific mutation was determined (Hamamy et al., 2006), the couple went successfully through preimplantation diagnosis in Saudi Arabia with the birth of an unaffected baby (Hellani et al., 2009). PGD is expected to expand and to be much utilized by high-risk couples to minimize the birth of affected children whenever this option is affordable.

The considerable challenge posed by genetic and congenital disorders among Arabs, including dysmorphic syndromes, calls for the development of prevention through strategies that should be carefully selected to match the unique demographic, cultural, and religious characteristics of the population, and take into consideration the priorities set and the resources available

There could be a considerable number of rare dysmorphic syndromes that prevail among Arabs, many of which are not hitherto published and their underlying molecular pathology not yet defined. Arab countries should thus strive to deploy DNA diagnostics and to build research capability around local priorities (Alwan & Modell, 2003). A *Nature Genetics* Editorial in October 2007 emphasizes the importance of attracting international collaborators, while at the same time Arab countries should put their own priorities first, so that international collaborators must commit to return benefits, training, and information to the developing region as the price of collaboration and, ideally, DNA should be genotyped and sequenced locally (Editorial, 2007)

Furthermore, the characterization of the prevailing genetic disorders in each geographic location, together with their mutations, is needed to plan for appropriate screening and testing protocols (Al-Gazali & Eapen, 2005).

REFERENCES

Abdelrahman S, Bin-Abbas B, Al-Ashwal A. (2000). Wolcott-Rallison syndrome in a Saudi infant. *Curr Pediatr Res* 4:51–54.

Abu Safieh L, Aldahmesh M, Shamseldin H, et al. (2010). Clinical and molecular characterization of Bardet-Biedl syndrome in consanguineous populations: the power of homozygosity mapping. *J Med Genet* 47(4):236–241.

Afifi AK, Der Kaloustian VM, Bahuth NB, Mire-Salman J. (1974). Concentrically laminated membranous inclusions in myofibres of Dyggve-Melchior-Clausen syndrome. *J Neurol Sci* 21(3):335–340.

Afzal AR, Rajab A, Fenske CD, et al. (2000). Recessive Robinow syndrome, allelic to dominant brachydactyly type B, is caused by mutation of ROR2. *Nature Genetics* 25:419–422.

Agarwal AK, Sinha V, Oral EA, et al. (2003). Phenotypic and genetic heterogeneity in congenital generalized lipodystrophy. *J Clin Endocrinol Metab* 88(10):4840–4847.

Al Aqeel AI. (2007). Islamic ethical framework for research into and prevention of genetic diseases. *Nat Genet* 39(11):1293–1298

Al Aqeel A, Al Sewairi W, Edress B, Gorlin RJ, Desnick RJ, Martignetti JA. (2000). Inherited multicentric osteolysis with arthritis: a variant resembling Torg syndrome in a Saudi family. *Am J Med Genet* 93:11–18.

Al-Arrayet S. (1987). Congenital anomalies in Bahrain *Bahrain medical bulletin* 9: 70–73.

Al-Awadi SA, Farag TI, Teebi AS, et al. (1985a) Primary hypogonadism and partial alopecia in three sibs with mullerian hypoplasia in the affected females. *Am J Med Genet* 22(3):619–622.

Al-Awadi SA, Teebi AS, Farag TI, Naguib KM, el-Khalifa MY. (1985b). Profound limb deficiency, thoracic dystrophy, unusual facies, and normal intelligence: a new syndrome. *J Med Genet* 22:36–38.

Alazami AM, Shaheen R, Alzahrani F, et al. (2009). FREM1 Mutations Cause Bifid Nose, Renal Agenesis, and Anorectal Malformations Syndrome. *Am J Hum Genet* 85:414–418.

Alazami AM, Al-Saif A, Al-Semari A, et al. (2008) Mutations in C2orf37, encoding a nucleolar protein, cause hypogonadism, alopecia, diabetes mellitus, mental retardation, and extrapyramidal syndrome. *Am J Hum Genet* 83:684–691.

Alders M, Hogan MH, Salehi F, et al. (2009). Mutations in CCBE1 cause generalized lymph vessel dysplasia in human. *Nat Genet* 41(12):1272–1274.

Alfi OS, Chang R, Azen SP. (1980). Evidence for genetic control of nondisjunction in man. *Am J Hum Genet* 32:477–483.

Al-Gazali L, Hamamy H, Al-Arrayad S. (2006): Genetic disorders in the Arab world. *BMJ* 333:831–834.

Al-Gazali L, Eapen V. (2005). Genetics: global challenge, regional focus. Advances in community and preventive genetics. *Community Genet* 8(1):5–6

Al-Gazali LI, Alwash R, Abdulrazzaq YM. (2005): United Arab Emirates: communities and community genetics. *Community Genet* 8:186–196.

Al-Gazali LI, Bakir M, Hamid Z, et al. (2003a). Birth prevalence and pattern of osteochondrodysplasias in an inbred high risk population. *Birth Defects Res A Clin Mol Teratol* 67:125–132.

Al-Gazali LI, Hertecant J, Ahmed R, Khan NA, Padmanabhan R. (2003b). Further delineation of Hennekam syndrome. *Clin Dysmorphol* 12:227–232.

Al-Gazali LI, Ravenscroft A, Feng A, Shubber A, Al-Saqqaf A, Haas D. (2003c). Stuve-Wiedemann syndrome in children surviving infancy: clinical and radiological features *Clin Dysmorphol* 12:1–8.

Al-Gazali LI, Jehier K, Nazih B, Abtin F, Haas D, Sadagahatian R. (2003d). Further delineation of Raine syndrome. *Clin Dysmorphol* 12(2):89–93.

Al-Gazali LI, Bakir M, Hamud OA, Gerami S (2002). An autosomal recessive syndrome of nasal anomalies associated with renal and anorectal malformations. *Clin Dysmorphol* 11:33–38.

Al-Gazali Li, Sztriha L, Skaff H, Haas D. (2001). Gerodermia osteodysplastica and wrinkly skin syndrome: are they the same? *Am J Med Genet* 101:213–220.

Al-Gazali LI, Sztriha L, Dawodu A, Bakir M, Varghese M, Varady E, Scorer J, Abdulrazzaq YM, Bener A, Padmanabhan R (1999a). Pattern of central nervous system anomalies in a population with a high rate of consanguineous mattiages *Clin Genet* 55:95–102.

Al-Gazali LI, Bakir M, Sadaghatian MR, Nath R, Haas D. (1999b). Anterior segment anomalies of the eye associated with multiple skeletal abnormalities and early lethality: confirmation of an autosomal recessive syndrome. *Clin Dysmorphol* 8:87–92.

Al-Gazali LI, Bakir M, Dawodu A, Haas D. (1999c). Recurrence of fibrochondrogenesis in a consanguineous family *Clin Dysmorphol* 8:59–61.

Al-Gazali LI, Bakalinova D, Bakir M, Dawodu A (1997). Fibrochondrogenesis: clinical and radiological features. *Clin Dysmorphol* 6:157–163.

Al-Gazali LI, Varghese M, Varady E, Al Talabani J, Scorer J, Bakalinova D. (1996a). Neonatal Schwartz-Jampel Syndrome: A common AR syndrome in the UAE population. *J Med Genet* 33:203–211.

Al-Gazeli LI, Aziz SA, Bakalinova D. (1996b). Desbuquois syndrome in an Arab Bedouin family. *Clin Genet* 50(4):255–259.

Al-Gazali LI, Bakalinova D, Sztriha L. (1996c). Spondylo-meta-epiphyseal dysplasia, short limb, abnormal calcification type. *Clin Dysmorphol* 5:197–206.

Al-Gazali LI, Al-Talabani, J (1996d). Setleis syndrome: autosomal recessive or autosomal dominant inheritance? *Clin Dysmorph* 5: 249–253

Al-Gazali LI, Dawodu AH, Sabrinathan K, Varghese M. (1995a). The profile of major congenital abnormalities in the UAE population. *J Med Genet* 32(1):7–13.

Al-Gazali LI, Makia S, Azzam A, Hall CM. (1995b). Wolcott-Rallison syndrome. *Clin Dysmorphol* 4(3):227–233.

Al-Gazali LI, Al Talabani J, Mosawi A, Lytle W. (1994). Anterior segment anomalies of the eye, clefting and skeletal abnormalities in two sibs of consanguineous parents: Michels syndrome or new syndrome?. *Clin Dysmorphol* 3:238–244.

Al-Gazali LI, Khalil M, Devadas K. (1993). A syndrome of insulin resistance resembling leprechaunism in five sibs of consanguineous parents. *J Med Genet* 30(6):470–475.

Ali BR, Akawi NA, Chadid F, et al. (2010). Molecular and clinical analysis of Ellis van Creveld syndrome in the UAE. *BMC Med Genet* 11:33.

Al-Kaissi A, Ammar C, Ben Ghachem MB, Hammou A, Chehida FB. (2003). Facial features and skeletal abnormalities in Larsen syndrome—a study of three generations of a Tunisian family. *Swiss Med Wkly* 133(45–46):625–628.

Al Kaissi A, Nessib N, Ghachem MB, Hammou A, Guiddana N, Kozlowski K. (2005). A novel syndrome resembling Desbuquois dysplasia. *Am J Med Genet A* 132A(1):68–75.

AlKuraya FS, Kilani RA. (2001). Attitude of Saudi families affected with hemoglobinopathies towards prenatal screening and abortion and the influence of religious ruling (Fatwa). *Prenat Diagn* 21(6):448–451.

Allanson JE, Biesecker LG, Carey JC, Hennekam RC. (2009). Elements of morphology: introduction. *Am J Med Genet A* 149A:2–5.

Allanson J, Austin W, Hecht F. (1986). Congenital cutis laxa with retardation of growth and motor development: a recessive disorder of connective tissue with male lethality. *Clin Genet* 29(2):133–136.

Al-Mane K, al-Dayel F, McDonald P. (1998). Intracranial calcification in Raine syndrome: radiological pathological correlation. *Pediatr Radiol* 28(11):820–823.

Al-Mayouf SM, Majeed M, Hugosson C, Bahabri S. (2000). New form of idiopathic osteolysis: nodulosis, arthropathy and osteolysis (NAO) syndrome. *Am J Med Genet* 93:5–10.

Albaramki J, Akl K, Al- Muhtaseb A, Al-Shboul M, Mahmoud T, El-Khateeb M, Hamamy H. Sanjad Sakati Syndrome: a case series from Jordan. (Manuscript accepted in EMHJ).

Akawi N, Al-Gazali L, Ali BR. Clinical and molecular analysis of UAE Fibrochondrogenesis patients expands the phenotype and reveals two COL11A1 homozygous null mutations. *Clin Genetics June 11 Epub ahead of print*

Al Talabani J, Shubbar AI, Mustafa KE. (1998). Major congenital malformations in United Arab Emirates (UAE): need for genetic counselling. *Ann Hum Genet* 62:411–418.

Al-Torki NA, al-Awadi SA, Cindro-Heberie L, Sabry MA. (1997). Gerodermia osteodysplastica in a Bedouin sibship: further delineation of the syndrome. *Clin Dysmorphol* 6(1):51–55.

Alwan A, Modell B. (2003). Recommendations for introducing genetics services in developing countries. *Nat Rev Genet* 4(1):61–68

Alwan A, Modell B. (1997). Community control of genetic and congenital disorders *EMRO Technical Publication Series 24*: WHO Regional Office for the Eastern Mediterranean Region, Egypt

Bargal R, Cormier-Daire V, Ben-Neriah Z, et al. (2009). Mutations in DDR2 gene cause SMED with short limbs and abnormal calcifications. *Am J Hum Genet* 84:80–84.

Bartels CF, Bukulmez H, Padayatti P, et al. (2004) Human skeletal growth is regulated by transmembrane natriuretic peptide receptor NPR-B. *Am J Hum Genet* 75:27–34.

Basaran N, Cenani A, Sayli BS, et al. (1992). Consanguineous marriages among parents of Down patients. *Clin Genet* 42: 13–15.

Biels SL, Silhavy JL, Brancati F, et al. (2009). Mutations in the inositol polyphosphate-5-phosphatase E gene link phosphatidyl inositol signaling to the ciliopathies. *Nat Genet* 41(9):1032–1036.

Bin-Abbas B, Al-Mulhim A, Al-Ashwal A. (2002). Wolcott-Rallison syndrome in two siblings with isolated central hypothyroidism. *Am J Med Genet* 111(2):187–190.

Bonthron DT, Dunlop N, Barr DG, El Sanousi AA, Al-Gazali LI. (1998). Organisation of the human PAX4 gene and its exclusion as a candidate for the Wolcott-Rallison syndrome. *J Med Genet* 35(4):288–292.

Boutheina BR, Aida M, Lamia S, et al. (2000). Lethal uropathies: prenatal diagnosis and feto-pathologic aspects. *Tunis Med* 78(2):120–124.

Borochowitz Z, Langer LO Jr, Gruber HE, Lachman R, Katznelson MB, Rimoin DL. (1993). Spondylo-meta-epiphyseal dysplasia (SMED), short limb-hand type: a congenital familial skeletal dysplasia with distinctive features and histopathology. *Am J Med Genet* 45(3):320–326.

Brickwood S, Bonthron DT, Al-Gazali LI, et al. (2003). Wolcott-Rallison syndrome: pathogenic insights into neonatal diabetes from new mutation and expression studies of EIF2AK3 *J Med Genet* 40:685–689.

Chaabouni-Bouhamed H. (2008). Tunisia: communities and community genetics. *Community Genet* 11(6):313–323.

Chaabouni H, Chaabouni M, Maazoul F, et al. (2001). Prenatal diagnosis of chromosome disorders in Tunisian population. *Ann Genet* 44(2):99–104.

Chan AKJ, Levin AV, Teebi AS. (2000). Craniofacial dysmorphism, agenesis of the corpus callosum and ocular colobomas: Temtamy syndrome?. *Clin Dysmorphol* 9:223–226.

Chandler KE, Kidd A, Al-Gazali L, et al. (2003). Diagnostic criteria, clinical characteristics, and natural history of Cohen syndrome. *J Med Genet* 40(4):233–241.

Christianson A, Howson C, Modell B. (2006) *Global Report on Birth Defects The Hidden Toll of Dying and Disabled Children.* White Plains, New York: March of Dimes Birth Defects Foundation.

Clarkson JHW, Homfray T, Heron CW, Moss AL. (2004). Catel-Manzke syndrome: a case report with severely malformed hands and feet. An extension of the phenotype or a new syndrome? *Clin Dysmorphol* 13:237–240.

CTGA (Catalogue of Transmission Genetic Diseases among Arabs). Available at: http://www.cags.org.ae/ctga_search.html.

Dagoneau N, Benoist-Lasselin C, Huber C, et al. (2004a). ADAMTS10 mutations in autosomal recessive Weill-Marchesani syndrome. *Am J Hum Genet* 75(5):801–806.

Dagoneau N, Scheffer D, Huber C, et al. (2004b) Null Leukemia Inhibitory Factor Receptor (LIFR) mutations in Stuve-Wiedemann/Schwartz-Jampel 2 syndrome. *Am J Hum Genet* 74:298–305.

Delepine M, Nicolino M, Barrett T, Golamaully M, Lathrop GM, Julier C. (2000). EIF2AK3, encoding translation initiation factor 2-alpha kinase 3, is mutated in patients with Wolcott-Rallison syndorme. *Nature Genetics* 25(4):406–409.

Der Kaloustian VM, Kronfol NM, Takla R, Habash A, Khazin A, Najjar SS. (1971). Leprechaunism. A report of two new cases. *Am J Dis Child* 122(5):442–445.

Desnick R J, Tukel T, Diaz N, et al. (2004). Homozygous nonsense mutations in TWIST2 gene cause Setleis syndrome. *Am Hum Genet* Meeting, Toronto.

Dixon-Salazar T, Silhavy JL, Marsh SE, et al. (2004) Mutations in the AHI1 gene, encoding Jouberin, cause Joubert syndrome with cortical polymicrogyria. *Am J Hum Genet* 75:979–987.

Donnai D, Read AP. (2003). How clinicians add to knowledge of development. *Lancet* 362:477–484.

Draz SA. (2006). Sanjad Sakati Syndrome. *Palestinian Med J* 2(1):10–14.

Editorial. (2007). Developing genetics for developing countries. *Nat Genet* 39(11):1287.

El Ghouzzi V, Dagoneau N, Kinning E, et al. (2003) Mutations in a novel gene Dymeclin (FLJ20071) are responsible for Dyggve-Melchoir-Clausen Syndrome. *Hum Mol Genet* 12:357–364.

El-Sawy MA, El-Kholy MS, El-Sedfy HH, Sleem MN. (2004). Sanjad-Sakati syndrome, case report. *The Egyp J Med Genet* 5(2):61–66.

Faivre L, Le Merrer M, Al-Gazali LI, et al. (2003). Homozygosity mapping of a Desbuquois dysplasia locus to chromosome 17q25.3 *J Med Genet* 40:282–284.

Faiyaz-Ul-Haque M, Zaidi SHE, Al-Ali, M, et al. (2004). A novel missense mutation in the galactosyltransferase-I (B4GALT7) gene in a family exhibiting facioskeletal anomalies and Ehlers-Danlos syndrome resembling the progeroid type. *Am J Med Genet* 128A: 39–45.

Farag TI, Teebi AS. (1989). High incidence of Bardet-Biedl syndrome among the Bedouin. *Clin Genet* 36:463–464.

Gillessen-Kaesbach G, Meinecke P, Ausems MG, et al. (1995). Desbuquois syndrome: three further cases and review of the literature. *Clin Dysmorphol* 4(2):136–144.

Gripp KW, Donnai D, Clericuzio CL, et al. (1997). Diaphragmatic hernia-exomphalos-hypertelorism syndrome: a new case and further evidence of autosomal recessive inheritance. *Am J Med Genet* 68(4):441–444.

Gül D, Özata M, Mergen H, Odabasi Z, Mergen M.(2000) Woodhouse and Sakati syndrome (MIM 241080): report of a new patient. *Clin Dysmorphol* 9:123–125.

Hafez M, et al. (1983). Demographic trends of Down's syndrome syndrome in Egypt. *Journal of the Egyptian Medical Association* 66:495–507.

Hall BD. (1993). The state of the art of dysmorphology. *Am J Dis Child* 147:1184–1189.

Hamamy H, Bittles AH. (2009). Genetic Clinics in Arab Communities: Meeting Individual, Family and Community Needs. Public Health Genomics. 12:30–40

Hamamy H, Al-Hait S, Alwan A, Ajlouni K. (2007a). Jordan: communities and community genetics. Community Genet 10:52–60.

Hamamy HA, Masri AT, Al-Hadidy AM, Ajlouni KM. (2007b). Consanguinity and genetic disorders. Profile from Jordan. *Saudi Med J* 28:1015–1017.

Hamamy HA, Teebi AS, Oudjhane K, Shegem NN, Ajlouni KM. (2007c). Severe hypertelorism, midface prominence, prominent/simple ears, severe myopia, borderline intelligence, and bone fragility in two brothers: new syndrome? *Am J Med Genet A*143(3):229–234.

Hamamy H, Saleh N, Oldridge M, Al-Hadidy A, Ajlouni K. (2006). Brachydactyly type B1: report of a family with de novo ROR2 mutation. *Clin Genet* 70(6):538–540.

Hamamy H, Masri A, Ajlouni K. (2005). Wrinkly skin syndrome. *Clin Exp Derm* 30:590–592.

Hamamy H, Alwan A. (1994). Hereditary disorders in the Eastern Mediterranean Region. *Bull World Health Organ* 72:145–154.

Hamamy HA, al-Hakkak ZS, al-Taha S. (1990). Consanguinity and the genetic control of Down syndrome. *Clin Genet* 37:24–29.

Han X-D, Cox VA, Slavotinek A. (2006). Atrioventricular block and wiry hair in Teebi hypertelorism syndrome. *Am J Med Genet* 140A:1960–1964.

Hattab FN, Yassin OM, Sasa IS. (1998). Oral manifestations of Ellis-van Creveld syndrome: report of two siblings with unusual dental anomalies. *J Clin Pediatr Dent* 22(2):159–165.

Heathcote K, Rajab A, Magre J, et al. (2002). Molecular analysis of Berardinelli-Seip congenital lipodystrophy in Oman: evidence for multiple loci. *Diabetes* 51(4):1291–1293.

Hellani A, Abu-Amero K, Azouri J, Al-Sharif H, Barblet H, El-Akoum S. (2009). Pregnancy after preimplantation genetic diagnosis for brachydactyly type B. *Reprod Biomed Online* 18(1):127–131.

Hellani A, Aqueel A, Jaroudi K, et al. (2004). Pregnancy after preimplantation genetic diagnosis for Sanjad-Sakati syndrome. *Prenat Diagn* 24:302–306.

Hennekam RCM, Geerdink RA, Hamel BCJ, et al. (1989). Autosomal recessive intestinal lymphangiectasia and lymphedema, with facial anomalies and mental retardation. *Am J Med Genet* 34:593–600.

Hennies H C, Kornak U, Ahang H, et al. (2008). Gerodermia osteodysplastica is caused by mutations in SCYL1BP1, a Rab-6 interacting golgin. *Nat Genet* 40:1410–1412.

Hennies HC, Rauch A, Seifert W, et al. (2004). Allelic heterogeneity in the COH1 gene explains clinical variability in Cohen syndrome. *Am J Hum Genet.* 75(1):138–145.

Hermanns P, Unger S, Rossi A, et al. (2008). Congenital joint dislocations caused by carbohydrate sulfotransferase 3 deficiency in recessive Larsen syndrome and humero-spinal dysostosis. *Am J Hum Genet* 82:1368–1374.

Hershkovitz E, Parvari R, Diaz GA, Gorodischer R. (2004). Hypoparathyroidism-retardation-dysmorphism (HRD) syndrome - a review. *J Ped Endocr Metab* 17:1583–1590.

Hoffmann K, Muller JS, Stricker S, et al. (2006). Escobar syndrome is a prenatal myasthenia caused by disruption of the acetylcholine receptor fetal gamma subunit. *Am J Hum Genet* 79(2):303–312.

Hone J, Acillin D, Al-Gazali LI, Lestringant GT, Orban, Taylor S. (1994) Homozygosity for a new mutation. Ile 119 - met in the insulin receptor gene in 5 siblings with familial insulin resistance. *J Med Genet* 9:715–717.

Horn D, Krebsova A, Kunze J, Reis A. (2000). Homozygosity mapping in a family with microcephaly, mental retardation, and short stature to a Cohen syndrome region on 8q21.3-8q22.1: redefining a clinical entity. *Am J Med Genet* 92(4):285–292.

Hosny GA, Fabry G. (1998). Treatment of hip subluxation in Dyggve-Melchior-Clausen syndrome. *J Pediatr Orthop B* 7(1):32–34.

Huber C, Oules B, Bertoli M, et al. (2009). Identification of *CANT1* mutations *in* desbuquois dysplasia. *Am J Hum Genet* 85(5):706–710.

Jarbhou H, Hamamy H, Al-Hadidy A, Ajlouni K. (2008). Cenani-Lenz syndactyly with facial dysmorphism, hypothyroidism, and renal hypoplasia: a case report. *Clin Dysmorphol* 17(4):269–270.

Jones K, (ed.) (2006) *Smith's Recognizable Patterns of Human Malformation*, 6th edition. Philadelphia, PA: Elsevier Saunders.

Kantarci S, Al-Gazali L, Hill RS, et al. (2007). Mutations in LRP2, which encodes the multiligand receptor megalin, cause Donnai-Barrow and facio-oculo-acoustico-renal syndromes. *Nat Genet* 39(8):957–959.

Kayserili H, et al. (2010). Hamamy syndrome (Fronto-cardio-frontal; Dysosteocraniofrontal syndrome, the first human clinical phenotype linked to a very early expressed transcription factor gene family. 14th Manchester Dysmorphology Conference, October 11–14, 2010, Genetic Medicine and Manchester Biomedical Research Center, Manchester, UK.

Knoblauch H, Urban M, Tinschert S. (1999). Autosomal recessive versus autosomal dominant inheritance in Larsen syndrome: report of two affected sisters. *Genet Couns* 10(3):315–320.

Kornak U, Reynders E, Dimopoulou A, et al. (2008): Impaired glycosylation and cutis laxa caused by mutations in the vesicular H(+)-ATPase subunit ATP6V0A2. Nature Genet 40: 32–34152.

Li Y, Laue K, Temtamy S, et al. (2010a). Temtamy preaxial brachydactyly syndrome is caused by loss-of-function mutations inchondroitin synthase 1, a potential target of BMP signaling. *Am J Hum Genet* 87(6):757–767.

Li Y, Pawlik B, Elcioglu N, et al. (2010b). LRP4 mutations alter Wnt/beta-catenin signaling and cause limb and kidney malformations in Cenani-Lenz syndrome. *Am J Hum Genet* 86(5):696–706.

Li J, Shivakumar S, Wakahiro M, et al. (2007). Agenesis of the corpus callosum, optic coloboma, intractable seizures, craniofacial and skeletal dysmorphisms: an autosomal recessive disorder similar to Temtamy syndrome. *Am J Med Genet* 143A:1900–1905.

Machado-Paula LA, Guion-Almeida ML. (2003). Teebi hypertelorism syndrome: Additional cases. *Am J Med Genet* 117A:181–183.

Marafie MJ, Redha MA, Al-Naggar RL. (2004). Wolcott-Rallison syndrome in a Bedouin boy. *Ann Saudi Med* 24(6):476–479.

Maroteaux P, Frezal J, Cohen-Solal L. (1986). The differential symptomatology of errors of collagen metabolism: a tentative classification. *Am J Med Genet* 24:219–230.

Martignetti JA, Aqeel AA, Sewairi WA, et al. (2001). Mutation of the matrix metalloproteinase 2 gene (MMP2) causes a multicentric osteolysis and arthritis syndrome. *Nature Genet* 28:261–265.

Megarbane A, Mustapha M, Bleik J, Waked N, Delague V, Loiselet J. (2000). Exclusion of chromosome 15q21.1 in autosomal-recessive Weill-Marchesani syndrome in an inbred Lebanese family. *Clin Genet* 58(6):473–478.

Mégarbané A, Haddad S, Berjaoui L. (1998). Prenatal ultrasonography: Clinical and radiological findings in a boy with fibrochondrogenesis. *Am J Perinatol* 15:403–407.

Mir NA, Galczek WC, Soni A. (1992). Easily identifiable congenital malformations in children: Survey of incidence and pattern in 32,332 live born neonates. *Ann Saudi Med* 12:366–371.

Mochida GH, Rajab A, Eyaid W, et al. (2004). Broader geographical spectrum of Cohen syndrome due to COH1 mutations. *J Med Genet* 41(6):e87.

Mokhtar MM, Kotb SM, Ismail, SR. (1998). Autosomal recessive disorders among patients attending the genetic clinic in Alexandria. *East Mediterr Health J* 4(3):470–479.

Morgan NV, Brueton LA, Cox P, et al. (2006). Mutations in the embryonal subunit of the acetylcholine receptor (CHRNG) cause lethal and Escobar variants of multiple pterygium syndrome. *Am J Hum Genet* 79(2):390–395.

Mostafa MI, Temtamy SA, El-Gammal MA, Mazen IM. (2005). Unusual pattern of inheritance and orodental changes in the Ellis-van Creveld syndrome. *Genet Couns* 16(1):75–83.

Murthy S, Malhotra A, Mani S, et al. (2007). Incidence of Down syndrome in Dubai, UAE. *Med Princ Pract* 16(1):25–28.

Naffah J, Taleb N. (1974). Deux nouveaux cas de syndrome de Dyggve-Melchior-Clausen avec hypoplasie de l'apophyse odontoide et compression spinale. *Arch Fr Pediatr* 31(10):985–992.

Nanda A, Alsaleh QA, Al-Sabah H, et al. (2008). Gerodermia osteodysplastica/wrinkly skin syndrome: report of three patients and brief review of the literature. *Pediat* 25:66–71.

Nazer H, Gunasekaran TS, Sakati NA, Nyhan WL. (1990). Concurrence of Robinow syndrome and Crigler-Najar syndrome in two offspring of first cousins. *Am J Med Genet* 37(4):516–518.

Nicolino PM, Dupin H, Macebeo V, Treppoz S, Chatelain PG. (1998). Wolcott-Rallison syndrome (diabetes mellitus and spondyloepiphyseal dysplasia): a plausible existence of a gene(s) important for the mutation of neonatal pancreatic beta cell function. *Hormone Res* 50(Suppl 3):77.

Oerter KE, Friedman TC, Anderson HC, Cassorla FG. (1992). Familial syndrome of endocrine and neuroectodermal abnormalities. *Am J Med Genet* 44:487–491.

Parvari R, Hershkovitz E, Grossman N, et al. (2002). Mutation of TBCE causes hypoparathyroidism-retardation-dysmorphism and autosomal recessive Kenny-Caffey syndrome. *Nature Genetics* 32:448–452.

Raas-Rothschild A, Goodman RM, et al. (1988). Pathological features and prenatal diagnosis in the newly-recognised limb/pelvis-hypoplasia/aplasia syndrome. *J Med Genet* 25:687–697.

Rajab A, Kornak U, Budde BS, Hoffmann K, Jaeken J, Nürnberg P, Mundlos S. (2008). Geroderma osteodysplasticum hereditaria and wrinkly skin syndrome in 22 patients from Oman. *Am J Med Genet A* 146A(8):965–976.

Rajab A, Bappal B, Al-Shaikh H, Al-Khusaibi S, Mohammed AJ. (2005a). Common autosomal recessive diseases in Oman derived from a hospital-based registry. *Community Genet* 8:27–30.

Rajab A, Hoffmann K, Ganesh A, Sethu AU, Mundlos S. (2005b). Escobar variant with pursed mouth, creased tongue, ophthalmologic features, and scoliosis in 6 children from Oman. *Am J Med Genet A* 134(2):151–157.

Rajab A, Kunze J, Mundlos S. (2004). Spondyloepiphyseal dysplasia Omani type: a new recessive type of SED with progressive spinal involvement. *Am J Med Genet* 126A:413–419.

Ramadani HM, Nasrat HA. (1992). Prenatal diagnosis of recurrent Meckel syndrome. *Int J Gynaecol Obstet* 39(4):327–332.

Ramocki MB, Dowling J, Grinberg I, et al. (2003). Reciprocal fusion transcripts of two novel Zn-finger genes in a female with absence of the corpus callosum, ocular colobomas and a balanced translocation between chromosomes 2p24 and 9q32. *Hum Mol Genet* 11:527–534.

Rauch A, Theil CT, Schindler D, et al. (2008). Mutations in the pericentrin (PCNT) gene cause primordial dwarfism. *Science* 319:816–819.

Reddy DV, Madenlioglu M. (1967). Chondroectodermal dysplasia (Ellis-van Creveld syndrome). Report of two cases from Saudi Arabia. *Clin Pediatr (Phila)* 6(1):51–56.

Reisner SH, Seelenfreund M, Ben-Bassat M. (1971). Cutis laxa associated with severe intrauterine growth retardation and congenital dislocation of the hip. *Acta Paediatr Scand* 60(3):357–360.

Reversade B, Escande-Beillard N, Dimopoulou A, et al. (2009) Mutations in PYCR1 cause cutis laxa with progeroid features. *Nature Genet* 41:1016–1021.

Riad W, Abouammoh M, Fathy M. (2006). Anesthetic characteristics and airway evaluation of patients with Weill-Marchesani syndrome. *Middle East J Anesthesiol* 18(4):725–731.

Sakati NO, Nyhan WL, Shear CS, et al. (1983). Syndrome of cutis laxa, ligamentous laxity, and delayed development. *Pediatrics* 72:850–856.

Sawardekar KP. (2005). Profile of major congenital malformations at Nizwa Hospital, Oman 10-year review. *J Paediatr Child Health* 41: 323–330.

Sbihi A, Benharbit R, Lahrichi A, Lahrech MT. (1980). [Dyggve-Melchior-Clausen dysplasia (apropos of a case)] *Maroc Med* 2(4):367–374.

Schorr S, Legum C, Ochshorn M, et al. (1977). The Dyggve-Melchio-Clausen syndrome. *AJR Am J Roentgenol* 128(1):107–113.

Shalev SA, Shalev E, Reich D, Borochowitz ZU. (1999). Osteosclerosis, hypoplastic nose, and proptosis (Raine syndrome): further delineation. *Am J Med Genet* 86(3):274–277.

Simpson MA, Hsu R, Keir LS, et al. (2007). Mutations in *FAM20C* are associated with lethal osteosclerotic bone dysplasia (Raine syndrome), highlighting a crucial molecule in bone development. *Am J Hum Genet* 81:906–912.

Smaoui N, Chaabouni M, Sergeev YV, et al.(2006). Screening of the eight BBS genes in Tunisian families: no evidence of triallelism. *Invest Ophthalmol Vis Sci* 47:3487–3495.

Smith UM, Tee JT, Consugar M, et al. (2006). Meckelin (MKS3), a novel transmembrane protein is mutated in Meckel-Gruber syndrome and the *wpk* rat. *Nature Genetics* 38:191–196.

Soliman AT, Rajab A, Alsalmi I, Bedair SM. (1998). Recessive Robinow syndrome: with emphasis on endocrine functions. *Metabolism* 47(11):1337–1343.

Soliman AT, El-Nawawy AA, El-Azzoni OO, Amer SA, Rajab A. (1995). Seip-Berardinelli Lipodystrophy: report of three cases and their endocrine functions. *Annals Saudi Med* 15(5):501–505.

Souraty N, Noun P, Djambas-Khayat C, et al. (2007). Molecular study of six families originating from the Middle-East and presenting with autosomal recessive osteopetrosis. *Eur J Med Genet* 50:188–199.

Stoetzel C, Laurier V, Davis EE, et al. (2006). BBS10 encodes a vertebrate-specific chaperonin-like protein and is a major BBS locus. *Nat Genet* 38:521–524.

Taban M, Memoracion-Peralta DSA, Wang H, Al-Gazali L, Traboulsi EI. (2007). Cohen syndrome: report of nine cases and review of the literature with emphasis on ophthalmic features. *J AAPOS* 11:431–437.

Tadmouri GO, Nair P, Obeid T, Al Ali MT, Al KN, Hamamy HA. (2009). Consanguinity and reproductive health among Arabs. *Reprod Health* 6:17.

Talisetti A, Forrester SR, Gregory D, Johnson L, Schneider MC, Kimonis VE. (2003). Temtamy-like syndrome associated with translocation of 2p24 and 9q32. *Clin Dysmorphol* 12:175–177.

Teebi AS, Teebi SA. (2005). Genetic diversity among the Arabs. *Community Genet.* 8(1):21–26.

Teebi A, Farag T. (1997). Genetic Disorders Among Arab Populations. *Oxford Monographs on Medical Genetics No30.* New York: Oxford University Press.

Teebi AS. (1994). Autosomal recessive disorders among Arabs: an overview from Kuwait. *J Med Genet* 31:224–233.

Teebi AS, al Saleh QA, Odeh H. (1992). Meckel syndrome and neural tube defects in Kuwait. *J Med Genet* 29:140.

Teebi AS, Al Awadi SA. (1991). Kuwait type faciodigitogenital syndrome (Letter). *J Med Genet* 28:805.

Teebi AS, Daoud AS. (1990). Multiple pterygium syndrome: a relatively common disorder among Arabs. *J Med Genet* 27:791.

Teebi AS. (1990). Autosomal recessive Robinow syndrome. *Am J Med Genet* 35(1):64–68.

Teebi AS, Naguib KK, Al-Awadi SA, Al-Saleh QA. (1988). New autosomal recessive faciodigitogenital syndrome. *J Med Genet* 25:400–406.

Teebi AS. (1987). New autosomal dominant syndrome resembling craniofrontonasal dysplasia. *Am J Med Genet* 28:581–591.

Temtamy SA. (2005). Catel-Manzke digitopalatal syndrome or Temtamy preaxial brachydactyly hyperphalangism syndrome? *Clin Dysmorphol* 14:211.

Temtamy SA, Abdel Meguid N, Mazen I, Ismail SR, Kassem NS, Bassiouni R. (1998a). A genetic epidemiological study of malformations at birth in Egypt. *East Mediterr Health J* 4:252–259.

Temtamy SA, Meguid NA, Ismail SI, Ramzy MI. (1998b). A new multiple congenital anomaly, mental retardation syndrome with preaxial brachydactyly, hyperphalangism, deafness and orodental anomalies. *Clin Dysmorphol* 7:249–255.

Temtamy SA, Salam MA, Aboul-Ezz EHA, Hussein HA, Helmy SAH, Shalash BA. (1996). New autosomal recessive multiple congenital abnormalities/mental retardation syndrome with craniofacial dysmorphism absent corpus callosum, iris colobomas and connective tissue dysplasia. *Clin Dysmorphol* 5:231–240.

Thiele H, Sakano M, Kitagawa H, et al. (2004). Lossof chondroitin 6-O-sulfotransferase-1 function results in severe human chondrodysplasia with progressive spinal involvement. *Proc Nat Acad Sci USA* 101:10155–10160.

Thong M-K, Chan L-G, Ting H-S. (2005). Further delineation of Al-Gazali syndrome (multiple skeletal abnormalities with anterior segment anomalies of the eye and early lethality) in a Malaysian family *Clin Dysmorphol* 14:1–5.

Tian J, Ling L, Shboul M, et al. (2010). Loss of CHSY1, a secreted FRINGE enzyme, causes syndromic brachydactyly in humans via increased NOTCH signaling. *Am J Hum Genet* 87(6):768–778.

Topley JM, Varady E, Lestringant GG. (1994). Larsen syndrome in siblings with consanguineous parents. *Clin Dysmorphol* 3(3):263–265.

Tsai AC, Robertson JR, Teebi AS. (2002). Teebi hypertelorism syndrome: Report of a family with previously unrecognized findings. *Am J Med Genet* 113:302–306.

Valente EM, Silhavy JL, Brancati F, et al. (2006). Mutations in the CEP290 gene, encoding a putative centosomal protein, cause pleotropic forms of Joubert Syndrome. *Nat Genet* 38:623–625.

Van Maldergem L, Magre J, Khallouf TE, et al. (2002). Genotype-phenotype relationships in Berardinelli-Seip congenital lipodystrophy. *J Med Genet* 39(10):722–733.

Wahab AA, Bener A, Teebi AS. (2006). The incidence patterns of Down syndrome in Qatar. *Clin Genet* 69:360–362.

Woodhouse NJY, Sakati NA. (1983). A syndrome of hypogonadism, alopecia, diabetes mellitus, mental retardation, deafness, and ECG abnormalities *J Med Genet* 20:216–219.

Woods CG, Stricker S, Seemann P, et al. (2006). Mutations in WNT7A cause a range of limb malformations, including Fuhrmann syndrome and Al-Awadi/Raas-Rothschild/Schinzel phocomelia syndrome. *Am J Hum Genet* 79:402–408.

Zankl A, Pachman L, Poznanski A, et al. (2007). Torg syndrome is caused by inactivating mutations in MMP2 and is allelic to NAO and Winchester syndrome. *J Bone Miner Res* 22:329–333.

Zankl A, Bonafé L, Calcaterra V, Di Rocco M, Superti-Furga A. (2005). Winchester syndrome caused by a homozygous mutation affecting the active site of matrix metalloproteinase 2. *Clin Genet* 67:261–266.

Zlotogora J. (1997a). Autosomal recessive diseases among Palestinian Arabs. *J Med Genet* 34: 765–766.

Zlotogora J. (1997b). Genetic disorders among Palestinian Arabs: 1. Effects of consanguinity. *Am J Med Genet* 68: 472–475.

53.

THE POPULATION GENETICS AND HEREDITARY DISEASES OF BAHRAIN

Shaikha Salim Al-Arrayed

INTRODUCTION—THE LAND OF BAHRAIN

The name *Bahrain* means the land of life, or the sacred land, as mentioned in the Sumerian legends, and it was so named for the presence of water and agriculture, a rarity in the Arabian Gulf. The state of Bahrain is made up of an archipelago of 40 islands in the Arabian Gulf. The Kingdom of Saudi Arabia is on the west, and Qatar on the East. The total land area of the kingdom is 720.14 km^2. The largest island among these 40 islands is the island of Bahrain, which represents 85% of the total land area (Central Statistics Organization, 2006).

THE PEOPLES OF BAHRAIN

The geographical, religious, and sociocultural aspects of Bahrain play an important part in the influence of genetic association with environmental factors on the people of Bahrain. History clearly shows that from the Stone Age until the advent of Islam, Bahrain was influenced by the Babylonian, Assyrian, and Greek civilizations, among other ancient civilizations. The eastern coast of the Arabian Peninsula was the one most influenced by these cultures; the Greek influence extended to the southernmost part of the Arabian Gulf, profoundly affecting Bahrain. The Sumerian legends gave a clear mention of the Bahrainis as "the people of Dilmun." However, around 500 BC, Greek and Roman geographers referred to Bahrain as "Tylos." It is maintained that it was basically the pearls and precious stones of Bahrain that gave the island its name and fame.

The Arabian Gulf was responsible for carrying out sea trade, and from the dawn of time Bahrain became an important commercial center on the major trade routes between the East and West. With geographic explorations and acquisition of overseas colonies by the Europeans, Bahrain became a battlefield and was occupied by the Portuguese in 1907. At the time when India became an important part of the British Empire, Bahrain was strategically located on the route to India. With the passage of time, its location contributed to an increased exchange with various parts of the world. Rapid economic progress with the advent of independence in 1971 made Bahrain a diversified industrial base and a major banking center, with a number of reputable international and offshore banks. Bahrain is now (1) an important communication center between the East and West, and has been (2) an important military base from the Second World War till today. All these factors highlight an infiltration of foreign cultures and civilization through time, giving a cosmopolitan blend of genes mingling with the local population.

THE HEALTH SERVICES OF BAHRAIN

The foundation of the modern health services dates back to 1900 when the first hospital was established. The first government doctor was appointed in 1925 and a full-fledged government clinic was set up 11 years later. In 1968, a comprehensive health services development plan was drawn up by the government in cooperation with the World Health Organization (WHO), and this was put into operation in 1970. The present government provides free comprehensive healthcare to all the residents of Bahrain, citizens and noncitizens, in a two-tier system. Primary Health Care Centers are drained into various clinical setups such as Salmaniya Medical Complex (SMC).

A Public Health Directorate is responsible for the prevention and control of diseases. Comprehensive programs for general sanitation, water purification, hygiene, health, and education have led to a decreasing trend in various contagious diseases. Presently, the health system delivery maintains a partnership between both the government and private sectors. In line with this, the Ministry of Health (MOH) has concomitantly played a major role in the provision, improvement, and sustenance of quality healthcare services. Health facilities have improved rapidly during the past five years. This can be witnessed clearly through the remarkable evolution with regard to the type and quality of the services offered at Salmaniya Medical Complex (the major hospital in Bahrain).

The construction work on the new King Hamad General Hospital in Al Muharraq Governorate was initiated in February 2006. This hospital is now a huge medical training center for the adjacent medical university (The Royal College of Surgeon Bahrain). The expansion of the services in Bahrain was not limited to secondary healthcare, but also included primary healthcare, in order to maximize capacities and accessibility to the services in primary healthcare. Private hospitals are a new addition to Bahrain's health care facilities that have started their services during the past 10 years

DEMOGRAPHIC DATA

Bahrain has been subjected to various influences from a number of different civilizations, due to its geographic location. Mingling of populations over generations has led to various social and cultural customs in the Bahraini people. An inherent desire of this small country to preserve its identity, as well as its religious and social culture, may have planted the seeds of consanguineous marriages, leaving behind genetic footprints on the sands of time (Central Statistics Organization, 2006; Bahrain Health Statistic Reports, 1985–2005).

The estimated population is 1,039,297 in 2010. In 2007, it was 646,551 based on the 2001census data, with the percentage of Bahraini nationals being 50.7% as against the non Bahrainis 49.3%. The majority of Bahraini Arabs are originally from Arabian Peninsula. The crude birth rate was estimated to be 15.4 per 1000 population, whilst the infant mortality rate is 83/1000 live births in 2007.

The infant mortality rate of live births was 8.9/1000. The percentage of people under 15 years was estimated as 40.3%, while in the 15–64 age groups it was 56%. However, in the age group of above 65, it was 3.4%.

GENETIC EPIDEMIOLOGY

Noncommunicable diseases, especially genetic diseases, are assumed to be a major cause of morbidity and mortality. Genetic disorders are a significant burden on healthcare delivery systems. Their chronic nature requires lifelong medical attention, expensive supportive and symptomatic therapy, and specialized care. Effective control of these diseases requires that their natural history, frequency, and distribution be studied (Bahrain Health Statistics Reports, 2005).

In 1983, a genetic unit was established at the Salmaniya Medical Complex (SMC). A national committee for the control of hereditary diseases was established in 1993. The aim of the committee was to conduct population studies on the prevalence of genetic diseases within the country, and to improve management and treatment standards of patients suffering from these diseases. During these years, many studies have been carried out.

HEREDITARY DISEASES IN BAHRAIN

CONSANGUINITY IN BAHRAIN

Consanguineous marriage has been a historically longstanding practice among the different social classes of Bahrain. Several studies were performed to determine the rate, and to study the trend to see if the frequency of cousin marriages has changed over time.

In 1990, the first study was performed on a group of 500 young married Bahraini women, in which each participant was asked to complete a standard questionnaire that included questions about the family relationship between the husband and the wife, and the relationship between their parents. The questionnaire thus provided information about 1000 couples back to the grandparents' generation. The rate of cousin marriage was estimated to be 39.4% in the present generation (fathers and mothers) and 45.5% in the previous generation (grandfathers and grandmothers), indicating a high rate of consanguinity with a significant decrease over time. The rate of first cousin marriage was evaluated as 21% (Al-Arrayed, 1995).

In 1998, through a National Family Health Survey, a total of 4,166 households were interviewed, including 26,723 participants. The overall consanguinity rate was estimated at 32%, including 24% for first cousin marriages, hence marking a significant gradual decline in the rate of marriages between relatives (Tawfeeq et al., 2000).

The third study was a neonatal screening for hemoglobinopathies performed in 2002, including a questionnaire with the intent to gather information about consanguinity. Marriage between relatives occurred at the rate of 20%, while the rate of first cousin marriages was 12.5%. Marriages between distant relatives comprised a total of 7% (Al-Arrayed, 2005).

The fourth figure came from the premarital counseling study conducted in 2006, where 500 couples participated in the study, which revealed the consanguineous marriage rate to be 20%, with 11.4% being marriages among first cousins and 8.6% among distant relatives (Table 53-1). About 80% of couples were nonrelated; hence, a gradual decline in the consanguinity rate was observed (Al-Arrayed & Al-Hajry, unpublished observations).

GENETIC BLOOD DISORDERS

Bahrain has been one of the first countries in the region to tackle the issue of genetic blood disorders, since 1986.

TABLE 53-1 PREVALENCE OF CONSANGUINEOUS MARRIAGES IN BAHRAIN

Year Type of Study Sample Size Relationship	1990 Previous Generation (G. Fathers & G. Mothers) n = 1000	1990 Current Generation n = 500	1998 Family Study n = 26,723	2002 Neonatal Sreening n = 2000	2006 Premarital Counseling n = 500
Not related	45.5%	60.6%	68%	80%	80%
Related	4535%	39.4%	32%	20%	20%
First cousins	24.5%	21.0%	24%	12.5%	11.4%
Second cousins	7.9%	7.8%	–	–	–
Distance relatives	7.1%	6.8%	8%	7%	8.6%

Genetic blood diseases are frequent in Bahrain, as in all the Middle Eastern countries. A history of malaria characterized the Gulf region, where it was endemic until eradicated in 1970, resulting in frequent manifestation of malaria-related diseases.

SICKLE CELL DISEASE (SCD)

Previous neonatal screening in 1984–1985 showed that the birth prevalence of SCD was estimated to be 2.1%, with Sickle Cell trait (SCT) being 11% and glucose-6-phosphate dehydrogenase (G6PD) deficiency being 25% (Mohammed et al., 1992; Nadkarni et al., 1991). In 2004, SCD was found to be the first cause of admission to SMC. The number of hospital admissions due to SCD gradually increased, with 2600 patients admitted during 2005, which formed 6.5% of the total admissions to SMC (Bahrain Health Statistics Reports, 2004–2005).

Several studies have been conducted on genetic blood disorders in Bahrain. One such study depicted the nature of SCD among the population. The finding revealed exposure to cold to be the principal precipitating factor of painful crisis (45% of cases), with fever (35%) and exhaustion (35%) being the next most common. Moreover, the most common symptoms were pain and fever. The main signs were anemia (92%), hepatomegaly (64%), jaundice (64%), and urinary tract infection (30%; Al-Arrayed, 1994; Al-Arrayed & Haites, 1995).

Another study sought to evaluate the clinical presentations and management of 200 Bahraini SCD patients who attended the SMC accident and emergency department for vaso-occlusive crisis during the period January–March 1994. The male:female ratio was estimated to be 2:1, with 60% of the group in the age range 15–30 years. Extremity pain was the most common presenting feature (86% of patients), followed by pain in the abdomen/generalized body ache (71%). Most patients (83%) responded to treatment with hydration, narcotic analgesics, or nonsteroidal anti-inflammatory drugs, and were discharged from the hospital. An analysis of the spectrum of manifestations suggested that the Bahrainis have a mild form of SCD (Al-Arrayed & Hamza, 1995). Molecular genetic studies were undertaken to determine the haplotypes of chromosomes carrying the sickle cell allele in Bahraini patients, and hence to consider the possible origin of these alleles (Al-Arrayed, 1995). The beta S gene was found to be linked to the Asian haplotype in 90%, and to the S2 haplotype in 5% of the cases. The Bantu haplotype was found in few patients (only 2.5%), and the haplotype was found in association with beta-thalassemia chromosomes in only 2.5% of the cases. The study showed that the Asian haplotype, which is linked to a benign form of sickle cell disease, is predominant in Bahrain (Al-Arrayed et al., 1995). Jassim and Al-Arrayed (2006) reported on the molecular basis of the benign form of sickle cell/beta-thalassemia syndrome in two healthy Bahraini patients. The aim was to study the different molecular determinants that might cause an extremely mild form of sickle cell/beta-thalassemia syndrome among the population. Two healthy Bahraini students belonging to two unrelated families with normal clinical pictures were noticed to have the sickle cell/beta-thalassemia syndrome through hemoglobin electrophoresis. Different molecular genetic techniques were employed to study blood samples from these girls. Three different molecular determinants were found in these students: compound heterozygosity for the sickle cell mutation, and beta-thalassemia mutation. Their corresponding haplotypes were (1) the Saudi-Indian haplotype for the sickle cell mutation and (2) –88 (C-to-A) beta-thalassemia mutation. (3) Alpha-globin gene mapping revealed homozygosity for the rightward deletion (–α3.7/–α3.7/) for both the students. The study concluded that different molecular determinants were found in association with the mild form of sickle cell/beta-thalassemia disease; namely, inheritance of a mild beta+-thalassemia mutation, an HbS haplotype-associated high (hemoglobin F) HbF expression, and coinheritance of alpha-thalassemia. All of these modulators were responsible for such a mild state of sickle cell disease in these patients.

Beta-Thalassemia

Studies have indicated that the incidence of beta-thalassemia among Bahraini students is 0.02%, while the trait rate remains between 2.5% and 3.5% (Al-Arrayed et al., 2003). The number of Bahraini babies born with beta-thalassemia has shown a decline from four babies annually to two during the same period of 20 years. Jassim and colleagues (2000) studied a total of 80 Bahraini individuals, of whom 35 were transfusion-dependent beta-thalassemia major patients, 37 presented with the trait, and 8 were sickle cell/beta-thalassemia patients. The application of different molecular techniques on 67 beta-thalassemia alleles identified 12 different mutations. However, IVS-I-3' end (-25bp), CD 39 (C-to-T), IVS-II-5 (G-to-C), IVS-II-2 (G-to-C), and IVS-II-2 (G-to-A) accounted for more than 80% of the total studied alleles. Other mutations common both to the Mediterranean basin [IVS-I-1 (G-to-A), IVS-II-1 (G-to- A), IVS-I-110 (G-to-A)] and the Indian subcontinent [CD8/9 (+G), CD 15 (G-to- A), CD 41/42 (-TCTT)] were also reported.

Alpha-Thalassemia

Newborn screening in 1985 showed a high frequency of alpha-thalassemia in Bahrain, at about 24% (Mohammed et al., 1992; Nadkarni et al., 1991), but only a few patients presented with manifestation of hemoglobin H (HbH) disease. There were no reported cases of hydrops fetalis. Jassim et al. (1999, 2001) reported on the molecular basis of alpha-thalassemia in Bahrain. Various polymerase chain reaction (PCR)-based methodologies were involved, namely differential PCR amplification, PCR-restriction fragment length polymorphism (PCR-RFLP), and direct PCR-amplified genomic DNA sequencing. Five alpha-thalassemia determinates were identified, which included three deletional types—the rightward 3.7 kilobase (kb) deletion, the leftward 4.2 kb deletion, and the penta nucleotide deletion in 5' splice donor side of intron I in the alpha 2-globin gene (GGTGAGG GG**GGTGAGG→GG-----**)—and two non-deletional alpha-thalassemia determinants, the Saudi type polyadenylation (polyA) signal mutation (AATAAA AATAAG) and the Turkish type polyA signal mutation (AATAAA AATGAA) in the alpha 2-globin gene. The study concluded that the three alpha-thalassemia mutations—the Saudi type polyA signal mutation, the pentanucleotide deletion, and the rightwards 3.7 kb deletion—accounted for 97% of all alpha-thalassemia determinants in Bahrain. Clearly, the interaction between the deletion alleles and the non-deletion alleles provides a complex picture in the phenotype.

Al-Mukharraq (1999) studied HbH disease in 26 Bahraini patients, followed by an assessment of the severity of anemia. The mean age at diagnosis was estimated to be 6.7 years. The mean hemoglobin was 8.1 g/dl, and the mean HbH was 17.4%. Fifty percent of these patients required blood transfusion, 17 children had average growth, while 9 were below average. There were no significant thalassemic bone changes or splenomegaly observed, thus depicting a mild to moderate clinical picture.

GLUCOSE-6-PHOSPATE DEHYDROGENASE DEFICIENCY

Glucose-6-phosphate dehydrogenase (G6PD) deficiency, a genetic defect underlying a biochemical red cell abnormality, occurs worldwide with variable incidence and clinical presentation among different ethnic groups. The disease may express itself as a drug-induced hemolytic anemia with neonatal jaundice, hemolytic anemia following infection, chronic nonspherocytic hemolytic anemia, favism, or may be asymptomatic. G6PD deficiency is an X-linked disorder. The incidence of G6PD deficiency has been observed to be higher in the Arab World, and is in the range of 8% to 50% in different populations. The frequency in Bahrain is between 20%–23%, but with few clinical problems (Bhagwat & Bapat, 1987; Mohammed et al., 1992; Al-Arrayed et al., 2003). Al Momen et al. (2004) reported on the molecular homogeneity of G6PD deficiency in Bahrain. DNA extraction was done for 83 G6PD-deficient subjects and 80 normal controls. Combinations of PCR-RFLP and PCR-DGGE procedures were employed to uncover the sequence variations at nt 563 and nt 1311 in subjects with deficient and with normal G6PD activity. The study showed that nearly 90% (93/102) of the X-chromosomes from G6PD-deficient subjects had nt 563 (C-to-T; G6PD Med) mutation, whereas ~9% of the X-chromosomes from G6PD-deficient subjects might have had other G6PD variant(s) or normal X-chromosomes in heterozygote females. Ninety-six percent (89/93) of the G6PD Med-bearing X-chromosomes showed thymine (T) at nucleotide position 1311. In contrast, 70% (82/117) of the normal X-chromosomes showed cytosine (C) at nucleotide position 1311, while it was thymine (T) in 30% (35/117) of the normal X-chromosomes. The study concluded that the vast majority (91%) of X-chromosomes from G6PD-deficient subjects in Bahrain are harboring the nt 563 (C-to-T) mutation. The G6PD Med variant in Bahrain is in tight linkage disequilibrium with thymine (T) at nt 1311. These data revealed a high molecular homogeneity of G6PD deficiency in Bahrain.

HEMOPHILIA

Hemophilia is an X- linked condition common in some Arab countries, but not common in Bahrain. Currently, only 10 patients are under treatment at SMC.

Frequencies of ABO Blood Groups and Rh Types

Al-Arrayed and colleagues (2001) estimated the frequencies of ABO and Rh phenotypes along with the respective gene frequencies in two study groups: (1) 5675 Bahraini school students in the aged 16–20 years, and (2) 7362 adult Bahraini blood donors. The frequencies of ABO groups in both study groups showed Group O>Group B>GroupA>Group AB, where group O contributed to nearly 50% of the total. Both study groups also showed >90% frequency of the RhD positive phenotype. In conclusion, the frequencies of ABO and Rh phenotypes in Bahrain were similar to those reported from other countries in the Arabian Gulf region. These frequencies appear to be intermediate between the frequencies seen in Europe and Southeast Asia.

CAMPAIGN TO CONTROL HEREDITARY BLOOD DISEASES

Bahrain successfully implemented a campaign to control hereditary blood diseases. This campaign included education and public awareness events, as well as antenatal, carrier, student, premarital, newborn, and prenatal screening methods followed by genetic counseling. Information booklets were prepared and distributed widely in schools and to the public, in an attempt to increase awareness. In 1991, the Bahrain Hereditary Anemia Society was established to further strengthen these efforts. In 1993, a premarital counseling (PMC) service was started, and in 1998, a student screening project was initiated (Al-Arrayed et al., 1997). Additionally, attempts aiming at screening Bahraini newborns for these genetic diseases will soon be put into practice. Also, a molecular genetic laboratory has been established to study the genotype of difficult cases. As a consequence of all these efforts, which have continued for about 20 years, evidence has shown a decline in the incidence of sickle cell disease by 50%–60%, from 2.1% to 0.9%.

PREMARITAL COUNSELING SERVICES

Premarital counseling is intended to identify couples at high risk, by requiring all couples intending to marry to undergo screening through history taking, physical examination, and further laboratory investigations. In 1993, a voluntary premarital counseling (PMC) service was established, which successfully contributed toward reducing the number of affected newborns, thereby reducing morbidity and mortality rates. More recently a law (Government Gazette, 2004) has been passed by the Bahraini Government that requires all Bahraini couples planning to marry to undergo a mandatory premarital counseling. Any citizen about to get married, even if the spouse is non-Bahraini, must now undergo a premarital checkup, which includes hereditary, infectious, and other diseases, according to the regulation issued by the Minister of Health. According to the results of the checkup, couples are given advice to undergo further investigation if needed, including treatment, health education, and counseling. Couples at risk of hemoglobinopathies are referred to the genetic department of SMC.

In addition, couples receive counseling regarding high-risk behaviors, including those related to HIV, Hepatitis B, and other infectious diseases, which facilitates early detection and treatment of some sexually transmitted diseases, and helps providing immunizations as required. It also promotes awareness regarding reproductive health, family planning, and healthy lifestyles, and provides couples with medical, social, and psychological support when needed. After successful counseling of both partners, certificates are issued, therefore leaving the decision of marriage to the couple even if they are not inclined to take the advice with respect to the genetics aspect. At-risk couples are followed up, if they decide to marry, and all sorts of medical care is extended to their children. The development of this law included wide consultation with all stakeholders to ensure that sociocultural mores, theological issues, and aspects of human rights had been considered. The possibility of issuing a law making premarital counseling obligatory was raised by the Ministry of Health with the Shura Council (a consultative body) after a thorough discussion, with the subsequent approval of the ministerial cabinet, and was then forwarded to the parliament to pass into law in 2004. In view of its high degree of sensitivity, the draft law was subjected to extensive consultation involving a wide range of stakeholders, and subsequently it received positive support and backing from the Islamic clergy. The WHO guidelines and recommendations (WHO, 1983) were followed at all stages of the project. In addition, the recommendations of the Genomics and World Health Report (WHO, 2002)—providing guidance on the ethical, legal, and social implications (ELSI) in genetic screening, while laying stress on the key points of the informed consent, confidentiality, stigmatization and discrimination—were carefully considered prior to implementation of the new law.

During 2005, which was the first year of the implementation of the law, 9107 clients underwent observation. Out of these, 97 (1%) cases presented with sickle cell disease and 1176 (12.9%) were carriers of the disease. A beta-thalassemia major status was reported in 19 patients (0.002%) and 239 individuals (0.03%) were carriers for beta-thalassemia. In addition, G6PD deficiency was reported in 2037 patients (22%). These frequency figures were similar to those obtained from a student screening program (Bahrain Health Statistics Reports, 2005; Al-Arrayed, 2006).

STUDENT SCREENING

The student screening for genetic blood diseases started in 1999. Covering about 50,000 students to date, the aim of the program is to raise awareness among the

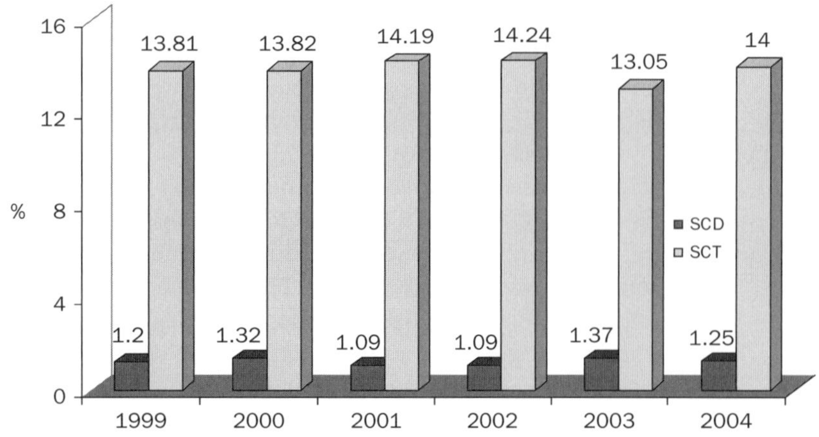

Figure 53-1 Prevalence of sickle cell disease (SCD) and sickle cell trait (SCT). Data obtained from the student screening program for genetic blood disease (1999–2004).

younger generation about these diseases through a comprehensive educational campaign, followed by screening. It also identifies carriers, thereby empowering them to make informed reproductive decisions in the future. The project operates for about six months on average every year for planning, education sessions, blood collection, laboratory testing and data processing, distribution of report cards, data analysis, and delivery of results.

The public was initially resistant to the screening program because they feared that girls identified as carriers would not be able to find husbands. The education campaigns, therefore, stressed that (a) all people carry some defective genes; (b) carriers could avoid the risk of giving birth to an affected child if they married a non-carrier; and (c) no one, including carrier–carrier couples, would be prevented from marriage. Broad-based public campaigns were also required to encourage parents and students to agree to undergo screening. The screening is equally applicable for both male and female students. Testing is stated to be voluntary and permission of the parents is sought before screening, or directly from the students in case they are 18 years or older. So far, approximately 80%–85% of parents and students have voluntarily agreed to be tested each year. In order to maintain privacy, test results are returned to students on a standardized official medical report card and are delivered via their school in a sealed envelope. In order to further protect their privacy and confidentiality, students are encouraged to open their envelopes at home with their families. Certainly, this system also helped them in choosing their appropriate future partners.

Each school receives reports on the prevalence of the three diseases in their student body. At the same time, the program also ensures that the Ministry of Health obtains accurate and updated statistics, which assists them in planning for future services. The Ministry of Education also observes the prevalence of these diseases among students, enabling them to take necessary steps to ensure improved health of the students. The study also yields important information about the frequencies of different types of abnormal hemoglobin levels among the population (Al-Arrayed et al., 2003; Figure 53-1). The study showed that the frequencies of HbD A = 0.66%, HbD D = 0.02%, HbE E = very rare, and HbE A = 0.13%. Surely, the benefits and effects of these measures will be seen in the improved genetic health of present and future generations (Al-Arrayed, 2005).

CONGENITAL ABNORMALITIES

The incidence of congenital abnormalities in Bahrain was observed in two different periods. The first study was carried out to determine the incidence of these anomalies in Bahrain by referring to the Ministry of Health statistics for the 11-year period, 1980 to 1990, while the second study looked at the same abnormalities in the period 1991 to 2000 (Table 53-2 and 53-3).

TABLE 53-2 AVERAGE ANNUAL INCIDENCE RATES OF CONGENITAL ANOMALIES (PER 1000 BIRTHS) IN BAHRAIN FOR THE PERIODS BETWEEN 1980–1990 AND 1991–2000.

Anomaly	1980–1990	1991–2000
Neural tube defects	0.95	2.68
Coronary heart defects	1.32	3.35
Respiratory	0.12	0.26
Cleft palate	0.35	0.9
Gastrointestinal	0.09	1
Genitourinary	1.93	2
Chromosomal	0.9	1.2
Musculoskeletal	2.07	3
Skin	–	0.3
Others	–	1.6
Total	27 per 1000	16 per 1000

TABLE 53-3 ANNUAL INCIDENCE RATES (PER 1000 BIRTHS) OF DIFFERENT ANOMALIES OBSERVED BETWEEN 1991 AND 2000 (ADAPTED FROM THE BAHRAIN HEALTH STATISTICS REPORTS, 1991-2000).

Anomaly	1991	1992	1993	1994	1995	1996	1997	1998	1999	2000	
Neural tube defects	3.3	3.2	2.2	2.2	2.0	2.6	2.1	3.6	2.6	2.68	
Congenital heart	2.7	2.25	3.7	3.1	1.8	4.2	3.6	4.8	2.4	3.35	
Respiratory	0.4	0.2	0.4	0.16	0.15	0.23	0.46	0.23	0.06	0.6	
Cleft palate	1.3	1.5	0.98	0.8	0.7	0.6	1.3	0.9	0.2	0.9	
Gastrointestinal	1.7	1.1	1.4	0.9	0.7	0.4	0.8	0.7	1.1	1.6	1
Genitourinary	2.2	3.1	2.3	2.24	3.3	1.6	1.7	2.4	1.3	0.96	2
Chromosomal	1.3	1.4	1.2	1.9	1.1	0.86	1.7	1.3	0.9	1.2	1.2
Musculoskeletal	5	4.3	3.2	2	3.5	1.9	2.3	4	3.7	0.73	3
Skin	0.5	0.1	0.2	0.5	1.1	0.4	0.3	0.2	0.2	0.14	0.35
Others	1.7	0.1	0.7	1.3	1.5	3	1.9	5.2	0.3	0.2	0.16

Total 16.4 per 1000.

The incidence of congenital anomalies in the first study was found to be 2.7% of live births, while it was found to be 1.7% in the second study. A study of each anomaly separately showed that anomalies of the musculoskeletal system had the highest incidence (2.28 per 1000 and 3/1000), followed by the genitourinary system (2.13 per 1000 births and 2/1000). The incidence of chromosomal disorders in the first study was 0.9 per 1000 births, while in the second it was 1.2/1000, (Al-Arrayed, 1995).

METABOLIC DISEASES

Al-Arrayed (1995) highlighted the frequency of metabolic diseases. In line with this, a neonatal metabolic screening pilot study was performed on 1000 neonates, in which electrospray tandem mass spectrometry was applied. Blood samples were collected on Guthrie cards from the heels of infants on the 3rd to 5th days of their birth. The dried samples were mailed to the mass spectrometry laboratory at King Faisal Specialist Hospital, where they were analyzed for more than 20 metabolic diseases, such as amino acids, organic acids, and carnitine esters. The results showed a high incidence of metabolic diseases.

Annually, approximately 100 neonates are expected to be affected by metabolic diseases in Bahrain, such as maple syrup urine disease, medium-chain acyl-Co-A dehydrogenase deficiency, primary carnitine deficiency, methylmalonic acidemia, and methylene tetrahydrofolate reductase deficiency. Many of these affected babies died within the first or second week, even before the diagnosis could be established. Al Jishi (unpublished observations) found an approximate prevalence of 1/10,000 for PKU, propionic acidemia, and fatty acid oxidation defect (CPTII); 2/10,000 for MSUD, isovaleric acidemia, 3-methylglutaric acidemia, and glutaric aciduria type 2; and 6/10,000 for methylmalonic acidemia, glyceric acidemia, and argino-succinic aciduria. Some other metabolic diseases that are frequently observed in Bahrain are mucopolysacharidosis (MPS), Neimann-Pick disease, GM2 gangliosidosis, Sanfilippo disease, I-cell disease, glycogen storage disease, biotinidase deficiency, and mitochondrial disorders, but further studies are needed to determine their prevalence (Al-Arrayed et al., 1999). These diseases have also been reported from other Arab countries. Ozand et al. (1990) reported that certain lysosomal storage diseases, including MPS type 1VA, Morquio disease, multiple sulphatase deficiency, Niemann-Pick disease type B, GM1 gangliosidosis type "0" (Sandhoff disease), and ceroid lipofuscinosis (Jansky-Bielschowsky and Batten-Spielmeyer-Vogt syndrome) are more frequent in Saudi Arabia. Another study (Benson et al., 1988) found that MPS type II (I-cell disease) was found to occur more frequently in Arabs than in any other ethnic group. Circumstances demand keeping track of the frequencies of these diseases, since most may be diagnosed in the prenatal period, and affected individuals could potentially receive treatment by bone marrow transplantation (Whitley et al., 1980; Schaison et al., 1989; Teebi et al., 1987; Wraith et al., 1987).

CHROMOSOMAL ABNORMALITIES

The incidence of chromosomal abnormalities was found to be relatively low in Bahrain. The incidence of Down syndrome among Bahrainis in 1993 was found to be 0.9 per 1000, compared with 1.4 per 1000 internationally, while in 1991–2000 it was found to be 1.2 per 1000 (Al-Arrayed, 1996). The number of confirmed trisomy 21 cases in the cytogenetic laboratory during 2004 and 2005 among 15,000 newborns was 12 and 17 respectively, indicating a prevalence of 0.8/1000, and 1.1/1000, respectively.

Al-Arrayed and Rajab (1995) studied the data, collected retrospectively, of 104 Bahraini patients presenting with Down syndrome from hospital records during the period 1989–1993, which included all patients suffering from Down syndrome admitted to the main hospital in Bahrain during the 5-year period. Cytogenetic analysis had been performed on 89 patients.

The mean age of patients admitted was 5 years, with 60% of patients under 1 year of age; the oldest patient was 31 years old. The most common complications were chest infections, congenital heart disease, increased susceptibility to all types of infection, anemia, and ear, nose, and throat, and eye complications. Karyotype analysis showed that 97% of those studied had free trisomy, 2% had translocation, and one patient had a mosaic karyotype. A slight increase in trisomy 13 and trisomy 18 births has been noticed during the last few years. Fragile X syndrome was reported in five Bahraini families (Al-Arrayed, 1990). Al-Arrayed (1996) also reported on cytogenetic studies on 500 Bahraini patients suspected of having chromosomal abnormalities on the basis of physical and/or developmental clinical features. Almost 27% of these patients were found to have abnormal karyotypes. Numerical abnormalities (including trisomy 21) were found in 19% of patients, whereas structural abnormalities manifested in 7%. Al-Arrayed (2003) studied inheritable chromosomal abnormality causing recurrent fetal wastage and abnormalities in children. Cytogenetic studies were performed on Bahraini patients suspected of having chromosomal abnormalities. Ten families inheriting different chromosomal abnormalities were reported.

The first family studied had one child and had six abortions. The chromosomal analysis showed interesting results. Both wife and husband presented with different chromosomal abnormalities. The wife had 45, XX, der (13; 14) (q10; q10), whereas the husband had 46, XY, inv (3) (p25q21). However, their child was phenotypically normal, though he inherited abnormal chromosomes from both parents. The second family had three mentally retarded children with Trisomy 21. The third family had a male child with 46, XY, rec (8), dup p, inv (8) (P23. 1 q 24. 2). He inherited the abnormal chromosome from his mother. In the fourth family, the patient was married to her first cousin. She had 4 p 3 D 2 A1L1. Both the wife and the husband had the same balanced reciprocal translocation, t (6; 10) (q 15; q 21.2). Their retarded baby girl presented with the same chromosomal abnormality. The fifth family had an encephalic stillbirth and an abortion. The husband had a pericentric inversion of the Y chromosome. The sixth family had two children with trisomy 21; the mother presented with mild Down syndrome features. The seventh family had four children, all mentally retarded with fragile X syndrome. The father was declared as the transmitting parent. The eighth family had a mentally retarded child with fragile X syndrome. The mother and five of her sisters were carriers for fragile X, and the grandfather was observed to be the transmitting parent. In the ninth family, the father's karyotype was 46, XY, inv. per. (2) (p12:q14). He had a balanced pericentric inversion of chromosome 2. His first child was abnormal with trisomy 13. In the tenth family, the father had 46, XY/46, XY, rea (2). The family had three abnormal mentally retarded children and had two abortions. This indicates that inherited chromosomal abnormality is not a rare condition, and that cytogenetic studies should be conducted for any family with abnormal children and with recurrent fetal wastage.

Al-Arrayed (1990, 2003) reported on a consecutive series of fifty couples with history of fetal wastage. They were cytogenetically studied with current banding techniques. Fetal wastage was defined to be occurring in the couples who had had more than two early abortions, stillbirth(s), or live birth(s) with multiple congenital anomalies. One couple was found to be carriers for a balanced reciprocal translocation. A woman was found to have Robertsonian Translocation, while another woman was found to be a mosaic (46, XX, 47, XXX). Moreover, one husband was found to have pericentric inversion of Y chromosome. These findings reveal that parental chromosome abnormalities account for fetal wastage in 8% of couples having such a history.

CYSTIC FIBROSIS

Cystic fibrosis (CF) is a hereditary multisystem disease transmitted as an autosomal recessive disorder. It leads to chronic pulmonary diseases, pancreatic enzyme deficiency, and abnormally high concentrations of electrolytes in sweat (Khan et al., 1985). Al-Arrayed and Abdullah (1996) performed an intensive retrospective search for patients with cystic fibrosis from clinical data and hospital records from SMC, with the aim of determining the prevalence of cystic fibrosis in Bahrain. The survey included 27 patients confirmed as having cystic fibrosis, born during the period 1978–1994. Almost 200,000 children were born in Bahrain during this period. Diagnosis was established by the presence of high sodium and chloride (≥ 70 mmol/L) concentrations in sweat.

The mean incidence during this period was found to be 1 in 7700. All cases were diagnosed during the first year of life, and 60% were diagnosed in the first 3 months of life. The male:female ratio was 14:13. The incidence of meconium ileus was 16%. Mortality in the neonatal period was 60%. The first cousin marriage rate among these families was 63%. Another study in 1998 reported the incidence of CF as 1/5800. The gene frequency seems to be low in Bahrain, while it is high in Caucasian population where the carrier frequency is about 1:20 to 1:25 (Warwick, 1978). Eskandarani (2002) undertook a genotypic study to characterize the cystic fibrosis transmembrane regulator (CFTR) gene mutations in the Bahraini cystic fibrosis population using a polymerase chain reaction-based direct gene test to search for 15 common CF mutations among Arabs. During the period October 2000 to May 2001, 19 patients from 13 families were recruited in the study. Patients were diagnosed as having CF, based on a typical clinical picture and sweat chloride levels (> 60 mmol/l). Eight mutations were detected in 21 of the 26 alleles examined; of the eight mutations detected, four were common among Bahrainis (2043delG > 548A-to-T > 4041C-to-G = F508, in order of decreasing frequency), accounting for 66% of the Bahraini CF alleles. However, four different heterozygous mutations were also detected; namely 1161delC, 1756G-to-T, 3120+1G-to-A, and 3661Ato- T, accounting for 16% of the Bahraini CF alleles.

CONGENITAL BLINDNESS

Congenital cataract is the most common eye condition leading to low vision and blindness in Bahrain (Al-Arrayed, 1992; Ahmed & Selvyn, 1988; Al-Alawi et al., 1998), followed by glaucoma and corneal diseases. The following causes of congenital blindness were reported in Bahrain: retinal dystrophy, congenital glaucoma, congenital cataract, retinopathy, macular degeneration, mesodermal dysgenesis, optic nerve dystrophy, retinitis pigmentosa, and Leber's optic dystrophy, while oculocutaneous albinism was observed to cause poor vision among some families. Ahmed and Selvyn (1988) found that consanguinity rate was positive in 42% of families with children with visual disability, but the consanguinity rate in the general population at that period was estimated to be 40%.

HEREDITARY DEAFNESS

Jamal examined 196 children presenting with deafness and speech delay, who were referred to a hearing-impaired children's clinic at SMC for over a period of 10 years from 1989–1996. The incidence of profoundly deaf children was estimated as 1.1 per 1,000 live births. Children with positive history of deafness in the family due to genetic causes constituted 37.8%. A review of the consanguinity rate among parents found that 48.5% were first cousins, whereas 12.2% were distantly related, which indicated high rate of consanguinity among parents. In one family, connexin 26 was found to be the responsible gene. However, further molecular studies are needed in this field (Jamal et al., 2000).

REVIEW OF THE SALMANIYA MEDICAL COMPLEX RECORDS (1988–2005)

The Salmaniya Medical Complex (SMC) is the main hospital for secondary and tertiary care in Bahrain. All rare, chronic, and difficult cases are referred to this hospital. The records were studied of patients admitted to SMC and diagnosed with genetic diseases. The record system was based on the WHO International Classification of Disease (ICD) 9 until 2002, when it was upgraded to ICD 10. The results highlight the magnitude of morbidity caused by each of these genetic diseases. The number of patients admitted into SCD during this period was estimated to be around 7000. On the other hand, the number of patients presenting with thalassemia did not exceed 100, whereas other genetic diseases occurred at lower frequencies and were recorded in dozens or less.

During this 17-year period, 14 patients were admitted with congenital hypothyroid, and the estimated prevalence for this disease was 1/2000. Patients with hemolytic anemia due to ABO were evaluated to be 21 in number. About 19 patients with leukodystrophy, 511 with disorders of lipoprotein metabolism, 138 with congenital hydrocephaly, 106 with anomalies of the spine, 10 with Alzheimer's, 15 with cerebral lipidosis, 65 with Huntington's disease, and 103 patients with multiple sclerosis were reported. The numbers of cases presented does not accurately reflect the epidemiology of these disorders in Bahrain, as there is no referral basis that ensures the investigation of all possible genetic diseases. It does, however, give an idea of the common genetic disorders among Bahrainis. Some other disorders reported include neurofibromatosis, epidermolysis bullosa, ichthyosis, G syndrome, polycystic kidney, polycystic ovaries, Parkinsonism, osteogenesis imperfecta, porphyria, hereditary 1,25-dihydroxyvitamin D-resistant rickets, and malignant hyperthermia (Khan et al., 1987, 1996; Al Jishi et al., 2004; Al-Arrayed, 1999; Aldeen et al., 1999; Kooheji, 1988; Zayyani, 1987; Mohammad & Ebrahim, 1985; Al-Ansari, 1984).

FREQUENTLY DIAGNOSED SYNDROMES IN THE GENETIC DEPARTMENT

The following diseases were frequently seen among patients attending the genetics clinic.

SPINOMUSCULAR ATROPHY

A significantly increasing number of families were found to present with spinomuscular atrophies. Four families were reported with spinomuscular atrophy (SMA), of which three had acute infantile type (Werdnig-Hoffmann, or SMA1), while one presented with chronic childhood SMA. Apparently, there was an increased frequency of SMA among Bahrainis, as three to four deaths occurred every year in the pediatric wards due to SMA1. One of the mothers had neuromuscular disorder manifested in diplopia and hoarseness of voice. Her younger brother and three of her male children had died in infancy. SMA1 carrier frequency in the Caucasian population was shown to be 1:80, and that of chronic childhood SMA was 1:90 (Pearn, 1973). However, the frequency of carriers is very high in some communities (Schap, 1985; Moosa & Dawood, 1990; Czeizel & Hamula, 1989; Czeizel, 1991). In the Occupied Territories, Fried and Mundel (1977) have found high incidence of SMA1 in the Egyptian Karaite community, where the frequency of heterozygotes was about 1:20. Chronic childhood SMA has now been mapped to chromosome 5q11.2-13.3m, with the gene of SMA1 located in the same general area (Brzustowicz et al., 1990; Melki et al., 1990; Malki, 1991).

NOONAN SYNDROME

Noonan syndrome (NS) is an autosomal dominant dysmorphic syndrome characterized by hypertelorism, a downward eye slant, and low-set, posteriorly rotated ears. Other features include short stature, a short neck with webbing or redundancy of skin, cardiac anomalies, epicanthic folds, deafness, motor delay, and a bleeding diathesis. Ten cases of NS have been reported from Bahrain. Four of them were Bahraini boys, and six were Pakistani girls (3 alive and 3 deceased). One of the Bahraini boys was a sporadic case, while the other three were siblings of one family. It was conjectured that the father could be the transmitting parent, as he had ptosis, but was normal mentally. A changing phenotype with age was found to manifest in this family (Allanson et al., 1985). The 2-year-old boy had more severe edema, ptosis, and a round face, whereas the eldest boy had EEG abnormality and seizures. Lemmi et al. (1983) reported that 71% of NS patients had electroencephalogram-graphic abnormalities, and that 82% had some type of neurological anomalies. In the cases of NS reported in the Pakistani family, all the affected girls had pulmonary stenosis. The mother had short stature of about 142 cm, but she was normal mentally, and could be the transmitting parent.

LAWRENCE-MOON BARDET-BIEDL SYNDROME

Two consanguineous Bahraini families with Lawrence-Moon Bardet-Biedl syndrome (LMBBS) were also reported from Bahrain. Out of these two families, one presented with Bardet-Biedl syndrome (BBS), with two siblings manifesting the typical clinical picture of this syndrome (Al-Arrayed & Al-Arrayed, 1991). The other family had Laurence-Moon syndrome (LMS). In 1977, Toledo et al. suggested the splitting of LMBBS into LMS and BBS as different but interrelated autosomal recessive disorders. Apparently, there is an increased frequency of this autosomal recessive syndrome among Arabs. Kalbian (1956) reported the first case of this condition in an Arab boy. Other cases were reported by Abdel Aziz in 1972, and Temtamy and Shalash in 1975. In Jerusalem, Ehrenfeld et al. (1970) reported 18 cases of BBS. These cases were descended from 12 families, with 3 of them being Arabs and 9 being Jewish. Farag and Teebi (1988, 1989) reported high incidence of BBS among Bedouins in Kuwait. They estimated the incidence in a mixed Arab population of Kuwait to be as high as 1:36,000, which was much higher than the prevalence of this disease in Switzerland (Buyse, 1979; Klien & Amman, 1969). Among 36 cases, 24 were Bedouins. They estimated the prevalence among Bedouins to be 1:13,000. The rest of the cases were from Saudi Arabia, Iraq, Syria, and Palestine (Temtamy & McKusick, 1978; Pagon et al., 1982; Bauman & Hogan, 1973; Hurley et al., 1975; Massof et al., 1979).

CARPENTER SYNDROME

Two of the families had Carpenter syndrome, with one family from Egypt and the other one from Iran. In both the families, the affected individuals were the products of consanguineous marriages.

The affected presented with acrocephaly, peculiar facies, brachydactyly, and syndactyly in the hands, and preaxial polydactyly and syndactyly of the toes.

GOLDENHAR SYNDROME

Two other families had babies with Goldenhar syndrome. One of them was Pakistani. The mother in the other family was Bahraini and the father Palestinian. The etiology of this syndrome is unknown; it is usually sporadic, but some familial cases have been reported. The risk for first degree relatives is about 2% (Burck, 1983; Wilson, 1983).

OTHER OBSERVED SYNDROMES IN BAHRAIN

Brachmann-de Lange syndrome

Smith-Lemli-Opitz syndrome

Seckel syndrome

Retinoblastoma

Achondroplasia

Wolman disease

NONCOMMUNICABLE DISEASES

Health problems in Bahrain are those generally experienced by countries passing through a stage of transition from developing to developed nations. Over the years, communicable diseases have shown a declining trend as the major causes of mortality and morbidity. They are being replaced by noncommunicable ones such as cardiovascular diseases, cancer, metabolic diseases, congenital anomalies, and accidents, as compared to spontaneous abortion/miscarriages that were the most common complications in pregnancies throughout the world. The National Family Health Survey (NFHS) in 1995 reported a dramatic increase in noncommunicable diseases (NCDs), such as cardiovascular disease (CVD), diabetes, and cancer in Bahrain. Cardiovascular diseases and cancer were the leading causes of death.

Diabetes prevalence was estimated to be between 20% and 30% depending on the study. This represented a common global trend, which affected the Arabian Gulf and the eastern Mediterranean region (EMR) exponentially. The share of noncommunicable diseases of the region's disease burden is expected to rise to 60% by the year 2020. The incidence of hypertension in the EMR area is 26%, and diabetes ranges between 7%–25%. The burden in terms of suffering and health costs is also immense. NFHS also reported that the most frequently reported chronic conditions for both men and women were high blood pressure, joint disease, and cardiac disease. Prevalence of these conditions rose steadily with age, among those aged 50 and over. Hypertension was the most common (15% for men and 27% for women), followed by diabetes (14% for men and 19% for women) and joint disease (7% for men and 18% for women), while heart disease prevalence was 7% for both men and women.

The Ministry of Health statistics for 2005 reported also on mental health. It showed that mental illnesses are increasingly becoming a universal problem in their distribution, due to the rapid changes in the country's socioeconomic status, education, and lifestyle. Although Bahrain is a very small country, it has its own share of problems in terms of frequency distributions of sociodemographic characteristics, patterns trends, and their relationships. A study was conducted on the data of inpatients at the Psychiatric Hospital over the past five years. The Psychiatric Hospital is the only hospital in the country that provides mental health care. In 2005, the distribution of discharges (1,159) according to the principal diagnosis showed that 24.2% of patients admitted were diagnosed as schizophrenics, 23% were depressed, 22% were drug dependent, and 30.8% suffered from other mental disorders. About 67% of all the discharges were male patients. It was clear that throughout the selected four age groups, males of age group 15–44 years represented the bulk of the total discharges, at 44.2%.

These statistics indicate that the noncommunicable diseases such as cardiovascular diseases, cancer, and diabetes are leading causes of morbidity and mortality in Bahrain.

DIABETES

Studies on the epidemiology of diabetes in Bahrain reported a prevalence of about 25% in those aged 20 years and over in a primary-care-based study (Zorba, 1996) and 30% in people aged 40–69 years in a population-based study (Al Mahroos, 1998). On the other hand, the study of Bahrain Family Health Survey, showed a total prevalence of 5.5% of diabetes mellitus and 6.3% of hypertension.

CARDIOVASCULAR DISEASES

Cardiovascular disease (CVD) is considered the leading identifiable cause of death in Bahrain. These diseases were reported to be responsible for 28.24% of deaths in 2003. The Health Statistics Report for 2005 states that cardiovascular diseases constitute the highest single cause of mortality in Bahrain, representing 69.9 per 100,000 of the population, accounting for more than 22% of total deaths at Salmaniya Medical Complex. The rate of deaths from circulatory disease distributed by gender showed 60.6 mortalities per 100,000 males, and 59.1 mortalities per 100,000 females.

This trend, however, has not changed dramatically since the seventies. Coronary heart disease (CHD; seems to be the predominant type of cardiopathy encountered in Bahrain. CHD was also found to be the fourth leading cause of admission from SMC in 2003. Coronary heart disease (CHD) and stroke are the predominant type of CVD encountered in clinical practice.

Known risk factors for CVD, such as smoking and raised blood cholesterol, and risk factors, such as lack of physical activity, obesity, and alcohol consumption, are expected to have increased in Bahrain over the last two decades. In addition to that, the continuing rise in the incidence of cardiovascular diseases is also associated with the rise in the size of the population over-65 years of age, who represent nearly 2.5% of the total population.

CANCER

Cancer is the third leading cause of death in Bahrain, accounting for 11.9% of all deaths (Annual Health Statistic Reports, 2004). The age-standardized rate for incidence of cancer in Bahrain is 167 and 141 per 100,000 males and females, respectively. However, these rates are considerably lower than the corresponding figures reported in industrialized countries. More than 300 individuals are diagnosed with cancer annually, with an estimated incidence rate of 66/100,000 population. Lung cancer is the leading type of cancer among males, accounting for 14.5% of registered cases in 1998, and 17.25% of Bahraini patients. The crude incidence rate was 6/100,000 of the population, and among Bahrainis it was 11.6/100,000. Breast cancer is the leading cancer among females, accounting for 30% of all new cases diagnosed in 1998 and 27% of Bahraini patients. The crude incidence was 17/100,000 of the female population and 18/100,000 of Bahraini females.

In general, healthcare in Bahrain has improved during the last two decades, and gradually the problems of genetic disorders are becoming a subject of interest. We conclude that much has to be done to study and reduce the rate of these genetic disorders among Bahrainis.

REFERENCES

Abdel Aziz M. (1972). Laurence Moon Beidel syndrome. *Bull Ophthalmol Soc Egypt 62*, 233–235.

Ahmed A, Selvyn D. (1988). Causes of low vision and blindness in Bahrain. *Bahrain Med Bull 10*(3), 142–144.

3-Al Jishi, Adel A, Sreekantaswamy. (2004). Fatal neurological manifestations of acute intermittent porphyria. *Bahrain Med Bull 26*, 67–69.

Al Mahroos FA. (1998). Obesity, physical activity and prevelance of diabetes in Bahraini Arab population. *Bahrain Med Bull 20*, 114–118.

Al-Alawi E, Ahmed A, Fateha B. (1998). Causes of visual disability at Al-Noor Institute for Blind. *Saudi J Ophthalmol 2*(2), 101–105.

Al-Ansari A. Gilles de la Tourette's syndrome. (1984). *Bahrain Med Bull 6*(3), 109–110.

Al-Arrayed H. (1992). Cataract blindness: A world health problem for 20 million people. *Bahrain Med Bull 14*(1), 3–4.

Al-Arrayed S, Hamza A. (1995). A survey of patients with sickle cell presentation to accident and emergency department of SMC in Bahrain. *J Bahrain Med Soc 7*(2), 105–112.

Al-Arrayed S. (2005). Campaign to control genetic blood diseases in Bahrain. *Comm Genet 8*, 52–55.

Al-Arrayed S. (1990). Cytogenetics of fetal wastage in Bahrain. *Bahrain Med Bull 12*(1), 21–24.

Al-Arrayed SS, Shome DK, Hafadh N, et al. (2001). ABO blood group and Rhd Phenotypes: Results of screening school children and blood donors. *Bahrain Med Bull 23*(3), 112–125.

Al-Arrayed SS, Abdullah F. (1996). Incidence of cystic fibrosis in Bahrain. *J Bahrain Med Soc 8*, 157–160.

Al-Arrayed SS, Al-Arrayed H. (1991). Bardet Beidel syndrome in a Bahraini family. *Bahrain Med Bull 13*(1), 38–40.

Al-Arrayed SS, Al-Hajry A. (2006). Premarital counseling services. J Health, Social Environ Issues 6(2).

Al-Arrayed SS, Hafadh N, Al-Serafi S. (1997). Premarital counseling: an experience from Bahrain. *East Mediterr Health J 13*(3), 415–419.

Al-Arrayed SS, Hafadh N. Amin S, Al-Mukhareq H, Sanad H. (2003). Student screening for inherited blood disorders in Bahrain. *East Mediterr Health J 9*(3), 344–351

Al-Arrayed SS, Haites N. (1995). Features of sickle-cell disease in Bahrain. East Mediterr Health J 1(4), 112–119.

Al-Arrayed SS, Rajab E. (1995). Morbidity of Down syndrome among the hospital population of patients in Bahrain. *J Bahrain Med Soc 7*(1), 13–16.

Al-Arrayed SS. (1995). Beta globin gene haplotype in Bahraini patients with sickle cell anemia. *Bahrain Med Bull 17*(1), 15–20.

Al-Arrayed SS. (1996). Chromosomal abnormalities in 500 referred cases in Bahrain. *Bahrain Med Bull 18*(1), 2–4.

Al-Arrayed SS. (1995). Epidemiology of congenital abnormalities in Bahrain. *East Mediterr Health J 1*(2), 248–252.

Al-Arrayed SS. (2003). Familial chromosomal abnormality. 1st GCC Genetics Conference.

Al-Arrayed SS. (1994). Nature of sickle-cell disease in Bahrain. *J Bahrain Med Soc 6*(3), 125–130.

Al-Arrayed SS. (1999). Review of the spectrum of genetic diseases in Bahrain. *East Mediterr Health J 5*(6), 114–120.

Al-Arrayed SS. (1995). The frequency of consanguineous marriages in the state of Bahrain. *Bahrain Med Bull 17*(2), 63–66.

Aldeen HAZ, Al-Hilli FA, Mohammed AM, Jamsheer NS, Ras Romani KH. (1999). Wolman's disease: The first two cases from Bahrain. *Bahrain Med Bull 21*(1), 24–26.

Allanson JE, Hall JG, Hughes HE, Preus M, Witt RD. (1985). Noonan syndrome: the changing phenotype. *Am J Med Genet 21*(3), 507–514.

Al-Momen NJ, Al-Arrayed SS, Al-Alawi AA. (2004). Molecular homogenity of G6PD deficiency. *Bahrain Med Bull 26*(4), 139–142.

Al-Mukharraq HJ. Hemoglobin H disease in Bahrain. (1999). *J Bahrain Med Soc 11*(3), 7–11.

Bahrain Health Statistic Reports, 1985–2005

Bauman ML, Hogan GR. (1973). Laurence Moon Biedl syndrome, report of two untreated children. *Am J Dis Child 123*, 119–126.

Benson FE, Fahmy NA, Fenson AH. (1988). Is there an ethnic predisposition to mucolipidosis type II? *Saudi Med J 9*(4), 381–388.

Bhagwat GP, Bapat JP. (1987). Glucose-6-phosphate dehydrogenase deficiency in Bahrain blood donors. *Bahrain Med Bull 9*(3), 120–122.

Brzustowicz LM, Lehner T, Castilla LH. (1990). Genetic mapping of chronic childhood onset spinal muscular atrophy to chromosome 5q11.2-13.3. *Nature 344*, 540–541.

Burck U. (1983). Genetic aspects of hemifacial microsomia. *Hum Genet 64*, 291–296.

Buyse M. (1979). Laurence Moon Biedl syndrome. In: Bergsma D, ed. *Birth Defect Compendium*. The National March of Dimes. New York, Alan R Liss.

Cancer Statistics, Ministry of Health. (2006). http://intranet.health.gov.bh/Cancer/.

Central Statistics Organization (2006). Available at www.bahrain.gov.bh.

Czeizel A, Hamula J. (1989). A Hungarian study on Werding Hoffmann disease. *J Med Genet 26*(12), 761–763.

Czeizel A. (1991). High incidence of acute infantile spinal atrophy in Hungary. *Hum Genet 86*(5), 539.

Ehrenfeld EN, Rowe H, Auerbach E. (1970). Laurance Moon Biedel syndrome in Israel. *Am J Opthalmol 70*, 524–532.

Eskandarani HA. (2002). Cystic fibrosis transmembrane regulator gene mutations in Bahrain. *J Trop Pediatr 48*(6), 348–350.

Farag TI, Teebi AS. (1988). Bardet Biedl and Lawrence Moon Biedl in a mixed Arab population. *Clin Genet 33*, 78–82.

Farag TI, Teebi AS. (1989). High incidence of Bardet Biedl syndrome among the Bedouin. *Clin Genet 36*(6), 463–464.

Fried K, Mundel G. (1977). High incidence of spinal muscular atrophy type 1 (Werding –Hoffmanns disease) in the Karaite community in Israel. *Clin Genet 12*, 250–251.

Government Gazette. (2004). Issue 2640, June 23.

Hurley RM, Dery P, Norady MB, Drummond KN. (1975). The renal lesion of the Laurence-Moon-Biedl syndrome. *J Pediatr 87*(2), 206–209.

Jamal A, Tammam NH, Buallai AK. (2000). Etiology of sensory neural hearing loss in children in Bahrain: a 10 years study. *J Bahrain Med Soc 1*(3), 147–150.

Jassim N, Al-Arrayed S, Gerard N, et al. (1999). A mismatched-primer polymerase chain reaction/restriction fragment length polymorphism strategy for rapid screening of the polyadenylation signal mutation alpha(T-Saudi) (AATAAA-->AATAAG) in the alpha2-globin gene. Hemoglobin 23(3), 213–220.

Jassim N, Al-Arrayed SS, Gerard N, et al. (2001). Molecular basis of alpha thalassemia in Bahrain. *Bahrain Med Bull 23*(1), 3–7.

Jassim N, Al-Arrayed SS. (2006). Molecular basis of the benign form of sickle cell-beta thalassemia syndrome. *Bahrain Med Bull, in press.*

Jassim NM, Al-Arrayed SS, Al-Mukhareq H, Merghoub T, Krishmamoorthy R. (2000). Spectrum of beta thalassemia in Bahrain. *Bahrain Med Bull 22*(1), 8–12.

Kalbian VV. (1956). Laurence Moon Biedl syndrome in an Arab boy: Familial incidence. *J Clin Endocr 14*, 1622–1625.

Khan IM, Al-Hermi B, Al-Abbasi AJ, Al-Ruffai M, Al-Arrayed SS. (1996). Hereditary 1,25-Dihdroxyvitamin D resistant rickets: a report of several cases in a kindred. *Bahrain Med Bull 18*(1), 28–30.

Khan IM, Al-Mukhareq H, Mohammed AM. (1987). G syndrome: A case report. *Bahrain Med Bull 9*(2), 87–90.

Khan IM, Mohammad AM, Akbar M. (1985). Mucoviscidosis: cystic fibrosis of the pancreas in Bahrain, Arabian Gulf. *Bahrain Med Bull 7*(1), 17–23.

Klien D, Amman F. (1969). The syndrome of Laurence Moon Biedl and allied disease in Switzerland. *J Neurol 9*, 479–513.

Kooheji AJ. (1988). Malignant hyperthermia. *Bahrain Med Bull 10*(3), 148–151.

Lemmi H, Summitt RL, Wilory RS, Duenas D. (1983). Electroencephalographic findings in the noonan syndrome. *Clin Electroencephalaog 4*, 4–8.

Malki J. (1991). Localization of the spinal muscular atrophy gene by reverse genetic methods. Prospect of a gene on chromosome 5. *Rev Prat 41*(18), 1677–1679.

Massof RW, Finkelstein D, Starr SJ. (1979). Bilateral symmetry of vision disorders in typical retinitis pigmentosa. *Br J Opthalmol 63*, 90–96.

Melki J, Abdelhak S, Sheth P. (1990). Gene for chronic proximal spinal muscular atrophies maps to chromosome 5q. *Nature 344*(6268), 767–768.

Ministry of Health. (2004). *Guidelines for Premarital Counseling Services*. The Bahrain Ministry of Health, Manama

Mohammad AM, Ebrahim AHK. (1985). Maple syrup urine disease in Bahrain. *Bahrain Med Bull 7*(3), 114–117.

Mohammed A, Al-Hilli F, Nadkarni K, Bhagwat G, Bapat J. (1992). Hemoglobinopathies and glucose-6-phosphate dehydrogenase deficiency in hospital birth in Bahrain. *Ann Saudi Med 12*, 536–539.

Moosa A, Dawood A. (1990). Spinal muscular atrophy in African children. *Neuropediatr 21*(1), 27–31

Nadkarni K, Al-Arrayed SS, Bapat J. (1991). Incidence of genetic disorders of haemoglobins in the hospital population of Bahrain. *Bahrain Med Bull 13*(1), 19–23.

Ozand PT, Gascon G, al Aqeel A, Roberts G, Dhalla M, Subramanyam SB. (1990). Prevalence of different types of lysosomal storage diseases in Saudi Arabia. *J Inherit Metab Dis 13*(6), 849–861.

Pagon RA, Haas JE, Bunt AH, Rodaway KA. (1982). Hepatic involvement in the Bardet-Biedl syndrome. *Am J Med Genet 13*(4), 373–381.

Pearn J H. (1973). The gene frequency of acute Warding Hoffmann disease (SMA). A total population survey in North East England. *J Med Genet 10*(3), 260–265.

Rasromani KH, Kamath MB. (1989). The incidence of correctable anomalies in Bahrain. *Bahrain Med Bull 11*(2), 64–67.

Schaison G, Bordigoni P, Leverger G. (1989). Bone marrow transplantation for genetic and metabolic disorders. *Nouv Rev Fr Hematol 31*(2), 119–123.

Schap T. (1985). Warding Hoffman disease on Reunion Island: A founder effect. *Clin Genet 27*, 617–619.

Tawfeeq N, et al. (2000). Bahrain National Family Health Survey, BFHS, 1995: Principle report. Manama: Ministry of Health, ISBN 9960-15-100-X, 83-96. http://www.moh.gov.bh/PDF/hs2005/hs2005_e.html.

Teebi AS, Al-Awadi SA, Farag TI, Naguib KK, El-Khalifa MY. (1987). Phenylketonuria in Kuwait and Arab countries. *Eur J Pediatr 146*(1), 59–60.

Temtamy SA, McKusick VA. (1978). The genetics of hand malformations. *Birth Defects Orig Artic Ser 14*(3), i-xviii, 1–619.

Temtamy SA, Shalash BA. (1975). The Laurance Moon Biedl syndrome in sibs. *Birth Defect X1 2*, 402–407.

Toledo SP, Medeiros GA, Knobel M. (1977). Evaluation of the hypothalamic pituitary gonadal function in the Bardet Biedl syndrome. *Metabolism 26*, 1277–1291.

Warwick WJ. (1978). The incidence of cystic fibrosis in Caucasian populations. *Helv Paediatr Acta 33*(2), 117–125.

Whitley CB, Draper KA, Dutton CM, Brown PA, Severson SL, France LA. (1980). Diagnostic test for mucopolysacharidosis 11. Rapid quantification of glycosaminglycan in urine samples collected on a paper matrix. *Clin Chem 35*(10), 2074–2081.

WHO. (1982). Hereditary anaemias: genetic basis, clinical features, diagnosis, and treatment. WHO working group. *Bull World Health Organ 60*(5), 643–660.

WHO. (1983). Community control of hereditary anaemias: memorandum from a WHO meeting. *Bull World Health Organ 61*(1), 63–80.

WHO. (2002). *ACHR recommendations on ELSI of human genetics*. Available at: http://www.who.int/genomics/elsi/recommendations/en/.

Wilson GN. (1983). Cranial defects in the Goldenhar syndrome. *Am J Med Genet 14*, 435–443.

Wraith JE, Rogers JG, Danks DM. (1987). The mucopolysacharidosis. *Aust Paediatr 23*(6), 329–334.

Zayyani NR. (1987). Familial Mediterranean fever in the Arabian Peninsula. *Bahrain Med Bull 9*(3), 130–136.

Zorba FI. (1996). Prevalence of diabetes mellitus in Bahrain. *Bahrain Med Bull* 18, 44–51.

54.

GENETIC DISEASES IN THE SULTANATE OF OMAN

Anna Rajab and Michael A. Patton

INTRODUCTION

Oman is situated in the southeast of the Arabian Peninsula along the east coast of the Persian Gulf. It has its borders with United Arab Emirates to the north, Saudi Arabia to the west, and Yemen to the southwest. It is the second largest territory in the Arabian Peninsula, with an area of 120,000 square miles and a coastline length of nearly 1,000 miles. Oman possesses a rich and varied terrain with rocky mountains, deserts, flat plains, and green oases. The coastal plains built up over the centuries by the outwash from the many *wadis* (valleys), which descend from the mountains. This forms the principle agricultural and date-growing area, and is also the most densely populated. The highest mountain range is Jebel Akhdar ("Green Mountain"), which rises to 10,000 feet.

After the accession of His Majesty Sultan Qaboos Bin Said in 1970, Oman opened the doors to the modern world after a long period of decline, debt, and restrictions. Since 1970 progress has been extremely rapid, and facilities have been brought to the remotest regions (Hanley, 1977). The income from oil in Oman is smaller than that of most of its neighbors in the Arabian Peninsula, and other sources of income come from fisheries, agriculture, industry, and tourism.

The Sultanate of Oman is administratively divided into five regions and four governorates with 61 *wilayats*. These regions are Ad Dakhliyah, Ash Sharqiyah, Al Batinah, Adh Dhahirah, and Al Wusta, and the governorates are Muscat, Dhofar, Musandam, and Al Buraymi. The regions of Ash Sharqiyah and Al Batinah have each been further subdivided into two Health Regions, as shown on Figure 54-1.

DEMOGRAPHIC FEATURES

In 2003, the second General Census of Population was carried out in the Sultanate of Oman, revealing a total population figure of 2,340,815, of which 23.9% are expatriates. Oman has a young population: about 11.58% and 36.2% of the population are under 5 years and under 15 years, respectively, and only 3.7% are 60 years and over (See Figure 54-2). More than a quarter (28.1%) of the total Omani population are females in the reproductive age group (15–49 years). The fertility rate of Omani women was estimated from 1993 census data to be 6.9, declined to 3.56 according to the 2003 census, and has further declined to 3.13 during 2007. The crude birth rate is estimated to be 25.0 per 1000 Omani population during 2007. At the present time, the Omani population is characterized by a rapid rate of growth, large family size, consanguineous marriages, and the presence of genetic isolates and semi-isolates.

GENETIC SERVICES IN THE SULTANATE OF OMAN

In the past thirty years, Oman has witnessed remarkable social and economic growth, which is best reflected in the well-organized and efficient healthcare system. With these achievements the country has had a shift in the pattern of disease. There has been a significant decrease in the incidence of communicable diseases and in the mortality and morbidity rates of infants and children under 5 years of age (Figure 54-3). In the past, the scale of the problem of congenital/genetic disorders was hidden in the high infant mortality rate because most affected infants died without being diagnosed. Now the majority are diagnosed and provided with the best possible treatment. As a result, the number of surviving affected children increases alongside the annual birth cohort, causing a considerable burden on the healthcare services (Alwan & Modell, 1997).

The Ministry of Health in the Sultanate of Oman is wishing to promote the potential benefits of genomic advances for the Omani population, and recognizing the urgent need for the application of genomics in the Sultanate of Oman. A specialized National Genetic Health Center is under construction in the capital city of Muscat, and is expected to open in 2012. The national strategies for the prevention and management of genetic and

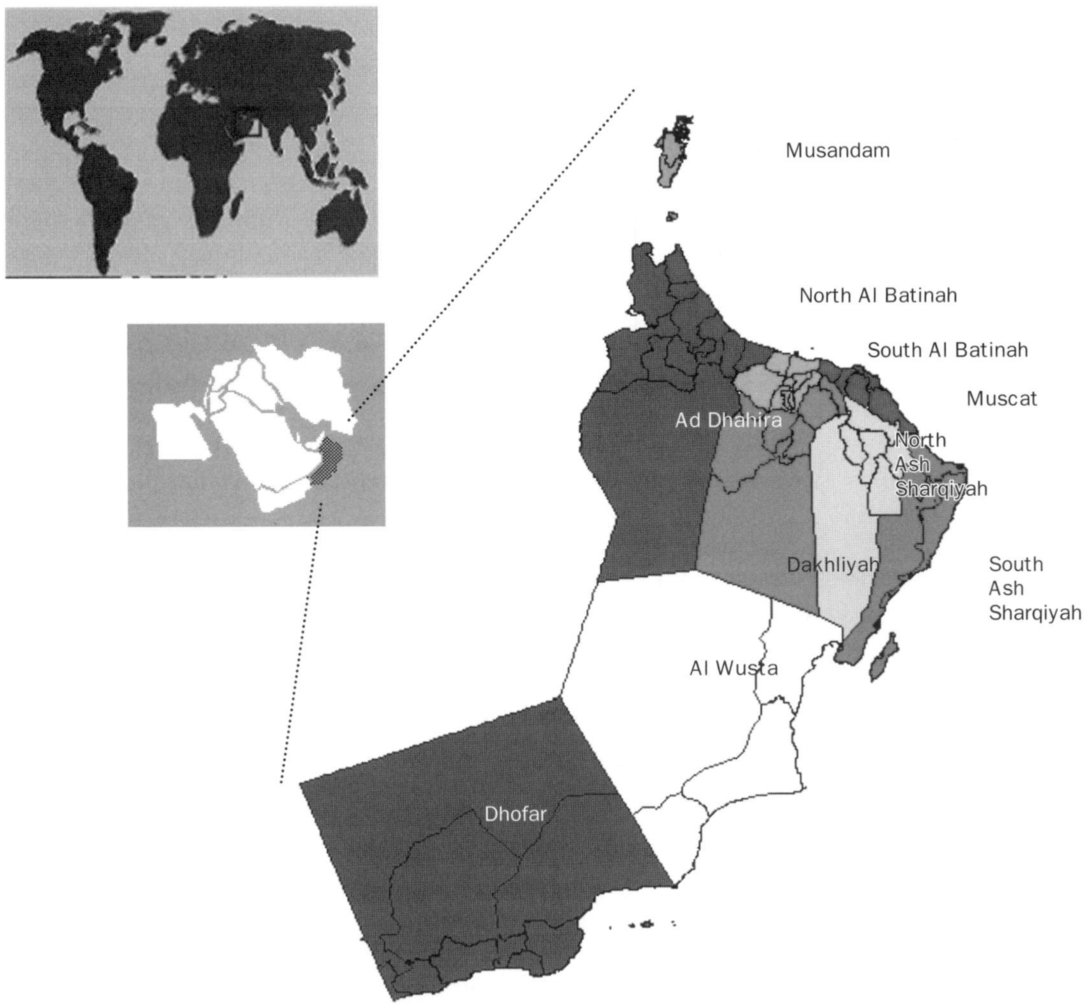

Fig. 54-1 Regional Map of Oman. Division of Oman Territory into Regions (Manataq). Refer color figure.

congenital disorders are defined with population structure, and with local needs and priorities where resources are available. The emphasis is on community genetic services that combine the skills of community medicine and medical genetics, and this approach is the preferred policy for the Sultanate of Oman. This includes the early identification and prevention of genetic risk, applied to the whole population with the empowerment of individuals to apply the relevant knowledge of genetics for their own benefit. In addition, the genetic services are in the process of being incorporated into primary healthcare. Educational packages and advocacy sessions are delivered regularly to increase the population's genetic knowledge. The first community genetic program was established in 1999 as the National Program for the Control of Genetic Blood Disorders, and is now integrated into the service. The clinical genetic service was established in 1993, the cytogenetic laboratory was established in 2000, and the molecular genetic laboratory was established in 2007.

The strategies developed in the Sultanate of Oman have been ethically compatible with the cultural and social makeup and religious beliefs of the population, and the legal system of the country. A prevention program is credible, and will enlist the confidence and cooperation of the people, only if it is accompanied by visible commitment to the care of affected people.

Understanding social beliefs, respecting the traditions set up in the Islamic community, and understanding the psychological difficulties faced by families affected by genetic disease, are essential for planning care and prevention (Rajab & El-Hazmi, 2007). A professional setting where geneticists, health educators, counselors, and community support groups all work together as a team in a friendly and positive manner creates a solid base for successful disease prevention.

COMMUNITY AND FAMILY STRUCTURE: PAST AND PRESENT

The preservation of a tribal structure in the community, coupled with isolation, has produced unique and favorable circumstances for building genealogical records and the study of genetic diseases.

The interior of Oman in the past had a tradition of being a closed, isolated, self-sufficient community, tribal

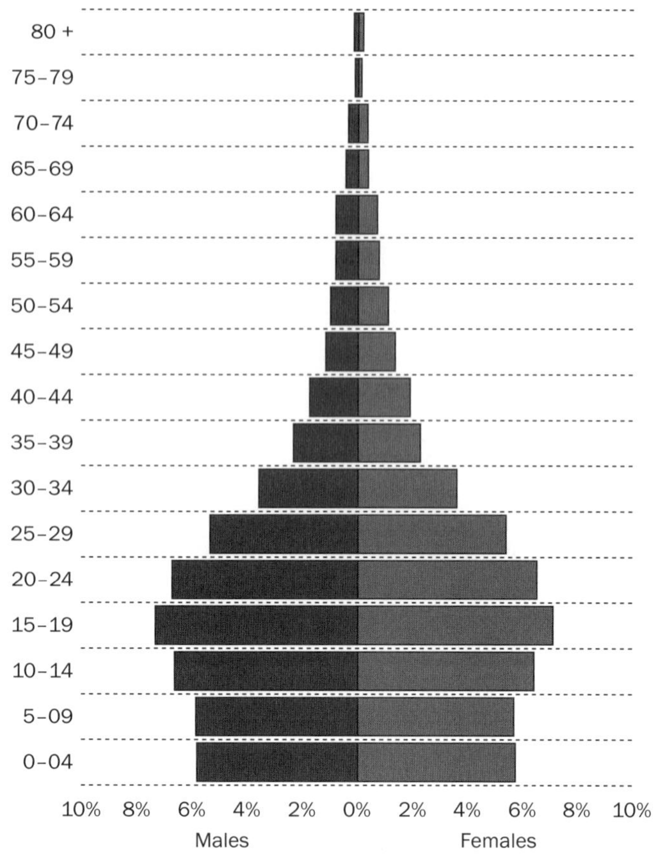

Figure 54-2 Population Pyramid for 2007 (Mid-Year Population)

in its organization, governed by an elected Imam (religious leader; Anthony, 1976; Wilkinson, 1987). In the classical Arab division between town and desert-dwelling people, the city dwellers are known as *Hadr*, and those who live in the desert are *Bedu*, or Bedouins (nomads).

In the past, the Bedouins represented not only the most numerous, but also politically and economically, the most important group of autonomous populations. They were believed to be the indigenous inhabitants of the country. The way of life and economy of the Bedouins is characterized by a nomadic lifestyle, which allows them to change their dwelling place frequently and to cover great distances to look for fresh pastures. A typical picture of the Bedouin way of life is found in the prevalence of a small camp and travel groups. The small nomadic group represents an optimal adaptation to the prevailing ecological conditions, where limited pastures will be quickly consumed by larger numbers of livestock, thereby necessitating a frequent change of location. It is often a household of a father and his married sons which make up a travel or camp group (Jansen, 1986).

The main strands of internal political history in Oman in the last two centuries have centered around the Bani Hina (Sons of Hina), which originate from southern Arabian stock (Yemenite or Qahtan tribes) and northwest Arabian descendants of Bani Ghafir (sons of either Ghafir, Nizar, or Adnan tribes; Miles, 1919). Thus, Bani Hina and Bani Ghafir have been opponents and have been competing for territories as pastures and watering places for the last two centuries. This fact is interesting from a genealogical point of view, as it was most unlikely that marriages took place between the members of the two clans.

Historically the survival of the tribes largely depended on the water sources available (Wilkinson, 1987). Water has been the most important factor for survival in the desert, and crucial for agriculture and animal breeding. It was vital for the survival of the tribes to control and protect their water sources and pastures, and traditionally a tribe controlled the territory of the *wadi* (valley) named after the tribe.

The common characteristics of a tribal unit in history were the existence of a leader (sheikh), a wide measure of political and military autonomy, and a well-defined territory with pastures and watering places. The tradition of tribal organization and the sheikh's leadership of the tribe is still preserved in Oman, and has been directed toward the welfare of the tribe rather than historical intertribal politics. Family and tribal pedigrees have great importance and respect in Arab culture, and are kept in many families. Present hospital records, as well as national identity cards, include four names (first, father's, grandfather's, and tribal name). This provides a unique opportunity to study genetic disorders among the tribal populations.

Oman had relatively small population in the past. There were less than quarter of a million inhabitants in the beginning of the nineteenth century (Lorimer, 1908). Until 1971, the lifestyle had remained unchanged for many centuries. No modern transportation means, such as airlines, asphalt roads, or electricity were available until 1970. Under these conditions, tribal structure was largely preserved in Oman. Links with the outside world were also limited before 1970. Sailing ships and

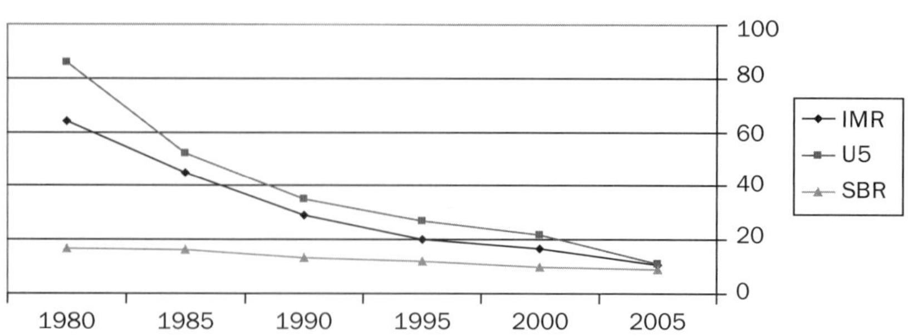

Figure 54-3 Changes in infant mortality rates (IMR), mortality under 5 years (U5) and stillbirth (SBR) at 5-year intervals observed in the past 25 years (1980–2005) from Ministry of Health Information System data.

camel caravans were the only means of trade and communication. The Indian subcontinent could be reached by sea in a month, and was the main trading link. East Africa could be reached in monsoon season in 3–4 weeks, and Zanzibar has been an Omani protectorate and trading post for centuries. In the southern Dhofar province, the geographic isolation was even greater, as 600 miles of desert lies between Dhofar and Muscat.

With increasing urbanization, many Omani citizens moved to Muscat where jobs and other modern facilities were available. The main tribe will still be found in their place of origin and be visited by many of its younger members during holidays.

There are a number of unique factors in Oman that make population analysis especially useful for genetic studies. Firstly, there are historical records of the population structure (Lorimer, 1908; Miles, 1919; Wellsted 1837, 1886); secondly, there has been a cultural preservation of tribal structures; and thirdly, there is an inclusion of the tribal names in hospital records and national identity cards.

As the population in Oman has remained constrained by tribal boundaries, it is likely that genetic diseases will also show similar constraint and have a clustering within tribes or genetic isolates. It follows, therefore, that the study of tribal origins and structures will be valuable in understanding the distribution of the genetic disorders, and ultimately may lead to efficient strategies for dealing with genetic disorders with economic spending on prevention programs.

Tribal structure has been clearly documented by several travelers in the last century (Lorimer, 1908; Miles, 1919; Wellsted, 1837, 1886). Miles provided a good description of Omani tribes and tribal politics, but the most detailed description of tribes and their sections was found in Lorimer's *The Gazetteer of the Persian Gulf, Oman and Central Arabia*, published in 1908. It included information about tribes as well as a description of tribal subsections, their location, and the political faction to which they belonged. The information about tribes of the southern province of Dhofar was derived from works of Anthony (1976).

In applying this to genetic research, the historical data was organized and entered on to a database using Microsoft Excel. It is in the form of a register that lists the names of the tribes in alphabetical order, their residence, and the numerical size of the tribe in the beginning of the century. The total number of tribal names registered is approximately 700. This includes the names of all tribes, tribal sections, and subsections (Rajab & Patton, 1999a). However, this number is reduced to 198 if tribal sections and subsections are incorporated within the main tribal name. The information on tribal size and names was subsequently omitted in the working database for confidentiality, as tribal information remains a matter of great sensitivity.

There are some advantages to studying the tribal populations of Oman. When an autosomal recessive disorder is found within an extended tribe, it may well be associated with single mutation, and so provide an opportunity to use homozygosity mapping to identify the gene concerned. In small populations of isolated villages, where an inter-village mating is a rare event, there is an increased risk for a genetic founder effect. With small isolated tribal units, the frequency of different alleles may change dramatically due to genetic drift and the effect of small numbers in the population. Due to this, the frequency of single-gene disorders can vary considerably in different parts of the country. The identification of an individual to a tribe, and vice versa, is important in terms of studying the incidence of single-gene diseases in particular population groups. It provides a unique opportunity to study whether there are any differences in the frequency of genetic disorders and birth defects among the tribal populations. The environmental, dietary, or other factors can be studied in the tribes with high prevalence of a particular disorder. This approach to population genetics would be expected to apply in neighboring Gulf cooperation countries, and other countries of the Middle East sharing similar historical and religious backgrounds.

CONSANGUINITY

The custom of consanguineous marriages in Muslim communities is deeply rooted in Arab culture (Al Awadi et al., 1985; Hamamy & Al-Hakkak, 1989; Bittles, 1995, 2003; El-Hazmi et al., 1995). The balance of opinion in the Middle East remains in favor of consanguinity, irrespective of increased risk of autosomal recessive diseases, congenital malformations, and mental retardation (Alwan & Modell, 1997; Al-Gazali et al., 2006).

Most studies on consanguinity in the Middle East have been based on hospital populations and have not tried to provide a comprehensive national overview. The study by Rajab and Patton (2000) in Oman is perhaps unique in this way, and may provide a model for others undertaking a study of consanguinity for health planning. The study was designed in two parts. The first part was a questionnaire survey conducted through all antenatal clinics in the country over a 3-year period (referred to as "the population study"). The second part involved collection of a full and detailed pedigree in a small subgroup attending other outpatient clinics (referred to as "the pedigree study"). The purpose of the second part was to validate the questionnaire, to determine the more complex patterns of intermarriage, and to determine the coefficient of inbreeding.

In the first part (population study) information was collected between October 1994 and January 1997, by introducing questions on consanguinity into the antenatal registration cards and transferring these data to the delivery registration books maintained in all such units in Oman. Data were collected on 60,635 couples of childbearing age, which represents around 20% of the population in this age range.

In the second part (pedigree study), a sample of 500 pedigrees was collected during the second year of the project from couples in different regions of Oman attending outpatient clinics for either accidental injuries or normal deliveries. In order to avoid bias, couples attending the genetic or pediatric clinics were not recruited, as it was believed that they might be more likely to be affected by, or be carriers of, a consanguinity-associated genetic disorder. The consanguinity was assessed first by the same questionnaire used during the larger study and, following this, a complete pedigree was collected. Detailed pedigrees included records of the extended pedigree as far back as recalled by the respondents. The pedigrees were analyzed using the Cyrillic-2 software program (Cherwell Scientific Publishing Ltd., Oxford). The coefficient of inbreeding (F) was estimated for each couple, and the mean coefficient of inbreeding (α) calculated for all couples.

During the population study, 7100 couples were interviewed in 1994, 28,480 couples in 1995, and 25,315 couples in 1996. These figures represent 15.5%, 57.0%, and 60.0% of all deliveries in Oman in 1994, 1995, and 1996 respectively. No major differences were observed between the consanguinity levels in 1994, 1995, and 1996, and so the results for all three years were pooled.

Data analysis confirmed first and second cousin marriages constituted 24.1% and 11.8% of all marriages respectively (Table 54-1). In addition, a further 20.4% of marriages were contracted within specific tribal groupings. By definition, many of these marriages would also be consanguineous but with a degree of relatedness less than that of second cousins ($F<0.0156$). Fewer than half (43.5%) of all marriages were between unrelated partners. The average inbreeding coefficient calculated for the Omani population from this survey was $\alpha = 0.0176$.

TABLE 54-1 CONSANGUINITY FIGURES IN OMAN (1994–1996)

Type of union	No. of couples studied, % of total	Coefficient of inbreeding
Double first cousins	2945 (4.8%)	0.006
First cousins	10,205 (16.8%)	0.0104
First cousins once removed	1575 (2.6%)	0.0008
Second cousins	7186 (11.8%)	0.0018
Marriages within the same tribe	12,459 (20.4%)	0.0008
Unrelated	26,525 (43.6%)	–
Total	60,895 (100%)	0.0198

Rajab and Patton (2002).

In the second part of the study, the pedigrees of 500 couples were recorded in full. Of the 500 couples interviewed with full pedigrees, 4.8% were double first cousins, 15.2% were first cousins, 1.2% were first cousins once removed, and 10.6% were second cousins (Rajab & Patton, 2000a). A further 18.4% of couples belonged to the same tribe, but their degree of consanguinity could not be accurately determined. The coefficient of inbreeding was calculated for each pedigree, and an α value of 0.0204 derived for the pedigree-based data as a whole. This was significantly higher than the value of $\alpha = 0.0176$ calculated from the large clinic-based survey. This showed that the population questionnaire had underestimated the consanguinity, as specific information on the prevalence of double first cousins and other complex relationships had not been collected.

Differences were noted between the consanguinity rates in the various regions of Oman, with a higher proportion of cousin marriages observed in rural and isolated areas, e.g., in the interior mountainous region (Nizwa). In the Empty Quarter (Al Wustah), the pedigrees obtained from twelve Bedouin families provided details of consanguineous unions for up to 5–6 previous generations, with all marriages contracted within their sub-tribes.

The study confirms that, in common with other Middle Eastern states, in Oman there has been a high proportion of consanguineous marriages in past generations and this practice continues to the present day. This information is relevant in terms of public health strategy, as it is likely to be associated with an increased pattern of specific genetic disorders. Studies in Oman have shown that consanguinity does indeed play a role in the frequency of hemoglobinopathies (Rajab & Patton 1997b), hydrocephalus (Rajab et al., 1998), Hirschsprung's disease (Rajab et al., 1997a), posterior urethral valves (Rajab et al., 1996), and deafness (Khabori & Patton, 2008)

In terms of its coverage of the population, the study is one of the largest and most comprehensive so far conducted. The 1994 national census in the Sultanate of Oman reported a total of 761,930 females, among whom 292,840 were of childbearing age (15–44 years old) and 64.3% of these were married. By interviewing 60,895 couples, representing 20.8% of the total female population of childbearing age or 32.3% of married women in Oman, a more accurate picture of the total population was provided than could have been expected from a study based in a single hospital.

The survey showed that 35.9% of marriages were between first or second cousins, a figure which is similar to the rates seen in other parts of the Middle East. Nonetheless, it probably is an underestimate of the total level of consanguinity in the population. Firstly, there were a further 20.4% of marriages contracted between members of the same tribe, which by definition were related to some degree. Secondly, there may have been consanguineous relationships in previous generations that were unknown to the couples interviewed and so were not recorded.

It would be expected that consanguinity over many generations would exert a cumulative effect on the degree of inbreeding, but this could not be demonstrated in the study. Ultimately, it would be interesting to compare estimates of inbreeding levels based on these results with those derived from molecular techniques.

The approach used in the study has demonstrated an inexpensive method of collecting data on consanguinity from a significant proportion of the Omani population of childbearing age. For greater precision, the questionnaire should be modified to define the specific types of marriage where more complex relationships are involved, such as double first cousin. However, even in this form the data can be applied in the study of inherited diseases, and will be of value in future healthcare planning in Oman. More recent preliminary data suggests that there is a reduction in the frequency of consanguineous marriages in urban areas in Oman.

CONGENITAL ANOMALIES

With the improvement in obstetric and neonatal care in Oman, there has been a steady decline in neonatal mortality from respiratory disease, infections, and asphyxia, and this has led to the emergence of congenital anomalies as an important cause of perinatal and neonatal morbidity and mortality.

Studies on congenital abnormalities in Oman started with those where there were obvious external malformations, or where there was likely to be referral to the only specialist pediatric surgery unit, as this was the best way to achieve complete national ascertainment. The conditions studied were neural tube defects, posterior urethral valves, facial clefts, and Hirschprung's disease. Later studies were made of congenital heart disease.

NEURAL TUBE DEFECTS

An analysis of the Delivery Registers data, which cover 80% of all national births, over the 1989–1995 period revealed a total incidence of neural tube defects at 1.25 per 1000. This total includes anencephaly (0.69 per 1000), myelomeningocele (0.45 per 1000), encephalocele (0.078 per 1000) and Meckel-Gruber syndrome (0.028 per 1000; (Rajab et al., 1998; Freeman & Rajab, 1999)). Consanguinity rates in the families of children affected with neural tube defects and congenital hydrocephalus were found to be much higher than in the general population, where the consanguinity rate is 56%. All (100%) cases of encephalocele and congenital hydrocephalus came from consanguineous marriages, and 75% of myelomeningocele came from consanguineous marriages.

There was tribal aggregation of neural tube defects cases in Oman in two tribes, but there was no geographical clustering within tribes of higher incidence. It makes it unlikely that a single environmental factor is responsible for the higher incidence.

FACIAL CLEFTING

Multiple sources of data were used to study facial clefts in Oman. The first source of the data was retrieved from hospital-based records of centralized plastic surgery services for cases born between 1989–1995. It revealed the incidence of facial clefting to be 1.5 per 1000 overall, with 0.62 per 1000 of combined cleft lip and palate, 0.34 per 1000 of cleft lip and 0.54 per 1000 of isolated cleft palate (Rajab & Thomas, 2001).

The second source of retrospective data was the national delivery registers of 1989–1985. The results were comparable to the figures for cleft lip, and cleft lip and palate from the first source, but underestimated the number of isolated cleft palates.

A further prospective study of 177 cases revealed associated congenital anomalies in 38.4% of cases, and recognized malformation syndromes were present in a quarter of cases studied. There was a family history of clefting in 23% of cases, and there was also a higher coefficient of inbreeding noted in families with clefting (0.0294) compared to general population figures (0.0198).

HIRSCHPRUNG DISEASE

A total of 110 cases of Omani children with histologically proven Hirschprung disease have been treated at Royal Hospital Pediatric Surgical Unit from 1987 to 1994. The frequency of Hirschsprung disease in Oman is 1 in 3070 (0.3/1000), which was considerably greater than in the Western studies (Rajab et al., 1997a).

There were significant variations in the birth incidence of Hirschsprung disease in different regions. The capital area, with 335,000 population, had a frequency of 1 in 4200, while North Batna and North Sharqiya regions had frequency 1 in 2000 and 1 in 1800 live births, respectively. A positive family history was recorded in 6 families out of 85 cases (7%). The rate of consanguineous marriages was higher, and specific autosomal recessive forms associated with piebaldism were noted. There was tribal aggregation of Hirschsprung disease, with 18.8% of cases occurring in two tribes.

POSTERIOR URETHRAL VALVES (PUV)

The pediatric surgery unit records at the Royal Hospital in Muscat were reviewed, and the case files of all patients treated for PUV for years (1987–1994) were extracted.

Only the definite cases where urethral vales were seen and ablated under direct vision were included in the study. Cases of anterior urethral valves and cases involving expatriate children have been excluded from the study. The incidence figures were calculated from the period 1989 to 1994. There were 56 cases of PUV from a total population of 261,000 newborns, or 133,000 newborn males (51%) in the Sultanate of Oman. This gave an incidence of 1 in 2375 males (Rajab et al., 1996). This is three times higher than that reported from the United Kingdom. After the completion of the study there were 9 additional cases of PUV who presented late (4 were born in 1989–1994 and 5 in 1995). This raised the overall incidence to 1 in 2100 males (Rajab et al., 1996).

The regional incidence of PUV showed a marked variation, with highest incidence in the interior mountainous district and the capital (Muscat) district, viz. 1 in 1700 males, and 1 in 2000 males respectively. In only 6 (13%) children, parents were unrelated; in 25 (54.4%) parents were first cousins, in 5 (10.9%) parents were second cousins, and in 11 (24%) parents came from the same tribe. Thus, 87% of the parents were consanguineous, and more than half of the children with PUV came from first cousin marriages. This shows a much higher frequency of first cousin marriages in the PUV cohort. The concentration of 18 patients in two of the tribes suggests there may be a genetic predisposition, but no specific genetic syndrome was identified.

CONGENITAL HEART DISEASE

The incidence and spectrum of congenital heart disease (CHD) was analyzed in the Sultanate of Oman from 1994 to 1996. CHD was detected in 992 of 139,707 live births (incidence 7.1/1000 live births). The common malformations were ventricular septal defect (24.9%), atrial septal defect (14.4%), and patent ductus arteriosus (10.3%). The frequency of atrioventricular septal defects (5.9%) was higher than reported from other countries. The incidence and pattern of CHD in Oman were similar to those reported from developed countries in Europe and America, except for a higher frequency of atrioventricular septal defects (Subramanyan et al., 2000). The fact that the rates are similar to the developed countries is also indicative of the high standards for neonatal and pediatric diagnosis.

CHROMOSOMAL ABNORMALITIES

The national cytogenetic laboratory was established in the year 2000, and chromosome abnormalities have been recorded since then. Chromosomal abnormalities were seen in 28.3% out of 1800 pediatric samples referred (Goud et al., 2005) with trisomies, tetrasomies, monosomies, and translocations occurring.

The birth prevalence of Down syndrome seems to be high in Middle Eastern countries. From 2000 to 2008, ascertainment of all Down syndrome cases in Oman has been completed. Analysis yielded an unusually high prevalence of 2.88/1,000 (1 in 350 live births). In almost 90% of cases, the cytogenetic diagnosis was performed within 6 months after birth. Based on a case-control study, increased maternal age was identified as a significant risk factor for Down syndrome, but not sufficient to explain its high prevalence. In all informative cases (N = 95) the nondisjunction error occurred during oogenesis. Consanguineous marriages were higher among parents of Down syndrome children than in the Omani population in general. The identification of additional risk factors is a challenge, and of great relevance with respect to primary prevention and adequate genetic services. The high prevalence of trisomy 21 may reflect the longer reproductive period in older women due to limited use of family planning and birth spacing. However, around 40% of mothers giving birth to Down syndrome children in Oman are younger mothers, which suggests the existence of an additional risk factors.

Due to recent improvements in healthcare, more Down syndrome patients will now survive to adult ages. Clearly, the Omani healthcare system and the Omani society in general should be prepared for this new situation. There is an urgent need for further epidemiological studies of trisomy 21 in Oman. Knowledge of potential hazards is essential from a public health perspective, under the aspect of risk avoidance and primary prevention. So far, maternal age, mosaicism, and altered recombination are the only other established risk factors. It is also possible that environmental factors affecting the grandmother of the Down syndrome child have to be studied, as the error may have occurred during meiotic prophase of oogenesis in the fetal period in the mother. On the other hand, there is convincing evidence that environmental factors around the time conception takes place can increase the rate of nondisjunction (Czeizel et al., 1993; Sperling et al., 1994). So far, no environmental factors have been identified in Oman. Clearly these aspects are of great importance from the perspective of primary prevention of trisomy 21.

SINGLE GENE DISORDERS

Hospital-based data on genetic disease has been collected in Oman from 1990 and presented in Tables 54-2, 54-3, 54-4, and 54-5. The number of diagnosed conditions has been growing over the years in parallel with improvements in diagnosis. It is apparent that before comprehensive care became available, the illness and death of these children may have been attributed to other causes.

TABLE 54-2 BIRTH PREVALENCE OF β-CHAIN DISORDERS IN OMAN

Region	Total birth 89–92	AS (%)	No. AS	β-thal trait (%)	No. β-thal trait	Total % β – thal trait and AS	Total No. β – thal trait and AS
Musandam	2174	14.3	311	5.4	117	19.7	428
Dakhliya	22,107	14.3	3161	3.3	730	17.6	3891
North Sharqiya	15,076	13.3	2005	2.4	362	15.7	2367
Dhahir	13,712	11.0	1508	3.3	452	14.3	1961
Capital	30,339	9.1	2761	4.2	1274	13.3	4035
South Bathna	26,551	10.0	2655	2.2	584	12.2	3239
South Sharqiya	14,059	8.5	1195	3.5	492	12	1687
North Batna	37,347	5.6	2091	4.5	1681	10.1	3772
Dhofar	12,053	0	0	0	0	0	0
Wustah	2110	0	0	0	0	0	0
Total	175,538	8.9	15,688	3.2	5692	12.2	21,380

Rajab & Patton (1999b).

TABLE 54-3 COMMONLY ASCERTAINED AUTOSOMAL RECESSIVE DISEASES IN THE SULTANATE OF OMAN AMONG 420,000 LIVE BIRTHS (1993–2002)

Autosomal Recessive Diseases	No of patients born 1993–2002	Observed incidence
Spinal Muscular Atrophy (Wernig-Hoffman disease)	56	1 in 10,000 births
Congenital Adrenal Hyperplasia	55	1 in 10,000 births
Polycystic kidneys	34	1 in 12,000 births
Cystic Fibrosis	32	1 in 15,000 births
Primary Microcephaly	31	1 in 15,000 births
Renal Tubular Acidosis	28	1 in 20,000 births
Congenital Nephrotic Syndrome (Finish type)	25	1 in 20,000 births
Nesidoblastosis	24	1 in 20,000 births
Apple-peel bowel syndrome	21	1 in 20,000 births
Zellweger Syndrome	19	1 in 20,000 births
Metachromatic Leukodystrophy	18	1 in 25,000 births
Congenital Generalized Lipodystrophy	18	1 in 25,000 births
Ellis-Van Creveld Syndrome	18	1 in 25,000 births
Scwartz-Jampel Syndrome	15	1 in 30,000 births
Bardet-Biedl Syndrome	14	1 in 30,000 births
Robinow syndrome	12	1 in 35,000 births
Oculocutaneous Albinism	14	1 in 30,000 births
Epidermolysis Bullosa	15	1 in 30,000 births
Galactosialidosis	9	1 in 50,000 births
Cerebro-Oculo-Musculo-Sceletal Syndrome	9	1 in 50,000 births
Meckel-Gruber Syndrome	9	1 in 50,000 births
Carbohydrate Deficient Glycoprotein Syndrome	8	1 in 50,000 births
Mucopolysacharidoses	8	1 in 50,000 births

Rajab et al. (2005). Community Genet 8, 27–30. Copyright permission from S. Karger AG, Basel 28.05.2005.

TABLE 54-4 AUTOSOMAL RECESSIVE DISORDERS (HOSPITAL-BASED DATA)

I. INBORN ERRORS OF METABOLOSM	OMIM	V. HANDICAPPING GENETIC DISORDERS WITH MENTAL RETARDATION	OMIM
A. Metabolic endocrinopathies		*A. Nonsyndromic mental retardation*	249500
Congenital adrenal hyperplasia	201910, 202110, 107910	*syndromes with microcephaly*	249500
Metabolic rickets	277440, 264700, 600785, 241520	Primary microcephaly	608716, 251200
Apparent mineralocorticoids excess Type 2	207765	Lethal microcephaly	Rajab et al., 2007
Nesidioblastosis	256450	Microcephaly with pontocerebellar hypoplasia	Rajab et al., 2003
		Cohen microcephaly	216550
B. Disorders of amino acids metabolism			
Maple syrup urine disease	248600	Microcephalic primordial dwarfism	210710
Tyrosinaemia	276700	*B. Syndromes with neuronal migration defects and brain malformations*	
Dihydropterin synthetase deficiency	261640	Walker-Warburg syndrome	236670
Homocystinuria	236200	Pena-Shokeir syndrome	214150
Glutaric aciduria	231670, 231680	Meckel-Gruber syndrome	249000
C. Organic acidurias		Fryns syndrome	220950
Propionic	606054	Jouber syndrome	213300
Isovaleric	243500	Miller-Dieker lissencephaly	247200
D. Lysosomal storage disorders			
Mucopolysaccharidoses	607014, 607015, 252300	*C. Conditions with regression/neurologic deficit/ seizures*	
Multiple sulphatase deficiency	272200	Baten's disease	256730
Mucolipidosis I and II	256500, 256550	Tay-Sach's disease	272800
Glycogen storage diseases	232200, 230800, 232400, 232300, 232700	Niemann-Pick's disease	257220
Galactosialidosis	256540	Leukodystriophies	251100, 254200, 264090
GM1–Gangliosidodis	230500	Progressive myoclonic epilepsy	310370
E. Long-chain fatty acid oxidation defects		Hereditary spastic paraplegia	270800
Zellweger syndrome	214100	Hyperekplexia	149400
Refsum disease	266500	Genetic form of cerebral palsy	Rajab et al., 2006

(Continued)

TABLE 54-4 CONTINUED

I. INBORN ERRORS OF METABOLOSM	OMIM	V. HANDICAPPING GENETIC DISORDERS WITH MENTAL RETARDATION	OMIM
F. Mitochondrial diseases		Olivopontocerebellar atrophy	Chand et al., 1996
Kearns-Sayre syndrome	530000	VI. NEUROMUSCULAR DISORDERS	
Congenital lactic acidosis	Bappal et al, 2001	Spinal Muscular Dystrophies (Koul R et al., 2007; Haider et al, 2001)	253300, 253400, 253900
Pyruvate carboxylase deficiency	266150	Congenital myopathies	255310, 161100, 601462
Pyruvate dehydrogenase deficiency	208800	Charcot-Marie-Tooth neuropathy	214400
G. Abnormality of fatty alcohol metabolism		Facioscapulohumeral dystrophy	158900
Sjogren-Larsen syndrome	270200	Limb girdle muscular dystrophies	608099, 160500, 253601
II. SKELETAL DYSPLASIAS AND DISEASES AFFECTING BONE STRUCTURE		**VII. IMMUNODEFICIENCIES AND CHROMOSOMAL INSTABILITY SYNDROMES**	
Asphyxiating thoracic dystrophy	208500	Ataxia-telangiectasia	208900
EllisVan-Creveld syndrome	225500	Bloom syndrome	210900
Hypochondrogenesis	200610	Chronic granulomatous disease	233700
Grebe chondrodysplasia	200700	Common variable immunodeficiency	240500
Robinow syndrome	268310	Cyclic neutropenia	610738
Spondyloepiphyseal dysplasia Omani type	608637	Hemophagocytic lymphohistiocytosis type 1	267700
Three M dwarfism (Dolichospondylic dysplasia)	273750	Fanconi anemia (Kannan K et al., 2003)	227650
Osteogenesis imperfecta	610195	Nijmegen syndrome	251260
Pycnodysostosis	265800	SCID	102700
Bamatter syndrome	231070	Xeroderma pigmentosusm	278700
Osteopetrosis	259700		
III. DISEASES AFFECTING KIDNEYS, LIVER AND GUT		**VIII. DISEASES AFFECTING SKIN**	
Apple-peel bowel syndrome	243600	Epidermolysis bullosa	226700, 226670
Infantile polycystic kidneys	263200	Hypohydrotic ectodermal dysplasia	224900
Focal segmental Glomerulosclerosis	603278	Lamellar ichthyosis	242300
Finnish type congenital nephrosis	256300	Ichthyosis congenita	242500
Steroid-resistant nephrotic syndrome	600995		
Nephronophtysis	256100, 602088	IX. MISCELLANEOUS RECESSIVE DISORDERS	
Cystinosis	219800	Bardet-Biedl syndrome	209900
Hyperoxaluria	259900	Congenital insensitivity to pain with anhydrosis	256800
Distal renal tubular acidosis	602722	Congenital generalized lipodystrophy	269700, 608594, 608154
Bartter syndrome	607364	Cutis Laxa with growth and developmental delay	219200

(Continued)

TABLE 54-4 CONTINUED

III. DISEASES AFFECTING KIDNEYS, LIVER AND GUT		VIII. DISEASES AFFECTING SKIN	
Familiar intrahepatic cholestasis	211600	Escobar syndrome	265000
Crigler-Najjar syndrome type I	218800	Long QT syndrome	152427
Wilson disease	277900	Neonatal progeroid syndrome	271900
IV. CONGENITAL BLINDNESS AND DEAFNESS		Oculocutaneous albinism	606952
Optic atrophy	258500	Schwartz-Jampel syndrome	255800
Nonsyndromic sensorineural deafness	220290	Cystic fibrosis (Frossard et al., 1996, 1998; Romey et al., 1999)	219700
Leber congenital amaurosis	204000		
Usher syndrome	276900		

Autosomal recessive conditions are by far commonest single-gene diseases, and remain the major contributors to childhood morbidity, mortality, and handicap (Table 54-3). In Oman an estimated 7% of children are born with a congenital or genetic disorder, compared with about 4.4% in Europe (EUROCAT Working Group). The difference is due to a high prevalence of inherited hemoglobinopathies, advanced maternal age, and customary consanguineous marriage. Registration of individuals with congenital or genetic disorders has just been started, and will provide more exact figures in due course.

In the past, the majority of genetic disorders caused early death. With modern medical care, about 17% remain lethal, about 36% are cured (mainly congenital malformations corrected by pediatric surgery), and about 47% cause lifelong disability. Because of a higher proportion of chronic inherited disorders in Oman, about 3% of children are expected to suffer disability for as long as they live, compared with about 1.3% in Europe (Modell, 2003). The crucial problem is that, to the extent that modern medical care improves survival, the total number of children surviving with chronic disorders rises year after year. Even with reduced family size these numbers are set to multiply about three times in the next 20 years unless prevention can be introduced. The birth prevalence of various recessive disorders is largely unknown. The figures of recessive disorders frequently ascertained in pediatric practice, derived from

TABLE 54-5 AUTOSOMAL DOMINANT AND X-LINKED DISORDERS OBSERVED IN OMANI POPULATION

Autosomal Dominant		X-linked disorders	
Achondroplasia	100800	Anhydrotic Ectodermal Dysplasia	305100
Adult polycystic kidney disease	601313	Alport syndrome	301050
Alligille syndrome	118450	Bruton agammaglobulinemia	300300
Craniosynostoses, Acrocephalosyndactily	101400	Hemophilia A	306700
Deafness congenital nonsyndromic	600965	Hemophilia B	306900
Epidermolysis bullosa simplex	131800	Duchenne muscular dystrophy	310200
Fanconi renotubular syndrome	134600	Lissencephaly (LIS 4A)	300121
Hereditary angioedema	106100	OTC deficiency	311250
Huntington's chorea	143100	Testicular feminization Syndrome	300068
Hyper- IgE syndrome	147060	X-linked forms of mental retardation	300624, 312750, 301040
Neurofibromatosis	162200		
Sotos syndrome	11755		0
Spinocerebellar ataxia	117360		
Tuberous sclerosis	191100		

1993–2002 hospital-based data, is presented in Table 54-2 (Rajab et al., 2005a). From this data it has been possible to provide some geographic mapping of the areas with the most frequent cases of the single-gene disorders (Figure 54-4).

HEMOGLOBINOPATHIES AND RED CELL DISORDERS

The hemoglobinopathies are one of the most prevalent single-gene disorders in the Middle East. At present there are around 400 patients with thalassemia major, and 3,000 sickle cell disorders in the Sultanate. In the Sultanate of Oman the birth prevalence of infants with hemoglobin disorders is 3.5–4.7/1,000 (Daar et al., 1998, 2000; Rajab & Patton, 1997b; Rajab et al., 2000b; White et al., 1993). Around 10% of Omani nationals are carriers of the gene of sickle cell anemia (OMIM 6039030), 2%–3% carry the gene of β-thalassemia (OMIM 141900), and 45% are carriers of α-thalassemia (OMIM 141800). The high frequency in some areas may reflect natural selection due to advantage for survival in the heterozygous state against malaria, and genetic drift in small isolated populations. The prevalence of the hemoglobinopathies shows some striking variations within the country. The birth prevalence of hemoglobin disorders is presented in Table 54-2.

There have been three original hemoglobin variants described in Oman. Hemoglobin Dhofar is linked to the codon 29 C-->T (IVS-1 nt-3) splice mutation, which causes β+ thalassemia (Williamson et al., 1995), hemoglobin Muscat (alpha 2 beta (2) 32 (B14) Leu→ Val) (Ramachandran et al., 1992), and hemoglobin S Oman (Glu6Val and Glu121Lys) (Langdown et al., 1989; Nagel et al., 1998).

The other red cell abnormality that is common in Oman is G6PD deficiency, which is found in 28% of males and 12% of females (Daar & Partare, 2006). In a majority of individuals the G6PD deficiency is benign (Daar et al., 1996; White et al., 1993), and asymptomatic forms A+ and B+ G6PD types predominate. However, it may still be a cause of neonatal jaundice or hemolytic crises seen in a small proportion of infants, especially in the northern parts of Oman.

INBORN ERRORS OF METABOLISM

There have been a number of hospital-based studies of metabolic disease in Oman (Bappal et al., 1999, 2001; Joshi et al., 2002; Joshi & Venugopalan, 2007). A wide variety of metabolic disorders have been observed; e.g., disorders of amino acids, organic acids, long-chain fatty acids, lysosomal storage disorders, mitochondrial disorders, and endocrinopathies. The range of metabolic disorders is summarized in Table 54-4.

NOVEL PHENOTYPES, GENE MAPPING, AND SPECIFIC MUTATIONS DESCRIBED FROM OMAN

With genetic isolation and consanguinity, certain novel mutations or clinical phenotypes will become markedly

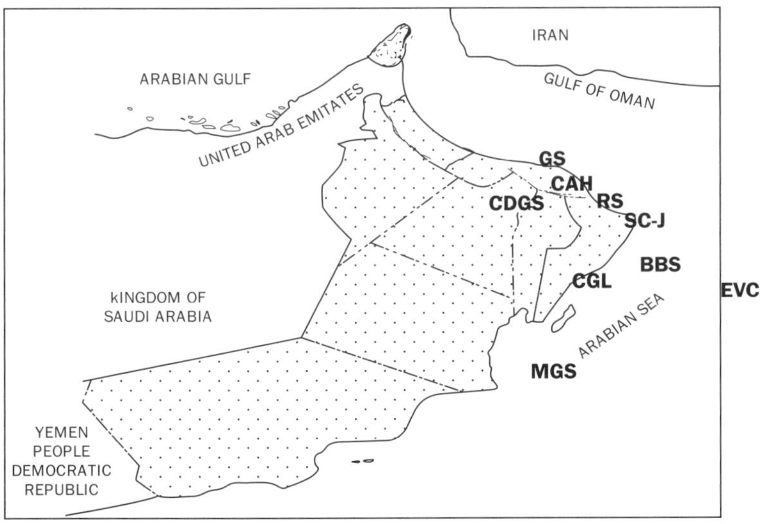

RS = Robinow Syndrome
CAH = Congenital Adrenal Hyperplasia
CGL = Congenital Generalised Lipodystrophy
CDGS = Carbohydrate Deficient Glycoprotein Syndrome
MGS = Meckel-Gruber Syndrome
EVC = Ellis Van Creveld Syndrome
SC-J = Schwartz Jampel Syndrome
GS = Galactosialidosis
BBS = Bardet-Biedl Syndrome

Figure 54-4 Geographical Areas of Oman Representing High Density of (90% of cases) Autosomal Recessive Conditions.

SOURCE: From Rajab et al. (2005). Community Genet (8), 27–30. Copyright permission from S. Karger AG, Basel 28.05.2005.

more prevalent and may be isolated to certain tribes or regions. This generalization seems to apply to the majority of the specific malformations studied in the Sultanate of Oman, and is expected to be relevant to communities in the Middle East. Below are listed novel recessive phenotypes and genotypes described in Omanis.

i. *Robinow syndrome.* This autosomal recessive disorder was first mapped through a number of related families in the North Batnah region of Oman. It causes "fetal facies," mesomelic limb shortening, skeletal abnormalities, and genital hypoplasia. The locus was mapped to chromosome 9 and the ROR2 gene identified in 2000 (Afzal et al., 2000a,b; Soliman et al., 1996). It is allelic to dominant brachydactyly type B.

ii. *Congenital generalized lipodystrophy.* A novel form of lipodystrophy has been described in Omani families with rippling of skeletal muscle and nonskeletal muscle involvement, mapped to 9q (Rajab et al., 2002; Heathcote et al., 2002). Apart from generalized lipodystrophy, there were reduced exercise tolerance, percussion myotonia, and nonskeletal muscle hypertrophy; e.g., hypertrophic pyloric stenosis, cardiomyopathy, prominent veins (phlebomegaly), hypertrophy of ureter, and macroglossia.

iii. *Congenital generalized lipodystrophy with deafness* is another lipodystrophy phenotype and was described in three Omani children with low birth weight, short stature, retarded bone age, tendency to fractures, striated metaphyses, sensorineural deafness, and delayed cognitive development (Rajab et al., 2003a).

iv. *Lethal microcephaly.* A new lethal form of microcephaly with a prenatal onset was described in 4 siblings in an Omani family (Rajab et al., 2007a). The features included low birth weight, microcephaly, apnea, seizures, and unusual facial features. Brain imaging revealed a simplified gyral pattern with normal to slightly thinned cortical gray matter, thin corpus callosum, and mild brainstem and cerebellar hypoplasia. All patients died within hours to weeks after birth following severe apnea attacks and central hypoventilation. No gene locus has been identified.

v. *Cohen syndrome.* This syndrome is characterized by microcephaly, mental retardation, unique facial features, neutropenia, and ophthalmologic findings. A novel mutation of the causative COH1 gene in Omani families, e.g., Hennies et al. (2004) and Mochida et al. (2004) show that Cohen syndrome has a wide geographic distribution.

vii. *Hereditary spastic paraplegia with microcephaly and mental retardation.* An autosomal recessive form of complex spastic paraplegia has been reported in two interrelated sibships presenting with mental retardation and spasticity. (Rajab et al., 2006). Genetic studies currently underway suggest this is a novel locus for hereditary spastic paraplegia.

viii. *Familial brain calcifications.* Autosomal recessive familial brain calcification with short stature and developmental delay was described in 8 children (Rajab et al., 2009). Brain imaging revealed extensive scattered calcifications of basal ganglia and cortex, similar to Aicardi-Goutieres syndrome or "Coats' Plus" syndrome, but the clinical features diverge substantially from these two syndromes. Gene mapping indicates a genetic locus for this disorder on chromosome 2 with a LOD score of 6.17 and suggests it is a new genetic entity.

ix. *Pontocerebellar hypoplasia.* A novel form of pontocerebellar hypoplasia has been mapped to chromosome 7q11-21. It is associated with postnatal microcephaly, mental retardation, optic atrophy, seizures, spasticity, and growth failure, and on brain imaging there was a simplified frontal gyral pattern with cerebellar hypoplasia, corpus callosum, and midbrain hypoplasia (Rajab et al., 2003b).

x. *Paroxysmal nonkinesigenic dyskinesia.* A novel mutation (c.20C>T: A7V; c.26C>T: A9V) of myofibrillogenesis regulator 1 gene (MR-1) was described in this disorder (Hempelmann et al., 2006).

xi. *Variant of Escobar syndrome.* A new syndrome with pursed mouth, creased tongue, scoliosis, and ophthalmic abnormalities was described from Oman and found to be due to mutations in the acetylcholine receptor pathway (Rajab et al., 2005b; Michalk et al., 2008).

xii. *Spondyloepiphyseal dysplasia Omani type.* This was described in nine individuals from two consanguineous Omani families. A missense mutation in C6ST-1 gene R304Q was found (Rajab et al., 2004; Tiele et al., 2004). The features include a normal length at birth with progressive growth delay and a final height of 110–130 cm, rhizomelic limbs, genu valgum, cubitus valgus, mild brachydactyly, camptodactyly, microdontia, severe shortening of the upper segment due to severe progressive kyphoscoliosis, severe progressive arthritic changes with joint dislocations, and normal intelligence. Spinal involvement resulted in physical handicap in early adulthood, and female patients were disabled earlier than males.

xiii. *Schwartz-Jampel syndrome.* The inclusion of an Omani family in the original gene-mapping study helped to identify the locus on chromosome 1p36, and led to the discovery of perlecan, which is a major proteoglycan of basement membranes (Nicole et al., 1995).

xiv. *Geroderma osteodysplastica.* This inherited skin disorder was described in three patients from Oman (Rajab et al., 2008), and subsequent analysis identified homozygosity for a 2-bp deletion in exon 1 of the SCYL1BP1 gene with loss of function (Hennies et al., 2008).

xv. *Wrinkly skin syndrome (autosomal recessive cutis laxa).* An international multicenter study included patients from Oman (Rajab et al., 2008b) and found impaired glycosylation caused by mutations in the vesicular H+-ATPase subunit ATP6V0A2 (Kornak et al., 2008).

xvi. *Three M syndrome.* An international study on this short-stature disorder included patients from Oman in the identification of the mutation (Q1469R) in the causative gene on 6p 21.1 (Huber et al., 2009).

xvii. *Grebe acromesomelic dysplasia.* Mutations were studied in an Omani family with identification of a novel deletion of G1144 in the CDMP1 gene (Al-Yahyaee et al., 2003).

xviii. *Hemophagocytic lymphohistiocytosis type 1.* Genetic studies performed in Oman have discovered a range of novel mutations in the Omani population: FHL1 gene (Del 9q 21.5-22; Pro 89→Thr, Arg225→Pro; trp374→stop codon; 50delT; mutation L17X; 12 bp inframe deletion (codon 284–287); Pro188→leu; Thr173→met change) (Ohadi et al., 1999; Al-Lamki et al., 2003; Muralitharan et al., 2005, 2007).

xix. *Hypopituitarism.* An Omani Family with three affected children with early onset hypopituitarism and hypoglycemia were found to have novel LHX3 mutation (Rajab et al., 2008a).

xx. *Apparent mineralocorticoid excess (AME).* Nine children with hypertension and hypokalemia from four Omani families were found with five different mutations in the HSD11B2 gene. Sequence data demonstrate the previously described L114Delta6nt mutation in exon 2, and new mutations in exon 3 (A221V), exon 5 (V322ins9nt), and, for the first time, in exon 1 (R74G and P75Delta1nt) of the HSD11B2 gene. These additional mutations provide further insight into AME and the function of the 11beta-HSD2 enzyme (Quinkler et al., 2004). The prevalence of monogenic forms of hypertension such as AME remains uncertain. However, our data suggests AME may be a relevant cause of hypertension in certain ethnic groups, such as the Omani population.

xxi. *Bardet-Biedl syndrome.* A number of families with Bardet-Biedl syndrome have been enrolled in international studies with identification of the mutation FLJ23560 in BBS10 on chromosome 12q21.2 (White et al., 2007). The Omani phenotype is more severe than many previous descriptions in the literature.

xxii. *Alopecia universalis congenital (AR).* This disorder has been reported in Oman with a mutation at 2776+1 (G>A) in exon 12 of the hairless gene (Cichon et al., 1998).

xxiii. *Familial factor X deficiency.* A new mutation of the gene regulating factor X function was found in an Omani family with a severe bleeding disorder. Familial CRM+ Factor X deficiency was found to be due to impaired prothrombinase activity of factor X (Gly381Asp Factor X) (Pinotti et al., 2003).

xxiv. *Long QT syndrome.* A malignant form of congenital Long QT syndrome was described in Omani families with multiple sibs affected (Supramayan & Venugopalan, 2002).

OMAN'S CONTRIBUTION TO GENETIC RESEARCH

The previous section demonstrates the opportunities for genetic research in a country like Oman, where there are large families with a tribal structure and a consanguineous population. The research has taken collaboration with several outside centers, from St. Georges University London (UK), Welch Laboratory, Harvard (USA), Weill Cornell College (USA), Humboldt University, Berlin (Germany), and Leiden University (Holland).

FUTURE PLANS AND CHALLENGES

The comprehensive healthcare systems and improvements in the quality of life have resulted in better survival of children with disadvantages. This has led to a progressive increase in the prevalence of genetic diseases and disabled in the community. This will lead to new priorities in healthcare. The Ministry of Health in Oman is addressing this issue with the development of new genetic facilities in the country. The delivery of genetic services ensures that such services are provided throughout the country by a community-based facility, but the laboratory facilities are concentrated in the main hospital. Cultural traditions are being observed in providing the service.

The planning of the genetic service will be guided by population research in genetics. This research has demonstrated the level of consanguinity, and the geographical variation of genetic disease. In developing laboratory services the identification specific disease mutations in an isolated population may make genetic testing easier. The role of basic research in this planning process is well proven.

ACKNOWLEDGMENTS

We would like to thank Professor Bernadette Modell for reviewing the manuscript and helpful advice.

REFERENCES

Afzal AR, Rajab A, Fenzke CD, et al. (2000a). Recessive Robinow syndrome, allelic to dominant brachydactyly type B, is caused by mutation of ROR2. *Nature Genetics 25*(4), 419–422.

Afzal AR, Rajab A, Fenske C, et al. (2000b). Linkage of recessive Robinow syndrome to a 4 cM interval on chromosome 9q22. *Hum. Genet 106*(3), 351–354.

Al-Lamki Z, Wali YA, Pathare A, Ericson KG, Henter JL. (2003) Clinical and Genetic studies of familiar hemophagocytic lymphohistiocytosis in Oman: need for early treatment. Pediatric *Hemtol. Oncol 20*(8), 603–609.

Alwan A, Modell B. (1997). Community control of genetic and congenital disorders. *EMRO Technical Publications* Series (24), Geneva.

Anthony JD. (1976). Historical and Cultural Dictionary of the Sultanate of Oman and the Emirates of the Eastern Arabia. The Scarecrow Press Inc., Metuchen, N.Y.

Al Awadi SA, Naguib KK, Moussa MA, et al. (1985). Consanguinity among the Kuwaiti population. Clin Genet *27*, 483–486.

Al-Gazali L, Hamamy H, Al-Arrayad S. (2006). Genetic disorders in the Arab world *BMJ 333*(7573), 831–834

Al-Yahyaee SAS, Al-Kindi MN, Habbal O, Kumar DS. (2003). Clinical and Molecular Analysis of Grebe Acromesomelic Dysplasia in an Omani Family. *Am J Med Genet 121A*, 9–14

Bappal B, Raghupathy P, de Silva V, Khusaiby SM. (1999). Permanent neonatal diabetes mellitus: clinical presentation and epidemiology in Oman. *Arch Dis Child, Fetal Neonatal Ed. 80*(3), F209–F212.

Bappal B, George M, Nair R, Khusaiby SA, De Silva V. (2001). Factitious hypoglycemia: a tale from the Arab world. *Pediatrics 107*(1), 180–181.

Bittles A. (1989) The Role and Significance of Consanguinity as a Demographic Variable. *Hum Hered 39*, 271–275.

Bittles AH. (2003). Consanguineous marriage and childhood health. *Dev Med Child Neurol* 45(8), 571–576. Review

Cichon S, Anker M, Vogt IR, et al. (1998). Cloning, genomic organization, alternative transcripts and mutational analysis of the gene responsible for autosomal recessive universal congenital alopecia. *Hum Mol Genet 7*(11), 1671–1679.

Czeizel AE, Elek C, Gundy S, et al. (1993). Environmental trichlorfon and cluster of congenital abnormalities. *Lancet 341*, 539–542.

Chand RP, Tharakan JK, Koul RL, Kumar SD. (1996). Clinical and radiological features of juvenile onset olivopontocerebellar atrophy. *Clin Neurol Neurosurg 98*(2), 152–156.

Daar S, Hussain HM, Gravell D, Nagel RL, Krishnamoorthy R. (2000). Genetic epidemiology of HbS in Oman: multicentric origin for the betaS gene. *Am J Hematol 64*(1), 39–46.

Daar S, Hussein HM, Merghoub T, Krishnamoorthy R. (1998). Spectrum of beta-thalassemia mutations in Oman. *Ann N Y Acad Sci 850*, 404–406.

Daar S, Vulliamy TJ, Kaeda J, Mason PJ, Luzzatto L. (1996). Molecular characterization of G6PD deficiency in Oman. *Hum Hered 46*(3), 172–176.

El-Hazmi MAF, Al-Swailem AR, Warsy AM, Sulaimani R, Al-Meshari AA. (1995). Consanguinity among the Saudi Arabian population. *J Med Genet 32*, 623–626.

EUROCAT Working Group. (eds.). Report No 9. Surveillance of congenital anomalies in Europe. (EUROCAT Central Registry, Room 1F08, University of Ulster, Newtownabbey, Co Antrim, Northern Ireland BT37 0QB.

Frossard PM, Girodon E, Dawson KP, et al. (1998). Identification of cystic fibrosis mutations in the United Arab Emirates. Mutations in brief No 133 *Online Hum Mutat 11*, 412–413.

Frossard P, Dawson KP, Girodon E, Gossens M. (1996). Characterisation of cystic fibrosis mutations in the United Arab Emirates. *European Journal of Human Genetics 4* (Suppl 1), 59.

Freeman N.V, Rajab A. (1999). Neural tube defects. *Tropical Doctor July*, 1–3.

Goud MT, Harassi S, Al-Khalili S, Salmani KK, Al-Busaidi SM, Rajab A. (2005) Incidence of chromosomal abnormalities in the Sultanate of Oman. *Saudi Med J 26*(12), 1951–1957.

Haider MZ, Moosa A, Dalal H, Habib Y, Reynold L. (2001). Gene deletion patterns in spinal muscular atrophy patients with different clinical phenotypes. *J Biomed Sci 8*(2), 191–196.

Hanley D. (1977). Oman and its Renaissance. 1st Ed. Stanley International, London.

Hamamy HA, Al-Hakkak ZS. (1989). Consanguinity and Reproductive Health in Iraq. *Hum Hered 39*, 271–275.

Heathcote K, Rajab A, Magre J, et al. (2002). Molecular analysis of Berardinelli-Seip congenital lipodystrophy in Oman: evidence for multiple loci. *Diabetes 51*(4), 1291–1293.

Hennies HC, Rauch A, Seifert W, et al. (2004). Allelic Heterogeneity in the COH1 Gene Explain Clinical Variability in Cohen Syndrome. *Am J Hum Genet 75*, 138–145.

Hennies HC, Kornak U, Zhang H, et al. (2008). Gerodermia osteodysplastica is caused by mutations in SCYL1BP1, a Rab-6 interacting golgin. *Nat Genet 40*(12), 1410–1412.

Hempelmann A, Kumar S, Muralitharan S, Sander T. (2006). Myofibrillogenesis regulator 1 gene (MR-1) mutation in an Omani family with paroxysmal nonkinesigenic dyskinesia. *Neurosci Lett 402*(1–2), 118–120.

Huber C, Delezoide AL, Guimiot F, et al. (2009) A large-scale mutation search reveals genetic heterogeneity in 3M syndrome. *Eur J Hum Genet 17*(3), 395–400.

Hoffmann K, Müller SJ, Stricker S, et al. (2006) Escobar syndrome is a prenatal myasthenia caused by disruption of the acetylcholine receptor fetal subunit. *Am J Hum Genet 79*, 303–312.

Jansen J. (1986). *Nomads in the Sultanate of Oman. Tradition and Development in Dhofar.* 1st ed. Westview Press.

Joshi SN, Hashim J, Venugopal P. (2002). Pattern of inborn errors of metabolism in an Omani population of the Arabian Peninsula. *Ann Trop Paediatr 22*(1), 93–96.

Joshi SN, Venugopalan P. (2007). Clinical characteristics of neonates with inborn errors of metabolism detected by Tandem MS analysis in Oman. *Brain Dev 29*(9), 543–546.

Kannan K, Dennison D, Nusrat N. (2003). Fanconi's anaemia. *Br J Haematol 122*(4), 521.

Khabori M Patton MA. (2008). Consanguinity and deafness in Omani children. *Int J Audiol 47*, 30–33.

Kornak U, Reynders E, Dimopoulou A, et al. (2008). Impaired glycosylation and cutis laxa caused by mutations in the vesicular H(+)-ATPase subunit ATP6V0A2. *Nature Genet 40*, 32–34.

Koul R, Al Futaisi A, Chacko A, et al.(2007). Clinical and genetic study of spinal muscular atrophies in Oman. *J Child Neurol 22*(10), 1227–1230.

Langdown JV, Williamson D, Knight CB, Rubenstein D, Carrell RW. (1989). Case report – a new doubly substituted sickling haemoglobin: HbS-Oman. *Br J Haematol 71*, 443–444.

Lorimer JG. (1908). *Gazetteer of the Persian Gulf, Oman, and Central Arabia. Vol. II. Geographical and Statistical*. Calcutta: Superintendent Government Printing.

Marengo-Rowe AJ, Lorkin PA, Gallo E, Lehmann N. (1968). Haemoglobin Dhofar – A New Variant from Southern Arabia. *Biochimica et Biophysica Acta 168*, 58–63.

Miles SB. (1919). *Countries and the Tribes of the Persian Gulf*. Second Edition, 1966. London: Frank Cass & Co. Ltd.

Michalk A, Stricker S, Becker J, et al. (2008). Acetylcholine receptor pathway mutations explain various fetal akinesia deformation sequence disorders. *Am J Hum Genet 82*(2), 464–476.

Mochida GH, Rajab A, Eyaid W, et al. (2004). Broader geographical spectrum of Cohen syndrome due to COH1 mutations. *J Med Genet 41*(6), e87.

Modell B. (2003) Possibilities for the control of congenital and genetic disorders in Oman. *Assignment Report to WHO*.

Muralitharan S, Al-Lamki Z, Dennison D, et al. (2005). An inframe perforin gene deletion in familial hemophagocytic lymphohistiocytosis is associated with perforin expression. *Am J Hematol 78*(1), 59–63.

Muralitharan S, Wali YA, Dennison D, et al. (2007). Novel spectrum of perforin gene mutations in familial hemophagocytic lymphohistiocytosis in ethnic Omani patients. *Am J Hematol 82*(12), 1099–1102.

Nagel RL, Daar S, Romero JR, et al. (1998). HbS-oman heterozygote: a new dominant sickle syndrome. *Blood 92*(11), 4375–4382

Nicole S, Ben Hamida C, Beighton P, et al. (1995). Localization of the Schwartz-Jampel syndrome (SJS) locus to chromosome 1p34-p36.1 by homozygosity mapping. *Hum Mol Genet 4*(9), 1633–1636.

Ohadi M, Lalloz MR, Sham P, et al. (1999). Localization of a gene for familial hemophagocytic lymphohistiocytosis at chromosome 9q21.3-22 by homozygosity mapping. *Am J Hum Genet 64*(1), 165–171.

Pinotti M, Camire RM, Baroni M, Rajab A, Marchetti G, Bernardi F. (2003). Impaired prothrombinase activity of factor X Gly381Asp results in severe familial CRM+ FX deficiency. *Tromb Haemost 89*, 243–248.

Quinkler M, Bappal B, Draper N, et al. (2004). Molecular basis for the apparent mineralocorticoid excess syndrome in the Oman population. *Mol Cell Endocrinol 217*(1–2), 143–149.

Rajab A, Freeman NV, Patton MA. (1996). The frequency of posterior urethral valves in Oman. *Brit J Urol 77*, 900–904.

Rajab A, Freeman NV, Patton MA. (1997a). Hirschprung's disease in Oman. *J Pediatr Surgery 32*(5), 724–727.

Rajab A, Patton MA. (1997b) Major factors determining frequencies of hemoglobinopathies in Oman. Letter to the Editor. *Am J Med Genet 71*, 240–242.

Rajab A, Vaishnav A, Freeman NV, Patton MA. (1998). Neural tube defects and congenital hydrocephalus in Oman. *J Trop Pediatr 44*, 300–303.

Rajab A, Patton M A. (1999a). Analysis of the population structure in Oman. *Community Genetics* 2, 23–25.

Rajab A, Patton MA. (1999b). Development and use of a national hemoglobinopathy register in Oman. Letter to the Editor. *Community Genetics 2*, 47–48.

Rajab A, Patton MA. (2000a). A study of Consanguinity in the Sultanate of Oman. *Annal Hum Biol 27*, 321–326.

Rajab A, Patton MA, Modell CB. (2000b). Study of haemoglobinopathies in Oman through a national register. *Saudi Med J 21*(12), 1168–1172.

Rajab A, Thomas C. (2001) Facial clefts in the Sultanate of Oman. *European Journal of Plastic Surgery 24*, 230–233.

Rajab A, Heathcote K, Joshi S, Jeffery S, Patton M. (2002) Heterogeniety or congenital generalised lipodystrophy in seventeen patients from Oman. *Am J Med Genet 110*, 219–225.

Rajab A, Khaburi M, Spranger S, Kunze J, Spranger J. (2003a). Congenital generalized lipodystrophy, mental retardation, deafness, short stature, and slender bones: A newly recognized syndrome? *Am J Med Genet 121A*, 271–276.

Rajab A, Mochida GH, Hill A, et al. (2003b). A novel form of pontocerebellar hypoplasia maps to chromosome 7q11-21. *Neurology 60*, 1664–1667.

Rajab A, Kunze J, Mundlos S. (2004). Spondyloepipheseal Dysplasia Omani type: A New Recessive Type of SED with Progressive Spinal Involvement. *Am J Med Genet 126A*, 413–419.

Rajab A, Bappal B, Al-Shaikh H, Al-Khusaibi S, Mohammed AJ. (2005a) Common autosomal recessive diseases in Oman derived from a hospital-based registry. *Community Genetics 8*, 27–30.

Rajab A, Hoffmann K, Ganesh A, Sethu AU, Mundlos S. (2005b). Escobar variant with pursed mouth, creased tongue, ophthalmologic features and scoliosis in 6 children from Oman. *Am J Med Genet 134A*, 151–157.

Rajab A, Seung-Yun Y, Abdulgalil A, et al. (2006). An autosomal recessive form of spastic cerebral palsy (CP) with microcephaly and mental retardation. *Am J Med Genet 140A*, 1504–1510.

Rajab, Manzini, Mochida, Walsh, Ross. (2007a). A Novel Form of Lethal Microcephaly with Simplified Gyral Pattern and Brain Stem Hypoplasia. *Am J Med Genet (A) 143*(23), 2761–2767.

Rajab A, El-Hazmi M. (2007b). The Gulf Cooperation Countries Genetic services. Understanding individuals, families, and community needs. Editorial. *Saudi Med J* 28(9), 1321–1323.

Rajab A, Kelberman D, de Castro SC, et al. (2008a). Novel mutations in LHX3 are associated with hypopituitarism and sensorineural hearing loss. *Hum Mol Genet 17*(14), 2150–2159.

Rajab A, Kornak U, Budde BS, et al. (2008b). Geroderma osteodysplasticum hereditaria and wrinkly skin syndrome in 22 patients from Oman. *Am J Med Genet 146A*, 965–976.

Rajab A, Neitzel H, Näthe J, et al. (in preparation) Down syndrome in Oman I: Epidemiological and molecular studies.

Rajab A, Aldinger KA, El-Shirbini HA, Dobyns WB, Ross ME. (2009) Recessive developmental delay, small stature, microcephaly and brain calcifications with locus on chromosome 2. *Am J Med Genet 149A*(2), 129–137.

Ramachandran M, Gu LH, Wilson GB, et al. (1992). A new variant, HB Muscat (alpha 2 beta (2) 32 (B14) Leu→ Val) observed in association with B S in an Arabian family. Hemoglobin *16*(4), 259–266.

Romey MC, Guittard C, Chazalette JP, et al. (1999). Complex allele (-102T>A+S549R(T>G)) is associated with milder forms of cystic fibrosis than allele S549R(T>G) alone. *Hum Genet 105*(1–2), 145–150.

Sawardekar KP. (2004). Genetic Analysis of lethal congenital malformations causing perinatal mortality at Nizwa Hospital, Oman. *Clin Genet 66*(3), 239–241.

Sperling K, Pelz J, Wegner RD, Dörries A, Grüters A, Mikkelsen M. (1994). Bewertung eines Trisomie 21 Clusters. *Med Genetik 6*, 378–385.

Subramanyan R, Joy J, Venugopalan P, Sapru A, al Khusaiby SM. (2000). Incidence and spectrum of congenital heart disease in Oman. *Ann Trop Paediatr 20*(4), 337–341.

Subramanyan R, Venugopalan P. (2002). Malignant familial long QT syndrome. *Saudi Med J 23*(6), 738–742.

Tiele H, Sakano M, Kitagawa K, et al. (2004). Loss of chonroitin 6-0-sulfotransferase-1 function result in severe human chondrodysplasia with progressive spinal involvement. *PNAS 101*(27), 10155–10160.

Wellsted JR. (1837). *Travels in Arabia*. (Printed 1840). London, Collbaren.

Wellsted JR. (1886) *Travels to the City of Caliphs, Along the Shores of the Persian Gulf and the Mediterranean*. London, Westview Press.

Williamson D, Brown KP, Langdown JV, Baglin TP. (1995). Haemoglobin Dhofar is linked to the codon 29 C-->T (IVS-1 nt-3) splice mutation which causes beta+ thalassaemia. *Br J Haematol 90*(1), 229–231.

Wilkinson JC. (1987). The *Immamate Tradition of Oman*. 1st Edition. Cambridge: Cambridge University Press.

White JM, Christie BS, Nam D, Daar S, Higgs DR. (1993). Frequency and clinical significance of Erythrocyte genetic abnormalities in Omanis. *J Med Genet 30*, 396–400.

White DR, Ganesh A, Nishimura D, et al. (2007). Autozygosity mapping of Bardet-Biedl syndrome to 12q21.2 and confirmation of FLJ23560 as BBS10. *Eur J Hum Genet 15*(2), 173–178.

Woolf DA, Leonard JV, Trembath RC, Pembrey ME, Grant DB. (1991). Nesidioblastosis: evidence for autosomal recessive inheritance. *Arch Dis Child 66*, 529–530.

55.

GENETIC DISORDERS IN LEBANON: CHALLENGES AND OPPORTUNITIES

Issam Khneisser, Salim M. Adib, and Andre Megarbane

INTRODUCTION

Lebanon is a small country (10,500 km^2) lying on the eastern shores of the Mediterranean Sea. Historically, it has always been a favored crossroads between Asia, Africa, and Europe, submitting it to many ethnic migratory currents. In the Lebanese population, currently estimated at about 4 million, coexist 17 officially recognized ethno-religious communities. For more than a millennium, most of these various communities have cohabitated in relatively well-defined regions of this mostly mountainous country. Within these sociological niches, endogamy and even consanguineous marriages were often the norm. Marriages among first cousins were frequent up till the 1980s, reaching 30% in certain reports (Der Kaloustian et al., 1980). The sociological transition with increased urbanization, population mixing, and women's educational and professional participation has greatly reduced these figures since then in most communities and regions. Most recent reports indicate that first cousin marriages have decreased to 15%, with the lowest rate found in the capital city of Beirut (CAS, 2004). These historical patterns have favored a high incidence of both common and rare genetic diseases. At least 184 different genetic diseases have been reported specifically from Lebanon (CAGS, 2008). In this chapter, we will review the evolution of genetic services in Lebanon, with a discussion of challenges and opportunities. We will also present the epidemiological and genetic aspects of some of those diseases, with an emphasis on the most frequently diagnosed ones.

RESEARCH CENTER HISTORY AND BACKGROUND

Diagnostic services and research on genetic disorders in Lebanon remained limited for years. Diagnosis and screening had been possible since the early 1960s in a few university-affiliated centers for a limited number of conditions such as G6PD deficiency or some hemoglobinopathies (Najjar et al., 1963; Der Kaloustian, 1986). All this started changing in the 1990s, following the end of a protracted period of civil wars and foreign invasions (1975–1991). Major centers such as the American University of Beirut Medical Center (AUBMC) and Saint-Joseph University Medical Genetics Unit (USJ-MGU) began acquiring state-of-the-art equipment and kits for cytogenetic defect research (Zahed & Megarbane, 1998), for molecular genetic investigation of neurosensory deafness (Chaib et al., 1996a; Saouda et al., 1998) and other conditions, and for neonatal screening (Daher et al., 2003; Khneisser et al., 2007). In addition to those two major centers, the Chronic Care Center (CCC), a multidisciplinary not-for-profit foundation dedicated to the treatment of some chronic diseases of childhood, was established in 1994. To date, CCC remains the largest diagnostic center in Lebanon for childhood hemoglobinopathies. Currently, projects for cytogenetic and/or molecular diagnostic laboratories are under development in some other academic centers in Lebanon.

EPIDEMIOLOGICAL STATUS AND CLINICAL CHALLENGES ASSOCIATED WITH SELECTED GENETIC CONDITIONS IN LEBANON

CHROMOSOMAL DEFECTS

Many cytogenetic chromosomal defects have been described in Lebanon in the past few years (Delague et al., 2006; Mahfouz et al., 2001; Megarbane et al., 1997a,b, 2000a, 2001a,b, 2002; Souraty et al., 2007a; Zahed et al., 2004a,b, 2005, 2007, 2008; Zahed & Megarbane, 1998). Most frequently diagnosed are entities leading to mental retardation such as Down syndrome.

About 100 new Down syndrome cases are diagnosed annually in Lebanon in a caseload of newborns estimated at about 60,000. Most cases are diagnosed now within the first month of age, up from 47.3% in 1998. Earlier diagnosis has become possible because of increased availability of cytogenetic centers and the disposition of physicians to use those centers. Practically all cases are diagnosed now either at the AUBMC or USJ-MGU laboratories in Beirut. Early diagnosis has improved the outcome of genetic counseling with families, and earlier targeted medical and educational care for the index case. The male to female ratio has been estimated at 1.66 (Zahed & Megarbane, 1998).

Secondary prevention of mental retardation cases involves behavioral management techniques, specialized schools, community support for the family, and careful medical follow-up. All these services are currently lacking in quality and insufficient to respond to the needs of patients in Lebanon. Most of those services are provided by NGOs, and there are no apparent plans for significant public provision of care on these issues in the coming few years. Although no curative treatment is yet available for these diseases, early diagnosis is essential to provide the family with adequate genetic counseling. The most critical point is still "to convince the family of the need for such an evaluation and to relieve the parents of any feeling of guilt" (Zafatayeff et al., 1998).

During the past years, new techniques have been used to better diagnose chromosomal defects (Delague et al., 2006). Other new cytogenetics technologies are still needed in Lebanon, such as microarrays-CGH, currently being developed in cooperation with external international laboratories (Zahed et al., 2007).

MOLECULAR DEFECTS

The most frequent molecular defects in Lebanon are believed to be those associated with hearing losses in children (about 1/1000 newborn children). The vast majority of prelingual forms of isolated deafness are highly genetically heterogeneous, with autosomal recessive modes of inheritance. Genetic variants responsible for this severe progressive sensorineural hearing loss have already been reported in Lebanon: DFNB1 (Mustapha et al., 2001; Medlej-Hashim et al., 2002a), DFNB6 and DFNB12 (Chaib et al., 1996a,b), DFNB9 (Yasunaga et al., 1999), DFNB13 and DFNB14 (Mustapha et al., 1998a,b), DFNB21 (Mustapha et al., 1999) DFNB31 (Mustapha et al., 2002a).

Other forms of deafness belong to the Usher syndrome (USH), an autosomal recessive disease in which neurosensory deafness is accompanied by progressive retinitis pigmentosa. Genetic defect variants associated with USH and described in Lebanon have included: USH1B (Mouglabey et al., 1998), USH1C (DeAngelis et al., 2001), USH1G (Mustapha et al., 2002b), USH2A (Saouda et al., 1998). The rarest form of hearing loss in Lebanon is the Pendred syndrome, an autosomal recessive disease characterized by congenital sensorineural deafness and goiter (Mustapha et al., 1998c).

Newborn hearing screening using a special audiometer is recommended before discharge from the hospital. In Lebanon, newborn hearing screening is currently available in only one university hospital in Beirut, affiliated with the Balamand University. The relative neglect surrounding this issue so far has prompted the recent creation of a civil coalition for persons suffering from hearing losses, dedicated to advancing the agenda for research, detection, and medical support.

Familial paroxysmal polyserositis, or familial Mediterranean fever, a molecular disease often found around the Mediterranean Sea, has been described in Lebanon (Der Kaloustian, 1986). The prevalence of the disease is not estimated but several case series had been reported in previous years in the Armenian community (Khachadurian & Armenian, 1974; Armenian, 1982), as well as in practically all other ethnic communities in Lebanon (Mansour et al., 2001; Medlej-Hashim et al., 2001, 2002b, 2004). Early diagnosis allows an early start of lifelong treatment with colchicine, which reduces the frequency and severity of acute crises of fever and serosal inflammation and prevents renal amyloidosis, the most severe complication of FMF.

The fragile-X syndrome is the most commonly inherited and one of the most devastating form of mental retardation worldwide. The genetics of the fragile-X syndrome is now understood, and the disorder is preventable through appropriate genetic counseling. Unfortunately, that issue in Lebanon has been discussed in only one paper, which described the clinical features of a series of 5 cases (Zafatayeff et al., 1998).

Other, rarer molecular defects have been described in local or familial clusters of genetic disorders. These have included, for example, oculocutaneous albinism type1 OCA1 (Zahed et al., 2005), ataxia (Delague et al., 2001, 2002; Megarbane et al., 2001c), progressive pseudorheumatoid dysplasia (PPD) (Delague et al., 2005), Charcot-Marie-Tooth neuropathy CMT4F (Delague et al., 2000; Guilbot et al., 2001) & CMT4H (De Sandre-Giovannoli et al., 2005; Delague et al., 2007), osteopetrosis (Souraty et al., 2007b), Bardet-Biedl syndrome (Laurier et al., 2006), odonto-onycho-dermal dysplasia syndrome (Adaimy et al., 2007; Megarbane et al., 2004), and Brown-Vialetto-Van Laere syndrome (Megarbane et al., 2000b). Congenital adrenal hyperplasia has been described in 25 unrelated Lebanese families from different religious communities presenting at least one affected child (Delague et al., 2000). Cystic Fibrosis (CF) was first reported in five individual cases (Bitar and Idriss, 1969), and later in 20 families with at least one CF-affected child (Desgeorges et al., 1997).

GENETIC DISEASES IN NEWBORN SCREENING PROGRAMS

The epidemiological features of disorders most frequently screened at birth in Lebanon are now relatively well known. Up till 2008, 150,000 newborn have been screened at USJ-MGU. Among those, the following cases were detected: 93 (60 per 100,000) of congenital hypothyroidism (CH); 830 (1% among males) of G6PD deficiency; 16 (10 per 100,000) of phenylketonuria (PKU) variants, and 6 (4 per 100,000) of congenital galactosemia variants (Khneisser et al., 2008b).

Among 18,000 newborn samples tested by MS/MS tandem spectrometry, 15 were confirmed to have an inborn error of metabolism other than PKU. Three of those 15 were methylmalonic cases (15 per 100,000); the remaining cases were 1 case each of tyrosinemia type II, methionine adrenyl transferase deficiency type III, arginino-succinate deficiency, isovaleric aciduria, propionic aciduria, carnitine uptake deficiency, hydratase methylgluconate, biotidinase deficiency, 3-oxothiolase deficiency, glutaric aciduria I, very long chain acyl-CoA dehydrogenase deficiency (VLCAD), and carnitine translocase deficiency. (Khneisser et al., 2008a).

Historically, CH has been one of the earliest described newborn diseases in Lebanon (Najjar et al., 1963). However, the most prevalent genetic disorder remains G6PD deficiency. Other diseases, such as GAL, homocystinuria, maple syrup urine disease, citrullinemia, and hyperornithinemia (Alexander, 1986; Mamo et al., 1971; Khneisser et al., 2008b) remain extremely rare.

Individual cases of fatty acids disorders have been reported in the past. They have included: *SCAD* (Mikati et al., 2007) and *VLCAD* (Touma et al., 2001; Khneisser et al., 2008b). Errors of carnitine metabolism such as CACT and CUD have also been reported in one child each (Khneisser et al., 2008b). Equally rare have been cases of organic acid disorders: methylmalonic aciduria (Alexander, 1986; Khneisser et al., 2008b), isovaleric aciduria, glutaric aciduria type I, and 3-methylglutaconylCoA-hydratase deficiency (Khneisser et al., 2008b). One case of mitochondrial acetoacetyl-CoA thiolase (T2) deficiency has been described (Zhang et al., 2004), and one case of 2-methylbutyryl-coenzyme A dehydrogenase deficiency (Sass et al., 2008).

Currently around 50% of the Lebanese newborn infants do not benefit from any coverage for screening. Even as newborn screening services have become available and accessible, their cost, though modest compared to potential benefits, may remain an obstacle for a large sector of the Lebanese population. Marketing efforts to promote the benefits of newborn screening among practitioners and in the public are conducted, essentially, by the concerned laboratories and affiliated academic centers. While goodwill has been expressed by national health authorities, a mandatory and publicly funded national screening program does not seem likely to become a reality soon.

HEMOGLOBINOPATHIES

Hemoglobinopathies are the largest group of genetic disorders in Lebanon, with a population incidence rate estimated at more than 2%. Major hemoglobinopathies detailed here are sickle cell anemia and the thalassemias.

Sickle cell anemia is known to be clustered in two geographic areas in north and south Lebanon. Results of a large-scale national survey published in 2007 described 387 patients with either sickle cell anemia (SS) and/or sickle beta-thalassemia (ST) mostly from these two areas. Fifty percent of the patients were males, and 56% were offspring of consanguineous parents. The SS/ST gene mutation distribution was 3:1. These survey results indicate that preventive efforts should target the northern and southern regions of Lebanon to decrease the number of new offspring afflicted with this disease (Inati et al., 2007).

Patients with thalassemia major (TM) or intermedia (TI) fare better than those with sickle cell anemia. Thalassemic patients have a unique opportunity of qualifying for comprehensive follow-up at the Chronic Care Center (CCC). To date, 452 patients aged 2–68 years have been followed at that center. Sixty-seven percent of patients are below 20 years. All affiliated patients undergo a complete assessment by the multidisciplinary team at least twice a year. Sixty-four percent have TM, and 36% TI. Fifty-seven percent are males. Geographically, patients are homogeneously distributed all over Lebanon, in contrast to patients with sickle cell disease. The CCC follow-up has remarkably reduced the mortality rate of affiliated thalassemia patients. Since 1994, only 12 TM patients have died from complications, mostly from heart failure. In addition, the current population of TM patients is relatively younger, which may cause more reduction in the mortality rate (Inati et al., 2006). While standards of care are the same regardless of religious or cultural background, the only demographic variable that seems to significantly affect TM patients' compliance to treatment is the parental level of education. Among parents of thalassemic patients, roughly 70% have not reached high school (Chehab et al., 1987).

OTHER CHRONIC HEMATOLOGICAL DISORDERS

The incidence of G6PD deficiency in Lebanon is estimated at 10 per 1000 in men and 0.4 per 1000 in women. Before newborn screening became available, G6PD deficient patients had a 77.8% risk to suffer at least one severe anemia crisis in their childhood following the ingestion of raw green fava beans.

(Khneisser et al., 2007). With the availability of newborn screening in Lebanon, that risk estimation had decreased to 4%. This is largely attributable to patients' awareness of their disease status, which leads them to avoid fava bean food products (Khneisser et al., 2006).

Hemophilia A (HA) is an X-linked recessive hereditary bleeding disorder affecting one in 5000 men, resulting from mutations in the F8 gene. A recent case series described 79 HA patients from 55 unrelated families in Lebanon (Djambas-Khayat et al., 2008). While relatively rare, this disease poses an economic challenge, as the treatment of recurrent bleeding episodes is very expensive. At the time of this writing, treatment was partially subsidized by the Ministry of Public Health (MOPH), but most patients still have to find philanthropic sources of funding to be able to obtain adequate care. Guidelines for control and prevention of HA in Lebanon are under review. They should include promoting the detection of female carriers in the affected family, extensive patient education, preventive physiotherapy, and rapid care following bleeding episodes.

CONCLUDING REMARKS: THE FUTURE OF GENETIC DISORDERS IN LEBANON

Any relevant program for the control and prevention of genetic disorders involves various types of genetic testing, depending on the family pedigree and on clinical signs. In a country like Lebanon, where high rates of consanguineous marriages and village endogamy have existed for centuries, a national mandatory newborn screening program should be put in place without further delay. The cost-effectiveness of such a program has been demonstrated repeatedly in Lebanon (Khneisser et al., 2007) and elsewhere. The evidence is convincing enough to warrant increased lobbying for its implementation.

The situation concerning genetic screening is different with adults. Regardless of the type of disease, the decision to take a genetic test remains a personal choice. It should be encouraged only where a multidisciplinary team exists to provide useful genetic counseling and advice on the various practical possibilities. In our small country, we are fortunate that at least two such multidisciplinary centers exist already and are accessible and culturally accepted. More are being planned and/or implemented at this time. Nevertheless, parents should be aware of whether or not there is a probability of a sick newborn, given that they have already had one. More ideally, this probability should be obtained for most frequently described diseases even before a first child is conceived in high-risk families: those with a known history of genetic disorders or those with consanguineous marriages.

To date, the only mandatory requirement vaguely associated with pregnancy outcomes in Lebanon is the premarital testing that was imposed in 1994. While screening for several infectious diseases with potential teratogenic effects are included, practically no genetic disorders are considered. The exception is thalassemia, which is indirectly targeted by requesting a hemoglobin mean corpuscular volume (MCV) as part of a complete blood count.

Some discussion started in 2008 to create a steering committee on genetic screening, which would lobby for mandatory neonatal screening and modifications of the premarital tests required in families at higher risk. The steering committee would also elaborate a national consensus on which diseases should be screened. Consensus is particularly important as new tools become available, and there is a temptation to use them in large-scale screenings even for rare diseases. This situation generally results in confusing, costly, and emotionally taxing false-positive results. In a small country like Lebanon, the loss of credibility from low positive predictive value of the method of choice can negatively affect the willingness of physicians to prescribe genetic screening tests and counseling, and the compliance of the public with prescribed services.

The genetic centers in Lebanon have by now acquired vast experience in providing services to the local population. They stand ready to absorb increased local demands as legal mandates are put in place and/or public awareness increases for early detection. Several centers have established ties with prestigious international centers, to provide continued training and education to local staff on state-of-the-art technologies. In turn, Lebanese centers desire, and indeed have already started, to provide collaboration and technology transfer to regional centers and colleagues in the larger Middle East and North African areas that may have difficulties accessing international centers (Mattit et al., 2006; Medlej-Hashim et al., 2000, 2002a, 2005; Souraty, 2007b).

REFERENCES

Adaimy L, Chouery E, Megarbane H, et al. (2007). Mutation in WNT10A is associated with an autosomal recessive ectodermal dysplasia: the odonto-onycho-dermal dysplasia. *Am J Hum Genet 81*, 821–828.

Bitar J, Idriss H. (1969). Cystic fibrosis of the pancreas. *J Med Liban 22*, 181–185.

Alexander DR. (1986). The study of inborn errors of metabolism in Lebanon. *J Med Libanais 36*, 54–61.

Armenian HK. (1982). Genetic and environmental factors in the aetiology of familial paroxysmal polyserositis. An analysis of 150 cases from Lebanon. *Trop Geogr Med 34*, 183–187.

CAGS. Center of Arab genetics studies (www.cags.org.ae). Accessed July 31, 2008.

CAS-Lebanon. Family health survey 2004, p. 46. Available at: http://www.socialaffairs.gov.lb/files/papfamreporten.pdf.

Chaib H, Place C, Salem N, et al. (1996a). A gene responsible for a sensorineural nonsyndromic recessive deafness maps to chromosome 2p22-23. *Hum Mol Genet 5*, 155–158.

Chaib H, Christophe Place C, Salem N, et al. (1996b). Mapping of *DFNB12*, a gene for a non-syndromal autosomal recessive deafness, to chromosome 10q21-22. *Hum Mol Gen 5*, 1061–1065.

Chehab FF, Der Kaloustian V, Khouri FP, et al. (1987). The molecular basis of beta-thalassemia in Lebanon: application to prenatal diagnosis. *Blood* 69, 1141–1145.

Daher R, Beaini M, Mahfouz R, et al. (2003). A neonatal screening in Lebanon: Results of five years' experience. *Ann Saudi Medicine 23*, 16–19.

De Sandre-Giovannoli A, Delague V, Hamadouche T, et al. (2005). Homozygosity mapping of autosomal recessive demyelinating Charcot-Marie-Tooth neuropathy (CMT4H) to a novel locus on chromosome 12p11.21-q13.11. *J Med Genet 42*, 260–265.

DeAngelis MM, McGee TL, Keats BJ, et al. (2001). Two families from New England with usher syndrome type IC with distinct haplotypes. *Am J Ophthalmol 131*, 355–358.

Delague V, Jacquier A, Hamadouche T, et al. (2007) Mutations in FGD4 encoding the Rho GDP/GTP exchange factor FRABIN cause autosomal recessive Charcot-Marie-Tooth type 4H. *Am J Hum Genet 81*, 1–16.

Delague V, Bareil C, Tuffery S, et al. (2000). Mapping of a new locus for autosomal recessive demyelinating Charcot-Marie-Tooth disease to 19q13.1-13.3 in a large consanguineous Lebanese family: exclusion of MAG as a candidate gene. *Am J Hum Genet 67*, 236–243.

Delague V, Souraty N, Khallouf E, et al (2000) A Mutational analysis in Lebanese patients with congenital adrenal hyperplasia due to a deficit in 21-hydroxylase. *Horm Res 53*, 77–82.

Delague V, Bareil C, Bouvagnet P, et al. (2002). A new autosomal recessive non-progressive congenital cerebellar ataxia associated with mental retardation, optic atrophy, and skin abnormalities (CAMOS) maps to chromosome 15q24-q26 in a large consanguineous Lebanese Druze family. *Neurogenetics 4*, 23–27.

Delague V, Bareil C, Bouvagnet P, et al. (2001). Nonprogressive autosomal recessive ataxia maps to chromosome 9q34-9qter in a large consanguineous Lebanese family. *Ann Neurol 50*, 250–253.

Delague V, Chouery E, Corbani S, et al. (2005). Molecular study of WISP3 in nine families originating from the Middle-East and presenting with progressive pseudorheumatoid dysplasia: identification of two novel mutations, and description of a founder effect. *Am J Med Genet A 138*, 118–126.

Delague V, Souaid M, Chouery E, et al. (2006). Screening for subtelomeric rearrangements using automated fluorescent genotyping of microsatellite markers: a Lebanese study. *Eur J Med Genet 49*, 117–126

Der Kaloustian VM, Naffah J, Loiselet J. (1980). Genetic diseases in Lebanon. *Am J Med Genet 7*, 187–203.

Der Kaloustian VM. (1986). Genetic diseases in Lebanon. *J Med Libanais* 36, 65–68.

Desgeorges M, Mégarbané A, Guittard C, et al. (1997). Cystic fibrosis in Lebanon: distribution of CFTR mutations among Arab communities. *Hum Genet 100*, 279–283.

Djambas-Khayat C, Salem N, Chouery E, et al. (2008). Molecular analysis of F8 in Lebanese haemophilia A patients: novel mutations and phenotype-genotype correlation. *Haemophilia 14*, 709–716.

El-Zahabi LM, Makarem J, Habbal Z, et al. (2007). Gaucher disease: different clinical manifestations associated with a rare mutation (R48W) in a Lebanese family. *Mol Genet Metab 91*, 402–404.

Guilbot A, Williams A, Ravisé N, et al. (2001). A mutation in periaxin is responsible for CMT4F, an autosomal recessive form of Charcot-Marie-Tooth disease. *Hum Mol Genet 10*, 415–421.

Habbal ZM, Touma EH. (1995). Positive interference from homocystinuria urine in a spot test for molybdenum cofactor deficiency. *Clin Chem 41*, 1056–1057.

Inati A, Jradi O, Tarabay H, et al. (2007). Sickle cell disease: the Lebanese experience. *Int J Lab Hematol 29*, 399–408.

Inati A, Zeineh N, Isma'eel H, et al. (2006). Beta-thalassemia: the Lebanese experience. *Clin Lab Haematol 28*, 217–227.

Khachadurian AK, Armenian HK. (1974). Familial paroxysmal polyserositis (familial Mediterranean fever); incidence of amyloidosis and mode of inheritance. *Birth Defects Birth Artic Ser 10*, 62–66.

Khlat M, Halabi S, Khudr A, Der Kaloustian VM. (1986). Perception of consanguineous marriages and their genetic effects among a sample of couples from Beirut. *Am J Med Genet 25*, 299–306.

Khneisser I, Adib SM, Loiselet J, Mégarbané A. (2006). Prevalence of G6PD deficiency and knowledge of diagnosis in a sample of previously unscreened Lebanese males: clinical implications. *J Med Screen 13*, 26–28.

Khneisser I, Adib SM, Loiselet J, Megarbane A. (2007). Cost-benefit analysis of G6PD screening in Lebanese newborn males. *J Med Liban 55*, 129–132.

Khneisser I, Adib SM, Lukacs Z, Megarbane A. (2008a). *Newborn screening by tandem mass spectrometry in Lebanon: High incidence of methylmalonic aciduria*. Paper presented at the 2008 Newborn Screening and Genetic Testing Symposium; San Antonio, TX, November 3–6, 2008.

Khneisser I, Adib SM, Megarbane A, Lukacs Z. (2008b). International cooperation in the expansion of a newborn screening programme in Lebanon: a possible model for other programmes. *J Inherit Metab Dis 31*(2), 441–446.

Laurier V, Stoetzel C, Muller J, et al. (2006). Pitfalls of homozygosity mapping: an extended consanguineous Bardet-Biedl syndrome family with two mutant genes (BBS2, BBS10), three mutations, but no triallelism. *Eur J Hum Genet 14*, 1195–1203.

Mahfouz R, al-Oreibi G, Darwiche N, et al. (2001). Constitutional chromosome abnormalities among patients referred for blood karyotype analysis: a 5-year study at the AUBMC. *J Med Liban 49*, 6–12.

Mamo JG, Tabbara KF. (1971). Homocystinuria in Lebanon : A report of five cases. *J Med Liban 24*, 473–482.

Mansour I, Delague V, Cazeneuve C, et al. (2001). Familial Mediterranean fever in Lebanon: mutation spectrum, evidence for cases in Maronites, Greek orthodoxes, Greek Catholics, Syriacs and Chiites and for an association between amyloidosis and M694V and M694I mutations. *Eur J Hum Genet 9*, 51–55.

Mattit H, Joma M, Al-Cheikh S, et al. (2006). Familial Mediterranean fever in the Syrian population: gene mutation frequencies, carrier rates and phenotype-genotype correlation. *Eur J Med Genet 49*, 481–486.

Medlej-Hashim M, Rawashdeh M, Chouery E, et al. (2000). Genetic screening of fourteen mutations in Jordanian familial Mediterranean fever patients. *Hum Mutat 15*, 384.

Medlej-Hashim M, Petit I, Adib S, et al. (2001). Familial Mediterranean Fever: association of elevated IgD plasma levels with specific MEFV mutations. *Eur J Hum Genet 9*, 849–854.

Medlej-Hashim M, Mustapha M, Chouery E, et al. (2002a). Non-syndromic recessive deafness in Jordan: mapping of a new locus to chromosome 9q34.3 and prevalence of DFNB1 mutations. Eur *J Hum Genet 10*, 391–394.

Medlej-Hashim M, Salem N, Chouery E, et al. (2002b). Familial Mediterranean fever: the potential for misdiagnosis of E148V using the E148Q usual RFLP detection method. *Clin Genet 61*, 71–73.

Medlej-Hashim M, Delague V, Chouery E, et al. (2004). Amyloidosis in familial Mediterranean fever patients: correlation with MEFV genotype and SAA1 and MICA polymorphisms effects. *BMC Med Genet 5*, 4.

Medlej-Hashim M, Serre JL, Corbani S, et al. (2005). Familial Mediterranean fever (FMF) in Lebanon and Jordan: a population genetics study and report of three novel mutations. *Eur J Med Genet 48*, 412–420.

Megarbane A, Bejjani BA, Shaffer LG, et al. (2002). Craniofacial anomalies, deafness, brachydactyly, short stature, and moderate mental retardation due to a cryptic 6p;11q translocation. *Am J Med Genet 108*, 69–74.

Megarbane A, Gosset P, Souraty N, et al. (2001a). Chromosome 10p11.2-p12.2 duplication: report of a patient and review of the literature. *Am J Med Genet 104*, 204–208.

Megarbane A, Le Lorc'H M, Elghezal H, et al. (2001b). Pure partial 7p trisomy including the TWIST, HOXA, and GLI3 genes. *J Med Genet 38*, 178–182.

Megarbane A, Delague V, Ruchoux MM, et al. (2001c). New autosomal recessive cerebellar ataxia disorder in a large inbred Lebanese family. *Am J Med Genet 101*, 135–141.

Megarbane A, Gosset P, Souraty N, et al. (2000a). Chromosome 7q22-q31 duplication: report of a new case and review. *Am J Med Genet 95*, 164–168.

Megarbane A, Desguerres I, Rizkallah E, et al. (2000b). Brown-Vialetto-Van Laere syndrome in a large inbred Lebanese family: confirmation of autosomal recessive inheritance? *Am J Med Genet 92*, 117–121.

Megarbane A, Souraty N, Prieur M, et al. (1997a). Interstitial duplication of the short arm of chromosome 2: report of a new case and review. *J Med Genet 34*, 783–786.

Megarbane A, Souraty N, Theophile D, et al. (1997b). Trisomy for the distal segment of the short arm of chromosome 17 in a boy with mild mental retardation and some dysmorphic features. *Ann Genet 40*, 55–59.

Mikati MA, Chaaban HR, Karam PE. (2007). Brain malformation and infantile spasms in a SCAD deficiency patient. *Pediatr Neurol 36*, 48–50

Mouglabey YB, Nimri S, Sayegh F, et al. (1998). Map refinement of the Usher syndrome type 1B gene, MYO7A, relative to 11q13.5 microsatellite markers. *Clin Genet 54*, 155–158.

Mustapha M, Chardenoux S, Nieder A, et al. (1998a). A sensorineural progressive autosomal recessive form of isolated deafness, DFNB13, maps to chromosome 7q34-q36. *Eur J Hum Genet 6*, 245–250.

Mustapha M, Salem N, Weil D, et al. (1998b). Identification of a locus on chromosome 7q31, DFNB14, responsible for prelingual sensorineural non-syndromic deafness. *Eur J Hum Genet 6*, 548–551.

Mustapha M, Azar ST, Moglabey YB, et al. (1998c). Further refinement of Pendred syndrome locus by homozygosity analysis to a 0.8 cM interval flanked by D7S496 and D7S2425. *J Med Genet 35*, 202–204.

Mustapha M, Weil D, Chardenoux S, et al. (1999). An alpha-tectorin gene defect causes a newly identified autosomal recessive form of sensorineural pre-lingual non-syndromic deafness, DFNB21. *Hum Mol Genet 8*, 409–412.

Mustapha M, Chouery E, Chardenoux S, et al. (2002a). DFNB31, a recessive form of sensorineural hearing loss, maps to chromosome 9q32-34. *Eur J Hum Genet 10*, 210–212.

Mustapha M, Chouery E, Torchard-Pagnez D, et al. (2002b). A novel locus for Usher syndrome type I, USH1G, maps to chromosome 17q24-25. *Hum Genet 110*, 48–50.

Mustapha M, Salem N, Delague V, et al. (2001). Autosomal recessive non-syndromic hearing loss in the Lebanese population: prevalence of the 30delG mutation and report of two novel mutations in the connexin 26 (GJB2) gene. *J Med Genet 38*, E36.

Najjar SS, Younoszai K, Derkaloustian V. (1963). Hypothyroidism in children; a review of 47 cases. *J Med Liban 16*, 181–196.

Saouda M, Mansour A, BouMoglabey Y, et al. (1998). The Usher syndrome in the Lebanese population and further refinement of the USH2A candidate region. *Hum Genet 103*, 193–198.

Sass JO, Ensenauer R, Röschinger W, et al. (2008). 2-Methylbutyryl-coenzyme A dehydrogenase deficiency: functional and molecular studies on a defect in isoleucine catabolism. *Mol Genet Metab 93*, 30–35.

Souraty N, Sanlaville D, Chédid R, et al. (2007a). Cytogenetic investigation of a child with a mosaic isochromosome 18q and ring 18q. *Eur J Med Genet 50*, 379–385.

Souraty N, Noun P, Djambas-Khayat C, et al. (2007b). Molecular study of six families originating from the Middle-East and presenting with autosomal recessive osteopetrosis. *Eur J Med Genet 50*, 188–199.

Touma EH, Rashed MS, Vianey-Saban C, et al. (2001). A severe genotype with favourable outcome in very long chain acyl-CoA dehydrogenase deficiency. *Arch Dis Child 84*, 58–60.

Yamak AA, Bitar F, Karam P, Nemer G. (2007). Exclusive cardiac dysfunction in familial primary carnitine deficiency cases: a genotype-phenotype correlation. *Clin Genet 72*, 59–62.

Yasunaga S, Grati M, Cohen-Salmon M, et al. (1999). A mutation in OTOF, encoding otoferlin, a FER-1-like protein, causes DFNB9, a nonsyndromic form of deafness. *Nat Genet 21*, 363–369.

Zafatayeff S, Zahed L, Souraty N, Megarbane A. (1998). Retrospective study of nine Lebanese families with fragile X syndrome and review of the literature. *J Med Liban 46*, 317–320.

Zahed L, Megarbane A. (1998). A cytogenetic register of down syndrome in Lebanon. *Community Gen 1*, 84–89.

Zahed L, Oreibi G, El-Amine H, et al. (2004a). A new patient with pure trisomy 4p resulting from isochromosome formation and whole arm translocation. *Am J Med Genet 128*, 60–62.

Zahed L, Oreibi G, Azar C, et al. (2004b). Ring chromosome 18q and jumping translocation 18p in an adult male with hypergonadotrophic hypogonadism. *Am J Med Genet 129*, 25–28.

Zahed L, Zahreddine H, Noureddine B, et al. (2005). Molecular basis of oculocutaneous albinism type 1 in Lebanese patients.*J Hum Genet 50*, 317–319.

Zahed L, Pramparo T, Farra C, et al. (2007). A patient with duplication (7)(p22.1pter) characterized by array-CGH. *Am J Med Genet 143*, 168–171.

Zahed L, Sismani C, Ioannides M, et al. (2008). Molecular and clinical description of a girl with a 46,X,t(Y;4)(q11.2;p16)/45,X,der(4)t(Y;4)(q11.2;p16) karyotype and a small cryptic 4p subtelomeric deletion. *Am J Med Genet 146*, 893–898.

Zhang GX, Fukao T, Rolland MO, et al. (2004). Mitochondrial acetoacetyl-CoA thiolase (T2) deficiency: T2-deficient patients with "mild" mutation(s) were previously misinterpreted as normal by the coupled assay with tiglyl-CoA. *Pediatr Res 56*, 60–64.

56.

GENETIC DISEASE IN PALESTINE AND PALESTINIANS

Bassam Abu-Libdeh, Peter D. Turnpenny, and Ahmed Teebi*

INTRODUCTION: THE PALESTINIAN PEOPLE

Palestinians are an indigenous people who either live in, or originate from, historical Palestine, which formed the land bridge between the continents of Europe, Asia and Africa. The great ancient trade routes crossed its territory, and it was therefore a natural meeting place, as well as a battleground, for culture and religious influences from Syria, Mesopotamia, Egypt, and Asia Minor. Control of the region was strategic and it became one of the most fought-over parts of the world, a situation that continues to this day.

Canaanites were the earliest known inhabitants of Palestine, as far back as the third millennium B.C., and the Hebrews, Semitic tribes from Mesopotamia, began to steadily migrate there after the fourteenth century B.C. Around 1125 B.C., more than 150 years after Moses led his people out of serfdom in Egypt, they finally defeated the Canaanites—but the Philistines, who had established an independent state on the southern coast of Palestine and controlled the Canaanite town of Jerusalem, proved a more formidable opponent. They defeated the Israelites around 1050 B.C., but approximately 50 years later were themselves defeated by united Hebrew tribes under King David.

After David's death in 922 B.C. the kingdom was divided into two—Israel, which fell to Assyria in 722 B.C., and Judah, which fell to Babylonia in 586. Jerusalem was destroyed and most of the Jews exiled. Not long afterward, however, Cyrus the Great of Persia conquered Babylonia and allowed the Jews to return, permitting them significant autonomy. Persian rule was eventually ended by Alexander the Great in 333 B.C. His successors attempted to impose Hellenistic (Greek) culture and religion, but the region subsequently came under Roman rule. The Jews revolted three times between 141 B.C. and A.D. 135; the last of these insurrections resulted in Jews being banned from Jerusalem. Judea was renamed Syria Palistina. As Christianity spread, and was adopted by the Romans, a time of prosperity followed in the Byzantine era.

Roman rule ended when Arab armies invaded Palestine and captured Jerusalem in A.D. 638, beginning 1,300 years of Muslim presence in what became known as Filastin. Palestine was holy to the Muslims because it was their first qabla (the direction they face when praying) and because the prophet Mohammad was believed to have been transported there from Mecca nocturnally, and to have consequently ascended to heaven from the area of Solomon's Temple, where the Dome of the Rock was later built. Although the Muslims guaranteed security and allowed religious freedom to all inhabitants of the region, the majority converted to Islam and adopted Arab culture. Palestine enjoyed, along with the rest of the Muslim empire at the time, a golden age of science, art, philosophy, and literature. This continued until the empire declined under the Mamelukes. It was under the Mamelukes that the period of greatest turbulence overtook Palestine. Between the eleventh and thirteenth centuries, conflict in the region was dominated by the Crusades from Western Europe, which began with religious motives but later included the expansion of commerce and trade. The era was effectively brought to an end by the conquests of Saladin.

This chapter is dedicated to Dr. Ahmad Said Teebi, born July 22, 1949, who passed away on July 22, 2010 from complications of cancer, at Trillium Hospital in Mississauga, Canada. He is survived by his loving family, including wife Amal Qudsi, sons Saeed and Basel, daughters Asil and Asma, brothers Essam, Mazen and Hani, sisters Layla, Samira and Rana, sons-in-law Saad Arif and Omar Soliman, daughters-in-law Jennifer Hamilton and Mais Abu Afifeh, and his grandson Leith. He was a proud Palestinian who truly loved his adopted home of Canada. Dr. Teebi had a distinguished career as a trailblazing geneticist and pediatrician. Having started his career in the country of Kuwait, he subsequently worked at Yale University, and was a professor at McGill University, the University of Toronto, and Cornell University. He published more than 200 scientific papers. His research focused primarily on the field of clinical genetics and dysmorphology, in which he was considered one of the world's great authorities, having described nearly 40 new disorders and syndromes. The second edition of his landmark textbook, *Genetic Disorders Among Arab Populations*, was published posthumously in 2010 by Springer. Dr. Teebi was the pride of his family, a wonderful father, brother, grandfather, and pillar of the Arab and Muslim communities.

1. Chief of Pediatrics & Genetics, Makassed Hospital, Jerusalem, and Associate Professor of Pediatrics & Genetics, Al-Quds Medical School, Jerusalem.

2. Clinical Genetics Department, Royal Devon and Exeter Hospital, Exeter, UK, and Honorary Senior Clinical Lecturer, Peninsula Medical School, Exeter, UK.

The Ottoman Turks defeated the Mamelukes in the thirteenth century and ruled Palestine until 1917. The country was divided into districts, the administration of which was placed mainly under the command of Arab Palestinians. The Christian and Jewish communities were allowed a large measure of autonomy. After a period of stagnation under the Ottoman Empire, the region was revitalized economically and socially in the nineteenth century by growing European interest in new markets and land.

The rise of European nationalism, and especially of anti-Semitism during the 1880s, encouraged Jews living in Europe to seek reinstitution in Palestine, or "the Promised Land." Zionism was thus born. Aided by the Arabs, the British captured Palestine from the Turks in 1917–1918. In the Balfour Declaration of 1917, Britain promised the Jews, whose help it needed in World War I, a Jewish "national home" in Palestine, even though Palestinian Arabs made up more than 95% of the population at the time. Helped by large-scale immigration between the two world wars, the state of Israel was established on May 14, 1948. The West Bank, including East Jerusalem, was put under the control of neighboring country Jordan, while Egypt took command of the Gaza Strip. However, these two territories were both captured by Israel in 1967 (Figure 56-1). The war resulted in the displacement of 60%–70% of the Palestinian population, whose numbers in diaspora are now estimated to be approximately 5 million; they scattered to neighboring countries, mainly Jordan, and later elsewhere, particularly the Arabian Gulf states and North America. Many Palestinians have maintained their identity and the desire to return to their homeland. The Oslo accord of 1993 resulted in the establishment of Palestinian autonomy in the Palestinian Territories of the West Bank and Gaza. According to the Statistical Abstract of Palestine No. (5) of the Palestinian Central Bureau of Statistics, the global population of Palestinians at the end of 2003 was estimated to be 9.6 million, with 4.8 million in diaspora, 1.1 million persons in Israel, and 3.7 million in the Palestinian Territories (Palestinian Central Bureau of Statistics, 2003).

Figure 56-1 The land of historical Palestine, now divided into Israel and the Occupied Palestinian Territories of the West Bank and Gaza.

CONSANGUINITY AND FAMILY PATTERN

Consanguinity is generally very high among Palestinians (Palestinian Central Bureau of Statistics, 2009), and higher than that of surrounding Arab nations (Rahim et al., 2009). In general, almost 50% of all marriages are consanguineous, officially 43.4% and 49.3% in the West Bank and Gaza Strip respectively. There has been no significant decline in this rate between the years 2000–2006 in spite of a public education and media campaign to discourage this kind of marriage (Figure 56-2). This can be attributed to political, economic, and social factors that play a major role in deciding the type of marriage.

A study from Israel (Freundlich and Hino, 1984), conducted among the Arab rural population in the western Galilee, showed unusually high overall figures (39%). They were highest in the Druze population (49%), lower among Muslims (40%), and still lower among Christians (29%). The most common type of consanguineous marriage was between first cousins, particularly paternal first cousins. A study in 1994 (Jaber et al., 1994) from Israel found a 44.3% consanguinity rate, with an average family inbreeding coefficient of 0.0192. The study group included urban, suburban, and rural communities with an average inbreeding coefficient of 0.01625, 0.01794, and 0.01958, respectively. First cousin marriage constituted more than 50% of all consanguineous marriages. In the village of Taybe near Tel-Aviv, one of the largest and most developed Arab villages in Israel, consanguineous mating was apparent in 236 out of 610 randomly chosen families (Jaber et al., 1992). Most of these were first cousin marriages (170

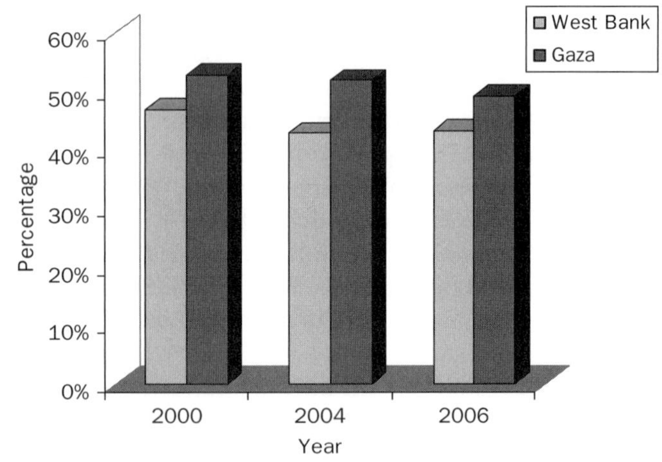

Figure 56-2 The rate of consanguineous marriages in the West Bank and Gaza, 2000–2006.

families), and the study highlighted a significant increase in the incidence of malformation, particularly in relation to the closeness of the parental relationship. Based on index cases, the incidence of major malformations was 5.8% in the offspring of inter-village marriages, 8.3% in intra-village nonrelated marriages, 15.1% in a distant consanguineous group, and up to 15.8% in the progeny of first cousin marriages ($p < 0.001$).

Many Palestinian families have large numbers of children. The average number of offspring per woman in the West Bank and Gaza was 7.8 in 1993 (Courbage and Khlat, 1993). The close family unit (parents and children) maintains strong ties with extended family, from which they obtain support and pride. The typical Palestinian village includes several clans (called *hamula*, pl. *hamayil*) with patrilineal structure. Each individual carries and uses at least four names: the first name, the father's name, the paternal grandfather's name, and the family name. It is usually possible to trace kinship using the father's first name (Lewitter et al., 1983). There is an excess of popular names that relate to Islamic (Mohammad, Ahmad, etc.) or Christian (Jiryis, Hanna, Butros, etc.) traditions.

GENETIC MARKERS AND POLYMORPHISMS

The molecular genetic profile of Palestinians was studied comprehensively for the first time in 2001 (Arnaiz-Villena et al., 2001) by using human leukocyte antigen (HLA) gene variability and haplotypes. Comparison with other Mediterranean populations, using neighbor-joining dendrograms and correspondence analyses, revealed that Palestinians are genetically very close to Jews and other Middle East populations, including Turks (Anatolians), Lebanese, Egyptians, Armenians, and Iranians. The closeness of these relationships was also studied using haplotypes constructed from Y-chromosome markers to trace the paternal origins of the Jewish diaspora (Hammer et al., 2000). A set of 18 biallelic polymorphisms was genotyped in 1,371 males from 29 populations, including 7 Jewish (Ashkenazi, Roman, North African, Kurdish, Near Eastern, Yemenite, and Ethiopian) and 16 non-Jewish groups from similar geographic locations. They concluded that Jewish and Middle Eastern non-Jewish populations share a common pool of Y-chromosome biallelic haplotypes. Mitochondrial DNA haplotypes were studied among Jews and Arabs (Ritte et al., 1993), and the results showed significant similarities, with only Ethiopian Jews showing distinct patterns. The remaining groups were genetically similar to Europeans.

The genetic similarity of Palestinians to the Jordanians is well illustrated in studies of mid-digital hair (Omari, 1986a), taste reaction to phenylthiourea (Omari, 1986b), ridge count and other dermatoglyphic parameters (Omari, 1985, 1991, 1992, 1993), and serum protein polymorphisms (Cleve et al., 1992; Nevo et al., 1993). Color vision, and tongue curling and folding have also been studied. No significant differences were found between Jordanians and Palestinians (West Bank versus East Bank of Jordan River) in regard to blood groups. Data from this study is compared to a study on the Druze community from northern Israel (Nevo, 1988; Table 56-1).

The blood group markers are similar among Jews and Arabs and have little, if any, African admixture (Banerjee et al., 1981). Of note is the high incidence of the O allele (Saha and Banerjee, 1986). Gc allele frequencies were also similar for Arab Druze and Muslims; the gene frequencies reported were as follows: Gc1F 0.1212 and 0.223, Gc1S 0.602 and 0.544, and Gc2 0.186 and 0.231, respectively (Cleve et al., 1978; Nevo and Cleve, 1983). While the Druze

TABLE 56-1 SELECTED ALLELE FREQUENCIES OF BLOOD GROUPS IN PALESTINIANS AND JORDANIANS IN GENERAL, AND COMPARISON WITH THE DRUZE COMMUNITY OF NORTHERN ISRAEL

	Druze [Nevo, 1988]	Jordanians/Palestinians [Omari, 1986a]
I^A	0.2543	0.2391
I^B	0.923	0.1378
I^O	0.6535	0.6230
Kell K	0.0170	0.0251
k	0.9830	0.9749
Duffy Fy^a	0.2929	0.3166
Fy^b + Fy	0.7071	0.6834
m	0.6350	0.5745
n	0.3649	0.4255
Rh D		0.7377
Rh d		0.2623

are believed to have separated from other Muslims some eight centuries ago, the gene frequencies suggest that there has been no significant genetic divergence between the two communities. The frequency of alleles of α-1 antitrypsin in Palestinian was found to be different from that reported in European populations (Nevo et al., 1982). The latter work also includes a description of a new allele, PiV-S.

GENETIC DISORDERS

Several population epidemiological studies have been conducted into the incidence of congenital malformations and genetic disorders among Palestinians. Similar to other Arab groups, the Palestinians have increased frequencies of congenital malformation and autosomal recessive disorders (Teebi, 1994). These have assumed greater healthcare significance as infectious diseases and nutritional problems have declined.

There is also an apparent high frequency of new autosomal recessive disorders among Palestinians. These are largely explained by the high rate of consanguinity, which also explains the high frequency of monogenic disorders in mentally retarded individuals from Palestine (Janson et al., 1990). In this study, parental consanguinity in cases of severe mental retardation was 67.5% compared to 50% in the general population. In the mixed Arab population of Kuwait, genetic causes accounted for over 50% of cases of mental retardation (Farag et al., 1993).

An unpublished study of births at Al-Makassed Hospital in Jerusalem (A. Dudin, personal communication) provides a glimpse of the health impact of genetic disorders. Stillbirths, neonatal deaths, and early neonatal deaths accounted for 2.6%, 2.4%, and 2.0% of new births, respectively, giving a total of 7%. Congenital anomalies (including neural tube defects) accounted for 23% of stillbirths, 40% of neonatal deaths, and 39% of early neonatal deaths. Further studies are needed to document the incidence of congenital anomalies in the general population, but it is clear that genetic disease has a significant health impact on this community.

COMMON AUTOSOMAL RECESSIVE DISORDERS

Several autosomal recessive disorders are reported or observed to be highly prevalent in the general population. Hemoglobinopathies, mainly β-thalassemia (β-thal), are common among Palestinians and represent a potentially preventable national health problem. The frequency of the β-thal trait among Muslim and Christian Palestinians, and Bedouins, is variable, ranging between 3%–4%, but >10% in some areas (Filon et al., 1994, 1995). Assuming a carrier frequency of 4%, the birth incidence of homozygous β-thal in the general population will be ~1:2,500. This incidence has been declining over the last decade, mainly because of the mandatory premarital thalassemia screening that was adopted on May 2000 (Younis, 2006). One additional factor for the decline was the introduction of an active antenatal diagnosis program for thalassemia at Al-Makassed Hospital in Jerusalem. During the period 1999–2005, 25 cases of homozygous β-thal were diagnosed antenatally, with subsequent termination of the pregnancy according to the wishes of the parents (Ayesh et al., 2005a). In 1990, there were 21 new cases of thalassemia, rising to 33 cases in 1995. In the year 2000, the number was 16, falling to 7 in 2003, and 5 in 2004 (Younis, 2006). Marked genetic heterogeneity was observed among Palestinians, in whom 17 mutations were detected (Darwish et al., 2005).

Most of the mutations detected are also found in neighboring Mediterranean countries. The 4 most common mutations, with their relative frequencies, were as follows: IVSI-110 (22.2%), IVSI-6 (13.6%), Cd37 (12%), and IVSI-I (9.7%) (Ayesh et al., 2005a). Darwish et al. (2005) reported that IVSI-6 was the most common mutation, with a frequency of 28.7% among their patient population. However, their sample was not representative of the whole population (Dr. Suheil Ayesh, personal communication). Other studies have shown slight differences in the relative frequencies of the most common mutations, with IVSI-110 being the most common and accounting for 37.5% of all β-thal alleles in Gaza (Filon et al., 1995). The IVS1-110 mutation is the most prevalent mutation in Lebanon and Egypt, accounting for 22% of the β-thal alleles in Israel, and is present in both Kurdish Jews and Arabs (Filon et al., 1994). In Bedouins, most of the β-thal genes were found in those living in malarial regions near the Huleh and Jezreel valleys in northern Israel.

A unique poly-A deletion (AATAAAA – A), accounting for 2.5% of all mutations in Gaza, has not been reported in any other country (Filon et al., 1995). A new β-hemoglobin variant, Hb Taybe, named after the Arab village of Taybe, was reported in 1994 (Galacteros et al., 1994). This hemoglobin (α 38 or 39 THR deleted) is silent in the heterozygote state and produces severe hemolytic anemia in the homozygous form.

HbS and HbO-Arab are present in several large Palestinian kindreds in the Arab-Israeli village of Jesser El-Zarka (Rachmilewitz et al., 1985). This village traces its ancestry to migrants from Jordan, the West Bank city of Nablus, and the south of Sudan. One Jewish family and eight Palestinian Arab families were found to have the Benin haplotype of the β-globin gene, and the ninth family had the CAR (Central African Republic) haplotype (Rund et al., 1990).

One other relatively common disease is familial Mediterranean fever (FMF), and the incidence among

Palestinians was estimated to be at least 1:2,000 (Barakat et al., 1989; Majeed and Barakat, 1989; Said et al., 1992). Such a high incidence is close to that observed in Armenians and Sephardic Jews (Sohar et al., 1967). The spectrum of mutations and genotypes in the pyrin (*MEFV*) gene in Palestinian FMF patients is similar to that among other patients in neighboring countries. The 5 most frequent mutations are M694V (49%), V726A (16.7%), M694I (11.9%), E148Q (8.5%), and M680I (4.4%), accounting for 90% of all detected mutations. The remaining 9 mutations (P369S, R408Q, A744S, M680Ib, R653H, 695R, E167D, F479L, and R761H) were much less frequent (Ayesh et al., 2005b; El-Shanti et al., 2006).

Cystic fibrosis (CF) is another relatively common genetic disorder. Reports from Israel suggest a high incidence among Palestinians (Katznelson, 1978, 1982). Observations from Kuwait indicate apparent high frequencies of CF among Palestinians and Jordanians (Issa et al., 1988). Data from Israel showed that the Phe508del (ΔF508) mutation accounts for 22%–25% of CF chromosomes in Palestinians versus 32.5% in Ashkenazi Jews; these can be compared to the 70% figure in northern Europeans (Lerer et al., 1990; Shoshani et al., 1992). Four mutations were found to be common in Arabs (Phe508del, G542X, W1282X, and N1303K), accounting for 55% of CF alleles (Abeliovich et al., 1992).

The incidence of classic phenylketonuria (PKU) in Kuwait was 1:6,479 live births, which is high compared to the general incidence of 1:11,000 (Teebi et al., 1987a). Two-thirds of cases detected in Kuwait's PKU screening project were Palestinians or Jordanians, who constituted about 22% of the whole population at that time. At the genetics clinic in Kuwait, Palestinian and Jordanian families with PKU represented more than 50% of all families identified. In a study from Israel, 36 Palestinian families with various hyperphenylalaninemias were included (Kleinman et al., 1994). Mutations and polymorphisms at the phenylalanine hydroxylase (PAH) locus were studied (Kleinman et al., 1994). Four mutations previously identified in Europe were found among Palestinians, indicating that gene flow from Europe to Palestine may have occurred early in history. In addition, three new PAH mutations unique to Palestinians were identified (Kleinman et al., 1992a, 1992b, 1993, 1994). A study from Kuwait, examining Kuwaiti, Palestinian, and Egyptian patients, showed the presence of four common European haplotypes and three unclassified ones (Bender et al., 1994).

In Kuwait, the minimum prevalence of Bardet-Biedl syndrome (BBS) in the general population was found to be 1:50,000 based on ascertained cases (Farag and Teebi, 1988). The prevalence was three times higher among Bedouins (Farag and Teebi, 1989). Four of 13 Arab families with BBS patients seen in Kuwait were Palestinians (Farag and Teebi, 1988). Reports from Israel indicate a high frequency of BBS with clustering among some Bedouin tribes (Kalbian, 1956; Ehrenfeld et al., 1970; Kwitek-Black et al., 1993). In a large Bedouin family, significant linkage was shown with chromosome 3 locus (Sheffield et al., 1994). Linkage was confirmed by homozygosity mapping. In another Bedouin family, the BBS locus showed linkage to 16q21 (Kwitek-Black et al., 1993). The most informative marker was homozygosity for the same allele in seven of nine affected individuals. This data indicated locus heterogeneity of BBS among Palestinians. Non-allelic genetic heterogeneity is further documented by finding two more BBS loci on chromosomes 11 and 15 (Leppert et al., 1994; Carmi et al., 1995). Phenotypic variability was evident comparing 3 unrelated Bedouin kindred, which were used for linkage mapping of BBS loci to chromosomes 3, 15, and 16 (Carmi et al., 1995).

Meckel (or Meckel-Gruber) syndrome is a rare malformation syndrome that was found to be highly prevalent in Kuwait, with an incidence of 1:3,500 births in the general population (Teebi et al., 1992). Many of the ascertained cases were Palestinians. A study from an Arab hospital in Jerusalem showed an incidence of 1:2,000 (Dudin, 1994). Data from Israel also indicate a high frequency among Arabs (J. Zlotogora, personal communication). Similar to a phenotype seen in Kuwait, Meckel syndrome without polydactyly was reported in a Palestinian family (Juabeh et al., 1987).

Metachromatic leukodystrophy was found to be common among Palestinians. Ten Muslim and Christian families with affected children have been found, three in the Jerusalem region and seven in a small area in lower Galilee (Zlotogora et al., 1994a; Heinisch et al., 1995). While multiple mutations accounted for the high frequency of metachromatic leukodystrophy among Muslim and Christian Palestinians living in a small geographic area (Heinisch et al., 1995), a single origin for the most frequent mutation was found among both Muslim and Christian patients (Zlotogora et al., 1994a), suggesting it may have been introduced to the region during the Crusades.

Familial hydrocephalus of prenatal onset was reported in Israel in 8 out of 14 families in which more than one child was diagnosed with this condition. All the Arab patients had consanguineous parents (Zlotogora et al., 1994b). A large Palestinian family with four affected children was also reported from Kuwait (Teebi and Naguib, 1988). The finding suggests that this entity is relatively frequent in Palestinians.

Classic congenital adrenal hyperplasia (CAH) due to 21-hydroxylase deficiency has an estimated incidence of at least 1:7,000 live births in the Arabs in Kuwait (Lubani et al., 1990). In this study, Palestinians and Jordanians represented 26.6% of all cases. Rare forms of CAH have also been reported (Rosler et al., 1996; Rosler, 2006).

Glanzmann's thrombasthenia was found to be the second most frequent bleeding disorder in Jordan (Awidi,

1983). One of us (AT) described 12 patients in nine families. Coller et al. (1987) found that immunoblot patterns of glycoprotein IIIa could distinguish the defect present in most Iraqi Jewish patients from that in the Palestinian patients. At the molecular level the mutations in the two populations are different (Newman et al., 1991).

Nesidioblastosis of the pancreas was found to be very frequent in Saudi Arabia, with an incidence of 1:2,675 live births (Mathew et al., 1988). From Israel, a large Bedouin family and an Arab Muslim family were reported (Glaser et al., 1990). In Jordan, this condition is frequently diagnosed among the newborns with hypoglycemia.

Wilson's disease (hepatolenticular degeneration) has an estimated incidence of 1:4,000 among Druze and 1:10,000 in Palestinian Arabs, compared to 1:3,000 in Turkish Jews and 1:40,000 in Ashkenazi Jews and Jews from the Maghreb countries (Bonné-Tamir et al., 1990).

In addition to the common disorders mentioned above, several disorders have been found to be very common in large kindreds or tribes restricted to a small geographic region, suggesting a founder effect.

NEW DISORDERS FIRST REPORTED AMONG PALESTINIANS AND JORDANIANS

A relativity large number of new disorders, most of which appeared to be autosomal recessive, have been reported among Palestinians. Many of these were observed and further characterized elsewhere. Examples of the newly described disorders among Palestinians are limb/pelvis-hypoplasia/aplasia syndrome; microcephaly with normal intelligence; Grebe-like chondrodysplasia; macrophthalmia, lethal; trigonobrachycephaly syndrome with bulbous nose and severe mental retardation; and hypogonadotrophic hypogonadism with mental retardation, obesity, and minor skeletal anomalies.

RARE DISORDERS

Several extremely rare disorders and unusual associations have been reported. Despite the fact that most of these disorders were reported only once, some of them may not be necessarily rare among Palestinians. In some instances the association is fortuitous, or due to closely linked genes, or simply represents a previously unrecognized syndrome. Examples include the congenital dyserythropoietic anemia type 1 associated with chronic recurrent osteomyelitis and Sweet's syndrome (Majeed et al., 1989), infantile osteopetrosis and Hirschsprung's disease (Dudin and Rambaud-Cousson, 1993), autosomal dominant syndactyly type IV/hexadactyly of feet associated with unilateral absence of the tibiae (Rambaud-Cousson et al., 1991), myelomeningocele and cloacal duplication (Telmesani, 1994). Also, Menetrier's disease with early childhood onset and possible autosomal recessive inheritance (Marcus and Verp, 1993), disordered pigmentation, spastic paraplegia neuropathy (Abdallat et al., 1980), Laron syndrome (Laron, 1993), Jancer syndrome (Zlotogora and Glick, 1993), autosomal recessive Robinow syndrome (Teebi, 1990), rhizomelic short-limbed dwarfism with abnormal facies (Robinow-like; Turnpenny and Thwaites, 1992), dyssegmental dwarfism (Svejcar, 1993), autosomal recessive nephrogenic diabetes insipidus due to deletion in aquaporin-2 gene (van Lieburg et al., 1994), familial thyroglossal duct syndrome (Klin et al., 1993), autosomal dominant Holt-Oram-like skeletal deformity (Boehme and Shotar, 1989), Krause-Kivlin syndrome (Frydman et al., 1991), and cerebro-oculo-facio-skeletal syndrome (Gershoni-Baruch et al., 1991).

OTHER OBSERVATIONS

Several other unifactorial and multifactorial disorders have been reported. The following are particularly notable observations:

- The incidence of neural tube defects among Palestinians was reported to be high (Dudin, 1994). The incidence for spina bifida, encephalocele, and anencephaly were 1.73, 0.24, and 1.96:1,000 births, respectively. This high incidence may be largely attributed to socioeconomic factors, including unbalanced nutrition.
- Clinical and molecular studies have been conducted into the significant problems of genetic/hereditary visual (Bandah et al., 2009; Elder and de Cock, 1993) and hearing (Shahin et al., 2002; Walsh et al., 2006) impairment.
- Multiple sclerosis (MS) was found to be 2.5 times more frequent among Palestinians than among Kuwaitis (Al-Din et al., 1990). The association with frequencies of genetic determinants (eye color, HLADR, HLA-DQW, blood group distributions) in Kuwaitis and Palestinians was also different. Among Palestinians, the higher frequency of MS and association with HLA-DR2 and HLA-DQW1 are similar to Europeans. It is possible that these characteristics are found because of admixture with Europeans.
- Familial aggregation of high blood pressure is also seen in some communities (Hurwich et al., 1982).
- Shoveling of the incisors is probably a polygenic trait, sometimes associated with invagination of affected incisor. Five to six percent of Palestinians/Jordanians have shoveling and 11% of those have invagination (Kharat et al., 1990).

- The twinning rate among Palestinians was about 12/1,000 gestations with a ratio of 1 monozygotic to 3.9 dizygotic twins (Gedda et al., 1992).
- A total of 38 individuals in four generations of one Palestinian family were affected with epidermolysis bullosa herpetiformis (Dowling Maera type), a presumed autosomal dominant condition (Hacham-Zadeh et al., 1988). Epidermolysis bullosa dystrophica triggered by sun exposure was also reported (Shubailat et al., 1983).
- Tay-Sachs disease exists but is rare (Navon et al., 1981). A Palestinian family was reported to have a specific mutation that was originally described in a British patient (Drucker and Navon, 1993).
- Homozygous protein C deficiency, resulting in massive venous thrombosis in newborns, was reported in Palestinians (Seligsohn et al., 1984). Heterozygotes had partial protein C deficiency but no thrombosis.
- Other disorders reported in Jordanians and Palestinians include apple peel syndrome (Farag and Teebi, 1989; Farag et al., 1993), acrodermatitis enteropathica (Majeed and Barakat, 1976), torsion dystonia (Besisso et al., 1987), erythropathogocytic lymphohistiocytosis, muscular dystrophies (Farag and Teebi, 1990; Mahjneh et al., 1992), Kabuki makeup (Niikawa-Kuroki) syndrome (Gillis et al., 1990), X-linked recessive hydrocephalus (Al-Awadi et al., 1984), pyruvate kinase deficiency (Karadsheh, 1993), Wolman's disease (Mahdi and Al-Nassar, 1991), cyclopia/holoprosencephaly (Bustami and Amr, 1986), and others (Al-Awadi et al., 1985; Der Kaloustian et al., 1980; Dudin and Thalji, 1991; Kohn et al., 1988; Raas-Rothschild et al, 2009; Teebi, 1991, 1993, 1994, 2000, 2005; Teebi and Al-Saleh, 1989a; Teebi et al., 1986a, 1986b, 1987b, 1989b; Teebi and Teebi, 2005; Yadav and Reavey, 1988).
- Besides the conditions mentioned, the following disorders have also been seen among Jordanians and Palestinians: split hand/split foot, retinitis pigmentosa (AR, AD, XLR), AD Parkinson's disease, 17,20-desmolase deficiency, Waardenburg syndrome, asymmetric crying facies syndrome (Caylor syndrome), asphyxiating June thoracic dystrophy, congenital ichthyosiform erythroderma, and wrinkly skin syndrome.

CYTOGENETIC ABNORMALITIES

The only cytogenetics laboratory in Palestine is located at Al-Makassed Hospital in Jerusalem. It was established in 1996, and between 1996 and 2008, 3704 peripheral blood samples were processed for chromosome analysis for different indications. The most frequent abnormality detected was trisomy 21, followed by sex reversal, Turner syndrome, and Klinefelter syndrome. Table 56-2 summarizes the data (Dr. Abu-Libdeh, unpublished data).

TABLE 56-2 PATTERNS OF CYTOGENETIC ABNORMALITIES ON PERIPHERAL BLOOD SAMPLES OF PALESTINIANS UNDERGOING KARYOTYPING AT THE CYTOGENETIC LABORATORY AT AL-MAKASSED HOSPITAL IN JERUSALEM, 1996–2008.

	Total samples	Abn.	Tri. 21	Tri. 18	Tri. 13	Trans. DS	KF. Synd.	TS	Sex revers.	Bal. Trans.	Others
1996	169	31	17	1	2	1	1	0	6	1	2
1997	232	46	27	1	3	1	3	2	2	2	5
1998	268	50	28	1	2	0	5	5	5	2	2
1999	257	53	35	2	1	1	3	2	4	1	4
2000	265	48	24	5	0	0	5	4	7	0	3
2001	240	40	22	0	1	2	1	4	5	2	3
2002	196	39	23	1	0	1	3	2	7	2	0
2003	279	56	33	4	0	0	1	9	8	1	0
2004	291	61	36	2	1	0	4	6	5	2	5
2005	384	82	44	4	2	1	3	6	20	2	0
2006	365	56	38	3	3	1	3	2	0	2	4
2007	350	57	39	2	0	0	2	2	4	3	5
2008	408	72	60	3	3	1	2	1	0	1	1
Total	3704	691	426	29	18	9	36	45	73	21	34

Abbreviations: Abn. = abnormal samples; Trio 21 = Trisomy 21; Trio 18 = Trisomy 18; Trio 13 = Trisomy 13; Trans. DS = Translocation Down Syndrome; KF. Synd. = Klinefelter syndrome; TS = Turner Syndrome; Sex revers. = Sex reversal; Bal. Trans = Balanced Translocation.

INBORN ERRORS OF METABOLISM

Inborn errors of metabolism (IEM), as with other autosomal recessive disorders, are diagnosed with increasing frequency among Palestinians compared to Western populations, due largely to consanguinity, as discussed. They have also been extensively characterized in many cases, and this review cannot do justice to all the details, so a few examples only are cited (Bach et al., 1993; Bargal et al., 2006; Zlotogora et al., 1985, 1991). Table 56-3 summarizes the patterns of IEM diagnosed at Al-Makassed Hospital in Jerusalem between 2004 and 2009 (Dr. Dweikat, unpublished data).

NEW GENES IDENTIFIED THROUGH MAPPING STUDIES IN PALESTINIAN KINDREDS

The increasing power of molecular techniques and availability of the complete human genome sequence have revolutionized discovery of the genetic basis of rare Mendelian disorders, especially through homozygosity mapping in large inbred kindreds. The involvement of Palestinian families in these research efforts has been

TABLE 56-3 INBORN ERRORS OF METABOLISM DIAGNOSED AT AL-MAKASSED HOSPITAL, JERUSALEM, 2004–2009

Disease	No of cases
Aminoacidopathies	38
Phenylketonuria	5
Alkaptonuria	6
Hereditary Tyrosinemia type 1	2
MSUD	6
Homocystinuria	2
Nonketotic hyperglycinemia (NKH)	7
Hypermethioninemia	1
Citrin deficiency	2
EPEMA	7
Urea cycle disorders	7
Ornithine transcarbamylase (OTC) def	1
Citrullinemia	3
Lysinuric protein intolerance (LPI)	2
Hyperornithinemia-hyperammonemia-homocitrullinemia (HHH) syndrome	1
Organic acidemias	*55*
Isovaleric acidemia	14
Biotinidase deficiency	2
3-methylglutaconic aciduria	1
β-ketothiolase deficiency	4
3-hydroxy 3-methylglutaric aciduria (HMG-CoA lyase deficiency)	2
Methylmalonic acidemia (MMA)	15
Propionic acidemia	11
5-Oxoprolinuria	1
3-OH-isobutyric aciduria	1
Glutaric aciduria	4
Glycogenosis	*31*
GSD Type I a	1
GSD Type I b	3
GSD Type III	26
GSD Type VI	1
Lysosomal storage disorders	39
GM1 Gangliosidosis	4
Niemann-Pick Disease	10
Krabbe	5
MLD	3
Mucolipidosis type II (I cell disease)	6
Hurler	3
Hunter	3
Sanfilippo	2
Morquio	1
Sandhoff	1
Galactosialidosis	1
Peroxisomal disorders	8
Mitochondrial disorders	
A. FATTY ACIDS OXIDATION DEFECTS	14
Carnitine palmityl transferase 1 (CPT) deficiency	1
Primary carnitine deficiency	1
VLCAD deficiency	2
LCHAD deficiency	2
MCAD deficiency	5
SBCAD deficiency	3
B. MITOCHONDRIAL RESPIRATORY CHAIN DISORDERS	22
Complex I def	6
Complex II def	1
Complex I and IV def	2
Cytochrome c oxidase def	2
Pyruvate DH E3 def	7
E2 (lipoamide DH) def.	1
Complex 1, 3, 4 deficiency – ?mtDNA depletion	3
Total	214

pioneering, and advances are likely to continue for the foreseeable future.

The identification of the cathepsin K gene (*CTSK*) which, when mutated gives rise to pycnodysostosis, the skeletal dysplasia that is believed to have been the cause of Toulouse-Lautrec's short stature, is owed to the availability of a Palestinian family in Galilee, Israel, with 12 affected individuals (Gelb et al., 1995, 1996). This was one of the first conditions to yield its secrets through autozygosity mapping. Many others have followed. The gene for molybdenum cofactor deficiency was localized to chromosome 6 in 1998 (Shalata et al., 1998), and the first Mendelian form of spondylocostal dysostosis, due to mutations in *DLL3*, was mapped in 1999 (Bulman et al., 2000; Turnpenny et al., 1991, 1999) through a large Arab kindred in Northern Israel. The causative *DLL3* mutation, 953insGCGGT, has been found in other ethnic groups on different haplotype backgrounds, and is therefore recurrent.

More recent examples of success using these now-standard techniques include spondylometaepiphyseal dysplasia, short limb-hand type, due to mutated *DDR2* (Bargal et al., 2009), and a focal epilepsy syndrome with intellectual impairment (Corbett et al., 2010).

MOLECULAR GENETIC TESTING

In addition to a few molecular genetic tests provided by private laboratories and usually sent for testing outside Palestine, the molecular genetics laboratory at Al-Makassed Hospital in Jerusalem represents the main site for performing such tests. The range of tests performed includes:

1. Hemoglobinopathies
2. Pyrin (*MEFV*) gene mutations (FMF/Behcet disease)
3. Thrombophilia mutations
4. Fragile-X study
5. Sex determination and Y deletion mutations
6. CML (9:22) translocation
7. Spinal muscular atrophy type 1

CONCLUSIONS AND FUTURE DEVELOPMENTS

The burden of genetic disease among Palestinians is large, and the rate of consanguinity very high, even when compared to the rate in surrounding Arab nations (Rahim et al., 2009). Healthcare planning and provision needs to take this into account, even though genetic services are often regarded as a luxury when resources are limited. Healthcare provision will become more aware of genetic causes and of their impact on human health. Even with the paucity of genetic data, it is evident that there are unique features in Palestinians. Some factors must be taken into consideration in developing genetic services, including the high consanguinity rate, large family size, and cultural and religious sensitivities.

In the Palestinian territories, there is still much to be done in the field of genetics. In particular, there is a need for:

1. More population studies to survey genetic diseases;
2. Recognition of the specialty of clinical genetics with appropriate specialized training programs at medical schools (currently available only in Israel);
3. Expansion of laboratory services in the fields of biochemical, cyto-, and molecular genetics;
4. Public awareness and education programs with emphasis on existing problems;
5. Promotion of genetic counseling services tailored to the needs of the community and to make preventive measures available;
6. Promotion of research into genetic disorders among Palestinians.

These developments require political will, allocation of resources, and organization, at a local level. However, external partnerships and support from well-established genetic centers can greatly accelerate the pace of change in addressing the burden of healthcare posed by genetic disease in Palestine. Many individuals and agencies have made contributions to clinical care, service development, and medical education in the Palestinian Territories over many years. One successful ongoing initiative with links to medical genetics is the Palestinian Child Health Diploma, a partnership between the U.K.'s Royal College of Pediatrics and Child Health, Medical Aid for Palestinians, and academic departments and pediatricians in Palestine (Waterston et al., 2009). There is no reason why similar initiatives specifically aimed at the development of genetic services—clinical, laboratory and counseling—should not be successful.

REFERENCES

Abdallat A, Davis SM, Farraage N, McDonald WI. (1980). Disordered pigmentation, spastic paraparesis and peripheral neuropathy in three siblings: a new neurocutaneous syndrome. *J Neurol Neurosurg Psychiatry* 43:962–966.

Abeliovich D, Lavon IP, Lerer I, et al. (1992). Screening for five mutations detects 97% of cystic fibrosis (CF) chromosomes and predicts carrier frequency of 1:29 in the Jewish Ashkenazi population. *Am J Hum Genet* 51:951–956.

Al-Awadi SA, Teebi AS, Farag TI, Naguib KM, El-Khalifa MY. (1985). Profound limb deficiency, thoracic dystrophy, unusual facies, and normal intelligence: a new syndrome. *J Med Genet* 22:36–38.

Al-Awadi SA, Farag TI, Naguib K, Teebi AS, El-Khalifa MY, Yassin S. (1984). X-linked hydrocephalus (Bickers-Adam-Edwards syndrome). *J Kwt Med Assoc* 18:187–190.

Al-Din AS, Khogali M, Poser CM, et al. (1990). Epidemiology of multiple sclerosis in Arabs in Kuwait: a comparative study between Kuwaitis and Palestinians. *J Neurol Sci* 100:137–141.

Arnaiz-Villena A, Elaiwa N, Silvera C, et al. (2001). The origin of Palestinians and their genetic relatedness with other Mediterranean populations. *Hum Immunol* 62:889–900.

Awidi AS. (1983). Increased incidence of Glanzmann's thrombasthenia in Jordan as compared with Scandinavia. *Scand J Haematol* 30:218–222.

Ayesh SK, Al-Sharef WA, Nassar SM, Thawabteh NA, Abu-Libdeh BY. (2005a). Prenatal diagnosis of beta-thalassemia in the West Bank and Gaza. *Saudi Med J* 26:1771–1776.

Ayesh SK, Nassar SM, Al-Sharef WA, Abu-Libdeh BY, Darwish HM. (2005b). Genetic screening of familial Mediterranean fever mutations in the Palestinian population. *Saudi Med J* 26:732–737.

Bach G, Moskowitz SM, Tieu PT, Matynia A, Neufeld EF. (1993). Molecular analysis of Hurler syndrome in Druze and Muslim Arab patients in Israel: multiple allelic mutations of the IDUA gene in a small geographic area. *Am J Hum Genet* 53:330–338.

Bandah D, Merin S, Ashhab M, Banin E, Sharon D. (2009). The spectrum of retinal diseases caused by NR2E3 mutations in Israeli and Palestinian patients. *Arch Ophthalmol* 127:297–302.

Bargal R, Cormier-Daire V, Ben-Neriah Z, et al. (2009). Mutations in DDR2 gene cause SMED with short limbs and abnormal calcifications. *Am J Hum Genet* 84:80–84.

Banerjee B, Saha N, Daoud ZF, Khalaf FH, Qudah H. (1981). A genetic study of the Jordanians. *Hum Hered* 31:65–69.

Barakat MH, Karnik AM, Majeed HWA, El-Sokbi NI, Fenech FF. (1989). Familial Mediterranean fever (recurrent heredity polyserositis) in Arabs. *Q J Med* 233:837–847.

Bargal R, Zeigler M, Abu-libdeh B, et al. (2006). When mucolipidosis III meets mucolipidosis II. GNPTA gene mutations in 24 patients. *Mol Genet Metab* 88:359–363.

Bender C, Buchler A, Schmidt-Mader B, et al. (1994). Haplotype analyses and a new Mspl-polymorphism at the phenylalanine hydroxylase gene in the Arabian population. *Eur J Pediatr* 153:392.

Besisso MS, Neubauer DH, Al-Rashied AA. (1987). Torsion dystonia in two Arab siblings. *J Kwt Med Assoc* 21:214–217.

Boehme DH, Shotar AO. (1989). A complex deformity of appendicular skeleton and shoulder with congenital heart disease in three generations of a Jordanian family. *Clin Genet* 36:442–450.

Bonné-Tamir B, Frydman M, Agger MS, et al. (1990). Wilson's disease in Israel: a genetic and epidemiological study. *Ann Hum Genet* 54:155–168.

Bulman MP, Kusumi K, Frayling TM, et al. (2000). Mutations in human Delta homologue DLL3 cause axial skeletal defects in spondylocostal dysostosis. *Nat Genet* 24:438–441.

Bustami FMF, Amr S. (1986). Cyclopia: a case study and review. *Jordan Med J* 21:283–297.

Carmi R, Elbedour K, Stone EM, Sheffield VC. (1995). Phenotypic differences among patients with Bardet-Biedl syndrome linked to three different chromosome loci. *Am J Med Genet* 59:199–203.

Cleve H, Patutschnick W, Nevo S, Wendt GG. (1978). Genetic studies of the Gc subtypes. *Hum Genet* 44:117–122.

Cleve H, Koller A, Patutschnick W, Rodewald A, Nabulsi A. (1992). Genetic serum protein polymorphisms in Jordanian Arabs: a pilot study of the systems AHSG, BF, FX111B, GC, PI, PLG and TF. *Gene Geogr* 6:31–40.

Coller BS, Seligsohn U, Little PA. (1987). Type 1 Glanzmann's thrombasthenia patients from the Iraqi-Jewish and Arab populations in Israel can be differentiated by platelet glycoprotein: IIIa immunoblot analysis. *Blood* 69:1696–1703.

Corbett MA, Bahlo M, Jolly L, et al. (2010). A focal epilepsy and intellectual disability syndrome is due to a mutation in TBC1D24. *Am J Hum Genet* 87:371–375.

Courbage Y, Khlat M. (1993). Structure et Accroissement de la population dans le Monde Arabe: Tendances Recentes. In: *Conference Arabe sur la population, Amman 4–8 April, 1993*, Institut National d'Etudes Demographiques, Paris.

Darwish HM, El-Khatib FF, Ayesh S. (2005). Spectrum of beta-globin gene mutations among thalassemia patients in the West Bank region of Palestine. *Hemoglobin* 29:119–132.

Der Kaloustian VM, Naffah J, Loiselet J. (1980). Genetic diseases in Lebanon. *Am J Med Genet* 7:187–203.

Drucker L, Navon R. (1993). Tay-Sachs disease in an Israeli Arab family: Try26 Stop in the alphasubunit of hexoaminidase A. *Hum Mut* 2:415–417.

Dudin AA, Thalji A. (1991). Diaphragmatic hernia and epidermolysis bullosa in two sibs. *Am J Med Genet* 39:498–499.

Dudin AA, Rambaud-Cousson A. (1993). Syndrome of infantile osteopetrosis and Hirschsprung disease in seven children born to four consanguineous unions in two families. *Am J Med Genet* 47:1083–1085.

Dudin A. (1994). Neural tube defects (NTD) among Palestinians living in the Southern part of the West Banks: a hospital based study. Report for Université Catholique de Louvain, Bruxelles.

Ehrenfeld E, Rowe H, Auerbach E. (1970). Laurence-Moon-Bardet-Biedl syndrome in Israel. *Am J Opthalmol* 70:524–532.

El-Shanti H, Majeed HA, El-Khateeb M. (2006). Familial Mediterranean fever in Arabs. *Lancet* 367:1016–1024.

Elder MJ, De Cock R. (1993). Childhood blindness in the West Bank and Gaza Strip: prevalence, aetiology and hereditary factors. *Eye* 7:580–583.

Farag TI, Teebi AS. (1988). Bardet Biedl and Laurence-Moon syndromes in a mixed Arab population. *Clin Genet* 33:78–82.

Farag TI, Teebi AS. (1989). High incidence of Bardet-Biedl syndrome among the Bedouin. *Clin Genet* 36:463–465.

Farag TI, Teebi AS. (1990). Duchenne-like muscular dystrophy in the Arabs. *Am J Med Genet* 37:290 (letter).

Farag TI, Al-Awadi SA, El-Badramary MH, et al. (1993). Disease profile of 400 institutionalized mentally retarded patients in Kuwait. *Clin Genet* 44:329–334.

Filon D, Oron V, Krichevski S, et al. (1994). Diversity in β-globin mutations in Israeli ethnic groups reflects recent historic events. *Am J Hum Genet* 54:836–843.

Filon D, Oron V, Shawa R, et al. (1995). Spectrum of b-thalassemia mutations in the Gaza area. *Hum Mutat* 5:351–353.

Freundlich E, Hino N. (1984). Consanguineous marriage among rural Arabs in Israel. *Isr J Med Sci* 20:1035–1038.

Frydman M, Weinstock AL, Cohen HA, Savir H, Varsana I. (1991). Autosomal recessive Peter's anomaly, typical facial appearance, failure to thrive, hydrocephalus, and other anomalies: further delineation of the Krause-Kivlin syndrome. *Am J Med Genet* 40:34–40.

Galacteros F, Girodon E, M'Rad A, et al. (1994). HB Taybe (a38 or 39 THR deleted): an α globin defect, silent in the heterozygous state and producing severe haemolytic anemia in the homozygous. *C R Acad Sci Paris, Life Sci* 317:437–444.

Gedda L, Martins M, Parisi P. (1992). Twinning in the Palestinian population of Jerusalem and the West Bank. *Acta Genet Med Gemellol (Roma)* 41:1–3.

Gelb BD, Edelson JG, Desnick RJ. (1995). Linkage of pycnodysostosis to chromosome 1q21 by homozygosity mapping. *Nat Genet* 10:235–237.

Gelb BD, Shi G-P, Chapman HA, Desnick RJ. (1996). Pycnodysostosis, a lysosomal disease caused by cathepsin K deficiency. *Science* 273:1236–1238.

Gershoni-Baruch R, Rm L, Lichtig C, Sojov P, Machoul I. (1991). Cerebro-oculo-facio-skeletal syndrome: further delineation. *Am J Med Genet* 41:74–77.

Gillis R, Klar A, Gross-Kieselstein E. (1990). The Niikawa-Kuroki (Kabuki make-up) syndrome in a Moslem Arab child. *Clin Genet* 38:378–381.

Glaser B, Phillip M, Carmi R, Lieberman E, Landau H. (1990). Persistent hyperinsulinemic hypoglycemia of infancy (nesidioblastosis): autosomal recessive inheritance in 7 pedigrees. *Am J Med Genet* 37:511–515.

Hacham-Zadeh S, Rappersberger K, Livshin R, Konrad K. (1988). Epidermolysis bullosa herpetiformis Dowling-Meara in a large family. *J Am Acad Dermatol* 18:702–706.

Hammer MF, Redd AJ, Wood ET, et al. (2000). Jewish and Middle Eastern non-Jewish populations share a common pool of Y-chromosome biallelic haplotypes. *Proc Natl Acad Sci* 97:6769–6774.

Heinisch U, Zlotogora J, Kafert S, Giselmann V. (1995). Multiple mutations are responsible for the high frequency of metachromatic leukodystrophy in a small geographic area. *Am J Hum Genet* 56:51–57.

Hurwich BJ, Rosner B, Nubani N, Kass EH, Lewitter FI. (1982). Familial aggregation of blood pressure in a highly inbred community, Abu Ghosh, Israel. *Am J Epidemiol* 115:646–656.

Issa ARA, Teebi AS, Issa MA, Shaabani IS, Ramadan DJ. (1988). Metabolic alkalosis in cystic fibrosis: atypical presentation in Kuwait. *Ann Trop Paediatr* 8:271–272.

Jaber L, Merlob P, Bu XD, Rotter JI, Shohat M. (1992). Marked parental consanguinity as a cause for increased major malformations in an Israeli Arab community. *Am J Med Genet* 44:1–6.

Jaber L, Bailey-Wilson JE, Haj-Yehia M, Hermandez J, Shohat M. (1994). Consanguineous matings in an Israeli-Arab community. *Arch Pediatr Adolesc Med* 148:412–415.

Janson S, Jayakoddy A, Abulaban A, Gustavson KH. (1990). Severe mental retardation in Jordanian children. A retrospective study. *Acta Pediatr Scand* 79:1099–1104.

Juabeh II, Dudin AA, Thalji A. (1987). Meckel-Gruber syndrome in one of non-identical twins: short case report. *Acta Genet Med Gemellol* 36:571–572.

Kalbian V. (1956). Laurence-Moon-Bardet-Biedl syndrome in an Arab boy: familial incidence. *J Clin Endocrinol* 16:1622–1625.

Karadsheh NS. (1993). Pyruvate kinase (PK) deficiency in Jordan. *Haematologica* 78:180–183.

Kharat DU, Saini TS, Mokeem S. (1990). Shovel-shaped incisors and associated invagination in some Asian and African populations. *J Dent* 18:216–220.

Katznelson D. (1978). Cystic fibrosis in Israel: clinical and genetic aspects. *Isr J Med Sci* 14:204–207.

Katznelson D. (1982). Cystic fibrosis in the Middle East. *Lancet* 1:112.

Kleinman S, Bernstein J, Schwartz Eisensmith CR, Woo SLC, Shiloh Y. (1992a). A defective splice site at the phenylalanine hydroxylase gene in phenylketonuria and benign hyperphenylalaninemia among Palestinian Arabs. *Hum Mutat* 1:340–343.

Kleinman S, Schwartz G, Akawi Y, Woo SLC, Shiloh Y. (1992b) A 22-bp deletion at the phenylalanine hydroxylase gene causing phenylketonuria in an Arab family. *Hum Mutat* 1:344–346.

Kleinman S, Li J, Schwartz G, Eisensmith RC, Woo SLC, Shiloh Y. (1993). Inactivation of phenylalanine hydroxylase by a missense mutation, R2705, in a Palestinian kinship with phenylketonuria. *Hum Mol Genet* 2:605–606.

Kleinman S, Avigad S, Vanagaite L, et al. (1994). Origins of hyperphenylalaninemia in Israel. *Eur J Hum Genet* 2:24–34.

Klin B, Serour F, Freid K, Efrati Y, Vinograd T. (1993). Familial thyroglossal duct cyst. *Clin Genet* 43:101–103.

Kohn G, El Shawwa R, El Ravves E. (1988). Isolated "clinical anophthalmia" in an extensively affected Arab kindred. *Clin Genet* 33:321–324.

Kwitek-Black AE, Carmi R, Duyk GM, et al. (1993). Linkage of Bardet-Biedl syndrome to chromosome 16q and evidence for non-allelic genetic heterogeneity. *Nat Genet* 5:392–396.

Laron Z. (1993). An update on Laron syndrome. *Arch Dis Child* 68:345–346.

Leppert M, Baird L, Anderson KL, Otterud B, Lupski JR, Lewis RA. (1994). Bardet-Biedl syndrome is linked to DNA markers on chromosome 11q and is genetically heterogenous. *Nat Genet* 6:108–112.

Lerer I, Cohen S, Chemke M, et al. (1990). The frequency of the DF508 mutation on cystic fibrosis chromosomes in Israeli families: correlation to CF haplotypes in Jewish communities and Arabs. *Hum Genet* 85:416–417.

Lewitter FI, Hurwich BJ, Nubani N. (1983). Tracing kinship through father's first name in Abu Ghosh, an Israeli Arab patrilineal society. *Hum Biol* 55:375–381.

Lubani MM, Issa ARA, Bushnaq R, et al. (1990). Prevalence of congenital adrenal hyperplasia in Kuwait. *Eur J Pediatr* 149:391–392.

Mahdi AH, Al-Nassar M. (1991). Wolman's disease in a Jordanian infant. *Ann Trop Paediatr* 11:305–308.

Mahjneh I, Vannelli G, Bushby K, Marconi GP. (1992). A large inbred Palestinian family with two forms of muscular dystrophy. *Neuromuscul Disord* 2:277–283.

Majeed H, Barakat M. (1989). Familial Mediterranean fever (recurrent hereditary polyserositis) in children: analysis of 88 cases. *Eur J Pediatr* 148:3636–3641.

Majeed HA, Barakat M. (1976). Acrodermatitis enteropathica – successful treatment with oral zinc. *J Kwt Med Assoc* 10:169–179.

Majeed HA, Kalaawi M, Mohanty D, et al. (1989). Congenital dyserythropoietic anemia and chronic recurrent multifocal osteomylitis in three related children and the association with Sweet syndrome in two siblings. *J Pediatr* 115:730–734.

Marcus P, Verp M. (1993). Menetrier disease in a child of a consanguineous union. *Am J Med Genet* 47:1231–1232.

Mathew PM, Young JM, Abu-Osba YK, et al. (1988). Persistent neonatal hyperinsulinism. *Clin Pediatr* 27:148–151.

Navon R, Nutman J, Kopel R, et al. (1981). Hereditary heat-labile hexosaminidase B: its implication for recognizing Tay-Sachs genotypes. *Am J Hum Genet* 33:907–915.

Nevo S. (1988). Genetic blood markers in Arab Druze of Israel. *Am J Phys Anthropol* 77: 183–190.

Nevo S, Cleve H. (1983). Gc subtypes in the Middle East: report on an Arab Moslem population from Israel. *Am J Phys Anthropol* 60:587–603.

Nevo S, Kanaaneh H, Cleve H. (1982). A genetic study of alpha-antitrypsin in an Israeli Arab population, with a new allele: Piv-s. *Isr J Med Sci* 18:891–893.

Nevo S, Cleve H, Koller A, et al. (1993). Serum protein polymorphisms in Arab Moslems and Druze in Israel: BF, F13B, AHSG, GC, PLG, PI, and TF. *Hum Biol* 64:587–603.

Newman PJ, Seligsohn U, Lyman S, Coller BS. (1991). The molecular genetic basis of Glanzmann thrombasthenia in the Iraqi-Jewish and Arab populations in Israel. *Proc Natl Acad Sci USA* 88:3160–3164.

Omari YI. (1985). A preliminary study on the fingerprint patterns of Jordanians. *Dirasat (Jordan University Publications)* 12:53–63.

Omari YI. (1986a). Distribution of mid-digital hair in Jordanian populations. *Broteria Genet* 7:133–144.

Omari YI. (1986b). Taste deficiency of phenylthiouria in Jordanian population. *Iraqi J Biol Sci* 71:253–265.

Omari YI. (1991). A dermatoglyphic study of Jordanian populations. *Broteria Genet* 7:151–160.

Omari YI. (1992). A dermatoglyphic study of Jordanians: main-line index and transversality. *Rev Brasil Genet* 15:183–189.

Omari YI. (1993). A dermatoglyphic study of Jordanian populations. Part II: palmar configurations. *Broteria Genet* 14:161–166.

Palestinian Central Bureau of Statistics. (2003). The total number of Palestinian population in the world for the end year 2003. *Statistical Abstract of Palestine* No (5).

Palestinian Central Bureau of Statistics. (2009). The final census report of 2007 statistics, issued February 2009.

Raas-Rothschild A, Dijkhuizen T, Sikkema-Raddatz B, et al. (2009). The8q22.1 microdeletion syndrome or Nablus mask-like facial syndrome: report on two patients and review of the literature. *Eur J Med Genet* 52:140–144.

Rachmilewitz EA, Tamari H, Liff F, Ueda Y, Nagel RL. (1985). The interaction of hemoglobin O Arab with Hb S and beta-thalassemia among Israeli Arabs. *Hum Genet* 70:119–125.

Rahim HF, Wick L, Halileh S, et al. (2009). Maternal and child health in the occupied Palestinian territory. *Lancet* 373:967–977.

Rambaud-Cousson A, Dudin AA, Zuaiter AS, Thalji A. (1991). Syndactyly type IV/hexadactyly of feet associated with unilateral absence of the tibia. *Am J Med Genet* 70:144–145.

Ritte U, Neufeld E, Prager EM, et al. (1993). Mitochondrial DNA affinity of several Jewish communities. *Hum Biol* 65:359–385.

Rosler A. (2006). 17 beta-hydroxysteroid dehydrogenase 3 deficiency in the Mediterranean population. *Pediatr Endocrinol Rev* 3(Suppl 3):455–461.

Rosler A, Silverstein S, Abeliovich D. (1996). A (R80Q) mutation in 17 beta-hydroxysteroid dehydrogenase type 3 gene among Arabs of Israel is associated with pseudohermaphroditism in males and normal asymptomatic females. *J Clin Endocrinol Metab* 81:1827–1831.

Rund D, Kornhendler N, Shalev O, Oppenheim A. (1990). The origin of sickle cell alleles in Israel. *Hum Genet* 85:521–524.

Saha N, Banerjee B. (1986). A study of some blood genetic characteristics of Bedouin and non-Bedouin Arabs of Jordan. *Hum Hered* 35:276–280.

Said R, Hamzeh Y, Said S, Tarawneh M, Al-Khateeb M. (1992). Spectrum of renal involvement in familial Mediterranean fever. *Kidney Int* 41:414–419.

Seligsohn U, Berger A, Abend M, Rubin L, Attias D, Zivelin A, Rapaport SI. (1984). Homozygous protein C deficiency manifested by massive venous thrombosis in the newborn. *N Eng J Med* 310:559–562.

Shahin H, Walsh T, Sobe T, et al. (2002). Genetics of congenital deafness in the Palestinian population: multiple connexin 26 alleles with shared origins in the Middle East. *Hum Genet* 110:284–289.

Shalata A, Mandel H, Reiss J, et al. (1998). Localization of a gene for molybdenum cofactor deficiency, on the short arm of chromosome 6, by homozygosity mapping. *Am J Hum Genet* 63:148–154.

Sheffield VC, Carmi R, Kwitekblack A, et al. (1994). Identification of a Bardet-Biedl syndrome locus on chromosomes 3 and evaluation of an efficient approach to homozygosity mapping. *Hum Mol Genet* 3: 1331–1335.

Shoshani T, Augarten A, Gazit E, et al. (1992). Association of a nonsense mutation (W 1282 X), the most common mutation in the Ashkenazi Jewish cystic fibrosis patients in Israel, with presentation of severe disease. *Am J Hum Genet* 50:222–228.

Shubailat G, Oumeish OY, Amr SS. (1983). Epidermolysis bullosa dystrophica triggered by sun exposure. *Ann Plast Surg* 10:239–243.

Sohar E, Gafni J, Pras M, Heller H. (1967). Familial Mediterranean fever: a survey of 470 cases and review of literature. *Am J Med Genet* 43:120–121.

Svejcar J. (1993). Biochemical abnormalities in connective tissue of osteodyplasty of Melnick-Needles and dyssegmental dwarfism. *Clin Genet* 23:369–375.

Teebi AS, Al-Awadi SA, Farag TI, Naguib KK. (1986a). Hypogonadotropic hypogonadism, mental retardation, obesity, and minor skeletal abnormalities: another new autosomal recessive syndrome from the Middle East. *Am J Med Genet* 24:373–378.

Teebi AS, Al-Awadi SA, Opitz JM, Spranger J. (1986b). Severe short limb dwarfism resembling Grebe chondrodysplasia. *Hum Genet* 74:386–390.

Teebi AS, Al-Awadi SA, Farag TI, Naguib KK, El-Khalifa MY. (1987a). Phenylketonuria in Kuwait and Arab countries. *Eur J Pediatr* 146:59–60.

Teebi AS, Al-Awadi SA, White AG. (1987b). Autosomal recessive nonsyndromal microcephaly with normal intelligence. *Am J Med Genet* 26:355–359.

Teebi A, Naguib K. (1988). Autosomal recessive nonsyndromal hydrocephalus. *Am J Med Genet* 31:467–470.

Teebi AS, Al-Saleh QA. (1989a). Nonsyndromal microphthalmia. *Clin Genet* 35:311–312

Teebi AS, Al-Saleh QA. Hassoon MM, Farag TI, Al-Awadi SA. (1989b). Macrosomia, microphthalmia, cleft palate, and early infant death: a new autosomal recessive syndrome. *Clin Genet* 36:174–177.

Teebi AS. (1990). Autosomal recessive Robinow syndrome. *Am J Med Genet* 35:64–68.

Teebi AS. (1991). Trigonobrachycephaly, bulbous bifid nose, mascrostomia, micrognathia, acral anomalies, and hypotonia in sibs. *Am J Med Genet* 38:529–531.

Teebi AS, Al-Saleh QA, Odeh H. (1992). Meckel syndrome and neural tube defects in Kuwait regional hospital. *J Med Genet* 29:140.

Teebi AS. (1993). Limb/pelvis/uterus-hypoplasia/aplasia syndrome. *Am J Med Genet* 30:797.

Teebi AS. (1994). Autosomal recessive disorders among Arabs: an overview from Kuwait. *J Med Genet* 31:224–233.

Teebi AS. (2000). Nablus mask-like facial syndrome. *Am J Med Genet* 95:407–408.

Teebi AS, Teebi SA. (2005). Genetic diversity among the Arabs. *Community Genet* 8:21–26.

Telmesani A. (1994). A rare association of myelomeningocele with cloacal duplication malformation. *Ann Trop Paediatr* 14:253–256.

Turnpenny PD, Thwaites RJ. (1992). Dwarfism, rhizomelic limb shortness, and abnormal face: a new short stature syndrome sharing some features with the Robinow Syndrome. *Am J Med Genet* 42:724–727.

Turnpenny PD, Thwaites RJ, Boulos FN. (1991). Evidence for variable gene expression in a large inbred kindred with autosomal recessive spondylocostal dysostosis. *J Med Genet* 28:27–33.

Turnpenny PD, Bulman MP, Frayling TM, et al. (1999). A gene for autosomal recessive spondylocostal dysostosis maps to 19q13.1-13.3. *Am J Hum Genet* 65:175–182.

van Lieburg AF, Verdijk MAJ, Knoers VVAM, et al. (1994). Patients with autosomal nephrogenic diabetes insipidus homozygous for mutations in the aquaporin 2 water-channel gene. *Am J Hum Genet* 55:648–652.

Walsh T, Abu Rayan A, Abu Sa'ed J, et al. (2006). Genomic analysis of a heterogeneous Mendelian phenotype: multiple novel alleles for inherited hearing loss in the Palestinian population. *Hum Genomics* 2:203–211.

Waterston T, Halileh S, Odeh J, Rudolf M, Hamilton P. (2009). Teaching child health in the occupied Palestinian territory. *Lancet* 373:878–880.

Yadav GC, Reavey PC. (1988). Aminoacidopathies: a review of three years' experience of investigations in a Kuwaiti hospital. *J Inherit Metab Dis* 11:277–284.

Younis K. (2006). Premarital testing for b-thalassemia in Palestine. *Thalassemia International Federation (TIF) Magazine* 49:23.

Zlotogora J, Glick B. (1993). Jancar syndrome: mental retardation, spasticity, and distal transverse limb defects. *Am J Med Genet* 47:89–90.

Zlotogora J, Regev R, Zeigler M, Iancu TC, Bach G. (1985). Krabbe disease: increased incidence in highly inbred community. *Am J Med Genet* 21:765–770.

Zlotogora J, Levy-Lahad E, Legum C, Ianca TC, Zeigler M, Bach G. (1991). Krabbe disease in Israel. *Isr J Med Sci* 31:196–198.

Zlotogora J, Furman-Shaharabani Y, Harris A, Barth ML, von Figura K, Gieselmann V. (1994a). A single origin for the most frequent mutation causing late infantile metachromatic leukodystrophy. *J Med Genet* 31:672–674.

Zlotogora J, Sagi M, Cohen. T (1994b). Familial hydrocephalus of prenatal onset. Am *J Med Genet* 49:202–204.

57.

GENETIC SKIN DISEASES IN THE ARAB WORLD

Mourad Mokni, Cherine Charfeddine, and Sonia Abdelhak

INTRODUCTION

Arab countries are characterized by genetic diversity. Major factors that contributed to their diversity include the migrations of Semitic tribes from the Arabian Peninsula, the Islamic expansion in the seventh century AD, the Crusade wars, and recent migration dynamics. These events have resulted in the admixture of the original Arabs with other populations extending from east and south Asia to Europe and Africa (Teebi and Teebi, 2005). Their demographic features include high rates of consanguinity in the range of 20% to 70% (Teebi et al., 2006), a large family size, and a rapid population growth. Consanguineous marriage leads usually to an increased birth prevalence of infants with recessive disorders (Teebi, 1994; Jaouad et al., 2009), and homozygosity for autosomal dominant congenital malformations, morbidity, and mortality (Abdulrazzaq et al., 1997). Several reports indicate that genetic disorders are frequent in Arab populations, but few detailed epidemiologic studies have been conducted so far to determine the exact frequency of each of the genetic defects.

Thousands of genetic diseases can associate skin features or defects, and hundreds of genodermatoses have mainly a cutaneous involvement with or without systemic defects. A review of these diseases in the Arab world cannot be exhaustive. In addition, the lack of precise epidemiological reports data (Souissi et al., 2007), the rarity of genetic studies, and genetic units contribute to underestimate the report and the burden of such diseases. In this review, we exclusively report literature data of genodermatosis described and genetically characterized in Arab patients.

EPIDERMOLYSIS BULLOSA

Epidermolysis bullosa (EB) is a group of heritable mechanobullous disorders characterized by fragility of the skin within the cutaneous basement membrane zone (BMZ; Fine et al., 2008). The severity of skin manifestations can be highly variable, so that at one end of the spectrum the skin findings may be a relatively minor blistering tendency, primarily on the hands and feet, whereas at the other end of the spectrum the compromised integrity of skin can result in early postnatal demise of the affected individual. Traditionally, EB has been divided into three broad categories: (a) the simplex forms demonstrate tissue separation within the basal keratinocytes; (b) the junctional forms depict tissue separation within the lamina lucida of the dermal–epidermal basement membrane; and (c) the dystrophic forms show tissue cleavage below the lamina densa within the upper papillary dermis. Epidermolysis bullosa (EB) encompasses a large group of inherited blistering skin disorders caused by mutations in at least 10 genes.

Numerous studies, mainly performed in European and U.S. families with EB, have revealed a number of characteristic epidemiological and genetic features, which form the basis for current diagnostic and counseling strategies. However, little is currently known about the molecular epidemiology of EB in Middle East populations (Nakano et al., 2002; Ayoub et al., 2005; Abu Sa'd et al., 2006).

In the Eastern Province of Saudi Arabia, a review of 49,902 dermatology cases revealed 16 cases of epidermolysis bullosa, 10 (62.5%) of which were of the dermolytic type (epidermolysis bullosa dystrophica), and 4 (25%) of the epidermolytic type (epidermolysis bullosa simplex; see Abahussein et al., 1993).

In North Africa, epidemiological data were available only in Tunisia. According to this study, 46% of the EB Tunisian patients have the dystrophic form, 18% the junctional, and 13% the simplex form. Only the dystrophic form, the most frequent, was studied on molecular level (Ouragini et al., 2010). The dystrophic epidermolysis bullosa (DEB, MIM 131750, 226600) is caused by mutations in the *COL7A1* gene encoding type VII collagen, the major component of anchoring fibrils. On the basis of the mode of inheritance and the clinical manifestations, DEB is classified into two major subtypes: one dominant (DDEB) and one recessive (RDEB). Ouragini et al. (2010) revealed six different mutations, three of which were family-specific mutations (c.1507+1G>A, c.4406_4415del and p.R1978X); one represented a worldwide recurrent mutation (c.7344G>A); and two were frequently found only in Tunisian patients (p.R1763X and p.G1483D). Molecular and clinical investigations in Tunisian patients showed large interfamilial and intrafamilial heterogeneity occurring for identical *COL7A1* genotype (Ouragini et al., 2009, 2010).

HEREDITARY DISORDERS OF KERATINIZATION

DARIER DISEASE

Darier Disease (DD; MIM 124200) is a rare dominantly inherited disorder that predominantly affects the skin. It is characterized by warty papules and plaques in seborrheic areas, palmoplantar pits, and distinctive nail dystrophy (Burge and Wilkinson (1992). The DD locus has been mapped to chromosome 12q23-24.1; mutations within the *ATP2A2* gene have been shown to be responsible for DD. Bchetnia et al. (2009a,b) reported the first study of DD in the North African population, and the results of clinical and genetic investigation of 8 Tunisian families with DD. Phenotypic variation was observed within and between the families, but none had severe DD. Linkage analysis to the *ATP2A2* gene showed a haplotype heterogeneity that was confirmed by mutational screening. Seven distinct mutations, among which 4 were novel, have been described. For only one family, no mutation was found; however, a rare variant of single nucleotide polymorphism (rs35235621) was identified. Mutational heterogeneity was also observed, owing to the richness of the genetic background of Tunisian population. The mutation spectrum identified in Tunisian families included 3 heterozygous missense mutations (G217A, D702N, and missense L900R), one nonsense mutation (R677X), two frameshift deletions (1747–1749 del 2T and 119–120 delAG), and one frameshift insertion (2772–2779 ins C). No obvious genotype–phenotype correlation was established. Differences in the expression of DD could be explained not only by environmental factors but also by modifier loci.

Histological investigations have also been reported in 6 unrelated Tunisian families with Darier's disease mutations (including 15 patients) among the 8 families mentioned above. The typical histological features of DD have been observed in all patients. Variable histological features have been observed among Tunisian patients ranging from mild to moderate lesions of Darier's disease. A significant correlation has been observed between the clinical presentation of the Darier's disease (mild or moderate) and the intensity of the histological features. Isolated acral form of Darier's disease was seen in one case. Two distinct original associations have been observed: DD/pemphigus vulgaris in one patient and DD/ichthyosis in the other patient. All these findings confirmed the clinical heterogeneity of Darier's disease on the basis of histological study. The intensity of the histological features could be closely correlated to the severity of Darier's disease clinical presentation (Kassar et al., 2009).

Recently, 3 new Tunisian cases with moderate forms of DD with "seborrheic" pattern have been reported (Bchetnia et al., 2009c).

The genetic investigations allowed the identification of 2 new heterozygous mutations (1713–1714del2A and R559Q). To date, 9 mutations have been identified in the *ATP2A2* gene in Tunisian populations (Bchetnia et al., 2009a,b,c).

ICHTHYOSES

Ichthyosis comprises a clinically and genetically heterogeneous group of disorders characterized by a skin desquamation. Different classifications have been proposed based on the mode of inheritance, the physiopathology, and the genetics of the different existing forms. The clinical observations and the etiology of the disease allowed distinguishing between the genetic forms, which generally appears like a keratinization disorder and nongenetic forms associated with cutaneous xerosis (DiGiovanna and Robinson-Bostom, 2003). The genetic forms have an inherited mode of transmission, and the age of onset and clinical observations may help the diagnostic. In fact, it is easier to distinguish the vulgaris form (IV), which appears generally 6 months after birth. It is the most common type of ichthyosis with an incidence of 1/250 is associated with keratinization disorders, and may appear later (Smith et al., 2006). The X-linked recessive form of ichthyosis begins with birth, but patients are not born as collodion babies. It is relatively frequent, with an incidence of 1/2000 (Arndt et al., 2000). The diagnosis of the other genetic forms of ichthyosis is much more problematic, because the etiology and clinical observations do not allow further specification. In fact, a collodion baby may evolve into congenital ichthyosiform erythroderma (CIE) or lamellar ichthyosis (LI), which are two different forms of autosomal recessive ichthyosis (ARCI). Their incidences are, respectively, 1/100,000 and 1/300,000 (Lorette et al., 1999).

The clinical observations have shown limits in the differentiation of those two forms. The molecular approach has been another way to help the diagnostic. But the genetic heterogeneity of the ARCI, which are induced by a mutation in 6 possible genes, has not helped the diagnostic; on the contrary, it has added a new difficulty to it. The Arab populations are a good field for the development of the genetic of orphan diseases such as autosomal recessive ichthyosis (ARCI), whose incidence is increased by the high rate of consanguinity in those populations (Al-Zayir et al., 2006; Al Aboud et al., 2007; Kharfi et al., 2008). Up to now one-third of the mutations inducing the ARCI forms of ichthyosis have been identified in the *TGM1* gene (Huber et al., 1995). Those mutations were found in Tunisian and Egyptian Arab populations, among others (Hennies et al., 1998; Shawky et al., 2004).

Seven homozygous mutations including five missense and two deletions were identified in the cytochrome *P450* gene in 12 consanguineous families from Algeria and Lebanon, with lamellar ichthyosis type 3 (Lefèvre et al., 2006). Another finding identified six homozygous mutations, including one nonsense and five missense mutations, in the *Ichthyin* gene mapped to chromosome 5q33 in23, in patients from 13 consanguineous families from Algeria, Syria, and Turkey (Lefèvre et al., 2004). Ichthyosis may be associated with vast number of extracutaneous pathologies such as neurological disorders, heart disease, and renal anomalies. Those syndromic forms are different clinically, and relatively easier to diagnose.

The SLS (Sjögren-Larsson syndrome) is induced by a mutation identified in the *FALDH* gene mapped to chromosome17. Mutations in this gene were found in patients from Morocco (Rafai et al., 2008) and Lebanon (Pigg et al., 1999). This syndrome is a rare neurocutaneous disorder (paralysis of the lower limbs, mental retardation, and convulsions) caused by a deficiency of the microsomal enzyme fatty aldehyde dehydrogenase (FALDH; see Rizzo et al., 2008).

Chanarin-Dorfman syndrome (CDS) is a rare association between triglyceride accumulation as lipid droplets in various organ's tissues and congenital ichthyosiform erythroderma. This syndrome is due to mutations in the *GCI-58* gene mapped to chromosome 3p21 (Baala et al., 2002). The syndrome was reported in the Tunisian population (Venencie et al., 1993).

The Harlequin fetus is a severe lethal form; clinical observations are clear and give a sure diagnosis. This form of congenital ichthyosiform erythroderma is associated with mutations in the *ABCA12* gene mapped to chromosome 2q34 (Kelsell et al., 2005; Thomas et al., 2006; Akiyama et al., 2005). This syndrome was detected in 3 Tunisian patients (Mkaouar, 1998).

PALMOPLANTAR KERATODERMA

Hereditary palmoplantar keratodermas (PPK) comprise a clinically and genetically heterogeneous group of genodermatoses, which share impaired epidermal differentiation resulting in prominent palmoplantar hyperkeratosis. Classically, keratodermas have been separated according to their clinical appearance into diffuse, focal, and as a feature of ectodermal dysplasias and many other syndromes. Since molecular genetic analyses have helped characterize the underlying genetic defects in an increasing number of hereditary PPK over the last two decades, a pathophysiological separation seems more reasonable. Today PPK can be classified based on defects in keratins, loricrin, desmosomes, connexins, and cathepsins. Although these proteins have different structures and functions, all of them influence epidermal differentiation and cornified envelope assembly. Depending on tissue distribution and location of mutation within a certain gene, the clinical spectrum of PPK ranges from a pure palmoplantar restricted skin abnormality to a complex combination of symptoms with dental anomalies, deafness, progressive cardiomyopathy, and even cancer. Solely for those reasons, a correct diagnosis based on molecular genetic analyses is mandatory, although a causal therapy is still not available. Instead, several therapeutic modalities including topical ointments, surgical interventions, and systemic retinoids help to reduce the patients' symptoms.

MAL DE MELEDA

Mal de Meleda (MDM, MIM 248300), also referred to as keratosis palmoplantaris transgrediens of Siemens, is a rare autosomal recessive skin disorder with a prevalence in the general population of 1 in 100,000. It was first described on the Adriatic island of Meleda, where it was relatively common. The main clinical characteristics are transgressive palmoplantar keratoderma, hyperhidrosis, and perioral erythema, but there are also associated features such as brachydactyly, nail abnormalities, and lichenoid plaques. Linkage analysis of large consanguineous families from Algeria localized the responsible gene to 8qter (*ARS* [component B]; see Fischer et al., 1998; Bouadjar et al., 2000).

This gene encodes for the SLURP-1 protein (Ly-6/uPAR related protein-1). New mutations were also described in patients from Tunisia (Marrakchi et al., 2003; Charfeddine et al., 2003), Palestine, Germany and Turkey (Eckl et al., 2003).

Genetic heterogeneity of MDM is suspected in a Tunisian family with recessive transgressive PPK closely resembling the MDM phenotype that excludes linkage to the *ARS* gene (Charfeddine et al., 2006). We demonstrated that some of the heterozygous female carriers had mild hyperkeratotic symptoms (Mokni et al., 2004).

BUSCHKE-FISCHER-BRAUER'S DISEASE

Buschke–Fischer–Brauer's disease, or the type I of punctate PPK (MIM 148600), is a rare hereditary skin disease that is usually inherited as an autosomal dominant trait. Punctate PPK is characterized by multiple punctate keratoses over the entire palmoplantar surfaces. The molecular basis of punctate PPK is still unknown. Recently, two punctate PPK loci were found to map to 15q22-15q24 and to 8q24.13-8q24.21. Recently, genetic study identified a candidate region reduced to 5.06 cM on 15q22.2-15q22.31. Clinical, histological, and genetic findings have been reported in 5 Tunisian families with a total of 15 patients (6 women and 9 men). The explored patients exhibited classical PPK phenotype, and a positive family history was observed that confirmed the autosomal mode of transmission of the disease in informative families—although several groups have reported association of punctuate PPK.

No PPK with malignancies has been associated with type I punctate PPK in the examined Tunisian patients. Special clinical findings were observed in the Tunisian studied families: there was occurrence of punctate PPK with moderate hearing impairment, and also the association of punctuate PPK with ichthyosis vulgaris. Recently, association of punctate PPK with other dermatological manifestations (psoriasis) was reported, but association with ichthyosis has not been previously described. This co-occurrence could be fortuitous because of the high prevalence of this type of ichthyosis worldwide. Genetic and haplotype analysis of Tunisian families confirmed linkage to the previously identified critical 15q22-15q24 interval (Bchetnia et al., 2009c).

NAXOS DISEASE

Naxos disease (MIM 601214) is a rare hereditary disorder characterized by palmoplantar keratoderma, woolly hair and cardiomyopathy. It is caused by the (Pk2157del2) mutation in the plakoglobin gene, which was identified in Greek families from Naxos Island where the disease had originally been described. Absence of the Pk2157del2 frameshift in an Arab Saudi affected child proved that Naxos disease in this Saudi Arab family is not caused by this mutation. Furthermore, heterozygosity for the IVS11+22G/A polymorphism provided evidence for exclusion of the plakoglobin gene (Stuhrmann et al., 2004).

VOHWINKEL SYNDROME

Vohwinkel syndrome (VS, MIM 124500), or keratoderma hereditarium mutilans, is a rare, diffuse, honeycombed, palmoplantar keratosis usually accompanied by pseudoainhum near the distal interphalangeal creases. The analysis of the *GJB2* gene in an Egyptian family with Vohwinkel syndrome showed a dominant missense mutation at a highly conserved residue, p.Gly130Val (Snoeckx et al., 2005).

PAPILLON-LEFEVRE SYNDROME

Papillon-Lefèvre syndrome (PLS, MIM 245000) is an autosomal recessive disorder characterized by palmoplantar keratoderma and severe, early-onset periodontis leading to premature loss of both the deciduous and permanent teeth. Mutations within the cathepsin C gene (*CTSC*) are responsible for the disease, leading to deficiency of cathepsin C activity. Nursier et al. (2002) presented a consanguineous Jordanian family with the IVS3-1G –> A mutation segregating within *CTSC* gene.

STRIATE KERATODERMA

Striate keratoderma (PPKS, MIM 148700) is characterized by palmoplantar keratoderma typified by streaking hyperkeratosis along each finger and extending onto the palm of the hand. Direct DNA sequencing analysis in families from Iran and Syria revealed a heterozygous C-to-A transversion at nt 395 of the *DSG1* gene. This mutation converted a serine residue (TCA) in exon 5 to a nonsense mutation (TAA) designated S132X. The mutation identified in this study is a novel mutation in the *DSG1* gene (Kljuic et al., 2003).

GENODERMATOSES ASSOCIATED WITH METABOLIC DISEASE

There are also forms of genodermatosis, in particular PPK, that have been associated with metabolic disorder.

TYROSINEMIA TYPE II

Tyrosinemia type II, or Richner-Hanhart syndrome (RHS, MIM 276600), is an autosomal recessive disorder characterized by keratitis, palmoplantar keratosis, mental retardation, and elevated blood tyrosine levels. The disease is due to a deficiency of hepatic cytosolic tyrosine aminotransferase (TATc), an enzyme involved in the tyrosine catabolic pathway. RHS is characterized by interfamilial and intrafamilial phenotypic variability. A large spectrum of mutations within *TATc* gene on chromosome 16q21.1-q22.3 has been shown to be responsible for RHS (Hühn et al., 1998). Recently, two novel missense mutations were identified (C151Y and L273P) within exon 5 and exon 8, respectively, in three unrelated consanguineous Tunisian families including 7 patients with confirmed biochemical diagnosis of tyrosinemia type II (Charfeddine et al., 2006).

ACRODERMATITIS ENTEROPATHICA

Acrodermatitis enteropathica (AEZ, MIM 201100) is a rare autosomal recessive disease that manifests as an inability of the affected individual to absorb intestinal zinc, and therefore patients have nutritional zinc deficiency. This disease is also characterized by periorificial and acral dermatitis, alopecia, and diarrhea. Mutations in the *SLC39A4* gene are responsible for acrodermatitis enteropathica. This gene encodes one member of a human zinc/iron-regulated transporter-like protein, also known as ZIP4, and consists of 12 exons.

Mutations are observed throughout the gene, and no obvious phenotype–genotype correlation can be made. However, there is no denying that the ethnic origin of patients could give some indications and orientation for a genetic diagnosis, especially in the case of Tunisian families. Indeed, among the small number of reported mutations in the Tunisian population, two mutations are far more prevalent: deletion c.123_1227 del and nonsense mutation p. Leu 48 X were found respectively in 5 and 3 unrelated Tunisian families (Küry et al., 2003; Meftah et al., 2006). It seems clear that a founder effect occurred in Tunisia.

BIOTINIDASE DEFICIENCY

Biotinidase deficiency (BD, OMIM 253260) is an autosomal recessively inherited disorder of biotin metabolism. The worldwide incidence of the disease is 1/60,000 (Wolf, 1991). Due to the high rate of consanguinity in Arab countries, the incidence is estimated about 1/14,800 in Turkey (Pomponio et al., 2000a); in Syria, and in Saudi Arabia (Pomponio et al., 2000b). Patients with biotinidase deficiency develop seizures, hypotonia, ataxia, developmental delay, hearing loss, optic abnormalities, alopecia, and skin rash, as well as ketolactic acidosis and organic aciduria (Pomponio et al., 1997). Those affected with complete biotinidase deficiency have enzymes that produce <10% of mean normal activity, while those affected with partial deficiency have enzymes that produce 10%–30% of normal activity (Hymes et al., 2001).

The human biotinidase gene (BTD MIM 609019) is localized to chromosome 3p25, consisting of 4 exons with a total length of 1629 bp, encoding a mature protein of 543 amino acids and resulting in a molecular mass of 56,771 Da (Knight et al., 1998).

More than 100 mutations causing biotinidase deficiency have been described (Wolf, 2005). The two most common mutations that cause profound BD in symptomatic children are: 98G:del7ins3 and R538C within exon 2 and exon 4 respectively (Pomponio et al., 1995, 1997). In children ascertained by newborn screening, the two most common mutations are a missense mutation Q456H and a double mutation A171T and D444H (Norrgard et al., 1997, 1998). Partial BD is due to the D444H missense mutation in one allele, in combination with a genetic lesion that leads to profound BD in the other allele (Swango et al., 1998). Several mutations were described in Turkey: Y57TER, R79C, C186Y, T196R, G310E and T532M (see Pomponio et al., 2000a,b); D543H (Muèhl et al., 2001); 393delC (Wolf, 2002); and Y454C (Wolf, 2005). Wolf (2002, 2005) described a deletion 393delC in two Palestinian children and a missense mutation (G114V) in Syrian patient. In Morocco, the most common mutation of two deletions and one insertion—98G:del7ins3, 1049delC, 1052delC and c.1264_1265insC—were described (Wolf, 2002; Iqbal et al., 2010).

XERODERMA PIGMENTOSUM

Xeroderma pigmentosum (XP, MIM 278700–278780) is a rare autosomal recessive disease characterized by hypersensitivity to UV rays, which occurs with a high frequency in the Middle East and North Africa. In Tunisia, although no precise epidemiological data are available, XP is suspected to be relatively frequent because of the high rate of consanguinity and endogamy. The frequency of the disease has been estimated at 1/10,000 (Zghal et al., 2006). The incidence of the disease in Libya has been estimated at 15–20 per million (Khatri et al., 1999). The disease is genetically and clinically heterogeneous, with 8 genes known to be responsible for the disease (*XPA, ERCC3, XPC, ERCC2, DDB2, ERCC4, ERCC5* and *POLH*; see Wood, 1996). The XP group C (XP-C) is the most prevalent type in Western Europe and in the United States. A survey of DNA repair characteristics among Egyptians has identified representatives of the 3 major complementation groups A, C, and variant. Group A patients showed very early onset of sun sensitivity and development of skin cancers, along with microcephaly and mental retardation. Group C patients showed an early onset of sun sensitivity (6 months), and most of them have XP without neurological abnormalities (Kraemer and DiGiovanna, 1993). The residual repair in XP-C cells facilitates more recovery of DNA synthesis than in other groups. This may contribute to the higher in-vitro survival in culture and milder clinical symptoms in group C as compared to group A (Cleaver et al., 1981; Hashem et al., 1980).

A high prevalence of the nonsense mutation R228X was reported in exon 6 of the *XPA* gene in Tunisian xeroderma pigmentosum group A patients, showing an intermediate phenotype (Nishigori et al., 1993). Haplotype analysis confirmed the hypothesis of a founder effect (Messaoud et al., 2010a). Although a common mutation is shared between almost all XP-A Tunisian patients, a particular phenotype characterized by a primarily neurological expression and caused by a novel mutation in the *XPA* gene was also described (Messaoud et al., 2010b).

In North Africa, the *XPC* V548A fs X572 mutation has been identified in Algerian and Moroccan patients (Daya-Grosjean et al., 1987; Li et al., 1993; Chavanne et al., 2000; Khan et al., 2006). This mutation has been also found in Tunisian XP-C patients at a high frequency with a common ancestor. Among them a complete phenotype–genotype correlation has been described. Indeed, all XP-C patients with a severe clinical form and without neurological abnormalities had this mutation (Ben Rekaya et al., 2009).

DISORDERS OF CONNECTIVE TISSUE

CUTIS LAXA

Cutis laxa (CL, MIM 123700; MIM 219100) comprises a heterogeneous group of inherited and acquired disorders characterized by loose, sagging, and redundant skin, which often leads to an appearance of premature aging (Uitto and Pulkkinen, 2002). Autosomal dominant forms (MIM 123700) are relatively mild, and may be caused by mutations in the elastin gene (*ELN*), whereas the more severe recessive forms have been associated with mutations in the fibulin 4 (FBLN4) and fibulin 5 (FBLN5) genes, as well as in a vesicular ATPase subunit (Zhang et al., 1999; Markova et al., 2003). The first autosomal recessive form of CL, caused by homozygous recessive mutations in exon 12 of the elastin gene (p.P211S), was reported in two related consanguineous Syrian families. Furthermore, the presence of a polymorphism in the fibulin 5 gene in one of the patients seems to modify the phenotype, producing more severe symptoms (Mégarbané et al., 2009). In a consanguineous family from Lebanon with autosomal recessive transmission, a homozygous missense mutation (c.649T -> C; p.C217R) in the fibulin-5 gene (*FBLN5*) was identified (Claus et al., 2008).

EHLERS-DANLOS SYNDROME

Abdul Wahab et al. (2003) described the clinical spectrum of anomalies of a new type of Ehlers–Danlos syndrome in 32 patients from a large interrelated extended family in Qatar. Anomalies included a variable degree of skin hyperextensibility, hypermobility of small and large joints, and tortuous systemic arteries. Peculiar facial features included epicanthic folds, flat saggy cheeks, elongated faces, and micrognathia. These patients are at risk for potentially catastrophic arterial rupture. Linkage to the major loci involved in Ehlers–Danlos syndrome and other connective tissue disorders, such as Cutis laxa, familial aneurysm, and osteogenesis imperfecta, was excluded by using specific DNA markers, confirming the uniqueness of this disorder.

ANHIDROTIC ECTODERMAL DYSPLASIA

Anhidrotic ectodermal dysplasia (HED, MIM 224900) is clinically characterized by sparse hair, absent eyebrows and eyelashes, missing teeth, decreased sweating, dry, thin skin, periorbital wrinkling and hyperpigmentation, prominent lips, and a saddle-shaped nose. Mutation in the ectodysplasin anhidrotic receptor gene (*EDAR*) or in the *EDARADD* gene can cause autosomal recessive hypohidrotic ectodermal dysplasia (EDA/HED). Mégarbané et al reported an 18-year-old woman, born to first cousin parents, presenting with a severe form of anhidrotic ectodermal dysplasia (EDA/HED). She had sparse hair, absent limb hair, absent sweating, episodes of hyperpyrexia, important hypodontia, and hyperconvex nails. She also showed unusual clinical manifestations such as an absence of breasts, a rudimentary extranumerary areola and nipple on the left side, and marked palmoplantar hyperkeratosis. Light microscopy of skin biopsies showed orthokeratotic hyperkeratosis and absence of sweat glands. A novel homozygous splice-site mutation (IVS9 + 1G > A) in the *EDAR* gene was identified. This mutation results in a total absence of EDAR transcripts, and consequently of the EDAR protein, which likely results in abolition of all ectodysplasin-mediated NF-kappaB signaling (Mégarbané et al., 2008).

CONCLUSION

In conclusion, although many programs to control genetic diseases have been adopted in Arab countries, most are poorly developed. A shift in public, political, and professional attitudes is needed to establish comprehensive services (Bayoumi and Yardumian, 2006).

Finally, as pointed by an anonymous editorial of Nature Genetics [No authors listed. (2006). The germinating seed of Arab genomics. *Nat Genet 38*(8), 851]: "The study of their constituent regional populations can form a new research resource from which their scientists can bring fresh insights to the world genomics community". As for several genetic diseases, large consanguineous families allowed identification of several genes involved in genetic skin disorders (Romdhane and Abdelhak, 2011). Taking into account particular familial structure and common genetic background, a global strategy for diagnosis and prevention of genodermatoses in the Arab world should be adopted (see Chapter 34).

ACKNOWLEDGEMENT

We would like to thank our colleagues who contributed to this chapter: A. S. Ben Brick, A Tiar, O. Messaoud, M. Ben Rekaya, M. Nagara, H. Ouragini, H. Azaiez and M. S. Boubaker.

REFERENCES

Abahussein AA, al-Zayir AA, Mostafa WZ, Okoro AN. (1993). Epidermolysis bullosa in the eastern province of Saudi Arabia. *Int J Dermatol* 32(8), 579–581.

Abdul Wahab A, Janahi IA, Eltohami A, Zeid A, Ul Haque NF, Teebi AS. (2003). A new type of Ehlers–Danlos syndrome associated with tortuous systemic arteries in a large kindred from Qatar. *Acta Paediatr* 92(4), 456–462.

Abdulrazzaq YM, Bener A, al-Gazali LI, al-Khayat AI, Micallef R, Gaber T. (1997). A study of possible deleterious effects of consanguinity. *Clin Genet* 51(3), 167–173.

Abu Sa'd J, Indelman M, Pfendner E, et al. (2006). Molecular epidemiology of hereditary epidermolysis bullosa in a Middle Eastern population. *J Invest Dermatol* 126(4), 777–781.

Akiyama M, Sugiyama-Nakagiri Y, Sakai K et al. (2005) Mutations in lipid transporter ABCA12 in harlequin ichthyosis and functional recovery by corrective gene transfer. *J Clin Invest* 115(7), 1777–1184.

Al Aboud D, Al Aboud K, Ramesh V, Kumar J. (2007). Lamellar ichthyosis in a Saudi kindred. *Skinmed* 6(1), 40–41.

Al-Zayir AA, Al-Amro Al-Alakloby OM. (2006). Clinico-epidemiological features of primary hereditary ichthyoses in the Eastern province of Saudi Arabia. *Int J Dermato* 45(3), 257–264.

Arndt T, Pelzer M, Nenoff P, et al. (2000). Lipoprotein and apolipoprotein electrophoresis in X-chromosome recessive ichthyosis. *Hautarzt 51,* 490–495.

Ayoub N, Tomb R, Charlesworth A, Meneguzzi G. (2005). Junctional epidermolysis bullosa. Identification of a new mutation in two Lebanese families. *Ann Dermatol Venereol* 132(6–7 Pt 1), 550–553.

Baala L, Hadj-Rabia S, Hamel-Teillac D, et al. (2002). Homozygosity mapping of a locus for a novel syndromic ichthyosis to chromosome 3q27-q28. *J Invest Dermatol* 119(1), 70–76.

Bayoumi RA, Yardumian A. (2006). Genetic disease in the Arab world. *BMJ* 333(7573), 819.

Bchetnia M, Benmously R, Ben Brick AS, et al. (2009a). New mutations of Darier disease in Tunisian patients. *Arch Dermatol Res* 301(8), 615–619.

Bchetnia M, Charfeddine C, Kassar S, et al. (2009b). Clinical and mutational heterogeneity of Darier disease in Tunisian families. *Arch Dermatol* 145(6), 654–656.

Bchetnia M, Charfeddine C, Kassar S, et al. (2009c). Clinical, histological and genetic investigation of Buschke-Fischer-Brauer's disease in Tunisian families. *J Dermatol Sc* 54(1), 54–56.

Ben Rekaya M, Messaoud O, Talmoudi F, et al. (2009). High frequency of the V548A fs X572 XPC mutation in Tunisia: implication for molecular diagnosis. *J Hum Genet* 54(7), 426–429.

Bouadjar B, Benmazouzia S, Prud'homme JF, Cure S, Fischer J. (2000). Clinical and genetic studies of 3 large, consanguineous, Algerian families with Mal de Meleda. *Arch Dermatol* 136(10), 1247–1252.

Burge SM, Wilkinson JD. (1992). Darier-White disease: a review of the clinical features in 163 patients. *J Am Acad Dermatol* 27(1), 40–50.

Charfeddine C, Mokni M, Ben Mousli R, et al. (2003). A novel missense mutation in the gene encoding SLURP-1 in patients with Mal de Meleda from northern Tunisia. *Br J Dermatol* 149(6), 1108–1115.

Charfeddine C, Mokni M, Kassar S, et al. (2006). Further evidence of the clinical and genetic heterogeneity of recessive transgressive PPK in the Mediterranean region. *J Hum Genet* 51(10), 841–845.

Charfeddine C, Monastiri K, Mokni M, et al. (2006). Clinical and mutational investigations of tyrosinemia type II in Northern Tunisia: identification and structural characterization of two novel TAT mutations. *Mol Genet Metab* 88(2), 184–1891.

Chavanne F, Broughton BC, Pietra D, et al. (2000). Mutations in the XPC gene in families with xeroderma pigmentosum and consequences at the cell, protein, and transcript levels. *Cancer Res* 60(7), 1974–1982.

Claus S, Fischer J, Mégarbané H, et al. (2008). A p.C217R mutation in fibulin-5 from cutis laxa patients is associated with incomplete extracellular matrix formation in a skin equivalent model. *J Invest Dermatol* 128(6), 1442–1450.

Cleaver JE, Zelle B, Hashem N, El-Hefnawi MH, German J. Xeroderma pigmentosum patients from Egypt: II (1981). Preliminary correlations of epidemiology, clinical symptoms and molecular biology. *J Invest Dermatol* 77(1), 96–101.

Daya-Grosjean L, James M, Drougard C, Sarasin A. (1987). An immortalized xeroderma pigmentosum, groupC, cell line which replicates SV40 shuttle vectors. *Mutat Res* 183(2), 185–196.

DiGiovanna JJ, Robinson-Bostom L. (2003). Ichthyosis: etiology, diagnosis, and management. *Am J Clin Dermatol* 4(2), 81–95.

Eckl KM, Stevens HP, Lestringant GG, et al. (2003). Mal de Meleda (MDM) caused by mutations in the gene for SLURP-1 in patients from Germany, Turkey, Palestine, and the United Arab Emirates. *Hum Genet* 112(1), 50–56.

Fine JD, Eady RA, Bauer EA, et al. (2008). The classification of inherited epidermolysis bullosa (EB): Report of the Third International Consensus Meeting on Diagnosis and Classification of EB. *J Am Acad Dermatol* 58(6), 931–950.

Fischer J, Bouadjar B, Heilig R, Fizames C, Prud'homme JF, Weissenbach J. (1998). Genetic linkage of Meleda disease to chromosome 8qter. *Eur J Hum Genet* 6(6), 542–547.

Hashem N, Bootsma D, Keijzer W, et al. (1980). Clinical characteristics, DNA repair, and complementation groups in xeroderma pigmentosum patients from Egypt. *Cancer Res* 40(1), 13–18.

Hennies HC, Raghunath M, Wiebe V, et al. (1998). Genetic and immunohistochemical detection of mutations inactivating the keratinocyte transglutaminase in patients with lamellar ichthyosis. *Hum Genet* 102(3), 314–318.

Huber M, Rettler I, Bernasconi K, et al. (1995). Mutations of keratinocyte transglutaminase in lamellar ichthyosis. *Science 267,* 525–528.

Hühn R, Stoermer H, Klingele B, et al. (1998). Novel and recurrent tyrosine aminotransferase gene mutations in tyrosinemia type II. *Hum Genet* 102(3), 305–313.

Hymes J, Stanley CM, Wolf B. (2001). Mutations in BTD causing biotinidase deficiency. *Hum Mutat* 18(5), 375–381.

Iqbal F, Item CB, Vilaseca MA, et al. (2010). The identification of novel mutations in the biotinidase gene using denaturing high pressure liquid chromatography (dHPLC). *Mol Genet Metab* 100(1), 42–45.

Jaouad IC, Elalaoui SC, Sbiti A, Elkerh F, Belmahi L, Sefiani A. (2009). Consanguineous marriages in Morocco and the consequence for the incidence of autosomal recessive disorders. *J Biosoc Sci* 41(5), 575–581.

Kassar S, Tounsi-Kettiti H, Charfeddine C, et al. (2009). Histological characterization of Darier's disease in Tunisian families. *J Eur Acad Dermatol Venereol* 23(10), 1178–1183.

Kelsell DP, Norgett EE, Unsworth H, et al. (2005). Mutations in ABCA12 underlie the severe congenital skin disease harlequin ichthyosis. *Am J Hum Genet* 76(5), 794–803.

Khan SG, Oh KS, Shahlavi T, et al. (2006). Reduced XPC DNA repair gene mRNA levels in clinically normal parents of xeroderma pigmentosum patients. *Carcinogenesis* 27(1), 84–94.

Kharfi M, El Fekih N, Ammar D, Khaled A, Fazaa B, Kamoun MR. (2008). Hereditary ichthyosis in Tunisia: epidemiological study of 60 cases. *Tunis Med* 86(11), 983–986.

Knight HC, Reynolds TR, Meyers GA et al. (1998). Structure of the human biotinidase gene. *Mamm Genome* 9(4), 327–330.

Khatri ML, Bemghazil M, Shafi M, Machina A. (1999). Xeroderma pigmentosum in Libya. Int J Dermatol 38(7), 520–524.

Kljuic A, Gilead L, Martinez-Mir A, et al. (2003). A nonsense mutation in the desmoglein 1 gene underlies striate keratoderma. Exp Dermatol 12(4), 523–527.

Kraemer KH, DiGiovanna JJ. (1993). In: Pagon RA, Bird TD, Dolan CR, Stephens K, editors. GeneReviews [Internet]. Seattle (WA): University of Washington, Seattle.

Küry S, Kharfi M. Kamoun R et al. (2003). Mutation spectrum of human SLC39A4 in a panel of patients with acrodermatitis enteropathica. *Hum Mutat* 22(4) 37-38.

Lefèvre C, Bouadjar B, Ferrand V, et al. (2006). Mutations in a new cytochrome P450 gene in lamellar ichthyosis type 3. Hum Mol Genet 15(5), 767–776.

Lefèvre C, Bouadjar B, Karaduman A, et al. (2004). Mutations in ichthyin a new gene on chromosome 5q33 in a new form of autosomal recessive congenital ichthyosis. Hum Mol Genet 13(20), 2473–2482.

Li L, Bales ES, Peterson CA, Legerski RJ. (1993). Characterization of molecular defects in xeroderma pigmentosum group C. *Nat Genet* 5(4), 413–417.

Lorette C, Arbeille B, Grangeponte MC, Vaillant L, et al. (1999). Ichtyoses. Encycl Med Chir (Elsevier, Paris), *Dermatologie* 12, 98–195, A10.

Markova D, Zou Y, Ringpfeil F, et al. (2003). Genetic heterogeneity of cutis laxa: a heterozygous tandem duplication within the fibulin-5 (FBLN5) gene. *Am J Hum Genet* 72(4), 998–1004.

Marrakchi S, Audebert S, Bouadjar B, et al. (2003). Novel mutations in the gene encoding secreted lymphocyte antigen-6/urokinase-type plasminogen activator receptor-related protein-1 (SLURP-1) and description of five ancestral haplotypes in patients with Mal de Meleda. *J Invest Dermatol* 120(3), 351–355.

Meftah SP, Kuivaniemi H, Tromp G, et al. (2006). A new mutation in exon 3 of the SCL39A4 gene in a Tunisian family with severe acrodermatitis enteropathica. *Nutrition* 22(10), 1067–1070.

Mégarbané H, Cluzeau C, Bodemer C, et al. (2008). Unusual presentation of a severe autosomal recessive anhydrotic ectodermal dysplasia with a novel mutation in the EDAR gene. *Am J Med Genet A* 146A(20), 2657–2662.

Mégarbané H, Florence J, Oliver Sass J, et al. (2009). An autosomal-recessive form of cutis laxa is due to homozygous elastin mutations, and the phenotype may be modified by a heterozygous fibulin 5 polymorphism. *J Invest Dermatol* 129(7), 1650–1655.

Messaoud O, Ben Rekaya M, Cherif W, et al. (2010a). Genetic homogeneity of mutational spectrum of group-A xeroderma pigmentosum in Tunisian patients. *Int J Dermatol* 49(5), 544–548.

Messaoud O, Ben Rekaya M, Kefi R, et al. (2010b). Identification of a primarily neurological phenotypic expression of xeroderma pigmentosum complementation group A in a Tunisian family. *Br J Dermatol* 162(4), 883–886.

Mkaouar M. (1998). Les ichtyoses héréditaires: étude épidémiologique et clinique à propos de 86 cas. Thèse de médecine, Sfax, 1998.

Mokni M, Charfeddine C, Ben Mously R, et al. (2004). Heterozygous manifestations in female carriers of Mal de Meleda. *Clin Genet* 65(3), 244–246.

Muèhl A, Moeslinger D, Item BC, et al. (2001). Molecular characterisation of 34 patients with biotinidase deficiency ascertained by newborn screening and family investigation. *Eur J Hum Genet* 9, 237–243.

Nakano A, Lestringant GG, Paperna T, et al. (2002). Junctional epidermolysis bullosa in the Middle East: clinical and genetic studies in a series of consanguineous families. *J Am Acad Dermatol* 46(4), 510–516.

Nishigori C, Zghal M, Yagi T, Imamura S, Kamoun MR, Takebe H. (1993). High prevalence of the point mutation in exon 6 of the xeroderma pigmentosum group A-complementing (XPAC) gene in xeroderma pigmentosum group A patients in Tunisia. *Am J Hum Genet* 53(5), 1001–1006.

Norrgard KJ, Pomponio RJ, Swango KL, et al. (1997). Mutation (Q456H) is the most common cause of profound biotinidase deficiency in children ascertained by newborn screening in the United States. *Biochem Mol Med* 61, 22–27.

Norrgard KJ, Pomponio RJ, Swango KL, et al. (1998). Double mutation (A171T and D444H) is a common cause of profound biotinidase deficiency in children ascertained by newborn screening in the United States. *Hum Mutation* 11, 410–414.

Nusier M, Zhang Y, Yassin O, Hart TC, Hart PS. (2002). Demonstration of altered splicing with the IVS3-1G -> a mutation of cathepsin C. *Mol Genet Metab* 75(3), 280–283.

Ouragini H, Cherif F, Brick SA, et al. (2010). Mutational survey of recessive dystrophic epidermolysis bullosa in Tunisian families unveils a spectrum of private, ethnic specific and world wide recurrent mutations. *J Dermatol Sci.* 57(2), 144–146.

Ouragini H, Cherif F, Kassar S, et al. (2009). Dystrophic epidermolysis bullosa phenotypes in a large consanguineous Tunisian family. *J Dermatol Sci* 54(2), 114–120.

Pigg M, Annton-Lamprecht I, Braun-Quentin C, et al. (1999). Further evidence of genetic homogeneity in Sjögren-Larsson syndrome. *Acta Derm Venereol* 79(1), 41–43. Pomponio RJ, Coskun T, Demirkol M, et al. (2000a). Novel mutations cause biotinidase deficiency in Turkish children. *J Inherit Metab Dis* 23(2), 120–128.

Pomponio RJ, Hymes J, Reynolds TR, et al. (1997). Mutations in the human biotinidase gene that cause profound biotinidase deficiency in symptomatic children: molecular, biochemical, and clinical analysis. *Pediatr Res* 42(6), 840–848.

Pomponio RJ, Ozand PT, Al Essa M, Wolf B. (2000b). Novel mutations in children with profound biotinidase deficiency from Saudi Arabia. *J Inherit Metab Dis* 23(2), 185–187.

Pomponio RJ, Reynolds TR, Cole H, et al. (1995). Mutational hotspot in the human biotinidase gene causes profound biotinidase deficiency. *Nat Genet*11, 96–98.

Rafai MA, Boulaajaj FZ, Seito A, et al. (2008). Sjögren-Larsson syndrome: a novel mutation in a Moroccan child. *Arch Pediatr* 15(11), 1648–1651.

Rizzo WB, Craft DA, Somer T, Carney G, Trafrova J, Simon M. (2008). Abnormal fatty alcohol metabolism in cultured keratinocytes from patients with Sjögren-Larsson syndrome. J Lipid Res 49, 410–419.

Romdhane L, Abdelhak S; Research Unit on Molecular Investigation of Genetic Orphan Diseases; Collaborators. (2011). Genetic diseases in the Tunisian population. *Am J Med Genet A* 155A(1), 238–267.

Shawky RM, Sayed NS, Elhawary NA. (2004). Mutations in transglutaminase 1 gene in autosomal recessive congenital ichthyosis in Egyptian families. *Dis Markers* 20(6), 325–332.

Smith FJ, Irvine AD, Terron-Kwiatkowski A, et al. (2006). Loss-of-function mutations in the gene encoding filaggrin cause ichthyosis vulgaris. *Nat Genet* 38, 337–342.

Snoeckx RL, Hassan DM, Kamal NM, Van Den Bogaert K, Van Camp G. (2005). Mutation analysis of the GJB2 (connexin 26) gene in Egypt. *Hum Mutat* 26(1), 60–61.

Souissi A, Zeglaoui F, Zouari B, Kamoun MR. (2007). A study of skin disease in Tunis. An analysis of 28.244 dermatological outpatient cases. *Acta Dermatoven* 16(3), 111–116.

Stuhrmann M, Bukhari IA, El-Harith el-HA. (2004). Naxos disease in an Arab family is not caused by the Pk2157del2 mutation. Evidence for exclusion of the plakoglobin gene. *Saudi Med J* 25(10), 1449–1452.

Swango KL, Demirkol M, Hüner G, et al. (1998). Partial biotinidase deficiency is usually due to the D444H mutation in the biotinidase gene. *Hum Genet* 102, 571–575.

Teebi AS. (1994). Autosomal recessive disorders among Arabs: an overview from Kuwait. *J Med Genet* 31(3), 224–233.

Teebi AS, Teebi SA. (2005). Genetic diversity among the Arabs. *Community Gene* 8(1), 21–26.

Thomas AC, Cullup T, Norgett EE, et al. (2006). ABCA12 is the major harlequin ichthyosis gene. *J Invest Dermatol* 126(11), 2408–2413.

Uitto J, Pulkkinen L. (2002). Heritable diseases affecting the elastic tissues: cutis laxa, pseudoxanthoma elasticum and related disorders. In: Rimoin DL, Connor JM, Pyeritz RE, Korf BR (eds) Emery & Rimoin's Principles and Practice of Medical Genetics, 4th ed. Churchill Livingstone, NewYork.

Venencie PY, Pauwels C, Rekik A. (1993). Ichtyose avec accumulation de lipides neutres Syndrome de Dorfman Chanarin: à propos d'une observation familiale. *Ann Dermatol Venereol* 120, 758–760.

Wolf B. (1991). Worldwide survey of neonatal screening for biotinidase deficiency. *J Inherit Metab Dis* 14(6), 923–927.

Wolf B, Jensen KP, Barshop B, et al. (2005). Biotinidase deficiency: novel mutations and their biochemical and clinical correlates. *Hum Mutat* 25, 413.

Wolf B, Jensen K, Hüner G, et al. (2002) Seventeen novel mutations that cause profound biotinidase deficiency. Mol Genet Metab. 2002 77, 108–111.

Wood RD. (1996). DNA repair in eukaryotes. *Annu Rev Biochem* 65, 135–167.

Zghal M, Fazaa B, Kamoun M-R. (2006). Xeroderma pigmentosum. EMC (Elsevier SAS, Paris), *Dermatologie* 98-660-A-10.

Zhang MC, He L, Giro M, Yong SL, Tiller GE, Davidson JM. (1999). Cutis laxa arising from frameshift mutations in exon 30 of the elastin gene (ELN). *J Biol Chem* 274(2), 981–986.

58.

HEREDITARY DISEASES OF THE NERVOUS SYSTEM IN ARABS

Nisrine Bissar-Tadmouri, Mohamad Al Homssi, and Pratiba Nair

INTRODUCTION

Neuroscience in Arab populations dates back to the ninth century, when many Muslim physicians made major contributions to the body of knowledge about neuroscience and developed major philosophical concepts of human understanding (Martín-Araguz et al., 2002). In his book, *Sustenance for Body and Soul*, Ahmed ibn Sahl al-Balki (850–934 CE) was the first to discuss psychosomatic diseases with an emphasis on mind and body (Deuraseh and Abu Talib, 2005). Al-Balki and Muhamed ibn Zakariya al-Razi were the first known physicians to describe psychotherapy (Syed, 2002). Ibn al-Haytham was the first scientist to argue that vision occurs in the brain rather than in the eyes, in his *Book of Optics* (Steffens, 2006). Ali ibn Abbas al-Majusi (982 CE) was the first person to discuss in detail such mental disorders as sleeping sickness, memory loss, coma, meningitis, vertigo, epilepsy, and hemiplegia (Haque, 2004). Avicenna was a pioneer of neuropsychiatry, as he first described numerous neuropsychiatric conditions including insomnia, mania, hallucinations, nightmare, dementia, epilepsy, stroke, paralysis, vertigo, melancholia, and tremors (Abbasi et al., 2007). Later on, in the thirteenth century, Maimonides described rabies and belladonna intoxication (Martín-Araguz et al., 2002). Avicenna also discovered the cerebellar vermis and the caudate nucleus (Aydin, 2001). Ibn Rushd suggested the existence of Parkinson's disease and attributed photoreceptor properties to the retina. Ibn Zuhr described meningitis, intracranial thrombophlebitis, mediastinal tumors, and made contributions to modern neuropharmacology (Martín-Araguz et al., 2002). Neurosurgery also flourished in that period, as Abulcasis (also known as al-Zahrawi) developed material and technical designs that are still used in neurosurgery today. During his era, neurosurgery in the Islamic world became a respected specialty practiced by reputable physicians (Mohamed, 2008). Later on, an Egyptian physician, Ibn al-Nafis, performed the earliest-known dissections on the human brain. Moreover, he corrected some of the incorrect theories of Galen and Avicenna on brain anatomy (Oataya, 1982).

Nowadays, studies suggest that at least 250 million people are affected by neurological disorders in the developing countries (Owolabi et al., 2007). The burden of neurological disease is likely to increase in the Arab world with the rise of the aging population (Benamer and Shakir, 2009). However, the available information on the pattern and frequency of neurological diseases in the Arab world is scanty. Data from community-based epidemiological studies are lacking, which makes it difficult to have an accurate idea about the burden of neurological illnesses in the Arab countries, a fact which hinders planning, organization, and delivery of neurology services. The paucity of reports on neurological disorders may not reflect a low frequency among Arabs but can be due to many reasons, among which is the limited number of neurologists practicing in the Arab world. It was reported by the WHO and World Federation of Neurology that the majority of Arab countries have less than one neurologist and one neurosurgeon per 100,000 population with almost no existence of neurological nurses and neuropediatricians (WHO, 2004). The scarcity of neurological investigations, including computerized tomography, magnetic resonance imaging, electroencephalography, and electromyography in many Arabic countries is also a problem (Bower and Zenebe, 2005). In addition, the paucity of formal neurological postgraduate specialist training programs in the Arab world is another problem.
Despite of these, inherited diseases of the nervous system in the Arab world have attracted major interest and occupied a considerable proportion of the literature in this field. This is mainly due to the structure of Arabic families that enhance the molecular studies of genetic diseases.

The Arab population is characterized by large family size, high maternal and paternal age, and a high level of inbreeding with consanguinity rates in the range of 25%–60%. Consequently, autosomal recessive disorders are widely spread. Data from the CTGA database on genetic disorders in Arab populations indicate that, in contrast to international databases, the overwhelming proportion of these disorders follow a recessive mode of inheritance (63%) compared to the smaller proportion of dominantly inherited traits (27%; Tadmouri et al., 2009). In fact, a considerable number of genes causing autosomal recessive conditions have been structurally and functionally determined at the molecular level through the joint collaboration of international and Arab scientists (Tadmouri et al., 2009). This chapter gives an overview of the inherited diseases of the nervous system in Arab populations, and the contribution of Arab families to the body of knowledge related to gene identification and characterization of inherited neurological diseases, especially the autosomal recessive forms (Box 58-1).

BOX 58-1

- Neuroscience in Arab populations dates back to the ninth century.
- Muslim physicians made major contributions to knowledge about neuroscience.
- Information on the pattern and frequency of neurological diseases in the Arab world is scanty.
- The Arab population is characterized by large family size, high maternal and paternal age, and high consanguinity rates.
- Autosomal recessive disorders are widely spread in Arabs.
- Arab families contributed to identification and characterization of inherited neurological disorders.

DEGENERATIVE DISEASES OF THE NERVOUS SYSTEM

DEMENTIA

Dementia presents clinically as progressive memory loss and cognitive and behavioral problems, and results from progressive destruction of neurons. It has several subtypes: dementia of Alzheimer type (DAT); vascular dementia (VaD); mixed (DAT & VaD); Parkinson's disease with dementia (PDD); dementia due to other medical reasons, or secondary dementia (SD; Hamad et al., 2004), and early-onset dementia (EOD; Chouery et al., 2008). Table 58-1 shows the distribution of these subtypes in various Arab countries.

No accurate prevalence studies in Arabs are available in the literature. The hospital frequency of dementia in a university hospital in Saudi Arabia is 19.3 per 100,000 patients (Ogunniyi et al., 1998). In the United Arab Emirates, the rate of institutionalization is 7–14 per 1000 people aged 65 or older, the female:male ratio is 1.6, and 89% have dementia. No further stratification of dementia is known in this country (Margolis and Reed, 2001). In Qatar, DAT is more prevalent than the other forms of dementia (Hamad et al., 2004). In Oman, 53% of 116 patients with dementia have DAT, 24% have VaD, and 10% have frontotemporal lobar degeneration, while early-onset dementia (EOD) constitutes 45% and potentially reversible dementia 9% (Shelley and Al Khabouri, 2007). Both dementia and depression are found to be present in more than half of the nursing home residents in Lebanon (Chahine et al., 2007). In neighboring Wadi Ara in Palestine, the prevalence of DAT is 20% of 821 population aged 60 or older (Bowirrat et al., 2001). In this population, the annual incidence rate of DAT in age-related cognitive decline (ARCD) is 4%. Studies indicated the presence of an association between VaD, illiteracy, and hypertension. Homocysteine levels are significantly higher than found in other studies. Old age, female gender, and illiteracy are risk factors for development of DAT. However, ApoE ε4 allele is found to be uncommon in this population with high DAT prevalence (Bowirrat et al., 2002). In Assiut (Egypt), the prevalence ratio is low for all forms of dementia (Farrag et al., 1998).

TABLE 58-1 DEMENTIA IN ARABS

Country	DAT	VaD	DAT & VaD	PDD	SD	EOD	All	References
Saudi Arabia	51.9%	18.2%	15.6%	7.8%	5.2%		19.3[1]	Ogunniyi et al., 1998
UAE							89%[2]	Margolis and Reed, 2001
Qatar	29%	22%	15%	5%	5%			Hamad et al., 2004
Oman	52.6%	24.1%	9.5%[3]		8.6%	45%		Shelley & Al Khabouri, 2007
Lebanon							>50%[4]	Chahine et al., 2007
Palestine	20.5%	22%						Bowirrat et al., 2001
Egypt	2.2%[3]	0.95%[3]	0.55%[3]			0.45%[3]		Farrag et al., 1998

Abbreviations: Dementia of Alzheimer type (DAT); vascular dementia (VaD); mixed (DAT & VaD), Parkinson's disease with dementia (PDD); dementia due to other medical reasons or secondary dementia (SD); early-onset dementia (EOD); United Arab Emirates (UAE)

[1]Of the institutionalized people aged 65 or older

[2]Frontotemporal lobar degeneration

[3]Prevalence ratio

[4]of nursing home residents

Despite the limited epidemiological studies on dementia in the Arab world, there are a few interesting studies dealing with the molecular basis of this disease. Tunisian patients with Alzheimer's disease (AD) have decreased amyloid β (1-42) levels and increased t-tau. This is thought to be useful in discriminating cases at risk for AD, such as ApoE ε4 allele carriers (Smach et al., 2008). Autosomal dominant early-onset Alzheimer's disease was studied in a Moroccan family that had six affected members. Three of them died at 34 and 40 years of age because of severe dementia. The disease resulted from a *presenilin 1* gene mutation as identified in one of the sisters living in France (Asri et al., 2003). On the other hand, a novel mutation leading to pure early-onset dementia (EOD) without bone cysts in *TERM2* (triggering receptors expressed on myeloid cells-2) gene at chromosome 6p21.2 locus was reported in three members of a Lebanese family (Chouery et al., 2008).

BOX 58-2

- Dementia has several subtypes: dementia of Alzheimer type, vascular dementia, mixed dementia, Parkinson's disease with dementia, secondary dementia, and early-onset dementia.
- All types of dementia have been reported in Arabs.
- A *presenelin 1* gene mutation was identified in Tunisia, while a *TERM2* gene was identified in a Lebanese family.

PARKINSON'S DISEASE

Parkinson's disease (PD) is an age-associated extrapyramidal neurodegenerative movement disorder caused by environmental and genetic factors. Its main pathological features are loss of dopaminergic neurons and presence of intraneuronal Lewy body inclusions. The vast majority of PD cases are sporadic, and less than 10% are familial cases (Mizuno et al., 2004). PD occurs at a prevalence of 27 per 100,000 in Saudi Arabia (Al-Rajeh et al., 1993), 31 per 100,000 persons in Libya (Ashok et al., 1986), and 43 per 100,000 persons in Tunisia (Attia et al., 1993). Its clinical features in Arabs are similar to those reported in other populations (Al-Bunyan, 2000).

Most studies on PD in the Arab states relate to the role of genes in the etiology of the disease, mainly in North African Arabs (Benamer et al., 2008). Gouider-Khouja and colleagues (2000) studied the inheritance patterns of familial PD in 21 Tunisian families. At least two inheritance patterns were identified (autosomal recessive and autosomal dominant with reduced penetrance). Linkage of autosomal recessive juvenile onset Parkinsonism to the PARK2 type locus on chromosome 6 was demonstrated in one Algerian family studied along with ten European families (Tassin et al., 1998). Similarly, linkage to the same locus was reported in 15 families including one from Saudi Arabia (Jones et al., 1998). Gouider-Khouja and colleagues (2003) later reported the clinical, pathologic, and genetic findings in a Tunisian kindred with autosomal recessive juvenile onset Parkinsonism caused by a mutation in the *Parkin* gene associated with PARK2 type.

A study on the autosomal dominant forms of Parkinson's disease evaluated the occurrence of the common G2019S mutation of the leucine-rich repeat kinase 2 (*LRRK2*) gene (PARK 8) in 198 probands from families including 17 families from Algeria, Tunisia, and Morocco, and one Lebanese family. Surprisingly, the frequency of the mutation in North African families was greater than in those from Europe (Lesage et al., 2005a) These results were confirmed by a later study including a larger cohort of North African Arabs (Lesage et al. 2006). Haplotype studies on patients of European origin, Ashkenazi Jews, and North African Arabs who carry this mutation concluded that patients from these populations shared a common Middle Eastern founder (Lesage et al., 2005b). The G2019S *LRRK2* mutation is also described in patients with apparently sporadic PD in North Africans of Berber origin (Hulihan al., 2008). A recent study including 226 autosomal dominant PD families, 182 from France, and 14 from North Africa, analyzed the 51 coding exons of *LRRK2* and identified 12 novel pathogenic variants in addition to three known variants (Lesage et al., 2009).

PTEN-induced putative kinase 1 (*PINK1*) gene, involved in PARK6 type, was reported in Saudi (Chishti et al., 2006), Sudanese (Leutenegger et al., 2006), and Algerian families (Ibanez et al., 2006). A novel mutation, A217D, caused autosomal recessive early-onset Parkinsonism in eight patients from two generations of a Sudanese family (Leutenegger et al., 2006), and a homozygous T313M *PINK1* mutation occurred in one Saudi family with four affected members (Chishti et al., 2006).

Kufor-Rakeb disease was first described in four patients of a consanguineous marriage in a large Jordanian family. Patients had Parkinsonian features associated with pallido-pyramidal degeneration, up-gaze paresis, and dementia. Linkage studies identified a locus on 1p36 nomenclatured as PARK9 (Hampshire et al., 2001). Later homozygous and compound heterozygous mutations of a P-type ATPase gene, *ATP13A2,* at the same locus showed association with this phenotype (Ramirez et al., 2006).

BOX 58-3

- Most studies of Parkinson's disease in the Arab states relate to the role of genes in the etiology of the disease in North Africa.

- The frequency of mutation of *LRRK2* gene in North African families is greater than in those from Europe.
- Patients of European origin, Ashkenazi Jews, and North African Arabs who carry the G2019S mutation share a common Middle Eastern founder.
- Parkinsonism associated with pallido-pyramidal degeneration, up-gaze paresis, and dementia was linked to 1p36 in a Jordanian family.

HEREDITARY ATAXIAS

The hereditary ataxias are a heterogeneous group of genetic disorders characterized by cerebellar symptoms, such as uncoordinated gait often associated with poor coordination of hands, speech, and eye movements, in association with other neurological features. The clinical course ranges from benign and/or nonprogressive to rapidly fatal (progressive neurodegenerative). Different modes of inheritance have been reported (Harding, 1983; Campanella et al., 1992).

Autosomal Dominant Cerebellar Ataxias (ADCAs)

ADCAs are a heterogeneous group of neurodegenerative disorders caused by genes that are numbered SCA1, SCA2, SCA3, and others. Most mutations are caused by abnormal expansion of an unstable polymorphic CAG trinucleotide repeat, encoding glutamine, in the corresponding gene (Crum and Josephs, 2006). SCA2 is the second most common form of ADCAs in ethnically diverse populations after SCA3 (Tang et al., 2000). In Tunisia, a family with ADCA was linked to the SCA2 locus (Belal et al., 1994). Similarly, linkage to SCA2 was reported in an Egyptian family with marked anticipation and integration instability among the generations. In one proband, infantile-onset SCA2 and intermediate CAG repeat expansion were striking (Abdel-Aleem and Zaki, 2008). Moreover, the unique presentation of polyphagia and obesity at the mid-phase of the disease, in this family, highlighted previous reports on the contribution of ataxin 2 protein, encoded by SCA2 gene, in the development of obesity (Kiehl et al., 2006).

Autosomal Recessive Cerebellar Ataxias (ARCAs)

Autosomal recessive cerebellar ataxias (ARCAs) are a heterogeneous group of rare neurological disorders involving both central and peripheral nervous systems and, in some cases, other systems and organs (Harding, 1993). The most prevalent autosomal recessive ataxia is Friedreich's ataxia (FA). Other recessive ataxias include ataxia telangiectasia, ataxia with ocular motor apraxia, ataxia with vitamin E deficiency, and ataxia with CoQ10 deficiency.

As for the majority of hereditary disorders in Arabs, autosomal recessive cerebellar ataxias tend to be more widely spread and more thoroughly studied than autosomal dominant forms. On the other hand, rare forms of ataxias have been reported in the Arab world. Mégarbané and colleagues (2001) reported five Druze Lebanese children of the same kindred with a new autosomal recessive cerebellar ataxia. The disease is characterized by severe mental retardation, short stature, speech defect, microcephaly, optic atrophy, and cerebellar atrophy. The disease locus in this family demonstrated linkage to chromosome 9q34-9qter (Delague et al., 2001). In Saudi Arabia, the disease demonstrated linkage to a new locus on chromosome 16q21-q23 in a consanguineous family with childhood onset, slowly progressive autosomal recessive spinocerebellar ataxia associated with epilepsy, and mental retardation (Gribaa et al., 2007). Another study identified a putative new locus on chromosome 22q11.2 that may be responsible for some cases of autosomal recessive cerebellar ataxia with developmental delay and failure to thrive (Baris et al., 2005).

Additionally, in a large Saudi Arabian family with spinocerebellar ataxia and axonal neuropathy (SCAN1), linkage mapping and positional candidate approach revealed the presence of a mutation in *TDP1* gene. The gene encodes tyrosyl-DNA phosphodiesterase 1 enzyme, which is involved in repairing covalent topoisomerase I-DNA single-strand or double-strand breaks (Takashima et al., 2002).

Friedreich's Ataxia (FA)

Friedreich's ataxia is a progressive autosomal recessive disorder affecting both the central and peripheral nervous systems, with onset occurring usually around puberty and signs characteristic of cerebellar dysfunction and peripheral neuropathy. Disease symptoms include ataxia of gait, muscle weakness, sensory disturbances, loss of incidence of tendon reflexes in the lower limbs, and dysarthria. The incidence of the disease is about 1/50,000 in European populations (Romeo et al., 1983; Skre et al., 1985).

A study of 112 families with established diagnosis of FRDA, including five from Tunisia, fine-mapped the FRDA gene locus in 9q13-q21 (Rodius et al., 1994). The frataxin gene, encoding a mitochondrial iron-chaperone involved in iron-sulfur cluster (ISC), was later identified in this locus (Campuzano et al., 1996). The most common mutation in Friedreich's ataxia is an expanded GAA trinucleotide repeat in intron 1 of the frataxin gene (Campuzano et al., 1996). Another study reported the occurrence of FA in a large inbred Arab family living near Jeddah, Saudi Arabia. The disease gene was localized to 9q13-q21.1, and the affected members showed typical features of FA except for two patients who showed retained tendon reflexes in the lower limbs, 13 and 19 years after onset of symptoms (Scrimgeour et al., 1996).

In Tunisia, a study including four consanguineous families indicated the presence of Friedreich's ataxia GAA repeat expansion in one family with early-onset cerebellar ataxia (EOCA). EOCA is clinically distinguished from FA by the fact that tendon reflexes are retained, and is characterized by less severe upper limb ataxia and less reduction in joint position sense than FA patients. These findings are indicative of considerable clinical heterogeneity in the clinical spectrum of FA in Arab families (Marzouki et al., 2001).

FRDA and ataxia with selective vitamin E deficiency (AVED) share very similar clinical phenotypes. The first report on AVED came from a study on two large inbred Tunisian families. The families did not show any linkage to the 9q13-q21.1 locus, and analysis of the patients revealed low levels of serum vitamin E (Ben Hamida et al., 1993a). Subsequently, homozygosity mapping in these families localized AVED to chromosome 8q (Ben Hamida et al., 1993b). A linkage study on 14 families, seven of which were from North Africa (2 from Morocco and 5 from Tunisia), refined the locus and revealed the frequent occurrence of the disease in this geographical area (Doerflinger et al., 1995). The identification of mutations in the alpha-tocopherol transport protein gene (*α-TTP*) confirmed that vitamin E deficiency is the cause of the neurological symptoms in AVED (Ouahchi et al., 1995). Vitamin E supplementation was found to improve symptoms and prevent disease progress (Gordon, 2001).

A study on 15 Moroccan families presented a clinical comparison between 16 patients with GAA expansion in the frataxin gene and 13 patients with the 744 del (A) mutation in the α-TTP gene. In FA, the intrafamilial variability in age at onset, severity, and associated symptoms was higher than expected for an autosomal recessive disease, but this is due in part to the variability of the GAA expansion. On the other hand, AVED resulting from 744 del (A) mutation was characterized by head titubation, less frequent neuropathy, and slower disease progression, as well as decreased visual acuity and retinitis pigmentosa (RP; Benomar et al., 2002).

In an attempt to smooth the progress of genetic counseling for Moroccan families with AVED, Bellayou and colleagues (2009a) made available to neurologists molecular analysis of the 744 del (A) mutation, the most widespread, and possibly a founder, mutation in Morocco. They were able to discover five new cases with homozygous mutation resulting from a consanguineous marriage. Consequently, these cases received preventative treatment by vitamin E supplementation that would prevent or report the symptoms of ataxia. This was a good example of a disease in which genetic counseling really plays its role in prevention of the disease, since the molecular analysis is very simple on a predominant mutation and there is a preventative treatment.

Ataxia Telangiectasia

Ataxia telangiectasia (AT) is an autosomal recessive disorder characterized by cerebellar ataxias and progressive neuromotor degeneration, immune deficiency, and the appearance of dilated blood vessels in the eyes and face (telangiectasias). Patients also manifest growth retardation, premature aging of skin and hair, chromosomal instability, lymphoreticular malignancies, and acute sensitivity to ionizing radiation (Gatti et al., 1999). The protein coding region of the AT gene (*ATM*) contains 66 coding exons spread over 146 Kb of genomic DNA (Savitsky et al., 1995). The ATM protein is a member of an expanding family of large phosphatidylinositol-3 kinase-related proteins that are involved in eukaryotic cell cycle control, DNA repair, and recombination (Zakian, 1995). An extremely wide spectrum of mutations in the *ATM* gene has been observed in various ethnic groups, the majority of which cause premature protein truncation (Gilad et al., 1996a). The large size and genomic structure of the AT gene complicates the process of screening for all possible sequence variations. Therefore, the identification of founder effects in specific communities would provide such an assay for that group. In line with this, Gilad and colleagues (1996b) reported a founder effect mutation in 17 Jewish AT families originating from North Africa, mainly Morocco. Similarly, Fares and colleagues (2004) reported two AT founder mutations in four large Druze clans originating from Lebanon, Syria, and Jordan.

On the other hand, preimplantation genetic diagnosis (PGD) was performed for a Saudi family with three AT-affected children. The family was screened to detect the mutation involved, a single cycle of PGD was performed, and an ongoing pregnancy was obtained (Hellani et al., 2002).

A 5-year retrospective study on 24 AT patients from southern Tunisia reported some typical AT clinical features, chromosomal instability, balanced chromosomal rearrangements, and development of neoplasia (lymphoma and leukemia) in two patients of the same family, confirming the role of ATM protein in tumor suppression (Triki et al., 2000).

BOX 58-4

- SCA2 resulting in autosomal dominant cerebellar ataxias (ADCAs) with unstable polymorphic CAG trinucleotide repeat was reported in Tunisia and Egypt.
- Autosomal recessive ataxias include Friedreich's ataxia, ataxia telangiectasia, ataxia with ocular motor apraxia, ataxia with vitamin E deficiency, and ataxia with CoQ10 deficiency.
- Rare forms of ataxias have been reported in the Arab world.
- *TDP1* gene was identified in a large Saudi Arabian family with spinocerebellar ataxia and axonal neuropathy (SCAN1).

- Friedreich's ataxia was localized to 9q13-q21 in Tunisian families.
- The most common mutation in Friedreich's ataxia is an expanded GAA trinucleotide repeat in intron 1 of the frataxin gene.
- Ataxia with selective vitamin E deficiency (AVED) was localized to chromosome 8q in Tunisian families.
- Genetic counseling played its role in prevention of the AVED following molecular analysis of the most common mutation in Morocco.
- A founder mutation causes ataxia telangiectasia in Jewish families originating from North Africa.
- A preimplantation genetic diagnosis (PGD) was performed for a Saudi family with three AT-affected children.

DEGENERATIVE DISEASES OF MOTOR NEURONS

Amyotrophic Lateral Sclerosis

Amyotrophic lateral sclerosis (ALS), also known as Lou Gehrig's disorder, after an affected American baseball player, is an adult-onset neurodegenerative disorder affecting the motor neurons. The condition progresses rapidly, from muscle weakness to disability and ultimately to death, sometimes within a span of 5 years after symptoms first begin to appear. Most studies on ALS seem to indicate the involvement of a complex genetic–environmental interaction in the pathogenesis of the disorder (Shaw, 2005; Siddique and Siddique, 2008). It is known, for instance, that about 10% of all cases of ALS are familial, being transmitted mostly in an autosomal dominant and sometimes recessive fashion. More than 12 independent genes or genetic loci have been implicated in these familial forms of ALS. The most well known of these genes is the copper zinc superoxide dismutase (*SOD1*) gene, mutations in which are responsible for up to 20% cases of the dominant form of ALS. Other genes involved include the *Alsin* gene on 2q33, *Senataxin* gene on 9q34, *VAPB* on 20q13.3, and *TDP43* on chromosome 1 (Wijesekera and Leigh, 2009).

In the Western countries, ALS shows an average point prevalence of 5.2 per 100,000 (Worms, 2001). Within the Arab world, ALS has been reported mainly from North African countries. A study from Libya reported the presence of 17 ALS patients with motor neuron disease (Radhakrishnan et al., 1986). A similar study from Sudan uncovered 19 cases of ALS (Abdulla et al., 1997).

Interestingly, most Arab patients affected with ALS have presented with nonclassical forms of the condition. Multiple members in a large Tunisian kindred were identified with a juvenile variant of ALS (OMIM: 205100), characterized by a very early age of onset, a slow progression of disease, and an autosomal recessive mode of inheritance (Ben Hamida et al., 1990a; Hentati et al, 1992). Linkage studies performed on this kindred enabled the mapping of this form of ALS, ALS2, to a 1.7 cM region on chromosome 2q33-35 (Hentati et al., 1994a; Hosler et al., 1998). Later, homozygous mutations in the ALS2 gene on this locus were identified in these Tunisian patients (Hadano et al., 2001; Yang et al., 2001). The alsin protein, product of the ALS2 gene, is expected to function as an activator of small GTPases that play major roles in cytoskeletal organization, membrane trafficking, and intracellular transport. In four separate Tunisian families, Hentati and colleagues (1998) detected the presence of a rare autosomal recessive form of ALS that mapped to a novel locus, ALS5, on chromosome 15q15-q22. This variant also had an early age of onset, but was accompanied by a rapid progression of the disease course. With four families affected, this form is likely to be the most common form of ALS in Tunisia.

Phenotypically variant disease forms were also discovered in affected Arab patients. A study on the course of disease progression among patients with ALS in the Occupied Territories demonstrated that affected Jews of North African origin had a significantly younger age, indicating a shorter survival rate. A plausible explanation emphasizes a genetic basis for this difference in survival. Interestingly, non-Jewish Arab patients in the same study were found to have the lowest mean age of onset, but had a relatively longer survival (Drory and Artmonov, 2007).

Spinal Muscular Atrophy

Spinal muscular atrophy (SMA) is another form of motor neuron disease characterized by degeneration of either the neuronal axons or the entire motor neurons themselves, resulting in weakness and atrophy of the voluntary muscles, especially of the limbs. Most cases of SMA are due to mutations in the survival motor neuron (*SMN*) gene. This form of the condition ranges in severity from the highly severe SMA type I (Werdnig Hoffman's disease) with infantile onset, type II SMA of moderate severity, juvenile onset type III SMA (Wolhlfart-Kugelberg-Welander disease), with a less severe manifestation, and the adult-onset type IV SMA, which is the least severe of all. Apart from mutations in the SMN gene, patients with both SMA type I and II have been found to carry mutations in the neuronal apoptosis inhibitory protein (*NAIP*) gene as well.

In Saudi Arabia, where a fairly good amount of research on SMA has been performed, a very high prevalence of the condition was discovered. In a study of children with genetic neurodegenerative disorders in the country, SMA was found to be one of the most encountered diagnoses (Mahdi, 1991). Another study, based on hospital incidence rates, estimated the prevalence rate of type I SMA to be 0.92 per 10,000 of the Saudi population; a prevalence level higher than that reported in Western populations (Al-Rajeh et al., 1992). This high prevalence in Saudi Arabia was also reflected in a carrier screening study where the population showed a carrier frequency of 2 in 50, higher that that seen in other epidemiological studies (Majumdar et al. in 2005). Similarly, in the first of its kind large-scale carrier screening program for SMA in an isolated Palestinian Arab community, 13% of women tested were found to be carriers of SMA (Basel-Vanagaite et al., 2008). These high prevalence levels in Arab populations have been thought to be due to the high level of consanguinity in these populations.

In Oman, Aithala and colleagues (1995) described twins with SMA in association with arthrogryposis multiplex congenita. They suggested that such patients may actually constitute a separate subgroup of SMA (Aithala et al., 1995).

Molecular studies on SMA among Arabs have also led to some distinctive findings. In multiple studies, dual deletion of both *SMN1* and *NAIP* were found to be more common among Arab patients with SMA than in other affected populations (Samilchuk et al., 1996; Al-Rajeh et al., 1998; Haider et al., 2001; Al-Jumah et al., 2003; Mrad et al., 2006; Essawi et al., 2007). In most of these studies, *NAIP* gene deletion was found to correlate with a more severe form of the disease. On the other hand, a study on a Moroccan patient population deduced that *NAIP* gene deletions are unimportant, since patients with dual mutations showed a mild phenotype (Bouhouche et al., 2003). In the same study, a hybrid *SMN* gene with exon 7 of *SMN1* and exon 8 of *SMN2* occurred in eight SMA patients, suggesting a founder mutation in this population. In 33 Egyptian patients, Shawky and colleagues (2001) detected homozygous deletions in *SMN1* in only 55% of the patients, prompting the need to look for other mutations in this population.

Hereditary Spastic Paraplegias

Hereditary spastic paraplegias (HSP), also known as Strumpell-Lorrain syndrome, are a group of heterogeneous disorders characterized by progressive axonal degeneration and resulting in progressive spasticity of the lower limbs. HSPs are classified clinically as complex or simple, based on the involvement, or lack thereof, of other signs such as ataxia, dementia, seizures, and neuropathy. Simultaneously, genetically the condition is classified on the basis of its mode of inheritance. To date, more than 25 different genetic loci have been implicated in the pathogenesis of these HSPs.

One of the earliest reports of HSP among Arabs originates from Libya, where an intensive search over a two-year period for spinocerebellar degeneration identified 24 patients from 10 families with HSP (Sridharan et al., 1985). Yet, this study could not identify any phenotypically native forms of the disease. In neighboring Tunisia, a study of 43 patients with hereditary motor disease showed 11 with familial spastic paraplegia and peroneal muscle atrophy (Ben Hamida et al., 1990a). More recently, however, a study on nine Lebanese patients identified variable and nonspecific MRI findings, including corpus callosum atrophy, white matter lesions, and abnormal T2 signal intensity in the posterior limb of the internal capsule (Hourani et al., 2009).

Most of the familial spastic paraplegias noticed in Arab populations show an autosomal recessive mode of inheritance, possibly due to the high consanguinity levels in these communities. A number of genetic linkage studies have been conducted on Arab patients with HSPs, and several of these studies have led to the identification of novel SPG loci. In Tunisia, linkage studies in four families with HSP mapped the disease to chromosome 8 (Hentati et al., 1994b). A similar study in 10 Algerian families showed that only one family linked to chromosome 8, indicating genotypic heterogeneity in this population (Coutinho et al., 1999). New loci identified first in Arab patients include SPG24 in Saudi Arabia (Hodgkinson et al., 2002), SPG23 in Palestinian Arabs (Blumen et al., 2003), SPG28 in Moroccans (Bouslam et al., 2005), SPG30 in Algerians (Klebe et al., 2006), and SPG18 and SPG35 in Omanis (Al-Yahyaee et al., 2006; Dick et al., 2008).

HSP associated with thin corpus callosum (TCC) is considered a rare autosomal recessive form of complicated HSP. Lossos and colleagues (2006) were the first to report this association among Arabs in three affected consanguineous families of which two mapped to the SPG11 locus. In a later study, the remaining family was localized to the SPG15 locus (Elleuch et al., 2007). In another study on eight Tunisian families with 19 patients affected with HSP-TCC, five families mapped to SPG11, while two of them mapped to SPG15. The two categories of patients show very little difference at the clinical level. Within the SPG11 patients, this study also identified two recurrent mutations, indicating a founder effect (Boukhris et al., 2008). One of these mutations had earlier been reported in a Moroccan family (Stevanin et al., 2007).

BOX 58-5

- Within the Arab world, ALS has been reported mainly in North Africans and presents with nonclassical forms.
- ALS5 is likely to be the most common form of ALS in Tunisia.
- North African Jews with ALS have a shorter survival rate than non-Jewish Arabs.
- A high prevalence of SMA is reported in Saudi Arabia and in the Palestinian Arab community due to the high level of consanguinity.

- SMA in association with arthrogryposis multiplex congenita is reported in Oman.
- Dual deletion of both *SMN1* and *NAIP* are more common among Arab patients with SMA than in other affected populations.
- A hybrid *SMN* gene occurred in Moroccan SMA patients.
- HSP among Arabs was reported in Libya, Tunisia, and Lebanon, mostly in autosomal recessive forms.
- New HSP loci identified first in Arabs include SPG24 in Saudi Arabia, SPG23 in Palestinian Arabs, SPG28 in Moroccans, SPG30 in Algerians, and SPG18 and SPG35 in Omanis.
- The rare HSP associated with thin corpus callosum (TCC) maps in Arabs to SPG11 and SPG15.

EPISODIC AND PAROXYSMAL DISORDERS

EPILEPSY

Epilepsy refers to a range of neurological disorders characterized by the presence of recurring seizures due to abnormal neuronal activity. Reported prevalence rates of epilepsy vary among Arab countries. Benamer and Grosset (2009) have provided a detailed review of the epidemiology of epilepsy in the Arabian region. In Libya, an epidemiological study of epileptic patients over 15 years of age revealed a prevalence of 2.3 per 1000 (Sridharan et al., 1986). More than half of these were of the primary generalized form of epilepsy. In Saudi Arabia, the hospital frequency rate of epilepsy is 8 per 1000 (Al-Rajeh et al., 1990). Interestingly, 23% of these patients have a positive family history, while 53% documented parental consanguinity. Later, a door-to-door survey of Saudi nationals set the prevalence rate of epilepsy in the Saudi population to 6.54 per 1000 (Al-Rajeh et al., 2001). The clinical features of epileptic patients in Saudi Arabia are similar to those reported from Western countries. Most of these patients have ages of onset below one year, and show generalized epileptiform activity with normal cranial CT findings (Al-Sulaiman and Ismail, 1999). Generalized seizures are also the most common type among Tunisian patients with infantile status epilepticus (Tabarki et al., 2001). In this group, the mortality rate stands at about 16%. Recently, a comparative study of the clinical characteristics of Jewish and Palestinian Bedouin children with febrile seizures revealed that although all other characteristics were same, the source of fever differed between the two groups. Among the Bedouin population, pneumonia was the most common cause of fever, while upper respiratory tract infections were common among the Jews (Shimony et al., 2009).

Several studies have also been performed on elucidating the genetic basis of epilepsy in Arab patients. Mendelian inheritance, as well as chromosomal aberrations, exists among 32 Egyptian patients with symptomatic generalized epilepsy (Hashishe, 1997). On the other hand, patients with idiopathic generalized epilepsy tend to show a non-Mendelian inheritance. Molecular genetic studies in Arab epileptic patients have turned up interesting results. Kuwaiti patients with epilepsy are rarely found to have any of the mutations in the *CHRNA4* and *SCN1B* genes that are frequently observed in Caucasian populations (Haider et al., 2005). Similarly, in two unrelated multigenerational Tunisian families with generalized epilepsy with febrile seizure plus (GEFS+), none of the known mutations could be detected in the *SCN1B*, *SCN1A*, or *GABRG2* genes. Instead, one study detected a new mutation in the *SCN1A* gene (Mrabet et al., 2007). On the other hand, Fendri-Kriaa and colleagues (2009) found linkage to *GABRG2* and *SCN1A* genes in 2 different Tunisian families with GEFS+, in addition to identifying a known mutation in the *SCN1A* gene in another family. It is clear, therefore, that considerable genotypic variability exists among Arab epileptic patients.

Specific subtypes of epilepsy have also been genetically studied among Arab populations, Lafora disease and Unverricht-Lundborg disease (ULD) being the most studied forms. The same novel homozygous deletion in *EPM2A* gene was detected in three Arab families affected with Lafora disease. Interestingly, in spite of this genotypic similarity, the age of onset of the condition varied, even within the same family (Gomez-Abad et al., 2007). However, in an Omani tribal isolate affected with Lafora epilepsy, a single mutation in the *EPM2B* gene (c.468-469delAG) results in uniform age of onset and death (Turnbull et al., 2008). At the same time, Arab families affected with ULD from the Maghreb region, Palestine, and Oman all map to the locus, 21q22.3, where the *EPM1* gene is located (Gouider et al., 1998; Mazarib et al., 2001; Santoshkumar et al., 2008). Dodecamer repeat expansions in the gene occur in both Palestinian and Omani families. However, in another Palestinian family, ULD maps to a new locus, designated as EPM1B, on the pericentromic region of chromosome 12 (Berkovic et al., 2005). Another Jordanian family with patients presenting with ataxia, juvenile-onset progressive action tremors, and atonic seizures also maps to the same region of chromosome 12. Despite the clinical differences between both of these families, it is speculated that they probably represent a single genetic entity (El-Shanti et al., 2006).

From the abundant studies done on the social consequences of epilepsy in Arabs, the lack of awareness about the disease as well as its social stigma is clear (Bener et al., 1998; Al-Adawi et al., 2003; Baker et al., 2005). On the positive side, surgery for epilepsy was found in a Lebanese population to significantly improve the quality of life (Mikati et al., 2004).

BOX 58-6

- Prevalence rates of epilepsy vary among Arab countries including Libya, Saudi Arabia, and Tunisia.
- Mendelian inheritance, as well as chromosomal aberrations, exists among Egyptian patients with symptomatic generalized epilepsy, while idiopathic generalized epilepsy shows a non-Mendelian inheritance.
- Kuwaiti patients with epilepsy are rarely found to have mutations in the genes that are frequently observed in Caucasian populations.
- The same novel homozygous deletion in *EPM2A* gene, with variable age of onset, was detected Arab families affected with Lafora disease.
- Unverricht-Lundborg disease from the Maghreb region, Palestine, and Oman maps to the *EPM1* gene locus.
- The lack of awareness about epilepsy as well as its social stigma is clear in the Arab world.

CEREBRAL PALSY

Cerebral palsy (CP) refers to a range of nonprogressive neurological disorders that have their origin in embryonic life, infancy, or early childhood. The main manifestation of the condition is in the form of abnormalities in body movement and muscle coordination, brought about by defects within the brain itself.

Most epidemiological studies of CP in the Arab world come from Saudi Arabia. A 10-year retrospective study for the years (1984–2003) recorded an incidence of CP as high as 0.41% of all live births in a single hospital in Riyadh (Al-Asmari et al., 2006). However, the same study showed a decreasing trend of CP in the country over the 10 years. In earlier studies on Saudi patients with CP, perinatal causes were found to be the most common cause of pathogenesis of the condition (Taha and Mahdi, 1984), while the most common specific etiological factors were found to be toxemia, followed by prematurity (El-Rifai et al., 1984a and b). Family history and consanguinity, along with other antenatal factors, seem to be a significant risk factor for the development of CP (Al-Rajeh et al., 1991).

Apart from spasticity, epilepsy is one of the most common complications of CP. An electroencephalographic (EEG) study of children with CP in Saudi Arabia showed that 76% of the patients have abnormalities (Al-Sulaiman, 2001a,b). Most common abnormalities are epileptiform, followed by slow waves. Most of the patients also have seizures, mostly of the generalized type. The principal abnormality noted on CT scan is cerebral atrophy. A similar study in the UAE pointed out that generalized tonic-clonic seizures are the most common kind of seizures among children with CP, and that the onset of these seizures is neonatal in the majority (Gururaj et al., 2003). Epilepsy is not the only complication of CP. A study on 50 adult Iraqi males with CP revealed that such patients are at risk of several secondary complications, including musculoskeletal disorders, mental retardation, and speech disorders, apart from epilepsy (Mezaal et al., 2009). In Jordan, children with spastic CP are found to have decreased physical growth parameters, especially in the quadriplegic form (Ibrahim and Hawamdeh, 2007).

Since the early 1990s, intramuscular injections of Botulinum toxin type-A (BT-A) have been used to relax contracting muscles in patients with CP. Studies in Egypt and Jordan have shown that BT-A injections significantly improved gross motor function in patients with spastic CP (El-Etribi et al., 2004; Hawamdeh et al., 2007), and that the best effect was obtained when the injections were given at both the gastrocnemius and adductors muscles (Ibrahim et al., 2007). In addition, Alrefai et al. (2009) demonstrated the effectiveness of BT-A in the management of drooling in children with CP.

The molecular pathology leading to CP in Arabs was only studied in a single family from Oman. The family showed a new autosomal recessive syndrome, characterized by spastic CP with microcephaly and mental retardation. The disease maps to a locus distinct from that on chromosome 2q, where a similar sort of condition had mapped to in a Pakistani family (Rajab et al., 2006).

BOX 58-7

- A high incidence CP is reported in Saudi Arabia, with epilepsy as a common complication.
- In UAE, generalized tonic-clonic seizures, mostly neonatal, are very common among children with CP.
- Iraqi males with CP are at risk of several secondary complications apart from epilepsy.
- In Jordan, children with spastic CP have decreased physical growth parameters.
- BT-A injections significantly improve gross motor function in patients with spastic CP
- An Omani family shows a new syndrome, characterized by spastic CP with microcephaly and mental retardation.

OTHER DISORDERS OF THE CENTRAL NERVOUS SYSTEM

IDIOPATHIC INTRACRANIAL HYPERTENSION

Idiopathic intracranial hypertension (IIH) is a relatively rare condition typified by headaches and nausea, experienced mostly by obese women of childbearing age. The condition is characterized by an increase in the intracranial pressure, in the absence of a tumor or any other underlying cause.

An epidemiological study in the Libyan population showed that the incidence of the condition (at 4.3 per 100,000 females) is three to four times higher than that reported in the United States (Radhakrishnan et al., 1993). In spite of this high incidence, reports of studies on IIH from the Arab world are scanty. Mezaal and Saadah (2005) studied the characteristics of IIH among 23 Arab patients in Dubai, United Arab Emirates. Most patients in their study had a relatively benign course. They also found obesity to be the most common factor associated with the condition, while the most common presenting signs were headache and double vision.

Lumbar puncture is one of the most effective ways to reduce the intracranial pressure in IIH. An 11-year-old girl from Oman with IIH was reported to develop cerebrospinal fluid hypovolemia following this procedure (Koul et al., 2002a). Meanwhile, in Egypt, a study was performed to assess the efficacy of laparoscopic surgery for both the implantation of a lumboperitoneal (LP) shunt , as well as to evaluate its patency. Laparoscopy-assisted LP shunting demonstrated more efficiency and was associated with more rapid postoperative recovery (Hay et al., 2004). In Saudi Arabia, Obeid and colleagues (2000) reported a patient with severe radiculopathy as a manifestation of IIH.

LEUKODYSTROPHY

The word *leukodystrophy* has its origin in the Greek root words of *leukos*, meaning white and *dystrophy*, meaning lack of growth. As the name suggests, the condition refers to a group of disorders, wherein imperfect growth or development of the myelin sheath result in progressive degeneration of the white matter of the brain. Included within this group are diseases such as metachromatic leukodystrophy, Krabbe disease, Canavan disease, vanishing white matter disease, and Alexander disease.

Several clinical descriptions of Arab patients with leukodystrophy are found in literature. A large Druze kindred demonstrated a very high incidence of Krabbe disease (Zlotogora et al., 1985, 1991). Within a study period of 3 years, Ozand and colleagues (1990) reported a fairly large group of 12 Saudi patients with Canavan disease, all of whom demonstrated defective aspartoacylase activity. Ben-Zeev and colleagues (2001) described nine patients from five Jewish-Libyan families who were diagnosed with vacuolating megalencephalic leukoencephalopathy. Considering the high level of consanguinity in the community, the authors suggested an autosomal recessive mode of inheritance of the disease in these families. In Syria, a novel syndrome of leukodystrophy with oligodontia was reported in a kindred with four affected patients. Again, the parents were consanguineous, suggesting an autosomal recessive mode of inheritance (Atrouni et al., 2003). Other reports of leukodystrophy in Arab patients include adrenoleukodystrophy in Lebanon (Sawaya et al., 1999) and Saudi Arabia (Al-Essa et al., 2000a), Canavan disease in Jordan (Masri and Hamamy, 2006), and Krabbe disease in the UAE (Al Talabani et al., 1998), Saudi Arabia (Al-Essa et al., 2000b), and Palestine (Korn-Lubetzki et al., 2003).

Molecular descriptions of leukodystrophy in Arab patients originate mostly from Palestinian families living in the Occupied Territories. In this group, a high incidence of late infantile metachromatic leukodystrophy occurred in 13 affected families (Zlotogora et al., 1994a). All patients from the Jerusalem region are homozygous for the same splice-site mutation in the *ARSA* gene. Interestingly, Muslim and Christian Arab patients originating from Jerusalem, as well as Caucasian patients, have the same haplotype in complete linkage disequilibrium with the splice-site mutation, prompting a suggestion that the mutation was probably introduced to the region around the time of the Crusades (Zlotogora et al., 1994b). In a later study, all 218 carriers of the Q190H mutation in the *ARSA* gene in a Palestinian Arab village were found to be descendants of a single couple (Zlotogora et al., 2007). In complete contrast, Arab patients with metachromatic leukodystrophy from the Lower Galilee region displayed five different mutations; an unexpected and surprising result for a population from such a small region (Heinisch et al., 1995).

BOX 58-8

- The incidence of idiopathic intracranial hypertension (IIH) in Libya is three to four times higher than that reported in the United States.
- In UAE, obesity is the most common factor associated with IIH, with presenting signs of headache and double vision.
- In Saudi Arabia, a patient with severe radiculopathy as a manifestation of IIH is reported.
- A large Druze kindred demonstrated a very high incidence of Krabbe disease.
- Saudi patients with Canavan disease demonstrated defective aspartoacylase activity.

- Jewish Libyan families diagnosed with vacuolating megalencephalic leukoencephalopathy are reported.
- A novel autosomal recessive syndrome of leukodystrophy with oligodontia is reported in Syria.
- Adrenoleukodystrophy, Canavan disease, and Krabbe disease were reported in other Arab states.
- A high incidence of late infantile metachromatic leukodystrophy occurred in Palestinian families sharing the same *ARSA* gene mutation.

DISEASES OF THE PERIPHERAL NERVOUS SYSTEM

Data on the incidence and prevalence of inherited peripheral neuropathies in the Arab world are very scarce. A study from Oman reported cases of childhood neuropathies from the records at Sultan Qaboos University Hospital. In these records, 82 children were found to have peripheral neuropathy, 17 of whom (20.7%) had hereditary motor and sensory neuropathy and 2 (2.4%) had hereditary sensory and autonomic neuropathy (Koul et al., 2002b). Another epidemiological study from Libya estimated the presence of 421 patients with hereditary sensory and motor neuropathy in Libya (Benamer, 2007).

HEREDITARY MOTOR AND SENSORY NEUROPATHIES (HMSN)

Neuropathies are common neurological diseases with a broad clinical spectrum. One disease subgroup in which tremendous progress has been made in terms of underlying molecular mechanisms is the hereditary motor and sensory neuropathies (HMSN), also called Charcot-Marie-Tooth (CMT) disease. CMT is a heterogeneous group of inherited motor and sensory neuropathies affecting ~ 1 of every 2500 individuals and characterized by chronic distal weakness with progressive muscular atrophy and sensory loss in the distal extremities (Dyck, 1975). Electrophysiological studies distinguish axonal forms (CMT2), with motor nerve conduction velocities (MNCV) greater than 40 m/s, from demyelinating forms (CMT1 and CMT4) with MNCV less than 35 m/s (Harding and Thomas, 1980; Bouche, 1983). All CMT forms are characterized as Mendelian traits and usually show complete penetrance, although the severity and extent of the disease can vary even between affected members of the same family. In each type, there are several modes of inheritance, and many loci including more than 30 genes have been identified (Nave et al., 2007; Zuchner and Vance, 2005). Progress in CMT genetics has led to a shift in the classification of disease subtypes, as DNA testing became the basis for detailed classification of the disease.

The demyelinating forms of CMT are characterized by the reduction in nerve conduction velocities and a reduction in compound muscle action potential amplitudes, which correlates with the clinical impairment (Krajewski et al., 2000). Clinical features of CMT1, the major form of the disease, include progressive distal muscle weakness and atrophy, often with *pes cavus* deformity. These signs are accompanied by mild sensory loss and absence of deep tendon reflexes. Segmental de myelination and re-myelination with typical onion-bulb formation are evident on nerve biopsies of CMT1 patients (Dyck, 1975). Clinical onset is usually during the second or third decade of life. CMT4, the autosomal recessive form of demyelinating CMT, is clinically similar to CMT1, but more severe and has an earlier average age of onset (Dyck et al., 1993).

Autosomal recessive CMT neuropathies are, in general, less common than the dominant forms. Skre (1974) estimated the frequency of autosomal recessive CMT to be 1.4 per 100,000 in western Norway. However, in communities with a high prevalence of consanguineous marriages, autosomal recessive inheritance is likely to account for 30%–50% of all forms of CMT and may even be the most common form (Martini et al., 2000). In line with these expectations, the great majority of CMT reports from the Arab world cover the autosomal recessive subtypes of the disease. This is probably due to the high rate of consanguinity and the structure of families in Arab societies, which played a major role in the identification of a number of autosomal recessive CMT loci and genes. A summary on the contribution of Arab families to the knowledge on autosomal recessive CMT is provided in this section.

Charcot-Marie-Tooth Type 4A (CMT4A)

In an attempt to localize the first subtype of CMT4, CMT4A, Ben Othmane and colleagues (1993) selected a subgroup of 4 families from a series of Tunisian CMT4 families according to specific criteria, mainly severe symmetric weakness and atrophy of the foot, slow motor nerve conduction velocity, severe hypomyelination upon nerve biopsy with basal lamina onion bulbs, and no myelin outfoldings. Disease onset was by one year of age, and some affected persons became wheelchair-dependent and/or developed kyphosis. Significant evidence for linkage was found for several markers from chromosome 8q13-21.1. Through additional positional cloning, Baxter and colleagues (2002) identified the gene encoding ganglioside-induced differentiation-associated protein-1 (*GDAP1*) as a causing gene in this locus. They found three *GDAP1* mutations in four Tunisian families. Interestingly, the same locus was found to be involved in an autosomal recessive axonal form of CMT in a large Tunisian family with 13 affected patients (Barhoumi et al., 2001). The disease was characterized by onset during the first decade, a progressive course, and distal atrophy in all four limbs, associated with pyramidal syndrome.

The nerve biopsy in two patients showed severe axonal neuropathy. These findings raised the question of whether mutations in the same *GDAP1* gene can cause both demyelinating and axonal neuropathic phenotypes. The answer came from two small and one large inbred family with Spanish ancestry, where an axonal CMT phenotype associated with hoarse voice and vocal cord paresis was found to be associated with *GDAP1* mutations (Cuesta et al., 2002).

In line with these findings, a number of *GDAP1* gene mutations were found in several Arab families in association with both demyelinating and axonal phenotypes. An S194X mutation was described in one of the Tunisian axonal CMT families reported by Barhoumi et al. (2001), and one of the Spanish families reported by Cuesta et al. (2002). The same mutation was reported later in a Moroccan family with compound heterozygote patients for the S194X and the R310Q mutation, and associated with an axonal form of the disease. The phenotype included hoarse voice and paralysis of the diaphragm (Azzedine et al., 2003b). On the other hand, a Moroccan family whose affected members were homozygous for the S194X mutation manifested a severe phenotype and earlier age of onset (Birouk et al., 2003). Electrophysiologic findings and morphologic data were consistent with an axonal form of CMT disease. Later, a novel GDAP1 mutation (P78L) was identified in three unrelated Moroccan families with early-onset demyelinating CMT (ARCMT1). Interestingly, the mutation was associated with a common haplotype suggesting a Moroccan founder mutation (Bouhouche et al., 2007a).

Charcot-Marie-Tooth Type 4B1

The locus of CMT4B1 was originally reported in an Italian family with early age of onset and remarkable proximal and distal weakness of the lower limbs and reduced motor nerve conduction velocities (Quattrone et al., 1996). The locus was later confirmed and refined in a Saudi family with similar clinical and neurophysiological features. Yet, the family showed some peculiar pathologic features; mainly, irregular and redundant loops associated with folding of the myelin sheaths (Salih et al., 2000). The gene responsible for this type was later identified to be *MTMR2* (Bolino et al., 2000).

Charcot-Marie-Tooth Type 4B2

The second locus of CMT4B, CMT4B2, was mapped to chromosome 11p15 in an inbred Tunisian family, with six affected individuals. Diagnostic criteria for CMT4B in this family included slowly progressive, unequivocal distal muscle weakness and atrophy, absent lower limb deep tendon reflexes, distal sensory loss, and *pes cavus* deformity. Sural nerve biopsies revealed the characteristic aspect found in CMT4B of abnormal folding of myelin sheaths (Othmane et al., 1999). Consequently, two large consanguineous families from Tunisia and Morocco with demyelinating CMT, characterized by myelin outfoldings on nerve biopsies and early-onset glaucoma, were mapped to the same locus (Azzedine et al., 2003a). A different nonsense mutation in the *MTMR13* gene was found to segregate with the disease phenotype in each of these families, identifying *MTMR13* as the causative gene for CMT4B2.

CMT4C

A study performed on 17 consanguineous families with autosomal recessive demyelinating CMT identified a locus for another subtype, termed CMT4C, on chromosome 5q23-q33. The families included two Algerian, one Tunisian, one Moroccan, and two French families (Guilbot et al., 1999). Mutations in the gene that code for SH3TC2 (*KIAA985*) are associated with this subtype in Turkish, Iranian, Greek, Italian, and German families (Senderek et al., 2003). In 10 Algerian families, SH3TC2 mutations, 6 of which are in exon 11, are associated with an age of onset between 2 and 10 years (Azzedine et al., 2006). Scoliosis or kyphoscoliosis and foot deformities are found in almost all patients, and are often inaugural. Patients can walk without help, indicating that CMT4C is less severe than other subtypes of autosomal recessive CMT, but spine deformities and giant axons on biopsies are a hallmark of CMT4C.

CMT4F

The only report of CMT4F from the Arab world involves a large inbred Lebanese family, which demonstrated linkage to a locus on chromosome 19q13.1-13.3 (Delague et al., 2000). The family was later found to carry a nonsense mutation in the gene encoding the periaxins (*PRX*). Histopathological and immunohistochemical analysis of a sural nerve biopsy of one patient revealed common features with the peraxin-null mice and the absence of L-periaxin from the myelin sheath (Guilbot et al., 2001).

CMT4H

The CMT4H locus on chromosome 12p11.21-q13.11 was described in two consanguineous Arab families, one Lebanese and one Algerian (De Sandre-Giovannoli et al., 2005). Detailed analysis depicted the *FGD4* gene to be responsible for this subtype in the same families (Delague et al., 2007).

CMT Type 2

Axonal CMT, or CMT2, appears to be less common than CMT1. It is characterized by progressive weakness and atrophy of distal muscles, hypoactive or absent deep tendon reflexes, distal sensory abnormalities, and foot deformities (Harding and Thomas, 1980). Sural nerve biopsy shows thinly myelinated fibers. Sufficient neurons remain normal so that conduction velocity of nerves is not much affected (Dyck et al., 1993). The normal NCV in this type makes the diagnosis entirely clinical for most family studies, thereby increasing diagnostic error and reducing the ability to narrow down the DNA region that contains the disease gene (Zuchner and Vance, 2005). Autosomal recessive CMT2 has been reported in Arab families. Of these recessive types is the CMT2B1, which was first described and mapped at chromosome 1q21.2-q21.3 in one large consanguineous Moroccan family (Bouhouche et al. 1999). By studying three consanguineous Algerian families, the disease-causing gene was localized to the same locus and a unique 892C>T (R298C) missense mutation was identified in the *LMNA* gene (De Sandre-Giovannoli et al., 2002). In addition, all CMT2B1 patients homozygous for this mutation originate from a restricted region of northwest Algeria and eastern Morocco, and carry a common homozygous ancestral haplotype at the *LMNA* locus, indicating a founder mechanism for this mutation. The phenotype in these families is unusual in terms of rapidity of evolution, involvement of the proximal muscles, and severity (Chaouch et al., 2003). Moreover, a study on seven Algerian families with the same mutation reported marked variability among the patients in the age of onset and course of disease (Tazir et al., 2004).

In a study on 13 Moroccan families with autosomal recessive axonal CMT (ARCMT2), four families had the *LMNA* 892C>T mutation and 6 families had the *GDAP1* S194X mutation. Founder effects for both mutations were suggested by the analysis of the microsatellite markers flanking the genes. The three remaining families did not link to these loci (Bouhouche et al., 2007b).

BOX 58-9

- Data on the incidence and prevalence of peripheral neuropathies in the Arab world are very scarce.
- The great majority of CMT reports from the Arab world cover the autosomal recessive subtypes of the disease.
- CMT4A was localized and the causing gene, *GDAP1*, was identified in Tunisian families.
- *GDAP1* gene mutations cause both demyelinating and axonal neuropathies and are reported in Tunisia and Morocco.
- The CMT4B1 locus was confirmed and refined in a Saudi family.
- CMT4B2 was mapped in an inbred Tunisian family, and the causing gene, *MTMR13,* was later identified in Tunisian and Moroccan Families.
- CMT 4C was first localized in Algerian, Tunisian, and Moroccan families.
- CMT4F was localized and the causing gene *PRX* identified in a Lebanese family.
- The CMT4H locus was described in Lebanese and Algerian families.
- CMT2B1 was first described and mapped in a Moroccan family, and the causing gene, *LMNA*, was identified in Algerian families.

GIANT AXONAL NEUROPATHY

Giant axonal neuropathy (GAN) is a rare hereditary disorder characterized by a progressive motor and sensory neuropathy with central nervous system involvement, mental retardation, and electroencephalographic abnormalities. Axonal loss and the presence of giant axonal swellings filled with neurofilaments on nerve biopsies are the hallmarks of this neurodegenerative disorder (Igisu et al, 1975; Ben Hamida et al., 1990b). The GAN gene was localized to chromosome 16q24.1 in three unrelated Tunisian families (Ben Hamida et al., 1997). It encodes a ubiquitously expressed protein, gigaxonin, which is implicated in the cytoskeletal functions and dynamics (Bomont et al., 2000). In an Algerian family with a peculiar clinical form of GAN and linkage to 16q24.1, patients presented with a CMT2-like phenotype, central nervous system involvement occurring 10 years later, and spastic paraplegia 16 years after onset (Zemmouri et al., 2000). Two years later, a report described a consanguineous Algerian family with a clinical picture different from the classical GAN and absence of linkage to 16q24.1, indicating a genetic heterogeneity in the disease (Tazir et al., 2002). Another study on six Algerian families with GAN reported variation in severity and clinical phenotypes associated with different mutations demonstrating obvious clinical heterogeneity (Tazir et al., 2009).

BOX 58-10

- The locus for giant axonal neuropathy was identified in unrelated Tunisian families.
- Clinical and genetic heterogeneity is demonstrated in Algerian GAN families.

HEREDITARY SENSORY AND AUTONOMIC NEUROPATHIES

Hereditary sensory and autonomic neuropathies (HSANs) are characterized by loss of pain sensation, in combination with other sensory and/or autonomic abnormalities, due to degeneration and/or abnormal development of peripheral nerves (Hilz, 2002). A number of loci and genes have been associated with HSAN, of which *HSN2* gene causes HSN type II in families from Quebec and Newfoundland (Lanfreniere et al., 2004). *HSN2* is a single-exon gene located within intron 8 of the *PRKWNK1* gene, and is transcribed from the same strand as *PRKWNK1*. A report on four affected members of a Lebanese family with HSAN type II phenotype revealed homozygosity for a frameshift mutation in the *HSN2* gene. This observation led to the conclusion that HSAN type II exists outside Quebec and Newfoundland, and supported the hypothesis that this type is caused by loss of function of the *HSN2* gene (Rivière et al., 2004).

HEREDITARY MOTOR NEUROPATHY

Hereditary motor neuropathy (HMN), or spinal muscular atrophy (SMA), groups a number of genetic disorders predominantly affecting the peripheral motor neuron and its axonal ramifications. Clinically, these disorders are characterized by weakness and wasting of distal and proximal muscles, the absence of significant sensory symptoms, absent or reduced deep tendon reflexes, and skeletal deformities (Dyck and Lambert, 1968). The distal forms of HMN (dHMN) have been classified into seven types, based on their clinical and genetic features. Three of these forms are of autosomal recessive inheritance. In a cluster of villages in the Jerash region of Jordan, nine families demonstrated with a novel form of autosomal recessive distal HMN. Distinctive clinical features in these families include the presence of pyramidal signs within the early stages of the disease, with persistence of knee hyperreflexia. This disease was termed the "Jerash type" or hereditary motor neuropathy (HMN-J). In these families, HMN-J maps to chromosome 9p21.1-p12 (Middleton et al., 1999).

Another study on an autosomal dominant type of dHMN, reported three families of Algerian Sephardic Jewish origin with a missense mutation (G526R) in the glycyl-tRNA synthetase (*GARS*) gene. The age of onset of the disease in these families varied widely, but the clinical and electrophysiologic presentation was uniform and progressed slowly (Dubourg et al., 2006).

BOX 58-11

- (HMN-J) is a novel form of autosomal recessive distal HMN characterized by pyramidal signs within the early stages of the disease, with persistence of knee hyperreflexia.
- Autosomal dominant dHMN was reported in Algerian Sephardic Jewish families with a *GARS* gene mutation.

MUSCULAR DYSTROPHIES

Inherited muscle diseases in the Arab World have been reported in Morocco, Algeria, Tunisia, Libya, Egypt, Sudan, Palestine, Jordan, Lebanon, Saudi Arabia, Kuwait, and Oman. The earliest reports came from Tunisia (Ben Hamida and Marakchi, 1980), Sudan (Salih et al., 1983), and Lebanon (Afifi et al., 1983). These reports presented many severe muscular dystrophy cases in a number of families. Later, small epidemiological studies were conducted in Jordan, Egypt, and Libya. In Jordan, 75 patients were diagnosed with muscle disease between 1990 and 1997. Fifty five (73%) of them had muscular dystrophy (MD), 28 (50.9%) had congenital muscular dystrophy, 11 (20%) had Duchenne muscular dystrophy (DMD), 9 (16.4%) had Becker muscular dystrophy (BMD), 4 (7.3%) had myotonic dystrophy, 2 (3.6%) had limb-girdle muscular dystrophy (LGMD), and 1 (1.8%) had facioscapulohumeral dystrophy (Al-Qudah and Tarawneh, 1998). In Assiut, Egypt, 40 patients with muscular disorders were identified, and the lifetime prevalence per 100,000 was 26.8 for muscular dystrophy, 11.49 for myotonia, 11.49 for myositis, 17.24 for systemic myopathy, and 9.57 for myasthenia gravis (El-Tallawy et al., 2005). In Benghazi, Libya, a 3-year search identified 34 DMD patients (prevalence 6/100,000) and 19 LGMD patients (3.7/100,000; Radhakrishnan et al., 1987). Two types of progressive MD were identified in Tunisia: typical DMD (Ben Hamida et al., 1992) and Duchenne-like muscular dystrophy (DLMD; Hamida et al., 1994).

DUCHENNE MUSCULAR DYSTROPHY

DMD is one of the most common human genetic diseases, accounting to 1/3300 male births (Hoffman and Kunkel, 1989). It presents as progressive proximal muscle weakness early in childhood, and follows an X-linked recessive inheritance. Becker muscular dystrophy (BMD) is a milder form of the disease with a later age of onset and a slower clinical progression. DMD and BMD are caused by partial deletions in the *dystrophin* gene in 60% of the cases, while point mutations and small deletions/insertions account for the remaining cases (Forrest et al., 1988). A recent study on 10 boys with DMD from Morocco reported the contribution of pathology and molecular genetics in the diagnostic strategy, and in the differentiation between DMD and BMD in Morocco, where sarcoglycanopathy DMD-like is the most frequent muscular dystrophy (Bellayou et al., 2009b). A study on the deletion frequency in 26 Kuwaiti and 16 Egyptian DMD patients documented a very high deletion detection rate (86%) in Arabs compared to other ethnic groups (Haider et al., 1998). In Saudi Arabia, a frequency of 63% for *dystrophin* gene deletions was reported in DMD and BMD patients (Al-Jumah et al., 2002). Screening of *dystrophin* gene deletions in 100 Egyptian families with DMD/BMD found that the deletion frequency was 55%. Of these, 55% had multiple deletions, the majority of them covering the major hot spot between exons 44 and 52, and 45% had single deletions (Effat et al., 2000). In addition, clinical, biochemical, and immunohistopathological analysis revealed that patients with double deletion within the *dystrophin* gene have a milder phenotype than patients harboring single deletions at either major or minor hotspots of the gene (El-Harouni et al., 2003). A later study from Egypt confirmed a deletion percentage of 61%, mostly in the major hotspot region (El Sherif et al., 2007). On the other hand, a third study detected frameshift deletion mechanisms in 51.3% of 152 unrelated Egyptian Duchenne and Becker muscular dystrophy patients (Elhawary et al., 2004).

Autosomal recessive Duchenne-like muscular dystrophy (DLMD), also known as severe childhood autosomal recessive muscular dystrophy (SCARMD) was recognized in communities with high consanguinity rates in Sudan (Salih et al., 1983; Salih et al., 1984), Tunisia (Ben Hamida and Marakchi, 1980; Ben Hamida et al., 1983), Kuwait (Zatz et al., 1990), Saudi Arabia (Salih et al., 1996), Libya, and Qatar (Dubowitz, 1980). The disease was localized to the pericentromeric region of chromosome 13q in three highly inbred families from Tunisia (Ben Othmane et al., 1992). DLMD linked to the same locus in Algerian families (Azibi et al., 1993) and Moroccan families (El Kerch et al., 1994) is associated with specific deficiency of adhalin, the 50-kDa dystrophin-associated glycoprotein (50DAG). Homogeneity of linkage data in the three Maghreb countries suggests a common gene, and possibly the same mutation in their populations.

LIMB-GIRDLE MUSCULAR DYSTROPHIES (LGMDS)

Limb-girdle muscular dystrophies (LGMDs) are a genetically heterogeneous group of muscle diseases inherited by either autosomal dominant (LGMD1) or recessive (LGMD2) pattern, and characterized by a primary and progressive muscle degeneration of the shoulder and pelvic girdles, with a range of clinical severity. Autosomal recessive LGMD have a particular high frequency in Arab states, especially Tunisia (Ben Hamida et al., 1983). Several types of autosomal recessive LGMDs (LGMD2A-I) have been identified, of which LGMD2C, LGMD 2D, LGMD 2E, and LGMD 2F are referred to as *sarcoglycanopathies*. Each type is due to mutations in the gene that encodes one of the sarcoglycan subunits. LGMD2B, also known as Miyoshi myopathy, was mapped to chromosome 2p12 in 12 families including a Tunisian family (Bejaoui et al., 1995), and is associated with mutations in the *dysferlin* gene (Bashir et al., 1998). The disease was reported in Moroccan (Mouton et al., 1999) and Saudi patients (Cupler et al., 1998). A cluster of LGMD2B is also reported in Libyan Jews, and is due to a founder *dysferlin* 1624delG mutation at about 10% with a prevalence of 1 per 1300 (Argov et al., 2000). LGMD2C is the most frequent LGMD2 in Tunisia. It constitutes 81% of sarcoglycanopathies and 75% of all LGMD2 in the country, and the majority of patients share the same founder γ-sarcoglycan (*γ-SGCG*) gene mutation (Kefi et al., 2003). Other autosomal recessive LGMD forms have been rarely reported in Arab populations. LGMD2D, associated with *α-SGCA* gene mutations, is described with LGMD2C in different members of the same consanguineous Tunisian family. One patient is homozygous for a γ-SGCG mutation, while two other patients were homozygous for a *α-SGCA* mutation (Fendri et al., 2006). LGMD2E is described in a Tunisian family with a homozygous *α-SGCB* gene missense mutation and severe phenotype (Bönnemann et al., 1998). LGMD2I was described and mapped to chromosome 19q13.3 in a large consanguineous Tunisian family (Driss et al., 2000). This locus was further narrowed down and a mutation in fukutin-related protein (*FKRP*) gene was identified in the same family (Driss et al., 2003). Recently, a second Tunisian family with LGMD2I associated with a compound heterozygous *FKRP* gene mutation was reported. All patients presented with the LGMD phenotype, but with variable cardiac involvement (Kefi et al., 2008). A novel missense mutation in exon 4 of the *FKRP* gene with marked phenotypic variability is identified in a large consanguineous LGMD2I Bedouin family from Upper Galilee (Harel et al., 2004).

EMERY-DREIFUSS MUSCULAR DYSTROPHY (EDMD)

Emery-Dreifuss muscular dystrophy (EDMD) is a humeroperoneal muscular dystrophy associated with contractures and cardiomyopathy (Emery, 2000). The X-linked type is caused by mutations in the *Emerin* gene, and is slightly less severe than the autosomal dominant type associated with mutations in the *Lamin A/C* gene (Bonne et al., 2002). A large Bedouin kindred with X-linked EDMD, and carrying a known *Emerin* gene insertion mutation, presented electrophysiological and morphological features suggestive of a neurogenic atrophy rather than clear-cut myopathy. Some of the affected members have an abnormal incremental response to high-frequency repetitive nerve stimulation, suggestive of a comorbid, mild congenital myasthenic syndrome (Ifergane et al., 2007).

CONGENITAL MUSCULAR DYSTROPHIES (CMD)

Congenital muscular dystrophies (CMD) are another clinically and genetically heterogeneous group of autosomal

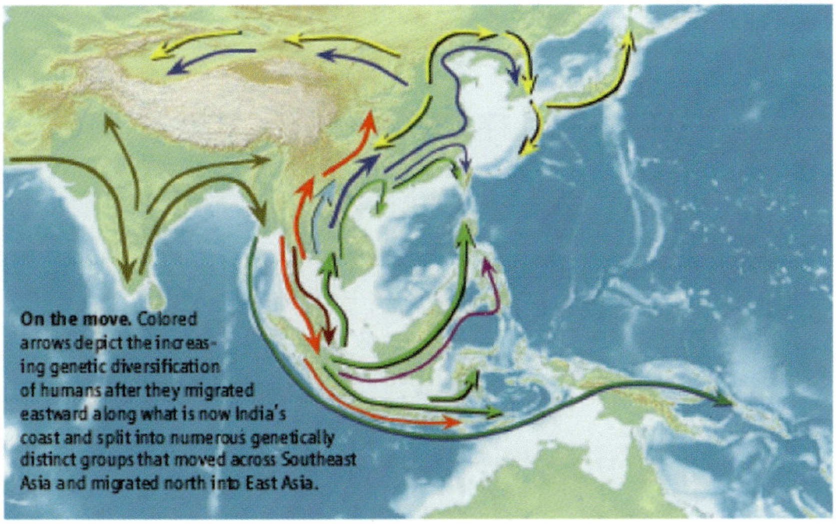

Figure 4-1 The migration of man and peopling in Asia (Adapted with permission from *Science,* December 11, 2009, page 1470).

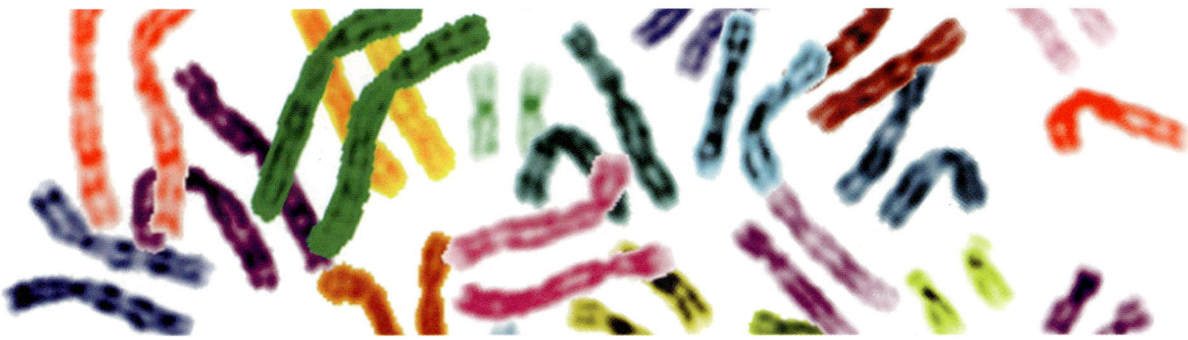

Figure 4-2 Heterogeneous human genomic regions showing minimal or no genetic differentiation.

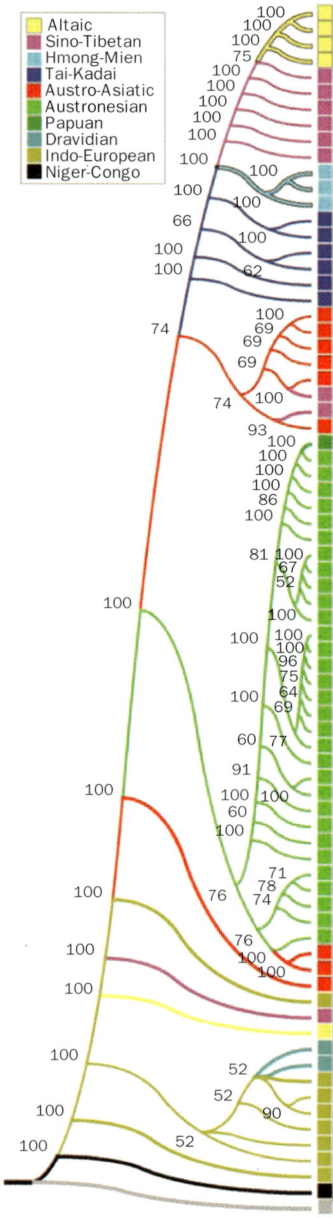

ID	Location	Latitude	Longitude	Ethnicity	Language	size
JP-RK	Japan	26.5	127.9	Ryukyuan	Okinawan	49
JP-ML	Japan	35.7	139.8	Japanese	Japanese	71
JPT	Japan	35.7	139.8	Japanese	Japanese	44
KR-KR	Korea	36.9	127.5	Korean	Korean	90
CHB	China	40.0	116.4	Han	Chinese	45
CN-SH	China	31.2	121.5	Han	Chinese	21
TW-HA	Taiwan	25.0	121.5	Han	MinNan	48
TW-HB	Taiwan	25.0	121.5	Han	Hakka	32
SG-CH	Singapore	1.4	103.8	Han	MinNan	30
CN-GA	China	23.3	113.5	Han	Cantonese	30
CN-HM	China	26.3	108.7	Hmong	Hmong	26
TH-HM	Thailand	18.6	98.1	Hmong	Hmong	20
TH-YA	Thailand	20.0	100.2	Yao	Iu-Mien	19
CN-CC	China	24.4	110.2	Zhuang	Zhuang	26
CN-JI	China	18.9	109.8	Jiamao	Jiamao	31
TH-TL	Thailand	19.2	100.9	Tai Lue	Lue	20
TH-TY	Thailand	18.4	98.9	Tai Yong	Tai Yong	18
TH-TK	Thailand	18.6	98.9	Tai Kern	Tai Kern	18
TH-TU	Thailand	19.0	99.0	Tai Yuan	Tai Yuan	20
TH-MA	Thailand	18.7	100.5	Mlabri	Mlabri	18
TH-TN	Thailand	19.1	100.9	H'Tin	Mal	18
TH-PP	Thailand	20.4	99.9	Plang	Blang	18
CN-WA	China	22.8	100.2	Wa	Wa	56
TH-LW	Thailand	18.4	98.1	Lawa	Lawa	19
TH-KA	Thailand	18.0	98.4	Karen	Karen	20
CN-JN	China	22.0	101.0	Jinuo	Jinuo	29
TH-PL	Thailand	19.9	99.2	Palong	Palong	18
AX-ME	Pacific	-5.8	155.1	Melanesian	Nasioi	5
ID-AL	Indonesia	-8.3	124.7	Alorese	Alor	19
ID-LE	Indonesia	-8.3	124.7	Lembata	Lembata	19
ID-LA	Indonesia	-8.3	123.0	Lamaholot	Lamaholot	20
ID-SO	Indonesia	-8.6	120.1	Manggarai	Manggarai	19
ID-RA	Indonesia	-8.7	120.5	Manggarai	Manggarai	17
ID-SB	Indonesia	-9.8	120.0	Kambera	Kambera	20
PI-AG	Philippines	13.7	123.3	Negrito	Agta	8
PI-AE	Philippines	14.9	120.2	Negrito	Aeta	8
PI-MW	Philippines	9.7	125.6	Negrito	Mamanwa	19
PI-IR	Philippines	13.0	121.1	Negrito	Iraya	9
PI-AT	Philippines	11.9	122.0	Negrito	Ati	23
AX-AM	Taiwan	23.7	121.4	Ami	Ami	10
AX-AT	Taiwan	24.6	121.4	Atayal	Atayal	10
PI-UB	Philippines	17.2	121.9	Urban	Ilocano	20
PI-UN	Philippines	14.6	121.0	Urban	Tagalog	19
PI-UI	Philippines	6.9	122.1	Urban	Visaya	20
PI-MA	Philippines	8.2	125.9	Manobo	Manobo	18
ID-MT	Indonesia	-0.3	98.4	Mentawai	Mentawai	15
ID-TR	Indonesia	-4.7	119.7	Toraja	Toraja	20
ID-ML	Indonesia	-3.0	104.7	Malay	Malay	12
ID-KR	Indonesia	1.5	100.0	Batak Karo	Batak Karo	17
ID-TB	Indonesia	2.3	99.1	Batak	Batak Toba	20
ID-DY	Indonesia	1.2	116.7	Dayak	Benuak	12
MY-MN	Malaysia	2.8	102.2	Malay	Minangkabau	20
SG-MY	Singapore	1.4	103.8	Malay	Malay	30
MY-KN	Malaysia	5.3	102.0	Malay	Malay	18
ID-JA	Indonesia	-6.2	106.7	Javanese	Javanese	34
ID-JV	Indonesia	-7.3	110.4	Javanese	Javanese	19
ID-SU	Indonesia	-6.2	106.7	Sundanese	Sunda	25
MY-BD	Malaysia	1.4	110.2	Bidayuh	Jagoi	50
MY-TM	Malaysia	2.9	102.1	Proto-Malay	Temuan	49
MY-JH	Malaysia	5.4	101.1	Negrito	Jehai	50
MY-KS	Malaysia	5.7	100.9	Negrito	Kensiu	30
TH-MO	Thailand	18.5	98.9	Mon	Mon	19
IN-NI	India	30.4	79.2	Tharu	Pahari	20
IN-TB	India	34.7	76.5	Ladakhi	Spiti	23
CN-UG	China	37.1	86.6	Uyghur	Uyghur	26
IN-DR	India	15.3	77.8	Upper-caste	Telugu	24
SG-ID	Singapore	1.4	103.8	India origin	Tamil	30
IN-WI	India	26.7	74.0	Bhil	Bhili	25
IN-EL	India	23.0	88.2	Upper-caste	Bengali	16
IN-SP	India	29.1	76.5	Upper-caste	Hindi	23
IN-WL	India	19.7	75.9	Upper-caste	Marathi	14
IN-NL	India	26.8	81.4	Upper-caste	Hindi	15
IN-IL	India	26.7	74.0	Upper-caste	Hindi	15
CEU	USA	51.5	-0.1	European	English	60
YRI	Nigeria	7.4	3.9	Yoruba	Yoruba	60

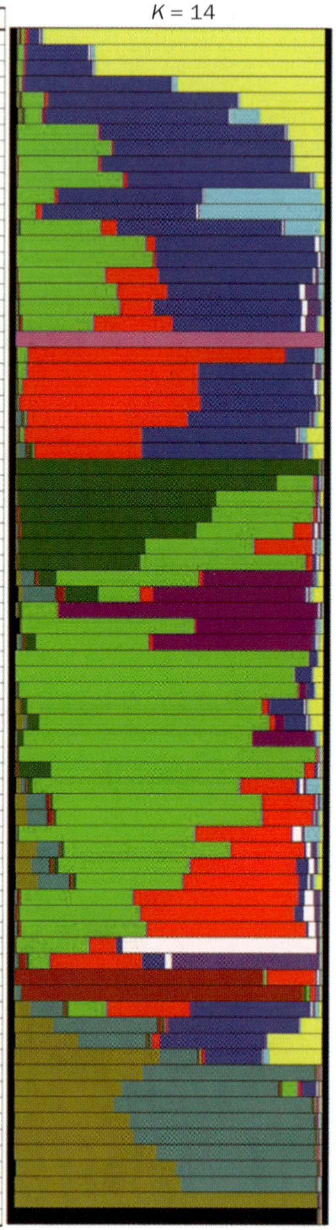

Figure 4-4 The geophysical and linguistic categories of the Asian populations analyzed by the HUGO Pan Asian SNP Consortium (adapted with permission from *Science*, September 17, 2009, page 1542). All population IDs except the four HapMap samples are denoted by four characters. The first two letters indicate the country where the samples were collected or (in the case of Affymetrix) genotyped, according to the following convention: AX, Affymetrix; CN, China; ID, Indonesia; IN, India; JP, Japan; KR, Korea; MY, Malaysia; PI, the Philippines; SG, Singapore; TH, Thailand; and TW, Taiwan. The last two letters are unique IDs for the population. To the right of the table, an averaged graph of results from STRUCTURE is shown for K = 14.

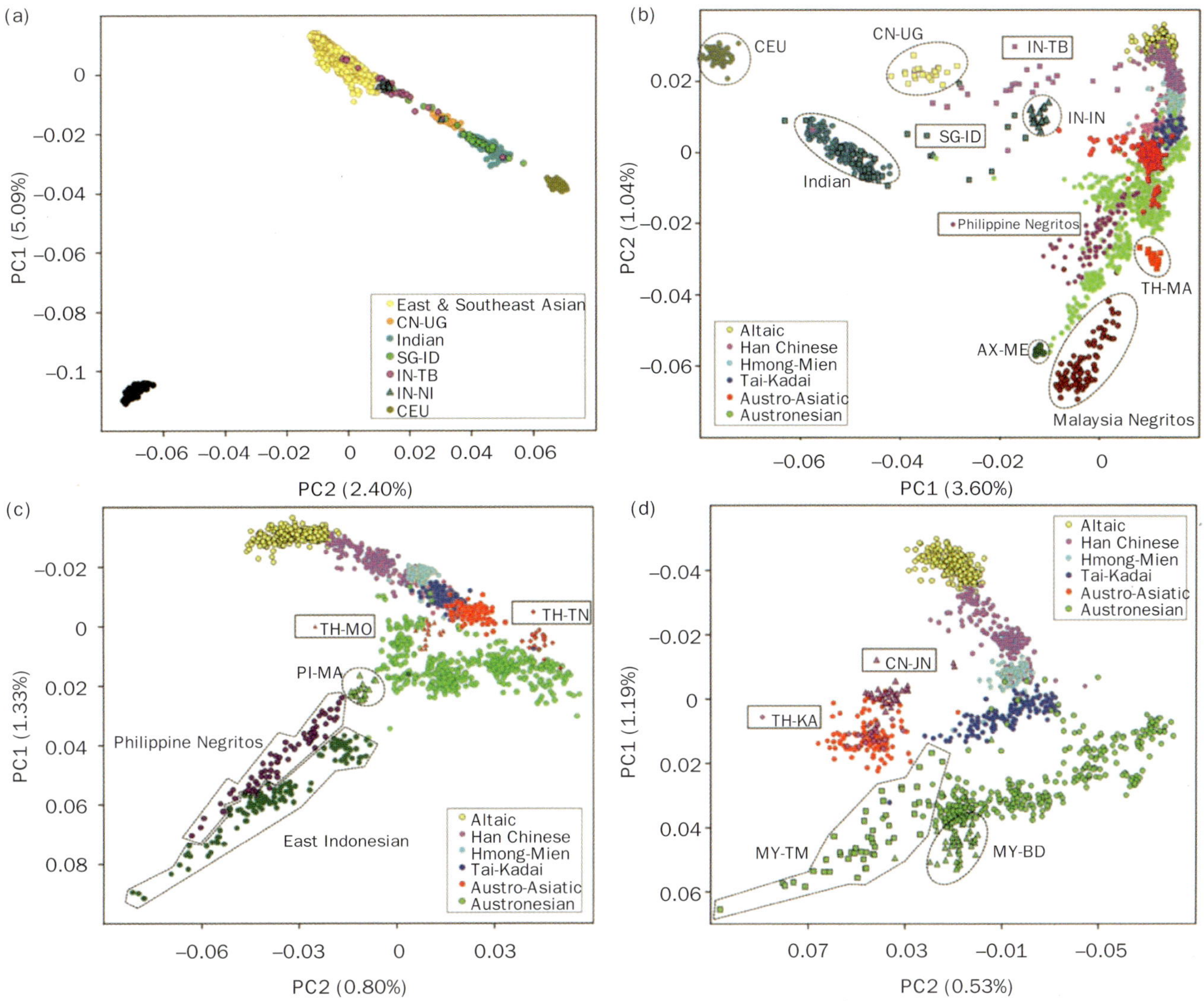

Figure 4-5 The population admixture of Asian population groups based on the principal component analysis (adapted with permission from *Science*, September 17, 2009, page 1543).

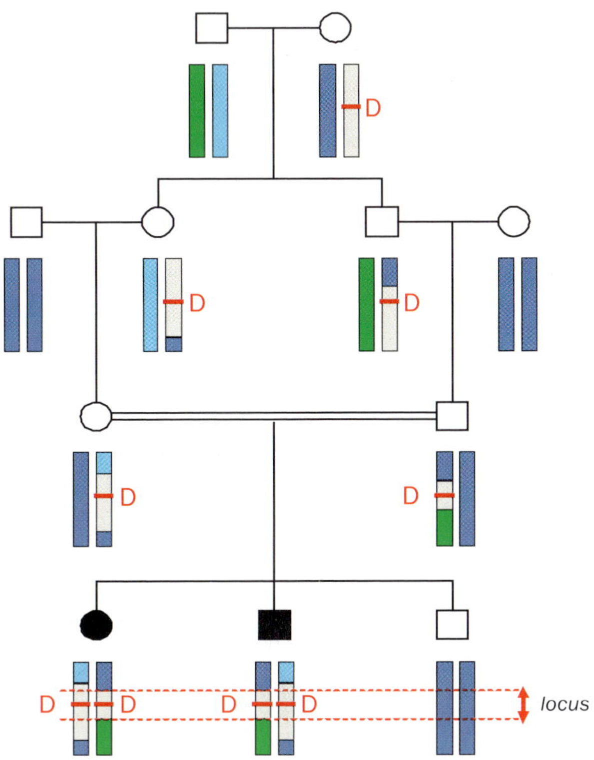

Figure 12-1 The principle of autozygosity mapping: a schematic representation of the inheritance of recessive disease gene in an idealized pedigree of a first cousin consanguineous family, leading to two individuals affected with a hypothetical recessive condition in the fourth (IV) generation. The great-grandmother in the first (I) generation is a carrier for one copy of the disease gene ("D"), indicated in red. Each of her children has a 50% chance of inheriting the disease allele, as well as the chromosomal segment or haplotype flanking the gene, shown in grey. Both children in generation II inherit the faulty gene, and then pass it on to their own children. As the disease gene is transmitted through the generations, the haplotype surrounding it becomes smaller due to meiotic recombination events. The parents in the third (III) generation are first cousins and carriers, each inheriting a copy of the disease gene from their grandmother, with a 1 in 4 risk that each of their children will inherit both copies of the disease gene. The two affected children (generation IV) in this pedigree are therefore *homozygous by descent* for this gene, but both are also homozygous for the chromosomal segment surrounding the disease gene. These shared homozygous regions can be identified by genotyping genetic markers spaced regularly throughout the genome in a genome-wide linkage scan.

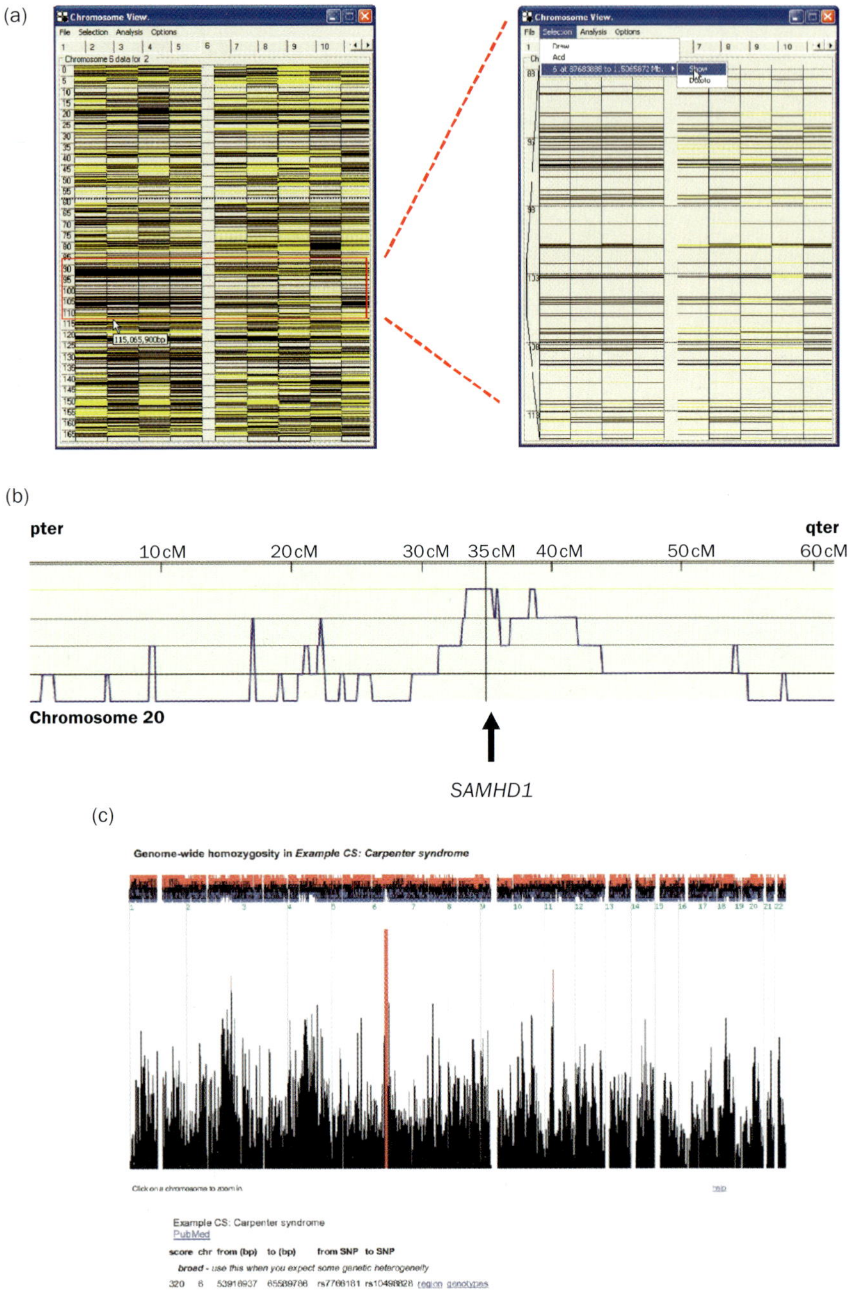

Figure 12-2 Programs to visualize chromosome and genome-wide SNP genotyping data for autozygosity mapping projects. (A) Regions of homozygosity on chromosome 6 displayed by "AutoSNPa" (Carr et al., 2006). The data is shown for one consanguineous family with affected and unaffected individuals on the left and right, respectively, in each "chromosome view" display panel. The extent of the chromosome is shown in the vertical axis, with physical location shown on the left. Homozygous genotypes are displayed in black, whereas heterozygous ones are in yellow. A clear shared region of homozygosity is seen in affected individuals, but is absent in unaffected individuals (red box), which can be expanded to show the region of interest in more detail, in this case chromosome 6, 876838888 to 115065872 bp. (B) "IBDFinder," a sequential SNP-counting program (Carr et al., 2009a), displays scores of homozygous genotypes along a chromosome to enable regions of homozygosity to be easily visualized in unrelated singleton cases. In the screenshot, the length of chromosome 20 is shown, marked with genetic distances, with a peak score identifying the causative gene, SAMHD1, for the AGS5 Aicardi-Goutieres syndrome locus (Rice et al., 2009). (C) Genome-wide homozygosity displayed by "Homozygosity Mapper" (Seelow et al., 2009). The screenshot shows the genome-wide homozygosity scores produced from SNP genotyping data in a Carpenter syndrome study (Jenkins et al., 2007). The scores are plotted as a bar chart with red bars indicating regions of interest, with chromosomes represented as bins along the horizontal axis. Above the bar chart, the excess or deficit of homozygous genotypes in cases versus controls is depicted. The program also automatically generates direct links to the most interesting regions, in this case on chromosome 6 between SNP markers rs7766181 and rs10498828 (53916937 to 65589786 bp).

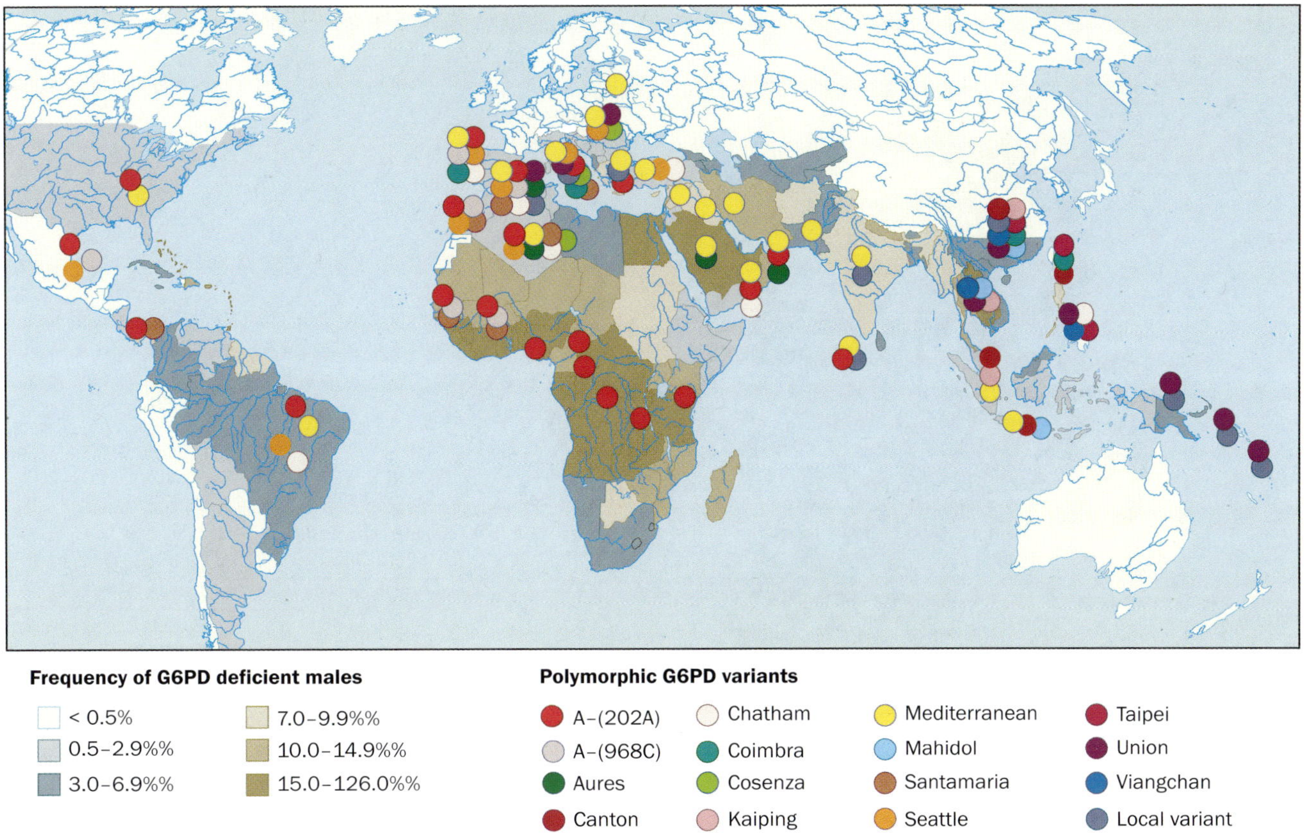

Figure 14-2 *World distribution of G6PD deficiency.* Shadings on the map indicate approximate frequencies of male G6PD deficiency frequencies in individual countries. Differently colored circles indicate where individual mutant G6PD alleles mostly contribute to the overall frequency of G6PD deficiency. (From Luzzatto & Notaro, 2001).

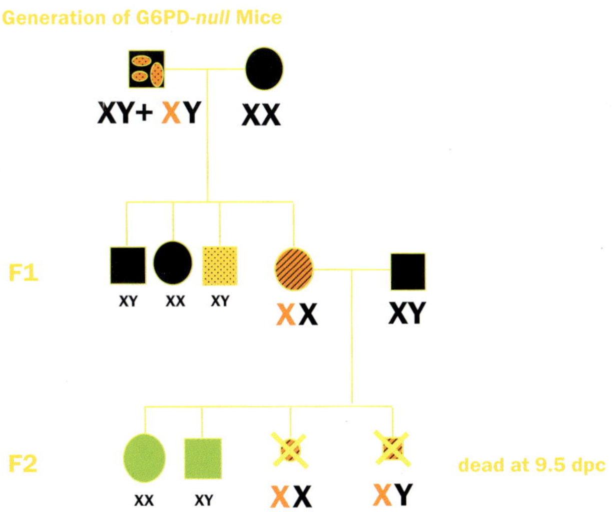

Figure 14-3 *A G6PD-null mutation is an embryonic lethal.* Chimeric mice were obtained by injecting into blastocysts G6PD knockout embryonic stem cells. Mating of a male chimeric mouse with a normal female (top line: the X-chromosome with the G6PD null mutation is in red) produced (F1), apart from normal offspring (in black), females heterozygous for the G6PD null mutation, but no hemizygous males, proving that G6PD null is an embryonic lethal. When a heterozygous female was mated with a hemizygous male, in the F2 both hemizygous male embryos and heterozygous female embryos died: the latter because of placental failure, since in the extraembryonic tissues, the normal paternal X is known to be selectively inactivated. (From Longo et al., 2002.)

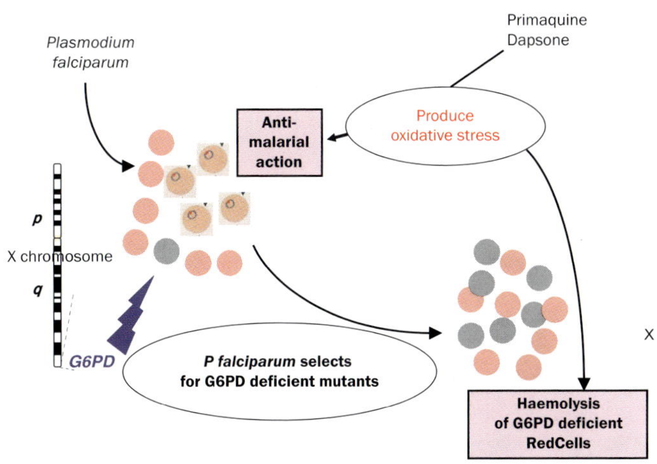

Figure 14-4 *Genetics and pharmacogenetics of G6PD deficiency.* Mutations of the *G6PD* gene arise spontaneously, like in any other gene. Some mutations entail deficiency of G6PD, and G6PD-deficient red cells confer protection against *P. falciparum*. As a result, genetically determined G6PD deficiency tends to become prevalent in areas where *P. falciparum* is endemic. Primaquine and dapsone are potent antimalarials, but unfortunately they are also potently hemolytic for G6PD-deficient red cells.

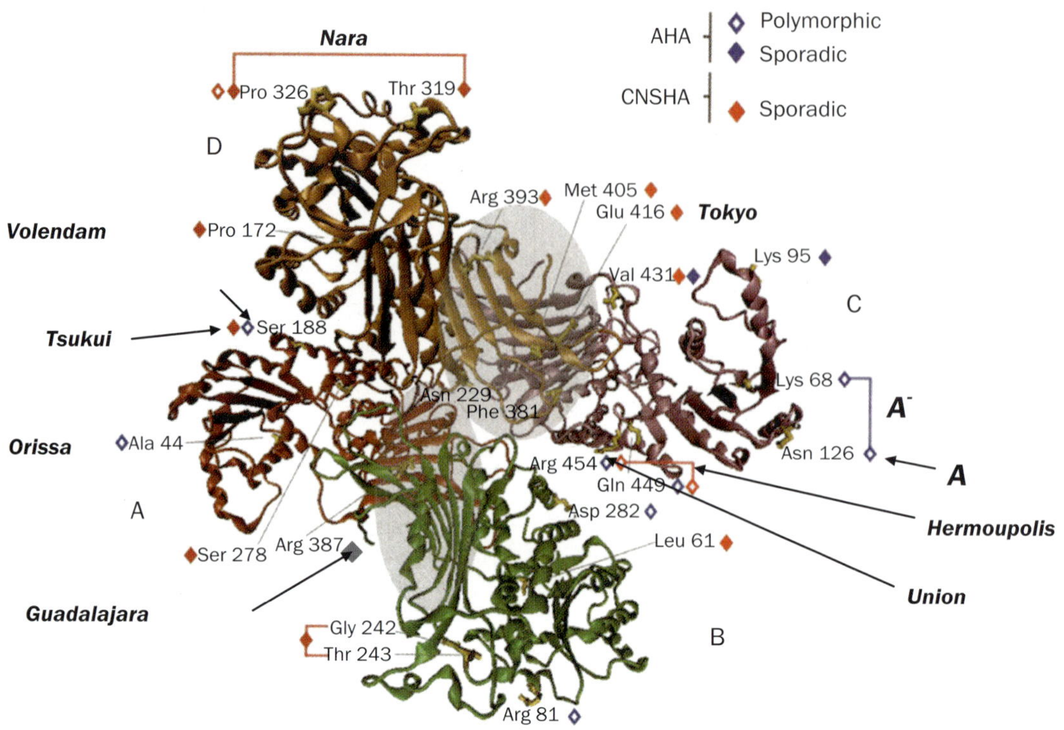

Figure 14-5 A three-dimensional view of G6PD.

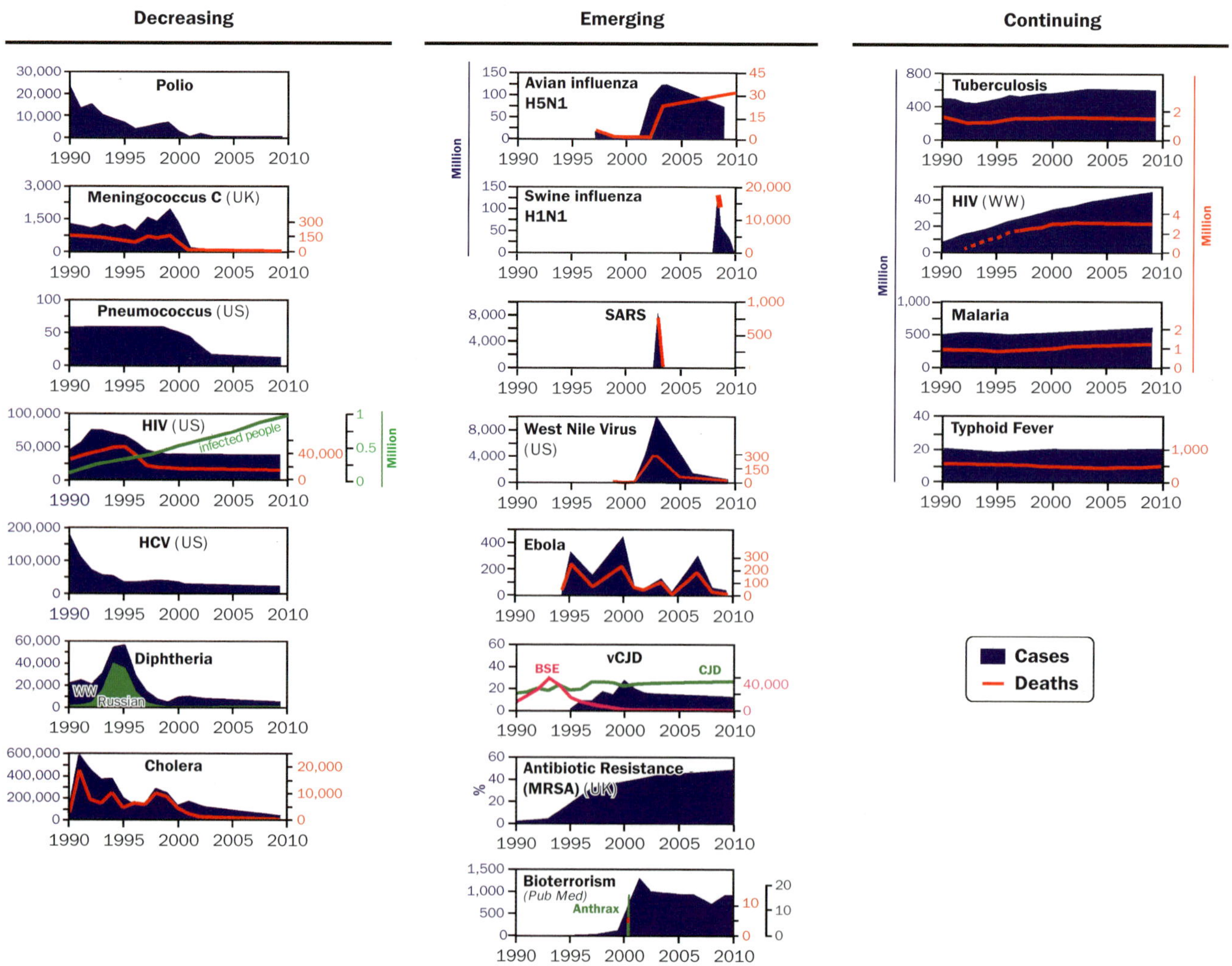

Figure 16-1 Major changes in infectious diseases during the last decade.

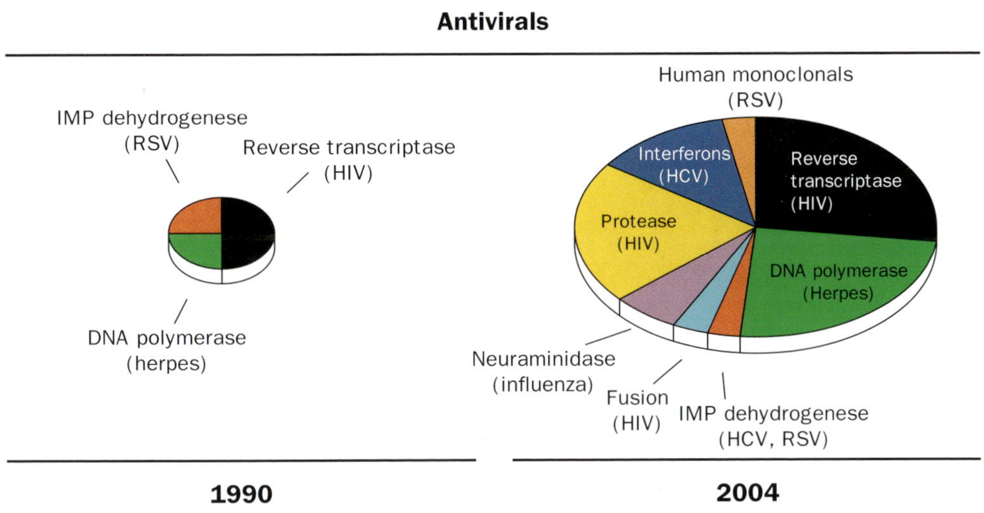

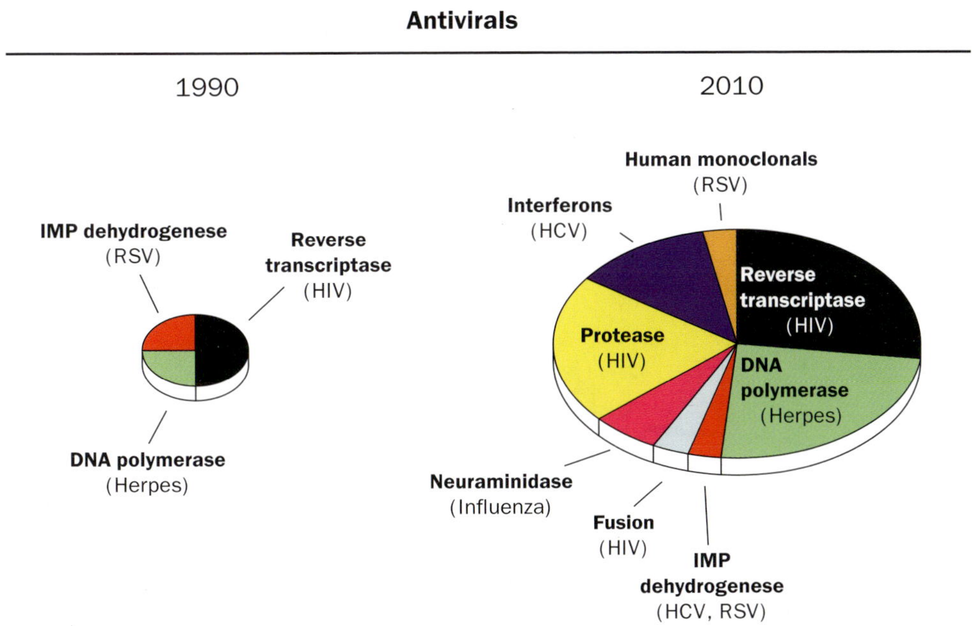

Figure 16-3 Antivirals approved by FDA in 1990 and 2004.

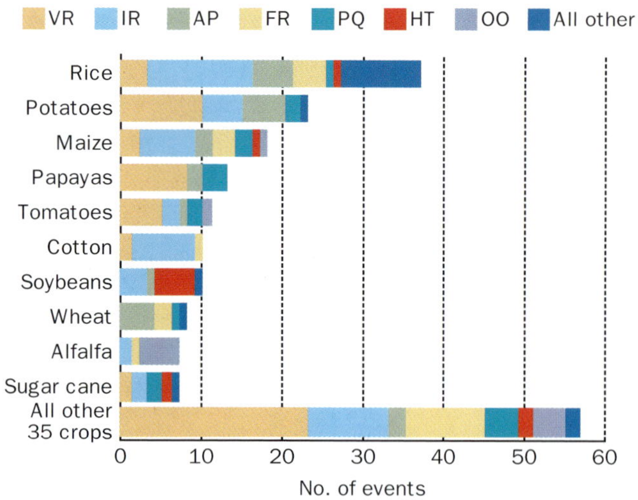

Figure 19-2 Phenotype characteristics sorted by number of transformation events among the top 10 crops in the study data set. AP, agronomic properties; BR, bacterial resistance; FR, fungal resistance; HT, herbicide tolerance; IR, insect resistance; OO, Other; PQ, product quality; VR, virus resistance.

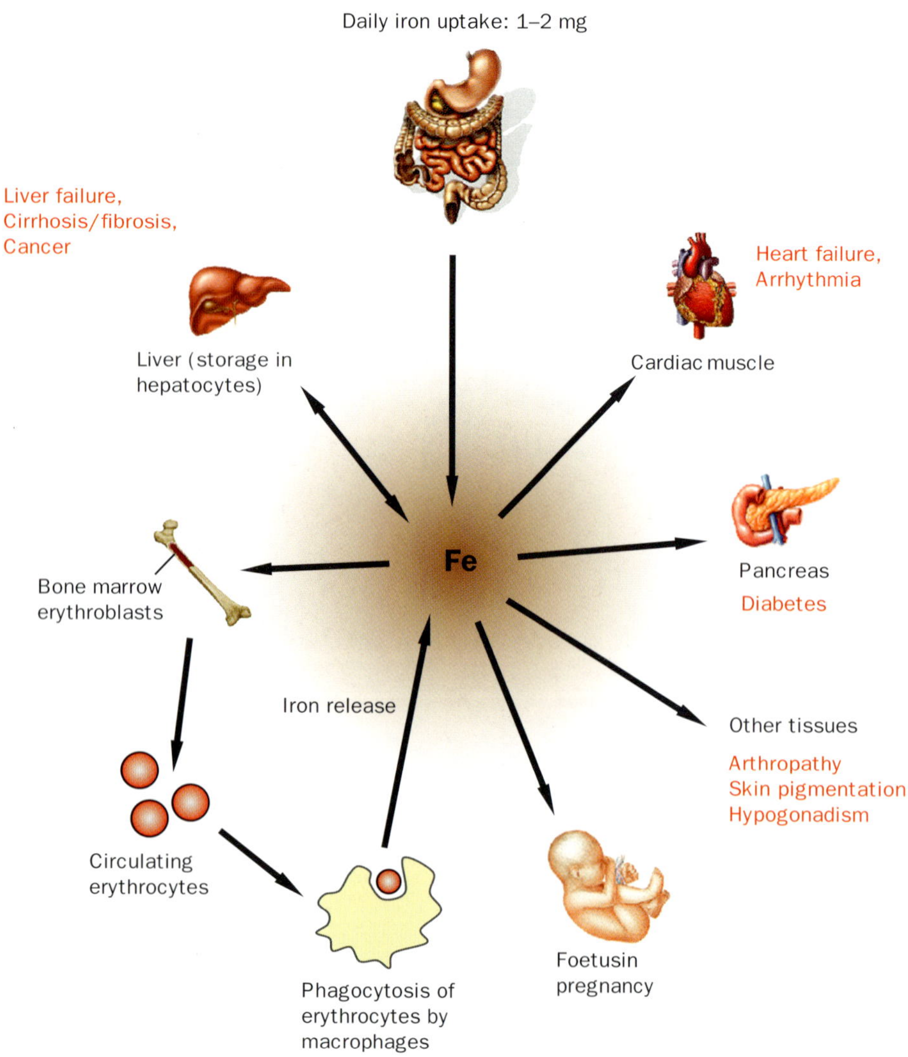

Figure 21-1 The Distribution of Iron

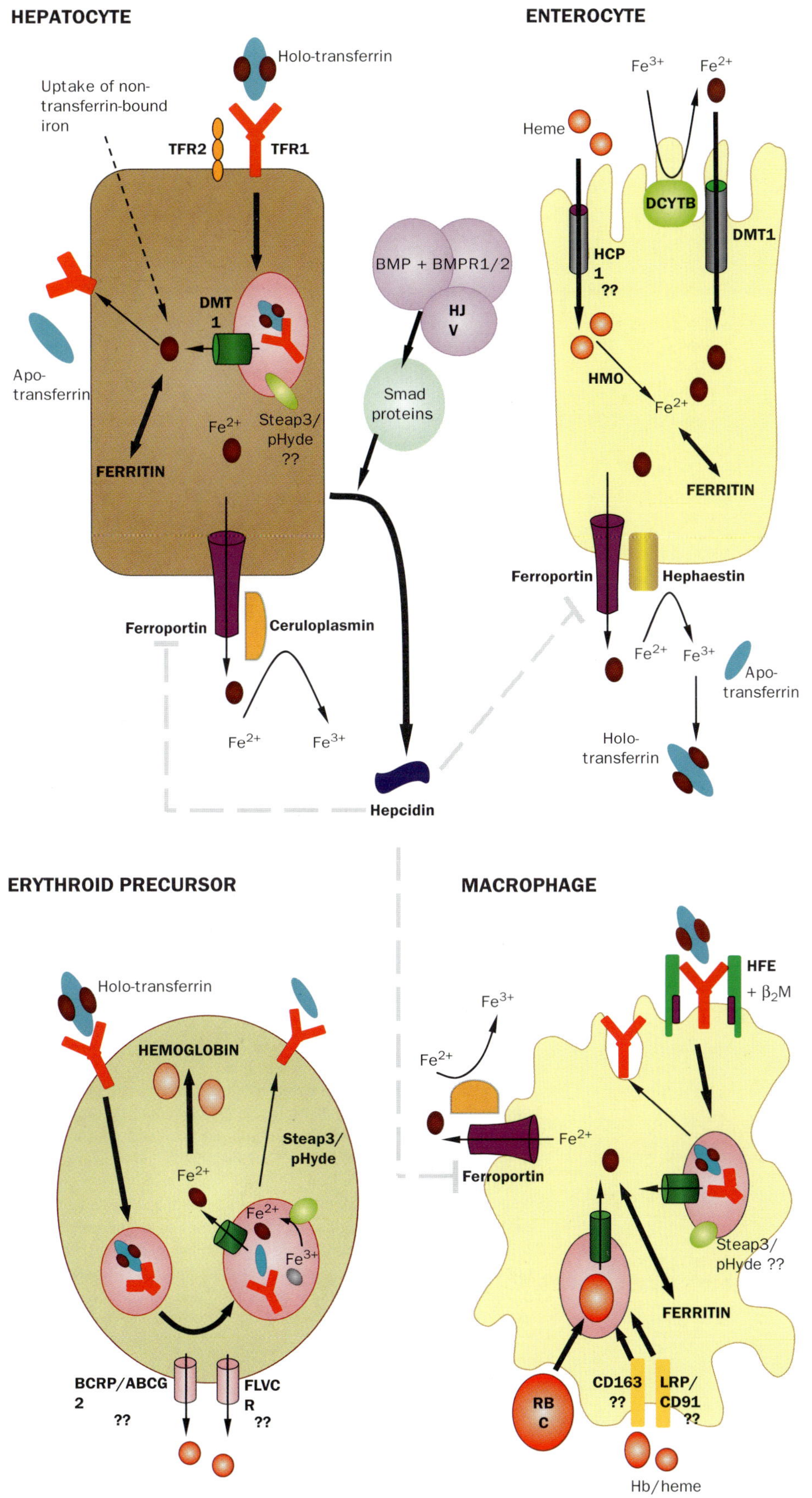

Figure 21-2 Schematic representation of iron transport and utilization by hepatocytes, enterocytes, erythroid cells, and macrophages. Hemojuvelin can upregulate hepcidin synthesis in the liver through the BMP-SMAD signaling pathway. When iron levels are high, increased hepcidin reduces iron uptake and availability by inhibitory binding to ferroportin. Different cell types are known to have different mechanisms of iron uptake. TFR1 levels in hepatocytes are low. It is likely that a non-transferrin-bound iron uptake pathway exists in the hepatocytes. In rat liver, it appears that *Hfe* is expressed predominantly in hepatocytes (Zhang et al., 2004). Arrows show the movement of iron, or the direction/activation of reactions/proteins. BCRP/ABCG2, breast cancer resistance protein/ABC transporter; β_2M, β_2-microglobulin; FLVCR, feline leukemia virus C receptor; HCP1, heme carrier protein 1; HMOX, heme oxygenase; LRP, low-density lipoprotein receptor-related protein; Steap3, six-transmembrane epithelial antigen of the prostate 3.

```
                         C80Y ↓                 G99R ↓
Human HJV          GGVGSGGLCRALRSYALCTRRTARTCRGDL 101
Mouse HJV          -GLASGGLCRALRSYALCTRRTARTCRGDL 94
Zebrafish HJV      -------YCSALRSYALCTQQTARACRGDL 92
Human RGMA         -------FCAALRSYALCTRRTARTCRGDL 97
Mouse RGMA         -------FCAALRTYALCTRRTARTCRGDL 98
Zebrafish RGMA     -------FCTALRAYNSCVRRTARTCRGDL 74
Zebrafish RGMB     -------FCKALRAYSACTQRTAKSCRGNL 94
                          * ***:*  *.::**::***:*

                      P192L ↓ ↓L194P
Human HJV          TCRVQGAWPLLDNDFLFVQATSSPMALGAN 213
Mouse HJV          TCRVQGAWPLLDNDFLFVQATSSPVSSGAN 206
Zebrafish HJV      TCAVQGAWPLIDNQYLYIQATSSPTRESSD 199
Human RGMA         TCKVQGAWPLIDNNYLNVQVTNTPVLPGSA 209
Mouse RGMA         TCKVQGAWPLIDNNYLNVQVTNTPVLPGSA 210
Zebrafish RGMA     TCKVEGAWPLIHNKYLSVQVTNTPVVVGSS 192
Zebrafish RGMB     TCKVEGAWPLIDNNYLSVQVTNVPVVYGSS 204
                   ** *:*****:.*.:* :*.*. *   .:

                                           ↓G320V
Human HJV          DVAMAFS--AEQDLQLCVGGCPPSQRLSRS- 330
Mouse HJV          DVARAFS--AEQDLQLCVGGCPPSQRLSRS- 323
Zebrafish HJV      AIVESYT--PEQDLQLCVWGCPISQRLEML- 316
Human RGMA         EVVNAVEDWDSQGLYLCLRGCPLNQQIDFQ- 328
Mouse RGMA         EVVNAVEDRDSQGLYLCLRGCPLNQQIDFQ- 329
Zebrafish RGMA     EVVNSVEDQDNQDLYLCLHGCPANQRIDFR- 311
Zebrafish RGMB     ELAMAFD--ETQDLQLCMNGCPTSERIDQEG 321
                    :. :      *.* **: ***  .:::.

                           ↓A343PfsX23
Human HJV          IDTARRLCKEGLPVEDAYFHSCVFDVLISG 369
Mouse HJV          IDTARRLCKEGLPVEDAYFQSCVFDVSVSG 362
Zebrafish HJV      FDPAYTHCSSLFPGRDVYFQACLFDVQVTG 350
Human RGMA         YETAVAKCKEKLPVEDLYYQACVFDLLTTG 386
Mouse RGMA         YETAVAKCKEKLPVEDLYYQACVFDLLTTG 385
Zebrafish RGMA     YQSAMAKCKERLPVEDLYFQSCVFDLLSSG 366
Zebrafish RGMB     LESASRRCRDQLEVKDIYFHSCVFDLLTTG 381
                    :.*    *  . :   .* *:::*:**:  :*
```

Figure 21-3 Multiple sequence alignment of hemojuvelin protein sequence. Alignment of selected regions of human hemojuvelin with the mutations, against the mouse and zebrafish hemojuvelin, and human, mouse, and zebrafish RGMs. Identified mutations are indicated by arrows. Red indicates small, hydrophobic, and aromatic residues. Blue indicates acidic residues. Magenta indicates basic residues. Green indicates hydroxyl, amine, and basic residues. * indicates identical residues in the alignment; colon (:), conserved substitutions; period (.), semi-conserved substitutions of residues. (Alignment produced using ClustalW version 1.83.)

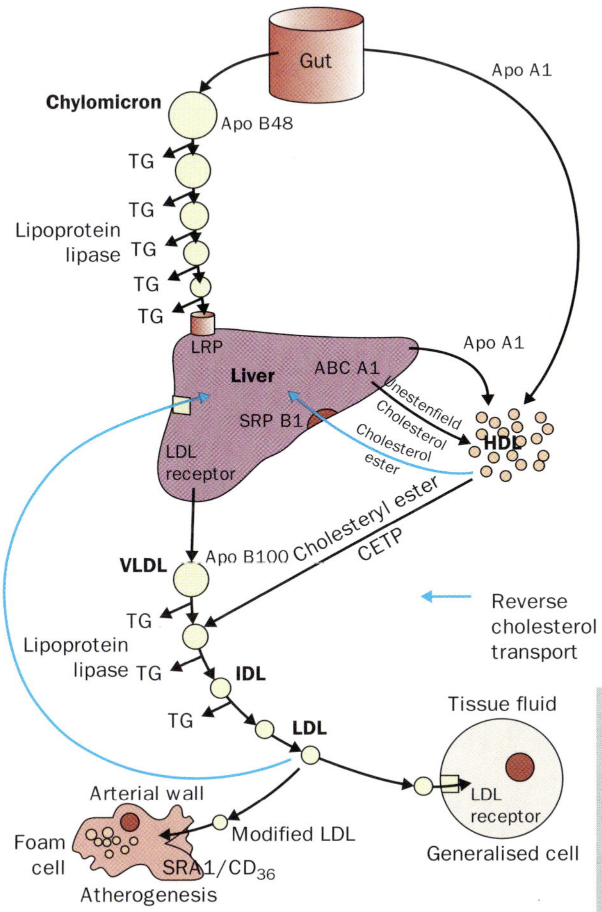

Dietary cholesterol absorbed in the gut is transported to the liver in chylomicrons, which are rich in triglyceride. Cholesterol is also synthesised de novo in the liver, gut, and central nervous system. Cholesterol in the liver is packaged along with triglycerides and secreted as VLDL, which is then converted into LDL in the circulation. LDL delivers cholesterol to most tissues, and the uptake of cholesterol is facilitated by LDL receptors. Excess cholesterol is removed from cells by HDL and is then either transferred to LDL and VLDL, whence it can return to the liver via hepatic LDL receptors or be taken up directly from HDL via the hepatic class B scavenger receptors. cholesterol is also excreted in bile both as free cholesterol and as bile acids, which are both subject to enterohepatic circulation. Most of the plasma cholesterol is present in LDL, with smaller amounts in VLDL and HDL. Total plasma cholesterol mainly reflects LDL cholesterol. In the fasting state, plasma triglyceride generally indicates the VLDL concentration.

LDL – low density lipoprotein
HDL – High density lipoprotein
VLDL – Very low density lipoprotein
IDL – Intermediate density lipoprotein
CETP – cholesteryl ester transfer protein
ABCAI – ATP binding cassette transporter 1
SRB 1 – class B scavenger receptors
LRP – lipoprotein related receptor
TG – triglyceride
Apo – apolipoprotein
SRA1 – steroid receptor RNA activator
$CD_{3/4}$ – thrombospondin receptor

Figure 24-1 The Role of Cholesterol in the Metabolism of Lipoprotein.
SOURCE: Adapted, with permission, from Bhatnagar et al., 2008

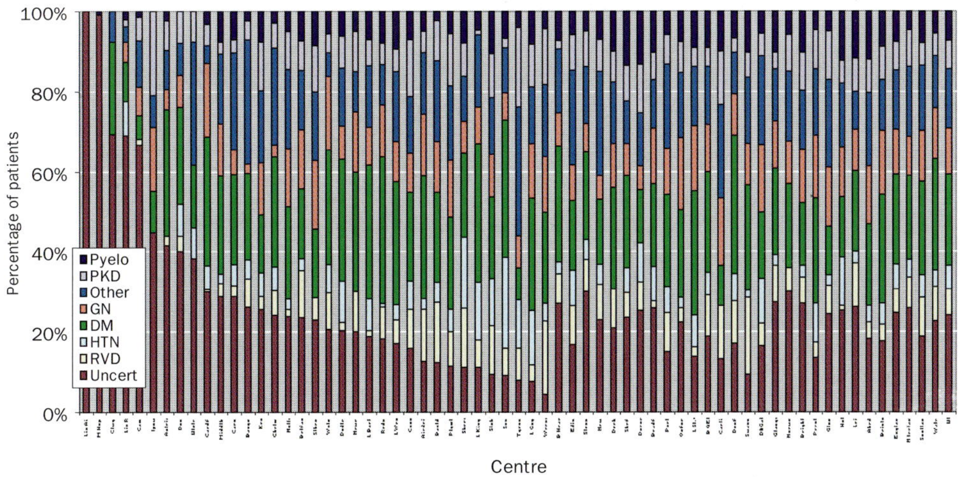

Figure 25-1 Proportion of patients with primary renal disease based on UK Renal Registry 2009 (Byrne, 2010).

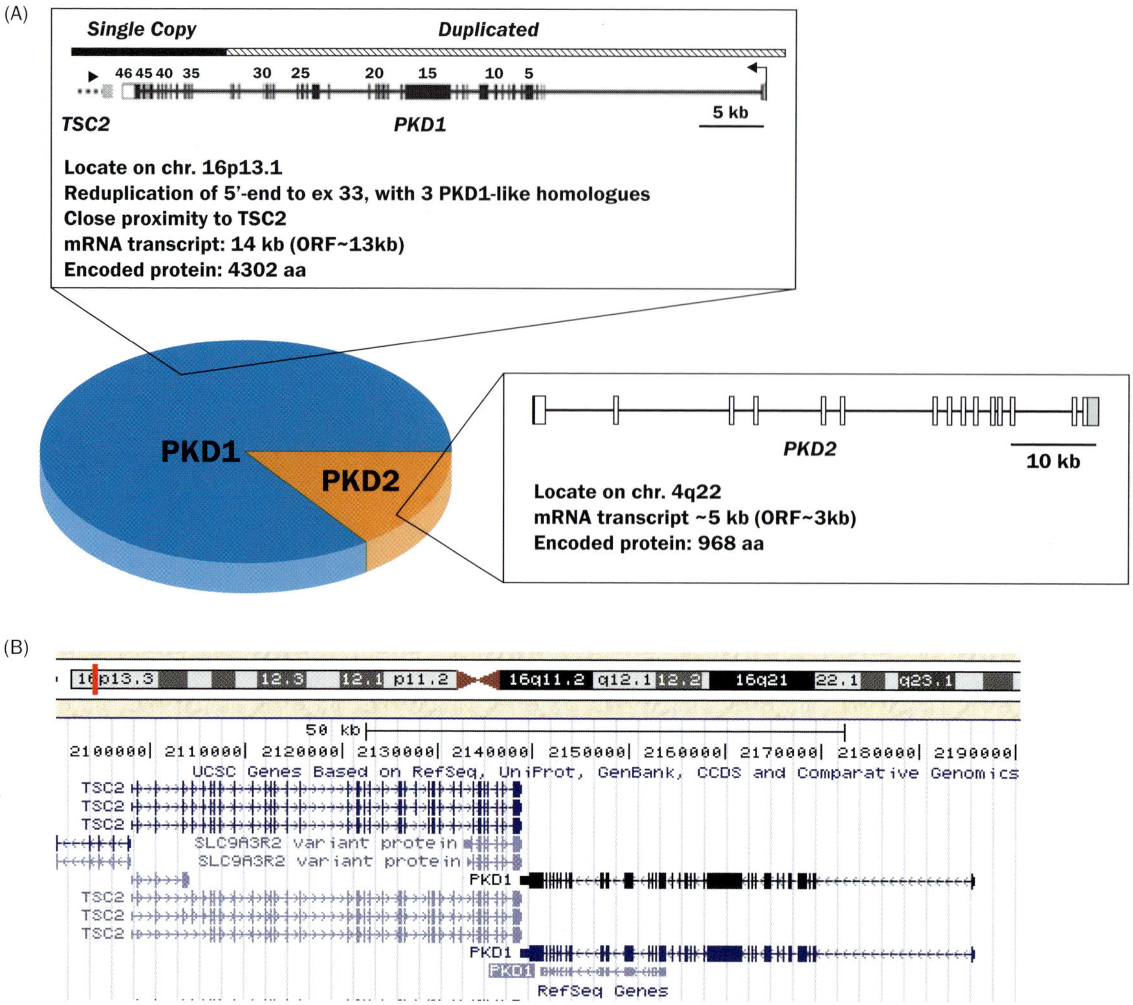

Figure 25-4 a) The two genes (PKD1 & PKD2) causing ADPKD and the contiguous gene TSC2 in close proximity to PKD1. b) Detailed molecular genetics of TSC2 and PKD1 on chromosome 16p.

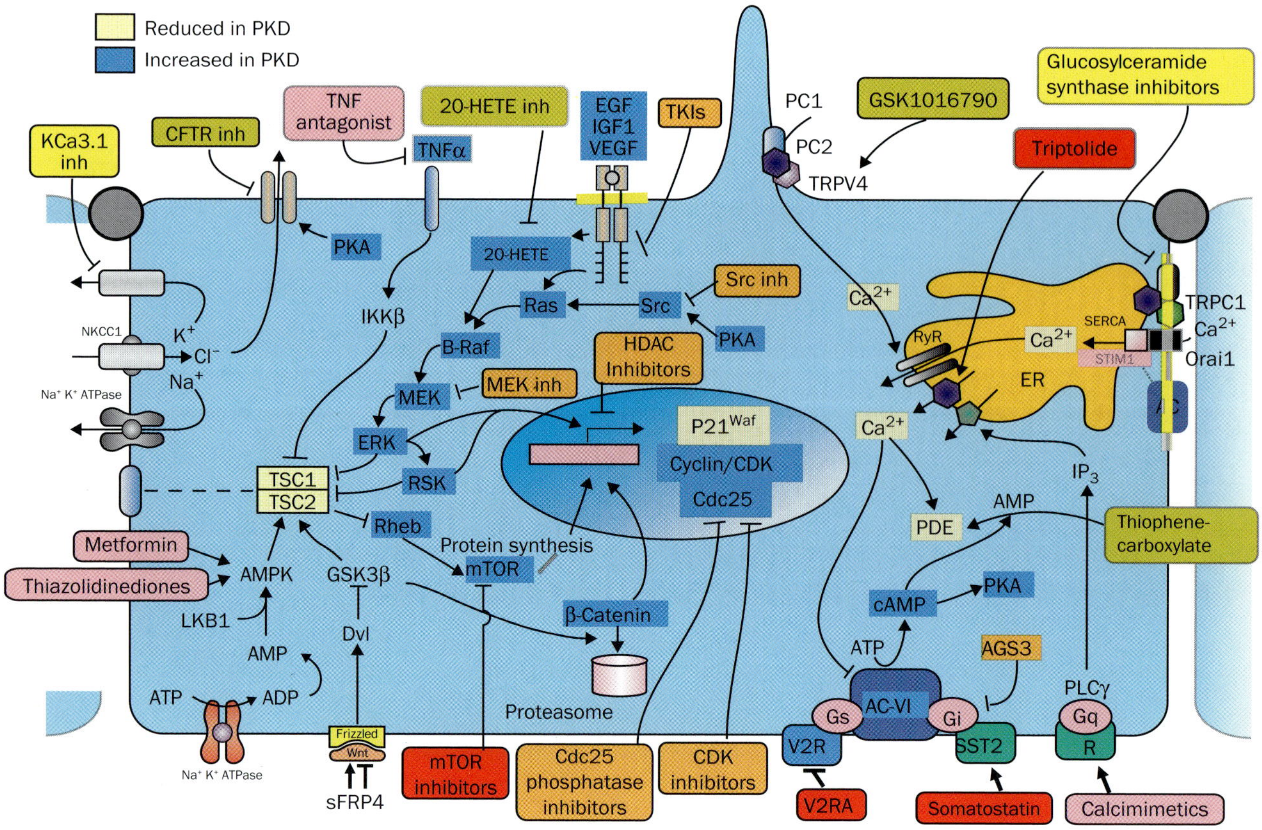

Figure 25-5 Possible sites for drug development shown in colored boxes (Adapted from Torres et al., 2007).

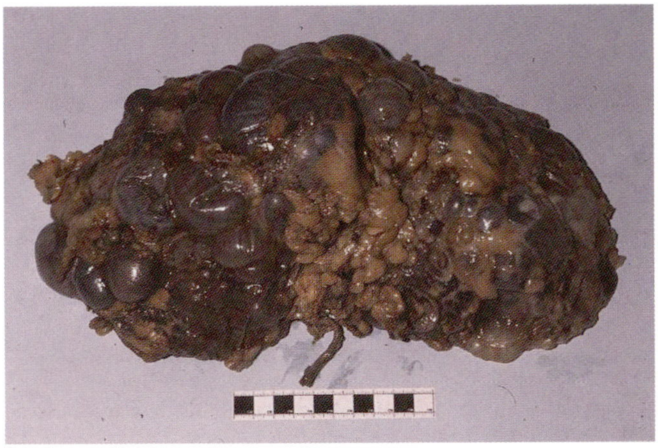

Figure 25-6 Typical appearances of an enlarged cystic kidney removed from a patient with ADPKD.

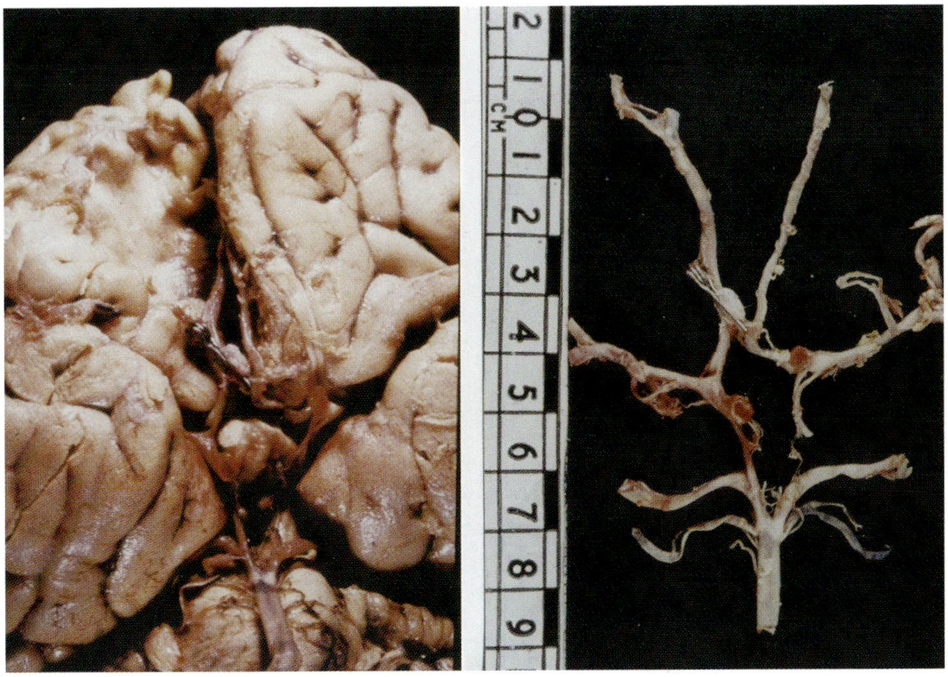

Figure 25-12 Aneurysm rupture with clipping showing (left panel).

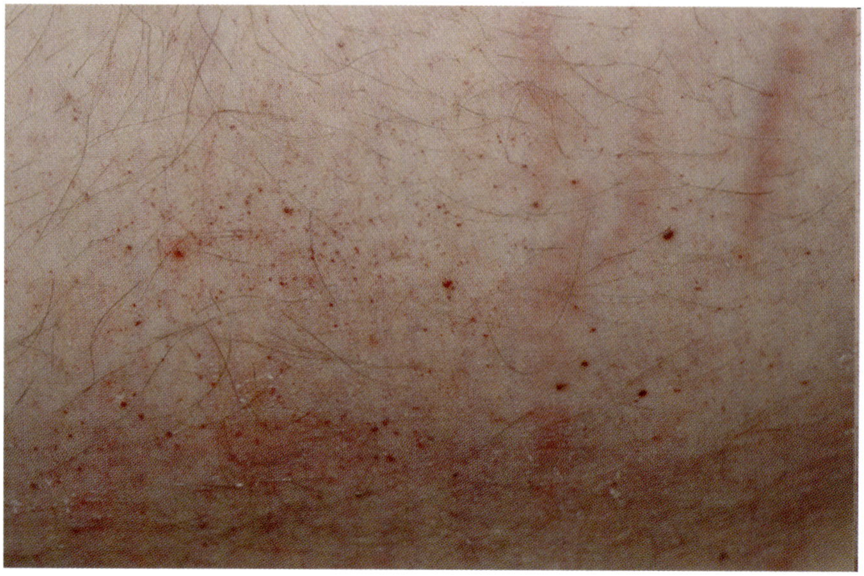

Figure 25-13 Angiokeratomas of the skin seen in a patient with Fabry disease caused by deposition of globotriaosylceramide (GL-3) in skin capillaries. Lesions are dark red to blue-black and punctate, they appear in clusters in the superficial layers of the skin. The lesions are often symmetrical, and most dense between the umbilicus and the knees. Mucosal areas are commonly involved.

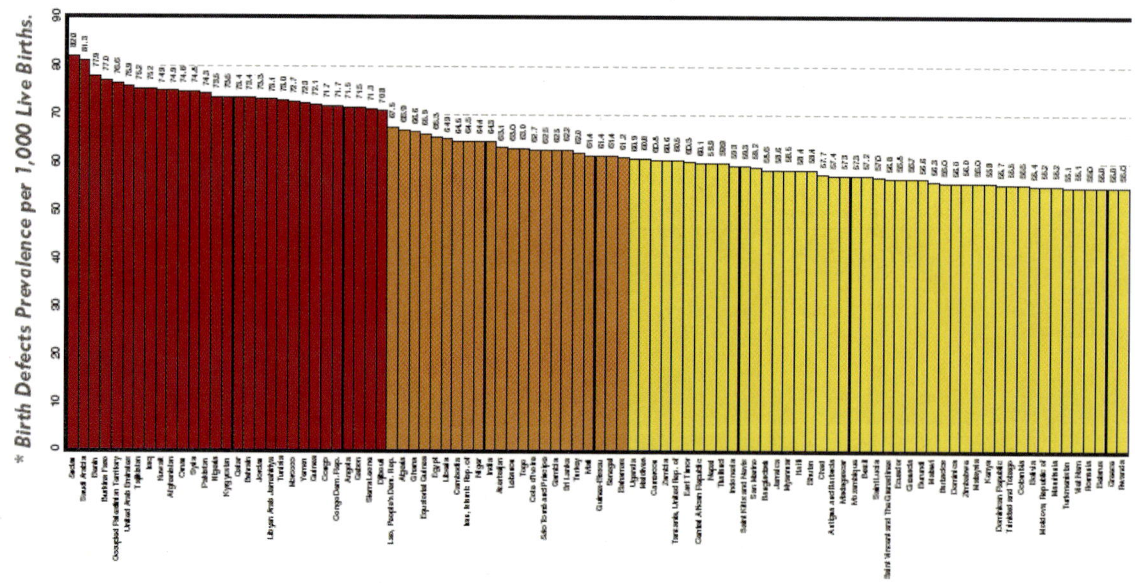

Figure 26-1 The Prevalence of "Birth Defects" in Developing Countries
SOURCE: adapted from Executive Summary, March of Dimes Global Report on Birth Defects, 2006

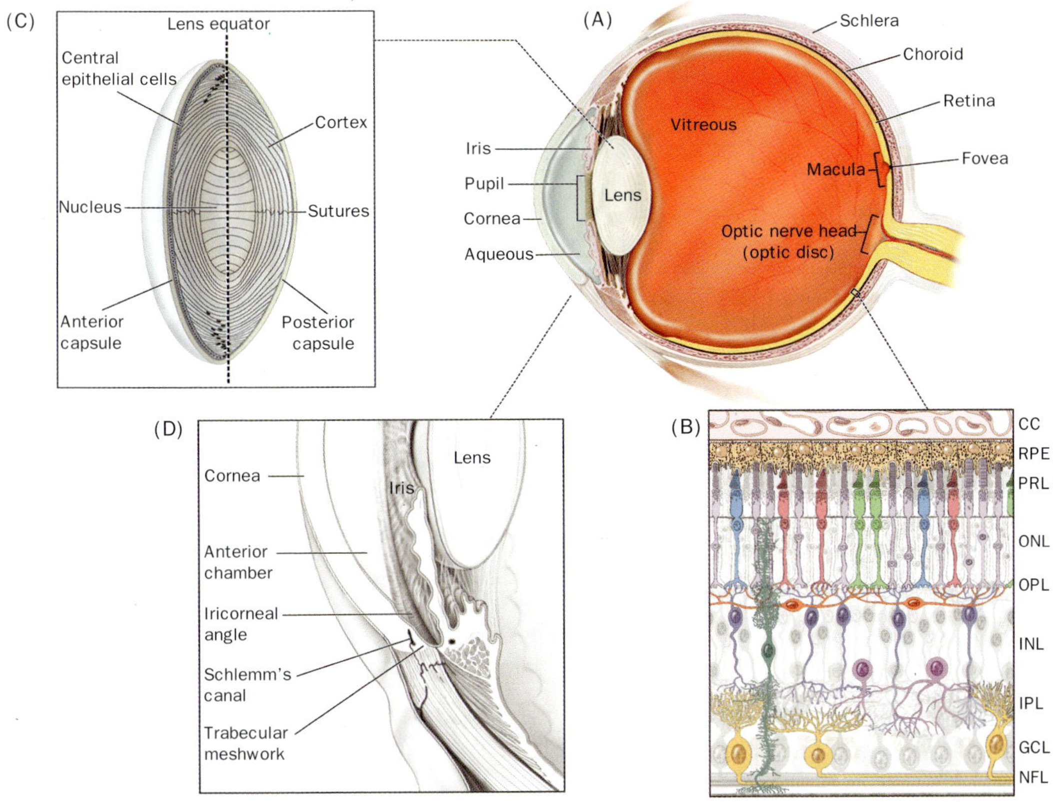

Figure 27-1 Schematic representation of the eye and anatomical structures affected in disease. A. Sagittal view of the eye structure; B. Layers and cell types in the retina; C. Histological organization of the lens; D. Magnification of the irido-corneal angle and trabecular meshwork. CC = choriocapillaris; RPE = retinal pigment epithelium; PRL = photoreceptor layer; ONL = outer nuclear layer; OPL = outer plexiform layer; INL = inner nuclear layer; IPL = inner plexiform layer; GCL=ganglion cell layer; NFL=neural fiber layer.

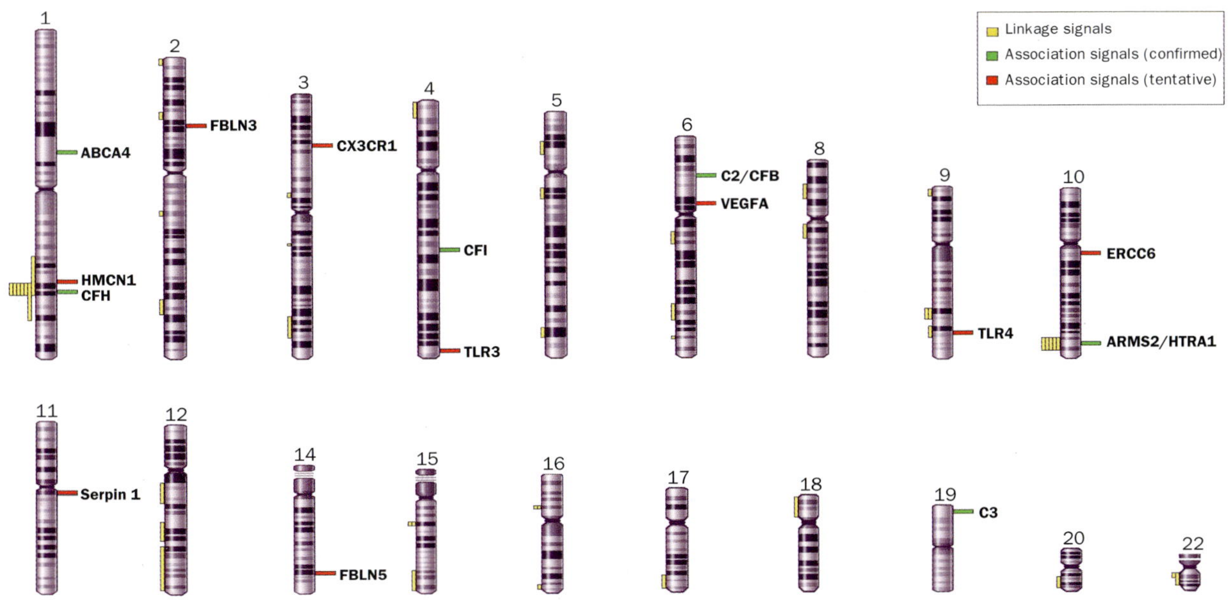

Figure 27-3 Summary of regions of linkage (yellow) and association signals (confirmed in green and tentative in red) for AMD susceptibility loci/genes.

SOURCE: Reproduced from ref. (Swaroop et al, 2009).

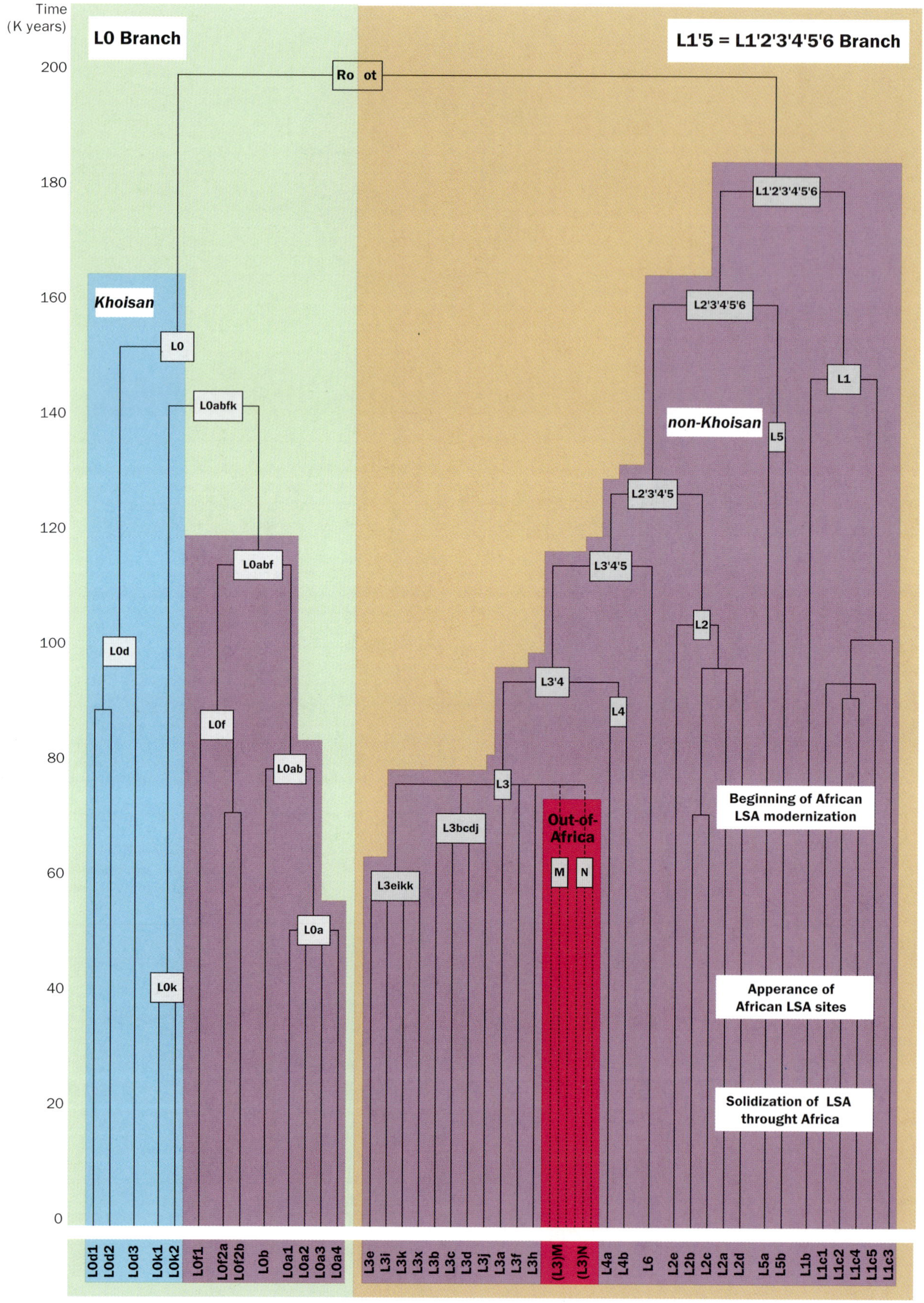

Figure 33-1 Tree showing global mtDNA macro-haplogroups. (Behar et al., 2008). Note: non-African mtDNA haplogroups trace ancestry to haplogroups M and N on this tree.

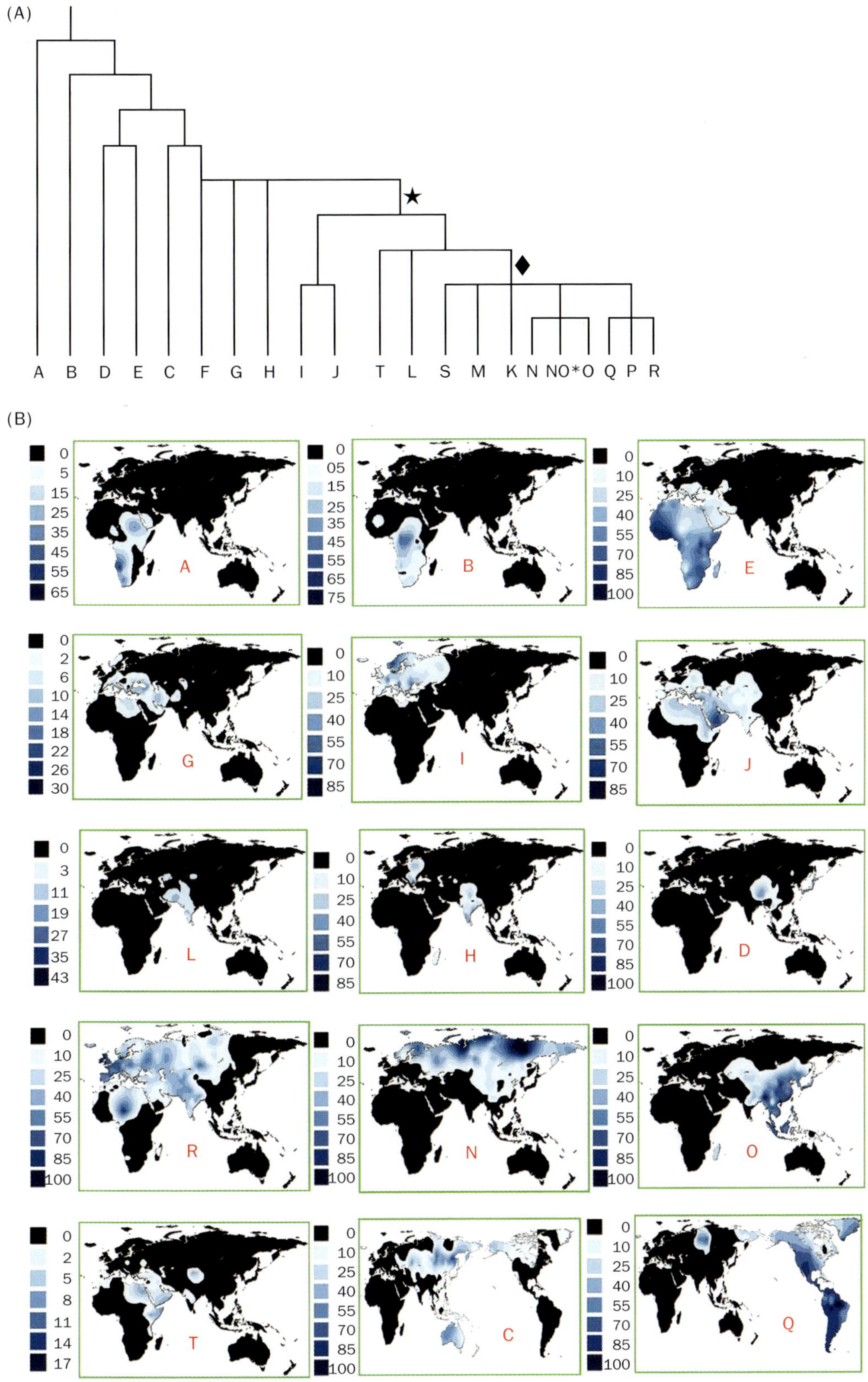

Figure 33-2 Y chromosome haplogroup tree showing the relationship of 20 major Y haplogroups (a) and the geographic frequency maps of some haplogroups (b) (Chiaroni et al., 2009). Haplogroups A, B and E are the common ones found in Africa.

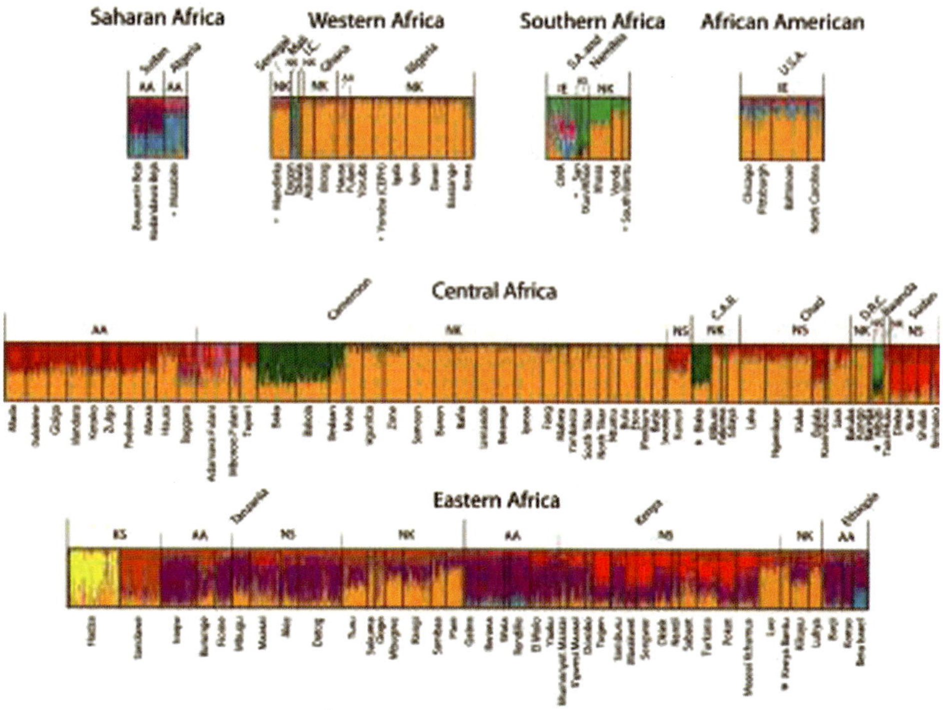

Figure 33-3 Geographic and genetic structure of populations within Africa. (Tishkoff et al., 2009).

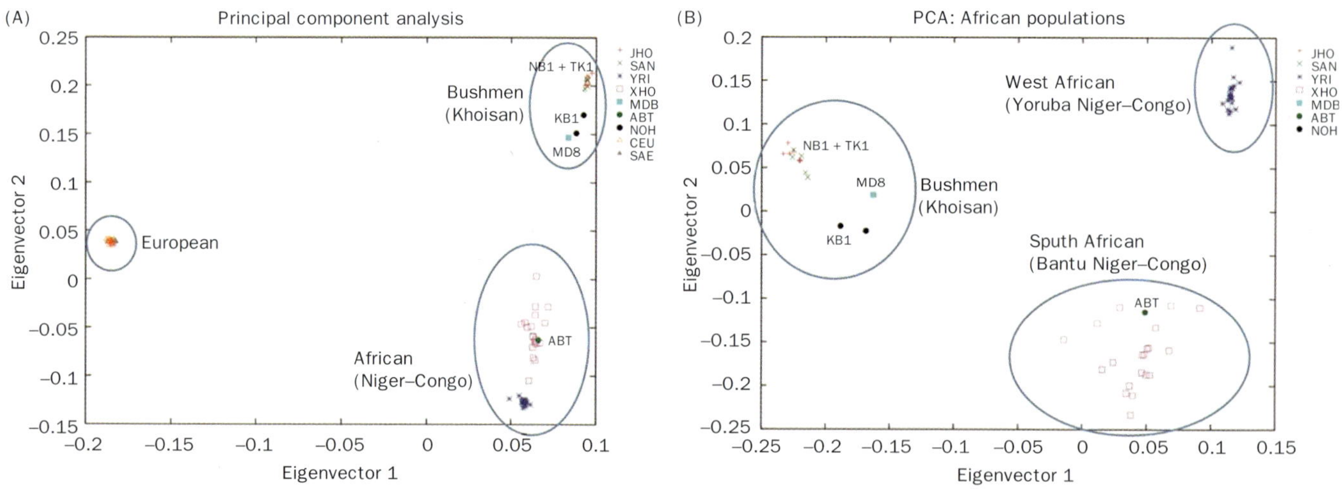

Figure 33-4 Three-way population structure based on 174,272 autosomal SNPs using PCA. (Schuster et al., 2010).

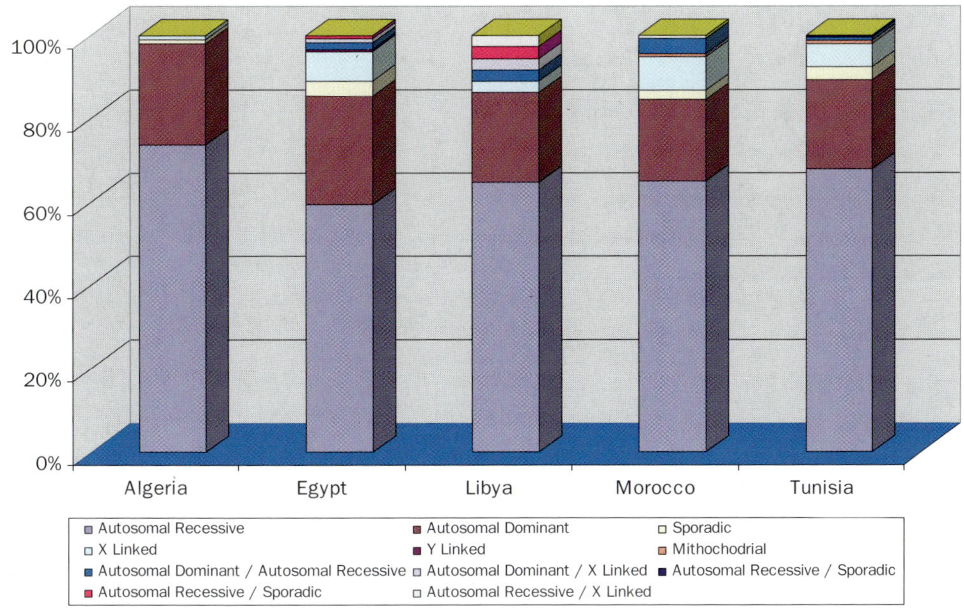

Figure 34-2 Mode of Inheritance of Genetic Disorders in North African Countries

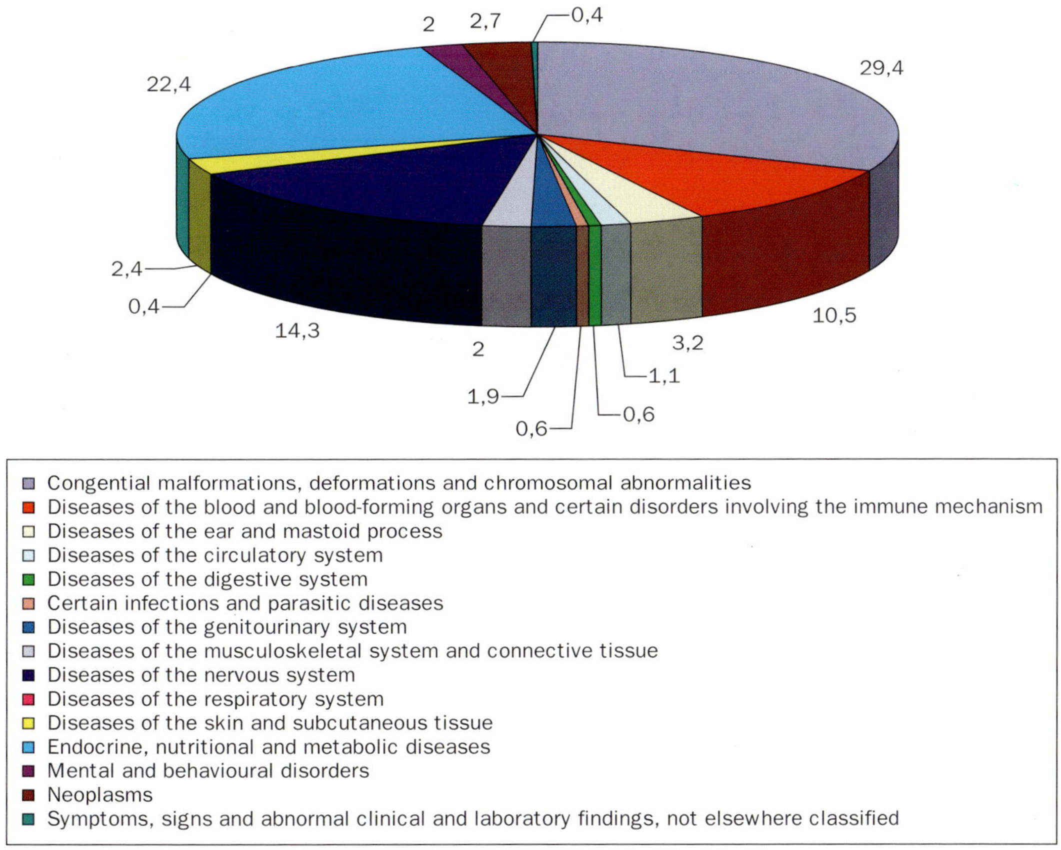

Figure 34-4 WHO-ICD10 Classification of Genetic Disorders in North Africa

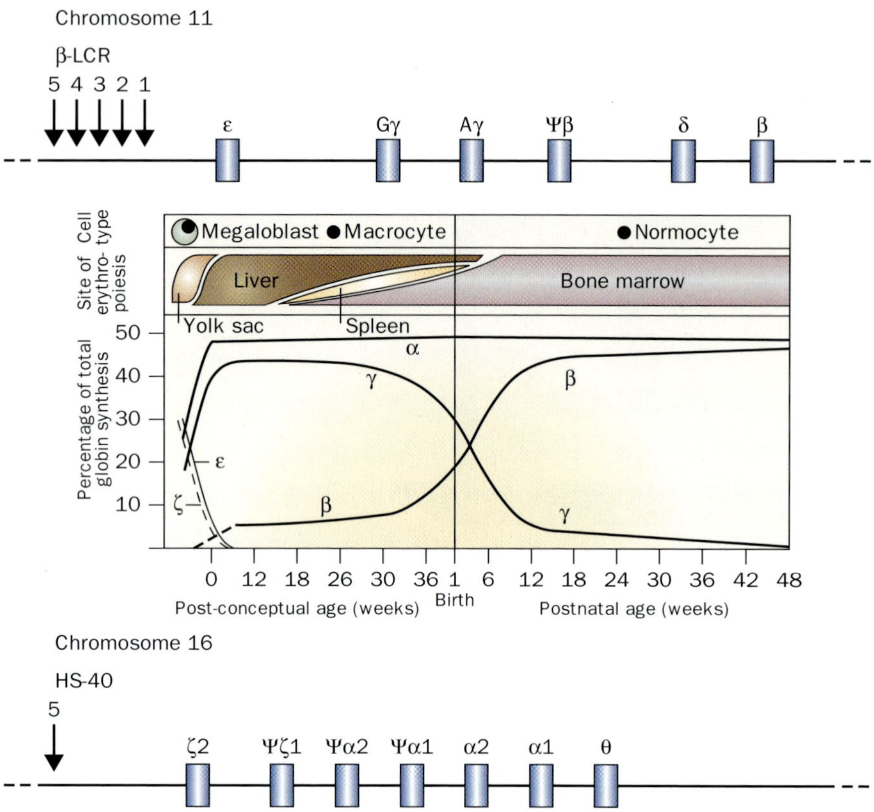

Figure 36-2 The structure of human hemoglobin (Hb) changes from embryonic to fetal and adult life. (Weatherall, 2001) Refer color figure.

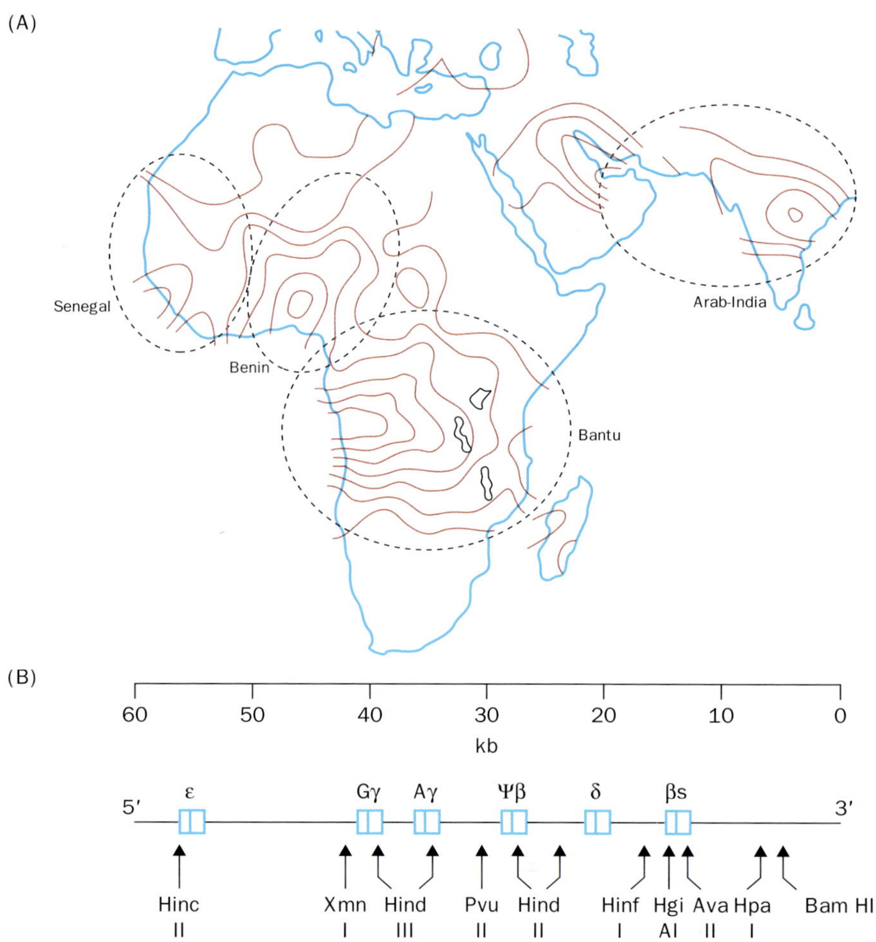

Figure 36-3 ORIGIN OF THE BETA S-GLOBIN GENE. (A). The S Hb gene variant appeared independently in three sites on the African continent, as illustrated by the 3 Hb gene cluster haplotypes that have been described on the continent (Senegal, Benin Bantu). (B). B-globin haplotypes are generated by linkage patterns of the restriction fragment length polymorphisms (RFLPs). Refer color figure.

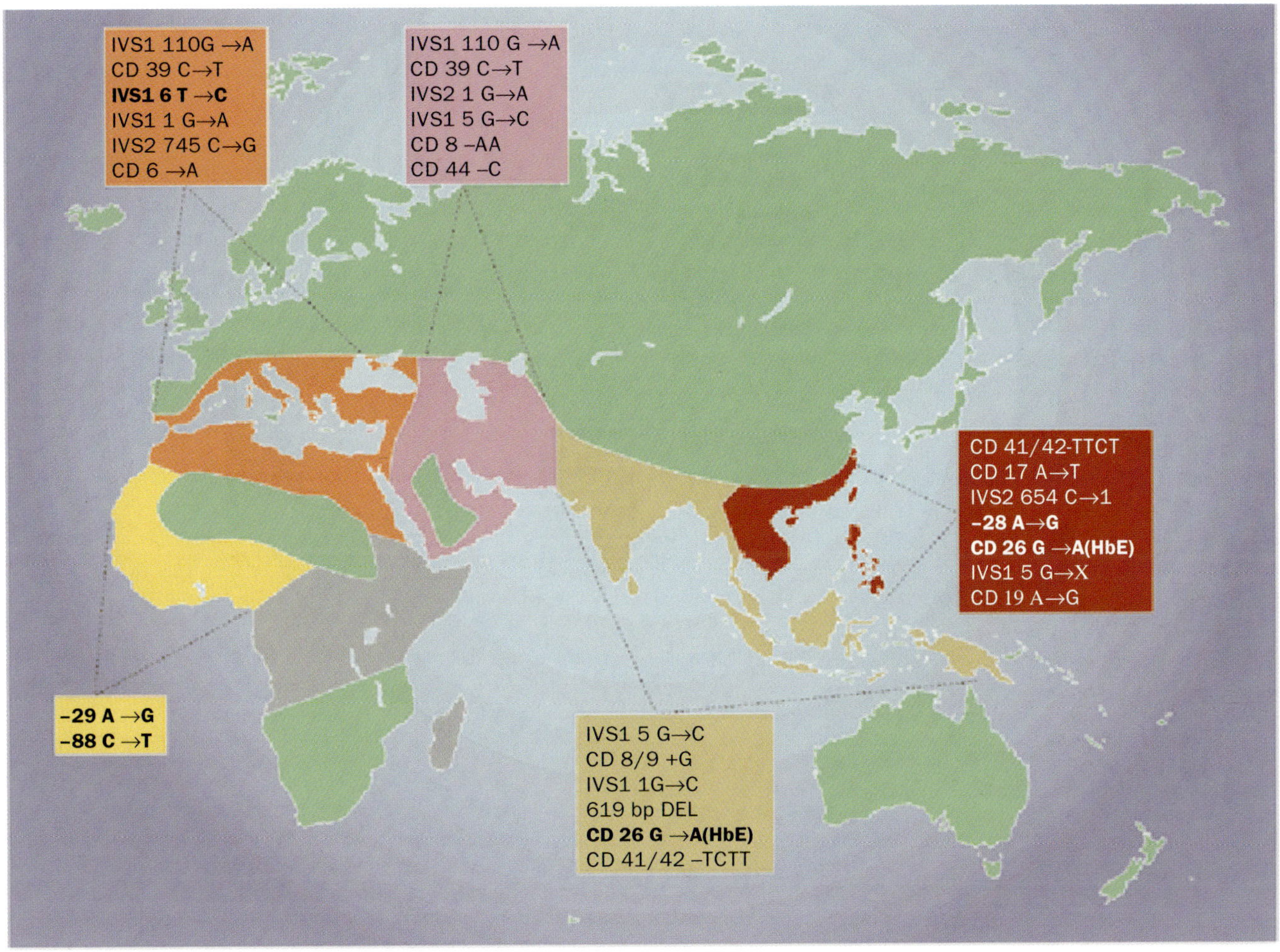

Figure 36-4 Global distribution of β-thalassemia mutations (Weatherall, 2001).
source: While β-thalassemia is not endemic in southern Africa, patients with β-thalassemia are managed in the hematology clinic. They originate predominantly from two areas, the Mediterranean (Greece and Cyprus) and the Indian subcontinent. The local mutation profiles differ somewhat from those of the parent population to local founder effects. For example, in the local Asian Indian population, the codon 8–9 mutation has never been observed, while the IVS1nt5 mutation predominates (Krause, 1994; see Table 36-1). Rare β-thalassemia patients from other population groups are also seen.

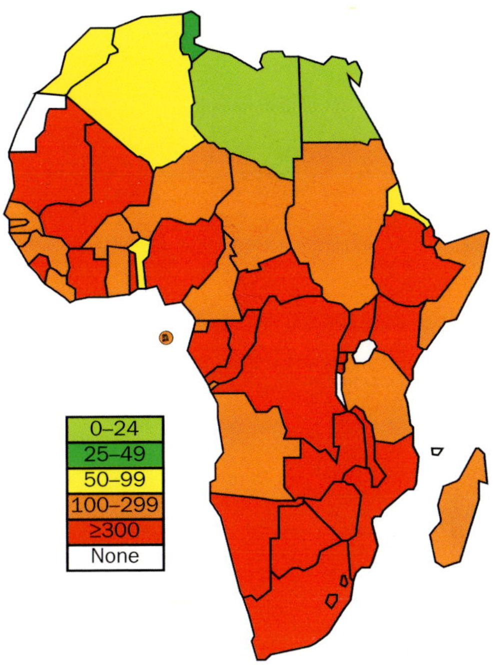

Figure 43-1 Estimated new TB cases (all forms) in African countries per 100,000 population (World Health Organization, 2009).

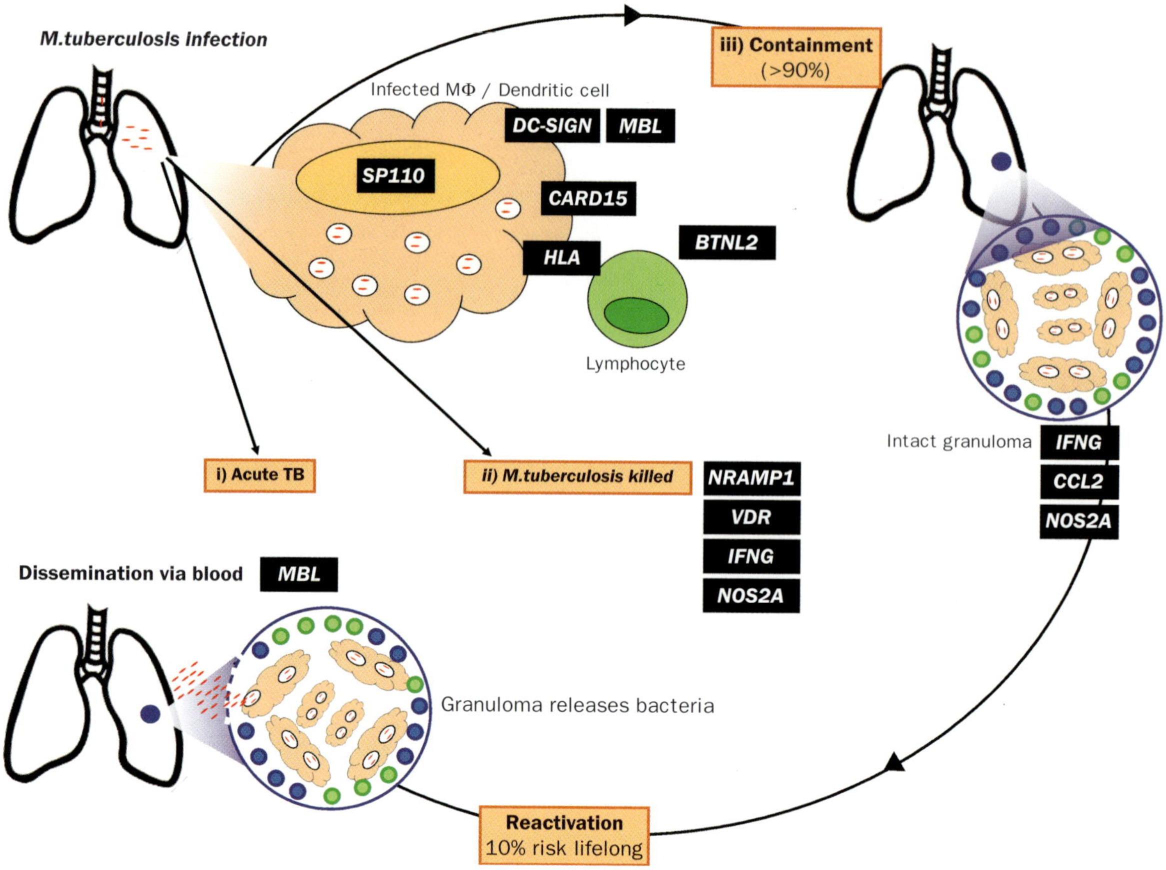

Figure 43-2 Genetic involvement in the TB disease process. A simplified representation of different outcomes after *M. tuberculosis* infection, and some of the genes that may be involved at various stages. The bacteria enter the respiratory system of the host via inhaled droplets and are engulfed by macrophages (MΦ) and dendritic cells. There are three potential outcomes after inhalation of *M. tuberculosis*: (i) infection advances to active TB, (ii) *M. tuberculosis* is instantly killed by the pulmonary immune system, or (iii) infection does not develop into active disease, because the bacteria are controlled in granulomas. This containment can last for decades, or the lifetime of the individual. *M. tuberculosis* can disseminate from granulomas, causing active TB (reactivation disease) (Adapted from Kaufmann & McMichael, 2005).

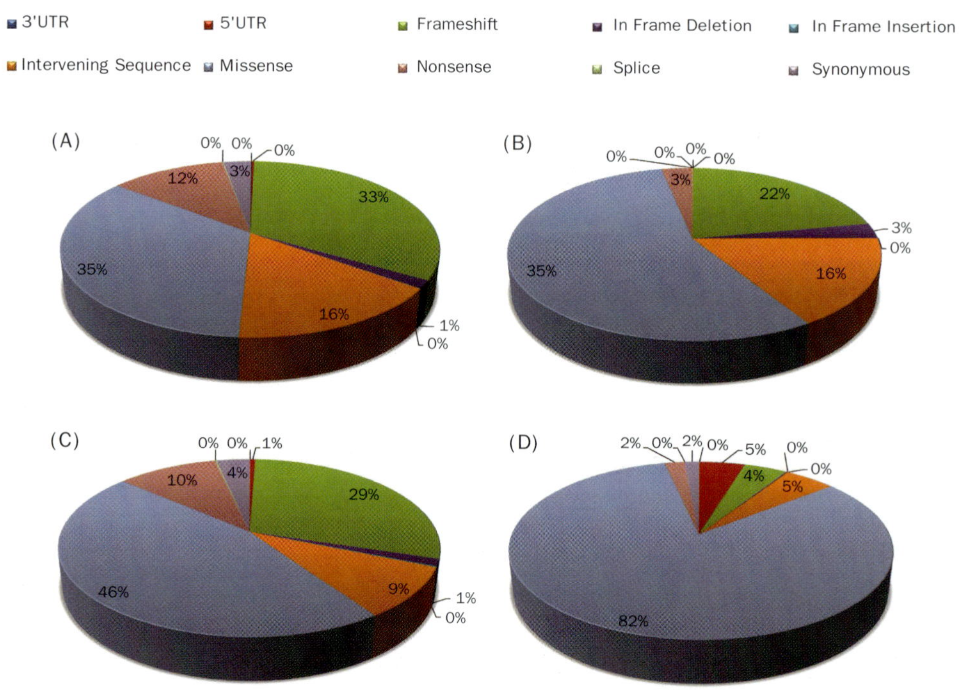

Figure 44-1 Pie charts of distinct mutations, polymorphisms and variants in *BRCA1* and *BRCA2*. Data were retrieved from Breast Cancer Information Core (BIC): http://research.nhgri.nih.gov/bic/). (A) *BRCA1*, worldwide; (B) *BRCA1*, Africa only; (C) *BRCA2*, worldwide; and (D) *BRCA2*, Africa only.

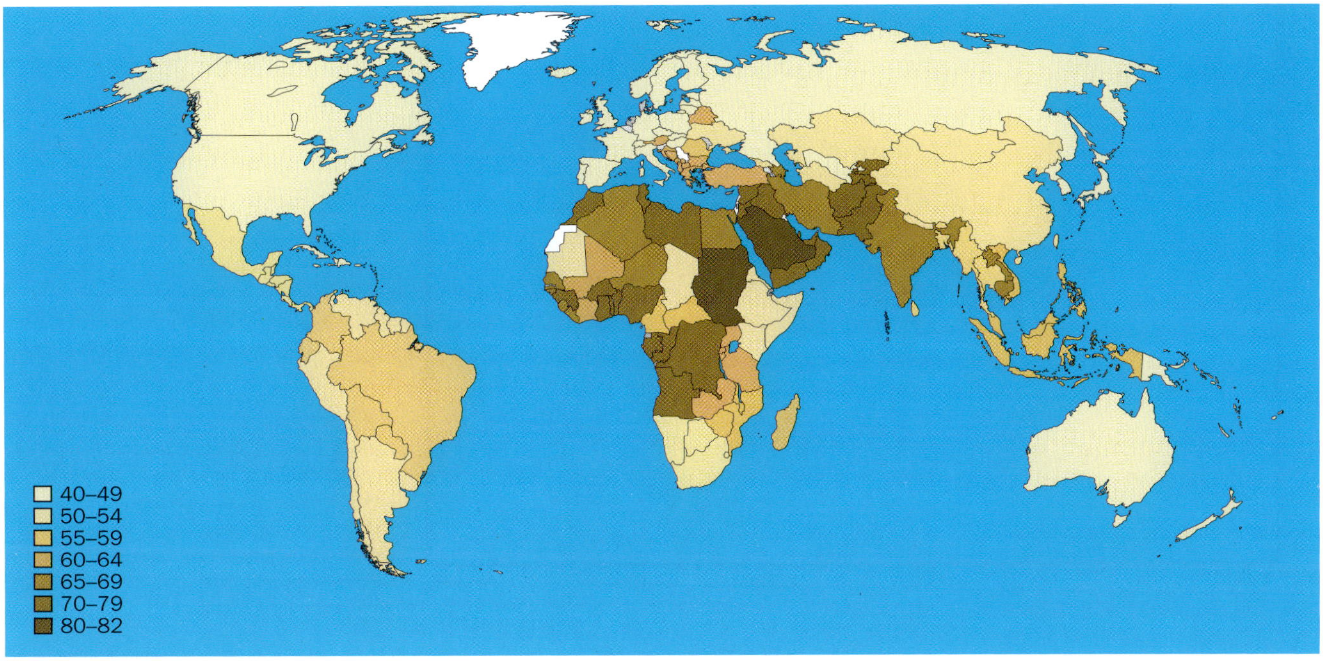

Figure 50-1 Global birth defects prevalence per 1000 live births (adapted from Christianson et al., 2006).

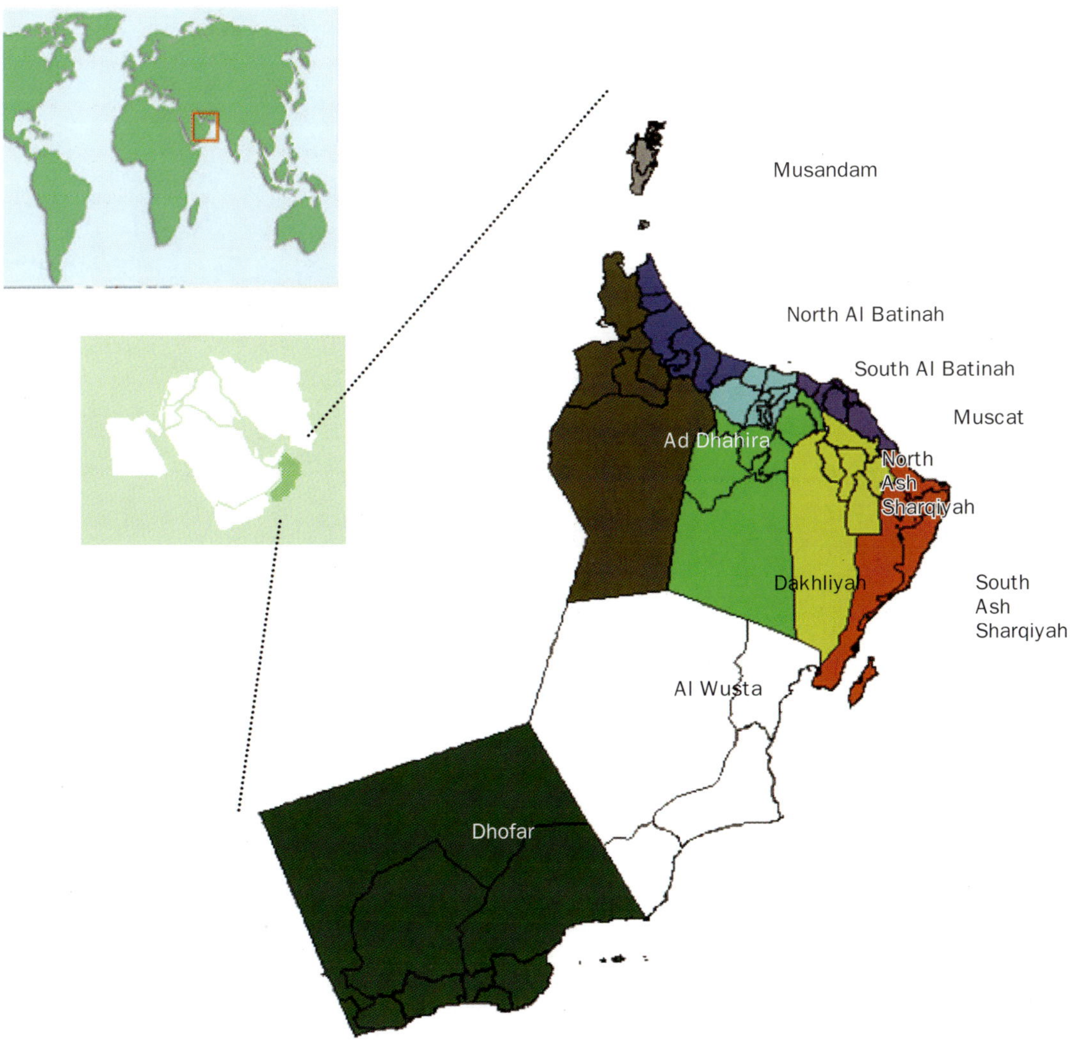

Fig. 54-1 Regional Map of Oman. Division of Oman Territory into Regions (Manataq).

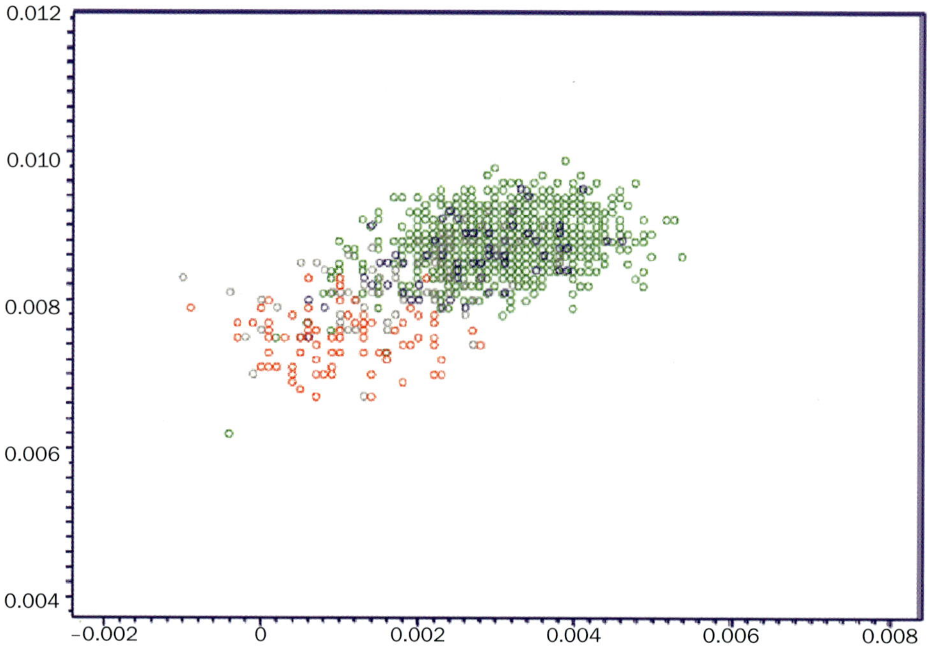

Figure 72-1 Principle component analysis of the Chinese and Japanese populations, using three Han Chinese groups: Han Chinese Beijing (grey circle), Han Chinese Denver (blue circle) and Taiwan Han Chinese (green circle), and the Japanese population (red circle).

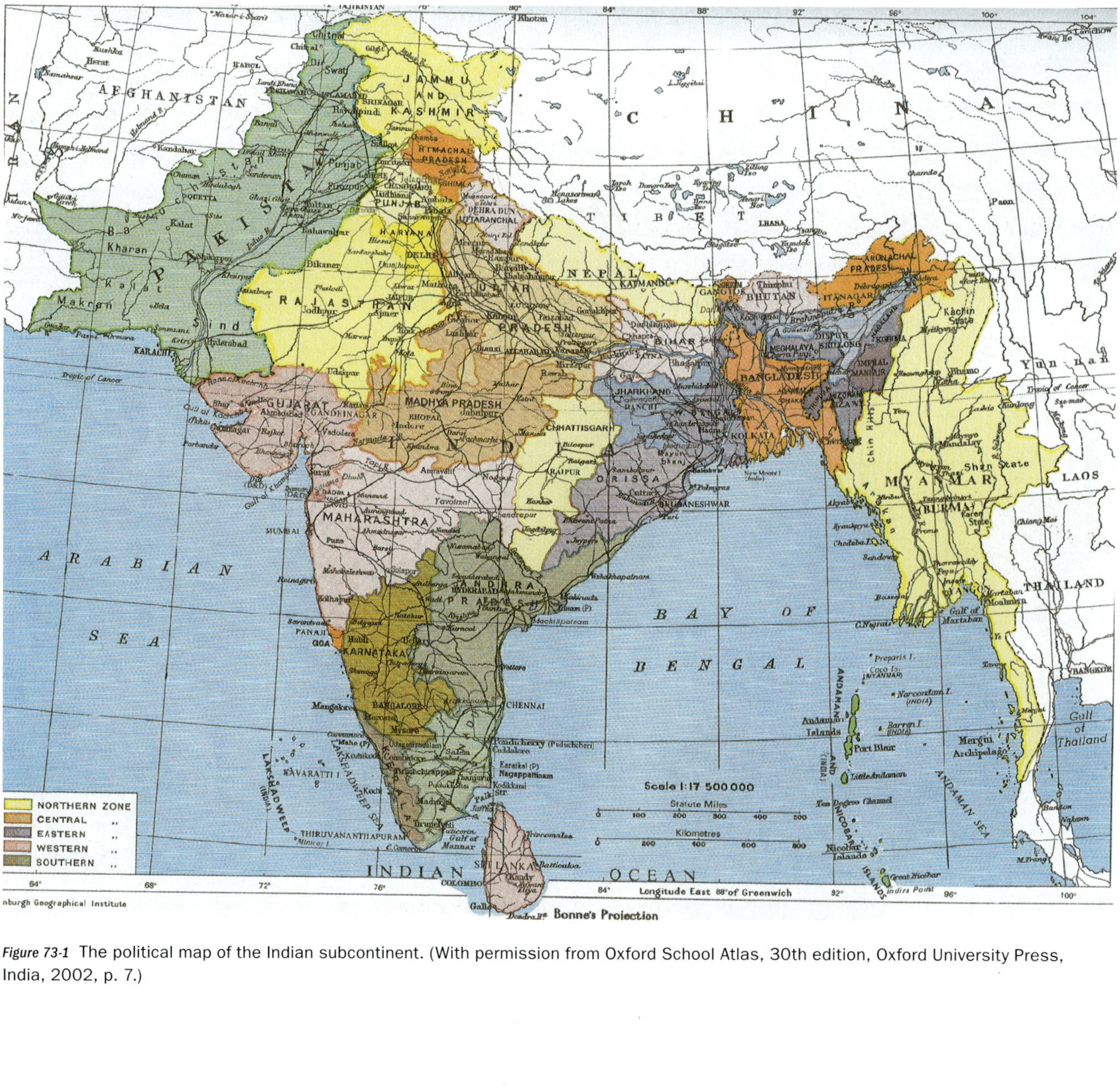

Figure 73-1 The political map of the Indian subcontinent. (With permission from Oxford School Atlas, 30th edition, Oxford University Press, India, 2002, p. 7.)

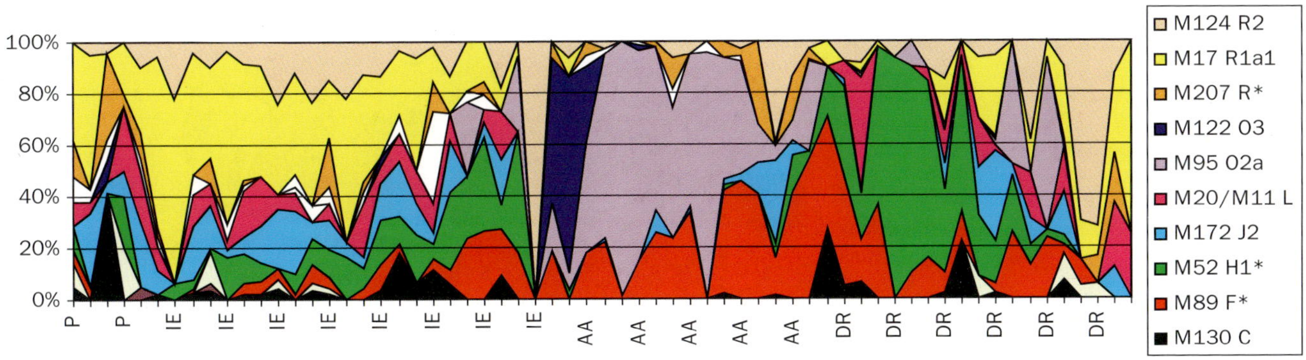

Figure 74-1 Picture of 100% stacked area of NRY HG allelic composition of 58 Indian and 5 Pakistani populations (totaling 2447 samples) arranged according to the language they speak at present. A clear trend of various alleles in different language speakers was discernible. Note the relative distributions of R1a1-M17 (yellow color), O2a-M95 (violet), H1-M52 (green); L-M20 (pink) F*-M89 F* (red) and C-M130 (black) in various language speakers. P = Pakistan; IE = Indo-European speakers; TB = Tibeto-Burmese speakers; AA = Austro-Asiatic speakers; and DR = Dravidian speakers, all from India. For exact population caste / tribe names and their references, refer to Table 74-1.

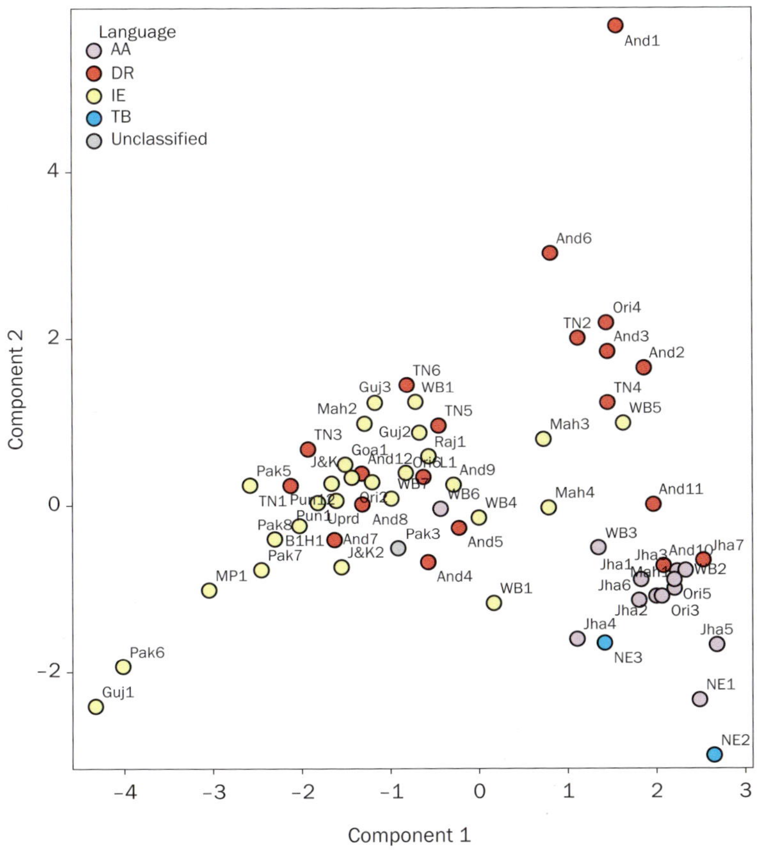

Figure 74-2 The principal component analysis (PCA) of NRY/HG data of Indian populations available in literature.
NOTE: The population Hazara from Pakistan was removed from analysis, because it showed very high variance, making differentiation among the rest of the populations difficult.

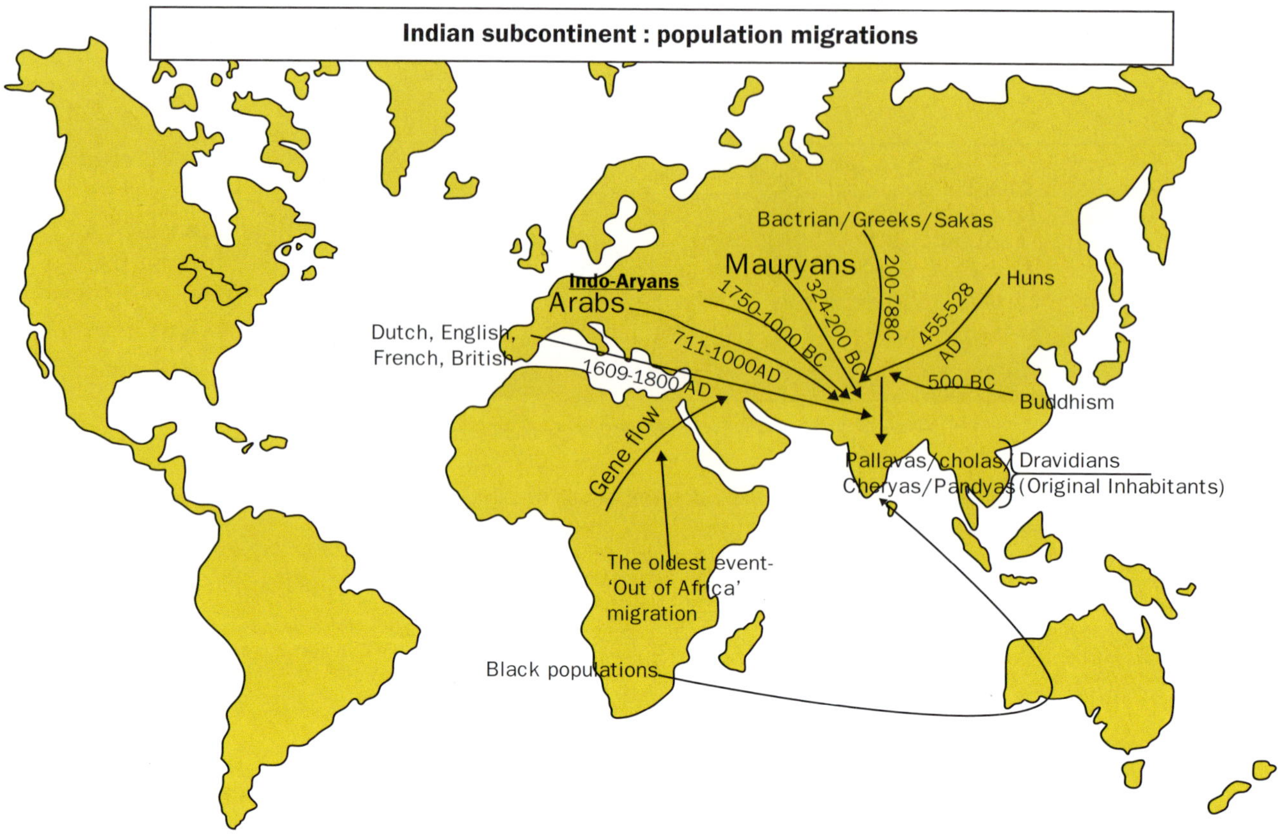

Figure 75-1 Pictorial representation of multiple waves of migration of various population groups that amalgamated with the people of India. Arrows point toward the direction of migration at particular periods in history.

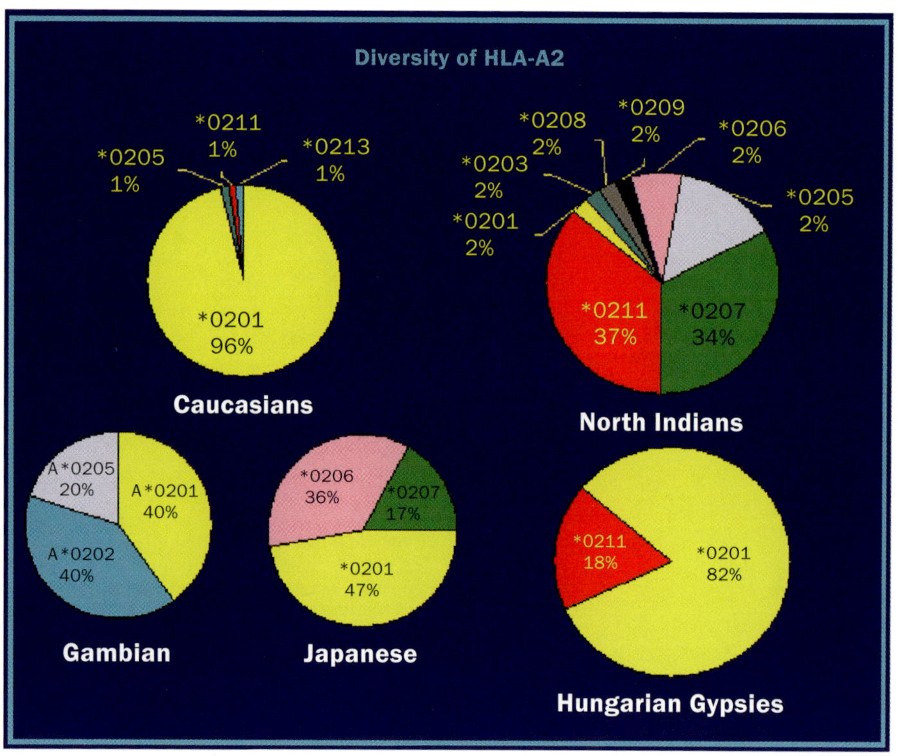

Figure 75-2 Distribution of HLA-A2 subtypes in Caucasians, Gambian, Japanese, North Indians, and Hungarian gypsies. Note the extreme heterogeneity in HLA-A2 in the Asian Indians, and similarity in the occurrence of unique A*0211 in Asian Indians and the Hungarian gypsies.

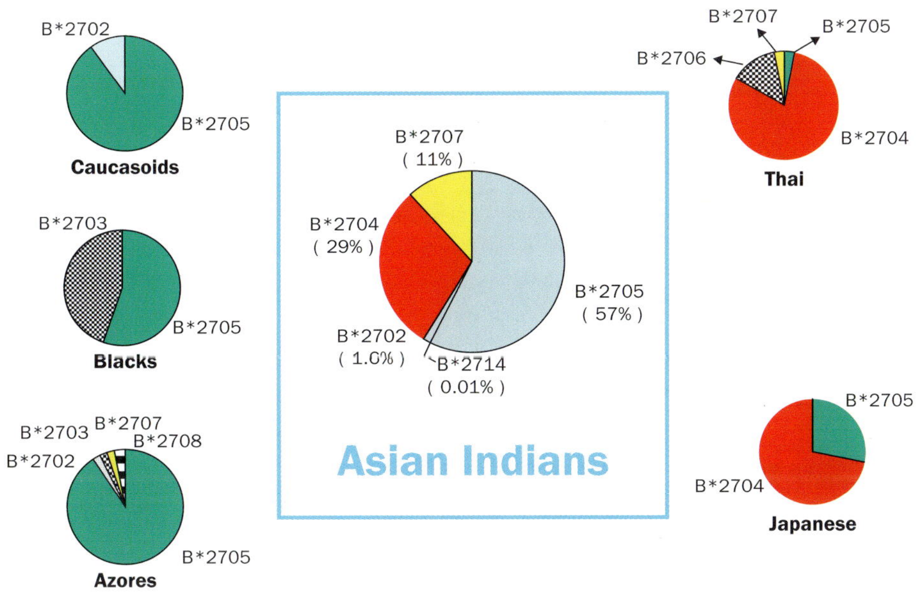

Figure 75-3 Distribution of HLA-B27 subtypes in Asian Indians and their comparison with other world population groups.

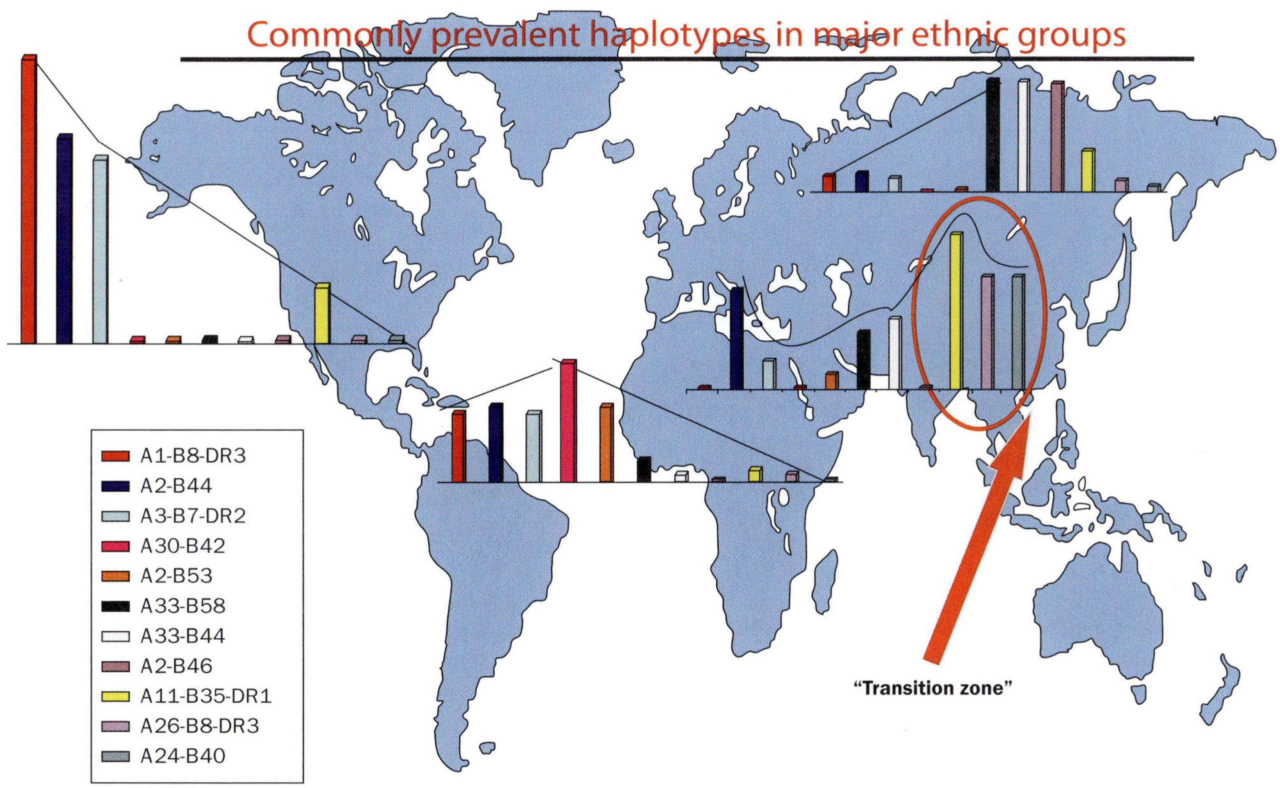

Figure 75-5 Worldwide comparison of common MHC haplotypes with those prevalent in the Indian population. The distribution pattern suggests that the Indian subcontinent represents a "transition zone" between Caucasoids and Negroids on one hand, and Mongoloids and Australoids on the other.

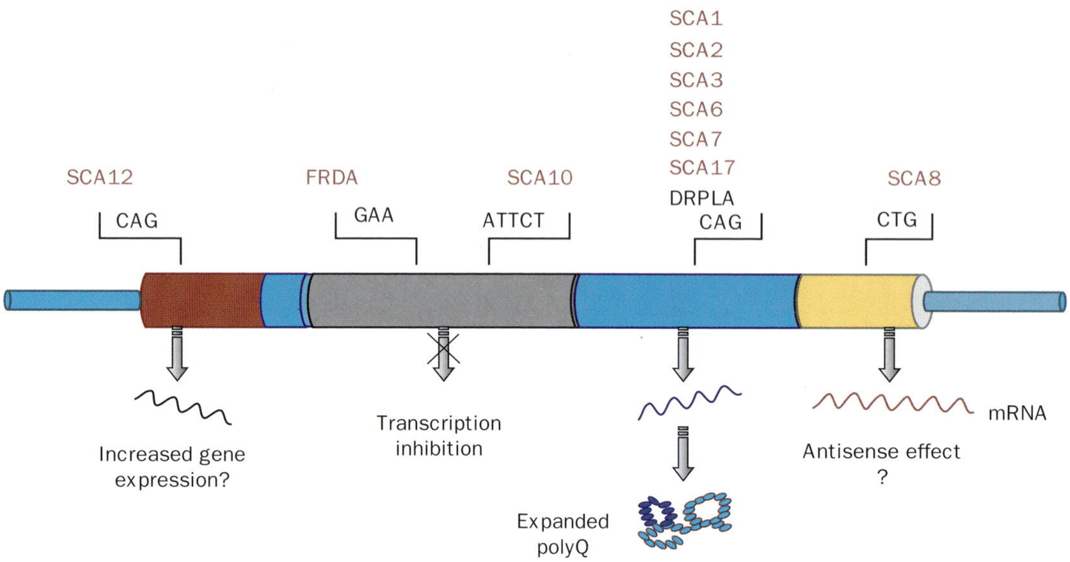

Figure 88-1 Location of repeats with respect to gene and possible consequences of repeat expansion in different ataxias.

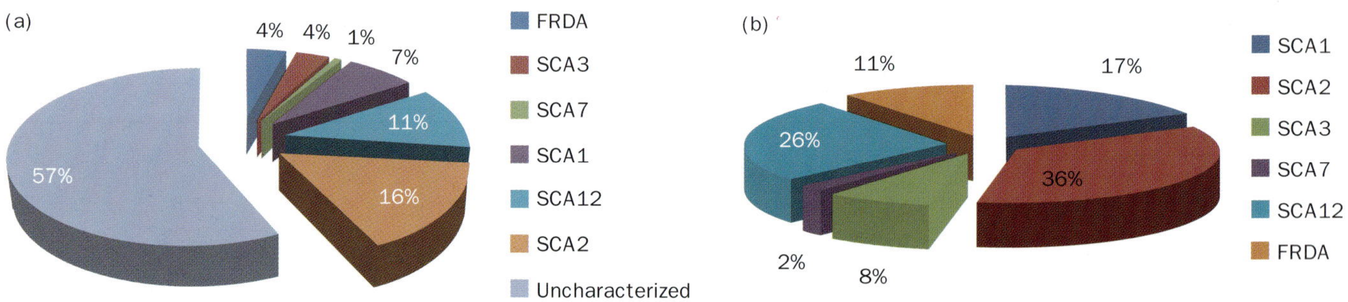

Figure 88-2 Prevalence of different SCAs in North Indian populations. (a) Frequencies of all SCA cases. (b) Relative frequencies of known SCAs. Uncharacterized SCA group is expansions-negative at SCA1, SCA2, SCA3, SCA6, SCA7, SCA8, SCA12, DRPLA, SCA17, and FRDA loci.

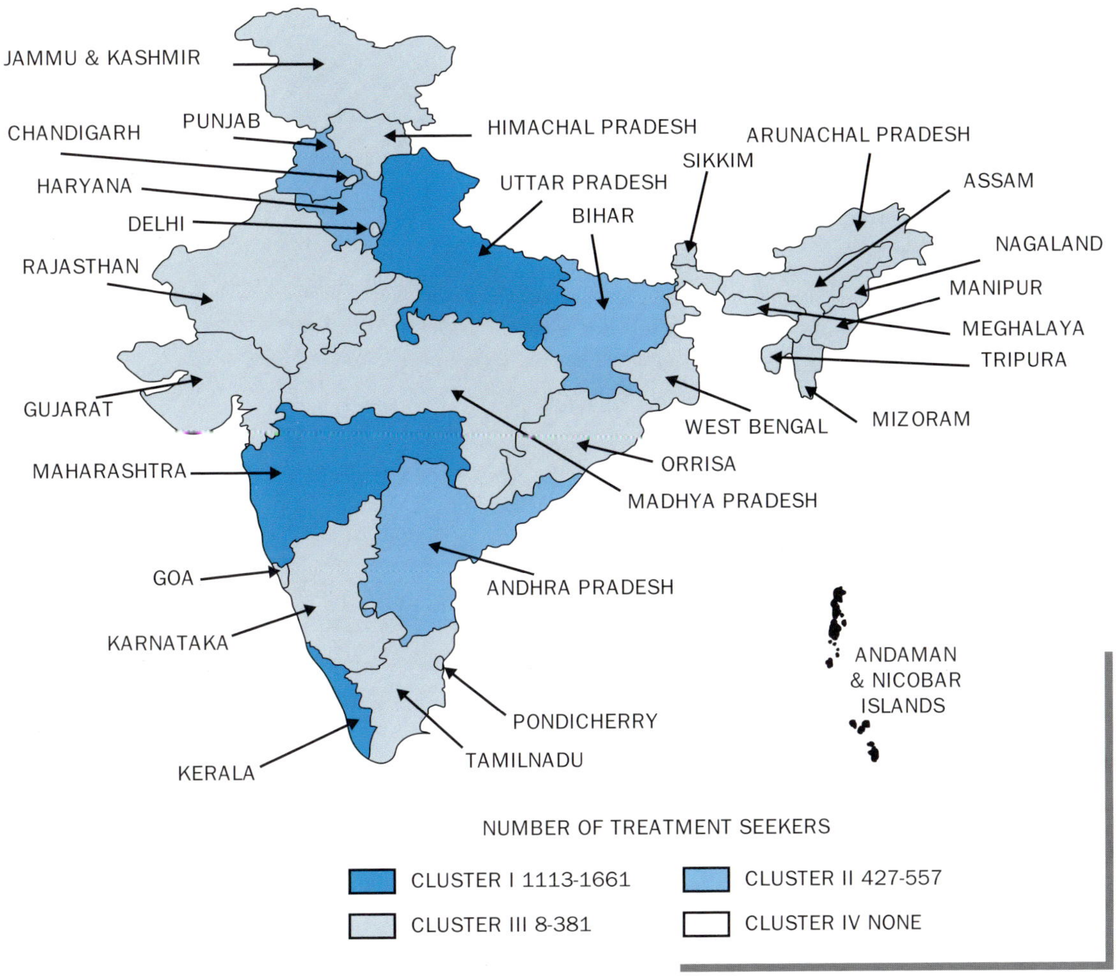

Figure 94-1 Alcohol abuse in various Indian states (N=8857).

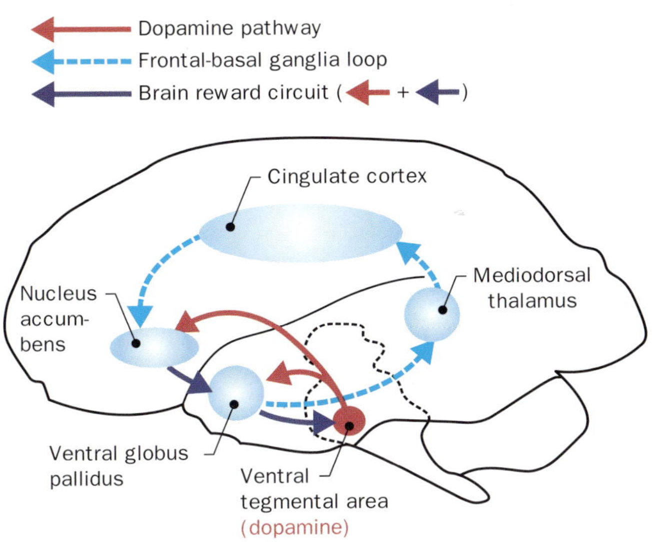

Figure 94-3 Neurobiological pathways of alcohol dependence and addiction.

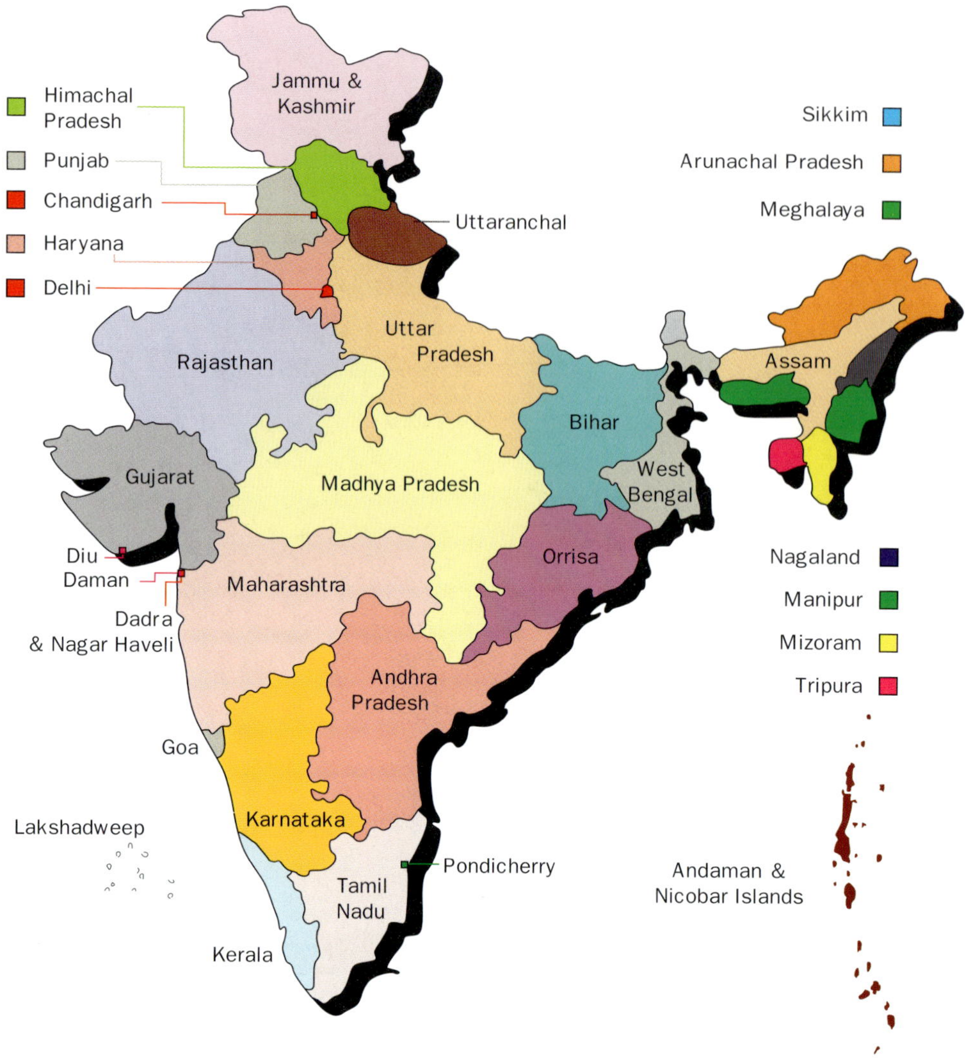

Figure 98-1 The political map of India showing 28 states and 7 union territories.

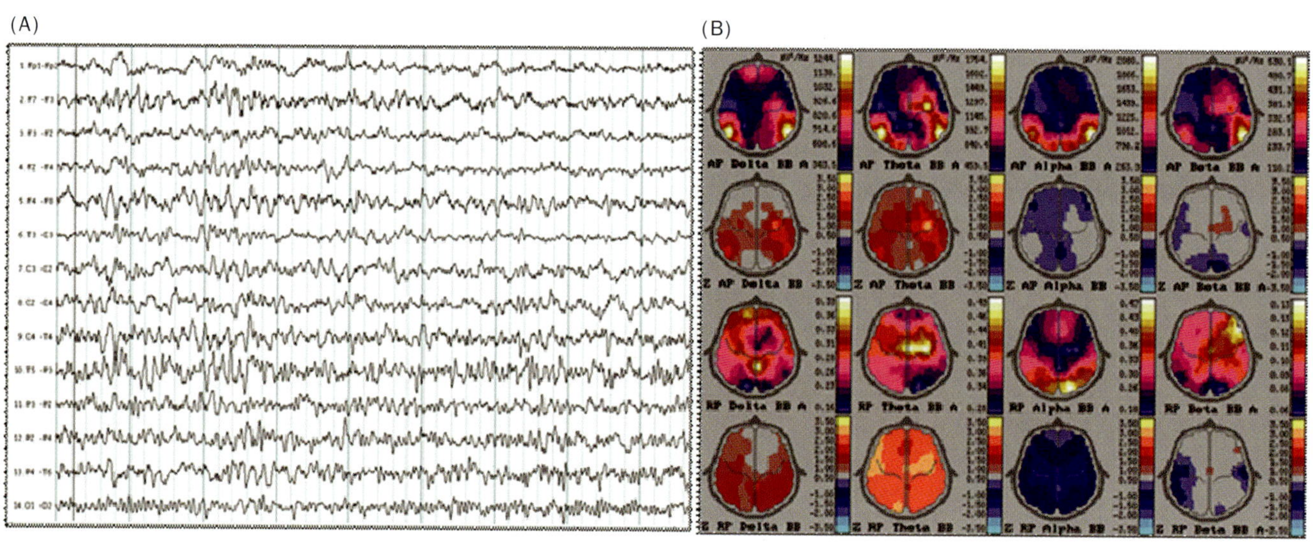

Figure 106-5 Conventional (A) and quantitative (B) EEG from a SCA2 patient 40 years old, with 10 years of disease duration, 40 CAG repeats in the SCA2 gene, and ataxia score of 17 units.

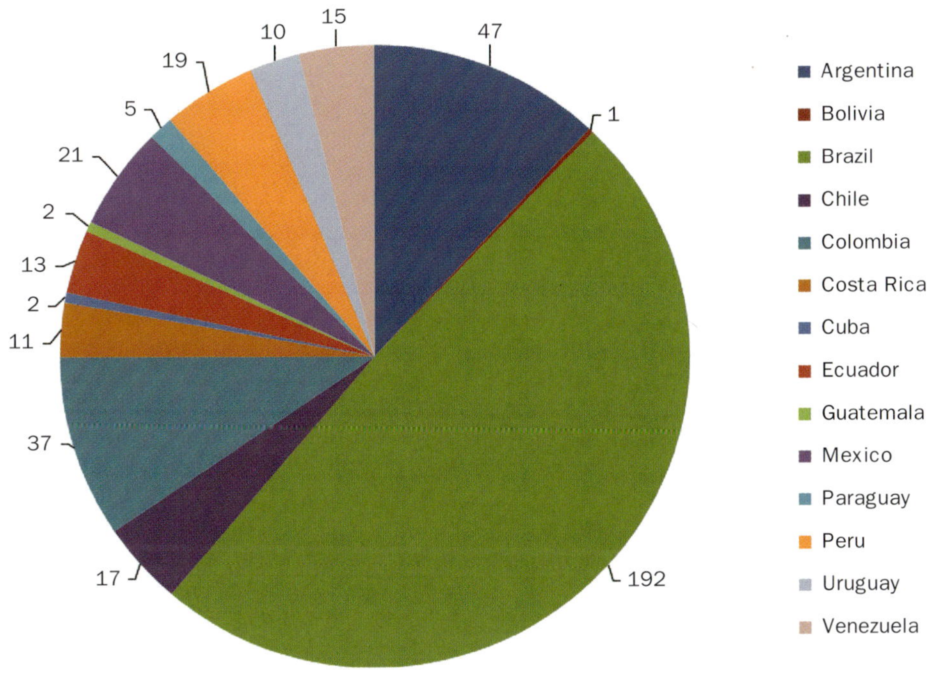

Figure 107-1 Number of students from different countries that have attended the ELAG course on its first 7 editions.

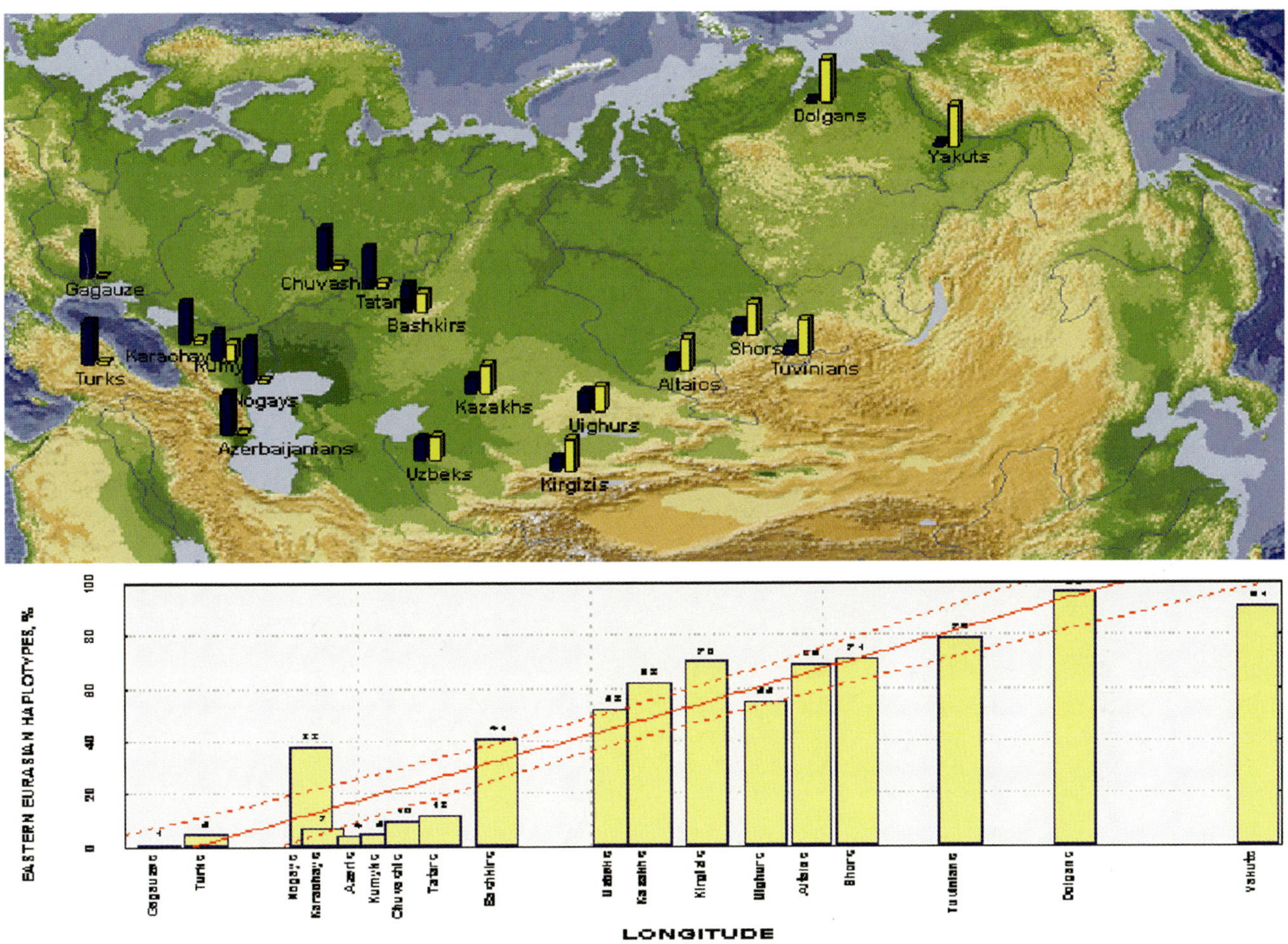

Figure 108-1 West to east frequency gradient of East Asian mtDNA lineages among 18 Turkic-speaking populations of Eurasia.

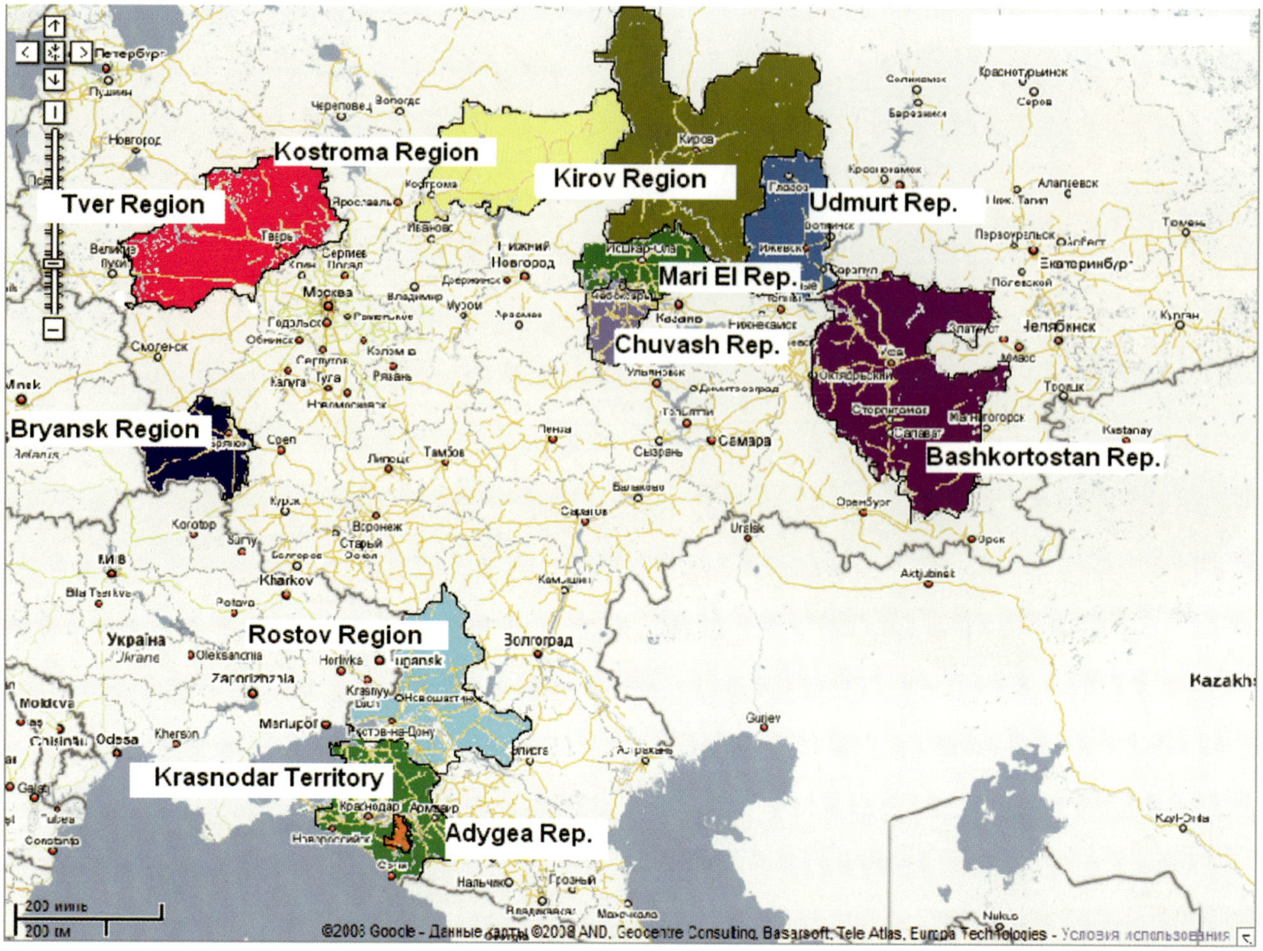

Figure 110-1 The geographical location of investigated populations

recessive neuromuscular disorders of early onset (Fardeau et al., 1996). The major clinical manifestations consist of an early-onset hypotonia and a muscle weakness, early and severe joint contractures, and a static or slowly progressive course (Voit, 1998). MDC1A, caused by mutations in *LAMA2* gene encoding the laminin α2 chain (merosin), represents 50% of CMD cases (Hillaire et al., 1994; Dubowitz, 1996). A 63 amino acid inframe deletion mutation in the *LAMA2* gene was described in two Saudi CMD siblings from a consanguineous family presenting a slight reduction in the expression of laminin α2-chain (Allamand et al., 1997). On the other hand, a patient from Tunisia was reported to have severe congenital muscular dystrophy and total deficiency of the laminin α2 chain due to a splicing mutation in the *LAMA2* gene, resulting in exon skipping and significant decrease of mRNA level (Siala et al., 2007). Interestingly, a later study reported another severe CMD with lumbar scoliosis and respiratory complications in a Tunisian patient, who died at the age of 10. *LAMA2* gene analysis in this patient revealed two *trans* mutations generating nonsense mRNA elicited to degradation (Siala et al., 2008a). A study on prenatal diagnoses of MDC1A followed by postnatal follow-up, in two consanguineous families from South Tunisia, reports phenotypic variability in two siblings carrying the same *LAMA2* gene out-of-frame mutation. These results are suggestive of the implication of modifier genes, especially that the dystrophic muscle undergoes a complex remodeling involving degeneration, regeneration, myogenesis, intracellular trafficking, and proteolysis (Siala et al., 2008b).

A different form of autosomal recessive CMD, characterized by proximal muscle weakness, muscle hypertrophy, and early respiratory failure, was described in a consanguineous family from the United Arab Emirates, and linkage to *LAMA2* was excluded (Muntoni et al., 1998). Linkage analysis of this family mapped this form of the disease to chromosome 1q24 (Brockington et al., 2000). Merosin-deficient CMD with mental retardation and cerebellar cysts, unlinked to any of the known CMD loci, was reported in three unrelated Tunisian patients, indicating a new CMD entity (Triki et al., 2003).

A new autosomal dominant form of L-CMD has been associated with de novo mutations in *LMNA* gene encoding lamins A/C in Arab patients (Quijano-Roy et al., 2008). Germinal mosaicism for *LMNA* mimicking autosomal recessive CMD has been reported in two children from a consanguineous Algerian family (Makri et al., 2009).

A number of other reports describe congenital myopathies in different Arab states. CMD is described in Kuwait, as the clinical phenotypes of 21 Arab CMD children with presence or absence of laminin α2 (merosin) in the muscle biopsies was reported (Habeeb et al., 2006). An interesting report describes the clinical, genetic, and histopathological features of 19 patients from a large inbred Palestinian family. The family includes 10 cases with CMD and nine other cases with LGMD in two generations (Mahjneh et al., 1992).

MYOTONIC DYSTROPHY (DM)

Myotonic dystrophy is an autosomal dominant adult form of muscular dystrophy characterized by myotonia weakness, muscle wasting, frontal baldness, cataract, hypogonadism, and electrocardiographic changes (Harper, 1989). The causative mutation is an expansion of an unstable tandem repeat of the CTG sequence in the 3' untranslated region of the *myotonin* gene on chromosome 19q13.3 (Harley et al., 1992). In unaffected individuals, the $(CTG)_n$ repeat number is polymorphic, ranges from 5 to 38 repeats, and is stably inherited (Brook et al., 1992). A study on the CTG repeat polymorphism in healthy Jews from different origins revealed the fact that North African and Yemenite Jews have significantly more CTG repeats in the normal range than Ashkenazi Jews and Muslim Arabs. This explains the higher frequency of DM in North African and Yemenite Jews (Mor-Cohen et al., 1997). A similar study on $(CTG)_n$ repeat number in healthy Kuwaiti individuals showed a statistically significant low frequency of alleles greater than 18 repeats, which explains the rare occurrence of DM in the Kuwaiti population (Alfadhli et al., 2004).

OTHER DISEASES OF MYONEURAL JUNCTION AND MUSCLE

Congenital fiber-type disproportion myopathy with type I fiber predominance and type II fiber smallness and atrophy is reported in three Omani siblings, with clinical features of congenital myopathy (Rao et al., 2005). Autosomal dominant hyaline body myopathy is reported in four Saudi patients. The causing gene, *MYH7*, was identified in the family and the mutation H1904L is thought to affect the slow myosin heavy chain rod domain (Bohlega et al., 2003; Bohlega et al., 2004). The clinical profile and disease course of juvenile dermatomyositis (JDM) were reported in 25 patients from Saudi Arabia (Shehata et al., 1999) and a case from Egypt presenting with extreme generalized edema “anasarca” (Zedan et al., 2008). Twenty of 130 dermatomyositis (DM) cases reported in Tunisia were associated with cancer; most commonly were breast cancer and nasopharyngeal cancer (Mebazâa et al., 2003). Myasthenia gravis (MG) was reported in 30 Iraqi patients and 104 Saudi patients, and was found to present at a younger age; thymectomy conferred good outcome (Al-Mahdawi and Al-Talib 2002; Al-Moallem et al., 2008).

BOX 58-12

- Inherited muscle diseases in the Arab World have been reported in Morocco, Algeria, Tunisia, Libya, Egypt, Sudan, Palestine, Jordan, Lebanon, Saudi Arabia, Kuwait, and Oman.
- A very high dystrophin deletion mutation frequency is documented in Arabs compared to other ethnic groups.
- Frameshift deletion mechanisms are detected in 51.3% of Egyptian Duchenne and Becker muscular dystrophy patients.
- DLMD was recognized in communities with high consanguinity rates in Sudan, Tunisia, Kuwait, Saudi Arabia, Libya, and Qatar.
- DLMD was mapped to chromosome 13q in Tunisian families and same locus was also reported in Algerian and Moroccan families.
- Autosomal recessive limb-girdle muscular dystrophies (LGMD) have a particularly high frequency in Arab states, especially Tunisia.
- LGMD2B, Miyoshi myopathy, was mapped to chromosome 2p12 in 12 families including a Tunisian family, and reported in Moroccan, Saudi, and Libyan Jewish patients.
- LGMD2C is the most frequent LGMD2 in Tunisia, where patients share the same founder *γ-SGCG* mutation.
- LGMD2I was described and mapped in a Tunisian family in which the causing gene *FKRP* was identified.
- A large Bedouin kindred with X-linked EDMD presented features suggestive of a neurogenic atrophy suggestive of a comorbid, mild congenital myasthenic syndrome.
- MDC1A caused by *LAMA2* gene is described in Saudi, Tunisian, and Kuwaiti CMD patients.
- A new autosomal recessive CMD was described and linked to a new locus in a family from the United Arab Emirates.
- North African and Yemenite Jews and Kuwaiti individuals have significantly more *myotonin* gene CTG repeats in the normal range than in Ashkenazi jews and Muslim Arabs, explaining the low frequency of myotonic dystrophies in these populations.

REFERENCES

Abbasi SS, Brasilience LBC, Workman RK (2007). The fate of medical knowledge and neurosciences during the time of Genghis Khan and the Mongolian Empire. *Neurosurgical Focus* 23:E13, 3.

Abdel-Aleem A, Zaki MS (2008). Spinocerebellar ataxia type 2 (SCA2) in an Egyptian family presenting with polyphagia and marked CAG expansion in infancy. *Journal of Neurology* 255:413–419.

Abdulla MN, Sokrab TE, el Tahir A, Siddig HE, Ali ME (1997). Motor neurone disease in the tropics: findings from Sudan. *East African Medical Journal* 74:46–48.

Afifi Bergman RA, Kazan W, Cortas S, Saad S, Hajj G (1983). Survey of neuromuscular disorders in Lebanon. I. Profile of the genetically determined muscular dystrophies. *Middle East Journal of Anesthesiology* 7:175–192.

Aithala GR, Chand P, Koelma I, El-Bulay MS (1995). Acute infantile spinal muscular atrophy with arthogryposis multiplex congenital in identical twins. *Emirates Medical Journal* 13:53–55.

Al-Rajeh S, Awada A, Bademosi O, Ogunniyi A (2001). The prevalence of epilepsy and other seizure disorders in an Arab population: a community-based study. *Seizure* 10:410–414.

Al-Rajeh S, Bademosi O, Ismail H, et al. (1993). A community survey of neurological disorders in Saudi Arabia: the Thugbah study. *Neuroepidemiology* 12:164–178.

Al-Rajeh S, Majumdar R, Awada A, Adeyokunnu A, Al Jumah M, Al Bunyan M, Snellen A (1998). Molecular analysis of the SMN and NAIP genes in Saudi spinal muscular atrophy patients. *Journal of Neurological Sciences* 158:43–46.

Al Talabani J, Shubbar AI, Mustafa KE (1998). Major congenital malformations in United Arab Emirates (UAE): need for genetic counselling. *Annals of Human Genetics* 62:411–418.

Al-Adawi S, Al-Salmy H, Martin RG, et al. (2003). Patient's perspective on epilepsy: self-knowledge among Omanis. *Seizure* 12:11–18.

Al-Asmari A, Al Moutaery K, Akhdar F, Al Jadid M (2006). Cerebral palsy: incidence and clinical features in Saudi Arabia. *Disability Rehabilitation* 28:1373–1377.

Al-Bunyan MA (2000). Parkinson's disease. Clinical and electrophysiological evaluation. *Saudi Medical Journal* 21:72–75.

Al-Essa MA, Bakheet SM, Patay ZJ, Powe JE, Ozand PT (2000)b. Clinical and cerebral FDG PET scan in a patient with Krabbe's disease. *Pediatric Neurology* 22:44–47.

Al-Essa MA, Sakati NA, Bakheet SM, et al. (2000)a. X-linked adrenoleukodystrophy. The Saudi experience. *Saudi Medical Journal* 21:61–71.

Alfadhli S, Elshafey AE, Bastaki L, Al-Awadi S (2004). CTG repeat number at the myotonic dystrophy locus in healthy Kuwaiti individuals: possible explanation of why myotonic dystrophy is rare in Kuwait. *Archives of Neurology* 61:895–898.

Al-Jumah M, Majumdar R, Al-Rajeh S, et al. (2003). Molecular analysis of the spinal muscular atrophy and neuronal apoptosis inhibitory protein genes in Saudi patients with spinal muscular atrophy. *Saudi Medical Journal* 24:1052–1054.

Al-Jumah, Majumdar R, Al-Rajeh S, et al. (2002). Deletion mutations in the dystrophin gene of Saudi patients with Duchenne and Becker muscular dystrophy. *Saudi Medical Journal* 23:1478–14882.

Allamand V, Sunada Y, Salih MA, et al. (1997). Mild congenital muscular dystrophy in two patients with an internally deleted laminin alpha2-chain. *Human Molecular Genetics* 6:747–752.

Al-Mahdawi AM, Al-Talib NM (2002). Outcome of thymectomy in patients with Myasthenia Gravis. *Saudi Medical Journal* 23:39–43.

Al-Moallem MA, Alkali NH, Hakami MA, Zaidan RM (2008). Myasthenia gravis: presentation and outcome in 104 patients managed in a single institution. *Annals of Saudi Medicine* 28:341–345.

Al-Qudah AA, Tarawneh M (1998). Congenital muscular dystrophy in Jordanian children. *Journal of Child Neurology* 13:383–386.

Al-Rajeh S, Abomelha A, Awada A, Bademosi O, Ismail H (1990). Epilepsy and other convulsive disorders in Saudi Arabia: a prospective study of 1,000 consecutive cases. *Acta Neurologica Scandinavia* 82:341–345.

Al-Rajeh S, Bademosi O, Awada A, Ismail H, al-Shammasi S, Dawodu A (1991). Cerebral palsy in Saudi Arabia: a case-control study of risk factors. *Developmental Medicine and Child Neurology* 33:1048–1052.

Al-Rajeh S, Bademosi O, Gascon GG, Stumpf D (1992). Werdnig Hoffman's disease (spinal muscular atrophy type I): A clinical study of 25 Saudi nationals in Al-Khobar. *Annals of Saudi Medicine* 12:67–71.

Al-Rajeh S, Majumdar R, Awada A, al-Jumah M (1999). Application of DNA-based tests for diagnosis of spinal muscular atrophy in Saudi Arabia. *East Mediterranean Health Journal* 5:1225–1229.

Alrefai AH, Aburahma SK, Khader YS (2009). Treatment of @@@@ sialorrhea in children with cerebral palsy: a double-blind placebo controlled trial. *Clinical Neurology and Neurosurgery* 111:79–82.

Al-Sulaiman A (2001a). Electroencephalographic findings in children with cerebral palsy: a study of 151 patients. *Functional Neurology* 16:325–328.

Al-Sulaiman AA (2001b). Epilepsy in Saudi children with cerebral palsy. *Saudi Medical Journal* 22:19–21.

Al-Sulaiman AA, Ismail HM (1999). Clinical pattern of newly-diagnosed seizures in Saudi Arabia: a prospective study of 263 children. *Child's Nervous System* 15:468–471.

Al-Yahyaee S, Al-Gazali LI, De Jonghe P, et al. (2006). A novel locus for hereditary spastic paraplegia with thin corpus callosum and epilepsy. *Neurology* 66:1230–1234.

Argov Z, Sadeh M, Mazor K, Soffer D, et al. (2000). Muscular dystrophy due to dysferlin deficiency in Libyan Jews. Clinical and genetic features. *Brain* 123:1229–1237.

Ashok PP, Radhakrishnan K, Sridharan R, Mousa ME (1986). Epidemiology of Parkinson's disease in Benghazi, North-East Libya. Clinical *Neurology and Neurosurgery* 88:109–113.

Asri F, Moussaoui D, Kadri N (2003). Autosomal dominant Alzheimer's disease. Study of a Moroccan family. *Encephale* 29:254–258.

Atrouni S, Daraze A, Tamraz J, Cassia A, Caillaud C, Megarbane A (2003). Leukodystrophy associated with oligodontia in a large inbred family: fortuitous association or new entity? *American Journal of Medical Genetics* 118:76–81.

Attia Romdhane N, Ben Hamida M, Mrabet A, et al. (1993). Prevalence study of neurologic disorders in Kelibia (Tunisia). *Neuroepidemiology* 12:285–299.

Aydin IH (2001). Avicenna and modern neurological sciences. *Journal of Academic Researches in Religious Sciences* 1:1–4.

Azibi K, Bachner L, Beckmann JS, et al. (1993). Severe childhood autosomal recessive muscular dystrophy with the deficiency of the 50-kDa dystrophin-associated glycoprotein maps to chromosome 13q12. *Human Molecular Genetics* 2:1423–1428.

Azzedine H, Bolino A, Taïeb T, et al. (2003a) Mutations in MTMR13, a new pseudophosphatase homologue of MTMR2 and Sbf1, in two families with an autosomal recessive demyelinating form of Charcot-Marie-Tooth disease associated with early-onset glaucoma. *American Journal of Human Genetics* 72:1141–1153.

Azzedine H, Ravisé N, Verny C, et al. (2006). Spine deformities in Charcot-Marie-Tooth 4C caused by SH3TC2 gene mutations. *Neurology* 67:602–606.

Azzedine H, Ruberg M, Ente D, et al. (2003b). Variability of disease progression in a family with autosomal recessive CMT associated with a S194X and new R310Q mutation in the GDAP1 gene. *Neuromuscular Disorders* 13:341–346.

Baker GA, Jacoby A, Gorry J, et al. (2005). Quality of life of people with epilepsy in Iran, the Gulf, and Near East. *Epilepsia* 46:132–140.

Barhoumi C, Amouri R, Ben Hamida C, et al. (2001). Linkage of a new locus for autosomal recessive axonal form of Charcot-Marie-Tooth disease to chromosome 8q21.3. *Neuromuscular Disorders* 11:27–34.

Baris H, Legum C, Levin L, et al. (2005). A putative new locus for an autosomal recessive cerebellar ataxia syndrome on chromosome 22q11. *Clinical Genetics* 68:185–187.

Basel-Vanagaite L, Taub E, Drasinover V, et al. (2008). Genetic carrier screening for spinal muscular atrophy and spinal muscular atrophy with respiratory distress 1 in an isolated population in Israel. *Genetic Testing* 12:53–56.

Bashir R, Britton S, Strachan T, et al. (1998). A gene related to *Caenorhabditis elegans* spermatogenesis factor fer-1 is mutated in limb-girdle muscular dystrophy type 2B. *Nature Genetics* 20:37–42.

Baxter RV, Ben Othmane K, Rochelle JM, et al. (2002). Ganglioside-induced differentiation-associated protein-1 is mutant in Charcot-Marie-Tooth disease type 4A/8q21. *Nature Genetics* 30:21–22.

Bejaoui K, Hirabayashi K, Hentati F, et al. (1995). Linkage of Miyoshi myopathy (distal autosomal recessive muscular dystrophy) locus to chromosome 2p12–14. *Neurology* 45:768–772.

Belal S, Cancel G, Stevanin G, et al. (1994).Clinical and genetic analysis of a Tunisian family with autosomal dominant cerebellar ataxia type 1 linked to the SCA2 locus. *Neurology* 44:1423–1426.

Bellayou H, Dehbi H, Bourezgui M, Slassi I, Nadifi S (2009a). Ataxia with vitamin E deficiency (AVED); an example of the contribution of research in molecular genetic to counselling in Morocco. *Pathologie Biologie (Paris)* 57:425–426.

Bellayou H, Hamzi K, Rafai MA, et al. (2009b). Duchenne and Becker muscular dystrophy: contribution of a molecular and immunohistochemical analysis in diagnosis in Morocco. *Journal of Biomedicine and Biotechnology* 2009:325210.

Ben Hamida C, Cavalier L, Belal S, et al. (1997). Homozygosity mapping of giant axonal neuropathy gene to chromosome 16q24.1. *Neurogenetics* 1:129–133.

Ben Hamida C, Doerflinger N, Belal S, et al. (1993b). Localization of Friedreich ataxia phenotype with selective vitamin E deficiency to chromosome 8q by homozygosity mapping. *Nature Genetics* 5:195–200.

Ben Hamida M, Belal S, Sirugo G, et al. (1993a). Friedreich's ataxia phenotype not linked to chromosome 9 and associated with selective autosomal recessive vitamin E deficiency in two inbred Tunisian families. *Neurology* 43:2179–2183.

Ben Hamida M, Fardeau M, Attia N (1983). Severe childhood muscular dystrophy affecting both sexes and frequent in Tunisia. *Muscle and Nerve* 6:469–480.

Ben Hamida M, Hentati F, Ben Hamida C (1990a). Hereditary motor system diseases (chronic juvenile amyotrophic lateral sclerosis): conditions combining a bilateral pyramidal syndrome with limb and bulbar amyotrophy. *Brain* 113:347–363.

Ben Hamida M, Hentati F, Ben Hamida C (1990b). Giant axonal neuropathy with inherited multisystem degeneration in Tunisian kindred. *Neurology* 40:245–250.

Ben Hamida M, Marakchi D (1980). Etude genetique des myopathies en Tunisie. *Union Medicale du Canada* 109:1–8.

Ben Hamida M, Miladi N, Turki I, Zaiem H (1992). Duchenne muscular dystrophy in Tunisia: a clinical and morphological study of 77 cases. *Journal of Neurological Sciences* 107:60–64.

Ben Hamida, Fardeau M, Attia N (1983). Severe childhood muscular dystrophy affecting both sexes and frequent in Tunisia. *Muscle and Nerve* 6:469–480.

Ben Othmane K, Ben Hamida M, Pericak-Vance MA, et al. (1992). Linkage of Tunisian autosomal recessive Duchenne-like muscular dystrophy to the pericentromeric region of chromosome 13q. *Nature Genetics* 2:315–317.

Ben Othmane K, Hentati F, Lennon F, et al. (1993). Linkage of a locus (CMT4A) for autosomal recessive Charcot-Marie-Tooth disease to chromosome 8q. *Human Molecular Genetics* 2:1625–1628.

Benamer HT (2007). Neurological disorders in Libya: an overview. *Neuroepidemiology* 29:43–49.

Benamer HT, de Silva R, Siddiqui KA, Grosset DG (2008). Parkinson's disease in Arabs: a systematic review. *Movement Disorders* 23:1205–1210.

Benamer HT, Grosset DG (2009). A systematic review of the epidemiology of epilepsy in Arab countries. *Epilepsia* 50:2301–2304.

Benamer HTS and Shakir RA (2009). The Neurology Map of the Arab World. *Journal of Neurological Sciences* 285:10–12.

Bener A, al-Marzooqi FH, Sztriha L (1998). Public awareness and attitudes towards epilepsy in the United Arab Emirates. *Seizure* 7:219–222.

Benomar A, Yahyaoui M, Meggouh F, et al. (2002). Clinical comparison between AVED patients with 744 del A mutation and Friedreich ataxia with GAA expansion in 15 Moroccan families *Journal of Neurological Sciences* 198:25–29.

Ben-Zeev B, Gross V, Kushnir T, et al. (2001). Vacuolating megalencephalic leukoencephalopathy in 12 Israeli patients. *Journal of Child Neurology* 16:93–99.

Berkovic SF, Mazarib A, Walid S, et al. (2005). A new clinical and molecular form of Unverricht-Lundborg disease localized by homozygosity mapping. *Brain* 128:652–658.

Birouk N, Azzedine H, Dubourg O, et al. (2003). Phenotypical features of a Moroccan family with autosomal recessive Charcot-Marie-Tooth disease associated with the S194X mutation in the GDAP1 gene. *Archives of Neurology* 60:598–604.

Blumen SC, Bevan S, Abu-Mouch S, et al. (2003). A locus for complicated hereditary spastic paraplegia maps to chromosome 1q24–q32. *Annals of Neurology* 54:796–803.

Bohlega S, Abu-Amero SN, Wakil SM, et al. (2004). Mutation of the slow myosin heavy chain rod domain underlies hyaline body myopathy. *Neurology* 62:1518–1521.

Bohlega S, Lach B, Meyer BF, et al. (2003). Autosomal dominant hyaline body myopathy: clinical variability and pathologic findings. *Neurology* 61:1519–1523.

Bolino A, Muglia M, Conforti FL, et al. (2000). Charcot-Marie-Tooth type 4B is caused by mutations in the gene encoding myotubularin-related protein-2. *Nature Genetics* 25:17–19.

Bomont P, Cavalier L, Blondeau F, et al. (2000). The gene encoding gigaxonin, a new member of the cytoskeletal BTB/kelch repeat family, is mutated in giant axonal neuropathy. *Nature Genetics* 26:370–374.

Bonne G, Capeau J, De Visser M, et al. (2002). 82nd ENMC international workshop, 5th international Emery-Dreifuss muscular dystrophy (EDMD) workshop, 1st Workshop of the MYO-CLUSTER project EUROMEN, September 15–16, 2000, Naarden, The Netherlands. *Neuromuscular Disorders* 12:187–194.

Bönnemann CG, Wong J, Ben Hamida C, Hamida MB, Hentati F, Kunkel LM (1998). LGMD 2E in Tunisia is caused by a homozygous missense mutation in beta-sarcoglycan exon 3. *Neuromuscular Disorders* 8:193–197.

Bouche P (1983). How to interpret the electromyogram (in neuromuscular diseases). *La Revue du Praticien* 33:549–555.

Bouhouche A, Benomar A, Birouk N, et al. (2003). High incidence of SMN1 gene deletion in Moroccan adult-onset spinal muscular atrophy patients. *Journal of Neurology* 250:1209–1213.

Bouhouche A, Benomar A, Birouk N, et al. (1999). A locus for an axonal form of autosomal recessive Charcot-Marie-Tooth disease maps to chromosome 1q21.2-q21.3. *American Journal of Human Genetics* 65:722–727.

Bouhouche A, Birouk N, Benomar A, Ouazzani R, Chkili T, Yahyaoui M (2007a). A novel GDAP1 mutation P78L responsible for CMT4A disease in three Moroccan families. *Canadian Journal of Neurological Sciences* 34:421–426.

Bouhouche A, Birouk N, Azzedine H, et al. (2007b). Autosomal recessive axonal Charcot-Marie-Tooth disease (ARCMT2): phenotype-genotype correlations in 13 Moroccan families. *Brain* 130:1062–1075.

Boukhris A, Stevanin G, Feki I, et al. (2008). Hereditary spastic paraplegia with mental impairment and thin corpus callosum in Tunisia: SPG11, SPG15, and further genetic heterogeneity. *Archives of Neurology* 65:393–402.

Bouslam N, Benomar A, Azzedine H, et al. (2005). Mapping of a new form of pure autosomal recessive spastic paraplegia (SPG28). *Annals of Neurology* 57:567–571.

Bower JH and Zenebe G (2005). Neurologic Services in the nations of Africa. *Neurology* 64:412–415.

Bowirrat A, Friedland RP, Farrer L, Baldwin C, Korczyn A (2002). Genetic and environmental risk factors for Alzheimer's disease in Israeli Arabs. *Journal of Molecular Neuroscience* 19:239–245.

Bowirrat A, Treves TA, Friedland RP, Korczyn AD (2001). Prevalence of Alzheimer typedementia in an elderly Arab population. *European Journal of Neurology* 8:119–123.

Brockington M, Sewry CA, Herrmann R, et al. (2000). Assignment of a form of congenital muscular dystrophy with secondary merosin deficiency to chromosome 1q42. *American Journal of Human Genetics* 66:428–435.

Brook JD, McCurrach ME, Harley HG, et al. (1992). Molecular basis of myotonic dystrophy: expansion of a trinucleotide (CTG) repeat at the 3' end of a transcript encoding a protein kinase family member. *Cell* 69:385.

Campanella G, Filla A, De Michele G (1992). Classifications of hereditary ataxias. A critical overview. *Acta Neurologica (Napoli)* 14:408–419.

Campuzano V, Montermini L, Moltò MD, et al. (1996). Friedreich's ataxia: autosomal recessive disease caused by an intronic GAA triplet repeat expansion. *Science* 271:1423–1427.

Chahine LM, Bijlsma A, Hospers AP, Chemali Z (2007). Dementia and depression among nursing home residents in Lebanon: a pilot study. *International Journal of Geriatric Psychiatry* 22:283–285.

Chaouch M, Allal Y, De Sandre-Giovannoli A, et al. (2003). The phenotypic manifestations of autosomal recessive axonal Charcot-Marie-Tooth due to a mutation in Lamin A/C gene. *Neuromuscular Disorders* 13:60–67.

Chishti MA, Bohlega S, Ahmed M, et al. (2006). T313M PINK1 mutation in an extended highly consanguineous Saudi family with early-onset Parkinson disease. *Archives of Neurology* 63:1483–1485.

Chouery E, Delague V, Bergougnoux A, Koussa S, Serre JL, Mégarbané A (2008). Mutations in TREM2 lead to pure early-onset dementia without bone cysts. *Human Mutation* 29:E194–E204.

Coutinho P, Barros J, Zemmouri R, et al. (1999). Clinical heterogeneity of autosomal recessive spastic paraplegias: analysis of 106 patients in 46 families. *Archives of Neurology* 56:943–949.

Crum BA, Josephs KA (2006). Varied electrophysiologic patterns in spinocerebellar ataxia type 2. *European Journal of Neurology* 13:194–197.

Cuesta A, Pedrola L, Sevilla T, et al. (2002). The gene encoding ganglioside-induced differentiation-associated protein 1 is mutated in axonal Charcot-Marie-Tooth type 4A disease. *Nature Genetics* 30:22–25.

Cupler EJ, Bohlega S, Hessler R, McLean D, Stigsby B, Ahmad J (1998). Miyoshi myopathy in Saudi Arabia: clinical, electrophysiological, histopathological and radiological features. *Neuromuscular Disorders* 8:321–326.

De Sandre-Giovannoli A, Chaouch M, Kozlov S, et al. (2002). Homozygous defects in LMNA, encoding lamin A/C nuclear-envelope proteins, cause autosomal recessive axonal neuropathy in human (Charcot-Marie-Tooth disorder type 2) and mouse. *American Journal of Human Genetics* 70:726–736.

De Sandre-Giovannoli A, Delague V, Hamadouche T, et al. (2005). Homozygosity mapping of autosomal recessive demyelinating Charcot-Marie-Tooth neuropathy (CMT4H) to a novel locus on chromosome 12p11.21-q13.11. *Journal of Medical Genetics* 42:260–265.

Delague V, Bareil C, Bouvagnet P, et al. (2001). Nonprogressive autosomal recessive ataxia maps to chromosome 9q34-9qter in a large consanguineous Lebanese family. *Annals of Neurology* 50:250–253.

Delague V, Bareil C, Tuffery S, et al. (2000). Mapping of a new locus for autosomal recessive demyelinating Charcot-Marie-Tooth disease to 19q13.1-13.3 in a large consanguineous Lebanese family: exclusion of MAG as a candidate gene. *American Journal of Human Genetics* 67:236–243.

Delague V, Jacquier A, Hamadouche T et al. (2007). Mutations in FGD4 encoding the Rho GDP/GTP exchange factor FRABIN cause autosomal recessive Charcot-Marie-Tooth type 4H. *American Journal of Human Genetics* 81:1–16.

Deuraseh N and Abu Talib M. (2005). Mental health in Islamic medical tradition. *The International Medical Journal* 4:76–79.

Dick KJ, Al-Mjeni R, Baskir W, et al. (2008). A novel locus for an autosomal recessive hereditary spastic paraplegia (SPG35) maps to 16q21-q23. *Neurology* 71:248–252.

Doerflinger N, Linder C, Ouahchi K, et al. (1995). Ataxia with vitamin E deficiency: refinement of genetic localization and analysis of linkage disequilibrium by using new markers in 14 families. *American Journal of Human Genetics* 56:1116–1124.

Driss A, Amouri R, Ben Hamida C, et al. (2000). A new locus for autosomal recessive limb-girdle muscular dystrophy in a large consanguineous Tunisian family maps to chromosome 19q13.3. *Neuromuscular Disorders* 10:240–246.

Driss A, Noguchi S, Amouri R, et al. (2003). Fukutin-related protein gene mutated in the original kindred limb-girdle MD 2I. *Neurology* 60:1341–1344.

Drory VE, Artmonov I (2007). Earlier onset and shorter survival of amyotrophic lateral sclerosis in Jewish patients of North African origin. A clue to modifying genetic factors? *Journal of Neurological Sciences* 258:39–43.

Dubourg O, Azzedine H, Yaou RB, et al. (2006). The G526R glycyl-tRNA synthetase gene mutation in distal hereditary motor neuropathy type V. *Neurology* 66:1721–1726.

Dubowitz V (1980). Rapidly progressive limb girdle muscular dystrophy in childhood. In: Angelini D, Danieli GA, Fontanari D (eds) *Muscular Dystrophy Research: Advances and New Trends*, pp. 129–133. Amsterdam, Exerpta Medica.

Dubowitz V (1996). 41st ENMC International Workshop on Congenital Muscular Dystrophy Naarden, The Netherlands. *Neuromuscular Disorders* 4:295–306.

Dyck PJ, Chance P, Lebo R, Carney JA (1993). Hereditary motor and sensory neuropathies. In: Dyck PJ, Thomas PK, Griffin JW, Low PA and Poduslo JF (eds). *Peripheral Neuropathy*, pp. 1094–1136. Saunders, Philadelphia.

Dyck PJ, Lambert EH (1968). Lower motor and primary sensory neuron diseases with peroneal muscular atrophy. Neurologic, genetic, and electrophysiologic findings in hereditary polyneuropathies. *Archives of Neurology* 18:603–618.

Dyck PJ (1975). Inherited neuronal degeneration and atrophy affecting peripheral motor, sensory, and autonomic neurons. In: Dyck PJ, Thomas PK, Griffin JW, Low PA and Poduslo JF (eds). *Peripheral Neuropathy*, pp. 825–867. Saunders, Philadelphia.

Effat LK, El-Harouni AA, Amr KS, El-Minisi TI, Abdel Meguid N, El-Awady M (2000). Screening of dystrophin gene deletions in Egyptian patients with DMD/BMD muscular dystrophies. *Disease Markers* 16:125–129.

El Kerch, Sefiani A, Azibi K, et al. (1994). Linkage analysis of families with severe childhood autosomal recessive muscular dystrophy in Morocco indicates genetic homogeneity of the disease in North Africa. *Journal of Medical Genetics* 31:342–343.

El Rifai MR, Ramia S, Moore V (1984a). Cerebral palsy in Riyadh, Saudi Arabia: I. Aetiological factors. *Annals of Tropical Paediatrics* 4:7–12.

El Rifai MR, Ramia S, Moore V (1984b). Cerebral palsy in Riyadh, Saudi Arabia: II. Associations between gestational age, birthweight and cerebral palsy. *Annals of Tropical Paediatrics* 4:13–17.

El Sherif RM, Fahmy NA, Nonaka I, Etribi MA (2007). Patterns of dystrophin gene deletion in Egyptian Duchenne/Becker muscular dystrophy patients. *Acta Myologica* 26:145–150.

El-Etribi MA, Salem ME, El-Shakankiry HM, El-Kahky AM, El-Mahboub SM (2004). The effect of botulinum toxin type-A injection on spasticity, range of motion and gait patterns in children with spastic diplegic cerebral palsy: an Egyptian study. *International Journal of Rehabilitation Research* 27:275–281.

El-Harouni AA, Amr KS, Effat LK, et al. (2003). The milder phenotype of the dystrophin gene double deletions. *Acta Neurologica Scandinavia* 107:400–404.

Elhawary NA, Shawky RM, Hashem N (2004). Frameshift deletion mechanisms in Egyptian Duchenne and Becker muscular dystrophy families. *Molecules and Cells* 18:141–149.

Elleuch N, Bouslam N, Hanein S, et al. (2007). Refinement of the SPG15 candidate interval and phenotypic heterogeneity in three large Arab families. *Neurogenetics* 8:307–315.

El-Shanti H, Daoud A, Sadoon AA, et al. (2006). A distinct autosomal recessive ataxia maps to chromosome 12 in an inbred family from Jordan. *Brain and Development* 28:353–357.

El-Tallawy, Khedr EM, Qayed MH, Helliwell TR, Kamel NF (2005). Epidemiological study of muscular disorders in Assiut, Egypt. *Neuroepidemiology* 25:205–211.

Emery AE (2000). Emery-Dreifuss muscular dystrophy - a 40 year retrospective. *Neuromuscular Disorders* 10:228–232.

Essawi ML, Effat LK, Shanab GM, Al-Ettribi GM, El-Haronui AA, Karim AM (2007). Molecular analysis of SMN1 and NAIP genes in Egyptian patients with spinal muscular atrophy. *Bratislava Medical Journal* 108:133–137.

Fardeau M, Tomé FM, Helbling-Leclerc A, et al. (1996). Congenital muscular dystrophy with merosin deficiency: clinical, histopathological, immunocytochemical and genetic analysis. *Revue Neurologique* 152:11–19.

Fares F, Axelord Ran S, David M, et al. (2004). Identification of two mutations for ataxia telangiectasia among the Druze community. *Prenatal Diagnosis* 24:358–362.

Farrag A, Farwiz HM, Khedr EH, Mahfouz RM, Omran SM (1998). Prevalence of Alzheimer's disease and other Dementing Disorders: Assiut-Upper Egypt Study. *Dementia and Geriatric Cognitive Disorders* 9:323–328.

Fendri K, Kefi M, Hentati F, Amouri R (2006). Genetic heterogeneity within a consanguineous family involving the LGMD 2D and the LGMD 2C genes. *Neuromuscular Disorders* 16:316–320.

Fendri-Kriaa N, Kammoun F, Rebai A, et al. (2009). Genetic screening of two Tunisian families with generalized epilepsy with febrile seizures plus. *European Journal of Neurology* 16: 697–704.

Forrest SM, Cross GS, Flint T, Speer A, Robson KJ, Davies KE (1988). Further studies of gene deletions that cause Duchenne and Becker muscular dystrophies. *Genomics* 2:109–114.

Gatti RA, Tward A, Concannon P (1999). Cancer risk in ATM heterozygotes: A model of phenotypic and mechanistic differences between missense and truncating mutations. *Molecular Genetics and Metabolism* 68:419–423.

Gilad S, Bar-Shira A, Harnik R, et al. (1996b). Ataxia-telangiectasia: founder effect among North African Jews. *Human Molecular Genetics* 5:2033–2037.

Gilad S, Khosravi R, Shkedy D, et al. (1996a). Predominance of null mutations in ataxia-telangiectasia. *Human Molecular Genetics* 5:433–439.

Gomez-Abad C, Afawi Z, Korczyn AD, et al. (2007). Founder effect with variable age at onset in Arab families with Lafora disease and EPM2A mutation. *Epilepsia* 48:1011–1014.

Gordon N (2001). Hereditary vitamin-E deficiency. *Developmental Medicine & Child Neurology* 43:133–135.

Gouider R, Ibrahim S, Fredj M, et al. (1998). Unverricht-Lündborg disease: clinical and electrophysiologic study of 19 Maghreb families. *Revue Neurologique*154:503–507.

Gouider-Khouja N, Belal S, Hamida MB, Hentati F (2000). Clinical and genetic study of familial Parkinson's disease in Tunisia. *Neurology* 54:1603–1609.

Gouider-Khouja N, Larnaout A, Amouri R, et al. (2003). Autosomal recessive parkinsonism linked to parkin gene in a Tunisian family. Clinical, genetic and pathological study. *Parkinsonism & Related Disorders* 9:247–251.

Gribaa M, Salih M, Anheim M, et al. (2007). A new form of childhood onset, autosomal recessive spinocerebellar ataxia and epilepsy is localized at 16q21-q23. *Brain* 130:1921–1928.

Guilbot A, Kessali M, Ravisé N, et al. (1999). The autosomal recessive form of CMT disease linked to 5q31-q33. *Annals of the New York Academy of Sciences* 883:453–456.

Guilbot A, Williams A, Ravisé N, et al. (2001). A mutation in periaxin is responsible for CMT4F, an autosomal recessive form of Charcot-Marie-Tooth disease. *Human Molecular Genetics* 10:415–421.

Gururaj AK, Sztriha L, Bener A, Dawodu A, Eapen V (2003). Epilepsy in children with cerebral palsy. *Seizure* 12:110–114.

Habeeb YK, Al-Bloushi MA, Al-Jumah ES, De Souza TM, Moosa A (2006). Congenital muscular dystrophy in Arab children. *Journal of Child Neurology* 21:400–405.

Hadano S, Hand CK, Osuga H, et al. (2001). A gene encoding a putative GTPase regulator is mutated in familial amyotrophic lateral sclerosis 2. *Nature Genetics* 29:166–173.

Haider MZ, Bastaki L, Habib Y, Moosa A (1998). Screening 25 dystrophin gene exons for deletions in Arab children with Duchenne muscular dystrophy. *Human Heredity* 48:61–66.

Haider MZ, Habeeb Y, Al-Nakkas E, et al. (2005). Lack of an association between candidate gene loci and idiopathic generalized epilepsy in Kuwaiti Arab children. *Journal of Biomedical Sciences* 12:815–818.

Haider MZ, Moosa A, Dalal H, Habib Y, Reynold L (2001). Gene deletion patterns in spinal muscular atrophy patients with different clinical phenotypes. *Journal of Biomedical Sciences* 8:191–196.

Hamad AI, Ibrahim MA, Sulaiti EM (2004). Dementia in Qatar. *Saudi Medical Journal* 25:79–82.

Hamida CB, Soussi-Yanicostas N, Bejaoui K, Butler-Browne GS, Hentati F, Ben Hamida M (1994). Expression of myosin isoforms and of desmin, vimentin and titin in Tunisian Duchenne-like autosomal recessive muscular dystrophy. *Journal of Neurological Sciences* 123:114–121.

Hampshire DJ, Roberts E, Crow Y, et al. (2001). Kufor-Rakeb syndrome, pallido-pyramidal degeneration with supranuclear upgaze paresis and dementia, maps to 1p36. *Journal of Medical Genetics* 38:680–682.

Haque A (2004). Psychology from Islamic Perspective: Contributions of early Muslim scholars and challenges to contemporary Muslim psychologists. *Journal of Religion and Health* 43:357–377.

Harding AE (1983). Classification of the hereditary ataxias and paraplegias. *The Lancet* 321:1151–1155.

Harding AE, Thomas PK (1980). The clinical features of hereditary motor and sensory neuropathy types I and II. *Brain* 103:259–280.

Harding, A (1993) Inherited, degenerative, and developmental cerebellar, and spinocerebellar disorder. In: Walton J (ed). *Brain's Disease of the Nervous System*, pp. 432–442. Oxford University Press, New York.

Harel T, Goldberg Y, Shalev SA, Chervinski I, Ofir R, Birk OS (2004). Limb-girdle muscular dystrophy 2I: phenotypic variability within a large consanguineous Bedouin family associated with a novel FKRP mutation. *European Journal of Human Genetics* 12:38–43.

Harley HG, Brook JD, Rundle SA, et al. (1992). Expansion of an unstable DNA region and phenotypic variation in myotonic dystrophy. *Nature* 355:545–546.

Harper PS (1989). *Myotonic Dystrophy.* 2nd edition. Philadelphia: WB Saunders Co.

Hashishe MM (1997). Genetic study of epilepsy. *Journal of Egyptian Public Health Association* 72:507–526.

Hawamdeh ZM, Ibrahim AI, Al-Qudah AA (2007). Long-term effect of botulinum toxin (A) in the management of calf spasticity in children with diplegic cerebral palsy. *Europa Medicophysica* 43:311–318.

Hay SA, Hay AA, Moharram H, Salama M (2004). Endoscopic implantation and patency evaluation of lumboperitoneal shunt: an innovative technique. *Surgical Endoscopy* 18:482–484.

Heinisch U, Zlotogora J, Kafert S, Gieselmann V (1995). Multiple mutations are responsible for the high frequency of metachromatic leukodystrophy in a small geographic area. *American Journal of Human Genetics* 56:51–57.

Hellani A, Laugé A, Ozand P, Jaroudi K, Coskun S (2002). Pregnancy after preimplantation genetic diagnosis for Ataxia Telangiectasia. *Molecular Human Reproduction* 8:785–788.

Hentati A, Bejaoui K, Pericak-Vance MA, et al. (1994a). Linkage of recessive familial amyotrophic lateral sclerosis to chromosome 2q33-q35. *Nature Genetics* 7:425–428.

Hentati A, Bejaoui K, Pericak-Vance MA, et al. (1992). The gene locus for one form of juvenile amyotrophic lateral sclerosis maps to chromosome 2. *American Journal of Human Genetics* 51(suppl.):A33.

Hentati A, Ouahchi K, Pericak-Vance MA, et al. (1998). Linkage of a commoner form of recessive amyotrophic lateral sclerosis to chromosome 15q15-q22 markers. *Neurogenetics* 2:55–60.

Hentati A, Pericak-Vance MA, Hung WY, et al. (1994b). Linkage of 'pure' autosomal recessive familial spastic paraplegia to chromosome 8 markers and evidence of genetic locus heterogeneity. *Human Molecular Genetics* 3:1263–1267.

Hillaire D, Leclerc A, Fauré S, et al. (1994). Localization of merosin-negative congenital muscular dystrophy to chromosome 6q2 by homozygosity mapping. *Human Molecular Genetics* 3:1657–1661.

Hilz MJ (2002). Assessment and evaluation of hereditary sensory and autonomic neuropathies with autonomic and neurophysiological examinations. *Clinical Autonomic Research* 12 Suppl 1:I33–43.

Hodgkinson CA, Bohlega S, Abu-Amero SN, et al. (2002). A novel form of autosomal recessive pure hereditary spastic paraplegia maps to chromosome 13q14. *Neurology* 59:1905–1909.

Hoffman EP, Kunkel LM (1989). Dystrophin abnormalities in Duchenne/Becker muscular dystrophy. *Neuron* 2:1019–1029.

Hosler BA, Sapp PC, Berger R, et al. (1998). Refined mapping and characterization of the recessive familial amyotrophic lateral sclerosis locus (ALS2) on chromosome 2q33. *Neurogenetics* 2:34–42.

Hourani R, El-Hajj T, Barada WH, Hourani M, Yamout BI (2009). MR imaging findings in autosomal recessive hereditary spastic paraplegia. *American Journal of Neuroradiology* 30:936–940.

Hulihan MM, Ishihara-Paul L, Kachergus J, et al. (2008). LRRK2 Gly2019Ser penetrance in Arab-Berber patients from Tunisia: a case-control genetic study. *Lancet Neurology* 7:591–594.

Ibáñez P, Lesage S, Lohmann E, et al.; French Parkinson's Disease Genetics Study Group (2006). Mutational analysis of the PINK1 gene in early-onset Parkinsonism in Europe and North Africa. *Brain* 129:686–694.

Ibrahim AI, Hawamdeh ZM (2007). Evaluation of physical growth in cerebral palsied children and its possible relationship with gross motor development. *International Journal of Rehabilitation Research* 30:47–54.

Ibrahim AI, Hawamdeh ZM, Al-Qudah AA (2007). Functional outcome of botulinum toxin injection of gastrocnemius and adductors in spastic hemiplegic cerebral palsied children. *Europa Medicophysica* 43:13–20.

Ifergane G, Al-Sayed I, Birk O, Harel T, Wirguin I (2007). Comorbidity of Emery-Dreifuss muscular dystrophy and a congenital myasthenic syndrome possibly affecting the phenotype in a large Bedouin kindred. *European Journal of Neurology* 14:305–308.

Igisu H, Ohta M, Tabira T, Hosokawa S, Goto I (1975). Giant axonal neuropathy. A clinical entity affecting the central as well as the peripheral nervous system. *Neurology* 25:717–721.

Jones AC, Yamamura Y, Almasy L, et al. (1998). Autosomal recessive juvenile Parkinsonism maps to 6q25.2-q27 in four ethnic groups: detailed genetic mapping of the linked region. *American Journal of Human Genetics* 63:80–87.

Kefi M, Amouri R, Chabrak S, Mechmeche R, Hentati F (2008). Variable cardiac involvement in Tunisian siblings harboring FKRP gene mutations. *Neuropediatrics* 39:113–115.

Kefi M, Amouri R, Driss A, et al. (2003). Phenotype and sarcoglycan expression in Tunisian LGMD 2C patients sharing the same del521-T mutation. *Neuromuscular Disorders* 13:779–787.

Kiehl TR, Nechiporuk A, Figueroa KP, Keating MT, Huynh DP, Pulst SM (2006). Generation and characterization of Sca2 (ataxin-2) knockout mice. *Biochemican and Biophysical Research Communications* 339:17–24.

Klebe S, Azzedine H, Durr A, et al. (2006). Autosomal recessive spastic paraplegia (SPG30) with mild ataxia and sensory neuropathy maps to chromosome 2q37.3. *Brain* 129:1456–1462.

Korn-Lubetzki I, Dor-Wollman T, Soffer D, Raas-Rothschild A, Hurvitz H, Nevo Y (2003). Early peripheral nervous system manifestations of infantile Krabbe disease. *Pediatric Neurology* 28:115–118.

Koul R, Chacko A, Javed H, et al. (2002b). A profile of childhood neuropathies at a university hospital in Oman. *Saudi Medical Journal* 23:450–456.

Koul R, Chacko A, Javed H, Jain R, Ganesh A, Srinivasan S (2002a). Syndrome of cerebrospinal fluid hypovolemia following lumbar puncture cerebrospinal fluid leak in a patient with idiopathic intracranial hypertension. Journal of Child Neurology 17:77–79.

Krajewski KM, Lewis RA, Fuerst DR, et al. (2000). Neurological dysfunction and axonal degeneration in Charcot-Marie-Tooth disease type 1A. *Brain* 123:1516–1527.

Lafreniere RG, MacDonald ML, Dube MP, et al. (2004). Study of Canadian Genetic Isolates. Identification of a novel gene (HSN2) causing hereditary sensory and autonomic neuropathy type II through the Study of Canadian Genetic Isolates. *American Journal of Human Genetics* 74:1064–1073.

Lesage S, Condroyer C, Lannuzel A, et al. (2009). Molecular analyses of the LRRK2 gene in European and North African autosomal dominant Parkinson's disease. *Journal of Medical Genetics* 46:458–464.

Lesage S, Dürr A, Tazir M, et al.; French Parkinson's Disease Genetics Study Group (2006). LRRK2 G2019S as a cause of Parkinson's disease in North African Arabs. *New England Journal of Medicine* 354:422–423.

Lesage S, Ibanez P, Lohmann E, et al.; French Parkinson's Disease Genetics Study Group (2005a). G2019S LRRK2 mutation in French and North African families with Parkinson's disease. *Annals of Neurology* 58:784–787.

Lesage S, Leutenegger AL, Ibanez P, et al.; French Parkinson's Disease Genetics Study Group (2005b). LRRK2 haplotype analyses in European and North African families with Parkinson disease: a common founder for the G2019S mutation dating from the 13th century. *American Journal of Human Genetics* 77:330–332.

Leutenegger AL, Salih MA, Ibáñez P, et al. (2006). Juvenile-onset Parkinsonism as a result of the first mutation in the adenosine triphosphate orientation domain of PINK1. *Archives of Neurology* 63:1257–1261.

Lossos A, Stevanin G, Meiner V, et al. (2006). Hereditary spastic paraplegia with thin corpus callosum: reduction of the SPG11 interval and evidence for further genetic heterogeneity. *Archives of Neurology* 63:756–760.

Mahdi AH (1991). Genetically determined neurodegenerative disorders: experiences in Saudi Arabia. *Annals of Tropical Paediatrics* 11:17–23.

Mahjneh I, Vannelli G, Bushby K, Marconi GP (1992). A large inbred Palestinian family with two forms of muscular dystrophy. *Neuromuscular Disorders* 2:277–283.

Majumdar R, Rehana Z, Al Jumah M, Fetaini N (2005). Spinal muscular atrophy carrier screening by multiplex polymerase chain reaction using dried blood spot on filter paper. *Annals of Human Genetics* 69:216–221.

Makri S, Clarke NF, Richard P, et al. (2009). Germinal mosaicism for LMNA mimics autosomal recessive congenital muscular dystrophy. *Neuromuscular Disorders* 19:26–28.

Margolis SA, Reed RL (2001). Institutionalized older adults in a health district in the United Arab Emirates: health status and utilization rate. *Gerontology* 47:161–167.

Martín-Araguz A, Bustamante-Martínez C, Fernández-Armayor Ajo V, Moreno-Martínez JM (2002). Neuroscience in Al Andalus and its influence on medieval scholastic medicine. *Revista de Neurologia* 34:877–892.

Martini R, Berciano J, Van Broeckhoven C (2000). 5th Workshop of the European CMT Consortium, 69th ENMC International Workshop: Therapeutic approaches in CMT neuropathies and related disorders 23–25 April 1999, Soestduinen, The Netherlands. *Neuromuscular Disorders 10:69–74.*

Marzouki N, Belal S, Benhamida C, Benlemlih M, Hentati F (2001). Genetic analysis of early onset cerebellar ataxia with retained tendon reflexes in four Tunisian families. *Clinical Genetics* 59:257–262.

Masri A, Hamamy H (2006). Canavan disease: a first case from Jordan. *Journal of Pediatric Neurology* 4:143–146.

Mazarib A, Xiong L, Neufeld MY, et al. (2001). Unverricht-Lundborg disease in a five-generation Arab family: instability of dodecamer repeats. *Neurology* 57:1050–1054.

Mebazâa A, Boussen H, Nouira R, et al. (2003). Dermatomyositis and malignancy in Tunisia: a multicenter national retrospective study of 20 cases. *Journal of the American Academy of Dermatology* 48:530–534.

Mégarbané A, Delague V, Ruchoux MM, et al. (2001). New autosomal recessive cerebellar ataxia disorder in a large inbred Lebanese family. *American Journal of Medical Genetics* 101:135–141.

Mezaal M, Saadah M (2005). Idiopathic intracranial hypertension in Dubai: nature and prognosis. *Acta Neurologica Scandinavia* 112:298–302.

Mezaal MA, Nouri KA, Abdool S, Safar KA, Nadeem AS (2009). Cerebral palsy in adults consequences of nonprogressive pathology. *The Open Neurology Journal* 3:24–26.

Middleton LT, Christodoulou K, Mubaidin A, et al. (1999). Distal hereditary motor neuronopathy of the Jerash type. *Annals of the New York Academy of Sciences* 883:65–68.

Mikati MA, Comair Y, Ismail R, Faour R, Rahi AC (2004). Effects of epilepsy surgery on quality of life: a controlled study in a Middle Eastern population. *Epilepsy & Behavior* 5:72–80.

Mizuno Y, Hattori N, Mochizuki H (2004). Etiology of Parkinson's disease. In Watts RL, Koller WC (eds). *Movement disorders: neurologic principles & practice*, pp. 209–232. McGraw-Hill.

Mohamed W (2008). History of Neuroscience: Arab and Muslim contributions to modern neuroscience. IBRO History of Neuroscience *http://www.ibro.info/Pub/Pub_Main_Display.asp?LC_Docs_ID=3433.*

Mor-Cohen R, Magal N, Gadoth N, Shohat T, Shohat M (1997). Correlation between the incidence of myotonic dystrophy in different groups in Israel and the number of CTG trinucleotide repeats in the myotonin gene. *American Journal of Medical Genetics* 71:156–159.

Mouton P, Maisonobe T, Behin A, Cohen L, Pierrot-Deseilligny C (1999). Distal Miyoshi muscular dystrophy in a Moroccan patient. *La Presse Medicale* 28:476–478.

Mrabet H, Belhedi N, Bouchlaka S, El Gaaied A, Mrabet A (2007). GEFS+ is not related to the most common mutations of SCN1B, SCN1A and GABRG2 in two Tunisian families. *Neurological Sciences* 28:311–314.

Mrad R, Dorboz I, Ben Jemaa L, et al. (2006). Molecular analysis of the SMN1 and NAIP genes in 60 Tunisian spinal muscular atrophy patients. *La Tunisie Medicale* 84:465–469.

Muntoni F, Taylor J, Sewry CA, Naom I, Dubowitz V (1998). An early onset muscular dystrophy with diaphragmatic involvement, early respiratory failure and secondary alpha2 laminin deficiency unlinked to the LAMA2 locus on 6q22. *European Journal of Paediatric Neurology* 2:19–26.

Nave KA, Sereda MW, Ehrenreich H (2007). Mechanisms of disease: inherited demyelinating neuropathies; from basic to clinical research. *Nature Clinical Practice Neurology* 3:453–464.

Oataya S (1982). Ibn al-Nafis has dissected the human body. Symposium on Ibn al-Nafis. Second International Conference on Islamic Medicine, Islamic Medical organization, Kuwait, Encyclopedia of Islamic World. *http://www.islamset.com/isc/nafis/oataya.html.* Retrieved 18/10/2009.

Obeid T, Awada A, Mousali Y, Nusair M, Muhayawi S, Memish S (2000). Extensive radiculopathy: a manifestation of intracranial hypertension. *European Journal of Neurology* 7:549–553.

Ogunniyi A, Daif AK, Al-Rajeh S, et al. (1998). Dementia in Saudi Arabia: experience from a university hospital. *Acta Neurologica Scandinavia* 98:116–120.

Othmane KB, Johnson E, Menold M, et al. (1999). Identification of a new locus for autosomal recessive Charcot-Marie-Tooth disease with focally folded myelin on chromosome 11p15. *Genomics* 62:344–349.

Ouahchi K, Arita M, Kayden H, et al. (1995). Ataxia with isolated vitamin E deficiency is caused by mutations in the alpha-tocopherol transfer protein. *Nature Genetics* 9:141–145.

Owolabi MO, Bower JH, Ogunniyi A (2007). Mapping Africa's way into prominence in the field of Neurology. *Archives of Neurology* 64:1696–1700.

Ozand PT, Gascon GG, Dhalla M (1990). Aspartoacylase deficiency and Canavan disease in Saudi Arabia. *American Journal of Medical Genetics* 35:266–268.

Quattrone A, Gambardella A, Bono F, et al. (1996). Autosomal recessive hereditary motor and sensory neuropathy with focally folded myelin sheaths: clinical, electrophysiologic, and genetic aspects of a large family. *Neurology* 46:1318–1324.

Quijano-Roy S, Mbieleu B, Bönnemann CG, et al. (2008). De novo LMNA mutations cause a new form of congenital muscular dystrophy. *Annals of Neurology* 64:177–186.

Radhakrishnan K, Ashok PP, Sridharan R, Mousa ME (1986). Descriptive epidemiology of motor neuron disease in Benghazi, Libya. *Neuroepidemiology* 5:47–54.

Radhakrishnan K, el-Mangoush MA, Gerryo SE (1987). Descriptive epidemiology of selected neuromuscular disorders in Benghazi, Libya. *Acta Neurologica Scandinavia* 75:95–100.

Radhakrishnan K, Thacker AK, Bohlaga NH, Maloo JC, Gerryo SE (1993). Epidemiology of idiopathic intracranial hypertension: a prospective and case-control study. *Journal of Neurological Sciences* 116:18–28.

Rajab A, Yoo SY, Abdulgalil A, et al. (2006). An autosomal recessive form of spastic cerebral palsy (CP) with microcephaly and mental retardation. *American Journal of Medical Genetics* 140:1504–1510.

Ramirez A, Heimbach A, Gründemann J, et al. (2006). Hereditary parkinsonism with dementia is caused by mutations in ATP13A2, encoding a lysosomal type 5 P-type ATPase. *Nature Genetics* 38:1184–1191.

Rao TV, Koul RL, Inuwa IM (2005). Congenital fiber-type disproportion myopathy with type I fiber predominance and type II fiber smallness and atrophy—a sterological analysis. *Clinical Neuropathology* 24:26–31.

Rivière JB, Verlaan DJ, Shekarabi M, et al. (2004). A mutation in the HSN2 gene causes sensory neuropathy type II in a Lebanese family. *Annals of Neurology* 56:572–575.

Rodius F, Duclos F, Wrogemann K, et al. (1994). Recombinations in individuals homozygous by descent localize the Friedreich ataxia locus in a cloned 450-kb interval. *American Journal of Human Genetics* 54:1050–1059.

Romeo G, Menozzi P, Ferlini A, et al. (1983). Incidence of Friedreich ataxia in Italy estimated from consanguineous marriages. *American Journal of Human Genetics* 35:523–529.

Salih MA, Ekmejian A, Ibrahim M, Omer A (1984). Respiratory insufficiency in a severe autosomal recessive form of muscular dystrophy. *Annals of Tropical Paediatrics* 4:45–48.

Salih MA, Mahdi AH, al-Rikabi AC, et al. (1996). Clinical and molecular pathological features of severe childhood autosomal recessive muscular dystrophy in Saudi Arabia. *Developmental Medicine and Child Neurology* 38:262–270.

Salih MA, Maisonobe T, Kabiraj M, et al. (2000). Autosomal recessive hereditary neuropathy with focally folded myelin sheaths and linked to chromosome 11q23: a distinct and homogeneous entity. *Neuromuscular Disorders* 10:10–15.

Salih MAM, Omer MIA, Bayoumi RA, Karrar O, Johnson M (1983). Severe autosomal recessive muscular dystrophy in an extended Sudanese kindred. *Developmental Medicine and Child Neurology* 25:43–52.

Samilchuk E, D'Souza B, Bastaki L, al-Awadi S (1996). Deletion analysis of the SMN and NAIP genes in Kuwaiti patients with spinal muscular atrophy. *Human Genetics* 98:524–527.

Santoshkumar B, Turnbull J, Minassian BA (2008). Unverricht-Lundborg progressive myoclonus epilepsy in Oman. *Pediatric Neurology* 38:252–255.

Savitsky K, Sfez S, Tagle DA, et al. (1995). The complete sequence of the coding region of the ATM gene reveals similarity to cell cycle regulators in different species. *Human Molecular Genetics* 11:2025–2032.

Sawaya RA, Mourad FH, Najjar S, Mikati M, Raymond GV (1999). Adrenoleukodystrophy associated with vitiligo and ulcerative colitis. *European Neurology* 42:169–172.

Scrimgeour EM, Krishna AG, Gasim A, et al. (1996). Friedreich's ataxia, with retained lower limb tendon reflexes, in a Saudi Arabian family. *Clinical Neurology and Neurosurgery* 98:8–11.

Senderek J, Bergmann C, Stendel C et al. (2003). Mutations in a gene encoding a novel SH3/TPR domain protein cause autosomal recessive Charcot-Marie-Tooth type 4C neuropathy. *American Journal of Human Genetics* 73:1106–1119.

Shaw PJ (2005). Molecular and cellular pathways of neurodegeneration in motor neurone disease. *Journal of Neurology, Neurosurgery & Psychiatry* 76:1046–1057.

Shawky RM, Abd el-Aleem K, Rifaat MM, Moustafa A (2001). Molecular diagnosis of spinal muscular atrophy in Egyptians. *East Mediterranean Health Journal* 7:229–237.

Shehata R, al-Mayouf S, al-Dalaan A, al-Mazaid A, al-Balaa S, Bahabri S (1999). Juvenile dermatomyositis: clinical profile and disease course in 25 patients. *Clinical and Experimental Rheumatology* 17:115–118.

Shelley BP, Al Khabouri J (2007). The spectrum of dementia: frequency, causes and clinical profile. A national referral hospital-based study in Oman. *Dementia and Geriatric Cognitive Disorders* 24:280–287.

Shimony A, Afawi Z, Asher T, Mahajnah M, Shorer Z (2009). Epidemiological characteristics of febrile seizures—comparing between Bedouin and Jews in the southern part of Israel. *Seizure* 18:26–29.

Siala O, Louhichi N, Triki C et al. (2007). Severe MDC1A congenital muscular dystrophy due to a splicing mutation in the LAMA2 gene resulting in exon skipping and significant decrease of mRNA level. *Genetic Testing* 11:199–207.

Siala O, Kammoun Feki F, Louhichi N, et al. (2008b). Molecular prenatal diagnosis of muscular dystrophies in Tunisia and postnatal follow-up role. *Genetic Testing* 12:581–586.Siala O, Louhichi N, Triki C, Morinière M, Fakhfakh F, Baklouti F (2008a). LAMA2 mRNA processing alterations generate a complete deficiency of laminin-alpha2 protein and a severe congenital muscular dystrophy. *Neuromuscular Disorders* 18:137–145.

Siddique N, Siddique T (2008). Genetics of amyotrophic lateral sclerosis. *Physical Medicine and Rehabilitation Clinics of North America* 19:429–439.

Skre H (1974). Genetic and clinical aspects of Charcot-Marie-Tooth's disease. *Clinical Genetics* 6:98–118.

Skre H, Haugstad TS, Berg K (1985). The hereditary ataxias. *Progress in Medical Genetics* 6:123–240.

Smach MA, Charfeddine B, Lammouchi T, et al. (2008). CSF beta-amyloid 1-42 and tau in Tunisian patients with Alzheimer's disease: the effect of APOE epsilon4 allele. *Neuroscience Letters* 440:145–149.

Sridharan R, Radhakrishnan K, Ashok PP, Mousa ME (1985). Prevalence and pattern of spinocerebellar degenerations in northeastern Libya. *Brain* 108:831–843.

Sridharan R, Radhakrishnan K, Ashok PP, Mousa ME (1986). Epidemiological and clinical study of epilepsy in Benghazi, Libya. *Epilepsia* 27:60–65.

Steffens B (2006). *Ibn al-Haytham: First Scientist*. Morgan Reynolds Publishing, Greensboro, NC.

Stevanin G, Santorelli FM, Azzedine H, et al. (2007). Mutations in SPG11, encoding spatacsin, are a major cause of spastic paraplegia with thin corpus callosum. *Nature Genetics* 39:366–372.

Syed IB (2002). Islamic Medicine: 1000 years ahead of its times. *Journal of Islamic Medical Association* 2:2–9.

Tabarki B, Yacoub M, Selmi H, Oubich F, Barsaoui S, Essoussi AS (2001). Infantile status epilepticus in Tunisia. Clinical, etiological and prognostic aspects. *Seizure* 10:365–369.

Tadmouri GO, Nair P, Obeid T, Al Ali MT, Al Khaja N, Hamamy H (2009). Consanguinity and reproductive health among Arabs. *Reproductive Health* 6:17

Taha SA, Mahdi AH (1984). Cerebral palsy in Saudi Arabia: a clinical study of 102 cases. *Annals of Tropical Paediatrics* 4:155–158.

Takashima H, Boerkoel CF, John J, et al. (2002). Mutation of TDP1, encoding a topoisomerase I-dependent DNA damage repair enzyme, in spinocerebellar ataxia with axonal neuropathy. *Nature Genetics* 32:267–272.

Tang B, Liu C, Shen L, et al. (2000). Frequency of SCA1, SCA2, SCA3/MJD, SCA6, SCA7, and DRPLA CAG trinucleotide repeat expansion in patients with hereditary spinocerebellar ataxia from Chinese kindreds. *Archives of Neurology* 57:540–544.

Tassin J, Dürr A, de Broucker T, et al. (1998). Chromosome 6-linked autosomal recessive early-onset Parkinsonism: linkage in European and Algerian families, extension of the clinical spectrum, and evidence of a small homozygous deletion in one family. The French Parkinson's Disease Genetics Study Group, and the European Consortium on Genetic Susceptibility in Parkinson's Disease. *American Journal of Human Genetics* 63:88–94.

Tazir M, Azzedine H, Assami S, et al. (2004). Phenotypic variability in autosomal recessive axonal Charcot-Marie-Tooth disease due to the R298C mutation in lamin A/C. *Brain* 127:154–163.

Tazir M, Nouioua S, Magy L, et al. (2009). Phenotypic variability in giant axonal neuropathy. *Neuromuscular Disorders* 19:270–274.

Tazir M, Vallat JM, Bomont P, et al. (2002). Genetic heterogeneity in giant axonal neuropathy: an Algerian family not linked to chromosome 16q24.1. *Neuromuscular Disorders* 12:849–852.

Triki C, Feki I, Meziou M, Turki H, Zahaf A, Mhiri C (2000). Clinical, biological and genetic study of 24 patients with ataxia telangiectasia from southern Tunisia. *Revues de Neurology (Paris)* 156:634–637.

Triki C, Louhichi N, Méziou M, et al.(2003). Merosin-deficient congenital muscular dystrophy with mental retardation and cerebellar cysts, unlinked to the LAMA2, FCMD, MEB and CMD1B loci, in three Tunisian patients. *Neuromuscular Disorders* 13:4–12.

Turnbull J, Kumar S, Ren ZP et al. (2008). Lafora progressive myoclonus epilepsy: disease course homogeneity in a genetic isolate. *Journal of Child Neurology* 23:240–242.

Voit T (1998). Congenital muscular dystrophies: 1997 update. *Brain & Development* 20:65–74.

Wijesekera LC, Leigh PN (2009). Amyotrophic lateral sclerosis. *Orphanet Journal of Rare Diseases* 4:3.

World Health Organization, World Federation on Neurology (2004). Atlas: country resources for neurological disorders. First edition.

Worms PM (2001). The epidemiology of motor neuron diseases: a review of recent studies. *Journal of Neurological Sciences* 191:3–9.

Yang Y, Hentati A, Deng HX, et al. (2001). The gene encoding alsin, a protein with three guanine-nucleotide exchange factor domains, is mutated in a form of recessive amyotrophic lateral sclerosis. *Nature Genetics* 29:160–165.

Zakian VA (1995). ATM-related genes: what do they tell us about functions of the human gene? *Cell* 82:685–687.

Zatz M, Passos-Bueno MR, Rapaport D (1990). Duchenne-like muscular dystrophy in the Arabs. *American Journal of Medical Genetics* 37:289–290.

Zedan M, El-Ayouty M, Abdel-Hady H, Shouman B, El-Assmy M, Fouda A (2008). Anasarca: not a nephrotic syndrome but dermatomyositis. *European Journal of Pediatrics* 167:831–834.

Zemmouri R, Azzedine H, Assami S, et al. (2000). Charcot-Marie-Tooth 2-like presentation of an Algerian family with giant axonal-neuropathy. *Neuromuscular Disorders*10:592–598.

Zlotogora J, Furman-Shaharabani Y, Harris A, Barth ML, von Figura K, Gieselmann V (1994b). A single origin for the most frequent mutation causing late infantile metachromatic leucodystrophy. *Journal of Medical Genetics* 31:672–674.

Zlotogora J, Gieselman V, von Figura K, Zeigler M, Bach G (1994a). Late infantile metachromatic leukodystrophy in Israel. *Biomedicine & Pharmacotherapy* 48:347–350.

Zlotogora J, Hujerat Y, Barges S, Shalev SA, Chakravarti A (2007). The fate of 12 recessive mutations in a single village. *Annals of Human Genetics* 71:202–208.

Zlotogora J, Levy-Lahad E, Legum C, Iancu TC, Zeigler M, Bach G (1991). Krabbe disease in Israel. *Israel Journal of Medical Sciences* 27:196–198.

Zlotogora J, Regev R, Zeigler M, Iancu TC, Bach G (1985). Krabbe disease: increased incidence in a highly inbred community. *American Journal of Medical Genetics* 21:765–770.

Zuchner S, Vance J (2005). Mechanisms of Disease: a molecular genetic update on hereditary axonal neuropathies. *Nature Clinical Practice* 2:45–53.

59.

PREIMPLANTATION GENETIC DIAGNOSIS IN SAUDI ARABIA

Serdar Coskun, Wafa Qubbaj, and Aida I. Al-Aqeel

INTRODUCTION

Advances in assisted reproductive technology and molecular genetics have made it possible to remove a single cell (blastomere) from an early embryo and analyze the genome of that single cell, giving rise to preimplantation genetic diagnosis (PGD). This procedure can be offered to families who are at risk of transmitting genetic diseases to their offspring. Essentially, PGD is a technique to determine the genotype of an embryo before implantation takes place, to avoid implantation of diseased embryos. PGD is an early form of prenatal diagnosis, offering an alternative to the classical prenatal diagnosis. Families are spared the difficult decision of pregnancy termination after having an affected fetus following classical prenatal diagnosis. On the other hand, they have to undergo an invasive and expensive infertility treatment, even though they may be fertile. This requires adherence to a strict family planning and effective contraceptive strategy.

The first successful PGD was performed by gender detection for two families who were at risk of transmitting X-linked diseases, adrenoleukodystrophy, and X-linked mental retardation (Handyside et al., 1990). Indications for PGD now cover almost any inherited disease with known molecular genetic mutations, in addition to human leukocyte antigen (HLA) testing of embryos. It is currently estimated that PGD is utilized in prevention of over 100 different diseases (Preimplantation Genetic Diagnosis International Society, 2008). Moreover, PGD can be used in detecting chromosomal aberrations of the embryos, either to improve the reproductive outcome or prevent aneuploidies in patients with certain chromosomal rearrangements (Kearns et al., 2005). However, this indication of PGD will not be discussed in this chapter.

A PGD cycle should be preceded with comprehensive genetic testing and counseling. Following these, couples should make an informed decision to continue with PGD. Once a decision to proceed is made, the cycle can be initiated. There are three aspects of a PGD cycle: first is the generation of embryos, then subsequently, the removal of cells from the embryo through micromanipulation, and finally, the genetic diagnosis. Embryos are generated through standard in-vitro fertilization procedures. Cleavage stage embryo biopsy is easily and most commonly performed using an inverted microscope with attached micromanipulators. The genetic diagnosis is usually done using a polymerase chain reaction (PCR) assay.

At present, PGD can be performed, in principle, for any genetic condition for which genetic alterations have been identified. Based on this information, special primers or chromosome-specific probes may be developed for accurate selection of mutation-free embryos, to ensure the establishment of a pregnancy with healthy embryos. Because PGD provides a valuable option for genetically disadvantaged couples to control the outcome of their reproductive planning at an early stage, PGD is currently applied worldwide (Verlinsky and Kuliev, 2000).

COUNSELING AND PATIENTS SELECTION

Genetic counseling is a process in which an individual or a family obtains information and advice about a genetic condition that may affect the individual, his progeny, his relatives, or the family as a whole. Based on this knowledge he/she can make the pertinent decision regarding marriage, reproduction, abortion, and health management. Counseling of families in different parts of the world must be done in accordance with the cultural and religious guidelines in each country, in order to appropriately initiate treatment and to prevent genetic disorders. Counseling can often include other family members, since they are also impacted by these disorders due to recurrence risk (AlBar, 1999, 2002; Al-Aqeel, 2005, 2007).

It is estimated that at least 20% of the global population live in communities with a preference for consanguineous marriage (i.e., first-cousins), and that at least 8.5% of children worldwide have consanguineous parents. This is linked with an increased birth prevalence of infants with severe recessive disorders, but has a relatively small effect on the prevalence of dominant and X-linked disorders, and its role in complex disorders is still uncertain (Modell and Darr, 2002).

Counseling of couples undergoing PGD should include its indications, technique, risks, and outcomes. An important role of the counselor is to support and inform the autonomous choices of women/couples: deciding whether to transfer "carrier" embryos within the goal of creating a "healthy" child, and understanding the possibility of embryos with other genetic abnormalities that are not tested during the PGD cycle (Ehrich, et al., 2007). The wife should be of suitable reproductive age, confirmed with her menstrual cycle and reproductive hormone measurements. In PGD, multiple embryos are required to increase the chances of having transferable embryos—this may be compromised with an advanced reproductive age.

Inherited disease in fetuses and offspring has long been a major problem for many couples in attempting to conceive a child. Today, this risk has lessened through medical science, which has at the same time raised significant ethical concerns. In many cases, the risk to a particular couple can be predicted from their family histories. If they wish to guard against a crippling inherited condition in their potential child, couples can choose to use prenatal diagnosis of their fetus at mid-pregnancy. Amniocentesis or chorionic villus sampling in mid-pregnancy enables cells and tissues to be collected from the fetus for diagnosis of familial or other forms of genetic diseases. Ultrasound examination of the fetus can also be used to detect morphological indicators of genetic disease. Parents can be advised by their physicians to have infusions of donor hemopoietic cells into the fetus to alleviate serious forms of immunological and hematological deficiencies. Whichever approach is used, parents are placed in a difficult and complex position, where they will have to decide whether to continue the pregnancy or abort their fetus if it is found to carry a severe genetic disease. Sadly, the great majority of inherited disorders as yet remain incurable. Parents must therefore make crucially important decisions about their fetus midway through gestation if prenatal diagnosis reveals a serious affliction. Counseling is essential to help them come to terms with this complex situation (Edwards, 2000). An alternative approach that has emerged is PGD. It is now routinely applied successfully to the detection of more and more severe forms of inherited disease (Edwards, 2000).

INDICATIONS

Although the most common use of PGD is aneuploidy screening for patients undergoing IVF treatment, we focus here on those with single gene disorders. Single gene disorders are a major health concern in the developing world, since the rate of consanguineous marriages are usually high, around 20%–70% (Al-Aqeel, 2007). Consanguinity, in combination with large family size and advanced maternal age, presents unique challenges and opportunities. Resulting pedigrees typically have multiple affected individuals. This helps in homozygosity gene mapping of a multitude of novel Mendelian diseases (Al-Aqeel, 2000; 2005; Martignetti et al., 2001). The tradition of consanguineous unions is likely to persist, and so the successful management of these disorders in affected children can incur a high financial cost. For example, the available injectable enzymes for Gaucher disease cost ~$30,000 per year for an infant. Consequently, it is a great burden on the healthcare system, the family, and the society, and carries a certain amount of stigmatization. Therefore, prevention is preferred, either through PGD or prenatal diagnosis in accordance with established ethical and religious guidelines (Al-Aqeel, 2005; 2007). Ideally, prevention of genetic diseases should start with community screening, possibly leading to PGD (Figure 59-1). PGD is considered one of the most powerful prevention tools for married couples who are at risk of transmitting genetic diseases.

PGD FOR SINGLE GENE DISORDERS

Single gene disorders are triggered by mutations in specific genes transmitted according to a predictable inheritance model, which can be dominant, recessive, or X-linked. Known single gene disorders number over 12,000, and they could be observed in 1/300 births (http://www.cdc.gov/ncbddd/single_gene/). In principle, PGD could be used in any of these disorders as long as disease-causing mutations are identified. Table 59-1 shows the list of single gene disorders for which PGD was performed in King Faisal Specialist Hospital and Research Center as examples of such disorders.

PGD FOR LATE-ONSET DISEASES AND CANCER SUSCEPTIBILITY GENES

Although ethically controversial, PGD application can be extended to diseases which have late onset, or mutations that merely predispose to certain cancers. In case of late-onset diseases, the penetrance is high and the presence of mutation is an almost certain indication that the disease will present later in life. PGD for certain Huntington's and Alzheimer disease gene mutations have been successfully performed (Sermon et al., 1998, Verlinsky et al., 2002). For genes linked to predisposition for cancer, the presence of a mutation does not always result in the occurrence of the malignancy. However, there is an increasing demand to utilize PGD for such conditions. PGD is performed for adenomatous polyposis of the colon (Ao et al., 1998), tumor-suppressor gene p53, Von Hippel–Lindau syndrome, retinoblastoma, rhabdoid tumor (Verlinsky et al., 2001; Rechitsky et al., 2002), neurofibromatosis type 2 (Abou-Sleiman et al., 2002), and BRCA1 (Spits et al., 2007).

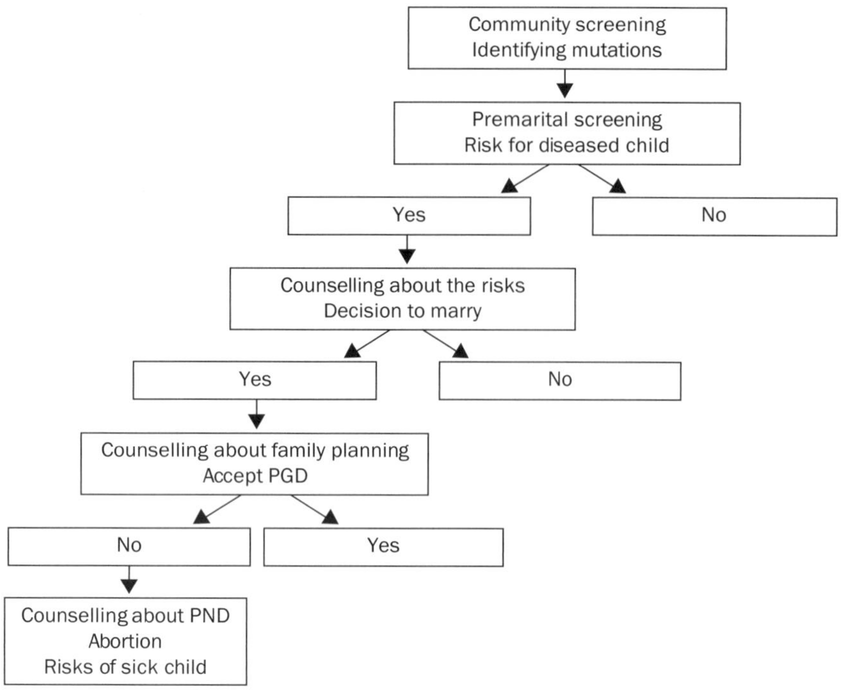

Figure 59-1 A flowchart of prevention methodologies of genetic diseases. PGD is one of the most effective preventive tools of genetic diseases among married couples.

TABLE 59-1 A LIST OF 51 DISEASES FOR WHICH PGD WAS PERFORMED IN KING FAISAL SPECIALIST HOSPITAL AND RESEARCH CENTER.

Achondroplasia	Maple syrup urine disease
Adrenoleukodystrophy (ADL)	Metachromatic leukodystrophy
Albinism, oculocutaneous type IA	Microphthalmia/anophthalmia
Alpha-thalassemia/mental retardation syndrome	Mucopolysaccharidosis, type I (Hurler syndrome)
Argininosuccinic acidemia	Mucopolysaccharidosis, type II (Hunter syndrome)
Ataxia telangiectasia mutated	Mucopolysaccharidosis, type IVA, Morquio syndrome A
Biotinidase deficiency	Mucopolysaccharidosis, type VI (Maroteaux-Lamy syndrome)
Bosley-Salih-Alorainy syndrome	Niemann Pick disease (NPD)
B-Thalassemia	Nonketotic hyperglycemia
Canavan disease	Osteogenesis Imperfecta Type IV
Carnitine-acylcarnitine translocase deficiency	Phenylketonuria (PKU)
Chronic granulomatous disease (CGD)	Progressive familial intrahepatic cholestasis 1
Citrullinemia	Propionic acidemia
Congenital adrenal hyperplasia (CAH)	Sanfilippo type B (MPS-IIIB)
Cystic fibrosis	Sanjad Sakati syndrome
Duchenne muscular dystrophy (DMD)	Severe combined immunodeficiency disease
Ehlers–Danlos syndrome	Sickle cell
Fragile-X	Smith Lemli Opitz syndrome; SLOS
Galactosemia	Spinal muscular atrophy/Werdnig Hoffmann
Glucose-6-phosphate dehydrogenase deficiency	Sulfite oxidase deficiency
Glutaric acidemia type 1	Tyrosinemia type I
glycogen storage disease type II, Pompe disease	Very long-chain acyl-CoA dehydrogenase deficiency
GM1-Gangliosidosis	Wiskott-Aldrich syndrome
Hereditary nonsyndromic sensorineural deafness	X-linked hydrocephalus
Hyperinsulinemia	Zellweger syndrome
Joubert syndrome (JBTS-3)	

HLA TYPING ("DESIGNER BABIES")

This indication of PGD selects the embryos based on their HLA types to be a matched stem cell donor for an affected sibling. These embryos are popularly known as "designer babies." Families requiring this procedure have a genetic disease in the family that affects hematopoietic cells, and the affected members are usually in need of stem cell transplantation. When the couple has decided to have another child, PGD offers a good option for them to have a child who is free of the disease and to be a stem cell matched donor for the living sibling. The chance of being a normal embryo in autosomal recessive diseases is 3 in 4, and likelihood of being HLA compatible is 1 in 4, which gives an overall likelihood of 3 in 16 embryos that would be suitable for transfer. The first PGD for this indication was applied to a family with Fanconi anemia (Verlinsky et al., 2001), and later extended to Wiskott-Aldrich syndrome, thalassemia, X-linked adrenoleukodystrophy, and hyperimmunoglobulin M syndrome (Rechitsky et al., 2004).

TECHNICAL ASPECTS

The PGD procedure has three technical aspects: (i) generation of embryos, (ii) removal of the cell, and (iii) genetic diagnosis. Embryos are generated through in-vitro techniques.

IN-VITRO FERTILIZATION

Although most of the patients who undergo PGD are fertile, they start a PGD cycle with infertility treatment to increase the number of available embryos to increase the chance of having transferable embryos for each cycle. After proper counseling and work-up, patients start with controlled ovarian hyperstimulation on day 3 of the menstrual cycle with daily gonadotropin injections with proper pituitary downregulation. The follicular development in response to the medication is monitored with regular ultrasound scanning to determine length of the stimulation, which may last a total of 7–15 days. When there are a good number of mature follicles, human chorionic gonadotropin (hCG) is given to induce final oocyte maturation. Oocytes are retrieved through ultrasound guided transvaginal aspiration 36 hours following hCG injection. Fertilization is achieved by injecting a single spermatozoon into the oocytes, which is known as intracytoplasmic sperm injection (ICSI), to avoid any genetic contamination arising from excess spermatozoa. The developments of embryos are followed daily until the diagnosis is made, and the normal embryos are transferred. Embryo transfers are usually performed on days 4–6 post retrieval at morulae/blastocyst stages.

BIOPSY

The removal of cells can be performed at different developmental stages. Biopsy is done under an inverted microscope with micromanipulator attached. The embryo is held with a holding pipette and cells are removed with an aspiration needle (Figure 59-2). The opening of a hole in the zona pellucida can be achieved by acid

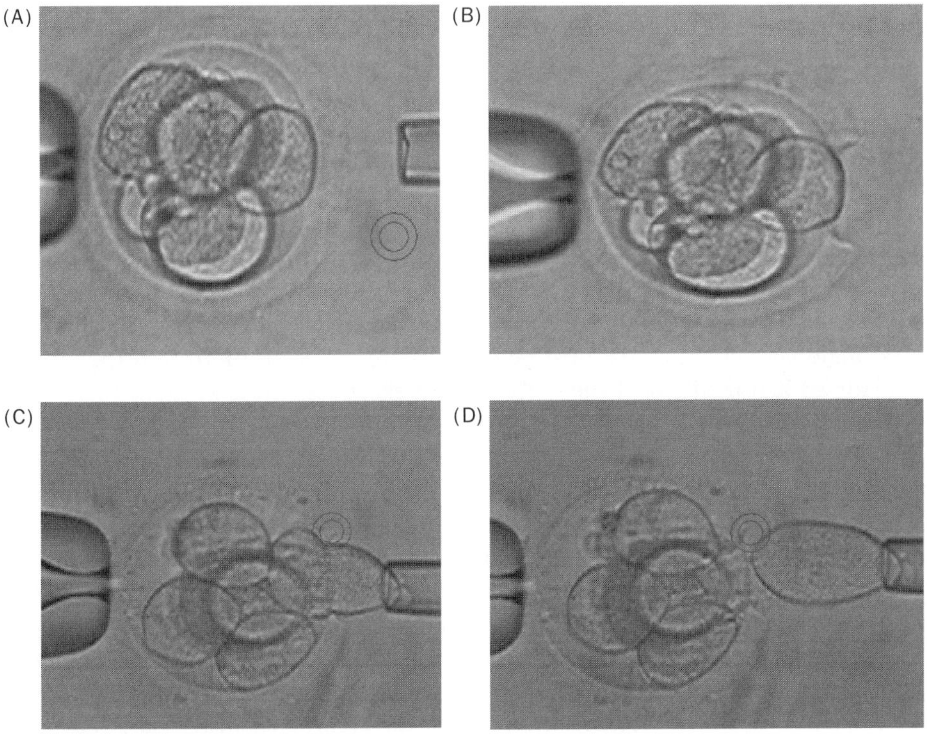

Figure 59-2 Blastomere biopsy. (A) an 8-cell embryo was held with a holding pipette; (B) a hole was made with the laser source; (C) the exposed blastomere was aspirated with a blastomere biopsy pipette; and (D) the blastomere was extracted from the embryo by applying gentle suction.

treatment, mechanical slitting or laser (De Vos and Van Steirteghem, 2001).

The most common biopsy stage is the removal of one or two blastomeres when the embryo is at the 6–10 cell stage on the third day following fertilization. A recent alternative is to perform the biopsy at the blastocyst stage on day 5 following fertilization. The blastomere removal has the advantage of providing more time for diagnosis, since the embryos have to be transferred by day 6. The blastocyst-stage biopsy, on the other hand, provides more cells for genetic diagnosis; however, the time is very limited for the diagnosis. Some centers cryopreserve blastocysts following biopsy to transfer the normal embryos in a replacement cycle (Magli et al., 2006). For this procedure, the center should have a good methodology and proven success with the blastocyst stage freezing.

First or second polar body biopsy are also performed. Polar bodies are the expelled genetic materials during the course of meiosis. They only provide information about the maternal genotype.

GENETIC ANALYSIS OF SINGLE CELLS IN SINGLE GENE DISORDERS

PCR is a powerful molecular technique as a part of PGD, for amplifying a particular DNA fragment to a level that it can be analyzed and a specific mutation can be detected (Saiki et al., 1985). However, it is still challenging to obtain a reliable diagnosis on a single cell where amplification failure (AF), allele dropout (ADO), and extraneous DNA contamination are still major problems. Along with direct mutation testing, the co-amplification of several informative linked markers such as short tandem repeats (STRs) and single nucleotide polymorphisms (SNPs) located very close to (extragenic) or within (intragenic) the disease-causing gene must be tested during the single cell diagnosis to identify ADO, to detect the presence and the source of any contamination.

AMPLIFICATION FAILURE (AF)

Problems with amplification failure in single cell PCR was observed when a PGD misdiagnosis for an X-linked disorder was reported (Hardy and Handyside, 1992), due to amplification failure of Y-chromosome-specific sequences leading to a false diagnosis as a female fetus. The reasons for amplification failure in single cell PCR are still unclear and occur frequently. However, many recommendations have been reported to minimize its risk, one of them being to ensure that the sampled cells do not lyse before they have been placed in the PCR tube, and that they should possess a visible nucleus (Sermon et al, 1995; Kontogianni et al, 1996).

ALLELE DROPOUT (ADO)

ADO is a phenomena unique to PCR with low abundance DNA, leading to preferential amplification (PA) of one allele, leaving the second allele under the detectable level (Findlay et al, 1995; Ray and Handyside, 1996; Rechitsky et al., 1996, 1998). ADO is a potential cause of misdiagnosis in the analysis of autosomal dominant disease, and recessive disorders when patients carry different mutations (compound heterozygotes). ADO of one of the mutant alleles in compound heterozygotes, or the mutant allele in autosomal dominant, results in a misdiagnosis as a heterozygote in the former case, and as healthy embryo in the latter, subsequently leading to transfer of an affected embryo.

If both parents carry the same mutation, then ADO of the mutant allele could result in an embryo diagnosis as normal, when in fact it is heterozygote, which imposes no serious misdiagnosis in returning back such an embryo (benign misdiagnosis). Yet, if ADO occurred to the normal allele, the embryo will be diagnosed as being affected, and consequently reduces the number of healthy embryos available for transfer.

ADO occurs in ~0.8–36% of single cell PCR amplifications (Ray and Handyside, 1996; Wells and Sherlock, 1998; Rechitsky et al., 1998). There are several possible causes of ADO, including DNA degradation, inefficient cell lysis, or imperfect PCR conditions. To minimize the risk of any misdiagnosis as a result of ADO, it is advisable to test two cells from each embryo if possible, since the probability of ADO affecting the same allele in both cells is low (Kontogianni et al., 1996; Thornhill and Snow, 2002). Additionally, using an informative linked polymorphism helps to detect any ADO in the gene of interest (Kuliev et al., 1998; Qubbaj et al., 2008).

CONTAMINATION

Peculiar DNA contamination from sperm, maternal cumulus cells, or any carryover contamination from previously amplified DNA fragments are also problems associated with misdiagnosis. Therefore, all cumulus cells must be carefully removed from around the oocyte and the embryo, and excess sperm is avoided by the use of the ICSI method. Biopsied cells (blastomeres) should be washed in a medium that has been tested for any contamination before being transferred to the PCR tubes. A separate room with filtered air under positive pressure (pre-PCR room) should be used for single cell PCR preparation and for cell transfer to the tube. To avoid any PCR template carryover, equipment and reagents for single cell PCR should be dedicated for this purpose, should not be in contact with previously amplified DNA, and reagents should be tested before a clinical case. All reagent preparations should take place in a laminar flow hood. Aliquoting of reagents into several tubes is important to avoid repeated freezing and thawing, and they should be discarded after a few uses. Staff working in the pre-PCR room should be required to wear clean gowns, gloves, and shoe covers.

WHOLE-GENOME AMPLIFICATION (WGA)

In PGD, only a single amplification reaction can be performed on a single cell. Optimizing various PCR conditions on a single cell level is costly and a time-consuming method. Additionally, testing more than one locus by including STR in multiplex PCR requires more extensive optimizations. In multiplex PCR, preferential amplification and ADO still occurs at a significantly higher rate (Findlay et al., 2001).

WGA is a technique to specifically increase the DNA quantities which originate from samples with limited DNA contents, from a starting template of 7 pg of DNA in a single cell, to a DNA yield of 35–40 μg (Cheung and Nelson, 1996; Hellani et al., 2004, Spits et al., 2006). WGA potentially can overcome many problems associated with single cell PCR. Molecular genetic testing protocols are the same as for genomic DNA; testing more than one locus can be achieved by multiplex PCR or single PCR.

Whole-genome amplification from a single cell can be achieved by primer extension pre-amplification (PEP; Zhang et al., 1992), and degenerate oligonucleotide primed PCR (DOP-PCR; Telenius et al., 1992). However, the amplification product from these techniques is short, characterized by strong biases, with artifacts, have incomplete genomic coverage, and inefficient microsatellite amplification (Cheung and Nelson, 1996; Paunio et al., 1996; Wells et al., 1999). Whole-genome amplification has been used in PGD cases for β–thalassemia by PEP-PCR (Jiao et al., 2003).

Recently, a new WGA method called *multiple displacement amplification* (MDA) has been introduced. MDA utilizes bacteriophage phi29 DNA polymerase and random hexamer primers under isothermal conditions and is not a PCR-based method (Blanco et al., 1989; Dean et al., 2002). MDA is found to overcome certain obstacles, including bias amplification, irreproducibility, and misdiagnosis (Hosono et al., 2003), and has been widely applied in PGD clinical cases (Hellani et al., 2005; Burlet et al., 2006; Coskun and Alsmadi, 2007; Renwick et al., 2006; Qubbaj et al., 2008).

The GenomePlex WGA method has also been recently introduced into the single cell field. It is a whole genome amplification method utilizing a proprietary amplification technology based upon random fragmentation of genomic DNA and conversion of the resulting small fragments to PCR-amplifiable OmniPlex Library molecules flanked by universal priming sites (Barker et al., 2004). The OmniPlex library is then PCR-amplified using universal oligonucleotide primers in a limited number of cycles. It has been successfully used in PGD for β-thalassemia and HLA typing (Chen et al., 2008).

ACCURACY OF PREIMPLANTATION GENETIC DIAGNOSIS AND CAUSES OF MISDIAGNOSIS

Prior to a PGD cycle, patients should be counseled and provide consent regarding the recommended PGD strategy and the likelihood of misdiagnosis. Misdiagnosis can be considered to be either adverse or benign. Adverse misdiagnoses are acknowledged due to the birth of an affected child or termination of an affected pregnancy, following a supposedly unaffected embryo transfer. Benign misdiagnoses resulted in the birth of an unaffected child carrying a mutation when the transferred embryo was thought to be homozygous normal. Misdiagnosis rates have been estimated at <1% for PCR cases (Wilton et al., 2009).

This percentage is usually estimated from the published misdiagnosis cases among the embryos having been transferred or the number of pregnancies. However, the true level of misdiagnosis in PGD may remain unknown for several reasons: not all embryos transferred resulted in pregnancy or a child birth; a proportion of misdiagnoses remained undetected; benign misdiagnoses were not considered; and some centers may not publish all such incidences.

During patient counseling it is recommended that both published and in-house estimates (individual misdiagnosis rates) should be available to allow informed consent for PGD (Wilton et al., 2009).

EMERGING (EXPANDING) PREIMPLANTATION GENETIC DIAGNOSIS TECHNOLOGIES

As molecular biology techniques advance, they often find an application in PGD. For example, WGA is becoming a preferred technique over direct single cell PCR, and the availability of excess DNA following WGA is providing new opportunities in single cell diagnosis like preimplantation genetic haplotyping (PGH) and microarray chip analyses.

PREIMPLANTATION GENETIC HAPLOTYPING

Testing of multiple markers to increase the efficiency of genetic diagnosis in single cells is a routine method. With the introduction of WGA in PGD, the diagnosis by using many linked markers was recently utilized and termed *preimplantation genetic haplotyping* (PGH;

Renwick et al., 2006). PGH is a novel approach in PGD to apply a rapid and straightforward application for any mapped single gene disorders, even in the absence of an established mutation. Several PGH cycles were carried out for cystic fibrosis (CF), Duchenne, and Becker muscular dystrophy, resulting in birth of babies in the PGD cycles for CF (Renwick et al., 2006). PGH requires careful study of the pedigree with at least one affected family member, and identification of several informative linked markers for the mutated gene.

MICROARRAY TECHNOLOGY

Although DNA microarrays are being used for detecting aneuploidies in embryos (Wells et al., 2008), genome-wide SNP microarrays hold great promise to be used in single gene disorders. SNPs are polymorphic sites spread throughout the entire genome of each individual. The information obtained from SNP analysis can be also used for DNA fingerprinting of embryos (Treff et al., 2010). Moreover, SNP microarrays can be instrumental in detecting the mutation-carrying gene by linkage analysis (Handyside et al., 2008). The proof of principle for this novel approach has been shown in a family with known mutation of cystic fibrosis by using genomic DNA from cell cultures (Handyside et al., 2008). If applied to the single cell, microarrays can revolutionize PGD by providing whole-genome information from a single chip, including chromosomal aneuploidy status, single gene disorder status, and HLA genotyping information. It even permits the assessment of multiple genetic loci diseases like cardiovascular diseases and diabetes (Handyside et al., 2008).

OUTCOME OF PREIMPLANTATION GENETIC DIAGNOSIS, AND RISK AND COMPLICATIONS

The European Society of Human Reproduction and Embryology (ESHRE) PGD Consortium was established in 1997, with data being collected from participating centers. The results are published annually in the journal, *Human Reproduction*.

The main risk of PGD involves misdiagnosis, which is having an affected embryo implantation following the transfer of supposedly normal embryos. The ESHRE PGD consortium reported a total of 12 misdiagnoses following PCR-based PGD from 912 pregnancies, or a misdiagnosis rate of ~1.3% (Wilton et al., 2009). The PGD procedure only covers the specific disease and the mutation previously identified in the family. Babies might also be born with other abnormalities that are not tested during PGD. Another complication related to infertility treatment, which might arise during stimulation, is ovarian hyperstimulation syndrome. Additionally, pregnancy following PGD is not guaranteed. Of ESHRE PGD consortium data collected since 1999, nine PGD datasets were published (data collection I-IX). Data were collected from 57 collaborating centers, comprising a total of 3530 PGD cycles to oocyte retrieval (OR) performed for monogenic disorders using PCR. Of these, 2775 Cycles to embryo transfer (ET) resulted in 769 pregnancies and more than 400 live born babies. Success rates for clinical pregnancy were quoted as 22% per cycle (OR) and 28% per embryo transfer procedure. To date, 12 cases of misdiagnosis for monogenic disorders using PCR were reported (0.36% per ET, 1.3% per pregnancy rate and 2.5% per delivered babies). As previously mentioned, it is essential that couples be counseled on all these issues before starting a PGD cycle.

ETHICAL CONSIDERATIONS

Although there is no single ethical issue that unifies the field of genetics and PGD, informed consent, confidentiality, the potential for social harm, and psychological distress are issues that physicians involved with testing should understand and address (Al-Aqeel, 2005).

The European Society of Human Reproduction and Embryology (ESHRE) published a report addressing ethical concerns related to PGD (ESHRE Task Force, 2001), which included the destruction of affected preimplantation embryos (and sometimes normal embryos due to technical limitations) and the risks of abuse for nonmedical reasons. The destruction of these embryos is not an issue within an Islamic context, since it is still performed in pre-ensoulment stage. With regard to the latter issue, information and consent of the couple, public transparency, and respect of professional guidelines would help limit risk of abuse.

Other indications for PGD that involve ethical concerns include cancer predisposition mutations and late-onset diseases, which may require nondisclosure testing. In nondisclosure PGD, which has been described for Huntington's disease but could also be applied to other late-onset diseases, patients do not wish to know their carrier status but want to have disease-free offspring. Their embryos are tested without revealing any of the details of the cycle or diagnosis (Stern, et al., 2002). Because nondisclosure testing puts practitioners in a difficult ethical position, such as having to undertake PGD cycles even when the results of previous cycles preclude the individual as a carrier, or having to do mock transfers if no embryos are available. The ESHRE ethics task force discourages nondisclosure testing (Shenfield, et al., 2003; Sermon, et al., 2002). PGD for late-onset genetic diseases and cancer predisposition genes has been subject to ethical debate, since they do not pose an immediate danger to the newborn. In some cases, an individual may not reach to an age to manifest the disease. These two groups are distinct in pathology. The late-onset diseases have full penetrance and are often not curable, whereas cancer predisposition syndromes do not have full penetrance and can often be treated. The value of the PGD procedure in these cases needs to be demonstrated. The ethical arguments with respect to PGD for these groups of diseases are quite similar. Many believe PGD is ethically acceptable for these indications because of the heavy burden imposed on patients who are carriers of one of these diseases, be it because of the burden imposed by the eventual fate in late-onset diseases, or because of the burden on the quality of life imposed by far-reaching preventive measures in cancer predisposition syndromes (Shenfield, et al., 2003; Robertson, 2003).

Another indication for PGD involves the selection of embryos according to their HLA type, in order to produce a stem cell donor for a sick sibling. In such cases, PGD is undertaken to select embryos not only being disease-free, but also having the same HLA type as the affected sibling. This application of PGD is controversial. The main ethical arguments against are that a child is being used as a treatment, hence breaching the Kantian imperative that a person should never be used as a means without consent. Furthermore, how will the child feel when they are told that they were made for the purpose of saving a sibling? Pennings and colleagues (2002), however, argue that the Kantian imperative is not violated because the future donor child is not solely made for the purpose of donating stem cells, but will also be loved and cared for by the family as an individual. Careful counseling by a psychologist should be arranged before PGD is used for this purpose, to ascertain the real motivations of the prospective parents. Pennings and coworkers (2002) also note that a child who is told that they were made to save a sibling will not experience feelings of low self-esteem but will, on the contrary, be proud to have contributed in such an important way to the well-being of a sibling.

Furthermore, giving cord blood at birth could be considered something one family member would do naturally for another. Finally, it is argued that parents would not hesitate to volunteer an existing child as a bone marrow donor for a sick sibling, and that it is ethically acceptable to make a child for a reason if it is acceptable to use an existing child for the same reason. Several other authors have reached the same conclusion (Shenfield, et al., 2003; Robertson, 2003). From an Islamic perspective, saving life is a priority in Islam; such an indication is encouraged by Islamic scholars, as the born child will also be loved and well cared for by his parents and family (Al-Aqeel, 2007).

Finally, several reports have been published on the use of sexing for social reasons, and have provoked mixed feelings. Opponents argue that sex selection for nonmedical purposes is sexist and tantamount to discrimination against women. Those in favor argue that sex selection at the embryonic stage of development is preferable to sex selection after prenatal diagnosis, or even after birth. Moreover, they claim that overall sex distribution will not be skewed, because only a few people will use PGD to choose their child's sex. A distinction should be made, however, between sex selection for convenience and for so-called family balancing. In the latter instance, the family should have at least one or two children of one sex before they can ask for a child of the opposite sex (Ray et al., 2003). In family balancing, the issues of sex discrimination and skewed sex ratios are avoided, giving the family an increased autonomy without conflicts with other ethical principles (Sermon et al., 2002; Robertson, 2003; Pennings, 2002; Ray et al., 2003).

A workshop organized by the International Islamic Center for Population Studies and Research (Al-Azhar University, Cairo) recognized the importance of PGD, but was guarded about its use on nonmedical grounds such as sex selection or family balancing, considering that each case should be treated on its own merits. The medical application of PGD was seen as marking progress in the field of ART, and as a welcome alternative to prenatal diagnosis that results in abortion. Sex-selection technologies have been condemned on the ground that their application is to discriminate against female embryos and fetuses, so perpetuating prejudice against the girl child and social devaluation of women (Serour and Dickens, 2001). Therefore, from an Islamic perspective, the Islamic Jurisprudence Council of the Organization of Islamic Countries did not approve social sexing. However, they allowed sex selection by PGD only for sex-linked genetic disorders. Applying the fourth Shari'ah principle (permissiveness in useful matters), this technique is ideal, as it is carried out in the pre-ensoulment (prior to the inspiration of the soul) stage at 48–72 hours of gestation (Al-Aqeel, 2007).

CONCLUSIONS

Diagnostic methods for PGD are based on DNA technologies. These techniques are advancing at a frantic pace. The methods are so swift that diagnosis on embryos can be made within a day, and normal embryos can be replaced or cryopreserved for future transfer if the parents desire.

Major advances in IVF have made their own unique contributions to these diagnostic studies. ICSI avoids the possible contamination of PCR by the numerous sperm heads adhering to the zona pellucida, as in a typical IVF insemination case. Embryo cryopreservation is becoming simpler and more routine, so embryos can be stored for later transfer. New understanding of the form of growth of human embryos, and improved culture medium to grow blastocysts in vitro, may soon permit a single embryo to be replaced in a mother with a very high probability of successful pregnancy. Indeed, the typical low rate of human implantation, in vitro or in vivo, has been a major problem in applying preimplantation genetic diagnosis, since it necessitated transfer of several normal embryos to achieve a single implantation. Moreover, the biopsy at the blastocyst stage could provide more genetic material for more accurate diagnosis.

The information presented briefly illustrates that PGD has become an established procedure in assisted reproduction practices and genetic services. For genetically disadvantaged couples, PGD makes it possible to plan for healthy progeny without facing the need for prenatal diagnosis and termination of unwanted pregnancies.

REFERENCES

Abou-Sleiman PM, Apessos A, Harper JC, Serhal P, Winston RM, Delhanty JD. (2002) First application of preimplantation genetic diagnosis to neurofibromatosis type 2 (NF2). *Prenat Diagn* 22:519–524.

Al-Aqeel A, Al Sewairi W, Edress B, Gorlin RJ, Desnick RJ, Martignetti JA. (2000) Inherited multicentric osteolysis with arthritis: a variant resembling Torg syndrome in a Saudi family. *Am J Med Genet* 93:11–18.

Al Aqeel AI. (2005) Ethical guidelines in genetics and genomics. An Islamic perspective. *Saudi Med J* 26:1862–1870.

Al Aqeel AI. (2007) Islamic ethical framework for research into and prevention of genetic diseases. *Nat Genet* 39:1293–1298.

Albar MA. (1999) Counselling about genetic disease: an Islamic perspective. *East Mediterr Health J* 5(6):1129–1133.

Albar MA. (2002) Ethical considerations in the prevention and management of genetic disorders with special emphasis on religious considerations. *Saudi Med J* 23(6):627–632.

Ao A, Wells D, Handyside AH, Winston RM, Delhanty JD. (1998) Preimplantation genetic diagnosis of inherited cancer: familial adenomatous polyposis coli. *J Assist Reprod Genet* 15:140–144.

Barker DL, Hansen MS, Faruqi AF et al. (2004) Two methods of whole-genome amplification enable accurate genotyping across a 2320-SNP linkage panel. *Genome Research* 14:901–907.

Blanco L, Bernad A, Lazaro JM, Martin G, Garmendia C, Salas M. (1989) Highly efficient DNA synthesis by the phage phi 29 DNA polymerase. Symmetrical mode of DNA replication. *J Biol Chem* 264:8935–8940.

Burlet P, Frydman N, Gigarel N, et al. (2006) Multiple displacement amplification improves PGD for fragile X syndrome. Mol *Hum Reprod* 12:647–652.

Chen SU, Su YN, Fang MY, et al. (2008). PGD of beta-thalassaemia and HLA haplotypes using OmniPlex whole genome amplification. *Reprod Biomed Online* 17:699–705.

Cheung VG, Nelson SF. (1996) Whole genome amplification using a degenerate oligonucleotide primer allows hundreds of genotypes to be performed on less than one nanogram of genomic DNA. *Proc Natl Acad Sci USA* 93:14676–14679.

Coskun S, Alsmadi O. (2007) Whole genome amplification from a single cell: a new era for preimplantation genetic diagnosis. *Prenat Diagn* 27:297–302.

De Vos A, Van Steirteghem A. (2001) Aspects of biopsy procedures prior to preimplantation genetic diagnosis. *Prenat Diagn* 21:767–780.

Dean FB, Hosono S, Fang L, et al. (2002) Comprehensive human genome amplification using multiple displacement amplification. *Proc Natl Acad Sci USA* 99:5261–5266.

Edwards RG. (2000). The resounding clinical opportunities and the severe ethical issues in preimplantation genetic diagnosis. In: Verlinsky Y, Kuliev A, eds. *An Atlas of Preimplantation Genetic Diagnosis*, pp. 3–5. New York, London: The Parthenon Publishing Group.

Ehrich K, Williams C, Farsides B, Sandall J, Scott R. (2007) Choosing embryos: ethical complexity and relational autonomy in staff accounts of PGD. *Sociology of Health & Illness* 29:1091–1106.

ESHRE Task Force on Ethics and Law. (2001) The moral status of the pre-implantation embryo. *Human Reprod* 16:1046–1048.

Findlay I, Ray P, Quirke P, Rutherford A, Lilford R. (1995) Allelic drop-out and preferential amplification in single cells and human blastomeres: implications for preimplantation diagnosis of sex and cystic fibrosis. *Hum Reprod* 10:1609–1618.

Findlay I, Matthews PL, Mulcahy BK, Mitchelson K. (2001) Using MF-PCR to diagnose multiple defects from single cells: implications for PGD. *Mol Cell Endocrinol* 183 Suppl 1:S5–S12.

Handyside AH, Kontogianni EH, Hardy K, Winston RM. (1990) Pregnancies from biopsied human preimplantation embryos sexed by Y-specific DNA amplification. *Nature* 344:768–770.

Handyside AH, Thornhill AR, Affara NA, Harton GL, Mariani BD, Griffin DK. (2008) Recombination mapping: a universal molecular karyotyping method for preimplantation genetic diagnosis of inherited disease. *Fertil Steril* 90(Suppl 1):S24.

Hardy K, Handyside AH. (1992) Biopsy of cleavage stage human embryos and diagnosis of single gene defects by DNA amplification. *Arch Pathol Lab Med* 116:388–392.

Hellani A, Coskun S, Benkhalifa M, et al. (2004) Multiple displacement amplification on single cell and possible PGD applications. *Mol Hum Reprod* 10:847–852.

Hellani A, Coskun S, Tbakhi A, Al Hassan S. (2005) Clinical application of multiple displacement amplification in preimplantation genetic diagnosis. *Reprod Biomed Online* 10:376–380.

Hosono S, Faruqi AF, Dean FB, et al. (2003) Unbiased whole-genome amplification directly from clinical samples. *Genome Res* 13:954–964.

Jiao Z, Zhou C, Li J, et al. (2003) Birth of healthy children after preimplantation diagnosis of beta-thalassemia by whole-genome amplification. *Prenat Diagn* 23:646–651.

Kearns WG, Pen R, Graham J, et al. (2005) Preimplantation genetic diagnosis and screening. *Semin Reprod* Med 23:336–347.

Kontogianni EH, Griffin DK, Handyside AH. (1996) Identifying the sex of human preimplantation embryos in X-linked disease: amplification efficiency of a Y-specific alphoid repeat from single blastomeres with two lysis protocols. *J Assist Reprod Genet* 13:125–132.

Kuliev A, Rechitsky S, Verlinsky O, et al. (1998) Preimplantation diagnosis of thalassemias. *J Assist Reprod Genet* 15:219–225.

Magli MC, Gianaroli L, Grieco N, Cefalu E, Ruvolo G, Ferraretti AP. (2006) Cryopreservation of biopsied embryos at the blastocyst stage. *Hum Reprod* 21:2656–2660.

Martignetti JA, Aqeel AA, Sewairi WA, et al. (2001) Mutation of the matrix metalloproteinase 2 gene (MMP2) causes a multicentric osteolysis and arthritis syndrome. *Nat Genet* 28:261–265.

Modell B, Darr A. (2002) Science and society: genetic counselling and customary consanguineous marriage. *Nat Rev Genet* 3:225–229.

Paunio T, Reima I, Syvanen AC. (1996) Preimplantation diagnosis by whole-genome amplification, PCR amplification, and solid-phase minisequencing of blastomere DNA. *Clin Chem* 42:1382–1390.

Pennings G. (2002) Personal desires of patients and social obligations of geneticists: applying preimplantation genetic diagnosis for non-medical sex selection. *Prenat Diagn* 22:1123–1129.

Pennings G, Schots R, Liebaers I. (2002) Ethical considerations on preimplantation genetic diagnosis for HLA typing to match a future child as a donor of haematopoietic stem cells to a sibling. *Hum Reprod* 17:534–538.

Preimplantation Genetic Diagnosis International Society (PGDIS). (2008) Guidelines for good practice in PGD: programme requirements and laboratory quality assurance. *Reprod Biomed Online* 16:134–147.

Qubbaj W, Al Aqeel AI, Al Hassnan Z, et al. (2008) Preimplantation genetic diagnosis of Morquio disease. *Prenat Diagn* 28:900–903.

Ray PF, Handyside AH. (1996) Increasing the denaturation temperature during the first cycles of amplification reduces allele dropout from single cells for preimplantation genetic diagnosis. *Mol Hum Reprod* 2:213–218.

Ray P, Munnich A, Nisand I, Frydman R, Vekemans M, Viville S. (2003) Sex selection by preimplantation genetic diagnosis: should it be carried out for social purposes? Is preimplantation genetic diagnosis for "social sexing" desirable in today's and tomorrow's society? *Hum Reprod* 18: 463–464.

Rechitsky S, Freidine M, Verlinsky Y, Strom CM. (1996) Allele dropout in sequential PCR and FISH analysis of single cells (cell recycling). *J Assist Reprod Genet* 13:115–124.

Rechitsky S, Strom C, Verlinsky O, et al. (1998) Allele dropout in polar bodies and blastomeres. *J Assist Reprod Genet* 15:253–257.

Rechitsky S, Verlinsky O, Chistokhina A, et al. (2002) Preimplantation genetic diagnosis for cancer predisposition. *Reprod Biomed Online* 5:148–155.

Rechitsky S, Kuliev A, Tur-Kaspa I, Morris R, Verlinsky Y. (2004) Preimplantation genetic diagnosis with HLA matching. *Reprod Biomed Online* 9:210–221.

Renwick PJ, Trussler J, Ostad-Saffari E, et al. (2006) Proof of principle and first cases using preimplantation genetic haplotyping--a paradigm shift for embryo diagnosis. Reprod Biomed Online 13:110–119.

Robertson JA. (2003) Extending preimplantation genetic diagnosis: the ethical debate. Ethical issues in new uses of preimplantation genetic diagnosis. *Hum Reprod* 18:465–471.

Saiki RK, Scharf S, Faloona F, et al. (1985) Enzymatic amplification of beta-globin genomic sequences and restriction site analysis for diagnosis of sickle cell anemia. *Science* 230:1350–1354.

Sermon K, Lissens W, Nagy ZP, Van Steirteghem A, Liebaers I. (1995) Simultaneous amplification of the two most frequent mutations of infantile Tay-Sachs disease in single blastomeres. *Hum Reprod* 10:2214–2217.

Sermon K, Goossens V, Seneca S, et al. (1998) Preimplantation diagnosis for Huntington's disease (HD): clinical application and analysis of the HD expansion in affected embryos. *Prenat Diagn* 18:1427–1436.

Sermon K, De Rijcke M, Lissens W, et al. (2002) Preimplantation genetic diagnosis for Huntington's disease with exclusion testing. *Eur J Hum Genet* 10:591–598.

Serour GI, Dickens BM. (2001) Assisted reproduction developments in the Islamic world. *Int J Gynaecol Obstet* 74:187–193.

Shenfield F, Pennings G, Devroey P, Sureau C, Tarlatzis B, Cohen J. (2003) Taskforce 5: preimplantation genetic diagnosis. *Hum Reprod* 18:649–651.

Spits C, Le Caignec C, De Rycke M, et al. (2006) Optimization and evaluation of single-cell whole-genome multiple displacement amplification. *Hum Mutat* 27:496–503.

Spits C, De Rycke M, Van Ranst N, et al. (2007) Preimplantation genetic diagnosis for cancer predisposition syndromes. Prenat Diagn 27:447–456.

Stern HJ, Harton GL, Sisson ME, et al. (2002) Non-disclosing preimplantation genetic diagnosis for Huntington disease. *Prenat Diagn* 22:503–507.

Telenius H, Carter NP, Bebb CE, Nordenskjold M, Ponder BA, Tunnacliffe A. (1992) Degenerate oligonucleotide-primed PCR: general amplification of target DNA by a single degenerate primer. *Genomics* 13:718–725.

Thornhill AR, Snow K. (2002) Molecular diagnostics in preimplantation genetic diagnosis. *J Mol Diagn* 4:11–29.

Treff NR, Su J, Tao X, Miller KA, Levy B, Scott RT Jr. (2010) A novel single-cell DNA fingerprinting method successfully distinguishes sibling human embryos. *Fertil Steril* 94:2017–2021.

Verlinsky Y, Kuliev A. (2000) *An Atlas of Preimplantation Genetic Diagnosis*. New York, London: The Parthenon Publishing Group.

Verlinsky Y, Rechitsky S, Schoolcraft W, Strom C, Kuliev A. (2001) Preimplantation diagnosis for Fanconi anemia combined with HLA matching. *JAMA* 285:3130–3133.

Verlinsky Y, Rechitsky S, Verlinsky O, et al. (2001) Preimplantation diagnosis for p53 tumour suppressor gene mutations. *Reprod Biomed Online* 2:102–105.

Verlinsky Y, Rechitsky S, Verlinsky O, Masciangelo C, Lederer K, Kuliev A. (2002) Preimplantation diagnosis for early-onset Alzheimer disease caused by V717L mutation. *JAMA* 287(8):1018–1021.

Wells D, Alfarawati S, Fragouli E. (2008) Use of comprehensive chromosomal screening for embryo assessment: microarrays and CGH. *Mol Hum Reprod* 14:703–710.

Wells D, Sherlock JK. (1998) Strategies for preimplantation genetic diagnosis of single gene disorders by DNA amplification. *Prenat Diagn* 18:1389–1401.

Wells D, Sherlock JK, Handyside AH, Delhanty JD. (1999) Detailed chromosomal and molecular genetic analysis of single cells by whole genome amplification and comparative genomic hybridisation. *Nucleic Acids Res* 27:1214–1218.

Wilton L, Thornhill A, Traeger-Synodinos J, Sermon KD, Harper JC. (2009) The causes of misdiagnosis and adverse outcomes in PGD. *Hum Reprod* 24:1221–1228.

Zhang L, Cui X, Schmitt K, Hubert R, Navidi W, Arnheim N. (1992) Whole genome amplification from a single cell: implications for genetic analysis. *Proc Natl Acad Sci USA* 89:5847–5851.

ASIA-PACIFIC

60.

GENOMICS FOR UNIVERSAL HEALTHCARE IN THAILAND[1,2]

Béatrice Séguin, Billie-Jo Hardy, Peter A. Singer, and Abdallah S. Daar

INTRODUCTION

One potential outcome of investing in genomic medicine is the provision of tools for creating a more cost-effective healthcare system. Partly with this aim in mind, Thailand has launched two genotyping initiatives: the Thai SNP Discovery Project and the Thai Center for Excellence in Life Sciences Pharmacogenomics Project. Together, these projects will help Thailand understand the genomic diversity of its population, and explore the role that this diversity has in drug response and disease susceptibility in its population. This chapter discusses challenges faced by Thailand to integrate genomic medicine in its relatively young universal healthcare system.

One of Thailand's distinguishing features as an emerging economy is the universal healthcare coverage for its population—a feature that can present both an opportunity and a challenge in the context of applying genomic medicine. An objective of Thai universal healthcare coverage is to ensure local access to essential medicines. However, as with most universal healthcare schemes, financial resources are limited and government providers must focus health expenditures on those medicines that are most likely to achieve the greatest health benefit at present. It is therefore legitimate to ask if drugs that have been developed and tested in Caucasian populations of industrialized countries would have the same therapeutic index (benefit/risk ratio) in the Thai population. Policymakers and scientists are aware that genomic diversity research presents an opportunity to understand local patterns of complex disease risk and variation in drug response, susceptibility to adverse drug reactions (ADRs), as well as patterns of infectious disease resistance (Marsh et al., 2006). This knowledge should translate into better diagnostics, better-informed therapeutic choices, and eventually perhaps new therapeutics and improved lifestyle choices that promote health and prevent disease.

The Thai SNP Discovery Project—a collaboration between Mahidol University's Faculty of Medicine, Ramathibodi Hospital, Oracle Co., Ltd (Thailand), the National Center for Genetic Engineering and biotechnology (BIOTEC; Thailand) and Centre National de Génotypage (CNG; France)—has collected samples from the Thai population and created an SNP database containing allele frequency and linkage disequilibrium (LD) block patterns for all annotated human genes and their regulatory regions. Initiated as Thailand's contribution to the Human Genome Organization (HUGO) Pan-Asian SNP Consortium, the database will also contain other information generated by the principal investigators of the Thai SNP Discovery Project, such as genomic sequences, genomic structure, primer sequences, and functional genomic information. A second major genotyping project in Thailand is the Thailand Center of Excellence in Life Sciences (TCELS)'s Pharmacogenomics Project, which is performing SNP genotyping of genes involved in drug responses in the Thai population. Although the two projects were initiated separately, the participating research scientists collaborate extensively on several projects. Applications resulting from these initiatives in the context of universal healthcare will require a long-term approach. The hope is that, as a result of these research initiatives, costs to Thailand's healthcare system will be reduced through selective use of safe and efficacious drugs. The cost of drugs in Thailand is an issue of major national concern; it is one of a few countries that have issued compulsory licenses in order to access existing therapeutics at a more affordable price (Stone, 2007). Compulsory licenses are issued by a government under certain conditions, such as a health emergency, to manufacture a patented product without the permission of the patent holder, and with adequate compensation (see the world Trade Organization's FAQs on compulsory licensing of pharmaceuticals). Thailand is thus at a crossroads, balancing its efforts between addressing the short-term problems of drug access while investing in long-term

1. Reprinted from Séguin B, Hardy B-J, Singer PA, Daar AS. (2008). Universal health care, genomic medicine and Thailand: investing in today and tomorrow. Copyright 2008 by Nature Publishing. Reproduced with permission of Nature Publishing.

2. The research presented in this chapter was conducted between 2006 and 2008, it has not been updated to reflect the many changes in the field.

research aimed at increasing the chances of meeting its own local health needs through local research and development (R&D) and innovation.

Using previously described qualitative methods (Séguin et al, 2008a,b) we performed 15 in-depth semi-structured interviews with key informants, including scientists and managers at the Thai SNP Discovery Project and the TCELS' Pharmacogenomic Project, as well as regulatory and government officials, a policy analyst, and experts in the area of genomics/pharmacogenetics and ethics from developed and developing countries. Below we explore these results within the context of a universal healthcare system in an emerging economy.

ADOPTION OF GENOMIC MEDICINE IN THAILAND

POLITICAL WILL

Thai stakeholders from various institutions, including local universities, BIOTEC, the National Science and Technology Development Agency (NSTDA), and the Thai Ministry of Public Health, participated in an Asia-Pacific Economic Cooperation (APEC)-wide foresight study in 2003. The meeting culminated in a report entitled: "DNA Analysis for Human Health in the Post-Genomic Era," which included a chapter authored by Thai stakeholders exploring opportunities and implications associated with the adoption of genomic medicine for member nations, titled "Issues Paper." In response, Thailand has promulgated its national policy, and in the National Biotechnology Policy Framework, 2004–2009, genomics and bioinformatics were identified as key strategy areas for improving the health of the Thai people, developing new bio-businesses as well as creating a self-sufficient economy. The second major project is aimed at a more downstream and potentially more immediate application of studies of human genomic variation. The establishment of TCELS, which initiated the Thai Pharmacogenomics Project by a royal decree of the King of Thailand in June 2004, attests to the importance of this project, as does the initial investment of 120 million bhat (approximately US$3.9 million) from the Thai government. This initiative is consistent with the Thai biotechnology policy framework. Although some of our key informants realistically expected that the fruits of their research will take time to have an impact on both local healthcare and the economy, they are aware that the applications of genomic medicine are already being applied clinically in ways that are relevant to Thailand. For example, Abacavir, an HIV antiretroviral, is associated with a hypersensitivity reaction that is associated with the allelic variant *HLA-B*5701*, the frequency of which varies across populations (Hughes et al., 2004; Hughes et al, 2008; Sadiq & Pakianathan, 2007; Watson et al., 2004). Prospective screening for this genotype can result in decreased occurrence of hypersensitivity reactions, because individuals with this variant are not prescribed Abacavir (Rauch et al., 2006; Zucman et al., 2007).

LOCAL HEALTH BENEFITS

With a population of ~64 million, Thailand spends 3.5% of its gross domestic product (GDP) on health care (see the World Health Organization's Thailand overview). Ninety-five percent of the Thai population has healthcare coverage, with 72% accessing it through the Thai universal health scheme (International Labor Office, 2004). Aiming toward cost-effective, local healthcare solutions, the Thai SNP Discovery Project will provide a reference database for spin-off projects that look specifically at disease associations. Such projects will include understanding genetic susceptibility to clinical malaria, identifying genetic factors involved in the clinical severity of thalassemia–hemoglobin E disease, and understanding the development of Dengue hemorrhagic fever. The TCELS' Pharmacogenomics Project, on the other hand, has collected 1,500 samples from patient populations or healthy Thais from throughout Thailand for five pharmacogenomics projects: an HIV pharmacogenomics project (studying Nevirapine-induced skin rash); drug allergy to allopurinol and carbamazepine; pharmacogenomics in childhood acute lymphoblastic leukemia; studies into oncology chemotherapy (using fluorouracil, 5-FU); and investigation of thalassemia–hemoglobin E disease. Researchers from the TCELS' Pharmacogenomics Project and the Thai SNP Discovery Project are focusing on diseases that are common in Thai patients. One key informant emphasized the value of conducting research for local benefit. He pointed out that given the small size of the Thai market, there is little to no incentive for multinational corporations to ensure that therapeutics have a good response rate with few adverse effects within the Thai population. The same, in fact, applies to many emerging economies and developing countries. One application that would be amenable to public health interventions is the screening of HIV patients for susceptibility to nevirapine-induced skin rash, an antiretroviral treatment that is used in Thailand. Such an application would have important public health implications. According to a document issued by the Thai Ministry of Public Health and the National Health Security Office, approximately 20% of patients using nevirapine-based triple antiretrovirals, such as GPO-vIr, develop the adverse reaction (National Health Security, Thailand, 2007; Phanuphak et al., 2007; Kiertiburanakul et al., 2007). Nevirapine can be replaced with an alternative therapeutic that does not cause the ADR—Efavirenz-based triple antiretroviral. However, the cost of this treatment is more than twice that of GPO-vIrs (National

Health Security, Thailand, 2007) and rational use of Efavirenz. In addition, TCELS has a particular interest in addressing a locally relevant mental health problem: it is collaborating with scientists at RIKEN Institute, Japan, to perform genome-wide scans on 3,000 samples from tsunami victims that have experienced post-traumatic stress disorder (PTSD), a disease with negative socioeconomic and health consequences on children and adults (Stone, 2005). This project provides a good example of north–south collaboration (that is, between developed and developing economies) and further supports the argument that genotyping projects build scientific capacity.

GENOMIC SOVEREIGNTY

At the onset of our study of large-scale genotyping projects in emerging economies, we found that in Mexico there was a desire to capture the value of investments in large-scale human genotyping projects and to protect Mexican biological samples from foreign exploitation. Key informants in Mexico spoke of protecting "genomic sovereignty" (Séguin et al., 2008a,c). The concept of genomic sovereignty did not appear as explicitly in interviews with our key Thai informants, but we heard of a need to protect Thai DNA samples. One key informant puts it thus: "the export of the DNA from the developed country, like the U.K. or Japan, is very difficult. But if the DNA is coming out from Thailand, it's very easy; we should have a clear regulation or law about how and who should authorize the transfer of the material abroad." Others expressed concerns that inappropriately tough guidelines and/or legislation pertaining to genomic sovereignty might impede international collaborations and partnerships, which Thai researchers consider valuable because they contribute toward building local scientific capacity. At present, there are institutional guidelines (within universities and hospitals) but no central unified national legislation on the export of human DNA samples. For example, there are guidelines from the Ministry of Public Health (MoPH) to restrict the export of DNA samples without proper justification (for instance, benefits to Thailand must be specified, and the MoPH requires that samples be accompanied by a local researcher), but these only apply to researchers who are funded through the Ministry. Thus, even if legislation in Thailand were not to go as far as to introduce a genomic sovereignty law similar to the one recently passed in Mexico (Becerril, 2008), it would benefit from a standardized central system that would, at a minimum, provide Thailand with a better understanding of where its DNA samples were going and why. This is an important consideration if Thailand is to ensure a self-sufficient economy and a thriving bio-business, as stated in the National Biotechnology Policy Framework. Given the recent controversies over Thailand issuing compulsory licenses for antiretrovirals, ensuring that knowledge generated from human genomic variation studies is protected through legislation or intellectual property could avoid similar future scenarios with respect to products and services derived from its own resources and research.

KNOWLEDGE-BASED ECONOMY

As part of a broader indication that innovation and commercialization of local research are a priority for the Thai government, it has made an initial public investment of $175 million in the Thailand Science Park (Thailand Science Park, 2007), which is intended to attract start-up as well as existing firms, including those focused on biotechnology and genomics. This type of support from the Thai government can be leveraged by organizations such as TCELS. The TCELS' Pharmacogenomics Project, by virtue of its association with TCELS, is able to leverage available in-house administrative systems for managing its own intellectual property and translating basic research.

Although TCELS does promote commercialization of research, no results have been commercialized so far. We therefore sought private firms that might be investing in human genomics, wanting to understand what types of bottlenecks are currently in place with respect to early commercialization of knowledge stemming from human genomic variation studies. We interviewed chief executive officers (CEOs) of two recently established start-up private companies that provide genotyping services to detect predispositions to common diseases. These services are different from a large-scale project such as the one initiated by Avesthagen in India (Séguin et al., 2008a; Jayaraman, 2007). They are more comparable to ones offered by health and wellness companies such as Sciona or 23andMe. Delphi Health Services, Ltd, provides genotyping services for complex diseases such as type II diabetes, psoriasis, and eczema, through physicians in private hospitals and to medical tourists. HeartGenetics is mainly focused on providing genotyping services for cardiovascular disease; mutations in the low-density lipoprotein receptor gene and the cholesterol ester transfer gene are screened for. The company offers its genotyping services to hospitals, but also directly to consumers via its website. Both companies are at an early stage in their development; hence it is too early to analyze their capabilities, business models, and economic benefit. Interestingly, however, key informants from the Thai private sector suggested that initial health benefits from genomics will be primarily directed toward the medical tourism industry and private Thai hospitals. One CEO stated that: "the initial market that we and others are interested in are the medical tourists. Thailand serves as an excellent base of operation for developing these tests [for example, SNP genotyping] because field labor is so cheap here." According to the Thailand

Board of Investment (BOI), typical labor costs for researchers and technicians are approximately $571 and $286 per month, respectively, making the cost of establishing local businesses very attractive and conducive to a medical tourism industry, which is already thriving in Thailand (see Thailand's Advantages on the BOI website). This feature is unique among the genotyping projects in the emerging economies we studied. We found that the efforts supported by TCELS might eventually lead to diagnostics and/or therapeutics that would be amenable to local public health, as well as economic benefits. The private sector firms we studied have so far not used these data because their focus, at present, is different.

INSTITUTIONAL LEADERSHIP

Key informants acknowledge that the Thai population needs to understand the potential behind genomic medicine to encourage its adoption. Various institutions in Thailand have shown leadership in the general area of public engagement. For example, science institutions like BIOTEC and the National Health Foundation promote public meetings to discuss general bioethics issues. One key informant stated that, at this stage, issues that concern the public relate more to stem-cell technologies, reproductive technologies, and cloning, which they perceive as more contentious issues than human genomics. TCELS representatives have made appearances in the media, published materials in local newspapers, encouraged companies to produce educational material about DNA and pharmacogenomics, and conducted tours of local pharmaceutical firms accompanied by journalists, in order to promote general public awareness on the importance of R&D in life sciences. In conjunction with local scientists, TCELS have released books and reports including "Pharmacogenomics for laymen" and "Pharmacogenomics in a resource-limited country, Thailand: building a better quality of life for Thai people." Written in Thai, they have been generated as part of an effort to gain support from the public, decision makers, and funding bodies. Such public engagement initiatives are fully supported by our informants, even though they might be misunderstood. One informant put it this way: "Someone even accused TCELS that 'you spend too much money on the advertisement' but I have a different idea. We need to educate people until they understand. And at that day, whether they will accept or reject it is up to their background. Not just based on something that they don't know."

GENOMIC MEDICINE IN THAILAND

DELIVERY AND ACCESS

Thailand has universal healthcare coverage, yet we did not come across any existing program to implement genomic medicine into the public healthcare system in the short term. Potential benefits are clear. For instance, one interviewee claimed that patients undergoing kidney transplant, if they were covered, would benefit from pharmacogenetic testing to prevent adverse reactions to azathioprine, a commonly used drug to stop organ rejection. As he later stated, "in order to get the benefit of health, of genomics, universal health insurance is probably a prerequisite." However, Thai universal health insurance coverage is new and only covers basic services. The addition of genetic tests as part of a public health initiative might be seen as an additional financial strain on the system, an issue common to all health systems globally. At present, such testing is within reach of only the wealthier portions of Thai society.

EDUCATION OF HEALTHCARE PROFESSIONALS

Adoption of genomic medicine will require a well-trained workforce. Although local genomics researchers have produced handbooks in Thai language to educate physicians on how to "maximize the effectiveness of HIV-1 treatment by using molecular techniques, bioinformatics, and pharmacogenomics" (Chantratita, 2008), others revealed that one of the main barriers to the adoption of genomic medicine in a public health setting is indeed the formal training of human resources in this field, including the lack of genetic counselors. This is a challenge facing most countries, including the industrially developed ones (SACGHS, 2008).

INNOVATION FRAMEWORK

Critics of the Thai innovation system suggest there is too much emphasis upon supporting research and capacity in the public sector, at the expense of providing similar support to domestic industry (Intarakumnerd & Chaminade, 2007). Accordingly, one participant pointed out that vaccine production is often outsourced to foreign companies, which he saw as obstructive toward long-term capacity building in science and technology in Thailand. This lack of support poses an additional challenge to developing newer fields such as genomics. In addition, Thailand is confronted by numerous capacity-building challenges. For instance, despite the National Health Foundation's "reverse brain drain" project, one key informant stressed that there are still few human genomics researchers in Thailand, and felt that many of their skilled PhD graduates continue to leave the country.

REGULATORY FRAMEWORKS

Thailand has been noteworthy internationally because of its policy of compulsory licensing of essential medicines (for example, antiretrovirals; see Ford et al., 2007; Seim, 2007; Steinbrook, 2007; Supakankunti et al., 2001). Thailand is therefore keen to innovate locally to reduce its dependence on other countries. One of the questions we heard repeatedly during our research in this and other countries was whether drugs developed in industrialized nations will have the same effect in different populations and circumstances. The absence of a requirement for adequate clinical trials for all imported drugs in Thailand makes this a difficult question to answer. For example, we were told that in many instances, bioequivalence studies alone are considered sufficient. The importance of this issue was reflected by one Thai participant who pointed out that there is a debate on whether or not products such as the Roche Amplichip apply to Thai SNP variation patterns. As reaffirmed by one Thai key informant, "a lot of drug companies will not have this research [population genomic variation] before launching their product, or they have but they are not including our group of people, which are different from European, Western, African. So we need these things so that they can be used for our people." We were told that, currently, systematic collection of genomic and pharmacogenomic data is not being requested by the Thai Food and Drug Administration during clinical studies. This is similar to other emerging economies we have studied (such as Mexico, India, and South Africa) and is also common in developed countries. Nevertheless, the fact that we found two private firms in Thailand who offer genotyping services through health professionals and directly to consumers indicates there is consumer demand for personalized genomic medicine and, therefore, the need to regulate such services (and products that derive from genomics) might be more imminent than expected. Given the above, the challenges posed by genomic medicine might lead to the rethinking of the global regulatory architecture.

COLLABORATION

The Pharmacogenomic Project at TCELS collaborates with the RIKEN Institute, the Craig Venter Institute, the Beijing Genomics Institute, and researchers at Yale University. The Thai SNP Discovery Project also has extensive collaborations, the most notable being with the HUGO Pan-Asian SNP Consortium. Nevertheless, during our interviews we identified local collaboration challenges. Key informants in Thailand stressed the lack of collaboration between the academic and hospital research institutions and the local private sector. For example, individuals in the private sector stated that academic and hospital research institutions are unsupportive of commercialization efforts, owing to "cultural" differences between the academic and private sectors. As one key informant stated, "People say that it is very unethical to commercialize research projects like this [genomics]." Another suggested that the fear of giving up one's own chances at academic publication is a factor. Within the private sector, although we identified two local firms that are developing genotyping services, they were not aware of each other. We have shown that, in general, successful firms in emerging economies proactively pursue collaborations and partnerships with the public and private sectors (Thorsteinsdottir et al., 2004; Frew et al., 2007).

RACE AND ETHNICITY

The issue of the use of race and ethnicity in biomedical research has received a lot of attention, especially in the United States (Burchard et al., 2003; Lee 2007; Tang 2006; Risch et al., 2002; Bowser 2004; Séguin et al., 2008d). This topic is also relevant in emerging economies that are investing in population genotyping projects. We found that, although it was generally acknowledged that there are probably differences between Caucasian and Thai populations, differences within Thai subgroups were considered more contentious: "If you set up a project saying you would like to see the ethnic difference between the northernmost province and the southernmost province, you are in trouble, immediately." One key informant explained that the word *Thai* refers to cultural identity, and that it is preferable not to emphasize any potential genetic differences in research projects. This issue is not likely to dissipate in the near future, and it will need to be considered within the cultural context of each country. For example, in a recent press release from the Indian Genome Variation (IGV) Consortium, IGV researchers have admitted to having "intense debates" on whether or not to reveal the names of the communities that have been sampled because of the social ramifications (IGVC, 2008; Mudur, 2008).

CONCLUDING REMARKS

Thailand is making efforts to meet its urgent healthcare needs as a result of infectious disease (for example, HIV) through universal healthcare coverage and compulsory licensing of essential medicines. At the same time, chronic disease rates are on the increase in Thailand, and elsewhere in the developing world, creating a double burden and often eclipsing the rate of infectious disease (Daar et al., 2007; Boutayeb, 2006). As a result, Thailand will soon face an additional financial burden on their universal health coverage system. Accordingly, Thailand must also make the appropriate investments in its science and technology sector to generate innovation in order to reduce its reliance upon compulsory licensing—a short-term solution at best.

The Thai SNP Discovery Project and TCELS' Pharmacogenomic Project reflect the vision of its government in addressing these challenges. Together, these initiatives provide an insight into how Thailand is harnessing genomic variation in the Thai population, and conducting basic research toward discovery. Although they are in their early stages, these projects can result in economic benefits through stimulation of the bio-economy, and might result in a new generation of locally developed and commercialized healthcare products and therapeutics. Through TCELS, the Thailand government has already begun to develop the essential infrastructure required for their effective translation. However, to meet this goal, Thailand will need to address the concerns raised here regarding the commercialization and collaboration gaps, and the lack of regulatory oversight in this sector. Implementing genomic medicine into public healthcare systems will require innovative models—the National Institute for Genomic Medicine in Mexico (INMEGEN) provides one, although Mexico lacks an equivalent universal health scheme (Séguin et al., 2008a,c). INMEGEN, however, is part of the Mexican National Institutes of Health, whereas TCELS' Pharmacogenomics Project is integrated within a not-for-profit government-funded entity with commercialization in life sciences as its mandate. In Thailand, many identified genomic medicine, such as diagnostics, as too costly for implementation within the universal healthcare system. For now, testing seems to be only within reach of the wealthier portions of Thai society and medical tourists. Thus, Thailand has a choice. Will it adopt a trickle-down model of incorporating genomic medicine into its healthcare system, or will it instead demonstrate leadership among other developing nations and emerging economies, and leverage its universal healthcare system toward this goal, providing equal access and local health benefits to all of its citizens?

ACKNOWLEDGMENTS

This chapter is based on the article "Universal health care, genomic medicine and Thailand: investing in today and tomorrow" published in *Nature Reviews-Genetics*, Special supplement October 2008: Genomic Medicine in Developing Countries; S14–s19. Authors and the Chief Editor are grateful to Nature Publishing Group for permission.

We gratefully acknowledge M. Maliakkal for her contributions to this project. This project was funded by Genome Canada through the Ontario Genomics Institute. The McLaughlin–Rotman Centre for Global Health, Program on Life Sciences, Ethics and Policy is primarily supported by Genome Canada through the Ontario Genomics Institute, the Ontario Research Fund, and the Bill and Melinda Gates Foundation. Other matching partners are listed the McLaughlin–Rotman Centre for Global Health web site. A.S.D. and P.A.S. are supported by the McLaughlin Centre for Molecular Medicine. P.A.S. is supported by a Canadian Institutes of Health Research Distinguished Investigator award.

REFERENCES

Becerril A. (2008). Aprueba el Senado sanciones a quienes trasladen tejido humano fuera de México. *La Jornada* [online], http://www.jornada.unam. mx/2008/03/28/index.php?section=sociedad&article=046n1soc.

Bowser R. (2004). Race as a proxy for drug response: the dangers and challenges of ethnic drugs. *DePaul Law Rev 53,* 1111–1126.

Boutayeb A. (2006). The double burden of communicable and non-communicable diseases in developing countries. *Trans R Soc Trop Med Hyg 100,* 191–199.

Burchard EG, et al. (2003). The importance of race and ethnic background in biomedical research and clinical practice. *N Engl J Med 348,* 1170–1175.

Chantratita W. (2008). In Maximize the Effectiveness of HIV-1 Treatment by Using Molecular Techniques, Bioinformatics, and Pharmacogenomics. 275 (Graphic Hut Co., Ltd, Bangkok.

Daar AS, et al. (2007). Grand challenges in chronic noncommunicable diseases. *Nature 450,* 494–496.

Ford N, Wilson D, Costa Chaves G, Lotrowska M, Kijtiwatchakul K. (2007). Sustaining access to antiretroviral therapy in the less-developed world: lessons from Brazil and Thailand. *AIDS 21*(Suppl 4), S21–S29.

Frew SE, et al. (2007). India's health biotech sector at a crossroads. *Nature Biotechnol 25,* 403–417.

Hughes CA, et al. (2008). Abacavir hypersensitivity reaction: an update. Ann. *Pharmacother 42,* 387–396.

Hughes AR, et al. (2004). Association of genetic variations in HLA-B region with hypersensitivity to abacavir in some, but not all, populations. *Pharmacogenomics 5,* 203–211.

Hughes AR, et al. (2008). Pharmacogenetics of hypersensitivity to abacavir: from PGx hypothesis to confirmation to clinical utility. *Pharmacogenomics J 11* March 2008 (doi: 10.1038/tpj.2008.3).

IGVC [Indian Genome Variation Consortium] (2008). Genetic landscape of the people of India: a canvas for disease gene exploration. *J Genet 87,* 3–20.

International Labour Office. (2004). A Technical Note to the Government: Financing Universal Health Care in Thailand. [online], http://www.ilo.org/public/english/protection/secsoc/downloads/publ/995sp1.pdf.

Intarakumnerd P, Chaminade C. (2007). Strategy versus practice in innovation systems policy: the case of Thailand. *Asian Journal of Technology Innovation 15,* 197–213.

Jayaraman KS. (2007). Database targets Parsi genes. Nature 446, 475.

Kiertiburanakul S, Khongnorasat S, Rattanasiri S, Sungkanuparph S. (2007). Efficacy of a generic fixed-dose combination of stavudine, lamivudine and nevirapine (GPO-VIR) in Thai HIV-infected patients. *J Med Assoc Thai 90,* 237–243.

Lee SS. (2007). The ethical implications of stratifying by race in pharmacogenomics. *Clin Pharmacol Ther 81,* 122–125.

Ministry of Public Health and The National Health Security Office Thailand (2007). Facts and evidences on the 10 burning issues related to the government use of patents on three patented essential drugs in Thailand. [online], <http://www.moph.go.th/hot/White%20Paper%20CL-EN.pdf>

Marsh S, Van Booven DJ, McLeod HL. Global (2006). pharmacogenetics: giving the genome to the masses. *Pharmacogenomics 7,* 625–631.

Mudur GS. (2008). Stamp on Tagore's India Genetic map blurs lines. *The Telegraph* (Calcutta, India) [online], <http://www.telegraphindia.com/1080425/jsp/frontpage/story_9186161.jsp>.

Phanuphak N, et al. (2007). Nevirapine-associated toxicity in HIV-infected Thai men and women, including pregnant women. *HIV Med 8,* 357–366.

Rauch A, et al. (2006). Prospective genetic screening decreases the incidence of abacavir hypersensitivity reactions in the Western Australian HIV cohort study. *Clin Infect Dis 43,* 99–102.

Risch N, Burchard E, Ziv E, Tang H.(2002). Categorization of humans in biomedical research: genes, race and disease. *Genome Biol 3,* comment2007.1–200712.

Secretary's Advisory Committee on Genetics, Health, and Society (2008). Realizing the potential of pharmacogenomics: opportunities and challenges. [online], http://www4.od.nih.gov/oba/SACGHS/reports/SACGHS_PGx_Report.pdf.

Sadiq ST, Pakianathan M. (2007). Uncertainties of routine HLA B*5701 testing in black African HIV cohorts in the UK. *Sex Transm Infect 83,* 181–182.

Séguin B, Hardy B, Singer PA, Daar AS.(2008a) Genomic medicine and developing countries: creating a room of their own. *Nature Rev Genet 9,* 487–493.

Séguin B, Hardy B, Singer PA, Daar AS. (2008b). Human genomic variation initiatives in emerging economies and developing countries *Nature Rev Genet 9* (Suppl), S3–S4.

Séguin B, Hardy B, Singer PA, Daar AS. (2008c) Genomics, public health and developing countries: the case of the Mexican National Institute of Genomic Medicine (INMEGEN). *Nature Rev Genet* 9(Suppl), S5–S9.

Séguin B, Hardy B, Singer PA, Daar AS. (2008d). Bidil: recontextualizing the race debate. *Pharmacogenomics J 8,* 169–173.

Seim L. (2007). Thailand: government issues compulsory licences for HIV/AIDS drugs. HIV AIDS Policy. *Law Rev 12,* 28–29.

Steinbrook R. (2001). Thailand and the compulsory licensing of efavirenz. *N Engl J Med 356,* 544–546 (2007).

Supakankunti S, et al. (2001). Impact of the World Trade Organization TRIPS agreement on the pharmaceutical industry in Thailand. Bull. *World Health Organ 79,* 461–470.

Stone R. (2007) Mongkol na Songkhla interview. Thai health minister defends controversial drug-patent policy. *Science 316,* 1408.

Stone R. (2005). Indian Ocean tsunami. In the wake: looking for keys to posttraumatic stress. *Science 310,* 1605.

Tang H. (2006). Confronting ethnicity-specific disease risk. *Nature Genet* 38, 13–15.

Thailand Science Park. (2007). Friendly cooperation agreement signed by Tsinghua Science Park and Thailand Science Park. Tsinghua Science Park [online], <http://www.thsp.com.cn/english/news/20070926Thailand/news.htm>.

Thorsteinsdottir H, Quach U, Daar AS, Singer PA. (2004). Conclusions: promoting biotechnology innovation in developing countries. *Nature Biotechnol 22*(Suppl), DC48–DC52.

Watson ME, Pimenta JM, Spreen WR, Hernandez JE. (2004). HLA-B*5701 and abacavir hypersensitivity. *Pharmacogenetics 14,* 783–784.

Zucman D, Truchis P, Majerholc C, Stegman S, Caillat-Zucman S. (2007). Prospective screening for human leukocyte antigen-B*5701 avoids abacavir hypersensitivity reaction in the ethnically mixed French HIV population. *J Acquir Immune Defic Syndr 45,* 1–3.

61.

SCREENING NEWBORNS IN THE ASIA-PACIFIC REGION

Carmencita D. Padilla and Bradford L. Therrell

INTRODUCTION

Newborn dried bloodspot screening (NDBS) began in the early 1960s with the work of Robert Guthrie.(Guthrie, 1962; Guthrie and Susi, 1963). He originated a process for collecting blood on special absorbent filter paper, transporting it to a testing laboratory, and analyzing the dried blood specimen (DBS) for phenylalanine concentration to detect phenylketonuria (PKU). This proved to be a cost-effective way to detect and reduce the adverse effects of PKU, including the institutionalization of affected patients. Over the next few years, many economically developed countries began NDBS as a public health strategy for early identification and treatment/management of newborns with PKU and other inborn errors of metabolism. Today, the Guthrie NDBS protocol is used to detect dozens of congenital conditions around the world, including metabolic and infectious diseases (Therrell, 2001), although his specific testing methodology is rapidly being replaced by more modern technologies. In particular, tandem mass spectrometry (MS/MS) has been embraced as the method of choice for simultaneous detection of over 20 metabolic disorders (Millington et al., 1990; Chace et al., 2003), and its use has become widespread including some screening in developing countries.

Early identification of screenable conditions is important, since timely intervention leads to significant reductions of morbidity, mortality, and associated disabilities in affected infants (American Academy of Pediatrics, 2000). In addition to traditional biochemical screening, other newborn screening processes can occur in the newborn nursery. For example, screening newborns for congenital hearing loss is a public health priority in many economically developed healthcare systems. Congenital heart disease (Valmari, 2007; Mahle, 2008), neonatal jaundice (Bertini and Rubaltelli, 2002; Mishra et al., 2009), and vision screening (American Academy of Pediatrics et al., 2003) are also receiving increased interest.

In countries with depressed and developing economies, such as many in Africa and Asia, NDBS and other newborn screening is either not yet a priority or is just emerging as a priority. In some countries, such as Australia and Japan, NDBS has a long history, while in others like Bangladesh and Indonesia it is only beginning. In some areas of the region there is either no program or screening information is not readily available. We will focus on NDBS where most data are available, acknowledging that hearing and other types of newborn screening are expanding rapidly in some cases, and may utilize NDBS infrastructure in order to speed their implementation.

Our objective is to provide an update of the status of NDBS programs within Asia and the Pacific Islands, building on a survey we previously conducted in 2006. We include selected national demographic data and other accumulated information about the impact of certain critical aspects of NDBS programs: operations, party(ies) responsible for advocacy, future plans, and major obstacles to full population coverage. For our purposes here, we will define the Asia-Pacific region broadly, and information from the following countries will be included: Australia, Bangladesh, China (Hong Kong and the rest of China), India, Indonesia, Japan, Korea, Malaysia, Mongolia, Myanmar, New Zealand, Pakistan, Philippines, Singapore, Sri Lanka, Taiwan, Thailand, Vietnam, and limited information from other Pacific Islands. Information sources include published papers, direct communication with country newborn screening managers, and output from a workshop conducted in the Asia-Pacific region in 2008 (Padilla and Therrell, 2007).

THE ASIA-PACIFIC NEWBORN SCREENING PROGRAM: BACKGROUND

Of approximately 136 million babies born annually in the world, 122 million are born in developing countries and 68 million are born in Asia (UNICEF, 2009). Countries within the region vary widely in size and birth populations. There are small countries (Singapore, New Zealand, Palau, etc.) and extremely large and populous ones (China and India). While there are many economically developed countries in the Asia-Pacific region (Japan, Taiwan, Korea,

Singapore, Australia, New Zealand, parts of China), there are also many countries with poor economies (most of the rest of the Asia-Pacific Region; see Padilla, 2008). Where poverty is abundant, the healthcare system is challenged and births often occur outside of hospital settings, sometimes as high as 80% (Padilla et al., 2010).

While NDBS has existed in parts of the Asia-Pacific region since the mid-1960s, there are currently more developing than developed NDBS programs there. The implementation barriers that face developing programs are in many ways similar to those of developed programs, and they are some of the same barriers faced in implementing other public health programs. These barriers (or challenges) include low literacy, language differences, cultural and religious beliefs, geographic/transportation extremes (many islands and mountains), unstable governments, and general absence of technical support. Hopefully, the lessons learned over time from developed programs can aid in more rapid implementation of NDBS in the developing world.

NDBS is a more complex *system* today than 45 years ago. From its early beginnings as a laboratory program testing for a single rare congenital condition, NDBS has evolved into a multidimensional system that requires dedicated individuals in order to succeed. Its sustainability requires smooth integration of the system's components, and it must succeed despite geographic, economic, and political constraints. The NDBS system includes: (1) *education* (of health professionals, parents, and politicians); (2) *screening* (specimen collection [timing and transport] and laboratory testing and reporting); (3) *short-term follow-up* (including out-of-range result notification and patient tracking that ensures confirmatory testing); (4) *diagnosis* (specialist consultation and evaluation); (5) treatment/*management* (counseling, compliance monitoring and long-term follow-up) and (6) *evaluation* (through systemwide quality assurance and outcome monitoring) (American Academy of Pediatrics, 2000; Pass et al., 2000; Therrell, 2001; McCabe et al., 2002).

As NDBS systems develop, they must not only create a sustainable infrastructure, but they must also assure quality screening services. While government involvement through the health ministry is not essential in the early stages of system development, ultimately its involvement is critical for sustainability. In addition to program oversight, the public health ministry must be active in assuring that there is coordination and government cooperation with academic centers, private facilities (confirmatory laboratories, private hospitals), insurers, and other nongovernment organizations. The health ministry can play a vital role in assuring system quality. Measurable indicators for monitoring must be identified both in the laboratory and in other parts of the screening system (Therrell and Hannon, 2006). An extensive listing of possible quality indicators has been developed for U.S. NDBS systems (Therrell et al, 2010; self-assessment tool available at: http://genes-r-us.uthscsa.edu/NBS_PEAS.htm), and developing programs have found these helpful in developing their own monitoring process (Padilla et al., 2009). Establishing the necessary government collaborations has proven to be one of the biggest challenges in establishing successful and sustainable NBS in many parts of the region (Padilla and Therrell, 2007).

THE ASIA-PACIFIC NEWBORN SCREENING PROGRAM: OUTCOMES

The history of the various Asia-Pacific programs naturally divides into four primary periods (eras) of activity. The first and second eras involved primarily the more developed countries, the third and fourth eras involved the developing countries.

FIRST ASIA-PACIFIC SCREENING ERA

Initial NDBS activities in the Asia-Pacific followed very closely the first NDBS activities in the United States in the early 1960s (Guthrie and Susi, 1963). At that time, many countries around the world, including some in the Asia-Pacific, either had or were beginning newborn PKU screening using a commercially available ferric chloride urine screening test, Phenistix (Ames Laboratories, USA). These included New Zealand (Houston and Veale, 1971), Japan (Nozue et al., 1983), and Australia (Wilcken et al., 1980). Urine PKU screening was prone to false negative tests, and proper specimen collection was not always easy or timely, so many physician researchers considered bloodspot testing for PKU as an alternative soon after Guthrie's report (Guthrie and Susi, 1963) and its adoption as a primary screen began to spread.

In 1965, Guthrie spent nine months traveling around the world advocating for his bloodspot procedure (Guthrie, 1992; Koch, 1997). His travels to New Zealand resulted in the creation of one of the first *national* NBS programs in 1966. Soon after, in 1967, programs also developed in Australia (Veale and Houston, 1978) and Japan (Nozue et al., 1983; see Table 61-1). A newborn screening pilot project to detect glucose-6-phosphate dehydrogenase (G6PD) deficiency from cord blood also began in Singapore in 1965, and became national in 1970. This cord blood screening program later evolved into a broader newborn screening initiative that included congenital hypothyroidism (CH).

New Zealand

In April 1966, the New Zealand Ministry of Health (MOH) began screening for PKU with Guthrie's bacterial

TABLE 61-1 PROGRAM DEMOGRAPHICS AND SCREENED CONDITIONS IN ASIA-PACIFIC NEWBORN SCREENING PROGRAMS

	Program Demographics							
Jurisdiction	Population (x 1000) 2008[a]	Births (x 1000) 2008[a]	Infant Mortality Rate (<1 yr) 2008[a]	Life expectancy at birth 2008[a]	Year NBS Began	Reference For Date Screening Began	Newborn Coverage Reported in 2006	Screening Cost Paid By
Australia	21,074	267	5	82	1967	Wilcken, 1999	100%	Gov
Bangladesh	160,000	3,430	54	64	1999	Moslem et al., 2003	<1%	Gov
Cambodia	14,562	361	69	61	-	-	?	?
China	1,337,441	18,134	18	73	1981	Chen and Guo, 1983	40%	Family
Hong Kong (China)[b]	7.055	52	2.9	82	1984	Lo and Lam, 1995	99%	Gov
India	1181412	26,913	52	64	1980	Devi and Naushad, 2004	<1%	Family
Indonesia	227,345	4,220	31	71	1999	Rustama et al., 2003	<1%	Family
Japan	127,293	1,034	3	83	1967	Irie et al., 1975	>99%	Gov
Korea (South)	48,152	452	5	79	1991	Lee, 1994	94%	Gov
Korea (North)	23,819	327	42	67	-	-	?	?
Laos	6,205	170	48	65	2008	Padilla and Therrell, 2008	?	?
Malaysia	27,014	551	6	74	1980	Singh, 2003	95%	Gov/Private
Mongolia	2,641	50	34	67	2000	Erdenechimeg, 2003	<1%	Grant
Myanmar	49,563	1020	71	62	2000	Thein, personal communication, 2006	<1%	Gov
Nepal	28,810	732	41	67	-	-	?	?
New Zealand	4,230	58	5	80	1966	Veale and Houston, 1976	100%	Gov
Palau	20	350	13	-	2009	Reyes, personal communication, 2009	0	Gov
Pakistan	156,952	5,337	72	67	2000	IAEA Report, 2002	<1%	?
Philippines	90,384	2,236	26	72	1996	Padilla and Domingo, 2002	10%	Family/Insur
Singapore	4,615	37	2	80	1965	Wong, 1965	>99%	Family 40%
Sri Lanka	20,061	365	13	74	2005	Nanayakkara 2007	<1%	Gov
Taiwan[c]	22,974	206	5.3	78	1985	Chen, 1993	>99%	Family
Thailand	67,386	977	13	69	1992	Charoensiriwatana et al., 1995	97%	Gov
Vietnam	87,096	1,494	12	74	2000	IAEA Report, 2002	<1%	Gov
Totals	3,544,580	66,561	25	69				
World Statistics	6,449,371	133,449	30	68				

[a] Source: UNICEF: The State of the World's Children – Special Edition Celebrating 20 Years of the Convention on the Rights of the Child (Statistical Tables) November 2009. New York: UNICEF, 1–52. Available at http://www.unicef.org/rightsite/sowc/pdfs/statistics/SOWC_Spec_Ed_CRC_Statistical_Tables_EN_111809.pdf.

[b] Hongkong – https://www.cia.gov/library/publications/the-world-factbook/geos/hk/html

[c] Taiwan – https://www.cia.gov/library/publications/the-world-factbook/geos/tw.html

inhibition assay (BIA). The New Zealand NDBS program was one of the first to include additional tests; i.e., homocystinuria (HCY), maple syrup urine disease (MSUD), tyrosinemia (TYR), histidinemia (HIS) and galactosemia (GAL; see Veale and Houston, 1978). New Zealand was also one of the early programs researching cystic fibrosis (CF) screening (Crossley et al., 1979; Lyon et al., 1983). Additionally, New Zealand provided screening for nine Pacific Island nations through a collaboration established by Guthrie and Veale in 1968–1969 (Veale and Houston, 1976; Guthrie, 1992).

The New Zealand program was also one of the first to attempt automated NDBS. Veale worked with California inventor Bob Phillips to pilot his automated filter paper punching machine, which simultaneously punched four samples for testing preparation (Houston and Veale, 1971). Later, when Dussault developed (Dussault and Laberge, 1973) and demonstrated (Dussault et al., 1975) the feasibility of a micro method for detecting CH from dried blood spots, it was soon added to the New Zealand screening panel (Lyon, 1983).

The New Zealand program was also a leader in developing and improving external laboratory quality control. In 1986, the New Zealand program began an internal quality assurance (QA) program which expanded regionally and eventually became an international blinded QA program (Webster, 1991). In July 2005, the National Screening Unit assumed funding and governance responsibility for the Newborn Metabolic Screening Program. Policies used in the New Zealand NDBS program are currently developed in conjunction with the Australasian Society of Human Genetics, which also includes the Australia NDBS program. In December 2006, New Zealand screening expanded to include tandem mass spectrometry (MS/MS), and now lists 28 conditions on its NDBS panel (Available at: http://www.nsu.govt.nz/current-nsu-programmes/2097.aspx). The MOH and the New Zealand Police have signed a memorandum of understanding (MOU) relating to the storage and use of NDBS samples effective on February 28, 2006. This MOU regulates requests from the police to the MOH for access to samples, and clarifies the circumstances under which such requests may be granted. To date, stored NDBS cards have been used by police once or twice a year to identify dead or missing persons, mostly with parental consent, e.g., when a child dies in a house fire. Stored NDBS cards are used only when all other avenues for identifying the person are not practicable or have failed (Available at: http://www.nsu.govt.nz/Publications/1454.asp).

Australia

During 1967–1968, NDBS for PKU began in the 6 states of Australia. In New South Wales, NDBS screening replaced a urine screening program that had been in place since 1964 (Wilcken et al., 1980). By 1968, all Australian states were offering NDBS for PKU (Szeinberg, 1971). About 10 years later, between 1977 and 1982, NDBS for CH started in the various other Australian states (Wilcken, 1999). CF screening began in New South Wales in July 1981, making it one of the first CF newborn screening programs in the world (Wilcken and Chalmers, 1985; Wilcken and Brown, 1987; Wilcken et al., 1995).

When newborn screening with MS/MS was suggested in the early 1990s, the screening program in New South Wales became one of the first NBS programs in the world to use electrospray MS/MS screen for inherited metabolic diseases. Reports of successes in the Australia program, which formally began MS/MS screening in 1998, coupled with those of the other pioneer programs, paved the way for others (Carpenter and Wilcken, 1999, Wiley et al., 1999; Wilcken et al., 2003; Wilcken and Gaskin, 2007).

Currently, all Australian screening programs are voluntary and fully publicly funded. There are five centralized screening laboratories providing testing—Western Australia, South Australia (also covering Tasmania and part of the Northern Territory), Victoria, New South Wales (also covering the Australian Capital Territory), and Queensland (also covering part of the Northern Territory). As with New Zealand, the policies governing the Australian NDBS system are developed as part of the work of a joint subcommittee of the Genetics Society of Australasia and the Division of Pediatrics of the Royal Australasian College of Physicians. The NDBS panels are state-based but generally include PKU, CH, CF, and over 20 other disorders of amino acid, organic acid, and fatty acid metabolism by MS/MS. GAL is screened for in all states except Victoria. Screening is currently not recommended for the following conditions: adenosine deaminase deficiency, Duchenne muscular dystrophy, familial hypercholesterolemia II, G6PD deficiency, hemochromatosis, lysosomal storage disorders, neuroblastoma, and toxoplasmosis (Available at: http://hgsa.com.au).

Japan

A urine screening program for PKU began in Japan in 1963. In 1967, the Japanese Ministry of Health and Welfare organized a "working party" that recommended adoption of the Guthrie NDBS method for PKU screening. In October1977, nationwide NDBS began for 5 inborn errors of metabolism—PKU, MSUD, HIS, HCY, and GAL. In the beginning, there were 52 screening laboratories covering 47 prefectures and 9 large cities. During the first 6 months of NDBS, coverage expanded from 15% to about 50%, reaching full population coverage by the end of 1978 (Naruse, 1980).

As newborn screening progressed, a professional society, the Japanese Society for Mass Screening (JSMS), emerged as a professional advocacy group. The JSMS was instrumental in getting the federal government to eventually implement a national QA system for NDBS (Naruse et al., 2003). The Japanese laboratory QA program became the responsibility of the Japanese Public Health Association. Upon its establishment in 1977, the Japanese QA program was the first national NDBS multi-laboratory QA program in the world (Naruse, 1980; Amino, 1987). The QA system has continued to evolve until, at present. monthly proficiency testing specimens are distributed from an officially designated Quality Control (QC) Center for Mass Screening in Japan. Guidelines for computerized reporting systems are also in place and monitored by the QC Center (Naruse et al., 2003).

Throughout the 1970s, Irie and his colleagues researched the feasibility of NBS for CH, and the Ministry of Finance eventually provided funds for the addition of CH in the late 1979 (Irie et al., 1975; Irie and Naruse, 1980). Congenital adrenal hyperplasia (CAH) was added in 1988 (Fujieda et al., 1987; Suwa et al., 1987). The feasibility and effectiveness of adding other NBS disorders were also studied, including urea cycle disorders (Suzuki et al., 1983; Fujimoto et al., 1987), MSUD (Matsuda et al., 1987), GAL (Kawamura, 1987), and biotinidase deficiency (Yamaguchi et al., 1987). Pilot screening with MS/MS was begun in six laboratories in 1997—Sapporo, Tokyo, Fukui, Osaka, Shimane, and Kumamoto. A total of 606,380 babies were screened from 1997–2007, and 65 cases of metabolic disorders were detected: 33 organic acidemias, 18 fatty acid oxidation disorders, and 14 amino acidemias. This translates to an approximate overall incidence of 1:9,330, which is lower than Europe or the United States (about 1:4,000–5,000; see Yamaguchi, 2008).

NDBS in Japan is currently considered the responsibility of local governments, and there is full population screening for PKU, MSUD, HCY, GAL, CH, and CAH in 47 prefectures and 15 designated large cities (Aoki, 2003). NDBS is part of the national health system, and there is no additional charge to patients receiving the typical screening panel. Reorganization to include MS/MS is ongoing. Medical education and support for NDBS expansion includes a newly published guidebook of targeted diseases detectable in MS/MS screening, and there is consideration of establishing consultation centers.

Diagnostic support includes the proposed creation of a network of several laboratories in which GC/MS, enzyme determination, or DNA tests will be available. A more reliable and quicker result reporting system is also being organized to include relocating screening laboratories so that at least 30,000 samples annually are analyzed in each laboratory. A continuing system for clinical follow-up and feedback to patients is being organized with patients consolidated in the National Center for Child Health and Development of Japan. Periodic publications are being used for communication between doctors, researchers and patients to provide more appropriate and up-to-date information. Finally, there is ongoing effort to ensure the quality of measurements and intensify efforts to solve NDBS related ethical issues (Yamaguchi, 2008).

Singapore

In the mid-1960s, early diagnosis of G6PD deficiency was made by determining enzyme activity in cord blood (Wong, 1965, 1975; Joseph et al., 1999; Joseph, 2003). G6PD symptoms range from mild hemolysis to hyperbilirubinemia leading to kernicterus. At that time in Singapore, nearly half of the cases of severe neonatal jaundice requiring exchange transfusion or terminating in kernicterus were associated with G6PD deficiency and were triggered from clothes stored with moth balls or various herbs (Wong, 1965, 1975). As a result, mass newborn screening for G6PD deficiency was started in Singapore in 1965, and Singapore became one of the first countries to have newborn cord blood screening for G6PD deficiency. The observed incidence of G6PD deficiency from screening was about 1:50 newborns. Newborn screening for G6PD deficiency has been responsible for essentially eradicating kernicterus secondary to G6PD deficiency, and no newborn deaths have been observed in the last 20 years (Joseph, 2003).

Building on the G6PD screening infrastructure, a pilot newborn screening program for CH using cord blood was started at the Kandang Kerbau Maternity Hospital in December 1981. This hospital accounted for about half of all births in Singapore at the time. The 18-month pilot provided data that supported universal CH screening, and in 1990, the G6PD screening program was expanded to include CH nationwide (Yeo et al., 1983; Joseph et al., 1999).

The Singapore NBS program recently expanded to include metabolic screening using MS/MS (31 conditions included) using NDBS collection cards. It is a voluntary program reaching about 50% (~ 90% in the public sector and <20% in the private sector). Newborn screening tests are routinely performed in hospital laboratories, with MS/MS testing performed in a single central laboratory. The government subsidizes 20%–60% of the cost of screening in public hospitals, approximately half of the births. Newborn screening is not mandated by a national policy, but is instead considered part of basic medical care (Joseph, 2003; Personal communication with R. Joseph, 2010).

SECOND ASIA-PACIFIC SCREENING ERA

The second Asia-Pacific screening era started in the 1980s. Limited newborn screening began in Malaysia, 1980 (cord blood for G6PD deficiency); India, 1980;

China, 1981; Taiwan, 1984; Hong Kong (then an independent country), 1984 (cord blood for G6PD deficiency); and Pakistan, 1987.

Malaysia

Pilot studies in Kuala Lumpur, Petabling Java, and Malacca led to cord blood screening for G6PD deficiency as part of routine newborn care in Malaysia in 1980 (Robinson et al., 1976; Choo et al., 1994). Approximately 10 years later, in June 1991, a national meeting of government pediatricians at Ipoh Hospital commissioned a report on the feasibility of a national screening program for CH. From 1991–1997, five separate pilot CH screening studies in various parts of the country revealed incidences ranging from 1:2400 to 1:3666, with an average CH incidence of 1:3026. In 1997, the MOH organized a national committee to consider implementation of CH NBS, and commissioned a systemic review (health technology assessment) of the benefits of CH screening. The result was a national CH newborn cord blood screening program that began in October 1998 with one district and three government hospitals participating. Slightly less than 6,000 newborns were screened during the first year of the project (approximately 1.8% of the newborn population). The program has continued to increase over the years, with approximately 45% of Malaysian newborns screened in 2003 (Singh, 2003).

The Institute for Medical Research included inborn errors of metabolism (IEM) as a focus during the 1990s, with the result that various analytical capabilities were initiated to aid in diagnosis. Upon its reorganization in 2001, the biochemistry unit became a function of the Specialized Diagnostic Center, with primary responsibility to aid in diagnosing various IEMs. It currently has MS/MS capability, as does a commercial metabolic screening service for newborns that was opened in April 2007 at the Universiti Sains Malaysia's Doping Control Center, serving both private and public hospitals for a fee (available at: http://www.imr.gov.my/org/biochem_r.htm; http://www.fodsupport.org/documents/Simpletesttocheckfor25disordersinnewborns.pdf). A third center at the University of Malaya has recently obtained MS/MS equipment and is expected to initiate extended newborn screening on a pilot basis in early 2012 (Personal communication with MK Thong, 2011).

Currently, the 1998 MOH program remains officially unchanged, although the results of an IMR study on MS/MS during the period 2005–2006 are being analyzed and may have a significant impact on the program in the future. Funding issues for expansion continue to exist. The current program is funded by the government for births in government hospitals. In private hospitals, screening is paid by insurance or private pay. Human resources remain a challenge, with an insufficient availability of clinical geneticists and metabolic dieticians. Confirmatory capacity is also limited. Cord blood remains the specimen of choice in the public hospitals and the CH coverage in government hospitals is reported to be 98.7% (Personal communication with MK Thong, 2009).

Taiwan

An official pilot study to initiate NBS with PKU and CH began in 1981.(Hwu et al., 2003) In 1983, the Taiwan Department of Health took over the studies and continued funding for two principal NBS screening laboratories. The National Science Council was responsible for beginning NBS in two institutes in Taipei. They established two medical genetic counseling centers in 1984 to act as clinical networks for inborn errors (National Taiwan University Hospital and V.A. General Hospital), and national NBS officially began in 1985. Both cord blood and heelstick specimens were accepted from 1987–1991, but cord blood specimens were discontinued after they were found to give less accurate results. Responsibility for specimen collection and follow-up resided with the city and country health bureaus, who referred possible cases to designated referral hospitals (Chen, 1993).

G6PD deficiency is common in Taiwan, and 20%–40% of patients with neonatal hyperbilirubinemia have G6PD deficiency. As a result, screening for G6PD deficiency was included as a part of Taiwan NBS from the beginning. Based on confirmed G6PD results from NBS, the incidence was found to be about 1:50 (essentially identical to Singapore). Kernicterus and exchange transfusions due to neonatal hyperbilirubinemia decreased dramatically when screening for G6PD was introduced; however, occasional deaths from environmental triggers continued to occur, emphasizing the need for better parent education (Hsiao et al., 1991).

NBS currently reaches about 97% of all Taiwan newborns, although it is not required by law (Hwu et al., 2003). There are 3 screening centers (one public and two private hospitals) and 18 referral hospitals, including outlying islands. The referral hospitals are geographically distributed and provide confirmatory tests, medical care, and genetic counseling as follow-up to positive screens. The Department of Health is responsible for screening and for the screening centers. Five conditions are included in the national NBS package—PKU, CH, HCY, GAL, and G6PD deficiency. The cost for NBS is $17, and an additional $9 is charged for shipping and handling. The government provides a $3 supplement for each screen. A monitoring system has existed since the onset of screening, and information is collected through annual records review or electronic submissions. Critical data are obtained on all cases, including compliance with medications and adverse outcomes (Padilla and Therrell, 2008).

CAH has been included in the screening package of one of the screening centers since 2000, reaching almost half of their newborn population. This center also began offering expanded MS/MS metabolic screening in 2001, covering approximately 30% of their births (Huang et al., 2006). A second center began CAH screening in 2001 and MS/MS screening in 2002. Parents have the option to obtain screening beyond the government-subsidized program, but an additional charge is incurred and consent is required. Currently, seven conditions are routine in the MS/MS panel and there is pilot testing for other conditions. The additional cost for MS/MS screening is US$10, which is partially subsidized by the government (Personal communication with W. Hwu, 2009).

The first reported large-scale newborn screening pilot program in the world for Pompe disease occurred in Taiwan between October 2005 and March 2008. Screening involved measuring acid alpha glucosidase (GAA) activity in DBS from approximately 45% of all newborns. The unscreened population was monitored as a control. Since the study period, all newborns in Taiwan have GAA screening available on a voluntary basis. Six of 206, 088 newborns screened in the pilot tested positive and were treated for Pompe disease (Chien et al., 2009). It is now routinely available at an additional cost to the parents (Personal communication with W. Hwu, 2009).

From July 2006 through June 2008, a pilot newborn screening program for Fabry disease was conducted. Of the 171,977 consecutive male and female newborns screened for α-Galactosidase A (α-Gal A) deficiency, by measuring DBS α-Gal A activities and β-galactosidase/α-Gal A ratios, the pilot study revealed a high frequency of Fabry disease (~1:1250) in males, of which the majority (86%) had the later-onset phenotype (~1:1390 males; Hwu et al., 2009).

Hong Kong

G6PD deficiency has been known to be a major genetic disorder in Hong Kong for decades, accounting for considerable mortality and morbidity. In 1984, the Central Genetic Neonatal Screening Unit was established as part of the Clinical Genetic Service in Hong Kong, and NBS for G6PD deficiency officially began in March 1984 using cord blood. Cord blood screening for CH was also implemented, and babies born at government hospitals and maternal and child health clinics (about 90% of newborns) routinely received both screening tests (Hau and Lam, 1994; Lo et al., 1995). By 1995, NBS was part of the routine neonatal services in 10 hospitals administered under the Hospital Authority of Hong Kong and six government maternity homes. Taken together, this represented the majority of all newborns delivered in the public sector. Coverage today is almost 99.9%. Cord blood continues as the specimen of choice (Lam and Cheng, 2003).

Both mortality and morbidity from neonatal hyperbilirubinemia have greatly decreased through the decades, down from over 300 cases annually to about one per year at present (Hau and Lam, 1994; Lo and Lam, 1995; Lo et al., 1995; Lam and Cheng, 2003; Personal communication with S. Lam, 2010). Currently, there is increasing interest in the capabilities of MS/MS as a newborn screening tool. Altogether, five spectrometers have been installed in Hong Kong, but only for diagnostic purposes. There is presently no plan to implement a territory-wide newborn screening program. Part of the reason is that the incidence of typical PKU and MCAD seems to be extremely low in the population. A pilot study is under development to investigate the possibility of MS/MS newborn screening (Personal communication with S. Lam, 2010).

China

A collaborative, integrated screening program was started in Shanghai in October 1981, with assistance from Guthrie (USA) and Naruse (Japan). This represented the first NBS screening in China. The NBS test panel included CH, PKU, and GAL (Chen and Guo, 1983). Eight years later, in 1989, NBS was started in Beijing (Xi, 1994), and from 1992–1993, the World Health Organization (WHO) and the Ministry of Public Health sponsored a cooperative NBS pilot project in seven major cities. In 1994, a survey of 17 cities in China showed that about 1% of babies born in the country were being screened. However, in Shanghai, Beijing, Tianjin, and Guangzhou, NBS had been systematically started and coverage was 80%–90% (Ying et al., 1996).

In October 1994, Presidential Order 33 was promulgated related to newborn screening and was added to the People's Republic of China law on maternal and infant healthcare, and it became enforceable on June 1, 1995. Article 24 of this Order stated: "Medical and health institutions shall... gradually develop medical and healthcare service such as the screening of the newborn babies..." This was interpreted by most provincial health departments as authorizing the development of newborn screening (Ying et al., 1996). While PKU and CH have been commonly included in provincial screening programs, optional screening for GAL and HIS were reported in 1999 (Gu et al., 1999) A more recent survey indicated that in 2007, 185 laboratories were providing screening for CH and PKU, with 35 of these also screening for G6PD, 9 for CAH, 2 for thalassemia, and one for GAL (Cao et al, 2009). Metabolic screening with MS/MS occurs in some parts of the country (Personal communication with Xue Fan Gu, 2009). NDBS is not routinely included in the healthcare services provided by the government, and parents must pay a nominal fee of about US$7.50 for PKU and CH (Gao et al., 2009).

In 2009, the MOH required all local bureaus to offer newborn screening, raising the coverage to almost 50% of the 17 million annual births. There are currently 124 newborn screening laboratories and 7 metabolic centers based in the cities of Shanghai, Guangzhou, and Hangzhou, supported by research grants. While families must pay the newborn screening fee, local citizens can obtain refunds from the government in Shanghai (since 2007), and Beijing (since January 1, 2010; also see http://www.interfax.cn/news/7467; Personal communication with Xue Fan Gu, 2009).

A work plan from the MOH has been developed aimed at increasing the early detection of diseases in newborns. According to the plan, the government will set up provincial newborn screening networks across China beginning in 2012, to be completed by 2015. Targets for newborn screening coverage in eastern China, central China and western China by 2012 are 90%, 50%, and 40% respectively, increasing to 95%, 80%, and 60% by 2015. The main diseases to be screened remain PKU and CH. Local government agencies have been asked to formulate plans for NDBS centers, to establish reasonable fees for NDBS services, and to allocate sufficient funds to carry out the NDBS program. Local governments have also been asked to upgrade the information system for NDBS and to set up national and provincial NDBS training centers for medical staff (Available at: http://www.interfax.cn/news/11954).

In order to accurately calculate incidences and coverage, accurate birth records are necessary, but a national birth registration system does not yet exist in China. As a result, and because an effective integrated newborn screening information system is not yet present, numbers of newborns screened are mainly derived from specimen records collected from delivery hospitals, and resident records are not separated from those of migrants, which are high in some provinces (Cao et al, 2009). From 2000 to 2007, a total of 17,961,826 were screened for PKU, with 1,527 cases identified (1:11,763). A total of 18,284,745 were screened for CH with 8,918 cases identified (1:2050). In the MS/MS pilot project in Shanghai, 116,000 newborns were screened for various inborn errors of metabolism with an overall incidence of 1:5800 (Gu et al., 2008).

India

With approximately 26 million births annually, India has the largest population of newborns in the world. A number of newborn screening pilot projects have been conducted since the 1980s in several cities and states, including Bangalore and Karnataka (South India), Hyderadad (North India), and Chattisgarh state (Rao et al., 1988; Devi and Naushad, 2004; Kaur et al., 1994; Bhatara et al., 2002). The incidence of conditions detected is reported to be: amino acid disorders, 1:3771; CH, 1:1985; CAH, 1:2600; GAL, 1:13,900; and G6PD deficiency, 1:2200 (Devi and Naushad, 2004).

Problems encountered with newborn screening programs in India have included: (1) time of sampling (traditionally after 48–72 hours); (2) early hospital discharge; (3) home deliveries; (4) other healthcare priorities; (5) not considered cost-effective; (6) lack of availability; (7) cost of treatment; (8) lack of feasibility for screening a huge population; (9) sample transport difficulties; (10) and lack of laboratories and trained personnel. In recent years, there has been increasing interest in establishing NBS, particularly in private laboratories (Padilla and Therrell, 2008).

In 2005, the Indian Council for Medical Research (ICMR) created a task force comprised of clinicians, pediatricians, geneticists, and laboratory scientists. This task force is currently coordinating a pilot study of over one million newborns from Mumbai, New Delhi, Cochin, and Hyderabad. The ICMR pilot project is the first coordinated nationwide screening effort, and is expected to be completed in two years. The objectives of the pilot studies are: (1) to evaluate the feasibility of newborn screening for the different geo-ethnic regions of the Indian population; (2) to attempt to define the incidence of the selected inborn metabolic errors (i.e., CH and CAH); and (3) to develop the capability of diagnosing and treating metabolic disorders using selective screening. This project will form the basis for national NDBS implementation. Plans include consideration of other disorders like hemoglobinopathies, hearing loss, and other IEMs (Padilla and Therrell, 2008).

MS/MS screening has recently become available in the private sector, primarily as part of high-risk screening for a fee. The Goa government recently initiated a state-funded NBS program that includes MS/MS. The ICMR high-risk screening project also offers free screening. The ICMR panel includes about 29 conditions. There are two private and two government laboratories included in the project (Personal communication with M. Kabhra, 2010).

Pakistan

CH screening started at Aga Khan University Hospital (AKUH) in March 1987. In the first two years of screening, all newborns delivered at AKUH and four other birthing facilities (maternity homes) affiliated with AKUH were targeted: approximately 5,000 live births. In subsequent years, due to logistical reasons, CH screening was limited to babies born only at AKUH. From 1989 to October 2008, 53,619 newborns were screened (Afroze et al., 2008).

In recent years, in an attempt to assist in revitalizing screening efforts in Pakistan, the International Atomic Energy Agency (IAEA), in partnership with the Pakistan

Atomic Energy Commission, supported two pilot newborn screening pilot projects: (1) Nuclear Medicine, Oncology and Radiotherapy Institute (NORI) Hospital, Islamabad, September 2006–August 2007, 2000 newborns screened; and (2) Institute of Nuclear Medicine Oncology (INMOL) Hospital Lahore, 3000 newborns screened (IAEA, 2002).

The problems encountered in the pilot studies included lack of awareness due to low literacy, accessibility in remote areas, general unavailability of services, lack of political will (including absence of policy on NDBS), out-of-hospital births, early hospital discharge (as early as 3–6 hours after birth), and lack of public sector funding. Some strategies identified to address the problems include public–private partnerships, public–public partnership with the Pakistan Atomic Energy Commission, and launching NDBS in the public sector, including teaching hospitals. The plan is to start newborn screening on a small scale with gradual expansion to include the entire population, guided by the lessons learned from the recent pilot studies (Padilla and Therrell, 2008).

THIRD ASIA-PACIFIC SCREENING ERA

A third era of NDBS in the Asia-Pacific occurred during the 1990s, after CH screening had been started in much of the economically developed world (Table 61-1). Because of the cost benefits realized from CH screening, and an increased CH incidence in iodine-deficient areas, there was considerable emphasis placed on screening newborns for CH in developing countries. NDBS projects were begun in Korea and iodine-deficient areas of Thailand, as pilots in 1991 and 1992, respectively, and in the Philippines in 1996. The Thailand pilot project became a national project in 1996, and the Korean pilot became a national program in 1997. The Philippine project became a national program in 2004.

Korea

In Korea, the Ministry of Health and Social Affairs adopted newborn screening for low-income families in 1991, with funding from the Social Welfare Fund. Six conditions were initially included in the screening panel - CH, PKU, GAL, MSUD, HCY, and HIS. In 1992, three screening centers were established and there was a cost increase. Central government subsidies helped to support the program. The number of laboratories increased to 21 by 1994. Local governments also began to assist in program funding (Lee, 1994; Han et al., 2003).

In 1995, the screening panel was reduced from 6 conditions to 2 conditions—PKU and CH—because the others were rarely detected and screening was not cost-effective. Funding in 1995 included only the central and local governments. In 1997, the program was expanded to include all newborns regardless of financial position. From 1998 onward, the cost was shared 40% from the central government and 60% from the local government. Follow-up support from the government for treatment items such as infant formula applied only to low-income families. Screening for GAL, MSUD, HCY, and HIS were optional if the parents agreed to an additional cost. Program participation over the years grew from about 35% in 1994 to about 90% in 2000 (Lee, 1994; Han et al., 2003).

While NDBS program coverage has increased steadily over the years, program management challenges still exist. There is no single organization responsible for program coordination or control of national screening services. The number of screening laboratories rose to 76 laboratories in 2000, with some laboratories processing small numbers of specimens. The amount of government support limits NDBS tests to PKU and CH unless parents choose to pay for others. Follow-up support services are considered inadequate. Recovering funds from the government is complicated and necessitates completion of a number of administrative processes (Han et al., 2003).

In 2006, the government added 4 diseases (CAH, HCY, MSUD, GAL) to the 2 diseases already screened (PKU and CH). Currently, the number of newborn screening laboratories has decreased to 12. Coverage for routine newborn screening was 96.8% in 2008. MS/MS was also available for 43 diseases at an additional fee. Screening uptake for MS/MS in 2008 was 70.9% of newborns (Personal communication with D. Lee, 2010).

Thailand

In 1992, a pilot project to establish the incidence of CH and PKU was conducted in 13 provinces in rural Northern Thailand. The survey showed a high incidence of CH, especially in iodine-deficient areas. Additionally, one case of PKU was detected. These data provided the basis for phased implementation of nationwide neonatal screening services. In 1995, a public health policy emphasized the importance of neonatal screening for CH and PKU, and in 1996, NDBS was declared a national project. The country was divided into four regions—North, East, Central, and South. Regional laboratories were established, along with a central oversight/QA laboratory—one central screening laboratory at NIH, and three regional medical-sciences centers. Extensive public relations materials in the Thai language now exist, along with a comprehensive data reporting system and a program website. Approximately 97% of the country now has screening available for CH and PKU.(Charoensiriwatana et al., 1995, 2003; In addition to the national program, a hospital-based screening program serves Siriraj Hospital, a large birthing hospital in Bangkok, screening for CH and PKU (Liammongkolkul, 2009).

In order to better assess the incidence of PKU in Thailand, since the initial pilot study in 1992–1995 provided only a single case, the results of NDBS for PKU in Thailand have twice been assessed. From October 1992–March 2001, the apparent incidence, from 5 PKU cases in 1,062,676 newborns screened was 1:221,535 (Pangkanon et al., 2003). A larger cohort of 5,243,841 screened between 1996 and 2006 with 16 confirmed cases gave a PKU incidence of 1:327,740 (Pangkanon et al., 2009).

A pilot screening project to determine the incidence and feasibility for screening for CAH was conducted beginning in 1999, in response to large numbers of clinical cases apparently diagnosed. Over 58,000 newborns were screened, and of the 144 with abnormal results only 68 were retested. The apparent incidence was reported to be about 1:19,500, but several babies died before retesting could occur. The study confirmed that without rapid follow-up and an understanding of the critical nature of NDBS for CAH, proper implementation would not be possible (Janejai et al., 2003).

Philippines

In 1996, a pilot project was initiated by a newborn screening study group of interested pediatricians and obstetricians in 24 hospitals in metro Manila. The pilot program sought to establish preliminary incidence data for five conditions (CH, CAH, GAL, PKU, HCY) in order to provide a recommendation to the government for policy adoption. The program proceeded in four phases: (1) 1996 – screening for CH, CAH, GAL, PKU, HCY; (2) 1998 – pilot screening for G6PD deficiency; (3) 2000 – evaluation for cost effectiveness and policy changes; and (4) 2004 – integration of NBS into the public health delivery system with the enactment of Republic Act 9288, or Newborn Screening Act of 2004. This act institutionalized the "National NBS System," which shall ensure the following: (a) that every baby born in the Philippines is offered NBS; (b) the establishment and integration of a sustainable NBS system within the public health delivery system; (c) that all health practitioners are aware of the benefits of NBS and of their responsibilities in offering it; and (d) that all parents are aware of NBS and their responsibility in protecting their child from any of the disorders (Padilla, 2003; Padilla, 2009). The result was a recommendation for nationwide screening for five conditions (excluding HCY, adding G6PD deficiency), with samples collected after 24 hours of age (Padilla and Domingo, 2002; Padilla, 2003). The next year, the DOH created a task force for NDBS. By year 2000, the National Technical Working Group was instituted to oversee nationwide implementation of the NDBS program. In 2003, the Philippine president proclaimed the first week of October as "newborn screening week." As a show of support for the NDBS program, the congress of the Philippines adopted the Newborn Screening Act of 2004 (available at: http://www.newbornscreening.ph), which requires NDBS to be offered to parents of all newborns.

In 2006, several steps were undertaken to further the goals of the Philippine NDBS program. These included requiring hospitals to provide NDBS before they are licensed or accredited, and addition of NBS to the national health insurance payment plan. Currently the screening program continues to grow, with more than 2500 hospitals offering screening and newborn coverage increased to 28% in 2009. The biggest challenge to the program is reaching the large number of home-delivered babies.

FOURTH ASIA-PACIFIC SCREENING ERA

Late in the 1990s, NDBS expansion in the Asia-Pacific received a boost from the IAEA, and the fourth era of screening in the region began. The IAEA expanded its support to countries in East Asia through a regional technical cooperation project: RAS 6/032, "Regional Screening Network for Neonatal Hypothyroidism." The project began in May 1999, ran for almost 10 years, and included countries in the region at different levels of newborn screening program development. Representatives from more developed programs shared their experiences and advice with emerging programs. As part of annual planning meetings, an internationally renowned newborn screening expert participated to assist in knowledge sharing. The following countries were involved: Bangladesh, China (Tianjin), Indonesia, Korea, Malaysia, Mongolia, Pakistan, Philippines, Thailand, and Vietnam. The original project was dedicated to CH screening, with the vision of using the same infrastructure in expanding newborn screening to other disorders. A guidance book for developing countries that were beginning CH screening was an important output of this project (Therrell and Padilla, 2005).

Bangladesh

Bangladesh began NDBS in 1999 as part of the IAEA East Asia Regional CH Project. IAEA funds initially supported the purchase of equipment, filter papers, reagents, training, and expert services. Since 1999, two pilot projects for CH screening have been completed with approximately 30,000 newborns screened and 16 confirmed with CH. Efforts during the past few years led the government to include a project on NDBS in its Annual Development Program (ADP) beginning in July 2006. Under the program, about 200,000 newborns from Dhaka City and from 11 districts in different parts of the country are being screened for CH, and laboratory facilities are being increased. The project is expected to cost 104 million taka (about US$1.5 million) and will run until June 2010. The end goal is to integrate NDBS into national and regional health policies in order to gradually include all newborns, to develop an efficient network throughout the country and region, to increase the number of diseases in screening, and to form a basis for research, particularly for the more common congenital anomalies.(Moslem et al., 2003; Hasan et al., 2003)

The current project has had to overcome several internal barriers. There has been a lack of involvement of active professionals, a lack of coordination between the centers, and, to date, integration into national health policy has not been achieved. At the local screening level, the following challenges have been identified: high rate of home deliveries; lack of information; insufficient knowledge and training; low national healthcare priority; lack of coordination between existing screening procedures; sociocultural patterns; and lack of motivation (Padilla and Therrell, 2008).

Indonesia

The IAEA introduced NDBS in Indonesia in 1999. After a year of planning, a regional working group was created and a feasibility study was conducted. In 2000, a pilot study was launched to determine the local incidence of CH. Results from the 5-year pilot study revealed an incidence of 1:4109 for CH. Several problems were identified: unsatisfactory blood samples (which resulted in high recall rates); delayed arrival of samples in the screening lab, and slow tracking of babies for further sampling; uncooperative health professionals who saw screening as an additional job burden; difficulty in obtaining consent; and early hospital discharge policies (Rustama et al., 2003).

In 2003, the Department of Health (DOH) offered financial support for reagents. During 2005–2006, the DOH conducted a series of workshop and training seminars in the capital cities of seven provinces. In addition, the Health Technology Committee (under the DOH) approved newborn hearing screening and newborn screening for CH, to be implemented nationwide with budget support from the Provincial Health Department. The additional challenges identified in expanding the program include the need for a policy statement from the Ministry of Health, inadequate understanding of newborn screening among policymakers at the provincial level, absence of well-developed infrastructure, high prevalence of home deliveries, inadequate number of healthcare professionals, and the fact that 40% of the population cannot afford the test fee (Padilla and Therrell, 2008).

Laos

In an effort to start NDBS in Laos, a strong newborn screening proponent attended the Japanese International Coooperation Agency (JICA) training course in Sapporo, Japan, in 1996 (Fukushi, 2007). Due to financial, technical, data (lack of sufficient incidence data), and public health priority challenges (polio, tuberculosis, and other infectious diseases), NDBS could not be started upon his return. In July 2008, a collaboration with the University of Dusseldorf in Germany allowed development of a pilot screening program for CH. The project was intended to screen 10,000 newborns. Of 7000 newborns actually screened, 2 were found to have CH (in line with international incidence predictions). The Ministry of Health has agreed to support the setting up of a newborn screening laboratory (Personal communication with Dr. Saysanasongkham, 2011).

Mongolia

In 2000, with assistance from IAEA, a pilot screening project for the detection of CH was developed. NBS was started at the Central Clinical Hospital and two maternity houses in Ulan Bataar. In the first year, 4% of the population was screened with no positive CH cases. In the following year, 15.2% of the newborn population was screened with one confirmed CH case. Another CH patient was confirmed in 2002. From 2000–2002, several educational activities were organized for physicians (pediatricians, obstetricians, and endocrinologists), nurses, and medical technologists. In 2002, the National IDD program was approved by the Government of Mongolia, which included the implementation of a screening program for CH (Erdenechimeg, 2003).

The following problems were identified during the pilot phase: lack of financial support for the program, lack of incentives for nurses to collect samples, and transportation and delivery of collected samples. In November 2007, a core management group for CH screening was established, and in the subsequent year, an operational plan was developed to expand the pilot program. The objectives were to define the incidence of CH in Ulaanbaatar, develop a national NBS program, establish NDBS at the MCHRC and Maternity House N1, develop training materials and handouts, conduct trainings on NDBS for nurses and physicians, and follow up children with positive screening test results (Padilla and Therrell, 2008).

Sri Lanka

Newborn screening was begun by efforts of the nuclear medicine unit (NMU), Faculty of Medicine, Peradeniya Medical School, with IAEA assistance in August 2005. Screening there has been sustained with government support and a few dedicated persons. Computerized records are maintained for all screened babies, which aids in the recall and follow-up process (Nanayakkara, 2007). While there is government support for CH screening, an internal war in the country has contributed to lack of finances and national screening policy development. There are discussions with the Health Ministry concerning regional efforts, and planning is ongoing (Personal communication with Damayanthi Nanayakkara, 2010).

Currently, NMU Perdaeniya provides CH screening tests for all births in two government hospitals in the central province of Sri Lanka. There is government funding for this effort and there is no charge for screening, confirmatory testing, treatment, and follow-up. However, specimen collection is not yet routine. From August 2005 – December 2009, 19,088 newborns were screened and 11 cases of CH were confirmed (incidence 1:1468). An additional 9 cases of transient CH were identified. The teaching hospital Karapitiya in a southern province also began a screening program, but encountered problems in sample collection; 1756 newborns were screened. This hospital recently received a grant to implement CH screening in the southern province; however, money has not yet been provided (Personal communication with Damayanthi Nanayakkara, 2010).

One of the major difficulties in NDBS implementation has been the reliability of the kit used for laboratory testing. The initial kit suffered from instability problems and, as a result, the program lost credibility among many physicians who were submitting specimens. This kit manufacturer was subsequently replaced, but restoring full confidence in the program has been an issue. As in some other developing countries, equipment service and maintenance have also been problematic in Sri Lanka (Personal communication with Damayanthi Nanayakkara, 2010).

Vietnam

Financial and technical support for pilot NDBS screening projects in Vietnam was initially provided by IAEA. Since 1999, several meetings and expert visits have been conducted. A blood collection form is currently in use, and guidelines have been written to communicate data. NDBS became a national program in 2006 with an action plan to increase coverage significantly by 2010. After several years of lobby efforts, the Ministry of Health approved a project entitled, "Increasing population quality through establishment and expansion of prenatal and neonatal screening system in 24 provinces in Vietnam up to 2010." The Prenatal Diagnosis Center – National Hospital for Obstetrics and Gynecology has been assigned to conduct the project in 12 northern provinces, and the Prenatal Diagnosis Center and Genetic Counseling – Tu Du Hospital, Ho Chi Minh City, is responsible for conducting the project in 12 southern provinces. This project has a budget of 32 Billion VND (equivalent to $US2 million; see Padilla and Therrell, 2008).

The government project aims to establish project management units from central to provincial levels. The following activities have been planned: design various educational and public relations materials; initiate communications campaigns (mass media, counseling); conduct training workshops on blood sampling, specimen transport, and program management; and begin training pediatricians on counseling and treatment (using training CD/DVD) of G6PD deficiency (567 clinical cases; 1% incidence) and CH (1:5000 cases). The government has invested $US2 million for both the prenatal screening and neonatal screening programs. A protocol for CAH screening expansion for 2009–2011 has been prepared (Padilla and Therrell, 2008).

The following disorders are currently being screened: CH, G6PD deficiency, and CAH in Ho Chi Minh City, with samples analyzed at the Tu Du Hospital; and CH and G6PD deficiency in Hanoi, with sample tested at the National Obstetric Hospital. Newborn screening for CH and G6PD deficiency began in the north in 2009 (Personal communication with Vu Chi Dung, 2009).

Pacific Islands

The Asia-Pacific region includes a large number of island countries with small populations and low numbers of births. However, taken together, they represent a significant number of babies that should be included in screening. During the mid-1960s, the New Zealand screening laboratory was active in receiving newborn screening specimens from many of these areas, as noted previously. Specimens were received from at least 10 island nations and occasionally others (Veale and Houston, 1978). Over time, the information from these programs has been limited, with only small numbers of specimens reported received by the New Zealand screening laboratory. Countries with island territories generally include screening for those babies in their national or regional program(s). So, for example, Australia receives samples from Norfolk Island and Lord Howe Island (Personal communication with B. Wilcken, 2007), Taiwan provides services for nearby islands (Chiang et al., 2003), Guam and Saipan send samples to the Oregon and Colorado programs in the United States (Personal communication with B. Therrell, 2009). Recently, Palau has entered into an arrangement to utilize the screening program in the Philippines and has now begun to screen their 350 newborns using the Philippine screening panel, excluding G6PD deficiency screening (Personal communication with E. Reyes, 2009). However, for the vast majority of the smaller island nations and territories in the Asia-Pacific there is no available information about newborn screening services.

QUALITY ASSURANCE PROGRAMS

Laboratory quality assurance is a critical aspect of NDBS services, and several options exist. Most notably, the

Centers for Disease Control and Prevention (CDC) in the United States provides free QA and proficiency testing (PT) materials for newborn screening worldwide. Most analytes are available as part of the CDC program, with the exception of G6PD, which is available from Taiwan. Because a huge influx of developing laboratory programs might overwhelm the CDC program, they routinely collaborate with newly established programs to provide whatever services may be needed, including technical consultation. Some examples of programs collaborating with CDC and their means of doing so are discussed below.

Thailand

The national screening program is housed at the Thailand NIH. In addition to their routine NDBS activities as one of the regional laboratories, selected personnel there are also responsible for supplying reagents for affiliated laboratories and providing NDBS QA. The Thai QA program was developed with the assistance of the CDC, provides challenge samples modeled on the CDC Newborn Screening Quality Assurance Program (NSQAP) proficiency testing (PT), and serves as a model for other Asian programs.

China

The CDC has assisted in developing a quality assurance program for NBS laboratory testing in China, and provides supplemental PT services in addition to those provided by the Chinese government. Similar to efforts in Thailand and other developing countries, the CDC has sought to identify partners in China who may develop the ability to provide their own PT program. In Shandong Province, at one point, PT materials were supplied to a central authority for distribution for distribution to all local screening laboratories. The testing results were reported back to the central authority, who in turn reported the results to the CDC for inclusion in their PT results assessment. Final performance results were then returned to the central authority for follow-up. The long-range plan was for the central authority to develop the capability to prepare specimens and manage their use for QA/PT (Wang et al., 2003). To date, this program has not yet accomplished its goal. In the meantime, a centralized PT program has been established in Beijing and is working to improve laboratory quality throughout the country.

Taiwan

In order to help NDBS provide quality G6PD testing, an external QA program has existed in Taiwan since January 1988, when it was developed to assess the reliability of confirmatory testing there. In 1999, an external QA/PT program was developed to assess the reliability of the G6PD screening in Taiwan and other interested countries. This program is the central source of G6PD QA/PT for most NDBS around the world. PT challenges containing 10 samples of unknown concentration of G6PD are submitted to participating laboratories at 1–2 month intervals. Results are reported and compared to a quantitative reference value from the Taiwan PT laboratory. Laboratories identified as needing technical assistance are contacted and counseled to improve their performance. This international PT program includes participants from Taiwan, the Philippines, Thailand, Lebanon, Vietnam, Macau (via Shanghai), Germany (Hamburg), the United States, and China (Chiang et al., 2003; Personal communication with K-J Hsiao, 2007).

CONCLUSIONS

Eighty percent of the births in the Asia-Pacific region occur in five countries – China, Indonesia, Bangladesh, India, and Pakistan (see Table 61-1). All countries with an infant mortality rate (IMR) of less than 10/1000 live births have achieved better than 90% screening coverage of their newborn population. Most are approaching full coverage, although the number of conditions screened varies widely. Of the remaining countries with higher IMRs, Thailand (IMR = 13/1000) is the only one that has achieved a high rate of newborn coverage (~97%). Among the countries that lack total coverage, the following obstacles are most often cited: poor economies, insufficient health education, lack of government support, early hospital discharge, and large numbers of out-of-hospital births.

The following items have been identified as integral to the success of full population newborn screening: (1) government prioritization; (2) full or partial government financing; (3) public education and acceptance; (4) health practitioner cooperation/involvement; and (5) government participation in institutionalizing a newborn screening system. Integration into the national healthcare delivery system was cited as the single most critical challenge. While most of the developed programs in the region successfully accomplished healthcare integration without legislative mandates, at least two of the developing programs (China and the Philippines) have found national legislation to be helpful. While the Chinese law is permissive and allows gradual system development, the Philippine law is more aggressive and requires that every newborn be given access to newborn screening by the health practitioner who delivers or assists in the delivery of a newborn. Government and physician responsibilities are also clearly outlined in the Philippine law.

Many of the developed programs of the Asia-Pacific region have been active in advancing newborn screening. New Zealand was one of the first to have a national

TABLE 61-2 SCREENED CONDITIONS IN ASIA-PACIFIC NEWBORN SCREENING PROGRAMS

	Screened Conditions												
									Groups Detected by MS/MS			Others	
Jurisdiction	Congenital Hypothyroidism	Phenylketonuria	Galactosemia	Maple Syrup Urine Disease	Congenital Adrenal Hyperplasia	Homocystinuria	Cystic Fibrosis	G6PD Deficiency	Other Amino Acid Disorders	Fatty Acid Oxidation Disorders	Organic Acid Disorders	Lysosomal Storage Disorders	Thalassemia
Australia	·	·	·	·		·	·		·	·	·	‡	
Bangladesh	·												
Cambodia	No Data Available												
China	·	·	‡	‡	‡			‡	‡	‡	‡		‡[a]
Hong Kong (China)	†							†					
India	‡	‡	‡	‡	‡	‡		‡	‡	‡	‡		
Indonesia	‡												
Japan	·	·	·	·	·	·			‡	‡	‡		
Korea (South)	·	·	·	·	·	·			‡	‡	‡		
Korea (North)	No Data Available												
Laos	‡												
Malaysia	·							·	‡	‡	‡		
Mongolia	‡												
Myanmar	‡												
Nepal	No Data Available												
New Zealand	·	·	·	·	·	·	·		·	·	·		
Palau	·	·	·		·								
Pakistan	‡												
Philippines	·	·	·		·			·					
Singapore	†							†	·	·	·		
Sri Lanka	‡												
Taiwan	·	·	·		·	·		·	·	·	·	‡	
Thailand	·	·			‡				‡	‡	‡		
Vietnam	‡				‡			‡					

· Offered to full population being screened – for MS/MS indicates screening and interpretations available for all detectable conditions

† Cord blood screening

‡ Pilot testing or select population screening – not full population screening

[a] Recent report (Cao et al., 2009) notes some thalassemia screening but confirmation unavailable

newborn screening program; Japan initiated the first national multi-laboratory QA program, and Australia was one of the first to utilize MS/MS. Both New Zealand and Australia were leaders in initiating CF NDBS.

Based on the information available, Table 61-2 provides an overview of current activities within the region. In several developing programs, screening for CH is still in the pilot phase. In some of the developed programs, MS/MS screening has been implemented but the number of disorders targeted for detection varies. When program representatives were asked the status of MS/MS screening in their programs, almost all indicated caveats in how their MS/MS results were interpreted (i.e., whether they were performing full-profile screening/interpretation). In Table 61-2, this lack of full-profile screening/interpretation has been indicated as "restricted interpretation." While MS/MS testing is routine in Australia and New Zealand, consensus does not exist regarding which conditions should be screening targets. In Singapore, MS/MS screening is routine in government hospitals, and detection is aimed at 25 metabolic conditions, but there is variation between hospitals as to whether a parent must opt in or opt out of this supplemental screening. In Malaysia and Thailand, some MS/MS screening exists in a limited number of hospitals. In Japan, conditions included in MS/MS pilot testing vary, despite having the same routine core screening panels in all prefectures and big cities. Pilot MS/MS testing is ongoing with support from the Japanese Ministry of Health, Labor and Welfare. In Hong Kong, MS/MS equipment is available for diagnostic testing in some facilities, but there is not yet acceptance of this technology for newborn screening. Limited MS/MS pilots exist in the rest of China.

The genetic nature of most of the screened conditions provides an interesting mix of screening conditions across the region. The universally important condition is CH, which has a similar incidence across all countries in the region, with the exception of pockets of iodine deficiency where it is much higher. These pockets most often occur in the developing countries, and so CH is usually the first condition considered for screening. The exception is G6PD deficiency, which is known to be of high incidence in the Asian population, in some countries approaching 5% in males (Lo et al., 1995). Therefore, several countries in the region began screening for G6PD before considering other conditions. Because cord blood samples were readily available, several programs began their G6PD or CH screening using cord blood. This program variability extends to MS/MS screening/interpretation, where programs vary in whether to use full scan MS/MS screening. It is not likely that these debates will soon end.

The developed programs within the region continue to refine and expand their NBS programs, and many research projects exist. Of particular interest is a pilot screening program for Pompe disease in Taiwan (Chien et al., 2009), and other lysosomal storage disease screening in Australia (Meikle et al., 2006). Hearing screening is also expanding throughout the region, with a number of national projects underway. The developing CH programs continue to show progress, and there are indications of screening interest in some of the countries currently without screening programs. The challenges to developing screening in some parts of the region are formidable, but they can be overcome with dedication, ingenuity, and support from developed programs and health advocacy organizations.

ACKNOWLEDGMENTS

We are grateful to the following individuals for providing country data: Mizanul Hasan (Bangladesh), Xue Fan Gu (China), Stephen Lam (Hong Kong) Madhulika Kabra (India), Diet Rustama (Indonesia), Shohei Harada (Japan), Dong Hwan Lee (Korea), Saysanasongkham Bounnack (Laos) Thong Meow Keong (Malaysia), Dianne Webster (New Zealand), Roy Joseph (Singapore), Damayanthi Nanayakkara (Sri Lanka), Wuh-Liang Hwu (Taiwan), Wiyada Charoensiriwatana (Thailand), and Vu Chi Dung (Vietnam).

REFERENCES

Afroze B, Humayun KN, Qadir M. (2008). Newborn screening in Pakistan – Lessons from a hospital-based congenital hypothyroidism screening programme. *Ann Acad Med Singapore 37*(Suppl 3), 114–116.

American Academy of Pediatrics, Newborn Screening Task Force. (2000). Serving the family from birth to the medical home – newborn screening: a blueprint for the future. *Pediatrics 106*(Suppl), 383–427.

American Academy of Pediatrics, Committee on Practice and Ambulatory Medicine, Section on Ophthalmology; American Association of Certified Orthoptists; American Association for Pediatric Ophthalmology and Strabismus; American Academy of Ophthalmology. (2003). Eye examination in infants, children, and young adults by pediatricians. *Pediatrics 111*, 902–907.

Amino Y. (1987). Quality control system in Japan. In: Therrell BL, ed. *Advances in Neonatal Screening,* pp. 531–535. Amsterdam: Elsevier Science Publishers.

Aoki K. (2003). Newborn screening in Japan. *Southeast Asia J Trop Med Pub Health 34*(Suppl 3), 80.

Bertini G, Rubaltelli FF. (2002). Non-invasive bilirubinometry in neonatal jaundice. *Semin Neonatol 7*, 129–133.

Bhatara V, Sankar R, Jurgen U, Peabody J. (2002). A review of the case for neonatal thytropin screening in developing countries: The example of India. *Thyroid 12*, 591–598.

Cao Y, Yuan P, Wang YP, Mao M, Zhu J. (2009). The profile of newborn screening coverage in China. *J Med Screen 16*, 163–166.

Carpenter KH, Wilcken B. (1999). Neonatal diagnosis of long-chain 3-hydroxyacyl-CoA dehydrogenase deficiency and implications for newborn screening by tandem mass spectrometry. *J Inherit Metab Dis 22*, 840–841.

Chace DH, Kalas TA, Naylor EW. (2003). Use of tandem mass spectrometry for multianalyte screening of dried blood specimens from newborns. *Clin Chem 49*, 1797–1817.

Charoensiriwatana W, Janejai N, Boonwanich W, Krasao P, Nakpresit S. (1995). Establishment of national neonatal screening programme in Thailand. In: Therrell B, Aldis B, eds. *Proceedings of 11th national Neonatal Screening Symposium, September 12-16, 1995, Corpus Christi, Texas*, pp. 1-4. Washington, DC: Association of State and Territorial Public Health Laboratory Directors.

Charoensiriwatana W, Janejai N, Boonwanich W, Krasao P, Chaisomechit S, Waiyasilp S. (2003). Neonatal screening program in Thailand. *Southeast Asia J Trop Med Pub Health 34*(Suppl 3), 94–100.

Chen JS. (1993). The organization of neonatal screening programs and international cooperation systems in Taiwan. In: Takasugi N, Naruse H, eds. *New Trends in Neonatal Screening*, pp. 33–35. Sapporo: Hokkaido University Press.

Chen R, Guo Di. (1983). The initiation of neonatal programmed in Shanghai, China. In: Naruse H, Irie M, eds. *Neonatal Screening*, pp. 483–484. Amsterdam: Excerpta Medica.

Chiang S-H, Wu K-F, Liu T-T, Wu, S-J, Hsiao K-J. (2003). Quality assurance program for neonatal screening of glucose-6-phosphate dehydrogenase deficiency. *Southeast Asia J Trop Med Pub Health 34*(Suppl 3), 130–134.

Chien Y-H, Lee N-C, Thurberg BL, Chiang SC, ZhangXK, Keutzer J, et al. (2009). Pompe disease in Infants: Improving the prognosis by newborn screening and early treatment. *Pediatrics 124*, e1116-e1125.

Choo KE, Ariffin WA, Normah J, Oppenheimer SF, Selamah G. (1994). Cord Blood G6PD Activity in Malay Neonates. In: Takasugi N, Naruse H, eds. *New Trends in Neonatal Screening, Screening*, pp. 7–10. Sapporo: Hokkaido University Press.

Crossley JR, Elliott RB, Smith PA. (1979). Dried-blood spot screening for cystic fibrosis in the newborn. *Lancet 1*, 472–474.

Devi AR, Naushad SM. (2004). Newborn Screening in India. *Indian J Pediatr 71*, 157–160.

Dussault JH, Laberge C. (1973). Thyroxine (T4) determination in dried blood by radioimmunoassay: A screening method for neonatal hypothyroidism. *Union Med Can 102*, 2062–2064.

Dussault JH, Coulomb P, Laberge C, Letarte J, Guyda H, Khoury K. (1975). Preliminary report on a mass screening program for neonatal hypothyroidism. *J Pediatr 86*, 670–674.

Erdenechimeg S. (2003). National newborn hypothyroid screening program in Mongolia. *Southeast Asia J Trop Med Pub Health 34*(Suppl 3), 85–86.

Fujieda K, Matsuura N, Takasugi N, Fukushi M, Arai O. (1987). *Five Years Experience of Newborn Screening Program for Congenital Adrenal Hyperplasia in Sapporo*. In: Therrell BL, ed. *Advances in Neonatal Screening*, pp. 281–286. Amsterdam: Elsevier Science Publishers.

Fujimoto A, Ohtake H, Miyagi T, et al. (1987). Blood spot enzyme multiple auxotroph screening for urea-cycle disorders in Osaka. In: Therrell BL, ed. *Advances in Neonatal Screening*, pp. 215–216. Amsterdam: Elsevier Science Publishers.

Fukushi M. (2007). An international training and support programme for the establishment of neonatal screening in developing countries. *J Inherit Metab Dis 30*(4), 593–595.

Gao Y, Yuan P, Wang YP, Mao M, Zhu J. (2009). The profile of newborn screening coverage in China. *J Med Screen 16*, 163–166.

Gu XF, Jun Y, Chen R-G. (1999). Neonatal screening of phenylketonuria and congenital hypothyroidism in China. *Southeast Asia J Trop Med Pub Health 30*(Suppl 2), 17–19.

Gu XF, Wang Z, Ye J, Han L, Qiu W. (2008). Newborn screening in China: Phenylketonuria, congenital hypothyroidism and expanded screening. *Ann Acad Med Singapore 37*(Suppl 3), 107–110.

Guthrie R. (1962). Screening for "inborn errors of metabolism" in the newborn infant – a multiple test program. *Birth Defects Original Article Series IV*, 92–98.

Guthrie R, Susi A. (1963). A simple phenylalanine method for detecting phenylketonuria in large populations of newborn infants. *Pediatrics 32*, 338–343.

Guthrie R. (1992). The origin of newborn screening. *Screening 1*, 5–15.

Han YJ, Lee D-H, Jong WK. (2003). Newborn screening in Korea. *Southeast Asia J Trop Med Pub Health 34*(Suppl 3), 81–84.

Hasan M, Nahar N, Ahmed A, Moslem F. (2003). Screening for congenital hypothyroidism – a new era in Bangladesh. *Southeast Asia J Trop Med Pub Health 34*(Suppl 3), 162–164.

Hau W, Lam TS. (1994). 20 years experience in neonatal screening in Hong Kong. In: *Abstracts 5th Asia Pacific Regional Meeting of the International Society for Neonatal Screening, September 14-17, 2004*, p. 31. Shanghai, China.

Houston IB, Veale AMO. (1971). Screening for inborn errors of metabolism. *Lab Manag 9*, 30–32, 53.

Hsiao K-J, Chiang SH, Chang TT, Liew DG, Chao YY. (1991). Experience of neonatal G6PD deficiency screening in Taiwan. In: Wilcken B, Webster D, eds. *Neonatal Screening in the Nineties*, pp. 217–218. Leura, NSW: The Kelvin Press.

Huang HP, Chu KL, Chien YH, Wei M, Wu S, Wang S, et al. (2006). Tandem mass neonatal screening in Taiwan – report from one center. *J Formos Med Assoc 105*, 882–886.

Hwu WL, Huang AC, Chen JS, Hsiao K-J, Tsai W-Y. (2003). Neonatal screening and monitoring system in Taiwan. *Southeast Asia J Trop Med Pub Health 34*(Suppl 3), 91–93.

Hwu W-L, Chien Y-H, Lee N-C. (2009). Newborn Screening for Fabry Disease in Taiwan Reveals a High Incidence of the Later-Onset Mutation c.936+919G>A (IVS4+919G>A). *Hum Mutat 30*(10), 1397–1405.

IAEA Report of East Asia Project Coordination and Planning Management Meeting (2002). *Workshop on Congenital Hypothyroidism: Quality Assurance and Validation (RAS6032), Tianjin, China*, pp. 10–12. Vienna: International Atomic Energy Agency.

Irie M, Enomoto K, Naruse H. (1975). Measurement of thyroid-stimulating hormone in dried blood spot. *Lancet 2*, 1233–1234.

Irie M, Naruse H. (1980). Mass screening for neonatal hypothyroidism in Japan. In: Bickle H, Guthrie R, Hammersen G, eds. *Neonatal Screening for Inborn Errors of Metabolism*, pp. 247–251. Berlin: Springer-Verlag.

Janejai N, Krasao P, Phansang J, Pankarnjanato R, Charoensiriwatana W. (2003). Congenital adrenal hyperplasia: should nationwide screening be implemented in Thailand? *Southeast Asia J Trop Med Pub Health 34*(Suppl 3), 170–173.

Joseph R, Ho LY, Gomez JM, Rajdurai VS, Sivasankaran S, Yip YY. (1999). Mass Newborn Screening for Glucose-6-phosphate dehydrogenase deficiency in Singapore. *Southeast Asia J Trop Med Pub Health 30*(Suppl 2), 70–71.

Joseph R, Ho LY, Gomez JM, Rajdurai VS, Sivasankaran S, Yip YY. (1999). Newborn Screening in Singapore. *Southeast Asia J Trop Med Pub Health 30*(Suppl 2), 23–24.

Joseph R. (2003). Mass newborn screening in Singapore. *Southeast Asia J Trop Med Pub Health 34*(Suppl 3), 89–90.

Kaur M, Das GP, Verma IC. (1994). Inborn errors of aminoacid metabolism in North India. *J Inherit Metabol Dis 17*, 230–233.

Kawamura M. (1987). Neonatal screening for galactosemia. In: Therrell BL ed. *Advances in Neonatal Screening*, pp. 227–230. Amsterdam: Elsevier Science Publishers.

Koch JH. (1997). *Robert Guthrie: The PKU Story*. Pasadena: Holt Publishing House.

Lam STS, Cheng ML. (2003). Neonatal screening in Hong Kong and Macau. *Southeast Asia J Trop Med Pub Health 34*(Suppl 3), 73–75.

Lee DH. (1994). Neonatal Screening in Korea. In: Takasugi N, Naruse H, eds. *New Trends in Neonatal Screening*, pp. 3–5. Sapporo: Hokkaido University Press.

Liammongkolkul S. (2009). Newborn screening of congenital hypothyroidism and hyperphenlalaninemia: A 3 years experience (2005–2008) at Siriraj Hospital, Thailand. *Japanese J Inherit Metab Dis 25*(2), 225(abstract).

Lo KK, Lam TS. (1995). Neonatal screening programme for congenital hypothyroidism in Hong Kong. In: Lam TS, Pang CP, eds.

Neonatal and Perinatal Screening: The Asian Pacific Perspectives, pp. 145–148. Hong Kong: The Chinese University Press.

Lo KK, Chan ML, Lo IFM, et al. (1995). Neonatal screening for glucose-6-phosphate dehydrogenase deficiency in Hong Kong. In: Lam TS, Pang CP, eds. *Neonatal and Perinatal Screening: The Asian Pacific Perspectives*, pp. 33–35. Hong Kong: The Chinese University Press.

Lyon IC, Crossley JR, Smith PA. (1983). Screening for cystic fibrosis. *New Zealand Medl J 96*, 673–675.

Lyon I. (1983). Screening for hypothyroidism: The New Zealand Experience. In: Naruse H, Irie M, eds. *Neonatal Screening*, pp. 111–112. Amsterdam: Elsevier Science Publishers.

Mahle WT. (2008). Physical examination and pulse oximetry in newborn infants: Out with the old, in with the new? *J Pediatr 152*, 747–748.

Matsuda I, Endo Y, Edo F, Akaboshi I. (1987). Clinical and biochemical observation of maple syrup urine disease patients found by neonatal screening. In: Therrell BL, ed. *Advances in Neonatal Screening*. Amsterdam: Elsevier Science Publishers, 217–218.

McCabe LL, Therrell BL, McCabe ERB. (2002). Newborn screening: Rationale for a comprehensive, fully integrated public health system. *Mol Genet Metab 77*, 267–273.

Meikle PJ, Grasby DJ, Dean CJ, Lang DL, Bockmann M, Whittle AM, et al. (2006). Newborn screening for lysosomal storage disorders. *Mol Genet Metab 88*, 307–314.

Millington DS, Kodo N, Norwood DL, Roe CR. (1990). Tandem mass spectrometry: A new method for acylcarnitine profiling with potential for neonatal screening for inborn errors of metabolism. *J Inherit Metab Dis 13*, 321–324.

Mishra S, Chawla D, Agarwal R, Deorari AK, Paul VK, Bhutani VK. (2009). Transcutaneous bilirubinometry reduces the need for blood sampling in neonates with visible jaundice. *Acta Paediatr. 98*(12), 1916–1919. Epub 2009 Oct 7.

Moslem F, Yasmeen S, Hasan M, Karim MA, Nahar N, Ahmed A. (2003). Newborn screening in Bangladesh. *Southeast Asia J Trop Med Pub Health 34*(Suppl 3), 71–72.

Nanayakkara DKK. (2007). The Practice and Limitations of Screening for Congenital Hypothyroidism in Selected Hospitals in Sri Lanka. Doctoral Dissertation, The University of Peradeniya Sri Lanka, April 2007.

Naruse H. (1980). System of neonatal screening for inborn errors of metabolism in Japan. In: Bickle H, Guthrie R, Hammersen G, eds. *Neonatal Screening for Inborn Errors of Metabolism*. Berlin: Springer-Verlag, 299–305.

Naruse H, Tsuji A, Suzuku E, Watanabe N, Maeda M, Fukushi M, et al. (2003). Quality control system for mass screening in Japan. *Southeast Asia J Trop Med Pub Health 34*(Suppl 3), 30–35.

Nozue G, Moriyama Y, Gomibuchi M, et al. (1983). Propagation and results of a neonatal screening program concerning inborn errors of metabolism in Japan. In: Naruse H, Irie M, eds. *Neonatal Screening*, pp. 477–480. Amsterdam: Elsevier Science Publishers.

Padilla C. (2003). Newborn Screening in the Philippines. *Southeast Asia J Trop Med Pub Health 34*(Suppl 3), 87–88.

Padilla C. (2008). Towards Universal Newborn Screening in Developing Countries: Obstacles and the Way Forward. *Ann Acad Med Singapore 37*(Suppl 3), 6–9.

Padilla C, Domingo C for the Newborn Screening Study Group. (2002). Implementation of Newborn Screening in the Philippines. *Philipp J Pediatr 51*, 2–10.

Padilla CD, Krotoski D, Therrell BL. (2010). Newborn screening progress in developing countries – overcoming internal barriers. *Semin Perinatol 34*(2), 145–155.

Padilla C, Therrell BL. (2007). Newborn screening in the Asia Pacific region. *J Inherit Metabol Dis 30*, 490–506.

Padilla C, Therrell B. (2008). *Consolidating Newborn Screening Efforts in the Asia Pacific Region* (Technical Report). Available at: www.newbornscreening.ph, Accessed December 31, 2009.

Padilla C, Basilio JA, Therrell BL. (2009). A performance evaluation and assessment scheme (PEAS) for improving the Philippine newborn screening program. *Acta Medica Philipp 49*, 58–63.

Padilla CD, Basilio J, and Oliveros Y. (2009). Newborn Screening: Research to Policy. *Acta Medica Philippina 43*(2), 6–14.

Pangkanon S, Ratrisawadi V, Charoensiriwatana W, et al. (2003). Phenylketonuria detected by the neonatal screening program in Thailand. *Southeast Asia J Trop Med Pub Health 34*(Suppl 3), 179–181.

Pangkanon S, Charoensiriwatana W, Janejai N, Boonwanich W, Chaisomchit S. (2009). Detection of phenylketonuria by the newborn screening program in Thailand. *Southeast Asia J Trop Med Pub Health 40*, 525–529.

Pass KA, Lane PA, Fernhoff PM, et al. (2000). U.S. Newborn Screening system guidelines II: Follow-up of children, diagnosis, management and evaluation – statement of the Council of Regional Networks for Genetics Services (CORN). *J Pediatr 137*(4 Suppl), S1–S46.

Rao NA, Devi AR, Savithri HS, Rao SV, Bittles AH. (1988). Neonatal screening for amino acidaemias in Karnataka, South India. *Clin Genet 34*, 60–63.

Robinson MJ, Lau KS, Lin HP, Chan GL. (1976). Screening for G6PD Deficiency. *Med J Malaysia 30*, 287–290.

Rustama DS, Fadil MR, Harahap ER, Primadi A. (2003). Newborn screening in Indonesia. *Southeast Asia J Trop Med Pub Health 34*(Suppl 3), 76–79.

Singh A. (2003). Malaysia country report – newborn screening. Hanoi, Vietnam: IAEA 2003 RAS 6/32 Project Report.

Suwa S, Shimozawa K, Titagawa T, et al. (1987). Collaborate study on regional neonatal screening for congenital adrenal hyperplasia in Sapporo. In: Therrell BL, ed. *Advances in Neonatal Screening*, pp. 279–280. Amsterdam: Elsevier Science Publishers.

Suzuki K, Tochizawa K, Suzuki R, Owada M, Sakayam T, Kitagawa T. (1983). A trial of the screening for urea cycle disorders. In: Naruse H, Irie M, eds. *Neonatal Screening*, pp. 372–373. Amsterdam: Elsevier Science Publishers.

Szeinberg A. (1971). Neonatal screening for phenylketonuria and some other metabolic disorders. In: Cohen BE, Rubin MI, Szeinberg A, eds. *Proceedings of the International Symposium on Phenylketonuria and Allied Disorders. Tel Aviv, June 15–22, 1969*. pp. 47–52. Tel Aviv: Translators' Pool.

Therrell BL. (2001). U.S. newborn screening policy dilemmas for the twenty-first century. *Mol Genet Metab 74*, 64–74.

Therrell BL, Padilla CD. (2005). *Screening of Newborns for Congenital Hypothyroidism Guidance for Developing Programmes*, p. 15. Vienna: International Atomic Energy Agency.

Therrell BL, Hannon WH. (2006). National evaluation of U.S. newborn screening system components. *Ment Retard Dev Disabil Res Rev 12*, 236–245.

Therrell BL, Schwartz M, Southard C, et al.; PEAS Organizing and Working Groups. (2010). Newborn Screening Performance Evaluation Assessment Schemes (PEAS), *Semin Perinato 34*, 105–120.

UNICEF. (2009). *The State of the World's Children – Special Edition Celebrating 20 Years of the Convention on the Rights of the Child (Statistical Tables)*, pp. 1–52. New York: UNICEF. Available at http://www.unicef.org/rightsite/sowc/pdfs/statistics/SOWC_Spec_Ed_CRC_Statistical_Tables_EN_111809.pdf, Accessed December 22, 2009.

Valmari P. (2007). Should pulse oximetry be used to screen for congenital heart disease? *Arch Dis Child Fetal Neonatal Ed. 92*(3), F219–F224.

Veale AMO, Houston IB. (1976). Administration and organization of the New Zealand screening program. In: Rubin MI, Jeljaszewicz J, Cabalska B, eds. *Proceedings of the International Symposium on Laboratory Screening Techniques for Inborn Errors of metabolism in Newborn and Selected High-Risk Infants, Warsaw: Poland September 3–6, 1972*, pp. 221–224. Warsaw: National Center for Scientific, Technical and Economic Information.

Veale AMO, Houston IB. (1978). Mass screening for inborn errors of metabolism. *Patient Management* 20–24.

Wang W, Chen X, Zang J, HE Y Zhang P, Shen F. (2003). Development of a newborn screening laboratory quality assurance system in

Shandong, China. *Southeast Asia J Trop Med Pub Health 34*(Suppl 3), 36–38.
Webster D. (1991). Why blind quality assurance – the HGSA QAP. In: Wilcken B, Webster D eds. *Neonatal Screening in the Nineties*, pp. 278–279. Leura, NSW: The Kelvin Press.
Wilcken B, Smith A, Brown DA. (1980). Urine screening for aminoacidopathies: Is it beneficial? Results of a long-term follow-up of cases detected by screening one million babies. *J Pediatr 97*, 492–497.
Wilcken B, Chalmers G. (1985). Reduced morbidity in patients with cystic fibrosis detected by neonatal screening. *Lancet 2*(8468), 1319–1321.
Wilcken B, Brown A. (1987). Screening for cystic fibrosis in the New South Wales, Australia: Evaluation of the results of screening 400 000 babies. In: Therrell BL, ed. *Advances in Neonatal Screening*, pp. 385–390. Amsterdam: Elsevier Science Publishers.
Wilcken B, Wiley V, Sherry G, Bayliss U. (1995). Neonatal screening for cystic fibrosis: A comparison of two strategies for case detection in 1.2 million babies. *J Pediatr 127*, 965–970.
Wilcken B. (1999). Neonatal Screening in Australia. *Southeast Asia J Trop Med Pub Health 30*(Suppl 2), 41–42.
Wilcken B, Wiley V, Hammond J, Carpenter K. (2003). Screening newborns for inborn errors of metabolism by tandem mass spectrometry. *N Engl J Med 348*, 2304–2312.
Wilcken B, Gaskin K. (2007). More evidence to favour newborn screening for cystic fibrosis. *Lancet 369*, 1146–1147.
Wiley V, Carpenter K, Wilcken B. (1999). Newborn screening with tandem mass spectrometry: 12 months' experience in NSW Australia. *Acta Paediatr Suppl 432*, 48–51.
Wong HB. (1965). Singapore kernicterus – the position in 1965. *J Singapore Paediatr Soc 7*, 35–43.
Wong HB. (1975). A surveillance system to prevent kernicterus in Singapore infants. *J Singapore Paediatr Soc 17*, 1–9.
Xi YS. (1994). Newborn screening in Beijing China. In: Takasugi N, Naruse H, eds. *New Trends in Neontal Screening*, pp. 19–21. Sapporo: Hokkaido University Press.
Yamaguchi A, Fukushi M, Mizushima Y, Sato Y, Takasugi N. (1987). A simple method for quantification of biotinidase activity in dried blood spot and its application to screening for biotinigase deficiency. In: Therrell BL, ed. *Advances in Neonatal Screening*, pp. 319–321. Amsterdam: Elsevier Science Publishers.
Yamaguchi S. (2008). Newborn Screening in Japan: Restructuring for the New Era. *Ann Acad Med Singapore 37*(Suppl 3), 13–17.
Yeo PP, Joseph R, Chng KP, Chua D, Thai AC, Ng CSA, et al. (1983). Screening programme for congenital hypothyroidism in Singapore. In: Naruse H, Irie M, eds. *Neonatal Screening*, pp. 113–114. Amsterdam: Elsevier Science Publishers.
Ying D, Chen R, Shen Y, Ye J, Huang X. (1996). Neonatal screening in Mainland China: Current status and future plans and proposals. In Lam S, Pang C, eds. *Neonatal and Perinatal Screening*, pp. 21–23. Hong Kong: The Chinese University Press.

62.

CLINICAL DYSMORPHOLOGY AND BIRTH DEFECTS REGISTRY IN THE ASIA-PACIFIC REGION

Meow Keong Thong

INTRODUCTION

The Global Burden of Disease Project showed birth defects as one of the leading causes of disability (Murray and Lopez, 1998).

The practice of assessing a population's health from the mortality data may lead to misallocation of health resources and underestimate the needs of patients with chronic disabilities (Modell and Kuliev, 1998). Congenital anomalies or birth defects are one of the most common causes of disability in developed and developing countries. Data on birth defects from population-based studies originating from developing countries in the Asia-Pacific region are lacking (Penchaszadeh, 2000). The March of Dimes estimated 7.9 million infants are born each year with a serious birth defect; of these births, 94% occur in middle- and low-income countries (Christianson et al., 2006). The estimated birth prevalence ranges from 40/1000 live births (high-income countries) to 82/1000 live births in low-income countries (WHO-MOD, 2006). Annually, over 3.3 million children under age five die from birth defects. This chapter deals with birth defects and related health issues in the Asia-Pacific region, which includes countries in East Asia and Southeast Asia with a combined population of two billion, accounting for one-third of the global population.

In many developing countries in the Asia-Pacific region, many genetic tests and chromosome studies are not available routinely or easily. Therefore, the main diagnostic tool in clinical dysmorphology in developing countries rests on clinical evaluation (Christianson and Modell, 2004). Often this leads to the second major issue in clinical dysmorphology in developing countries— the paucity of clinical photographs and information on dysmorphology in the diverse population in the Asia-Pacific region. The majority of patient photographs in dysmorphology databases and textbooks contain patient data largely derived from Western or Caucasian populations (Hall et al., 1989; Jones, 1997; Gorlin et al., 2001). Therefore, subtle dysmorphic features in Asians are often difficult to interpret, and many dysmorphologists and clinical geneticists in this region often have to rely on personal experience and clinical acumen to make the syndromic diagnosis (Thong et al., 2005). Furthermore, in addition to the limited expertise in the field of clinical dysmorphology, there are limited resources for investigations and management of these conditions.

A major birth defect (also known as congenital disorder) is an abnormality of prenatal origin that, if uncorrected or uncorrectable, significantly impairs normal physical or social function or reduces normal life expectancy. This may be structural defect, i.e., related to problems of the body parts or functional birth defects, such as disorders involving the central nervous system (learning and behavioral difficulties, speech or language disorders, seizures and movement disorders), inborn errors of metabolism, and degenerative conditions. The etiology for birth defects is heterogeneous (ICBDSR, 2007; Table 62-1). A minor birth defect is an abnormality of prenatal origin that is present at birth and does not have surgical, medical, or cosmetic importance at the time of examination. Examples are preauricular pits and umbilical hernia. The presence of three or more minor birth defects may suggest the possibility of a syndrome. (Leppig et al., 1987)

Multiple birth defects are found in 1% of all newborns and the majority are associated with dysmorphic features. Of these, about 30%–40% can be diagnosed as having a specific or recognizable syndrome, while the remainder have unknown entities that will require more time and further testing before they may be delineated (Figure 62-1). The study of human birth defects or congenital malformations, in particular with its emphasis on structural abnormalities of development, was first known as *dysmorphology* as described by Dr. David Smith (Jones, 1997). Clinical dysmorphology refers to the study and care of individuals with multiple birth defects and their families. There are nearly 300 well-described syndromes in *Smith's Recognizable Patterns of Human Malformation* (Jones, 1997) and close to

TABLE 62-1 ETIOLOGY AND EXAMPLES OF BIRTH DEFECTS

Chromosomal (5%) Unbalanced chromosomal rearrangements, e.g., duplication, deletion, trisomy, or monosomy will result in severe structural or developmental defects. Examples: Trisomy 21 (Down syndrome), trisomy 18 (Edward syndrome), trisomy 13 (Patau syndrome), chromosome 22q11 deletion (DiGeorge syndrome), monosomy X (Turner syndrome)
Single Gene Defects (10%) Some may be isolated involving one organ body system, while others may present with multiple birth defects or syndromal conditions. Examples: *Isolated*: alpha-thalassemia (AR), polydactyly (AD), infantile polycystic kidney disease (AR), osteogenesis imperfecta (AR/AD/X-linked), G6PD deficiency (X-linked) *Multiple/syndromal*: Williams syndrome (elastin gene), neurofibromatosis (neurofibromin gene), Marfan syndrome (fibrillin gene)
Multifactorial (30%) Occurs as a result of interplay between genetic predisposition or susceptibility and environmental factors. Empiric risks are derived from large family studies to provide the likelihood of recurrence for siblings and future offspring. Examples: Most of these birth defects are isolated: Congenital heart disease (most common birth defects), neural tube defects, such as spina bifida and anencephaly (association with folic acid deficiency), cleft lip/palate (exposure to tobacco products in utero) and talipes equinovarus.
Environmental (5%–10%) A teratogen is an agent (chemical, physical, or biological) that causes a birth defect by interfering with normal embryonic or fetal development in the first trimester. Examples: **Drugs/Chemical** Thalidomide, retinoids, valproic acid, alcohol, warfarin. **Biological** Infective agents and microbials such as TORCHES organisms. Maternal illnesses such as lupus, insulin-dependent diabetes mellitus **Physical** Ionizing radiation and prolonged maternal hyperthermia.
Unknown (50%) Many birth defects do not have a readily identifiable cause. Some of these may be due to submicroscopic chromosomal changes, genomic imprinting errors, mosaicism, trinucleotide repeat disorders, and new dominant conditions. It is expected that more causes will be identified in the near future. Examples: *Isolated*: imperforate anus, diaphragmatic hernia, facial clefts, hypospadia, hydrocephalus, microcephaly. *Multiple*: VACTERL association, MURCS association.

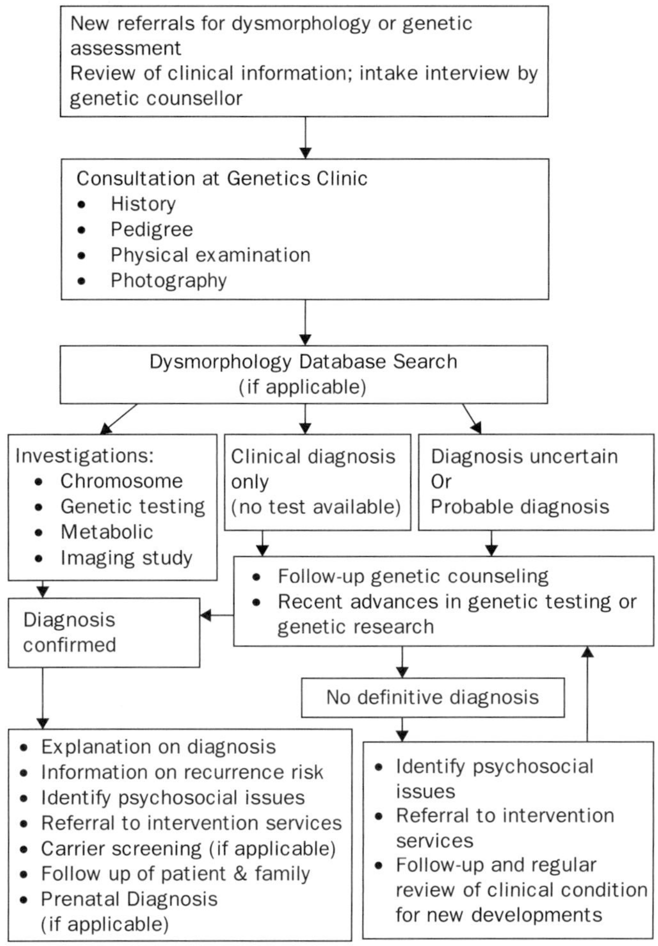

Figure 62-1 Approach to Dysmorphology Consultation—Flowchart.

2000 unique syndromes in dysmorphology databases, such as "Pictures of Standardized Syndromes and Undiagnosed Malformations" (POSSUM; www.possum.net.au) and the Baraitser-Winter dysmorphology databases (www.lmdbases.com). Unfortunately, while anecdotal publications and case reports of various rare and unique syndromes from the Asia-Pacific regions are available, many of these are published not in the English language. In addition, there are relatively few publications featuring larger series of patients with confirmatory or mutation analyses. Long-term and outcome studies and their relationship to health economics are also uncommon.

Delineation of a syndrome diagnosis is important because the natural history, prognosis, treatment options, screening for possible complications, inheritance pattern, risk of recurrence, and reproductive options for the family can be provided during genetic counseling. Not having a definitive syndromic diagnosis can be distressing for some individuals or families, and may result in constant searching for information or specialists to "confirm" the diagnosis.

Assigning an erroneous diagnosis can have adverse outcome—information provided will be inappropriate, particularly the natural history, treatment implications, recurrence risk, and prenatal testing. This may lead to serious negative implications; for example, offering palliative care to an infant wrongly thought to have a "lethal" skeletal dysplasia. Undoing a misdiagnosis or reassigning another diagnosis at a later time may be difficult for the family, and can have devastating consequences—both psychologically and socially.

It is appropriate to inform families that no specific or definitive syndromic diagnosis could be made, when the diagnosis is not certain or not proven by tests or investigations. Early intervention programs and surgical management can be performed for specific problems, and empiric genetic counseling provided even without an overall definitive diagnosis.

The clinical approach to assessment in clinical dysmorphology in developing countries should include the following:

BOX 62-1. I. HISTORY

- Pedigree (family tree) analysis. This is crucial, as many Asian families are large with frequent polygamous and polyandric relationships.
- Paternal and maternal age at conception.
- Consanguinity. This is culturally acceptable in many Asian cultures, including arranged marriages, and therefore this history must be taken with great care and sensitivity.
- Family history of inherited conditions. Often, accurate information on this aspect is not available as medical records are incomplete and postmortems are not frequently performed.
- Obstetric history, e.g., abortions, intrauterine growth restriction.
- Exposure to teratogens, e.g., infective agents, substance use, drugs.

BOX 62-2 II. ANTHROPOMETRY

- Weight, height and occipitofrontal circumference
- Body segments and limb length discrepancies
- Other facial and body measurements

BOX 62-3 III. PHYSICAL EXAMINATION

- All measurements must be plotted on appropriate normograms.
- A clear description of all the facial features and all major and minor anomalies must be made, using the appropriate terminology.
- Abnormal external features, e.g., cleft lip and palate, scalp defects, absent digits, skin changes must be identified, while internal features—e.g., congenital heart defects, dislocated joints, coloboma—must be sought.
- Facial measurement and craniofacial evaluation, examples using direct facial anthropometric measurements, photo anthropometric methods, and cephalometrics are used in many research environments. However, these techniques are not easily available for routine use and require considerable expertise.
- Photographic or videographic documentation is important. Ask for old family photographs or portraits. Serial photography may reveal mild but definite changes over a period of time. Photography is also important when a comparative dysmorphology search is performed using commercial databases. Standard views include the frontal and lateral shots of the face and head, whole body, hands and feet (dorsal and plantar views), and close-ups of unique findings of the skin or any feature. Parental consent should be obtained and the photographs must be labeled with the date and hospital numbers for future reference.

INVESTIGATIONS OF AN INDIVIDUAL WITH MULTIPLE BIRTH DEFECTS

There are no specific "gene tests" for all the syndromes known to man, although this list is increasing. Many clinicians trained in developing countries have limited exposure to this "new genetics," and to the use of genetic technology in the clinical field. For example, many are not aware that a "normal" report ("no mutation detected" or "no chromosomal abnormalities found") does not rule out a syndromic diagnosis. A mutation detected may be falsely regarded as "pathogenic" when they may in fact be benign (silent mutations, variants, or polymorphism). Therefore, genetic tests are performed only after genetic counseling is given by a clinical geneticist or genetic counselor. The test is performed in a laboratory experienced in the condition. Many clinicians are not aware of the limitations of genetic testing and the ethical, legal, and social issues associated with it. Often, results obtained from research projects are used in the management of patient care. Lastly, there exists an inequitable access to genetic testing in many developing countries. Many communities are unable to afford genetic testing or prenatal diagnosis, and there is a risk of individuals who are genetically handicapped becoming a mark of social class. Some of the modalities for investigations in clinical dysmorphology are listed below:

- Chromosomal study for patients with multiple birth defects
- Skeletal survey and bone age in suspected bone dysplasia
- Imaging studies: cranial ultrasonogram (infants) or magnetic resonance imaging for structural brain abnormalities, ultrasonogram for renal abnormalities, echocardiogram for cardiac defects. In many developing countries, some of these imaging facilities may not be available.

- Metabolic studies and endocrine tests, for example, thyroid function test and urinary metabolic and mucopolysaccharidosis "screen," can be performed in most developing countries. For many other inborn errors of metabolism with dysmorphism, such as peroxisomal disorders (e.g., Zellweger syndrome), lysosomal disorders (e.g., oligosaccharidosis), sterol synthesis disorders (e.g., Smith-Lemli-Opitz syndrome, Menkes disease), congenital disorders of glycosylation, and homocystinuria may require specialized laboratory testing. These tests may not be routinely available in many developing countries.
- Other investigations that are increasingly made available in developing countries include fluorescence in-situ hybridization (FISH) study in microdeletion or microduplication syndromes, hematological and genetic studies for Barts hydrops fetalis, and serological studies for intrauterine infections.
- Molecular genetic studies or mutation analyses are performed if the diagnosis has medical consequences for the patient (e.g., confirmation of Prader-Willi syndrome) or use in prenatal diagnosis for future offspring.

APPROACH TO CLINICAL DYSMORPHOLOGICAL DIAGNOSIS

There are a number of approaches that are used in making a dysmorphology diagnosis. (Aase, 1990). In developing countries, two practical approaches are often used to make a dysmorphological diagnosis.

A. FACIAL GESTALT ("SPOT" DIAGNOSIS)

Many syndromes have specific or distinctive facial appearance, thus enabling "spot" diagnosis to be done. However, this approach may be difficult in developing countries with limited expertise in clinical dysmorphology. Mild expression of facial surgery may make it difficult to diagnose the condition confidently. In making a facial gestalt diagnosis, the following general principles and caveats are useful to follow:

- Familial features or ethnic traits must be taken into consideration. Therefore, an experienced doctor familiar with the common facial features of the specific ethnic group will be helpful.
- Minor anomalies – individual congenital anomalies that have no serious medical or cosmetic effects to the patient, and that are found in up to 14%–20% of the newborn population. However, they may be indicators of abnormal morphogenesis, or clues in the diagnosis of a specific pattern of malformations, especially if three or more are found in an individual.
- Evolving phenotype – some syndromes exhibit longitudinal changes over time in the same individual, e.g., Noonan syndrome.
- Variable expressivity – variance in the extent of abnormality among individuals with the identical syndrome, e.g., severity of mental handicap or growth failure in chromosomal conditions, and in many autosomal dominant conditions, e.g., neurofibromatosis.
- Heterogeneity of the condition – modification due to genetic or environmental factors; for example, teratogenic effects of diabetic embryopathy shares many features with other phenocopies. Insulin-dependent diabetes mellitus and poorly controlled gestational diabetics are common in many developing countries.
- Lumping and splitting phenomenon – similar phenotypes arising from different etiologies (Marfan syndrome and homocystinuria), or different phenotypes from the same etiology (GLI3 mutations).

B. DYSMORPHOLOGY DATABASES

Commercial electronic dysmorphology databases, e.g., POSSUM (Pictures of Standardized Syndromes and Undiagnosed Malformations), Baraitser-Winter (previously London) Dysmorphology Database, and SYNDROC are available (Pelz et al., 1996). While these computer programs may appear easy to use, they require some experience in identifying the significant birth defects or definitive features (the "handle") as the traits selected for analysis. Once the list of differential diagnoses is generated, the user is expected to peruse the list of syndromes provided and to consider the few "most likely" diagnoses for further evaluation. They do not provide a definitive diagnosis. For the Asia-Pacific region, there are no commercial electronic dysmorphology databases that have originated from this region, although a few, such as the "Computerized Dysmorphology Database Mackay (CDDM)," have been reported in the literature (Hsu et al, 1994). In addition, many of these commercial dysmorphological databases are costly and/or require licensing fees, and these may not be affordable to many hospital users in developing countries. The users of these electronic databases are also encouraged to attend dysmorphology meetings to familiarize themselves with the use of these databases. In using these databases, the general principles for gestalt diagnosis apply:

- Familiarize yourself with the common syndromes. A good place to start is *Smith's Recognizable Patterns of Human Malformation,* or any dysmorphological publications.

- Selection of traits to be entered in the computer database should be accurate and based on unique features found in the patient.
- Features such as "learning disability" and "failure to thrive" are found in a great number of conditions. Therefore, the list of possible syndromes generated will be long and lack sensitivity.
- Conversely, even a "small" finding, e.g., cutis aplasia (scalp defect), transverse defects of limbs, and cutis marmorata will rapidly narrow the list to Adams Oliver syndrome. Similarly, the findings of hypoplastic alae nasi, malabsorption, and deafness in a child will narrow the list to Johansson-Blizzard syndrome and congenital disorder of glycosylation.
- Scan through all the photographs available on the particular syndrome. In this way, you develop a sense of the relevant "salient" features of that syndrome.
- Often, the diagnosis may not be achieved. In this case, you may wish to consult an experienced person in dysmorphology by sending a detailed history, findings, and photographs of the patient for a second opinion.

MANAGEMENT OF CHILDREN WITH "SYNDROME" DIAGNOSIS

The affected individual and family are usually referred to a genetic service for confirmation of diagnosis and to receive genetic counseling. The myriad of manifestations, prognoses and complications for each syndrome mean that the care and health and medical surveillance are individualized for each respective syndrome.

In some developing countries with antenatal ultrasonography, the diagnosis may have been made already (Ho et al., 2006). Ideally, the parents should be counseled appropriately on the condition, as well as some decision made regarding management in the postnatal period. If the condition was not known prior to the birth of the child, resuscitation of the newborn and management of immediate medical/surgical problems should be attended to. After stabilization of the patient's condition, steps must be taken to confirm the diagnosis. Referral to a genetic service may be required. Definitive diagnosis may not be made in some patients during the neonatal period or first visit. This may require regular reviews for new changes and health monitoring.

Genetic counseling should be provided, so that parents may be informed of the genetic basis of the condition; e.g., mode of inheritance, recurrence risks, burden of disease, and future reproductive options and so on. More importantly, helping parents to come to terms with the diagnosis is the most difficult step. In many developing countries, first cousin marriages are culturally acceptable. Some societies have low literacy rates and discussion with these clients have to be tailored to their understanding of the condition. In many Asian or oriental cultures, "Eastern" cultural values—such as importance of cohesion and stability of society; consensual decision-making process; doing things only if it benefits family and society; change only for the better; life concepts such as fate or karma; time being relative and non-linear—are sometimes difficult to grasp for those familiar only with "Western" or Christian-Judeo beliefs.

Palliative care should be considered for infants with severe birth defects incompatible with life, or facing futility of care, along with grief and bereavement support for neonatal death or loss of a "normal child" as well as providing psychosocial counseling and support. Rehabilitation and early intervention programs should be instituted. A referral to support groups may be helpful. The family should be provided with the appropriate resources including a listing of support organizations, for example the National Organization for Rare Diseases (NORD; www.rarediseases.org). Unfortunately, in many developing countries, these services are lacking and understaffed. Therefore long-term health and medical surveillance may be suboptimal. Reproductive options and prenatal diagnosis in future pregnancies should be discussed, and the options made available if required.

PREVENTION OF BIRTH DEFECTS IN DEVELOPING COUNTRIES

Strategies for the prevention of birth defects are well known and evidence-based (WHO-MOD Report, 2006). Many of these low-technology approaches can be adopted in developing countries easily. In some countries, prenatal diagnosis and termination of pregnancy are sensitive issues and must be handled professionally. Some of the preventive strategies are listed:

- Increase public awareness and professional education regarding birth defects
- Genetic counseling for at-risk families
- Rubella vaccination
- Population screening for genetic disease, for example, thalassemia trait
- Periconceptional folic acid supplementation 400ug daily at least one month before conception through the first 3 months of pregnancy
- Treat and control insulin-dependent diabetes and other maternal illnesses

- Early antenatal visit (first trimester) and fetal ultrasound scanning (second trimester)
- Maternal serum screening tests and diagnostic procedures for prenatal diagnosis

BIRTH DEFECTS REGISTRY

The systematic collection and analysis of data on major birth defects is an integral part of a clinical genetic service (Thong et al., 2005; Suresh et al., 2005). Epidemiological data on birth defects and risk factors from population-based studies in developing countries are urgently required. With recent advances in healthcare, many babies with major birth defects are surviving. Some live with disabilities with significant medical, psychological, and socioeconomic challenges to family and community. Up to 70% of birth defects could be prevented or treated effectively (Christianson et al., 2006). This requires the strengthening of medical genetic services in low- and middle-income countries (WHO-MOD Report, 2006).

Therefore, reliable data on the toll of mortality and morbidity due to birth defects relies on the ongoing surveillance of the types, birth prevalence, severity, and outcomes of children with birth defects. A Birth Defects Register (BDR) may alert observers to the geographic and temporal variations in the frequencies of different birth defects and possible etiological or teratogenic agents, as well as enabling them to audit the effectiveness of public health interventions. Groups that may contribute data information to the BDR include health professionals, special treatment centers, private hospitals and healthcare practices, autopsy services, diagnostic services, and medical records departments from all hospitals. The notification of birth defects may be done on a voluntary basis or under the authority of legislation. There is a need to obtain ethical approval to review patient case files from all hospitals and healthcare facilities to ensure that patient privacy is protected and confidentiality of information assured.

USES OF A BIRTH DEFECTS REGISTER

The functions of a BDR may include the following:

- To establish local prevalence rates for birth defects, so as to facilitate and determine accurate baseline incidence and detection of trends of birth defects. This is to provide basic population data on the different types of birth defects, as different communities may have different prevalence rates for the various conditions (Forrester and Merz et al., 2006).
- To monitor the occurrence of defects over time and by geographical area, to allow earlier detection of the emergence of any new teratogen and further investigations of suspected teratogens. This is particularly important in developing or industrializing countries, where new teratogenic agents may be introduced into the environment or food production. (Shi et al., 2002; Hwang et al., 2008).
- To utilize local prevalence rates to plan healthcare facilities and allocate appropriate health resources required for the design and implementation of various preventative programs to reduce the occurrence of birth defects (Tan et al., 2007). As birth defects are reportedly found in 3%–5% of the population, adequate resources must be made available for prevention, treatment, care, and rehabilitation programs, in line with the ethical tenet of distributive justice.
- To perform epidemiological studies on the causation of birth defects, and to be used as an accurate diagnostic index for clinical research. This is important to identify specific risk factors that may be reduced via various preventive programs (Li et al., 2003). The use of folic acid fortification to reduce the occurrence of neural tube defects is an example of such research.
- To train medical/health staff in the skills to recognize major birth defects within the framework of public health. In many developing countries, healthcare resources and professional manpower are limited. However, many paramedical staff may be trained to identify major birth defects, and early intervention and treatment may be provided with such an approach (Thong et al., 2005).
- To establish a referral system and information network in the acute and long-term management of babies/children with major birth defects. Similarly, logistical difficulties may be an important factor, and a BDR may be helpful in establishing a suitable referral system in the various communities.
- To facilitate access to genetic counseling services. Genetic counseling is not easily available in most health services, and a BDR may serve as a nucleus to train providers of genetic counseling, population screening, and public education. With such amenities, prenatal diagnosis for specific genetic disease can be achieved for at-risk families.
- To increase community knowledge about birth defects through education. Empowerment of the members of the public is essential to ensure success of any preventive programs.

SETTING UP A BIRTH DEFECTS REGISTER

The setting up of a Birth Defects Register requires careful and detailed planning, extensive consultation with various experts, and open collaboration with many professional groups and healthcare services. The staff of the BDR should include several specialists and professionals including clinical geneticist, pediatrician, obstetrician, representative from the maternal and child health service, epidemiologist and biostatistician, data manager, nursing staff, and clerical assistants.

It also requires assured sources of funding, as the health returns from a BDR may not be evident for many years. Frequently, experts from the public health and community genetics work closely together to design a program that will be the most suitable for the local situation. A BDR may be population-based or hospital-based. A population-based dataset is more representative of the epidemiology of birth defects, as it will include information from the private sector as well as milder forms of birth defects that may not be available from public hospitals (Tan et al., 2005). A hospital-based BDR is relatively easier to manage and low-cost, and clinical information may be easier to verify. However, it may provide a biased pattern of birth defects, particularly if the hospital is a tertiary referral center (Noraihan et al., 2005).

A BDR may need to decide whether to include all major birth defects or just sentinel birth defects. Ideally, all major birth defects should be included in a BDR. However, some major birth defects, for example structural malformation of the central nervous system, cardiovascular system, and renal tract abnormalities, may not be evident on external examination. This is particularly so in many developing countries where medical imaging services are limited. Therefore, some BDRs have advocated sentinel defects only, and these are usually birth defects that are evident at birth on inspection. The disadvantage of this is that major birth defects arising in certain populations or following a teratogenic effect may be missed, if reporting is based on sentinel findings that are not involved.

Due to logistical difficulties, some BDRs may opt to report birth defects in certain regions or localities only, rather than cover the entire population. Some BDRs may choose to recruit subjects up to births only, or include children up to a specified age only. Again, this may not provide accurate data and may miss important findings. Another important aspect is that the BDR must operate within the framework of the current health system and services. It is multidisciplinary, as it may involve the public health, maternal and child health services, pathological services, medical imaging facilities, and medical record services. A BDR is usually governmental in origin, although nongovernmental organizations may be active participants in the program. Reporting may be voluntary or legislated. A voluntary service may result in incomplete data collection and errors of ascertainment, while mandatory reporting means that government funding must be provided to ensure the program is managed to the highest standards. An important factor for a BDR to consider is to provide timely and up-to-date feedback to stakeholders in the program. This may include members of the medical profession, healthcare planners, and hospitals and services that have contributed to the BDR.

Examples of Birth Defects Register include the International Clearinghouse for Birth Defects Surveillance & Research (ICBDSR); in South America, there is the Latin American Collaborative Study of Congenital Malformations (ECLAMC), which is a hospital-based multinational group; the European Surveillance of Congenital Anomalies (EUROCAT); the Metropolitan Atlanta Congenital Defects Program in the United States; the JAOG (Japan Association of Obstetricians & Gynecologists), and the BDSS (Birth Defects Surveillance System) in Beijing, China (WHO/ICBD/EUROCAT, 2003).

LIMITATIONS OF A BIRTH DEFECTS REGISTER

It is to be expected that there will be a number of difficulties and limitations arising from the planning and maintenance of a BDR. Many medical practitioners are bewildered by the various terms used in a BDR. For example, the terms "congenital abnormalities," "congenital disorders," "congenital anomalies or malformations" and "birth defects" have been used interchangeably in some centers, although specific definitions are available for all the above. The March of Dimes uses "birth defects," and these include abnormalities in structure or function, including metabolism, which are present from birth. The World Health Organization (WHO) Human Genetics Program (HGN) prefers "congenital disorder," which is defined as any potential pathological condition arising before birth, including all disorders caused by environmental, genetic, or unknown factors, whether they are evident at birth or become manifest later in life. The WHO Burden of Disease Unit uses "congenital anomalies" defined as macroscopic morphological anomalies present from birth. This excludes functional birth defects, disability, common single gene disorders such as cystic fibrosis and glucose-6-phosphate dehydrogenase deficiency, and inborn errors of metabolism. In a recent report of a joint WHO-March of Dimes meeting in 2006, it was decided that the term *birth defect* is synonymous with the term *congenital disorder* as defined and used by the HGN (WHO-MOD, 2006).

Most BDRs were established to *monitor* birth defects, although *surveillance* of birth defects is preferred, as it denotes an active process of data collection, identification of risk factors, with possibly intervention to prevent birth defects. Another difficult issue has been ascertainment of infants with multiple birth defects. This may require the assessment of a clinical geneticist to make a syndromic diagnosis, or to perform the appropriate investigations. Increasingly, prenatal diagnosis and termination of pregnancies are performed in many communities. This may reduce the births of children with birth defects, but the true prevalence remains the same. Hence, many BDRs include termination of pregnancies in the data collection. Unfortunately, data collection and ascertainment of diagnosis may be made more difficult in this situation.

There have been few studies on the cost-effectiveness of BDRs, and long-term benefits are uncertain (Thong et al, 2005). Many medical or healthcare planners are reluctant or unable to commit public funds for medical registers due to the lack of definitive evidence of cost-effectiveness. There may be underutilization of the data in the BDR due to lack of expertise or research funding, for example, to maintain a teratogen warning system. There is a need to increase public and professional awareness on birth defects and to facilitate research collaboration, locally and regionally. Collaboration with the public health sector, or "public health genetics," may result in specific targeted preventive programs for birth defects. It is expected that with these developments, the issue of stigmatization and discrimination of individuals with birth defects may be further reduced.

FOLLOW-UP AND OUTCOME

In the Asia-Pacific region, there are a number of established BDRs (ICBDSR, 2007). The Japan Association of Maternal Welfare (later known as Japan Association of Obstetricians and Gynecologists [JAOG] in 1994) and the Chinese Birth Defects Monitoring Network (CBDMN) programs began in 1972 and 1984, respectively, and became full members of the International Clearinghouse for Birth Defects Surveillance and Research (ICBDSR) in 1988 and 1987, respectively. The JAOG program is based on reports from 270 hospitals throughout Japan, representing 9% of Japanese births. Stillbirths of 22 weeks or more gestation are included. The BDR is a research program acknowledged by the Ministry of Welfare and supported by JAOG. Reports are obtained from delivery units and pediatric clinics, and detailed information on exposure to various teratogens such as drugs, X-ray, and viral infections are included. The top five birth defects were cleft lip and palate, Down syndrome, small intestinal atresia/stenosis, hydrocephaly, and anorectal atresia.

Meanwhile, the CBDMN was first known as Chinese Birth Defects Program of Sichuan Province, China, but this was changed to Chinese Birth Defects Monitoring Network in 1994. The program is hospital-based and covers 260,000 births annually in 31provinces. Participation is voluntary, while funding is from local health authorities and the Chinese Ministry of Health. Reports are obtained from delivery units, pediatric clinics, and pathology departments of the participating hospitals. Maximum age of diagnosis is 7 days, while stillbirth is defined as 28 weeks gestation. The main birth defects were cleft lip and palate, polydactyly, limb reduction defects, spina bifida, and hypospadia. In contrast, the Birth Defect Surveillance System in Thirty Counties of Four Provinces, People's Republic of China (BDSS-Beijing) which began in 1992, became a full member of the ICBDMS in 1997. The program is a population-based monitoring system. Reports were obtained from all hospitals and village health stations, which together cover all geographically defined populations. Total population in these areas is around 17 million and total number of births per year is around 150,000. Funding is from China's Ministry of Health and local health authorities, while the sources of ascertainment are from delivery units, pediatric clinics, ultrasound departments, pathology departments, and perinatal healthcare departments of different-level hospitals, MCH institutes, and village health stations in the participating counties and cities. Exposure information is obtained from the perinatal healthcare surveillance system (PHCSS) in the same areas for all women and their babies, from premarital examination till six weeks after birth. BDSS data is linked with PHCSS data by using an ID number assigned to each woman. The main defects reported were cleft lip and palate, polydactyly, hydrocephaly, microtia, and anaencephaly.

In Southeast Asia, only Singapore has a national BDR, which was established in 1992. It is population-based and funded by the government. It collects all major and minor birth defects but not inborn errors of metabolism. The common birth defects are congenital heart defects, cleft lip and palate, gastrointestinal defects, neural tube defects, and limb defects. A number of publications have been reported in the literature (Tan et al., 2005, 2007). In Malaysia, several hospital-based birth defects registers were started. More than 20% of medically certified infant deaths in Malaysia were classified as due to birth defects in 1990. As the data collected were hospital-based, a pilot study to set up a population-based BDR using research grants was initiated in 2000. In Malaysia, a collaborative project between the University of Malaya, Royal College of Medicine Perak and the Ministry of Health of Malaysia was performed at the Kinta district, with the base located at the Ipoh Hospital (Thong et al, 2005). The aims of the project were to ascertain the epidemiology of major birth defects in Malaysian births using a population-based birth defects register, and to identify adverse factors associated with major birth defects.

All major birth defects in infants weighing 500 g (or more) or infants delivered at 22 weeks of gestation till one week of life were included. Details of the study have been published (Thong et al., 2005; Ho et al., 2006). The results showed the birth prevalence of birth defects were one in seventy births, and several risk factors were identified. It recommended prenatal and antenatal screening for insulin-dependent diabetes, genetic counseling, investigating causes for previous abortions, and public education on avoidance of teratogens. It also indicated that periconceptional folic acid supplementation or food fortification to reduce birth defects required further investigation. Of concern was that only 15% of the birth defects were detected during the antenatal period (Ho et al., 2006). The main birth defects were chromosomal disorders, congenital heart disease, cleft lip and palate, club feet, and central nervous system defects.

The population-based study was the first such study in Malaysia. Useful information on the baseline incidence of birth defects and their associated risk factors were identified. The approach and algorithm for the recognition, reporting, and management of birth defects by paramedical staff were evaluated. A dysmorphology database was maintained. The Malaysian government announced the fortification of flour with folic acid in 2008. As a long term follow-up, a section on birth defects was included in the Malaysian National Neonatal Registry, beginning from 2005. The preliminary data on the project is awaited.

In the Philippines, several pilot projects on congenital anomalies were started (Padilla et al., 2003). In 2008, a Birth Defects Surveillance Project was announced and will be coordinated from the Institute of Human Genetics, National Institutes of Health, with funding provided by the Department of Health, University of Philippines, and March of Dimes Foundation. It will be both hospital-based and community-based, and only physical defects will be included. The major birth defects found in the pilot studies were ankyloglossia, multiple birth defects (not elsewhere classified), cleft lip and palate, talipes equinovarus, and anencephaly. In Thailand, there is no BDR yet (Sirinavin, 1995). Plans are afoot to commence pilot studies consisting of 3-year retrospective study of birth defects in 2–3 medical schools, as well as prospective study in major provincial hospitals for 2 years, followed by expansion to 76 provincial hospitals in the third year of study with integration into a national policy in 2011. This project will be undertaken by the Ministry of Public Health and supported by professional societies such as the Pediatric Genetics Society of Thailand, The Perinatal Society of Thailand, The Royal College of Obstetricians and Gynecologists of Thailand, and The Royal College of Pediatricians of Thailand (Wasant, personal communication).

CONCLUSIONS

Clinical dysmorphology is an emerging medical discipline in the Asia-Pacific region. An accurate dysmorphological diagnosis is crucial for appropriate management and genetic counseling, as well as to institute preventive strategies. The major difficulties have included lack of expertise in this field, low level of awareness on the importance of birth defects by public and medical professionals, inadequate clinical and laboratory resources, as well as various cultural and social issues that have an impact on provision of clinical genetics services in the region. The systematic collection and analysis of data on major birth defects (Birth Defects Register) is an integral part of any clinical genetic service. It is essential to the planning and developing of care and prevention services at local and national levels. Increasingly, the surveillance of birth defects has shaped up to become a major thrust of public health genetics. A Birth Defects Register must function in a systematic and scientific manner, with clearly defined objectives, methodology, and adherence to ethical principles. Awareness of its uses and limitations are essential in the planning of the Register.

ACKNOWLEDGMENTS

The Ministry of Science, Technology and Innovation, Malaysia (IRPA grant number: 06-02-03-0586) and all collaborators in the Malaysian Kinta District Birth Defects Study, including the Ministry of Health of Malaysia, Ipoh Hospital, Royal College of Medicine Perak and, in particular, Dr. Jackie Ho, Dr. Khatijah Nurani, and Dr. Amar Singh. The author would also like to thank Dr. Carmencita Padilla (Institute of Human Genetics, NIH Manila Philippines), Dr. Angeline Lai (KK Women's and Children's Hospital, Singapore) and Dr. Pornswan Wasant (Sriraj Hospital, Mahidol University, Bangkok, Thailand) for their input.

REFERENCES

Aase JM. (1990). *Diagnostic Dysmorphology.* Plenum Medical Book Company, New York.

Christianson AL, Modell B. (2004). Medical genetics in developing countries. *Annual Reviews in Genomics & Human Genetics* 5:219–265.

Christianson AL, Howson CP, Modell B. (2006). The hidden toll of dying and disabled children. *March of Dimes Global Report on Birth Defect* (pp. 2–4). White Plains, New York: March of Dimes Birth Defects Foundation.

Forrester MB, Merz RD. (2006). Rates for specific birth defects among offspring of Japanese mothers, Hawaii, 1986–2002. *Congenit Anom (Kyoto)* 46(2):76–80.

Gorlin RJ, Cohen MM Jr, Hennekam RCM. (2001). *Syndromes of the Head and Neck.* 4th Edition, Oxford University Press, New York.

Hall JG, Froster-Iskenius UG, Allanson JE. (1989). *Handbook of Normal Physical Measurements.* Oxford Medical Publications, New York.

Ho JJ, Thong MK, Nurani NK. (2006). Prenatal detection of birth defects in a Malaysian population: Estimation of the influence of termination of pregnancy on birth prevalence in a developing country. *Australian & New Zealand J of Obstetrics & Gynaecology* 46 (1):55–57.

Hsu CH, Lin SP, Huang FY. (1994). Computerized Dysmorphology Database Mackay (CDDM): preliminary report. *Zhonghua Min Guo Xiao Er Ke Yi Xue Hui Za Zhi.* 35(1):19–26.

Hwang BF, Jaakkola JJ, Guo HR. (2008). Water disinfection by-products and the risk of specific birth defects: a population-based cross-sectional study in Taiwan. *Environ Health* 7:23–27.

ICBDSR [International Clearinghouse for Birth Defects Surveillance and Research]. (2007). *Annual Report 2007 with data for 2005.* Rome: The International Centre on Birth Defects.

Jones KL. (ed.) (1997). *Smith's Recognizable Patterns of Human Malformations,* 5th Edition. W.B. Saunders Co., Philadelphia.

Leppig KA, Werler MM, Cann CI, et al. (1987). Predictive value of minor anomalies: association with major malformations. *J Pediatr.* 1120:531–537.

Li S, Moore CA, Li Z, et al. (2003). A population-based birth defects surveillance system in the People's Republic of China. *Paediatr Perinat Epidemiol.* 17(3):287–293.

Murray CJL, Lopez AD. (1998). Global Burden of Diseases and Injury Series. *Volume 1: The Global Burden of Disease.* Boston: Harvard School of Public Health.

Modell B, Kuliev A. (1998). The history of community genetics: the contribution of the haemoglobin disorders. *Community Genetics* 1998: 1:3–11.

Noraihan MN, See MH, Raja R, Baskaran TP, Symonds EM. (2005). Audit of birth defects in 34,109 deliveries in a tertiary referral center. *Med J Malaysia.* 60(4):460–468.

Padilla- CD, Cutiongco EM, Sia JM. (2003). Birth defects ascertainment in the Philippines. *Southeast Asian J Trop Med Public Health.* 34(Suppl 3):239–243.

Pelz J, Arendt V, Kunze J. (1996) Computer assisted diagnosis of malformation syndromes: an evaluation of three databases (LDDB, POSSUM, and SYNDROC). *Am J Med Genet.* 63(1):257–267.

Penchaszadeh VB. (2000). Delivery of Genetic Services in Developing Countries. In: Khoury MJ, Burke W, Thompson EJ, eds. *Genetics and Public Health in the 21st Century: Using Genetic Information to Improve Health and Prevent Disease.* New York: Oxford University Press.

Suresh S, Thangavel G, Sujatha J, Indrani S. (2005). Methodological issues in setting up a surveillance system for birth defects in India. *Natl Med J India.* 18(5):259–262

Shi LM, Chia SE, Chan OY, Chew SK, Foong BH. (2002). Prevalence of birth defects and parental work in Singapore live births from 1994 to 1998: a population-based study. *Occup Med (Lond).* 52(6):325–331.

Sirinavin C. (1995). Medical genetics in Thailand. *Southeast Asian J Trop Med Public Health.* 26(Suppl 1):26–33.

Tan KH, Tan TY, Tan J, Tan I, Chew SK, Yeo GS. (2005). Birth defects in Singapore: 1994–2000. *Singapore Med J.* 46(10):545–552.

Tan KB, Tan SH, Tan KH, Yeo GS. (2007). Anencephaly in Singapore: a ten-year series 1993–2002. *Singapore Med J.* 48(1):12–15.

Thong MK, Ho JJ, Noor Khatijah N. (2005). A population-based study of birth defects in Malaysia. *Annals of Human Biology.* 32 (2): 180–187.

WHO/ICBD/EUROCAT. (2003). *World Atlas of Birth Defects,* 2nd Edition. Geneva: World Health Organization.

WHO-MOD. (2006). Report of a Joint WHO-March of Dimes Meeting: *Management of Birth Defects and Haemoglobin Disorders.* Geneva: World Health Organization.

63.

PREVENTION AND CONTROL OF THALASSEMIA IN THAILAND

Chanane Wanapirak, Jintana Pattanapongthorn, Pranee Winichagoon, and Suthat Fucharoen

INTRODUCTION

Thalassemia refers to a group of inherited blood disorders. They occur most commonly among Asian people, especially of Southeast Asian, Southern Asian, and Mediterranean ancestry. The fundamental abnormality in thalassemia is impaired production of the globin chain. Thalassemia can be classified into α- and β-thalassemias by their clinical manifestations and their genetic background. Children with β-thalassemia major appear healthy at birth, but during the first decade of life they become pale, listless, and fussy with poor appetite. They grow slowly and have persistent jaundice. Without proper treatment, the spleen, liver, and heart become greatly enlarged. Bones turn thin and brittle, and face bones become distorted. Heart failure and infection are the leading causes of death among untreated patients.

Complications of thalassemia can be prevented when children with thalassemia major are treated with regular blood transfusions to keep their hemoglobin level around 10 g/dl. This "hypertransfusion" method enhances the child's growth, and prevents heart failure and bone deformities, with better quality of life. Unfortunately, regular blood transfusions may lead to certain complications, mostly related to blood safety issues, especially blood-transmitted diseases such as hepatitis, AIDS, malaria, and syphilis. Moreover, frequent blood transfusions also lead to an accumulation of iron in the body, which can damage the heart, liver, and other organs. Iron chelator helps the body to eliminate excess iron and prevent or delay problems related to iron overload. Desferrioxamine is the only iron chelator used during the last 40 years. It is usually administered daily via a mechanical pump that administers the drug subcutaneously while the patient is sleeping, for 10–12 hours at a dose of 40 mg to 60 mg/kg/day, at least 5 days/week, in patients on regular blood transfusion.

Patients with thalassemia major who are treated with regular blood transfusions and iron chelation live longer and are spared much of the suffering endured by untreated patients. During these last few years, there have been many research initiatives for oral iron chelators. Deferiprone, L1, is the first oral iron chelator to be widely used in many countries. Deferasirox, ICL670, is a new oral iron chelator that has been recently approved by the U.S. Food and Drug Administration. These new oral iron chelators permit better quality of life because the patient does not have to undergo long hours of subcutaneous injection every day. The long-term efficacy and side effects need careful monitoring and evaluation.

Thalassemia may be cured by stem cell transplantation. However, this treatment is possible only for a minority of patients who have a suitable HLA-matched donor and can afford the costly treatment. The transplant procedure is risky and some complications may occur. Some patients benefit from splenectomy, especially those with very large spleen and some evidence of hypersplenism.

The other treatments for thalassemia are symptomatic and supportive care for infection, osteoporosis, and other endocrine complications. While conventional therapy involving blood transfusions and iron chelation are intolerably expensive, prenatal diagnosis and selective abortion are more economical. However, these should not be carried out indiscriminately. This approach points to the need of accurate diagnosis and good understanding of the disease. It is time for international health organizations such as the World Health Organization to consider thalassemias as the international public health problems also in need of good health planning, if health for all is to be achieved.

THALASSEMIA IN THAILAND

The types and frequencies of thalassemias are heterogeneous in Thailand. Both α-thalassemia and β-thalassemia and some abnormal hemoglobins (Hb), such as Hb E and Hb Constant Spring, are prevalent in Thailand. The frequencies are 20% to 30% for α-thalassemia, 3% to 9% for β-thalassemia, 10% to 50% for Hb E, and 1% to 8% for Hb Constant Spring (Wasi et al., 1969; Wasi 1981; Fucharoen and Winichagoon1987).

These abnormal genes in different combinations lead to over 60 thalassemia syndromes. Different thalassemia genotypes have greatly variable severity; even patients with apparently identical genotypes can have remarkably different severity. The two major alpha-thalassemic diseases are Hb Bart's hydrops fetalis, or homozygous α-thalassemia 1 (homozygous α^0-thalassemia), and Hb H disease that occurs from the interaction between α-thalassemia 1 (α^0-thalassemia) gene and α-thalassemia 2 gene (α^+-thalassemia), or between α^0-thalassemia 1 gene and Hb CS gene. Almost all Hb Bart's hydrops fetuses die intra-utero or a few minutes after birth, and their mothers often suffer from obstetrics complications such as preeclampsia, dystocia, postpartum hemorrhage due to an enlarged placenta, and psychological burden of carrying a nonviable fetus to term. Interaction between β-thalassemia genes or β-thalassemia and Hb E genes leads to homozygous β-thalassemia and β-thalassemia/Hb E disease that are major beta-thalassemic syndromes in this region. In β-thalassemia/Hb E disease, although the patients have identical genotype, the degree of anemia varies greatly with hemoglobin levels ranging from 3–13 g/dl (Winichagoon et al., 2000).

Due to the large number of thalassemic patients and limited medical service resources, it is not possible to give optimal blood transfusions and iron chelating agents to the majority of patients. The estimated direct cost for the management of ONE thalassemia major patient who lives until 10–30 years old is about 1.3 to 6.6 millions Baht (US $39,394–$200,000; Table 63-1). The majority of thalassemia major patients receive no or minimal blood transfusions and no iron chelators because it is too expensive. These masses of untreated thalassemic patients develop a multitude of complications as previously described. Although we are prepared to use all available cures, they are either not yet available or available to the very few. Moreover, we need to develop human resources and infrastructure to cope with the sheer number of patients. We are thus left to face taking care of a large number of patients with complications not now seen in the West. The best approach to cope with thalassemia in developing countries, including Thailand, is to prevent birth of new cases with major thalassemic disease.

PREVENTION AND CONTROL PROGRAM FOR THALASSEMIA IN THAILAND

Effective prevention programs for thalassemia have been demonstrated in many countries where carriers of abnormal genes are very high, such as Cyprus, Greece, and Italy. Premarital screening for thalassemia is standard practice and national audit data are available. Most at-risk couples are identified early for prenatal diagnosis in the first pregnancy, and the majority uses this service and produces healthy offspring. Screening programs need to be supported by public education and regulatory structures empowering individuals to make informed decisions and to ensure that people are protected against discrimination as a result of their test results (Angastiniotis et al., 1995). The thalassemia carriers are detected mainly through the expanded family study of cases with thalassemic diseases. Laboratory progress during these last few years has enabled us to do mass screening for beta-thalassemia and Hb E carriers with accurate diagnosis by using the automatic high-performance liquid chromatography (HPLC) system (Angastiniotis et al., 1995). In recent years, a nationwide program has commenced to prevent and control homozygous β-thalassemia, compound heterozygote β-thalassemia/Hb E, and Hb Bart's hydrops fetalis, with encouraging results.

Recently, Wanapirak et al. (2004) reported the prevalence of thalassemia in pregnant women in northern area of Thailand, with the overall prevalence of thalassemia trait (heterozygote) being 25.4%; those classified as α-thalassemia 1 trait were 6.6%, β-thalassemia trait 3.7%, Hb E trait 11.6%, homozygous hemoglobin E 0.8%. The combination of α-thalassemia 1 trait and β-thalassemia trait was 1.2%, and combination of α-thalassemia

TABLE 63-1 ESTIMATED DIRECT COST FOR THE MANAGEMENT OF ONE BETA-THALASSEMIA MAJOR PATIENT FOR 30 YEARS (NUMBER IN THAI BAHT, US$1 = 33 BAHT).

Cost (Thai Baht)	1–10 years	11–20 years	21–30 years
1. Blood transfusion plus filter/month	2,500 (1 unit)	5,000 (2 units)	7,500 (3 units)
2. Desferrioxamine Vial/month	5,000 (20 vials)	10,000 (40 vials)	15,000 (60 vials)
3. Hospital care 1 day/month	2,000	2,000	2,000
4. Other: Lab tests/month	1,000	1,000	1,000
5. Total cost/month	10,500	18,500	26,500
6. Total cost/year	126,000	222,000	318,000
7. Total cost/10 years	1,260,000	2,220,000	3,180,000

1 and hemoglobin E traits were 1.5%. They also found β-thalassemia/hemoglobin E disease at 0.2%. This high prevalence of thalassemia trait leads to many births of severe thalassemia, as previously described. Before the prevention and control program was initiated in 1994, our department encountered about 20 cases of homozygous β-thalassemia, 30–40 of β-thalassemia/Hb E, and 20–30 of Hb Bart's hydrops each year, from total births of about 6000–8000. Therefore, these three severe thalassemia syndromes need to be controlled, especially with a prenatal approach.

CHIANG MAI STRATEGY FOR THE PREVENTION AND CONTROL OF THALASSEMIA

Concerning the prenatal approach to prevention and control of new cases of severe thalassemia syndrome, several studies on fetal DNA analysis have been reported (Fucharoen et al., 1991; Lee et al., 1998; Liu et al., 1997; Yong et al., 1999). The major advantage of this approach is the possibility of an early prenatal diagnosis from 10–12 weeks of gestation onward, by analyzing chorionic villi or amniocytes. However, there are some drawbacks, such as the requirement of considerable technical effort, high cost, and the impossibility of wide availability. To date, fetal DNA analysis has not been proven to be effective in prevention of severe thalassemia in larger areas, especially in developing countries, whereas delayed first visits of pregnant women to a doctor are yet another problem, apart from the economic issues. Since 1998, the Chiang Mai team developed a more feasible method and have had great success in control of severe thalassemia, simply and at much lower cost than the fetal DNA analysis approach, and has modified the original strategy to be more accurate and effective (Wanapirak et al., 1998). The strategy includes:

1. Genetic counseling
2. Identification of pregnancies at risk by:
   - 2.1 Retrospective screening (history review of known risk)
   - 2.2 Prospective screening for asymptomatic women

   Screening Test

   2-minute OF (2-min. osmotic fragility test)

   HbE screening test

   Diagnostic Test (if both partners of the couple are positive according to the screening test)

   HbA2 level and PCR for α-thal 1 if 2-min. OF is positive

   HbA2 level (negative 2-min. OF but positive HbE)
3. Prenatal diagnosis for pregnancy at risk
   - 3.1 Prenatal counseling
   - 3.2 Cordocentesis at 16–22 weeks' gestation
   - 3.3 Fetal blood analysis with high-performance liquid chromatography (HPLC)
4. Termination of the affected pregnancy

The prospective screening consisted of 2-minute osmotic fragility (OF) and Hb E screening tests in women with no risk, and testing the husbands of the women with a positive result. Subsequently, the OF test was replaced by mean corpuscular volume (MCV) when the automated cell counters became available nationwide. If both partners of the couple had a positive result, the further diagnostic test (HbA2 level and PCR for α-thalassemia1 genotype) for the carrier was needed. A pregnancy in which both partners of the couple were carriers was considered a couple at risk (CAR), and further in-detail counseling and cordocentesis was offered for prenatal diagnosis of the severe thalassemia syndrome (Figures 63-1 and 63-2; Tongsong et al., 2000).

Since the program was started the number of new cases of thalassemia at Chiang Mai Medical School is gradually decreasing as shown in Figure 63-3 and Table 63-2 (Tongsong et al., 2001; Maternal Fetal Medicine Unit, 1994–2008).

In the first few year of program, we evaluated the cost-effectiveness of the program and found that among the total pregnant population of 21,000 individuals that we screened, 80 affected fetuses had been found and terminated. The total cost of the prevention program was about US$257,140, and the cost of management of these affected cases, if they had been born, would be US$7,200,000. The cost-benefit ratio is 1:28, which is a very highly cost-effective project.

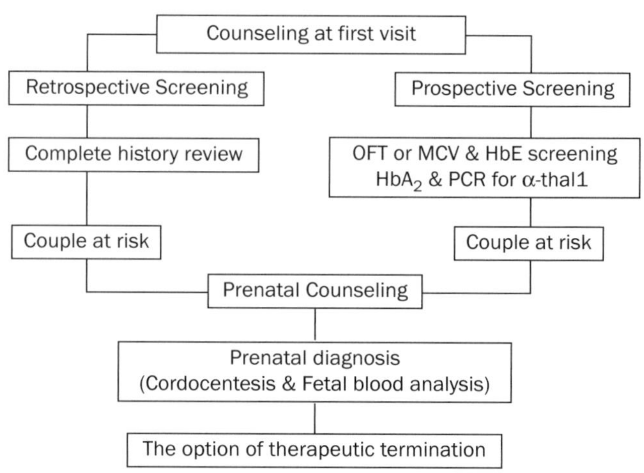

Figure 63-1 Main scheme for the screening of thalassemia carriers, Chiang Mai strategy.

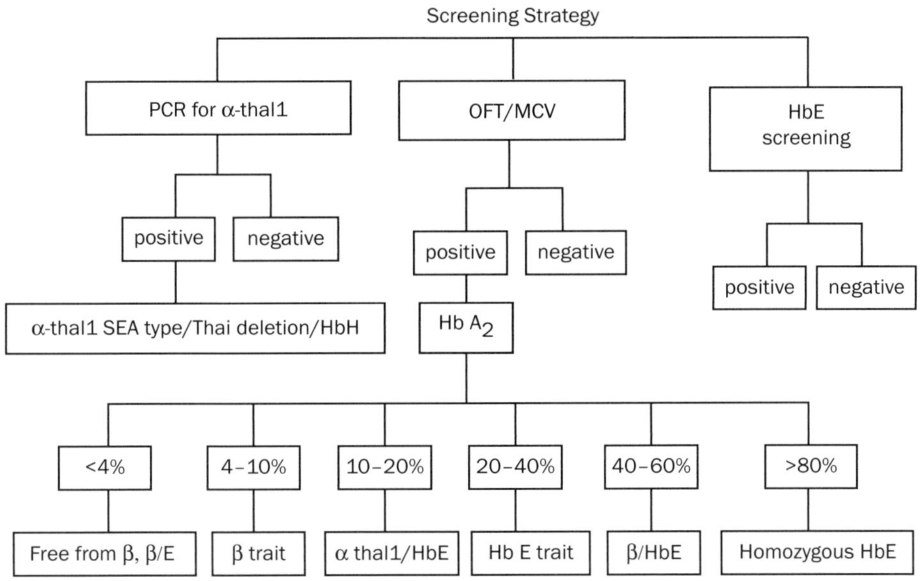

Figure 63-2 Outline of thalassemia screening techniques.

RECENT PROGRESS

The increasing problem of thalassemia and sickle cell anemia were recognized by the WHO and reported in *Genomics and World Health* (WHO, 2002). Among the recommendations of the report it was suggested that for the more effective control and management of the thalassemias, the principle of the further development of North/South partnerships should be pursued, and that the possibility of South/South networks should also be examined. Under the support of Professor Prawase Wasi, Thailand, Sir David Weatherall, U.K., and Professor Alan Bittles, Australia, the Asian Network for the Control of Thalassemia was established during the conference on *Genetics and Population Health* held in Fremantle, Australia, 2004. Representatives from a number of Asian countries together with workers in the thalassemia field from the United Kingdom, Canada, and Australia will work together toward the prevention and control of thalassemia in Asia. It was agreed that the network should focus on fact-finding regarding the extent of the problem in individual countries, together with an account of the facilities that exist for the diagnosis and management of the different forms of thalassemia in each country. The health burden of the thalassemias in Asia should be translated into disability-adjusted life years (DALYs). Using this approach it is possible to compare the health burden of the thalassemias with other health problems in Asia. Although the preliminary data indicate that thalassemia will pose a health burden comparable with some of the major communicable diseases, far more data are required, particularly gene frequency and accurate costing of both prevention and treatment regimes.

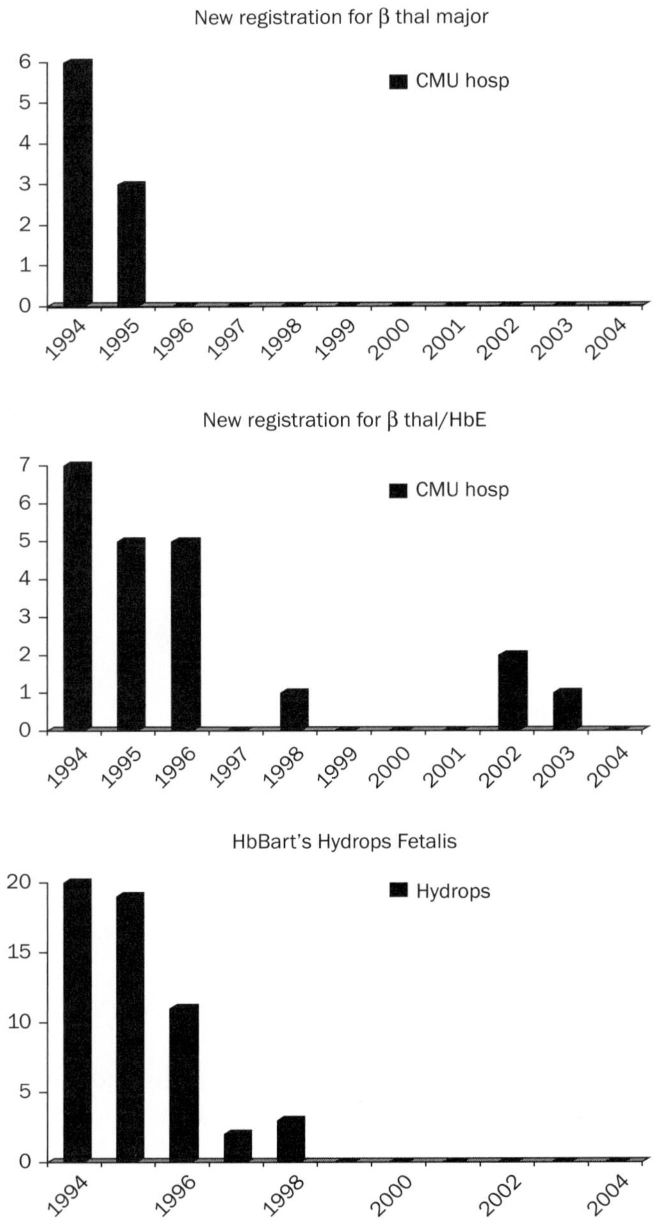

Figure 63-3 New case of β-thalassemia major, β-thalassemia/HbE and Hb Bart's disease at Chiang Mai university hospital

TABLE 63-2 EFFECTIVENESS OF THE PREVENTION AND CONTROL PROGRAM AT FACULTY OF MEDICINE, CHIANG MAI UNIVERSITY, BETWEEN 2004–2010.

Year	2003	2004	2005	2006	2007	2008	2009	2010	Total
Total screened	2152	1765	1737	1504	1563	1580	1400	1269	12970
Positive cases	634	697	697	487	598	633	403	383	4532
Couple at risk	41	43	47	31	31	32	25	23	273
# Hb Bart's	13	12	12	10	12	12*	10	8	89*
# β thal major	7	8	6	3	5	4	2	3	38
# HbE/β thal	21	23	29	18	14	17*	13	12	147*
Affected	5	7	10	10	7	8	5	6	58

	Number
Total screened	12970
Couple at risk	273
% risk couple	2.11 (2.4%)

	Number
Total PND	211/273
Affected	58
% affected	27.5 (25%)

273-211=62 →

- ✓ Refuse PND 15
- ✓ Missing data 28
- ✓ Over GA 9
- ✓ Abortion 10

Estimated couple at risk 2.4% (311 couples)
Missing couple at risk (CAR) 38 (12%)
Efficacy of CAR detection 88%

The first full meeting of the Network was held in Bangkok in July 2005. A draft outline of the Network program and guidelines for the prevention and control of thalassemia in Asia are as follows:

OBJECTIVES AND PRIORITIES OF THE NETWORK

a. To ensure that each country in the Network has a National Thalassemia Center for providing expertise in the diagnosis, control, and management of the thalassemias.

b. To provide the training in the diagnosis and management of the thalassemias required to develop these thalassemia centers.

c. To determine the most cost-effective and accurate approaches to population screening for thalassemia.

d. To establish a population screening program within each country to determine more accurately the frequency of the thalassemias and hence the likely health burden for each country in the future.

OUTLINE OF APPROACH TO ACHIEVE THESE OBJECTIVES

a. Establish a thalassemia centers in countries in which they do not exist.

b. Establish at least one and possibly two major education and training workshops in the diagnosis and management of thalassemia in the first year of the program.

c. In order to sustain the Network, involve local partnerships between countries in which the thalassemia programs are well established, and those in which facilities are not yet available. A firm and ongoing partnership of this type would probably be the most ideal way to sustain the future of the Network.

d. Organize at least one annual meeting of the representatives of each country to assess progress and to evolve joint research programs. The evolution of a population screening program should be the top research priority, and hence the way in which this is done should be comparable for each center.

e. A start should be made in incorporating help from health economists toward defining the true cost of the control and management of thalassemia in each country.

f. Through the efforts of the Network, and with the help of the WHO, Thalassemia International Federation (TIF), and other agencies, governments of countries with a high frequency of thalassemia should be persuaded to identify the importance of developing national programs for achieving these aims.

In May 2006, the World Health Organization and the March of Dimes Birth Defects Foundation held a joint meeting in Geneva. Afterwards, the WHO had adopted a resolution on sickle cell anemia (WHO, 2006a) and made recommendations to the forthcoming WHA to support national and international activities on the control of sickle cell anemia worldwide. Subsequently, similar discussion on sickle cell disease was held at the WHA/59 on thalassemia and other hemoglobinopathies at the EB/118, and similar resolutions have been adopted (WHO, 2006b).

SUMMARY

Thalassemia and abnormal hemoglobin are the most common inherited disorders in Thailand. The standard treatment is regular blood transfusion and iron chelation in severely anemic patients with iron overload. In the last 20 years there has been much progress in diagnosing, preventing, and managing thalassemia. This has led to success in prevention and control of thalassemia in many Mediterranean countries such as Cyprus, Italy, and Greece. It is expected that the same approaches may be applied for the control of thalassemia in developing countries, including Thailand, where thalassemias are very prevalent.

CONCLUSION

Thalassemia is one of the major public health problems in Asia and Thailand. The strategy for a control program consists of treatment and prevention; i.e., treatment of existing patients in the most cost-effective way, and reducing the birth of new cases. The specific target is to set up the core of an effective treatment, prevention, and training service in each country. This should define the methods of treatment and prevention appropriate to each individual country. To achieve success in the prevention and control of thalassemia needs continuity and a holistic approach. It is expected that with optimal collaboration and support, effective prevention and control can be achieved. This will lead to a healthier and better quality of life for new generations.

ACKNOWLEDGMENT

This work is supported by the Office of the Higher Education Commission and Mahidol University under the National Research University Initiative.

REFERENCES

Angastiniotis M, Modell B, Englezos P, Boulyjenkov V. (1995). Prevention and control of haemoglobinopathies. *Bull WHO 73*, 375–386.

Fucharoen S, Winichagoon P, Thonglairoam V, et al. (1991). Prenatal diagnosis of thalassemia and hemoglobinopathies in Thailand: experience from 100 pregnancies. *Southeast Asian J Trop Med Public Health 22*, 16–29.

Fucharoen S, Winichagoon P. (1987). Hemoglobinopathies in Southeast Asia. *Hemoglobin 11*, 65–88.

Lee AC, Ha SY, Wong KW, et al. (1998). Prevention of beta thalassemia major by antenatal screening in Hong Kong. *Pediatr Hematol Oncol 15*, 249–254.

Liu TC, Lin SF, Yang TY, Lee JP, Chen TP, Chang JG. (1997). Prenatal diagnosis of thalassemia in the Chinese. *Am J Hematol 55*, 65–68.

Maternal Fetal Medicine Unit. (YEAR) Statistics Annual Report. Department of Obstetrics and Gynecology, Faculty of Medicine, Chiang Mai University 1994–2010.

Tongsong T, Wanapirak C, Sirivatanapa P, et al. (2000). Prenatal control of severe thalassemia: Chiang Mai strategy. *Prenat Diagn 20*, 229–234.

Tongsong T, Wanapirak C, Sirivatanapa P, et al. (2001). Prenatal eradication of Hb Bart's hydrops fetalis. *J Reprod Med 46*, 18–22.

Wanapirak C, Muninthorn W, Sanguansermsri T, Dhananjayanonda P, Tongsong T. (2004). Prevalence of thalassemia in pregnant women at Maharaj Nakorn Chiang Mai Hospital. *J Med Assoc Thai 87*, 1415–1418.

Wanapirak C, Tongsong T, Sirivatanapa P, Sanguansermsri T, Sekararithi R, Tuggapichitti A. (1998). Prenatal strategies for reducing severe thalassemia in pregnancy. *Int J Gynaecol Obstet 60*, 239–244.

Wasi P, Na-Nakorn S, Pootrakul S, et al. (1969). Alpha and beta thalassemia in Thailand. *Ann NY Acad Sci 165*, 60–82.

Wasi P. (1981). Haemoglobinopathies including thalassemia. Part I: Tropical Asia. *Clin Haematol 10*, 707–729.

WHO [World Health Organization]. (2002). *Genomics and World Health: Report of the Advisory Committee on Health Research.* Newton, MA: Digital Design Group.

WHO. (2006a). Executive Board Resolution on Sickle Anemia. EB117. R3. Geneva, Switzerland: WHO. [www.who.int/gb].

WHO. (2006b). Executive Board Resolution on Thalassaemia and other Hemoglobinopathies. EB118.R1. Geneva, Switzerland: WHO. [www.who.int/gb].

Winichagoon P, Fucharoen S, Chen P, Wasi P. (2000). Genetic factors affecting clinical severity in beta-thalassemia syndromes. *J Pediatr Hematol/Oncol 22*, 573–580.

Yong KN, Wadaworth D, Langlois S, Yong SL, Wilson RD. (1999). Thalassemia carrier screening and prenatal diagnosis among the British Columbia (Canada) population of Chinese descent. *Clin Genet 55*, 20–25.

64.

INHERITED METABOLIC DISORDERS IN THAILAND AND THE ASIA-PACIFIC

Pornswan Wasant

INTRODUCTION

The study of inherited metabolic disorders (IMD) in Thailand is in its infancy when compared with developed countries. Prior to 1987, the majority of disorders were clinically diagnosed, since there were only a handful of clinicians and scientists with expertise in inborn errors of metabolism, and a lack of well-equipped laboratory facilities and government support. In developing countries, inherited metabolic disorders are not considered a priority due to the prevalence of infectious diseases, such as HIV infection and congenital infections. A multicenter survey conducted in 1994 and 2001 revealed the existence of numerous IMDs from all over the country. Case reports and publications on IMD in Thai (and international) medical journals in the past 20 years had undoubtedly raised awareness among Thai pediatricians and scientists. In 2001, the "Genetic Metabolic Center," the first of its kind in Thailand, was first established at Siriraj Hospital Faculty of Medicine. Numerous new cases of IMD have been identified since then.

HISTORICAL BACKGROUND: MILESTONES OF IMD IN THAILAND

The study of inherited metabolic disorders in Thailand started in 1987.

1. **The First Decade (1987–1997):** The majority of IMDs were clinically diagnosed, since there were no laboratory facilities available in Thailand. In May 1990, the 5th International Congress of Inborn Errors of Metabolism (ICIEM) was held in Asilomar, California, and only one Thai delegate (the author) attended that meeting. In June 1993, the first Asia-Pacific Regional Meeting (APRM) of the International Society on Neonatal Screening (ISNS) was held in Sapporo, Japan. Experiences from these two international meetings made a great impact on the development of IMD in Thailand. In July 1994, the first Asia-Pacific Conference on Medical Genetics was held in Bangkok, Thailand, supported by Mahidol University and the International Center for Medical Research (ICMR) at Kobe University School of Medicine, Japan. Subsequently, a multicenter study on IMD was initiated (1994 to1998) to collect data and explore the prevalence of IMD.

From 1993 to1997 a pilot project on newborn screening was started at Siriraj Hospital, Mahidol University in Bangkok. In November 1995, the 2nd APRM of ISNS was held in Hong Kong. The following year, 1996, the Department of Medical Science, Ministry of Public Health (MOPH), Thailand initiated a pilot project in newborn screening in Thailand.

2. **The Second Decade (1997–2007):** In 1998, the 3rd APRM of the ISNS was organized in Chiang Mai, Thailand, and was attended by more than 200 people. The objectives of the meeting were to raise awareness of IMD, to educate Thai physicians (pediatricians & obstetricians), and to help nurses understand the importance of newborn screening.

From 1998–2000, there were collaborations with Japanese scientists/experts in which gas-liquid chromatography/mass spectrometry (GC/MS) was first introduced to Thailand. In 2001 the Genetic Metabolic Center was established in the Division of Medical Genetics, Department of Pediatrics, Faculty of Medicine Siriraj Hospital, Mahidol University with support from the faculty/hospital (HPLC) and JICA (Japanese Intergovernmental Cooperation Agency) (GC/MS) including transfer of technology to Thai scientists. The first Thai textbook on IEM—a collection of data on IEM gathered from a 15-year experience of IEM referrals to Siriraj Hospital—was completed in 2003, with 2 volumes and 958 pages. (@@Wasant, 2003).

The 9th International Congress on IEM (ICIEM) was organized in Brisbane, Australia in 2003; two papers on urea cycle disorders were presented at the satellite meeting, and "IEM in Thailand—Siriraj Experience" was also presented. In 2005, the First Genetic Metabolic Symposium was held in Bangkok to raise awareness among Thai pediatricians.

The 10th International Congress on IEM (ICIEM) was organized in Chiba, Japan in September 2006. The author was invited to be one of the local organizing committee and an invited speaker in the symposium, "IEM in Asia." Twenty-five posters on IEM in Thailand, all first reported cases, were presented. The 6th ISNS meeting was held at Awaji Island, Tokushima, Japan, and the title of the invited lecture was "Newborn Screening in Thailand: Challenges and Opportunities." A poster, "PKU Infant from Newborn Screening at Siriraj Hospital," was presented. In November 2007, the second Genetic Metabolic Symposium was held in Bangkok with support from the Asia-Pacific Society of Human Genetics.

RESEARCH COLLABORATIONS

Research collaborations with U.S. and Japanese experts/scientists were initiated beginning in 1989.

1. Research collaborations with the United States (1987–1997) include: (1) Prof. Edwin Kolodny (NYU) – lysosomal enzyme assays; (2) Dr. Edwin Naylor (McGeeWomen's & Children) – tandem mass spectrometry; (3) Dr. George Thomas (Johns Hopkins) – amino acid analysis (AA); (4) Prof. Hugo Moser (Kennedy Krieger Institute) – peroxisomal disorders; (5) Dr. Robert Guthrie – newborn screening; (6) Prof. Harvey Levy (Boston, MA) – newborn screening; (7) Prof. Saul Brusilow (Johns Hopkins) – urea cycle disorders; (8) Dr. Holmes Morton (Lancaster, PA) – MSUD; (9) Prof. Vivian Shih (Boston, MA) – amino acid disorders. (Wasant et al., 1995, 1999, 2002)
2. Research collaborations with Japan (1997–2007) include: (1) Dr. Toshiaki Oura (Osaka) – newborn screening; (2) Dr. Hiroshi Naruse (Tokyo) – newborn screening; (3) Prof. Isamu Matsumoto (Kanazawa) – GC/MS; 4) Prof. Seiji Yamaguchi (Shimane) – tandem mass spectrometry; (5) Prof. Keiko Kobayashi (Kagoshima) – mutation analysis (UCD); (6) Assoc. Prof. Toshihiro Shinka (Kanazawa) – GC/MS; (7) Assoc. Prof. Kenji Hara (Fukuoka) – GC/MS. (Wasant et al., 1999, 2002)

ESTABLISHMENT OF GENETIC METABOLIC CENTER IN THAILAND

Since 1998, research collaboration with Chulabhorn Research Institute in Bangkok has also helped us in the area of amino acid analysis, enzyme assays, and mutation analysis, from which almost 20 publications arose, representing this pioneering work on IMD in Thailand.

In 2001, the "Genetic Metabolic Center" was established at Siriraj Hospital Faculty of Medicine, the first of its kind in Thailand, with assistance from JICA (Japanese Intergovernmental Cooperation Agency) which provided Gas-liquid Chromatography/Mass Spectrometry (GC/MS) and technology transfer, together with funding from the Chaofa-Maha Chakri Pediatric Building for high-performance liquid chromatography (HPLC). Numerous new IMD cases have been identified since (Wasant et al., 1995, 1999, 2002, 2008).

MULTICENTER STUDY OF IEM (2001–2004)

The following IMDS were identified: (1) *Carbohydrate disorders* – galactosemia, glycogen storage diseases (GSD type I, GSD type II, GSD type III), fructose-1,6-bisphosphatase deficiency; (2) *Amino acid disorders* – PKU, hyperphenylalaninemia, tyrosinemia type I, MSUD, homocystinuria, albinism, NKH (Wasant et al., 2005); (3) *Urea Cycle disorders* – ALD, OTC, ASD, unidentified UCD (Wasant et al., 2005); (4) *Organic acid disorders* – IVA, MMA, PA, alkaptonuria, multiple carboxylase deficiency (MCD) (Wasant et al., 1999); (5) *Mitochondrial disorders* – MCAD, translocase deficiency, carnitine deficiency (Wasant et al., 1999); (6) *Peroxisomal disorders* – RCDP, Zellweger, primary hyperoxaluria type I; (7) *Lipidosis* – Niemann-Pick type I, Gaucher, Sandhoff, GM_1 gangliosidosis; (8) *Mucopolysaccharidosis* – Hurler, HurlerScheie, Scheie, Hunter, Sanfilippo, Morquio, Maroteaux-Lamy, Sly, unidentified MPS; (9) *Disorder of Purine Metabolism* – LeschNyhan; (10) *Disorder of Copper Transport* – Menkes; (11) Leucodystrophies – X-ALD, others; (12) *Others* – lipoprotein lipase deficiency, hyperlipoproteinemia, porphyria, cystinuria, methemoglobinuria, HMG CoA lyase deficiency.

NEWBORN SCREENING PROGRAM IN THAILAND

A national neonatal screening program has been implemented into the public health infrastructure by the Ministry of Public Health (MOPH) since 1996 (Charoensiriwatana W et al., 2003). At present, approximately 90% of all newborns are being screened for congenital hypothyroidism (CH) and phenylketonuria (PKU) (Pangkanon S et al., 2003, 2005, 2006). However, many problems still exist due to poor organization (lack of adequate follow-up and genetic counseling, problems with medical management and noncompliance) including the lack of a systematic and holistic approach.

TABLE 64-1 DATA ON INHERITED METABOLIC DISORDERS IN THE ASIA-PACIFIC (2006)

A. Developed countries	Common IMD	Other IMDs	Metabolic lab	Confirmation	Consanguinity	Newborn Screening	Data given by
1. Australia	Cystic fibrosis,	MCAD, GALT, MMA, MSUD,	HPLC, GC/MS,	Australia	small percentage	CH, PKU, CF, GALT,	Bridget Wilcken
(Pop. 20 million)	phenylketonuria	NKH, OTC, CBS	MS/MS	overseas		MSUD	
2. Japan							
(Pop. 125 million)	CH, CAH, PKU	LSD, OA, Wilson disease	HPLC,	Japan &	small percentage	CH, CAH, GALT,	Seiji
		UCD, citrin deficiency	GC/MS, MS/MS	overseas	no data	PKU, MSUD,	Yamaguchi
						homocysteinemia	
3. Korea	PKU, CH,	PKU, MPS, Gaucher, OTC,	HPLC,	Korea &	no data	PKU, CH	Dong Hwan Lee
(Pop. 48 million)	Wilson disease	citrullinemia	GC/MS, MS/MS	overseas		(since 1991)	
4. Hong Kong (Pop. 7 million)	G6PD	AA, OA, FAOD, UCD	HPLC,	Hong Kong &	no data	G6PD, CH	Stephen TS
			GC/MS, MS/MS	overseas		(since 1984)	Lam
5. Singapore	OA, AA, UCD	MPS, carbohydrate disorders	HPLC,	Singapore &	<1%	G6PD, CH, Hearing,	Ivy Ng
(Pop. 4.2 million)			GC/MS, MS/MS	Australia		TMS	
6. Taiwan	G6PD, PKU,	numerous	HPLC,	Taiwan,	no data	CH, G6PD, PKU,	Wuh Liang
(Pop. 20 million)	GSD, MPS		GC/MS, MS/MS	(USA, Japan)		GALT, CAH,	Hwu
						homocystinuria	

B. Developing countries	Common IMD	Other IMDs	Metabolic lab	Confirmation	Consanguinity	Newborn Screening	Data given by
1. China	PKU, CH	OA, MMA, FAOD,	HPLC, GC/MS,	China & overseas	no data	CH, PKU, G6PD,	Xue Fan Gu
(Pop. 1.3 billion)		MCAD, tyrosinemia,	MS/MS			CAH	
		homocystinuria					
2. Indonesia	G6PD, unknown	GSD, MPS, LSD, MMA,	developing	overseas	no data	CH (only in Bandung,	Damayanti R Sjarif
(Pop. 230 million)		mitochondrial disorders		(Netherlands,		Jakarta)	
				Japan, Australia)			

(Continued)

TABLE 64-1 (CONTINUED)

B. Developing countries	Common IMD	Other IMDs	Metabolic lab	Confirmation	Consanguinity	Newborn Screening	Data given by
3. Malaysia	G6PD, UCD	MSUD, MMA,	HPLC, GC/MS	Malaysia, Australia	1.2%	G6PD, TSH	Meow Keong Thong
(Pop. 24 million)		tyrosinemia,				(Cord blood)	
		mitochondrial disorders					
4. Philippines	G6PD	MSUD, MMA, GALT,	HPLC, GC/MS	Australia	no data	CH, CAH, GALT,	Carmencita Padilla,
(Pop. 80 million)		OTC, ALD, Gaucher				PKU, G6PD	Mary Anne Chiong
5. Thailand	G6PD, CH, PKU,	LSD, FAOD,	HPLC, GC/MS	Thailand,	no data	PKU, CH	Pornswan Wasant
(Pop. 64 million)	MSUD, OA, MMA,	mitochondrial disorders		(overseas – USA,	(probably fair	(since 1996)	
	UCD, MPS			Japan)	percentage)		

A national committee has recently been established in 2006 and problems are being solved.

Pilot project on newborn screening for congenital hypothyroidism and phenylketonuria (PKU) with funding from Mahidol University was started from 1993 to 1996 (Wasant et al., 1999). In 1998, the 3rd Asia-Pacific Regional Meeting of the International Society on Neonatal Screen (ISNS) was successfully organized in Chiang Mai, northern Thailand with about 200 participants. Routine newborn screening program was established at Siriraj Hospital in 2005 with 9,000 newborns screened annually. Between 2006–2007, working with Department of Maternal & Child Health, Ministry of Public Health to educate medical professionals – pediatricians and nurses in the 4 regions (North, Northeast, South and Central Thailand). In 2006, the educational program for medical personnels at Siriraj Hospital Annual Scientific Meeting was provided for 300 participants. In 2009, the educational program was provided for 13 Bangkok Metropolitan hospitals with 200 participants. Since 2009, the expanded program has covered 45,000–50,000 newborns annually. Future plan is coverage of 100,000 newborns per year in Bangkok metropolitan area with support from the National Health Security Office (NHSO).

ORGANIZATION OF IEM PATIENT GROUP MEETINGS

In November 2004, an IEM and LSD (lysosomal storage disorders) patient group meeting, the first of its kind in Thailand, was organized at Siriraj Hospital to raise awareness among Thai pediatricians and Ministry of Public Health personnels, in order to provide better care for IEM and LSD patients in Thailand. It was a multicenter effort, since there were only a handful of IEM experts in Thailand.

The 8th LSD Annual Asia Symposium was organized in Bangkok in November 2005, with support from a commercial company. The objectives were an exchange of ideas and knowledge, including collaboration among local and international experts in the field. The First Genetic Metabolic Symposium was also organized as a post-conference satellite symposium.

In October 2007, the second Genetic Metabolic Symposium was held in Bangkok, and coincided with the 20th year celebration of my work in Thailand at Siriraj Hospital, with support from the Asia-Pacific Society of Human Genetics.

Most recently, in March 2009, we organized the first LSD patient group meeting at Siriraj hospital; there were 40 families attending, and it was the beginning of the establishment of a foundation for LSD patients in Thailand.

INBORN ERRORS OF METABOLISM IN THE ASIA-PACIFIC

The study of inherited metabolic disorders (IMD) has received well-deserved attention in developed countries; e.g., the United Kingdom, United States, and European countries, since the early 1900s. By contrast, these disorders have received much less or no attention in the Asia-Pacific countries due to (1) prevalence of other diseases/problems in the region, e.g., malnutrition and infectious diseases (HIV in particular); (2) other, more common genetic disorders such as thalassemia; and (3) economic conditions in different countries.

Attention to the inherited metabolic disorders in developed countries in the Asia-Pacific region such as Japan, Korea, and Australasia, had been well established since 1960s, particularly in Australia and Japan. As for other developing countries such as Indonesia, the Philippines, and Thailand, the study of IMD is still in its infancy due to (1) lack of government support, (2) only a handful of clinicians and scientists with expertise in IMD, and (3) lack of or only a few well-equipped laboratory facilities. Existing data on common IMDs in different countries, the existence or non-existence of metabolic laboratories, including confirmation inside or outside of each country, and the existence of newborn screening are presented in Table 64-1.

SUMMARY

There has been quite a remarkable progress of work and research in inherited metabolic disorders in Thailand in the past 20 years. Numerous publications, both at the national and international levels, have been published. Organization of services for IMD has improved the quality of life for Thai patients in the past several years.

ACKNOWLEDGMENTS

The author would like to acknowledge much valuable assistance from several collaborators. From the United States, the author would like to particularly acknowledge Professor Edwin H. Kolodny (NYU), who assisted us in performing lysosomal enzyme assays in many patients, for example in Scheie & Sly syndrome, and gave us advice about patients with neurogenetic problems. Dr. Edwin Naylor (Pittsburgh, Pennsylvania) who helped us with tandem mass spectrometry on high-risk newborns and patients with delayed development on a research basis for many years; Dr. George Thomas (Johns Hopkins) for performing plasma amino acid analysis when we did not have any such instrument in our laboratory; and other collaborators who provided laboratory support in the early stages prior to the establishment of our genetic metabolic center.

From Japan, the author would like to acknowledge Dr. Toshiaki Oura (Osaka) for helping us in setting up newborn screening at Siriraj Hospital; also Dr. Hiroshi Naruse, who gave support in organizing the 3rd Asia-Pacific Regional meeting of ISNS in Chiang Mai in 1998. The author would like to express our gratitude to late Professor Isamu Matsumoto (Kanazawa) who helped us a great deal in the establishment of Genetic Metabolic Center at Siriraj Hospital in Bangkok in 2001. We would also like to express our thanks to Associate Professor Toshihiro Shinka (Kanazawa) who helped us in confirmation of diagnosis of many organic acid disorders, and Professor Seiji Yamaguchi (Shimane) who helped us in performing tandem mass spectrometry in high-risk patients with IEM.

The author would also like to thank Professor Jisnuson Svasti, Chulabhorn Research Institute in Bangkok and his team for helping us in amino acid analysis, enzyme assays and molecular study in our patients with lysosomal storage disorders (Hurler & Hunter syndrome). (Ketudat Cairns JR et al., 2005; Keeratichamroen S et al., 2007, 2008)

Lastly, I would like to thank our Asia-Pacific Society of Human Genetics group who provided us with invaluable data in the Asia-Pacific region; e.g., Professors Bridget Wilcken, Seiji Yamaguchi, Dong Hwan Lee, Wuh Liang Hwu, Xue Fan Gu, Meow Keong Thong, Carmencita Padilla, and Drs. Stephen TS Lam, Ivy Ng, Damayanti Sjarif, and Mary Anne Chiong.

REFERENCES

Charoensiriwatana W, Janejai N, Boonwanich W, et al. (2003). Neonatal screening program in Thailand. *Southeast Asian J Trop Med Public Health.* 34(Suppl 3):94–100.

Keeratichamroen S, Cairns J, Sawangareetrakul P, et al. (2007). Novel mutations found in two genes of Thai patients with isolated methylmalonic acidemia. *Biochemical Genetics* 45(5–6):421–430.

Keeratichamroen S, Ketudat Cairns JR, et al. (2008). Molecular analysis of the iduronate-2-sulfatase gene in Thai patients with Hunter syndrome. *J Inherit Metab Dis.* 31(Suppl 2):S303–S311.

Ketudat Cairns JR, Keeratichamroen S, Sukcharoen S, et al. (2005). The Molecular basis of mucopolysaccharidosis type I in two Thai patients. *Southeast Asian J Trop Med Public Health.* 36(5):1308–1312.

Pangkanon S, Ratrisawadi V, Charoensiriwatana W, et al. (2003). Phenylketonuria detected by the neonatal screening program in Thailand. *Southeast Asian J Trop Med Public Health.* 34(Suppl 3):179–181.

Pangkanon S, Charoensiriwatana W, Techasena W, Srisomsap C. (2005). Phenylketonuria: a report of 18 cases in Thailand. *Japanese Journal for Inherited Metabolic Diseases* 21(2):163.

Pangkanon S, Charoensiriwatana W, Liamsuwan S. (2006). 6-pyruvoyltetrahydropterin synthase deficiency two-case report. *J Med Assoc Thai.* 89(6):872–877.

Wasant P (ed.). (2003). *Inherited Metabolic Disorders in Children: Thai Textbook of Pediatrics*, 1st ed. Bangkok: P.A. Living.

Wasant P, Naylor E, Matsumoto I, Liammongkolkul S. (2002). Mitochondrial fatty acid oxidation disorders in Thai infants: a report of 3 cases. *J Med Assoc Thai* 85(Suppl 2):S710–S719.

Wasant P, Wattanaweeradej S, Raksadawan N, Kolodny EH. (1995). Lysosomal storage disorders – Siriraj experience. *Southeast-Asian J Trop Med Public Health* 26(Suppl 1):S54–S58.

Wasant P, Liammongkolkul S, Srisomsap C, Svasti J, Naylor EW, Matsumoto I. (1999). Inherited metabolic disorders in Thailand – Siriraj experience. *Southeast Asian J Trop Med Public Health.* 30(Suppl 2):124–137.

Wasant P, Liammongkolkul S, Srisawat C, et al. (1999). Neonatal screening for congenital hypothyroidism and phenylketonuria at Siriraj hospital, Mahidol University, Bangkok, Thailand – A pilot study. *Southeast Asian J Trop Med Public Health.* 30(Suppl 2):33–37.

Wasant P, Naylor EW, Liammongkolkul S. (1999). Detection of inherited metabolic disorders via tandem mass spectrometry in Thai infants. *Southeast Asian J Trop Med Public Health* 30(Suppl 2):154–159.

Wasant P, Matsumoto I, Liammongkolkul S. (1999). Detection of inborn errors of metabolism in Thai infants via gas chromatography and mass spectrometry. *Southeast Asian J Trop Med Public Health* 30(Suppl 2):160–165.

Wasant P, Vatanavicharn N, Pipopmongkol C, et al. (2005). Retrospective study of patients with suspected inborn errors of metabolism at Siriraj Hospital, Bangkok, Thailand. (1997–2001). *J Med Assoc Thai.* 88(6):746–753.

Wasant P, Viprakasit V, Srisomsap C, et al. (2005). Argininosuccinate synthetase deficiency: mutation analysis in 3 Thai patients. *Southeast Asian J Trop Med Public Health.* 36(3):757–761.

Wasant P, Liammongkolkul S, Kuptanon C, et al. (2008). Organic Acid disorders detected by urine organic acid analysis: twelve cases in Thailand over three-year experience. *Clinica Chimica Acta.* 392(1–2):63–68.

65.

GENETIC SKELETAL DYSPLASIAS IN THAILAND

Pornswan Wasant

INTRODUCTION

Genetic skeletal dysplasias are a heterogeneous group of disorders associated with abnormalities in the size and shape of the limbs, trunk, and/or skull that frequently result in disproportionate short stature. There are well over 100 distinct skeletal dysplasias that have been classified primarily on the basis of their clinical or radiographic characteristics (Rimoin & Lachman, 2007; Lachman, 2007).

The Genetic Skeletal Dysplasia Clinic was established at the Division of Medical Genetics, Department of Pediatrics, Faculty of Medicine Siriraj Hospital, Mahidol University in Bangkok in 1992—the one and only clinic of its kind in Thailand. The objectives are: (1) to diagnose and provide genetic counseling to patients and their families; (2) to educate pediatric and orthopedic residents in training; (3) to perform research in genetic skeletal dysplasias, which are rare disorders with heavy burden to patients and their families; and (4) to be a model for holistic care and a multidisciplinary approach. There are numerous cases identified (Table 65-1): achondroplasia, hypochondroplasia, pseudoachondroplasia (Wasant et al., 1994), atelosteogenesis, pycnodysostosis, spondyloepiphyseal dysplasia congenita, spondylometaepiphyseal dysplasia, osteogenesis imperfecta, chondroectodermal dysplasia (Ellis van Creveld syndrome) (Wasant et al., 1993), cleidocranial dysostosis, thanatophoric dysplasia, rhizomelic chondrodysplasia punctata (Wasant & Moser, 2006), trichorhinophalangeal syndrome, mucopolysaccharidoses, (Hurler, Hunter, Sanfilippo, Morquio, Maroteaux-Lamy, Sly), osteopetrosis, campomelic dysplasia, Stickler syndrome, Kniest dysplasia, nail-patella syndrome, diastrophic dysplasia, short-rib polydactyly syndrome, Larsen syndrome, multiple epiphyseal dysplasia, metaphyseal dysplasia, and hypophosphatemic rickets. The diagnoses are established at clinical, radiological, and biochemical levels. There are limitations in confirmation at the molecular level in developing countries. A multidisciplinary approach and genetic counseling are provided. We present here some of the common and rare genetic skeletal dysplasias.

ACHONDROPLASIA (MIM# 100800)

Clinical manifestation: the most common genetic skeletal dysplasia, short-limbed dwarfism, large head with prominent forehead, saddle nose, gibbus in thoracolumbar region, bowed legs and trident hands, spinal stenosis, normal intelligence, neurologic complications (Figures 65-1 & 65-2).

Radiography: large calvarium, short base and small foramen magnum, decreased lumbosacral interpedicular distance, short pedicles and short, broad limb bones.

Mode of inheritance: autosomal dominant

OSTEOGENESIS IMPERFECTA (MIM# 166210, 259400)

OSTEOGENESIS IMPERFECTA TYPE I (MIM# 166200, 166240):

Clinical manifestation: a hereditary skeletal condition leading to susceptibility to fracture and severe skeletal deformity, with extreme variability resulting from genetic and biochemical heterogeneity; osseous fragility, dentinogenesis imperfecta characterized by excessive bone fragility associated with osteoporosis, distinctly blue sclera, susceptibility to conductive hearing loss, frequency 1:30,000 live births (Figure 65-3).

Radiography: generalized osteopenia, evidence of old fractures and callus formation at the site of recent fractures, angulation, bowed-leg deformity and Wormian bones

Genetics: autosomal dominant with variable expressivity

Molecular pathology: reduction type I collagen synthesis (≈ 50%)

TABLE 65-1 GENETIC SKELETAL DYSPLASIA CLINIC, FACULTY OF MEDICINE SIRIRAJ HOSPITAL, MAHIDOL UNIVERSITY: TWENTY-YEAR EXPERIENCE (WASANT ET AL., 1993, 1994, 1995A,B; SURASERANIVONGSE ET AL., 2003; CAIRNS ET AL., 2005; WASANT & LIAMMONGKOLKUL, 2006; WASANT ET AL., 2006; WASANT & MOSER (2006); KEERATICHAMROEN ET AL., 2008).

Osteochondrodysplasias	Mode of inheritance	Present at birth	Number of patients
1. **Achondroplasia group**			
Thanatophoric dysplasia, Type I	AD	+	2
Achondroplasia	AD	+	>35
Hypochondroplasia	AD	–	5
2. **Metatropic dysplasia group**			
Metatropic dysplasia	AD	+	1
3. **Short-rib dysplasia (SRP)**			
(with or without polydactyly) group			
SRP type?	AR	+	1
Chondroectodermal dysplasia			
(Ellis-van-Creveld dysplasia)	AR	+	4
4. **Diastrophic dysplasia group**			
Diastrophic dysplasia	AR	+	4
Atelosteogenesis type?	AR	+	1
5. **Type II collagenopathies**			
Achondrogenesis	AD	+	1
Hypochondrogenesis	AD	+	1
Kniest dysplasia	AD	+	1
Spondyloepiphyseal dysplasia (SED) congenita	AD	+	3
Spondyloepimetaphyseal dysplasia (SMED)	AD	+	7
Spondyloepimetaphyseal dysplasia (SMED),			
Strudwick type	AD	+	2
Stickler dysplasia	AD	+	2
6. **Other spondyloepi-(meta)-physeal SE(M)D dysplasias**			
Spondyloepiphyseal dysplasia with joint laxity	AR	+	2
7. **Multiple epiphyseal dysplasias & pseudoachondroplasia**			
Pseudoachondroplasia	AD	–	2
Multiple epiphyseal dysplasia (MED)	AD	–	2
8. **Chondrodysplasia punctata (stippled epiphyses group)**			
Rhizomelic type	AR	+	1
Zellweger syndrome*	AR	+	1
9. **Metaphyseal dysplasias**			
Jansen type	AD	+	1
McKusick type (Cartilage-hair hypoplasia)	AR	+	4
10. **Mesomelic dysplasias**			
Dychondrosteosis (Leri-Weill)	AD	–	1
11. **Acromelic and acromesomelic dysplasias**			
Trichorhinophalangeal dysplasia, type II	AD	+	1
(Langer-Giedion)			
Pseudohypoparathyroidism		–	
(Albright Hereditary Osteodystrophy)			
12. **Dysplasias with prominent membranous bone**			
involvement			
Cleidocranial dysplasia	AD	+	5
Osteodysplasty, Melnick-Needles	XLD	–	1
13. **Bent-bone dysplasia group**			
Campomelic dysplasia	AD	+	2
14. **Multiple dislocations with dysplasias**			
Larsen syndrome	AR	+	>7
Pseudodiastrophic dysplasia	AR	+	2
15. **Dysostosis multiplex group**			
Mucopolysaccharidosis IH	AR	–	3
Mucopolysaccharidosis IH-S	AR	–	1
Mucopolysaccharidosis IS	AR	–	1
Mucopolysaccharidosis II	XLR	–	16
Mucopolysaccharidosis III (A, B, C, D)	AR	–	7
Mucopolysaccharidosis IV (A, B)	AR	–	17
Mucopolysaccharidosis VI	AR	–	4

(Continued)

TABLE 65-1 (CONTINUED)

Osteochondrodysplasias	Mode of inheritance	Present at birth	Number of patients
Mucopolysaccharidosis VII	AR	–	1
GM 1 Gangliosidosis	AR	+	1
Sialidosis	AR	+/–	1
Mucolipidosis	AR	+/–	1
16. Dysplasias with decreased bone density	AD	+/–	12
Osteogenesis imperfecta I			
Osteogenesis imperfecta II	AD, AR	+	6
Osteogenesis imperfecta III	AD, AR	+	27
Osteogenesis imperfecta IV	AD	+	1
Cole-Carpenter dysplasia	SP	+	1
17. Dysplasias with defective mineralization			
Hypophosphatasia—perinatal lethal and infantile form	AR	+	1
Hypophosphatemic rickets	XLD	–	2
18. Increased bone density without modification of bone shape			
Osteopetrosis	AR	–	1
Pycnodysostosis	AR	+	2
19. Increased bone density with diaphyseal involvement			
Caffey dysplasia	AD, AR	–	1
20. Disorganized development of cartilaginous and fibrous components of the skeleton			
Enchondromatosis, Ollier	SP	–	1
Fibrous dysplasia (McCune-Albright)	SP mosaic	–	1
Fibrodysplasia ossificans progressiva	AD	+	1
21. Patella dysplasias			
Nail-patella dysplasia	AD	–	1
22. Others			
Recessive Larsen syndrome and Humero-spinal dysostosis	AR	+	1
Thanatophoric dysplasia with clover-leaf skull	AD	+	1
Tattoo Acrolaryngeal dysplasia	unknown	+	1
Multiple exostosis	AD	–	4
Spondylocostal dysostosis	AD	+	1
23. Unidentified skeletal dysplasia			>10

Jaruratanasirikul S, Vasiknanont P, Saetung P, Mitanun W. (1995). Zellweger syndrome: First reported case in Thailand and literature review. Proceedings, Asia-Pacific Conference in Medical Genetics. *Southeast Asian J Trop Med Public Health 26*(Suppl 1), 47–51.

From: Lachman RS. (1998). International nomenclature and classification of the osteochondrodysplasias (1997). *Pediatr Radiol* 28, 737–744.

(A) (B)

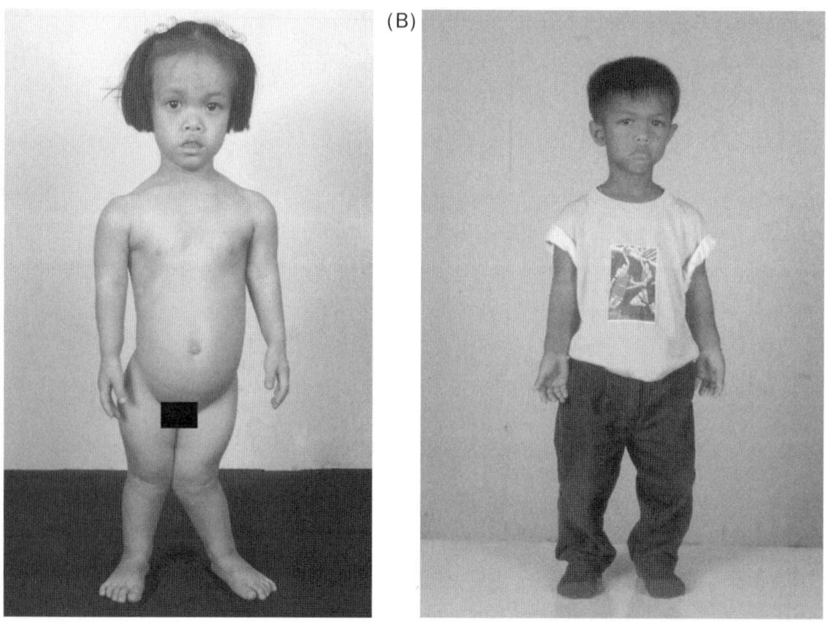

Figure 65-1 Achondroplasia (Wasant et al., 1995a)

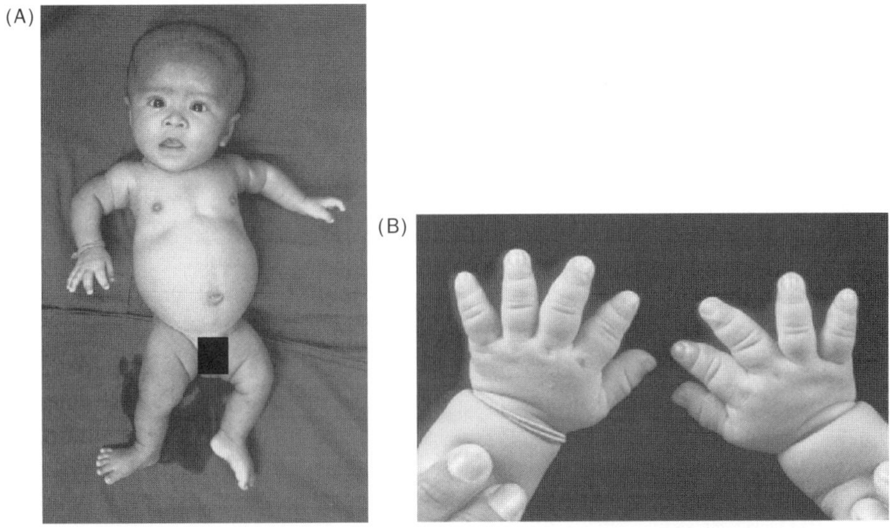

Figure 65-2 Achondroplasia (Wasant et al., 1995a)

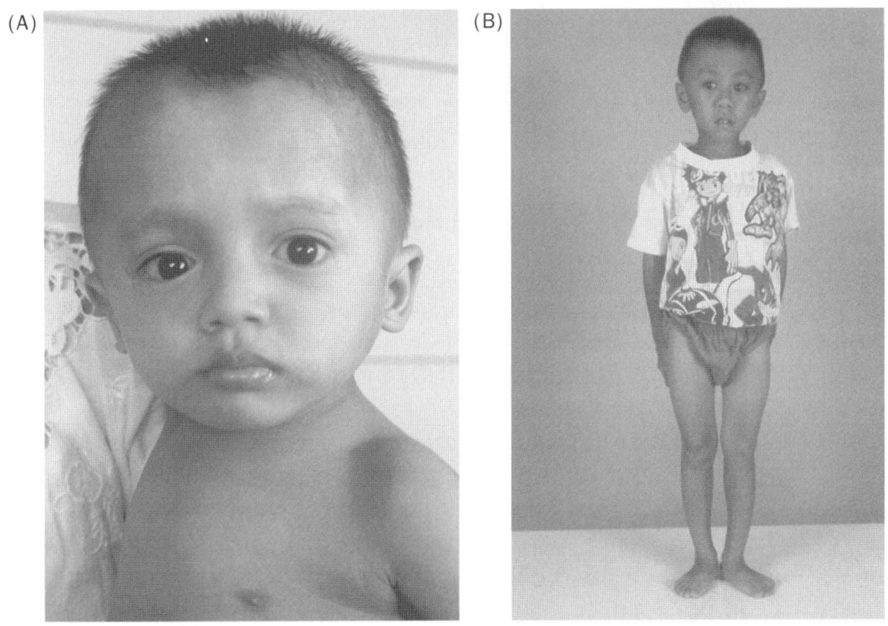

Figure 65-3 Osteogenesis imperfecta type I (Wasant et al., 1995a)

OSTEOGENESIS IMPERFECTA TYPE II (MIM# 166210, 610854)

Clinical manifestation: (perinatal/lethal type) characterized by extreme bone fragility, leading to intrauterine or early infant death, clinically and biochemically heterogeneous there are 3 subgroups (A,B,C; see Figure 65-4).

Group A – broad crumpled long bones (accordion-like, beaded ribs)

Group B – broad crumpled long bones, no beaded rib, fractures, survival possible

Group C – thin fractured long bones, thin mildly beaded ribs, very rare type

Radiography: small thorax with shortened ribs (beading contour), broad and hypoplastic femora, demineralized long and short tubular bones, angulated tibias, flattened vertebrae and Wormian bones

Genetics: parental germline mosaicism, autosomal recessive inheritance

Molecular pathology: diminished production of type I collagen

OSTEOGENESIS IMPERFECTA TYPE III (MIM# 259420)

Clinical manifestation: nonlethal OI with severe bone fragility leading to progressive deformity of the skeleton; fractures are present in most cases at birth and occur frequently during childhood; progressive kyphoscoliosis, severely affected limb deformities at birth, poor growth, “triangular facies” with frontal bossing; blue sclerae at birth (Figure 65-5).

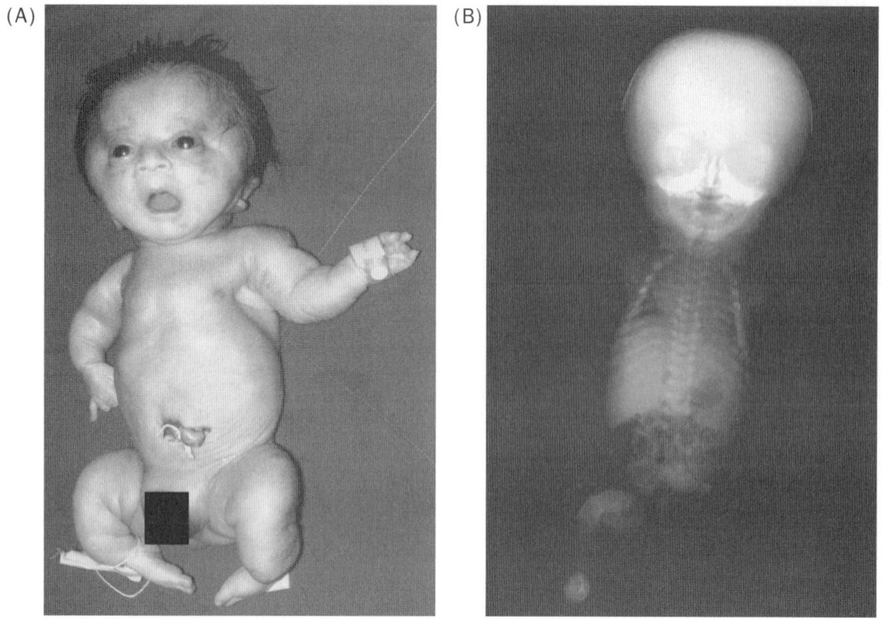

Figure 65-4 Osteogenesis imperfecta type II (Wasant et al., 1995a)

Radiography: generalized osteopenia and multiple fractures, undermodeling of shafts of long bones producing a "broad-bone" appearance; multiple Wormian bones

Mode of inheritance: parental germline mosaicism, autosomal recessive, rare disorder

OSTEOGENESIS IMPERFECTA TYPE IV (MIM# 166220)

Clinical manifestation: osseous fragility (mild to moderate); (1) newborn fractures (25%), (2) childhood-fractures (maximal)

Radiography: similar to type I

Mode of inheritance: autosomal dominant, normal sclerae

Group A – normal teeth

Group B – dentinogenesis imperfecta

DIASTROPHIC DYSPLASIA (MIM# 222600)

Clinical manifestation: present at birth, micromelic dwarfism, clubfoot, symphalangism of feet, deformed great toes, ulnar deviation of hand, abduction of hypermobile and proximally inserted thumb (hitchhiker thumb), flexion contracture of joints,

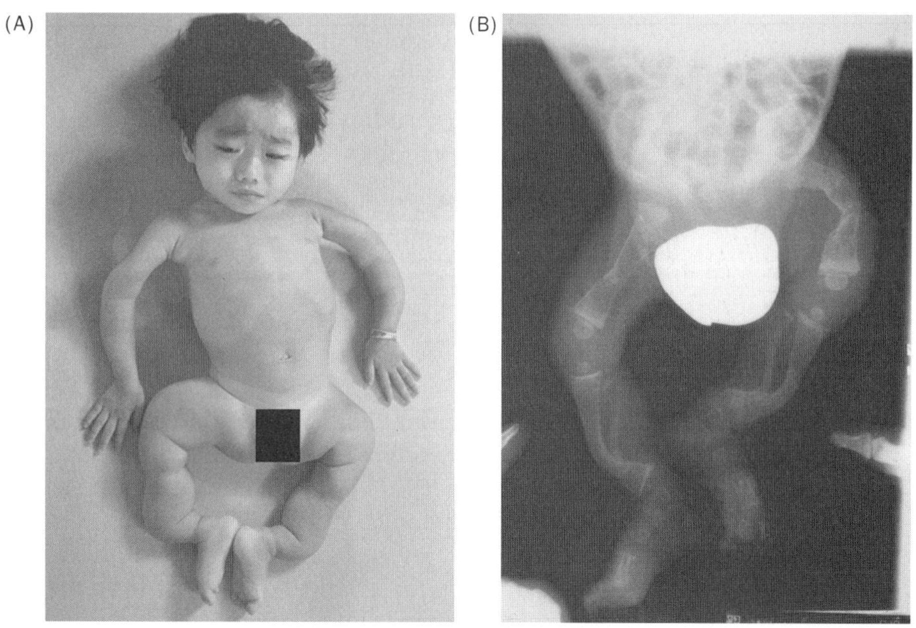

Figure 65-5 Osteogenesis imperfecta type III (Wasant et al., 1995a)

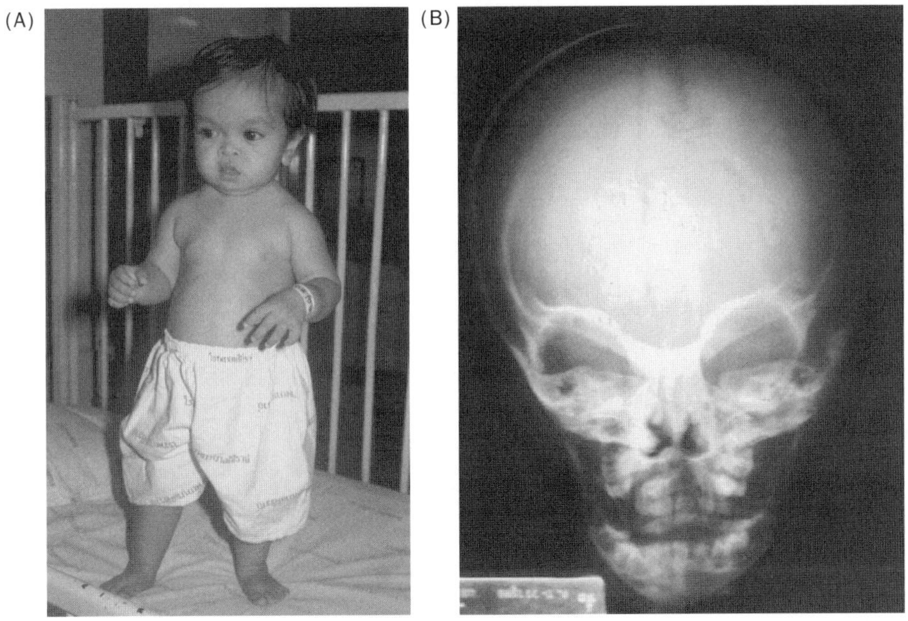

Figure 65-6 Pycnodysostosis (Wasant et al., 1995a)

progressive postnatal dorsal lordosis and scoliosis, acute swelling of pinnae of ears in infancy, later cauliflower ears, cleft or high-arched palate; short, thick and clubbed tubular bones

Radiography: ossified ear pinnae, precocious ossification of costal cartilage, scoliosis and lumbar interpedicular narrowing, short limb bones with broad metaphyses, delayed epiphyseal ossification, short, often ovoid first metacarpal with proximally located thumb, short limbs

Mode of inheritance: autosomal recessive; frequency: common, more than 300 cases reported to 1988, 0.2 per 100,000 births

PYCNODYSOSTOSIS (MIM# 265800)

Clinical manifestation: short-limbed dwarfism, prominent calvaria with smallness of facial structures, micrognathia, dental abnormalities, fractures, shortness of fingers, rare generalized hyperostotic bone disease recognized by short-limbed short stature, characteristic facies and wide anterior fontanelle persistent into adult life, large skull with frontal and occipital bossing, short and broad hands and feet; bluish sclerae and tendency to fractures with minimal trauma (Figure 65-6).

Differential diagnosis: cleidocranial dysplasia, osteopetrosis and other dense bone disorders

Mode of inheritance: autosomal recessive trait; frequency: uncommon, approximately 125 cases to 1988; increased frequency of parental consanguinity

Radiography: increased bone density, characteristic findings in the skull are wide sutures, Wormian bones, and small mandible

Mode of inheritance: autosomal recessive trait, cathepsin K deficiency

ELLIS VAN CREVELD SYNDROME (MIM# 225500)

Clinical manifestation: chondroectodermal dysplasia, extremely rare autosomal recessive syndrome, short distal limbs, postaxial polydactyly, nail hypoplasia, short ribs, natal teeth, oral frenulae and other features of an ectodermal dysplasia, 50% with cardiac defects typically common atrium, large atrial septal defect, meso/acromelic limb shortening. and small chest (Figure 65-7).

Radiography: normal skull, short ribs, normal vertebrae, square ilia with hooklike spurs at acetabulae, acromesomelic shortening with cone epiphyses, hamate–capitate fusion

Mode of inheritance: autosomal recessive, EVC protein gene mapped to 4p16 with unknown function

X-LINKED HYPOPHOSPHATEMIC RICKETS (MIM# 307800)

Clinical manifestation: recognized in infancy in boys and girls with bowing of legs, radiographic evidence of rickets resistant to vitamin D therapy,

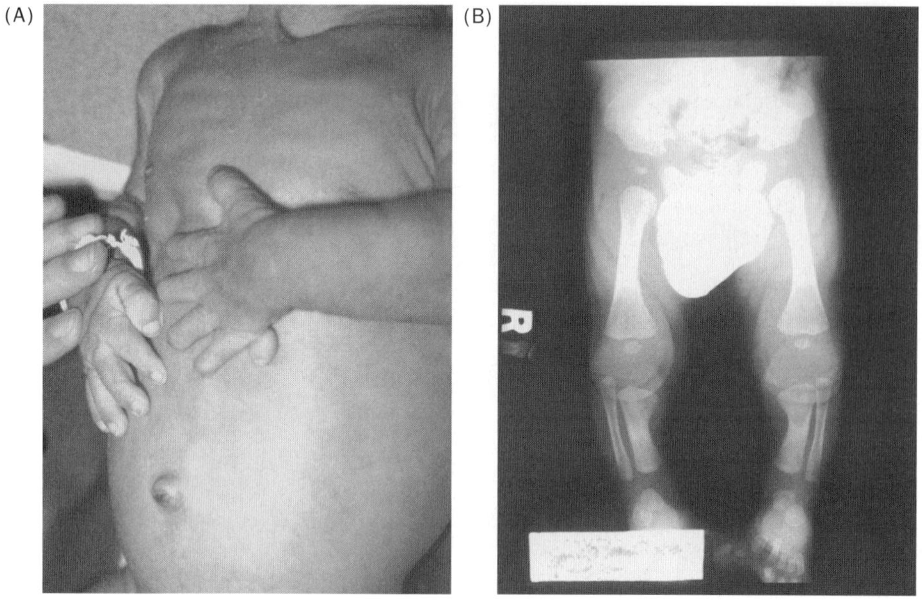

Figure 65-7 Ellis van Creveld syndrome (Wasant et al., 1995a)

and characteristically low serum phosphate (Figures 65-8 & 65-9).

Mode of inheritance: X-linked dominant trait

THANATOPHORIC DYSPLASIA (MIM# 187600)

Clinical manifestation: severe short-limbed dwarfism, narrowed chest, neonatal death, macrocephaly, depressed nasal bridge, curved short femurs, metaphyseal flaring, H-shaped flattened vertebrae, occasional clover-leaf skull (Figure 65-10).

Radiography: curved short long bones with metaphyseal spikes, short ribs and inverted U-shaped vertebrae in the lumbar area.

RHIZOMELIC CHONDRODYSPLASIA PUNCTATA (RCDP) (MIM# 215100)

Clinical manifestation: peroxisomal disorder, short humeri and femurs, wide coronal cleft in vertebrae, punctate epiphyseal mineralization, prenatal-onset growth deficiency, severe flat face, depressed nasal bridge and tip of nose, proximal shortening, cataracts, icthyosiform erythroderma, joint contractures, midface hypoplasia (Figure 65-11).

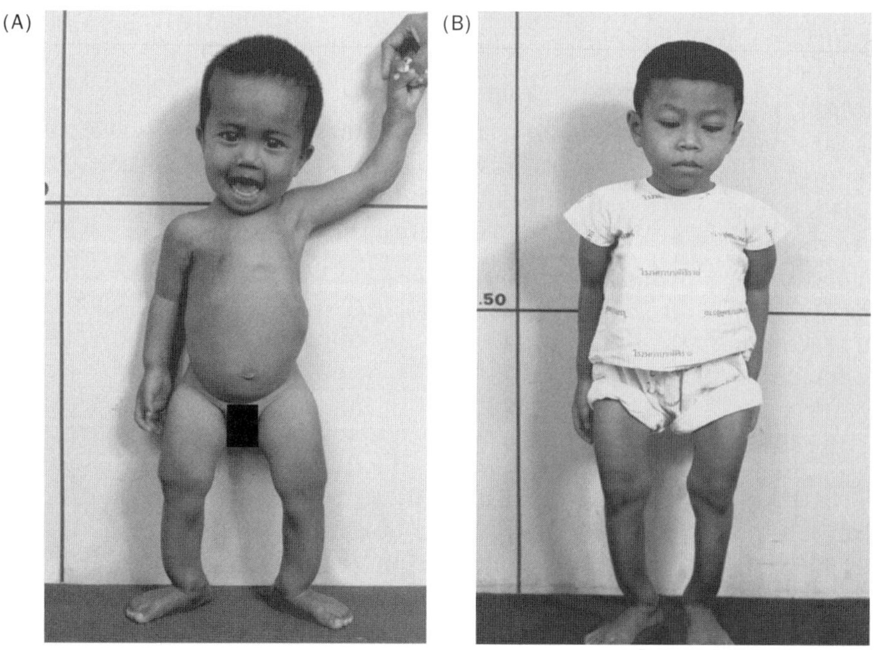

Figure 65-8 Hypophosphatemic rickets (Wasant et al., 1995a)

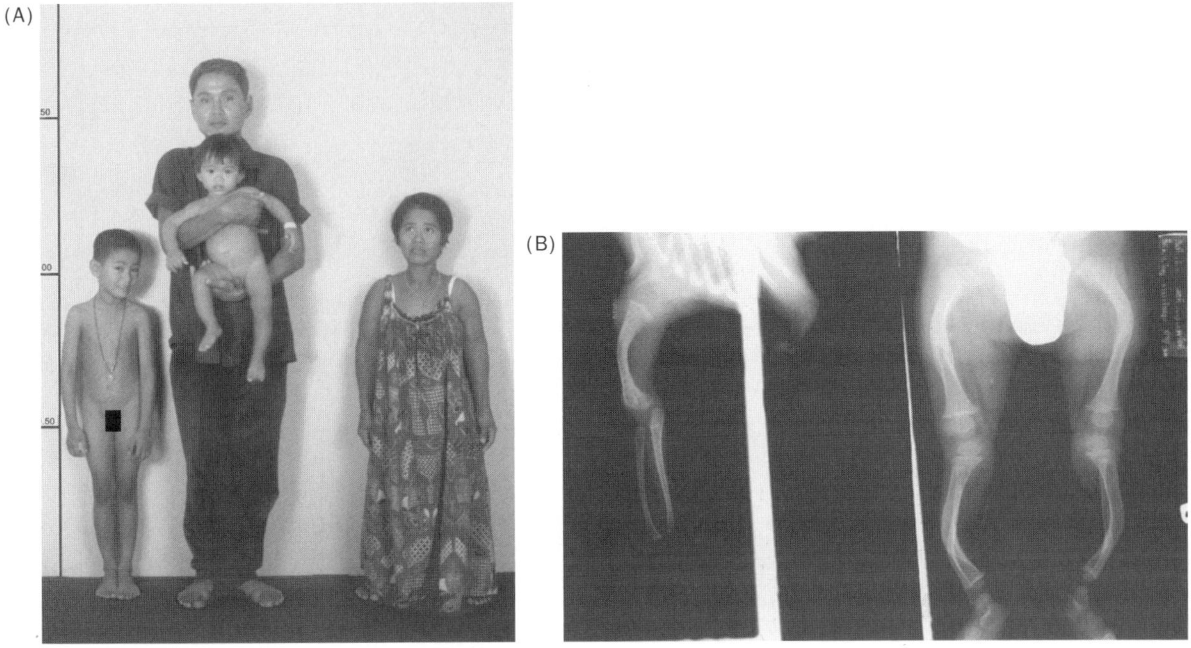

Figure 65-9 Hypophosphatemic rickets (Wasant et al., 1995a)

Radiography: short femurs and humeri, stippled calcification in epiphyses and periarticular areas

Mode of inheritance: autosomal recessive

HYPOCHONDROPLASIA (MIM# 146000)

Clinical manifestation: rhizomelic shortening of extremities, short broad hands, limited extension at elbow, mild short stature, muscular appearance, normal head to slight prominence of forehead, normal–mild lumbar lordosis, normal intelligence

Radiography: normal skull, normal to slightly flared ribs, lumbosacral interpedicular narrowing

Mode of inheritance: autosomal dominant

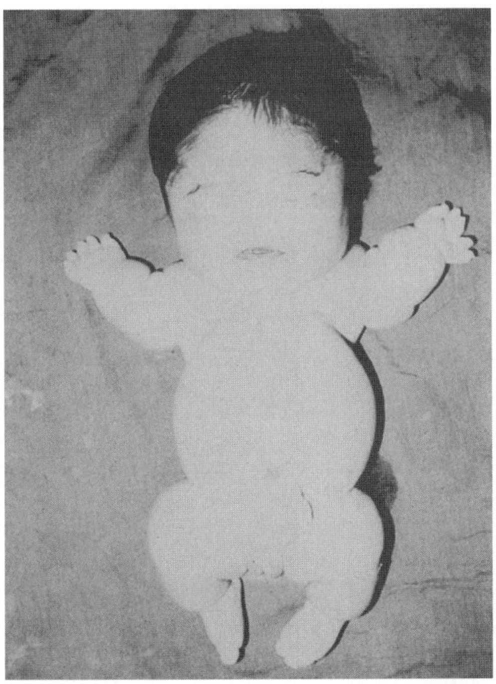

Figure 65-10 Thanatophoric dysplasia (Wasant et al., 1995a)

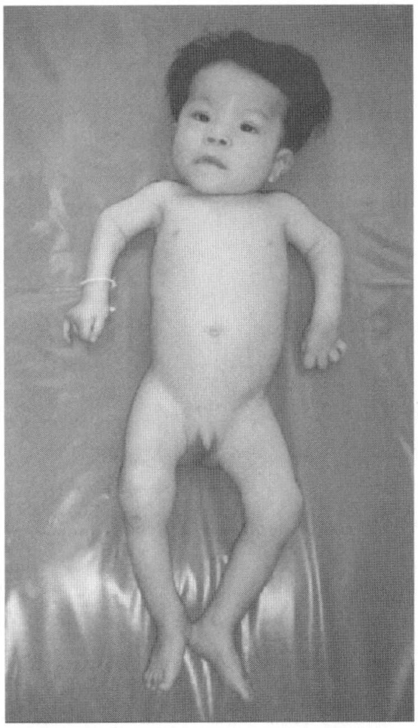

Figure 65-11 Rhizomelic dysplasia punctata (RCDP) (Wasant et al., 1995a)

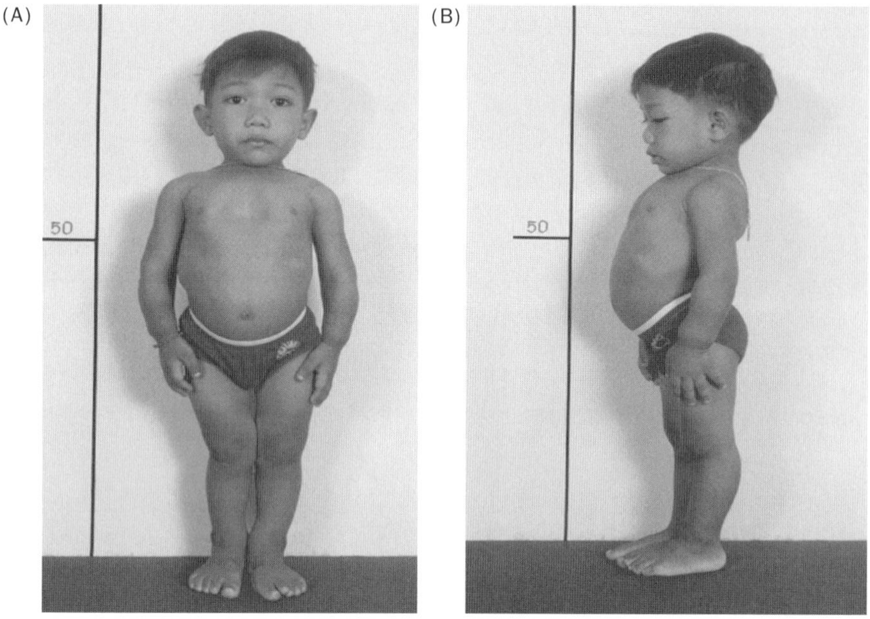

Figure 65-12 Spondyloepiphyseal dysplasia (SED) (Wasant et al., 1995a)

PSEUDOACHONDROPLASIA (MIM# 177150)

Clinical manifestation: normal skull and face, trunk appears disproportionately long, very short limbs with genu valgum or varum, hypermobility of joints, small broad hands, not manifested until at least 2 years of age

Radiography: normal skull and ribs, platyspondyly with anterior tongue-like protrusion, acetabular irregularity, generalized epiphyseal and metaphyseal dysplasia, short tubular bones with cupped metaphyses

Mode of inheritance: autosomal dominant

SPONDYLOEPIPHYSEAL DYSPLASIA (MIM# 183900, 184100, 313400)

Clinical manifestation: characterized by predominant involvement of vertebral bodies and epiphyses of the proximal joints, considerable phenotypical and genetic heterogeneity; classical severe form of SED congenita (SEDC) inherited as autosomal dominant trait, recognized at birth, osteoarthropathy of the hip joint, myopia (Figure 65-12).

Radiography: degenerative osteoarthropathy of the hip joints

Mode of inheritance: variable

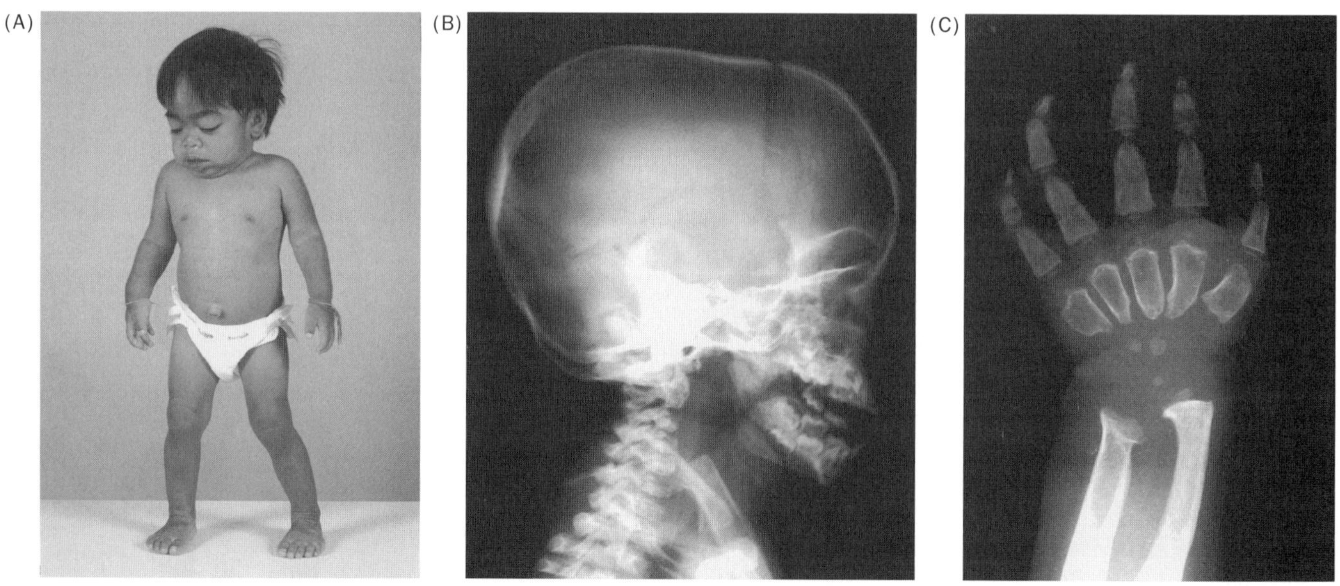

Figure 65-13 Hurler syndrome (MPS I) (Cairns et al., 2005; Wasant et al., 2006)

METABOLIC BONE DISEASES (MUCOPOLYSACCHARIDOSES)

HURLER SYNDROME (MUCOPOLYSACCHARIDOSIS I) (MIM#607014)

Clinical manifestation: coarse facies, severe mental retardation, coarse hair, hirsutism, corneal opacification, dwarfism, thoracolumbar gibbus, flexion contractures of joints, claw – hand deformity, harsh voice, hepatosplenomegaly, excess storage of acid mucopolysaccharides in tissues and excess excretion in urine (Figure 65-13).

Radiography: craniofacial-macrocephaly, J-shaped sella turcica, thoracolumbar gibbus, thorax, pelvis and hips, claw – hand deformity, "dysostosis multiplex congenita"

Mode of inheritance: autosomal recessive, defective enzyme alpha-L-iduronidase enzyme

HUNTER SYNDROME (MUCOPOLYSACCHARIDOSIS II) (MIM# 309900)

Clinical manifestation: Coarse facies, growth deficiency, stiff joints, clear cornea, hepatosplenomegaly, progressive deafness, mental retardation (Figure 65-14).

Radiography: Dysostosis multiplex congenita, ovoid-shaped vertebral bodies, hypoplastic iliac bones, shallow acetabular fossa and flared iliac wings, bullet-shaped proximal and middle phalanges, coarse bone trabeculation and thin cortices

Mode of inheritance: X-linked recessive

MORQUIO SYNDROME (MUCOPOLYSACCHARIDOSIS IV) (MIM# 253000)

Clinical manifestation: characteristic osteochondrodysplasia caused by the defective intralysosomal degradation of keratan sulfate, mild coarse facies, short-trunk dwarfism with normal intelligence, severe kyphoscoliosis, genu valgum, odontoid hypoplasia, corneal clouding, hearing loss, hepatomegaly, aortic and mitral valve stenosis. Most important complication: spinal cord compression at the upper cervical level due to atlantoaxial instability (Neufeld & Muenzer, 1995; Figure 65-15).

Radiography: dysostosis multiplex congenita, characteristic spondyloepiphyseal bone dysplasia with platyspondyly, anterior hypoplasia of the vertebral bodies at the thoracolumbar junction, odontoid hypoplasia, and shortening of the tubular bones

Mode of inheritance: autosomal recessive

MANAGEMENT OF PATIENTS WITH SKELETAL DYSPLASIA

- Effective management of these children includes precise diagnosis, prompt recognition of specific skeletal and nonskeletal complications, appropriate orthopedic and rehabilitative care, emotional support, psychological and genetic counseling.
- Prenatal diagnosis is now available for many of the skeletal dysplasias.

(A) (B)

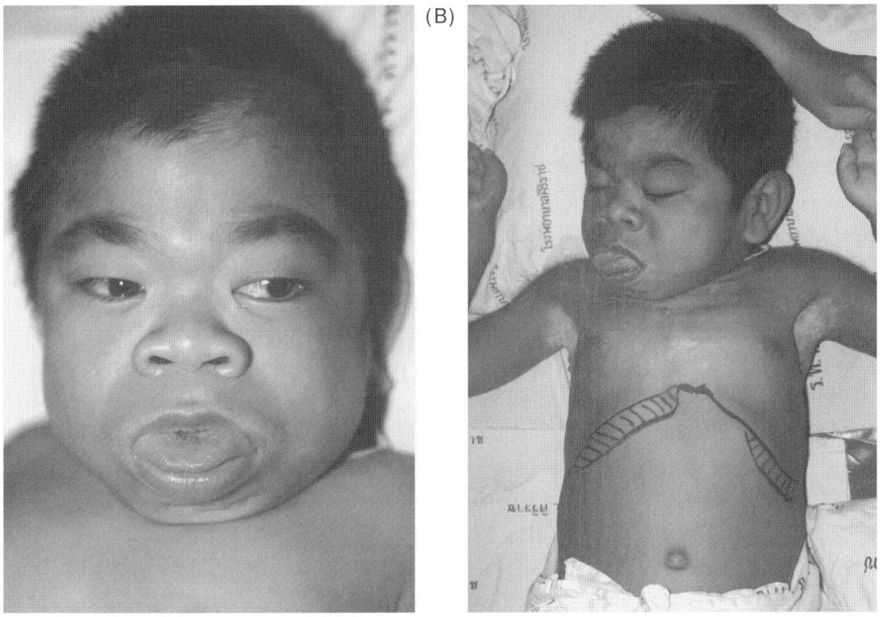

Figure 65-14 Hunter syndrome (MPS II) (Wasant et al., 1995b; Wasant & Liammongkolkul, 2006; Keeratichamroen et al., 2008)

(A) (B)

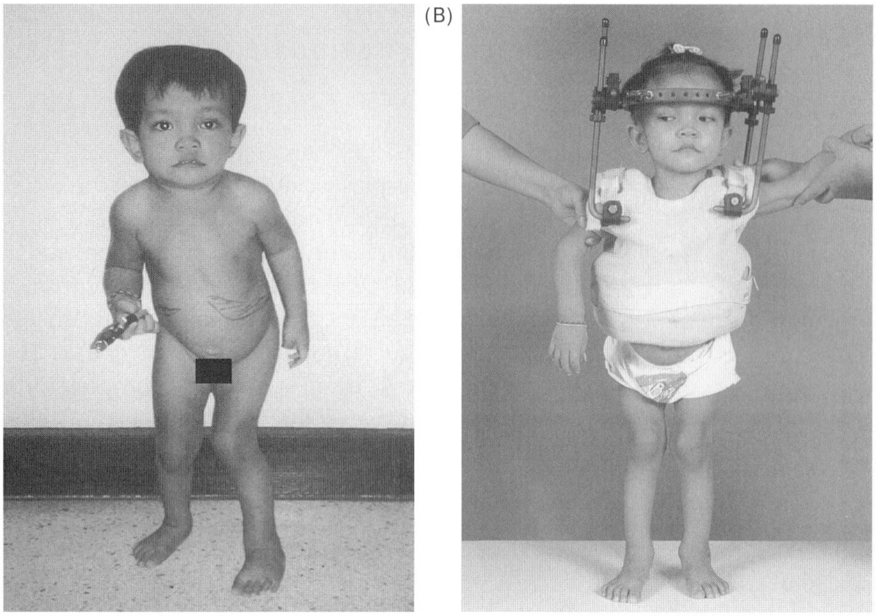

Figure 65-15 Morquio syndrome (MPS IV) (Suraseranivongse et al., 2003; Wasant & Liammongkolkul, 2006; Wasant & Liammongkolkul, 2006)

- Treatment with human growth hormone has been tried in achondroplasia, and a short-term beneficial effect has been reported.
- Surgical limb lengthening has been successfully applied for many years to patients with achondroplasia in Israel and Europe.
- It is important to realize, and to convey to the parents, that at this time there is no known cure for any of these conditions.
- Orthopedic, neurologic, other systemic signs and symptoms must be recognized as early as possible in order to maintain function and minimize morbidity.

GENETIC SKELETAL DYSPLASIA IN THE DEVELOPING COUNTRIES

Questionnaires were sent to the clinical geneticists in the Asia-Pacific countries (twice, in 2009 and 2010); e.g. Indonesia, Hong Kong (China), Malaysia, Philippines, Taiwan, Singapore. However, we received the only response from Malaysia (Meow Keong Thong). The author believes genetic skeletal dysplasia can be identified by any clinical geneticist in any clinical genetics clinic in any country. It usually depends on the special interest of the clinical geneticists and his/her training whether they will focus on this group of genetic disorders or not.

Common obstacles in the developing countries are: lack of genetic laboratories for cytogenetics, molecular diagnosis for confirmation of diagnosis, lack of qualified clinical geneticists and genetic counselors, and insufficient funding for treatment of genetic diseases. The solution should be increased government funding for genetic laboratories, and overseas training for clinical geneticists and genetic counselors. We also need increased public and professional awareness about inherited and genetic diseases, more research funding, and publication data on Asian patients with genetic disorders (Personal communication: Meow Keong Thong, Malaysia).

SUMMARY

Clinical evaluation of the skeletal dysplasias requires a wide variety of clinical, radiographic, and pathological tools. Diagnosis of the specific form of skeletal dysplasias can be of great importance in the prognosis, prevention, and treatment of these disorders, and in the provision of accurate genetic counseling.

ACKNOWLEDGMENT

The author would like to acknowledge the late Professor Emeritus Victor A. McKusick, the Moore Clinic, The Johns Hopkins Hospital, Baltimore, Maryland, U.S.A., who provided the author the opportunities to learn (1975–1977) as a postdoctoral fellow in medical genetics, and inspired the author during the training years, to appreciate and be knowledgeable about genetic skeletal dysplasia, which are a group of genetic disorders with great challenge, both in diagnosis and treatment.

The author would also like to thank Professor David L. Rimoin and Professor Ralph Lachman (Cedars Sinai Hospital, Los Angeles, U.S.A.) who provided several confirmations of the diagnoses of rare genetic skeletal dysplasia in our clinic.

Lastly, the author would like to thank Professor Jisnuson Svasti and his staffs at Chulabhorn Research Institute in Bangkok, who performed enzyme assays for patients with Hurler and Hunter syndromes; Khun Somporn Liammongkolkul, our chief biochemical scientist, who developed the thin-layer chromatography for identification of all patients with mucopolysaccharidoses; and Drs. Nitiwat Vattanavicharn and Achara Sathienkijkanchai, who helped take care of the patients.

REFERENCES

Cairns JRK, Keeratichamroen S, Sukcharoen S, et al. (2005). The molecular basis of mucopolysaccharidosis type I in two Thai patients. *Southeast Asian J Trop Med and Public Health 36*(5), 1308–1312.

Keeratichamroen S, Ketudat Cairns JR, et al. (2008). Molecular Analysis of the iduronate-2-sulfatase gene in Thai patients with Hunter syndrome. *JIMD Short Report #107* (Online).

Lachman RS. (2007). *Taybi and Lachman's Radiology of Syndromes, Metabolic Disorders, and Skeletal Dysplasias*, 5th edition. Mosby.

Rimoin DL, Lachman RS. (2007). Chondrodysplasias. In: *Emery and Rimoin's Principle and Practice of Medical Genetics*, 5th edition. Churchill Livingstone.

Suraseranivongse S, Wasant P, Sathienkijkanchai A, Thanapipatsiri S. (2003). Cervical cord decompression and posterior cervical immobilization following anterior c1-c2 subluxation in a patient with MPS Type IV (Morquio syndrome): a team approach. *Interdepartmental Conference, Siriraj Hosp Gaz 55*(4), 240–244.

Wasant P, Tritrakarn A, Kamolsilp M, et al. (1993). Ellis-Van Creveld syndrome (chondroectodermal dysplasia). Case report and review of literature. *Siriraj Hosp Gaz 1993; 45*: 850–854.

Wasant P, Tritrakarn A, Channgarm P, et al. (1994). Pseudoachondroplastic dysplasia (spondylometa- epiphyseal dysplasia). Case report and review of literature, *Thai Journal of Pediatrics 33,* 69–75.

Wasant P, Waeteekul S, Rimoin DL, Lachman RS. (1995a). Genetic Skeletal Dysplasia – Siriraj Experience. *Southeast Asian J of Trop Med Public Health 26*(Suppl 1), 59–67.

Wasant P, Wattanaweeradej S, Raksadawan N, Kolodny E H (1995b). Lysosomal Storage Disorder in Thailand: The Siriraj Experience. *Southeast Asian J of Trop Med Public Health 26*(Suppl 1), 54–58.

Wasant P, Liammongkolkul S, Vatanavicharn N. (2006). Mucopolysaccharidosis type I (Hurler syndrome) in Thai children – report of 4 cases (MPS IH, IH/S, IS). *Journal of Inherited Metabolic Disease 29*(Suppl 1), 154.

Wasant P, Liammongkolkul S. (2006). Mucopolysaccharidosis Type II (Hunter syndrome) in Thai children – report of 16 cases. *Journal of Inherited Metabolic Disease 29*(Suppl 1), 154.

Wasant P, Liammongkolkul S. (2006). Mucopolysaccharidosis Type IVa (Morquio syndrome) in Thai children – report of 3 cases. *Journal of Inherited Metabolic Disease 29*(Suppl 1), 155.

Wasant P, Liammongkolkul S. (2006) Mucopolysaccharidosis Type IVb (Morquio syndrome) in Thai children – report of 4 cases. *Journal of Inherited Metabolic Disease 29*(Suppl 1), 155.

Wasant P, Moser H. (2006). Rhizomelic chondrodysplasia punctata (RCDP) type I in 2 Thai infants-first reported case. *Journal of Inherited Metabolic Disease29*(Suppl 1), 139.

66.

MEDICAL GENETICS SERVICES IN THE PHILIPPINES

Carmencita Padilla, Eva Maria Cutiongco-de la Paz, and Catherine Lynn Silao

INTRODUCTION

The Philippines is an archipelago of more than 7000 islands with a population of almost 90 million. Of the 1.7 million babies born annually, the infant mortality rate is 23 per 1000; the neonatal infant mortality rate is 15 per 1000; the under-5 mortality rate is 28 per 1000; and the life expectancy is 72 years old (UNICEF, 2009). Infectious diseases have dominated the top ten causes of morbidity in the past seven decades and, thus, government efforts are primarily focused toward their eradication. Awareness of the impact of genetics on health is currently limited to impacts of newborn screening on the health of children. Congenital anomalies have been among the top 10 causes of infant mortality in the past seven decades, but there are no concrete programs responding to this problem. The Department of Health is currently identifying creative strategies to address the early identification, diagnosis, management, and referral of patients.

THE BEGINNINGS OF GENETICS IN THE PHILIPPINES

The beginnings of genetics were evident in the 1960s and 1970s upon the return of Filipinos who had received genetics training overseas. As a response to increased awareness and recognition of the value of genetic testing, cytogenetic laboratories were opened by the following: Domiciana Dauis-Lawas in 1963; Flora Pascasio and Bayani-Sioson in 1969; Angelina Tantengco in 1970; Teresita Briones in 1975; Florencio Isagani Medina in 1974; and the University of the Philippines Los Baños, Institute of Plant Breeding, in 1975. A cytogenetics laboratory specializing in tumor cells was opened at the Veterans Memorial Hospital in 1978. The Philippine Society of Human Genetics, Inc. was founded in 1980 by Briones, Pascasio, Tantengco, Palo-Garcia, and Medina. (Padilla, 1992).

A series of postgraduate courses were conducted in 1981 and 1982 on the following topics: Teaching Medical Genetics (Corazon Gonzales); Patterns of Inheritance (Fe Palo-Garcia); Management of Genetic Disorders (Flora Pascasio); Genetic Counseling (Teresita Briones); Cytogenetic and Cytological Analysis (Florencio Isagani Medina); Cytogenetic Disorders (Angelina Tantengco); Molecular Genetics (Marita Reyes); and Inborn Errors of Metabolism (Teresita Briones). In 1983, the topics were genetic growth problems and newborn screening. In 1984, urine metabolic screening was offered at the University of the Philippines Department of Biochemistry and, shortly thereafter, the Philippine Children's Hospital and the Makati Medical Center offered similar services. Also in 1984, a prenatal diagnosis clinic was opened at the University of Santo Tomas Hospital. The first edition of the local book, *Applied Human Genetics,* was launched in 1984. This heightened interest in genetics was dampened by the untimely demise of Briones and Pascacio in 1985 and 1986, respectively, as well as the retirement of Dauis-Lawas (Padilla, 1992). As a result, both clinical and laboratory genetic services slowed down.

During the 1980s, four units provided genetic counseling services, namely: (1) the Department of the Pediatrics in the University of the Philippines – Philippine General Hospital Medical Center, headed by Dr. Teresita Briones; (2) the Department of Obstetrics and Gynecology in the University of the Philippines – Philippine General Hospital Medical Center, headed by Dr. Fe Palo-Garcia; (3) the Department of Internal Medicine in the University of the Philippines – Philippine General Hospital Medical Center, headed by Dr. Flora Pascasio; and (4) the Department of Pathology and Microbiology in the University of the East – Ramon Magsaysay Memorial Medical Center, headed by Dr. Angelina Tantengco (Briones et al., 1984).

The Flora M. Pascasio, M.D. Research Foundation Inc. was launched on September 23, 1987 to promote and sponsor research and training projects in the disciplines of genetics and endocrinology. The foundation has conducted several postgraduate courses, with the aim of: (1) developing awareness of the importance and application of genetics in medicine and allied fields; (2) providing training through a practicum to demonstrate the practical application of genetics and related fields; and (3) stimulating and encouraging research in genetics, particularly its application in the Philippine setting (Reyes et al., 2001).

GENETICS BEYOND THE 1990s

The next wave of developments in medical genetics started with the setting up of genetic services in 1990 by the College of Medicine – University of the Philippines Manila (see below).

Although it was in the mid-1970s when the Philippine Nuclear Research Institute (PNRI) started to perform chromosomal analysis, it was only on field rats for research projects. Subsequently, chromosomal analysis for humans was added as a service. At present, the center focuses on providing services to occupationally exposed workers, endocrinology cases, and confirmatory testing for newborns and infants (personal communication with Mrs. Gregorio, 2010).

St. Luke's Medical Center (SLMC), a private hospital-based biomedical research facility in the country, contributed to the development of genetics in the Philippines through the establishment of its Research and Biotechnology Division (RBD) in July 1995 (Natividad, 2001). The facility is equipped with state-of-the-art equipment for conducting basic and medical research in the fields of molecular medicine, including cytogenetics, microbiology, biotechnology, and stem cell research. SLMC has five research centers within the RBD—the Center for Stem Cell Research, Center for Human Genome Research, Center for Drug Discovery and Development, Center for Clinical Trials, and Center for Biostatistics and Molecular Epidemiology. Through these centers, SLMC has been able to create integrated, multidisciplinary research programs in partnership with clinicians. In 2004, SLMC established a molecular diagnostics department. Among others, they have pathogen detection services utilizing various DNA techniques, molecular cytogenetic services for chromosomal analysis, cancer testing, chromosomal fragility testing, and bone marrow transplantation. The center also offers molecular immunology, electron microscopy, and other customized services for detection of gene mutations for various cancers and other genetic diseases and biomarkers, and use sequence-based technology for detection of unknown pathogens (personal communication with Dr. Natividad, 2010).

GENETIC SERVICES AT THE UNIVERSITY

GENETIC SERVICES AT THE UNIVERSITY OF THE PHILIPPINES DILIMAN

The Natural Science Research Institute (NSRI) – University of the Philippines Diliman established the DNA Analysis Laboratory in 1996. The mandate of the lab is to pioneer the development of forensic DNA technology in the country. The center has three main thrusts: research, extension, and training. NSRI aims to establish a population database for forensic applications. NSRI conducts DNA testing in cases upon request from individuals, private and government agencies, or as ordered by the court. They also provide training in DNA forensics to various stakeholders. NSRI also performs paternity testing and DNA profiling for the resolution of civil and criminal cases, and for the identification of disaster victims (personal communication with Dr. De Ungria, 2010).

Aside from these centers, the Philippine National Police (PNP) and the National Bureau of Investigation (NBI) also perform DNA testing, mainly for forensic applications. (De Ungria et al., 2008)

GENETIC SERVICES AT THE UNIVERSITY OF THE PHILIPPINES MANILA

The University of the Philippines (UP) Manila is the forerunner in the establishment of clinical and laboratory genetic services in the Philippines. Recognizing the need to provide comprehensive genetic services in the country, the Medical Genetics Unit was created in 1990 as a special project at the Office of the Dean of College of Medicine. It was then housed at the Department of Pediatrics of the Philippine General Hospital (PGH). A dysmorphology clinic, more recently known as the Clinical Genetics Clinic, was then opened at the outpatient department of the Philippine General Hospital (PGH) in 1991. Cytogenetics laboratory services were also offered in the Pediatric Research Laboratory in the same year. In 1997, the laboratory genetic services were transferred to the National Institutes of Health (NIH). By 1998, the Medical Genetics Unit became the Institute of Human Genetics (IHG), a component institute of the NIH. The IHG is the largest provider of genetic services in the country today, providing services in clinical genetics, cytogenetics, newborn screening, molecular genetics, and biochemical genetics, committed to make genetic testing available and accessible to Filipinos all over the country (Padilla, 2008).

Due to the continued growth of the services being offered, the institute is now composed of various professionals—clinical geneticists, biochemical geneticists, endocrinology/metabolic physicians, molecular geneticists, molecular biologists, biochemists, chemists, medical technologists, nurses, genetics fellows, university researchers, and technical and administrative support staffs—from what started with a single staff member in 1990.

CLINICAL GENETICS SERVICES

The Clinical Genetics Unit provides comprehensive clinical and laboratory diagnostic services to families and

TABLE 66-1 TOP 5 DIAGNOSES FOR OUT-PATIENT CLINIC CONSULTATIONS AND IN-PATIENT REFERRALS (PADILLA, 2008)

Out-patient clinic consultations

1. Down Syndrome
2. Maple Syrup Urine Disease (MSUD)
3. Non-syndromic Global Developmental Delay/Mental Retardation
4. Phenylketonuria (PKU)
5. Multiple congenital anomalies

In-patient referrals

1. Down syndrome
2. MSUD
3. Multiple congenital anomalies
4. VACTERL
5. Failure to thrive

individuals with, or at risk for, an inherited disease. The clinics offer information on the etiology, diagnosis, and treatment of specific genetic disorders, as well as genetic counseling. Tables 66-1 and 66-2 list the common conditions seen Clinical Genetics Unit (Padilla, 2008).

The unit, together with the Department of Pediatrics in PGH, offers a two-year fellowship program in Clinical Genetics. It is designed to provide broad clinical exposure to areas of dysmorphology, biochemical genetics, cytogenetics, molecular genetics, and neonatal screening programs. Some important components of the training program are genetics and metabolic clinics, pediatric ward rounds, consultations regarding familial disorders and birth defects, and participation in regularly scheduled pre-clinic and post-clinic conferences. Experience in genetic counseling and training in laboratory procedures for the diagnosis of genetic disorders are likewise included (Padilla, 2008). In response to the Department of Health's goal of providing genetic services, IHG, together with the Department of Pediatrics PGH is currently offering a two-year Master in Science in Genetic Counseling (Padilla, 2008).

The Clinical Genetics Unit also holds a metabolic clinic twice a month to take care of the diagnostic and therapeutic needs of patients with metabolic disorders. The metabolic team undertakes enzyme replacement therapy in patients with lysosomal storage diseases—three cases of Gaucher disease and one case of Pompe disease (Chiong et al., 2009). Referrals of patients from other departments of PGH and other hospitals nationwide for confirmation of IEMs are also accepted by the clinic.

Recently, the Clinical Genetics Unit added a mucopolysaccharidosis (MPS) clinic, which is held every six months. It is a multidisciplinary clinic where MPS patients are seen by cardiologists, pulmonologists, otorhinolaryngologists, ophthalmologists, developmental specialists, and geneticists.

TABLE 66-2 TOP DIAGNOSES FOR PRENATAL COUNSELING CONSULTATIONS (PADILLA, 2008)

1. Hydrocephalus
2. Anencephaly
3. Conjoined twinning
4. Gastroschisis

CYTOGENETIC SERVICES

The Cytogenetics Unit performs diagnostic procedures for patients with multiple birth defects, mental retardation, and abnormal sexual development, couples with infertility or multiple miscarriages, and patients with malignancies and hematologic disorders, using samples from peripheral blood cells, solid tissues, and bone marrow. It also offers high-resolution banding and fluorescence in-situ hybridization (FISH) analysis in testing for Angelman syndrome, Prader Willi syndrome, Di George/Velo-cardio-facial syndrome, William syndrome, chronic myelogenous leukemia (CML), and breast cancer.

The cytogenetics laboratory provides cytogenetic analysis of peripheral blood, cord blood, bone marrow, and skin/tissue samples throughout the country. In a 16-year review of the chromosomal analyses, 1.8% of samples were skin and tissues, 8.5% were cord blood, 14.9% were bone marrow, and the remaining 74.8% were peripheral blood. The majority of samples were from patients referred by physicians from government or private hospitals and private health practitioners, and some were research samples on the cytogenetic effects of environmental toxins (i.e., pesticides, etc.) on high-risk populations (Padilla et al., 2009).

In a retrospective study of the results of chromosomal analyses done on peripheral blood from 1991 to 2007, the top ten reasons for referral for cytogenetic studies were confirmation of a chromosomal diagnosis, cytogenetic effects of environmental toxins, recurrent miscarriages/poor obstetric history, multiple congenital anomalies, possible chromosomal findings in syndromes, ambiguous genitalia/CAH suspect, developmental delay, parents and siblings of patients with chromosomal aberrations, cytogenetic effects of leukemia, and hydrops fetalis. Results showed numerical chromosome abnormalities were more common than structural chromosome abnormalities. Full trisomy 21 was the most common aneuploidy seen. Classic Turner syndrome was the most frequent sex chromosome abnormality identified. Deletions, additions, and translocations were the most common structural chromosome abnormalities ascertained (Table 66-3; Padilla et al., 2009).

BIOCHEMICAL GENETICS SERVICES

The Biochemical Genetics Unit provides diagnostic testing and physician-assisted consultative services to clinicians to promote adequate diagnosis and management of inherited metabolic disorders. The array of biochemical tests includes urine metabolic screen by high-voltage electrophoresis,

TABLE 66-3 DISTRIBUTION OF CHROMOSOMAL ABNORMALITIES SEEN AT THE CYTOGENETICS LABORATORY, IHG-NIH-UP (1991–2007) (PADILLA ET AL., 2009)

Chromosome abnormality		Number cases				
Numerical						**2368**
Autosomal abnormalities					2136	
	Trisomy 21			1858		
		Full	1640			
		Mosaic	134			
		Translocation type	62			
		Other variation	22			
	Trisomy 18			228		
		Full	217			
		Mosaic	7			
		Translocation type	1			
		Other variation	3			
	Trisomy 13			44		
		Full	38			
		Translocation type	3			
		Mosaic	3			
	Trisomy 8			1		
	Trisomy 9			1		
	Trisomy 11			1		
	Trisomy 19			1		
	Monosomy 21			1		
	Trisomy 22			1		
Sex chromosome abnormalities					231	
	Turner syndrome (45, X)			185		
		Classic	72			
		Mosaic with normal cell line	47			
		Mosaic with i(X)	28			
		Mosaic with Trisomy X	3			
		Other variation	35			
	Klinefelter (47, XXY)			13		
	Trisomy X			11		
		Full	10			
		Mosaic	1			
	Monosomy Xp			8		
	48, XXXY			4		
	49, XXXXY			4		
	47, XYY			3		
	Monosomy Xq			2		
	48, XXYY			1		
Triploidy					1	

(Continued)

TABLE 66-3 (CONTINUED)

Chromosome abnormality		Number cases				
Structural						**366**
Deletions					112	
	Terminal			98		
	Mosaic			12		
	Interstitial			2		
Translocation					99	
	Two-break			87		
	Robertsonian			10		
	Three-break			2		
Addition					64	
Marker chromosomes					44	
Insertions					12	
Ring chromosomes					9	
Fragile sites					9	
Inversions					7	
Duplications					4	
Derivative chromosomes					3	
Isochromosomes					2	
Dicentric chromosome					1	

* Does not include research data and results from bone marrow and solid tissue.

quantitative amino acid analysis by high-performance liquid chromatography, and organic acid analysis by gas chromatography-mass spectrometry (Padilla, 2008).

From January 2002 to September 2011, there were a total of 3445 patients referred to the biochemical genetics laboratory for a diagnostic investigation. The majority were infants from 29 days to 24 months old (52.64%), followed by children from 24 months to 6 years old (20.96%) and neonates from 0–28 days old (13.82%). The top three clinical indications for referral were neurologic disorders (34.04%), neurodevelopmental disorders (25.5%), and infectious disorders (5.76%). MSUD was the most common inborn error of metabolism (IEM) detected (Table 66-4).

MOLECULAR GENETICS SERVICES

The goals of the Molecular Genetics Unit are to understand the molecular, genetic etiology of a range of single gene disorders, syndromes, and complex traits in humans; to develop new tests for diagnosing and monitoring these conditions; and to translate research findings into improved care for children, adults, and families.

Research undertakings in the unit focus on identification of changes in the DNA of candidate genes, and on whether these changes contribute toward the clinical manifestations, onset, and severity of diseases. Although specific research areas across the unit are disorder-diverse, the common goals are reflected within all the research groups. In collaboration with local and international academic clinicians and scientists and with specialist groups, the unit is involved in several interdisciplinary research studies of the genetics of monogenic and complex genetic conditions in the Filipino population. Using relevant technology, the unit also undertakes studies on the genetic basis of common conditions such as cardiovascular disease, cancer, and diabetes.

GENETIC CONDITIONS BEING STUDIED

All the efforts by the Molecular Genetics Unit in recent years have been aimed at more effectively understanding the mechanisms leading to a wide variety of disorders, their diagnosis and treatment, as well as prevention. Rapid advancement is expected to continue with further research in the pipeline on cancer, cardiovascular diseases, and metabolic disorders, to name a few.

METABOLIC DISORDERS

Studies on the characterization of common mutations of inborn errors of metabolism have been performed in

TABLE 66-4 REGISTRY OF IEM CASES AT BIOCHEMICAL GENETICS UNIT* (BIOCHEMICAL GENETICS UNIT, IHG)

IEM Case	Number of cases
Maple Syrup Urine Disease (MSUD)	101
Phenylketonuria	17
Mucopolysccharidosis	36
Galactosemia (classical)	5
Galactosemia (non-classical)	18
X-linked Adrenoleukodystrophy	11
Methylmalonic Acidemia	15
Gaucher Disease	10
Hyperphenylalanemia	10
Pompe Disease	3
Carbamoyl Phosphate Synthase Deficiency	4
Ornithine Transcarbamoylase Deficiency	3
6-Pyruvoyltetrahydropterin Synthase Deficiency	4
Argininosuccinic lyase deficiency	1
Holocarboxylase Synthase Deficiency	1
Citrullinemia	1
Tay Sach's Disease	1
Tyrosinemia Type I	1
HeterozygousCystinuria	9
Homozygous cystinuria	3
Classical Homocystinuria	1
Multiple Sulfatase deficiency	1
Lowe Syndrome	1
Neimann Pick Syndrome	1
Menkes	1
Complex 1 deficiency (Mitochondrial respiratory chain disorder)	1

*As of September 2011

the country, particularly on maple syrup urine disease (MSUD), phenylketonuria (PKU), methylmalonic aciduria (MMA), and glucose-6-phosphate dehydrogenase (G6PD) deficiency.

More than 100 cases of MSUD are recorded in the registry of the Biochemical Genetics Unit of the Institute of Human Genetics, National Institutes of Health, as of February 2010. There has been a steady increase noted in the number of patients diagnosed in recent years (Padilla et al., 2001). Studies on the most common gene deletions—particularly the 4.6 kb deletion of the dihydrolipoamide branched chain transacylase E2 (DBT) gene—have been conducted, and results were found to be beneficial for the early diagnosis and determination of the proper course of treatment of the patients, in order to prevent permanent neurological impairments (Silao et al., 2004; 2008).

From 1996 to 2010, 16 out of 2,389,959 newborns were diagnosed with phenylketonuria (PKU) in the Philippines. Filipino PKU patients have been found to have previously identified p.I65T, p.R413P, p.Ex6-96A>G, p.R243Q mutations in the PAH gene. This significantly aided in the medical management and genetic counseling of patients and their families (Silao et al., 2009).

Since MMA is one of the diagnostic challenges for pediatricians, sequence analysis of the MUT gene has been done. The molecular genotyping of Filipino MMA patients showed compound heterozygosity for c.1595G>A, c.2011A>G, c.322C>T, c.982C>T and c.1280G>A, allowing directed mutational analyses for the local population (Silao et al., 2009).

Another common genetic disorder, CAH, was found to have an estimated crude incidence higher than what is reported in most populations. Studies that used a method of combined differential PCR and amplification created restriction site approach, and direct probing for the presence of known mutations in the CYP21 and CYP21P genes of Filipino CAH patients showed that a majority of cases had a premature splicing error mutation. Determination of frequent alleles facilitates rapid screening for mutations in the 21-OH gene and leads to a definitive diagnosis and possible prenatal intervention (Cutiongco-de la Paz et al., 2009).

Based on the newborn screening data as of December 2010, 44,273 babies have been diagnosed with G6PD deficiency in the country (NSRC), and thus genetic studies have been undertaken. Using multiplex polymerase chain reaction (PCR) with multiple tandem forward primers and a common reverse primer (MPTP), G6PD variants were detected with the G6PD Viangchan being the most common, followed by the G6PD Union, G6PD Vanua Lava, G6PD Chatham and G6PD Canton (Padilla et al., unpublished).

CANCER

A molecular epidemiologic study that attempted to associate environmental factors and specific genetic polymorphisms with cancer susceptibility was done. The study was able to identify genetic polymorphisms that could be potential biomarkers for breast, colorectal, and oral cavity cancer in the Filipino population. Of these, only one—the GSTP1c. 313A>G homozygote genotype in oral cavity cancers—was found to be associated with cancer risk when adjusted for epidemiologic risk factors identified in the study (Cutiongco-de la Paz et al., unpublished). In the same study, epidemiological risk factors that were identified to be significantly related to the cancers include:

cigarette smoking for lung and oral cavity cancer; passive smoking, tobacco chewing, and inverted cigarette smoking for oral cavity cancer; family history of cancer for lung and breast cancer; and increasing age at first pregnancy for breast cancer. These findings are consistent with existing literature from other countries, and are all believed to be preventable, reinforcing the need to always include public awareness and continuing healthcare education in cancer prevention programs in the country (Ngelangel et al., 2009).

The highest breast cancer incidence rates in Southeast Asia have been reported by the Manila Cancer Registry in the Philippines. A case-control study on Filipinos revealed that BRCA mutations had a prevalence of 5.1%, with a prevalence of 4.1% for solely BRCA2 mutations. This may be due to the presence of 2 common founder mutations in BRCA2 among Filipino women with breast cancer (De Leon Matsuda et al., 2002).

Another study on cancer, specifically of the colon and rectum, revealed that the incidence of *ras* mutations among Filipinos is lower than other populations. This data implies that the detection of this mutation is a highly sensitive but not a specific diagnostic tool for colorectal cancer among Filipinos (Carillo et al., 2009).

CARDIOVASCULAR DISEASES

Studies that aim to identify genes among Filipinos that place the population at risk for cardiovascular disease have also been done. One of these studies involves the association of mutations on the low-density lipoprotein-receptor (LDL-R) gene that is related to familial hypercholesterolemia (FH). There were 6 novel mutations identified, and it was found that there is a significant association between mutations and number of clinical variables, including family history of dyslipidemia, FH score, and LDL-cholesterol level (Punzalan et al., 2005).

Another study aimed to associate cholesterol ester transfer protein (TaqIB) polymorphism with HDL-C levels among Filipinos. The frequencies of polymorphisms were 40% for B1B1, 50% for B1B2, and 10%for B2B2. Additionally, B1B1 is found to be associated with low HDL-C levels. With the identification of these polymorphisms, causes of low HDL-C other than the traditional causes have been established (Sy R et al., 2007).

PAN-ASIAN SNP

The Institute participated in a large-scale study that looked at the autosomal variations of the Asian population. Results showed considerable relatedness and prevalent gene flow among the populations involved, and may indicate a predominant south-to-north migration of East Asian populations. Data suggested that there is a common origin for all East Asian and Southeast Asian populations studied; the majority of the gene pool—both of the Negrito and non-Negrito populations—in Asia was derived from a single initial entry of modern humans into the continent (HUGO Pan-Asian SNP Consortium, Science, 2009).

STUDY GROUPS, SURVEILLANCE PROJECTS, AND REGISTRIES

Aside from the clinical and diagnostic services offered by the IHG, it is also home to several study groups (as listed in Table 66-5). The Institute is actively involved in the conduct of research studies that are ethically sound and culturally sensitive. The studies aim to understand the genetic basis of common but complex disorders found in the population. The results are focused on the development of intervention strategies at the level of the doctor and patient. The study groups also target the formulation of policies at the national level to further improve the standards of genetic health care in the country (IHG Executive Summary, 2010).

IHG also conducts the Philippine Birth Defects Surveillance Project, a product of the Philippine Birth Defects Surveillance Study Group, which is a joint effort with the Department of Health (DOH). Congenital malformation of the heart ranks 6th, and other congenital anomalies rank 8th among the top ten causes of infant mortality (Philippine Health Statistics 2004, DOH). The causes are unknown and risk factors and knowledge on preventive measures are not yet established. The surveillance of Filipino newborns with birth defects was initiated in September 2008 under the guidance of the March of Dimes Global Network for Maternal and Infant Health. This project aims to provide statistics that will be significant in policy and program planning, and also to address other urgent public health concerns, i.e., prenatal care, genetic counseling, and management of preventable causes and risk factors. From this project, the Philippine

TABLE 66-5 IHG STUDY GROUPS (IHG EXECUTIVE SUMMARY)

1. Genetics of Diabetes Study Group
2. Cardiovascular Genetics Study Group
3. Philippine Cancer Genetics Study Group
4. Bipolar Research Consortium of the Philippines
5. Philippine Schizophrenia Study Group
6. Philippine Autism Study Group
7. Philippine Thalassemia Research Study Group
8. Philippine Kawasaki Disease Study Group
9. Philippine Subacute Sclerosing Panencephalitis (SSPE) Study Group
10. Philippine Birth Defects Surveillance Study Group
11. Philippine Oral Cleft Study Group and Phenotype and Genetics in Oral-facial Cleft in Families Study Group

Birth Defects Registry was created in order to establish the incidence and prevalence of birth defects in the country (Table 66-6).

Another registry of the IHG is the Philippine Oral Cleft Registry (POCR), which was launched in May 2003. It is under the initiative of the Philippine Oral Cleft Registry Study Group (POCRSG), composed of the IHG-NIH-UP Manila, the Philippine Association of Plastic, Reconstructive and Aesthetic Surgeons, Inc. (PAPRAS), the Philippine Society of Otolaryngology-Head and Neck Surgery (PSO-HNS), the Operation Smile Philippines Foundation, Inc. (OSPF), and the Philippine Band of Mercy (PBM). The objective of the group is to implement a registry program for patients with oral clefts (cleft lip and/or palate) in the Philippines. Specifically, the POCSG aims to determine the incidence of oral clefts, the different types of oral clefts, and identify possible risk factors for oral clefts in the country. The registry also collects data on maternal history and demographic profile to better understand oral cleft occurrences. The POCSG aims to cover all identified cases of oral clefts in the country.

From May 2003 to December 2006, covering all cases of oral clefts in the country both from community-based surgical missions and from hospitals, the POCR recorded a total of 2324 cases representing a prevalence of 0.42 per 1000 live births (or 1:2367 live births) using an estimated birth rate of 1.5 million babies a year, assuming full ascertainment. This is an underestimate of the true prevalence of oral clefts in the country, since there are other groups or organizations that are doing oral cleft missions (Padilla et al., 2008).

Of the 2324 cases reported, the majority were males (60.11%) with an overall female to male ratio of 1:1.5. This slight male preponderance has also been reported by previous studies. The combination of cleft palate with cleft lip accounted for the highest number of cases (56.8%) followed by cleft lip only (25.3%) and cleft palate only (17.9%; see Padilla et al., 2008).

TABLE 66-6 TOP BIRTH DEFECTS REGISTERED AT THE BIRTH SURVEILLANCE PROJECT AS OF AUGUST 2011

Rank	Birth Defect
1.	Cleft lip and palate (CLAP)
2.	Limb deformities
3.	Skin tags
4.	Ankyloglossia
5.	Multiple congenital anomalies (MCA)
6.	Neural tube defects (NTD)
7.	Undescended testis (Cryptorchidism)
8.	Down Syndrome
9.	Hydrocephalus
10.	Nevus

Birth Defect
Oral and facial cleft (cleft lip, clect palate, facial cleft)
Limb deformities/limb reduction
Multiple congenital anomalies (MCA)
Neural tube defects (NTD)
Undescended testis (Cryptorchidism)
Down Syndrome
Hydrocephalus
Imperforate Anus

NEWBORN SCREENING

In 1996, the Philippine Newborn Screening Project was launched in collaboration with 24 private and government hospitals in Metro Manila. Newborn screening (NBS) in the Philippines started in 1996 as a pilot project initiated by the Newborn Screening Study Group, composed of pediatricians and obstetricians in 24 hospitals (listed in Table 66-7) in Metro Manila (Padilla, 2003). The pilot project established the incidence of five metabolic conditions—namely, congenital hypothyroidism (CH), congenital adrenal hyperplasia (CAH), galactosemia (GAL), phenylketonuria (PKU), and homocystinuria (HCY)—which were used as the basis for recommendations to the government for policy adoption. The program proceeded in four phases: (1) 1996, screening for CH, CAH, GAL, PKU, HCY; (2) 1998, pilot screening for glucose-6-phosphate dehydrogenase (G6PD) deficiency; (3) 2000, evaluation of the program for cost effectiveness and recommended policy changes; and (4) 2004, integration of NBS into the public health delivery system with the enactment of Republic Act 9288, or Newborn Screening Act of 2004. This act institutionalized the "National NBS System," which shall ensure the following: (a) that every baby born in the Philippines is offered NBS; (b) the establishment and integration of a sustainable NBS system within the public health delivery system; (c) that all health practitioners are aware of the benefits of NBS and of their responsibilities in offering it; and (d) that

TABLE 66-7 PREVALENCE OF THE FIVE CONDITIONS SCREENED IN THE NEWBORN SCREENING PROGRAM, AS OF DECEMBER 2010 (NSRC)

Condition	Screened	Confirmed	Prevalence
CH	2,389,959	719	1:3,324
CAH	2,389,959	253	1:9,446
GAL	2,389,959	22	1:108,635
PKU	2,389,959	16	1:149,372
G6PD Deficiency	2,312,144	44,273	1:52

all parents are aware of NBS and their responsibility in protecting their child from any of the disorders (NSRC). The highlights of the law and its implementing rules and regulations are:

- DOH is the lead agency tasked with implementing this law;
- Any health practitioner who delivers or assists in the delivery of a newborn in the Philippines shall, prior to delivery, inform parents or legal guardians of the newborns of the availability, nature, and benefits of NBS;
- Health facilities shall integrate NBS in their delivery of health services;
- Creation of a Newborn Screening Reference Center at the NIH to serve as the technical partner of the Department of Health;
- Establishment and accreditation of NBS centers equipped with a NBS laboratory and recall/follow-up program;
- Provision of NBS services as a requirement for licensing and accreditation by the DOH and the Philippine Health Insurance Corporation (PHIC); and
- Inclusion of the cost of NBS in insurance benefits (Padilla et al., 2009).

With the implementation of the Newborn Screening Act of 2004 (Republic Act 9288), the advocacy and policy component of newborn screening was transferred to the Newborn Screening Reference Center (NSRC), which is also housed at the NIH (Padilla, 2008). Today, five conditions are included in the screening panel: CH, CAH, PKU, GAL and G6PD deficiency (Padilla, 2008). From 24 hospitals in 1996, there are now more than 3000 newborn screening facilities throughout the country as of December 2010 with 35% national newborn screening coverage.

PRENATAL DIAGNOSIS

Currently, prenatal diagnosis is practiced in a very restricted way in the Philippines. The most extensively used prenatal diagnostic procedure is prenatal ultrasonography, which utilizes 2-dimensional ultrasound for congenital anomaly screening during the second trimester of pregnancy. This can be complemented by fetal echocardiography when indicated, such as in cases of congenital heart defects detected after a congenital anomaly scan. First-trimester ultrasound screening has been introduced in some urban medical centers. In this procedure, fetal nuchal translucency is measured and the presence or absence of the nasal bone is assessed, both of which screen for the most common chromosomal abnormality, Down syndrome. More advanced techniques of 3-dimensional and 4-dimensional ultrasonography have also become popular lately, and have been included in routine prenatal diagnosis. Color Doppler ultrasound is also used by most perinatology centers in the country as a prenatal diagnostic technique to detect intrauterine growth retardation (Cutiongco-de la Paz, 2006).

Maternal serum screen, whether using single or multiple markers, as a noninvasive form of prenatal diagnosis, has not been customarily offered in the country. For the results of this screening to be valid, a population-based analysis of the serum markers must be conducted, due to the existence of race-ethnicity variation (Cutiongco-de la Paz, 2006).

Amniocentesis has been done in some cases in the past, not for the purpose of termination of pregnancy but more to provide guidance for obstetricians and pediatricians as they plan for anticipatory care in preparation for the birth of an affected child. Likewise, chorionic villi sampling has not been normally offered in any institution. Other important issues that need to be addressed include the processing and analysis of samples from amniotic fluid and chorionic villi, since there are only a few genetic laboratories in the country able to perform such procedures. Additionally, although cost is one of the major issues deterring the general population from having invasive prenatal diagnosis procedures, the more prevailing concern is that termination of pregnancy is not allowed by law in the country (*Article II, Section 12 of the 1987 Constitution of the Philippines*) (Cutiongco-de la Paz, 2006).

The Committee for the Formulation of Ethics and Practice Guidelines in ART and IUI of the Philippine Society of Reproductive Endocrinology and Fertility published a handbook entitled "Guidelines on the Ethics and Practice of Assisted Reproductive Technology and Intrauterine Insemination." These guidelines aim to direct and regulate the development or promotion of novel assisted reproductive procedures. One of the pertinent issues mentioned in the handbook is preimplantation genetic diagnosis, in anticipation of the ethical issues that may arise in the use of techniques such as PGD in assisted reproductive technology. The guidelines state that "PGD is an acceptable procedure when carried out to identify specific genetically transmissible abnormalities to help couples avoid the possibility of having abnormal children." It was also emphasized that PGD utilized for embryonic gender identification is allowable only in cases when there is a strong family history of sex-linked genetic disorders in a couple undergoing assisted reproductive procedures (Cutiongco-de la Paz, 2006).

Noninvasive techniques of prenatal ultrasonography are still the most commonly used procedures for prenatal diagnosis in the Philippines. Due to the increased use of reproductive technology at present, the trend may lean toward the use of newer technologies for prenatal diagnosis, such as preimplantation genetic diagnosis. More invasive techniques in prenatal diagnosis, however, will only be useful in providing anticipatory care and guidance (Cutiongco-dela Paz, 2006).

ADVOCACY PROGRAMS

In collaboration with the government and other health organizations, the IHG participates actively in tri-media campaigns for better understanding of genetic disorders and proper referral of patients with genetic conditions. It also assists in setting up family support groups for rare genetic conditions. It participates in continuing education activities and medical missions of the Down Syndrome Association of the Philippines, Inc. (since 1994) and the Philippine Society of Orphan Disorders, Inc. (since 1996), the MSUD Parent Support Group (since 2006) and the Cornelia de Lange Parent Support Group (since 2006; see Padilla, 2008).

The Institute has also spearheaded the establishment of the Volunteer Youth Leaders for Health – Philippines (VYLH-Philippines) in July 2009. The VYLH-Philippines is part of an international effort to establish the Global Network for Maternal and Infant Health participated in by youth counterparts in China and Lebanon, and linked by the common interest of volunteerism and public service to improve birth outcomes worldwide through advocacy. The VYLH-Philippines is a network of leaders from different youth organizations based in universities and communities in the country, and was organized with the aim of mobilizing the youth toward health work. Currently, the network is doing advocacy and promotional work in their respective schools and communities, with a focus on (1) the significance of folic acid supplementation in the prevention of birth defects, (2) the promotion of newborn screening; and (3) lobbying for public support for the urgent passage of the Rare Disease Act (*The Philippine Star*, 2009).

INTERNATIONAL COOPERATION

The IHG is recognized as one of the leading genetic centers in Asia. The Newborn Screening Center receives foreign laboratory and program coordinators for short-term training, and the Clinical Genetics Unit accepts foreign physicians for short-term observerships at the PGH. The faculty members of the IHG also have key roles in the Asia-Pacific Society of Human Genetics and the International Society for Neonatal Screening. The Institute has also been organizing workshops and hosting meetings for local and regional participants (Padilla, 2008).

CHALLENGES

With the increasing awareness of the Filipino people of different genetic disorders, the demand for genetic services has also increased throughout the Philippines. Provision of basic genetic healthcare services to every region remains the biggest challenge in the country. While it would be preferable to have at least one geneticist and one genetic counselor in each region, this is currently not possible. There are only a few geneticists with clinical practices, available only in the major areas such as Manila, Cebu, and Davao. In response to this critical lack of specialists, the Department of Health and the Newborn Screening Reference Center offer scholarships for fellowships in clinical genetics for pediatricians committed to practicing clinical genetics in regions currently without services.

ACKNOWLEDGMENTS

The authors would like to thank the following:

Dr. Filipinas Natividad, Vice President, Research and Biotechnology Division, St. Luke's Medical Center (RBD-SMLC)

Mrs. Juana Gregorio, Senior Science Research Specialist, Philippine Nuclear Research Institute (PNRI)

Ms. Maria Corazon De Ungria, University Researcher III and Head of the DNA Laboratory, Natural Science Research Institute, University of the Philippines Diliman (NSRI-UPD)

Dr. Mary Anne Chiong, Medical Specialist III and Head of the Biochemical Genetics Laboratory, Institute of Human Genetics, National Institutes of Health, University of the Philippines Manila (IHG-NIH, UPM)

Drs. Kathryn Ty and Ma-Am Joy Tumulak, University Researcher II, Institute of Human Genetics, National Institutes of Health, University of the Philippines Manila (IHG-NIH, UPM)

REFERENCES

Briones TK, Tantengco AT, Pascasio FM, Garcia FP, Reyes, MT. (1984). Genetic Counseling. In: *Applied Human Genetics* (pp. 145–154). Quezon City, Philippines: JMC Press, Inc.

Carillo MCO, Chao F-I, Silao CLT, Cutiongco-de la Paz E, David-Padilla C. (2009). Evaluation of *ras* mutations in Filipino colorectal patients. *ACTA Medica Philippina*. 42:17–21.

Chiong MAD, Silao CLT, Lee JY, Abarquez CG, Estrada SC. (2009). Enzyme replacement therapy in Filipino patients with Gaucher Disease and Pompe Disease. *Acta Medica Philippina*. 43(1):33–38.

Cutiongco-de la Paz EM. (2006). Prenatal diagnosis and its role in reproductive risk screening, prevention and treatment of genetic diseases: is the Philippines ready for it? *Acta Medica Philippina*. 40(2):54–57.

Cutiongco-de la Paz EM, Abaya EC, Silao CLT, Capistrano-Estrada S, Padilla CD. (2009). Mutations of the steroid 21-hydroxylase gene among Filipino patients with congenital adrenal hyperplasia. *Acta Medica Philippina.* 43(2):32–35.

De Leon Matsuda ML, Liede A, Kwan E, et al. (2002). BRCA1 and BRCA2 mutations among breast cancer patients from the Philippines. *Int J Cancer.* 98(4):596–603.

De Ungria MCA, Calacal GC, Delfin FC, et al. (2008). Overview of Forensic Services by the UP-NSRI DNA Laboratory: a 4-year Report. *Acta Medica Philippina.* 42(2): 34–38.

Department of Health (2004). Philippine Health Statistics. Available at *www.doh.gov.ph.* Accessed on September 23, 2009.

HUGO Pan-Asian SNP Consortium (2009). Mapping Human Genetic Diversity in Asia. *Science.* 326(5959):1541–1545.

Institute of Human Genetics, National Institutes of Health, University of the Philippines. Biochemical Genetics Unit Registry (Metabolic Clinic Census) (2001–2010). Unpublished data.

Institute of Human Genetics. (2010). National Institutes of Health, University of the Philippines. Executive Summary.

Natividad FF. (2001). Cytogenetics and the Molecular Laboratory: The St. Luke's Model. In: Reyes MVT, Lee LV, Tañega AL. (eds.), *Gene Talk* (pp. 33–35). Flora Pascasio Foundation Inc. 2001.

Newborn Screening Reference Center (NSRC), National Institutes of Health, University of the Philippines Manila. Available at *www.newbornscreening.ph.* Accessed on May 19, 2009.

Ngelangel CA, Javelosa MAU, Cutiongco-de la Paz EM, Philippine Cancer Genetics Study Group. (2009). Epidemiological Risk Factors for Cancers of the Lung, Breast, Colon-rectum & Oral cavity: A case-control study in the Philippines. *Acta Medica Philippina.* 43(4):29–34.

Padilla C. (1992). Evolution of Genetics in the Philippines: Genetics Then and Now (unpublished)

Padilla CD. (2003). Newborn screening in the Philippines. *Southeast Asia J Trop Med Pub Health.* 34(Supplement 3):87–88.

Padilla CD. (2008). Overview of Genetic Services at UP Manila. *Acta Medica Philippina.* 42(2):7–10.

Padilla CD, Basilio J, Oliveros Y. (2009). Newborn Screening: Research to Policy. *Acta Medica Philippina.* 43(2):6–14.

Padilla CD, Cutiongco-de la Paz EM, Lucero F, Villafuerte C, et al. (2008). Profile of oral cleft cases reported in the Philippine Oral Cleft Registry from May 2003 to December 2006. *Acta Medica Philippina.* 42(2):27–33.

Padilla CD, Cutiongco-de la Paz EM, Cadag NS, Salonga EAG, Chiong MAD. (2009). A review of the results of chromosomal analyses done at the National Institutes of Health from 1991 to 2007. *Acta Medica Philippina.* 43(1): 4–6.

Padilla CD, Silao CLT, Lee YL. (2001). Maple syup urine disease: a report of 26 cases in the Philippines. In: Poh San L, Yap EP. (eds.), *Frontier in Human Genetics Diseases and Technologies* (1st ed., pp. 185–193). Singapore: World Scientific Publishing Co., 2001.

Padilla CD, Silao CLT, de la Paz EM, et al. (n.d.). Frequency of Glucose-6-phosphate dehydrogenase deficiency among Filipino newborns. (unpublished data).

Prevalence update. Philippine Newborn Screening Program. Institute of Human Genetics. National Institutes of Health. University of the Philippines, Manila. February 2004.

Punzalan FE, Sy RG, Santos RS, Cutiongco, et al. (2005). Low density lipoprotein-receptor (LDL-R) gene mutations among Filipinos with familial hypercholesterolemia. *J Atheroscler Thromb.* 12(5):276–283.

Reyes MVT, Lee LV, Tañega AL. (eds.), (2001). Gene Talk (Foreword). xiii-xv. Flora M. Pascasio, MD Research Foundation Inc.

Sy RG, Cutiongco EM, Punzalan FE, Santos RS, Geronimo FR, Tangco RV. (2007). Human clholesteryl ester transfer protein (TaqIB) polymorphism among Filipinos with cardiovascular risk factors. *J Atheroscler Thromb.* 14(3):116–121.

Silao CLT, Canson DM, Hernandez KN, Chiong MAD, Capistrano-Estrada S, Padilla CD. (2009). Mutations of the phenylalanine hydroxylase (PAH) gene in Filipino patients with phenylketonuria. *Acta Medical Philippina.* 43(2):36–39.

Silao CLT, Hernandez KN, Canson DM, et al. (2009). Molecular analysis of the MUT gene in Filipino patients with methylmalonic acidemia. *ACTA Medica Philippina.* 43:29–32.

Silao CLT, Padilla CD, Matsuo M. (2004). A novel deletion creating a new terminal exon of the dihydrolipoyl transacylase gene is a founder mutation of Filipino maple syrup urine disease. *Mol Genet Metab.* 81(2):100–104.

Silao CLT, Padilla CD, Matsuo M. (2008). Early based diagnosis of maple syrup urine disease using polymerase chain reaction-based mutation detection. *Pediatr Int.* 50(3):312–314.

Philippine Cancer Genetics Study Group. (2008). Evaluation of Genetic Polymorphisms as Biomarkers of Cancer Risk Among Filipinos (unpublished data).

Philippine Star. Genetics research towards better health for the Filipinos. March 22, 2009. Page 21.

UNICEF [United Nations Children's Fund]. (2009). *The State of the World's Children 2009: Maternal and Newborn Health.* New York: UNICEF. Available at: http://www.unicef.org/sowc09/docs/SOWC09-FullReport-EN.pdf. Accessed on July 14, 2010.

CHINESE SUBCONTINENT

67.

COMMUNITY AND PUBLIC HEALTH GENETICS IN CHINA

Manshu Song, Youxin Wang, Lijuan Wu, Yun Wang, and Wei Wang

DEMOGRAPHICS AND MIGRATION IN CHINA

The demographics of China are uneven, exaggerated by the imbalanced economy, by the large number of highly migrant populations, and by the rapid aging. China is the third largest country in the world, lying in the eastern part of Asia, on the western coast of the Pacific Ocean, with a total land area of 9.6 million square kilometers. The total population numbers were over 1.3 billion on the mainland in 2007 (Zhao, 1999; Year Book China, 2008). About 56% of the population lives in the countryside, although a migration trend is happening. The current life expectancy for males is 70.0 years and for females 74.0 years. The average life expectancy of Chinese is 73.0 years (Year Book China, 2008, p. 79). Compared with developed countries, China's birth rate is still high; however, there has been a declining trend in the past decade. In particular, after establishing a basic national policy of family planning, the birth rate dropped from 36% in 1949 (Year Book China, 1991) to 12.10‰ in 2007 (Year Book China, 2008, p. 315). Meanwhile, the mortality declined from 20‰ in 1949 (Year Book China, 1991) to 6.93‰ in 2007 (Year Book China, 2008, p. 315). As a result, China's population has an aging trend. The United Nations predicts that more than 453 million Chinese will be older than 60 years by 2050 (United Nations, 2006).

China is a country with 56 officially recognized ethnicities distributed in different areas, and all of them have their own lifestyles. However, there are about 1.2 billion people of the Han ethnic group, accounting for 91.6% of the total population in China. They use Chinese characters as their written language, and most speak Mandarin. While the other 55 ethnic groups account for only 8.4% of the total population in China, they use 6 distinct languages and speak over 50 dialects.

China uses the register or *Hukou* system to manage its population. The Hukou system was established in Chinese cities in 1951, and was extended to the rural areas in 1955. It was formalized as a permanent system in 1958. Despite significant modifications since the early 1980s, the system remains unchanged in its nature today.

According to the *Hukou* system, China was once a country with a very low rate of labor mobility. Since 1978, the large-scale rural labor (*nongmingong* in Chinese) migration has become a striking phenomenon in the process of China's social and economic development with the deepening of reform and open-door policy, and the improvement of labor market and related policies. The rural population has provided a vast reservoir of people willing to work for low wages in factories, at construction sites, and wherever another pair of hands is needed. Thus, China's subsequent exceptionally high rates of internal migration are both the consequence and the cause of economic development. Huge internal migration from rural to urban areas was estimated at about 140 million in 2005, or 10% of the total population. Three-quarters of this migration was within provinces (Hu et al., 2008).

Most domestic migrations are due to rural labor migrants, who are almost entirely without any medical benefits compared to most urban residents, and they do not enjoy the state subsidies granted permanent urban residents (Wang & Zuo, 1999). They also tend to live in crowded, low-quality housing, often at the work site (Roberts,1997; Wang et al., 2002; Shen & Huang, 2003). All these factors aggravate health dangers such as maternal and infant diseases (Hu et al., 2008; Yang et al., 2005; Zhan et al., 2002).

CHINA'S FAMILY PLANNING

China instituted a "one family, one child" policy in 1979, and a family planning law in 2002. In the one-child policy, couples were encouraged to marry late, usually in their mid-20s, and allowed to have only one child. This policy brought a conspicuous birth rate decline from 37.88‰ in 1965 to 18.21‰ in1980 (Flaberty et al., 2007). In 1985, the birth rate had a rebound to 21.04‰ because of the "1.5 child" policy of 1984, which permitted peasants whose first child was a girl to have a second child after a suitable period. Since then, the birth rate has been declining gradually. Now, the birth rate is under 15‰. To date, China's one-child policy has contributed

200–400 million less people to the total growth of the world population (King, 2005).

The one-child policy, insofar as it limits couples to having 1 or 2 children, leads to more attention to the diagnosis of birth defects and greater involvement of parents in child care, which is named "Healthy Birth and Child Care" (Short et al., 2001). The Law on Maternal and Infant Health (implemented in 1994) requires physicians to recommend a postponement of marriage if either member of a couple has a genetic disease. If one spouse has a serious hereditary disease, the couple may only marry if they agree to use long-term contraception or to undergo sterilization. If prenatal tests reveal that a fetus has a serious hereditary disease or serious deformity, the physician should advise the pregnant woman to have an abortion, and the law states that the pregnant woman is supposed to follow this recommendation. A survey in genetic research and practice was done in 402 genetic service providers in China, using a Chinese version of an internationally circulated survey questionnaire on ethics and genetics. In all, 255 participants completed questionnaires (63%). The majority of the respondents (89%) reported that they agreed with the current Chinese laws and regulations on termination of pregnancy for genetic abnormalities, on the basis of considerations of population control and family planning (Mao & Wertz, 1997).

HEALTHCARE STRUCTURE IN CHINA

In 2005, the fund allocated to the health sector was 60,150 million RMB, about 1.77% of the total government budgetary expenditure in China. In 2006 there were 308,969 health institutions in urban China, and 609,128 village clinics in rural China (Health Statistical Year Book, 2007). In China, there are two significant differences in the healthcare systems between urban and rural areas. Both urban and rural health services deliver through 3 levels: primary, secondary, and tertiary.

URBAN HEALTH CARE SYSTEM

In the urban system, the Chinese Center for Disease Control and Prevention (CDC) operates at provincial, city or district, and community levels. Meanwhile, there are three types of hospitals in the urban medical care system, which are tertiary, secondary, and primary hospitals (Figure 67-1). A tertiary hospital has at least 500 beds, and is operated by the provincial government. A secondary hospital has 100–500 beds, offering complete medical services to a fixed population of several communities. A primary hospital has basic facilities and fewer than 100 beds, offering prevention, sanitation, health education, and treatment services for a specific community.

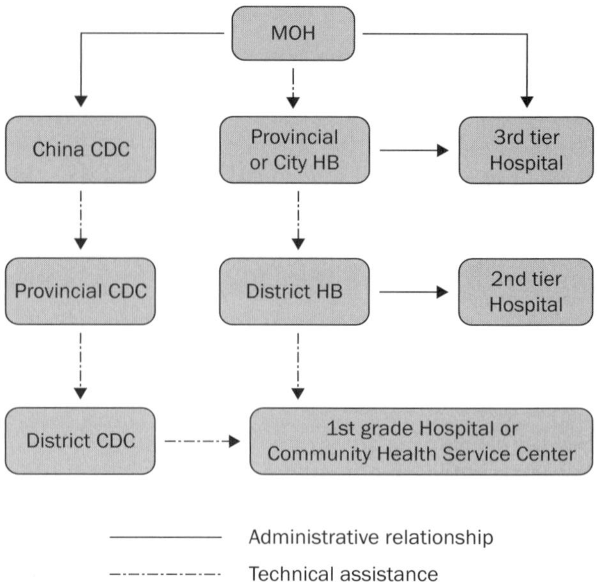

Figure 67-1 The health service system in urban areas in China. MOH: Ministry of Health; HB: Health Bureau.

The process of the urban healthcare payment system has two stages. Before 1994, the Government Insurance Scheme and the Labour Insurance Scheme were the mainstream sources of insurance, and only the employees of government agencies, public institutions, and state-owned enterprises were covered by the two schemes—which also partly covered the cost of healthcare for the dependants of employees. According to a national survey in 9 provinces in 1986, less than 14% of the urban population was not covered by any health insurance or plan (Ministry of Health, 1989). After 1994, a new Urban Employee Basic Health Insurance Scheme—which covered more of the urban population, including the employees of institutions mentioned above, foreign-funded enterprises, individual enterprises, and those who were urban inhabitants but had no stable jobs—was developed to replace the previous two schemes. In this scheme, government-run insurance schemes have been decreasing, while non-mainstream insurance (e.g., commercial schemes) have been increasing.

RURAL HEALTHCARE SYSTEM

In the rural system, there is a three-tier healthcare network based on the administrative relationship of between the county and town (Figure 67-2). The county and townshiptiers play a communication role between the higher tier and the lower tier. The village clinics deliver all the health services including diagnosis, treatment, prevention, health education, vaccination, and women's and children's healthcare directly to rural populations.

The payment mechanism in the rural healthcare system has changed a lot since the People's Republic of China

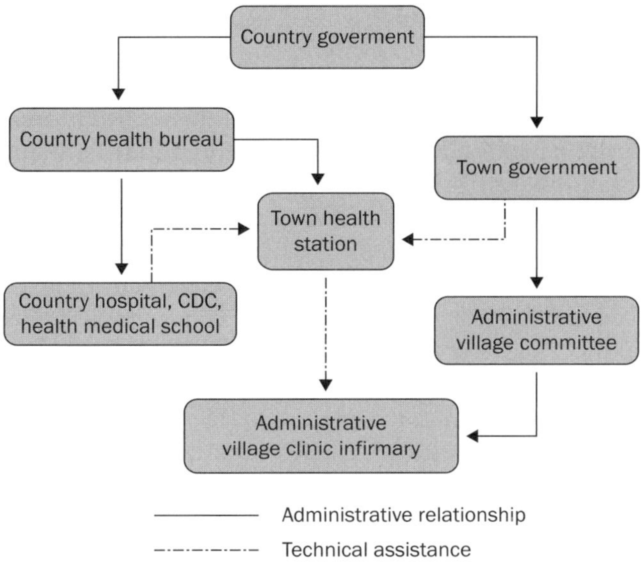

Figure 67-2 The health service system in rural areas in China.

was founded in 1949. Before the economic reform was launched in 1978, over 90% of the rural population was covered by the Cooperative Medical Scheme, but from the late 1980s to 2000, when the rural economic system changed a lot but the Cooperative Medical Scheme was not improved in a timely way, less than 10% of the rural population was covered by insurance schemes (Liu et al., 1998). Between 1993 and 1998, the proportion of people having to pay out-of-pocket was estimated as high as 90% in rural areas (Tang, et al., 2008). In 2002, a new rural healthcare system was established by the Chinese government. The new system is administered by the central and territorial governments. The funds are collected from private entities, local government, and the central government. Until 2007 there were about 685 million peasants (about 66% of the total rural population) who had health insurance.

In addition, commercial health insurance (also called "private" health insurance), which is a kind of health insurance paid for by organizations other than the government, is used in both urban areas and rural areas. It is usually paid by the client's employer, by a union, by the client and employer sharing the cost, or by the client alone. Commercial health insurance can overcome the deficiencies of the rural and urban healthcare system in China. Everyone can choose this insurance according to his or her health or economic condition. Until now, commercial insurance schemes are less than 2% of all health insurance schemes in China.

CHINA'S COMMUNITY-RELATED HEALTH PROGRAMS

There is no nationwide approach to public health genomics in China. However, the opening of the State Key Laboratory of Medical Genetics in 1991 established regulations and procedures for public health genomics-related programs and services.

In addition, the National Center for Women's and Children's Health (China WCH) of the China CDC is a state-level professional organization for women's and children's health, under the aegis of the China CDC. The China WCH provides the administrative hub of the China Women's and Children's Health Network (http://www.chinawch.com/) to facilitate coordination of relevant activities, prenatal screening, and congenital anomalies surveillance.

To date—except Qinghai, Xinjiang, Ningxia, and Tibet—22 provinces, 2 autonomous regions, and 4 municipalities in mainland China have established their local Internet websites/networks providing information on the laws, policies, research counseling, education, and surveillance relevant to women's and children's healthcare.

PRENATAL SCREENING

The general protocol at present for prenatal screening is predominately focused on those diseases with relatively high prevalence in China, such as screenings for Down syndrome, open neural tube defect, and also for trisomy 18 in some cases, using multiple serum markers. This practice varies and depends on a specific request by the doctor, generally prompted by the age, ethnicity, or family history of the pregnant woman. Prenatal screening usually is strongly recommended to those women of advanced reproductive age, or to those of advanced paternal age, since they are at an increased risk of having babies with a birth defect. A survey of 1,416 new mothers about their willingness to accept prenatal screening for DS (after they were provided with information on prenatal screening for DS) showed that 91.2% of all mothers would accept it (E et al., 2008).

In the metropolitan territories and regions with advanced economics in China, almost every hospital with a department of gynecology and obstetrics can provide regular prenatal screening services. Pregnant women will be guided by the relevant medical staff (e.g., obstetricians) to register and set up their own healthcare documents. Thereby, the healthcare handbooks for the prenatal period are offered to pregnant women as long as they come to the hospitals with a request for prenatal screening. Suggestions on medical termination of pregnancy will be offered by the doctor if the pregnancy has defects with high probability confirmed by subsequent DNA diagnostic testing. Screening may also be performed pre-pregnancy.

In China, the coverage rates of prenatal screening are varied because of the imbalanced economic development. For example, the coverage rates in Wenzhou city and Hangzhou city, which had enjoyed good economic development, were 63.15% and 60%, respectively in 2007, while in Gansu province, with poor economic development, the coverage rate was only 31.47% in 2006.

Fees for these screening tests vary from province/hospital to province/hospital. In some advanced cities of China, such as Beijing, Tianjin, Shanghai, and Shenzhen, pregnant women with urban census register/Hukou are given the privilege of having the special prenatal screening free of charge, and the fees are paid by their insurance or by the public medical care service. In order to reduce the rate of birth defects, some city governments with relatively good economics began to provide free special prenatal screening to rural pregnant women, for whom usually no medical insurance or free medical care is provided by the publicly funded service. For example, the government of Tianjin invested over 2 million RMB for such purposes in 2006. Thereby, 80% of rural pregnant women could take advantage of the service, and 40,000 rural newborns would benefit from such policy in Tianjin.

Although the Chinese government has been providing more services in prenatal screening, not everyone can benefit from it. For example, most pregnant migrant workers are aware of neither prenatal screening nor other essential healthcare, as they are usually of a lower social and economic status with little opportunity to take advantage of public medical care services. Dangerous birth delivery conditions and birth defects usually happen in such communities.

NEWBORN SCREENING

There are about 20 millions births each year in China. For hyperphenylalaninemia (including phenylketonuria) alone, there are 1,600–1,800 new cases each year. A report (2003–2004) from the National Working Committee on Children and Women under the State Council of China showed that there were 800–1,200 newborns with birth defects, about 4%–6% of all newborns.

Newborn screening in China was first introduced in 1981, when screenings for phenylketonuria, congenital hypothyroidism, and galactosemia were first implemented in Shanghai. Both the Law on Maternal and Infant Health (implemented in 1994) and its "Action Program" (launched in 2000) mandated newborn screening for congenital hypothyroidism and phenylketonuria throughout the country. Since then, variations in provision have emerged across provinces and territories. The Guangdong and Guangxi provinces added a glucose-6-phosphatase dehydrogenase (G6PD) screening program due to their local disease spectrum. Some territories in Nanjing, Wuxi, and Shanghai have implemented congenital adrenal cortical hyperplasia screening. In addition, some territories of Shanghai have implemented tandem mass spectrometry based screening for amino, organic, and fatty acid metabolic disorders. Now, the Chinese Ministry of Health has targeted screening for hearing deficiency and piloted certain territories such as Shanghai. Until 2003, screening for hearing deficiency covered 90% of newborns in Shanghai.

In 2003, newborn screening in Beijing, Shanghai, and Guangdong covered 95% newborns, but only 20% newborns were screened throughout the whole country; namely, extensive areas of the countryside, small villages, and towns remained blank.

A total of 133 laboratories for newborn screening had been established in 2003. In China, congenital hypothyroidism and hyperphenylalaninemia (phenylketonuria) have become two essential screening items implemented in all provinces and territories. To date, over 30 genetic disorders including amino acid disorders (e.g., phenylketonuria), organic acid disorders (e.g., propionic acidemia), and fatty acid oxidation disorders (e.g., carnitine transporter deficiency) can be screened in advanced provinces and territories in China using a variety of assays, including high-performance liquid chromatography (HPLC) and tandem mass spectrometry. The infants diagnosed with a condition will be followed up and provided treatment and counseling accordingly, by the services associated with the newborn screening services, in each province or territory within 10–20 days after birth.

Provision and fees for these screening tests vary from provinces/territories to provinces/territories, as well as from hospital to hospital, and most of them are paid by out-of-pocket expenses.

SEX SELECTION IN CHINA

In traditional Chinese ideology, the man has stronger power to support a family, and thus the woman has a lower social position. In the social context, sex selection in China is evident through different social aspects, as women are not allowed to play roles similar to those of men. Sex selection oppresses women in their autonomy. In some parts of China, female infanticide is common and abandoned newborn girls can be found in the countryside. This situation has changed in recent years. In 1994, China implemented a "Maternal and Infant Healthcare Law." This law aims to improve maternal and child health, and empowers physicians to recommend postponement of marriage if either member of the couple has an infectious, genetic, or psychiatric disease. If one spouse has a serious hereditary disease, the couple may not marry. If prenatal tests reveal that a fetus has a serious hereditary disease or serious physical deformity, the physician must advise the pregnant woman to undergo an abortion, and the law states that the pregnant woman is supposed to follow this recommendation. This statute also bans determining the sex of a fetus through the use of technology unless such tests are medically necessary. This ban is the reaction to the combination of China's one-child policy and the technological ability to predict the sex of a fetus, which has led to a change in China's sex ratio (Lai-wan et al., 2006). Despite the law, in some areas of China sex selection is regularly performed and female fetuses are aborted. This has led to alarming imbalance in the male:female birth ratio, to around 130:100.

Ethical, legal and social issues in genetic study and practice were investigated in 402 genetic services providers from 30 provinces and autonomous regions in China. In total, 255 participants completed questionnaires (63%). The majority of the respondents (89%) reported that they agreed with the current Chinese laws and regulations on termination of pregnancy for genetic abnormalities and nonmedical indications, on the basis of considerations of population control and family planning. More than half the respondents opposed sex selection by prenatal diagnosis in the absence of an X-linked disorder. However, most of them (86%) preferred directive counseling (Xia et al., 2004; Mao & Wertz, 1997). More than half agreed to disclose genetic information to relatives at risk, and would permit third parties such as law enforcement agencies, spouse/partner, blood relatives, employers, life insurers and health insurers to access stored DNA without prior agreement. In recent years, more and more sons moved out of home towns for business, leaving daughters to support elderly parents. A girl in China is thought to be a "warm coat" for parents, and now more couples wish to have daughters. Even so, there are now an estimated 80 million missing females in China. The large cohorts of "surplus" males now reaching adulthood are predominantly of low socioeconomic class, and concerns have been expressed about their lack of ability to marry. It is widely feared that their consequent marginalization in the Chinese society may lead to antisocial behavior and violence, threatening societal stability and security. Measures to reduce sex selection must include strict enforcement of existing legislation, the ensuring of equal rights for women, and public awareness campaigns about the dangers facing China as a result of sex selection in the last decade (Hesketh & Xing, 2006).

THE "ONE-CHILD" POLICY

To date the number of people inhabiting China has already surpassed 1.3 billion. Most of them live in the rural areas, although recently more people have migrated from rural to urban areas. In 1970s, the Chinese government realized the importance of controlling population growth. A government superstructure called the "National Family Planning Committee" was established and provincial committees were set up. The China adopted "one family, one child" policy in 1979 and "family planning" legislation was passed in 2002. These measures helped in reducing the size of the population tremendously; however, the number of people is still increasing remarkably. To ensure the quality of a single child, the government launched a program to improve maternal and child health (MCH). Now, in some minorities, one couple can have more children. Many positive steps were taken to benefit the one-child family. For example, a single child may receive free primary education. The government keeps the public informed about the implications of uncontrolled population. Family planning workers can be found in almost all villages, neighborhood committees, factories, and military units (Shen,1984). There is little doubt that China has made major improvements in controlling the growth of its population during the last three decades. However, the balance may be difficult to maintain due to a rising aged population with its low productivity. The government of China is trying to reform its "one couple, one child" policy to reflect its growing economy. Enforcement of the law is now less rigid, as some minority residents and farmers have ignored the law and had more children.

GENETIC TESTING

In China, some genetic testing laboratories have been founded. The genetic tests mainly deal with diseases such as breast cancer, carcinoma of the large intestine, heart rate abnormality, and nerve/muscle defects, and tests are undertaken as a kind of auxiliary examination for early prevention, early diagnosis, and early treatment.

The State Key Lab of Medical Genetics of China was founded in 1984 in Hunan province and opened in 1991. It is the only designated organization to authenticate the identification of novel abnormal chromosomal karyotypes in China. It has been maintaining the Database of the Family Collection of Genetic Diseases in China, and has collected novel human abnormal chromosomal karyotypes identified in China for the past 30 years. The database has assembled 2,144 novel human abnormal karyotypes identified in China (excluding those in Tibet, Taiwan, Hong Kong, and Macau) and their related disease information. The karyotypes include 4,510 chromosome breakages and 57 balanced translocations between the X-chromosome and autosomes. The database provides also statistical information, chromosome breakage frequencies, abnormal karyotype descriptions, and related clinical information (http://skl.host4.90i.cn/index.html).

Some tests are paid by insurance providers in advanced cities (e.g., a test for human papilloma virus/HPV in Beijing), but most of them require out-of-pocket payment.

PROJECTS RELATED WITH THE HUMAN GENOME IN CHINA

China has proposed or entered several projects related to the human genome in the past decades. The earliest project started in 1991 when the Human Genome Diversity Project (HGDP) was initiated. The Chinese Human Genome Diversity Project (CHGDP) has collected cell lines from the 56 Chinese ethnic groups and tested the DNA samples. In 1999 China entered the international Human Genome Project (HGP) and undertook the sequencing of 1% of the human genome working draft (i.e., the region 3pter-D3S3610 containing 30 million bp). China is also responsible for about 10% of the International HapMap Project, which was launched in 2002 jointly by the United States, United Kingdom, Japan, China, and Canada. This project aims to guide the design and analysis of genetic association studies, shed light on structural variation and recombination, and identify loci that have been subject to natural selection during human evolution. In January 2008, the "Yanhuang (abbreviated as YH)" (Emperor Yan) Project was generated by the Beijing Genomics Institute (BGI) which aims to sequence the entire genome of 100 Chinese individuals over a 3-year period. To date, the YH Database (http://yh.genomics.org.cn/index.jsp) has been set up, and raw sequences, alignments, consensus genome, variants of theYH diploid genome and relevant tools are freely available to download, which accelerates the discovery of disease gene and mutation in Asian population.

GENETICS EDUCATION IN CHINA

Genetics education in China started fairly late due to the education disaster of the Cultural Revolution in 1966–1976. In the 1980s, the colleges began to offer genetics courses, but only the basics of human genetics were emphasized in the regular colleges and universities. Now, the main form of Chinese genetics education for the public is genetic counseling in clinics. Since the Law on Maternal and Infant Health was passed in 1994, genetic counseling in clinics has been developed in the advanced hospitals, mainly as a unit attached to departments of gynecology and obstetrics. However, compared with developed countries, there are still many problems for genetic counseling in China. Firstly, genetic counseling clinics are almost all located in big cities, while over 50% of the population is living in rural areas where people have little medical knowledge and more children in one family. Secondly, genetic counseling in China now is not offered by professionally trained genetic counselors, but clinicians such as pediatricians or obstetricians (Ren et al., 2002). Thirdly, there is no official genetic counseling program in China (Zhang & Zhong, 2006).

In 2007 the first website, named China Genetic Counseling Network (CGCN), on genetic counseling and genetic education was established in China (http://www.gcnet.org.cn). It introduces basic knowledge of genetic diseases, their clinical symptoms, diagnosis, prevention, and treatment.

CHINA'S UNMET NEEDS IN HEALTH

China has a big population spread over a large land mass, and is going through an imbalanced economic development period. Although China has invested in the basic research of genome science, it is urgent for the country to develop public health genomic programs and clinical genetic services that meet the needs of the Chinese people, particularly residents in the rural areas. A standard framework for evaluating genetic/genomic testing and related interventions in China is also needed. What is more, the Chinese government should implement genomic education into the Chinese healthcare system. China needs an easy way for the public to get genetic information on innovation, education, and prevention.

ACKNOWLEDGMENTS

The project was partially supported by National Natural Science Foundation of China (No. 30901238), Research Fund for the Doctoral Program of Higher Education of China (No. 20050025002; 20091107120017), Major State Basic Research Program of China-973 (No. 2011CB503806), Major State Applied Research Program-863 (2006AA02Z434), National Key Technologies R&D Program of China (2008JX10001–003) and Funding Project for Academic Human Resources Development in Institutions of Higher Learning under the Jurisdiction of Beijing Municipality (PHR201008393).

REFERENCES

E Q, Tang ZL, Huang JY, Han P, Tian D, Li CL, Chen YY. (2008). A survey on the willingness to accept prenatal screening for Down's syndrome. *Chinese Primary Health Care 8*, 33–35.

Flaberty JH, Liu ML, Dong B, Ding QF, Li X, Xiao SF. China: the aging giant. (2007). *J Am Geriatr Soc 55*, 1295–1300.

Hesketh T, Xing ZW. (2006). Abnormal sex ratios in human populations: Causes and consequences. *PNAS 103*(36), 13271–13275.

Health Statistical Year Book. (2007). Beijing: Ministry of Health. Available at: www.moh.gov.cn/publicfiles/business/htmlfiles/zwgkzt/ptjnj/200807/37168.htm.

Hu XJ, Cook S, Salazar MA. (2008). International migration and health in China. *Lancet 372*, 1717–1719.

King M. (2005). China's infamous one-child policy. *Lancet 365*, 215–216.

Lai-wan CC, Blyth E, Hoi-yan CC. (2006). Attitudes to and practices regarding sex selection in China. *Prenat Diagn 26*(7), 610–613.

Liu Y, Rao K, Fei J. (1998). Economic transition and health transition: comparing China and Russia. *Health Policy 44*, 103–122.

Mao X, Wertz DC. (1997). China's genetic services providers' attitudes towards several ethical issues: a cross-cultural survey. *Clin Genet 2*, 100–109.

Ministry of Health. (1989). Urban medical service study in China—survey data on medical care demand and utilization from nine provinces and municipalities. Beijing: Ministry of Health.

Ren AG, Wang L, Zhao P, Li Z. (2002). Current status of genetic counseling in maternal and child health institutions. *Chinese J Reprod Health 13*, 131–134.

Roberts KD. (1997). China's tidal wave of migrant labor: what can we learn from Mexican undocumented migration to the United States? *Int Migr Rev 31*, 249–293.
Shen G. (1984). A brief introduction to China's family planning. *China Popl News 1*(3), 4–10.
Shen JF, Huang YF. (2003). The working and living space of the 'floating population' in China. *Asia Pac Viewp 44*, 51–62.
Short SE, Zhai FY, Xu SY, Yang ML. (2001). China's one-child policy and the care of children: an analysis of qualitative and quantitative data. *Soc Forces 79*, 913–943.
Tang SL, Meng QY, Chen L, Bekedam H, Evans T, Whitehead M. (2008). Tackling the challenges to health equity in China. *Lancet 372*, 1493–1501.
Wang F, Zuo XJ. (1999). Inside China's cities: institutional barriers and opportunities for urban migrants. *Am Econ Rev 2*, 276–280.
Wang F, Zuo XJ, Ruan DC. (2002). Rural migrants in Shanghai: living under the shadow of socialism. *Int Migr Rev 36*, 520–545.
United Nations. (2006). *World population prospects: the 2004 revision population database.* New York: United Nations Population Division.
Xia DY, Liao SS, He QY, Liao JF, Wang XC, Wu QH. (2004). A questionnaire-based survey on attitude and behavior of sex among rural women in Hainan province. *Chinese Journal of Epidemiology 25*(7), 586–589.
Yang YM, Li Y, Liu CR. (2005). The situations and measures of infant birth qualities from 2000 to 2004. *Chinese Primary Health Care 6*, 10–11.
Year Book China. (1991). Beijing, National Bureau of Statistics.
Year Book China (2008). Beijing, National Bureau of Statistics (pp. 79–80).
Zhan SK, Sun ZW, Erik B. (2002). Economic transition and maternal health care for internal migrants in Shanghai, China. *Health Policy Plan 17*, 47–55.
Zhang YZ, Zhong N. (2006). Current genetic counseling in China. *J Peking Univ (Health Sciences) 1*, 33–34.
Zhao J. (1999). *China Geography.* Beijing: Higher Education Press.

68.

GENETIC EPIDEMIOLOGY IN CHINA

Peng-Gao Li, Yan He, and Wei Wang

Epidemiology is usually defined as the study of the distribution, determinants, and control of health-related states and events in populations. Genetic epidemiology could be regarded as a discipline closely allied to traditional epidemiology that focuses on the familial, and in particular genetic, determinants of disease and the joint effects of genes and nongenetic determinants. Traditionally, genetic diseases include chromosomal, monogenic, and polygenic/multifactorial diseases.

MONOGENIC DISEASE

Monogenic disease is an inherited disease controlled by a single pair of genes. It results from modifications in a single gene occurring in all cells of the body. Such disorders are inherited (passed on from one generation to another) in a simple pattern according to Mendel's laws. As such, these disorders are often referred to as *Mendelian disorders*. Though relatively rare, they affect millions of people worldwide. Scientists currently estimate that over 10,000 human diseases are known to be monogenic. Pure genetic diseases are caused by a single error in a single gene in the human DNA. The nature of disease depends on the functions performed by the modified gene. The single-gene or monogenic diseases can be classified into three main categories:

- Dominant
- Recessive
- X-linked

To date, except Qinghai, Xinjiang, Ningxia, and Tibet, 22 provinces, 2 autonomous regions, and 4 municipalities in mainland China have established their local Internet websites/networks providing information on the laws, policies, research counseling, education, and surveillance relevant to women's and children's healthcare including newborn screening for common genetic diseases (Table 68-1).

POLYGENIC DISEASE

A genetic disease caused by the combined actions of nongenetic factors and several genes with low penetrance. Examples of polygenic diseases include coronary heart disease, diabetes mellitus, common cancers, hypertension, and peptic ulcers. Because such disorders depend on the simultaneous presence of several genes, they are not inherited in simple Mendelian patterns.

Polygenic diseases, also called multifactorial or complex diseases, are conditions that are influenced by the actions of multiple genes, their interactions with each other and with the environment. Genetic epidemiologic studies include family, twin, case-control and adoption studies, as well as methods for quantifying the strength of the genetic influences on a disease. A major focus will be the identification of specific gene effects using both linkage and association analysis and their variants.

NUTRITIONAL FACTORS INFLUENCING GENETIC DISORDERS IN CHINESE POPULATIONS

The phenotype in several monogenic and polygenic conditions is influenced by several nongenetic factors. Interactions between specific genes and environmental factors (Tiret, 2002) in most cases are dynamic, beginning at conception and continuing through adulthood. Because the environmental factors are complex and keep changing, the majority of patients affected with genetic disorders require meticulous multidisciplinary treatment throughout their lives. As a consequence, there is a strong imperative for a better understanding of the etiology of these diseases, through which exposures to modifiable risk factors can be eliminated or reduced, resulting in elevated life quality or prevention of these malformations in the next generation.

In this respect, food intake is one of the important "modifiable" and "environmental" factors to which we

TABLE 68-1 CONDITIONS INCLUDED IN THE NEWBORN SCREENING PROGRAMS IN CHINA

Disease	Incidence
Congenital hypothyroidism	1:4000
Phenylketonuria	1:11,000
Glucose-6-phosphatase dehydrogenase	3.6:100
Congenital adrenal cortical hyperplasia	1:15,000

are all exposed continuously, from conception to death. Thus, the interactions between nutritional factors and genetic disorders are recognized and extensively studied. The concept of "gene–diet" interaction describes the modulation of the effect of a dietary component on a specific phenotype (plasma lipid concentrations, obesity, glycemia, etc.) by a specific gene mutation or a genetic polymorphism. Alternatively, this notion refers to the dietary modification of the effect of a genetic variant on a phenotypic trait. The potential benefits of harnessing the power of genomics for dietary prevention of disease are enormous, and this approach is considered the future of nutritional research in the postgenomic era (Mensink et al., 2002).

In China, the prominent role of diet in the etiology of disease has been recognized for a long time in traditional Chinese Medicine (TCM; see Veith, 1971). TCM has been used for monogenic diseases (e.g., phenylketonuria, lactose intolerance, and favism) in which diet plays a determinant role in the final phenotype, and also for multifactorial disorders (e.g., cardiovascular disease and osteoporosis) usually associated with the aging process. Currently, considering that gene modification in humans is neither technically feasible nor ethically permitted (Guttmacher & Collins, 2003), using genomics-based knowledge to recommend personalized behavioral changes in relation to dietary habits, which should mean more effective disease prevention and treatment, is a feasible strategy.

THE INFLUENCE OF NUTRITIONAL FACTORS ON GENETIC DISORDERS

Dietary factors exert their effects on genetic disorders at different levels. First, fetal development and the "in utero" conditions would be essential to produce the first gene–nutrient interactions. At this stage, molecular biological processes are modified by nutrients at the cellular level by influencing the expression and silencing of the genes implicated in embryogenesis. In this respect, it is important to point out that the embryonic nutritional status is fully dependent on maternal food intake and metabolism. However, due to increased needs, inadequate intake, decreased absorption, disturbances in extra-embryonic transfer, or underlying genetic aberrations in the mother and/or embryo, maternal nutritional deficiencies during early pregnancy may occur and significantly affect the nutritional status and gene expression of the embryonic tissues (Natsume et al., 1999). On the other hand, ingestion of specific food or nutrient supplements inappropriately, such as vitamin A and zinc, may cause nutrient deficiency, which also can affect the gene expression of the embryonic tissues.

Secondly, in some conditions, as in the case of the inborn errors of metabolism, nutrition in the first years of life becomes a determinant key of health or disease manifestation. For instance, in phenylketonuria (PKU), galactosemia, lactose intolerance, and celiac disease, diet plays a major role in phenotypic expression. Dietary modification has been largely used to prevent the development of the disease manifestation. Although in some of these diseases, the specific gene(s) and/or the particular mutations have not been well characterized, restricting the harmful nutrient in affected subjects has long been used and proved effective. In this regard, nutritional genomics will be crucial to improving the prevention or treatment of these diseases by the early identification of specific mutations or haplotype combinations that modulate dietary response in affected subjects, and then to provide a more individualized and specific dietary restriction for the disease in the future.

Thirdly, for multifactorial diseases such as atherosclerosis and cancer, a long period of exposure to the same dietary pattern would be necessary to develop the disease phenotype (Leong et al., 2003). And, it is the same with monogenic diseases. Although genetic diseases have been classified as monogenic or multifactorial, it has been recognized that this traditional classification is an oversimplification and the reality is far from clear-cut, because most monogenic traits share some of the features found in the multifactorial diseases. One must consider the dramatic phenotypic diversity of the so-called "classical monogenic diseases" reflecting the heterogeneity of mutations at the major locus, the action of some secondary and tertiary modifiers, and the influence of a wide range of environmental factors (Scriver et al., 1999).

Therefore, diet may be the most influential environmental factor modulating the phenotypic expression of both monogenic and multifactorial diseases in a patient's life.

MECHANISMS THROUGH WHICH NUTRIENTS AFFECT GENETIC DISORDERS

Nutrients affect diseases in several ways. It is widely accepted that nutrients, either in excess or deficiency, may cause damage in animals and humans. Although the specific mechanisms through which each nutrient affects genetic disorders are, as yet, largely unknown, evidences are rapidly accumulating. Briefly, nutrients exert their effects on organisms through the following approaches. First, nutrients are involved in the molecular biological processes as substrates, cofactors, and ligands. Therefore, they are essential in normal biochemical and physiological processes, as well as in disease convalescence. Second, studies reveal that nutrients at the cellular level can influence gene expression either directly by the regulation of relevant gene expression, or indirectly through epigenetic events (e.g., niacin, folate; see Berry et al., 1999) or by influencing genomic stability (e.g., magnesium, folate, and zinc; see Oommen et al., 2005), and thereby control the gene and protein expressions by posttranscriptional and posttranslational alterations. In this aspect, mechanisms of various nutrients differ from each other and are quite complex. For example, it is difficult to interpret transcriptional profiling data alone, because dimerization, posttranslational modifications, relative quantities of co-activators and co-repressors, and the formation of multiple transcriptional complexes binding to common promoter regions ultimately determine the binding activation potential of transcription factors. Nevertheless, clustering analysis of differentially regulated genes can identify numerous genes known to be regulated by specific transcription factors, thereby enabling several transcription factor signatures to be established. These signatures then provide clues concerning the various pathways through which nutrients may affect gene transcription. Moreover, in the multifactorial etiology of genetic disorders, a deficiency of (e.g., folate) and/or excess of (e.g., retinoic acid) nutrients could also play a role indirectly, through other genes such as growth factors. Furthermore, as for the epigenetic modulation, it is known that the methyl cycle supplies 1-carbon units critical for a variety of methylation reactions, essential for proper gene expression and maternal and paternal imprinting by methylated DNA (Fenech, 2003). Growth factor genes are commonly imprinted in this manner, and these methylation patterns can be altered by nutrition (Razin et al., 2005). For instance, faulty embryonic methylation of DNA due to abnormal folate levels or improper methyl cycle gene expression at a critical developmental juncture could inappropriately silence growth factors necessary for proper organ development, resulting in congenital diseases. Also, differences in the cellular concentration of the nutrients and their intermediates may result in nutrient-related hormonal changes—which could also be a major determinant of the gene–nutrient interaction—are especially important in women's health, and could be the basis for future gender-specific and age-specific recommendations based on genetic makeup. Finally, differences in the cellular concentration of the nutrients and their intermediates can also cause alterations in the intracellular signaling pathways, such as the homocysteine pathway, in which riboflavin (Waterland et al., 2003), folate (McNulty et al., 2002), pyridoxine (McNulty et al., 2002), cobalamin (McNulty et al., 2002), and zinc (Verhoef et al., 2005) play roles as cofactors or substrates, and the oxidative pathway, in which the balance is determined by oxidants such as glucose and homocysteine and antioxidants such as ascorbic acid and glutathione, as well as the hematopoiesis pathway, which is influenced by iron, cobalamin, and folate (Peariso et al., 1998).

NONGENETIC FACTORS AFFECTING GENETIC DISORDERS

PHENYLKETONURIA (PKU)

Phenylketonuria (PKU) is an autosomal recessive genetic disorder characterized by a deficiency in the enzyme phenylalanine hydroxylase (PAH), which converts phenylalanine to tyrosine. The result is a buildup of phenylalanine in the blood to a level where it causes brain damage (mental retardation) in infants. Excess phenylalanine is converted into several metabolites, which are excreted into urine where they cause a mousy smell. Understanding the condition has led to an effective treatment. Patients must be placed on a diet with very little phenylalanine until they are teenagers. The diet must be started within a few weeks of birth, the earlier the better.

From 1985 to 2001, the Group of Neonatal Screening, consisting of the Chinese Society of Child Health Care, the Chinese Preventive Medical Association, and the Center for Neonatal Screening Quality Control Laboratory at the National Center for Clinical Laboratories (NCCL), conducted a neonatal screening for PKU at the 39 neonatal screening centers all over the country. PKU was detected by the bacterial inhibition assay of heel-pricked blood samples 72 hours postnatal, or by fluorometric determination of the blood phenylalanine concentration. A total of 5,817,280 newborns were screened for PKU. Among them, 522 cases of PKU were detected with an incidence of 1/11,144 (Gu & Wang, 2004). Because of the increasing screening coverage to test for the presence of this disease in China since the year 1981, more and more children with this disease have been saved from its damages.

GLUCOSE-6-PHOSPHATE DEHYDROGENASE DEFICIENCY AND FAVISM

Favism is an abnormal sensitivity to fava beans that causes a hemolytic anemia in patients. Even to smell the flowers of the bean may be enough to trigger an attack. This disease is more common in the Mediterranean region, and was known to Pythagoras who exhorted his pupils to avoid beans. The disease is caused by deficiency of the enzyme glucose-6-phosphate dehydrogenase (G6PD). G6PD deficiency is the most common enzymopathy of man that is of clinical significance (Carter & Gross, 2005). It is thought to affect more than 400 million people worldwide (Noori-Daloii et al., 2004). The highest prevalence of G6PD deficiency is found in tropical Africa, the Middle East, tropical and subtropical Asia, Papua New Guinea, and various Mediterranean locations [19]. Only a proportion of G6PD-deficient sufferers are prone to favism, and this proportion is variable among populations. A study in Sardinia, where there is a high prevalence of G6PD deficiency (7.5% of all males are hemizygotes), found 508 cases of favism over a 9-year period (Meloni et al., 1992). It therefore appears to be a relatively rare manifestation of a common genetic polymorphism.

Doctor Du Shun De (1952) was the first Chinese scientist who discovered favism in China in the 1950s and gave it the name *Can Dou Bing* in Chinese. Later studies indicated that favism was relatively more common in southern China than in the north (Song, 1994). Now, favism is included in the neonatal screening programs of many southern provinces such as Guangdong, Yunnan, and Guangxi.

To prevent the symptoms of hemolytic anemia due to favism, avoidance of ingestion of broad beans is very important in patients known to be Glucose-6-phosphate dehydrogenase (G6PD)-deficient, or those who have suffered previous episodes of favism. Genetic counseling and screening may be useful where there is a family history of G6PD deficiency, to allow diagnosis before exposure to hemolytic precipitants. Population screening and health education programs in areas of high prevalence of G6PD deficiency have been shown to reduce the incidence of favism in the at-risk population (Song, 1994).

NEURAL TUBE DEFECTS

Neural tube defects (NTDs) are birth defects of the brain and spinal cord. The two most common neural tube defects are spina bifida and anencephaly. In spina bifida, there is usually nerve damage that causes at least some paralysis of the legs. In anencephaly, much of the brain does not develop. Babies with anencephaly are either stillborn or die shortly after birth.

Several studies have revealed that China has the highest known rate of NTDs in the world. For example, a collaboration project conducted by Peking University (PU) and the National Center on Birth Defects and Developmental Disabilities (NCBDDD) at the Centers for Disease Control and Prevention (CDC) of the United States since 1983 has revealed that NTDs affect 5 in 1,000 newborns in Northern China (http://www2.cdc.gov/od/gharview/GHARciomain). It was also reported that in Shanxi Province, one of the areas in China that has the highest prevalence of NTDs, the prevalence during January 2000 to December 2004 was 49.77/10,000 (730 in 130,783) births (Zhang & Ma, 2006). However, in some rural areas of Shanxi, the prevalence was reported as high as 140/10,000 births in 2002 (Zhang et al, 2006).

Fifty to seventy percent of NTDs can be prevented when women supply their diet with folic acid, a water-soluble B vitamin. Since 1991, DBDDD staff have collaborated with Chinese colleagues and conducted a population-based prospective cohort study on the benefits of periconceptional use of folic acid (FA) in preventing NTDs. This program documented that maternal use of 400 μg of FA taken without other vitamins could reduce the prevalence of NTDs by up to 85% in high-rate areas, and that maternal FA consumption before and during early pregnancy was safe and did not increase the likelihood for miscarriage or twinning. Now, the Chinese Centers for Disease Control (Chinese CDC) recommends that all women of childbearing age eat a diet high in folic acid or take a multivitamin with 0.4 mg of folic acid each day, especially one month prior to conception and through the first three months of pregnancy. However, women who have had a previous NTD pregnancy are recommended to take an even higher dosage of folic acid prior to planning a pregnancy. They should increase the daily dose of folic acid from 0.4 mg to 4.0 mg, from one month prior to conception through the first three months of pregnancy. Of course, the 4.0 mg tablets of folic acid should only be obtained through a prescription from a doctor.

DOWN SYNDROME (DS)

Down syndrome, or trisomy 21, was once considered a disorder of genetic origin whose abnormalities are present at birth, and it was assumed that nothing could be done about it. Now, however, researchers and physicians are discovering that all potential damage is not present at birth. It actually can progress as the child grows—unless the molecular damage is halted. Thankfully, this is possible.

As we know, Down syndrome is caused by the presence of all or part of an extra 21st chromosome. This means the presence of extra genes and an overexpression of their functions, resulting in the production of too much of some enzymes and an overusage of others, causing elevated levels of hydrogen peroxide that combines with iron, producing free radicals, while depleting the antioxidant selenium. Fewer antioxidants and more free radicals lead to cell death in the brain. Other enzyme-related problems are higher risks for vascular disease, mental retardation, and other physical abnormalities.

Understanding of these underlying mechanisms has led to some solutions; for example, supplementation of folic acid and B vitamins for would-be mothers. Several recent studies have reported that folate metabolism is abnormal in mothers of children with DS (Coppede et al., 2006, 2007; Martínez-Frías et al., 2006; Devlin et al., 2006; Scala et al., 2006). This can lead to a lack of DNA methylation and abnormal chromosomal segregation during meiosis. Moreover, a deficiency in folic acid can lead to an increase in homocysteine levels, a risk factor for vascular problems commonly seen in DS patients. This imbalance can be corrected by supplementing with higher levels of folic acid, vitamin B6, and vitamin B12.

Other deficiencies and allergies may also be involved in DS, so it is important to do thorough physical, nutritional, and biochemical assessments upon its diagnosis. The sooner one corrects any imbalances, the greater the chance of preventing further molecular damage to the brain and the rest of the body. These immediate treatments with nutrition, antioxidants, and amino acids may lead to remarkable improvements, not only in mental functioning but also in physical appearance.

CONCLUSIONS

Nutritional factors play an important and complex role in the pathogenesis of genetic disorders. Food intake is one of the most important modifiable "environmental" factors that exert their influences on populations. Meanwhile, future possibilities to develop preventive nutritional strategies are of great interest and great prospect.

So far, however, although insight is increasing into the functions of relevant genes, their proteins and metabolites, and their interactions with specific nutrients, knowledge about the underlying biological mechanisms of genetic disorders affected by nutrients is limited in relation to the 25,000 or so genes of the human genome, and the more than 3 million common variants that lurk within those genes. Add the fact that food is full of hundreds of bioactive compounds, each of which varies depending on where plants are grown or animals are raised. To achieve a more comprehensive knowledge of genetic disorders and their relationships with nutrients, we have still got quite a lot of work to do in the future.

REFERENCES

Berry RJ, Li Z, Erickson JD, et al. (1999). Prevention of neural-tube defects with folic acid in China. China-U.S. collaborative project for neural tube defect prevention. *New England Journal of Medicine 341*(20), 1485–1490.

Carter S, Gross S. (2005). Glucose-6-Phosphate Dehydrogenase Deficiency. *eMedicine*. Available at: http://www.emedicine.com/med/topic900.htm.

Cheng N, Bai Y, Hu X, et al. (2003). A base-line survey on birth defects in Gansu province, West China. Annals of Tropical Paediatrics 23, 25–29.

Coppede F, Colognato R, Bonelli A, et al. (2007). Polymorphisms in folate and homocysteine metabolizing genes and chromosome damage in mothers of Down syndrome children. *American Journal of Medical Genetics 143*, 2006–2015.

Coppedè F, Marini G, Bargagna S, et al. (2006). Folate gene polymorphisms and the risk of Down syndrome pregnancies in young Italian women. *American Journal of Medical Genetics 140,* 1083–1091.

Devlin AM, Clarke R, Birks J, Evans JG, Halsted CH. (2006). Interactions among polymorphisms in folate-metabolizing genes and serum total homocysteine concentrations in a healthy elderly population. *American Journal of Clinical Nutrition 83*, 708–713

Du SD. (1952). Favism in west China. *Chinese Medical Journal 70*, 17–26.

Fenech M. (2003). Genomic stability: a new paradigm for recommended dietary allowances (RDAs). *Forum of Nutrition 56*, 97–100.

Gao Q, Tao J, Tian D, Lu J, Cai Y. (2006). Screening for Down's syndrome in over 16000 pregnant women diagnosis analysis by over 500 amniotic fluid examination. *Chinese Journal of Birth Health & Heredity 14*, 26–34.

Gu XF, Wang ZG. (2004). Screening for phenylketonuria and congenital hypothyroidism in 5.8 million neonates in China. *Chinese Journal of Preventive Medicine 38*, 99–102.

Guttmacher AE, Collins FS. (2003). Welcome to the genomic era. *New England Journal of Medicine 349*, 996–998

Leong NM, Mignone LI, Newcomb PA. (2003). Early life risk factors in cancer: the relation of birth weight to adult obesity. *International Journal of Cancer 103*, 789–791.

Martínez-Frías ML, Pérez B, Desviat LR, et al. (2006). Maternal polymorphisms 677C-T and 1298A-C of MTHFR, and 66A-G MTRR genes: is there any relationship between polymorphisms of the folate pathway, maternal homocysteine levels, and the risk for having a child with Down syndrome? *American Journal of Medical Genetics 140*, 987–997.

McNulty H, McKinley MC, Wilson B. (2002). Impaired functioning of thermolabile methylenetetrahydrofolate reductase is dependent on riboflavin status: implications for riboflavin requirements. *The American Journal of Clinical Nutrition 76*, 436–441.

Meloni T, Forteleoni G, Meloni GF. (1992). Marked decline of favism after neonatal glucose-6-phosphate dehydrogenase screening and health education: the northern Sardinian experience. *Acta Haematologica 87*, 29–31.

Mensink RP, Plat J. (2002). Post-genomic opportunities for understanding nutrition: the nutritionist's perspective. *The Proceedings of the Nutrition Society 61*, 401–404.

Natsume N, Sugimoto S, Yoshida K. Kawai T. (1999). Influence of maternal anaemia during early pregnancy on the development of cleft palate. *British Journal of Oral & Maxillofacial Surgery 37*, 330–331.

Noori-Daloii MR, Najafi L, Mohammad Ganji S, Hajebrahimi Z, Sanati MH. (2004). Molecular identification of mutations in G6PD gene in patients with favism in Iran. *Journal of Physiology & Biochemistry 60*, 273–277.

Oommen AM, Griffin JB, Sarath G, Zempleni J. (2005). Roles for nutrients in epigenetic events. *Journal of Nutritional Biochemistry 16*, 74–77.

Peariso K, Goulding CW, Huang S, Matthews RG, Penner-Hahn JE. (1998). Characterization of the zinc binding site in methionine synthase enzymes of Escherichia coli: the role of zinc in the methylation of homocysteine. *Journal of the American Chemical Society 120*, 8410–8416.

Razin A, Cedar H. (2005). DNA methylation in epigenetic control of gene expression. *Progress in Molecular & Subcellular Biology 38*, 151–167.

Ren AG, Wang L, Zhao P, Li Z. (2002). Current status of genetic counseling in maternal and child health institutions. *Chinese Journal of Reproductive Health 13*, 131–134.

Scala I, Granese B, Sellitto M, et al. (2006). Analysis of seven maternal polymorphisms of genes involved in homocysteine/folate metabolism and risk of Down syndrome offspring. *Genetics in Medicine 8*, 409–416.

Scriver CR, Waters PJ. (1999). Monogenic traits are not simple: lessons from phenylketonuria. *Trends in Genetics 15*, 267–272.
Song WX. (1994). *Anemia: Basic and Clinical Concerns*. Tianjin: Tianjin Science & Technology Publishing House. pp. 173–174.
Tiret L. (2002). Gene-environment interaction: a central concept in multifactorial diseases. *The Proceedings of the Nutrition Society 61*, 457–463.
Veith I. (1971). *The Yellow Emperor's Classic of Internal Medicine*. Berkeley, CA: University of California Press.
Verhoef P, de Groot LC. (2005). Dietary determinants of plasma homocysteine concentrations. *Seminars in Vascular Medicine 5*, 110–123.
Waterland RA, Jirtle RL. (2003). Transposable elements: targets for early nutritional effects on epigenetic gene regulation. *Molecular & Cellular Biology 23*, 5293–5300.
Zhang L, Ren A, Li Z, Hao L, Tian Y, Li Z. (2006). Folate concentrations and folic acid supplementation among women in their first trimester of pregnancy in a rural area with a high prevalence of neural tube defects in Shanxi, China. Birth defects research Part A: *Clinical & Molecular Teratology 76*, 461–466.
Zhang XJ, Ma NL. (2006). Dynamic monitoring of neural tube defects in Shanxi during 2000 to 2004. *Journal of Shanxi Medical University 3*, 187–192.
Zhang YZ, Zhong N. (2006). Current genetic counseling in China. *Journal of Peking University (Health Sciences) 1*:33–34.
Zhu F, Zhu L, Guo X, Xiao G. (2005). Analysis of Down's syndrome screening and antenatal diagnosis of 3195 cases in the middle period of pregnancy. *China Journal of Modern Medicine 15*, 3079–3081, 3084.

69.

TRADITIONAL CHINESE MEDICINE IN THE AREA OF GENOMICS

Hong-Min Yun, Manshu Song, and Wei Wang

INTRODUCTION

Traditional Chinese Medicine (TCM) is a range of medical practices used in China for more than four millennia. It was created with the birth of Chinese civilization when Emperor Yan ruled China, and an integrated theory system has been formed through years of pharmacological and clinical practice. As an important natural resource, TCM and its pharmacological actions have received more and more attention worldwide over the last decade (Yu et al., 2006). Contemporary research highlights the process of the modernization of TCMs by focusing on combining genomic medicine with TCM and making a canonical, more quantitative and accurate description for its concepts and theories.

APPLING GENETIC TECHNIQUES IN THE AUTHENTICATION OF TCM

The constituents in TCMs are influenced by three principal factors: heredity (genetic composition), ontogeny (stage of development), and environment (climate, associated flora, soil, method of cultivation). Genetic makeup inspections provide a definitive answer to the botanical identity of the TCM constituent, as the DNA profiling of herbal species does not vary with their physical form, or physiological and external conditions (Hon et al., 2003). In light of the advances in molecular biotechnology in the past few decades, genetic tools are able to provide more standardized and reliable methods for authentication of herbal materials at the DNA level. Different DNA manipulation techniques have been developed and used in TCM authentication, such as random amplified polymorphic DNA (RAPD), DNA fingerprinting using multi-loci probes, restriction fragment length polymorphic DNA (RFLP), amplified fragment length polymorphism (AFLP), and microsatellite marker technology (Hon et al., 2003).

Ginseng has long been used to maintain physical vitality in China and the Far East. Ginsenosides can inhibit early antigen activation of Epstein-Barr virus, and also show anticarcinogenic effects in a two-stage mouse skin model with 9,10-dimethyl-1,2-benzanthracene (DMBA), and in lung carcinogenesis induced by 4-nitroquinolin-1-oxide (4NQO) (Konoshima et al., 1996). The DNA probe method for the identification of host-specific DNA fragments is employed in DNA fingerprinting assay of ginseng, and generates a distinctive banding pattern with similarity index of 0.55 between Chinese and American ginseng (Ho & Leung, 2002). In another investigation, RAPD technique has been applied on the authentication of Panax species and their adulterants, and the distinct RAPD fingerprints of American and Chinese ginseng have been obtained, which are consistent irrespective of sources and ages (Shaw & But, 1995). RFLP is also applied to ginseng authentication based on ribosomal ITS1-5.8S-ITS2 region and 18S rRNA gene, and shows promising results on identification of Chinese and American Panax species (Fushimi et al., 1997; Ngan et al., 1999).

USING GENETIC TECHNIQUES TO REVEAL THE RELATIONSHIP BETWEEN FORMULA AND COMPONENTS

Most Chinese herbal formulae consist of several individual herbal components. In a research study of the herbal formula of San-Huang-Xie-Xin-Tang (SHXXT), microarray technique was used to analyze the putative mechanism of SHXXT and to define the relationship between SHXXT and its individual herbal components. Gene expression profiles of HepG2 cells treated with SHXXT's components are obtained by DNA microarray. Gene set enrichment analysis indicates that SHXXT's components display a unique anti-proliferation pattern via p53 protein through DNA damage signaling pathways in HepG2 cells. In addition, hierarchical clustering analysis shows that Rhizoma Coptis, the principle herb, shares a similar gene expression profile with SHXXT. These findings may explain why Rhizoma Coptis exerts the major effect in the herbal formula of

SHXXT. This is one of the examples of revealing the relationship between formulae and their herbal components in TCM using microarray and bioinformatics tools (Cheng et al., 2008).

UNDERSTANDING THE MECHANISMS OF TCM WITH A MOLECULAR APPROACH

A study on airway inflammation shows that Bakumondo-to has anti-inflammatory, anti-allergic, immunomodulatory, secretory-modulating and metabolic regulatory actions, and all of its actions are based on the restoration of damaged cellular functions through DNA transcriptional regulation (Miyata, 2007). Research studies on royal jelly (RJ) demonstrate that it has diverse nutritional and pharmacological functions in humans, such as vasodilative and hypotensive, or antihypercholesterolemic and antitumor activity (Fujii, 1995). RJ competes with 17β-estradiol for binding to the human estrogen receptors α and β. Treatment of MCF-7 cells with RJ enhances proliferation, and concomitant treatment with tamoxifen blocks this effect (Mishima et al., 2005). A reporter gene assay shows that RJ enhances transcription of the luciferase gene through the estrogen-responsive element in MCF-7 cells. Subcutaneous injection of RJ can restore the expression of vascular endothelial growth factor gene in the uteri of ovariectomized rats (Suzuki et al., 2008)].

Berberine is a well-known component of the Chinese herb medicine Huanglian (Coptis chinensis), and is capable of inhibiting growth and endogenous platelet-derived growth factor (PDGF) synthesis in vascular smooth muscle cells (VSMCs) after in-vitro mechanical injury. Berberine can significantly suppress PDGF-stimulated Cyclin D1/D3 and Cyclin-dependent kinase (Cdk) gene expression. Moreover, berberine increases the activity of AMP-activated protein kinase (AMPK), which leads to phosphorylation activation of p53 and increased protein levels of the Cdk inhibitor p21Cip1. These observations offer a molecular explanation for the anti-proliferative and anti-migratory properties of berberine (Liang et al., 2008).

Many kinds of Chinese herb have been confirmed to have antitumor probability. Research on such Chinese herbals shed light on the mechanisms and provide clues for searching for new drugs. A research on the anticancer activity of artemisinin shows that cytotoxicity of its derivative, artesunate, is associated with inhibition of inducible nitric oxide synthase (iNOS). The fact that a number of genes are involved in nitric oxide (NO) signaling and are significantly upregulated or downregulated by artesunate means that artesunate does not only inhibit iNOS, but also affects other NO-related pathway genes. Microarray analysis also shows that the Wnt/β-catenin signaling pathway, which plays an important role in colon cancer, is regulated by artesunate, and colon cancer cell lines are most sensitive toward artesunate among all solid tumor types. Artesunate might also attenuate the growth of human colorectal carcinoma by inhibition of the Wnt/β-catenin pathway (Konkimalla et al., 2008).

Nuclear factor-κB (NF-κB) is critically important for tumor cell survival, growth, angiogenesis, and metastasis. One of the key events in the NF-κB signaling is the activation of inhibitor of NF-κB kinase (IKK) in response to stimuli of various cytokines. The root of E. fischeriana Steud has been used as a traditional Chinese medicine for more than 2,000 years. 17-acetoxyjolkinolide B (17-AJB) is one of the constitutes of E. fischeriana Steud, and a novel small-molecule inhibitor of IKK.17-AJB effectively inhibits tumor necrosis factor-A–induced NF-κB activation and induces apoptosis of tumor cells. Detailed analysis revealed that 17-AJB keeps IKK in its phosphorylated form irreversibly to inactivate its kinase activity, leading to its failure to activate NF-κB. The effect of 17-AJB on IKK is specific and has no effect on other kinases such as p38, p44/42, and JNK. The effects of 17-AJB on apoptosis also correlate with inhibitions of expressions of the NF-κB-regulated genes. It is suggested that 17-AJB is a novel type NF-κB pathway inhibitor, and its unique interaction mechanism with IKK may render it a strong apoptosis inducer of tumor cells and a novel type anticancer drug candidate (Yan et al., 2008).

APPLYING GENOMIC MEDICINE TO INVESTIGATE HERB–DRUG INTERACTIONS

In view of the increasing use of herbal medicines all over the world, concerns have been raised about herb–drug interactions. Accumulating evidence has demonstrated that concurrent administration of herbal remedies may alter the pharmacokinetic or pharmacodynamic behaviors of drugs, and severe adverse effects may occur at extreme conditions (Fugh-Berman, 2000).

Tian Xian is a traditional Chinese herbal anticancer remedy that activates human pregnane X receptor (PXR) in cell-based reporter gene assays. Tian Xian products are herbal dietary supplements manufactured in China by the China-Japan Feida Union Co., Ltd. that are distributed worldwide and aggressively marketed as anticancer herbal therapy. These products are also marketed as herbal therapies that alleviate the unpleasant side effects associated with Western-medicine anticancer treatments. Activation of PXR in the liver regulates the expression genes encoding proteins that are intimately involved in the hepatic uptake, metabolism, and elimination of toxic compounds from the body. PXR-mediated herb–drug interactions can have undesirable effects in patients receiving combination therapy. Tian Xian can alter the strength of interaction between the human PXR protein and transcriptional cofactor proteins. It can increase expression of Cyp3A11 in primary cultures of rodent hepatocytes, and induce expression of CYP3A4 in primary cultures of human hepatocytes. Those data indicate that coadministration of Tian Xian is probably contraindicated in patients undergoing anticancer therapy with conventional chemotherapeutic agents (Lichti-Kaiser & Staudinger, 2008).

COMBINING PHENOME AND GENOME WITH TRADITIONAL CHINESE MEDICINE THEORIES

Zang-Fu theory is one of the foundational principles of TCM, and *Jing* is one of the most important concepts in *Zang-Fu* theory. The *Jing* concept has probably referred to the genetic material as well as its activities, as *Jing* is thought to be "the substance essential for development, growth, and maturation" and "conception is made possible by the power of *Jing*, growth to maturity is the blossoming of *Jing*, and the decline into old age reflects the weakening of the *Jing*" (Kaptchuck, 1983). An injured *Jing*, in the form of various genetic mutations, may thereby result in abnormal development that primarily distributes along the four *Jing-Mai*, connected with the kidneys, which significantly overlap with the body's developmental midline (Li-Ling, 2001). The *Jing-Luo, Zang-Fu* and *Wu-Xing* theories, which constitute the fundamental concepts of TCM, have summarized the physical, functional, and spiritual aspects of human phenotypes. Deciphering the function of a gene often has to rely on the precise identification of its phenotypes. Delineation of the concepts of TCM may significantly improve the understanding of human disease and facilitate ongoing genomic research.

SUMMARY

Traditional Chinese Medicine (TCM) encompasses a mass of human pharmacological information that couples with clinical experience, having existed in China as an integrated theory system for over several thousand years. Genomic medicine is playing an important role in the modernization of TCM by combining TCM theory with modern biological and genetic concepts, e.g., to elucidate the active components of TCM herbal medications and their pharmacodynamic mechanisms at a molecular level. *Zang-Fu* theory is the foundational principles of TCM, and *Jing* is the most important concepts in *Zang-Fu* theory. The *Jing* could be referred to the genetic material and its activities, as *Jing* is thought to be "the substance essential for development, growth, and maturation" and "conception is made possible by the power of *Jing*, growth to maturity is the blossoming of *Jing*, and the decline into old age reflects the weakening of the *Jing*."

ACKNOWLEDGMENTS

The project was partially supported by National Natural Science Foundation of China (30771193, 30600690, 30901238), Research Fund for the Doctoral Program of Higher Education of China (20050025002), Beijing Municipal Commission of Education (KZ200610025014), Beijing Natural Science Foundation (5072007), Major State Basic Research Program-973 of China (2005CB522804), National High Technology Research and Development Program-863 of China (2006AA02Z434), and Beijing Municipal Commission of Science and Technology Foundation (D0906005040391).

REFERENCES

Cheng WY, Wu SL, Hsiang CY, Li CC, Lai TY, Lo HY, Shen WS, Lee CH, Chen JC, Wu HC, Ho TY. (2008). Relationship Between San-Huang-Xie-Xin-Tang and its herbal components on the gene expression profiles in HepG2 cells. *Am J Chin Med* 36:783–797.

Fugh-Berman A. (2000). Herb-drug interactions. *Lancet* 355:134–138.

Fujii A. (1995). Pharmacological effect of royal jelly. *Honeybee Science* 16:97–104.

Fushimi H, Komatsu K, Isobe M, Namba T. (1997). Application of PCR-RFLP and MASA analyses on 18S ribosomal RNA gene sequence for the identification of three Ginseng drugs. *Biol Pharm Bull* 20:765–769.

Ho IS, Leung FC. (2002). Isolation and characterization of repetitive DNA sequences from Panax ginseng. *Mol Genet Genomics* 266:951–961.

Hon CC, Chow, YC, Zeng FY, Leung FC. (2003). Genetic authentication of ginseng and other traditional Chinese medicine. *Acta Pharmacol Sin* 24:841–846.

Kaptchuk TJ. (1983). *Chinese Medicine, A Web That Has No Weaver.* London: Rider, pp. 62–114.

Konkimalla VB, Blunder M, Korn B, et al. (2008). Effect of artemisinins and other endoperoxides on nitric oxide-related signaling pathway in RAW 264.7 mouse macrophage cells. *Nitric Oxide* 19:184–191.

Konoshima T, Takasaki M, Tokuda H. (1996). Anti-tumor-promoting activities of the roots of notoginseng. *Nat Med* 50:158–162.

Liang KW, Yin SC, Ting CT, et al. (2008). Berberine inhibits platelet-derived growth factor-induced growth and migration partly through an AMPK-dependent pathway in vascular smooth muscle cells. *Eur J Pharmacol* 590:343–354.

Lichti-Kaiser K and Staudinger JL. (2008). The traditional Chinese herbal remedy tian xian activates pregnane X receptor and induces CYP3A gene expression in hepatocytes. *Drug Metab Dispos* 36:1538–1545.

Li-Ling J. (2001). Connections between traditional Chinese medicine and congenital syndromes. *Am J Med Genet* 103:257–262.

Mishima S, Suzuki KM, Isohama Y, et al. (2005). Royal jelly has estrogenic effects in vitro and in vivo. *J Ethnopharmacol* 101:215–220.

Miyata T. (2007). Pharmacological basis of traditional medicines and health supplements as curatives. *J Pharmacol Sci* 103:127–131.

Ngan F, Shaw P, But P, Wang J. (1999). Molecular authentication of Panax species. *Phytochemistry* 50:787–791.

Shaw PC, But PP. (1995). Authentication of Panax species and their adulterants by random-primed polymerase chain reaction. *Planta Med* 61:466–469

Suzuki KM, Isohama Y, Maruyama H, et al. (2008). Estrogenic activities of Fatty acids and a sterol isolated from royal jelly. *Evid Based Complement Alternat Med* 5:295–302.

Yan SS, Li Y, Wang Y, Shen SS, et al. (2008). 17-Acetoxyjolkinolide B irreversibly inhibits IkappaB kinase and induces apoptosis of tumor cells. *Mol Cancer Ther* 7:1523–1532.

Yu F, Takahashi T, Moriya J, et al. (2006). Traditional Chinese medicine and Kampo: a review from the distant past for the future. *J Int Med Res* 34:231–239.

70.

INHERITED FACTORS IN CARDIOVASCULAR DISEASE IN CHINA

Ling Zhang, Ying Dai, and Wei Wang

INTRODUCTION

Cardiovascular diseases (CVDs) commonly include a number of disorders, such as rheumatic fever, hypertension, ischemic heart disease, cardiopulmonary disease, cerebrovascular disease, atherosclerosis, structural heart conditions, and many less common conditions. Among the common CVDs, coronary heart disease (CHD), stroke, heart failure (HF), and arterial fibrillation (AF) are major causes of morbidity and mortality. China is rapidly going through a transforming stage that features GDP and drastic changes in people's lifestyles. With its high morbidity and mortality rate and occurrence in a younger and younger age group, CVD is now a serious public health issue that imposes a heavy burden on the Chinese economy and society.

Even though the Sino-MONICA project (Monitoring Trends and Determinants in Cardiovascular Disease) suggested that the incidence and mortality rate of coronary heart disease in China were lower than the world average from 1985 to 1993, the incidence and mortality rate were high (Wu Z et al., 2001a). It is estimated that annually, 2 million new cases of stroke are added with a prevalence of about 7 million patients with stroke, two thirds of these being either disabled or deprived of work capacity to different degrees (NCCD, 2006). Each year, 500,000 people in China present as new cases of myocardial infarction (MI). Statistics indicate that in China there are probably 2 million people with MI, 4,2 million people with heart failure, and 5 million with cardiopulmonary disease (NCCD, 2006). The mortality and proportional mortality rate for CVD is highest among other diseases in China (Ministry of Health, 2006). From 1990 to 2005, the mortality rate fluctuated from 160 to 240/100,000 in rural areas and 210 to 240/100,000 in urban areas. It is estimated that 3 million people die of CVD each year, accounting to more than one third of the total death in China (NCCD, 2006]. The data shows that CVD is an issue with severe outcomes that Chinese can not afford to ignore.

In China, since the prevalence of CVD is rapidly increasing in rural areas, the gap between urban and rural areas is narrowing significantly. It is alarming to note that CVD is now increasingly diagnosed a younger age, probably due to increased fat intake and less physical exercise, which is a common phenomenon in China.

CVD is a complex of different medical and surgical conditions, resulting from environmental causes and the interaction of genetic predispositions and environmental factors. Environmental triggers are primarily those related to changes in lifestyle including high fat intake, overweight, smoking, and lack of physical exercise, all of which are considered to be the major risk factors for CVD in China. In contrast to monogenic disorders, the phenotype for a complex disorder such as CHD is controlled by several genes with small effect (polygenic), and mechanisms involving complex biological pathways (Liu L, 2007). Identifying genetic variants that contribute to CVD may not only provide better understanding of the pathophysiology of the disease, but also may elucidate the biochemical and physiological pathways that link various risk factors to CVD. Furthermore, it will help us to understand the confounding gene–gene interactions and gene–environment interactions that are keys to the causation of CVD. Currently there are few English publications on the status of Chinese CVD research, and reliable statistics information on the prevalence, incidence, and mortality are not available. In this chapter, the epidemiology and etiology of CVD in China are reviewed with emphasis on inherited factors.

EPIDEMIOLOGY

INCIDENCE, PREVALENCE, AND MORTALITY RATE OF CVD

The prevalence of CVD is high in China. A nationwide survey in 2006 revealed 200 million people with hypertension, 200 million with dyslipidemia, 420,000 with heart failure, 2 million with myocardial infarction, 7 million with stroke, and 5 million with cardiopulmonary disease. Around 3 million people die of CVD each year, accounting

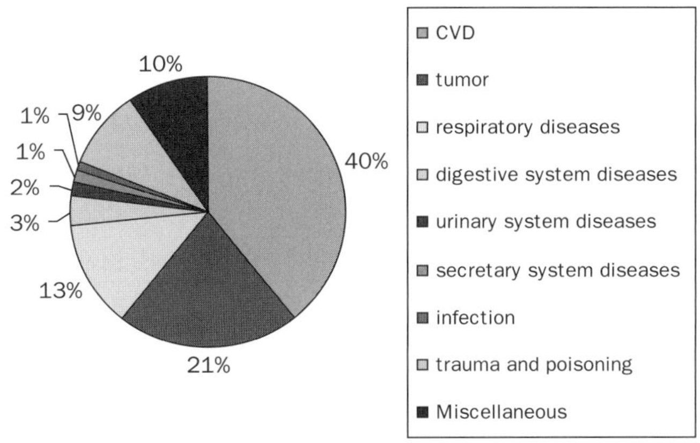

Figure 70-1 Proportional mortality rate of different diseases of urban areas in China in 2005.

for more than one-third of the total annual mortality in China. CVD is now the primary cause of death in both rural and urban areas (Figure 70-1, Figure 70-2; NCCD, 2006).

Among other cardiovascular conditions, CHD, stroke, and hypertension are the most common with severe outcomes. According to the Sino-MONICA project (1983 to 1993), the highest incidence of CHD was 108.7 per 100,000, and the lowest was 3.3 per 100,000 for men aged 35 to 64 years of age, with a 33-fold difference (Wu Z et al., 2001a). In 2003, the Third National Health Care Survey showed the prevalence of CHD was 12.4% in urban areas and 2.0% in rural areas (NCCD, 2006). In general, the mortality rate is higher in urban areas than rural areas (Table 70-1) and it grows exponentially with increasing age in both sectors. Similarly, the incidence and mortality rate for cerebrovascular disease (stroke) were high compared with world averages, ranging from 553.3/100,000 (1987 to 1989) to 33.0/100,000 for men aged 35 to 64 years, a 17-fold difference (Wu Z et al., 2001a). In 2002, the Chinese National Nutrition and Health Survey (CNNHS) reported that the prevalence of hypertension was 18.8% among those 18 years or older, in a survey of 272,023 people from 132 counties (districts and cities) of 31 provinces (including autonomous regions and municipalities) of China, indicating more than 160 million people suffering from hypertension in China.

AGE AND GENDER DISTRIBUTION

The report on cardiovascular disease in China in 2006 showed that the mortality rate for CHD in urban areas grew exponentially with increased age in both males and females (NCCD, 2006). The Sino-MONICA project and Report on Cardiovascular Disease in China in 2006 indicated that males generally had higher incidence and mortality rate of CHD than females (Table 70-1, Figure 70-2). In rural areas, the situation was similar, except that women had the higher mortality rate among those aged 85 and above. A significant difference of incidence and mortality rate of stroke in gender also can be seen (Table 70-1, Figure 70-3). In most Chinese epidemiological studies, the prevalence of hypertension has increased with age in both men and women. Men have a stronger association with high blood pressure than women. It was reported that the highest prevalence of hypertension occurs in the people aged 60–64 years, with 58.9% in men and 56.6% in women. The prevalence is 6-fold and 16-fold higher than those in the age group of 25–29 years. A significant difference of prevalence in gender can be seen (Figure 70-4).

GEOGRAPHICAL CHARACTERISTICS OF CVD

Generally, urban areas have higher incidence and mortality rate of CVD than rural areas. The Sino-MONICA project showed significant geographic variations in CVD rates, with higher rates in the north and lower rates in the south (Tables 70-2 and 70-3). The highest incidence of CHD among men and women was found in Shandong (108.7/100,000 for men and 34.0/100,000 for women) and the lowest in Anhui (3.3/100,000 for men and 0 for women). The incidence varied 33-fold among men. The highest incidence of stroke events among men and women was found in Heilongjiang (553.3/100,000 for men

Figure 70-2 The incidence of CHD in people aged between 35 to 74 in Beijing district from 1984 to 1997 (adjusted by age).

TABLE 70-1 CHANGES IN THE MORTALITY RATES OF CEREBROVASCULAR DISEASE AND CORONARY HEART DISEASE

		City				Country			
Disease	Sex	2003	2004	2005	2006	2003	2004	2005	2006
Cerebrovascular disease (1/100,000)	Total	105.40	100.90	116.63	93.69	89.89	74.95	111.74	105.48
	Men	109.08	107.75	116.63	97.28	87.49	79.97	119.81	112.62
	Women	101.72	93.7	105.19	83.96	72.11	63.89	128.53	81.91
Coronary heart disease (1/100,000)	Total	76.23	99.36	98.22	90.72	45.53	67.18	107.11	84.94
	Men	75.01	100.36	99.49	95.95	62.15	70.74	116.46	87.86
	Women	67.10	98.30	96.88	91.36	60.95	92.96	106.11	90.12
Cardiovascular disease (1/100,000)	Total	181.64	200.26	214.85	184.41	135.42	142.13	218.85	190.42
	Men	184.09	208.11	216.12	193.23	149.64	150.71	236.27	200.48
	Women	168.82	192.00	202.07	175.32	133.06	156.85	234.64	172.03

Data based on the death rate of 10 main diseases in certain regions in China (2003–2006), Chinese Health Statistical Digest, Ministry of Health of the People's Republic of China.

and 337.7/100,000 for women), and the lowest in Anhui among men (33.0/100,000) and in Fujian among women (29.7/100,000). The incidence varied 17-fold among men and 11-fold among women in the populations studied (Liu L, 2007). According to the data of the 1991 CNBPS, the highest prevalence of hypertension was in Beijing at 20.58%, followed by Tibet, Tianjin, Xinjiang, Zhejiang, Sichuan, and Yunnan at 16.89%. Gansu had the lowest prevalence at 7.65% (http://cmbi.bjmu.edu.cn/epidemic/epi-hyper/chyper1.htm). The difference in the incidence and mortality rates of CVD between north and south may attribute to climate and diet patterns. Economic status and life styles mainly explain the contrast of incidence and mortality rate between rural and urban areas.

RESEARCH AND MEASUREMENTS TAKEN TO CONTROL CVD IN CHINA

CVD has been a serious issue in developed countries for a long time. In many industrialized countries, CVD claims more lives each year than any other disease. Heart disease and stroke are the first and third leading causes of death in the United States (Liu J et al., 2004). However, since it started the Framingham Heart Study in 1948, the United States has reported many advances in CVD research and made significant achievements in CVD prevention and control (Dawber et al., 1959; Kannel et al., 1961, 1964). According to the Sino-MONICA project, both the incidence and mortality rate of stroke in China was higher than world average. The incidence and mortality rate of stroke events was 2- to 6-fold higher than that of CHD. Even though the incidence of coronary disease was relatively lower than world average, the number of patients is still enormous due to China having largest population in the world. Even more disturbing is that since 1993, which marks the end of the Sino-MONICA project, the incidence and mortality rate of CVD have been trending upward in China. This distressing situation calls for urgent efforts toward research and control of CVD in China.

The Sino-MONICA is a WHO collaborative project that lasted 10 years, involving approximately 5 million people of 17 regions in China. It elucidated the epidemiology and the main risk factors of CVD. Although it represents research of the largest scale on CVD in China up to now, due to its cross-sectional nature, its conclusions

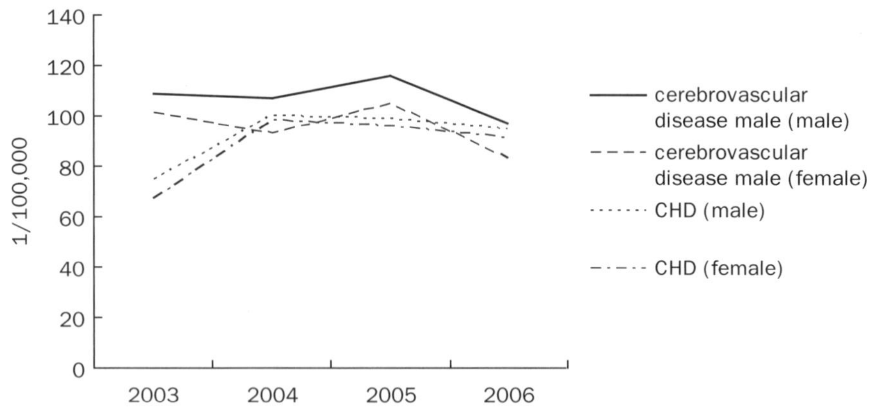

Figure 70-3 Mortality rates of cerebrovascular disease and CHD in urban areas in China. from 2003 to 2006. Data based on *Report On Cardiovascular Disease In China,* Encyclopedia of China Publishing House, Beijing.

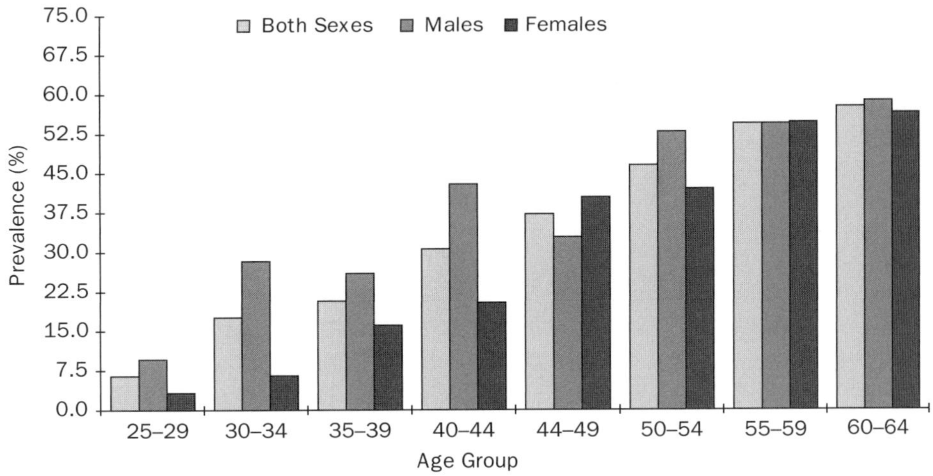

Figure 70-4 The increasing prevalence of hypertension by age. Data from WHO Global InfoBase Online: Source Survey Detail Viewer InfoBase, ref. #:101219a2. (http://www.who.int/infobase/report-viewer.aspx?rptcode=ALL&surveycode=101219a2&dm=18). Hypertension: SBP≥140mmHg and/or DBP≥90mmHg and/or under antihypertensive medication.

TABLE 70-2 AGE-STANDARDIZED INCIDENCE AND MORTALITY OF CORONARY EVENTS DURING 1987–1993 BY POPULATION, 35 TO 64 YEARS OF AGE

	Incidence of CHD (1/100,000)			Mortality of CHD (1/100,000)		
Population	1987–1989	1990–1991	1992–1993	1987–1989	1990–1991	1992–1993
MALE						
Beijing	70.3	87.5	78.0	38.0	50.0	43.5
Shijiazhuang	51.0	59.5	52.5	38.7	52.5	37.0
Huhehaote	52.3	79	.	334.0	79.0	.
Shenyang	49.7	47.5	49.5	32.7	31.0	30.0
Anshan	88.7	82.0	73.0	44.0	45.5	32.5
Jilin	40.5	49.5	48.0	21.5	13.5	8.0
Daqing	95.7	94.5	96.0	41.0	19.5	39.0
Shanghai	7.3	6.0	3.5	3.7	3.5	3.5
Haimen	8.3	5.0	1.5	5.3	1.5	1.5
Chuzhou	3.3	8.5	0	3.3	2.0	0
Nanchang	53.0	.	.	4.3	.	.
Qingdao	12.0	.	.	3.5	.	.
Zhengzhou	108.7	.	.	58.0	.	.
Zhengzhou	40.7	.	.	33.7	.	.
Zhanjiang	59.7	.	.	34.3	.	.
Mianyang	14.0	20.5	11.5	9.3	6.5	5.5
Wulumuqi	79.7	48.0	70.0	25.3	16.0	9.5
FEMALE						
Beijing	31.3	28.0	31.5	12.0	22.0	24.0
Shijiazhuang	15.7	14.0	23.0	13.3	12.0	23.0
Huhehaote	20.0	40.0	.	14.3	35.0	.
Shenyang	16.0	18.0	21.0	10.3	13.0	19.0
Anshan	27.0	26.5	38.0	22.0	17.5	20.0
Jilin	13.0	10.5	10.0	9.5	2.5	2.0
Daqing	39.7	30.0	39.0	27.0	13.5	17.5

(Continued)

TABLE 70-2 (CONTINUED)

Population	Incidence of CHD (1/100,000) 1987–1989	1990–1991	1992–1993	Mortality of CHD (1/100,000) 1987–1989	1990–1991	1992–1993
Shanghai	1.7	3.0	1.0	1.7	3.0	1.0
Haimen	3.0	0.5	0	3.3	0.5	0
Chuzhou	0	2.0	0	0	2.0	0
Nanchang	39.7	.	.	4.3	.	.
Qingdao	0	.	.	0	.	.
Zhengzhou	34.0	.	.	20.0	.	.
Zhengzhou	21.7	.	.	18.0	.	.
Zhanjiang	12.0	.	.	9.7	.	.
Mianyang	1.3	6.0	0	1.3	4.0	0
Wulumuqi	12.0	23.0	9.5	7.0	16.5	9.5

were inevitably confounded by some biases (Wu Z et al., 2001a). Since Sino-MONICA, there have been projects of different scales; the most important one, however, is the Chinese Multi-Provincial Cohort Study (CMCS), a prospective cohort study on cardiovascular disease incidence in 11 provinces of China that lasted 10 years and involved approximately 30,000 subjects aged 35 to 64. It suggested that CVD and its risk factors were common in the Chinese population, and the most important risk factors were hypertension, smoking, and hyperglycemia (Wu Z et al., 2001b). The results of this study were of importance in CVD prevention and control in China.

TABLE 70-3 AGE-STANDARDIZED INCIDENCE AND MORTALITY OF STROKE EVENTS DURING 1987–1993 BY POPULATION, 35 TO 64 YEARS OF AGE

Population	Incidence of CHD (1/100,000) 1987–1989	1990–1991	1992–1993	Mortality of CHD (1/100,000) 1987–1989	1990–1991	1992–1993
MALE						
Beijing	255.3	248.0	251.5	64.0	71.0	61.0
Shijiazhuang	184.3	232.0	259.0	68.0	81.5	98.0
Huhehaote	191.7	165.5	.	61.7	65	.
Shenyang	246.3	263.0	255.0	79.0	85.5	91.5
Anshan	171	335.5	257.0	37.7	83.0	55.0
Jilin	268	444.0	496.0	75.5	53.0	68.0
Daqing	553.3	705.0	751.0	107.3	104.0	98.5
Shanghai	11.3	103.0	78.5	57.0	49.5	46.5
Haimen	70.0	90.0	73.7	29.0	42.5	55.0
Chuzhou	33.0	51.5	75.0	27.0	24.0	40.5
Nanchang	101.7	.	.	43.3	.	.
Qingdao	85.5	.	.	32.5	.	.
Zhengzhou	146.7	.	.	40.7	.	.
Zhengzhou	234.5	.	.	109.0	.	.
Zhanjiang	231.0	.	.	64.3	.	.
Mianyang	90.0	112.0	99.0	43.7	53.5	49.5
Wulumuqi	227.0	154.0	.	45.3	36.5	.

(Continued)

TABLE 70-3 (CONTINUED)

Population	Incidence of CHD (1/100,000)			Mortality of CHD (1/100,000)		
	1987–1989	1990–1991	1992–1993	1987–1989	1990–1991	1992–1993
FEMALE						
Beijing	178.0	179.5	182.5	50.3	51.0	50.0
Shijiazhuang	133.3	143.5	148.5	56.0	59.0	62.0
Huhehaote	120.3	159.0	.	61.7	56.0	.
Shenyang	123.0	122.0	131.0	52.3	51.0	54.0
Anshan	154.0	170.5	203.0	50.7	35.0	51.0
Jilin	134.5	221.5	250.0	50.0	48.0	49.0
Daqing	337.7	360.0	405.5	85.0	63.5	81.5
Shanghai	93.0	95.0	65.5	43.3	30.5	29.5
Haimen	49.0	65.5	69.0	27.3	37.0	41.5
Chuzhou	48.0	46.0	40.5	36.0	21.0	21.0
Nanchang	29.7	.	.	14.0	.	.
Qingdao	88.0	.	.	33.0	.	.
Zhengzhou	108.7	.	.	44.0	.	.
Zhengzhou	150.0	.	.	49.0	.	.
Zhanjiang	145.0	.	.	46.3	.	.
Mianyang	68.3	86.0	62.0	45.3	34.5	22.5
Wulumuqi	238.7	147.5	.	39.3	32.5	.

RISK FACTORS OF CVD:

Hypertension, smoking, dyslipidemia, hyperglycemia, overweight, and obesity are now widely acknowledged as the main risk factors of CVD in China (NCCD, 2006). CMCS indicated that hypertension, smoking, diabetes, hypercholesterolemia were the independent risk factors for CHD, and hypertension, diabetes, hypercholesterolemia, smoking, and obesity were independent risk factors for stroke events (NCCD, 2006).

CMCS also suggested that the incidence of CVD grew with the levels of risk factors (see Table 70-4; Wu Z et al., 1999). The prospective cohort study of 17,330 subjects aged from 35 to 59 suggested the main risk factors of CVD were hypertension, smoking, hypercholesterolemia (≥5.70mmol/l), and hyperglycemia. All of them contribute significantly to relative risk of cardiovascular disease (CHD and stroke). Among them, hypertension is of the most significance. Furthermore, estimated by the prevalence of the population baseline, the population proportional attributable rate (PPAR) was up to 80%, that is to say, these risk factors were responsible for 80% of the ischemic vascular disease. The PPAR of hypertension and smoking were highest at 34.5% and 31.9%, respectively (see Table 70-5A & 70-5B; Zhou B, 2005).

HYPERTENSION

Not only a manifestation of CVD, hypertension is also a trigger that induces a set of complications such as CHD and stroke. Cohort studies confirm that hypertension is the most important independent risk factor for CHD.

A cohort study to evaluate the relationship between blood pressure and CVD risk, carried out in 11 provinces (1992–2002), suggests that 36.1% of all CVD, 44.0% of acute stroke, and 23.7% of acute CHD were associated with hypertension (Wang W et al., 2004). This study, which spanned 10 years, included about 31,728 representative Chinese persons aged 35–64 years and showed that there was a stepwise close relationship between blood pressure and CVD, and that it was more strongly related to stroke than CHD. Among stroke types, blood pressure was more closely related to ischemic stroke than hemorrhage stroke (Figure 70-5).

An epidemiological study of cardiovascular and cardiopulmonary disease risk factors in four populations in China, by the PRC–USA Collaboration Group, indicated that compared with groups whose SBP<110 mmHg, as SBP grew every 10 mmHg, the multifactor-adjusted relative risk of CHD and stroke grew significantly. The abovementioned cases of CHD and stroke

TABLE 70-4 INCIDENCE OF CHD AND STROKE EVENT WITH DIFFERENT LEVELS OF RISK FACTORS

	Incidence of CHD (1/100,000)		Incidence of Stroke Event (1/100,000)	
Risk factors	Male	Female	Male	Female
BLOOD PRESSURE (MMHG)				
Normal	41.1	57.5	118.2	48.7
Higher-normal	95.9	23.5	206.9	332.3
Diagnosed hypertension	196.0	50.1	633.2	480.3
CHOLESTEROL (MMOL/L)				
<=4.14	54.6	26.3	162.2	143.1
4.15–5.16	55.3	69.4	209	126.8
≥5.17	112.7	40.5	304.8	172.8
SMOKING (CIGARETTE)				
0	42.6	52.3	238.4	142.1
1–10	80.2	0	251.3	320.7
≥10	121.7	0	197.3	0
GLUCOSE (MMOL/L)				
<6.10	69	30.7	234.7	149.1
6.10–7.75	106.6	149.1	140.2	180.2
≥7.76	112.7	186.7	349.4	0
BMI (KG/M^2)				
<25	49.9	35.4	189	97.7
25–29	122	88	287.2	199.6
≥30	101	0	421.8	332.7

Blood pressure: Normal (SBP<140 mmHg and DBP<90mmHg); Diagnosed hypertension (SBP≥160 mmHg and/or DBP≥95); Normal-higher (blood pressure between normal and diagnosed hypertension)

among the 120–139 mmHg group accounted for 33.5% and 32.9% of the total cases, while the 140–159 mmHg group accounted for 24.0% and 25.7%. More than half of the total incidence was induced by higher-than-normal blood pressure, or Class I hypertension (see Table 70-6; Zhou B, 2005). This suggested that more attention should be paid to preventing and controlling hypertension.

TABLE 70-5A RR (RELATIVE RISK) AND PPAR (POPULATION PROPORTIONAL ATTRIBUTABLE RISK) OF RISK FACTORS OF CVD OF PEOPLE IN 14 REGIONS

Risk factors	Incidence (%)	RR (95%CI)	PPAR (%)
Hypertension	22.6	3.4 (2.3–4.9)	34.5
Smoking	35.6	2.3 (1.3–3.0)	31.9
Hypercholesterolemia	12.7	2.0 (1.5–3.7)	11.4
Diabetes	3.3	1.9 (1.1–3.6)	3.0

Adjusted by age, gender, and alcohol intake.

Definition of risk factors: Hypertension: SBP≥140mmHg and/or DBP≥90mmgHg or under treatment for hypertension. Smoking: have more than one cigarette per day within one year. Hypercholesterolemia: cholesterol in serum ≥5.70mmol. Diabetes: fasting blood glucose ≥126mg.

There are papers reporting that other risk factors of CVD grew alongside that of hypertension. Compared to people with normal blood pressure, the levels of TC, TG, glucose, BMI, and waist circumference were higher with statistical significance (Zhao D et al., 2006). Thus, along with treating the patients proactively, public health education that effectively and economically changes unhealthy lifestyles would prove beneficial in developing countries, especially in China, which has a large population of CVD patients (Zhou B et al., 2002).

SMOKING

CMCS showed that smoking contributed to 19.9% of the incidence of CHD and 11% of the incidence of acute stroke event in the 30,000 subjects aged between 35 to 64, in a 10-years follow-up study. Compared with nonsmokers the risk for smokers to develop CHD, acute ischemic stroke, and hemorrhage stroke was 1.75-fold, 1.37-fold, and 1.21-fold of that for nonsmokers. The study of 10,000 subjects by PRC-USA group confirmed that the incidence of 31.9% of ischemic vascular diseases was attributable to smoking

TABLE 70-5B RR AND PERCENTAGE OF TOTAL ABOVE-QUOTED CASES OF CHD IN DIFFERENT SBP GROUPS

SBP (mm Hg)	Number of survival person-years	Cases	Multifactor adjusted RR	Above-quoted cases	Percentage of the total above-quoted cases
<110	120,650	20	1	0	–
110–	106,055	117	1.7	44.8	8.1
120–	72,022	139	2.78	90	16.2
130–	37,353	118	4.08	92.6	16.7
140–	17,070	72	5.02	60.4	10.9
150–	10,183	89	9.93	82.1	14.8
160–	5576	69	14.31	65.2	11.8
170–	3109	53	19.71	50.9	9.2
180–	2807	70	29.49	68.1	12.3

using multifactor analysis. Male and female smokers stood a 1-fold and 0.59-fold higher risk respectively of ischemic vascular diseases (RR=2.04, 95%CI 1.43–2.92 for males and RR=1.59, 95%CI 1.10–2.30 for females; see NCCD, 2006).

According to a national survey in 2002, the smoking rate in the population aged above 15 was 35.8%–66.0% for males and 3.1% for females. There were 350 million smokers and 540 million secondhand smokers in China in 2002. Even gloomier is that the smoking rate is trending upward among teenagers aged 15 to 19. Health education and public campaigns for smoking cessation would be of great significance in China.

DYSLIPIDEMIA

Dyslipidemia is a complex of disorders including hypercholesterolemia (TC≥5.72mmol/L), hypertriglyceridemia (TG≥1.70mmol/L) and lipoproteinemia of low HDL-C (HDL-C<1.04mmol/L). Hypertriglyceridemia is the most frequent type of dyslipidemia in China (Wang W et al., 2001). The China National Nutrition and Health Survey (CNNHS) in 2002 indicated the prevalence of hypercholesterolemia (TC≥5.72mmol/L), hypertriglyceridemia (TG≥1.70mmol/L), and lipoproteinemia of low HDL-C (HDL-C<1.04mmol/L) was 2.9%, 11.9%, and 7.4%

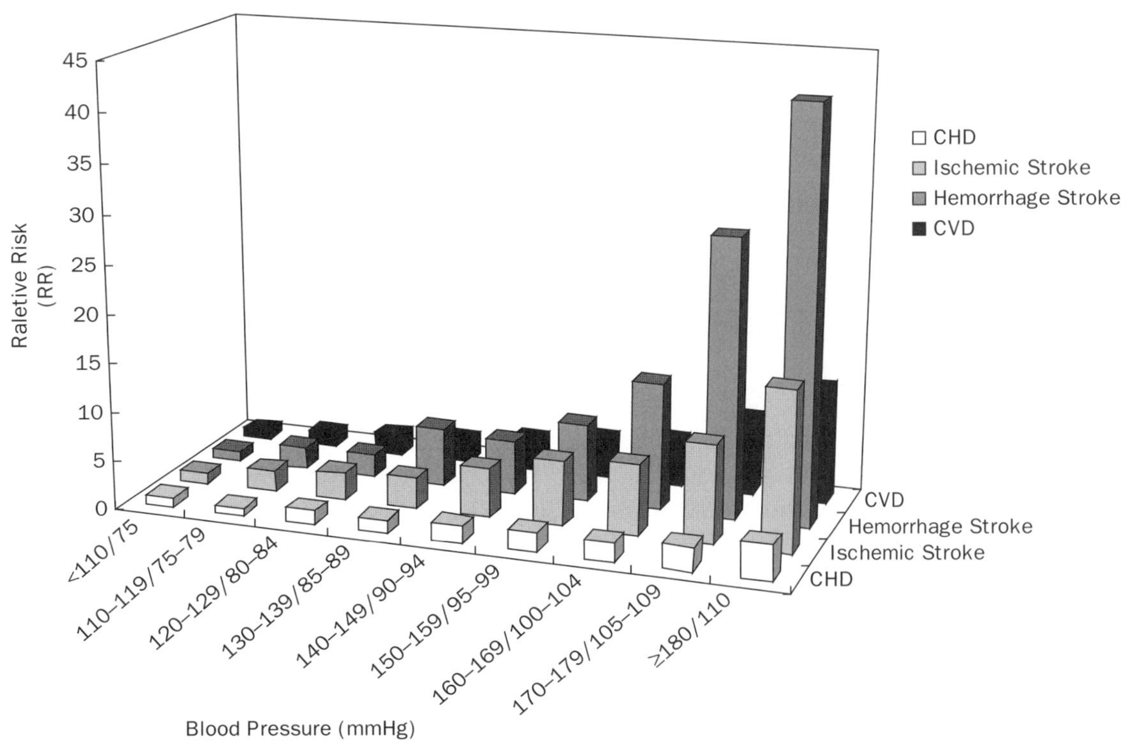

Figure 70-5 Different blood pressure levels and the relative risk of CVD. Data based on a cohort study carried out in 11 provinces to evaluate the relationship between blood pressure and CVD risk (1992–2002).

TABLE 70-6 CANDIDATE GENES SUPPOSED CORRELATED WITH CVD BY STUDIES IN CHINA

Gene Name	Polymorphic allele/Genotype	Association	Reference
LIPID METABOLISM PATHWAY			
ApoE	ε4 allele	Higher risk to develop CHD	Song & Stamfer, 2004; Cao Wei et al.,1999
ApoA I	–75A and +83T allele	Decrease the risk of CHD	Zhou et al., 2003
ApoA V	–1131C allele	Risk factor of CHD	Bi et al., 2005; Tang et al., 2005
ApoB	3' VNTR genotype	Risk factor of CHD	Zhang YX, 2006
	M+ allele	Promotes atherosclerosis and CHD	
ApoA(a)	PNR TTTTA 5 allele	Risk factor of CHD	Hu Bo et al., 2003
ENDOTHELIAL FUNCTION AND OXIDATIVE STRESS PATHWAYS			
e-NOS	–786C allele and 894T allele	Increases risk of cardiovascular events	Li D et al., 2004
	intron4 polymorphism genotype (A allele, four repeats)	Correlates with ischemic stroke	Hou et al., 2001
Glutathione peroxidase1 (GPX1)	198Pro/leu variant genotype	Associates with coronary artery disease	Tang NP et al., 2008
HEMOSTATIC/HEMOLYTIC PATHWAYS			
PAI-1	4G/4G genotype	Risk factor for acute cerebral infarction	Zhang et al., 2001
GPIbαKozak	–5C allele	Risk factor for transient ischemic attack	Song et al., 2006
MSP I	A353G(G→A)	Risk factor of myocardial infarction	Cai et al., 2000
FXIIIA	Pro564Leu genotype	Increases the fatalness of cerebral hemorrhage	Tang Y et al., 2008
β-fibrinogen	–A455 allele	Risk factor of ischemic stroke	Dong & Zhang, 2004; Liu Y, 2004
HOMOCYSTEINE METABOLISM			
Methylenetetrahydrofolate reductase (MTHFR)	677T allele	Risk factor of CHD and stroke	Zhou et al., 1999; Zhang et al., 2004; Qin et al., 2002
CYTOKINE			
TNF-alpha	–238 A allele	Decreases risk of CHD	Hou et al., 2008
Interleukin-10 (IL-10)	–592C/C genotype	Protects against early atherosclerosis through anti-inflammatory pathways	Xie et al., 2008; Xie et al., 2008;
CRP	–717A allele	Risk factor of CHD	Shang et al., 2005; Chen et al., 2005
RENIN ANGIOTENSIN SYSTEM (RAS)			
Angiotensin-converting enzyme (ACE) gene	Intron 16 deletion genotype	Risk factor for cerebral infarction, peripheral arterial occlusive disease and increases the risk for left ventricular hypertrophy	Zhang et al., 2001; Chen et al., 2002; Shen & Daiwen, 1998
Angiotensin II type I receptor (AT1R)	1166A allele	Protective factor of CAD	Yang T et al., 2008

respectively. Based on this, it was estimated there were 200 million people with dyslipidemia in 2006. With the population growth and aging trends of China, the number is expected to rise (Zhao W et al., 2005).

Cohort studies have confirmed that the rise of TC, TG, and LDL-C levels, and decline of HDL levels, increase the incidence of acute CHD and ischemic stroke event significantly (Zhao D et al., 2006; Wang W et al., 2006; Yue H et al., 2004; Pan C, 2004). CMCS showed the incidence of ischemic cardiovascular disease (CHD and ischemic stroke) adjusted by age was on a rising trend with the increased levels of TC (Li G, Yan & Wang W, 2006), and incidence of acute CHD and ischemic stroke rose significantly among populations with HDL-C lower than 1mmol/L. The relative risk for CHD and stroke was 1.39 (95%CI 1.00–1.92) and 1.45 (95%CI 1.15–1.83; see Wang W et al., 2006). A cohort study lasting 20.8 years indicated that compared with subjects whose TC<4.7mmol/L, the risk for incidence of myocardial infarction of subjects whose TC was 4.7–5.1mmol/L, 5.2–5.6mmol/L,

5.7–6.1mmol/L, and ≥6.2mmol/L increased 0.7-fold, 0.95-fold, 1.76-fold, 2.69-fold, respectively (adjusted by age, blood pressure, and smoking; Yue H et al., 2004). Another study by the PRC-USA group that analyzed the growth cutoff points of TC or LDL-C and the incidence ischemic cardiovascular disease found that with the rise of TC and LDL-C, the relative risk of ischemic cardiovascular disease rose correspondingly (Pan C, 2004). A 14-group cohort study (follow-up of 17,330 subjects for 6.8 years) further confirmed that subjects with TC≥5.70 had 1-fold higher risk of ischemic cardiovascular disease (RR=2.0 95%CI 1.5–3.7; see Zhao D et al., 2006).

According to a sampling survey of 20 provinces from 2001–2002, the rate of awareness, treatment, and control of subjects with TC≥6.22mmol/L or under medication was 21.3%, 14.0%, and 11.3% for males and 18.1%, 11.6%, and 9.5% for females. Of subjects with TC≥5.18mmol/L the number was even smaller, at 8.8%, 3.5%, and 1.9% for males and 7.5%, 3.4%, and 1.5% for females (He J et al., 2004). As an important independent risk factor of CVD, dyslipidemia with its high prevalence and low rate of awareness, treatment, and control would be a formidable obstacle for China to preventing and controlling CVD if insufficient attention is paid by government and medical institutes.

DIABETES

Diabetes is an important risk factor for atherosclerosis (Zhao D et al., 2006). CMCS presented data showing that incidence of both CHD and stroke event rose with the rise of blood glucose levels (see Table70-4; Wu Z et al., 1999). A report of the PRC-USA group suggested that among groups of normal fasting blood glucose (< 6.1mmol/L), abnormal fasting blood glucose (6.0–7.0mmol/L), and diabetes (≥7.1mmol/L), the incidence of CHD and stroke event adjusted by age varied from 8.2 per 100,000 to 30.2 per 100,000, a 3.7-fold difference for males, and varied from 3.6 per 100,000 to 19.8 per 100,000, a 5.5-fold difference for females (Zhou B et al., 2003). The relative risk of CHD for the group of diabetes was 3.22 (95%CI 1.37–7.56) compared with the group of normal blood glucose. And the relative risk of ischemic stroke for diabetes was 3.05 (95%CI 1.37–7.56).

OVERWEIGHT AND OBESITY

For Chinese adults the normal body mass index (BMI) is 19–24 kg/m^2; overweight is BMI≥24 kg/m^2; and obesity is BMI≥28 kg/m^2. Waist circumference ≥85cm in males and ≥80cm in females is considered abdominal obesity.

A lot of large-scale cohort studies found that with the rise of BMI, the risk for CVD grows significantly. The risk factors for stroke event in the group of overweight (BMI 24.0–27.9) and obesity (BMI≥28.0) subjects was 1.47 (95%CI 1.06–2.03) and 2.48 (95%CI 1.58–3.89), respectively (Zhao L et al., 2002). There was a paper indicating that the relative risk of ischemic cardiovascular disease was 1.43 (95%CI 0.897–2.272) for males (waist circumference 85–90cm) and for females (waist circumference 80–95cm). For males and females with waist circumference ≥95 and ≥90cm, the relative risk was up to 2.36 (95%CI 1.394–3.994; see Zhao D et al., 2006).

Like hypertension, overweight and obesity are the triggers of many disorders. The study by the PRC-USA group found that BMI was positively related to blood pressure. The 10-group study confirmed that the average level of BMI baseline was positively related to incidence of diagnosed hypertension (Wu X et al., 1996). Furthermore, following large-scale studies it was reported that obesity, including abdominal obesity, was not only significantly related to blood pressure, blood fat, blood glucose respectively, but also related to a clustering of hypertension, hypercholesterolemia, hyperglycemia, and low high density lipoprotein cholesterol (Wu Y et al., 1996; Zhao L et al., 2003a,b).

Due to high fat intake and lack of physical exercise, the rate of overweight and obesity have been growing during the last 20 years. Compared with the data from the Third China National Nutrition and Health Survey (CNNHS) in 1992, the rate of overweight and obesity increased by 38.6% and 80.6%, to 17.6% and 5.6% (Huang Z et al., 1994). Other large-scale studies also confirm this trend. There were 200 million people of overweight and 60 million people of obesity in 2002. Calculated in terms of the size of population, there were 240 million people of overweight and 70 million people of obesity who were vulnerable to CVD, or already with CVD in China in 2006 (NCCD, 2006). It would be wise to change people's lifestyles by health education that encourages people to lose weight scientifically and effectively.

ETIOLOGY

There have been three stages in the process of controlling CVD in China. The first was to study whether the incidence and mortality of CVD decline with risk factors. The second aimed to prevent and control CVD in populations of large scale through healthcare services and public health education. While some traditional ("Framingham") risk factors such as older age, male gender, family history, and cigarette smoking contribute to disease, as in the general population, others, including hypertension, diabetes/insulin resistance, and dyslipidemia occur with a higher prevalence and/or severity in patients with CVD. Furthermore, in the more prominent occurrences of some of the traditional risk factors in CVD, the extent and severity of cardiovascular complications is not necessarily proportionate to the underlying risk factor profile. Thus, we are now in the third stage, which emphasizes research of the etiology of CVD at the molecular level (Hu Y& Zhao D, 2003).

GENETIC RISK IN CVD, GENOME-WIDE ASSOCIATION APPROACH, CANDIDATE GENE APPROACH AND PATHOPHYSIOLOGIC MECHANISMS OF CVD

Many risk factors for CVD, such as lipid levels, hypertension, obesity, diabetes, and the metabolic syndromes, show significant heritability in family and population-based epidemiologic studies. In contrast to Mendelian or monogenic disorders, where linkage analysis may be successful in mapping a disease locus, the phenotype for a complex disorder such as CVD and its complications is controlled by multiple genes (polygenic, locus heterogeneity) and involved in multiple biological pathways. The contribution of each gene is small (low genotypic relative risk). Phenotypic expression is often modified by gene–environment interactions and gene–gene interactions.

GENOME-WIDE ASSOCIATION APPROACH

Recent advances in genomics have ushered in a genome-wide association study (GWAS), an approach in which a substantial fraction of common genetic variation is tested for its role in determining phenotypic variation (Wang L et al., 2003). GWAS has been successfully applied in many studies to identify novel genetic loci related to several medical phenotypes, and the Framingham Heart Study Project is a good example.

The study, based on 1345 participants from 310 pedigrees, provided results of genome-wide association for 4 CVD outcomes with great public health impact: major CVD, major CHD, AF, and HF. No associations attained genome-wide significance in their analyses using GEE (generalized estimating equations) or FBAT (family-based association tests) additive genetic models. Even though designed and conducted under strict standards, the study had its own limitations, which led to negative results of candidate gene analyses that might have underestimated associations for genes that are incompletely covered by the SNPs used in that investigation (Liu J et al., 2004). Replication studies are now performed in different regions to confirm variants that associate with disease outcomes. For instance, Broeckel et al. (2002) located a candidate gene for myocardial infarction on chromosome 14 by GWAS on 1406 participants from 513 pedigrees. In China, however, study of matched scale involving so many ideal nuclear families is now rare, and reliable data on genome-wide association is not available.

With dramatic declines in the cost of high-throughput genotyping, selective genotyping of SNPs with suggestive evidence of association can be considered after GWAS. Up to now, two-stage approaches—genome-wide association followed by selective genotyping—have been adopted as a practical and efficient strategy for pursuing initial genome-wide results (Skol et al., 2006; Wang H et al., 2006).

CANDIDATE GENE APPROACH

Single nucleotide polymorphisms (SNPs) are the most common form of DNA variation in the human genome and are biallelic (only two choices exist at a given site). Other types of polymorphisms include insertion/deletion and mini- and microsatellites (di-, tri-, and tetranucleotide repeats). The approach to studying complex disorders has classically been gene association studies using the candidate gene approach, where cases of diseases or traits are compared with appropriate controls from the same population, with the objective of detecting small effects of a candidate gene on the phenotype. The candidate gene is selected on the basis of biological plausibility that allows an a priori hypothesis about its etiologic role in disease, and is supported by evidence from functional in vitro and in vivo experiments including expression studies, genetic mapping experiments, and protein assays, which individually or collectively describe the intermediate phenotype (Rao et al., 2005).

PHENOTYPES OF CVD AND ITS PATHOPHYSIOLOGIC MECHANISMS

The spectrum of CVD covers a wide range of phenotypes, broadly classified as the outcomes associated with atherosclerosis (CHD, cerebrovascular disease, peripheral vascular disease), myocardial muscle involvement (left ventricular hypertrophy [LVH] and heart failure), and arterial disease (arteriosclerosis). Phenotype definition is one of the critical caveats in the process of ensuring validity in gene association studies. Despite overlap to different degrees, phenotypes tend to cosegregate with pathophysiologic processes, metabolic pathways, and therefore candidate genes. Specific phenotypes are the final expression of interactions among environmental, metabolic, and genetic factors. Atherosclerosis, for instance, is a time-dependent, multistep process involving multiple pathways—lipoprotein metabolism, oxidative stress, coagulation, inflammation, and endothelial function. Similarly, LVH is the culmination of the effects of hypertension (systolic and diastolic), volume overload, anemia, age, and chronic dialysis, with activation of the renin-angiotensin system (RAS) and cardiac remodeling mediated by cytokines and matrix proteins (Rao et al., 2005) (Figures 70.6 and 70.7).

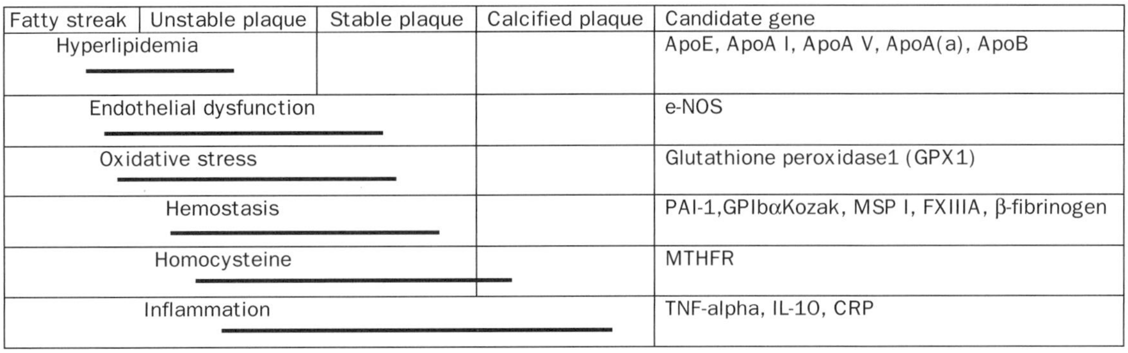

Figure 70-6 The stages of evolution in the atherosclerosis lesion from fatty streak to calcified plaque, with the injury risk processes and risk factors operating during these stages. Candidate genes with a possible relationship to disease at each stage are enumerated (Rao et al., 2005).

OVERVIEW OF SNP ASSOCIATION STUDIES IN CVD

SNPS IN LIPID METABOLISM PATHWAYS

No definite relationship has been elucidated between lipoprotein and CVD up to now, but researches of recent years have reported that the level of lipoprotein in serum and its gene polymorphisms are correlated with the incidence and severity of CVD.

APOE GENE

ApoE is the most studied lipoprotein. ApoE regulates lipid levels in serum by interacting with ApoE receptors in the liver and LDL receptors to transport cholesterol. The ApoE gene is located at 19q13.2 with 3 introns and 4 exons. Gene polymorphisms of ApoE (ε2, ε3, ε4) lead to 3 isomers (E2, E3, E4). The difference in phenotypes of ApoE in populations is decided by different distribution of allele frequencies. Even though there are a lot of reports about the effect of ApoE on CHD and stroke, some conclusions are conflicting. Kosunen et al. (1995) and Contois et al. (1996) reported that ApoE was a risk factor for CHD, and Xie Jie Yonghong et al. (1989) supported their points based on their research in the Chinese population. However, there are contrary results presented by Kurz et al. (1995). Pedro-Botet et al. (1992) found ε4 was the risk factor for ischemic cerebrovascular disease, and Conderc et al. (1993) claimed ε3/ε3 might prevent cerebrovascular disease, while ε2 was a risk factor. The study by Cao Wei et al. (1999) in the Chinese northeastern population showed there was no significant difference among allele frequencies of ApoE in CVD patients and the control group. Song et al. (2004) analyzed 48 papers from 1966 to 2004 concerning ApoE and CVD and found the ε4 carriers had a 42% higher risk to develop CHD than the ε3/ε3 carriers. Jiang Xin et al. (2006) reached similar conclusions by meta-analysis of CHD and ApoE in the Chinese population and concluded that allele ε4 was a risk factor for CHD in China.

APOA I GENE

The ApoA I gene is located at 11q ter with 3 introns and 4 exons; –75G/A in the promoter region and + 83 C/T (or +84G/A) at intron 1 are the most reported polymorphisms. Shioji et al. (2004) found that +84 A allele was a risk factor for myocardial infarction in the Japanese population, and that it would decrease the level of HDL-C. Chhabra et al. (2005) reported that –75A allele would decrease the level of HDL-C and ApoA I, and was significantly related to severity of CHD in northern Indians. The research of Zou Yangchun et al. (2003), on the other hand, showed that both –75A and +83T alleles would increase the level of HDL-C and ApoA I, thus decreasing the risk of CHD in the Han Chinese ethnicity. The conflicts among different studies are attributed to differences of races studied, definitions of CVD phenotype, and sizes of samples.

APOA V GENE

The ApoA V gene is located at 11q 23 with 3 introns and 4 exons. ApoA V plays an important role in TG and lipoprotein

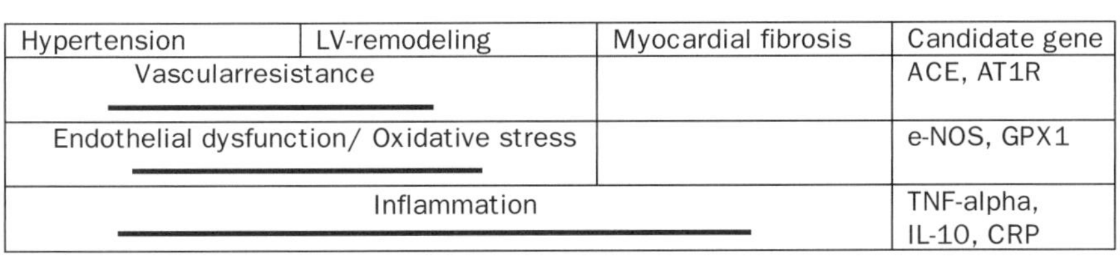

Figure-70-7 Pathophysiologic phase in the left ventricular remolding. Candidate genes with possible relationship to disease at each stage are enumerated (Rao et al., 2005).

level regulation. There are up to 17 SNPs related to ApoA V reported, and –1131T/C and 56C/G are the most studied polymorphisms. Talmud et al. (2005) studied a British population and found that the TG levels of –1131T/T carriers were significantly higher than that of –1131C/C carriers, but there was no relationship between the polymorphism and CHD. They also reported that the 56G allele was a risk factor in atherosclerosis. Bi Nan et al. (2005) found –1131C was a risk factor for CHD by case-control study of the Han Chinese ethnicity; Tang Y et al. (2005) supported their point.

APOB GENE

The ApoB gene is located at 2p23 with 28 introns and 29 exons. The most studied polymorphisms are 3' VNTR, Msp I, and X ba I. Studies in China and abroad all find 3' VNTR to be a risk factor in CHD. Studies in Chinese populations and other races report that X ba I was a genetic marker of CHD, but the mechanisms are not known yet. Studies in China and abroad also confirm that polymorphism of Msp I is related to atherosclerosis, and the M+ allele promotes atherosclerosis and CHD (Zhang YX, 2006).

APO (A) GENE

Epidemiological studies show that the growth of Apo (a) level is an independent risk factor of atherosclerosis. The most studied polymorphism is TTTTA at the upstream of promoter, which is one kind of PNR (pentanucleotide repeat). Studies abroad showed that big size of PNR was a risk factor of CHD (Volleova et al, 1997; Kalina et al., 2001). Hu Bo H et al. (2003) reported that Apo (a) PNR polymorphism is closely related to Apo (a) levels and that the PNR TTTTA 5 allele might be a risk factor in CHD in the Chinese population.

LDL-RECEPTOR (LDL-C) GENE

The LDL-C gene is located at 19q13.1-p13.3 with 17 introns and 18 exons. Most of its variations are located at the promoter and exons. The polymorphisms lead to dysfunction of LDL-R and are responsible for the growth of TC levels. Wang C et al. (2004) reported that Ava II + carriers who drink and smoke had higher risk for CHD in China. One study found that the N+ allele was closely related to myocardial infarction (Zang B et al., 2005).

SNPS IN ENDOTHELIAL FUNCTION AND OXIDATIVE STRESS PATHWAYS

Synthesis of nitric oxide from L-arginine by endothelial nitric oxide synthase (eNOS) plays a key role in preventing atherosclerosis. The gene encoding eNOS is located on chromosome 7q35-36 (13). The C allele at–786 was associated with an increased risk of cardiovascular events in Han Chinese (Li D et al., 2004) and significant association with hypertension was found in Chinese Mongolia (Wang C et al., 2006). The T allele at 894 was confirmed to be associated with cardiovascular disease in both Han Chinese and in Mongolians (Li D et al., 2004; Zhang C et al., 2001). According to study of Hou et al. (2001), the intron 4 polymorphism (A allele, four repeats) was related to ischemic stroke in Chinese populations.

The reduced form of nicotinamide adenine dinucleotide phosphate (NADPH) oxidase (p22phox) is an important source of superoxide in blood vessels, which increases oxidative stress and reduces nitric oxide bioactivity. The C242T polymorphism may determine enzyme function, the T allele being associated with reduced NADPH oxidase activity, and has been explored as a candidate gene for ischemic stroke (Mata-Balaguer et al., 2004).

Glutathione peroxidase 1 (GPX1), the key antioxidant enzyme in vascular endothelial cells, has been shown to exert a protective effect against the presence of coronary artery disease (CAD). There was study providing evidence that GPX1 198Pro/leu variant genotypes are significantly associated with CAD risk in this Chinese population (Tang N et al., 2008).

SNPS IN HEMOSTATIC/HEMOLYTIC PATHWAYS

For young people who rarely have other complications pf CVD, the imbalance of hemostatic/hemolytic system plays an important role in inducing stroke. Gene polymorphisms are risk factors for stoke occurring at an early age for people aged under 50.

Plasminogen activator inhibitor (PAI)-1 is produced by the vascular endothelium, and is a major inhibitor of tissue-type plasminogen activator (tPA) and therefore intravascular fibrinolysis. An insertion/deletion polymorphism within the PAI-1 locus (4G/5G) affects gene expression; the 4G/4G PAI-1 gene polymorphism that is associated with higher PAI-1, and therefore low fibrinolytic activity, was associated with fatal CVD. The results of study in China suggest that PAI-1 gene polymorphism may be a susceptible factor for acute cerebral infarction in Chinese, and 4G allele homozygous genotype may be a major risk factor for acute cerebral infarction; it may be especially an independent risk factor for cerebral infarction in female patients (Zhang C et al., 2001).

Blood coagulation factors and platelet are the two most important systems for hemostasis.

There are three kinds of glycoprotein receptors on the surface of platelets responsible for triggering the aggregation of platelets. They are GP Ia/IIa, GP Ib/Xb, and GP IIb/IIIa. According to a study of Song Yuqiang et al. (2006) in Qingdao, China, the −5T/C polymorphism of the platelet glycoprotein GPIbα Kozak gene (one polymorphism of GP Ib/Xb) is implicated in the pathogenesis of transient ischemic attack (TIA). Also, the C allele is an independent risk factor for TIA in Chinese populations.

The result of a study by Cai Qiangjun et al. (2000) supported that in China, the role of plasma factor VII coagulant (VII c) activity as a risk factor for myocardial infarction, and VII c was influenced by its gene MSP I polymorphism. There is a study in Guangxi, China, suggesting that coagulation factor XIII (FXIIIA) Pro564Leu polymorphism might have some relationship with cerebral hemorrhage; people with FXIIIA Pro564Leu polymorphisms could experience increased fatality of cerebral hemorrhage. The prevalence of FXIIIA Pro564Leu in the Guangxi population might be lower than that in the white race populations (Tang Y et al., 2008).

Several prospective and case-control studies have shown that an elevated level of fibrinogen is an independent risk factor for cardiovascular diseases (Wilhelmsen et al., 1984; Ernst & Koenig, 1992). And, there are studies in China that indicated plasma fibrinogen expression was affected by the β-fibrinogen gene-455A/G polymorphism, so the −A455 allele might be a risk factor for ischemic stroke in Chinese populations (Dong Q & Zhang C, 2004; Liu Y et al., 2002).

SNPS IN HOMOCYSTEINE METABOLISM

Homocysteine is an amino acid that accelerates atherosclerosis by increasing lipoprotein oxidation and endothelial injury. Methylenetetrahydrofolate reductase (MTHFR) is a key enzyme in the remethylation of homocysteine to methionine, and reduces total homocysteine levels. Studies in China found that the frequency of homozygous and heterozygous MTHFR mutation in stroke and CHD patients was higher than that in controls, and that T alleles in stroke and CHD patients was significantly more frequent than that in control subjects. Plasma Hcy levels in stroke and CHD patients were higher than that in normal subjects. These results suggested that polymorphisms of MTHFR gene and hyperhomocysteinemia might be an independent risk factor for stroke and CHD (Zhou X et al., 1999; Zhang C et al., 2004; Qin Q et al, 2002). But there are also contrary conclusions in China. For instance, Huang Yining et al. (2002) reported that even though homozygote of mutant MTHFR gene C677T caused hyperhomocysteinemia, and there was a significant association between hyperhomocysteinemia and ischemic cerebrovascular disease, the phenotype of MTHFR had no relation with stroke.

SNPS IN CYTOKINE GENES

Interleukin (IL)-6 and tumor necrosis factor (TNF)-α are proinflammatory cytokines, and IL-10 inhibits proinflammatory cytokine production. Several promoter region polymorphisms have been identified that influence cytokine production and therefore the inflammatory state.

A study by Hou L et al. (2008) identified a significant interaction between TNF-alpha gene and smoking status. Both single locus and haplotype analyses indicated that A allele of the −238G>A polymorphism decreased the risk of CHD among nonsmokers in Chinese Han populations.

According to Xie G et al. (2010), interleukin-10 (IL-10) −592C/C genotype was significantly associated with decreased risk of early carotid atherosclerosis compared to −592 A/A genotype. This observed association was significantly modified by known atherogenic risk factors, namely smoking and obesity. In people with normal cholesterol (<200 mg/L), −592 C/C genotype was also associated with lower levels of highly-sensitive C-reactive protein. The findings generated an interesting hypothesis that the IL-10 −592C/C genotype might exert its protection against early atherosclerosis through anti-inflammatory pathways. The results also confirmed that gene–risk factor interaction played an important role in pathogenesis of early atherosclerosis in Chinese populations. A study by Xie G (2008) also supported their conclusion.

Given the central role of CRP in the inflammatory response, and its strong associations to both outcome and complications in CVD, it may be useful to study the polymorphisms of CRP. The CRP polymorphisms widely acknowledged concerning CVD are rs1800947, rs2794521, and rs309124. Among them, rs1800947 and rs2794521 polymorphisms were reported to be correlated to CRP levels, and −717A allele was a risk factor for CHD in China (Shang X et al., 2005; Chen J et al., 2005).

SNPS IN THE RENIN ANGIOTENSIN SYSTEM (RAS)

It is reported that the angiotensin-converting enzyme (ACE) gene carries an insertion/deletion polymorphism on intron 16 and accounts for approximately 50% of interindividual variation in plasma ACE levels. Study indicated that ACE gene deletion polymorphism might serve as an independent risk factor for cerebral infarction in the Han Chinese nationality, while circulation ACE activity might be related to gene deletion polymorphism (Zhang X et al., 2001). According to Chen Jin et al. (2002), ACE gene deletion polymorphism was also a susceptible gene to the presence of peripheral arterial occlusive disease (PAOI). There are also studies in China suggesting that a deletion polymorphism of ACE gene is probably an important hereditary factor of brain infarction's morbidity, and increases the risk for left ventricular hypertrophy in Chinese hypertensive patients (Shen D & Ha D, 1998a,b).

A study by Yang Tao et al. (2008) indicated that incidence of coronary arterial disease was correlated with polymorphism of angiotensin II type I receptor (AT1R), and SNP1166A was a protective factor in CAD. Although there is a report that A1166C polymorphism of AT1R did not influence the risk of MI in China (Pan X et al., 2005), research by Qiu C et al. (2003) showed ACE deletion polymorphism should be an independent risk factor for CHD and myocardial infarction in Chinese people. The AT1R –C allele might increase the relative risk of myocardial infarction and CHD associated with deletion polymorphism of ACE; there was a synergistic interaction between ACE –I/D and AT1R –C gene polymorphism.

It should be noted that although genetic factors (SNPs) that modulate individual susceptibility are common, they generally have a modest effect at an individual level. However, because of their frequency in the population, the genetic variants can be associated with a high attributable risk if large enough populations are examined.

GENE–GENE INTERACTION AND GENE–ENVIRONMENT INTERACTION

Instead of an isolate factor that solely decides the development and prognosis of CVD, genes modify CVD by gene–gene interaction and gene–environment interaction.

GENE–GENE INTERACTION

Genes modify CVD with synergistic and antagonistic effects. The cytokine network is highly complex, containing interactive cascades of gene activation and suppression. It is widely acknowledged that the expression of CRP is regulated at the transcription level by IL-6 promoting its translation, with IL-1 playing a synergistic role. Thus, the IL-1 and IL-6 gene polymorphisms are associated with CRP gene expression. And, in the RAS system, there is also a report in China that showed a synergistic interaction between ACE –I/D and AT1R –C gene polymorphism (Qiu C et al., 2003).

GENE–ENVIRONMENT INTERACTION

Study in China showed that the effect of ApoE gene polymorphism on fasting glucose and insulin was modified by overweight and obesity. Obese ε4 carriers tended to have higher fasting glucose and insulin levels (Liu Q et al., 2008). There is also report in China suggesting that the carriers of ε3, ε4 (ApoE gene polymorphisms) or Ava II+ (LDL-C receptor polymorphism) alleles would have a higher risk of suffering from CHD if they drank alcohol or smoked heavily (Wang CH et al., 2004).

The interesting "thrifty genotype" hypothesis, proposed by Neel in 1962, is now more and more acknowledged with further studies on CVD (Neel, 1999). The hypothesis suggested that thrifty genes evolved with humans struggling to survive in adversity. They are responsible for energy-saving when food is deficient, and would eventually lead to the so-called "urbanized diseases" such as CVD when food is sufficient. The hypothesis highlighted the interaction of genes and environment. Individuals with CVD are carriers of the wild genotype, while individuals without phenotypes of CVD are the variant group, which is the result of successful adjustment to an adverse environment characterized by high energy intake and sedentary work in the modern world. Though not quite mature, this hypothesis might shed light on the research and prevention of CVD in the future.

Different races may have different gene polymorphism frequencies, inhabit various environments, and share diverse lifestyles; thus they show different genotype frequencies. It is understandable that studies in different countries show conflicting results concerning the effect of the same gene polymorphism on CVD. Ruling out the biases, such as different phenotype definition, sampling standards, and racial differences, can contribute a lot to understanding the conflicts.

Given the high prevalence and mortality of CVD in China, it is of high priority to study its etiology at the molecular level based on Chinese populations. Even though there is a data pool available for association study based on research of other countries, racial differences should always be considered when trying to reach conclusions concerning the effects of a certain gene on CVD in Chinese populations.

REFERENCES

Bi N, Yan S, Li G, et al. (2005). Polymorphisms in the apolipoprotein A5 gene and apolipoprotein C3 gene in patients with coronary artery disease. *Chinese Journal of Cardiology* 33(2): 116–121.

Broeckel U, Hengstenberg C, Mayer B et al. (2002). A Comprehensive Linkage Analysis for Myocardial Infarction and Its Related Risk Factors. *Nat Genet* 30:210–214

Cai Q, Chen J, Yuan J, et al (2000). Study on the association of coagulation factor VII and its gene MsPI polymorphism with the risk of myocardial infarction. *Chinese Circulation Journal* 15(2):75–77.

Cao W, Chen F, Teng L, Wang S, Fu S, Zhang G. (1999). The relationship between apolipoprotein E gene polymorphism and coronary heart disease and arteriosclerotic cerebral infarction. *Chinese Journal of Medical Genetics* 16(4): 249–251.

Chen J, Zhao J, Gu D, et al. (2005). -717A>G polymorphism of human C-reactive protein gene associated with coronary heart disease in ethnic Han Chinese: the Beijing atherosclerosis study. *Journal of Molecular Medicine* 83(1):72.

Chen J, Li X, Li J, et al. (2002). The Relationship between Angiotensin Converting Enzyme Gene Polymorphism and Peripheral Occlusive Disease. *Molecular Cardiology of China* 2(6):7–10.

Chhabra S, Narang R, Lakshmy R, et al. (2005). ApoA I-75G to a substruction associated with severe forms of CAD, lower level of HDL and ApoA I among Northern Indians. *Dis Markers* 21(4):169–174.

Contois JH, Anamani DE, Tsongalis GJ. (1996). The underlying molecular mechanism of apolipoprotein E polymorphism: Relationships to lipid disorders. *Clin Lab Med* 16(1):105–123.

Couderc R, Mahieux F, Bailleul S, et al. (1993). Prevalence of apolipoprotein E phenotypes in ischemic cerebrovascular disease. A case-control study. *Stoke* 25(11):2296–2297.

Dawber TR, Kannel WB, Revotskie N, et al (1959). Some Factors Associated With the Development of Coronary Heart Disease; Six years' Follow-up Experience in the Framingham Study. *Am J Public Health* 49: 1349–1356.

Dong Q, Zhang C. (2004). Association between the polymorphism of β-fibrinogen gene-455G/A and ischemic stroke. *Chinese Journal of Medical Genetics* 21(3):274–276.

Ernst E, Koenig W. (1992). Fibrinogen and cardiovascular risk. *Vasc Med* 2:115–125.

He J, D Gu, et al. (2004). Serum total and lipoprotein cholesterol levels and awareness, treatment, and control of hypercholesterolemia in China. *Circulation* 110(4): 405–411.

Hou L, Huang J, Lu X., Wang L, Fan Z, Gu D. (2008). Polymorphisms of tumor necrosis factor alpha gene and coronary heart disease in a Chinese Han population: Interaction with cigarette smoking. *Thrombosis Research* 123(6):822–826.

Hou L, Osei-Hyiaman D, et al. (2001). Association of a 27bp repeat polymorphism in e NOS gene with ischemic stroke in a Chinese population. *Neurology* 56(4):490–496.

Hu B, Zhou X, Ye SX, Liang MJ, Li ZX. (2003). Expression of pentanucleotide repeat polymorphism of apolipoprotein(a) in Chinese Han and the association with coronary heart disease and serum lipid level. *Chinese Journal of Clinical Rehabilitation* 7 (21):2899–2901.

Hu Y, Zhao D. (2003). New features of cardiovascular disease prevention and control in China. *Chinese Journal of Epidemiology* 24(12):1069.

Huang Y, Zhao YL, Li S. (2002). Hyperhomocysteine, methylenetetrahydrofolate reductase gene, and other risk factors in ischemic stroke. *Chinese Medical Journal* 82(2):119–122.

Huang Z, Wu X, et al. (1994). A north-south comparison of blood pressure and factors related to blood pressure in the People's Republic of China: a report from the PRC-USA Collaborative Study of Cardiovascular Epidemiology. *Journal of Hypertension* 12(9): 1103–1112.

Jiang X, Dong S, Luo L, et al. (2006). Apolipoprotein E genotypes and risk for coronary heart disease in Chinese population: A meta-analysis. *Chinese Circulation Journal* 21 (2):117–120.

Kalina A, Csazar A, Fust G, et al. (2001). The association of serum lipoprotein(a) levels, apoliprotein (a) size and (TTTTA)n polymorphisms with coronary heart disease. *Chin Chim Acta* 309(1):45–51.

Kannel WB, Dawber TR, Friedman GD, et al. (1964). Risk Factors In Coronary Heart Disease: An Evaluation of Several Serum Lipids As Predictors of Coronary Heart Disease. *Ann Intern Med* 61:888–889

Kannel WB, Dawber TR, Kagan A, et al (1961). Factors of Risk In the Development of Coronary Heart Disease-Six Year Follow-up Experience: the Framingham Study. *Ann Intern Med* 55:33–50.

Kosunen O, Talasniemi S, Lehtovirta M, et al. (1995). Relation of coronary atherosclerosis and apolipoprotein E genotypes in Alzhemier patients. *Stroke* 26(5):743–748.

Kurz A, Egensperger R, Lautenschlager N, et al. (1995). The apolipoprotein E gene and Alzheimer disease phenotype. *Z Gerontol Geriatr* 28(3):195–199.

Li D, Hua Q, Pi L. (2004). Association of G894T polymorphism of endothelial nitric oxide synthase gene and essential hypertension. *Chinese Journal of Hypertension* 12(4): 326–330.

Li D, Hua Q, Pi L. (2004). Association of T786C polymorphism of endothelial nitric oxide synthase gene with essential hypertension and the effect of valsartan. *Chinese Journal of Hypertension* 12(4):331–334.

Li G, Yan Q, Wei L. (2006). Relationship between serum leptin concentration and lipids & body mass index in patients with chronic congestive heart failure. *Chinese Journal of Cardiology* 34(2):169–173.

Liu J, Zhao D, Wang W, et al. (2004). Comparison between the results from the Chinese Multi-provincial Cohort Study and those from the Framingham Heart Study. *Chinese Journal of Cardiology* 32:167–172.

Liu L. (2007). Cardiovascular diseases in China. *Biochemistry and Cell Biology*. 85: 157–163.

Liu Q, Zho D, Liu J, Sun JY. (2008). Obesity modulates the association among ApoE genotype, insulin and glucose. *Molecular Cardiology of China* 8(4):217–222.

Liu Y, Pan J, Wang S, Li X, Huang Y. (2002). β-fibrinogen gene -455A/G polymorphism and plasma fibrinogen level in Chinese stroke patient. *Chinese Medical Journal* 115(2):214–216.

Mata-Balaguer T, de la Herran R, Ruiz-Rejon C, Ruiz-Rejon M, Garrido-Ramos MA, Ruiz-Rejon F. (2004). Angiotensin-converting enzyme and p22(phox) polymorphisms and the risk of coronary heart disease in a low-risk Spanish population. *Int J Cardio* l95:145–151.

NCCD [National Center for Cardiovascular Disease, Ministry of Health of People's Republic of China]. (2006). *2006 Report On Cardiovascular Disease In China* Encyclopedia of China Publishing House, Beijing

Neel JV. (1999). Diabetes mellitus: a "thrifty" genotype rendered detrimental by "progress?" *Bull World Health Organ* 77:694–703.

Pan C. (2004). Metabolism: A group with high risk of CHD. *Chinese Journal of Cardiology* 32(7):643–647.

Pan XH, Fan HX, Hu JQ, et al. (2005). Relationship between angiotensin II type1 receptor gene polymorphism and risk factors of coronary heart disease. *Molecular Cardiology of China* 5(5): 717–720.

Pedro-Botet J, Senti M, Nogues X, et al. (1992). Lipoprotein and apolipoprotein profile in men with ischemic stroke. Role of lipoprotein(a), triglyceride-rich lipoproteins, and apolipoprotein e polymorphism. *Stroke* 23(11):1556–1562.

Qin Q, Zhao BR, Cui RZ, et al. (2002). The association of the abnormal homocysteine metabolism and coronary disease. *Molecular Cardiology of China* 2(4): 16–19.

Qiu CG, Fan XH, Fan HX, et al. (2003). Association between angiotensin-converting enzyme and angiotensin II Type1 receptor gene polymorphism and the coronary heart disease. *Molecular Cardiology of China* 3(4): 202–206.

Rao M, Jaber BL, et al. (2005). Gene polymorphism association studies in dialysis: cardiovascular disease. *Semin Dial* 18(3): 217–225.

Shang X, Zhao B, Dai X. (2005). The clinical significance of level and genotypes of plasma c-reactive protein among different kinds of inflammatory disease on the development of coronary heart disease. *Section of Internal Medicine, Foreign Medical Science* 32(6):369.

Shen D, Ha D. (1998). The relationship between angiotensin-converting enzyme gene polymorphism and brain infarction in Chinese hypertensives. *Chinese Journal Of Medical Genetics* 15(3):136–138.

Shen D, Ha D. (1998). The relationship between angiotensin-converting enzyme gene polymorphism and left ventricular hypertrophy in Chinese hypertensives. *Chinese Journal Of Medical genetics* 15(6):364–366.

Shioji K, Mannann T, Kokubo Y, et al. (2004). An association analysis between apoA I polymorphisms and the high-density lipoprotein (HDL) cholesterol level and myocardial infarction (MI) in Japanese. *J Hum Genet* 49(8):453–439.

Skol AD, Scott LJ, Abecasis GR, Boehnke M. (2006). Joint Analysis is More Efficient than Replication-based Analysis for Two-stage Genome-wide Association Studies. *Nat Genet* 38:209–213.

Song Y, Stamfer MJ. (2004). Meta-analysis: apolipoprotein E genotypes and risk for coronary heart disease. *Ann Intern Med* 141(2):137–147.

Song Y, Zou H, Zhang C. (2006). Correlation between the polymorphism of platelet glycoprotein GPIbαKozak gene and transient ischemic attack. *Chinese Journal of Medical Genetics* 23(5):561–564.

Talmud DJ, Martin S, Taskinen MR, et al. (2005). ApoA5 gene variants, lipoprotein particle distribution and progression of coronary heart disease: Results from the LOCAT study. *J Lipid Res* 45(4):750–756.

Tang NP, Wang LS, Yang L, et al. (2008). Genetic variant in glutathione peroxidase 1 gene is associated with an increased risk of coronary artery disease in a Chinese population. *Clin Chim Acta* 395:89–93.

Tang Y, Qin P, Huang Q, et al. (2008). Correlation between coagulation factor FXIIIVA Pro564Leu polymorphisms and cerebral hemorrhage. *Journal of Apoplexy and Nervous Disease* 125(5):537–539.

Tang YB, Sun P, Guo DP, et al. (2005). Association between apolipoprotein A5-1131T>C polymorphism and susceptibility of coronary artery disease in Chinese. *Chinese Journal of Medical Genetics* 22(3): 281–283.

Volleova MV, Vasruna VI, et al. (1997). Comparative analysis of Apo(a) gene-alleles: Distribution of pentanucleotide repeats in position -1373 and C/T transition in position +93 among patients with myocardial infarction and a cohort group in St Petersburg Russia. *Biochem Mol Med* 61(2):208–213.

Wang C, Sun G, Yan X, Ding Y. (2006). Association of polymorphisms G894T and T786C in the endothelial nitric oxide synthase with the risk of essential hypertension in Mongolia population. *Molecular Cardiology of China* 6(2):81–84.

Wang C, Zhou X, Zhou G, Han D, Zheng F. (2004). Interaction of ApoE and LDL-R gene polymorphisms and alcohol drinking and smoking on coronary heart disease. *National Medical Journal of China* 84(7): 554–558.

Wang H, Thomas DC, Pe'er I, Stram DO. (2006). Optimal Two-stage Genotyping Designs for Genome-wide Association Scans. *Genet Epidemiol* 30:356–368

Wang L, Fan C, Topol SE, Topol EJ, Wang Q. (2003). Mutation of MEF2A in an inherited disorder with features of coronary artery disease. *Science* 302:1578–1581.

Wang W, Zhao D, Liu J, et al. (2004). A Prospective study of relationship between blood pressure and 10-year cardiovascular risk in a Chinese cohort aged 35–64 years. *Chinese Journal of Internal Medicine.* 43: 730–734.

Wang W, Zhao D, Liu J, et al. (2006). Serum Total Cholesterol and 10-year Cardiovascular Risk in a Chinese Cohort Aged 35–64 year. *Chinese Journal of Cardiology* 34(2):169–173.

Wang W, Zhao D, Sun JY, et al. (2006). Risk factors comparison in Chinese patients developing acute coronary syndrome, ischemic or hemorrhagic stroke: A multi-provincial cohort study. *Chinese Journal of Cardiology* 34(12):1133–1137.

Wang W, Zhao D, Wu Z, et al. (2001). Study of triglyceride distribution aged 35–64, and their association with other cardiovascular disease risk factors in 11 provinces. *Chinese Journal of Epidemiology* 22(1):22–29.

Wilhelmsen L, Svardsudd K, Korsan-Bengtsen K, et al. (1984). Fibrinogen as a risk factor for stroke and myocardial infarction. *New England Journal of Medicine* 311:501–550.

Wu X, Wu Y, Zhou B. (1996). The incidence of hypertension and associated factors in 10 population groups of China. *Natl med J China* 76(1):24–29.

Wu Y, et al. (1996). Body mass index relationship with clustering of major cardiovascular risk factors. *Chinese Journal of Prevention and Control of Chronic Non-Communicable Diseases* 4(5):208–215.

Wu Z, Yao C, Zhao D, et al. (1999). A prospective cohort study on cardiovascular disease incidence in 11 provinces in China Part I. Association between risk factor level and the cardiovascular disease incidence. *Chinese Journal of Cardiology* 27(1): 5–8.

Wu Z, Yao C, Zhao D, et al. (2001a). Sino-MONICA Project: A collaborative study on trends and determinants in cardiovascular diseases in China, Part I: Morbidity and mortality Monitoring. *Circulation* 103:462–468.

Wu Z, Yao C, Zhao D, et al. (2001b). A prospective cohort study on cardiovascular disease incidence in 11 provinces in China Part II. Associations between individual risk factor aggregation and cardiovascular disease incidence. *Chinese Journal of Cardiology* 29(4):246–250.

Xie G, Myint PK, Zhao L, et al. (2010). Relationship between -592A/C polymorphism of interleukin-10 (IL-10) gene and risk of early carotid atherosclerosis. *Int J Cardiol* 143(1):102–104.

Xie Y, et al. (1989). Analysis of apo E phenotype of 107 CHD patients. *Journal of Clinical Cardiology* 5:35–37.

Yang T, Cheng J, Shi Y, et al. (2008). Ang II Type I receptor polymorphism in association with coronary artery disease. *Molecular Cardiology of China* 8(2): 91–94.

Yue H, Gu D, Wu X, et al. (2004). A 20-year prospective study on risk factors for myocardial infarction of 5137 men in Capital Steel and Iron Company. *Chinese Journal of Preventive Medicine* 38(1): 43–46.

Zang B, Li X, Guo J, Pan L, Wu K. (2005). Effects of Nco I polymorphisms at the LDL receptor gene on myocardial infarction in elderly patients. *Chinese Journal of Geriatrics* 24(7):517–519.

Zhang C, Li J, Li L, Luo B. (2001). The study of PAI-1 promoter region gene polymorphism in cerebrovascular disease. *Chinese Journal of Medical Genetics* 18(5):383–387.

Zhang C, Ye J, Hui R. (2004). Association of polymorphism of MTHFR gene and plasma homocysteine with stroke in Chinese elderly people. *Molecular Cardiology of China* 4(1):3–7.

Zhang YX. (2006). Polymorphisms of genes involving in lipid metabolism and coronary heart disease. *Molecular Cardiology of China* 6:350–353.

Zhang X, Xia J, Jin D, et al. (2001). The relationship between angiotensin converting enzyme gene polymorphism and risk factors for cerebral infarction. *Chinese Journal of Epidemiology* 22(6):435–438.

Zhao D, Li C, Wang W, et al. (2006). The 10-year cardiovascular risk in a Chinese cohort aged 35–64 years with high-normal blood pressure. *Chinese Journal of Geriatric Heart Brain Vessel Disease* 8(11):730–733.

Zhao L, Li Y, Wu Y, et al. (2003b). Relation of body mass index and waist circumference to the level of high density lipoprotein cholesterol (HDL-C) and low HDL-C prevalence. *Chinese Journal of Cardiology* 31(4):302–305.

Zhao L, Wu Y, Zhou B, Li Y, Yang J. (2003a). Mean level of blood pressure and rate of hypertension among people with different levels of body mass index and waist circumference. *Chinese Journal of Epidemiology* 24(6):471–475.

Zhao L, Zhou B, Wu Y, Li Y, Yang J. (2002). A prospective study on body mass index and mortality. *Chinese Journal of Epidemiology* 23 (1):24–27.

Zhao W, Zhang J, You Y, et al. (2005). Epidemiologic characteristics of dyslipidemia in people aged 18 years and over in China. *Chinese Journal of Preventive Medicine* 39(5):306–310.

Zhou B, Liu X, Wu Y, et al. (2002). Pulse pressure in prediction of incidence of cardiovascular disease in middle-aged Chinese cohort. *Chinese Journal of cardiology* 30(11):687–691

Zhou B, Liu X, Wu Y, et al. (2003). Predictive value of diabetes and impaired fasting glucose to incidence of cardiovascular disease in middle-aged Chinese. *Chinese Journal of Cardiology* 31(3):226–230.

Zhou B. (2005). Cohort study of features of risk factors of cardiovascular disease in China. *Chinese Journal of Epidemiology* 26(1): 58–61.

Zhou X, Hu A, Hui R, et al. (1999). The relationship of polymorphisms of MTHFR gene and plasma homocysteine levels with stroke. *Chinese Journal of Cardiology* 27(2):121–123.

Zhou Y, Hu D, Yang X, et al. (2003). Relationship among apolipoprotein A I gene polymorphism, lipid levels and coronary atherosclerosis disease. *Chinese Medical Journal (English)* 116(5):665–668.

71.

FAMILIAL AND GENETIC CANCER IN CHINA

Fen Liu and Wei Wang

INTRODUCTION

The development and progress of cancer is a highly complicated biological process. From the macroscopic aspect, this process involves a series of interactions and reactions between environment and host, and especially the genetic factors of the host; however, from the microcosmic view, the development of a tumor actually reflects the result of the influence of carcinogenic agents upon the heredity factors. Therefore, the main concern of genetic researchers has focused on the function of genetic factors in the development and progress of tumors. Although cancer epidemiological investigations show that more than 80% of tumors are caused by various environmental factors, the statistical data also indicates that those who suffer from cancer are minority among those who are in the same living conditions, and even in some high-risk regions the incidence rate is only between 0.1% and 0.2%. So the function of the internal factors of host, especially the genetic factors, in the development of cancer is important. A history of cancer in the family, familial carcinoma, and family aggregation of the common malignant tumors are still major contributing factors.

The mapping and incidence rate of the various cancers show that the Chinese people are different from other racial/ethnic groups around the world. For example, more than half of esophageal cancers in the world take place in the Chinese people, and the incidence rates of nasopharyngeal carcinoma and liver cancer among Chinese people are also higher than those of other races (Lin, 2003). The differences in the prevalence of these cancers in different populations are closely associated with the differences in people's lifestyles and the environmental exposures of different ethnic groups. At the same time, the biological differences among races are also important determining factors.

The Chinese population accounts for about 22% of the population of the world, encompassing 56 nationalities (*minzu*) and 205 languages that have been identified (Xu and Du, 1999). There are extremely abundant and precious resources of population in Chinese nation. Thus, the epidemiological and genetic research pertaining to high-incidence tumors, and the characterization of the population-based tumor research in China, are important subjects for life sciences research in China and in the world.

It has not been our purpose in this chapter to provide the reader with an exhaustive review of all the cancers in the populations of Chinese. Rather, we have tried to report on the more specific, thoroughly studied cancers and to illustrate them through selected examples.

RETINOBLASTOMA

Retinoblastoma (RB; MIM# 180200) is the most common and severe form of malignant tumor of the eyes that affects infants and young children. It can be familial or sporadic, with an incidence of 1 in 15,000–20,000 live births in most populations. There is no evidence that frequencies are lower in Chinese than in Caucasian populations (Chiu et al., 1997; Zhang et al., 1997; Harbour, 1998).

It was the first tumor for which Knudson's hypothesis of a two-mutation event was initially proposed, and largely confirmed (Knudson, 1971). In the familial forms the first mutation is germinal. This mutation, present in all the somatic cells, predisposes the individual to retinoblastoma when a second mutation, somatically acquired, modifies the second allele of the same gene. Thus, susceptibility is inherited as a dominant trait. Inactivation of *RB1* (the retinoblastoma susceptibility gene mapped to chromosome band 13q14) occurs in both familial and sporadic retinoblastoma (Bookstein and Lee, 1991). The trait is transmitted in an autosomal-dominant fashion with 80% to 90% penetrance (Vogel, 1979; Naumova and Sapienza, 1994). No "hotspot" mutation has been identified. About 80% of known *RB1* sequence mutations are nonsense mutations, which abrogate the two globular (A and B) domains required for binding to Rb1-associated proteins. Missense mutations and small inframe deletions represent approximately 10% of all mutations. Large chromosomal alterations are found in about 7% of patients (Lohmann, 1999).

In China, among the early studies on Retinoblastoma, Shen et al. (Li, 1998) reported 108 RB patients (affected in one or both eyes) diagnosed by clinical and pathology examination in Shanghai between 1972 and 1981. Retrospective family investigation and genetic epidemiological analysis showed that the majority of patients were affected in one eye (76%), with both eyes less (24%), and with a sex ratio of 1:1.25 (M:F). The youngest patient had a family history of RB in both eyes (average age was 5.2 months). The family data also suggested that being affected in both eyes was not related to the birth order however, the single-eye sufferers were influenced by the birth order, which means the single-eye sufferers were more easily influenced by the environmental factors compared with patients afflicted in both eyes.

There are several RB research teams in China, which mainly concern themselves with clinical and basic ophthalmology research rather than epidemiological investigation. These experimental research projects have included identifying the abnormality of the Rb gene at the chromosome and the molecular levels (Huang et al., 1997; Chen et al., 1995; Li et al., 1998; Shen et al., 2003; Du et al., 2002), establishing several RB tumor cells (Jin et al., 1991; Feng et al., 1994), producing RB monoclone antibodies, and developing immunity guides and gene therapy for RB (Li et al., 1997a; Zhong et al., 2002; Ying et al., 2003). It is hoped that this research will make an important contribution, together with more clinical data, to achieving a breakthrough in cancer treatments in the future.

NASOPHARYNGEAL CARCINOMA

Nasopharyngeal carcinoma (NPC) is one of the most common malignant tumors in China, particularly in the southern provinces. Guangdong province is the area with the highest rate of NPC in the world. Based on the cancer mortality information of years 1973–1975, from the first Chinese national survey, the age-standardized mortality rates of NPC are 2.94/100,000 (men) and 1.67/100,000 (women), whereas the rates are up to 10.82/100,000 (men) and 5.44/100,000 (women) in Guangdong province. There is a slight decrease in NPC mortality from the 1970s to the 1990s, with a mortality of NPC at 2.34/100,000 (men), 1.10/100,000 (women) in the sampling regions, with Guangdong province at 10.19/100,000 (men) and 3.78/100,000 (women) (Dai et al., 1999). It is estimated that the mortality will decrease to 1.35/100,000 in China in the twenty-first century.

The geographic characteristics of NPC death are obvious; the incidence of NPC is the highest in southern China and southeastern Asia. The World Health Organization (WHO) estimates that about 80% of NPC occurs in the south part of China. In China the highest rates are seen in Zhaoqing county, Foshan city, and Guangzhou areas in Guangdong province and in the Wuzhou area in the eastern Guangxi province (Hong et al., 2003). In Guangdong province, the rate of NPC is especially high in the Cantonese-speaking population, moderate in those who speak the *Minnan* dialect, and lowest in the *Hakka* dialect population (Parkin et al., 1992).

The cancer high-risk family is the first research target in genetic epidemiology. The high incidence of NPC in the Chinese population is clustered in families. It is a common observation that 2–4 NPC patients are from a family constellation (Jia, et al, 2004). For example, a family in Nanhai city of Guangdong province is a representative high-risk family on whom complete data have been gathered. The husband and wife both are farmers; they did not suffer from the nasopharyngeal carcinoma. There were 15 NPC patients among their offspring. Among them, 3 patients are from the next generation (11 people), 10 from the third generation (38 people), and 2 from the fourth generation (58 people). In the fourth generation, the ratio of NPC patients to unaffected family members decreased, which may be related to the younger age of this fourth generation, who have not yet reached the high-risk age range. Moreover, in addition to 15 NPC patients there are two people suffering from hepatocellular carcinoma and one person suffering from breast cancer. For details, see the family tree in Figure 71-1. It suggests that predisposing genes play an important role in NPC etiology.

Epstein-Barr virus (EBV) infection is closely associated with NPC. The downstream target genes of the EBV-encoded latent membrane protein 1 (LMP1) was cloned and identified by suppression subtractive hybridization in an LMP1 expressing NPC cell line, CNE-2, at the University of Hong Kong in 2001 (Kwok Fung Lo et al., 2001). In addition to this, there are predisposing genetic factors that are also closely related to the incidence of NPC. The linkage analysis of NPC families collected from mainland China, Singapore, Hong Kong, and Malaysia has proven the existence of these predisposing genes, special haplotypes of human leukocyte antigen (HLA) domain (Hong et al., 2003; Lu et al., 1990). In addition, the frequencies of sister chromatid exchanges (SCE) in peripheral blood lymphocytes, glutathione S-transferase (GST), and glucose-6-phosphate dehydrogenase (G6PD) are also common markers identified in the research of susceptibility to NPC. Furthermore, there are reports that chromosomal abnormalities, activation of oncogene, and inactivation of tumor-suppressing genes are associated with increasing risk of NPC (Shao et al., 2002).

ESOPHAGEAL CARCINOMA

The main epidemiological regions for esophageal cancer is in the developing countries, and geographic variation in incidence is very striking. The highest risk area is

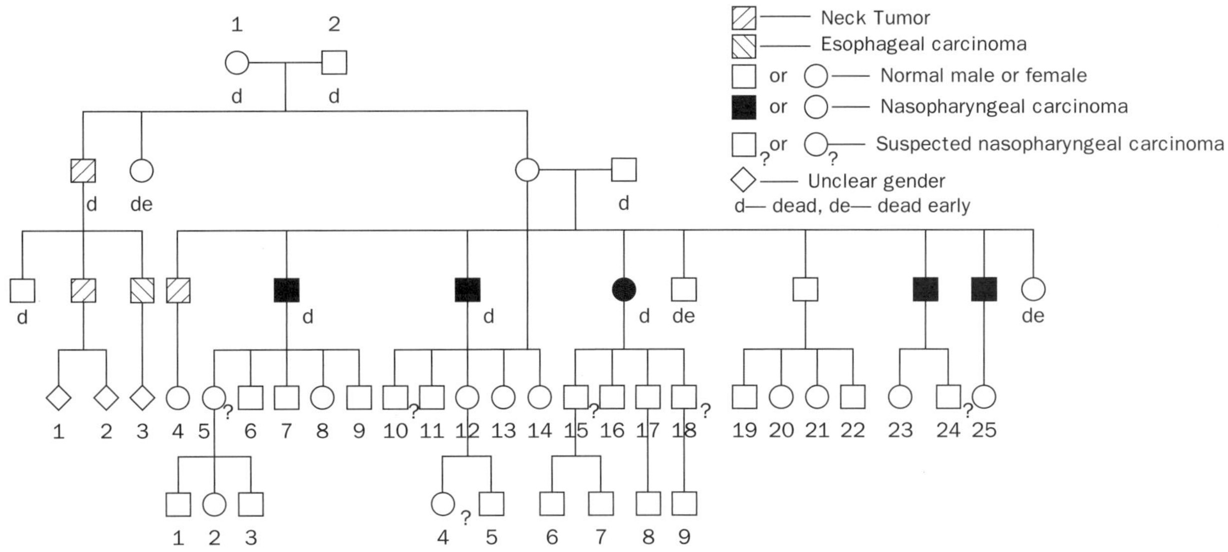

Figure 71-1 The family tree of a nasopharyngeal carcinoma high-risk family (From Lu et al., 1998; *Cancer Genetics*. Science Press)

in the Asian "esophageal cancer belt" (an area stretching from northern Iran through the central Asian republics to northern-central China), with incidence rates as high as 200/100,000 (Parkin, 2001). Mortality of esophageal cancer is higher in China, and the new cases are about 54% of those in the world. The high-risk regions are: (1) Taihang mountain area along the borderlines of Shanxi, Henan, and Hebei provinces in northern China; (2) the northern region of Sichuan province; (3) northern region of Jiangsu province; (4) Dabie mountain area along the borderline between Hubei and Anhui provinces; (5) coastland areas on the border between Fujian and Guangdong provinces; and (6) among Kazak residents in the Xinjiang Uygur Autonomous Region. There is usually an obvious high-risk center in each of these high-incidence regions, and both morbidity and mortality gradually decrease toward the outlying areas. Incidence and mortality of esophageal cancer gradually increases along with age; and the group from 50–70 years old accounts for three-quarters of total deaths from esophageal cancer. Incidence and mortality of esophageal cancer in males are usually higher than in females (Dong, 2004).

The research in esophageal cancer is mainly focused on esophageal squamous cell carcinoma (ESCC), beginning from the initial stages in China in the 1970s. Over the past three decades, many studies have been conducted on the etiologic and epidemiologic factors associated with esophageal cancer in China (Carter et al., 1992). Environmental factors that have been implicated as causally related include consumption of moldy foodstuffs, pickled vegetables, and steaming hot gruel; consumption of high levels of nitrosamines or various trace elements; nutritional deficiencies; the occurrence of fungal infections; and poor personal and oral hygiene.

Epidemiologic surveys have also disclosed a strong familial aggregation in the occurrence of esophageal cancer cases (Shi et al., 2000; Carter et al., 1992). The segregation analysis of 221 high-risk nuclear families from the Yaocun Commune, Linxian, China, reveals the effects of an alleged Mendelian recessive gene that is present (at a frequency of 19%) in the population, causing 4% of this population to be predisposed to develop esophageal cancer (Carter et al., 1992). Sister chromatid exchanges (SCE), chromosomal aberration at fragile sites, and endoreduplication in the cultured lymphocytes may play an important role in esophageal carcinogenesis in the high-risk cancer families (Wu, 1986; 1990; 1991).

In addition, a large number of case-control studies were conducted to identify genetic polymorphisms and markers for ESCC risk, such as methylene tetrahydrofolate reductase (MTHFR; Wang et al., 2005; Li et al., 2008; Qin et al., 2008), cytochrome P450 2E1 (CYP2E1; Lin et al., 1998b; Tan et al., 2000; Guo et al., 2008; Qin et al., 2008), glutathione S-transferase (GST; Lin et al., 1998a; 1998b; Tan et al., 2000), cyclooxygenase-2 (COX-2; Zhang et al., 2005), thymidylate synthase (TS; Wang et al., 2005), xeroderma pigmentosum group C (XPC; Zhou et al., 2006), ALDH2 (Guo et al., 2008), ADH1B (Guo et al., 2008), DNA repair gene Mlh3 (Liu et al., 2006a) and hoGG1 (Xing et al., 2001), cell cycle-related gene TP53 (Hu et al., 2003) and MDM2 (Hong et al., 2005), esophageal cancer related gene 2 (ECRG2; Yue et al., 2004), tumor suppressor gene PTEN (Ge et al., 2008), and P53 (Lu et al., 2004; Li et al., 2002). Overall, the odds ratios (ORs) are 2.0–10.0 between these genetic polymorphisms markers with ESCC risk. Furthermore, mitochondrial DNA (mtDNA) haplogroups (D4a and D5a) might be candidate genetic markers for screening populations susceptible to esophageal cancer (Li et al., 2007).

There are also some reports concerning the serum markers of ESCC; for example, serum pepsinogen I (PGI) and pepsinogen I/pepsinogen II ratio (PGI/II ratio; see Ren et al., 2009), soluble interleukin-2 receptor alpha (sIL-2Ralpha; Wang et al., 2000; Liu et al., 2006b), p53 antibody (Wang, et al., 2004a; Liu et al., 2006b), matrix metalloprotease-9 (MMP-9; Liu et al., 2006b), tissue inhibitor of matrix metalloprotease21 (TIMP21; Liu et al., 2006b), copper/zinc superoxide dismutase (Cu/ Zn SOD; Liu et al., 2006b), MG7-Ag (Ren et al., 2000), hepatocyte growth factor/scatter factor (HGF/SF; Ren et al., 2005), CDC25B (Liu et al., 2008) and CYFRA21-1 (Wang et al. 2001). These may be useful biomarkers in ESCC screening and diagnosis.

The development, progress and metastasis of esophageal cancer may involve several gene changes, and these changes can be on the level of gene DNA, mRNA, or at the protein level. Although some important progress has already been made in research to identify genes related to esophageal cancer, the mechanisms of development and progress of this disease are still unclear. For early warning and diagnosis, and effective therapy of esophageal cancer, unremitting hard work is still needed to identify which genes are responsible for the beginning of cancerization, especially in the cancer precursor stage, and the key genes causing cancer development and progress.

HEPATOCELLULAR CARCINOMA

Primary hepatocellular carcinoma (HCC), i.e., liver cancer, ranks as the third most common tumor-caused death, and 81% of cases occur in the less developed countries (54% in China; Parkin, 2001). Since the 1990s, liver cancer ranks second among tumor-caused deaths in China. The mortality of liver cancer in males is about threefold higher than that in female patients, with a crude death rate of 29.1/100,000 and 11.21/100,000, respectively (Li et al., 1997b).

In China, the incidence of liver cancer varies by region, with higher incidence in the southeast and coastlands and lower rates in the northwest and inland areas. The highest incidence is reported in Qidong county of Jiangsu province, Fusui county of Guangxi province, and Shanghai city, with a mortality of 25~50/100,000/year. Moreover, Taiwan and Hong Kong also have high incidence. The city areas have lower mortality (28.26/100,000 in men and 10.11/100,000 in women) than rural regions (29.28/100,000 in men and 11.59/100,000 in women; Rubinfeld et al., 1993). Liver cancer can develop at any age, but occurs more often in middle-aged people (90% in those 40~60 years old; You, 2005).

Mortality of liver cancer is overall consistent in China, and in the regions that have better prevention and treatment measures, such as Shanghai, a decreasing trend is already evident. The standardized mortality in men decreased from 34.1/100,000 to 26.5/100,000 (22.2%), and in women decreased from 12.0/100,000 to 9.4/100,000 (21.4%) during 1972~1994 (Jin et al., 1999). But the survival rate of liver cancer has changed little per year worldwide (Dong, 2004). The mortality rates of liver cancer in migrant groups are inconsistent. Among male Chinese living in San Francisco, mortality is 22.1/100,000—lower than that of local people at 29.0/100,000, but still sevenfold higher than the local white population. Mortality of Chinese in Singapore is the same as the people in mainland China (Dong, 2004).

The true pathogenesis of liver cancer is still not clear, and many related risk factors have been discovered, such as viral hepatitis and aflatoxin-polluted food and water. More than 60% of liver cancer is associated with B-type hepatovirus (HBV) infection. Currently, studies have already confirmed that hepatitis B vaccine inoculation can effectively prevent HBV infection, thus raising the possibility of liver cancer prevention.

From a genetic viewpoint, liver cancer is a polygenic hereditary disease. The occurrence of liver cancer is the result of interactions between environmental carcinogenic factors and heredity (Long et al., 2006; Chen et al., 2006a). Familial aggregations of liver cancer have been seen repeatedly (Wu, 2004). Studies of high-risk areas in Qidong in Jangsu province, and Fusui in Guangxi Zhuang autonomous region, revealed that liver cancer cases have obvious familial aggregations. And, interestingly, hepatitis type B, which was thought to be one of the most important and related causes of liver cancer, also showed a certain extent of familial aggregation (Wu, 2004). The study found that patients with Au homozygote (Australia antigen), after infection with HBV, easily display autosomal recessive inheritance with the virus. The genetic predisposition to HBV is identified as an additive factor of quantitative trait of polygenic inheritance of primary hepatic carcinoma. There are also reports that genetic predisposition is due to the interaction of HBV infection and a recessive gene (Shen et al., 1991) in Chinese families.

In China, the extensive research about human liver carcinogenesis has been undertaken primarily in Shanghai Cancer Institute. In 1984, Gu's laboratory identified N-*ras* as a transforming gene in human HCC; then they found that at least four oncogenes, or related genes, were involved in human HCC, such as N-*ras*, *c-myc*, *est*-2 and IGF-II(Gu, 1988). They also reported that a p53 gene mutation at the third base of codon 249 was detected in HCC samples from Qi-Dong county, Chin. A correlation between aflatoxin Bl (AFB1) exposure and p53 aberrations was also observed (Li et al., 1993). In 2001, there was a report on a comprehensive characterization of gene expression profiles of hepatitis B virus-positive HCC, through the generation of a large set of 5'-read expressed sequence tag (EST) clusters (11,065 in total) from HCC and noncancerous liver samples, which then were applied to a cDNA microarray system containing 12,393 genes/

ESTs and compared with a public database. The commercial cDNA microarray, which contains 1176 known genes related to oncogenesis, was used also for profiling gene expression. Integrated data from the above approaches identified 2253 genes/ ESTs as candidates with differential expression. A number of genes related to oncogenesis and hepatic function/ differentiation were selected for further semi-quantitative reverse transcriptase–PCR analysis in 29 paired HCC/noncancerous liver samples. Many genes involved in cell cycle regulation, such as cyclins, cyclin-dependent kinases, and cell cycle negative regulators, were deregulated in most patients with HCC. Aberrant expression of the Wnt-β-catenin pathway and enzymes for DNA replication also could contribute to the pathogenesis of HCC. The alteration of transcription levels was noted in a large number of genes implicated in metabolism, whereas a profile change of others might represent a status of dedifferentiation of the malignant hepatocytes—both considered as potential markers of diagnostic value. Notably, the altered transcriptome profiles in HCC could be correlated to a number of chromosome regions with amplification or loss of heterozygosity, providing one of the underlying causes of the transcription anomaly of HCC [Xu et al., 2001]. Further research on the mechanisms of this cancer, of utmost clinical and biological significance, is still needed.

LUNG CARCINOMA

Lung cancer is the most common cancer in the world, both in terms of incidence (1.2 million new cases) and mortality (1.1 million deaths; Parkin, 2001). Investigations of the cause of death in a 1/10 sampling population showed that 16.2% of total cancer death was due to lung cancer, which ranked second in China in 1990~1992. The world standard population mortality is 29.7/100,000 in men and 11.7/100,000 in women in China (Chen et al., 2006b).

There are great differences in incidence and mortality of lung cancer in different areas: 52% of new cases occur in more developed countries; in less developed countries, the highest rates are seen in the Middle East and China (Parkin, 2001). Incidence and mortality of lung cancer are also characterized by regional differences within China. Surveys in parts of the cities and counties in China over the period from 1998 to 2002 showed that the regions with incidence and mortality over 50.0/100,000 were Shanghai (incidence 58.8/100,000 and mortality 53.6/100,000), Dalian (incidence 55.0/100,000 and mortality 51.2/100,000), and Anshan (incidence 50.5/100,000 and mortality 51.7/100,000), while the lower-risk regions were Yangcheng county and Linzhou city, with both incidence and mortality of 10.0/100,000 (Chen et al., 2006b). These statistics show that lung cancer incidence for men in industrialized cities is clearly higher than in the remote areas.

Incidence and mortality of lung cancer are different among different populations. The investigation of death causes in China showed that mortality due to lung cancer is different among different minorities; Chinese Koreans have a higher rate of lung cancer than the national average (7.09/100,000); the ethnic Mongolians and the Kazak people have slightly lower rates; and the other minorities have the lowest mortality due to lung cancer (Liao, 2003). Incidence and mortality of lung cancer increases with age in almost all nations and regions, and both are higher in males than in females (Parkin, 2001).

In 1988, the adjusted mortality rate of lung cancer in men was 32.77/100,000 in China, and rose to 41.20/100,000 in 1998, an increase of 25.72% over ten years. The adjusted mortality rate in women was 14.72/100,000 in 1988, and rose to 17.77/100,000 in 1998, an increase of 20.72% (Liao, 2003). Compared with the Western developed countries, China has showed a rising trend in incidence and mortality since the 1970s, while in Western developed countries there has been little change in incidence during that period.

Etiology studies of lung cancer show that risk factors are smoking, occupation, environmental pollution, and hereditary susceptibility. It is estimated that 80%~90% of lung cancer may attribute to smoking. However, among smokers there are only 10%~15% attacked by lung cancer, and 10% ~15% of lung cancer cases occur in nonsmokers. Obviously, these individual differences indicate some hereditary susceptibility to lung cancer. Recent research showed the involvement of metabolic enzyme polymorphisms, sensitivity of mutagenic agents, DNA damage, tumor suppressor gene mutation, and chromosome instability in the carcinogenesis (Dong, 2004; Tan and Zhao, 2002; Hu et a., 2006).

COLORECTAL CARCINOMA

Worldwide, colorectal cancer ranked third in incidence (945,000 new cases) and mortality (492,000 deaths) in 2000, with similar chances of occurrence in men and women (1.1 to 1; Parkin, 2001). Colorectal cancer is one of the common cancers in Chinese, ranking fifth in malignant tumor–caused deaths both in men and women. Incidence of colorectal cancer is comparatively low in China, but has a rising trend. Based on a survey representing one-tenth of the population of the whole country in 1990~1992, mortality of colorectal cancer is 5.30/100,000 in China, which represented an increase of 32.2% when compared with the first nationwide survey in 1970s (Li et al., 1997c).

The geographic distribution of incidence and mortality of large-bowel cancer in China is characterized by a higher incidence in the eastern coastland than northwest inland, and the highest incidence was observed in the middle-downstream region of Yangtze River. The distribution of large-bowel cancer is associated with regional economic development, lifestyles, and diet. The national survey in the 1970s showed that incidence (25.6/100,000) and mortality (20.6/100,000) of large-bowel cancer are highest in Jiashan county of Zhejiang province (Dong, 2004).

The peak age incidence of large-bowel cancer is between 40 and 65 years, descending gradually after 65–70 years. The incidence is higher in men than in women generally. The age adjusted mortality rate of large-bowel cancer is highest among Mongolians (4.03/100,000), while lower in other ethnic groups and lowest in Yi people (2.04/100,000; Dong, 2004). Incidence and mortality of large-bowel cancer in Chinese migrants who move to the United States is much higher than in mainland Chinese residents (Muir, 1987). The similar finding is also seen in Singapore Chinese, due to the changes of lifestyle and diet (Lee et al., 1998).

Incidence and mortality of colorectal cancer rise by 2% annually, with obvious regional differences including high incidence in medium-size and large cities (You, 2005). Beijing and Shanghai rank third in incidence of malignant tumors, and Tianjin city ranks fourth. For example, in Shanghai, incidence of colorectal cancer increased 87.0% from 1973 to 1993, with an average increase of 4.2% annually. Along with the change of people's lifestyles and dietary habits, the incidence of colorectal cancer is likely to further increase.

Epidemiological research of large-bowel cancer indicates that environmental factors and lifestyles, such as high fat, high protein, high salt, and low cellulose food, are related to the development of large-bowel cancer (Dong, 2004). Moreover, it is estimated that genetic factors may contribute about 20% to the etiology of large-bowel cancer (Eddy, 1987). Family adenomatous polyposis (FAP) and hereditary nonpolyposis colorectal cancer (HNPCC) are examples of such cases. Prevalence of FAP is around 1.5/100,000 in China, with 1% of patients developing large-bowel cancer (Liu et al., 1997). The onset of FAP is young; the occurrence of adenoma usually takes place around the age of 20 years, with cancer progression at around 40~50 years of age. FAP is associated with the mutations identified in APC [Rubinfeld et al., 1993]. HNPPC is an autosomal dominant disease, accounting for 5%~15% of all large-bowel cancers. The mutations of mismatch-repaired genes (hMLH1, hMSH2, hPMS1, hPMH2; Fishel et al., 1994; Yuan et al., 1999) are responsible for HNPPC.

STOMACH CARCINOMA

Stomach cancer is the fourth most frequent cancer, with 876,000 new cases and 647,000 deaths worldwide in 2000. Almost two-thirds of these cases occurred in less developed countries (Parkin, 2001). Stomach cancer ranks as the second of cause of death among the total of malignant tumor cases in China. There were around 220,000 Chinese who died of stomach cancer in 1985, and an estimated 360,000 in 2000, which together accounted for 40% of the number of deaths from stomach cancer in the world during the same period (Li et al., 1997).

Age-standardized incidence of stomach cancer is high in China—9.8/100,000 in 1973~1992, and 21.8 /100,000 in 1990~1992 (Li et al., 1997). The high-incidence areas for stomach cancer in China are Liaodong Peninsula, Shandong Peninsula, eastern China coastland (Shanghai city, Jiangsu, Zhejiang and Fujian provinces), and inland regions (Ningxia Hui Autonomous Region, Gansu and Shanxi provinces). The rates are low in provinces of southern China, such as Hunan and Guangdong provinces, Guangxi Zhuang autonomous region, and Sichuan and Yunnan provinces (You, 2006). The highest mortalities in men (120.5/100,000) and women (15.1/100,000) were observed in Changle county of Fujian province and Yangzhong county of Jiangsu province, respectively [You, 2006].

Mortality of stomach cancer increases progressively with aging in a log-linear trend, displays as a cumulative curve. Mortality of stomach cancer is usually low in young people under 35 years old, increases quickly after 40 years of age, and highly concentrates among people over 55 years old, accounting for 70% of the total deaths with a sex ratio of 1.5:2.5 (M: F). Mortalities of stomach cancer are different among different ethnic groups; it is high in the Kazak, Hui, Korean, and Mongolian people, whereas it is low among minorities such as the Miao and Yi in Yunnan and Guizhou provinces. The differences in mortality detected among the ethnic groups could be explained by the differences in their genetic backgrounds, cultural practices, diets, and lifestyles (You, 2006).

The mortality of stomach cancer is clearly decreasing in most countries. The national data of dynamic time has shown that mortality of stomach cancer experienced a small increase or stable rates since 1970 in China, with an increasing tendency in the villages, where it increased from 19.37/100,000 in 1973~1975 to 24.36/100,000 in 1990~1992. However, mortality of stomach cancer had a decreasing tendency in the cities, decreasing from 20.3/100,000 in 1973~1975 to 15.34 /100,000 in 1990~1992. The declining rates of mortality in stomach cancer in some nations has mainly resulted from a reduction of cancers localized to the antrum and the angulus of stomach (Dong, 2004).

Stomach cancer is mainly divided into intestinal and suffused types by histological type (according to Lauren's grouping). The onset age of the intestinal type is late, and the cancer occurs more often in the antrum of stomach, which is caused mainly by environmental carcinogenic agents. The onset age of the suffused type is young, with hereditary susceptibility. The studies showed that 25%

of hereditary diffuse gastric cancers (HDGC) are caused by the mutation of embryonic tumor suppressor gene E-cadherin (CDH1; Guilford et al., 1998). Recently, a study in a high-risk population of China revealed that genetic polymorphisms (–347*del/del* genotype) in the *E-cadherin* promoter may increase susceptibility to sporadic gastric carcinoma (SGC) among males (Zhang, et al., 2008). Overall, intestinal gastric carcinoma is more common in clinical practice, and has a relatively defined developmental process.

A series of studies of gastric cancer and its precursors in a high-risk population of Linqu County in China revealed that Helicobacter pylori *(H pylori)* infection, high intake of sour pancakes, salted foods, cigarette smoking, and low consumption of fresh fruits and vegetables, along with family stomach cancer history, are all associated closely with stomach cancer development (You, et al., 1988). In 1994, the International Agency for Research on Cancer (IARC) classified infection with *H pylori* as carcinogenic to human beings, with a relative risk over 1.8–3.6 (You, 2006).

In addition to environmental factors, genetic variations may play a role in process of gastric cancer carcinogenesis and is another avenue of research to investigate. The genetic polymorphisms of *cytochrome P450 (CYP), glutathione-S-transferase (GST), Lewis and interleukins (IL)* and *Cyclooxygenase (COX)-2* are all associated with the risk of gastric cancer and play a synergistic role with environmental exposures (You et al., 2005; Jiang et al., 2003; Lu et al., 2005; Liu et al., 2006c). The development of stomach cancer is the long-term outcome of an interaction between individual hereditary susceptibility and the surrounding environment.

BREAST CANCER

Breast cancer is the second most frequent cancer in the world (1.05 million cases in 2000), and is by far the most common malignant disease in women (22% of all new cancer cases; Parkin, 2001). Incidence of breast cancer is low in the Chinese, but is increasing rapidly and continuously, and has become the malignant tumor with the highest prevalence in Chinese females (You, 2005). There were 470,000 new breast cancer cases and 130,000 deaths from breast cancer in 2005 (Yang et al., 2006).

There are also big differences among different regions in China; i.e., the prevalence in developed regions is higher than in less developed regions, and cities' rates are higher than those in villages (Dong, 2004). Data from Tianjin and Shanghai showed that from 1982 to 1992, incidence of breast cancer increased 38.9% in urban areas and 39.7% in the villages. Since 1978, breast cancer has become the malignant tumor with the highest prevalence in females residing in Beijing, and still keeps rising with a 2.4% increase each year, to a level of 45/100,000 in 2005 (Dong, 2004). It is estimated that along with rapid developments in economic and living conditions, incidence of breast cancer in Chinese women will catch up with the Western developed countries quickly (Dong, 2004). The age-group peak of breast cancer is between 40 and 49 years old in Chinese women (Hu, 2001). Migrant epidemiology studies show that once women from low-incidence countries or regions move to high-incidence countries or regions, the incidence of breast cancer becomes higher than that in their birthplaces, such as in Asian females who migrated to the United States, whose incidence of breast cancer increased, and incidence in their children is similar to that of the local American females, indicating that environmental factors play a role in the carcinogenesis of breast cancer (Wu, 2004).

Many high risk factors, such as longtime exposure to estrogenic hormones, i.e., prolonged menstruating years, late procreation, and no breastfeeding are associated with breast cancer. Family history is also an important risk factor. Individuals whose first-degree relatives (mother, sister, or daughter) are breast cancer patients will have two- to threefold higher risk of developing breast cancer compared to the normal population. It has been proved recently that development of familial breast cancer (especially early-onset breast cancer) is associated with the BRCA1 gene on chromosome 17, the BRCA2 gene on chromosome 13, and/or a gene deletion on chromosome 8 (Dong, 2004; Song et al., 2006). Also, gene polymorphism (vascular endothelial growth factor, VEGF) might be a susceptible factor for breast cancer among Chinese women (Kataoka et al., 2006). These candidate genes provide good markers for genetic screening in the early diagnosis and prevention of breast cancer.

CERVICAL CANCER

This is the second most common cancer in women worldwide (471, 000 annual cases, 233, 000 deaths; Parking, 2001). There are 10,000 new cases of cervical cancer annually in China, accounting for one-fifth of total new cases in the world (Qiao et al., 2004). Although in the past 20 years the mortality rate has obviously decreased in China, there are still high morbidity and mortality rates in the western regions, and have been for decades. For example, in Wudu county in Gansu province and Yangcheng county in Shanxi province, the mortality rate is 36/100,000 (Li et al., 1997), which is 10 times higher than the rest of the country.

The onset age of cervical cancer reported varies in different countries. In China, the onset age is usually later than 35, and highest incidence rate is between ages 45–49. Investigations among minority ethnic groups in China showed the highest mortality rates in the Uygur minority, moderate rates in Mongoloid and Hui minorities, and lower rates in the Zang, Miao, and Yi minority populations. Generally, the morbidity rate is high in women with low income and low educational levels. People living in the villages have higher rates than those living in cities in China. In addition, women with multiple sex partners and who have migrated into the city are also a high-risk population (Qiao et al., 2004).

Incidence and mortality in cervical cancer have decreased quite substantially, particularly in more developed countries, where there are long-standing screening programs (Parkin, 2001). Decreases are also seen in some less developed countries, particularly in China where the estimated age-standardized incidence rate in 2000 was 5.2, compared with an estimated 17.8 in 1985 (Parkin et al., 1993). For example, a spinning and weaving department in Shanghai, Jing'an County in Jiangsu province, and other mandatory screening regions all have seen noticeable effects. In recent years, the number of younger patients has increased in some parts of China, as they have in other countries (Zhang et al., 1999).

Epidemiology and biological research has suggested that the risk factors for cervical cancer mainly include the following: first is behavioral factors, such as early sexual behavior, more than one sex partner, parity and contraceptives, low socioeconomic level, and malnutrition; second is biological factors, including bacterial and viral infections, etc. Human papillomavirus (HPV) is now understood to be the most important cause of cervical cancer (IARC, 1995). Now, there are only a few studies showing that cervical cancer may have family aggregation.

PROSTATE CARCINOMA

With 543,000 new cases, prostate cancer is the sixth most common cancer in the world, and third in importance in men (10.2% of new cancer cases—16% in more developed countries, and 4.5% in less developed countries; Parkin, 2001). Incidence rates of prostate cancer are comparatively lower in China, with 2.1/100,000 in 1992 (Parkin et al., 1992), but are increasing in recent years and it has become the third most common malignant tumor in Chinese men (Sun, 2004). There are great differences in incidence of prostate cancer in different countries and different racial/ethnic groups. Overall, the incidence in Western countries is clearly higher than in developing countries. Epidemiology investigations find that the Chinese who migrated to the United States are more likely to have prostate cancer than those who live in their homeland (Reiter and deKernion, 2002).

With the numbers of older people increasing, China has been at the leading edge of aging societies. The number of people above age 60 was 13 million in 2000, representing one-fifth of the older population in the world, with about a 3% annual increase (Wang et al., 2004b). Men between ages 50–70 are at the highest risk of developing prostate cancer. Because of the increasing of numbers of elderly, and the change of diet, the number of prostate cancer patients has a tendency to increase year after year in China.

The risk factors of prostate cancer are related to age, genetics, diet, androgen, etc. This is a malignant disease of the elderly—78% cases are in men over age 65 (Parkin, 2001). Men are at risk for early onset of the disease if there is a family history (Goldgar et al., 1994). Different races have different incidences of prostate cancer. Mortality rates and incidence rates are low in Asian populations, and high in Europe and America (Parkin, 2001). This suggests that genetic determinants are important in the pathogenesis of prostate cancer. Polymorphisms in genes controlling androgen metabolism seem to provide at least part of the explanation (Beemsterboer et al., 1999). Among migrants from mainland China to Hong Kong and the United States, the incidence rates increase 5 and 16 times, respectively (Muir, 1996). This change is due partly to the environment, possibly including diet and Westernization of lifestyles, including increasing obesity and increased consumption of meat and fat (Parkin, 2001). There are also other latent risk factors, such as vasoligation, drinking, cadmium exposure, excess intake of vitamin D and vitamin A, and other dietary factors (Morgan et al., 1996; Graham et al., 1983). The screening of prostate-specific antigen (PSA) is a technology to detect prostate cancer patients earlier.

CONCLUSION

In addition to the relatively common cancers mentioned above, there are numerous rare conditions in the literature, and an exhaustive discussion of all of them is beyond the scope of this chapter. Even though this presentation of various cancers in the populations of Chinese is incomplete and fragmentary, it provides an overview of human genetics in terms of its anthropological and medical implications.

REFERENCES

Beemsterboer PM, Kranse R, de Koning HJ, et al. (1999). Changing role of 3 screening modalities in the European randomized study of screening for prostate cancer (Rotterdam). *Int J Cancer* 84(4):437–441.

Bookstein R, Lee WH. (1991). Molecular genetics of the retinoblastoma suppressor gene. *Crit Rev Oncog* 2:211–227.

Carter CL, Hu N, Wu M, Lin PZ, Murigande C, Bonney GE. (1992). Segregation analysis of esophageal cancer in 221 high-risk Chinese families. *J Natl Cancer Inst* 84:771–776.

Chen DN, Li AD, Huang Q, Li AR, Gu JR, Luo CR.*(1995)*. The study of gene biological function in retinoblastoma. *Zhonghua Yan Ke Za Zhi*. [Chinese Journal of Ophthalmology] 31:87–91.

Chen X, Wang H, Xie W, et al. (2006a). Association of CYP1A2 genetic polymorphisms with hepatocellular carcinoma susceptibility: a case-control study in a high-risk region of China. *Pharmacogenet Genomics* 16:219–227.

Chen WQ, Zhang SW, Li LD, et al. (2006b). An analysis of lung cancer incidence and mortality from 30 cancer registries in China, 1998–2002. *Bull Chin Cancer* 15:570–574.

Chiu LL, Lai PS, Low PS, et al. (1997). Molecular studies of loss of heterozygosity in retinoblastoma. *Ann Acad Med Singap* 26:315–319.

Dai XD, Li LD, Lu FZ, et al. (1999). 20 years change tendency and prediction analysis in the near future of mortality rate of nasopharyngeal carcinoma in China. *J Pract Oncol* 13:1–5.

Dong ZW. (2004). *The study progress of cancer in China* (Vol. VII). Peking University Medical Press, Beijing, China. pp. 9–182.

Du Q, Jiang YH, Gallie BL. (2002). Low-penetrance retinoblastoma due to exons 24 and 25 deletions in the Rb1 gene. *Zhonghua Yi Xue Yi Chuan Xue Za Zhi.* [Chinese Journal of Medical Genetics] 19:370–374.

Eddy DM. (1987). Screening for colon cancer in a high risk population. *Gastroenterology* 92:682.

Feng GG, Li YP, Yi YZ, Zhen JL, Zhen HL, Lin JX. (1994). Heterotransplantation of the SO-Rb50 cell line into athymic "nude" mice. *Zhonghua Yan Ke Za Zhi.* [Chinese Journal of Ophthalmology] 30:47–49.

Fishel R, Ewel A, Lee S, et al. (1994). Binding of mismatched microsatellite DNA sequences by the human MSH2 protein. *Science* 266:1403–1405.

Ge H, Cao YY, Chen LQ, et al. (2008). PTEN polymorphisms and the risk of esophageal carcinoma and gastric cardiac carcinoma in a high incidence region of China. *Dis Esophagus* 21(5):409–415.

Goldgar De, Easton DF, Cannon-Albright LA, et al. (1994). Systematic population-based assessment of cancer risk in first-degree relatives of cancer probands. *J Natl Cancer Inst* 86:1600–1608.

Graham S, Harighey B, Marshall J, et al. (1983). Diet in the epidemiology of prostate cancer. *J Natl Cancer Inst* 70:687.

Gu JR. (1988). Molecular aspects of human hepatic carcinogenesis. *Carcinogenesis* 9(5):697–703.

Guilford P, Hopkins J, Harraway J, et al. (1998). E-cadherin gremlin mutations in familial gastric cancer. *Nature* 392(26):402.

Guo YM, Wang Q, Liu YZ, Chen HM, Qi Z, Guo QH. (2008). Genetic polymorphisms in cytochrome P4502E1, alcohol and aldehyde dehydrogenases and the risk of esophageal squamous cell carcinoma in Gansu Chinese males. *World J Gastroenterol* 14:1444–1449.

Harbour JW. (1998). Overview of RB gene mutations in patients with retinoblastoma. Implications for clinical genetic screening. *Ophthalmology* 105:1442–1447.

Hong MH, Guo X, Min HQ. (2003). *Nasopharyngeal carcinoma.* Chinese Medical & Pharmaceutical Technology Publication, Beijing. pp. 16.

Hong Y, Miao X, Zhang X, et al. (2005). The role of P53 and MDM2 polymorphisms in the risk of esophageal squamous cell carcinoma. *Cancer Res* 65(20):9582–9587.

Hu N, Li WJ, Su H, et al. (2003). Common genetic variants of TP53 and BRCA2 in esophageal cancer patients and healthy individuals from low and high risk areas of northern China. *Cancer Detect Prev* 27:132–138.

Hu YS. (2001). *Modern image diagnostics of breast cancer,* p. 215. Beijing: Science Publication.

Hu Z, Shao M, Yuan J, et al. (2006). Polymorphisms in DNA damage binding protein 2 (DDB2) and susceptibility of primary lung cancer in the Chinese: a case-control study. Carcinogenesis 27:1475–1480

Huang Q, Dryja TP, Yandell DW. (1997). The characteristics of Rb gene mutations in 108 cases of retinoblastoma. *Zhonghua Yan Di Bing Za Zhi.* [Chinese Journal of Ocular Fundus Diseases] 13:12–16.

IARC. (1995). *Monographs on the evaluation of carcinogenic risks to humans.* Volume 64, human papillomavirus. Lyon, France: International Agency for Research on Cancer.

Jia WH, Feng BJ, Xu ZL, et al. (2004). Familial risk and clustering of nasopharyngeal carcinoma in Guangdong, China. *Cancer* 101:363–369.

Jiang JM, Pan KF, Ma JL, et al. (2003). Analysis of Lewis gene polymorphism in high and low incidence area of gastric cancer in Shandong province. *Yi Chuan* 25:258–260.

Jin F, Ddevesa SS, Chow WH, Zheng W, Ji BT, Fraumeni JF Jr, Gao YT. (1999). Cancer incidence trends in urban Shanghai, 1972–1994: An update. *Int J Cancer* 83:435–440.

Jin J, Yi YZ, Zheng JL, et al. (1991). Establishment and characterization of a human retinoblastoma cell line SO-Rb50. *Zhongshan Da Xue Xue Bao (Yi Xue Ke Xue Ban).* [Journal of Sun Yat-sen University (Medical Sciences)] 12:173–177.

Kataoka N, Cai Q, Wen W et al. (2006). Population-based case-control study of VEGF gene polymorphisms and breast cancer risk among Chinese women. *Cancer Epidemiol Biomarkers Prev* 15:1148–1152.

Knudson AG. (1971). Mutation and cancer: statistical study of retinoblastoma. *Proc Natl Acad Sci USA* 68:820–823.

Kwok Fung Lo A, Liu Y, et al. (2001). Identification of downstream target genes of latent membrane protein 1 in nasopharyngeal carcinoma cells by suppression subtractive hybridization. *Biochim Biophys Acta* 1520:131–140.

Lee HP, Duffy SW, Day NE, Shanmugaratnam K. (1998). Recent trends in cancer incidence among Singapore Chinese. *Int J Cancer* 42:159.

Li D, Cao Y, He L, Wang NJ, Gu JR. (1993). Aberrations of p53 gene in human hepatocellular carcinoma from China. *Carcinogenesis* 14(2):169–173.

Li D, Diao Y, Li H, Fang X, Li H. (2008). Association of the polymorphisms of MTHFR C677T, VDR C352T, and MPO G463A with risk for esophageal squamous cell dysplasia and carcinoma. *Arch Med Res* 39:594–600.

Li H, Feng GG, Fang Y, Li YP, Zheng JL, Yi YZ. (1998). An investigation on chromosome aberration of SO-Rb50cloned cell strains. *Yan Ke Xue Bao.* [Eye Science] 14:220–223.

Li YP, Feng GG, Ning B, et al. (1997a). Studies of monoclonal antibody against retinoblastoma cells. *Zhonghua Yan Di Bing Za Zhi.* [Chinese Journal of Ocular Fundus Diseases] 13:3–5.

Li LD, Zhang SW, Lu FZ, et al. (1997b). Characteristic study of death notation and classification composition of malignant tumor in China. *Chin J Oncol* 19:323–328.

Li LD, Lu FZ, Zhang SW, et al. (1997c). 20 years change tendency and prediction analysis in the near future of malignant tumor in China. *Chin J Oncol* 19:3–9.

Li ML. (1998). Cancer and genetic diseases. In: Lu ZB, Zhao SY, et al., eds. *Cancer genetics*, pp. 80–. Beijing: Science Press.

Li T, Lu ZM, Guo M, et al. (2002). p53 codon 72 polymorphism (C/G) and the risk of human papillomavirus-associated carcinomas in China. *Cancer* 95(12):2571–2576.

Li XY, Su M, Huang HH, Li H, Tian DP, Gao YX. (2007). mtDNA evidence: genetic background associated with related populations at high risk for esophageal cancer between Chaoshan and Taihang Mountain areas in China. *Genomics* 90(4):474–481.

Liao ML. (2003). *Lung Cancer,* pp. 1–7. Beijing: Chinese Medical & Pharmaceutical Technology Press.

Lin D, Tang YM, Lu SX, Kadlubar FF. (1998a). Glutathione S-transferase M1, T1 genotypes and the risk of esophageal cancer: a case-control study. *Zhonghua Liu Xing Bing Xue Za Zhi.* [Chinese Journal of Epidemiology] 19:195–199.

Lin DX. (2003). The genetic susceptibility of cancer. In: Zeng YX, ed. *Oncology* (2nd ed., p. 50). Beijing: People's Health Publication.

Lin DX, Tang YM, Peng Q, Lu SX, Ambrosone CB, Kadlubar FF. (1998b). Susceptibility to esophageal cancer and genetic polymorphisms in glutathione S-transferases T1, P1, and M1 and cytochrome P450 2E1. *Cancer Epidemiol Biomarkers Prev* 7:1013–1018.

Liu HX, Li Y, Jiang XD, et al. (2006a). Mutation screening of mismatch repair gene Mlh3 in familial esophageal cancer. *World J Gastroenterol* 12:5281–5286.

Liu B, Ciu JF, Liu XF, et al. (2006b). Serum markers as predictors of esophageal squamous dysplasia and early cancer. *Cancer Res Prev Treat* 33(8):551–554.

Liu F, Pan KF, Zhang XM, et al. (2006c). Genetic polymorphisms of cyclooxygenase-2 (COX-2), expression and risk of gastric cancer and its precursors in a Chinese population. *Gastroenterology* 130:1975–1984.

Liu J, Zheng S, Feng YZ. (1997). The initial report and registration of family survey familial adenomatous polyposis. *Chin J Oncol* 24:756–759.

Liu WL, Zhang G, Wang JY, et al. (2008). Proteomics-based identification of autoantibody against CDC25B as a novel serum marker in esophageal squamous cell carcinoma. *Biochem Biophys Res Commun* 375(3):440–445.

Lohmann DR. (1999). RB1 gene mutations in retinoblastoma. *Hum Mutat* 14:283–288.

Long XD, Ma Y, Wei YP, Deng ZL. (2006). The polymorphisms of GSTM1, GSTT1, HYL1*2, and XRCC1, and aflatoxin B1-related hepatocellular carcinoma in Guangxi population, China. *Hepatol Res* 36:48–55.

Lu SJ, Day NE, Degos L, et al. (1990). Linkage of nasopharyngeal carcinoma susceptibility locus to the HLA region. *Nature* 346:470–471.

Lu WL, Pan KF, Zhang L, Lin DX, Miao XP, You WC. (2005). Genetic polymorphisms of interleukin(IL)-1B, IL-1RN, IL-8, IL-10 and tumor necrosis factor α and risk of gastric cancer in a Chinese population. *Carcinogenesis* 26:631–636.

Lu XM, Zhang YM, Lin RY, et al. (2004). p53 polymorphism in human papillomavirus-associated Kazakh's esophageal cancer in Xinjiang, China. *World J Gastroenterol* 10(19):2775–2778.

Morgan MS, Griffiths K, Blacklock N. (1996). The preventive role of diet in prostatic disease. *Br J Urol* 77:481–493.

Muir C. (1987). *Cancer incidence in five continents.* Lyon, France: IARC Scientific Publication, Vol. V.

Muir CS. (1996). Epidemiology of cancer in ethnic group. *Br J Cancer* 29(Suppl):S12–16.

Naumova A, Sapienza C. (1994). The genetics of retinoblastoma, revisited. *Am J Hum Genet* 54:264–273.

Parkin DM, Muir CS, Whclan SL, Gao YT, Ferlay J, Powell J. (1992). *Cancer incidence in five continents.* IARC Scientific Publication, Vol. VI.

Parkin DM, Pisani P, Ferlay J. (1993). Estimates of worldwide incidence of eighteen major cancers in 1985. *Int J Cancer* 54:594–606.

Parkin DM. (2001). Globocal cancer statistics in the year 2000. *Lancet Oncol* 2:533–543.

Qiao YL, Zhang TH, Zhao FH, Pan QJ, Li L (2004). Screening and early diagnosis and therapy to the cervical cancer. In: Dong ZW, ed. *The study progress of cancer in China* (Vol. VII, pp. 212–217). Beijing: Peking University Medical Press.

Qin JM, Yang L, Chen B, Wang XM, Li F, Liao PH, He L. (2008). Interaction of methylenetetrahydrofolate reductase C677T, cytochrome P4502E1 polymorphism and environment factors in esophageal cancer in Kazakh population. *World J Gastroenterol* 14:6986–6992.

Reiter RE, deKernion JB. (2002). Epidemiology, etiology, and prevention of prostate cancer. In: Walsh PC, Retik AB, Vaugh ED Jr, et al. (eds.), *Campell Urology* (8th ed.). Philadelphia: Saunders.

Ren J, Chen Z, Juan SJ, Yong XY, Pan BR, Fan DM. (2000). Detection of circulating gastric carcinoma-associated antigen MG7-Ag in human sera using an established single determinant immuno-polymerase chain reaction technique. *Cancer* 88(2):280–285.

Ren JS, Kamangar F, Qiao YL, et al. (2009). Serum pepsinogens and risk of gastric and esophageal cancers in the General Population Nutrition Intervention Trial cohort. *Gut* 58:636–642.

Ren Y, Cao B, Law S, et al. (2005). Hepatocyte growth factor promotes cancer cell migration and angiogenic factors expression: a prognostic marker of human esophageal squamous cell carcinomas. *Clin Cancer Res* 11(17):6190–6197.

Rubinfeld B, Souza B, Albert I, et al. (1993). Association of the APC gene product with beta-catenin. *Science* 262:1731–1734

Shao JY, Zeng WF, Zeng YX. (2002). The molecular genetics progress of nasopharyngeal carcinoma. *Cancer* 21:1–10.

Shen FM, Lee MK, Gong HM, Cai XQ, King MC. (1991). Complex segregation analysis of primary hepatocellular carcinoma in Chinese families: Interaction of inherited susceptibility and hepatitis B viral infection. *Am J Hum Genet* 49:88–93.

Shen HX, Zhang QJ, Xiao XS, Li SQ, Guo XM, Wu ZY. (2003). A missense mutation outside the large pocket of the retinoblastoma protein. *Zhongguo Bing Li Sheng Li Xue Za Zhi.* [Chinese Journal of Pathophysiology] 19 (3):314–316.

Shi Q, Xu D, Sun C. (2000). Study on family aggregation of esophageal cancer in Linzhou city. *Zhonghua Yu Fang Yi Xue Za Zhi* [Chinese Journal of Preventive Medicine] 34:269–270.

Song CG, Hu Z, Wu J, et al. (2006). The prevalence of BRCA1 and BRCA2 mutations in eastern Chinese women with breast cancer. *J Cancer Res Clin Oncol* 132:617–626.

Sun YH. (2004). The current status of prostate cancer in China. *Zhonghua Mi Niao Wai Ke Za Zhi.* [Chinese Journal of Urology] 25(2):77–81.

Tan JZ, Zhao GM. (2002). *History of genetics in China*, p. 322. Shanghai: Shanghai Science and Technology Press.

Tan W, Song N, Wang GQ, et al. (2000). Impact of genetic polymorphisms in cytochrome P450 2E1 and glutathione S-transferases M1, T1, and P1 on susceptibility to esophageal cancer among high-risk individuals in China. *Cancer Epidemiol Biomarkers Prev* 9:551–556.

Vogel F. (1979). Genetics of retinoblastoma. *Hum Genet* 52:1–54.

Wang MH, Li YP, Wang XD, Zhang H, Xiong LH, Huang HZ. (2004a). Correlation between serum p53 antibody and chemosensitivity in patient with esophageal cancer. *Ai Zheng* [Chinese Journal of Cancer] 23(4):467–470.

Wang JY, Zhu G, Zhu SC, Zhang YG, Liu M. (2004b). The screening and diagnosis and therapy early in prostate cancer. In: Dong ZW, ed. *The study progress of cancer in China* (Vol. VII, pp. 241–245). Beijing: Peking University Medical Press.

Wang LD, Guo RF, Fan ZM, et al. (2005). Association of methylenetetrahydrofolate reductase and thymidylate synthase promoter polymorphisms with genetic susceptibility to esophageal and cardiac cancer in a Chinese high-risk population. *Dis Esophagus* 18:177–184.

Wang LS, Chow KC, Li WY, Liu CC, Wu YC, Huang MH. (2000). Clinical significance of serum soluble interleukin 2 receptor-alpha in esophageal squamous cell carcinoma. *Clin Cancer Res* 6(4):1445–1451.

Wang T, Zhang W, Liu Y. (2001). Clinical significance of the novel tumor marker CYFRA21-1 in patients with esophageal cancer. *Zhonghua Yi Xue Za Zhi* [Chinese Journal of Medicine] 81(22):1390–1391.

Wu H. (2004). *Cancer Genetics*, p. 338. Beijing: Science Publication.

Wu YQ. (1986). Genetic etiology of esophageal cancer–III. The spontaneous and mitomycin-C (MMC) induced sister chromatid exchanges (SCE) and chromosome aberrations in cultured lymphocytes from members of "high risk" and "low risk" esophageal cancer families in Linxian County. *Zhonghua Zhong Liu Za Zhi* [Chinese Journal of Oncology] 8:23–25.

Wu YQ. (1990). Genetic etiology of esophageal cancer. V. Fragile sites on chromosome of cultured lymphocytes from members of high risk cancer families in Linxian County. *Zhonghua Zhong Liu Za Zhi* [Chinese Journal of Oncology] 12:174–178.

Wu YQ. (1991). Genetic etiology of esophageal cancer. VI. Significance of endoreduplication in the cultured lymphocytes from members of high risk cancer families in Linxian County. *Zhonghua Zhong Liu Za Zhi* [Chinese Journal of Oncology]. 13:273–275.

Xing DY, Tan W, Song N, Lin DX. (2001). Ser326Cys polymorphism in hOGG1 gene and risk of esophageal cancer in a Chinese population. *Int J Cancer* 95:140–143.

Xu LP, Du RF. (1999). New progress of human genome diversity project. *Zhonghua Yi Xue Yi Chuan Xue Za Zhi* [Chinese Journal of Medical Genetics] 16:108–109.

Xu XR, Huang J, Xu ZG, Qian BZ, et al. (2001). Insight into hepatocellular carcinogenesis at transcriptome level by comparing gene expression profiles of hepatocellular carcinoma with those of corresponding noncancerous liver. *Proc Natl Acad Sci USA* 98(26):15089–15094.

Yang L, Li LD, Chen YD, DM Parkin. (2006). Time trends, estimates and projects for breast cancer incidence and mortality in China. *Chin J Oncol* 28:438–440.

Ying FW, Li YP, Tang S, et al. (2003). Separation of SO-Rb50 cells antigen corresponding to the monoclonal antibody of anti-retinoblastoma. *Zhonghua Yan Di Bing Za Zhi.* [Chinese Journal of Ocular Fundus Diseases] 19:152–155.

You WC, Blot WJ, Chang YS, et al. (1988). Diet and high risk of stomach cancer in Shandong, China. *Cancer Res* 48:3518–3523.

You WC, Hong JY, Zhang L, et al. (2005). Genetic polymorphisms of CYP2E1, GSTT1, GSTP1, GSTM1, ALDH2, and ODC and the risk of advanced precancerous gastric lesions in a Chinese population. *Cancer Epidemiol Biomarkers Prev* 14:451–458.

You WC. (2006). *Stomach cancer*, pp. 1–34. Beijing: Chinese Medical and Pharmaceutical Technology Press.

You WC. (2005). The epidemic information of common malignant tumor. *China Med News* 20:22.

Yuan Y, Huang J, Zheng S. (1999). Mutation of human mismatch repair genes in hereditary nonpolyposis colorectal cancer (HNPCC) families. *Zhonghua Zhong Liu Za Zhi.* [Chinese Journal of Oncology] 21(2):105–107.

Yue CM, Bi MX, Tan W, et al. (2004). Short tandem repeat polymorphism in a novel esophageal cancer-related gene (ECRG2) implicates susceptibility to esophageal cancer in Chinese population. *Int J Cancer* 108(2):232–236.

Zhang BZ, Pan KF, Liu ZJ, et al. (2008). Genetic polymorphisms of the E-Cadherin promoter and risk of sporadic gastric carcinoma in Chinese populations. *Cancer Epidemiol Biomarkers Prev* 17(9):2402–2408.

Zhang Q, Minoda K, Zeng R, et al. (1997). Exon-by-exon screening for RB germline mutations using Heteroduplex-SSCP analysis. *Yan Ke Xue Bao* [Eye Science]. 13:5–11.

Zhang WH, Bai P, Wu LY, Ma SH. (1999). Carcinoma of the cervix in younger women (≤35years). *Zhongguo Zhong Liu Lin Chuang Yu Kang Fu* [Chinese Journal of Clinical Oncology and Rehabilitation] 6(6):39–41.

Zhang XM, Miao XP, Tan W, et al. (2005). Identification of functional genetic variants in cyclooxygenase-2 and their association with risk of esophageal cancer. *Gastroenterology* 129:565–576.

Zhong XF, Li YP, Ning B, et al. (2002). Preparation of anti-human retinoblastoma monoclonal antibody by solid tumor cell immunization and study on characteristics of its antigen. *Zhonghua Yan Ke Za Zhi.* [Chinese Journal of Ophthalmology] 38:691–694.

Zhou RM, Li Y, Wang N, Zhang XJ, Dong XJ, Guo W. (2006). Correlation of XPC Ala499Val and Lys939Gln polymorphisms to risks of esophageal squamous cell carcinoma and gastric cardiac adenocarcinoma. *Ai Zheng* [Chinese Journal of Cancer] 25:1113–1119.

72.

GENOMIC MEDICINE IN TAIWAN: RESEARCH AND DEVELOPMENTS

Ming Ta Michael Lee, Chien-Hsiun Chen, and Yuan-Tsong Chen

BACKGROUND ON TAIWAN

Taiwan, also known as Formosa ("beautiful island" in Portuguese), is located in eastern Asia off the southeastern coast of China, separated by the 180-kilometer-wide Taiwan Strait and surrounded by east China, the Philippine Sea, and the South China Sea. Taiwan is a small island (35,801 km^2, approximately the size of Holland) with a population of 23 million. It is made up almost entirely of ethnic Han Chinese, with a small proportion (2%) of the Taiwanese population being of aboriginal ancestry.

THE HAN CHINESE POPULATION IN TAIWAN

Most of the Han Chinese ancestors came from two southeast coast provinces of China, Fujian and Guangdong, due to the overpopulation of the two provinces in the past few centuries and the close geographic distances between the two provinces and Taiwan. The first Taiwan census taken in 1905 showed that 95.1% of the total Taiwan population was of Han Chinese origin, of which 86.2% was of Fujian origin and 13.7% was of Guangdong origin. The ratios did not change until the end of World War II. Right after World War II, a major migration wave from mainland China increased the HanChinese to 98% of the total population (Chen, 1997).

People with Fujian origins are often referred as the "Minnan" people (*min* is an abbreviation of Fujian and *nan* means *south* in Chinese), since most of their ancestors were from the southern counties of Fujian. On the other hand, people with Guangdong origins are referred as the "Hakka" people (*hakka* means *guest* in Chinese) since their ancestors were newcomers or minorities in the Han Chinese population back in the old days. All the other Han Chinese people are referred as "mainlanders" since most of them came from Mainland China after World War II and registered themselves with origins from other providences. A survey conducted by the Council for Hakka Affairs Executive Yuan, Republic of China (Taiwan) in 2004 showed that 73.3% of the Taiwan Han Chinese self-reported as having Fujian origins, 13.4% with Hakka origins, 8.0% as mainlanders, and 1.9% as Taiwan aborigines.

The Minnan and Hakka people share similar genetic backgrounds (Yang et al., 2006). Genetically they show affinities to southern Asian populations, as determined by phylogenetic trees and correspondence analysis calculated from HLA allele frequencies. This corresponds historically with the fact that they are the descendants of the southeast coastal indigenous population (Yueh) of China (Lin et al., 2001).

It has been argued that many of the Han Chinese intermarried with the preexisting indigenous tribes after arrival, mainly with the plains tribes but also with the mountain tribes. However, the degree of contribution of these tribes to the Han Chinese gene pool is at present unknown (Lin et al., 2000). A recent study using genome-wide autosomal SNPs traced the genetic origins of Asian populations in 73 Southeast Asian and East Asian populations and inferred that there was a single major inflow of human migration from Southeast Asia into other parts of Asia (Abdulla et al., 2009). The findings may in part interoperate the genetic sharing between Taiwan Han Chinese and Autronesian populations. Further studies for more accurate ancestry of the Taiwan Han Chinese are required.

Even though there are several subpopulations of Han Chinese in Taiwan, a recent study using SNPs located in the HLA region (Yang et al., 2006) showed that these groups are basically identical. Thus, this strongly supported the homogeneity of the Taiwan Han Chinese. As described above, the Han Chinese in Taiwan are mostly from the southern populations of Han Chinese; they are more closely related to the HapMap's Han Chinese Denver population (Figure 72-1) and farther from the Han Chinese Beijing and the Japanese populations.

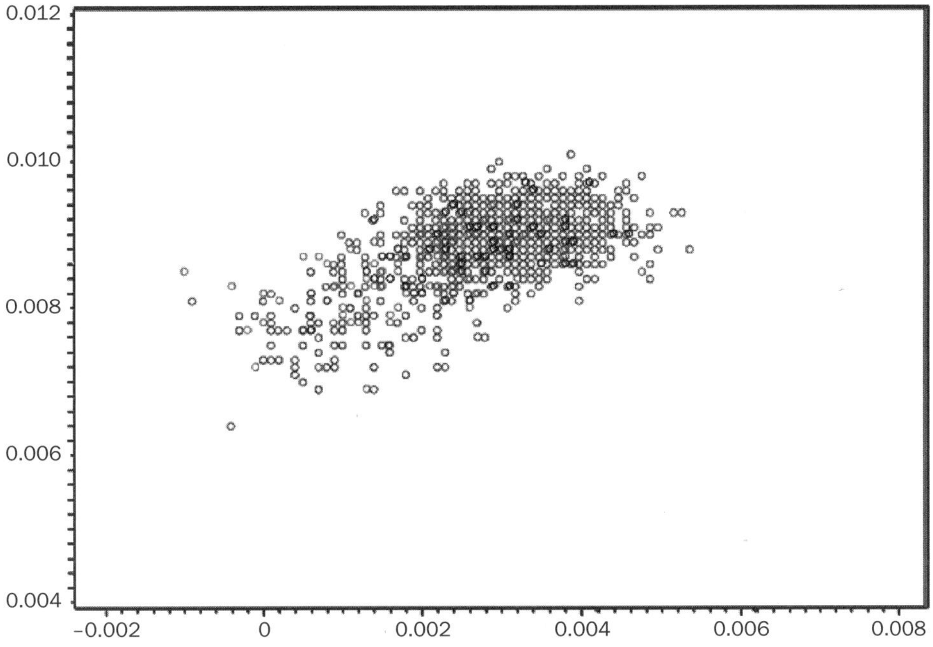

Figure 72-1 Principle component analysis of the Chinese and Japanese populations, using three Han Chinese groups: Han Chinese Beijing (grey circle), Han Chinese Denver (blue circle) and Taiwan Han Chinese (green circle), and the Japanese population (red circle). Refer color figure.

THE ABORIGINAL POPULATIONS IN TAIWAN

There are more than 490,000 indigenous people in Taiwan, about 2% of the total Taiwan population. Currently, 14 tribes have been recognized by the government. They are the Amis, Atayal, Paiwan, Bunun, Puyuma, Rukai, Tsou, Saisiyat, Yami, Thao, Kavalan, Truku, Sakizaya, and Sediq, each with their own distinct language, cultural features, traditional customs, and social structure (http://www.apc.gov.tw/main/index_en.jsp). Studies have been conducted to search for Asian roots for the aboriginal Taiwanese tribes in the context of modeling the peopling of the Pacific; however, conflicting conclusions were made based on different types of evidence (Gibbons, 2001).

A main hypothesis concerning Polynesian origins called the "express train to Polynesia" (often referred as "out of Taiwan"), supported by archaeology (red-slipped pottery) and linguistic evidence, has been suggested (Diamond, 1988; Bellwood, 1997; Blust, 1999). Under the express train model, Neolithic pottery-making and rice-growing farmers from South China settled in Taiwan, spread to the Philippines, eastern Indonesia, and Melanesia, and then set sail to the Pacific (Diamond and Bellwood, 2003). In particular, a linguistic analysis of the relationship between 1200 Austronesian languages showed that all of the languages were clustered into 10 subgroups, among which nine were spoken only by the aboriginal Taiwanese tribes (Blust, 1999). In addition, a language family tree based on 77 Austronesian languages with 5,185 lexical items was found to have a root at the aboriginal Taiwanese languages, and was highly compatible with the express train model (Gray and Jordan, 2000). Genetic studies analyzing mitochondrial DNA (mtDNA) variations also supported that the aboriginal Taiwanese tribes held an ancestral position in the spread of mtDNAs throughout Southeast Asia and Oceania (Melton et al., 1995; Sykes et al., 1995). Further studies of mtDNA and nuclear DNA suggested that the indigenous Taiwanese tribes have temporally deep roots in South China (Melton et al., 1998; Trejaut et al., 2005).

An alternative model, called the "entangled bank" (Terrell and Welsch, 1997), argued that the Polynesians evolved locally in Melanesia, within the voyaging corridor between island Southeast Asia and Melanesia (Terrell and Welsch, 1997). Another model, called the "slow boat" (Oppenheimer and Richards, 2001), proposed that the ancestors of Polynesians originated in Asia but moved slowly through Melanesia and mixed with local settlers on the way to the Pacific. Genetic evidence also supported these models.

HEALTHCARE IN TAIWAN

Taiwan has established one of the best healthcare systems in Asia with the institution of National Health Insurance (NHI), a compulsory medical insurance plan that offers equal access to healthcare and had reached 99% coverage by 2007 The NHI also maintains an extensive medical database (http://w3.nhri.org.tw/nhird//en/index.htm), which records information such as indications (using ICD9 codes) and drugs prescribed. This database has been very valuable to healthcare research in Taiwan. There are over 500 public and private hospitals in Taiwan; however, the referral system has not been well established, so the patients tend to go to major medical centers for their medical needs, even for common colds. Thus, more than 80% of patient care is covered by Taiwan's sixteen major medical centers. Population homogeneity, good medical recordkeeping, advanced medical infrastructure and concentrated medical centers make Taiwan an ideal place to recruit patients for medical research.

Medical research in Taiwan, like almost other countries, focuses on diseases most prevalent and has an important impact on public health in Taiwan. Malignant tumor has been the number one cause of death in Taiwan for more than 27 years (Table 72-1), with lung and hepatic cancer being the most prevalent.

THE EARLY YEARS

Genomic research in Taiwan can be roughly divided into two eras divided by the year 2002. Before 2002, there was no infrastructure in Taiwan dedicated to genomic research. Research was mostly carried out by individual labs, and most research focused on epidemiology and molecular biology. Genetic studies were also conducted; however, they were mostly replication studies. No significantly novel genetic discovery was identified during this era in Taiwan.

TAIWAN'S SUCCESS IN THE IT INDUSTRY

Taiwan always has been known as an international powerhouse in the IT industry, with virtually all electronics having parts made in Taiwan. This was made possible when the Taiwan government decided to focus on semiconductors in order to upgrade its industry in the late 1970s. The government provided enormous support not only for academics but also industry, hoping that important discoveries made, or newly devised techniques with commercial potential, could be quickly developed into products. The best known example is the establishment of Hsinchu Science Park (also known as Taiwan's Silicon Valley, http://www.sipa.gov.tw/english/index.jsp). It was established in 1980 with US$1679 million of investment by the government in park facilities and infrastructure, It is located near two of Taiwan's science and engineering powerhouses, National Chiao Tung University and National Tsing Hua University, with close academic–industry ties, and houses the largest semiconductor manufacturing company in the world, Taiwan Semiconductor Manufacturing Company and United Microelectronics Corporation, in addition to more than 350 companies involved in semiconductors, computers, telecommunications, and optoelectronics. The park's production value has reached more than NT$ 1 trillion (US$30.76 billion) and has been hailed as one of the most successful science parks in the world.

TABLE 72-1 TOP 10 CAUSES OF DEATH IN TAIWAN IN 2008

Rank	Cause	% of Total Deaths
1	Malignant Tumor	27.3%
2	Cardiac Disease	11.1%
3	Cerebrovascular Disease	7.5%
4	Pneumonia	6.1%
5	Diabetes	5.6%
6	Accidental Injury	5%
7	Chronic Lower Respiratory Diseases	3.8%
8	Chronic Liver Disease and Cirrhosis	3.5%
9	Suicide	2.9%
10	Nephritis, Nephritic Syndrome and Nephrosis	2.8%

THE NATIONAL RESEARCH PROGRAM FOR GENOMIC MEDICINE ERA

However, with decreasing profit margin for the semiconductor and computer industry, the Taiwan government again started to look for new areas to focus on, in order to bring Taiwan's industry to a new level and to compete globally. With the Human Genome project published in year 2001, the Taiwan government realized the importance of genomic medicine and the potential it can offer. As a result of this, The National Research Program for Genomic Medicine (NRPGM) was launched as a national priority program by the National Science Council (NSC) of Taiwan. Three government agencies—NSC, Department of Health, and the Ministry of Economic Affairs—joined forces to fund this national program. The goal of NRPGM is to perform basic research to better understand human diseases, capitalize on these discoveries to develop new paradigms for diagnosis and treatment, and ultimately foster the development of biotechnology industries and build a robust economy through technology transfer and commercialization. The NRPGM is divided into two phases: the aim of the first phase (2002–2005) was to establish infrastructure required for genomic research and to develop the foundation for research in several common diseases. This included genomics, proteomics, structural biology, bioinformatics, and ELSI (ethical, legal and social issues) related to biomedical technology. Phase II (2006–2010) of NRPGM consists of four core parts: national core facilities, basic research programs, industrial/academic collaboration, and international collaboration. The scope for this phase is to integrate the various components for genomic research and to focus on disease-oriented projects and accelerate the development of biotechnology industries. It is hoped that phase II of the program will identify discoveries for the prevention, diagnosis, and treatment of the three major diseases in Taiwan: cancer, infectious disease, and highly heritable diseases.

THE NATIONAL CORE FACILITIES

The National Core Facilities of NRPGM were formed in order to establish infrastructures required for genomic medicine research. Genomic medicine research often requires sophisticated instrumentation, which demands great amount of resources to acquire and maintain; therefore, the National Core Facilities were also set up to integrate and centralize management and resources. The facilities are not only responsible for offering high quality, high-throughput services using the state-of-the-art instruments, but also to offer consulting and analytical services to their users. In addition, the cores are encouraged to develop multidisciplinary technologies and participate in collaborative research projects, especially at international level. They are also responsible for promoting and educating the researchers in Taiwan on the technologies possessed by the cores. A total of 17 core facilities were established by the end of 2009, and covered 5 major areas: A. animal models, phenotyping, and imaging facilities for diseases; B. clinical sample banks for genomics; C. facilities for basic genomics; D. facilities for proteomics and structural genomics; and E. facilities for bioinformatics.

Facilities related to genomic research established by NRPGM are as follows.

Clinical Sample Banks for Genomics

Two cores have been established in this field. The National Clinical Core for Genomic Medicine is to aid in the development of the nation's genetic medicine research by training genetic counseling professionals, assisting researchers with all aspects of genetic counseling, as well as conducting genetic studies such as preparation of case report forms, informed consent forms, phenotype assessment, pedigree assessment, and biological sample collection. The core also can provide statistical consultations with respect to sample size estimation, statistical power evaluation, interpretation of the statistical analysis, and maintaining a Good Clinical Practice compliance data capture and storage system. Many studies in Taiwan are still being conducted using paper-based questionnaires with data managed via Excel. However, errors such as data inconsistencies between different questions and missing data often occur with this method. To prevent these errors, the core has implemented "iQuestion," a centralized clinical study management software for questionnaire design, data capture, data management, dataset export, and clinical report management (http://iquestion.ngc.sinica.edu.tw/ecrf/do/home) in all the studies managed by the core. Since the implementation of iQuestion, errors in clinical data ascertainment have dropped considerably and this has resulted in the reduction of the time required for data cleaning as well as data analysis. This core was also responsible for the release of the Han Chinese cell and genome bank DNA described above.

Facilities for Basic Genomics

Five facilities were set up for basic genomic study needs. **The High-Throughput Genome Analysis Core Facility** offers high-throughput sequencing for viral, bacterial, and human genomes using the next-generation high-throughput genome sequencers. The core also offers small RNA identification, ChIP sequencing, and methylation profiling for transcriptome analysis. **The National Genotyping Center** aims to help researchers identify disease-causing and drug-susceptible genes from the vast human genome by providing high-throughput SNP (single nucleotide polymorphism) and STRP (short tandem repeat polymorphism) genotyping services. The BioIT core within the center also has established extensive bioinformatics tools for SNP research, which are provided as free online services to all the researchers. **The Integrated Core Facility for Functional Genomics** has established microarray technology platforms for cDNA microarray, in-house spotting oligo microarray, both in membrane and glass formats, oligo microarray, real-time quantitative PCR, array CGH, and microRNA quantification. The core also has developed a comprehensive expression algorithm provided as a free online tool. RNA interference (RNAi) has become a very efficient reverse genetics technology for functional genomics. **The RNAi Core Facility** was established to distribute and maintain RNAi resources (lentiviral-based RNAi libraries targeting 15,000 human and 15,000 mouse transcripts), in addition to its platform for RNAi screening, assay, and analysis.

Facilities for Bioinformatics

Two facilities are currently in operation for the bioinformatics. **The Bioinformatics Core for Genomic Medicine and Biotechnology Development** is composed of five units: statistical genetics, computational proteomics, comparative genomics and interactomes, structural bioinformatics, and applied medical genetics in specific areas in bioinformatics. The integration of manpower and expertise has been emphasized to provide comprehensive and in-depth services to investigators in genomics research. Several dozen kinds of analysis tools and network service database are offered on its website. **The Advanced Bioinformatics Core Facility** consists of comparative bioinformatics, functional bioinformatics, genomic statistics, and information technology, with the services provided as online web services and collaboration. Both of the bioinformatics cores aim to assist researchers to take advantage of the available bioinformation and to use bioinformatics tools, to become the information center of the NRPGM in order to integrate and share the information generated by this national program, and to develop next-generation bioinformatics tools to increase our competitiveness.

Other infrastructures established by NRPGM are described below.

Facilities for Proteomics and Structural Genomics

Five facilities have been built to serve all proteomics and structural genomics studies.

The Core Facilities for Proteomics and Glycomics, the Core Facility for Protein Production and X-Ray Structural Analysis, and the National Synchrotron Radiation Research Center were all set up to meet the needs of proteomic and structural studies.

Animal Models, Phenotyping, and Imaging Facilities for Diseases

Many human diseases cannot be studied directly and thus many researchers rely heavily on animal models, especially mutagenic mouse. Thus, the National Mouse Mutagenesis Program Facility, Transgenic Mouse Model Core Facility, the Functional and Micro-Magnetic Resonance (MR) Imaging Center, the Molecular and Genetic Imaging Core, and the Taiwan Mouse Clinic were established to create mutant mice (either knockout or ENU mutagenesis mouse) and are phenotyping these mice.

Thus, NRPGM has established very extensive arrays of core facilities to virtually cover all areas of genomic medicine. Five major activities—service, collaboration, R&D, dissemination, and training—are the goals for each national core facility. These national core facilities have very well met the needs of NRPGM projects. Basic genomics core facilities provide high-throughput screening tools—such as whole-genome sequencing and microarray platforms for expression and SNP profiling for the identification of disease susceptibility genes—as well as providing bioinformatics tools to help researchers with data management and analysis. Proteomics core facilities help to express the identified genes, characterize their proteins, and prepare for future drug screening. Imaging and animal core facilities provide platforms for researchers to use MRI, micro-PET, and microscopic imaging equipment in order to understand the genes and the associated diseases in animal models constructed or produced by transgenic mice and ENU mutagenesis core facilities. Bioinformatics core facilities provide analysis tools and an integrated network to help researchers analyze the data generated from genomic, proteomic, and imaging services. The National Clinical Core Facility helps to train genetic counselors and design future genetic studies, as well as ensuring the quality of the biological samples collected. From 2002–2006, they served more than 1358 PIs/laboratories (Figure 72-2). There are more than 192 reported collaborative projects, and the core facilities contribute to more than hundreds of research publications each year, as well as many patents and technology transfers. More than 275 conference and training courses have been hosted by the core facilities and have attracted more than 13,216 scientists. These are remarkable achievements in just 4 years. However, the importance and impact of these cores go beyond what can be measured by papers, patents, or service charges. The core program has also trained a large number of highly skilled technical personnel who will aid development of the biotech industry in Taiwan. The cores have become an integral and essential component not only for genomic research, but also for all life sciences in Taiwan.

BASIC RESEARCH PROGRAM

The Basic Research Program component of NRPGM focuses on diseases important for public health in Taiwan, as well as developing new technologies for the diagnosis, prevention, and treatment of these diseases. Projects funded under NRPGM usually receive more resources than those funded by the National Science Council (the main funding body in Taiwan), and the NRPGM tends to grant high-risk projects. The additional funding allows researchers to explore areas that have not been tapped in Taiwan previously. One of the aims of the research program is also to identify Taiwan's own niche in biomedical research. As a small nation with limited resources, Taiwan needs to identify unique areas in order to compete on the

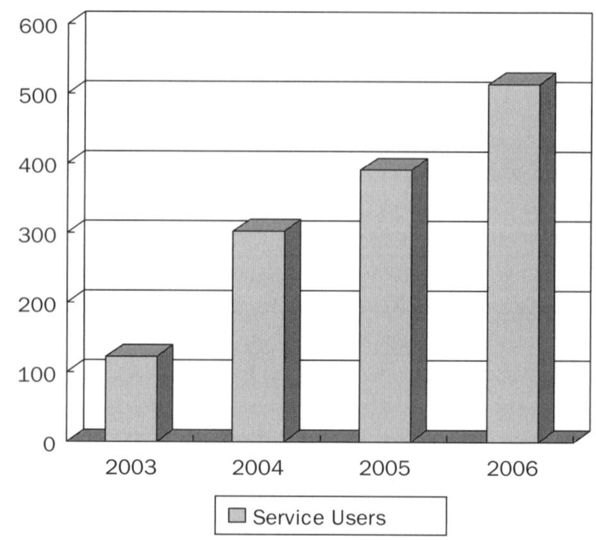

Figure 72-2 NRPGM core facilities service users.

international stage. The basic research programs focus on 4 major diseases, innovative research, and ethical, legal, and social implications of human genomics (ELSI).

1. **Liver Cancer**: The liver cancer program aims to identify pathological mechanisms of liver cancers, and to develop reagents and equipment for early diagnosis and treatment. Since hepatitis B and hepatitis C viral infections are highly prevalent, and among the major causes of liver cancer in Taiwan, this program also aims to integrate medical centers in Taiwan to build a clinical and epidemiological database and specimen bank of HBV- and HCV-related hepatocellular carcinoma.
2. **Lung Cancer**: Lung and liver cancers are the most prevalent malignant tumors in Taiwan. Therefore, understanding of the genetic epidemiology of lung cancer in Taiwan and elucidation of the molecular pathogenesis of lung cancer are high priorities for NRPGM. This program will also focus on understanding gene–environment interactions, as well as pharmacogenomic studies with the ultimate goal of personalized diagnosis and therapy for lung cancer.
3. **Infectious Diseases**: Taiwan is also endemic for several infectious diseases, thus making it a top priority for the NRPGM research programs to understand the pathogenesis of these pathogens and hope that new preventive and treatment strategies can be developed. Pathogens important to Taiwan's public health include hepatitis virus (HBV, HCV and HDV), Dengue virus, enterovirus, *Klebsiella pneumoniae*-associated pyogenic liver abscess, and *Mycobacterium tuberculosis,* and these will be the focus for NRPGM.
4. **Highly Heritable Diseases**: Many diseases, whether common, complex, or rare, have a genetic component associated with them (especially rare diseases). Thus, this program intends to identify disease susceptibility genes for diseases such as mental disorders (e.g., schizophrenia and bipolar I disorder), type II diabetes, congenital heart disease, and metabolic syndromes. It is the goal of this program that genetic research on these diseases will lead to a better understanding of pathogenesis and help to identify potential drug targets.
5. **Innovative Research**: This research program focuses on novel research and promotes interdisciplinary collaboration using genomics, and development of new tools and technologies with a goal to identify new niches where Taiwan can excel. Areas of focus will include pharmacogenomics, genomic medicine, bioinformatics, proteomics and structural biology, stem cell research, and Chinese herbal medicine.
6. **Ethical, Legal and Social Implications (ELSI):** Genomic research on human diseases often requires recruiting patients with the disease, and collecting medical information as well as biological samples from them. Therefore, to ensure the research is conducted properly, and that the patients' rights and confidentiality are not violated, ELSI guidance needs to be established. This program will establish a national infrastructure for ELSI research in Taiwan, and a network for exchange and collaboration of international research, conduct and analyze opinion surveys on topics related to ethical, legal, and social implications of genomic medicine research, initiate ELSI forums on genomic medicine, and educate citizens on ELSI-related topics through lecture series and publications accessible to the public.

INDUSTRIAL/ACADEMIC COLLABORATION

In order to increase the economic value of basic research and to increase vigor among biotech companies, NRPGM encourages interactions and collaborations between academia and industry. The "Bridging Program," in collaboration with the National Science and Technology Program for Biotechnology and Pharmaceuticals (NSTPBP), will identify research projects with industry potential, provide help in patent and marketing analyses, and further assist principal investigators to apply for intellectual property rights and develop business plan. The Center for Drug Evaluation (CDE) will also offer consultation services related to medical regulation, so as to expedite the process of commercialization.

INTERNATIONAL COLLABORATION

NRPGM encourages international collaboration in order to upgrade the quality of research in the field of genomic medicine and enhance the international visibility of Taiwan. It is also expected to create opportunities of cooperation with international pharm/biotech companies through those interactions. The research project applied for can only be funded in a situation where the corresponding project from the other country is funded by its funding agency.

GENETIC EPIDEMIOLOGICAL STUDIES CONDUCTED IN TAIWAN

After the blueprint for the core facilities and research programs was finalized, and the funding to carry out these plans was secure, Taiwan was in a good position to do good science. Almost all the cores described above were established within two years of the start of the program, and they have attracted numerous users across the island. The successful implementation of the core facilities laid the foundations for success of several major genetic epidemiological projects described below.

ESTABLISHMENT OF A MASTER CONTROL DATABASE

The first milestone for the research program was the successful completion of the Han Chinese Cell and Genome Bank in Taiwan (Pan et al. 2006). Recruited for this project were 3380 Han Chinese in Taiwan, to serve as the genetic representation of the Taiwan population. A stratified, 3-staged, probability clustering sampling scheme was employed to ensure the representativeness of the samples collected. Approximately 278 male and female individuals respectively were recruited in each of 6 age groups (20–, 30–, 40–, 50–, 60–, 70–); this allowed a sufficient number of sex/age matched controls for various diseases. Plasma, DNA, and B-lymphocytes were collected and banked. In addition, clinical measurements, such as anthropometric parameters, basic blood chemistry, and blood pressure were also measured (Table 72-2). A questionnaire about ethnicity, disease and medication history, lifestyles, and cognitive function (for the elderly) was also administered by trained nurses. Questions regarding ancestral origin of parents and grandparents were included, to exclude the aboriginal population. The advantage of this is that this cohort could serve as master controls for many genetic studies for Han Chinese in Taiwan (and it has), so the investigators would not need to recruit controls; they could simply apply DNA for the controls. This would save investigators lots of time and resources, as patient recruitment can be costly and laborious. More than 1000 of the controls were also genotyped with high-density oligonucleotide arrays (both Affymetrix and Illumina platforms), and these data were also used in various genetic studies (see below). The creation of EBV transformed B-lymphoblastiod cell lines, a longer-term supply of DNA materials for the next generation of researchers in Taiwan.

The creation of the master controls, for the first time in Taiwan, also allowed researchers to conduct detailed studies on the genetic similarities among the Han Chinese subgroups (Yang et al. 2006). Analysis showed that all the subgroups are highly similar, and no stratification is needed when conducting genetic studies using the Han Chinese in Taiwan.

THE TAIWAN BIOBANK PROJECT

With the advancement in biomedical research and the completion of human genome sequencing, medical geneticists and genetic epidemiologists are now able to provide biomedical researchers with a wider variety of disease information than ever before, further aiding researchers in more in-depth research on disease-causing mechanisms. Numerous countries have already begun active participation in the construction of their own biobanks, with the goal of creating a large population-based cohort database, allowing combinations of genetics, epidemiology, and other medical information to explore disease etiology. Given unique

TABLE 72-2 PHENOTYPE INFORMATION COLLECTED BY THE HAN CHINESE CELL AND GENOME BANK IN TAIWAN

Questionnaire Information	Physical Exam Items	Clinical Chemistry
Birthday	Blood pressure	Cholesterol
Gender	Pulse rate	HDL-C
Education	Body weight	Uric acid
Ancestral ethnic origin of the parents and grand parents	Body height	Electrolytes (Na^+, Cl^-)
Medical history	Waist circumference	BUN
Familial disease history	Hip circumference	Creatinine
Medication history for the past six months	Peak flow rate	SGOT
Life style (smoking, alcohol consumption and betel nut chewing)		SGPT
Short portable Mental Status for the elderly		HbA1C
		Antibody for hepatitis C
		Antibody for hepatitis B-surface antigen
		Antibody for hepatitis B-core protein

features of risk-factor profiles and genetic background, it is therefore highly beneficial for Taiwan to construct a biobank specific to the Taiwanese population, as it would enable large-scale cohort studies to be carried out for common diseases occurring locally in Taiwan. There are several unique features that characterize the biobank in Taiwan. The Taiwan biobank aims to understand descriptive epidemiological features and to obtain background risk profiles of common diseases in Taiwan, to explore gene-risk factor/gene–gene interactions associated with common diseases in Taiwan, to establish essential bioinformatic networks of biobanks, leading to the development of a bio-IT industry, and to explore critical issues related to ethnic, legal, and social issues of genomic medicine in Taiwan. The project is currently in the logistical evaluation phase, which will recruit 15,000 individuals to test the feasibility of the biobank project. If this phase is successful, the project aims to recruit 250,000 individuals with regular follow-up. It is hoped that this project will lead to greater understanding of the driving forces behind common diseases in Taiwan, and allow introduction of more effective disease-prevention and treatment strategies, ultimately leading to reduced consumption of medical resources and improving overall health for the population in Taiwan.

Identification of Disease-Causing Genes in Mendelian Disorders and Rare Diseases

The most notable achievement is the identification of type II collagen for the avascular necrosis of the femoral head (ANFH; Liu et al., 2005). Three families with well-defined ANFH phenotypes were recruited and genome-wide linkage scans using microsatellites were performed to localize the region harboring the disease-causing gene. It was eventually narrowed down to a 5 Mb region on chromosome 12, which contained 21 known genes and 25 predicted genes. This is usually the difficulty associated with linkage analysis, as researchers are often left with numerous genes to be sequenced and the even more daunting task to establish which of the hundreds of changes could alter function and link with the disease. Therefore, careful prioritization of the candidate genes is required to efficiently identify the culprit. The ANFH researchers did, and perhaps with a striking piece of luck, identify a G→A transition in exon 50 of COL2A1 which predicted the replacement of glycine with serine at codon 1170 in a GXY repeat of type II collagen. This was a very interesting discovery. Type II collagen is the most abundant protein in the cartilage, and numerous mutations in this gene have been linked to genetic diseases of the skeleton. It forms long fibers that are assembled from three chains more than 1000 amino acids long, with every third amino acid in the sequence being glycine. Any substitution of this glycine would destroy the long fiber structure. This identification would be useful, as one could screen ANFH patients with early-onset symptoms for type II collagen mutation. Early diagnosis might improve the results of ANFH therapies.

Animal Models of Human Diseases and Metabolic Diseases

As described above, the ENU mutagenesis core was established to unravel genes and pathways associated with human diseases and to generate animal models for these human diseases. This was accomplished by a genome-wide, large-scale production of mutant mice by N-ethyl-N-nitrosourea (ENU, a potent chemical mutagen), followed by phenotype screening and genotyping to identify the disease genes. The ENU program has mouse models of neurological diseases (brain atrophy, necrotizing encephalopathy, hydrocephalus, pain insensitivity, phenobarbital resistance), cardiovascular diseases (aortic stenosis, cardiac arrhythmia), renal disease (hydronephrosis), hematology (lymphopenia) and most notably, metabolic diseases. Metabolic diseases are an important group of clinical disorders. Most metabolic disorders cause serious medical problems (complications) resulting in multiple hospitalizations and lifelong medical treatment. A comprehensive analysis of small metabolites and metabolic flux in biological fluids and tissues will allow a comprehensive assessment of the whole metabolic pathway of health and disease states in the body. Tandem mass spectroscopy (MS/MS) analysis of blood plasma, urine, and tissues has become a powerful tool to detect metabolic diseases in Taiwan, the United States, and some European countries. Using the same technique, researchers in Taiwan used MS/MS to screen ENU mice for abnormal amino acid and fatty acid metabolism. They were able to identify an ENU mouse line with elevated blood branched-chain amino acids, ketoaciduria, and clinical features closely resembling human maple syrup disease, which is a severe genetic metabolic disorder caused by deficiency of branched-chain α-keto acid dehydrogenase (BCKD). However, BCKD gene and enzymatic functions were normal in these mice. Eventually, sequencing revealed a mutation that caused a splicing defect in the mitochondrial form of branched-chain aminotransferase (Kao et al. 2006). This work is important, as mice and humans share similar metabolism; thus, it is likely that humans with similar clinical manifestations and biochemical profiles will also have the mutation.

The same approach was also used to identify the mutation for an ENU-mouse model exhibiting hair loss; severe osteoporosis; cachexia, a wasting syndrome characterized by loss of weight weakness and fatigue often seen in patients with chronic diseases such as cancer, AIDS, or tuberculosis; and systemic amyloidosis, a deadly protein misfolding disease in which misfolded proteins destroy vital organs such as the heart, kidneys and liver. It also caused early death (Saleem et al. 2010).

These phenotypes were caused by defective palmitoylation resulting from a mutation in the Zdhhc13 gene, which codes for palmitoyl acyltransferase, the enzyme that catalyzes the reaction of protein palmitoylation. The multiorgan/system failure of mice with mutated Zdhhc13 established a direct link between protein palmitoylation and the regulation of many important diverse physiological functions, and indicates that the absence of Zdhhc13 can result in profound disease pathologies.

The mouse model developed will be useful for further investigation of the mechanisms by which improper palmitoylation leads to disease, including the identification of target proteins of ZDHHC13. The elucidation of these proteins would be an exciting first step to understanding the molecular mechanisms underlying human alopecia, osteoporosis, and many neurodegenerative diseases caused by protein misfolding and amyloidosis.

Pharmacogenomics and Personalized Medicine

The greatest achievement of genetic epidemiology study in Taiwan is perhaps in the area of pharmacogenetics and its applications in personalized medicine. Adverse drug reactions (ADRs) are important causes of morbidity, which are potentially life threatening and thus are considered to be a major public health issue. One of the most severe ADRs is toxic epidermal necrosis (TEN, also known as Lyell's syndrome) and Stevens-Johnson Syndrome (SJS), a rare and severe mucocutaneous blistering disorder. Even though the incidence of TEN and SJS is rare (less than two patients per million per year), they have high mortality and can cause severe disability in surviving patients. Over 100 different drugs have been documented to associate with TEN and SJS, but the most associated drug groups in Taiwan are carbamazepine (CBZ) and allopurinol. The incidence for CBZ-induced SJS/TEN was estimated as 1.4 for every 10,000 new users in Germany, 2 in the United States, 5 in France and 9 in the UK. However, the incidence is much higher in Southeast Asian countries such as Taiwan (59 cases per 100,000 CBZ patients per year). However, physicians have had no reliable tests to predict the occurrence of SJS and TEN when they prescribed the drugs. The first major breakthrough in searching for genetic markers associated with TEN came in 2004, in the study of CBZ-induced SJS and TEN. Chung et al. (2004) reported a 100% association between HLA-B*1502 and 44 Han Chinese residing in Taiwan with CBZ-induced SJS and TEN, compared with the observed frequency of 3% (3/101) of CBZ-tolerant patients [P value = 3.1×10^{-27}, odds ratio (OR) = 2505, 95% confidence interval (CI) = 126–49522]. The HLA-B*1502 allele was present in 8.6% of the healthy controls without prior history of CBZ use. In their follow up study, 59 of 60 (98.3%) CBZ-induced SJS/TEN patients carried the HLA-B*1502 allele, while only 6 of the 144 CBZ-tolerant controls (4.16%) were tested positive for the allele (P value = 1.6×10^{-41}, OR = 1357, 95% CI = 193.4–8838.3; Hung et al. 2006). This is perhaps one of the strongest associations identified for ADRs. The same association was also observed in Southeast Asian populations. However, the HLA-B*1502 allele is not associated with Caucasian, African, Japanese, and Korean populations due to low HLA-B*1502 frequency. Because of the strong genetic evidence, the U.S. Food and Drug Administration (FDA) changed the label for CBZ and allopurinol by including genetic information, and recommends genetic testing.

Another strong association was also identified for allopurinol-induced SJS and TEN. Using 21 SJS/TEN patients, 135 allopurinol-tolerant and 93 healthy individuals, HLA-B*5801 was tested positive in 100% (51/51) of the allopurinol-induced SJS, TEN, or HSS patients compared to 15% (20/135) allopurinol-tolerant controls (P value = 4.7×10^{-24}, OR = 580.3, 95% CI = 34.4–9780.9) and in 20% (19/93) of the population controls (P value = 8.1×10^{-18}, OR = 393.5, 95% CI = 23.2–6665.26; see Hung et al., 2005). Unlike HLA-B*1502, this allele was found to associate with allopurinol-induced SJS and TEN in Caucasian as well as Japanese populations.

Warfarin, a commonly prescribed anticoagulant, exhibited large interindividual and interethnic differences in the dose required for its anticoagulation effect. This makes prescribing the correct dose difficult. Incorrect warfarin dose might cause adverse events such as bleeding, the most serious complication of warfarin treatment. A functional promoter polymorphism of vitamin K epoxide reductase complex subunit 1 (VKORC1 –1639A>G) was found to be associated with interindividual and interethnic differences in warfarin dosage, and can be used to guide warfarin treatment (Yuan et al., 2005). Using our data, along with data provided by other research groups, the U.S. FDA changed the label for warfarin by including genetic information and recommended genetic testing.

Clinical trials have been performed for the abovementioned projects to assess the usefulness of DNA biomarkers in preventing adverse drug reaction (Wen et al., 2008). Because of these trials, infrastructures for genetic testing have been established in all major medical centers. These tests also have been licensed to local biotech companies. The government has started to pay for genetic testing, as well. The HLA-B*1502 test was recently picked up by the Bureau of National Health Insurance in Taiwan. Thus, these studies have established Taiwan as the leader in personalized medicine.

Genome-Wide Association Studies on Complex Diseases

It can be said that the latter half of the first decade in the twenty-first century is all about genome-wide association studies (GWAS). There is at least one GWAS study for every major disease. However, most of the GWAS were performed on Caucasian populations; findings from these studies may not be applicable to the Han Chinese population. In order to identify genetic determinants for complex diseases, researchers in Taiwan also carried out several GWAS studies.

Type 2 diabetes is a chronic metabolic disorder caused by inadequate response to insulin by the body. Both environmental factors, such as diet and exercise, and genetic factors affect the age of disease onset and symptom progress. Type 2 diabetes currently affects approximately 360 million people, or about 6% of the world population; by 2025 this figure is expected to rise to around 25%.

Researchers in Taiwan, led by NRPGM's National Genotyping Center, analyzed genome-wide data of 2798 patients with type 2 diabetes using high-density genotyping technology. They identified two genes, named protein *tyrosin phosphatase receptor* (PTPRD, P = 8.54610210; odds ratio [OR] = 1.57; 95% confidence interval [CI] = 1.36–1.82) and *serine racemase* (SRR, P = 3.0661029; OR = 1.28; 95% CI = 1.18–1.39) that were associated with type 2 diabetes in ethnic Chinese (Tsai et al., 2010). Variation in PTPRD may affect the body's resistance to insulin and, therefore, influence the body's blood glucose metabolism. SRR gene variations may alter glutamate signaling in the pancreas, thus regulating insulin and/or glucagon secretion. The study also confirmed that the KCNQ1 gene, which was previously identified in a study of type 2 diabetes in Japanese, also plays an important role in the pathogenesis of type 2 diabetes in Han Chinese.

Bipolar I disorder is a mood disorder defined by the presence of recurrent episodes of abnormally elevated mood (mania). A proportion of patients who experience manic episodes also experience depressive episodes, or mixed episodes in which features of both mania and depression are present at the same time. The disease often seriously interferes with patients' personal lives and socio-occupational function, as well as their families' quality of life. Heritability of bipolar I disorder has been estimated to be around 80%, and a polygenic inheritance model is favored by scientists. However, the exact cause of the disease remains to be elucidated. Bipolar I disorder currently affects approximately 1% of the world population, and has been a major public health concern.

For this GWAS study, Taiwanese researchers analyzed the genes of 1409 patients with bipolar I disorder and 1000 normal individuals using high-density genotyping technology. They identified four genes, named SP8, ST8SIA2, CACNB2 and KCTD12, that were associated with bipolar I disorder in the Han Chinese population (Lee et al., 2011). Three of them (SP8, CACNB2 and KCTD12) were identified for the first time. Both SP8 and ST8SIA2 are involved in the development of the brain, and thus strongly support the neurodevelopmental model for the pathogenesis of bipolar I disorder. This study also increases evidence supporting a common mechanism for bipolar I disorder and schizophrenia, as ST8SIA2 has previously been identified to be associated with both disorders. CACNB2 and KCTD12 are genes regulating calcium and potassium ion channels. In addition, this study also replicated another gene, ANK3, that regulates sodium ion channels and was previously found in Caucasian patients. All of these findings strongly suggest ion channelopathy (or ion channel disorder) in the development of bipolar I disorder. These genes may be related to the pharmacological effect of mood stabilizers, including lithium salt and some anticonvulsants.

One of the major implications of the research findings is to show that common and different disease genes exist even for the same disease in different populations. The findings from these studies could help scientists to understand more about the disease progress of type 2 diabetes and bipolar I disorder, as well as to design and develop new drugs for these diseases.

Bioinformatics

In the post-genomic era, genetic studies usually involve hundreds, if not thousands of samples, especially for genome-wide association studies (GWAS). It is virtually impossible to record clinical information, track the whereabouts of biological samples, and store and analyze large amounts of data of with pen and paper. Even data recorded with commonly used software such as Excel or Access are plagued with errors and inconsistencies, as they cannot reflect the real-time movements of the samples collected and record any changes to the database; thus, samples can often get lost and the recorded data could be altered without anyone knowing. In addition, one of the big problems with Excel or Access is that it these programs are unable to perform realtime data cleaning for illogical answers and inconsistencies. Another hurdle in analyzing genetic information is the overwhelming amount of information generated that the researchers have to handle, as all the current high-density oligonucleotide chips contain more than 1 million single nucleotide polymorphisms (SNPs). The researchers might even have to work with hundreds of SNPs showing associations after statistical analysis. To help the researchers to overcome these problems, the National Genotyping Center has established an array of bioinformatics tools that cover sample collection (SamPortal), clinical and phenotype information ascertainment (iQuestion), and data analysis (GenePipe, http://genepipe.ncgm.sinica.edu.tw/). GenePipe (Tsai et al. 2007; Chen et al., 2008; Liu et al., 2008) is a collection of tools that will help users to determine the location of the SNPs with rsIDs, list any known genes located in the regions of the SNPs, and extract gene information, location, structure, function, tissue specificity, disease, subcellular location, pathway, and oncology associated with these genes. This will save researchers' time significantly, as they do not need to open multiple windows just to extract the above information; all that is required is a simple mouse click. It also allow users to design sequencing primers simply by enter the name of the gene, or even chromosome regions simply by entering the physical position, and check the quality of the designed primers with built-in UCSC in-silico PCR and Blast. These programs are freely open to all the researchers and have helped many of the research accomplishments described in this chapter. The National genotyping Center also created Totibase (http://totibase.ngc.sinica.edu.tw/totibase/index_en.html#), a laboratory resource management system which allows researchers to inventory all their laboratory items, such as chemicals, plasmids, and bacterial strains.

International Collaboration

With many diseases exhibiting ethnic differences, it is important that each population conducts its own study to identify genetic associations with diseases unique to each population. However, it is also important to have international collaboration, as it might identify genetic associations common to all the populations. Thus, several major research studies also participated in major international collaborations.

As described above, the National Genotyping Center identified polymorphisms within the VKORC1 gene that play an important role in determining warfarin dosage. This association was also found to be true in all the major populations examined by various research groups throughout the world. Thus, 21 research groups from nine countries joined together to form the International Warfarin Pharmacogenetics Consortium (IWPC) to work collaboratively to further understanding of the pharmacogenetics of warfarin. Using the collected data, which consisted of anonymous data from about 5,700 people across the globe that were taking stable doses of warfarin, researchers were able to develop a pharmacogenetic dosing algorithm for the prediction of warfarin dose that is outlined in the new article. This algorithm would improve warfarin dosage prediction in the major races (Klein et al., 2009).

The study revealed that when the predictions of ideal dose included genetic information, they were more accurate, especially for patients at the low or high ends of the dosing range. This is significant, as nearly half of those who take warfarin are at the extremes of the range, and these people are usually at the greatest risk for excessive bleeding or clotting. By quickly optimizing dosages for these patients, doctors can minimize dangerous complications and improve the effectiveness of warfarin treatment.

The National Genotyping Center, with over 90 scientists from China, India, Japan, South Korea, Singapore, and Malaysia formed the Human Genome Organization's (HUGO's) Pan-Asian Single Nucleotide Polymorphism (PASNP) Consortium and traced the genetic origins of Asian populations in 73 Southeast Asian and East Asian populations. They concluded that genetic ancestry was highly correlated with ethnic and linguistic groups, there was a clear increase in genetic diversity from northern to southern latitudes, and there was a single major inflow of human migration from Southeast Asia into other parts of Asia. These findings challenge the current predominant theory that presumes there had been multiple inflows of human migration along both southern and northern routesx (Abdulla et al., 2009).

This genetic mapping of Asia may have important implications, especially in the further understanding of human migratory patterns in human history, and the study of genetics and diseases.

Breast cancer is one of the most common cancers in women, and its risk factors strongly suggest that hormones are involved in the development of the disease. Early onset of menstrual periods, late menopause, and having no history of pregnancy have been shown to significantly increase the risk of developing breast cancer. However, only one-half of the women who develop the disease fit into these currently recognized at-risk groups. This prompted Taiwanese scientists to form the Taiwan Breast Cancer Study, together with The Breast Cancer Association Consortium (BCAC), to search for genes that may help increase understanding of the cancer. Using genome-wide scan (GWS), the scientists uncovered multiple genomic sites that may harbor breast cancer genes on chromosomes two (2q35, rs13387042, per-allele OR = 1.12, 95% confidence interval [CI] = 1.09 to 1.15; $P_{trend} = 1.0 \times 10^{-19}$), three (3p24, rs4973768: per-allele OR = 1.11, 95% CI = 1.08–1.13, $P = 4.1 \times 10^{-23}$) and seventeen (17q23.2, rs6504950: per-allele OR = 0.95, 95% CI = 0.92–0.97, $P = 1.4 \times 10^{-8}$). (See Easton et al. 2007; Ahmed et al. 2009; Milne et al. 2009; Udler et al. 2009). The discoveries could result in new anticancer drug targets and new preventive, therapeutic, and diagnostic approaches. This study is a very good example of studying a disease with multiple populations, as it allowed the comparison of genetic profiles between Taiwanese breast cancer and breast cancer in other countries and populations, and has identified the commonalities as well as differences in the patients.

THE FUTURE

Most of the important infrastructures for genomic study have been established in the first decade of the twenty-first century, and Taiwan has made use of these facilities to make several important discoveries. The next stage will focus on how to translate important findings into clinical use. Taiwan already has such experience—the pharmacogenetics of warfarin and Stevens-Johnson Syndrome are prime examples of this, and laboratory findings are already being used clinically. In addition to translational medicine, a major focus will also be on bioinformatics to anticipate the needs for next-generation sequencing. With the knowhow and the infrastructure generated from the first decade, these will ensure Taiwan will continue to contribute to genomics in the years to come.

REFERENCES

Abdulla, MA, Ahmed I, et al. (2009). Mapping human genetic diversity in Asia. *Science* 326(5959):1541–1545.

Ahmed S, Thomas G, et al. (2009). Newly discovered breast cancer susceptibility loci on 3p24 and 17q23.2. *Nat Genet* 41(5):585–590.

Bellwood P. (1997). *Prehistory of the Indo-Malaysian Archipelago*. Honolulu, Hawaii, University of Hawaii Press.

Blust R. (1999). Subgrouping, circularity and extinction: some issues in comparative Austronesian linguistics. *Symposium Series of the Institute of Linguistics, Academica Sinica*, Taipei, Taiwan.

Chen C, Sun T, Tsai H. (1997). *Taiwan Population*. Taipei, SMC Publishing Inc.

Chen YH, Liu CK, et al. (2008). GenoWatch: a disease gene mining browser for association study. *Nucleic Acids Res* 36(Web Server issue):W336–W340.

Chung WH, Hung SI, et al. (2004). Medical genetics: a marker for Stevens-Johnson syndrome. *Nature* 428(6982):486.

Diamond J, Bellwood P. (2003). Farmers and their languages: the first expansions. *Science* 300(5619):597–603.

Diamond JM. (1988). Express train to Polynesia. *Nature* 336(6197):307–308.

Easton DF, Pooley KA, et al. (2007). Genome-wide association study identifies novel breast cancer susceptibility loci. *Nature* 447(7148):1087–1093.

Gibbons, A. (2001). The peopling of the Pacific. *Science* 291(5509):1735–1737.

Gray RD, Jordan FM. (2000). Language trees support the express-train sequence of Austronesian expansion. *Nature* 405(6790):1052–1055.

Hung SI, Chung WH, et al. (2006). Genetic susceptibility to carbamazepine-induced cutaneous adverse drug reactions. *Pharmacogenet Genomics* 16(4):297–306.

Hung SI, Chung WH, et al. (2005). HLA-B*5801 allele as a genetic marker for severe cutaneous adverse reactions caused by allopurinol. *Proc Natl Acad Sci USA* 102(11):4134–4139.

Kao HJ, Cheng CF, et al. (2006). ENU mutagenesis identifies mice with cardiac fibrosis and hepatic steatosis caused by a mutation in the mitochondrial trifunctional protein beta-subunit. *Hum Mol Genet* 15(24):3569–3577.

Klein TE, Altman RB, et al. (2009). Estimation of the warfarin dose with clinical and pharmacogenetic data. *N Engl J Med* 360(8):753–764.

Lee MT, Chen CH, et al. (2011). Genome-wide association study of bipolar I disorder in the Han Chinese population. *Mol Psychiatry* 16(5):548–556.

Lin M, Chu CC, et al. (2001). The origin of Minnan and Hakka, the so-called Taiwanese, inferred by HLA study. *Tissue Antigens* 57(3):192–199.

Lin M, Chu CC, et al. (2000). Heterogeneity of Taiwan's indigenous population: possible relation to prehistoric Mongoloid dispersals. *Tissue Antigens* 55(1):1–9.

Liu CK, Chen YH, et al. (2008). Functional analysis of novel SNPs and mutations in human and mouse genomes. BMC *Bioinformatics* 9 (Suppl 12):S10.

Liu YF, Chen WM, et al. (2005). Type II collagen gene variants and inherited osteonecrosis of the femoral head. *N Engl J Med* 352(22):2294–2301.

Melton T, Clifford S, et al. (1998). Genetic evidence for the proto-Austronesian homeland in Asia: mtDNA and nuclear DNA variation in Taiwanese aboriginal tribes. *Am J Hum Genet* 63(6):1807–1823.

Melton T, Peterson R, et al. (1995). Polynesian Genetic Affinities with Southeast-Asian Populations as Identified by mtDNA Analysis. *Am J Hum Genet* 57(2):403–414.

Milne RL, Benitez J, et al. (2009). Risk of estrogen receptor-positive and -negative breast cancer and single-nucleotide polymorphism 2q35-rs13387042. *J Natl Cancer Inst* 101(14):1012–1018.

Oppenheimer SJ, Richards M. (2001). Polynesian origins – Slow boat to melanesia? *Nature* 410(6825):166–167.

Pan WH, Fann CS, et al. (2006). Han Chinese cell and genome bank in Taiwan: purpose, design and ethical considerations. *Hum Hered* 61(1):27–30.

Saleem AN, Chen YH, et al. (2010). Mice with alopecia, osteoporosis, and systemic amyloidosis due to mutation in Zdhhc13, a gene coding for palmitoyl acyltransferase. *PLoS Genet* 6(6):e1000985.

Sykes B, Leiboff A, et al. (1995). The origins of the Polynesians: an interpretation from mitochondrial lineage analysis. *Am J Hum Genet* 57(6):1463–1475.

Terrell JE, Welsch RL. (1997). Lapita and the temporal geography of prehistory. *Antiquity* 71(273):548–572.

Trejaut JA, Kivisild T, et al. (2005). Traces of archaic mitochondrial lineages persist in Austronesian-speaking Formosan populations. *PLoS Biol* 3(8):e247.

Tsai FJ, Yang CF, et al. (2010). A genome-wide association study identifies susceptibility variants for type 2 diabetes in Han Chinese. *PLoS Genet* 6(2):e1000847.

Tsai MF, Lin YJ, et al. (2007). PrimerZ: streamlined primer design for promoters, exons and human SNPs. *Nucleic Acids Res* 35(Web Server issue):W63–65.

Udler MS, Meyer KB, et al. (2009). FGFR2 variants and breast cancer risk: fine-scale mapping using African American studies and analysis of chromatin conformation. *Hum Mol Genet* 18(9):1692–1703.

Wen MS, Lee M, et al. (2008). Prospective study of warfarin dosage requirements based on CYP2C9 and VKORC1 genotypes. *Clin Pharmacol Ther* 84(1):83–89.

Yang HC, Lin CH, et al. (2006). A comparison of major histocompatibility complex SNPs in Han Chinese residing in Taiwan and Caucasians. *J Biomed Sci* 13(4):489–498.

Yuan HY, Chen JJ, et al. (2005). A novel functional VKORC1 promoter polymorphism is associated with inter-individual and inter-ethnic differences in warfarin sensitivity. *Hum Mol Genet* 14(13):1745–1751.

INDIAN SUBCONTINENT

73.

THE INDIAN SUBCONTINENT AND ITS PEOPLE

Dhavendra Kumar

INTRODUCTION

The Indian subcontinent is a vast region inhabited by heterogeneous populations. Geographically it is often referred to as "Southwest Asia." The people from this region are often collectively referred to as "South Asians." However, inappropriate use of the term *Asian* can be misleading and damaging (Bhopal et al. 1991). For example "South Asians" may be confused with "Southeast Asians," who are entirely different. The term *Asian Indian* is preferably used in North America in the sociopolitical context to refer to people originating from the Indian subcontinent (Fisher, 1996). It helps to differentiate them from other populations with the "Indian" suffix, for example, American Indian, Brazilian Indian, Mexican Indian, etc. The Indian subcontinent is one of the largest geopolitical regions of the Asian continent. The geographic and political boundaries include India, Pakistan, Bangladesh, Sri Lanka, Nepal, Bhutan, and other small islands of the Indian Ocean. There is marked cultural and ethnic diversity with considerable overlap. The admixture of various populations is evident in neighboring states stretching from Afghanistan in the northwest to Burma (Myanmar) in the northeast. The combined population of the whole region probably exceeds that of the Chinese subcontinent (www.censusindia.net).

THE PHYSICAL MAP

The Indian subcontinent is a large region from north to south and east to west. Within these horizons are a number of rivers, mountains, deserts, basins, and areas prone to natural disasters. The vast and dense Himalayan range stretches from the Hindu Kush in the northwest to the Karakoram range in the north, which descends into the vast plateau of Tibet. The vast terrain of the Himalayas covers the northern sector and extends to the east, providing a long border to the north of India, and includes the sovereign Himalayan states of Nepal and Bhutan. The highest point of Mount Everest lies between Nepal and Tibet, which is now included in China.

Below the Himalayan range of mountains lies the Shiwalik range. The mainland is divided between the plains of the Indus river in the northwest and the large plain of the River Ganges, in most of the northern and eastern part of the Indian subcontinent. The River Yamuna and other rivers in the surrounding region join the River Ganges. The River Brahmaputra serves the northeastern mountainous region and the wooded plains of Bengal. The land served by these rivers is rich in natural resources and provides the livelihood of millions of people. The central part of India includes the Great Indian Desert and a large, densely populated region. It is separated from the southern part of India, called "Deccan," by the Vindhya mountain range stretching from east to west. Two other mountainous ranges stretch along the western coastal region (Western Ghats) and the eastern coast (Eastern Ghats). These two ranges include southern Indian states. Several rivers and their tributaries serve this vast area. The region is densely populated and has a rich culture.

The political map of the subcontinent has seen many changes over the last 1000 years. The land was divided between several hundreds of small and large states or kingdoms. These were included in major empires, among them that of the Ashoka, Maurya, the Moghuls, and lastly, the British Empire. The present political map (Figure 73-1) includes the sovereign nations of Afghanistan, Pakistan, India, Nepal, Bhutan, Bangladesh, Sri Lanka, and The Maldives (Oxford School Atlas, 2002). Although the political boundaries are clearly laid out, there remains some controversy surrounding the northern states of Jammu and Kashmir between India and Pakistan. There is an ongoing border dispute between China and India over Ladakh on the northeastern side of the border. This book has not addressed these differences, but has concentrated on population heterogeneity, social-cultural variation between peoples, and genetic factors affecting health.

The Hindu Kush is the gateway to the northwest frontier region of Pakistan and Afghanistan. Historically, it has been one of the major entry routes to the Indian mainland for foreign invaders. Numerous invasions, spanning over several hundreds of years, brought people from the Middle

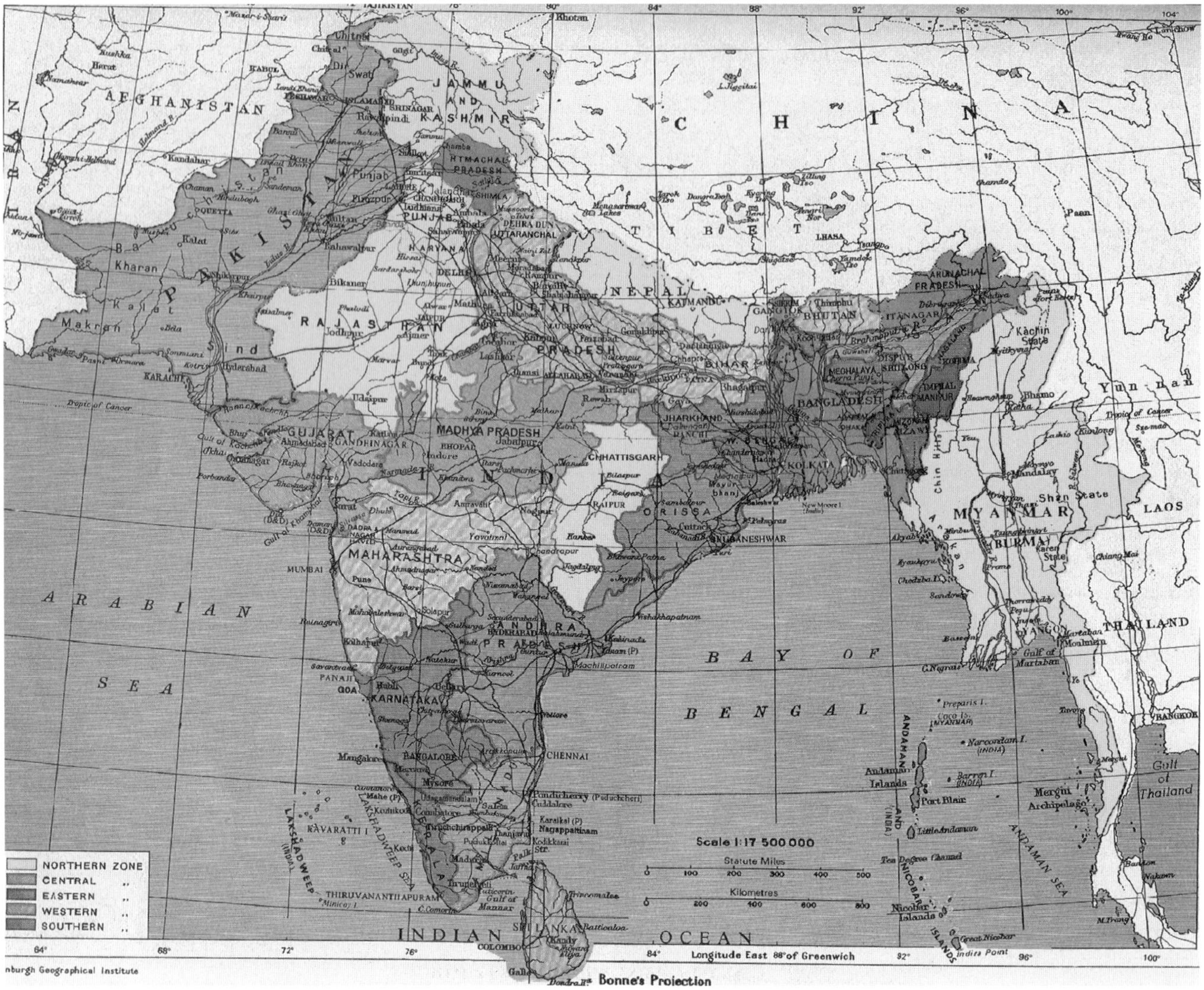

Figure 73-1 The political map of the Indian subcontinent. (With permission from Oxford School Atlas, 30th edition, Oxford University Press, India, 2002, p. 7.) Refer color figure.

East, Central Asia, and Mongolia. These peoples, from various corners of Asia and Central and Eastern Europe, laid the foundation for a large heterogeneous population. The native people of the Indian subcontinent were gradually outnumbered and now remain confined to the deep south or scattered throughout in small tribal populations. The Dravidians in South India are probably the oldest authentic, distinct ethnic group of the Indian subcontinent.

THE PEOPLE OF THE INDIAN SUBCONTINENT

The name of the subcontinent is derived from the river Indus, or *Sindhu*. This river has attracted hundreds of historians and archaeologists. The Indus Valley civilization is believed to have existed in this region about five thousand years ago (Nehru, 1946). Proof has been provided by the excavations of Harappa in the west Punjab (Pakistan) and Mohanjeo-daro in the Sind. The existence of a highly organized and developed ancient civilization has been the subject of much interest and debate. Professor Childe (1943) writes, “The Indus civilization represents a very perfect adjustment of human life to a specific environment that can only have resulted from years of patient effort. And it has endured; it is already specifically Indian and forms the basis of modern Indian culture.” It is astonishing that any culture or civilization should have continued for five thousand years and more. There is evidence that during this period there was constant and intimate contact with people from other continents, including Persians (Iranians), Egyptians, Greeks, Chinese, Arabs, Central Asians, and peoples from the Mediterranean. This resulted in a multilateral population and cultural admixture reflected in physical and cultural heterogeneity (Nehru, 1946).

The Indus Valley civilization was highly developed and must have taken thousand of years to evolve. It was a predominantly secular civilization and the religious element, though present, did not exert a strong influence. It was also clearly the precursor of later cultural periods. Among several important observations about the Indus civilization, comments of Childe (1943) provide evidence that the cities were governed by a well-planned and administrated system. This is reflected in well-planned streets and a regularly maintained, magnificent system of drains. The municipal government had town planning bylaws and a maintenance system for streets, lanes, and properties, rendered necessary by recurrent floods.

This civilization was probably not restricted to the northwestern regions. There can be little doubt that many other cities and places lie buried together with the remains of a rich ancient cultural heritage. Evidence of this civilization has been found as far apart as Kathiawar in the western Indian state of Gujarat and the Ambala district of the state of Haryana in India. There is good reason to believe that it spread to the Gangetic valley. It is thus clear that the Indus Valley civilization evolved over many millennia. This civilization is equally important to that of Persia, Mesopotamia, and Egypt.

The people of the Indus Valley had much contact with the Sumerian civilization of that period, and there is some evidence for an Indian merchant colony in Akkad. Childe (1943) adds that produce from the Indus cities reached as far as the markets on the Tigris and Euphrates. Also, produce from these places reached the Indus Valley, as shown by presence of Sumerian artifacts and Mesopotomian toilet sets. Trade was not confined to raw materials and luxury articles. Fish was regularly imported from the Arabian seacoasts, augmenting the food supplies of the Indus Valley cities. Who were these people who inhabited the Indus Valley? This question has been debated by several eminent anthropologists and population geneticists. It is possible that the whole valley came to an abrupt end due to some natural catastrophe, such as an earthquake or major floods. The Indus River is well known for catastrophic flooding, which could easily have washed away cities and villages. It is postulated that the Indus Valley culture was an indigenous culture, since its roots and offshoots may be found in southern India. Some scholars find an essential similarity between these people and the Dravidian races and culture of south India. It is likely that migrations may have taken place several thousand years before the development of the Indus Valley civilization. This is in accordance with the "out of Africa" concept of man's migration across the Arabian Sea, or by land through the Middle East (Figure 73-2), and which is believed to have taken place about 50–60 thousand years ago.

The people inhabiting the region after the end of the Indus Valley civilization are referred to as Aryans. The word *aryan* in ancient Sanskrit script means *noble*. The Aryan migrations are believed to have taken place about a thousand years after the Indus Valley period. These people probably came from Central Asia and the European areas of the Baltic Sea, the Black Sea, and the Caspian Sea. This race is also called "Indo-European" and is recognised by its Indo-European class of languages. It is now firmly believed that the first great cultural synthesis and fusion took place between the incoming Aryans and the Dravidians, who were probably the true and original representatives of the Indus Valley civilization.

Out of this synthesis and fusion grew the modern races and cultures of the Indian subcontinent, which

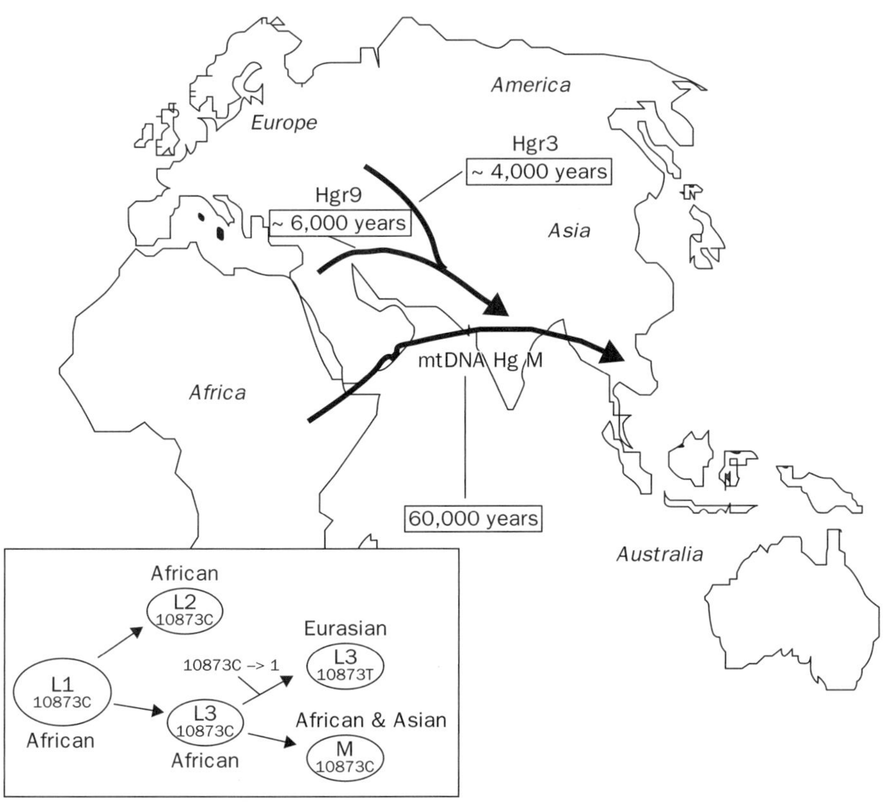

Figure 73-2 "Out-of-Africa" migration of man to the Indian subcontinent—the land route. (With kind permission from Dr. Ken McElreavey, Pasteur Institute, Paris, France.)

have distinctive elements of both races. Over the following hundreds of years, a number of other races were assimilated into the Aryan-Dravidian race. These included Iranians, Greeks, Parthians, Bactrians, Scythians, Huns, Turks (before Islam), early Christians, Jews, and Zoroastrians. They came, left their mark, and were absorbed like a drop of water in the ocean (Nehru, 1946). It is thus not surprising to find evidence of population heterogeneity in recent reports using mitochondrial DNA (mtDNA) polymorphisms and Y-chromosome linked genetic polymorphisms (McElreavey and Quintana-Murci, 2002).

Population geneticists are keen to establish the origin of the human populations inhabiting the Indian subcontinent. Population movements can be dated using inherited mutated alleles (Jobling and Tyler-Smith, 2000). Studies using mtDNA polymorphisms (Quintana-Murci et al., 1999) and Y-linked DNA polymorphisms (Quintana-Murci et al., 2001) provide evidence of the origin of various population groups through maternal and paternal lineages. There is now data available in support of major population movements having taken place from east Africa to western and southern Asia around 60,000 "years before present" [YBP]. Migrations of populations to India from the Fertile Crescent in southwestern Iran date back to 8,000 YBP, and of the Indo-European speakers from Central and West Asia to approximately 5,000 YBP. Chapter 73 of this book provides evidence for the sea route of migrations from Africa to the Indian subcontinent dating back to approximately 70,000 years.

The analysis of two Y-chromosome variants, Hgr9 and Hgr3, provides interesting data (Quintan-Murci et al., 2001). Microsatellite variation of Hgr9 among Iranians, Pakistanis, and Indians indicate an expansion of populations to around 9,000 YBP in Iran, and then to 6,000 YBP in India. This migration originated in what was historically termed *Elam* in southwest Iran, to the Indus valley, and may have been associated with the spread of Dravidian languages from southwest Iran (Quintana-Murci et al., 2001). The same study has shown the distribution of another haplogroup, termed Hgr3, among the central and eastern European populations. This marker is seen with decreasing frequency toward west Europe and east Asia.

The distribution of Hgr3 in Iran shows marked differences between the western Caspian province (3%) and the eastern provinces (31%), with a decreasing frequency cline toward India. This suggests a population movement from central Eurasia into modern Iran and, following an eastern route, into India (Figure 73-2). The estimation of Hgr3 expansion times determined from populations at different geographical locations suggests that the migration dates back to approximately 4,000 YBP, a date which coincides with the disappearance of the Harrapan culture in the Indus valley and the appearance of Indo-European languages in the Indian subcontinent (Quintan-Murci et al., 2001).

The diversity of the Indian subcontinent is tremendous, both in physical appearance and sociocultural practices. All these people are unique by way of their racial origin and traits. The Pathans of Afghanistan and Tamils of south India and Sri Lanka apparently have little in common; they differ in physical appearance, style of clothing, food habits, and language, but even so, they share common strands. There is ample evidence of a Central Asian presence in the northwestern province and in Kashmir. For several thousands of years, Afghanistan has had close links with the states and people of the Indian subcontinent, of which it geographically is a part. Historically, this part of the subcontinent enjoyed great popularity for its famous University of Taxila, which was an important place of learning about two thousand years ago.

The Pathans and Tamils are probably two extreme examples, with the majority of the people somewhere in between. All of them have their distinctive features. It is fascinating to see how the Pathans, the Sindhis, the Punjabis, the Kashmiris, the Rajputs, the Hindustani-speaking people of north and central India, the Bengalis, the Gujratis, the Tamils, the Andhras, the Oriyas, the Assamese, the Canarese, the Malyalis, and the Tamils share, and have maintained, unique, distinct, similar characteristics for hundreds of years spanning over at least five millennia. These people share a common heritage. There is something living and dynamic about this heritage, which is reflected in lifestyles and philosophical attitudes to life and its problems.

Ancient India, like ancient China and Egypt, was a world unto itself, with a culture, civilization, and heritage shaped by numerous invasions and external influences.

REFERENCES

Bhopal RS, Phillimore P, Kohli HS. (1991). Inappropriate use of the term 'Asian': an obstacle to ethnicity and health research.' *J Public Health* 13:244–246.

Gordon C. (1943). *What Happened in History*. Pelican Books, London. p. 112.

Fisher NL. Ed. (1996). Cultural and ethnic diversity: A guide for genetics professionals. Johns Hopkins University Press, Baltimore and London.

Jobling MA, Tyler-Smith C. (2000). New uses for new haplotypes of the human Y chromosome in disease and selection. *Trends Genet* 16:356–362.

McElreavey K, Quintana-Murci L (2002): Understanding inherited disease through human migrations: A south-west Asian perspective. *Community Genetics* 5:153- 156.

Nehru J (1946): '*The Discovery of India*', Oxford University Press, New Delhi, India, 1999. pp. 69–74.

Oxford School Atlas, Oxford University Press, New Delhi, India, 2002. pp. 6–7.

Quintana-Murci L, Semino O, Bandelt HJ, Passarino G, McElreavey K, Santachiara-Benerecetti AS. (1999). Genetic evidence of an early exit of *Homo sapiens sapiens* from Africa through eastern Africa. *Nature Genet* 23:437–441.

Quintana-Murci L, Krausz C, Zerjal T, Sayar SH, Hammer ME, Mehdi SQ, Ayub Q, Qamar R, Mohyddin A, Radhakrishna U, Jobling MA, Tyler-Smith C, McElreavey K. (2001). Y-chromosome lineages trace diffusion of people and languages in southwestern Asia. *Am J Hum Genet* 68:537–542.

74.

GENOMIC PERSPECTIVES OF PEOPLING AND LANGUAGES OF THE INDIAN SUBCONTINENT

Ganesh Prasad Arunkumar, Varatharajan Santhakumari Arun, Adhikarla Syama, Valampuri John Mary Selvam Kavitha, and Ramasamy Pitchappan

INTRODUCTION

"Genomics," the study of the genome of a species, is the buzz word of the twenty-first century, thanks to the Human Genome project that ushered in this new era of biology (Baltimore, 2001). The genomic tools are simple, straightforward and more accurate, thus making biology a more exact science, similar to physics and chemistry. Nonetheless, the purpose of the studies and the design of the experiments become more critical in such a venture due to the enormous diversity and complexity of the biological phenomena that operate in evolutionary processes. The Indian subcontinent, the second successful home of humankind, is special in this evolutionary process due to her population's long history, many migrations, isolation, divergence, and cultural evolution since the first emigration of man (Wells et al., 2001). The impact of natural selection that has operated on these disparate gene pools in an alien environment is a matter of intense scrutiny, since no parallel for such longstanding and sympatrically isolated populations exists in other parts of the world, apart from the birthplace of mankind in Africa. Many of these isolations seem to have occurred prior to language developments. The geographical subsistence and cultural isolations presumably lead to different language developments in various parts of India. We attempt here to interpret the Non Recombinant Y (NRY) chromosome polymorphisms of India in the context of migrations and origin of languages.

In 1901, Karl Landsteiner, the discoverer of the human ABO blood group and Nobel laureate, first provided direct evidence for the existence of genomic diversity in human populations. In 1919, Hirszfeld and Hirszfeld found ABO gene variations among human populations. The B blood group was unique and most prevalent in South Asia, particularly in southern Indian tribal populations (Cavalli-Sforza et al., 1994). During the 1950s and 1960s, more systemic analysis of variation in genes and proteins became possible with the detection by Pauling et al. (1949) of blood protein polymorphisms in hemoglobin. The 1980s were a transitional period from the analysis of gene polymorphisms to protein polymorphism (Sanghvi et al., 1981), to the studies of DNA sequence polymorphisms in the form of Human Genome and other variome projects (Baltimore, 2001). To better understand the origin of this genomic diversity, one may need to study population-level forces such as migration and miscegenation, which play major roles in creating diversity. It has been proposed that genomic differentiation in populations is mostly due to "fission" followed by independent evolution (Cavalli-Sforza, 1997). Mutations, natural selection, and drift play important roles in deciphering diversity at the population level. While mutations supply raw material for genomic diversity by introducing new alleles, their survival and expansion is dependent on their fitness and functional importance. The study of polymorphism at the single nucleotide (SNP) level in introns (noncoding region) or exons (coding region), or at the microsatellite level, becomes a powerful tool in studying genomic diversity in both health and disease. Recent literature on NRY chromosomes makes them ideal candidates to study population diversity. NRY is evolutionarily a neutral marker, thus permitting us to reconstruct a population's history. The distribution of NRY variations of various linguistic states of India becomes more interesting. The analyses throw better light on the population migrations, language development, and its spread.

RECENT AFRICAN ORIGINS

The recent African origin and spread of anatomically modern humans suggested that *Homo sapiens sapiens*, our species, evolved from a small African population that had subsequently colonized the whole world, supplanting former hominids, ~120–200 thousand years ago (kya) around the time of the first appearance of anatomically modern humans (Cann et al., 1987). This replacement model, now widely accepted, has been later called the "Out of Africa," or "Recent African Origin" (RAO) model, in contrast to the earlier Multi-Regional Evolution Model (MRE; see Wolpoff et al., 1984). Molecular evidence favors the RAO model. Older populations evolved for longer must have had more time to accumulate genomic diversity. The excess African diversity can thus be explained by older onset of population demographic expansion in Africa, combined with higher effective population size, population size fluctuations, and also periodic extinctions of populations outside Africa or positive selection through adaptation to new environments outside Africa (Eller, 2001; Aquadro et al., 2001). The non-African patterns of genetic variation are indeed a subset of African ones. Microsatellite studies also showed a gradual reduction of diversity with increasing distance from Africa, and linkage disequilibrium values, which reflect the lower ages of haplotypes in non-African populations (Tishkoff et al., 1996). The Indian subcontinent, the second to be occupied by man, thus attracts our attention to investigate further in these directions. The RAO model proposes one, two, or multiple migrations using various routes over a period of time. Two routes have been proposed: the first is the "northern route" over Sinai, leading to eastern Asia through the steppes of central Asia and southern Siberia, and the second is the "southern route" over southern Arabia, followed by migration along the coastline of India. While the northern route model could explain the peopling of the whole of Eurasia by a single migration from Africa, the southern route model is interpreted as implying at least two separate late Pleistocene dispersal events, one leading to the northwest and the other to the east of Eurasia (Cavalli-Sforza et al., 1994).

INDIAN CORRIDOR

Being positioned at the tri-junction of African, northern Eurasian and oriental realms, India has served as a major corridor for the dispersal of modern humans (Cann, 2001) and attracted many streams of people since the Paleolithic, starting with the Late Pleistocene as supported by archaeological evidence (Paddaya, 1982; Misra, 2001; Petraglia et al , 2010). Though the modern anthropology tends to reject the somatoscopic and anthropological measurements, there is a revival of interest in deciphering skin color genes and studying their genome with modern tools (Yuldasheva et al., 2002). Sanghvi and Karve, distinguishing various castes of Tamil Nadu, India, have deciphered that nose shape and skin color are the most discriminative (Sanghvi et al., 1981). Even today, the Indian physical anthropologists consider these ancient classifications, identifying four different morphological groups in India (Bhasin, 2006). These are: (i) "negritos," characterized by dwarf stature and frizzy hair, who are common in Nilgiri hills of Tamil Nadu (Paniya, Irula, and Kadar tribes) and the Andaman islands: we see them nowadays in many caste populations, including Brahmins; (ii) "proto-Austroloids," characterized by long head, dark skin, and broad nose, found in central and southern India and speaking Dravidian languages/dialects; (iii) "Mongoloids," characterized by broad face, medium stature, yellow skin, and slightly obliquely set eyes, exclusively found in sub-Himalayan and northeastern regions, speaking Austro-Asiatic (AA) or Tibeto-Burman (TB) languages; and (iv) "Dinaric" type (Mediterranean element) with medium to light pigment, hook nose, acrocephalic and round heads, found in Bengal and Orissa. The "Caucasoids" or the "Nordic", with blond hair and long heads and speaking Indo-European (IE) languages is most common in the north and northwestern regions of India. The four major language families of India seem to have their own non-overlapping geographic clines. It will be interesting to compare the distribution of the NRY markers and the origin of these languages, and answer whether these could have arisen through fission and a long process of isolation in various regions of India.

NRY PHYLOGENY IN INDIA

AFRICAN ROOT

The roots of Y phylogeny roots in Africa have been dated around 100 kya (Underhill, 2003), characterized by HG A-M91 and HG B-M60 NRY–SNP haplogroups (HGs), and restricted to Africa. These migrations and subsequent mutations formed the scaffold on which all other Y- chromosome diversification with geographical cline has occurred. The majority of Y lineages across the globe are composed of a tripartite assemblage consisting of (1) HG C-M130, (2) HG D-M174 and HG E-M96, and (3) overarching HG F-M89, which defines the internal node of all remaining HGs, G-M201 through R-M207 (Underhill et al., 2001; Wells, 2007).

OUT OF AFRICA EMIGRATIONS

The HG C-M130, not seen in any African populations presumably originated somewhere in Asia on an M168 lineage sometime after an early departure event (Capelli et al., 2001; Underhill et al., 2001; Table 74-1). This clade (C-M130) characterizes the first migrants into India: the descendants we could identify near Madurai (Wells et al., 2001). This clade has many sublineages displaying irregular geographic patterning consistent with diversification and northward migration of this HG C-M130, since the last ice age, with the westernmost limit in India: thus HG C3-M217, a transversion mutation, is common in East Asia and Siberia, with representatives in North America (Karafet et al., 2001; Lell et al., 2002), and eastern and central parts of central Asia, while C5-M356 is common in India (Sengupta et al., 2006). This lineage is absent in Indonesia, Oceania (Kayser et al., 2000), and Yunnan, China (Karafet et al., 2001).

The ancestors who accumulated HG D-M174 and HG E-M96 mutations could have arisen in Africa or Asia (Underhill, 2003). HG E-M96 lineages are the most frequent in Africa, and display subsequent binary and microsatellite diversification. Conversely, Asian haplogroup D-M174 occurs at low frequencies throughout eastern Asia, except in remote and isolated locations like Tibet, Japan, and the Andaman islands (Underhill et al., 2001; Thangaraj et al., 2003).

The third major and most successful subclade of M168 lineages, characterized by super-haplogroup F-M89, defines the root from which all others (HGs G-M201 through R-M207) originated and have evolved outside Africa (Kivisild et al., 2003). HG F-M89 diversified into many branches with region-specific markers—the Middle East showing HGs G-M201 and J-M304, Europe with HG I-M170, and India with F-M89 and H-M69 lineages, seldom observed elsewhere (Table 74-1).

EXPANSION IN INDIA

HG F*-M89* is the most paraphyletic subcluster (unclassified derivative) of M168 lineages, ubiquitous but found with lesser frequency in various parts of India. Many tribal populations of southern India possess higher frequencies of F*-M89* with high STR variance, particularly Dravidian speaking groups of Tamil Nadu and Koya of Orissa (Kavitha, 2008; Wells et al., 2001; Kivisild et al., 2003; Cordaux et al., 2004a; Table 74-2). The high STR variance of HG F*-M89* from Tamil Nadu and Andhra Pradesh has suggested a deep time depth of 45,000 YBP (Sengupta et al., 2006).

TABLE 74-1 THE AGES OF THE NRY HAPLOGROUPS AND THEIR DISCRIMINATING ALLELES PREVALENT IN INDIA AND NEARBY REGIONS.

NRY HG	Marker	Estimated Age of the mutation YBP*	Distribution	Reference
C	M130	50,000	India, Australia, Central Asia America,	Genographic#
F	M89	45,000	India	Genographic#
G	M201	30,000		Genographic#
H	M69	20,000–30,000		Genographic#
H1	M52	25,000	India	Genographic#
J	M304	31,700	Middle East	Semino et al. (2004)
J2	M172	15,000–20,000		Hammer et al. (2000)
K	M9	40,000		Genographic#
L	M20	30,000	India	Genographic#
L1	M27/M76	9100		Sengupta et al. (2006)
O	M175	35,000	Orissa, North East, South East Asia,	Genographic#
O2a	M95	11,700	Orissa	Sengupta et al. (2006)
O3	M122	10,000	North East, South East Asia, China	Genographic#
P	M45	40,000	North Asia	Wells et al. (2001)
Q	M242	15,000–18,000		Seielstad et al. (2003)
R	M207	30,000		Genographic#
R2	M124	25,000	India	Genographic#
R1a1	M17	15,000	Caucus, Europe, India	Wells et al. (2001)

*The estimated ages have all been determined based on the available data: if it was not a representative sampling, then the age may vary. Ascertainment bias is possible due to smaller samples and sampling errors; hence, some of these ages may need to be considered with caution.

#www.nationalgeographic.com/genographic website.

TABLE 74-2 NRY HG ALLELE DISTRIBUTION DATA USED FOR COMPUTING PAN-ASIAN PCA

Serial No	Population	Country / Region	Province / State	Language families	Population Codes in Pan-Asian PCA, Fig 2	Sample size	C-M130	D-M174	E*-M96	G-M201	I-M170	F*-M89
1	Pathan	Pakistan		IE	Pak7	21	4.762	0	0	9.524	0	4.762
2	Sindhi	Pakistan		IE	Pak8	21	0	0	0	0	0	4.762
3	Hazara	Pakistan		IE	Pak4	25	40	0	0	0	4	0
4	Kalash	Pakistan		IE	Pak5	20	0	0	0	20	0	0
5	Makrani	Pakistan		IE	Pak6	20	0	0	5	0	0	0
6	Konka Brahmin	India West	Goa	IE	Goa1	43	2.326	0	0	0	0	4.651
7	Gujarat	India West	Gujarat	IE	Guj3	29	17.24	0	0	0	0	3.448
8	Gujarat Brahmin	India West	Gujarat	IE	Guj1	64	3.125	0	3.125	10.94	0	0
9	Bhils	India West	Gujarat	IE	Guj2	22	9.091	0	0	0	0	18.18
10	Desasath Brahmin	India West	Maharashtra	IE	Mah2	16	6.25	0	0	0	0	0
11	Kathari	India West	Maharashtra	IE	Mah3	19	0	0	0	0	0	26.32
12	Maratha	India West	Maharashtra	IE	Mah5	36	5.556	0	0	0	0	5.556
13	Punjab	India West	Punjab	IE	Pun2	66	3.03	0	0	0	0	1.515
14	Punjab Brahmin	India North	Punjab	IE	Pun1	49	4.082	0	0	4.082	0	4.082
15	Kashmir Gujars	India North	Jammu Kashmir	IE	J&K1	49	2.041	0	0	0	0	4.082
16	Kashmiri Pandits	India North	Jammu Kashmir	IE	J&K2	51	1.961	0	0	1.961	0	3.922
17	Rajput	India North	Rajasthan	IE	Raj1	29	3.448	0	0	0	0	10.34
18	Uttar Pradesh Brahmin	India North	Uttar Pradesh	IE	Uprd	31	0	0	0	0	0	0
19	Bihar Brahmins	India Central	Bihar	IE	Bih1	56	1.786	0	0	0	0	0
20	Madhya Pradesh Brahmins	India Central	Madhya Pradesh	IE	MP1	42	0	0	0	0	0	2.381
21	Halba	India Central	Maharashtra	IE	Mah4	21	0	0	0	0	0	23.81
22	Karan	India Central	Orissa	IE	Ori1	18	0	0	0	0	0	0
23	Oriya Brahmin	India Central	Orissa	IE	Ori2	24	0	0	0	0	0	4.167

TABLE 74-2 (CONTINUED)

H1*-M52	J2-M172	L-M20/M11	K-M9	N-M231	O*-M175	O2a-M95	O3-M122	P-M74/ M45	Q*-M242/ P36	R*-M207	R1a1-M17	R2-M124	Reference
9.524	0	9.524	0	0	0	0	0	0	9.524	14.29	38.1	0	Sengupta et al., 2006
0	28.57	4.762	0	0	0	0	0	0	4.762	0	52.38	4.762	Sengupta et al., 2006
0	4	0	0	0	0	0	8	0	8	32	0	4	Sengupta et al., 2006
20	10	25	0	0	0	0	0	0	0	5	20	0	Sengupta et al., 2006
0	25	20	0	0	0	0	0	0	5	10	25	10	Sengupta et al., 2006
6.977	13.95	18.6	2.326	0	0	0	0	0	0	0	41.86	9.302	Kivisild et al., 2003
10.34	20.69	10.34	3.448	0	0	0	0	6.897	0	0	24.14	3.448	Kivisild et al., 2003
1.563	15.63	7.813	3.125	3.125	0	0	0	0	0	9.375	32.81	9.375	Sharma et al., 2009
9.091	18.18	18.18	0	0	0	0	0	0	0	0	9.091	18.18	Sharma et al., 2009
18.75	12.5	12.5	0	0	0	0	0	0	0	0	43.75	6.25	Sahoo et al., 2006
36.84	5.263	5.263	0	0	0	0	0	5.263	0	5.263	15.79	0	Sahoo et al., 2006
30.56	19.44	11.11	0	0	0	0	0	0	0	0	13.89	13.89	Sengupta et al., 2006, Sahoo et al., 2006
3.03	21.21	12.12	0	0	0	0	0	7.576	0	0	46.97	4.545	Kivisild et al., 2003
0	22.45	6.122	0	0	0	0	0	0	0	0	34.69	24.49	Sharma et al., 2009
10.2	6.122	16.33	8.163	0	0	0	0	0	2.041	2.041	40.82	8.163	Sharma et al., 2009
9.804	9.804	5.882	9.804	0	0	0	0	0	5.882	17.65	19.61	13.73	Sharma et al., 2009
17.24	13.79	6.897	0	0	0	0	3.448	0	0	0	31.03	13.79	Sengupta et al., 2006
16.13	3.226	3.226	0	0	0	0	0	0	6.452	0	67.74	3.226	Sharma et al., 2009
0	8.929	8.929	3.571	0	0	0	0	0	3.571	3.571	64.29	5.357	Sahoo et al., 2006, Sharma et al 2009
7.143	23.81	7.143	0	2.381	0	0	0	2.381	4.762	0	38.1	11.9	Sharma et al., 2009
23.81	0	0	0	0	0	28.57	0	0	4.762	0	19.05	0	Sengupta et al., 2006
16.67	5.556	0	0	0	0	0	0	0	0	0	55.56	22.22	Sahoo et al., 2006
8.333	4.167	20.83	0	0	0	0	0	4.167	0	4.167	41.67	12.5	Sahoo et al., 2006

(Continued)

TABLE 74-2 (CONTINUED)

Serial No	Population	Country / Region	Province / State	Language families	Population Codes in Pan-Asian PCA, Fig 2	Sample size	C-M130	D-M174	E*-M96	G-M201	I-M170	F*-M89
24	Lambadi	India South	Andhra Pradesh	IE	And9	53	11.32	0	0	0	0	3.774
25	W.Bengal	India East	West Bengal	IE	WB1	31	3.226	0	0	3.226	0	6.452
26	Karmali	India East	West Bengal	IE	WB4	16	0	0	0	0	0	0
27	Kora	India East	West Bengal	IE	WB5	17	0	0	0	0	0	17.65
28	WB. Brahmin	India East	West Bengal	IE	WB7	49	0	0	0	0	0	0
29	Garo	India East	Meghalaya	TB	NE2	33	0	0	0	0	0	18.18
30	Jamatia	India East	West Bengal	TB	NE3	30	0	0	0	0	0	0
31	Korku	India West	Maharashtra	AA	Mah1	59	0	0	0	0	0	15.25
32	Asur	India Central	Jharkand	AA	Jha1	55	0	0	0	0	0	25.45
33	Birjia	India Central	Jharkand	AA	Jha2	24	0	0	0	0	0	0
34	Korwa	India Central	Jharkand	AA	Jha3	42	0	0	0	0	0	33.33
35	Savar	India Central	Jharkand	AA	Jha4	47	0	0	0	0	0	40.43
36	Kharia	India Central	Jharkand	AA	Jha5	46	2.174	0	0	0	0	39.13
37	Munda	India Central	Jharkand	AA	Jha6	60	0	0	0	0	0	23.33
38	Juang	India Central	Orissa	AA	Ori3	59	0	0	0	0	0	1.695
39	Ho	India Central	Orissa	AA	Ori5	116	0	0	0	0	0	22.41
40	Mahali	India Central	West Bengal	AA	WB3	38	0	0	0	0	0	39.47
41	Khasi	India East	Meghalaya	AA	NE1	92	0	0	0	0	0	17.39
42	Mudi	India East	West Bengal	AA	WB2	37	0	0	0	0	0	45.95
43	Lodha	India East	West Bengal	AA	WB6	71	1.408	0	0	0	0	14.08

TABLE 74-2 (CONTINUED)

H1*-M52	J2-M172	L-M20/M11	K-M9	N-M231	O*-M175	O2a-M95	O3-M122	P-M74/ M45	Q*-M242/ P36	R*-M207	R1a1-M17	R2-M124	Reference
5.66	3.774	11.32	3.774	0	0	0	0	33.96	0	11.32	13.21	1.887	Sahoo et al., 2006, Kivisild et al., 2003
9.677	6.452	0	0	0	3.226	0	0	6.452	0	0	38.71	22.58	Kivisild et al., 2003
0	0	0	0	0	0	0	0	0	0	0	0	100	Sahoo et al., 2006
47.06	0	0	0	0	0	29.41	0	0	0	0	5.882	0	Sahoo et al., 2006
6.122	0	0	0	0	0	0	0	0	0	0	71.43	22.45	Sharma et al., 2009, Sengupta et al., 2006
0	0	0	0	0	3.03	18.18	54.55	0	0	6.061	0	0	Kumar et al., 2007
3.333	0	0	0	0	0	6.667	76.67	0	0	0	6.667	6.667	Sengupta et al., 2006
0	0	0	0	0	0	81.36	1.695	1.695	0	0	0	0	Kumar et al., 2007
0	9.091	0	0	0	0	63.64	0	0	0	1.818	0	0	Kumar et al., 2007
0	0	0	0	0	0	95.83	0	4.167	0	0	0	0	Kumar et al., 2007
0	2.381	0	0	0	0	59.52	0	0	0	0	0	4.762	Kumar et al., 2007
0	12.77	0	0	0	0	14.89	0	0	0	31.91	0	0	Kumar et al., 2007
2.174	2.174	0	0	0	2.174	45.65	0	0	0	6.522	0	0	Kumar et al., 2007, Sahoo et al., 2006
0	0	0	1.667	0	0	50	0	6.667	0	11.67	0	6.667	Kumar et al., 2007, Sahoo et al., 2006
0	0	0	0	0	0	98.31	0	0	0	0	0	0	Sahoo et al., 2006, Kumar et al., 2007
0	0.862	0	0	0	0	71.55	0	2.586	0	0	0	2.586	Sengupta et al., 2006, Kumar et al., 2007, Sahoo et al., 2006
13.16	5.263	0	5.263	0	0	7.895	0	0	0	15.79	0	13.16	Kumar et al., 2007, Sahoo et al., 2006
0	0	0	0	0	2.174	41.3	29.35	4.348	0	5.435	0	0	Kumar et al., 2007
0	2.703	0	0	0	0	43.24	0	2.703	0	2.703	0	2.703	Kumar et al., 2007
5.634	30.99	0	2.817	0	0	5.634	0	0	0	0	1.408	38.03	Sengupta et al., 2006, Kumar et al., 2007

(Continued)

TABLE 74-2 (CONTINUED)

Serial No	Population	Country / Region	Province / State	Language families	Population Codes in Pan-Asian PCA, Fig 2	Sample size	C-M130	D-M174	E*-M96	G-M201	I-M170	F*-M89
44	Oraon	India Central	Jharkand	DR	Jha7	100	0	0	0	0	0	54
45	Muria	India Central	Orissa	DR	Ori4	20	0	0	0	0	0	10
46	Koraga	India South	Andhra Pradesh	DR	And1	33	0	6.061	0	0	0	0
47	Koya	India South	Andhra Pradesh	DR	And2	41	0	0	0	0	0	36.59
48	Yerava	India South	Andhra Pradesh	DR	And3	41	26.83	0	0	0	0	43.9
49	Kappu naidu	India South	Andhra Pradesh	DR	And4	18	0	0	0	5.556	0	0
50	Komati	India South	Andhra Pradesh	DR	And5	20	0	0	0	5	0	10
51	Naikpod Gond	India South	Andhra Pradesh	DR	And6	18	22.22	0	0	0	0	11.11
52	Raju	India South	Andhra Pradesh	DR	And7	19	0	0	0	0	0	0
53	Yerkual	India South	Andhra Pradesh	DR	And8	18	0	0	0	0	0	0
54	Konda Reddy	India South	Andhra Pradesh	DR	And10	30	0	0	0	0	0	23.33
55	Koya Dora	India South	Andhra Pradesh	DR	And11	27	0	0	0	0	0	25.93
56	Andh	India South	Andhra Pradesh	DR	And12	54	1.852	0	0	0	0	3.704
57	Iyer	India South	Tamil Nadu	DR	TN1	29	6.897	0	0	10.34	0	3.448
58	Kurumba	India South	Tamil Nadu	DR	TN2	19	0	0	0	0	0	15.79
59	Iyengar	India South	Tamil Nadu	DR	TN3	47	0	0	0	8.511	0	0
60	Irula	India South	Tamil Nadu	DR	TN4	40	5	0	0	0	0	42.5
61	Pallan	India South	Tamil Nadu	DR	TN5	44	2.273	0	0	0	0	6.818
62	Kallar	India South	Tamil Nadu	DR	TN6	93	6.452	0	0	0	0	16.13
63	Sinhalese	SriLanka	SriLanka	DR	SL1	39	0	0	0	0	0	12.82
64	Burushaki	Pakistan		unclassified	Pak3	20	5	0	0	5	0	0

Abbreviations of language families: IE=Indo European, AF=Afro-Asiatic, DR=Dravidian, TB=Tibeto-Burman, ST=Sino-Tibetan, AA=Austro-Asiatic, AT=Altaic, BR=Brusaki.

H1*-M52	J2-M172	L-M20/M11	K-M9	N-M231	O*-M175	O2a-M95	O3-M122	P-M74/ M45	Q*-M242/ P36	R*-M207	R1a1-M17	R2-M124	Reference
3	0	0	0	0	0	35	0	1	0	4	0	3	Kumar et al., 2007, Sahoo et al., 2006
80	0	0	0	0	0	10	0	0	0	0	0	0	Sengupta et al., 2006
87.88	0	0	0	0	0	0	0	0	0	0	0	6.061	Cordaux et al., 2004a
60.98	0	0	0	0	0	0	0	0	0	0	2.439	0	Kivisild et al., 2003
19.51	0	0	0	0	0	0	0	0	0	0	9.756	0	Cordaux et al.,2004a
0	0	0	0	0	0	0	0	0	0	11.11	11.11	72.22	Sahoo et al., 2006
0	0	0	0	0	0	0	0	0	0	0	15	70	Sahoo et al., 2006
61.11	0	5.556	0	0	0	0	0	0	0	0	0	0	Sahoo et al., 2006
0	10.53	21.05	15.79	0	0	0	0	0	0	15.79	26.32	10.53	Sahoo et al., 2006
0	0	11.11	55.56	0	0	0	0	0	0	0	33.33	0	Sahoo et al., 2006
3.333	0	0	0	0	0	66.67	0	0	0	0	6.667	0	Sengupta et al., 2006
22.22	3.704	0	0	0	0	48.15	0	0	0	0	0	0	Sengupta et al., 2006
16.67	35.19	1.852	0	0	0	1.852	1.852	0	0	0	31.48	5.556	Thanseem et al., 2006
3.448	17.24	17.24	0	0	0	0	0	0	0	3.448	27.59	10.34	Sengupta et al., 2006
68.42	0	5.263	0	0	0	0	0	0	0	0	0	10.53	Sengupta et al., 2006
23.4	19.15	19.15	0	0	0	0	0	0	0	0	23.4	6.383	Sengupta et al., 2006, Sahoo et al., 2006
35	2.5	7.5	0	0	0	0	0	0	0	0	0	7.5	Sengupta et al., 2006, Sahoo et al., 2006
29.55	9.091	11.36	9.091	0	0	0	0	0	0	2.273	15.91	13.64	Sengupta et al., 2006, Sahoo et al., 2006
18.28	1.075	44.09	0	0	0	1.075	0	1.075	0	0	3.226	8.602	Wells et al., 2001, Sahoo et al., 2006
7.692	10.26	17.95	0	0	0	0	0	0	0	0	12.82	38.46	Kivisild et al., 2003
15	5	15	5	0	0	0	5	0	0	30	0	15	Sengupta et al., 2006

NRY HG H1-M52, a derivative of M69, has been reported in higher frequencies in southern India, cutting across the caste and tribal boundaries (Wells et al., 2001; Kivisild et al., 2003). A few other studies have found HG H1-M52 with high STR variance in Maharashtra (Sengupta et al., 2006) and western India (Trivedi et al., 2008). Thanseem et al. (2006) have suggested that M52 originated in the Indian subcontinent immediately after the Late Pleistocene settlements. The available samples in literature show an estimated age of 25,000 years (Table 74-1).

NEOLITHIC CATTLE KEEPERS

The J-M172 clade implicated in agricultural expansion through Neolithic cattle keepers is thought to have arisen in the Caucusus and Anatolia and spread to southwestern Europe (Cavalli-Sforza et al., 1994; Semino et al., 2004; Hammer et al., 1998; Rosser et al., 2000); Bedouin and Palestinian Arabs possess the highest frequency of this mutation (66%—55%) followed by Sephardic Jews and Muslim Kurds (40%; see Semino et al., 2004). The J HG is divided into two sub-haplogroups, J1-M267 and J2-M172, with the former showing an ancestral Y-STR haplotype 14-16-23-11-12, for loci DYS19-DYS388-DYS390-DYS392-DYS393 and the latter 14-15-23-11-12 (Giacomo et al., 2004). A one-step mutated haplotype of this J2, viz. 15-15-23-11-12, is the common clade in India: the Thodas of Nilgiris possess this J2 in higher frequencies, correlating with their pastoral buffalo cult life (Kavitha, 2008). It has been proposed that earlier migrations brought agriculture and Dravidian speakers into India, while another, much later one brought rice cultivators from Southeast Asia (Diamond and Bellwood, 2003; Fuller, 2003). HG J2-M172 has shown a high STR diversity in Dravidian tribal populations, but to hypothesize that this HGJ2-M172 is a part of a Neolithic expansion would require more evidence (Thanseem et al., 2006). The available datasets thus do not correlate well with these two major events. The archaeology once again does not support this contention.

Some studies have found HG J2-M172 at higher frequencies in Dravidian and Indo-European castes than in tribes (Sengupta et al., 2006, Cordaux et al., 2004a). It is absent in East Asia, and typically present in Central Asia at frequencies of 10%–20%, leading Cordaux et al. to interpret that Indian HG J2-M172 originated from Central Asia rather than West Asia. The data listed in Table 74-2 shows the presence of J2 in a wide variety of populations across India. Neolithic markers of early farmers—HGs E3-M35 and G-M201, that are prevalent in Europe, Anatolia, the southern Caucusus, and Iran—are, however, sporadic in Indians (Semino et al., 2000; Underhill et al., 2001).

CENTRAL ASIAN EXPANSION

An expansion of HG F-M89 lineages toward Central Asia or the Caucusus also gave rise to a founder that acquired the HG K-M9 mutation, defining another major bifurcation in the phylogeny. Distinctive HG K-M9 sublineages have been observed in India, the Middle East, and Europe, while some HG K-M9 and HG M-M186 lineages are restricted to Oceania. Three major lineages of K-M9, HG P-M45 are characteristic of North Asia, while HG Q-M242 is found in Siberia and North America, and the westward-expanding HG R-M207 in Eurasia. K-M9 has given rise to two offshoots, one HG L-M20 prevalent in the Indian subcontinent to become L1-M76 in southern India, and another HG O-M175 found in eastern Asia, the whole of oriental populations including the Chinese, and also in the Austro-Asiatic speakers and Tibeto-Burmese speakers of India. The genomic evidence further supports this (HUGO Pan-Asian SNP Consortium, 2009).

ORIGIN OF L AND DRAVIDIAN SPEAKERS

The NRY HG L-M20 is virtually absent in Europe, but found irregularly and at low frequencies in populations of the Middle East and southern Caucusus (Nebel et al., 2001). It occurs at a frequency of 4.3% in Pakistan and 13.5% in Central Asia (Qamar et al., 2002; Semino et al., 2000; Wells et al., 2001). At a resolution of six STR loci, four Chenchu tribal individuals from Andhra Pradesh shared a widespread common haplotype 14-12-22-10-14-11; DYS19- DYS388- DYS390-DYS391- DYS392- DYS393. This is shared by Lambadis, Punjabis, and Iranians. An Armenian haplotype 15-12-23-10-13-11, commonly found in their HG L-M20, is a three-step mutation (Weale et al., 2001). These differences indicate two distinct founders and independent expansions: more data is required to identify the antiquity of these populations. The hitherto available L subtyping data shows the presence of HG L1-M76 in many northwestern states and Dravidian-speaking southern belts of India (Trivedi et al., 2008). Sengupta et al. (2006) found a subtype of HG L-M20 to be the most common haplogroup in India, and proposed its early diversification in Dravidian speakers and subsequent expansion toward peripheral regions, suggesting an Indian origin of Dravidian speakers. The Brahmin populations from Tamil Nadu have been considered as Dravidian speakers, to prove their argument. However, Sahoo et al. (2006) observed absence of HG L-M20 in IE speakers from Bihar, Orissa, and West Bengal, and has concluded that distribution of NRY HGs in India was associated with geography rather than linguistics. Among Austro-Asiatic (AA) speakers of India, as mentioned earlier, HG O2-M95 is predominantly a Southeast Asian marker (Basu et al., 2003) and virtually absent in central

Asia (Wells et al., 2001). Thanseem et al. (2006) have found HG O2-M95 at the highest frequency in AA tribes (52%), and a deeper coalescence age (68,000 YBP) that does not fit with the history of other NRY clades. Non-AA castes and tribes have a frequency of this marker of 6.3%, and the scenario has suggested a footprint of earlier AA settlers carrying this defining mutation. HG O3-M122, and its sublineage HG O3e-M134 that spread through East Asia (Su et al., 2000) showed the highest frequency among Tibeto-Burman (TB) speakers of North East India , while the caste groups of the region possess only 3% (Trivedi et al., 2006). Further, since the coalescence age for HGs C-M130, H-M69, and R2-M124 was deeper compared to HG O-M175, they concluded that AA speakers could not have been the earliest settlers of India. More recent data in Table 74-1, however, suggest an estimated age of 35,000 years for HGO-M175 and 11,700 years for HG O2a-M95.

R2 RESTRICTED TO INDIA AND ITS NEIGHBORS

HG R2-M124, a last major clade of significance to appear in India, is restricted to India, Pakistan, Iran, and southern Central Asia (Kivisild et al., 2003); however, this has been seen with highest frequency (53%) among Sinte Romani (Gypsy) (Wells et al., 2001). Cordaux et al. (2004a) have suggested that this HG R2 originated in India; this conclusion was based on the presence of this clade in both Dravidian and Indo-European speakers. Within India it is predominant in the east coast and southern India (Sahoo et al., 2006). Network analysis of available data has depicted that a large number of haplotypes were shared between populations of South India, while the populations of eastern India harbored more discrete haplotypes, originating in situ.

THE ENIGMA OF R1A1

Contrary to R2, the widespread northern Indian clade among Brahmin-related groups, HG R1a1-M17, has been linked with the recent spread of Kurgan culture originating in southern Russia/ Ukraine and dispersing to Europe, Central Asia, and India between 3000–1000 BCE (Passarino et al., 2001; Quintana-Murci et al., 2001; Wells et al., 2001). In a global analysis, a deeper Palaeolithic time depth of ~15,000 YBP for HG R1a1-M17 mutation has been suggested (Semino et al., 2000; Wells et al., 2001). Further, two region-specific Y-STR allele patterns have been associated with HG R1a1-M17 among Europeans (Passarino et al., 2002): allele 15 at DYS19 and alleles 19 and 21 at locus YCA IIa,b against the background of HG R1a1-M17 characterize populations of Western Europe, while alleles 16 for DYS19 and 19,23 for YCA IIa,b characterize Eastern European populations.

Interestingly, the high frequency of HG R1a1-M17 is concentrated around the elevated terrain of central and western Asia, and is present at a relatively low frequency in Caucusus and Middle East. In Central Asia, its frequency is highest in the highlands among Tajiks, Kyrgyz, and Altais (>50%) and drops down to <10% in the plains among the Turkmenians and Kazakhs (Wells et al., 2001; Zerjal et al., 2002). In contrast to the above, other studies have observed a high HG frequency in Central Asians and lower average STR diversity than in Indian castes and tribes. This has been attributed to a founder effect from southern and western Asia during the early Holocene expansion, contributing HG R1a1-M17 chromosomes to both Central Asian and South Asian tribes prior to the arrival of the Indo-European speakers (Kivisild et al., 2003; Thanseem et al., 2006; Trivedi et al., 2008). Zerjal et al. (2002), however, attributed the low Y- STR diversity to a bottleneck effect in Central Asian populations. Some authors also propose an Indian origin for the HG R1a1-M17 based on the high frequency and associated STR variance in India (Sharma et al., 2009), while others attribute the origin to C.Asia (Wells et al., 2001). While extensive subclades and subtypes have been identified for NRY HG R1b, the R1a1 has been the least studied (Underhill et al., 2010) for lack of new markers; we await more data from our genographic project (www.nationalgeographic.com/genographic), in order to further decipher the early population-movements scenario.

LANGUAGE CORRELATES OF NRY DISTRIBUTION—A PRINCIPAL COMPONENT ANALYSIS

The overall NRY diaspora based on the hitherto available data suggests a pattern of peopling of India. While HG R1a1 is prevalent in the northern Indian belt, the HG O and its derivatives are predominantly seen in east-central and northeastern regions of India, mostly among tribals. HG L is restricted to various Dravidian speaking populations of India and some populations of Pakistan (Figure 74-2). The data in Table 74-2 and Figure 74-1, showing the stacked areas of various alleles in different populations, reveals a striking correlation between NRY composition and the languages they speak. This is further brought out by the principal component analysis (PCA) (Figure 74-2) of the data in Table 74-2. The first two components of the PCA account for more than 50% of the total variance (Figure 74-3).

The geographical distribution of the NRY HGs described in the previous paragraph is clearly brought out in the PCA plot of various NRY HG frequency distribution in these populations The three language speakers, i.e., Indo-European (IE), Dravidian (DR) and Austro-Asiatics (AA), irrespective of tribes or castes, are seen to be influenced by

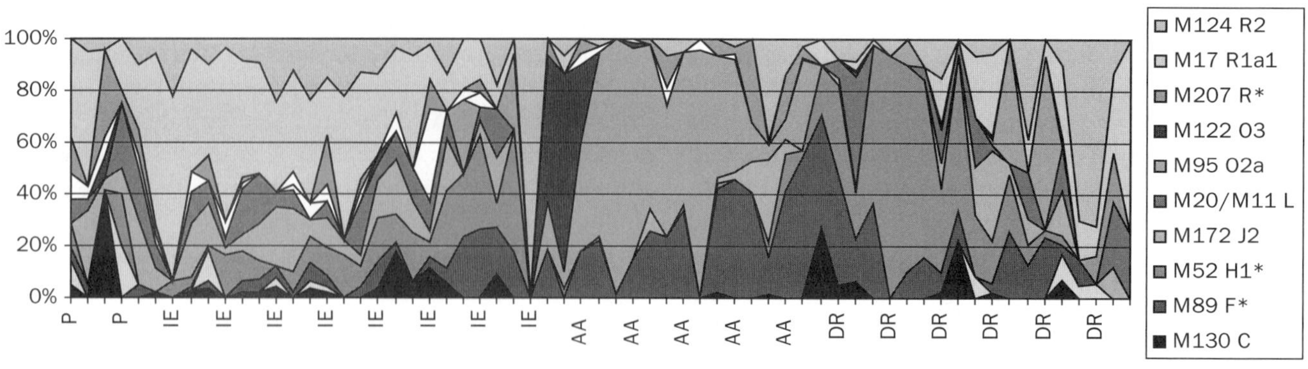

Figure 74-1 Picture of 100% stacked area of NRY HG allelic composition of 58 Indian and 5 Pakistani populations (totaling 2447 samples) arranged according to the language they speak at present. A clear trend of various alleles in different language speakers was discernible. Note the relative distributions of R1a1-M17 (yellow color), O2a-M95 (violet), H1-M52 (green); L-M20 (pink) F*-M89 F* (red) and C-M130 (black) in various language speakers. P = Pakistan; IE = Indo-European speakers; TB = Tibeto-Burmese speakers; AA = Austro-Asiatic speakers; and DR = Dravidian speakers, all from India. For exact population caste/tribe names and their references, refer to Table 74-1. Refer color figure.

various eigenvectors: thus the Dravidian speakers, mostly tribals, are distributed in upper right quadrangle, while the Orissa, West Bengal, and northeastern tribal populations speaking AA languages cluster on the right bottom quadrangle of the plot. Many Brahmin and other populations of northern India, speaking IE languages, are clustered on the left bottom quadrangle of the plot. The overlap between IE speakers and DR speakers seen in the middle of the plot can be attributed to either a confluence of two ancestors, miscegenation, or founders to varying degrees, or to language replacement. The populations found at the extremes, with highest Eigen in one direction, possessed the highest frequencies of one or another NRY allele. This can be attributed to a small founder or bottleneck effect, and uninterrupted expansion without any foreign gene flow. The terrain and climate of the eastern central India and northeastern India favors such a population expansion. However, the absence or low frequencies of many other NRY clades in the AA speakers, and the concentration of these tribal populations in huge numbers in the

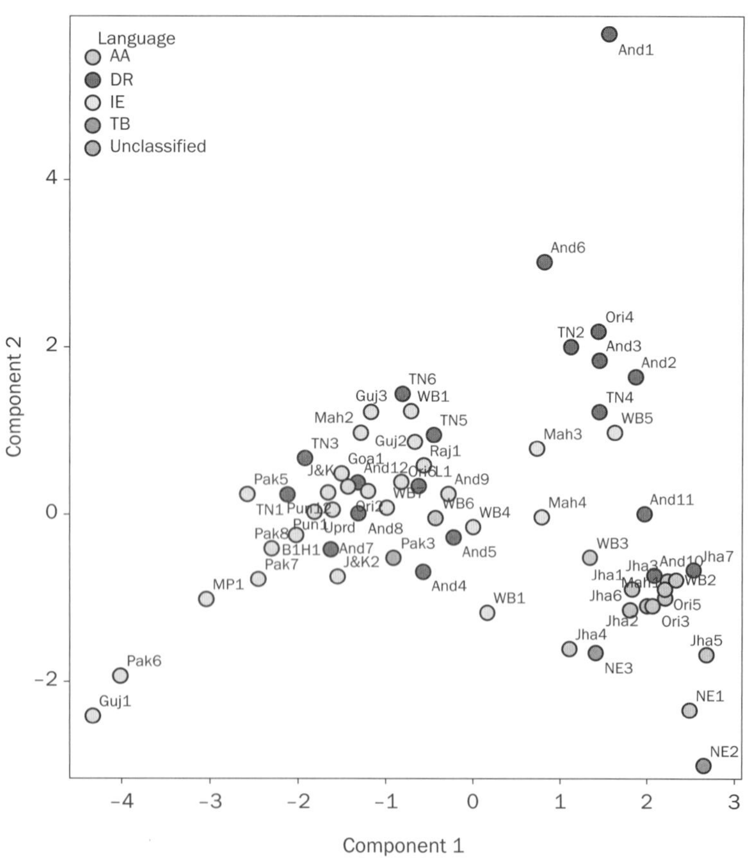

Figure 74-2 The principal component analysis (PCA) of NRY/HG data of Indian populations available in literature. Refer color figure.

NOTE: The population Hazara from Pakistan was removed from analysis, because it showed very high variance, making differentiation among the rest of the populations difficult.

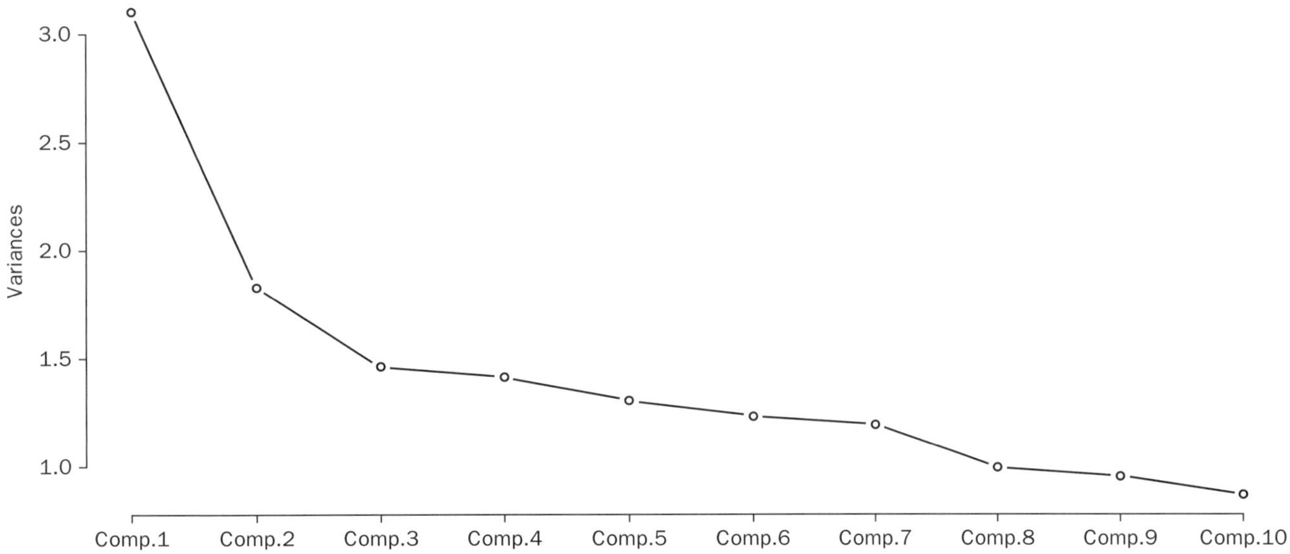

Figure 74-3 The Scree Plot of the principal component analysis of NRY HG data of Indian populations available in literature.

eastern central India/Orissa belt and the northeastern hilly tracts of India suggests a concomitant origin of these NRY clades and their language, and a spurt of huge expansion from the small founder. This is reiterated by the observation that all these populations possessed very little of other parallel and later-derived NRY HGs.

This proposition is further supported by the data that the Chinese and other oriental populations with the derivatives of O3 have are decedents from a common ancestor somewhere from the northeast/Myanmar region (Shi et al., 2005). Basu et al. (2003) have suggested that the AA and TB speaking tribal groups might have entered India first from a northwest corridor and, much later, some through a northeast corridor. Contrary to this, Cordaux et al. (2004b) have proposed that northeast India acted as a barrier. Kumar et al. (2007), in an extensive recent study, identified a strong genetic link among sublinguistic groups of Indian AA-speaking populations and has suggested an origin of AAs in India who later spread to Southeast Asia. The analysis of the present study, in light of the population size and the extent of distribution of these AA-speaking tribal groups, reiterates the concomitant origin of these clades and their language. The time of origin of the clade O2a, i.e., 11,700 years ago (Sengupta et al., 2006; Table 74-1) fits well with the assumptions that spoken language originated ~10,000 years ago. A very interesting observation was the higher frequencies (50%–90%) of O2a in half of the AA-speaking populations hitherto available in literature (Table 74-2).

CONCLUSION

India, the second continent to be successfully occupied by modern man, is heterogeneous in itself, in terms of geography, climate, and populations. The whole of India thus cannot be considered as a single gene pool. The migratory history as revealed by NRY shows definitive pathways, origin, and autochthonous expansion of various NRY clades and populations in different parts of the country. Many of these populations are ancient than the languages they speak. Thus, as various languages developed, presumably in small founders, the population expansion and language spread must have taken place concomitantly: hence, we see a good correlation between languages and NRY in India.

SUMMARY

Modern man (*Homo sapiens sapiens*), originating in Africa, first emigrated ~70,000 years ago, walked through the coasts of India (southern coastal route model) and reached Australia. Since then many migrations, settlements, and expansions have taken place in various parts of India. The island model of human settlements and expansions may explain the origin of settled communities and languages in India.NRY chromosome markers help to unravel the details of early migrations of man into South Asia. The analysis of the literature thus suggests the origin and expansion of languages, superimposed by the genomic data: the data implies small founders, autochthonous origin (mutation) of new NRY markers, nuclear origins, and uninterrupted expansion / dispersal of populations and languages in India as exemplified by Austro-Asiatic (AA) speakers. The better communication means and the language presumably led to the settlements (founders), rapid expansion, formation of culture and societies, and their dispersal to newer horizons and territories. South Asia and Southeast Asia thus seem to be the cradle of many new founders, and autochthonous civilizations commensurate with languages that are preserved till today.

REFERENCES

Aquadro CF, DuMont BB, Redd FA. (2001). Genome wide variation in human and fruitfly: a comparison. *Curr Opinion Genet Dev* 11:627–634.

Baltimore D. (2001). Our genome unveiled. *Nature* 409:814–816.

Basu A, Mukherjee N, Roy S, et al. (2003). Ethnic India: A genomic view, with special reference to peopling and structure. *Genome Res* 13:2277–2290.

Bhasin MK. (2006). Genetics of castes and tribes of India: Indian population milieu. *Int J Hum Genet* 6(3):233–274.

Cann RL, Stoneking M, Wilson AC. (1987). Mitochondrial DNA and human evolution. *Nature* 325:31–36.

Cann RL. (2001). Genetic clues to dispersal of human populations: Retracing the past from the present. *Science* 291:1742–1748.

Capelli C, Wilson JF, Richards M, et al. (2001). A predominantly indigenous paternal heritage for the Austronesian speaking peoples of insular South East Asia and Oceania. *Am J Hum. Genet* 68:432–443.

Cavalli-Sforza LL, Menozzi P, Piazza A. (1994). The history and geography of human genes. Princeton, New Jersey: Princeton University Press, p. 1088.

Cavalli-Sforza LL. (1997). Genes, peoples, and languages. *Proc Natl Acad Sci U S A* 94:7719–7724.

Cordaux R, Aunger R, Bentley G, Nasidze I, Sirajuddin SM, Stoneking M. (2004a). Independent origins of Indian caste and tribal paternal lineages. *Curr Biol* 14:231–235.

Cordaux R, Weiss G, Saha N, Stoneking M. (2004b). The northeast Indian passageway: a barrier or corridor for human migrations. *Mol Biol Evol* 21:1525- 1533.

Diamond J, Bellwood P. (2003). Farmers and their languages: The first expansions. *Science* 300:597–603.

Eller E. (2001). Estimating relative population sizes from simulated data sets and the question of greater African effective size. *Am J Phys Anthropol* 116:1–12.

Fuller D. (2003). An agricultural perspective on Dravidian historical linguistics: archaeological crop packages, livestock and Dravidian crop vocabulary. In Bellwood P, Renfrew C, eds. *Examining The Farming/Language Dispersal Hypothesis*. Cambridge: McDonald Institute for Archaeological Research, pp. 191–213.

Giacomo Di F, Luca F, et al.(2004). Y chromosome haplogroup J as a signature of the post-neolithic colonization of Europe. *Hum Genet* 115:357–371.

Hammer MF, Karafet MT, Rasanayagam A, et al. (1998). Out of Africa and back again: nested cladistic analysis of human Y chromosome variation. *Mol Biol Evol* 15:427–441.

Hammer MF, Redd AJ, Wood ET, et al. (2000). Jewish and Middle Eastern non-Jewish populations share a common pool of Y-chromosome biallelic haplotypes. *Proc Natl Acad Sci U S A* 97(12):6769–774.

HUGO Pan-Asian SNP Consortium. (2009). Mapping Human Genetic Diversity in Asia. *Science* 326(5959):1541–1545.

Karafet TM, Xu L, Du R, et al. (2001). Paternal population history of East Asia: Sources, patterns, and microevolutionary processes. *Am J Hum Genet* 69:615–628

Kavitha VJ. (2008). Studies on the Genomic Diversity of Southern Indian Breeding Isolates. PhD Thesis. Madurai Kamaraj University, India.

Kayser M, Brauer S, Weiss G, et al. (2000). Melanesian origin of Polynesian Y chromosomes. *Curr Biol* 10:1237- 1246.

Kivisild T, Rootsi S, Metspalu M, et al.(2003). The genetic signatures of earliest settlers persist in Indian tribal and caste populations. *Am J Hum Genet* 72:313–332.

Kumar V, Reddy AN, Babu JP, et al. (2007). Y-chromosome evidence suggests a common paternal heritage of Austro Asiatic populations. *BMC Evol Biol* 28;7:47.

Lell JT, Sukernik RI, Starikovskaya YB, et al. (2002). The duel origin and Siberian affinities of Native American Y chromosomes. *Am J Hum Genet* 70:192–198.

Misra VN.(2001). Prehistoric colonisation of India. *J Biosci* 26:491–531.

Nebel A, Filon D, Brinkmann B, Majumder PP, Faerman M, Oppenheim A. (2001). The Y chromosome pool of Jews as part of the genetic landscape of the Middle East. *Am J Hum Genet* 69:1095–1112.

Paddaya K. (1982). The Transition from Lower to Middle Paleolithic and the Origin of Modern Man. Ronen A, ed. *British Archaeological Reports International series,* Oxford, U.K.: vol 151, pp. 257–264.

Passarino G, Semino O, Magri C, et al. (2001). The 49 af haplotype 11 is a new marker of the EU19 lineage that traces migrations from northern regions of the Black Sea. *Hum Immunol* 62:922–932.

Passarino G, Cavalleri GL, Cavalli-Sforza LL, Børresen-Dale A-L, Underhill PA. (2002). Different genetic components in the Norwegian population revealed by the analysis of mtDNA and Y chromosome polymorphisms. *Eur J Hum Genet* 10:521–529.

Pauling L, Itano HA, Singer SJ and Wells IG. (1949). Sickle-cell anemia, a molecular disease. *Science* 110:543–548.

Petraglia MD, Haslam M, Fuller DQ, Boivin N. (2010). The southern dispersal route and the spread of modern humans along the Indian Ocean rim: New hypotheses and evidence. *Annals of Human Biology* 37(3):288–311

Qamar R, Ayub Q, Mohyuddin A, et al. (2002). Y chromosomal DNA variation in Pakistan. *Am J Hum Genet* 70:1107–1124.

Quintana-Murci L, Krausz C, Zerjal T, et al.(2001).Y-chromosome lineages trace diffusion of people and languages in Southwestern Asia. *Am J Hum Genet* 68:537–542.

Rosser ZH, Zerjal T, Hurles ME, et al.(2000). Y Chromosomal Diversity in Europe Is Clinal and Influenced Primarily by Geography, Rather than by Language. *Am J Hum Genet* 67:1526–1543.

Sahoo S, Singh A, Himabindu G, et al. (2006). Prehistory of Indian Y chromosomes: Evaluating demic diffusion scenarios. *Proc Natl Acad Sci U S A* 103(4):843–848.

Sanghvi LD, Balakrishnan V, Karve I. (1981). Biology of the people of Tamil Nadu. Pune: Indian Society of Human Genetics and Calcutta: Indian Anthropological Society.

Seielstad M, Yuldasheva N, Singh N, et al. (2003) A novel Y-chromosome variant puts an upper limit on the timing of first entry into the Americas. *Am J Hum Genet* 73(3):700–705.

Semino O, Passarino G, Oefner PJ, et al.(2000). The genetic legacy of Paleolithic *Homo sapiens sapiens* in extant Europeans: a Y chromosome perspective. *Science* 290:1155–1159.

Semino O, Magri C, Benuzzi G, et al.(2004). Origin, Diffusion and Differentiation of Y-chromosomal Haplogroups E and J: Inferences on Neolithization of Europe and later migratory events in the Mediterranean Area. *Am J Hum Genet* 74:1023–1034.

Sengupta S, Zhivotovsky LA, King R, et al.(2006). Polarity and temporality of highresolution Y chromosome distributions in India identify both indigenous and exogenous expansions and reveal minor genetic influence of central Asian pastoralists. *Am J Hum Genet* 78:202–221.

Sharma S, Rai E, Sharma P, et al. (2009). The Indian origin of paternal haplogroup R1a1* substantiates the autochthonous origin of Brahmins and the caste system. *J Hum Genet* 54(1):47–55

Shi H, Dong YL, Wen B, et al. (2005). Y-chromosome evidence of southern origin of the East Asian-specific haplogroup O3-M122. *Am J Hum Genet* 77(3):408–419.

Su B, Xiao C, Deka R, Seielstad MT, et al. (2000). Y chromosome haplotypes reveal prehistorical migrations to the Himalayas. *Hum Genet* 107:582–590

Thangaraj K, Singh L, Reddy AG, et al. (2003). Genetic affinities of the Andaman Islanders, a vanishing human population. *Curr Biol* 13:86–93

Thanseem I, Thangaraj K, Chaubey G, et al. (2006). Genetic affinities among the lower castes and tribal groups of India: inference from Y chromosome and mitochondrial DNA. *BMC Genet* 7:42.

Tishkoff SA, Dietzsch E, Speed W, Pakstis AJ, Kidd JR. (1996). Global patterns of linkage disequilibrium at the CD4 locus and modern human origins. *Science* 271:1380–1387.

Trivedi R, Sitalakshmi T, Baneerjee J, Singh A, Sircar PK and Kasyap VK. (2006). Molecular insights into the origins of the Shompen a declining population of the Nicobar archipelago. *J Hum Genet.* 51:217–226.

Trivedi R, Sahoo S, Singh A, et al.(2008). Genetic imprints of Pleistocene origin of Indian Populations: A comprehensive phylogeographic sketch of Indian Y-Chromosomes. *Int J Hum Genet* 8(1–2):1–20

Underhill PA, Passarino G, Lin AA, et al. (2001). The phylogeography of Y chromosome binary haplotypes and the origins of modern human populations. *Ann Hum Genet* 65:43–62

Underhill PA. (2003). *Inferring Human History: Clues from Y-Chromosome Haplotypes.* Cold Spring Harbor Symposia on Quantitative Biology, Cold Spring Harbor Laboratory Press. Volume LXVIII, pp. 487–493.

Underhill PA, Myres NM, Rootsi S, et al. (2010). Separating the post-Glacial coancestry of European and Asian Y chromosomes within haplogroup R1a. *Eur J Hum Genet* 18(4):479–484

Weale ME, Yepiskoposyan L, Jager RF, et al. (2001). Armenian Y chromosome haplotypes reveal strong regional structure within a single ethno-national group. *Hum Genet* 109:659–674.

Wells RS, Yudlasheva N, Ruzibakiev R, et al.(2001). The Eurasian heartland: a continental perspective of y-chromosome diversity. *Proc Natl Acad Sci USA* 98:10244–10249.

Wells S. (2007). Deep Ancestry: Inside The Genographic Project. National Geographic Society

Wolpoff MH, Wu X and Throne A. (1984). Modern *Homo Sapiens* origins: A general theory of hominid evolution involving the fossil evidence from East Asia. In: Smith FH, Spencer F, eds. *The Origins of Modern Humans: A World Survey of the Fossil Evidence.* New York: Liss AR. pp. 465–467.

Yuldasheva N, Wells RS, Ruzibakiev RM, et al. (2002). High Polymorphism at the MC1R locus is consistent with phenotype diversity in Eurasian populations. Paper presented in *The Human Genome Organisation*, Shanghai, China.

Zerjal T, Wells R, Yuldasheva N, Ruzibakiev R, Tyler-Smith C. (2002). A genetic landscape reshaped by recent events: Y chromosomal insights into central Asia. *Am J Hum Genet* 71:466–482.

75.

GENOMIC DIVERSITY OF HLA IN THE INDIAN SUBCONTINENT

Narinder K. Mehra and Gurvinder Kaur

INTRODUCTION

The post-genomics era has ushered in a renewed interest in genomic diversity studies, particularly with regard to definition of polymorphisms and their functional significance. An understanding of the functionality of these large sequences and their variability between populations and individuals appears to be the next logical step to provide meaning to the voluminous human genome sequences. Population-based description of polymorphisms, especially in functionally relevant gene complexes such as the major histocompatibility complex (MHC), has gained immense importance, particularly with regard to the impact of genomics in biology and medicine. Because of the large degree of polymorphism and extensive linkage of its various loci, the gene complex within the human leukocyte antigen (HLA) region, which is the human equivalent of MHC, serves as a good mini-genome model for anthropological investigations of genetic relatedness and migrations of different world populations. Three outstanding features of the human MHC— namely the extraordinarily high polymorphism, tight linkage among the loci, and the nonrandom association of its various alleles—make the system of particular interest from the perspective of population genetics.

THE INDIAN SUBCONTINENT

The Indian subcontinent is a land of enormous cultural, geographical, religious, and linguistic diversity. It harbors appreciably greater genetic diversity than any other comparable global region after Africa. The population comprises various tribal groups, ranked caste, and other ethnic groups, several of which are largely endogamous. Why and how this genetic diversity arose in India is not clear. The "Out of Africa" migration model of modern humans suggests that a small population, most probably from East Africa, expanded throughout Africa around 100 kya. This was followed by a second wave of expansion between 60 and 40 kya into Asia and from there to other continents (Cavalli-Sforza & Feldman, 2003). Currently, it is accepted that the modern human evolved in Africa, migrated out of the continent, and one of these major waves came to India, and this served as an important platform for the dispersal of modern humans to other geographical regions, including Southeast Asia.

Genetic diversity of the Indian population has had immense influence due to successive waves of migration and invasions from the Middle East and Central Asia, as well as Mongolia, all of which contributed to the available gene pool (Figure 75-1). Starting from the fourth century BC onward, India—particularly the northern region—experienced several waves of migration that continued for almost 2000 years. History testifies that Greeks, Pathans, Sakas, and Pahlavas, including the Kushans, were the first to come after the Indo-Aryan civilizations. On the west coast, Jews and Parsis came after fleeing from their own homelands. Similarly, Arabian and Persian Muslims, Turks, and Afghans, whose total number was very large, invaded India from time to time. The Muslim immigration into India began even before the Arab invasion of Sind, quite early in the eighth century AD (Bhasin & Walter, 2001), and ended with the establishment of the Mughal Empire in the sixteenth century. This was the last major movement that produced perceptible change in the composition and culture of the indigenous population. The Muslims did not accept the Hindu religion, and this led to large-scale conversions. The Europeans, Portuguese, Dutch, French, and British started their intrusions from the fifteenth century AD onwards, and the Indo-European community grew from these infusions. The Mongolians entered the eastern part of India from Southeast Asia. The Indian subcontinent therefore became the melting pot of the various racial groups who largely halted here and intermingled with the local population.

As a result of the evolutionary history and the prevalent practice of endogamy and consanguinity, the populations of the subcontinent demonstrate high genetic differentiation and extensive substructuring

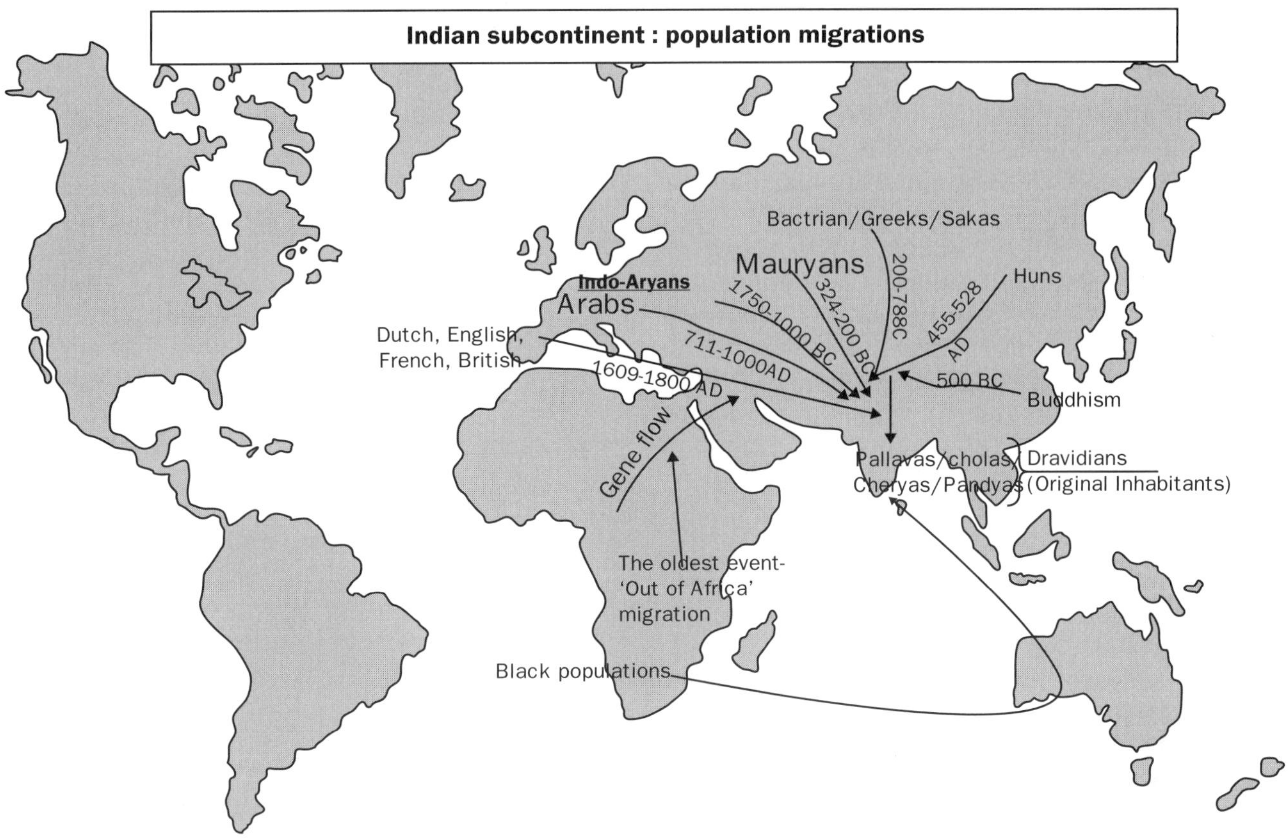

Figure 75-1 Pictorial representation of multiple waves of migration of various population groups that amalgamated with the people of India. Arrows point toward the direction of migration at particular periods in history. Refer color figure.

(Majumdar, 2001). A recent study involving 25 well-defined ethnic groups in India, along with the Europeans and Chinese populations, concluded that the ancestry is unique to India (Reich et al., 2009). The authors suggested that Indian groups have inherited different proportions of ancestry from the "ancestral North Indians" and the "ancestral South Indians." Hence, consistent with social history, northern regions show closer affinities with Indo-European speaking populations of central Asia as compared to those inhabiting southern regions, which may have been derived from early colonizers arriving from Africa along the southern exit route.

An historical perspective of the racial admixture following multiple waves of migration into the Indian subcontinent, overlaid with evolutionary pathogen- cum environmental-driven natural selection, have established remarkable heterogeneity in MHC genes in this population on the one hand, and origin of "novel alleles" and "unique HLA haplotypes" on the other. Accordingly, the distribution of HLA genes varies significantly among and between various ethnic groups in this part of the world. Such variations may occur at three levels. First, common HLA genes can occur in all populations, but at different frequencies. Second, certain HLA genes may get restricted to a specific population with complete absence or rare occurrence in others; e.g., HLA-A*36 is novel to the African population, while HLA-A*0211 is the most prevalent HLA-A2 subtype among Asian Indians and occurs with the least frequency among Caucasians. Finally, certain haplotypes, held together through strong linkage disequilibrium at various loci, remain population-restricted. The best example is that of the autoimmune-favoring HLA-DR3 positive haplotypes, as described in a later section of this chapter.

NOVEL HLA ALLELES IN ASIAN INDIANS

Molecular studies on the distribution of HLA alleles and haplotypes on a representative sample of northern India indicate an appreciable Caucasoid and Oriental influence in the generation of allelic diversity in the MHC class I and class II regions. This is supported by the data that Oriental-specific alleles like the A*0206, A*0207 and A*3303 are more commonly found in the Indian population than the typical Caucasian alleles A*0201, A*2901, A*3002 and A*3004 (Jaini et al., 2002b; Mehra et al., 2001). Similarly, Oriental haplotypes, such as A*3303-B*4403, A*3101-B*51xx, including some that are disease associated (A*3303-B*5801) occur with an appreciable frequency in this population.

DIVERSITY IN HLA CLASS I REGION

HLA-A2

HLA-A2 is a ubiquitous antigen that occurs with a frequency of ~30%–35% in most populations around the world. Molecular analyses indicate at least 373 different subtypes or alleles of this antigen that vary in their distribution in various populations. For example, the Caucasoid populations are characterized by the presence of only one subtype, namely A*0201, while the North Indian population shows extensive heterogeneity (Mehra et al., 2001). The most common HLA-A2 subtype in North India is HLA*0211 (representing more than one-third of all HLA-A2+ve individuals) while alleles specific to Oriental (A*0206 and A*0207) and Negroid groups (A*0205) are also represented here with appreciable frequency (Figure 75-2). On the other hand, A*0201 is rarely observed among the Asian Indians. Sequence analysis of HLA-A*0211 indicates that it has a unique peptide binding site, with critical differences at amino acid residues T73I and H74D in the α1 domain. The amino acid differences in peptide binding pockets could result in different specificities of peptide binding and presentation. These studies have important implications in our understanding of antigen presentation and designing peptide-based immunological therapeutic approaches.

The origin of gypsies in Europe and elsewhere has been a subject of some debate. They represent a closed population with significant inbreeding and genetic identity that is very similar to that of the Indian population. In a controlled study carried out by us on the Hungarian gypsy populations, we observed a unique genetic heritage shared by them with Indians. For example, the unique HLA-*0211 allele that is known to be specific to the Asian Indians was found with a significant frequency (18%) among the gypsies (Figure 75-2). In contrast, this allele was almost completely absent in the Hungarian Magyars living in the same geographic location. Linguistic evidence suggests that the gypsies are of diverse Indian origin, more like a collection of genetically isolated founder populations. Shared presence of the unique HLA alleles in these geographically distant populations strongly indicates common genetic lineage, by way of a common ancestor or migration of one from the other. Therefore, it is expected that the gypsies might have conserved at least some of their genetic component bits that were originally derived from Indians, but then evolved under the influence of different driving forces.

HLA-A19

The HLA-A19 is another interesting multi-gene family, where HLA-A*33 is the most prevalent and polymorphic subgroup observed among North Indians. Of the four known subtypes of A*33, the Caucasians are characterized by the predominant occurrence of A*3301, while the Orientals have only the A*3303 subtype. We observed the latter allele (rather than the former) as the most predominant A*33 subtype in the Indian population (Jaini et al., 2002b). Further, we identified another novel allele

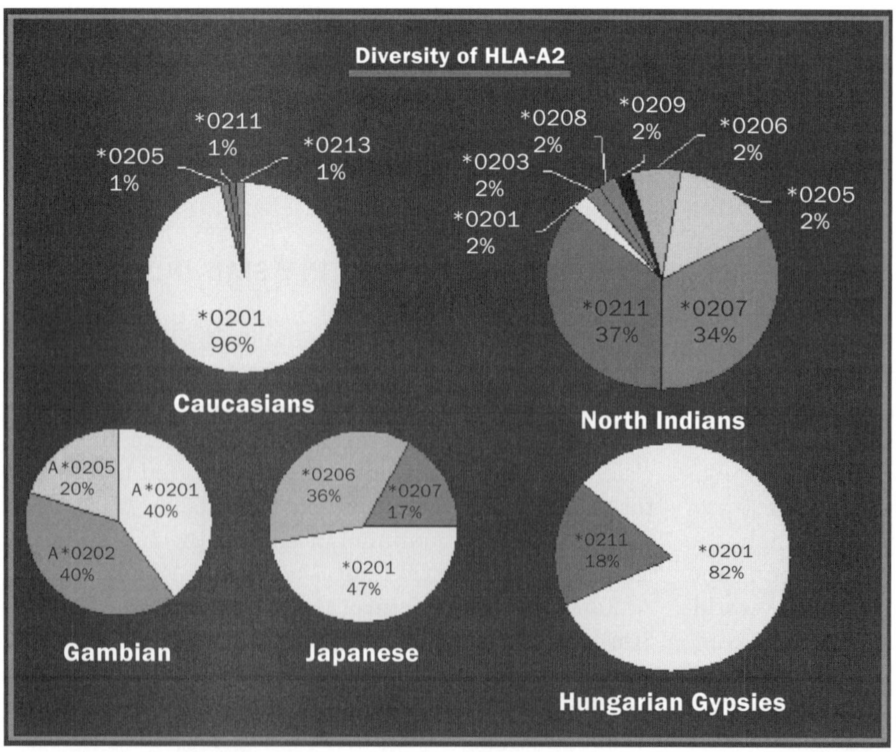

Figure 75-2 Distribution of HLA-A2 subtypes in Caucasians, Gambian, Japanese, North Indians, and Hungarian gypsies. Note the extreme heterogeneity in HLA-A2 in the Asian Indians, and similarity in the occurrence of unique A*0211 in Asian Indians and the Hungarian gypsies. Refer color figure.

designated as A*3306 (accession no. HWS 10000697) in the Indian population that differs from A*3303 at position 228 [Ile 52 Met] (Rozemuller et al., 2002). Our studies provide evidence that the novel A*3306 allele may have arisen from A*3303 by a single nucleotide mutation, and thus represents Oriental influence in the generation of HLA diversity.

The predominant presence of such novel alleles as A*3303 and A*0211 among North Indians is significant and indicates a natural course of positive selection of such alleles and their selective advantage in this population. These results emphasize the uniqueness and high heterogeneity of the MHC repertoire among Indians.

HLA-B27

Although HLA-B27 occurs with a population frequency of 4% to 6% in most populations, its frequency rises to >90% in patients with ankylosing spondylitis and other related spondyloarthropathies. Of the currently known 90 molecular subtypes (alleles) of HLA-B27, we observed the occurrence of both B*2704 (an Oriental subtype) and *2705 (common in Caucasians) in North Indians with an appreciable frequency, followed by *2707, *2702 and others (Figure 75-3). The Indian population indeed reveals much greater heterogeneity of the HLA-B27 gene as compared to Caucasoids and other population groups.

It has been hypothesized that the B27 subtypes that are disease-associated (for example *2704, *2705) differ from those that show negative association (for example *2706 in the Thais) at critical amino acid positions in the peptide binding groove, particularly in exon 3 (His114Asp and Asp116 Tyr) or at positions 77 and 152. These differences lead to alteration of the peptide binding ability of the B27 molecules in such a way that influences accommodation of "pathogenic peptides" and their presentation to cytotoxic T-cells.

HLA-C AS LIGANDS FOR KIRS

The course of natural selection has been one of the major driving forces of genetic evolution, and this can be traced through the extent of HLA diversity at the population level. This can be best illustrated by the study of tribal groups, because of their limited conserved habitat, custom of consanguineous marriages, and occurrence of frequent epidemics of infections. The Andaman and Nicobar islands on the east coast of India are home to two broad racial groups, namely the Negritos (constituting tribes like Great Andamanese, Onges, Jarawas, and the Sentinelese) and the Mongoloids (i.e., Nicobarese and the Shompens). The Negrito race is numerically and physically the smallest human race known. People from mainland India that are primarily descendents of ex-convicts and settlers from Bangladesh and Sri Lanka make up the population of these islands. Very little is known about these classical isolated tribal groups except that their physical features resemble those of the African pygmies and that they are currently facing extinction due to challenges posed by an alarmingly high rate of outbreaks of infections such as viral hepatitis. The Great Andamanese constitute a diminishing population of barely 36 individuals inhabiting the Strait Island. Likewise, the current census of Jarawas stands at ~250 in number.

The HLA-Cw alleles serve as ligands for killer immunoglobulin receptors (KIRs) and aid in natural killer (NK) cell-mediated function. Of the 960 known alleles of the HLA-C locus, the common allele families, such as Cw*02/04/05/06, serve as ligands for KIR2DL1, while Cw*01/03/07 serve as ligands for KIR2D2. Studies carried

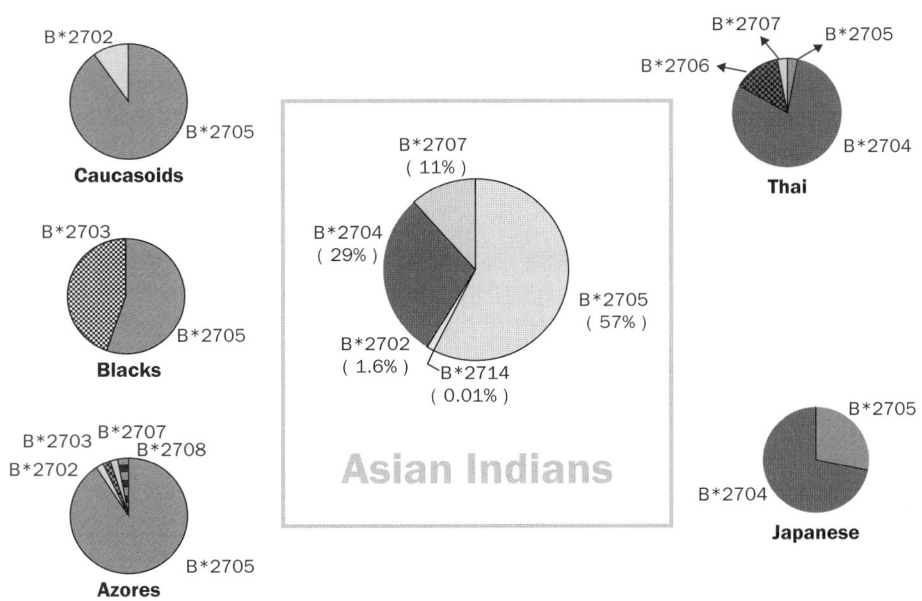

Figure 75-3 Distribution of HLA-B27 subtypes in Asian Indians and their comparison with other world population groups. Refer color figure.

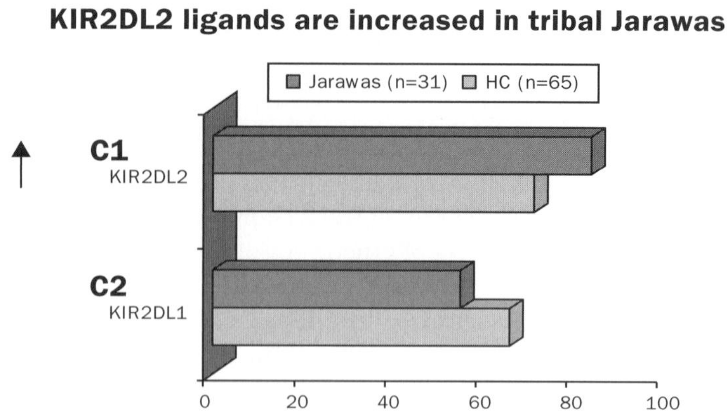

	KIR2DL1 C2	KIR2DL2 C1
Jarawas (n=31)	54.83	83.87
HC (n=65)	65.77	71.03

C2 = Cw*02/04/05/06 = ligands for KIR2DL1
C1 = Cw*01/03/07 = ligands for KIR2DL2

Figure 75-4 Comparison of KIR2DL1 ligands C1 and C2 among the diminishing tribes of Andaman and Nicobar islands and the mainland people of the Indian peninsula. Note that the C1 includes HLA-Cw*01/03/07 and represents ligands for KIR2DL2. Similarly, C2 group includes Cw*02/04/05/06 i.e. ligands for KIR2DL1.

out by us showed predominant occurrence of Cw*04 and Cw*07 allele families in the Jarawas as compared to the healthy mainland population of the Indian subcontinent. On the other hand, frequencies of Cw*06, Cw*03 and Cw*12 were significantly reduced in this tribal population (Figure 75-4). These results suggest that KIR2DL2 ligands are significantly increased in the Jarawa population as compared to the mainland Indian population, which has more of the KIR2DL1 ligands. This is an important observation because of its direct implications in susceptibility to infectious diseases and cancer, and their translational significance in diminishing tribes.

DIVERSITY IN HLA CLASS II REGION

Significant genetic diversity has also been noted in HLA class II allelic groups in the Indian population. Molecular analysis conducted by us revealed that the distribution of HLA-DR and DQ alleles in this population broadly follows the Caucasoid pattern, with predominant occurrence of DRB1*15/16 and DRB1*07 (43% and 28.2%) along with their associated DQA1 and DQB1 alleles. Similarly, the distribution patterns of subtypes of DR*15/16 and DR*11/12 allele families in the Indian population is similar to that observed in Caucasians. By contrast, the distribution of HLA-DR3 and DR4, both of which are incidentally disease-associated, reveals interesting differences. For example, the occurrence of HLA-DR4 family of alleles in India reveals more of an Oriental influence, with predominant prevalence of DRB1*0403 and DRB1*0405 subtypes (Jaini et al., 2002a). The former allele occurs more frequently in the Japanese (~13%) but rarely in the Caucasians (~3%). Moreover, more than 10% of the DR4 haplotypes occur in typical Oriental combination with DQB1*0401, while others show an association with DQB1*0302 (70%). Extensive studies done by our group have shown that HLA-DR4 is closely associated with rheumatoid arthritis (RA), plausibly due to the occurrence of cumulative differences corresponding to residues 67, 70, 71 and 74 in pocket 4, Val/Gly dimorphism at residue 86 in pocket 1, and others in pockets 6 and 9. Molecular subtyping studies have shown a strong association of RA with DRB1*0405 in the Asian Indian population, just as it occurs in the Orientals, Polynesians, Spaniards, Greeks, and Israeli Jews. While *0405 imparts high risk in the Indian RA patients, followed by *0401, the allele *0403 acts more as a protective allele. This is in contrast to the situation in Caucasoids, in whom the major RA susceptibility allele is DRB1*0401 followed by *0404. Hence, shared motifs in DR4 molecules may have different consequences for recognition of different peptides in different populations.

It is thus conceivable that a greater Oriental influence has been favored in the phylogenesis of the HLA class I region alleles in the Indian population, while for HLA class II region genes, both Oriental and Caucasian influence is visible. These observations suggest that Asian Indians have shared parallel natural-selection driving forces and racial admixing with the Caucasian and Oriental populations, as their footprints can be found in the Indian MHC genealogy.

UNIQUE FEATURES OF HLA-DR3 +VE HAPLOTYPES IN ASIAN INDIANS

The population of the Indian subcontinent, placed as it is between the Caucasoids and Negroids on one hand and Australoids and Mongoloids on the other, provides a rich source of both diverse and unique MHC haplotypes (Figure 75-5). While the most frequent Caucasoid haplotypes occur with a reasonable frequency in Asian Indians, those found predominantly in other ethnic groups (e.g., Australian Aborigines, populations of Oceania, China,

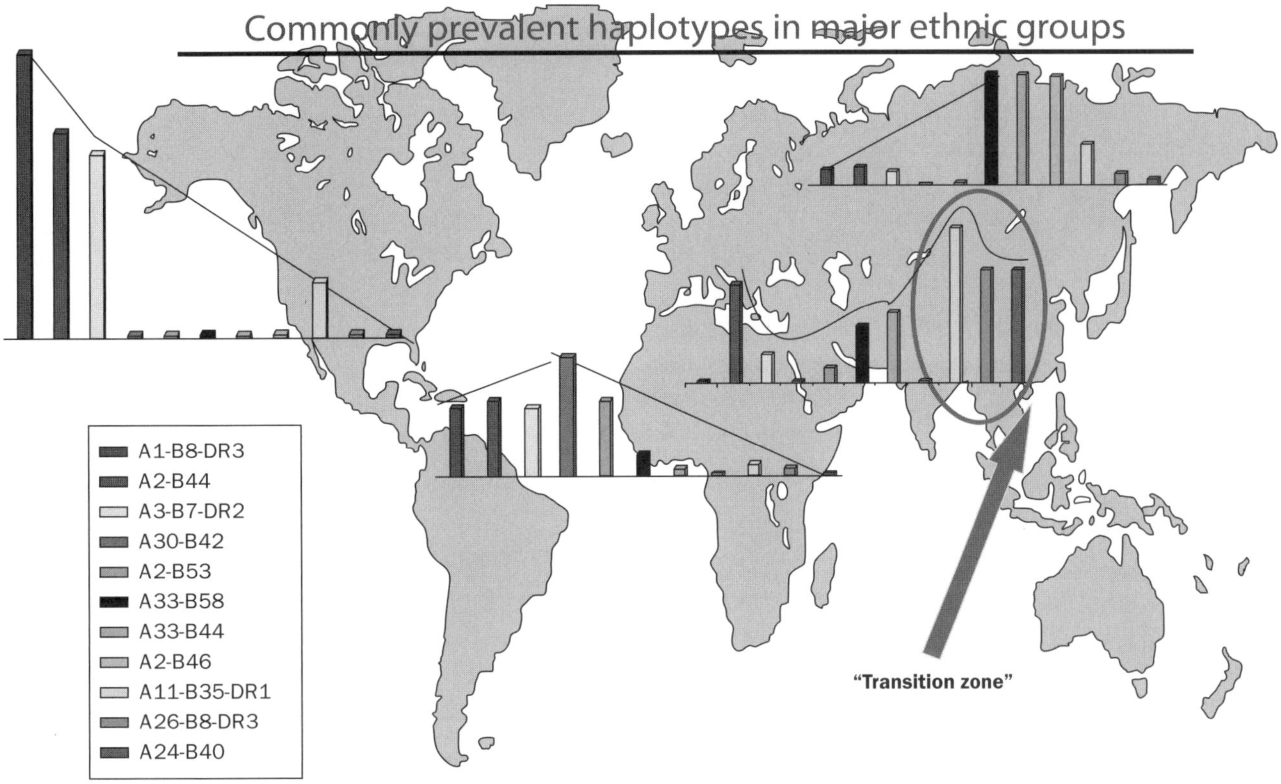

Figure 75-5 Worldwide comparison of common MHC haplotypes with those prevalent in the Indian population. The distribution pattern suggests that the Indian subcontinent represents a "transition zone" between Caucasoids and Negroids on one hand, and Mongoloids and Australoids on the other. Refer color figure.

and Japan) are also detected. Of these, the three most common HLA haplotypes that occur in North India are HLA-A33-B44-DR7, HLA-A33-B58-DR3, and HLA-A2-B50-DR3. Incidentally, HLA-A33-B44-DR7 is also observed in most other Asian countries like Malaysia, Thailand, South Korea, Bangladesh, Pakistan, and China, but not reported in rest of the world. Similarly, HLA-A33-B58-DR3 is one of the commonest haplotypes in the Chinese population, and it occurs with a significantly high frequency in the Indian subcontinent, suggesting racial admixture from the East. On the other hand, haplotype HLA-A2-B50-DR3 is observed exclusively in India, particularly in patients with type 1 diabetes, and is therefore "unique" to this population.

The extended haplotype that favors type 1 diabetes among western Caucasians is HLA-A1-B8-DR3-DQ2, often referred to as *ancestral haplotype 8.1* or AH8.1. This haplotype occurs with a significantly reduced frequency in the Indian population and has been compensated by another, related haplotype that represents a variant form differing from the Caucasian AH8.1 at several gene loci. We have designated this haplotype as AH8.1v. On the other hand, the most common DR3 positive haplotype in the Indian population is HLA-A26 B8 DR3, designated as *ancestral haplotype 8.2* or AH8.2 (Kaur et al., 2008). A few additional HLA-DR3 haplotypes, HLA-A24B8DR3 (AH8.3), A3B8DR3 (AH8.4), and A31B8DR3 (AH8.5), also occur in this population and interestingly, all of these are disease-associated (Figure 75-6).

Fine molecular analyses revealed that AH8.1 and AH8.1v might have coevolved from a common ancestor. The preferential divergence of AH8.2 over AH8.1 leading to survival advantage might have been driven by vigorous pathogenic challenges encountered by the Indian population (Kumar et al., 2009). It is interesting to note that certain haplotypes continue to remain conserved in different populations, despite the proportion of polymorphism possessed within them. Mechanisms underlying this are not fully clear, although pressures exerted by environmental factors, microbiota, and racial admixtures, acting individually or in unison, are relevant. Similarly, it is not clear how HLA genes orchestrate with non-HLA genes and contribute toward predisposition or resistance to a given disease. Answers to these questions are needed for understanding the molecular basis of MHC-mediated immune surveillance mechanisms and may serve as paradigms for future perspectives in molecular vaccine design.

IMPLICATIONS OF MHC HETEROGENEITY

HLA-based studies carried out in the Indian population have highlighted the unique and extensive MHC diversity of the Indian population (Figure 75-7). Our experience indicates

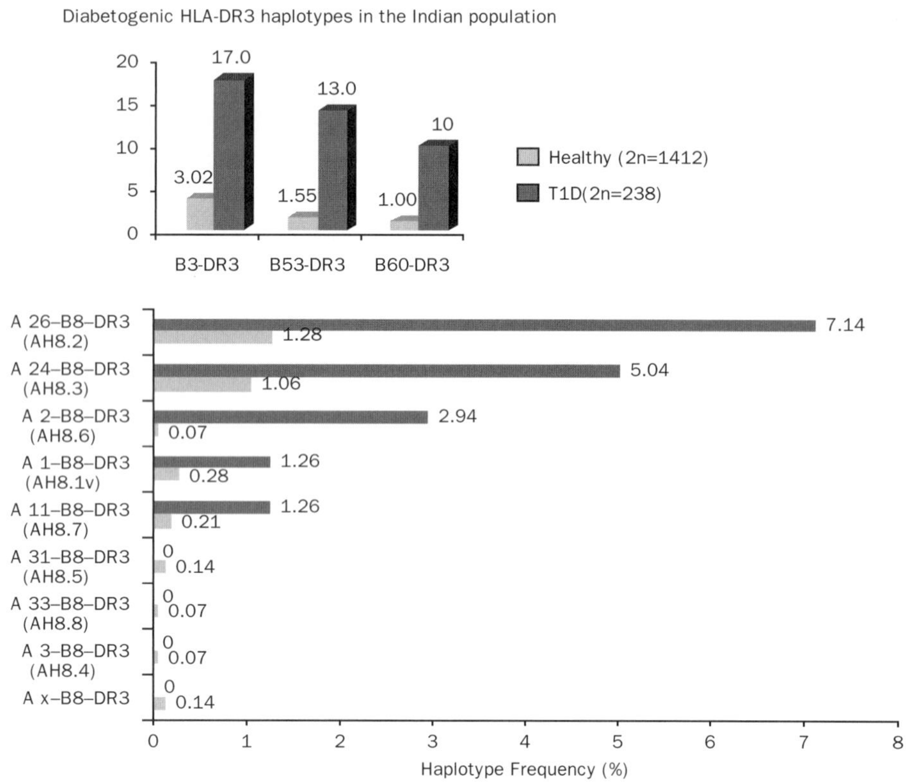

Figure 75-6 Distribution of multiple and diverse DR3+ve haplotypes in the Indian healthy population and patients with type 1 diabetes.

that the probability of finding an individual with homozygous HLA haplotype varies between 0.01% to 0.8% in the Indian population, depending on the ethnic group under study. It is conceivable that establishment of such homozygous HLA stem cell banks could have immense applications in the field of hematopoietic stem cell transplantation.

Such diversity ensures variable composition of the peptide-binding pockets, and their cumulative combinations in a population confer the potential to bind a broad spectrum of peptides. With an insight into the stereochemical properties and solvation entropies of these molecules, it is now possible to predict the peptide-binding specificities of the peptide-binding groove using computer-based algorithms, and extrapolate the findings to vaccine design strategies. Selective predominance of different HLA alleles and haplotypes in different populations poses hindrances in the generation of universal peptide-based vaccines. A dedicated screening effort is required at the global level to determine specificities of various peptide-presenting molecules followed by presentation to T-cells, and thus gain an insight into the subsequent effector immune response.

PERSPECTIVES

It is evident that HLA-based anthropological studies in different ethnic groups have important implications in the field of population genetics, disease association studies, and molecular vaccine design modalities. Such data

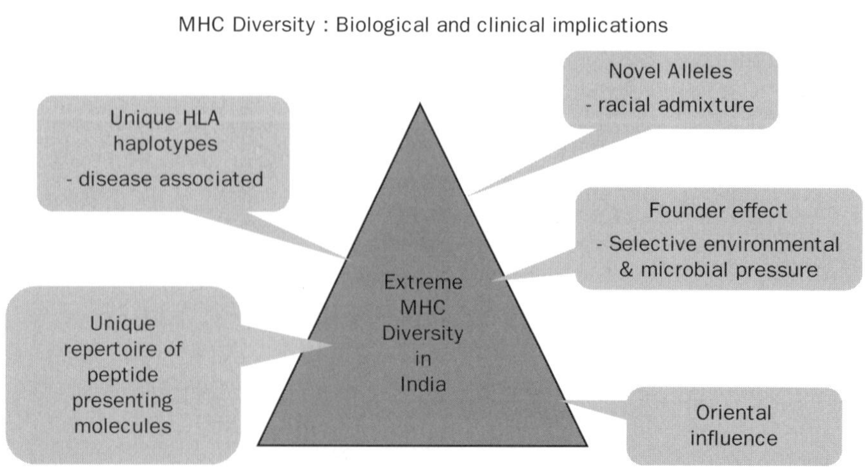

Figure 75-7 Diagrammatic representation highlighting extreme MHC diversity observed in the Indian population alongside its biological and translational implications.

are also necessary in the translational context for donor selection during organ and bone marrow transplantation, identification of permissible mismatches, and of "risk genes" for various diseases.

Major lessons drawn from these studies are summarized as follows:

i. Most populations in the Indian subcontinent follow the Caucasoid pattern of HLA allelic distribution with variable admixture of genes from the Mongoloid and Aryan races;

ii. The observed extreme diversity in the HLA class I and class II region genes in these populations is due primarily to the existence of several "novel alleles" and "unique HLA haplotypes" that could have arisen as a consequence of the founder effect, racial admixture, and geophysical or socioeconomic barriers contributing to the genetic drift;

iii. There is compelling evidence to suggest that selection pressure could have directed the generation of such novel alleles and unique haplotypes.

iv. The high level of variance in the repertoire of peptide presenting MHC molecules implies altered antigen presentation, immune responses, and HLA genetic profiles in relation to disease susceptibility and resistance.

v. The extreme degree of polymorphism may pose limitations for the development of potentially global vaccines and may require development of novel vaccination protocols for effective therapeutic use in extremely diverse populations like those in the subcontinent.

vi. The existence of a large variety of alleles and haplotypes, both unique and representative of other ethnic groups in the Indian subcontinent, poses additional challenges in the transplantation context, particularly with regard to the hematopoietic stem cell transplantation.

In view of the importance of HLA in defining genetic architecture of the human population, understanding the demographic history and population kinship, along with unique constitution of MHC in ethnically well-defined populations, has important implications for immunobiology. This requires the use of advanced molecular technologies determining HLA diversity at high-resolution levels, use of appropriate statistical methods for measuring the selection pressure (balancing or directional, etc.), maximum likelihood analysis, Ewans-Watterson homozygosity, and strict selection of sampling algorithm to avoid sampling errors for better representation of a population. The data generated in the Indian subcontinent on diverse population groups could become an invaluable genomic resource for assessing the implications of new polymorphisms in the generation of immune response, defining susceptibility/resistance factors in diseases, survival in varying environmental conditions, design of universal drugs or vaccines, and finally, to generate an insight into the extent and comparative effect of different evolutionary forces (founder effect, isolation, divergence, genetic admixture, mutations, and environmental selection) on the generation and maintenance of polymorphism.

ACKNOWLEDGMENTS

The authors wish to thank the ICMR and DBT, Government of India, for their generous support toward HLA-based studies. Contributions made by Drs. Ritika Jaini and Neeraj Kumar for anthropological studies, and of Shekhar Neolia toward HLA typing, Drs. KM Murhekar and SC Sehgal for sample collection from Jarawas, and of Drs. Gyozo Petranyi and Katalin Rajczy for collecting gypsy DNA samples, are gratefully acknowledged.

REFERENCES

Bhasin MK, Walter H. (2001). *Genetics of Caste and Tribes of India.* Delhi: Kamla-Raj Enterprises.

Cavalli-Sforza LL, Feldman MW. (2003). The application of molecular genetic approaches to the study of human population. *Nat Genet* 33:266–275.

Jaini R, Kaur G, Mehra NK. (2002a). Heterogeneity of HLA-DRB1*04 and its associated haplotypes in the North Indian population. *Hum Immunol* 63(1):24–29.

Jaini R, Naruse T, Kanga U, et al. (2002b). Molecular diversity of the HLA-A*19 groups of alleles in North Indians: Possible oriental influence. *Tissue Antigens* 59:487–491.

Kaur G, Kumar N, Szilagyi A, et al. (2008). Autoimmune-associated HLA-B8-DR3 haplotypes in Asian Indians are unique in C4 complement gene copy numbers and HSP-2 1267A/G. *Hum Immunol* 69(9):580–587.

Kumar N, Kaur G, Mehra N. (2009). Genetic determinants of type 1 diabetes: immune response genes. *Biomark Med* 3(2):153–173.

Majumdar PP. (2001). Ethnic population of India as seen from an evolutionary perspective. *J Biosci* 26:533–545.

Mehra NK, Jaini R, Balamurugan A, Kaur G. (2001). HLA-A*02 diversity in North Indians. Predominance of HLA-A*0211. *Tissue Antigens* 57:502–507.

Reich D, Thangaraj K, Patterson N, Price AL, Singh L. (2009). Reconstructing Indian population history. *Nature* 461:489–494.

Rozemuller EH, Mehra N, van der Zwan AW, Jaini R, Tilanus MG. (2002). HLA-A*3306, a novel variant in the Indian population. *Tissue Antigens* 59(5):421–423.

76.

GENOMIC APPLICATIONS IN MEDICINE AND HEALTH IN INDIA

Billie-Jo Hardy, Béatrice Séguin, Peter A. Singer, Mitali Mukerji, Samir K. Brahmachari, and Abdallah S. Daar

INTRODUCTION

India currently has the world's second-largest population along with a fast-growing economy and significant economic disparity. It also continues to experience a high rate of infectious disease and increasingly higher rates of chronic diseases. However, India cannot afford to import expensive technologies and therapeutics, nor can it, as an emerging economy, emulate the health delivery systems of the developed world. Instead, to address these challenges it is looking to biotechnology-based innovation in the field of genomics. The Indian Genome variation (IGv) consortium, a government-funded collaborative network among seven local institutions, is a reflection of these efforts. The IGv has recently developed the first large-scale database of genomic diversity in the Indian population that will facilitate research on disease predisposition, adverse drug reactions, and population migration.

THE INDIAN GENOME VARIATION

With one of the world's largest populations, a high burden of infectious disease, and an increasing rate of chronic disease (Boutayeb, 2006), India cannot afford to adopt existing and expensive Western healthcare models to address its local health needs. Instead, India needs to look to innovative, cost-effective models, including adopting emerging health technologies. Genomic sciences and related technologies can add value to India's local healthcare systems by emphasizing prediction and prevention, and possibly decreasing the cost of healthcare through better diagnosis, early detection, and improved treatment and management. Moreover, by investing in these emerging genomic health technologies India could advance its health product innovation system, building capacity in science and technology (S&T) to provide it with a competitive edge in the knowledge-based economy, resulting in economic stimulation. Here we present our findings on a case study of the Indian Genome Variation (IGV) database, a collaborative network initiative that recognized both the richness and the significance of human genomic diversity within the Indian population, as well as the need for a pan-Indian data resource. The resulting database captures essential data about disease predisposition, adverse drug reactions (ADRs) and population migration within India. The IGVdb is a collaboration between the Council of Scientific and Industrial Research (CSIR) Institutes, the Institute of Genomics and Integrative Biology (IGIB), the Center for Cellular and Molecular biology, the Indian Institute of Chemical biology, the Central Drug Research Institute, the Industrial Toxicological Research Center, the Institute of Microbial Technology, and the Indian Statistical Institute (ISI), Kolkata. The IGV data bank contains samples from up to 15,000 unrelated individuals from various subpopulations in India, on the basis of their different geographical zone (northern, southern, eastern, western, central and northeastern) and linguistic categories (Tibeto-Burman, Dravidian, Indo-European, and Austro-Asiatic; Indian Genome Variation Consortium, 2005) Using previously described qualitative methods (Séguin et al., 2008b, 2008c), we performed 15 in-depth semistructured interviews with 17 key informants representing scientists and managers at the IGV, and key informants from diverse backgrounds, such as regulatory and government officials, one policy analyst, and experts in the areas of genomics/pharmacogenetics and ethics from developed and developing countries, to understand the motivation for and the process behind the creation of an initiative such as the IGV database.

KNOWLEDGE-BASED ECONOMY OF INDIA

Economic benefits were a key driver behind the creation of the IGV—a reflection of the Indian government's broader attempt to support economic development through investment in S&T–based innovation. Participants explained to us that emerging innovative genomic-based initiatives such as the IGV contribute toward scientific capacity building, developing and retaining valuable human resources, and providing Indian scientists and the growing domestic private sector with a competitive edge in the global market. In a step toward this last goal, members of the IGV have proposed developing a predictive population therapy database. The database, either solely using IGV data or, potentially, in combination with additional clinical data, would enable various clients to select clinical trial participants in order to stratify clinical trials. Multinationals have been increasingly conducting early-stage clinical trials in India because of the cost savings and the readily accessible, largely drug-naive population resource. India has thus positioned itself as a global hub for conducting clinical trial research by investing in capacity and infrastructure (Nundy & Gulhati, 2005; Maiti, 2007; Singh, 2006; Thiers et al., 2008). Some domestic companies subsidize their research and development (R&D) platforms by providing contract services for clinical trials to multinational and foreign companies (Frew et al., 2007). Accordingly, a "predictive population database" could help maintain India's competitive edge by improving the selection specificity through stratification of the test population, thereby further reducing the time and cost associated with conducting clinical trials in India. Members of the IGV are also exploring innovative applications in niche markets. For instance, a new field, termed *ayurgenomics*, is investigating the potential relationship between human genomic variation, pharmacogenomics, personalized therapeutics, and the traditional Indian medicine, Ayurveda. Individual variation was described in India by Charaka over 4,000 years ago, and it continues to be the basic principle of the Ayurveda, which is still widely practiced across India (Patwardhan, 2003; Patwardhan et al., 2004). The IGV also hopes to contribute to the Indian biotechnology sector, which experienced a growth rate of 30% in 2007 led by the biopharma sector (Indo-Asian News Service, 2007). For instance, it is making efforts to protect its basic research discoveries. Currently, the IGV provides the global academic community with access to its database through an Internet portal under a click wrap agreement (an online agreement prompting users to read and agree to the terms of use prior to access). We were told by participants, however, that the discoveries made by the IGV will be protected through patents and will subsequently need to be licensed through the IGV. The IGV has also begun to provide its researchers with commercialization support in order to encourage translation of basic research to potential products and therapeutics. It has partnered with a not-for-profit, public–private technology transfer laboratory—The Center for Genomic Applications (TCGA). TCGA is a partnership between the Institute for Molecular Medicine, the IGIB, the Council for Scientific and Industrial research (CSIR) and The Chatterjee Group. In partnership with IGV researchers, TCGA will provide the technological expertise and capacity to conduct the proof-of-concept and Phase I clinical studies that are needed to encourage domestic and foreign industry investment in early-stage commercial opportunities in India.

The development of therapeutic products or services from the IGV database is unlikely in the near future, given its early stage of development. Thus, in terms of more immediate commercialization opportunities from human genomic diversity knowledge, we tried to gauge the involvement of the domestic private sector. Generally, we found it to be limited; however, Avesthagen Ltd., an Indian life sciences company, announced in March 2007 that it will be initiating a large-scale genotyping project of the Parsi population—the Avesthagenome Project (Jayaraman, 2007). Estimated to take up to 5 years to complete and costing approximately US$32 million, the project proposes to genotype India's entire Parsi population (~69,000 people) and develop a database containing genomic information as well as linkage to family history and medical records. The Parsis are an ethnic minority in India who are considered genetically homogenous and are feared to be at risk of extinction owing to their religion, which prohibits marriages outside of the community (Jayaraman, 2007). –The aim is to determine the link between genes, disease, and environmental factors, to better understand disease and to develop new therapies and diagnostics with a focus on chronic diseases, such as cancer and central nervous system disorders, that can be used to directly benefit the Parsi population, with an additional potential to be marketed globally where appropriate. The IGV is looking to position the Indian research community as a global knowledge partner, as opposed to a service provider that facilitates foreign research. Crucial to their success are their exploration of niche market opportunities (for example, ayurgenomics), their efforts to encourage patent protection of research discoveries, and the establishment of strategic partnerships to foster the translation of basic research. In addition to these approaches, the IGV needs to consider mechanisms to provide support to researchers wishing to out-license or create companies on the basis of their technology. They will also need to further pursue strategic opportunities with the domestic private sector. For instance, how can the complementarities between the IGV and the Avesthagenome project be best leveraged? Until much of the IGV data is validated, the existing initiatives will provide the consortium with the momentum required to get started.

GENOMIC SOVEREIGNTY

Key informants in several developing countries have told us that in order to generate the potential health and economic benefits expected from human genomic diversity studies, they will need to prevent unapproved exports of genomic materials and protect the data contained therein through legislation (Séguin et al., 2008b). India is no exception. One-fifth of the global population lives in India, and thus a significant amount of human genomic variation can be found there (Séguin et al., 2008a). Accordingly, key informants in this study asserted that the availability of India's large population resource, with its multigenerational endogamous families and well-maintained genealogical records, provides them with a unique research advantage. Participants advised us that, in the past, Indian physicians were provided with incentives (for example, vacations and/or equipment) by foreign researchers in order to encourage them to share their patient samples. These physicians were often not aware that by doing so, they might inadvertently be contributing to foreign patents on discoveries made using Indian patient samples that would subsequently limit local access to resulting therapeutics. Many of the participants we interviewed also cited repeated conflicts between foreign and local researchers regarding benefit sharing and publications. Thus, in November 1997, the Guidelines for Exchange of Human Biological Material for Biomedical Research Purposes were updated by the Indian Council of Medical Research (ICMR) so that government permission is required for the export of human tissue samples acquired in India. By preventing their export, these guidelines have provided Indian researchers with a mechanism to protect their discoveries. However, several participants pointed out that these guidelines are not enforceable owing to a lack of human resources. Moreover, multinational pharmaceutical companies conducting research within India are able to bypass these guidelines and, although they might be providing training and employment, there are no guarantees that the Indian population will gain significant long-term economic or health benefits from this research. Interviewees thus recommended that the Indian government establish legislation to enforce the guidelines and extend them to address the collection and use of genomic data by multinationals conducting research within the country. However, if the Indian government were to establish genomic sovereignty legislation, such as that recently enacted in Mexico (Becerril, 2008), it is not clear how it would be enforced. Moreover, recent concerns have arisen regarding biodiversity legislation in India—in its current form it impedes international collaboration, reducing opportunities for Indian scientists to share their data and contribute to larger research projects (Prathapan et al., 2008). The debate is an important one, as the knowledge generated from these national data banks might need to be interpreted in larger international collaborative efforts before it can contribute to generating practical health benefits.

POLITICAL WILL

In 2001, the Indian Ministry of Health pledged $20 million in public funding toward medical genomics (Mudur, 2001). Consequently, in addition to the IGV, which received $5 million from the CSIR, there are several genotyping and genomic-based research initiatives that receive public funding for basic research from the ICMR, the Department of Biotechnology (DBT) and the CSIR. Moreover, in 2007, the Indian government approved the National Biotechnology Development Strategy (also known as "the Strategy"), a policy document prepared by the DBT. The DBT falls under the Ministry of Science and Technology, and is linked with the commercial development of biology and biotechnology in India. The Strategy outlines proposed actions for boosting the translation of life-sciences knowledge, such as creating investment capital, boosting technology transfer and intellectual property regulations, addressing regulatory issues, and enhancing public understanding. Earlier drafts of the Strategy also highlighted genomic medicine as a strategic direction and proposed that India could act as a global hub for genomic medicine R&D. In addition to strategic policy, ethical guidelines for biobanking in India have been established by the ICMR (see the Ethical Guidelines for Biomedical Research on Human Participants). These types of guidelines can stimulate R&D because they provide a framework with parameters that scientists and industry can work within, while also ensuring transparency and public accountability. In order to ensure that the research generated by the IGV is leveraged toward economic and health benefits, it is crucial that the political will in place now is maintained, and that basic research is supported throughout development, commercialization, and delivery. The approval of these strategies and guidelines provides the momentum necessary to do so. However, the Indian government will need to develop success measures for the Strategy, and focus on developing future policy guidelines for regulatory and delivery issues to ensure public acceptance. For instance, it is thought that the Genetic Information Nondiscrimination Act (GINA), recently passed by the U.S. Congress, will help spur public acceptance of the genomic medicine industry in the United States (Stein, 2008).

LOCAL HEALTH BENEFITS

India has a history of investing in genetic research to improve local health. Key informants told us that genetic research in India has resulted in tangible public health benefits, such as blood banks and screening registries. For instance, the DBT has established 16 units throughout India that provide genetic diagnosis and genetic counseling services for prenatal diagnosis and genetic disorders, such as beta-thalassemia, Duchenne muscular dystrophy, hemoglobinopathies, and neurogenetic disorders that are prevalent in the country (see The Human Genome and Human Genetics and Genome Analysis, produced by the DBT). We were also told that genetic research in India has resulted in social benefits, such as reduced stigmatization previously associated with genetic disease. Accordingly, the IGV has been publicly funded with the expectation that, in the long term, it will promote local health savings through advances in understanding predisposition to disease, mechanisms of disease, pharmacogenomics, and personalized medicine. Members of the IGV have recently published their results on 405 SNPs in 75 disease genes and a 5.2 Mb region of chromosome 22 in 1871 individuals from 55 population groups in India (Séguin et al., 2008a). Using these SNPs, members of the IGV are currently conducting population-based screens for susceptibility to asthma, diabetes, neuropsychiatric disorders, high-altitude sickness, retinitis pigmentosa, malaria, and infectious diseases, as well as variation in drug metabolism. Members of the IGV also told us of an application they are developing that might reduce healthcare costs in India—a predictive population therapy database that investigates drug response variability in the Indian population. One participant provided the basis for why a database could be beneficial when he told us that "[up to] 13% of the Indian population does not respond to up to 30 very essential drugs in northern India...." As a result, low-cost therapeutics for which there are high rates of nonresponse and ADRs are often substituted with newer, more expensive therapeutics. If the nonresponder and ADR cases can be explained by genomic variation, these expensive substitutions can be superseded with better tailored treatments. As a proof of concept, researchers at the IGV have demonstrated an association between a panel of SNPs located at the beta(2)-adrenergic receptor (ß2Ar) locus and the response to salbutamol, a low-cost asthma therapeutic, in Indian asthma patients. These results could be used to design a diagnostic that would identify patients that are resistant to salbutamol (Séguin et al., 2008a; Kukreti et al., 2005; Bhatnagar et al., 2006). Although this method can contribute to reducing individual healthcare costs and ultimately contribute to improving public health, given the size of India's population and disparate levels of education and poverty across the country, there are significant challenges in ensuring the adoption and delivery of the benefits associated with genomic medicine. For instance, in contrast to the National Institute of Genomic Medicine (INMEGEN) in Mexico (Séguin et al., 2008d), the IGV is not linked with the Indian Ministry of Health, creating a gap between R&D and the eventual delivery of local health benefits.

GENOMIC MEDICINE CHALLENGES IN INDIA

Academic Links.

Strong links and synergies between institutions provide opportunities for knowledge flow and the development of expertise, and are part of a successful health biotechnology innovation system (Thorsteinsdottir et al., 2004). The IGV is a unique example in this regard, and has set the trend for many new consortia initiatives in India. Nevertheless, according to our key informants, coordination between academic institutions in India remains weak. One government official pointed out how this lack of coordination can result in barriers to progress and can prevent industry involvement. Consequently, public research institutions in India need to develop a national foresight map for large-scale genotyping projects, human genomic variation studies, and disease association studies that will help leverage existing data networks and create opportunities for collaboration between institutions.

Public–Private Sector Links.

Local private sector involvement, to develop and commercialize knowledge, is necessary for sustainable development in a country's health-related biotechnology sector (Thorsteinsdottir et al., 2004; Frew et al., 2006). Although partnerships between academic institutions and industry continue to strengthen in India, to date they remain insubstantial (Frew et al., 2007). Policy efforts exist to promote these partnerships. For instance, the CSIR provides public–private partnership grants (see the New Millennium Indian Technology Leadership Initiative). In addition, the National Biotechnology Development Strategy recently proposed increased spending on public–private partnership initiatives, and a biotechnology industry partnership program for advanced technologies.

Although these programs are a good start, their robustness remains to be tested. Key informants told us that data sharing often created a barrier to these partnerships; negotiations regarding data sharing can be affected by disagreements over publication timelines, intellectual property rights, and return-on-investment strategies. Moreover, our informants maintained that the local private sector is either uninterested or not prepared to foray into genomic medicine, partly owing to their interest in short-term gain and financial restraints, as well as their lack of R&D capacity. However, the announcement of Avesthagenome might signal a burgeoning interest by the domestic private sector.

Commercialization in the Public Sector.

Commercialization of research associated with the IGV, as well as other Indian institutions involved in human genomic variation studies, is generally seen as a long-term goal. Researchers point out that human genomic variation studies are in their early stages, require further validation, and currently have few obvious commercial applications. As a result, Indian researchers generally view peer-reviewed publications as a more important goal than research commercialization. Perhaps reflecting this lack of interest in commercialization, or merely the need for clinical validation, we were told that some institutions are currently providing local clinics with their findings on drug response variation to enable physicians to account for these differences when treating their patients. Some members of the IGV have addressed this challenge by entering into a partnership with TCGA, thereby ensuring that commercialization of basic research is encouraged. However, additional incentives, such as the proposals made in the recent National Biotechnology Development Strategy, will also need to be implemented if India is to become a hub for R&D in genomic medicine.

The Need for Regulatory Frameworks

The Indian Ministry of Health regulates all drugs and related products, as well as clinical trials in collaboration with the ICMR and the DBT, through the Central Drugs Standard Control Organization (CDSCO). The CDSCO has not yet established guidelines for clinical trials or therapeutics that rely on population differences, genomic sciences, population-based therapeutics, or pharmacogenomics. We were told by a regulatory official that regulations were not necessary, as there were no available products as yet. Instead, he suggested that the regulatory framework will evolve as these products become available. By contrast, researchers felt that the lack of regulatory guidelines in India will serve as a barrier to the adoption of genomic sciences for health benefits because there is no incentive to report clinical data or develop products. Key informants recommended that imported therapeutics that show population variance in their efficacy should undergo clinical trials in India prior to approval, and that all clinical trials should submit genomic variation data. For instance, one participant mentioned that "a lot of vaccine trials are going to happen in India in the near future; [it would be] very important to look at the pharmacogenomics of that." But even if the CDSCO were to implement guidelines, voluntary or otherwise, the resources needed to initiate and monitor adherence to regulatory guidelines are not in place. As one interviewee pointed out, "Indian systems are very proactive today. They really want to do things. So the leadership is very clear. But what the leadership is lacking are people under them."

Delivery requires innovative solutions

It has been suggested that the benefits of genomic medicine will not reach the poor owing to the expected high cost of delivery (Holm, 2006). Thirty-four percent of India's population continues to exist on less than $1 per day (see India Country Overview 2007 from the World Bank website). Moreover, there is significant inequality between the rural and urban regions, with approximately 72.2% of India's population living in rural areas (see the Census of India Rural–Urban Distribution). The suggestion that these rural poor make India a potentially poor market for genomic medicine is now being challenged by models that suggest that these populations are more receptive to innovative technologies. Indeed, the rural poor are thought to make up part of a potentially multitrillion dollar market, with suggestions that their informal economy accounts for 40%–60% of the economic activity in the developing world (Prahalad & Hart, 2002).

These lessons are directly applicable in India. For example, we learned of a partnership being negotiated between a private company and an Indian bank for a rural model for delivery, through micro-franchising, of functional foods (processed foods with added nutritional claims of health promotion and/or disease prevention) and nutraceuticals. In the long term, similar delivery models could be extended to genomic medicine: cheap diagnostics could be provided at the level of the village to identify those patients who would need to travel to the urban regions for treatment. Large healthcare units, banks, and industry could develop partnerships to cut costs of treatment and improve their ability to harness this market by tapping into the existing infrastructure. Clearly, genomic medicine is in its early stages and cheap diagnostics have yet to be developed. However, if local health benefits are expected to reach the whole population, it is appropriate to begin examining models of how to best harness these markets. In order to move such proposals forward, stakeholders will need to address the concerns associated with the involvement of the private sector, insurance providers, and banks in the delivery of healthcare.

Physicians' Understanding

Physicians' understanding of genomic medicine and its applications is limited in India. Consequently, integration of genomic medicine into the clinic will pose a challenge. Physicians will require instruction on human genomic variation and how to use this variation for the benefit of medical practice, including how drug efficacy and safety can be affected by common genetic polymorphisms in drug-metabolizing enzymes. However, as one researcher said, "... our medical curricula don't have structure for pharmacogenomics or for understanding how genetics would be applied to medicine." In an effort to bridge this gap between research and physicians, we were told of nongovernmental physician education programs, such as the Moving Academy of Medicine, and the government website, Genetics India, which provide information on the use of genetics in medicine. Moreover, institutes such as the Manipal Life Sciences Center at Manipal

University provide presentations on genomics and potential applications to doctors. However, for effective physician education to take place, standardized programs on genomics and genomic-based applications will need to be implemented within the medical curricula at universities, such as those that are being implemented by INMEGEN (Séguin et al., 2008d).

Public Awareness

Public awareness of human genomic variation, genotyping initiatives, and genomic medicine in India remains unclear. The informants we interviewed stated that public awareness of these subjects was low because of the lack of public engagement. One interviewee pointed out that scientists and researchers rarely engage with the public on scientific issues because "... we have problems in our country with respect to how much information you would divulge because of the status of education." Even so, there is evidence of nongovernmental organizations spearheading efforts to inform the public about medical biotechnology and its potential applications. The Jan Swasthya Abhiyan group, a national coordination committee, distributes a pamphlet entitled "New Technologies in Public Health—Who Pays and Who Benefits?" that provides the general public with the People's Health Movement views on the ethical and social implications of genomics and health. However, some participants felt that because of the public's familiarity with Ayurveda, a traditional form of personalized medicine, extending the personalized approach to include genomics would meet with little resistance. One interviewee told us: "Of course that is something different about the way we look at pharmacogenomics and the genotyping and the phenotyping correlation, but that concept had existed." But the interviewees were clear that without the support of the public, the adoption of genomic medicine into the healthcare setting will be difficult. Thus, interviewees recommended the development and implementation of public engagement programs to increase public awareness and understanding and to increase public support.

CONCLUDING REMARKS

India is spearheading efforts to benefit from genotyping projects and associated technologies, including pharmacogenomic and genome-wide disease association studies. Here, we documented this trajectory by exploring the IGV.

The IGV is a reflection of the desire to fuel economic development through investment in S&T innovation, the belief that there will be significant public health benefits and savings to the healthcare system, and the need to ensure that genomic variation contained within the Indian population is harnessed to its benefit. The IGV was initiated, despite limited resources, by instituting a vertically integrated network of existing research institutions and public–private partnerships with service providers. The IGV is currently validating its data and, given its early stage of development, whether it meets any of its long-term goals remains to be seen. Before the IGV can become a sustainable initiative with concrete benefits, such as health applications derived from the knowledge of genomic variation, there are a number of challenges that will need to be addressed within the Indian life-sciences innovation framework. In some cases, the IGV and the DBT have addressed these challenges, as exemplified by the IGV partnership with TCGA, and the DBT's biotechnology strategy that address challenges to commercialization and collaboration. However, in the area of regulatory guidelines, physician understanding, and public engagement, there has been little or no activity. These important challenges need to be addressed in the near future as other genotyping and pharmacogenomic initiatives are underway in India, both in the private and public sector. A few of these challenges are being effectively addressed by similar initiatives in Mexico (Séguin et al., 2008d; INMEGEN), South Africa (Hardy et al., 2008; Africa Genome Education Institute) and Thailand (Séguin et al., 2008b; Thailand Center for Excellence in Life Sciences), to which India could look for guidance; others are being experienced internationally, providing India with the opportunity to develop innovative solutions and become a global leader in this sector.

REFERENCES

Becerril A. (2008). Aprueba el Senado sanciones a quienes trasladen tejido humano fuera de México. *La Journada* [online]: http://www.jornada.unam.mx/2008/03/28/index.php?section=sociedad&article=046n1soc (in Spanish).

Bhatnagar P, Guleria R, Kukreti R. (2006). Variable therapeutic response in asthma: a genetic perspective. *Personalized Medicine* 3:61–78.

Boutayeb A. (2006). The double burden of communicable and non-communicable diseases in developing countries. *Trans R Soc Trop Med Hyg* 100:191–199.

Frew SE, Sammut SM, Siu WW, Daar AS, Singer PA. (2006). The role of the domestic private sector in developing countries for addressing local health needs. *Int J Biotechnology* 8:91–102.

Frew SE, et al. (2007). India's health biotech sector at acrossroads. *Nature Biotechnol* 25:403–417.

Hardy B, Séguin B, Ramesar R, Singer PA, Daar AS. (2008). South Africa: from species cradle to genomic applications. *Nature Rev Genet* 9 (Suppl):S19–S23.

Holm, S. (2006). In: Cohen JC, Illingworth P, Schuklenk U, eds. *The Power of Pills: Social, Ethical and Legal Issues in Drug Development, Marketing and Pricing*. pp. 98–109. Ann Arbor, Michigan: Pluto Press.

Indian Genome Variation Consortium. (2005). The Indian Genome Variation database (IGVdb): a project overview. *Hum Genet* 118:1–11.

Indo-Asian News Service. (2007). Indian biotech sector crosses $2 billion in FY 2007. *Twocircles.net* [online]: http://www.twocircles.

net/2007jun07/indian-biotech-sector-crosses-2-billion-fy-2007.html.

Jayaraman KS. (2007). Database targets Parsi genes. *Nature* 446:475.

Kukreti R, et al. (2005). Beta(2)-adrenergic receptor polymorphisms and response to salbutamol among Indian asthmatics. *Pharmacogenomics* 6:399–410.

Maiti R, Raghavendra M. (2007). Clinical trials in India. *Pharmacol Res* 56: 1–10.

Mudur G. (2001). India invests heavily in genomics research. *BMJ* 322:576.

Nundy S, Gulhati CM. (2005). A new colonialism?—Conducting clinical trials in India. *N Engl J Med* 352:1633–1636.

Patwardhan B. (2003). Ayurgenomics: integration for customized medicine. *Indian J Nat Products* 19:16–23.

Patwardhan B, Vaidya A, Chorghade M. (2004). Ayurveda and natural product drug discovery. *Current Sci* 86:789–799.

Prahalad CK, Hart SL. (2002). The Fortune at the Bottom of the Pyramid. *Strategy and Business Magazine*. [online]: http://www.cs.berkeley.edu/~brewer/ict4b/Fortune-BoP.pdf.

Prathapan K, et al. (2008). Death sentence on taxonomy in India. *Current Sci* 94:170–171.

Séguin B, Hardy B, Singer PA, Daar AS. (2008a). Indian Genome Variation Consortium: Genetic landscape of the people of India: a canvas for disease gene exploration. *J Genet* 87:3–20.

Séguin B, Hardy B, Singer PA, Daar AS. (2008b). Genomic medicine and developing countries: creating a room of their own. *Nature Rev Genet* 9:487–493.

Séguin B, Hardy B, Singer PA, Daar AS. (2008c). Human genomic variation initiatives in emerging economies and developing countries. *Nature Rev Genet* 9 (Suppl):S3–S4.

Séguin B, Hardy B, Singer PA, Daar AS. (2008d). Genomics, public health and developing countries: the case of the Mexican National Institute of Genomic Medicine (INMEGEN). *Nature Rev Genet* 9 (Suppl):S5–S9.

Séguin B, Hardy B, Singer PA, Daar AS. (2008e). Universal health care, genomic medicine and Thailand: Investing in today and tomorrow. *Nature Rev Genet* 9(Suppl): S14–S19.

Singh S. (2006). India Bullish on Global Clinical Trial Business Opportunity. *Science and Marketplace*

Stein L. (2008). Congress Passes Bill Barring Genetic Discrimination. *Scientific American* [online], http://www.scientificamerican.com/article.cfm?id=bill-bars-genetic-discrimination.

Thiers F, Sinskey A, Berndt E. (2008). Trends in the globalization of clinical trials. *Nature Rev Drug Discov* 7:13–14.

Thorsteinsdottir H, Quach U, Daar AS, Singer PA. (2004). Conclusions: promoting biotechnology innovation in developing countries. *Nature Biotechnol* 22 (Suppl): DC48–DC52.

FURTHER INFORMATION

McLaughlin-Rotman Centre for Global Health: http://www.mrc-global.org

Africa Genome education institute: http://www.africagenome.com/

Census of India Rural–Urban Distribution: http://www.censusindia.gov.in/census_Data_2001/India_at_glance/rural.aspx

Ethical Guidelines for Biomedical Research on Human Participants: http://www.icmr.nic.in/ethical_guidelines.pdf

Guidelines for Exchange of Human Biological Material for Biomedical Research Purposes: http://icmr.nic.in/min.htm

Human Genetics and Genome Analysis: http://india.gov.in/sectors/science/index.php?id=34

National Biotechnology Development Strategy: http://dbtindia.nic.in/biotech_strategy.htm

National institute of Genomic Medicine (INMEGEN), Mexico: http://www.inmegen.gob.mx

New Millennium Indian Technology Leadership Initiative: http://www.csir.res.in/external/Heads/collaborations/Nmitli.htm

Thailand Center for Excellence in Life sciences: http://www.tcels.or.th/

The Human Genome: http://pib.nic.in/feature/feyr2001/fsep2001/f120920011.html

77.

EPIDEMIOLOGY OF GENETIC DISEASES IN THE INDIAN SUBCONTINENT

Ishwar C. Verma and Dhavendra Kumar

INTRODUCTION

The Indian subcontinent includes several developing countries—India, Pakistan, Sri Lanka, and Bangladesh. The infant mortality rate (per 1000 infants under 1 year) in these countries is gradually declining but remains high compared to developed countries in the west (WHO, 2010). In all these countries there is an accelerating demographic switch to noncommunicable diseases, particularly among the urban populations (Verma and Bijarnia, 2002). This is most apparent in Sri Lanka, with its low mortality rate and it is not surprising, therefore, that noncommunicable diseases of adults are receiving great attention in this country. In all the four countries, a gradually increasing prevalence of complex diseases due to mutifactorial/polygenic inheritance, such as diabetes mellitus, coronary artery disease, hypertension, etc., has been recorded.

There are number of factors that significantly increase the prevalence of genetic disorders in the Indian subcontinent. Consanguineous marriages are a major factor (see Chapters 11 & 80). This custom is common, accounting for almost 30%–40% of marriages among south Indians and Muslims of north India, and a smaller percentage among other communities like the Sikhs and Maharashtrians. Almost two-thirds of marriages in Pakistan (Afzal et al., 1994) and Bangladesh are between close relatives. Several studies have attributed this custom to be an important factor in mortality and morbidity among children as well as adults (Bittles, 2001). In a series of papers by Yaqoob and colleagues (1993, 1995) a significant harmful effect of inbreeding was observed in children in Pakistan. Much information on the increased morbidity and mortality among the consanguineous families from Pakistan and Bangladesh has come from England, where there are a large number of migrants from these countries (Balarajan et al., 1989). The study of consanguineous families from these countries has led to the identification of many new genes (Roberts et al., 2002; Ahmed et al., 2003). An international consortium on consanguinity provides regular updates on various aspects of consanguinity (www.consang.net).

The large population of these countries, especially India, Pakistan, and Bangladesh, and a high birth rate results in many patients with genetic disorders (Verma, 2000). Another important factor is the relatively improved medical and diagnostic facilities that allow early identification and management of genetic disorders, which hitherto were submerged by infections and malnutrition (Verma and Bijarnia, 2002).

In India, congenital anomalies and genetic disorders have become important causes of morbidity and mortality. A significant proportion of the perinatal and infant mortality in cities is related to major congenital anomalies and potentially fatal genetic disorders. For example, New Delhi's All India Institute of Medical Sciences Hospital recorded fewer perinatal deaths due to congenital malformations in the 1970s compared to the 1980s and 1990s, when the trend had reversed and malformations had surpassed immaturity as a major factor in perinatal mortality (Verma, 2000). A somewhat similar trend has been observed in other urban metropolitan cities of India. In a major Mumbai (previously Bombay) hospital serving a predominantly poor population, congenital malformations were the third commonest cause (13.2%) of perinatal mortality (Mehta, 1998). The neonatal–perinatal database of the National Neonatology Forum, India, surveyed 1460 stillbirths and 1400 neonatal deaths (National Neonatology Forum, 1995). The forum recorded congenital malformations as the second commonest cause (9.9%) for mortality among stillbirths, and the fourth commonest cause (9.6%) of neonatal mortality.

The present chapter discusses prevalence of genetic diseases with reference to the Indian population. The data from other countries is limited; however, chapters on Pakistan, Bangladesh, Nepal, and Sri Lanka provide some epidemiological data on genetic diseases.

THE PREVALENCE OF BIRTH DEFECTS AND GENETIC DISORDERS

There are almost 24 million births per year in India. A large number of infants with genetic disorders are born annually. The data from population census and published frequencies of disorders is summarized in Table 77-1. The estimates for glucose-6-phosphate dehydrogenase (G6PD) deficiency are for males only, although G-6PD deficiency is also observed in females, since due to the high prevalence of the gene, both the father and the mother may possess the mutant gene. Sickle cell disease occurs predominantly among tribal communities, although it is observed in nontribal communities living in these areas.

CONGENITAL MALFORMATIONS

The frequency of congenital malformations in India is derived from various sources: a meta-analysis of 23 studies carried from 1969 to 1982 (Verma, 1978), a hospital-based prospective study (Verma et al., 1990), and a multicenter study sponsored by the Department of Atomic Energy, India (Verma et al., 1998). All these studies provide a prevalence figure between 2% and 3%. The latter study included 94,610 newborns. It provided a prevalence figure of 2.03%, employing rigorous diagnostic criteria. The prevalence figures (per 10,000 births) for various malformations, based on the three-center study (Verma et al., 1998), are provided in Table 77-2.

NEURAL TUBE DEFECTS

Analysis of 496,891 births from various parts of India revealed a frequency of neural tube defects at 4.2 per 1000, while anencephaly and spina bifida had a frequency of 3.07 and 2.0 per 1000 births, respectively (Verma, 1978). The prevalence was extremely high in north India (more than about 1 in 250 to 285 births per year), especially in the states of Punjab, Haryana, Delhi, and Rajasthan, with a decreasing trend toward the east (Bihar and West Bengal) and south of the country (Tamil Nadu and Kerala). In India, the Indian Council of Medical Research (1999) carried out a multicentric trial that demonstrated a significant protective effect of periconceptional vitamin therapy to mothers with previously affected children. In this trial, multivitamins were used rather than folic acid alone, on the premise that Indian women have greater deficiencies of vitamins than women in the West. In South India, Kulkarni and associates (1997), in a much smaller study, showed that folic acid prevented neural tube defects in South India.

TABLE 77-1 PREVALENCE OF GENETIC DISEASE AT BIRTH IN INDIA

Disorder	Incidence	Estimated cases/year
Congenital malformations	1 in 50	495,096
G6PD deficiency	1:35 to 1:10 (M)	390,000
Down syndrome	1:1139	21,412
β-thalassemia	1:2700	9,000
Sickle cell disease	1:1200	5,200
Amino acid disorders	1: 2347	9,760
Other metabolic disease	1:2500	9,000
Duchenne muscular dystrophy	1: 5000 (M)	2,510
Spinal muscular atrophy	1:10,000	2,510

Modified from Verma and Bijarnia, 2002.

TABLE 77-2 PREVALENCE OF COMMON MALFORMATIONS IN INDIA (VERMA ET AL., 1998)

Malformations	Cases per 10,000	Estimated births
Neural tube defects	36.3	88,532
Talipes	14.5	35,364
Polydactyly	11.6	28,364
Hydrocephalus	9.5	23,169
Cleft lip and/or palate	9.3	22,681
Congenital heart disease	7.1	17,316
Hypospadias	5.0	12,194
Tracheoesophageal fistula	3.7	9,023
Diaphragmatic hernia	2.6	6,341
Anorectal atresia/stenosis	2.4	5,853
Microcephaly	2.2	5,365
Cleft palate alone	1.7	4,146
Intersex and cryptorchidism	1.6	3,902
Intestinal atresia/stenosis	1.2	2,926
Anophthalmia/microphthalmia	1.0	2,438

DOWN SYNDROME

A meta-analysis of various studies (75,103 births) on the prevalence of malformations in India revealed 82 cases (frequency of 1 in 916 births) of Down syndrome (Verma, 1988). However, ascertainment was carried out using only the clinical criteria. Another survey from three centers indicated a frequency of 1 in 1,150 births (Verma et al., 1998). This study used cytogenetic methods to confirm the diagnosis suspected on clinical grounds. Each baby included in the study was assessed by a doctor with special training in dysmorphology.

FRAGILE-X SYNDROME

Fragile-X syndrome is the second commonest cause of inherited mental retardation. In a study on mentally retarded persons, the prevalence was 1.8% among patients of both sexes, and 2.8% in males only (Elango and Verma, 1996). However, 5.8% males with nonspecific mental retardation were found to have fragile-X syndrome. Another study revealed the fragile-X syndrome in 4.5% of children referred to hospitals with mental retardation, and in 2.8% of children attending special schools (Jain et al., 1998). Sharma et al. (2001) determined the frequency of fragile-X A and E syndromes in an institutionalized population (n = 130) in New Delhi, India. Eligibility criteria for inclusion of subjects in the study were mild/moderate mental retardation, with/without family history, and the fragile-X clinical phenotype. Screening by Southern hybridization revealed an overall frequency of 7.7% of the syndrome in the sample population.

METABOLIC DISORDERS

A multicenter survey on inherited metabolic diseases indicated a wide spectrum of metabolic disorders including mucopolysaccharidoses, lysosomal storage disorders (Gaucher's disease, Niemann-Pick disease), Pompe's disease, Wilson disease, and galactosemia (Bharucha et al., 1994). A screen of 112,269 neonates for amino acid disorders in the southern state of Karnataka showed four common metabolic diseases—tyrosinemia, maple syrup urine disease, and phenylketonuria. A combined frequency of 1 per 2,495 births was recorded for these conditions. The frequency for generalized aminoaciduria was higher at 1 per 1,605 births (Appaji Rao et al., 1988; Appaji Rao, 1991). In another hospital-based study (Kaur et al., 1994), biochemical screening of 4400 cases of mental retardation revealed abnormalities in 256 individuals (5.7%). The most frequent of these included hyperglycinemia, homocystinuria, alkaptonuria, and maple syrup urine disease. In a study at Sir Ganga Ram Hospital extending over 5 years, we found the commonest disorders to be amino acid disorders, followed by metachromatic leukodystrophy, Krabbe's disease, and mucopolysaccharide disorders (Verma et al., 2003).

LATE-ONSET GENETIC DISORDERS

Major tertiary teaching hospitals across the Indian subcontinent look after several families with late-onset dominantly inherited neurological diseases. However, there are no reliable prevalence figures, due to lack of published data in peer-reviewed periodicals. Some reports provide molecular data in families with Huntington's disease, Machado-Joseph disease and spinocerebellar ataxias (see Chapter 88). Molecular studies on fragile X-syndrome are well established at University of Delhi, south campus (Sharma et al., 2001), and at Sir Ganga Ram Hospital, while other triplet repeat disorders are tested at Delhi, Kolkatta, and Hyderabad (Murgod et al., 2001; Basu et al., 2000; Saleem et al., 2000, 2003. At Sir Ganga Ram Hospital, triplet repeat disorders were seen among 4.6% of all patients referred for molecular studies (Verma et al., 2003).

TABLE 77-3 BURDEN OF GENETICALLY RELATED DISORDERS IN ADULT LIFE

Disorder	No. in millions
Coronary artery disease (> 30 yr)	10.6
Hypertension (20–60 yr)	19.4
Diabetes mellitus (>15 yr)	11.1
Mental illness (schizophrenia, manic depressive psychosis)	40.0
Epilepsy	26.7
Mental retardation	25.6
Dyslexia	22.6
Total	156.0

COMMON MULTIFACTORIAL MEDICAL DISORDERS

Clinical studies from India, Pakistan, Sri Lanka, and the United Kingdom indicate that there is a high prevalence of polygenic/multifactorial disorders, such as ischemic heart disease, oro-pharyngeal cancer (see Chapter 85), diabetes mellitus (see Chapter 87), neuro-psychiatric illnesses (including Parkinsonism, schizophrenia and manic depressive psychoses—Chapter 93) and bronchial asthma in adulthood (see Chapter 82). Although the available data (Table 77-3) is based on selected populations, nevertheless the alarming rise in the frequency of these disorders is disconcerting and requires careful examination by the public health authorities in these countries. Studies on coronary heart disease and diabetes mellitus among the immigrants from the Indian subcontinent to United Kingdom provide useful medical evidence, but fall short of providing reasonable prevalence data (McKeigue et al., 1988, 1989).

SUMMARY

The Indian subcontinent has a very large population with high birth rate. A conservative estimate provides a relatively higher incidence and prevalence of genetic diseases.

However, no reliable countrywide published data exists. This is compounded by higher numbers of consanguineous marriages among the Muslim communities, as well as in South Indian communities. An estimated 495,000 infants with congenital malformations, 390,000 with G6PD deficiency, 21,400 with Down syndrome, 9000 with β-thalassemia, 5200 with sickle cell disease, and 9760 with amino acid disorders are born each year in India alone. A similar proportion of these cases probably occurs in other countries of the Indian subcontinent. The prevalence of late-onset multifactorial/polygenic disorders (including hypertension, coronary heart disease, diabetes mellitus, schizophrenia, bipolar affective disorder, and bronchial asthma) is also high. The burden of these disorders is heavy on the affected individual or family, as the cost of therapy and management falls on the individual, while the rural areas have limited diagnostic and management resources.

REFERENCES

Afzal M, Ali SM, Siyal HB. (1994). Consanguineous marriages in Pakistan. *Pak Dev Rev 33*(4 Pt 2), 663–674.

Ahmed ZM, Morell RJ, Riazuddin S, et al. (2003). Mutations of MYO6 are associated with recessive deafness, DFNB37. *Am J Hum Genet 72*(5), 1315–1322.

Appaji Rao N (1991). Genetic consequences of inbreeding in a large human population. *Proc Nat Sci Acad B 57*, 361–368.

Appaji Rao N, Rama Devi R, Savithri HS, Venkat Rao S, Bittles AH. (1988). Neonatal screening for aminoacidemias in Karnataka, South India. *Clin Genet 34*, 60–63.

Balarajan R, Soni Raleigh V, Botting B. (1989). Mortality from congenital malformations in England and Wales: variations by mother's country of birth. *Arch Dis Child 64*(10), 1457–1462.

Basu P, Gangaopadhaya PK, Mukherjee SC, et al. (2000). Molecular anatomy of CTG expansion in myotonin protein kinase gene among myotonic dystrophy patients from eastern India. *Hum Mutat 16*(4), 372.

Bharucha BA, Kher AS, Naik GG, et al. (1994). Spectrum of inherited metabolic disease in India – A 15-year experience (Abstract). In: Chaudhary P, Sachdev HPS, Puri RK, et al., eds. *Proceedings Asian Congress of Pediatrics*, p. 98. New Delhi: Jaypee Publishers.

Bittles AH (2001). Consanguinity and its relevance to clinical genetics. *Clin Genet 60*, 89–98.

Elango R, Verma IC. (1996). Fragile X syndrome among children with mental retardation. *Indian J Pediatr 63*, 533–538.

Jain U, Verma IC, Kapoor AK. (1998). Prevalence of fragile X(A) syndrome in mentally retarded children at a genetic referral centre in Delhi, India. *Indian J Med Res 108*, 12–16.

Kaur M, Das GP, Verma IC. (1994). Inborn errors of amino acid metabolism in North India. *J Inherit Metab Dis 17*, 230–233.

Kulkarni ML, Jose S. (1997). Folic acid prevents neural tube defects in high prevalence area. *Indian Pediatr 34*, 561–562.

McKeigue PM, Marmot MG, Syndercombe G, et al. (1988). Diabetes, hyperinsulinaemia, and coronary risk factors in Bangladesh in east London. *Br Heart J 60*(5), 390–396.

McKeigue PM, Miller Marmot MG (1989) Coronary heart disease in south Asians overseas: A review. *J Clin Epidemiol 42*(7), 597–609.

Mehta AC. (1998). Neonatal-perinatal data of Nowrosjee Wadia Maternity Hospital, Mumbai. *Bull Natl Neonatal Forum (New Delhi) 12*, 12–15.

Murgod UA, Saleem Q, Anand A, et al. (2001). A clinical study of patients with genetically confirmed Huntington's disease from India. *J Neurol Sci 190*(1–2), 73–78.

National Neonatology Forum (1995). *National Neonatal-Perinatal Database*. New Delhi: All India Institute of Medical Sciences.

Roberts E, Hampshire DJ, Pattison L, et al. (2002). Autosomal recessive primary microcephaly: An analysis of locus heterogeneity and phenotypic variation. *J Med Genet 39*(10), 718–721.

Saleem Q, Choudhry S, Mukerji M, et al. (2000). Molecular analysis of autosomal dominant hereditary ataxias in the Indian population: High frequency of SCA2 and evidence for a common founder mutation. *Hum Genet 106*(2), 179–187.

Saleem Q, Roy S, Murgood U, et al. (2003). Molecular analysis of Huntington's disease and linked polymorphisms in the Indian population. *Acta Neurol Scand 108*(4), 281–286.

Sharma D, Gupta M, Thelma BK. (2001). Expansion mutation frequency and CGG/GCC repeat polymorphism in FMR1 and FMR2 genes in an Indian population. *Genet Epidemiol 20*(1), 129–144.

Verma IC. (1978). Genetic disorders in India. In: Verma IC (ed). *Medical Genetics in India,* Pondicherry, Auroma Enterprises, Vol. 1, pp. 5–18.

Verma IC. (1978). High frequency of neural tube defects in India. *Lancet i*, 879–880.

Verma IC. (1988). Genetic causes of mental retardation. In: Niermeijer M, Hicks E, eds. *Mental Retardation, Genetics and Ethical Considerations*. Amsterdam: Reidel, pp. 99–106.

Verma IC. (2000). Burden of genetic disorders in India. *Indian J Pediatr 67*, 893–898.

Verma IC, Anand NK, Modi UJ, Bharucha BA. (1998). *Study of malformations and Down syndrome in India – A multi-centre study.* Bhabha Atomic Research Centre, Tromaby, Department of Atomic Energy, India.

Verma IC, Bijarnia S. (2002). The burden of genetic disorders in India and a framework for community control. *Community Genet 5*, 192–196.

Verma IC, Elango R, Mehta L. (1990). Monitoring reproductive and developmental effects of environmental factors in India – A review. (Abstract). Indo-US symposium on effects of environmental and genetic factors on pregnancy outcome, New Delhi. pp. 63–73. All India Institute of Medical Sciences.

Verma IC, Saxena R, Lall M, Sharma RL. (2003). Genetic counseling and prenatal diagnosis in India – experience at Sir Ganga Ram Hospital. *Indian J Pediatr 70*, 293–297.

WHO (2010). *Birth defects*—Report of the secretariat, 63rd World Health Assembly, April 1, 2010.

Yaqoob M, Gustavson K-H, Jalil F, et al. (1995). Early childhood health in Lahore, Pakistan, II: Inbreeding. *Acta Pediatr 390*(suppl), 17–26.

Yaqoob M, Bashir A, Tareen K, et al. (1993). Severe mental retardation in 2-24 month children in Lahore, Pakistan, a prospective cohort study. *Acta Pediatr 84*, 267–272.

78.

GENETIC SERVICES IN INDIA: A MODEL FOR DEVELOPING COUNTRIES

Ratna D. Puri and Ishwar C. Verma

INTRODUCTION

Genetic disorders occur in two waves, the first in infancy and childhood and the second in adulthood. The disorders in childhood are birth defects due to various malformations and single gene disorders. Their frequency has been estimated to be 6.43% of total births worldwide (MOD report, 2006). This figure is applicable to all countries, with some adjustments related to the frequency of consanguineous marriages and the maternal ages at conception. The genetic disorders in adulthood are commoner and affect a very large population. Indeed, the developing countries are witnessing an epidemic of these late-onset disorders, such as the metabolic syndromes related to the changing lifestyle. The list includes disorders such as diabetes mellitus, hypertension, coronary artery disease, mental illness, etc. After the Human Genome Project was completed, the genetic component in these disorders attracted great attention from governments in the developing countries. In India, a separate budget was allocated for their control, especially for diabetes mellitus; and numerous large -scale studies are in progress to examine the role of genetic markers in these disorders.

Based on the March of Dimes report, it is estimated that the annual birth prevalence of congenital defects in India is 1.7 million. In spite of the high frequency of these disorders and the huge numbers, these have been neglected by successive health administrations and the medical establishment. This indifference is partly related to their low frequency compared to communicable disorders, nutritional problems, and perinatal asphyxia. This has resulted in ill representation of the subject at all levels of medical education, lack of awareness of genetic disorders among the medical students and doctors, and thus relatively tardy progress in the development of genetics as a specialty. As a result, facilities for diagnosis, treatment, and prevention of genetic disorders are inadequate. This translates to ignoring the potential of genetic disorders as a public health problem with lack of easy accessibility and high costs of genetic tests (Agarwal SS, 2009). This burden is further compounded by absence of screening programs and birth defects registers. On the other hand, there are reports of improper utilization of the knowledge of genetics even in developed countries (Schmidtke et al., 2006).

The government of India does not lay emphasis on providing medical services for genetic disorders. Nevertheless, the need for these services is increasing, especially in the cities. In most states, the infant mortality rate has reduced to 53 per 1000 live births (SRS Bulletin, 2011), reflecting improvement in nutrition, prevention of communicable diseases, and generally better living conditions. However, congenital malformations and childhood genetic conditions remain significant in childhood morbidity and mortality. Planning for provision of the genetic services in a developing country requires careful analysis and carrying out "needs assessment." This can be looked at under the following headings:

1. Burden of genetic disorders in a particular region or country
2. Causes of perinatal mortality contributing to infant mortality
3. Causes of morbidity and mortality in childhood as well as adulthood
4. Evaluation of cases referred for genetic counseling
5. Evaluation of the load of genetic disorders in primary and tertiary care facilities

BURDEN OF GENETIC DISORDERS

The development of any health policy or program for services in a country requires data on epidemiology and burden of specific disorders, for example genetic conditions. An Institute of Medicine report (Institute of Medicine, 1994) indicated that in developing countries about 5% of all pregnancies result in the birth of a child with a significant genetic disorder, congenital malformation, or disability. An estimated 43% of cases of severe learning difficulties (IQ <50) are caused by single genes or chromosomal abnormalities (Institute of Medicine, 1994). In developing nations, genetic disorders account for about 15% to 25% of perinatal and infant mortality (Verma and Singh, 1989; Penchaszadeh, 1993a,b). Based on the March of Dimes (MOD, 2006) report, over 3.3 million children under 5 years of age die each year from birth defects, and the majority occur in developing countries. Many noninfectious diseases are now known to have a genetic component. In India and other developing countries, genetic diseases and congenital malformations are emerging at an alarming rate, in addition to infections, which continue to flourish. Accordingly, in university and city hospitals, congenital malformations and genetic disorders are reported as important causes of morbidity and mortality (Verma, 1986a, 2000; Verma & Bijarnia, 2002).

This shift is well illustrated by the change as the predominant cause of perinatal mortality in India. For example, New Delhi's All India Institute of Medical Sciences (AIIMS) recorded sepsis as the predominant cause for perinatal mortality in the 1970s; thereafter, in the 1980s, preterm birth was the leading cause, while in the 1990s congenital malformations surpassed both preterm birth and neonatal sepsis as the leading cause. (Verma, 2000). This trend is not confined to AIIMS only, but is also reflected in the causes of perinatal mortality in other major hospitals across India (Mehta, 1998). In the report of the National Neonatal Perinatal Database (2002–2003), congenital malformations contributed to 9.2% of neonatal mortality, after birth asphyxia, infections, and hyaline membrane disease. In a recent study from a rural population of Uttar Pradesh, 6.5% of neonatal deaths were due to congenital malformations, 25.7% due to infections, 14.1% as a result of birth asphyxia, and 26.9% were preterm births (Baqui et al., 2006). A retrospective analysis from 1995 to 2003 of infant mortality in urban slums in Vellore, in South India, reports that congenital malformations were responsible for 9.6% infant deaths (Vaid et al., 2007).

The population of India is multiethnic and multicultural, and many genetic disorders occur in varying frequencies in different communities. To state a few examples, sickle cell anemia is predominantly observed in the tribal communities inhabiting the states of central India (Maharashtra, Andhra Pradesh, Gujarat, Orissa, Bihar, Chattisgarh) and the Nilgiri hills in Tamil Nadu and Wayanad district in Kerala. Of 427 tribal communities in India, at least 300 are known to have the sickle cell (SC) gene. It is estimated that there are about 5 million carriers of the SC gene and about 200,000 affected persons (homozygotes). These communities live in remote regions of the country with poor access to healthcare and have little income, making the establishment of a control program more difficult. Similarly, one survey among schoolchildren demonstrated a carrier frequency of β-thalassemia at 5% in Delhi, 4% in Mumbai, and 8% in Kolkatta (ICMR study, 1991).

The Agarwal community in India harbors a number of disorders that have founder mutations. For example, megalencephalic leukoencephalopathy with subcortical cysts (MLC) occurs in India predominantly among this community, and they all harbor a common mutation (320insC) in the MLC1 gene, suggesting a founder effect. (Gorospe JR et al., 2004). Spinocerebellar ataxia type 12 is also predominantly reported in this community in India. In contrast, there are some disorders that are fairly prevalent in all communities, and patients seek confirmation of diagnosis, management and prevention of recurrence in subsequent pregnancies, as well as carrier screening for at-risk individuals in the family. Common examples of these disorders are beta-thalassemia, spinal muscular atrophy, Duchenne muscular dystrophy, and hemophilia A, to name a few. Incidence of these disorders is not known in India; however, evaluation of reports of diagnosis in genetic counseling centers of tertiary care hospitals does show that these disorders are not uncommon.

NEEDS ASSESSMENT FOR GENETIC SERVICES

Though genetic disorders are individually rare, collectively they still affect a large number of people. Genetic diseases are commoner in India than in the West due to consanguineous marriages, a vast population, and increased frequency of β-thalassemia. To evaluate the adequacy of a genetic program it is important to assess the relationships between genetics and health for the traditional single gene disorders, as well as noncommunicable diseases like diabetes mellitus, hypertension, etc. Needs assessment addresses the availability of quality genetics resources in the community, the appropriateness of genetics technologies offered to the community, the accessibility of clinical and laboratory services, the cost benefit of using genetics technology, and the community's knowledge of the use of genetics to improve health. The assessment of these factors alerts health officials and others to areas in which policies should be developed and for which better provision of services is needed.

In India, the first genetic clinic and register started at the All India Institute of Medical Sciences in New Delhi in 1969 (Verma et al., 2003). This has recently developed into a center of excellence. Reddy et al. (1986) started a genetic clinic at Guntur and then Hyderabad, although unfortunately this languished after his death. In his analysis, disorders of the central nervous system were commonest, followed by chromosomal disorders. The role of consanguinity was highlighted in a later study from

TABLE 78-1 GENETIC COUNSELING AT SIR GANGA RAM HOSPITAL, NEW DELHI, INDIA (2002–2006; N = 9770)

During pregnancy	6252 (64%)
Chromosomal disorder	720 (7.4%)
Metabolic disorders	1057 (10.8%)
Neuromuscular – DMD, SMA	900 (9.2%)
Mental retardation	270 (2.7%)
Malformations and NTDs	190 (1.9%)
Miscellaneous (triple repeat/skeletal dysplasias, short stature, etc.)	381 (3.9%)

Pondicherry that showed a higher incidence of autosomal recessive disorders both in the clinic and in the general population. (Verma, 1986b) More recent data (2002–2006) from analysis of cases provided with genetic counseling at Sir Ganga Ram Hospital, New Delhi, shows that referrals are commonest for pregnancy-related events (Table 78-1). This suggests that the severe burden of disease compels patients to prevent recurrence in a subsequent pregnancy. Compared to 38.4% of patients who sought advice during pregnancy a decade ago, 56% of patients were counseled for this indication in 2009.

The remarkable improvements in genetic technology, and increased awareness of genetic disorders through the media and Internet, have helped many families to prevent recurrence of single gene disorders and chromosomal disorders (Table 78-2).

Genetic data from the medical genetics outpatients at Sanjay Gandhi Postgraduate Institute (SGPGI) shows an increase in the number of patients referred for genetic counseling from 1998–2003 (Table 78-3). The maximum referrals were during pregnancy, similar to the trend noted at Delhi. Genetic hemolytic disorders, mental retardation, skeletal dysplasias, and neuromuscular disorders were some of the other common indications for counseling (Phadke, 2004)

TABLE 78-2 INDICATIONS OF GENETIC REFERRALS DURING PREGNANCY, SIR GANGA RAM HOSPITAL, NEW DELHI, INDIA (N = 6252)

Abnormal Triple/AMA	33.4%
Abnormal ultrasound	14.6%
Previous child with Down syndrome	5.7%
Single gene disorders	14.5%
Metabolic diseases	1.8%
Recurrent miscarriage/infertility	9.1%
Previous pregnancy with malformations	5.2%
TORCH/Varicella infections	5.2%
Drugs during pregnancy	1.6%
Miscellaneous (previous child with MR, family history of genetic disease, etc.)	8.8%

The severe burden of genetic disease compels couples to seek antenatal diagnosis to make sure that the fetus is not abnormal. In fact, many couples having a child with a birth defect seek genetic counseling only when the wife is again pregnant, so as to avoid the birth of another affected child. An ICMR-sponsored multicentric study on causes for genetic counseling, carried out by the Task Force of Human Genetics from 1992 to 1995, also showed that events related to pregnancy were the commonest referrals to a genetic center. The other common disorders for which referrals were sought were identifiable syndromes (12.1%), chromosomal disorders (11.3%), and mental retardation (11%; Report of an ICMR Task Force Study, 1996). Another recent multicentric ICMR study concluded that 14% of pregnant women were at risk of having an adverse pregnancy outcome/genetic disorder. In this study, of the 42 variables used for screening an increased risk in pregnant women, past history of spontaneous abortion, stillbirth, or previous child with NTD/congenital malformation were found to be important risk factors for adverse outcome of pregnancy (ICMR, 2009). This study provided evidence for the need and utility of preventive genetic services in the National Family Welfare Program.

HEALTHCARE SYSTEM IN INDIA

Health is a state subject in India, and individual states are given primary authority over healthcare, including public health and hospitals. An estimated 80% of public healthcare funding comes from the states. However, the central government formulates comprehensive health policies and plans under the direction of the Ministry of Health and Family Welfare. The center funds many important health projects like the Rural Health Mission, control programs for cancer, iodine deficiency disorders, etc. (Ministry of Health Annual report). Areas of national importance like medical education and the medical profession are the shared responsibility of the central government and individual states (Garg, 2001). India has a two-tiered healthcare system that funnels more difficult cases into urban hospitals, while attempting to provide routine medical care to the vast majority in the rural areas. There is a network of village health centers, subcenters, primary centers, community health centers, district hospitals, and medical colleges. Government hospitals provide medical treatment either free or at a very nominal charge. However, with only 3.6% of the gross domestic product (GDP) allocated to health, the limitations faced in this sector are inadequate staff and facilities. In addition, recently there has been a major growth in the private healthcare industry in different parts of the country and this sector too plays a major role in health.

TABLE 78-3 GENETIC COUNSELING DATA: SANJAY GANDHI POSTGRADUATE INSTITUTE OF MEDICAL SCIENCES, LUCKNOW, UTTAR PRADESH, INDIA

Clinical Presentation	1998–1999 (n = 599)	2000–2001 (n = 516)	2002–2003 (n = 632)
Mental retardation/developmental delay	28	40	68
Abnormalities in sexual development & differentiation	46	34	14
Poor reproductive outcome/recurrent fetal losses	93	79	121
Neuromuscular disorders	63	55	43
Skeletal dysplasias	15	16	18
Genetic hemolytic anemia	104	107	111
Genetic bleeding disorders	18	25	21
Multiple malformation syndromes	78	64	87
Neural tube defects (NTDs) (pre/postnatal)	43	31	14
Isolated congenital malformations	23	32	16
Exposure to teratogen during pregnancy	9	3	6
Miscellaneous	82	62	67

The challenge for India is providing healthcare to a vast population with three distinct categories—rural, urban, and urban poor (slums). The rural population has limited access to medical facilities and has a resource crunch. The urban poor cannot afford health facilities of prevention and care of children with birth defects. In the cities, the urban population who can pay, obtain healthcare for birth defects at the best of hospitals and institutions. The majority in this group can cover the expenses themselves. Only 10% of the population has health insurance, which is usually inadequate and very often does not cover genetic disorders. Recognizing the importance of health in development, the government of India has recently launched the National Rural Health Mission to help in improving basic healthcare delivery systems. The goal of the mission is to improve the availability of and access to quality healthcare by people, especially those residing in rural areas, the poor, women, and children.

PREVALENT GENETIC SERVICES IN INDIA

Genetics as a service encompasses two essential components: facilities for a sound clinical diagnosis, and establishing the laboratory tests for confirmation of diagnosis. Both these components, the genetic counseling centers and laboratories for genetic testing, need to be developed as a composite unit for efficient genetic services (Agarwal, 2009). This allows for appropriate utilization of the ever-evolving technologies in genetics for good patient care. The biggest challenge in developing countries is establishing the laboratory infrastructure for diagnosis of genetic disorders inclusive of molecular genetics, cytogenetics, biochemical genetics, and the HLA/immunogenetics. Genetics departments are present in many universities doing research in human or medical genetics. A PubMed search with "genetics" and "India" as keywords lists about 19,000 articles. However, not all these centers extend their services to patients. Professor Jai Rup Singh attributes this to the absence of a government mandate to extend research benefits to patient care (Singh et al., 2010; see Chapter 98). There is limited networking between centers of research and patient care, in addition to absence of a structured online website to generate information of the ongoing research in most centers. This may also be as a result of limited awareness of genetics among medical doctors, as this subject forms a rudimentary part of medical curriculum in India. Professor Jai Rup Singh lists 47 centers in India that undertake genetic investigations, spread out in 18 states. Among these, cytogenetics is better represented, followed by molecular diagnostics. With only about 26 centers for biochemical investigations and prenatal diagnosis, he observes that these services are grossly inadequate for the 1.15 billion population in India. (Jai Rup Singh et al., 2010). An alarming absence of medical genetic education and genetic services is recorded in about 90% of the medical colleges in India (Ghosh and Mohanty, 2002). The apparent lack of opportunities, and a perceived notion that monetary return from practice of genetics is poor, discourage aspiring physicians from taking up any branch of genetics as a career goal. Interestingly, medical students are attracted to study genetics because they read about the new discoveries in genetics frequently in the newspapers.

Most of the genetic counseling, at least for common disorders, is provided by pediatricians and physicians who learn the rudiments of genetic medicine through practice and while faced with questions being raised by the patient families. In most developing countries, physicians are good at diagnosing genetic disorders because the recognition of the common syndromes is easy. However, often in a busy medical practice insufficient time is available for obtaining a proper family history, and counseling of these patients. Lack of awareness about the exact nature of genetic mechanisms and knowledge about the availability of tests for diagnosis, and prenatal diagnosis for prevention, often results in imparting wrong information to the couple, resulting in the birth of more affected children.

However, with increasing awareness among patients and physicians about genetic disorders, this is now undergoing a change. Clinical genetic services are becoming more prevalent, their utilization is increasing, and patients are referred to specialized genetic centers for counseling in complex cases. The average number of patients seen at the Center of Medical Genetics at Sir Ganga Ram Hospital is approximately 4000 in a year. An encouraging observation in the last decade is the creation of many new and composite genetic units in some government and private hospitals. They have also established genetic tests for common disorders like thalassemia, spinal muscular atrophy, Duchenne muscular dystrophy, hemophilia, etc. Unfortunately there is no database or organized system that lists the genetic centers in India and the tests they perform. A beginning is being made to establish a web-based directory of genetic services in India supported by the Indian Council of Medical Research, on the pattern of "Genetests" in the United States. This is likely to be operational by December 2011. In India the nature of genetic counseling may vary to some extent from that in the developed countries. Patients are mostly referred from the primary care physician with a possibility of a genetic disorder and without any diagnosis. In India, there is no structured social healthcare system or medical health insurance coverage. A diagnosis is to be established with as few investigations as possible, as the cost of testing in India is usually borne by the patient himself or his family. Most families would prefer to bear the cost of prenatal diagnosis. Most Indian patients often desire more directive genetic counseling, and this appears justified because of the poor knowledge in the vast majority of people about biology and genetics. Certainly the final choice of the course of action is left to the patient, as the threat of litigation is increasing for wrong advice and/or care.

In a large country like India, there is a need to regionalize the genetic services, so that molecular diagnosis for the common genetic disorders is available at many places. Each center should, however, establish the capabilities for molecular diagnosis of a selected number of somewhat rare disorders so that, among these centers, the whole spectrum of disorders is covered. Prenatal diagnostic facilities should be an essential component of these centers (Verma, 2000). An initiative on these lines has been taken up by the Indian Council of Medical Research, and a multicentric project on lysosomal storage disorders (LSDs) is underway. Three major centers in North India—All India Institute of Medical Sciences, Sir Ganga Ram Hospital, and Maulana Azad Medical College—will collaborate and set up facilities for testing all major LSDs.

In the last decade there has been a remarkable change, in that many genetic diagnostic centers have been established. These provide inexpensive molecular tests at a fraction of the cost that is charged in the Western developed nations. Some centers in India receive samples for testing form neighboring countries like Bangladesh, Pakistan, Sri Lanka, and United Arab Emirates (UAE). Cytogenetics is better established in India, with many laboratories offering chromosomal analysis in peripheral blood. All large cities have at least 3–4 of such laboratories. A major innovation has been made possible by the 3–4 large private laboratories, who have established collecting centers even in the smaller towns, so that the only limitation is the awareness that a certain test should be done to help a particular family.

Many of the older geneticists had a dream that India should establish molecular diagnostics of common disorders like thalassemia and Duchenne muscular dystrophy. This dream has been realized, as many centers now exist that provide molecular diagnosis such that quite an extensive list of disorders is covered (Nadkarni et al., 2009; Singh JR et al., 2010).

For biochemical disorders, diagnostic leukocyte enzyme and amino acid studies are available in a limited number of centers. Though advanced diagnostic facilities are available in a few private and government centers, these are grossly inadequate for the vast number of children affected with biochemical disorders. With availability of enzyme replacement therapy for some children of LSDs, there is a need to make these diagnostics widely available.

NEW GENETIC INITIATIVES IN INDIA

GENETIC EDUCATION

A study conducted by Chattopadhyay (2005) on understanding of genetic information in higher secondary students in northeast India, reports the poor foundation and knowledge of this subject. A leading report cautioned that the level of genetics education in medical colleges in India is unstructured and mostly rudimentary. This is a cause for concern and needs to be addressed immediately. Inadequate comprehension of genetics has them at a disadvantage to appreciate modern medicine. The government has recently established a new body to oversee medical education in the country, and it is hoped that the new curriculum will contain more genetic teaching.

Young medical students' interest in genetics is kindled by reports of new gene discoveries and articles in most top journals on the genetic basis of common diseases. However, what is lacking are enough establishments to train these students. At present there is only one center in India at Lucknow that offers a postgraduate doctoral program in Medical Genetics since 1990, and now a few other centers have started a genetic fellowship, confirming the increasing awareness and demand for this specialty. The government sector is now awakening to the requirement of incorporating genetics as a specialty in hospitals. This is well illustrated by a recent collaboration between Nizam's Institute of Medical Sciences (NIMS), and the Center for DNA Fingerprinting and Diagnostics (CDFD), Hyderabad in Andhra Pradesh, to establish a "Medical Genetics Unit". The former is a hospital involved in patient care, training, and research activities, and, the latter an autonomous institute of Department of Biotechnology, Government of India, Over the last few years, increasing numbers of continuing medical education programs are being held in different regions of the country, with good participation from specialists and medical students. A landmark in genetics in India was an international conference on "Birth Defects and Disabilities in the Developing Countries," which was hosted in Delhi in 2009 and had excellent participation. This illustrates the increasing awareness and the urgent need to introduce medical genetics education in the country. This serious shortage of geneticists needs to be addressed. We need to create more departments of genetics in medical colleges, and more jobs in this discipline. Since the completion of the Human Genome Project, the understanding and practice of medicine has been revolutionized. Diagnostic tests and new forms of therapy based on better understanding of disease pathogenesis are available. For optimum utilization of modern day medicine, we need to integrate molecular biology with clinical genetics for good patient care, diagnosis, and treatment. The government of India is also creating centers of teaching and education integrated with research. The prime minister of India has declared 2010–2020 as the *Decade of Innovation*. Students are being encouraged toward research in basic sciences with better funding, government and private sector partnership, and provision of infrastructure and technology.

NEWBORN SCREENING PROGRAMS

Of the 25 million babies born in India every year, it is estimated that 1 in every 2000 newborns suffers from a metabolic disorder. The incidence varies from community to ethnicity to religion to location. For instance, while hypothyroidism is prevalent in sub-Himalayan areas, sickle cell disease is common in the tribal populations of Maharashtra, Gujarat, Orissa, Madhya Pradesh, Kerala, and Tamil Nadu, including the nontribal population from Chattisgarh state, according to various studies.

Currently there is no universal newborn screening program in India. Some private laboratories provide facilities for screening tests, and individual private hospitals have their own policies for newborn screening. About five to six private labs are screening newborns for genetic disorders at a cost. We at Sir Ganga Ram Hospital screen newborns for three disorders; i.e., hypothyroidism, congenital adrenal hyperplasia, and G6PD deficiency. All newborns are also screened for hearing loss. Extended metabolic screening by mass tandem spectrometry is offered to high-risk symptomatic newborns. The problem of an unstructured screening program, as prevalent in India, is that many newborns with treatable disorders are missed. There is lack of standardization of the laboratory services being provided, absence of quality control, as well as inadequate and improper surveillance.

Recently the Indian Council of Medical Research has initiated the first nationwide mass screening program that is free. It is a multicenter project to assess the feasibility of newborn screening for congenital hypothyroidism and congenital adrenal hyperplasia in five centers all over India. This is a step ahead, and it is hoped that it will create the infrastructure to offer a nationwide screening program.

CARRIER SCREENING FOR THALASSEMIA

In India, the most spectacular progress in recent years has been in the molecular analysis, prenatal diagnosis, and community control of hemoglobin disorders and thalassemia. A few common mutations account for > 95% of severe β-thalassemia cases. The carrier rate for β-thalassemia varies from 3%–17%. Hb E (hemoglobin E) is an abnormal hemoglobin prevalent in northeastern states. There are many reports in PubMed of the different prevalent mutations in various communities in India. This knowledge has been utilized for molecular analysis in families. The common mutations are first checked by simpler techniques, like ARMS PCR, Reverse DOT Blot. It is only in a small percentage of patients that gene sequencing is required. This makes molecular testing less laborious and cheaper. There are many centers working in the field of thalassemia in Delhi, Chandigarh, Lucknow, Vellore, Mumbai, Ahmedabad, Calcutta, Hyderabad, etc. Many government and nongovernment organizations are working for a thalassemia-free India. Ahmedabad Red Cross Society is one such example of pioneering work, in the state of Gujarat in thalassemia control.

GENETIC SUPPORT GROUPS IN INDIA

THALASSEMICS INDIA is a nongovernmental organization and was established in 1986. It is working zealously across the country, operating in close association with doctors, drug/equipment companies throughout the country and abroad, hospitals, as well as state thalassemia associations and centers. In addition to its main goals of providing good medical facilities for treatment of affected children, it also works to create awareness among the public, and facilitate screening and prevention programs for control of thalassemia. Other support groups like those for hemophilia, the Lysosomal Storage Disorders Support Society of India, and Down syndrome support groups play an important role in creating awareness of genetic disorders, as well as the need for organized counseling services and centers for diagnosis and treatment.

BIRTH DEFECTS REGISTRY OF INDIA

Though there are many reports of children with malformations from the Indian subcontinent, there have been no uniform methods for the surveillance of birth defects and recording of national data. However, recently there has been a step forward in this direction and a Birth Defects Registry was started in 2001 at a private institution, Fetal Care Research Foundation in Chennai. Over the years it has expanded to include all states of India, and aims to systematically collect information on the incidence of birth defects and assess their prevalence in India, as well as evolve management strategies for birth defects all over India. The ICMR has also initiated a multicenter task force program for a national birth defects registry in North India.

PUBLIC–PRIVATE PARTNERSHIPS

Analysis of available data shows extremely uneven health and development progress in various parts of the country. The large, widespread health infrastructure that has been set up throughout the country is unable to meet newer health challenges. Public sector hospitals have insufficient funds to keep pace with technological advances. In keeping with the developments of the Indian economy, there has been an unprecedented growth of the private sector all over the country. As the demand for high-quality and up-to-date medical services increase, many genetic units are now being established in the private sector, where procurement of expensive equipment and investments for expertise are possible. Public–private partnerships (PPP) in genetic services offers a solution to the challenge faced by the public sector, due to lack of resources and infrastructure, to provide complex, expensive genetic services. To address the issue of equity, we at Sir Ganga Ram Hospital charge those who can afford to pay, and subsidize costs for those less able. This policy allows all sections of society to utilize the expertise and services available.

A well-implemented PPP process allows improved access to care, reduced costs, and enhanced effectiveness of service provision. This is an effective means of providing rational, affordable, and comprehensive care to the entire population, provided that the private partners are chosen with care Public–private partnerships can be utilized to support large randomized placebo controlled trials that can have a major impact on health related to genetic disorders. In India, the National Rural Health Mission advocates PPP for delivery of tertiary medical care (NHRM – mission document). It is clear that utilizing non-state abilities in delivery of healthcare can be a powerful tool.

DEVELOPMENT OF CAPACITY IN HUMAN RESOURCES AND TRANSFER OF TECHNOLOGY—"THE NORTH-SOUTH PARTNERSHIP"

This is about developing networks and partnerships between countries to share knowledge, expertise, and transfer of technology between developed and developing countries for control and management of disease. A fine example in India is the progress toward the care and prevention of thalassemia (Weatherall, 2005). Through North–South partnerships, good control programs for thalassemia control have developed. Sustained efforts and networking, and some support from international health organizations, will create awareness for control and prevention and a platform for discussion with government agencies. Without these concerted efforts, genetic disorders will remain neglected as a public health priority.

ACCREDITATION

Accreditation of laboratories is not mandatory in India. However, recently the central government has implemented a voluntary accreditation program. The National Accreditation Board for Testing and Calibration Laboratories (NABL) is an autonomous body under the Department of Science and Technology, Government of India, and is the sole government-authorized accreditation body for laboratories. The task ahead is to ensure that all genetic laboratories enter the accreditation program, as well as other quality control programs.

THE WAY AHEAD

Dr. Mohanty (Ghosh and Mohanty, 2002), in her editorial, outlines how functioning genetic services in the medical colleges can be established while government policies are being laid down on this matter. The existent manpower should be spared and trained in the present centers of excellence across the country. They then set up genetic units in their respective colleges, and repeated short-term trainings in genetics can be offered to improve their skills in clinical and laboratory work. Funds can be obtained through research projects from different funding authorities. In fact, this is already in place and many projects are funded by government bodies like the Indian Council of Medical Research, Department of Biotechnology, etc. For best utilization of resources, medical colleges in the same city can prevent duplication of services. As most genetic tests are expensive and require special expertise, networking of the existing genetic centers is important to provide optimal services in the present set-up. Pooling of our clinical and laboratory services in a collaborative network is the way ahead.

A major part of care and prevention of birth defects is to be implemented through the primary healthcare programs in India. We should start with common birth defects for care and prevention in the health programs, like neural tube defects, cleft lip/palate, talipes equino varus, chromosomal disorders like Down syndrome, congenital heart disease, and embryopathies due to intrauterine infections and drugs. The strategy to be implemented may have to differ in different regions of the country. In the rural areas and urban slums, we have to strengthen our preconception care—good maternal nutrition, fortification of food with folic acid, and prevention of maternal infections. Food fortification has been started in Chandigarh, and this can serve as a model to be implemented at the national level with government assistance. Training of medical and paramedical staff is essential to increase the awareness of genetic disorders. Training in prenatal ultrasound to detect fetal malformations and soft markers for chromosomal disorders appears more cost-effective than biochemical screening. For the urban paying sector, ultrasound screening for NTD and other malformations, first and second trimester biochemical tests for aneuploidies, infections, and fetal echocardiography are in place. Preconception and antenatal care, along with postnatal management of children with birth defects, is followed at our center. What more is required is education of doctors, especially obstetricians and pediatricians who are the first point of contact with these patients.

CONCLUSION

India, like other developing counties, is facing an accelerating demographic switch to noncommunicable diseases. Recently an increased awareness among patients and physicians has been observed. India is one of the first few among the developing countries to have recognized the importance of biotechnology in the health sector. Funding agencies like the Department of Biotechnology, Indian Council of Medical Research, Council of Scientific and Industrial Research (CSIR), and Department of Science and Technology (DST) have provided major funding to strengthen and develop capabilities in genetics and genomics research. This is well illustrated by the recent report of the success of an Indian CSIR scientist in decoding the 3.1 billion basepair sequence of an Indian using next-generation sequencing technology. A large Indian Human Genome Variation project has been completed. India has joined in the International Cancer Genome Project. Another important initiative is the Open Source Drug Discovery (OSDD) project, with global partnerships and a vision to provide affordable healthcare to the developing world. Four thousand top scientists, including 10% from the United States, have joined hands to solve the problem of tuberculosis. The entire intellectual property will be shared. Out of 4000 scientists, 700 are from India. A new institute of translational medicine has been set up by the Department of Biotechnology. A National Institute of Genomics and Genetics has been established in Kolkatta.

However, many challenges lie ahead. The limited resources, poor infrastructure, and inadequate manpower in genetics cannot address the huge burden of genetic disorders in India. Improvement in the teaching of genetics at all levels of education is required. Translational research should receive a high priority. Greater numbers of jobs in genetics in both private and public sectors are required. A multilevel targeted program is the need of the hour. Integration of genetic services into the existing health structure is essential. Improvement of facilities in existing centers, coordination, and networking of all the centers are essential components to reap the benefits of the new genetics and provide them to the people.

REFERENCES

Agarwal SS. (2009). Medical genetics in India – What needs to be done? *Indian J Med Res 130*, 354–356.

Andrews LB, Fullarton JE, Holtzman NA, Motulsky AG. (eds.). (1994). Assessing genetic risks: implications for health and social policy. Washington, D.C.: National Academy Press.

Baqui et al. (2006). Rates, timing and causes of neonatal deaths in rural India: implications for neonatal health programmes. *Bulletin of World Health Organization 84*(9), 685–764.

Chattopadhyay A. (2005). Understanding of Genetic Information in Higher Secondary Students in Northeast India and the Implications for Genetics Education. *Cell Biology Education 4*, 97–104.

Christianson A, Howson CP, Modell B. (2006). March of Dimes Global Report on Birth Defects, The Hidden Toll of Dying and Disabled Children. White Plains, NY: March of Dimes Birth Defects Foundation.

Garg CC. (2001a). A conceptual framework for national health accounts in India: a pilot study on Punjab. New Delhi, India. *Institute of Economic Growth.*

Ghosh K, Mohanty D (2002) Teaching of medical genetics in the medical colleges of India -way ahead. *Indian Journal of Human Genetics 8*(2), 43–44

Indian Council of Medical Research, New Delhi, India. (2009). Feasibility of introducing genetic services in the National Family Welfare Programme in India. *Indian J Med Res 130*, 404–412.

Indian Council of Medical Research. (1996). Genetic counseling and antenatal diagnosis of common genetic disorders. Report of an ICMR task force study. Indian Council of Medical Research, New Delhi.

Mehta AC. (1998). Neonatal-perinatal data of Nowrosjee Wadai Maternity Hospital, Mumbai. Maulana Azad Medical College, New Delhi. *Bulletin National Neonatology Forum 12*(1), 12–15.

Nadkarni A, Gorakshakar A, Surve R, et al (2009) Hematological and molecular analysis of novel and rare beta-thalassemia mutations in the Indian population. *Hemoglobin* 33(1), 59–65.

National Neonatal-Perinatal Database, NNPD 2002–2003. NNPD Network. http://www.newbornwhocc.org/pdf/nnpd_report_2002-03.PDF.

National Rural Health Mission – Ministry of Health and Family Welfare mohfw.nic.in/NRHM/Documents/Mission_Document.pdf.

Penchaszadeh VB. (1993a). Reproductive genetic testing from an international perspective: impact on women in underdeveloped countries. *Fetal Diagn Ther 8*(Suppl 1), 202–209.

Penchaszadeh VB. (1993b). Reproductive health and genetic testing in the third world. *Clin Obstet Gynecol 36*, 485–495.

Phadke SR. (2004). Genetic counseling. *Indian Journal of Pediatrics 71*(2), 151–156.

Reddy YR, Ahuja YR, Sujatha M, et al. (1986). Genetic registers in India. In: Verma IC, ed. *Genetic Research in India*. (pp. 5–10). New Delhi: Sagar Printers and Publishers.

Schmidtke J, Paul Y, Nippert I. (2006). Education in medical genetics for physicians: Germany. *Community Genet 9*, 235–239.

Singh JR, Singh AR, Singh AR (2010). Directory of Human Genetic Services in India - 2007. *Int J Hum Genet, 10*(1–3), 187–192.

SRS bulletin Jan 2011 - Census of India Website. Census and Vital Statistics Website on the Internet: http://www.censusindia.gov.in.

Vaid A, et al. (2007). Infant mortality in an urban slum. *Indian J Pediatr 74*(5), 449–453.

Verma IC. (1986a). Genetic disorders need more attention in developing countries. *World Health Forum 7*, 69–70.

Verma IC. (1986b) Genetic counseling and control of genetic diseases in India. In: Verma IC. (ed.), *Genetic Research in India* (pp. 21–37). New Delhi: Sagar Printer and Publishers.

Verma IC. (2000). Burden of Genetic disorders in India. *Indian J Pediatr 67*, 893–898.

Verma IC, Singh B. (1989). Ethics and medical genetics in India. In: Wertz DC, Fletcher JC, eds. *Ethics and Human Genetics: A Cross-Cultural Perspective* (pp. 250–270). Berlin: Springer-Verlag.

Verma IC, Bijarnia S. (2002). The burden of genetic disorders in India and a framework for their control. *Community Genet 5*, 192–196.

Verma IC, Saxena R, Lall M, Bijarnia S, Sharma R. (2003). Genetic counseling and prenatal diagnosis in India—experience at Sir Ganga Ram Hospital. *Indian J Pediatr 70*(4), 293–297.

Weatherall DJ. (2005). The global problem of genetic disease. *Annals of Human Biology 32*(2), 117–122

WHO. (1999). Prevention and care of genetic diseases and birth defects in developing countries. Geneva, WHO/HGN/GL/WAOPBD/99.1.

79.

THE PRACTICE OF GENETIC COUNSELING IN INDIA

Meenakshi Bhat

INTRODUCTION

Genetic medicine is a branch of medical science which deals with genetic factors in human health and disease as well as the diagnosis and management of hereditary disorders. Major advances in the field have brought with them a whole host of new screening strategies, diagnostic tools, medical therapies, and prenatal diagnostic procedures. All this has also brought along new bioethical, legal, and social issues. There is probably no other branch of medicine in which transmission of medical information in lay terms to a family seeking advice is as important.

Genetic counseling, which is an important component of genetic medicine, differs from other medical branches in that it is aimed at the family unit rather than the individual. The process of communication with the family about the details of the genetic disorder, and the options available to them in dealing with it, is the art and science of genetic counseling. Of vital importance is the underlying theory of non-directiveness (Clarke A, 1997). This mandates that the final decision about the best intervention for a particular condition should be made by the family. Although genetic counseling has been in practice over many years the world over, very few definitions of the process are available in the literature. A useful definition is that it is:

> *An educational process that seeks to assist affected and/or other at-risk individuals to understand the nature of the genetic disorder, its transmission and the options open to them in management and family planning*
>
> (Kelly, 1986; Harper, 2004).

Knowledge of recent advances in genetic testing and emerging bioethical issues is not the only requirement in providing counseling services. It is equally important to take into consideration local or regional sociocultural beliefs. The aim of this chapter is to focus on some of the commonest sociocultural practices prevalent in Indian society, and the special challenges posed by them in the genetics clinic.

The first clinics offering genetic counseling as we understand it today began in the 1940s. The term *genetic counseling* was coined by Sheldon Reed in 1947. The earliest Indian published reports on genetic counseling appeared in the late 1960s, some with directive, eugenic messages (Kumar, 1968).

Unlike in the Western countries, clinical genetics in India is still a specialty available only in a few tertiary medical institutions. The reasons for this may include a dearth of medical genetic training programs and the prioritization of limited government health funding to national programs in prevention and treatment of infectious diseases. Only recently has the government of India approved the first pilot project on screening of newborns for inherited metabolic disorders (ICMR, 2007). Understandably, specialized branches such as genetic counseling and quality diagnostics are available only in a few major cities, and largely in privately funded institutions. Structured long-term genetic counseling courses, finishing with a certification or qualification, are not nationally available.

WHO IS QUALIFIED TO DO GENETIC COUNSELING?

Although there are no national norms, most of the genetic counseling done in the majority of Indian centers is by medical doctors with some genetics training. In other countries, postgraduates in biological sciences and qualified nurses also are eligible to undergo training in genetic counseling and become full-fledged counselors. Arguments against the abilities of nonmedical counselors in genetics include limitation of understanding of dysmorphology, or appreciation of subtle clinical features that may underlie major genetic disorders. Counterarguments to this may include the training and salary expenses of a medical doctor in this field. A two-tier system, with both medical and nonmedical counselors, may be the solution where limited funding is a perennial issue.

PROCESS OF GENETIC COUNSELING IN INDIA

Referrals to a genetics clinic are most often by medical specialists treating the individual or the family, and only very occasionally initiated by the "at-risk" individual or family. Most clinics consist of walk-in patients, often without a referral letter to outline the reasons for a genetic consultation. As a result, the numbers of individuals attending each clinic, or the waiting time for families, are very difficult to streamline. Many clinics therefore adopt a first-come-first-served approach to seeing patients. Pre-appointment or post-appointment home visits by counselors or coworkers are rarely done in India for a variety of reasons, most notably a shortage of staff.

RECORDING THE PEDIGREE

As in other societies, a genetic counseling session begins with a detailed pedigree collection from the family seeking advice. Some of the difficulties faced at the level of recording the pedigree are:

- In the absence of written records or a general practitioner, as in the National Health Service (NHS) of the United Kingdom, it is often difficult to confirm or verify medical conditions of other family members of the proband.
- Female siblings are often not acknowledged or accounted for unless specifically asked about.
- Previously deceased siblings, miscarriages, or other pregnancy losses are also not mentioned unless specifically enquired about.
- Detailed postal addresses and accurate dates of birth are very difficult to obtain from rural and uneducated families.

However, in our center, due to the widespread availability of mobile telephones and the low cost, we have found it very useful to obtain and record mobile phone numbers to ensure rapid follow-up and communication with the family. Most families do not regard this as an intrusion of their privacy, and many actually prefer this as a means of maintaining contact.

CLINICAL EVALUATION

The first appointment includes a detailed examination of the proband, recording of all measurements, and arranging genetic and other tests that may help arrive at a diagnosis. Written informed consent is obtained whenever blood tests and/or clinical photographs are requested. On many occasions, the request for signed consent for taking photographs is met with surprise and sometimes suspicion, as it is not common practice locally for consent to be obtained even for minor surgical procedures.

Some of the issues encountered in the clinic are:

- Islamic women in *burqa* (full veil) and orthodox Hindu women prefer not to reveal their faces or undress if a male attendant or medical colleague is present.
- Many families object to having the genital region photographed.
- Some families regard taking pictures of newborns as inauspicious.
- It is not always easy to obtain childhood photographs of dysmorphic individuals; photographic records in the absence of an important celebration in the family, such as a religious event or a wedding, are rare in poor families.
- In many cases, such pictures show elaborate make-up on the face, hair hidden under floral hair decorations, or touching up with special effects, often masking the dysmorphology we set out to record.

IMPARTING RESULTS

A second appointment is usually arranged when results of the genetic tests become available. Invariably, all or many members of the family attend a counseling session. The Western concept of privacy of information is usually not considered to be an issue in Indian society. In all but a few situations, asking only the individual concerned or immediate family into the clinic separately is viewed with suspicion by the rest of the family. This is particularly so if the wife alone is asked to come in for prenatal diagnostic counseling.

With prenatal issues in a couple, and with congenital malformations in a child, the wider family is seen in accompaniment, more out of curiosity than any other type of involvement. We have found it useful to point to the small size of most counseling rooms or lack of adequate number of chairs in the room to tactfully limit the number of people present during consultation. It is not unusual for grandparents, uncles and aunts, or even friends of the family to come separately after the consultation to ask for information about the medical details in the index family. We usually refer them back to the family members who have just been counseled.

As in other societies, the counselor has to depend on visual aids such as diagrams, models, and pictures of karyotypes to facilitate understanding (Greenwood Genetics Center, 2002). This may be very useful in creating an image in the mind of a lay person of the chromosome and gene being described. Occasionally this may "backfire," as in a case counseled by the author. A mother who was shown a karyotype to explain trisomy 21 in her child wanted to know how the extra "worm" could be removed from the body. A colleague's experience after counseling parents regarding a trisomy was also amusing. The father came back later to inquire if the extra chromosome seen in their child could belong to the mother's former boyfriend. Having a dialog with the family and enquiring about their understanding at periodic intervals during the counseling session may help gauge their comprehension of the genetic facts and to avoid confusion.

GENETIC COUNSELING

The practice of genetic counseling combines technical knowledge of the clinical disorder in question, the ability to communicate the scientific details in easily understandable terms, and empathizing with the family seeking advice without compromising on the accuracy of information imparted. As with other professions that demand skill in communication, experience in the field is invaluable in judging what and how much information should be imparted in each session of contact with the family seeking advice.

It is always useful first to ascertain what issues concern the affected individual or parents most, and to address them. Very often, the parents of an affected child are most concerned about life limitation or future marriage prospects, rather than the name of the medical condition (Girisha et al., 2007).

It is usually safe to assume that counseling in all cases should start at the beginning, with a complete explanation of the genetic disorder. Even if the parents are medical personnel, they may not always be conversant with principles of clinical genetics and Mendelian patterns of inheritance, simply because the subject is either not taught or has rudimentary representation in the syllabi of undergraduate medical education. An example is an eminent and eloquent head of a software institution who had downloaded reams of facts about neurofibromatosis type 1 (NF1), which affected his wife and daughter. He wished to be informed of all therapeutic clinical trials happening in different countries. During counseling it was noted that he was astonished to learn that it was as likely that a male could inherit NF1 as a female.

The most commonly asked question in clinics where children are seen with congenital malformations is: "Why did this happen?" What should be said to a family has to be delicately balanced with what is not to be said in the context of the likely consequences of a well-understood session. The perception that the mother is somehow responsible for any disorder detected in a child is widely prevalent. Explaining X-linked inheritance is therefore tricky, and most experienced counselors will have their own ways of approaching this. In many situations, when a family is told that "the mother is the carrier of a faulty gene" that may be responsible for the occurrence of the particular genetic disorder in male offspring, the mother in question ends up facing social difficulties, including abandonment with the affected child or children, divorce, or other forms of social ostracization (Phadke and Agarwal, 2001).

A similar scenario is often witnessed in the case of autosomal recessive disorders. An illustration of this is a consanguineous couple seeking genetic advice after the loss of two children to an undiagnosed severe neuro-metabolic disorder. To complicate matters, the wife had delivered by Caesarian section both times, which limited the number of further surgical deliveries possible. An enthusiastic explanation of autosomal recessive inheritance by the author, and the inability to provide definitive prenatal testing to the couple in the next pregnancy, resulted in the appearance of a new wife in a couple of months. Following the Hippocratic Oath of "first do no harm" takes on a special meaning in such situations. Another nonconsanguineous couple's first child, a female baby with ambiguous genitalia, was identified as having congenital adrenal hyperplasia. While counseling, autosomal recessive inheritance was discussed with the family. The very dominant mother-in-law of the woman, who insisted on being part of the consultation, immediately accused the wife of incest with a male relative, as "nothing is wrong in *our* family." Ridiculous as these accusations may seem, various versions of these scenarios are played out in numerous counseling situations or subsequently, when a latent blaming of the mother becomes an overt accusation. What is worrying is that this is not restricted only to the illiterate or the poor.

CONTINUED SUPPORT

In the West, following a consultation the genetic counselor summarizes the discussion that has ensued, and all results, into a letter to the family with a copy to the referring doctor. This is probably the most time-consuming part of the counselor's job. In most centers in India, the ability to follow these good practices rarely exists owing to limitation of time and resources. This often leads to a discontinuity of genetic care at the community level, as genetic information discussed in the clinic is either only partially understood or completely forgotten. In the author's center, where an average of 20–25 families are seen every week in the genetics clinic by one specialist, the same time and resource restrictions are faced. This is overcome in part by making two handwritten copies of the consultation, one filed in the medical records and then numbered and stored serially in the clinic, and the second copy given to the family. Usually, a student or nurse who is sitting in on the consultation copies details in the medical record while the specialist obtains the family details and writes in the patient's copy. In the absence of another person in the clinic, a photocopy of the consultation details is made. In the patient's copy of the handwritten consultation, the provisional or definite genetic diagnosis, test results, recurrence risks, and key follow-up instructions for the family or the treating doctor are also written in simple English or the regional language. As mentioned earlier, it has been very useful to record mobile telephone numbers, which helps us to keep in touch with families requiring follow-up.

For common disorders such as Down's syndrome, certain microdeletion syndromes, such as chromosome 22q11 micro-deletion and William's syndrome, which are commonly seen in our clinic, a standard prepared computer format akin to an information leaflet about the condition forms the template of the letter to the family. While the family waits after the consultation, a brief paragraph of the individual's details are added to this, the letter printed, and handed over to them. The letter is usually reinforced by simple verbal instructions to the family, such as "please telephone or email us as soon as your wife is pregnant." With this strategy, it has been possible for our center to keep in touch with approximately 80% of the families seen in our clinic, irrespective of the educational status or geographical distance of the family.

Regular systematic review of undiagnosed dysmorphic syndromes or families with chromosomal rearrangements is seldom possible for a host of reasons. Follow-up of families who have been counseled in the clinic is at best patchy, and depends largely on the family, who may or may not contact the center again for a fresh query. Distance may be one reason for this, as the family may return after the first consultation to their town or village, where medical genetics services may not be available.

Support groups for specific genetic conditions are only occasional and often initiated by families of affected children. Many of these support groups are limited by the number of affected individuals they reach, mainly because of lack of funding or wider dissemination of information. Most counselors resort to putting families with rare disorders in Internet contact with support groups in other countries.

Very few registries of genetic disorders exist, and many of them contain mainly regional rather than national data.

NATIONAL ISSUES IN GENETIC MEDICINE AND COUNSELING

POPULATION

In a vast and diverse country of over a billion people, absolute numbers of individuals and families at risk of a genetic disorder are enormous. In general, awareness of the availability of the genetic counseling services is low. However, paucity of centers offering genetic counseling means that most counselors see between 10 and 20 families in a day's clinic, and the time available for counseling is therefore limited. In most centers the average time spent in each counseling session is a fraction of that in Western countries.

HEALTHCARE SYSTEM

India has a two-tier healthcare system, with a state-run health service responsible for training of the majority of medical professionals, and treatment of the underprivileged and poorly educated sections of society. The smaller second tier, consisting of privately run medical schools and hospitals, provides expensive medical care paid for by the patient. The former caters to the vast majority of the rural and the urban poor.

COST OF GENETIC TESTS

Unlike the situation with the National Health Service of the United Kingdom and the medical insurance coverage of other developed countries, in the majority of Indian cases all genetic tests are paid for by the family. Although the costs of common genetic tests are a fraction of their costs abroad, they are often unaffordable for most families. Even where families have private health insurance, most policies restrict coverage to inpatient rather than outpatient investigations and treatment. In India, cost of testing is one of the main obstacles to thorough investigation of a genetic disorder. With the limited availability of genetic diagnostic facilities nationally, and the prohibitive costs of investigations available abroad, there is a constant, daily struggle to appropriately investigate all families for genetic disorders. In this context, sequential cascade screening of family members for single gene disorders and chromosomal rearrangements becomes even more difficult to achieve.

SPECIALIZED EDUCATION IN GENETICS

Although there are over 270 medical schools in India, only rudimentary medical genetics is taught in the undergraduate curriculum. There are very few centers in India offering formal, specialized training in medical genetics to doctors. In particular, there are no long-term, specialized, state-recognized courses in genetic counseling, despite the huge need for this specialty. As a result, most genetic counseling is usually undertaken by clinicians in other medical specialties, only partially trained or without formal training in genetic counseling. The advice given to families often reflects the doctor's own limited understanding of the genetic disorder, along with his or her own personal beliefs. It is still common to hear of couples with one or more children with a genetic disorder having been advised "not to have any further children." Specialized concepts in counseling, such as predictive testing in late-onset disorders and cancer genetics, counseling in imprinted disorders and mitochondrial inheritance, genetic counseling and testing in children, and ethical issues of consent and confidentiality are not familiar to many.

There is also a real issue about "intellectual inbreeding." Individuals trained in the few institutions in the country that offer medical genetics training often become trainers and teachers in the same institutions. It is the author's belief that genetic counseling is best experienced and learned as an apprenticeship, and not only from the written word. In the absence of external training and techniques, the practice of genetic counseling is limited to local experience. With little exposure to methods of genetic counseling in other countries and societies, new ideas and concepts in the practice of counseling are hard to come by. Ways of overcoming these limitations may include having nationally designed genetic counseling courses complemented by visiting experts contributing from their experience.

COMMON CONDITIONS SEEN IN INDIAN CLINICS

Down syndrome and other chromosomal anomalies, mental retardation syndromes, congenital malformation syndromes, beta-thalassemia, glucose-6-phosphate dehydrogenase (G6PD) deficiency, sickle cell anemia and hemophilias, congenital adrenal hyperplasia, Gaucher's disease, mucopolysaccharidoses and other lysosomal storage disorders, spinal muscular atrophy, neural tube defects, Duchenne and Becker muscular dystrophy, spinocerebellar ataxia, recurrent miscarriages, and poor obstetric history are among the commonest referrals in most genetics clinics in the country (Verma and Bijarnia, 2002; Verma et al., 2003; Colah et al., 2005; Kumta, 2005; Ghosh and Shetty, 2008). It was estimated that in the last decade, among the 24 million births occurring annually, 780,000 had congenital malformations at birth, 340,000 were diagnosed with G6PD deficiency, 20,800 with inborn metabolic disorders, 21,000 with Down syndrome, 10,400 with congenital hypothyroidism, 9,000 with beta-thalassemia and 5,200 with sickle cell anemia (Verma, 2000). These mind-boggling numbers speak of the volume of work and the need for genetic and counseling services.

GENETIC SCREENING PROGRAMS

The government of India has recently approved the first national pilot project of newborn screening (ICMR, 2007). In this pilot project, 100,000 newborns will be tested for congenital hypothyroidism and congenital adrenal hyperplasia, two disorders reported to be common in Indian society, so that early remedial therapy can be instituted. Regional screening programs for prevalence of birth defects, beta-thalassemia, sickle cell anemia, and retinoblastoma have been reported. Antenatal screening for Down syndrome by nuchal translucency scans and maternal serum screening is available in the larger cities only, and to a small minority of pregnant mothers.

GENETIC COUNSELING IN INDIA: SOME SPECIAL ISSUES

LANGUAGE

The practice of genetic counseling in India can be likened to the complexity of using an imaginary single norm for the entire continent of Europe. Each Indian state, like each country in Europe, has a different language, culture, customs, food habits, and population profile. India has 18 official languages (71st Amendment to the Constitution of India) and over 2000 dialects. The 2001 census recorded 29 individual languages as having more than a million native speakers. This linguistic diversity makes it impossible for counseling to be practiced in any one language throughout India.

On an average day in a tertiary hospital dealing with patients from various regions of the country, it is not unusual to be expected to counsel in five or six different languages. Many educated Indians are trilingual, and this helps in overcoming some of the language problems in a counseling session. Professional translators are rarely available to overcome language barriers. Sometimes relatives or friends accompanying the family assist in translation; this then raises the issue of confidentiality.

GENETIC TERMINOLOGY

More importantly, although the concept of hereditary traits is widely appreciated, there is no easily understood word for *gene* in any Indian language. Counselors usually resort to the English equivalent or a functional translation which may be better understood by most families. The closest Hindi translation for gene is *paitrika,* which literally translates to "paternal." In Kannada (the official language of the state of Karnataka, the capital of which is Bangalore), *jivanu* is the dictionary translation of *gene.* As very few individuals understand this technical term, the author usually uses the word *huttu-guna*, which literally translates to "birth traits."

RELIGION AND SOCIAL PRACTICES

The major religions in India are Hinduism, Islam, Christianity, Sikhism, Buddhism, and Jainism. Approximately

80% of Indians are Hindus. Marriages between individuals of different religions are still the exception rather than usual. Religious sentiments as well as traditional beliefs have to be taken into consideration while counseling. Some examples are:

- Tuesdays are considered inauspicious by many Hindus, who prefer not to have an appointment to receive genetic results or have an elective Caesarian section on the day.
- Fridays are not preferred for hospital appointments by Muslims as it is a special prayer day.
- A solar or lunar eclipse is considered inauspicious for pregnant women. Widely held beliefs include occurrence of congenital birth defects if a pregnant woman ventures outdoors during an eclipse.
- Many parents paint black spots on the faces of young children to ward off an "evil eye." Branding of children to ward off evil spirits or as part of tradition is also common in some societies, and thereby adds to acquired dysmorphology.
- In certain North Indian states, pre-lacteal feeds containing opiates are fed to newborns, causing constipation and constricted pupils as a side effect. It is important to inquire about such feeding before embarking upon investigation of an organic etiology.

SOCIAL FACTORS IN MARRIAGE, REPRODUCTION AND PRENATAL DIAGNOSIS

Marriage is a universal concept in Indian society. It is unusual for men, and especially women, to remain unmarried by choice in the absence of medical or economic reasons. Single parenthood, parenthood outside marriage, and having multiple partners are still unusual. In some sections of Indian society, in the event of a wife's early death, the husband usually marries the woman's sister.

Childlessness is also regarded as a social aberration, with very few married couples choosing to remain childless in the absence of a medical reason. In fact, the overwhelming desire for biological children in Indian society has led to a flourishing practice for the many assisted reproduction therapy (ART) centers in the country.

Most rural marriages and many urban ones are still "arranged" marriages. The alliance is usually agreed upon by the parents and other elders of the family, rather than the couple themselves. The process of getting a couple married is a complicated one. Matching of horoscopes of the two individuals concerned, belonging to the same religion and caste, social and economic background, absence of any major physical defects or diseases in the woman, and family approval are all major factors in arriving at the decision. While the support and approval of the extended family are invaluable in contributing toward longevity of the marriage, the converse may also be true. When a genetic condition results in infertility, repeated pregnancy loss, or birth of affected children, the tendency to blame the wife is still widely prevalent. Often, even if the couple in question may have no major differences between themselves, the extended family may exert pressure on the couple, influencing the course of the marriage in the pursuit of a perfect child and inheritor.

According to Hindu scriptures, life starts at the time of conception. Therefore, discontinuing a pregnancy at any stage of pregnancy is against Hindu religious beliefs. Legally, in India termination of pregnancy is permitted until twenty weeks of gestation where "there is a substantial risk that if the child were born, it would suffer from such physical or mental abnormalities as to be seriously handicapped" or "the continuance of the pregnancy would involve a risk to the life of the pregnant woman or of grave injury to her physical or mental health" (Medical Termination of Pregnancy Act, 1971). In contrast are modern societal trends, where many Indians resort to discontinuation of pregnancy as a method of limiting family size. Although exact numbers are not known, the number of abortions performed in India each year is estimated to be between five and six million (Mohan and Goyal, 2000). This may be one of the reasons why advice regarding discontinuation of pregnancy as an option in case of a serious birth defect or genetic disorder is not usually met with much resistance by a majority of counseled couples.

In case of couples with an unborn child at risk of a genetic disorder, the husband usually leads the discussion and states decisions; the wife usually is the "silent" partner; even when asked directly she usually acquiesces to the husband's decision. This may reflect one or more of several reasons:

- Educational status of the wife, who may be young at the time of marriage and may not have completed her education or may be educationally less qualified than the husband
- Social conditioning to accede always to the husband's decision
- Economic dependence of many women, even educated ones, who choose to stop work after marriage or opt for a career subordinate to that of the husband.

Decisions regarding having an invasive test for prenatal diagnosis or discontinuing a pregnancy are usually taken in consultation with family "elders," who may include parents of the couple and other relatives. Women visiting prenatal diagnostic clinics are more often accompanied by female relatives than by the husband. A Dutch study reported that the most important factors governing reproductive planning after genetic counseling included a strong desire to have children and lack of personal experience of the disorder, rather than the magnitude of genetic risk or availability of prenatal diagnosis (Frets et al., 1990). A similar but smaller study from the northern states of India, of couples who were provided genetic counseling, showed no change in reproductive planning based on the risk of recurrence, but change based on the presence and number of normal live children in the family (Phadke et al., 2004).

PREFERENCE FOR A MALE CHILD

A major factor that influences reproductive decision making is the desire for a male child. Culturally, the desire for a male child in Hindus is a major factor not just in reproductive planning, but also in the choice of prenatal diagnostic tests. Widespread use of noninvasive and invasive sex determination for selective termination of the girl child in the last two decades has led to an increasingly skewed female to male sex ratio in many parts of India (George, 2006; Sharma et al., 2007; Garg and Nath, 2008). The 2001 population census reported an all-India average of less than 930 women for every 1000 men, with the lowest ratios from North Indian states. Government legislation, in the form of the Pre-conception and Pre-natal Diagnostic Techniques Act of 1994 (amended in 2002) and National Policy for Empowerment of Women, 2001, is attempting to prevent discrimination against and exploitation of women.

Even after birth, it is usually the male child for whom the family, if circumstances require it, is willing to spend on genetic testing and treatment, when available. In one case in the author's experience, a couple with two affected children with Wilson's disease preferred to pay for d-penicillamine therapy for the male child in preference to the female child, as they could only afford treatment of one child. Numerous such examples of discrimination against the female in various situations have been reported from Indian counseling centers (Naveed et al., 1992; Phadke, 2004). In disorders of sexual development, the preference of most parents to raise their child as a male is seen often, even when appropriately counseled otherwise. One example is of a couple with a male child born after two daughters; the third child had complete agenesis of the penis and a male karyotype. The parents were overwhelmingly in favor of raising the child as a male, preferring to consider assisted reproductive technology to enable the child to have biological offspring in the future. An international study that surveyed attitudes of medical geneticists from India about their willingness to perform prenatal diagnosis for selection of a male fetus reported that 52% had answered in the affirmative (Wertz and Fletcher, 1993). One of only a few centers in India offering preimplantation genetic diagnosis has claimed results in achieving exclusively male pregnancies mainly for "family balancing" (Malpani et al., 2002).

CONSANGUINITY

Consanguineous marriages are common in many Indian communities. The incidence of consanguineous marriages ranges from 0.5% in northeast India (Mizoram) to 38.2% in the southern state of Tamil Nadu (Bittles, 2002). Even among Hindus, social customs differ widely between the northern and the southern parts of India. In North India, marriage between two individuals who are related through seven generations on the male side, or five generations on the female side, is not permitted. In several communities of South India, uncle–niece and first cousin marriages are the norm. Traditionally, a girl is expected to marry her maternal uncle, unless she does not have one. Among Muslims, marriage between first cousins is common.

In families where the parents are consanguineously married, multi-generation consanguinity is often seen noted. In a recent survey of 105 individuals with inborn errors of metabolism from Bangalore, we found the prevalence of close consanguinity to be 60% in affected individuals and 20% in matched controls (unpublished data). This gives rise to difficulties at two levels in genetic counseling. First, accurate risk calculation for the occurrence of an autosomal recessive genetic disorder gets complicated with multi-generational consanguinity. Secondly, it is not uncommon to see two separate autosomal recessive disorders, or an autosomal recessive disorder and a separate dominant disorder, segregating within the same kinship. Needless to say, genetic counseling and prenatal and postnatal testing of individuals at risk in these situations are difficult for both the counselor and the family.

Beta-thalassemia is believed to be the commonest serious single gene disorder in India. The carrier rate is estimated at 4%, with higher rates among Sindhis, Punjabis, and Gujaratis (Agarwal et al., 2003). At present, healthy consanguineous couples from these communities seeking preconception counseling are offered carrier testing for beta-thalassemia because of the increased carrier rate for this disease.

DISORDERS OF SEXUAL DEVELOPMENT (DSD)

Congenital adrenal hyperplasia (CAH), complete and partial androgen insensitivity, 5-alpha-reductase deficiency, and mixed gonadal dysgenesis are the commonest DSDs encountered in India. Genital ambiguity at birth can be considered a genetic emergency, as there is need for early sex assignment. However, in many Indian families this may not come to light till much later, as genital ambiguity is viewed by many families as a matter of shame and secrecy with a social stigma against the affected child. In a society where marriage and reproduction are imperatives, the difficulty encountered by these individuals in finding partners for marriage and the inability to reproduce makes it difficult for them to be integrated adequately in society. Counseling of these individuals requires skill and psychological support from trained counselors familiar with societal expectations.

Females with masculinized genitalia due to CAH may be accurately diagnosed and commenced on appropriate treatment. However, parents have been noted to discontinue treatment, jeopardizing the life of their child. This usually arises from the unspoken fear that their children may grow up to be "hijras." This fear must be raised directly and addressed in a genetic counseling session, as the parents or guardians of the child may not voice these concerns at a first meeting with the counselor.

Hijras are a distinct group of transgender individuals common in Indian society. Although most individuals who become part of a *hijra* community are believed to have been genetically male at birth, they tend to dress as females. A large number of such men are known to undergo surgery for removal of penis and both testes before complete integration into the community. Most members of this community tend to live together in groups, although they are not related. They earn by begging or performing at auspicious ceremonies such as marriages and births. The description of several venerated transgender characters in Hindu scriptures (Shikhandi and Shiva-shakti) makes their presence at Hindu religious ceremonies acceptable. Many also make a living as sex workers engaged in prostitution.

British laws in nineteenth century India, such as the Criminal Tribes Act of 1871 and Section 377 of the Indian Penal Code (IPC) of 1861, were responsible for providing a legal basis for discrimination against individuals with transgender issues. The more recent Immoral Trafficking Prevention Act (IPTA) of 1986 inadvertently also encourages discrimination against *hijras* and other transgender individuals.

GENDER IDENTITY ISSUES

While homosexuality may not be a direct concern of genetic medicine, it is important to discuss it in the context of DSD. Many parents are anxious that children with DSD may grow up with homosexual tendencies. This often comes in the way of optimal genetic counseling and management of affected children. The difficulty of many parents to accept reasons for homosexuality may be because of the history and laws of the country. Under Section 377 of the Indian Penal Code (1861), homosexuality was until recently regarded as a criminal offense in India, punishable by ten years imprisonment. Ancient Hindu codes of conduct as proscribed by the *Manu Smriti* also prohibit homosexuality, with more severe punishment outlined for females than for males engaging in homosexuality. In traditional Indian society, the institution of marriage and the family are an important part of culture, and sex outside marriage meets strong societal disapproval. Although many governmental, legal, and social groups have recently come out in favor of decriminalizing homosexuality, widespread disapproval continues to prevail (PUCL-K report, 2003). Prevalence of homosexuality in Indian society is therefore shrouded in secrecy and not openly discussed.

ROLE OF THE MEDIA

Much of what is reported in the national and regional Indian media regarding genetic disorders are reports of sensational individual cases with rare diseases or serious congenital malformations, often with graphic photographs. Education about genetic disorders and availability of local diagnostic or therapeutic services involves a very small part of the print and audiovisual media. More commonly, medical advances in other countries are reported in a manner that suggests availability of new therapies being more imminent than scientific research would permit. As a result, it is not unusual to see several referrals a month in any Indian genetics clinic seeking "stem cell therapy" for alleviation of various ailments. The medical fraternity must also shoulder part of the blame, as dialog with media colleagues is often limited to issuing reports highlighting individual or institutional achievements in the diagnosis of esoteric genetic disorders, rather than dealing with the wider issue of public education.

SUGGESTIONS FOR DEVELOPMENT OF GENETIC COUNSELING SERVICES IN INDIA

First and foremost, quality diagnostic laboratories with affordable genetic testing will have to be the backbone of a unit offering genetic counseling and support. Formation of a national committee of experts to design and lay down national guidelines for certification of courses in genetic counseling is obviously necessary. The role of national governmental organizations such as the Indian Council of Medical Research and Department of Biotechnology in organizing, monitoring, and certifying these courses will be essential to promote quality and standards nationally.

Specialized counseling courses designed for medical specialties where a genetic counselor is an integral part of the multidisciplinary management team will have to be considered: for example, counselors specializing in cancer, hematology, and ophthalmic genetic disorders. Creation of job opportunities for genetic counselors in medical schools and hospital departments, and education of medical colleagues to utilize their services, will also be necessary.

As many states in India are limited by linguistic boundaries, training in counseling should be made available in all states with emphasis on the specific terminologies, customs, and traditions prevalent in the geographical region. A three-language formula may be essential, and all counselors should be able to provide counseling in English (widely spoken by the educated), Hindi (the national language), and the regional language of the state.

Current practices in genetic counseling target only a very small minority of the affected or at-risk individuals and families who seek these services. Looking to the future, genetic diagnosis and counseling in India will have to reach the wider community in a proactive manner if it is to have an impact on national health goals.

ACKNOWLEDGEMENTS

I would like to gratefully acknowledge Professor H. Sharat Chandra, Director, Centre for Human Genetics, and Mr. Prakash for their constructive criticism and editorial help with this manuscript.

Preparation of this review has also been aided by grants to the Centre for Human Genetics from the Indian Council of Medical Research and the Department of Biotechnology, Government of India.

REFERENCES

Agarwal S, Gupta A, Gupta UR. (2003). Prenatal diagnosis in beta-thalassemia: an Indian experience. *Fetal Diagn Ther* 18(5):328–332.

Bittles AH. (2002). The impact of consanguinity on the Indian population. *Indian J Hum Genet* 8:45–51.

Clarke A. (1997). The process of genetic counselling: beyond non-directiveness. In: Harper PS, Clarke A, eds. *Genetics, Society and Clinical Practice*. Oxford: Oxford BIOS, pp. 179–200.

Colah R, Surve R, Nadkarni A, et al. (2005). Prenatal diagnosis of sickle syndromes in India: dilemmas in counseling. *Prenat Diagn* 25(5):345–349.

Frets PG, Duivenvoorden HJ, Verhage, F, et al. (1990). Factors influencing the reproductive decision after genetic counseling. *Am J Med Genet* 35(4):496–502.

Garg S, Nath A. (2008). Female feticide in India: issues and concerns. *J Postgrad Med* 54(4):276–279.

George SM. (2006). Millions of missing girls: from fetal sexing to high technology sex selection in India. *Prenat Diagn* 26(7):604–609.

Ghosh K, Shetty S. (2008). Quality and easy availability of medical support overwhelmingly determines the attitude and belief towards medical resting in haemophilia. *Haemophilia* 14(2):392–393.

Girisha KM, Sharda SV, Phadke SR. (2007). Issues for counseling for Down syndrome. *Indian Pediatr* 44(2):131–133.

Greenwood Genetic Center. (2002). *Genetic counseling Aids*, 4th ed., Greenwood SC: Greenwood Genetic Center.

Harper PS. (2004). Introduction to genetic counseling. In Harper PS. (ed.), *Practical Genetic Counseling* (6th ed.). London: Hodder Headline Group. ISBN 0340 81196 X.

ICMR [Indian Council of Medical Research]. (2007). Task force on Inborn Metabolic Disorders. *www.icmrmetbionetindia.org.*

Kelly TE. (1986). *Clinical Genetics and Genetic Counseling.* Chicago: Year Book.

Kumar P. (1968). Genetic counseling in family planning. *Antiseptic* 65 (11):831–834.

Kumta NB. (2005). Inborn Errors of Metabolism (IEM) – An Indian Perspective; *Indian J Pediatr* 72(4):325–332.

Legislation: legislation dealing with the use of languages, Constitution of India. Articles 29, 30, 120, 210, 343–351, as amended in the 21st and 71st Amendments.

Malpani A, Malpani A, Modi D (2002). Preimplantation for sex selection for family balancing in India. *Hum Reprod* 17:11–12.

Medical Termination of Pregnancy Act. (1971). Act No. 34 of 1971, Government of India.

Mohan R, Goyal RS. (2000). The changing pattern of medical termination of pregnancies in Rajasthan and India: a comparative analysis. *The Journal of Family Welfare* 46(1):39–46.

Naveed M, Phadke SR, Sharma A, Agarwal SS. (1992). Sociocultural problems in genetic counseling. *J Med Genet* 29(2):140.

Phadke SR, Agarwal SS. (2001). Adverse effects of genetic counseling on women carriers of disease: the Indian perspective. *Natl Med J India* 14(1):47–49.

Phadke SR, Pandey A, Puri RD, Patil SJ. (2004). Genetic counseling: the impact in Indian milieu. *Indian J Pediatr* 71(12):1079–1082.

Phadke SR. (2004). Genetic counseling. *Indian J Pediatr* 71(2):151–156.

Pre-Conception and Pre-natal Diagnostic Techniques (prohibition of sex selection) Act (PNDT), 1994. Amended in 2002.

PUCL-K [Peoples' Union for Civil Liberties, Karnataka]. (2003). Human rights violations against the transgender community, a study of Kothi and Hijra sex workers in Bangalore, India. Sept 2003 Report.

Sharma BR, Gupta N, Relhan N. (2007). Misuse of prenatal diagnostic technology for sex-selected abortions and its consequences in India. *Public Health* 121(11):854–860.

Verma IC, Bijarnia S. (2002). The burden of genetic disorders in India and a framework for community control. *Community Genet* 5(3):192–196.

Verma IC, Saxena R, Lall M. et al. (2003). Genetic counseling and pre-natal diagnosis in India- experiences at Sir Ganga Ram hospital. *Indian J Pediatr* 70(4) 293–297.

Verma IC. (2000). Burden of genetic disease in India. *Indian J Pediatr* 67(12):893–898.

Wertz DC, Fletcher JC. (1993). Prenatal diagnosis and sex selection in 19 nations. *Soc. Sci Med* 37(11):1359–1366.

80.

SOCIAL AND CULTURAL ASPECTS OF CUSTOMARY CONSANGUINEOUS MARRIAGES IN SOUTH ASIA

Rafat Hussain

INTRODUCTION

This chapter provides an overview of consanguineous unions with a particular emphasis on regions where such marriages are part of the social and cultural ethos. Consanguineous marriages are common in many parts of North Africa, West Asia and South Asia. There are also a substantial number of immigrants in many Western countries from these regions. However, the focus of this chapter is on patterns and correlates of cousin marriages in the Middle East and South Asia, and a discussion of social customs and cultural norms associated with consanguinity. The terms *consanguinity* and *cousin marriages* are used interchangeably throughout the chapter. Consanguineous unions are conventionally defined as marriages between individuals up to the third cousin level. However, in most regions of the world where consanguineous unions are common, marriages are predominantly among close relatives such as first cousins. Most of the studies on consanguinity have assessed the prevalence of consanguineous relationship over one generation. However, in biological terms the coefficient of inbreeding may be higher due to multiple relationships across generations. Moreover, the practice of cousin marriage needs to be viewed within the larger cultural continuum of endogamy.

REGIONAL VARIATIONS IN PREVALENCE AND PATTERNS OF COUSIN MARRIAGES

Among the African countries covered in this section include Tunisia, Algeria, Morocco, Libya, Egypt, Sudan, and Mauritania. Consanguineous unions are common in Tunisia (Ben Arab et al., 2004) and Kerkeni et al. (2007) reported from a hospital-based study that a quarter (24.8%) of all live births were to consanguineously married parents. In Algeria analysis of data from the 1994 Maternal and Child Health Survey showed that 36% of marriages in the sample population were cousin marriages (Jurdi & Saxena, 2003). Similar estimates are reported by Zaoui and Beimont (2002) from the Tlemcen region of West Algeria. In Morocco, the prevalence of consanguineous unions was 28% in the rural population resident in the Atlas mountains (Cherkaoui et al., 2005). A more recent study which included both urban and rural regions of Tunisia reported lower prevalence at around 15% (Jaouad et al., 2009). The Libyan Maternal and Child Health Survey reported the prevalence of consanguineous unions at around 46.5% (General People's Committee for Health and Social Insurance [Libya] 1996, cited in Jurdi & Saxena, 2003). In Egypt, prevalence figures for consanguineous unions range from 21%–33% across different studies (Hafez et al., 1983; Mokhtar & Abdel-Fattah, 2001; Mohamed, 1995) and 39% as determined by the nationally representative 1988 Egyptian Demographic and Health Survey (Bittles, 1991). Studies from Sudan report prevalence estimates ranging from 44%–63% (Ahmed, 1979; Saha & Sheikh, 1988; Saha et al., 1990; Jurdi & Saxena, 2003). Recent information from Mauritania is limited to one study, which reported a prevalence figure of around 47% (Hamamy et al., 2005).

In contrast to North Africa, there are a large number of studies that have assessed the prevalence of consanguineous unions in the Middle East. A study from Saudi Arabia based on cross-sectional surveys across all 13 regions of the country during 2004–2005 reported that 56% of all marriages in the survey population were consanguineous (El-Mouzan et al., 2007). The authors found considerable regional variation within the kingdom, with the prevalence being as high as 67% in Madina compared to 42% in the much smaller region of Al-Baha. The findings of the study by El-Mouzan et al. are comparable to earlier studies from Saudi Arabia. The national prevalence of consanguinity was estimated to be around 58% (El-Hazmi et al., 1995), while localized studies conducted in Riyadh (Al Hussain & Al Bunyan, 1997) and Dammam (Al-Abdulkareem & Ballal, 1998) reported prevalence rates of 51.3% and 52% respectively. In Jordan, a number of studies have investigated the prevalence and correlates of consanguineous unions. A wide variation in prevalence figures was found from 26%–64%, with prevalence being higher in smaller, rural-based studies (Khoury & Massad, 1992; Al-Salem & Rawashdeh, 1993; Nabulsi, 1995; Sueyoshi & Ohtsuka, 2003,; Hamamy et al., 2005). Obeidat et al. (2008) report the prevalence of 49% from a cross-sectional hospital-based study in 2007 in the north of Jordan.

The prevalence of consanguineous unions in Lebanon has been reported to be between 12.8%–42%. Results of the Maternal and Child Health Survey in Lebanon found the prevalence to be 21% (Ministry of Public Health [Lebanon] 1998, cited in Jurdi & Saxena, 2003). In another study from Lebanon, the overall prevalence was 35.5% (Barbour & Salameh, 2009). There has been a marked decline in the popularity of cousin marriages in large urban centers such as Beirut, from close to a third of all marriages being consanguineous prior to the 1940s to 18% by 1983 (Khlat, 1997). In studies based on data collected through the National Collaborative Perinatal Neonatal Network, on pregnancy and birth histories for a large cohort from the Greater Beirut region between 1998–2001, the prevalence of consanguinity was reported to be around 13%–14% (Tamim et al., 2003; Mumtaz et al., 2007). Information from the Maternal and Child survey in Syria reported that 38% of the marriages in the survey sample were consanguineous (Central Bureau of Statistics [Syria] 1995, cited in Jurdi & Saxena, 2003). In Syria, a large-scale study conducted across several urban and rural regions found 35% of marriages were consanguineous unions; prevalence varied from 67.5% in Al Raqa province to 22.1% in Latakia province. In general, prevalence of consanguineous unions was lower in western and northwestern provinces compared to central, northern, and southern provinces (Othman & Saadat, 2009). In the Palestinian Territories, a secondary analysis of 2 rounds of data (1995 and 2004) of women of reproductive age (usually defined as 15–49 years) found that by 2004, consanguineous unions comprised 45% of all marriages (Assaf & Khawaja 2009). Estimates of studies conducted on Arab Muslim population residents in areas comprising Israel show prevalence rates to be between 17%–63% (Jaber et al., 1994; Zlotogora et al., 2002; Vardi-Saliternik et al., 2002; Raz & Atar, 2004).

In the United Arab Emirates, the overall prevalence of consanguineous marriages has been reported to be between 40%–54% (Fahmy et al., 1993; Al Ghazali et al., 1997). Similar prevalence figures are reported for other countries in the region, such as Qatar [54%] (Abdulrazzaq et al., 1997; Bener & Alali, 2006) and Bahrain [39%–45%] (al-Arryed, 1995), Kuwait [54%] (Al-Awadi et al., 1985), and 35.9% in the Sultanate of Oman (Rajab & Patton, 2000). In Yemen, the overall prevalence has been reported to range between 40%–47% (Jurdi & Saxena, 2003; Gunaid et al., 2004).

Findings of the 2002 Iranian Fertility Transition Survey, conducted in the four southeastern provinces of Gilan, Sistan and Baluchistan, Yazd, and West Azarbaijan, show a high level of consanguineous unions. There was a wide variation in the regional prevalence ranging from 23% in Gilan to 78% in Sistan and Baluchistan (Abbasi-Shavazi et al., 2008). A large-scale study by Saadat et al. (2004) investigated the prevalence of consanguineous marriages in 12 ethnic/religious populations: the Persians (Shi'a and Sunni), Kurds (Shi'a and Sunni), Lurs, Azaris, Baluchis, Zabolis, Turkamans, Bakhtiaris, Ghashghais, and Arabs. The overall rate of consanguineous marriage was 38.6%. Other recent studies that have assessed the prevalence of consanguineous unions include Hashemipour et al. (2007), who found the prevalence of 28% among parents of live born infants in Isfahan. There is paucity of information from Afghanistan. However, a recent population-based study conducted in 2010 in 8 of the 34 provinces reported the overall prevalence of consanguinity to be 46.2%, ranging from 38.2% in Kabul to 51.2% in Bamyan province (Saify & Saadat, 2011).

In India, at the national level the prevalence of cousin marriages was 14% as determined in the 1992–1993 National Family Health Survey (IIPS, 1995). But the overall figure obscures the considerable regional and religious differences, as the preference of cousin marriage is not only predicated on religious differences but also on the geographical divide between North and South India. For example in Tamil Nadu, a south Indian state, 47% of all marriages were consanguineous, compared to 25% in Maharashtra (west India), and less than 10% of marriages in the northern and eastern states of India were consanguineous (IIPS, 1995). Among the minority Muslim population in India, which numbered over 100 million at the time of the national survey, the overall prevalence was estimated to be 27%, and in the Jammu region of the state of Jammu and Kashmir, the prevalence was 40.4% (Bittles & Hussain, 2000).[1] Unfortunately, similar data was not collected in the two additional rounds of the Indian NFHS conducted in 1998–1999 and 2005–2006. Data from a recent small-scale study from Mysore in southern India shows that approximately 28% of all participants were consanguineously married. However, the type of cousin marriages were not specified (Rao et al., 2009). The prevalence of consanguineous marriages in Pakistan, as estimated by a nationally representative survey of ever-married women aged 15–49 years, was reported to be around 63% (Ahmed et al., 1992). Smaller regional studies show a wide variation in the practice, with prevalence figures ranging from 30%–70% (Shami et al., 1990; Hashmi, 1997; Hussain & Bittles, 1998; Fikree et al., 2006; Badshah et al., 2008). A recent study reported a very high level of consanguinity (72.7%) among parents of children admitted to a large public tertiary care hospital in Karachi (Akram et al., 2008).

For other countries of South Asia, no data is available on national prevalence, although indirect information obtained from studies point towards the presence of cousin marriages do occur in Bangladesh. Information from nonpublished sources indicates that the prevalence of consanguineous marriages was about 17% around the Teknaf region in southern Bangladesh (Khan et al., unpublished 1994). Similar figures are reported from the study by Do et al. (2006) using information from the 1996 Matlab Health and Socioeconomic Survey comprising data on 5083 married men and 6068 married women. The Matlab region, though only 130 km from the capital city of Dhaka, is relatively isolated from communication and transportation and has a predominant agrarian economy. Of all the female respondents in the sample, 10% were married to a first cousin and 8% were married to a relative other than first cousin. A similar figure of nearly 18% is reported by a recent study on causes of causes of blindness in Bangladeshi children. The majority of the study subjects were residents in rural regions (Muhit et al., 2007). Anthropological studies undertaken by Fricke and his colleagues in the Timling region of Nepal (along the Tibet and Burmese borders) shows that cross-cousin marriages are common, and part of the social structure that is strongly based on kin alliance with reciprocal rights and obligations (Fricke et al., 1993; Fricke, 1995). There is no recent information on the prevalence of consanguineous unions for other countries of South Asia—Sri Lanka, Bhutan and Maldives—they share two common features: populations that predominantly follow Buddhism and/or Hinduism, and sociocultural similarities with other South Asian countries. Information from Thailand, Malaysia, and the Philippines shows that consanguineous marriages do take place among the resident non-Muslim population, albeit at a much lower rate than the Muslim population (Morgan et al., 2002), and there is little reason to believe that this is unlikely to be the case for countries in the South Asian region on whom currently no data is available.

TYPES OF COUSIN UNIONS

The most common type of consanguineous unions across the regions where cousin marriages are customary includes first cousin unions, with marriage to a second cousin or distant relative forming only a small proportion. The pattern of first cousin unions varies by region and culture. Among Muslims, all four types of cousin marriages (i.e., matrilateral and patrilateral, parallel and cross-cousin marriages) are contracted, and the type of marriage depends both on the availability of appropriate potential spouse, as well the degree of social closeness among the immediate families (Hussain & Bittles, 1998). A study from the southern Iran reported that among the 2686 clients referred for premarital, preconception, prenatal, and postnatal genetic counseling advice, the overwhelming majority of consanguineous couples (married/engaged to be married) were first cousins: 73% of the total sample, and 86% of the consanguineous unions (Fathzadeh et al., 2008). There was no particular dominant pattern of first cousin unions, with patrilateral parallel cousin unions almost as frequent (36.6%) as cross-cousin unions (34.7%), and matrilateral parallel cousin marriages being a close third (28.6%). Additionally, a small proportion of the clients were double first cousins (1.5%). A figure of 1%–5% is reported for double first cousin unions for almost all the countries in the region. In the recent Afghan study discussed earlier, 55% of all consanguineous unions were among first cousins whereas 17% were double first-cousins (Saify & Saadat, 2011).

Another marriage pattern that deserves some discussion is that of direct exchange marriages between offspring of two siblings. An anthropological study of the Janjua Rajput tribe in a small village in the mountainous region of Salt Ranges in northern Pakistan found that almost all the households in the village had at least one pair of siblings married in a direct exchange relationship—*watta satta*—and some households had such practice dating back to three generations (Alvi, 2007). Jacoby and Mansuri (2007) make similar observations about exchange marriages in their study from southern Pakistan. The most common form of such an exchange marriage is between offspring of a brother and a sister, whereby the brother's daughter is married to the sister's son in exchange for marriage of a sister's daughter to the brother's son, and vice versa. However, exchange marriages can involve two sets of parallel cousins, as well. The underlying principle governing such exchange marriages in case of cross-cousin marriages is the desire to unite the affinal and kin groups (Alvi, 2007). However, equally important is the rationale of risk minimization, which applies to both parallel and cross-cousin marriages, whereby the risk of mistreatment of the daughter is counterbalanced by the threat of retaliation through mistreatment of the daughter-in-law.

The Aryan traditions among the north Indian Hindus proscribes marriage within seven generations on the father's side and five generations on the maternal side. Hence, marriages among Hindus in North India are organized more around ethnic, caste, and class boundaries, and a large proportion of marriages are endogamous. In contrast, among Hindus in South India, with their predominantly Dravidian ancestry, both endogamy and consanguinity exist and overlap substantially. Among the Hindus in South India, the common pattern is cross-cousin marriage, especially matrilateral cross-cousin marriages. The prevalence of uncle–niece marriages, once common among the Hindu population of South India, has declined markedly. Such a decline can be attributed both to changing social norms as well as shrinking pool of appropriately aged matches. The 1992–1993 NFHS data shows that only 0.6% of marriages in Karnataka were uncle–niece unions. However, in the same national survey, 7.4% of marriages in Tamil Nadu and 4.2% in Andhra Pradesh were uncle–niece unions (IIPS, 1995).

Cultural norms can and do at times transcend religious proscriptions for consanguineous unions. For example, although both uncle–niece marriage and double first cousin have an equivalent coefficient of inbreeding, the former is not allowed by Islam, whereas the latter is. Thus, while double first cousin unions do occur among Muslims, although not as commonly as first cousin unions, uncle–niece marriages are not only unheard of, but would be considered incestuous under current laws in most countries. In India, however, 0.4% of all marriages among Muslims were uncle–niece marriages (IIPS, 1995). Similarly, there is considerable anecdotal evidence to suggest that first cousin unions were common among upper-caste Hindus living in western regions of undivided Punjab (which now constitutes Punjab province in Pakistan), while such marriages were largely proscribed in northern India. Consanguineous unions do occur in isolated tribal population subgroups in north India, such as the Dhankut, a small endogamous group in the state of Uttar Pradesh in northern India, who practice both parallel and cross-cousin marriages (Chantia, 2008).

Despite the contemporary unfavorable view of both Roman and Protestant churches about first cousin unions (Bittles, 2001; 2008), such marriages do occur among the Christian population, though the prevalence is much lower than among the corresponding Muslim population from the same setting/country. In the Middle East, consanguineous unions have been reported among Christian populations in Lebanon, Jordan, and Israel (Khlat, 1997; Khoury & Massad, 1992; Vardi-Slaternik et al., 2002). A recent study from Lebanon based on hospital births in the greater Beirut region found that the prevalence of cousin unions in the Christian sample was 2% compared to 13% in the Muslim sample (Mumtaz et al., 2007). In India, the proportion of consanguineous marriages among the Christian population included in the 1992 national-level survey was 13%, and among the Buddhist around 19% (IIPS, 1995). In Pakistan, while national-level figures are not available, a study conducted among low-income groups in Karachi reported that 38% of marriages among Christians, both Roman Catholics and Protestants, were consanguineous (Hussain & Bittles, 1998). Although first cousin marriages are common among Christians in Pakistan, one particular type of union, the patrilateral parallel cousin marriage, is avoided as offspring of brothers are considered to share the same blood and any such union would be akin to incest (Hussain, 1999; 2002).

SOCIAL AND CULTURAL FACTORS INFLUENCING MARRIAGE CHOICE

A common feature across most parts of the world where cousin marriages are customary is the social structure that regards marriage decision making largely a preserve of parents or family elders. Within this overarching decision-making process, marriage choice is largely restricted not only within religious and class affiliation, but also ethnic boundaries (Hussain, 2004). Many of the studies have only investigated prevalence and type of consanguineous unions, but without taking into account the societal structures that favor endogamy. Among the Arab populations in the Middle East, belonging to the same tribe, also commonly known as *hamula* or *kabeela*, is a cultural tradition (Tadmouri et al., 2009). Furthermore, the traditional rationale of "cousin-right" is widely accepted as part of the marriage norms in many Arab societies where cousin marriages are customary (Al-Rifai & Woody, 2007). Similar cultural perceptions exist in other communities where cousin marriage is customary. In countries such as Pakistan, the custom of cousin marriage emanates largely from the strong *biraderi* system, which is steeped in conventions of reciprocal rights and obligations.[2] A study conducted in major cities and towns of Punjab, where in addition to 46.5% of marriages being consanguineous, an additional 37.7% of marriages had been contracted within the *biraderi* or clan grouping (Shami, Grant & Bittles, 1994). Badshah et al. (2008, p. 197) paraphrase a common cultural perception, especially in rural areas: "first and second cousin marriage is characterized as gold and silver, whereas nonconsanguineous union is considered worthless."

Marriage patterns in the parental generation can also be a key influencing factor for type of marriage in the net generation.

A study from the northern Iranian city of Mashad reports that of the 500 youth interviewed as part of premarital counseling program, 47% of those who had consanguineously married parents were going to marry a cousin, compared to less than a quarter (23%) of the sample whose parents were not married to a cousin (Hasanzadeh-Nazarabadi et al., 2006). Similar results have been reported from Qatar, where the prevalence of consanguineous unions in the parents was reported to be 42% and approximately a third (36%) of all marriages among the parents-in-law was consanguineous (Bener & Hussain, 2006). Moreover, attitudes toward cousin marriage are heavily influenced by marriage patterns among siblings. In the Iranian study discussed above, consanguinity was viewed favorably by 22% of the sample whose siblings were not married to a cousin, whereas the corresponding proportion of the sample with 1–2, 3–4, or more than 4 siblings married to a cousin was 31%, 39%, and 83% respectively (Hasanzadeh-Nazarabadi et al., 2006).

A recent study in Karachi, Pakistan, found that among the grandparental generation, 64.1% of maternal and 60.3% of paternal grandparents were consanguineously married. Most of the marriages in the grandparental generation were first cousins: 77% of maternal grandparents and 83% of paternal grandparents (Akram et al., 2008). Thus, in many communities across the Middle East and South Asia, consanguinity and endogamy are part of the intergenerational pattern of marriage customs.

A number of demographic correlates of consanguineous unions have been observed in a large number of studies across diverse settings in the regions where cousin marriages are customary. These correlates include lower educational attainment for females, younger age at marriage, lower socioeconomic status measured variedly as employment status, family income and assets, type of housing, and number of family members. A persistent urban–rural differential is seen across studies (Hussain & Bittles, 1999; Sallam et al., 2001; Bittles, 2001, 2008; Bittles & Black 2010a; Tamim et al., 2003; Mumtaz et al., 2007). Fertility levels for women in consanguineous unions are reported to be higher both due to younger age at marriage and lower contraceptive use, and some increase in reproductive wastage (abortions and stillbirths and early offspring mortality; Bittles et al., 2002; Hussain & Bittles, 2004; Bittles & Black 2010b). Taken together, these correlates point to a portrait of individuals and communities that have more conservative and traditional cultural norms and practices.

Since the custom of cousin marriage exists in regions that predominantly have a Muslim population, it is somewhat erroneously believed that the high prevalence of cousin marriages among Muslim populations is due to some religious preference. Many scholars have pointed to the marriage of Fatima, the only daughter of the Prophet Mohammad, to his paternal first cousin Ali as an example that the prophet favored close cousin marriages. However, other scholars have pointed to the fact that endogamy was widely practiced in the Arab world before the advent of Islam (Teebi & El-Shanti, 2006). Not only is there no particular indication to this effect in the Koran, but also, according to Jaber et al. (1992), the Prophet Mohammad discouraged close cousin marriages. The underlying rationale for this perhaps was to ensure wider marriage alliances, which could facilitate the spread of Islam through Arabia. Similarly, Al Aqeel (2007), in his commentary on the Islamic ethical framework for research into and prevention of genetic diseases, states that Islamic teachings discourage the practice of close cousin unions. The second caliph Omer Ibn Al-Khatab noticed that the tribe of *Bani Althaa'b* had become weak and unhealthy because of intermarriage between cousins and advised avoidance of close cousin marriages and seeking of spouses from tribes further afield (Albar 1999, cited by Al Aqeel, 2007, p. 1294). However, for some Muslim communities maintaining the purity of descent can be an important factor in marriage choice, but in such subgroups marriage choice is not restricted to cousin marriage, but marriage within the endogamous clan boundaries. The main clan groups in Pakistan that practice endogamy (though not always close consanguineous marriage) for religious reasons are *Sayyeds* (who claim to have lineage associated directly with offspring of the Prophet) and *Qureshis* (who claim to have lineage associated with the patriline of the Prophet), and therefore are concerned with keeping the lineage "pure" (Hussain, 1999, 2004).

There is a general perception that cousin marriages are preferential because the economic costs are lower. Among the more affluent socioeconomic groups, retaining land and wealth within the family could be a factor that influences the choice of cousin marriages; however, among the general populace the reasons stem more from cultural norms. This is largely because there is little wealth to share in the first place. Moreover, women's subordinate social position often requires them to willingly forego their share of patrimony in order to retain or enhance their links with their natal kin, on whom they may need to heavily depend in case of desertion or widowhood. The one aspect where the economic argument does play a role is in societies, such as most parts of South Asia, where dowries are the cultural norm. Relatives are less likely to have an expectation of a large dowry, and the bride and her family are less likely to be "penalized" for any perceived shortcomings in the expected dowry (Caldwell et al., 1983; Reddy, 1988; Hussain, 1999). An Indian study found that marriage payments (both bridewealth and dowry) were much lower in consanguineous unions (Badaruddoza & Afzal, 1995). In a similar vein, a recent paper by Do et al. (2006) from Bangladesh postulates that in societies where the practice of providing large dowries for daughters is commonplace, consanguineous unions are a rational response to a marriage market failure rather than simply a consequence of culture, religion, or cultural preference. In such a situation, marriage to a relative with reduced or non-obligatory dowry payments provides an alternative to high dowry costs, and allows families to avoid the social repercussions associated with delayed marriage of daughters. However, the study found only a third of the sample women received any dowry. Moreover, after controlling for other important attributes at the time of marriage, such as age at marriage, years of schooling, religion, birth order, and number of brothers and sisters, women in consanguineous unions were on average 6%–7% less likely to bring a dowry at marriage, and on average 4% less likely to receive any inheritance. The results are not surprising. The study was conducted in a rural region of Bangladesh where, although agriculture is the main occupation, a third of the households were landless. In other parts of South Asia, such as rural areas in Nepal where farming and manual labor are the predominant occupations, cross-cousin marriages also have an economic value, as the boundaries of endogamy are shaped by family allegiances and reciprocal rights and obligations (Fricke, 1995).

Though differentials in bride-wealth or dowry provision could be an important reason for consanguineous unions rather than sociocultural customs, the decision for cousin marriages is based on a complex set of attributes that come into play in marriage choice for an offspring. Among the important factors that determine consanguineous marriages are the perceived social benefits of such unions including parents feeling reassured that there are no hidden character 'flaws' or unfavorable personal attributes, and relatively easier adjustment and integration of daughters into the new household through greater social and cultural compatibility between the husband and the wife, and the bride and the mother-in-law. As divorce still carries considerable social stigma, the presence of common relatives who can intercede with both spouses in times of marital conflict is perceived to be a tremendous advantage that such marriages have over marriages to non-cousins (Hussain, 1999, 2004; Bittles & Black, 2010a). A study from Damascus reported that thorough knowledge of the future spouse and the family is a prerequisite for marriage in Syrian society (Al-Rifai & Woody, 2007). Such in-depth knowledge is an important qualifier for many societies where consanguineous unions are customary. It serves as a fundamental starting point onto which other factors of the multilayered process of marriage decision making are added.

The perception of greater social support and better treatment of women in consanguineous marriages is widely held across most communities in the Middle East and South Asia. Although, as pointed out by Tadmouri et al. (2009), no direct empirical evidence is available that compares divorce rates among consanguineous and nonconsanguineous marriages, some recent studies have included consanguinity as an important correlated factor in domestic violence. A study aimed at determining frequency and severity of domestic violence prior to and during pregnancy, reported that Pakistani women married to close relatives had a significantly lower risk of being victims of domestic violence (Fikree et al., 2006). The authors conclude that "premarital family relationships appear to a play a significant protective role on an abusive relationship. Women whose spouses were distantly related were nearly four times as likely to be in an abusive relationship as compared to women married to their first cousins (p. 256)." Similarly, a recent study from reproductive health clinics in Jordan found that consanguinity was protective against violence during pregnancy (Clark et al., 2009). Thus, it seems that maintaining family and spousal harmony is viewed by many parents as one of the key reasons for continuing with the custom of cousin marriages. It has also been postulated that, although in most societies the custom of cousin marriage exists despite many such unions resulting in women having to relocate from home villages and communities, as the residential pattern in the early years of marriage is to live in extended household with the husband's parents and his siblings, stronger kinship relationships lead to better social networks for women, which can be of considerable value in dealing with child ill-health and disability (Robinson, 2007; Balhotra et al., 2010; Tadmouri et al., 2009).

SUMMARY

In communities with a strong preference for consanguinity, such unions will continue to be a common feature, due to the social honor associated with remaining within the ethnic boundaries and/or to retain family wealth. However, many studies show a decline in the prevalence of cousin marriages across most of the Middle Eastern and Asian countries, both from changing social norms and issues associated with family size. The ongoing fertility transition in many regions will lead to greater non-availability of the appropriate spouses, as a small number of children combined with an increase in female education, a small but significant increase in age at marriage for both young women and men, and other factors associated with modernization, such as increasing urbanization and monetization, are likely to have a negative impact on prevalence of customary cousin unions. On the other hand, social inequity is on the rise in some communities, with many parents already finding it difficult to find the resources to pay bride-wealth or dowry and associated marriage costs. Therefore, it is likely that parents would prefer to arrange marriage of at least one or more of their children within the family where possible. The combination of demographic and social changes is likely to lead to a situation where cousin marriages will be part of the marriage customs of most communities for the next generation or more.

NOTES

1 The survey was not conducted in large parts of Kashmir, which has a predominant Muslim population, due to civil unrest.

2 The term is derived from the Persian word *birader* and implies brotherly or fraternal relationship, and often extends to include nonrelatives belonging to the same social caste.

REFERENCES

Al-Abdulkareem AA, Ballal SG. (1998). Consanguineous marriages in an urban area of Saudi Arabia: rates and adverse health effects on the offspring. *Journal of Community Health 23*(1), 75–83.

Abdulrazzaq YM, Bener A, Al-Ghazali LI, Al-Khayat AI, Micallef R, Gaber T. (1997). A study of possible deleterious effects of consanguinity. *Clinical Genetics 51*, 167–173.

Ahmed T, Ali SM, Aliaga A, Arnold F, Ayub M, Bhatti MH. (1992). *Pakistan Demographic and Health Survey 1990/91*. National Institute of Population Studies and Macro International Inc., Islamabad and Columbia, MD.

Ahmed AH. (1979). Consanguinity and schizophrenia in Sudan. *British Journal of Psychiatry 134*, 635–636.

Akram DS, Arif F, Fayyaz J. (2008). How frequent are consanguineous marriages? *Journal of Dow University of Health Sciences 2*(2), 76–79.

Alvi A. (2007). India and the Muslim Punjab: A unified approach to South Asian kinship. *Journal of the Royal Anthropological Institute 13*(3), 657–678.

Assaf S, Khawaja M. (2009). Consanguinity trends and correlates in the Palestinian Territories. *Journal of Biosocial Science 41*(1), 107–124.

Al Aqeel A. (2007). Islamic ethical framework for research into and prevention of genetic diseases. *Nature Genetics 39*(11), 1293–1298.

Al-Awadi SA, Moussa MA, Naguib KK, Frag TI, Teebi ES, el-Khalifa M. et al. (1985). Consanguinity among the Kuwaiti population. *Clinical Genetics* 27, 483-486.

Al Ghazali LI, Bener A, Abdulrazzaq YM, Micallef R, Al-Khayat AI, Gaber T. (1997). Consanguineous marriages in the United Arab Emirates. *Journal of Biosocial Science 29*, 491–497.

Al Hussain M, Al Bunyan M. (1997). Consanguineous marriages in a Saudi population and the effect of inbreeding on perinatal and postnatal mortality. *Annals of Tropical Paediatrics 17*, 155–160.

Abbasi-Shavazi MJ, McDonald P, Hosseini-Chavoshi M. (2008). Modernization or cultural maintenance: The practice of consanguineous marriage in Iran. *Journal of Biosocial Science 40*(6), 911–933.

Al-Rifai MT, Woody RC. (2007). Marriage patterns and pediatric neurologic disease in Damascus, Syria. *Pakistan Journal of Neurological Sciences 2*(3), 136–140.

Al-Salem, M, Rawashdeh, N. (1993). Consanguinity in North Jordan. *Journal of Biosocial Science 25*(4), 553–556.

Badaruddoza A, Afzal M. (1995). Effects of inbreeding on marriage payment in North India. *Journal of Biosocial Science 27*(3), 333–337.

Badshah S, Mason L, McKelvie K, Payne R, Lisboa PJG. (2008). Risk-factors for low birthweight in the public-hospitals at Peshawer, NWFP-Pakistan. *BMC Public Health 8*, 197. Full text available from http://www.biomedcentral.com/1471-2458/8/197.

Balhotra S, Valente C, van Soest A. (2010). The puzzle of Muslim advantage in child survival in India. *Journal of Health Economics 29*(2), 191–204.

Barbour B, Salameh P. (2009). Consanguinity in Lebanon: prevalence, distribution and determinants. *Journal of Biosocial Science 41*(4), 505–517.

Bener A, Alali KA. (2006). Consanguineous marriage in a newly developed country: the Qatari population. *Journal of Biosocial Science 38*(2), 239–246.

Bener AB, Hussain R. (2006). Consanguineous unions and child health in the state of Qatar. *Paediatric and Perinatal Epidemiology* 20, 372–378.

Ben Arab S, Masmoudi S, Beltaief N, Hachicha S, Ayadi H. (2004). Consanguinity and endogamy in Northern Tunisia and its impact on non-syndromic deafness. *Genetic Epidemiology 27*(1), 74–79.

Bittes AH. (2001). Consanguinity: A major variable in the study of North African reproductive behaviour, morbidity and mortality? In Moore S. (ed.), Proceedings of the Demographic and Health Surveys World Conference, Macro International Inc., Columbia, pp. 321–341.

Bittles AH. (2008). A Community Genetics Perspective on Consanguineous Marriage. *Community Genetics 11*(6), 324–330.

Bittles AH. (2001). Consanguinity and its relevance to clinical genetics. *Clinical Genetics 60*(2), 89–98.

Bittles AH, Hussain R. (2000). An analysis of consanguineous marriage in the Muslim population of India at regional and state levels. *Annals of Human Biology 27*(2), 163–171.

Bittles AH, Black M. (2010a). Consanguinity, human evolution, and complex diseases. *Proceedings of the National Academy of Sciences 107*(1), 1779–1786.

Bittles AH, Black M. (2010b). Consanguineous marriage and human evolution. *Annual Reviews in Anthropology 39*, 193–207.

Bittles AH, Grant JC, Sullivan SG, Hussain R. (2002). Does inbreeding lead to decreased human fertility? *Annals of Human Biology 29*(2), 111–131.

Caldwell JC, Reddy PH, Caldwell P. (1983). The causes of marriage change in South India. *Population Studies 37*(2), 343–361.

Chantia A. (2008). Effects of inbreeding on health among Dhankut – An endogamous group of Bahraich, Uttar Pradesh. *Anthropologist 10*(3), 219–223.

Cherkaoui M, Baali A, Larrouy G, Sevin A, Boetsch G. (2005). Consanguinity, fertility of couples and mortality of children in the high Atlas population (commons of Anougal and Azgour, Marrakesh, Morocco). *International Journal of Anthropology 20*(3–4), 199–206.

Clark CJ, Hill A, Jabbar K, Sliverman JG. (2009). Violence during pregnancy in Jordan: its prevalence and associated risk and protective factors. *Violence Against Women 15*(6), 720–735.

Do QT, Iyer S, Joshi S. (2006). *The Economics of Consanguineous Marriages* (2006). World Bank Policy Research Working Paper No. 4085. Full text available from http://ssrn.com/abstract=950455.

El-Hazmi MAF, Al-Swailem AR, Warsy AS, Al-Swailem AM, Sulaimani R, Al-Meshari AA. (1995). Consanguinity among Saudi Arabian population. *Journal of Medical Genetics 32*, 623–626.

El-Mouzan MI, Al-Salloum AA, Al-Herbish AS, Qurachi MM, Al-Omar AA. (2007). Regional variations in the prevalence of consanguinity in Saudi Arabia. *Saudi Medical Journal 28*(12), 1881–1884.

Fahmy NA, Benson PF, Al-Garrah DB. (1993). Consanguinity in UAE: Prevalence and analysis of some risk factors. *Emirates Medical Journal 1*, 39–41.

Fathzadeh M, Bigi MAB, Bazrgar M, Yavarian M, Tabatabacc HR, Akrami SM. (2008). Genetic counselling in southern Iran: consanguinity and reason for referral. *Journal of Genetic Counselling 17*(5), 472–479.

Fikree FF, Sadiqua NJ, Korejo R, Afshan A, Durocher JM. (2006). Intimate partner violence before and during pregnancy: experiences of postpartum women in Karachi, Pakistan. *Journal of Pakistan Medical Association 56*, 252–257.

Fricke T. (1995). Marriage change and moral change: culture, value and demographic transition. Paper presented at the conference, *The Continuing Demographic Transition: The J.C. Caldwell Seminar*, August 14–17, Canberra.

Fricke T, Axinn AG, Thornton A. (1993). Marriage, social inequality and women's contact with their natal families in alliance societies: two Tamang examples. *American Anthropologist 95*(2), 395–419.

Gunaid AA, Hummad NA, Tamim KA. (2004). Consanguineous marriage in the capital city Sana'a, Yemen. *Journal of Biosocial Science 36*(1), 111–121.

Hafez M, El-Tahan H, Awadalla M, El-Khayat H, Abdel-Gafar A, Ghoneim M. (1983). Consanguineous matings in the Egyptian population. *Journal of Medical Genetics 20*, 58–60.

Hamamy A, Chalbi N, Ben AM, Elgazzeh M. (2005). Effects of consanguinity and social factors on mortality and fertility in Mauritania. *Tunis Med 83*, 221–226.

Hamamy H, Jamhawi L, Al-Darawsheh J, Ajlouni K. (2005). Consanguineous marriages in Jordan: why is the rate changing with time? *Clinical Genetics 67*, 511–516.

Hasanzadeh-Nazarabadi N, Rezaeetalab GH, Dastfan F. (2006). Study of youths' knowledge, behaviour, and attitude towards consanguineous marriages. *Iranian Journal of Public Health 35*(3), 47–53.

Hashmi MA (1997) Frequency of consanguinity and its effect on congenital malformations: a hospital-based study. *J Pak Med Assoc 47*(3), 75–78.

Hashemipour M, Amini M, Talaie M, et al. (2007). Parental consanguinity among parents of neonates with congenital hypothyroidism in Isfahan. *Eastern Mediterranean Health Journal 13*(3). Full text available from http://www.emro.who.int/Publications/EMHJ/1303/article9.htm

Hussain R, Bittles AH. (1998). The prevalence and demographic characteristics of consanguineous marriages in Pakistan. *Journal of Biosocial Science 30*(2), 261–279.

Hussain. R. (1999). Community perceptions of reasons for preference for consanguineous marriages in Pakistan. *Journal of Biosocial Science 31*(4), 449–461.

Hussain R. (2002). Lay perceptions of genetic risks attributable to inbreeding in Pakistan. *American Journal of Human Biology 14*(2), 264–274.

Hussain R, Bittles AH. (2004). Assessment of association between consanguinity and fertility in Asian populations. *Journal of Health, Population & Nutrition 22*(1), 1–12.

Hussain, R. (2004). Consanguinity, cultural variation, ethical and psychosocial factors, in: D. Kumar, ed. *Genetic Disorders of the Indian Subcontinent*. The Netherlands: Kluwer Academic Publishers, pp. 125–136.

IIPS. (1995). *National Family Health Survey 1992-93*. International Institute for Population Sciences, Bombay.

Jaber L, Merlob P, Bu X, Rotter JI, Shohat M. (1992). Marked parental consanguinity as a cause for increased major malformations in an Israeli Arab community. *American Journal of Medical Genetics 44*(1), 1–6.

Jurdi R, Saxena PC. (2003). The prevalence and correlates of consanguineous marriages in Yemen: Similarities and contrasts with other Arab countries. *Journal of Biosocial Science 35*(1), 1–13.

Saify K, Saadat M. (2011). Consanguineous marriages in Afghanistan. *Journal of Biosocial Science 00*, 1–9, doi10.1017/S0021932011000253

Kerkeni E, Monastiri K, Saket B, Guediche MN, Cheikh HB. (2007). Interplay between socio-economic factors, consanguinity, fertility and offspring mortality in Monastir, Tunisia. *Croatian Medical Journal 48*, 701–707.

Khlat M. (1997). Endogamy in Arab countries, in: Teebi, A, Farag, TI (eds). *Genetic Disorders Among Arab Populations*, New York: Oxford University Press, pp. 63-80.

Khoury SA, Massad D. (1992). Consanguineous marriage in Jordan. *American Journal of Medical Genetics 43*(5), 769–775.

Jaber L, Bailey-Wilson JE, Haj-Yehia M, Hernandez J, Shohat M. (1994). Consanguineous matings in an Israeli-Arab community. *Archives of Pediatrics & Adolescent Medicine 148*(4), 412–415.

Jacoby HG, Mansuri G. (2007). *Watta Satta: Bride Exchange and Women's Welfare in Rural Pakistan*. World Bank Policy Research Working Paper No. 4126. Full text available from http://ssrn.com/abstract=960383

Jaouad IC, Elalaoui SC, Sbiti A, Elkerh F, Belmahi L, Sefiani A. (2009). Consanguineous marriages in Morocco and the consequence for the incidence of autosomal recessive disorders. *Journal of Biosocial Science 41*(5), 75–81.

Mohamed MS. (1995). An epidemiological study on consanguineous marriage among urban population in Alexandria. *Journal of Egyptian Public Health Association 70*(3), 293–305.

Mokhtar MM, Abdel-Fattah MM. (2001). Consanguinity and advanced maternal age as risk factors for reproductive losses in Alexandria, Egypt. *European Journal of Epidemiology 17*(6), 559–565.

Morgan SP, Stash S, Smith HL, Mason KO. (2002). Muslim and non-Muslim difference in female autonomy and fertility: evidence from four Asian countries. *Population & Development Review 28*(3), 515–537.

Muhit MA, Shah SP, Gilbert CE, Foster A. (2007). Causes of severe visual impairment and blindness in Bangladesh: a study of 1935 children. *British Journal of Opthalmology 91*(8), 1000–1004.

Mumtaz G, Tamim H, Kanaan M, et al. (2007). Effect of consanguinity on birth weight for gestational age in a developing country. *American Journal of Epidemiology 165*(7), 742–752.

Nabulsi A. (1995). Mating patterns of the Abbad tribe in Jordan. *Social Biology 42*(3–4), 162–174.

Obeidat BR, Khader YS, Amarin ZO, Kassawneh M, Al OM. (2008). Consanguinity and adverse pregnancy outcomes: the North of Jordan experience. *Maternal & Child Health Journal 4*(2), 283–289.

Othman H, Saadat M. (2009). Prevalence of consanguineous marriages in Syria. *Journal of Biosocial Science 41*(5), 685–692.

Rao TSS, Prabhakar AK, Rao KSJ, et al. (2009). Relationship between consanguinity and depression in a South Indian population. *Indian Journal of Psychiatry 51*(1), 50–52.

Rajab A, Patton MA. (2000). A study of consanguinity in the Sultanate of Oman. *Annals of Human Biology 27*, 321–326

Raz AE, Atar M. (2004). Cousin marriage and premarital carrier matching in a Bedouin community in Israel: Attitudes, service development and educational intervention. *Journal of Family Planning & Reproductive Health Care 30*(1), 49–51.

Reddy PG. (1988). Family structure and consanguineous marriage among three caste groups of Andhra Pradesh, India. *International Journal of Anthropology 3*(4), 355–359.

Robinson R. (2007). Indian Muslims: The varied dimensions of marginality. *Economic and Political Weekly 42*(10), 839–843.

Saadat M, Ansari-Lari M, Farhud DD. (2004). Consanguineous marriage in Iran. *Annals of Human Biology 31*(2), 263–269.

Saha N, el Sheikh FS. (1988). Inbreeding levels in Khartoum. *Journal of Biosocial Science 20*(3), 333–336.

Saha N, Hamad RE, Mohamed S. (1990). Inbreeding effects on reproductive outcome in a Sudanese population. *Human Heredity 40*(44), 208–212.

Sallam SA, Mahfouz AA, Dabbous NI. (2001). Reproductive health of married adolescent women in squatter areas in Alexandria, Egypt. *Eastern Mediterranean Health Journal 7*(6), 935–942.

Shami SA, Grant JC, Bittles AH. (1994). Consanguineous marriage within social/occupational class boundaries in Pakistan. *Journal of Biosocial Science 26*(1), 91–96.

Shami SA, Schmitt LH, Bittles AH. (1990). Consanguinity, spousal age at marriage and fertility in seven Pakistani Punjab cities. *Annals of Human Biology 17*(2), 97–105.

Sueyoshi S, Ohtsuka R. (2003). Effects of polygyny and consanguinity on high fertility in the rural Arab population in South Jordan. *Journal of Biosocial Science 35*, 13–526.

Tadmouri GO, Nair P, Obeid T, Al Ali MT, Al Khaja N, Hamamy HA. (2009). Consanguinity and reproductive health among Arabs. *Reproductive Health 6*(1), 17. Full text available from http://www.reproductive-health-journal.com/content/6/1/17.

Tamim H, Khogali M, Beydoun H, Melki I, Yunis K. (2003). Consanguinity and apnea of prematurity. *American Journal of Epidemiology 158*, 942–946.

Teebi AS, El-Shanti HI. (2006). Consanguinity, implications for practice, research and policy. *The Lancet 367*(9515), 970–971.

Vardi-Saliternik R, Friedlander Y, Cohen T. (2002). Consanguinity in a population sample of Israeli Muslim Arabs, Christian Arabs and Druze. *Annals of Human Biology 29*(4), 422–431.

Zaoui S, Biemont C. (2002). Frequency of consanguineous unions in the Tlemcen area (West Algeria). *Sante 12*, 289–295.

Zlotogora J, Habiballa H, Odatalla A, Barges S. (2002). Changing family structure in a modernizing society: A study of marriage patterns in a single Muslim village in Israel. *American Journal of Human Biology 14*(5), 680–682.

81.

MEDICAL GENETICS AND GENOMICS IN SRI LANKA

Vajira H.W. Dissanayake

EARLY DEVELOPMENTS AND THE CURRENT STATE OF MEDICAL GENETIC SERVICES IN SRI LANKA

The origins of scientific study of heredity and the study of genes and their function in the medical field in Sri Lanka can be traced back to the 1960s and 1970s, when Sri Lankan academics and clinicians trained in the United Kingdom and returned to the country to live and work here.

The first major milestone toward formal organization of delivery of genetics as a clinical service, however, was the establishment of the Human Genetics Unit (HGU) of the Faculty of Medicine, University of Colombo, in October 1983 [http://www.hgucolombo.org]. This single-handed effort of Professor Rohan W. Jayasekara—a U.K. trained geneticist—resulted in the establishment of the first genetic diagnostic and genetic counseling service in the country. In addition, he also formalized basic medical genetics and clinical genetics curricula in Sri Lanka, which were incorporated into the undergraduate medical curriculum of two of the four medical schools in the country at the time—Colombo and Galle (Ruhuna). He also introduced genetics to postgraduate curricula and today, genetics is being taught in many postgraduate courses conducted by the Postgraduate Institute of Medicine of the University of Colombo, which is the national center for postgraduate medical training in the country (http://www.cmb.ac.lk/pgim/). These early developments had a positive impact on shaping the views of the medical community in Sri Lanka toward incorporating genetics in their day-to-day clinical practice (Jayasekara, 1989).

Interest in genetics started to grow in the early 1990s, with molecular genetic techniques becoming increasingly important in infectious disease diagnostics and in forensic sciences, in addition to genetic disease diagnostics. This resulted in academic scientists in medical faculties within the university system developing an interest in genetics. The Department of Biochemistry of the Faculty of Medicine, University of Colombo was one of the first centers in the country to introduce molecular genetic testing in a research setting, for genetic disorders (in collaboration with the HGU) and for infectious disease diagnostics (Welihinda et al., 1993). They conducted several successful research projects. None of these, however, grew into regular molecular genetic diagnostic services. In early 1990s the Department of Biochemistry came to be renamed the Department of Biochemistry and Molecular Biology—the first academic department in the country with "molecular biology" in its name. The department also had the distinction of offering the first master level course in molecular biology in the country. The academics of this department in 2004 formed the Institute of Biochemistry, Molecular Biology and Biotechnology, of the University of Colombo (http://www.ibmbb.lk/). The institute now has a wide-ranging research and training portfolio, but it does not provide genetic diagnostic or genetic counseling services.

With Sri Lanka engaged in a protracted civil conflict from the early 1980s, funding for genetic diagnostic services was hard to come by in the public sector. This was a major limitation for the development of genetic services in the country. Against this background, in 2005 the HGU made a strategic alliance with the largest private sector hospital chain in Sri Lanka—Asiri Group of Hospitals—to establish the first fully fledged state-of-the-art genetic diagnostic service in the country. This laboratory, which has now grown into the Asiri Centre for Genomic and Regenerative Medicine (ACGRM; http://www.asiri.lk/acgrm), is the largest provider of genetic diagnostic services in the country. It provides cytogenetic and molecular genetic diagnostics for a wider range of genetic disorders, as well as infectious disorders. Tests for all the common genetic disorders found in the country are available at ACGRM. The services provided include prenatal diagnosis. ACGRM is the only private sector center in Sri Lanka providing both genetic diagnostics as well as genetic counseling services.

In addition to these major centers, several other organizations provide genetic services. These include Genetech, a private sector diagnostic and research laboratory with a special interest in DNA fingerprinting and paternity testing (http://www.genetechsrilanka.

com); the Molecular Medicine Unit at the Faculty of Medicine, University of Kelaniya, a public sector university unit specializing in infectious diseases diagnostics and DNA typing for human identification and forensic case work (http://174.132.189.92/~medkel/medkel.kln/dept/templete/tepl.php?dept=MMU); the Department of Physiology, Faculty of Medicine, University of Kelaniya, a public sector university department that has a genetic clinic run by a clinical geneticist; and the Department of Pathology, Durdans Hospital, Colombo, a private sector hospital laboratory that provides limited genetic diagnostic services, especially for infectious diseases.

There are several other academic departments in universities in Sri Lanka engaged in genetic research projects, but they do not provide diagnostic or counseling services. As such, the major genetic diagnostic and counseling service provider in Sri Lanka is the alliance of HGU and ACGRM. In spite of these developments, the genetic community in Sri Lanka is still very small, and much more has to be done in the country to understand the genetic basis of disease affecting the Sri Lankan population, and to improve the availability of diagnostic and clinical genetic services.

The remaining sections of this chapter review the contemporary scientific literature based on work carried out on Sri Lankans living in Sri Lanka as well as abroad. It has been strengthened with unpublished data obtained from the abovementioned centers. The Sri Lankan work prior to the year 2000 is not covered in this review unless relevant to the discussion, as this data has already been reviewed in a previous book chapter (De Silva, 2004).

THE POPULATION STRUCTURE AND VARIATION

Sri Lanka is a unique island in the Indian Ocean. Its 65,610 sq. kilometers is rich in its diversity. The 21 million population consists of Sinhalese at 73.9% (an indigenous population that had mixed with migrants from the Bengali region of India over 2000 years ago), Sri Lankan Tamils at12.3% (an indigenous population that had mixed with migrants from Southern India), Indian Tamils at 5.5% (recent migrant tea estate workers from Southern India), Moors at7.1% (Arab traders who married indigenous women and settled down in the country), Malays at 0.3% (traders from the east who married indigenous women and settled down in the country), and Burghers at 0.3% (a hybrid population of Europeans and Sri Lankans) [Source: The total population is the current estimate for 2009. The percentages are based on Census Data for 1981 – http://www.statistics.gov.lk]. A small aboriginal group called the *Vaddhas* is widely believed to be the descendants of the original inhabitants of the country who were separated from other communities due to the geographic isolation of the part of the country in which they lived. There are no major differences observed in the pattern of disease affecting any of these ethnic groups. There is, however, clustering of thalassemia, which is the commonest single gene disorder in the country, in the northwestern, north central, central, and Uva provinces (De Silva et al., 2000).

The genetic differences between these ethnic groups and their relationship to other populations, especially in neighboring India, have been widely studied (Roychoudhury and Nei, 1985; Harihara et al., 1988; Saha, 1988; Juneja et al., 1994; Kshatriya, 1995; Papiha et al., 1996a, 1996b; Mastana and Papiha, 2001; Soejima and Koda, 2005; Gutala et al., 2006; Malavige et al., 2007). These studies, for the most part, confirm the close genetic affinity of different ethnic groups living in the country and provide little evidence in support of historical accounts of the Sinhalese originating from the Bengali region of eastern India.

During the past few years the researchers at the HGU have undertaken some studies on the differences in allele frequencies of selected single nucleotide polymorphisms (SNPs) in the three major ethnic groups in the country—Sinhalese, Sri Lankan Tamils, and Moors. This data, as well as other SNP frequency data obtained from published literature and validated unpublished data provided by genetic laboratories in Sri Lanka, is catalogued in the Sri Lankan Genome Variation Database (SLGVD). This database is an open-access online database maintained by the HGU (http://hgucolombo.org/SLGVD.aspx). There have not been any large-scale projects to catalog allele frequencies of SNPs in the Sri Lankan populations. Sri Lanka is a partner in the MalariaGen Project (http://www.malariagen.net). In this project, a large number of SNPs are expected to be genotyped in 1100 Sinhalese. Another project wherein a large number of people of Sri Lankan origin were genotyped was a genome-wide association study of skin pigmentation in a population of south Asian ancestry in the United Kingdom (Stokowski et al., 2007). Once publicly released, SNP data from these projects, too, would be a useful resource for future genetic work in the country.

Although there has been considerable interest on DNA fingerprinting for forensic work, data on allele frequencies and statistical parameters of forensic interest in different ethnic groups in the Sri Lankan population is still scarce. A recent report fills this void and may serve as a useful starting point for further work in this important area of genetics in the country (Illeperuma et al., 2009).

CHROMOSOMAL DISORDERS

Antenatal screening for congenital abnormalities by maternal multiple-marker screening and ultrasound scanning has been in practice in Sri Lanka for several decades. Antenatal genetic testing for chromosomal abnormalities and other single gene disorders, however, has become available only recently (Nanayakkara and Dissanayake, 2006, 2008). Termination of pregnancy for fetal indications is not legally permitted in Sri Lanka and as such, the uptake of antenatal screening is low and many children with chromosomal abnormalities are born in the country. However, in the absence of systematic collection of epidemiological data about these conditions, reliable statistics are not available on their incidence.

The following were the reasons for referral for cytogenetic testing to the two cytogenetic testing laboratories in the country (HGU and ACGRM) in 2006 and 2007: Down syndrome – 234 (21%); Turner syndrome – 78 (7%); sex determination in cases of ambiguous genitalia – 60 (5%); other dysmorphic babies – 78 (7%); family history of a congenital abnormality/dysmorphic syndrome – 57 (5%); primary amenorrhea/other problems of secondary sexual development – 111 (10%); subfertility/spontaneous pregnancy loss – 320 (31%), hematological malignancies – 154 (14%), other indications – 27 (2%). There were a total of 1119 tests performed in these two years (Dissanayake et al., 2008).

There has been considerable attention on Down syndrome because it continues to be the commonest disorder seen in genetic clinics. In a series of 615 patients with Down syndrome tested at HGU and ACGRM over a period of 6 years, 89% were due to nondisjunction, 10% were due to mosaicism, and the rest due to translocations (Dissanayake et al., 2008). Recently, one family each, with a child with Down syndrome born to women with mosaic trisomy 21, have been reported from the two centers. One of these families had had two children with Down syndrome.

The cytogenetic abnormalities seen among women with primary amenorrhea, too, have been interesting. In a series of 118 tested at the HGU, the following abnormal karyotypes were seen: 45,X in 10 (8.5%) [Turner syndrome]; 45, X/46,XX in 20 (17%) [mosaic Turner syndrome]; 46XX/46XY in 2 (1.7%) [sex chromosome mosaicism]; 45,X/46,XY in 2 (1.7%); and 47,XXX/45,X/46,XX in 2 (1.7%). The others had a normal female karyotype (Wetthasinghe et al., 2006).

Among the individuals undergoing cytogenetic testing at the HGU is an interesting series of five patients, including a pair of twins, with true hermaphroditism, a rare condition which is genetically heterogeneous. In this series—unlike most other cases of true hermaphroditism reported in the scientific literature, where the 46,XX karyotype seems to be the predominant one—four had the 46,XY karyotype and one was a mosaic 46,XX/46,XY (Andraweera et al., 2008).

Among the rare cytogenetic syndromes that have been detected at ACGRM recently have been a child with ambiguous genitalia and developmental delay who had a karyotype of 49,XXXXY, confirming the 49,XXXXYsyndrome; a child with dysmorphic features who had a karyotype of 46,XX,del(11)(q24-q25), confirming Jacobsen syndrome; a child with cat-cry syndrome and other dysmorphic features, who had a karyotype of 46,XX,der(5)add(8)(q13)del(5)(p13)pat, which had been inherited from a balanced translocation carrier father; a child with micropenis and delayed secondary sexual development, who had a karyotype of 46,XY,t(9;11)(p24;q22); a child with congenital heart disease and other dysmorphic feature, who had 22q11.2 deletion syndrome; and the first case of translocation Edward syndrome detected in the country, who had a karyotype of 46,XX,t(18,20) (q21.1;p13; see Dissanayake et al., 2008; Pedurupillay et al., 2008).

With fertility rates in the country almost below replacement levels, testing subfertile coupes and couples experiencing recurrent pregnancy loss is fast becoming the main workload of cytogenetic laboratories. At ACGRM, in a series of 442 patients (213 couples and 16 individuals) tested for these indications, the detection rate of chromosomal abnormalities and polymorphic chromosomal variants that may interfere with fertility was reported to be as high as 9.5%. This rate is almost doubled to 19.4% when couples are tested following failed attempts at assisted reproduction (Dissanayake et al., 2009). There is also a demand from the clinicians caring for couples experiencing recurrent pregnancy loss to seek a genetic answer to their clients' problem by genetic evaluation of the abortus material for chromosomal abnormalities (Nanayakkara and Dissanayake, 2008).

COMMON MENDELIAN DISORDERS

THALASSEMIAS

Thalassemia is the commonest monogenic disorder in Sri Lanka. Both α and β thalassemia is found in the country, with β-thalassemia by far being the commoner of the two. The estimated 2000 thalassemics in the country are said to consume 7.5% of the national health budget annually for transfusion and iron chelation–related hospital costs. In the absence of a national program for prevention of thalassemia, annually about 80 new children are added to this number (source: Ministry of Health, Sri Lanka). Genetic testing for thalassemia is available as a clinical genetic service in the country (Nanayakkara and Dissanayake, 2008). Data from research (De Silva et al., 2000; Fisher et al., 2003) and clinical testing implicate three common mutations that account for more than 70% of the β-thalassemia cases. They are IVS1-5 (G→C), IVS1-1 (G→A), and CD 26 (G→A). The last mutation, which causes HbE disease, interacts with other mutations to cause HbE β-thalassemia, which accounts for up to 30% of the cases in some parts of the country (Premawardhena et al., 2005a). α-thalassemia in Sri Lanka is mainly associated with the –α3.7 and –α4.2 deletions (De Silva et al. 2000).

Chronic hyperbilirubinemia and gallstone formation are common complications in patients with HbE β-thalassemia. The UGT*1 genotype may be important in the etiology of these complications (Premawardhena et al., 2001). The high prevalence of this disease-associated genotype in the Sri Lankan population suggests that these patients in Sri Lanka would be at a much higher risk of developing hyperbilirubinemia and gallstones (Premawardhena et al., 2003).

Ongoing studies on different forms of thalassemia in Sri Lanka has begun to shed light on different aspects of the disease, which may redefine treatment protocols and help reduce the massive expenditure in caring for these patients (Premawardhena et al., 2004; 2005a, 2005b, 2008). In addition, sequence-based genetic testing available locally is also helping to further clarify the genotype–phenotype correlation of these patients (Nanayakkara and Dissanayake, 2008).

HEMOPHILIA

Thalassemia is followed by hemophilia in terms of the numbers of people affected. The Hemophilia Association has 260 registered patients with hemophilia, although the estimate for the country exceeds 1000 cases. Of this number, approximately 90% have hemophilia A, and the rest hemophilia B (10%). [Source: National Hemophilia Association of Sri Lanka]. Research into the genetics of hemophilia has not been undertaken in Sri Lanka. Although genetic testing is available locally at ACGRM on a clinical basis, there have been no referrals for testing.

HEREDITARY IRON OVERLOAD DISORDERS

Although iron overload secondary to β-thalassemia is common in the country, hereditary iron overload disorders, such as hereditary hemochromatosis caused by HFE gene mutations, are uncommon. In fact, although rare, the p.H63D mutation in the HFE gene is commoner than the p.C282Y mutation in the Sri Lankan population as is the case in other Asian populations (Rochette et al., 1999; Nanayakkara and Dissanayake, 2008). Against that background, the report of a Sri Lankan woman with iron overload due to the c.V162del mutation in the ferroportin 1 gene (causing type 4 hemochromatosis, or ferroportin disease) is interesting, because this suggests that analysis of this and other ferroportin 1 mutations may be useful in the study of iron overload disorders, as it may be the basis for hitherto unexplained cases of iron overload in Sri Lanka (Wallace et al., 2005).

NEUROGENETIC DISORDERS

The classic neurogenetic disorders, too, are not rare in the Sri Lankan population. Although prevalence data is not available, Duchene muscular dystrophy (DMD) is the main condition seen in children. In fact, genetic testing for DMD was the first molecular genetic test to be introduced in the country (Welihinda et al., 1993). The availability of newer genetic techniques locally has made it possible to define the complete deletion/duplication profile of Sri Lankan children with DMD (Nanayakkara and Dissanayake, 2008). In the adult population, ataxias and Huntington's disease are the main conditions encountered by clinicians (Dissanayake et al., 2006). Genetic testing for ataxias has just been introduced to the country at the time of writing; therefore, information on subtypes is not available. Genetic testing for Huntington's disease has been available in the country for some time. Data from these studies suggest that the low prevalence of the condition in Sri Lanka compared to that in the West is possibly due to the shorter CAG repeats in normal alleles of the IT15 gene in our population (Nanayakkara and Dissanayake, 2008).

OTHER RARE DISORDERS

There has been very little investigation into other single gene disorders in the Sri Lankan population. However, there are a couple of interesting case studies in the scientific literature. The first is a report of a mutation in the growth hormone releasing hormone receptor gene (GHRHR) which causes dwarfism (Netchine et al., 1998). This appears to be a new form of dwarfism due to isolated growth hormone (GH) deficiency, secondary to inactivating mutations in the GHRHR gene, which has been described in three families from the Indian subcontinent (Baumann 1999). The second is a report of factor X (FX) deficiency in a French patient of Sri Lankan ancestry caused by the Gly366Ser mutation in the FX gene, the first such report (Isshiki et al., 2005).

ACGRM, which offers sequencing-based genetic testing, reports several rare mutations. The first is a CD50G>A homozygous mutation in the G6P transporter (SLC37A4) gene found in a child suspected to have glycogen storage disease (GSD), confirming the diagnosis to be GSD type 1b. The second is a heterozygous P406L mutation in the TYR gene found in a child with oculocutaneous albinism. In this case, sequence analysis failed to identity the second mutation, compound heterozygosity for which, with the P406L mutation, would have caused the disease.

MITOCHONDRIAL DISORDERS

There appears to be only one report of a mitochondrial disorder in the Sri Lankan population. This report deals with maternally inherited diabetes and deafness (MIDD), which is a disorder caused by mutations in several mitochondrial genes. The commonest of these is the mt3243A>G in the mitochondrial DNA-encoded tRNA (Leu, UUR) gene. This mutation is reported to be prevalent among 0.9% of young diabetic subjects in the Sri Lankan population. However, a maternal family history of diabetes and either a personal and/or family history of deafness could only distinguish half of patients with MIDD from Sri Lankan subjects with young-onset diabetes (Katulanda et al., 2008).

COMPLEX NONCOMMUNICABLE DISEASES

GENETIC THROMBOPHILIAS

Since the first description of the genetic thrombophilic polymorphisms in the mid-1990s, there has been considerable attention on the prevalence of these polymorphisms in the Sri Lankan population. The initial work, in fact, was in Sri Lankans living in London (Alagratnam et al., 2000), and in Sri Lankan samples that were a part of a worldwide collection at Oxford (Schneider et al., 1998). These studies established that the methylenetetrahydrofolate (MTHFR) 677C>T polymorphism was polymorphic in the Sri Lankan population. The HGU developed an interest in genetic thrombophilias in the early part of this decade. This resulted in a considerable amount of work that established the allele frequencies of the variant alleles of the functionally significant polymorphisms in these genes in different ethnic groups in the Sri Lankan population (Dissanayake et al., 2006, 2007, 2008) and examined the association of thrombophilic polymorphisms, including factor V (F5) Leiden, Factor II (F2) 20210G>A, and MTHFR 677C>T polymorphisms with preeclampsia and recurrent pregnancy loss in Sinhalese women. The latter work failed to find an association for these polymorphisms with these two conditions (Dissanayake et al., 2004, 2008).

The HGU began testing patients with a history of thromboembolism for the presence of variant alleles of the F5 Leiden, Factor II 20210G>A and MTHFR 677C>T polymorphisms as a clinical genetic testing service in 2005 (Andraweera et al., 2007; Dissanayake et al., 2008). The prevalence of the variant alleles of these polymorphisms in different groups of patients tested at HGU, as part of research and clinical testing, is summarized in Table 81-1.

REPRODUCTIVE DISORDERS

The special interest of the HGU in genetics of reproductive disorders—preeclampsia, recurrent pregnancy loss, and fetal growth restriction—has resulted in a considerable amount of genetic work in this area (Dissanayake, 2009), some of which is also mentioned in the previous section. The most noteworthy result of this work has been the discovery of an association between polymorphic variants in the maternal epidermal growth factor gene with preeclampsia (Dissanayake et al., 2009) and the weight of babies at birth (Dissanayake et al., 2007). The latter finding has been replicated in two white Western European populations in the United Kingdom. The researchers at the HGU attempted to replicate a recent European report, reporting an association for microdeletions in the Y-chromosome with recurrent pregnancy loss in affected Sinhalese couples, and found that none of the male partners of women who experienced recurrent pregnancy loss had Y-chromosome microdeletions (Wetthasinghe et al., 2008).

PARKINSON'S DISEASE

Parkinson's disease is an important problem in the elderly population in Sri Lanka. A small study of Parkinson's disease conducted in Colombo reports that as much as 10.3% of cases are familial; this is a much higher proportion than that reported in other series (Samarasinghe et al., 2006).

BREAST CANCER

Breast cancer is by far the commonest cancer among Sri Lankan women. The first large-scale study to report mutations in the BRCA1 gene, published recently, report two novel pathogenic nonsense mutations in two Sri Lankan families with breast cancer: c.3086delT and c.5404delG (De Silva et al., 2008). The latter mutation has also been described in a second family tested at ACGRM. There seems to be some confusion on the assignment of the nucleotide position of this mutation. It appears that it should be designated c.5409delG (Dissanayake et al., 2008). A third, possibly pathogenic, novel missense mutation has also been described in this population (c.856T>G; se De Silva et al., 2008).

The only study that attempted to functionally characterize breast cancer in Sri Lankan women reports the prevalence of estrogen and progesterone receptor expression and Her/2neu amplification in carcinoma of the breast among Sri Lankans, as well as the relationship of these to each other, age at diagnosis, and grade of tumor, to be similar to that described in other parts of the world (Ratnatunga and Liyanapathirana, 2007).

TABLE 81-1 THE FREQUENCY/PERCENTAGE OF GENOTYPES IN GENETIC THROMBOPHILIC POLYMORPHISMS IN THE FACTOR V (F5), FACTOR II (F2), AND METHYLENETETRAHYDROFOLATE REDUCTASE (MTHFR) GENES IN 711 PATIENTS WITH THROMBOEMBOLIC CONDITIONS, TESTED IN THE HUMAN GENETICS UNIT, COLOMBO.

	F5 1691G>A (Leiden)			F2 20210G>A		MTHFR 677 C>T		
	GG	GA	AA	GG	GA	CC	CT	TT
Indication for Testing	n (%)	n (%)	n (%)	n (%)	n (%)	n (%)	n (%)	n (%)
Cerebrovascular Accidents	104 (100)	0 (0)	0 (0)	103 (99)	1 (1)	85 (82)	16 (15)	3 (3)
Cerebral Venous Thrombosis	12 (86)	2 (14)	0 (0)	14 (100)	0 (0)	13 (93)	1 (7)	0 (0)
Transient Ischaemic Attacks	2 (100)	0 (0)	0 (0)	2 (100)	0 (0)	2 (100)	0 (0)	0 (0)
Myocardial Infarction	19 (90)	2 (10)	0 (0)	21 (100)	0 (0)	15 (71)	6 (29)	0 (0)
Ischaemic Heart Disease	4 (100)	0 (0)	0 (0)	4 (100)	0 (0)	1 (25)	3 (75)	0 (0)
Pulmonary Hypertension	4 (100)	0 (0)	0 (0)	4 (100)	0 (0)	3 (75)	1 (25)	0 (0)
Deep Vein Thrombosis	68 (99)	1 (1)	0 (0)	68 (99)	1 (1)	49 (71)	19 (28)	1 (1)
Pulmonary Embolism	9 (100)	0 (0)	0 (0)	9 (100)	0 (0)	9 (100)	0 (0)	0 (0)
Thrombosis in Veins in Other Sites	15 (94)	1 (6)	0 (0)	16 (100)	0 (0)	14 (88)	2 (13)	0 (0)
Thrombosis in Arteries in Limbs	5 (100)	0 (0)	0 (0)	5 (100)	0 (0)	4 (80)	1 (20)	0 (0)
Recurrent Pregnancy Loss	249 (98)	4 (2)	0 (0)	253 (100)	0 (0)	198 (78)	51 (20)	4 (2)
Preeclampsia	168 (96)	7 (4)	0 (0)	175 (100)	0 (0)	136 (78)	36 (21)	3 (2)
Other Pregnancy Complications	4 (100)	0 (0)	0 (0)	4 (100)	0 (0)	2 (50)	2 (50)	0 (0)
Presymptomatic Family Screening	10 (59)	5 (29)	2 (12)	16 (94)	1 (6)	14 (82)	3 (18)	0 (0)
Miscellaneous Indications	13 (93)	1 (7)	0 (0)	14 (100)	0 (0)	11 (79)	2 (14)	1 (7)
Total	686	23	2	708	3	556	143	12

ORAL CANCER

Oral cancer is the commonest cancer among males in Sri Lanka. Betel quid chewing has been implicated strongly in the etiology of the condition, including inducing the development of specific p53 mutations (Chiba et al., 1998) and/or hypermethylation of p14, p15, and p16 in precancerous lesions (Takeshima et al., 2008).

It is evident, however, that inherited genetic variations also play a role in the development of oral cancer among Sri Lankan men, by the fact that deficient CYP2A6 activity due to genetic polymorphism in the gene encoding for it seems to reduce oral cancer risk in this population (Topcu et al., 2002). A recent microarray-based study of oral squamous cell carcinomas in the Sri Lankan population has provided valuable early insights for the selection of biomarkers that can be used in early detection of oral cancer (Suhr et al., 2007).

HEMATOLOGICAL MALIGNANCIES

With the detection of chromosomal abnormalities and expression of fusion genes coming into the forefront of diagnosing, treating, and prognosticating hematological malignancies, the demand for genetic testing for these conditions has grown rapidly in Sri Lanka (Athapaththu et al., 2008). In addition to cytogenetic evaluation of blood and bone marrow, the molecular genetic tests that are available in the country include testing for the prognostically significant translocations in acute lymphoblastic leukemia, detection and quantification of BCR/ABL transcripts in patients with chronic myeloid leukemia, JAK2 V617F mutation and exon 12 mutation testing in patients with polycythemia, detection of the FIP1L1-PDGFRA fusion gene in hyper-eosinophilic syndrome, as well as detection of high affinity hemoglobin variants (Athapaththu et al., 2008).

The body of knowledge that is being collected as a result of the availability of these tests is providing unique insights into the etiology of these conditions. These include differences in the pattern of chromosomal translocations (Hidellage et al., 2008), especially those seen in children with acute lymphoblastic leukemia (Athapaththu et al., 2008), and the higher prevalence of aberrant BCR/ABL transcripts in adults with chronic myeloid leukemia (Opanayake et al., 2008).

NEUROBLASTOMA

The recognition of the importance of detection of MYCN amplification in the management of neuroblastoma has created a demand for genetic analysis of these tumors. The early data from analyzing these tumors for MYCN amplification and other chromosomal abnormalities, using the multiplex ligation-probe amplification method, suggest that the tumor characteristics in Sri Lankan patients may be different from those in the West (Nanayakkara and Dissanayake, 2008).

HOST SUSCEPTIBILITY TO INFECTIOUS DISORDERS

Sri Lanka is a tropical country with its share of infectious disorders. Diseases such as malaria, which were responsible for widespread morbidity and mortality in the past, have largely been eradicated, and other diseases such as dengue, chikungunya, and cutaneous leishmaniasis have raised their heads. A review of the molecular genetics of these pathogens is beyond the scope of this chapter. The aim of this section is to review the host genetic factors that confer susceptibility to infections.

Host genetic factors conferring susceptibility to malaria and cutaneous leishmaniasis have been the subject of research in the country. One study attempted to dissect the variation in the incidence of malarial infection and the clinical intensity of the disease, into its component causes, by pedigree analysis of longitudinal data. It reported that overall, the hosts show consistent repeatable differences among their susceptibility to clinical disease, about half of which could be attributed to host genes (Mackinnon et al., 2000). One study that took the candidate gene association approach to study this genetic contribution reports that the odds in favor of carrying the TNFα*2 allele, mainly of the heterozygous genotype (TNFα*1,*2), are two to three times greater among individuals with severe malaria compared to healthy controls, or to those with uncomplicated malarial infections (Wattavidanage et al., 1999).

In a study of host factors in cutaneous leishmaniasis, another group, while reporting a possible genetic susceptibility to cutaneous leishmaniasis among the Sinhalese (Samaranayake et al., 2008), could not find an association with the condition for variations in both TNFα and TNFβ genes (Samaranayake et al., 2008). However, they report that allele frequencies of the TNFβ gene among Sinhalese were significantly different from that of Sri Lankan Tamils and Moors (Samaranayake et al., 2008).

PHARMACOGENOMICS

Genetic testing for VCORK1 and CYP2C9 variants to determine the correct dosing for warfarin has been available in Sri Lanka since the U.S. Food and Drug Administration recommended it in mid-2008. The uptake for genetic testing, however, has been low, and none of the patients thus far tested have shown the alleles associated with slow metabolism (Nanayakkara and Dissanayake, 2008).

There has also been some interest in the past on the effect of variations in the cytochrome P450 system on drug metabolism among a group of Sri Lankan pharmacologists (Weerasuriya et al., 1994). Research has been initiated recently by the same group, in collaboration with the HGU, to take a fresh look at this system, taking into consideration advances in genetics and genomics and focusing on variations in the CYP2D6 gene. This study reports that a considerable proportion of Sri Lankans carry enzyme inactivation alleles in the CYP2D6 gene, which would mean that they require smaller doses of drugs metabolized through the CYP2D6 enzyme (Jayakody et al., 2009).

GENETIC TOXICOLOGY

The only study in the field of genetic toxicology in the Sri Lankan population is a study on metabolism of arsenic. Arsenic (+3 oxidation state) methyltransferase (AS3MT), an S-adenosyl-l-methionine-dependent enzyme, catalyzes the methylation of arsenite. Monomethylarsonous acid is more cytotoxic and genotoxic than arsenate and arsenite. The 287T variant of the AS3MT gene has been shown to increase the concentration of monomethylated arsenic (MMA) in urine of those with arsenic exposure. This variant is reported to be nonexistent or present in very low frequencies in Tamils and Sinhalese (Fujiharaa et al., 2008).

ETHICAL, LEGAL, AND SOCIAL ASPECTS

There are no special laws governing genetic research or the delivery of genetic services in Sri Lanka. At the beginning of this century, a study group convened by the National Science and Technology Commission of Sri Lanka (http://www.nastec.lk) drew up "Draft Guidelines for New Genetics and Assisted Reproductive Technologies in Sri Lanka." These guidelines are now being incorporated into draft legislation that is being drawn up by a subcommittee on bioethics of the National Science Foundation, Sri Lanka. It will be some time into the future, however, when such legislation might be enacted.

Sri Lanka is one of the first countries in the region to have had formal ethics review mechanisms (Dissanayake et al., 2006). The Forum for Ethics Review Committees in Sri Lanka (FERCSL) has been working tirelessly over the past few years to further strengthen these mechanisms in the country (http://www.fercsl.net).

ACCESS TO AND UTILIZATION OF GENETIC SERVICES

As genetic services expand in the country, it is necessary to identify barriers to utilization of these services and take measures to overcome them. There have been some positive developments in this regard among the public, with several patient and family groups such as the Parents Association for the Special Child, the Thalassemia Association, the Hemophilia Association, and a support group for families with Duchene muscular dystrophy being formed and becoming active.

At present, one of the main limitations to accessing and utilizing genetic services in the country is the fact that all the major genetic centers are located in and around the capital Colombo. Therefore, those who require such services have to travel to Colombo. The HGU has come up with an innovative solution to overcome this problem. In 2005, the government of Sri Lanka began establishing Internet centers named *Nanasalas* in many parts of the country to provide Internet access to rural people. So far, over 575 of these centers have been set up. Some of them are in the most remote parts of the country. The HGU has made use of the opportunity created by this and linked up with the Nanasala network to offer online genetic consultations via videoconferencing using Skype. This service is in its infancy still, and it is hoped that it might help solve problems of accessibility to those who need assistance most. Briefly, the service involves the following: When a patient is referred to the HGU, the patient can contact a Nanasala in his/her area and hand over the referral letter; the Nanasala operator scans and emails the letter to the HGU, the HGU gives an appointment for the consultation, and at the appointed time, the consultand arrives at the Nanasala and the online link is established with the HGU. This link is established using Skype software, which allows free videoconferencing. Following the consultation, an email letter is sent to the Nanasala, addressed to the referring physician; this email is printed and given to the consultand, who takes it to the referring physician. Testing, if necessary, can be arranged by sending samples via courier without the need for the patient to travel to Colombo.

While there have been no studies aimed at understanding the factors that influence the health-seeking behavior of individuals and families affected with genetic disorders in Sri Lanka (although such behavior is bound to be largely influenced by the fact that genetic testing in Sri Lanka is not available freely through the public health system, and the issues of accessibility mentioned above), the views of doctors and medical students have been surveyed in some detail. These studies report high levels of support for use of the new genetic technologies (Dissanayake et al., 2002; Simpson et al., 2005) and support for a change in law in favor of liberalizing pregnancy termination when a genetic defect is detected antenatally (Simpson et al., 2003).

These favorable attitudes would augur well for the future of medical genetics and genomics in Sri Lanka, as Sri Lanka looks to a future of rapid development in all spheres including medical genetics and genomics, with the war situation that gripped the country for nearly 30 years coming to an end.

REFERENCES

Alagratnam D, Wierzbicki AS, Swaminathan R, Turner C, Wickramasinghe SN. (2000). Serum homocysteine, folate and thermolabile variant of MTHFR in healthy Sri Lankans living in London. *Atherosclerosis 149*, 221–222.

Andraweera PH, Gammulla CG, Weerasekara LY, Dissanayake VHW, Jayasekara RW. (2007). The prevalence of thrombophilic genetic variants in patients with a history of thromboembolism. *Ceylon Medical Journal 52*, 44.

Andraweera PH, Wetthasinghe TK, Dissanayake VHW, Jayasekara RW. (2008). A case series of true hermaphrodites. *Ceylon Medical Journal 53*, 17.

Athapaththu AMMH, Hidellage NT, Opanayake DJS, Pedurupillay CRJ, Ratnayake A, Dissanayake VHW. (2008). Prognostically significant translocations in Sri Lankan children with Acute Lymphobastic Leukaemia (ALL). *Ceylon Medical Journal 53*, 40.

Athapaththu AMMH, Nanayakkara HLV, Opanayake DJS, et al. (2008). An audit of tests performed for hematological malignancies in a Genetic Laboratory. *Proceedings of the 9th Annual Academic Sessions of the Sri Lanka College of Haematologists 27.*

Baumann G. (1999). Mutations in the growth hormone releasing hormone receptor: A new form of dwarfism in humans. *Growth Hormone and IGF Research 9*(Suppl B)24–29; discussion 29–30.

Chiba I, M. Muthumala, Y. Yamazaki, et al. (1998). Characteristics of mutations in the p53 gene of oral squamous-cell carcinomas associated with betel-quid chewing in Sri Lanka. *International Journal of Cancer 77*, 839–842.

De Silva D. (2004). Genetic Disorders and Medical Genetics in Sri Lanka. In: Kumar D, ed. *Genetic Disorders of the Indian Subcontinent*, pp. 519–540. Dordrecht: Kluwer Academic Press.

De Silva S, Fisher CA, Premawardhena A, et al. (2000). Thalassaemia in Sri Lanka: Implications for the future health burden of Asian populations. Sri Lanka Thalassaemia Study Group. *Lancet 355*, 786–791.

W. De Silva, E. H. Karunanayake, K. H. Tennekoon, et al. (2008). Novel sequence variants and a high frequency of recurrent polymorphisms in BRCA1 gene in Sri Lankan breast cancer patients and at risk individuals. *BMC Cancer 8*, 214.

Dissanayake VHW. (2009). Genetics of Pre-eclampsia and the weight of babies at birth – Clinical and Genetic Studies in Sri Lanka. *Ceylon Medical Journal 54*(3), 90–92.

Dissanayake VHW, Athapaththu AMMH, Opanayake DJS, Hidellage NT, Wijetunge K, Pedurupillay CRJ (2009). Chromosomal abnormalities in patients with recurrent spontaneous pregnancy loss and sub-fertility. *Sri Lanka Journal of Obstetrics and Gynecology 31*, 84–87.

Dissanayake VHW, Gammulla CG, Jayaweera SLD, Broughton Pipkin F, Morgan L, Jayasekara RW. (2007). Genetic polymorphisms in the methylenetetrahydrogolate reductase, factor V and prothrombin genes in the Sri Lankan population. *Ceylon Medical Journal 52*, 44.

Dissanayake VHW, Giles V, Jayasekara RW, et al. (2009). A study of three candidate genes for pre-eclampsia in a Sinhalese population from Sri Lanka. *Journal of Obstetrics and Gynaecology Research 35*, 234–242.

Dissanayake VHW, Jayasekara RW. (2008). Cytogenetic testing in paediatrics: Some aspects of the Sri Lankan scenario. *Sri Lanka Journal of Paediatrics 37*, 38–41.

Dissanayake VHW, Jayasekara RW, Senevirathne HR, Kalsheker N, Broughton Pipkin F, Morgan L. (2006). The prevalence of the methylenetetrahydrofolate reductase 677>T polymorphism in the Sri Lankan population. *Ceylon Medical Journal 51*, 8–9.

Dissanayake VHW, Kariyawasam C, Perera BM, Sandamal LYS, Weerasekera LY, Jayasekara RW. (2008). Genetic testing for the Factor V gene polymorphisms – should we be testing for both Leiden and R2? *Ceylon Medical Journal 53*, 16.

Dissanayake VHW, Lanerolle RD, Mendis N. (2006). Research ethics and ethical review committees in Sri Lanka: A 25 year journey. *Ceylon Medical Journal 51*, 110–113.

Dissanayake VHW, Mallawarachchi PH, Wettasinghe KT, et al. (2008). Cytogenetic diagnostic services in Sri Lanka 2006–2007. *Ceylon Medical Journal 53*, 58.

Dissanayake VHW, Morgan L, Broughton PF, Jayasekara RW, Seneviratne HR, Kalsheker N. (2004). A study of the association of the methylene-tetrahydrofolate reductase 677C>T polymorphism and pre-eclampsia in Sinhalese women in Sri Lanka. *Hypertension in Pregnancy 23*, 20.

Dissanayake VHW, Nanayakkara NV, Weeratunga R. (2008). A novel BRCA1 mutation in a Sri Lankan Sinhalese Family with Breast Cancer. *Proceedings of the Annual Scientific Sessions of the Sri Lanka College of Surgeons*, College of Surgeons, Sri Lanka, 45.

Dissanayake VHW, Perera WMDTN, Govindapala DS, Gamage R, Perera S, Jayasekara RW. (2006). An audit of the Neurogenetics Clinic at the Institute of Neurology (ION), National Hospital of Sri Lanka (NHSL). *Ceylon Medical Journal 51*, 14.

Dissanayake VHW, Simpson B, Jayasekara RW. (2002). Attitudes towards the new genetic and assisted reproductive technologies in Sri Lanka: A preliminary report. *New Genetics and Society 21*, 65–74.

Dissanayake VHW, Tower C, Broderick A, et al. (2007). Polymorphism in the epidermal growth factor gene is associated with birthweight in Sinhalese and white Western Europeans. *Molecular Human Reproduction 13*, 425–429.

Dissanayake VHW, Weerasekare LY, Gammnulla GC, Andraweera PH, Jayasekara RW. (2008). Thrombophilic genetic mutations and recurrent pregnancy loss – A case control study. *Sri Lanka Journal of Obstetrics and Gynaecology 30*, 98.

Dissanayake VHW, Weerasekera LY, Gammulla CG, et al. (2008). Genetic thrombophilic polymorphisms and pregnancy complications. *Ceylon Medical Journal 52*, 30.

Fisher CA, Premawardhena A, De Silva S, et al. (2003). The molecular basis for the thalassaemias in Sri Lanka. *British Journal of Haematology 121*, 662–671.

Fujiharaa J, Soejimab M, Kodab Y, Kunitoc T, Takeshitaa H. (2008). Asian specific low mutation frequencies of the M287T polymorphism in the human arsenic (+3 oxidation state) methyltransferase (AS3MT) gene. *Mutation Research 654*, 158–161.

Gutala R, Carvalho-Silva DR, Jin L, et al. (2006). A shared Y-chromosomal heritage between Muslims and Hindus in India. *Human Genetics 120*, 543–551.

Harihara S, Saitou N, Hirai M, et al. (1988). Mitochondrial DNA polymorphism among five Asian populations. *American Journal of Human Genetics 43*, 134–143.

Hidellage N, Athapaththu A, Opanayake D, Pedurupillay C, Dissanayake V. (2008). Recurrent balanced reciprocal chromosomal translocations seen in haematological malignancies. *Ceylon Medical Journal 53*, 16.

Illeperuma RJ, Mohotti SN, De Silva TM, Fernandopulle ND, Ratnasooriya WD. (2009). Genetic profile of 11 autosomal STR loci among the four major ethnic groups in Sri Lanka. *Forensic Science International Genetics 3*, e105–e106.

Isshiki I, Favier R, Moriki T, et al. (2005). Genetic analysis of hereditary factor X deficiency in a French patient of Sri Lankan ancestry: In vitro expression study identified Gly366Ser substitution as the molecular basis of the dysfunctional factor X. *Blood Coagulation & Fibrinolysis 16*, 9–16.

Jayakody RL, Galappatthy P, Risama MF, et al. (2009). Genetic variants in the Cytochrome P4502D6 (CYP2D6) gene in the Sri Lankan population. *Ceylon Medical Journal 54*, 35.

Jayasekara RW. (1989). The attitudes of doctors and students towards a genetic service in an Asian country: Sri Lanka. *Asia Oceania Journal of Obstetrics and Gynaecology 15*, 267–270.

Juneja RK, Saha N, Tay JS, Low PS, Gahne B. (1994). Distribution of plasma alpha-1-B-glycoprotein (A1BG) polymorphism in several populations of the Indian subcontinent. *Annals of Human Biology 21*, 443–448.

Katulanda P, Groves CJ, Barrett A, et al. (2008). Prevalence and clinical characteristics of maternally inherited diabetes and deafness caused by the mt3243A > G mutation in young adult diabetic subjects in Sri Lanka. *Diabetes Medicine 25*, 370–374.

Kshatriya GK. (1995). Genetic affinities of Sri Lankan populations. *Human Biology 67*, 843–866.

Mackinnon MJ, Gunawardena DM, Rajakaruna J, Weerasingha S, Mendis KN, Carter R. (2000). Quantifying genetic and nongenetic contributions to malarial infection in a Sri Lankan population. *Proceedings of the National Academy of Science U S A 97*, 12661–12666.

Malavige GN, Rostron T, Seneviratne SL, et al. (2007). HLA analysis of Sri Lankan Sinhalese predicts North Indian origin. *International Journal of Immunogenetics 34*, 313–315.

Mastana SS, Papiha SS. (2001). D1S80 distribution in world populations with new data from the UK and the Indian sub-continent. *Annals of Human Biology 28*, 308–318.

Nanayakkara HLV, Dissanayake VHW. (2006). Antenatal genetic screening for fetal aneuploidy in Sri Lanka: Preliminary results. *Sri Lanka Journal of Obstetrics and Gynaecology 28*, 21.

Nanayakkara HLV, Dissanayake VHW. (2008). Aneuploidy screening of abortus material by Multiplex-ligation Probe Amplification (MLPA) provides useful information for counselling couples with spontaneous pregnancy loss. *Ceylon Medical Journal 53*, 30.

Nanayakkara HLV, Dissanayake VHW. (2008). Pharmacogenomic testing for warfarin therapy in the era of personalised medicine. *Ceylon Medical Journal 53*, 58.

Nanayakkara HLV, Dissanayake VHW. (2008). Preliminary results of a genetic testing service for thalassaemia in Sri Lanka – new opportunities for clinical phenotyping. *Ceylon Medical Journal 53*, 60.

Nanayakkara HLV, Rodrigo N, Dissanayake VHW. (2008). Prenatal genetic screening in Sri Lanka. *Ceylon Medical Journal 53*, 60.

Nanayakkara HLV, Weerathunga RP, Dissanayake VHW. (2008). Alpha and beta thalassaemia mutations detected in a clinical genetic testing laboratory in Sri Lanka. *Proceedings of the 9th Annual Academic Sessions of the Sri Lanka College of Haematologists*, Sri Lanka College of Haematologists, 23.

Nanayakkara HLV, Weerathunga RP, Dissanayake VHW. (2008). The CAG repeat length in Sri Lankan patients with Huntington Disease. *Proceedings of the 41st Annual Academic Sessions of the Ceylon College of Physicians*, Ceylon College of Physicians, 50.

Nanayakkara HLV, Weerathunga RP, Dissanayake VHW. (2008). Genetic testing for Duchene Muscular Dystrophy by Multiplex-Ligation

Probe Amplification. *Proceedings of the 11th Annual Scientific Congress of the Sri Lanka College of Paediatricians*, Sri Lanka College of Paediatricians, 65.

Nanayakkara HLV, Weerathunga RP, Gunasekara NCA, Dissanayake VHW. (2008). Mutations in the HFE gene and HFE-Associated Hereditary Hemochromatosis (HFE-HHC) in Sri Lanka. *Proceedings of the 41st Annual Academic Sessions of the Ceylon College of Physicians*, Ceylon College of Physicians, 66.

Nanayakkara HLV, Weerathunga RP, Peiris DC, Dissanayake VHW. (2008). Genetic classification of Sri Lankan children with neuroblastoma. *Proceedings of the 11th Annual Scientific Congress of the Sri Lanka College of Paediatricians*, Sri Lanka College of Paediatricians, 62.

Netchine I, Talon P, Dastot F, Vitaux F, Goossens M, Amselem S. (1998). Extensive phenotypic analysis of a family with growth hormone (GH) deficiency caused by a mutation in the GH-releasing hormone receptor gene. *Journal of Clinical Endocrinology and Metabolism 83*, 432–436.

Opanayake DJS, Athapaththu AMMH, Hidellage NT, Pedurupillay CRJ, Ratnayake A, Dissanayake VHW. (2008). Are aberrant BCR-ABL transcripts more common in Sri Lankan adults with Chronic Myeloid Leukaemia (CML)?. *Ceylon Medical Journal 53*, 59.

Papiha SS, Mastana SS, Jayasekara R. (1996a). Genetic variation in Sri Lanka. *Human Biology 68*, 707–737.

Papiha SS, Mastana SS, Purandare CA, Jayasekara R, Chakraborty R. (1996b). Population genetic study of three VNTR loci (D2S44, D7S22, and D12S11) in five ethnically defined populations of the Indian subcontinent. *Human Biology 68*, 819–835.

Pedurupillay CRJ, Athapaththu AMMH, Opanayake DJS, Dissanayake VHW. (2008). A case series of rare cytogenetic abnormalities. *Ceylon Medical Journal 53*, 59.

Premawardhena A, Arambepola M, Katugaha N, Weatherall DJ. (2008). Is the beta thalassaemia trait of clinical importance? *British Journal of Haematology 141*, 407–410.

Premawardhena A, De Silva S, Arambepola M, et al(2004). Thalassemia in Sri Lanka: A progress report. *Human Molecular Genetics 13*(Spec No. 2), R203–R206.

Premawardhena A, De Silver S, Arambepola M, et al. (2005a). Hemoglobin E-beta-thalassemia: Progress report from the International Study Group. *Annals of the New York Academy of Sciences 1054*, 33–39.

Premawardhena A, Fisher CA, Fathiu F, et al. (2001b). Genetic determinants of jaundice and gallstones in haemoglobin E beta thalassaemia. *Lancet 357*, 1945–1946.

Premawardhena A, Fisher CA, Liu YT, et al. (2003). The global distribution of length polymorphisms of the promoters of the glucuronosyltransferase 1 gene (UGT1A1): Hematologic and evolutionary implications. *Blood Cells, Molecules and Diseases 31*, 98–101.

Premawardhena A, Fisher CA, Olivieri NF, et al. (2005). Haemoglobin E beta thalassaemia in Sri Lanka. *Lancet 366*, 1467–1470.

Premawardhena A, Fisher CA, Olivieri NF, et al. (2005). A novel molecular basis for beta thalassemia intermedia poses new questions about its pathophysiology. *Blood 106*, 3251–3255.

Ratnatunga N, Liyanapathirana LV. (2007). Hormone receptor expression and Her/2neu amplification in breast carcinoma in a cohort of Sri Lankans. *Ceylon Medical Journal 52*, 133–136.

Rochette J, Pointon JJ, Fisher CA, et al. (1999). Multicentric origin of hemochromatosis gene (HFE) mutations. *American Journal of Human Genetics 64*, 1056–1062.

Roychoudhury AK, Nei M. (1985). Genetic relationships between Indians and their neighboring populations. *Human Heredity 35*, 201–206.

Saha N. (1988). Blood genetic markers in Sri Lankan populations–reappraisal of the legend of Prince Vijaya. *American Journal of Physical Anthropology 76*, 217–225.

Samaranayake TN, Dissanayake VH, Fernando SD. (2008). Clinical manifestations of cutaneous leishmaniasis in Sri Lanka - possible evidence for genetic susceptibility among the Sinhalese. *Annals of Tropical Medicine and Parasitology 102*, 383–390.

Samaranayake TN, Fernando SD, Jayasekara RW, Dissanayake VHW. (2008). Inherited susceptibility to Cutaneous Leishmaniasis in Sri Lanka *International Journal of Infectious Diseases 12*, e386.

Samaranayake TN, Fernando SD, Jayasekara RW, Dissanayake VHW. (2008). Inherited variability in tumour necrosis factor production and potential susceptibility to Cutaneous Leishmaniasis in the Sri Lankan population. *International Journal of Infectious Diseases 12*, e386.

Samarasinghe HD, K. Mayurathan, Dissanayake VHW, R. Gamage, Jayasekara RW. (2006). A preliminary study of familial Parkinson's Disease in Sri Lanka. *Ceylon Medical Journal 51*, 13.

Schneider JA, Rees DC, Liu YT, Clegg JB. (1998). Worldwide distribution of a common methylenetetrahydrofolate reductase mutation. *American Journal of Human Genetics 62*, 1258–1260.

Simpson B, Dissanayake VH, Wickramasinghe D, Jayasekara RW. (2003). Prenatal testing and pregnancy termination in Sri Lanka: Views of medical students and doctors. *Ceylon Medical Journal 48*, 129–132.

Simpson B, Dissanayake VHW, Jayasekara RW. (2005). Contemplating choice: Attitudes towards intervening in human reproduction in Sri Lanka. *New Genetics and Society 24*, 99–117.

Soejima M, Koda Y. (2005). Denaturing high-performance liquid chromatography-based genotyping and genetic variation of FUT2 in Sri Lanka. *Transfusion 45*, 1934–1939.

Stokowski RP, Pant PV, Dadd T, et al. (2007). A genomewide association study of skin pigmentation in a South Asian population. *American Journal of Human Genetics 81*, 1119–1132.

Suhr ML, Dysvik B, Bruland O, et al. (2007). Gene expression profile of oral squamous cell carcinomas from Sri Lankan betel quid users. *Oncology Reports 18*, 1061–1075.

Takeshima M, Saitoh M, Kusano K, et al. (2008). High frequency of hypermethylation of p14, p15 and p16 in oral pre-cancerous lesions associated with betel-quid chewing in Sri Lanka. *Journal of Oral Pathology and Medicine 37*, 475–479.

Topcu Z, Chiba I, Fujieda M, et al. (2002). CYP2A6 gene deletion reduces oral cancer risk in betel quid chewers in Sri Lanka. *Carcinogenesis 23*, 595–598.

Wallace DF, Browett P, Wong P, Kua H, Ameratunga R, Subramaniam VN. (2005). Identification of ferroportin disease in the Indian subcontinent. *Gut 54*, 567–568.

Wattavidanage J, Carter R, Perera KL, et al. (1999). TNFalpha*2 marks high risk of severe disease during Plasmodium falciparum malaria and other infections in Sri Lankans. *Clinical and Experimental Immunology 115*, 350–355.

Weerasuriya K, Jayakody RL, Smith CA, Wolf CR, Tucker GT, Lennard MS. (1994). Debrisoquine and mephenytoin oxidation in Sinhalese: A population study. *British Journal of Clinical Pharmacology 38*, 466–470.

Welihinda J, Karunanayake EH, Jayasekara R, Peiris JB, Pettersson U, Wadelius C. (1993). Deletion screening of Sri Lankan Duchenne muscular dystrophy patients using the polymerase chain reaction. *Annals of Tropical Paediatrics 13*, 83–86.

Wetthasinghe TK, Dissanayake VHW, Jayasekara RW. (2008). A descriptive study on the role of Y chromosome microdeletions in spontaneous recurrent pregnancy loss. *Sri Lanka Journal of Obstetrics and Gynaecology 30*, 98.

Wetthasinghe TK, Mallawarachchi PHR, De Saram EMNR, Perera VS, Dissanayake VHW, Jayasekara RW. (2006). Chromosomal abnormalities in Sri Lankan women with primary amenorrhoea. *Sri Lanka Journal of Obstetrics and Gynaecology 28*, 36–37.

82.

INHERITED FACTORS IN OBSTRUCTIVE LUNG DISEASES IN INDIA

Amrendra Kumar Gupta, Balaram Ghosh, and Anurag Agrawal

INTRODUCTION

Chronic obstructive pulmonary diseases (COPD) have been a scourge of public health across the world (Eder et al., 2006). As can be intuitively understood from their name, they relate to obstruction to the normal flow of air during breathing, characterized by either increased resistance of the conducting airways or dynamic airflow limitation. These defects are either variable or episodic, such as in asthma, or relatively fixed and chronic; e.g. chronic bronchitis and emphysema. Together and individually, obstructive lung diseases are a source of significant disability and burden to society in terms of healthcare expenditure, man-hours of work lost, and reduction in productivity (Bergman, 1992). While a full discussion of the classification of obstructive lung diseases is outside the scope of this chapter, it is important to realize that any understanding of inherited factors for a disease is relevant only in context of the definition of the disease (Barnes, 2008). A useful clinical classification is based on the reversibility and variability of the airflow limitations. Asthma is characterized by variable and reversible airway obstruction, as judged by clinical history and pulmonary function tests, while COPD is characterized by a relatively fixed airflow limitation with poor response to bronchodilators. There is further heterogeneity within asthma and COPD. In asthma, subphenotypes such as atopic asthma, non-atopic asthma, eosinophilic asthma, aspirin-sensitive asthma, exercise-induced asthma, etc., have been defined based on etiological, pathological, or symptomatological considerations. In COPD, subphenotypes can be based on reversibility of airflow limitation (partly reversible or poorly reversible), environmental etiology (smoking-related or not), and pathology (bronchitis or emphysema), among others. These subphenotypes are nontrivial in terms of molecular pathology, may reflect distinct diseases, and must be carefully considered during interpretation of studies. Nevertheless, both longitudinal and cross-sectional epidemiological studies point toward significant roles played by inherited factors (Vercelli, 2008). By now it is quite well accepted that genetic factors are predisposition factors that are modulated by environmental triggers to bring about phenotypes characteristic of asthma, COPD, or other obstructive lung diseases. While efforts to fish out inherited factors for asthma or COPD had started a long time back, major advances in biomolecular technology in the last two decades have led to accelerated progress. In the following discussion on the current understanding of inherited factors in obstructive lung diseases in India, we will try to present data within the context of specific study design, unique clinical considerations relevant to the Indian subcontinent, and the global scenario. Importantly, population genetics of complex disorders in India is still in its infancy compared to extensively studied populations in the West, and faces unique challenges. India is a melting pot of civilizations with striking ethnic variations across its lands. A recent project, the Indian Genome Variation Database (IGVdb, 2008), found that different geographical regions of India correspond to different reference populations from the Human Genome Project, mostly in accordance with anthropological observations and recorded history. This study has provided a framework for appropriate comparisons between genetic studies in India and abroad (Hardy et al., 2008).

BRONCHIAL ASTHMA

The prevalence of asthma in India has been variously estimated to be between 2.3% to 16.6% and is rising (Jindal, 2007a). While the highest estimates are from urban schoolchildren cohorts in Bangalore in South India, the lower estimates are from adult populations across the country. Even after accounting for differences in methodology, there is a large variation in asthma prevalence across geographical regions. In a large urban and semi-urban population survey from Delhi, India, over 11% of adult subjects had asthma and 60% of subjects with asthma had a family history of asthma (Gaur et al., 2006). Atopy or family history of atopy was strongly associated with asthma, with 70% of asthmatics also suffering from allergic rhinitis. In a larger study across entire India, sponsored by the Indian Council of Medical Research, the prevalence of asthma in adults was only 2.4%, but a similar pattern regarding family history and allergy was seen. The odds for a first-degree relative having asthma were sixfold higher for asthmatics than non-asthmatics. The odds ratio for atopy was 12.3. Gender and age importantly determined risk for asthma, similar to what has been seen elsewhere. While boys were at higher risk for asthma than girls in school cohorts, adult asthma was more commonly seen in women. Lifestyle was also an important determinant. Thus, the Indian asthma scene is dominated by atopic asthma running through families, suggesting an important role for inherited factors. Being a vast country with marked ethnic and racial differences, these factors are unlikely to be constant throughout. Also, the increasing burden of urban asthma highlights the important role of environment in asthma pathogenesis. The marked variations in asthma prevalence across the country may relate to genetic racial differences, environmental differences related to ethnic and sociocultural factors, differential gene–environment interactions, and epigenetic influences. It is believed that asthma is a complex genetic disorder and that there is significant gene–environment interaction (McLeish & Turner, 2007).

More than a hundred genes have been implicated in asthma so far. Prior to discussing the data, we will briefly review the methodology. More extensive reviews can be found elsewhere (Burke, 2003; Illig & Wjst, 2002; Vercelli, 2008). Population genetics/genomics studies are either hypothesis-driven, as in the case of candidate gene association studies, or without any prior hypothesis, such as linkage studies or genome-wide associations. Linkage studies are particularly useful in large extended family samples, and statistical tests are carried out to detect if the phenotypes shows linkage to any genetic markers. Once a region has been identified to be linked to the phenotypes under consideration, it could be fine-mapped to detect the susceptibility genes that reside in that area. Since asthma is a complex disorder, with multiple factors interacting to orchestrate the disease pathogenesis, there are no single factors with sizable effect, leading to high false negative discovery rates in linkage studies. Additionally, linkage studies require large extended families with data from three generations, making it cumbersome to collect samples. For candidate gene association studies, genes are selected from biochemical/molecular pathways expected or shown to a play role in disease pathogenesis, followed by detection of polymorphisms, genotyping, and statistical analysis. Association studies could either be case-control studies or family-based association studies. In case-control studies, ethnically matched cases and controls are recruited to see if there is any statistically significant difference of allele/genotype frequency between cases and controls. Family-based association studies usually carry out tests to identify if there is any association of loci with disease phenotypes in the presence of linkage. Any deviation of transmission of alleles/genotypes from the expected, between affected and unaffected offspring, results in statistically significant transmission disequilibrium. Candidate gene studies or association studies are quite sensitive in detecting loci with minor effect on phenotypes. However, case-control studies are prone to high false positive discovery rates, and family-based study designs have high false negative discovery rates; therefore, neither of the approaches is foolproof in terms of specificity and sensitivity. Ideally, case-control association studies are validated using family-based association studies, or vice-versa. Also, replication studies using ethnically different populations provide more confidence to the observations.

Discovery of microarray-based platforms have led to the development of genome-wide scan with coverage of single nucleotide polymorphism (SNPs) from almost the entire genome. Genome-wide association studies using these polymorphisms offer the opportunity to discover novel genes/loci with high sensitivity and specificity. Moreover, the tasks of genotyping large number of samples for these kinds of studies are becoming simpler and cheaper with recent sequencing and genotyping technologies. Unfortunately, most reports of association fail to replicate across different populations. Although methodological issues, such as sample collection criteria, phenotype definitions, etc., might be responsible for such discrepancies, ethnically different populations might have significantly different genetic compositions due to varying evolutionary history and environment. This necessitates identification of the genetic factors relevant to a particular ethnic group, since this could have pharmacological implications. The Indian subcontinent, in that respect, is unique in its composition, although several populations show close affinity to HapMap populations. Although, as mentioned earlier, population genetic studies of asthma in Indian populations have not been as enormous and comprehensive as those in Western populations, there have been significant discoveries like identification of phosphatidylinositol 4-phosphatase (INPP4a), a novel candidate gene in asthma (Sharma et al., 2008), which has also been replicated in another study (Rogers et al., 2009). Most of the genetic studies in the Indian subcontinent have adopted a candidate gene approach. Linkage studies are absent so far, even though India has very well structured families, often co-inhabiting together as joint families or large extended families, especially in the rural areas. In our experience, there is poor compliance of family members, which makes it difficult to collect a sufficient number of three-generation large extended families. There are no reports of genome-wide association studies either, as of now, probably due to previous budgetary constraints. Ongoing efforts are addressing these lacunae.

The group led by one of the authors (AA) has made efforts to identify genes important in the pathogenesis of bronchial asthma in India. Only a few other investigators have been involved in asthma population genetics in India; consequently, most of the genetic studies reports are from the author's laboratory. The main focus has been on atopic asthma, since it is the dominant subphenotype in India. Blood samples were collected following the guidelines of the American Thoracic Society for inclusion and exclusion of individuals with atopic asthma and matched healthy controls. For case-control studies, a homogeneous population belonging to the Indo-Aryan ethnic group was collected mostly from northern, central and northwestern parts of India. For family-based association tests, samples from almost all over the country were utilized; however, the southern part of the country was inadequately represented. The case-control study samples were tested for stratification using neutral markers as yet unlinked to asthma or any associated phenotypes. Using candidate gene approaches, we identified around 15–20 genes that might play a crucial role in asthma pathogenesis in the Indian population. Candidate genes include T helper cell development and differentiation into various functional subsets, eosinophil recruitment, and innate immune system.

INTERFERON γ

Interferon γ (IFNG) is a critical player of T helper cell differentiation (Kumar & Ghosh, 2008). Also it regulates many features critical to asthma pathogenesis (Teixeira et al., 2005). IFNG leads to polarization of T helper cell differentiation toward Th1 response, which is protective since it counterbalances Th2-mediated responses that underlie the atopic state. Failure to produce adequate amounts of IFNG may predispose to Th2-mediated asthma, since higher levels of IFNG are found in children who outgrow asthma compared to those who do not (Smart et al., 2002). The IFNG gene lies on chromosome 12q21, which is one of the most widely replicated loci found to be associated with asthma across various world populations (Raby et al., 2003). In the Indian population, a CA repeat in intron I of this gene showed association with asthma (Nagarkatti et al., 2002). Yet, a family-based study to revalidate the result failed to show any association (Kumar & Ghosh, 2008), perhaps because this locus is multiallelic and the study may have been underpowered. One SNP (rs1861494, A→G) in intron 3 of this gene was found to be significantly associated with asthma in both case-control and family-based studies. The G allele was negatively associated with asthma, implying that it might have a protective role. The two alleles showed differential binding to a nuclear factor(s) which remains unidentified.

INTERLEUKIN-4

Interleukin-4 (IL-4) is secreted by Th2 cells and is one of the vital cytokines involved in the pathology of bronchial asthma (Renauld, 2001). It is located on chromosome 5q31-33 in close proximity to other loci shown to be linked to asthma across various populations (Vercelli, 2008). IL-4 interacts with IL4R, leading to dimerization and nuclear translocation of STAT6 (see below). This results in expression of IL-4 responsive genes, including Cε immunoglobulin gene, regulating IgE levels (Nelms et al., 1999). In a case-control cohort, a CA repeat in intron 2 of IL-4 was shown to have statistically significant association with bronchial asthma (Nagarkatti et al., 2004a). However, this association was not replicated in a family-based study design (unpublished data). In addition, the functional role of this CA repeat has not been elucidated. There is not enough data to suggest significant association of IL-4 gene polymorphism with asthma in the northern Indian population; however, the possibility of false negatives cannot be ruled out.

SIGNAL TRANSDUCER AND ACTIVATOR OF TRANSCRIPTION 6

Signal transducer and activator of transcription 6 (STAT6) is an important candidate for asthma. It is the major downstream effector molecule for IL-4 and IL-13, both of which are critical Th2 cytokines (Yang et al., 2001). STAT6 knockout mice (STAT6 –/–) that are sensitized and challenged with allergens fail to develop an asthma-like phenotype, unlike wild type mice (Finkelman et al., 2000; Takeda et al., 1996). A novel CA repeat in the proximal promoter of this gene was identified and studied in a case-control cohort along with another CA repeat in 5'UTR (Nagarkatti et al., 2004b). A $(CA)_{16}$ repeat in the promoter was found to be associated with both asthma and serum total IgE levels. The 15 and 16 CA repeats at the 5' UTR locus were positively and negatively associated with asthma, respectively. The haplotype analysis suggested that combination of alleles at the two loci may have larger effects on asthma compared to either of them alone. Several other polymorphisms in this gene show association with allergy, asthma, or related phenotypes in other populations (Chen et al., 2007).

CLARA CELL SECRETORY GENE

Clara cell secretary gene CC16 was selected as a candidate gene for asthma due to its chromosomal location (11q13) that consistently showed linkage to bronchial asthma (Hay et al., 1995). CC16 is one of the major anti-inflammatory or immunomodulatory molecules, and plays an important part in the secretion/expression pattern in bronchoalveolar lavage (BAL; see Bernard et al., 1992; Broeckaert & Bernard, 2000; Mansur et al., 2002). An intragenic microsatellite repeat and a +38 A/G polymorphism that have been shown to be associated with asthma in other populations were investigated in our study population (Sharma et al., 2004b). The microsatellite allele with repeat size 18 showed significant association with asthma, in both case-control and family-based studies. The haplotype A_18 (allele A at +38A/G and repeat size 18) was associated with asthma as well as serum total IgE.

HIGH AFFINITY IGE RECEPTOR

The high affinity IgE receptor (FcεRI), is highly expressed on the surface of mast cells and basophils (Kinet, 1999). IgE plays an important role in the development of bronchial asthma, with enhanced production in atopic asthma. The cross-linking of IgE antibodies bound to FcεRI by multivalent allergens induces the activation of these cells, resulting in secretion of allergic mediators and induction of cytokine gene transcription. The number of IgE receptors (FcεRI) per basophil, and FcεRI-mediated histamine release from basophils, is higher in allergic individuals as compared to normal controls. Association studies suggest an important role for (G/A) polymorphism in intron 2, a (C/T) polymorphism in the 3'-untranslated region (3'-UTR), and a dinucleotide $(CA)_n$ repeat in intron 5 of the *FcεRIβ* gene in the pathogenesis of bronchial asthma (Sharma et al., 2004c). A significant association was observed with asthma and allele A in intron 2, C allele in the 3'-UTR, and the $(CA)_{16}$ repeat. On the other hand, the $(CA)_{18}$ allele was found to be negatively associated with asthma and may be protective. In order to investigate the combined association of the intron 2, $(CA)_n$ and 3'-UTR polymorphisms, three locus haplotypes were generated. Two major haplotypes, *A_16_C* and *G_18_T,* were found to be positively and negatively associated with asthma, respectively. This further correlated with histamine release and the expression level of the FcεRIβ transcripts in the basophils (Sharma et al., 2006a). Thus, in a North Indian population, an individual with FcεRIβ A_16_C haplotype in homozygous condition is at increased risk to develop atopic asthma (Sharma et al., 2004c, 2006a). Since this has been functionally validated, it could be useful for identifying individuals predisposed to atopic asthma.

CD14

CD14 is the co-receptor of LPS, leading to signaling via the toll-like receptor 4 on antigen presenting cells (Zdolsek & Jenmalm, 2004). It is located on 5q31 along with a cytokine gene cluster that harbors other genes, such as IL-4, IL-5, IL-13. It seems to be deficient in allergy, making it an important candidate gene for asthma (Marsh et al., 1994). Polymorphisms in this gene have variably shown association or no association with asthma (Baldini et al., 2002; LeVan et al., 2001; Sharma et al., 2004a). A −159 C/T polymorphism was shown to be associated with asthma, with elevated serum total IgE levels in both case-control and family-based studies. Allele C was found at a higher frequency in asthmatics than normal controls. This polymorphism has been shown to affect Sp protein binding and transcriptional activity (LeVan et al., 2001) that could have important implications since CD14 is a critical regulator of Th1/Th2 polarization.

TRANSFORMING GROWTH FACTOR β-1

Transforming growth factor β 1 (TGFβ1) is a pleiotropic cytokine that has been shown to have both proinflammatory and anti-inflammatory roles (Sanjabi et al., 2009). Its role in airway remodeling is well documented (Minshall et al., 1997; Ohno et al., 1996). It is implicated in fibrosis to account for sub-epithelial fibrosis developing in patients with severe asthma. On the other hand, it prevents inflammatory responses to allergen through inhibition of basophil and eosinophil proliferation (Alam et al., 1994). Several polymorphisms have been identified in this gene, and studies in different populations have had variable results (Nagpal et al., 2005). The author's laboratory found a novel repeat located 24.9 kb upstream of this gene, and two single nucleotide polymorphisms (−800 G/A, and −509 C/T) significantly associated with asthma in case-control and family studies. In addition, another repeat (−509 C/T) polymorphism showed association with serum total IgE and serum TGFβ1 levels. The haplotype 23_G_T was found to be the risk haplotype, while 22_G_C was found to be the protective haplotype. Individuals with the 22_G_C haplotype had lower serum TGFβ1 levels compared to those with the 23_G_T haplotype. A reporter gene assay, with the luciferase construct containing the −509 C/T polymorphism, showed evidence of altered gene expression, providing functional validation (unpublished).

INTERLEUKIN 10

Interleukin 10 (IL-10) downregulates the production of proinflammatory cytokines from monocytes and macrophages (Moore et al., 2001). In models of asthma, engineering of T-cells to produce IL-10 reduces the development of airway hyperresponsiveness (Oh et al., 2002). Three polymorphisms, namely −1082A/G, −819C/T, and −592C/A, have been found in the promoter region and have been studied in different populations (Kube et al.,

1995). In the Indian population, an association with the −1082 A/G polymorphism in case-control and family studies is reported (Chatterjee et al., 2005). Three major haplotypes (GCC, ACC, and ATA) had been described in different populations; in our study we identified another haplotype, ATC. ATA was found to be the risk haplotype, while ATC and GCC may be protective. While the functional basis of this is unclear, all these polymorphisms lie in putative transcription binding sites and may influence IL-10 gene regulation.

CHEMOKINE RECEPTOR 5

Chemokine receptor 5 (CCR5) drives a Th1 type immune response upon its interaction with RANTES, MIP 1α, and MIP 1β (Loetscher et al., 1998). A 32bp deletion has been identified that shows protection against HIV infection (Samson et al., 1996). This 32bp deletion results in premature stop codon, which might lead to aberration of CCR5-mediated signaling and inflammation. This deletion has also been investigated in relation to asthma in many populations and was found to be protective against asthma in Caucasian children (Hall et al., 1999). In the Indian population, CCRΔ32 frequency is low, and insufficient numbers of homozygotes were found to form any firm conclusions (Batra et al., 2005). Interestingly, heterozygosity was present with higher frequency in asthmatics than controls, indicating that this mutation may be positively affecting asthma pathogenesis, perhaps through influencing susceptibility to infections. However, this is speculative.

MAST CELL CHYMASE 1

Mast cell chymase 1 (CMA1) is stored in high amounts in secretory granules of mast cells. It is a chymotrypsin-like protease that upon its release activates a number of proinflammatory molecules (Hart, 2001; Tomimori et al., 2002). It modulates airway remodeling by formation of collagen fibrils from type I procollagen, and by activation of MMP1, MMP9, and intestinal procollagenase (He et al., 2004; Lazaar et al., 2002). These features make it an important candidate gene for asthma pathogenesis. We genotyped one previously known SNP (1903G/A) and a novel (TG)n(GA)m repeat, 254bp downstream of this gene, in case-control cohorts (Sharma et al., 2005). There was no significant association between the −1903G/A polymorphism and asthma, although there was an association with serum Total IgE levels. The repeat was associated with asthma, and combined haplotypic analysis showed significant association with both asthma and serum total IgE levels. Haplotype G_43 was found to be potentially protective. The functional basis of this is unknown.

N-ACETYLTRANSFERASE 2

The N-acetyltransferase 2 (NAT2) gene has been found to be associated with asthma in various studies (Bosse & Hudson, 2007; Vercelli, 2008). NAT enzymes are involved in detoxification of a number of drugs, such as hydrazines and arylamines (Hein et al., 1993, 2000). NAT2 is one isoform of this enzyme that has high affinity for N-acetylation of most aromatic compounds, resulting in their detoxification. This enzyme is highly polymorphic, resulting in segregations of human populations as rapid, intermediate, and slow acetylators (Hein, 2000). These variants have also been studied in relation to various malignancies, and other diseases like rheumatoid arthritis and diabetes mellitus. The observation that NAT 2 mRNA is present in the epithelial cells lining of respiratory bronchioles suggests that this enzyme may be involved in metabolizing the inhalants and other molecules in the lungs, thus affecting susceptibility to respiratory diseases. In a Russian study, the slow acetylation genotype showed a protective effect on bronchial asthma (Makarova et al., 2000). We studied the possible effect of NAT2 variants on asthma susceptibility in India (Batra et al., 2006). A +481C/T polymorphism showed significant association with asthma. The CC genotype was more frequently present in controls than asthmatics. This polymorphism also showed marginal association with total serum IgE, as well as percentage blood eosinophil counts. The +282C/T polymorphism showed association with total serum IgE but not with asthma.

TUMOR NECROSIS FACTOR α

Tumor necrosis factor α (TNFA) and lymphotoxin α (LTA) are located on chromosome 6p21.3. TNFA is a pleiotropic inflammatory cytokine, levels of which are elevated in sputum, bronchoalveolar lavage, and biopsy sample from asthmatics (Kips, 2001). It is also shown to cause airway hyperresponsiveness upon inhalation. Lymphotoxin α has a role in homing of lymphocytes, formation of spleen and lymph nodes, B-cell proliferation, and IgE synthesis (Worm et al., 1998). TNFA −863C/A polymorphism was found to be associated with asthma in our study population (Sharma et al., 2006b). Allele A was found to be associated with reduced serum TNFA levels, reduced serum IgE levels, and reduced risk of asthma. This polymorphism influences the binding of nuclear factors at this locus, as determined by gel shift assays. Allele A has lesser affinity for the nuclear factor than allele C. Another polymorphism, denoted as LTA_NcoI, also showed marginal association with asthma. Replication in family-based studies would provide more confidence to these observations. In another Indian study on TNFA, a −308 G/A polymorphism was associated with asthma in a North

Indian population. The inclusion criteria for this study were physician-diagnosed asthma, active clinical symptoms, and >15% reversibility. The frequency of GG genotype was higher in controls, whereas GA and AA were higher in patients. This polymorphism had also shown marginal association with asthma in our study (Sharma et al., 2006b). Since allele A is known to be correlated with high plasma TNF levels, the findings have a functional basis (Wilson et al., 1992).

EOTAXIN

Chromosomal region 17q11-17q21 shows linkage to asthma and related traits across various populations and harbors a number of genes which play important roles in inflammatory processes (Koppelman et al., 2002). Some of the important genes in this chromosomal region include RANTES, MCP1, MCP3, iNOS, and Eotaxin. Eotaxin is considered to be one of the predominant chemoattractants for eosinophils, which play a vital role in asthma pathology by secreting a number of mediators like granular protein, reactive nitric oxide, chemokines and cytokines, etc. (Rothenberg & Hogan, 2006) Eotaxin signals through chemokines receptor 3 (CCR3), which is also expressed by other cell types such as Th2 lymphocytes, basophiles, dendritic cells, mast cells, etc., thereby modulating effector function of various other cell types besides eosinophils (Rothenberg, 1999). Some polymorphisms in this gene may have an important role in the pathogenesis of bronchial asthma in the Indian population (Batra et al., 2007b). One hexanucleotide repeat (GAAGGA)n 10.9 kb upstream of this gene, and three SNPs (+67G/A, –384A/G and –426C/T) were investigated in case-control and family-based association studies. The repeat showed very significant association with asthma, total serum IgE, and plasma eotaxin levels. The +67G/A polymorphism showed association with asthma and plasma eotaxin levels, but not with serum IgE. On the other hand, SNP +384A/G showed association with serum total IgE levels and plasma eotaxin levels, but not asthma. Haplotypic analysis showed very strong association with asthma, serum total IgE levels, and plasma eotaxin levels. 8_G_A_C was identified as the risk haplotype. Since these polymorphisms show correlation with eotaxin levels, they could be directly or indirectly affecting its transcription or translation. Further functional studies are warranted.

NITRIC OXIDE SYNTHASE

Nitric oxide synthase (NOS) is an enzyme that catalyzes the production of nitric oxide from semi-essential amino acid L-arginine (Dweik, 2005). Various isoforms of this gene have been identified that are referred to as *constitutive* NOS (cNOS) and *inducible* NOS (iNOS). Constitutive NOS is supposed to generate small amounts of nitric oxide in response to physiological stimuli, while inducible NOS production is triggered by inflammatory cytokines during inflammation. iNOS is expressed predominantly by inflammatory cells (T-cells and macrophages, etc.) and epithelial cells. It has been shown that nitric oxide in exhaled breath is primarily from the iNOS activity. Higher levels of iNOS are associated with asthma pathology. A genetic association study of iNOS gene polymorphisms and asthma (and related phenotypes) in the Indian population (Batra et al., 2007a) revealed four repeat polymorphisms: (CCTTT)n in promoter, intron 2 (GT)n repeat, intron 4 (GT)n repeat, and intron 5 (CA) n. The intron 4 (GT)n allele with repeat size 15 showed significant association with asthma, asthma severity, and percentage blood eosinophil counts. Importantly, individuals with this genotype had significantly higher serum nitric oxide levels in contrast to other individuals. Since it has been shown that variable repeat size could modulate gene expression, it would be interesting to see if this repeat polymorphism has some regulatory role in gene expression, especially in the context of various stimulatory signals.

INTERLEUKIN 21

Interleukin 21 belongs to the IL-2 family of cytokines which binds to the γ_c receptor along with its specific receptor IL-21α (Leonard & Spolski, 2005). Upon its interaction with the receptor complex, it activates JAK1 and JAK3, that in turn activates STAT1 and STAT3. IL-21 could modify inflammatory responses by interacting with receptors present on B-cells, T-cells, natural killer cells, and dendritic cells (Nurieva et al., 2007; Strengell et al., 2003). IL-21 is secreted by Th17 cells, which have been recently identified to play critical roles in development of autoimmune disorders (Korn et al., 2007). IL-21 has also been shown to modulate IgE synthesis and production by B-cells. So far, genetic association studies involving candidate genes involved in TH17 development, differentiation, or its effector function and asthma, are lacking, even though some evidence indicates that it could play important roles in asthma pathogenesis (Bullens et al., 2006). From our lab we had provided the first evidence of association of an exonic variant in this gene with asthma, serum total IgE, and IL-21 levels (Chatterjee et al., 2009). Four single nucleotide polymorphisms (1455C/T, 1472G/T, 5250C/T, and 8381C/T) and one (GAAT)n repeat—1183bp upstream of this gene—were analyzed in case-control and family-based studies. 5250C/T was found to be significantly different between cases and controls. Also, there was significantly different transmission to the affected and unaffected offspring in an independent family-based analysis. This SNP also showed association with serum IL-21 levels. While bioinformatic analysis suggests that this polymorphism might alter the mRNA structure, it remains to be experimentally validated.

ACID MAMMALIAN CHITINASE

Acid mammalian chitinase (CHIA) has been found to promote Th2-mediated inflammation (Elias et al., 2005). It is believed that chitinase enzymes could be polarizing Th2 response by eradicating chitin containing pathogens, or by acting as chemotactic agents. One exonic variant has previously been associated with asthma in a German pediatric cohort (Bierbaum et al., 2005). CHIA polymorphisms might influence the asthma risk in Indian populations (Chatterjee et al., 2008). Two SNPs (rs3806448 G/A and rs10494132 T/C) were found to be associated with total serum IgE. Two locus haplotype analyses showed that AT was the risk haplotype, and GT the protective haplotype. Reporter gene assays showed increased transcription activity of the AT haplotype, indicating that these polymorphisms have functional roles in gene regulation. It is quite possible that SNPs in this gene might be playing roles in the regulation of CHIA gene expression and function, and thereby acting as susceptibility factors for asthma.

β-2 AGONISTS

β-2 agonists (β_2-agonist) are widely used as bronchodilators and recommended as first-line anti-asthma drugs. β_2-agonist receptor (β_2AR) is the key target of β_2-agonist drugs, and widely expressed throughout the body, especially in smooth muscle cells of bronchi. It is a G protein coupled receptor that mediates the action of catecholamines. It is encoded by intronless gene at 5q31-5q32 (Drysdale et al., 2000). It has many polymorphisms, and it is believed that they could be potential modifiers of asthma or might be responsible for interindividual variation of responses to β_2-agonist drugs. The results of genetic association studies with asthma and drug responsiveness are contradictory across various studies (Kukreti et al., 2005). In the Indian population, sequencing of a 2193bp covering various polymorphic sites resulted in identification of ten polymorphisms in this region. One of these, leading to substitution of glycine for arginine in position 16, was found to associate with differences in response to therapy with salbutamol, a β agonist. Subjects homozygous for Arg16 were poor responders to salbutamol, in terms of increase in forced expiratory volume at 1 second (FEV1), an established objective measure of lung function, in comparison to Gly16. Interestingly, in initial studies on this subject, asthmatic children homozygous for arginine at codon 16 were shown to have significantly greater (>fivefold) bronchodilator response to albuterol than individuals homozygous for glycine residues, and the Arg16 variant was found to be more sensitive in vitro (Martinez et al., 1997). Yet, the finding of our study was in agreement with increased sub-sensitivity to β agonists noted in other studies (Lee et al., 2004), as well as with poorer asthma control in asthmatic subjects with Arg16 homozygosity on regular doses of salmeterol (Palmer et al., 2006). Since regular use of these drugs has been reported to have detrimental effects on symptoms and lung function in double-blinded placebo-controlled studies, it is possible that the sub-sensitivity and poorer control of asthma noted by us and others is a result of increased initial sensitivity coupled with chronic use (Israel et al., 2004). In support of this, Palmer et al. (2006) only found Arg16 variant to be associated with poorer control of asthma during regular use, but not intermittent use. In another study, adult asthmatic patients on regular use were studied following withdrawal of bronchodilators up to 24 hours prior to the study (Kukreti et al., 2005). Thus, it is possible that asthma patients homozygous for the arginine residue are potentially good responders in the native state, who have been overtreated, resulting in a poorer response phenotype as in our study. They may benefit from minimizing the use of both short-acting and long-acting β_2-agonists. We believe that while further study is necessary before coming to firm conclusions regarding management, genotyping at this locus is of potential benefit to asthmatic patients (Wechsler & Israel, 2005).

PHOSPHOINOSITIDE 3-KINASE

Phosphoinositide 3-kinase (PI3K) participates in a number of inflammatory events, and dysregulation of the signaling mechanism involving PI3K leads to various immune disorders (Deane & Fruman, 2004; Rommel et al., 2007). PI3K-γ and PI3K-delta are the two isoforms shown to be critical regulators in immune cell signaling. PI3K is involved in phosphorylation of phosphatidylinositols (PtdIns) on the third carbon of the inositide ring, leading to generation of second messengers like PtdIns-3-P, PtdIns-3, 4, 5-P3, PtdIns-3, 4-P2, etc., which mediate its effector functions. A number of phosphatases have been identified that dephosphorylate these lipid messengers and downregulate PI3K-mediated signaling. These messengers have also been shown to modulate asthma features (Kumar & Ghosh, 2009). SHIP (src homology 2-containing inositol phosphatase) is 5' phosphatase and it downregulates mast cell degranulation upon IgE crosslinking (Huber et al., 2002). PTEN (phosphatase and tensin homologue), which is 3' phosphatase, has been shown to downregulate IL-4, IL-5, and eosinophilic cationic protein that are expressed in ovalbumin-challenged mice (Kwak et al.,

2003). In the Indian population using bioinformatics, genetics, and molecular biology techniques, another phosphatase, phosphatidylinositol 4-phosphatase (INPP4A), was found to be associated with asthma (Sharma et al., 2008). INPP4A has higher affinity for PtdIns 3,4-P2 and removes 4-phosphate from it. Since PtdIns 3, 4-P2 has been shown to interact with PDK, Akt, etc., and mediate inflammation, its dephosphorylation by INPP4A is potentially an important homeostatic mechanism, downregulation of which may predispose to inflammation. In an initial screen, the microsatellite repeat rs3217304 (+99097) in intron 11 of INPP4A showed association with atopic asthma. INPP4A was then fine-mapped to identify other polymorphisms, to further study its role in asthma pathogenesis. Since an SNP (+92031 A/T) and the microsatellite repeat (+99097) showed significant association with asthma in both case-control and family-based studies, the intervening genomic region was sequenced in 16 individuals, leading to identification of another polymorphism, +110832A/G, which was 18.8 and 11.7 kb from +92031A/T and +99097 (CA)n, respectively. This polymorphism was in the PEST sequence of the α_3 splice variant, and was a nonsynonymous SNP that leads to Thr/Ala substitution. Allele +110832A was found to be overtransmitted to asthmatic offspring. Frequency of genotype GG was higher in controls in comparison to asthmatics. Our haplotypic studies using +92031A/T, +99097 (CA)n, and +110832A/G further revealed significant difference in distribution of haplotypes between normal controls and asthmatics. Haplotype A_2_A was found to be positively associated with asthma, while T_1_G was negatively associated with asthma. Bioinformatic analysis indicated that Thr/Ala substitution results in altered protein stability, probably via altered recognition by calpain proteases. To demonstrate this experimentally, INPP4A protein levels were measured in platelets of individuals with different genotypes for +110832A/G SNP. Individuals with AA genotype had lesser INPP4A protein in comparison to other genotypes. Further, upon stimulation of platelets with Ca+ ionophore and ionomycin, it was observed that protein from AA individuals was more susceptible to degradation. +110832A/G also showed positive association with RANTES levels, such that AA individuals had higher levels in comparison to AG or GG individuals. Recently, we have also analyzed freely available gene expression datasets, looking for pathways that may be differently regulated in asthmatics compared to the normal subjects in conditions of cellular stress. In this independent analysis, we have found that key components of the inositol signaling pathway are upregulated during stress in normal individuals, but show failure to upregulate in asthmatics (unpublished data). This provides additional support for our finding of genetic susceptibilities to asthma being associated with INPP4A. Further work dissecting this pathway's relationship to asthma is underway.

CHRONIC OBSTRUCTIVE PULMONARY DISEASE

Similar to asthma, chronic obstructive pulmonary disease (COPD) in India has been recognized and investigated mainly through small individual investigator-led studies focused in a few geographical areas (Jindal, 2007b). In these, COPD prevalence was estimated to range from 2% to 22% in men and 1.9–19% in women. In a review of three decades of literature on COPD in India, the median prevalence was 5% in men and 2.7% in women. Smoking was associated with the great majority of disease, with other factors being domestic pollution. A recent large multicentric study, sponsored by the Indian Council of Medical Research (ICMR), showed 5% prevalence in men and 3.2% in women. Smoking was the principal associated risk, with smokers outnumbering nonsmokers by 2.65:1. Alternate types of smoking unique to the Indian subcontinent such as *hukkah* (smoke filtered through water) and *bidi* (rolled leaves) were associated with greater risk for COPD than conventional cigarette smoking. Smoking per se is the most important risk factor for development of COPD; however, it has been shown that only 10% to 15% of chronic smokers develop the disease, suggesting that genetic factors might be modifiers modulating the susceptibility. Unfortunately, none of the large cohorts was part of a genetic study, and all available literature on genetics of COPD in India derives from small cohorts led by individual investigators, focused almost exclusively on smoking-related COPD (Jindal, 2007b). The methods to fish out susceptibility genes for COPD are similar to asthma. In COPD, the environmental influence is supposedly more than asthma, which might be the reason that globally the progress in gene identification for COPD is lagging behind. It is expected that in future, environmental factors will be accounted for in genetic studies, which might be helpful in detection of genetic factors. There are very scanty data from India with respect to COPD genetics. There are no family-based association studies so far, and only case-control candidate gene studies have been reported. Also, only smoking-related COPD has been investigated, despite nonsmoking COPD being prevalent in India. There are studies from the laboratory of Q. A. Pasha, also at our institute, comparing smokers with or without COPD (Ahsan et al., 2004; Arif et al., 2007a,b; Vibhuti et al., 2007). Ethical and geographical matching was used to get rid of the population stratification. COPD was diagnosed by clinical features, and chest X-ray and spirometry were also performed. Patients had forced expiratory volume in 1s (FEV_1) <80% and FEV_1/FVC (force vital capacity) <70%. Patients with evidence of reversibility in airway obstruction >200 ml and >12% increase in FEV_1 after nebulization of salbutamol were excluded. Patients with bronchiectasis, tuberculosis, coronary artery disease, valvular heart disease, systemic hypertension, primary myocardial disease, diabetes mellitus, etc., were also excluded. Control subjects had no family history of COPD.

RENIN-ANGIOTENSIN-ALDOSTERONE AND KALLIKREIN-KININ SYSTEMS

Renin-Angiotensin-Aldosterone system (RAAS) and Kallikrein-Kinin system (KKS) influence pulmonary vasoconstriction and pulmonary vascular remodeling, which are features associated with COPD (Morrell et al., 1997). Angiotensin converting enzyme (ACE) and endothelial nitric oxide synthase (eNOS) modulate RAAS and KKS. Ahsan et al. (2004) investigated the association of three polymorphisms (one in ACE gene and two polymorphisms in eNOS gene) with COPD, and plasma ACE and NO levels. The genotype distribution and allelic distribution for the three polymorphisms studied did not vary between the cases and controls. However, they observed that ACE activity was higher in patients. Subjects with insertion genotype in the ACE gene, and GG genotype at a –894 G/T SNP in the eNOS gene, had lower ACE activity and high NO levels. Since the two enzymes interact in several pathways, the interactions observed at the genetic level could be relevant. This study had a very low sample size, and replication in family studies or in a large case-control cohort should be informative. In another study, –786T/C and –922A/G polymorphisms of the eNOS gene were found to be associated with COPD (Arif et al., 2007a). Correlation studies with plasma nitrite levels, plasma malonaldehyde levels, and FEV1 also showed associations of these polymorphisms. Since these polymorphisms show association not only with the dichotomous trait, i.e., COPD, but also with relevant subphenotypes, further study is warranted.

DETOXIFICATION ENZYMES

Enzymes involved in the detoxification of xenobiotics have shown to be involved in pathogenesis of obstructive lung diseases such as asthma and COPD (Teramoto et al., 2005). Since smoking is one of the most significant risk factors for COPD, involvement of candidates from this metabolic pathway was studied in relation to COPD (Arif et al., 2007c). Detoxification of xenobiotics takes place in two phases: phase I, where xenobiotics undergo activation, usually by cytochrome P450 (CYPP450) enzymes; followed by conversion of the activated compound into inactivated hydrophilic compounds by phase II enzymes, such as glutathione S-transferase (GST) and NAT. A total of nine polymorphisms in CYP2E1 (-1053C/T, –1293G/C, 7632T/A and 9893C/G), CYP2D6 (1934G/A), and NAT2 {481C/T (NAT*4/4), 590G/A (NAT*5/5), 803A/G (NAT*6/6), 857G/A(NAT*7/7)} genes were investigated. –1053C/T and –1293G/C were in complete LD, and found to be associated with COPD. –1293C and –1053T have been shown to alter the transcription of CYP2E1, where these polymorphic alleles have higher transcriptional activity, protein levels, and enzymatic activity in comparison to the wild types (i.e., –1293G and –1053C) respectively. Increased enzymatic activity of CYP2E1 implies more activation of the xenobiotics, which may lead to increased damage to the lungs (Tsutsumi et al., 1994). CYP2D6 polymorphism 1934G/A did not show any association in their study cohort. This polymorphism has also been shown to alter enzyme activity (Smith et al., 1992). NAT2 polymorphisms showed significant difference in both the groups. There was overrepresentation of the slow acetylator allele in patients, implying that slow acetylation or detoxification of xenobiotics might enhance its damaging effect. Slow acetylation alleles have also been shown to be overrepresented in patients with asthma (Nacak et al., 2002).

GLUTATHIONE-S-TRANSFERASE

Glutathione-S-transferase (GST) is a member of membrane-associated proteins in the eicosanoid and glutathione (MAPEG) family of transmembrane proteins, which catalyze the conjugation of reduced glutathione via the sulfhydryl group to the electrophilic center on a wide variety of substances (Hayes & Strange, 2000). This group of enzymes acts as antioxidants and detoxifying agents by degrading xenobiotics, etc. GSTP1 is the isoform, which is expressed more widely by the airway cells and tissues such as alveoli, alveolar macrophages, respiratory bronchioles, etc. Amino acid substitutions due to nonsynonymous SNPs in the exonic region of this gene have been shown to regulate enzyme activity. Vibhuti A et al. (2007) genotyped two nonsynonymous SNPs—I105V (rs1695; A/G in exon 5) and A114V (rs1138272; C/T in exon6)—of this gene, looking for associations with COPD. Biochemical parameters such as MDA, plasma GST levels, and GPx activity were also reported. V105V and V114V were significantly higher in patients than controls. Further, I105V+V105V and A114V+V114V combinations were higher in patients than normal controls. Two locus haplotypes, estimated using a maximum likelihood procedure, showed significant difference between the two groups. Haplotype 105V-114V was more frequent in patients, and haplotype 105I-114A was more frequent in controls. Although there are variable findings with respect to GSTP1 gene polymorphisms in COPD and other diseases, this study indicates that the genotypes and haplotypes associated with risk for COPD may result from lower enzymatic activity, and thereby incomplete catabolization of the toxicants. In support of this, the 105V and 114V alleles found to be associated with COPD in this study have been shown to result in lower enzymatic activities (Johansson et al., 1998; Sundberg et al., 1998).

MICROSOMAL EPOXIDE HYDROLASE

Microsomal epoxide hydrolase (mEPHX) catalyzes the hydrolysis of a wide range of exogenous arenes and aliphatic epoxides, to form water-soluble dihydroidals (Watabe et al., 1981). Since this enzyme is another potential detoxifying agent, Vibhuti et al. (2007) have studied two nonsynonymous SNPs (rs1051740 T/C in exon 3 and rs2234922 A/G in exon 4) with COPD, MDA levels, GSH levels, and GPx activity. These nonsynonymous SNPs lead to amino acid substitutions that change the enzymatic activity (Y113H and H139R, respectively). They found H113H and H139H to be more frequent in patients than controls. H113H was associated with decreased GSH levels and GPx activity. When genotype combinations from both the loci were analyzed, H113H+H139H showed association with decreased FEV1, increased MDA levels, and decreased GSH levels and GPx activity.

CYCLOOXYGENASE

Cyclooxygenase (COX), which acts as deoxygenase and peroxidase, is one of the key enzymes in the biosynthesis of prostanoids (prostaglandins, prostacyclins, thrombaxanes; Mitchell et al., 1995; Simon, 1999). It has constitutive as well an inducible isoforms, which vary in their tissue distribution and regulation of expression. COX-1 is the isoform's constitutive form, while COX-2 is the inducible isoform. COX-2 is found abundantly in activated macrophages and other cells at the site of inflammation, thereby playing a critical role in inflammatory events. In asthma and COPD, immunoreactivity of COX-2 is increased significantly. Protein-53 or tumor protein-53 (p53) is a transcription factor that regulates cell cycles and is known to be a tumor suppressor. p53 is found to be frequently mutated in the epithelium of smokers, and is suggested to be regulating pulmonary vascular epithelial remodeling in COPD. COX-2 is induced by p53 through the ERK/MAPK pathway, thereby regulating inflammation (Han et al., 2002). Polymorphisms in these genes were studied in relation to COPD (Arif et al., 2008). It was found that the C allele at the –765 G/C SNP in COX2, and the nonsynonymous praline allele at p53 Pro72Arg allele, were more frequent in patients. The authors also performed pair-wise gene–gene interaction tests and identified the interaction between –765G/C and 72Pro/Arg. Functional studies should make these findings more interesting.

α-1 ANTITRYPSIN

While there are no primary classical genetic studies on established genetics risks like α1 antitrypsin mutations in emphysematous COPD, low population prevalence for this genotype is reported for West Asia and India (Jindal, 2007b). There is a single report of association of PIM3 (wild type of α_{1-}AT or α-1 antitrypsin) variants with COPD in one case-control cohort in the north Indian population from our center (Gupta et al., 2005). This cohort consisted of COPD patients who were smokers, as well as nonsmokers. Physician-diagnosed COPD patients were enrolled, but the sample size (n=73 total; 37 COPD patients and 37 normal controls) in this study was very low. Deficiency of α_{1-}AT is a major risk factor for emphysematous COPD. It has many allelic variants that have differential activity. PIM3 is the wild type allele of this gene, and its variants are supposed to have normal α_{1-}AT serum levels. However, some of the PIM3 variants have been shown to be associated with diseases like asthma (Gaillard et al., 1994). In their study, Gupta et al. (2005) observed that PIM3 allele (Glu376GAA→ AspGAC) has increased prevalence in COPD patients as compared to controls. Validation of these findings in a larger cohort would be informative, as well as delineation of any functional role of these polymorphisms.

SUMMARY

In summary, genetics of obstructive lung diseases is an emerging field in India. As it spreads beyond select institutions, more data covering a larger spectrum of the Indian population is accumulated. We expect that, given the striking differences in lifestyles and rich genetic diversity in India, much remains to be done before inherited factors contributing to the obstructive lung disease burden in India are fully understood. One of the important challenges will be that these diseases are truly syndromes, which are clinically defined in terms of similarities at the level of symptomatology and treatment. These features are only the tip of the iceberg, and at a molecular pathology level, there may be marked heterogeneity with a strong environmental component. For example, it was recently shown that the C159T polymorphism in the CD14 gene had opposite effects on the risk of asthma in rural and urban areas, because of varying levels of endotoxin in the environment. While the TT genotype was protective in urban populations with low endotoxin load, it was associated with increased risk in rural communities where the endotoxin load was higher (Baldini et al., 2002; Martinez, 2007; McLeish & Turner, 2007). This is one reason why, despite larger and larger studies from around the globe, there has been only limited progress in identifying the genetic basis of complex diseases. Developments in technology make it easier for us to sequence ever-increasing numbers of samples at previously unimaginable levels of detail, but simply increasing sample size has never been and will not be a viable strategy, as has been elegantly discussed elsewhere (Vercelli & Martinez, 2006). We need to make efforts to subphenotype and to interpret genetic data in the context of detailed phenotypic data as well as environmental data. While a wealth of biological understanding is now within our reach, we must make sure that it does not get lost in translation.

REFERENCES

Ahsan A, Ram R, Baig MA, Pasha MA. (2004). ACE I allele and eNOS G allele crosstalk may have a role in chronic obstructive pulmonary disease. *Clin Biochem 37*(11), 1037–1040.

Alam R, Forsythe P, Stafford S, Fukuda Y. (1994). Transforming growth factor β abrogates the effects of hematopoietins on eosinophils and induces their apoptosis. *J Exp Med 179*(3), 1041–1045.

Arif E, Ahsan A, Vibhuti A, et al. (2007a). Endothelial nitric oxide synthase gene variants contribute to oxidative stress in COPD. *Biochem Biophys Res Commun 361*(1), 182–188.

Arif E, Vibhuti A, Alam P, et al. (2007c). Association of CYP2E1 and NAT2 gene polymorphisms with chronic obstructive pulmonary disease. *Clin Chim Acta 382*(1–2), 37–42.

Arif E, Vibhuti A, Alam P, et al. (2007b). Association of CYP2E1 and NAT2 gene polymorphisms with chronic obstructive pulmonary disease. *Clin Chim Acta 382*(1–2), 37–42.

Arif E, Vibhuti A, Deepak D, Singh B, Siddiqui MS, Pasha MA. (2008). COX2 and p53 risk-alleles coexist in COPD. *Clin Chim Acta 397*(1–2), 48–50.

Baldini M, Vercelli D, Martinez FD. (2002). CD14: An example of gene by environment interaction in allergic disease. *Allergy 57*(3), 188–192.

Barnes PJ. (2008). Immunology of asthma and chronic obstructive pulmonary disease. *Nat Rev Immunol 8*(3), 183–192.

Batra J, Pratap ST, Mabalirajan U, Sinha A, Prasad R, Ghosh B. (2007a). Association of inducible nitric oxide synthase with asthma severity, total serum immunoglobulin E and blood eosinophil levels. *Thorax 62*(1), 16–22.

Batra J, Rajpoot R, Ahluwalia J, et al. (2007b). A hexanucleotide repeat upstream of eotaxin gene promoter is associated with asthma, serum total IgE and plasma eotaxin levels. *J Med Genet 44*(6), 397–403.

Batra J, Sharma M, Chatterjee R, Sharma S, Mabalirajan U, Ghosh B. (2005). CCR5 Delta32 deletion and atopic asthma in India. *Thorax 60*(1), 85.

Batra J, Sharma SK, Ghosh B. (2006). Arylamine N-acetyltransferase gene polymorphisms: Markers for atopic asthma, serum IgE and blood eosinophil counts. *Pharmacogenomics 7*(5), 673–682.

Bergman AB. (1992). An economic evaluation of asthma in the United States. *N Engl J Med 327*(8), 571–572.

Bernard A, Marchandise FX, Depelchin S, Lauwerys R, Sibille Y. (1992). Clara cell protein in serum and bronchoalveolar lavage. *Eur Respir J 5*(10), 1231–1238.

Bierbaum S, Nickel R, Koch A, et al. (2005). Polymorphisms and haplotypes of acid mammalian chitinase are associated with bronchial asthma. *Am J Respir Crit Care Med 172*(12), 1505–1509.

Bosse Y, Hudson TJ. (2007). Toward a comprehensive set of asthma susceptibility genes. *Annu Rev Med 58*, 171–184.

Broeckaert F, Bernard A. (2000). Clara cell secretory protein (CC16): Characteristics and perspectives as lung peripheral biomarker. *Clin Exp Allergy 30*(4), 469–475.

Bullens DM, Truyen E, Coteur L, et al. (2006). IL-17 mRNA in sputum of asthmatic patients: Linking T cell driven inflammation and granulocytic influx? *Respir Res 7*, 135–141.

Burke W. (2003). Genomics as a probe for disease biology. *N Engl J Med 349*(10), 969–974.

Chatterjee R, Batra J, Das S, Sharma SK, Ghosh B. (2008). Genetic association of acidic mammalian chitinase with atopic asthma and serum total IgE levels. *J Allergy Clin Immunol 122*(1), 202–208.

Chatterjee R, Batra J, Ghosh B. (2009). A common exonic variant of interleukin21 confers susceptibility to atopic asthma. *Int Arch Allergy Immunol 148*(2), 137–146.

Chatterjee R, Batra J, Kumar A, et al. (2005). Interleukin-10 promoter polymorphisms and atopic asthma in North Indians. *Clin Exp Allergy 35*(7), 914–919.

Chen W, Khurana Hershey GK. (2007). Signal transducer and activator of transcription signals in allergic disease. *J Allergy Clin Immunol 119*(3), 529–541

Deane JA, Fruman DA. (2004). Phosphoinositide 3-kinase: Diverse roles in immune cell activation. *Annu Rev Immunol 22*, 563–598.

Drysdale CM, McGraw DW, Stack CB, et al. (2000). Complex promoter and coding region β 2-adrenergic receptor haplotypes alter receptor expression and predict in vivo responsiveness. *Proc Natl Acad Sci U S A 97*(19), 10483–10488.

Dweik RA. (2005). Nitric oxide, hypoxia, and superoxide: The good, the bad, and the ugly! *Thorax 60*(4), 265–267.

Eder W, Ege MJ, von Mutius E. (2006). The asthma epidemic. *N Engl J Med 355*(21), 2226–2235.

Elias JA, Homer RJ, Hamid Q, Lee CG. (2005). Chitinases and chitinase-like proteins in T(H)2 inflammation and asthma. *J Allergy Clin Immunol 116*(3), 497–500.

Finkelman FD, Morris SC, Orekhova T, et al. (2000). Stat6 regulation of in vivo IL-4 responses. *J Immunol 164*(5), 2303–2310.

Gaillard MC, Zwi S, Nogueira CM, et al. (1994). Ethnic differences in the occurrence of the M1(ala213) haplotype of α-1-antitrypsin in asthmatic and non-asthmatic black and white South Africans. *Clin Genet 45*(3), 122–127.

Gaur SN, Gupta K, Rajpal S., Singh AB, Rohtagi A. (2006). Prevalence of bronchial asthma and allergic rhinitis among urban and rural adult population of Delhi. *Indian J Allergy Asthma Immunol 20*(2), 90–97.

Gupta J, Bhadoria DP, Lal MK, et al. (2005). Association of the PIM3 allele of the α-1-antitrypsin gene with chronic obstructive pulmonary disease. *Clin Biochem 38*(5), 489–491.

Hall IP, Wheatley A, Christie G, McDougall C, Hubbard R, Helms PJ. (1999). Association of CCR5 delta32 with reduced risk of asthma. *Lancet 354*(9186), 1264–1265.

Han JA, Kim JI, Ongusaha PP, et al. (2002). P53-mediated induction of Cox-2 counteracts p53- or genotoxic stress-induced apoptosis. *EMBO J 21*(21), 5635–5644.

Hardy BJ, Seguin B, Singer PA, Mukerji M, Brahmachari SK, Daar AS. (2008). From diversity to delivery: The case of the Indian Genome Variation initiative. *Nat Rev Genet 9*(Suppl 1), S9–S14.

Hart PH. (2001). Regulation of the inflammatory response in asthma by mast cell products. *Immunol Cell Biol 79*(2), 149–153.

Hay JG, Danel C, Chu CS, Crystal RG. (1995). Human CC10 gene expression in airway epithelium and subchromosomal locus suggest linkage to airway disease. *Am J Physiol 268*(4 Pt 1), L565-L575.

Hayes JD, Strange RC. (2000). Glutathione S-transferase polymorphisms and their biological consequences. *Pharmacology 61*(3), 154–166.

He SH, Xie H, Zhang XJ, Wang XJ. (2004). Inhibition of histamine release from human mast cells by natural chymase inhibitors. *Acta Pharmacol Sin 25*(6), 822–826.

Hein DW. (2000). N-Acetyltransferase genetics and their role in predisposition to aromatic and heterocyclic amine-induced carcinogenesis. *Toxicol Lett* 112–113, 349–356.

Hein DW, Doll MA, Fretland AJ, et al. (2000). Molecular genetics and epidemiology of the NAT1 and NAT2 acetylation polymorphisms. *Cancer Epidemiol Biomarkers Prev 9*(1), 29–42.

Hein DW, Doll MA, Rustan TD, et al. (1993). Metabolic activation and deactivation of arylamine carcinogens by recombinant human

NAT1 and polymorphic NAT2 acetyltransferases. *Carcinogenesis 14*(8), 1633–1638.

Huber M, Kalesnikoff J, Reth M, Krystal G. (2002). The role of SHIP in mast cell degranulation and IgE-induced mast cell survival. *Immunol Lett 82*(1–2), 17–21.

IGVdb (2008). Genetic landscape of the people of India: A canvas for disease gene exploration. *J Genet 87*(1), 3–20.

Illig T, Wjst M. (2002). Genetics of asthma and related phenotypes. *Paediatr Respir Rev 3*(1), 47–51.

Israel E, Chinchilli VM, Ford JG, et al. (2004). Use of regularly scheduled albuterol treatment in asthma: Genotype-stratified, randomised, placebo-controlled cross-over trial. *Lancet 364*(9444), 1505–1512.

Jindal SK. (2007a). Bronchial asthma: The Indian scene. *Curr Opin Pulm Med 13*(1), 8–12.

Jindal SK. (2007b). The burden of chronic airflow obstruction in India. *J Assoc Physicians India 55*, 742.

Johansson AS, Stenberg G, Widersten M, Mannervik B. (1998). Structure-activity relationships and thermal stability of human glutathione transferase P1-1 governed by the H-site residue 105. *J Mol Biol 278*(3), 687–698.

Kinet JP. (1999). The high-affinity IgE receptor (Fc epsilon RI): From physiology to pathology. *Annu Rev Immunol 17*, 931–972.

Kips JC. (2001). Cytokines in asthma. *Eur Respir J Suppl 34*, 24S-33S.

Koppelman GH, Stine OC, Xu J, et al. (2002). Genome-wide search for atopy susceptibility genes in Dutch families with asthma. *J Allergy Clin Immunol 109*(3), 498–506.

Korn T, Bettelli E, Gao W, et al. (2007). IL-21 initiates an alternative pathway to induce proinflammatory T(H)17 cells. *Nature 448*(7152), 484–487.

Kube D, Platzer C, von Knethen A, et al. (1995). Isolation of the human interleukin 10 promoter. Characterization of the promoter activity in Burkitt's lymphoma cell lines. *Cytokine 7*(1), 1–7.

Kukreti R, Bhatnagar P, Rao C, et al. (2005). β(2)-adrenergic receptor polymorphisms and response to salbutamol among Indian asthmatics. *Pharmacogenomics 6*(4), 399–410.

Kumar A, Ghosh B. (2008). A single nucleotide polymorphism (A -> G) in intron 3 of IFNγ gene is associated with asthma. *Genes Immun 9*(4), 294–301.

Kumar A, Ghosh B. (2009). Genetics of asthma: A molecular biologist perspective. *Clin Mol Allergy 7*(7), 7.

Kwak YG, Song CH, Yi HK, et al. (2003). Involvement of PTEN in airway hyperresponsiveness and inflammation in bronchial asthma. *J Clin Invest 111*(7), 1083–1092.

Lazaar AL, Plotnick MI, Kucich U, et al. (2002). Mast cell chymase modifies cell-matrix interactions and inhibits mitogen-induced proliferation of human airway smooth muscle cells. *J Immunol 169*(2), 1014–1020.

Lee DK, Currie GP, Hall IP, Lima JJ, Lipworth BJ. (2004). The arginine-16 β2-adrenoceptor polymorphism predisposes to bronchoprotective subsensitivity in patients treated with formoterol and salmeterol. *Br J Clin Pharmacol 57*(1), 68–75.

Leonard WJ, Spolski R. (2005). Interleukin-21: A modulator of lymphoid proliferation, apoptosis and differentiation. *Nat Rev Immunol 5*(9), 688–698.

LeVan TD, Bloom JW, Bailey TJ, et al. (2001). A common single nucleotide polymorphism in the CD14 promoter decreases the affinity of Sp protein binding and enhances transcriptional activity. *J Immunol 167*(10), 5838–5844.

Loetscher P, Uguccioni M, Bordoli L, et al. (1998). CCR5 is characteristic of Th1 lymphocytes. *Nature 391*(6665), 344–345.

Makarova SI, Vavilin VA, Lyakhovich VV, Gavalov SM. (2000). Allele NAT2*5 determines resistance to bronchial asthma in children. *Bull Exp Biol Med 129*(6), 575–577.

Mansur AH, Fryer AA, Hepple M, Strange RC, Spiteri MA. (2002). An association study between the Clara cell secretory protein CC16 A38G polymorphism and asthma phenotypes. *Clin Exp Allergy 32*(7), 994–999.

Marsh DG, Neely JD, Breazeale DR, et al. (1994). Linkage analysis of IL4 and other chromosome 5q31.1 markers and total serum immunoglobulin E concentrations. *Science 264*(5162), 1152–1156.

Martinez FD. (2007). CD14, endotoxin, and asthma risk: Actions and interactions 3. *Proc Am Thorac Soc 4*(3), 221–225.

Martinez FD, Graves PE, Baldini M, Solomon S, Erickson R. (1997). Association between genetic polymorphisms of the β2-adrenoceptor and response to albuterol in children with and without a history of wheezing. *J Clin Invest 100*(12), 3184–3188.

McLeish S, Turner SW. (2007). Gene-environment interactions in asthma. *Arch Dis Child 92*(11), 1032–1035.

Minshall EM, Leung DY, Martin RJ, et al. (1997). Eosinophil-associated TGF-β1 mRNA expression and airways fibrosis in bronchial asthma. *Am J Respir Cell Mol Biol 17*(3), 326–333.

Mitchell JA, Larkin S, Williams TJ. (1995). Cyclooxygenase-2: Regulation and relevance in inflammation. *Biochem Pharmacol 50*(10), 1535–1542.

Moore KW, de Waal MR, Coffman RL, O'Garra A. (2001). Interleukin-10 and the interleukin-10 receptor. *Annu Rev Immunol 19*, 683–765.

Morrell NW, Danilov SM, Satyan KB, Morris KG, Stenmark KR. (1997). Right ventricular angiotensin converting enzyme activity and expression is increased during hypoxic pulmonary hypertension. *Cardiovasc Res 34*(2), 393–403.

Nacak M, Aynacioglu AS, Filiz A, et al. (2002). Association between the N-acetylation genetic polymorphism and bronchial asthma. *Br J Clin Pharmacol 54*(6), 671–674.

Nagarkatti R, Kumar R, Sharma SK, Ghosh B. (2004a). Association of IL4 gene polymorphisms with asthma in North Indians. *Int Arch Allergy Immunol 134*(3), 206–212.

Nagarkatti R, Rao B, Rishi JP, et al. (2002). Association of IFNG gene polymorphism with asthma in the Indian population. *J Allergy Clin Immunol 110*(3), 410–412.

Nagarkatti R, Rao C, Vijayan V, Sharma SK, Ghosh B. (2004b). Signal transducer and activator of transcription 6 haplotypes and asthma in the Indian population. *Am J Respir Cell Mol Biol 31*(3), 317–321.

Nagpal K, Sharma S, Rao C, et al. (2005). TGFβ1 haplotypes and asthma in Indian populations. *J Allergy Clin Immunol 115*(3), 527–533.

Nelms K, Keegan AD, Zamorano J, Ryan JJ, Paul WE. (1999). The IL-4 receptor: Signaling mechanisms and biologic functions. *Annu Rev Immunol 17*, 701–738.

Nurieva R, Yang XO, Martinez G, et al. (2007). Essential autocrine regulation by IL-21 in the generation of inflammatory T cells. *Nature 448*(7152), 480–483.

Oh JW, Seroogy CM, Meyer EH, et al. (2002). CD4 T-helper cells engineered to produce IL-10 prevent allergen-induced airway hyperreactivity and inflammation. *J Allergy Clin Immunol 110*(3), 460–468.

Ohno I, Nitta Y, Yamauchi K, Hoshi H, et al. (1996). Transforming growth factor β 1 (TGF β 1) gene expression by eosinophils in asthmatic airway inflammation. *Am J Respir Cell Mol Biol 15*(3), 404–409.

Palmer CN, Lipworth BJ, Lee S, Ismail T, Macgregor DF, Mukhopadhyay S. (2006). Arginine-16 β2 adrenoceptor genotype predisposes to exacerbations in young asthmatics taking regular salmeterol. *Thorax 61*(11), 940–944.

Raby BA, Silverman EK, Lazarus R, Lange C, Kwiatkowski DJ, Weiss ST. (2003). Chromosome 12q harbors multiple genetic loci related to asthma and asthma-related phenotypes. *Hum Mol Genet 12*(16), 1973–1979.

Renauld JC. (2001). New insights into the role of cytokines in asthma. *J Clin Pathol 54*(8), 577–589.

Rogers AJ, Raby BA, Lasky-Su JA, et al. (2009). Assessing the reproducibility of asthma candidate gene associations, using genome-wide data. *Am J Respir Crit Care Med 179*(12), 1084–1090.

Rommel C, Camps M, Ji H. (2007). PI3K delta and PI3K γ: Partners in crime in inflammation in rheumatoid arthritis and beyond? *Nat Rev Immunol 7*(3), 191–201.

Rothenberg ME. (1999). Eotaxin. An essential mediator of eosinophil trafficking into mucosal tissues. *Am J Respir Cell Mol Biol 21*(3), 291–295.

Rothenberg ME, Hogan SP. (2006). The eosinophil. *Annu Rev Immunol 24*, 147–174.

Samson M, Libert F, Doranz BJ, et al. (1996). Resistance to HIV-1 infection in caucasian individuals bearing mutant alleles of the CCR-5 chemokine receptor gene. *Nature 382*(6593), 722–725.

Sanjabi S, Zenewicz LA, Kamanaka M, Flavell RA. (2009). Anti-inflammatory and pro-inflammatory roles of TGF-β, IL-10, and IL-22 in immunity and autoimmunity. *Curr Opin Pharmacol 9*(4), 447–453.

Sharma M, Batra J, Mabalirajan U, et al. (2004a). Suggestive evidence of association of C-159T functional polymorphism of the CD14 gene with atopic asthma in northern and northwestern Indian populations. *Immunogenetics 56*(7), 544–547.

Sharma M, Batra J, Mabalirajan U, et al. (2008). A genetic variation in inositol polyphosphate 4 phosphatase a enhances susceptibility to asthma. *Am J Respir Crit Care Med 177*(7), 712–719.

Sharma S, Ghosh B. (2004). Association of an intragenic microsatellite marker in the CC16 gene with asthma in the Indian population. *J Hum Genet 49*(12), 677–683.

Sharma S, Nagarkatti R, Rao C, et al. (2004b). A_16_C haplotype in the FcepsilonRIβ gene confers a higher risk for atopic asthma in the Indian population. *Clin Genet 66*(5), 417–425.

Sharma S, Rajan UM, Kumar A, Soni A, Ghosh B. (2005). A novel (TG) n(GA)m repeat polymorphism 254 bp downstream of the mast cell chymase (CMA1) gene is associated with atopic asthma and total serum IgE levels. *J Hum Genet 50*(6), 276–282.

Sharma S, Nagpal K, Mabalirajan U, Kumar A, Ghosh B. (2006a). Correlation of FCERIB gene haplotypes with histamine release from basophils in atopic asthma. *J Allergy Clin Immunol 118*(4), 960–963.

Sharma S, Sharma A, Kumar S, Sharma SK, Ghosh B. (2006b). Association of TNF haplotypes with asthma, serum IgE levels, and correlation with serum TNF-α levels. *Am J Respir Cell Mol Biol 35*(4), 488–495.

Simon LS. (1999). Role and regulation of cyclooxygenase-2 during inflammation. *Am J Med 106*(5B), 37S–42S.

Smart JM, Horak E, Kemp AS, Robertson CF, Tang ML. (2002). Polyclonal and allergen-induced cytokine responses in adults with asthma: Resolution of asthma is associated with normalization of IFN-γ responses. *J Allergy Clin Immunol 110*(3), 450–456.

Smith CA, Moss JE, Gough AC, Spurr NK, Wolf CR. (1992). Molecular genetic analysis of the cytochrome P450-debrisoquine hydroxylase locus and association with cancer susceptibility. *Environ Health Perspect 98*, 107–112.

Strengell M, Matikainen S, Siren J, et al. (2003). IL-21 in synergy with IL-15 or IL-18 enhances IFN-γ production in human NK and T cells. *J Immunol 170*(11), 5464–5469.

Sundberg K, Johansson AS, Stenberg G, et al. (1998). Differences in the catalytic efficiencies of allelic variants of glutathione transferase P1-1 towards carcinogenic diol epoxides of polycyclic aromatic hydrocarbons. *Carcinogenesis 19*(3), 433–436.

Takeda K, Tanaka T, Shi W, et al. (1996). Essential role of Stat6 in IL-4 signalling. *Nature 380*(6575), 627–630.

Teixeira LK, Fonseca BP, Barboza BA, Viola JP. (2005). The role of interferon-γ on immune and allergic responses. *Mem Inst Oswaldo Cruz 100*(Suppl 1), 137–144.

Teramoto S, Ishii T, Yamamoto H, Yamaguchi Y, Matsuse T. (2005). Xenobiotic enzymes and genetics of COPD. *Chest 127*(1), 408–409.

Tomimori Y, Tsuruoka N, Fukami H, et al. (2002). Role of mast cell chymase in allergen-induced biphasic skin reaction. *Biochem Pharmacol 64*(7), 1187.

Tsutsumi M, Wang JS, Takase S, Takada A. (1994). Hepatic messenger RNA contents of cytochrome P4502E1 in patients with different P4502E1 genotypes. *Alcohol Alcohol Suppl 29*(1), 29–32.

Vercelli D. (2008). Discovering susceptibility genes for asthma and allergy. *Nat Rev Immunol 8*(3), 169–182.

Vercelli D, Martinez FD. (2006). The Faustian bargain of genetic association studies: Bigger might not be better, or at least it might not be good enough. *J Allergy Clin Immunol 117*(6), 1303–1305.

Vibhuti A, Arif E, Deepak D, Singh B, Qadar Pasha MA. (2007). Genetic polymorphisms of GSTP1 and mEPHX correlate with oxidative stress markers and lung function in COPD. *Biochem Biophys Res Commun 359*(1), 136–142.

Watabe T, Kanai M, Isobe M, Ozawa N. (1981). The hepatic microsomal biotransformation of delta 5-steroids to 5 α, 6 β-glycols via α- and β-epoxides. *J Biol Chem 256*(6), 2900–2907.

Wechsler ME, Israel E. (2005). How pharmacogenomics will play a role in the management of asthma. *Am J Respir Crit Care Med 172*(1), 12–18.

Wilson AG, di Giovine FS, Blakemore AI, Duff GW. (1992). Single base polymorphism in the human tumour necrosis factor α (TNF α) gene detectable by NcoI restriction of PCR product. *Hum Mol Genet 1*(5), 353.

Worm M, Ebermayer K, Henz B. (1998). Lymphotoxin-α is an important autocrine factor for CD40 + interleukin-4-mediated B-cell activation in normal and atopic donors. *Immunology 94*(3), 395–402.

Yang M, Hogan SP, Henry PJ, et al. (2001). Interleukin-13 mediates airways hyperreactivity through the IL-4 receptor-α chain and STAT-6 independently of IL-5 and eotaxin. *Am J Respir Cell Mol Biol 25*(4), 522–530.

Zdolsek HA, Jenmalm MC. (2004). Reduced levels of soluble CD14 in atopic children. *Clin Exp Allergy 34*(4), 532–539.

83.

INHERITED METABOLIC DISEASE IN INDIA

Bal N. Apte

INTRODUCTION

In 1899, Sir Archibald Garrod described alkaptonuria, a disease that he classified as a lifelong congenital chemical alteration (Garrod, 1899). Later in 1908, he described other diseases, such as albinism, cystinuria, porphyria, and pentosuria, which he called "inborn errors of metabolism" (Garrod, 1908). Garrod's conclusions were completely correct in relation to the genetic basis of metabolic disorders and the gene–enzyme concept (Beadle and Tatum, 1948). The importance of Garrod's observation was that he implied that the inborn errors of metabolism (IEM) are manifestations of biochemical individuality. This constitutes the earliest important medical application of the laws of Mendel (Mendel, 1865). IEM results from the absence or dysfunction of a gene product or its cofactor, leading to either accumulation or the deficiency of a specific metabolite. IEM thus reflects homeostatic derangement in the patient. Gene products in IEMs are usually enzymes. It is, however, being increasingly recognized that receptors, transport proteins, or regulators of other gene products or cofactors may play a role in the pathogenesis of IEMs. IEMs are individually rare, but as a group are relatively common.

CLASSIFICATION OF INHERITED METABOLIC DISEASES

A simple-minded classification of IEMs would be to group them on the basis of diseases of small and large molecules (Stacey, 2005). Diseases of small molecules would include disorders of intermediary metabolism, defects of energy generation, and defects of neurotransmitters. Diseases of large molecules would include storage disorders in which proteins, lipids, carbohydrates, or combinations of these accumulate within cellular organelles and disrupt cellular function. Diseases of small molecules typically present as life-threatening catastrophic events in the neonatal period, resulting from low or absent enzyme levels involved in their metabolism. Episodic dysfunction usually occurs in the setting of higher levels of residual enzyme activity in the face of catabolic stress. Diseases of large molecules typically present as progressive neurodegenerative disorders with extra-neurological manifestations as a frequent accompaniment. However, there are exceptions to this rule; patients with small molecule disorders may present with non-specific and non-progressive developmental delay and mild mental retardation (Stacey, 2005). The classical example is that of phenylketonuria, in which persistently raised phenylalanine levels are neurotoxic and, in the absence of phenylalanine restricted diet, this may result in significant intellectual handicap (Krause et al., 1985). On the other hand, glycogen storage disease type I is a disease of a large molecule that may present acutely with neonatal hypoglycemia (Stacey, 2005).

This classification of human disease derives from observational correlation between pathological analysis and clinical syndromes. Characterizing diseases in this way establishes a classification scheme that has served clinicians well to the current time, relying on observational skills to define the syndromic phenotype. Throughout the last century, this approach became more objective; as the molecular underpinnings of many disorders were being identified and definitive laboratory tests became an essential part of the overall diagnostic paradigm. Yet, this classic diagnostic strategy has widely recognized shortcomings that reflect both a lack of sensitivity in identifying preclinical disease and a lack of specificity in defining disease unequivocally (Whitfield et al., 2004).

The importance of redefining human disease in this post-genomic era cannot be overemphasized. Several examples serve to prove this point well. Subcategorizing histologically similar cancers by differences in surface biomarkers, transcription profiling, or proteomic analysis is currently being applied to several malignancies, including lymphomas (Dave et al., 2006) and adenocarcinoma of the breast (Hedenfalk et al., 2001; Hall et al., 2006), in an effort to provide better information about prognosis and response to therapy. This approach defines the expanding field of molecular pathology, in which molecular signatures replace histopathology to diagnose disease and predict outcome.

As an example of a disease whose genetic basis not only is felt to be much simpler than that of malignancies, but also is affected by host genomic and environmental complexities, consider sickle cell disease (Sebastiani et al., 2005). This classic Mendelian disease is, by definition, viewed as a monogenic disorder, in which all affected individuals have a single point mutation at position 6 of the beta-chain of hemoglobin, leading to a substitution of valine for glutamic acid. This single mutation changes the oxygen affinity of hemoglobin, and leads to its ability to form polymers under hypoxic conditions, which, in turn, deform the erythrocyte into the characteristic sickle shape. Yet, despite this well-defined mutation, and its biochemical and physiological consequences, the genotype simply cannot invariably predict the phenotype of patients with the disease. Patients with this mutation are not at all homogeneous in their clinical presentations: some develop principally painful crises with or without bony infarcts; others are prone to hemolytic crises; some develop vasoocclusive crises, including stroke; still others develop acute chest syndrome, while many are phenotypically normal, except for mild anemia. There are many reasons for these different clinical phenotypes, including the presence of disease-modifying genes (Sebastiani et al., 2005) and environmental influences (ambient oxygen concentration, infection, dehydration), which can interact to yield different phenotypes (Kato et al., 2007). This example points out that our true understanding of even the most straightforward of genetic disorders is quite limited. Indeed, several examples could be listed that show this pattern (Sebastiani et al., 2005; Dave et al., 2006; Hedenfalk et al., 2001; Hall et al., 2006; Kato et al., 2007; Farber and Loscalzo, 2004; Seidman and Seidman, 2001; Kim, 2006; Kitano, 2004; Loscalzo et al., 2007).

SCENARIO IN THE POST-GENOMIC ERA

With the complete sequence of the human genome a reality, and with a growing body of transcriptomics, proteomics, and metabolomics sets of data in health and disease, we are now in a unique position in the history of medicine to define human disease precisely, uniquely, and unequivocally, with optimal sensitivity and specificity (Whitfield, et al., 2004). These advances have been driven by the novel technologies that can analyze global sets of gene products. Transcriptomics defines the population of mRNA species in a cell at a specific time and set of conditions. Proteomics addresses the challenging problem of defining changes in protein expression, protein dynamics, and posttranslational modifications. The upcoming field of metabolomics measures changes in populations of low molecular weight metabolites under a given set of conditions (Fiehn, 2001). Low molecular weight metabolites represent the end products of cellular processes, and as such, advertise the response of human systems to a variety of genetic and environmental influences (Fiehn, 2002). Metabolic work-up of a patient, therefore, gives us a global view of the functioning of his various systems. Our strategy of carrying out investigations for aminoacidopathies, organic acidopathies, and enzymopathies precisely results in giving us this global view of the patient and enables us in the diagnosis of the disease (Apte, 2006).

India, like all developing countries, is going through an accelerated demographic switch to noncommunicable diseases. In the cities, congenital malformations and genetic disorders are important causes of morbidity and mortality. India has a high birth rate, because of which a large number of infants with genetic disorders are born every year— about half a million with malformations and 21,000 with Down syndrome (Verma, 2000). The primary causes of referral for genetic investigations and counseling in India have been recurrent abortions, multiple congenital anomalies, infertility, Down syndrome, muscle dystrophy, convulsions, and mental retardation in the previous issue (Kumta, 2005).

A study carried out in three centers (Mumbai, Delhi, and Baroda) on 94,610 newborns by using a uniform pro forma, showed congenital malformation frequency of 2.03%, the commonest malformations being neural tube defects and musculoskeletal disorders. The frequency of Down syndrome among 94,610 births was 1 per 1150. Screening of 112,269 newborns for amino acid disorders showed three disorders to be the commonest-- tyrosinemia, maple syrup urine disease, and phenylketonuria. A correlation of patients with mental retardation related to amino acid metabolism revealed hyperglycinemia, homocystinuria, alkaptonuria, and maple syrup urine disease to be the commonest metabolic disorders. However, the frequency of other genetic diseases, such as beta-thalassemia at birth is 1:2700, accounting for approximately 9,000 cases of thalassemia major born every year, including around 5,200 infants with sickle cell disease born annually (Verma, 2000).

Disorders of the amino acid metabolism constitute the largest group of IEM. Most of these disorders cause brain damage leading to mental retardation. According to the World Health Organization, 3% of the world population suffers from mental retardation (Swarna et al., 2004). Biochemical investigations of cases of mental retardation in AIIMS, Delhi, and KEM Hospital, Mumbai, demonstrated that common metabolic disorders involve mucopolysaccharides, lysosomal enzymes, Wilson's disease, glycogen storage disease, and galactosemia (Verma, 2000).

Consanguineous marriages are strongly favored among some populations of South India. In a study conducted on 407 infants and children, 35 genetic diseases were diagnosed in 63 of them. Forty-four showed single gene defect, twelve had polygenic disorders, and seven were with Down's syndrome. The coefficient of inbreeding of the whole study group was F=0.0414, which was significantly higher than that for the general population, F=0.0271. Autosomal recessive disorders formed a single largest disease category. The results suggest that long-term inbreeding may not have resulted in appreciable elimination of recessive lethals and sublethals from the gene pool (Devi et al., 1987; Devi and Naushad, 2004).

In a screening program of urine and plasma for the IEMs of amino acids, carried out on 2560 cases over a period of six years, 2.4% showed eleven different Mendelian disorders including homocystinuria, alkaptonuria, MSUD, and nonketotic hyperglycinemia. Two cases each had ornithinemia and PKU. Generalized aminoaciduria was found in about 3.52%. This frequency distribution of aminoacidopathies in North India is remarkably different from that observed in the West (Kaur et al., 1994).

We had conducted two camps, one in 2007 at the Spandan Holistic Institute, Mumbai, and the other earlier in 2005 in a village in the interior of Maharashtra (Apte and Gogate, 2005). The aim of conducting these camps was to examine and carry out biochemical investigations of previously undiagnosed pediatric cases of mental retardation and autism, for various inborn errors of metabolism (IEM). Children from special schools for autism and mental retardation participated in the camps, along with their parents. These children were being given training in special schools in occupational therapy, physiotherapy, speech therapy, social development, etc.

Patients' documentation included detailed history, pedigree, main complaints, birth history, developmental milestones, family history, dietary habits, anthropometry, phenotypic abnormalities, organomegaly, behavioral patterns, etc. We examined 120 children, out of which parents of 37 children came forward for detailed biochemical investigations. This low response could be because of two reasons: (a) the cost of the comprehensive genetic investigations involving a broad screen, and (b) the future prospect of spending on the management procedure, together with the skepticism and the ignorance as to whether any diagnosis of the disease will be done. These two factors certainly contribute, to a large extent, to why such data has not been collected in India, if screening programs are to be conducted without any government funding. All children were in the age group 0.5–12 years, and of Asian ethnicity. As mentioned before, they were being treated for their symptoms and were being given supportive therapy like physiotherapy, occupational therapy, speech therapy, etc. But the basic defect was never even investigated.

Children with chromosomal abnormalities, which are known to cause mental retardation such as Down syndrome and fragile-X syndrome, were not included in this screening exercise. Out of 37 children, 29 were males and 7 females.

This pilot study, done on 37 children with features of mental retardation and autism, has revealed some interesting findings. This exercise certainly gives us the trend of events in the society. The results of the biochemical investigations of one of the two camps are shown in Table 83-1 and Figure 83-1 (Datar and Apte, 2008). Fifty-four percent of children were given definitive diagnosis; 37.5% had MPS and formed the largest single group. Twenty-five percent of children had biotinidase deficiency, and another 25% had homocystinuria; 12.5% had both biotinidase deficiency and were homocystinuric. We had examined 150 patients in the second camp, conducted in 2005. The results of this camp were very similar to the one presented here, and hence we will not describe them here. The important finding of the camps is that more than 50% of the children having mental retardation or autism have some diagnosable metabolic defect.

Children with manageable IEMs, such as biotinidase deficiency, homocystinuria, and those having both disorders together, were instituted management by dietary restrictions and nutritional supplementation. A significant number of them have shown remarkable improvement in their general behavior, and have now been admitted to normal schools.

The sample size is indeed small and, therefore, the outcome is not really statistically significant. More such efforts need to be done on larger sample sizes to give us better epidemiological data on IEMs in schools for children with mental retardation. This will also help in bringing out a fair number of children languishing in special schools to give them a gift of normal, independent life. Our efforts continue in that direction.

A recent study carried out to detect causes of mental retardation in 2000 children with developmental delay, by a multidisciplinary team comprising a pediatrician, cytogeneticist, biochemist, psychologist, and speech and occupational therapist, ascribed the causes to genetics, environment, and idiopathic. Definitive diagnosis was possible in 1192 (60%) who were classified as cases due to genetic factors, environmental factors, and chromosomal abnormalities. Mucopolysaccharidosis, followed by amino acidopathies such as PKU, tyrosinemia, and alkaptonuria, were the most common causes of IEMs (Bhatt et al., 2008). Our experience of the past 25 years corroborates well with these findings (Apte and Mhatre, 2011).

TABLE 83-1 DISTRIBUTION OF PATIENTS DIAGNOSED WITH IEM (54%)

MPS	37.5%
Biotinidase deficiency	25%
Homocystinuria	25%
Biotinidase deficiency +	12.5%

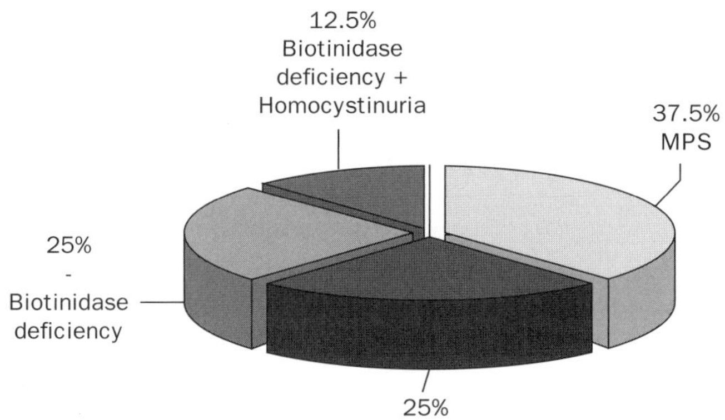

Figure 83-1 Distribution of patients diagnosed with IEM.

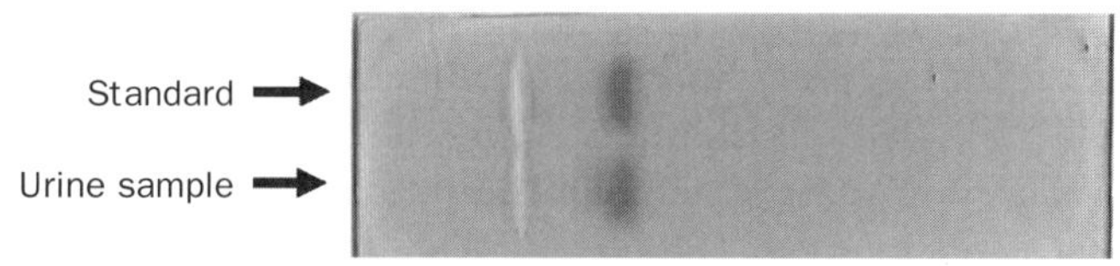

Figure 83-2 A small piece of filter paper is impregnated with 5L sample of patient's urine, and another piece is impregnated with 5L sample containing 10 nanograms of standard heparin sulfate. A strong band in the urine indicates the excretion of mucopolysaccharides in urine. The type of MPS is diagnosed by quantitatively measuring the leukocyte enzymes related to MPS.

A substantial proportion of genetic diseases due to IEM can be managed by dietary restrictions and nutritional supplementation. However, the key to the management of patients is the correct diagnosis of the genetic disease.

We have developed a simple and rapid procedure for the diagnosis of mucopolysaccharidosis (Apte, 2009). Mucopolysaccharidoses (MPS) are hereditary, progressive disorders with an insidious onset, which are caused by the excessive intralysosomal accumulation of glycosaminoglycans (acid mucopolysaccharides or GAGs) in various tissues. GAGs are long-chain, complex carbohydrates consisting of a variety of uronic acids, amino sugars, and neutral sugars. They are usually linked to proteins to form proteoglycans, which form the major constituents of the ground substance of connective tissue. They are present in all membranes of the cell organelles (Apte, 2009).

DIAGNOSTIC SERENDIPITY

We describe below our rapid and sensitive method for the diagnosis of MPS. This is based on the electrophoretic analysis of urine samples on agarose gel, prepared on a regular 7.5 × 2.5 cm microscope slide. The procedure has been described in detail elsewhere (Apte, 2009). The presence of a band of MPS provides definite, positive, and strong evidence for carrying out further analysis of the patient's leukocyte enzymes related to MPS, so that the type of MPS of the patient can be diagnosed. The procedure offers high sensitivity and rapidity, and has enabled us to diagnose a large number of asymptomatic cases with mucopolysaccharidosis. The result of the simple urine analysis provides a strong guideline for undertaking more detailed enzyme investigations of the patient. Figure 83-2 shows a slide with a band of MPS in a patient's urine sample, together with a band of heparin sulfate used as a standard.

The importance of a simple electrophoretic analysis of urine provides an excellent benchmark as a lead to the diagnosis of MPS. The method can detect as low as 20 nanograms of MPS, giving us a clear visible band. The spot tests and the turbidity tests that are being used today in several laboratories require at least ten times more MPS concentration to give a positive result. Both spot tests and turbidity test procedures are very much subjective and, therefore, the result will be totally dependent on personal judgment. We have screened now more than 1000 urine samples each of carriers and normal individuals for the excretion of MPS. In none of them have we seen a band of MPS, thus making our electrophoretic screening procedure full-proof for the detection of MPS only in patients.

NOVEL PROCEDURE FOR ENZYME ESTIMATION

I would now like to describe examples of enzyme deficiencies in patients leading to genetic disease. Our laboratory has developed a novel procedure for the definitive diagnosis of enzymopathies (Apte, 2006, 2009). The procedure clearly delineates a patient from a carrier and from a normal individual. We have been applying this procedure routinely in all our enzyme estimations. The procedure has generality of application to any pathological sample—plasma, serum, leukocyte extract, amniotic fluid, urine, tissue extract, etc. We will describe the procedure again, in brief, as applied to the estimation of biotinidase enzyme in serum.

Biotin is found in a covalently bound form in nature. Biotinidase enzyme is involved in releasing biotin from its bound form and making it available as free biotin. Biotin acts as a cofactor for several carboxylases that are involved in the metabolism of carbohydrates, proteins, and fats (Hymes and Wolf, 1996). Biotinidase has also been shown to have biotinyl transferase activity, resulting in the transfer of biotin from biocytin to nucleophilic acceptors such as histones. Defective biotinidase activity causes multiple carboxylase deficiency (Sander, et al., 1980). This results in ataxia, muscular hypotonia, seizures, alopecia, periorificial cutaneous eruptions, and episodic metabolic acidosis. The enzyme biotinidase is estimated in serum. Blood sera of a 6-month-old patient, P, the patient's father, FP, and the mother, MP, as carriers, and a normal 6-month-old child, NP, were collected.

TABLE 83-2 DECAY OF SERUM BIOTINIDASE ACTIVITY

SUBJECT	% ENZYME ACTIVITY RETAINED NUMBER OF DAYS			
	1	2	3	4
NP	100 (6.4)	90 (5.6)	77 (4.93)	63 (4.03)
FP	100 (3.2)	82 (2.6)	57 (1.8)	35 (1.12)
MP	100 (2.5)	80 (2.0)	55 (1.38)	32 (0.8)
P	100 (3.9)	70 (2.73)	43 (1.68)	5 (0.19)

Normal range of serum biotinidase = 4.3–7.5 units (nmoles of PABA liberated/ml/min). All values in parentheses are actual readings. NP = Normal person; FP = Father of the patient; MP = Mother of the patient; P = Patient

The enzyme activity was measured using synthetic substrate (Wolf et al., 1983; Galjaard, 1980) and followed for three consecutive days. The samples were stored at 10C after the assay each day. The percentage of the activity retained on each day was calculated assuming the activity on the first day of assay as 100% in that sample. The results of the quantitative estimation of biotinidase in all the samples are shown in Table 83-2, and graphically in Figure 83-3.

The enzyme activities in both carriers of the disease, FP and MP, the father and the mother, are lower than the patient, P, to start with. However, the patient retains only 5% of his initial enzyme activity on the fourth day of assay, whereas the patient's father and mother retain 35% and 32% of their initial enzyme activity, respectively, on the fourth day. The normal person, NP, retains 63% of his initial biotinidase activity. In this particular situation, both father and mother would have been diagnosed as patients, and the patient himself would have been classified as a carrier, if the diagnosis had been based on a single point assay readings of these samples.

TABLE 83-3 DECAY OF SERUM BIOTINIDASE ACTIVITY

SUBJECT	% ENZYME ACTIVITY RETAINED NUMBER OF DAYS			
	1	2	3	4
NP	100 (6.4)	90 (5.6)	77 (4.93)	63 (4.03)
P	100 (3.2)	0 (0)	0 (0)	0 (0)

NP–Normal person; P–Patient

Our invention has thus become an asset in such ambiguous situations, providing us with a definitive diagnostic procedure in all our enzyme assays. The procedure was successfully applied to the prenatal diagnosis of MPS by following the decay of enzyme activity in the amniotic fluid (Apte and Bhingarde, 2000).

We have a patient who was absolutely normal on all counts till the age of 35 years. He started developing weakness in the lower limbs with progressive dementia, severe depression, and difficulty in getting up from sitting posture, with reducing muscle tone. The patient finally came down to using a wheel chair. He was a successful engineer with his own expanding business. All his outdoor activities came to a halt. The patient had no history of convulsions, no family history of similar symptoms, no history of diabetes, hypertension, or cardiac problems. His genetic biochemical investigations were revealing. He had low biotinidase activity. The decay kinetics shown in Table 83-3 and Figure 83-4 is striking. Biotinidase activity came down to zero on the

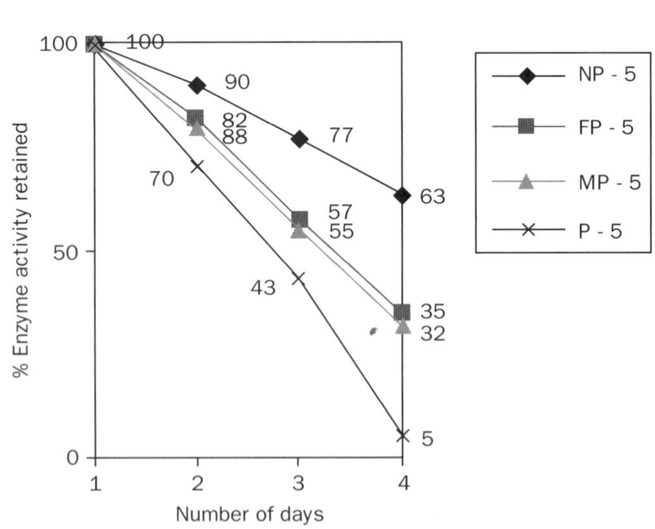

Figure 83-3 Decay of serum biotinidase activity over a period of four days. The enzyme activity obtained on the first day was considered 100%. The enzyme activities obtained on the 2nd, 3rd and 4th days are plotted as percentage of the original enzyme activity retained on that day.

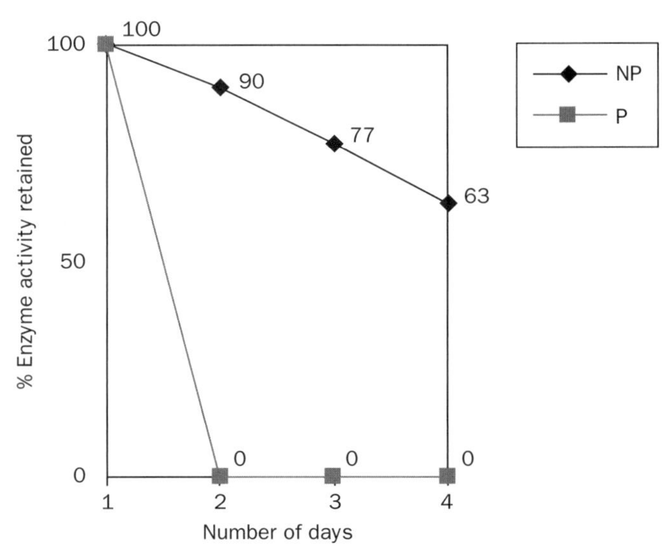

Figure 83-4 Decay of serum biotinidase enzyme activity over a period of four days. The enzyme activity obtained on the first day was considered 100%. The enzyme activities obtained on the 2nd, 3rd, and the 4th days are plotted as percentage of the original enzyme activity retained on that day.

TABLE 83-4 PATIENTS DIAGNOSED WITH APTE PROCEDURE

SERIAL NO.	DISEASE	ENZYME DEFICIENCY	NO. OF PATIENTS
1	Hurler Syn. (MPS I)	α – Iduronidase	27
2	Morquio B Syn. (MPS IV B)	β – Galactosidase	35
3	Maroteux Lamy Syn. (MPS VI)	Arylsulphatase B	112
4	Sly Syn. (MPS VII)	β–Glucuronidase	117
5	Metachromatic Leukodystrophy (MLD)	Arylsulphatase A	93
6	Sandhoff Disease	Hexosaminidase Total	17
7	Tay-Sachs Disease	Hexosaminidase A	21
8	Gaucher Disease	β – Glucosidase	1
9	Pompe Disease	α – Glucosidase	4
10	Glycogen Storage Disease I	Glucose–6–Phosphatase	3
11	Multiple Carboxylase Deficiency (MCD)	Biotinidase	138

second day itself. The normal individual had 90% activity on the second day and retained 63% activity on the 4th day.

We would have missed this patient and several others if we had not applied the Apte procedure for the assay of his biotinidase. This patient was put on biotin (10 mg BD) and has responded very well for the past six years on biotin supplementation and dietary restrictions. The patient is now up and about.

We have encountered several examples of patients with fulminant clinical manifestations of the disease, whose disease-related enzyme activities fall well within the normal range. If we had based our diagnosis on a single point enzyme assay result, these subjects would have been classified as normal subjects. Similarly a large number of carriers would have been diagnosed as patients by the same token as they had enzyme activities below normal range

Table 83-4 shows a list of patients (N = 568) with a variety of enzyme deficiencies who were diagnosed in the last ten years by the application of Apte procedure. A substantial number of them had enzyme activities which were in the normal range. We have now adopted this novel procedure, and follow the kinetics of decay of enzyme activity over a period of four days in all our routine enzyme estimations.

A list of carriers (N = 1129) diagnosed in the last ten years by this procedure is also shown in Table 83-5, in which there were about 30% of individuals who had either very low enzyme activities or enzyme activities that were well within the normal range.

Enzyme assays carried out using Apte procedure gave us a clear picture and a great confidence in the results. This enabled us to delineate between patients, carriers, and normal individuals, and has enabled us to unambiguously diagnose patients and carriers.

TTABLE 83-5 CARRIERS DIAGNOSED WITH THE APTE PROCEDURE

NO.	DISEASE	ENZ DEFICIENCY	NO. OF PATIENTS
1	Hurler Syn. (MPS I)	α – Iduronidase	52
2	Morquio B Syn.(MPS IV B)	β – Galactosidase	67
3	Maroteux Lamy Syn. (MPS VI)	Arylsulphatase B	227
4	Sly Syndrome (MPS VII)	β–Glucuronidase	231
5	Metachromatic Leukodystrophy (MLD)	Arylsulphatase A	207
6	Sandhoff Disease	Hexosaminidase Total	30
7	Tay-Sachs Disease	Hexosaminidase A	33
8	Gaucher Disease	β – Glucosidase	2
9	Pompe Disease	α – Glucosidase	2
10	Glycogen Storage Disease I	Glucose–6–Phosphatase	1
11	Multiple Carboxylase Deficiency (MCD)	Biotinidase	277

SUMMARY

Correct diagnosis of IEM is the key to the successful management of patients with genetic diseases. Our pilot study, conducted on a group of children from schools for mentally retarded and autistic children, has revealed metabolic derangements in more than 50% of the cases. Despite the well-known fact that mental retardation can be caused by defective metabolic processes, investigations to detect them are rarely ordered. Early diagnosis of these disorders can lead to effective management in a substantial proportion of cases. An awareness of genetic entities in the causation of a variety of genetic diseases, especially lysosomal storage disorders as a cause of joint abnormalities in children, is essential for diagnosing the disorders and intervening before permanent musculoskeletal damage occurs. New therapies are available, but their effect on the diseases of bone and connective tissue is slow, so they need to be instituted very early to maximize their effect (Ponder and Haskins, 2007; Rutecki, 2010). The Human Genome Project has provided a great impetus to the science of medical genetics. Development of advanced molecular, cytogenetic, and biochemical techniques could be regarded as spinoff benefits of this project. Metabolomics is an emerging post-genomic science, which is being employed to reinvestigate gene function and the pathogenesis and pathophysiology of the genetic disease states. Metabolomics promises to play a major role in nutritional science for defining dietary regimens for patients with IEM.

Human genetics is in a revolutionary phase worldwide, emerging as a core area of biological sciences and human health. The modern genetics is going to find its way to every nook and corner of the practice of medicine, human evolution, history, anthropology, forensic science, etc. This is going to revolutionize clinical practice in the coming years. Every few weeks, new study reports of a possible link between a gene and human behavior appear. The Indian population, which represents one-sixth of the world population, provides a goldmine for genetic studies.

A broad-based, comprehensive program will diminish IEM-related morbidity, mortality, and handicap to an acceptable level. Even an orthodox country like Saudi Arabia has realized the importance of screening programs, and has now made it mandatory for couples to undergo genetic investigations before conception (El–Hazmi et al., 1995; El-Hazmi, 2006). At the medical level, services developed should have a general appeal. A number of newly developed tests and modalities have much wider applications in the field of IEMs, and physicians should find them instrumental in diagnostic work-up of their patients. Clinical practice in genetics is an intellectually intensive and psychologically rewarding activity. A number of IEMs are still not included in the present screen that is used in our genetic centers. The possibilities are enormous (Apte, 2006). The field is wide open. India provides a variety of clinical material for diagnosis, management, genetic counseling, and research. The future is inviting.

ACKNOWLEDGMENTS

I am thankful to Dr. Rahul S. Kamat, of Vidyasagar Institute of Genetic Studies & Research, for providing me with logistic facilities. I am also thankful to Ms. Samhita Mhatre for her help in the preparation of this manuscript.

REFERENCES

Apte BN. (2009). A novel approach to the definitive diagnosis of enzymopathies. *J Clin & Diagnostic Res 3*, 1381–1388.

Apte BN. (2009). A simple and rapid method for the diagnosis of mucopolysaccharidoses (MPS). *J Clin & Diagnostic Res 4*, 416–420.

Apte BN, Gogate SG. (2005). Population screening for genetic disorders – Jalna Camp Experience. *Bombay Hosp J 47*(2), 217–219.

Apte BN. (2006). Private practice in genetics: Future scope in Mumbai. *Bombay Hosp J 48*(1), 1–7.

Apte BN. (2004). The Management of Inherited Metabolic Diseases in India. In Dr. Dhavendra Kumar (ed.), Genetic Disorders of the Indian Subcontinent (pp. 207–222). Dordrecht, Germany, London, U.K. and Boston, USA: Kluwer Academic Publishers.

Apte BN, Bhingarde S. (2000). Prenatal diagnosis of Maroteaux – Lamy syndrome. *Bombay Hosp J 43*(2), 327.

Apte BN, Mhatre SS. (in preparation). Repertoire of genetic diseases – Our experience of the past twenty years at our genetic clinic.

Beadle GW, Tatum EL. (1948). Genetic control of biochemical mutations in Neurospora. *Proc Natl Acad Sci 27*, 499–503.

Bhatt C, Misra Z, Goyel N. (2008). Detection of inherited metabolic diseases in children with mental handicap. *Ind J Clin Bio 23*(1), 10–16.

Datar CA, Apte BN. (2008). Screening of previously undiagnosed paediatric cases of mental retardation and autism for specific metabolic disorders. *Bombay Hosp J 50*, 560–565.

Dave SS, Fu K, et al. (2006). Molecular diagnosis of Burkitt's lymphoma. *N Eng J Med 354*, 2431–2442.

Devi AR, Rao NA, Bittles AH. (1987). Inbreeding and the incidence of childhood genetic disorders in Karnataka, South India. *J Med Genet, 24*, 362–365.

Devi ARR, Naushad SM. (2004). Newborn screening in India. *Ind J Ped 71*, 157–161.

El-Hazmi et al. (1995). Consanguinity among the Saudi Arabian population. *J Med Genet 32*(8), 623–626.

El-Hazmi MA. (2006). Pre-marital examination as a method of prevention from blood genetic disorders. Comm views. *Saudi Med J*, 27, 1291–1295.

Farber H, Loscalzo J. (2004). Pulmonary hypertension. *N Engl J Med 351*, 1655–1665.

Fiehn O. (2001). Combining genomics metabolome analysis and biochemical modeling to understand metabolic networks. *Comp Funct Genom* 2, 155–168.

Fiehn, O. (2002). Metabolomics: The link between genotypes and phenotypes. *Plant Mol. Biol.* 48, 155–171.

Galjaard H. (1980). Genetic metabolic diseases – Early diagnosis and prenatal analysis. New York: Elsevier/Oxford: North Holland Biomedical Press.

Garrod AE. (1899). A contribution to the study of alkaptonuria. *Proc R Med Chir Soc 2*, 130–135.

Garrod AE. (1908). Inborn errors of metabolism (Croonian lectures). *Lancet 1*(73), 142–214.

Hall P, et al. (2006). Hormone-replacement therapy influences gene expression profiles and is associated with breast cancer prognosis: A cohort study. *BMC Med 4*, 16–24.

Hedenfalk I, et al. (2001). Gene-expression profiles in hereditary breast cancer. *N Eng J Med 344*, 539–548.

Hymes J, Wolf B. (1996). Biotinidase and its role in biotin metabolism. *Clin Chem Acta 255*(1), 1–11.

Kato GJ, Gladwin MT, Steinberg MH. (2007). Deconstructing sickle cell disease: Reappraisal of the role of hemolysis in the development of clinical subphenotypes. *Blood Rev 21*, 37–47.

Kaur M, Das GP, Verma IC. (1994). Inborn errors of amino acid metabolism in North India. *J Inher Metab Dis 17*, 230–233.

Kim J. (2006). Emergence: Core ideas and issues. *Synthese 151*, 547-559.

Kitano H. (2004), Biological Robustness, *Nat Rev Genet* 5, 826–837.

Krause W, Halminski M, McDonald L, et al. (1985). Biochemical and neuropsychological effects of elevated plasma phenylalanine in patients with treated phenylketonuria. *J Clin Invest 75*, 40–46.

Kumta NB. (2005). Inborn errors of metabolism (IEM) – An Indian perspective. *Ind J Ped 72*(4), 325–332.

Loscalzo J, Kohane I, Barabasi AL. (2007). Human disease classification in the postgenomic era: A complex systems approach to human pathobiology. *Mol Syst Biol 3*, 124–132.

Martins AM. (1999). Inborn errors of metabolism: A clinical overview. *Sao Paulo Med J 117*(6), 251–265.

Mendel G. (1865). Experiments in Plant Hybridization. Versuche über Pflanzen-Hybriden. Verh. Naturforsch. Ver. Brünn 4, 3–47 (in English in 1901, *J Royal Hortic Soc* 26, 1–32).

Ponder KP, Haskins ME (2007). Gene therapy for mucopolysaccharidosis. *Expert Opin Biol Ther 7*(9), 1333–1345.

Rutecki GW. (2010). Pre-prenatal care: A primary care primer on the future. *Consult 50*(3), 1–2.

Sander JE, Malamud N, Cowan MJ, Packman S, Amman AJ, Wara DW. (1980). Intermittent ataxia and immunodeficiency with multiple carboxylase deficiency: A biotin responsive disorder. *ANN Neurol 8*, 544–550.

Sebastiani P, et al. (2005). Genetic dissection and prognostic modeling of overt stroke in sickle cell anemia. *Nat Genet 37*, 435–440.

Seidman JC, Seidman C. (2001). The genetic basis for cardiomyopathy from mutation identification to mechanistic paradigms. *Cell 104*, 557–567.

Stacey KHT. (2005). Who is afraid of metabolic diseases? A guide to neurometabolic disorders. National University of Singapore: Dept. of Pediatrics Publication.

Swarna M, Jyothy, Rani PU, et al. (2004). Amino acid disorders in mental retardation: A two-decade study from Andhra Pradesh. *Biochem Genet 42*, 85–98.

Verma IC. (2000). Burden of genetic disorders in India. *Ind J Ped 67*, 893–898.

Verma IC, Saxena R, Lall M, et al. (2003). Genetic counseling and prenatal diagnosis in India–Experience at Sir Ganga Ram Hospital. *Ind J Ped 70*, 293–297.

Verma IC. (2003). Implications of human genome project for pediatric practice. *Ind Ped 40*, 481–485.

Whitfield PD, German AJ, Noble PJM. (2004). Metabolomics: An emerging post-genomic tool for nutrition. *Br J Nutrition 92*, 549–555.

Wolf B, Greer RE, Allen RJ, Goodman SI, Kien CL. (1983). Biotinidase deficiency: The enzymatic defect in late onset multiple carboxylase deficiency. *Clin Chem Acta 131*, 273–281.

84.

GENETIC VARIATION AND GENETIC EPIDEMIOLOGY OF SRI LANKA

Deepthi C. de Silva

INTRODUCTION

Sri Lanka is a lower-middle-income country that has achieved control over its population growth and maintained better health and education indices compared with her neighbors in South Asia. There is a health service that is nominally free at the point of delivery, but underfunding is a problem. The ending of a crippling civil war has resulted in optimism for an economic recovery and improvements in healthcare funding but it remains to be seen if this optimism is fulfilled. Genetics is becoming recognized as an important cause for predisposition to disease, but the level of public understanding about it is limited. Genetic counseling is available both in the state and private sectors. Genetic testing is funded by the patient. There is a problem in funding genetic testing due to the relative high cost, especially for families already economically disadvantaged by having a chronically sick child. The penal code does not allow termination of pregnancy for genetic disease or fetal abnormalities, which poses ethical problems regarding the use of prenatal diagnosis. This chapter aims to review the current state of medical genetics in Sri Lanka, deficiencies that have not been addressed, and the likely future developments in this field.

HISTORY OF SRI LANKA AND THE ORIGINS OF ITS POPULATION

Sri Lanka is an Indian Ocean island of around 62,000 square kilometers. Its close proximity to India and position in the middle of the sea routes between Southeast Asia and Africa explains much of its social and political history.

Historical evidence suggests that the earliest human remains from this country date to around 37,000 BC, with the discovery of the evidence of an anatomically modern man referred to as the "Balangoda man" (www.lankalibrary.com). One of the earliest records of the history of this island is to be found in the *Mahavamsa*, written by Buddhist monks around the sixth century AD. It documents that two tribes existed around the fifth century BC when Prince Vijaya landed on the island. The *Mahavamsa* offers a mythical account of the Indian colonization of the island and suggests that the Sinhalese, the majority ethnic group in Sri Lanka, are the descendants of Prince Vijaya and his companions (De Silva, 1981).

Modern historians believe that the Indo-Aryan colonization of the island occurred in waves, with both warriors and traders sailing down the coast of India. These settlers probably absorbed or chased away the indigenous tribes as they established settlements around the coastal regions.

The Dravidians from South India also had strong links with the island. The multiethnic society that was established in the country was to result in strong social and cultural links between the different indigenous tribes and the Indian colonizers.

The Veddas, a distinct group of Sri Lankans, are possibly the descendants of the original inhabitants of the island (Ellepola and Wickramanayake, 1986). Their hunter-gatherer life style is gradually being changed as a result of assimilation into modern society.

Trading, cultural and social links had also existed between the island and other countries in Asia and Arabia. Sri Lankan Moors and Malays can trace their ancestry to these links.

Over the next ten centuries, the island developed a strong cultural and religious identity with the Anuradhapura and Polonnaruwa kingdoms. There were conflicts, especially from Chola invaders from southern India, and the setting up of a Tamil kingdom in northern parts of the island. The center of power moved toward the hilly central parts of the island and led to the development of the Kandyan kingdom.

European colonization of the island started with the Portuguese, followed by the Dutch and British. Each of these races left an indelible mark on the country. Portuguese names are common among current Sri Lankans, as is the Catholic faith, introduced with much fanaticism to the country. The Dutch introduced the Roman Dutch law and initiated registration of births, marriages, and deaths. Their architecture is still to be found in the forts located in coastal towns and a system of transportation canals. The British were the only invaders to rule both the coastal areas and the previously independent kingdom of

Kandy. Their influence includes the English language, spoken widely in the country, tea estates, and an abiding passion for cricket!

RECENT HISTORY AND CURRENT GOVERNMENT

The island gained its independence in 1948 and became a republic in 1972 on the model of a parliamentary democracy. In 1978, the constitution was altered to create the post of an executive president and a legislature (parliament), both elected by a system of universal suffrage and proportional representation (www.priu.gov.lk). Power is further devolved to eight provinces, each containing a number of districts. The provinces are governed by a locally elected parliament and a governor appointed by the president. Each district has further administrative divisions depending on the size of the population: large towns have municipal councils; smaller towns urban councils; the smallest administrative unit is the *Pradeshiya Sabha*. These three units also have elected representatives. The process of provincial devolution of power is, theoretically, to improve and democratize government.

Following independence, there was rapid change in many aspects of the country. The election in 1956 was a landmark one: the government's policy of "Sinhala only" as the language of government was popular among the Sinhala-speaking majority but unacceptable to the English-educated elite and ethnic minorities. The recent history of the country has been marred by violence, both political and ethnic. There have been two armed, Maoist insurrections supported by Sinhalese youth (in 1971 and 1987–1990) and a much publicized ethnic conflict between the Liberation Tigers of Tamil Eelam (LTTE) and the Sri Lankan government. The civil war by the LTTE from 1978 to 2009, has resulted in the loss of more than 70,000 lives and has caused the disablement of larger numbers of adults and children. This period has also given rise to a large number of refugees within Sri Lanka and abroad. The end of the war was marked by a humanitarian crisis with 250,000 civilians trapped in the war zone. Reconstruction and resettlement of displaced people is currently in progress.

ECONOMIC SITUATION

Since 1978, Sri Lanka has been an open economy with a current GDP of US$49.5 billion and a growth rate of 6% (www.data.wordbank.org). Services including tourism, financial, trade, and transport make up 60% of the GDP, while industry (garment and leather goods, gems and jewellery, food processing, and construction) and agriculture (rice, tea, rubber, and spices) make up 28% and 12% of the GDP respectively. The economy has a high public debt load (around 86% of GDP) with donors including the World Bank and countries such as Japan, Iran, and China. The Asian tsunami in 2004 affected the country with the loss of more than 30,000 lives but the expected impact of a downturn in the economy did not take place. In 2008, the world economic crisis associated with hikes in food and fuel prices resulted in an economic crisis necessitating a loan from the International Monetary Fund (IMF) conditional on economic targets (www.imf.org/external/country).

Sri Lanka is classified as a lower-middle-income country with a per capita GDP of US$1780 (www.wordbanksl.stats). There is a disparity in wealth with a Gini coefficient of 0.4 (a measure of 0 indicates total equality while 1 indicates total inequality) and 23% of the population lives below a government-defined poverty line (www.data.wordbank.org). The disparity in wealth is especially marked between the different parts of the country with the areas affected by the civil war and the estate sector being more affected than the rest of the country. The unemployment rate is 5.2% (www.statistics.gov.lk). The ending of the costly civil war with aid-funded rehabilitation and infrastructure development programs is expected to enable continued economic growth.

The total health expenditure is 4.9% of total expenditures and 1.2% of GNP. This is around US$21 per capita of government expenditure on health. A significant proportion of this expenditure was for salaries (Annual Health Bulletin, 2007).

DEMOGRAPHY OF SRI LANKA

ETHNIC GROUPS AND RELIGIOUS AFFILIATIONS OF SRI LANKANS

Sri Lanka has a population of 20.01 million with a ratio of 97 males to 100 females (Annual Health Bulletin, 2007). It is a densely populated country with around 309 people per square kilometer. Sri Lanka is a multiethnic and multireligious society, and most of the main cities have all ethnic groups living in them. In spite of the very real ethnic tensions that have resulted in international publicity and criticism, at an individual level the ethnic groups are able to live in relative harmony.

Census data divides the population into different ethnic categories. The Sinhalese make up around 76.4% of the total and include both low country and Kandyan people (the two categories were separated in previous census data). They are the majority ethnic group in most parts of the country except the north and east. The Tamils are separated into Sri Lankan Tamils (11.1%) and Indian Tamils (1.17%). The Indian Tamils are the descendants of workers brought to the island during British rule as manual labor, and they continue to work mainly on the tea estates. Sri Lankan Moors constitute the third category, making up around 9.1% of the population. The descendants of Europeans include Burghers (descended from the Dutch) and Eurasians, and they comprise 0.72% of the population. No census data is available on the current Vedda population, whose numbers and traditions are dwindling (www.statistics.gov.lk).

Most of the world's religions are practiced in the country. Buddhism is the most widely practiced, with around 15 million (70% of population) adherents. This includes many Sinhalese and some Tamils. The Buddhists follow the Theravada doctrine. Hinduism is practiced by around 9% of the population, the Tamils being the main ethnic group following this religion. Islam is practiced by 11% of the population and is the main religion among the Sri Lankan Moors and Malays. Christianity is practiced by around 10% with all the main denominations represented. Most Christians are either Sinhalese or Tamils. Christian denominations include a long-established Catholic and Protestant tradition and a more recent increase in the numbers of evangelical Christian sects (www.statistics.gov.lk).

POPULATION GROWTH AND ITS CONTROL

Sri Lanka is the first South Asian country to achieve replacement level fertility (RLF), when the total fertility rates (TFR) are between 2.1–2.2/woman, with a total fertility rate of 1.9 for the period 1995–2000 (Annual Health Bulletin, 2000). The current population growth rate is 1.1%. Fertility control has been an important part of health policy since the early 1970s, as it was recognized that the high birth rate would reduce the general level of health of the population. The initial decline in fertility was related to postponement of the age at marriage and, later, to the use of contraceptives to increase the gap between children and to limit family size (De Silva, 1994). Although abortion is illegal (except when the mother's physical health is at risk) there is evidence of its widespread use (Rajapakse, 2002).

The reduction in TFR is associated with an expected rise in the proportion of the older population in the twenty-first century. A rise in the incidence of chronic degenerative diseases and neoplasms is also predicted.

The reduction in the TFR and improvement of health indices is attributed partly to higher levels of education, especially of women. Sri Lanka has a high literacy rate with 92.2% males and 89.2% females over the age of 10 years being able to read and write at least one language. Free education is available to the tertiary level, and there are a large number of private schools providing primary and secondary education in the main cities. There is an increase in the number of private institutions offering tertiary education including a private medical school. Under 10% of the population has never been to school (7.6% of males and 4.6% of females), while nearly 22% have not gone beyond primary education (Annual Health Bulletin, 2007).

MEDICAL STATISTICS IN SRI LANKA

In spite of its size and relatively low gross national product, Sri Lanka has some of the best health indices among the developing countries (Annual Health Bulletin, 2007). The life expectancy at birth is 76.4 and 71.7 for females and males, respectively. Its crude birth and death rates are declining (18.9 and 5.8/1000) as are the neonatal and infant death rates (8.4 and 11.17/1000 live births). The maternal mortality rate is 14.3/100,000 live births.

Primary healthcare is well organized with high attainment of immunization targets. In infants, diphtheria, pertussis, tetanus (DPT), polio, and BCG immunization rates are 99%. All pregnant mothers have immunity to tetanus (Demographic health survey, 2007, www.statistics.gov.lk). There have been no recorded cases of polio and diphtheria since 1996 (www.epid.gov.lk).

Around 87% of all infants up to 3 months of age are exclusively breastfed. Malnutrition remains a problem, with 15.7% of children under 5 years being affected by chronic malnutrition (height for age less than –2SD below expected); 15.8% are acutely malnourished (weight for height <-2SD than expected) and 25.3% have evidence of acute and chronic malnutrition (weight for age is low).

MORBIDITY AND MORTALITY IN SRI LANKA

The changing demographic pattern and improvements in primary healthcare have resulted in a reduction in morbidity and mortality rates. The death rates from infectious diseases have declined from 23.9 per 100,000 to 12.3 per 100,000 between 1980 and 2007 (Annual Health Bulletin, 2007). The decline in deaths from preventable and easily treatable infections has been replaced by increases in the incidence of other infections such as dengue hemorrhagic fever. The incidence of malaria has fallen from 678.9 to 5.2 cases per 100,000 between 1990 and 2007. Morbidity and mortality from vascular diseases and neoplasms has increased. Ischemic heart disease is the commonest cause of death (10%) and cerebrovascular disease accounts for 9% of all adult deaths. Malignant neoplasms cause 7.5% of deaths in adults. There has been an increase in alcohol-related cirrhosis and all smoking-related diseases, mainly among males. Endocrine and metabolic disease-related morbidity is also rising. The deaths related to congenital malformations, deformations and chromosome abnormalities have risen from 32.3 to 63.9/100,000 population. Among infants under one year of age, perinatal problems are the commonest cause of death (799/100,000) while congenital malformations, deformities, and chromosomal abnormalities account for 89.5/100,000 deaths. Among older teenagers and young adults, external injury (including war injuries) and self-harm are the most important and preventable causes of death and the death rates are again higher among males than females (Annual Health Bulletin, 2000 and 2007).

HEALTH POLICY AND THE ORGANIZATION OF THE HEALTHCARE SYSTEM

The broad aim of the health policy is to "increase life expectancy and improve quality of life." The main priorities include controlling preventable diseases and health promotion. They also intend to address health problems like "inequalities in health service provision, care of the elderly and disabled, noncommunicable diseases, accidents and suicides, substance abuse and malnutrition" (Annual Health Bulletin, 2000).

Sri Lanka boasts that free, Western-type health care is available for any individual within a 4.8 km distance from his or her home. Inpatient care is mainly in state hospitals, but there is a large and growing private sector in most urban areas. Around 60% of patients seek outpatient care in the private sector, while 10%–15% seek inpatient care. The private hospitals have a bed capacity of 8000 but probably need to expand by another 3000 to deal with the demand (http://www.dailynews.lk). There is an increasing market of private health insurance, both self funded and employer funded.

There are also other local government and armed forces hospitals providing Western medicine. General practitioners (all self-employed and independent of the state) probably handle around a quarter of the primary care consultations per year. Indigenous medicine, especially Ayurveda, remains popular, and there are both state funded and private Ayurvedic hospitals. Other healthcare systems available include Unani, Siddha, and homeopathy.

There are three levels of curative care:

1. Central dispensaries, maternity homes, rural and district hospitals as primary care institutions.
2. Base and provincial hospitals as secondary care institutions
3. Teaching hospitals and specialist hospitals (maternity, pediatric, and cancer) as tertiary care institutions

The tertiary care institutions are managed by the central government, while local government is responsible for primary and secondary care hospitals. There is no referral system and, as a result, there is underutilization of some hospitals while others (especially the tertiary care institutions) remain overcrowded.

Health promotion and prevention of disease is organized by around 250 Medical Officers of Health (MOH) and Divisional Directors of Health Services (DDHS). Family health workers provide care for individual families, especially targeting women and infants. They act as a link between the community and primary care clinics.

Most of the births occurring in the country are within hospitals, and around 94% of all births are in the state sector. There are a few specialist maternity hospitals in the country, and these and the other teaching hospitals account for up to a third of all deliveries. There are two specialist pediatric hospitals, but separate pediatric wards are available in all the teaching and provincial hospitals. There is an expansion of facilities for the care of neonates as well as other specialties. (Annual Health Bulletin, 2007).

MEDICAL AND PARAMEDICAL STAFF TRAINING

There are eight Medical Faculties in the country: Colombo, Kelaniya, and Sri Jayawardenapura (Western Province), Peradeniya (Central Province), Ruhuna (Southern Province), Jaffna (Northern Province) and two new Faculties in Anuradhapura (North Central Province) and Batticaloa (Eastern Province). Around 1000 doctors graduate each year. Genetics forms an integral part of the undergraduate curriculum, usually in the preclinical years. There are ten nursing schools providing training for staff nurses (3-year course) and family healthcare workers (1-year course). A recent development has been the commencement of undergraduate degrees in nursing. There is only one dental school. The government also trains medical laboratory technicians, pharmacists, physiotherapists, occupational therapists, radiographers, and public health inspectors (Annual Health Bulletin, 2007). Degree courses are also run by some universities for paramedical training. A few private-sector hospitals also train nurses for their institutions.

There are around 55.1 doctors and 6.6 dental surgeons per 100,000 people. The number of nurses and public health midwives is 157.3 and 30.8 per 100,000 people, respectively. There is a shortage of nursing and paramedical staff in most hospitals across the country. There is misdistribution of specialist services with some of the larger cities being well provided, while there is a paucity of specialists in other parts of the island. There is a problem of many doctors and nurses leaving the country to work abroad (Annual Health Bulletin, 2007).

MEDICAL GENETICS IN SRI LANKA—RESEARCH AND CLINICAL APPLICATIONS

NEED FOR GENETICS AND LIMITATIONS TO ITS DEVELOPMENT

Sri Lanka has the infrastructure to control many common causes of morbidity and mortality observed in other developing countries. As a result, the patterns of morbidity and mortality are also in transition, with a rise in congenital anomalies and noncommunicable diseases associated with a decrease in preventable or easily treatable diseases. The emphasis of health policy is likely to be targeted to the common causes of neonatal and infant deaths including perinatal insults and congenital abnormalities. In adults, the emphasis will be on the common noncommunicable diseases.

It is likely that there will be a need for research and application of current genetic knowledge to identify the role of genes and their interaction with environmental risk factors in the causation of these common problems in Sri Lanka.

Data on the prevalence of genetic diseases in Sri Lanka is scanty. No individual genetic disease is classified in the morbidity and mortality figures. Research data is also very limited, partly due to the difficulty of tracing notes within hospitals and incomplete classification of diseases following discharge. Disease databases are rare: there is a cancer registry in the main cancer hospital in Maharagama, but this is still not complete (de Silva, 1999). More recently, a register of patients with thalassemia has been started in the Kurunegala (de Silva, 1996), and this has also been extended to other parts of the country.

Population-based data collection using the family health workers' (FHW) records is also fraught with difficulty, as most genetic disorders are not common and ascertainment may not be complete, especially in conditions where there is marked variability in clinical phenotype or in those which present in adult life. Population-based prevalence studies are possible for genetic conditions where there is a known high prevalence in a particular area (for example, thalassemia in the town of Kurunegala and its surrounding areas) or the genetic disorder is common or relatively easy to recognize (as is the case in Down syndrome). The developments in the care of thalassemics in Kurunegala have been stimulated by research on the genetics of the topic, and similar studies may be of benefit in the future.

The current state of research and its application for the diagnosis and treatment of patients in Sri Lanka will be reviewed.

POPULATION GENETICS

Research on Genetic Polymorphisms in Sri Lanka

The genetic variations in the different ethnic groups have been investigated using hematological, biochemical, and DNA polymorphisms. The nearly extinct race of Veddas has also been a source of interest for geneticists.

The initial interest in population genetic polymorphisms was in the study of blood groups. This is of significance for the blood transfusion service but the data generated was also used for paternity testing. De Zoysa (1985) reported on the prevalence of the ABO and rhesus groups in the country. She observed that there were variations in the prevalence of the ABO groups among the ethnic groups.. Group O was present in 46%, 39%, and 42%, group B was present in 26%, 33%, and 28%; Group A in 22%, 21%, and 23%, and lastly, group AB, the rarest was present in 5%, 6.8%, and 6.1% respectively of Sinhalese, Tamils, and Moors. There was less variation within the population with regard to Rhesus status, and 4.66% were Rhesus negative.

Genetic polymorphisms in the Vedda population indicate that in spite of the morphological similarity with the hunter-gatherers of Australia and the Far East, genetically they are closer to tribes in South India (Roychoudry, 1984; Saha, 1988, 1995). Comparing the Vedda and Sinhalese populations, Ellepola and Wickramanayake (1986) studied 16 polymorphic and 6 monogenic red cell and plasma loci and performed morphological studies. They suggest that genetic divergence between these two groups is likely to have occurred between the populations around 30,000 years ago.

The genetic variations between the Sinhalese and Tamils have also been studied. (Jayasekera et al., 1994, Mastana et al., 1993). Using serum protein markers, Jayasekera et al. (1994) and Mastana et al. (1993) reported that there is heterogeneity between the ethnic groups of Sri Lanka. Papiha and colleagues (1996a), using DNA polymorphisms, found little difference between the Sinhalese and Tamil populations. The study found that the Sinhalese had a slight affinity to the population of the northwest of India, but this was not as high as the affinities between the different groups from within India. Within the population of Sri Lanka, the Malays and Burghers show more similarities to the local Sinhalese and Tamil genetic polymorphisms than to their population of origin (Papiha et al., 1996b). Kshatriya (1995) also noted that the Sinhalese and the Tamils are closer genetically to each other and to the populations of South India. In particular, there was little genetic affinity between the Sinhalese and the Gujarati, Punjabi, or Bengali. There also was little affinity to the Vedda population.

HLA subtyping has revealed that the Sinhalese showed a preponderance of the HLA-A*0211 allele in a reported frequency similar to the north Indian population, and this in conjunction with other HLA frequencies (A, B, C, DR and DQ alleles) was interpreted as indicating a north Indian origin for the origin of the Sinhalese (Malavige, 2007).

Genetic Polymorphisms with Significance to the Practice of Medicine

A study of thalassemic patients in Sri Lanka has reported that gallstones and hyperbilirubinemia were particularly common in HbE/thalassemia individuals. They noted that the promoter sequence for the enzyme UGT1A1 involved in hepatic conjugation of bilirubin had alleles of differing sizes. Individuals who were homozygous for a 7 base pair repeat (compared with the 6 bp wild type or compound heterozygotes) were more likely to be affected with jaundice and gallstones. They also noted that this haplotype had a high frequency in Sri Lanka, India, and Bangladesh, but was rare in Southeast Asia, with Europeans and Africans having an intermediate frequency (Premawardhena, 2001, 2003).

Rochette et al. (1999), studying the origins of the genetic hemochromatosis gene, found that in Sri Lanka, the H63D mutation was present in three new haplotypes and the C282Y in one new haplotype, indicating an independent origin for both mutations in the island. The significance of this finding with regard to genetic hemochromatosis in the population (which is rare) remains to be elucidated.

The gene CYP2D6 has also been studied with regard to the slow and fast metabolizers of debrisoquine and mephenytoin (Weerasuriya et al., 1994). The mutant allele was present in around 9% of the population and the rate was half compared to that reported in Europeans. Similarly, acetylator phenotype studies have shown that 63% are slow acetylators, a figure similar to but slightly lower than that reported among Indians (Perera and Weerasinghe, 1987).

Polymorphisms of the epidermal growth factor gene were found to be associated with fetal growth restriction among the infants of a group of Sinhalese and Western European women (Dissanayake, 2007). Homozygosity for the EGF 2566G>A polymorphism was associated with increasing the risk of preeclampsia (Dissanayake 2009)

Dissanayake et al. (2009a) studied three genes (methylene tetrahydrofolate reductase, Factor V, and prothrombin) where polymorphisms have been associated with thrombophilia in other populations. Studying 240 Sri Lankans, they report the frequencies of the polymorphisms in this group and suggest that this would be useful for planning future tests offered by a clinical genetics service.

The use of short tandem repeat (STR) polymorphisms in autosomal loci have been reported by two groups studying its forensic uses among the different ethnic groups in Sri Lanka (Illeperuma, 2008; Manamperi 2009). A minimum of 9 STR loci (CSF1PO, TPOX, THO1, D16S539, D7S820, D13S317, vWA, FESFPS, and F13B) were found to have a matching probability of 3.67×10^{-10} and a power of discrimination of $1–3 \times 10^{-10}$ (Manamperi, 2009). Illeperuma (2008), using 2 additional STR polymorphisms, reported that the combined power of exclusion was greater than 0.99999 in all four ethnic groups that were studied, and that a combined genetic database could be used for DNA based identification of all ethnic groups in the country.

SINGLE GENE DISORDERS

The commonest single gene disorder that has been extensively investigated in Sri Lanka is thalassemia. The commonest form is beta-thalassemia, but around 15% have alpha globin mutations causing alpha + thalassemia. Up to 40% are estimated to be compound heterozygotes for HbE and a beta-thalassemia mutation, with a resulting phenotype being variable (de Silva et al., 2000a). The distribution of thalassemia in Sri Lanka varies, with most cases reported in the northwestern, north central, and Uva provinces, which also happen to be areas where malaria is endemic. It is estimated that there are around 2000 current cases and a birth rate of around 100 new cases each year (de Silva et al., 2000a).

A random population screening study among school children has reported a carrier rate varying between 1%–5% for β-thalassemia and 0%–2% for HbE, with an overall prevalence of thalassemia carriers of 2.2% and HbE carriers of 0.5% (de Silva, 2000a; Perera, 2000). In Kurunegala, a carrier rate of 12% was reported using a screening test, but ascertainment bias is possibly responsible for the high rate of carriers (de Silva, 1996). There have been 23 beta-thalassemia mutations described in the Sri Lankan population (de Silva, 2000b) but of these, two are common, with 70% being homozygotes or compound heterozygotes for IVS1-5(G→C) or IVS1-1(G→A). Individuals who are affected with these mutations develop a transfusion-dependent type of thalassemia. The third common mutation was the codon 26(G→A) mutation causing HbE. In combination with another beta globin mutation, HbE can cause a more variable thalassemic phenotype.

The management of thalassemic patients in the Kurunegala has been given great impetus by the associated research project on the genetics of this disorder. The management has been streamlined with registration of all cases, a dedicated area within the hospital for the patients, a patient-friendly atmosphere in the wards, and the creation of a donor roster. As a result of this, the patients have regular transfusions, their growth and psychological states have improved, and the stigmatization of the affected child and his family has been reduced (de Silva, 1996).

In spite of all that can be achieved by a dedicated team, concern has been expressed about the burden this condition will pose the healthcare sector in the country (de Silva, 2000a; Perera, 2000). The cost of transfusions and treatment with iron chelating therapy is high. There is increasing publicity regarding thalassemia, and attempts are being made to increase public awareness of this condition. The present strategy is to identify carriers (using cascade screening in affected families and voluntary screening for individuals from high-risk areas). Screening is performed using high-performance liquid chromatography (HPLC), and this is intended to target individuals before marriage so that they can be offered counseling regarding their carrier status. Advice is given by medical personnel regarding avoiding marrying carriers (http://www.sundayobserver.lk/2009/12/20/spe40.asp). Both the carrier testing and counseling are offered free. The uptake of cascade screening is low. The minister of health has stated that he supports compulsory screening and prevention of marriage between carriers; this is unlikely to be implemented.

There are no prevalence figures in Sri Lanka for any other single gene disorder. In a series of 70 consecutive cases referred to a genetic clinic (de Silva et al., 1998a) a single gene disorder was queried in 11 (16%) patients. This low rate may reflect the fact that some common disorders like thalassemia or hemophilia are unlikely to be referred. Referred conditions included muscular dystrophies, myopathies, possible cases of cystic fibrosis (rare in Sri Lanka), Fanconi anemia, and an unknown peroxisomal disorder.

A study of dystrophin gene deletions in 24 patients with X-linked muscular dystrophy reported that deletions were identified in 62.5% of cases and that 82% of the deletions were on the deletion hot spots of dystrophin (Welihinda, 1993).

The only study regarding Huntington disease (HD) is presented in an abstract from a poster presentation (Nanayakkara, 2008a). Mutation test findings from 20 HD patients were reported [9 (43%) male, mean age 45 years (range 15 to 64)]. An expanded allele (>39 CAG repeats) was reported in 18 cases, while two had expanded reduced penetrance alleles (36–39 repeats). The testing was performed at a private laboratory in Colombo and no more clinical information was available in this abstract.

Nanayakkara (2008b) also presented data from testing for fragile-X syndrome using FMR1 methylation. They reported that of the 12 males clinically suspected of fragile-X syndrome, only 2 were found to be positive. No clinical data were given to evaluate these findings.

A study of 257 patients referred for immunological assessment (de Silva, 2007a) reported that 20 had an identifiable immune disorder (8 cases of agammaglobulinaemia, 2 cases each of ataxia telangiectasia (AT), IgA deficiency and hyper IgE syndrome, 3 cases of common variable immune deficiency, and one case each of Griscelli syndrome, hyper IgM syndrome, and X-linked severe combined immunodeficiency). Apart from reporting that there were two familial cases of agammaglobulinemia, no other genetic information was given. They do not report that genetic testing had been performed to confirm AT.

Yu Lok (2009), investigating iron overloading among Asians, reported that HFE gene related hemochromatosis is rare. In a larger report of mutations in other iron-overloading associated genes, a 17-year-old Sri Lankan male was found to be homozygous for a hemojuvelin gene mutation causing diabetes, hyperpigmentation, and abnormal liver function.

There are a number of reports of single gene disorders, mainly consisting of a single case or family.

CHROMOSOME DISORDERS

Data reported from the Colombo Medical Faculty laboratory for the period 1983 to 1987 have been published (Jayasekera, 1988). Out of 409 referrals, 58 were found to have chromosomal anomalies. These include 54 (71%) with nondisjunction Down syndrome (DS) and 2 with translocation DS (one each with a 13/21 and 14/21 translocation). There were also single cases of trisomy 13 and trisomy 18 in this series. Sex chromosome anomalies reported included 9 (11.9%) cases of Turner syndrome (TS), and 4 cases of Klinefelter syndrome (KS). There were also two cases of Turner syndrome and one case of mosaic Klinefelter syndrome on lymphocyte analysis. Data from this study also report that 93% of cases of nondisjunction Down syndrome had parents who were under 35 years of age.

The same laboratory (Dissanayake and Jayasekara, 2008) has reported their data regarding 615 cases of Down syndrome identified in the period 2002–2007. This consisted of 89% cases with nondisjunction, 10% with mosaicism, and 0.7% with translocation (4 cases of a 14;21 translocation and 1 case of 21;21 translocation).

In a study of 442 cases with >2 pregnancy losses, subfertility, and failed assisted reproduction, Dissanayake et al. (2009b) identified 42 (9.5%) with chromosome anomalies. This included mosaic aneuploidies of the autosomes or sex chromosomes (found interestingly in the peripheral blood of adults), polymorphisms of chromosomes (9qh+, 21ps+, Yqh+, 13ps+, 14ps+, 15ps+, 22ps+), which were reported to be commoner among this group, and structural anomalies (deletions of chromosome 1[q42-44], chromosome 12[q13-q14], duplication of chromosome 4[q12-21], two translocations, and an inversion) in the presumably otherwise healthy adult cases who were investigated.

Y-chromosome microdeletions affecting the azoospermia factor a (AZFa), b (AZFb), and c (AZFc) were studied in oligozoospermic and azoospermic males in Sri Lanka (Fernando, 2006). Seven out of a total of 96 infertile males had deletions affecting one or more of these genes, reflecting the pattern reported from other countries.

In a series of 70 patients referred for genetic counseling, 59 required cytogenetic investigations to confirm or exclude a chromosome anomaly (de Silva, 1998a). Of the ten samples that were tested as part of a research project, 4 cases of nondisjunction DS and two siblings with a 13/21 translocation, and a case of Wolf Hirschorn syndrome were confirmed.

Buccal smears have been used to determine the presence or otherwise of an X chromatin body in buccal cells. De Mel et al. (1992) attempted to use this test to diagnose Turner syndrome among a group of girls with severe short stature, identified by screening a cohort of school entrants. Buccal smear was performed in 61 out of 97 selected girls (16,100 school entrants had their height assessed). Nine children were found to have 30% or more cells with an absent X chromatin body, but subsequent karyotyping was normal in all these cases and no mosaicism was noted on lymphocyte analysis. The low prevalence rate of Turner syndrome may indicate poor case selection for testing, overdependence on the percentage of chromatin negative cells, or a true low prevalence.

Wickramanayake (1992) attempted to identify X-chromosome anomalies using the buccal smear in a group of moderate-to-severe intellectually impaired children (n=104, out of whom 10% could not be tested), hospital-based psychiatric patients (n=484, 233 males) and a group of medical students (n=143). There were no anomalies in the intellectually impaired group or in the undergraduates. Among the psychiatric patients, one male patient was found to have a positive test suggesting Klinefelter syndrome, and one female each was found to have either absence (Turner syndrome) or a double positive test (indicating triple X syndrome).

SYNDROME DIAGNOSIS

As with the single gene disorders, there is no prevalence data, even on syndromes that are easily recognized. This includes conditions that are commonly found and diagnosed on clinical grounds; for example, Noonan syndrome, or more investigation-based diagnoses like 22q11 microdeletion associated velo-cardio-facial syndrome.

CONGENITAL ANOMALIES

Data on congenital anomalies is more easily available but, again, prevalence data is lacking. There were low rates of rubella immunity in women of the reproductive age group, with 43% of women in the reproductive age group being non-immune (Mendis, 1989). The incidence of congenital rubella is likely to reduce as the current immunization program does include measles and rubella vaccination for all children at 3 years and repeated for all girls between 10 and 15 years. Non-immune women between 15–44 years are also given rubella immunization. (www.epid.gov.lk). There were 8 cases nationally of congenital syphilis reported in 2000 (Annual Health Bulletin). This is partly due to at-risk women not receiving antenatal care and poor coverage with serum screening. Folate supplementation is offered to women attending antenatal care, but often this is not given during the periconception period.

The Birth Defects Research Unit based in Kandy has examined 20,000 consecutive births in two teaching hospitals of Kandy during September 1996 to February 1998 (de silva, 1998b). They report the incidence of central nervous system malformations to be 3.7/1000 live births. These include 17 anencephaly, 5 encephalocele, 1 iniencephaly, 7 microcephaly, 21 hydrocephaly and 15 spina bifida cases. Limb defects were noted in 8/1000 live births, genitourinary abnormalities in 5/1000 live births, and cleft lip and/or palate or other gastrointestinal anomalies in 2.6/1000 live births. Risk factors for congenital anomalies recorded include parental consanguinity, low socioeconomic status (also associated with consanguineous marriage) and non-use of folate supplements in the periconception period.

Gunasekera and Gunasekera (1998) reported a series of 3745 consecutive births (including 55 stillbirths) at a teaching hospital and noted that congenital anomalies were present in 50 cases (including 4 stillbirths). Central nervous system anomalies (24%), gastrointestinal tract anomalies including cleft lip and/or palate (20%), dysmorphic syndromes (14%), and cardiovascular anomalies (6%) were found.

Congenital heart defects (CHD) are an obvious abnormality to study, as these cause significant morbidity and mortality. In a study of admissions to the Professorial Paediatric Unit at the Lady Ridgeway Hospital for Children, 102 admissions with CHD were studied (Wickramasinghe, 2000). Of these children, 52% were male and 69 had acyanotic CHDs. Nineteen cases died during the course of the study, while only 12 cases had operative correction. The lack of adequate resources to offer prompt therapy is likely to have contributed to this death rate. Twenty-two cases in this study were considered to be syndromic, on the basis of the presence of other abnormalities. They included 8 patients (7.8%) with Down syndrome and 3 cases of congenital rubella. The delay in identification and treatment causes CHDs to be an important cause of morbidity and mortality in adults. Jayasinghe and Walloopillai (1972) reported that there were 1296 CHD cases over the age of 40 years presenting between 1966 and 1972. Septal defects and PDA were the commonest problems but others included tetralogy of Fallot and pulmonary stenosis. The presence or otherwise of other congenital anomalies or a family history were not recorded. Thenabadu (1982) presented a larger review but the contribution from syndromic CHD or the presence or otherwise of a family history was not recorded.

Amaratunga and Chandrasekera (1989), in a study on 51,542 live births and 5263 stillbirths, reported an incidence of cleft lip and/or palate to be 0.83 per 1000 births. Cleft palate alone accounted for 0.19 per 1000 births. There was a positive family history in 19.6% of cleft lip/palate, while the rate was lower for cleft palate alone (9%). The prevalence was higher among Moors, but no other associated socioeconomic factors were identified. These rates are probably underestimates due to the cases being selected from a tertiary center.

The incidence of congenital dislocation of the hip was reported to be low in a questionnaire survey of pediatricians, general surgeons and orthopedic surgeons, although some comment that carrying babies on the mothers' side with the hips abducted (which is thought to reduce this condition) is not common in Sri Lanka (Corea, 1992).

In a survey of 226 visually impaired or blind children, 25% were found to have microphthalmos and 17% had cataracts, but no mention is given to the possible etiology or estimation of the proportion likely to be related to genetic factors (Eckstein et al., 1995).

COMMON MEDICAL DISORDERS: LIKELY GENETIC CONTRIBUTION IN THE SRI LANKAN POPULATION

Asthma

In a case control study on childhood asthma (300 cases and age-matched controls), Karunasekera (2001) reported that there was a significant association of this condition with either a father or sibling who had asthma or a family history of atopy. A relative risk ratio was not given.

Ischemic Heart Disease and Other Vascular Diseases

Mendis and Ekanayake (1994) estimated the prevalence of ischemic heart disease among 975 males between 35–49 years. ECG changes and symptoms were present in 16/1000, history suggestive of disease 54/1000, and ECG changes only 32/1000. Interestingly, Mendis (1991) reported absence of ischemic heart disease and a low prevalence of hypertension in the Vedda population.

Polymorphisms in the cholesteryl ester transferase protein, apolipoprotein and Apo B gene have been studied with no significant association found (Mendis, 1990, 1991). Hyperhomocysteinemia has been associated with an increased risk of hypertension and coronary artery disease (Mendis, 1999, 2002).

Diabetes Mellitus

In Sri Lanka, the prevalence of diabetes is between 2.5% to 5.02% in the general population (Fernando et al., 1994; Illangasekera, 1993). Impaired glucose tolerance is also reported to be between 5.27% and 8%. Higher figures are reported from the urban and suburban areas compared to the rural areas.

A questionnaire survey of 1000 type 2 diabetics (502 males, mean age of onset 47 years; de Silva et al., 2002) reported that 37% had a parent with diabetes. In those with an affected parent, the age of onset was younger, there was a significant excess of maternal transmission, and other siblings were more likely to be affected.

A study investigated 994 diabetics <45 years for the presence of the mitochondrial 3243A>G mutation causing maternally inherited diabetes and deafness (MIDD). Nine were found to have this with >5% heteroplasm giving a prevalence of 0.9%. The identified cases had a younger age of onset (25.9 +/– 4.8y compared with those without a mutation whose mean age was 31.9y +/– 5. 6) and they were more lean with a body mass index (BMI) of 18.7 +/– 2.7 kg/m^2 compared with the mutation negative cases whose BMI was 24.7 +/– 4.0 kg/m^2. A maternal family history of diabetes, and a personal or maternal family history of deafness was found in 55% of the identified cases (Katulanda, 2008).

Psoriasis

Psoriasis is reported to have a 0.4% prevalence rate in the population, and a lower age of manifestation is associated with a positive family history of the condition (Gunawardene, 1978).

Cancer

The mortality rate from cancer has risen from 27.9/100,000 in 1985 to 43.6/100,000 in 2003, with 45% of all diagnosed cases being in males. Among males the commonest malignancy is that of the lip, oral cavity, and pharynx (20%), followed by bronchial (11%) and esophageal (8%) malignancies. Colorectal cancer accounts for 6.8%, while prostate cancer accounts for 5% of all malignancies in males. Among females, the commonest malignancy is of the breast (25%) followed by cervix (12%), ovary (8%), and colorectum (4.8%).

There is a cancer registry that collects data from all cancer units in the country, and there will be an attempt to make this a population-based database in the future. There are no national cancer screening programs, but private hospitals and others are now offering such services. Such interventions are targeted to the relatively affluent and are unlikely to reduce the national morbidity or mortality figures. There is an attempt to offer cervical cancer screening with a vaginal inspection to detect an early cancer (rather than a premalignant lesion) until the infrastructure is available to develop a PAP smear based screening program (Kumarasinghe, 1999). Mammography and colonoscopy are now available in most teaching hospitals, where they are used mainly for diagnostic purposes. Targeted screening for breast or colorectal cancer is likely to be practiced in some state and private hospitals for people who are perceived to be at high risk because of a past personal or family history of cancer.

Population prevalence data and assessments of lifetime risk are not available for most common cancers. There is also no information about the significance of the family history and its impact on altering the lifetime risk if a close relative is affected with cancer. Data used to assess high-risk individuals are from Northern European or American populations. Case control studies should be possible, but there may be underreporting of a family history, and case verification is difficult as record tracing is nearly impossible in most hospitals. Death certificate data may not be entirely accurate, as the underlying diagnosis is often not recorded.

There are some reports that attempt to assess details of early-onset cancer cases. De Silva, (2000b) reports a case control study of 60 colorectal cancers in patients less than 40 years of age, and 245 colorectal cancers from patients over 40 years of age. There were no differences in the duration of symptoms prior to diagnosis, presenting features, site of lesion, or Dukes stage at diagnosis. There were significant differences in histology, with 13.3% having mucoid and 5% having a signet ring appearance. Two out of the sixty cases (3.3%) gave a family history of a similar cancer, but the details of the relationship and whether other cancers were found in the patient or close relatives is not given. Around 3.3% of the young cancer cohort had a history of ulcerative colitis. Around 19% of colorectal carcinomas in Sri Lanka are found in the < 40 year age group according to this data. This figure is similar to that observed in other developing countries, but higher than in developed countries: this could be due to a greater proportion of the population being young, but may also indicate a genetic predisposition to the condition. Further epidemiological and molecular research is required to define this more clearly.

There are no estimates in the Sri Lankan population of the proportion of breast cancers likely to be related to an inherited predisposition. One group (Hettiaratchi, 2000) have attempted to sequence the BRCA1 gene but report no mutations from a small number of breast cancer patients whose samples were analyzed for the common mutations of exon 2 (185delAG) or exon 20 (5382insC). The case selection criteria were not given.

De Silva W et al. (2008a) report data from BRCA1 mutation screening using SSCP and sequencing of abnormal SSCP products, as well as direct sequencing of exon 11. They used 130 cases with breast cancer (half with a positive family history and half without) and, in addition, screened 70 unaffected individuals with a family history and 40 controls with no family history of cancer. Ethical approval had been obtained from their institutional review board and written consent had been given by all the subjects. This study identified 2 novel frame shift mutations (c.3086delT and c.5404delG) among affected cases.

They also reported the data from testing of unaffected family members (possibly including children) of the case with the c5404delG mutation. This family history included premenopausal breast cancer in the proband and her mother, and uterine cancer in a maternal aunt, oral cancer in a maternal uncle, one cousin with premenopausal breast cancer (daughter of the proband's unaffected maternal aunt), and a niece diagnosed with a salivary cancer aged 13 years. None of the other affected breast cancer cases could be tested, as they were deceased. The researchers had tested 19 other relatives without cancer and two were found to have the mutation. Two other sequent variants were classified as possibly pathogenic (c.856T>G and IVS17-2A>T). They also reported 15 BRCA1 variants, of which 6 were missense polymorphisms, 4 were silent, and 4 unclassified. A mutation of possible pathogenic significance was identified in a patient diagnosed with breast cancer at 45 years. Her mother and sister were also affected at the ages respectively of 55 and 35. Here again, unaffected first degree relatives were tested and found to be negative for this possible mutation.

There are published reports on p53 mutations in oral carcinomas associated with betel chewing. Ranasinghe (1993) demonstrated nuclear activity of the p53 protein using immunochemistry, in 11% of the 38 tumors. Further sequence analysis of these 5 tumors did not reveal any p53 mutations in exon 5–9 that they sequenced. Chiba (1999) also studied squamous carcinomas in betel chewers and found p53 mutations in 10 out of 23 carcinomas that were studied. The betel chewers, homozygous for a CYP2A6 deletion polymorphism, had a lower rate of oral squamous carcinoma than those carrying the wild type alleles (Topcu, 2002).

In a study of 15 cases of squamous cell carcinoma of the cervix, human papilloma virus 16 was identified in 73%, while type 18 was identified in 20% (de Silva et al., 2006).

Psychiatric and Neurodegenerative Diseases

Using a mini mental stage examination for screening a group individuals of >65 years, and following up those with a score of <17 with further dementia-related investigations confirmed that 3.98% of this population had dementia (95% CI 2.6–5.7). Of these, 71% were diagnosed as probable Alzheimer's disease, 14.4% with vascular dementia, and 7% with mixed vascular and Alzheimer's disease. Rare causes included Lewy body dementia and syphilis (de Silva et al., 2003). A study of 23 patients with Alzheimer's disease (AD) and 21 unaffected controls reported that the mean total homocysteine levels were higher among AD cases compared with controls. The AD cases also had a higher likelihood of having an ApoE4 allele (the adjusted odds ratio for one or more of these alleles compared with none was 10.39%). Medial temporal lobe atrophy was also present in AD cases compared with controls (de Silva et al., 2005).

A case with cerebral autosomal dominant arteriopathy with subcortical infarcts and leukoencephalopathy (CADASIL) associated with a NOTCH 3 mutation (c260G) has been reported, and the authors suggest that this condition is underrecognized in the country (de Silva et al., 2009).

Susceptibility to depression was studied among a population-based twin registry. There was a higher genetic contribution in females (61%) than males (4%). The authors concluded that at least among females, the genetic contribution to depression was as strong as in higher-income countries (Ball, 2009).

Infectious Diseases

The use of molecular techniques for diagnosis of infectious diseases will be discussed later in the chapter.

Identifying the genetic basis for the variation in the severity of infectious diseases has also been assessed. Chan et al. (1984) reported that in elephantiasis, complicating filarial infection was higher among individuals with HLAB15 than controls. Severe malaria was more likely in patients who were heterozygous for TNFα2/TNFα1 compared to those found to be homozygous for the TNFα1 allele (Wattavidanage, 1999).

A mathematical model was used to assess malaria transmission in Sri Lanka (MacKinnon, 2000). This reported that variations in the susceptibility of malaria and the severity of illness could be attributed to host genetic factors.

Polymorphisms of the TNF, LTA and SLC11A1 genes were studied for an association with an increased risk for cutaneous Leishmaniasis but no such was identified (Samaranayake, 2010).

GENETIC SERVICES IN SRI LANKA

GENETIC COUNSELING

This is offered at the Human Genetics Unit, Medical Faculty, Colombo, and the Physiology Department, Medical Faculty, Ragama. Both centers are in the western province. Most pediatricians are skilled at counseling for common genetic disorders, thus avoiding large numbers of referrals for genetic counseling.

Jayasekera (1986) reported that there was acceptance for a genetic service in Sri Lanka among the majority of medical students and doctors. They felt that on principle, genetic testing and prenatal diagnosis was acceptable. De Silva et al. (1997) also reported a similar acceptance for the process of nondirective counseling.

No data exists about the population attitudes towards genetic testing, prenatal diagnosis, and termination of pregnancy.

CYTOGENETIC SERVICE

The first cytogenetic laboratory was started in 1983 at the Colombo Medical Faculty, and since 2006, a private laboratory has also been offering cytogenetic investigations. Both laboratories charge the patients directly for tests. A private molecular laboratory also offers postnatal aneuploidy testing.

The private cytogenetic laboratory offers prenatal and postnatal diagnosis of chromosome anomalies (www.asiri.lk), as well as blood and marrow investigations for malignancies. Chromosome fragility testing is also offered.

There is a need to expand the cytogenetic laboratories to other parts of the country while attempting to reduce the cost to make it more easily affordable to poor families. There is also a need to develop less expensive molecular techniques in order to more accurately define deletion and microdeletion syndromes. Microarray testing is not yet available for clinical use but may be of use diagnostically, although the cost of testing may make it prohibitive for many families. The cost of genetic testing, although much lower than in laboratories in developed countries, remains an issue for many families already burdened with the care of a sick child (de Silva et al., 2007).

MOLECULAR GENETIC SERVICE

There are private and state institutions performing molecular testing as well as a number of university-based institutions performing human genetics research.

Rapid diagnostic methods for infectious diseases are becoming available and widely used, either as part of research studies or for rapid diagnosis. The diagnostic investigations are mainly funded by the patient, and these include tuberculosis, dengue hemorrhagic fever, hepatitis B and C, malaria, leishmaniasis, and filariasis. Research projects are currently underway to investigate the genome of pathogens and to develop targets for therapy or immunization.

DNA fingerprinting is currently one of the most widely used molecular investigations in Sri Lanka. Indications for DNA fingerprinting include cases of disputed paternity, criminal cases, and to identify decomposed bodies.

Genetic testing for specific human genetic diseases is now available for a large number of conditions in the private sector with some tests being performed in laboratories abroad and some as research projects.

One private laboratory offers an extensive list of investigations. These include 36 separate molecular investigations for diagnosing infectious diseases, 6 syndrome diagnostic tests, metabolic diseases (9 different conditions), 7 common single gene disorders, 16 cancer syndromes including BRCA1/2, 21 blood disorders, MODY, cardiomyopathies, and 5 pharmacogenetic investigations. They also offer karyotyping (see above), prenatal diagnosis, and testing aborted material for aneuploidy using MLPA kits. Another private laboratory offers DNA fingerprinting, postnatal aneuploidy testing, and testing for the common thalassemia mutations, Duchenne deletion screening, and hemochromatosis (www.genetechlanka.org). The third private laboratory offers molecular tests for infectious diseases (www.durdans.com). There is an Institute of Biochemistry Molecular Biology and Biotechnology, and molecular laboratories at other universities, some of which offer diagnostic testing. Other planned developments include a thalassemia testing laboratory based in a university, and a molecular laboratory associated with the national twin registry.

There is a cancer screening center, which is a subsidiary of an insurance company in Sri Lanka. It aims to offer a state-of-the-art level of care from initial risk assessment to appropriate screening for cancer. Genetics is part of the offered package. The whole unit functions in collaboration with the Washington Cancer Institute (http://www.ceylincohealth.com).

PRENATAL DIAGNOSIS

Prenatal diagnosis for chromosome anomalies is available in the private sector. Nanayakkara et al. (2008c) report their experience with 89 cases of prenatal diagnosis using amniotic fluid (n=88) and cord blood (n=1). They offered aneuploidy screening to 85 cases (indications including maternal age >35, bad obstetric history, fetal anomalies on ultrasound scanning, positive fetal marker screening, and a family member with congenital anomaly/mental retardation). They used the MLPA P095 kit (MRC-Holland) for an aneuploidy screen. They do not specify if the family history of congenital anomaly or mental retardation was found to be related to a chromosome aneuploidy and if it was confirmed in an affected family member. They report the detection of trisomy 21 in 7 cases (7%), XYY in 1 case and monosomy X in one case. There is no discussion available in this abstract regarding the counseling that was offered and the outcome of the pregnancies diagnosed with an aneuploidy, as well as those that were not. They also do not state if karyotyping was performed on the samples. They also tested for thalassemia and Duchenne muscular dystrophy in families where the parents (or mother in the Duchenne cases) were "confirmed or suspected" carriers of these conditions. They report the detection of two thalassemia carriers in the prenatal samples but do not state if the mutations in the affected family members were known prior to offering prenatal diagnosis.

It is unclear if such testing that risks the life of the fetus and costs Rs 17,300 (around \$170; www.asiri.lk) is justifiable in a setting where there is no provision in the law for termination of pregnancy for fetal anomalies. Service providers would argue that the parents are given information regarding their pregnancy to justify the invasive investigations and costs of prenatal diagnosis. They may also seek termination either locally or abroad.

Senanayake (2006) reported a study where interviews were conducted among 10 women found to have pregnancies complicated by lethal fetal anomalies. Eight of them felt that the identification of the fetal anomaly prepared them for the eventual birth (and death) of their child, but two would have preferred not to have found this information. All of them coped by taking part in religious activities, which may have raised unrealistic expectations but also may have given them some comfort. All of them supported the availability of termination of pregnancy for lethal fetal anomalies.

Termination of pregnancy for fetal genetic disease has widespread acceptance among healthcare workers in Sri Lanka. De Silva et al. (2008b) reported results from a questionnaire survey of current and future doctors and nurses. Acceptability of prenatal diagnosis was 94%, 91%, 86%, and 71% respectively for lethal recessive diseases, Down syndrome, hemophilia, and late-onset neurodegenerative diseases. Termination of pregnancy (TOP) was acceptable for Down syndrome (84%), and lethal recessive disease (82%), and these rates were higher compared with hemophilia (65%) and neurodegenerative disease (53%). There was willingness to consider termination for self/spouse for Down syndrome (79%), lethal recessive disease (78%), hemophilia (60%), and neurodegenerative disorder (54%). The proportions willing to participate in a pregnancy termination (Down syndrome 54%, lethal recessive disease 51%, hemophilia 38%, and neurodegenerative disorder 38%) were lower. Religious affiliation influenced attitudes regarding TOP, with Christians being more opposed than Buddhists.

THE FUTURE OF GENETICS IN SRI LANKA

FUTURE DEVELOPMENTS

It is likely that genetics will continue to interest scientists and doctors as a research area, as well as a potential source of funds to develop their institutions. It is unlikely that state health funds will be used immediately to fund molecular diagnostic methods, as it is essential to develop more routine diagnostic services such as hematology, clinical chemistry, microbiology, pathology or radiology.

The development of collaborations between the academics and their foreign counterparts, or with the private sector locally, seems a likely way forward and this is taking place at present. The private sector may take over many of the genetic testing services, although long-term feasibility and cost effectiveness of such testing needs to be assessed. Equity of access to genetic testing is a major problem.

At present, only one of the laboratories offering genetic testing is accredited. The Sri Lanka accreditation board was started in collaboration with the Ministry of Science and technology following a parliamentary act in 2005. It has recently started the process of voluntary accreditation of state and private laboratories offering diagnostic investigations in the country. The aim of this is to endorse the laboratory's competence, credibility, impartiality, and integrity in carrying out their activities. Obtaining accreditation would enable building up of trust between the organization and its customers, get international recognition, minimize risk, and demonstrate credibility. The process of accreditation follows the methods that are practiced in other countries (www.slab.lk).

REGULATION OF GENETIC RESEARCH AND SERVICES

Regardless of who offers genetic investigations, it is essential to have a clear set of nationally accepted guidelines for research as well as diagnostic genetic testing. Currently, most research projects get approval from a local research ethics committee, but a central research ethics committee has recently been commenced. There are no guidelines regarding genetic testing for confirmation of the diagnosis of a genetic disease, identification of carriers or presymptomatic individuals, or prenatal testing. This issue is being addressed at present. As indicated by some research data, international guidelines are not always being followed.

More research is needed to assess the acceptability and psychosocial effects of genetic testing, especially with regard to predictive tests for cancer susceptibility, late-onset neurodegenerative disorders, for carrier testing and prenatal diagnosis.

Sri Lanka is a small country with a limited number of experts in this field. The same individuals may find themselves involved in offering counseling, organizing and running laboratories, offering and reporting genetic investigations, performing and reviewing research and drawing up guidelines regarding future genetic practice. This is likely to cause potential conflicts of interests.

LEGAL ASPECTS

There has been discussion about the legalization of abortion for fetal anomalies. The laws regarding abortion are from the nineteenth century and do not take into account the availability of new diagnostic methods for fetal anomalies and genetic disorders. An amendment to the law was attempted in the 1990s but withdrawn due to religious objections. It is likely that at some point in the future, the law will allow terminations for fetal anomalies incompatible with life or causing severe abnormalities after birth. Even in these cases, ethical guidelines should be developed to protect all those involved from criticism regarding unethical conduct.

FUNDING OF GENETIC SERVICES

It is unlikely that many Sri Lankans will be able to afford genetic testing, as the costs are likely to be prohibitive for most individuals with a low income and especially those families who are financially disadvantaged because of a chronic illness in a family member. The availability of testing locally will make genetics more accessible, as this will reduce costs compared with tests performed abroad. It is imperative that the individuals (or the parents of a child) undergoing genetic testing are fully informed of the advantages and disadvantages of a genetic test before testing, and this can only happen with more education of referring doctors and the general public.

It is most likely that the recent advances in diagnosis will be utilized by the more affluent and influential minority, or those whose disease is likely to be of interest as a research project.

It will be necessary to gather information about the cost effectiveness of genetic testing in Sri Lanka before they are offered widely. This should be in addition to evaluating the benefits and disadvantages of genetic testing. Assumptions regarding any of these issues should not be made based on data gathered in the West.

ACKNOWLEDGMENT

The author of this chapter contributed a chapter entitled "Genetic disorders and medical genetics in Sri Lanka" in "*Genetic Disorders of the Indian Subcontinent*" (Ed. Dhavendra Kumar), published by Kluwer/Springer, Dordrecht, 2004. The author and chief editor of the book acknowledge that some data and information may appear similar to the previous chapter. The Springer-Verlag Science+Media B.V. (Dordrecht) has given permission for the use of this data and information in the current chapter.

REFERENCES

Amaratunga A, Chandrasekera A. (1989). Incidence of cleft lip and palate in Sri Lanka. *J Oral Maxillofac Surg* 47(6):559–561.

Annual Health Bulletin, 2000 and Annual Health Bulletin 2007, Medical Statistics Unit, Department of Health, Ministry of Healthcare & Nutrition, Sri Lanka.

Ball HA, Sumathipala A, Siribaddana SH, et al. (2009). Genetic and environmental contributions to depression in Sri Lanka. *Br J Psychiatry* 195(6):504–509.

Chan S, Dissanayake D, Mak J, et al. (1984). HLA and filariasis in Sri Lankans and Indians. *Southeast Asian J Trop Med Public Health* 15(3):281–286.

Chiba I, Muthumala M, Yamazaki Y, et al. (1998). Characteristics of mutations in the p53 gene of oral squamous cell carcinoma associated with betel chewing in Sri Lanka. *Int J Cancer* 77(6):839–842.

Corea J. (1992). Is congenital dislocation of the hip rare in Sri Lanka? *Ceylon Med J* 37:96.

de Mel T, Jayasekera R, Warnasooriya N. (1992). Screening for Turner syndrome: how useful is the buccal smear test? *Ceylon Med J* 37:83–84.

de Silva D, Jayasekera K, Rubasinghe N, De Silva D. (1997). Attitudes towards genetic counseling and testing among medical students and newly qualified doctors. *Ceylon Med J* 42:129–132.

de Silva D, de Silva DG. (1998a). Referrals for syndrome diagnosis and implications for development of a clinical genetics service in Galle, Sri Lanka. *Proceeding of the Sri Lanka College of Paediatricians*, Colombo, Sri Lanka.

de Silva D. (2007). Emerging issues in genetic testing in Sri Lanka. *Indian J Med Ethics* 15(3):133.

de Silva DC, Jayawardana P, Hapangama A, et al. (2008b). Attitudes toward prenatal diagnosis and termination of pregnancy for genetic disorders among healthcare workers in a selected setting in Sri Lanka. *Prenat Diagn* 28(8):715–721.

de Silva HA, Gunatilake SB, Smith AD. (2003). Prevalence of dementia in a semi-urban population in Sri Lanka: report from a regional survey. *Int J Geriatr Psychiatry* 18(8):711–715.

de Silva HA, Gunatilake SB, Johnston C, Warden D, Smith AD. (2005). Medial temporal lobe atrophy, apolipoprotein genotype, and plasma homocysteine in Sri Lankan patients with Alzheimer's disease. *Exp Aging Res* 31(3):345–354.

de Silva KR, Gamage R, Dunuwille J, et al. (2009). Cerebral autosomal dominant arteriopathy with subcortical infarcts and leukoencephalopathy (CADASIL): a patient from Sri Lanka. *J Clin Neurosci* 16(11):1492–1493.

de Silva KM (1981). Colonisers and settlers. In: *A History of Sri Lanka*, pp. 3–16. Oxford University Press, Delhi.

de Silva M, Fernando M, Fernando D. (2000b). Comparison of some clinical and histological features of colorectal carcinoma occurring inpatients below and above the age of 40 years. *Ceylon Med J* 45(4):166–168.

de Silva MVC, Fernando MS, Gooneewardene SAS. (1999). Prostatic carcinoma in Sri Lanka—is it more common than cancer register statistics? *Ceylon Med J* (44):192.

De Silva PHD. (1998b). A review of 20,000 births. *Birth Res News* 3(3): 2–3.

de Silva R, Gunawardena S, WickramasingheG, Ranasinghe B, Namasivayam Y. (2007a) Primary immunodeficiency among patients with recurrent infections. *Ceylon Med J* 52(3): 83–86.

de Silva R, Karunaratne K, Mendis LN, Ramesh R, Chow V. (2006a). PCR detection and typing of human papilloma virus DNA in squamous carcinoma of the cervix in a cohort of Sri Lankan women. *Ceylon Med J* 51(3):114–117.

de Silva S. (1996). The control of thalassaemia in the Kurunegala district: a dauntless challenge. *Ceylon J Child Health* 1(25):16–24.

de Silva S, Fisher CA, Premawardhena A, et al. (2000a) Thalassaemia in Sri Lanka: implications for the future health burden of Asian populations. *Lancet* 355(9206):786–791.

de Silva S, Weerasuriya N, De Alwis N, De Silva M, Fernando D. (2002). Excess maternal transmission and familial aggregation of type 2 diabetes in Sri Lanka. *Diabetes Res Clin Pract* 58(3):173–177.

de Silva W, Karunanayake EH, Tennekoon KH, et al. (2008a). Novel sequence variants and a high frequency of recurrent polymorphisms in BRCA1 gene in Sri Lankan breast cancer patients and at risk individuals. *BMC Cancer* 8:214.

de Silva WI. (1994). Ahead of achievement of replacement level fertility in Sri Lanka before the Year 2000. *Asia Pac Popul J* 9(4):3–22.

de Zoysa NS. (1985). The distribution of ABO and Rhesus (Rh) blood groups in Sri Lanka. *Ceylon Med J* 30:37–41.

Dissanayake VHW, Tower C, Broderick A, et al. (2007). Polymorphism in the epidermal growth factor gene is associated with birth weight in Sinhalese and white Western Europeans. *Molecular Human Reproduction* 13(6):425–429.

Dissanayake VHW, Jayasekara RW. (2008). Cytogenetic testing in paediatrics: Some aspects of the Sri Lankan scenario. *Sri Lanka Journal of Paediatrics,* 37: 38–41

Dissanayake VHW (2009). Genetics of pre-eclampsia and the weight of babies at birth—clinical and genetic studies in Sri Lanka. *Ceylon Med J* 54(3): 90–99

Dissanayake VHW, Gammulla CG, Jayaweera LSD, Jayasekara RW. (2009a). Prevalence of genetic thrombophilic polymorphisms in the Sri Lankan population—implications for association study design and clinical genetic testing services. *Experimental and Molecular Pathology* 87(2):159–162.

Dissanayake VHW, Maheshi A, Athapaththu H, et al. (2009b). *Sri Lanka Journal of Obstetrics and Gynaecology* 31:84–87

Eckstein M, Foster A, Gilbert C. (1995). Causes of childhood blindness in Sri Lanka: results from children attending six schools for the blind. *Br J Ophthalmol* 79(7):624–625.

Ellepola SB, Wickramanayake ER. (1986). A genetic study of the Veddas and the Sinhalese. *Ceylon J Med Sci* 29(1):1–21.

Fernando D, Siribaddana S, De Silva D. (1994). Impaired glucose tolerance and diabetes mellitus in a suburban Sri Lankan community. *Postgrad Med J* 70(823):347–349.

Fernando L, Gromoll J, Weerasooriya TR, Nieschlag E, Simoni M. (2006). Y-chromosomal microdeletions and partial deletions of the Azoospermia Factor c (AZFc) region in normozoospermic, severe oligozoospermic and azoospermic men in Sri Lanka. *Asian J Androl* 8(1):39–44.

Gunasekera DP, Gunasekera PC. (1998). Prevalence of congenital abnormalities in a teaching hospital obstetric unit. *Proceeding of the Sri Lanka College of Paediatricians* July 1–17, 1998, Colombo, Sri Lanka.

Gunawardene D, Gunawardene K, Vasanthanathan N, Gunawardene J. (1978). Psoriasis in Sri Lanka—a computer analysis of 1366 cases. *Br J Dermatology* 98(1):85–96.

Hettiaratchi U, Karunanayake E, Wijesundera W, Angumawela P. (2000). Preliminary study on BRCA1 mutation analysis of Sri Lankan breast cancer patients. *Proceedings of SLAAS.* November 27–December 1, 2000, Colombo, Sri Lanka.

Illangasekera U, Nugegoda D, Perera L. (1993). Prevalence of diabetes mellitus and impaired glucose tolerance in a rural Sri Lankan community. *Ceylon Med J* 38(3):123–126.

Illeperuma RJ, Mohotti SN, de Silva TM, Fernandopulle ND, Ratnasooriya WD. (2008). Genetic profile of 11 autosomal STR loci among the four major ethnic groups in Sri Lanka. *Forensic Sci Int Genet* 3(3):e105–e106. Epub November 28, 2008.

Jayasekera R. (1988). Chromosome anomalies in Sri Lanka: a cytogenetic profile. *Ceylon J Med Sci* 31(1):1–5.

Jayasekera R. (1986). The attitudes of doctors and students towards a genetic service in an Asian country, Sri Lanka. *Asia Oceania J Obs and Gynaecology* 15(3):267–270.

Jayasekera R, Mastana S, Papiha S. (1994). Distribution of group specific component and transferring subtypes in the populations of Sri Lanka. *Gene Geogr* 8(2):151–156.

Jayasinghe MS, Wallooppillai NJ. (1972). Congenital heart disease in the over 40 age group in Sri Lanka. *Ceylon Med J* 17(4):204–206.

Kumarasinghe M. (1999). An effective and affordable cervical carcinoma screening programme for Sri Lanka. *Ceylon Med* J 44:156–158.

Karunasekera K, Jayasinghe J, Alwis L. (2001). Risk factors for childhood asthma: a Sri Lankan study. *J Trop Paediatr* 47(3):142–145.

Katulanda P, Groves CJ, Barrett A, et al. (2008). Prevalence and clinical characteristics of maternally inherited diabetes and deafness caused by the mt3243A > G mutation in young adult diabetic subjects in Sri Lanka. *Diabet Med* 25(3):370–374. Kshatriya GK. (1995). Genetic affinities of the Sri Lankan populations. *Hum Biol* 67(6):843–866.

MacKinnon M, Gunwardena D, Rajakaruma J, Weerasinghe S, Mendis K, Carter R. (2000). Quantifying genetic and non-genetic contributions to malarial infection in a Sri Lankan population. *Proc Natl Acad Sci (USA)* 97(23):12661–12666.

Malavige GN, Rostron T, Seneviratne SL, et al. (2007). HLA analysis of Sri Lankan Sinhalese predicts North Indian origin. *Int J Immunogenet* 34(5):313–315.

Manamperi A, Hapuarachchi C, Silva Gunawardene N, Bandara A, Dayanath D, Abeyewickreme W. (2009). STR polymorphisms in Sri Lanka: evaluation of forensic utility in identification of individuals and parentage testing *Ceylon Med J* 54(3):85–89.

Mastana S, Jayasekera R, Fisher P, Sokol R, Papiha S. (1993). Genetic polymorphisms of orosomucoid in populations of the United Kingdom, Indian subcontinent and Cambodia. *Jpn J Hum Genet* 38(3):289–296.

Mendis L. (1989). Susceptibility to Rubella virus among Sri Lankan women. *Ceylon Med J* 34:73–75.

Mendis S, Shepherd J, Packard C, Gaffney D. (1990). Genetic variations in the cholesteryl ester transfer protein and apolipoprotein A-I genes and its relation to coronary heart disease in the Sri Lankan population. *Atherosclerosis* 83(1):21–27.

Mendis S, Shepherd J, Packard C, Gaffney D. (1991) Restriction fragment length polymorphisms in the ApoB gene in relation to coronary heart disease in a southern Asian population. *Clin Chim ACTA* 196(2–3):107–117.

Mendis S. (1991). Coronary heart disease and coronary risk profile in a primitive population. *Trop Geogr Med* 43(1–2):199–202.

Mendis S, Ekanayake E. (1994). Prevalence of coronary heart disease and cardiovascular risk factors in middle-aged males in a defined population in central Sri Lanka. *In J Cardiol* 46(2):135–142.

Mendis S, Athauda S, Naser M, Takahashi K. (1999). Association between hyperhomocysteinaemia and hypertension in Sri Lankans. *J Int Med Res* 27(1):38–44.

Mendis S, Ranatunga P, Jayatilake M, Wanninayake S, Wickremasinghe R. (2002). Hyperhomocystinaemia in Sri Lankan patients with coronary artery disease. *Ceylon Med J* 47(3):89–92.

Nanayakkara HLV, Weerathunga HLV, Dissanayake VHW. (2008a). The CAG repeat length in Sri Lankan patients with Huntington Disease. *2nd Annual Scientific Sessions of the Sri Lankan Association of Neurologists,* November 21–23, 2008, Colombo, Sri Lanka.

Nanayakkara HLV, Weerathunga RP, Dissanayake VHW. (2008b). Fragile X syndrome detected in Sri Lankan patients due to methylation. *2nd Annual Scientific Sessions of the Sri Lankan Association of Neurologists*, November 21–23, 2008, Colombo, Sri Lanka.

Nanayakkara HLV, Weerathunga RP, Rodrigo N, Dissanayake VHW. (2008c). Prenatal genetic screening in Sri Lanka. *CMJ* 53(Suppl. 1). Poster presentation 28. Sri Lanka Medical Association Annual Scientific Sessions March 2008.

Papiha SS, Mastana SS, Purandare CA, Jayasekara R, Chakraborty R. (1996a). Population genetic study of three VNTR loci (D2S44, D7S22 and D12S11) in five ethnically defined populations of the Indian subcontinent. *Hum Biol* 68(5):819–835.

Papiha SS, Mastana SS, Jayasekara R. (1996b). Genetic variation in Sri Lanka. *Hum Biol* 68(5):707–737.

Perera D, Gunasekera D, Wijekoon A., Petrou M. (2000) Thalassaemia control in developing countries—the Sri Lankan perspective. *Ceylon Med J* 45(1):12–16.

Perera PL, Weerasinghe WM. (1987). Determination of acetylator phenotype in Sri Lanka. *Ceylon Med J* 32:29–34.

Premawardhena AP, Fisher CA, Fathiu F, et al. (2001). Determinants of jaundice and gallstones in haemoglobin E beta thalassaemia. *Lancet* 357(9272):1945–1946.

Premawardhena AP, Liu YT, Fisher CA, Clegg JB, Weatherill DJ. (2003). The worldwide prevalence of the UGT-1A1 promotor polymorphism as a contributor for phenotypic variability in thalassaemia. *Proceedings of the Sri Lanka Medical Association*; March 26–29, 2003, Colombo, Sri Lanka.

Rajapakse LC. (2002). Estimates of induced abortion in urban and rural Sri Lanka. *J College Commun Phys Sri Lanka* 7:10–15.

Ranasinghe A, Warnakulasuriya K, Johnson N. (1993). Low prevalence of expression of p53 oncoprotein in oral carcinomas from Sri Lanka associated with betel and tobacco chewing. *Eur J Cancer B Oral Oncol* 29B(2):147–150.

Rochette J, Pointon J, Fisher C, et al. (1999). Multicentric origin of the haemochromatosis gene (HFE) mutations. *Am J Hum Genet* 64:1056–1062.

Roychoudry AK. (1984). Genetic relationship between Indian tribes and Australian Aboriginals. *Hum Hered* 34(5):314–320.

Saha N, Mark JW, Tay JS, et al. (1995). Population genetic study among Asli (Semai Senoi) of Malaysia: Malayan aborigines. *Hum Biol* 67(1):37–57.

Saha N. (1988). Blood genetic markers in Sri Lankan populations—reappraisal of the legend of Prince Vijaya. *Am J Anthropol* 76(2):217–225.

Samaranayake TN, Fernando SD, Dissanayake VH. (2010). Candidate gene study of susceptibility to cutaneous leishmaniasis in Sri Lanka. *Trop Med Int Health.* 15(5):632–638.

Senanayake H, de Silva D, Premaratne S, Kulatunge M. ((2006). Psychological reactions and coping strategies of Sri Lankan women carrying fetuses with lethal congenital malformations. *Ceylon Med J* 51(1):14–17.

Thenabadu N. (1982). Congenital heart disease in Sri Lanka. *Ceylon Med J* 27:100–115.

Topcu Z, Chiba I, Fujieda M, et. al. (2002). *Carcinogenesis* 23(4): 595–598.

Wattavidanage J, Carter R, Perera K, et al. (1999). TNFalpha 2 marks high risk of severe disease during plasmodium falciparum malaria and other infections in Sri Lankans *Clin Exp Immun* 115(2):350–355.

Weerasuriya K, Jayakody RL, Smith CA, Wolf CR, Ticker GT, Lennard MS. (1994). Debrisoquine and mephenytoin oxidation in Sinhalese: a population study. *B J Clin Pharmacol* 38(5):466–470.

Welihinda J, Karunanayake E, Jayasekera R, Peiris J, Pettersson U, Wadelius C. (1993). Deletion screening of Sri Lankan Duchenne muscular dystrophy patients using the polymerase chain reaction. *Ann Trop Paediatr* 13(1):83–86.

Wickramanayake E. (1992). X chromosome number and mental health status in Sri Lanka *Ceylon J Med Sci* 35:35–38.

Wickramasinghe P, Lamabadasuriya S, Narenthiran S. (2000). Prospective study of congenital heart disease in children. *Ceylon Med J* 46(3):96–98.

Yu Lok C, Merryweather-Clarke A, Viprakasit V, et al. (2009). Iron overload in the Asian community. *Blood* 114(1):20–25.

ON-LINE RESOURCES

www.asiri.lk/acgrm/acgrm3.html
www.ceylincohealth.com
www.dailynews.lk/2010/02/22/bus01.asp
www.durdans.com
www.epid.gov.lk
www.genetechlanka.org
www.imf.org
www.lankalibrary.com
www.slab.lk
www.statistics.gov.lk
www.sundayobserver.lk/2009/12/20/spe40.asp
www.sundaytimes.lk
www.worldbank.org/lk

85.

FAMILIAL CANCER AND CANCER GENETICS IN THE INDIAN SUBCONTINENT

Dhavendra Kumar

INTRODUCTION

Cancer rates in India and neighboring countries of the Indian subcontinent are lower than those seen in Western countries, but are rising with increasing migration of rural population to the cities, increase in life expectancy, and changes in lifestyles. In India, rates for oral and esophageal cancers are some of the highest in the world. In contrast, the rates for colorectal, prostate, and lung cancers are among the lowest.

Cancer is global and undoubtedly one of the major causes of mortality and morbidity irrespective of geographic and ethnic origin. It equally applies to people living in the Indian subcontinent as well as the migrant groups from various countries of the subcontinent (Kamath et al., 1999; Winter et al., 1999). It is estimated that 75,000 new breast cancer cases occur in Indian women every year (Chopra, 2001). However, a full appreciation of its implications on health and living are lacking due to the overwhelming burden of environmentally related diseases in the Indian subcontinent, such as malnutrition, infection, and pollution (Chopra, 2001). Some cancers are attributed to smoking and chronic exposure to toxic substances. An alarming excess of oral cancer is a good example, probably related to the social habit of tobacco chewing.

THE EPIDEMIOLOGY OF CANCER

There are number of epidemiological reports from various cancer centers in India (Malhotra et al., 2001; Sen et al., 2002) and Pakistan (Bhurgri et al., 2002). The study by Malhotra et al. (2001), reported from a major city of the state of Punjab in India, looked at hospital-based cancer records. Patients came from all over the state, as well as neighboring states of Himachal Pradesh and Haryana in north India. Over a ten-year period, 4730 cancer cases were referred for histopathological examination. The incidence of malignant tumors was estimated to be 8.36%. Females outnumbered males, with male to female ratio of 1:1.09. Most of the cancers were seen in the age group of 41–50 years. In females the two most common cancer sites were breast (21.07%) and cervix (19.4%), while in males hypopharynx/larynx (13.94%) and prostate (9.65%) were the most common sites of cancer. No instance of familial occurrence of any cancer was cited.

The first report of the Kolkata (previously known Calcutta) cancer registry (Sen et al., 2002) includes 11,700 cases over a two-year period (January 1, 1998 to December 31, 1999). The overall age-adjusted (world population) incidence rates were 102.1/100,000 males and 114.6/100,000 females. The most frequently reported malignancies in males were lung cancer (16.3%) followed by cancers of the oral cavity (7.1%), pharynx (5.7%), and larynx (5.7%). In females, the most frequently reported malignancies were breast (22.7%) followed by uterine cervix (17.5%), gall bladder (6.4%) and ovary (5.8%). The data reported by the Kolkata cancer registry provide information on the cancer profile in eastern India. However, this report does not give any indication of proportion of familial cancers.

The report of Bhurgri et al. (2002) reports cancer incidence and patterns in Karachi, the largest city of Pakistan. The data was collected from various hospitals within the city over the period 1998–1999. All cancers were reported and registered with the Karachi Cancer Registry. The age-standardized incidence rate (ASR) of cancer at all sites was 132.4/100,000 for the males. Cancer of the lung (10.8%; ASR 17.3) was the most frequently recorded malignancy, followed by oral cavity (10.5%; ASR13.2) and larynx (5.0%; ASR 7.4). The ASR in females for all cancers at all sites was 133.0/100,000. Cancer of the breast (32.0%; ASR 40.7) was the most frequently recorded malignancy, followed by oral cavity (8.1%; ASR 11.7) and gall bladder (3.6%; ASR 5.5). This study concluded an underestimate of about 15%–20%. Tobacco-associated cancers were estimated to be 38.3%; two principal cancers, breast and oral cavity, accounted for 40% of all cancers in females. In addition, a high incidence of gall bladder cancer in the females was also found. It was difficult to deduce any specific etiological factors from the study. No examples of familial occurrence were cited.

The above two reports and many others list incidence and prevalence of different cancers from the Indian subcontinent. Some of these reports have included possible etiological factors. The list of potential etiological factors is constantly increasing. Not surprisingly, the majority of cancers across the Indian subcontinent is directly or indirectly linked to one or more environmental factors. Within the bulk of environmentally related cancers lie a small proportion of cancers that might be associated with inherited predisposition. However, the data is either not available or lost in the middle of vast amounts of epidemiological data. Another factor is possibly the lack of scientific evidence to demonstrate specific molecular or genetic changes.

Winter et al. (1999) examined cancer incidence among South Asians in the United Kingdom (ethnic origins from India, Pakistan and Bangladesh) and compared it to non-South Asians and Indian subcontinent rates. English South Asian cancer incidence rates for all cancer sites combined (lung cancer in males, breast cancer in females, and lymphoma in both sexes) were significantly lower than non-South Asian rates but higher than the indigenous Indian subcontinent rates. This report highlights a transition from a lower cancer risk in the country of ethnic origin to that of the country of residence. This observation probably reflects changes in lifestyle and other environmental exposures to the migrant Indian subcontinent populations.

One of the possible factors could be exogenous hormonal changes as reported by Kamath et al. (1999). This study examined various hormonal profiles (serum total 17 beta-estradiol [E2], sex hormone-binding globulin-bound E2, available E2, estrone [E1], and dehydroepiandrosterone sulfate), reproductive history, family history of cancer, body composition/size, dietary intake, physical activity, and excretion of isoflavones, between two distinct ethnic groups (migrant Asian Indians and local women) of premenopausal women residing in the United States. The sex hormone profiles were significantly higher among American women compared to the Indian and Pakistani women (Asian Indians). It was concluded that difference in circulating sex hormone concentrations could be an important risk factor for breast cancer.

The concept of genes and relationship to oncogenesis is not new. The fact that cancer is in itself a genetic change, wholly at the cellular level, indicates that whatever the etiology of cancer in the Indian subcontinent, genetic factors should be important in its etiopathogenesis. Although there are occasional reports indicating somatic genetic changes in various cancers reported from India, reports indicating associated germ line mutations are extremely limited. However, anecdotal and unpublished data provide examples of several instances of familial occurrence of breast and colorectal cancers. Given the large volume of population inhabiting the Indian subcontinent, and equally a sizable migrant population living abroad, a conservative estimate of familial cancer could be very high. The present chapter provides limited information on genetic aspects of common cancers, with particular reference to familial cancers. Any information related to known environmental factors associated with commonly occurring cancers is excluded. This includes cancer of the lung and cervix. Information on these cancers is available through other medical and public health resources.

ORAL CANCER

The majority of oral cancers are squamous cell carcinomas (OSCC). The frequency varies considerably in different geographical regions, being highest in the Indian subcontinent (Hodgson and Maher, 1999).

Uncommon association of oropharyngeal malignancy is known to complicate autosomal dominant pachyonychia congenita (OMIM 167200) and the X-linked recessive condition, dyskeratosis congenita (OMIM 305000). Oropharyngeal cancers are prominent among solid cancers arising in patients with Fanconi's anemia (OMIM 227650), not uncommon among offspring of consanguineous Pakistani parents. However, reports of oral malignancy related to Mendelian disorders are relatively infrequent.

The cancer of the oral cavity across the Indian subcontinent is widespread (Babu, 2001). Oral cancers in India constitute a major proportion of all cancers. With a population of over one billion, the huge burden of cancer poses many challenges. Its incidence among the migrant population outside the subcontinent (1%) is considerably lower compared to that of the Indian subcontinent (40%). This observation has provided sufficient evidence in favor of one or more environmental factors. Cultural and dietary risk factors, including some ethnic variation, have been attributed to high incidence of and prevalence of oral cancer in India (Zain, 2001). Etiological association with poor oral hygiene and the custom of tobacco/betel quid/areca nut chewing has been the subject of several studies (Balaram et al., 2002). There is overwhelming evidence to support a direct link with these environmental factors. The habit of tobacco/betel nut chewing is prevalent across a wide social spectrum throughout India.

The study from South India (Balaram et al., 2002) attributed 35% of oral cancers in men to the combination of the social habits of smoking and alcohol drinking, About half (49%) was found to be related to tobacco chewing. The same study found 95% of oral cancers among women directly linked to the social habit of tobacco chewing and poor oral hygiene. In a majority of the studies there is no correlation of age at diagnosis and associated risk factors. The study by Iype et al. (2001) reports lack of associated factors in 40% of oral cancer patients under the age of 35 years. This study probably reflects non-environmental etiological factors associated with oral squamous cell cancers. Submucosal fibrosis and lichen planus appear to be premalignant lesions (Yaacob et al., 2002). Association with celiac disease is also reported (Ferguson and Kingstone, 1996). There are no reports to indicate a high incidence of oral cancer associated with these diseases among people from the Indian subcontinent. Genetic susceptibility to oral squamous cell carcinoma remains an important question, to explain the etiology and molecular pathology of oral cancer in the Indian subcontinent. A number of reports highlight possible genetic associations with oral cancer in India (Saranath et al., 1993; Sreelekha et al., 2001).

Polymorphism in glutathione *S*-transferase (*GST*) genes, causing variations in enzyme activities, may influence susceptibility to oral cancer and leukoplakia in smokers and/or smokeless tobacco users (Buch et al., 2002; Sikdar et al., 2003). Buch et al. (2002) studied the influence of genetic polymorphism at *GSTM1*, *GSTM3* and *GSTT1* gene loci on oral cancer risk among Indians habituated to the use of smokeless tobacco, bidis, or cigarettes. DNA was analyzed in 297 oral cancer patients and 450 healthy controls using the polymerase chain reaction (PCR) and PCR-restriction fragment length polymorphism (RFLP) methods. Lifetime tobacco exposure was evaluated as a risk factor in relation to the polymorphism at the GST gene loci using logistic regression analysis. There was no significant difference in the distribution of the *GSTM3* and *GSTT1* genotypes between oral cancer patients and controls. However, a significant threefold increase in risk was seen for patients with the *GSTM1* null genotype (age adjusted OR = 3.2, 95% CI 2.4–4.3). The impact of the *GSTM1* null genotype on oral cancer risk was also analyzed in separate groups of individuals with different tobacco habits. The odds ratio associated with the GSTM1 null genotype was 3.7 (95% CI 2.0–7.1) in tobacco chewers, 3.7 (95% CI 1.3–7.9) in bidi smokers, and 5.7 (95% CI 2.0–16.3) in cigarette smokers. Furthermore, increased lifetime exposure to chewing tobacco appeared to be associated with a twofold increase in oral cancer risk in *GSTM1* null individuals. The results suggest that the *GSTM1* null genotype is a risk factor for development of oral cancer among Indian tobacco habitués.

The study by Sikdar et al. (2003), consisting of 109 leukoplakia and 256 oral cancer patients and 259 controls, report genotype frequencies at *GSTM1*, *GSTT1*, *GSTM3*, and *GSTP1* determined by polymerase chain reaction-restriction fragment length polymorphism methods, and analyzed by multiple logistic regression to determine the risks of the diseases. There were no significant differences in the distributions of *GSTM1*, *GSTM3*, and *GSTT1* genotypes in patients and controls when all individuals were compared. However, frequencies of *ile/ile* genotype at codon 105 and variant *val-ala* haplotype of *GSTP1* was significantly higher (OR = 1.5; 95% CI = 1.0–2.0) and lower (OR = 1.4; 95% CI = 1.0–1.9) in oral cancer patients compared to controls, respectively. The impacts of all genotypes on risks of oral cancer and leukoplakia were also analyzed in patients with different tobacco habits and doses. Increased risks of cancer and leukoplakia were observed in tobacco smokers with *GSTM3* (*A/A*) genotype (OR = 2.0, 95% CI = 1.0–4.0; OR = 2.0, 95% CI = 1.0–4.4, respectively). The study observed that *GSTM3* (*A/A*) genotype could become one of the markers for assessing the premalignant potential of leukoplakia. Heavy tobacco chewing (>124 chewing-years) increased the risk of cancer in individuals with *GSTT1* homozygous null genotype (OR = 3.0; 95% CI = 1.0–9.8). The study inferred that the lifetime exposure to tobacco smoking (>11.5 packs per year) increased the risk of leukoplakia in individuals with *GSTM1* homozygous null genotype (OR = 2.4; 95% CI = 1.0–5.7). Similar to the study of Buch et al. (2002), Sikdar et al. (2003) also suggest that polymorphisms in *GSTP1*, *GSTM1*, *GSTM3*, and *GSTT1* genes regulate risk of cancer and leukoplakia differentially among different tobacco habitués.

Infrequent loss of heterozygosity (LOH) at adenomatous polyposis coli (APC) gene locus (6%) is reported in Indian oral cancer patients (Tandle et al., 2000). This study concluded that the infrequent LOH at the polymorphic RsaI locus in exon 11 of the APC gene was probably associated with tobacco-associated Indian oral cancer, unlike tobacco/alcohol associated oral cancers from Western countries.

Genomic instability as reflected by fragile sites and microsatellite alterations in specific target regions is shown to be an important characteristic of oral squamous cell carcinoma (OSCC). Subhadra et al. (2003) reported genetic susceptibility to oral cancer associated with common fragile sites. In a study including 100 patients, the expression of bleomycin-induced fragile sites (FS) in the blood lymphocytes of 150 individuals (100 oral cancer patients and 50 age and sex matched controls), was significantly higher compared to controls. FS expression was site-specific in oral cancer patients. Chromosome 5 was the most affected. Enhanced expression of FS at the centromeric region was observed in the patient group. The study emphasized the role of FS in the genetic susceptibility to Indian oral cancer. The study by Mahale and Saranath (2000) reports a high incidence (62%) of microsatellite alterations on 9p21023 in Indian OSCC patients. This included 27% with microsatellite instability (MSI) and 52% LOH. The authors concluded that increased incidence of MSI and LOH at 9p21 involved the p16 (CDKN2) tumor suppressor gene in a subset of tobacco-induced Indian OSCC. However, no association with the tumor suppressor gene ING1 on 13q34, involved in p53 signaling pathway, was found in another study on Indian OSCC patients (Krishnamurthy et al., 2001).

Association of Indian OSCC and mutations in the p53 gene, and changes in the p53 product, has attracted several research centers in India (Kaur et al., 1994; Ravi et al., 1996; Ralhan et al., 2001). The study by Kaur et al. (1994) reports overexpression of p53 protein in betel- and tobacco-related human oral dysplasia and OSCC in India. The same center has reported a positive correlation with the serological changes in the p53 protein and somatic p53 missense mutations (Ralhan et al., 2001). This study recommends serological detection of p53 antibodies as the surrogate marker for p53 gene alterations in premalignant and malignant oral lesions. The study reported by Ravi et al. (1996) indicated overexpression of p53 protein in both premalignant and malignant oral lesions, including over expression of the bcl-2 anti-apoptotic protein. However, it is not clear from any studies whether oral cancer patients had germ line p53 mutations.

Heinzel et al. (1996) studied a cohort of Indian patients from Singapore diagnosed with OSCC. Mutations and polymorphisms in the p53, p21, and p16 genes were analyzed. The study was based on tumor analysis alone. No association with p21 was found. However, p53 and p16 genes were frequently mutated. A positive correlation with the personal habit of betel quid chewing was observed. A high mutation rate in the H-ras and p53 genes is also reported from Japan in oral cancers associated with tobacco chewing. The study concluded association of both H-ras and p53 mutational events were independent and mutually exclusive (Munirajan et al., 1998). Mutations in H- and K-ras genes were observed at a frequency of 28% and 33% respectively in oral tumor specimens from eastern Indian populations (Das et al., 2000). The clinical significance of p53 gene mutations in OSCC is unclear. One study has indicated its prognostic significance in conjunction with overexpression of P-glycoprotein (Pgp) encoded by the MDR1 gene. The p53 gene is shown to activate the MDR1 gene promoter. The study points out a positive correlation between overexpression of p53 and Pgp and aggressive nature of the tumor and poor prognosis.

HEPATOBILIARY CANCER (INCLUDING PANCREATIC CANCER)

Most cases of hepatocellular carcinoma (HCA) occur in adults, and are usually sporadic. There is considerably higher incidence of HCA in the Indian subcontinent, approaching 40 per 100,000 per year. The majority are attributed to environmental factors, such as hepatitis B viral infections and aflatoxin. Association with hepatitis C viral infection appears to be valid, but lacks support of convincing data. Familial occurrence of hepatocellular carcinoma is rare, probably accounting for exposure to common environmental factors.

Hepatocellular cancer is known to occur in some of the Mendelian disorders including hemochromatosis, tyrosinemia, Fanconi's anemia, Wilson's disease, glycogen storage disease (type I and IV), and α1-antitrypsin deficiency. These recessively inherited disorders should be considered in any patient belonging to one of the Indian subcontinent population groups. This will be important in cases of parental consanguinity. Liver cancer may occur as a complication of chronic liver disease and cirrhosis of a variety of causes.

The glutathione *S*-transferases (GSTs) are a polymorphic supergene family of detoxification enzymes that are involved in the metabolism of numerous potential carcinogens. Several allelic variants of polymorphic GSTs show impaired enzyme activity and are suspected to increase the susceptibility to various cancers, notably oral cancer. Pandey et al. (2006) report an association with gall bladder cancer. The study determined the distribution of polymorphisms in the GST family of genes (*GSTT1*, *GSTM1*, *GSTP1*, and *GSTM3*) that were studied in 106 cancer patients and 201 healthy controls. Genotypes were analyzed by polymerase chain reaction (PCR) and PCR-restriction fragment length polymorphism (RFLP). The frequencies of *GSTM1* null and *GSTM3*BB* genotypes did not differ between patients and controls. The overall frequency of *GSTT1* null was lower in cases as compared with controls ($p = 0.003$, Odds ratio (OR) = 0.2, 95% confidence interval (CI), 0.1–0.6). After sex stratification, the *GSTT1* null frequency was reduced only in female patients ($p = 0.008$, OR = 0.2, 95% CI = 0.1–0.6). However, the *GSTP1*, *ile/val* genotype and the *val* allele were significantly higher in cases than controls ($p = 0.013$, OR = 1.9, 95% CI = 1.1–3.1; $p = 0.027$, OR = 1.5, 95% CI = 1.0–2.1, respectively). To study gene–gene interactions, a combined risk of gallbladder cancer due to *ile/val* or *val/val* was calculated in combination with null alleles of *GSTM1* and *GSTT1* or the *B allele of *GSTM3*, but there was no enhancement of risk. Gallstones were present in 57.5% of patients with gallbladder cancer, but there were no significant differences between allelic/genotype frequencies of the studied *GST* genes polymorphisms between patients with or without gallstones. The study has put forward an observation showing *ile/val* genotypes and *val* allele of *GSTP1* in association with gallbladder cancer. Further studies are clearly warranted for clinical application, specifically for gallbladder cancer risk estimation.

Pancreatic cancer is uncommon and is rarely reported from the Indian subcontinent (Chari et al., 1994). Familial clustering is uncommon, but documented in the Western literature. Such reports are extremely uncommon from the Indian population groups. However, association of adenocarcinoma of the pancreas with recurrent familial pancreatitis is well described. It is an uncommon autosomal dominant condition, manifesting from early childhood with attacks of recurrent pancreatitis (Kattwinkel et al., 1973). The condition has been mapped to chromosome 7q, and mutations in cationic trypsinogen gene have been detected in affected individuals (Whitcomb et al., 1996). There is one common mutation, R117H in exon 3, which has been described in several affected families. The incidence of cancer in gene carriers may approach 25%. The condition is probably similar to recurrent hereditary tropical calcific pancreatitis, which is now increasingly diagnosed in Bangladesh and adjoining parts of the Indian subcontinent. It is suggested that risk of pancreatic cancer in the tropical variety may be similar to that of non-calcific hereditary pancreatitis (Chari et al., 1994).

BREAST AND OVARIAN CANCER

Although breast and ovarian cancers account for a significant proportion of all cancers in the Indian subcontinent, reports on familial occurrence are relatively infrequent. The age standardized (world) rate (ASR) of breast cancer in the Asian continent is less than 40 per 100,000 per year, except for Manila in the Philippines (47.7/100,000) and Pakistan (51.7/100,000; see Parkin et al., 1997). Breast cancer rates reported from India are comparatively lower, ranging from 8.7/100,000 to 28.2 per 100,000 per year (Parkin et al., 1997). Higher ASR for breast cancer is also reported from cancer registries in Pakistan (Malik et al., 1992; Bhurgri et al., 2000). Affected women in Pakistan are typically young and often present with advanced disease (Usmani et al., 1996; Ahmed et al., 1997). Although the pattern of age-specific breast cancer rate is broadly similar to that of Western countries, relatively rapid premenopausal growth is seen in Pakistani women after 45 years of age. This difference in the slope of the age-specific incidence curves in Eastern and Western populations indicates that premenopausal breast cancer is relatively more common among the Indian subcontinent populations.

Carcinoma of the ovary is the most common cancer of gynecologic origin in Pakistan (Bhurgri et al., 2000). The ASR of ovarian cancer in Karachi, Pakistan (10.2 per 100,000 per year), is comparable to that of Ontario, Canada (10.7 per 100,000 per year) (Parkin et al., 1997; Bhurgri et al., 2000). In contrast, ASRs for ovarian cancer in India range from 1.2/100,000 to 7.2/100,000 per year (Parkin et al., 1997).

It is not clear why Pakistan has one of the highest ASRs for both breast and ovarian cancers. Lifestyle and reproductive factors may influence the incidence of breast and ovarian cancer, but the specific factors have not been identified. It is possible that a proportion of the breast and ovarian cancers may be linked to mutations in the breast and ovarian cancer susceptibility genes BRCA1 (MIM113705) and BRCA2 (MIM 600185). Further, it is stipulated that higher cancer incidence may reflect parental consanguinity. Pakistan has one of the highest rates of consanguinity in the world (Hashmi, 1997). The overall frequency of consanguineous marriages is 60%–76% in Pakistan, and that frequency is 75% among Pakistanis living in the United Kingdom (Darr and Modell, 1988; Hashmi, 1997; Hussain and Bittles, 1998). Inbreeding is known to increase the risk of diseases caused by homozygosity of deleterious recessive genes. An excess of childhood cancers was also reported among children of consanguineous marriages in Britain (Powell et al., 1995). There is little information on the possible role that recessive genes play in adult cancer. One study from Pakistan has described an association between consanguinity and the risk of breast cancer (Shami et al., 1991). Further studies are needed to explore the contribution of consanguinity in the incidence of breast and ovarian cancers.

There are only a few studies reported on the contribution of BRCA1 and BRCA2 gene mutations to breast and ovarian cancers in the Indian subcontinent populations (Kumar et al., 2002; Liede et al., 2002; Pestonjamasp and Mittra, 2000; Kannan et al., 2000; Saxena et al., 2002; Varamathi et al., 2003). The incidence of breast cancer in North India is reported to have increased from 17.9 to 24.9 per 100,000 during the period from 1965 to 1985 (Saxena et al., 2002). The same study analyzed a cohort of 20 breast cancer patients with either a family history of breast and/or ovarian cancer (2 or more affected first-degree relatives) or early age of onset (<35 years). Two novel splice site variants (331+1G>T; 4476+2T>C) in BRCA1 (10%) were identified. In addition, two BRCA2 missense variants were each identified in more than one parent (two unaffected individuals each), and likely represent population-specific polymorphisms.

Another study from India (Kumar et al., 2002) reports coding sequence of BRCA1 gene in 14 breast cancer patients with a positive family history of breast and/or ovarian cancer. Mutation analysis was carried out using the conformation-sensitive gel electrophoresis (CSGE) followed by sequencing. Three mutations (21%) in the BRCA1 gene were identified. Two of them were considered to be novel, of which one is a missense in exon 7 near the ring finger domain, while the other was a one base pair deletion in exon 11 resulting in protein truncation. The third mutation, 185 delAG, was similar to that described in the Ashkenazi Jewish families. The study concluded a lower prevalence of BRCA1 germ line mutations in hereditary breast and/or ovarian cancer among Indian women.

The study by Varamathi et al. (2003) reports germ line BRCA1 mutations in three breast and ovarian cancer families (3 or more affected relatives under 45 years of age). Ten sporadic breast cancer patients and 50 normal control subjects were also studied, making a total of 90 cases as the study sample. Mutation analysis was carried out using the polymerase chain reaction-mediated site directed mutagenesis (PSM) and single stranded conformation polymorphism (SSCP), followed by sequencing of selected variant bands. No mutation was identified in the sporadic cases. Eight point mutations were identified in the familial group. Two deleterious pathogenic, protein truncating nonsense mutations were detected in exon 11 (E1250X) and exon 20 (E1745X), and six novel and unique amino acid substitutions (F1734S, D1739Y, V1741G, Q1747H, P1749A, R1753K). One complex missense mutation in exon 20 (V1741G; E1754X) was observed in only one family. These complex mutations exist only in breast cancer families but not in control populations of women. Three splice site variants (IVS20+3A>C, IVS20+4A>T, IVS20+5A>T) and two intronic variants (IVS20+21_22insG, IVS20+21T>G) were also detected. This study and the other two Indian studies provide useful data on the prevalence of BRCA1 germ line mutations in families with breast and/or ovarian cancer.

The study of Pakistani and Canadian collaborators (Liede et al., 2002) included a cohort of 341 case subjects with breast cancer, 120 case subjects with ovarian cancer, and 200 female control subjects from Karachi and Lahore, the two major cities of Pakistan. The prevalence of BRCA1 or BRCA2 mutations among case subjects with breast cancer was 6.7% (95% confidence interval [CI] 4.1%–9.4%), and that among case subjects with ovarian cancer was 15.8% (95% CI 9.2%–22.4%). Mutations of the BRCA1 gene accounted for 84% of the mutations among case subjects with ovarian cancer, and 65% of mutations among case subjects with breast cancer. The majority of the reported mutations are unique. Five BRCA1 mutations (2080insA, 3889delAG, 4184del4, 4284delAG, and IVS14-1A->G) and one BRCA2 mutation (3337->T) were found in multiple case subjects and represent founder mutations. The penetrance of deleterious mutations in BRCA1 and BRCA2 is comparable to that of Western populations. The cumulative risk of cancer to age 85 years in female first-degree relatives of BRCA1 mutation-positive case subjects was 48%, and was 37% for the first-degree relatives of the BRCA2-mutation-positive case subjects. A higher proportion of case subjects with breast cancer than control subjects were the progeny of first-cousin marriages (P = .001). The effects of consanguinity were significant for case subjects with early-onset cancer (age <40 years) (P = .0008) and case subjects with ovarian cancer (P = .002).

The authors of the Ontario-Pakistani study excluded the major nongenetic risk factors (dietary, smoking, alcohol, and exogenous hormones) as potential risk factors. The study concluded that genetic factors play a significant role in breast and ovarian cancer incidence in Pakistan. The prevalence and penetrance estimates for Pakistan suggest that dominant BRCA1 and BRCA2 mutations are significant contributors to breast and ovarian cancer incidence in this population. The association between parental consanguinity and breast cancer risk suggests that recessive genes may play a role in the etiology of breast cancer in young women. These data are also pertinent to immigrant women from Pakistan, who may seek genetic counseling for cancer risk in North America and Europe. This would probably equally apply to other women from other neighboring countries (India, Bangladesh, and Afghanistan) of the Indian subcontinent.

GASTROINTESTINAL CANCERS

Unlike oral cancer, there is no consistent epidemiologic data on esophageal, gastric, and colorectal cancers in the Indian subcontinent. Some studies on oral cancer have included esophageal cancers and utilize the same laboratory methods and biomarkers in establishing causal or preventive potential. Cancer genetic researchers and molecular biologists across India favor the glutathione S-transferases (GSTs) as one of the major detoxification enzymes that provide critical defense against carcinogens. This includes the hypothesis that altered frequencies of GST genotypes and environmental exposures might be associated with increased susceptibility for the development of esophageal cancer (Jain et al., 2006). A leading medical genetics center in north India, analyzed GST gene polymorphisms in a cohort of 100 esophageal cancer patients and 137 age and gender matched healthy controls (Jain et al., 2006). Frequencies of GSTT1 null, GSTM1 null and GSTP1 genotypes did not differ between patients and controls. However, a twofold risk was observed for GSTM1 null genotype in esophageal adenocarcinoma (OR 2.1; 95% CI = 0.53–8.6). The study design also included gene–environment interactions with smoking and alcohol in esophageal cancer. In patients with smoking habits, GSTM1 null and GSTP1 *ile/ile* genotype were at higher risk for esophageal cancer (OR 1.5; 95% CI = 0.50–4.4 and OR 1.3; 95% CI = 0.40–3.5, respectively). A moderate risk for cancer was observed from alcohol usage along with GSTM1 null (OR 1.3; 95% CI = 0.50–3.6) and GSTP1 *val/val* genotypes (OR 1.2; 95% CI = 0.20–5.7). Interaction of GST genotypes with occupational exposure did not affect the risk for esophageal cancer. These findings suggest that genetic polymorphisms of *GSTT*1, *GSTM*1, and *GSTP*1 are not associated with higher risk of esophageal cancer. However, interaction of smoking or alcohol with *GSTM*1 null or *GSTP*1 *ile/ile* moderately increases the risk for esophageal cancer in north Indian populations.

Cancer of the colon and rectum is one of the commonest causes of death (Hodgson and Maher, 1999). Its incidence among the Asian populations (<2 per 100,000 per year) is comparatively lower than in Africa (2.8 per 100,000 in Nigeria) and Europe (38 per 100,000 in the United Kingdom). Almost 98% of colorectal cancers are adenocarcinomas. Adenomas predispose to malignancy and coexist in about 75% of cases (Morson, 1966). Increased risk of colorectal cancer is associated with inflammatory bowel disease (ulcerative colitis and Crohn's disease). Kochhar et al. (1992) reported 1.8% incidence of bowel malignancy in patients with idiopathic ulcerative colitis. This study recommended colon cancer surveillance in all patients with inflammatory bowel disease (IBD) from the Indian subcontinent. A higher colon cancer incidence with IBD was also reported among Indian patients living in South Africa (Pillay et al., 1981).

Genetic factors are important in the pathogenesis of colorectal cancer, which is a multistep process (Fearon and Vogelstein, 1990). Mendelian types of colorectal cancer follow autosomal dominant inheritance with marked variable expression and incomplete penetrance. Familial adenomatous polyposis coli (FAPC) and hereditary nonpolyposis colorectal cancer (HNPCC) remain the major hereditary causes of colorectal cancer. However, dominant hereditary factors account for only 5% of colorectal cancers. The majority of the colorectal cancers follow multifactorial inheritance models. Empiric recurrence risk estimates are available for advising family members on their increased lifetime risk. Identification of high-risk family members is important to offer screening, which is shown to reduce morbidity and mortality associated with colorectal cancer (Hodgson and Maher, 1999).

Incidence of digestive tract malignancy including colorectal cancers (CRC) among the Indian subcontinent populations is not clear. However, it is estimated to be lower than the Western populations (Jussawalla and Gangadharan, 1977). Low incidence is likely to be due to generous intake of starch in the diet and the presence of natural antioxidants such as curcumin in Indian cooking (Mohandas and Desai, 1999). Nevertheless, it is recognized to be a major health burden in tertiary hospitals (Hart et al., 1992; Rawal et al., 1995; Mohandas and Jagannath, 2000) in the Indian subcontinent and also among the migrant Indian subcontinent ethnic groups.

Hart et al. (1992) collected data from 37 missionary establishments in India, Nepal, Pakistan, Bangladesh, and Bhutan. Information was collected on the frequency of both gastric cancers and CRCs. During the year of 1980, about 500,000 patients attended various hospitals. A total of 291 gastric tumors and 169 CRCs were diagnosed. In India and Pakistan there was no difference in number between gastric and CRC cases. The relative risk of developing gastric cancer rather than CRC in Bangladesh was 8 (95%, confidence limits 4.5–14.2) and in Nepal the relative risk was 4 (95%, confidence limits 2.0–7.0). A significant variation in the occurrence of cancer was observed between countries. Nepal had the highest and Pakistan the lowest numbers of both gastric and colorectal tumors. The family history was not reported with any tumor. Difference in relative incidence was attributed to local environmental factors, probably dietary.

Rawal et al. (1995) looked at hospital records of a major tertiary hospital in Delhi, India. Over the period from 1984 to 1993, a total of 1751 gastrointestinal malignancies (GIM) were confirmed, providing a 2.1% incidence of all GIMs. The relative distribution of malignancy according to site was esophagus 36.0%, stomach 19.9%, liver secondaries 13.9%, colon 9.8%, periampullary 5.6%, gallbladder 4.5%, duodenum 3.0%, malignant ascites 2.6%, hereditary colon cancer (HCC) 2.3%, and pancreas 1.1%. Mean age for cancer of esophagus was 53.5 +/– 11.4 years, stomach 51.8 +/– 12.9, colon 49.1 +/– 16.7, duodenum 45.3 +/– 11.4, malignant ascites 51.8 +/– 13.1, pancreas 46.9 +/– 15.3 and HCC 52.5 +/– 12 years, with an overall age of onset for all GIMs being 49.7 +/– 13.4 years. All malignancies were common in males except for gallbladder. Mohandas and Jagannath (2000) have provided population estimates for various GIMs. The incidence of cancer of esophagus and stomach is gradually declining. However, the incidence of cancers of colon, pancreas, liver, and gallbladder is gradually rising. This is attributed to rapid urbanization and changes in lifestyle including dietary habits. Although no examples of any hereditary form of GIM are cited, screening for secondary prevention is recommended in high-risk individuals including relatives of patients with familial bowel cancer.

However, there are occasional families living in the Western hemisphere with a hereditary form of CRC. Wallis et al. (1992) described germ line mutation in familial adenomatous polyposis coli (FAP) in an Indian family from Britain. There are a few families of an Indian subcontinent ethnic background, living in Britain, with multiple affected relatives with hereditary nonpolyposis colon cancer (HNPCC). Reports of familial CRC from India, Pakistan, or any other country of the region are uncommon. The large kindred reported by Bamezai et al. (1984) could be one of several families in India with FAP or HNPCC. It is likely that these are not properly accounted for due to lack of formal cancer registry systems. The few cancer registries in India and Pakistan do not offer any clear statistics on CRC (Sen et al., 2002; Bhurgri et al., 2002). Nevertheless, it is possible that there are families scattered across the subcontinent that are either not accounted for or remain undiagnosed. A recent report from south India describes application of mutation analysis and secondary prevention in a family with FAP (Savithri et al., 2000). A positive family history of colon cancer was present in 3.3% of young Sri Lankan patients (under 40 years) diagnosed to have CRC (de Silva et al., 2000). Increased clinical awareness for existence of familial CRC in Sri Lanka is recommended using the FAP model for predictive genetic testing, genetic counseling, and secondary prevention by colonoscopy and prophylactic surgery (de Silva and Fernando, 1998).

UROLOGICAL CANCERS

Prostate cancer is not widespread across the Indian subcontinent. The true age-related prevalence is unknown and the available figures are not representative of the total burden. Due to enormous sociocultural influences, serious life-threatening or life-limiting illnesses like prostate cancer are underreported, or often not reported at all. Similar to other populations, prostate cancer may also occur in a family sharing inherited predisposition. There are no structured epidemiological studies reported from the Indian subcontinent. However, a few studies have been reported focusing on genetic associations with selected candidate genes.

Among the selected candidate genes, glutathione-S-transferase (GST) gene has been the focus of a few studies in patients with sporadic prostate cancer (Srivastava et al., 2004). The GSTs are active in detoxifying a wide variety of endogenous or exogenous carcinogens, and genetic polymorphisms of GSTM1, GSTT1 and GSTP1 have been assessed to evaluate the relative risk of various cancers including prostate cancer. One study from north India (Srivastava et al., 2004) studied 127 patients with sporadic prostate cancer and 144 age- and population-matched controls. The GSTT1 and GSTM1 null genotypes were identified by multiplex polymerase chain reaction (PCR) in peripheral blood DNA samples, and GSTP1-313 A/G polymorphism was determined by PCR/restriction fragment length polymorphism. There was a significant association in null alleles of the GSTM1 (odds ratio 2.239, 95% confidence interval 1.37–3.65, $P = 0.001$) and GSTT1 (1.891, 1.089–3.282, $P = 0.026$) with prostate cancer risk, and in the –313 G alleles (Val) of the GSTP1 gene (2.48, 1.51–4.08, $P < 0.001$). The combined analysis of these three genotypes showed a further increase in the risks of prostate cancer (7.23, 2.42–22.6, $P < 0.001$). The study concluded that the GSTP1-313 G polymorphism, and null alleles of GSTM1 and GSTT1, are strong predisposing risk factors for sporadic prostate cancer in north India.

The prevalence of genetic polymorphism in the *CYP2D6*, *GSTM1*, and *GSTT1* genes has been investigated to find their association with risk of bladder cancer (Sobtia et al., 2005). The study consisted of 100 patients (97 males and 3 females) suffering from bladder cancer and 76 matching controls. The maximum number of patients in this study was in the age group of 61–70 years. While there was no association between the heterozygous (HEM) genotype of the *CYP2D6* gene with the risk of bladder cancer (OR = 1.00; 95% CI = 0.46–2.16), it was 1.5-fold with poor metabolizers (PM) genotype. When stratified according to different grades of bladder cancer, a significant association was found with an OR = 3.54 (95% CI = 0.89–13.98) in grade II, 3.3 (95% CI = 0.12–20.6) in grade III, and 1.67 (95% CI = 0.15–18.45) in grade IV. When stratified in relation to smoking status, significant association of the disease was found in heavy smokers with an OR = 2.13 (95% CI = 0.71–6.43). Subjects with the null genotype for *GSTM1* had a slightly significant association with the bladder cancer risk, and the risk increased to twofold with the *GSTT1* null genotype. Smoking status also revealed an impact on the prevalence of bladder cancer in the individuals with *GSTM1* and *GSTT1* null genotypes. The study by Sobtia et al. (2005) estimates a threefold increase in risk of developing bladder cancer in the presence of one copy of the variant *CYP2D6* (HEM) allele and null *GSTT1*.

NUTRITIONAL FACTORS AND CANCER PREVENTION

India is a developing country with one of the most diverse populations and diets in the world. Studies of Indian immigrants in Western societies indicate that rates of cancer and other chronic diseases, such as coronary heart disease and diabetes, increase dramatically after a generation in the adopted country. Change of diet is among the factors that may be responsible for the changing disease rates. Diet in India encompasses a diversity unknown to most other countries, with many dietary patterns emanating from cultural and religious teachings that have existed for thousands of years. However, the role of the Indian diet in causation or prevention of cancer is unclear. Increasingly, more attention is being focused on certain aspects of the Indian diet, such as vegetarianism, spices, and food additives. Of particular interest for cancer prevention is the role of turmeric (curcumin), one of the many ingredients in the popular Indian curry spice. Other ingredients of interest include cumin, chilies, kalakhar, Amrita Bindu, and various plant seeds. There is widespread belief that some of these spices possess anticancer properties for both cancer development and treatment. Few prospective studies, however, have been conducted to investigate the role of the Indian diet and its various components in prevention of cancer. From a public health perspective, there is an increasing need to develop cancer prevention programs responsive to the unique diets and cultural practices of the people of India (Sinha et al., 2003). This field has huge potential for basic and translational research, in particular with exponential growth in new generation genomic technologies and nutrigenomic database and bioinformatic tools.

SUMMARY

Clinical cancer genetics is a relatively new subspecialist field within the broad sphere of human genetics. The application and practice of human genetics in the Indian subcontinent is a new start. It is thus not unusual to face difficulty in finding adequate material to review in a specialized field such as cancer genetics. However, there are several reports and research publications describing wide-ranging methods and results related to human cancer. Not surprisingly, the bulk of the material is on oral cancer, which is a major health problem in India as well as in other neighboring countries. There is enormous data to link this with environmental factors, including sociocultural practices, notably chronic exposure to smoking, alcohol drinking, and the habit of tobacco and betel nut (and its products) chewing. Although several reports indicate chromosomal fragility and somatic mutations in p53 tumor suppressor genes, there is not much evidence to support a clear genetic susceptibility. However, it remains to be shown by further research.

The glutathione *S*-transferases (GSTs) are a polymorphic supergene family of detoxification enzymes involved in the metabolism of numerous potential carcinogens. Several allelic variants of polymorphic GSTs show impaired enzyme activity, and are suspected to increase the susceptibility to various cancers, notably oral, esophageal, gallbladder, prostate, and bladder. Most studies were carried out on north Indian cohorts and need to be replicated in other populations of the Indian subcontinent.

Familial occurrence of breast and colorectal cancers is occasionally reported. Recent reports describing various germ line BRCA1 and BRCA2 mutations in the Indian and Pakistani cohorts with breast and ovarian cancer provide evidence that a small, but nevertheless significant, proportion of these cancers are etiologically related to mutations in the breast/ovarian cancer susceptibility genes. Even a small percentage of incidence associated with germ line mutations would reflect in large numbers of families, given that over one billion people inhabit the Indian subcontinent. Although the incidence of colorectal cancer in the Indian subcontinent is lower compared to the Western populations, a small number of families should be spread across with either FAP or HNPCC. Further epidemiological work supported by core molecular analysis is urgently needed to work out the incidence of gastrointestinal malignancies associated with genetic susceptibility. A similar approach might help in studying relatively uncommon cancers, including hepatocellular carcinoma and malignant pancreatic tumors.

REFERENCES

Ahmed R, Shaikh H, Hasan SH. (1997). Is carcinoma breast a different disease in Pakistani populations? *J Pak Med Assoc* 47:114–116.

Babu KG. (2001). Oral cancers in India. *Semin Oncol* 28(2):169–173.

Balaram P, Srdhar H, Jajkumar T, et al. (2002). Oral cancer in southern India: The influence of smoking, drinking, paan-chewing and oral hygiene. *Int J Cancer* 98(3):440–445.

Bamezai R, Singh G, Khanna NN, Singh S. (1984). Genetics of site specific colon cancer: A family study. *Clin Genet* 26(2):129–132.

Bhurgri Y, Bhurgri A, Hasan SH, et al. (2000). Cancer incidence in Karachi, Pakistan: first results from Karachi Cancer Registry. *Int J Cancer* 85:325–329.

Bhurgri Y, Bhurgri A, Hasan SH, et al. (2002). Cancer patterns in Karachi division (1998–1999). *J Pak Med Assoc* 52(6):244–246.

Buch SC, Notani PN, Bhisey RA. (2002). Polymorphism at GSTM1, GSTM3 and GSTT1 gene loci and susceptibility to oral cancer in an Indian population. *Carcinogenesis* 23(5):803–805.

Chari ST, Mohan V, Pitchumoni CS, Viswanathan M, Madanagopalan N, Lowenfels AB. (1994). Risk of pancreatic carcinoma in tropical calcifying pancreatitis: an epidemiologic study. *Pancreas* 9(1):62–66.

Chopra R. (2001). The Indian scene. *J Clin Oncol* 19(Suppl 18):106S–111S.

Darr A, Modell B. (1988). The frequency of consanguineous marriage among British Pakistanis. *J Med Genet* 25:186–190.

Das N, Majumdar J, DasGupta UB. (2000). Ras gene mutations in oral cancer in eastern India. *Oral Oncol* 36(1):76–80.

De Silva MV, Fernando MS, Fernando D. (2000). Comparison of some clinical and histological features of colorectal carcinoma occurring in patients below and above 40 years. *Ceylon Med J* 45(4):166–168.

De Silva DC, Fernando R. (1998). Familial adenomatous polyposis. *Ceylon Med J* 43(2):99–105.

Fearon ER, Vogelstein B. (1990). A genetic model for colorectal tumorigenesis. *Cell* 61:759–767.

Ferguson A, Kingstone K. (1996). Coeliac disease and malignancies. *Acta Paediatr Suppl* 412:78–81.

Hart AR, Mann R, Mayberry JF. (1992). Gastric and colorectal cancer in the rural Indian subcontinent: A survey of patients attending mission hospitals. *Digestion* 51(2):110–114.

Hashmi M. (1997). Frequency of consanguinity and its effect on congenital malformation – a hospital based study. *J Pak Med Assoc* 47:75–78.

Heinzel PA, Balaram P, Bernard HU. (1996). Mutations and polymorphisms in the p53, p21 and p16 genes in oral carcinomas of Indian betel quid chewers. *Int J Cancer* 68(4):420–423.

Hodgson SV, Maher E. 1999. A practical guide to Human cancer genetics. Cambridge: Cambridge University Press.

Hussain R, Bittles AH. (1998). The prevalence and demographic characteristics of consanguineous marriages in Pakistan. *J Biosoc Sci* 30:261–275.

Iype EM, Pandey M, Mathew A, Thomas A, Sebastian P, Nair MK. (2001). Oral cancer among patients under the age of 35 years. *J Postgrad Med* 47(3):171–176.

Jain M, Kumar S, Rastogi N, Lal P, Ghoshal UC, Tiwari A, Pant MC, Baiq MQ, Mittal B.(2006). GSTT1, GSTM1 and GSTP1 genetic polymorphisms and interaction with tobacco, alcohol and occupational exposure in esophageal cancer patients from North India. *Cancer Lett*. 242(1):60–67.

Jussawalla DJ, Gangadharan P. (1977). Cancer of the colon: 32 years of experience in Bombay, India. *J Surg Oncol* 9(6):607–622.

Kamath SK, Murillo G, Chatterton RT Jr, et al. (1999). Breast cancer risk factors in two distinct ethnic groups: Indian and Pakistani vs. American premenopausal women. *Nutr Cancer* 35(1):16–26.

Kannan K, Krishnamurthy J, Feng J, Nakajima T, Tsuchida N, Shanmugan G. (2000). Mutation profile of the p53, fhit, p16INK4a/p19ARF and H-ras genes in Indian breast carcinomas. *Int J Oncol* 17(5):1031–1035.

Kattwinkel J, Lapey A, Di Sant'Agnese PA, Edwards WA. (1973). Hereditary pancreatitis: three new kindreds and a critical review of the literature. *Pediatrics* 51(1):55–69.

Kaur J, Srivastava A, Ralhan R. (1994). Over expression of p53 protein in betel- and tobacco-related human oral dysplasia and squamous-cell carcinoma in India. *Int J Cancer* 58(3):340–350.

Kochhar R, Goenka MK, Kaushik SP, Gupta NM, Nagi B, Mehta SK. (1992). Colorectal carcinoma in Indian patients with idiopathic ulcerative colitis. *Eur J Cancer Prev* 1(4):293–296.

Krishnamurthy J, Kannan K, Fend J, Mohanprasad BK, Tsuchida N. (2001). Mutational analysis of the candidate tumor suppressor gene ING1 in Indian oral squamous cell carcinoma. *Oral Oncol* 37(3):222–224.

Kumar BV, Lakhotia, Ankthli R, et al. (2002). Germline BRCA1 mutation analysis in Indian breast/ovarian cancer families. *Cancer Biol Ther* 1(1):18–21.

Liede A, Imtiaz A, Malik ZA, de los Rios P, Kwan E, Narod S. (2002). Contribution of BRCA1 and BRCA2 mutations to breast and ovarian cancer in Pakistan. *Am J Hum Genet* 71:595–606.

Mahale A, Saranath G. (2000). Microsatellite alterations on chromosome 9 in chewing tobacco-induced oral squamous cell carcinomas from India. *Oral Oncol* 36(2):199–206.

Malhotra V, Shah BS, Sabharwal S. (2001). Pattern of cancer in Dayanand Medical College & Hospital, Ludhiana (a ten year retrospective study). *Indian J Pathol Microbiol* 44(1):27–30.

Malik IA, Mubarak A, Luqman M, et al. (1992). Epidemiological and morphological study of breast cancer in Pakistan. *J Environ Pathol Toxicol Oncol* 11:353.

Mohandas KM, Desai DC. (1999). Epidemiology of digestive tract cancers in India. V. Large and small bowel. *Indian J Gastroenterol* 18(3):118–121.

Mohandas KM, Jagannath P. (2000). Epidemiology of digestive tract cancers in India. VI. Projected burden in the new millennium and the need for primary prevention. *Indian J Gastroenterol* 19(2):74–78.

Morson, BC (1996). Factors influencing the prognosis of early cancer of the rectum. *Proc. R. Soc. Med.* 15:607–612.

Munirajan AK, Mohanprasad BK, Shanmugam G, Tsuchida N. (1998). Detection of a rare mutation at codon 59 and relatively high incidence of H-ras mutation in Indian oral cancer. *Int J Oncol* 13(5):971–974.

Pandey SN, M. Jain, P. Nigam, G. Choudhuri and B. Mittal (2006). Genetic polymorphisms in *GSTM1, GSTT1, GSTP1, GSTM3* and the susceptibility to gallbladder cancer in North India. *Biomarkers* 11(3):250–261.

Parkin DM, Whelan SL, Ferlay J, Raymond L, Young J, eds. (1997). *Cancer incidence in five continents.* Lyon, France: IARC Scientific Publications No. 143, Vol. 7.

Pestonjamasp PH, Mittra I. (2000). Analysis of BRCA1 involvement in breast cancer in Indian women. *J Biosci* 25(1):19–23.

Pillay SP, Moshal MG, Spitaels JM, et al. (1981). Etiology of strictures in South African black and Indian patients. *Dis Colon Rectum* 24(2):107–113.

Powell JE, Kelly AM, Parkes SE, Cole TRP, Mann JR. (1995). Cancer and congenital abnormalities in Asian children: a population-based study from the West Midlands. *Br J Cancer* 72:1563–1569.

Ralhan R, Agarwal S, Nath N, Mathur, Wasylyk B, Srivastava A. (2001). Correlation between p53 gene mutations and circulating antibodies in betel-and tobacco-consuming North Indian population. *Oral Oncol* 37(3):243–250.

Ralhan R, Swain RK, Agarwal S, et al. (1999). P-glycoprotein is positively correlated with p53 in human oral pre-malignant lesions and is associated with poor prognosis. *Int J Cancer* 84(1):80–85.

Ravi D, Nalinakumari KR, Rajaram RS, Nair MK, Pillai MR. (1996). Expression of programmed cell death regulatory p53 and bcl-2 proteins in oral lesions. *Cancer Lett* 105(2):139–146.

Rawal KK, Jain M, Kumar N, Broor SL, Malhotra V. (1995). Spectrum of gastrointestinal malignancy over a period of 10 years at a tertiary referral center of India. *Trop Gastroenterol* 16(3):34–37.

Saranath D, Bhoite LT, Deo MG. (1993). Molecular lesions in human oral cancer: The Indian scene. *Eur J cancer B Oral Oncol* 29B(2):107–112.

Saranath D, Bhoite LT, Deo MG, et al. (1994). Detection and cloning of potent transforming tobacco-related human oral carcinomas. *Eur J cancer B Oral Oncol* 30B(4):268–277.

Saxena S, Szabo CI, Chopin S, et al. (2002). BRCA1 and BRCA2 in Indian breast cancer patients. *Hum Mutat* 20(6):473–474.

Savithri HS, Venkatesha Murthy HS, Baskaran G, et al. (2000). Predictive testing for familial adenomatous polyposis in a rural South Indian community. *Clin Genet* 58(1):57–60.

Sen U, Sankaranarayanan R, Mandal S, Ramanakumar AV, Parkin DM, Siddiqi M. (2002). Cancer patterns in eastern India: The first report of the Kolkata cancer registry. *Int J Cancer* 100(1):86–91.

Shami SA, Qaisar R, Bittles AH. (1991). Consanguinity and adult morbidity in Pakistan. *Lancet* 338:954–955.

Sikdar N, Ranjan Rashmi Paul, Bidyut Roy (2003). Predictive Markers and Cancer Prevention – Glutathione *S*-transferase *M3* (*A/A*) genotype as a risk factor for oral cancer and leukoplakia among Indian tobacco smokers. *International Journal of Cancer* 109(1):95–101.

Sinha R1, DE Anderson, SS McDonald, P Greenwald. (2003). Cancer Risk and Diet in India. *Journal of Postgraduate Medicine* 49(3):222–228.

Sreelekha TT, Ramadas K, Pandey M, Thomas G, Nalinakumari KR, Pillai MR. (2001). Genetic polymorphism of CYP1A1, GSTM1 and GSTT genes in Indian oral cancer. *Oral Oncol* 37(7):593–598.

Srivastava, DSL, Mandhani A, Mittal B, Mittal RD. (2004). Genetic polymorphism of glutathione S-transferase genes (GSTM1, GSTT1 and GSTP1) and susceptibility to prostate cancer in Northern India. *BJU International* 95(1):170–173.

Sobtia RC, Al-Badrana AI, Sharmaa S, Sharmab SK, Krishanc A, Mohand H. (2005). Genetic polymorphisms of *CYP2D6, GSTM1*, and *GSTT1* genes and bladder cancer risk in North India. *Cancer Genetics and Cytogenetics* 156(1):68–73.

Subhadra NV, Sundareshan TS, Satyanarayana M. (2003). Genetic susceptibility to oral cancer and the expression of common fragile sites: A study of 100 patients. *Cancer Genet Cytogenet* 140(1):70–72.

Tandle A, Sanghavi V, Saranath D. (2000). Infrequent loss of heterozygosity at adenomatous polyposis coli gene locus in Indian oral cancers. *Cancer Lett* 157(2):155–160.

Usmani K, Khanum A, Afzal H, Ahmad N. (1996). Breast carcinoma in Pakistani women. *J Environ Pathol Toxicol Oncol.* 15(2–4):251–253.

Varamathi MTAA, Deo SS, Shukla NK, Das SN. (2003). BRCA1 germline mutations in Indian familial breast cancer. *Hum Mutat* 21(1):98–99.

Wallis Y, MacDonald F, Rindi PM, et al. (1992). Germline APC mutation familial adenomatous polyposis coli in an Indian family. *Lancet* 340(8826):1035.

Whitcomb DC, Gorry MC, Preston RA, et al. (1996). Hereditary pancreatitis is caused by a mutation in the cationic trypsinogen gene. *Nat Genet.* 14(2):141–145.

Winter H, Cheng KK, Cummins C, Maric R, Silcocks P, Varghese C. (1999). Cancer incidence in the South Asian population of England (1990–1992). *Br J Cancer* 79(3–4):645–654.

Yaacob HB, Tan PL, Ngeow WC. (2002). Malignancy in oral lichen planus: A review of a group from the Malaysian population. *J Oral Sci* 44(2):65–71.

Zain RB. (2001). Cultural and dietary risk factors of oral cancer precancer–a brief overview. *Oral Oncol* 37(3):205–210.

86.

INHERITED DISORDERS OF THE EYE IN THE INDIAN SUBCONTINENT

Govindasamy Kumaramanickavel

அரிது அரிது மானிடராய் பிறத்தல் அரிது
அதனினும் அரிது கூண் குருடு செவிடு பேடு நீங்கி பிறத்தல் அரிது
பேடு நீங்கி பிறப்பினும் அறிவுடை மாந்தராதல் அரிது.

ஔவையார்

Rare, rare
It is rare to be born as a human being
It is very rare to be born without any handicap like physical deformities, blindness and deafness
Finally, it is RARE to be born as a normal, intelligent human being.

AVVAIYAR
An ancient Tamil female poet, India

INTRODUCTION

Ophthalmology and genetics have a very close association in the history of ophthalmology. The field of ophthalmology is referred to as a "playground for genetics." It was through ophthalmology that some of the fundamental issues in genetics were recognized and understood. One of the genetic diseases that was identified very early in human history, in 1796, was protanopia (color blindness). The first genetic disease to be mapped (associated) to the X-chromosome, in 1911, was color vision deficiency. In 1947 it was proved for the first time that one disease (hemophilia) and another (color vision deficiency) were linked together, which meant that the genes were lying close to each other in the X-chromosome. In 1963, for the first time a human disease was linked to an autosome, cataracta centralis pulverulenta, was linked to the Duffy blood group in chromosome 1. In 1995, it was shown that human disease could occur due to digenic inheritance; in retinitis pigmentosa, two genes could cause one genetic disease instead of two. Additionally, the concept of tumor suppressor genes in the field of cancer biology was established through the fundamental knowledge that was gained in retinoblastoma.

Medical genetics is *the study of inheritance of diseases in man*. Professor Victor A. McKusick of John Hopkins is the father of medical genetics, and his catalog *On-line Mendelian Inheritance in Man* (OMIM) is a compilation of 19,995 single gene disorders and mitochondrial diseases (see Chapter 1). Genetic abnormalities can be classified into four major categories: chromosomal abnormalities, single gene (monogenic or Mendelian) disorders, complex (multifactorial or polygenic) disorders, and mitochondrial disorders. We inherit 21,000 genes from each of our parents, and all these dictate what should be our skin and iris colors, height, weight, morphological features, etc. All these genes reside somewhere in the 23 pairs of chromosomes. From start of the first chromosome to the end of the last, we possess 3.2 billion letters of ATGC nucleotides. A major worldwide, concerted collaborative biological scientific effort was undertaken to genetically map and sequence the human genome, called the *Human Genome Project*. Nearly 100,329 mutations in 3739 disease genes have now been identified, including ocular genes.

Understanding human genetics and drafting the human genome has resulted in our currently possessing massive information—the potentials of this are unfolding remarkably. Currently we can identify defective genes in unborn children (prenatal diagnosis), and also predict to a prospective couple what diseases their children could carry (premarital or heterozygote screening); in addition, in late-onset diseases like cancers, presymptomatic diagnosis can be made, and technology exists now where we could screen a fertilized ovum for a genetic disease (preimplantation genetic diagnosis). With all this, a new medical field is emerging called *predictive medicine*.[1] The present chapter reviews the global problem of blindness in the genetics context. The genetic diseases of the eye are discussed as a major cause of blindness in a developing country like

India. Discussion of various nongenetic causes of blindness is beyond the scope of this chapter, and can be found in any ophthalmology text. The main focus is on relevant aspects of ophthalmology and medical genetics.

THE VISUAL SYSTEM

The genetic disorders of the eye involve both its complex structure and function. It is expected that the reader will have some basic understanding of the anatomy and physiology of the eye (see "Further Reading"). This section outlines the various components of the anatomy and physiology of the eye relevant to important genetic disorders.

The eye is situated in the orbit, and it is protected from outside by the upper and lower eyelids and eyelashes. The outer surface is kept moist and is lubricated by the lacrimal duct system. The pupil divides the globe into anterior and posterior segments (Figure 86-1). The anterior segment comprises the iris and is filled with the aqueous humor. Several genes regulate development of various structures of the anterior chamber. The paired-box axial system gene 6 [PAX6] is one of the important genes. The posterior chamber contains the lens and vitreous humor, surrounded by the choroid and retina. The retina has several neural layers sandwiched between the choroid and the retinal pigment epithelium (Figure 86-1). The lens and vitreous humor are made of collagen tissue regulated by the expression of COL2A1, COL11A1 and COL11A2 genes. Several complex gene systems are associated with the development, physiology, and biochemistry of the choroid, retina, and retinal pigmentary epithelium. Perception of the vision could be divided into two major biochemical cascades of reactions that can be broadly classified into (a) phototransduction pathway, and (b) visual cycle. The former takes place in the photoreceptors and the latter in the retinal pigment epithelium.

PHOTOTRANSDUCTION PATHWAY

The complex chemical reaction in the photoreceptors triggered by a photon is called *phototransduction* (Figure 86-2). A rod-specific opsin, bound to a photosensitive chromophore 11-cis-retinal, is the rhodopsin molecule. The chromophore is triggered by the photon. Initially, the chromophore undergoes a conformational change, and then a single rhodopsin catalyzes the activation of hundreds of G-protein transducins. The transducin stimulates cyclic guanosine monophosphate phosphodiesterase (cGMP PDE) to hydrolyze cyclic guanosine monophosphate (cGMP). The consumption of cGMP in the rod outer segment (ROS), and the consequent fall in the concentration, leads to the closure of the cGMP-gated cation channels (CNCG) in the photoreceptor plasma membrane.

The electrical equilibrium of ROS is maintained by inflow of sodium and calcium through CNCG channels and outflow of sodium and calcium through another channel. Hence, closure of the CNCG channels due to phototransduction events (but the other channel continues cation outflow) causes hyperpolarization due to sodium depletion that inhibits neurotransmission to the brain. Recovery of the photoreceptor cell and deactivation of rhodopsin is triggered by decreased calcium concentration. Firstly, rhodopsin is phosphorylated by rhodopsin kinase, followed by the capture of rhodopsin by the protein arrestin, which prevents further activation of transducin and releases the 11-trans-retinal from rhodopsin. This change prevents further activation of cGMP PDE; in addition, the concentration of cGMP is increased by the synthesis of cGMP by a retinal guanylate cyclase (RetGC) and also by a family of calcium-binding proteins, including recoverin (inhibits activity of rhodopsin kinase) and guanylyl cyclase activating protein [GCAP]. Due to the decrease of calcium in the phototransduction pathway, it is removed from recoverin and this releases the inhibition of rhodopsin kinase.

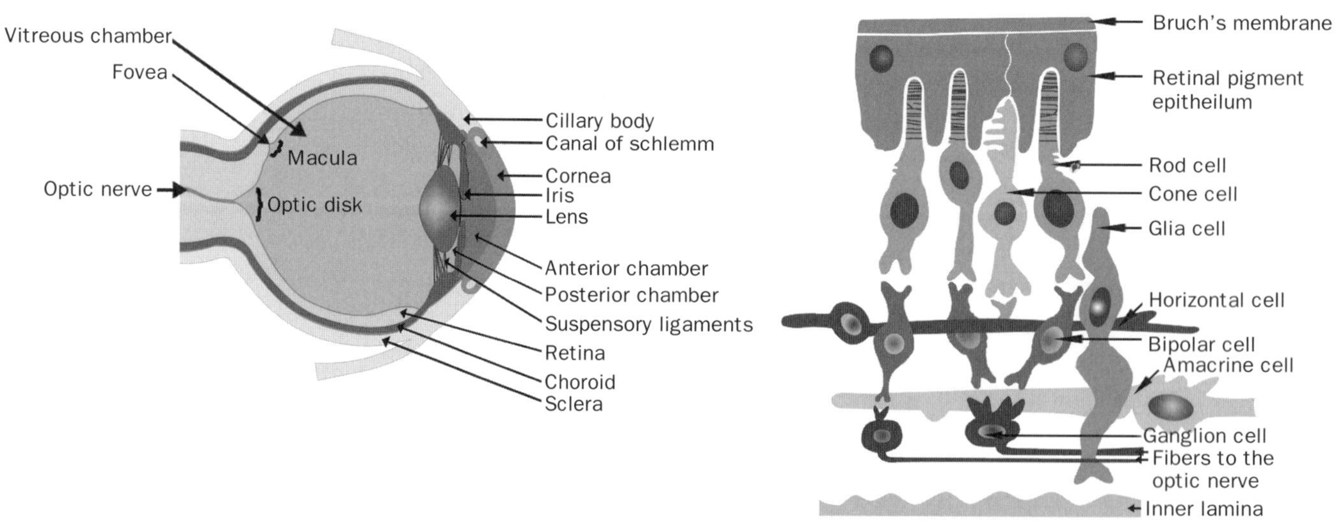

Figure 86-1 The structure of the eye (left) and the neural layers of the retina.

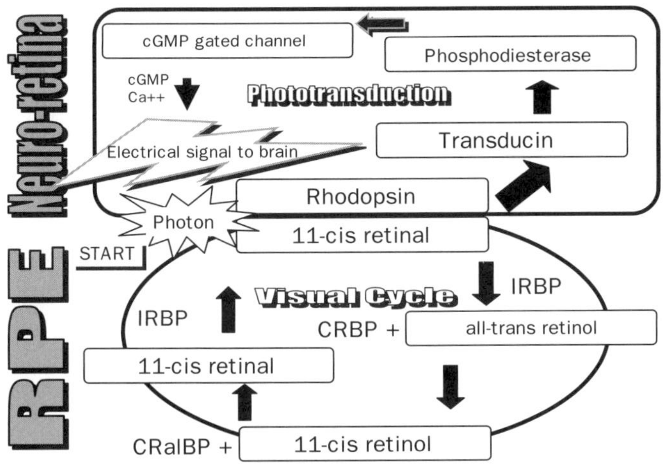

Figure 86-2 Phototransduction pathway and the visual cycle.

THE VISUAL CYCLE

In the rod outer segment (ROS), 11-*cis*-retinal is isomerized to all trans-retinal, and then is transported to the RPE by the interphototrector retinoid-binding protein (IRBP). Entering the RPE cell, all-trans-retinol is transferred to the cellular retinoid-binding protein (CRBP) and here, a specific isomerase converts all-trans-retinol to 11-*cis*-retinol. Now 11-*cis*-retinol gets bound to the cellular retinaldehyde-binding protein [CRalBP] and then is oxidized to 11-*cis*-retinal. The 11-*cis*-retinal released from the RPE cells is retransported to the ROS once again by the IRBP, where it helps in the regeneration of rhodopsin. One other key protein involved in this process is the RPE65 protein, constituting 10% of the total amount of proteins in the RPE cells.

In the three-dimensional structure, molecules obtain a particular structure that enables them to effectively capture quantum and conformational alterations, leading to unmasking of receptor sites that react with the cytoplasmic proteins. The 346-aminoacid-rhodopsin polypeptide folds into seven predominantly α-helices and embeds in the disk bilayer. The 11-*cis*-retinal chromophore is attached to lysine residue (codon 296) at the midpoint of the seventh helix, and the rest of the helices provide a cage-like structure for the chromophore. Photoactivation of 11-*cis*-retinal triggers isomerization of the chromophore to all-trans form, and there is conformational change in rhodopsin, exposing a binding site (residues 231 to 252) for G-protein transducin on the cytoplasmic surface, kicking off the phototransduction cascade pathway. In the termination of the phototransduction pathway and initiation of recycling of the rhodopsin molecule, first the serines and threonines on the cytoplasmic surface of the carboxy terminal of the activated rhodopsin (R*) are phosphorylated by rhodopsin kinase. On phosphorylation of R*, arrestin binds and prevents further binding of transducin. Two highly conserved cysteines on the intradiscal surface are important to structural stability of rhodopsin. In the hydrophobic regions or nearby, the oligosaccharide attachment may interfere with initial folding and glycosylation that would disturb membrane incorporation of rhodopsin. Transduction would be reduced or prolonged if there are mutations in the C-terminal of rhodopsin.

THE PROBLEM OF BLINDNESS

The leading causes of blindness and low vision in the world are cataract, trachoma, glaucoma, onchocerciasis, and xerophthalmia (Table 86-1). Worldwide today, it is

TABLE 86-1 GLOBAL DISTRIBUTION OF BLINDNESS BY CAUSE AND BY REGION

	Number of blind (x 1000)				
Region	Cataract	Trachoma	Glaucoma	Onchocerciasis	Other
Established market economies*	84	–	180	–	2,136
Former socialist economies of Europe	91	–	74	–	935
India	5,120	865	1,141	–	1,774
China	2,166	1,174	1,514	–	1,846
Other Asian countries and islands	2,314	1,362	973	–	1,151
Sub-Saharan Africa	3,101	1,380	853	358.5	1,407.5
Latin America and the Caribbean	1,326	158	183	1.5	631.1
Middle-Eastern Crescent†	1,627	927	205	–	841
Total (%)	15,820 (41.8)	5,866 (15.5)	5,123 (13.5)	360 (0.9)	10,722 (283)

*Western Europe, North America, Australia, Japan, and New Zealand. † Plus newly independent states in Central Asia.

estimated that about 180 million people have visual disability. According to the World Health Organization, around 40 to 45 million persons are blind and cannot walk about unaided, and if decisive public health intervention measures are not undertaken this figure is expected to double in the next 25 years.

Blindness due to cataract and glaucoma in India and China affects 6261 and 3680 million people, respectively; India has 42.3% more blind people than China. In India, 4 million people develop cataract each year. Astonishingly, 4 million cataract surgeries a year are performed by private practice ophthalmologists, private ophthalmic institutions, and the state governments, together with nongovernmental agencies. It is remarkable to note that the district of Coimbatore in the southern state of Tamil Nadu, India, is free of cataract-related blindness.

WHO data show that glaucoma is one of the leading causes for bilateral blindness. Recent prevalence estimates conclude that glaucoma is the third leading cause of blindness worldwide after cataract and trachoma. WHO estimates 105 million suspected cases of glaucoma (Thylefors et al., 1995; WHO, 1997). Approximately 80% of these are blind and live in the developing world. Studies on the prevalence rate of glaucoma in developed countries have shown it to vary from 1.2% to 2.1% (Tielsch et al., 1995). India is estimated to have one-fourth of the world's visually impaired population, and among them 9.1 million people have primary open-angle glaucoma (Dandona et al., 1999).

GENETIC EYE DISORDERS AND INDIA

In India, genetic disorders of the eye are widely prevalent; however, the data are scarcely retrievable. For example, in our country there are exclusive ocular genetic departments in Naryana Nethralaya (Bangalore), Sankara Nethralaya (Chennai), Arvind Eye Hospital (Madurai) and L.V. Prasad (Hyderabad). In a tertiary ophthalmic clinic in India, Vijaya et al. (1997) observed predominantly retinitis pigmentosa patients, followed by retinoblastoma, congenital stationary night blindness including Oguchi disease, aniridia, anophthalmos, chromosomal disorders (Down, Turner, Klinefelter syndromes), cone dystrophy, glaucoma, Laurence-Moon/Bardet-Biedl syndrome, Leber congenital amaurosis, myopia, oculocutaneous albinism, and others (Table 86-2).

In addition, uncommon disorders are also occasionally seen. For example, two unrelated male patients presented with retinitis pigmentosa (RP); one patient had sickle cell disease (SCD), while the other had situs inversus totalis (SIT). The combination of RP with SCD and RP with SIT has not hitherto been reported in literature (Madhavan et al., 2001). On average, every year, 1000 eye genetic disorder patients attend the clinic requiring a genetic diagnosis and counseling. Cytogenetic and molecular genetic laboratories support the clinic.

TABLE 86-2 GENETIC EYE DISORDERS IN AN OPHTHALMIC GENETIC CLINIC IN INDIA (VIJAYA AND KUMARAMANICKAVEL, 1997).

Disorder	Percentage
Retinitis pigmentosa	67.2%
Stargardts disease	6.6%
Retinoblastoma	6.9%
Congenital stationary night blindness including Oguchi disease	3.2%
Aniridia, Anophthalmos, Chromosomal disorders (Down, Turner, Klinefelter syndromes) Cone dystrophy, Glaucoma, LMBB syndrome, Leber congenital amaurosis, Myopia, Oculocutaneous albinism, Others	16.1%

CHROMOSOMAL DISORDERS

Down syndrome is the commonest chromosomal disorder seen in the clinic. In addition they come across patients with Turner and Klinefelter syndromes. Turner syndrome, however, is more common compared to Klinefelter syndrome; the latter was diagnosed incidentally in an RP patient. The common presenting ophthalmic problems in Down syndrome cases include squint and cataract. Structural chromosomal abnormalities are seen in patients with retinoblastoma, blepharophimosis, and aniridia—13q14, 3q22, and 11q15 deletions respectively. Overall, the chromosomal abnormalities constitute about 1.5% of the total caseload of the eye genetic clinic. The remaining 98.55% of cases belong to the single gene (Mendelian) or complex disorders. Predominantly, however, they diagnose and offer genetic counseling for Mendelian diseases.

CONSANGUINITY AND OCULAR GENETIC DISEASES

Consanguinity is widely practiced in Asia and Africa, and about 20% to 50% of all marriages in this part of the world are consanguineous (Bittles et al., 1991). Muslims in Pakistan, Bangladesh, Sri Lanka, and India, and Hindus in south India—particularly Andhra Pradesh and Tamil Nadu—generally practice close-relative marriage.

In a study conducted in south India, it has been found that 28.82% of patients had parents of consanguineous marriage (Kumaramanickavel et al., 2002a). This rate is particularly high compared to only 0.5% of consanguineous marriages in North America and Western Europe

(Bittles et al., 1991). Due to migration of rural people to the urban areas, consanguinity continues to be practiced in all areas. Despite conversion of a small section of the population to Roman Catholicism, consanguinity is still practiced, although the Catholic Church forbids this practice. Migration of such communities to the developed world has not stopped them from such practices, which is a cause of concern for health services in the Western countries, and particularly in the United Kingdom.

The most common type of consanguineous marriage is among first cousins (53.26%), followed by uncle–niece (26.3%) marriages, and 85.3% of them belong to the southern states of India and comprising Hindus (83.06%) and Muslims (14.86%). The coefficient of inbreeding in 673 families was 0.021 (Kumaramanickavel et al., 2002a). In these 673 families, 63.89% had retinitis pigmentosa; Stargardt's disease and cone dystrophy were diagnosed in 4.9%, and Leber congenital amaurosis in 4.3% of the cases. Other diagnoses included Lawrence-Moon/Bardet-Biedl syndrome (3.26%), Usher syndrome (1.9%), oculocutaneous albinism (1.6%), and Oguchi disease (1.33%).

SINGLE GENE DISORDERS AND THE EYE

Rapid progress has been made in the field of Mendelian genetics with the advent of polymerase chain reaction (PCR) and powerful linkage analysis software. Since 1989, the Human Genome Project has helped in mapping and cloning of several genes. Blindness was one focus of this effort, and therefore many eye-related genes were rapidly identified. The retina had a special place in this research. There are now several hundreds of these genes identified, which are listed in the database and are regularly updated (Tables 86-3 to 86-7; Genome Data Base, Human Genome

TABLE 86-3 CORNEAL DYSTROPHY GENES AND THEIR PHENOTYPES

Phenotype	Gene symbol	Gene	Chromosome
Corneal dystrophy, lattice type 1 Corneal dystrophy, Corneal dystrophy 3 Avellino type, Corneal dystrophy 4 Reis-Bucklers type, Corneal dystrophy lattice type IIIA, Corneal dystrophy, epithelial basement membrane 1, Groenouw type 1 granular corneal dystrophy	TGFB1	Transforming growth factor, beta induced (BIGH3)	5q31
Corneal dystrophy Fuch's endothelial 1, Corneal dystrophy 2 posterior polymorphous	COL8A2	Collagen, type VIII, alpha 2	1p34.3
Corneal dystrophy associated with teen-age perceptive deafness, corneal dystrophy Fuchs endothelial 4	SLC4A11	Solute carrier family 4, sodium bicarbonate transporter-like, member 11	20p13
Corneal dystrophy juvenile epithelial 1 of Meesmann	KRT3	Keratin 3	12q13
Pachyonychia congenita type 2	KRT6B	Keratin 6B	12q13.1-q13.2
Corneal dystrophy, juvenile epithelial 2 of Meesmann	KRT12	Keratin 12	17q21.1
Pachyonychia congenita type II	KRT17	Keratin 17	17q21-q22
Macular corneal dystrophy, type 1	CHST6	Carbohydrate(N-acetylglucosamine 6-O) sulfotransferase 6	16q22
Corneal dystrophy, gelatinous drop-like	TACSTD2	Tumor-associated calcium signal transducer 2 (membrane component, chromosome 1, surface marker 1, M1S1)	1p32-p31
Familial amyloid polyneuropathy IV	GSN	Gelsolin	9q33.3
Corneal fleck dystrophy	PIP5K3	Phosphatidylinositol-3 phosphate/phosphatidylinositol 5-kinase, type III	2q35
Congenital stroma corneal dystrophy	DCN	Decorin	12q21.33
Corneal dystrophy, Fuchs endothelial 6	ZEBI	Zinc finger E-box binding homeobox 1	10p11.2
Fish-Eye disease	LCAT	Lecithin-cholesterol acyltransferase	16q22.1
Keratosis follicularis spinulosa decalvans	SAT1	Spermidine/spermine N1-acetyltransferase 1	Xp22.13
Posterior polymorphous corneal dystrophy	VSX1	Visual system homeobox 1	20p11.21
Posterior polymorphous corneal dystrophy 3	ZEB1	Zinc finger E-box binding homeobox 1	10p11.22
Schnyder crystallin corneal dystrophy	UBIAD1	UbiA prenyltransferase domain containing 1	1p36.2

TABLE 86-4 CATARACT GENES AND THEIR PHENOTYPES

Phenotype	Gene symbol	Gene	Chromosome
Cataract punctate progressive, lamellar, aculeiform	CRYGD	Crystallin, gamma D	2q33–35
Cataract coppock-like	CRYGC	Crystallin, gamma C	2q33-q35
Cataract, zonular nuclear pulverulent	GJA8	Gap junction protein, alpha 8, 50 kDa (Connexin 50)	1q21-q25
Cataract progressive juvenile onset	BFSP2	Beaded filament structural protein 2, phakinin	3q21-q22
zonular pulverulent	GJA3	Gap junction protein, alpha 3, 46 kDa (connexin 46)	13q11-13
posterior polar congenital	PITX3	paired-like homeodomain transcription factor 3	10q25
posterior polar congenital	CRYAB	crystallin, alpha B	11q23.3-24.2
Polymorphic punctate lamellae cataract	AQPO	aquaporin family	12q12-14.1
Nuclear lamellar with sutural component	CRYBA3	Crystallin beta A3	17q11-q12
Cerulean (nuclear and cortical)	CCA1		17q24
Cortical, pulverulent, presenile cataract	LIM2	Lens intrinsic membrane protein 2	19q13.4
cataract, congenital, Marner type	HSF4	Heat shock transcription factor 4	16q22.1
Congenital zonular central Nuclear (microcornea)	CRYAA	Crystallin alpha A	21q22.3
Coppock (nuclear lamellar)	CRYBB2	Crystallin beta B2	22q11.2
Pulverulant	CRYBB1	Crystallin beta B1	22q11.2
Nuclear	CRYBB3	Crystallin beta B1	22q11.2
Congenital lamellar	CRYBB4	Crystallin beta B1	22q11.2

Mutation Database, GeneClinic, Genatlas, GeneCards, RetNet electronic databases).

DISEASES OF THE ORBIT AND LID

Ptosis is quite often seen at our clinic, along with some complicating squint disorders like Duane syndrome. Both are complex disorders, but familial cases are uncommonly seen.

A large Indian family with 17 affected members presented with ophthalmic abnormalities (ptosis and blepharophimosis), craniostenosis, limb, and external ear abnormalities (Maw et al., 1996). The inheritance appears to be autosomal dominant and all the affected members had nonsense *TWIST* gene mutation in chromosome 7p. (Dollfus et al., 2001, 2002)

DISEASES OF THE CORNEA

The cornea has several layers. Anterior to the posterior are epithelium, stroma, and endothelium. BIGH3 gene is involved in causing corneal dystrophies-lattice type

TABLE 86-5 GLAUCOMA GENES AND THEIR PHENOTYPE

Phenotype	Gene symbol	Gene	Chromosome
Peters' anomaly, Glaucoma primary congenital (buphthalmos), open angle	CYP1B1	Cytochrome P450, subfamily 1 (dioxin inducible), polypeptide 1	2p21
Axenfeld-Rieger anomalies with glaucoma, Peter's anomaly	FOXC1	Forkhead box C1 Gene	6p25
Glaucoma 1, open angle	MYOC	Myocilin, trabecular meshwork inducible glucocorticoid response (TIGR)	1q23-q24
Aniridia, Cataract, Secondary glaucoma, Peters' anomaly Cataract, with nystagmus Nystagmus, Ectopia pupillae Foveal hypoplasia, isolated Ocular anterior segment anomaly Keratitis, autosomal dominant WAGR syndrome	PAX6	Paired box homeotic gene 6	11p13
Exfoliation syndrome	LOXL1	Lysyl oxidase-like 1	15q24.1
Glaucoma, primary open-angle 1E	OPTN	Optineurin	10p14-p13

TABLE 86-6 RETINAL DEGENERATIVE GENES AND THEIR PHENOTYPES

Retinal Diseases	Locations known, genes yet to be identified	Genes identified	Some of the genes
Retinitis pigmentosa (AD, AR, XL) Cone or cone-rod dystrophy (AD, AR, XL) Congenital stationary night blindness (AD, AR, XL) (Above three retinal diseases have been grouped together, as many genes in the phototransduction pathway or visual cycle could cause any of these three disorders)	8 4 0	36 10 9	Retinal degeneration slow (RDS), Retinitis pigmentosa GTPase regulator, Rhodopsin (RHO), Retinal pigment epithelium specific protein (RPE65), Crumbs drosophila homologue of 1 (CRB1), Phosphodiesterase 6B cGMP-specific rod beta (PDE6B), Retinal outer segment membrane protein 1 (ROM1), C-mer proto-oncogene tyrosine kinase (MERTK), Tubby-like protein 1(TULP1), Retinitis pigmentosa GTPase regulator interacting protein 1 (RPGRIP1), Retinaldehyde binding protein (RLBP1), Neural retina leucine zipper (NRL), Phosphodiesterase 6A, cGMP-specific, rod, alpha (PDE6A), Nuclear receptor subfamily 2 group E3 (NR2E3), human homolog of yeast pre-mRNA splicing factor 3 (PRPF3), human homolog of yeast pre-mRNA splicing factor C8 (PRPF8), human homolog of yeast pre-mRNA splicing factor 31 (PRPF31), small nuclear ribonucleoprotein 200kDa (U5) (SNRNP200), semaphorin 4A (SEMA4A), RP1 protein (RP1), RP9 protein or PIM1-kinase associated protein 1 (RP9), topoisomerase I binding arginine/serine rich protein (TOPORS), Ceramide kinase-like protein (CERKL), Rod cGMP-gated channel alpha subunit (CNGA1), Rod cGMP-gated channel beta subunit (CNGB1), Lecithin retinol acyltransferase (LRAT), prominin 1 (PROM1)
Bardet-Biedl syndrome, autosomal recessive	0	12	McKusick-Kaufman syndrome protein, ADP-ribosylation factor-like 6, BBS proteins
Leber congenital amaurosis (AD, AR)	1	9	Cone-rod otx-like photoreceptor homeobox transcription factor (CRX), Inosine monophosphate dehydrogenase 1 (IMPDH1), Retinol dehydrogenase 12 (RDH12), Retinal-specific guanylate cyclase (GUCY2D), Retinal pigment epithelium specific protein (RPE65), Spermatogenesis associated protein 7 (SPAT7)
Macular degeneration (AD, AR)	6	8	Bestrophin 1 (BEST1), Elongation of very long fatty acids protein (EELOVL4), Retinal fascin homolog 2, actin bundling protein (FSCN2), Hemicentin 1 (fibulin 6) (HMCN1), Tissue inhibitor of metalloproteinases-3 (TIMP3), ATP-binding cassette transporter – retinal (ABCA4), Retinitis pigmentosa GTPase regulator (RPGR)
Optic atrophy (AD, AR, XL)	4	2	OPA1 protein, Transmembrane protein 126A (TMEM126A), Inner mitochondrial membrane translocase 8 homolog A (TIMM8A)
Usher syndrome (AR)	2	9	Cadherin-like gene 23 (CDH23), Whirlin (DFNB31), Myosin VIIA (MYO7A), harmonin (USH1C), Human homolog of mouse scaffold protein containing ankyrin repeats and SAM domain (USH1G), usherin (USH2A)
Retinopathy (mitochondrial)	1	7	Several mitochondrial proteins (KSS), complex I, III or IV proteins (LHON),
Total		202*	

* Includes other retinopathies not listed in the table.

TABLE 86-7 COLOR GENES AND THEIR PHENOTYPE

Phenotype	Gene symbol	Gene	Chromosome
Color blindness, total	CNGA3	Cyclic nucleotide gated channel alpha 3	2q
Tritanopia	OPN1SW	Opsin 1 (cone pigments), short-wave-sensitive (color blindness, tritan)	7q
Blue cone monochromatism Cone degeneration, x-linked Visual dichromacy	OPN1LW	Opsin 1 (cone pigments), long-wave-sensitive (color blindness, protan) RCP	Xp28
Blue cone monochromatism Cone degeneration, x-linked Visual dichromacy	OPN1LW	Opsin 1 (cone pigments), medium-wave-sensitive (color blindness, deutan) GCP	Xp28

1, Avellino, Reis-Bucklers, granular, lattice type 3A and Groenouw type 1 (Munier et al., 1997). It is a transforming growth factor β-induced gene, and codes for keratoepithelin. Macular dystrophies type 1 and 2 are caused by carbohydrate (N-acetylglucosamine 6-O) sulfotransferase 6 gene. Tumor-associated calcium signal transducer 2 causes gelatinous drop-like corneal dystrophy. However, mutations in genes encoding collagen type VIII, alpha 2, and keratin 3 & 12 cause Meesmann corneal dystrophy. Defects in the gelsolin gene caused corneal dystrophy 2 amyloidosis (Maury, 1990).

The following gives an account of various molecular genetic studies in India. Paliwal et al. (2009) have demonstrated a novel VSX1 gene c.525G>C, glutamine175histidine mutation, in an Indian patient with keratoconus. Paliwal et al. (2009), in another study, have shown a TGFBI c. 418G>A, arginine124 histidine mutation in two families with Avellino corneal dystrophy. Hemadevi et al. (2008) have established 10 different mutations in the SLC4A11 gene, including a complete deletion in patients with congenital hereditary endothelial dystrophy (CHED); for the same disease, Sultana et al. (2007) have established 27 different mutations; Ramprasad et al. (2007) showed 9 mutations in CHED2, including compound heterozytes, and Kumar et al. (2007) have also shown mutations in CHED2. In two families with severe granular corneal dystrophy (GCD), an arginine 555 tryptophan mutation was identified in the TGFB1 gene. Chakravarthi et al. (2005) have reported several mutations in the same gene causing both lattice and GCD. In macular corneal dystrophy (MCD) Sultana et al. (2003) have shown carbohydrate sulfotransferase-6 (CHST6) mutations as causative.

MYOPIA

Myopia is a major public health problem in Singapore and Hong Kong. In the races of Chinese origin, myopia seems to be widely prevalent. In these countries increased "nearwork" has brought this problem into an epidemic proportion. More than 60% of middle school children have myopia in these two countries. High myopia, a refraction error of equal to or below – 6 diopters in each eye, can be inherited as an autosomal dominant or recessive trait.

Three chromosomal loci have been identified (MYP1, 2, 3) for myopia (Young et al., 1998a,b). Gene locus for autosomal dominant myopia in seven families is mapped to chromosomal loci 18p11.31 (MYP2). Further analysis by transmission disequilibrium test suggested that the gene for 18p11.31-linked high myopia is most proximal to marker D18S52, with a likely interval of 0.8 cM between markers D18S63 and D18S52. The second locus for autosomal dominant myopia was mapped to chromosomal loci 12q21.3-q22 and named as MYP3, with a maximum LOD score of 3.85 by 2-point linkage analysis at a recombination fraction of 0.0010, for markers D12S1706 and D12S327 (Paluru et al., 2003). Decorin and lumican, which map to 12q23 and 12q21.3-q22 regions respectively, are suggested as possible candidate genes. These are members of the small interstitial proteoglycan family of proteins that are expressed in the extracellular matrix of various tissues. Both interact with collagen and limit the growth of fibril diameter. Two autosomal loci are mapped to chromosome 7q36 and 12 (Young et al., 1998b). One another locus, MYP1, is mapped to Xq28 in myopia with X-linked inheritance (Metlapally et al., 2009).

GLAUCOMA

The prevalence of glaucoma from various Western population-based studies varies between 1%–3% (Tielsch et al., 1995). Studies on the prevalence rate of glaucoma in developed countries indicate it to vary from 1.2% to 2.1% (Tielsch et al., 1995). Recent epidemiological studies in India have reported a higher prevalence in our population, ranging from 3.6% in the Andhra Pradesh Eye Diseases Study (APEDS) to 4.73% in the Vellore Eye Study (Thomas et al., 2003).

The commonest type of glaucoma worldwide is primary glaucoma, which is present in 2% of those over the age of 40 years, and the prevalence increases with age. India is estimated to have one-fourth of the world's visually impaired population. According to a worldwide survey on the prevalence of glaucoma, India is estimated to have 5.6 million people with primary open-angle glaucoma (POAG; see World Health Organization, 1998). In a study on the prevalence of glaucoma in the south Indian population, Dandona et al. (1999, 2000) have reported a prevalence rate of 2.56% for POAG and 1.08% of angle-closure glaucoma among people 40 years of age or older. The familial nature of glaucoma was first recognised in the mid-1800s. It was observed that glaucoma develops in 16% of first-degree relatives of those with POAG compared to <2% of the general population. High concordance of POAG was observed between monozygotic twins, and 13%–47% of POAG patients are reported to have positive family history (Leighton, 1976).

The mode of inheritance for POAG varies from autosomal dominant to multifactorial inheritance. Juvenile open-angle glaucoma, an early-onset form of POAG, is an autosomal dominant disorder with high penetrance. The first gene identified by positional cloning approach for POAG is in the GLC1A locus (1q21-31). This locus was first mapped to juvenile open-angle glaucoma (JOAG), an early-onset POAG, by Sheffield et al. in 1993. The GLC1A locus maps to the trabecular meshwork inducible glucocorticoid response (TIGR) gene in a study reported by Stone et al. (1997). The GLC1A gene spans about 20kb and has 3 exons of size 604, 126, and 782 bp. Sequence analysis predicted that GLC1A encodes an approximately 58-kD-504-amino acid protein, expressed as a 2.3kb transcript, sharing 40% homology in the carboxyl terminal domain with olfactomedin, a major component of the extracellular matrix of the olfactory neuroepithelium. The protein is expressed in cells of the trabecular meshwork (TM), a major site of aqueous humor outflow and the presumed site of the altered fluid dynamics found in patients with open-angle glaucoma. Mutations in the TIGR gene are shown to be present in ~4.6% of the randomly screened patients with adult forms of POAG, and include both missense and nonsense mutations It has been shown that accumulation of myocilin in the extracellular matrix might contribute to the obstruction of the outflow pathway and result in an elevated IOP (Fautsch et al., 2000).

About 199 sequence variants have been reported so far in the TIGR gene within its structural domain, out of which more than 30 are pathogenic. Among them, 16 variants are shown to have strong association with POAG in different ethnic populations (Fingert et al., 1999; Suzuki et al., 1997; Mansergh et al., 1998; Lam et al., 2000; Alward, 2000). Some of the mutations have been found to result in accumulation of the truncated protein in the cells, and to reduce the secretion of endogenous MYOC. Among the mutations in TIGR gene reported so far, the Gln368stop mutation in exon 3 of the gene is by far the prevalent mutation in different studies. This mutation was confined in the patients with glaucomatous optic atrophy due to high IOP, and also in patients with normal tension glaucoma. The Tyr 437 His and Ile 477 Asn mutation was found in every clinically affected member screened in a study. These mutations are reported to be associated with a form of glaucoma for which the average age at diagnosis is almost four decades earlier than that of the Gln368stop mutation, and in which the intraocular pressure is high. The Ile477Asn mutation is associated with such high IOP. The P370L mutation is consistently seen with earlier age at onset, and accompanying high pressures and tendency to be unresponsive to medications. By contrast, the E323K mutation is not associated with such an extremely high IOP, and seems more responsive to medications. A few groups in India have studied the role of MYOC gene in POAG (Kanagavalli et al., 2003; Markandaya et al., 2004; Sripriya et al., 2004; Bhattacharjee et al., 2007; Rose et al., 2007). It has been shown that Glycine48Histidine is the most common MYOC change that is observed in the Indian population (Chakrabarti et al., 2005). PCG and CYP1B1 mutation associations have been shown earlier, although the molecular mechanism is unknown.

In India, the pathology of primary congenital glaucoma (PCG) has been identified in 20 members while screening the CYP1B1 gene, and by observing five mutations in eight patients. five of them being novel (Panicker et al., 2002). Many more mutations were also identified (Reddy et al, 2004). Tanwar et al. (2009) have shown some novel mutations causing PCG in a profile of north Indian PCG patients. Chakrabarti et al. (2009) have shown FOXC1 gene as causative for PCG. Bhattacharjee et al. (2008) have shown the association of a CYP1B1 variant with POAG. Chakrabarti et al. (2007) have shown CYP1B1 haplotypes to be sharing mutations with POAG and PACG. Ramprasad et al. (2007) have applied linkage analysis and counseling in a family with CYP1B1 mutation. Kumar et al. (2007) interestingly analyze the role of CYP1B1, MYOC, OPTN, and OPTC genes in Indian POAG patients, and conclude that none of these genes have any significant role. The OPTN gene role in POAG is not significant (Mukhopadhyay et al., 2005; Sripriya et al., 2006). Ramprasad et al. (2008) established a strong association of LOXL1 gene rs3825942 variant in patients with exfoliation syndrome and glaucoma.

DISEASES OF THE LENS

One of the fascinating organs of the body is the lens of the eye. Its transparent nature is well maintained until middle age; however, it can become opaque and have unique morphological characters, which are called *cataracts*. The cataracts that can appear at birth, young age, or old age, are called congenital, developmental, or senile cataracts, respectively. Senile cataracts constitute 81% of blindness of all causes and affect 4 million people each year in India

(Vijaya et al, 1997). Environmental and genetic reasons are the causative agents for cataract.

So far, 18 different genetic factors have been shown to cause cataract; however, nothing is known about the genetic aspect of age-related cataracts. Cataract is one of the abnormalities in some syndromes like Peter's anomaly. The PAX6 gene, involved in causing cataract, also causes aniridia, secondary glaucoma, Peters' anomaly, cataract with nystagmus, nystagmus, ectopia pupillae, foveal hypoplasia, isolated ocular anterior segment anomaly, keratitis, and autosomal dominant WAGR syndrome. The PAX6 protein has a pan-effect at various parts of the eye, particularly the anterior segment; hence, a defect in the gene leads to broad consequences. The only posterior segment disorder that happens in the PAX6 mutation is foveal hypoplasia; why it is isolated, however, is not clearly understood. In a sutural cataract family from India, Kannabiran et al. (1998) have reported a splice site mutation in the crystalline β A1 gene in chromosome 17q.

DISEASES OF THE RETINA: RETINAL DEGENERATIVE DISEASES AND RETINITIS PIGMENTOSA

The knowledge of retinitis pigmentosa (RP) is about two and a half centuries old. In 1744, Ovelgun was the first person to have described familial night blindness and mentioned the hereditary nature of the disease. Nettleship in 1908 showed the hereditary nature of RP in a study involving 976 families, which was later confirmed by Usher in 1914. RP is the most common cause of inherited blindness in humans. It usually manifests in adolescence with night vision problems and loss of the midperipheral visual field. The diagnosis of RP can be made by demonstrating bilateral and symmetrical involvement, loss of peripheral vision, and rod dysfunction, evidenced by dark adaptometer or electroretinogram (ERG). In most patients the disease is progressive, and by adulthood there is considerable reduction in peripheral vision and macular function. At an advanced stage, the retinal vessels are attenuated, black bone spicule pigment deposition is distinctly observed in the retina, and the optic nerve head appears pale and waxy. Attenuated blood vessels further augment degenerative processes in the photoreceptor and neuroretina.

RP is inherited in all the three Mendelian forms: autosomal dominant (adRP), autosomal recessive (arRP) and X-linked recessive (xlRP). In the genetic forms, arRP is more rapidly progressive than adRP, whereas xlRP, although least common, is the most severe form (Kar et al., 1995). Among various inherited groups, arRP constitutes 60%, adRP 10% to 25%, and xlRP 5% to 18% (Kar et al., 1995). The most common are sporadic or isolated cases, where there is no history of any other family member with the disease. The worldwide prevalence of RP is 1 in 3500, and presently there should at least be 1.5 million affected individuals around the world. Countrywise, the prevalence rates of RP are: USA – 1/3700, UK – 1/3195, Switzerland – 1/7000, Israel – 1/4500, China – 1/4016, Norway – 1/4440, and Slovenia – 1/6023. The xlRP is predominant in the United Kingdom, adRP in North America, and arRP in consanguinity-practicing countries like Switzerland, Israel, and southern India. Sporadic RP [sRP] cases form approximately half of the cases reported in several studies. Genetic counseling could be difficult for the sRP group of patients and family members. The majority of sRP will be either arRP or xlRP.

Understanding RP at the molecular level started with linkage shown by Bhattacharya et al. (1984) using the RFLP DNA probe L1.28 in xlRP. In 1989, McWilliam et al. (1989) showed linkage of adRP to chromosome 3, and later Dryja et al. (1995) demonstrated the mutation in the rhodopsin gene at the same region. Later, several genes were linked and mutation identified by using brilliant and powerful techniques like linkage mapping, positional cloning, and molecular analysis of candidate genes. Presently there are 36 known or predicted RP genes—19 different gene mutations have been identified, but in 17 only a location has been determined. Currently the genes implicated in RP are RHO, PDE6A, PDE6B, CNGA1, SAG, RPE65, RLBP1, ABCA4, RGR, RDS, ROM1, PROML1, NRL, CRX, RP1, RP2, RPGR, CRB1, and TULP1. So far more than 80 mutations in the rhodopsin gene and 40 in the peripherin/RDS gene have been reported. The rhodopsin gene mutations account for 30% of adRP disease (al-Maghtheh et al., 1993). Unlike codon 347, which is identified all over the world, codon 23 mutations have been identified only in the United States and could be due to a "founder effect." Interestingly, mutations in rhodopsin also cause congenital stationary night blindness (CSNB; see al-Jandal et al., 1999). The hypothesis that these mutations might ultimately lead to apoptosis in the human retina, causing degeneration, has gained considerable importance.

The puzzle of ABCR/RIM protein not expressed in cones suggests that there could be an analogous equivalent in the cones. It has been observed that mutation in codon 258 with termination causes clinical presentation similar to adult vitelliform macular dystrophy, and mutations at codons 140 and 167 cause pattern dystrophy. Peripherin is made up of 346 amino acids, and has four transmembrane hydrophobic domains and two putative N-linked glycosylation sites. Peripherin/rds maintains the structural stability of ROS, and this is done by non-covalently binding to ROM1. Though puzzling, it has been shown that mutations do not directly cause photoreceptor cell death but actually trigger an apoptotic reaction (Marigo, 2007). Mutations in ABCR/RIM protein cause RP and cone-rod dystrophy. Mutations in peripherin/rds cause butterfly-shaped pigment dystrophy, Bull's Eye and adult vitelliform macular dystrophy, cone-rod pattern dystrophy, and fundus flavimaculatus. It is likely that a combination of L185P mutation in peripherin/rds and a null mutation in its homologous subunit ROM-1 causes a digenic form of adRP (Kajiwara et al, 1994). Cone rod homeobox [CRX] is emerging as a key gene causing retinal degenerations. CRX protein is member of the Otx/Otd subfamily of paired-like homeodomain transcription factors, which binds to the promoter regions of many photoreceptor-specific genes and transactivates expression of rhodopsin, IRBP, β-phosphodiesterase, and arrestin (Zack et al, 1999). CRX and the neural retina leucine zipper (Nrl) protein together modulate the expression of rhodopsin. Mutation in CRX gene causes Leber's congenital amaurosis, RP, and approximately 5% of cone-rod dystrophy. Mutation in the NRL gene causes adRP. The cellular retinaldehyde-binding protein [CRALBP], which is expressed predominantly in RPE and Mueller cells, when mutated causes arRP.

Van Soest and colleagues (1999) have classified all the mutations into three groups: mutations affecting (a) the regeneration and shedding of the photoreceptor outer segment, (b) visual transduction cascade, and (c) the visual cycle. Of the 100 rhodopsin mutations so far reported, except for a few with deletion, all result in an abnormal protein causing adRP (van Soest et al., 1999). The Pro23His mutation is observed in 12% of the population in the United States; most other mutations have been reported only in very few patients.

Transgenic mice studies have shown that there is defective folding and transport of mutated rhodopsin to the ROS and embedding in the membrane in adRP-like Pro23His mutations. In in-vitro studies, transducin is continuously activated in 296-codon rhodopsin mutation, the 11-*cis*-retinal binding site. However, when the same mutation is expressed in mice, rhodopsin was permanently phosphorylated and inactivated by arrestin; hence, instead of continuous activation, shutoff of the cascade sets the pace for photoreceptor degeneration. Interestingly, mutations proximal to this site cause congenital stationary night blindness and not RP. This could possibly be due to some aberration or increased incidence of spontaneous isomerization; however, in both situations there is no degeneration. Although more than 100 mutations have been described, surprisingly only two of them have arRP (Kumaramanickavel et al., 1994).

Peripherin/RDS mutations do not affect the photoreceptors in early stages; later, there is functional loss and appearance of lipofuscin-containing yellow deposits in the RPE, and finally there is degradation of outer segments, burdening the RPE cell metabolism. The affected phenotype is quite varied, including RP, retinitis punctata albescens, fundus flavimaculatus, pattern dystrophy of the RPE, or cone and cone-rod dystrophies (van Soest et al, 1999; Zack et al, 1999). Mice missing peripherin/RDS in homozygosity have increased outer segment dysplasia and accelerated photoreceptor cell death compared to heterozygous loss. ROM1 mutation separately does not cause disease; however, when it combines with peripherin/RDS mutation it has been shown to cause RP in four cases. In all the cases, the peripherin/RDS mutation was Leu185Pro, but ROM1 was three different mutations; proteins of these genes interact and form heterotertramers.

The cGMP phosphodiesterase (cGMP PDE) has four subunits (αβγγ). Both γ-subunits cover the α- and β-subunits catalytic binding sites, but activated transducin uncovers the binding site by releasing the inhibitory γ-subunits. Subsequently, cGMP PDE hydrolyzes and depletes cGMP in the ROS, which finally results in the closure of CNCG. Mutations in both the αβ subunits disrupts the function of cGMP PDE, resulting in high concentrations of cGMP leading to degeneration of photoreceptor cells by apoptotic mechanisms, causing adRP, and patients are night blind with constricted visual fields. Though γ-subunit is a key candidate gene for RP, to date no mutations have been reported. The rod CGMP-gated cation channel (CNCG) is composed of three subunits (αβγ). Mutations in the α subunit make the transduction pathway ineffective, resulting in constant activation of the process, in turn resulting in degeneration of the retina.

In India, a few mutations have been identified in the rhodopsin gene, but those too in a recessive manner (Kumaramanickavel et al., 1994)—Oguchi disease and arrestin, retinal pigment epithelium 65 protein and early-onset RP, and frameshift mutation in prominin (mouse)-like 1 protein mutation and RP (Maw et al., 2000). Gandra et al. (2008) screened RHO, PRPF31, RP1, and IMPDH1 genes and found the frequency to be 4% in isolated RP and 2% in autosomal dominant retinitis pigmentosa. The same group screened the RPE65 gene in Leber congenital amaurosis patients and found it to be less frequent (Mamatha et al., 2008). Singh et al. (2006) have shown various mutations in autosomal recessive RP (arRP) by screening TULP1, RLBP1, ABCA4, RPE65, and RP1 genes. Dikshit and Agarwal (2001) looked specifically at codons 345 and 347 of the rhodopsin gene in patients with RP.

Suganthalakshmi et al. (2007) reported RS1 mutations in patients with X-linked retinoschisis. Nallathambi et al. (2006) reported mutations in the FZD4 gene in patients with familial exudative vitreoretinopathy. In a patient with autosomal recessive retinal dystrophy, Singh et al. (2006) demonstrated a homozygous mutation in the ABCA4 gene. Kumar et al. (2004) showed a MYO7A gene mutation in a four-generation Usher syndrome family.

As it has been across the globe, Tyr302His was associated to age-related macular degeneration in the Indian population (Kaur et al., 2006). In the Indian population the same CFH SNP variation was shown to be associated with rs547154 (C2) and rs641153 (CFB) as reduced risk alleles (Kaur et al., 2010). The same group also showed that LOC387715 and HTRA1 SNPs were risk alleles for AMD in our population (Kaur et al., 2008).

RETINOBLASTOMA

The retinoblastoma (RB) gene is a tumor suppressor, one that regulates cell division, and any derangement due to two hits in it can cause malignancy (Knudson, 1971). Both alleles of the RB gene have to be inactivated by two-steps in the embryonal stem cells to result in a tumor (see also Chapters 44 & 85). Retinoblastoma gene inactivation could be due to chromosomal deletion, single nucleotide alteration, microdeletion, loss of heterozygosity (LOH), or methylation of the promoter region. Usually, alleles with premature termination of codons have complete penetrance and bilateral retinoblastoma, while missense alterations, substitutions in the promoter region, and some splice site mutations are associated with incomplete penetrance and reduced expressivity. In bilateral retinoblastoma patients, most of the mutations would result in premature termination of the protein, and such patients on average have more than three tumors in each eye (Harbour, 1998). Single base substitutions (40%–50%) and small length mutations (25%–30%) are the majority of mutations observed in retinoblastoma. Loss of constitutional heterozygosity at the intragenic loci was observed in 71% of the tumors from retinoblastoma patients, and of these, 91% had another mutation (Harbour, 1998).

Madhavan et al. (2009) showed overexpression of KIF14 and E2F3 in a large cohort of retinoblastoma tumors. They also correlated it with the older age at diagnosis and high expression of KIF14. In a high-throughput gene expression study Chakraborty et al. (2007) showed 1004 genes were upregulated and 481 were downregulated and studied few genes by RT-PCR.

Ramprasad et al. (2007) concluded that loss of heterozygosity could play a role in RB tumor formation. Joseph et al. (2006) could not establish a genotype–phenotype correlation in their study cohort. The same group tried to establish a diagnostic model for the region, including cytogenetic studies (Harini et al., 2001; Joseph et al., 2005), and they showed an epigenetic mechanism of methylation in the tumors at 6.6% (Joseph et al., 2004). They also established that genetic screening was cheaper than conventional clinical screening (Joseph et al., 2004). A fluorescence in-situ hybridization study Amare Kadam et al. (2004) showed that 12.3% of RB patients had a cytogenetic deletion; the usual observed worldwide is 6%–8%. Two groups in India have shown mutations in the RB gene as causative of RB (Kumaramanickavel et al., 2003; Kiran et al., 2003; Ata-ur-Rasheed et al., 2002).

DIABETIC RETINOPATHY

Diabetic retinopathy (DR) is one of the leading causes of visual morbidity. Prolonged hyperglycemia is a major risk factor for DR, and strict control of glycemia helps delay or prevents the onset of complications in diabetic patients. The Wisconsin Epidemiological Study on Diabetic Retinopathy (WESDR) suggests that genetic factors could promote the severity and rapidity of onset of retinopathy in diabetic patients. In general, hyperglycemia, poor metabolic control, cytokines, vasoactive hormones, and insulin resistance are responsible for the progression of diabetic microvascular diseases. Many genetic factors influence the onset of complications in diabetes. The potential candidate genes and gene products include nitric oxide, aldose reductase, tumor necrosis factor β; and receptors for advanced glycation end product (*ALR2*) (Chakravarthy et al., 1995; Kumaramanickavel et al., 2001a,b, 2002b,c, 2003).

A signal transduction molecule, nitric oxide, when modulated causes various diseases including diabetic retinopathy. In Northern Ireland a study has shown that allelic polymorphism of the inducible nitric oxide synthase (iNOS) gene is associated with retinopathy (Warpeha & Chakravarthy, 2003). In the Asian Indian population, eleven alleles, 175–225 bp, were identified, and allele 210 bp was significantly associated with retinopathy (Kumaramanickavel et al., 2002b). Allele 210 bp of the iNOS gene is a high-risk allele for developing retinopathy, and allele 220 bp protects an individual from developing further retinal complications. In the eye, NO is possibly produced by the iNOS in the Muller cells and the retinal pigment epithelium. The iNOS produced in the eye is believed to be involved in pathogenesis of diabetic retinopathy. Nitric oxide thus formed may also act as a potent angiogenic mediator by stimulating the growth-promoting effect of vasodilator peptides, including vascular endothelial growth factor (Uthra et al., 2008).

Upstream to the promoter region of the aldose reductase gene (ALR2), a $(CA)_n$ repeat polymorphism has been shown to be associated with retinopathy in different ethnic populations (Kumaramanickavel et al., 2003). We identified 13 alleles in our diabetic population, and the Z-2 allele also showed a significant association with those DR patients who had proliferative retinopathy and maculopathy. The Z-2 allele is, therefore, a high-risk allele for diabetic retinopathy in Asian Indian patients. *ALR2* is a rate-limiting enzyme in the polyol pathway that has been implicated in the development of long-term complications in diabetes for more than a quarter-century. *ALR2*, through the polyol pathway, converts glucose to sorbitol, which is finally converted to fructose. This pathway is accelerated in chronic hyperglycemia, producing increased metabolic and vascular abnormalities that may lead to tissue ischemia affecting eyes, kidneys, and the nervous system. In Australia and Britain, for example, type 1 diabetes patients with retinopathy have been shown to have increased frequency of the *ALR2* Z-2 allele that has $(CA)_{23}$ repeats. Similarly, in type 2 diabetes, Chinese and Japanese retinopathy patients have been shown to have an increased association with the Z-2 allele (Demaine et al., 2000).

The binding of advanced glycation end products to the receptor induces cellular oxidative stress and vascular dysfunction, and this is implicated in the pathogenesis of diabetic retinopathy. Our result suggests that Ser82 allele in the receptor for advanced end products gene is a low-risk allele for developing diabetic retinopathy in the Asian Indian patients who have type 2 diabetes. Among several etiopathological mechanisms proposed in diabetic retinopathy, advanced glycation end products (AGE) formed due to non-enzymatic glycation of proteins is one of the key components causing microvascular complications (Kumaramanickavel et al., 2002b,c).

The association of tumor necrosis factor beta (TNFβ) with diabetic retinopathy has been described previously (Hawrami et al., 1996). The TNFβ gene, otherwise known as lymphotoxin α (LTα), is shown to be one of the candidate genes involved in the predisposition to diabetic retinopathy. Individuals inheriting certain alleles have varying levels of expression of the gene, suggesting a functional link between genotype and phenotype.

In one of the earliest studies, Hawrami et al. (1996) showed the association of a tumor necrosis factor (TNF) beta-related polymorphism with insulin and non-insulin-dependent diabetes mellitus. Kumaramanickavel et al. (2001) showed low-risk association of TNF gene polymorphism allele 4 with DR, and allele 8 as a risk allele for proliferative diabetic retinopathy (PDR). The same group showed that the 210 allele of inducible nitric oxide synthase gene polymorphism is a high-risk allele for DR, whereas alleles 200 and 220 were protective (Kumaramanickavel et al., 2002c). They also showed that the serine 82 allele in the receptor for the advanced glycation end products (RAGE) gene is a low-risk allele for DR (Kumaramanickavel et al., 2002c). However, Ramprasad et al. (2007) showed a modest association of –347/A allele with RAGE. Kumaramanickavel et al. (2003) also showed that the Z-2 allele of the aldose reductase gene is strongly associated with DR; this is an allele that is much more uniformly associated across many populations.

Uthra et al. (2010) have demonstrated lack of association of protein kinase C beta 1 alleles and vascular endothelial growth factor (VEGF). However, Suganthalakshmi et al. (2006) have shown association of another polymorphism of VEGF to DR. The insulin growth factor-1 gene 18 repeats polymorphism is a susceptible genotype for DR (Uthra et al., 2007).

COLOR VISION DEFECT

Total color blindness is caused by cyclic nucleotide gated channel alpha 3 gene. Tritanopia is caused by the opsin-1 gene, which is short-wave sensitive. Blue cone monochromatism cone degeneration and X-linked visual dichromacy are caused by mutations in the opsin-1 gene; both disorders could be either medium- or long-wave sensitive. We rarely see patients with color vision defect, who come for genetic counseling worried about their profession and risk in future generations.

SUMMARY

In the history of medicine, ophthalmology and genetics have strong links. India stands ahead of China in world blindness with a record 42.3% more blind people. Cataract, glaucoma, and diabetic retinopathy are growing disorders in India due to the aging population. Eye genetics is a nascent field in India and is fast emerging as an exclusive sub-specialist field of ophthalmology. Retinitis pigmentosa and retinoblastoma are some of the common eye genetic disorders seen in a tertiary hospital. It could be due to the visual morbidity that these diseases get the most attention. Consanguinity is one of the main cases for autosomal recessive eye diseases, commonly encountered in genetic clinics.

As in the United Kingdom and other developed countries, India should focus on establishing more genetic clinics, as the large population of the country requires prenatal and presymptomatic diagnostic services. China, though equally populous, has embarked on a one-child formula, whereas India, being a democratic country, believes in family planning through acceptance. Hence, India would need such services for the future.

With the help of the Human Genome Project, several genes for varied single and complex eye disorders have been mapped. It is now possible to confirm and characterize specific disorders at the molecular level. Identification of a specific mutation is essential for presymptomatic diagnosis. The future applications may include gene therapy.[2] It is thus important that genetic laboratory services be developed to support eye genetic clinics. This would be required by major tertiary hospitals over the next decade across India. These facilities would need trained clinical and scientific personnel.

NOTES

1. Gene Therapy and Ophthalmology
Therapeutics in the next decade will be different. Gene therapy will be one of the major forms of treatment. By definition gene therapy is the "replacement of a deficient gene product or correction of an abnormal gene." Genetic disorders compared to heart or lung diseases are rare in a population, but when 14,729 single gene disorders and various other complex or multifactorial diseases are put together, they form a major group in health and disease. Gene therapy might not restrict itself to treating single gene disorders, but could also be used in treating complex or multifactorial disorders like diabetes, hypertension, glaucoma, psychiatric disorders, and cancer.

In 1986, in the United States, medical history was made when adenosine deaminase (ADA) enzyme was given to Ms. Laura Cay Boren, who had severe combined immune deficiency (SCID) disease. This triggered off an avalanche of research projects in gene therapy for both treating genetic diseases and finding techniques for the delivery of genes.

Gene therapy can be divided into *somatic cell therapy* and *germ line therapy*. In somatic cell therapy the somatic cells like brain, kidney, liver, or muscle are addressed for therapeutics. In this form of treatment the phenotype (the clinical expression) alone is addressed for correction, and this means that it would result in these people reaching reproductive age and having children who in turn would be transmitting the disease to future generations. Nevertheless, this would be the form of treatment in the next few decades. Germ line therapy will address the genetic problem at the inheritance level, hence correcting the gene forever in the gamete or fertilized ovum, and would not transmit the disease to future generations. Presently this method is considered to be morally and ethically unacceptable and therefore is prohibited in many countries.

In the mechanics of gene therapy, delivery vehicles at the molecular level have to be used to introduce genes into the somatic cells. The delivery vehicles or vectors are viruses, but other modes like liposome transfer, chemical, electroporation, microinjection, and particle bombardment are also used. Gene therapy can be delivered via the following methods: (i) Ex vivo: Cells are corrected outside the body and then reimplanted; such attempts have been made in treating various disorders including familial hypercholesterolemia and SCID. (ii) In situ: Vectors are introduced to a particular part of the body. In malignant melanoma, for example, genes are inserted into the melanoma patch in the skin. (iii) In vivo: Vectors with the desired gene are introduced into the venous blood, as in sickle cell anemia or hemophilia, or cystic fibrosis.

The tissues gene therapy targets are stem cells, endothelium skin, muscle, liver, lungs, nerve tissue, and eye. Gene therapy experiments for retinoblastoma, gyrate atrophy, and retinitis pigmentosa are in progress. Gene therapy for ocular inflammation, including autoimmune disease and infection, are also being looked into. In endophthalmitis, genes preventing apoptosis (*apoptosis* means cells that are unable to correct accumulating genetic defects undergo programmed cell death) of retinal elements, or genes that can influence neuroregenerative activity in the retina, like nerve growth factors, could be introduced, and this could stabilize the vision of RP patients. The Briard dog, a natural model of Leber congenital amaurosis, has been treated successfully, indicating directions for human treatment. Ribozymes are watchdogs in a cell that destroys defective mRNA, and rats have been treated with this method successfully for RP. In RPE 65 mice, vitamin A substitute has restored vision. The eye is an ideal organ for gene therapy, as such therapy would not leave the eye due to the blood-eye barrier and thus would avoid systemic effects. Furthermore, the eye is an easily accessible organ. For example, gene eye drops could be used to treat inherited corneal dystrophies!

Treating genetic disorders has now moved from speculation to reality. Dr. Robin Ali and his team at University College in London have successfully treated patients, for the first time, with the blinding disease LCA. Though there was an untoward incident initially in gene therapy, this work of scientists at London and few others in the United States has paved the way for gene therapy to be a mode of treatment in medicine. Presently, gene therapy clinical trials are ongoing in pulmonary emphysema, prostate cancer, age-related macular degeneration, painful diabetic nephropathy, hemophilia B, thyroid tumors, oral and maxillofacial malignant tumors, head and neck malignant tumors, Alzheimer's disease, non-small-cell lung cancer, Tay-Sachs, cancer and rheumatoid arthritis, cystic fibrosis, HIV, Huntington's disease, muscular dystrophy, Parkinson's disease, and severe combined immunodeficiency syndrome.

2. Predictive Medicine and Ophthalmology
Predictive medicine (PM) is an emerging medical specialty. The dawn of the Human Genome Project has lent further importance to PM. It is anticipated that PM might emerge as a major discipline in the field of clinical medicine. At present, using microarray molecular profiling techniques, it might be possible to predict clinical prognosis—and such a prediction has been reported in clear cell renal cell carcinoma. Similarly, in breast cancer there has been an association of amplification of erbB-2 receptor with poor clinical outcome, but the same association has also been shown to have good response for adriamycin-based therapy. Predicting the clinical status of human breast cancer is now possible by using gene expression profiles. Shortly, it is also predicted that genotype profile of a patient and drug response for a disease could be investigated in advance, using computer software, well before drugs are prescribed. Hence it is anticipated that the practice of medicine in the coming decades is going to be entirely different. Some of the pharmaceutical companies involved in PM research and service are Becton Dickinson, Bristol-Myers Squibb, Celera Genomics, Health Screen America, and Millennium Predictive Medicine.

Predictive testing using association and mutation data would need mandatory precounseling prior to the test, where the pros and cons of the test would be explained to the consultand (the person who reports for counseling, who could be affected or normal) and the family. The test would involve collection of few ml. of blood from the consultand to obtain DNA. The test would be performed by screening the possible genes involved or, in certain situations, blood would be drawn from family members to establish, for example, a linkage to a locus for glaucoma. Single nucleotide polymorphism (SNP) or allelic polymorphism could also be tested for association or linkage disequilibrium with glaucoma.

A dense SNP map of DNA sequence is being developed for the human genome, and presently studies are ongoing to show its association with various human diseases. The single nucleotide variation in the human genome is a rich and abundant polymorphism compared to microsatellite polymorphism. In SNP there is at least one change every 200 bases, and hence it is more significant and powerful in detecting linkage disequilibrium for disease loci. The potential of this variation would be exploited in the application of disease associations, developing diagnostics, studying therapeutic variations, and designing preventive strategies. All this can be applied and would be fine-tuned for complex diseases like diabetes, hypertension, glaucoma, coronary artery disease, or even cancers and obesity. Now, several disease susceptibility alleles and even SNP haplotypes have been shown to be associated with diseases. Constructing haplotypes for SNPs and their associations to diseases will have good predictive advantage with less chance for error. For successful practice, predictive medicine would need expertise from various disciplines such as human genetics, population genetics, biostatistics, and computational biology. As widely believed, disease susceptibility alleles and mutations should be tested only when there are preventive benefits, availability of lifestyle modifying strategies, and early treatment modalities; otherwise, these tests could do more harm than good. Predictive tests for

Alzheimer's disease has led susceptible but presymptomatic individuals to commit suicide, as there are no preventive or lifestyle modifying strategies or treatment modalities available for this disease. However, glaucoma is an ideal disease where such tests could help to prevent or reduce the morbidity of the disease.

The efficacy to breakdown and effect of toxicity of drugs is a complex mechanism, driven by multiple genes. The science, which deals with the genetic basis for variable response to drugs among individuals, is called *pharmacogenetics* and *pharmacogenomics*. Individuals with homozygous CYP2D6 null alleles are poor metabolizers of drugs like debrisoquine, metoprolol, nortriptyline, and propafone. Likewise, individuals with null allele for P450 isoform CYP2C19 have metabolizing difficulty with drugs like omeprazole, diazepam, propranolol, mephenytoin, amitriptyline, hexobarbital, and other drugs. The psychiatric illness depression is estimated to cost the United States US$53 billion each year, and the market for antidepressant drugs in 1998 was $8 billion. The effect of antidepressant drugs varies considerably from person to person, and this area is being extensively investigated for personalized therapy. Out of the 3.1 billion prescriptions given in the United States, approximately 2.1 million develop adverse reactions, 1 million need hospitalization, and more than 100,000 may die. The individual variation in drug response can be classified by accurate genotyping of SNPs and showing association to drug response. Good fundamental understanding in this science would ensure drug prescriptions to be safe and personalized. Prescription of drugs would move from a "one-drug-fits-all" concept to personalized and safe therapy. Some of the companies involved in this research area are Celera Genomics, diaDexus, DNAPrint, deCODE Genetics, Gemini Genomics, Genaissance Pharmaceuticals, Gene Logic, Genentech, Genome Therapeutics, Genomica, Genset, Golden Helix, GlaxoSmithKline, Genzyme Genetics, Incyte Genomics, Millennium Pharmaceuticals, Myriad Genetics, Novartis, Oxagen, Quintiles, and Variagenics.

REFERENCES

al-Jandal N, Farrar GJ, Kiang AS, et al. (1999). A novel mutation within the rhodopsin gene (Thr-94-Ile) causing autosomal dominant congenital stationary night blindness. *Hum Mutat* 13(1):75–81.

al-Maghtheh M, Gregory C, Inglehearn C, Hardcastle A, Bhattacharya S. (1993). Rhodopsin mutations in autosomal dominant retinitis pigmentosa. *Hum Mutat* 2(4):249–255. Review..

Alward WL. (2000). The genetics of open-angle glaucoma: the story of GLC1A and myocilin. *Eye (London)* 14(Pt 3B):429–436. Review.

Amare Kadam PS, Ghule P, Jose J, et al. (2004). Constitutional genomic instability, chromosome aberrations in tumor cells and retinoblastoma. *Cancer Genet Cytogenet* 150(1):33–43.

Ata-ur-Rasheed M, Vemuganti G, Honavar S, Ahmed N, Hasnain S, Kannabiran C. (2002). Mutational analysis of the RB1 gene in Indian patients with retinoblastoma. *Ophthalmic Genet* 23(2):121–128.

Bhattacharya SS, Wright AF, Clayton JF, Price WH, Phillips CI, McKeown CME, Jay M, Bird AC, Pearson PL, Southern EM, Evans HJ. (1984). Close genetic linkage between X-linked retinitis pigmentosa and a restriction fragment length polymorphism identified by recombinant DNA probe L1.28. *Nature* 309:253–255.

Bhattacharjee A, Acharya M, Mukhopadhyay A, et al. (2007). Myocilin variants in Indian patients with open-angle glaucoma. *Arch Ophthalmol* 125(6):823–829.

Bhattacharjee A, Banerjee D, Mookherjee S, Acharya M, Banerjee A, Ray A, Sen A, Variation Consortium TI, Ray K. (2008). Leu432Val polymorphism in CYP1B1 as a susceptible factor towards predisposition to primary open-angle glaucoma. *Mol Vis* 14:841–850.

Bittles AH, Mason WM, Greene J, Rao NA. (1991). Reproductive behavior and health in consanguineous marriages. *Science* 252:789–794. [www.consang.net].

Chakrabarti S, Devi KR, Komatireddy S, et al. (2007). Glaucoma-associated CYP1B1 mutations share similar haplotype backgrounds in POAG and PACG phenotypes. *Invest Ophthalmol Vis Sci* 48(12):5439–4445.

Chakrabarti S, Kaur K, Komatireddy S, et al. (2005). Gln48His is the prevalent myocilin mutation in primary open angle and primary congenital glaucoma phenotypes in India. *Mol Vis* 11:111–113.

Chakrabarti S, Kaur K, Rao KN, et al. (2009). The transcription factor gene FOXC1 exhibits a limited role in primary congenital glaucoma. *Invest Ophthalmol Vis Sci* 50(1):75–83.

Chakraborty S, Khare S, Dorairaj SK, Prabhakaran VC, Prakash DR, Kumar A. (2007). Identification of genes associated with tumorigenesis of retinoblastoma by microarray analysis. *Genomics* 90(3):344–353.

Chakravarthi SV, Kannabiran C, Sridhar MS, Vemuganti GK. (2005). TGFBI gene mutations causing lattice and granular corneal dystrophies in Indian patients. *Invest Ophthalmol Vis Sci* 46(1):121–125.

Chakravarthy U, Stitt AW, McNally J, Bailie JR, Hoey EM, Duprex P. (1995). Nitric oxide synthase activity and expression in retinal capillary endothelial cells and pericytes. *Curr Eye Res* 14:285–294.

Chand A, Kumaramanickavel G, Abraham M, Gahlot DK, Apte BN, Denton MJ. (1991). Indian recessive pedigrees with recessive retinitis pigmentosa: A potential for homozygosity mapping. In: Humphries P, Battacharya S, Birrd A. (eds.), *Degenerative Retinopathies: Advances in Clinical and Genetic Research* (pp. 35–40). Boca Raton, FL: CRC Press.

Dandona L, Dandona R, Mandal P. (2000). Angle-closure glaucoma in an urban population in southern India. The Andhra Pradesh eye disease study. *Ophthalmology* 107:1710–1716.

Dandona L, Dandona R, Naduvilath TJ, et al. (1999). Burden of moderate visual impairment in an urban populaion in southern India. *Ophthalmology* 106:497.

Demaine A, Cross D, Millward A. (2000). Polymorphisms of the aldose reductase gene and susceptibility to retinopathy in type 1 diabetes mellitus. *Invest Ophthalmol Vis Sci* 41(13):4064–4068.

Dikshit M, Agarwal R. (2001). Mutation analysis of codons 345 and 347 of rhodopsin gene in Indian retinitis pigmentosa patients. *J Genet* 80(2):111–116.

Dollfus H, Biswas P, Kumaramanickavel G, et al. (2002). Saethre-Chotzen Syndrome Notable intrafamilial phenotypic variability in a large family with Q28X TWIST mutation. *Am J Med Genet* 109(3):218–225.

Dollfus H, Kumaramanickavel G, Biswas P, et al. (2001). Identification of a new TWIST mutation (7p21) with variable eyelid manifestations supports locus homogeneity of BPES at 3q22. *J Med Genetics* 58:470–471.

Dryja TP, Finn JT, Peng YW, McGee TL, Berson EL, Yau KW. (1995). Mutations in the gene encoding the alpha subunit of the rod cGMP-gated channel in autosomal recessive retinitis pigmentosa. *PNAS* 92(22):10177–10181.

Dryja TP, Li T (1995). Molecular genetics of retinitis pigmentosa. *Hum Mol Genet* 4:1739–1743.

Fautsch MP, Bahler CK, Jewison DJ, Johnson DH. (2000). Recombinant TIGR/MYOC increases outflow resistance in the human anterior segment. *Invest Ophthalmol Vis Sci* 41(13):4163–4168.

Finckh U, Xu S, Kumaramanickavel G, et al. (1998). Homozygosity mapping of autosomal recessive retinitis pigmentosa locus (RP22) on chromosome 16p12.1-p12.3. *Genomics* 48:341–345.

Fingert JH, Héon E, Liebmann JM, et al. (1999). Analysis of myocilin mutations in 1703 glaucoma patients from five different populations. *Hum Mol Genet* 8(5):899–905.

Harbour JW. (1998). Overview of RB gene mutations in patients with retinoblastoma: implications for clinical genetic screening. *Ophthalmology* 105:1442–1447.

Gandra M, Anandula V, Authiappan V, et al. (2008). Retinitis pigmentosa: mutation analysis of RHO, PRPF31, RP1, and IMPDH1 genes in patients from India. *Mol Vis* 14:1105–1113.

Genetic glossary is available at http://www.nhgri.nih.gov/glossary/.

Goldberg I. (2001). Glaucoma: diagnosis, screening, and management. *Asian Journal of Ophthalmology* 3:1.

Gu S, Kumaramanickavel G, Srikumari CR, Denton MJ, Gal A. (1999). Autosomal recessive retinitis pigmentosa locus RP28 maps between D2S1337 and D2S286 on chromosome 2p11-p15 in an Indian family. *J Med Genet* 36(9):705–707.

Gu SM, Thompson DA, Srikumari CR, et al. (1997). Mutations in RPE65 cause autosomal recessive childhood-onset severe retinal dystrophy. *Nature Genetics* 17:194–197.

Harbour JW. (1998). Overview of RB Gene Mutations in patients with retinoblastoma. *Ophthalmology* 105:1442–1447.

Kajiwara K, Berson EL, Dryja TP. (1994). Digenic retinitis pigmentosa due to mutations at the unlinked peripherin/RDS and ROM1 loci. *Science* 264(5165):1604–1608.

Harini R, Ata-ur-Rasheed M, Shanmugam MP, Amali J, Das D, Kumaramanickavel G. (2001). Genetics of retinoblastoma: A cytogenetic and segregation analyses profile of 81 patients from a referral hospital in southern India. *Ind J Ophthalmol* 49:37–42.

Hawrami K, Hitman GA, Rema M, et al. (1996). An association in non-insulin-dependent diabetes mellitus subjects between susceptibility to retinopathy and tumour necrosis factor polymorphism, *Human Immunol* 46:49–54.

Hayashi N, Ido E, Yuji O, Ueno H. (1999). An experimental application of gene therapy for human retinoblastoma. *Invest Ophthalmol Vis Sci* 40:265–272.

Hemadevi B, Veitia RA, Srinivasan M, et al. (2008). Identification of mutations in the SLC4A11 gene in patients with recessive congenital hereditary endothelial dystrophy. *Arch Ophthalmol* 126(5):700–708.

Inglehearn CF. (1998). Molecular genetics of human retinal dystrophies. Eye 12:571–579.

Jacob A, et al. (1998). Prevalence of Primary Glaucoma in an Urban South Indian Population. *Indian J Ophthalmol* 46:81–86.

Joseph B, Anuradha S, Soumittra N, et al. (2002). RPE 65 gene: Multiplex PCR and Mutation screening in patients from India having retinal degenerative diseases. *Journal of Genetics* 81:1.

Joseph B, Madhavan J, Mamatha G, Ramprasad VL, Gopal L, Kumaramanickavel G. (2006). Retinoblastoma: A diagnostic model for India. *Asian Pac J Cancer Prev* 7(3):485–488.

Joseph B, Mamatha G, Raman G, Shanmugam MP, Kumaramanickavel G. (2004). Methylation status of RB1 promoter in Indian retinoblastoma patients. *Cancer Biol Ther* 3(2):184–187.

Joseph B, Paul PG, Elamparithi A, et al. (2005). Karyotyping in retinoblastoma–a statistical approach. *Asian Pac J Cancer Prev* 6(4):468–471.

Joseph B, Raman R, Uthra S, et al. (2006). Genotype-phenotype correlation analysis in retinoblastoma patients from India. *Asian Pac J Cancer Prev* 7(4):619–622.

Joseph B, Shanmugam MP, Srinivasan MK, Kumaramanickavel G. (2004). Retinoblastoma: Genetic testing versus conventional clinical screening in India. *Mol Diagn* 8(4):237–243.

Kanagavalli J, Krishnadas SR, Pandaranayaka E, Krishnaswamy S, Sundaresan P. (2003). Evaluation and understanding of myocilin mutations in Indian primary open angle glaucoma patients. *Mol Vis* 9:606–614.

Kannabiran C, Rogan PK, Olmos L, Basti S, Rao GN, Kaiser-Kupfer M, Hejtmancik JF. (1998). Autosomal dominant zonular cataract with sutural opacities is associated with a splice mutation in the ßA3/A1-crystallin gene. *Mol Vision* 4:21.

Kannabiran C, Sridhar MS, Chakravarthi SK, Vemuganti GK, Lakshmipathi M. (2005). Genotype-phenotype correlation in 2 Indian families with severe granular corneal dystrophy. *Arch Ophthalmol* 123(8):1127–1133.

Kar B, John S, Kumaramanickavel G. (1995). Retinitis pigmentosa in India: A genetic and segregation analysis. *Clinical Genetics* 47: 75–79.

Kaur I, Hussain A, Hussain N, et al. (2006). Analysis of CFH, TLR4, and APOE polymorphism in India suggests the Tyr402His variant of CFH to be a global marker for age-related macular degeneration. *Invest Ophthalmol Vis Sci* 47(9):3729–3735.

Kaur I, Katta S, Hussain A, et al. (2008). Variants in the 10q26 gene cluster (LOC387715 and HTRA1) exhibit enhanced risk of age-related macular degeneration along with CFH in Indian patients. *Invest Ophthalmol Vis Sci* 49(5):1771–1776.

Kaur I, Katta S, Reddy RK, et al. (2010). The involvement of complement factor B and complement component C2 in an Indian cohort with age-related macular degeneration. *Invest Ophthalmol Vis Sci* 51(1):59–63.

Kiran VS, Kannabiran C, Chakravarthi K, Vemuganti GK, Honavar SG. (2003). Mutational screening of the RB1 gene in Indian patients with retinoblastoma reveals eight novel and several recurrent mutations. *Hum Mutat* 22(4):339.

Klintworth GK, Sommer JR, Obrian G, et al. (1998). Familial subepithelial corneal amyloidosis (gelatinous drop-like corneal dystrophy): Exclusion of linkage to lactoferrin gene. *Mol Vis* 31(4):31.

Knudson AG Jr. (1971). Mutation and cancer: statistical study of retinoblastoma. PNAS 68(4): 820–823.

Kumar A, Babu M, Kimberling WJ, Venkatesh CP. (2004). Genetic analysis of a four generation Indian family with Usher syndrome: a novel insertion mutation in MYO7A. *Mol Vis* 10:910–916.

Kumar A, Basavaraj MG, Gupta SK, et al. (2007). Role of CYP1B1, MYOC, OPTN, and OPTC genes in adult-onset primary open-angle glaucoma: Predominance of CYP1B1 mutations in Indian patients. *Mol Vis* 13:667–676.

Kumar A, Bhattacharjee S, Prakash DR, Sadanand CS. (2007). Genetic analysis of two Indian families affected with congenital hereditary endothelial dystrophy: two novel mutations in SLC4A11. *Mol Vis* 13:39–46.

Kumaramanickavel G, Joseph B, Narayana K, et al. (2003). Molecular-genetic analysis of two cases with retinoblastoma: Benefits for disease management. *J Genet* 82(1–2):39–44.

Kumaramanickavel G, Joseph B, Vidhya A, Arokiasamy T, Shetty NS. (2002a). Consanguinity and ocular genetic diseases in South India: Analysis of a five-year study. *Community Genet* 5:182–185.

Kumaramanickavel G, Ramprasad VL, Sripriya S, Upadyay NK, Paul PG, Sharma T. (2002b). Association of Gly82Ser polymorphism in the RAGE gene with diabetic retinopathy in type II diabetic Asian Indian patients. *J Diabetes Complications* 16(6):391–394.

Kumaramanickavel G, Sripriya S, Vellanki RN, et al. (2002c). Inducible Nitric Oxide Synthase gene and Diabetic Retinopathy in Asian Indian patients. *Clinical Genetics* 61(5):344–348.

Kumaramanickavel G, Maw MA, Denton MJ, et al. (1994). Missense rhodopsin mutation in a family with recessive RP. *Nature Genetics* 8:10–11.

Kumaramanickavel G, Sripriya S, Ramprasad VL, Upadyay NK, Paul PG, Tarun S (2003). Z-2 aldose reductase allele and diabetic retinopathy in India. *Ophthalmic Genetics* 24(1): 41–48.

Kumaramanickavel G, Sripriya S, Vellanki RN, et al. (2001a). Tumor necrosis factor allelic polymorphism with Diabetic Retinopathy in India. *Diabetes Research and Clinical Practice* 54:89–94.

Kumaramanickavel G, Ramprasad VL, Sripriya S, Upadyay NK, Paul PG, Sharma T. (2001b). Association of *Gly82Ser* polymorphism in the RAGE gene with Diabetic Retinopathy in Type II diabetic Asian Indian patients. *J Diabetes and its complications* 16:391–394.

Lam DSC, Leung YF, Chua JKH, et al. (2000). Truncations in the *TIGR* Gene in Individuals with and without Primary Open-Angle Glaucoma. *Invest Ophthalmol Vis Sci* 41: 1386–1391.

Leighton DA. (1976). Survey of the first-degree relatives of glaucoma patients. *Trans Ophthalmol Soc UK* 96(1):28–32.

Lewin AS, Drenser KA, Hauswirth WW, et al. (1998). Ribozyme rescue of photoreceptor cells in a transgenic rat model of autosomal dominant retinitis pigmentosa. *Nature Medicine* 4:967–971.

List of Genes causing Retinal degenerative diseases: RetNet, http://utsph.sph.uth.tmc.edu/www/utsph/RetNet/home.htm.

McWilliam P, Peter G, Farrar J, Kenna P, Bradley DG, Humphries MM, Sharp EM, McConnell DJ, Lawler M, Sheils D, Ryan C, Stevens K,

Daiger SP, Humphries P. (1989). Autosomal dominant retinitis pigmentosa (ADRP): Localization of an ADRP gene to the long arm of chromosome 3. Genomics 5(3): 619–622.

Madhavan C, Bhende P, Lekha G. Vasanthi SB, Kumaramanickavel G. (2001). Retinitis pigmentosa patients with sickle cell disease and dextrocardia and situs inversus syndrome. *Indian Journal of Ophthalmology* 49:193–195.

Madhavan J, Coral K, Mallikarjuna K, et al. (2007). High expression of KIF14 in retinoblastoma: association with older age at diagnosis. *Invest Ophthalmol Vis Sci* 48(11):4901–4906.

Madhavan J, Mitra M, Mallikarjuna K, et al. (2009). KIF14 and E2F3 mRNA expression in human retinoblastoma and its phenotype association. *Mol Vis* 15:235–240.

Mansergh FC, Kenna PF, Ayuso C, Kiang AS, Humphries P, Farrar GJ. (1998). Novel mutations in the TIGR gene in early and late onset open angle glaucoma. *Hum Mutat* 11(3):244–251.

Mamatha G, Srilekha S, Meenakshi S, Kumaramanickavel G. (2008). Screening of the RPE65 gene in the Asian Indian patients with leber congenital amaurosis. *Ophthalmic Genet* 29(2):73–78.

Marigo V. (2007). Programmed cell death in retinal degeneration: targeting apoptosis in photoreceptors as potential therapy for retinal degeneration. *Cell Cycle* 6(6):652–655. Review.

Markandaya M, Ramesh TK, Selvaraju V, et al. (2004). Genetic analysis of an Indian family with members affected with juvenile-onset primary open-angle glaucoma. *Ophthalmic Genet* 25(1):11–23.

Maury CPJ. (1990). Isolation and characterization of cardiac amyloid in familial amyloid polyneuropathy type IV (Finnish): Relation of the amyloid protein to variant gelsolin. *Biochim Biophys Acta* 1096:84–86.

Maw MA, Corbeil D, Koch J, et al. (2000). A frameshift mutation in prominin (mouse)-like 1 causes human retinal degeneration. *Hum Mol Genet* 9:27–34.

Maw MA, John S, Jablonka S, et al. (1995). Oguchi disease: Suggestion of linkage to markers on chromosome 2q. *J Medical Genetics* 32:396–398.

Maw MA, Kar B, Biswas J, et al. (1996). Linkage of Blepharaophimosis syndrome in a large Indian pedigree to chromosome 7p. *Human Molecular Genetics* 5:2049–2054.

Maw MA, Kumaramanickavel G, Kar B, John S, Bridges RJ, Denton MJ. (1998). Two Indian siblings with Oguchi disease are homozygous for an arrestin mutation encoding premature termination. *Human Mutations* 1:S317-S319.

Metlapally R, Michaelides M, Bulusu A, et al. (2009). Evaluation of the X-linked high-grade myopia locus (MYP1). with cone dysfunction and color vision deficiencies. *Invest Ophthalmol Vis Sci* 50(4):1552–1558.

Mukhopadhyay A, Komatireddy S, Acharya M, et al. (2005). Evaluation of Optineurin as a candidate gene in Indian patients with primary open angle glaucoma. *Mol Vis* 11:792–797.

Munier FL, Korvatska E, Djema A, et al. (1997). Kerato-epithelin mutations in four 5q31-linked corneal dystrophies. *Nat Genet* 15(3):247–251.

Naiglin L, Gazagne C, Dallongeville F, et al. (2002). A genome wide scan for familial high myopia suggests a novel locus on chromosome 7q36. *J Med Genet* 39:118–124.

Nallathambi J, Shukla D, Rajendran A, Namperumalsamy P, Muthulakshmi R, Sundaresan P. (2006). Identification of novel FZD4 mutations in Indian patients with familial exudative vitreoretinopathy. *Mol Vis* 12:1086–1092.

National Human Genome Research Institute: http://www.nhgri.nih.gov.

Nussenblatt RB, Csaky K. (1997). Perspectives on gene therapy in the treatment of ocular inflammation. *Eye* 11:217–222.

Paliwal P, Gupta J, Tandon R, Sharma A, Vajpayee RB. (2009). Clinical and genetic profile of Avellino corneal dystrophy in 2 families from north India. *Arch Ophthalmol* 127(10):1373–1376.

Paliwal P, Singh A, Tandon R, Titiyal JS, Sharma A. (2009). A novel VSX1 mutation identified in an individual with keratoconus in India. *Mol Vis* 15:2475–2479.

Paluru P, Ronan SM, Heon E, et al. (2003). New locus for autosomal dominant high myopia maps to the long arm of chromosome 17. *Invest Ophthalmol Vis Sci* 44:1830–1836.

Panicker SG, Reddy AB, Mandal AK, et al. (2002). Identification of novel mutations causing familial primary congenital glaucoma in Indian pedigrees. *Invest Ophthalmol Vis Sci* 43:1358–1366.

Ramprasad S, Radha V, Mathias RA, Majumder PP, Rao MR, Rema M. (2007). Rage gene promoter polymorphisms and diabetic retinopathy in a clinic-based population from South India. *Eye (London)* 21(3):395–401.

Ramprasad VL, Ebenezer ND, Aung T, et al. (2007). Novel SLC4A11 mutations in patients with recessive congenital hereditary endothelial dystrophy (CHED2). *Hum Mutat* 28(5):522–523.

Ramprasad VL, George R, Soumittra N, Sharmila F, Vijaya L, Kumaramanickavel G. (2008). Association of non-synonymous single nucleotide polymorphisms in the LOXL1 gene with pseudoexfoliation syndrome in India. *Mol Vis* 14:318–322.

Ramprasad VL, George RJ, Sripriya S, Nirmaladevi J, Vijaya L, Kumaramanickavel G. (2007). Molecular genetic analysis of a consanguineous south Indian family with congenital glaucoma: Relevance of genetic testing and counseling. *Ophthalmic Genet* 28(1):17–24.

Ramprasad VL, Madhavan J, Murugan S, et al. (2007). Retinoblastoma in India: microsatellite analysis and its application in genetic counseling. *Mol Diagn Ther* 11(1):63–70.

Reddy AB, Kaur K, Mandal AK, et al. (2004). Mutation spectrum of the CYP1B1 gene in Indian primary congenital glaucoma patients. *Mol Vis* 10:696–702.

Rose R, Karthikeyan M, Anandan B, Jayaraman G. (2007). Myocilin mutations among primary open angle glaucoma patients of Kanyakumari district, South India. *Mol Vis* 13:497–503.

Sheffield VC, Stone EM, Alwar WLM, Drack AV, Johnson AT, Streb LM, Nichols BE. (1993). Genetic linkage of familial open angle glaucoma to chromosome 1q21–q31. *Nat Genet* 4: 47–50.

Singh HP, Jalali S, Hejtmancik JF, Kannabiran C. (2006). Homozygous null mutations in the ABCA4 gene in two families with autosomal recessive retinal dystrophy. *Am J Ophthalmol* 141(5):906–913.

Singh HP, Jalali S, Narayanan R, Kannabiran C. (2009). Genetic analysis of Indian families with autosomal recessive retinitis pigmentosa by homozygosity screening. *Invest Ophthalmol Vis Sci* 50(9):4065–4071.

Sripriya S, Nirmaladevi J, George R, et al. (2006). OPTN gene: profile of patients with glaucoma from India. *Mol Vis* 12:816–820.

Sripriya S, Uthra S, Sangeetha R, et al. (2004). Low frequency of myocilin mutations in Indian primary open-angle glaucoma patients. *Clin Genet* 65(4):333–337.

Stone EM, Fingert JH, Alward LM, Nguyen DT, Polansky JR, et al. (1997). Identification of a gene that causes primary open angle glaucoma. *Science* 275:668–670.

Suganthalakshmi B, Anand R, Kp R, et al. (2006). Association of VEGF and eNOS gene polymorphisms in type 2 diabetic retinopathy. *Mol Vis* 12:336–341.

Suganthalakshmi B, Shukla D, Rajendran A, Kim R, Nallathambi J, Sundaresan P. (2007). Genetic variations in the hotspot region of RS1 gene in Indian patients with juvenile X-linked retinoschisis. *Mol Vis* 13:611–617.

Sultana A, Garg P, Ramamurthy B, Vemuganti GK, Kannabiran C. (2007). Mutational spectrum of the SLC4A11 gene in autosomal recessive congenital hereditary endothelial dystrophy. *Mol Vis* 13:1327–1332.

Sultana A, Sridhar MS, Jagannathan A, Balasubramanian D, Kannabiran C, Klintworth GK. (2003). Novel mutations of the carbohydrate sulfotransferase-6 (CHST6) gene causing macular corneal dystrophy in India. *Mol Vis* 9:730–734.

Suzuki Y, Shirato S, Taniguchi F, Ohara K, Nishimaki K, Ohta S. (1997). Mutations in the TIGR gene in familial primary open-angle glaucoma in Japan. *Am J Hum Genet* 61(5):1202–1204.

Tanwar M, Dada T, Sihota R, Dada R. (2009). Identification of four novel cytochrome P4501B1 mutations (p.I94X, p.H279D, p.Q340H, and p.K433K) in primary congenital glaucoma patients. *Mol Vis* 15:2926–2937.

Tanwar M, Dada T, Sihota R, Das TK, Yadav U, Dada R. Mutation spectrum of CYP1B1 in North Indian congenital glaucoma patients. *Mol Vis* (2009). Jun 13;15:12009.

Tielsch JM, Katz J, Quigley HA, Javitt JC, Sommer A. (1995). Diabetes, intraocular pressure, and primary open-angle glaucoma in the Baltimore Eye Survey. Ophthalmology. 102(1):48–53.

Thomas R, George R, Parikh R, Muliyil J, Jacob A. (2003). Five year risk of progression of primary angle closure suspects to primary angle closure: a population based study. *Br J Ophthalmol* 87(4):450–454.

Thylefors B, Nigrel AD, Pararajasegaram R, Dadzie KY. (1995). Global data on blindness. *Bull World Health Organ* 73(1):115–121. Review.

Uthra S, Raman R, Mukesh BN, et al. (2010). Protein kinase C beta (PRKCB1) and pigment epithelium derived factor (PEDF) gene polymorphisms and diabetic retinopathy in a south Indian cohort. *Ophthalmic Genet* 31(1):18–23.

Uthra S, Raman R, Mukesh BN, et al. (2007). Diabetic retinopathy and IGF-1 gene polymorphic cytosine-adenine repeats in a Southern Indian cohort. *Ophthalmic Res* 39(5):294–299.

Uthra S, Raman R, Mukesh BN, et al. (2008). Association of VEGF gene polymorphisms with diabetic retinopathy in a south Indian cohort. *Ophthalmic Genet* 29(1):11–15.

van Soest S, Westerveld A, de Jong PT, Bleeker-Wagemakers EM, Bergen AA. (1999). Retinitis pigmentosa: defined from a molecular point of view. *Surv Ophthalmol* 43(4):321–334. Review.

Vijaya R, Gupta R, Panda G, Ravishankar K, Kumaramanickavel G. (1997). Genetic analysis of adult-onset cataract in a city-based ophthalmic hospital. *Clinical Genetic* 52:427–431.

Warpeha KM, Chakravarthy U. (2003). Molecular genetics of microvascular disease in diabetic retinopathy. *Eye* 17:305–311.

World Health Organization. (1997). *The world health report 1997. Conquering suffering, enriching humanity.* Geneva: WHO, 68:1997.

World Health Organization. (1998). World health report. Geneva: WHO.

Young TL, Ronan SM, Alvear AB, et al. (1998a). A second locus for familial high myopia maps to chromosome12q. *Am J Hum Genet* 63:1419–1424.

Young TL, Ronan SM, Drahozal LA, et al. (1998b). Evidence that a locus for familial high myopia maps to chromosome 18p. *Am J Hum Genet* 63:109–119.

Zack DJ. (1993). Ocular gene therapy. *Arch Ophthalmol* 111:1477–1478.

Zack DJ, Dean M, Molday RS, et al. (1999). What can we learn about age-related macular degeneration from other retinal diseases? *Mol Vis* 5:30. Review.

87.

TYPE 2 DIABETES MELLITUS IN INDIA

Radha Venkatesan, Kanthimathi Sekar and Viswanathan Mohan

INTRODUCTION

Type 2 diabetes (T2D) is a disease characterized by insulin resistance and impaired pancreatic beta-cell function, affecting about 170 million people worldwide (Wild et al., 2004). According to the latest Diabetes Atlas, India currently has 50.8 million people with diabetes and these figures are predicted to increase to 87.0 million by 2030 (Sicree et al., 2009). The prevalence of T2D is reaching epidemic proportions in Indian populations. The increase in diabetes prevalence is seen in all age classes, but is proportionally highest among the younger age groups.

GENETIC EPIDEMIOLOGY OF TYPE 2 DIABETES

T2D is a complex heterogeneous group of conditions characterized by elevated levels of plasma glucose, caused by impairment in both insulin secretion and insulin action. The etiopathogenesis of T2D involves interplay of both genetic and environmental factors. Although the recent increase in diabetes prevalence reflect the effects of changing environmental factors, multiple lines of evidence support the view that genetic factors are equally important:

- The prevalence of T2D varies widely among populations. The observation that the disease prevalence varies substantially among ethnic groups that share a similar environment supports the idea that genetic factors contribute to predisposition to the disease (Diamond, 2003).
- Familial aggregation of the disease is another source of evidence for a genetic contribution to the disease, although admittedly families also share common environmental traits. The odds ratio for offspring of a single affected parent is 3.5 compared to those with no parental diabetes history, and this increases to 6.1 if both parents are affected (Meigs et al., 2000; Radha and Mohan, 2007).
- The high concordance in monozygotic twins (over 80%) compared to 50% concordance in dizygotic twins provides compelling evidence for a genetic component in etiology of T2D (Vaag et al., 1995; Poulsen et al., 1999).
- Data from various studies are in support of a genetic basis for measures of both insulin sensitivity and insulin secretion (Gerich, 1998; Elbein et al., 1999).

Genetic epidemiology seeks to elucidate the role of genetic factors and their interactions with environmental factors in the occurrence of disease in different populations. The advantages of genetic epidemiology over conventional strategies in epidemiology include: (i) genetic markers are stable indicators of host susceptibility; (ii) case-control designs are facilitated, thus allowing the study of uncommon endpoints; and (iii) gene effects lead directly to genetic mechanisms (Kaprio, 2000). These methods apply equally to assessing the presence or absence of disease, to assessing genetic influences on quantitative or semi-quantitative traits and also to study individual families or populations.

HERITABILITY OF DIABETES AMONG INDIANS

Studies done in the 1980s in a group of 135 Asian Indian and 146 European diabetic patients attending a diabetic clinic in the United Kingdom showed that 36% of the Europeans had a first-degree relative with diabetes compared to 45% of Asian Indians (Mohan et al., 1986a). Of greater interest was the excess of parental diabetics, with nearly 10% of Asians compared to 1% of Europeans having two diabetic parents. Another study from our group measured the prevalence of diabetes among the offspring of two T2D parents in India. This study showed that 60% of offspring had diabetes or impaired glucose tolerance (Viswanathan et al., 1985). This figure was considerably higher than the figures reported for prevalence of diabetes in offspring of European diabetic parents. It was also

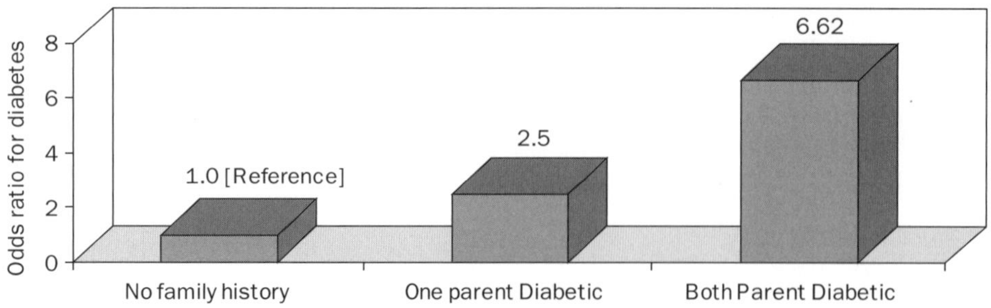

Figure 87-1 Family history of diabetes and risk for diabetes – The Chennai Urban Population Study [CUPS] (Mohan et al., 2003).

observed that the age at onset of diabetes among the offspring was much earlier compared to their parents (Meigs et al., 2000; Sargeant et al., 2000).

In a population-based study conducted by the Chennai Urban Population Study (CUPS) on 1262 individuals in Chennai (formerly Madras) in southern India, we assessed the influence of family history on prevalence of T2D. The prevalence of diabetes was higher among subjects who had a positive family history of diabetes (18.2%) compared to subjects without a family history of diabetes (10.6%, p = 0.0015). Moreover, 9.3% of subjects with family history of diabetes had impaired glucose tolerance (IGT—a pre-diabetic stage) compared to 5.0% of subjects without a family history (p = 0.016). The overall prevalence of glucose intolerance (diabetes + IGT) among subjects with two diabetic parents was significantly higher (55%) than those who had diabetic a parent (22.1%, p =0.005) or those with two nondiabetic parents (15.6%, p<0.0001). The odds ratios of the risk for diabetes among subjects with one diabetic parent was 2.5, and this increased to 6.62 in subjects who had both parents affected by diabetes (Figure 87-1; Mohan et al., 2003).

FAMILY HISTORY OF DIABETES

A review described the heterogeneity of Type 2 diabetes with four different types of inheritance (Stern, 2002). It was suggested by this study that there could be polygenic inheritance with no major genes, or oligogenic inheritance with several major genes. Heterogenity in type 2 diabetes arises because it is part of the metabolic cluster, the insulin resistance syndrome, which comprises glucose intolerance (diabetes, impaired glucose tolerance), dyslipidemia (hypertriglyceridemia, low HDL levels), hypertension, and abdominal obesity.

In the Chennai Urban Population Study (CUPS), the subjects with both diabetic parents had higher prevalence rate of all the components of the insulin resistance syndrome compared to subjects with one diabetic parent and those with no family history of diabetes (Table 87-1; Mohan et al., 2003).

Very recently, heritability of quantitative traits associated with type 2 diabetes in large multiplex families from South India was estimated (Mathias et al., 2009). Heritability estimates were calculated for all quantitative traits at the univariate level, and bivariate analyses were done to determine the correlation in genetic and environmental control across these quantitative traits. The study revealed strong familial aggregation of quantitative traits that are typically associated with type 2 diabetes, lending credence to the role of genetic factors in type 2 diabetes.

GENETIC STUDIES IN TYPE 2 DIABETES MELLITUS

The genetics of T2D can be categorized in two broad groups: genetics of monogenic forms of diabetes, where a single gene is causal in the development of the disease, and genetics of polygenic forms of diabetes, where a number of genes are responsible for the susceptibility of the disease.

TABLE 87-1 PREVALENCE OF COMPONENTS OF INSULIN RESISTANCE SYNDROME IN SUBJECTS WITH AND WITHOUT FAMILY HISTORY OF DIABETES – DATA FROM CUPS STUDY (MOHAN ET AL., 2003)

Variables	No family history	One parent diabetic	Both parents diabetic
Glucose intolerance (WHO consulting group criteria)	15.6%	27.4%	55.0%
Obesity (WHO guidelines for Asians)	24.8%	40.3%	50.0%
Abdominal Obesity (WHO guidelines for Asians)	26.6%	43.2%	52.5%
Hypertension (JNC VI criteria)	21.7%	23.8%	30.0%
Hypercholesterolemia (NCEP guidelines)	21.7%	29.8%	37.5%
Hypertriglyceridemia (NCEP guidelines)	21.3%	24.1%	35.0%
Low HDL levels (NCEP guidelines)	57.6%	56.9%	70.0%

Patterns of inheritance suggest that T2D is both polygenic and heterogeneous, that is, multiple genes are involved, and different combinations of genes play a role in different subsets of individuals. Exactly how many genes and what their relative contributions are remains ambiguous. Earlier studies have shown that non-insulin-dependent diabetes mellitus is likely to be a polygenic disease (Hitman et al., 1992).

To unravel the genetics of T2D, the following approaches have been adopted:

- To focus on forms of T2D transmitted with a Mendelian dominant pattern of inheritance and/or other specific clinical features, which resulted in the discovery of genes involved in maturity-onset diabetes of the young (MODY), several syndromes of severe insulin resistance, neonatal diabetes, mitochondrial diabetes, and other rare genetic syndromes.
- To search for genetic variants in candidate genes that might be associated with the common T2D. These studies have focused on functional candidate genes—i.e., genes whose products are known to play a role in glucose homeostasis—or positional candidate genes, i.e., genes located in chromosomal regions that had been identified in linkage studies.
- To perform microarray gene expression analysis in an attempt to define genetic alterations in T2D.
- To perform high-throughput genome-wide association studies (GWAS), which is expected to accelerate the speed of gene identification in T2D.

MONOGENIC DIABETES

Some of the most compelling evidence that inherited variations can cause glycemic dysregulation comes from the clinical and genetic description of monogenic diabetes, including maturity-onset diabetes of the young (MODY), insulin resistance syndromes, mitochondrial diabetes, and neonatal diabetes. Although rare, these syndromes provide a framework for understanding and investigating the complex genetics of T2D.

MATURITY-ONSET DIABETES OF THE YOUNG

MODY was first described in 1975 as a unique type of non-insulin-dependent, characterized by β-cell dysfunction, autosomal dominant diabetes in thin, young adults (usually <25 years of age), who are not prone to ketoacidosis (Tattersall and Fajans, 1975; Fajans et al., 2001). All patients clinically designated as MODY satisfied the following criteria of Tattersall and Fajans (1975):

a. Age at onset of diabetes is <25 years.

b. Correction of hyperglycemia for a minimum period of 2 years without insulin.

c. Absence of ketonuria at any time.

d. Evidence of autosomal dominant inheritance with three-generation transmission of diabetes.

This syndrome accounts for approximately 1%–2% of diabetes cases worldwide, although MODY has been reported as responsible for as many as 5% of European diabetic cases (Hattersley et al., 1998). Six different MODY genes have been identified to date, although ~10% of MODY patients have no identifiable genetic mutation in any of them, possibly due to the presence of other undiscovered gene(s): the MODY "X."

- MODY 1 results from mutations in *HNF4A* on chromosome 20, which affects the development and function of human pancreatic β cells. MODY 1 is the third most common MODY subtype, with 31 mutations in 40 families described to date (Ellard and Colclough, 2006).
- MODY 2 results from mutations in the glucokinase gene (*GCK*) on chromosome 7p (Froguel et al., 1992; Hattersley et al., 1992) and accounts for ~20% of MODY cases. Inactivating mutations of *GCK* result in mild, stable, lifelong fasting hyperglycemia because those affected lack an effective glucose sensor. Drug treatment is rarely required, and microvascular disease is unusual (Frayling et al., 2001).
- MODY 3 results from mutations in *HNF1A* (*TCF1*), which impairs key steps of glucose transport, metabolism, and mitochondrial metabolism (Yamagata et al., 1996b). It is the most common MODY subtype, with more than 300 different mutations. Retinopathy and nephropathy are common in MODY 3; however, macrovascular disease is not (Bellanne-Chantelot et al., 2008; Ellard and Colclough, 2006).
- MODY 4 results from heterozygous mutations in the gene that encodes insulin promoter factor 1 (*IPF1* or *PDX1*), a transcription factor that regulates insulin gene transcription as well as islet and pancreatic development. Homozygous mutations in exon 1 of *IPF1* give rise to pancreatic agenesis (Stoffers et al., 1997).
- MODY 5, a more common MODY variant than originally suspected, is unique in that it is associated with renal anomalies. *HNF1B* is highly expressed in the pancreas, liver, and kidney. Diabetes results from both hepatic insulin resistance and β cell loss. β cell dysfunction is more severe in MODY 5 than in

MODY 3, and ketoacidosis has also been reported. These patients typically require insulin therapy and do not respond to sulfonylureas. *HNF1B* mutations have been associated with renal dysfunction, genitourinary problems, abnormal liver function, and hyperuricemia (Horikawa et al., 1997).

- MODY 6 has been shown to result from mutations in *NEUROD1* (*BETA2*), which is important for pancreatic development and insulin gene transcription. Patients develop moderately severe and progressive β cell dysfunction. MODY 6 is extremely rare, with only a few cases reported in the literature (Malecki et al., 1999)

Among these, mutations in the glucokinase gene cause a relatively mild form of diabetes. Mutations in the hepatocyte nuclear factor-1 beta gene may lead to a severe form of diabetes associated with renal and genital defects. Heterozygous mutations in insulin promoter factor (IPF-1) cause MODY phenotype, while the homozygous mutations produce pancreatic aplasia or hypoplasia.

Earlier studies reported on the high prevalence of MODY (using the clinical criteria used at that time) in South Indians (4.8%; Mohan et al., 1985). The insulin responses in MODY and the beta-cell response in the offspring of MODY indicated an increased insulin resistance compared to classical Indian type 2 diabetic subjects (Mohan et al., 1985).

Recently, our work has revealed important insights into the genetics of MODY in south India. We identified 9 novel variants comprising 7 mutations (one novel mutation, −538G→C at promoter region, and six novel coding region mutations) and 2 polymorphisms in the *HNF1A* gene. Functional studies revealed reduced transcriptional activity of the *HNF1A* promoter for two of the promoter variants. One of the novel mutations, Arg263His, was identified in a family of 30 individuals. The mutation cosegregated with diabetes in this family, and this mutation was not seen in nondiabetic members in the family, thus providing evidence for the mutation to be involved in causing MODY (Radha et al., 2009). Further, 3 novel MODY 1 mutations have also been identified in *HNF4 A* gene in the same population (Anuradha et al., 2010). Thus, about 9% of the clinically diagnosed MODY subjects are MODY 3, and 3% are MODY 1 in south India.

MODY3 is the commonest form of MODY worldwide (Fajans et al., 2001). These patients are of normal BMI, often present with severe hyperglycemia, and have microvascular complications. The prevalence of MODY3 in patients with early-onset T2D varies from 2.5% to 36% (Kaisaki et al., 1997; Lehto et al., 1999; Aguilar-Salinas et al., 2001; Ng et al., 2001; Owen et al., 2003).

Common variations in the *HNF1A* gene have also been associated with impaired insulin secretion, and I27L and A98V polymorphisms in the MODY3 gene (*TCF1*) are associated with increased risk of T2D in overweight individuals (Holmkvist et al., 2006). An association of Val 98 allele with South Indian patients presenting with MODY phenotype has been reported by Anuradha et al. (2005), and this allele was also associated with an earlier age at onset of T2D.

We also found that SNPs rs4810424 and rs3212198 in the *HNF4A* gene showed a modest association with early-onset T2D. This is in line with findings from some previous studies (Love-Gregory et al., 2004; Silander et al., 2004). The −30G/A polymorphism in the *GCK* gene β-cell promoter has been associated with increased fasting plasma glucose (fPG) and reduced β-cell function (Stone et al., 1996). In addition, it has been shown to affect birth weight (Hattersley et al., 1998). Carriers of the −30G/A polymorphism in the *GCK* gene have a sustained increase in fasting glucose concentrations that is not translated into an increased risk of future T2D. Even though the −30G/A polymorphism has been associated with diabetes-related traits, its contribution to T2D is less well known. However, in a recent meta-analysis it was shown to have a modest but significant effect on T2D risk (Winckler et al., 2007).

Taken together, the results indicate that common variants in the *HNF1A*, *HNF4A*, and *GCK* genes are associated with a modestly increased risk of future T2D (Holmkvist et al., 2008).

NEONATAL DIABETES

Neonatal diabetes, which usually presents in the initial days or months of life, can be transient or permanent. The transient type of neonatal diabetes (TNDM), resolves at a median of 12 weeks, usually remits in childhood, but is associated with an increased risk of T2D in adulthood. Permanent neonatal diabetes mellitus (PNDM), traditionally treated with insulin, is associated with activating mutations in the *KCNJ11* gene, which encodes the islet ATP-sensitive potassium channel Kir6.2 (Gloyn et al., 2004).

Activating mutations in *KCNJ11* result in constitutive opening of the inwardly rectifying potassium channel, hyperpolarization of the β cell membrane, and subsequent hypoinsulinemic diabetes. Conversely, loss-of-function polymorphisms have been associated with persistent hyperinsulinemia of infancy (Shyng et al., 1998). Several novel heterozygous missense activating mutations in *KCNJ11* have now been identified in TNDM. The severity of the mutation correlates with the clinical phenotype—some *KCNJ11* defects result in PNDM, and even more severe genetic mutations result in the DEND syndrome (developmental delay, epilepsy, neonatal diabetes; see Hattersley and Ashcroft, 2005). Mutations in this gene appear to be the most common cause of neonatal diabetes. It is interesting to note that the closely associated sulfonylurea receptor (*SUR1*) also plays an important role in neonatal diabetes. The dominant mutation results in neonatal hyperinsulinemia and subsequent hypoglycemia, and the heterozygous form may increase adult susceptibility to common T2D (Babenko and Bryan, 2003).

In contrast to the rarer monogenic forms of diabetes, as described above, the more common T2D is a polygenic disorder. The complex individual susceptible and protective genes are more difficult to identify, given the limited individual impact of a single genetic locus. Indeed, a full understanding of the complex gene–gene and gene–environment interactions at play in this disease has proven quite challenging thus far.

The susceptibility to complex forms of T2D is associated with frequent polymorphisms that create amino acid variants in exons, or influence the expression of genes in the regulatory pathways (McCarthy and Froguel, 2002). Alleles of these polymorphisms are present in both healthy individuals and T2D patients, although with statistically significant differences in the frequencies. The sequence variants are associated with an increase in the risk of developing the disease and are considered susceptibility variants, but not causative factors that unequivocally determine the disease.

The candidate gene approach focuses on the search for an association between T2D and sequence variants in or near biologically defined candidate genes, which have been chosen based on their known physiological function. The importance of these variants or other nearby variants is tested by comparing the frequency in T2D patients and normal glucose-tolerant subjects. A second approach is the genome-wide linkage scan strategy (Whittemore, 1996), in which regularly spaced markers are traced in families and sibling pairs for segregation with T2D. No prior knowledge of gene or gene effects is necessary, but the genetic locus must have sufficient impact on the disease susceptibility in order to be detectable.

The starting point for the candidate gene approach is the potential implication that either altered expression and/or function of a particular gene product (conferred by noncoding or coding genetic variants) may have on a biological function or on disease.

Extending the analysis of genes implicated in monogenic forms of diabetes has proved successful also for T2D, as exemplified by *HNF4A*, *HNF1A* and *KCNJ11* genes. Common variants of *HNF4A* (MODY1) have been associated with T2D in both Finnish and Ashkenazi Jewish populations (Love-Gregory et al., 2004; Silander et al., 2004).

SPECIFIC GENETIC ASSOCIATIONS IN THE INDIAN SUBCONTINENT

India is presently facing an epidemiological transition, with more urbanization and a rapid shift in the economy. Consequent to this, changes in lifestyle occur with a Western dietary pattern leading to increased consumption of dietary fat and calories. This has led to a shift in health burden from communicable diseases to noncommunicable diseases, particularly diabetes. The influence of urbanization on diabetes has been clearly shown in the "CUPS" study where the prevalence of diabetes and impaired glucose tolerance (IGT) was twice as high among the middle-income group compared to the low-income group (Mohan et al., 2001). Subjects who followed a sedentary lifestyle had three times higher risk of developing diabetes compared to the more physically active (Mohan et al., 2003). Furthermore, the study also clearly showed that family history and lifestyle factors like physical activity and high socioeconomic status have a synergistic effect in the development of diabetes. It is thus clear that a combination of genetic and environmental factors contribute to diabetes.

Sharp et al. (1987) and Mohan et al. (1986a), showed that insulin resistance and the prevalence of hyperinsulinemia are higher in Indians compared to matched groups of Europeans. Though body mass index (BMI), an indicator of obesity, is lower among Indians, the waist-to-hip ratio for any given BMI was higher among Indians compared to other ethnic groups (Joshi, 2003). Furthermore, for any BMI, Indians also had higher body fat and for any given body fat, Indians had higher insulin resistance compared to other ethnic groups (Chandalia et al., 1999). Studies on low birth weight and insulin resistance in Indian neonates have shown that Indians have hyperinsulinemia and adiposity even at birth compared to Caucasians (Yajnik et al., 2002). These studies suggest that Indians seem to be genetically more prone to diabetes and insulin resistance.

Various genetic studies using the candidate gene approach have pointed at the role of different genes in the susceptibility to type 2 diabetes.

The genetic study on plasma cell glycoprotein 1 (*PC-1*) gene polymorphism K121Q (ENPP1 K121Q), to evaluate the role of this polymorphism in prediction of T2D in three study groups, included 679 South Indians living in Chennai, India (223 with T2D); 1,083 migrant South Asians living in Dallas, Texas (121 with T2D); and 858 Caucasians living in Dallas, (141 with T2D). The prevalence of subjects carrying the polymorphic ENPP1 121Q allele in the nondiabetic and diabetic groups were 25% and 34% for South Asians living in Chennai ($p < 0.01$), 33% and 45% ($p < 0.01$) for the South Asians living in Dallas, and 26% and 39% ($p < 0.003$) for the Caucasians. This study supports the hypothesis that ENPP1 121Q predicts genetic susceptibility to T2D in both South Asians and Caucasians (Abate et al., 2005).

Another genetic study in collaboration with the same populations was conducted on peroxisome proliferator activated receptor-G (*PPARG*) gene polymorphism, one of the main candidate genes that is implicated in adipogenesis, insulin resistance, and T2D. The frequencies of the common Pro 12 Ala polymorphism in South Indians living in Chennai was found to be 19%, while that in South Asians in Dallas was 18% and in Caucasians in Dallas it was 20%. The study concluded that despite the frequency of the Ala allele at the PPAR-Pro12Ala locus being the same in individuals of South Asian descent as in Caucasians, this particular polymorphism does not appear to improve insulin sensitivity or decrease risk for T2D in South Asians (Asian Indians), as it does in Caucasians, thereby supporting the Pro12Ala polymorphism as protective against diabetes in Caucasians but not in South Asians (Radha et al., 2006). Further, the 3 variants (–1279G/A, Pro12Ala, and His478His) in the *PPARG* gene were studied for their association with T2D and its related traits in a South Indian population. Although –1279G/A, Pro12Ala, and His478His variants of the *PPARG* gene were not associated with T2D mellitus, the Pro12Ala and –1279G/A promoter variant in combination showed increased susceptibility to T2D mellitus (Vimaleswaran et al., 2010a). In continuation of this study, the association of Pro12Ala polymorphism with adiponectin levels was studied in Asian Indians, which concluded that the Pro12Ala polymorphism of the *PPARG* gene is not associated with serum adiponectin levels in Asian Indians (Radha et al., 2007).

A genetic study was conducted to examine whether variants in the regulatory regions of the adiponectin gene contribute to T2D in Asian Indians. It was found that among Asian Indians there is an association of +10211T/G polymorphism in the first intron of the adiponectin gene with T2D, obesity, and hypoadiponectinemia (Vimaleswaran et al., 2008).

In one study, the A allele of Thr394Thr (G-A) polymorphism of the peroxisome proliferator-activated receptor-γ coactivator-1A (*PGC-1A*) gene, a cofactor of *PPARG* gene involved in adaptive thermogenesis, fatty acid oxidation, and gluconeogenesis, was found to be associated with T2D and also with total, visceral, and subcutaneous body fat among Asian Indians (Vimaleswaran et al., 2005; 2006). Another study showed that the Thr394Thr and Gly482Ser variant genotypes provide protection against T2D mellitus in two North Indian population groups (Bhat et al., 2007). A study on K121Q polymorphism of plasma cell glycoprotein 1 (*PC-1*) gene supports the hypothesis that 121Q predicts genetic susceptibility to T2D in both South Asians and Caucasians (Abate et al., 2005).

A study on three polymorphisms—namely, –3826A→G, an A→C transition in the 5′-untranslated region (UTR) and Met229Leu of the gene encoding for uncoupling protein-1 (UCP1)—showed that they were not associated with T2D independently. However, the frequency of the *A-C-Met* haplotype was significantly associated with the increased genetic risk for developing T2D in Asian Indians (Vimaleswaran et al., 2010b). A similar study on Ala55Val and –55C/T polymorphisms at the uncoupling protein 3-2 (*UCP3-2*) loci showed association with a significantly reduced risk of developing T2D in Asian Indians (Vimaleswaran et al., 2011).

Genetic study on the PPARG coactivator1 alpha (PPARGC1A), peroxisome proliferator activated receptor gamma (PPARG), and uncoupling protein1 (UCP1) gene polymorphisms, showed lack of association with metabolic syndrome in Asian Indians (Vimaleswaran et al., 2007).

Insulin receptor substrate (*IRS*)-2 plays an important role in insulin signaling, and its disruption results in diabetes in mice. In humans, the *IRS-2* Gly1057Asp substitution was associated with lower risk of T2D in lean individuals, but with a higher risk in obese individuals. A study on this polymorphism in Asian Indians found that the DD genotype increases susceptibility to T2D by interacting with obesity (Bodhini et al., 2007a).

The most exciting and promising gene associated with T2D is the transcription factor 7 like 2 gene (*TCF7L2*) gene, since positive replication has been reported by virtually all the studies conducted so far (Grant et al., 2006; Zhang et al., 2006; Groves et al., 2006; Florez et al., 2006; Chandak et al., 2007; Horikoshi et al., 2007, Sanghera et al., 2008a). In a study conducted in a South Indian population, the T allele of the rs12255372(G/T) and rs7903146(C/T) polymorphisms of TCF7L2 gene confer susceptibility to T2D in Asian Indians. Although the exact role of *TCF7L2* is unknown, it appears to act through the Wnt pathway (Bodhini et al., 2007b).

Earlier studies on rs9939609 T→A and rs7193144 C→T variants of the intron 1 of the fat mass and obesity-associated (*FTO*) gene showed an association with T2D which was independent of BMI in Asian Indians (Sanghera et al., 2008b; Yajnik et al., 2009). A recent association study on six variants—the rs9940128, rs7193144, rs8050136 (intron 1), rs918031, rs1588413 (intron 8), and rs11076023 (3'UTR)—across three regulatory regions of the *FTO* gene with obesity and T2D in a South Indian population showed that the rs9940128 A/G, rs1588413 C/T, and rs11076023 A/T variants of the *FTO* gene were associated with T2D but not with obesity. The rs8050136 C/A variant was associated with obesity, and its association with T2D was also mediated through obesity. The haplotype "ACCTCT" confers a lower risk of T2D in this South Indian population (Ramya et al., 2010).

GENETIC STUDIES IN OTHER POPULATIONS

Mutations in some genes have been found to account for a small percentage of type 2 diabetes cases. Among these, the glucokinase gene on chromosome 20 (Gelding et al., 1994) and beta-3adrenergic receptor gene (beta 3AR;

Silver et al., 1996) have attracted some attention. However, genes accounting for the majority of genetic susceptibility to T2D remain to be identified (Raffel and Rotter, 2002).

There is some evidence from the Swedish population that the genetic susceptibility to T2D might be related to a major autosomal recessive gene (Iselius et al., 1982). This observation has also been confirmed in both Mexican-Americans and non-Hispanic Caucasians (Haffner et al., 1996). The evidence for this hypothesis is based on marked intrafamilial correlation for decreased insulin secretion in response to oral glucose load, as well as with fasting insulin levels. This hypothesis is likely to be relevant to the Indian subcontinent populations, particularly where the practice of consanguinity is very common. There is some evidence that certain late-onset disorders, for example breast and ovarian cancers, might be related to recessively inherited factors as a result of parental consanguinity. So far, there is no evidence to support this argument. However, it could be one of the many inherited factors predisposing to T2D in the Indian subcontinent populations.

An association of the islet amyloid polypeptide (IAPP) gene (protein amylin) was excluded as the likely candidate gene in the pathogenesis of T2D among South Indians (Dravidians) compared to Europeans with type 2 diabetes (McCarthy et al., 1992). Similarly, no evidence was found for the APOC3-482C>T variant on insulin, glucose, and triglyceride concentrations in South Asian diabetics compared to Caucasians with T2D (Waterworth et al., 2001).

Polymorphic variations of genes encoding the glucose transporters glycoproteins (GLUT1 snf GLUT4) are associated with genetic susceptibility to T2D (Pontiroli et al., 1996). There was no difference in various population groups including the Asian Indians. However, higher genetic susceptibility was inferred for overweight/obese women. Mutations in the coding region of the human obese gene (OB) did not show any positive correlation in Japanese and Asian Indian obese patients with T2D (Niki et al., 1996). This is likely to be due to mutations in another gene manifesting with obesity-associated T2D.

The single nucleotide polymorphism (SNP) E23K of *KCNJ11* has now been convincingly associated with T2D. Although initial smaller studies failed to replicate the association of the E23K polymorphism with T2D, large-scale studies and meta-analyses have consistently associated the lysine variant with T2D, with an OR of 1.15 (Moore and Florez, 2008).

One of the earliest diabetes genes identified was *CAPN10*, which encodes an intracellular calcium-dependent cysteine protease that is ubiquitously expressed and can hydrolyze substrates important in calcium-regulated signaling pathways (Croall and DeMartino, 1991). Although subsequent studies have not produced a consistent relationship between variation in *CAPN10* and metabolic phenotypes, recent meta-analyses indicate that variants in *CAPN10* may indeed confer a modest increase in T2D susceptibility. Interestingly, calpain inhibitors increase insulin secretion and increase the metabolism of muscle glucose to glycogen in animal models (Moore and Florez, 2008). A recent study showed a positive correlation of calpain-10 expression on insulin secretory function in healthy human isolated pancreatic islets, whereas it may enhance lipotoxic activation of apoptosis in T2D islets effecting overall insulin secreting function (Ling et al., 2009).

One of the main candidate genes that is implicated in adipogenesis, insulin resistance, and T2D is the peroxisome proliferator activated receptor-γ (*PPAR-γ*) gene. Within a unique domain of *PPAR-γ2* gene that enhances ligand-independent activation, a common Pro12Ala polymorphism has been identified (Altshuler et al., 2000). Deeb et al. (1998) reported that the Ala allele of this polymorphism was associated with increased insulin sensitivity and decreased risk of T2D in a Finnish and second-generation Japanese cohort. The risk of T2D conferred by this SNP has been studied prospectively in the Finnish Diabetes Prevention Study and the larger Botnia Prevention Study. In the Finnish study of 500 subjects with impaired glucose tolerance, the relative risk of developing diabetes was doubled in alanine carriers, contradicting prior evidence that the alanine allele was protective. In the larger Botnia study, comprising >2000 subjects, proline homozygotes were 1.7 times more likely to develop diabetes than alanine carriers (Lyssenko et al., 2005).

The K121Q polymorphism of the *ENPP1/PC-1* gene is associated with insulin resistance/atherogenic phenotypes, including earlier onset of T2D and myocardial infarction (Bacci et al., 2005). Prudente et al. (2007) suggested that the Q121 allele is a gene variant with pleiotropic deleterious effects on insulin resistance, obesity, and T2D. Other studies showed that this polymorphism has no significant impact on insulin sensitivity and is not a critical determinant for either diabetes or obesity (González-Sánchez et al., 2003; Seo et al., 2008).

Stefan et al. (2003) suggest that the association of homozygosity for the Asp1057 allele in *IRS-2* with T2D in Pima Indians may be mediated by interaction of the polymorphism with obesity on several diabetes-related traits. In a Japanese study, Asp1057 allele and haplotype pairs were not associated with the risk of diabetes. However, T2D patients, particularly obese patients carrying the Asp1057 allele and the CA haplotype, were associated with insulin resistance. These results strongly suggest that the Gly1057Asp polymorphism in insulin receptor substrate-2 is not associated with ß-cell dysfunction (Okazawa et al., 2005).

Single nucleotide polymorphisms (SNPs) of *ADIPOQ* gene have been genotyped in large datasets from various ethnic groups, and several SNPs associated with hypoadiponectinemia, obesity, and T2D have been identified (Menzaghi et al., 2002; Vasseur et al., 2003; Gibson and

Froguel, 2004; Berthier et al., 2005; Heid et al., 2006). Adiponectin has been associated with low diabetes risk. The metabolic effects of adiponectin are mediated by adiponectin receptors 1 (ADIPOR1) and 2 (ADIPOR2). In a study of six polymorphisms in *ADIPOR1* and 16 polymorphisms in *ADIPOR2*, a significant association between *ADIPOR1* haplotypes and diabetes risk was observed (Qi et al., 2007).

A genome-wide association (GWA) study for T2D in a U.K. population revealed a novel locus associated with body mass index (BMI) – the fat mass and obesity associated (*FTO*) gene on chromosome 16 (Frayling et al., 2007). The representative SNP rs9939609 was confirmed to be associated with elevated BMI after replication in over 38,000 study participants of European ancestry. In addition, adiposity appeared to mediate the association between the *FTO* variant and the risk of T2D (Scott et al., 2007; Zeggini et al., 2007). Several other studies have also observed associations between *FTO* variants and obesity-related traits in various populations (Dina et al., 2007, Chang et al., 2008).

GWA STUDIES FOR TYPE 2 DIABETES GENES

Understanding the role of genetics in T2D has been done through epidemiological studies, studies of candidate genes, and genetic linkage in families, for decades. While this has provided important insights into some rare monogenic forms of diabetes, understanding the genetics of common T2D remains a major challenge. By examining the entire genome it is possible to identify, in an unbiased manner, the genetic variation that is associated with specific diseases. Over the past year, a number of studies were done based on high-throughput genome-wide association (GWA) studies. The results of five independent GWA screens for T2D genes have been published (Sladek et al., 2007; Zeggini et al., 2007; Scott et al., 2007; Saxena et al., 2007; Steinthorsdottir et al., 2007), and these were followed by five smaller GWA studies (Salonen et al., 2007; Rampersaud et al., 2007; Hayes et al., 2007; Florez et al., 2007; Hanson et al., 2007). The five large studies were all conducted using a two-stage strategy consisting of a GWA screen in an initial cohort of unrelated cases and controls, followed by replication of the most significant findings in additional sets (Sanghera et al., 2008b; 2009). Advances in genotyping techniques and the availability of large patient cohorts have made it possible to identify common genetic variants associated with T2D through genome-wide association studies. So far, genetic variants on 19 loci have been identified (Lyssenko and Leif, 2009). The GWA studies suggest that from a genetic perspective, T2D represents a very large number of different disorders.

The GWA association studies continue to open our eyes to the broad nature of molecules that might contribute to the pathogenesis of T2D. Clearly, many of the genes identified in both GWA and positional cloning studies would not be considered typical "candidate" genes. It is especially important to keep in mind that most of the observed associations are in large domains of strong linkage disequilibrium in noncoding regions of the genome (Vavouri et al., 2006; Carroll et al., 2005; Birney et al., 2007). Thus, the true T2D genes may be placed at some distance from the association signals. With regard to prediction of disease, most of the single loci identified by GWA mark only a 10%–20% increase in risk of disease, and even the strongest single association, *TCF7L2*, represents only about a doubling of disease risk in its homozygous state. Higher relative risks can be generated by combining multiple markers together. Indeed, about 97% of the population carry between 9 and 20 risk alleles at the 15 T2D loci identified thus far (Cauchi et al., 2008b). Thus, at present, even a panel of multiple genes has uncertain usefulness in

TABLE 87-2 ALLELE FREQUENCY IN DIFFERENT POPULATIONS

				MDRF Study (South Indian)		Hong Kong		West Africa		French		FUSION		WTCCC	
S. NO	Gene	SNP	Risk Allele	Cases	control	Cases	control	Cases	control	Cases	control	Cases	control	Cases	control
1	CDKAL1	rs7756992 A/G	G	0.267	0.218	0.517[1]	0.462	0.612[1]	0.625	0.295[1]	0.258				
2	CDKAL1	rs7754840 G/C	C	0.252	0.207	0.420[2]	0.358					0.406[3]	0.372	0.35[3]	0.31
3	CDKAL1	rs6931514 A/G	G	0.263	0.215										
4	CDKN2A/B	rs7020996 C/T	C	0.088	0.13										
5	HHEX	rs7923837 A/G	G	0.411	0.458	0.205[2]	0.176			0.335[1]	0.377				
6	BAZ1B	rs12056034 A/G	A	0.062	0.088										

[1] Steinthorsdottir et al., 2007

[2] Scott et al., 2007

[3] Ng et al., 2007

genetic prediction, and a more complete identification of markers needs to be done to understand its usefulness.

A recent meta-analysis on three genome-wide association (GWA) scans identified six loci (NOTCH2, THADA, ADAMTS9, JAZF1, CDC123/CAMKID, and TSPAN8/LGRS) highly associated with T2D in Caucasians. Sanghera et al.(2009) genotyped highly significant variants from each locus in a case-control cohort consisting of 680 T2D cases and 637 normoglycemic (NG) controls in Khatri Sikh diabetics of North India. The findings suggest that CDC123/CAMKID could be a major risk factor for the development of T2D in Sikhs by affecting b-cell function. This is the first study reporting the role of recently emerging loci in this high-risk population from the South Asian subcontinent.

A total of 45 single nucleotide polymorphisms (SNPs) from 15 genes and 13 un-annotated loci identified from recent GWAS T2D studies were genotyped on 926 unrelated T2D and 812 normal glucose tolerant (NGT) subjects of Southern India (Table 87-2). Only 6 of 45 SNPs studied were replicated in this South Indian population. Three SNPs—rs7756992 ($p = 0.007$), rs7754840 ($p = 0.015$), and rs6931514 ($p = 0.029$) of the *CDKAL1* gene, rs7020996 ($p = 0.003$) of the *CDKN2A/B* gene, rs7923837 ($p = 0.038$) of the *HHEX* gene, and rs12056034 ($p = 0.033$) of the *BAZ1B* gene—were associated with T2D in our population.

TABLE 87-3 SUMMARY OF RECENT GENETIC STUDIES IN TYPE 2 DIABETES

1	*PPAR g* gene (Pro12Ala)	Lyssenko et al., 2005	Pro12Ala polymorphism protective in Caucasians
2	*PPAR g* gene (Pro12Ala)	Radha et al., 2006	Pro12Ala polymorphism not protective in South Asians
3	*PGC-1 α* gene (Thr394Thr, Gly482Ser)	Bhat et al., 2007	Associated with type 2 diabetes
4	*PGC-1 α* gene (Thr394Thr)	Vimaleswaran et al., 2005; 2006	Associated with type 2 diabetes and with body fat.
5	*PGC-1 α* gene (Gly482Ser)	Liang et al., 2006	Associated with the development of insulin resistance and T2D
6	*PGC-1 α* gene (Gly482Ser)	Ek et al., 2001	Associated with relative risk of type 2 diabetes in a European population
7	*PC-1* gene (K121Q)	Bacci et al., 2005	Associated with insulin resistance/atherogenic phenotypes
8	*PC-1* gene (K121Q)	Abate et al., 2005	Associated with type 2 diabetes
9	*PC-1* gene (K121Q)	González-Sánchez et al., 2003	Has no significant impact on insulin sensitivity
10	*IRS-2* gene (Gly1057Asp)	Bodhini et al., 2007a	D1057D genotype susceptible to diabetes by interacting with obesity
11	*IRS-2* gene (Gly1057Asp)	Okazawa et al., 2005	Is not associated with ß-cell dysfunction
12	*TCF7L2* gene (rs7903146)	Sanghera et al., 2008a	Associated with type 2 diabetes in Asian Indians
13	*TCF7L2* gene (rs7903146)	Chandak et al., 2007	Associated with type 2 diabetes
14	TCF7L2 gene (rs12255372; rs7903146)	Bodhini et al., 2007b	Associated with type 2 diabetes in Asian Indians
15	TCF7L2 gene (rs7903146)	Grant et al., 2006	T allele of rs7903146 associated with an increased risk of T2D
16	Adiponectin gene	Vimaleswaran et al., 2008	+10211T→G Associated with type 2 diabetes
17	Adiponectin gene	Stumvoll et al., 2002	SNP 45 is associated with obesity in a German population
18	Adiponectin gene (+45T/G; +276G/T)	Hara et al., 2001; Menzaghi et al., 2002; Vasseur et al., 2002	Significantly associated with T2D and adiponectin level in Japanese population and with insulin resistance in some Caucasian populations
19	*FTO* gene	Yajnik et al., 2009	Associated with type 2 diabetes in south Asian Indians
20	*FTO* gene	Frayling et al., 2007	Associated with body mass index
21	*FTO* gene	Dina et al., 2007, Chang et al., 2008	Associated with obesity-related traits
22	*FTO* gene	Ramya et al., 2010	Associated with T2D and obesity in South Asian Indians

Large-scale studies are needed in this population to validate the findings (Chidambaram et al., 2010).

In summary, the conventional and recent high-throughput technologies have led to the identification of a few genes playing a role in T2D (Table 87-3). The validation and characterization of these genes and gene loci is underway.

REFERENCES

Abate N, Chandalia M, Satija P, et al. (2005). ENPP1/Pc-1 K121Q polymorphism and genetic susceptibility to type 2 diabetes. *Diabetes* 54:1207–1213.

Aguilar-Salinas CA, Reyes-Rodriguez E, Ordonez-Sanchez ML, et al. (2001). Early-onset type 2 diabetes: Metabolic and genetic characterization in the Mexican population. *J Clin Endocrinol Metab* 86:220–226.

Altshuler D, Hirschhorn JN, Klannemark M, et al. (2000). The common PPAR-gamma Pro12Ala polymorphism is associated with decreased risk of Type 2 diabetes. *Nat Genet* 26:76–80.

Anuradha S, Radha V, Deepa R, Hansen T, Carstensen B, Pederson O, Mohan V. (2005). A prevalent amino acid polymorphism at Codon 98 (Ala98Val) of the Hepatocyte Nuclear Factor-1α is associated with maturity onset diabetes of the young and younger age at onset of Type 2 diabetes in Asian Indians. *Diabetes Care* 28:2430–2435.

Anuradha S, Radha V, Mohan V. (2010). Association of novel variants in the hepatocyte nuclear factor 4A gene with maturity onset diabetes of the young and early onset type 2 diabetes. *Clin Genet.* doi: 10.1111/j.1399-0004.2010.01577.x. [Epub ahead of print].

Babenko AP, Bryan J. (2003). SUR domains that associate with and gate KATP pores define a novel gatekeeper. J Biol Chem 278:41577–41580.

Bacci S, Ludovico O, Prudente S, et al. (2005). The K121Q polymorphism of the ENPP1/PC-1 gene is associated with insulin resistance/atherogenic phenotypes, including earlier onset of type 2 diabetes and myocardial infarction. *Diabetes* 54(10):3021–3025.

Bellanne-Chantelot C, Carette C, Riveline JP, et al. (2008). The type and the position of HNF1A mutation modulate age at diagnosis of diabetes in patients with maturity-onset diabetes of the young (MODY)-3. *Diabetes* 57:503–508.

Berthier MT, Houde A, Cote M, et al. (2005). Impact of adiponectin gene polymorphisms on plasma lipoprotein and adiponectin concentrations of viscerally obese men. *J Lipid Res* 46:237–244.

Bhat A, Koul A, Rai E, Sharma S, Dhar MK, Bamezai RNK. (2007). PGC-1αThr394Thr and Gly482Ser variants are significantly associated with T2D in two North Indian populations: A replicate case-control study. *Hum Genet* 121:609–614.

Birney E, Stamatoyannopoulos JA, Dutta A, et al. (2007). Identification and analysis of functional elements in 1% of the human genome by the ENCODE pilot project. *Nature* 447:799–816.

Bodhini D, Radha V, Deepa R, et al. (2007a). The G1057D polymorphism of IRS-2 gene and its relationship with obesity in conferring susceptibility to type 2 diabetes in Asian Indians. (CURES-19). *International Journal Obesity* 31:97–102.

Bodhini D, Radha V, Monalisa Dhar, Narayani N, Mohan V. (2007b). The rs12255372 (G/T) and rs7903146(C/t) polymorphisms of the TCF7L2 gene are associated with type 2 diabetes mellitus in Asian Indians. *Metabolism Clinical and Experimental* 56:1174–1178.

Carroll JS, Liu XS, Brodsky AS, et al. (2005). Chromosome-wide mapping of estrogen receptor binding reveals long-range regulation requiring the forkhead protein FoxA1. *Cell* 122:33–43.

Cauchi S, Meyre D, Durand E, et al. (2008b). Post genomewide association studies of novel genes associated with type 2 diabetes show gene-gene interaction and high predictive value. *PLoS ONE* 3, e2031.

Chandak GR, Janipalli CS, Bhaskar S, et al. (2007). Common variants in the TCF7L2 gene are strongly associated with type 2 diabetes mellitus in the Indian population. *Diabetologia* 50:630–667.

Chandalia M, Abate N, Garg A, Stray-Gundersen J, Grundy SM. (1999). Relationship between generalized and upper body obesity to insulin resistance in Asian Indian men. *J Clin Endocrinol Metab* 84:2329–2335.

Chang YC, Liu PH, Lee WJ, et al. (2008). Common variation in the fat mass and obesity-associated (FTO) gene confers risk of obesity and modulates BMI in the Chinese population. *Diabetes* 57:2245–2252.

Chidambaram M, Radha V, Mohan V. (2010). Replication of recently described type 2 gene variants in a South Indian population. *Metabolism* 59(12):1760–1766.

Croall DE, DeMartino GN. (1991). Calcium-activated neutral protease (calpain) system: Structure, function, and regulation. *Physiol Rev* 71:813–847.

Deeb SS, Fajas L, Nemoto M, et al. (1998). A Pro12Ala substitution in PPAR-gamma2 associated with decreased receptor activity, lower body mass index and improved insulin sensitivity. *Nat Genet* 20:284–287.

Diamond J. (2003). The double puzzle of diabetes. *Nature* 423:599–602.

Dina C, Meyre D, Gallina S, et al. (2007). Variation in FTO contributes to childhood obesity and severe adult obesity. *Nat Genet* 39:724–726.

Ek J, Andersen G, Urhammer SA, et al. (2001). Mutation analysis of peroxisome proliferator-activated receptor-g coactivator-1 (PGC-1) and relationships of identified amino acid polymorphisms to Type II diabetes mellitus. *Diabetologia* 44, 2220–2226.

Elbein SC, Hasstedt SJ, Wegner K, Kahn SE. (1999). Heritability of pancreatic beta-cell function among nondiabetic members of Caucasian familial type 2 diabetic kindreds. *J Clin Endocrinol Metab* 84:1398–1403.

Ellard S, Colclough K. (2006). Mutations in the genes encoding the transcription factors hepatocyte nuclear factor 1 alpha (HNF1A) and 4 alpha (HNF4A) in maturity-onset diabetes of the young. *Hum Mutat* 27(9):854–869.

Fajans SS, Bell GI, Polonsky KS. (2001). Molecular mechanisms and clinical pathophysiology of maturity-onset diabetes of the young. *N Engl J Med* 345:971–980.

Florez JC, Jablonski KA, Bayley N, et al. (2006). TCF7L2 polymorphisms and progression to diabetes in the Diabetes Prevention Program. *N Engl J Med* 355:241–250.

Florez JC, Manning AK, Dupuis J, et al. (2007). A 100K genome-wide association scan for diabetes and related traits in the Framingham Heart Study: Replication and integration with other genome-wide datasets. *Diabetes* 56:3063–3074.

Frayling TM, Evans JC, Bulman MP, et al. (2001). Beta-cell genes and diabetes: Molecular and clinical characterization of mutations in transcription factors. *Diabetes* 50(Suppl 1):S94–100.

Frayling TM, Timpson NJ, Weedon MN, et al. (2007). A common variant in the FTO gene is associated with body mass index and predisposes to childhood and adult obesity. *Science* 316:889–894.

Froguel P, Vaxillaire M, Sun F, et al. (1992). Close linkage of glucokinase locus on chromosome 7p to early-onset non-insulin-dependent diabetes mellitus. *Nature* 356:162–164.

Gelding SV, Andres C, Nithyanathan R, Richmond W, Gray IP, Johnston DG. (1994). Serum lipoprotein levels and plasma concentrations of insulin, intact and 32,33 split proinsulin in normoglycaemic relatives of patients with type 2 diabetes. *Diabetes Med* 11(8):748–754.

Gerich JE. (1998). The genetic basis of type 2 diabetes mellitus: Impaired insulin secretion versus impaired insulin sensitivity. *Endocr Rev* 19:491–503.

Gibson F, Froguel P. (2004). Genetics of the ADIPOQ locus and its contribution to Type 2 diabetes susceptibility in French Caucasians. *Diabetes* 53:2977–2983.

Gloyn AL, Pearson ER, Antcliff JF, et al. (2004). Activating mutations in the gene encoding the ATP-sensitive potassium-channel subunit Kir6.2 and permanent neonatal diabetes. *N Engl J Med* 350:1838–1849.

González-Sánchez JL, Martínez-Larrad MT, Fernández-Pérez C, Kubaszek A, Laakso M, Serrano-Ríos M. (2003). K121Q PC-1 Gene Polymorphism Is Not Associated with Insulin Resistance in a Spanish Population. *Obesity Research* 11:603–605.

Grant SF, Thorleifsson G, Reynisdottir I, et al. (2006). Variant of transcription factor 7-like 2 (TCF7L2) gene confers risk of type 2 diabetes. *Nature Genetics* 38(3):320–320.

Groves CJ, Zeggini E, Minton J, et al. (2006). Association analysis of 6,736 U.K. subjects provides replication and confirms TCF7L2 as a type 2 diabetes susceptibility gene with a substantial effect on individual risk. *Diabetes* 55:2640–2644.

Haffner SM, Miettina H, Stern MP. (1996). Insulin secretion and resistance in nondiabetic Mexican-Americans and non-Hispanics whites with a parental history of diabetes. *J Clin Endocrinol Metabol* 81(5):1846–1851.

Hanson RL, Bogardus C, Duggan D, et al. (2007). A search for variants associated with young-onset type 2 diabetes in American Indians in a 100K genotyping array. *Diabetes* 56:3045–3052.

Hara K, Mori Y, Dina C, et al. (2001). Genetic variation in the gene encoding adiponectin is associated with increased risk of type 2 diabetes. *Diabetes* 50(Suppl. 1):A244.

Hattersley AT, Turner RC, Permutt MA, et al. (1992). Linkage of type 2 diabetes to the glucokinase gene. *Lancet* 339:1307–1310.

Hattersley AT, Beards F, Ballantyne E, Appleton M, Harvey R, Ellard S. (1998). Mutations in the glucokinase gene of the fetus result in reduced birth weight. *Nature Genetics* 19:268–270.

Hattersley AT, Ashcroft FM. (2005). Activating mutations in kir6.2 and neonatal diabetes: New clinical syndromes, new scientific insights, and new therapy. *Diabetes* 54:2503–2513.

Hayes MG, Pluzhnikov A, Miyake K, et al. (2007). Identification of type 2 diabetes genes in Mexican Americans through genome-wide association studies. *Diabetes* 56:3033–3044.

Heid IM, Wagner SA, Gohlke H, et al. (2006). Genetic architecture of the APM1 gene and its influence on adiponectin plasma levels and parameters of the metabolic syndrome in 1,727 healthy Caucasians. *Diabetes* 55:375–384.

Hitman GA, McCarthy MI, Mohan V, Viswanathan M. (1992). The genetics of non-insulin-dependent diabetes mellitus in South India: An overview. *Ann Med* 24:491–497.

Holmkvist J, Cervin C, Lyssenko V, et al. (2006). Common variants in HNF-1_ and risk of type 2 diabetes. *Diabetologia* 49:2882–2891.

Holmkvist J, Almgren P, Lyssenko V, et al. (2008). Common variants in maturity-onset diabetes of the young genes and future risk of type 2 diabetes. *Diabetes* 57(6):1738–1744.

Horikawa Y, Iwasaki N, Hara M, Furuta H, Hinokio Y, Cockburn B, Lindner T, Yamagata K, Ogata M, *et al.* (1997). Mutation in hepatocyte nuclear factor-1b gene (TCF2) associated with MODY. *Nature Genetics* 17:384–385.

Horikoshi M, Hara K, Ito C, Nagai R, Froguel P, Kadowaki T. (2007). A genetic variation of the transcription factor 7-like 2 gene is associated with risk of type 2 diabetes in the Japanese population. *Diabetologia* 50:747–751.

Iselius L, Lindstein J, Morton NE et al. (1982) Evidence for an autosomal recessive gene regulating the persistence of insulin response to glucose in man. *Clin Genet* 22:180–194.

Joshi SR. (2003). Metabolic syndrome–emerging clusters of the Indian phenotype. *J Assoc Physicians India* 51:445–446.

Kaisaki PJ, Menzel S, Lindner T, et al. (1997). Mutations in the hepatocyte nuclear factor-1α gene in MODY and early-onset NIDDM. Evidence for a mutational hotspot in exon4. *Diabetes* 46:528–535.

Kaprio J. (2000). Science, medicine, and the future. Genetic epidemiology. *BMJ* 320:1257–1259.

Lehto M, Wipemo C, Ivarsson SA, et al. (1999). High frequency of mutations in MODY and mitochondrial genes in Scandinavian patients with familial early-onset diabetes. *Diabetologia* 42:1131–1137.

Liang H, Ward WF. (2006). PGC-1α: A key regulator of energy metabolism. *Adv Physiol Educ* 30:145–151.

Ling C, Groop L, Guerra SD, Lupi R. (2009). Calpain-10 Expression Is Elevated in Pancreatic Islets from Patients with Type 2 Diabetes. *Plosone* 4(8):E6558.

Love-Gregory LD, Wasson J, Ma J, et al. (2004). A common polymorphism in the upstream promoter region of the hepatocyte nuclear factor-4α gene on chromosome 20q is associated with type 2 diabetes and appears to contribute to the evidence for linkage in an Ashkenazi Jewish population. *Diabetes* 53:1134–1140.

Lyssenko V, Almgren P, Anevski D, et al. (2005). Genetic Prediction of Future Type 2 Diabetes. *PLoS Med* 2(12):e345.

Lyssenko V, Leif G. (2009). Genome-wide association study for type 2 diabetes: clinical applications. *Current Opinion in Lipidology* 20(2):87–91.

Malecki MT, Jhala US, Antonellis A, Fields L, Doria A, Orban T, Saad M, Warram JH, Montminy M, Krolewski AS. (1999). Mutations in NEUROD1 are associated with the development of Type 2 diabetes mellitus. *Nature Genetics* 23(3):323–328.

Mathias RA, Deepa M, Deepa R, Wilson AF, Mohan V. (2009). Heritability of quantitative traits associated with type 2 diabetes mellitus in large multiplex families from South India. *Metabolism* 58(10):1439–1445.

McCarthy MI, Hitman GA, Mohan V, Ramachandran A, Snehlatha C, Viswanathan M. (1992). The islet amyloid polypeptide gene and non-insulin-dependent diabetes mellitus in south Indians. *Diabetes Res Clin Pract* 18(1):31–34.

McCarthy MI, Froguel P. (2002). Genetic approaches to the molecular understanding of type 2 diabetes. *Am J Physiol Endocrinol Metab* 283:E217–E225.

Meigs JB, Cupples LA, Wilson PW. (2000). Parental transmission of type 2 diabetes: The Framingham Offspring Study. *Diabetes* 49:2201–2207.

Menzaghi C, Ercolino T, Di Paola R, et al. (2002). A haplotype at the adiponectin locus is associated with obesity and other features of the insulin resistance syndrome. *Diabetes* 51:2306–2312.

Mohan V, Ramachandran A, Snehalatha C, Mohan R, Bharani G, Viswanathan M. (1985) High prevalence of maturity onset diabetes of the young (MODY) among Indians. *Diabetes Care* 8:374–374.

Mohan V, Sharp PS, Aber VR, Mather HM and Kohner EM. (1986a). Family histories of Asian Indian and Europeans non-insulin-dependent diabetic patients. Practical *Diabetes* 3:254–256.

Mohan V, Shanthi Rani S, Deepa R, Premalatha G, Sastry NG, Saroja R. (2001). Intra urban differences in the prevalence of the metabolic syndrome in southern India – The Chennai Urban Population Study (CUPS). *Diabet Med* 18:280–287.

Mohan V, Shanthirani CS, Deepa R. (2003). Glucose intolerance (diabetes and IGT) in a selected south Indian population with special reference to family history, obesity and life style factors – The Chennai Urban Population Study (CUPS 14). *Journal of Association of Physicians of India* 51:771–777.

Moore AF, Florez JC. (2008). Genetic Susceptibility to Type 2 Diabetes and Implications for Antidiabetic Therapy. *Annual Review of Medicine* 59:95–111.

Ng MC, Lee SC, Ko GT, et al. (2001). Familial early-onset type 2 diabetes in Chinese patients: Obesity and genetics have more significant roles than autoimmunity. *Diabetes Care* 24:663–671.

Ng MC, Park KS, Oh B, et al. (2007). Implication of genetic variants near TCF7L2, SLC30A8, HHEX, CDKAL1, CDKN2A/B, IGF2BP2, and FTO in type 2 diabetes and obesity in 6,719 Asians. *Diabetes* 57(8):2226–2233.

Niki T, Mori H, Tamori Y, et al. (1996). Human obese gene: Molecular screening in Japanese and Asian Indian NIDDM patients associated with obesity. *Diabetes* 45(5):675–678.

Okazawa K, Yoshimasa Y, Miyamoto Y, et al. (2005). The haplotypes of the IRS-2 gene affect insulin sensitivity in Japanese patients with type 2 diabetes. *Diabetes Res Clin Pract* 68(1):39–48.

Owen KR, Stride A, Ellard S, Hattersley AT. (2003). Etiological investigation of diabetes in young adults presenting with apparent type 2 diabetes. *Diabetes Care* 26:2088–2093.

Pontiroli AE, Capra F, Veglia F, et al. (1996). Genetic contribution of polymorphism of the GLUT1 and GLUT4 genes to the susceptibility to type 2 (non-insulin-dependent) diabetes mellitus in different populations. *Acta Diabetol* 33(3):193–197.

Poulsen P, Kyvik KO, Vaag A, Beck-Nielsen H. (1999). Heritability of type II (non-insulin-dependent) diabetes mellitus and abnormal glucose tolerance – a population-based twin study. *Diabetologia* 42:139–145.

Prudente S, Chandalia M, Morini E, et al. (2007). The Q121/Q121 Genotype of ENPP1/PC-1 Is Associated with Lower BMI in Non-diabetic Whites. *Obesity* 15, 1–4.

Qi L, Doria A, Giorgi E, Hu FB. (2007). Variations in adiponectin receptor genes and susceptibility to type 2 diabetes in women: A tagging-single nucleotide polymorphism haplotype analysis. *Diabetes* 56(6):1586–1591.

Radha V, Vimaleswaran KS, Babu HNS, et al. (2006). Role of genetic polymorphism peroxisome proliferator – Activated Receptor – 2 Pro 12Ala on ethnic susceptibility to diabetes in south Asian and Caucasian subjects. *Diabetes Care* 29:1046–1051.

Radha V, Mohan V. (2007). Genetic predisposition to type 2 diabetes among Asian Indians. *Indian J Med Res* 125:259–274.

Radha V, Vimaleswaran KS, Babu S, et al. (2007). Lack of association between serum adiponectin levels and the Pro12Ala polymorphism in Asian Indians. *Diabet Med* 24(4):398–402.

Radha V, Ek J, Anuradha S, Hansen T, Pedersen O, Mohan V. (2009). Identification of novel variants in the hepatocyte nuclear factor 1 alpha gene in South Indian patients with maturity onset diabetes of young, *J Clin Endocrine & Metabolism* 94(6):1959–1965.

Raffel LJ, Rotter JI. (2002). Diabetes Mellitus. In: Rimoin DL, Connor JM, Pyeritz RE, Korf BR, eds. *Emery and Rimoin's Principles and Practice of Medical Genetics*. Edinburgh: Churchill Livingstone, pp. 2231–2276.

Rampersaud E, Damcott CM, Fu M, et al. (2007). Identification of novel candidate genes for type 2 diabetes from a genome-wide association scan in the Old Order Amish: Evidence for replication from diabetes-related quantitative traits and from independent populations. *Diabetes* 56:3053–3062.

Ramya K, Radha V, Ghosh G, Majumder PP, Mohan V. (2011). Genetic variations in the fto gene are associated with type 2 diabetes and obesity in South Indians. *Diabetes Technol Ther* 13(1):33–42.

Salonen JT, Uimari P, Aalto JM, et al. (2007). Type 2 diabetes whole-genome association study in four populations: The DiaGen consortium. *Am J Hum Genet* 81:338–345.

Sanghera DK, Nath SK, Ortega L, et al. (2008a). TCF7L2 polymorphisms are associated with type 2 diabetes in Khatri Sikhs from North India: Genetic variation affects lipid levels. *Ann Hum Genet* 72(4):499–509.

Sanghera DK, Ortega L, Han S, et al. (2008b) Impact of nine common type 2 diabetes risk polymorphisms in Asian Indian Sikhs: PPARG2 (Pro12Ala), IGF2BP2, TCF7L2 and FTO variants confer a significant risk. *BMC Med Genet* 9:59.

Sanghera DK, Been L, Ortega L, et al. (2009). Testing the association of novel meta-analysis-derived diabetes risk genes with type II diabetes and related metabolic traits in Asian Indian Sikhs. *Journal of Human Genetics* 54:162–168.

Sargeant LA, Wareham NJ, Khaw KT. (2000). Family history of diabetes identifies a group at increased risk for the metabolic consequences of obesity and physical inactivity in EPIC-Norfolk: A population-based study. The European Prospective Investigation into Cancer. *Int J Obes Relat Metab Disord* 24:1333–1339.

Saxena R, Voight BF, Lyssenko V, et al. (2007). Genome-wide association analysis identifies loci for type 2 diabetes and triglyceride levels. *Science* 316:1331–1336.

Scott LJ, Mohlke KL, Bonnycastle LL, et al. (2007). A genome-wide association study of type 2 diabetes in Finns detects multiple susceptibility variants. *Science* 316:1341–1345.

Seo HJ, Kim SG, Kwon OJ. (2008). The K121Q Polymorphism in ENPP1 (PC-1) Is Not Associated with Type 2 Diabetes or Obesity in Korean Male Workers. *J Korean Med Sci* 23:459–464.

Sharp PS, Mohan V, Levy JC, Mather HM, Kohner EM. (1987). Insulin resistance in patients of Asian Indian and European origin with non-insulin dependent diabetes. *Horm Metab Res* 19:84–85.

Shyng SL, Ferrigni T, Shepard JB, et al. (1998). Functional analyses of novel mutations in the sulfonylurea receptor 1 associated with persistent hyperinsulinemic hypoglycemia of infancy. *Diabetes* 47:1145–1151.

Sicree R, Shaw J, Zimmet P. (2009). Diabetes and impaired glucose tolerance. In: Gan D, ed. *Diabetes Atlas*. 4th ed, pp. 1–105. Brussels: International Diabetes Federation.

Silander K, Mohlke KL, Scott LJ, et al. (2004). Genetic variation near the hepatocyte nuclear factor-4alpha gene predicts susceptibility to type 2 diabetes. *Diabetes* 53:1141–1149.

Silver K, Walston J, Wang Y, Dowse G, Zimmet P, Shuldiner AR. (1996). Molecular scanning for mutations in the beta 3-adrenergic receptor gene in Nauruans with obesity and noninsulin dependent diabetes mellitus. *J Clini Endocrinol* 81(11):4155–4158.

Sladek R, Rocheleau G, Rung J, et al. (2007). A genome-wide association study identifies novel risk loci for type 2 diabetes. *Nature* 445:881–885.

Stefan N, Kovacs P, Stumvoll M, et al. (2003). Metabolic effects of the Gly1057Asp polymorphism in IRS-2 and interactions with obesity. *Diabetes* 52(6):1544–1550.

Steinthorsdottir V, Thorleifsson G, Reynisdottir I, et al. (2007). A variant in CDKAL1 influences insulin response and risk of type 2 diabetes. *Nat Genet* 39:770–775.

Stern MP. (2002). The search for type 2 diabetes susceptibility genes using whole-genome scans: An epidemiologist's perspective. *Diabetes Metab Res Rev* 18:106–113.

Stoffers DA, Ferrer J, Clarke WL, Habener JF. (1997). Early-onset type-II diabetes mellitus (MODY4) linked to IPF1. *Nature Genetics* 17:138–139.

Stone L, Kahn S, Fujimoto W, Deeb S, Porte D. (1996). A variation at position -30 of the betacell glucokinase gene promoter is associated with reduced beta-cell function in middle-aged Japanese-American men. *Diabetes* 45:422–428.

Stumvoll M, Tschritter O, Fritsche A, et al. (2002). Association of the T-G polymorphism in adiponectin (exon 2) with obesity and insulin sensitivity: Interaction with family history of type 2 diabetes. *Diabetes* 51:37–41.

Tattersall RB, Fajans SS. (1975) A difference between the inheritance of classical juvenile-onset and maturity-onset type diabetes of young people. *Diabetes* 24:44–53.

Vaag A, Henriksen JE, Madsbad S, Holm N, Beck-Nielsen H. (1995). Insulin secretion, insulin action, and hepatic glucose production in identical twins discordant for non-insulin-dependent diabetes mellitus. *J Clin Invest* 95:690–698.

Vasseur F, Lepretre F, Lacquemant C, Froguel P. (2003). The genetics of adiponectin. *Curr Diab Rep* 3:151–158.

Vavouri T, McEwen GK, Woolfe A, Gilks WR, Elgar G. (2006). Defining a genomic radius for long-range enhancer action: Duplicated conserved non-coding elements hold the key. *Trends Genet* 22:5–10.

Vimaleswaran KS, Radha V, Ghosh S, et al. (2005). Peroxisome proliferators-activated receptor- co-activator-1 (PGC-1) gene polymorphisms and their relationship to Type 2 diabetes in Asian Indians. *Diabetic Medicine* 22:1516–1521.

Vimaleswaran KS, Radha V, Anjana M, et al. (2006). Effect of polymorphisms in the PPARGC1A gene on body fat in Asian Indians. *International Journal of Obesity* 30:884–891.

Vimaleswaran KS, Radha V, Deepa R, Mohan V. (2007). Absence of Association of Metabolic Syndrome with PPARGC1A, PPARG and UCP1 Gene Polymorphisms in Asian Indians. *Metab Syndr Relat Disord* 5(2):153–162.

Vimaleswaran KS, Radha V, Ramya K, et al. (2008). A novel association of a polymorphism in the first intron of adiponectin gene with type 2 diabetes, obesity and hypoadiponectinemia in Asian Indians. *Human Genetics* 123(6):599–605.

Vimaleswaran KS, Radha V, Jayapriya MG, et al. (2010a). Evidence for an association with type 2 diabetes mellitus at the PPARG locus in a South Indian population. *Metabolism* 59(4):457–462.

Vimaleswaran KS, Radha V, Ghosh S, Majumder PP, Rao MR, Mohan V. (2010b). A haplotype at the UCP1 gene locus contributes to genetic risk for type 2 diabetes in Asian Indians. *Metab Syndr Relat Disord* 8(1):63–68.

Vimaleswaran KS, Radha V, Ghosh S, Majumder PP, Rao MR, Mohan V. (2011). Uncoupling protein 2 and 3 gene polymorphisms and their association with type 2 diabetes in Asian Indians. *Diabet Technol Therapeut* 13(1):19–25.

Viswanathan M, Mohan V, Snehalatha C, Ramachandran A. (1985). High prevalence of type 2 (Non insulin dependent) diabetes among offspring of conjugal diabetic parents in India. *Diabetologia* 28:907–910.

Waterworth DM, Talmud PJ, Humphries SE, et al. (2001). Variable effects of the APOC3-482C>T variant on insulin, glucose and triglyceride concentrations in different ethnic groups. *Diabetologia* 44(2):245–248.

Whittemore AS. (1996). Genome scanning for linkage: An overview. *Am J Hum Genet* 59:704–716.

Wild S, Roglic G, Green A, Sicree R, King H. (2004). Global prevalence of diabetes: Estimates for the year 2000 and projections for 2030. *Diabetes Care* 27:1047–1053.

Winckler W, Weedon MN, Graham RR, et al. (2007). Evaluation of common variants in the six known maturity-onset diabetes of the young (MODY) genes for association with type 2 diabetes. *Diabetes* 56:685–693.

Yajnik CS, Lubree HG, Rege SS, et al. (2002). Adiposity and hyperinsulinemia in Indians are present at birth. *J Clin Endocrinol Metab* 87:5575–5580.

Yajnik CS, Janipalli CS, Bhaskar S, et al. (2009). FTO gene variants are strongly associated with type 2 diabetes in South Asian Indians. *Diabetologia* 52(2):247–252.

Yamagata K, Oda N, Kaisaki PJ, et al. (1996b) Mutations in the hepatic nuclear factor 1alpha gene in maturity-onset diabetes of the young (MODY3). *Nature* 384:455–458.

Zeggini E, Weedon MN, Lindgren CM, et al. (2007). Replication of genome-wide association signals in UK samples reveals risk loci for type 2 diabetes. *Science* 316:1336–1341.

Zhang C, Qi L, Hunter DJ, et al. (2006). Variant of transcription factor 7-like 2 (TCF7L2) gene and the risk of type 2 diabetes in large cohorts of U.S. women and men. *Diabetes* 55:2645–2648.

88.

HEREDITARY SPINOCEREBELLAR ATAXIAS IN INDIA

Mohammed Faruq, Uma Mittal, Achal K. Srivastava, and Mitali Mukerji

INTRODUCTION

Spinocerebellar ataxias (SCAs) are a group of rare progressive neurodegenerative disorders characterized by symptoms that include gait ataxia, dysarthria, peripheral neuropathy, oculomotor disorder, etc. (Cummings and Zoghbi, 2000). SCAs have been classified clinically by Harding into four broad categories of autosomal dominant cerebellar ataxias (ADCA I-IV) based on some unique features as well as occurrence of other associated symptomatology due to noncerebellar neuronal degeneration (Harding, 1993; Table 88-1). The majority of the SCAs belong to ADCA1 category of Harding's classification system. Members of this group have prominent cerebellar ataxias and associated one or more features like peripheral neuropathy, muscular atrophy, pyramidal, extrapyramidal, and one or more signs eye movement abnormalities. ADCA-II is clinically distinguishable by presence of retinal pigmentary degeneration in addition to all the features of ADCA-I. ADCA-III cases exhibit pure cerebellar features, though extracerebellar involvement has also been reported in some studies. ADCA-IV presents with seizure along with cerebellar involvement (Everett and Wood, 2004; Harding, 1993). This classification provides some clues for clinical identification of SCAs, which helps in determining genetic investigatory directions, follow-up, as well as prognosis of these incurable disorders. One of the unique features of SCAs is the phenomenon of anticipation, wherein there is an increase in severity and decrease in the age of onset of disease in successive generations. Once the disease onsets, the rate of progression is variable both among and within the same SCAs. In the initial stages of the disease, each of the SCAs to some extent can be clinically distinguished. However, as the disease progresses there is a significant overlap of clinical features between members of ADCA groups (Table 88-2).

MOLECULAR GENETICS

Nearly 30 loci have been identified to be associated with ataxia, which include both candidate linkage regions as well as well characterized genes (Table 88-1). Expansion and instability of repetitive sequences has been implicated in one-third of the genes associated with ataxia. Furthermore, expansion of triplet repeat has also been reported in a few other neurodegenerative disorders like Huntington's disease and Friedreich's ataxia (Cagnoli et al., 2006; Chamberlain et al., 1988; Chung et al., 2003; Devos et al., 2001; Holmes et al., 1999; Imbert et al., 1996; Kawaguchi et al., 1994; Knight et al., 2004; Koide et al., 1994; Koob et al., 1999; Orr et al., 1993; Pulst et al., 1996; Schelhaas et al., 2004; Storey et al., 2009; Swartz et al., 2002; van Swieten et al., 2003; Verbeek et al., 2004; Vuillaume et al., 2002; Waters et al., 2006; Yu et al., 2005). This has also been the underlying molecular cause for the observed non-Mendelian phenomenon of anticipation, as the repeats are unstable during transmission and the length of the repeat correlates with both increase in severity and earlier age of onset. The repeats—a majority being triplets, especially CNG (N is any nucleotide) or GAA—are located either in coding or noncoding regions of the respective disease genes (Figure 88-1). These repeats are polymorphic with respect to both length and interruption pattern in the normal population. However, the extent of polymorphism varies at different loci. In some cases there is an overlap between normal and expanded alleles, whereas in others there is a distinct gap between the two ranges (Figure 88-3). These repeats become unstable once they cross a particular threshold, leading to disease manifestation (Cummings and Zoghbi, 2000). The pathological threshold varies in these diseases depending on whether the repeat is coding or noncoding. In polyglutamine (polyQ) disorders (caused by expansion of glutamine-encoding CAG repeats), it is generally 30–40 triplets, with the exception of spinocerebellar ataxia 6 (SCA6), wherein as few as 20 triplets are sufficient to cause the disease. The expansions of coding repeats are generally small, bearing no more than 100 repeats in contrast to the massive expansions to over thousands observed in the noncoding repeats. In expansions involving noncoding repeats, a small pool of cryptic, intermediate-size expansions or premutations generally expands to the affected range following germline transmission (Table 88-1).

TABLE 88-1 MOLECULAR GENETICS OF SPINOCEREBELLAR ATAXIAS

SCA type	Gene	Chromosome	Mutation	SN	LN	Unstable	Reduce penet	Pathog	References
ADCA-I (CEREBELLAR ATAXIA, PERIPHERAL NEUROPATHY, OCCULAR MOVEMENT ABNORMALITY, PYRAMIDAL AND EXTRAPYRAMIDAL INVOLVEMENT)									
SCA1	ATXN1	6p22.3	(CAG)n	6–29	>30	36–38	39	40–82	Orr et al., 1993
SCA2	ATXN2	12q24.13	(CAG)n	13–22	>22	NR	NR	32–200	Imbert et al., 1996; Pulst et al.,1996
SCA3	ATXN3	14q32.12	(CAG)n	13–25	26–36	NR	45–51	61–84	Kawaguchi et al., 1994
SCA8	ATXN8OS	13q21	(CTG)n	15–50	NR	NR	71–80	>80	Koob et al., 1999
SCA12	PPP2R2B	5q32	(CAG)n	4–32	NR	NR	40–41	>45	Holmes et al., 1999
DRPLA	ATN1	12p13.31	(CAG)n	6–20	20–35	NR	NR	>48	Koide et al., 1994; Takano et al., 1998
SCA4	PLEKHG4	16q22.1	PM						Ishikawa et al., 2005
SCA13	KCNC3	19q13.33	PM						Waters et al., 2006
SCA27	FGF4	13q33.1	PM						van Swieten et al., 2003
SCA18	M	7q31-q32	Uk						Brkanac et al., 2002
SCA19*	M	1p21-q21	Uk						Schelhaas et al., 2004
SCA20	M	11p13-q11	Uk						Knight et al., 2004
SCA21	M	7p21.3-p15.1	Uk						Vuillaume et al., 2002
SCA22*	M	1p21-q23	Uk						Schelhaas et al., 2004
SCA23	M	20p13-p12.2	Uk						Verbeek et al., 2004
SCA24	M	1p36	Uk						Swartz et al., 2002
SCA25	M	2p21-p13	Uk						Stevanin et al., 2004
SCA28	M	18p11.22-q11.2	Uk						Cagnoli et al., 2006
SCA29	M	3p26	Uk						Dudding et al., 2004
SCA30	M	4q34.3-q35.1	Uk						Storey et al., 2009
ADCA-II (CEREBELLAR ATAXIA, RETINAL DEGENERATION, PYRAMIDAL AND EXTRAPYRAMIDAL INVOLVEMENT)									
SCA7	ATXN7	3p14.1	(CAG)n	<19	20–28	28–33	34–36	37–306	David et al., 1997
ADCA-III (PURE CEREBELLAR FORM)									
SCA6	CACNA1A	19p13.13	(CAG)n	<18	NR	NR	19	>20–29	Zhuchenko et al., 1997
SCA5	SPTBN2	11q13.2	Del, PM						Ikeda et al., 2006
SCA9*	Uncharacterized								
SCA11	TTBK2	15q14-q21.3	Ins, Del						Houlden et al., 2007
SCA14	PRKCG	19q13.42	PM						Chen et al., 2003
SCA15	ITPR1	3p24.2-3ptr	Del.						Hara et al., 2004
SCA16	M	8q23-q24.1	Uk						Miyoshi et al., 2001
SCA26	M	19p13.3	Uk						Yu et al., 2005
ADCA-IV (CEREBELLAR ATAXIA + SEIZURES, DEMENTIA AND PSYCHOMOTOR DISTURBANCES IN SCA17)									
SCA10	ATXN10	22q13.31	(ATTCT)n	10–29	NR	NR	280–370	500–4500	Matsuura et al., 2000
SCA17	TBP	6q27	(CAG)n	25–39	40–42	NR	43–48	>49	Nakamura et al., 2001

* reserved; PM, point mutation; M, mapped; Uk, Unknown; Del, deletion; Ins, Insertion; NR, not reported;

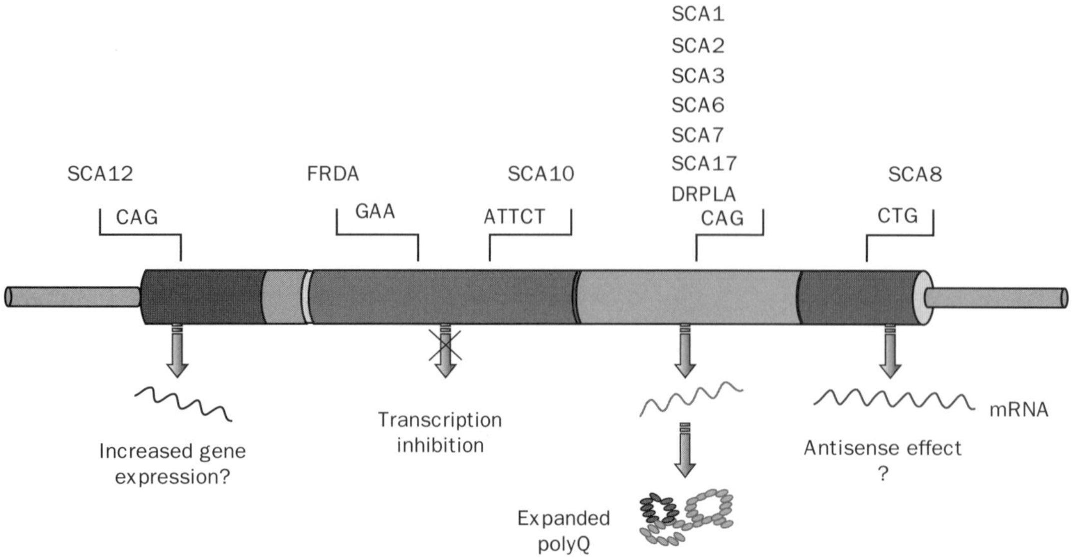

Figure 88-1 Location of repeats with respect to gene and possible consequences of repeat expansion in different ataxias. Refer color figure.

NEUROPATHOLOGY AND CLINICAL CORRELATIONS

Gross cerebellar and brainstem atrophy have been observed in the commonest subtypes of SCAs, such as SCA1, SCA2 and SCA3, which correlates with cerebellopontine phenotypes of these diseases as well as with the severity of illness. (Estrada et al., 1999; Pang et al., 2002; Rolfs et al., 2003; Rub et al., 2005; Yamada et al., 2008) Besides this, variable neuronal loss has also been documented in different regions of brain that in some cases could exhibit clinical symptoms beyond cerebellar ataxic syndromes. For instance, basal ganglia involvement in SCA3 and SCA17, macular degeneration in SCA7, and frontotemporal involvement in SCA2 could explain dystonia and rigidity, visual failure, and early psychomotor disturbance, respectively, in these subclasses (Durr et al., 1995; Estrada et al., 1999; Genis et al., 1995; Pang et al., 2002). But sparing of motor cortex and higher spinal cord nuclei does not correlate with the pyramidal features observed in SCAs (Schols et al., 2004). This could be due to extensive loss of cerebellar connections to various regions of the brain, leading to abnormal neural discharge from these tracks without their direct involvement.

PATHOLOGICAL CONSEQUENCES OF REPEAT EXPANSION

Various mechanisms at the DNA, RNA, and protein levels have been proposed for disease pathogenesis due to repeat expansion, based on cell culture and model system studies. A common feature of these disorders, the triplet repeat expansion and cell specific degeneration has allowed an insight into underlying pathological cellular consequences in SCAs, i.e., gain-of-function mechanisms at the RNA or protein level. (Bates, 2005; Gatchel and Zoghbi, 2005; Figure 88-2). CAG repeats expansion in a coding region translate into abnormally long stretches of glutamine residues in mutant protein, which has the propensity to aggregate either in cytosolic or nuclear compartments (Ross and x, 2004). However, the exact role of these long polyQ stretches in cellular cytotoxicity is unclear. Cellular/murine model-based studies have demonstrated that expanded polyQ-containing proteins lose

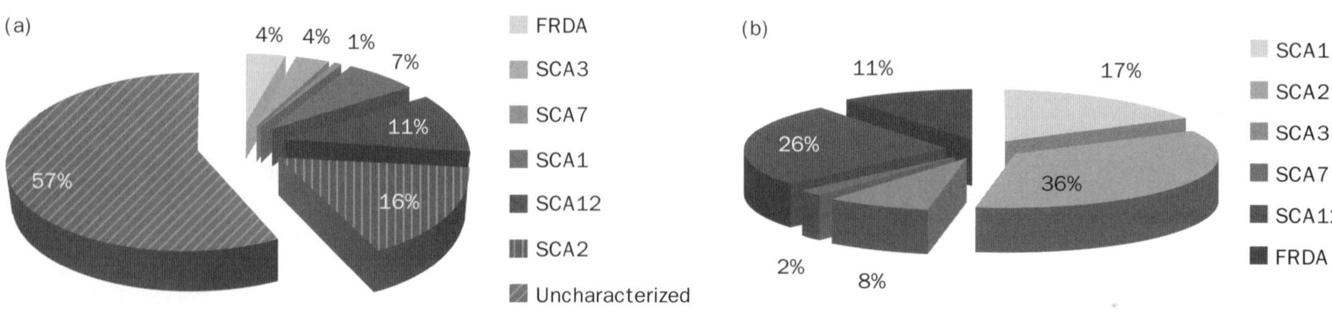

Figure 88-2 Prevalence of different SCAs in North Indian populations. (a) Frequencies of all SCA cases. (b) Relative frequencies of known SCAs. Uncharacterized SCA group is expansions-negative at SCA1, SCA2, SCA3, SCA6, SCA7, SCA8, SCA12, DRPLA, SCA17, and FRDA loci. Refer color figure.

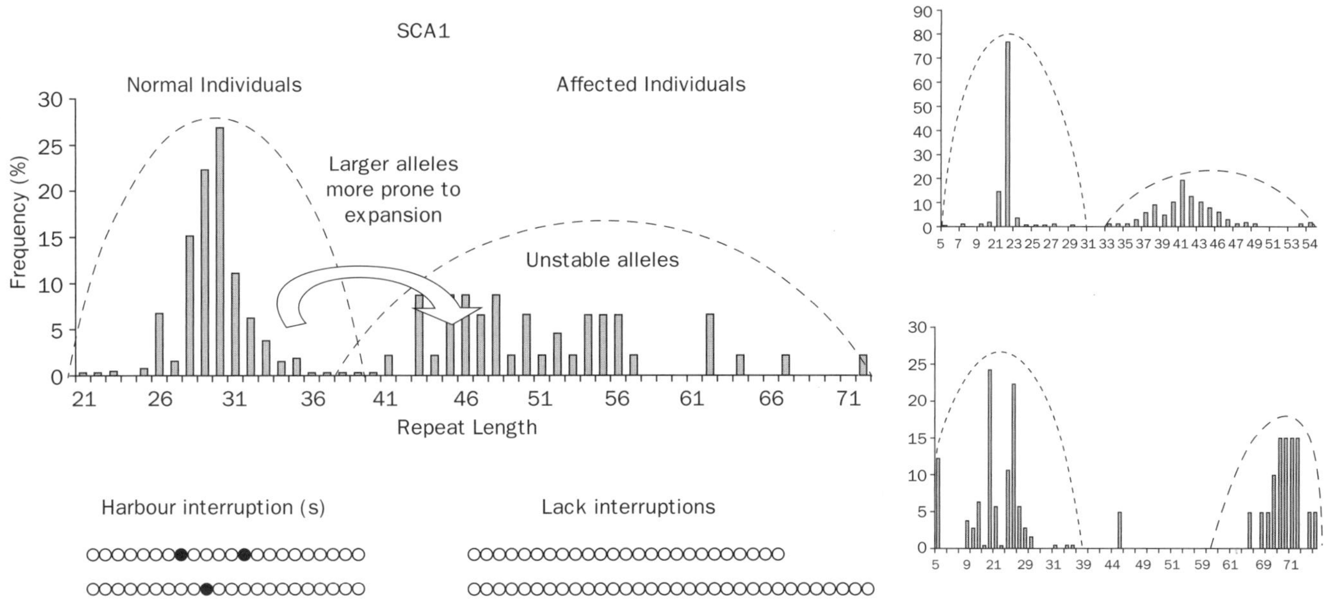

Figure 88-3 Spectrum of normal and expanded repeat alleles with respect to length and interruptions. The frequencies of normal and expanded alleles at respective loci are plotted on the basis of their frequencies in the Indian population. X-axes and Y-axes represent the repeat length and the frequencies (%) respectively.

their native conformation and are misfolded which are substrates for cellular UBQ-proteosome mediated clearance machinery and formation of cellular inclusion particles. Inclusion bodies containing various sequestrated proteins—e.g, cellular defense proteins, various transcription factors, etc.—have been implicated in pathogenesis (Cummings et al., 1998; McCampbell et al., 2000; Schmidt et al., 2002). These could be the result of aberrant interactions of toxic proteins, or failure of clearance machinery. The UPS machinery seems to play a major role in cellular toxicity exerted by these misfolded proteins, as evidenced by the following observations: (i) various components of these clearance mechanisms have been isolated from inclusion bodies (Cummings et al., 1998); (ii) altering expressions of UPS could modify the formation of cellular aggregates in in vitro studies (Jana et al., 2001); (iii) ataxin-1 and ataxin-3 interact with UPS machinery and alter its function by decreasing its activity, and acting as a deubiquitinating component of UPS, respectively (Nicastro et al., 2005; Park et al., 2005); and (iv) the role of UPS-associated E3 ligases, i.e, CHIP (c-terminus of HSP70 interacting protein), E6-AP, and parkin (RING-finger type E3 ligase; see Cummings et al., 1999; McDonough and Patterson, 2003; Tsai et al., 2003) has been implicated in different polyglutamine-mediated neurodegenerative disorders. Some evidence also questions the direct implication of polyQ toxicity mechanisms in SCAs as a sole pathology as indicated by some cellular model studies of SCA1. It is observed that the neighboring protein context of polyQ, containing mutant protein and its interacting partners, also plays a major role in their cellular toxicity. The phosphorylation status at Ser 776 site and binding of various interacting molecules at AXH domain of atxn-1 modulates mediates the toxic consequences of mutant proteins (Emamian et al., 2003; Lam et al., 2006; Mizutani et al., 2005; Serra et al., 2006; Tsuda et al., 2005). Cellular toxicity directly by abnormal short fragments of toxic molecules containing polyQ have also been implicated.

In noncoding repeat expansion, a gain-of-function mechanism has also been demonstrated at the RNA level in an SCA8 model in which CUG expansion containing mRNAs at the 3'UTR have altered binding with various RNA binding proteins (Liquori et al., 2001; Mutsuddi et al., 2004; Philips et al., 1998; Worth et al., 2000). In SCA12, where CAG repeat expansion has been reported at the 5' nonprotein coding region of the PPP2R2B gene, the role of CAG expansion in pathogenesis is still unclear (Holmes et al., 1999). However, irrespective of the mechanisms underlying these disorders, all of them lead to disruptions of cellular homeostasis resulting in neurodegeneration (Chou et al., 2006; Lipinski and Yuan, 2004; Wang et al., 2006).

The direct role of mitochondrial mutations has also been implicated in various early-onset neurodegenerative disorders, i.e., Leber's hereditary optic neuropathy (LHON), Leigh's disease, Parkinsonism, etc. Similarly, mutations in nuclear encoded mitochondrial proteins have been well documented in Friedreich's ataxia (GAA expansion-FXN), MIRAS (POLG mutations). The interplay of calcium homeostasis and mitochondrial mediated apoptotic pathways regulating cellular cytotoxicity could determine the extent and variable nature of disease spectrum observed in SCAs. Thus, the role of mitochondrial mutations as a direct causal event in uncharacterized

SCAs, or their effect on modulating disease phenotype in known ataxias, requires further exploration.

DISEASE PREVALENCE: GEOGRAPHICAL AND ETHNIC DIFFERENCES

The relative frequencies of different subtypes of autosomal dominant cerebellar ataxias (ADCAs) vary between different ethnic and geographical populations. The majority of them are prevalent in most of the studied populations, whereas a few are restricted to specific populations. For example, SCA3/MJD (Machado-Joseph Disease) seems to be the major subtype worldwide, representing more than 70% of the cases in Portugal (Gaspar et al., 2001). On other hand, dentato-rubro- pallidoluysian atrophy (DRPLA) is mostly reported from Japan (Sasaki et al., 2003), and SCA12 from India (Bahl et al., 2005; Srivastava et al., 2001). The prevalence is probably accounted for by regional founder effects, as evidenced through linkage disequilibrium studies using flanking markers at various loci such as SCA1, SCA2, SCA3, SCA6, SCA12, DRPLA, and FRAXA in different populations (Bahl et al., 2005; Basu et al., 2000; Chiurazzi et al., 1996; Mittal et al., 2005b, 2005c; Rubinsztein et al., 1994; Terasawa et al., 2004).

GENETIC STUDIES IN SCAS: THE INDIAN SCENARIO

Clinically, SCAs were first described as early as in 1971 by Wadia and Swami et al., with the identification of 9 families with spinocerebellar degeneration and slow eye movements (Wadia and Swami 1971). Later, in 1986, two groups (Bharucha et al., 1986; Jain and Maheshwari, 1986) identified three families of Machado Joseph disease in India. Subsequently, with the discovery of triplet repeat expansions, occurrence of SCAs were reported from different regions of the country. Due to the rarity of the disease, studies on ataxias have been carried out by only a few groups from North India: the All India Institute of Medical Sciences, Southern India-National Institute of Mental Health and Neurosciences (NIMHANS), Saha Institute of Nuclear Physics (Kolkata), Vivekananda Institute of Medical Sciences (Kolkata), Bangur Institute of Neurology (Kolkata), and Advanced Diagnostic Centre in Ranchi, India, in the eastern part of the country. These studies have provided estimates of prevalence of ataxias in different parts of the country. The extended pedigree of Indian SCA families has also allowed discovery as well as identification of founder chromosomes for a few ataxias.

PREVALENCE OF ATAXIA IN INDIA

Among repeat associated ADCAs, there are differences in prevalence of SCA subtypes in different geographical populations and ethnically diverse populations the world over. Similarly, in India, frequencies of recognized SCAs vary in subregions; however, SCA1, SCA2, and SCA3 are commonly observed varieties, and less commonly observed cases of SCA12 and SCA7 are found only in particular ethnic groups (Table 88-2). There is only one earlier reported isolated case of SCA6 (Gangopadhyay et al., 2001) and a recent description of two kindreds of SCA17 has been observed (Hire et al., 2010). SCA6, SCA8 and SCA10, have not been experienced in India so far. SCA2 is commonest type in north and eastern India, whereas SCA1 is found to be more common among South Indian cohorts of SCAs. Overall, SCA2 is the commonest SCA identified as reported from various studies in the country. The prevalence of SCA12 has remained confined to one ethnic community, ancestries of which belong to Haryana state, and it is second to SCA2 in prevalence in north India.

Understanding the mechanism(s) of repeat instability is a challenging task, and there have been major breakthroughs in this field. Several distinct hypotheses (Jodice et al., 1997; Saleem et al., 2001; Webster et al., 2002) have

TABLE 88-2 FREQUENCY DISTRIBUTION OF SCAS STUDIED BY VARIOUS GROUPS IN INDIA

	Region	N	SCA1	SCA2	SCA3	SCA6	SCA7	SCA12	DRPLA	FRDA	UC-SCA
(Faruq et al., 2009)	Delhi (North)	400	27	60	14	–	3	45	–	17	234
(Krishna et al., 2007)	Bangalore (South)	284	39	32	18	–	–	–	–	–	202
(Alluri et al., 2007)	Hyderabad (South)	124	6	1	1	–	1	–	1	–	114
(Rengaraj et al., 2005)	Chennai (South)	17	17	–	–	–	–	–	–	–	–
(Wadia et al., 1998)	Mumbai (South)	6/31*	–	6	–	–	–	–	–	–	–
(Basu et al., 2000; Chattopadhyay et al., 2003),(Sinha et al., 2004),(Chakravarty and Mukherjee, 2002)	Kolkata (East)	99	12	30	12	1	–	–	–	6	47

*6 families were taken for genetic studies out of 31 identified families of cerebellar ataxia and slow saccades; UC-SCA, Unclassified SCAs.

been pursued, with a majority of them based on studies in model systems (Cleary and Pearson, 2003). Nevertheless, none of these models can satisfactorily explain various aspects of these human disorders. Intergenerational instability associated with these disorders is not observed in all the murine models of repeat instability. Furthermore, humans show a proclivity toward expansions once a repeat stretch exceeds the threshold length, whereas in model systems, contractions are more common; thereby, these models fail to recapitulate the prevalence of expansions. In addition, there is a disparity in the behavior of homologous repeats in different mouse models (Savouret et al., 2003). These limitations of the model systems could be explained by the requirement for specific cis- or trans-acting elements as contributing factors of repeat instability that is suggested by genetic studies in human patients and model systems. However, the model systems of these disorders also fail to exhibit the same devastating phenotypes observed in humans. The likely explanation for this has been ascribed to the short life span of a mouse when compared to humans. It is therefore crucial to establish a correlation between the data derived from in vitro studies or from model systems and the etiology of repeat expansions in humans. Many reviews have covered comprehensively the insights gained from these studies (Cleary and Pearson, 2003; Lahue and Slater, 2003; Lenzmeier and Freudenreich, 2003; Usdin and Grabczyk, 2000).

We felt that in the post–human genome sequencing era, it would be possible to complement model system studies with a population polymorphism approach to understanding the mechanism of repeat instability in the genomic context. In this approach, utilizing association and linkage disequilibrium information, one can correlate the specific allele(s) of the variant(s) to the disease locus in a population. This would not only allow identification of the predisposed chromosomal backgrounds, but would also help elucidate the role of various genetic factors in determining the extent of repeat instability and heterogeneity in the prevalence of these disorders among populations. Comparisons of the disease variant and the associated polymorphisms in different human populations, as well as with nonhuman primates, would also provide valuable insights into the mutational history and spread of various diseases. In the process of this study we have built a considerable resource of ataxia families that includes ~900 families from AIIMS and NIMHANS.

In the following section, we provide an overview of different studies on ataxia in India, and demonstrate how, solely using a population polymorphism approach, a number of mechanisms related to trinucleotide repeat instability have been elucidated in the Indian population. We have also compared patterns of prevalence in India and other populations. We have discussed it in the purview of the following questions:

- What factors determine the repeat instability?
- What are the mechanisms and origin of repeat instability?
- What factors determine the prevalence of a disorder in a population?
- Is there a universal mechanism of repeat instability?

MECHANISM OF ORIGIN AND EXPANSION OF TRIPLET REPEAT

SPINOCEREBELLAR ATAXIA TYPE 1 (SCA 1)

The causative mutation for SCA1 [MIM 164400] is the expansion of a polyglutamine encoding CAG repeat in the coding region of SCA1 gene located on chromosome 6p23 (Orr et al., 1993). The repeat stretch is highly polymorphic and varies in length from 6 to 44 repeats in normal individuals, and is interrupted by 1 to 3 CAT (histidine) triplets when the length exceeds 21 repeats (Chung et al., 1993; Orr et al., 1993; Sobczak and Krzyzosiak, 2004). In the expanded alleles, only pure, uninterrupted stretches of 39 to 83 CAG repeats have been reported (Figure 88-2). The repeat lengths in the range of 39–41 triplets overlap between the normal and affected individuals, and therefore are either transmitted stably or manifest the phenotype depending on the presence or absence of the CAT interruptions respectively (Chong et al., 1995; Zuhlke et al., 2002).

There is a wide heterogeneity in the prevalence of SCA1 between different ethnic and geographical groups, which has been correlated with the frequency of large normal (LN) alleles and/or the interruption pattern in different populations (Mittal et al., 2005b; Sobczak and Krzyzosiak, 2004; Takano et al., 1998; Zhou et al., 2001). In the Siberian population, though the frequency of the LN alleles of SCA1 is very rare, the disease prevalence is extremely high. In this population, the high prevalence is attributed to the very high frequency (65%) of normal chromosomes with single interruptions (Zhou et al., 2001). In contrast, in the Caucasian population the frequency of LN correlates with the prevalence of SCA1 among ADCAs (Takano et al., 1998), and most of the normal chromosomes harbor more than 1 interruption (Sobczak and Krzyzosiak, 2004).

Haplotype analyses using microsatellite markers (D6S288 and D6S274), as well as SNPs in world populations, have revealed multiple origins of SCA1. These studies have been carried out primarily in the Japanese, South African, and Indian populations (Mittal et al., 2005b; Ramesar et al., 1997; Wakisaka et al., 1995). In a recent study, using markers linked to the disease locus in normal and affected human chromosomes in SCA1, we suggested both repeat length and interruption pattern as the determinants of prevalence of SCA1 in the Indian population (Mittal et al., 2005b). In addition to using population polymorphism analyses in humans and nonhuman primates, we also showed that expanded alleles at the SCA1 locus in the Indian population might have arisen through two distinct events (Mittal et al., 2005b)—one an ancestral event, and the other a human-specific event. In the first case, simultaneous lengthening of the ancestral chromosomes during evolution might have resulted in the origin of LN and expanded chromosomes. This is because the ancestral haplotype is present in ~44% of the affected chromosomes, which are also associated with LN alleles. However, all the LN alleles sharing this haplotype with the expanded chromosomes have accumulated protective repeat stabilizing CAT interruptions, whereas the expanded alleles have pure repeat stretches. This pool of expanded chromosomes might have escaped selection and persisted in the population due to a late age of onset of the disease in individuals with smaller expanded alleles. In the second case, haplotype analysis suggested that expanded alleles could have arisen from small normals (SNs) harboring single interruptions, through a reciprocal recombination mechanism between two singly interrupted SN alleles. This could have led to loss of interruption in one case, and gain of two interruptions in the other case, resulting in expanded and LN alleles, respectively. As expanded alleles have originated simultaneously with the LNs, the frequency of the LNs might reflect the prevalence of the disease. The direct correlation between the disease prevalence and singly interrupted alleles, as observed in the Siberian and the Indian populations, might be attributed to the mechanism of origin of expanded alleles from more expansion-prone, singly-interrupted alleles (Figure 88-4).

Thus, in SCA1 there seem to be two distinct mutational mechanisms. One set of expanded alleles have arisen through gradual lengthening of an ancestral allele by polymerase slippage. The other involves a more recent human-specific event of recombination involving singly interrupted alleles. Both events have also led to generation of large normal alleles with protective interruptions. The frequencies of both LN alleles, as well as alleles with single interruptions, are predictors of prevalence of SCA1 in a population.

SPINOCEREBELLAR ATAXIA TYPE 2 (SCA 2)

SCA2 (MIM 183090) is caused by polyglutamine repeat expansion in the SCA2 gene located on chromosome 12q24

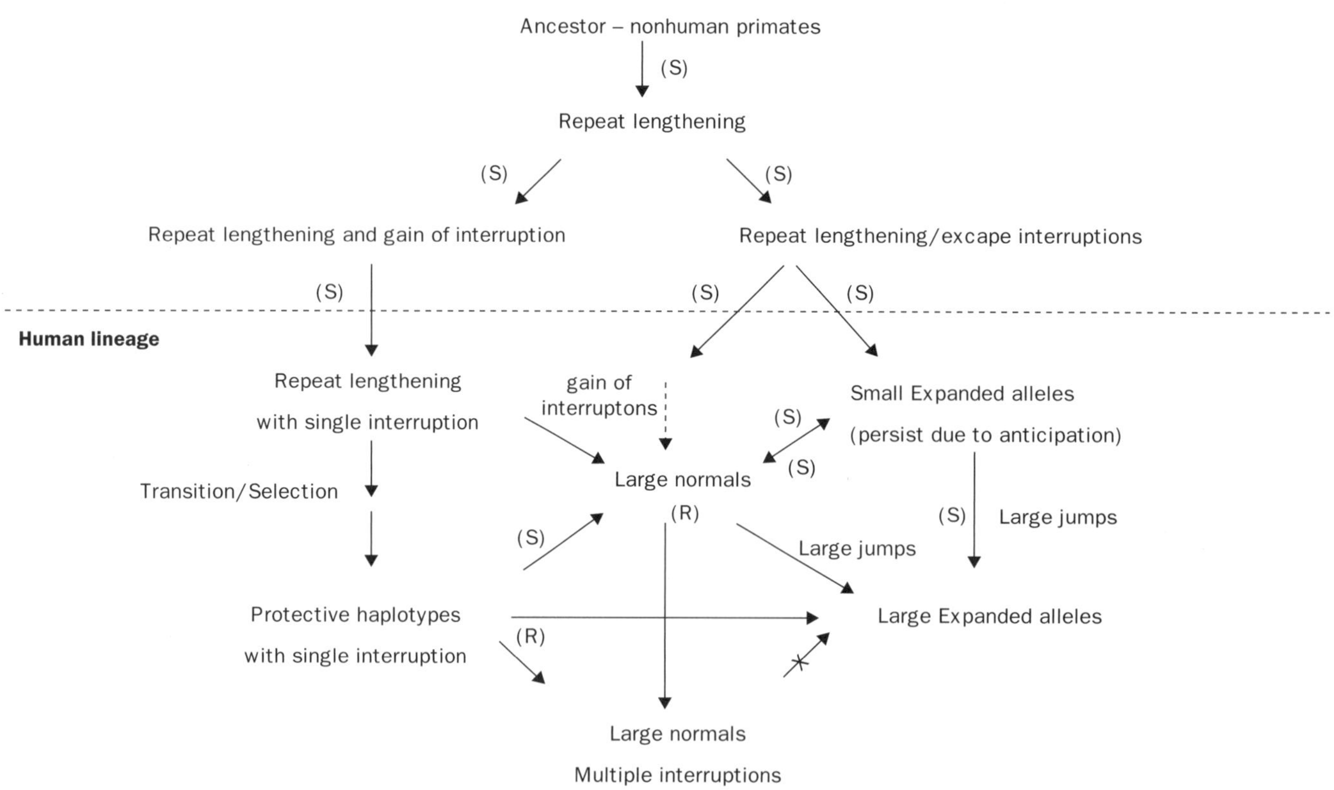

Figure 88-4 Course of normal and expanded repeat alleles with respect to length and interruptions in primates. We propose that large normal alleles with multiple interruptions are not predisposed to expansion. (S) – Slippage, (R) – Recombination. The figure is compiled on the basis of various studies discussed in the text (Choudhry et al., 2001; Dombrowski et al., 2002; Eichler et al., 1994; Mittal et al., 2005c; Mittal et al., 2005b; Saleem et al., 2000; Sobczak and Krzyzosiak, 2004).

(Imbert et al., 1996; Pulst et al., 1996; Sanpei et al., 1996). As in SCA1, the repeat tracts are cryptic, having one or more CAA interruptions in normal individuals. However, in contrast to SCA1, repeats at the SCA2 locus in the normal chromosomes are nearly monomorphic with respect to length (Figure 88-2), with ~95% of the alleles having 22 repeats (Choudhry et al., 2001; Saleem et al., 2000; Sobczak and Krzyzosiak, 2004). There is no gap between the normal and the diseased range, because of the presence of rare larger alleles. The alleles at the higher end of the normal range do not harbor interruptions and could be responsible for very late age-of-onset of SCA2 (Matsumura and Futamura, 2001). Even though the repeats have less heterozygosity with respect to length, there is wide heterogeneity in repeat interruption patterns in various populations. Population polymorphism analysis using associated SNPs provided insights into the possible mechanism for repeat instability at this locus (Choudhry et al., 2001; Sobczak & Krzyzosiak, 2004). In this case, unlike SCA1, an exclusive CC haplotype in the expanded alleles, which was present in all nonhuman primates and is a minor allele in normal human chromosomes, was observed in the Indian population (Choudhry et al., 2001b). A majority of the normal alleles with this haplotype are either pure or without the most proximal 5′ CAA interruption. This suggested that frequency of alleles with longer pure CAG stretch, rather than LNs, determines the prevalence of this disease in a population. The other haplotype, GT, which is human-specific, was present only in the normal chromosomes, with a majority of them having two or more stabilizing interruptions (Figure 88-4). This is similar to SCA1, wherein repeats in the ancestral chromosomes increase in length with gain of stabilizing interruptions, whereas those that escape interruptions persist as expanded alleles in the population. The analysis at the SCA2 locus also revealed the polarity effect in repeat instability, with most of the changes in the CAG repeat tract occurring at the 5′ end. This suggested the effect of the orientation of repeat with respect to the replication origin on repeat mutability. In SCA2 we also observed a horizontal increase in repeat length within a generation, wherein the younger siblings have larger length of repeats (Saleem et al., 2000). This suggested a prezygotic expansion in the germinal cell of the transmitting parent, which was substantiated by an observation of similar repeat lengths in a pair of dizygotic twins. A horizontal increase in repeat length also suggested a role for age of the transmitting parent in repeat instability. Further, repeat length increase by one triplet in majority of the transmissions also suggested a slippage-mediated expansion at SCA2 locus. The possibility of hairpin-mediated slippage was discounted at this locus, as a minimum of three triplet repeat expansions should have been observed in such a case.

Thus, in SCA2 the frequency of alleles with single interruptions in the normal population might determine prevalence. As in SCA1, the ancestral alleles that escape interruptions give rise to expanded alleles, and those that gain interruptions are stabilized normal alleles. Gain of interruption is protective and a human-specific event. Repeat lengthening is polar, occurring at the 5' end of the repeat stretch, and could be mediated through a simple polymerase slippage (Figure 88-4).

MACHADO-JOSEPH DISEASE (MJD)

MJD/SCA3 (MIM 109150), which was originally described in a Portuguese-Azorean population, is the most common ADCA, caused by CAG repeat expansion in the coding region of the MJD1 gene located on chromosome 14q32.1 (Kawaguchi et al., 1994). Normal individuals have 12–40 CAG repeats, expanding to 61–86 repeats in affected individuals (Kawaguchi et al., 1994; Maciel et al., 1995; Maruyama et al., 1995). Unlike SCA1 and SCA2, MJD has a unique spectrum of alleles (Figure 88-2). There is a wide gap between the normal and the expanded alleles, although a few cases of intermediate alleles of MJD in the range of 46–56 CAG repeats have been reported worldwide (Padiath et al., 2005; Takiyama et al., 1997; van et al., 2001). Additionally, loss of interruptions does not seem to play any role in repeat expansion at the MJD locus, as disease-associated repeats remain interrupted even after expansion.

Haplotype analysis has revealed that ~72% of the MJD families worldwide share a common A-C-A haplotype that has been traced to a Portuguese–Azorean ancestry (Gaspar et al., 2001). The prevalence of MJD in different ethnic groups has been correlated to the frequency of large normal alleles (Takano et al., 1998). But unlike other cases, the disease-associated haplotype was not found to be significantly associated with the large normals (Maciel et al., 1999), indicating other than Portuguese–Azorean origin of the founder haplotype. In a recent study by our group, we observed the same A-C-A haplotype associated with the expanded alleles in the Indian population (Mittal et al., 2005c). The surprising finding was that this haplotype is much more common in the normal Indian population (45%), in contrast to its extremely rare occurrence in the Portuguese population. We therefore suggested a founder event on a chromosomal background that has a South Asian origin, which is also substantiated by historical evidence of population admixture during colonization. Furthermore, this haplotype is very ancient in the Indian population, as it is present in small and ancient isolated populations of India (Chattopadhyay et al., 2003). Unlike SCA1 and SCA2, the founder haplotype for MJD mutation is not ancestral, as nonhuman primates have non-A-C-A haplotype (Mittal et al., 2005c).

Through tracing haplotypes by including an SNP within the repeat tract, we suggested that expansion at the

MJD locus is mediated through a gene conversion/recombination event (Mittal et al., 2005c). This is because we observed a rare intermediate allele that shared the same flanking haplotype (A-C-A) with the large normals and expanded alleles, but a different allele of the SNP within the repeat tract. This allele is also present in the small normals, suggesting that a gene conversion event within the repeat tract could have led to the lengthening of the repeat tract. If the two alleles involved in recombination have large repeat numbers, a large expanded allele is expected, and if one of the alleles happens to be smaller, then an intermediate repeat allele would be formed. Since the mechanism of expansion mostly involves expansion-prone large normals, haplotypes associated with the expanded alleles might depend on the frequencies of the haplotypes associated with large normals in a population. This might also explain why different haplotypes are associated with expanded alleles at the MJD locus in other populations, as well as dependence of prevalence on frequency of large normal alleles (Maciel et al., 1999). It is worthwhile to validate our observations in the other populations. A recombination mechanism might also explain the gap between normal and expanded alleles. Like SCA1, expansion by recombination at the MJD locus is a human-specific event.

At the MJD locus, the expanded alleles arise through a process of recombination. This might explain the large gap observed between the normal and expanded alleles, as well as the presence of limited founders. Recombination leading to generation of an expanded allele seems to be a human-specific event.

SCA7

Spinocerebellar ataxia type 7 (SCA7) is a rare and severe form of progressive autosomal dominant cerebellar ataxia (ADCA), characterized by cerebellar ataxia associated with progressive macular degeneration. The disease is caused by CAG repeat expansion in the coding region of the atxn-7 gene on chromosome 3p14.1 (David et al., 1997). SCA7 alleles contain 4–35 CAG repeats in normal healthy individuals, whereas pathological alleles contain 36–400 CAG repeats (Lebre and Brice, 2003). Intermediate alleles (IAs) in the range of 28–35 CAG repeats can be meiotically unstable when transmitted paternally, and de novo expansions occur in these large, but still normal, repeat sequences (Giunti et al., 1999; Stevanin et al., 1998). Despite marked anticipation and large jump, affecting subsequent generations' reproductive capacities, recurrent mutations occur on IAs; as a result, SCA7 persists with varying prevalence from 1.0% to 11.7% of genetically diagnosed ADCAs in different populations (Michalik et al., 2004), except in Sweden and Finland, where SCA7 is the most prevalent (Jonasson et al., 2000). Different haplotypes have been reported to be associated with the diseased alleles, indicating multiple origins of the SCA7 mutation (Stevanin et al., 1999).

The prevalence of SCA7 is very low in India populations (Alluri et al., 2007; Mittal et al., 2005a). In our one study we found only two SCA7 families with three affected individuals out of 235 SCA families showing characteristic features of the ADCA-II category. In one of the families we identified a de novo mutation in a subject with no family history, and genetic testing revealed that the unaffected father had CAG repeats in the intermediate range (31 repeats), which underwent pathological expansion to 59 repeats in the son. Haplotype analysis using microsatellite markers (D3S3566, D3S3698, D3S1600, and D3S3635) showed that the two affected pedigrees had different genetic backgrounds, indicating that different ancestral mutations were involved in SCA7 in the Indian population (Mittal et al., 2005a), as reported previously for other populations (Stevanin et al. 1998). We further analyzed CAG repeat at the SCA7 locus in a sperm sample of the transmitting parent with mutable IA, using SP-PCR and reconstruction experiments to assess the impact of the allele competition during PCR, which revealed absence of sperm cells carrying the expanded allele. Since no expanded alleles were observed in SP-PCR analysis of sperms, it was unlikely that a rare sperm cell with an expanded repeat, undetectable in the father's sperms, contributed to the zygote formation. Additionally, no IA was observed in somatic cells of the offspring using SP-PCR with ~10 haploid genomes. We suggested that the de novo expansion might have occurred during an early stage of embryogenesis as a post-zygotic event. This was corroborated by an earlier study that reported probability of repeat instability to be highest during early post-zygotic cell divisions (Wohrle et al., 1993). In the other family, DNA analysis of the transmitting parent revealed CAG repeat mosaicism in sperm samples compared to blood. This was in corroboration with the finding in SCA7 that gonadal mosaicism is greater than that observed in blood (David et al., 1998). This study, therefore, emphasizes the necessity of identification of modifier loci that govern the onset of repeat expansion.

HUNTINGTON'S DISEASE (HD)

HD (MIM 143100) is caused by polyglutamine repeat expansion encoded by CAG triplets in the IT15/Huntington gene present on chromosome 4p16.3 (1993). The repeat numbers vary from 6 to 35 in the normal population. In patients, the number of repeats can vary from 36 to an upper range of ~250 repeats reported in the case of juvenile Huntington's disease (The Huntington's Disease Collaborative Research Group,1993; Nance et al., 1999; Rubinsztein et al., 1996). The prevalence of HD is very heterogeneous around the world, with relatively high prevalence reported from most European populations (Harper, 1992). Lower rates of the disease have been reported from Japan, China, and among African Blacks (Chang et al., 1994; Harper, 1996; Leung et al., 1992). These observations suggested that the original HD mutation had its origin in Western Europe, which subsequently spread around the world largely through migrations linked to U.K. colonization (Harper, 1996; Hayden et al., 1981). Haplotype analysis at the HD locus demonstrated the presence of multiple founders in several populations, thereby suggesting that multiple mutation events underlie this disorder (Andrew et al., 1993; Saleem et al., 2002, 2003; Squitieri et al., 1994).

The Δ2642 D allele of del/ins polymorphism is overrepresented in the expanded HD chromosomes but is very rare in normal chromosomes (Rubinsztein et al., 1995). Moreover, in the normal chromosomes, the D allele is only associated with larger alleles. The proportion of LN and expanded chromosomes associated with D allele is similar, which explained the overrepresentation of D allele in the HD chromosomes. Analysis of this allele in nonhuman primates suggested that the deletion event might have happened on the normal alleles at the upper end in the human lineage (Rubinsztein et al., 1995; Saleem et al., 2002). Haplotype analysis also suggested that that the deletion-bearing chromosomes are not more mutable than non-deletion chromosomes, and it might only represent an apparent founder for HD mutation.

Multiple founders of Huntington's disease suggest the presence of cis factors, which might govern recurrent instability in different populations. This is further corroborated by the association of a deletion polymorphism with the predisposed alleles. Large normals are more prone to instability, and hence their frequency determines the prevalence of the disease.

TESTING THE HYPOTHESIS FOR A REPEAT EXPANSION MECHANISM FROM POPULATION POLYMORPHISM STUDIES

Friedreich's ataxia (FRDA; MIM 606829) is an autosomal recessive disorder caused by expansion of GAA repeats in the intron of the frataxin gene on chromosome 9q13. In normal individuals, the length of the repeat ranges from 7–38 triplets which, through intermediate alleles, expand to thousands in affected individuals. With insights gained through in vitro, in vivo, comparative genomics, and population-based approaches, it is possible to identify loci sharing similar STR dynamics. This is exemplified by GAA repeats that expand to more than a thousand in FRDA, and that has its origin at the 3' end of the poly "A" stretch within an Alu repeat (Chauhan et al., 2002). From a nonpolymorphic short stretch in nonhuman primates, the GAA repeat length has increased in humans with a concomitant shortening of the poly "A" to a nonpolymorphic one. A search for similar repeats in the entire human genome revealed that 60% of GAA repeats have arisen from Alu repeats, and the length of the GAA repeat is inversely correlated to the "A" stretch. Given that both kinds of repeats can form a triplex structure that allows slippage during replication, it is possible that relative stability of either of the tracts might determine polymorphism and thereby their instability. We examined the polymorphism in both tracts in normal individuals at different loci, harboring variable ratios of A to GAA lengths, to test this hypothesis, and observed that a pure stretch of "A" was more polymorphic if it was longer than the GAA repeat, and vice versa; at an intermediate range, neither repeats were polymorphic. Thus, the relative length of the flanking "A" stretch may determine the instability of GAA repeats. We further suggested that larger GAA alleles with a relatively shorter stretch of "A" repeats are prone to expansion, corroborating the observation that haplotypes of expanded chromosomes at the FRDA locus are similar to LN alleles (Cossee et al., 1997). This also explains why large normal GAA repeats are predictors of prevalence of FRDA in the population.

FOUNDER STUDIES AND CONFOUNDING OBSERVATIONS

All founder haplotypes may not be equally informative in all populations, but might give insights into the historical events that could lead to origin of the disease. Haplotype analysis at the HD locus, along with historical evidence, suggested spread of HD mutation through U.K. colonization in the Indian population (Saleem et al., 2002). Similarly, haplotype analysis has suggested that DRPLA founder mutation has occurred in an ancient haplotype of Asian origin, and these expanded chromosomes were directly introduced in the Portuguese population (Martins et al., 2006). Haplotype analysis as well as the historical evidence suggested that the expansion-prone large normal alleles with A-C-A haplotype at the MJD locus could have been introduced into the Portuguese population through admixture with South Asian populations (Mittal et al., 2005c).

SCA12: AN ATAXIA UNIQUE TO INDIA

Spinocerebellar ataxia type 12 (SCA12) is a rare and unique subtype of SCAs falling under the ADCA-I category caused by noncoding CAG repeat expansion in the 5' promoter, encompassing the region of the *PPP2R2B* gene on chromosome 5q31-5q32. SCA12 is so far observed worldwide only in two ethnic groups. Almost all the cases of SCA12 have been identified in a single endogamous ethnic group from Haryana state in northern India, except for one large American family of German descent (Bahl, et al., 2005; Srivastava, et al., 2001; Holmes, et al., 1999). SCA12 at present accounts for ~16% of the ADCA cases in the North Indian population. A PPP2R2B gene product is one of the regulatory subunits of PP2A holoenzyme, which is serine threonine phosphatase. A role of PP2A has been implicated in the regulation of protein phosphorylation, cell growth and differentiation, DNA replication, regulation of kinase cascades, ion channel function neurotransmitter release, microtubule assembly, and apoptosis. Normal repeat length at SCA12 loci ranges from 7–41, with 10 triplets being the most common allele length in all the populations studied. SCA12 pathology results when repeats expand beyond a length of 51 (51–78).

STUDIES CARRIED OUT SO FAR IN SCA12 RESEARCH

In an earlier study (Fujigasaki et al., 2001), we had reported expansion in the PPP2R2B gene in 6 patients and 3 asymptomatic at-risk individuals from 5 Indian families, which suggested that SCA12 is not rare in India (Srivastava, et al., 2001). In a following study, we identified 15 additional families in India harboring this mutation and, surprisingly, all pedigrees were part of only one endogamous population of Haryana. We used a population polymorphism approach, and analyzed intragenic SNPs (rs7446269, rs10302810, rs6897368, and rs10302811) and repeats to look for the presence of an ancestral founder for SCA12 in the Indian population. Haplotype analysis using these polymorphic markers revealed a single haplotype (A-C-A-G) to be associated with all the expanded chromosomes in the Indian SCA12 pedigrees, as compared to ethnically matched unrelated normal individuals (18.5%), suggesting the presence of a common founder for the SCA12 mutation in the Indian population.

We had also genotyped an American SCA12 family which did not share the A-C-A-G haplotype, signifying that the mutation in this family arose from a different founder. This finding is additional evidence that the expansion is causative, and not a marker in linkage disequilibrium with yet another undetected mutation. We observed considerable LD spanning a region of ~137kb downstream of the CAG repeat, marked by the 4 polymorphic SNPs in the recently originated founder population (~5000 years ago). This observation supports the use of endogamous Indian populations for fine-mapping of disease genes by linkage disequilibrium mapping (Bahl et al., 2009).

MITOCHONDRIA AND NEURODEGENERATION

Several mitochondrial-DNA encoded genes and nuclear encoded mitochondrial genes have been implicated directly or indirectly in etiopathogenesis of various neurological disorders, ranging from juvenile-onset Leigh disease, MERRF, etc., to adult or late-onset Parkinson's disease, Friedreich's ataxia, Huntington's disease, ALS (amyotrophic lateral sclerosis). and Alzheimer's disease. Despite enormous efforts to determine the role of mitochondria in disease causation of late-onset neurological disorders, is not straightforward due to high mutation rates and dosage effect of mitochondrial copies in the cytoplasm. In a very recent study we have tried to understand the cumulative role of different mitochondrial DNA variations in a group of 92 sporadic cerebellar ataxia cases against a control based on mtSNP scores. We compared the entire mitochondrial genome, and 324 variations were observed across the mitochondrial genome in these ataxia patients. Finally, from an initial set of 324 variations we prioritized 8 potentially disease-associated candidates. (Bhardwaj et al.,2009).

GENOTYPE-PHENOTYPE CORRELATION: CREATION OF SCA-LSVD TO UNDERSTAND THE DISEASE SPECTRUM ACROSS GLOBAL POPULATIONS

There is wide clinical heterogeneity among and between subtypes of SCAs, along with differences in prevalence and clinical manifestation of the same SCAs across different ethnic groups. In addition to this enormous clinical heterogeneity, the rarity and incurable nature of the disease preempts participation of clinicians from large number of primary and tertiary hospitals. Thus, a true estimate of the disease burden and prevalence of SCAs is lacking in most of the populations. In this scenario, we felt it imperative to develop an integrated locus-specific variation database (SCA-LSVD) for ataxia, cataloging variations with associated clinical features of ataxia from different global populations (http:\\miracle.igib.res.in\ataxia; also see Faruq et al., 2009). The database at present houses information on genes that have repeat involvement (SCA1, 2, 3, 6, 7, 8, 12, 17, FRDA, DRPLA). We initiated this activity by cataloging variations in 400 ataxia families that we have studied over the last 10 years.

The purpose of this database would also be to disseminate and increase the awareness of the phenotypic spectrum of different SCAs that might aid community participation in implementing genetic testing of SCAs. We aim to update this database with associated haplotypes in disease gene region, additional micro-phenotypic information in concordance with the international cooperative ataxia rating scale (ICARS), and data from other research groups working in ataxia. This would facilitate researchers in genotype-to-phenotype (G2P) studies, and also provide a helpful resource for tracing founder chromosomes and discovery of novel mutations. SCA-LSVD would also allow cross-comparisons between different cohorts of SCA patients and help in understanding the molecular correlates of phenotypes in ataxia, a multilocus disease that converges to overlapping phenotypes probably due to related molecular mechanisms.

SUMMARY

Spinocerebellar ataxias (SCAs) are a group of rare hereditary neurodegenerative disorders with overlapping clinical spectrum. Globally, trinucleotide repeat expansion has been implicated in the etiopathogenesis in a majority of the SCA-associated genes. Prevalence of SCAs varies widely across different ethnic populations. In this review, we have covered various aspects of genetic studies on ataxia that have been undertaken in Indian populations. We discuss how variability information from normal and affected individuals has revealed factors determining repeat instability and disease prevalence, as well as mechanisms and origins of repeat instability. We also highlight the development of a new locus-specific variation database for spinocerebellar ataxias (SCA-LSVD) that aims to catalog and integrate genotype–phenotype information on different SCAs from India. This provides a framework for understanding the molecular correlates of phenotypes in ataxia where related molecular mechanisms converge to overlapping phenotypes.

CONCLUDING REMARKS

Based on the analysis of population polymorphism studies described above, the following general themes in repeat instability emerge (Figure 88-4). From an ancestral state of short and possibly uninterrupted repeats, there is a progressive lengthening of repeats that are prone to expansion through slippage, with a concomitant gain of protective interruptions in normal individuals. A few alleles that escape interruptions might progressively increase in length to become expanded alleles. Since there is an inverse correlation of the repeat length to age at onset, the smaller expanded alleles may not lead to disease symptoms during reproductive age, and might escape selection and persist in the population. These studies further suggest that the loss of interruption may not be a prerequisite for expansion, since the positively associated haplotypes have multiple interruptions in normal individuals. Instead, the gain of interruption is a protective event in the human lineage. It is possible that recombination or gene conversion between normal alleles can concomitantly generate larger normals with multiple interruptions and uninterrupted expanded alleles, which is well supported by in vitro studies. This may also explain the presence of large gaps and jumps observed between the normal and expanded range in a few loci. So far, no study has reported loss of interruption during transmission as a triggering event in expansion. However, other cis- or trans-acting factors that determine the age at onset and timing of repeat expansion remain to be elucidated. The evidence of functional consequences of repeat length variability is now being examined.

In conclusion, population polymorphism could become an important functional link between naturally occurring polymorphism, the polymorphic structures they adopt, and the consequent polymorphic functions. As natural variant sequences are available, it is not necessary to undertake studies with extensive site-directed mutagenesis. Furthermore, insights from these approaches may help identify expansion-prone candidate loci, as well as provide insights into the mechanisms that modulate onset and progression of ataxia. Concerted efforts are needed for community participation toward integration of variation and related phenotype information from different ethnic populations and SCAs. The SCA-LSVD provides a platform in this direction.

REFERENCES

Alluri RV, Komandur S, Wagheray A, et al. (2007) Molecular analysis of CAG repeats at five different spinocerebellar ataxia loci: correlation and alternative explanations for disease pathogenesis. *Mol Cells* 24:338–342.

Andrew S, Theilmann J, Almqvist E, et al. (1993) DNA analysis of distinct populations suggests multiple origins for the mutation causing Huntington disease. *Clin Genet* 43:286–294.

Bahl S, Ahmed I, Mukerji M. (2009) Utilizing linkage disequilibrium information from Indian Genome Variation Database for mapping mutations: SCA12 case study. J Genet 88:55–60.

Bahl S, Virdi K, Mittal U, et al. (2005) Evidence of a common founder for SCA12 in the Indian population. Ann Hum Genet 69:528–534.

Basu P, Chattopadhyay B, Gangopadhaya PK, et al. (2000) Analysis of CAG repeats in SCA1, SCA2, SCA3, SCA6, SCA7 and DRPLA loci in spinocerebellar ataxia patients and distribution of CAG repeats at the SCA1, SCA2 and SCA6 loci in nine ethnic populations of eastern India. Hum Genet 106:597–604.

Bates GP. (2005) History of genetic disease: the molecular genetics of Huntington disease – a history. Nat Rev Genet 6:766–773.

Bharucha NE, Bharucha EP, Bhabha SK. (1986) Machado-Joseph-Azorean disease in India. Arch Neurol 43:142–144.

Bhardwaj A, Mukerji M, Sharma S, et al. (2009) MtSNPscore: A combined evidence approach for assessing cumulative impact of mitochondrial variations in disease. *BMC Bioinformatics* 10(Suppl 8):S7.

Cagnoli C, Mariotti C, Taroni F, et al. (2006) SCA28, a novel form of autosomal dominant cerebellar ataxia on chromosome 18p11.22-q11.2. *Brain* 129:235–242.

Chakravarty A, Mukherjee SC. (2002) Autosomal dominant cerebellar ataxias in ethnic Bengalees in West Bengal – an Eastern Indian state. *Acta Neurol Scand* 105(3):202–208.

Chamberlain S, Shaw J, Rowland A, et al. (1988) Mapping of mutation causing Friedreich's ataxia to human chromosome 9. *Nature* 334:248–250.

Chang CM, Yu YL, Fong KY, et al. (1994) Huntington's disease in Hong Kong Chinese: epidemiology and clinical picture. *Clin Exp Neurol* 31:43–51.

Chattopadhyay B, Basu P, Gangopadhyay PK, et al. (2003) Variation of CAG repeats and two intragenic polymorphisms at SCA3 locus among Machado-Joseph disease/SCA3 patients and diverse normal populations from eastern India. *Acta Neurol Scand* 108:407–414.

Chauhan C, Dash D, Grover D, Rajamani J, Mukerji M. (2002) Origin and instability of GAA repeats: insights from Alu elements 835. *J Biomol Struct Dyn* 20:253–263.

Chiurazzi P, Genuardi M, Kozak L, et al. (1996) Fragile X founder chromosomes in Italy: a few initial events and possible explanation for their heterogeneity. *Am J Med Genet* 64:209–215.

Chong SS, McCall AE, Cota J, et al. (1995) Gametic and somatic tissue-specific heterogeneity of the expanded SCA1 CAG repeat in spinocerebellar ataxia type 1 952. *Nat Genet* 10:344–350.

Chou AH, Yeh TH, Kuo YL, et al. (2006) Polyglutamine-expanded ataxin-3 activates mitochondrial apoptotic pathway by upregulating Bax and downregulating Bcl-xL. *Neurobiol Dis* 21:333–345.

Choudhry S, Mukerji M, Srivastava AK, Jain S, Brahmachari SK. (2001) CAG repeat instability at SCA2 locus: anchoring CAA interruptions and linked single nucleotide polymorphisms. *Hum Mol Genet* 10:2437–2446.

Chung MY, Lu YC, Cheng NC, Soong BW. (2003) A novel autosomal dominant spinocerebellar ataxia (SCA22) linked to chromosome 1p21-q23. *Brain* 126:1293–1299.

Chung MY, Ranum LP, Duvick LA, Servadio A, Zoghbi HY, Orr HT. (1993) Evidence for a mechanism predisposing to intergenerational CAG repeat instability in spinocerebellar ataxia type I. *Nat Genet* 5:254–258.

Cleary JD, Pearson CE. (2003) The contribution of cis-elements to disease-associated repeat instability: clinical and experimental evidence. *Cytogenet Genome Res* 100:25–55.

Cossee M, Schmitt M, Campuzano V, et al. (1997) Evolution of the Friedreich's ataxia trinucleotide repeat expansion: founder effect and premutations 76. *Proc Natl Acad Sci U S A* 94:7452–7457.

Cummings CJ, Mancini MA, Antalffy B, DeFranco DB, Orr HT, Zoghbi HY. (1998) Chaperone suppression of aggregation and altered subcellular proteasome localization imply protein misfolding in SCA1. *Nat Genet* 19:148–154.

Cummings CJ, Reinstein E, Sun Y, et al. (1999) Mutation of the E6-AP ubiquitin ligase reduces nuclear inclusion frequency while accelerating polyglutamine-induced pathology in SCA1 mice. *Neuron* 24:879–892.

Cummings CJ, Zoghbi HY. (2000) Trinucleotide repeats: mechanisms and pathophysiology. *Annu Rev Genomics Hum Genet* 1:281–328.

David G, Abbas N, Stevanin G, et al. (1997) Cloning of the SCA7 gene reveals a highly unstable CAG repeat expansion. *Nat Genet* 17:65–70.

David G, Durr A, Stevanin G, et al. (1998) Molecular and clinical correlations in autosomal dominant cerebellar ataxia with progressive macular dystrophy (SCA7). *Hum Mol Genet* 7:165–170.

Devos D, Schraen-Maschke S, Vuillaume I, et al. (2001) Clinical features and genetic analysis of a new form of spinocerebellar ataxia. *Neurology* 56:234–238.

Dombrowski C, Levesque S, Morel ML, Rouillard P, Morgan K, Rousseau F (2002) Premutation and intermediate-size FMR1 alleles in 10572 males from the general population: loss of an AGG interruption is a late event in the generation of fragile X syndrome alleles. Hum Mol Genet 11:371–378.

Durr A, Smadja D, Cancel G, et al. (1995) Autosomal dominant cerebellar ataxia type I in Martinique (French West Indies). Clinical and neuropathological analysis of 53 patients from three unrelated SCA2 families. *Brain* 118:1573–1581.

Eichler EE, Holden JJ, Popovich BW, et al. (1994) Length of uninterrupted CGG repeats determines instability in the FMR1 gene. *Nat Genet* 8:88–94.

Emamian ES, Kaytor MD, Duvick LA, et al. (2003) Serine 776 of ataxin-1 is critical for polyglutamine-induced disease in SCA1 transgenic mice. *Neuron* 38:375–387.

Estrada R, Galarraga J, Orozco G, Nodarse A, Auburger G. (1999) Spinocerebellar ataxia 2 (SCA2): morphometric analyses in 11 autopsies. *Acta Neuropathol* 97:306–310.

Everett CM, Wood NW. (2004) Trinucleotide repeats and neurodegenerative disease. *Brain* 127:2385–2405.

Faruq M, Scaria V, Singh I, Tyagi S, Srivastava AK, Mukerji M. (2009) SCA-LSVD: a repeat-oriented locus-specific variation database for genotype to phenotype correlations in spinocerebellar ataxias. *Hum Mutat* 30:1037–1042.

Fujigasaki H, Verma IC, Camuzat A, et al. (2001) SCA12 is a rare locus for autosomal dominant cerebellar ataxia: a study of an Indian family. *Ann Neurol* 49:117–121.

Gangopadhyay PK, Ghosh B, Roy T, Basu N, et al. (2001) Spinocerebellar ataxia—type 6. *J Assoc Phys India* 49:658–659.

Gaspar C, Lopes-Cendes I, Hayes S, et al. (2001) Ancestral origins of the Machado-Joseph disease mutation: a worldwide haplotype study. *Am J Hum Genet* 68:523–528.

Gatchel JR, Zoghbi HY. (2005) Diseases of unstable repeat expansion: mechanisms and common principles. *Nat Rev Genet* 6:743–755.

Genis D, Matilla T, Volpini V, et al. (1995) Clinical, neuropathologic, and genetic studies of a large spinocerebellar ataxia type 1 (SCA1) kindred: (CAG)n expansion and early premonitory signs and symptoms. *Neurology* 45:24–30.

Giunti P, Stevanin G, Worth PF, David G, Brice A, Wood NW. (1999) Molecular and clinical study of 18 families with ADCA type II: evidence for genetic heterogeneity and de novo mutation. *Am J Hum Genet* 64:1594–1603.

Harding AE. (1993) Clinical features and classification of inherited ataxias. *Adv Neurol* 61:1–14.

Harper PS. (1996) Epidemiology of Huntington's disease. *Hum Genet* 89:365–376.

Hayden MR, Berkowicz AL, Beighton PH, Yiptong C. (1981) Huntington's chorea on the island of Mauritius. *S Afr Med J* 60:1001–1002.

Hire R, Katrak S, Vaidya S, et al. (2010) Spinocerebellar ataxia type 17 in Indian patients: two rare cases of homozygous expansions. *Clin Genet* 10-0004.

Holmes SE, O'Hearn EE, McInnis MG, et al. (1999) Expansion of a novel CAG trinucleotide repeat in the 5' region of PPP2R2B is associated with SCA12. *Nat Genet* 23:391–392.

Imbert G, Saudou F, Yvert G, et al. (1996) Cloning of the gene for spinocerebellar ataxia 2 reveals a locus with high sensitivity to expanded CAG/glutamine repeats. *Nat Genet* 14:285–291.

Jain S, Maheshwari MC. (1986) Joseph disease in India–report of two families. J *Neurogenet* 3:61–73.

Jana NR, Zemskov EA, Wang G, Nukina N. (2001) Altered proteasomal function due to the expression of polyglutamine-expanded truncated N-terminal huntingtin induces apoptosis by caspase activation through mitochondrial cytochrome c release. *Hum Mol Genet* 10:1049–1059.

Jodice C, Giovannone B, Calabresi V, Bellocchi M, Terrenato L, Novelletto A. (1997) Population variation analysis at nine loci containing expressed trinucleotide repeats. *Ann Hum Genet* 61:425–438.

Jonasson J, Juvonen V, Sistonen P, et al. (2000) Evidence for a common Spinocerebellar ataxia type 7 (SCA7) founder mutation in Scandinavia. *Eur J Hum Genet* 8:918–922.

Kawaguchi Y, Okamoto T, Taniwaki M, et al. (1994) CAG expansions in a novel gene for Machado-Joseph disease at chromosome 14q32.1. *Nat Genet* 8:221–228.

Knight MA, Gardner RJ, Bahlo M, et al. (2004) Dominantly inherited ataxia and dysphonia with dentate calcification: spinocerebellar ataxia type 20. *Brain* 127:1172–1181.

Koide R, Ikeuchi T, Onodera O, et al. (1994) Unstable expansion of CAG repeat in hereditary dentatorubral-pallidoluysian atrophy (DRPLA). *Nat Genet* 6:9–13.

Koob MD, Moseley ML, Schut LJ, et al. (1999) An untranslated CTG expansion causes a novel form of spinocerebellar ataxia (SCA8). *Nat Genet* 21:379–384.

Krishna N, Mohan S, Yashavantha BS, Rammurthy A, Kiran Kumar HB, Mittal U et al. (2007) SCA 1, SCA 2 & SCA 3/MJD mutations in ataxia syndromes in southern India. *Indian J Med Res* 126(5):465–470.

Lahue RS, Slater DL. (2003) DNA repair and trinucleotide repeat instability. *Front Biosci* 8:s653–s665.

Lam YC, Bowman AB, Jafar-Nejad P, et al. (2006) ATAXIN-1 interacts with the repressor Capicua in its native complex to cause SCA1 neuropathology. *Cell* 127:1335–1347.

Lebre AS, Brice A. (2003) Spinocerebellar ataxia 7 (SCA7). *Cytogenet Genome Res* 100:154–163.

Lenzmeier BA, Freudenreich CH. (2003) Trinucleotide repeat instability: a hairpin curve at the crossroads of replication, recombination, and repair. *Cytogenet Genome Res* 100:7–24.

Leung CM, Chan YW, Chang CM, Yu YL, Chen CN. (1992) Huntington's disease in Chinese: a hypothesis of its origin. *J Neurol Neurosurg Psychiatry* 55:681–684.

Lipinski MM, Yuan J. (2004) Mechanisms of cell death in polyglutamine expansion diseases. *Curr Opin Pharmacol* 4:85–90.

Liquori CL, Ricker K, Moseley ML, et al. (2001) Myotonic dystrophy type 2 caused by a CCTG expansion in intron 1 of ZNF9. *Science* 293:864–867.

Maciel P, Gaspar C, DeStefano AL, et al. (1995) Correlation between CAG repeat length and clinical features in Machado-Joseph disease. *Am J Hum Genet* 57:54–61.

Maciel PF, Gaspar CF, Guimaraes LF, et al. (1999) Study of three intragenic polymorphisms in the Machado-Joseph disease gene (MJD1) in relation to genetic instability of the (CAG)n tract. *Eur J Hum Genet* 7(2):147–156.

Martins S, Calafell F, Wong VC, Sequeiros J, Amorim A. (2006) A multistep mutation mechanism drives the evolution of the CAG repeat at MJD/SCA3 locus. *Eur J Hum Genet* 14:932–940.

Maruyama H, Nakamura S, Matsuyama Z, et al. (1995) Molecular features of the CAG repeats and clinical manifestation of Machado-Joseph disease. *Hum Mol Genet* 4:807–812.

Matsumura R, Futamura N. (2001) Late-onset SCA2: 33 CAG repeats are sufficient to cause disease. *Neurology* 57:566.

McCampbell A, Taylor JP, Taye AA, et al. (2000) CREB-binding protein sequestration by expanded polyglutamine. *Hum Mol Genet* 9:2197–2202.

McDonough H, Patterson C. (2003) CHIP: a link between the chaperone and proteasome systems. *Cell Stress Chaperones* 8:303–308.

Michalik A, Martin JJ, Van BC. (2004) Spinocerebellar ataxia type 7 associated with pigmentary retinal dystrophy. *Eur J Hum Genet* 12:2–15.

Mittal U, Roy S, Jain S, Srivastava AK, Mukerji M. (2005a) Post-zygotic de novo trinucleotide repeat expansion at spinocerebellar ataxia type 7 locus: evidence from an Indian family. *J Hum Genet* 50:155–157.

Mittal U, Sharma S, Chopra R, et al. (2005b) Insights into the mutational history and prevalence of SCA1 in the Indian population through anchored polymorphisms. Hum Genet 118:107–114.

Mittal U, Srivastava AK, Jain S, Jain S, Mukerji M. (2005c) Founder haplotype for Machado-Joseph disease in the Indian population: novel insights from history and polymorphism studies. *Arch Neurol* 62:637–640.

Mizutani A, Wang L, Rajan H, et al. (2005) Boat, an AXH domain protein, suppresses the cytotoxicity of mutant ataxin-1. *EMBO J* 24:3339–3351.

Mutsuddi M, Marshall CM, Benzow KA, Koob MD, Rebay I. (2004) The spinocerebellar ataxia 8 noncoding RNA causes neurodegeneration and associates with staufen in Drosophila. *Curr Biol* 14:302–308.

Nance MA, Mathias-Hagen V, Breningstall G, Wick MJ, McGlennen RC. (1999) Analysis of a very large trinucleotide repeat in a patient with juvenile Huntington's disease. *Neurology* 52:392–394.

Nicastro G, Menon RP, Masino L, Knowles PP, McDonald NQ, Pastore A. (2005) The solution structure of the Josephin domain of ataxin-3: structural determinants for molecular recognition. *Proc Natl Acad Sci U S A* 102:10493–10498.

Orr HT, Chung MY, Banfi S, et al. (1993) Expansion of an unstable trinucleotide CAG repeat in spinocerebellar ataxia type 1. *Nat Genet* 4:221–226.

Padiath QS, Srivastava AK, Roy S, Jain S, Brahmachari SK. (2005) Identification of a novel 45 repeat unstable allele associated with a disease phenotype at the MJD1/SCA3 locus. *Am J Med Genet B Neuropsychiatr Genet* 133B:124–126.

Pang JT, Giunti P, Chamberlain S, et al. (2002) Neuronal intranuclear inclusions in SCA2: a genetic, morphological and immunohistochemical study of two cases. *Brain* 125:656–663.

Park Y, Hong S, Kim SJ, Kang S. (2005) Proteasome function is inhibited by polyglutamine-expanded ataxin-1, the SCA1 gene product. *Mol Cells* 19:23–30.

Philips AV, Timchenko LT, Cooper TA. (1998) Disruption of splicing regulated by a CUG-binding protein in myotonic dystrophy. *Science* 280:737–741.

Pulst SM, Nechiporuk A, Nechiporuk T, et al. (1996) Moderate expansion of a normally biallelic trinucleotide repeat in spinocerebellar ataxia type 2. *Nat Genet* 14:269–276.

Ramesar RS, Bardien S, Beighton P, Bryer A. (1997) Expanded CAG repeats in spinocerebellar ataxia (SCA1) segregate with distinct haplotypes in South african families. *Hum Genet* 100:131–137.

Rengaraj R, Dhanaraj M, Arulmozhi T, Chattopadhyay B, et al. (2005) High prevalence of spinocerebellar ataxia type 1 in an ethnic Tamil community in India. *Neurol India* 53(3):308–310.

Rolfs A, Koeppen AH, Bauer I, et al. (2003) Clinical features and neuropathology of autosomal dominant spinocerebellar ataxia (SCA17). *Ann Neurol* 54:367–375.

Ross CA, Poirier MA. (2004) Protein aggregation and neurodegenerative disease. *Nat Med* 10(Suppl):S10–S17.

Rub U, Brunt ER, Gierga K, et al. (2005) Spinocerebellar ataxia type 7 (SCA7): first report of a systematic neuropathological study of the brain of a patient with a very short expanded CAG-repeat. *Brain Pathol* 15:287–295.

Rubinsztein DC, Leggo J, Amos W, Barton DE, Ferguson-Smith MA. (1994) Myotonic dystrophy CTG repeats and the associated insertion/deletion polymorphism in human and primate populations. *Hum Mol Genet* 3:2031–2035.

Rubinsztein DC, Leggo J, Coles R, et al. (1996) Phenotypic characterization of individuals with 30–40 CAG repeats in the Huntington disease (HD) gene reveals HD cases with 36 repeats and apparently normal elderly individuals with 36–39 repeats. *Am J Hum Genet* 59:16–22.

Rubinsztein DC, Leggo J, Goodburn S, Barton DE, Ferguson-Smith MA. (1995) Haplotype analysis of the delta 2642 and (CAG)n polymorphisms in the Huntington's disease (HD) gene provides an explanation for an apparent 'founder' HD haplotype. *Hum Mol Genet* 4:203–206.

Saleem Q, Anand A, Jain S, Brahmachari SK. (2001) The polyglutamine motif is highly conserved at the Clock locus in various organisms and is not polymorphic in humans. *Hum Genet* 109:136–142.

Saleem Q, Choudhry S, Mukerji M, et al. (2000) Molecular analysis of autosomal dominant hereditary ataxias in the Indian population: high frequency of SCA2 and evidence for a common founder mutation. *Hum Genet* 106:179–187.

Saleem Q, Muthane U, Verma IC, Brahmachari SK, Jain S. (2002) Expanding colonies and expanding repeats. *Lancet* 359:895–896.

Saleem Q, Roy S, Murgood U, et al. (2003) Molecular analysis of Huntington's disease and linked polymorphisms in the Indian population. *Acta Neurol Scand* 108:281–286.

Sanpei K, Takano H, Igarashi S, et al. (1996) Identification of the spinocerebellar ataxia type 2 gene using a direct identification of repeat expansion and cloning technique, DIRECT. *Nat Genet* 14:277–284.

Sasaki H, Yabe I, Tashiro K. (2003) The hereditary spinocerebellar ataxias in Japan. *Cytogenet Genome Res* 100:198–205.

Savouret C, Brisson E, Essers J, et al. (2003) CTG repeat instability and size variation timing in DNA repair-deficient mice 1518. *EMBO J* 22:2264–2273.

Schelhaas HJ, Verbeek DS, van de Warrenburg BP, Sinke RJ. (2004) SCA19 and SCA22: evidence for one locus with a worldwide distribution. *Brain* 127:E6.

Schmidt T, Lindenberg KS, Krebs A, et al. (2002) Protein surveillance machinery in brains with spinocerebellar ataxia type 3: redistribution and differential recruitment of 26S proteasome subunits and chaperones to neuronal intranuclear inclusions. *Ann Neurol* 51:302–310.

Schols L, Bauer P, Schmidt T, Schulte T, Riess O. (2004) Autosomal dominant cerebellar ataxias: clinical features, genetics, and pathogenesis. *Lancet Neurol* 3:291–304.

Serra HG, Duvick L, Zu T, et al. (2006) RORalpha-mediated Purkinje cell development determines disease severity in adult SCA1 mice. *Cell* 127:697–708.

Sinha KK, Worth PF, Jha DK, Sinha S, et al. (2004) Autosomal dominant cerebellar ataxia: SCA2 is the most frequent mutation in eastern India. *J Neurol Neurosurg Psychiatry* 75(3):448–452.

Sobczak K, Krzyzosiak WJ. (2004) Patterns of CAG repeat interruptions in SCA1 and SCA2 genes in relation to repeat instability. *Hum Mutat* 24:236–247.

Squitieri F, Andrew SE, Goldberg YP, et al. (1994) DNA haplotype analysis of Huntington disease reveals clues to the origins and mechanisms of CAG expansion and reasons for geographic variations of prevalence. *Hum Mol Genet* 3:2103–2114.

Srivastava AK, Choudhry S, Gopinath MS, et al. (2001) Molecular and clinical correlation in five Indian families with spinocerebellar ataxia 12. *Ann Neurol* 50:796–800.

Stevanin G, David G, Durr A, et al. (1999) Multiple origins of the spinocerebellar ataxia 7 (SCA7) mutation revealed by linkage disequilibrium studies with closely flanking markers, including an intragenic polymorphism (G3145TG/A3145TG). *Eur J Hum Genet* 7:889–896.

Stevanin G, Giunti P, Belal GD, et al. (1998) De novo expansion of intermediate alleles in spinocerebellar ataxia 7. *Hum Mol Genet* 7:1809–1813.

Storey E, Bahlo M, Fahey M, Sisson O, Lueck CJ, Gardner RJ. (2009) A new dominantly inherited pure cerebellar ataxia, SCA 30. *J Neurol Neurosurg Psychiatry* 80:408–411.

Swartz BE, Burmeister M, Somers JT, Rottach KG, Bespalova IN, Leigh RJ. (2002) A form of inherited cerebellar ataxia with saccadic intrusions, increased saccadic speed, sensory neuropathy, and myoclonus. *Ann N Y Acad Sci* 956:441–444.

Takano H, Cancel G, Ikeuchi T, et al. (1998) Close associations between prevalences of dominantly inherited spinocerebellar ataxias with CAG-repeat expansions and frequencies of large normal CAG alleles in Japanese and Caucasian populations. *Am J Hum Genet* 63:1060–1066.

Takiyama Y, Sakoe K, Nakano I, Nishizawa M. (1997) Machado-Joseph disease: cerebellar ataxia and autonomic dysfunction in a patient with the shortest known expanded allele (56 CAG repeat units) of the MJD1 gene. *Neurology* 49:604–606.

Terasawa H, Oda M, Morino H, et al. (2004) A novel haplotype of spinocerebellar ataxia type 6 contributes to the highest prevalence in western Japan. *Neurosci Lett* 358:107–110.

The Huntington's Disease Collaborative Research Group. (1993) A novel gene containing a trinucleotide repeat that is expanded and unstable on Huntington's disease chromosomes.. *Cell* 72:971–983.

Tsai YC, Fishman PS, Thakor NV, Oyler GA. (2003) Parkin facilitates the elimination of expanded polyglutamine proteins and leads to preservation of proteasome function. *J Biol Chem* 278:22044–22055.

Tsuda H, Jafar-Nejad H, Patel AJ, et al. (2005) The AXH domain of Ataxin-1 mediates neurodegeneration through its interaction with Gfi-1/Senseless proteins. *Cell* 122:633–644.

Usdin K, Grabczyk E. (2000) DNA repeat expansions and human disease. *Cell Mol Life Sci* 57:914–931.

van Swieten JC, Brusse E, de Graaf BM, et al. (2003) A mutation in the fibroblast growth factor 14 gene is associated with autosomal dominant cerebellar ataxia. *Am J Hum Genet* 72:191–199.

van AN, Sinke RJ, Zwarts MJ, et al. (2001) Intermediate CAG repeat lengths (53,54) for MJD/SCA3 are associated with an abnormal phenotype. *Ann Neurol* 49:805–807.

Verbeek DS, Piersma SJ, Hennekam EF, Ippel EF, Pearson PL, Sinke RJ. (2004) Haplotype study in Dutch SCA3 and SCA6 families: evidence for common founder mutations. *Eur J Hum Genet* 12:441–446.

Vuillaume I, Devos D, Schraen-Maschke S, et al. (2002) A new locus for spinocerebellar ataxia (SCA21) maps to chromosome 7p21.3-p15.1. *Ann Neurol* 52:666–670.

Wadia NH, Swami RK (1971) A new form of heredo-familial spinocerebellar degeneration with slow eye movements (nine families). Brain 94(2):359–374.

Wadia N, Pang J, Desai J, Mankodi A, et al. (1998) A clinicogenetic analysis of six Indian spinocerebellar ataxia (SCA2) pedigrees. The significance of slow saccades in diagnosis. Brain 121(Pt 12):2341–2355.

Wakisaka A, Sasaki H, Takada A, et al. (1995) Spinocerebellar ataxia 1 (SCA1) in the Japanese in Hokkaido may derive from a single common ancestry. *J Med Genet* 32:590–592.

Wang HL, Yeh TH, Chou AH, et al. (2006) Polyglutamine-expanded ataxin-7 activates mitochondrial apoptotic pathway of cerebellar neurons by upregulating Bax and downregulating Bcl-x(L). *Cell Signal* 18:541–552.

Waters MF, Minassian NA, Stevanin G, et al. (2006) Mutations in voltage-gated potassium channel KCNC3 cause degenerative and developmental central nervous system phenotypes. *Nat Genet* 38:447–451.

Webster MT, Smith NG, Ellegren H. (2002) Microsatellite evolution inferred from human-chimpanzee genomic sequence alignments. *Proc Natl Acad Sci U S A* 99:8748–8753.

Wohrle D, Hennig I, Vogel W, Steinbach P. (1993) Mitotic stability of fragile X mutations in differentiated cells indicates early post-conceptional trinucleotide repeat expansion. *Nat Genet* 4:140–142.

Worth PF, Houlden H, Giunti P, Davis MB, Wood NW. (2000) Large, expanded repeats in SCA8 are not confined to patients with cerebellar ataxia. *Nat Genet* 24:214–215.

Yamada M, Sato T, Tsuji S, Takahashi H. (2008) CAG repeat disorder models and human neuropathology: similarities and differences. *Acta Neuropathol* 115:71–86.

Yu GY, Howell MJ, Roller MJ, Xie TD, Gomez CM. (2005) Spinocerebellar ataxia type 26 maps to chromosome 19p13.3 adjacent to SCA6. *Ann Neurol* 57:349–354.

Zhou YX, Qiao WH, Gu WH, et al. (2001) Spinocerebellar ataxia type 1 in China: molecular analysis and genotype-phenotype correlation in 5 families. *Arch Neurol* 58:789–794.

Zuhlke C, Dalski A, Hellenbroich Y, Bubel S, Schwinger E, Burk K. (2002) Spinocerebellar ataxia type 1 (SCA1): phenotype-genotype correlation studies in intermediate alleles. *Eur J Hum Genet* 10:204–209.

89.

MALFORMATION SYNDROMES IN INDIA

Shubha R. Phadke

INTRODUCTION

Congenital malformations are present at birth. Major malformations may need special care soon after birth or during the neonatal period. However, many multiple malformation syndromes associated with subtle dysmorphism and minor malformations may not become clinically relevant till the latter part of infancy or childhood when associated with developmental delay or mental retardation. Hence, incidence of multiple malformation syndromes obtained from examination of the newborn may be an underestimate. The studies on Indian populations by Kolah et al. (1967) and Kulkarni and Kurian (1990) provide a prevalence of congenital malformation syndromes at 0.34 per 1000 births and 2.4 per 1000 births, respectively. The latter study (Kulkarni and Kurian, 1990) covered the South Indian population with very high rates of consanguinity. The study by Kolah et al. (1967) reflects the period when information about syndromes and diagnostic facilities were scarce. The study of 7015 neonates born in a tertiary care hospital in New Delhi showed a 2.3% incidence of congenital malformations (Singh et al., 1991). The authors observed that that recording of multiple malformations was inadequate, as in most of the cases only the major malformations were recorded and the associated minor defects were not mentioned. Dr. Suresh's group in South India has started a registry of birth defects during last few years (personal communication). They have established the Birth Defects Registry of India (BDRI), which is a hospital-based, descriptive surveillance program, and passively collects (voluntary reporting) data on congenital malformations from hospitals of defined geographic regions. A detailed discussion about the methodology has been published (Suresh et al., 2005). There have been 46 registries established between 2001–2009, and they have been functional in several districts distributed all over India. During 2001–2009 there were 766,390 births reported from the member registries, of which 97.5% were live born. There were 7687 cases with birth defect(s), of which 44.1% were medically terminated. The overall crude birth prevalence is 100.3/10,000. Central nervous system anomalies (39.9/10,000) are leading, followed by musculoskeletal anomalies (28.9/10,000) during the past 9 years of data analyzed. The crude birth prevalence of cardiovascular system anomalies was 17.3/10,000, gastrointestinal tract anomalies were 13.0/10,000, and genitourinary tract anomalies were 14.1 per 10,000. Among the major anomalies, neural tube defect (NTD; 28.6/10,000) is the highest, followed by congenital talipes equinovarus (CTEV; 9.3/10,000).

Though a limited number of hospitals are included in the registry, systematically and prospectively collected data in this registry will provide reliable data about prevalence of malformations.

With improving perinatal care, the relative contribution of malformations to neonatal mortality and morbidity is increasingly being recognized in India, as is the situation in developed countries (Verma, 1994). In addition, preference of a small nuclear family unit in the urban educated working class of people has made each birth more precious for a family. Thus, the need for the care of malformed children and genetic counseling for such families after a genetic diagnosis is now increasingly sought by lay people, as well as by health professionals engaged in providing primary and secondary healthcare. However, the number of specialized tertiary centers providing genetic services is still very limited and confined to few large metropolitan cities. New genetic centers have been established in India over the last decade. But the number of trained clinicians with special interest in genetics and clinical geneticists is still very limited. There has been some interest expressed in dysmorphology among pediatricians. This is reflected from the case reports of multiple malformation syndromes published from India. Currently there is no separate journal devoted to dysmorphology published from any Indian subcontinent country. Two important and indexed pediatrics journals published from India are *Indian Pediatrics* and *Indian Journal of Pediatrics*. These journals play an important role in documenting the rare syndromes, creating awareness about diagnosis and providing information on genetic counseling and prenatal diagnosis (Bijarnia et al., 2003; Gupta et al., 2003; Gulati et al., 2003; Phadke et al., 1997). The number of case reports of malformations and malformation syndromes published per year is more than 12 each journal. In addition, dysmorphologists from India have published case reports in peer reviewed international journals. Case series of malformations also have been reported (Thakur et al., 2004a; Dalal et al., 2006). The London Dysmorphology Database has more than 40 references of papers published by Indian authors. Online Mendelian Inheritance in Man (OMIM) has also cited a few papers from India. OMIM also has included new malformation disorders reported for the first time from the Indian subcontinent (Phadke & Gautam, 1999; Verma et al., 1975). A spondyloepiphyseal dysplasia, highly prevalent in a district from South India, has also been included in OMIM (Agarwal et al., 1994).

KNOWN MALFORMATION SYNDROMES

NONCHROMOSOMAL SYNDROMES

Some well-known multiple malformation syndromes reported from India are listed in Table 89-1. Though there are no prevalence data for any syndrome, the various case reports document their presence in the Indian population and indicate the prevalence may not be greatly different from that observed in other countries. The facial phenotypes of these dysmorphic syndromes are similar to those of cases reported from other ethnic groups. These include typical malformation syndromes, including some syndromes of metabolic etiology and with skeletal dysplasias. The distinction among these is often subtle and arbitrary. Dalal and Phadke (2007) have studied the utility of morphometric analysis of the face in dysmorphology.

KNOWN SYNDROMES WITH UNUSUAL FEATURES

Additional features that are not known to be the part of the phenotype of a syndrome are commonly observed in many cases. Similarly, cases with overlapping features of two syndromes are also not uncommon (Sharma et al., 1992; Phadke et al., 1999). Cases with such unusual features are reported from India (Ghosh et al., 2007; Gambhir et al., 2010). These include the following cases.

Asphyxiating thoracic dystrophy with hydrocephalus (Singh et al., 1988)
Neu Laxova syndrome with spina bifida and cryptorchidism (Naveed et al., 1990)
OEIS syndrome with craniofacial abnormalities (Halder et al., 1994)
Oro-facio-digital syndrome with features overlapping with Ellis-van Creveld syndrome (Phadke et al., 1999)

TABLE 89-1 SYNDROMES REPORTED FROM INDIA

Aarskog syndrome	Adams Oliver syndrome
Aicardi syndrome	Amniotic band syndrome
Anhidrotic ectodermal dysplasia	Aniridia Wilms tumor
Apert syndrome	Asphyxiating thoracic dystrophy
Branchio oto facial syndrome	Cerebro oculo facial syndrome
CHARGE syndrome	Chundley McCullough syndrome
Cleidocranial dysplasia	Cockayene syndrome
Congenital contractural arachnodactyly	Cornelia de Lange syndrome
Costello syndrome	Crouzon syndrome
Cutis laxa syndrome	Dellaman syndrome
Denys Drash	Ectrodactyly
Ellis van Creveld syndrome	Femoral hypoplasia unusual facies syndrome
Fraser syndrome	Goltz syndrome
Gorlin syndrome	Grebbe syndrome
Hajdu Cheney syndrome	Hallerman Streiff syndrome
Incontinentia pigmenti	Johansson Blizzard syndrome
Kabuki make up syndrome	KID syndrome
Klippel Trenauney Weber syndrome	Larsen syndrome
Laurence Moon syndrome	Lenz microphthalmia syndrome
Mandibuloacral dysostosis	Marden Walker syndrome
Marfan syndrome	Marshal Smith syndrome
Meckel Gruber syndrome	Moebius syndrome
MURC association	Nagar craniofacial dysostosis
Nail Patella syndrome	Noonan syndrome
OMENS plus	Oro facio digital syndromes
Parault syndrome	Peter syndrome
Poland anomaly	Popliteal pterygium syndrome
Reiger syndrome	Russel Silver syndrome
Seckel syndrome	Shah Waardenburg syndrome
Short rib polydactyly syndrome	Spondylocostal syndrome
Stickler syndrome	Towne Brocks syndrome
Tricho rhino phalangeal syndrome	VACTERAL
Van der Woude syndrome	Waardenburg syndrome
Weil Marchesani v	Wildervank syndrome
Walker Warburg syndrome	

Cardiofacial syndrome with anotia and facial palsy (Girisha and Phadke, 2005)
Tel Hashomer camptodactyly with hirsutism in a case with Patel and Adhia, (2004)
Adams Oliver syndrome with delayed bone age and abnormal terminal phalanx (Baskar et al., 2009)

(A)

(B)

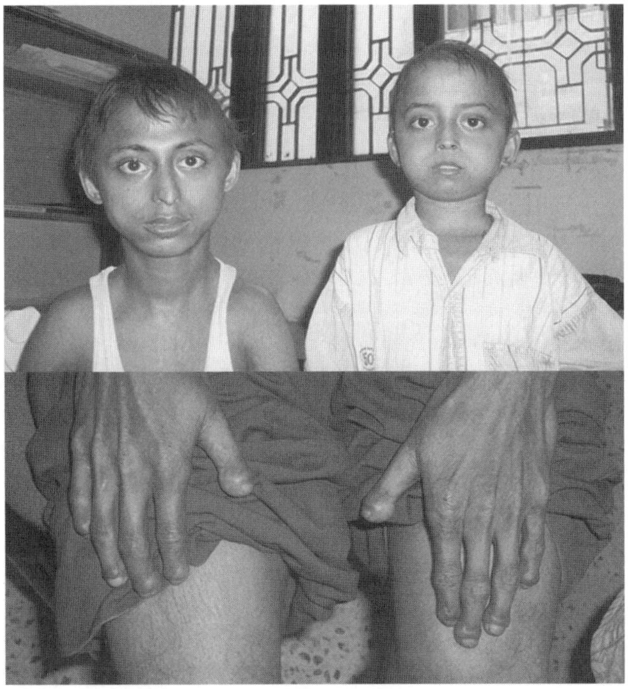

Figure 89-1 Two siblings from a consanguineous Muslim family with acro-mandibulo-facial dysostosis and osteolysis with homozygous R527H (1680 G>A) *MMP2* mutation; note marked osteolysis and pseudoclubbing in fingers.

misdiagnosed as progeria (Biswas & Reddy, 1997, Prakash et al, 1990, Cusana and Scarano, 1991). In two siblings with mandibuloacral dysplasia, homozygous mutation R527H (1580 G>A) was identified (Figure 89-1).

NEW SYNDROMES FROM NORTH INDIA

With an increasing population of clinical geneticists and increasing awareness about genetic disorders, any new cases with unusual constellations of malformations are observed in clinics and some are published. Due to its large population and high prevalence of consanguinity, India is a mine for rare genetic disorders. With improved availability of technological advancements and genetic expertise, these families with rare disorders are a good substrate for gene-mapping research. Identification of a new gene for microcephaly and a new locus for nonsyndromic deafness in recent years have paved the way for gene-mapping studies in other genetic disorders.

CAMPTOSYPOLYDACTYLY, COMPLEX DISORGANIZATION TYPE (OMIM 607539).

Phadke and Gautam (1999) described a complex type of camptopolydactyly in a firstborn child of a consanguineous Muslim family. The hand was completely disorganized, with syndactyly and camptodactyly of some fingers (Figure 89-2). Major abnormalities included two fingers arising from the dorsum of hands, dysplastic nails, and syndactyly of toes. The growth and development of the girl was normal. There was no facial dysmorphism or malformations of other systems.

The family had recurrence of a similar malformation disorder in the next pregnancy (Phadke et al., 2003). The hands of this male fetus were exactly similar to the proband. The feet showed soft tissue syndactyly. There was a fully formed digit arising from the dorsum of right foot. The recurrence and parental consanguinity supports an autosomal recessive mode of inheritance.

SHORT RIB POLYDACTYLY SYNDROME (SRP) TYPE III (OMIM 263510; VERMA-NAUMOFF SYNDROME)

The case of a lethal skeletal dysplasia with polydactyly, reported by Verma et al. (1975), is listed as SRP type III in OMIM. The most important distinguishing features of SRP type III are found in the skull. The cranial base is short, with bulging forehead and a depressed nasal bridge and flat occiput. The long bones show a distinct corticomedullary demarcation, with widened metaphases and marked longitudinal spurs. Dagoneau et al. (2009) identified homozygosity or compound heterozygosity for missense or nonsense mutations in the DYNC2H1 gene in families with SRP III.

MENTAL RETARDATION, PTOSIS AND POLYDACTYLY: A NEW AUTOSOMAL RECESSIVE SYNDROME?

Three affected siblings in a consanguineous family had short stature, severe mental retardation, down-slanting

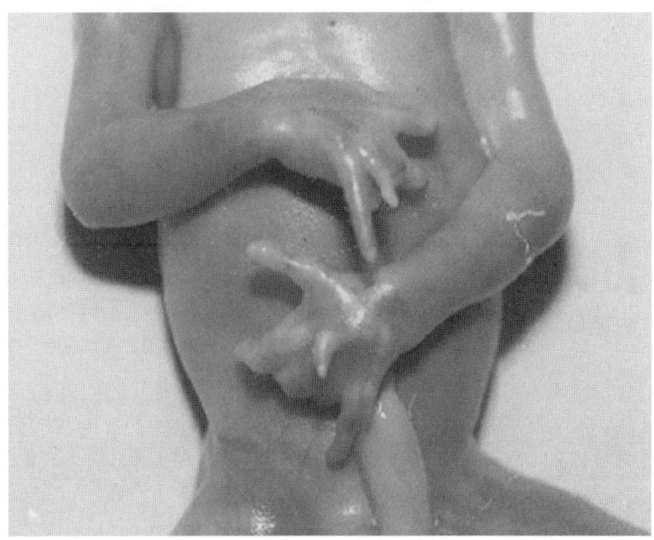

Figure 89-2 Complex camptosynpolydactyly with disorganization – note fingers arising from dorsum of hands (OMIM *607539).

palpebral fissures, and ptosis (Panigrahi et al., 2002). Two of them had postaxial polydactyly. One affected male and one affected female in the postpubertal age group had normal secondary sexual characteristics. None of them had obesity, retinopathy, or renal involvement.

MICROCEPHALY, MICROPENIS: A MONOGENIC SYNDROME

Two male siblings of a consanguineous Muslim couple were reported to have severe retardation microcephaly, micropenis, and bilateral cryptorchidism (Pradhan et al., 1999). Neuroimaging by magnetic resonance (MRI) showed paucity of gyri, suggestive of pachygyria. However, further review of the MRI by an expert pediatric neuroradiologist suggested findings of microcephaly with simplified gyral pattern. Consanguinity in this family favors autosomal recessive etiology for the microcephaly associated with hypogenitalism. However, the possibility of X-linked recessive mode of inheritance cannot be ruled out.

VAN DEN ENDE-GUPTA SYNDROME OF BLEPHAROPHIMOSIS, CONTRACTUAL ARACHNODACTYLY, AND CHARACTERISTIC FACIES

Phadke et al. (1998) reported 2 unrelated Indian girls with blepharophimosis, arachnodactyly, digital contractures, beaked nose, everted lips, prominent ears and elbow deformity. The features were similar to previously reported cases, one of which was from Brazil (Van Den Ende et al., 1992, Gupta et al., 1995). Mutations in SCARF 2 gene were identified recently in some cases of Van den Ende Gupta syndrome (Anastasio et al., 2010).

PREAXIAL BRACHYDACTYLY WITH ABDUCTION OF THUMBS AND HALLUX VARUS

In this family, a father and his daughter presented with a unique pattern of preaxial brachydactyly and facial appearance reminiscent of Rubinstein-Taybi syndrome (Sharma et al., 1994). Both of them had short, broad, abducted thumbs and hallices. The second digits were medially angulated and short. The father had normal mental function, and the child also had normal development for her age. These cases had some similarities with Rubinstein-Taybi syndrome, otopalatodigital syndrome type I, and brachydactyly type D and E. But they are different in many ways and appear to be distinct from the syndromes described to date.

HANDIGODU DISEASE (HD)

This is a unique type of spondylo-epi-metaphyseal skeletal dysplasia (Figure 89-3) described in the Chanangi and Chaluvadi communities of the Shimoga district of Karnataka, a state in the southern part of India (Agarwal et al., 1994). The disease is equally common among males

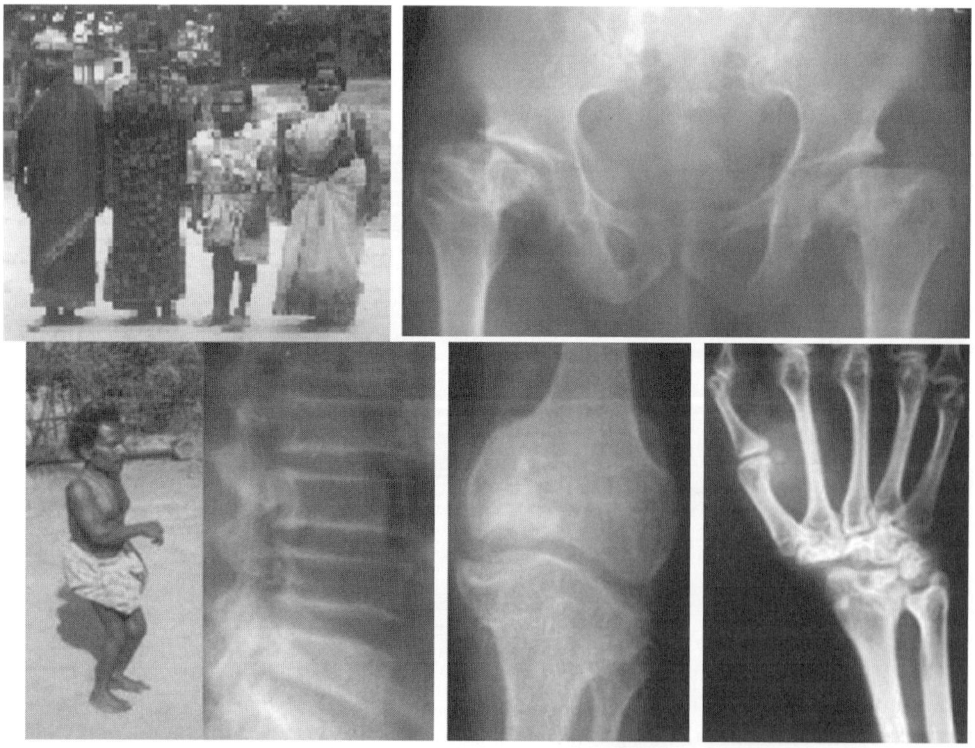

Figure 89-3 Handigodu disease: 2 persons on the right are affected with HD and 2 women on the left are normal. The affected male member of the family (lower left) showing typical phenotype, and radiographs of the pelvis, spine, knee, and hands showing radiological changes in HD.

and females, but there is great variability in the severity of the disease among the affected population. HD is a progressive disease of the skeletal system, predominantly involving the hip joints and spine and presenting in late childhood and in young adults. The radiological changes in the hips and spine are mainly dysplastic, while osteoarthritic changes predominate in some cases. Clinical presentation is early-onset osteoarthritis of hips and/or short stature.

On the basis of anthropometric and radiological findings, three main groups of clinical presentations have been identified (Table 89-3). The first consists of average-build individuals with predominantly osteoarthritic changes or dysplasia of the hips. The second group consists of slightly short-statured individuals with an arm span exceeding their height, indicating disproportionate shortening of the trunk. In these cases, dysplastic changes in the hips and spine predominate on radiological examination. The third group presents clinically with significant short stature with involvement of spine, hips, pelvis, and epiphyses, and other long bones as well. Some of them also have osteoarthritis of knees and shortening of digits. All these phenotypes co-segregate in the families. Table 89-1 shows distribution of 234 cases according to their presentations.

The osteoarthritis is, in many cases, severe enough to need hip joint replacement. A majority of the young affected persons with HD become physically handicapped due to lack of facilities for hip joint replacement. The disease has remarkable similarity with spondyloepiphyseal dysplasia reported from South Africa (Agarwal et al., 1997). The molecular studies to map the gene for HD by linkage to candidate genes are in progress (personal communication).

A NEWLY RECOGNIZED SYNDROME WITH DOUBLE UPPER AND LOWER LIP, HYPERTELORISM, EYELID PTOSIS, BLEPHAROPHIMOSIS, AND THIRD FINGER CLINODACTYLY

Parmar and Muranjan (2004) reported a 21-year-old male with double upper and lower lip, hypertelorism, unilateral ptosis, blepharophimosis, and broad nose with broad nasal tip, highly arched palate, and bilateral third finger clinodactyly. The disorder differs from Ascher syndrome and appears to represent a newly recognized syndrome.

TABLE 89-3 PHENOTYPE OF HANDIGODU DISEASE

Hip joint involvement	64
Osteoarthritis	36
Dysplastic changes	28
Dysplasia spine only	26
dysplasia hips and spine with or without	
Involvement of other bones	144
Total	234

HANDLESS FOOTLESS SYNDROME

Phadke et al. (2006) reported a fetus with symmetrical absence of hands and feet with normal middle and proximal segments of limbs. The limb reductions were too symmetrical to ascribe to amniotic band disruption. From the data of animal experiments, the authors suggested the possibility of genetic defects in AER-FGF.

SHORT STATURE, ULNAR DEVIATION OF HANDS WITH ABSENT CARPALS AND JOINT CONTRACTURES: A NEW SYNDROME

Phadke and Dalal (2007) described a 7-year-old boy with mild motor delay, short stature, joint contractures, ulnar deviations of hands, and mild facial dysmorphism. Similar association of malformations was not found in the literature.

GENES AND MUTATIONS IN DYSMORPHOLOGY

At present, genes for many malformation syndromes have been identified. But the basis of dysmorphology diagnosis continues to remain clinical. Some data is becoming available about mutations in these syndromes in Indian patients. A case of Costello syndrome with severe cutis laxa was found to be mosaic for HRAS G12S (c.34G>A) mutation (personal communication). This is the commonest mutation, and somatic mosaicism for HRAS mutation has been reported only once before in the past. HOX D 13 mutations and genotype–phenotype correlation has been reported in an Indian family with synpolydactyly (Malik et al., 2007). Tamhankar et al. (2010) reported an A353V mutation in DKC1 gene, and this is the commonest mutation worldwide responsible for an X-linked variety of dyskeratosis congenita. There are many ophthalmological centers active in research in genetics. Nallathambi et al. (2007) reported a nonsense mutation (c.273C>G/p.Y91X) in the FOXL2 gene in all affected members of a blepharophimosis epicanthus inversus family. The same center has reported PAX 6 mutations in Indian patients with aniridia and optic nerve anomalies (Neethirajan et al., 2006; Nallathambi et al., 2006). They reported 4 novel mutations—c.1174delTG (in exon 10), c.710delC (exon 6), c.406delTT (exon 5), and c.393insTCAGC (exon 5)—and one novel mutation (c.1080C>T in exon 9). In one of our aniridia family, a novel mutation (IVS12+2 T>C) was identified (Aggarwal et al., 2011). Vanita et al. (2007a,b) have described mutations in various types of cataracts and a novel FBN1mutation (R240C) in autosomal dominantly inherited ectopia. The cataract mutations include a novel, heterozygous c.134G>C change that resulted in the substitution of a highly conserved tryptophan by serine (p.W45S) in a family with autosomal dominantly inherited jellyfish-like cataract and microcornea (Vanita et al., 2008a), a heterozygous c.262C>A change in the gene for connexin 50 (GJA8) in a case with balloon-like cataract (Vanita et al., 2008b), a heterozygous c.176G>A change in gamma crystallin gene, and a novel heterozygous 98G>T change in connexin 46 gene in two families with autosomal dominantly inherited congenital cataract (Guleria et al., 2007; Vanita et al., 2009, 2007). For rare monogenic disorders, an attempt is made to look for mutations in candidate genes (Girisha et al., 2010b, 2011) or to study molecular mechanisms of the disease (Dalal et al., 2010). Identification of a mutation in a syndrome helps in prenatal diagnosis (Jagadeesh et al., 2009). Some groups in India have participated in gene mapping studies (Alazami et al., 2008; Radhakrishna et al., 1999, 2006) and are a part of large consortia for study of genetic disorders (Brancati et al., 2007).

Data about mutations in Indian patients with genetic bone disorders have been discussed in the previous section already.

MULTIPLE MALFORMATION SYNDROMES AND CONSANGUINITY

Many multiple malformation syndromes are inherited in autosomal recessive mode, and hence are more likely to occur in consanguineous marriages. Consanguinity is common among Muslims and South Indian Hindus. Kulkarni and Kurian (1990) reported increased incidence of malformations in consanguineous populations of South India as compared to nonconsanguineous populations. Parental consanguinity was present in all of the eight autosomal recessive syndromes in their study. Our experience also showed that most of the rare and new syndromes reported had parental consanguinity, and many of them were from the Muslim community (Table 89-4). A high level of consanguinity is also present in Hindus in South India, where marriages between maternal uncle and niece are common.

Efforts of mapping of genes for rare syndromes are limited by the unavailability of many large families with such diseases. Availability of consanguineous families with uncommon dysmorphic disorders partly bypasses this hurdle by allowing the use of autozygosity (homozygosity) mapping. With new microarray-based linkage analysis and exome sequencing, even small families with

TABLE 89-4 COMPARISON OF PREVALENCE OF CONSANGUINITY IN PATIENTS WITH BETA-THALASSEMIA HOMOZYGOTES, WHICH IS A COMMON AUTOSOMAL RECESSIVE (AR) DISORDER, WITH THAT OF RARE AUTOSOMAL RECESSIVE DISORDERS (UNPUBLISHED DATA).

	Beta Thalassemia homozygotes	Other AR disorders
Consanguineous	10.7%	26.6%
Nonconsanguineous	89.3%	73.4%

rare disorders can be used for gene mapping. Hence, reporting of such rare cases, and storing the DNA samples and setting up the cell lines from affected individuals of these families, is essential. It can play an important role in mapping genes for multiple malformation syndromes, and thus helpful in understanding the underlying developmental genetic pathology. Lack of curative treatment and cost of investigations are important problems limiting complete work-up of these rare cases.

MALFORMATION SYNDROMES BEFORE BIRTH AND IN STILLBIRTHS

Many malformations are not compatible with life and are seen in higher frequency in stillbirths. The clinicians are increasingly appreciating the importance of autopsy of stillbirths and its utility in genetic counseling (Pahi et al., 1998). This is obvious from reports of lethal disorders. Routine ultrasonography during the antenatal period is increasingly becoming common. With availability of good resolution machines and expertise, many malformation syndromes are diagnosed prenatally. (Sharma et al., 1994; Sharma &Phadke, 1997; Shivarajan et al., 2003; Suresh et al., 1995; Kulkarni et al., 1995; Phadke et al., 2003). Prenatal diagnosis of associated malformations helps greatly in counseling families regarding prognosis of the fetus with malformation (Phadke et al., 2003). The study by Sankar and Phadke (2006a) compared postmortem findings with prenatally detected malformations and found that the results were similar to those of other international studies, indicating good levels of prenatal diagnosis of malformations in India. Autopsy of fetuses terminated after prenatal diagnosis of malformations also helps in delineation of new syndromes. Phadke et al. (2007) identified a severe form of a neurodegenerative disorder of fetal onset with congenital cerebral and cerebellar atrophy. Prenatal diagnosis of malformations has become better and more easily available. Treatable malformations, mild malformations, and some abnormalities of uncertain significance get identified in prenatal ultrasonography, sometimes in the latter part of pregnancy, and this raises many ethical and emotional dilemmas for the families. The society and medical fraternity need to evolve some guidelines regarding severity of malformations that justify termination of pregnancy (Phadke et al., 2007, 2003).

SUMMARY

In this review an attempt is made to give an overview of malformation syndromes from India. It gives an idea about the occurrence of various dysmorphic disorders in the Indian population, and the availability of diagnostic and genetic services in India. As compared to this chapter in the previous edition of the book, there have been many changes. These are due to rapid developments in the field of clinical genetics in India. In 2010, the terms "malformation" and "India" in PubMed gave more than 1000 references, out of which more than 700 were after the year 2000 and more than 500 were after the year 2005. There is increased availability of molecular work-up of the cases and research in medical genetics. Although it is not much. there is data about mutations in the cases of malformation syndromes, and this has been useful in genetic counseling and prenatal diagnosis. In addition to the syndromes described above, there are many more published cases from India in the literature. Not surprisingly, there must be still many more in the population. Because of the large heterogeneous population, and ethnic groups with high rates of consanguinity, many different types of multiple malformation syndromes possibly exist in India. The majority of these are not even brought to medical attention and remain or perish undiagnosed. This is complicated by the extreme lack of diagnostic facilities and limited number of clinical geneticists or pediatricians with an interest in dysmorphology. Identification of these cases and correct diagnosis can go a long way in helping the families by way of providing accurate genetic counseling and helping in making informed reproductive choices, including prenatal diagnosis. It will also help in providing valuable resources for research in carrying out clinical and molecular studies for mapping of genes and identifying mutations.

ACKNOWLEDGMENTS

I thank Dr. M.L. Kulkarni (Davangere), Dr. Madhulika Kabra (New Delhi), Dr. Dala (Hyderabad), Dr. Girisha (Manipal), Dr. Patil (Bangalore), Dr. R. Dua Puri (New Delhi), Dr. Shenoy (Manglore), Dr. Sankar (Trivendrum), and Dr. Archana Kher (Mumbai) for providing their publications related to the subject. Special thanks to Dr. Suresh, Dr. Sujatha, and the group (Chennai) for sharing their unpublished data. My sincere thanks are due to the editors of *Indian Pediatrics* and *Indian Journal of Pediatrics* for giving me the opportunity to review the manuscripts. It has helped me to keep myself informed about the Indian scenario in the field of malformations. I also thank my senior residents, who enthusiastically helped in work-up of the cases and providing reviews of the literature.

REFERENCES

Agarwal SS, Phadke S, Fredund V, Viligeon D, Beighton P. (1997). Mselini and Handigodu familial osteoarthropathies: Syndromic identity. *Am J Med Genet* 72:435–439.

Agarwal SS, Phadke SR, Phadke RV, et al. (1994). Handigodu disease: A radiological study. A new variety of spondyloepimetaphyseal dysplasia of autosomal dominant type. *Skeletal Radiol* 23:611–619.

Aggarwal S, Jinda W, Limwongse C, Atchaneeyasakul LO, Phadke SR. (2011) Run-on mutation in the PAX6 gene and chorioretinal degeneration in autosomal dominant aniridia. *Mol Vis.* 17:1305–1309.

Al Aqueel A, Al Sewairi W, Edress B, Gorlin RJ, Desnick RJ, Martignetti JA. (2000). Inherited multicentric osteolysis variant resembling Torg syndrome in a Saudi Family. *Am J Med Genet* 93:11–18.

Alazami AM, Al-Saif A, Al-Semari A, et al. (2008). Mutations in C2orf37, encoding a nucleolar protein, cause hypogonadism, alopecia, diabetes mellitus, mental retardation, and extrapyramidal syndrome. *Am J Hum Genet* 83(6):684–691.

Anastasio N, Ben-Omran T, Teebi A, Ha KCH, Lalonde E, Ali R, Almureikhi M, Der Kaloustian VM, Liu J, Rosenblatt DS, Majewski J, Jerome-Majewska LA. (2010). Mutations in SCARF2 are responsible for **van den Ende-Gupta** syndrome. *Am. J. Hum. Genet.* 87:553–559.

Awasthi A, Das R, Varma N, et al. (2005). Hematological disorders in Down syndrome: Ten-year experience at a Tertiary Care Centre in North India. *Pediatr Hematol Oncol* 22(6):507–512.

Baskar S, Kulkarni ML, Kulkarni AM, Vittalrao S, Kulkarni PM. (2009). Adams-Oliver syndrome: Additions to the clinical features and possible role of BMP pathway. *Am J Med Genet A* 149A(8):1678–684.

Bhatia MS, Kabra M, Sapra S. (2005). Behavioral problems in children with Down syndrome. *Indian Pediatr* 42(7):675–680.

Bhattacharyya R, Sanyal D, Roy K, Saha S. (2009). A study of cluster behavioral abnormalities in Down syndrome. *Indian J Med Sci* 63(2):58–65.

Bijarnia S, Baijal A, Verma IC. (2003). Genetic Counseling in Acrocallosal syndrome. *Indian Pediatrics* 70(2):169–171.

Biswas AC, Reddy A. (1997). Images in clinical practice: Progeria. *Indian Pediatrics* 34(10):945–946.

Brancati F, Barrano G, Silhavy JL, et al. 2007). CEP290 mutations are frequently identified in the oculo-renal form of Joubert syndrome-related disorders. *Am J Hum Genet* 81(1):104–113.

Chaddha V, Agarwal S, Phadke SR, Halder A. (2003). Low level of mosaicism in atypical Prader Willi syndrome: Detection using fluorescent in situ hybridization. *Indian Pediatr* 40(2):166–8.

Chelvan HT, Malathi N, Kailasam V, Ponnudurai A. (2009). Cleidocranial dysplasia: A family report. *J Indian Soc Pedod Prev Dent* 27(4):249–252.

Cusana F, Scarano G. (1991). Familial progeria or Mandibuloacral dysplasia. *Am J Med Genet* 41:139.

Dalal A, Mehrotra RN. (2009). Hypertrichosis, hyperkeratosis and mental retardation syndrome: Further delineation of phenotype. *Clin Dysmorphol* 18(2):83–84.

Dalal A, Phadke SR. (2005). Hemihyperplasia with Ehlers Danlos syndrome like skin changes. *Clin Dysmorphol* 14(4):207–208.

Dalal A, Phadke SR. (2007). Morphometric analysis of face in dysmorphology. *Comput Methods Programs Biomed* 85(2):165–172.

Dalal A, Sarkar A, Padma Priya T, Madhusudan N. (2010). Giuffre – Tsukahara syndrome-Evidence for X linked dominant inheritance and review. *Am J Med Genet A* 152A:2057–2060.

Dalal AB, Phadke SR, Pradhan M, Sharda S. (2006). Hemihyperplasia syndromes. *Indian J Pediatrics* 73(3):609–615.

Datta V, Sinha A, Saili A, Nangia S. (2005). Bruck syndrome. *Indian J Pediatr* 72(5):441–442.

Danda S, Phadke SR, Agarwal SS. (1997). Acromesomelic dwarfism: Report of a family with two affected siblings. *Indian Pediatr* 34(4):1127–1130.

Dagoneau, N., Goulet, M., Genevieve, D., Sznajer, Y., Martinovic, J., Smithson, S., Huber, C., Baujat, G., Flori, E., Tecco, L., Cavalcanti, D., Delezoide, A.-L., Serre, V., Le Merrer, M., Munnich, A., Cormier-Daire, V. (2009). DYNC2H1 mutations cause asphyxiating thoracic dystrophy and short rib-polydactyly syndrome, type III. *Am. J. Hum. Genet.* 84:706–711,

Eckstein MB, Brown DW, Foster A, Richards AF, Gilbert CE, Vijayalakshmi P. (1996). Congenital rubella in south India: diagnosis using saliva from infants with cataract. *BMJ* 312(7024):161.

Gambhir PS, Gambhir SP, Bankar SM. (2010). Ophthalmoacromelic syndrome: Two further cases expanding the phenotype. *Clin Dysmorphol* 19(2):91–94.

Garg P. (2007). Experience in counselling Down's syndrome. *Indian Pediatr* 44(8):623.

Ghosh S, Setty S, Sivakumar A, Pai KM. (2007). Report of a new syndrome: Focus on differential diagnosis and review of Ellis-van Creveld, Curry-Hall, acrofacial dysostosis, and orofacial digital syndromes. *Oral Surg Oral Med Oral Pathol Oral Radiol Endod* 103(5):670–676.

Girisha KM, Cormier-Daire V, Heuertz S, Phadke RV, Phadke SR. (2008). Novel mutation and atlantoaxial dislocation in two siblings from India with Dyggve-Melchior-Clausen syndrome. *Eur J Med Genet* 51(3):251–256.

Girisha KM, Ganesh HK, Rao L, Srilatha PS. (2010a). Massive cranial osteolysis, skin changes, growth retardation and developmental delay: Gorham syndrome with systemic manifestations? *Am J Med Genet A* 152A(3):759–763.

Girisha KM, Lewis LE, Phadke SR, Kutsche K. (2010b). Costello syndrome with severe cutis laxa and mosaic HRAS G12S mutation. *Am J Med Genet A*. 152A(11):2861–2864.

Girisha KM, Phadke SR, Khan F, Agrawal S. (2006). S252W mutation in Indian patients with Apert syndrome. *Indian Pediatrics* 43(80):733–735.

Girisha KM, Phadke SR. (2005). Anotia and facial palsy: Unusual features of cardiofacial syndrome. *Indian J Pediatr* 72:525–526.

Girisha KM, Sharda SV, Phadke SR. (2007). Issues in counseling for Down syndrome. *Indian Pediatr.* 44(2):131–133.

Girisha KM, Vasudevan T, Shah H, Satyamoorthy K. (2011). Hypoplasia/aplasia of pelvis, femora, fibula, ulna, digits and nails: Fuhrman syndrome with WNT7A mutations. *Clin Dysmorphol.* 20(4):205–209.

Gulati S, Menon S, Kabra M, Kalra V. (2003). Schinzel Acrocallosal syndrome. *Indian J Pediatrics* 70(2):173–176.

Guleria K, Sperling K, Singh D, Varon R, Singh JR, Vanita V. (2007). A novel mutation in the connexin 46 (GJA3) gene associated with autosomal dominant congenital cataract in an Indian family. *Mol Vis* 13:1657–1665.

Gupta A, Hall CM, Ransley YF, Murday VA. (1995). A new autosomal recessive syndrome of characteristic facies, joint contractures, skeletal abnormalities and normal development: Second report with further clinical delineation. *J Med Genet* 32:809–812.

Gupta A, Phadke SR. (2001). Bowen Conradi Syndrome in an Indian infant. first non-Hutterite case. *Clinical Dysmorphology* 10:155–156.

Gupta A, Thakur S, Phadke SR. (2003). Diagnosing Acrocallosal syndrome. *Indian J Pediatrics* 70:177–179.

Gupta N, Kabra M, Das CJ, Gupta AK. (2008). Pyle metaphyseal dysplasia. *Indian Pediatr* 45(4):323–325.

Halder A, Fauzdar A, Kabra M, Saxena A. (2005). Detection of 22 q11.2 hemizygous deletion by interphase FISH in a patient with features of CATCH 22 syndrome. *Indian Pediatr* 42(12):1236–1239.

Halder A, Pahi J, Pradhan M, Pandey A, Gujral R, Agarwal SS. (1998). Iniencephaly: A report of 19 cases. *Indian Pediatrics* 35:891–896.

Halder A, Sharma AK, Phadke SR, Jain A, Agarwal SS. (1994). OEIS complex with craniofacial anomalies defect of blastogenesis? *Am J Med Genet* 53:21–23.

Halder A, Sharma AK, Phadke SR, Jain A, Agarwal SS. (1994). OEIS complex with craniofacial anomalies defect of blastogenesis ? *Am J Med Genet* 53:21–23.

Jagadeesh S, Bhat L, Suresh I, Muralidhar SL. (2009). Prenatal diagnosis of restrictive dermopathy. *Indian Pediatr* 46(4):349–351.

Jyothy A, Kumar KS, Rao GN, et al. (2000). Cytogenetic studies of 1001 Down syndrome cases from Andhra Pradesh, India. *Indian J Med Res* 111:133–137.

Kochar GS, Choudhary A, Gadodia A, Gupta N, Simpson MA, Crosby AH, Kabra M.(2010). Raine syndrome: a clinical, radiographic and genetic investigation of a case from the Indian subcontinent. *Clin Dysmorphol.* 19(3):153–156.

Kolah PJ, Master PA, Sanghri LD. (1967): Congenital malformations and perinatal mortality in Bombay. *Am J Obstet Gynecol* 97:400–406.

Kulkarni ML, Baskar K, Kulkarni PM. (2007). A syndrome of immunodeficiency, autoimmunity, and spondylometaphyseal dysplasia. *Am J Med Genet A* 143(1):69–75.

Kulkarni ML, et al. (1995). Antenatal diagnosis of Grebe syndrome in a twin pregnancy by ultrasound. *Indian Pediatrics* 32:1007–1011.

Kulkarni ML, Kulkarni BM, Nasser PU. (1995). Antenatal diagnosis of Grebe syndrome in a twin pregnancy by ultrasound. *Indian Pediatr* 32(9):1007–1011.

Kulkarni ML, Kurian M. (1990). Consanguinity and its effects on fetal growth and development: A south Indian study. *J Med Genet* 27:348–352.

Kulkarni ML, Matadh PS, Praveen Prabhu SP, Kulkarni PM. (2005a). Fibrochondrogenesis. *Indian J Pediatr* 72(4):355–357.

Kulkarni ML, Samuel K, Bhagyavathi M, Suresh Kumar C. (1994). Skeletal dysplasias in hospital in Southern India. *Indian Pediatrics* 32:657–665.

Kulkarni ML, Samuel K, Bhagyavathi M, Sureshkumar C. (1995). Skeletal dysplasias in a hospital in southern India. *Indian Pediatr* 32(6):657–665.

Kulkarni ML, Zaheeruddin M, Naveen RB, Kulkarni PM. (2005b). Langer-Giedion syndrome with renal cyst. *Indian J Pediatr* 72(2):181.

Kulkarni ML, Zaheeruddin M, Shenoy N, Vani HN. (2006). Fetal valproate syndrome. *Indian J Pediatr* 73(10):937–939.

Kumar S, Tullu MS, Muranjan MN, Kamat JR. (2005). Rabson-Mendenhall syndrome. *Indian J Med Sci* 59(2):70–73.

Kurian K, Shanmugam S, Harsh Vardah T, Gupta S. (2007). Chondroectodermal dysplasia (Ellis van Creveld syndrome): A report of three cases with review of literature. *Indian J Dent Res* 18(1):31–34.

Lakshminarayana P, Ibrahim S, Venkataraman P, Jagatheesan T, Kamala KG. (1991). KAP study on mothers of children with Down syndrome. *Indian Pediatr* 28(9):997–1001.

Lewis LE, Ramesh Bhat Y, Naik P, Sethi K, Girisha KM. (2010). Opsismodysplasia. *Indian J Pediatr* 77(5):567–568.

Malik S, Girisha KM, Wajid M, et al. (2007). Identical HOXD 13 polyalanine expansion mutations in syndactyly type II families of Pakistani and Indian origin. *BMC Med Genet* 8(1):78.

Mandal K, Boggula VR, Borkar M, Agarwal S, Phadke SR. (2009). Use of multiplex ligation probe amplification (MLPA) in screening of subtelomeric regions in children with idiopathic mental retardation. *Indian J Pediatrics* 76(10):1027–1031.

Mandal K, Phadke SR, Kalita J. (2008). Congenital swan neck deformity of fingers with syndactyly. *Clin Dysmorphol* 17(2):109–111.

Martignetti JA, Aqeel AA, Sewairi WA, Boumah CE, Kambouris M, Mayouf SA, Sheth KV, Eid WA, Dowling O, Harris J, Glucksman MJ, Bahabri S, Meyer BF, Desnick RJ.(2001). Mutation of the matrix metalloproteinase 2 gene (MMP2) causes a multicentric osteolysis and arthritis syndrome. *Nat Genet.* 28(3):261–265.

Girisha KM, Ganesh HK, Rao L, Srilatha PS. (2010) Massive cranial osteolysis, skin changes, growth retardation and developmental delay: Gorham syndrome with systemic manifestations? *Am J Med Genet A.* 152A(3):759–763.

Murugesan C, Kumar P, Muralidhar K. (2009). A new association of multiple congenital anomalies/mental retardation syndrome with bradycardia-tachycardia syndrome: A case report. *J Med Case Reports* 3:9309.

Nahar R, Puri R, Saxena R, Kohli S, Verma IC. (2010). Utility of Molecular studies in Skeletal Dysplasias. *Genetic Clinics* 3(4):12–15.

Nahar R, Saxena R, Kohli S, Puri R, Verma IC. (2009). Molecular studies of achondroplasia. *Indian J Orthop* 43(2):194–196.

Nallathambi J, Neethirajan G, Shashikant S, Vijayalakshmi P, Sundaresan P. (2006). PAX6 missense mutations associated in patients with optic nerve malformation. *Mol Vis* 12:236–242.

Nallathambi J, Neethirajan G, Usha K, Jitendra J, De Baere E, Sundaresan P. (2007). FOXL2 mutations in Indian families with blepharophimosis-ptosis-epicanthus inversus syndrome. *J Genet* 86(2):165–168.

Nampoothiri S, Lakshman LR, Anilkumar A, Thampi MV. (2008). Partial trisomy 9q due to maternal 9q 17q translocation. *Indian Pediatr* 45(7):595–598.

Nanda A, Kaur S, Bhakoo ON, Kapoor MM, Kanwar AJ. (1989). Fetal hydantoin syndrome: a case report. *Pediatr Dermatol.* 6(2):130–133.

Naveed, Manjunath CS, Sreenivas V. (1990). New manifestations of Neu-Laxova syndrome. *Am J Med Genet* 5(1):55–59.

Neethirajan G, Nallathambi J, Krishnadas SR, et al. (2006). Identification of novel mutant PAX6 alleles in Indian cases of familial aniridia. *BMC Ophthalmol* 6:28.

Pahi J, Phadke SR, Halder A, Gupta A, Pandey R, Agarwal SS. (1998). Does autopsy of antenatally diagnosed malformed fetus aid in genetic counseling. *National Med J India* 11:169–170.

Pangrazio A, Pusch M, Caldana E, et al. (2010). Molecular and clinical heterogeneity in ClCN7-dependent osteopetrosis: report of 20 novel mutations. *Hum Mutat* 31(1):E1071–1080.

Panigrahi I, Das RR, Sharda S, Marwaha RK, Khandelwal N. (2010). Response to zolendronic acid in children with type III osteogenesis imperfecta. *J Bone Miner Metab* 28(4):451–455.

Panigrahi I, Phadke SR, Agarwal SS. (2002). Mental retardation, ptosis and polydactyly: A new autosomal recessive syndrome? *Clinical Dysmorphology* 11:289–292.

Parmar RC, Muranjan MN, Kotvaliwale S, Sharma S, Bharucha BA. (2003). Ring chromosome 12 with variable phenotypic features: Clinical report and review of the literature. *Am J Med Genet A* 117A(3):275–277.

Parmar RC, Muranjan MN. (2004). A newly recognized syndrome with double upper and lower lip, hypertelorism, eyelid ptosis, blepharophimosis, and third finger clinodactyly. *Am J Med Genet A* 124A(2):200–201.

Patel ZM, Adhia RA. (2004). Tel-Hashomer camptodactyly syndrome with hirsuitism in an Indian family. *J Assoc Physicians India* 52:837–838.

Patil SJ, Banerjee M, Phadke SR, Mittal B. (2009). Mutation analysis in Indian children with achondroplasia – utility of molecular diagnosis. *Indian J Pediatr* 76(2):147–149.

Patil SJ, Bhat M, Rao S, Krishnan RS. (2009). Spondylocarpotarsal synostosis: A rare case of vertebral segmentation defect. *Indian J Pediatr* 76(4):417–419.

Patil SJ, Phadke SR. (2007). Pericentric inversion causing duplication and deletion of chromosome region 13q22 --> qter in the offspring. *Am J Med Genet A* 143(1):82–824.

Phadke SR, Agarwal S, Puri RD. (2003). Recurrence of complex camptopolydactyly in a sibling suggestive of autosomal recessive mode of inheritance. *Am J Med Genet* 116A:94–96.

Phadke SR, Dalal A. (2007). Short stature, ulnar deviation of hands with absent carpals and joint contractures: A new syndrome. *Clin Dysmorphol* 16(1):55–57.

Phadke SR, Dua Puri Ratna, Phadke RV. (2007). Severe form of congenital cerebral and cerebellar atrophy: a neurodegenerative disorder of fetal onset. *J of Clinical Ultrasound* 35(6):347–350.

Phadke SR, Fischer B, Gupta N, Prajnya R, Kabra M, Kornak U. (2010b). Mutation spectrum in Indian patients with autosomal recessive infantile malignant osteopetrosis: Report of novel mutations. *Indian J Med Res* 131:508–514.

Phadke SR, Gautam P. (1999). Complex camptopolydactyly: An unusual hand malformation. *Am J Med Genet* 83:191–192.

Phadke SR, Girisha KM, Dalal A. (2006). Handless footless fetus. *Clin Dysmorphology* 15:233–234.

Phadke SR, Girisha KM. (2007). Issues in counseling for Down syndrome: Indian scenario. *Indian Pediatrics* 44:131–133.

Phadke SR, Gulati R, Agarwal SS. (1998). Further delineation of a new (Van Den Ende- Gupta) syndrome of blepherophimosis, contractual arachnodactyly and characteristic face. *Am J Med Genet* 77:16–18.

Phadke SR, Gupta A, Pahi J, Pandey A, Gautam P, Agarwal SS. (1999). Malignant recessive osteoporosis. *Indian Pediatrics* 36:69–72.

Phadke SR, Gupta N, Girish KM, et al. (2010a). Rhizomelic chondrodysplasia punctata type 1– Report of 3 cases from India. *Journal of Applied Genetics* 51(1):107–110.

Phadke SR, Pahi J, Pandey A, Agarwal SS. (1999). Oral-facial digital syndrome with acromelic short stature: A new variant overlap with Ellis Van Creveld syndrome. *Clinical Dysmorphology* 8(3):185–188.

Phadke SR, Pahi J, Phadke RV, Pradhan S, Agarwal SS. (1997). Importance of etiologic diagnosis of hydrocephalus as illustrated by a case of Walker Warburg Syndrome. *Indian Pediatrics* 34:1037–1038.

Phadke SR, Patil SJ, Niraj Kumari, Krishnani N. (2007). Spondylothoracic dysplasia: Prenatal diagnosis and the problems of nosologic overlap. *Am J Med Genet A* 143(8):899–902.

Phadke SR, Puri RD, Agarwal S, Thakur S. (2003). Counseling of prenatally detected malformations of uncertain prognosis: A dilemma. *Perinatology* 5(3):132–137.

Phadke SR, Ramirez M, Difeo A, Martignetti JA, Girisha KM.(2007). Torg-Winchester syndrome: Lack of efficacy of pamidronate therapy. *Clin Dysmorphol* 16(2):95–100.

Phadke SR, Sharda S. (2008). A report of a patient with interstitial deletion of 15q22: Further delineation of a new micro deletion syndrome. *Am J Med Genet A* 146A(15):1999–2000.

Phadke SR, Sharma AK, Halder A, Pandey R, Bhatia VL, Agarwal SS. (1994). GAPO syndrome in a child without dermal hyaline deposits. *Am J Med Genet* 51:191–193.

Phadke SR, Thakur S. (2002). Prenatal diagnosis of iniencephaly and alobar holoprosencephaly with trisomy 13 mosaicism: A case report. *Prenatal Diagnosis* 22:1240–1241.

Pradhan M, Phadke SR, Jain S, Agarwal SS. (1999). Pachygyria/Hypogenitalism: A monogenic syndrome. *Am J Med Genet* 87:254–257.

Prakash H, Sidhu SS, Raghavan R, Deshmukh RN. (1990). Brief clinical report: Hutchinson Guilford Protegra: Familial occurrence. *Am J Med Genet* 36:431–433.

Puri RD, Phadke SR. (2003). Catel-Manzke syndrome without cleft palate – a case report. *Clinical Dysmorphology* 12(4):279–281.

Puri RD, Thakur S, Verma IC. (2007). Spectrum of severe skeletal dysplasias in North India. *Indian J Pediatr* 74(11):995–1002.

Purohit SK, Kumta SM, Rao PP, Thatte RL. (1989). An interesting case of aglossia-adactyly syndrome. *Br J Plast Surg.* 42(2):228–229.

Radhakrishna U, Bornholdt D, Scott HS, et al. (1999). The phenotypic spectrum of GLI3 morphopathies includes autosomal dominant preaxial polydactyly type-IV and postaxial polydactyly type-A/B; No phenotype prediction from the position of GLI3 mutations. *Am J Hum Genet* 65(3):645–655.

Radhakrishna U, Ratnamala U, Gaines M, et al. (2006). Genomewide scan for nonsyndromic cleft lip and palate in multigenerational Indian families reveals significant evidence of linkage at 13q33.1-34. *Am J Hum Genet* 79(3):580–585.

Sankar VH, Phadke SR. (2006a). Clinical utility of autopsy and comparison with prenatal ultrasound findings. *J of Perinatology* 26:224–229.

Sankar VH, Phadke SR. (2006b). Ring chromosome 13 in an infant with ambiguous genitalia. *Indian Pediatr* 43(3):258–260.

Shah I, Johari A. (2007). Oral bisphosphonate treatment for osteogenesis imperfecta—an Indian perspective. *Ann Trop Paediatr* 27(1):39–43.

Shankar VN, Ajila V, Kumar G. (2010). Osteoglophonic dysplasia: A case report. *J Oral Sci* 52(1):167–171.

Sharma AK, Halder A, Phadke SR, Agrawal SS. (1994). Preaxial brachydactyly with abduction of thumbs and hallus varus: A distinct entity. *Am J Med Genet* 49:274–277.

Sharma AK, Jain A, Phadke SR, Srivastava S. (1999). Prenatal diagnosis of Robert syndrome. *Indian Pediatr* 31:1261–1264.

Sharma AK, Phadke SR, Chandra K, et al. (1992). Overlap between Majewski and Hydrolethalus syndromes: A report of two cases. *Am J Med Genet* 43:949–953.

Sharma AK, Phadke SR, Halder A, Agarwal SS. (1994). Jarcho-Levin Syndrome. *Indian Pediatrics* 31:707–708.

Sharma AK, Phadke SR. (1994). Another case of spondylocostal dysplasia and severe anomalies: A diagnostic and counseling dilemma. *Am J Med Genet* 50:383–384.

Sharma AK, Upreti M, Kamboj M, et al. (1994). Incidence of neural tube defects at Lucknow over a 10 year period from 1982–1991. *Indian J Med Res* 99:223–226.

Shenoy RD, Kamath N. (2010). Dandy-Walker malformation, occipital meningoencephalocele, meso-axial polydactyly and bifid hallux. *Clin Dysmorphol* 19(3):166–168.

Sheth J, Joshi R, Sheth F. (2007). Ring chromosome 9 in a dysmorphic child. *Indian J Pediatr* 74(5):507–508.

Shivanand G, Jain V, Lal H. (2007). Progressive pseudorheumatoid chondrodysplasia of childhood. *Singapore Med J* 48(5):e151–153.

Shivarajan MA, Suresh S, Jagadeesh S, Lata S, Bhat L. (2003). Second trimester diagnosis of Neu Laxova syndrome. *Prenat Diagn* 23(1):21–24.

Singh M, Ray D, Paul VK, Kumar A. (1988). Hydrocephalus in asphyxiating thoracic dystrophy. *Am J Med Genet.* 29(2):391–395.

Singh M, Deorari AK, Khajuria RC, Paul VK. (1991). A four year study on neonatal morbidity in a New Delhi hospital. *Indian J Med Res.* 94:186–192.

Singh K, Phadke SR, Agarwal SS. (1999). Mandibuloacral dysplasia: Indian patient with severe bony changes. *J Assoc Physn India* 47: 833–834.

Sneth KV, Al Eid W, Dowling O, Harvis J, Gluksman MJ, Bahabris, Meyer BF, Desnick RJ. (2001).: Mutation of the matrix metalloproteinase 2 gene (MMP2) causes a multicentric osteolysis and arthritis syndrome. *Nature Genetics* 28:261–265.

Superti-Furga A, Unger S. (2007). Nosology and classification of genetic skeletal disorders: 2006 revision. *Am J Med Genet A* 143(1):1–18.

Suresh S, Rajesh K, Suresh I, Raja V, Gopisn D, Gnanasoundari S. (1995). Prenatal diagnosis of orofaciodigital syndrome Mohr type. *J Ultrasound Med* 14(11):863–866.

Suresh S, Thangavel G, Sujatha J, Indrani S. (2005). Methodological issues in setting up a surveillance system for birth defects in India. *Natl Med J India* 18(5):259–262.

Suri M, Kabra M, Verma IC. (1994). Blepharophimosis, telecanthus, microstomia, and unusual ear anomaly (Simosa syndrome) in an infant. *Am J Med Genet* 51:222–223.

Tamhankar PM, Zhao M, Kanegane H, Phadke SR. (2010). Identification of DKC1 gene mutation in an Indian patient. *Indian Journal of Pediatrics* 77(3):310–312.

Tandon A, Bhargava SK, Goel S, Bhatt S. (2008). Pseudoachondroplasia: A rare cause of rhizomelic dwarfism. *Indian J Orthop* 42(4):477–479.

Thakur S, Paul L, Phadke SR. (2004b). Lethal arthrogryposis with icthiosis: Overlap with Neu Laxova syndrome, restrictive dermopathy and Harlequin fetus. *Clinical Dysmorphology* 13(2):117–119.

Thakur S, Singh R, Pradhan M, Phadke SR. (2004a). Spectrum of holoprosencephaly. *Indian J Pediatr* 71:593–597.

Van Den Ende JJ, Van Bever Y, Rodini ESO, Richieri-Costa A. (1992). Marden-Walker like syndrome without psychomotor retardation. Report of a Brazilian girl born to consanguineous parents. *Am J Med Genet* 42:467–469.

Vanita V, Singh JR, Singh D, Varon R, Robinson PN, Sperling K. (2007). A recurrent FBN1 mutation in an autosomal dominant ectopia lentis family of Indian origin. *Mol Vis* 13:2035–2040.

Vanita V, Singh JR, Singh D, Varon R, Sperling K. (2008a). A mutation in GJA8 (p.P88Q) is associated with "balloon-like" cataract with Y-sutural opacities in a family of Indian origin. *Mol. Vis.* 17(14):1171–1175.

Vanita V, Singh JR, Singh D, Varon R, Sperling K. (2008b). A novel mutation in GJA8 associated with jellyfish-like cataract in a family of Indian origin. *Mol Vis* 14:323–326.

Vanita V, Singh JR, Singh D, Varon R, Sperling K. (2009). Novel mutation in the gamma-S crystallin gene causing autosomal dominant cataract. *Mol Vis* 15:476–481.

Verma IC, Bhargava S, Agarwal S. (1975). An autosomal recessive form of lethal chondrodystrophy with severe thoracic narrowing, rhizoacromelic type of macromelia, polydactyly and genital anomalies. *Birth Defects Orig Art Ser XI* (6):167–174.

Verma IC. (1994). The challenge of genetic disorders in India: Molecular genetics and gene therapy. The new frontier. In: *Proceedings of First Annual Ranbaxy Science Foundations Symposium.* New Delhi, pp. 11–20.

Vijayalakshmi P, Kakkar G, Samprathi A, Banushree R.(2002). Ocular manifestations of congenital rubella syndrome in a developing country. *Indian J Ophthalmol.* 50(4):307–311.

Weimer J, Shivakumar S, Danda S, et al. (2007). Premature ovarian failure in a patient with a complex chromosome rearrangement involving the critical region Xq24, characterized by analysis using fluorescence in situ hybridization by chromosome microdissection. *Fertil Steril* 88(6):1677.

90.

THALASSEMIAS AND ABNORMAL HEMOGLOBINS OF THE INDIAN SUBCONTINENT

John Old

INTRODUCTION

The hemoglobinopathies are a heterogeneous group of autosomal recessive disorders caused by defects in the human globin genes. The disorders consist of the thalassemias, characterized by the reduced synthesis of one or more of the globin chains, and the hemoglobin variants, characterized by the synthesis of a structurally abnormal globin (Weatherall and Clegg, 2001). They form the most common single gene disorder in the world, and are found at high frequencies in many populations as a result of positive selection pressure due to falciparum malaria. Individuals with the carrier state are easily identifiable, permitting the control of the serious hemoglobinopathies by a program of carrier screening, counseling, and prenatal diagnosis (Petrou and Modell, 1995). The most important disorders for which prenatal diagnosis is considered are β^{o}-thalassemia, β-thalassemia, sickle cell anemia, and the various compound heterozygous states that result in a clinically significant disease.

The mutations causing the hemoglobinopathies are regionally specific, with four main regional groups: African, Mediterranean, southeast Asian, and Asian Indian. Within each region, each local population has its own characteristic spectrum thalassemia mutations and hemoglobin variants. The majority of the globin gene mutations have now been characterized at the molecular level, and the particular mutations and their frequencies found in most at-risk ethnic groups have been published (Huisman et al., 1997). The majority of the molecular defects can be diagnosed using a variety of techniques involving the analysis of DNA amplified by the polymerase chain reaction (PCR); see Old, 1996). PCR-based techniques provide a quick and sensitive approach to fetal DNA diagnosis, and normally enable a result to be obtained in two to three days.

This chapter describes the thalassemia mutations and abnormal hemoglobins found on the Indian subcontinent, and summarizes the molecular diagnostic approaches for screening and prenatal diagnosis of these mutations. The spectrum and gene frequencies of thalassemia mutations reported for other developing countries has recently been reviewed elsewhere (Henderson et al., 2009).

α-THALASSEMIA

CLINICAL PHENOTYPES

α-thalassemia is the most common type of thalassemia found in the Indian subcontinent. It is characterized by a deficiency of α-globin chain synthesis, with defective gene expression occurring in either one of the two globin genes (called α-2 or α^{+}-thalassemia), or in both genes (α-1 or α^{o}-thalassemia; see Weatherall and Clegg, 2001). Thus, there are four clinical states, depending upon the number of functional α-globin genes (3, 2, 1 or 0). The most severe form of α-thalassemia is the homozygous state for α^{o}-thalassemia, known as Hb Bart's hydrops fetalis syndrome. This condition results from a deletion of all four globin genes, and an affected fetus cannot synthesize any α-globin to make Hb F or Hb A. Fetal blood contains only the abnormal hemoglobin Bart's (γ_4) and a small amount of Hb Portland. The resulting severe fetal anemia leads to asphyxia, hydrops fetalis, and stillbirth or neonatal death. Individuals with one functional α-gene have Hb H disease, and are compound heterozygotes for α^{+} and α^{o}-thalassemia. They have a moderately severe hypochromic microcytic anemia, and produce large amounts of Hb H (β_4) as a result of the excess β-chains in the reticulocyte. Patients with deletional Hb H disease suffer from fatigue, general discomfort, and splenomegaly, but they rarely require hospitalization and lead a relatively normal life. Individuals with two normal and two nonfunctional α-genes have α^{o}-thalassemia trait, or homozygous α^{+}-thalassemia, and those with three normal and one nonfunctional gene have α^{+}-thalassemia trait. Finally, some nondeletion α^{+}-thalassemia point mutations result in an unstable variant. These variants, while barely symptomatic in the heterozygous state, may result in Hb H disease and/or a thalassemia intermedia syndrome in the homozygous state or compound heterozygous state with α^{o}-thalassemia, and some cases are even born with a hydrops fetalis syndrome (Wajcman et al., 2008).

GENOTYPES

The α-thalassemia mutations reported to occur in Asian Indian individuals are listed in Table 90-1. There are very few reports of Hb H disease patients of Asian Indian origin. Thus, α^{o}-thalassemia trait is extremely uncommon in the Indian subcontinent. One α^{o}-thalassemia allele ($—^{SA}$) has been reported to occur in Indian patients. The 23kb deletion was first discovered in the "Cape coloured" population of South Africa (Vandenplas et al., 1987), and later characterized in an individual from the Gujurat region of India (Drysdale and Higgs, 1988). A different α^{o}-thalassemia deletion mutation has recently been observed in a fetus diagnosed with Hb Bart's hydrops fetalis syndrome born to a British couple of Pakistani origin (Old, personal communication).

α^{+}-thalassemia has been reported in many different areas of India (Brittenham, 1981). Early studies using electrophoresis to measure the presence of Hb Bart's in newborns suggested a carrier frequency of 1%. However, more sensitive methods applied to cord blood samples from Mumbai showed that 15% had α-thalassemia (Desai et al., 1997). This frequency has been confirmed by DNA analysis studies of British Asians of Bengali, Gujarati, Pakistani, and Punjabi ethnic origin (Hassal et al., 1998). A carrier frequency of 15% was seen in individuals of Pakistani origin, 11% in Gujaratis, and 13% in Punjabis. The predominant mutation was the $-\alpha^{3.7}$ allele (80%), the rest being $-\alpha^{4.2}$allele. A similar gene frequency for α^{+}-thalassemia has been observed in Sri Lanka. Both the $-\alpha^{3.7}$ and $-\alpha^{4.2}$ alleles were found, with an overall gene frequency of 14% for the $-\alpha^{3.7}$ and 2% for the $-\alpha^{4.2}$ allele (de Silva et al., 2000). DNA studies have revealed even higher gene frequencies for α^{+}-thalassemia in the Indian tribal populations.

A study of individuals with sickle cell trait from Orissa found a gene frequency of the $-\underset{.}{\alpha}$ deletion alleles to be 0.29, giving a carrier frequency of 50% (Kulozik et al., 1988). The predominant form was the $-\alpha^{3.7}$ allele, with the $-\alpha^{4.2}$ allele constituting about 30%.

A study of the Valmiki, Khonda Reddi, and Koya Dora tribal groups revealed a remarkably high gene frequency of 0.58 (Brittenham et al., 1980). This high figure implies that 80% of individuals are carriers or homozygotes for α^{+}-thalassemia. Similarly high gene frequencies have been detected in tribals from the Kolam (Fodde et al., 1991), the Baiga (Reddy et al., 1995) and the Gond tribe (Gupta et al., 1991). In all the studies, the gene frequency of the $-\underset{.}{\alpha}$ deletion alleles in nontribals from the same regions was found to be much lower (0.07–0.14). Finally, the highest gene frequency recorded for the $-\underset{.}{\alpha}$ deletion alleles is 0.8, reported for the Tharu people of the Terai region of southern Nepal (Modiano et al., 1991).

A few rare α^{+}-thalassemia alleles have been characterized in individuals of Asian Indian origin. A 3.5 kb deletion allele ($-\alpha^{3.5}$) has been described in two Indian patients (Kulozik et al., 1988), and a deletion of 7.9 kb ($-\alpha^{7.9}$) has recently been characterized in two individuals

TABLE 90-1 LIST OF α-THALASSEMIA MUTATIONS

Mutations	Phenotype	Origin	References
DELETION			
$-\alpha^{4.2}$	α^{+}	Asian-Indian	Embury et al., 1980
$-\alpha^{3.7}$	α^{+}	Asian-Indian	Embury et al., 1980
$-\alpha^{7.9}$	α^{+}	Surinam-Indian	Harteveld et al., 2003
$-\alpha^{3.5}$	α^{+}	Indian	Kulozik et al., 1988
$—^{SA}$	α^{0}	Indian	Vandenplas et al., 1987
$—^{PAK}$	α^{0}	Pakistani	Old, personal communication
NON-DELETION			
α1 CD 14 TGG→CGG (Trp→Arg) Hb Evanston	α^{+}	Indian	Bernini & Harteveld., 1998
α1 IVS1-117 G→A (GCAGGA→GCAAGA)	α^{+}	Indian	Cürük et al., 1993
α2 CD 104 TGC→TAC (Cys→Tyr) Hb Sallanches	α^{+}	Asian-Indian	Khan et al., 2000; Dash et al., 2006
α2 CD 130 GCT→CCT (Ala→Pro) Hb Sun Prairie	α^{+}	Indian/Pakistani	Harkness et al., 1990
α2 CD 142 (TAA→TCA) Hb Koya Dora	α^{+}	Indian	De Jong et al., 1975
α2 Poly A (AATAAA→AATA—)	α^{+}	Indian	Harteveld et al., 1994

of Suriname-Indian origin (Harteveld et al., 2003). Several nondeletional α^+-thalassemia alleles have been described in Indians, including the α2-globin gene chain termination variant Hb Koya Dora (De Jong et al., 1975) and the mRNA processing mutant α1 IVS-117 (G→A) (Cürük et al, 1993). In Pakistan, a rare nondeletional α^+-thalassemia gene with a severe phenotype has been reported for the first time. The α2-globin chain variant, Hb Sallanches, was observed to cause transfusion-dependent HbH disease in three homozygous patients (Khan et al., 2000), and has also now been observed in an Indian child (Dash et al., 2006).

β-THALASSEMIA

CLINICAL PHENOTYPES

β-thalassemia is characterized by a deficiency of β-globin chain synthesis. The majority of individuals homozygous for β-thalassemia have the transfusion-dependent condition called *β-thalassemia major.* This condition results from the homozygous state for a β^o or a severe β^+-thalassemia mutation, or, more commonly, the compound heterozygous state for two different β^o or β^+ type mutations. At birth, β-thalassemia homozygotes are asymptomatic because of the high production of Hb F, but as this declines, affected infants present with severe anemia during the first or second year of life. Treatment is by frequent blood transfusion to maintain a hemoglobin level above 10g/dl, coupled with iron chelation therapy to control iron overload; otherwise, death results in the second or third decade from cardiac failure. This treatment does not cure β-thalassemia major, although some patients have now reached the age of 40 in good health and have married and produced children. With the prospects for gene therapy remaining as distant as ever, the only cure for β-thalassemia for the foreseeable future is bone marrow transplantation. Although this form of treatment has proved successful when carried out in young children, it is limited by the requirement of an HLA-matched sibling or relative.

Some individuals homozygous for β-thalassemia have a milder clinical condition called *thalassemia intermedia.* Such patients present later in life relative to those with thalassemia major, and are capable of maintaining a hemoglobin level above 6g without transfusion. Thalassemia intermedia is caused by a wide variety of genotypes and covers a broad clinical spectrum. Patients with a severe condition present between 2 and 6 years of age, and although they are capable of surviving with an Hb level of 5 to 7g/dl they will not develop normally and are treated with minimal blood transfusion. At the other end of the spectrum are patients who do not become symptomatic until they reach adult life, and remain transfusion independent with Hb levels of 8 to 10g/dl. However, even these milder patients tend to accumulate iron with age, and many thalassemia intermedia patients develop clinical problems relating to iron overload after the third decade.

β-thalassemia is an important genetic disease on the Indian subcontinent. The average reported incidence of β-thalassemia trait is 3.3%, with 1–2 per 1000 couples being at risk of having an affected child each year. In India, 6000 children with thalassemia major are born annually (Modell and Petrou, 1983). Prevention of the disease by genetic counseling and prenatal diagnosis has a particularly important role in controlling the disease in developing countries where there are limited resources for the medical care of affected individuals. In order to develop a first trimester prenatal diagnosis program by fetal DNA analysis, the molecular basis of the β-thalassemia mutations in the population at risk needs to be defined. This was first done for the Indian population in the United Kingdom (Old et al., 1990), and now many prenatal diagnosis programs have been established in various diagnostic centers on the Indian subcontinent.

In Mumbai, Gorakshakar et al. (1997) reported their first trimester prenatal diagnosis experience for 55 couples using PCR and denaturing gradient gel electrophoresis. The following year, Saxena et al. (1998a) in New Delhi published their experience of prenatal diagnosis in 415 pregnancies using the amplification refractory mutation system (ARMS). The ARMS technique has also been used by another Mumbai diagnostic center, as reported by Thakur et al. (2000) for the analysis of 494 fetuses by DNA analysis, and who also diagnosed 293 fetuses in the second trimester by globin chain synthesis. Prenatal diagnosis is also feasible and acceptable in some Muslim countries, as the reported experience of Ahmed et al. (2000) in Rawalpindi, Pakistan demonstrates. They reported that 47 out of 53 women who were diagnosed as carrying an affected fetus opted for the termination of pregnancy before 120 days of gestation, as permitted by Islamic scholars if the fetus is affected with a serious genetic disorder.

GENOTYPES

β-thalassemia has been shown to be caused by 170 point mutations or small insertions/deletions of DNA sequence in and around the β-globin gene, and by 13 large deletions of 44 nucleotides or greater (Huisman et al, 1997). The mutations either reduce the expression of the β-globin gene (β^+-type) or result in the complete absence of β-globin (β^o-type). However, for each country there is usually just a small number of mutations that account for the majority of the β-thalassemia alleles, making molecular diagnosis simple and feasible.

The mutations are regionally specific, and can be grouped into Mediterranean, Asian Indian, Chinese, and African regions, with each country within a region having its own characteristic spectrum of 10–30 mutations.

Table 90-2 lists the β-thalassemia mutations found in countries of the Indian subcontinent. Except for codon 41/42 (-TCTT), they are all classified as Asian Indian mutations and only are very rarely found in individuals from the other three regions. The relative allele frequencies of the mutations listed in Table 90-2 have been compiled from several reports of large studies of thalassemic patients by molecular analysis. Thalassemia mutations are listed for India (Parikh et al., 1990; Varawalla et al., 1991a, 1991b; Baysal et al., 1994; Old et al., 2001; Bashyam et al., 2004; Nadkarni et al., 2009; Agarwal S et al., 2010), for Pakistan (Varawalla et al., 1991a; Khan and Riazuddin, 1998; Old et al., 2001) for the Maldives (Furuumi et al., 1998), for Sri Lanka (de Silva et al., 2000; Old et al., 2001; Fisher et al., 2003) and for Mauritius (Kotea et al., 2000; Old et al, 2001).

In India and Pakistan, five mutations account for the majority of the β-thalassemia alleles. These are IVSI-5 (G→C), IVSI-1 (G→T), codon 41/42 (-TCTT), codon 8/9 (+G), and the 619bp deletion. In India, these five mutations account for 92% of the alleles. The most frequently observed mutation is IVSI-5 (G→C), as is also the case for Sri Lanka, Mauritius, and the Maldives. India and Pakistan also share a number of less common mutations, such as codon 15 (G→A), codon 30 (G→C) and codon 47/48 (+ATCT) (Saxena et al., 1998b), and then there are a number of reports of rare mutations specific for each country. For example, the mutations codon 44 (-C) and IVSII-848 (C→A) have only been observed in the Pakistani population (Khan et al., 2000). The rare splice acceptor site mutation IVSI-130 (G→C) has only been reported in Indians (Bandyopadhyay et al., 2001), as has the 7bp deletion at codons 22/23/24, –87 (C→A) (Nadkarni et al., 2009), and the splicing defect IVSII-837 (T→G) (Bashyam et al, 2004). A rare initiation codon mutation, (ATG→ACG), has been reported in both Pakistani and Indian individuals (Khan et al., 2000; Gupta et al., 2002).

A number of novel mutations have been described in the Indian population. Four small deletion mutations have been reported in Indians: a 8bp deletion at codons 36–39, a 22bp deletion at codons 81–88, a 17bp deletion at codons 121–131, and a 7bp deletion in exon 2 (Shaji et al., 2002a, 2002b; Nadkarni et al., 2002; Gupta et al., 2003). Similarly, for two novel point mutations CD 16 (C→T) and IVSII-613 (C→T) (Nadkarni et al., 2009) and a novel 3 base substitution mutation at codons 4–6 (TCCTG→ATCTT) (Saxena et al., 2000).

In Sri Lanka, just two mutations, IVSI-5 (G→C) and IVSI-1 (G→A), have been shown to comprise 83% of all the β-thalassemia mutations (de Silva et al., 2000; Old et al., 2001; Fisher et al., 2003). The finding that IVSI-1 (G→A) was the second commonest allele at a relative frequency of 27% is interesting, as this is a mutation from the Mediterranean group and is extremely uncommon in other Asian Indian countries (Varawalla et al., 1991b). A further 12 mutations were identified, including three novel mutations not described before in any at-risk population: CD55 (-A), CD 6 (-13bp), and IVSI-129 (A→C). The CD55 (-A) mutation has since been discovered in the Indian population (Nadkarni et al., 2002).

A small number of rare β^+-thalassemia mutations have a milder phenotype than the majority of the β^+ and β^o-thalassemia mutations. The homozygous state for these mutations is associated with the milder condition of thalassemia intermedia. The mild mutations observed in Asian Indians include two transcription mutations located in the promoter region upstream of the β-globin gene, –88 (C→T) and –28 (A→G) (Thein et al., 1988; Pawar et al., 1997), the CAP+1 (A→C) site mutation, and the Poly A (AAT→AAC) mutation. The latter mutation, together with CAP+1 (A→C), are associated with an Hb A_2 level of 3.4%–4.0%, which is only just above the normal range of 0%–3.5%. These two mutations are examples of the type classed as normal Hb A_2 β-thalassemia mutations, and carriers of these mutations may be misdiagnosed as having β-thalassemia trait.

REGIONAL VARIATION

The relative gene frequencies for the common β-thalassemia mutations have now been determined for different regions of India and Pakistan. A study of β-thalassemia mutations in homozygous patients in six different ethnic groups of Pakistan revealed a total of 22 different β-thalassemia mutations (Ahmed et al., 1996; Khan and Riaduzzin, 1998). The combined list of mutations for the six ethnic groups is presented in Table 90-3.

The four most common mutations were IVSI-5 (G→C), CD8/9 (+G), 619bp deletion and IVSI-1 (G→T), accounting for 79% of the alleles. There were notable differences between the six ethnic groups, and also differences between the four provinces of Pakistan. Notably, IVSI-5 (G→C) was found to be more prevalent in Sindh and Baluchistan (bordering with India in the southeast and Iran in the southwest), while CD8/9 (+G) was more common in Punjab and the Northwest Frontier Province (bordering with India in the northeast and Afghanistan, respectively). The 619bp deletion was the most common mutation in Gujarati and Memon (46%).

Several studies on the regional distribution of β-thalassemia mutations in India have been published (Parikh et al., 1990; Varawalla et al., 1991b; Venkatesan et al., 1992; Dastidar et al., 1994; Verma et al., 1997; De et al., 1997; Madan et al., 1998; Agarwal et al., 2000; Kukreti et al., 2002). The IVSI-5 (G→C) mutation is the most common

TABLE 90-2 β-THALASSEMIA MUTATIONS (%) IN THE INDIAN SUBCONTINENT

Mutation	India *(674 alleles)*	Pakistan *(678 alleles)*	Sri Lanka *(1240 alleles)*	Mauritius *(55 alleles)*	Maldives *(78 alleles)*
-88 (C→T)		0.3			
-87 (C→A)	0.1				
-28 (A→G)				3	
CAP+1 (A→C)	1	1			
Initiation CD (T→C)	0.1	0.3			
CD4/5/6	0.1				
CD5 (-CT)		4			
CD6 (-13bp)			0.1		
CD8/9 (+G)	12	21	2	4	
CD15 (G→A)	0.7	4	1	4	
CD15 (-T)	0.5	0.2	1		
CD16 (-C)	0.5	2.0	1		
CD16 (C→T)	0.1				
CD22/23/24 (-7bp)	0.1				
CD30 (G→A)		0.2			
CD30 (G→C)	1	2	1	4	15
IVSI-1 (G→A)		1	27		9
IVSI-1 (G→T)	23	9			
IVSI-5 (G→C)	30	37	56	83	74
IVSI-129 (A→C)	0.1		0.2		
IVSI-130 (G→C)	0.1		1		
IVSI-130 (G→A)			0.7		
CD36-39 (-8bp)	0.1				
CD39 (C→T)		0.3			
CD41/42 (-TCTT)	7	3	2	3	2
CD44 (-C)		0.2			
CD47/48 (+ATCT)	0.7	0.2			
CD55 (-A)	0.1		0.2		
CD88 (+T)		0.2			
CD110 (T→C)	0.5				
IVSII-1 (G→A)		1			
IVSII-613 (C→T)	0.1				
IVSII-837 (T→G)	0.1				
IVSII-848 (C→A)		0.3			
PolyA (T→C)			0.8		
619bp deletion	20	12	0.2		
7bp del	0.1				
22bp del	0.1				
Unknown	4		7		

TABLE 90-3 PREVALENCE OF THE β-THALASSEMIA MUTATIONS IN INDIA AND PAKISTAN

	IVSI-5 (G→C)	CD8/9 (+G)	619 bp del	IVSI-1 (G→T)	CD41/42 (-TCTT)	CD5 (-CT)	CD15 (G→A)
INDIAN							
Gujarati	41	4	26	17	9	–	0
Punjabi	38	13	15	11	13	–	1
Sindhi	12	11	54	18	4	–	0
Tamil Nadu	81	3	1	0	10	–	0
Northern India	23	13	35	20	10	–	0
Eastern India	55	2	2	0	11	–	7
PAKISTANI							
Gujarati	16	11	46	17	4	0	1
Punjabi	28	38	4	6	6	2	5
Sindhi	42	11	14	14	4	2	4
Balochi	73	7	1	4	1	2	4
Urdu speaking	50	18	6	6	4	2	0
Mohajir	41	12	16	4	6	1	4
Pashtoon	14	50	3	3	6	9	4

mutation in the Gujarati, Punjabi Tamil Nadu, and the eastern region of India, reaching the highest relative frequency of 81% in southern India. In the Sindhi and northern India, the 619bp deletion is the most frequently observed mutation, but in southern India it is extremely uncommon. Another example of the regional variation of mutations is the mutation codon 15 (G→A), which is the third commonest mutation in eastern India at a frequency of 7%, but in all the other regions it is extremely rare or nonexistent.

THE δβ-THALASSEMIAS AND HPFH DISORDERS

δβ-thalassemia and the deletion types of hereditary persistence of fetal hemoglobin (HPFH) are characterized by the complete absence of Hb A and Hb A_2 in homozygotes and an elevated level of Hb F in heterozygotes. Both conditions are caused by large DNA deletions involving the β-globin gene cluster affecting the β and δ-genes, but leaving either one or both of the γ-globin genes intact. More than fifty different deletion mutations have been identified (Hardison et al., 2002), and they can be classified into the $(\delta\beta)^o$ and $(^A\gamma\delta\beta)^o$-thalassemias, HPFH conditions, fusion chain variants Hb Lepore and Hb Kenya, and the $(\varepsilon\gamma\delta\beta)^o$-thalassemias.

$(\delta\beta)^o$-THALASSEMIA AND $(^A\Upsilon\delta\beta)^o$-THALASSEMIA

The $(\delta\beta)^o$-thalassemias are characterized by the presence of Hb F consisting of both $^G\gamma$ and $^A\gamma$-globin chains, as both γ-globin genes remain intact in these conditions. Heterozygotes for $(\delta\beta)^o$-thalassemia have normal levels of Hb A_2 and a raised Hb F level usually in the range of 2%–15%. For the majority of mutations, the Hb F is heterogeneously distributed in the red cells. The red cells are hypochromic and microcytic, indicating that the production of γ-chains is insufficient to fully compensate for the lack of β-globin. Homozygotes for this condition have thalassemia intermedia. One type of $(\delta\beta)^o$-thalassemia has been described in Asian Indians (Mishima et al., 1989). The Indian $(\delta\beta)^o$-thalassemia mutation extends for about 32.6 kb from a point between the $^A\gamma$-gene and the ψβ-gene (Gilman et al., 1992).

The $(^A\gamma\delta\beta)^o$-thalassemias are characterized by the Hb F containing only $^G\gamma$-globin chains, and heterozygotes produce 6%–23% Hb F. Apart from this distinction, the phenotype of the heterozygous and homozygous states are identical to those for $(\delta\beta)^o$-thalassemia. One type of $(^A\gamma\delta\beta)^o$-thalassemia has also been discovered in Asian Indians. The Indian $(^A\gamma\delta\beta)^o$-thalassemia mutation involves a complex deletion/inversion rearrangement of the β-globin gene cluster (Jones et al., 1981). There is a loss 834bp from the $^A\gamma$-globin gene and a deletion of 7.46 kb removing parts of the δ and β-globin genes, combined with an inversion of the sequence between the two deletions (Jennings et al., 1985).

$(\varepsilon\Upsilon\delta\beta)^o$-THALASSEMIA

The $(\varepsilon\gamma\delta\beta)^o$-thalassemias are rare conditions that result from several different long deletions that start upstream of the ε-gene and remove all of the β-globin gene cluster or, in two cases, the deletion ends between the δ and β-genes, thus sparing the β-globin gene—but in both cases, no β-globin synthesis occurs. This is because the deletions remove the β-globin gene cluster locus control region (LCR) located 50 kb upstream of the ε-gene. In adult life, heterozygotes for this condition have a similar hematological picture to β-thalassemia trait with a normal Hb A_2 level. The homozygous condition is presumed to be incompatible with fetal survival. Six different deletions have been characterized, but none have been described in Asian Indians to date.

HEREDITARY PERSISTENCE OF FETAL HEMOGLOBIN (HPFH)

The deletional HPFH conditions can be regarded as a type of δβ-thalassemia in which the reduction in β-globin chain production is almost completely compensated for by the increased γ-globin chain production. Homozygous individuals have 100% F comprised of both $^{A}\gamma$ and $^{G}\gamma$-globin chains but, in contrast to $(\delta\beta)^{o}$-thalassemia, homozygotes are clinically normal. Heterozygotes have an elevated Hb F level of 17%–35%, higher than that found in δβ-thalassemia heterozygotes, and the Hb F is distributed uniformly (pancellular) in red cells with near normal MCH and MCV values. Six different HPFH deletions have been identified by gene mapping studies, one of which is found in Asian Indians. The HPFH-3 deletion, known as Indian HPFH, has been described in more than six Indian families (Kutlar et al., 1984; Wainscoat et al., 1984) and extends for about 48.5kb from a point between the $^{A}\gamma$-gene and the ψβ-gene (Henthorn et al., 1986). The HPFH-6 allele, (the Vietnamese/Chinese 27kb deletion) has also been diagnosed in five cases from India (Nadkarni et al., 2008).

Finally, there is a group of conditions called non-deletion HPFHs, caused by a point mutation in the promoter region of the $^{A}\gamma$ or $^{G}\gamma$-globin genes. Heterozygous individuals have normal red cells and no clinical abnormalities, and an elevated Hb F level, ranging from 2%–40%. At least twelve different mutations have been described in many different ethnic groups, although none have been recorded in Asian Indian individuals as yet.

ABNORMAL HEMOGLOBINS

More than 800 hemoglobin variants have been characterized to date, many of which can cause clinical disorders. These can be divided into variants that alter the oxygen-carrying properties of hemoglobin, unstable variants that produce a hemolytic anemia, a few that are ineffectively synthesized and thus result in the phenotype of thalassemia, and the variants that cause sickling disorders.

Although most are not very common, three (Hb S, Hb C, and Hb E) have reached gene frequencies high enough to cause major public health problems in many countries. Both Hb S and Hb E are found at high gene frequencies in parts of the Indian subcontinent and are discussed in more detail below. Table 90-4 lists all the abnormal hemoglobins that have been characterized in individuals of Asian Indian origin. After Hb S and Hb E, the most frequently observed ones are Hb D-Punjab and Hb Q-India (Abraham et al., 2003; Moiz et al., 2008).

HB S

Hb S is the commonest abnormal hemoglobin in the world, occurring in equatorial Africa, Mediterranean countries, the Middle East, and parts of India. It has reached high gene frequencies in some populations as a result of malarial selection.

Children heterozygous for Hb S suffer from malaria less frequently and less severely than normal individuals, and are consequently more likely to survive and pass on their genes. The loss of Hb S genes through death of homozygotes was more than compensated for by the survival of heterozygotes, and over time the sickle cell trait became very common in malarial areas.

Clinical Phenotypes

The sickling disorders include the heterozygous state, called sickle cell trait (AS), the homozygous condition, called sickle cell disease (SS) and the compound heterozygous states for the Hb S gene in association with β-thalassemia or the β-globin chain variants such as Hb C, Hb D-Punjab and Hb O-Arab. Sickle cell disease is characterized by a lifelong hemolytic anemia, the occurrence of acute exacerbations called crises, and a variety of complications resulting from an increased propensity to infection and the deleterious effects of repeated vaso-occlusive episodes. With active management, the proportion of patients expected to survive to 20 years of age is approximately 90%. The course of the illness is very variable, even within individual sibships, let alone different racial groups.

The clinical course of sickle cell β-thalassemia is also very variable, ranging from a disorder identical with sickle cell anemia to a completely asymptomatic condition. The heterogeneity is mostly due to the type of β-thalassemia mutation that is co-inherited. It tends to be very mild in Africans because of the likelihood of the co-inheritance of one of the mild β^{+} mutations commonly found in this racial group. However, those patients who inherit a β^{o}-thalassemia allele exhibit a clinical disorder very similar to sickle cell anemia.

TABLE 90-4 ABNORMAL HEMOGLOBINS REPORTED IN ASIAN INDIAN INDIVIDUALS

Name	Codon	Amino acid change	Presumed mutation	Phenotype
Hb Karachi	α 5	Ala →Pro	GCC→CCC	Normal
Hb Dunn	α 6	Asp→Asn	GAC→AAC	Increased oxygen affinity
Hb J-Parisl	α 12	Ala→Asp	GCC→GAC	Normal
Hb I	α 16	Lys→Glu	AAG→GAG	Normal
Hb Mexico	α 54	Gln→Glu	CAA→GAA	Normal
Hb G_Waimanalo	α 64	Asp→Asn	GAA→AAC	Normal
Hb Persepolis	α 64	Asp→Tyr	GAC→TAC	Not recorded
Hb Q-India	α 64	Asp→His	GAC→CAC	Normal
Hb Q-Iran	α 75	Asp→His	GAC→CAC	Normal
Hb Luton	α 89	His→Leu	CAC→CTC	Increased oxygen affinity
Hb J-Rajappen	α 90	Lys→Thr	AAG→ACG	Normal
Hb Rampa	α 95	Pro→Ser	CCG→TCC	Increased oxygen affinity
Hb Godavari	α 95	Pro→Thr	CCG→ACG	Increased oxygen affinity
Hb J-Meerut	α 120	Ala→Glu	GCG→GAG	Normal
Hb Sun Prairie	α 130	Ala→Pro	GCT→CCT	Unstable
Hb Natal	α 140/1	Tyr +Arg	TAC→TAA	Increased oxygen affinity
Hb Constant Spring	α 142	Term→Gln	TAA→CAA	Unstable
Hb Koya Dora	α 142	Term→Ser	TAA→TCA	Unstable
Hb S	β 6	Glu→Val	GAG→GTG	Sickling
Hb D-Agri	β 9 β 121	Ser→Tyr Glu→Gln	TCT→TAT GAA→CAA	Normal
Hb D-Iran	β 22	Glu→Gln	GAA→CAA	Normal
Hb E	β 26	Glu→Lys	GAG→AAG	Unstable
Hb Leeds	β 56	Gly→Cys	GGC→TGC	Normal
Hb M-Ratnagiri	β 63	His→Tyr	CAT→TAT	Cyanosis
Hb Headington	β 72	Ser→Arg	AGT→AGA	Increased oxygen affinity
Hb Chandigarh	β 94	Asp→Gly	GAC→GGC	Not recorded
Hb Showa-Yakushiji	β 110	Leu→Pro	CTG→CCG	β-thalassemic
Hb Harrow	β 118	Phe→Cys	TTT→TGT	Normal
Hb D-Punjab	β 121	Glu→Gln	GAA→CAA	Normal
Hb A_2-Canada	δ 99	Asp→Gly	GAT→AAT	Increased oxygen affinity
Unnamed	δ 116	Arg→Leu	CGC→CTC	Not recorded
Hb F-Kuala Lumpur	Aγ 22	Asp→Gly	GAT→GGT	Not recorded

(Continued)

TABLE 90-4 (CONTINUED)

Name	Codon	Amino acid change	Presumed mutation	Phenotype
Hb Lepore	δβ	Hybrid	Hollandia,	β-thalassemic
Hb Lepore	δβ	Hybrid	Washington-Boston	β-thalassemic
Hb P-India	βδ	Hybrid	β to 87; δ from 116	Not recorded

The interaction of the Hb S gene with hemoglobin variants is also variable. The interaction with Hb C (hemoglobin SC disease) is normally mild, though it may be associated with ocular, CNS and bone complications, and thrombotic problems in pregnancy. The interactions with Hb D-Punjab and Hb O-Arab are more severe, resulting in both cases in a disease indistinguishable from homozygous sickle cell anemia.

Genotype

The sickle cell gene mutation is a single base change (A→T) at codon 6 of the β-globin gene, changing the nucleotide triplet of the genetic code for valine (GAG) to glutamic acid (GTG). The mutation is thought to have arisen independently at least four times in Africa and once in Asia because of its association with five completely different β-globin gene haplotypes (Kulozik et al., 1986). Such studies have shown that the sickle gene found in Mediterranean individuals is of African origin—the Benin haplotype. The haplotypes have been assigned the name of the geographic area where they are most frequently found. The other haplotypes most frequently found are the Central African Republic (the CAR, or also termed the Bantu haplotype), Senegal, and Cameroon. The fifth haplotype is the Arab-India haplotype, found in both Saudi Arabia and India. Because of the great variability in the phenotypic expression of the disease, it is not possible to predict accurately how severe the disease will be in any one individual. However, epidemiological studies have shown that the CAR haplotype is associated with the most clinically severe condition (Powars, 1991).

The haplotype associated with the mildest course of the disease is the Arab-India haplotype. SS patients with this haplotype exhibit high levels of Hb F, because this particular β^S mutation is linked to the –158 C→T mutation 5' to the ${}^G\gamma$-globin gene (the *Xmn*1 polymorphism), which results in enhanced γ-chain production in response to anemic stress. The other factor known to modify the disease is the co-inheritance of α-thalassemia (Higgs et al., 1982). In individuals of Asian Indian origin this is always the α^+ type. A study of Hb S heterozygotes in a tribal group and a nontribal group of western India revealed a mean Hb S expression of 27.9% in the tribal group compared to 35.5% in the nontribal group, with the lower Hb S expression caused by the co-inheritance of α^+-thalassemia. Hb SS patients in both groups had the same amount of Hb F (15%) and the same haplotype (the Arab-Indian), but the tribal Hb SS patients appeared to have a mild clinical phenotype whereas the nontribals had a more severe phenotype (Mukherjee et al., 1997). Hb SS patients homozygous for α^+-thalassemia have reduced levels of hemolysis and a higher hemoglobin level compared to Hb SS patients with a normal α-genotype.

Hb S was first described in India in the Nilgiri district of Madras, observed at a frequency of up to 20% (Lehman and Cutbush, 1952). It has since been reported in many other areas of India, including Orissa (up to 25%), Madhya Pradesh (up to 20%), Andhra Pradesh (up to 17%), Maharashtra (up to 10%), Gujarat (up to 30%), and Assam (Livingstone, 1985), both in tribal groups as well as scheduled Hindu castes. Hb S is unevenly distributed across India, forming a patchwork distribution. Haplotype data collected on three widely separated populations showed that 91% of the β^S alleles examined had the same haplotype, the Arab-Indian haplotype (Kulozik et al., 1986; Labie et al., 1989).

The finding of the same haplotype in tribal populations living in distant regions strongly suggests that the mutation arose once in a single population at a time when these tribal populations were in close contact. The strong association of the sickle mutation and one haplotype points to a recent origin of the mutation. This is because it is unlikely that the same mutation arose independently in different groups on the same haplotype, and the linkage of mutation to haplotype has not been broken to any extent by recombination. Anthropological and archaeological evidence has suggested that the Indian carriers of the sickle gene are descendants of the Harappa culture, which flourished in the region of the Indus river (Kennedy, 1981). Between 2000 and 4000 years ago this group dispersed after invasions from the north by peoples who form the majority of present-day India's population. Thus, it appears the sickle mutation arose in one location in India about 5000 years ago, spreading to other regions of India in populations which then became isolated from each other by mainstream Indians that do not carry the β^S-globin gene.

Hb S is also found at a significant frequency in individuals of Mauritius (Kotea et al., 1995). Studies showed that all Hb S carriers of Indian origin had the Arab-Indian haplotype. Thus, the haplotype data indicates that Indians of tribal origin were among those who migrated to the Mauritius islands. At much lower gene frequencies, Hb S has also been observed in Sri Lanka and the Maldives.

HB E

Hb E is the second commonest abnormal hemoglobin in the world. It is found in Asia, from Assam in northern India through to China, and is most concentrated at the border of Thailand, Laos, and Cambodia. For Hb E (and also Hb C) there is not such a good case that the high gene frequencies found in some populations are due to malarial selection. Although there is a good geographical correlation of Hb E and malaria-infected regions, the homozygous state is a mild disorder and is not lethal, as is the case with Hb C, and it is not clear how heterozygosity for Hb E or Hb C confers resistance against malaria.

Clinical Phenotypes

The heterozygous and homozygous states for Hb E are associated with no clinical disability. The importance of Hb E lies in its interaction with β-thalassemia. Compound heterozygotes have a variable clinical picture, ranging from a disorder indistinguishable from β-thalassemia major to a mild form of β-thalassemia intermedia. Again, compound heterozygotes with a β^{o}-thalassemia allele exhibit the most severe disorder, while those with a β^{+} allele may sometimes have a milder disorder, although the co-inheritance of α-thalassemia also plays a significant role as an ameliorating factor in Hb E/β-thalassemia (Verma et al., 2007).

Genotype

Hb E is caused by a single base change (G→A) at codon 26 of the β-globin gene. This point mutation activates a cryptic splice site located between codons 24 and 27, resulting in two forms of β-globin mRNA. The normally spliced mRNA containing the β^{E} mutation is produced at a low level compared to the abnormally spliced mRNA, and thus there is a deficiency of β^{E}-globin because the abnormally spliced mRNA does not produce a recognizable β-globin, resulting in a mild β-thalassemia phenotype.

Hb E is most commonly observed in the eastern regions of the Indian subcontinent, where a high preponderance of Hb E/β-thalassemia is observed. For example, in Bengalis, the carrier frequency of Hb E approaches 3.5%, similar to that for β-thalassemia (3.7%; see Ajmani et al., 1976). Studies on patients with Hb E/β-thalassemia from West Bengal and Uttar Pradesh have shown that the majority have the genotype β^{E}/IVSI-5 (G→C) (Agarwal et al., 1997; De et al., 1997). Further to the north, in Assam, Hb E reaches much higher gene frequencies. Hb E has been reported to occur at carrier frequencies of up to 50% in the Karachi people in Assam (Hundrieser et al., 1988).

Hb E has also been found in Sri Lanka, where a recent study of patients affected with thalassemia has shown that 40% have Hb E/β-thalassemia (de Silva et al., 2000). A survey of schoolchildren from different parts of Sri Lanka by these authors revealed a gene frequency of Hb E of 1%, compared to 2% for β-thalassemia. Similarly, in Bangladesh both Hb E and β-thalassemia are quite common, resulting in many cases of Hb E/β-thalassemia (Weatherall and Clegg, 2001).

FUSION HEMOGLOBINS

Fusion chain variants are abnormal hybrid hemoglobins resulting from unequal crossing over between different globin genes. Hb Lepore is a hybrid globin chain comprised of δ and β gene sequences, and Hb Kenya is comprised of γ and β gene sequences. The former results in a β-thalassemia phenotype with little or no increase in Hb F, while the latter has an HPFH phenotype. Hb Lepore homozygotes and compound heterozygotes with β-thalassemia have a variable phenotype, ranging from to thalassemia major to thalassemia intermedia.

Hb Kenya has only been observed in the heterozygous state, and is similar to heterozygous HPFH, with individuals having 5%–10% Hb F, normal red cell morphology, and balanced globin chain synthesis. Most cases of Hb Kenya appear to originate from east Africa, while Hb Lepore is more widespread, occurring in individuals from Mediterranean countries through to Southeast Asia. A study of the spectrum of thalassemia major and intermedia patients in the Indian population has revealed both Hb Lepore-Hollandia and Hb Lepore-Washington (Shaji et al., 2003). Hb P-India, the antigene of the Hb Lepore-Boston-Washington type, consisting of β chains up to codon 87 and δ chains from codon 116, has also been found in an Indian individual (Prehu et al., 1994).

OTHER ABNORMAL HEMOGLOBINS

More than 30 other abnormal hemoglobins have been described in individuals of Indian and Pakistani origin. These include 18 alpha chain variants, 14 beta chain variants, two delta chain variants, and one gamma chain variant, as described in the Hb Var database (http://globin.cse.psu.edu; Patrinos et al., 2004). Most of these variants are very rare and are associated with no significant clinical manifestations. The exceptions are a Hb M-Ratnagiri, a rare Hb M variant which causes congenital methemoglobinemia (Kedar et al., 2005), and Hb Showa-Yakushiji, a hyper-unstable β-chain variant which is associated with a dominant β-thalassemia phenotype (Edison et al., 2005).

DIAGNOSTIC METHODS

There are many different PCR-based techniques that can be used to diagnose the globin gene mutations, including dot blot analysis, reverse dot blot analysis, the amplification refractory mutation system (ARMS), denaturing gradient gel electrophoresis, gap PCR, and restriction endonuclease analysis. Each method has its advantages and disadvantages, and the particular one chosen by a laboratory to diagnose point mutations depends not only on the technical expertise available in the diagnostic laboratory, but also on the type and variety of the mutations likely to be encountered in the individuals being screened.

DIAGNOSTIC APPROACHES

The DNA analysis approach used depends on which disorder or combination of disorders is targeted (α-thalassemia, β-thalassemia, δβ-thalassemia/HPFH, Hb variants) by the results of hematological screening tests (Old, 2007). Different molecular diagnostic techniques are used for each type of disorder, but the approach used is usually the same, i.e., to screen for the mutations known to be common in the local population and then, if required, to screen again for rarer known mutations or to switch to a method that will identify unknown mutations. The ones used in my laboratory for known point mutations are ARMS-PCR and restriction endonuclease analysis of amplified product (RE-PCR) and gap-PCR. Direct DNA sequencing is used for rare/unknown point mutations, and multiplex ligation-dependent probe amplification (MLPA) for rare/unknown deletion mutations.

α-THALASSEMIA

Gap-PCR provides a quick diagnostic test for the common α^+-thalassemia and α^o-thalassemia deletion mutations but requires careful application for prenatal diagnosis. Primer sequences have been published for the multiplex gap-PCR diagnosis of five α^o-thalassemia deletions and two α^+-thalassemia deletions (Liu et al., 1999). The α^o-thalassemia deletions diagnosable by PCR are: the $—^{SEA}$ allele, found in Southeast Asian individuals; the $—^{MED}$ and $-(\alpha)^{20.5}$ alleles found in Mediterranean individuals; the $—^{FIL}$ allele, found in Filipino individuals; and finally the $—^{THAI}$ allele, found in Thai individuals. The two α^+-thalassemia deletion mutations are 3.7 kb and the 4.2 kb single α-gene deletion mutations, designated $-\alpha^{3.7}$ and $-\alpha^{4..2}$.

Amplification of sequences in the α-globin gene cluster is technically more difficult than that of the β-globin gene cluster, requiring more stringent conditions for success due to the higher GC content of the breakpoint sequences and the considerable sequence homology within the α-globin gene cluster, resulting occasionally in unpredictable reaction failure and the problem of allele dropout. Hence, great care must be taken when using gap-PCR for prenatal diagnosis of homozygous α^o-thalassemia, and confirmation by MLPA is recommended.

MLPA is used for the diagnosis of rare or unknown α^o and α^+-thalassemia mutations, which cannot be diagnosed by gap-PCR because their breakpoint sequences have not been determined (Harteveld et al., 2005). The method does not provide a definitive diagnosis for any particular deletion mutation, as it simply detects the deletion of DNA sequence between two probe locations, and thus only a presumed identification of a deletion mutation can be made.

α^+-thalassemia is also caused by more than 40 point mutations in either of the two α-globin genes. These non-deletion alleles can be detected by selective amplification of each α-globin gene followed by DNA sequence analysis, or, for a few specific known mutations such as the unstable α-globin chain variant Hb Constant Spring, by the use of ARMS-PCR or RE-PCR (Old, 2010).

β-THALASSEMIA

The majority of the defects are single nucleotide substitutions, small insertions, or deletions, and the strategy depends on knowing the spectrum of common and rare mutations in the ethnic group of the individual being screened. The common ones are analyzed first, using a PCR method designed to detect specific mutations simultaneously. This approach will identify the mutation in more than 80% of cases for most ethnic groups. A further screening of the known rare mutations will identify the defect in another 10%–15% of cases if necessary. Mutations remaining unidentified at this stage are then characterized by DNA sequencing.

A widely used technique for detecting common known Indian and Pakistani mutations is the amplification refractory mutation system (ARMS). This technique fits the main requirements of a PCR technology, i.e., speed, cost, convenience, and the ability to test for multiple mutations simultaneously, as it provides a quick cheap screening method without any form of labeling of primers or amplified DNA, and can be multiplexed (Old et al., 2001; Baig, 2007). Other approaches used in developing countries are denaturing gradient gel electrophoresis (DGGE) and allele-specific oligonucleotide hybridization by reverse dot blotting (RDB). For example, some laboratories in India have reported results of comprehensive mutation screening surveys using a combination of ARMS-PCR and RDB to detect the common mutations, followed by DNA sequencing wherever necessary (Bashyam et al., 2004; Baig et al., 2006; Colah et al., 2009).

δβ-THALASSEMIA AND HPFH

A number of these deletion mutations have had their breakpoint sequences characterized and are easily diagnosed using gap-PCR, which produces a characteristic breakpoint product to identify the deletion mutation. A recent study of δβ-thalassemia and HPFH by gap-PCR in the Indian population characterized three different alleles, although in nine cases (16%) the deletion allele remained uncharacterized (Nadkarni et al., 2008), showing that other types of δβ-thalassemia and HPFH remain to be identified in the Indian population. The recently developed technique of MLPA can now be used to screen for all types of δβ-thalassemia or HPFH deletion alleles (Gallienne et al, 2010). However, the technique does not give a definitive identification of the allele; it only signals the presence of a deletion allele and its approximate deletion size.

HB VARIANTS

More than 700 hemoglobin variants have been described to date, most of which were identified by protein analysis and have never been characterized at the DNA level. Hb variants are usually first observed in a screening program using HPLC or gel electrophoresis methods, and only a presumptive diagnosis can be made. A definitive diagnosis requires DNA analysis or mass spectroscopy to reveal the underlying DNA sequence or amino acid change. For the vast majority of variants, positive identification at the DNA level is achieved by selective globin gene amplification and DNA sequence analysis. Once a variant has been characterized by DNA sequencing, more specific techniques such as ARMS-PCR can be developed for its detection, as shown for Hb Q-India (Abraham et al., 2003).

For all the clinically important variants—Hb S, Hb C, Hb E, Hb D-Punjab and Hb O-Arab—simple DNA analysis techniques such as dot blotting or ARMS-PCR can be used. All except Hb C can also be diagnosed by restriction enzyme digestion of amplified product (RE-PCR), providing a second method confirmation of a prenatal diagnosis. The sickle cell gene mutation abolishes a *Dde*I recognition site at codon 6, and diagnosis by *Dde*I digestion of amplified product remains the simplest method of DNA analysis for sickle cell disease. Similarly, the mutations giving rise to Hb D-Punjab and Hb O-Arab abolish an *Eco*RI site at codon 121, and the Hb E mutation abolishes a *Mnl*I site at codon 26 (Old, 2010).

SUMMARY

The hemoglobinopathies are the most common clinically serious single gene disorders in the world. For example, in India, the average incidence of β-thalassemia trait is 3.3%, with 1–2 per 1000 couples being at risk of having an affected child, and approximately 6000 children with thalassemia major are born annually. The disorder represents a major public health problem for the treatment of affected individuals, and prevention programs for β-thalassemia by genetic counseling and prenatal diagnosis by DNA analysis have been established.

At least 27 different mutations have been identified to cause β-thalassemia on the Indian subcontinent. The regional distribution and gene frequencies of the mutations have now been extensively studied. α^+-thalassemia has also been found to occur extensively throughout the Indian subcontinent at high gene frequencies. The most common alleles are the 3.7 kb and 4.2 kb single alpha gene deletion mutations. These have been reported in to occur at a frequency of 10%–15% in the general population. However, in some of the Indian tribal populations, much higher gene frequencies have been found by DNA studies.

Also observed on the Indian subcontinent are regionally specific mutations causing δβ-thalassemia, and hereditary persistence of fetal hemoglobin (HPFH). A number of abnormal hemoglobins have also been characterized, most occurring at very low gene frequencies, and have no clinical significance. However, two clinically significant abnormal hemoglobins are also found—Hb S in many tribal populations, and Hb E in the eastern parts of the Indian subcontinent. Hb S results in sickle cell anemia in the homozygous state, and also interacts with Hb D-Punjab and β-thalassemia trait to produce a severe form of sickle cell disease. Hb E interacts with β-thalassemia trait to produce a severe β-thalassemia disorder. All the different β-thalassemia, α-thalassemia, and abnormal hemoglobins that have been characterized in the Asian Indian population are discussed or summarized in this chapter.

REFERENCES

Abraham R, Thomas M, Britt R, Fisher C, Old J. (2003). Hb Q India: an uncommon variant diagnosed in three Punjabi patients with diabetes is identified by a novel DNA test. *J Clin Pathol 56*, 0–3.

Agarwal S, Gulati R, Singh K. (1997). Hemoglobin E-beta thalassaemia in Uttar Pradesh. *Indian Pediatr 34*, 287–292.

Agarwal S, Pradhab M, Gupta UR, Sarwai S, Agarwal SS. (2000). Geographic and ethnic distribution of beta thalassaemia mutations in Uttar Pradesh, India. *Hemoglobin 24*, 89–97.

Agarwal S, Tamhanker PM, Kumar R, Dalal A. (2010). Clinical and haematological features in a compound heterozygote (HBB:c.92+5G>C/HBB:c.93-2A>C) case of thalassaemia major. *Int J Lab Hematol 32*, 369–372.

Ahmed S, Petrou M, Saleem M. (1996). Characterisation of beta-thalassaemia mutations in Pakistan; A basis for prenatal diagnosis. *British Journal of Haematology 57*, 476–482.

Ahmed S, Saleem M, Sultana N, et al. (2000). Prenatal diagnosis of beta-thalassaemia in Pakistan: experience in a Muslim country. *Prenatal Diagnosis 20*, 378–383.

Ajmani M, Sharma A, Talukder G, Bhattacharya DK. (1976). Beta thalassaemia trait in West Bengal–a methodological study. *Curr Sci 45*, 476.

Baig SM, Azhar A, Hassan H, et al. (2006). Spectrum of beta-thalassemia mutations in various regions of Punjab and Islamabad, Pakistan: Establishment of prenatal diagnosis. *Hematologica 91*, ELT02.

Baig SM. (2007). Molecular diagnosis of beta-thalassemia by multiplex ARMS-PCR: A cost effective method for developing countries like Pakistan. *Prenat Diagn 27*, 580–581.

Bandyopadhyay A, Bandyopadhyay S, Dasgupta UB, Chandra S, Da MK. (2001). Detection of rare splice acceptor site mutation (IVSI nt 130 G→C) of the beta globin gene in 3 patients of Eastern India. *Am J Haematol 67*, 149.

Bashyam MD, Bashyam L, Savithri GR, Gopikrishna M, Sangal V, Devi AR. (2004). Molecular genetic analysis of beta-thalassemia in South India reveals rare mutations in the beta globin gene. *J Hum Genet 49*(8), 408–413.

Baysal E, Sharma S, Wong SC, Jogessar VB, Huisman THJ. (1994). Distribution of β-thalassemia mutations in three Asian Indian populations with distant geographical locations. *Hemoglobin 18*, 201–209.

Bernini LF, Harteveld CL. (1998). β-Thalassaemia. In: Rodgers GP, ed. *Clinical Haematology: Sickle Cell Disease and Thalassaemia*, pp. 53–90. London: Bailliere Tindall.

Brittenham G, Lozoff B, Harris JW, Kan YW, Dozy AM, Nayudu NVS. (1980). Alpha globin gene number: Population and restriction enzyme studies. *Blood 55*, 706–708.

Brittenham GM. (1981). Hemoglobinopathies in India. *Hemoglobin 5*, 751–766.

Colah R, Gorakshakar A, Nadkarni A, et al. (2009). regional heterogeneity of beta-thalassemia mutations in the multiethnic Indian population *Blood Cells Mol Dis* 42: 241–246.

Cürük MA, Baysal E, Gupta RB, Huisman THJ. (1993). An IVS-I-117 (G→A) acceptor splice site mutation in the β_1-globin gene is a non-deletional β-thalassaemia-2 determinant in an Indian population. *British Journal of Haematology 85*, 148–152.

Dash S, Harano K, Menon S. Hb Sallanches [alpha104(G11)Cys→Tyr, TGC→TAC (alpha2)]: An unstable haemoglobin variant found in an Indian child. (2006). *Hemoglobin 30*, 393–396.

Dastidar DG, Dutta RN, Gupta P, Old J. (1994). detection of beta thalassaemia mutations in eastern Indian population by polymerase chain reaction. *Indian J Med Res 100*, 111–114.

De M, Das SK, Bhattacharya DK, Talukder G. (1997). The occurrence of beta thalassaemia mutations and its interaction with hemoglobin E in the eastern India. *Int J Hematol 66*, 31–37.

De Jong WW, Khan PM, Bernini LF. (1975). Hemoglobin Koya Dora: High frequency of a chain termination mutant. *Am J Hum Gen 27*, 81–90.

de Silva S, Fisher CA, Premawardhena A, et al. (2000). Thalassaemia in Sri Lanka: Implications for the future health burden of Asian populations. Sri Lankan Thalassaemia Study Group *Lancet 355*, 786–791.

Desai S, Cohal R, Gupte S, Mohanty D. (1997). Is cellulose acetate electrophoresis a suitable technique for detection of Hb Bart's at birth? *Hum Hered 47*, 181–184.

Drysdale HC and Higgs DR. (1988). α-thalassaemia in an Asian Indian *Br J Haematol 68*, 264–264.

Edison ES, Shaji RV, Devi SG, Kumar SS, Srivastava A, Chandy M. (2005). Hb Showa-Yakushiji [beta110(G12)Leu→Pro] in four unrelated patients from west Bengal. *Hemoglobin* 29; 19–25.

Embury SH, Miller JA, Dozy AM, Kan YW, Chan V, Todd D. (1980). Two different molecular organizations account for the single α-globin gene of the α-thalassaemia-2 genotype. *Journal of Clinical Investigation 66*, 1319–1325.

Fisher CA, Premawardhena A, de Silva S, et al. (2003). The molecular basis for the thalassaemias in Sri Lanka. *Br J Haematol 121*, 662–671.

Fodde R, Losekoot M, van den Broek MN, et al. (1991). Prevalence and molecular heterogeneity of alpha thalassaemia in two tribal populations from Andhra Pradesh, India. *Hum Genet 80*, 157–160.

Furuumi H, Firdous N, Inoue T, et al. (1998). Molecular basis of beta-thalassaemia in the Maldives. *Hemoglobin 22*, 141–151.

Gallienne A, Dreau H, McCarthy J, et al. (2009). Multiplex ligation-dependent probe amplification identification of 17 diffrerent β-globin deletions (including 4 novel mutations) in the UK population by multiplex ligation-dependent probe amplification *Hemoglobin 33*, 406–416.

Gilman JG, Brinson EC, Mishima N. (1992). The 32.6 kb Indian δβ-thlassaemia deletion ends in a 3.4 kb L1 element downstream of the β-globin gene. *Br J Haematol 82*, 417–421.

Gorakshakar AC, Lulla CP, Nadkarni AH, et al. (1997). Prenatal diagnosis of beta-thalassaemia among Indians using denaturing gradient gel electrophoresis. *Hemoglobin 21*, 421–435.

Gupta RB, Tiwary RS, Pande Pl, et al. (1991). Hemoglobinopathies among the Gond tribal groups of Central India; interaction of α- and β-thalassaemia with β chain variants. *Hemoglobin 15*, 441–458.

Gupta A, Hattori Y, Agarwal S. (2002). Initiation mutation in an Asian Indian family. *Am J Haematol 71*, 134–136.

Gupta A, Hattori Y, Gupta UR, et al. (2003). Molecular genetic testing of beta-thalassemia patients of Indian origin and a novel 8-bp deletion mutation at codons 36/37/38/39. *Genetic Test 7*, 163–168.

Hardison RC, Chui DHK, Giardine B, et al. (2002). HbVar: a relational database of human hemoglobin variants and thalassemia mutations at the globin gene server. *Human Mutation 19*, 225–233.

Harkness M, Harkness DR, Kutlar F, et al. (1990). Hb Sun Prairie or α-2130(H13)Ala→Pro-β-2: A new unstable variant occurring in low quantities. *Hemoglobin 14*, 479–490.

Harteveld CL, Losekoot M, Haak H, Heister GA, Giordano PC, Bernini LF. (1994). A novel polyadenylation signal mutation in the α2-globin gene causing α thalassaemia. *British Journal of Haematology 87*, 139–143.

Harteveld CL, van Delft P, Wijermans PW, et al. (2003). A novel 7.9 kb deletion causing α^+-thalassaemia in two independent families of Indian origin *Br J Haemat 120*, 364–366.

Harteveld Cl, Voskamp A, Phylipsen M et al. (2005). Nine unknown rearrangements in 16p13.3 and 11p15.4 causing alpha- and beta-thalassaemia characterised by high resolution multiplex ligation-dependent probe amplification. *J Med Genet 42*, 922–931.

Hassal OW, Tillyer ML, Old JM. (1998). Prevalence and molecular basis of alpha thalassaemia in British South Asians. *J Med Screen 5*, 31–33.

Henthorn PS, Mager DK, Huisman THJ, Smithies O. (1986). A gene deletion ending within a complex array of repeated sequences 3' to the human beta globin gene cluster. *Proc Natl Acad Sci USA 83*, 5194–5198.

Henderson S, Timbs A, McCarthy J, et al. (2009). Incidence of haemoglobinopathies in various populations – The impact of immigration. *Clinical Biochemistry 42*, 1745–1756.

Higgs DR, Aldridge BE, Lamb J, et al. (1982). The interaction of alpha-thalassemia and homozygous sickle-cell disease. *New England Journal of Medicine 306*, 1441–1446.

Huisman THJ, Carver MFH, Baysal E, eds. (1997). *A Syallabus of Thalassemia Mutations*. Augusta, GA: The Sickle Cell Anemia Foundation.

Hundrieser J, Deka R, Gogol BC. (1998). B-globin gene associated DNA haplotypes and frameworks in the Karachi population of Assam (India). *Human Heredity 38*, 240–245.

Jennings MW, Jones RW, Wood WG, Weatherall DJ. (1985). Analysis of an inversion within the human beta globin gene cluster. *Nucl Acids Res 13*, 2897–2906.

Jones RW, Old JM, Trent RJ, Clegg JB, Weatherall DJ. (1981). Major rearrangement in the human beta-globin gene cluster. *Nature 291*, 39–44.

Kedar PS, Nadkarni AH, Phanasgoankar S, et al. (2005). Congenital methaemoglobinaemia caused by Hb-MRatnagiri (beta-63CAT→TAT, His→Tyr) in an Indian family *Am J Haematol 79*, 168–170.

Kennedy KAR. (1981). Skeletal biology: When bones tell tales. *Archeology 34*, 17–24.

Khan SN, Riazuddin S. (1998). Molecular characterisation of β-thalassaemia in Pakistan. *Hemoglobin 22*, 333–345.

Khan SN, Butt FI, Riazuddin S, Galanello R. (2000). Hb Sallanches [alpha104(G11)Cys→Tyr]: A rare alpha 2-globin chain variant found in the homozygous state in three members of a Pakistani family. *Hemoglobin 24*, 31–35.

Khan SN, Riazuddin S, Galanello R. (2000). Identification of three rare β-thalassaemia mutations in the Pakistan population. *Hemoglobin 24*, 15–22.

Kotea N, Baligadoo S, Surran SK, et al. (1995). Bicentric origin of sickle hemoglobin among the inhabitants of Mauritius Island. *Blood 86*, 407–408.

Kotea N, Ramsawawmy R, Lu CY, et al. (2000). Spectrum of beta thalassaemia mutations and their linkage to beta- globin gene haplotypes in the Indo-Mauritians. *Am J Hematol 63*, 11–15.

Kukreti R, Dash D, Vineetha KE, et al. (2002). Spectrum of beta thalassaemia mutations and their association with allele sequence polymorphisms at the beta globin gene cluster in an eastern Indian population. *Am J Haematol 70*, 269–377.

Kulozik AE, Wainscoat JS, Serjeant GR, et al. (1986). Geographical survey of βs-globin gene haplotypes: Evidence for an independent Asian origin of the sickle-cell mutation. *American Journal of Human Genetics 39*, 239–244.

Kulozik AE, Kar BC, Serjeant BE, Weatherall DJ. (1988). The molecular basis of α-thalassaemia in India; its interaction with the sickle gene. *Blood 71*, 467–472.

Kutlar A, Gardiner MB, Headlee MG, et al. (1984). Heterogeneity in the molecular basis of three types of hereditary persistence of fetal hemoglobin and the relative synthesis of the $^{G}\gamma$ and $^{A}\gamma$ types of chains. *Biochem Genet 22*, 21- 35.

Labie D, Srinivas R, Dunda O, et al. (1989). Haplotypes in tribal Indians bearing the sickle gene: evidence for the unicentric origin of the βs mutation and the unicentric origin of the tribal populations of India. *Hum Biol 61*, 479–491.

Lehmann H, Cutbush M. (1952). Sickle-cell trait in Southern India. *Br Med J 1(4755)*, 404–405.

Livingstone FB. (1985). *Frequencies of hemoglobin variants*. New York: Oxford University Press.

Livingstone FB. (1989). Who gave who hemoglobin S: The use of restriction site haplotype variation for the interpretation of the evolution of the β^{S}-globin gene. *Am J Hum Biol 1*, 289–302.

Liu YT, Old JM, Fisher CA, Weatherall DJ, Clegg JB. (1999). Rapid detection of a-thalassaemia deletions and a-globin gene triplication by multiplex PCRs. *Br J Haematol 108*, 295–299.

Madan N, Sharma S, Rusia U, Sen S, Sood SK. (1998). Beta thalassaemia mutations in northern India (Delhi). *Indian J Med Res 107*, 134–141.

Mishima N, Landman H, Huisman THJ, Gilman JG. (1989). The DNA deletion in an Indian δβ-thalassaemia begins one kilobase from the $^{A}\gamma$-globin gene and ends in a L1 repetitive sequence. *Br J Haematol 73*, 375–379.

Modell BC, Petrou M. (1983). The problem of haemoglobinopathies in India. *Indian Journal of Haematology 1*, 5–16.

Moiz B, Moatter T, Hashmi MR, et al. (2008). Identification of haemoglobin Q India (alpha 1-64 Aap-His) through ARMS-PCR. First report from Pakistan. *Ann Hematol 87*, 385–389.

Modiano G, Morpurgo G, Terranato L, et al. (1991). Protection against malaria morbidity: near fixation of the α-thalassaemia gene in a Nepalese population. *Am J Hum Genet 48*, 390–397.

Mukherjee MB, Lu CY, Ducrocq R, et al. (1997). Effect of alpha-thalassaemia on sickle-cell anemia linked to the Arab-Indian haplotype in India *Am J Haematol 55*, 104–109.

Nadkarni A, Sakaguchi T, Takaku H, et al. (2002). A novel beta0-thalassaemia mutation at codon 55 (-A) and a rare 17 bp deletion at codons 126-131 in the Indian population. *Hemoglobin 26*, 41–47.

Nadkarni A, Wadia M, Gorakshakar A, Kiyama R, Colah RB, Mohanty D. (2008). Molecular characterisation of delta beta-thalassemia and hereditary persistence of fetal hemoglobin in the Indian population. *Hemoglobin 32*, 425–433.

Nadkarni A, Gorakshakar A, Surve R, et al. (2009). Haematological and molecular analysis of novel and rare beta-thalassemia mutations in the Indian population. *Hemoglobin 33*, 59–65.

Old JM, Varawalla NY, Weatherall DJ. (1990). Rapid detection and prenatal diagnosis of β-thalassaemia: Studies in the Indian and Cypriot populations in the UK. *Lancet 336*, 834–837.

Old J. (1996). Haemoglobinopathies. *Prenat Diag 16*, 1181–1186.

Old J, Khan S, Verma I, et al. (2001). A multi-centre study in order to define further the molecular basis of β-thalassemia in Thailand, Pakistan, Sri Lanka, Mauritius, Syria, and India, and to development a simple molecular diagnostic strategy by ARMS-PCR. *Hemoglobin 25*, 397–407.

Old JM. (2007). Screening and genetic diagnosis of haemoglobinopathies. *Scand J Clin Lab Invest 67*, 71–86.

Old JM. (2010). Prenatal Diagnosis of the Hemoglobinopathies. In: Milunksy A, Milunsky J, eds. *Genetic Disorders and the Fetus*, 6th edition, pp. 646–679. Oxford: Blackwell Publishing.

Parikh P, Cotton M, Boehm C, Kazazian HH Jr. (1990). Ethnic distribution of beta-thalassaemia in the Indian subcontinent. *Lancet 336*, 1006–1006.

Patrinos GP, Giardine B, Riemer C, et al. (2004). Improvements in the HbVar database of human hemoglobin variants and thalassemia mutations for population and sequence variation studies. *Nucleic Acids Res 32*(Database issue), D537-D541.

Pawar AR, Colah RB, Mohanty D. (1997). A novel beta+-thalassemia mutation (codon 10 GCC —> GCA) and a rare transcriptional mutation (-28A —> G) in Indians. *Blood 89*, 3888–3889.

Petrou, M, Modell B. (1995). Prenatal screening for haemoglobin disorders. *Prenatal Diagnosis 15*, 1275–1295.

Powars DR. (1991). β_s-gene cluster haplotypes in sickle cell anemia. *Hematology/Oncology Clinics of North America 5*, 475–493.

Prehu MO, Prehu C, Goossens M, Galacteros E, Wajcman H. (1994). A new anti-Lepore hemoglobin Hb P India (β87-δ105), found in coincidence with a C→G substitution at position 162 of IVS2 in both the δ and δβ genes, questions on the genetic mechanisms leading to Hbs Lepore and anti-Lepore. *Blood 83*, 261a.

Reddy PH, Petrou M, Reddy PA, Tiurory RS, Modell B. (1995). Hereditary anaemias and iron deficiency in a tribal population (the Baiga) of central India. *Eur J Haematol 55*, 103–109.

Saxena R, Jain PK, Thomas E, Verma IC. (1998a). Prenatal diagnosis of beta-thalassaemia: experience in a developing country. *Prenatal Diagnosis 18*, 1–7.

Saxena R, Thomas E, Verma IC. (1998b). Detection of beta-thalassaemia mutation insertion ATCT at codon 47/48 by ARMS technique for screening and prenatal diagnosis. *Ind J Med Res 105*, 275–277.

Saxena R, Moi L, Demurtas M, Rosatelli MC, Cao A, Verma IC. (2000). A beta-thalassaemia allele with 3 base substitutions in codons 4/5 & 6 (ACT CCT GAG→ACA TCT TAG) detected by denaturing gradient gel electrophoresis & sequencing. *Indian J Med Res 111*, 24–27.

Shaji RV, Gerard N, Krishnamoorthy R, Srivastava A, Chandy M. (2002a) Co-existence of a novel beta-globin gene deletion (codons 81-8) with the codon 30 (G→C) mutation in an Indian patient with beta0-thlassaemia. *Hemoglobin 26*, 49–57.

Shaji RV, Srivastava A, Krishnamoorthy R, Chandy M. (2002b) A novel beta-thalassaemia mutation in an Asian Indian. *Hemoglobin 26*, 237–243.

Shaji RV, Edison ES, Krishnamoorthy R, Chandy M, Srivastava A. (2003). Hb Lepore in the Indian population. *Hemoglobin 27*, 7–14.

Thakur (Mahadik) C, Vaz F, Bannerjee M, et al. (2000). Prenatal diagnosis of beta-thalassaemia and other haemoglobinopathies in India. *Prenatal Diagnosis 20*, 194–201.

Thein SL, Hesketh C, Wallace RB, Weatherall DJ. (1988). The molecular basis of thalassaemia major and thalassaemia intermedia in Asian Indians: Application to prenatal diagnosis. *British Journal of Haematology 70*, 225–232.

Vandenplas S, Higgs DR, Nicholls RD, Bester AJ, Mathew CGP. (1987). Characterization of a new α^{o} thalassaemia defect in the South African population. *British Journal of Haematology 66*, 539–542.

Varawalla NY, Old JM, Weatherall DJ. (1991a) Rare β-thalassaemia mutations in Asian Indians. *Brit J Haemat 79*, 640–644.

Varawalla NY, Old JM, Sarkar R, Venkatesan R, Weatherall DJ. (1991b). The spectrum of β thalassemia mutations on the Indian subcontinent: The basis for prenatal diagnosis. *Brit J Haemat 78*, 242–247.

Venkatesan R, Sarkar R, Old JM. (1992). β-thalassaemia mutations and their linkage to β-haplotypes in tamil Nadu in Southern India. *Clin Genet 42*, 251–256.

Verma IC, Saxena R, Thomas E, Jain PK. (1997). Regional distribution of β-thalassemia mutations in India. *Human Genetics 100*, 109–113.

Verma IC, Kleanthous M, Saxena R, et al. (2007). Multicenter study of the molecular basis of thalassaemia intermedia in different ethnic populations. *Hemoglobin 31*(4), 439–452.

Wainscoat JS, Old JM, Wood WG, Trent RJ, Weatherall DJ. (1984). Characterisation of an Indian $(\delta\beta)^{o}$ thalassaemia. *Br J Haematol 58*, 353–360.

Wajcman H, Traeger-Synodinos J, Papassotiriou I, et al. (2008). Unstable and thalassemic α chain Hb variants: A cause for Hb H disease and thalassemia intermedia. *Hemoglobin 32*(4), 327–349.

Weatherall DJ, Clegg JB, eds. (2001). *The Thalassemia Syndromes,* 4th edition. Oxford: Blackwell Scientific Publications.

91.

THE MANAGEMENT OF INHERITED BLOOD DISEASES IN THE INDIAN SUBCONTINENT

Reena Das

INTRODUCTION

Beta-thalassemia syndromes are the commonest symptomatic monogenic disorder in the Indian subcontinent. Thalassemia arises from reduced or absent synthesis of globin chains of the hemoglobin molecule, and the hemoglobinopathies are due to the qualitative difference of the globin chains causing an abnormal hemoglobin molecule. The inheritance pattern of β-thalassemia syndromes and hemoglobinopathies is autosomal recessive. The presentation of thalassemia has a wide clinical spectrum, ranging from the asymptomatic carrier state (beta-thalassemia trait; βTT) constituting the reservoir of the disease in the community, to thalassemia major (TM), with an early presentation in infancy and frequent three-weekly intervals of blood transfusion to sustain a reasonable quality of life. If not treated with regular blood transfusions, patients with TM die within their first few years. In between these two clinical end points is thalassemia intermedia (TI), which refers to a diverse collection of phenotypes in which the clinical features lie between TM and βTT (Weatherall and Clegg, 2001).

The globin chains depending on their composition are designated alpha (α), beta (β), gamma (γ), delta (δ), etc. The normal adult hemoglobin consists of two α-chains and two β-chains. Since the β-chain is only fully activated after birth, β-thalassemia—that is, the more severe form—does not appear till six months of age, when the γ-chains that form fetal hemoglobin are switched off. In contrast, α-thalassemia manifests both in fetal and adult life, as the α-chains are shared by both fetal and adult hemoglobins. Thalassemias are classified, depending on which globin chain is defective or absent, as α-thalassemia, β-thalassemia, δ-thalassemia, δβ-thalassemia, and εγδβ. Based on the quantitative output of the beta chains, two main types of β-thalassemia exist—β^0-thalassemia, in which no β-globin is produced, and β^+-thalassemia, where some β-globin is produced in reduced amounts. Less severe forms of β-thalassemia are designated β^{++} to indicate that the defect is mild. Because of reduced β-chains, there is an excess of α-chains that precipitate and cause damage to both the developing red cell precursors as well as to the mature red cells. Thus, the central pathophysiological mechanism in β-thalassemia is α-chain excess and the damage it causes at every stage of erythropoiesis in the bone marrow (Schrier, 2002).

Though much is known about thalassemias and hemoglobinopathies, due to clinical and molecular research over the last five decades, it has been noted that the disorders are heterogeneous in their manifestations in different regions of the world where they are common. Molecular characterization has revealed that the genetic defects are diverse for the same clinical phenotype and a small number of genetic defects, many unique to the population, account for the majority of the cases. Widespread identification of the asymptomatic carriers of the disease has led to screening of partners, who, if detected to be carriers, are offered prenatal diagnosis of at-risk pregnancies (WHO, 1982). Increased awareness regarding the thalassemia syndromes has resulted in a concern among the medical fraternity and social workers to decrease the frequency of the disease by use of screening programs.

The population heterogeneity in the Indian subcontinent is also because of numerous castes and tribal units characterized by cultural isolation and inbreeding within the population. The problems faced include: *(i)* a vast population, low socioeconomic status, low health budgets, and almost no or minimal health insurance schemes to look after the patients. The numbers of government hospitals are also inadequate to look after the increasing numbers of patients reaching hospitals in the hope of availing health care. *(ii)* Patients diagnosed as TM require regular blood transfusions and hospitals are struggling to get adequate and safe blood. *(iii)* Iron chelation is necessary to remove the extra iron deposited, since the excess iron deposition is toxic and damages various organs; this forms the third difficult aspect of tackling the thalassemia syndromes. However, due to the limited health resources and cost involved, only 10% of patients are being optimally treated with regular blood transfusions with safe blood, and iron chelation in India (Colah et al., 2008). The rural population in India comprises approximately 70% of the total population, and most of the data on population frequency is based on the urban population. If data has to be extrapolated from the limited population frequency data available for the country, then we are dealing with an astonishing figure of 35.8 million beta-thalassemia traits in the country, as per the 2001 census of India, with >10,000 to 15,000 new cases of thalassemia major born each year. The total number of newborns each year with any hemoglobinopathies is estimated to be >22,500 in India, which includes sickle cell disease and double heterozygotes of beta-thalassemia and Hb E disease. The data emphasizes the immediate the need for a widespread community control program to reach out to the urban and rural populations.

The reason why the disease appears to be thriving in this region is also partly due to the traditional system of marriage within castes and subcastes, and the practice of consanguinity in certain populations, leading to higher frequencies of the disease than expected. The spectrum of surviving patients with hemoglobinopathies is difficult to estimate in India and the neighboring countries, since the survival of the affected patients is dependent on the level of medical care that the patients receive. In this chapter we will take a glimpse at the frequency data, molecular spectrum characterized, and available preventive measures taken to tackle the disease in this part of the world.

HEMOGLOBINOPATHIES IN INDIA

A majority of the mutations of beta-thalassemia encountered in India are the β^0 type of mutation. Most patients with β-thalassemia are compound heterozygotes, having inherited two different mutations, one from each parent. In some families, the patient is true homozygous, in that the two mutations are identical. This happens particularly in families with consanguineous marriages or when marriages occur within close communities, or with mutations that are common in a given population. Genotype of β-thalassemia can be complicated by co-inheritance of either α-thalassemia or structural hemoglobin variants that can influence the clinical phenotype, as well as affect genetic counseling.

Structural variants with a β-thalassemia phenotype are often called "thalassemic hemoglobinopathies." They comprise distinctive mutations of hemoglobin gene, causing amino acid substitutions, additions, and deletions leading to structural variants of the hemoglobin molecule. The structural hemoglobins most commonly found in India are HbS (beta 6(A3)Glu>Val), HbE (beta 26 Glu>Lys), and HbD Punjab (beta 121(GH4)Glu>Gln). Of these, only HbS—either in a homozygous state leading to sickle cell disease, or interacting with β-thalassemia—and double heterozygosity for HbE and β-thalassemia lead to symptomatic disease, both of which have a restricted regional distribution in India. The other uncommon structural hemoglobin variants, both in the alpha- and beta-globin genes in India are HbQ India (alpha1 64(E13) Asp>His), Hb J Meerut (alpha1 or alpha2 120 Ala>Glu), Hb D Iran (beta 22(B4)Glu>Gln) and Hb Lepore (Shaji et al., 2003). However, HbC (beta 6(A3)Glu>Lys), which is common in African Blacks, is not seen in Asian Indians. Conditions giving rise to high fetal hemoglobin, such as δβ thalassemia trait, are found in a prevalence of 0.02% to 0.7%. The prevalence of HPFH trait has been found to be 0.1% to 0.4%.

PREVALENCE OF β-THALASSEMIA IN DIFFERENT REGIONS OF INDIA

The diagnostic hallmark of heterozygous β-thalassemia (βTT) is an elevated HbA_2. Individuals with βTT may have a mild anemia with characteristic hypochromic microcytic red blood cells, elevated red cell counts, and increased levels of HbA_2 (>4%) and variable levels of Hb F. However, in the "silent" carriers the quantity of HbA_2 may be normal. The distribution of β-thalassemia trait is heterogeneous in India, and the incidence varies from 1%–15% in different parts of the country. The heterogeneity pattern in different communities is as a result of the practice of arranged marriages in India, resulting in a higher frequency of hemoglobinopathies in some communities. Numerous studies undertaken subsequently have shown that the prevalence of β-thalassemia varies from 1% to 17% in different population groups, with an overall frequency of 3%–4%. In a multicentric collaborative study by the Indian Council of Medical Research (1993), the prevalence of βTT was found to be 2.7% in Mumbai, 6.6% in Delhi, and 10.2% in Kolkata (Sukumaran, 1975; Sood et al., 1993). UNICEF in 1996 estimated that there are 29.7 million carriers of β-thalassemia trait in India, and about 7000 infants with homozygous β-thalassemia born every year. In the northern region of India we have found the prevalence of βTT to be 3.5% among almost 2000 voluntary blood donors of Punjabi origin (Garewal and Das, 2003). A screening project based on 56,814 college students and pregnant women recruited in the states of Maharashtra, Gujarat, Punjab, Karnataka, West Bengal, and Assam indicated a carrier rate of 2.78% (Mohanty et al., 2008). A WHO update on β-thalassemia in India indicated a similar overall carrier frequency of 3%–4%, which, given the current national population, would translate to between 35.1 and 46.8 million carriers of the disorder nationwide (WHO, 2008; PRB, 2009). With a population of over one billion, implementing preventive measures for β-thalassemia in India remains the main method of controlling the disease. Immense health resources would be required to control the disease, and it has been felt by all the clinical and laboratory hematologists and scientists working in this field that a national preventive program is the need of the hour, with constant training to be imparted to the health professionals. Also, health insurance schemes are in their infancy in India; the onus of treatment of an affected child falls on the parents or on the government. Educating and increasing awareness of β-thalassemia in the community are required, so that increasing numbers of young individuals opt for voluntary screening programs.

To identify high-risk groups in the northern region, a caste structure analysis was undertaken of 780 consecutive individuals with βTT. Almost half of these individuals belonged to the Khatri–Arora castes of Punjab (47.8%), with the remaining 52.2% distributed among 13 other caste groups (Garewal and Das, 2003). The Khatri-Arora castes form 10%–15% of the Punjabi population, and a screening program specifically directed to this community would be beneficial. A recent study on 446 healthy, unrelated Sindhi individuals before marriage or before reproduction revealed the prevalence of βTT in Sindhis of Nagpur (central India) to be 16.81% (Mulchandani et al., 2008). A study to estimate βTT across the five major Sindhi subcastes varied, and Larkana Sindhis had the highest (17%), whereas Dadu Sindhis had the lowest (8%) frequency of the beta-thalassemia allele (Jawahirani et al., 2007). These studies highlight that for micromapping the frequencies in India, it is important to analyze the caste structure, and the finding of higher frequencies of β–thalassaemia in certain communities is not surprising, as majority of marriages in India are arranged by the parents, who look for a prospective spouse within their own caste.

Preventive strategies include: (a) population screening at either premarital or preconception stage; (b) genetic counseling of carriers; and (c) prenatal diagnosis (Petrou and Modell, 1995). These strategies have been successfully carried out on a voluntary basis by couples of childbearing age in the populations of Sardinia, Cyprus, and Greece. Earlier screening for βTT was done in the general population to determine the prevalence in the general population. However, following up with genetic counseling does not yield productive results, since many years may elapse between conveying the report and planning the birth of a child. Presently, a consensus has evolved that screening antenatal women during their first antenatal visit is the most cost-effective measure of screening the population. All women who are found to be carriers are advised to get their spouse screened. This strategy has been found to be most effective, as the concern to have a normal, healthy baby is a universal phenomenon. The other group where targeted screening yields a good return is "cascade or extended family screening" in family members who have an index case of thalassemia major or severe thalassemia intermedia. This subgroup is also initiated to the process of screening, and the frequency of finding βTTs is much higher than in general population screenings.

The laboratory tests for screening include complete automated blood cell counts followed by HPLC analysis for quantitating the types of hemoglobin fraction. Many studies on the usefulness of the naked eye single tube red cell osmotic fragility test (NESTROFT) as a screening test in resource-restricted settings have been carried out. Iron deficiency anemia also results in a positive NESTROFT test, and since it is a common problem in the subcontinent, the usefulness of the test is compromised. The other problem faced is that certain beta-thalassemia mutations may remain truly "silent," where red cell indices are not suggestive of a βTT. Such individuals may be missed in a screening program, and it is important to keep this fact in mind during the process of genetic counseling.

β-GENE MUTATIONS IN INDIANS

Though approximately 250 mutations in the beta-globin gene are described, it is well recognized that in a particular population only a handful of mutations prevail (Giardine et al., 2007). Most of the mutations are point mutations, and few deletions have been described in the β-gene worldwide. In India there are five common Asian Indian mutations—i.e., IVS I-5 (G-C), IVS I-1 (G-T), Fr 8/9 (+G), Fr 41/42 (-TTCT), 619 bp deletion—that account for a majority of the mutations in various studies (Table 91-1). Figure 91-1 shows the basic structure of the beta-gene with the positions of the common Indian mutations. The genomic DNA is analyzed, and the methodologies followed to determine the specific mutations are the amplification refractory mutation system (ARMS; Old et al., 1990) or the reverse dot blot hybridization (RDB; Colah et al., 1997)—and, when uncharacterized, DNA sequencing must be carried out. Kazazian et al. (1984) were the first to characterize seven new mutations in Asian Indians in 44 chromosomes, with 18.1% of mutations remaining uncharacterized. Since then, many studies have been carried out to determine the regional spectrum of mutations. Most of the studies from India show a frequency of the common five Indian mutations to be between 80%–90% (Edison et al., 2008). A total of 63 mutations have been characterized in the Indian population so far. Most of the studies were published from either west or north India, and only recently publications from southern and eastern India are generating data to show heterogeneity of the mutations across different regions of the country. Table 91-1 shows the type and list of Indian mutations observed

TABLE 91-1 LIST OF INDIAN MUTATIONS OF BETA-THALASSEMIA

	Class of Mutation	Type	Distribution
I	**NONFUNCTIONAL MRNA**		
a.	**Nonsense mutants**		
	Codon 15 (G→A)	β^0	Widespread
	Codon 17 (A→C)	β^0	Sporadic, Assam
b.	**Frame shift mutants**		
	–2 Codon 5 (-CT)	β^0	AP, Kerala, Punjab*
	+1 Fr 8/9 (+G)	β^0	Widespread, highest in North, West
	–2 Codon 8 (-AA)	β^0	Sporadic, Maharashtra
	–1 Codon 16 (-C)	β^0	Widespread
	–4 Fr 41/42 (-TTCT)	β^0	Widespread
	–2 Codon 5 (-CT)	β^0	AP, Kerala, North
	–1 Codon 15 (-T)	β^0	Sporadic, MP, WB
	–1 Codon 16 (-C)	β^0	South, Maharashtra, WB, North
	–7 Codon 22-24	β^0	Sporadic, Kerala, Maharashtra
	+1 Codon 27/28 (+C)	β^0	Sporadic, UP
	–1 Codon 36/37 (-T)	β^0	Sporadic, One Muslim patient
	–8 Codon 36-39	β^0	Sporadic
	–1 Codon 41 (-C)	β^0	Sporadic, Meghalaya, Manipur, WB
	–1 Codon 44 (-C)	β^0	Sporadic, Maharashtra
	+4 Fr 47/48 (+ATCT)	β^0	Restricted to Sikhs
	+1 Fr 54/55 (+A)	β^0	Sporadic, Maharashtra
	–1 Codon 55 (-A)	β^0	Sporadic, Maharashtra
	+1 Codon 55 (+A)	β^0	Sporadic, Maharashtra
	–7 Codon 62-64	β^0	Sporadic, WB
	–22 Codon 81-87	β^0	Sporadic, WB
	+1 Codon 57/58 (+C)	β^0	Restricted to Sikhs
	+1 Codon 88 (+T)	β^0	Sporadic
	+1 Codon 106/107 (+G)	β^0	Sporadic, Kerala
	–1 Codon 111 (-C)	β^0	Sporadic, Maharashtra
	–17 Codon 126-130	β^0	Sporadic, Maharashtra
II	**RNA PROCESSING MUTATIONS**		
a.	**Splice junction changes**		
	IVS I-1 (G→T)	β^0	#North, West
	IVS I-1 (G→A)	β^0	North*
	IVS I, 3'-end-25 bp deletion	β^0	Sporadic, Kerala, Maharashtra
	IVS I 130 (G→A)	β^0	Karnataka, Maharashtra
	IVS I 130 (G→C)	β^0	Tamil Nadu,
b.	**Consensus changes**		
	Codon 30 (G→A)	β^0	AP, Tamil Nadu, Gujarat, Punjab*
	Codon 30 (G→C)	β^0	East, Andhra Pradesh
	IVS I-5 (G→C)	β^+	Widespread, 54% to 80.5%
	IVS I-129 (A→C)	β^0	Sporadic, WB, Himachal*
	IVS II-1 (G→A)	β^0	Sporadic, South
	IVS II-837 (T→G)	β^0	Goa, South, Maharashtra
c.	**Internal IVS change**		
	IVS I-128 (T→G) [IVS I 3'-end (T→G)]	β^+	Punjab
	IVS I-110 (G→A)	β^+	Sporadic, AP, Kerala
	IVS II-591 (T→C)	β^+	Sporadic
	IVS II-613 (C→T)	β^+	Sporadic, Tamil Nadu, WB
	IVS II-654 (C→T)	β^+	Sporadic, Maharashtra
	IVS II-745 (C→G)	β^+	Sporadic, Maharashtra
d.	**Coding regions substitutions affecting processing**		
	Codon 26 (G→A)	β^+	East and North East
III	**TRANSCRIPTIONAL MUTANTS IN PROMOTER REGION**		
	–90 (C→T)	β^+	Sporadic, WB, Kerala
	–80 (C→T)	β^+	Sporadic, Andhra Pradesh
	–87 (C→A)	β^+	Sporadic, Orissa
	–88 (C→T)	β^{++}	Restricted to North
	–88 (C→A)	β^{++}	North*
	–28 (A→G)	β^{++}	South
	–29 (A→G)	β^{++}	Sporadic, Tamil Nadu
	–25 (A→G)	β^{++}	Sporadic, Karnataka
IV	**CAP SITE MUTANTS**		
	Cap +1 (A→C)	β^{++}	Mostly from North & Sporadic cases
V	**DELETION**		
	619 bp del	β^0	#North, West
VI	**MISCELLANEOUS**		
	Initiation Codon 1 T→C	β^0	Sporadic, Tamil Nadu
	Codon 4 (T→A) + Codon 5 (G→T) + Codon 6 (G→T)in cis form	β^0	Sporadic, UP
	Codon 5 (-CT) + Codon 13 (C→T) + Codon 26 (C→G) + Codon 26/27 (+C)	β^0	Sporadic, UP
	Codon 13 (C→T) + Codon 26 (G→A) + Codon 26/27 (-C)	β^0	Sporadic, UP

(Continued)

TABLE 91-1 (CONTINUED)

	Class of Mutation	Type	Distribution
	Codon 8 (A→G)	?	Sporadic, UP
	Codon 13 (C→T)	?	Sporadic
	Codon 16 (C→T)		Sporadic, Maharashtra
	Poly A (T→C)	β^+	South, Maharashtra, WB

AP: Andhra Pradesh, MP: Madhya Pradesh, WB: West Bengal, UP: Uttar Pradesh

*Unpublished #Higher in migrants from Pakistan

in India, along with the regional distribution, based on publications. (Old et al., 1990; Varawalla et al., 1991; Venkatesan et al., 1992; Agarwal et al., 1994, 1997, 2000; Verma et al., 1997; Madan et al., 1998; Vaz et al., 2000; Kukreti et al., 2002; Bashyam et al., 2004; Edison et al., 2008; Colah et al., 2009).

Most of the analysis gives data as percentage prevalence, which results in skewed data as the number of cases studied from various regions is different. However, the major use of the data generated has been the direct application of the mutations for antenatal diagnosis. There are 8 to 10 centers in major cities for prenatal diagnosis (Gorakshakar et al., 1997; Saxena et al., 1998; Thakur et al., 2000; Garewal and Das, 2003; Garewal et al., 2005a,b; Colah et al., 2009). Unfortunately, most patients from the rural parts of the country are neither diagnosed nor treated optimally.

COMMON FIVE ASIAN INDIAN MUTATIONS

The IVS-1,5 (G-C) mutation is the most widespread of the Indian mutations, with a prevalence of 30%–35% in northern India to 65%–75% in the eastern region and 80%–90% in the south. Fr 8/9 (+G) and Fr 41/42 (-TTCT) are two common mutations with a prevalence of 2.8% to 11% and 2% to 9.4%, respectively. They are both β^0 and do not have predilection for any caste or region. IVS-1, 1 (G-T) mutation is mainly seen in Gujaratis, Sindhis, and Punjabis. In fact, very few alleles if any have been reported outside these communities. Among the Punjabis, it has been observed that this mutation is seen predominantly in the Khatri-Arora caste. Its importance lies in the fact that it is almost invariably linked with the Xmn1$^G\gamma$ –158 polymorphism. As a result, homozygotes for this mutation present as thalassemia intermedia, or may confer a milder phenotype. Because of this fact, the IVS 1, 1 (G-T) mutation was more prevalent in the thalassemia intermedia group (28.5%) as compared the thalassemia major group (8.5%; our observation). An analysis by Varawalla et al. in 1991 has shown that 99% of individuals with the 619 bp deletion, which is seen only in Asian Indians, originally come from the neighboring regions of Sindh, Gujarat, and Punjab. Moreover, the mutation is associated with a single haplotype consistent with a common origin.

UNCOMMON MUTATIONS

The CAP+1 (A-C) mutation is a mild β^{++} mutation affecting the initiation codon and is exclusively seen in Asian Indians, and in homozygotes it presents as βTT. However, unlike the IVS 1,1 (G-T) and 619 bp deletion, it has a wide caste distribution and is not restricted to any caste groups. Its prevalence in Punjabis is 3.2%, and the importance of this mutation lies in the fact that it is hematologically "silent" in about half the βTT cases and is likely to be missed in a screening program. Individuals with this mutation usually have red cell indices in the borderline range or normal, and also often have normal HbA_2. There is every possibility that such individuals will be missed in a screening for BTT during pregnancy. A study on the clinical significance of interactions of CAP+1 (A→C) with other beta thalassemia mutations and globin gene modifiers showed a spectrum of phenotypic severity, and a majority (18/30; 60%) of compound heterozygotes were transfusion-dependent and phenotypically severe enough to necessitate appropriate therapy and counseling. Phenotype was modified with co-inheritance of α–thalassemia or *Xmn*-1 $^G\gamma$ polymorphism. The –88 (C-T) mutation occurs in the promoter region of the β-globin gene and is classified as a mild mutation (β^{++}). It has been reported in African Blacks and in Asian Indians, where its reported prevalence has been less than 1%.

Another β^{++} mild promoter mutation, –88 (C→T), is found in a very high prevalence (46%: 41/88) in the "Jat Sikh" subcaste of Punjabis, and 16 individuals presenting

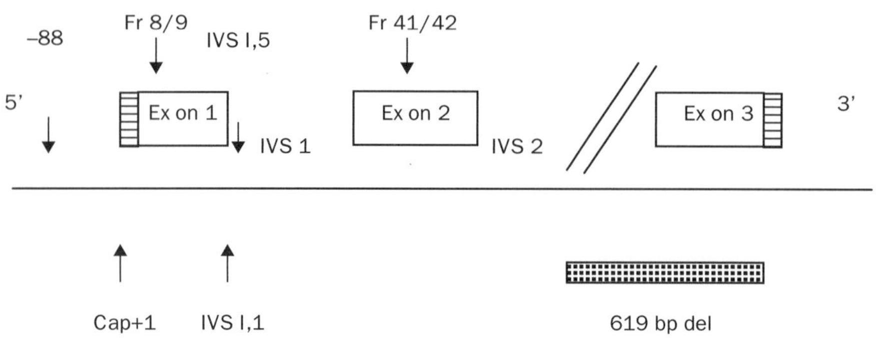

Figure 91-1 Beta-globin gene structure showing the common Asian Indian mutations.

in homozygous form were clinically mild. Untransfused patients had characteristic hematological findings: high Hb F (38.1%–68.6%), high Hb A_2 (5.7%–9.8%), and the rest adult hemoglobin. The 19 subjects with compound heterozygosity for –88 (C→T) and another β-gene mutation presented both as thalassemia intermedia (4 cases) and thalassemia major (15 cases). One of the four patients with milder phenotype had a second mild mutation, CAP+1(A→C). In the other three cases with milder phenotype, the second mutation was associated with the presence of *Xmn*-1 $^{G}\gamma$ polymorphism. Haplotype analysis showed a single distinct haplotype (+ – – – – + –) in all but two individuals, which is different from the description in the Africans with homozygous –88 (C-T), showing that the mutation in our population occurred independently (Garewal et al., 2005a,b). The Fr 47/48 (+ATCT) mutation was first described by Garewal et al. (1994) in a Punjabi Sikh family, and has subsequently been reported primarily in this population. Fr 57/58 (+C) has also seen restricted to Sikhs. In Punjab it forms 1% of the alleles. Codon 15 (G→A), Codon 16 (-C) and Codon 5 (-CT) mutations are noted to be present in lower frequencies throughout the country. CD15(G→A) is the second most common mutation in Maharashtra in the west (17.2%) and in the neighboring southern state of Karnataka (11.2%). IVS II-837(T→G) has been shown to be the most common (53.8%) mutation in Goa. It has also been identified in the neighboring coastal areas in Karnataka (4%) and occasionally in Maharashtra (0.3%). The highest frequency of polyadenylation site mutation, poly A (T/C), is in Karnataka (14.41%), and the lowest in Andhra Pradesh (2.94%). Codon 30 (G/C) has a frequency of 6.62% in Andhra Pradesh, which makes it one of the most common mutations in that state alone. The frequency was much lower (0%–0.85%) in the other southern states. Most of the other mutations have been seen in small number of cases, or as sporadic cases in different studies from India (Gupta et al., 2002, 2003).

PREVENTION OF THALASSEMIA

Population screening combined with premarital/preconception/antenatal screening, followed by appropriate genetic counseling, constitutes a rational approach for prevention. In case both the husband and wife are carriers, prenatal diagnosis should be offered. Mutation detection should be performed before fetal sampling and DNA analysis from chorionic villous biopsy at 10–12 weeks, or amniocytes after 14 weeks, is done. Precaution needs to be exercised to exclude maternal contamination by examining the variable number tandem repeats of DNA of the mother and fetal samples. Problems to keep in mind are occasional failure of DNA amplification or mispaternity.

INTERACTION OF β-THALASSEMIA WITH STRUCTURAL HB VARIANTS

HBE/β-THALASSEMIA

An example of a mutation causing activation of the cryptic splice site is HbE (β 26 Gln →Lys). HbE production is quantifiably reduced, and therefore it is also known as a "thalassemia hemoglobinopathy." The compound heterozygous state, HbE/β–thalassemia, results in a clinical picture closely resembling homozygous β–thalassemia, ranging from severe anemia and transfusion dependency to thalassemia intermedia. In eastern regions of India (Assam, Meghalaya, Bengal, Bihar, and Orissa) Bangladesh, and Sri Lanka, Hb E and β-thalassemia traits coexist. The genetic factor for these differences is not yet fully elucidated. However, some factors like the total amount of HbF and alpha gene interactions can interact and influence the phenotype. By itself, HbE is clinically asymptomatic both in the heterozygous and homozygous state.

HBS/β-THALASSEMIA

HbS is found all over the country, and seen mostly distributed among the various tribal groups and schedule castes in different areas. Sporadic cases of HbS in other communities are not uncommon. The average frequency of sickle cell allele is 4.3% based on 308 populations studied in India (Kar, 1991; Rao, 1998; Chhotray et al., 2004; Balgir, 2005). In the tribal and schedule caste communities, carrier prevalences of up to 40% have been reported due to high endogamy. Though heterozygous HbS is clinically silent, the homozygous state results in sickle cell disease. Its interaction with β-thalassemia results in moderate to severe clinical phenotype. A study from western India, on 29 sickle cell anemia patients with associated alpha-thalassemia, showed that the alpha-thalassemia decreases the clinical symptoms and hematological severity of sickle cell anemia in this region (Mukherjee et al., 1998).

HBD/β-THALASSEMIA

HbD Punjab was originally described in a Sikh from Punjab. Its prevalence in Punjabis is approximately 3%. The double heterozygous state of HbD/β-thalassemia is usually asymptomatic. Its importance is that if the other partner is heterozygous for β-thalassemia, there are chances of having an affected thalassemia major offspring. The interaction of HbD with HbS results in a clinical phenotype of moderate thalassemia intermedia.

OTHER HB VARIANTS

Other Hb variants found along with β-thalassemia in India are HbQ India (an alpha-chain variant) and HbJ Meerut, Hb Lepore. Conditions giving rise to high fetal hemoglobin, such as δβ thalassemia trait, are found in a prevalence of 0.02% to 0.7%. The prevalence of HPFH trait has been found to be 0.1% to 0.4%. By themselves, even in the homozygous state they do not cause significant abnormality. However, in the combination of one allele with a β^+ or β^0 allele, they can lead to severe thalassemia major or intermedia phenotypes. For γ-chain analysis, PCR-restriction enzyme digestion for *Xmn*-$1^G\gamma$ polymorphism and specific PCR-based tests for HPFH, δβ-thalassemia are done.

α-GLOBIN GENOTYPE IN INDIA

In many populations in which β–thalassemia is prevalent, α–thalassemia also occurs at a high frequency. Homozygotes or compound heterozygotes for β-thalassemia who co-inherit α-thalassemia have less redundant α-globin and tend to have a less severe condition of thalassemia intermedia. At the other extreme, patients who have co-inherited HbH (the equivalent of only one functioning α gene) and homozygous β-thalassemia also have thalassemia intermedia. Conversely, the presence of increased α-globin product (ααα/αα or αααα/ αα) in β-thalassemia traits tips the globin chain imbalance further and crosses the threshold, converting a typically clinical asymptomatic state to that of thalassemia intermedia. Alpha-gene gross deletions were identified using α-globin probes for Southern Blotting, which has been replaced by Gap-PCR in the last decade. MLPA analysis and DNA sequencing is being done for point mutations causing alpha-thalassemia.

An ICMR Task Force study in New Delhi, Bombay, and Calcutta in 1993 determined the prevalence of α-thalassemia by cord blood screening and found an overall prevalence of 1.1%. Alpha genotyping done by Southern Blot hybridization of *BamH*1 and *Bgl*ll digests, or GAP PCR, shows the prevalence of single alpha gene deletion (–α/αα) to be about 12%–13%, and α gene triplication (ααα/αα) to be 3% (Trehan et al., 2000; Sankar et al., 2006; Nadkarni et al., 2008). Two alpha gene deletions (–α/–α) is low at 0.9%. A study on tribals of the Gond group showed gene frequency of α–thalassemia ($-\alpha^{3.7}$, $-\alpha^{4.2}$, and Hb Koya Dora) to be 0.54 (Gupta et al., 1991). A study from eastern India on various tribes of West Bengal, Arunachal Pradesh, and Assam found alpha-globin gene deletion alleles in 18% of subjects from West Bengal, 3.9% from Arunachal Pradesh, and 3.84% from Assam tribesmen (Sen et al., 2005). The molecular basis for α^0-thalassemia from India was defined by Shaji et al. in 2003, and they found the most common molecular defect is α^0 of the South African (SA) deletion involving 23 kb that removes both the α1 and α2 genes, as first described from South Africa in an individual from Gujarat. They found this deletion in 17/21 patients of HbH disease. They also documented one case each of—MED and—SEA deletion, and two cases of homozygous polyadenylation signal mutation in the α_2-globin gene. Sarkar et al. (2005a,b) reported alpha zero deletion,—KOL, in the heterozygous state, encompassing the embryonic zeta2-globin and the duplicated alpha-globin genes extending from c. 1150 bp upstream of the zeta2-globin gene to c. 960 bp downstream of the theta1-gene. Other deletions present in 120 unrelated, eastern Indian, putative alpha-thalassemia patients were –3.7 kb (16.25%), –4.2 kb (5%) and —SEA (3.33%). Bhattacharya et al. (2008) detected alpha-globin polyadenylation (poly A) signal mutation, AATAAA>AATA-, which produces HbH disease in the homozygous state and occurs frequently in the Indian population.

Non-deletional alpha-thalassemias have also been reported from India. From Madhya Pradesh, Gupta et al. (1991) described Hb Koya Dora mutation and Cürük et al. (1993) described a G>A mutation at IVS-I-117 of the alpha 1-globin gene (acceptor splice site), which makes this gene nonfunctional. Its frequency was established at approximately 6%, which raises the total frequency of alpha-thalassemia determinants in this population to approximately 60%. One case involving a young Punjabi female with persistent hypochromic microcytic red cell indices, in whom iron deficiency and BTT had been ruled out, revealed a non-deletion type of α-thalassemia, i.e., Hb Evanston (alpha 14 (A12) Try→Arg; Das et al., 2005). Sarkar et al. in 2005a,b described Hb Sun Prairie (alpha2-globin cd130, GCT→CCT, Ala→Pro) in three unrelated chromosomes, in association with a C→T transition in the 5'-untranslated region (UTR), two bases upstream from the translation start site, which led to clinically severe anemia.

HEMOGLOBINOPATHIES IN PAKISTAN

Pakistan has a large population—172,800,000 in a census in July 2008—and the overall prevalence of β-thalassemia trait is 5.5%. Punjab is the largest province of the country, having more than 50% of the population. The state of beta-thalassemia is alarming, and it is the most common inherited disorder in Pakistan. Consanguinity is very high (>81%) and the literacy rate is low in south Punjab. A thalassemia prevention program is the need of the hour in this part of Pakistan. There are very inadequate treatment facilities for the over 4000 homozygotes born each year. Pathans showed significantly ($P < 0.02$) higher prevalence rate (7.96%) than Punjabis (3.26%; Khattak and Saleem, 1992). A study by Ghani et al. in 2002 placed the prevalence of βTT at 13%, with 5.1% sickle cell disease, 0.76% HbD Punjab, 0.32% HbC, and 0.22% HPFH. The practice of consanguinity is high among the Muslims of Pakistan, and a study screening for beta-thalassemia in 10 large Pakistani families with a history of a hemoglobin disorder showed that 183/591 (31%) were carriers. Seventeen of 214 married couples (8%) consisted of two carriers. All carriers have used the information provided in the testing and counseling process: carriers married to carriers with two or more healthy children have avoided further pregnancy, and most such couples with one or no healthy children have used prenatal diagnosis. They concluded that testing of extended families is a feasible way of deploying DNA-based genetic screening in communities in which consanguineous marriage is common (Ahmed at al., 2002).

The analysis of the spectrum of mutations found in Pakistan shows the mutations to be common, as found in Punjabis from north India, as there was population admixture in undivided India and lot of migration took place during the time of partition. Molecular analysis of 602 alleles from six ethnic groups in Pakistan showed that 99.2% of alleles were characterized. Nineteen different mutations in all ethnic groups were found, and the four most common mutations were IVS-I-5 (G→C) (37.7%), codons 8/9 (+G) (21.1%), the 619 bp deletion (12.4%), and IVS-I-1 (G→T) (9.5%), accounting for 80.7% of the alleles. There were differences between ethnic groups and also between provinces. In the four provinces of Pakistan, the IVS-I-5 (G→C) mutation is more prevalent in Sindh and Balochistan, bordering India in the south and Iran in the southwest, while the codons 8/9 (+G) mutation is more common in the Punjab and the Northwest Frontier province, bordering India in the northeast and Afghanistan, respectively. The 619 bp deletion is high (46%) in Gujaratis and Memons residing in the province of Sindh, neighboring Gujarat in India. An analysis from Rawalpindi on caste distribution showed that thalassemia was most frequent in Rajputs followed by Jatts, Arain, Sheikhs, and Pathans. Consanguinity was common, and 56.7% of the couples were first cousins and 19.8% were relatives. The commonest mutations were Fr 8-9 at 33.5%, IVS 1-5 at 17.2%, Fr 4142 at 8%, IVS 1-1 at 5.2%, deletion 619 at 4.2%, and codon 5 at 4.2% (Khan et al., 1995; Khan & Riazuddin, 1998). Three rare beta-thalassemia mutations were identified by single strand conformation polymorphism analysis followed by direct sequencing. Two mutations, IVS-II-848 (C→A) and initiation codon (ATG→ACG), were found in the homozygous condition in patients belonging to Balochi and Sindhi ethnic groups of Pakistan, together with heterozygous and homozygous alpha (–3.7) deletions, respectively (Khateeb et al., 2000). A frameshift mutation at codon 44 (–C) was identified in a patient belonging to the Gujarati ethnic group, together with IVS-I-1 (G→T) and a normal complement of four α-globin genes. Haplotype analysis was performed to identify the chromosomal background associated with these mutations, and for tracing their origin and spread. (Khan et al., 2000). Centers in Lahore, Rawalpindi, Faisalabad, and Karachi are offering prenatal diagnosis based on the data generated of the mutations in the region (Baig et al., 2006; Ahmed, 2007; Hafeez et al., 2007; Tasleem et al., 2007). Prenatal diagnostic service was introduced in the country in 1994. Most of the couples are open to the idea of prenatal testing and a few couples opt not to be tested (Ahmed et al., 2000). However, over 97% of the couples who requested prenatal testing already had an affected child. An important factor contributing to high incidence of thalassemia is the unawareness among people about the available diagnostic facilities for prenatal diagnosis in Pakistan. Strict implementation of collective measures, including carrier identification, genetic counseling, and prenatal diagnosis, are required for preventing beta-thalassemia in the country.

Various hemoglobinopathies have been reported from Pakistan, such as HbD Punjab, HbS, HbD Iran, HbQ India, HbC, δβ-thalassemia, and HPFH. An analysis of HPLC tracings of 11,008 subjects over a 13-month period identified 10 individuals with hemoglobin Q, showing the alpha-chain variant to be only 0.9%. HPLC is a useful screening tool, especially in developing countries where PCR facilities may not be accessible, to identify and characterize various hemoglobinopathies (Moiz et al., 2008). A study from Armed Forces Institute of Pathology, Rawalpindi, from January 1994 to April 2004, found 13 individuals from six unrelated Pakistani families with a hematological diagnosis of delta beta-thalassemia (Ahmed and Anwar, 2006). The mutation analysis revealed that all 13 individuals had the same Inv/Del $^{G}\gamma(^{A}\gamma\delta\beta)$. They concluded that δβ-thalassemia is a rare disorder in Pakistan and the clinical picture is very mild, but its combination with beta-thalassemia trait can produce a severe transfusion-dependent thalassemia.

Alpha-thalassemia has also been studied in Pakistan. A study was carried out to find out the prevalence of the alpha-thalassemia gene in the population of northern Pakistan by estimating the concentration of Hb Bart's in 500 cord blood samples during the period 1986–1987 at AFIP Rawalpindi. Hb Bart's was detected in 12 neonates, thus indicating a rate of 2.4% of general population as carrier of alpha-thalassemia gene (Rehman et al., 1991). To determine the prevalence and molecular basis of alpha-thalassemia in the British South Asian population, and the implications for genetic screening for hemoglobin disorders, 397 South Asian subjects were screened for hemoglobinopathy. Alpha-thalassemia was found to be the commonest hemoglobinopathy in British South Asian subjects. They recommended that screening and counseling should focus on subjects with more marked red cell hypochromia (MCH < 25 pg), thus concentrating resources on subjects of different ethnic origins who may carry the α^0-thalassemia gene (Hassall et al., 1998).

An analyses of 204 individuals from the normal population of Lahore revealed the presence of only the $-\alpha^{3.7}$ allele with an overall frequency of 8.3%, 0.2% of $-\alpha^{4.2}$ and 0.9% of triplicated $\alpha\alpha\alpha^{anti3.7}$ Five patients with triplicated alpha genes were homozygous for either the β^{+} or the β^{0} genotype. The $-\alpha^{4.2}$ allele was found only in Sindhis, while triplicated $\alpha\alpha\alpha^{anti3.7}$ was present in Punjabis, Sindhis, and Balochis. Ethnic differences were statistically significant for Pashtoon vs. Balochi ($p < 0.0005$) and Pashtoon vs. Sindhi ($p < 0.002$; Khan et al., 2003).

HEMOGLOBINOPATHIES IN SRI LANKA

The thalassemias pose an increasing burden for healthcare services in many Asian countries including Sri Lanka where the population estimate is 19,238,575 (see also Chapters 81 and 84). In order to conserve rare resources, it is essential to determine the reasons for the remarkable phenotypic heterogeneity and natural history of these disorders so that the most cost-effective methods for their control and management can be established. A long-term observational study of patients with different forms of thalassemia in Sri Lanka suggests that in addition to the well-defined primary, secondary, and tertiary genetic modifiers, environmental factors, particularly malaria, and variation in the ability to adapt to the profound anemia that characterizes these conditions, may play a significant role in determining their clinical severity. These findings may have important implications for the control and management of thalassemia in Asian populations (Perera et al., 2000; de Silva et al., 2000; Premawardhena et al., 2004, 2005a,b). Both beta-thalassemia and HbE are common disorders in Sri Lanka. The interaction of hemoglobin E and beta-thalassemia results in a wide spectrum of clinical disorders, some indistinguishable from thalassemia major, and some milder and not transfusion-dependent.

Partially as a result of this wide range of phenotypes, clear guidelines for approaches to transfusion and to iron-chelating therapy for patients with HbE beta-thalassemia have not been developed. By contrast, data that have accumulated during the past 10 years in patients with beta-thalassemia permit a quantitative approach to the management of iron overload, and provide guidelines for the control of body iron burden in individual patients treated with iron-chelating therapy. These guidelines may be applicable to patients with HbE beta-thalassemia. Preliminary evidence of iron loading in affected patients with HbE beta-thalassemia in Sri Lanka suggest that this disorder may be associated with variable but accelerated gastrointestinal iron absorption, and that the iron loading associated with chronic transfusions in patients with HbE beta-thalassemia is similar to that observed in patients with beta-thalassemia. These data, in the only cohort of patients with HbE beta-thalassemia to have undergone quantitative assessment of body iron burden, suggest that the principles that guide assessment of iron loading and initiation of chelating therapy in patients with beta-thalassemia may be generally applicable to those with HbE beta-thalassemia. Further quantitative studies in both nontransfused and transfused patients will be necessary to permit firm conclusions (Olivieri et al., 2000).

Chronic hyperbilirubinemia, gallstone formation, and gall bladder disease are unusually common in people with hemoglobin E beta-thalassemia in Sri Lanka. To determine whether this has a genetic basis we compared the bilirubin levels and frequency of gallstones in patients with different alleles of the UGT*1 gene. There was a significantly higher bilirubin level in those with the 7/7 genotypes compared with 6/6 and 6/7 genotypes (p=0.032 and 0.0015, respectively), who also appeared more prone to gallstone formation. These results suggest that the UGT*1 genotype is of importance in the genesis of gallstones in this population of patients (Premawardhena et al., 2001).

Blood samples from 703 patients with beta-thalassemia and from 1600 schoolchildren showed that the thalassemia mutations were unevenly spread. Although 23 different beta-thalassemia mutations were found, three accounted for the thalassemia phenotype in about 70% of the patients, most whom were homozygotes or compound heterozygotes for IVS1-5 (G→C) or IVS1-1 (G→A). The third common mutation, codon 26 (G→A), which produces HbE, interacts with one or other of these mutations to produce HbE beta-thalassemia; this comprises 13.0%–30.9% of cases in the main centers. Samples from 472 patients were analyzed to determine the alpha-globin genotype. Overall, 15.5% patients were carriers for deletion forms of alpha+-thalassemia. Average gene frequencies showed that there will be more than 2000 patients requiring treatment at any one time, in the future, of whom those with HbE beta-thalassemia will account for about 40%. Three novel mutations were identified—codon 55 (-A) and IVS-I-129 (A→C) and a 13-bp deletion removing the last nucleotide in CD6 to CD10 inclusively (Fisher et al., 2003). Two beta-thalassemia mutations were found to coexist in one beta-globin gene: Sri Lankan patients homozygous for codon 16 (-C) frameshift were also homozygous for the beta+ codon 10 (C→A) mutation. Studies of Sri Lankan, Pakistani, and Indian carriers suggest the codon 10 (C→A) mutation is just a rare polymorphism on an ancestral allele, on which the $beta^{0}$ codon 16 (-C) mutation has arisen (Old et al., 2001).

The allele frequency of alpha+-thalassemia was 6.5% and 1.1% for $-\alpha^{3.7}$ and $-\alpha^{4.2}$ deletions, respectively. Nondeletion alpha-thalassemia was not observed. Triplicate or quadruplicate alpha-globin genes were unusually common. In 1.5% of cases it was impossible to identify beta-thalassemia alleles, but in Kurunegala, detailed family studies led to an explanation for the severe thalassemia phenotype in every case, including a previously unreported instance of homozygosity for a quadruplicated alpha-globin gene together with beta-thalassemia trait. These findings have implications for the control of thalassemia in high-frequency populations with complex ethnic histories (Olivieri et al., 2008).

HEMOGLOBINOPATHIES IN BANGLADESH

Bangladesh is also a populous country with a current estimate of total population being 150,448,339. A study of 735 school children from all six divisions of Bangladesh was carried out to determine the frequency of βTT and HbE trait (Table 91-2). The overall frequency of βTT was 4.1% and HbE trait was 6.1%, which also is a considerable health burden for the country (Khan et al., 2005). Table 91-2 shows the frequency of βTT and HbE trait in various districts. Calculated value of the expected births of beta-thalassemia major annually in Bangladesh is 1040, and HbE thalassemia is 6443. Prevalence of both beta-thalassemia trait and HbE trait is significantly high in Bangladesh, and thalassemia is a major genetic problem in the country. Limited genotyping data available on 17 thalassemia patients revealed that IVS 1-5 (G-C) is the most common beta-thalassemia mutation, which was seen in 13 patients (76.5%). Most of the patients were transfusion-dependent (Khan et al., 1999).

HEMOGLOBINOPATHIES IN THE MALDIVES

The Maldives comprises 1192 islands covering a land mass that amounts to under 1% of the total geographical territory of the country. The population of 369,031, as per an estimate in July 2007, is dispersed across 200 isolated communities, with an average of 1000 people per community. Recent progress in health terms include a reduction in the infant mortality rate from 62 in 1992 to 14 in 2003, and 95% coverage in child immunization. In 1992, SHE, a nongovernmental organization, established that the beta-thalassemia prevalence rate was 18.1% (1 in 5) and, on the basis of the result, launched a nationwide awareness and population screening program, visiting each island in the Maldives every 5 years and targeting 12–35-year-olds. Screening of 100 cord blood samples indicated a 28% incidence of alpha-thalassemia. Screening results highlighted significantly high incidence of more than one hemoglobinopathy on individual islands. This is of particular importance given the norm of intra-island marriages. Specific mutation analysis showed that three mutations accounted for more than 95% of the thalassemia genes, ensuring a high detection rate and cost effectiveness of a prenatal diagnosis program. Outcomes of the screening program include; screening of more than 25% of the target population; the establishment of a Government National Thalassemia Center; inclusion of thalassemia into the school curriculum; the legal requirement for screening prior to marriage; legalization of prenatal diagnosis and medical termination of pregnancy; and the commencement of prenatal diagnostic services. The program successes include effective advocacy, resource mobilization, motivation for screening, voluntary blood donation, and thalassemia becoming a household word in the country (Firdous, 2005).

TABLE 91-2 FREQUENCY OF βTT AND HB E TRAITS IN VARIOUS BANGLADESHI DISTRICTS

District	Frequency of βTT (%)	Frequency of HbE (%)
Barishal	8.1	8.1
Rajshakhi	5.5	16.5
Sylhet	5.2	4.2
Dhaka	3.2	5.2
Chitagong	2.9	2.9
Tribal school	4.2	41.7
Khulna	2.4	2.4

The molecular basis of beta-thalassemia in the Maldives on a systematic analysis from 41 unrelated patients, including 37 beta-thalassemia homozygotes (three with beta-thalassemia/Hb E, and one with beta-thalassemia/Hb S) was carried out. Four different beta-thalassemia mutations were detected in 78 alleles. These are the IVS-I-5 (G→C), codon 30 (AGG→ACG) [also indicated as IVS-I (-1)], IVS-I-1 (G→A), and codons 41/42 (-TTCT) mutations. The distribution of the beta-thalassemia mutations in the Maldives is 58 alleles (74.3%) with the IVS-I-5 (G→C) mutation, 12 (15.4%) with the codon 30 (AGG→ACG) mutation, seven (9%) with the IVS-I-1 (G→A) mutation, and one with the codons 41/42 (–TTCT) mutation. The first three mutations account for 98.7% of the total number of beta-thalassemia chromosomes studied. These mutations are clustered in the region spanning 6 bp around the junction of exon 1 and the first intervening sequence of the beta-globin gene. These observations have significant implications for setting up a thalassemia prevention and control program in the Maldives. Analysis of haplotypes and frameworks of chromosomes bearing each beta-thalassemia mutation suggested that the origin and spread of these mutations were reflected by the historical record (Modell et al., 1990; Furuumi et al., 1998).

HEMOGLOBINOPATHIES IN NEPAL

Though thalassemia and hemoglobinopathies are described in sporadic estimates from Nepal, no proper data based on population screening is available. The population estimate is 26,469,569 based on a survey in July 2003. Among four ethnic groups in a lowland area of Nepal, the prevalences of abnormal hemoglobin, thalassemia, glucose-6-phosphate-dehydrogenase (G6PD) deficiency, hereditary Southeast Asian ovalocytosis (SAO) and Duffy blood-group antigen Fy/Fy were determined and related to each group's habitat. The group that has lived for many decades in a malaria-endemic lowland area, the Danuwar, was found to have a high prevalence of alpha+-thalassemia (79.4%) and low prevalences of hemoglobin E and G6PD deficiency. Much lower prevalences of alpha+-thalassemia were observed in the Newar (20.5%), Parbate (16.5%), and Tamang (8.8%), who, until the 1950s, all spent their hot-season nights in malaria-free areas at higher altitudes. No subjects with any other identified abnormal hemoglobin, beta-thalassemia, SAO, or Fy/Fy were detected (Suzuki et al., 2007).

In a report on protection against malaria morbidity, near-fixation of the alpha-thalassemia gene in a Nepalese population was noted. It was found that the Tharu people of the Terai region in southern Nepal have an incidence of malaria about sevenfold lower than that of sympatric non-Tharu people. In order to find out whether this marked resistance against malaria has a genetic basis, in vitro cultures of *Plasmodium falciparum* in both Tharu and non-Tharu red cells were performed. The significant finding was relatively low and variable frequencies of beta-thalassemia, beta S, G6PD (-), and Duffy (a-b-) in different parts of the Terai region. The average in vitro rate of invasion and of parasite multiplication did not differ significantly in red cells from Tharus versus those from non-Tharu controls. By contrast, the frequency of alpha-thalassemia is uniformly high in Tharus, with the majority of them having the homozygous alpha-/alpha-genotype and an overall alpha-thalassemia gene frequency of 8. It was suggested that holoendemic malaria has caused preferential survival of subjects with alpha-thalassemia, and that this genetic factor has enabled the Tharus as a population to survive for centuries in a malaria-holoendemic area. From the data it was estimated that the alpha-thalassemia homozygous state decreases morbidity from malaria by about tenfold. This is an example of selection evolution toward fixation of an otherwise abnormal gene (Sakai et al., 2000).

A microepidemiological study was carried out on thalassemia in two neighboring populations in Nepal, the Danuwar and the Tamang. Settlements of the Danuwar are located below the limit of the malarial zone (1200m in altitude), whereas those of the Tamang are found in malaria-free uplands. Three heterozygotes for HbE were observed in the Danuwars. Alpha deletion of $-\alpha^{3.7}$ in the Danuwars was found at a high gene frequency of 63%, while the gene frequency in the Tamangs was only 5%. Analysis of the alpha-globin gene cluster revealed that four different haplotypes were associated with the type of alpha+-thalassemia in the Danuwars (Modiano et al., 1991).

MAJOR KEY POINTS

Beta thalassemia syndromes are the commonest symptomatic monogenic disorder in the Indian subcontinent.

- The population heterogeneity in the Indian subcontinent is because of numerous castes and tribal units characterized by cultural isolation and inbreeding within the population.
- The problems faced include a vast population, low socioeconomic status, low health budgets, and almost no or minimal health insurance schemes to look after the patients.
- The numbers of government hospitals are also inadequate to look after the increasing numbers of patients reaching hospitals in the hope of availing healthcare.
- Patients diagnosed as thalassemia major require regular blood transfusions, and adequate and safe blood with optimal chelation is available to a minority.
- A national preventive program is the need of the hour in each country, with a focus on constant training to health professionals to develop manpower.
- Prevention is the key to controlling thalassemia, and antenatal screening and extended family screening is required.

CONCLUSIONS

The thalassemia syndromes and hemoglobinopathies are a common problem in the Indian subcontinent. The large population, financial constraints, and limited health resources make the disorder a substantial proportion of the world's hemoglobinopathy burden. It is important to get accurate frequency data from each country, so that the true numbers can be extrapolated to present the burden of the disease to the relevant government or funding agencies. Due to marked endogamy and the practice of consanguinity in certain populations, the frequency is increased in some communities. Optimal management is based on finances, and prevention is a cost-effective measure to decrease the burden of the disorder. Since the traits or carriers are asymptomatic, targeted screening programs and availability of genetic counseling is required urgently to be available in greater numbers in the subcontinent. A good referral system with the help of nongovernmental agencies is essential to direct couples seeking chorionic villous sampling and further DNA testing.

REFERENCES

Agarwal S, Pradhan M, Gupta UR, Sarwai S, Agarwal SS. (2000). Geographic and ethnic distribution of beta-thalassemia mutations in Uttar Pradesh, India. *Hemoglobin* 24:89–97.

Agarwal S, Wei I, Hattori Y, Agarwal S. (1997). Codon 4 ACT→ACA, codon 5 CCT→TCT and codon 6 GAG→TAG mutations in cis position: a form of thalassemia trait. *Am J Hematol* 56:187–188.

Agarwal S, Naveed M, Gupta UR, Kishore P, Agarwal SS. (1994). Characterization of beta-thalassaemia mutations in 57 beta-thalassaemia families seen at Lucknow. *Indian J Med Res* 100:106–110.

Ahmed S, Saleem M, Sultana N, et al. (2000). Prenatal diagnosis of beta-thalassaemia in Pakistan: experience in a Muslim country. *Prenat Diagn* 20(5):378–383.

Ahmed S, Saleem M, Modell B, Petrou M. (2002). Screening extended families for genetic hemoglobin disorders in Pakistan. *N Engl J Med* 347(15):1162–1168.

Ahmed S, Anwar M. (2006). Haematological and genetic features of deltabeta-thalassaemia in Pakistan. *J Coll Physicians Surg Pak* 16:19–22.

Ahmed S. (2007). Prenatal diagnosis of beta-thalassemia: 12 years' experience at a single laboratory in Pakistan. *Prenat Diagn* 27:1224–1227.

Baig SM, Azhar A, Hassan H, et al. (2006). Spectrum of beta-thalassemia mutations in various regions of Punjab and Islamabad, Pakistan: Establishment of prenatal diagnosis. *Haematologica* 91:ELT02.

Baig SM, Azhar A, Hassan H, et al. (2006). Prenatal diagnosis of beta-thalassemia in Southern Punjab, Pakistan. *Prenat Diagn* 26(10):903–905.

Balgir RS. (2005). The spectrum of hemoglobin variants in two scheduled tribes of Sundargarh district in north-western Orissa, India. *Ann Hum Biol* 32:560–573.

Bashyam M, Bashyam L, Savithri G, Gopikrishna M, Sangal V, Devi A. (2004). Molecular genetic analyses of beta-thalassemia in South India reveals rare mutations in the beta-globin gene *Hum Genet* 49:408–413.

Bhattacharya G, Sarkar AA, Banerjee D, Chandra S, Das M, Dasgupta UB. (2008). Polymerase chain reaction-based search for two alpha-globin gene mutations in India. *Hemoglobin* 32:485–490.

Chhotray GP, Dash BP, Ranjit M. (2004). Spectrum of hemoglobinopathies in Orissa, India. *Hemoglobin* 28:117–122.

Colah R, Gorakshakar A, Nadkarni A, et al. (2009). Regional heterogeneity of β-thalassemia mutations in the multi ethnic Indian population. *Blood Cells Molecules and Diseases* 42:241–246.

Colah RB, Gorakshakar AC, Lu CY, et al. (1997). Application of covalent reverse dot blot hybridization for rapid prenatal diagnosis of the common Indian thalassaemia syndromes. *Indian J Haematol Blood Transfusion* 15:10–13.

Colah R, Surve R, Wadia M, et al. (2008). Carrier screening for β thalassemia during pregnancy in India: A 7 year evaluation. *Genetic Test* 12:181–185.

Cürük MA, Baysal E, Gupta RB, Sharma S, Huisman TH. (1993). An IVS-I-117 (G→A) acceptor splice site mutation in the alpha 1-globin gene is a nondeletional alpha-thalassaemia-2 determinant in an Indian population. *Br J Haematol* 85:148–152.

Das R, Garewal G, Shaji RV, Ahluwalia J, Bali HK, Varma S. (2005). Investigation of persistent hypochromic microcytosis unmasks hemoglobin Evanston [alpha 14 (A12) Try→Arg] in a patient of cyclic thrombocytopenia preceding Takayasu's disease. *Hematology* 10:427–429.

de Silva S, Fisher CA, Premawardhena A, et al. (2000). Thalassaemia in Sri Lanka: implications for the future health burden of Asian populations. Sri Lanka Thalassaemia Study Group. *Lancet* 355:786–791.

Edison ES, Shaji RV, Devi SG, et al. (2008). Analysis of beta globin mutations in the Indian population: presence of rare and novel mutations and region-wise heterogeneity. *Clin Genet* 73:331–337.

Firdous N. (2005). Prevention of thalassaemia and haemoglobinopathies in remote and isolated communities – the Maldives experience. *Ann Hum Biol* 32(2):131–137.

Fisher CA, Premawardhena A, de Silva S, et al. (2003). The Sri Lanka Thalassaemia Study Group. The molecular basis for the thalassaemias in Sri Lanka. *Br J Haematol* 121(4):662–671.

Furuumi H, Firdous N, Inoue T, et al. (1998). Molecular basis of beta-thalassemia in the Maldives. *Hemoglobin* 22(2):141–151.

Garewal G, Fearon CW, Warren TC, et al. (1994). The molecular basis of β thalassemia in Punjabi and Maharastrian Indians includes a multilocus aetiology involving triplication β-globin loci. *British Journal of Haematology* 86:372–376.

Garewal G, Das R. (2003). Spectrum of β thalassemia mutations in north India. *International Journal of Human Genetics* 3:217–219.

Garewal G, Das R, Kaur J, Marwaha RK, Gupta I. (2005a). Establishment of prenatal diagnosis for beta-thalassaemia: a step towards its control in a developing country. *Ann Hum Biol* 32:138–144.

Garewal G, Das R, Ahluwalia J, Marwaha RK, Varma S. (2005b). Nucleotide −88 (C-T) promoter mutation is a common β-thalassemia mutation in the Jat Sikhs of Punjab, India. *Am J of Haematology* 79:252–256.

Garewal G, Das R, Awasthi A, Ahluwalia J, Varma N, Marwaha RK. (2007). The clinical significance of the spectrum of interactions of CAP+1 (A→C), a silent beta globin gene mutation, with other beta thalassemia mutations and globin gene modifiers in North Indians. *Eur J Haematol* 79:417–421.

Ghani R, Manji MA, Ahmed N. (2002). Hemoglobinopathies among five major ethnic groups in Karachi, Pakistan. *Southeast Asian J Trop Med Public Health* 33(4):855–861.

Ghosh Dastidar D, Dutta RN, Gupta P, Old JM. (1994). Detection of beta-thalassaemia mutations in eastern Indian population by polymerase chain reaction. *Indian J Med Res* 100:111–114.

Giardine B, Van Baal S, Kaimakis P, et al. (2007). HbVar database of human hemoglobin variants and thalassemia mutations: 2007 update. *Hum Mut* 28:206.

Gorakshakar AC, Lulla CP, Nadkarni AH, et al. (1997). Prenatal diagnosis of beta-thalassemia among Indians using denaturing gradient gel electrophoresis *Hemoglobin* 21: 421–435.

Gupta A, Hattori Y, Agarwal S. (2002). Initiation codon mutation in an Asian Indian Family. *Am J Hematol* 71:134–136.

Gupta A, Hattori Y, Gupta UR, Sarwai S, Nigam N, Singhal P, Agarwal S. (2003). Molecular genetic testing of beta thalassemia patients of Indian origin and a novel 8 bp deletion mutation at codon 36/37/38/39. *Genet Test* 7:163–168.

Gupta RB, Tiwary RS, Pande PL, et al. (1991). Hemoglobinopathies among the Gond tribal groups of central India; interaction of alpha- and beta-thalassemia with beta chain variants. *Hemoglobin* 15:441–458.

Hafeez M, Aslam M, Ali A, Rashid Y, Jafri H. (2007). Regional and ethnic distribution of beta thalassemia mutations and effect of consanguinity in patients referred for prenatal diagnosis. *J Coll Physicians Surg Pak* 17(3):144–147.

Hassall OW, Tillyer ML, Old JM. (1998). Prevalence and molecular basis of alpha thalassaemia in British South Asians. *J Med Screen* 5:31–33.

Jawahirani A, Mamtani M, Das K, Rughwani V, Kulkarni H. (2007). Prevalence of beta-thalassaemia in subcastes of Indian Sindhis: Results from a two-phase survey *Public Health* 121:193–198.

Kar BC. (1991). Sickle cell disease in India. *J Assoc Physicians India* 39(12):954–960.

Khan WA, Alam MA, Sarwardi G, et al. (1999). Genotypic study of Bangladeshi thalassemia patients and transfusion requirements. *DS (Child) H J* 115:37–39.

Khan WA, Banu A, Amin SK, et al. (2005). Prevalence of Beta thalassemia trait and Hb E trait in Bangladeshi school children and health burden of thalassemia in our population. *DS (Child) H J* 21:1–7.

Khateeb B, Moatter T, Shaghil AM, Haroon S, Kakepoto GN. (2000). Genetic diversity of beta-thalassemia mutations in Pakistani population. *J Pak Med Assoc* 50:293–296.

Khattak MF, Saleem M. (1992). Prevalence of heterozygous beta-thalassemia in northern areas of Pakistan. *J Pak Med Assoc* 42:32–34.

Khan SN, Zafar AU, Riazuddin S. (1995). Molecular genetic diagnosis of beta thalassemia in Pakistan. *J Pak Med Assoc* 45:66–70.

Khan SN, Riazuddin S. (1998). Molecular characterization of beta-thalassemia in Pakistan. *Hemoglobin* 22:333–345.

Khan SN, Riazuddin S, Galanello R. (2000). Identification of three rare beta-thalassemia mutations in the Pakistani population. *Hemoglobin* 24:15–22.

Kukreti R, Dash D, E VK, et al. (2002). Spectrum of beta-thalassemia mutations and their association with allelic sequence polymorphisms at the beta-globin gene cluster in an Eastern Indian population. *Am J Hematol* 70:269–277.

Khan SN, Hasan F, Sollaino C, Perseu L, Riazuddin S. (2003). Molecular characterization of alpha-thalassemia in Pakistan. *Hemoglobin* 27:161–166.

Madan N, Sharma S, Rusia U, Sen S, Sood SK. (1998). Beta-thalassaemia mutations in northern India (Delhi). *Indian J Med Res* 107:134–141.

Modell B, Razzak A, Hindley N. (1990). Thalassaemia in the Maldives. *Lancet* 335(8698):1169–1170.

Modiano G, Morpurgo G, Terrenato L, et al. (1991). Protection against malaria morbidity: near-fixation of the alpha-thalassemia gene in a Nepalese population. *Am J Hum Genet* 48(2):390–397.

Mohanty D, Colah R, Gorakshakar A, eds. (2008). Report of Jai Vigyan S & T Mission Project on Community Control of Thalassaemia Syndromes – Awareness, Screening, Genetic Counseling and Prevention. A National Multicentre Study of ICMR (2000 – 2005). Indian Council of Medical Research, New Delhi.

Moiz B, Moatter T, Hashmi MR, et al. (2008). Identification of hemoglobin Q India (alpha 1-64 Asp-His) through ARMS-PCR. First report from Pakistan. *Ann Hematol* 87(5):385–389.

Mukherjee MB, Surve R, Tamankar A, et al. (1998). The influence of alpha-thalassaemia on the haematological & clinical expression of sickle cell disease in western India. *Indian J Med Res* 107:178–181.

Mulchandani DV, Fulare MB, Zodpey SP, Vasudeo ND. (2008). Prevalence and some epidemiological factors of beta thalassaemia trait in Sindhi community of Nagpur City, India. *Indian J Public Health* 52(1):11–15.

Nadkarni A, Phanasgaonkar S, Colah R, Mohanty D, Ghosh K. (2008). Prevalence and molecular characterization of alpha-thalassemia syndromes among Indians. *Genet Test* 12(2):177–180.

Old JM, Varawalla NY, Weatherall DJ. (1990). Rapid detection and prenatal diagnosis of beta-thalassaemia: studies in Indian and Cypriot populations in the UK. *Lancet* 336:834–837.

Old JM, Khan SN, Verma I, et al. (2001). A multi-center study in order to further define the molecular basis of beta-thalassemia in Thailand, Pakistan, Sri Lanka, Mauritius, Syria, and India, and to develop a simple molecular diagnostic strategy by amplification refractory mutation system-polymerase chain reaction. *Hemoglobin* 25:397–407.

Olivieri NF, De Silva S, Premawardena A, et al. (2000). Iron overload and iron-chelating therapy in hemoglobin E-beta thalassemia. *J Pediatr Hematol Oncol* 22:593–597.

Olivieri NF, Muraca GM, O'Donnell A, Premawardhena A, Fisher C, Weatherall DJ. (2008). Studies in haemoglobin E beta-thalassaemia. *Br J Haematol* 141:388–397.

Perera DM, Gunasekara D, Wijekoon A, Petrou M. (2000). Thalassaemia control in developing countries – the Sri Lankan perspective. *Ceylon Med J* 45:12–16.

Petrou M, Modell B. (1995). Prenatal screening for haemoglobin disorders. *Prenat Diagn* 15:1275–1295.

PRB. (2009). World population data sheet. Washington, DC: Population Reference Bureau.

Premawardhena A, Fisher CA, Fathiu F, de Silva S, Perera W, Peto TE, Olivieri NF, Weatherall DJ. (2001). Genetic determinants of jaundice and gallstones in haemoglobin E beta thalassaemia. *Lancet* 357(9272):1945–1946.

Premawardhena A, De Silva S, Arambepola M, et al. (2004). Thalassemia in Sri Lanka: a progress report. *Hum Mol Genet* 13(Spec No 2):R203–R206.

Premawardhena A, De Silver S, Arambepola M, et al. (2005a). Hemoglobin E-beta-thalassemia: Progress report from the International Study Group. *Ann N Y Acad Sci* 1054:33–39.

Premawardhena A, Fisher CA, Olivieri NF, et al. (2005b). Haemoglobin E beta thalassaemia in Sri Lanka. *Lancet* 366(9495):1467–1470.

Rao VR (1988). Genetics and epidemiology of sickle cell anaemia in India. *Indian Journal of Medical Sciences* 18:218–221.

Rehman Z, Saleem M, Alvi AA, Anwar M, Ahmed PA, Ahmad M. (1991). Alpha-thalassaemia: prevalence and pattern in northern Pakistan. J Pak Med Assoc 41(10):246–247.

Sakai Y, Kobayashi S, Shibata H, et al. (2000). Molecular analysis of alpha-thalassemia in Nepal: correlation with malaria endemicity. *J Hum Genet* 45(3):127–132.

Sankar VH, Arya V, Tewari D, Gupta UR, Pradhan M, Agarwal S. (2006). Genotyping of alpha-thalassemia in microcytic hypochromic anemia patients from North India. *J Appl Genet* 47:391–395.

Sarkar AA, Mukhopadhyay C, Chandra S, Banerjee S, Das MK, Dasgupta UB. (2005a). Co-inheritance of the Hb Sun Prairie mutation with a point mutation at 5'-UTR in the eastern Indian population. *Br J Haematol* 129(2):282–286.

Sarkar AA, Banerjee S, Chandra S, et al. (2005b). A novel 33.3 kb deletion (– –KOL) in the alpha-globin gene cluster: a brief report on deletional alpha-thalassaemia in the heterogeneous eastern Indian population. *Br J Haematol* 130:454–457.

Saxena R, Jain PK, Thomas E, Verma IC. (1998). Prenatal diagnosis of β-thalassemia: Experience in a developing country. *Prenat Diagn* 18:1–7.

Schrier SL. (2002). Pathophysiology of thalassemia. *Current Opinion in Hematology* 9:123–126.

Sen R, Chakrabarti S, Sengupta B, et al. (2005). Alpha-thalassemia among tribal populations of Eastern India. *Hemoglobin* 29(4):277–280.

Shaji RV, Edison ES, Krishnamoorthy R, Chandy M, Srivastava A. (2003). Hb Lepore in the Indian population. *Hemoglobin* 27:7–14.

Shaji RV, Eunice SE, Baidya S, Srivastava A, Chandy M. (2003). Determination of the breakpoint and molecular diagnosis of a common alpha-thalassaemia-1 deletion in the Indian population. *Br J Haematol* 123:942–947.

Sood SK, Madan N, Colah R, Sharma S, Apte SV, eds. (1993). Collaborative Study of Thalassemia: Report of an ICMR Task Force Study. New Delhi: Indian Council of Medical Research.

Sukumaran PK. (1975). Abnormal hemoglobins in India. In: Sen NN, Basu AK, Chatterjea JB Memorial Committee, eds. *Trends in Haematology*. Calcutta: J.B. Chatterjea Memorial Volume, pp. 225–261.

Suzuki A, Hamano S, Shirakawa T, et al. (2007). The distribution of hereditary erythrocytic disorders associated with malaria, in a lowland area of Nepal: A micro-epidemiological study. *Ann Trop Med Parasitol* 101(2):113–122.

Tasleem S, Tasleem H, Siddiqui MA, Adil MM, Rashid Y. (2007). Prenatal diagnosis of beta-thalassaemia by chorionic villous sampling. *J Pak Med Assoc* 57(11):528–531.

Thakur MC, Vaz E, Banerjee M, et al. (2000). Prenatal diagnosis of beta thalassemia and other hemoglobinopathies in India. *Prenat Diagn* 20:194–201.

Trehan U, Garewal G, Kaul D, Das R. (2000). Alpha thalassemia and alpha gene triplications in Punjabis, with and without beta thalassemia trait. *Hematology* 6:153–160.

UNICEF. (1996). *The State of the World's Children*. Oxford: Oxford University Press, p. 80.

Varawalla NY, Old JM, Venkatesan R, Weatherall DJ. (1991). The spectrum of β-thalassemia mutations on the Indian subcontinent: The basis for prenatal diagnosis. *British Journal of Haematology 78*, 242–247.

Vaz FE, Thakur CB, Banerjee MK, Gangal SG. (2000). Distribution of beta-thalassemia mutations in the Indian population referred to a diagnostic center. *Hemoglobin* 24:181–194.

Venkatesan R, Sarkar R, Old JM. (1992). Beta-Thalassaemia mutations and their linkage to beta-haplotypes in Tamil Nadu in southern India. *Clin Genet* 42:251–256.

Verma IC, Saxena R, Thomas E, Jain PK. (1997). Regional distribution of β-thalassemia mutations in India. *Human Genetics* 100:109–113.

Weatherall DJ, Clegg JB. (2001). *The Thalassaemia Syndromes*. Oxford: Blackwell Science, 4th ed.

WHO Working Group Report. (1982). Community control of hereditary anemias. *Bull WHO* 60:643–660.

WHO. (2008). Joint WHO-TIF meeting on management of haemoglobin disorders (2nd: 2008: Nicosia, Cyprus) Geneva: World Health Organization. (NLM classification: WHO 190).

92.

CONVENTIONAL AND MOLECULAR CYTOGENETICS IN INDIA

Prochi F. Madon

INTRODUCTION

Cytogenetics, the study of chromosomes, is carried out by the conventional method of karyotyping and the comparatively recent molecular cytogenetic method of fluorescence in-situ hybridization (FISH). Among the genetic diagnostic services currently available in India, cytogenetics, especially karyotyping, is the most routinely carried out genetic test, though the FISH test is also being widely used now. In 1980, Dr. I.C. Verma compiled the first directory of human genetic services in India (Verma, 1980). More recently, Professor Dr. Jai Rup Singh and colleagues published a detailed directory of 54 centers in India from 18 states and union territories that undertake any type of genetic diagnostic investigations (see Chapter 98). Cytogenetic investigations were carried out in 45 of these centers (Singh et al., 2007a), although today the number is estimated to be higher. The state of Maharashtra in western India has the most centers, while there seems to be a dearth of genetic facilities in northeastern India (Singh et al., 2007b). However, more updated information regarding the total number of centers offering various types of genetic services across the country is not available, though a list with contacts of 47 centers for genetic services in India, most of which include cytogenetics, is mentioned in the article by Singh et al. (2010), which can be accessed freely (www.krepublishers.com).

The cities in India where human cytogenetic facilities are currently available include Delhi, Mumbai, Chennai, Kolkata, Bangalore, Hyderabad, Amritsar, Lucknow, Chandigarh, Vellore, Kochi, Edappal, Manipal, Trivandrum, Coimbatore, Mysore, Visakhapatnam, Pune, Ahmedabad, Baroda, Surat, Bhopal, Jammu-Tawi, Nashik, Nagpur, Kolhapur, Jaipur, Aligarh, Faridabad, Gurgaon, Siliguri, Bambolim, Rohtak, Kurukshetra, Aligarh, Patiala, and probably many more.

THE HISTORY OF HUMAN CYTOGENETICS IN INDIA

The chromosome number of man was established correctly as 46, by Tjio and Levan (1956). Within the next few years, many abnormal karyotypes were discovered and techniques for karyotyping were refined. The earliest publications in medical cytogenetics in India were by Chaudhuri and Chaudhuri (1963), and Chaudhuri (1964). This was followed shortly by a study of 226 patients investigated in a private laboratory at Kolkata (Chaudhuri & Chaudhuri, 1966), a case of translocation Down syndrome (Chaudhuri et al., 1966) and others (Khanna, 1966; Sadasivan & Ebenezer, 1967; Solanki et al., 1967; Grover et al., 1968; Gupta, 1970; Verma & Ghai, 1970; Verma et al., 1971). Professor Sharat Chandra started the Institute for Genetic Studies in Bangalore in 1967 and was interested in X-chromosome inactivation (Brown & Chandra, 1973). Dr. G.S. Mutalik, who was in the Division of Medical Genetics, Johns Hopkins University, in the 1960s (Murphy & Mutalik, 1969), started the genetics division at B.J. Medical College, Pune. Other genetic laboratories were set up in India mainly in the late 1960s and early 1970s.

The Indian Society of Human Genetics (ISHG) was established in 1973, with efforts of the founder members, Drs. L.D. Sanghvi, S.R. Das, G.S. Mutalik, A.J. Baxi, H.M. Bhatia, K.C. Das, K.C. Malhotra, D.P. Mukherjee, P.K. Sukumaran, J V. Undevia, and others (Sanghvi, 1974). The 1st Annual Conference of ISHG was held in Mumbai in 1974, by which time there were about a dozen centers in India where work in human cytogenetics was being carried out. These included the All India Institute of Medical Sciences (AIIMS) in Delhi, Institute for Research in Reproduction (IRR, under ICMR – Indian Council of Medical Research), Cancer Research Institute (CRI), and Sir J.J. Group of Hospitals & Grant Medical College (GMC) in Mumbai B.J. Medical College in Pune, St. Johns Medical College in Bangalore, and Punjab University in Chandigarh, from where papers were presented or published (Das, 1974; Indravathi, 1974; Master & Sukumaran, 1974; Pavri, 1976; Pavri et al., 1974; Peter, 1974; Varma et al., 1974; Mutalik et al., 1975; Verma & Jacob, 1976; Ghosh et al., 1978; Joseph & Thomas, 1982). At the 2nd Annual Conference of ISHG, held in Kolkata in 1975, the number of published abstracts in cytogenetics almost doubled, with papers presented from Hyderabad,

Kolkata, and Waltair as well as the others (Ahuja et al., 1975; Bhasin, 1975a, 1975b; Ghosal, 1975; Ghosh et al., 1975; Kucheria et al., 1975; Navalulkar & Shukla, 1975; Phadke et al., 1975b; Sadasivan, 1975; Saxena, 1975; Sobti, 1975, Verma et al., 1975). A symposium on "Structural and Functional Aspects of Chromosomes" held at the Bhabha Atomic Research Centre in 1975 was inaugurated by their emeritus scientist, Dr. A.R. Gopal-Ayengar, and organized by Dr. N.K. Notani, both of whom had a keen interest in cytogenetics. A seminar on medical genetics was also held in 1975 at Osmania University, Hyderabad, where Professor Allan Emery from the University of Edinburgh took an active part, and Dr. O.S. Reddi was the main organizer. A majority of papers in cytogenetics presented at the seminar were on antenatal diagnosis (Kucheria & Keswani, 1975; Thomas, 1975; Verma, 1975). An international workshop on medical genetics was organized by Dr. I.C. Verma at Pondicherry in October 1977. The papers presented were compiled into a book entitled *Medical Genetics in India*. Topics in cytogenetics included abnormal sex phenotypes (Pavri et al., 1978), hypogonadism (Kucheria et al., 1978), Robertsonian translocations (Thomas et al., 1978), the role of NOR's in nondisjunction (Singh & Ghosh, 1978), and satellite association in Down syndrome (Murthy & Ambani, 1978). The first PhD students (Bamezai, Kadam-Amare, Kolluri, Moghe-Kulkarni, Pavri-Madon, Peter, Potti-Agustus, Sarkar, etc.) working on topics in human cytogenetics (Kolluri et al., 1978; Madon, 1979) were awarded doctorates in the early 1980s. At that time, dermatoglyphics in cytogenetic disorders was widely studied (Mutalik & Lokhandwala, 1968; Saksena & Kumar, 1968, Kher et al., 1971). This was the study of dermal ridges such as arches, loops, whorls, ridge count, Simian crease, etc. on the palms and soles, since there was a very strong association of a specific dermatoglyphic pattern in Down syndrome. Dermatoglyphic studies in cases of sex chromosome mosaicism showed an increased frequency of arches and a low total finger ridge count (Pavri et al., 1977b).

The *Indian Journal of Pediatrics* started publishing some of the first articles on cytogenetics in India (Chaudhuri & Chaudhuri, 1963). Cytogenetic data from Indian laboratories was published in a regularly updated International Registry (Borgaonkar et al., 1993), right from its first listing in 1975, and in a subsequent book (Borgaonkar, 1997). The *Bulletin of Human Genetics*, a former newsletter of ISHG, was initiated in 1975 and was converted in 1995 to the *Indian Journal of Human Genetics,* which was published biannually. Since 2005, this journal has published three issues per year (www.ijhg.com). The inaugural issue of the quarterly *International Journal of Human Genetics* (KRE publishers, Delhi) was published in March 2001 (http://www.krepublishers.com/02-Journals/IJHG/IJHG-00-0-000-000-2001-Web/IJHG-00-0-000-000-2001-1-Cover.htm). The biannual newsletter of the Indian Society of Prenatal Diagnosis and Therapy (ISPAT), called *Fetal Medicine and Surgery,* started around 1998 and was upgraded in 2010 as the *Journal of Prenatal Diagnosis and Therapy* (www.jpdt.org). All these journals are accessible with free full text online at the websites given above. A recent addition is *Genetic Clinics*, a free quarterly newsletter of the Genetics Chapter of the Indian Academy of Pediatrics (geneticsiap@gmail.com) from SGPGI, Lucknow.

The annual conferences of ISHG have seen an increasing participation, even of postgraduate students, mainly from the zoology departments of various universities, and many of their research publications are in the journals mentioned above. A "young scientist award in cytogenetics," named after Dr. K.H. Marimuthu, was instituted by ISHG in the 1970s together with a few other awards in different disciplines, and papers are presented and judged at the annual conferences. The 35th Annual Conference of ISHG recently held at SGPGIMS in Lucknow, and a few earlier ones, have now shown an increasing trend toward research in molecular genetics and pharmacogenomics. ISHG 2011 at Manipal focused on genomics, genetic diseases, and diagnostics.

The first private cytogenetics laboratory in India was started in Mumbai in the 1980s by Dr. Hema Purandare and Dr. Amit Chakravarty, followed shortly by another one run by Dr. Saroj Kotwaliwale, all of whom were from Grant Medical College and Sir J.J. Group of Hospitals, Mumbai. A number of new laboratories were also opening up in government, municipal, and private hospitals, medical colleges, research institutions, and zoology departments of various universities. The genetics unit at IRR (now known as NIRRH) in Mumbai was upgraded as the Genetic Research Centre (GRC), headed by Dr. Zarine Patel, and was shifted to Bai Jerbai Wadia Childrens' Hospital in the mid-1980s, though it has now moved back to NIRRH. Another ICMR institute (NIIH) in the neighboring campus of KEM Hospital also carries out research and diagnostics in cytogenetics. Many laboratories, such as those run by SN GeneLab, Lal Path Lab, Metropolis, Perkin Elmer, Piramal Diagnostics, OncQuest, Preventine, Quest Diagnostics, Reliance Life Sciences, Super Religare Laboratories, Thyrocare, Vimta Lab, etc., also carry out cytogenetic diagnostic tests or have affiliated with or taken over existing laboratories, and have well-established logistics for sample collection from all over India.

MOLECULAR CYTOGENETICS

FLUORESCENCE IN SITU HYBRIDIZATION (FISH)

The molecular cytogenetic technique of fluorescence in situ hybridization (FISH) was standardized and tested in India in a few laboratories in the late 1990s (Jobanputra et al., 1998; Amare et al., 2001). Before the advent of commercially available, easy-to-use labeled FISH probes, a few institutes in India, such as Tata Memorial Hospital, Mumbai, and SGPGI, Lucknow, had standardized the indirect labeling technique using biotin-streptavidin and a combination of antigen-antibody complexes, and conducted workshops in the 1990s. Diagnostic FISH of specific tests was carried out for a limited group of patients. After the availability of directly labeled FISH probes, Jaslok Hospital and Research Centre in Mumbai was one of the first to open up routine FISH diagnostics on a wide variety of samples. They have been accepting prenatal, postnatal, and leukemia samples from all over India since 2001, and were also the first to successfully introduce preimplantation genetic diagnosis (PGD) in India for aneuploidies and translocations by FISH (Parikh et al., 2001a; Madon, 2006a, Athalye et al., 2007, Madon et al., 2010a,b,c).

There have been a number of Indian publications on the diagnostic application of the FISH technique, particularly for rapid prenatal diagnosis of common aneuploidies in high risk cases (Ankathil et al., 1992; Jobanputra et al., 2001, 2002; Athalye et al., 2002; Parikh et al., 2002; Modi & Bhartiya, 2002; Madon et al., 2010c), diagnosis of leukemias (Kucheria et al., 1998, 2003; Amare, 2002; Berde et al., 2002; Madon et al., 2003a, 2007b; Ganguly et al., 2007a), breast cancer (Panjwani et al., 2010), bladder cancer (Madon et al., 2006b), microdeletion syndromes (Chaddha et al., 2003; Halder et al., 2005a; Gawde et al., 2006; Madon et al., 2010a), spontaneous abortions (Madon et al., 2001; Rao et al., 2002; Kokate et al., 2010), arrested embryos (Fauzdar et al., 2008, Madon et al., 2010c), low-grade mosaicism (Madon et al., 2004b; Patki et al., 2004; Parikh et al., 2006) and other cases (Rajangam et al., 1993; Rao & Mohanty, 2005; Gowda et al., 2010). Kokate et al. (2010) carried out aneuploidy screening by FISH on placental tissue/fetal skin of 2272 cases of spontaneous abortions and detected common aneuploidies in 16% cases, the most common being monosomy X followed by triploidy and trisomy 21, seen mainly in male fetuses. Sequential FISH was also carried out on previously G-banded metaphases to delineate cryptic abnormalities (Madon et al., 2003b). A recent publication on microdeletion syndromes (Madon et al., 2010a) detected 73 positive cases out of 374 studied by FISH. These included 29 cases of Angelman syndrome, 24 of Williams, 16 of Prader-Willi and 4 cases of deletion 22q, although the clinical features of diGeorge syndrome were not present in all, some of them having other features of the CATCH-22 spectrum. The diagnosis of Phelan-McDermid syndrome caused by deletion 22q13 (control region of the diGeorge probe) was confirmed by FISH after a suspicion on karyotyping with high-resolution banding in a child with mild features of the autism spectrum disorder (personal unpublished data). Advanced techniques, such as M-FISH, m-BAND, CGH, and array-CGH, have been applied in a few publications from India (Halder et al., 2005b; Bakshi et al, 2006; Sheth et al., 2009, 2010; Chakravarty & Chakravarty 2010).

CLINICAL CYTOGENETICS

DOWN SYNDROME

There have been a number of cytogenetic studies in Down syndrome (DS) in India. Studies from Western countries initially focused on the prevalence of DS in pregnancies of older mothers, but studies in India found that a majority of such children, even of the common free trisomy 21 type, were born to young mothers under the age of 30 (Murthy & Ambani, 1978; Verma et al., 1979; Jyothy et al., 2000; Kaur et al., 2003; Kava et al., 2004; Malini & Ramachandra, 2010) or 35 (Kothare et al., 2002). It was also observed that the presence of an increased frequency of satellite association, which could cause nondisjunction, was seen more often in a parent of a child with trisomy 21 (Murthy & Ambani, 1981; Vishwanath et al., 1996; Yashwanth et al., 2010). Kochupillai et al. (1976) demonstrated a significantly high frequency of mental retardation and Down syndrome in an area of high-level natural radiation from the sands of Kerala. However, Cheriyan et al. (1999) karyotyped 16,169 newborns and found that the frequencies of total autosomal and sex aneuploidies, as well as structural anomalies, were comparable between the high-level and normal-level natural radiation areas, though they reported the presence of rogue cells, the rarely occurring metaphases with a high level of chromosomal damage, in newborns. Murthy et al. (1987) reported 60 cases of DS from Ahmedabad. Shah, et al., (1990), karyotyped 144 cases suspected of various chromosomal abnormalities and detected cytogenetic anomalies in 39.58% cases, of which DS was found in 25.7% and sex chromosome abnormalities in 13.2%. Lakshminarayana (1990) karyotyped 500 DS children and detected 15 translocations, of which 9 were de novo. Verma et al. (1991) carried out cytogenetic analysis in 645 cases of Down syndrome and compared data on 2410 cases from different studies in India, showing the mean frequency of different types as free trisomy 21 (92%), translocations (4%), and mosaics (4%). An unusual case of Down syndrome in four sibs due to a balanced reciprocal translocation 2;21in the mother was published (Kotwaliwale et al., 1991). Rajangam et al. (1995) reported a rare case of a trisomy 21 female with mosaicism for a paternally inherited small supernumerary marker chromosome comprising the short arms of acrocentrics. A large multicentric "study of malformations and Down syndrome in India" (SOMDI)—funded by the Board of Research in Nuclear Sciences (BRNS), Department of Atomic Energy (DAE), Government of India—published data pooled from Mumbai (Bharucha,

1998), Baroda (Modi et al., 1998), and Delhi (Verma et al., 1998), showed that from a total of 97,446 consecutive births from selected hospitals, there were 59 cytogenetically proven cases of Down syndrome, giving an incidence of 0.61 per 1000 births. All these had free trisomy 21 except one case that had a Robertsonian translocation t(13;21). Jyothy et al. (2000) studied 1001 cases of DS from Andhra Pradesh over a 20-year period and found that the origin of the extra chromosome 21due to meiotic nondisjunction was maternal in 79.24% cases. In a study of 8551 consecutive births in Goa (Vaz & Shyama, 2005), cytogenetic analysis was carried out in all cases of congenital malformations and the incidence of DS was 1.52 per 1000 births. Mosaicism was seen in 2/13 cases of trisomy 21. Unusual cases of Down syndrome have also been reported (Cyrus et al., 2005, 2006; Sheth et al., 2007; Vashist et al., 2009). Data on 874 karyotyped cases of DS collected over a period of 35 years at a medical college in Bangalore (Jayalakshamma et al., 2010) showed free trisomy in 86.9%, translocation in 8.8%, and mosaicism in 4.3%, with a preponderance of males (54.24%). Translocation t(14;21) formed 62.34% of all translocations. Another review of 1020 karyotyped cases in Chennai (Chandra et al., 2010) showed free trisomy in 83.82% and translocations in 5% cases (of which 50% were de novo), and mosaicism in 10.78%. A 14-year cytogenetic study of 1950 cases in Amritsar (Kaur & Singh, 2010) detected 222 cases of DS, of which 70% were males. Mandava et al. (2010) analyzed 1572 cases of Down syndrome detected at a large private laboratory in India, and reported cases of double aneuploidy, mosaic double aneuploidy, and novel translocations. Bansal et al. (2010) reviewed genetic counseling in chromosomal abnormalities, focusing on different types of Down syndrome.

IDIOPATHIC INTELLECTUAL DISABILITY

Walter et al. (1971) carried out cytogenetic studies in mental retardation and reported a case of the XXXXY variant of Klinefelter syndrome. Kulkarni et al. (1981) detected partial trisomy and partial monosomy for 5p in two sibs, whose father was a balanced translocation carrier. Moghe et al. (1981) studied the frequency of karyotype abnormalities in 74 mentally retarded patients excluding Down syndrome, selected from a total of 306 cases, and 14 of these had chromosome abnormalities including sex chromosome abnormalities. Structural autosome abnormalities, such as deletions 1q, 2q, 5p and 17p, duplication 5p and translocation t(13;14), were detected. Ambani et al. (1984) studied 709 subjects with mental retardation and found cytogenetic abnormalities in 248 (35%). Reddy and Thomas (1985) studied 200 such subjects, of which 39 (19.5%) showed cytogenetic abnormalities. In an ICMR multicentric task force study on 1314 subjects (Hegde, et al., 1989), cytogenetic abnormalities were detected in 312 (23.7%) cases. Gambhir et al. (1991) reported a translocation between chromosomes 1 and 10 in a boy with mental retardation and dysmorphic features. A neonate with multiple dysmorphic features and growth retardation was found to have partial trisomy 10q arising from a balanced translocation in the father (Suresh et al., 1995). Prabhakara et al. (2002) reported a case of partial trisomy 15q22, due to segregation of maternal t(10;15) reciprocal translocation. In a study of 150 female subjects with mental retardation from Kolkata (Dutta et al., 2009), besides common aneuploidies like trisomy 21 and Turner syndrome, detectable chromosome abnormalities were not seen, although marked polymorphisms such as large satellites were reported in two cases. Interesting cases of structural aberrations associated with mental retardation have also been reported (Thomas et al., 2001; Shetty et al., 2006; Shetty & Dave, 2010). Most of the PhD theses on cytogenetics in mental retardation are from NIMHANS, Bangalore (Latha, 1996; Bhaskar Rao, 1999; Venkatesh, 2006). Vallabhajosyula et al. (2009) reported a rare case of ring 21in a male child with dysmorphic features and mental retardation. Analysis of a 14-year study of 1760 cases of mental retardation karyotyped at a private laboratory in Mumbai (Dave & Shetty, 2010) was part of a PhD thesis of the University of Mumbai. Chetan et al. (2010) carried out cytogenetic investigations including FISH in 100 cases of idiopathic mental retardation, and found chromosomal anomalies in 12. These included cases of the microdeletion 15(q11q13), pericentric inversion 9 with different breakpoints, a paracentric inversion Xq, ring chromosomes, and translocations.

FRAGILE-X AND AUTISM

Manjunatha et al. (1988) reported the first case of fragile-X from India. This is the second most common genetic cause of mental retardation after Down syndrome (Chetan et al., 2002). Kar (1995) reported the expression of fragile-X in a patient with clinical features of deletion 9p syndrome. Elango and Verma (1996) screened 1111 patients with mental retardation and selected 55 for cytogenetic studies to detect folate-sensitive fragile sites, of which 20 were positive. The frequency of cytogenetically diagnosed fragile-X cases among 360 boys with mental retardation was found to be 6.38% in a study by Jain et al. (1998). Bhaskaran et al. (1999) reported spontaneous expression of the fragile site Xq27.3 in a fragile-X patient. Chetan et al. (2001) demonstrated induced fragile sites at the "qter" region of all C-group chromosomes, stressing the need for confirmation of fragile-X by subsequent G-banding. The frequency, association, and genetic implications of chromosomal fragile sites in 300 subjects with idiopathic mental retardation was compared to that of 100 controls

(Manjunatha et al., 2002) and 19 fragile sites were seen expressing in more than 4% cells in 28 subjects from 23 families. Fragile-X and other autosomal fragile sites have been the topic of a few recent PhD dissertations (Mehta, 2005; Athalye, 2008).

A study on 150 subjects with autism analyzed for various chromosomal fragile sites (Manjunatha et al., 2001) found 12 individuals positive for fragile-X chromosome. The percentage of expression varied from 4%–30% cells in different cultures. Other chromosomal fragile sites (3%–14%) were noticed among 25 autistic individuals who were negative for fragile-X. Genetic studies in children with intellectual disability and autistic spectrum disorders (ASD) from Mumbai concluded that children falling into the category of classical autism do not have associated chromosomal abnormalities, whereas those with autistic traits and PDD have chances of associated chromosomal abnormalities (Balasubramanian et al., 2009). They recommended that detailed cytogenetic evaluation should be carried out in all subjects with ASD, particularly if the subject additionally shows intellectual disability, abnormal EEG patterns or seizures, hypotonia, severe motor and gait problems, or dysmorphic features.

COUPLES WITH RECURRENT SPONTANEOUS ABORTIONS (RSA)

One of the earliest cytogenetic studies in India on couples with repeated abortions was by Rani et al. (1974). Singhania et al. (1985) carried out cytogenetic studies in repeated fetal loss. Gupta and Chakravarty (1995) karyotyped 225 couples with recurrent abortions, where clinical, hormonal, and biochemical factors were negative, and found chromosomal anomalies in 19% cases, of which 47% were mosaics. Some of these had only 10% abnormal metaphases. They also stressed the role of satellite association in recurrent abortion and congenital defects, which was supported by others (Anuradha et al., 2002). A bisatellited supernumerary marker chromosome shown to be an *idic* 14 or 22 was seen in the male partner of a couple with recurrent miscarriages (Thanemozhi et al., 1997) and the role of the marker chromosome in the etiology of RSA was discussed. The cytogenetic basis of recurrent abortions was the subject of the L.D. Sanghvi oration given by Manorama Thomas (1999) at the ISHG conference in Mumbai. Kalpana et al. (2000) compiled data on 17,047 aborters, where abnormalities were seen in 5.84% cases. Patel et al. (1996b, 2000) reported chromosome abnormalities and variants in couples with RSA from western India. Premature centromere division was also associated with habitual aborters (Anuradha et al., 2001; Kalpana et al., 2004). An explanation of the causes of RSA and other genetic disorders is given online by Jalan (http://jalananil.tripod.com/NIRMAN/id12.html).

The proper technique of specimen collection of aborted tissue is very important for detailed cytogenetic analysis (Madon, 2004a). Dubey et al. (2005) karyotyped 742 couples with a history of RSA and found chromosome abnormalities in 2% and variants in 3.2% cases. An overview of RSA (Meka & Reddy, 2006) describes various genetic and nongenetic causes within the Indian scenario. Nair et al. (2006) reported 4 cases of RSA with balanced translocations between chromosomes 12 and 15, with different breakpoints, from Kerala. Another study (Mangalgiri & Pathak, 2008) detected 3 Robertsonian translocations among 40 couples studied. A case of a de novo complex chromosomal rearrangement (CCR) 46,XX,t(7;15;13)(p15;q21;q31) in a female with an adverse obstetric history was reported (Iyer et al., 2009) with literature citations of other such cases involving 3–5 chromosomes in RSA. Cytogenetic analysis of 6750 couples with repeated miscarriages (Mandava & Das, 2010) showed the prevalence of major chromosomal abnormalities, including complex and double translocations and insertions, to be 3.3%. Tilak (2010) described phenotypic abnormalities in 58 cases of reciprocal translocations. Venkateshwari et al. (2010) reported a rare case of a 21/22 translocation from Hyderabad, while Kar & Thakkar (2010) reported another with t(14;15) and inv(9) from Chennai. Abilash et al. (2010) described a rare case of a female with 9 consecutive abortions, who had a Robertsonian translocation t(13;21) and the SRY gene translocated to her X-chromosome, and discussed the role of an interchromosomal effect. Dutta et al. (2011) studied 1162 couples with recurrent miscarriages in southern India and detected chromosome anomalies, mainly reciprocal translocations, in 78 cases.

INFERTILITY AND RELATED CONDITIONS

Some of the earliest reports of the scope and limitations of the sex chromatin test, also known as *Barr body analysis*, for clinical diagnosis and research were by Shah (1957) Mallik (1962), Shetty et al. (1966) and Malhotra et al. (1969), followed by others (Sen & Thombre, 1971; Verma et al., 1973; Chakravarty et al., 1978; Purandare & Chakravarty, 1980). This test was an easy alternative to karyotyping for the detection of numerical X-chromosome abnormalities such as Turner syndrome, Klinefelter syndrome, XXX, and others, though the percentage of mosaicism could not be determined. The Barr body seen in interphase nuclei from buccal smears represented the inactive X-chromosome; therefore, a normal female would have 1 Barr body, while it would be absent in a female with Turner syndrome and present in a male with XXY, etc. This test has now been replaced by FISH, which is an accurate quantitative test and can even help in detecting low-grade mosaicism.

Cytogenetic studies of numerical sex chromosome anomalies and common autosomal aneuploidies were carried out even before G-banding was routinely done in India (Pathak & Sharma, 1968; Chaudhuri, 1969; Grover & Shivde, 1971; Phadke et al., 1975a; Pavri et al., 1977a; Kelkar et al., 1978; Madon et al., 1978a, b; Verma et al., 1978). Later studies with banding techniques included cytogenetic abnormalities in male infertility (Peter et al., 1980; Ghosh et al., 1986), primary amenorrhea (Joseph & Thomas, 1982; Mulye et al., 1983; Roy & Banerjee, 1995; Shanker et al., 1995; Kalpana & Satyanarayana, 1997; Mondal et al., 2002; Kalavathi et al., 2010; Vijayalakshmi et al., 2010) and various disorders of sexual development including ambiguous genitalia (Sankar & Phadke, 2006), hypospadias, true hermaphroditism, pseudohermaphroditism, Turner syndrome, Klinefelter syndrome, and variants (Kannan & Ganpathy, 1982; Shah et al., 1982; Iyengar et al., 1986; Lata et al., 1986; Maji et al., 1990; Shreeniwas et al., 1994; Suri et al., 1995; Panthaki, 1998; Thangaraj et al., 1998; Rajangam et al., 1999; Raman et al., 2001; Sood & Trehan, 2001; Jyothy et al., 2002; Mukherjee & Pandit, 2002; Ganguly & Sahni, 2003; Madon et al., 2003b; Kaur et al., 2004; Movva et al., 2005; Nagvenkar et al., 2005; Goud et al., 2006; Rao et al., 2008b; Bertha et al., 2010; Kaur & Singh, 2010; Khan & Gangane, 2010; Kumar et al., 2010; Margaret et al., 2010; Rajasekhar, et al., 2010; Salaskar, 2010; Sangeetha et al., 2010).

Polymorphic variants of heterochromatin, such as 9qh+ and Yqh+, have been reported to be associated with infertility and various other disorders (Bakshi et al., 1997; Patel et al., 1997; Madon et al., 2005; Rao & Ghosh, 2005; Rao et al., 2006) from the early days, though they were also seen in the general population (Bhasin, 2005) and were considered as variations in "junk DNA" with no clinical significance. However, there is now renewed interest in the subject, as they may be associated with epigenetic factors that could play a role in infertility (Minocherhomji et al., 2010). A case-control study has shown a significantly higher frequency of 9qh+ and Yqh+ associated with infertility, when compared with a matched fertile control group (Minocherhomji et al., 2009).

PRENATAL DIAGNOSIS AND PRODUCTS OF CONCEPTION

In India, antenatal diagnosis with ultrasound started only in the latter part of the 1970s (Agarwal, 1974), diagnostic chorionic villous biopsy (CVB) and amniotic fluid karyotyping in the early 1980s, while cordocentesis started in the late 1980s (Mathur et al., 2002). Gogate et al. (1994) critically evaluated various methods of fetal tissue sampling and karyotyping for prenatal diagnosis. A case of holoprosencephaly with trisomy 13 was detected prenatally by USG (Patel et al., 1996a). Jobanputra et al. (2001) reported on prenatal diagnosis in 82 women with high-risk pregnancies by karyotyping and FISH. Verma et al. (2003) described their experiences in genetic counseling and prenatal diagnosis at a tertiary genetic center in India. Of the 3500 subjects who were provided genetic counseling, 28.7% were for prenatal diagnosis. Their data, compared with previous studies, showed a change toward increased prenatal diagnostic tests, the commonest indication for amniocentesis being advanced maternal age. Bansal et al. (2010) described the importance of genetic counseling in chromosome abnormalities and its role in prenatal diagnosis. Agarwal et al. (2010) reported prenatal diagnosis at 20 weeks in an interesting case where the triple test showed an increased risk for a neural tube defect. There was a history of a previous spontaneous abortion in the first trimester. The husband's niece had global developmental delay, and was found to have a deletion of the critical region for Angelman syndrome due to a reciprocal translocation between chromosomes 10 and 15. The husband was karyotyped and showed t(10;15)(p21;p11.3). Prenatal ultrasonography did not show any abnormality. Amniocentesis showed a microdeletion of 15q11-q13 by FISH, and an unbalanced chromosomal rearrangement 45,der(10)t(10;15)(q26;q15),-15 by karyotyping. Autopsy on termination of pregnancy showed no malformation except hypertelorism and micrognathia, highlighting the importance of genetic counseling. Gazala et al. (2010) described prenatal diagnosis in a case where a previously affected child was found to have a recombinant 6 resulting from maternal inv(6)(p21.2q25.3) on repeat karyotyping. The fetus had an inversion similar to the mother, and the pregnancy was continued.

Beula et al. (2010) reported on evaluation of cardiac defects in a fetus with an interstitial deletion of 8p. Lall et al. (2010) karyotyped 5500 amniotic fluid samples in high-risk cases and detected chromosome abnormalities in 2%, of which 59% were numerical and 41% structural anomalies. Nine pericentric inversions of chromosomes 2, 8, 9, 19, and Y were also detected, requiring careful genetic counseling. Sahetya et al., (2010) stressed the importance of prenatal karyotyping in high-risk cases to reduce the burden of stress, as a majority of cases are normal. Madon et al. (2010c) reported prenatal cytogenetic diagnosis in over 3100 cases, where 92% were from amniotic fluid, 6% CVS, and 2% cord blood. Aneuploidy was detected in only 2% of the high-risk cases, trisomy 21 being the most frequent (75%), followed by sex chromosome anomalies (13%) and trisomy18 (8%). The other anomalies included trisomy 9 mosaicism, partial trisomy 9 (Dandy Walker malformation), monosomy 13 and ring 13 mosaicism, trisomies 11, 2, small supernumerary markers, inherited balanced translocations, and polymorphisms (pericentric inversions 9 and Y). A few samples were checked for microdeletions such as Prader Willi/Angelman, Williams, and diGeorge, where there was a history of a previous affected child, and the subsequent pregnancy was normal in all cases. In contrast, cytogenetic analysis on 550 products of conception (mainly from miscarriages) showed abnormalities in 141 (25.64%) cases. Of these, autosomal and sex chromosome aneuploidies were present in an equal number (31.9% each). Cases of triploidy (15.6%), tetraploidy (12.8%), and structural anomalies (7.1%) including balanced and unbalanced Robertsonian translocations, a supernumerary isochromosome 12p (Pallister Killian syndrome), and pericentric inversion 9 variant (6 cases) were also observed. The 45 cases of autosomal aneuploidies included trisomies 13 (10/45 cases), 16 (8/45), 18 (7/45), 21 (6.45), 22 (6/45), and 1 case each of trisomy 3, 4, 6, and 15. Double aneuploidy was seen in 3 cases. The sex chromosome anomalies included cases of Turner (28/45), Klinefelter syndrome (3/45), and mosaicism.

PREIMPLANTATION GENETIC DIAGNOSIS (PGD)

PGD is a state-of-the-art technique where genetic analysis is done on a single cell from cleavage-stage embryos, and normal embryos are selected for transfer. This is an additional step in the IVF procedure. PGD by FISH for chromosome aneuploidies and translocations has been successfully established in India at Jaslok Hospital and Research Centre, Mumbai, by Dr. Firuza Parikh and her team (Parikh et al., 2001b; Parikh & Madon, 2004; Athalye et al., 2006; Madon et al, 2007a, 2010b,c). PGD by FISH is useful in couples with a history of recurrent miscarriages, when one of the partners is a carrier of a balanced reciprocal or Robertsonian translocation. PGD helps the couples to have a normal child faster than by natural conception. Preimplantation genetic screening (PGS) for common aneuploidies is also done by FISH. Multicolor probes are used in the first round. The same cell is washed and hybridized with different probes in 2–4 subsequent rounds to check for aneuploidy of 7–11 chromosomes, depending on the quality of the fixed nucleus. A lot of practice of biopsy, fixation, and FISH on arrested embryos is essential before undertaking PGD.

CYTOGENETIC EFFECTS OF OCCUPATIONAL HAZARDS

There are a number of studies on individuals exposed to various occupational hazards and other chemicals. Jaju et al. (1986) studied the cytogenetic effect of neomycin on human lymphocytes in vitro. Vegetable garden workers exposed to pesticides were screened and found to have a higher frequency of chromosomal aberrations and sister chromatid exchanges compared to controls (Rupa et al., 1988). Results of immunological and chromosome studies on the effects of exposure to the toxic gas methyl isocyanate in the Bhopal gas tragedy were published by Saxena et al. (1988). Rupa et al. (1991) studied the clastogenic effect of pesticides in lymphocytes of cotton field workers. Kaur et al. (1999) analyzed the frequency of sister chromatid exchanges in treated and untreated leprosy patients. A preliminary report by Krishnaja and Chauhan (2000) showed that quinacrine dihydrochloride, the female nonsurgical sterilant, induces dicentrics, rings, and marker chromosomes in peripheral blood lymphocytes treated in vitro. Chromosomal aberrations in the peripheral lymphocytes of workers exposed to diagnostic X-rays were studied by Maddileti et al. (2002) and Gadhia et al.(2004). Murugesan (2003) studied radiation-induced chromosomal hotspots at G_1 and G_2 stages of human lymphocytes in culture. Krishnaja and Sharma (2004) studied transmission of gamma ray–induced unstable chromosomal aberrations through successive mitotic divisions in human lymphocytes in vitro. Prabhu et al. (2004) studied the effect of 2-deoxy-D-glucose on the induction of chromosomal aberrations in lymphocytes exposed in vitro to gamma radiation at a dose rate of 1.0 Gy/minute. Sasikala et al. (2003) conducted a hemato-biochemical and cytogenetic study of active and passive smokers. Chromosome damage in lymphocytes of petrol pump workers (Yadav & Seth, 2001), stainless steel welders (Yadav et al., 2001a), rubber industry workers (Yadav & Chhillar, 2001), panmasala chewers (Yadav & Chadha, 2002a), aluminum foundry workers (Prasad et al., 2002), mint factory workers (Padmaja et al., 2002), traffic policemen (Kamboj & Sambyal, 2006), individuals exposed to vehicular pollution (Yadav & Chadha, 2002b), and cement dust (Jude et al., 2002) was studied. Chromosomal aberrations induced by ferric chloride (Poddar et al., 2004), cyproterone acetate (Siddique & Mohammad, 2004), and nickel and chromium, which was ameliorated by curcumin (Rao et al., 2008a), were also reported.

CANCER CYTOGENETICS

Some of the first Indian publications on cytogenetic studies in hematological malignancies and solid tumors were by Biswas and Chowdhury (1968), Sarkar and Chowdhury (1972), Manna (1974), and Parekh (1974 a, b). Khare et al. (1977) described a consistent chromosome abnormality of the "E" group chromosomes in acute leukemias. Bhisey et al. (1978) carried out cytogenetic studies in the blastic crisis of chronic myeloid leukemia (CML). Many cytogenetic and molecular cytogenetic studies in CML have been reported to date (Das et al., 1978; Bhisey et al., 1982; Chandra & Marimuthu 1987; Parikh et al., 1992; Chinnappan et al., 2000; Amare et al., 2002; Jacob et al., 2002; Madon et al., 2002; Talwar et al. 2002; Gadhia et al., 2005; Chavan et al., 2006). Among the other hematological malignancies, there have been reports of cytogenetic studies in acute leukemias (Pande et al., 1986; Gladstone et al., 1994; Shivakumar et al., 2002; Bakshi C et al., 2003;

Bakshi S et al., 2004; Ganguly et al., 2007b; Gujral et al., 2009; Trivedi et al., 2009), myelodysplastic syndromes (Nair et al., 1992; Kadam et al., 1995; Korgaonkar et al., 2008; Varma & Varma 2008), chronic lymphocytic leukemia (Lakshmaiah et al., 2010), and Hodgkin's disease (Kadam et al., 1990), among others.

Among the solid tumors, there have been studies in retinoblastoma (Rao et al., 1975; Harini et al., 2001), breast cancer (Bhattacharya et al., 2005; Kumari & Bhat, 2008; Kaur et al., 2009; Guleria & Sambyal, 2010), pleural and peritoneal effusions (Augustus et al., 1989) head and neck squamous cell carcinoma (Bhattacharya et al., 2003; Chakraborty et al., 2003), gastrointestinal tract cancer (Guleria & Sambyal, 2003; Jude et al., 2005; Kaur & Sambyal, 2008), bladder cancer using the Urovysion FISH test (Madon et al., 2006b), etc. Chromosome breakage syndromes such as Fanconi anemia and ataxia telangiectasia have also been studied (Athale et al., 1991; Chowdhury et al., 1996; Mathur et al, 2000; Madon et al., 2001).

REMARKS

An attempt has been made to include papers from a maximum number of cytogenetic centers in India and first authors by gathering publications from various Indian journals. Earliest publications and published abstracts have also been checked. Inadvertent omissions may be excused. A lot of data from many centers is still unpublished. Repetition of some cases in different publications from the same center is unavoidable. All laboratories involved in human cytogenetics are requested to send me their contact details to compile an updated listing, which will be informative.

ACKNOWLEDGEMENTS

Thanks are due to Mr. Dhananjay Kulkarni for assisting in compilation of references, and to my family for their patience during preparation of this manuscript.

REFERENCES

Abilash, V. G., Saraswathyand, R & Marimuthu, K. M. (2010). Inter chromosomal effect (ICE) resulting in increased abnormal pregnancies in an infertile female with a rare Robertsonian translocation (13;21)(p10;p10) and SRY gene carrier with 9 consecutive abortions. *International Journal of Human Genetics,* 10(1–3):169–174.

Agarwal, S. S. (1974). Antenatal diagnosis of genetic disorders. *Journal of the Association of Physicians of India,* 22(1):37–40.

Agarwal, M., Prajnya & Phadke, S., (2010). Prenatal diagnosis in a family with chromosomal translocation between chromosomes 10 and 15: A case report. *Journal of Prenatal Diagnosis and Therapy,* 1(1):45, PO-1.

Ahuja, Y. R., Nath, Y. R. & Geetha, L. (1975). Difficulties of human cytogenetic work in India and how to overcome them. In *Abstracts: 2nd Annual conference of the Indian Society of Human Genetics* (Abstract 42). ISI, Kolkata.

Amare, P., Baisane, C., Saikia, T., Nair, R., Gawade, H. & Advani, S. (2001). Fluorescence in situ hybridization: A highly efficient technique of molecular diagnosis and prediction for disease course in patients with myeloid leukemias. *Cancer Genetics and Cytogenetics,* 131:125–134.

Amare, P.S. (2002). Chronic myeloid leukemia: Cytogenetics and molecular genetics. *Indian Journal of Human Genetics,* 8(1):4–10.

Ambani, L.M., Patel, Z.M., Dhareshwar, S.S., Krishnamurthy, D. S., Moghe, M.S., Mulye, V.R., et al., (1984). Clinical, biochemical and cytogenetic studies in mental retardation. *Indian Journal of Medical Research,* 79:384–391.

Ankathil, R., Vijayakumar, T. & Nair, M. K.(1992). Molecular cytogenetics of human cancer. *National Medical Journal of India,* 5(2):63–68.

Anuradha, N., Satyanarayana, M. & Manjunatha, K.R. (2001). Premature centromeric division and abortions. *International Journal of Human Genetics,* 1(3):191–193.

Anuradha, N., Satyanarayana, M. & Manjunatha, K. R. (2002). Satellite associations in recurrent aborters. *International Journal of Human Genetics,* 2(1):61–64.

Athale, U. H., Rao, S. R., Kadam, P. R., Gladstone, B., Nair, C. N., Kurkure, P. A., et al. (1991). Fanconi's anemia: a clinico-hematological and cytogenetic study. *Indian Pediatrics,* 28(9):1003–1011.

Athalye, A. S., Madon, P. F., Dhumal. S., B., Saxena, S. G., Naik, N. J. & Parikh, F. R., (2002). The application of FISH in routine cytogenetic diagnosis of aneuploidies and malignancies. In Bhasin, M. K. & Kalla, K. K. (Eds) Proceeding volume of National Conference on Expanding Horizons of human Genetics, Delhi, pp. 9–13.

Athalye, A. S., Parikh, F. R., Naik, N. J. & Madon, P. F. (2006). Preimplantation Genetic Diagnosis. *Health Screen* (Thyrocare), 2(19):34–36.

Athalye, A. S., Madon, P. F. & Parikh, F. R., (2007). FISH – Introduction and setting up a FISH lab. In M. B. Agarwal (Ed.), *Advanced Techniques in Haematological Malignancies.* Ranbaxy Super Specialities, Mumbai, pp. 87–97.

Athlaye, A. S. (2008). Fragile X chromosome study at the site Xq27.3 by cytogenetic and molecular analysis. *Thesis submitted to University of Mumbai for Ph.D. in Applied Biology.* 118 pp.

Augustus, M., Visveshwara, R. N. & Bhargava, M. K. (1989). Chromosome analysis in effusions as a diagnostic adjunct to cytology. *Indian Journal of Cancer,* 26(2):67–75.

Bakshi, C., Amare-Kadam, P., Abhyankar, D., Baisane, C., Banavali, S. & Advani, S. (2003). Chromosomal rearrangement in Down syndrome with acute myeloid leukemia. *Indian Journal of Pediatrics,* 70(9):755–758.

Bakshi, S.R., Trivedi, A.H., Patel, R.K., Roy, S.K., Alladi, P.A., Bhatavdekar, J.M., Patel, D.D., Shah, P.M. & Rawal, U.M. (1997). Constitutive heterochromatin polymorphism in pediatric cancer patients. *Indian Journal of Human Genetics,* 3(2):89–94.

Bakshi, S. R., Roy, S.K., Trivedi, P. J., Brahmbhatt, M. M., Rawal, S. M., Kakadia, P.M., et.al. (2004). AML-M2 with der(18)t(1;18)(q2?;p11.3) in addition to t(8;21) and del(9q). *Indian Journal of Human Genetics,* 10(2):78–80.

Bakshi, S. R., Brahmbhatt, M. M., Trivedi, P. J. & Chudoba, I. (2006). Constitutional tetrasomy 18p. *Indian Pediatrics,* 43(4):357–360.

Balasubramanian, B., Bhatt, C. V. & Goyel, N. A. (2009). Genetic studies in children with intellectual disability and autistic spectrum of disorders. *Indian Journal of Human Genetics,* 15(3):103–107.

Bansal, V., Suresh, S., Suresh, I., Jagadeesh, S. & Fazal, G. J. (2010). Genetic counseling in chromosomal abnormalities. *Journal of Prenatal Diagnosis and Therapy,* 1:14–19.

Berde, P., Kopakka, N., Banerjee, M., Bhartiya, D. & Parikh, F. (2002). Fluorescent *in situ* hybridization and detection of microdeletion syndromes. In Bhasin, M. K. & Kalla, K. K. (Eds) Proceeding volume of National Conference on Expanding Horizons of human Genetics, Delhi, pp. 15–29.

Bertha, A., Danda, S., Tjandrajana, E., Srivastava, V., Subramanian, R. & Mathew, J. (2010). Klinefelter syndrome with systemic lupus erythematosus in an Indian man. *Lupus*, 19(7):870–871.

Beula, A. P. R., Hegde, S, Venkatesh, P. & Jayaprakash. (2010). Evaluation of cardiac defect in a fetus with 8p interstitial deletion. *International Journal of Human Genetics*, 10(1–3):183–186.

Bharucha, B.A. (1998). Study of malformations and Down Syndrome in India (SOMDI): Bombay region. *Indian Journal of Human Genetics*, 4(1):88–92.

Bhasin, M. K. (1975a). Cytogenetics study in primary amenorrhoea. In *Abstracts: 2nd Annual Conference of the Indian Society of Human Genetics* (Abstract 44). ISI, Kolkata.

Bhasin, M. K. (1975b). Giemsa banding technique and review of the literature on identifying human chromosomes. In *Abstracts: 2nd Annual conference of the Indian Society of Human Genetics* (Abstract 33). ISI, Kolkata.

Bhasin, M.K. (2005). Human population cytogenetics: A review. *International Journal of Human Genetics*, 5(2):83–152.

Bhaskaran, S., Tilak, P., Lincoln, S., Thomas, I M. & Brahmachari, V. (1999). Spontaneous expression of fragile site at Xq27.3 in a Fragile X patient. *Indian Journal of Human Genetics*, 5(1):37–40.

Bhaskar Rao, G. V., (1999). A systematic and comparative study on the association and significance of chromosomal aberrations in subjects with idiopathic mental retardation, congenital malformations and syndromes. *PhD Thesis, submitted to NIMHANS (Deemed University)*, Bangalore.

Bhattacharya, N., Tripathi, A., Dasgupta, S., Sabbir, M. G, Roy, A., Sengupta, A., et al., (2003). Association of deletion in the chromosomal 8p.21.3-23 region with the development of invasive head & neck squamous cell carcinoma in Indian patients. *Indian Journal of Medical Research*, 118:77–85.

Bhattacharya, N., Chunder, N., Roy, A., Alam, N., Roychoudhury, S. & Panda, C.K. (2005). Amplification of c-myc locus is independently associated with the deletions of chromosome 8p in breast carcinoma. *International Journal of Human Genetics*, 5(1):49–55.

Bhisey, A.N., Khare, A. G., Advani, S.H. & Ranadive, K. J. (1978). Cytogenetic studies in leukemias—blastic crisis of chronic myeloid leukemia. *Indian Journal of Medical Research*, 68:285–294

Bhisey, A. N., Advani, S. H. & Khare, A. G., (1982). Cytogenetic anomalies in a child born to a mother receiving busulfan for chronic myeloid leukaemia. *Indian Journal of Cancer*, 19(5):272–276.

Biswas, S. & Chowdhury, J. R., (1968). Evaluation of chromosomal morphology in human normal and cancer cervix cells in vitro. *Indian Journal of Medical Research*, 56(11):1595–1599.

Borgaonkar, D. S. (1997). *Chromosomal variation in man: A catalog of chromosomal variants and anomalies* (8th ed.). Wiley-Liss, New York.

Borgaonkar, D. S., Reisor N., Ahn, J., Caldwell R., Gahunia, H. Jr., Pasquenza, N., et al. (1993). *Repository of chromosomal variants and anomalies: International registry of abnormal karyotypes* (14th Listing). Medical Centre of Delaware, USA.

Brown, S. W. & Chandra, S. H. (1973). Inactivation system of the mammalian X chromosome. *Proceedings of the National Academy of Sciences of USA*, 70(1):195–199.

Chaddha, V., Agarwal, S., Phadke, S. R. & Halder, A. (2003). Low level of mosaicism in atypical Prader Willi syndrome: Detection using fluorescent in situ hybridization. *Indian Pediatrics*. 40(2):166–168.

Chakraborty, S.B., Sabbir, M.G., Roy, A., Sengupta, A. & Panda, C.K. (2003). Multiple deletions in chromosome 3p are associated with the development of head and neck squamous cell carcinoma. *International Journal of Human Genetics*, 3(2):79–87.

Chakravarty, A., Purandare, H., Mehta, L. & Chakravarty, B.P. (1978). Effect of synthetic and natural sex steroids on X-chromatin. *Indian Journal of Medical Research*. 68:785–789.

Chakravarty A. & Chakravarty, S. (2010). Chromosome structural variants: Impact on cytogenetic diagnosis and genetic medicine. *Journal of Prenatal Diagnosis and Therapy*, 1:5–13.

Chandra, N., Cyril, C., Lakshminarayana, P., Nallasivam, P., Ramesh, A., Gopinath, P. M., et al., (2010). Cytogenetic evaluation of Down syndrome: A review of 1020 referral cases. *International Journal of Human Genetics*, 10(1–3):87–93.

Chandra N. & Marimuthu, K. M., (1987). Correlation of cytogenetic findings with clinical manifestation in chronic phase of myeloid leukaemia. *Indian Journal of Medical Research*, 85:553–664.

Chaudhuri, A. & Chaudhuri, K. C. (1963). The chromosome number in man. *Indian Journal of Pediatrics*, 30:75–77.

Chaudhuri, A., (1964). The Chromosomes – An introduction – with some observations in Indian children. *Indian Journal of Pediatrics*, 31:175–186.

Chaudhuri, A. & Chaudhuri, K. C. (1966). A survey of patients investigated cytogenetically in a medical genetics unit. *Indian Journal of Pediatrics*, 33(9):267–274.

Chaudhuri, A., Patel, S. & Chaudhuri, K. C. (1966). G21/G translocation chromosome in a boy with mongolism. *Indian Journal of Pediatrics*, 33(224):288–292.

Chaudhuri A. (1969). 45,X-46,XX mosaicism in Turner's syndrome. *Indian Journal of Pediatrics*, 36(256):171–174.

Chavan, D., Ahmad, F., Iyer, P., Dalvi, R., Kulkarni, A., Mandava, S. et al. (2006). Cytogenetic investigation in chronic myeloid leukemia: study from an Indian population. *Asian Pacific Journal of Cancer Prevention*. 7(3):423–426.

Cheriyan, V. D., Kurien, C. J., Das, B., Ramachandran, E. N., Karuppasamy, C. V., Thampi, M. V. et al. (1999). Genetic monitoring of the human population from high-level natural radiation areas of Kerala on the Southwest coast of India. II. Incidence of numerical and structural chromosomal aberrations in the lymphocytes of newborns. *Radiation Research*, 152:S154–S158.

Chetan, G.K., Latha, P., Arathi, R., Padma, S., Rao, G.V.B., Girimaji, S.R., et.al.(2001). Cytogenetics of Fragile X chromosome: autosomal sites as potential markers for fragile X chromosome analysis. *International Journal of Human Genetics*, 1(2):91–95.

Chetan, G. K., Manjunatha, K. R., Arathi, R., Latha, P., Padma, S. & Bhakar Rao, G. V., (2002). Genetics of Fragile X syndrome: A systematic data from the Indian population. *International Journal of Human Genetics*, 2(1):69–72.

Chetan, G. K., Manjunatha, K. R., Venkatesh, H. N., Balu, S., Venkataswamy, E. & Roy, S. (2010). Cytogenetic studies of idiopathic mental retardation: A report. *International Journal of Human Genetics*, 10(1–3):33–39.

Chinnappan, D., Verma, I. C., Choudhry, V. P. & Arya, L. S., (2000). Cytogenetics investigation in childhood chronic myeloid leukemia. *Indian Journal of Pediatrics*, 67(2):107–112.

Chowdhury, M. R., Singh, G., Shukla, A. & Verma, I. C. (1996). Cytogenetic studies in ataxia telangiectasia and their use in prenatal diagnosis. *Indian Journal of Medical Research*, 103:155–158.

Cyrus, C., Chandra, N., Jegatheesan, T., Chandralekha, K., Gopinath, P.M, Marimuthu, K.M. et.al. (2005). Down syndrome child with 48,XXY,+21 karyotype. *Indian Journal of Human Genetics*, 11(1):47–48.

Cyrus, C., Teena, K., Paul, S.F.D., Chandra, N., Meena, J., Marimuthu, K.M., et.al. (2006). Familial Robertsonian translocation 13;21 in a Down syndrome patient with XYY/XY mosaicism. *International Journal of Human Genetics*, 6(4):291–295.

Das, K. C. (1974). Cytogenetic studies in two ethnic groups of population in India. In *Abstracts: 1st Annual conference of the Indian Society of Human Genetics* (pp. 24). Mumbai.

Das, K.C., Nayak, J., Mohanty, D. & Garewal, G. (1978). Cytology and cytogenetics in chronic myelogenous leukaemias. *Indian Journal of Medical Research*. 68:148–163.

Dave, U. & Shetty, D., (2010). Chromosomal abnormalities in mental retardation: Indian experience. *International Journal of Human Genetics*, 10(103):21–32.

Dubey, S., Chowdhury, MR., Prahlad, B., Kumar, V., Kumar, R., Kabra, M., Verma, IC., et al. (2005). Cytogenetic causes for recurrent spontaneous abortions – An experience of 742 couples. (1484 cases). *Indian Journal of Human Genetics*, 11(2):94–98.

Dutta, S., Shaw, J., Sinha, S. & Mukhopadhyay, K. (2009). Chromosomal abnormalities associated with mental retardation in female subjects. *Indian Journal of Human Genetics,* 15(1):28–31.

Dutta, U. R., Rajitha, P., Pidugu, V. K. & Dalal, A. B. (2011). Cytogenetic abnormalities in 1162 couples with recurrent miscarriages in Southern region of India: report and review. *Journal of Assisted Reproduction and Genetics,* 28(2):145–149.

Elango, R. & Verma, I.C. (1996). Fragile X syndrome among children with mental retardation. *Indian Journal of Pediatrics,* 63(4):533–538.

Fauzdar, A., Sharma, R. K., Kumar, A. & Halder, A. (2008). A preliminary study on chromosome aneuploidy and mosaicism in early pre-implantation human embryo by fluorescence in situ hybridization. *Indian Journal of Medical Research,* 128(3):287–293.

Gadhia, P.K., Shah, N., Nahata, S., Patel, S., Patel, K., Pithawala, M. & Tamakuwala, D. (2004). Cytogenetic analysis of radiotherapeutic and diagnostic workers occupationally exposed to radiations. *International Journal of Human Genetics*, 4(1):65–69.

Gadhia, P.K., Vaniawala, S. & Pithawala, M. (2005). Some observations on spontaneous sister chromatid exchange frequencies and cell cycle progression in stimulated lymphocytes of patients with different malignancies. *International Journal of Human Genetics,* 5(3):187–191.

Gambhir, P. S., Gole, L. A., Tembe, M. V., Saraph, A. A., Phadke, M. A. & Khedkar, V. A., (1991). A translocation between chromosome 1 and 10 in a boy with mental retardation and dysmorphic features. *Indian Pediatrics,* 28(3):289–291.

Ganguly, B. B. & Sahni, S. (2003). X chromosome abnormalities in Indian adolescent girls. *Teratogenesis, Carcinogenesis and Mutagenesis*, 23:245–253.

Ganguly, B. B., Kadam, N. N. & Agarwal, M. B. (2007a). Conventional and fluorescence *in situ* hybridization analysis of three-way complex BCR-ABL rearrangement in a chronic myeloid leukemia patient. *Journal of Cancer Research and Therapeutics*, 3(2):124–126.

Ganguly, B. B., Loher, Y. & Agarwal, M. B. (2007b). Translocation t(11;20)(p15;q11) detected in AML M0: A case report. *Atlas of Genetics and Cytogenetics in Oncology and Haematology.* URL: http://AtlasGeneticsOncology.org/Reports/1120GangulyID100028.html

Gawde, H., Patel, Z.M., Khatkhatey, M. I., D'Souza, A., Babu, S., Adhia, R., et al., (2006). Chromosome 22 microdeletion by FISH in isolated congenital heart disease. *Indian Journal of Pediatrics,* 73(10):885–888.

Gazala, J., Amithkumar, I. V., Praveena, K. K. & Sujatha, J. (2010). Prenatal diagnosis in a pericentric inversion 6. *International Journal of Human Genetics*, 10(1–3):175–178.

Gladstone, B., Kadam, P. R., Balsara, B. R., Pai, S. K., Gopal, R., Nair, C. N. et al., (1994). Cytogenetic studies on patients of acute lymphoblastic leukemia Burkitt's type with (8;14) & (14;18) translocations. *Indian Journal of Medical Research,* 99:264–266.

Gogate, S., Panthaki, M. H., Shah, H, Desai, K. K. & Thomas, S. P. (1994). Critical evaluation of various methods of fetal tissue sampling and karyotyping for prenatal diagnosis. In *Souvenir: 3rd National workshop on prenatal diagnosis and therapy* (pp. 34). Bangalore.

Ghosal, K. K. (1975). Chromosomal, hormonal and exfolitive vaginal cytological study of a case of bilateral ovarian transplantation. In *Abstracts: 2nd Annual conference of the Indian Society of Human Genetics* (Abstract 123). ISI, Kolkata.

Ghosh, P. K., Kenue, R. & Singh, I. P. (1975). Karyotypic polymorphism in normal individuals. In *Abstracts: 2nd Annual conference of the Indian Society of Human Genetics* (Abstract 12). ISI, Kolkata.

Ghosh, R., Shome, D., Chatterjee, B. & Ghosh, P. K. (1986). Chromosomal abnormalities in infertile males. *Indian Journal of Medical Research*, 83:408–416.

Ghosh, S. N., Shah, P. N. & Gharpure H. M. (1978). Absence of H-Y antigen in XY females with dysgenetic gonads. *Nature,* 276(5684):180–181.

Goud, K. I., Raina, V., Verma, S. & Chadha, G. (2006). Cytogenetics and genetic counseling of patients in North India. *JK Science,* 8(1):28–30.

Gowda, S. R., Gawde, H. M. Hyderi, A., Mysore, R. S., Krishnamurthy, B., Karat, S. C., et al. (2010). Chromosomal anomalies and congenital heart disease in Mysore, South India. *International Journal of Human Genetics*, 10(1–3):131–139. (FISH)

Grover, S., Solanki, B. K., Khandelwal, M.K. & Jaiswal, R. B. (1968). Chromosomal abnormalities in measles. *Indian Pediatrics,* 5(7):326–329.

Grover, S. & Shivde, J. V. (1971). An unusual variant of Turner's syndrome. *Journal of Postgraduate Medicine*, 17(4):195–197.

Gujral, S., Polampalli, S., Badrinath, Y., Kumar, A., Subramanian, P. G., Raje, G., et al., (2009). Clinico-hematological profile in biphenotypic acute leukemia. *Indian Journal of Cancer.* 46(2):160–168.

Guleria, K. & Sambyal, V. (2003). Chromosomal instability in peripheral blood leucocytes of oesophageal cancer patients. *International Journal of Human Genetics*, 3(3):179–186.

Guleria, K. & Sambyal, V. (2010). Spectrum of chromosomal aberrations in peripheral blood lymphocytes of gastrointestinal tract (GIT) and breast cancer patients *International Journal of Human Genetics* 10(1–3):147–158.

Gupta, A. D. (1970). Human genetics and the practitioner. *Journal of the Indian Medical Association.* 54(5):203–204.

Gupta, N.J. & Chakravarty, B.N. (1995). Chromosomal abnormality in cases of recurrent abortions. *Indian Journal of Human Genetics,* 1(2):111–115.

Halder, A., Fauzdar, A., Kabra, M. & Saxena, A. (2005a). Detection of 22 q11.2 hemizygous deletion by interphase FISH in a patient with features of CATCH 22 syndrome. *Indian Pediatrics,* 42(12):1236–1239.

Halder, A., Halder S. & Fauzdar, A., (2005b). A preliminary investigation of genomic screening in cervical carcinoma by comparative genomic hybridization. *Indian Journal of Medical Research.* 122(5):434–446.

Hegde, S., Kalyankar, G. D., Thomas, I. M., Ambani, L. M., Dhareshwar, S. S., Patil, S. M., et al., (1989). Multicentric task force study on identification of genetic causes of mental retardation. ICMR, New Delhi.

Indravathi, K. (1974). Cytogenetic studies in repeated spontaneous abortions. In *Abstracts: 1st Annual conference of the Indian Society of Human Genetics* (pp. 25). Mumbai.

Iyer, P.A., Vyas, J.C., Ranjan, P. & Saranath, D. (2009). A de novo complex chromosomal rearrangement of 46,XX,t(7;15;13)(p15;q21;q31) in a female with an adverse obstetric history. *International Journal of Human Genetics*, 9(2):139–143.

Iyengar, J.K., Rohatgi, M., Menon, P.S., Mathews, A.R., Verma, I.C. & Bhargava, S., (1986). Clinical, cytogenetic and hormonal profile in extreme hypospadias with bilaterally descended testes. *Indian Journal of Medical Research*, 83:604–609.

Jacob, R. T., Gayathri, K., Surath, A. & Rao, D. R. (2002). Cytogenetic profile of chronic myeloid leukemias. *Indian Journal of Cancer,* 39(2):61–65.

Jain, U., Verma, I. C. & Kapoor, A. K. (1998). Prevalence of fragile X(A) syndrome in mentally retarded children at a genetics referral centre in Delhi, India. *Indian Journal of Medical Research*, 108:12–16.

Jaju, M., Jaju, M. & Ahuja, Y. R. (1986). Cytogenetic effect of neomycin on human lymphocytes in vitro. *Indian Journal of Experimental Biology*, 24(9):595–598.

Jayalakshamma, Margaret, M., Amudha, S., Tilak, P., Devi, R. & Rajangam, S., (2010). Cytogenetic analysis in Down syndrome. *International Journal of Human Genetics,* 10(1–3):95–99.

Jobanputra, V., Kriplani, A., Choudhry, V. P. & Kucheria K. (1998). Detection of chromosomal abnormalities using fluorescence in situ hybridization (FISH). *National Medical Journal of India,* 11(6):295–263.

Jobanputra, V., Roy, K. K., Kriplani, A. & Kucheria K. (2001). Prenatal diagnosis of chromosomal abnormalities in women with high risk pregnancies. *Indian Journal of Medical Research,* 114:148–155.

Jobanputra, V., Roy, K. K. & Kucheria, K. (2002). Prenatal detection of aneuploidies using fluorescence in situ hybridization: a preliminary experience in an Indian set up. *Journal of Biosciences.* 27(2):155–163.

Joseph, A. & Thomas, I. M. (1982). Cytogenetic investigations in 150 cases with complaints of sterility or primary amenorrhoea. *Human Genetics,* 61(2):105–109.

Jude, A.L.C., Sasikala, K., Kumar, A., Sudha, S. & Raichel, J. (2002). Haematological and cytogenetic studies in workers occupationally exposed to cement dust. *International Journal of Human Genetics,* 2 (2):95–99.

Jude, A.L.C., Sasikala, K., Chandrasekar, T.S., Kumar, R. A., Sudha, S., Devi, M. V. & Balachandar, N. (2005). Cytogenetic findings in cancerous and non-cancerous lesions of the digestive system. *International Journal of Human Genetics,* 5(3):199–203.

Jyothy, A., Kumar, K. S., Rao, G. N., Rao, V. B., Swarna, M., Devi, B. U., et al. (2000). Cytogenetic studies of 1001 Down syndrome cases from Andhra Pradesh, India. *Indian Journal of Medical Research,* 111:133–137.

Jyothy, A., Kumar, K.S.D., Swarna, M., Raja Sekhar, M., Uma Devi, B. & Reddy, P.P.(2002).Cytogenetic investigations in 1843 referral cases of disordered sexual development from Andhra Pradesh, India. *International Journal of Human Genetics,* 2(1):55–59.

Kadam, P. R., Advani, S. H. & Bhisey, A. N. (1990). Cytogenetic studies in Hodgkin's disease. *Indian Journal of Medical Research,* 92:307–314.

Kadam, P., Dadabhoy, K., Bhisey, A., Athale, U., Nair, C., Nair, R. et al., (1995). Chromosome investigations & clinical outcome in patients with myelodysplastic syndromes. *Indian Journal of Medical Research,* 101:163–169.

Kalavathi, V., Chandra, N., Nambiar, R., Shanker, J., Sugunashankari, P, Meena, J., et al., (2010). Chromosomal abnormalities in 979 cases of primary amenorrhoea. *International Journal of Human Genetics,* 10(1–3):65–69.

Kalpana, V.L.& Satyanarayana, M. (1997). Cytogenetic analysis of primary amenorrhea cases. *Indian Journal of Human Genetics,* 3(2):95–100.

Kalpana, V.L., Satyanarayana, M. & Rao, G.V.B. (2000). Demographic and cytogenetic study in recurrent spontaneous aborters. *Indian Journal of Human Genetics,* 6(1):15–19

Kalpana, V.L., Satyanarayana, M., Prabhaker, S. & Sarvani, R. (2004). Chromosomal instability in recurrent spontaneous aborters. *International Journal of Human Genetics,* 4(1):31–36.

Kamboj, S.S. & Sambyal, V. (2006). Increased chromosomal aberrations in peripheral blood lymphocytes of traffic policemen of Amritsar city. *International Journal of Human Genetics,* 6(2):125–131.

Kannan, V. & Ganapathy, V., (1982). XXX infertility syndrome. *Journal of the Association of Physicians of India,* 30(12):910–911.

Kar, B. (1995). Expression of fragile Xq 27.3 in a patient with clinical features of deletion 9p syndrome. *Indian Pediatrics,* 32(7):801–803.

Kar, B. & Thakkar, N., (2010). Reproductive outcome following prenatal diagnosis: An important diagnostic tool in a female with recurrent pregnancy loss presenting with Robertsonian translocation der(14;15) and inversion inv(9). *Journal of Prenatal Diagnosis and Therapy,* 1(1):42, FP-7.

Kaur, A. & Singh, J.R. (2010). Chromosomal abnormalities: genetic disease burden in India. *International Journal of Human Genetics,* 10(1–3):1–14.

Kaur, A., Mahajan, S. & Singh, J.R. (2003). Cytogenetic profile of individuals with mental retardation. *International Journal of Human Genetics,* 3(1):13–16.

Kaur, A., Mahajan, S. & Singh, J.R. (2004). Cytogenetic analysis in cases with sex anomalies. *International Journal of Human Genetics,* 4(3):167–171.

Kaur, P. & Sambyal, V. (2008). Lymphocytic chromosomal instability in sporadic gastrointestinal tract (GIT) cancer patients and their first-degree relatives. *International Journal of Human Genetics,* 8(4):335–342.

Kaur, H., Kaur, M. G., Setia, K., Sudan, M., Uppal, M.S., Sambyal, V., et.al. (2009). Chromosomal instability in the lymphocytes of breast cancer patients. *Indian Journal of Human Genetics,* 15(1):13–18.

Kaur, P. & Sambyal, V. (2008). Lymphocytic chromosomal instability in sporadic gastrointestinal tract (GIT) cancer patients and their first-degree relatives. *International Journal of Human Genetics,* 8(4):335–342.

Kaur, S., Sambyal, V. & Rai, S.K. (1999). SCE analysis in treated & untreated leprosy patients. *Indian Journal of Human Genetics,* 5(2):11–13.

Kava, M. P., Tullu, M. S., Muranjan, M. N. & Girisha, K. M. (2004). Down syndrome: Clinical profile from India. *Archives of Medical Research,* 35:31–35.

Kelkar, A. V., Madon, P., Kolluri, R. V., Nair, S. & Kulkarni, R. D. (1978). Trisomy D and E: Case reports. Abstracts of *V Annual Conference of the Indian Society of Human Genetics,* Mumbai, Session B4.

Khan, A. & Gangane, S. D. (2010). Presence of multiple cell lines in short stature. *Journal of Prenatal Diagnosis and Therapy,* 1(1):47, PO-14.

Khanna, K. K. (1966). Chromosomes in health and disease. *Indian Journal of Pediatrics,* 33(224):307–311.

Khare, A. G., Advani, S. H., Bhisey, A. N. & Ranadive, K. J., (1977). A consistent chromosome abnormality involving "E" group chromosome in acute leukaemias. *Indian Journal of Cancer,* 14(1):81–86.

Kher, M. B., Indurkar, M. B. & Lokur, U. V. (1971). Dermatoglyphics in pediatric practice. *Indian Journal of Medical Sciences,* 25(9):618–622.

Kochupillai, N., Verma, I. C., Grewal, M. S. & Ramalingaswamy V. (1976). Down's syndrome and related abnormalities in the area of high background radiation in costal Kerala. *Nature,* 262:60–61.

Kokate, P., Shah, L., Yadav, A. & Mandava, S., (2010). Molecular cytogenetic study on a large cohort of aborted tissue samples. *Journal of Prenatal Diagnosis and Therapy,* 1(1):47, PO-12.

Kolluri, R. V., Mutalik, G. S., Madon, P. F, Nair, S, Kelkar, A. V. & Kulkarni, R. D. (1978). Cytogenetic and dermatoglyphic studies in Down syndrome. Abstracts of *V Annual Conference of the Indian Society of Human Genetics,* Mumbai, Session B4.

Korgaonkar, S., Babu, V. R., Kerketta, L. & Ghosh, K., (2008). Chromosomal breakage in myelodysplatic syndrome. *Asian Pacific Journal of Cancer Prevention,* 9(1):151–154.

Kothare, S., Shetty, N. & Dave, U. (2002). Maternal age and chromosomal profile in 160 Down syndrome cases – Experience of a tertiary genetic centre from India. *International Journal of Human Genetics,* 2(1):49–53.

Kotwaliwale, S. V., Dicholkar, V. V. & Motashaw, N. D. (1991). Maternal transmission of translocation 2;21 associated with Down's syndrome. *Journal of Medical Genetics,* 28:415–416.

Krishnaja, A. P. & Chauhan, P. S. (2000). Quinacrine dihydrochloride, the non-surgical female sterilant induces dicentrics, rings, and marker chromosomes in human peripheral blood lymphocytes treated in vitro: a preliminary report. *Mutation Research/Genetic Toxicology and Environmental Mutagenesis,* 466(1):43–50.

Krishnaja, A. P. & Sharma, N. K. (2004). Transmission of gamma ray induced unstable chromosomal aberrations through successive mitotic divisions in human lymphocytes in vitro. *Mutagenesis,* 19(4):299–305.

Kucheria, K. & Keswani, N. H. (1975). Antenatal diagnosis of chromosomal disorders. In Emery, A. E. H., Reddi, O. S., Sanjeeva Rao, M. (Eds.), *Proceedings of the seminar on medical genetics* (pp. 125–133). Osmania University, Hyderabad.

Kucheria, K., Mehra, N. K., Bamezai, R & Vaidya, M. C. (1975). Chromosome banding patterns (G-bands) and Autoradiography. In *Abstracts: 2nd Annual conference of the Indian Society of Human Genetics* (Abstract 28). ISI, Kolkata.

Kucheria, K., Kumar, V., & Bamezai, R. (1978). A family of male and female sibs with hypogonadism. In Verma, I. C. (Ed.), *Medical Genetics in India (Vol. 1),* Auroma Enterprises, Pondicherry, pp. 157–160.

Kucheria, K., Bose, S., Jhanwar, M. & Choudhry, V.P. (1998). Confirmation of cytogenetic diagnosis in Ph' positive and negative cases by Fluorescence In Situ Hybridization. *Indian Journal of Human Genetics,* 4(2):144–150.

Kucheria, K., Jobanputra, V., Talwar, R., Ahmed, M. E., Dada, R. & Sivakumaran, T. A. (2003). Human molecular cytogenetics: diagnosis, prognosis and disease management. *Teratogenesis, carcinogenesis and mutagenesis,* Suppl. 1, 225–233.

Kulkarni, M. V., Patel, Z. M., Peter, J. J. & Ambani, L. M., (1981). Partial trisomy and monosomy for short arm of chromosome number 5 in two sibs and a balanced translocation carrier father. *Indian Pediatrics,* 18(4):263–266.

Kumar, P., Gupta, A., Fotra, R., Upma, Raina, S., Sethi, S, et al. (2010). Chromosomal anomalies in referred cases with suspected genetic disorders: First report from Jammu and Kashmir. *International Journal of Human Genetics,* 10(1–3):41–47.

Kumari, U. & Bhat, G.K. (2008). Sensitivity of lymphocytes of patients with fibroadenoma and breast cancer to gamma-rays as evaluated by chromosomal aberrations. *International Journal of Human Genetics,* 8(4):351–356.

Harini, R., Ata-ur-Rasheed, M., Shanmugam, M. P., Amali, J., Das, D. & Kumaramanickavel, G. (2001). Genetic profile of 81 retinoblastoma patients from a referral hospital in southern India. *Indian Journal of Ophthalmology.* 49(1):37–42.

Lakshmaiah, K. C., Tejinder, S., Kumari, P., Gowri, M. M., Satheesh, C. T., Lokanatha, D., et al., (2010). Significance of t (8: 14) in CLL?. *Indian Journal of Human Genetics,* 16:(2), 102–103.

Lakshminarayana, P. (1990). Translocation Down's syndrome. *Indian Journal of Pediatrics,* 57:265–271.

Lall, M., Puri, R., Pandita, S., Chaudhary, P., Tiwari, L. & Verma, I. C. (2010). Chromosome inversions detected prenatally warrant careful genetic counseling. *Journal of Prenatal Diagnosis and Therapy,* 1(1):49–50, PO-25.

Lata, M., Meena, B. & Mehta, L. (1986). Phenotype genotype correlation in 46,X,i(Yq) and 46,X,i(Xq). *Journal of Obstetrics and Gynaecology of India,* 36(2):346–349.

Latha, P., (1996). Cytogenetics of autosomal fragile sites: Association and their implications in subjects with mental retardation. *Ph. D Thesis, submitted to NIMHANS (Deemed University).* Bangalore.

Maddileti, U., Padmaja, T., Prasad, M.H. & Reddy, P.P. (2002). Analysis of chromosomal aberrations in the peripheral lymphocytes of workers exposed to diagnostic x-rays. *International Journal of Human Genetics,* 2(4):265–268.

Madon, P. F., Mutalik, G. S., Nair, S., Kolluri, R. V., Kelkar, A. V. & Kulkarni, R. D. (1978a). Ambiguous genitalia and the syndrome of testicular feminization. Abstracts of *V Annual Conference of the Indian Society of Human Genetics,* Mumbai, session B4.

Madon, P. F., Nair, S., Patel, R. A., Jain, M. M. & Kolluri, R. V. (1978b). A male with a uterus in an inguinal hernia. Abstracts of *V Annual Conference of the Indian Society of Human Genetics,* Mumbai, session B4.

Madon, P. F., (1979). Studies in abnormal sex phenotypes with special reference to correlation between cytogenetic, dermatoglyphic and clinical aspects. *Thesis submitted to University of Bombay for Ph.D. in Applied Biology.* 226 pgs.

Madon, P.F, Athalye, A. S., Soonawala R. P. & Parikh, F. R. (2001). 46,XX/46,XY karyotype in a phenotypically female abortus. *Bulletin of the Jaslok Hospital and Research Centre,* 25(3):7–8.

Madon, P. F., Athalye, A. S., Lulla, C. P. & Parikh, F. R. (2001). Fanconi anemia with triphalangeal thumbs, syndactyly and contractures of the fingers in a 2 year old boy. *International Journal of Human Genetics,* 1:2, 87–90.

Madon, P. F., Athalye, A. S., Bandkar, V., Dhumal, S., Sopariwala, A. & Parikh, F.R. (2002). Cytogenetic analysis of hematological malignancies. In Bhasin, M. K. & Kalla, K. K. (Eds) Proceeding volume of National Conference on Expanding Horizons of human Genetics, Delhi, pp. 37–42.

Madon, P.F., Athalye, A.S., Bandkar, V., Dhumal, S., Sopariwala, A. & Parikh, F.R. (2003a). Fluorescence *in-situ* Hybridization (FISH) – a rapid and useful technique for diagnosis and management in leukemia. *International Journal of Human Genetics,* 3(2):115–119.

Madon P. F., Kopakka, N. T., Berde, P. P, Shewale, L. H., Sarangdhar, V. & Parikh, F. R., (2003b). 45,X,t(Y;13)(q11.2;q11),-der(13) with Y microdeletions in an azoospermic male. *Annales de Genetique,* 46(2–3):178.

Madon, P. F. (2004a). Guide to specimen collection for cytogenetic investigations. In A. K. Debdas (Ed.) *Screening and Diagnosis of Fetal Malformation – A practical guide.* B. I. Publications, India, pp. 173–178.

Madon, P. F., Athalye, A. S. & Parikh, F. R., (2004b). Fluorescence in-situ Hybridization – Application in Genetic Diagnostics. In P. Usha Rani, A. Jyothy and M. Hema Prasad (Eds.) *Recent Advances in Genetics and Health.* Osmania University, Hyderabad, pp 97–101.

Madon, P. F., Parikh, F. R. & Athalye, A. S. (2005). Polymorphic variants on chromosomes probably play a significant role in infertility. *Reproductive Biomedicine Online,* 11(6):726–732.

Madon, P. F., (2006a). Development of diagnostic genetic services in India: The experience of molecular cytogenetics. *Challenges of Genetics and Genomics for the Indian Subcontinent.* Precongress Satellite Meeting of the International Congress of Human Genetics, Brisbane.

Madon, P. F., Athalye, A. S., Parikh, F. R. (2006b). Urovysion – A FISH test for bladder cancer detection. Proceedings and abstracts of *APSSM & AGASSI* conference, pp 60.

Madon, P. F., Athalye, A. S., Naik, N. J. & Parikh, F. R. (2007a). Preimplantation genetic diagnosis. In: Telang, M., (Ed.). *Atlas of human assisted reproductive technologies.* Jaypee, New Delhi: pp. 167–174.

Madon, P. F., Athalye, A. S., Parikh, F. R. (2007b). Application of FISH in Hematological Malignancies. In R. Colah, K. Vasantha, S. Shetty (eds), *Proceedings of International Conference on Emerging Trends in Hematology and Immunohematology,* Mumbai, pp 245–250.

Madon, P. F., Athalye, A. S., Sanghavi, K & Parikh, F. R., (2010a). Microdeletion syndromes detected by FISH – 73 positive from 374 cases. *International Journal of Human Genetics,* 10(1–3):15–20.

Madon, P. F., Athalye, A.S., Naik, N. J., Naik, D. J. & Parikh, F. R. (2010b). PGD for a Robertsonian translocation by FISH: First successful pregnancy from India. *Journal of Prenatal Diagnosis and Therapy,* 1(1):20–22.

Madon, P., Athalye, A., Bandkar, V., Kawle, M., Sanap, R., Padyal, P., et al., (2010c). Cytogenetic analysis of prenatal samples, abortuses and arrested embryos at a private hospital in India. *Prenatal Diagnosis,* 30(Suppl. 1):S91, P2–46.

Maji, D., Mukhopadhyay, B., Chaudhuri, U. R., Sarkar, S. & Sadhukhan, B., (1990). True hermaphroditism: a report of two cases. *Journal of the Association of Physicians of India.* 38(11):868–870.

Malhotra, K. K., Deo, M. G., Kapur, M. M. & Shah, P.K. (1969). Chromatin-positive Klinefelter's syndrome. *Indian Journal of Medical Research,* 57(3):543–549.

Malini, S. S. & Ramachandra, N. B., (2010). Young mothers produce more chromosomal syndrome babies in Mysore, South India. *International Journal of Human Genetics,* 10(1–3):105–112.

Mallik, K. C. (1962). Sex chromatin. *Journal of the Indian Medical Association.* 38:430–431.

Mandava, S. & Das, B. R. (2010). Cytogenetic analysis of 6750 couples with repeated miscarriages and malformed children: SRL experience. Abstract FP-1, *Journal of Prenatal Diagnosis and Therapy,* 1(1):41.

Mandava, S., Koppaka, N., Bhatia, V. & Das, B. R. (2010). Cytogenetic analysis of 1572 cases of Down syndrome: A report of double aneuploidy and novel findings 47,XY, t(14;21)(q13;q22.3)mat,+21 and 45,XX,t(14;21) in an Indian population. *Genetic Testing and Molecular Biomarkers,* 14(4):499–504.

Mangalgiri A. S & Pathak, S.A. (2008). Cytogenetics in recurrent abortions. *People's Journal of Scientific Research.* 1:5–8.

Manjunatha, K.R., Rao, B.S., Narayanan, H.S., Girimaji, S.R., Shobha, S., Gandhi, D.H. et al. (1988). Fragile X syndrome: The first case report from India. *Genome,* 30(Suppl 1):205.

Manjunatha, K.R., Chetan, G.K., Arathi, R., Padma, H.N., Venkatesh, S., Srinath, S.R., Girimaji, S.R. & Sheshadri, S. (2001). Genetics of autism: association of chromosomal fragile sites. *International Journal of Human Genetics,* 1(4):293–299.

Manjunatha, K.R., Chetan, G.K., Arathi, R., Rao, G.V.B., Latha, P., Padma, S., et.al. (2002). Frequency, association and genetic implications of chromosomal fragile sites in mental retardation. *International Journal of Human Genetics*, 2(1):33–39.

Manna, G. K., (1974). Chromosomal polymorphism and cancer. *Indian Journal of Cancer,* 4(2):175–179.

Margaret, M., Tilak, P. and Rajangam, S. (2010). 45, X/47, X, i(X)(q10), i(X)(q10)/46, X, i(X)(q10). Isochromosome Xq in mosaic Turner syndrome. *International Journal of Human Genetics*, 10(1–3):77–80.

Master, H. R. & Sukumaran, P. K. (1974). A simple method for leukocyte cultures from peripheral blood for chromosome studies. In *Abstracts: 1st Annual conference of the Indian Society of Human Genetics* (pp. 28). Mumbai.

Mathur, R., Chowdhury, M. R. & Singh, G., (2000). Recent advances in chromosome breakage syndromes and their diagnosis. *Indian Pediatrics*, 37(6):615–625.

Mathur, R., Dubey, S., Hamilton, S., Singh, G., Deka, D., Kriplani, A., et al., (2002). Rapid prenatal karyotyping using foetal blood obtained by cordocentesis. *National Medical Journal of India,* 15(2):75–77.

Mehta, M., (2005). Correlation of fragile sites with genetic disease. *Thesis submitted to University of Mumbai for Ph.D. in Science.*

Meka, A. & Reddy, B.M. (2006). Recurrent spontaneous abortions: An overview of genetic and non-genetic backgrounds. *International Journal of Human Genetics,* 6(2):109–117.

Minocherhomji, S., Athalye, A., Madon, P., Kulkarni, D., Uttamchandani, S. & Parikh, F. (2009). A case-control study identifying chromosomal polymorphic variations as forms of epigenetic alterations associated with the infertility phenotype. *Fertility and Sterility,* 92:1, 88–95.

Minocherhomji, S., Madon, P. F. & Parikh, F. R. (2010). Epigenetic Regulatory Mechanisms Associated with Infertility. *Obstetrics and Gynecology International*, 2010:198709, 7 pages.

Modi, D. N. & Bhartiya, D. (2002). Clinical experience with the use of Fluorescence in Situ Hybridization on uncultured cells for prenatal diagnosis. *Indian Journal of Human Genetics, 8*(2):60–65.

Modi, U. J., Nayak, U., Aiyer, S., Bharani, S., Master, D. C., Shah, T. et al. (1998). Study of malformations and Down Syndrome in India (SOMDI): Baroda region. *Indian Journal of Human Genetics,* 4(1):93–98.

Moghe, M., Patel, Z. M., Peter, J. J. & Ambani, L. M. (1981). Cytogenetic studies in a selected group of mentally retarded children. *Human Genetics,* 58:184–187.

Mondal, S. K., Guha, D., Banerjee, D. & Sinha, S. K., (2002). Study of primary amenorrhoea with special reference to cytogenetic evaluation. *Indian Journal of Pathology & Microbiology,* 45(2):155–159.

Movva, S., Kumar, M., Najeeb, S., Murthy., Sreelatha, K. & Hasan, Q.(2005). A novel case of live born with 49, XXXY, + 10 karyotype: implications of autosomal trisomies other than 13, 18 and 21. *International Journal of Human Genetics*, 5(3):179–182

Mukherjee, S. & Pandit, K., (2002). Gonadal dysgenesis due to isochromosome formation, *Journal of the Association of Physicians of India*, 50:1090.

Mulye, V., Ambani, L., Shringi, M., Peter, J. & Motashaw, N. (1983). Cytogenetic studies in primary amenorrhoea. *Indian Journal of Medical Research.* 78:53–58.

Murphy, E. A. & Mutalik, G. S. (1969). The application of Bayesian methods in genetic counseling. *Human Heredity,* 19(2):126–151.

Murthy, D. S. K. & Ambani, L. M. (1978). Acrocentric chromosome association in Down syndrome patients and mothers. In Verma, I. C. (Ed.), *Medical Genetics in India (Vol. 1),* 171–175.

Murthy, D. S. K. & Ambani, L. M. (1981). Frequency and pattern of association of acrocentric chromosomes in patients of Down syndrome and their mothers. *The Indian Journal of Medical Research,* 73:53–60.

Murthy, S. K, Krishna Murthy, D. S., Shah, V. C. and Desai, A. B. (1987). Down syndrome in Ahmedabad: Karyotype analysis in 60 cases. *Indian Journal of Pediatrics,* 54(5):723–727.

Murugesan, R. (2003). Radiation-induced chromosomal hot spots at G 1 and G 2 stages of human lymphocytes in culture. *Indian Journal of Human Genetics*, 9(1):21–24.

Mutalik, G. S. & Lokhandwala, V. A. (1968). Application of dermatoglyphical studies in medical diagnosis. *Journal of the Association of Physicians of India*, 16(12):925–932.

Mutalik, G. S., Nair, S., Pavri, P. N., Chandorkar, B. R. & Shetty, V. V. (1975). A study of three sex chromosome mosaics. *In: Proceedings of the Symposium on Structural and Functional Aspects of Chromosomes, BARC, Mumbai*, 300–314.

Nagvenkar, P., Desai, K., Hinduja, I. & Zaveri, K., (2005). Chromosome studies in infertile men with oligozoospermia and non-obstructive azoospermia. *Indian Journal of Medical Research,* 122:34–42.

Nair, R., Athale, U. A., Iyer, R. S., Nair, C. N., Pai, S. K., Kurkure, P. A. et al., (1992). Childhood myelodysplastic syndromes: clinical features, cytogenetics and prognosis. *Indian Journal of Pediatrics,* 59(4):443–448.

Nair, S.B., Mukundan, G., Paul, B.M., Ramachandran, L., Gopinathan, K.K. & Joseph, S. (2006). Chromosome 12;15 rearrangements in patients with recurrent miscarriage. *Indian Journal of Human Genetics,* 12(3):133–139.

Navalulkar, P. K. & Shukla, R. N. (1975). Comparative study of leukemic and non-leukemic cells by cytogenetic analysis. In *Abstracts: 2nd Annual conference of the Indian Society of Human Genetics* (Abstract 73). ISI, Kolkata.

Padmaja, T., Devi, Ch.V.R. & Reddy, P.P. (2002). Analysis of chromosomal aberrations in mint factory workers. *International Journal of Human Genetics,* 2 (2):81–86.

Pande, G., Advani, S. H. & Bhisey, A. N. (1986). Cytogenetic studies on acute non-lymphatic leukaemia. *Indian Journal of Medical Research.* 83:179–186.

Panjwani, P., Epari, S., Karpate, A., Shirsat, H., Rajsekharan, P., Basak, R., et al. (2010). Assessment of HER-2/neu status in breast cancer using fluorescence *in situ* hybridization and immunohistochemistry: Experience of a tertiary cancer referral centre in India. *Indian Journal of Medical Research*, 132:287–294.

Panthaki, M. H. (1998). Prevention of genetic disorders. *Indian Journal of Medical Sciences*, 52(2):66–69.

Parekh, J. G., (1974a). Editorial: Genetic aspects of leukemia. *Journal of the Association of Physicians of India,* 22(12):920–922.

Parekh, J. G., (1974b). Editorial: Familial constitutional panmyelocytopathy, Fanconi's anemia (F.A.). *Journal of the Association of Physicians of India.* 22(7):552–553.

Parikh, F. R., Madon, P. F., Athalye, A. S., Naik, N. J., Gada, S. D., Ganla, K. N., et al. (2001a). Preimplantation genetic diagnosis of chromosomal abnormalities by multicolour fluorescence in situ hybridization. *Journal of the Indian Medical Association*, 99:441–444.

Parikh, F. R., Naik, N.K., Gada, S., Bhartiya, D., Athalye, A.S. & Madon, P.F. (2001b). Preimplantation genetic diagnosis for the better management of couples during assisted reproduction. *International Journal of Human Genetics*, 1(2):117–121.

Parikh, F. R., Madon, P. F., Athalye, A., S., Saxena, S. G. & Naik, N. J., (2002). FISH in infertility, prenatal diagnosis and PGD in India. *Cytogenetics Genome Research*, 97:275.

Parikh, F. R. & Madon, P. F., (2004). Preimplantation genetic diagnosis. *Modern Medicare,* 1:24–26.

Parikh, F. R., Athalye, A. S., Uttamchandani, S. A., Kawle, M. T., Dhumal, S. B. & Madon, P. F. (2006). Gonadal chromosome mosaicism in cells of the follicular fluid predicts reproductive failure in women undergoing *in vitro* fertilization. *Fertility and Sterility,* 86:3, S143.

Parikh, P. M., Iyer, R. S., Saikia, T. S., Gopal. R. & Advani, S. H. (1992). Philadelphia Chromosome. *Journal of the Association of Physicians of India*, 40(2):98–100.

Patel, R.K., Trivedi, A.H., Bakshi, S.R., Jalu, R.J., Bhatavdekar, J.M., Shah, P.M. & Patel, D.D. (1997). C-band heteromorphism in down syndrome patients. *Indian Journal of Human Genetics,* 3(2):83–87.

Patel, Z. M., Adhia, R. A. & Madon, P. F., (1996a). Prenatal diagnosis of holoprosencephaly associated with Trisomy 13 by ultrasonography. *Journal of Clinical Genetics and Tribal Research,* 1:3, 188–190, 1996.

Patel, Z. M., Madon, P. F., Adhia, R., et al. (1996b). Chromosomal rearrangements in couples with recurrent pregnancy losses – 874 cases. Abstract book of *VIII International Conference on Early Prenatal Diagnosis and Therapy,* Goa.

Patel, Z. M., Sanghavi, D. A. & Adhia, R. A. (2000). Chromosomal rearrangements amongst couples with recurrent abortions in Indian population. *The Journal of Clinical Genetics and Tribal Research,* 5:41–49.

Pathak, A. K. & Sharma, N. L. (1968). Turner's syndrome in a male child. Report of a case. *Indian Journal of Pediatrics,* 35(250):532–535.

Patki, A., Gada-Saxena, S., Madon, P. F., et al., (2004). Low-grade gonadal mosaicism detection by follicular fluid FISH. *Human Reproduction,* 19(suppl 1):i175, P-513.

Pavri, P. N., Mutalik, G. S. & Phadke, M. A. (1974). Giemsa banding technique for human chromosomes. *Journal of the Indian Medical Profession,* 20:8870.

Pavri, P. N. (1976). Chromosomes. *Bulletin of the Indian Medical Association,* 5:77–79.

Pavri, P. N., Mutalik, G. S., Nair, S., & Kulkarni, R. D. (1977a). Cytogenetic studies in cases of abnormal sexual development and multiple congenital anomalies. *Journal of the J.J.Group of Hospitals and Grant Medical College,* 22:159–164.

Pavri, P. N., Mutalik, G. S., Nair, S., Kolluri, R. V & Kulkarni, R. D. (1977b). Dermatoglyphics of patients with sex chromosome mosaicism. *Bulletin of the International Dermatoglyphics Association,* 6(1):7–8.

Pavri, P. N., Mutalik, G. S., Kolluri, R. V, Nair, S. & Kulkarni, R. D. (1978). Cytogenetic studies in abnormal sex phenotypes. In Verma, I. C. (Ed.), *Medical Genetics in India (Vol. 1).* Auroma Enterprises, Pondicherry, pp. 153–156.

Peter, J. (1974). Hot aceto-orcein for staining human chromosomes. *Journal of Genetics,* 61(3):251.

Peter, J. J., Vaidya, R. A., Rege, N. R., Gokarn, V. V., Phadke, A.G. & Ambani, L. M. (1980). Cytogenetic abnormalities in male infertility. *Indian Journal of Medical Research,* 71:225–231.

Phadke, M.A., Mutalik, G.S., Sainani, G.S., Kate, S.L. & Khedkar, V.A. (1975a). Turner's syndrome—a clinical and cytogenetic study. *Journal of the Association of Physicians of India.* 23(7):427–431.

Phadke, M. A., Mutalik, G. S., Sainani, G. S., Kate, S. L., Khedkar, W. A. & Arnikar, H. H. (1975b). A clinical and cytogenetic study of Down's Syndrome. In *Abstracts: 2nd Annual conference of the Indian Society of Human Genetics* (Abstract 114). ISI, Kolkata.

Poddar, S., Talukder, G. & Sharma, A. (2004). Chromosome damage induced by ferric chloride in human peripheral lymphocytes. *International Journal of Human Genetics,* 4(4):261–264.

Prabhakara, K., Dutta, U. & Ramadevi, A. R., (2002). Partial 15q22 trisomy due to segregation of maternal 10:15 reciprocal translocation. *Indian Pediatrics,* 39(11):1050–1054.

Prabhu, B.K., Gowri, B., Muthuvelu, K., Venkatachalam, P., Paul, S.F.D. & Jayanth, V. R. (2004). Effect of 2-Deoxy-D-Glucose on the induction of chromosomal aberrations in lymphocytes exposed *in vitro* to gamma radiation at a dose rate of 1.0 Gy/minute. *International Journal of Human Genetics,* 4(1):45–49.

Prasad, M.H., Devi, G.S., Prabhavathi, A., Padmavathi, P., Rao, M.S. & Reddy, P.P.(2002).Analysis of chromosomal aberrations in the peripheral blood lymphocytes of aluminium foundry workers. *International Journal of Human Genetics,* 2 (2):73–75.

Purandare H. & Chakravarty A. (1980). Buccal mucosal X-chromatin frequency in the female. *Indian Journal of Medical Research,* 71:790–794.

Rajangam, S., Lincoln, S., Hegde, S., Jayashree, Manjunath, N. A. & Thomas, I. M., (1993). FISH techniques in a Turner mosaic with ring X chromosome. *Indian Pediatrics,* 30(12):1451–1454.

Rajangam, S., Varghese, M., Lincoin, S., Iyer, U. & Thomas, I. M. (1995). Familial marker in a Down Syndrome Child. *Indian Journal of Human Genetics,* 1(1):42–45.

Rajangam, S., Lincoln, S., Tilak, P. & Thomas, I. M., (1999). Primary amenorrhea with Xq duplication. *Indian Journal of Medical Sciences,* 53(2):49–52.

Rajasekhar, M., Murugesan, R., Rekharao, Shetty, H., Jyothirao, Gopinath. P. M. & Satyamoorthy, K. (2010). Cytogenetic analysis of 1400 referral cases: Manipal experience. *International Journal of Human Genetics,* 10(1–3):49–55.

Raman, P. G., Shende, A., Chakraborty, D. & Akolekar, S. S. (2001). 46,XY,3q+ karyotype presenting as syndrome with obesity and hypogonadism. *Journal of the Association of Physicians of India,* 49:481–484.

Rani, K., Kotwani, B. G. & Rao, S. R. (1974). Cytogenetical studies on couples with repeated abortions. *Indian Journal of Experimental Biology,* 12(1):98–99.

Rao, B. V, Kerketta, L., Korgaonkar, S., Ghosh, K. & Mohanty, D. (2002). Fluorescence in situ hybridization analysis of balanced t (8;22)(q24.3;q12.2) in a female with recurrent spontaneous abortions. *Indian Journal of Human Genetics,* 8(1):32–33.

Rao, B.V. & Ghosh, K. (2005). Chromosomal variants and genetic diseases. *Indian Journal of Human Genetics,* 11(2):59–60.

Rao, B.V. & Mohanty, G.D. (2005). Nucleolar organising region evaluation using new NOR FISH probe. *Indian Journal of Human Genetics,* 11(1):44–46.

Rao, B. V., Kerketta, L., Korgaonkar, S. & Ghosh, K. (2006). Pericentric inversion of chromosome 9[inv(9)(p12q13)]: Its association with genetic diseases. *Indian Journal of Human Genetics,* 12(3):129–132.

Rao, K. S., Rao, C. R. & Radhakrishna, G. (1975). Cytogenetics in retinoblastomas. *Indian Journal of Ophthalmology.* 23(3):23–25.

Rao, M. V., Jhala, D. D., Patel, A. & Chettiar, S. S. (2008a). Cytogenetic alteration induced by nickel and chromium in human blood cultures and its amelioration by curcumin *International Journal of Human Genetics,* 8(3):301–305.

Rao, M. V., Hydrabadi, V. R. & Chandel, D. (2008b). Complete androgen insensitivity in three generations of a family. *International Journal of Human Genetics,* 8(4):361–363.

Reddy, K. S. & Thomas, I. M., (1985). Cytogenetic investigations in mentally retarded patients. *Indian Journal of Medical Research,* 82:164–170.

Roy, A. K. & Banerjee, D. (1995). Cytogenetic study of primary amenorrhoea. *Journal of the Indian Medical Association,* 93(8):291–292.

Rupa, D. S., Rita, P., Reddy, P. P. & Reddi, O. S. (1988). Screening of Chromosomal Aberrations and Sister Chromatid Exchanges in Peripheral Lymphocytes of Vegetable Garden Workers. *Human Experimental Toxicology,* 7(4):333–336.

Rupa, D. S., Reddy, P. P. & Reddi, O. S. (1991). Clastogenic effect of pesticides in peripheral lymphocytes of cotton-field workers. *Mutation Research/Genetic Toxicology,* 261(3):177–180.

Sadasivan, G. & Ebenezer, L.N. (1967). Abnormal X in a Turner mosaic. *Indian Journal of Medical Research,* 55(10):1060–1063.

Sadasivan, G. (1975). The chromosomes of habitual aborters. In *Abstracts: 2nd Annual conference of the Indian Society of Human Genetics* (Abstract 34). ISI, Kolkata.

Sahetya, A. R., Balsarkar, G. D., Allahbadia, S., Bandekar, P., Purandarey H. & Prabhu, V. (2010). Why a positive screening test should undergo invasive cytogenetic diagnosis. *Journal of Prenatal Diagnosis and Therapy,* 1(1):49, PO-23.

Saksena, P. N. & Kumar, N., (1968). Dermatoglyphics in Turner's syndrome. Analysis of patterns in 4 cases and their families. *Indian Journal of Pediatrics,* 35(248):429–435.

Salaskar, V., (2010). 163 cases of structural and numerical abnormalities reported in Metropolis. *Journal of Prenatal Diagnosis and Therapy,* 1(1):49, PO-24.

Sangeetha, R., Balachandar, V., Mohanna Devi, S., Velmurugan, P., Lakshman Kumar, B, Manikantan, P. et al., (2010). Cytogenetic study on sexual ambiguity in humans. *International Journal of Human Genetics,* 10(1–3):81–86.

Sanghvi, L. D. (1974). Mystery that is heredity. In *Souvenir: 1st Annual Conference of the Indian Society of Human Genetics* (pp. 1–13), Mumbai.

Sankar, V. H. & Phadke, S. R. (2006). Ring chromosome 13 in an infant with ambiguous genitalia. *Indian Pediatrics,* 43(3):258–260.

Sarkar S. & Chowdhury, J. R. (1972). Effect of hydroxyurea on the chromosomes of tumour cells. *Indian Journal of Cancer,* 9(2):132–137.

Sasikala, K., Rosalin, F.R., Jude, A.L.C., Kumar, R.A., Sudha, S., Devi, M.V., et.al. (2003). Active and passive smokers – a haematobiochemical and cytogenetic study. *International Journal of Human Genetics,* 3(1):29–32.

Saxena, M. B. (1975). Possible mechanisms of chromosomal evolution for the origin of the human karyotype. In *Abstracts: 2nd Annual conference of the Indian Society of Human Genetics* (Abstract 2). ISI, Kolkata.

Saxena, A. K., Singh, K. P., Nagle, S. L., Gupta, B. N., Ray, P. K., Srivastav, R. K., et al. (1988). Effect of exposure to toxic gas on the population of Bhopal: Part IV—Immunological and chromosomal studies. *Indian Journal of Experimental Biology.* 26(3):173–176.

Sen, S. C. & Thombre, D. P. (1971). Nuclear sex determination in infertile males. *Indian Journal of Medical Research.* 59(11):1761–1766.

Shah, PN. (1957). Sex chromatin test; its scope and limitations in clinical practice and research. *Indian Journal of Medical Sciences,* 11(9):679–686.

Shah, V. C., Krishna Murthy, D. S., Roy, S., Contractor, P. M. & Shah, A. V., (1982). True hermaphrodite 46,XX/46,XY: Clinical, cytogenetic and histopathological studies. *Indian Journal of Pediatrics,* 49(401):885–890.

Shah, V.C., Murthy, D.S. & Murthy, S.K. (1990). Cytogenetic studies in a population suspected to have chromosomal abnormalities. *Indian Journal of Pediatrics,* 57(2):235–43.

Shanker, J., Chandra, N., Lakshimanarayana, P and Gopinath, P. M. (1995). Cytogenetic evaluation of 194 patients with primary and secondary amenorrhoea. *Journal of Reproductive Biology and Comparative Endocrinology,* 7:104–111.

Sheth, F.J., Uppala, R., Morris, M.A., Blouin, J.L., Sheth, J.J., Multani, A., et.al. (2007). Cytogenetic, molecular and FISH analysis of an isodicentric chromosome 21 idic(21)(q22.3) in a mildly-affected patient with Down syndrome. *International Journal of Human Genetics,* 7(3):215–218.

Sheth, F. J., Ewers, E., Kosyakova, N., Weise, A., Sheth, J. Desai, M., et al. (2009). A small supernumerary marker chromosome present in a Turner syndrome patient not derived from X- or Y-chromosome: a case report. *Molecular Cytogenetics,* 2:22.

Sheth, F. J., Andrieux J. & Sheth, J., (2010). Supernumerary marker chromosome in a child with microcephaly and mental retardation. *Indian Pediatrics,* 7(3):277–279.

Shetty, D., Desai, K. & Dave, U. (2006). Trisomy 9p in a mentally retarded female inherited from maternal reciprocal translocation. *International Journal of Human Genetics,* 6(3):203–207.

Shetty, D. & Dave, U., (2010). Genetic syndromes with mental retardation: A challenge for prenatal diagnosis. *Journal of Prenatal Diagnosis and Therapy,* 1(1):49, PO-21.

Shetty, K. T., Sharma, N. L. & Wahal, K. M. (1966). Sex chromatin in newborns. *Journal of the Indian Medical Association.* 47(8):372–374.

Shivakumar, S., Poonkhuzhali, B., Gunasekaran, S., Srivastava, A. & Chandy M., (2002). Cytogenetic, fluorescent in situ hybridization & reverse transcriptase-polymerase chain reaction analysis in acute promyelocytic leukaemia patients. *Indian Journal of Medical Research,* 115:59–67.

Shreeniwas, Kher, A., Datta, S., Kanade, S., Vaidya, M., & Bharucha, B. A. (1994). Mixed gonadal dysgenesis with 45, X/46, XX/46, XY mosaicism. *Indian Journal of Pediatrics,* 61(4):445–446.

Siddique, Y.H. & Mohammad, A. (2004). Induction of chromosomal aberrations and sister chromatid exchanges by cyproterone acetate in human lymphocytes. *International Journal of Human Genetics,* 4(3):187–191.

Singh, I. P. & Ghosh, P. K. (1978). Role of nucleolar organizing chromosomes in the etiology of nondisjunctional error. In Verma, I. C. (Ed.), *Medical Genetics in India (Vol. 1).* Auroma Enterprises, Pondicherry, pp. 167–169.

Singh, J. R., Singh, A. R. & Singh, A. R. (2007a). Present scenario of genetic services. *In: Directory of human genetic services in India.* Satvik Media Pvt. Ltd., Amritsar, India, pp 9–10.

Singh, J. R., Singh, A. R. & Singh, A. R. (2007b). State-wise availability of genetic services – Chromosomal investigations. *In: Directory of human genetic services in India.* Satvik Media Pvt. Ltd., Amritsar, India, pp 10–12.

Singh, J. R., Singh, A. R. & Singh, A. R. (2010). Directory of human genetic services in India – 2007. *International Journal of Human Genetics,* 10(1–3):187–192.

Singhania, L., Talukder, G. & Sharma, A. (1985). Chromosomal factors in repeated foetal loss. *Journal of the Indian Medical Association,* 83(5):168–169

Sobti, R. C. (1975). Chromosomal evolution in acute leukemias. In *Abstracts: 2nd Annual conference of the Indian Society of Human Genetics* (Abstract 80). ISI, Kolkata.

Solanki, B. R., Grover, S. & Junnarkar, R. V., (1967). Cytogenetic studies in nutritional anaemia with special reference to megaloblastic anaemia. *Indian Journal of Medical Research,* 55(4):360–369.

Sood, M. & Trehan, A. (2001). Turner's syndrome in infancy. *Indian Pediatrics,* 38(5):559

Sudha, T. & Gopinath, P. M. (1990). Homologous Robertsonian translocation (21q21q) and abortions. *Human Genetics,* 85(2):253–255.

Suresh, S., Suresh, I., Lakshminarayana, P., Jabeen, G. & Rajesh, K. (1995). Partial trisomy of chromosome 10 inherited from a carrier father. *Indian Journal of Pediatrics,* 62(5):615–618.

Suri, M., Kabra, M., Jain, U, Sanders, V., Saxena, R., Shukla, A., et al. (1995). A clinical and cytogenetic study of Turner syndrome. *Indian Pediatrics,* 32:433–442.

Talwar, R., Choudhry, V. P., Jobanputra, V. & Kucheria, K. (2002). Evaluation of minimal residual disease in patients with chronic myeloid leukaemia on IFN-alpha 2b therapy using conventional cytogenetics and fluorescence in situ hybridization. *National Medical Journal of India,* 15(4):195–198.

Thanemozhi, G., Santhiya, S.T., Chandra, N., Martin, T.H., Hansmann, I. & Gopinath, P.M. (1997). A bisatellited supernumerary marker chromosome associated with recurrent miscarriages. *Indian Journal of Human Genetics,* 3(2):71–76.

Thangaraj, K., Gupta, N. J., Chakravarty, B. & Singh, L. (1998). XXY Female: An unique case of sex reversal. *The Lancet,* 352 (9134):1121.

Thomas, I.M. (1975). Risk prediction in chromosomal disorders. In Emery, A. E. H., Reddi, O. S., Sanjeeva Rao, M. (Eds.), *Proceedings of the seminar on medical genetics* (pp. 89–99). Osmania University, Hyderabad.

Thomas, I. M., Narayanan, H. S., Jayakumari, R. N. & Victor, R. (1978). D/D translocation: Report of two cases. In Verma, I. C. (Ed.), *Medical Genetics in India (Vol. 1).* Auroma Enterprises, Pondicherry, pp. 163–166.

Thomas, I. M. Cytogenetic basis of Recurrent abortions, *Perinatology* 1999, 1(4):181–187.

Thomas, I.M., Neelamma, M., Reddy, K.S., Gopalkrishna, B.G. & Rajangam, S. (2001). Translocation (9;10) Pat, in a male child leading to monosomy 9p13 →9pter and trisomy 10q24→10qter. *International Journal of Human Genetics,* 1(4):301–302.

Tilak, P., (2010). Effect of reciprocal translocations on phenotypic abnormalities. *International Journal of Human Genetics*, 10(1–3):113–119.

Tjio, J. B. & Levan, A. (1956). The chromosome number of man. *Hereditas*, 42:1–6.

Trivedi, P.J., Patel, P.S., Brahmbhatt, M.M., Patel, B.P., Gajjar, S.B., Bakshi, S.R. et.al. (2009). A new recurring chromosome 13 abnormality in two older patients with *de novo* acute myeloid leukemia: An Indian experience. *Indian Journal of Human Genetics*, 15(3):137–139.

Vallabhajosyula, R., Tilak, P. & Rajangam, S. (2009). Ring chromosome 21: A case report. *International Journal of Human Genetics*, 9(3–4):235–238.

Varma, K. P. S., Bhatia, P & Pai, R. A. (1974). Down's syndrome in Indian population. In *Abstracts: 1st Annual conference of the Indian Society of Human Genetics* (pp. 25). Mumbai.

Varma, N. & Varma, S., (2008). Proliferative indices, cytogenetics, immunophenotye and other prognostic parameters in myelodysplastic syndromes. *Indian Journal of Pathology and Microbiology*, 51(1):97–101.

Vashist, M., Neelkamal, Yadav, R. & Lal, M. (2009). *International Journal of Human Genetics*, 9(2):105–107.

Vaz, N. & Shyama, S. K. (2005). Numerical chromosomal abnormalities in the malformed newborns of Goa. *International Journal of Human Genetics*, 5(4):237–240.

Venkatesh, H. N., (2006). Genetics of mental retardation: Molecular cytogenetic studies among idiopathic mental retardation. *Ph.D. Thesis, submitted to NIMHANS (Deemed University)*, Bangalore.

Venkateshwari, A., Begum, A., Srilekha, A., Sujatha, M., Nallari, P. & Jyothy, A. (2010). A 21/22 translocation in a female with repeated abortions: A case report. *International Journal of Human Genetics*, 10(1–3):179–181.

Verma, I. C. & Ghai, O. P. (1970). Pediatrics and the human chromosomes. *Indian Journal of Pediatrics*, 37(264):28–43.

Verma, I. C., Ghai, O. P. & Chawla, V. (1971). Cytogenetic studies in Indian childhood cirrhosis. *Indian Journal of Pediatrics*, 38(278):125–127.

Verma, I. C., Bawa, B., Ghai, O. P. & Guha, D. K. (1973). Survey of X-chromatin aberrations in newborn babies in Delhi. *Indian Pediatrics*, 10(9):537–539.

Verma, I. C. (1975). Prenatal diagnosis in 65 cases in Delhi. In Emery, A. E. H., Reddi, O. S., Sanjeeva Rao, M. (Eds.), *Proceedings of the seminar on medical genetics* (pp. 183–190). Osmania University, Hyderabad.

Verma, I. C., Ahuja, M. M. S. & Sansi, P. K. (1975). A new form of familial male pseudohermaphroditism. In *Abstracts: 2nd Annual conference of the Indian Society of Human Genetics* (Abstract 66). ISI, Kolkatta.

Verma, I. C. & Jacob T.T. (1976). Down syndrome in India. *Lancet*, *i*:1200.

Verma, I. C., Prakhya, B. M. & Malhotra, A. K., (1978). Study of sex chromosomal aneuploidy in an Indian prison population. *Indian Journal of Medical Research*, 67:493–498.

Verma, I. C., Malhotra, A. K., Malik, G. R. & Ghai, O. P. (1979). Cytogenetic profile of Down's syndrome in India. *Indian Journal of Medical Research*, 69:147–151.

Verma, I. C. (1980). *Directory of human genetic services in India* (1st ed.,) New Delhi.

Verma, I. C., Mathew, S., Elango, R. & Shukla, A. (1991). Cytogenetic studies in Down syndrome. *Indian Pediatrics*, 28:991–996.

Verma, I.C., Anand, N.K., Kabra, M., Menon, P.S.N. & Sharma, N. (1998). Study of malformations and Down Syndrome in India (SOMDI): Delhi region. *Indian Journal of Human Genetics*, 4(1):84–87.

Verma, I. C., Saxena, R., Lall, M., Bijarnia, S., & Sharma, R. (2003). Genetic counseling and prenatal diagnosis in India—experience at Sir Ganga Ram Hospital. *Indian Journal of Pediatrics*, 70(4):293–297.

Vishwanath, A., David, S., Rajangam, S. & Thomas, M. (1996). Down syndrome: Acrocentric association. *Indian Journal of Human Genetics*, 2(2):131–134.

Vijayalakshmi, J., Koshy, T., Kaur, H., Mary, A., Selvi, R., Parvathi, D., et al., (2010). Cytogenetic analysis of patients with primary amenorrhoea. *International Journal of Human Genetics*, 10(1–3):71–76.

Walter, A., Mammen, K.C., Martin. B., Gopalakrishnamoorthy, H.R. & John. T.J. (1971). Cytogenetic studies in mental retardation with report of a case of XXXXY variant of Klinefelter's syndrome. *Indian Journal of Medical Research*, 59(7):1072–1076.

Yadav, J.S. & Seth, N. (2001). Cytogenetical damage in petrol pump workers. *International Journal of Human Genetics*, 1(2):145–150.

Yadav, J.S., Yadav, A.S. & Sharma, T. (2001a). Chromosome damage in lymphocytes of stainless steel welders. *International Journal of Human Genetics*, 1 (3):195–202.

Yadav, J.S. & Chhillar, A.K. (2001). Cytogenetic risk assessment in workers of rubber industry. *International Journal of Human Genetics*, 1(4):243–248.

Yadav, J.S. & Chadha, P. (2002a). Genotoxic studies in pan masala chewers: A high cancer risk group. *International Journal of Human Genetics*, 2(2):107–112.

Yadav, J.S. & Chadha, P. (2002b). Cytogenetic damage in individuals exposed to Vehicular pollution. *International Journal of Human Genetics*, 2 (2):113–117.

Yashwanth, R., Chandra, N. & Gopinath, P. M. (2010). Satellite associations in Down syndrome. *International Journal of Human Genetics*, 10(1–3):101–104.

93.

GENETIC STUDIES OF PARKINSON'S DISEASE IN INDIA

Jharna Ray

INTRODUCTION

Parkinson's disease (PD; MIM 168600) is the second most common progressive neurodegenerative disease after Alzheimer's disease (Nussbaum & Ellis, 2003). The disease severity increases over time, causing premature death. The first clinical description of the disease was published in a paper entitled "An Essay on the Shaking Palsy," by James Parkinson in 1817 (Parkinson, 1817).

PD is principally diagnosed by a few clinical symptoms of "Parkinsonism," designated by the acronym TRAP (Tremor, Rigidity, Akinesia, Postural instability). In addition to four cardinal signs of Parkinsonism, other features are gait difficulties, mask faces (expressionless face), micrographia, depression, and speech problems including hypophonia and dysarthria.

The progressive depigmentation and degeneration of dopaminergic neurons (DA) in the substantia nigra pars compacta (SNpc) and the presence of degenerating ubiquitin-positive neurofilaments and Lewy bodies (LB) in the affected brainstem regions are the characteristic pathological features of PD. The loss of striatal dopamine results in bradykinesia and rigidity (Kish et al., 1988). In addition, a heterogeneous neuron population, including catecholamine producing neuron serotonin producing neuron, is also lost, resulting in a substantial depletion of serotonin and a variable loss of acetylcholine. This supports some of the mood and cognitive changes associated with PD (Agid et al., 1987). Lewy body is an eosinophilic hyaline inclusion deposited in surviving neurons (Gibb and Lees, 1989) and is made up of misfolded and aggregated alpha-synuclein, ubiquitin, and Torsin A (Andersen, 2000; Lucking et al., 2000a; Cole & Murphy, 2002). Not all clinically diagnosed PDs have Lewy body accumulation; particularly, patients with parkin mutations suggesting Lewy bodies found in PD-affected brains, which are not a hallmark for diagnosis (Eriksen et al., 2003).

The majority of PD cases are idiopathic; only a small percentage is inherited. Prior to the discovery of genes underlying monogenic forms of PD, environmental and other factors have been implicated in idiopathic PD (Dauer & Przedborski, 2003; Greenamyre & Hastings, 2004; Cookson, 2005). There is ample evidence to suggest that mitochondrial dysfunction, oxidative stress, and impairment of the ubiquitin proteasome system (UPS) may represent the principal molecular mechanism of pathogenesis of both sporadic and familial forms of PD.

The exposure to environmental toxins was considered to be a significant cause of PD. This was supported by the fact that MPTP (1-methy-4-phenyl-1,2,3,6-tetrahydropyridine) causes PD. Later, MPTP has been extensively used to develop an experimental model for PD. The environmental toxins, like MPTP and rotenone, disrupt mitochondrial complex-I activity causing energy failure and cell death. The inhibition of complex-I can increase the generation of superoxide anions, resulting in increased dopamine oxidation (Ramsay and Singer, 1992; Lotharius & O'Malley, 2000). It appears that DA neurons are especially vulnerable to complex-I deficits, which are central to the pathogenesis of sporadic PD.

The genetic evidence also suggests that alterations in complex-I activity play a role in the pathogenesis of sporadic PD. The polymorphism Thr to Ala in NADH dehydrogenase 3 gene of complex-I leads to a significant reduced risk of developing PD in Caucasian populations (van der Walt et al., 2003). However, no specific disease-related mutations have been detected in sporadic PD in mitochondrial or nuclear genes that encode complex-I proteins, nor have any pathogenic mutations in mtDNA been identified.

EPIDEMIOLOGY

The prevalence of Parkinson's disease is notably diverse worldwide among different ethnic groups and in different geographical locations (Liat BenMoyal-Segal, 2006). The highest prevalence is found in countries with intensive agriculture; namely, Australia (a crude prevalence ratio of 414 per 100,000 (McCann et al., 1998) and Argentina (with crude prevalence 656.8 per 100,000 (Melcon et al., 1997). A cluster of relatively low PD prevalence is apparent in the Far East, where the lowest prevalence is found in mainland China (Wang et al., 1991). The prevalence of PD in India is less than that in many Western countries, but higher than that in African, Japanese, and some Chinese populations. However, recent prevalence data from Asian countries show an increasing trend.

In India the prevalence of PD varied markedly from one study to other (Table 93-1). There is no uniform, large-scale, nationwide epidemiological data on the incidence and prevalence of PD among Indians, mostly due to dissimilar ethnicity and age structure of the study populations. A crude prevalence rate of 14.1 per 100,000 is reported in a population of 63,645 from rural Kashmir in north India (Razdan et al., 1994). Similarly, low crude prevalence rates were observed from Bangalore, south India (27/100,000), and rural Bengal from eastern India, (16.1/100,000; Das & Sanyal, 1996; Gourie-Devi et al., 2004). In a study involving 52,377 subjects from the city of Kolkata, West Bengal, a crude prevalence rate has been reported to be 45.82 /100,000 (Das et al., 2006). Evaluation of PD in geriatric populations above age 60 has demonstrated a prevalence rate of 330/100,000 (Das et al., 2008). The occurrence of PD among Anglo-Indians of India demonstrated a lower risk of PD as compared to the Indians (Ragothaman et al., 2003). One of the interesting findings is that the Parsi community living in India (Bharucha et al., 1988) has a very high prevalence similar to that found in Europe and North America. The Parsis, descendants from Iran, migrated to India and form a closed community. These facts account for their high prevalence rates despite their Asian origin. The prevalence rate of PD in Bulgarian gypsies was found to be much less than Caucasians (Milnov et al., 2000). The origin of gypsies is thought to be in north India, a group who later moved to Balkans. No differences exist between the Caucasians and gypsies with respect to their living environment. Since the gypsies originated in India, the prevalence of PD is comparable to other Indian or Asian populations (Tanner et al., 1992).

RISK FACTORS

There are many factors that are considered to be risks for PD. These include age, gender, ethnicity, familial factors, exposure to environmental toxins such as pesticides and heavy metals, rural living and well water drinking, physical activity, mental stress, etc.

Increasing age is unequivocally associated with increasing risk for PD. About 0.3% of the general population and 3% of people older than 65 and above have the disease (Moghal et al., 1994). The incidence of PD in India is likely to increase with an increase in life expectancy. It is predicted that in 2012 the total Indian population would be around 1.2 billion, of whom 93 million would be over 60 years of age. Assuming that 1% of the population over 60 years has PD, one could estimate the incidence would be about 930,000 PD patients in 2010 in India (Singhal et al., 2003).

In most of the studies PD is more common in men than in women (1:3.5) irrespective of geographic location or race (Zhang & Roman, 1993). A study from NIMHANS, Bangalore, involving both urban and rural populations, reported that males had slightly higher rate than females (31 vs. 24 per 100,000; Gourie-Devi et al., 2004). An increased prevalence was reported in females in a rural population of Bengal (214.2 (female) vs. 164.2 (male) per 100,000; Saha et al., 2003). On the other hand, the female to male ratio was reported to be 1:3.96 in north India (Behari et al., 2001).

Some lifestyle elements, such as rural living, farming, well water drinking, etc., may act as risk factors for developing PD. In India a positive correlation was found with PD and well water drinking for more than 10 years duration (Behari et al., 2001). The dietary intake of antioxidants and caffeine confers protection and slows down the disease progression (de Rijk et al.,

TABLE 93-1 PREVALENCE RATES OF PARKINSON'S DISEASE FROM DIFFERENT STUDIES IN INDIA

Region	Period of study	Size of population surveyed (Rural/Urban)	Crude prevalence rate (per 100,000)	Reference
West India	1988	14,010	328.3	Bharucha et al., 1988
North India	1994	63,645 (R)	14.1	Razdan et al., 1994
South India	1982–84	39,926 (R & U)	7	Gouri-Devi et al., 2004
South India	1993–95	102,557 (R & U)	33	Gouri-Devi et al., 2004
East India	1989–90	37,286 (R)	16.09	Das et al., 1996
East India	1992	8763	80	Das et al., 2006
East India	1992–93	20,842 (R)	53	Saha et al., 2003
East India	2003–04	52,377 (U)	45.82	Das et al., 2008

1997; Michael et al., 2002). Antipsychotic drugs, which are used to treat schizophrenia and psychosis, can induce the symptoms of PD by lowering dopaminergic activity.

GENETICS OF PARKINSON'S DISEASE

PD is mostly sporadic; only a small portion (5%–10%) is familial, inherited either as an autosomal dominant or autosomal recessive trait. To date at least thirteen chromosomal loci (PARK1–PARK13, of which PARK1 and PARK4 are same) and nine causal genes have been identified for PD. Although these genes account for only a very small percentage of PD cases, the characterization of encoded protein in patients and controls helps to understand the molecular basis of pathogenesis, and also allows researchers to construct animal models to elucidate the pathogenic process in PD. The mode of transmission, chromosomal localization, and the identified causal genes are summarized in Table 93-2. Five genes—α-synuclein (SNCA), ubiquitin carboxy-terminal hydrolase L-1 (UCHL-1), leucine rich repeat kinase 2 (LRRK2), Grb10-interacting GYF protein 2 (GIGYF2), or trinucleotide repeat containing 15 (TNRC15), HtRA2—and two loci (on chromosomes 2p13 and 1p32) have been implicated in the pathogenesis of autosomal dominant Parkinson's disease, and the four genes, parkin, DJ-1, PTEN-induced kinase 1 (PINK1), and ATP13A2, have been implicated for autosomal recessive PD (Biskup et al., 2008). The genetic studies to understand the molecular basis of pathogenesis of PD in India has begun only in recent years (Nagar et al., 2001) and is summarized in Table 93-3.

PARK1 & PARK4 LINKED PD: α-SYNUCLEIN (SNCA; OMIM 163890)

Alpha-synuclein (SNCA), the first gene implicated in PD, is unequivocally associated with familial PD and is central to the pathophysiology of familial and sporadic PD. The SNCA protein is the major component of Lewy bodies and Lewy neurites in PD and other α-synucleinopathies.

SNCA contains six exons and encodes a 140 amino acid protein. Human alpha-synuclein consists of an N-terminal amphipathic region containing six imperfect repeats (with a KTKEGV motif), a hydrophobic central region (containing the non-amyloid β component, NAC domain) and an acidic C-terminal region (George, 2002). It has been estimated to account for as much as 1% of the total protein in soluble cytosolic brain fraction (Iwai et al., 1995). The physiological function of α-synuclein is unclear; however, a role in synaptic transport of vesicles or in synaptic plasticity has been suggested (Jensen et al., 1998).

A total of three missense mutations (A53T, A30P, and E46K) have been reported in SNCA (PARK1) and are very rare. The first mutation, A53T substitution, was reported in a large family from southern Italy with PD apparently inherited in autosomal dominant fashion (Polymeropoulos et al., 1997). The autopsy of 2 patients in these kindred showed the pathological changes typical of PD with Lewy

TABLE 93-2 LOCI AND GENES ASSOCIATED WITH FAMILIAL PD OR IMPLICATED IN PD

Locus	Location	Gene	Inheritance	Special Features	Reference
PARK 1	4q21	*α–Synuclein*	Autosomal dominant	Early onset, lower prevalence of tremor	Polymeropoulos et al., 1997
PARK 2	6q25.2-q27	*Parkin*	Autosomal recessive	Early onset, more frequent dystonia, levodopa-induced dyskinesia, slow progression	Kitada et al., 1998
PARK 3	2p13	Unknown	Autosomal dominant	Dementia in some individuals, rapid progression	Gasser et al., 1998
PARK 4	4q21	*α–Synuclein*	Autosomal dominant	Early onset, rapid progression, dementia, autonomic dysfunction, postural tremor	Singleton et al., 2003
PARK 5	4p14	*UCHL1*	Autosomal dominant	Classic PD	Leroy et al., 1998
PARK 6	1p35-37	*PINK1*	Autosomal recessive	Early onset, slow progression	Valente et al., 2004
PARK 7	1p36	*DJ-1*	Autosomal recessive	Early onset, slow progression	Bonifati et al., 2003
PARK 8	12p11.2-q13.1	*LRRK2*	Autosomal dominant	Classic PD	Funayama et al., 2002
PARK 9	1p36	*ATP13A2*	Autosomal recessive	Juvenile onset, dementia, supranuclear gaze paralysis	
PARK 10	1p32	Unknown	Dominant (?)	Classical PD	Hicks et al., 2002
PARK 11	2q36-37	*GIGYF2*	Autosomal dominant	Classical PD	Pankratz et al., 2003
PARK 12	Xq31	Not known	X-linked2	Classical PD	Marx et al., 2003
PARK 13	2p12	*Omi/HtrA2*	ND	Classical PD	Le et al., 2003

The PARK1 and PARK4 loci are same. ND: Not determined. All PD patients demonstrate typical Parkinsonian features, including tremor, bradykinesia, rigidity, and postural instability

TABLE 93-3 MUTATION PROFILE OF INDIAN PARKINSON'S DISEASE PATIENTS

PARK loci	Candidate Genes	Patient Analyzed	Mutations Screened	Mutations identified	Susceptibility variants	Reference
PARK1/PARK4	*SNCA*	Familial, juvenile, EOPD & sporadic	A30P, A53T	Not found	Not found	Nagar et al., 2001
PARK 2	*Parkin*	EOPD, juvenile, LOPD, familial & sporadic	Coding sequences	11 point mutations, 9 exon rearrangements, exon deletions [3&4; 8&9]	Homozygous Ser167 & Val380 increases risk for PD	Biswas et al., 2006 Biswas et al., 2007 Madegowda et al., 2005 Chaudhury et al.,2005
PARK 6	*PINK 1*	Familial & sporadic, YOPD & LOPD	Coding sequences	2 point mutations (Het)	Not found	J Ray (personal communication)
PARK 7	*DJ1*	One Indian male in a study in US population	Coding sequences	1 point mutations (Het)		
PARK 8	*LRRK 2*	Familial & sporadic, YOPD & LOPD	G2019S, I2012T, I2020T, R1441C, R1441G, R1441H	G2019S	Not found	Punia et al., 2006

bodies and L-DOPA responsive PD (Golbe et al., 1990). Among the reported mutations, A53T, the most frequent one, was found in at least 12 Mediterranean families, notably Greek and Italian, probably with a common ancestor (Spira et al., 2001). The A53T mutation was also reported in a Korean family with a different haplotype, suggesting two independent mutational events (Ki et al., 2007). Most patients with point mutations have prominent dementia and earlier onset than in sporadic PD (Cookson et al., 2005). In addition, duplications and triplications of the locus containing SNCA (initially PARK4; Singleton et al., 2003) suggest that overexpression of SNCA is toxic. Nine families with duplications and three with triplications of the SNCA locus give an overall mutation frequency of ~2% in familial Parkinsonism.

Genetic variability in the promoter is reported to be association with PD in Japanese populations but not with European populations (Spadafora et al., 2003; Ross et al., 2002). An association between a dinucleotide repeat (Repl) in the SNCA and sporadic PD was found in different populations (Parsian et al., 1998; Mizuta et al., 2006; Tan et al., 2004a).

Alpha-synuclein is highly expressed throughout the mammalian brain, and is enriched in presynaptic nerve terminals (Clayton et al., 1999). It has been estimated to account for as much as 1% of the total protein in soluble cytosolic brain fraction (Iwai et al., 1995). It is a major structural component of Lewy bodies. Mutation in alpha-synuclein causes a toxic gain of function. The genomic triplication results in approximately double expression of wild type protein, which causes the disease. The mutant proteins (A30P and A53T) display an increased propensity to self-aggregate, to form an oligomeric species and LB-like fibrils in vitro (Conway et al., 2000). Oligomers are the precursor of higher-ordered aggregates, such as amyloid-like fibrils, which precipitate as filamentous structures observed in the LBs and Lewy neuritis. The mechanism by which wild-type synuclein aggregates in sporadic PD is not clear. However, mitochondrial complex-I inhibitors, such as rotenone and Paraquat, lead to synuclein aggregation and accumulate in vitro and in animal models (Manning-Bog et al., 2002; Sherer et al., 2003).

The first genetic study of PD in Indian patients was reported in 2001 (Nagar et al., 2001). The entire exon 3 and 4 of SNCA was screened in 169 PD patients comprising 18 familial, 3 juvenile, 48 early-onset, and 100 sporadic cases. No mutation was detected, including two reported variants, G88C (Ala30P) in exon 3 and G209A (Ala53Thr) in exon 4. This suggests that mutation in the alpha-synuclein gene is a rare cause of PD in India.

PARK2-LINKED PD (PARKIN; MIM 602544)

Among the PD-causing genes, parkin is the most common, and an exonic deletion in the parkin gene was first reported in a Japanese family with autosomal recessive juvenile Parkinsonism (ARJP; Kitada et al., 1998). Mutations in the parkin gene have been found in patients from all over the world, and are the most frequent cause of autosomal recessive PD. Parkin is a large gene (~1.5 Mb), located on chromosome 6 and consisting of 12 exons. More than 195 different mutations have been identified, of which missense and gross deletions are the most common. Over 79 different point mutations and numerous exonic rearrangements, duplications, and deletions have been reported (HGMD; http://www.hgmd.cf.ac.uk/ac/gene.php?gene=PARK2). The most prevalent mutation worldwide is the deletion c.255A, which results in a truncated peptide of 82 amino acids. Other changes in the open reading frame that produce stop codons are 202-203delAG and 321-322insGT, and one nonsense mutation, Trp453Stop (Abbas et al., 1999; Mena et al.,

2008). Parkin mutations account for about 50% of familial cases and are found in about 70% of sporadic cases with age of onset <20 years (Lucking et al., 2000b; Periquet et al., 2003; Mata et al., 2004). Screening of sporadic young-onset cases (age of onset <45 years) gives a variable frequency between 2% to 18% depending on the ethnicity of the study populations (Lucking et al., 2000b; Periquet et al., 2003; Mata et al., 2004; Wu et al., 2005). Limited studies have suggested that parkin mutational frequency in late-onset PD is lower than in early-onset cases, accounting for 0% to 11% depending on the inclusion criteria (familial or sporadic; Mata et al., 2004; Foroud et al., 2003; Oliveri et al., 2001). In several early-onset PD cases a single heterozygous parkin mutation is found. No other mutations are detected by dosage analysis or sequencing of the promoter region in these patients. Moreover, the age of PD onset is later in these patients than in those with mutations identified in both parkin alleles (Kann et al., 2002; Hedrich et al., 2002; Lohmann et al., 2003; West et al., 2002). This raises the question of whether single heterozygous parkin mutations are sometimes sufficient to cause early-onset PD. Mechanisms to explain the disease in these cases might involve haploinsufficiency, dominant negative, or dominant gain-of-function mechanisms. Positron emission tomography (PET) studies showing that heterozygous carriers of parkin mutations have mild nigrostriatal dysfunctions also suggest that these mutations might be harmful, at least at the subclinical level (Hilker et al., 2001; Khan et al., 2002).

The parkin promoter lacks TATA or CAAT boxes and appears to share homology with the alpha-synuclein promoter. It has several putative *cis*- elements for various transcription factors (Asakawa et al., 2001; West et al., 2001). One variant, –258T/G located in the core promoter region, binds to nuclear protein from human substantia nigra in vitro, and it was found that the G-allele is associated with decreased gene expression (West et al., 2002). Several single nucleotide polymorphisms (SNPs) have been identified in the parkin gene, including missense changes.

The clinical phenotype of patient with parkin mutations is variable and, in most cases, related to the genotype (Mena et al., 2008). In general, the age at onset and the severity of the clinical syndrome depends on the kind of mutation, and is more severe with large deletions or with mutations that produce a truncated peptide. Patients with deletions of one or more exons or with mutations, which change the open reading frame of DNA and produce a truncated, nonfunctional protein, have a very early onset of symptoms, respond to L-DOPA or DA agonists, and have early development of dyskinesia.

Lewy bodies are absent in the majority of patients with parkin mutations, suggesting parkin might be an essential component of Lewy bodies (Chung et al., 2001). Interestingly, Lewy bodies have been found in a patient who is a compound heterozygote for parkin mutations, with 40 bp deletion in exon 3 and Arg275Trp. The brain of this patient had abundant ubiquitin and alpha-synuclein positive inclusions throughout the substantia nigra (Farrer et al., 2001). It has been demonstrated that the Arg275Trp mutation, located within the RING finger 1 domain, leads to an unusual cellular distribution of the protein, with large cytoplasmic and nuclear inclusions (Cookson et al., 2003). These observations help to explain Lewy body pathology in some patients with parkin mutations.

The parkin gene encodes a 465 amino acid protein containing several important functional domains—an N-terminal ubiquitin-like domain (Ub-L), two RING finger domains separated by an in-between-RING (IBR) domain in the C-terminal, and a linker domain connecting N- & C- terminals (Morett & Bork, 1999). Parkin acts as an ubiquitin ligase (E3 ligase) in the ubiquitin proteasome system (UPS; Shimura et al., 2000). The protein is ubiquitously expressed in neurons and glial cells. The E3 ligase controls the key step in the cycle of ubiquitin-mediated hydrolysis of damaged or misfolded proteins that are degraded via proteasome. Parkin mutations may result in an incorrect ubiquitination and an invalid targeting to the proteasome, leading to protein accumulation.

Recently, three studies reported from India, describing the mutation profile of parkin gene among southern, northern and eastern Indian PD patients (Table 93-3; Madegowda et al., 2005; Chaudhary et al., 2006; Biswas et al., 2006). Screening of 102 EOPD patients (20 familial and 82 sporadic; age of onset <50 years) and 105 normal controls from South India identified a homozygous missense mutation, Thr240Met, in exon 6, in one case, and homozygous deletions of exons 8 and 9 in another case. These constituted 2% of all early-onset PD, 10% of all familial early-onset PD, and 25% of all autosomal recessive early-onset PD patients. In addition, seven heterozygous exonic changes were identified among 23 sporadic cases. These may represent heterozygous pathogenic changes (Madegowda et al., 2005).

In another study, Chaudhary et al. (2006) evaluated parkin mutation(s) in 259 PD patients from India, both familial and sporadic young onset (YO; <40 years) and late onset (LO; >41 years). Among these 259 PD patients, 53 (32 YOPD, 3 LOPD, and 18 familial PD) were from the south, and 206 (141 YOPD, 48 LOPD and 17 familial PD) were from north India. Nine point mutations, including two homozygous (Ala31Asp and Arg334Cys), one compound heterozygous, and six heterozygous, were identified in 20 patients (7.7%). No homozygous exonic deletions were found, but heterozygous exon rearrangements were observed in 6 of 65 (9.2%) patients. Two mutations, Arg334Cys and Gln34Arg, were detected in a relatively high frequency: Arg334Cys in exon 9 in eight of the 20 patients (40%), and Gln34Arg in exon 2 in four probands (20%). A silent mutation (C917T in exon 7) was also seen in three patients (15%). All these mutations except Arg42Cys were screened in 55 controls and in the available family members. It was interesting to note that all nine point mutations were observed among the north Indian patients. However, only two mutations, Gln34Arg and Arg334Cys, were seen in two unrelated south Indian patients. Parkin mutations were notably higher among patients from South India (18.83%) compared to those from North India (7.76%). Also, the frequency of exon rearrangements was higher among south Indians than north Indian PD patients. Point mutations and heterozygous deletions taken together, the overall frequency of parkin mutations reported in this study was 8.5%.

We also screened parkin in 138 PD patients (mean age of onset, 47±14; age range, 5–77 years), and 100 controls from eastern India. Among the patients, 107 (77%) represent sporadic cases, 8 (6%) confirmed familial cases, and 23 (17%) have a history of other neurological problems in the family. A total of 18 nucleotide variants including six novel changes were detected. These include five missense mutations (Gln34Arg, Arg42Cys, Arg42His, Tyr143Cys, and Arg334Cys) detected in eight patients in the heterozygous condition, and a homozygous deletion encompassing exons 3 and 4 in two sibs affected with PD. Among eastern Indian PD patients, mutation in parkin was identified in 7.24% (10/138) cases.

In this study we tried to determine genotype–phenotype correlation. The patients bearing parkin mutations had an age range of 31–60 years, with mean age of onset 47±11. They responded well with levodopa treatment. The disease started with rest tremor in a majority of the patients, with slow progression except the patients harboring Arg334Cys. Two patients bearing Arg334Cys mutation were observed to have all four cardinal features of Parkinsonism including dyskinesia. Two sibs with the homozygous deletion at exons 3–4 had a strikingly similar phenotype. For example, both the patients presented with severe rest tremor of the whole body at 31 years of age, and gradually developed bradykinesia, rigidity, postural instability, and drug-induced dyskinesia. In addition, the older sib manifested several secondary features including slurring of speech and loss of memory. However, multiple patients sharing the Gln34Arg mutation were observed to have considerable variation in their phenotypes.

ASSOCIATION BETWEEN PARKIN POLYMORPHISMS AND PD

Parkin mutation causes autosomal recessive PD when familial; however, because of the complex etiology of PD, it is possible that single nucleotide polymorphisms (SNPs) in parkin could be associated with the disease and cause pathogenesis by co-participation of other genetic or environmental factors. Several studies have analyzed the association between common polymorphisms in parkin and PD, in different populations with inconsistent results (Eerola et al., 2002; Oliveri et al., 2001; Mata et al., 2002; Lincoln et al., 2003; Hu et al., 2000). We identified several nucleotide variants including two nonsynonymous SNPs—Ser167Asn (c.601G/A) in exon 4 and Val380Leu (c.1239G/C) in exon 10—in both the patients and the controls, and evaluated the potential for association of these two SNPs with PD by genotyping 138 eastern Indian PD patients (mean age of onset 47 ±14 years) and 141 ethnically matched controls (mean age 55 ± 10 years; Biswas et al., 2006, 2007). The association study revealed that for c.1239G/C SNP, the G-allele and (G,G) genotype were significantly overrepresented in PD patients compared to controls—the p value for both cases were 0.00. Similarly, the allelic distribution and genotype of c.601G/A SNP were observed to be differentially distributed between the case and control groups with p values of 0.001 in both cases. Both the cSNPs in parkin, within this cohort, were found to be significantly associated with PD independent of age of onset, sex, and the presence of other polymorphisms (Biswas et al., 2007).

Unlike many other populations of the world, India consists of ethnically, geographically, and genetically diverse populations, comprising more than a billion people of four major linguistic lineages, consisting of 4693 communities with several thousand endogamous groups (Indian Genome Variation Consortium, 2005). Because of this diversity, this study was extended to determine the frequency of Ser167Asn polymorphism in 1674 samples from 55 ethnic groups encompassing the entire country. The minor allele (c.601A, 167Asn) frequency (MAF) for the SNP varied significantly (0 to 0.467) among the four linguistic lineages and geographical zones of India, with the highest MAF observed in the Tibeto-Burman populations. Genotype frequency of Ser167Asn was also found to vary significantly between different populations of India. Interestingly, among 55 population groups, 15 had the major genotype (G,G) greater than 90% similar to our patient pool, thereby raising the possibility of higher prevalence of PD in these population groups, which could be tested for its validity on the availability of epidemiological data. A high prevalence of parkin mutations has also been reported in Europeans. Our data clearly indicates that homozygous Ser167 and Val380 are significantly associated with the disease, suggesting that the parkin polymorphisms are risk factors for PD, and their frequency is greatly influenced by ethnic origin. This study also provides an opportunity to correlate genetic variation with epidemiological data in Indian population.

UBIQUITIN CARBOXYL-TERMINAL HYDROLASE L1 (PARK5, UCH-L1)

The UCHL1 gene encodes a protein that belongs to a family of deubiquitinating enzymes and is abundant in neurons. It constitutes 1% to 2% of the total soluble brain protein (Wilkinson et al., 1989). It is present in LB of sporadic PD (Lowe et al., 1990). A mouse with a homozygous UCHL1 intragenic deletion develops neurodegeneration with ubiquitylated inclusions (Saigoh et al., 1999). It is therefore, considered a candidate gene for PD pathogenesis. So far, two nucleotide variants have been reported. A heterozygous missense change, Ile93Met, was identified in an affected sibling pair in a German family, which causes a partial loss of the catalytic activity of this thiol protease and may lead to aggregation of proteins (Leroy et al., 1998). This was not reported in any other populations. The second variant, a common nonsynonymous polymorphism (Ser18Tyr), was reported in a European cohort of cases and controls (Maraganore et al., 1999). Inconsistent reports have been obtained regarding association with PD in Caucasians. However, it was observed that the presence of 18Tyr allele and 18Tyr/Tyr homozygosity in the UCH-L1 gene is associated with a reduced risk for development of sporadic PD in a Japanese population. Studies in a Chinese population suggest that this polymorphism may be a weak protective factor against early-onset PD (Wang et al., 2002). No studies have been reported in Indian population.

PHOSPHATASE AND TENSIN (PTEN)-INDUCED KINASE1 (PARK6, PINK1, MIM 608309)

Mutations in PINK1 (PARK6 locus; MIM 605909) gene, encoding PTEN-induced kinase 1, are found to be associated with both familial (autosomal recessive) and sporadic early onset Parkinsonism. A missense (G309D) and a truncating (W437X) mutation in the PINK1 gene were discovered first in familial PD (PARK6) of Italian and Spanish origin (Valente et al., 2004). Subsequently, point mutations, frame shifts, and truncating mutations have been identified (Ibanez et al., 2006; Tan et al., 2006). PINK1 mutations account for approximately 1%–7% of early-onset or autosomal recessive PD (ARPD) in Caucasians, about 8.9% in Japanese ARPD, and 3.7% in sporadic and familial PD of Chinese origin (Healy et al., 2004; Valente et al., 2004; Tan et al., 2006). A total of 65 mutations have been reported in PINK1 (HGMD), the most frequent being Gln456Stop (Grunewald et al., 2007).

The PINK1 gene, spanning about 18.8 kb, contains 8 exons and encodes a protein of 581 amino acids. The protein contains a 34 amino acid mitochondrial targeting sequence (MTS) and a conserved serine/threonine kinase domain (amino acids 156 to 509; Silvestri et al., 2005). It is ubiquitously expressed in the human brain, and is considered to be neuroprotective as illustrated in flies with removed PINK1. Most of the reported mutations of PINK1 are indeed localized within the serine/threonine kinase domain, suggesting the loss of kinase activity, which invokes a link between mitochondria and neurodegeneration in PD.

A total of 44 PINK1 heterozygous variants have been reported so far. The prevalence of PINK1 heterozygous rare variants in PD has been reported to vary considerably among different populations, ranging from 0.3% to 2.3% (Healy et al., 2004; Abou-Sleiman et al., 2006; Toft et al., 2007). The heterozygous variants include 40 (91%) missense, two nonsense (W437X, Q456X), one splice site (g.15445_15467del23), and one 3 bp insertion (c.1602_1603insCAA) mutations. Therefore, it is arguable whether a single heterozygous PINK1 mutation is the causative factor, or merely a risk allele. On 18F-dopa PET imaging, a 20% to 30% reduction in the caudate and putamen in PARK-6 carriers could be demonstrated, providing some support that this type of mutation may increase the risk of PD (Khan et al., 2002). The PINK1 carriers showed typical pathological hallmarks of idiopathic PD—neuronal cell loss in the substantia nigra, and presence of Lewy body (Gandhi et al., 2006), suggesting that the heterozygous mutations may have a functional effect on the encoded protein as a result of haploinsufficiency.

Pathogenic PINK1 mutations have also been reported in patients of Asian origin (Hatano et al., 2004a, 2004b). PINK1 screening in a cohort of 40 ARPD patients (34 ethnic Chinese, 3 Malays and 3 Indians), identified a heterozygous missense mutation (C26G in exon 4) in an 82-year-old Chinese patient (age at onset, 73 years). This results an amino acid substitution (Leu268Val) in the conserved kinase domain (Tan et al., 2005). This heterozygous PINK1 mutation is considered to be pathogenic, as has been shown, in that 18F-dopa PET imaging has demonstrated a 20%–30% reduction in the caudate and putamen of some PARK6 carriers (Khan et al., 2002).

We analyzed PINK1 variants in Indian patients (250 PD cases and 205 ethnically matched controls). Two potentially pathogenic variants (Arg246Gln and Arg276Gln) were detected in the heterozygous state in 5 patients. In addition, 13 other variants were identified, including a known polymorphism (Ala340Thr), and a few synonymous or intronic changes, none of which are likely to be pathogenic. On comparison of the clinical data of PD patients harboring potential mutations of Arg246Gln and Arg276Gln, it was observed that 5 patients had some common features in addition to Parkinsonian features. These features included an age of onset < 50 years (age range 28–48 years), and common symptoms like pain, speech impairment, and impaired ocular movement. They also had very little tolerance to levodopa, and interestingly, all of these patients were exposed to environmental toxins (pesticides and insecticides) for about 2 to 20 years. Unlike the Chinese population, the Ala340Thr variant did not show any association with PD in Indian population (294 patients and 105 controls). Six single nucleotide polymorphisms (SNPs) selected from dbSNP were genotyped in 531 normal healthy individuals representing different ethnic groups of India. Most of the SNP markers were observed to be highly heterozygous among Indians, which could be used for segregation analysis of PINK1 alleles in familial PD cases (Biswas et al., 2010).

DJ-1 (PARK7; MIM 602533)

The DJ-1 gene (PARK7 locus) was reported to be involved in autosomal recessive early-onset Parkinsonism (van Duijn et al., 2001). The identification of homozygous point mutation (L166P) in an Italian family and a 14 kb deletion in a Dutch family provide the first evidence for the involvement of DJ-1 in PD (Bonifati et al., 2003). Mutations in the DJ-1 gene are very rare and are being implicated in about 1%–2% of cases (Lockhart et al., 2004). A total of 18 mutations have been reported in this gene, including deletions, missense, and splice site alteration (HGMD; Bonifati et al., 2004). The human DJ-1, ~25 cM away from PARK6 locus, is a 24 kb gene and contains eight exons. The first two exons are noncoding and subject to alternative splicing (Bonifati et al., 2003; Taira et al., 2001). The affected individuals with DJ-1 mutations have similar phenotype to those having parkin mutation. The DJ-1 gene encodes a highly conserved homo dimeric protein of 189 amino acids that belongs to the DJ-1/ThiJ/PfpI superfamily. DJ-1 is expressed in most mammalian tissues, including the brain (Bandopadhyay et al., 2004). It responds to oxidative stress (Mitsumoto & Nakagawa, 2001) and has a chaperone activity (which suppresses the aggregation of unfolding intermediates; Lee et al., 2003). DJ-1 does not appear in Lewy bodies of sporadic PD and other synucleinopathies, but co-localize with tau positive inclusions in a number of neurodegenerative disorders.

Mutations in the DJ-1 have been described in ARPD of European ancestry, young-onset (YOPD), Ashkenazi Jewish, and Afro-Caribbean patients. There is little information on the prevalence of DJ-1 mutations among Asian PD populations. In a study from Singapore, DJ-1 mutations were evaluated in YOPD and ARPD in a multiethnic cohort (Chinese, Malays, and Indians) of PD patients. However, no mutations were reported in these patients (Tan et al., 2004b).

In a study from the United States, DJ-1 was screened in 107 EOPD patients. Two heterozygous mutations (R98Q and A104T) and two mutations, predicted to alter substantially the protein product (IVS6-1G~C and c.56delC c.57G~A), were reported. Mutations IVS6-1G~C and c.56delC c.57G~A were identified in a single subject. The heterozygous R98Q mutation was found in three unrelated subjects; one of these patients is a 46-year-old Indian man with PD diagnosed at the age of 41 years. The mother of this patient was diagnosed with PD by history only (Hague et al., 2003). The heterozygous R98Q is considered to be a risk factor, since it forms a heterodimers with wild-type DJ-1, and also showed reduced sensitivity to H2O2-induced cell death (Takahashi-Niki et al., 2004).

The heterozygous R98Q mutation was also reported in a patient from a cohort of 386 PD (familial and sporadic) patients, mostly Asian, suggesting that DJ-1 mutation is very rare among patients with familial and sporadic Parkinsonism from Asian countries (Tomiyama et al., 2009).

LEUCINE-RICH REPEAT KINASE2 (LRRK2; PARK8; OMIM 607060)

Mutations in the LRRK2 (leucine-rich repeat kinase 2) gene at the PARK8 locus causes an autosomal dominant form of familial PD (Paisan-Ruiz et al., 2004, Zimprich et al., 2004b) and also, more frequently, sporadic PD (Lesage et al., 2005). It accounts for 10% of autosomal dominant familial PD, and 3.6% of sporadic PD (Lesage & Brice, 2009). The 144 kb LRRK2 contains 51 exons and encodes a multi-domain protein kinase comprising 2527 amino acids (Zimprich et al., 2004a). The protein domains include an armadillo (ARM) and ankyrin (ANK) repeat region, leucine-rich repeats (LRR), a Roc GTPase domain (followed by an associated C-terminal of Roc (COR) domain), a tyrosine kinase-like (TKL) domain, and a WD40 repeat. More than 40 different nucleotide variants, almost all missense, have been reported. The penetrance of LRRK2 mutation is age-specific; penetrance increases from 17% at 50 years to 85% at 70 years of age. LRRK2 mutations have also been found in asymptomatic individuals. Seven variants have been reported to be proven pathogenic mutations, which are conserved through evolution and are clustered within the functionally important regions of the protein. The most common mutation reported so far is the G2019S, found in 3%–7% of familial PD and 1%–3% of sporadic PD. The highest prevalence is reported in North African Arabs (39%) and Ashkanezi Jews (18%; Deng et al., 2005; Nichols et al., 2005; Aasly et al., 2005; Kachergus et al., 2005; Lesage et al., 2006). It is very rare in Asia, South Africa, and some European countries (Tan et al., 2005; Fung et al., 2006; Punia et al., 2006). In three independent large-scale studies involving more than 2000 Chinese PD patients and controls, no G2019S mutation was detected (Tan et al., 2005; Fung et al., 2006). The G2019S carriers have been demonstrated to share a haplotype, suggesting a common ancestral founder (Kachergus et al., 2005; Lesage et al., 2005). In a recent study, the six most commonly reported mutations in LRRK2 (G2019S, R1441C, R1441G, R1441H, I2012T, and I2020T) were investigated among 1012 (PD, 800; controls, 212) Indians; only one female patient showed a heterozygous G2019S mutation, which suggests that these mutations may be a rare cause of PD among Indians

(Punia et al., 2006). Gly2385Arg has been reported to be a risk factor for Japanese and Chinese populations, and thus suggests a risk factor for Asians (Funayama et al., 2007; Di Fonzo et al., 2006). However, analysis of 472 non-Chinese Asian subjects, including 66 Indian PD cases and 133 controls, did not identify the Gly2385Arg mutation among Indian PD patients, suggesting it is a genetic risk factor only to selected Asian races (Tan et al., 2007).

The LRRK2 protein, a multi-domain protein kinase, is involved in multiple functions, including substrate binding, protein phosphorylation, and protein–protein interaction (Biskup et al., 2006). It is expressed in most brain regions, including the substantia nigra, caudate nucleus, and putamen (Zimprich et al., 2004b), and regulates the length and branching of neuritis; these functions might be impaired by disease-causing mutations. Different studies suggest that mutations in the protein may result in loss of kinase activity, suggesting haploinsufficiency as a disease mechanism (Bonifati, 2007). A majority of the reported mutations lie within these functional domains, leading to pathology associated with LRRK2. The protein has also been shown to bind RING domains of parkin protein, in vitro, but the pathological significance of this needs to be investigated (Smith et al., 2005).

GENETIC SUSCEPTIBILITY FACTORS IN PARKINSON'S DISEASE

PD is a complex disease; monogenic forms represent only <10% of PD in most populations. The vast majority of PD results from complex interactions among genes, and between genes and environmental factors. Genetic variations may be risk factors or disease modifiers (Tables 93-4 and 93-5), affecting penetrance, age at onset, severity, and progression of the disease.

Are genes responsible for monogenic disorders also susceptibility factors? Screening of polymorphic variants of the parkin gene, Ser167Asn, and Val380Leu, in Indian PD patients and controls found a strong association with PD (see under "Parkin"; Biswas et al., 2007).

ROLE OF MAPT IN PD

Apart from the candidate genes that may cause PD, several other genes are also known to modify the disease. Microtubule-associated protein tau (MAPT) is one such gene involved in the pathogenesis of several neurodegenerative diseases, including PD. Tau encoded by the MAPT gene is enriched in the axons and is involved in the organization and integrity of the microtubules, thereby regulating axonal transport. The MAPT locus is genetically complex, owing to a sub-chromosomal inversion of ~900 kb that took place about 3–3.6 million years ago (Stefansson et al., 2005). Extensive genotyping across the interval identified two haplotypes; viz., H1 and H2, in almost complete linkage disequilibrium. The frequency of H2 varies considerably among different ethnic groups—it is almost absent in East Asians and Native Americans, rare in Africans and Indians (~5%), but frequent (20%–30%) in populations of Caucasian origin, and is under positive selection there (Das et al., 2009; Stefansson et al., 2005). The H1 haplotype has been found to be associated with increased risk of developing PD, In addition, several other MAPT polymorphisms have been investigated for an association with PD, and different populations show different risk polymorphisms or haplotypes; however, reports from the developing countries are rare. By genotyping rs1467967, rs242557, rs3785883, rs2471738, del-In9, and rs7521 of MAPT in a Taiwanese series, Fung et al. (2006) identified two risk haplotypes (GGGC + G and GAGT + G), while in a recent report from India that analyzed the same SNPs except rs3785883 (since it violated Hardy-Weinberg equilibrium) GAC + G was shown to be a risk for PD (OR = 1.563; 95% CI = 1.045—2.337; p = 0.03), AAC + A (OR = 2.787; 95% CI = 1.372—5.655; p = 0.004) for early-onset PD (age at onset <=40 years vs. young controls, age <=45 years), and AAC + G (OR = 2.233; 95% CI = 1.018—4.895; p = 0.045) for late-onset PD (age at onset > 40 years vs. young controls, age >45 years). This highlights the significance of rs7521 in modifying the age at onset of PD under a common haplotype background (AAC +). More precisely, in people who are predisposed to develop PD in their lifetime, the A allele of rs7521 lowers the age at onset of the disease, while the G allele delays the age at onset, indicating the involvement of different MAPT haplotypes on PD pathology in diverse ethnic groups. In addition, haplotype AGC + A was also found to be a risk for sporadic PD (OR = 2.773, 95% CI = 1.198—6.407, p = 0.016) (Table 93-5; Das et al., 2009).

The early evidence of MPTP-induced neurodegeneration suggests that the genes from xenobiotic metabolism pathways have been favorite candidates and this resulted in several association studies (Prasad et al., 1999). On the basis of biochemical, pharmacological, and pathophysiological evidence, variations in the genes from the dopaminergic oxidative stress pathways and mitochondria have been widely implicated (Ischiropoulos & Beckman, 2003; Kosel et al., 1999).

The depletion of dopamine levels in the brain due to degeneration of dopaminergic neurons of SNpc is a hallmark of PD. Therefore, the genes involved in its synthesis, storage, binding, transport, and metabolism have been considered as good candidates for PD pathogenesis. The cumulative contribution of genetic variations in genes from the dopaminergic pathway has been widely implicated to confer susceptibility to idiopathic PD.

An association study with PD samples from south (147 cases, 130 controls) and north (340 cases, 344

TABLE 93-4 ASSOCIATION STUDIES IN INDIA ON PD PATIENTS AND CONTROLS

Gene	Marker	Location	Sample	Result	Reference
DRD1	rs4532 (−48A>G)	5′UTR	147 PD cases and 130 controls from South India; replicated in 340 cases, 344 controls of North India		Juyal et al., 2006
	rs5330 (Ser50Arg)	Exon 2		Not associated	
	rs5331 (Ala199Ser)	Exon 2		Monomorphic	
	rs13306309 (Ala 229 Thr)	Exon 2		Monomorphic	
	rs686 (A>G)	3′UTR		Monomorphic	
DRD2	rs1799732 (−141ins/delC)	5′UTR	147 PD cases and 130 controls from South India; replicated in 340 cases, 344 controls of North India	Not associated	Juyal et al., 2006
	rs17294542 (G>A)	1 kb upstream from exon 2		Not associated	
	rs1800498 (T>C)	3.5 kb downstream from exon 2		Not associated	
	rs1801028 (Ser311Cys)			Not associated	
	rs2234689 (G>C)	1.9 kb downstream of exon 8		Not associated	
	rs1800497 (T>C)	10 kb downstream of exon 8		Not associated	
	rs1079597	5′UTR	70 PD cases and 100 control samples of North India	Not associated	Singh et al., 2008
	rs1800497 (T>C)	3′UTR		Significantly associated in South Indian population	
DRD3	rs324026 (T>C)	Intron 1	147 PD cases and 130 controls from South India; replicated in 340 cases, 344 controls of North India	Not associated	Juyal et al., 2006
	rs6280 (Ser9Gly)	Exon 2		Not associated	
	rs1503670 (G>A)	3′UTR		Not associated	
	rs905568 (G>C)	Intron 3		Not associated	
	STR	3′end		Not associated	
DRD4	120 bp duplication	1.2 kb upstream of initiation codon	147 PD cases and 130 controls from South India; replicated in 340 cases, 344 controls of North India	Significantly associated in both South and North Indian populations	Juyal et al., 2006
	−521C>T	Promoter		Not associated	
	48 bp VNTR	Exon 3		Not associated	
DAT	40-bp VNTR	3′UTR	147 PD cases and 130 controls from South India; replicated in 340 cases, 344 controls of North India	Not associated	Juyal et al., 2006
	1215A>G	Exon 9	70 PD cases and 100 control samples of North India	Not associated	Singh et al., 2008
CYP2D6	CYP2D6*3 (1 bp del)	Exon 5	70 PD cases and 100 control samples of North India	Not associated	Singh et al., 2008
	CYP2D6*4 (1934G>A)	Intron 3		Not associated	
CYP2E1	CYP2E1*5B (-1019C>T)	5′UTR	70 PD cases and 100 control samples of North India	Not associated	Singh et al., 2008
	CYP2E1*6 (7632T>A)	Intron 6		Not associated	
MAO-B	rs1799836 (A>G)	Intron 13	70 PD cases and 100 control samples of North India	Significantly associated in both male and female samples	Singh et al., 2008

(Continued)

TABLE 93-4 (CONTINUED)

Gene	Marker	Location	Sample	Result	Reference
GSTP1	313A>G	Exon 5	70 PD cases and 100 control samples of North India	Not associated	Singh et al., 2008
GSTT1	Wild type and null genotypes	-	70 PD cases and 100 control samples of North India	Not associated	Singh et al., 2008
MnSOD	Ala9Val		70 PD cases and 100 control samples of North India	Deviated from Hardy-Weinberg equilibrium in controls	Singh et al., 2008
Parkin*	rs1801474 (Ser167Asn)	Exon 4	138 PD cases and 141 controls of Eastern India	Significantly associated	Biswas et al., 2007
	rs1801582 (Val380Leu)	Exon 10		Significantly associated	
MAPT	rs1467967	Intron –1	301 PD cases and 243 controls of Eastern India	Not associated	Das et al., 2009
	rs242557	Intron –1		Not associated	
	rs3785883	Intron 3		Deviated from Hardy-Weinberg equilibrium	
	rs2471738	Intron 9		Not associated	
	del-In9	Intron 9		Not associated	
	rs7521	3' region		Not associated	

* This observation has been validated in a larger Indian cohort including 363 PD cases and 224 controls (unpublished observation – J Ray).

controls) India was reported, using a total of 20 markers including SNPs, STR, VNTR, duplication markers from five candidate genes—namely, dopamine receptors DRD1, DRD2, DRD3, and DRD4—and dopamine transporter DAT (Juyal et al., 2006). Of these, dopamine receptor genes DRD2 and DRD4 modulate motor behavior and activity of nigrostriatal neurons, and DRD1 responds to exogenous levodopa in PD patients. DRD3 receptors are localized to limbic areas, and are associated with cognitive, emotional, and endocrine functions. The dopamine transporter (DAT), present on the DA-synthesizing neurons, controls the intensity and duration of neurotransmission by rapid DA uptake into the presynaptic terminals. The allelic, genotypic, and haplotypic associations of these markers with PD were tested. Among the several markers tested, the 120 bp duplication marker of the DRD4 gene showed significant association with PD from both north and south India. A significant allelic association in south India (odds ratio, OR (95% confidence interval, CI)=0.67 (0.47–0.97) for 120 bp dup; 1.48 (1.03–2.13) for 120 bp WT) and genotypic association in south India (OR (95%CI)= 0.56 (0.35–0.91) for 120 bp dup/dup; 1.62 (0.99–2.64) for 120 bp dup/120 bp WT) and in north India (OR (95% CI)= 1.41 (1.03–1.93) for 120 bp dup/120 bp WT) was observed.

The association of polymorphism in the genes involved in toxication–detoxication are also considered as susceptibility factors for PD and were analyzed in Indian PD cases (Tables 93-4 and 93-5). The genes studied include cytochrome P4502D6, CYP2E1, GSTT1, GSTT1, GSTP1, MAO-B, and MnSOD (Singh et al. 2008). Glutathione S-transferases (GSTs) constitute a superfamily of ubiquitous, multifunctional enzymes, which play a key role in cellular detoxification, provide defense against toxic effects of xenobiotics or environmental toxins, and release oxidative stress. Among the various types of GSTs, GSTT, GSTM, and GSTP are the most prevalent.

Seventy PD patients and 100 healthy controls belonging to the same geographical location and same ethnicity were analyzed. Heterozygous genotypes of CYP2D6*4, CYP2E1*5B (RsaI) polymorphism, and homozygous mutant genotypes of CYP2E1*6 (Dral) were found to be overrepresented in PD cases when compared to the controls. No association was found in PD with manganese-superoxide dismutase (MnSOD), dopamine receptor-D2 (DRD2), and dopamine transporter (DAT). They reported an increased risk in patients carrying glutathione S-transferaseT1 (GSTT1) null, or homozygous variant genotypes of GSTP1. A similar study using 331 PD patients and 177 controls from eastern India resulted in no significant association with GSTT1, GSTM1, and GSTP1 (Biswas et al., 2011).

MAO-B is also a significant risk factor for PD. Monoamine oxidase-B (MAO-B) enzyme is responsible for the metabolism of biogenic amines, such as dopamine. MAO-B levels increase with age and in neurodegenerative diseases (Saura et al., 1994). MAO-B catalyzes dopamine-forming hydrogen peroxide (H2O2), which elicits inhibitory effects on mitochondrial complex-I activity, triggering neuronal cell death. In PD, MAO-B activity level doubled in the substantia nigra, suggesting a possible cause of dopaminergic cell death (Damier et al., 1996). A polymorphism (A > G) in intron 13 of the MAO-B gene was reported to show increased risk with the "G" allele

TABLE 93-5 HAPLOTYPES ASSOCIATION WITH PD SUSCEPTIBILITY

Gene	Markers						Comment
MAPT (Das et al., 2009)	rs1467967	rs242557	rs2471738	*del-In9*	rs7521		
	G	A	C	+	G		Risk for PD
	A	A	C	+	A		Risk for early-onset PD
	A	A	C	+	G		Risk for late-onset PD
	A	G	C	+	A		Risk for sporadic PD
DRD2 (Juyal et al., 2006)	rs1799732	rs17294542	rs1800498	Rs1801028	rs2234689	rs1800497	
	-C	G	C	C	C	C	Protective among South Indians
DRD3 (Juyal et al., 2006)	rs324026	rs6280	rs1503670	Rs905568			
	C	A	A	G			Risk for North Indian PD cases
DRD4 (Juyal et al., 2006)	120 bp duplication	−521C>T	48 bp VNTR				
	120 bp WT	T	467 bp				Risk for South Indian PD cases
	120 bp WT	T	611 bp				Risk for South Indian PD cases

in Caucasians (Costa et al., 1997). A significant association was also reported for MAO-B variant allele "G" and PD in an Indian PD population (Singh et al., 2008). Singh et al. also reported gene–gene interactions. Genotype combinations characterized by the presence of two variant genotypes on their corresponding loci revealed that four combinations of GSTT1 null and MnSOD (-9Val), or GST null and MAOB-G, or CYP2E1*5B and MAO-B-AG, or CYP2E1*5B and DRD2 (Taq1A-het) genotypes in the patients exhibited several fold higher and significant association with risk to PD. It was suggested that the genes involved in detoxification and dopamine regulation may modulate susceptibility to PD and could be important risk factors in the pathogenesis of PD.

CONCLUSION

Genes implicated in monogenic forms of PD have provided insight into the pathogenesis of both familial and sporadic forms of the disease. The clinical heterogeneity of Parkinsonism is probably the cumulative effect of different gene–environment and/or gene–gene interactions. Current literature suggests that the frequency of pathogenic gene mutations in PD may vary according to ethnicity and to the specific gene. The copy-number variation of SNCA can also affect the phenotypic expression of the disease. Penetrance of some of the recurrent mutations is incomplete and may vary with age. However, more studies are needed to determine the frequency of pathogenic gene mutations in different ethnic populations, genotype–phenotype correlation, and penetrance of recurrent mutations.

The molecular events that contribute to PD and related disorders result in the death of dopaminergic neurons in vulnerable brain regions, and are consequently reflected in the shared phenotype. However, known PD-causing genes account for only a small fraction of the monogenic forms. Robust high-density SNP genotyping technologies and data analysis programs, combined with the analysis of copy-number variation and large pathogenic genomic rearrangements, will identify novel loci. Large-scale studies and publicly available GWA databases, crucial for statistical power, require collaborative efforts using stringent clinical, genetic, and analytic methods.

PD prevalence, although low among Indians compared to Caucasians, has an increasing trend, which may be due to higher longevity among Indians. The PD prevalence in Parsis living in India is reported to be high. However, this observation may be due to overrepresentation of the elderly population in the study group. A true picture of the relative prevalence of PD among different population groups will emerge if the studies are done based on similar sets of criteria, including the age groups examined in the study cohort. The meta-analysis of all available published data by multiple investigators and further data sharing is showing a healthy trend in this direction—such study for PD may be helpful to gain a bigger picture of its occurrence.

Among the genes studied in Indian PD patients (parkin, PINK1, SNCA, LRRK2), parkin has the highest number of mutations—consistent with observations made for other populations. However, a few nucleotide variants have been found exclusively in Indian patients. Thus, screening those variants in a larger Indian cohort will shed light on the extent of occurrence and the necessity for molecular diagnostic tests to screen suspect PD cases. In addition, a few SNPs in parkin have been reported to be associated with PD, which varies in frequency in population groups studied under the Indian Genome Variation

Consortium. It would be useful to replicate the study in additional patient cohorts and, if validated, then attempt a correlation of the frequency of these genomic variants with future epidemiological studies in different population groups.

ACKNOWLEDGMENTS

The author would like to thank Mr. Arindam Biswas and Miss Gautami Das for providing technical help.

REFERENCES

Aasly JO, Toft M, Fernandez-Mata I, et al. (2005). Clinical features of LRRK2- associated Parkinson's disease in central Norway. *Ann Neurol* 57:762–765.

Abbas N, Lucking CB, Ricard S, et al. (1999). A wide variety of mutations in the parkin gene are responsible for autosomal recessive parkinsonism in Europe. French Parkinson's Disease Genetics Study Group and the European Consortium on Genetic Susceptibility in Parkinson's Disease. *Hum Mol Genet* 8:567–574.

Abou-Sleiman PM, Muqit MM, McDonald NQ, et al. (2006). Heterozygous effect for PINK1mutations in Parkinson's disease? *Ann Neurol* 60:414–419.

Agid Y, Javoy-Agid F, Ruberg M. (1987). Biochemistry of neurotransmitters in Parkinson's disease. In: Marsden CD, Fahn S, eds. *Movement Disorders* 2 London: Butterworths, pp. 166–230.

Andersen JK. (2000). What causes the build-up of ubiquitin-containing inclusions in Parkinson's disease? *Mech Ageing Dev* 118:15–22.

Asakawa S, Tsunematsu K, Takayanagi A, et al. (2001). The genomic structure and promoter region of the human parkin gene. *Biochem Biophys Res Commun* 286:863–868.

Bandopadhyay R, Kingsbury AE, Cookson MR, et al. (2004). The expression of DJ-1 (PARK7) in normal human CNS and idiopathic Parkinson's disease. *Brain* 127:420–430.

Behari M, Srivastava AK, Das RR, et al. (2001). Risk factors of Parkinson's disease in Indian patients. *J Neurol Sci* 190:49–55.

Bharucha NE, Bharucha EP, Bharucha AE, et al. (1988). Prevalence of Parkinson's disease in the Parsi community of Bombay. *India Arch Neurol* 45:1321–1323.

Biskup S, Moore DJ, Celsi F, et al. (2006). Localization of LRRK2 to membranous and vesicular structures in mammalian brain. *Ann Neurol* 60:557–569.

Biskup S, Gerlach M, Kupsch A, et al. (2008). Genes associated with Parkinson syndrome. *J Neurol* 255(Suppl 5):8–17.

Biswas A, Gupta A, Naiya T, et al. (2006). Molecular pathogenesis of Parkinson's disease: Identification of mutations in the Parkin gene in Indian patients. *Parkinsonism and Related Disorders* 12:420–426.

Biswas A, Maulik M, Das SK, et al. (2007). Parkin polymorphisms: Risk for Parkinson's disease in Indian population. *Clin Genet* 72:484–486.

Biswas A, Sadhukhan T, Majumder S, et al. (2010). Evaluation of PINK1 variants in Indian Parkinson's disease patients. *Parkinsonism and Related Disorders* 16:167–171.

Biswas A, Sadhukhan T, Bose K, et al. (2011). Role of glutathione S-transferase T1, M1 and P1 polymorphisms in Indian Parkinson's disease patients. *Parkinsonism and Related Disorders*. PMID: 21993019. doi:10.1016/j.parkreldis.2011.09.019.

Bonifati V, Rizzu P, van Baren MJ, et al. (2003). Mutations in the DJ-1gene associated with autosomal recessive early-onset parkinsonism. *Science* 299:256–259.

Bonifati V, Oostra BA, Heutink P. (2004). Linking DJ-1 to neurodegeneration offers novel insights for understanding the pathogenesis of Parkinson's disease. *J Mol Med* 82:163–174.

Bonifati V. (2007). LRRK2 low-penetrance mutations (Gly2019Ser) and risk alleles (Gly2385Arg)-linking familial and sporadic Parkinson's disease. *Neurochem Res* 32:1700–1708.

Chaudhary S, Behari M, Dihana M, et al. (2006). Parkin mutations in familial and sporadic Parkinson's disease among Indians. *Parkinsonism Relat Disord* 12:239–245.

Chung KK, Zhang Y, Lim KL, et al. (2001). Parkin ubiquitinates the alpha-synuclein-interacting protein, synphilin-1: Implications for Lewy-body formation in Parkinson disease. *Nat Med* 7:1144–1150.

Clayton DF, George JM. (1999). Synucleins in synaptic plasticity and neurodegenerative disorders. *J Neurosci Res* 58:120–129.

Cole NB, Murphy DD. The cell biology of alpha-synuclein: A sticky problem? Neuromolecular Med. (2002). 1:95–109.

Conway KA, Lee SJ, Rochet JC, et al. (2000). Acceleration of oligomerization, not fibrillization, is a shared property of both α-synuclein mutations linked to early-onset Parkinson's disease: Implications for pathogenesis and therapy. *Proc Natl Acad Sci USA* 97:571–576.

Cookson MR, Lockhart PJ, McLendon C, et al. (2003). RING finger 1 mutations in Parkin produce altered localization of the protein. *Human Molecular Genetics* 12:2957–2965.

Cookson MR. (2005). The biochemistry of Parkinson's disease. *Annu Rev Biochem* 74:29–52.

Cookson MR, Xiromerisiou G, Singleton A. (2005). How genetics research in Parkinson's disease is enhancing understanding of the common idiopathic forms of the disease. *Curr Opin Neurol* 18:706–711.

Costa P, Checkoway H, Levy D, et al. (1997). Association of a polymorphism in intron 13 of the monoamine oxidase B gene with Parkinson disease. *Am J Med Genet* 74:154–156.

Damier P, Kastner A, Agid Y, et al. (1996). Does monoamine oxidase type B play a role in dopaminergic nerve cell death in Parkinson's disease? *Neurology* 46:1262–1269.

Das G, Misra AK, Das SK, Ray K, Ray J. (2009). Microtubule-associated protein tau (*MAPT*) influences the risk of Parkinson's disease among Indians. *Neuroscience Letters* 460:16–20.

Das SK, Sanyal K. (1996). Neuroepidemiology of major neurological disorders in rural Bengal. *Neurology India* 14:47–58.

Das SK, Biswas A, Roy T, et al. (2006). A random sample survey for prevalence of major Neurological disorders in Kolkata. *Indian journal of Medical Research* 124:163–172.

Das SK, Biswas A, Roy J, et al. (2008). Prevalence of major neurological disorders among geriatric population in the metropolitan city of Kolkata. *J Assoc Physicians India* 56:175–181.

Dauer W. and Przedborski S. (2003). Parkinson's disease: Mechanisms and models. *Neuron* 39:889–909.

Deng H, Le W, Guo Y, et al. (2005). Genetic and clinical identification of Parkinson's disease patients with LRRK2 G2019S mutation. *Ann Neurol* 57:933–934.

de Rijk MC, Breteler MM, den Breeijen JH, et al. (1997). Dietary antioxidants and Parkinson disease. The Rotterdam Study. *Arch Neurol* 54:762–765.

Di Fonzo A, Wu-Chou YH, Lu CS, et al. (2006). A common missense variant in the LRRK2 gene, Gly2385Arg, associated with Parkinson's disease risk in Taiwan. *Neurogenetics* 7:133–138.

Eriksen JL, Dawson TM, Dickson DW, Petrucelli L. (2003). Caught in the act: Alphasynuclein is the culprit in Parkinson's disease. *Neuron* 40:453–456.

Eerola J, Launes J, Hellstrom O, et al. (2002). Apolipoprotein E (APOE), PARKIN and catechol-O-methyltransferase (COMT) genes and susceptibility to sporadic Parkinson's disease in Finland. *Neurosci Letters* 330:296–298.

Farrer M, Chan P, Chen R, et al. (2001). Lewy bodies and Parkinsonism in families with Parkin mutations. *Ann Neurol* 50:293–300.

Foroud T, Uniacke SK, Liu L, et al. (2003). Heterozygosity for a mutation in the parkin gene leads to later onset Parkinson disease. *Neurology* 60:796–801.

Fung HC, Chen CM, Hardy J, et al. (2006). Lack of G2019S LRRK2 mutation in a cohort of Taiwanese with sporadic Parkinson's disease. *Mov Disord* 21:880–881.

Funayama M, Li Y, Tomiyama H, et al. (2007). Leucine-rich repeat kinase 2 G2385R variant is a risk factor for Parkinson disease in Asian population. *Neuroreport* 18:273–275.

Gandhi S, Muqit MM, Stanyer L, et al. (2006). PINK1 protein in normal human brain and Parkinson's disease. *Brain* 129:1720–1731.

George JM. (2002). The synucleins. *Genome Biol* 3:REVIEWS 3002.1–3002.6

Gibb WR, Lees AJ. (1989). The significance of the Lewy body in the diagnosis of idiopathic Parkinson's disease. *Neuropathol Appl Neurobiol* 15:27–44.

Gourie-Devi M, Gururaj G, SatishChandra P, et al. (2004). Prevalence of Neurological disorders in Bangalore, India: A community based study with a comparison between urban and rural areas. *Neuroepidemiology* 23:261–268.

Golbe LI, Di Iorio G, Bonavita V, et al. (1990). Large kindred with autosomal dominant Parkinson's disease. *Ann Neurol* 27:276–282.

Greenamyre JT, Hastings TG. (2004). Biomedicine. Parkinson's-divergent causes, convergent mechanisms. *Science* 304:1120–1122.

Grunewald A, Breedveld GJ, Lohmann-Hedrich K, et al. (2007). Biological effects of the PINK1 c.1366C>T mutation: Implications in Parkinson disease pathogenesis. *Neurogenetics* 8:103–109.

Hague S, Rogaeva E, Hernandez D, et al. (2003). Early-onset Parkinson's disease caused by a compound heterozygous DJ-1 mutation. *Ann Neurol* 54:271–274.

Hatano Y, Li Y, Sato K, et al. (2004a). Novel PINK1 mutations in early-onset parkinsonism. *Ann Neurol* 56:424–427.

Hatano Y, Sato K, Elibol B, et al. (2004b). PARK6-linked autosomal recessive early-onset parkinsonism in Asian populations. *Neurology* 63:1482–1485.

Healy DG, Abou-Sleiman PM, Gibson JM, et al. (2004). PINK1 (PARK6) associated Parkinson disease in Ireland. *Neurology* 63:1486–1488.

Hedrich K, Marder K, Harris J, et al. (2002). Evaluation of 50 probands with early onset Parkinson's disease for Parkin mutations. *Neurology* 58:1239–1246.

Hilker R, Klein C, Ghaemi M, et al. (2001). Positron emission tomographic analysis of the nigrostriatal dopaminergic system in familial Parkinsonism associated with mutations in the parkin gene. *Ann Neurol* 49:367–376.

Hu CJ, Sung SM, Liu HC, et al. (2000). Polymorphisms of the parkin gene in sporadic Parkinson's disease among Chinese in Taiwan. *Eur Neurol* 44:90–93.

Ibanez P, Lesage S, Lohmann E, et al. (2006). Mutational analysis of the PINK1 gene in early onset parkinsonism in Europe and North Africa. *Brain* 129:686–694.

Indian Genome Variation database (IGVdb): A project overview. Indian Genome Variation Consortium. (2005). *Hum Genet* 118:1–11.

Ischiropoulos H, Beckman JS. (2003). Oxidative stress and nitration in neurodegeneration: Cause, effect, or association? *J Clin Invest* 111:163–169.

Iwai A, Masliah E, Yoshimoto M, et al. (1995). The precursor protein of non-A beta component of Alzheimer's disease amyloid is a presynaptic protein of the central nervous system. *Neuron* 14:467–475.

Jensen PH, Nielsen MS, Jakes R, et al. (1998). Binding of alpha-synuclein to brain vesicles is abolished by familial Parkinson's disease mutation. *J Biol Chem* 273:26292–26294.

Juyal RC, Das M, Punia S, et al. (2006). Genetic susceptibility to Parkinson's disease among South and North Indians: I. Role of polymorphisms in dopamine receptor and transporter genes and association of DRD4 120-bp duplication marker. *Neurogenetics* 7:223–229.

Kachergus J, Mata IF, Hulihan M, et al. (2005). Identification of a novel LRRK2 mutation linked to autosomal dominant parkinsonism: Evidence of a common founder across European populations. *Am J Hum Genet* 76:672–680.

Kann M, Jacobs H, Mohrmann K, et al. (2002). Role of parkin mutations in 111 community-based patients with early-onset parkinsonism. *Ann Neurol.* 51:621–625.

Khan NL, Brooks DJ, Pavese N, et al. (2002). Progression of nigrostriatal dysfunction in parkin kindred: An (18F) dopa PET and clinical study. *Brain* 125:2248–2256.

Kish SJ, Shannak K, Hornykiewicz O. (1988). Uneven pattern of dopamine loss in the striatum of patients with idiopathic Parkinson's disease. Pathophysiologic and clinical implications. *N Engl J Med* 318:876–880.

Kitada T, Asakawa S, Hattori N, et al. (1998). Mutations in the parkin gene cause autosomal recessive juvenile parkinsonism. *Nature* 392:605–608.

Kosel S, Hofhaus G, Maassen A, et al. (1999). Role of mitochondria in Parkinson disease. *Biol Chem* 380:865–870.

Lee SJ, Kim SJ, Kim IK, et al. (2003). Crystal structures of human DJ-1 and Escherichia coli Hsp31 that share an evolutionarily conserved domain. *J Biol Chem* 278:44552–44559.

Leroy E, Boyer R, Auburger G, et al. (1998). The ubiquitin pathway in Parkinson's disease. *Nature* 395:451–452.

Lesage S, Ibanez P, Lohmann E, et al. (2005). G2019S LRRK2 mutation in French and North African families with Parkinson's disease. *Ann Neurol* 58:784–787.

Lesage S, Dürr A, Tazir M, et al. (2006). LRRK2 G2019S as a cause of Parkinson's disease in North African Arabs. *N Eng J Med* 354:422–423.

Lesage S, Brice A. (2009). Parkinson's disease: From monogenic forms to genetic susceptibility factors. *Hum Mol Genet* 18:R48–R59.

Ki CS, Stavrou EF, Davanos N, et al. (2007). The Ala53Thr mutation in the alpha-synuclein gene in a Korean family with Parkinson disease. *Clin Genet* 71:471–473.

Liat Benmoyal-Segal, Soreq H. (2006). Gene-environment interactions in sporadic Parkinson's disease. *J Neurochem* 97:1740–1755.

Lincoln SJ, Maraganore DM, Lesnick TG, et al. (2003). Parkin variants in North American Parkinson's disease: Cases and controls. *Mov Disord* 18:1306–1311.

Lockhart PJ, Lincoln S, Hulihan M, et al. (2004). DJ-1 mutations are a rare cause of recessively inherited early onset parkinsonism mediated by loss of protein function. *J Med Genet* 41:e22.

Lohmann E, Periquet M, Bonifati V, et al. (2003). How much phenotypic variation can be attributed to parkin genotype? *Ann Neurol* 54:176–185.

Lotharius J, O'Malley K. (2000). The parkinsonism-inducing drug 1-methyl-4-phenylpyridinum triggers intracellular dopamine oxidation. *J Biol Chem* 275:38581–38588.

Lowe J, McDermott H, Landon M, et al. (1990). Ubiquitin carboxyl-terminal hydrolase (PGP 9.5) is selectively present in ubiquitinated inclusion bodies characteristic of human neurodegenerative diseases. *J Pathol* 161:153–160.

Lucking CB, Durr A, Bonifati V, et al. (2000). Association between early-onset Parkinson's disease and mutations in the parkin gene. *N Engl J Med* 342:1560–1567.

Madegowda RH, Kishore A, Anand A, et al. (2005). Mutational screening of the Parkin gene among South Indians with early onset Parkinson's disease. *J Neurol Neurosurg Psychiatri* 76:1588–1590.

Manning-Bog AB, McCormack AL, Li J, et al. (2002). The herbicide paraquat causes up-regulation and aggregation of α-synuclein in mice: Paraquat and α-synuclein. *J Biol Chem* 277:1641–1644.

Maraganore DM, Farrer MJ, Hardy JA, et al. (1999). Case-control study of the ubiquitin carboxy-terminal hydrolase L1 gene in Parkinson's disease. *Neurology* 53:1858–1860.

Mata IF, Alvarez V, Garcia-Moreira V, et al. (2002). Single-nucleotide polymorphisms in the promoter region of the PARKIN gene and Parkinson's disease. *Neurosci Letters* 329:149–152.

Mata IF, Lockhart PJ, Farrer MJ. (2004). Parkin genetics: One model for Parkinson's disease. *Hum Mol Genet* 13:R127–R33.

McCann SJ, LeCouteur DG, Green AC, et al. (1998). The epidemiology of Parkinson's disease in an Australian population. *Neuroepidemiology* 17:310–317.

Melcon MO, Anderson DW, Vergara RH, et al. (1997). Prevalence of Parkinson disease in Junin, Buenos Aires province, Argentina. *Movement Disorders* 12:197–205.

Mena MA, Rodriguez-Navarro JA, Ros R, et al. (2008). On the pathogenesis and neuroprotective treatment of Parkinson Disease: What have we learned from the genetic forms of this disease? *Current Medicinal Chemistry* 15:2305–2320.

Schwarzschild MA, Chen J-F, Ascherio A. (2002). Caffeinated clues and the promise of adenosine A2A antagonists in PD. *Neurology* 58:1154–1160.

Milnov I, Kmetski TS, Lyons KE, et al. (2000). Prevalence of Parkinson's disease in Bulgarian Gypsies. *Neuroepidemiology* 19:206–209.

Mizuta I, Satake W, Nakabayashi Y, et al. (2006). Multiple candidate gene analysis identifies a-synuclein as a susceptibility gene for sporadic Parkinson's disease. *Hum Mol Genet* 15:1151–1158.

Mitsumoto A, Nakagawa Y. (2001). DJ-1 is an indicator for endogenous reactive oxygen species elicited by endotoxin. *Free Radic Res* 35:885–893.

Moghal S, Rajput A, D'Arcy C, et al. (1994). Prevalence of *Movement Disorders* in elderly community residents. *Neuroepidemiology* 13:175–178.

Morett E, Bork P. (1999). A novel transactivation domain in parkin. *Trends Biochem Sci* 24:229–231.

Nagar S, Juyal RC, Chaudhary S, et al. (2001). Mutations in the alpha-synuclein gene in Parkinson's disease among Indians. *Acta Neurol Scand* 103:120–122.

Nichols WC, Pankratz N, Hernandez D, et al. (2005). Genetic screening for a single common LRRK2 mutation in familial Parkinson's disease. *Lancet* 365:410–412.

Nussbaum RL, Ellis CE. (2003). Alzheimer's disease and Parkinson's disease. *N Engl J Med* 348:1356–1364.

Oliveri RL, Zappia M, Annesi G, et al. (2001). The parkin gene is not involved in late-onset Parkinson's disease. *Neurology* 57:359–362.

Paisan-Ruiz C, Jain S, Evans EW, et al. (2004). Cloning of the gene containing mutations that cause PARK8-linked Parkinson's disease. *Neuron* 44:595–600.

Parkinson J. (1817). An essay on the shaking palsy. *Medical Classics* 10:964–997.

Parsian A, Racette B, Zhang ZH, et al. (1998). Mutation, sequence analysis and association studies of a-synuclein in Parkinson's disease. *Neurology* 51:1757–1759.

Periquet M, Latouche M, Lohmann E, et al. (2003). Parkin mutations are frequent in patients with isolated early-onset Parkinsonism. *Brain* 126:1271–1278.

Polymeropoulos MH, Lavedan C, Leroy E, et al. (1997). Mutation in the alpha-synuclein gene identified in families with Parkinson's disease. *Science* 276:2045–2047.

Prasad KN, Cole WC, Kumar B. (1999). Multiple antioxidants in the prevention and treatment of Parkinson's disease. *J Am Coll Nutr* 18:413–423.

Punia S, Behari M, Govindappa ST, et al. (2006). Absence/rarity of commonly reported LRRK2 mutations in Indian Parkinson's disease patients. *Neurosci Letters* 409:83–88.

Ramsay RR, Singer TP. (1992). Relation of superoxide generation and lipid peroxidation to the inhibition of NADH-Q oxidoreductase by rotenone, piericidin A, and MPP + . *Biochem Biophys Res Commun* 189:47–52.

Razdan S, Kaul RL, Motta A, et al. (1994). Prevalence and pattern of Major Neurological disorders in rural Kashmir (India) in 1986. *Neuroepidemiology* 13:113–119.

Ragothaman M, Murgod UA, Gururaj G, et al. (2003). Lower risk of Parkinson's disease in an admixed population of European and Indian origins. *Mov Disorders* 18:912–914.

Ross OA, Awayn NH, McWhinney D, et al. (2002). A novel polymorphic triplet repeat in intron five of the alpha-synuclein gene: No evidence of expansion or allelic association with idiopathic Parkinson's disease in the Irish. *Neuroreport* 13:1621–1625.

Saha SP, Bhattacharya S, Das SK, et al. (2003). Epidemiological study of neurological disorders in a rural population of Eastern India. *J Indian Medical Association* 101:299–300.

Saigoh K, Wang YL, Suh JG, et al. (1999). Intragenic deletion in the gene encoding ubiquitin carboxy-terminal hydrolase in gad mice. *Nat Genet* 23:47–51.

Saura J, Richards JG, Mahy N. (1994). Age-related changes on MAO in Bl/C57 mouse tissues: A quantitative radioautographic study. *J Neural Transm* 41(Suppl):89–94.

Sherer TB, Kim JH, Betarbet R, et al. (2003). Subcutaneous rotenone exposure causes highly selective dopaminergic degeneration and α-synuclein aggregation. *Exp Neurol* 179:9–16.

Shimura H, Hattori N, Kubo S, et al. (2000). Familial Parkinson disease gene product, parkin, is a ubiquitin-protein ligase. *Nat Genet* 25:302–305.

Silvestri L, Caputo V, Bellacchio E, et al. (2005). Mitochondrial import and enzymatic activity of PINK1mutants associated to recessive parkinsonism. *Hum Mol Genet* 14:3477–3492.

Singh M, Khan AJ, Shah PG, et al. (2008). Polymorphism in environment responsive genes and association with Parkinson disease. *Mol Cell Biochem* 312:131–138.

Singleton AB, Farrer M, Johnson J, et al. (2003). alpha-Synuclein locus triplication causes Parkinson's disease. *Science* 302:841.

Singhal B, Lalkaka J Sankhla C. (2003). Epidemiology and treatment of Parkinson's disease in India. *Parkinsonism Relat Disord* 9(Suppl 2):S105–S109.

Smith WW, Pei Z, Jiang H, et al. (2005). Leucine-rich repeat kinase 2 (LRRK2) interacts with parkin, and mutant LRRK2 induces neuronal degeneration. *Proc Natl Acad Sci USA* 102:18676–18681.

Spadafora P, Annesi G, Pasqua AA, et al. (2003). NACP-Rep1 polymorphism is not involved in Parkinson's disease: A case-control study in a population sample from southern Italy. *Neurosci Letters* 351:75–78.

Spira PJ, Sharpe DM, Halliday G, et al. (2001). Clinical and pathological features of a Parkinsonian syndrome in a family with an Ala53Thr alpha-synuclein mutation. *Ann Neurol* 49:313–319.

Stefansson H, Helgason A, Thorleifsson G, et al. (2005). A common inversion under selection in Europeans. *Nat Genet* 37:129–137.

Taira T, Takahashi K, Kitagawa R, et al. (2001). Molecular cloning of human and mouse DJ-1 genes and identification of Sp1-dependent activation of the human DJ-1 promoter. *Gene* 263:285–292.

Takahashi-Niki K, Niki T, Taira T, et al. (2004). Reduced anti-oxidative stress activities of DJ-1 mutants found in Parkinson's disease patients. *Biochem Biophys Res Commun* 320:389–397.

Tan EK, Chai A, Teo YY, et al. (2004a). Alpha synuclein haplotypes implicated in risk of Parkinson's disease. *Neurology* 62:128–131.

Tan EK, Tan C, Zhao Y, et al. (2004b) Genetic analysis of DJ-1 in a cohort Parkinson's disease patient of different ethnicity. *Neurosci Letters* 367:109–112.

Tan EK, Shen H, Tan LC, et al. (2005). The G2019S LRRK2 mutation is uncommon in an Asian cohort of Parkinson's disease patients. *Neurosci Letters* 384:327–329.

Tan EK, Yew K, Chua E, et al. (2006). PINK1 mutations in sporadic early-onset Parkinson's disease. *Mov Disord* 21:789–793.

Tan EK, Zhao Y, Tan L, et al. (2007). Analysis of LRRK2 Gly2385Arg genetic variant in non-Chinese Asians. *Mov Disord* 22:1816–1818.

Tanner CM, Thelen JA, Offord KP, et al. (1992). Parkinson's disease incidence in Olmsted County, MN: 1935–1988. *Neurology* 42:194.

Toft M, Myhre R, Pielsticker L, et al. (2007). PINK1 mutation heterozygosity and the risk of parkinson disease. *JNNP* 78:82–84.

Tomiyama H, Li Y, Yoshino H, et al. (2009). Mutation analysis for DJ-1 in sporadic and familial parkinsonism: Screening strategy in parkinsonism. *Neurosci Letters* 455:159–161.

Valente EM, Abou-Sleiman PM, Caputo V, et al. (2004). Hereditary early-onset Parkinson's disease caused by mutations in PINK1. *Science* 304:1158–1160.

van der Walt JM, Nicodemus KK, Martin ER, et al. (2003). Mitochondrial polymorphisms significantly reduce the risk of Parkinson disease. *Am J Hum Genet* 72:804–811.

van Duijn CM, Dekker MC, Bonifati V, et al. (2001). Park7, a novel locus for autosomal recessive early-onset parkinsonism, on chromosome 1p36. *Am J Hum Genet* 69:629–634.

Wang Y, Shi Y, Wu Z, et al. (1991). Epidemiologic Survey: Parkinson's disease in China. *Chinese Med J* 104:960–964.

Wang J, Zhao CY, Si YM, et al. (2002). ACT and UCH-L1 polymorphisms in Parkinson's disease and age of onset. *Mov Disord* 17:767–771.

West A, Farrer M, Petrucelli L, et al. (2001). Identification and characterization of the human parkin gene promoter. *J Neurochem* 78:1146–1152.

West A, Periquet M, Lincoln S, et al. (2002). Complex relationship between Parkin mutations and Parkinson disease. *Am J Med Genet* 114:584–591.

Wilkinson KD, Lee KM, Deshpande S, et al. (1989). The neuron-specific protein PGP 9.5 is a ubiquitin carboxyl-terminal hydrolase. *Science* 246:670–672.

Wu RM, Bounds R, Lincoln S, et al. (2005). Parkin mutations and early-onset parkinsonism in a Taiwanese cohort. *Arch Neurol* 62:82–87.

Zimprich A, Muller-Myhsok B, Farrer M, et al. (2004a). The PARK8 locus in autosomal dominant parkinsonism: Confirmation of linkage and further delineation of the disease-containing interval. *Am J Hum Genet* 74:11–19.

Zimprich A, Biskup S, Leitner P, et al. (2004b). Mutations in LRRK2 cause autosomal-dominant parkinsonism with pleomorphic pathology. *Neuron* 44:601–607.

Zhang ZX, Roman GC. (1993). Worldwide occurrence of Parkinson's disease: An updated review. *Neuroepidemiology* 12:195–208.

94.

THE GENETIC BASIS OF ALCOHOLISM IN INDIA

Meera Vasani

HISTORICAL BACKGROUND OF ALCOHOL USE IN INDIA

Alcohol use in India can be traced back to antiquity. The degree of cultural permissiveness toward alcohol has varied according to the changing historical periods. India, as a civilization, has a rich cultural history marked by enormous diversity and deep-rooted traditional value systems. There have been contemporary influences of industrialization and globalization in recent times. The attitudes, belief systems, and patterns of alcohol use have been modified with the changing religious, cultural, and sociopolitical influences. Some of these influences are reflected in alcohol use patterns in contemporary India. Alcohol distillation, possibly, originated in the Indus valley civilization as early as 2000 B.C. *Sura* was mostly derived from fermented rice, barley, jaggery, or *mahuwa* (flowers of *madhyaka* tree) and was relatively stronger than *soma*.

During the Vedic and Epic periods (Iron Age; 1700–500 BC), Hinduism was seen not only as a religion but a way of life. Drinking was totally prohibited by the *Rig Veda*. The great Indian epic *Ramayana,* in its description of society, has generally followed a bad/good or black/white dichotomy. In the Ancient period (500 BC–550 AD), a balanced view of drinking was promoted and moderation was counseled. Alcohol was thought to increase the mental principle *(pitta)*, while diminishing the physical and vitality principles (*kapha* and *vata).* Kautilya, the prime minister of India's first Great Emperor, Chandragupta Maurya, codified the rules and regulations for alcohol, pointing to its common usage among people in the regime. With the origin of Buddhism, which was spread by Emperor Ashoka (reigned ca. 273–232 BC) to several neighboring Asian countries, a movement for general abstention from alcohol was started. However, in the Medieval (550–1500 AD) and post-Medieval (1500–1800 A.D) periods, alcohol consumption became prevalent in a considerable section of the Hindu society, especially home-brewed low-ethanol beers (Prakash, 1961). The post-Medieval period was dominated by the Mughal regime. The holy book *Koran* prohibited the use of alcohol, referring to it as "great sin and harm." The Hindu society, in the face of the external Mughal regime, became somewhat more introverted and rigid in following the preexisting codes of behaviors and restrictive attitudes toward alcohol. In the Colonial Era (1800–1947 AD), the British administration noted the widespread production of toddy, arrack, and other traditional beverages all over India. In addition, the Portuguese monks in Goa had also developed distilled liquor called *fenny* from the cashew fruit. The British colonial administration, however, felt alcoholic beverages to be an excisable commodity (Achaya, 1998).

DRINKING CULTURE IN INDIA

India essentially has never accepted alcohol as a part of normal social discourse or eating behavior, and there is a virtual absence of "normative pattern" of drinking alcohol and attitudes toward it. India does not neatly fit into any of the conceptualizations of *abstinent* (condemnation of alcohol use in any form, e.g., Muslims and Protestants), *ambivalent* (negative attitudes coexist with attitudes idealizing intoxication, e.g., English and Scandinavian countries), *permissive* (moderate consumption accepted, e.g., Jews and Italians) or *over-permissive* (alcohol use widely accepted and drunkenness permitted in certain contexts, e.g., French and Japanese cultures (by Pittman, 1964). India displays the culture of purity and abstinence on one hand, and permissiveness toward drinking on the other. The pattern of drinking in India has undergone a change from occasional and ritualistic use, to being a social event with a common purpose to get drunk (Mohan et al., 2001). A nationwide survey reflects on the fact that those who drink alcohol tend to do so in heavy amounts (Ray et al., 2004).

ALCOHOL USE IN CONTEMPORARY INDIA

The changes brought by the Colonial Era are reflected in the post-Independence period and present India. The traditional drinks such as *toddy* or cottage-made *arrack* have been largely replaced by more expensive factory-made alcoholic beverages with greater alcohol content. This in turn has led to organized and unorganized production of illicit alcoholic beverages, in most parts of the country.

Most authentic figures for prevalence of alcohol use in the general population of India (Figure 94-1) are available from the National Survey on Extent Pattern and Trends of Drug Use in India (Ray et al., 2004). A general population survey (National Household Survey, NHS; Ray et al., 2004), was conducted in which a nationally representative sample of over 40,000 males was interviewed (age group of 12–60 years), and information on various aspects of alcohol use was obtained. About 53% were in the age range of 12–30 years (Figure 94-2). The prevalence of current alcohol use (defined as "use in preceding 30 days") was found to be 21.4%. Projecting this figure to the absolute number, it can be safely estimated that there are at least 62.5 million current users of alcohol in the country. The prevalence ranged from 7% in the western state of Gujarat (officially under complete prohibition) to 75% in the northeastern state of Arunachal Pradesh. The complexity of Indian culture is reflected in the fact that there is a wide variation among various regions, urban/rural setting, socioeconomic status, caste groups, and religious beliefs. An earlier study covering both urban and rural areas of east, north, south, and central India showed that alcohol use was higher among specific groups like the industrial workers, tea plantation workers, and tribals, compared to the general population. Alcohol in contemporary India, thus, reflects the same paradox as in other areas of development of the country, where travel by modern airlines is juxtaposed with the continuing use of oxcarts. On similar lines, the drinking in India also exhibits widely divergent patterns (Mohan, & Sharma, 1995).

WHAT IS ALCOHOL ADDICTION?

Alcohol addiction is a chronically relapsing disorder that is defined by two major characteristics: a compulsion to take the drug, with a narrowing of the behavioral repertoire toward excessive drug intake and a loss of control in limiting intake (APA, 1994). An important challenge for neurobiological research is to understand the neuroadaptive

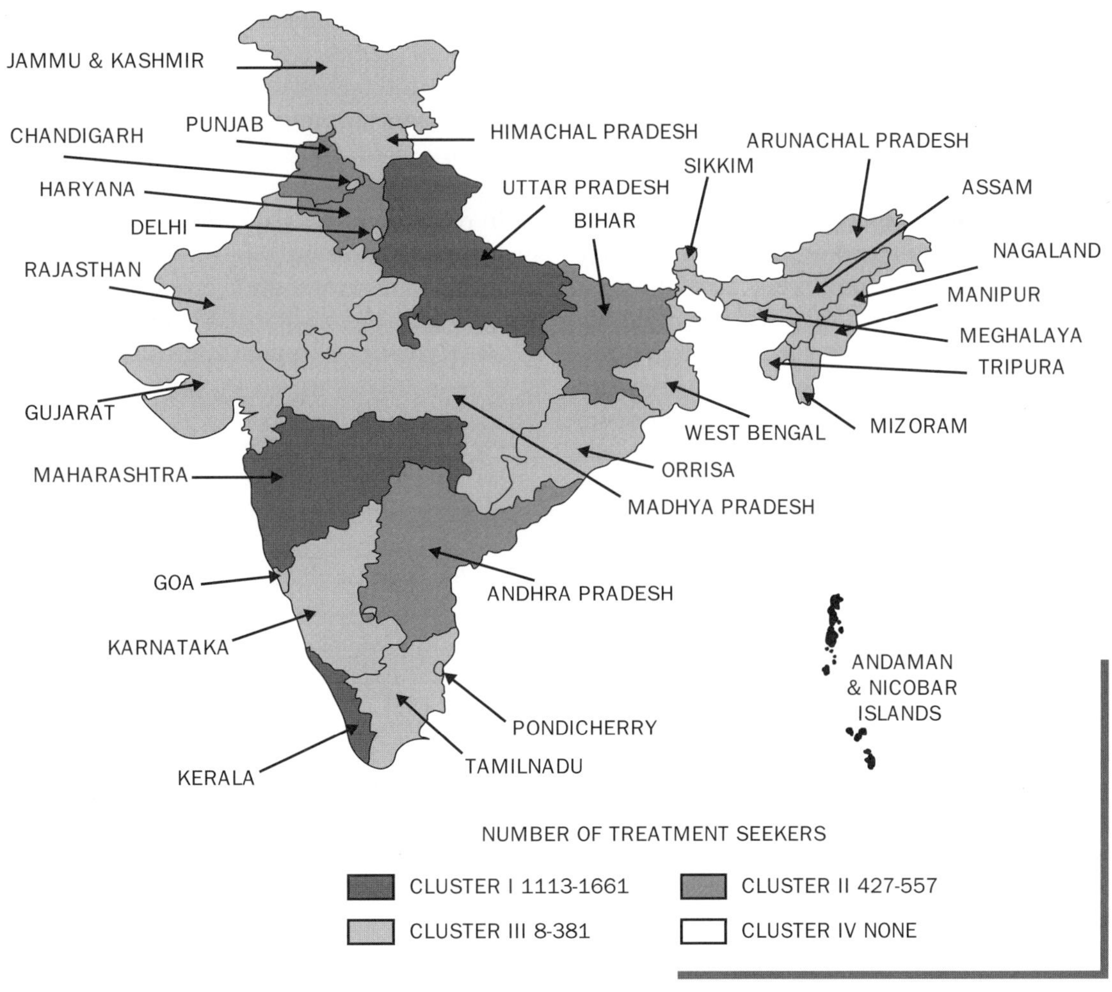

Figure 94-1 Alcohol abuse in various Indian states (N=8857). Refer color figure.

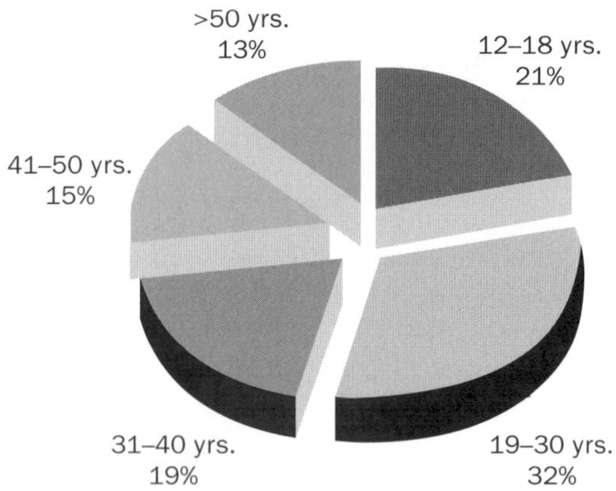

Figure 94-2 Alcohol abuse in India- Age distribution (years) N=40,697.

differences between controlled drug use and loss of control, and, by extension, the molecular, cellular, and system processes that lead to addiction (Koob & Le Moal, 1997). Advances in neuroscience research are rapidly unraveling the complexity of processes underlying addiction in general and of alcohol abuse in particular.

Genetic influences on human substance abuse (alcohol or any other drug) and vulnerability have been documented by using classical and molecular genetic approaches. More recently, research to find specific genes contributing to human vulnerability is being developed to define the limits of possible genetic and environmental factors in vulnerability (Uhl et al., 1995). From clinical and naturalistic approaches, it is generally agreed that preexisting intrinsic vulnerability does exist that leads from recreational use to abuse and dependence (Anthony et al., 1994). Why some individuals succumb and others do not is not a trivial question, but it has been generally neglected despite clinical evidence. Genetic interactions with compulsive alcohol use may be explained by an individual differences-centered approach to the problem. Substantial evidence exists for an important genetic contribution to the vulnerability to develop alcoholism. Twin studies in Scandinavia and the United States have shown consistently higher rates of alcoholism in monozygotic compared to dizygotic twins of male alcoholics (Koob & Le Moal, 2001).

NEUROBIOLOGICAL SYSTEMS IN ALCOHOL ADDICTION

The essential features of alcohol addiction are loss of control over consumption, compulsion to obtain the next drink, and continuation despite knowledge of negative health and social consequences. Processes of reward and reinforcement have great impact on addiction, after which long-lasting or permanent neuroadaptations occur, leading to tolerance and dependence (Enoch & Goldman, 2001).

The positive reinforcing effects of drugs, when described in the context of reward, often are equated with the pleasurable effects of drugs in the absence of a deficit state. Negative reinforcement, in contrast, occurs with alleviation of an existing aversive state or alleviation of a drug-generated aversive state (e.g., withdrawal). An important challenge is to understand the neuroadaptive differences between controlled drug use and loss of control, and, by extension, the molecular, cellular, and system processes leading to addiction. Advances in neuroscience research are rapidly unraveling the complexity of processes underlying addiction in general, and of alcohol abuse in particular (Koob & Le Moal, 2001). It is likely that genetic variation predisposes some individuals to increased craving and loss of control.

Alcohol affects a range of neurotransmitter systems in different regions of the brain (Figure 94-3). A pathway crucial to the action of alcohol is the mesolimbic dopamine system, associated with the ability to feel pleasure. This pathway originates in the ventral tegmental area (VTA) of the midbrain, and projects to the nucleus accumbens (NAC), the limbic system, and the orbitofrontal cortex. The amygdala, hippocampus, and medial prefrontal cortex send excitatory projections to the NAC (Koob et al., 1998). Serotonergic neurons originating in the dorsal and medial raphe nuclei project to mesolimbic structures, including the VTA and NAC, and may exert inhibitory control on mesolimbic dopamine neuron activity (Di Matteo et al., 1999). 5-HT3 is the only serotonin receptor that directly gates an ion channel; it plays an important role in reinforcement, and the effect of serotonin at this receptor is directly potentiated by ethanol (Grant, 1995). Enhanced gamma-aminobutyric acid (GABA), the brain's major inhibitory neurotransmitter,

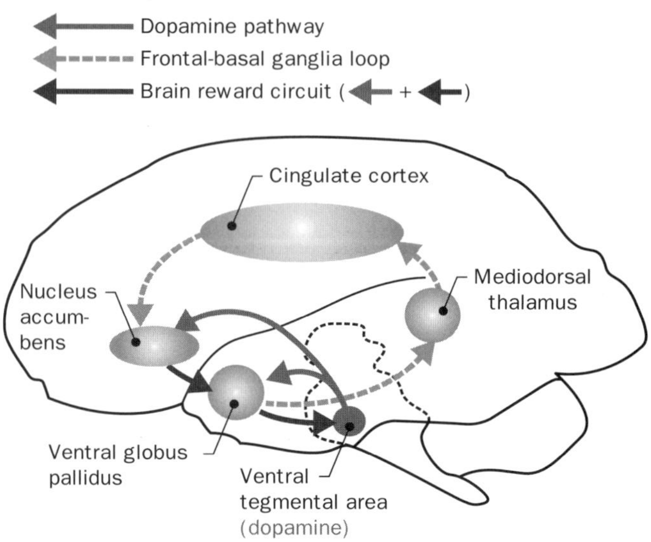

Figure 94-3 Neurobiological pathways of alcohol dependence and addiction. Refer color figure.

glutamate (an excitatory neurotransmitter), dopamine, opioid peptides, and serotonin neurotransmission have been associated with acute ethanol administration, and potentially mediate some of alcohol's reinforcing effects. The development of tolerance may be caused by adaptive changes in the $GABA_A$ receptor system. Ethanol's inhibition of the glutamatergic excitatory neurotransmitter pathways, especially at the level of the postsynaptic *N*-methyl-D-aspartate (NMDA) receptor, may be an important cause of its neurotoxic effects (Tsai & Coyle, 1998), particularly intoxication, withdrawal (Koob et al., 1998), and blackouts.

The midbrain dopamine systems have critical roles not only in the reward and motor systems, but also in higher-order functions, including cognition and memory. Positive reinforcement occurs when consumption of alcohol increases the probability of a response to obtain it again. The positive reinforcing effects of drugs, when described in the context of reward, often are equated with the pleasurable effects of alcohol in the absence of a deficit state. Negative reinforcement, in contrast, occurs with alleviation of an existing aversive state or alleviation of a drug-generated aversive state (e.g., withdrawal). Thus, dysregulation of reward neurotransmission is hypothesized to involve compromised function in the dopamine at the molecular and cellular levels in specific components of the extended amygdala, and constitutes part of the allostatic state of the brain reward system (Koob & Le Moal, 2001).

REWARD AND IMPULSE REGULATION IN ALCOHOLISM

A logical approach to understanding vulnerability to alcohol addiction is to identify variants in neurotransmitter genes implicated in ethanol use, even though individual allele effects are likely to be small. Of particular interest are the reward pathway and the impulse-control pathway.

SEROTONIN

Serotonin inhibits variety of behaviors. Low levels of serotonin appear to contribute to impulsivity and early-onset alcoholism (Linnoila et al., 1986). Normal levels of serotonin appear to be correlated with a sense of well-being, and lower levels with anxiety and dysphoria. The serotonin transporter regulates synaptic serotonin levels by reuptake of the neurotransmitter. At serotonin terminals, the 5HT1B receptor regulates serotonin release. Crabbe et al. (1996) observed that 5-HT1B knockout mice drink twice as much ethanol, are less intoxicated, and show enhanced aggression compared with normal mice.

GABAA RECEPTORS

Cross-tolerance of benzodiazepines with ethanol and their effectiveness in treating both anxiety and alcohol withdrawal suggests that $GABA_A$ receptors play a key role in alcoholism, particularly alcoholism comorbid with anxiety disorders. Differences in allele frequencies between alcoholics and control subjects have been found in $GABA_A$alpha3 (Parsian & Cloninger, 1997).

DOPAMINE

Dopamine is involved in arousal, reward, motivation, and cognition. The controversial association of a *DRD2* receptor polymorphism with alcoholism has been replicated in some case-control studies (Paterson et al., 1999). Goldman et al. (1997) did not find any such association with alcoholism. However, abnormality in dopaminergic neurotransmission seems to be one of the prime pathogenic mechanisms of alcoholism (Bowirrat & Oscar-Berman, 2005). The dopaminergic system influences brain reward mechanisms, and therefore is considered a strong candidate for alcohol dependence. Dopamine activity in mesolimbic regions is associated with the reinforcing effects of drugs (Volkow et al., 2004). Several studies suggest that when an individual has a hypodopaminergic trait, he or she is predisposed to increased release of dopamine by taking substances such as alcohol, creating a high-risk scenario for developing alcohol dependence (Addolorato et al., 2005a, 2005b). Current research supports the idea that initial exposure to alcohol activates the reward pathway by releasing DA in the nucleus accumbens, which in turn sends messages to the cortex to be coded as experiences and perhaps as memories. Once coded, these experiences can promote subsequent behavior such as further alcohol intake. In addition, these "memories" are linked to the effects of alcohol on the brain and its functioning.

REWARD PATHWAY

Further, dopamine exerts the above mentioned reward and reinforcing mechanisms by acting on multiple dopamine receptors (D1–D5) present in brain. Therefore, genetic polymorphisms in dopamine receptor genes are believed to influence the development and/or severity of alcoholism. Genetic association studies in several populations around the world have been carried out to determine susceptibility to alcohol dependence conferred by polymorphisms in candidate genes in the dopaminergic system; however, the association reports are inconsistent across different populations.

The D1 dopamine receptor has been involved in a number of brain functions, including motor control, inattentive symptoms, and reward and reinforcement mechanisms. The -48A>G marker of the DRD1 gene has been associated with a large set of addictive behaviors/phenotypes including alcohol dependence. Kim et al. (2007) examined the genetic effects of the dopamine receptor gene family (DRD1–DRD5) in the Korean population. Although the authors did not find any association of DRD1–DRD5 genes with the risk of alcoholism, severity of alcohol-related problems was significantly associated with 5' UTR polymorphism in the DRD1 (DRD1-48A>G) gene.

Of the various dopamine receptors, polymorphism in the DRD2 gene is considered most crucial in conferring risk to alcohol dependence. Previous analysis of the dopamine receptors (D1–D5) and transporter protein (DAT) has indicated that the dopamine receptor D2 (DRD2) is involved in susceptibility to alcoholism (Sasabe et al., 2007).

The genetic association of DRD2 gene with alcohol dependence draws support from clinical investigation whereby lower DRD2 receptor density and function in alcoholics may be responsible for craving for alcohol and subsequent relapse in these patients.

The DRD2 gene has three commonly investigated polymorphisms (-141C Ins/Del, TaqIB, and TaqIA), but results are thus far equivocal for their association with alcohol dependence (Ferguson & Goldberg, 1997; Goldman, 1995). The promoter polymorphism of the DRD2 gene involving the insertion (Ins)/deletion (Del) of a cytosine (–141C Ins/Del, rs1799732)) is related to receptor density (Arinami et al., 1997). SNP TaqI B is closer to the regulatory and structural coding regions (5' region) of the gene (Hauge et al., 1991), and is considered to play an important role in transcription regulation. Taq I A SNP is regarded as the most dramatic polymorphism of DRD2, for its highly inconsistent association with AD. The A1 allele of Taq I A SNP has been found to be associated with low D2 dopamine receptor availability in the striatum (Pohjalainen et al., 1998). Recently, it has become evident that the TaqI-A polymorphism, previously considered to be present in 3' UTR of DRD2, is actually located in a nearby novel gene (in reverse orientation) named *ankyrin repeat and kinase domain containing* (ANKK1), where it causes a missense substitution. TaqI-A SNP of the ANKK1 gene, is associated with increased striatal activity of aromatic L-amino acid decarboxylase, the final enzyme in the biosynthesis of dopamine (Laakso et al., 2005). SNP TaqI B is closer to the regulatory and structural coding regions (5' region) of the gene (Miller et al., 1988).

DRD1 AND DRD2 GENE POLYMORPHISMS: INDIAN DATA

Prasad, Ambekar and Vaswani (2010), in a population-based case-control study, investigated the association of DRD1 and DRD2 gene polymorphisms with alcohol dependence in male populations from north India. The results indicated that unlike the findings of Kim et al. (2007), no allelic or genotypic association of DRD1 -48A>G marker with alcohol dependence or severity (measured by AUDIT) score was observed. With regard to DRD2 gene polymorphisms (-141 Ins/Del, TaqIB and TaqIA), Prasad et al. (2010) observed a highly significant association ($p = 0.004$) of -141C Ins/Del SNP with alcohol dependence. SNP TaqIB did not show any allelic or genotypic association. For TaqI A polymorphism, genotype Taq A1A1 was present in negligible frequency in Indian populations. Significant association of genotype Taq A1/A2 (Fisher's exact p value = 0.04) with alcohol dependence was also seen. Though difference in allelic frequencies was not statistically significant, an overrepresentation of allele TaqA2 was found among alcohol-dependent subjects as compared to healthy control populations. A combined analysis involving two associated SNPs (-141C Ins/Del and Taq I A) of DRD2, and crucial laboratory parameters (SGOT, SGPT, GGT, and alcohol consumed/day) indicated that patients of AD with -141C ins/ins genotype (of -141C Ins/Del polymorphism) had significantly higher values ($P < 0.01$) of SGOT, SGPT, and GGT as compared to those with -141C Ins/Del and -141C Del/Del genotypes. Further, sub-analysis of the above group of patients highlighted that SGOT, SGPT, and GGT values were significantly higher among patients with -141C Ins/Ins + TaqA2A2 genotypic combination as compared to -141C Ins/Ins + TaqA1A2/A1A1 genotypic combinations ($P < 0.0001$). Results of independent association of SNPs were replicated in haplotypic association. As expected, haplotype containing alleles -141C InsC and Taq I A2 were found to be predisposing to alcoholism and conferred ≈2.5 times risk to patients with this haplotypic combination (-141C Ins-A-A2). On the other hand, haplotypic combination with -141C Del allele and Taq I A1 alleles (-141C Del-A-A1) seemed to confer protection from alcoholism.

GENETICS OF ALCOHOL METABOLISM

Presently, the genes known to have a major impact on alcoholism are two enzymes involved in ethanol metabolism. Alcohol dehydrogenase (ADH) metabolizes ethanol to acetaldehyde, a toxic intermediate, which is then converted to acetate by aldehyde dehydrogenase (ALDH). Approximately half the population of Southeast Asians has functional polymorphisms at four different genes: *ADH2*, *ADH3*, *ALDH1*, and *ALDH2*. The most important variants are *ALDH2*2* (Glu487-Lys487) and *ADH2*2* (Arg47-His47). *ALDH2*2* inactivates *ALDH2*, the mitochondrial enzyme responsible for most acetaldehyde metabolism in cells. *ADH2*2* is a super-active variant. *ADH2*2* and *ALDH2*2* raise levels of acetaldehyde by increasing rate of synthesis, by decreasing rate of metabolism, and thus interact additively but not synergistically (Enoch & Goldman, 2001). Ingestion of small amounts of ethanol in carriers of *ADH2*2* and *ALDH2*2* produces an unpleasant reaction characterized by facial flushing, headache, hypotension, palpitations, tachycardia, nausea, and vomiting. In an analogous fashion, disulfiram, used in the treatment of alcoholism, inhibits *ALDH2* and thereby causes a flushing reaction after alcohol consumption. Therefore, the *ALDH2*2* and *ADH2*2* genotypes can be regarded as protective and acting through a common mediator, acetaldehyde. Although the *ALDH2*2* allele dominantly inactivates the enzyme, the dominance is incomplete. This explains the nearly complete protective effect of the *ALDH2*2/2* homozygous genotype, and the partial protective effect of the heterozygous genotype. Because the *ADH3*1* allele is in linkage disequilibrium (population association) with *ADH2*2* (Enoch & Goldman 2001), it is also predictive of vulnerability, but only as a genetic marker.

GENES ENCODING ALCOHOL METABOLISM PATHWAY

Genetic polymorphisms, particularly those of alcohol metabolizing enzymes ADH and ALDH, have been largely implicated in development of alcohol dependence. The two outstanding examples of verified human "addiction genes" encode for enzymes that catalyze consecutive steps in alcohol metabolism: ADH1B and ALDH2 (Kuo et al., 2008). The functional variants of allele ADH1B*2 (His47) and ADH1B*3 (Cys369) have high enzyme activity. These variants have markedly different frequencies in different ethnic groups. ADH1B*2 allele is common in Asian populations (Eriksson et al., 2001) but is less common in Caucasians. ADH1B*3 has been found in African and Native American populations (Osier et al., 2002; Wall et al., 1997). Thus, protective effect seems to vary across environments, and the effects of genotypes are additive.

Several significant associations have been reported with alcohol dependence for subsets of alcohol metabolism genes (e.g., ADH1B, ADH1C, ALDH2) primarily in East Asian populations (Cheng et al., 2004; Higuchi et al., 2004). The roles of these genes in non-Asian populations are less clear. The ALDH2 gene is highly expressed in the liver and stomach, with a very high affinity for acetaldehyde, and plays a central role in human acetaldehyde metabolism. The ALDH1 gene also has high affinity for acetaldehyde. A functional polymorphism of the ALDH2 gene, ALDH2*2, has lower enzyme activity than the wild-type allele. This mutant allele is mainly present in East Asians (approximately 30%). Although homozygous individuals have no ALDH2 activity, heterozygous individuals maintain 30% to 50% activity. The deficient ALDH2*2 allele is associated with decreasing risk of alcohol dependence in Asian populations (Higuchi et al., 2004; Cheng et al., 2004; Zintzaras et al., 2006). In several eastern Asian countries, such as Japan, where both His47 and Lys487 are highly abundant, most of the population carries a heterozygous or homozygous genotype that is protective against alcoholism (Goldman et al., 2005). Overall, both ADH and ALDH enzymes exhibit genetic polymorphism and ethnic variation.

ALDH2, ADH1B, AND ADH1C GENE POLYMORPHISMS

Asian Data

Prevalence of *ALDH2*, *ADH1B*, and *ADH1C* alleles vary greatly across Asian ethnic groups. Approximately half of Chinese-American and Japanese samples and approximately one-third of Korean, Han Chinese, and Taiwanese studied carry at least one *ALDH2*2* allele. The prevalence of this allele is much lower (10%) in Thais, and almost no Filipinos. Similarly, *ADH1B*2* allele is found in 80% or more of Han Chinese and Taiwanese, Filipino, Japanese, Korean, and some Chinese and Taiwanese aborigine people but only in about 15% of Indians. *ADH1C*1* allele has been found in almost all Chinese and Korean people studied, but it has not been analyzed yet in other Asian ethnic groups.

Indian Data

Though prevalence of alcohol dependence (AD) in Asian populations is low, in India a sizable number of people are suffering from alcohol-use disorders (Ray et al., 2004). Genetic predisposition conferred by various candidate genes in AD differs across ethnic groups (Tu & Israel, 1995). This differential contribution leads to inconsistent genetic association and nonreplication of association between different populations. However, India, representing about one-sixth of the world's population, has very little data on genes/polymorphisms that confer susceptibility to alcohol dependence. Genetic polymorphisms, particularly those of alcohol metabolizing enzymes, alcohol ADH, and ALDH, have been largely implicated in development of alcohol dependence. The last two decades have seen growing recognition of the significance of genes encoding enzymes responsible for hepatic alcohol metabolism, namely ADH, ALDH, and CYP2E1, as protective factors in the development of alcoholism. Functional genetic polymorphisms in genes for these enzymes, along with other genes, can be envisaged to influence the organ-specific complications of alcoholism. Presence/absence of these polymorphisms in a population could modulate the probability of development and relapse of alcohol dependence, and confer protection/susceptibility to alcohol abuse and alcohol-related complications. ADH1B*2 (ADH1B*47His) allele codes for a higher activity enzyme as compared to the ADH1B*1 (ADH1B*47Arg) allele (Goldman et al., 2005; Osier et al., 2002; Whitfield, 2002) and is thus known to influence drinking behavior resulting in protection from alcoholism. A considerable variation in the allele frequency of this polymorphism has been observed among different ethnic groups. However, due to very low frequency of ADH1B*2 allele in Caucasian populations, association studies have mostly remained inconclusive (Chen et al., 1999). Reports on ADH1B*2 allele (Arg47His polymorphism) frequency in healthy Indian populations are inconsistent. Goedde et al. (1992) reported 9.9% ADH1B*2 allele frequency in heterogeneous samples from Indian populations, whereas another study on a Kachari population (Osier et al., 2002) reported the frequency to be 6.6%. However, Reddy et al. (2006), in a recent study involving 28 Indian tribal populations, found the polymorphism to be monomorphic (ADH1B*1/*1) with complete absence of ADH1B*2 allele in healthy people. Based on these findings, authors argued that Indians (mainly those comprising lower caste and the labor class) could have developed tolerance against aversive effects of alcohol, and thus the protection-conferring ADH1B*2 allele has been selected out from the population.

Mitochondrial aldehyde dehydrogenase 2 (ALDH2) plays a major role in ethanol metabolism. Clinical and enzyme-pharmacokinetic studies have indicated that individuals who show initial sensitivity to alcohol by virtue of their genetic make-up (i.e., presence of ALDH2*2 allele) are discouraged from drinking excessive alcohol (Bhaskar et al., 2007). Assanangkornchai et al. (2003) found that ALDH2*2/*2 subjects are at minimum risk for developing alcohol dependence and alcohol-related organ damage as compared to those with ALDH2*1/*2 and ALDH2*1/*1 genotypes. A significant variation in the allele frequency of Glu487Lys polymorphism has been observed across different populations worldwide, with almost complete absence of ALDH2*2 allele in Caucasians, Africans, and Americans (SNP database; NCBI) and relatively higher frequency in East Asians. Studies on Chinese and Korean populations found that alcoholics are less likely to have ALDH2*2 alleles than controls (Chen et al., 1999; Shen et al., 1997). Pertinent to East Asian populations, approximately 2%–12% of Chinese and Koreans are homozygous for the ALDH2*2 allele (ALDH2*2/*2 genotype) and appear to be almost completely protected. However, Glu487Lys polymorphism is found to be monomorphic in Indian healthy populations from Madhya Pradesh, Maharashtra, and Andhra Pradesh (Bhasker et al., 2007) (Figure 94-4).

Due to the paucity of Indian data with regard to ADH1B/ALDH2 gene polymorphisms, Vaswani et al. (2009) aimed to characterize the Arg47His and Glu487Lys polymorphisms in Indian subjects with alcohol dependence and to see the genotype/phenotype correlation (if any). The results indicated that ADH1B gene polymorphism was found to be largely monomorphic with minor allele frequency (ADH1B*2) =0.0014. For the ALDH2 Glu487Lys SNP, genotypic frequencies were 0.73 (2*1/*1), 0.16 (2*1/*2), and 0.11 (2*2/*2), with minor allele frequency (ALDH2*2) = 0.19. About 80% of subjects with ALDH 2*1/*2 and 38% with ALDH2*1/*1 reported a history of severe alcohol withdrawals (delirium) at some time in their drinking career. About 39% of ALDH2*1/2*1 and 29% of ALDH2*1/*2 drank almost every day. However, subjects drinking excessive amounts (>60 g/day) were higher in ALDH2*1/*1 & ALDH2*1/*2 as compared to ALDH2*2/*2. Two clinical parameters, namely, duration of alcohol dependence and age at onset of alcohol dependence, correlated well with the protection-conferring property of the ALDH2 Glu487Lys polymorphism. Duration of alcohol dependence was found to be significantly longer (p-value <0.01) among individuals with the ALDH2*1/*1 genotype as compared to subjects with ALDH2*2/*2. However, age at which patients developed alcohol dependence was significantly lower (p-value <0.01) in individuals with the ALDH2*1/*1 genotype as compared to subjects with ALDH2*2/*2, whereas the quantity of alcohol consumed was much higher in ALDH2*1/*1 carriers as compared to ALDH2*2/*2 (Figure 94-5). Significantly higher values of SGOT and SGPT in individuals with ALDH2*1/*1 genotype as compared to ALDH2*2/*2 is suggestive of severe liver damage due to higher amounts of alcohol consumption among ALDH2*1/*1 subjects (Figure 94-5). Association

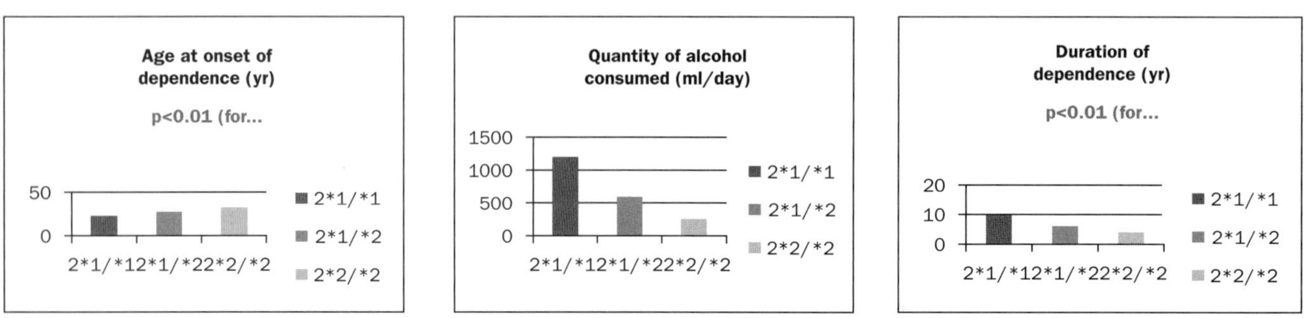

Figure 94-4 Association of ALDH polymorphisms and alcohol dependence.

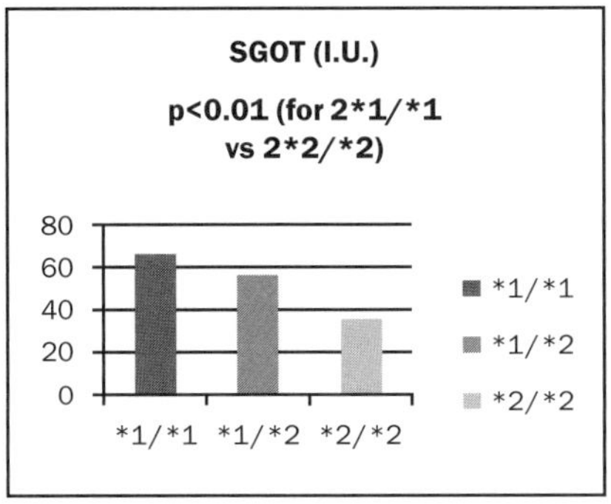

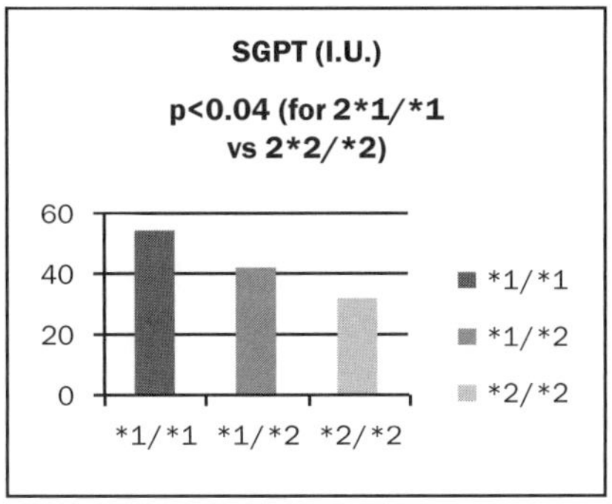

Figure 94-5 Association of ALDH polymorphisms with hepatotoxicity.

of clinical/laboratory parameters with ALDH2 polymorphism is indicated in Figure 94-6.

Contrasting to the reports from neighboring East Asian populations, Vaswani et al. (2009) reported comparatively higher frequencies of ALDH2*2/*2 in alcohol-dependent individuals. However, a recent study on healthy controls by Bhasker et al. (2007) from India, is equally exciting, as it found the SNP to be monomorphic (ALDH2*1/2*1). A higher frequency of inactive ALDH2 allele (Vaswani et al., 2009) in Indian subjects as compared to controls (Bhaskar et al., 2007; Goedde & Agarwal, 1986) provides evidence for a direct role of acetaldehyde in the development of alcohol addiction, despite its toxic actions.

Lower intake of alcohol in subjects carrying ALDH2*2/*2 polymorphism is supported by fact that although subjects with ALDH2*2/*2 genotype were classified alcohol-dependent by DSM IV, these individuals were drinking 3 to 5 restaurant portions (of 8g/drink in Indian liquor), which is not enough to produce ethanol intoxication, but rather acetaldehyde intoxication. However, the authors felt that although their findings of higher ALDH 2*2 frequency in AD is contradictory to the findings in East Asian populations, they are in agreement with Hahn et al.(2006). Thus, development of alcohol dependence in Indian subjects carrying ALDH2*2/*2 genotype with low levels of drinking but with presumably high levels of acetaldehyde strongly implies acetaldehyde in the causality of the dependence.

GENETICS OF ALCOHOLIC PANCREATITIS

Chronic pancreatitis (CP) is characterized by progressive and irreversible damage to both exocrine & endocrine components of the pancreas. Prevalence of CP is 10–15/million population with annual incidence of 3.5–4/million. Tropical pancreatitis is more common in India (70% of CP is of the tropical variety and 25% is alcoholic). However, in North India, the prevalence of alcoholic CP is higher (41%). The hypothesis states that pancreatitis develops as an acute process that progress to chronic damage. Risk of developing alcoholic pancreatitis in heavy drinkers (>60gms/day) over long duration is likely to be 2%–3%. However, the role of mild to moderate amounts of alcohol use in the development or progression of pancreatitis remains unclear. Multiple candidate genes (PRSS1, SPINK1, CFTR, ADH, ALDH) have been identified that can influence the pathogenesis in various forms of CP.

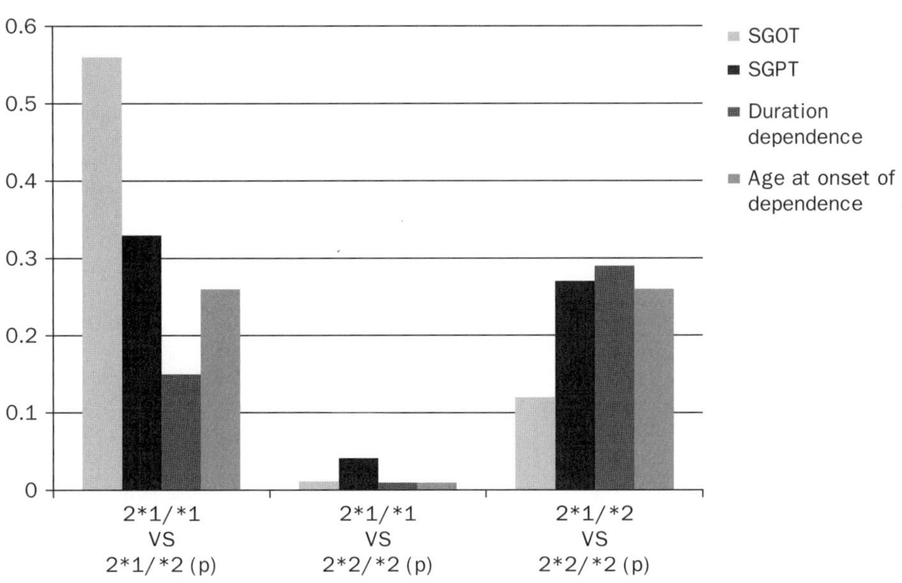

Figure 94-6 Correlation of hepatic toxic effects and ALDH gene polymorphisms.

Genetic variation at the ADH, ALDH, and CYP2E1 genes may influence the rate of elimination of ethanol and the rate of formation and elimination of acetaldehyde. It is also a known fact that ethnic populations differ in allele frequencies at these loci. Therefore, it could be hypothesized that specific genotype(s) with reference to alcohol metabolizing enzymes, combined with mutations in other specific genes (associated with pancreatic function) may influence, (at least in part) specific types of organ damage, specifically alcoholic pancreatitis. Considering the importance of genetic factors in alcoholic pancreatitis, Sarya et al. (personal communication) aimed to study the association of polymorphisms in the genes encoding alcohol metabolizing enzymes ADH2, ALDH2, and CYP2E1 in Indian patients suffering from alcoholic chronic pancreatitis (ACP), nonalcoholic pancreatitis (NAP), alcoholics (ALC), and unrelated controls(UR).

The ADH2 gene was monomorphic in all groups except one subject in ALC. All patients were found to carry a single genotype (ADH 2*1/*1). A considerable variation in the allele frequency of this polymorphism has been observed among different ethnic groups. ADH2*2 allele is found to be more common in the nonalcoholic group than in the alcoholic populations from East Asia, Taiwan, Spain, New Zealand, and Jews from Israel and the United States. However, due to very low frequency of ADH2*2 allele in Caucasian population, association studies have mostly remained inconclusive. Reports on ADH2*2 allele (Arg47His polymorphism) frequency in Indian populations have been found to be inconsistent (Goedde et al., 1992; Osier et al., 2002). In a recent report involving 28 Indian tribal populations, the polymorphism was found to be monomorphic (ADH1B*1/*1) with complete absence of ADH1B*2 allele (Reddy et al., 2006). For ADH3, patients with alcoholic chronic pancreatitis (ACP) were found to have Val allele. Three patients with ACP and one with nonalcoholic pancreatitis (NAP) were found to be heterozygous for G1259C mutation in CYP2E1, while none of the controls carried the mutated allele.

With regard to ALDH2 polymorphism, more than 70% of subjects (in all four groups) were homozygous for ALDH2-1/2-1 genotype, while 10% were homozygous for ALDH*2/*2 genotype. The presence of ALDH2-2/2-2 genotype impacted the mean age of patients. The age of onset of pancreatitis was significantly lower in patients carrying ALDH2-2/2-2 genotype as compared to those with ALDH2-1/2-1. Odds ratio analysis revealed significant association of ALDH 2-2 allele with ACP, and significant association of ALDH 2-1 allele with NAP. Significant association of ALDH 2*2/*2 in alcoholic pancreatitis indicated that it confers almost 3 times greater risk to develop alcoholic pancreatitis as compared to those carrying ALDH *2-*1 or ALDH*1*1 genotype. The results also indicated that the age at onset of pain was significantly higher in ACP than in NAP. Pseudocyst was more common in ACP than NAP, while the reverse pattern was seen in cases of steatorrhea in the two groups.

GENETIC EPIDEMIOLOGY OF ALCOHOLISM IN INDIA

According to National Health survey data (Ray et al., 2004), in India alcohol is the most common substance of abuse (excluding tobacco). The current prevalence rate in the past one month for alcohol is 21.4% and the prevalence is highest among the age group 41–50 yrs (32.2%), out of which only 2% actually sought help. However, very few genetic studies to unravel genetic predisposition/candidate genes in the Indian population have been carried out to date. To detect genes of small and moderate effect, large genetic samples are usually needed. The methods most widely used include the candidate gene approach, in which the co-inheritance of phenotypes and genetic markers for candidate gene(s) is followed by pedigree (genetic linkage study) and genetic association studies (GSAs) using a set of gene polymorphisms or single nucleotide polymorphisms (SNPs) in cases and matched controls. Only limited data is available on alcohol dependence from GSAs carried out in selected Indian cohorts.

CONCLUSIONS

Excluding tobacco, alcohol is the most common substance of abuse in India. Alcohol use and abuse reflect wide population diversity and wide-ranging sociocultural practices. The current prevalence of alcohol consumption varies from 21.4% (21–50 years age group) to 32.2% (age group 41–50 years).

- To date, very few genetic association studies to unravel genetic predisposition to tolerance (dependence and addiction) and intolerance (avoidance) in Indian populations have been carried out.
- Two polymorphisms, namely, -141C Ins/Del and Taq I A in *DRD2*, seem to have clinical implications in development of alcohol dependence in a selected Indian cohort. However, the results have to be viewed in the perspective of potential limitation posed by small sample size and lack of replication in larger sample sets.
- Genetic association studies using functional polymorphisms of gene encoding aldehyde dehydrogenase (ALDH) enzyme provide evidence for developing alcohol dependence with ALDH2*/2* polymorphism. Loss of or diminished ALDH function probably leads to poor metabolism of acetaldehyde to acetate, resulting in acetaldehyde toxicity and its various side effects.

REFERENCES

Achaya KT. (1998). *Indian Food: A Historical Companion*. Delhi: Oxford University Press.

Addolorato G, Abenavoli L, Leggio L, et al. (2005a). How many cravings? Pharmacological aspects of craving treatment in alcohol addiction: a review. *Neuropsychobiology* 51:59–66.

Addolorato G, Leggio L, Abenavoli L, et al. (2005b). Neurobiochemical and clinical aspects of craving in alcohol addiction: a review. *Addict Behav* 30:1209–1224.

American Psychiatric Association. (1994). *Diagnostic and Statistical Manual of Mental Disorders*, 4th ed. Washington DC: American Psychiatric Press.

Anthony JC, Warner LA, Kessler RC. (1994). Comparative epidemiology of dependence on tobacco, alcohol, controlled substances, and inhalants: Basic findings from the National Comorbidity Survey. *Exp Clin Psychopharmacol* 2:244–268.

Arinami T, Gao M, Hamaguchi H, et al. (1997). A functional polymorphism in the promoter region of the dopamine D2 receptor gene is associated with schizophrenia. *Hum Mol Genet* 6: 577–582.

Assanangkornchai S, Noi-pha K, Saunders JB, et al. (2003). Aldehyde dehydrogenase 2 genotypes, alcohol flushing symptoms and drinking patterns in Thai men. *Psychiatry Res* 118(1):9–17.

Bhaskar LV, Thangaraj K, Osier M. et al. (2007). Single nucleotide polymorphisms of the ALDH2 gene in six Indian populations. *Ann Hum Biol.* 34(6): 607–619.

Bowirrat A, Oscar-Berman M. (2005). Relationship between dopaminergic neurotransmission, alcoholism, and reward deficiency syndrome. *Am J Med Genet (Neuropsychiatr Genet)* 132:29–37.

Chen CC, Lu RB, Chen YC, et al. (1999). Interaction between the functional polymorphism of the alcohol metabolism genes in protection against alcoholism. *Am J Hum Genet* 65:795–807.

Cheng ATA, Gau SF, Chen THH, et al. (2004). A 4-year longitudinal study on risk factors for alcoholism *Arch Gen Psychiatry* 61:184–191.

Crabbe JC, Phillips TJ, Feller DJ, et al. (1996). Elevated alcohol consumption in null mutant mice lacking 5-HT1B serotonin receptors. *Nat Genet* 14:98–101.

Di Matteo V, Di Giovanni G, Di Mascio M, et al. (1999). SB 242084, a selective serotonin 2C receptor antagonist increases dopaminergic transmission in the mesolimbic system. *Neuropharmacology* 38:1195–1205.

Eng MY, Luczak SE, et al. (2007). ALDH2, ADH1B, AND ADH1C genotypes in Asians: A literature review. *Alcohol Res Health* 30(1):22–27.

Enoch MA, Goldman D. (2001). The genetics of alcoholism and alcohol abuse. *Current Psychiatry Reports* 3:144–151.

Eriksson CJP, Fukunaga T, Sarkola T, et al. (2001). Functional relevance of human ADH polymorphism *Alcohol Clin Exp Res* 25:157S–163S.

Ferguson RA, Goldberg DM. (1997). Genetic markers of alcohol abuse. *Clin Chim Acta* 257:199–250.

Goedde HW, Agarwal DP. (1986). Aldehyde oxidation: ethnic variations in metabolism and response. *Prog Clin Biol Res* 214:113–138.

Goedde HW, Agarwal DP, Fritze G, et al. (1992). Distribution of ADH1B and ALDH2 genotypes in different populations. *Hum Genet* 88:344–346.

Goldman D. (1995). Candidate genes in alcoholism. *Clin Neurosci* 3:174–81.

Goldman D, Urbanek M, Guenther D, et al. (1997). Linkage and association of a functional DRD2 variant [Ser311Cys] and DRD2 markers to alcoholism, substance abuse and schizophrenia in Southwestern American Indians. *Am J Med Genet (Neuropsychiatr Genet)* 74:386–394.

Goldman D, Oroszi G, Ducci F. (2005). The genetics of addictions: uncovering the genes. *Nat Rev Genet* 6:521–532.

Grant KA. (1995). The role of 5-HT3 receptors in drug dependence. *Drug Alcohol Depend* 38:155–171.

Hahn CY, Huang SY, Ko HC. (2006). Acetaldehyde involvement in positive and negative alcohol expectancies in Han Chinese persons with alcoholism. *Arch Gen Psychiatry* 63(7):817–823.

Hauge XY, Grandy DK, Eubanks JH, et al. (1991). Detection and characterization of additional DNA polymorphisms in the dopamine D2 receptor gene. *Genomics* 10:527–530.

Higuchi S, Matsushita S, Masaki T, et al. (2004). Influence of genetic variations of ethanol-metabolizing enzymes on phenotypes of alcohol-related disorders. *Ann N Y Acad Sci* 1025:472–480.

Kim DJ, Park BL, Yoon S, et al. (2007). 5' UTR polymorphism of dopamine receptor D1 (DRD1) associated with severity and temperament of alcoholism. *Biochem Biophys Res Commun* 357(4):1135–1141.

Koob GF, Le Moal M. (1997). Drug abuse: Hedonic homeostatic dysregulation. *Science* 278:52–58.

Koob GF, Roberts AJ, Schulteis G, et al. (1998). Neurocircuitry targets in ethanol reward and dependence. *Alcohol Clin Exp Res* 22:3–9.

Koob GF, Le Moal M. (2001). Drug addiction, dysregulation of reward, and allostasis. *Neuropsychopharmacology* 24:97–129.

Kuo PH, Kalsi G, Prescott CA, et al. (2008). Association of ADH and ALDH genes with alcohol dependence in the Irish Affected Sib Pair Study of alcohol dependence (IASPSAD) sample. *Alcohol Clin Exp Res* 32(5):785–795.

Laakso A, Pohjalainen T, Bergman J, et al. (2005). The A1 allele of the human D2 dopamine receptor gene is associated with increased activity of striatal L-amino acid decarboxylase in healthy subjects. *Pharmacogenet Genomics* 15:387–391.

Linnoila M, Virkkunen M, Roy A. (1986). Biochemical aspects of aggression in man. *Clin Neuropharmacol* 9(suppl 4):377–379.

Miller SA, Dykes DD, Polesky HF. (1988). A simple salting out procedure for extracting DNA from human nucleated cells. *Nucleic Acids Res* 16:1215.

Mohan D, Sharma HK. (1995). Alcohol and culture in India. In: Heath DB, ed. *International Handbook on Alcohol and Culture*. Westport, CT: Greenwood Press.

Mohan D, Chopra A, Ray R, et al. (2001). Alcohol consumption in India: a cross-sectional study. In: Demers A, Room R, Bourgault C. *Surveys of drinking patterns and problems in seven developing countries*. Geneva: World Health Organization.

Osier MV, Pakstis AJ, Soodyall, H et al. (2002). A global perspective on genetic variation at the ADH genes reveals unusual patterns of linkage disequilibrium and diversity. *Am J Hum Genet* 71:84–99.

Osier MV, Pakstis AJ, Goldman D, et al. (2002). A proline-threonine substitution in codon 351 of ADH1C is common in Native Americans. *Alcohol Clin Exp Res* 26:1759–1763.

Parsian A, Cloninger CR. (1997). Human GABAA receptor alpha 1 and alpha 3 subunits genes and alcoholism. *Alcohol Clin Exp Res* 21:430–433.

Paterson AD, Sunohara GA, Kennedy JL. (1999). Dopamine D4 receptor gene: novelty or nonsense? *Neuropsychopharmacology* 21:3–16.

Pittman D. (1964). Social and cultural factors in drinking patterns, pathological and non-pathological. In *Selected Papers Presented at 27th International Congress on Alcohol and Alcoholism,* Vol. 1, pp. 1–13. Laussane: International Council on Alcohol and Alcoholism.

Pohjalainen T, Rinne JO, Någren K, et al. (1998). The A1 allele of the human D2 dopamine receptor gene predicts low D2 receptor availability in healthy volunteers. *Mol Psychiatry* 3:256–260.

Prasad P, Ambekar A, Vaswani, M (2010). Dopamine D2 receptor polymorphisms and susceptibility to alcohol dependence in Indian males: a preliminary study. *BMC Medical Genetics.* 11:24; doi:10.1186/1471-2350-11-24.

Prakash O. (1961). *Food and drinks in Ancient India.* New Delhi: Munshi Ram Manohar Lal.

Pushplata P, Ambekar A, Vaswani M. (2010). Dopamine receptor D2 polymorphisms and susceptibility to alcohol dependence in Indian males: a preliminary study: in press. *BMC Med Genet.* 11:11–24.

Ray R, Mondal AB, Gupta K, Chatterjee A, Bajaj P. (2004). *The Extent, Pattern and Trends of Drug Abuse in India: National Survey.* New Delhi: United Nations Office on Drugs and Crimes & Ministry of Social Justice and Empowerment, Government of India.

Reddy BM, Reddy ANS, Nagaraja T, et al. (2006). Single nucleotide polymorphisms of the alcohol dehydrogenase genes among the 28 caste and tribal populations of India. *Int J Hum Genet* 6(4):309–316.

Sasabe T, Furukawa A, Matsusita S, et al. (2007). Association analysis of the dopamine receptor D2 (DRD2) SNP rs1076560 in alcoholic patients. *Neurosci Lett* 412:139–142.

Shen YC, Fan JH, Edenberg HJ, et al. (1997). Polymorphism of ADH and ALDH genes among four ethnic groups in China and effects upon the risk for alcoholism. *Alcohol Clin Exp Res* 21:1272–1277.

Tsai G, Coyle JT. (1998). The role of glutamatergic neurotransmission in the pathophysiology of alcoholism. *Ann Rev Med* 49:173–184.

Tu GC, Israel Y. (1995). Alcohol consumption by Orientals in North America is predicted largely by a single gene. *Behav Genet* 25(1):59–65.

Uhl GR, Elmer GI, LaBuda MC, Pickens RW. (1995). Genetic influences in drug abuse. In: Bloom FE, Kupfer DJ, eds. *Psychopharmacology: The Fourth Generation of Progress,* pp. 1793–1806. New York: Raven Press.

Vaswani M, Pushplata P, Suman K. (2009). Association of *ADH1B* and *ALDH2* gene polymorphisms with alcohol dependence: a pilot study from India. *Human Genomics* 3(3):213–220.

Volkow ND, Fowler JS, Wang GJ. (2004). The addicted human brain viewed in the light of imaging studies: brain circuits and treatment strategies. *Neuropharmacology* 47:3–13.

Wall TL, Garcia Andrade C, Thomasson HR, et al. (1997). Alcohol dehydrogenase polymorphisms in Native Americans: identification of the ADH2*3 allele. *Alcohol Alcohol* 32:129–132.

Whitfield JB. (2002). Alcohol dehydrogenase and alcohol dependence: variation in genotype-associated risk between populations. *Am J Hum Genet* 71:1247–1250.

Zintzaras E, Stefanidis L, Santos M, et al. (2006). Do alcohol-metabolizing enzyme gene polymorphisms increase the risk of alcoholism and alcoholic liver disease? *Hepatology* 43:352–361.

95.

THE GENETICS OF AUTISM IN INDIA

G.K. Chetan, K.R. Manjunatha, Sam Balu, H.N. Venkatesh, and S.C. Grimaji

INTRODUCTION

Autism (from Greek *autos*, meaning *self*) is not a single disorder but a name given to a distinctive group of behavioral phenotypes. First described by Leo Kanner, a child psychiatrist, in a 1943 report describing 11 children with "disturbances of affective contact" led him to coin the term *autism* to describe the "extreme aloneness" that he saw as its characteristic trait (Kanner, 1943). Autism therefore can be described as a severe neurodevelopmental disorder characterized by profound impairments in language and socio-emotional functioning, and restricted, repetitive interests and behavior (Losh et al., 2008). Also, there are other developmental disorders and syndromes that have similar features to those of autism; these are grouped into what are now known as the Autism Spectrum Disorders (ASDs), a spectrum of psychological conditions characterized by widespread abnormalities of social interactions and communication, as well as severely restricted interests and highly repetitive behavior (WHO, 2006). The ASDs include Asperger syndrome and pervasive developmental disorders (PDD). Autism has been associated disorders such as phenylketonuria (PKU), neurofibromatosis and tuberous sclerosis, and fragile-X syndrome (Folstein and Rutter, 1988; Gillberg and Forsell, 1984; Cohen et al., 1991; Gillberg et al., 1994; Manjunatha et al., 2001; Reddy, 2005). Epilepsy has the highest association with autism, reported in up to a third of individuals with an ASD by adulthood (Fombonne and du Mazaubrun, 1992).

The worldwide prevalence of autism is 7.1 per 10,000 in individuals under 18 years of age in countries such as the United States, United Kingdom, France, and Japan (Williams et al., 2006). Overall, boys are affected more often than girls, and the average male: female ratio is 3.8:1 (Fombonne, 1999). In recent times in the United States, population studies have suggested that the prevalence of autism has increased two to three times over the last three decades (Yeargin-Allsopp et al., 2003; Barbaresi et al., 2005). There is no specific data available from India of prevalence rates of autism in the general population. The authors of recent case series reports of 16 and 62 autistic children from tertiary hospitals in Chandigarh and New Delhi, respectively, have stated that autism is not uncommon in India; however, the diagnosis is frequently missed as there is tremendous lack of awareness and knowledge about the disorder among health professionals and the lay public alike (Singhi and Malhi, 2001; Kalra et al., 2005). Another systematic genetic study was carried out by Arathi (1998) for 150 autistic cases at the National Institute of Mental Health and Neuro Sciences (NIMHANS), Bangalore, India.

GENETIC ANALYSIS OF AUTISM

Autism has a strong underlying genetic component. It is a complex polygenic disorder with variable expression and heritability. About 5%–10% of individuals with autism have "secondary autism," in which an environmental agent, chromosomal abnormality, or a single gene disorder can be easily identified. The remaining 90% have idiopathic autism for which a major gene or group of genes have not yet been clearly identified (Reddy, 2005). The development of a variety of newer high-throughput genetic techniques and the sequencing of the human genome (Lander et al., 2001), as well as the HapMap project (International HapMap Consortium, 2005), have helped provide a better insight into the genetic factors that play a role it the etiology of this disorder. With the exception of Rett syndrome, which is caused in the majority of cases by de novo mutations or microdeletions of the methyl-CpG-binding protein 2 (MeCP2) gene on Xq28, there is no current evidence that the other DSM-IV subtypes of autism are linked to any particular genetic disorder (Amir and Zoghbi, 2000). Studies showed a link between autism and different autosomal fragile sites (Arrieta et al., 1996; Arathi, 1998). Results from linkage studies and the drop in the concordance rates between monozygotic and dizygotic twins suggests that the genetic etiology of autism is certainly heterogeneous (different genes in different families), polygenic (more than one affected gene per individual), with epigenetic influences and allelic heterogeneity (different variants in the same gene may lead to different patterns of genetic disease).

There are different approaches to identifying genetic loci, chromosomal regions likely to contain relevant genes: cytogenetic studies that may guide molecular studies by pointing to relevant inherited or de novo chromosomal abnormalities in affected individuals and their families; evaluation of candidate genes known to affect brain development in these significantly linked regions; or, alternatively, linkage of candidate genes selected a priori because of their presumptive contribution to the pathogenesis of autism; and whole-genome screens, searching for linkage of autism to shared genetic markers in populations of multiplex families (families with more than one family member affected) and screening for the genes contributing the different endophenotypes of this disorder (Muhle et al., 2004; Losh et al., 2008).

Numerous studies have shown the occurrence of chromosomal aberrations in subjects with autism (Table 95-1). In a study carried out by us (Manjunatha et al., 2001), 150 autistic subjects were analyzed for both autosomal fragile sites as well as the fragile site at Xq27.3 (that of the fragile-X linked mental retardation). Among the 150 subjects analyzed, 25 subjects had autosomal fragile sites, with expression from 3%–14% (these subjects were negative for fragile-X syndrome). The different types of fragile sites observed are given in Table 95-2a. Twelve individuals were positive for fragile-X (expression ranged from 4%–30% in their cells), and the details are given in Table 95-2b. Certain features common in both autism and fragile-X syndrome, like elongated face, large lop ears, and high arched palates, were noticed among autistic individuals with fragile-X chromosome. Apart from these features, macroorchidism, a very common finding in fragile-X syndrome patients, was observed in only 3 out of the 11 autistic males with fragile-X chromosome. Other features typical to fragile-X syndrome—e.g., hyperextensible joints, double-jointedness, pectus deformity, mitral valve prolapse, bushy eyebrows—were also not observed

TABLE 95-1 STUDIES SHOWING CHROMOSOMAL ABERRATIONS AMONG AUTISTIC SUBJECTS

Sl. No.	Chromosomal Aberrations	Study
1	Heteromorphic chromosomes-9qh+, long/short Y	Gillberg & Wahlström, 1985
2	del(15)(q12)	Kerbeshian et al., 1990
3	inv/dup (15)(pterq13)	Schinzel, 1990
4	mosaic, dup(18q)	Ghaziuddin et al., 1993
5	dup(15q11q13)	Baker et al., 1994
6	Monosomy 5p,del(17)(p11.2)	Vostanis et al., 1994
7	Partial tetrasomy 15	Hotopf and Bolton, 1995
8	del(2q); del Xp; del(18q); Sex chromosome aneuploidies	Gillberg, 1998
9	del(13)(q12q13)	Smith et al., 2002
10	t(5;18)(q33.1;q12.1)	Kroisel et al., 2004
11	Partial trisomy of 8p	Papanikolaou et al., 2006
12	dup(7)(q11)	Depienne et al., 2007
13	t (7;16) (p22.1; p11.2)	Bayou et al., 2008
14	del(1q42)	Williams et al., 2009
15	Micro-duplication of chromosome 15q11.2	van der Zwaag et al., 2009
16	del(4)(q35.1q35.2);del(8)(p23.2pter)	Chien et al., 2010

TABLE 95-2A AUTISTIC SUBJECTS ANALYZED THAT WERE POSITIVE FOR FRAGILE SITE AT XQ27.3 (ADAPTED FROM MANJUNATHA ET AL., 2001)

Sl No.	Gender	Age (yrs.)	Degree of Mental Retardation	Percentage Expression of Fragile Site at Xq27.3
1	Male	5	Nil	20
2	Male	6	Mild	18
3	Male	33	Moderate	10
4	Male	26	Mild	11
5	Male	9	Mild	7
6	Female	8	Moderate	5
7	Male	9	Mild	9
8	Male	7	Mild	7
9	Male	3.5	Mild	12
10	Male	17	Moderate	22
11	Male	12	Moderate	27
12	Male	12	Moderate	10

TABLE 95-2B AUTISTIC SUBJECTS ANALYZED AS POSITIVE FOR AUTOSOMAL FRAGILE SITES (ADAPTED FROM MANJUNATHA ET AL., 2001)

Sl. No.	Fragile Site	No. of Subjects	Type of Fragile Site	Percentage of Expression
1	2q31	4	Common	4–6
2	2q33	3	Common	5–6
3	4q31	2	Common	4–10
4	5q21	4	Common	4
5	5q31	3	Common	5–14
6	6p23	1	Folate Sensitive	12
7	6q26	1	Common	6
8	12q24	1	Folate Sensitive	5
9	13q21	1	Common	4
10	Xq22	3	Common	4–5
11	Xq26	2	Rare Site	3–4

in these individuals. Further, some features that are rare among fragile-X syndrome patients, like anteverted, posteriorly placed ears, macrocephaly, microcephaly, short stature, and low birth weight, were also observed in autistic subjects. Therefore, phenotypic correlations become very difficult to use in establishing the disease pattern. Our study indicated that both the disorders, in spite of having a few common behavioral characteristics, have certain variations in features suggesting that autism is distinct from fragile-X syndrome as far as phenotype–genotype correlations are concerned.

A similar study was carried out by Reddy in 2005, where 433 autistic subjects were screened for chromosomal aberrations, using both high-resolution banding techniques as well as fluorescent in situ hybridization (FISH). Chromosomal abnormalities were found in 14 of 421 cases (details in Table 95-3a), with 4/14 having supernumerary marker chromosomes (among these, 3 had markers for the critical region for Prader-Willi/Angelman syndromes), 4/14 cases had deletions, 3/14 subjects carried duplications, 3/14 subjects had inversions, 2/14 individuals carried translocations, and 1/14 had a chromosomal duplication. The FISH analysis was carried out in 23 cases excluding those which had an already identified aberration. All these were negative. In addition to this, PCR-based analysis for the size and methylation status of the CGG repeat at the fragile-X locus was also carried out; the details are given in Table 95-3b. In addition to the above-mentioned studies, several other studies have established a link between fragile-X syndrome and autism (listed in Table 95-4) and the association of other autosomal fragile sites (details given in Table 95-5).

CYTOGENETIC ANALYSIS, CHROMOSOMAL ABERRATIONS, FRAGILE SITES, FRAGILE-X SYNDROME

Genome-wide linkage analysis is a method that was earlier used to analyze the genetic factors involved in autism. Numerous studies were carried out, with varying results (Auranen et al., 2002; Alarcón et al., 2005; Schellenberg et al., 2006; Duvall et al., 2007; Szatmari et al., 2007). These studies reported numerous suggestive linkage peaks in

TABLE 95-3A DETAILS OF AUTISTIC SUBJECTS WHO WERE POSITIVE FOR CHROMOSOMAL ABERRATIONS (ADAPTED FROM REDDY, 2005)

Sl.	Age	Sex	Clinical Indication	Karyotype	FISH Details	Fragile-X Analysis
MARKER CHROMOSOMES 4/14= 28%						
1	6.1	M	Autism	47,XY+mar[27]/46,XY[3]	.ish der(2)(D2Z+)	NRM
2	5	F	Autism	47,XX,+mar	.ish der(15)(D15Z+,D15SS++,GABRB3+)	NT
3	3	M	Autism	47,XY,+mar de novo	.ish i(15)(q11.2)(rRNA++,D15Z,D15511-)	NT
4	5.1	F	Autism, MR	mos47,XY,+r [15]/46XX[15]	.ish der(15))(rRNA++,D15Z,D15S11-, GABRB3+)	NT
DUPLICATION 1/14= 7%						
5	16.8	F	Autism	46,XX,dup(15) (q11.2q13) de novo	.ish dup(15)(q11.2) (D15S11++, GABRB3++)de novo	NT
DELETION 1/14=7%						
6	6.5	F	Autism, DD, MR, macrocephaly	46,XX,del(2) (q37.7)	.ish del(2)(q37.7)(D2S447)	NT
7	6.6	F	Autism, DD	46,XX,del(3) (p25)de novo	ish del(3)(p25)	NT
8	11.7	M	Autism, DD, MR, MCA- h/o DS	46,XY,del(12) (q21.2q23.3)	.ish 12(wcp12x2)	NRM
9	3.6	M	Autism	46XY,del(13) (q13.2q14.1)de novo	.ish 13q13(D13356x2), 13q14(RBx2)	NT
INVERSION 3/14= 21%						
10	2.8	F	Autism	46,XX,inv(10)(p11.2q21.1)	NT	NRM
11	3.5	M	Autism	mos46,XY,inv(14)(q11.2q33) [3]/46,XY[17]	NT	NRM
12	3.3	M	Autism, Hypotonia, DD	46,XY,add(17)(q23) or inv(17) (q23q25) de novo	.ish inv(17)(q24.2q25.3) (wcp17x2,MPOx2,D17S928x2)	NT
TRANSLOCATION						
13	2.7	F	Autism	46,XX,t(1;14)(q25;q31.2)	NT	NT
14	3.3	M	Autism	46,XY,der(14;18)(q10;q10)	NT	NT

TABLE 95-3B DETAILS OF AUTISTIC SUBJECTS WHO WERE TESTED FOR FRAXA (CGG) REPEAT SIZE AND METHYLATION STATUS (ADAPTED FROM REDDY, 2005)

Sl.	Age in Years	Sex	Clinical Indication	FRAXA Result	Methylation Status	Karyotype
FULL MUTATION						
1	4.5	M	Autism	ABN Full Mutation	Abnormal	46,XY
2	5	M	Autism	ABN Full Mutation	Abnormal	46,XY
3	4.5	M	Autism, DD	ABN Full Mutation	Abnormal	46,XY
MOSAIC MUTATION						
4	2.3	M	Autism	ABN-size mosaic. Full mutation [800r]/ premutation [155r]	Normal & abnormal	46,XY
5	1.7	M	Autism	ABN-size mosaic. Full mutation [400r]/ premutation [150r]	Normal & abnormal	46,XY
6	8.5	M	Autism, DD	ABN-size mosaic. Full mutation [200–900r]/ premutation [30r]	Partial	46,XY
MOSAIC PREMUTATION CARRIER						
7	6	F	Autism	ABN-premutation mosaic [29,39,65,80r]	Absent	46XX

the human genome. The most consistent region based on linkage analysis appears to be on 7q, the region 7q22-23 being strongly implicated by meta- analysis (Trikalinos et al., 2006, Rossi et al., 2008). Though this technique has had great success in bringing about discoveries for more than 1600 Mendelian disorders (McKusick, 2007), it has limitations, especially for the results produced when dealing with oligogenic disorders, involving multiple risk alleles with smaller effects (Risch and Merikangas, 1996).

Candidate gene based studies, as the name implies, deal with the identification of genes that are involved in neurodevelopmental pathways or relevant molecular mechanisms. Some of the candidate genes analyzed were MET-a receptor tyrosine kinase, SLC6A4-a serotonin transporter, PTEN-tumor suppressor gene, TSC1, and TSC2-genes coding for the tuberous sclerosis complex (Campbell et al., 2006; McCauley et al., 2004; Sutcliffe et al., 2005; Guhathakurta et al., 2008; Butler et al., 2005; Asato and Hardan, 2004; Curatolo et al., 2004). In another candidate gene, REL (reelin, involved in regulation of neuronal migration), the 5' UTR region seems to play an important role in this aspect, as observed both in global and Indian studies (Ashley-Koch et al., 2007; Dutta et al., 2007b). However, the findings in another study on

TABLE 95-4 STUDIES SHOWING AUTISTIC SUBJECTS WHO WERE FRAGILE-X POSITIVE

Study	Autistic Subjects Studied	Fragile-X Positive	Percentage
Blomquist et al., 1985	102	13	13
Wahlström et al., 1986	143	16	12
Fisch et al., 1986	144	18	13
Brown et al., 1986	183	24	13
Matsuishi et al., 1987	38	2	5
Crowe et al., 1988	20	2	10
Payton et al., 1989	85	2	2
Wahlström et al., 1989	32	2	6
Cantú et al., 1990	67	1	1
Tranebjaerg and Kure, 1991	32	2	6
Cohen et al., 1991	344	24	7
Manjunatha et al., 2001	150	12	8
Reddy, 2005	316	7	2
Shen et al., 2010	933	4	<1

TABLE 95-5 VARIOUS AUTOSOMAL FRAGILE SITES REPORTED AMONG AUTISTIC SUBJECTS (ADAPTED FROM ARATHI, 1998)

Sl. No.	Fragile Sites	Study
1	Constitutive fragile sites like 3p14	Gillberg & Wahlstrom, 1985
2	16q23,6q26,Xp22,Xq27	Gillberg & Wahlström, 1985
3	1p31,3p14,3q21,3q27,6q26, 16q22	Goldfine et al., 1985
4	2q13	Jayakar et al., 1986
5	1p32,1q32,1q44,2q,23,2q37, 3p14, 6q21, 6q27,7q32, 8q22, 9q32, 10q22, 10q26, 12p12,14q23,14q24,16q24, 18q22, Xp22, Xq22	Li et al., 1993
6	2q13, 6q23,12q13	Arrieta et al., 1996
7	2p16,2q11,2q33,2q36,3p14,4q3 1,5p14,5q21,5q31,6q26,10q25, 11q14,12q21,12q24,13q32	Arathi, 1998
8	16q22-23	Kerbeshian et al., 2000

six single nucleotide polymorphisms (SNPs), of the reelin gene itself, suggested that these six SNPs of RELN are unlikely to be associated with ASD in the Indian population (Dutta et al., 2008). A similar result was also seen for the SNPs of the glutamate receptor 6 gene (GluR6 or GRIK2), when studied in the Indian population (Dutta et al., 2007a). The details of some more candidate genes identified from both global and Indian studies are listed in Table 95-6.

GENOME-WIDE LINKAGE ANALYSIS, CANDIDATE GENES

Genome-wide association studies (GWAS) compare genetic risk factors (in the form of specific genetic markers) in a case versus control manner. GWAS has only recently been applied to psychiatric disorders. In GWAS, markers are distributed throughout the genome rather than limited to candidate regions, thus providing a more unbiased canvassing of the genome. GWAS could be a more effective approach for genetic studies of complex disorders such as autism. In a study carried out by Liu et al. in 2008, 976 multiplex families from the Autism Genome Project consortium were selected, and genome-wide linkage analyses on two quantitative subphenotypes was performed. Two chromosomal regions, 11p15.4-p15.3 and15q13.3-q14 (which was previously linked to schizophrenia), had high logarithm of odds (LOD) scores. Another study by Bucan et al. (2009) carried out GWAS of exonic copy-number variants in 912 families for novel autism susceptibility genes. They identified over150 loci harboring rare variants in multiple unrelated probands, including exonic deletions at NRXN1 and whole-gene duplications encompassing UBE3A and several other genes in the 15q11-q13 region. Similarly, significant support was observed for previously unreported genes, such as BZRAP1, an adaptor molecule known to regulate synaptic transmission, and the lesser-known MDGA2 gene (encoded protein shows a very high similarity to Contactin 4). A GWAS of social and nonsocial autistic-like traits was carried out by Ronald et al., 2010. The allele frequency differences in approximately 430,000 autosomal SNPs were screened using pooled DNA on 500K SNP microarrays. Then, 22 and 20 SNPs among the social and nonsocial studies, respectively, were tested for QTL association. The study found 3 SNPs that were nominally significant.

TABLE 95-6 SOME OF THE CANDIDATE GENES IDENTIFIED FOR AUTISM [*-STUDY CARRIED OUT IN INDIAN POPULATION]

Sl.	Gene	Locus	Function	Study
1	MET	7q31	Hepatocyte growth factor receptor and encodes tyrosine-kinase activity	Campbell et al., 2006
2	SLC6A4	17q11.2	Solute carrier family 6 (neurotransmitter transporter, serotonin)	McCaulley et al., 2004; Sutcliff et al., 2005; Guhathakurta et al., 2008*
3	5-HTTLPR	17q11.2	5-HTT gene-linked polymorphic region of SLC6A4, is located approximately 1 kb upstream of the transcription initiation site and is composed of 16 repeat elements.	Guhathakurta et al., 2006*
4	RELN	7q22	Reelin-extracellular matrix serine protease that plays a role in layering of neurons in the cerebral cortex and cerebellum.	Ashley-Koch et al., 2007; Dutta et al., 2007*
5	PTEN	10q23	Tumor suppressor, a dual-specificity protein phosphatase	Butler et al., 2005
6	TSC1	9q34	Tuberous sclerosis 1 protein (Hamartin) – tumor suppressor, plays a role in the stabilization of tuberin protein.	Asato and Hardan, 2004
7	TSC2	16p13.3	Tuberous sclerosis 2 protein (Tuberin) – tumor suppressor, mutations lead to tuberous sclerosis disorder	Curatolo et al., 2004
8	NLGN3	17p13.2	Neuroligin 3– codes for neuronal cell surface proteins	Yan et al., 2005
9	NLGN4X	Xp22.33	Neuroligin 4, X-linked – codes for neuronal cell surface proteins	Laumonnier et al., 2004
10	NRXN1	2p16.3	Neurexin 1– codes for nervous system cell adhesion molecules and receptors	Kim et al., 2008
11	CNTNAP2	7q35	Contactin associated protein-like 2– member of the neurexin family, which functions in the vertebrate nervous system as cell adhesion molecules and receptors	Arking et al., 2008
12	SHANK3	22q13.3	SH3 and multiple ankyrin repeat domains – member of the Shank gene family – coding for multi-domain scaffold proteins of the postsynaptic densities	Durand et al., 2007; Moessner et al., 2007
13	EN2	7q36.2	Engrailed-2, Homeobox Gene – codes for homeodomain-containing proteins that have been implicated in the control of pattern formation during development of the central nervous system.	Sen et al., 2009*
14	HTR-2A	13q14-q21	5-hydroxytryptamine (serotonin) receptor 2A – codes for Serotonin receptor, type 2A.	Guhathakurta et al., 2009*

GENOME-WIDE ASSOCIATIONS, CANDIDATE GENES, ENDOPHENOTYPES

Endophenotypes are subclinical markers of disease (e.g., behavioral, physiological, neuropsychological) that are present among both affected and unaffected individuals. An endophenotype is very useful for simplifying disorders with complex etiology like autism, by breaking it down to simpler endophenotypes. The genetic analysis of such endophenotypes would yield more straightforward correlations within the analyzed population. Endophenotypes are measurable in both affected and unaffected individuals thereby affording analysis of larger sample sizes with greater power (Gottesman and Gould, 2003). The details

TABLE 95-7 VARIOUS AUTISTIC ENDOPHENOTYPES WITH THE ASSOCIATED GENETIC LOCI (ADAPTED FROM LOSH ET AL., 2008)

Endophenotype	Chromosome	Locus	Reference
1) Language, communication	1p	1p36.23-p36.13/1p21.2	Alarcon et al., 2005; Schellenberg et al., 2006
	2q	2q31.1/2q33.1	Buxbaum et al., 2001, Shao et al., 2002
	3q	3q13.31-q26.1	Alarcon et al., 2005
	5q	5p15.2p13.1/5q12.3	Alarcon et al., 2005; Schellenberg et al., 2006
	5p	5p15.2-p13.1	Alarcon et al., 2005
	6q	6q25.3	Schellenberg et al., 2006
	7q	7q21.2-q21.3/7q 36.2/7q35/7q36.1/7q36.3	Bradford et al., 2001; Alarcon et al., 2002; Alarcon et al., 2005; Alarcon et al., 2008; Schellenberg et al., 2006
	8q	8q23-24	Chen et al., 2006
	9q	9q34.2	Schellenberg et al., 2006
	10q	10q22.1-q23.22/10q22.3	Alarcon et al., 2002; Alarcon et al., 2005
	10p	10p14-p15.3	Alarcon et al., 2005
	11q	11q23.3-q24.1/11q25	Alarcon et al., 2002; Alarcon et al., 2005; Schellenberg et al., 2006
	13q	13q22.1	Bradford et al., 2001
	15q	15q26.3	Alarcon et al., 2005
	16q	16q24.1	Alarcon et al., 2005
	16p	16p12-p13	Chen et al., 2006
	17q	17q11.2/17q23.2-q25.1	Buxbaum et al., 2004; Alarcon et al., 2005
	18q	18q23	Schellenberg et al., 2006
	20p	20p11.22-q13.32	Alarcon et al., 2002; Alarcon et al., 2005
	22q	22q13.1	Schellenberg et al., 2006
2) Social responsiveness	4q	4q34.1	Duvall et al., 2007
	10q	10q11.23-q26.3	Duvall et al., 2007
	11p	11p15.4-q22.1	Duvall et al., 2007
	17p	17p13.3-q23.2	Duvall et al., 2007
3) OCD/Repetitive behavior	1q	1q42.2	Buxbaum et al., 2004
	7q	7q31	Sakurai et al., 2006
	15q	15q11-q13	Shao et al., 2003
	16p	16p13.3	Alarcon et al., 2005
	17q	17q11.2/17q23.2-q25.1	Brune et al., 2006; Alarcon et al., 2005
4) Savant skills	15q	15q11-q13	Nurmi et al., 2003
5) Developmental regression	7q	7q36.1	Molloy et al., 2005
	10p	10p15.3	Schellenberg et al., 2006
	14q	14q32.2	Schellenberg et al., 2006
	21q	21q21.1	Molloy et al., 2005
6) Rapid milestones	19p	19p13	McCauley et al., 2004
7) Head circumference	7p	7p15.2	Conciatori et al., 2004

of certain autistic endophenotypes and the associated genetic loci are given in Table 95-7 (Losh et al., 2008).

CONCLUSION: USING AN INTEGRATED APPROACH TO DECIPHER THE GENETICS OF AUTISM

Autism/ASD are developmental disorders with complex etiology and have considerable heterogeneity in their clinical manifestation. In developing countries such as India, the early detection of autism is still not possible in most places, especially in the rural regions. Also, the identification and categorization of these disorders may not entirely be accurate, as it is a difficult task and requires trained professionals. Hence, autism cases are by and large studied in specialized mental healthcare centers. In terms of genetic analysis, another important aspect is the ethnic diversity of the population. Genes identified within a certain community may not find straightforward correlations with the general population studies. Genes that are promising candidates from studies in other countries may not find any significant association with the Indian population, as in the case of RELN SNPs.

In order to achieve significant information about underlying genetic factors of such disorders, several steps should be taken. Firstly, improved training and education of healthcare professionals at various levels will insure that more cases will be reported and the requisite information (such as family history) will be accurately recorded. With this information, population studies (cytogenetic, linkage-based, GWAS, etc.) with larger sample sizes can be carried out with greater specificity, even more so when done using the autistic endophenotypes. The findings from such studies can then be correlated with studies carried out in particular subpopulations, in order to bring out both the common genetic factors for the whole population, as well as specific genetic markers for the subpopulation studied. With the information garnered from all the studies, a central database can be established, facilitating easier discovery of candidate genes, which can then be analyzed in detail. The outline for this integrated, multicentric approach has schematically represented in Figure 95-1. Thus, by using an integrated approach, the

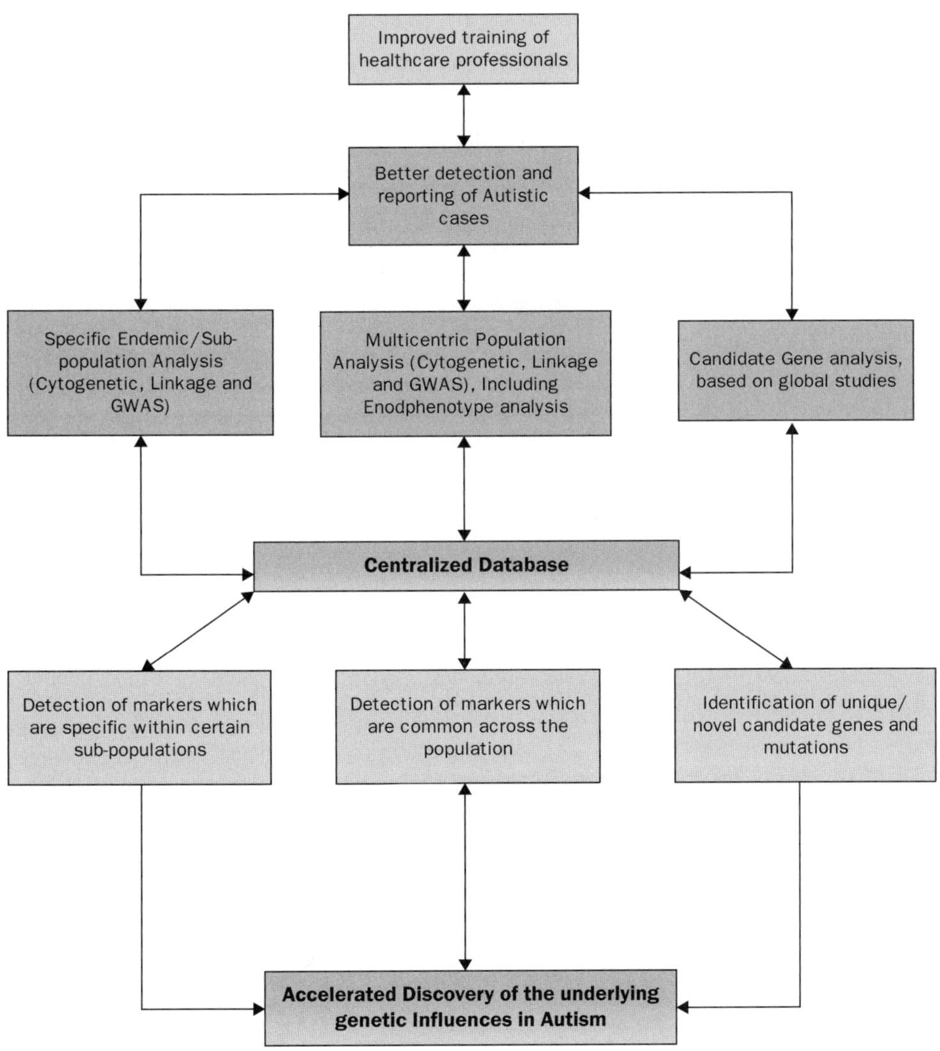

Figure 95-1 Outlines for a multi-tiered integrated approach for the genetic analysis of autism.

process of determining the underlying genetic influences on complex disorders such as autism can be truly successful.

ACKNOWLEDGMENTS

We are grateful to the ICMR, New Delhi and UGC, New Delhi for providing financial support for the studies carried out at our institute. We are grateful to the Director, NIMHANS, Bangalore, for providing the necessary infrastructure and financial support. Our sincere thanks to all the patients and their family members who participated in the genetic studies. We are thankful for all the support from the various departmental faculties and staff of the institute.

REFERENCES

Alarcón M, Cantor RM, Liu J, Gilliam TC, Geschwind DH. (2002). Evidence for a language quantitative trait locus on chromosome 7q in multiplex autism families. *Am J Hum Genet* 70:60–71.

Alarcón M, Yonan AL, Gilliam TC, Cantor RM, Geschwind DH. (2005). Quantitative genome scan and ordered-subsets analysis of autism endophenotypes support language QTLs. *Mol Psychiatry* 10:747–757.

Alarcón M, Abrahams BS, Stone JL, et al. (2008). Linkage, association, and gene-expression analyses identify CNTNAP2 as an autism-susceptibility gene. *Am J Hum Genet* 82:150–159.

Amir RE, Zoghbi HY. (2000). Rett syndrome: methyl-CpG-binding protein 2 mutations and phenotype-genotype correlations. *Am J Med Genet* 97(2):147–152.

Arathi R. (1998). Cytogenetic of Fragile-X Chromosome: Frequency, Association and their Implications in subjects with Autism. *PhD. Thesis*, NIMHANS, Banglaore, India.

Arking DE, Cutler DJ, Brune CW, et al. (2008). A common genetic variant in the neurexin superfamily member CNTNAP2 increases familial risk of autism. *Am J Hum Genet* 82:160–164.

Arrieta I, Nuñez T, Gil A, Flores P, Usobiaga E, Martinez B. (1996). Autosomal folate sensitive fragile sites in an autistic Basque sample. *Ann Genet* 39(2):69–74.

Asato MR, Hardan AY. (2004). Neuropsychiatric problems in tuberous sclerosis complex. *J Child Neurol* 19:241–249.

Ashley-Koch AE, Jaworski J, Ma de Q, et al. (2007). Investigation of potential gene-gene interactions between APOE and RELN contributing to autism risk. *Psychiatr Genet* 17:221–226.

Auranen M, Vanhala R, Varilo T, et al. (2002). A genomewide screen for autism-spectrum disorders: Evidence for a major susceptibility locus on chromosome 3q25-27. *Am J Hum Genet* 71:777–790.

Baker P, Piven J, Schwartz S, Patil S. (1994). Brief report: duplication of chromosome 15q11-13 in two individuals with autistic disorder. *J Autism Dev Disord* 24(4):529–535.

Barbaresi WJ, Katusic SK, Colligan RC, Weaver AL, Jacobsen SJ. (2005). The incidence of autism in Olmsted County, Minnesota, 1976–1997: Results from a population-based study. *Arch Pediatr Adolesc Med* 159:37–44.

Bayou N, M'rad R, Belhaj A, et al. (2008). De novo balanced translocation t (7;16) (p22.1; p11.2) associated with autistic disorder. *J Biomed Biotechnol* 2008:231904.

Blomquist HK, Bohman M, Edvinsson SO, et al. (1985). Frequency of the fragile-X syndrome in infantile autism. A Swedish multicenter study. *Clin Genet* 27(2):113–117.

Bradford Y, Haines J, Hutcheson H, et al. (2001). Incorporating language phenotypes strengthens evidence of linkage to autism. *Am J Med Genet (Neuropsychiatr Genet)* 105:539–547.

Brown WT, Jenkins EC, Cohen IL, et al. (1986). Fragile-X and autism: a multicenter survey. *Am J Med Genet* 23(1–2):341–352.

Brune CW, Kim SJ, Salt J, Leventhal BL, Lord C, Cook EH Jr. (2006). 5-HTTLPR genotype-specific phonotype in children and adolescents with autism. *Am J Psychiatry* 163:2148–2156.

Bucan M, Abrahams BS, Wang K, et al. (2009). Genome-wide analyses of exonic copy number variants in a family-based study point to novel autism susceptibility genes. *PLoS Genet.* 5(6):e1000536.

Buxbaum JD, Silverman JM, Smith CJ, et al. (2001). Evidence for a susceptibility gene for autism on chromosome 2 and for genetic heterogeneity. *Am J Hum Genet* 68:1514–1520.

Buxbaum JD, Silverman J, Keddache M, et al. (2004). Linkage analysis for autism in a subset families with obsessive-compulsive behaviors: Evidence for an autism susceptibility gene on chromosome 1 and further support for susceptibility genes on chromosome 6 and 19. *Mol Psychiatry* 9:144–150.

Butler MG, Dasouki MJ, Zhou XP, et al. (2005). Subset of individuals with autism spectrum disorders and extreme macrocephaly associated with germline PTEN tumour suppressor gene mutations. *J Med Genet* 42:318–321.

Campbell DB, Sutcliffe JS, Ebert PJ, et al. (2006). A genetic variant that disrupts MET transcription is associated with autism. *Proc Natl Acad Sci U S A* 103:16834–16839.

Cantú ES, Stone JW, Wing AA, Langee HR, Williams CA. (1990). Cytogenetic survey for autistic fragile-X carriers in a mental retardation center. *Am J Ment Retard* 94(4):442–447.

Chen GK, Kono N, Geschwind DH, Cantor RM. (2006). Quantitative trait locus analysis of nonverbal communication in autism spectrum disorder. *Mol Psychiatry* 11:214–220.

Chien WH, Gau SS, Wu YY, et al. (2010). Identification and molecular characterization of two novel chromosomal deletions associated with autism. *Clin Genet.* 78(5):449–456.

Cohen IL, Sudhalter V, Pfadt A, Jenkins EC, Brown WT, Vietze PM. (1991). Why are autism and the fragile-X syndrome associated? Conceptual and methodological issues. *Am J Hum Genet* 48(2):195–202.

Conciatori M, Stodgell CJ, Hyman SL, et al. (2004). Association between the HOXA1 A218G polymorphism and increased head circumference in patients with autism. *Biol Psychiatry* 55:413–419.

Crowe RR, Tsai LY, Murray JC, Patil SR, Quinn J (1988). A study of autism using X chromosome DNA probes. *Biol Psychiatry* 24(4):473–479.

Curatolo P, Porfirio MC, Manzi B, Seri S. (2004). Autism in tuberous sclerosis. *Eur J Paediatr Neurol* 8:327–332.

Depienne C, Heron D, Betancur C, et al. (2007). Autism, language delay and mental retardation in a patient with 7q11 duplication. *J Med Genet* 44(7):452–458.

Durand CM, Betancur C, Boeckers TM, et al. (2007). Mutations in the gene encoding the synaptic scaffolding protein SHANK3 are associated with autism spectrum disorders. Nat Genet 39:25–27.

Dutta S, Das S, Guhathakurta S, Sen B, et al. (2007a). Glutamate receptor 6 gene (GluR6 or GRIK2) polymorphisms in the Indian population: a genetic association study on autism spectrum disorder. *Cell Mol Neurobiol* 27(8):1035–1047.

Dutta S, Guhathakurta S, Sinha S, et al. (2007b). Reelin gene polymorphisms in the Indian population: a possible paternal 5'UTR-CGG-repeat-allele effect on autism. *Am J Med Genet B Neuropsychiatr Genet.* 144B(1):106–112.

Dutta S, Sinha S, Ghosh S, Chatterjee A, Ahmed S, Usha R. (2008). Genetic analysis of reelin gene (RELN) SNPs: no association with autism spectrum disorder in the Indian population. *Neurosci Lett* 15; 441(1):56–60.

Duvall JA, Lu A, Cantor RM, Todd RD, Constantino JN, Geschwind DH. (2007). A quantitative trait locus analysis of social responsiveness in multiplex autism families. *Am J Psychiatry* 164:656–662.

Fisch GS, Cohen IL, Wolf EG, Brown WT, Jenkins EC, Gross A. (1986). Autism and the fragile-X syndrome. *Am J Psychiatry* 143(1):71–73.

Folstein SE, Rutter ML. (1988). Autism: familial aggregation and genetic implications. *J Autism Dev Disord* 18(1):3–30.

Fombonne E, du Mazaubrun C. (1992). Prevalence of infantile autism in four French regions. *Soc Psychiatry Psychiatr Epidemiol* 27:203–210.

Fombonne E. (1999). The epidemiology of autism: A review. *Psychol Med* 29:769–786.

Ghaziuddin M, Sheldon S, Tsai LY, Alessi N. (1993). Abnormalities of chromosome 18 in a girl with mental retardation and autistic disorder. J Intellect Disabil Res. 37(Pt 3):313–317.

Gillberg C, Forsell C. (1984). Childhood psychosis and neurofibromatosis—more than a coincidence? *J Autism Dev Disord* 14(1):1–8.

Gillberg C. (1998). Chromosomal disorders and autism. *J Autism Dev Disord* 28(5):415–425.

Gillberg C, Wahlström J. (1985). Chromosome abnormalities in infantile autism and other childhood psychoses: a population study of 66 cases. *Dev Med Child Neurol.* 27(3):293–304.

Gillberg IC, Gillberg C, Ahlsén G. (1994). Autistic behaviour and attention deficits in tuberous sclerosis: a population-based study. *Dev Med Child Neurol* 36(1):50–56.

Goldfine PE, McPherson PM, Heath GA, Hardesty VA, Beauregard LJ, Gordon B. (1985). Association of fragile-X syndrome with autism. *Am J Psychiatry* 142(1):108–110.

Gottesman II, Gould TD. (2003). The endophenotype concept in psychiatry: etymology and strategic intentions. *Am J Psychiatry* 160(4):636–645.

Guhathakurta S, Ghosh S, Sinha S, et al. (2006). Serotonin transporter promoter variants: Analysis in Indian autistic and control population. *Brain Res* 1092(1):28–35.

Guhathakurta S, Sinha S, Ghosh S, et al. (2008). Population-based association study and contrasting linkage disequilibrium pattern reveal genetic association of SLC6A4 with autism in the Indian population from West Bengal. *Brain Res* 1240:12–21.

Guhathakurta S, Singh AS, Sinha S, et al. (2009). Analysis of serotonin receptor 2A gene (HTR2A): association study with autism spectrum disorder in the Indian population and investigation of the gene expression in peripheral blood leukocytes. *Neurochem Int* 55(8):754–759.

Hotopf M, Bolton P. (1995). A case of autism associated with partial tetrasomy 15. *J Autism Dev Disord* 25(1):41–49.

International HapMap Consortium. (2005). A haplotype map of the human genome. *Nature* 437(7063):1299–1320.

Jayakar P, Chudley AE, Ray M, Evans JA, Perlov J, Wand R. (1986). Fra(2) (q13) and inv(9) (p11q12) in autism: causal relationship? *Am J Med Genet.* 23(1–2):381–392.

Kalra V, Seth R, Sapra S. (2005). Autism - Experiences in a tertiary care hospital. *Indian J Pediatr* 72:227–230.

Kanner L. (1943). Autistic disturbances of affective contact. *Nerv Child* 2:217–250.

Kim HG, Kishikawa S, Higgins AW, et al. (2008). Disruption of neurexin 1 associated with autism spectrum disorder. *Am J Hum Genet* 82:199–207.

Kerbeshian J, Burd L, Randall T, Martsolf J, Jalal S. (1990). Autism, profound mental retardation and atypical bipolar disorder in a 33-year-old female with a deletion of 15q12. *J Ment Defic Res.* 2:205–210.

Kerbeshian J, Severud R, Burd L, Larson L. (2000). Peek-a-boo fragile site at 16d associated with Tourette syndrome, bipolar disorder, autistic disorder, and mental retardation. *Am J Med Genet* 96(1):69–73.

Kroisel PM, Windpassinger C, Wagner K, et al. (2004). De novo translocation t(5;18)(q33.1;q12.1) associated with autistic disorder. *Am J Med Genet A* 129A(1):98–100.

Lander ES, Linton LM, Birren B, et al. (2001). Initial sequencing and analysis of the human genome. *Nature* 409(6822):860–921.

Laumonnier F, Bonnet-Brilhault F, Gomot M, et al. (2004). X-linked mental retardation and autism are associated with a mutation in the NLGN4 gene, a member of the neuroligin family. *Am J Hum Genet* 74:552–557.

Losh M, Sullivan PF, Trembath D, Piven J. (2008). Current developments in the genetics of autism: from phenome to genome. *J Neuropathol Exp Neurol* 67(9):829–837.

Li SY, Chen YC, Lai TJ, Hsu CY, Wang YC. (1993). Molecular and cytogenetic analyses of autism in Taiwan. *Hum Genet* 92(5):441–445.

Liu XQ, Paterson AD, Szatmari P; Autism Genome Project Consortium. (2008). Genome-wide linkage analyses of quantitative and categorical autism subphenotypes. *Biol Psychiatry.* 64(7):561–570.

Manjunatha KR, Chetan GK, Arathi R, et al. (2001). Genetics of autism: association of chromosomal fragile sites. *IJHG* 1(4):293–299.

Matsuishi T, Shiotsuki Y, Niikawa N, et al. (1987). Fragile-X syndrome in Japanese patients with infantile autism. *Pediatr Neurol* 3(5):284–287.

McCauley JL, Olson LM, Dowd M, et al. (2004). Linkage and association analysis at the serotonin transporter (SLC6A4) locus in a rigid-compulsive subset of autism. *Am J Med Genet B Neuropsychiatr Genet* 127:104–112.

McKusick VA. (2007). Mendelian inheritance in man and its online version, OMIM. *Am J Hum Genet* 80:588–604.

Moessner R, Marshall CR, Sutcliffe JS, et al. (2007). Contribution of SHANK3 mutations to autism spectrum disorder. *Am J Hum Genet* 81:1289–1297.

Molloy CA, Keddache M, Martin LJ. (2005). Evidence for linkage on 21q and 7q in a subset of autism characterized by developmental regression. *Mol Psychiatry* 10:741–746.

Muhle R, Trentacoste SV, Rapin I. (2004). The genetics of autism. *Pediatrics* 113(5):e472–e486.

Nurmi EL, Dowd M, Tadevosyan-Leyfer O, Haines JL, Folstein SE, Sutcliffe JS. (2003). Exploratory subsetting of autism families based on savant skills improves evidence of genetic linkage to 15q11-q13. *J Am Acad Child Adolesc Psychiatry* 42:856–863.

Papanikolaou K, Paliokosta E, Gyftodimou J, et al. (2006). A case of partial trisomy of chromosome 8p associated with autism. *J Autism Dev Disord* 36(5):705–709.

Payton JB, Steele MW, Wenger SL, Minshew NJ. (1989). The fragile-X marker and autism in perspective. *J Am Acad Child Adolesc Psychiatry* 28(3):417–421.

Reddy KS. (2005). Cytogenetic abnormalities and fragile-X syndrome in autism spectrum disorder. *BMC Med Genet* 6:3.

Risch N, Merikangas K. (1996). The future of genetic studies of complex human diseases. *Science* 273:1516–1517.

Ronald A, Butcher LM, Docherty S, et al. (2010). A genome-wide association study of social and non-social autistic-like traits in the general population using pooled DNA, 500 K SNP microarrays and both community and diagnosed autism replication samples. *Behav Genet* 40(1):31–45.

Rossi E, Verri AP, Patricelli MG, et al. (2008). A 12Mb deletion at 7q33-q35 associated with autism spectrum disorders and primary amenorrhea. *Eur J Med Genet* 51(6):631–638.

Sakurai T, Ramoz N, Reichert JG, et al. (2006). Association analysis of the NrCAM gene in autism and in subsets of families with severe obsessive-compulsive or self-stimulatory behaviors. *Psychiatr Genet* 16:251–257.

Schellenberg GD, Dawson G, Sung YJ, et al. (2006). Evidence for multiple loci from a genome scan of autism kindreds. *Mol Psychiatry* 11:1049–1060.

Schinzel A. (1990). Autistic disorder and additional inv dup(15)(pter—q13) chromosome. *Am J Med Genet.* 35(3):447–448.

Sen B, Surindro Singh A, et al. (2009). Family-based studies indicate association of Engrailed 2 gene with autism in an Indian population. *Genes Brain Behav* 9(2):248–255.

Shao Y, Raiford KL, Wolpert CM, et al. (2002). Phenotypic homogeneity provides increased support for linkage on chromosome 2 in autistic disorder. *Am J Hum Genet* 70:1058–1061.

Shao Y, Cuccaro ML, Hauser ER, et al. (2003). Fine mapping of autistic disorder to chromosome 15q11-q13 by use of phenotypic subtypes. *Am J Hum Genet* 72:539–548.

Shen Y, Dies KA, Holm IA, et al. (2010). Clinical genetic testing for patients with autism spectrum disorders. *Pediatrics* 125(4):e727–e735.

Singhi P, Malhi P. (2001). Clinical and neurodevelopmental profile of young children with autism. *Indian J Pediatr* 38:384–390.

Smith M, Woodroffe A, Smith R, et al. (2002). Molecular genetic delineation of a deletion of chromosome 13q12-->q13 in a patient with autism and auditory processing deficits. *Cytogenet Genome Res* 98(4):233–239.

Sutcliffe JS, Delahanty RJ, Prasad HC, et al. (2005). Allelic heterogeneity at the serotonin transporter locus (SLC6A4) confers susceptibility to autism and rigid-compulsive behaviors. *Am J Hum Genet* 77:265–279.

Szatmari P, Paterson AD, Zwaigenbaum L, et al. (2007). Mapping autism risk loci using genetic linkage and chromosomal rearrangements. *Nat Genet* 39:319–328.

Tranebjaerg L, Kure P. (1991). Prevalence of fra(X) and other specific diagnoses in autistic individuals in a Danish county. *Am J Med Genet* 38(2–3):212–214.

Trikalinos TA, Karvouni A, Zintzaras E, et al. (2006). A heterogeneity-based genome search meta-analysis for autism-spectrum disorders. *Mol Psychiatry* 11:29–36.

van der Zwaag B, Staal WG, Hochstenbach R. (2009). A co-segregating microduplication of chromosome 15q11.2 pinpoints two risk genes for autism spectrum disorder. *Am J Med Genet B Neuropsychiatr Genet* 153B(4):960–966.

Vostanis P, Harrington R, Prendergast M, Farndon P. (1994). Case reports of autism with interstitial deletion of chromosome 17 (p11.2 p11.2) and monosomy of chromosome 5 (5pter-->5p15.3). *Psychiatr Genet* 4(2):109–111.

Wahlström J, Gillberg C, Gustavson KH, Holmgren G. (1986). Infantile autism and the fragile-X. A Swedish multicenter study. *Am J Med Genet* 23(1–2):403–408.

Wahlström J, Steffenburg S, Hellgren L, Gillberg C. (1989). Chromosome findings in twins with early-onset autistic disorder. *Am J Med Genet* 32(1):19–21.

Williams JG, Higgins JP, Brayne CE. (2006). Systematic review of prevalence studies of autism spectrum disorders. *Arch Dis Child* 91:8–15.

Williams JM, Beck TF, Pearson DM, Proud MB, Cheung SW, Scott DA. (2009). A 1q42 deletion involving DISC1, DISC2, and TSNAX in an autism spectrum disorder. *Am J Med Genet A* 149A(8):1758–1762.

World Health Organization. (2006). *International Statistical Classification of Diseases and Related Health Problems.* 10th ed. (ICD-10). World Health Organization.

Yan J, Oliveira G, Coutinho A, et al. (2005). Analysis of the neuroligin 3 and 4 genes in autism and other neuropsychiatric patients. *Mol Psychiatry* 10:329–332.

Yeargin-Allsopp M, Rice C, Karapurkar T, Doernberg N, Boyle C, Murphy C. (2003). Prevalence of autism in a US metropolitan area. *JAMA* 289:49–55.

96.

THE GENETICS OF LEPROSY IN INDIA

Mariakuttikan Jayalakshmi, Narayanan Kalyanaraman, and Ramasamy Pitchappan

INTRODUCTION

Sushruta samhita, an ancient Indian medical treatise, and Kautilya's *Arthashastra*, dated 6th century AD, have described leprosy, one of the oldest recorded diseases (Zysk, 1993; Jopling & McDougall, 1996; Bloomfield, 2004). The evidence of lepromatous leprosy in skeletal remains from Balathal, a Chalcolithic site (2300–1550 BC) in India (Misra, 1997; Robbins et al., 2009) further supports the existence of leprosy in prehistoric India. The disease is also known very much in biblical times, and in various civilizations (ancient Egypt, China, and India; see, e.g., Hulse, 1972; McLeod & Yates, 1981; Robbins et al., 2007).

Dr. Armauer Hansen of Norway discovered the lepra bacilli in 1873. With the discovery it became clear that Hansen's disease is a chronic infection caused by *Mycobacterium leprae*, an obligate intracellular bacillus. The bacillus normally infects macrophages and Schwann cells, and colonizes the low-temperature extremities, causing debilitating deformities and weeping wounds in some patients (lepromatous leprosy (LL), multibacillary), while in others it leads to nerve damage and resultant deformities (tuberculoid type (TT), paucibacillary). This broad clinical spectrum of the disease correlates to pathological and immunological spectra (Ridley & Jopling, 1966). It is interesting to note that in long-standing high endemic areas such as India, the prevalence of LL has decreased dramatically over the years, while new transmissions and TT leprosy are still occurring. Nonetheless, no definite vector or mode of transmission has yet been identified. It is assumed that person-to-person transmission occurs under circumstances such as crowding, inadequate housing, and lack of hygiene (Kerr-Pontes et al., 2004), though it is known to clinicians and epidemiologists that while a husband has leprosy, the wife may not show any symptoms of leprosy.

This paradox has been resolved by twin studies, sib pair studies and affected sib pair studies. All these have for a long time pointed toward a genetic basis for disease susceptibility (Mohammed Ali & Ramanujam, 1966; Chakravarti & Vogel, 1973; Fine et al., 1979; de Vries et al., 1980; van Eden et al., 1985; Dessoukey et al., 1996; Siddiqui et al., 2001; Mira et al., 2003). This interindividual variability, at the genetic level (in contracting or resisting the disease), seems to play a major role in the survival of the pathogen in the host. The immunogenetic dictum, "Not all the infected develop disease" is now better understood in this genomic era, with many candidate and whole-genome studies available.

The disappearance of the disease from the developed world is generally attributed to better environmental sanitation and personal hygiene, and essentially not due to medical intervention by chemotherapy alone. As a result, most of the cases now seen are in the developing world. Thus, there is not much incentive for active research in the Western world, though a few groups have continued to persistently work on the problem. The registered prevalence of leprosy globally from 121 countries was 213,036 cases in 2009 (WHO, 2009). The major chunk of this burden is from South-East Asian countries, accounting for 94% of global incidence, and within this region India represents 80.1%. Thus, the responsibility of finding their own solutions lies with developing countries; nonetheless, the global scientific community and the genomic era have ushered in a new age in science.

The human genome project and its outcome, delineating the entire genome sequence, is a milestone in the history of mankind. This was made all the more feasible by the availability of technology such as PCR, BDT sequencing, cloning, and informatics. All these technologies and understandings have been put to better use in investigating the whole of the biological scientific sphere and mycobacteriosis. The completion of the genomic sequence of *M. leprae* in 2001 also became a remarkable milestone in the history of microbiology. This has led to newer insights into the mechanisms of parasite evolution and the evasive mechanisms they adopt. The leprosy bacilli seem to have undergone extensive reductive evolution with considerable downsizing of its genome; namely, 3,268,203 bp (~3.3 Mb) compared with *Mycobacterium tuberculosis* at 4,411,532 bp (~4.4 Mb; Cole et al., 2001).

These developments in the field of genomics thus ushered in the new-generation sciences and a better understanding of mycobacterioses. We review here the literature and evidence that favor the idea of a host genetic basis for leprosy susceptibility. The genomic evidence corroborates with the age-old twin studies and candidate gene studies, reiterating the role of host genetics in the disease manifestation.

HERITABILITY ANALYSES

Heritability analysis is a population genetics tool to analyze the contributions of environmental and genetic factors in disease manifestation. The best of these methods is the twin study. In this approach, monozygotic and dizygotic twin pairs are compared to determine the degree of concordance for the phenotype of interest. Various twin studies from India have suggested a genetic influence in leprosy (Mohammed Ali & Ramanujam, 1966; Chakravarti & Vogel, 1973). A high concordance rate in monozygotic twins (60%–85%) compared with dizygotic twins (5%–20%) has been observed, highlighting the contribution of host genetic factors in leprosy manifestation (Chakravarti & Vogel, 1973), although in these early days no genetic tools were available.

CANDIDATE GENE ASSOCIATION STUDIES

Candidate gene association studies are the conventional, most extensively employed, and easiest approach to studying disease susceptibility. This employs either a population-based case-control design or a family-based design. The population-based case-control association study investigates the relationship between disease and genetic markers by comparing genotype frequencies from unrelated patients and controls, while the family-based association study makes use of pedigrees and evaluates the transmission of a marker allele from heterozygous parents to their affected children by using the transmission disequilibrium test (TDT; Spielman et al., 1993) or variation thereof (Sun et al., 1999).

HLA ASSOCIATION STUDIES

With the delineation of MHC (major histocompatibility complex), HLA (human leucocyte antigen) has been associated with immune response; there was a spurt of HLA association studies in various diseases. Leprosy is not an exception to this. The chronic nature of leprosy pertains to the immune response to the pathogen, and whether this immune response is adequate and appropriate seems to be determined by the IR (immune response) genes located within the HLA region (Zinkernagel et al., 1976; Benacerraf, 1980; Dausset, 1980; Snell, 1980).

The major histocompatibility complex consists of multiple loci situated on the short arm of the chromosome (C6p21.3). It spans about 3500 kb and includes many loci coding for immune response and inflammatory receptors and molecules. Of these, the HLA encodes a set of transmembrane proteins that bind and present antigenic peptides to T-cells. The MHC restriction phenomenon thus may determine the variance of susceptibility to a disease (Zinkernagel, 1996).

Class I Allele Associations

HLA A, B, C, the MHC class I alleles, function through effective CTL response in infection. (Tables 96-1a and 96-1b present their associations). It has been suggested that the host responses resulting in different leprosy phenotypes are under genetic control, especially under the influence of HLA (de Vries et al., 1976; de Vries, 1992; Soebono et al., 1997). With this background, HLA has been used extensively as a candidate marker in many of the case-control and family studies of mycobacterioses (Pitchappan, 1990).

HLA association with leprosy varies in different populations. Susceptibility is attributed to HLA-A*2 and HLA-B*7 in Thai populations (Greiner et al., 1978), HLA-A*9, HLA-A*10, HLA-A*32, B*5, B*21, Bw6, HLA-Cw1, and Cw2 in Turkish populations (Kocak et al., 2002), B*7 in Japanese (Izumi et al., 1982), and A*02, A*11, B*5110, and HLA-Cw*0703 in Indian populations (Shankarkumar et al., 2003). HLA-A*01, Cw*0401, and Cw*0602 in Indians (Shankarkumar et al., 2003), A*03, B*44, and B*49 in Turkish (Kocak et al., 2002), HLA-Bw40 in Thai (Greiner et al., 1978), and HLA-B*46 in Southern China (Wang et al., 1999) confer resistance to leprosy. Thus, there is no uniformity of HLA Class I alleles association in various populations.

HLA Class II Allele Associations

Of all the DRB1 alleles, HLA-DR2 and its subtypes are consistently associated with Mycobacterioses (Brahmajothi et al., 1991, Ravikumar et al., 1999, Tosh et al., 2006). HLA-DR2 associations with specific leprosy phenotypes, namely, multibacillary leprosy or lepromatous leprosy (Rani et al., 1992) and tuberculoid leprosy (Mehra et al., 1995; Visentainer et al., 1997) have also been reported. A study from South India has reported a stronger and significant association of HLA-DR2 with erythema nodosum

TABLE 96-1A LEPROSY SUSCEPTIBILITY: GLOBAL GENOMIC AND ALLELE ASSOCIATION STUDIES LISTED BY CHROMOSOME

	Genome/allele studied	Assay used	Place/Country	Chromosome	C/Allele/ Microsatellite marker	Patient sample size (Disease type)	Control Sample size	Pt (%)	Ctl (%)	LOD Score	RR	OR	CI	P value	Author	Year
1	IL - 12RB	Direct PCR sequencing	Japan	C 5	IL 12RB2 -1035G	176 (LL: 130, TT: 46)	68				3.24	3.97		0.001000	Ohyama et al.	2005
2	IL - 12RB	Direct PCR sequencing	Japan	C 5	IL 12RB2 -1023G	176 (LL: 130, TT: 46)	68				2.48	2.95		0.010000	Ohyama et al.	2005
3	IL - 12RB	Direct PCR sequencing	Japan	C 5	IL 12RB2 -464G	176 (LL: 130, TT: 46)	68				3.03	3.64		<0.01	Ohyama et al.	2005
4	Genome-wide scan	Microsatellite mapping	Southern Vietnam	C 6	C 6q25-27	86 F (PB: 90, MB:115)				4.31				0.000005	Mira et al.	2003
5	Genome-wide scan	Microsatellite mapping	Southern Vietnam	C 6	C 6q25	86 F (PB: 90, MB:115)				2.62				0.000250	Mira et al.	2003
6	HLA-DRB1/DQA1	Microarray	New Delhi, India	C 6	C 6p21: rs1071630	209	239	68.60	47.80			2.38	1.80–3.15	$8.5x10^{-10}$	Wong et al.	2010
7	HLA-DRB1/DQA1	Microarray	Kolkata, India	C 6	C 6p21: rs1071630	168	131	71.80	53.90			2.17	1.53–3.08	$1.1x10^{-05}$	Wong et al.	2010
8	HLA-DRB1/DQA1	Microarray	New Delhi, India	C 6	C 6p21: rs9270650	209	239	42.90	21.80			2.70	1.93–3.76	$6.4x10^{-10}$	Wong et al.	2010
9	HLA A locus	MLC	Turkish	C 6	HLA - A*32	80 (BL: 35, LL: 25, BT: 15, TT: 5)	120	6.30	0.80			7.50	0.892–63.005	0.027800	Kocak et al.,	2002
10	HLA A locus	MLC	Turkish	C 6	HLA - A*9	80 (BL: 35, LL: 25, BT: 15, TT: 5)	120	16.30	2.50			6.50	1.913–22.084	0.000400	Kocak et al.	2002
11	HLA A locus	MLC	Turkish	C 6	HLA - A*10	80 (BL: 35, LL: 25, BT: 15, TT: 5)	120	10.00	2.50			4.00	1.094–14.624	0.022600	Kocak et al,	2002
12	HLA A locus	PCR-SSOP	Maharashtra, India	C 6	HLA - A*1102	32 (LL:16, BL:16)	67	31.00	1.00			30.00		0.000010	Shankarkumar et al.	2003
13	HLA C locus	MLC	Turkish	C 6	HLA - Cw1	80 (BL: 35, LL: 25, BT: 15, TT: 5)	120	10.00	1.70			6.00	1.307–27.527	0.008000	Kocak et al.	2002
14	HLA DR locus	PCR-SSOP	Southern Brazilian	C 6	HLA - DRβ1*08	169 (TT: 41, LL: 63, IM: 4, BL: 49)	217	7.30	5.50			12.00	1.51–95.12	0.284700	da Silva et al.	2009
15	HLA DR locus	Line Probe Assay - RH	Brazil	C 6	HLA - DRβ1*10	578 (MB:371, TB: 207)	691					1.83		0.083000	Vanderborght et al.,	2007

(Continued)

TABLE 96-1A (CONTINUED)

	Genome/allele studied	Assay used	Place/Country	Chromosome	C/Allele/ Microsatellite marker	Patient sample size (Disease type)	Control Sample size	Pt (%)	Ctl (%)	LOD Score	RR	OR	CI	P value	Author	Year
16	HLA DR locus	PCR-SSP	Southern Vietnam	C 6	HLA - DRβ1*10	194 F (PB, MB)	691					2.03	1.20-4.05	0.040000	Vanderborght et al.,	2007
17	HLA DR locus	PCR-SSOP	Argentina	C 6	HLA - DRβ1*1401	71 (MB: 56, PB: 15)	81	9.15	0.62		14.83	16.22		0.001000	Borras et al.	2008
18	HLA DR locus	PCR-SSOP	Argentina	C 6	HLA - DRβ1*1406	71 (MB: 56, PB: 15)	81	9.15	0.62		14.83	16.22		0.001000	Borras et al.	2008
19	Linkage Studies	MLC	Egypt	C 6	HLA-DR2	15 F								<0.005	Dessoukey et al.	1996
20	HLA DR locus	MLC	New Delhi, India	C 6	HLA - DR2	118 (LL: 68, BL: 30, BB: 20)	237	42.37	19.41					0.000310	Rani et al.	1992
21	HLA DR locus	MLC	Japan	C 6	HLA - DR2	132 (LL: 84, TL: 28)	55	63.10	16.10		8.7			0.000000	Izumi et al.	1982
22	HLA DR locus	PCR-SSOP	North India	C 6	HLA - DRβ1*15	93 (LL: 41, BL: 25, TT: 28)	47	75.30	21.30		11.3			0.000010	Rani et al.	1993
23	HLA DR locus	PCR-SSOP	North India	C 6	HLA - DRβ1*1501~02	54 TT	44	69.00	32.00			4.70	2.0– 11	0.006300	Zerva et al.	1996
24	HLA DR locus	PCR-SSOP	Southern Brazil	C 6	HLA - DRβ1*1601-1602	169 (TT: 41, LL: 63, IM: 4, BL: 49)	217	7.30	3.20			2.52	1.26-5.01	0.010000	da Silva et al.	2009
25	HLA DQ Locus	MLC	Turkish	C 6	HLA - DQ1	80 (BL: 35, LL: 25, BT: 15, TT: 5)	120	30.00	10.80			2.77	1.500–5.111	0.000600	Kocak et al.	2002
26	HLA DQ Locus	MLC	Turkish	C 6	HLA - DQ3	80 (BL: 35, LL: 25, BT: 15, TT: 5)	120	25.00	9.20			2.73	1.382–5.379	0.002400	Kocak et al.	2002
27	TNF- α	PCR-SSOP	East India	C 6	TNF- α -308A	228 (LL: 121, TL: 107)	160					3.00		0.020000	Roy et al.,	1997
28	TNF- α	PCR-SSP	Egypt	C 6	TNF- α -308 (GG)	47 (PB: 17, MB: 29)	98	17.30	6.10			3.23	1.05–9.9	<0.05	Settin et al.,	2007
29	IL-1Ra	PCR-SSP	Egypt	C 6	IL-1Ra	47 (PB: 17, MB: 29)	98	84.10	58.80			3.60	1.5–9.0	<0.05	Settin et al.	2007
30	Whole Genome	Microsatellite mapping	South India	C 10	C 10p13	224 F (PB: 220, MB: 4)				4.09				0.000002	Siddiqui et al.	2001

31	Whole Genome	Microsatellite mapping	South India	C 10	D10S548 - 43.4	224 F (PB: 220, MB: 4)				2.02				0.000015	Siddiqui et al.,	2001
32	Whole Genome	Microsatellite mapping	South India	C 10	D10S582 - 46.7	224 F (PB: 220, MB: 4)				3.52				0.000010	Siddiqui et al.	2001
33	Whole Genome	Microsatellite mapping	South India	C 10	D10S1653 - 38.8	224 F (PB: 220, MB: 4)				2.54				0.000013	Siddiqui et al.	2001
34	Whole Genome	Microsatellite mapping	South India	C 10	D10S1660 - 47.2	224 F (PB: 220, MB: 4)				2.73				0.000015	Siddiqui et al.,	2001
35	Whole Genome	Microsatellite mapping	South India	C 10	D10S1661- 40.9	224 F (PB: 220, MB: 4)				2.02				0.000007	Siddiqui et al.	2001
36	Whole Genome	Microsatellite mapping	South India	C 10	D10S1662 - 46.2	224 F (PB: 220, MB: 4)				2.64				0.000008	Siddiqui et al.	2001
37	VDR	PCR-in-situ hybridization	East India	C 12	VDR Codon 352 CC	231 (TT: 107, LL: 124)	166					3.22	1.47–7.13	0.001000	Roy et al.	1999
38	VDR	LDR-PCR	Malawi	C 12	VDR Codon 352 CC	270 (MB: 26, PB: 254)	452					4.30	1.6–11.4	0.004000	Fitness et al.	2004
39	NOD2	Microarray	China	C 16	C 16q12: rs9302752	706	1225					2.28	1.70–3.06	1.42×10^{-9}	Zhang et al.(b)	2009b
40	NOD2	Microarray	China	C 16	C 16q12: rs7194886	706	1225					2.25	1.58–3.21	4.43×10^{-7}	Zhang et al.(b)	2009b
41	NOD2	Chip based MALDI -TOF	Nepal	C 16	C 16q12: rs12448797	933 (LL: 582, TL: 342)	101					2.18	1.05–4.52	0.310000	Berrington et al.	2010
42	Genome-wide scan	Microsatellite mapping	Brazil	C 17	C 17q11-q21 D17S1795	72 F (372)				2.67				0.004000	Jamieson et al.	2004
43	Genome-wide scan	Microsatellite mapping	Brazil	C 17	C 17q11-q21 D17S250	72 F (372)				2.34				0.010000	Jamieson et al.	2004
44	Genome-wide scan	Microsatellite mapping	Brazil	C 17	C 17q22 D17S1868	21F (192)				2.38				0.000500	Miller et al.	2004
45	IL -10	PCR-RFLP	Brazil	C 19	IL10: -819T	300 (MB: 210, PB: 90)	92					2.64	0.93-8.04	0.040000	Santos et al.	2002
46	IL -10	PCR-RFLP	Brazil	C 19	IL10: -3575T/-2849A/-2763C	297	283	5.00	2.00			2.37	1.04–5.39	0.027000	Moraes et al.	2004
47	IL -10	PCR - SSP	Egypt	C 19	IL10-1082 (GG)	47 (PB: 17, MB: 29)	98	26.20	5.10			6.60	2.1–20.5	<0.05	Settin et al.	2007

BB - Borderline Borderline; BL - Borderline Leprosy; BT - Borderline Tuberculoid; C - Chromosome; C13orf31 - Chromosome 13 open reading frame 31.; CCDC122 - Coiled-Coil Domain Containing 122; F - Family; HLA - Human Lecukocyte Antigen; I - Individuals; IL - Interleukin; IL-1Ra - Interleukin - 1 Receptor a; IL-12RB - Interleukin - 12 Receptor B; IM - Intermediate; LDR-PCR - Ligation Detection Reaction - PCR; LL - Lepramatous Leprosy; LRRK2 - Leucine-Rich Repeat Kinase 2; MALDI - TOF - Matrix Assisted Laser Desorption Ionization - Time of Flight; MB - Multibacillary; MHC - Major Histocompatibility Complex; MICA - MHC Class Chain Related Genes A; MICB - MHC Class Chain Related Genes B; MLC - Microlymphocytotoxicity Assay; MRC-1 - Mannose Receptor C-type Lectin-1; NOD2 - Nucleotide-Binding Oligomerization Domain Containing 2; PACRG - Parkin Coregulated Gene Protein; PARK2 - Parkinson Disease Gene; PB - Paucibacillary; PCR-RFLP - PCR - Restriction fragment Length Polymorphism; PCR-SSCP - PCR - Single Strand Conformational Polymorphism; PCR-SSOP - PCR - Sequence Specific Oligonucleotide Probes; PCR-SSP - PCR - Sequence Specific Polymorphisms; Res - Resistance; RIPK2 - Receptor-Interacting Serine–Threonine Kinase 2; Sus - Susceptibility; TB - Tuberculoid Bacillary; TL - Tuberculoid Leprosy; TLR - Toll Like Receptor; TNF - Tumor Necrosis Factor; TNFSF15 - Tumor Necrosis Factor (ligand) Superfamily Member 15; TT - Tuberculoid Type; U - Unclassified; VDR - Vitamin D Receptor

TABLE 96-1B LEPROSY RESISTANCE: GLOBAL GENOMIC AND ALLELE ASSOCIATION STUDIES LISTED BY CHROMOSOME

	Genome/ allele studied	Assay used	Place/Country	Chromosome	C/Allele/ Microsatellite marker	Patient sample size (Disease type)	Control Sample size	Pt (%)	Ctl (%)	LOD Score	RR	OR	CI	P value	Author	Year
1	Complement Receptor 1	LDR-PCR	Malawi	C 1	K1590E (GG)	270 (MB: 26, PB: 254)	452					0.30	0.1-0.8	0.020000	Fitness et al.,	2004
2	TLR1	Microarray	New Delhi, India	C 4	C 4p14: rs5743618	209	239	3.80	13.00			0.27	0.15-0.47	$1.3x10^{-06}$	Wong et al.,	2010
3	TLR1	Microarray	Kolkata, India	C 4	C 4p14: rs5743618	168	131	3.60	8.40			0.40	0.20-0.83	0.012000	Wong et al.,	2010
4	TLR1	Microarray	Turkey	C 4	C 4p14: rs5743618	90	57	26.30	42.80			0.48	0.29-0.80	0.004000	Wong et al.,	2010
5	TLR1	Microarray	Kumbakonam, India	C 4	C 4p14: rs5743618	189F		-	-			0.61	0.35-1.09	0.090000	Wong et al.,	2010
6	TLR2	Sequencing	Ethiopia	C 4	TLR2: C597T	216						0.34	0.17-0.68	0.002000	Bochud et al.,	2008
7	HLA A locus	PCR-SSOP	Maharashtra, India	C 6	HLA - A*0101	32 (LL:16, BL:16)	67	6.00	22.00			0.23		0.040000	Shankarkumar et al.,	2003
8	HLA A locus	MLC	Turkish	C 6	HLA - A*3	80 (BL: 35, LL: 25, BT: 15, TT: 5)	120	1.30	12.50			0.10	0.013-0.742	0.004000	Kocak et al.,	2002
9	HLA B locus	MLC	Turkish	C 6	HLA - B*44	80 (BL: 35, LL: 25, BT: 15, TT: 5)	120	1.30	8.30			0.15	0.019-1.149	0.031300	Kocak et al.,	2002
10	HLA C locus	PCR-SSOP	Maharashtra, India	C 6	HLA - Cw*04011	32 (LL:16, BL:16)	67	3.00	33.00			0.07		0.001000	Shankarkumar et al.,	2003
11	HLA C locus	PCR-SSOP	Maharashtra, India	C 6	HLA - Cw*0602	32 (LL:16, BL:16)	67	13.00	37.00			0.24		0.010000	Shankarkumar et al.,	2003
12	HLA DR locus	Line Probe Assay - RH	Brazil	C 6	HLA - DRβ1*04	578 (MB: 371, TB: 207)	691					0.61		0.008000	Vanderborght et al.,	2007

13	HLA DR locus	PCR-SSP	Southern Vietnam	C 6	HLA - DRβ1*04	194 F	691					0.48	0.24-0.96	0.030000	Vanderborght et al.,	2007
14	HLA DR locus	PCR-SSO	Argentina	C 6	HLA - DRβ1*0808	71 (MB: 56, PB: 15)	81	1.41	12.35		0.114	0.10		0.000600	Borras et al.,	2008
15	HLA DR locus	PCR-SSOP	Argentina	C 6	HLA - DRβ1*1103	71 (MB: 56, PB: 15)	81	1.41	12.96		0.109	0.10		0.004000	Borras et al.,	2008
16	HLA DR locus	PCR-SSOP	Indonesia	C 6	HLA - DRβ1*12	79 (TT/BT: 41, LL/ BL: 38)	50					0.33-0.35		0.05	Soebono et al.,	1997
17	HLA DQ Locus	MLC	Turkish	C 6	HLA - DQ5	80 (BL: 35, LL: 25, BT: 15, TT: 5)	120	2.50	10.00			0.25	0.057-1.087	0.041700	Kocak et al.,	2002
18	MRC1	Sequencing	Vietnam	C 10	MRC1 Exon 7	339F						0.76	0.60–0.96	0.035000	Alter et al.,	2010
19	VDR	PCR-in-situ hybridization	East India	C 12	VDR Codon 352 TC	231 (TT: 107, LL: 124)	166					0.58	0.38-0.89	0.008000	Roy et al.,	1999
20	IL -10	PCR-RFLP	Brazil	C 19	IL10: -3575A/-2849G/-2763C	297	283	1.00	4.00			0.35	0.13-0.91	0.005000	Moraes et al.,	2004
21	Genome wide scan	Microsatellite mapping	South India	C 20	C 20p12 - D20S115	233 F (PB: 223, MB: 10)				2.17				0.000030	Tosh et al.,	2002
22	Genome wide scan	Microsatellite mapping	South India	C 20	C 20p12 - D20S835	233 F (PB: 223, MB: 10)				3.48				0.021000	Tosh et al.,	2002

BB - Borderline Borderline; BL - Borderline Leprosy; BT - Borderline Tuberculoid; C - Chromosome; C13orf31 - Chromosome 13 open reading frame 31.; CCDC122 - Coiled-Coil Domain Containing 122; F - Family; HLA - Human Leukocyte Antigen; I - Individuals; IL - Interleukin; IL-1Ra - Interleukin - 1 Receptor a; IL-12RB - Interleukin - 12 Receptor B; IM - Intermediate; LDR-PCR - Ligation Detection Reaction - PCR; LL - Lepramatous Leprosy; LRRK2 - Leucine-Rich Repeat Kinase 2; MALDI - TOF - Matrix Assisted Laser Desorption Ionization - Time of Flight; MB - Multibacillary; MHC - Major Histocompatibility Complex; MICA - MHC Class Chain Related Genes A; MICB - MHC Class Chain Related Genes B; MLC - Microlymphocytotoxicity Assay; MRC-1 - Mannose Receptor C-type Lectin-1; NOD2 - Nucleotide-Binding Oligomerization Domain Containing 2; PACRG - Parkin Coregulated Gene Protein; PARK2 - Parkinson Disease Gene; PB - Paucibacillary; PCR-RFLP - PCR - Restriction fragment Length Polymorphism; PCR-SSCP - PCR - Single Strand Conformational Polymorphism; PCR-SSOP - PCR - Sequence Specific Oligonucleotide Probes; PCR-SSP - PCR - Sequence Specific Polymorphisms; Res - Resistance; RIPK2 - Receptor-Interacting Serine–Threonine Kinase 2; Sus - Susceptibility; TB - Tuberculoid Bacillary; TL - Tuberculoid Leprosy; TLR - Toll Like Receptor; TNF - Tumor Necrosis Factor; TNFSF15 - Tumor Necrosis Factor (ligand) Superfamily Member 15; TT - Tuberculoid Type; U - Unclassified; VDR - Vitamin D Receptor.

leprosum (ENL), suggesting a role in disease progression (Agrewala et al., 1988). Studies based on microlymphocytotoxicity assays have also proven HLA-DR, DQ allele associations. HLA-DR2 (Visentainer et al., 1997), HLA-DR9, DR10, DRw52, (Kocak et al., 2002), DR3 (van Eden et al., 1982), DQ1, DQ3 (Kocak et al., 2002), and DQW1 (Rani et al., 1992) have thus been associated with the disease. The most consistent association with leprosy as well as tuberculosis have been with HLA-DRB1*15 (HLA DR2) alleles in different world populations (Rani et al., 1993; Mehra et al., 1995; Zerva et al., 1996; Tosh et al., 2006; Vanderborght et al., 2007; Zhang et al., 2009a). However, different associations with leprosy susceptibility have been shown in Brazilian populations (HLA-DRB1*08, DRB1*10, DRB1*16; da Silva et al., 2009), Argentinean populations (DRB1*1401, DRB1*1406; Borras et al., 2008), and Vietnamese populations (HLA-DRB1*10; Vanderborght et al., 2007).

NON-HLA GENES

MICA & MICB Genes

The human MHC class I chain-related genes (MICA and MICB) are located within the HLA class I region of chromosome 6 (C6p21.3). Their organization, expression, and products differ considerably from classical HLA class I genes, and act as ligands for cells expressing a common activatory natural killer-cell receptor (NKG2D). A study from southern China has suggested that an exon 5 variant of the MICA gene may be associated with resistance to the multibacillary form of the disease. Our study on paucibacillary leprosy-affected sib-pair families from South India has identified significant association with a functional variant of the MICA gene, as well as a microsatellite in the flanking region of the MICB gene. The MICA*5A5.1 allele per se is associated with leprosy susceptibility (P = 0.017; Tosh et al., 2006).

Tumor Necrosis Factor (TNF)

Studies on non-HLA gene associations have largely been prompted by SNP markers. TNF, a proinflammatory cytokine, is implicated in mycobacterial infection, LL, and ENL patients, who have high levels of serum TNF (Parida et al., 1992; Barnes et al., 1992). TNF-α promoter -308GG is associated with susceptibility to leprosy in Egyptian populations (Settin et al., 2007), whereas SNP at promoter -308A (O.R. = 3.0) is associated with susceptibility to leprosy in Indian populations (Roy et al., 1997). Paradoxically, TNF-α promoter polymorphism -308A is associated with resistance in Brazilian populations (Santos et al., 2000). Some of these discrepancies can be attributed to varied linkage disequilibrium of implicated genes and susceptibility in different populations.

Interleukin – 10 (IL-10)

The IL-10 locus is located at 1q31-q32 and its product is classically considered as an anti-inflammatory molecule that counterbalances the effect of proinflammatory cytokines such as TNF-α and IL-1β. Also, IL-10 is an immunoregulatory cytokine involved in IL-4 producing T-cell expansion and B-cell proliferation and differentiation (Llorente et al., 1995). High TNF-α/IL-10 levels have been correlated with better prognosis in leprosy outcome among household contacts (Lima et al., 2000). A Brazilian study has reported a susceptibility phenotype IL-10 promoter polymorphism, -819T (Santos et al., 2002). While some extended haplotypes of IL-10 are associated with resistance, another haplotype is associated with susceptibility (Moraes et al., 2004; Malhotra et al., 2005; Pereira et al., 2009).

Interleukin – 12 (IL-12)

One of the functions mediated by IL-12 is to drive uncommitted T-cells into a Th1 or Th2 phenotype, which secretes cytokine typical of inflammatory or humoral responses, respectively (Trinchieri, 2003). There is a great range in outcome after mycobacterial infections such as TT and LL, and this might be attributed to immune responses influenced by the key cytokine regulator IL-12. Individuals with leprosy are less likely to have the 3' UTR genotype associated with lower IL-12B expression (P = 0.001; Morahan et al., 2007). SNPs within the 5' flanking region of IL-12RB2 affect the degree of expression of this gene, and may be implicated in individual differences in cell mediated immune (CMI) responsiveness to mycobacterial antigens, leading to LL or TL (Ohyama et al., 2005). Another study suggests the association of IL-12 p40 3'UTR 1188 A/C polymorphism with greater susceptibility to LL in patients from western Mexico (Alvarado-Navarro et al., 2008).

Vitamin D Receptor (VDR)

The nuclear hormone receptor VDR coding gene is located at 12q12-q14: it mediates many immunomodulatory effects of the active form of vitamin D, which includes suppression of cytokine synthesis, immunoglobulin production, and lymphocyte proliferation. Homozygotes for a silent T to C change in codon 352 of the VDR gene has shown an increased risk for developing leprosy (O.R. = 4.3, P = 0.004; Fitness et al., 2004). An Indian case-control study in a Bengali population has suggested a modulatory effect of the VDR polymorphism influence in susceptibility to leprosy by affecting the type and the strength of the host immune response (Roy et al., 1999).

Nucleotide-Binding Oligomerization Domain–Like Receptors (NOD)

The NOD locus is located at C16q21. These are a diverse set of 22 innate immune receptors that are involved in cytoplasmic detection of microbes via recognition of muramyl dipeptide and activation of inflammatory cascades against pathogens (Bourhis et al., 2007; Franchi et al., 2008). Genetic variant studies of the NOD2 gene in leprosy patients from Nepal have shown an association with susceptibility to leprosy and the development of leprosy reactive states. (Berrington et al., 2010).

PARK2/PACRG Gene

Parkinson's disease gene (PARK2) and its co-regulated genes (PACRG) are involved in the intracellular antigen processing pathway via the ubiquitin–proteasome system. These genes are located at loci 6q25.2-q27. Mira et al. (2004) have investigated this region in Vietnamese families and found a significant association between leprosy and 17 markers located in a block of ~80 kb overlapping the 50 regulatory regions shared by the PARK2 and PACRG. This finding has been repeated in a sample of 975 unrelated leprosy case controls from Brazil. Variants in the regulatory region shared by PARK2 and PACRG therefore seem to act as common risk factors for leprosy: However, no significant association has been identified in an Indian study. The noninvolvement of PARK2 and PACRG locus SNPs with leprosy susceptibility in Indian populations has been attributed to a differential effect of these SNPs in regulating genetic susceptibility to leprosy in different populations (Malhotra et al., 2006). Only a comparison of the whole-genome regions of PARK2/PACRG in various populations would resolve this puzzle.

Toll-Like Receptor (TLR)

Toll-like receptors are type 1 transmembrane proteins expressed in innate immune cells that play a critical role in the inflammatory response to microbial pathogens (Akira et al., 2006; Beutler, 2004). These genes are located in various chromosomes (C3[TLR9], C4 [TLR1,2,3,6], C5[TLR5], C9[TLR4], CX[TLR7,8]). Three polymorphisms in *TLR2* (597C>T, 1350T>C, and a microsatellite marker) analyzed in 431 Ethiopian patients with leprosy and 187 control subjects have suggested a significant association of TLR-2 with leprosy reversal reaction (Bochud et al., 2008).

Mannose Receptor C-type Lectin (MR)

MRC1 encodes the mannose receptor C-type lectin, a cell surface protein that belongs to a family of receptors for pathogen-associated molecular patterns (PAMPs) and is part of the innate arm of the immune system. The locus lies in chromosome 10p12.33. Three closely spaced amino acid changes encoded by exon 7 of MRC1 are associated with altered susceptibility to paucibacillary leprosy (Cooke & Hill, 2008). This SNP was genotyped in 490 simplex and 90 multiplex leprosy families comprising 704 patients from Vietnam, and the authors have observed significant undertransmission of the serine allele of the G396S polymorphism with leprosy per se, and multibacillary leprosy (protective). In a sample of 384 Brazilian leprosy cases and 399 healthy controls, Alter et al. (2010) have also observed similar association of the glycine allele of the G396S polymorphism with multibacillary leprosy.

WHOLE GENOME SCAN STUDIES IN LEPROSY

Genome-wide association (GWA) studies represent a powerful approach toward understanding the genetic basis involved in common human diseases and offer new avenues for exploring their pathophysiology. The use of genome scans has the ability to uncover a region and sequence not apparently associated with disease, and thus not possible to test for candidate genes. It is expensive and cost-intensive but has greater power tp identify the genome predisposing for disease (Davies et al., 1994). This is a great advantage in the study of complex diseases, such as leprosy, in which low mortality rates enable the collection of samples from several generations, which is necessary in this kind of test (Tosh et al., 2002). This gave us an opportunity to investigate disease susceptibility rather than clinical disease alone, and the pioneering genetic linkage scan study in leprosy came from our laboratories (Siddiqui et al., 2001). The study of 224 families from South India (Tamil Nadu), containing 245 independent affected sib pairs with leprosy, mainly of the paucibacillary type, in a two-stage genome screen using 396 microsatellite markers, has shown a significant linkage on chromosome 10p13 for a series of neighboring microsatellite markers. Thus it has provided evidence for a non-HLA–linked susceptibility for leprosy (Siddiqui et al., 2001).

In 2003, another study from southern Vietnam consisting of a panel of 86 families including 205 siblings affected with leprosy has confirmed verbatim our findings and further have shown an added association in multibacillary disease (Mira et al., 2003). A recent Brazilian case-control study consisting of 389 leprosy patients and 399 healthy subjects has identified the MRC1 gene (C10p13), encoding the human mannose receptor (MR), as a causative factor (Alter et al., 2010). Our further study on the Tamil Nadu families has further identified another locus in chromosome 20p12 with protection against leprosy, but this region did not map with leprosy susceptibility in a Vizag cohort (Tosh et al., 2002). Genome-wide scans conducted in 21 leprosy families (173 individuals) in Brazil have revealed three susceptibility regions (C6p21.32, C17q22, and C20p13; Miller et al., 2004). Another Brazilian study (Jamieson et al., 2004) has identified 17q11.2 as a susceptibility locus controlling leprosy in a cohort of 72 multicase leprosy families.

Studies in Han Chinese populations highlight that the variants of genes CCDC122, C13orf31, NOD2, TNFSF15, HLA-DR, RIPK2, and LRRK2, involved in the innate immune response, are risk factors for developing leprosy. Polymorphisms in the NOD2 gene, involved in the signaling pathway linked with innate immune responses, have shown significant association with susceptibility to infection with *M. leprae* (Zhang et al., 2009b). Though these studies suggest interpopulation genomic diversity as a probable cause of variations observed in associations, these studies have confirmed for the first time that susceptibility to even such a chronic infectious disease as leprosy is determined by host genetics.

Recently meta-analysis employing GWAs of more than 1500 individuals from different case-control and family studies, inclusive of our 2001 study (Kumbakonam cohorts) and from New Delhi and Kolkata, India, Malawi, Gambia, and United Kingdom, has brought out consistent associations between genetic variants of both *TLR1* and the *HLA-DR*B*1/DQA1* regions and susceptibility to leprosy. Further, the population differentiation analysis shows that the TLR1 locus is extremely differentiated, and the protective dysfunctional 602S allele is rare in Africa but expands to become the dominant allele among individuals of European descent. This supports the hypothesis that this locus may be under selection from mycobacterium or similar pathogens that are recognized by TLR1 and its co-receptors. The observations provide an insight into the long-standing host–pathogen relationship between human and mycobacterium, and highlight the key role of the TLR pathway in infectious diseases (Wong et al., 2010). It also reveals the importance of carrying out a genetic epidemiology approach, even in this genomic era.

CONCLUSION

Hitherto available literature has convincingly shown a role for host genetics in leprosy susceptibility. While the bacterium inflicts damage, it is the host response and immunopathology that determines the clinical manifestation. The protective (negative) associations show the involvement of beneficial immune response in nonmanifestation of the diseases. This may be true for most of the chronic and even acute infections, such as HIV, as has been shown already. Functional genomics may further throw light on the pathways involved in susceptibility and resistance. The need of the hour is well planned, multicentric whole genome studies. We may not need a vaccine to protect 1/10,000 (of prevalence) individuals while 9,999 individuals may not develop the disease due to their immunogenetics!

REFERENCES

Agrewala JN, Ghei SK, Sudhakar KS, Girdhar BK, Sengupta U. (1988). HLA antigens and erythema Nodosum leprosum (ENL). *Tissue Antigen* 33:486–487.

Akira S, Uematsu S, Takeuchi O. (2006). Pathogen recognition and innate immunity. *Cell* 124:783– 801.

Alter A, de Leseleuc L, Van Thuc N, et al. (2010). Genetic and functional analysis of common MRC1 exon 7 polymorphisms in leprosy susceptibility. *Hum Genet* 127(3):337–348.

Alvarado-Navarro A, Montoya-Buelna M, Munoz-Valle JF, Lopez-Roa RI, Guillen-Vargas C, Fafutis-Morris M. (2008). The 3'UTR 1188 A/C polymorphism in the interleukin-12p40 gene (IL-12B) is associated with lepromatous leprosy in the west of Mexico. *Immunol Lett* 118(2):148–145.

Barnes PF, Chatterjee D, Brennan PJ, Rea TH, Modlin RL. (1992). Tumor necrosis factor production in patients with leprosy. *Infect Immun* 60(4):1441–1446.

Benacerraf B. (1980). The role of MHC gene products in immune regulation and its relevance to alloreactivity. Nobel Lecture.

Berrington WR, Macdonald M, Khadge S, et al. (2010). Common polymorphisms in the NOD2 gene region are associated with leprosy and its reactive states. *J Infect Dis* 201(9):1422–1435.

Beutler B (2004). Inferences, questions and possibilities in Toll-like receptor signalling. *Nature* 430:257– 63.

Bloomfield M. (2004). *Hymns of the Atharva-Veda*. Whitefish, MT: Kessinger Publishing.

Bochud PY, Hawn TR, Siddiqui MR, et al. (2008). Toll-like receptor 2 (TLR2) polymorphisms are associated with reversal reaction in leprosy. *J Infect Dis* 197(2):253–256.

Borras SG, Cotorruelo C, Racca L, et al. (2008). Association of leprosy with HLA-DRB1 in an Argentinean population. *Ann Clin Biochem* 45(1):96–98.

Bourhis L, Benko S, Girardin SE. (2007). NOD1 and NOD2 in innate immunity and human inflammatory disorders. *Biochem Soc Trans* 35:1479–1484.

Brahmajothi V, Pitchappan RM, Kakkanaiah VN, et al. (1991). Association of pulmonary tuberculosis and HLA in south India. *Tubercle* 72(2):123–132.

Chakravarti, MR, Vogel F. (1973). A twin study in leprosy. In: Becker PE, ed. *Topics in Human Genetic*. Stuttgart: Georg Thieme Verlag, pp. 1–24.

Cooke G, Hill A. (2008). Tuberculosis, leprosy and other mycobacterial diseases. In: Kaslow RA, McNicholl JM, Hill AVS, eds. *Genetic susceptibility to infectious diseases*. New York: Oxford University Press, pp. 333–351.

Cole ST, Eiglmeier K, Parkhill J, et al. (2001). Massive gene decay in the leprosy bacillus. *Nature* 409(6823):1007–1011.

da Silva SA, Mazini PS, Reis PG, et al. (2009). HLA-DR and HLA-DQ alleles in patients from the south of Brazil: Markers for leprosy susceptibility and resistance. *BMC Infect Dis* 9:134.

Davies JL, Kawaguchi Y, Bennett ST, et al. (1994). A genome-wide search for human type 1 diabetes susceptibility genes. *Nature* 371(6493):130–136.

Dausset J. (1980). The Major Histocompatibility Complex in Man—Past, Present and Future concepts. Nobel Lecture.

Dessoukey MW, el-Shiemy S, Sallam T. (1996). HLA and leprosy: Segregation and linkage study. *Int J Dermatol* 35:257–264.

de Vries RR, Fat RF, Nijenhuis LE, van Rood JJ. (1976). HLA-linked genetic control of host response to Mycobacterium leprae. *Lancet* 2:1328–1330.

de Vries RR, Mehra NK, Vaidya MC, Gupte MD, Meera Khan P, Van Rood JJ. (1980). HLA-linked control of susceptibility to tuberculoid leprosy and association with HLA-DR types. *Tissue Antigens* 16:294–304.

de Vries RR (1992). HLA and disease: From epidemiology to immunotherapy. *Eur J Clin Invest* 22(1):1–8.

Fine PE, Wolf E, Pritchard J, et al. (1979). HLA-linked genes and leprosy: A family study in Karigiri, South India. *J Infect Dis* 140:152–161.

Fitness J, Floyd S, Warndorff DK, et al. (2004). Large-scale candidate gene study of leprosy susceptibility in the Karonga district of Northern Malawi. *Am J Trop Med Hyg* 71:330–340.

Franchi L, Park JH, Shaw MH, et al. (2008). Intracellular NOD-like receptors in innate immunity, infection and disease. *Cell Microbiol* 10(1):1–8.

Greiner J, Schleiermacher E, Smith T, Lenhard V, Vogel F. (1978). The HLA system and leprosy in Thailand. *Hum Genet* 42(2), 201–213.

Hulse EV. (1972). Leprosy and Ancient Egypt. *Lancet* 2:1024.

Izumi S, Sugiyama K, Matsumoto Y, Ohkawa S. (1982). Analysis of the immunogenetic background of Japanese leprosy patients by the HLA systems. *Vox Sang* 42(5):243–247.

Jamieson SE, Miller EN, Black GF, et al. (2004). Evidence for a cluster of genes on chromosome 17q11- q21 controlling susceptibility to tuberculosis and leprosy in Brazilians. *Genes Immun* 5(1):46–57.

Jopling WH, McDougall AC. (1996). *Handbook of Leprosy,* 5th ed. New Delhi, India: CBS Publishers and Distributors.

Kerr-Pontes LRS, Montenegro ACD, Barreto ML, Werneck GL, Feldmeier H. (2004). Inequality and leprosy in Northeast Brazil: An ecological study. *Int J Epi* 33: 262–269.

Kocak M, Balci M, Pence B, Kundakci N. (2002). Associations between human leukocyte antigens and leprosy in the Turkish population. *Clin Exp Dermatol.* 27(3):235–239.

Lima MC, Pereira GM, Rumjanek FD, et al. (2000). Immunological cytokine correlates of protective immunity and pathogenesis in leprosy. *Scand J Immunol* 51:419–428.

Llorente L, Zou W, Levy Y, et al. (1995). Role of interleukin 10 in the B lymphocyte hyperactivity and autoantibody production of human systemic lupus erythematosus. *J Exp Med* 181:839–844.

Malhotra D, Darvishi K, Sood S, et al. (2005). IL-10 promoter single nucleotide polymorphisms are significantly associated with resistance to leprosy. *Hum Genet* 118(2):295–300.

Malhotra D, Darvishi K, Lohra M, et al. (2006). Association study of major risk single nucleotide polymorphisms in the common regulatory region of PARK2 and PACRG genes with leprosy in an Indian population. *Eur J Hum Genet* 14(4):438–442.

McLeod K, Yates R. (1981). Forms of Chin Law: An Annotated Translation of the Feng-chen shih. *Harvard Journal of Asiatic Studies* 41(1):111–163.

Mehra NK, Rajalingam R, Mitra DK, Taneja V, Giphart MJ. (1995). Variants of HLA-DR2/DR51 group haplotypes and susceptibility to tuberculoid leprosy and pulmonary tuberculosis in Asian Indians. *Int J Lepr Other Mycobact Dis* 63(2):241–248.

Miller EN, Jamieson SE, Joberty C, et al. (2004). Genome-wide scans for leprosy and tuberculosis susceptibility genes in Brazilians. *Genes Immun* 5(1):63–67.

Mira MT, Alcais A, Van Thuc N, et al. (2003). Chromosome 6q25 is linked to susceptibility to leprosy in a Vietnamese population. *Nat Genet* 33(3):412–415.

Mira MT, Alcais A, Nguyen VT, et al. (2004). Susceptibility to leprosy is associated with PARK2 and PACRG. *Nature* 427(6975):636–640.

Misra VN. (1997). Balathal: A Chalcolithic Settlement in Mewar, Rajasthan, India: Aesults of First Three Seasons Excavations. *South Asian Archaeology* 13:251–273.

Mohamed Ali P, Ramanujam K. (1966). Leprosy in twins. *Int J Lepr* 34:405–407.

Moraes MO, Pacheco AG, Schonkeren JJ, et al. (2004). Interleukin-10 promoter single-nucleotide polymorphisms as markers for disease susceptibility and disease severity in leprosy. *Genes Immun* 5(7):592–595.

Morahan G, Kaur G, Singh M, et al. (2007). Association of variants in the IL12B gene with leprosy and tuberculosis. *Tissue Antigens* 69(1):234–236.

Ohyama, Ogata K, Takeuchi K, et al. (2005). Polymorphism of the 50 flanking region of the IL-12 receptor beta2 gene partially determines the clinical types of leprosy through impaired transcriptional activity. *J Clin Pathol* 58:740–743.

Parida SK, Grau GE, Zaheer SA, Mukherjee RO. (1992). Serum tumor necrosis factor and interleukin 1 in leprosy and during lepra reactions. *Clin Immunol Immunopathol* 63:23–27.

Pitchappan RM. (1990). Genetics of Tuberculosis Susceptibility. *Trop Med Parasitol* 51:355–356.

Pereira AC, Brito-de-Souza VN, Cardoso CC, et al. (2009). Genetic, epidemiological and biological analysis of interleukin-10 promoter single-nucleotide polymorphisms suggests a definitive role for -819C/T in leprosy susceptibility. *Genes Immun* 10(2):174–180.

Rani R, Zaheer SA, Mukherjee R. (1992). Do human leukocyte antigens have a role to play in differential manifestation of multibacillary leprosy: A study on multibacillary leprosy patients from north India. *Tissue Antigens* 40(3):124–127.

Rani R, Fernandez-Vina MA, Zaheer SA, Beena KR, Stastny P. (1993). Study of HLA class II alleles by PCR oligotyping in leprosy patients from north India. *Tissue Antigens* 42:133–137.

Ravikumar M, Dheenadhayalan V, Rajaram K, et al. (1999). Associations of HLA-DRB1, DQB1 and DPB1 alleles with pulmonary tuberculosis in South India. *Tuber Lung Dis* 79(5):309–317.

Ridley DS, Jopling WH. (1966). Classification of leprosy according to immunity. A five-group system. *Int J Lepr Other Mycobact Dis* 34(3):255–273.

Robbins G, Mushrif V, Misra VN, Mohanty RK, Shinde VS. (2007). Report on the Human Remains at Balathal. *Man and Environment* 31(2):50–65.

Robbins G, Mushrif Tripathy V, Misra VN, et al. (2009). Ancient Skeletal Evidence for Leprosy in India (2000 BC). *PLoS One* 4(5):e5669.

Roy S, McGuire W, Mascie-Taylor CG, et al. (1997). Tumor necrosis factor promoter polymorphism and susceptibility to lepromatous leprosy. *J Infect Dis* 176:530–532.

Roy S, Frodsham A, Saha B, Hazra SK, Mascie-Taylor CG, Hill AV.(1999). Association of vitamin D receptor genotype with leprosy type. *J Infect Dis* 179:187–191.

Santos AR, Almeida AS, Suffys PN, et al. (2000). Tumor necrosis factor promoter polymorphism (TNF2) seems to protect against development of severe forms of leprosy in a pilot study in Brazilian patients. *Int J Lepr Other Mycobact Dis* 68:325–327.

Santos AR, Suffys PN, Vanderborght PR. (2002). Role of tumor necrosis factor-α and interleukin-10 promoter gene polymorphisms in leprosy. *J Infect Dis* 186:1687–1691.

Settin A, Nassar S, Abdel-Latif A, et al. (2007). Association of cytokine gene polymorphism with susceptibility and clinical types of leprosy. *Int J Heal Sci* 1(1):25–33.

Shankarkumar U, Ghosh K, Badakere S, Mohanty D. (2003). Novel HLA Class I Alleles Associated with Indian Leprosy Patients. *J Biomed Biotechnol* (3):208–211.

Siddiqui MR, Meisner S, Tosh K, et al. (2001). A major susceptibility locus for leprosy in India maps to chromosome 10p13. *Nat Genet* 27(4):439–441.

Snell GD. (1980). Studies in Histocompatibility. Nobel Lecture.

Soebono H, Giphart MJ, Schreuder GM, Klatser PR, de Vries RR. (1997). Associations between HLA-DRB1 alleles and leprosy in an Indonesian population. *Int J Lepr Other Mycobact Dis* 65(2):190–196.

Spielman RS, McGinnis RE, Ewens WJ. (1993). Transmission test for linkage disequilibrium: The insulin gene region and insulin-dependent diabetes mellitus (IDDM). *Am J Hum Genet* 52(3):506–516.

Sun F, Flanders WD, Yang Q, Khoury MJ. (1999). Transmission disequilibrium test (TDT) when only one parent is available: The 1-TDT. *Am J Epidemiol* 150:97–104.

Tosh K, Meisner S, Siddiqui MR, et al. (2002). A Region of Chromosome 20 Is Linked to Leprosy Susceptibility in a South Indian Population. *J Infect Dis* 15:186(8):1190–1193.

Tosh K, Ravikumar M, Bell JT, Meisner S, Hill AV, Pitchappan R. (2006). Variation in MICA and MICB genes and enhanced susceptibility to paucibacillary leprosy in South India. *Hum Mol Genet* 15(19):2880–2887.

Trinchieri G. (2003). Interleukin-12 and the regulation of innate resistance and adaptive immunity. *Nat Rev Immunol* 3:133.

Vanderborght PR, Pacheco AG, Moraes ME, et al. (2007). HLA-DRB1*04 and DRB1*10 are associated with resistance and susceptibility, respectively, in Brazilian and Vietnamese leprosy patients. *Genes Immun* 8(4):320–324.

van Eden W, de Vries RR, D'Amaro J, Schreuder I, Leiker DL, van Rood JJ. (1982). HLA-DR-associated genetic control of the type of leprosy in a population from surinam. *Hum Immunol* 4(4):343–350.

van Eden W, Gonzalez NM, de Vries RR, Convit J, van Rood JJ. (1985). HLA-linked control of predisposition to lepromatous leprosy. *J Infect Dis* 151:9–14.

Visentainer JE, Tsuneto LT, Serra MF, Peixoto PR, Petzl-Erler ML. (1997). Association of leprosy with HLA-DR2 in a Southern Brazilian population. *Braz JMed Biol Res* 30(1):51–59.

Wang LM, Kimura A, Satoh M, Mineshita S. (1999). HLA linked with leprosy in southern China, HLA-linked resistance alleles to leprosy. *Int J Lepr Other Mycobact Dis* 67:403–408.

WHO (World Health Organization). (2009). Leprosy Fact Sheet: *http://www.who.int/mediacentre/factsheets/fs101/en/*

Wong SH, Gochhait S, Malhotra D, Pettersson FH, Teo YY, Khor CC, Rautanen A, Chapman SJ, Mills TC, Srivastava A, et al. (2010). Leprosy and the adaptation of human Toll-like receptor 1. *PLoS Pathog* 1(6):e1000979.

Zerva L, Cizman B, Mehra NK, et al. (1996). Arginine at positions 13 or 70–71 in pocket 4 of HLADRB1alleles is associated with susceptibility to tuberculoid leprosy. *J Exp Med* 183:829–836.

Zhang F, Liu H, Chen S, et al. (2009a). Evidence for an association of HLA-DRB1*15 and DRB1*09 with leprosy and the impact of DRB1*09 on disease onset in a Chinese Han population. *BMC Med Genet* 11(10):133.

Zhang FR, Huang W, Chen SM, et al. (2009b). Genome wide Association Study of Leprosy. *N Engl J Med* 361:2609–2618.

Zinkernagel RM, Dunlop MB, Blanden RV, Doherty PC, Shreffler DC. (1976). H-2 compatibility requirement for virus-specific T-cell-mediated cytolysis Evaluation of the Role of H-2I Region and Non-H-2 Genes in Regulating Immune Response. *J Exp Med* 144:519–532.

97.

THE CURRENT STATUS OF MEDICAL GENETICS IN INDIA

Neerja Gupta and Madhulika Kabra

INTRODUCTION

In the post-genomic era, medical genetics has become integral to all disciplines in medicine. The discipline of Human and medical genetics was introduced in India about forty years ago; however, the progress in this field has been slow, probably due to poor planning and organization. The progress in medical genetics in any country is evident by medical genetics education, availability of diagnostic and counseling services, extent of integration of services with the existing healthcare system, prevention programs, research in related areas, and important issues of ethical and social considerations. This chapter examines the present status of medical genetics in India and considers the scope and challenges that lie ahead.

EXTENT OF THE PROBLEM

With over one billion people, the burden of genetic disorders in India is enormous due to its high birth rate, consanguineous marriages, and widespread occurrence of hemoglobinopathies and other birth defects such as neural tube defects, and Down syndrome. Congenital and hereditary genetic diseases are becoming a significant healthcare burden in India, hence there is a need for adequate and effective genetic testing and counseling services (Verma, 2001). With better control of communicable diseases and improved antenatal and newborn care, there is a significant reduction in neonatal and infant mortality. In states like Kerala and Goa, the health indices are comparable to developed countries. In some urban areas, congenital malformations and genetic disorders are the third most common cause of mortality in newborns (WHO-SEAR, 2004). In many tertiary care centers, congenital malformations are emerging as a major cause of neonatal mortality and morbidity. According to a publication of the National Neonatal and Perinatal Database, a compilation of statistics for neonatal mortality and mortality from 18 major teaching hospitals, congenital malformations contribute to about 8% of neonatal deaths (NNPD, 2000). Currently, in India most middle-class urban families face socioeconomic constraints, and thus desire that every child born should be normal. The hopes of the people are high, and they expect and demand modern genetic tests. Increasingly, the importance of medical genetics is being recognized in the prevention and management of birth defects (Balgir, 1999a,b).

EDUCATION

In India, there are no structured teaching programmes in medical genetics in the undergraduate and postgraduate curriculum. There is lack of exposure to training in basic and clinical genetics. Principles of genetics are taught theoretically in the first few preclinical years of medical education in most medical colleges. In most postgraduate courses, for example in Pediatrics there is hardly any exposure to clinical genetics practice except in few referral centers. The reason for this could be twofold. One there are hardly any medical colleges in India with facilities for clinical genetic services, and second is the mindset of policymakers, who still consider that the time has still not come when priority should be given to medical genetics education.

Postgraduates in biochemistry, pharmacology, microbiology, pathology, etc., do learn basics of genetics and molecular biology—but again, it is theoretical and research-based. Absence of clinical genetics services in the majority of the medical colleges in India hardly encourages the medical students to think of medical genetics as a career goal. In a survey conducted among 94 doctors including junior residents, senior residents, practitioners, and medical college teachers, the knowledge of medical genetics was inadequate in about 70%, fair in 26%, and good in only about 4% of participants (Phadke et al., 2002).

In India, medical genetics exists as a super-specialty course (DM) only at one place. Other institutions are likely to offer similar courses in the future. There is also lack of any dedicated training courses in genetic counseling, which is evident by the shortage of trained genetic counselors. In most instances, genetic counseling is offered by clinicians. Recently, one of the universities in Hyderabad has started a structured genetic counseling course. MSc in Genetics and PhD in genetics are available at various universities, and generate qualified scientists that are involved in diagnostics and genetic research. The national funding agencies like Department of Biotechnology (DBT), Department of Science & Technology (DST), and Indian Council of Medical Research (ICMR funding short courses in clinical genetics and genetic counseling. The Medical Council of India (MCI) has also recently approached faculty involved in clinical genetics to revise the undergraduate and postgraduate curriculum, including relevant topics on medical genetics. Even if the new curriculum were implemented, there will be a lack of trained professionals to teach medical genetics in the majority of medical institutions (Ghosh and Mohanty, 2002). The Council should seriously consider establishing medical genetics divisions/departments in all medical colleges. To begin with, it can be a division of any of the existing major clinical departments, for example, medicine, pediatrics, or obstetrics (Agarwal, 2009). In the present era, there is a definite need to improve standards of medical education across the country with specific focus on the new genetics and molecular biology (Agarwal, 2009).

DIAGNOSTIC AND COUNSELING SERVICES

The healthcare system in India is multifaceted, which not only includes public and private facilities but also alternative systems of medicine. The private sector plays a major role in healthcare delivery, providing about 80% of ambulatory and 60% of inpatient care (Nandraj and Khot, 2003). The quality of care varies across public and private facilities.

There are very few establishments within the country that provide comprehensive clinical and laboratory services for Genetic disorders, either in public or private domains. In the last few years many laboratory facilities, mostly private, have been set up without proper counseling and clinical genetics facilities. There has been tremendous progress in the practice of fetal medicine, with exceptionally trained professionals in fetal ultrasonography. However, largely these facilities are not available to the masses. The awareness among doctors and the public is definitely increasing, as is evident from the increasing number of referrals to our facility over last few years.

The laboratory services available within the country include cytogenetic testing, both conventional and rapid, molecular testing for almost all common and many rare single gene disorders, and also facilities for metabolic and neurodegenerative disorders. The services are however very expensive, especially in private setups. Government setups also provide services at a cost, but these are far less than those in private services. There is also a provision for exemption of charges based on the socioeconomic situation of the family in public sector facilities. There is urgent need to organize and coordinate these services for better affordability and accessibility to most consumers. In most cases, clinical services are not integrated with the primary health care system. This is important for the benefit of the majority of people living in rural areas. There are a few basic interventions, which can be integrated with some effort. These include identification of high-risk cases, folic acid prophylaxis for neural tube defects, aneuploidy screening, and thalassemia screening (Rasaily, 2009).

Another important issue that needs urgent attention is that of quality control for Genetic testing since there is no universal accreditation system.

PREVENTIVE SERVICES

At the community level, the strength of medical genetics lies in its contribution to the prevention of disease. In India, national screening programs either for the newborn or prenatal screening, which have a potential to reduce morbidity and mortality are largely absent. A newborn screening program has been initiated by the ICMR as a pilot project at five major centers in the country with this vision. In Goa and Chandigarh (one hospital), newborn screening has recently been implemented as a state-funded program. In view of the high incidence, there is an urgent need to initiate newborn screening for congenital hypothyroidism in the country (Kapoor and Kabra, 2010). For disorders such as hemoglobinopathies, Down syndrome, and major congenital malformations such as neural tube defects, prevention programs at the national level are needed to decrease the incidence of these disorders.

Prenatal diagnosis using invasive techniques is available in India at selected centers. Both obstetricians and ultrasonologists are performing these procedures. There is a need to train more professionals in these techniques. Many institutions and professional organizations are making efforts in this direction and offer training courses in fetal medicine. Such professional organizations include the Indian Society of Human Genetics, Indian Society for Prenatal Diagnosis and Therapy, Society of Fetal Medicine, and the Genetics Chapter of the Indian Academy of Pediatrics.

The increasing number of facilities for prenatal diagnosis specially those using ultrasonography has raised concerns about misuse, primarily sex selection practices. The Prenatal Diagnosis and Therapy Act has been implemented in the country to take care of existing and emerging issues (PNDT Act, 2011).

TREATMENT FACILITIES

There are a few disorders for which treatment is available and is being provided to patients in India, though such facilities are not accessible to all. These disorders include hemoglobinopathies, where both therapeutic and supportive treatment is available at various centers in India. Similarly, coagulation disorders like hemophilia are also being managed, and Factor VIII and IX are provided free or at a very low cost. Patients and their parents are running thalassemia or hemophilia societies successfully. Similarly, for various lysosomal storage disorders (LSD) like Gaucher disease and mucopolysaccharidoses, enzyme replacement therapy is available through Genzyme India under the India Charitable Access Program (INCAP) program. Recently, a LSD support group has been formed. As these therapies are extremely expensive, government support is required to help these families. The government of India has special provisions for providing certificates for children with special needs, including those with learning difficulties, for railway concession, income tax exemption, etc.

ETHICAL AND SOCIAL ISSUES

Genetics as a specialty probably has the maximum number of ethical dilemmas being faced the world over. In our country, they are all the more important because of widespread poor educational and economic status, and related concerns to protect the rights of people and avoid exploitation. Furthermore, there is a need to have local guidelines for genetic testing and research, keeping in view our sociocultural scenarios. Various bodies in India such as the ICMR are coming up with guidelines for genetic research, international collaborations, stem cell research, etc. (DBT/ICMR, 2007).

KEY POINTS

Genetic disorders constitute a major burden in India.

- Education in medical genetics needs promotion at the undergraduate and postgraduate level, with emphasis on clinical applications.
- There has been significant improvement in clinical genetic services and diagnostics, but there is a need for integrating these services with the existing healthcare system and wider availability.
- There is an urgent need to improve the awareness about the subject among the medical and paramedical professionals and the public.

CONCLUSION

In the last few years, India has seen advances in all fields of medicine including, medical genetics, but there is a scope for improvement. To improve genetic education, services and research, we need to enhance awareness about genetics at the community level, and incorporate the medical genetics specialty in the core curricula of MBBS and postgraduate studies. There is a need to develop medical genetics in an organized manner as a medical school/hospital activity, rather than as a strictly research tool. The interest and expertise in genetics must percolate to all specialties and superspecialties. Services should be integrated with the existing health system, with preventive programs at a nationwide level. There is an urgent need for strengthening networking between clinical and laboratory services to ensure efficient use of specialized laboratory space, time, and expertise.

REFERENCES

Agarwal SS. (2009). Medical genetics in India—what needs to be done? *Indian J Med Res* 130:354–356.

Balgir RS. (1999a). Medical genetics in clinical practice in India. *Current Medical Trends* 3:567–572.

Balgir RS. (1999b). Control and prevention of the genetic load of haemoglobinopathies in India. *Natl Med J India* 12:234–238.

Department of Biotechnology [DBT] and Indian Council of Medical Research [ICMR]. (2007). *Guidelines for Stem Cell Research and Therapy.*

Ghosh K, Mohanty D. (2002). Teaching of medical genetics in the medical colleges of India—Way ahead. *Indian J Hum Genet* 8:43–44.

Kapoor S, Kabra M.(2010) Newborn screening in India: current perspectives. *Indian Pediatr* 47:219–224.

Nandraj S, Khot A. (2003). Accreditation system for health facilities: challenges and opportunities. *Economic and Political Weekly* 50:5251–5255.

NNPD. [National Neonatal Perinatal Database]. (2004). National Neonatology Forum of India Morbidity and mortality among outborn neonates at 10 tertiary care institutions in India during the year 2000. *J Trop Pediatr* 50:170–174.

Phadke SR, Agarwal S, Agarwal SS. (2002). Medical genetics education in India. Natl Med J India 15:363.

Pre-Conception and Pre- Natal Diagnostic Techniques (Prohibition of Sex Selection) Act. (2011). Ministry of Health and Family Welfare, Government of India.

Rasaily R. (2009). Feasibility of introducing genetic services in the national family welfare program in India. Indian Council of Medical Research (ICMR) Collaborating Centers & Central Technical Coordinating Unit, ICMR, New Delhi, India. *Indian J Med Res* 130:404–412.

Verma IC. (2001). Burden of genetic disorders in India. *Indian J Pediatr* 68:25.

WHO-SEAR [World Health Organization Regional Office for Southeast Asia]. (2004). Identifying regional priorities in the area of human genetics in SEAR: Report of an Intercountry Consultation, Bangkok, Thailand, 23–25 September 2003. New Delhi, (SEA-RES-121).

98.

THE INDIAN LANDSCAPE OF HUMAN/MEDICAL GENETICS RESEARCH AND SERVICE

Swati Naik, Arvind Rup Singh, and Jai Rup Singh

INTRODUCTION

India is geographically one of the largest countries in the Asian continent. Its population in July 2009 stood at 1.15 billion persons. With this, India became the second country in the world after China to cross the one billion mark, and is the second most densely populated country in the world.

It is likely to exceed China in 2030 by having a population of about 1.53 billion (Rosenberg, 2009). The nation has 28 states and 7 union territories; each state has its unique character and a unique native language (Figure 98-1). Most of the Indians speak the national language, Hindi, and many can speak English; hence, the language barrier is overcome relatively easily throughout India. India is well known worldwide for its economic boom, prestigious culture, and fantastic tourism.

HUMAN GENETICS

Medicine is currently in a state of transformation, created by the combination of two major aspects of advanced technology. First is the explosion in information technology and second, the rapidly expanding science of genetics. The potential outcome is that within the foreseeable future, we will see the introduction of a new kind of individualized medicine, designed uniquely to the personal and genetic needs of each patient. Human genetic services are advanced and well established in developed countries like United Kingdom, United States, Australia, and Germany. The awareness of genetics as a branch of medicine has increased in these countries, though it is not optimized yet. Grossly speaking, human genetics is a specialty of medicine and it deals with all the genetic aspects of humans as a species, including inherited diseases and disorders. It is a relatively new area of medicine, particularly in developing countries like India.

The advances in genetics have revolutionized the management of a vast majority of medical conditions. These have revealed the cause of a number of medical conditions, which has helped us understand the pathophysiology of the condition and hence also helped in the management of many medical disorders. These advancements, as related to genetic diseases, can be broadly divided into the areas of clinical genetics and laboratory-based genetics (Figure 98-2).

CLINICAL GENETICS

Clinical genetics is concerned with the diagnosis, prognosis, management, and prevention of genetic disorders. It also provides for genetic counseling and genetic services to individuals or families with, or at risk of, conditions that may have a genetic basis. The aim of genetic services is to help those affected by, or at risk of, a genetic disorder to live as normally as possible.

Genetic disorders can affect any body system and at any age. These include:

- Chromosomal abnormalities, which cause birth defects, mental retardation, and/or reproductive problems, etc.
- Single gene disorders such as cystic fibrosis, muscular dystrophy, Huntington's disease, sickle cell anemia, etc.
- Mitochondrial disorders due to dysfunction of the mitochondrial respiratory chain, such as Leber hereditary optic neuropathy.
- Multifactorial disease like diabetes, schizophrenia, multiple sclerosis, etc.
- Familial cancer and cancer-prone syndromes such as inherited breast or colorectal cancer, neurofibromatosis, etc.
- Birth defects with a genetic component, such as neural tube defects and cleft lip and palate.

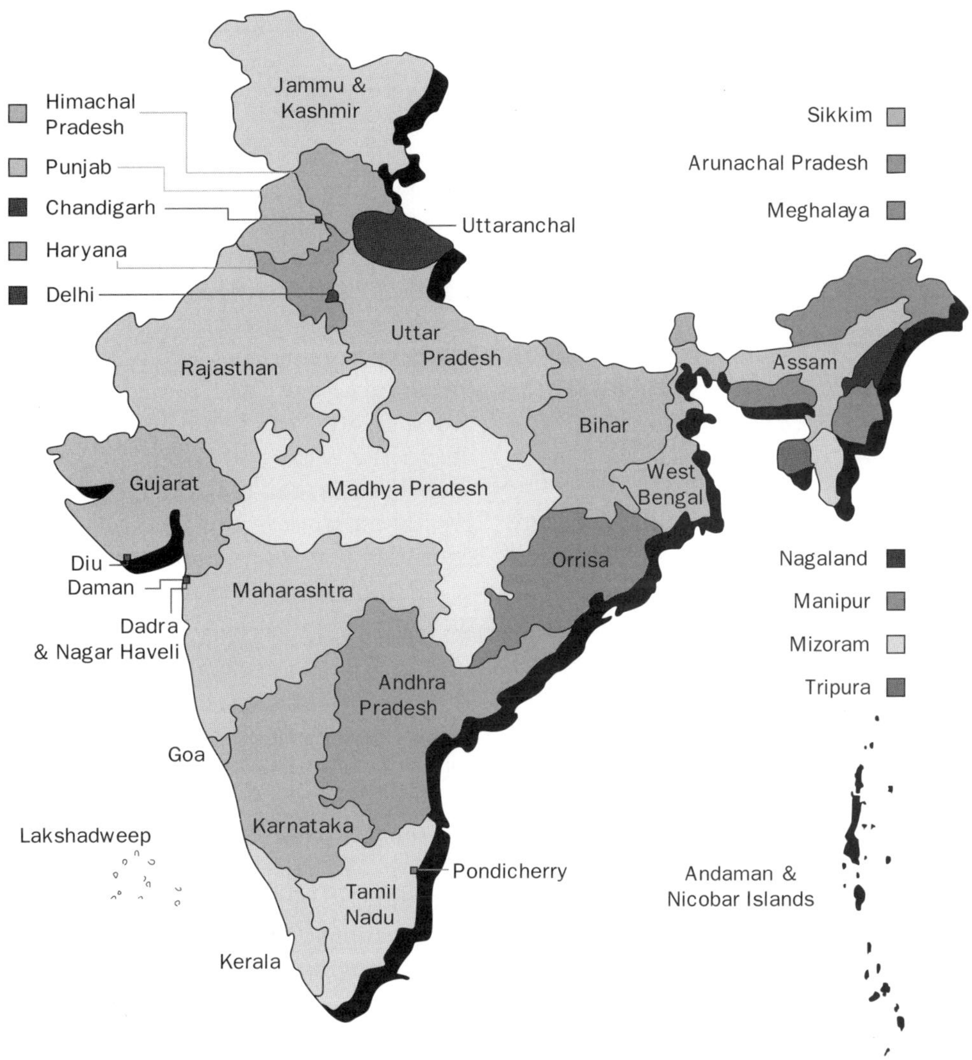

Figure 98-1 The political map of India showing 28 states and 7 union territories. Refer color figure.

Clinical genetics involves multidisciplinary work, including liaising with different medical specialties, mainly involving clinic or outpatient-based services. The genetic counselors with special interest in education can play an important role to spread the use of genetic services among general practitioners and midwives.

LABORATORY GENETICS

Clinical Laboratory Services

The laboratory services play an important role in conducting investigations such as karyotyping, mutation analysis, and DNA sequencing. For example, the availability of tests like "Array CGH" has made tremendous positive change in diagnosing variety of genetic conditions. Array CGH is sophisticated chromosome analysis that shows the chromosomes in fine detail, and gives information whether any part of the genome is missing or added.

Research in Genetics

Research is an integral part of genetics and improves our understanding of any genetic condition. It helps to reveal etiopathogenesis of many medical conditions that ultimately helps in their better management. Many times, the prevalence of various genetic conditions is different in different geographical parts of the world. Therefore, it is essential to establish research centers that may cater to general and specific local needs of particular regions.

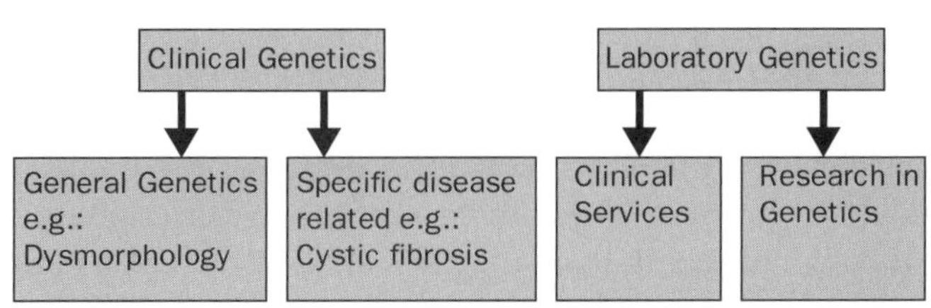

Figure 98-2 The commonly adopted model for genetic services.

CURRENT STATUS OF GENETIC SERVICES IN INDIA

The awareness of and use of genetic services is increasing among health professionals, and also in the general population in India, although at a slow pace. To our knowledge, currently there are more than 54 genetic centers in India. The genetic services are expanding in India, but still there is a need to rapidly establish new genetic centers to fulfill the demands of the given population size.

WHY IS IT IMPORTANT TO ESTABLISH GENETIC SERVICES IN INDIA?

Approximately, more than 50 million individuals are affected with single gene disorders in India. The exact proportion of individuals affected with chromosomal abnormalities, a variety of syndromes, inherited cancer, etc., in India is unknown. The rapidly evolving science of genetics will shed further light on genetic causes of common diseases such as diabetes and hypertension in the near future. Therefore, the total number of individuals who are at great risk of being affected with a genetic condition is enormous. This high number of genetically affected patients would warrant many well-established genetic service centers throughout India.

Infectious diseases form a major burden on the government as well as on the private health services in India. The control of infectious diseases is an important aspect of population health, although equally important is the timely and precise diagnosis of genetic disorders. A majority of the genetic disorders documented so far are only the tip of the iceberg of the actually afflicted individuals in the population. The vast majority of persons with genetic disorders in the Indian population do not seek medical attention, or remain undiagnosed due to unavailability of genetic services. Due to the perpetual recurring nature of genetic disorders, and lack of awareness of the available genetic services, their total but unrealized economic burden on the society is huge.

Human genetic services are developing in India, although at a slow pace. Therefore, at the present time genetic services are unavailable for the majority of the population in India. The public healthcare system organized by the government is underdeveloped, and there is no compulsory health insurance system, which could take care of expenses incurred by the affected individuals. In such a scenario, only a small number of persons, particularly the rich people, can think of paying for expensive genetic tests. For the majority of the population, this is unachievable in current times. For them, the birth of a congenitally malformed child or a child with a genetic disorder is the will of God. They have no option except to let such a child perish, and then go in for next unplanned pregnancy hoping that God will be merciful this time.

There is lack of awareness about genetic conditions among healthcare professionals and in the general population in India. Clinical genetics is relatively new and evolving in India, and there are few trained junior doctors available. Introducing awareness of medical genetics among trainee doctors and creating more trained geneticists will help to overcome this problem. The interest of public in genetic or inherited diseases has been enhanced by extensive media coverage given to the success of the Human Genome Project. There is now a greater realization about the diagnosis of genetic diseases, a better understanding of the pathogenesis of such diseases, and the power of predictive tests has been greatly enhanced. The medical professionals are now enthusiastic to know more about genetic diseases. They are also curious about prenatal and postnatal management of such disorders.

POSSIBLE PRIORITIES FOR HUMAN GENETIC SERVICES IN INDIA

Inherited Disorders in Families

Currently, there is less awareness of the availability of investigations for familial/inherited disorders, such as Huntington's disease or neurofibromatosis type 1. Predictive testing of these conditions can help in their better medical management. Prenatal testing for single gene disorders and other genetic diseases can ultimately help to decrease the incidence of such disorders and lead to a healthier society.

Prenatal Diagnosis

The dramatic developments in female education are changing the Indian society, particularly in urban populations. Women are planning pregnancies at a later age due to their aspirations for achievement of their career goals. This has resulted in an increase in maternal age at the time of conception. The increase in maternal age is associated with certain genetic/syndromic conditions such as Down syndrome. Prenatal testing for three common trisomies— Down syndrome, Patau syndrome (trisomy 13), and Edward syndrome (trisomy 18)—is a useful tool to screen for these conditions. The testing for these conditions is relatively easy and can be easily implemented by the government. If the pregnancy is affected, counseling about the real situation can help families cope in the decision making for termination of the pregnancy. The prenatal testing can reduce the incidence of certain genetic/syndromic conditions.

Chromosomal Imbalances

Chromosomal imbalances such as unbalanced translocations are common in the general population. Their exact figures for Indian populations are not known, due to lack of comprehensive studies. Chromosomal translocations result in recurrent miscarriages, infertility, and congenital abnormalities. Therefore, proper diagnosis of the chromosomal abnormalities can drastically improve their management.

Consanguinity

Consanguinity literally means *common blood* and indicates common ancestry. Consanguineous marriages are thus marriages among close relatives, e.g., with a cousin or an uncle, and these are quite common in many parts of India. The chance of children being affected with certain autosomal recessive genetic conditions is high in such marriages. The best examples of such conditions are found in metabolic diseases. Genetic diagnosis of these disorders can help to reduce their recurrence risk, and screening of extended family members will be helpful in their better management.

Population Screening for Genetic Disorders

There are certain genetic conditions that are differentially common in different populations of the world. For example, sickle cell anemia is common in Afro-Caribbean and Mediterranean populations; cystic fibrosis is common in Caucasian populations; and thalassemia and sickle cell anemia are common in the Indian population. Population screening for carrier detection of such genetic disorders can prevent or reduce the occurrence of these disorders. In our opinion, government as well as private healthcare organizations should establish and develop such screening programs.

SOME CONCERNS

The World Health Organization (WHO) has expressed great concern and stressed the need of early diagnosis, prevention, and management of genetic disorders in developing countries and has proposed detailed guidelines for it. There are obvious resource limitations in the implementation of WHO guidelines in India. To implement these guidelines, efforts need to be initiated for the development of human resources, opening of more diagnostic centers, broadening of the base for prenatal and postnatal screening, documentation of population-specific distribution of genetic diseases, and expanding the population screening strategies.

TABLE 98-1 TYPES OF GENETIC SERVICES OFFERED BY DIFFERENT CENTERS (TOTAL CENTERS: 54) (SINGH ET AL., 2007)

Type of service	Number of centers
Cytogenetic Investigations	45
Biochemical Investigations	26
Molecular Investigations	28
DNA Extraction and Storage	31
Genetic Counseling	45
Antenatal Genetic Counseling	26
Clinical Diagnosis of Genetic Disorders	31

Some institutes and universities have encouraged departments and research centers to undertake research in the area of human and medical genetics. A few of these provide limited diagnostic genetic services to patients. In addition to these, there are a number of institutions in India that have excellent infrastructure and facilities, created through government funding, and that are involved in research on different aspects of human genetics. However, a majority of these institutions do not extend

TABLE 98-2 POPULATION BEING SERVED BY EACH GENETIC CENTER (SINGH ET AL., 2007)

State/U.T.	Population (in millions)	No. of centers	No. of persons per center (in millions)
Andhra Pradesh	75.7	6	12.6
Chandigarh	0.90	1	0.9
Chhattisgarh	20.80	1	20.8
Gujarat	50.6	2	25.3
Haryana	21.1	1	21.10
Jammu and Kashmir	10.1	1	10.10
Karnataka	52.7	3	17.60
Kerala	31.8	1	31.80
Madhya Pradesh	60.4	3	20.10
Maharashtra	96.8	12	8.07
Manipur	2.4	1	2.40
Orissa	36.7	1	36.70
Punjab	24.3	1	24.30
Rajasthan	56.5	1	56.50
Tamil Nadu	62.1	8	7.76
Uttar Pradesh	166.0	2	83.00
West Bengal	80.2	3	26.70
Delhi	13.8	6	2.30

(As per census of India, 2001)

their facilities to patients for the diagnosis of genetic diseases. There is no mandate from the government funding agencies, sponsoring such patient-based research projects, to earmark a small component of these grants for genetic services including diagnostics. It is not possible to have meaningful or useful genetic services in a country like India without the government's support and commitment. Although the private sector is opening up new genetic centers, the total number of centers that are available across the country remains awfully inadequate. Accreditation of the centers providing genetic services is another important aspect. WHO guidelines recommend strict quality control for diagnostics. In India, at the moment, there is no regulatory control for these services. The need is to introduce appropriate accreditation of the centers for genetic services.

One point of concern is the absence of any link-up or tie-up between different genetic centers in India. Obviously, it is not feasible that each genetic center be self-sufficient for undertaking all types of genetic investigations required to diagnose all the genetic conditions. Therefore, it is pertinent that there should be information sharing and some sort of networking between various centers so that the material for relevant genetic investigations can be quickly sent to the appropriate centers.

In 1980, Professor I.C. Verma had made a maiden effort by bringing out a Directory of Genetic Services in India. Later on, in the year 2007, a more detailed Directory of Human Genetic Services was prepared by Professor Jai Rup Singh, Dr. Arvind Rup Singh, and Mr. Arsh Rup Singh. This is an authentic directory of genetic services available so far, and most of the contents published in this chapter are borrowed from it (see Tables 98-1, 98-2, and 98-3). The authors claimed that it is not a comprehensive directory, but they tried to get as much information as possible and set the ball rolling in the right direction. The data compiled is quite interesting and revealing. There are approximately 54 genetic centers throughout India. Among these, cytogenetic investigations are being

TABLE 98-3 TYPES OF GENETIC SERVICES AVAILABLE (SINGH ET AL., 2007) (STATE-WISE AND HEAD OF GENETIC SERVICE-WISE DISTRIBUTION)

State/U.T., and Head of Genetic Services	Clinical diagnosis	Chromosomal investigations	Molecular tests	DNA extraction	Genetic counseling	Pregnancy counseling	Management of genetic disorders
Andaman & Nicobar Islands (U.T.)							
Andhra Pradesh							
Ahuja, Y.R.		■	■	■	■		
Busi, Bhaskararao		■	■	■			■
Devi, Radha Rama	■	■	■	■	■		■
Hasan, Quarratulain		■	■	■	■		
Jyothi, A.		■	■		■	■	■
Rao, U.V. Krishna	■	■	■	■	■		■
Arunachal Pradesh							
Assam							
Bihar							
Chandigarh (U.T.)							
Das, Reema	■	■	■	■	■	■	
Mitra, Mitashree			■	■	■		■
Dadar & Nagar Haveli							
Daman & Diu (U.T.)							
Delhi (U.T.)							
Ghosh, R.		■	■				
Husain, Sayed Akhtar		■		■			
Kabra, Madhulika		■		■			
Kucheria, Kiran		■		■	■	■	
Nath, Surinder		■			■		

(Continued)

TABLE 98-3 (CONTINUED)

State/U.T., andHead of Genetic Services	Clinical diagnosis	Chromosomal investigations	Molecular tests	DNA extraction	Genetic counseling	Pregnancy counseling	Management of genetic disorders
Verma, I.C.							
Goa							
Gujarat							
Shah, Pankaj M.							
Sheth, Frenny							
Haryana							
Kushwaha, K.P.S.							
Himachal Pradesh							
Jammu & Kashmir							
Gupta, Subash							
Jharkhand							
Karnataka							
Bagai, Kamal Kant							
Hedge, Sridevi							
Rajangam, Sayee							
Kerala							
Ankathil, Ravindran							
Lakshwadeep (U.T.)							
Madhya Pradesh							
Choube, Ramesh							
Gupta, R.B.							
Jain, Sharad							
Maharashtra							
Amare, Pratibha S.							
Apte, B.N.							
Chopade, Dnyandeo							
Dani, Ashok							
Dave, Usha							
Gogate, Ganesh							
Jalan, Anil B.							
Khan, Abeda							
Lahiri, K.R.							
Malpani, Aniruddha							
Mohanty, Dipika							
Parikh, Firuzza R.							
Manipur							
Shah, S.A. Latif							
Meghalaya							
Mizoram							
Nagaland							
Orissa							

(Continued)

TABLE 98-3 (CONTINUED)

State/U.T., andHead of Genetic Services	Clinical diagnosis	Chromosomal investigations	Molecular tests	DNA extraction	Genetic counseling	Pregnancy counseling	Management of genetic disorders
Balgir, R.S.			■		■		■
Pondicherry (U.T.)							
Punjab							
Singh, Jai Rup	■	■	■	■	■	■	
Rajasthan							
Sharma, Nalinkash		■			■	■	■
Sikkim							
Tamil Nadu							
Gowrishankar, K.		■			■	■	■
Jayanth, V.	■	■		■	■	■	■
Kannan, J.		■	■		■	■	■
Kumaramanickavel		■	■	■	■		
Mammen		■	■	■	■		■
Rajkumar, T.	■			■	■		■
Sundari, Maya				■			
Suresh, S.		■	■	■	■	■	■
Tripura							
Uttar Pradesh							
Mohammad, A.		■			■	■	■
Phadke, Shubha		■	■	■	■	■	■
Uttaranchal							
West Bengal							
Benerjee, A.R.		■	■		■	■	■
Chaudhuri, T.K.		■	■	■	■	■	■
Dey, Subarata		■	■	■	■		

undertaken at 45 centers, biochemical investigations at 26 centers, molecular diagnostics at 28 places, prenatal diagnosis at 26 centers, and genetic counseling at 45 places.

We hope that this chapter will highlight the existing lacunae and the extreme shortage that exists for genetic services in India. Making these services available throughout India is a matter of great concern, and is crucial for Indian population.

CONCLUSION

In conclusion, human genetics services are slowly developing in India. These services should be established rapidly throughout India to fulfill the demands of the large population size of the country. An organized collaboration between the genetic centers is crucial and the government of India and the large private health organizations need to take responsible steps to establish genetic services throughout the country.

REFERENCES

Singh JR, Singh AR, Singh AR. (2007). Directory of Human Genetic Services in India. Satvic Media Pvt. Limited, Amritsar, India. pp. 155.

Rosenberg M. (2009). *India's Population.* Available at: http://geography.about.com.

LATIN AMERICA AND THE CARIBBEAN

99.

DESIGN AND IMPLEMENTATION OF A PLATFORM FOR GENOMIC MEDICINE IN MEXICO

Gerardo Jiménez-Sánchez, Julio Frenk, and Guillermo Soberón

INTRODUCTION

The spectacular progress in biochemistry, molecular biology, genetic engineering, and biotechnology that occurred during the second half of the twentieth century paved the way for the sequencing and mapping of the human genome. Genomes contain all of the information needed to determine the functions of an organism and maintain homeostasis, a state of equilibrium within an organism's internal environment. The healthy state depends on the correct functioning of the genome in interaction with the environment. Because of these homeostatic systems, organisms can dynamically adapt to changes in the environment. When this equilibrium is disrupted, disease occurs.

The Human Genome Project (HGP) represented a milestone in human history, not only because of its scientific and technological scope but also because of the great impact that this information will have on people's daily lives. Knowing how variations in the genome confer human individuality will be of great use in medicine, giving physicians the ability to prevent, diagnose, and treat common diseases (Childs et al., 2001). Moreover, the rise of genomic sciences has propelled the development of new technologies to sequence and study the human genome and that of many other species. These technologies have led to novel and more sophisticated opportunities for economic growth through the development of new products, the creation of new intellectual property, commercialization, and identification of new markets. However, these discoveries also pose important ethical, legal, social, political, and even religious challenges that demand adaptive changes in contemporary societies. Thus, it is of great importance that all members of society understand the opportunities and risks that this information generates, so that necessary actions can be implemented to allow equal access to medical applications while preventing inappropriate use of information, thus complying with universal principles related to human rights and respect for societal values (Jiménez-Sánchez, 1999).

The analysis of the results from the HGP has increasingly made evident important differences between traditional genetics, which studies individual genes or relatively small groups of genes, and genomics, which is dedicated to the comprehensive study of genomes with a global focus that covers all the genes in a species. Research in this field has become increasingly focused on understanding the function and regulation of each gene and its interactions with other genes and with the environment.

In February 2001, the first "draft" of the human genome was published. The results of the HGP, supported by the United States government, were published in *Nature* (Lander et al., 2001). Simultaneously, the sequence resulting from the private project by Celera Genomics was published in *Science* (Venter et al., 2001). In addition to mapping and sequencing of the human genome, these projects have produced the first analysis of the genome's contents. Thus, to systematically identify genetic variations in the human genome, a new project was developed: the "HapMap." Its main goal was to describe the most frequent genetic variations in the genome of three ancestral populations: European, Asian, and African. This project successfully concluded in 2006, providing a series of fundamental tools to contribute to the identification of the genetic bases of common diseases (HapMap, 2005). This and other subsequent projects, including ENCODE and "The 1000 Genomes" (Kaiser, 2008), have significantly improved our knowledge of genetic variation and have contributed to the identification of disease-associated genes.

The human genome contains variations in its sequence, the combinations of which confer individuality to each member of the human species. All the members of our species share around 99.8% of our genome. However, there are different kinds of genetic variation in the human genome. Among these, there are between 3 to 10 million single nucleotide polymorphisms. If we calculate the number of possible combinations of these, and other variations also available in the human genome, it will become clear why each individual has his or her own genetic combination, accounting for about 0.1% of the total sequence. The ability to uniquely identify each individual will have applications in different areas. In fact, in some countries, DNA banks are being developed to identify individuals based on currently known polymorphic variants.

In addition to the complete sequence of the human genome, a map that contains around 23,000 genes and a catalog of more than 1200 genes associated with diseases (Jiménez-Sánchez et al., 2001a) and genetic variations were published. These variations include single nucleotide polymorphisms (SNPs), copy-number variations (CNVs), duplications, and deletions (Iafrate et al., 2004; Wakeley et al., 2001).

IMPLICATIONS FOR HEALTHCARE

Genomic medicine, defined as the analysis of genetic variation to identify risks for common diseases, will lead to a more individualized, more predictive, and more preventive medical practice based on the genetic protection or risk conferred by each individual's genome. Thus, to develop specific applications to diagnosis and treatment of human disease, it is necessary to identify genetic variants associated with common disease risks (Jiménez-Sánchez et al., 2001a).

Our ability to analyze the whole human genome and associate genetic variations to common traits makes an important distinction relative to traditional human genetics approaches, although both disciplines can complement each other to achieve their own goals. Applications derived from the human genome will offer new forms of prevention, including the ability to develop presymptomatic identification of individuals at risk. This may lead to new strategies to influence individuals' lifestyles as part of their primary healthcare.

Overall, genomic medicine has the potential to prevent or delay the presence of common diseases in humans. As a result, it is not too bold to anticipate that, within a few years, clinical medicine will be increasingly immersed in the interpretation of genomic findings and in the design of programs to optimize healthcare of individuals based on those findings.

One of the benefits derived from genomics and proteomics is the ability to design new pharmaceutical drugs based on the identification of their molecular targets and their regulatory mechanisms. Furthermore, it will make it possible to identify those individuals likely to have unusual responses to certain drugs, leading to a safer use of them (Jiménez-Sánchez et al., 2002c). Thus, pharmacogenomics studies the response of individuals to pharmaceutical agents based on the characteristics of their genome (Evans & Relling, 1999; Emilien et al., 2000; Sadée, 1989). Its development potentially will lead to the design of products directed to population groups that share specific DNA sequences associated with better medical effects and less toxicity.

ETHICAL, LEGAL, AND SOCIAL IMPLICATIONS

The use of genomic information may have implications of great individual and social relevance. There are important ethical, legal, and social issues related to the use of genomic information, which contemporary societies need to address in a timely way (Jiménez-Sánchez & Lara Alvarez, 2007). Some of the most relevant challenges include prevention of stigmatization and discrimination of individuals based on their genomic makeup. This includes fair use of genetic information by insurance companies, employers, the legal system, schools, and other institutions. It is therefore important to develop regulatory frameworks that regulate access to this information, among other issues.

Thus, it is essential that the study of the human genome and the development of medical applications be undertaken under ethical principles that will allow societies to benefit from these new opportunities, as well as to avoid the risks of misuse, ensuring respect for the rights and dignity of individuals.

UNESCO produced a "Universal declaration on the human genome and human rights" during the 29th meeting of its General Assembly in November 1997. This declaration, unanimously approved by 186 United Nations member countries, indicates that "no research or research applications concerning the human genome... should prevail over respect for the human rights, fundamental freedoms and human dignity of individuals or, where applicable, of groups of people."

Research on this topic has received worldwide attention. In Mexico, investigators from the Center for Interdisciplinary Studies on Health and Law (http://www.juridicas.unam.mx/invest/areas/neisd/) at the Research Institute of the National Autonomous University of Mexico (UNAM) have worked on different aspects of the legal and social implications of genomic research and have made important contributions to this field (Muñoz de Alba, 2002).

In addition, Mexico has a National Commission of Bioethics that analyzes such challenges in the context of universal principles and national legal frameworks. This commission acquired the status of a decentralized agency of the Mexican Federal Government in September 2005. Recently, it has been provided with growing resources to undertake an ambitious program, which includes studying the ethical, legal, and social aspects of genomic medicine.

HUMAN GENETICS IN MEXICO

Mexico has a solid tradition of biomedical and health sciences, upon which it is possible to construct the basis for genomic medicine. The disciplines of human genetics, biochemistry, molecular biology, microbiology, immunology, biotechnology, neuroscience, pediatrics, internal medicine, and geriatrics are all mature, and they constitute indispensable elements to the development of genomic medicine research and its applications to public health.

The study of human genetics in Mexico is a relatively recent development; it was only at the end of the 1940s that the first publications of Dr. Mario Salazar Mallén described the distribution of blood types among the Mexican population. In the 1960s, the first genetic research groups were established in hospitals both in Mexico City and Guadalajara. Soon after, the Mexican Association of Human Genetics emerged, along with human genetics as a specialty of medicine at the National Medical Center of the Mexican Institute of Social Security (IMSS) and the Mexican Board of Human Genetics.

Since its foundation in Mexico, human genetics has achieved significant progress in different areas, such as clinical practice, cytogenetics, population genetics, biochemical genetics, neonatal screening and, more recently, molecular genetics. The most relevant contributions of Mexican geneticists are discussed in three publications compiled by Kofman-Alfaro et al. (1991), Salamanca and Armendares (1995), and Lisker and Carnevale (1995). The groups that were progressively established during the last decades have made contributions to scientific research, healthcare, and higher education.

Human genetics in Mexico is a highly decentralized discipline resulting from the establishment of groups of geneticists across the country. Most of these scientists have made important contributions for which they have received international recognition. In the last years, not only have clinical geneticists been trained, but also doctoral programs have produced graduates in the field.

GENOMIC SCIENCES IN MEXICO

During the last decade, the Mexican government has strengthened its commitment to improving competitiveness and innovation through science and technology. Although investment in this area has been consistently limited, from 1995 to 2005 the percentage of the gross domestic product (GDP) devoted to science and technology increased from 0.35% to 0.43%. The number of students registered in doctoral programs in science and technology increased from 488 in 1994 to 2009 in 2005. From 2001 to 2005, the number of researchers in the National System of Investigators increased by 62% (Triunfol, 2007). In addition, in 2002, the Chamber of Deputies of the Mexican Congress approved a new Law of Science and Technology, through which new funding sources, known as the Sectorial Funds for Science and Technology, were established. These funds are composed of equal contributions from the different secretaries (departments), states, districts, and the National Council of Science and Technology (CONACYT).

Mexico currently has more than 12,100 researchers registered in the National System of Investigators. Among them, 24.5% are established at UNAM, and the rest are at public state universities throughout the country. In medical research, the National Institutes of Health (NIH) have about 500 researchers (4.1%), the Mexican Institute of Social Security has around 294 (2.4%), and other areas in the health sector have 87 researchers (0.7%). The fields of biology, chemistry, medicine, and health sciences contribute with 26.7% of the total, reflecting a sustained increase during the past ten years (CONACYT, www.siicyt.gob.mx). Overall, the number of investigators continues to increase, as well as the number of areas in which they work.

In the past 15 years, Mexican investigators have made initial contributions to genomic sciences. These include participation in the genome project for *Escherichia coli* (Blattner et al., 1997) and the first large-scale sequencing project in Mexico, the sequencing of the genome of *Rhizobioum etli*, conducted by the Center of Genomic Sciences (CCG; www.ccg.unam.mx), formerly known as the Center for Research in Nitrogen Fixation (CIFN; Gonzalez et al., 2006), at UNAM. These and other projects have led to a series of genomic initiatives related to different organisms, including the parasite *Taenia solium*, whose genome is also currently being sequenced at UNAM (Aguilar-Diaz et al., 2006). Other projects include the sequencing of plant species, which is mainly being undertaken at the Center of Research and Advanced Studies (CINVESTAV) of the National Polytechnic Institute (www.cinvestav.mx). Recently, the National Laboratory of Genomics for Biodiversity (LANGEBIO) was created with the goal of carrying out important genomics projects in plants, such as the sequencing and functional analysis of the genome of maize. Additional infrastructure for genomic research is located in the states of Guanajuato, Nuevo León, Jalisco, Tamaulipas, and Yucatán, among others (Jiménez-Sánchez et al., 2008).

In 2004, an undergraduate program in genomic sciences (LCG) began at UNAM (www.lcg.unam.mx) at the CCG and the Biotechnology Institute in Cuernavaca, Morelos. The curriculum includes intensive education in mathematics, statistics, computational sciences, and biology (Palacios & Collad0-Vides, 2007). In recent years, two new scientific societies have been founded: the Mexican Society of Genomic Sciences (http://smcg.ccg.unam.mx/) and the Mexican Society of Genomic Medicine (www.somegen.org.mx), with the latter headed by Gerardo Jiménez-Sánchez as its founding president. These societies gather together the majority of scientists who work in the field of genomics in Mexico.

FROM HUMAN GENETICS TO GENOMIC MEDICINE

Mexico has a long tradition in the biomedical sciences and health. Currently, there are more than 250 professionals registered with the Mexican Association of Human Genetics, and more than 175 certified geneticists registered by the Mexican Board of Human Genetics. A variety of institutions together offer more than 50 formal, independent graduate programs in human genetics and molecular biology. As a result, independent research groups have been established in areas that include clinical genetics, cytogenetics, genetic epidemiology, population genetics, inborn errors of metabolism, neonatal screening, genetic toxicology, and molecular genetics (Salamanca & Armendares, 1995).

Historically, the majority of important contributions in genetic research by Mexican-Americans have occurred outside of Mexico, and some contributions have resulted from the collaboration with Mexican researchers. In Mexico, the majority of studies related to the genetic basis of complex diseases have primarily analyzed known variants in candidate genes. Initially, such studies included the association of the HLA-B27 allele with ankylosing spondylitis (Fraga et al., 1979) and the association of other HLA alleles with rheumatoid arthritis. Since then, researchers have described associations between HLA alleles and various complex diseases common among Mexicans. These diseases include type 2 diabetes mellitus and some of its complications (Perez-Luque et al., 2003), generalized lupus erythematosus (GLE; Bekker-Mendez et al., 1998), scleroderma (Vargas-Alarcon et al., 1995), and type 1 diabetes mellitus (Gorodezky et al., 1995).

Other complex diseases have also been studied using candidate gene approaches. Some examples include the analysis of SNPs in the genes encoding insulin, and others associated with different abnormalities in serum lipid levels (Sanchez-Corona et al., 2004), such as hepatic nuclear factor 4-alpha (Weissglas-Volkov et al., 2006), tumor necrosis factor-alpha (Parra-Rojas et al., 2006) and *ABCA1* (Villarreal-Molina et al., 2007). In addition, there have been associations described between GLE and polymorphisms in *PTPN22* (Baca et al., 2006) and *PCDC1* (Velazquez-Cruz et al., 2007) in children, as well as between GLE and polymorphisms in *PCDC1* in adult patients (Prokunina et al., 2002).

The 677C>T polymorphism in the methylenetetrahydrofolate reductase gene (*MTHFR*) is associated with diseases related to hyperhomocysteinemia and folate deficiency. It has been found that, among Mexicans, the 677C>T allele is weakly associated ($p = 0.05–0.01$) with hyperhomocysteinemia (Torres-Sanchez et al., 2006), anencephaly (Munoz et al., 2007) and gastric cancer (Lacasana-Navarro et al., 2006).

The analysis of genetic variations related to pharmacological responses has mainly focused on the genes that encode for metabolizing enzymes, such as *CYP2D6* (Contreras et al., 2011; Hidalgo-Miranga & Jiménez-Sánchez, 2009). In the case of this gene, the allelic frequencies show significant differences among the different populations of Mexico. For example, *CYP2D6*10* has a frequency of 12.45% among individuals in Mexico City (Lopez et al., 2005) and a frequency of only 2.3% among individuals in the north-central state of Durango (Sosa-Macias et al., 2006).

Clearly, Mexican geneticists have achieved important milestones; however, to identify novel genes associated to common diseases in Mexicans, it is important to implement modern research strategies such as genome-wide association studies (GWAS), dense genotyping, and resequencing approaches, among others, to achieve an emergence of research in genomic medicine.

Genomic analysis of cancer in Mexico is currently undergoing a transition from the classic genetic approaches to whole-genome analysis, including whole expression and copy-number variations analyses (Hidalgo et al., 2005; Valladares et al., 2006; Vazquez-Ortiz et al., 2007). This is transforming the area of molecular oncology in Mexico, particularly in cervicouterine and breast cancers.

To further develop human genetics and genomic medicine in Mexico, new experimental designs are required, including those considering the genetic structure of the Mexican population, supported by the collection of large numbers of disease-specific samples. In addition, training a higher number of health professionals in genomics and public awareness programs are needed so that Mexico takes full advantage of genomic medicine.

DEVELOPMENT OF GENOMIC MEDICINE IN MEXICO

Genomic medicine will allow for more individualized, thus effective, interventions. These will need to be incorporated into health policies to turn medical practice into a more individualized, predictive, and preventive discipline. Thus, scientific research must be a priority to identify disease-associated genetic variants in the Mexican population. Currently, an important number of genetic variants have been identified for various common diseases, including diabetes mellitus (Wellcome Trust, 2007; McCarthy, 2010), cardiac diseases (Wellcome Trust, 2007), Crohn's disease (Mathew, 2008), and various types of cancer (Easton et al., 2007; Zanke et al., 2007). This information, and its corroboration in different populations, has the potential to strengthen screening programs, diagnostic and prognostic strategies, and the identification of new therapeutic targets and novel molecules of pharmacological use. Such strategies would not only decrease chronic complications and their associated costs, but can also improve medical care and the patient's quality of life, and lead to a more rational use of health resources and better-supported decisions in public health (Jiménez-Sánchez et al., 2001b).

Many believe that the main challenges that Mexico faces can be divided into two categories: on one hand, overcoming painful lags in development that work against the dignity of its people, and on the other, the capacity to develop strategies to join emerging opportunities at the frontier of science and technology for the benefit of its people.

The challenges that a country must face in relation to its economic and social development are many and varied. There are pressing problems that will be overcome in the midterm and long term future. However, we have learned that challenges should not be dealt with in chronological order. Thus, it is not valid to suggest that Mexico should not embrace the emerging frontiers of science and technology to improve healthcare and wellbeing until previous, pressing challenges are solved. In fact, overcoming current hurdles and embarking upon promising new paths should be done simultaneously. As such, it is universally recognized that economic and social development must be pursued simultaneously.

In the case of healthcare, it is a false dilemma to put problems of underdevelopment such as malnutrition, infections, and pathologies related to reproduction, against the important opportunities that emerge in biotechnology, information technologies, and telecommunications—which can translate into greater possibilities for the diagnosis, treatment, control, and prevention of common diseases. This means that the Mexican health system should continue to improve its programs by incorporating new cost-effective interventions, increasing coverage for services that are already nearly universal, strengthening primary healthcare programs, promoting community medicine, and developing a health culture based in education. Furthermore, the National Health System should have the means to identify, evaluate and, when necessary, implement the rapidly emerging technological innovations that can strengthen healthcare in a variety of ways.

These thoughts were included in the 2002 declaration at the World Health Organization stating that Mexico cannot afford to be indifferent and passive about genomic medicine. The potential benefits that genomic medicine offers are significant, and it is absolutely essential that these opportunities be considered as part of the national health programs (Jiménez-Sánchez, 2003).

At that time, it was clear in Mexico that genomic medicine will translate into a new paradigm for healthcare, with a number of opportunities that would represent important benefits for the Mexican population. In addition, the solid infrastructure in genetics and clinical medicine, along with the enthusiasm to develop genomic sciences in Mexico, represented significant assets that could serve as the basis to develop genomic medicine. Moreover, it was recognized that the development of genomic medicine would require the analysis of genetic variations in the Mexican population, as well as a robust human and technological infrastructure to support/facilitate research and implementation programs. Thus, genomic medicine clearly was not an "out of the box one-fits-all" solution.

The available evidence indicated the important benefits that genomic medicine could represent for Mexico, as well as the risks if such development was delayed. These included lack of access to novel knowledge and applications to public health, and the risk of increasing Mexico's dependence on developed economies in sensitive areas such as health. This led to the decision of incorporating genomic medicine as a part of Mexico's National Health Plan. As a result, it stimulated the development of cutting-edge infrastructure to implement world-class genomic medicine, oriented to the development of early interventions for prevention, diagnosis, and treatment of common diseases such as hypertension, type II diabetes mellitus, asthma, acute myocardial infarction, and a number infectious diseases that are becoming increasingly relevant in Mexico as part of the epidemiological transition in progress (Jiménez-Sánchez, 2003). The sustained development of this strategy would lead to improvement of healthcare in Mexico, and makes it reasonable to predict an important economic impact related to the reduction of health-related costs, mainly those of chronic treatment and loss of productivity from individuals with common diseases (Jiménez-Sánchez et al., 2002a).

A major component of this strategy was oriented toward educational programs aimed at professionals and the general public. The development of a national platform in genomic medicine will lead to the creation of infrastructure to develop training programs at different levels. In addition, such a strategy would lead to new opportunities to increase participation in a knowledge-based economy, including the stimulation of the genomic industry as a means to contribute to the economic development of Mexico (WHO, 2002). Although the costs of this initial infrastructure seemed high, they were relatively moderate compared with the financial costs that result from the burden of disease or the dependence on other countries with expertise in such a significant strategic area.

Mexico has extraordinary opportunities to identify genetic variants associated with medical traits, due to the availability of isolated populations, which are highly inbred and genetically homogenous. We suggested that international collaborations should be stimulated; however, abusive relationships must be avoided in which samples from Mexicans are taken to more industrialized countries, impeding access to research and results. This has occurred in the past in the field of archaeology and in the study of Mexican fauna and flora. For these reasons, it is strategically important that Mexico takes advantage of the new knowledge and technologies generated by the HGP. To do this, it will be necessary to train a critical mass of researchers, professionals, and technicians who have the capacity to assimilate, develop, and apply genomics to medicine (Jiménez-Sánchez et al., 2008).

ANCESTRY OF THE MEXICAN POPULATION

Mexico is the fourteenth largest country in the world, with a total area of 1,972,550 km^2 and close to 112 million inhabitants (INEGI, 2005). Its varied topography results in a variety of climate conditions, from arid deserts in the north to rainy tropical climates in the south and along the southeast coast. In addition, geographically distant regions have different demographics. These differences arise from both the country's distinct ancestral components and from the demographic conditions that characterize each region (Gerhard, 1986). During the pre-Hispanic era, the majority of the population was concentrated in the center and south of Mexico. The ethnic groups that inhabited the north of Mexico did not have linguistic, religious, or political unity. It was not until two centuries after the Spanish conquest that the northern regions drew the attention of the Spaniards, mainly because of the silver deposits discovered in those regions (Gerhard, 1986). After the notable reduction of the Amerindian population as a consequence of epidemics between 1545 and 1548, African slaves were brought into Mexico. These slaves mixed both with indigenous inhabitants and with Mestizos, and many were transferred to other regions to work in mineral mines. The Yucatan Peninsula, located in the southeast region of Mexico, was populated by different Amerindian groups who were decimated by diseases, which reduced the original population by at least one-half (Gerhard, 1991). This unique history translates into a population that is derived from more than 60 groups of local Amerindians, Europeans, and, to a lesser degree, Africans. A survey of the language and geographic location of Amerindian groups in Mexico is currently being conducted. These groups have mixed throughout the past 500 years, leading to the Mestizo population that currently represents 80% of Mexicans (Gonzalez Burchard et al., 2005). Because of this unique demographic history, it is important to characterize the genetic composition of the Mexican population as an initial step toward the successful development of genomic medicine in Mexico.

STRATEGY TO ESTABLISH A NATIONAL INSTITUTE OF GENOMIC MEDICINE

As is the case for the majority of developing countries, Mexico faces demographic and epidemiological transitions that have important implications for the standards of disease, disability, and mortality. On one hand, it faces the unresolved problems of infections, malnutrition, and reproductive health, and on the other, the emerging challenges of chronic and degenerative diseases of the industrialized world. In Mexico, the adult population has a high prevalence of diabetes mellitus (7.0%), hypertension (30.8%), and obesity (29.4%; Olaiz, 2006). The two main causes of mortality are cardiovascular diseases (22.9%) and diabetes mellitus (15.3%; INEGI, 2005). Preventing these diseases is a key strategy to reducing their significant economic and health burden.

In 2000, an initial working group analyzed reasons why genomic medicine represented important opportunities for healthcare in Mexico. The group identified the following eight areas of opportunity: (1) contribution to a more individualized, predictive, and preventive medical practice; (2) strengthening of scientific research and technology in Mexico; (3) potential cost reduction of healthcare; (4) development of pharmacogenomics; (5) generation of novel products and services; (6) strengthening of the potential to participate in the knowledge economy; (7) timely development of an ethical and legal framework for genomic medicine in Mexico; and (8) the development of public education programs related to genomics and society.

The establishment of the National Institute of Genomic Medicine was preceded by multi-institutional efforts aimed at developing a national platform for genomic medicine in which this institution would have a central role. In 1999, the Mexican Health Foundation (FUNSALUD), then under the leadership of Dr. Guillermo Soberón, organized a working group composed of specialists from the NHIs and UNAM. The group analyzed the state of the art of genomic sciences at that time, as well as the possibilities to use such progress to improve healthcare for the Mexican population.

Simultaneously, the federal government of Mexico, then headed by President Ernesto Zedillo, began to generate the infrastructure that Mexico would need to coordinate public policies and actions regarding the human genome. In consequence, José Antonio González Fernández, Secretary of Health at that time, established the National Commission for the Human Genome on October 23, 2000. This organism was created by the action of an executive decision of the President of Mexico, which was co-signed by the secretaries of health and education. This Commission provided an ideal forum for discussion at the highest levels of decision making, and was instrumental in speeding the process of developing genomic medicine in Mexico.

In 2000, a strategic alliance was established to generate synergies for the project. This alliance was integrated by the Department of Health (SSA), UNAM, CONACYT, and FUNSALUD. Its goal was to evaluate the best way to create an institution that would coordinate the development of a national platform of genomic medicine. On August 27, 2001, with the assistance of specialists from the McKinsey consulting firm, this group produced a feasibility study to launch the initial steps for a national platform of genomic medicine in Mexico. This study recommended the creation of a Consortium to promote and perform executive and detailed studies, in order to create and develop the first National Health Institute in Latin America dedicated to genomic medicine, while the Mexican Congress analyzed a bill to create the National Institute of Genomic Medicine (INMEGEN). As a result, the Consortium for the Institute of Genomic Medicine was established on November 22, 2001.

The Consortium was designed with a very specific goal, namely to promote and perform executive and detailed studies to establish and develop INMEGEN (Jiménez-Sánchez et al., 2002b). Its organic structure was headed by a governing board, integrated by Julio Frenk, Secretary of Health; Juan Ramon de la Fuente, Rector of UNAM; Jaime Parada Avila, Director of CONACYT; and Antonio Lopez de Silanes, President of the Board of Directors of FUNSALUD. From it, a board of directors was formed that included representatives from each of the institutions of the alliance: Misael Uribe (SSA), Alfonso Serrano Perez-Grovas (CONACYT), Juan Pedro Laclette (UNAM), and Guillermo Soberón (FUNSALUD), along with a fifth person, Dr. Jorge Rosenkranz, who was included because of his experience in the fields of research and the industrial sector. Guillermo Soberón was appointed coordinator of this board. Subsequently, they elected Gerardo Jiménez-Sánchez as director of the Consortium, and Cuauhtemoc Valdes from FUNSALUD was invited to take care of administrative coordination.

The Consortium existed from January 2002 until May 2005, when it had successfully achieved its goal. The director of the Consortium periodically presented the work programs and reports for the consideration of the board of directors. The board monitored and supervised the activities and financial status. Furthermore, the Board of Governors was kept informed on an ongoing basis, and periodic meetings were held to formally present the appropriate progress reports.

Given the novel nature of genomic medicine, the Consortium dedicated part of its efforts to broadly disseminate information on the nature and the scope of the project to the academic community and the general public. Throughout the project, more than 100 lectures were held. Many of them were offered at the most distinguished academic institutions, including El Colegio Nacional, the National Academy of Medicine, and the Industrial's Club, the latter aimed at the business community. These conferences were delivered both by domestic and foreign speakers such as Barton Childs, Francis Collins, Aravinda Chakravarti, David Valle, Rick Ward, Julio Frenk, Juliana Gonzalez, Gerardo Jiménez-Sánchez, and Guillermo Soberon. In addition, the first National Congress of Genomic Medicine was celebrated in 2004 with participation of top international academic leaders in genomic medicine.

Antonio López de Silanes, a prominent businessman in Mexico and, at that time, president of the Board of Directors from FUNSALUD, organized a group of 37 FUNSALUD associates with experience in the pharmaceutical industry, health, and insurance services, along with other people interested in the topic. This group made it possible for FUNSALUD to contribute to the trust of the Consortium. Additionally, some companies (Laboratorios Silanes, GlaxoSmithKline, Merck Sharp & Dohme, and Novartis) provided support for specific projects and actions of the Consortium. One example of such support was the establishment of the Silanes-INMEGEN Award in genomic medicine in 2002, and the Silanes-FUNSALUD Chair of Genomic Medicine in 2004.

Since then, FUNSALUD has been disseminating information about genomic medicine on its own, as well as through joint events with the National Academy of Medicine and through written reports published in the *Gaceta Medica de Mexico*. An edition of *Cuadernos FUNSALUD* was dedicated to reviewing the achievements of genomic medicine and the opportunities for the pharmaceutical industry.

As has been mentioned, the Consortium was in charge of formulating the distinct projects and documents that served as the basis for the discussion of this initiative with the legislators at that time, mainly with those in the health and science and technology committees from each of the two chambers of the Mexican Congress, as well as with the Department of the Treasury to lay the foundations of the financial feasibility of the project.

The Consortium produced a number of national congresses, seminars, and courses. Among those, Gerardo Jiménez-Sánchez designed three courses that integrated genomic medicine and clinical practice; namely, "introduction to genomic medicine," "pediatric applications of genomic medicine" and "genomic applications in internal medicine," which were registered with the Faculty of Medicine at UNAM. In addition, an electronic portal was established to identify potentially interested candidates to pursue a career in the field. This portal produced extraordinary results: an enormous interest was seen among young doctors, biologists, chemists, engineers, and other professionals, many of them Mexicans receiving graduate training abroad. Several of them subsequently joined INMEGEN.

The research activities soon began, and it became necessary to construct an adequate space for the laboratory work and for highly specialized technological units. We established the initial headquarters of the Consortium on a 3000m^2 area at the Zafiro corporate tower in south Mexico City. In this space, we established the first laboratories and the following three core centers: sequencing, microarrays, and high-performance computing.

The results obtained by the Consortium were published in two separate reports (Jiménez-Sánchez, 2002, 2004b), which were approved by the Board of Governors and the Board of Directors of the Consortium. They were made publicly available and widely distributed to the institutions and public in Mexico.

THE FOUNDATION AND INITIAL DEVELOPMENT OF INMEGEN

In Mexico, the law of the National Institutes of Health indicates that NIHs are to be created by the Mexican Congress. The work done by the Consortium laid the foundations for discussion in the Mexican Congress regarding the need for the National Institute of Genomic Medicine. Thus, the process that began at the end of 1999 concluded in April 2004, following a significant number of actions to promote such legislation. On July 19 of the same year, a decree was published creating the National Institute of Genomic Medicine, in a ceremony headed by then President of Mexico, Vicente Fox, and in which then Secretary of Health, Julio Frenk, and the president of the Health Commission of the Chamber of Deputies, Jose Angel Cordoba Villalobos, participated. It is worth mentioning that Dr. Cordoba Villalobos followed Julio Frenk as Secretary of Health on December 1, 2006.

During the administration of President Felipe Calderon, INMEGEN received a visit from the First Lady of Mexico, Ms. Margarita Zavala, who inaugurated the first three High-Technology Core Units on the third anniversary of the founding of the institution. Later, on May 11, 2009, during the presentation of the results from the Mexican Genomic Diversity Project, the President of Mexico expressed his support of INMEGEN and urged it to continue working toward full development of genomic medicine in Mexico.

According to the Law of the National Institutes of Health, INMEGEN is the National Reference Center for matters related to the human genome and its medical applications. It has the following main objectives: (1) to perform clinical, epidemiological, experimental, technological, and basic studies and research in its areas of specialty for the understanding, prevention, diagnosis, and treatment of diseases and the rehabilitation of the ill; (2) to promote public health measures; and (3) to perform activities inherent to the National Institutes of Health, except for those related to providing medical services. The latter is relevant because in Mexico, all NIHs except for two (Public Health and Genomic Medicine) are highly specialized hospitals in addition to research institutions, whereas INMEGEN was planned as an institution where clinical research would be carried out using the clinical infrastructure of the rest of the NIHs. This feature, called the *horizontal dimension*, was designed during the feasibility study to develop robust synergies without diluting INMEGEN's research budget. In addition, the Institute was tasked with promoting projects to develop specialized technology based on specific projects involving technological innovation.

The same law defines the administrative bodies of the Institute. Thus, the Governing Board was established for the period from 2004–2009, in accordance with Chapter III (dedicated to the governing bodies), Article 14 of the cited law. From the first to the fifth years, the Governing Board consisted of the Secretary of Health, who presided over it; the General Coordinator of the National Institutes of Health; a representative of the Department of the Treasury; the President of the Board of Trustees of the Institute; a representative of the educational sector; and four invited members designated by the Secretary of Health from other institutions who were recognized for their moral quality, merits, prestige, and experience in their areas of expertise. These last members are invited for four years, with the possibility of being reappointed for one additional term; however, none of them were reappointed by the following federal administration. In addition, the Governing Board had a secretary and a deputy secretary.

During 2004, this Governing Board was presided over by the Secretary of Health, Dr. Julio Frenk Mora, and was composed of the following members: Dr. Jaime Sepulveda Amor, Deputy Chairman and General Coordinator the National Institutes of Health; Dr. Guillermo Soberon, President of the National Commission of Bioethics; Dr. Manuel H. Ruiz de Chavez, Executive President of the Mexican Health Foundation; Dr. Juliana Gonzalez Valenzuela, Emeritus Professor of the Department of Philosophy and Arts of UNAM; and Dr. Jorge Rosenkranz Weiner, distinguished Mexican scientist with special expertise in intellectual property in biotechnology; Mr. Sergio Montaño Fernandez, representative of the Department of Treasury, and Dr. Rene Santoveña Arredondo, representative of the educational sector and Rector of the Autonomous University of the State of Morelos; Mr. Carlos Eduardo Represas de Almeida, President of the Board of Trustees and Chair of the Governing Board of Nestlé Mexico, S.A. de C.V.; and Sergio Vazquez Cordova, as Commissary.

The members mentioned above formed the Governing Board until 2006. With the change of federal administration, beginning in 2007, Dr. Jose Angel Cordova Villalobos, Secretary of Health, presided over the organization, and Dr. Julio Sotelo Morales joined as Alternate Chairman and Trustee of the Coordinating Commission of the National Institutes of Health and High-Specialization Hospitals.

Several other members of the board changed, including the four invited members named.

In September 2008, given that no members were reappointed upon concluding their terms, the following new members joined: Adolfo Martinez Palomo, Emeritus Investigator of the Center for Research and Advanced Studies of the IPN; Dr. Ruben Lisker Yourkowitzky, Director of Research of the National Institute of Medical Sciences and Nutrition Salvador Zubiran; Dr. Misael Uribe Esquivel, Chief of the Department of Gastroenterology of the National Institute of Medical Sciences and Nutrition Salvador Zubiran; and Dr. Teresita Corona Vazquez, General Director of the National Institute of Neurology and Neurosurgery Manuel Velasco Suarez. Later, in July 2009, Dr. Jose Narro Robles, Rector of UNAM, was invited to represent the educational sector as a substitute for Dr. Fernando Bilbao.

It is worth mentioning that according to Article 16 of the cited law, this Governing Board had the following non-delegable duties conferred by the Federal Law of Public Enterprises: (1) approve the distribution of the definitive annual budget of the entity and the investment programming, according to the total amount authorized by its budget; (2) approve budget adjustments for its programs, which does not imply changes in total authorized amount, investment resources, projects financed by external credit, or in the fulfillment of the institutional objectives and goals; (3) establish guidelines for the application of self-generated resources; (4) authorize the use of spaces in the areas and facilities of the Institute that are not for hospital use; (5) approve and modify the basic structure of the institution according to the total budget authorized for personal services; (6) establish a system for continuing professional education for the personnel of the Institute; (7) determine the rules and the percentages of additional salary for the personnel who participate in extramurally- funded research projects, and the royalties that result from the application or exploitation of industrial and intellectual property rights that are derived from projects performed by the Institute; and (8) evaluate and approve candidates for executive posts presented by the General Director. In parallel, according to Article 17, the Governing Board of the Institute carried out ordinary sessions twice each year in addition to the extraordinary sessions proposed by its Chair (Jiménez-Sánchez, 2009).

Based upon the above, the Governing Board of INMEGEN chose Dr. Gerardo Jimenez-Sanchez as the founding General Director of the Institute. He had served as Director of the Consortium for the Institute of Genomic Medicine. In addition, the 2004–2009 Work Program for the National Institute of Genomic Medicine (Jiménez-Sánchez, 2004a) was prepared by Dr. Jiménez-Sánchez as part of the selection process for INMEGEN's General Director.

THE BOARD OF TRUSTEES

The Board of Trustees was established according to Article 21 of the Law of the National Institutes of Health. According to Article 23 of the Law, this board comprised a president, a secretary, a treasurer, and the members, all of whom were recognized for their reputations, were part of the social and private sector or of the community in general, and who had a dedication to service. The operations of the Board of Trustees and the tenure of its members in their posts were established by corresponding internal rules of operation. It is worth mentioning that the posts of the members were honorific, and they did not receive remuneration, emolument, or compensation, according to Article 24 of the law mentioned above.

The Board of Trustees of INMEGEN assisted the Governing Board and the General Director through the development of the following functions, according to Article 25 of the cited law: (1) support the activities of the Institutes and formulate suggestions aimed at improving performance; (2) contribute to the procurement of resources that promote the fulfillment of the objectives of the Institutes; and (3) complete other functions indicated by the Governing Board.

Within this legal framework, in the first ordinary session of the Governing Board celebrated on March 30, 2005, the integration of the Board of Trustees of INMEGEN was authorized, according to the current regulations; the Board of Trustees was included as an advisory and consulting body, which had the goal of supporting the work program of INMEGEN, mainly by encouraging excellence in scientific research, by training highly specialized human resources, and by facilitating the link with the business sector that will allow the medical applications derived from genomic medicine research to be translated into goods and services that contribute to the healthcare of Mexicans.

In this context, in August 2004, the Secretary of Health, Dr. Julio Frenk Mora, president of the Governing Board of INMEGEN, invited Mr. Carlos Eduardo Represas de Almeida to join and preside over the Board of Trustees and to join the Governing Board of INMEGEN as a member. Mr. Represas has distinguished himself as a leading member of the Mexican Health Foundation, having served as Vice President of Finance of that organization, and as a member of the Board of Directors on various occasions. His performance as chair of the Board of Trustees was of great value in directing the activities of the organization to consistently benefit the institution and improve its performance.

The Board of Trustees of the Institute was composed of distinguished members of the business sector, whose work was fundamental in the support of INMEGEN's program of work. This collegial body was composed of the following founding members: Carlos Eduardo Represas de Almeida (Chair), Jorge Arevalo Chavez (Secretary), Emilio Azcarraga Jean, Luis German Carcoba García, Henry S. Davis, Pierre Froidevaux Chavan, Marcos Martinez Gavica, Ernesto Rubio del Cueto, Jaime Serra Puche, and Nina Zambrano.

This board contributed in a creative way to add value to the development of the activities of INMEGEN, through a work program related to the one proposed by the General Director and approved by the Governing Board (Jiménez-Sánchez, 2004a). The board established a trust fund to receive funds on behalf of INMEGEN, and also performed various activities aimed at supporting the development of the Institute. Among these, there was the creation of an informative video on INMEGEN and the establishment of strategic partnerships, for example with Nestlé, a company headquartered in Switzerland that develops scientific research of the highest level, including studies related to genomics and proteomics. This relationship led to the establishment of two scholarships and a Nestlé Chair of Nutrigenomics to promote the development of this field in Latin America.

ALIGNMENT WITH FEDERAL PROGRAMS

The creation and operation of INMEGEN contributed to the development of one of the areas of interest indicated by the National Health Program of 2001–2006 (PRONASA); the creation of INMEGEN was considered to be within the scope of the tenth strategy: "Strengthen investment in human resources, research and infrastructure in health." In particular, this project was related to the objectives stated in action line 10.4 of the National Health Program: "Strengthen the research and development of health technology."

Because of the federal goals stated above, it is important to mention that INMEGEN's 2004–2009 Work Program was in the context of the 2001–2006 National Health Program, as well as the Health Research Action Program. In this context, the creation of INMEGEN responded to the challenge of improving health research quantitatively and qualitatively. In particular, the strategy was intended to generate scientific innovation and to make the development of genomic medicine in Mexico a reality through a multidisciplinary approach and a focus on sectorial collaboration and inter-sector liaison. With the creation of INMEGEN, the national health system and its public institutions were strengthened by the promise of research in genomic medicine and training of human resources.

Furthermore, to contribute to the objectives of the 2007–2012 National Health Program, INMEGEN aligned its efforts with the following strategies: fortify and integrate actions that promote health and prevent diseases; promote investment in systems, information technologies, and communications that improve the efficiency and the integration of the sector; strengthen research and training in the healthcare field; and support the ability to offer health services through the development of the necessary infrastructure and equipment.

THE 2004–2009 WORK PROGRAM FOR INMEGEN

The Governing Board of INMEGEN approved the Work Program for the period of 2004–2009 that was proposed by Dr. Gerardo Jiménez-Sánchez as part of the selection process of the General Director of the Institute. In this program, the mission, vision, and objectives of INMEGEN were established, which included priorities for the initial development of the Institute and which were grouped into nine strategies.

The mission of INMEGEN is "to contribute to the healthcare of Mexicans by developing excellent scientific research and providing high-level training for human resources, which will lead to the medical applications of genomic knowledge through an innovative culture, cutting-edge technology, and strategic alliances that adhere to universal ethical principles."

The vision of the Institute is "to be the undisputed leader in Latin America and one of the main research hubs worldwide for developing genomic medicine, undertaking scientific research, training human resources, encouraging technological innovation, and developing goods and services. The most important values to INMEGEN are excellence, honesty, creativity, responsibility, institutional pride, loyalty, and respect."

The Work Program included nine specific strategies with 44 action lines and 125 goals to achieve during the first five years of INMEGEN's existence. The strategies were as follows: (1) organizational design: "the INMEGEN system"; (2) high-level scientific research in genomic medicine; (3) excellence of training and teaching in genomic medicine; (4) cutting-edge genomic technology applied to medicine; (5) establishment of the initial infrastructure; (6) development of strategic alliances for the integral development of genomic medicine; (7) translation of scientific knowledge into health goods and services; (8) compliance with the ethical, legal, and social frameworks of genomic medicine; and (9) administration of the research and teaching services.

Currently, INMEGEN (www.inmegen.gob.mx) is one of the twelve NIHs of Mexico. INMEGEN constitutes the cornerstone of the Mexican strategy to develop a national program in genomic medicine. The organization is designed to grow as an autonomous institution, linked to the educational and health sectors across the entire country. Its main funding source comes from federal fiscal funds, although it also receives support in the form of donations, both national and international. During its first three years of operation, the Mexican government assigned more than 125 million dollars for its initial operations and infrastructure. This included modern facilities located in Mexico City that consisted of high-technology genomic units for sequencing, genotyping and expression analysis, supercomputers, proteomics, and biomarker discovery laboratories. In addition, a research center for the ethical, legal, and social implications of genomic medicine was established, along with a center for the incubation of businesses and technology transfer. Very soon, INMEGEN will open its new 60,000m^2 facilities on the NIHs campus, in the south of Mexico City (Figure 99-1).

The scientific agenda of INMEGEN includes the understanding of the genomic structure of the Mexican population and the study of its most complex health problems, which include diabetes, obesity, cardiovascular diseases, infections, autoimmune diseases, macular degeneration associated with age, and tumors of the thyroid, breast, prostate, and blood. The majority of the projects involve the identification of genes associated with common disease risk and the biomarkers for their early diagnosis and prognosis. Furthermore, pharmacogenomics is an important part of the scientific agenda, given that strategies are being proposed to create more effective and less toxic medicines for the Mexican population (Seguin et al., 2007).

MAIN ACHIEVEMENTS OF INMEGEN DURING 2004–2009

During its first five years of operation, INMEGEN laid the foundation necessary to develop a national platform for genomic medicine in Mexico. The institute recruited a group of talented professionals who were dedicated to the construction of an innovative institution, and who shared an ambitious vision for the future. This vision was one of developing genomic medicine oriented to ease national health problems in the context of the programs established by the federal government.

The organizational design of INMEGEN was the product of careful planning aimed at developing a modern institution whose contributions would not only have a direct impact on public health, but would also be governed by ethical guidelines that guarantee adherence to universal ethical principles. Thus, the Institute established its Code of Ethics and Conduct, in addition to 34 internal normative manuals and instruments, which were incorporated into the daily activities of the personnel through more than ten seminars about organizational culture.

The main research areas considered in INMEGEN's 2004–2009 Work Program were population genetics, metabolic diseases (obesity and diabetes mellitus), cardiovascular diseases, cancer, infectious diseases, and pharmacogenomics. Over time, scientific progress generated the opportunity to broaden these areas of research. So, beginning in 2007, new areas of research were added, including medical proteomics, functional genomics for cancer, nutrigenomics, and genomic eye diseases. Currently, INMEGEN carries out genome-wide association studies (GWAS) on age-related macular degeneration, obesity, systemic lupus erythematosus, and cardiovascular diseases. INMEGEN also analyzes the expression of leukemias, sarcomas, and breast (Hidalgo-Miranda & Jiménez-Sánchez, 2009), lung, thyroid, and prostate cancers both in humans and in experimental mouse models

Figure 99-1 The National Institute of Genomic Medicine in Mexico City. Located in the Campus of the National Institutes of Health.

(Mendoza-Villanueva et al, 2008). In the area of medical proteomics, INMEGEN is working to discover biomarkers for melanoma, lung, and breast cancer through an analysis of the proteomic profiles of blood, saliva, and other body fluids.

ACADEMIC ACTIVITIES AND TECHNOLOGICAL INFRASTRUCTURE

The selection and recruitment process for the academic personnel of INMEGEN ensures that each member of the team can contribute to specific areas of medical genomics. By August of 2009, INMEGEN had recruited 31 researchers organized in nine laboratories, each of them with cutting-edge technology. They conducted 42 scientific research projects (Table 99-1), 11 of which received external financing. The results included over 55 peer-reviewed publications, training of human resources, and the first two National Congresses of Genomic Medicine in Latin America. In addition, a laboratory of genomic diagnosis was established to develop genomic services, initially those related to pharmacogenomics of oral anticoagulants (Acenocoumarol and Warfarin) and the antiretroviral Abacavir.

The high-technology units established at INMEGEN focused on supporting specific research and teaching programs of the Institute and other affiliated institutions. These units included the genomic sequencing and polymorphism analysis unit, powered by Affymetrix, Illumina, and Applied BioSystems platforms, the genomic expression unit, the medical proteomics and biomarker identification unit, and the validation area, which has the capacity to use tissue arrays for the simultaneous and comparative analysis of proteins. All of these units have access to the most advanced technology in genomic medicine.

The high-technology units were connected to a supercomputer and an information technology system, which had a capacity of over 2.2 million operations per second. This was done for the purpose of meeting the analysis requirements of the research projects. Additionally, a storage capacity of 22 terabytes was implemented, which was essential for storing the large quantity of genomic data generated by the various research projects. This infrastructure included computer and telecommunication networks to support the projects of both local scientists and others located in various institutions around the world.

Training of human resources increased rapidly, as various courses were formally integrated into undergraduate and graduate programs. Over 20 courses were offered during this period, attended by 455 students from across the country. Furthermore, more than 42 graduate students from various institutions of higher education carried out their scientific research projects in the laboratories of INMEGEN, obtaining their respective degrees.

INMEGEN hosted more than 130 academic events of the highest level and has been honored with the participation of distinguished international speakers. These events included the first and second National Congresses of Genomic Medicine. The participation of the medical, academic, and scientific community of Mexico and other parts of the world was overwhelming, with more than 20,700 individuals participating.

The initial facilities of INMEGEN included a Center for Information and Documentation (CID), which specialized in the different areas of genomic medicine. The CID collection contains more than 1100 books and subscriptions to 167 specialized scientific journals. Moreover, the innovative culture of the Institute led to the establishment of an electronic system to provide more than ten different electronic services to both the personnel of INMEGEN and interested parties in other geographic locations.

TABLE 99-1 SELECTION OF RESEARCH PROJECTS DEVELOPED AT INMEGEN, 2004–2009.

1. Genomic diversity of Mexican populations.
2. GWAS study for essential hypertension in the Mexican population.
3. Genomic structure of the angiotensinogen gene.
4. Genomic variability associated with age-related macular degeneration.
5. Analysis of proteomic profiles in the saliva of smokers with lung adenocarcinoma.
6. Resequencing analysis of genes related to drug metabolism.
7. Pharmacogenomics of Acenocoumarol and Warfarin.
8. Pharmacogenomics of Abacavir.
9. Analysis of maternal and paternal lineages with mitochondrial markers and of the Y-chromosome in the Mestizo population.
10. Genomic risk analysis of morbid obesity in the Mexican population.
11. The regulatory role of *CEMP1* during the process of in vitro cementogenesis/osteogenesis.
12. Identification of early biomarkers for diabetic nephropathy.
13. Genomic basis of autoimmune diseases among the Mexican population.
14. Identification of polymorphisms among genes associated with asthma.
15. Genetic risk factors for obesity and metabolic syndrome.
16. Allelic variants associated with the oxidative stress response in health and disease.
17. Genetic association with susceptibility to asthma and treatment response.
18. The Nrf2-Keap1 signaling pathway in the susceptibility to myeloid leukemia.
19. Early molecular biomarkers of liver cancer.
20. Global expression analysis in a three-dimensional culture model in different types of tumors.
21. Regulation of the expression of Smac/DIABLO.
22. Sequence analysis of the amyloidogenesis of immunoglobulin light chains.
23. Proteomic biomarkers associated with melanoma.
24. Proteomic analysis of breast cancer and identification of cancer biomarkers.
25. Proteomic profiling in gastric cancer.
26. Ethical, legal, and social implications of genomic medicine.
27. Pan-American Initiative in Bioethics.

Since its creation, one of the priorities of INMEGEN has been the development of scientific, academic, and business interactions with various organizations in Mexico and abroad. During 2005–2009, 48 cooperation agreements were signed (Jiménez-Sánchez, 2009). Such agreements have contributed to the development of a national platform of genomic medicine, thus creating synergies that will assure permanent world-class competitiveness.

The strengthening of the links with the academic sector continued in 2006, through academic and scientific collaboration with the University of Guanajuato, the Autonomous University of Tamaulipas, the Autonomous University of the State of Morelos, the National Council of Science and Technology, and the College of Postgraduates. In 2007, INMEGEN's interaction with the Institute of Anthropological Research of UNAM led to the project, "Lineages of mitochondrial DNA and of the chromosome in the Mestizo and indigenous Mexican population."

Important alliances were established with the departments of health in various states of Mexico, an activity that began at the end of 2004 through an interaction with the state of Zacatecas, which established a program for visits and rotations for students. Another collaboration with the state of Yucatan led to a pilot project to study cardiovascular diseases in the Mestizo-Mayan population.

In 2005 and 2006, collaboration agreements were formalized with the states of Yucatan, Veracruz, Sonora, Guerrero, Zacatecas, Guanajuato, and Tamaulipas. These states participated in the project to develop a genomic map of Mexican populations. They also stimulated information exchange related to scientific progress in genomic medicine, academic activities, technological services, and publications, as well as training of human resources and development of research projects.

The link with the industrial sector became an ideal space to explore nutrigenomics, one of the areas with the greatest potential in the coming years. During the second semester of 2006, INMEGEN´s interaction with Nestlé, S.A. de C.V. began. The following year, the relationship was formalized with the signing of a collaboration agreement for the establishment of the "Nestlé Chair of Nutrigenomics" and the "Nestlé Fellowships in Nutrigenomics." The purpose of these positions was to support scientific research and training of human resources in nutrigenomics at INMEGEN. The positions also served to foster new ideas that could contribute to improve healthcare of the Mexican population.

The inauguration of the Nestlé Chair in Nutrigenomics was planned for a period of three years, with the goal of recruiting a senior researcher capable of conducting research on fundamental topics related to improving the health status of Mexicans through nutrition and genomics. Furthermore, two masters, doctoral, or postdoctoral positions were created under the name, "Nestlé Fellowship in Nutrigenomics." These positions were designed to encourage young Mexicans to develop proficiency in the area of nutrigenomics.

The selection process for the awardees of those fellowships began with the publication of a call for applications in the last quarter of 2007; three applications for study in nutrigenomics and two applications for the position of Chair of Nutrigenomics were received. By 2008, a review process was established by an ad-hoc group. Currently, Dr. Elizabeth Tejero holds the Chair in nutrigenomics.

The links with foreign institutions were of great importance, and the first contacts with organizations outside of Mexico began in 2004. These international organizations included the National Institutes of Health of France (INSERM), the National Institutes of Health in the United States, Johns Hopkins University, and the Broad Institute of Harvard University and the Massachusetts Institute of Technology. Vanderbilt University had an important role in developing collaborative efforts in the areas of ethical, legal, and social implications of genomic medicine, as well as in the areas of bioinformatics and pharmacogenomics.

In 2005, strategic alliances were established with the Translational Genomics Research Institute (TGen), under the direction of Dr. Jeff Trent, which turned into specific projects of academic and scientific collaboration. Furthermore, beginning on December 20, 2005, INMEGEN joined the "Public Population Projects in Genomics" (P3G) under the leadership of Professor Bartha Knoppers, a nonprofit international consortium with the main objective of promoting collaboration among researchers in population genomics through the creation of a public, open, accessible and transparent database.

In 2006, a collaborative agreement was established with the Interdisciplinary Center on Bioethics of the Pan-American Health Organization/World Health Organization for the purpose of training human resources, developing collaborative research projects on ethical aspects of genomic medicine, and exchange of information. Several publications resulted from this effort, particularly in the area of ethical, legal, and social aspects of genomic medicine.

In 2007, an academic and scientific collaboration agreement was signed with the Genome Institute of Singapore, represented by Dr. Edison T. Liu, Executive Director of the Institute. The purpose of this agreement was to strengthen and broaden cooperation efforts to establish research programs and other collaborative projects. Through this link, both institutions developed research activities in genomics, pharmacogenomics, educational opportunities, and ethical, legal, and social matters related to genomic medicine.

During the same year, INMEGEN's international collaborations were particularly productive. A collaboration with the United Nations University of Biotechnology for Latin America and the Caribbean was established. This agreement included the granting of funds to INMEGEN for the design and delivery of the course "Ethical, Legal and Social Matters in Genomic Medicine," which was the first of its kind in Latin America. Similarly, as a result of the agreement between the *Centre de Recherche en Droit Public* of the University of Montreal and INMEGEN, the HumGen Spanish portal (http://www.humgen.umontreal.ca/int/partenaires.cfm?&lang=3) was launched. This portal serves as a Spanish-language tool to explore the ethical, legal, and social aspects of genomic medicine. It could also serve as a communication bridge between Canada and Latin America on legal and social matters related to genomic medicine.

In addition, INMEGEN organized an "International Meeting of Leaders in Genomic Medicine: Emerging Regulatory Aspects" in collaboration with the Drug Information Association (DIA). The event had attendees and speakers from Mexico, Latin America, the United States, Canada, Switzerland, India, Singapore, Belgium, and Japan, among other countries.

INMEGEN created a Center for Research on the Ethical, Legal, and Social Aspects of Genomic Medicine, which brought together researchers from different institutions who together developed scientific studies on new topics in this field. The Intellectual Property Program was also developed, and a business incubator called ProGen was established (http://progen.inmegen.gob.mx). The work of ProGen was tightly linked to INMEGEN research projects and also included matters of copyright, associated with multiple brands and characters in comic books elaborated by INMEGEN.

The National Institute of Genomic Medicine has played a fundamental role in the dissemination of information about genomic medicine, and it has also earned national and international recognition. The methods used to disseminate information turned out to be very effective. Among them, the design of three widely distributed comics on genomic medicine stand out, as well as the design of a bilingual web portal that has become a key site to spread genomic medicine information. Over 53 million successful consultations were made through INMEGEN's portal between 2004 and 2009, and more than 45 million documents were downloaded by users from 140 countries.

In addition to INMEGEN's initial facilities, the construction of its permanent headquarters is now 80% complete (Figure 99-1). This new facility will provide a 17-fold increase in space and technological capacities that will be essential for the next phase of Mexico's construction of a national platform of genomic medicine (Hardy et al., 2008).

The results from INMEGEN's first administration were released to the public in 2009 in a four-volume report entitled: "Report on the 2004–2009 Activities of the National Institute of Genomic Medicine" (Jiménez-Sánchez, 2009). This documents reports the full implementation of INMEGEN's Work Program and was approved by the Governing Board of the Institute (Jiménez-Sánchez, 2004a).

THE MEXICAN GENOME DIVERSITY PROJECT

The characterization of the genetic structure in the Mexican population was one of the strategic areas of research at INMEGEN. Such information will be an essential tool for the development of genomic medicine for the Mexican population. The development of this line of research is of special relevance in the context of the initial phases of the International HapMap Project, which did not include any Latin American populations (HapMap, 2005).

Mexican Mestizos constitute a recently admixed population, mostly composed of European, Amerindian, and African lineages. The genetic heterogeneity of Mexicans is derived from a range of different demographic dynamics in geographically distinct regions. The first stage of the project included the genotyping of 110,356 SNPs from 300 unrelated individuals who identified themselves as Mestizos and originated from six geographically distant regions of Mexico (Silva-Zolezzi et al., 2009). Next, the genetic diversity, the patterns of linkage disequilibrium (LD), and the extent of common haplotypes were determined using data from these individuals along with information from the HapMap. The initial results of the project indicated that, even when there are regional genetic differences among Mexican subpopulations, these groups are sufficiently similar to be analyzed as one single group. However, the results of this study provided evidence of the population structure of Mexicans that should be considered when designing and analyzing genomic association to human disease. The initial analysis used a set of 2824 informative markers for ancestry derived from three populations of the HapMap. The analysis indicated that there are different proportions of mixes among Mestizos, and it showed a fourth ancestral component that is present in different proportions and that correspond to the Amerindian contribution (Silva-Zolezzi et al., 2009). The results of this project indicate that a haplotype map of Mexican Mestizos would improve the selection of tag-SNPs and make it possible to better capture common genetic variations in studies on the associations of diseases common among the Mexican population. This is the first genome-wide genotyping effort of a recently admixed Latin American population that is available to the general public (http://diversity.inmegen.gob.mx).

To generate a more comprehensive description of the common genetic variations and to adequately describe the genomic structure of these populations, the SNP density will be increased to approximately 1.5 million. In addition, the ensuing analysis will include Amerindian populations. The results derived from these efforts will provide the foundation for translating knowledge of the genetic structure of the Mexican population into a better understanding of common complex diseases. Additional studies on the genomic structure will include the systematic analysis of copy-number variations in the Mestizo and Amerindian populations.

The initial results of this project suggest that the genetic differences among Mestizos of various regions of Mexico are mainly due to differences in the ancestral contributions of European and Amerindian populations. Comparative analysis of the proportion of these contributions among the participating states demonstrated important differences among some of them. In the majority of the analysis, the samples of the central regions had a higher similarity to indigenous groups, while the samples from the northern regions were more similar to European populations. These findings are supported by the current and historical Amerindian population densities in these regions (Silva-Zolezzi et al., 2009).

An analysis of the shared alleles and haplotypes was performed to evaluate the degree of genetic variability that is shared among samples of Mexican Mestizos and the populations of the international HapMap project (Europeans, Africans, Chinese, and Japanese). The results indicated that Mexicans share 64% of the common haplotypes (present in more than 5% of the samples) with Africans, 75% with the population of Asia, and 80% with the populations of northern Europe. When the information is combined with the four populations from HapMap, the percentage of shared haplotypes rose to 96%. These results indicate that there is a percentage of genetic variability present only among Mexicans, and that this variability is derived from the Amerindian component of the Mestizo population. On the other hand, the results of the comparison among Mestizo populations indicate that a specific haplotype map for them would allow a better selection and would reduce the number of genetic markers necessary for association studies of complex diseases, making these studies more efficient and less expensive.

The analysis of private alleles was performed for the purpose of identifying genetic variants exclusive to Mexican populations that had a frequency greater than 5%, and that were absent from the populations examined in the initial phases of the HapMap Project (African, European, and Asian). Eighty-nine markers were identified, only in some of the Mexican populations analyzed. In addition, 86 of these markers were also found in the samples of the Zapoteca population, which indicates their Amerindian origin (Silva-Zolezzi et al., 2009).

The initial results of the project demonstrated the feasibility of generating a public genomic database of the Mexican population and a catalog of the most common variants (http://diversity.inmegen.gob.mx). This public resource will become an important source of information for the design of genomic studies aimed at finding genes associated with common diseases, not only in Mexico but also in many parts of Latin America.

The Mexican Genome Diversity Project does not include direct medical applications, but rather was designed as a public tool that will lay the foundation for clinical studies on genomic medicine that will be performed by researchers across the country and in other parts of the world. In addition, it will allow new areas of research to be explored that will help identify genetic variations associated with a predisposition to common diseases, and how individuals with those genes respond to medications. This project was funded with public resources assigned to INMEGEN, and developed with important donations from the Mexican Health Foundation and the Gonzalo Rio Arronte Foundation.

The Mexican Genome Diversity Project has represented important progress for genomic sciences in Mexico. In addition, it has stimulated the recruitment of a critical mass of researchers in various disciplines. Furthermore, the technological infrastructure created by this project has become a valuable asset. These accomplished milestones, combined with our highly qualified scientists and the strategic liaisons established by INMEGEN, will contribute to develop a robust national platform in genomic medicine.

CURRENT AND FUTURE CHALLENGES

Mexico has invested significant efforts and resources to develop genomic medicine as a means to improve healthcare for the Mexican population. The results obtained in the first ten years of work have exceeded anyone's expectations in establishing the foundations of a mid-term national project.

Developing genomic medicine in Mexico to the point where it can perform at a leadership level and truly improve the healthcare of the Mexican population requires more than just solid scientific work. This ambitious goal will demand several cultural changes, including serious team efforts from medical clinics and scholars of the basic sciences, and a focus on common objectives. These collaborative efforts will be needed to address national health problems such as diabetes mellitus or breast cancer. It is likely that disarticulated projects will not be as successful with problems of this magnitude. It is important, therefore, that the Mexican model fosters a real component of innovation. In general, it will be the younger generations that will better deal with these challenges.

There are other challenges, including the need for new financing mechanisms, new incentives to innovate, and new ways to register intellectual property. The current model dominating scientific research has a significant impact on the scientists' income and creates a legitimate urgency to publish results. Although it has been proven that such systems increase the number and quality of scientific publications, they do not stimulate innovation in areas in which the risk of failure is significantly greater. In those areas, results are usually published after registration of the intellectual property. In recent years, the Mexican government has established programs to stimulate innovation and to generate new business models based on scientific knowledge. These programs, located within CONACYT, reflect the awareness within the federal government of the strategic importance of innovation for economic growth. INMEGEN has established a synergy with CONACYT and the Mexican Institute of Industrial Property (www.impi.org.mx). The goal of this synergy is to develop mechanisms to incorporate innovation as a part of the local community, particularly in genomic medicine. Although there may be a number of genomic projects that will be developed by Mexican scientists alone, many genomic projects are generally the result of international collaborations, whereby Mexican scientists send samples to countries that have a more solid infrastructure. The goal of this collaboration is to combine forces and accelerate their research (Table 99-2A & 99-2B).

Even though international collaboration is desirable and encouraged, as the Mexican groups increase their critical scientific and technological mass, the collaborations tend toward a more equitable level of contribution, and groups are increasingly embarking on more independent research. Even though training programs in genomic medicine are recent, the number of applicants has significantly increased from dozens to hundreds at INMEGEN and the LCG. This clearly indicates that there is an opportunity to host new training programs in genomic sciences that respond to the needs of the current generations and contribute to the development of genomic medicine in Mexico. Unfortunately, "brain drain" continues to be a significant problem, mainly because of the lack of competitive salaries, the appropriate infrastructure, and funding opportunities for research throughout the country. The challenge of genomic medicine includes the recruitment and repatriation of qualified professionals, and the development of means to retain them for long enough to successfully develop their scientific contributions. In recent years, this situation has improved thanks to better infrastructure and consistent research programs that attract young researchers and bring them back to Mexico. CONACYT is facing this challenge by implementing a repatriation program, and a program aimed at retaining young scientists in Mexico. As a result, between 2001 and 2005, a total of 845 scientists were repatriated, some of which joined our genomic medicine program.

The limitations of sustained funding represent a challenge for science worldwide. Mexico's investment in research and development (R&D) is the lowest among member states of the Organization for Economic Cooperation and Development (OECD) and is equivalent to approximately a seventh of the average funding level of a member state of the OECD. It is significantly lower

TABLE 99-2A COLLABORATIVE AGREEMENTS SIGNED WITH INTERNATIONAL INSTITUTIONS 2004-2009

1. Translational Research Institute, Phoenix, AZ, USA
2. Bioethics Unit of the Pan American Health Organization – WHO.
3. Vanderbilt University, USA
4. McGill University, Genome Quebec, Canada
5. The State University of New York (SUNY), USA
6. United Nations University, UN
7. Genome Institute of Singapore, Singapore
8. Alcon Laboratories, USA
9. Centre de Recherche en Droit Public, Université de Montreal, Canada
10. Drug Information Association (DIA) USA
11. Foundation for the Popularization of Science and Technology and Institute of Science and Technology, Salamanca University, Spain

TABLE 99-2B COLLABORATIVE AGREEMENTS SIGNED WITH INSTITUTIONS FROM MEXICO 2004-2009

1. Yucatan Health Services; Autonomous University of Yucatan and FUNSALUD "Peninsular Chapter."
2. Zacatecas Health Services and Autonomous University of Zacatecas.
3. National Council of Science and Technology, MUO.
4. Sonora Health Services, the Autonomous University of Sonora and FUNSALUD.
5. Veracruz State Government and Veracruz University.
6. Ministry of Health of the State of Guerrero and Autonomous University of Guerrero.
7. National Institute of Public Health.
8. The State's Employees Social Security and Social Services Institute (ISSSTE).
9. National Institute of Public Health, Specific Agreement.
10. Guanajuato Health Services and Autonomous University of Guanajuato State.
11. Autonomous University of Morelos State.
12. Mexican Association of Human Genetics, MUO.
13. Mexican Association of Human Genetics, Specific Agreement.
14. Mexican Institute of Industrial Property.
15. "Gonzalo Río Arronte" Foundation.
16. Mexican Health Foundation, MUO.
17. Post-Graduate College (Agricultural sciences).
18. National Council of Science and Technology, HR Specific Agreement.
19. Tamaulipas Health Services and Autonomous University of Tamaulipas.
20. National Commission on Social Health Protection.
21. NESTLÉ, Mexico, MUO.
22. Ministry of Health of the State of Oaxaca; Autonomous University of Oaxaca "Benito Juárez" and Southeast Regional University.
23. Ministry of Health, Health Services of the State of Durango and State University of Durango.
24. Institute of Ophthalmology Conde de Valenciana Foundation.
25. Ministry of Health of the State of Campeche and Autonomous University of Campeche.
26. PROA Group Laboratories.
27. Federal District Government.
28. National Institute of Psychiatry "Ramón de la Fuente Muñiz."
29. National Autonomous University of Mexico.
30. Mexican Foundation for Promoting Education for Timely Prevention of Breast Cancer.
31. Genomi-k Laboratories.
32. Institute for Anthropological Research.
33. University Teaching and Research (Tec Milenio University).
34. Children's Interactive Museum (Papalote).
35. Hospital Juarez of Mexico.
36. Medix Products.
37. Nestlé México S.A.

than that of other emerging economies, such as China (0.7%), India (0.8%), or Brazil (>0.8%; OECD, 2006; Hardy et al., 2008). For comparison with other countries in Latin America, in 2004, Mexico invested 0.41% of its GDP in R&D, while Argentina invested 0.44%, Chile invested 0.68% and Brazil invested 0.91%. This represents a significant challenge for the successful development of a nascent field such as genomic medicine in Mexico. The Mexican government has not yet committed itself to investing in this promising field, but it is essential that the R&D situation be improved among public institutions in Mexico and that the technological infrastructure be kept updated. To reduce reliance on the federal government, various initiatives have been explored that provide additional support for genomic medicine. These include competition for international funding opportunities and increased private sector participation in R&D. Currently, around 35% of the spending on science and technology research in Mexico comes from the industrial sector (OECD, 2006). Even though this percentage is one of the lowest among member countries of the OECD, strategic alliances are being established between the industrial sector and public research institutions for specific projects, including those related to genomic medicine.

There are other challenges that play an important role in the successful development of genomic medicine. Some of these challenges involve the importation of equipment and chemical reagents from other countries. This is particularly important for the generation of new genomic technologies, because acquiring them involves significant costs from customs duties and tariffs, which must be paid for by the research budget. In addition, the excessive regulations and lengthy customs procedures damage many pieces of equipment and cause unfortunate delays during research projects.

As genomic medicine develops in Mexico, the need for modern legislation addressing both the ethical and social implications of genomic medicine will also increase. Questions of discrimination, confidentiality, equality in access to medical services, and financial and labor implications, among others, will require regulations based on Mexican laws.

We have learned valuable lessons during the planning and implementation of this strategy (Seguin et al., 2008). Some of them may be useful for similar emerging economies interested in developing genomic medicine. Careful planning is needed, along with the participation of the main organizations in the fields of health and education. Scientific and philanthropic organizations must work together on an equal level, which requires solid leadership at the highest levels to coordinate all of the efforts. It is important to determine the degree of academic, scientific, financial, and political capacity available before trying to establish feasible goals. It is also useful to take advantage of the local experiences of the past cases of both success and failure to design a more effective effort, to focus on ambitious but realistic goals, and to define specific challenges. These goals should include the development of programs for training, the development of infrastructure, and the selection of alliances. In addition, it is desirable that scientific research includes a component to translate knowledge into products and services, so that the programs are more attractive for public–private investment and contribute to the knowledge economy. Despite the large number of challenges ahead, Mexico has successfully begun to develop an expertise in genomic medicine. There is still much to do, but by carefully planning the commitments of the different actors in Mexican society, and by forming solid alliances, genomic medicine in Mexico will flourish and benefit the population through improved healthcare. Additionally, the country's scientific work, infrastructure, and commitment to innovation will have an effect throughout Latin America and will allow Mexico to participate in the worldwide transition toward a knowledge-based economy.

REFERENCES

Aguilar-Diaz H, et al. (2006). The genome project of Taenia soliu m. *Parasitol Int* 55(Suppl):S127–S130.

Baca V, et al. (2006). Association analysis of the PTPN22 gene in childhood-onset systemic lupus erythematosus in Mexican population. *Genes Immun* 7:693–695.

Bekker-Mendez C, et al. (1998). Haplotype distribution of class II MHC genes in Mexican patients with systemic lupus erythematosus. *Scand J Rheumatol* 27:373–376.

Blattner FR, et al. (1997). The complete genome sequence of Escherichia coli K-12. *Science* 277:1453–1474.

Childs B, Valle D, Jiménez-Sánchez G. (2001). The inborn error and biochemical individuality. In: Scriver CR, Beaudet AL, Sly WS, Valle D, eds. *The Metabolic and Molecular Basis of Inherited Disease*, Vol. I, Chapter 4, pp. 155–166. New York: McGraw Hill.

Contreras AV, et al. (2011). Resequencing, haplotype construction and identification of novel variants of CYP2D6 in Mexican Mestizos. *Pharmacogenomics* 12(5):745–756.

Easton DF, et al. (2007). Genome-wide association study identifies novel breast cancer susceptibility loci. *Nature* 447(7148):1087–1093.

Emilien G, Ponchon M, Caldas C, Isacson O, Maloteaux JM. (2000). Impact of genomics on drug discovery and clinical medicine. *QJM* 93:391–423.

Evans WE, Relling MV. (1999). Pharmacogenomics: Translating functional genomics into rational therapeutics. *Science* 286:487–491.

Fraga A, et al. (1979). HLA-B27 in Mexican patients with ankylosing spondylitis. *Arthritis Rheum* 22:302.

Gerhard P. (1986). *Geografia historica de la Nueva España, 1519—1821*. Mexico City: Universidad Nacional Autónoma de México (UNAM).

Gerhard P. (1991). *La frontera sureste de la Nueva España*. Mexico City: Universidad Nacional Autónoma de México (UNAM).

Gonzalez Burchard E, et al. (2005). Latino populations: A unique opportunity for the study of race, genetics, and social environment in epidemiological research. *Am J Public Health* 95:2161–2168.

Gonzalez V, et al. (2006). The partitioned Rhizobium etli genome: Genetic and metabolic redundancy in seven interacting replicons. *Proc Natl Acad Sci U S A* 103:3834–3839.

Gorodezky C, et al. (1995). [MHC-dependent molecular mechanisms of susceptibility and protection in type I diabetes in Mexicans]. *Gac Med Mex* 131:395–403.

HapMap [The International HapMap Consortium]. (2005). A haplotype map of the human genome. *Nature* 437:1299–1320.

Hardy BJ, et al. (2008). The next steps for genomic medicine: Challenges and opportunities for the developing world. *Nat Rev Genet* 9(Suppl 1):S23–S27.

Hidalgo A, et al. (2005). Microarray comparative genomic hybridization detection of chromosomal imbalances in uterine cervix carcinoma. *BMC Cancer* 5:77.

Hidalgo-Miranga A, Jiménez-Sánchez G. (2009). Genomic basis for breast cancer: Advances in personalized medicine. *Salud Publica Mex* 51(Suppl 2):S197–S207.

Iafrate AJ, et al. (2004). Detection of large-scale variation in the human genome. *Nat Genet* 36:949–951.

INEGI (I.N.d.E.G.e.I.). (2005). II Conteo de poblacion y vivienda 2005. *Mexico.*

Jiménez-Sánchez G. (1999). Implicaciones Médicas y Sociales del Genoma Humano en la Sociedad Mexicana. In: Musik-Asali GA, Medina-Gonzalez S, eds. *México 2020: Retos y perspectivas.* Mexico DF: CONACyT.

Jiménez-Sánchez G. (2002). *Hacia el Instituto Nacional de Medicina Genómica. Consorcio Promotor del Instituto de Medicina Genómica. Informe de Actividades 2002.* Mexico: [2002 Annual Report of the Consortium for the Institute of Genomic Medicine], ISBN 968-5018-77-5.

Jiménez-Sánchez G. (2003). Developing a platform for genomic medicine in Mexico. *Science* 300:295–296.

Jiménez-Sánchez G. (2004a). *Programa de Trabajo para dirigir el Instituto Nacional de Medicina Genómica 2004—2009.* Mexico DF: Fundación Mexicana para la Salud.

Jiménez-Sánchez G. (2004b). *Un paso mas hacia el establecimiento del INMEGEN.* Mexico: Fundación Mexicana para la Salud A.C.

Jiménez-Sánchez G. (2009). *Informe de Actividades 2004—2009 del Instituto Nacional de Medicina Genómica.* Mexico City: INMEGEN®.

Jiménez-Sánchez G, Childs B, Valle D. (2001a). Human disease genes. *Nature* 409:853–855.

Jiménez-Sánchez G, Childs B, Valle D. (2001b). The effect of Mendelian Disease on Human Health. In: Scriver CR, Beaudet AL, Sly WS, Valle D, eds. *The Metabolic and Molecular Basis of Inherited Disease,* Vol. I, Chapter 4, pp. 167–174. New York: McGraw Hill.

Jiménez-Sánchez G, Lara Alvarez C. (2007). Retos bioéticos de la medicina genómica. In: Brena-Sesma IC, ed. *Panorama Internacional en Salud y Derecho. Culturas y Sistemas Jurídicos Comparados*, pp. 391–410. Mexico City: Instituto de Investigaciones Jurídicas. UNAM.

Jiménez-Sánchez G, Silva-Zolezzi I, Hidalgo A, March S. (2008). Genomic medicine in Mexico: Initial steps and the road ahead. *Genome Res* 18:1191–1198.

Jiménez-Sánchez G, Valdes Olmedo J, Soberon G. (2002a). Desarrollo de la medicina genómica en México. *Este pais* 139:17–23.

Jiménez-Sánchez G, Valdes Olmedo J, Soberon G. (2002b). El Instituto Nacional de Medicina Genómica [The National Institute of Genomic Medicine]. *Este pais* 141:50–56.

Jiménez-Sánchez G, Valdes Olmedo J, Soberon G. (2002b). En el umbral de la medicina genómica. *Este pais* 138:21–30.

Kaiser J. (2008). DNA sequencing. A plan to capture human diversity in 1000 genomes. *Science* 319:395.

Kofman-Alfaro S, et al. (1991). La genética humana en México. *Gac Med Méx* 127:85–103.

Lacasana-Navarro M, Galvan-Portillo M, Chen J, Lopez-Cervantes M, Lopez-Carrillo L. (2006). Methylenetetrahydrofolate reductase 677C>T polymorphism and gastric cancer susceptibility in Mexico. *Eur J Cancer* 42:528–533.

Lander ES, et al. (2001). Initial sequencing and analysis of the human genome. *Nature* 409:860–921.

Lisker R, Carnevale A. (1995). Diagnóstico y perspectivas de la medicina. In: Fortes M, Gomez CW, eds. *Retos y perspectivas de la ciencia en México*, pp. 142–154. México: Academia de la Investigación Científica.

Lopez M, Guerrero J, Jung-Cook H, Alonso ME. (2005). CYP2D6 genotype and phenotype determination in a Mexican Mestizo population. *Eur J Clin Pharmacol* 61:749–754.

Mathew CG. (2008). New links to the pathogenesis of Crohn disease provided by genome-wide association scans. *Nat Rev Genet* 9:9–14.

McCarthy MI. (2010). Genomics, type 2 diabetes, and obesity. *N Engl J Med* 363:2339–2350.

Mendoza-Villanueva D, et al. (2008). Gene expression profile of cervical and skin tissues from human papillomavirus type 16 E6 transgenic mice. *BMC Cancer* 8:347.

Muñoz de Alba MC, ed. (2002). *Reflexiones en torno al derecho genómico.* Mexico DF: Instituto de Investigaciones Jurídicas, UNAM.

Munoz JB, et al. (2007). Methylenetetrahydrofolate reductase gene polymorphisms and the risk of anencephaly in Mexico. *Mol Hum Reprod* 13:419–424.

OECD. (2006). Main science and technology indicators 2006-2. http://www.oecd-ilibrary.org/science-and-technology/main-science-and-technology-indicators/volume-2006/issue-2_msti-v2006-2-en-fr.

Olaiz G. (2006). *Encuesta Nacional de Salud y Nutrición 2006.* Mexico DF: Instituto Nacional de Salud Pública y Secretaría de Salud.

Palacios R, Collado-Vides J. (2007). Development of genomic sciences in Mexico: A good start and a long way to go. *PLoS Comput Biol* 3:1670–1673.

Parra-Rojas I, Ruiz-Madrigal B, Martinez-Lopez E, Panduro A. (2006). Influence of the -308 TNF-alpha and -174 IL-6 polymorphisms on lipid profile in Mexican subjects. *Hereditas* 143:167–172.

Perez-Luque E, et al. (2003). Protective effect of DRB1 locus against type 2 diabetes mellitus in Mexican Mestizos. *Hum Immunol* 64:110–118.

Prokunina L, et al. (2002). A regulatory polymorphism in PDCD1 is associated with susceptibility to systemic lupus erythematosus in humans. *Nat Genet* 32:666–669.

Sadée W. (1989). Pharmacogenomics. *BJM* 319:1286–1289.

Salamanca F, Armendares S. (1995). The development of human genetics in Mexico. *Arch Med Res* 26(Spec No):S55–S62.

Sanchez-Corona J, et al. (2004). Polymorphisms in candidate genes for type 2 diabetes mellitus in a Mexican population with metabolic syndrome findings. *Diabetes Res Clin Pract* 63:47–55.

Seguin B, Essajee S, Jiménez-Sánchez G, Singer P, Daar A. (2007). Human Genomic Variation Studies and Pharmacogenomics are Critical for Global Health. In: Suarez-Kurtz G, ed. *Pharmacogenomics in Admixed Populations.* Bandes Bioscience.

Seguin B, Hardy BJ, Singer PA, Daar AS. (2008). Genomics, public health and developing countries: The case of the Mexican National Institute of Genomic Medicine (INMEGEN). *Nat Rev Genet* 9(Suppl 1):S5–S9.

Silva-Zolezzi I, et al. (2009). Analysis of genomic diversity in Mexican Mestizo populations to develop genomic medicine in Mexico. *Proc Natl Acad Sci U S A* 106:8611–8616.

Sosa-Macias M, et al. (2006). CYP2D6 genotype and phenotype in Amerindians of Tepehuano origin and Mestizos of Durango, Mexico. *J Clin Pharmacol* 46:527–536.

Torres-Sanchez L, et al. (2006). Dietary and genetic determinants of homocysteine levels among Mexican women of reproductive age. *Eur J Clin Nutr* 60:691–697.

Triunfol, ML. (2007). Latin American science moves into the spotlight. *Cell* 131:1213–1216.

Valladares A, et al. (2006). Genetic expression profiles and chromosomal alterations in sporadic breast cancer in Mexican women. *Cancer Genet Cytogenet* 170:147–151.

Vargas-Alarcon G, Granados J, Ibanez de Kasep G, Alcocer-Varela J, Alarcon-Segovia D. (1995). Association of HLA-DR5 (DR11) with systemic sclerosis (scleroderma) in Mexican patients. *Clin Exp Rheumatol* 13:11–16.

Vazquez-Ortiz G, et al. (2007). Differentially expressed genes between high-risk human papillomavirus types in human cervical cancer cells. *Int J Gynecol Cancer* 17:484–491.

Velazquez-Cruz R, et al. (2007). Association of PDCD1 polymorphisms with childhood-onset systemic lupus erythematosus. *Eur J Hum Genet* 15:336–341.

Venter JC, et al. (2001). The sequence of the human genome. *Science* 291:1304–1351.

Villarreal-Molina MT, et al. (2007). The ABCA1 R230C Variant Affects HDL-cholesterol levels and body mass index in the Mexican population: Association with obesity and obesity-related comorbidities. *Diabetes* 56:1881–1887.

Wakeley J, Nielsen R, Liu-Cordero SN, Ardlie K. (2001). The discovery of single-nucleotide polymorphisms – and inferences about human demographic history. *Am J Hum Genet* 69:1332–1347.

Weissglas-Volkov D, et al. (2006). Common hepatic nuclear factor-4-alpha variants are associated with high serum lipid levels and the metabolic syndrome. *Diabetes* 55:1970–1977.

Wellcome Trust Case Control Consortium. (2007). Genome-wide association study of 14,000 cases of seven common diseases and 3,000 shared controls. *Nature* 447:661–678.

WHO. (2002). *Genomics and world health. Report of the Advisory Committee on Health Research.* Geneva: World Health Organization.

Zanke BW, et al. (2007). Genome-wide association scan identifies a colorectal cancer susceptibility locus on chromosome 8q24. *Nat Genet* 39:989–994.

100.

ANTHROPOLOGICAL AND MEDICAL IMPLICATIONS OF GENETIC ADMIXTURE IN THE MEXICAN MESTIZO POPULATION

Ricardo M. Cerda-Flores and Augusto Rojas-Martínez

INTRODUCTION

Common complex diseases like cardiovascular disorders, cancer, asthma, diabetes, high blood pressure, schizophrenia, cleft lip and palate, etc., are the main focus of current genomic and epidemiological research. For most of these illnesses, ethnic background is suspected to play an important etiopathogenic role. Genome-wide association studies have been conducted to find genetic sequences or genes associated to several of these multifactorial disorders, but most of them have found a small effect size and too low a heritability to be valuable for genetic probability testing (Holtzman & Marteau, 2000). It is clear that in the near future, better planning of translational-focused genomics research and better bioinformatic approaches will improve our knowledge of gene–gene and gene–environment interactions, and genome tests will be better able to predict clinical outcomes (Hawkins et al., 2009). Most of these genome analyses have been conducted in subjects from particular ethnic groups, mainly Caucasians, Africans, and East Asians, for whom specific and defined panels of polymorphic markers have been developed after the conclusion of the HapMap project (http://hapmap.ncbi.nlm.nih.gov), but conducting studies in populations with ethnic admixtures represents a challenge, as well an opportunity to understand the genetic contributions to mixed racial phenotypes. Because of their recent history, the admixed populations of Latin America constitute an interesting case study for clinical genomics research.

Latin America contributes to about 8.3% of the global population (www.nationsonline.org/oneworld/world_population.htm). and is an active, growing population, not just in the subcontinent but in industrialized countries like the United States, Canada, and some European countries. The main ethnic population of this area is currently known as "Hispanic" in the medical literature, a very confusing word to describe a mixed population descending from at least three main parental groups: Native Americans, Caucasians from Spain, and Africans. In its most precise definition, *Hispanic* means native or belonging to Spain, but the term is used to define Spanish speakers, or subjects with Spanish-derived last names, although this can include Filipinos, a group clearly different from the groups populating Latin America. The Medical Subject Headings of the US National Library of Medicine (MeSH) define Hispanic Americans as: "any person living in the United States of Mexican (Mexican Americans), Puerto Rican, Cuban, Central or South American, or other Spanish culture or origin. The concept does not include Brazilian Americans or Portuguese Americans." (http://www.ncbi.nlm.nih.gov/mesh?term=hispanic). This concept was introduced in this index in 1978, probably based on a former concept of the US Census Bureau, which recently states that:

> US federal government agencies must adhere to standards issued by the Office of Management and Budget (OMB) in October 1997, which specify that race and Hispanic origin (also known as ethnicity) are two separate and distinct concepts. These standards generally reflect a social definition of race and ethnicity recognized in this country and they do not conform to any biological, anthropological, or genetic criteria. The standards include five minimum categories for data on race: "American Indian or Alaska Native," "Asian,'" "Black or African American," "Native Hawaiian or Other Pacific Islander," and "White." There are two minimum categories for data on ethnicity: "Hispanic or Latino" and "Not Hispanic or Latino." The concept of race reflects self-identification by people according to the race or races with which they most closely identify. (http://www.census.gov)

As reflected, *Hispanic* is still a vague, ill-defined concept, even for two agencies of the same government; but in the world medical literature, this definition has become operational in clinical and biomedical research. Although debatable, it may be preferable to use the Spanish term *Mestizo*, which describes the descendents of mating in the above described parental ethnic groups, to conclude that a large proportion of the Latin American population is constituted of Mestizos.

Proportions of original ethnic groups that arrived in the Americas and Mestizos differ in each Latin American country, and even within different regions inside some countries. For example, the Mexican population represents a particular intermingling of racial admixture with a predominant Mestizo population and a representative proportion of Native American language speakers above 5 years old (7.1% in year 2000) distributed in 62 linguistic groups that are unevenly distributed across the country (La Población Indígena en México. INEGI, 2004). Mestizos constitute the predominant group in the Mexican urban areas and are the main focus of healthcare coverage. Several anthropologic and genetic studies have shown a gradient distribution of the Mestizo to Native American populations along the north-south axis of the country, with relevant African contributions in communities of the Gulf and Pacific coasts. There is also some presence of descendants from other geographic areas, mainly from the Middle East, Central and South Europe, Portugal, and East Asia, most of whom arrived during the twentieth century.

ADMIXTURE STUDIES TO DEFINE THE GENETIC COMPONENTS OF THE MEXICAN MESTIZO

Initial work on the genetic structure of the Mexican population, based on blood groups, polymorphic proteins, hemoglobin variants, HLA, and DNA markers, were crucial to define the interracial variability— mainly among the Caucasian, Native American, and African components—within diverse Mexican communities (Lister, 1981); but over time, the question of ethnic admixture in the Mestizo populations started to generate interest among scholars (Lisker et al., 1990). An intriguing question for population geneticists concerns the homogeneity of the Mexican population, or, more precisely, the Mestizo Mexican population. The appropriate answer to this question is fundamental for the design of genome-based population studies, genome-wide association studies for complex disease, and pharmacogenomics, among others, because there is always a risk of structure sub-estimation.

To develop understanding of the topic of genetic homogeneity in the Mexican population, this chapter will focus on reports of open population genetic studies, preferably those carried out with anonymous genetic markers, to avoid confounding factors related to possible genetic markers associated with prevalent complex diseases or particular premorbid life styles. We have also discarded studies performed with non-randomly selected subjects, such as university students, groups with defined income, religious communities, etc. We also will follow a historic perspective of the described contributions in order to correlate results with advancements in the methodologies for genetic marker screening as applied to population studies.

Monterrey, Nuevo León, is the third largest urban conglomerate in Mexico and the main metropolitan area in the northeast of the country. Cerda-Flores et al. (1987) analyzed genetic admixture in a randomly selected sample of 400 inhabitants from Monterrey and the genetic distances from parental populations. Allele frequencies of the ABO, Rh, MNSs, Duffy, P, Kidd, and Lutheran blood group systems within populations from Monterrey were determined and grouped in accordance with the grandparents' place of birth. Data from parental populations was obtained from published data on Tlaxcaltecan Native Americans and Mestizos (Crawford et al., 1974), West Europeans (Spaniard, Portuguese, French, German, and Sephardic Jews; Mourant, 1976; Sandler et al., 1979), and Africans (Mourant, 1976). After estimations of genetic admixture and distances, it was found that the gene frequencies in inhabitants of the metropolitan area were intermediate to those found in their putative ancestral populations, confirming that they were Mestizo. The resulting genetic structure differed from that reported by Lisker, et al. (1980) for the Mestizo population from central Mexico, who found a predominant genetic contribution of Native Americans, but similar to the Mexican-American population from south Texas reported by Hanis et al. (1986). Cerda-Flores et al. also found that the European genetic contribution was higher in the subpopulation descending from grandparents born in northeastern Mexico than in the subpopulation descending from grandparents born in central or southern Mexico.

Studies of genetic admixture in the Mexican-American population of Starr County, Texas, are relevant contributions to understanding Mexican admixture. Mexican Mestizos account for 93.8% of the population of this county, which is located at the shared border between south Texas and northeast Mexico. In a study reported by Hanis et al. (1986), using a collection of 1004 randomly selected samples from Mestizo subjects of the county, it was originally established that the gene pool of this community had contributions of 61%, 31%, and 8% from the Spanish, Native American, and West African parental populations, respectively. In a subsequent study of subpopulations of Mexican-Americans divided by birth place (Mexico and Texas) using the same sample collection, Cerda-Flores et al. (1992) aimed at establishing the genetic diversity, and to estimate genetic admixture with 21 polymorphic erythrocyte enzymes and plasma proteins as genetic markers. They found that both subpopulations were genetically homogeneous and similar, with more than 99% of the total average gene diversity (H_T) attributed to individual variation within the population. Again, the genetic admixture analysis showed a predominant influence coming from the Spanish, followed by a lesser contribution from Amerindians, and a slight influence from the West Africans. Contributions from the ancestral populations to both subpopulations were similar, establishing that Mexican Americans of Starr County were similar to northeastern Mexicans. An additional study of genetic variation by birth cohorts in Mexican Americans of the same Texas County, grouped in three cohorts (1896–1925, 1926–1955, 1956–1985) and using 21 genetic markers, was conducted to ascertain the extent of birth cohort–related genetic variation within this population and the genetic differences from the Monterrey population. The three cohort groups were genetically indistinguishable. Gene diversity analysis attributed more than 99.8% of the total gene diversity to variation between individuals within the birth cohort populations. Subdivision by birth cohort contributed with only 0.18% to the total gene diversity. Once more, the genetic admixture analysis suggested a predominant influence from the Spaniards. The genetic structure of the Mexican-American population of Starr County was found again similar to the population of Monterrey (Cerda-Flores et al., 1994).

Studies performed by Lisker et al. (1981 and 1990) help to delineate the parental genetic contributions of the Mexican Mestizo, and to establish the pattern of the regional north-south gradient of the European and Native American contributions. In their study of admixture in Mexican Mestizos from four urban populations located across the country, 202 individuals from the city of León in the central state of Guanajuato, 228 from Merida, and 220 from Oaxaca City—from the southeastern states of Yucatán and Oaxaca, respectively—and 257 from Saltillo in the northeastern state of Coahuila (close to Monterrey), were analyzed to define the admixture components by genotyping ABO, MN, Rh, and Duffy blood groups, serum haptoglobin, albumin, and factor Bf types, and hemoglobin and glucose-6-phosphate dehydrogenase types. Admixture calculations with a maximum likelihood method were used to generate a tri-hybrid model for the parental contributions for the European, Native American, and African components. They found the following percentage contributions: 40.3%, 51.3%, and 8.4% in León: 42.9%, 51.2%, and 5.9% in Merida; 30.6%, 67.6%, and 1.8% in Oaxaca; and 38.0%, 54.7%, and 7.3% in Saltillo (Lisker et al., 1990). Figure 100-1 shows the results of this study and delineates the north-south gradient of the European contribution.

After the initial study of genetic admixture in the Monterrey area, a subsequent report was published by Cerda-Flores et al. (1991), which describes the genetic structure of the populations migrating to the Monterrey metropolitan area from the central states of San Luis Potosí and Zacatecas. A total of 4680 residents of Monterrey were grouped by generation and birthplace of the four grandparents (Monterrey, San Luis Potosí, and Zacatecas) to determine the extent of genetic variation within this population and the genetic distances among Monterrey natives and the immigrant populations from the above-mentioned states. Nine blood group systems were analyzed as genetic markers. The genetic distance analysis indicated that immigrants from San Luis Potosí and Zacatecas were similar to the Monterrey population, irrespective of birthplace and generation. More than 96% of the total gene diversity was attributed to individual variation within the population. The genetic admixture analysis suggested that populations from the three Mexican

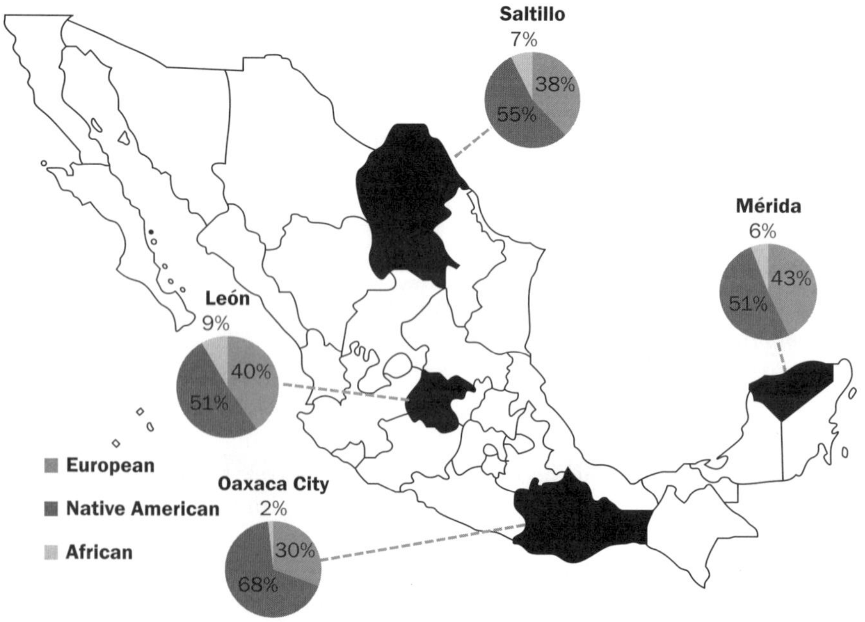

Figure 100-1 Contributions from parental populations to four Mestizo populations using blood groups, and polymorphic erythrocyte enzymes and plasma proteins as genetic markers. (Lisker, 1981; Lisker et al., 1990)

locations, stratified by birthplace and generation, received a predominant Spanish contribution (78.5%), followed by a Native American contribution (21.5%). An admixture analysis of the population from Nuevo León stratified by generation, indicated a substantial contribution from Monterrey (64.6%), followed by Zacatecas (22.1%) and San Luis Potosí (13.3%).

We studied 202 individuals from the city of Leon in Guanajuato state, 228 from Merida, Yucatan, 220 from Oaxaca, Oaxaca, and 257 from Saltillo, Coahuila, to learn the distribution of the ABO, MN, Rh, and Duffy blood groups, serum haptoglobin, albumin, and factor Bf types, and red cell hemoglobin and glucose-6-phosphate dehydrogenase types. With the gene frequencies obtained, we performed admixture measurements with a maximum likelihood method, obtaining a tri-hybrid model for black, Indian, and white ancestry with the following proportions: 0.084, 0.513, and 0.403 in Leon; 0.059, 0.512, and 0.429 in Merida; 0.018, 0.676, and 0.306 in Oaxaca; and 0.073, 0.547, and 0.380 in Saltillo. The general pattern has high Indian ancestry followed by white and black ancestry. This pattern is congruent with most other studies performed in Mexico, including the east coast, where Indian ancestry predominates despite a clear increase in the black contribution.

After the description of DNA profiling by Gill, Jeffreys, and Werrett (1985), a genotyping technique based in the analysis of polymorphic DNA sequences, classical serological genetic markers such as blood groups, HLA, and polymorphic proteins and enzymes, were to become obsolete, mainly because of their restricted informative power and the simplicity of the DNA genotyping techniques. Genotyping based on DNA polymorphic markers and improvements in marker screening methodologies impacted positively the understanding of the genetic structure of human populations, and these tools were rapidly assimilated for the study of admixture in Mexico. A genetic admixture study of three Mexican Mestizo populations (Nuevo León, Jalisco, and Mexico City) based on polymorphisms of the D1S80 and HLA-DQA1 loci (Cerda-Flores et al., 2002), showed that allele frequency distributions were relatively homogenous in the three samples, with only the Mexico City population showing minor differences of the HLA-DQA1 allele frequencies. Genetic admixture analyses demonstrated contributions ranging from 60%–70% from Spanish-European, 37%–49% from Native American, and 1%–3% from African parental populations. These estimates were virtually the same as those reported earlier from blood group loci.

In a further study of 143 Mestizos from northeast Mexican subjects in whom 13 short tandem repeat (STR) loci were genotyped (D3S1358, vWA, FGA, D8S1179, D21S11, D18S51, D5S818, D13S317, D7S820, CSF1PO, TPOX, TH01, and D16S539), maximum likelihood estimates of admixture components generated a tri-hybrid model with admixture proportions of 54.99% +/- 3.44, 39.99% +/- 2.57, and 5.02% +/- 2.82 for the Spanish, Native American, and African contributions, respectively (not different from those obtained using D1S80 and HLA-DQA1 loci). Figure 100-2 shows the admixture contributions to Mestizos from these locations.

Larger panels of DNA polymorphic markers resulted after the conclusion of the Human Genome Project, as well as new techniques for high-throughput genotyping, like the microarray genotyping and the next-generation sequencing platforms. A large study to understand the genomic diversity of the Mexican Mestizos was carried out by the National Institute of Genomic Medicine of Mexico (Silva-Zolezzi et al., 2009) in 300 samples of unrelated Mestizos from the Mexican states of the north

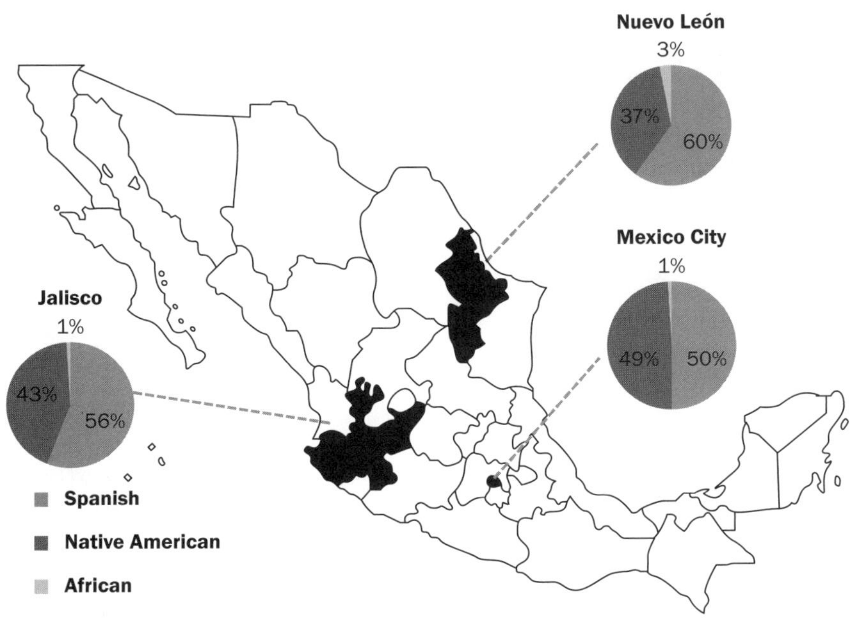

Figure 100-2 Contributions from parental populations to three Mestizo populations using 13 STR markers. (Cerda-Flores et al., 2002)

(Sonora and Zacatecas), central (Guanajuato), Gulf and Pacific coasts (Veracruz and Guerrero, respectively), and south (Oaxaca and Yucatán), and 30 Native Americans (Zapoteco from Oaxaca). A microarray platform was used for the screening of 99,953 single nucleotide polymorphisms (SNPs) and a subset of 1814 ancestry informative markers was generated to analyze genome-wide distribution and minimize linkage disequilibrium between SNPs. Data on ancestral contributions came from the International Hap Map Project (Northern Europeans, Africans, Chinese, and Japanese data sets; http://hapmap.ncbi.nlm.nih.gov). Mean ancestries were 55.2% ± 15.4% for Native American, 41.8% ± 15.5% for European, 1.8% ± 3.5% for African, and 1.2% ± 1.8% for East Asian. The highest and lowest estimates of mean European ancestry were 61.6% ± 8.5% for Sonora and 28.5% ± 12.0% for Guerrero. This study established that genetic differences among Mexican Mestizos are due to regional differences in the European and Native American parental contributions, and also describes an admixture gradient in which the European contribution predominates in the northern population, while populations in central states are closer to the studied Native American community. Some new private alleles are found in the study that clearly show a Native American origin. Despite the fact that the African contribution is low (<10%), its contribution is higher for the coastal states. As expected, the study shows linkage disequilibrium decay, which is similar to the observed in non-African populations, correlating with the reduced haplotype diversity found in Native American populations.

Although the previous study defines admixture informative markers with strong F_{st} values amenable for designing a Mestizo SNP panel for admixture mapping, it did not suggest a set of minimal useful markers for such a panel, since this is clearly out of the scope of this research. Admixture mapping is based in the long-range linkage disequilibrium generated by admixture among genetically different ancestral populations, and its purpose is to find genetic variants differing in frequency across populations, possibly associated to a particular trait or disease. In contrast with current methods for genome-wide association in case-control studies, it requires fewer markers for genome screening, is less affected by allelic heterogeneity, and can achieve higher mapping resolution. This method has also better statistical power than family-linkage studies (McKeigue, 2005). An ideal panel for admixture mapping must allow fine-mapping association, stratification control, and distinction between genetic and environmental interactions. In addition it should demonstrate sufficient imputing power to predict nonscreened genetic variants associated with a particular trait. No such Mestizo panel is available yet for admixture mapping, although there are some interesting initiatives under intensive investigation. While this panel is available, notable efforts have been made to define the genetic structure of the pre-Columbian ethnic groups and their contribution to the present-day Mexican Mestizos. Rubi-Castellanos et al. (2009) analyzed 13 combined DNA index system-short tandem repeats (CODIS-STRs) in 1984 unrelated Mestizos, representing 10 communities from different regions of Mexico (north, west, central, and southeast). The study included samples of Amerindian (Purepechas), European (Huelva, Spain), and African (Fang) origin. The analysis of molecular variance (AMOVA) test demonstrated low but significant differentiation among Mestizos from defined regions (F_{ST} = 0.34%; P = 0.0000). The spatial analysis of molecular variance (SAMOVA) predicted clustering of Mestizo populations into the four delimited regions. Once more, the most differences were observed between the northwest and the other regions, and a STRUCTURE analysis, performed to identify the three ancestral populations, demonstrated an increasing north-to-south gradient of Amerindian ancestry. This study showed discrepancies for parental contributions to the admixture in the state of Nuevo León in comparison to those reported by Cerda-Flores et al.: percentages of 43.3%, 38.2%, and 18.5% were found for the Native American, European and African parental groups, respectively. Interestingly, the distribution of Amerindian ancestry reflected the pre-Hispanic Native population density, particularly toward the Mesoamerican area.

In addition, some efforts are under development to define a set of admixture informative markers for Mexican Mestizos. Martinez-Marignac et al. (2007) tested a panel of 69 ancestral informative markers in 275 residents from Mexico City for prospective admixture mapping studies for type 2 diabetes mellitus. The average proportions of Native American, European, and West African admixture contributions were 65%, 30%, and 5%, respectively. Using a logistic model, they also found an association between higher educational status and European contribution. This association is probably maintained by socioeconomic stratification. The effective number of generations back to un-admixed ancestors was 6.7 (95% CI 5.7–8.0), and they estimate that genome-wide admixture mapping will require typing about 1400 evenly distributed ancestry informative markers to localize genes underlying disease risk between populations of European and Native American ancestry. An additional study by Martinez-Fierro et al. (2009) was carried out in 100 nonrelated residents of Monterrey, with a panel of 74 ancestral informative SNPs. They found contributions of 56%, 38%, and 6% from the Native American, European, and West African parental groups, respectively. A direct comparison of the admixture estimates for Mestizo populations in Mexico City and Monterrey was possible by comparing the data obtained from 54 overlapping markers used in

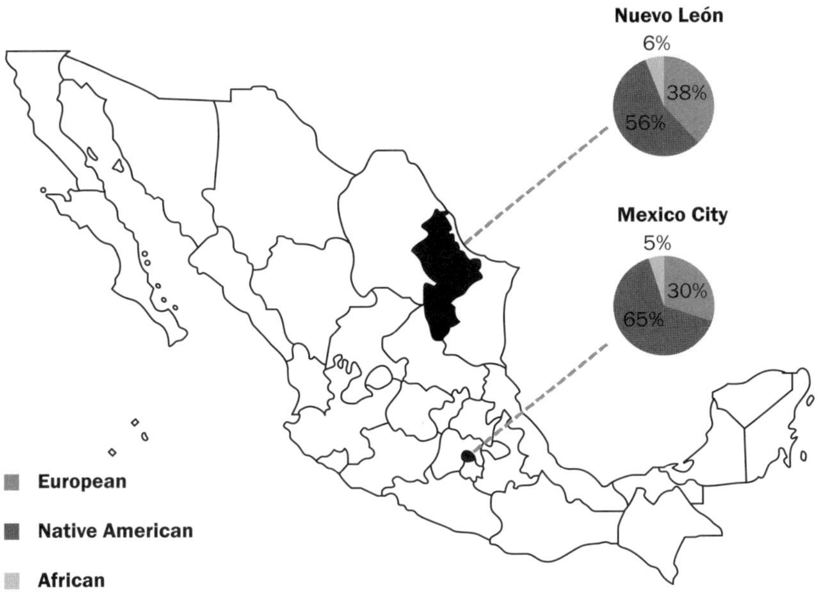

Figure 100-3 Contributions from parental populations to three Mestizo populations using 54 admixture informative markers. (Rubi-Castellanos et al., 2009; Martinez-Fierro et al., 2009)

the study by Martinez-Marignac and in this study. This comparison revealed that Mestizos from Monterrey have higher European ancestry (38% vs. 30%) and lower Native American ancestry (56% vs. 65%) than Mestizos from Mexico City (Figure 100-3). Both samples showed similar African contributions (6% vs. 5%), as reflected in previous studies. Parental contributions resemble those reported by Rubi-Castellanos et al. (2009) and Lisker et al. (1990), but differ from those found about 25 years earlier by Cerda-Flores et al. These differences can be explained by the type and number of genetic markers used to estimate admixture (as suggested by the evolution of the markers), the statistical methods, the number of individuals included in each study, by selection of the population sample, or by migratory dynamics, which has a great impact in modern-day Monterrey.

CONCLUSION

Although not perfect, the word *Mestizo* describes the current predominant population more appropriately than *Hispanic*. The genetic pool of the current Mestizo population of Mexico is mainly made of parental contributions from Native Americans, Spaniards, and to a lesser extent, from Africans. There is a north-south gradient in the European contribution, and, inversely, a south-north gradient of Native American contribution, but admixture homogeneity is observed in most Mestizo subpopulations. Although there is not a defined panel of markers for Mexican Mestizo studies yet, recent progress in the methodologies for admixture mapping, a powerful technology to discover genes associated to complex traits and diseases, will allow ascertaining genes causing some of the most prevalent chronic diseases, like cardiovascular disorders, diabetes, and cancer. In this sense, our Mestizo population has an interesting advantage over their parental ethnic groups.

REFERENCES

Cerda-Flores RM, Barton AS, Hanis CL, Chakraborty R. (1994). Genetic variation by birth cohorts in Mexican Americans of Starr County, Texas. *Am J Hum Biol* 6:669–674.

Cerda-Flores RM, Kshatriya GK, Barton SA, et al. (1991). Genetic structure of the populations migrating from San Luis Potosi and Zacatecas to Nuevo León in Mexico. *Hum Biol* 63:309–327.

Cerda-Flores RM, Kshatriya GK, Bertin TK, Hewett-Emmett D, Hanis CL, Chakraborty R. (1992). Gene diversity and estimation of genetic admixture among Mexican-Americans of Starr County, Texas. *Ann Hum Biol* 19:347–360.

Cerda-Flores RM, Ramirez-Fernandez E, Garza-Chapa R. (1987). Genetic admixture and distances between populations from Monterrey, Nuevo Leon, Mexico, and their putative ancestral populations. *Hum Biol* 59:31–49.

Cerda-Flores RM, Villalobos-Torres MC, Barrera-Saldaña HA, et al. (2002). Genetic admixture in three Mexican Mestizo populations based on D1S80 and HLA-DQA1 loci. *Am J Hum Biol* 14:257–263.

Crawford MH, Leyshon WC, Brown K, Lees F, Taylor L. (1974). Human biology in Mexico. II. A comparison of blood group, serum and red cell enzyme frequencies, and genetic distances of the Indian populations of Mexico. *Am J Phys Anthropol* 41:251–268.

Gill P, Jeffreys AJ, Werrett DJ. (1985). Forensic application of DNA 'fingerprints'. *Nature* 318:577–579.

Hanis CL, Chakraborty R, Ferrell RE, Schull WJ. (1986). Individual admixture estimates: disease associations and individual risk of diabetes and gallbladder disease among Mexican-Americans in Starr County, Texas. *Am J Phys Anthropol* 70:433–441.

Hawkins N, de Vries J, Boddington P, Kaye J, Heeney C. (2009). Planning for translational research in genomics. *Genome Med* 1:87.

Holtzman NA, Marteau TM. (2000). Will genetics revolutionize medicine? *N Engl J Med* 343:141–144.

Lisker R, Ramirez E, Briceño RP, Granados J, Babinsky V. (1990). Gene frequencies and admixture estimates in four Mexican urban centers. *Hum Biol* 62:791–801.

Lisker R. (1981). *Estructura Genética de la Población Mexicana. Aspectos Médicos y Antropológicos*. Mexico: Salvat Mexicana.

Martinez-Fierro ML, Beuten J, Leach RJ, et al. (2009). Ancestry informative markers and admixture proportions in northeastern Mexico. *J Hum Genet* 54:504–509.

McKeigue PM. (2005). Prospects for admixture mapping of complex traits. *Am J Hum Genet* 76:1–7.

Mourant AE. (1976) *The distribution of the human blood groups and other polymorphisms.* 2nd edition. New York: Oxford University Press.

Rubi-Castellanos R, Martínez-Cortés G, Muñoz-Valle JF, et al. (2009). Pre-Hispanic Mesoamerican demography approximates the present-day ancestry of Mestizos throughout the territory of Mexico. *Am J Phys Anthropol* 139:284–294.

Sandler SG, Kravitz C, Sharon R, Hermoni D, Ezekiel E, Cohen T. (1979). The Duffy blood group system in Israeli Jews and Arabs. *Vox Sang* 37:41–46.

Silva-Zolezzi I, Hidalgo-Miranda A, Estrada-Gil J, et al. (2009). Analysis of genomic diversity in Mexican Mestizo populations to develop genomic medicine in Mexico. *Proc Natl Acad Sci U S A* 106:8611-8616.

101.

MEDICAL AND HUMAN GENETICS IN ECUADOR

César Paz-y-Miño

INTRODUCTION

Currently, the field of genetics is widely known among societies, individuals, and populations. It has reached the law; it takes on life, death, health, and diseases, extending toward ecology along with truth and ethics. Basically, it has to do with everything including medicine. There is no particular field that can escape its knowledge and there is no physician that will not come across it, now more than ever since we live in the century of genetics. However, it is also the century of genomic medicine, which goes beyond genes and explores them in their true dimension as genes that produce proteins and determine functions.

Ecuador has had a slow development in medical and human genetics. Within Latin America, it is one of the few countries where the government and the public health system do not include genetic services in their hospitals. This has generated an anachronistic development of private genetic services. The insufficiency of official services for genetic health has generated a series of unattended, negligibly assisted, and poorly prevented genetic health problems.

Although the constitution of the Ecuadorian government guarantees the genetic health of the population, and the Ministry of Health has established a committee for the Ecuadorian Genome in 2001 with a genomic program planned for 2010, specific progress has not occurred and the attention to genetic health problems is overdue. In spite of this, various private health centers, military medicine, social security, and some universities possess genetic services or offer care, and somehow mitigate the genetic health needs of the community. The private services have introduced modern tests such as studies of metabolic diseases, and forensic and kinship tests. Also, there are molecular techniques like real-time PCR and sequencing being used to release scientific publications that slowly introduce our genetic activity to the world. The Ecuadorian scientific production in biomedicine represents less than 1% in Latin America, and 0.01% globally.

Health problems in Ecuador are those of a country in transition toward development. The official vital statistics show diarrheal and gastrointestinal problems as the first cause of morbimortality. Also, at least six of the 10 first causes of the aforementioned statistics are closely related to genetic problems and congenital defects. In ninth position is stomach cancer though overall cancer occupies the first position and is 6 times higher than the first cause of morbimortality.

ORIGIN OF GENETICS IN ECUADOR

The origin of genetics in Ecuador is interesting, because awareness of physical problems was already represented by archaeological remains of different ancient cultures. We can observe examples of this legacy in the Valdivia culture with its two-headed Venus (6000 years ago), the Machalilla culture with its figures representing ectrodactyly and polidactyly (3500 years ago), the Chorrera culture with a figure of achondroplastic dwarfism, the Jama-Coaque culture with representations of dwarves and giants (2500 years ago) and hermaphrodites, in addition to other cultures that also give us a look at genetics from a different, anthropological-historical and archaeological perspective (Paz-y-Miño, 1994, 1995).

In the development of biological sciences, and in genetics in particular, Ecuador was influenced by several thinkers. Perhaps the most outstanding was Charles Darwin, who arrived to the Galápagos Islands in 1835 and left a biological-evolutionary restlessness. His noticeable influence can be found in Father Solano's writings on biological subjects (1839; see Paz-y-Miño, 1994, 1995). His "genetic" subjects and ideas can be seen in interesting information where subjects like pathological inheritance were discussed in 1915. Also, anatomical anomalies and embryological development were subjects talked about at that time.

Issues related to inheritance were also considered by several authors from 1931 onwards. In 1946 and 1947, "The meaning of inheritance and possible applications" was written while the first Ecuadorian book (*Inheritance and Man*) on genetics was published in 1947.

In Ecuador, very precise contributions were made to the subject of genetics, including lessons on inheritance, inheritance of hematological disorders, and chromosomal studies in bone marrow for diagnosis of leukemia. There are also studies on Down syndrome, which represents one of the more frequent chromosomal pathologies in Ecuador and affects 1/527 live born babies. During 1947 to 1967, excellent studies on blood groups in Ecuadorian natives, anthropological evaluations and physical characteristics on skulls were carried out. These contributions have led to an even better knowledge of the Ecuadorian population. The same type of study was conducted by Eduardo Ruiz in Guayaquil only in blood groups (Paz-y-Miño, 1994, 1995, 2004).

Few genetic studies from the early 1970s until 1983 are found, and this is exactly at the time when development of genetics around the world took off. However, some of the most interesting publications around this time are studies on malformations. Here, investigations carried out by N. Dávila, C. Naranjo, C. Varas, S. Romero and G. Montalvo (Paz-y-Miño, 1994, 1995, 2004) are noteworthy.

GENETICS AS A SPECIALTY

Ecuador has been slowly introduced to genetics as a specialization in medical practice. Meanwhile, the organizations of geneticists formed a society. It is only in 1983 that genetics began to participate as a medical specialization in Ecuador at the Armed Forces hospital. Two years later, a genetic service was created at the Social Security Hospital. Later, several doctors specialized in genetics joined the hospital to serve and assist in healthcare. Curiously, the number of health practitioners specialized in genetics has not increased. There are only 10 medical geneticists in the country, and that number has remained the same since 10 years ago.

The first study of congenital malformations including chromosomal evaluation was conducted at the Isidro Ayora Maternity Hospital. This is a hospital where the highest numbers of births in the Ecuadorian mountain range region are handled. The study showed chromosomopathies are the third cause of congenital malformations in Quito (Paz-y-Miño, 1994).

The development of medical genetics has been intricate and difficult, since it does not count on government support. For example, not one public hospital yet exists that has a genetics laboratory or offers genetic services. Nevertheless, genetic health is partly covered in different centers with a variable number of geneticists—and on the other hand, only very few universities have contributed to healthcare and have committed resources to research and investigation.

When medical activity in genetics first started, the following areas were considered: clinical diagnosis, classic cytogenetics, prenatal cytogenetic diagnosis, and genetic advisory services. In 1989, the Ecuadorian Society of Human Genetics was formed and activities started to change along with its creation. Today, it can be noticed how tests and trials have increased, along with a few established groups of medical geneticists that carry out even more techniques than those used in the very beginning. These include molecular cytogenetics, metabolic and biochemistry genetics, molecular genetics (PCR, RT-PCR, sequencing), ontogenetics, filiations, and forensics (Paz-y-Miño, 2004, 1997a, 1997b, 1998).

The scientific production in medical genetics in Ecuador has little output at the international level, contributing with 0.01% to the world's achievements in biomedical knowledge. This is due to a series of problems, such as lack of money, a small critical group of investigators, nonexistent investigation policy, reduced groups of investigators, hardly any development in investigation, very little institutional support, and technological scientific dependency, among others. It has been found that the scientific production related to Ecuador it is greater (about 3600 works) than the scientific production from Ecuador, which accounts for the small number of 400 registered international publications since 1985. The scientific publications of international impact can be divided according to journal. *Clinical Neurology* has 53 published works done by several groups, *Clinical Parasitology* has 47 works achieved by several groups, and *Genetics* has 43 works contributed by two groups. Even though UDLA is responsible for 90% of these works, what it remains to be done for scientific production is a challenge.

Even after the number of geneticists in Ecuador increased and the establishment of the Ecuadorian Society of Human Genetics became a reality, there has not been much activity. It was only in July 2007 that the Society managed to carry out the First Ecuadorian Congress on Human Genetics, with a second one in July 2009.

After years of collecting information, we now know some important data on the genetics of Ecuadorians. We know that diagnosed cancer, which has its origin in at least four groups of genes (oncogenes, tumor suppressor genes, repair genes, and apoptosis genes) is the first cause of morbimortality in Ecuador. Palatal fissure lip, an illness of multifactorial or polygenetic inheritance, appears in first place of all the occurring malformations in Ecuador. Congenital hip dysplasia and microtia, both of polygenetic etiology, appear as the second and third most frequent congenital malformations. As compared to the rest of Latin America, the frequencies of the three aforementioned diseases are four to six times higher. Chromosomopathies are quite frequent, followed by cardiopathies (heart defects), sexual differentiation disorders, and mental retardation (Paz-y- Miño, 1994, 1995, 1997a,b, 1998, 2004; Paz-y- Miño et al., 2002c; Montalvo et al., 2006).

TABLE 101-1 Genetic, Chromosomal, and Multifactorial Alterations Found in Genetic Services

TYPE OF ALTERATION	No.
Chromosomopathies	1924
Dominant autosomal disorders	393
Recessive autosomal disorders	395
X-linked disorders	191
Multifactorial and polygenic disorders	583
Environment injuries	98
TOTAL	3584

Source: Biomedical Research Institute Database. C. Paz-y-Miño, 2010

TABLE 101-2 CONGENITAL MALFORMATION FREQUENCIES RELATED TO GEOGRAPHICAL HIGH ALTITUDE (DATA FROM THE ECLAMC, 2005)

			<2000m	>2000m
Cleft Lip +Palate H	11,08		20,18	
Hip Dysplasia		3,08		15,41
Down Syndrome	11,70		13,94	
Club Foot		9,85		9,54
Polydactyly		9,85		12,84
Microtia		3,80		11,74
Hydrocephaly		2,46		8,70
Spina Bifida		4,93		4,77
Cleft Palate		6,16		2,94
Esophageal Atresia	2,46		3,30	
Hypospadias		1,23		3,67
Anencephaly		1,23		4,40
Septal defects		0		4,40
Microphthalmia	2,46		2,94	
Trunk defects		0,62		2,94

Montalvo G, et al. ECLAMC-Ecuador, 2006.

The official statistics on adult mortality in Ecuador show that diabetes of polygenetic origin is placed third, while ischemic cerebral disease and hypertension, also of polygenetic origin, occupy fifth and sixth place. Also, malignant stomach tumor is placed ninth. In the child population, congenital cardiac malformations hold the eighth place and malformations in general occupy the ninth place. The data for child population is the result of genetic consultation statistics reported by the Baca Ortiz Children's Hospital in Quito, where children arrive mainly because of malformations (Paz-y-Miño, 2004; Paz-y-Miño et al., 2002c).

During these 20 years, we have evaluated 3800 patients with genetic and malformation problems. In Table 101-1, the distribution of consultation and pathologies can be observed.

A successful genetic research in Ecuador is the study and characterization of populations at high altitude. Much of the data compiled explains an association between altitude and certain genetic problems. It can be argued that the Ecuadorian mountain range populations, especially in Quito, belong to a special geographic and ethnic group where malformations and genetic problems appear in an unusual form. The following chart gathers data of evaluated malformation problems during studies conducted by the ECLAMC (Latin American Collaborative Study of Congenital Malformations; Montalvo et al., 2006; Castilla et al., 1975, 1999). The Laboratory I am in charge of participated in this study as a reference center for DNA testing (Table 101-2).

In order to genetically characterize the population of Ecuador, we began working with simple phenotype characteristics that account for simple Mendelian inheritance, such as nose and superior lip shape, hair implantation, blood type groups, etc. We detected that recessive characteristics prevail in the analyzed Ecuadorian population, reflecting important genetic crossing (Paz-y-Miño et al., 2000). We have further corroborated this data with DNA studies.

We also work with simple physiological and enzymatic characteristics, such as the presence or absence of catalase in individuals. We find that catalase deficiency is present in less than 1% of the Ecuadorian population, which is similar to data from the world's population (Paz-y-Miño, 1997b). We researched the ability to distinguish flavors, and the results showed that for sweet and salty flavors the concentrations needed to differentiate the studied substances are similar to the world's population, although, for bitter and sour flavors, the Ecuadorians needed 20 times higher concentrations to distinguish them (Paz-y-Miño, 1997b). Then, according to the cytogenetic studies carried out, we figured out that the chromosomopathies in Ecuadorians are similar to those of the world's population except for mosaic Down Syndrome (this is the coexistence of two or more cellular lines in the same individual, one normal and one altered). We have observed that 10% of the approximately 2000 evaluated patients and karyotypes display mosaicism (Paz-y-Miño, 1994, 2004). This would be an unusual number internationally speaking, since worldwide, mosaicism accounts for 2% of the Down syndrome population. This finding is significant because it generates questions to be looked into: Why does it appear in high-altitude populations? Does high altitude have some influence? Why does it appear in young mothers? Does a gene for non-mitotic disjunction exist? If so, the population of Quito would certainly display it, especially since mosaics are always post-zygotic (Tables 101-3, 101-4, and 101-5).

Subsequently, we also evaluated polymorphisms of chromosomes such as the long arm of chromosome Y,

TABLE 101-3 AUTOSOMAL CHROMOSOMOPATHIES

Type	N	%
TRISOMY21		
Simple	662	78,90
Mosaic	88	10,49
Isochromosomes	2	0,24
Simple translocation	13	1,55
Mosaic translocation	1	0,12
Undetected trisomy in peripheral blood	4	0,48
Trisomy 22	1	0,12
Trisomy 13	14	1,67
Mosaic trisomy 13	1	0,12
Trisomy 18	19	2,26
Mosaic trisomy 18	1	0,12
Trisomy 19	2	0,24
Mosaic trisomy 19	1	0,12
Mosaic trisomy 9	2	0,24
Trisomy 8	0	0,00
PARTIAL TRISOMY		
9p+	1	0,12
7p+	1	0,12
4p+	2	0,24
19p+	2	0,24
14p+	3	0,36
15p+	2	0,24
+7p	1	0,12
+21q-	1	0,12
22p+	2	0,24
Non-identified	0	0,00
Partial monosomy		0,00
18p-	1	0,12
12p-	1	0,12
5p-	3	0,36
13q-	2	0,24
13 del (q21->q22)	1	0,12
Non-identified	0	0,00
Isochromosome 17	1	0,12
Chromosome 1 ring	1	0,12
Chromosome 9 ring	1	0,12
Chromosome 15 ring	1	0,12
Inv (9) mosaic	1	0,12
TOTAL	839	100,00

TABLE 101-4 GONOSOMAL CHROMOSOMOPATHIES

Type	N	%
TURNER SYNDROME		
Monosomy X Simple	128	39,75
Mosaic	106	32,92
Isochromosome iXp-iXq	6	1,86
Isochromosome mosaic	8	2,48
Microchromosome mosaic	1	0,31
Ring mosaic	1	0,31
Trisomy X	3	0,93
Trisomy X mosaic	3	0,93
Xq.27 Fragile (Martin Bell syndrome)	6	1,86
Klinefelter syndrome XXY		0,00
Simple	15	4,66
Mosaic	4	1,24
Microchromosome mosaic	1	0,31
47, XYY	7	2,17
48, XXYY	4	1,24
49, XXXXY	1	0,31
Abnormal sexual differentiation		0,00
Female pseudo-hermaphroditism	6	1,86
Male pseudo-hermaphroditism	17	5,28
Mixed gonadal dysgenesis	2	0,62
Pure gonadal dysgenesis	1	0,31
Ambiguous genitalia	2	0,62
TOTAL	322	100,00

the short arms of acrocentric chromosomes, and heterochromatin regions, among others. We found out that their frequencies are similar to those of the world's population. What drew attention is that the pericentric inversion of chromosome 9 is a chromosomopathy present in up to 3% of the world's population, but it is present only in 0.11% of Ecuadorians, a percentage similar to the Asian population (Paz-y-Miño, 1994, 1995, 1997a,b, 1998, 2004).

After pioneering the implementation of molecular techniques in Ecuador in the year 1995, we began working with common diseases such as pancreas cystic fibrosis (FQ). By studying the delta F508 mutation of the CFTR gene, which is the most frequent mutation, we discovered that in Ecuador it only appears in 25% of people suffering from the disease, whereas in other places it can reach up to 95%. Even when the results are compared to other studies of Latin American populations, Ecuadorians show a smaller percentage (Paz-y-Miño et al., 1999). However, the study of the Ecuadorian coastal region's population showed different statistics. The delta F508 mutation is present in 37%, while the other five mutations of European origin are present in rather lower occurrences (Valle et al., 2007).

The mutations of the HFE gene related to hemochromatosis showed that the H63D mutation is more related to the disease than other mutations described around the

TABLE 101-5 Chromosome Variants and Polymorphisms

TYPE	N	%
Yq+	14	7,33
Yq-	3	1,57
9qh+	5	2,62
14p+	3	1,57
15p+	2	1,05
22p+	2	1,05
t (2;18)	2	1,05
t (4;11)	3	1,57
t (5;14)	5	2,62
t (6;7)	2	1,05
t (7;10)	1	0,52
t (8;X)	3	1,57
t (13;14)	3	1,57
t (13;15)	2	1,05
t (X;21)	1	0,52
Xq+	1	0,52
1qh+	2	1,05
16qh+	1	0,52
t (1;20)	1	0,52
t (14;21)	1	0,52
Chromosome fragility	134	70,16
TOTAL	191	100,00

world, such as the C282Y mutation in Caucasian populations (Leone et al., 2005).

In the world's population, gene polymorphisms have been related to susceptibility or resistance to diseases. There are many studies on this matter, for example on tuberculosis, malaria, leprosy, leishmaniasis, Chagas disease, filariasis, etc. We evaluated a non-positive HIV population and the relation to the gene CCR5, which has been shown to be involved in the infection or non-infection with the HIV virus and the development of AIDS. We found that the mutation percentage for gene CCR5 is rather low, with 1.2% in the general population and 0% in those affected with HIV. If we compare these data with the European population, we observe that the mutation reaches 20% of the general population, while it is extremely low in African populations (less than 0.5%). The CCR5 gene works so that the virus can attach itself to lymphocytes. If the gene has mutated, the virus does not attach itself. Moreover, it provides natural and genetic resistance in individuals that have the mutation (Paz-y-Miño et al., 2005).

One of the latest studies was on Alzheimer's disease. Here, we studied the genetic polymorphisms in apolipoprotein E (Apo E) and glutathione peroxidase 1 (GPX-1) genes. The study's main objective was to determine the prevalence of Apo E and GPX-1 polymorphisms and their influence on the development of Alzheimer's disease (AD) in the Ecuadorian population. We performed an analytic cross-sectional case-control study. The study group (n=39) was constituted of patients with Alzheimer's disease dementia. The control group (n=39) was composed of elderly adults who had not been diagnosed with dementia and had the same age and education as the study group. The inclusion period was from 2007 until 2008. Then, a blood sample was collected from them to apply standardized DNA protocols, after signing the respective consent forms and finishing a structural interview. We also performed PCR-RFLP techniques to determine the genotype of each individual. A positive association between ε4 and ε2 alleles of Apo E was found with an odds ratio (OR) of 2.40 and 2.58, respectively. The GPX1 gene showed a strong association to Leu allele with OR 5.05; meanwhile the Pro allele showed a negative association with OR 0.27. This concluded that Apo E is neither a risk factor nor a protective one for Alzheimer's disease, while the Leu allele of GPX1 is indeed a risk factor for the disease.

STUDIES OF THE GENETIC ORIGIN AND CHARACTERIZATION OF THE ECUADORIAN POPULATION

In order to complete the analysis of population studies, we evaluated repeated sequences of DNA, also known as STRs. These sequences aid in the identification of individuals in forensic genetics and paternity tests. We studied 17 polymorphic genetic markers in 404 individuals that allowed us to observe that out of all the 361 described polymorphic variants, Ecuadorians displayed only 205. If we compared these variants with populations from other countries we would find that they concur in frequencies with some populations and differ in others (Paz-y-Miño, 2001). When we calculate genetic distances between the studied populations, we can verify that most of the Ecuadorian population is racially mixed (Mestizos). A comparison between the Mestizo population and the indigenous population could show that we have more in common with the latter than with populations from the Iberian Peninsula. Additionally, the Mestizo population has genes from African-American populations, as well as from indigenous populations. Essentially, we confirm the high crossing index that Ecuadorians own. Ecuador is not formed by a sole, special ethnic group. Here, pure groups do not exist and all the analyzed groups share genes. Therefore, all the Ecuadorian population is polyhybrid, a fact supported by modern studies of the world's population, which show that all humans have more genes in common than genes that differentiate us (Paz-y-Miño, 2001, 2004). Other populations show similar conclusions to those found by our studies, becoming a good source of reference for our work (Gonzales-Andrade, 2006).

Through the study of mitochondrial DNA we can characterize populations and study their origins. We evaluated the CYP-B gene CYP-B and found 4 frequent variants in diverse populations. In Mestizos, the G15043A variant is the most frequent and it originated in central Asia. The C15040T variant is most frequent in the Quechua population, and comes from as far as Australia and India. The T15115C variant is located in the African-American population of the coastal province of Esmeraldas, and comes from Central Africa and the Dominican Republic. Additionally, the A14926G variant is more frequent on the coast of Ecuador. The Cholo group from the province of Guayas comes from East Asia, Jordan, and Arabia. These data represent a reliable source of the origins of Ecuadorian populations and are in accord with historical data (Paz-y-Miño et al., 2008).

Currently, we are associated to the global genetic characterization project of the population, whose purpose is to unravel the origins of the human race. This project is supported by National Geographic and IBM. It involves hundreds of investigators who work toward the same goal of evaluating 100,000 individuals, studying their Y-chromosomes as well as their mitochondria, and thus understand where we come from and how human migrations happened (Wells, 2008). The laboratory of UDLA is the reference center for the Ecuadorian study since 2009.

Other groups have been working on the genetic ethnography of Ecuador since approximately 10 years ago (Gonzales-Andrade, 2006). Their investigation team has worked with Ecuadorian ethnic groups. Ecuador has three main ethnic components, classified not by importance but by the number of inhabitants and their own ethnic self-identification. In the first place we find the urban population of Mestizos, followed by the American indigenous natives, who constitute more than 15 multicultural groups. At last, we find the African-Ecuadorian population that descends from African slaves brought during colonial times. Their investigations have analyzed molecular indicators of Mestizos, Amerindian Kichwas, Amerindian Waoranis and African-Ecuadorians. Randomly, they researched autosomal microsatellites, Y-chromosome microsatellites, and mitochondrial DNA, with medical-forensic and anthropological purposes.

Regarding mitochondrial DNA, the A2 haplotype was present in the Waoranis, showing also a new variant (sub-haplogroup) and ancestral mutation in position 1611. This finding is very interesting because it is the link to the Na-Dene, Chukchi, Esquimo-Aleutian populations, and a recently revealed population of El Salvador. This could explain the first migratory routes toward South America.

In relation to the Y-chromosome, Gonzalez-Andrade, et al. (2009) analyzed a total of 341 samples obtaining 277 different haplogroups divided into 99 groups in Mestizos, 91 in Kichwas, 97 in African-Ecuadorians, and 6 in Waoranis. They observed great haplogroup diversity that varied from 0.997 to 0.999 in all the groups except for the Waoranis, which showed a variation of 0.80 and also showed unique haplotypes not shared with any other groups. Continuing the studies, especially those on Y-chromosome SNPs on two Waorani communities (Bameno and Toñampari), it was found that there are haplogroups Q-M3 (xM19) with 8 different sub-haplogroups. Also by means of this Y-chromosome STR study, it was observed that African-Ecuadorians are the closest to Guineans, but even closer to the Kichwas and the Spaniards than to the Guineans. Between the Mestizos and the Kichwas there is a small distance, though the distance between Mestizos and Kichwas is clearly less than that among the Kichwas and the Spaniards.

Referring to autososomal chromosomes, 20 autosomal forensic markers were analyzed in 1438 different samples (González-Andrade et al., 2006). The findings are that Mestizos, Amerindian Kichwas, and African-Americans are in balance according to the Hardy-Weinberg equilibrium. It was also confirmed that Mestizos have a small genetic distance from Amerindian Kichwas, followed by Spaniards and Caucasian populations of the Iberian Peninsula and Europe. The Kichwas, clearly native, are closer to Mestizos than any other major urban populations from other Andean countries, mainly those places where an indigenous presence exists. The African-Ecuadorian population is close to the Guineans but is actually closer to the Kichwas and the Spaniards than to Africans. This type of study has led to the analysis of new and better markers with anthropological aims.

An interesting study is one of ethnic combination carried out with interest in autosomal microsatellites and the Y-chromosome of Mestizos, Kichwas, and African-Ecuadorians. Here, they found through autosomal STRs markers that Mestizos and African-Ecuadorians are trihybrid populations. The Mestizos contain ~73% of autosomal chromosomes of Amerindian origin, a ~19% presumed European and ~8% of African origin. The African-Ecuadorians can be considered ~57%African, ~28% European and ~15% Native American. It is observed that there is a greater contribution from European men and their Y-chromosome. In Mestizos, the presumed European lines reach ~70%; ~2% are of African origin and ~28% of Native American, data that concurs with the introduction of paternal lines during the Spanish conquest. In African-Americans, we can observe a ~44% African, ~31% European, and ~15% Native American mix. In the Kichwas, ~10% are of presumed European origin and ~78% of Native American origin. These findings go along to demonstrate combined genetics.

GENES AND CANCER

The strongest field of investigation in our laboratory is the study of the relation between genes and cancer. The purpose has been to evaluate genes that are or can be involved in the origin, progression, and tumor evolution. Considering the background described of genetic variability in the Ecuadorian population, we asked ourselves if the genes related to cancer also display variations. The studies of cancer distribution by population groups show that different incidence does exist according to sex, geographic region, and population group with common genetic characteristics. Studies conducted in Haiti, Bolivia, Russia, and the Ecuadorian population (Paz-y-Miño, 2002c; Gonzales-Andrade et al., 2006, 2009) show that the incidence of cancer is greater in the mountain population. A similar phenomenon occurs with tumors, such as the carotid body and its relation with altitude (Sandberg, 1990).

Evidence related to genetic factors show that cancer has its foundation in the origin, as well as in progression and evolution, according to family studies that showed higher predisposition to cancer development. Family studies have also displayed cancers with Mendelian inheritance patterns, development of congenital cancers, and a series of evidence that shows genome instability (e.g., chromosomal damage) reached spontaneously or induced by clastogenics, exposure to genotoxic agents like X-rays, ultraviolet rays, and pesticides. These are factors that trigger cancer and cause the presence of chromosomal instability in some genetic clinical syndromes (chromosomal fragility syndromes) as well as other rare associations with a genetic basis in people and populations (Paz-y-Miño et al, 2002c; Gonzales-Andrade et al., 2006, 2009; Vogelstein & Kinzler, 1998; Rodriguez-Cuevas et al., 1986).

We have worked in several types of cancers, including tumors of the central nervous system (meningiomas and neuromas), in which we studied the NF2 gene. Here, we detected three new mutations of this gene (Paz-y-Miño et al., 1997). We continued working with retinoblastoma and its Rb gene, detecting two new mutations of the gene Paz-y-Miño et al., 2002c). In breast, stomach, prostate, and bladder cancer we have evaluated specific genes, finding new polymorphisms.

We evaluated lung cancer and studied the incidence of L858R and G719S mutations of epidermal growth factor receptor oncogene (*EGFR*; Paz-y-Miño et al., 2010). In this factor, it is known that the tyrosine kinase domain's phosphorylation begins the downstream of intracellular signal transduction regulating apoptosis, angiogenesis, and cell proliferation. The presence of L858R and G719S mutations causes the overexpression of receptors, deregulating the intracellular signal system and leading to the inhibition of apoptosis, uncontrolled cell proliferation, and metastasis. In our research, the relationship between these mutations and lung cancer risks was evaluated. One hundred and eighty-nine (189) Ecuadorian individuals were analyzed. Eighty (80) samples of lung tissue of individuals with lung cancer were embedded in paraffin, and 109 samples of peripheral blood of healthy individuals were taken. DNA from these samples was extracted and their genotype was determined by PCR-RFLP technique. The statistical analyses were carried out by means of chi-square (χ^2) and odds ratio (OR) tests. Highly significant differences were found between healthy and affected individuals regarding the presence of the L858R variant ($P < 0.001$), and nonsignificant differences were found for the G719S variant ($P > 0.05$). The relative risk for the L858R variant individuals was OR = 5.9, 95% confidence interval (CI) 2.7–13.1, $P < 0.001$, and for the G719S variant was OR = 2.5, 95%CI 0.71–8.9, $P > 0.05$. The results obtained conclude that *EGFR* mutations have influence on the risk of lung cancer development. Therefore, a research treatment with tyrosine kinase inhibitors is suggested.

Studies on bladder cancer show the frequency of polymorphisms Pro198Leu in the *GPX-1* gene and Ile58Thr in the *MnSOD* gene in high-altitude Ecuadorian populations with this type of cancer. Bladder cancer represents 8% of all malignancies diagnosed in men, and 3% in women. The risk factors for developing bladder cancer, including the incidence rate, morbidity, and mortality, vary according to the ethnic group, exposure rates at work, age, gender, and tobacco consumption. Furthermore, there is a risk of developing this carcinoma due to dietary conditions, demonstrating that certain enzymes neutralize the oxidative compounds derived from carcinogens. If the carcinogens are not destroyed, they eventually accumulate in the body and destroy the epithelial cells in the bladder, thus increasing the risk of developing this disease. The detoxifying enzymes inactivate dangerous chemical compounds and anions—one reason why it is important to know if the polymorphisms Pro198Leu in *GPX-1* and Ile58Thr in *MnSOD* are associated with bladder cancer. In this study, 120 individuals were analyzed as controls, and 97 individuals with previously diagnosed bladder cancer. Highly significant differences were observed, and individuals with the Pro198Leu polymorphism presented a probability of developing bladder cancer 3.8 times higher than that of controls (OR = 3.8; CI 95% 2.16 – 6.78; $P < 0.001$). No significant differences for the *MnSOD* gene's Ile58Thr polymorphism occurred when compared to the control population. The results indicate that the polymorphism of *GPX-1* gene influences the risk of developing bladder cancer in the Ecuadorian population. It also suggests that more research on detoxifying genes in bladder cancer should be conducted.

We evaluated prostate cancer (Paz-y-Miño et al., 2009). Androgens are essential to normal prostate growth and development. It is therefore possible that polymorphisms in the androgen synthesis gene 5alpha-reductase type II (SRD5A2) may be involved in the progression of prostate tumors. We evaluated the relationship of two single-nucleotide polymorphisms, A49T and V89L, with prostate cancer risk in a case-control study. A total of 114 prostate cancer patients and 144 healthy control males were genotyped. We found highly significant differences between the two polymorphisms, the risk of developing prostate cancer, and some of the clinical-pathologic characteristics. Individuals who carry at least one V allele may have a higher risk of developing prostate cancer [odds ratio (OR) = 7.5, 95% confidence interval (CI) = 2.57–22.08, P<0.001]. In addition, individuals with LL genotype showed reduction in the progression to a higher tumor stage (OR = 0.10, 95%CI = 0.040–0.27, P<0.001). The A49T substitution was associated with a higher pTNM stage (OR = 2.87, 95%CI 1.14–7.21, P = 0.003) and elevated Gleason grade (OR = 3.14, 95%CI = 1.12–8.78; P = 0.004). Furthermore, the allelic frequencies of the A49T variant (33% controls and 45% cases) are the highest reported worldwide. These findings suggest that among the Ecuadorian populations, these polymorphisms influence the risk of developing prostate cancer.

Additionally we have evaluated DNA damage repair genes related to exposure to genotoxics such as the MSH2 gene and apoptosis genes like the BCL2 gene. We have found that there is some relation between these genes polymorphisms and a higher damage frequency of genetic material (Paz-y-Miño et al., 2002c). Applying the available techniques in Ecuador (PCR, SSCP, manual sequencing, direct study of mutation by RFLPs), we have found peculiar variants and new mutations of the investigated genes in these tumors. These variants are precisely listed in the International Registry of Genetic Variations (Gene Bank) and they lead us to think that the Ecuadorian population has a unique genetic behavior (Paz-y-Miño et al., 2002a).

Also, there are several works on leukemia demonstrating that chromosomal and gene restructure frequencies are different from those found in other countries. For example, the Philadelphia chromosome is present in 95% of myeloid chronic leukemias around the world, but in Ecuador it is only present in 86% of cases, as is also seen in other variants such as B3/A3 (Paz-y-Miño et al., 2002a,b,c). As previously mentioned, we do not represent a special ethnic group, but we do have interesting genetic variants that encourage us to continue studying the Ecuadorian population. We have data on acute leukemia (AL), an aggressive type of leukemia that occurs primarily in early age groups. To determine an effective treatment and to obtain a favorable prognosis it is important to know the frequency of the translocations and the expression of chimeric genes involved in this type of cancer. We analyzed 666 patients diagnosed with AL whose cytogenetic analysis involved the GTG banding technique. As far as the molecular analysis, RNA isolation and PCR for chimeric genes (BCR-ABL, MLL-AF4, AML-ETO and CBFβ-MYH11) was performed. Significant differences were found in the t(15;17)(q21;q11), t(8;21)(q22;q22) translocations and the 16 (p13;q22) inversion of children diagnosed with acute myeloid leukemia (AML). In adults with AML, the t(15;17)(q21;q11) had very low frequencies. The AML incidence was significantly different according to age groups. In patients with acute lymphoid leukemia (ALL), the t(9;22) (q34;q11) translocation was the most frequent. No t(1;19)(q23;p13) translocation was found in patients between the age of 0 to 50 years old, while the t(4;11)(q21;q23) translocation was not present in children between the age of 0 to 12 years old. We determined that there were some variations between the frequencies obtained in this study and those of prior investigations. The variations between revised frequencies and frequencies found in this study may occur due to the different genetic constitution of the studied population, especially because the individuals belong to a mixed population.

PROBLEMS IN MEDICAL GENETICS

Some problems that the field of genetics will have to solve in the years to come include those of government and authorities' carelessness regarding genetic activities and patients. Other problems are the limited number of personnel attending genetic patients, the limited interest of health professionals in genetic clinical and research areas, and the limited research activity on genetics. Also, investigation groups and personal activity from geneticists are almost nonexistent; there is a lack of genetic networks and collaborations, along with little support for genetic work on behalf of nongovernmental organizations and little social support for genetic activities. Moreover, there is limited technology development and absence of more sophisticated and expensive techniques, such as biomicrochips, robotized sequencing of DNA, pharmacogenetic product development, and more. Lastly, there is little incentive for medical students or those in compatible sciences, so that if they choose to specialize and work in medical genetics there is a lack of a job market in this area (Paz-y-Miño, 1994, 1995, 1997a,b, 1998, 2004).

Problems such as therapeutic abortion and its legalization, the study of stem cells, personalized therapy, and technology transference are yet to be approached from a bioethical perspective. These situations will have to be dealt with by medical genetics, although they could worsen depending on the type of attention received. If Ecuador and Latin America do not aim toward the development of genomic medicine and what it implies, all sorts of investment, investigation, formation, qualification, and quality control will be replaced by mere consumption of old genetic science.

OTHER ACHIEVEMENTS AND CHALLENGES FOR GENETICS IN THE 21ST CENTURY

Something important has been obtained since 1987, and it is the inclusion of genetics as a unit in medical programs at universities and in Bachelor's degrees, as well as in many of the postgraduate courses. As a result, we can now account for a better evaluation of patients and genetics can be used more efficiently (Paz-y-Miño, 1994, 1995, 1997a,b, 1998, 2004). Universidad de las Américas created the Institute of Biomedical Investigations, which is focused on lines of investigation related to human molecular and medical genetics.

According to the OMS-OPS recommendations, the requirement for geneticists in Latin American countries is of one geneticist for every one hundred thousand citizens. This means that there should be 24 in Ecuador, but the truth is that there are actually only 8 medical geneticists. Until we have a suitable number of specialists, a centralized system for patient care should be implemented, one that would make the existing services more functional.

Currently, the OPS leads a movement for genetic care rights. This care is quality controlled, accessible, cheap, and generalized. It also goes along with the new vision of socialized medicine, based on care rights and on the redress of rights violated due to care negligence, which is considered now as human or child abuse. This proposal also includes reproductive rights, the access to voluntary family planning, the use of contraceptives, the use of emergency contraceptives, the therapeutic interruption of pregnancy, the legalization of abortion, and the right to a healthy atmosphere. These issues have a bioethical background that involves medical genetics and medical actions (Paz-y-Miño, 1994, 1995, 1997a,b, 1998, 2004). Ethical discussions have increased due to the formation of ethic committees and the work of the Ecuadorian Society of Bioethics.

Some issues discussed in these fields include the exposure of people to toxic agents (Paz-y-Miño, et al., 2002c, 2007; Comisión Científica Ecuatoriana, 2007), restricted access to new technologies, the use of organizations of genetically disabled people to obtain money as charity, the responsibility of families or couples with children with genetic diseases, and the government′s evasion of this responsibility. Other subjects of discussion are the relegation of people to disabling environments (undernourished, poor, lack of health access, medicines, etc.), the limited access for genetic testing, the limitations on genetic treatments, the poor quality control in genetic activity, the financing of genetic research with therapeutic effectiveness, and the availability or restriction of genetic information (Paz-y-Miño, 1994, 1995, 1997a,b, 1998, 2004).

Genetics played an important role internationally when it was used to sustain scientific arguments of the Ecuadorian government regarding the damage that aerial dispersions with glyphosate produced on the populations in the area, as they were being carried out by Colombia at the northern border. Genetic and health studies conducted at the border demonstrated that the nontechnical preparations of this weed killer produced genetic damage with carcinogenic risk (Paz-y-Miño et al., 2002b,d, 2009).

Genetic counseling has been established in Ecuador since the beginning of the geneticists′ work, and it has served to guide patients in the best way possible. It helps them make their own decisions, freely and responsibly. With this guidance, the geneticist becomes a technician who offers information on the risks of recurrence, and the evolution and future of the disease. It is the patient who freely chooses the path, although the impartiality of genetic advising has been questioned. From my point of view it is not possible to be unbiased, seeing the current inequality of opportunities and medical attention. It is not possible to be unbiased when there is welfare injustice, or when we face the government's indolence regarding genetic problems that cause family and social conflicts. This is an idea currently supported by the Ecuadorian Society of Human Genetics.

REFERENCES

Castilla EE, et al. (1975). Incidence of congenital malformations in Latin America. Results of a cooperative study. *Rev Med Chile* 103:331–334.

Castilla EE, et al. (1999). Altitude as a risk factor for congenital anomalies. *Am J Med Genet* 86:9–14.

Comisión Científica Ecuatoriana. (2007). El Sistema de Aspersiones aéreas del plan Colombia y sus impactos sobre el ecosistema y la salud en la frontera Ecuatoriana. Ministerio de Relaciones Exteriores, Quito.

Gonzales-Andrade F. (2006). *Ensayos Médicos sobre Genética*. Quito: Imprenta Noción.

González-Andrade F, Roewer L, Willuweit S, Sánchez D, Martínez-Jarreta B. (2009). Y-STR variation among ethnic groups from Ecuador: Mestizos, Kichwas, Afro-Ecuadorians and Waoranis. Forensic Sci Int Genet. 3(3):83–91.

González-Andrade F, Sánchez-Q D, Martínez-Jarreta B. (2006). Genetic analysis of the Amerindian Kichwas and Afroamerican descendents populations from Ecuador characterised by 15 STR-PCR polymorphisms. *Forensic Sci Int* 160(2–3):231–235.

Leone PE, et al. (2005). Analysis of HFE gene mutations (C282Y, H63D, and S65C). in the Ecuadorian population. Annals Hematology 84(2):103–105.

Montalvo G, et al. (2006). Frecuencia de malformaciones congénitas en hospitales ecuatorianos de la red ECLAMC período junio 2001–junio 2005. *CAMBIOS Órgano Oficial de Difusión Científica H.C.A.M.* 5(9).

Paz-y-Miño C. (1994). Genética Humana: Conceptos Básicos e Investigaciones en el Ecuador. Fundación Simón Bolivar, Quito.

Paz-y-Miño C. (1995). Panorama Ecuatoriano en materia de estudios Genéticos en las enfermedades infantiles. En Bipatología Andina y Tropical Ecuatoriana, R. Benítez y G Ordoñez Ed., Academia Ecuatoriana de Medicina. Universidad Andina, Simón Bolívar, CCE. Vol II, Quito, pp. 721–735.

Paz-y-Miño C. (1997a). La genética médica en el Ecuador: Situación y perspectivas de desarrollo *Brazilian Journal of Genetics* 20(1):55–62.

Paz-y-Miño C. (1997b). Situación y Perspectivas de desarrollo de la Genética Humana en el Ecuador. Fundación Simón Bolívar, Quito.

Paz-y-Miño C. (1998). *Medical Genetics Services in Latin America: Ecuador.* World Health Organization, New York.

Paz-y-Miño C. (2001). Resultados preliminares de la frecuencia de STR´s en la población ecuatoriana. *Revista Nuestra Ciencia* 3:20–22.

Paz-y-Miño C. (2004). Genetic services in Ecuador. *Community Genetics* 7(2–3):137–141.

Paz-y-Miño C, Leone PE. (2000). Three novel somatic mutations in the *NF2* tumor suppressor gene [g816T>A; g1159A>G; gIVS11-1G>T] *Hum Mutat* 15(5):487. Mutation and Polymorphism # 116, Online.

Paz-y-Miño C, et al. (1995). Follow up study of chromosome aberrations in lymphocytes in hospital workers occupationally exposed to low levels of ionizing radiation. *Mutat Res* 335:245–251.

Paz-y-Miño C, et al. (1997). Comparative study of chromosome aberrations induced with Aphidicolin in women affected by breast cancer and cervix uterine cancer. *Cancer Genet Cytogen* 94:120–124.

Paz-y-Miño C, et al. (1999). The ΔF508 mutation in Ecuador, South America. *Human Mutation* 14:348–350.

Paz-y-Miño C, et al. (2000). Frecuencias alélicas de características mendelianas en población universitaria de la ciudad de Quito. *Revista de la Facultad de Ciencias Médicas de la Universidad Central del Ecuador* 25(2):23–26.

Paz-y-Miño C, Bustamante G, Sánchez ME, Leone PE. (2002a). Cytogenetic monitoring in a population occupationally exposed to pesticides in Ecuador. *Environ Health Persp* accepted No. 5435ART.

Paz-y-Miño C, et al. (2002b). BCR-ABL rearrangement frequencies in chronic myeloid leucemia and acute lymphoblastic leukemia in Ecuador, South America. *Cancer Genet Cytogenet* 132(1):65–67

Paz-y-Miño C, et al. (2002c). *Genética Toxicológica y Carcinogénesis.* Ed. Abya Yala, Quito.

Paz-y-Miño C, Pérez JC, Fiallo BF, Leone PE. (2002d). A polymorphism in the *hMSH2* gene (gIVS12-6T>C). associated with non-Hodgkin lymphomas. *Cancer Genet Cytogen* 133: 29–33.

Paz-y-Miño C, et al. (2003). B3/A3 Rearrangement in a patient with chronic myeloid leukemia. *Leukemia and Lymphoma* 44(2):375–376.

Paz-y-Miño C, et al. (2005). Study of *CCR5delta32, ccr2-641, and SDF1-3'A* polymorphisms related to resistance to HIV-1 infection and disease in the Ecuadorian population. *Human Biology* 77(4):521–526.

Paz-y-Miño C, et al. (2007). Evaluation of chromosome and DNA damage related with glyphosate mixture exposure in Ecuadorian population. *Genetics and Molecular Biology* 30(2):456–460.

Paz-y-Miño C, et al. (2008). Polymorphic variants of the human mitochondrial cytochrome b gene (CYB) in Ecuadorian Population. *Current Anthropology*, enviado en revision No. 250055.

Paz-y-Miño C, Witte T, Robles P, Llumipanta W, Díaz M, Arévalo M. (2009). Association among polymorphisms in the steroid 5alpha-reductase type II (SRD5A2). gene, prostate cancer risk, and pathologic characteristics of prostate tumors in an Ecuadorian population. *Cancer Genet Cytogenet* 189(2):71–76.

Paz-y-Miño C, López-Cortés A, Muñoz MJ, Cabrera A, Castro B, Sánchez ME. (2010). Incidence of the L858R and G719S mutations of the epidermal growth factor receptor oncogene in an Ecuadorian population with lung cancer. *Cancer Genet Cytogenet* 15;196(2):201–203.

Rodriguez-Cuevas H, Lau I, Rodriguez HP. (1986). High-altitude paragangliomas diagnostic and therapeutic considerations. *Cancer* 57:672–676.

Sandberg A. (1990). *The Chromosomes in Human Cancer and Leukemia.* (2nd edition). New York: Elsevier.

Valle P, et al. (2007) Analysis of CFTR gene mutations and Cystic Fibrosis incidence in the Ecuatorian population. *Investigación Clínica* 48(1):91–98.

Vogelstein B, Kinzler KW. (1998). *The Genetics Basis of Human Cancer.* New York: McGraw-Hill.

Wells S. (2008). Genographic Project. National Geographic & IBM. Available at: https://www.nationalgeographic.com/genographic/

102.

THE BRAZILIAN ANCESTROME

Sérgio D.J. Pena

INTRODUCTION

Modern humanity had a single and relatively recent origin, around 200,000 years ago, in Africa. From there he migrated and occupied progressively Asia, Europe, Oceania, the Americas, and all other corners of the earth (reviewed in Pena, 2007).

This diaspora was accompanied by morphological diversification, the result of adaptation to climatic and environmental conditions of different regions of the world. But the genome has remained the same, or, better said, equally different between individuals. Studies of modern molecular genetics have shown that over 90% of human genomic diversity occurs among individuals and not between populations of different continents (Bastos-Rodrigues et al., 2006 and references therein). In other words, from a biological standpoint, there are no human races.

Thousands of years after the first out-of-Africa human migration, people from Asia (Amerindians), Europe (Portugal) and Africa (slaves from West and Central Africa) met again in Brazil during the sixteenth century. From this confluence began a process of genetic mixing unusual in the history of mankind, creating the present-day Brazilian, which we can call, a little irreverently, *Homo brasilis*.

This chapter is a phylogeographic exercise around the question: who are the Brazilian people? Obviously there is no simple answer to this question. *Homo brasilis* is probably the most genetically diverse population on our planet—heterogeneous, complex, multidimensional, and far beyond any attempt at synthesis. What we want is simply to describe and celebrate its diversity.

In 1843, the then recently founded Instituto Histórico e Geográfico Brasileiro (presently called Instituto Brasileiro de Geografia Estatística – IBGE) instituted a prize for proposals on how to best describe the formation of the Brazilian people. The competition was won by the Bavarian physician and botanist Karl Friedrich Philipp von Martius, who had extensively traveled in Brazil. He proposed that the starting point should be the confluence of the three ancestral roots of Brazilians: the Amerindians, Europeans, and Africans, much in the way three tributaries coalesce to form a great river (Carvalho, 2003). We decided to use the same approach in this text.

Genome is the name given to the full DNA complement in a haploid set of chromosomes of a particular organism. From gen*ome* arose the "neo-suffix" *-ome* that refers to a *totality* of some sort; e.g., prote*ome* for all proteins of a tissue considered collectively. Likewise, we wish to give the name "Brazilian Ancestrome" to the totality of genetic characteristics of the three main founding populations of Brazil: Amerindians, Europeans, and Africans.

It is worthwhile to make a small parenthesis here, to remember that since modern humanity had a single origin in Africa circa 200,000 years ago (reviewed in Pena, 2007), Europeans and Asians are descendants from the relatively small group that migrated out of Africa roughly 60,000 years before present. Thus, they are genealogically directly related to Africans, rather than constituting biologically distinct groups (Hunley et al., 2009).

HISTORICAL FORMATION OF THE BRAZILIAN PEOPLE

Little is known about the number of indigenous people living in the area of what is now Brazil when the Portuguese arrived in 1500 (Vainfas, 2000), although a figure often cited is that of 2.5 million individuals (IBGE, 2000). The Portuguese-Amerindian admixture started soon after the arrival of the first colonizers and later became commonplace, being after 1755 even encouraged as a strategy for population growth and colonial occupation of the country (Salzano and Freire-Maia, 1970).

From the middle of the sixteenth century, Africans were brought to Brazil to work on sugarcane farms and, later, in the gold and diamond mines and on coffee plantations. Historical records suggest that between circa 1550 and 1850 (when the slave trade was abolished), around four million Africans arrived in Brazil (IBGE, 2000).

In reference to the European immigration, it is estimated that about 500,000 Portuguese arrived in the country between 1500 and 1808 (IBGE, 2000). From then on, the Brazilian ports were legally opened to all friendly nations.

Significantly, in the approximately one hundred-year period 1872–1975, Brazil received 5.5 at least million other immigrants from Europe and parts of the world (IBGE, 2000). These were, in decreasing order, 34% Italians, 29% Portuguese, 14% Spanish, 5% Japanese, 4% Germans, 2% Lebanese and Syrians, and 12% others. This phenomenon, which has been denominated the "whitening of Brazil" had complex economic and sociological causes, was tinged with racist ideology, and has been well discussed in the literature (Seyferth, 1985; Santos, 2002).

In such reunion we can roughly envisage three periods. Initially, the largest population component was represented by the indigenous Amerindians, who thus contributed heavily to the initial formation of Brazilians. The subsequent decrease in the number of Amerindians by the combined effect of guns, germs, and steel (Diamond, 1997), and the large influx of Africans from the slave trade, led to a second phase that lasted until 1850. The third period occurred after 1850, when African immigration stopped and the very prominent entry of Europeans occurred, leading to the "whitening" of Brazil (Seyferth, 1985; Santos, 2002).

GENETIC VARIATION IN BRAZILIANS

In the past few years we have been using several different molecular tools to try to characterize the ancestry and formation of the Brazilian people. We will describe briefly these studies, from which we could unravel evidence of genetic admixture in levels much higher than had previously been suspected. Unfortunately, size limitations do not allow us to review the whole field, especially the many important contributions of other research groups.

UNIPARENTAL GENETIC MARKERS IN BRAZILIANS

There are several types of genetic markers at the DNA level, and they can be classified according to their molecular nature and genomic localization. Because of extensive recombination, autosomal haplotypes are evanescent, constituting excellent individuality markers.

On the other hand, uniparental maternal (mitochondrial DNA/mtDNA) and paternal (non-recombinant region of the Y-chromosome/NRY) polymorphisms are excellent stable lineage markers because they are haploid and do not undergo recombination. As such, blocks of genes (haplotypes) transmitted to the next generations remain unaltered in the matrilineages and patrilineages until a mutation supervenes. The new mutations that have occurred and reached high frequencies after the dispersion of modern man from Africa can be specific to certain regions of the globe, and can serve as geographical markers.

The mitochondrial DNA and the NRY provide complementary information that can trace back to several generations in the past, and allow reconstitution of the history of a nation through the migrations of women and men respectively. At the individual level, lineage markers can also be employed, with the caveat that mtDNA and NRY will only provide information about, respectively, a single woman or man ancestor of an individual among thousands of ancestors, hence making up a small proportion of the genetic constitution of a person. However, when lineage markers are applied to sets of people, they provide reliable information about the range, composition, and proportions of ancestral roots (that we might call the *ancestrome*) of the groups.

We initially examined DNA polymorphisms in the nonrecombining portion of the Y-chromosome to investigate the contribution of distinct patrilineages to the present-day white Brazilian population. Y-chromosome polymorphisms were typed in 200 unrelated males from four geographical regions of Brazil, and in 93 Portuguese males. In our Brazilian sample, the vast majority of Y-chromosomes proved to be of European origin—only 2% of the Y-chromosome lineages were from sub-Saharan Africa (haplogroup E3a*), and none was Amerindian (haplogroup Q3*; see Pena et al., 2000; Carvalho-Silva et al., 2001).

There was no significant differentiation among the proportions of Y lineages of the four geographical regions of Brazil. Likewise, there were no significant differences when the haplogroup frequencies in Brazil and Portugal were compared by means of exact tests of population differentiation (Carvalho-Silva et al., 2006). Nevertheless, by typing with fast-evolving NRY markers, we later could uncover a higher within-population haplotype diversity in Brazil than in Portugal, explainable by the input of European Y-chromosomes of diverse origins (Carvalho-Silva et al., 2006).

To learn about the maternal counterpart, we analyzed mtDNA, which revealed a different reality. Considering Brazil as a whole, 33%, 39%, and 28% of matrilineages were of Amerindian, European and African origin, respectively (Pena et al., 2000; Alves-Silva et al., 2000). As expected, the frequency of different regions reflected their genealogical histories: most matrilineal lineages in the Amazonian region had Amerindian origin, while African ancestry was preponderant in the northeast (44%) and European haplogroups were prevalent in the south (66%). These data have since been amply confirmed by innumerous other research studies. For instance, we recently analyzed the mtDNA haplogroup structure of 242 self-identified white individuals from São Paulo and ascertained matrilineal proportions of 24% Amerindian,

22% African, and 54% European (HG Dornelas, SP Bydlowski, and SDJ Pena, unpublished results).

Next, for further confirmation, we studied mtDNA lineages in 120 Black individuals from the city of São Paulo (Gonçalves et al., 2008). The results, as expected, showed a mirror image of those previously found in white Brazilians: on one hand, 85% of the lineages originated in sub-Saharan Africa, 12% were from Amerindians, and only 3% were from Europe; on the other, only 48% of the Y-chromosome lineages originated from sub-Saharan Africa (the vast majority belonging to haplogroups E3a7 and E3a*). Studies in Black individuals from the cities of Rio de Janeiro and Porto Alegre (Hünemeier et al., 2007) produced very similar results.

Together, these numbers configure a picture of very strong directional mating between European males and Amerindian and African females, which agrees perfectly with the known history of the peopling of Brazil since 1500. These studies also reveal that the genomes of most Brazilians are mosaic, having mtDNA and NRY with different phylogeographical origins.

BIPARENTAL GENETIC MARKERS AND ANCESTRY IN BRAZILIANS

In Brazil, not withstanding relatively large levels of genetic admixture and a myth of "racial democracy," there exists a widespread social prejudice that seems to be particularly connected to the physical appearance of the individual (Harris and Kotak, 1963). Color (in Portuguese, *cor*) denotes the Brazilian equivalent of the English term race (*raça*) and is based on a complex phenotypic evaluation that takes into account, besides skin pigmentation, also hair type, nose shape, and lip shape (Telles, 2002). The reason why the word *color* is preferred to *race* in Brazil is probably because it captures the continuous aspects of phenotypes (Telles, 2002). In contrast with the situation in the United States, there appears to be no racial descent rule operational in Brazil, and it is possible for two siblings differing in color to belong to completely diverse racial categories (Harris and Kotak, 1963).

Based on the criteria of self-classification of the 2000 census of the Instituto Brasileiro de Geografia e Estatística (IBGE) the Brazilian population was then composed of 53.4% Whites, 6.1% Blacks, and 38.9% Brown (*pardos* in Portuguese). How do these numbers correlate with genomic ancestry?

INFERENCES ABOUT THE EUROPEAN AND AFRICAN GENOMIC ANCESTRAL ROOTS

Using a panel of genetic polymorphisms that display large differences in allelic frequencies (>0.40; these polymorphisms are called ancestry-informative markers, or AIMs for short) between Europeans and Africans, Parra et al. (1998) showed that, at a population level, it was possible to estimate with great precision the degree of European and African ancestry among North Americans. We decided to ascertain whether this same panel of markers would be capable of estimating, on an individual level, the degree of African ancestry in Brazilians. For that, we selected ten of the best AIMs used in the American study (Parra et al., 2003).

With the purpose of verifying the individual discrimination power of this set of 10 AIMs, we initially genotyped a small sample of individuals from the northern part of Portugal and from the island of São Tomé, located in the Gulf of Guinea, on the west coast of Africa. These population sources were chosen because they are geographically related to the European and African population groups that participated in the peopling of Brazil. A complete individual discrimination between the European and African genomes was obtained. It was thus clear that the 10-allele set of Parra et al. (2003) was highly efficient and provided reliable individual discrimination between European and African genomes.

Our initial Brazilian sample was composed of 173 individuals from a southeastern rural community, clinically classified according to their color (white, black, or intermediate) with a multivariate evaluation based on skin pigmentation in the medial part of the arm, hair color and texture, and the shape of the nose and lips. When we compared the African genomic ancestry values assessed for these individuals, we observed that the groups had much wider ranges than those of Europeans and Africans and that there was very significant overlap between them. This indicated that in Brazil there is significant dissociation of color and genomic ancestry; i.e., at an individual level it was not possible to infer the ancestry of a Brazilian from his/her color (Parra et al., 2003).

To corroborate these findings we undertook a second investigation based on data from 12 forensic microsatellites that had been utilized to estimate the personal genomic origin for each of 752 individuals from the city of São Paulo, belonging to different Brazilian color categories (275 Whites, 192 Browns and 285 Blacks; Pimenta et al., 2006). The genotypes permitted the calculation of a personal likelihood-ratio estimator of European or African ancestry. Again we observed great overlaps among color categories of Brazilians. This was confirmed quantitatively, using a Bayesian analysis of population structure that did not demonstrate significant genetic differentiation between the color groups. These results corroborate and validate our previous conclusions using ancestry-informative markers.

If we consider some peculiarities of Brazilian history and social structure, we can construct a model to explain why color should indeed be a poor predictor of African

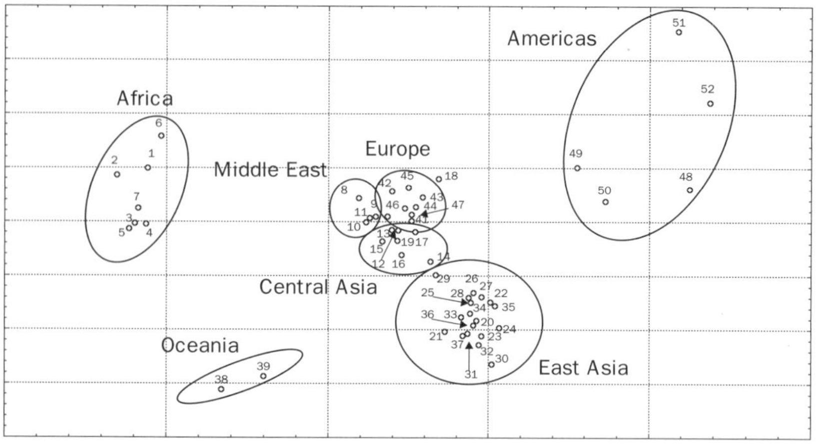

Figure 102-1 To test the discrimination power of our 40-indel set, we obtained a distance matrix of the 52 populations of the CEPH-HGDP international sample set (Cann et al., 2002) using the Reynolds genetic measure, which is based on the F_{ST} linearized for short divergence times (Reynolds et al., 1983). From the matrix we undertook a multidimensional scaling analysis (Kruskal and Wish, 1978) using the program *Statistica*. It is immediately apparent that the resolution power was excellent and that the individuals from the 52 populations aggregate into five widely separated clusters that correspond to Africa, Oceania, East Asia, America, and a Central Europe/Middle East/Central Asia group (reprinted with permission from Pena et al., 2009).

ancestry (Parra et al., 2003). Nowadays most Africans have black skin, genetically determined by a very small number of genes that were evolutionarily selected in adaptation to the tropical and subtropical climate. Thus, if we have a social race identification system based primarily on phenotype, such as occurs in Brazil, we classify individuals on the basis of the presence of certain alleles at a small number of genes that have impact on the physical appearance, while ignoring all of the rest of the genome. Assortative mating based on color, which has been shown by demographic studies to occur in Brazil, will produce strong associations among the individual components of color. Indeed, we detected the presence of such positive associations at highly significant levels in a southeastern Brazilian population (Parra et al., 2003). On the other hand, we expect that any initial admixture association between color and the AIMs will inevitably decay over time because of genetic admixture. It is easy to see how this combination of social forces could produce a population with distinct color groups, and yet with similar levels of African ancestry.

GENOMIC STUDIES OF THE AMERINDIAN, EUROPEAN, AND AFRICAN ANCESTRAL ROOTS

The two studies mentioned above did not take into account the Amerindian contribution to the Brazilian population. To achieve that, we needed new polymorphic markers that would be sensitive to all three ancestries.

We screened the database of 2000 human diallelic short indels characterized by Weber et al. (2002), and identified 40 polymorphisms that fulfilled the following criteria: widespread chromosomal location in the human genome, increasing amplicon sizes that allowed multiplex PCR amplification and electrophoretic analysis, and allele frequency close to 0.5 in the European population. We used these 40 indel markers to study worldwide human genome variation, namely all the samples in the CEPH-HGDP Diversity Panel (Cann et al., 2002), composed of 52 populations originated from five geographical regions: Americas, sub-Saharan Africa, East Asia, Oceania, and a cluster composed of Europe, the Middle East, and Central Asia (Bastos-Rodrigues et al., 2006). We obtained a distance matrix of the 52 populations using the Reynolds genetic measure, which is based on the F_{ST} linearized for short divergence times (Reynolds et al., 1983). On visualization using multidimensional scaling analysis (Kruskal and Wish, 1978) we obtained a very adequate graphical representation that showed five widely separated clusters corresponding to Africa, Oceania, East Asia, Americas, and a central Europe/Middle East-/Central Asia group (Figure 102-1).

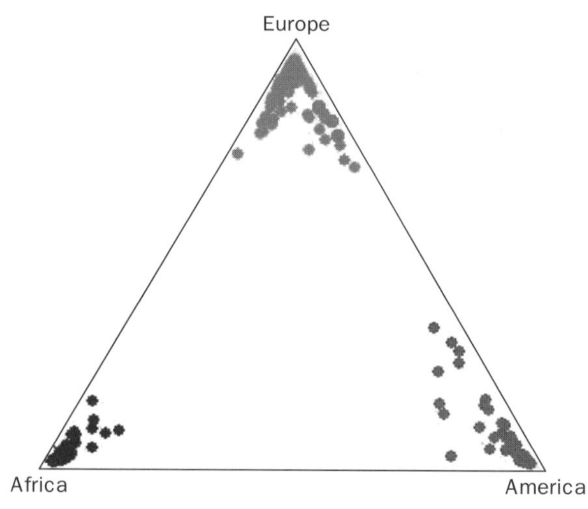

Figure 102-2 Triangular plot produced by the *Structure* program version 2.1 (Pritchard et al., 2000) on analysis of the genotypes of the European, Amerindian, and sub-Saharan African individuals of the CEPH-HGDP panel (the three Brazilian ancestral roots). *Structure* uses multi-local genotypes to group individuals on the basis of their genotypes. The run consisted of 100,000 burn -in steps, followed by 2.5×10^5 Markov Chain Monte Carlo iterations, without any prior information on the population origin of each sampled individual. We used the "admixture" model and assumed the allele frequencies of different populations to be correlated (reprinted with permission from Pena et al., 2009).

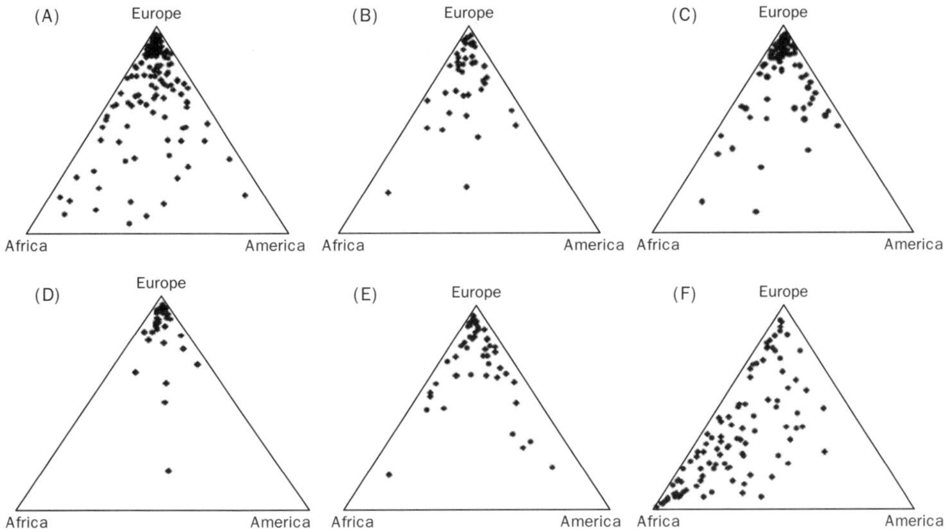

Figure 102-3 Triangular graphs of the genomic proportion of Amerindian, European, and African ancestry of Brazilian individuals, drawn using the Tri-Plot program (Graham and Midgley, 2000). We used our set of 40 insertion-deletion polymorphisms (Bastos-Rodrigues et al., 2006) and the *Structure* program to study 272 Brazilians self-defined as White (Carvalho-Silva et al., 2001, Parra et al., 2003).

Next, we submitted the genotypes of the Amerindians, Europeans, and sub-Saharan Africans of the CEPH panel (the three Brazilian ancestral roots) to *Structure* version 2.1 (Pritchard et al., 2000), a Bayesian software that uses multi-local genotypes to infer the structure of population and group individuals on the basis of their genotypes, even without any prior information on the population origin of each sampled individual. The program produced a triangular plot in which the three different populations clustered in different vertices with no overlap (Figure 102-2). European individuals had on average 94.6% European ancestry, sub-Saharan Africans had on average 96.5% sub-Saharan African ancestry, and Amerindians had on average 94.8% Amerindian ancestry (Pena et al., 2009).

INFERENCES ABOUT THE AMERINDIAN, EUROPEAN, AND AFRICAN GENOMIC ANCESTRAL ROOTS AMONG BRAZILIANS

As shown in the previous section, Amerindian, African, and European samples of CEPH can be used to define a triangular landscape with ancestry-specific vertices on which we can plot the results of *Structure* analyses of Brazilian samples (Pena et al, 2009). Figure 102-3 shows triangular plots for Brazilian self-declared White individuals from Minas Gerais in the southeast region, from the northern region (states of Amazonas, Acre, Rondônia, and Pará), the state of Pernambuco in the northeast region, the southern region (states of Rio Grande do Sul, Santa Catarina, and Paraná) and the city of São Paulo in the southeast region. The last graph in Figure 102-3 (F) shows 100 Brazilian self-declared Black individuals from the city of São Paulo.

We can observe from the figure that most of the White individuals of all the regions examined clustered in the "European" vertex of the triangle plot (Figure 102-3, A–E), although a proportion of them were scattered throughout

TABLE 102-1 AVERAGE VALUES OF THE GENOMIC PROPORTIONS OF AMERINDIAN, EUROPEAN AND AFRICAN ANCESTRY OF INDIVIDUALS FROM THE CEPH-HGDP PANEL (21) AND FROM THE BRAZILIAN POPULATION (SEE DESCRIPTION IN THE LEGEND OF FIGURE 102-3). ALL INDIVIDUALS WERE TYPED FOR ALL 40 INSERTION-DELETION POLYMORPHISMS (22) AND ANALYZED WITH THE *STRUCTURE* PROGRAM USING THE CEPH-HGDP POPULATIONS AS REFERENCES.

		Ancestry		
	Number	Amerindian	European	African
CEPH-HGDP POPULATION				
Europeans	161	0.028	0.946	0.026
Amerindians	105	0.948	0.037	0.014
Sub-Saharan Africans	126	0.011	0.023	0.965
BRAZILIAN WHITES				
Southeast (Minas Gerais)	142	0.131	0.708	0.161
North	45	0.121	0.742	0.137
Northeast	49	0.147	0.711	0.142
South	36	0.092	0.819	0.089
Southeast (São Paulo)	88	0.105	0.779	0.116
BRAZILIAN BLACKS				
Southeast (São Paulo)	100	0.132	0.371	0.497

A: 142 individuals from the State of Minas Gerais in the southern region;

B: 45 individuals from the north (States of Amazonas, Acre, Rondônia e Pará);

D: 36 individuals from the south (States of Rio Grande do Sul, Santa Catarina and Paraná); and

E: 49 individuals from the northeast (state of Pernambuco).

We also studied 88 White (C) and 100 Black (F) men from the city of São Paulo, Brazil, randomly drawn from a larger sample described in a previous publication (Pimenta et al., 2006). This figure is reprinted with permission from Pena et al. (2009).

the triangle area. As can be checked in Table 102-1, all average European contributions for White individuals were above 0.700, with a maximum of 0.819 in southern Brazil (a region of heavy European immigration) and a minimum of 0.709 in Minas Gerais. When we compared the regions pairwise using a Monte Carlo resampling strategy (Manly, 1997), however, we could not find statistical significance. We then pooled all the samples of Brazilian Whites and compared it with the sample of Europeans from the CEPH panel, now finding a significant difference ($P < 0.0001$).

Black individuals from the city of São Paulo showed very high individual variation in their biogeographical ancestry, as indicated by a large spread of data points in the triangular graph (Fig 102-3F). On average, they had an average degree of African ancestry slightly below 50% (Table 102-1), which was significantly different from White individuals of the same region and also from individuals from sub-Saharan Africa.

An interesting observation is that the extent of Amerindian ancestry is relatively low (range 0.092–0.147) and not statistically different among White individuals from different geographical regions, and also between White and Black individuals from São Paulo. Further results have been recently published by Pena et al. (2011).

CONCLUSIONS

Many authors have used historical, sociological, and anthropological methodology to analyze the origins of Brazilians: Paulo Prado in *Retrato do Brasil* (published in 1927), Sérgio Buarque de Holanda in *Raízes do Brasil* (published in 1933), Gilberto Freyre in *Casa Grande & Senzala* (published in 1933), and Darcy Ribeiro in *O Povo Brasileiro* (1995). We have used new molecular genetics tools for the same purpose.

The data presented in this review demonstrate that currently available DNA tests can provide an important molecular confirmation of the proposals of the authors mentioned above, and also are capable of providing new, valuable insights into the process of genetic formation and structure of the Brazilian people.

Studies with uniparental markers in both white and black Brazilians demonstrate strong directional mating between European males and Amerindian and African females, which agrees with the known history of the peopling of Brazil since 1500. These data reveal that the genomes of most Brazilians are mosaic, having mtDNA and NRY with different phylogeographical origins.

Studies with autosomal biparental markers reveal very elevated levels of genetic admixture between the three ancestral roots. However, it is also evident that there was an important population effect of the program of "whitening" of Brazil promoted through the immigration of circa six million Europeans in the roughly one-hundred-year period after 1872. This manifests itself both in a predominant (>70%) European genomic ancestry in Brazilian Whites, regardless of geographical region, and on a high average European genomic ancestry (37.1%) in Brazilian Blacks.

The correlation between color and genomic ancestry is very imperfect: at an individual level one cannot safely predict the skin color of a person from her level of European, African, and Amerindian ancestry, nor the opposite. Regardless of their skin color, the overwhelming majority of Brazilians have a high degree of European ancestry. Also, regardless of their skin color, the overwhelming majority of Brazilians have a significant degree of African ancestry. Finally, most Brazilians have a significant and very uniform degree of Amerindian ancestry!

The high ancestral variability observed in Whites and Blacks suggest that each Brazilian has a singular and quite individual proportion of European, African, and Amerindian ancestry in their mosaic genomes. Thus, the only possible basis to deal with genetic variation in Brazilians is not as members of color groups, but on a person-by-person basis, as 190 million human beings, with singular genome and life histories.

SUMMARY

We call "Brazilian Ancestrome" the totality of genetic characteristics of the three main founding populations of Brazil: Amerindians, Europeans, and Africans. We here review studies from our laboratories using different molecular tools to characterize such ancestrome. Initially we used uniparental DNA markers to investigate the contribution of distinct Y-chromosome and mitochondrial DNA lineages to present-day populations. High levels of genetic admixture and strong directional mating between European males and Amerindian and African females were unraveled. We next analyzed different types of biparental autosomal polymorphisms. Especially useful was a set of 40 insertion-deletion polymorphisms (indels) that, when studied worldwide, proved exquisitely sensitive in discriminating between Amerindians, Europeans, and sub-Saharan Africans. When applied to the study of Brazilians these markers confirmed extensive genomic admixture, but also demonstrated a strong imprint of the massive European immigration wave in the nineteenth and twentieth centuries. The high observed individual ancestral variability suggests that each Brazilian has a singular proportion of Amerindian, European, and African ancestries in their mosaic genomes. In Brazil, one cannot predict individually the color of persons from their genomic ancestry, nor the opposite. Brazilians should be assessed on a personal basis, as 190 million human beings, and not as members of color groups.

ACKNOWLEDGMENTS

This research was sponsored by Conselho Nacional de Desenvolvimento Científico e Tecnológico of Brazil (CNPq) and a Millenium Institute Project of the Ministério de Ciência e Tecnologia of Brazil on Genomic Ancestry (coordinator: F.M. Salzano).

REFERENCES

Alves-Silva J, da Silva Santos M, Guimarães PE, et al. (2002). The ancestry of Brazilian mtDNA lineages. *Am J Hum Genet* 67:444–461.

Bastos-Rodrigues L, Pimenta JR, Pena SD. (2006). The genetic structure of human populations studied through short insertion-deletion polymorphisms. *Ann Hum Genet* 70:658–665.

Cann HM, de Toma C, Cazes L, et al. (2002). A human genome diversity cell line panel. *Science* 296: 261–262.

Carvalho JM. (2003). Martius e a nossa história. *Nossa Hist* 1:96–102.

Carvalho-Silva DR, Santos FR, Rocha J, Pena SD. (2001). The phylogeography of Brazilian Y-chromosome lineages. *Am J Hum Genet* 68:281–286.

Carvalho-Silva DR, Tarazona-Santos E, Rocha J, Pena SD, Santos FR. (2006). Y chromosome diversity in Brazilians: switching perspectives from slow to fast evolving markers. *Genetica* 126:251–260.

Diamond J. (1997). *Guns, Germs, and Steel: The Fates of Human Societies*. New York: WW Norton & Co.

Freyre G. (2001) Casa Grande & Senzala. Rio de Janeiro, Livros do Brasil.

Gonçalves VF, Carvalho CMB, Bortolini MC, Bydlowski SP, Pena SDJ. (2008). The phylogeography of African Brazilians. *Hum Hered* 65:23–32.

Graham DJ, Midgley NG. (2000). Graphical representation of particle shape using triangular diagrams: an Excel spreadsheet method. *Earth Surf Proc Landf* 25:1473–1477.

Harris M, Kotak C. (1963). The structural significance of Brazilian categories. *Sociologia* 25:203–208.

Holanda SB. (1986). Raizes do Brasil. Belo Horizonte, J. Olympio Editora.

Hünemeier T, Carvalho C, Marrero AC, Salzano FM, Pena SDJ, Bortolini MC. (2007). Niger-Congo speaking populations and the formation of the Brazilian gene pool: mtDNA and Y-chromosome data. *Am J Phys Anthrop* 133:854–867.

Hunley KL, Healy ME, Long JC. (2009). The global pattern of gene identity variation reveals a history of long-range migrations, bottlenecks, and local mate exchange: implications for biological race. *Am J Phys Anthropol* 139:35–46.

IBGE. (2000). *Brasil- 500 anos de povoamento*. Rio de Janeiro: Instituto Brasileiro de Geografia Estatística – IBGE.

Kruskal JB, Wish M. (1978). *Multidimensional Scaling*. New York: SAGE Publications.

Manly BFJ. (1997). *Randomization, Bootstrap and Monte Carlo Methods in Biology*. New York: Chapman & Hall.

Parra EJ, Marcini A, Akey J, et al. (1998). Estimating African American admixture proportions by use of population-specific alleles. *Am J Hum Genet* 63:1839–1851.

Parra FC, Amado RC, Lambertucci JR, Rocha J, Antunes CM, Pena SDJ. (2003). Color and genomic ancestry in Brazilians. *Proc Natl Acad Sci U S A* 100:177–182.

Pena SDJ, Bastos-Rodrigues L, Pimenta JR, Bydlowski SP. (2009). DNA tests probe the genomic ancestry of Brazilians. *Braz J Med Biol Res* 42:870–876.

Pena SDJ, Di Pietro G, Fuchshuber-Moraes M, Genro JP, Hutz MH, Kehdy Fde S, Kohlrausch F, Magno LA, Montenegro RC, Moraes MO, de Moraes ME, de Moraes MR, Ojopi EB, Perini JA, Racciopi C, Ribeiro-Dos-Santos AK, Rios-Santos F, Romano-Silva MA, Sortica VA, Suarez-Kurtz G. (2011). The genomic ancestry of individuals from different geographical regions of Brazil is more uniform than expected. PLoS One. 16:e17063.

Pena SDJ, Carvalho-Silva DR, Alves-Silva J, Prado VF, Santos FR. (2000). Retrato molecular do Brasil. *Ciência Hoje* 27(159):16–25.

Pena SDJ. (2007). The evolution and structure of human genetic diversity. In: Suarez-Kurtz G, ed. *Pharmacogenomics in Admixed Populations*. (pp. 1–11). Austin: Landes Bioscience.

Pimenta JR, Zuccherato LW, Debes AA, et al. (2006). Color and genomic ancestry in Brazilians: a study with forensic microsatellites. *Hum Hered* 62: 190–195.

Prado P. (1999). Retrato do Brasil: Ensaio sobre a Tristeza Brasileira. São Paulo, Companhia das Letras.

Pritchard JK, Stephens M, Donnelly P. (2000). Inference of population structure using multilocus genotype data. *Genetics* 155:945–959.

Reynolds J, Weir BS, Cockerham CC. (1983). Estimation of the co-ancestry coefficient: basis for a short-term genetic distance. *Genetics* 105:767–779.

Ribeiro D. (1995). *O Povo Brasileiro: a formação e o sentido do Brasil*. São Paulo: Companhia das Letras.

Salzano FM, Freire-Maia N. (1970). *Problems in human biology: a study of Brazilian populations*. Detroit: Wayne State University Press.

Santos AS. (2002). Historical roots of the "Whitening" of Brazil. *Latin Am Persp* 29:61–82.

Seyferth G. (1985). A antropologia e a teoria do branqueamento de raça no Brasil: a tese de João Batista de Lacerda. *Rev Mus Paulista* 30:81–98.

Telles EE. (2002). Racial ambiguity among the Brazilian population. *Ethn Racial Stud* 25:415–441.

Vainfas, R. (2000). História indígena: 500 anos de despovoamento. In: IBGE, *Brasil: 500 anos de povoamento*. (pp. 37–59). Rio de Janeiro: Instituto Brasileiro de Geografia Estatística – IBGE.

Weber JL, David D, Heil J, Fan Y, Zhao C, Marth G. (2002). Human diallelic insertion/deletion polymorphisms. *Am J Hum Genet* 71:854–862.

103.

THE PRACTICE OF MEDICAL GENETICS IN BRAZIL

Dafne Dain Gandelman Horovitz, Antonia Paula Marques-de-Faria, and Victor Evangelista de Faria Ferraz

INTRODUCTION

BRAZIL – THE LAND AND THE PEOPLE

The Federative Republic of Brazil is a country of continental dimensions, whose territory covers more than 8.5 million km^2 and borders all the countries of South America except for Chile and Ecuador. After more than three centuries as a colony of Portugal, it became an independent empire in 1822 and a federal republic in 1889. It was under a military dictatorship during the second half of the last century and today, after six presidential elections, the country has emerged as a consolidated democracy and as a political and economic leader in the region. In fact, the Brazilian economy also acquired a strong position in the global economy in recent years. (IBGE, 2010; United Nations, 2009).

The federation is formed by the Federal District and 26 states, and represents the largest territory (8,514,876,599 km^2) and the largest population in South America. It is distinguished by an admirable national unity, based mainly in the Portuguese language, which is spoken in all regions, and by a magnificent biodiversity and natural resources. The agriculture is large and well developed, while mining, manufacturing, and service sectors are enjoying significant growth. The production is diversified and the Brazilian products are competitive in several sectors of the globalized world. Notwithstanding, the country still faces economic, political, and social problems; the highly unequal income distribution is of paramount concern (IBGE, 2010; United Nations, 2009).

In 2004, the Brazilian population surpassed 180 million and today is estimated in the 190 million range. In the last decades, the country underwent consistent changes in demography. The birth rate is around 3 million per year and the population growth rate declined by an average of 1.9% annually during 1980–1990, and in 2001 it was 1.4%, reflecting the effect of birth control programs developed by the Brazilian government during the 1990s. Eighty-two percent of the population are urbanized, reaching over 90% in some regions, such as the states of Rio de Janeiro and São Paulo, where there are such megalopolises as São Paulo, the main industrial and financial center of the nation, surrounded by other large cities connected by modern highways; in contrast, there are regions with small hidden villages, some of them only reached by precarious trails, plane, or boat (IBGE, 2010; United Nations, 2009).

It is a country predominantly young, with around 40% of the population under 19 years old, although the proportion of people over 35 years old has been increasing, reaching 34.5% in 2000 with a narrowing down in the base of the age group pyramid. The life expectancy at birth was increased to 72.8 years in 2008, an increase of 2.3 years when compared with 2000 (70.5 years). Besides gradual aging of the population, other demographic modifications are the marked decline of fertility, the changes in the causes of death and the reduction of mortality level, as well as the improving literacy (IBGE, 2010).

Concerning racial aspects, the country is a melting pot of different ethnic groups, although the majority is composed of bi- and tri-hybrids, mainly from the mixture of Caucasians and African descendants, in addition to Amerindians. Taking into account the ethnic characteristics obtained during the 2000 census, using the options White, Black, "*Pardo*" (mixed color or race; mulatto or Mestizo), Asiatic, and Native Brazilian Indian, the percentage of people that considered themselves White are 54%, while 39.9% are *Pardos*, 5.4% Black, 0.5% Asiatic, and 0.2% Native Indians. The distribution is irregular, since in southern region the White population reaches 84%, while it is 62% in the southeast. Inversely, there are most *Pardos* or mixed-color people in the populations of the north (63.5%) and northeast (59.8%) regions (IBGE, 2010; Marques-de-Faria et al., 2004).

BRAZIL – HEALTH SYSTEM AND SOME HEALTH INDICATORS

In 2008, the mortality rate of children under 5 years was 22 per 1000, which placed Brazil in 100th place among

193 countries and territories of the world (Unicef, 2009). The infant mortality rate was significantly reduced over the last decade and was 20.7 per 1000 in 2006, although it shows marked regional disparities ranging from 29.8 per 1000 in the less developed northeast region, to 13.3 per 1000 in the more industrialized regions of the south and southeast (Brasil, 2008). The leading causes of registered infant deaths in Brazil are perinatal conditions, which accounted for 46.5% of infant mortality in 1985–1987, and 56.8% in 1995–1997. This relative increase in proportional mortality was observed for all regions, with higher rates in the north and northeast than in the south and southeast regions (Victora and Barros, 2001).

The second leading cause of registered infant deaths during the last decade were congenital malformations (including chromosome abnormalities), which surpassed diarrhea and respiratory infections, increasing from 7.1% in 1985–1987 to 11.2% in 1995–1997 (of the total infant mortality rate). The national infant mortality rate due to malformations remained stable in the decade under study—4.4 (1985–1987) and 4.2 per 1000 (1995–1997)—which is a good indicator of data consistency (Victora and Barros, 2001).

There are more than 347,000 active physicians in the country (Conselho Federal de Medicina, 2011), with the majority living in the main cities, especially in state capitals. The national average corresponds to 1 physician per 594 inhabitants, ranging from 1 per 833 in the north to 1 per 359 inhabitants in the southeast region. More than 70% of all Brazilian medical professionals and medical schools are concentrated in the most developed regions of the south and southeast (Conselho Federal de Medicina, 2011).

The majority of the population is served by the public Unified Health System ("Sistema Único de Saúde," or SUS). It was created in conformity with the provisions of the 1988 Federal Constitution (Senado Federal, 1988), to ensure universal and equal access for promoting, protecting, and recovering the health of all citizens fairly. The system has a municipal basis, whose management is under state administrations with the support of the Federal Ministry of Health. One fundamental step has been the development of a national health information database within the Ministry of Health (DATASUS). The SUS has gradually become more efficient in improving health conditions nationwide.

In addition to the SUS, there is a thriving private health sector that serves 24% of the Brazilian population (around 46 million inhabitants). The quality of these services is variable, despite the efforts from the Ministry of Health to regulate and improve them. Finally, for a significant part of the population, it is still difficult to access either public or private services (Conselho Federal de Medicina, 2011).

In order to reduce the striking health disparities by strengthening primary healthcare, in 1981 the federal government instituted the Family Health Program ("Programa de Saúde da Família," or PSF) and the Community Health Agents Program. Under these programs, by 2010 there were over 30,600 PSF teams, each composed of 1 nurse, 2 nurse assistants, and 6 community health agents (usually recruited among the community) in 5565 municipalities throughout the country (Brasil, 2009g). Each team provides basic health services to between 600 and 1000 families. The aim of the program is not only treatment and prevention of diseases, but also to reinforce individual responsibility for one's own health and the health of one's family and community (United Nations, 2001).

Birth Defects in Brazil

In the past decades, with health improvement of the population in general, with more effective sanitation and better control of infectious and nutritional diseases, it can be stated that Brazil has been living through an epidemiological transition. Taking such a statement into consideration, birth defects are increasingly assuming a more important role in morbidity and mortality in Brazil, especially in large centers and reference hospitals (Horovitz et al., 2005).

When looking for information regarding congenital malformations in the Brazilian Ministry of Health's database, we can observe that the 2008 hospital mortality rate in the pediatric age range (one to nineteen years) was 4.7 times greater when the primary diagnosis for admission was included as given in Chapter XVII ("Congenital malformations, deformities and chromosomal abnormalities") of the International Statistical Classification of Diseases and Related Health Problems (ICD-10). We can also notice an increment of the permanency and hospitalization cost for this group of diseases (Brasil, 2009e). Besides the high morbidity and mortality, chronic diseases present a complex problem when considering that continuous treatment is necessary, resulting in lifelong high costs.

By processing information from the Ministry's database on mortality, the evolution of infant mortality causes from 1980 to 2006 can be observed (Figure 103-1; 2009f). When removing and proportionately redistributing all the nondefined causes, perinatal causes were the leading cause of infant mortality in 1980, corresponding to 38% of deaths under one year; at the same time, congenital anomalies occupied the fifth position, representing 5% of the total. Over time, such profiles have changed, and a proportional reduction in infectious, nutritional, and respiratory diseases can be noticed. In that way, congenital malformations became the second cause of deaths in children under age 1 year, representing 13% of the total in 2000. In 2006, congenital anomalies not only kept the second position, but also became accountable for 16.3%

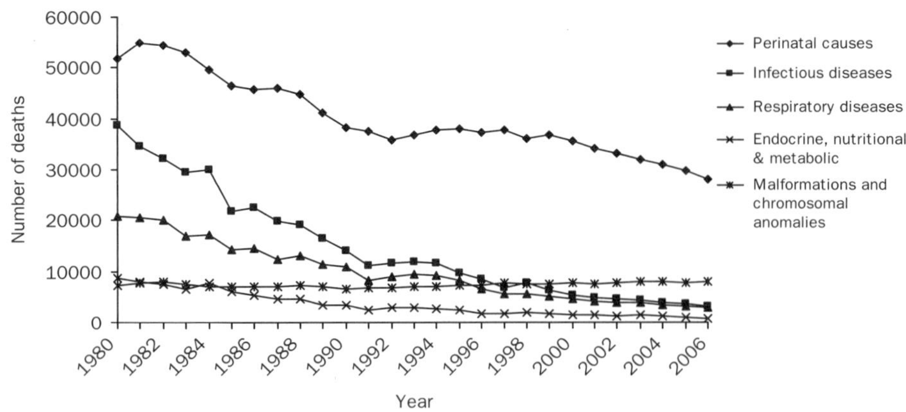

Figure 103-1 Graphical representation of the evolution of proportional infant mortality due to the five major causes of death between 1980 and 2006 in Brazil. Original data was obtained from the Mortality Information System, accessed through DATASUS (http://www.datasus.gov.br; 2009f).

of the deaths, exceeding the sum of the third and fourth causes (infectious and respiratory, respectively) (2009f). These statistics on birth defects and infant mortality in Brazil can be classified as surprising due to their magnitude, but they did not determine government policies or initiatives for the prevention and management of such a group of problems until 2004. In that year the process began for introducing medical genetics as a health policy of the Ministry of Health. This topic will be addressed after the presentation of some of the already established government programs related to birth defects in Brazil.

The Brazilian Unified Health System (Sistema Único de Saúde – SUS)

The Brazilian Unified Health System (Sistema Único de Saúde, or SUS) is one of the largest public health systems in the world. It includes most aspects of healthcare, from outpatient care to organ transplantation, and proposes to ensure full, universal, and free-of-charge medical access for the entire Brazilian population. The SUS was conceived during the discussions encompassing reformist health ideals, and bolstered by a broader concept of health, which led to the determination of the principles of universal coverage and service, and of equal access to services and health actions, in the 1988 Brazilian Constitution, Besides offering appointments, medical exams, and hospitalizations, the SUS also promotes immunization campaigns, prevention, and health monitoring, such as food control and medicament registration (Brasil, 2009b).

In spite of the early establishment of these principles in the 1988 Brazilian Constitution, the SUS was officially created only in September 1990, defining the need for a hierarchical network and delimiting powers at federal, state, and regional levels (Brasil, 1990b). The basic principles of SUS are (Brasil, 2006):

Participative management with health councils in all states and municipalities, national policies development, and funding of at least half of the expenses of public health in the country by the Union; the states and the municipalities are responsible for implementing the health policies.

Management of actions and health services by municipalities, including arrangements among municipalities/states that would allow the population's comprehensive care.

Reference-based system, with the basic (primary) care being its entry gate.

The supply of medicines, being the strategic, the exceptional, and the high-cost ones subsidized by the Ministry of Health.

According to the SUS operational rules, the municipalities considered "fully enabled for management" have autonomy in health management. This could complicate the structuring of a health services network, which needs to be regionalized or hierarchical in some situations. Complications could arise especially in more complex specialties such as medical genetics. In this chapter, the increasing importance and growing impact of birth defects and genetic diseases on health in Brazil has been already introduced. We will describe below some initiatives that are aimed at specific questions involving genetic diseases and, finally, the strategies proposed for the implementation of medical genetics in the SUS.

HEALTHCARE IN BRAZIL AND THE IMPACT OF BIRTH DEFECTS/GENETIC DISEASES

MEDICAL GENETICS AND HEALTHCARE IN BRAZIL

Most centers and care services related to the clinical genetics field are concentrated in the southeast and south, the most developed regions of the country. Such services are usually integrated into university hospitals and are responsible for the medical care of thousands of individuals and families annually. Furthermore, they are considered as references at regional or national levels. There are also hospitals that specialize in specific conditions, such as diseases and disorders of the locomotor system and craniofacial anomalies. Several of these institutions have laboratory infrastructure and are also capable of offering tests such as cytogenetic exams and several molecular biology techniques. However, the access to these services is limited. Although some of them are partly financed by the SUS, medical genetic services lack organization as a network and are not fully and officially integrated into the national health system (Marques-de-Faria et al., 2004).

ACTIONS IN BRAZIL RELATED TO BIRTH DEFECTS

Despite the lack, until 2009, of a comprehensive health policy specifically focused on birth defects, and that aimed to include genetics in the SUS, some government programs have emerged over the years. Sometimes such programs originated from the pressure of groups of patients/families who fought for diagnosis or treatment, thus emphasizing the absence of the government in their formulation. These programs show a timid presentation of the birth defects issue in the public policy agenda. However, they are expected to be incorporated in an integrated manner as part of the clinical genetics policy that hopefully will be implemented.

National Newborn Screening Program

Since 1990, laws and regulations regarding screening of newborns for genetic/congenital disorders were introduced in several parts of the country (Brasil, 1990a; Brasil, 1992). The goal of newborn screening would be to detect diseases that could lead to severe mental retardation if not treated early, whose diagnosis and early treatment would allow the proper development of the affected children.

The National Newborn Screening Program ("Programa Nacional de Triagem Neonatal," or PNTN) was established in 2001. It states that the diseases to be screened must follow criteria such as not presenting clinical signs at birth, as well as the possibility of early detection by safe and reliable tests. In addition to these criteria, the diseases included in the program must have an effective treatment, to be administered by programs with defined logistics, with an economically viable and socially acceptable cost-benefit relation (Brasil, 2001b; Brasil, 2001c). Based on these principles, screening has been recommended in Brazil for the following diseases: phenylketonuria (PKU), congenital hypothyroidism (CH), sickle cell disease and other hemoglobinopathies (SC), and cystic fibrosis (CF)[1]. This recommendation followed a rationale of progressive implementation phases (Phase I: CH and PKU, Phase II: PKU, CH, and SC; Phase III: PKU, CH, SC, and CF). The reference services for newborn screening were registered as "Newborn Screening Reference/Monitoring and Treatment of Congenital Diseases" and they should ensure not only the screening process but also diagnostic confirmation and appropriate monitoring and treatment of screened patients. By 2007, there were reference services registered by PNTN in all states of Brazil, which provided coverage of 80% of births[2] (Brasil, 2009d).

The newborn screening programs' experiences show important data, such as the possibility of good quality coverage, agility in the process, and also the integration capability between the primary healthcare and reference centers. As a main rule, newborn screening must be based on regional reference laboratories under state (or regional) supervision, and be also integrated into basic healthcare with the clinical support of reference centers. The global supervision is entrusted to the technical advisory group on newborn screening, established and coordinated by the healthcare department of the Ministry of Health[3] (Brasil, 2001c).

The analysis of the programs' results in various Brazilian regions is not only of epidemiological but also of economic interest to the country. Expanding the program to other diseases must be justified by epidemiologic data, and to diseases with real indication of traceability, if possible, according to ethnic regionalization. So far, the early detection of affected children and institution of therapy have resulted in the prevention of medical complications and severe mental retardation in many Brazilian children in the past decade.

Special Programs of the Ministry of Health for the Treatment of Genetic Diseases

Some genetic disease treatments are being undertaken by the Ministry of Health, and regulated by ministerial decrees. The process is not a simple one, especially due to issues such as the very high cost of some treatments. In addition to that, this high cost often involves legal actions or strong lobbying by families, who base such actions in the federal constitution, whereby everyone has the right to health and to healthcare that should be provided by the state (Senado Federal, 1988).

There are several arguments against the government's responsibility to take care of certain treatments, such as lack of federal funds, and the statement that many primary healthcare interventions could be implemented with such funds, instead of just treating a patient with a rare disease. If the unit "disability-adjusted life years" (DALY) was used, the investment in health activities with high cost per DALY gained would not be justified in developing countries (Mattos, 2003). On the other hand, the "Rule of Rescue" (Jonsen, 1986) could be used as a favorable argument, as nobody should be left on their own due to a treatment's high cost, especially if it is available and proven effective (Clarke et al., 2001). Not allowing a needy patient to receive effective treatment is also inconsistent with the principle of integrality, as well as with the constitutional principles of universality and equity in health (Mattos, 2003). Some of these treatment programs are described below:

Osteogenesis Imperfecta Program: Osteogenesis imperfecta (OI) is a disease caused by mutations in type I collagen, and is characterized by bone fragility and osteopenia. The patients present multiple fractures that can evolve into severe bone deformities and functional limitations. However, treatment with substances in the drug group of the bisphosphonates helps to decrease the bone reabsorption, to gain bone mass, to reduce the number of fractures, and to improve the patient's quality of life.

In December 2001, an ordinance was published instituting the treatment of OI within the Brazilian Health System (Brasil, 2001e). Instituto Fernandes Figueira/FIOCRUZ, a tertiary maternity and children's hospital that is one of the reference units of the Ministry of Health, was designated not only as a reference unit for the treatment of OI but also as the program coordinator in the country. Furthermore, approximately 150 patients were enrolled in that single center by 2008. Following the standards set by the ordinance, other centers were registered as reference centers for the treatment of OI. In this way, care was extended to other patients in all Brazilian regions. It is presently estimated that there are more than 700 OI patients under follow-up through this specific program in Brazil (Santos, 2009).

The Gaucher Disease Program: Gaucher disease is an inherited metabolic disease, classified as a lysosomal storage disease, and has as its main characteristics the enlargement of the liver and the spleen, hematological compromise, and bone lesions. It was the first metabolic disease to be treated by enzyme replacement therapy, introduced in clinical practice in the 1990s.

In 1995, the Brazilian Association of Gaucher disease, founded in 1994, won the first court injunction that forced the Ministry of Health to supply the enzymatic therapy. Nowadays, the medication is directly financed by the Ministry of Health. In 2002, an official protocol was established which set up state reference centers for evaluation, treatment, and monitoring of patients (Brasil, 2002). Currently, Imiglucerase is the most expensive drug in the country's extraordinary dispensation medicaments program. Moreover, the creation of the treatment program allowed an optimal access to therapy for more patients, not only due to the decrease in the individual medication's cost (direct negotiation between the Ministry of Health and the pharmaceutical industry), but also due to more organized treatment dynamics, as opposed to treatment on an individual basis. Despite the presence of more patients enrolled in the program, it is now more effective due to dosing rationalization, sharing of vials, and reduction of losses due to improper storage by families; in addition, tiring legal actions demanding access to treatment are no longer needed. The funding for the treatment of Gaucher disease has provided the means for the Brazilians who suffer from this rare disorder to have normal and productive lives.

The Cystic Fibrosis Program: Cystic fibrosis is a genetically determined disease, inherited in an autosomal recessive manner, and is characterized by chronic infections of the respiratory tract that lead to bronchiectasis, pancreatic insufficiency, intestinal malabsorption, sweat gland abnormalities, and urogenital dysfunction. Its incidence is estimated at 1:3000 live births among Caucasians, dropping to 1:17,000 among African Americans and 1:90,000 among Asians. In Brazil, the estimated incidence is around 1:10,000 but varies greatly among regions (its highest prevalence is in the south, due to European immigration). The disease involves very high morbidity and mortality, thus leading many patients never to reach adulthood. It is advocated that an early diagnosis and the institution of preventive measures would improve the quality of life and patients' survival. Newborn screening for cystic fibrosis is included in the Brazilian program for all the states that are in phase III (Brasil, 2001b; Brasil, 2001c). Besides the inclusion of this disease in the national screening program, an ordinance was published in 2001 that established a clinical protocol and guidelines for treatment, which also includes the supply of pancreatic enzymes for the disease (Brasil, 2001d).

Some aspects should be highlighted regarding the therapeutic programs for genetic diseases available via SUS. Treatments for other metabolic diseases are presently available (such as lysosomal diseases—Fabry, mucopolysaccharidosis types I, II, and VI, and Pompe disease, for which there are already enzyme replacement therapies—and aminoacidopathies benefited by special dietary treatment, among others) or are in final development stages. Treatment requests made by families and patient associations are not only part of everyday life, but they are also being routed through the law. Based on the federal constitution, the requests are almost always upheld by courts. It is essential that new programs follow well-established protocols such as those currently available. Along with the treatment programs, strategies for the prevention of new cases should be outlined, and the role of medical genetics and genetic counseling is essential in this process. Genetic counseling is one of the few possibilities of stabilizing the treatments' rising costs (WHO, 2000).

Birth Defects Registry: The Field Number 34 of the Liveborn Declaration

The liveborn declaration (LD) – official document issued by hospitals, without which the children's parents cannot perform their civil registration – had a new field introduced in 2000 (field number 34 – Brasil, 1999), partially modified in 2011 (fields 6 and 41 – CEInfo, 2011). Such LD allows congenital anomalies present at birth to be registered systematically. This information is not available in most countries; the data obtained may be an extremely valuable tool for monitoring and preventing birth defects in Brazil. Although the completion of the fields regarding birth defects is not mandatory, the way they were elaborated, with a yes/no question for the presence of congenital malformation plus an open field provided for the description and classification of the all the observed defects according to ICD-10 could provide, if properly filled in, a fairly complete record of abnormalities.

By analyzing data available through the DATASUS (the Ministry of Health's database) related to births in Brazil in 2006, only 0.6% of liveborn were registered as having a congenital anomaly (Brasil, 2009c), suggesting that such anomalies are being underreported. When comparing data from the deaths caused by a congenital malformation with data from the liveborn registry, Cunha et al. (2002) demonstrated an underreporting of 60.7% of birth defects on birth certificates. Guerra, who also suggested an underreporting, showed unsatisfactory reliability in the coding of the recorded anomalies by the Secretary of Health. These papers warn us about underreporting and underutilization of field number 34 (Guerra et al., 2008a; Guerra et al., 2008b). This item's addition to the LD will only be relevant for programs regarding birth defects if it is effectively implemented; in addition, its importance has to be recognized, not only as epidemiological data, but also as information that could help the planning of healthcare and actions to prevent birth defects.

THE "NATIONAL POLICY FOR COMPREHENSIVE CARE IN CLINICAL GENETICS AT SUS" (*POLITICA NACIONAL DE ATENÇÃO INTEGRAL EM GENÉTICA CLINICA NO SUS*)

The Brazilian Ministry of Health recently published a decree that proposes the creation of a "National Policy for Comprehensive Care in Clinical Genetics at SUS" (*Politica Nacional de Atenção Integral em Genética Clinica no SUS;* Brasil, 2009a). This is an important historical fact. At the time this chapter was written, however, no supplementary ordinance, which would be absolutely essential to organize and regulate this policy, had been published. At least, supported by the epidemiological transition which we are living, the need for an organized action in the area of medical genetics/birth defects in Brazil has been acknowledged. Some of the topics presented below will discuss some issues that justify the proposal of such policy, showing how it could work, and also pointing out the problems that could be encountered along the way, such as the training of qualified manpower and service capacitation.

Studies indicate that the overall incidence of birth defects in Latin America is not significantly different from that found worldwide (Penchaszadeh et al., 1999). In general, one can consider that no less than 5% of liveborns present some developmental abnormality, wholly or partly determined by genetic factors. The impact of such factors can be felt today in infant mortality in most of the world and even in Brazil.

In Brazil, medical practice in genetics has had a recent onset compared to other medical specialties. The Brazilian Genetics Society (Sociedade Brasileira de Genética, or SBG) was founded in 1955, and it brought together all people interested in issues related to genetics (Sociedade Brasileira de Genética, 2009). In the early 1980s, a group of physicians within the SBG, who worked in the area of clinical genetics, began to discuss the need for a society made up mostly of physicians. That would be essential to turn clinical genetics into a medical specialty, since a fully recognized residency program already existed at the time. That way, in 1986, the Brazilian Society of Clinical Genetics (Sociedade Brasileira de Genética Clínica, or SBGC) was created. Later it was renamed the Brazilian Society of Medical Genetics (Sociedade Brasileira de Genética Médica, SBGM), which endows the board certification in medical genetics through an agreement with the Brazilian Medical Association and the Federal Medicine Council, also regulating this medical specialty in the country (Brunoni, 1997).

INTRODUCTION AND HISTORICAL ASPECTS OF THE NATIONAL POLICY FOR COMPREHENSIVE CARE IN CLINICAL GENETICS AT SUS

The process that led the Ministry of Health to acknowledge the need to establish a national policy for care in clinical genetics began in 2001, and was partly influenced by the announcement of the sequencing of the human genome, a milestone in the history of science and humanity. The ethical, political, legal, and administrative matters related to the expectation of access to human genetic material became an issue in most countries. At this time, a committee to discuss the access and use of the human genome was created in the Ministry of Health (Brasil, 2001a). It comprised representation of various organs and entities of the ministry, and was coordinated by the Department of Science and Technology in Health (Departamento de Ciência e Tecnologia em Saúde; DECIT).

Moreover, one of the first work proposals of this committee was that every representative should map the activities and/or internal demands that were related to such topics in their own areas[4]. After that request was issued, a document that included the major endowments and duties related to the use of genomic technology in healthcare was compiled by the SAS (Health Care Department). Such a document, however, highlighted that given the existing demands and difficulties, most Brazilian regions were hardly prepared for clinical genetics practice, while the committee addressed issues relevant to genomic medicine[5,6].

At this point, it is worth mentioning that the emphasis on "genomics" as opposed to "genetics" could lead to the misconception that the former would involve different technologies that would be out of context in medical genetic services, or, even worse, that these technologies would have precedence over the development of appropriate and equitable healthcare for genetic diseases in general (WHO, 2002). Fortunately, the committee understood that there is no dichotomy between medical genetics and genomics. This is due to fact that the same molecular analysis tool, which allows the detection of a rare monogenic disease, can be used for determining the predisposition to common complex diseases. In other words, genomic technology can facilitate dealing with different diseases by elucidating the molecular mechanisms associated, thus contributing to the improvement of healthcare as a whole.

From this premise, the main conclusion was that there were basic problems of infrastructure and shortcomings in the area. The SAS would have the challenge of establishing a minimum organizational structure, from which strategic actions would be applied to ensure comprehensive care in genetics. Such actions would have to provide a universal, fair, and equitable access to tests and procedures in this area, including the forthcoming genomic technology.

After the impact of the Human Genome Project, the Ministry of Health finally began to pay attention to medical genetics and to consider the preparation of a National Policy for Comprehensive Care in Clinical Genetics at SUS. Such initiatives, however, transcended the initial attributions of the committee regarding the human genome and depended on a specific action of the SAS. This action was subjected to internal decisions related to agenda issues and the sector's priorities, and also took into consideration political aspects due to the change of government in the period. This delayed the process until a Working Group (WG) was formally established by an ordinance (Portaria GM/MS No. 2380) in October 28, 2004, to address the issue (Brasil, 2004).

WORKING GROUP (WG) ACTIVITIES AND STRUCTURING THE POLICY FOR CLINICAL GENETICS

The WG had as its designated task the development of the national policy for comprehensive care in clinical genetics at SUS, taking into consideration the lack of regulation and standardization, and also regional inequities in healthcare in clinical genetics in the country. The coordination of the WG was managed by the Department of Science and Technology (DECIT), of the Science, Technology and Strategic Inputs Secretary, and by the Department of Healthcare (SAS) of the Ministry of Health. It was formed by representatives of various agencies and entities of the Ministry of Health, of the Brazilian Society of Medical Genetics (SBGM), the Brazilian Society of Genetics (SBG), and of reference services in clinical genetics.

During the period 2004–2006, aiming to assess the situation of care in clinical genetics and to promote a debate on the various instances of the SUS, involving local healthcare managers and service providers, as well as to obtain grants for the development of the policy, there were several meetings and two regional workshops (south/southeast and north/northeast/midwest). Representatives of the Ministry of Health also participated in special sessions during clinical genetic conferences.

This process resulted in a proposal that ultimately led to ordinance No. 81 of the Ministry of Health in January 20, 2009, establishing the National Policy for Comprehensive Care in Clinical Genetics within the SUS (Brasil, 2009a), and also designated the strategies for actions that must be taken into account in its regulation. This was a long overdue landmark for medical genetics professionals in Brazil.

Care networks would be established based on the actions of the state health secretariats together with the multiple municipal health secretariats. The agreement between the state and municipal managers from each region would determine the organization, and would also provide the necessary conditions for the networks' operation. Such organization would have to follow specific guidelines proposed by the Ministry of Health; the ministry would also support the network organization by partially financing the system (CONASS, 2007).

The fundamental objective of such policy is the integration of specialized services with primary healthcare. Such integration would prioritize prevention, but with a strategy of comprehensive and integral healthcare with promotion, prevention, treatment, and rehabilitation, in order to involve all levels of healthcare by offering interdisciplinary and multiprofessional assistance. Building a comprehensive network focused on the citizens' needs in the area of clinical genetics, from the first appointment to the specialized procedures, is one of the goals of this policy. As the entrance gateway into the system, primary healthcare will have to be prepared to identify individuals and families with birth defects and genetic diseases, who will then receive adequate healthcare at the primary level and will be referred to specialized care whenever necessary, for evaluation or treatment. Specialized healthcare should be provided with infrastructure and trained professionals to take care of those referred both from primary healthcare and from other specialties. That means that targeted multidisciplinary monitoring and other foreseen procedures will be included in this level.

For the structuring of the network, Clinical Genetic Services (CGS) and Centers for Specialized Health Care in Clinical Genetics (CSHCCG) would be accredited. Most CSHCCG would be headquartered in teaching hospitals, and also be able to train professionals and support local health managers. On the other hand, CGSs would be responsible for specialized healthcare in accordance with the states′ regionalization master plans.

The implementation of such a policy will be progressive, taking into account local capacities of network organization. The major structuring axis for healthcare in clinical genetics will be genetic counseling, which must be guaranteed to any individual or family who may require it. Care guidelines should be established for investigating the most frequent conditions, such as mental retardation, congenital anomalies, and inborn errors of metabolism.

A thorough evaluation of the financial impact of the inclusion of specific genetic tests and of genetic counseling in health costs/financing is still needed. In addition, manuals and guidelines should be developed to guide professionals from primary and specialized care.

All these initiatives and other necessary measures for fully structuring, monitoring, and improving the policy will be empowered to the SAS, who will pursue these actions through the constitution of a technical advisory board (CONASS, 2007).

THE GENETICS NETWORK AVAILABLE IN BRAZIL: THE BACKBONE OF THE POLICY TO BE INTRODUCED

The most recent census work related to the mapping of medical genetics services in Brazil was completed in 2003 (Horovitz, 2003). This work aimed to locate public medical genetic services available to the Brazilian population, including those related to laboratories, whether integrated or not to clinical centers and to medical schools. At the time, clinical care in the specialty was offered in 48 institutions, out of which 33 were somehow integrated into a genetic laboratory. More recent data[7] show that new genetic services are being implemented in states previously devoid of such kind of assistance. Nevertheless, the following Brazilian states still have no specialized care in genetics: Amazonas, Amapa, Roraima, Rondônia, and Tocantins.

The greater lack of services is certainly in the northern regions. Regarding the northeast region, services are mainly clinical, without local laboratory support in some of the states; on the other hand, the number of yearly medical consultations is quite impressive in some of those services. The majority of medical genetic services in Brazil are located in the southeast region or in state capitals, with Sao Paulo state being a noticeable exception, where there are also services in the interior of the state.

According to data provided by services (Horovitz, 2003), approximately 56,000 outpatient consultations in clinical genetics are estimated annually in the country (including counseling of couples and prenatal genetic counseling). From this amount, fewer than 16,000 cases represent first-visit patients. And it is noteworthy that out of the total demand, more than half of them are performed in only nine facilities.

If it is considered that all newborns with birth defects need a genetic evaluation, with a minimally estimated incidence of malformations (2% of total births), at least 60,000 first-visit consultations per year would be necessary in the country[8]. The estimate has an even higher deficit if we consider the broadest definition of birth defects, including problems not necessarily detected at birth, along with developmental disorders (in that case, 5% of total births would represent 150,000 new cases per year in Brazil). It is also worth remembering that many medical consultations regarding birth defects are related to problems during pregnancy. Genetic counseling is recommended to parents of malformed stillborn babies, and such statistics are not even included in the estimates above. By analyzing these data, the number of appointments offered is far below the actual need.

Despite the estimated shortage in availability, the genetic services that already exist could form the "backbone" of any organized plan to implement appropriate diagnosis, prevention, and treatment of genetic diseases within the community.

MEDICAL GENETICS LABORATORIES AND LABORATORY NETWORK

Medical genetics is an interdisciplinary area, in essence, and also collaborative. Characteristically, this specialty deals with rare diseases, making it impossible to have "in-house" laboratory support for all diagnoses, even in the most comprehensive services. Having that stated, many laboratories around the world, including in Brazil, have been specializing in specific diagnostic tests.

Gathered data shows that the distribution of the medical genetic services in Brazil is uneven, with higher concentration in the south-center region and with a variable complexity range (Marques-de-Faria et al., 2004; Horovitz, 2003). In addition, the availability of diagnostic exams for genetic diseases follows similar distribution logic. Presently, there is no logical referenced flow for diagnostic tests, although clinical services around the country send and receive samples based on informal agreements between them.

A closer look at the latest research on genetic services in Brazil (Horovitz, 2003), updated with information not yet published[9], shows that diagnostic genetic tests were available in 47 out of 66 genetic services linked to the SUS (71%). Among these, 83% offered conventional cytogenetics, 55% high resolution cytogenetics, 32% fluorescent in situ hybridization, 36% tests for inborn errors of metabolism, and 32% prenatal diagnosis. Around 50% also performed investigations using molecular biology techniques for several groups of diseases, including mental retardation, dysmorphic syndromes, cancer predisposition, infertility, and metabolic diseases, among others. To date, there is no structured or published data on the number of genetic tests available in Brazil. Obviously, it is understood that this set of laboratories is insufficient to meet the needs of new annual cases, and also of the rising number of cases that still lack diagnosis in the country. Just for new Down syndrome patients, over 4000 cytogenetic tests would be needed annually; considering that about 50% of children with birth defects would also need this exam, we would have an annual need of 45,000 cytogenetic tests (assuming a rate of 3% birth defects). Such figures do not take into account infertility and other genetic disorders.

Genetic tests can be quite complex and expensive, which partly explains the need to organize a laboratory network to support diagnosis. This network is informally organized in Brazil by health professionals. Unpublished data show that the most popular e-mail topic in a medical genetics web-based discussion list[10] (about 21%) is requests for information about exams for genetic disorders (GENETICACLINICA discussion list, 2009).

The information regarding genetic testing in Brazil is not officially organized, nor does there exist a referral network in SUS. By 2009, the cost of tests sent outside of their units, even within the same state or to other states in Brazil, were not being paid to the executing laboratories in most cases. According to information from such laboratories, there are no official billing mechanisms to charge for outside exams (sent from other institutions/countries/states). Therefore, such tests are sometimes directly paid by the users, or, even worse, routine and necessary testing ends up being covered by research grants (Horovitz, 2003). Due to the absence of formal accounting mechanisms in Brazil, much of the clinical testing required by medical genetic services is performed by the research reference laboratories based mainly on personal connections and goodwill. It is expected that the implementation of a special policy for clinical genetics in SUS will change this scenario, stimulating the creation of regional laboratories and formalizing the informal network created by professionals working in public health.

A census to outline the laboratories available within the Brazilian public sector, which tests they already offer, and the development of a dynamic tool for updating and providing these and new tests, are important for the organization of a comprehensive network for care in genetics. Instruments that provide information about the exams and laboratories already exist in some countries, such as in Scotland (Kelly, 2002), as well as in the United States, with the GeneTests project. In May 2010, this database included data regarding tests in 578 laboratories, encompassing 2067 diseases (GENETESTS, 2010).

Obviously, besides the existing organization for genetic testing, laboratories at several centers distributed in Brazil should be funded to introduce diagnostic techniques for at least the most frequent diseases, such as karyotyping, screening for inborn errors of metabolism, and even molecular biology techniques, for example, for the diagnosis fragile-X mental retardation. The inclusion of new tests could be addressed by a standing committee of experts from the Ministry of Health, in a process similar to that in England, where the clinical and epidemiological impacts are considered (Kroese et al., 2007). Furthermore, due to the epidemiological characteristics of genetically determined diseases, sending samples abroad for diagnosing rare conditions in reference laboratories must also be contemplated.

MEDICAL GENETICS AND PRIVATE HEALTHCARE IN BRAZIL

About 25% of the Brazilian population has some sort of private health insurance coverage (private healthcare plans). The Supplementary National Health Agency (*Agência Nacional de Saúde Suplementar* – ANS) was founded in 2000 and is affiliated to the Ministry of Health. Its aim is to promote the public interest in supplemental healthcare. This agency regulates the relationship between providers, government, and clients, by determining the minimum and mandatory coverage requirements for plans under contract.

Genetic testing within the private sector is quite different. Due to the wide media coverage and publicity, there is more pressure on the private sector for laboratory diagnosis of genetic diseases. Since April 2008, a special regulation (RN no. 167, 10.1.2008), which updated the list of procedures in the private health sector, was introduced (ANS, 2009). In the resolution, several laboratory procedures in genetics (cytogenetics, biochemical, and molecular techniques including in situ hybridization and prenatal diagnostic tests) were included with mandatory coverage; some genetic tests have to be made available according to specific guidelines. At the same time, the resolution contemplates the possibility of conducting any molecular genetic test, if all other diagnostic possibilities have been unsuccessful and if a clinical geneticist orders such an exam.

Far from being a unanimous solution, this resolution shows that genetic tests are a reality in the country. An informal search of the Internet can show that. A large number of clinical laboratories now offer testing for genetic diseases by molecular biology techniques. In addition, several laboratories exclusively specialized in the diagnosis of genetic diseases already exist in Brazil. The distribution of these laboratories also follows the concentration pattern in the center-south region, although other laboratories throughout the country offer such testing, and function as blood drawing centers; the samples will be sent out and the actual testing will be performed in the reference laboratories within the center-south.

The new exam list is still being adjusted. However, what has been observed in practice since its implementation was an extension, although still modest, of the access to healthcare and clinical testing in genetics.

TRAINING OF HUMAN RESOURCES FOR COMPREHENSIVE HEALTHCARE IN MEDICAL GENETICS

Perhaps one of the most critical points for the success of a national policy based on comprehensive care in medical genetics is human resources. The evaluation and care of families with needs in this area requires training in topics poorly covered during higher education in medicine and in the healthcare field, such as nondirective genetic counseling, dysmorphology and syndromology, and also basic knowledge of genetic testing. Moreover, the training of specialists in medical genetics still has a low numerical impact. Adding to this the fact that specialists tend to settle within the major services in the center-south region of the country, it turns out that most Brazilian states have very few or no trained professionals in clinical genetics. Thus, the professional training demanded by the health needs in this field could be divided into genetics training within the undergraduate health courses, and specialist postgraduate training.

GENETICS TRAINING IN THE HEALTH CAREERS

Integrated healthcare in medical genetics cannot depend only on specialized training. Certainly, it should be encouraged and enhanced with a better distribution of specialization programs around the country. However, a key concern in human resources training is to provide the necessary conditions for health professionals in order for them to be competent in identifying problems, and even to manage the most common diseases. A key competence is to recognize reproductive issues as important when it comes to genetic diseases, and to adequately address recurrence risks through nondirective genetic counseling.

In the international literature several recommendations concerning the competence of health staff not specialized in genetic diseases can be found. The U.K. model is quite interesting, where a list of skills that the health team must have for healthcare in genetics was created, focusing on the needs of patients and their families (UK Workforce, 2007). Such skills involve recognition of genetics in their daily practice, recognition of individuals with health needs in this area, obtaining and analyzing family history, estimating genetic risks, recognizing the need for genetic testing, referring to specialists, and communicating genetic information to patients, family, and the healthcare team. The approach is not focused on professionals, but on the needed care, therefore focusing on the customers of such services. Thus, some of the responsibilities are assigned to professionals specifically trained in medical genetics, both physicians and nonphysicians, although the needs will have to be recognized by the whole team.

This model of care has an impact on the general education of health professionals, as well as in specialized training. A comprehensive proposal is made in a document (now in its third edition) by the National Coalition for Education for Health Professional Education in Genetics, who suggests minimum competencies for all health professionals, based on the acquisition of knowledge, skills, and attitudes in three broad guiding areas: (1) to examine their practice skills regularly, identifying areas where professional development related to genetics and genomics would be beneficial; (2) to understand that genetic information related to health has important social and psychological implications for individuals and families; and (3) to know how and when to refer an individual or family to a specialized professional in genetics. (Core Competencies Working Group, 2007).

For medical schools in particular, an important document is the proposal for a core curriculum in genetics by the American Society of Human Genetics. In this document, which indicates abilities, skills, and attitudes desirable for medical genetics, there are four guiding propositions: (1) genetics should be taught along the entire course (basic and specific); (2) there must be a specific discipline of medical genetics, because even though there are several areas of interrelationship, there are several specific contents to be addressed; (3) the responsibility of structuring the curricula in medical genetics will be conducted by someone trained in medical genetics; and (4) several methods of teaching and learning can be used, including clinical discussions and class medical practice (ASHG, 2001).

In Brazil, genetics is part of the curricula in several health-related graduate schools, either as a structured discipline or among one of the major themes such as cell biology. It is classically taught within the basic disciplines of the courses, with little, if any, integration with practice. A survey conducted by Porciuncula in 2004 shows great content heterogeneity in medical schools. There is a strong presence of genetics within the so-called basic cycle, but also a growing connection with professional practice, especially with the inclusion of specific disciplines of medical genetics in the clinical cycle (Porciuncula, 2004). When analyzing his data on the surveyed medical schools in comparison to national and international literature, the author proposes that an array of abilities, skills, and attitudes toward medical genetics be built, adapted to Brazilian needs (Porciuncula, 2004). Education in medical genetics in Brazil still suffers from the absence of such a document, and the strengthening of the contents of medical genetics in healthcare teaching, strongly anchored epidemiologically, is a fundamental part of the effort to better deal with these problems in Brazil today.

Postgraduate Training: The Srictu-Senso Postgraduate Programs

It is widely known that the strictu-senso postgraduate programs in genetics in Brazil have been collaborating with the training of professionals for teaching and research in medical genetics. Although the primary purpose is to train professionals for universities and research centers, a reasonable proportion eventually ends up being trained in specific areas of medical genetics, particularly laboratory investigation in cytogenetics, biochemistry, and molecular biology.

The traditional academic training in genetics in Brazil leads to a diverse panel of scientific research in medical genetics. A very brief analysis performed by Brunoni (2002), when assessing major themes of papers presented at the Brazilian national meetings of genetics and medical genetics in 2000 and 2001, also points to a large concentration in a few research centers, with high concentration in the center-south of the country.

The nonmedical professionals whose undergraduate or postgraduate training enables them to perform within medical genetics, or even those who enter the area through practice, may obtain qualification through their associations, although few opt do so. The Brazilian Society of Genetics (SBG) has been conducting certifications in human cytogenetics and in human molecular genetics since 1999, through an agreement with the Federal Councils of Biology, Biomedicine and Pharmacy. Until 2010, fifteen professionals had obtained such certification (Sociedade Brasileira de Genética, 2011).

Postgraduate Training: Medical Specialization in Medical Genetics

The first medical residency program in medical genetics in Brazil, recognized by the Ministry of Education, was installed in the hospital of the medical school of Ribeirão Preto, University of São Paulo, in 1977. Medical genetics was recognized as a medical specialty by the Brazilian Medical Association in 1983. Since then, several new residency programs were created, totaling 11 programs and 23 vacancies for new physician trainees yearly. The program, approved by the National Medical Residency Committee, is of direct access (after the end of undergraduate medical training) and lasts three years, with its shape governed by Resolution CNRM No. 02/2006 of 17 May 2006 (Table 103-1; Comissão Nacional de Residência Médica, 2006).

The residency programs in medical genetics in Brazil are highly concentrated in the center-south region of the country. Of the 11 programs, 8 are in the southeast (four in Sao Paulo, three in Rio de Janeiro, and one in Minas Gerais), two in the south (Rio Grande do Sul), and one in Brasília – Federal District (Secretaria de Educação Superior, 2009a,b). This distribution follows the already described logic of the high-complexity services, also heavily concentrated in these regions.

Expertise in a medical specialty in Brazil is not based solely on the titles coming from medical residencies and recognized by the Ministry of Education. Another form of professional recognition comes from board certification,

TABLE 103-1 MEDICAL TRAINING IN BRAZIL

First Year: (80%) programs in internal medicine and pediatrics, focusing primarily on general and community pediatrics, well-baby nursery, pediatric neurology, general practice (internal medicine), neurology, endocrinology

Second Year: Training in clinical genetics outpatient clinic (60%); dysmorphology (25%); minimum of 120 cases per resident per year metabolic diseases (15%), minimum 50 cases per resident per year; infertility (10%), minimum 20 cases per resident per year; genetic counseling (10%), minimum 20 cases per resident per year; consulting as specialist in other hospital clinics (10%); nursery/fetal pathology 10%; fetal medicine 10%

Third Year: Training in clinical genetics laboratory activities. (30%) cytogenetics–biochemical genetics–molecular genetics general outpatient clinic; (40%) outpatient related specialties; 10% cancer genetics, neuromuscular diseases, or other.

Elective training: Radiology and Diagnostic Imaging–Dermatology–Ophthalmology–Cardiology

Compulsory courses: - Basic Medical Genetics–Genetic Diseases and Birth Defects - Pathogenic Mechanisms in Genetic Diseases

The institution must have at least the following required infrastructure in facilities, equipment, and personnel to offer training in the laboratory of cytogenetics, biochemical genetics, and molecular genetics. The service must have a basic laboratory unit for biological sample preparation. The minimum number of procedures required per year per 2nd-year resident is, in addition to outpatient clinics, at least 5 autopsies (fetal pathology) and 10 evaluations with fetal medicine. The minimum number of laboratory activities required per year per 3rd-year resident is (a) biochemical genetics: a minimum of 15 tests for screening of inborn errors of metabolism and 15 chromatography tests; (b) cytogenetics: a minimum of 15 lymphocyte cultures and preparation of karyotypes; and (c) molecular genetics: a minimum of 15 DNA isolation procedures and 15 southern blot and/or PCR tests.

Teaching/scientific and complementary theoretical activities (20%): clinical case discussions, diagnostic exam sessions, scientific paper discussions.

awarded by the societies of medical specialties and recognized by the Brazilian Medical Association. Board certifications in medical genetics are held annually since 1993 by SBGM, and involve a theoretical test, analysis of curriculum, and interviews. This type of certification, often also pursued by those who were already certified by medical residency programs, also allows physicians with strictu-senso postgraduate training, and those with practical training for several years in medical genetic services, to be recognized as specialists. An interesting item regarding board certification is the need for continuous updating to keep it valid, by a scoring system for five-year periods, obtained from attendance at conferences, courses, etc. Thus, this complex system of training and qualification seems to be appropriate in format, but unfortunately has not yet come close to the fulfilling the professional needs for the country in the specialty.

It is also important to point out that despite many professionals being involved in genetics care, clinical evaluation and genetic counseling are delivered exclusively by physicians, with a few exceptions (nurses counseling patients with inherited cancer in some reference centers; genetic information regarding sickle cell trait after newborn screening given in primary care settings, for instance). Genetic counseling is not a recognized profession in Brazil.

Training of Human Resources – Conclusion

Professional training for health professionals in medical genetics is present in Brazil, although highly concentrated in central-south regions of the country. This would partly allow the training of human resources for the SUS network, in a continuing education strategy, enabling them to identify and appropriately manage the most common problems, while recognizing the need to refer to specialized services. Direct action on the undergraduate courses in the health-related areas, however, is long overdue. The medical schools, as well as other health-related courses, have been heterogeneously adapting to these new requirements, demanded by the changing epidemiological profile of Brazil. As for skilled professionals, they are few, most working in medical genetic services highly concentrated in the central-south area of the country. Data collected in 2010 by the Brazilian Medical Association and the National Commission of Medical Residency shows that around 200 physicians have been awarded board certification in medical genetics since 1981 in Brazil.

The educational apparatus for medical experts would have the means, in addition to expanding the number of specialists, to provide suitable environments for training other professionals in a multidisciplinary team. All this must be encouraged, and part of the incentive will be the placement of these professionals within the service network, which would involve the creation and implementation of a public health policy to deal with care in genetics.

FINAL CONSIDERATIONS – MEDICAL GENETICS IN BRAZIL

Contrary to the prevailing common sense, care in medical genetics does not necessarily involve high costs and high technology. Health actions in the area of genetics do not differ conceptually from other healthcare actions, and involve the various aspects of diagnosis, prevention, and treatment. In that sense, organizing a network would be nothing more than properly channeling the investments, and structuring the already existing services in an integrated and efficient manner, leading to a fair and more democratic access to healthcare.

In addition to formalizing and carrying out the organization of a network in clinical genetics in Brazil, other actions need to be implemented for the system not only to properly function, but also to be gradually expanded and adapted to the country's growing needs. Government initiatives that already exist should be *consolidated*, and nongovernmental programs may eventually be added and enrich the system. As examples of optimization and integration, informing city officials about the importance of the correct completion of "Field 34" in the newborn declarations should be encouraged. Such data will be of great relevance for monitoring incidences of birth defects, programming specific health actions, and evaluating the effectiveness of actions already implemented, such as folic acid fortification in flour. The importance of patient/parent organizations should also be reinforced; besides offering support and comfort to their members, such associations have as a goal the dissemination of information among lay people and also to physicians. These nongovernmental associations can also play the fundamental role of introducing new topics on the political agenda.

Finally, the issue of prevention must be addressed. We could have pages of discussion on what is acceptable and what would be considered ethical; for many situations in genetics, early detection (in some cases, during pregnancy) is the only way to prevent the birth of an affected child[11], It is important to note that professionals, geneticists or not, are often faced with issues of great ethical and legal complexity. In such instances, education is crucial. Complex decisions should only be made based on real understanding, as well as respecting people's rights. Clinical genetics advocates that families should have autonomy in their decisions.

The network configuration of services in clinical genetics (regionalized, hierarchical, and functional, as recommended in the creation of the SUS/Unified Health System in Brazil) will be a crucial item for care in medical genetics. Unfortunately, until November 2011, more than two years after its publication, the National Policy for Comprehensive Care in Clinical Genetics, whose inception was idealized and formatted by the Ministry of Health, has not yet been implemented in Brazil. The Federal Public Prosecution is contemplating the instigation of a public civil inquiry in order to identify (and if necessary, to punish) those responsible[12]. Considering the magnitude of the impact that birth defects and genetic diseases already have on health, even in a country like Brazil, it is recommended that healthcare in genetics truly becomes part of the Unified Health System.

NOTES

1. The movement that began in civil society organizations (families with cystic fibrosis) was essential for the creation of the National Newborn Screening Program in Brazil. For this reason, despite existing doubts regarding the validity of screening for that disease on a public health level in Brazil, and at the expense of not investigating for other, possibly more frequent diseases, or for diseases for which screening or treatment could be more effective, it was incorporated into the group of diseases to be investigated.
2. It must be remembered that such statistics comprise tests performed by PNTN compared to the *total* number of births. Newborns in whom screening was performed by private medical insurance are not included, possibly leading to a sub-registry of actual newborn screening coverage in the country.
3. The *Secretaria de Assistência a Saúde—SAS*, referenced in this text as the Health Care Department, is a branch of the Brazilian Ministry of Health that plans and coordinates integrated healthcare actions.
4. Proceedings of the Committee on Access and Use of the Human Genome's 1st Meeting, Ministry of Health, Department of Health Policy, Department of Science and Technology in Health, August 1, 2001.
5. Proceedings of the Committee on Access and Use of the Human Genome's 2nd Meeting, Ministry of Health, Department of Health Policy, Department of Science and Technology in Health, December 20, 2001.
6. Document presented by one of the authors of this chapter (Marques-de-Faria AP) in the 2nd meeting of the Committee on Access and Use of the Human Genome on December 20, 2001.
7. Such data was obtained by personal communications, annual conference proceedings of the Brazilian Society of Medical Genetics, and also from the society's member listings.
8. These figures consider an average of 3 million births per year in Brazil.
9. Information updates on services, including those nonexisting or not computed in 2003 (time of the survey); the information was obtained through personal communications or through published data in the annals of the Brazilian Society of Medical Genetics meetings between 2004 and 2009.
10. GENETICACLINICA - http://health.groups.yahoo.com/group/geneticaclinica/
11. In Brazil, abortion is only considered legal in cases of pregnancy resulting from rape, or if the continuation of the pregnancy is considered life-threatening for the mother (Criminal Code of 1940). From the 1990s, court orders that allow pregnancy terminations in cases of anencephaly and severe fetal malformations (mostly incompatible with life) have been obtained, although discussions on the decriminalization of abortion still have much to evolve.
12. Brazilian Society of Medical Genetics – Information sent to its members by electronic mail – April 8, 2010.

REFERENCES

American Society of Human Genetics. (2001). Medical school core curriculum in Genetics. Available at: http://www.ashg.org/pdf/Medical%20School%20Core%20Curriculum%20in%20Genetics.pdf. Accessed on 10/31/2011.

ANS (Agência Nacional de Saúde Suplementar). 2009. Resolução ANS nº 167, de 09 de janeiro de 2008. Atualiza o Rol de Procedimentos e Eventos em Saúde, que constitui a referência básica para cobertura assistencial nos planos privados de assistência à saúde, contratados a partir de 1º de janeiro de 1999, fixa as diretrizes de Atenção à Saúde e dá outras providências. Available at: http://www.ans.gov.br/texto_lei.php?id=1242. Accessed on 10/31/2011.

Brasil. (1990a). Lei nº. 8069 de 13 de julho de 1990. Dispõe sobre o estatuto da criança e do adolescente e dá outras providências. Inciso III, artigo 10. *Diário Oficial da União* Jul 16; p.13563.

Brasil. (1990b). Presidência da República. *Lei no. 8.080* de 19 de setembro de 1990. Dispõe sobre as condições para a promoção, proteção e recuperação da saúde, a organização e o funcionamento dos serviços correspondentes e dá outras providências.

Brasil. (1992). Ministério da Saúde, Gabinete do Ministro. Portaria GM/MS nº. 22, 1992 15 Jan. Trata do programa de diagnóstico precoce do Hipotireoidismo Congênito e Fenilcetonúria.

Brasil, Ministério da Saúde. (1999). Sistema de Informações sobre nascidos vivos. Manual de instruções para o preenchimento da declaração de nascido vivo, 3ed.

Brasil, Ministério da Saúde. (2001a). Portaria N.º 470/GM, de 6 de abril de 2001 – Institui a comissão sobre acesso e uso do genoma humano.

Brasil. (2001b). Ministério da Saúde, Gabinete do Ministro. Portaria GM/MS nº. 822, 2001 06 Jun. Institui, no âmbito do Sistema Único de Saúde, o Programa Nacional de Triagem Neonatal/PNTN.

Brasil. (2001c). Ministério da Saúde, Secretaria de Assistência à Saúde. Portaria SAS nº. 189, 2001 06 Jun. Institui, no âmbito da Secretaria de Assistência à Saúde – Ministério da Saúde, o Grupo Técnico de Assessoramento em Triagem Neonatal.

Brasil. (2001d). Programa Fibrose Cística. Available at: http://dtr2001.saude.gov.br/sas/PORTARIAS/Port2001/PT-263.htm - *Portaria SAS/Nº 263 de 18 de julho de 2001* - Aprova o PROTOCOLO CLÍNICO E DIRETRIZES TERAPÊUTICAS – FIBROSE CÍSTICA – Enzimas Pancreáticas.

Brasil. (2001e). Ministério da Saúde, Gabinete do Ministro. Portaria GM/MS nº. 2305, 2001 Dez 19. Aprova o Protocolo de Indicação de Tratamento Clínico da osteogenesis imperfecta com pamidronato dissódico no âmbito do Sistema Único de Saúde – SUS.

Brasil. (2002). Ministério da Saúde, Secretaria de Assistência à Saúde. Portaria SAS nº. 449, Jul 08. Aprova o protocolo clínico e diretrizes terapêuticas – doença de Gaucher - Imiglucerase. *Diário Oficial da União* Jul 09; 130(1).

Brasil, Ministério da Saúde. (2004). Portaria GM no. 2.380, de 28 de outubro de 2004 - Institui o Grupo de Trabalho de Genética Clínica, e dá outras providências. *Diário Oficial da União*, 209 (Oct 29, 2004).

Brasil. (2006). Ministério da Saúde. Entendendo o SUS. Available at: http://portal.saude.gov.br/portal/arquivos/pdf/cartilha_entendendo_o_sus_2007.pdf. Accessed on 10/31/2011.

Brasil. (2008). Ministerio da Saude, Secretaria Executiva, Departamento de Informatica do SUS. Rede Interagencial de Informações para a

Saúde. Indicadores e Dados Básicos. Available at: http://tabnet.datasus.gov.br/cgi/idb2008/c16.htm.

Brasil. (2009a). Ministério da Saúde, portaria GM no. 81, 20 de janeiro de 2009 – Institui, no ambito do Sistema Único de Saude (SUS), a Politica Nacional de Atenção Integral em Genetica Clinica. *Diário Oficial da União* 21/01/2009.

Brasil. (2009b). Ministério da Saúde. Portal da Saúde–portalsaude.saude.gov.br; principal (sobre o SUS) http://portalsaude.saude.gov.br/portalsaude/index.cfm/?portal=pagina.visualizarArea&codArea=347. Accessed on 10/31/2011.

Brasil. (2009c). Ministério da Saúde. SINASC (Sistema de Informações sobre nascidos vivos), data regarding the year of 2006. Available at: http://tabnet.datasus.gov.br/cgi/tabcgi.exe?sinasc/cnv/nvuf.def.

Brasil. (2009d). Ministério da Saúde - Programa Nacional de Triagem Neonatal: Indicadores. Fonte: Relatórios anuais enviados pelas Secretarias da Saúde dos Estados e Distrito Federal http://portal.saude.gov.br/portal/arquivos/pdf/INDICADORES_TRIAGEM_NEONATAL.pdf

Brasil. (2009e). Ministério da Saúde, Departamento de Informação e Informática do SUS – DATASUS, Sistema de Informações de Saúde. Sistema de Informações Hospitalares – SIH/SUS. Available at: http://tabnet.datasus.gov.br/cgi/tabcgi.exe?sih/cnv/nruf.def).

Brasil (2009f). Ministério da Saúde, Departamento de Informação e Informática do SUS – DATASUS, Sistema de Informações de Saúde. Sistema de Informações sobre Mortalidade (SIM). Available at: http://www2.datasus.gov.br/DATASUS/index.php?area=0205&VObj=http://tabnet.datasus.gov.br/cgi/deftohtm.exe?sim/cnv/obt10.

Brasil (2009g). Ministério da Saúde. Sala de Situação em Saúde do Ministério da Saúde. Available at: http://189.28.128.178/sage/?saude=http://189.28.128.178/sage/&botaook=OK&obj=http://189.28.128.178/sage/

Brunoni D (1997). Estado atual do desenvolvimento dos serviços de genética médica no Brasil. *Brazilian Journal of Genetics* 20(suppl):11–23.

Brunoni, D (2002). Aconselhamento Genético. *Ciência e Saúde Coletiva* 7(1):101–107.

CEInfo - Coordenação de Epidemiologia e Informação. (2011). Manual de preenchimento da Declaração de Nascido Vivo. São Paulo: Secretaria Municipal da Saúde São Paulo (cidade). Declaração de Nascido Vivo, 2011. 24p.

Clarke J, Amato D, Deber R (2001). Managing public payment for high-cost, high-benefit treatment: enzyme replacement therapy for Gaucher's disease in Ontario. *Canadian Medical Association Journal* 165:595–596.

Comissão Nacional de Residência Medica (2006). *Resolução CNRM nº 02, de 15 de maio de 2006.* Dispõe sobre requisitos mínimos dos Programas de Residência Médica e dá outras providências. Available at: http://portal.mec.gov.br/sesu/arquivos/pdf/cnrm/resolcnrm002_2006.pdf

CONASS [Conselho Nacional de Secretários de Saúde]. (2007). Política Nacional de Atenção Integral em Genética Clínica. *Nota Técnica n.º05/2007.* Available at: http://www.conass.org.br/admin/arquivos/NT05-07.pdf

CFM (2011) - Conselho Federal de Medicina. Available at: http://portal.cfm.org.br/index.php?option=com_content&view=article&id=22200:informacoes-sobre-a-saude-suplementar-no-brasil&catid=3 Accessed on 10/31/2011.

Core Competencies Working Group – National Coalition for Health Professional Education in Genetics. (2007). *Core Competencies in Genetics for Health Professionals* 3rd Edition. Available at: http://www.nchpeg.org/index.php?option=com_docman&task=doc_download&gid=44&Itemid=. Accessed on 10/31/2011.

Cunha J, Aerts d, Leite Jcl, Castilla ee. (2002). Defeitos congênitos em Porto Alegre: uma investigação da qualidade dos dados registrados na Declaração de Nascido Vivo. *Revista Brasileira de Epidemiologia* (suppl.):51.

DATASUS – Ministério da Saúde. http://www.datasus.gov.br

Discussion List GENETICACLINICA. (2009). Managed by Dr. Victor Evangelista de Faria Ferraz - Departamento de Genética da Faculdade de Medicina de Ribeirão Preto – Universidade de São Paulo, Brasil. Available at: http://health.groups.yahoo.com/group/geneticaclinica

GENETESTS (2010): Medical Genetics Information Resource (database online). Copyright, University of Washington, Seattle. 1993–2006. Available at: http://www.genetests.org

Guerra FA, Llerena Jr JC, Gama SGN, Cunha CB, Theme Filha MM. (2008a). Defeitos congênitos no Município do Rio de Janeiro, Brasil: uma avaliação através do SINASC (2000–2004) *Cad Saúde Pública* (Rio de Janeiro) 24(1):140–149.

Guerra FA, Llerena Jr JC, Gama SGN, Cunha CB, Theme Filha MM. (2008b). Confiabilidade das informações das declarações de nascido vivo com registro de defeitos congênitos no Município do Rio de Janeiro, Brasil, 2004 *Cad Saúde Pública* (Rio de Janeiro) 24(2):438–446.

Horovitz DDG, Llerena Jr JC, Mattos RA. (2005). Atenção aos Defeitos Congênitos no Brasil: Panorama Atual. *Cadernos de Saúde Pública* 21:1055–1064.

Horovitz DDG. (2003). Atenção aos defeitos congênitos no Brasil: propostas para estruturação e integração da abordagem no sistema de saúde [Tese de Doutorado]. Rio de Janeiro: Instituto de Medicina Social, Universidade do Estado do Rio de Janeiro.

IBGE – Fundação Instituto Brasileiro de Geografia e Estatística. (2010). Brasil em Síntese. Available at: http://www.ibge.gov.br/brasil_em_sintese/default.htm.

Jonsen AR. (1986). Bentham in a box: technology assessment and health care allocation. *Law, Medicine and Health Care* 14: 172–174.

Kelly KF. (2002). The Scottish molecular genetics consortium—15 years on. Health Bull (Edinb) 60(1):83–90.

Kroese M, Zimmern RL, Farndon P, Stewart F, Whittaker J. (2007). How can genetic tests be evaluated for clinical use? Experience of the UK Genetic Testing Network. *Eur J Hum Genet* 15: 917–21.

Marques-de-Faria AP, Ferraz VFE, Acosta AX, Brunoni D. (2004). Clinical Genetics in Developing Countries: The case of Brazil. *Community Genet* 7:95–105.

Mattos RA. (2003). Integralidade e a formulação de políticas específicas de saúde. In: Pinheiro R, Mattos RA. (Org.) Construção da Integralidade: cotidiano, saberes e práticas em saúde. Rio de Janeiro: UERJ, IMS: ABRASCO, 45–59.

Pagon RA, et al. (2002). GeneTests–GeneClinics: genetic testing information for a growing audience. *Human Mutation* 19:501–509.

Penchaszadeh VB, Christianson AL, Giugliani R, Boulyjenkov V, Katz M. (1999). Services for the prevention and management of genetic disorders and birth defects in developing countries. *Community Genet* 2:196–201.

Porciuncula CGG. (2004). Avaliação do Ensino de Genética Médica nos Cursos de Medicina no Brasil. Tese (Doutorado em Ciências Médicas) – Faculdade de Ciências Médicas, Universidade de Campinas, Campinas.

Santos FB. (2009). I Encontro Nacional dos Centros de Referência em Osteogênese Imperfecta (Personnal communication).

Senado Federal. (1988). Constituição da Republica Federativa do Brasil, Imprensa Oficial do Estado S.A. IMESP.

Secretaria de Educação Superior – Comissão Nacional de Residência Medica – Ministério da Educação. (2009a). Sistema CNRM – Instituição x Programas x Vagas – Programas de Residência em Genética Médica. Available at: http://mecsrv04.mec.gov.br/sesu/SIST_CNRM/APPS/inst_especialidades.asp

Secretaria de Educação Superior – Comissão Nacional de Residência Medica – Ministério da Educação. (2009b). Histórico. Available at: http://www.sbg.org.br/AtaFundacao.html.

Sociedade Brasileira de Genetica. (2011). *Relação dos Aprovados em Concurso de Título de Especialista*. Available at: http://www.sbg.org.br/relPremioTitEspecialista.html. Accessed on 10/31/2011.

Sociedade Brasileira de Genetica. (2007). UK Workforce Competences for Genetics in Clinical Practice for NonGenetics Healthcare Staff. Enhancing patient care by integrating genetics in clinical practice. Available at: http://www.geneticseducation.nhs.uk/media/16686/Competence_Framework.pdf. Accessed on 10/31/2011.

104.

POPULATION MEDICAL GENETICS IN BRAZIL

Lavinia Schuler-Faccini and Roberto Giugliani

INTRODUCTION

Population medical genetics is the area of medical genetics that aims at the study and medical care of the population, and not of the individual or the family, as is the normal practice in clinical or medical genetics (Castilla, 2005).

Population medical genetics combines different aspects of genetics: *clinical genetics*; *human population genetics*, which investigates populations according to micro-evolutionary parameters; *epidemiological genetics*, traditionally involved in the study of common chronic diseases of polygenic etiology, except for Mendelian diseases; and *sanitary* or *community genetics*, which stands in the interface with public health, giving support to preventive health measures.

This is a poorly developed science in the world as a whole, and still nonexistent in Brazil. The population structure and dynamics of communities in the northeast Brazil and, to a lesser extent, in other regions of the country have been submitted to thorough scientific evaluation during the past 50 years, with results that have benefited the international scientific community in terms of knowledge and new methods developed. (Krieger et al., 1965; Azevedo et al., 1969; Freire-Maia et al., 1975, 1978, 1981). These studies, however, have been restricted to normal genetics characteristics, such as blood groups and other markers, and not to mono-, oligo- or polygenic diseases and their medical aspects of diagnostics, prevention, treatment, and prognostics. Similarly, no program of community education has yet aimed to advise the population about their historic and present reproductive structure, with the final goal of primary prevention. This deficiency is possibly due to the complexity and multidisciplinary tasks involved, as well as to the poor level of priority given until recently by research funding agencies to education and scientific divulgation projects.

CONCEPTS

Conceptually, population medical genetics is an area of knowledge that studies and cares for populations with high prevalence of diseases with a strong genetic component in their etiology, and/or congenital anomalies (present at birth) even when the cause is mainly environmental. It is referred to by different expressions, including *population medical genetics*, or "medical genetics at the population level," or the study of populations with high prevalence of genetic and/or congenital diseases.

As such, medical population genetics is at the interface between clinical genetics in medicine, community genetics in public health, epidemiologic genetics in the population as a whole, and population genetics in the anthropological context, in the study of normal characteristics (blood groups, etc.) aiming at the investigation of the mechanisms responsible for microevolution.

Whereas the differences among some of these partially overlapping five branches of human genetics are clear and qualitative, among others they are less clear and sometimes quantitative, depending more on operational than on conceptual definitions. Abnormal hereditary characteristics may sometimes segregate with genetic markers or variants of normal characteristics, explaining the overlap of population genetics and population medicine. In another example, population medical genetics would not be different from genetic epidemiology if the latter had not concentrated on the study of common chronic diseases and with complex inheritance in populations at large, overlooking the study of monogenic (Mendelian) or oligogenic diseases, in isolated populations.

The convergence of different aspects of human genetics demands close collaboration between experts in these different areas. Although population medical genetics was initiated almost 50 years ago at the Johns Hopkins University in Baltimore, with the first studies by Victor A. McKusick and colleagues with medical genetics of more traditional Amish populations ("Old Order") of Maryland and Pennsylvania (Cross and McKusick, 1970; Khoury et al., 1987), it has not been established as such in the international community.

POPULATION MEDICAL GENETICS IN BRAZIL

In Brazil, several population isolates are being studied by professionals of medical and human genetics, who have serious difficulties in the collection and analysis of genealogical data in populations with thousands of individuals. These researchers have limited technical/scientific support, due to the lack of literature adequate to their needs (Salzano and Freire-Maia, 1967).

For centuries, Brazil received immigrants from a large and varied number of founder populations that brought with them, at times, characteristic parental mutations. The migratory event, in itself, may be responsible for modifications in the original gene frequencies, due to a sampling effect. In Rio Grande do Sul, descendants of the Azoreans have a high frequency (1:30,000) of a neurodegenerative disorder known as spinal cerebellar ataxia type III, or Machado-Joseph disease. These families are currently followed by the team of the proposing institution, but have not yet been evaluated through a populational approach. The same occurs with some forms of familial cancer that are also under study by members of our group.

Another situation with similar difficulties is the one presented by populational groups that are exposed or potentially exposed to mutagenic and teratogenic agents, such as thalidomide in the regions were leprosy disease is endemic, nuclear radiation in the region of Angra dos Reis, and contaminated industry waste in the region of Triunfo in the state of Rio Grande do Sul.

Finally, the tendency or susceptibility to infectious diseases is both strongly associated to environmental characteristics, as to genetic characteristics of the host and pathogen. It is possible that genes that influence resistance/susceptibility are associated to the immune response in various infectious diseases, so that their identification can produce breakthroughs in the prevention or cure of various infections. It should be emphasized the possible positive impacts diagnosis, treatment, and public health actions that this kind of study may cause, especially when dealing with tropical endemic diseases in Brazil. Such diseases have little or no commercial appeal to the large international pharmaceutical companies, which are responsible for development of almost all of the new alternative pharmacological therapies.

Considering these limitations, a nationwide project using populational genetics tools was initiated two years ago in Brazil. INAGEMP (National Institute for Population Medical Genetics) applies this knowledge to the study of Brazilian populations with high frequency of genetic and/or congenital diseases. This approach is expected to shed light upon mechanisms of disease that are still obscure, and whose understanding may have a significant impact on healthcare that goes well beyond genetics and rare diseases.

ONGOING POPULATION MEDICAL GENETICS ACTIONS

The INAGEMP activities are summarized on Table 104-1, and described below.

MONTE SANTO, BA

City with 56,970 inhabitants (2006 census), in the north of the state of BA, with 3273 km^2. The population presents a high consanguinity rate and high prevalence of several genetic diseases, including mucopolysaccharidosis VI, phenylketonuria, congenital hypothyroidism, and cystic fibrosis. The Sertão Museum and the Monte Santo Church are the main local entities of historical/cultural memory for the population. The Canudos War constitutes the main axis of the demographical history of the region. The Monte Santo Church holds baptism records since 1831, weddings since 1856, and deaths since 1874, as well as chrism records since 1958. A Genogram Analysis Pack (PAH, *Pacote de Análise de Heredograma*), is being developed as part of this project, and will be housed in the Sertão Museum.

TABULEIRO DO NORTE, CE

City with 33,000 inhabitants, located in the southeast of the Ceará state, preserves excellent historic-genealogical documentation since its beginning as Tabuleiro da Areia

TABLE 104-1 COMMUNITIES AND SPECIFIC PROBLEMS STUDIED BY THE NATIONAL INSTITUTE OF POPULATION MEDICAL GENETICS

Community	Characteristic
Monte Santo, Bahia State	Mucopolysaccharidosis VI
Tabuleiro do Norte, Ceará State	Type I Gaucher disease
São José do Pântano, São Paulo State	Malformations of the neural system
Cândido Godói, Rio Grande do Sul State	High twinning rate
Angra dos Reis, Rio de Janeiro State	Impact of the nuclear power plant
Leprosy endemic, North Region of Brazil	Thalidomide teratogenicity
Porto Velho, Rondônia State	Susceptibility to tropical infectious diseases
Triunfo, Rio Grande do Sul State	Environmental contamination
Azorean descendants, Rio Grande do Sul State	Machado-Joseph Disease (SCA 3)
South Region of Brazil	Familial breast cancer

farm, in the seventeenth century. The population has a high level of inbreeding and a high frequency of Gaucher disease (OMIM# 230800), a severe chronic disease of recessive autosomal inheritance. The PAH will be housed in the *Tabuleiro da Areia* museum.

SÃO JOSÉ DO PANTANO, MG

This is a rural neighborhood (4000 inhabitants) of the city Pouso Alegre (125,209 inhabitants), where the regional archdiocese is located. The population shows a high frequency of neural tube closure defects and hydrocephaly not related to neural tube defects. Cases of close consanguinity are evident, and remote inbreeding is also possible. The Metropolitan Cathedral of Pouso Alegre and Paróquia Bom Jesus have records of baptisms, weddings, deaths, and chrisms since the year 1811. Earlier records, since 1747, are available in Silvianópolis and Ouro Fino. The books have been recently bound in black and identified with golden letters in the cover. Some pages of the Baptism book of 1811 were removed and partially destroyed in the right inferior angle by moths and rodents, but are still legible. The more recent books have an alphabetical summary, but this information is not available for the earlier records.

CÂNDIDO GODÓI, RS

The etiology of twin births in humans is still largely unclear, and is the subject of a number of investigations. Twinning rates show a wide geographical and time variation, being extremely rare in Asian populations (5–6 per 1000 maternities) and more frequent in sub-Saharan populations (23 per 1000 maternities; Hoekstra et al., 2008). Previous studies show that twinning rates in European countries started to decline around 1900, but have increased steadily from the 1970s onward (Imaizumi, 1998). Even in a contiguous continent, such as Europe, variations on twinning rates between different countries are observed (Hall, 2003). The variability in twinning rates has been assigned to the variation in DZ twinning rates, as MZ twinning has a constant prevalence around the world and time (4/1000 maternities; Tong et al., 1997).

Although familial aggregation of DZ twins has been known for a long time, only recently a few of genes are being identified and implicated in the etiology of DZ twinning. Nongenetic factors such as advanced maternal age, increased parity, lactation, diet, higher maternal height, body mass index, and race are observed also as risk factors for DZ twinning (Hoekstra et al., 2008).

The use of assisted reproductive technology (ART) procedures is cited as the main cause for the increasing occurrence of DZ and MZ twins during the past two decades, but advanced maternal age is possibly also playing a role (Fauser et al., 2005; Martin et al., 2005). The higher occurrence of DZ twins after ART procedures is a result of ovarian stimulation protocols and multiple embryo transfers.

Cândido Godói is a small town in South Brazil (lat 27° 57' 07'; long 54° 45' 07') with approximately 6000 inhabitants, and it is known as the "Twins' Town" due to its high rate of twin births. According to data available from the Brazilian Ministry of Health, between 1994 to 2006 around 2% of live births in Cândido Godói were twins, compared to an average of 1% for the whole country. Noteworthy, however, is the fact that the births of twins are not equally distributed throughout the municipality but are mainly concentrated in one locality, named Linha São Pedro (LSP). This locality was founded at the beginning of the twentieth century by a few families of German ancestry, coming from other Rio Grande do Sul towns. Presently, the population of LSP is less than 600 inhabitants and most are of the Catholic religion.

In 1994, our group was called to investigate further the question of gemelarity in LSP. At that time, the twinning birth rate was estimated as 10% (Matte et al., 1996). More recently, a controversial theory was raised by an Argentinian journalist (Camarasa, 2008). He wrote a book alleging a possible link between the twinning phenomenon in Cândido Godói and supposed experiments of the Nazi physician, Joseph Mengele. According to Camarasa, Mengele could have lived at Cândido Godói at the beginning of the 1960s. Camarasa's suppositions, however, were not based on any actual historical records available.

After this event that brought the attention of worldwide media, the municipality of Candido Godoi asked INAGEMP to establish an investigation into the possible genetic reasons for the high incidence of twinning there.

ANGRA DOS REIS, RJ

A city with 144,000 inhabitants, located in the west coast of Rio de Janeiro state and with an electronuclear power plant established since 1985. Despite lack of scientific evidence of radiation as a cause of malformations in humans, with exception of a role in the interruption of neuronal migration in the brain (Brent, 1999), and in spite of the safety measures adopted, populations living near nuclear power stations are in constant fear of nuclear exposure, due to sub leak or possible accidents similar to the one at Chernobyl. Although the risk of congenital anomalies is similar to that of the general population (5% of births, Castilla and Orioli, 2004), several malformed infants are born every month. Each of these births renew the fear of a permanent, ongoing threat. This kind of situation may be greatly improved by a permanent program of community education, to avoid the worries caused by lack of information or, still worse, by misinformation.

Therefore, there is no need for special investigations, other than those already underway, to evaluate possible contributions of a "possible increase in radiation" to the incidence of congenital malformations. Furthermore, intrauterine radiation in the embryo phase is not expected to increase the frequency of congenital defects. There is undoubtedly an immediate need for special operational studies in all the cities with low incidence of congenital defects, since the most probable cause is lack of accuracy of existing methods, such as 34 of the Declaration of Liveborn, or the death cause in the Declaration of Stillborn. We are worried about the possible stigmatization of the population in Angra dos Reis due to alarmist articles published in the press. This has already happened with communities in Cubatão, in the 1980s, and in Goiânia in 1987. In 2007, ECLAMC and the Angra dos Reis Health Foundation started the implementation of educational actions among employees of the Electronuclear Plant and the population of Angra dos Reis and neighbor cities, in the states of Rio de Janeiro and São Paulo. This situation represents a model for any ecological agent wrongly and without scientific basis considered as teratogenic, with psychological and economic consequences for the community, as has already happened in Cubatão in 1990 (Monteleone Neto and Castilla, 1994) and Goiânia in 1987, as well as in less known situations such as Caçapava in 1988, among others (Castilla et al., 1997).

ENDEMIC AREAS FOR LEPROSY

In 1996, ECLAMC reported 32 cases of children seriously affected by thalidomide, born in Brazil after recognition of the drug as teratogenic and its removal from the market (Castilla et al., 1996).

The geographical distribution of the cases was superimposed to areas endemic for leprosy, suggesting a leak in the system of distribution control of health organizations. Ten years later, new cases were reported by ECLAMC (Schüler-Faccini et al., 2007), showing that the situation is still out of control. This situation affects a population of 33 million people who live in less favored areas in Brazil, mainly in the north and center-west regions, which are considered to be hyperendemic for leprosy, and demands coordinated and efficient education actions to alert the population about the dangers threatening them.

This situation represents a model for any environmental (nongenetic) avoidable agent, affecting several subpopulations, spread through several regions of the country. The education program to be developed may be used to control other teratogens, such as misoprostol, which has been widely used in Brazil (where abortion is an illegal procedure) as a popular abortifacient (Castilla and Orioli, 1994; Schuler et al., 1999, Orioli and Castilla, 2000).

MONTE NEGRO, RONDONIA

Municipality located about 260 km to the southwest of Porto Velho (10° 15' S, 63°18' W) with around 15,000 inhabitants, of which about 60% live in the rural area. In this locality, the sample was collected at random, following a complete ascertainment model of investigation, gathering data from 900 individuals, including about 250 nuclear families. Personal information was collected about the family and the health of individuals, followed by medical examination and blood collection. More details about the population and its demographic and epidemiological patterns can be found in Camargo et al. (2002). In that sample, surveys will be carried out related to infections, malaria, and leishmaniasis.

TRIUNFO, RS

The geographic area under investigation is contaminated by chemicals used for wood preservation, according to an environmental impact study FEPAM (Brazilian Federal Agency for Environmental Protection), including risk analysis by qualified teams. This region is at the left bank of the Taquari river and its flood plain near the confluence with the Jacuí river in the Barreto district, Triunfo, RS. The area includes a nearby dwelling hosting agricultural activities, retail marketplaces, a fishing and recreational park, and a shipyard. The Taquari right bank is mainly agricultural. Besides the city of Triunfo, General Câmara is also among the populations served by the ETAs of Corsan, with intakes near the area under study.

Activities were suspended in 2004, after 44 years of production, preservation, manufacturing, and trading in the national and international market of light posts and other wood-derived products. A surplus of chemicals used to treat wood was left behind from that active period, including Seeley's process of cold-warm baths using pentachlorophenol solution in oil and/or creosote oil. Since 1982, this activity was alternated with creosote and/or hydrosalt CCA (Cu-Cr-Ar). Pentachlorophenol was eliminated as a basic constituent in 1998, when prohibited by law, leaving hydrosalt as the only wood preserver.

Considering the described contaminated site under the FEPAM process, the study of disseminating paths, and those pollutants, along with the definition of the areas directly or indirectly influenced by the site, they all should be investigated to define the size of the exposed human population in view of the known toxicity of those pollutants. In this respect, a collaboration with the Department of Genetics UFRGS was already established in order to make an evaluation of the frequency of reproductive harm, including congenital malformations, and cancer.

COMMUNITIES OF AZOREAN ANCESTRY IN RIO GRANDE DO SUL

Rio Grande do Sul is the southernmost state of Brazil and received during the eighteenth century many immigrants from the Azores islands. Machado-Joseph disease, or spinocerebellar ataxia, is an autosomal dominant adult-onset ataxia, with high incidence in the Azorean population. The estimated prevalence of MJD is 3.8/100,000 inhabitants for Portugal (continental and isles); 36.9/100.000 in Açores; 27.1 in S. Miguel, and 835.2 in As Flores. On the other hand, the same estimate for the south of Brazil is 2/100,000 inhabitants (Jardim et al., 2001a).

In spite of the not-alarming prevalence rate in the south of Brazil, the psychosocial impact of MJD on patients and their families is very strong. This group of illnesses causes great disabilities and dependence, and during 10 years of evolution 50% of patients need support to stand and walk. Patients are left out of the working group from the start of this disease, and at a given stage of evolution will need assistance for the most basic physiological acts. The large number of patients in a given family greatly affects the family group having to deal with severe neurological impairments. A program of care and research considering molecular testing aimed at prediction could avoid eugenic attitudes, assuring complete access to information on these illnesses to patients and their families.

A multidisciplinary team is now evaluating the psychological status of those interested in predictive testing, providing the psychological support needed for a free and autonomous decision about testing, as well as support upon informing results, and life reprogramming of the affected person.

FAMILIAL BREAST CANCER IN SOUTHERN BRAZIL

For the systematic investigation of genetic risks for breast cancer in the state of Rio Grande do Sul, the state with the second largest gross incidence rate for this tumor in the country, a program of exploratory research (a pilot study named *Projeto Genética Núcleo Mama Porto Alegre – NMPOA*) was delineated in 2003 with the aim to determine the prevalence of individuals at higher risk for "hereditary" breast cancer, and the prevalence of germinal mutations for given predisposing genes, in a sample with positive family history of cancer.

A sample from a larger cohort study was used, including women aged 15 years or more at health stations of PSF, *Programa Saúde da Família* (Family Health Program) in Porto Alegre. This study, which will follow up the cohort of women from southern Porto Alegre for 10 years, includes now, at 24 months from inception, 9218 women. At recruitment, women are seen by health attendants at 19 PSF units, and a family history form is completed with 7 simple questions about cancer in first-degree relatives (FDR), plus a general history of family malignancies. This form was validated in the studied population, and for a threshold of just one positive answer, a sensitivity of 0.876, and a specificity of 0.564 for the identification of several hereditary forms of hereditary breast cancer was obtained. Women with at least one positive answer the questionnaire were selected for genetic risk evaluation by the multidisciplinary team at NMPOA. An analysis of the first 8665 enrolled women showed the diagnosis of cancer in at least one or at least two FDR, for 24.2% and 5.1% of them, respectively. The majority of those FDR were females, and the most frequent five tumor types were breast, uterus, colon, lung, and ovary for females, and lung, prostate, esophagus, head and neck, and gastric for males. The mean age at diagnosis of breast and ovary cancer in those FDR was less than 50 years, much lower than expected for the general population according to data from the National Cancer Institute (INCA). The presence of at least one risk factor for hereditary breast cancer was observed in 13% of patients. All women with at least one risk factor for hereditary breast cancer were sent for genetic evaluation, and 12.7% of those fulfill criteria for a hereditary syndrome of CM, including the syndrome for hereditary breast and ovary cancer (24%), Li-Fraumeni/Li-Fraumeni-like syndromes (SLF/LFL; 68%), and a hereditary breast and colon cancer syndrome (HBCC; 8%). There was no previous indication of such a high frequency of SLF/LFL syndromes for this population. SLF is produced by germ mutations in a tumor suppressor gene (*TP53*), it follows an autosomal dominant pattern, and predisposes the affected persons to several tumors in infancy, adolescence, or early adult life, including breast cancer, soft tissue sarcomas, leukemias, CNS tumors, and adrenocortical carcinomas, among others. This was the first Brazilian study to evaluate systematically the prevalence of family risk factors for cancer and breast cancer predisposing syndromes, searching in a population sample for clues to the elevated number of cases with breast cancer, at a very early age, in a geographical area with elevated mortality rates due to this tumor.

This study is working now toward the next objectives, which are: (1) to determine the prevalence of positive family history for cancer in first-degree relatives, and (2) the prevalence of positive risk factors for hereditary breast cancer, by means of a validated simple questionnaire in different regions of Brazil; (3) to identify high-risk populations for hereditary breast cancer; (4) to characterize population structure and dynamics (reproductive pattern, ancestry, genetic isolation, etc.)

for different populations, correlating findings with the risk for hereditary breast cancer; (5) to develop training programs for basic care health attendants, for the identification of persons and families in risk for hereditary cancer; and (6) to propose a strategy of intervention and handling of persons and families in risk for hereditary cancer, particularly in those Brazilian regions with high prevalences for these phenotypes.

This study will allow for the construction of plan for identification of patients at risk for breast cancer and other forms of hereditary cancer in the public health network, providing valuable information on the distribution of persons at risk and on the etiopathogenesis of hereditary cancer (breast and others), allowing for the referral of persons at risk to programs for screening and prevention of cancer.

CONCLUDING REMARKS

In a setting where several fields of science take the direction of ultra-specialization and follow divergent routes, an important group of investigators from the fields of population genetics, epidemiologic genetics, community genetics, and clinical genetics decide to concentrate their abilities, experiences, and vocations on an innovative proposal, whose backbone is convergence. Using populational genetics tools, solutions are being sought for health problems not yet understood or resolved. This approach is expected to shed light upon mechanisms of disease that are still obscure, and whose understanding may have a significant impact on healthcare that goes well beyond genetics and rare diseases.

REFERENCES

Azevedo E, Morton NE, Miki C, Yee S. (1969). Distance and kinship in northeastern Brazil. *Am J Hum Genet* 21:1–22.

Camarasa J. (2008). Joseph Mengele: O Anjo da Morte da na América do Sul. NORMA, Argentina.

Camargo LM, Moura MM, Engracia V, Pagotto RC, Basano SA, da Silva LH, Camargo EP, Beiguelman B, Krieger H. (2002). A rural community in a Brazilian Western Amazonian Region: some demographic and epidemiological patterns. *Mem Inst Oswaldo Cruz* 97:193–195.

Castilla EE, Adams JP. (1996). Genealogical information and the structure of rural Latin American populations: Reality and fantasy. *Human Heredity* 46:241–255.

Castilla EE, Ashton-Prolla P, Barreda-Mejia E, et al. (1996). Thalidomide, a current teratogen in South America. *Teratology* 54:273–277.

Castilla EE, Orioli IM. (2004). ECLAMC: The Latin American Collaborative Study of Congenital Malformations. *Community Genetics* 7:76–94, 2004.

Castilla EE. (2005). On being a medical geneticist. *American Journal of Medical Genetics* 138A:197–198.

Cross HE, McKusick VA. (1970). Amish demography. *Soc Biol* 18:28–39.

Fauser BC, Devroey P, Macklon NS. (2005). Multiple birth resulting from ovarian stimulation for subfertility treatment. *Lancet* 365:1807–1816.

Freire-Maia A, Freire-Maia N, Morton NE, Azevedo ES, Quelce-Salgado A. (1975). Genetics of acheiropodia (the handless and footless families of Brazil). VI. Formal genetic analysis. *Am J Hum Genet* 27:521–527.

Freire-Maia A, Laredo-Filho J, Freire-Maia N. (1978). Genetics of acheiropodia ('the handless and footless families of Brazil'): X. roentgenologic study. *Am J Med Genet* 2:321–330.

Freire-Maia, A. (1981). The extraordinary handless and footless families of Brazil—50 years of acheiropodia. *Am J Med Genet* 9:31–41.

Hall JG. (2003). Twinning. *Lancet* 362:735–743.

Hoekstra C, Zhao ZZ, Lambalk CB, Willemsen G, Martin NG, Boomsma DI, Montgomery GW. (2008). Dizygotic twinning. *Hum Reprod Update* 14:37–47.

Imaizumi Y. (1998). A comparative study of twinning and triplet rates in 17 countries, 1972–1996. *Acta Genet Med Gemellol (Roma)* 47:101–114.

Khoury MJ, Cohen BH, Diamond EL, Chase GA, McKusick VA. (1987). Inbreeding and prereproductive mortality in the Old Order Amish. I. Genealogic epidemiology of inbreeding. *Am J Epidemiol* 125:453–461.

Krieger H, Morton NE, Mi MP, Azevedo E, Freire-Maia A, Yasuda N. (1965). Racial admixture in northeastern Brazil. *Ann Hum Genet* 29:113–125.

Matte U, Le Roux MG, Bénichou B, Moisan JP, Giugliani R. (1996). Study of possible increase in twinning rate in a small village in south Brazil. *Acta Genet Gemellol* 45:431–437.

Monteleone Neto R, Castilla EE. (1994). Apparently normal frequency of congenital anomalies in the highly polluted town of Cubatão, Brazil. *Am J Med* Genet 52:319–323.

Orioli IM, Castilla EE. (2000). Epidemiological assessment of misoprostol teratogenicity. *British Journal of Obstetrics and Gynecology* 107:519–523.

Martin JA, Hamilton BE, Sutton PD, Ventura SJ, Menacker F, et al. (2005). Births: final data for 2003. *Natl Vital Stat Rep* 54:1–116.

Salzano FM, Freire-Maia N. (1967). *Populações Brasileiras: Aspectos Demográficos, Genéticos e Antropológicos.* São Paulo: Editora da Universidade de São Paulo.

Schüler L, Pastuszak A, Sanseverino MTV, et al. (1999). Pregnancy outcome after exposure to misoprostol in Brazil: a prospective, controlled study. *Reprod Toxicol* 13:147–151.

Schüler-Faccini L, Soares RCF, de Sousa ACM, et al. (2007). New cases of thalidomide embryopathy in Brazil. *Birth Defects Res A Clin Mol Teratol* 79:671–672.

Tong S, Caddy D, Short RV. (1997). Use of dizygotic to monozygotic twinning ratio as a measure of fertility. *Lancet* 349:843–845.

105.

MEDICAL GENETIC SERVICES IN LATIN AMERICA: FOCUS IN ARGENTINA

Victor B. Penchaszadeh

LATIN AMERICA: AN OVERVIEW OF HISTORY, ETHNICITY AND HEALTH CONDITIONS

Latin America is the part of the Americas that was conquered and colonized by Spain and Portugal in the sixteenth century and beyond, and extends north–south from Mexico in North America, through Central America and virtually all of South America, all the way to the most southern city of the world, Usuahia, in Argentina. In addition, the Spanish and French-speaking Caribbean are considered part of Latin America. The original aboriginal population, numbering several millions when the conquistadors arrived, was reduced to less than one-tenth in size by war, oppression, and inhumane forced labor conditions. The current population of Latin America numbers about 600 million people, with different degrees of admixture of the original indigenous population with the Spanish and Portuguese colonizers, the Africans brought as slaves in the seventeenth and eighteenth centuries, and the extensive European, and to a lesser degree Asian, immigration of the nineteenth and twentieth centuries. The extensive ethnic diversity is obviously accompanied by a very rich cultural diversity, notwithstanding which there are many common cultural and historical features that characterize modern Latin Americans, including their common roots, similar Spanish and Portuguese languages, common revolutionary wars against colonial powers and, more recently, a growing political and commercial integration. Amerindians continue to be a significant part of the population (average about 10%), in variable proportions according to sub-regions and countries.

Most Latin American countries have medium–low annual incomes (regional average $5000 per capita). Industrialization and urbanization is occurring at a fast pace, albeit with marked disparities between and within countries. The degree of socioeconomic inequality in the region is one of the highest in the world: in 2006, over half the population was living in poverty and 27.6% was earning less than $2 per day (World Bank, 2006). Average life expectancy is 73.4 years, infant mortality rate 20.2 per 1000, and maternal mortality 89.4 per 100,000 (with wide ranges according to country, median income, and social class; PAHO, 2007).

Due to epidemiological transition, the main burden of disease is that of noncommunicable diseases with genetic contribution, particularly cardiovascular diseases, cancer, hypertension, diabetes, and mental illness, while congenital anomalies rank high in the causes of infant mortality (PAHO, 2007). The prevalence of genetic diseases and congenital defects in Latin America is similar to other regions of the world (Penchaszadeh, 2004), with hemoglobinopathies posing a significant public health burden in countries with a high proportion of African ancestry (Brazil and the Caribbean).

LATIN AMERICA: HEALTH SYSTEMS AND MEDICAL GENETIC SERVICES

Most Latin American countries organized their health services in the mid-twentieth century under the responsibility of the state, with three main coexisting health subsystems. The *public system*, run directly by the state and funded from general revenues, comprises national, provincial, and municipal networks of healthcare services at primary, secondary, and tertiary levels, with largely free access (albeit with a number of deficiencies that require out-of-pocket expenses by patients). Although in theory any citizen can access the public system, in practice it covers 50%–60% of the population. A second subsystem is funded by employer–employee contributions, usually administered by the *social security system*, providing health services to employees and retired persons, either directly or through contracts with the private or the public sector, and covering about 35% of the population. The private sector caters to the 10% of the population with the highest income and is based in private, for-profit insurance. The average annual per capita health expenditure in

Latin America in 2006 was only USD 262 (PAHO, 2007), with wide variations by social class, marked inequities in distribution, and favoring high technology medicine in tertiary centers, rather than primary care–centered healthcare.

Access to family planning services and contraception is limited, and although induced abortion is largely illegal in almost all countries of the region, it is practiced extensively. Recently, national or local legislation in some countries began to allow pregnancy termination for severe fetal anomalies (Colombia, Argentina) and, exceptionally, outright elective abortion (Mexico City). In November 2008, the Uruguayan congress passed a law allowing elective abortion but the law was vetoed by the president.

Latin America has a skilled, well-trained and sizable workforce of approximately 600 physicians trained as clinical geneticists (i.e., 1 per million population; Penchaszadeh, 2004). While the distribution of this workforce is very unequal between and within countries, virtually all countries have national professional societies of human or medical genetics. Most clinical geneticists work part-time in the public sector as well as in private practice. The specialty of clinical genetics has been recognized in several countries, and genetic counseling is largely performed by clinical geneticists, since there are no training programs in genetic counseling for nonphysicians (Penchaszadeh, 2004).

Medical genetics services have had a fragmentary and uneven development in Latin America, being mostly concentrated in urban and wealthy areas. The public sector is the largest provider of genetic services in the region, but underfinancing, poor planning and coordination, and deficient regulations and quality assurance have hindered their quality and responsiveness to population needs. The social security system and the private sectors' main goal is profit, at the expense of quality of care. Departments of health of the public sector at city, provincial, and national levels have not been up to in the task of formulating and implementing policies and programs in genetic services, except for newborn screening, which tend to be mandated by national law (Penchaszadeh, 2004). Notably, however, Cuba, Costa Rica, and more recently Brazil have socialized healthcare systems with genetic services as an integral component (Heredero-Baute, 2004; de Céspedes et al., 2004, Marques-de-Faria et al., 2004; Horovitz et al., 2006).

ARGENTINA: AN OVERVIEW

Argentina is situated in the southern tip of South America, with a surface of 2,791,810 km^2 and a length of 3694 km from north to south, bordering to the west with Chile and to the north and northeast with Bolivia, Paraguay, Brazil, and Uruguay. To the east, Argentina faces the Atlantic Ocean, with a 4725 km-long coast. The estimated total population is 40 million, 90% of whom live in urban areas.

The territory into which the Spanish conquistadors arrived at the beginning of the sixteenth century, and was called "Argentina," was then inhabited by hundreds of thousands of Amerindians of at least 30 different ethnicities. In spite of extensive admixture occurring in the next two centuries, a sizable native population remained in the region. In the late nineteenth century, an extermination campaign run by the government decimated most remaining aboriginals, who currently number less than 400,000 persons concentrated in the northwest, northeast, and the south of the country, living in abysmal conditions. West Africans brought as slaves numbered in the tens of thousands at the beginning of the nineteenth century, but they virtually disappeared in the second half of the century, probably through admixture and migration to neighboring countries. In the second half of the nineteenth century and first half of the twentieth century, Argentina received a huge influx of immigration from virtually all countries of Europe, including large contingents of European Jews and, to a lesser extent, Middle Eastern peoples. In recent times immigration is coming primarily from neighboring countries (Paraguay and Bolivia) with strong Amerindian ancestry. Finally, immigration from South Korea and China has been the latest wave in the 1980s. It is estimated that about 50% of the population descend from Italian immigrants and 25%–30% from Spaniards. In the 1960s Jews numbered close to half a million people, being the third largest concentration of Jews in the world, after Israel and the United States.

Reproductive patterns among Argentines have been largely open, with a resulting extensive admixture. In Buenos Aires, the relative European, Amerindian, and West African genetic contributions to the gene pool have been estimated at 67.5%, 25.9%, and 6.5%, respectively (Martínez-Marignac et al., 2004). Average consanguinity is less than 1% except in some remote isolated areas (Liascovich et al., 2001). The official language is Spanish and the majority of the population is nominally of Catholic faith

ARGENTINA: POLITICAL, SOCIAL AND ECONOMIC CHARACTERISTICS

Argentina is a federal republic with a presidential system, where each province has its own constitution and elects its governing officials. In the twentieth century, Argentina went from being the 8th world economy to recurrent political and economic crises, which included a succession of military dictatorships, the last of which engaged in the most brutal repression the country had ever seen: between 1976 and 1983, 30,000 persons were "disappeared" by the military (Argentina's Commission on Disappeared People, 1986), while hundreds of thousands fled the country to escape repression. During the dictatorship the country incurred in a huge foreign debt, much of the national industry was dismantled, and most strategic state enterprises and services, including health services, were privatized. Civilian governments that followed were not able or willing to alter the wave of privatizations and concentration of the economy in few hands, resulting in huge unemployment and poverty, and an economic meltdown in 2001, with a 300% depreciation of the currency and a default in the foreign debt repayment. Annual per capita income dropped from USD 7470 in 2000 to USD 3580 in 2004, and later recovered to a current of USD 5000). In 2002, unemployment and extreme poverty climbed to over 25%, affecting drastically the living conditions of vast population sectors. The economic crisis was eventually brought under control by 2004, and the annual economic growth rate for the past few years increased to a healthy 9% (World Bank, 2011).

ARGENTINA: HEALTH CONDITIONS AND SERVICES

The burden of disease is contributed largely by noncommunicable disorders, although diseases of poverty (infections and malnutrition) are still a major problem, particularly among the disadvantaged (PAHO, 2007a; Penchaszadeh, 2008). Maternal mortality is very high (40 per 100,000), and 31% of it is due to complications of unsafe and illegal induced abortion. Infant mortality is relatively low by regional standards (12.9 per 1000), and its main causes are perinatal conditions (52%) and congenital anomalies (27%; PAHO, 2007; Argentina, 2011). Genetic and congenital disorders have a similar prevalence that in developed countries (Penchaszadeh and Beiguelman, 1998), with variations in individual conditions due to founder effects and consanguineous isolates, including some clusters of rare single gene disorders in isolated rural areas, such as Sandhoff disease (Dodelson de Kremer et al., 1987) and oculocutaneous albinism (Castilla and Adams, 1996).

The modern health system of Argentina dates from 1945–1955, a period of economic bonanza characterized by industrialization, rapid urbanization, and active unionism. The national Ministry of Health is responsible for the development of general health objectives, policies and regulations, executing national health programs (such as vaccinations, campaigns against tobacco, cancer, obesity, etc.), overseeing the general operation of health services, and negotiating the coverage and fees of health insurance plans, all in coordination with provincial health authorities. However, its leverage to influence decisions at the provincial level is weak, as each province has the authority and the funds to manage and deliver their own healthcare. This results in wide variations in coverage and quality of care between jurisdictions (PAHO, 2007; Penchaszadeh et al., 2010).

Annual total expenditure in health is approximately 8.0% of GNP, which translates into about USD 450 per capita. Health care delivery is fragmented in three sectors (PAHO, 2007; Penchaszadeh et al., 2010):

a. The public system, funded through general taxation, maintains public hospitals and ambulatory health centers throughout the country and provides free healthcare. The public system spends 22% of total health expenditures to serve 48% of the population, mostly the poor and those without insurance.
b. The social security system, funded by employer and employee contributions, spends 29% of total health expenditures to serve 47% of the population, largely blue-collar workers, employees, and retired persons, and mostly contracting with the private sector.
c. The private sector is largely based on a for-profit insurance model that spends 49% of total health expenditures and serves 7.5% of the population (upper middle classes).

The Argentine health system has an overall good level of spending and benefits from good human and structural resources; however, it suffers from significant segmentation, fragmentation and inefficiency, financing inequalities, and inequities in access to and quality of care (PAHO, 2007; Penchaszadeh et al., 2010).

MEDICAL GENETIC SERVICES

Clinical genetic services developed in Argentina in the late 1960s in a small number of public hospitals, led by a few physicians trained abroad. In 1970 the Ministry of Health created a National Center of Medical Genetics, to conduct epidemiological research on genetic conditions and training in medical genetics. In the mid-1970s, many medical geneticists left the country because of the military dictatorship. Currently, about 120 clinical geneticists are certified as such by the Argentine Society of Genetics (certification recognized by the Ministry of Health). These clinical geneticists staff about 41 clinical genetics units throughout the country, mainly in pediatric and maternity public hospitals of major cities, providing genetic consultations and counseling, mostly for single gene disorders and congenital defects in children and newborns. Understaffing and long waiting lists are a problem. (Penchaszadeh, 2008). A recent survey of eight of the most busy genetic units, in pediatric hospitals in six different jurisdictions, reported 18,138 first-time visits in a 5-year period. Of those, 40% were affected with a single gene disorder, 23% with congenital defects and/or mental retardation, and 13% with a chromosome abnormality (Barreiro, 2007). Statistics recently compiled by the Ministry Health show that about 20% of pediatric patients in tertiary care centers present a condition caused by a single gene disorder or a chromosome abnormality. These patients are rarely diagnosed in the primary or secondary levels of care (Alba et al., 2009).

Cytogenetic laboratories exist in a few major hospitals, and they perform mainly standard chromosome analysis, with few of them equipped for molecular cytogenetics. DNA testing for diagnostic purposes is provided in a few public hospitals and academic centers, for cystic fibrosis, fragile-X, congenital adrenal hyperplasia, muscular dystrophy, beta thalassemia, hemophilia, breast cancer (BRCA1 and BRCA2), thyroid cancer, MEN 1 and 2, FAP, Huntington disease, and spinal cerebellar ataxias (Penchaszadeh, 2008). While genetic services in the public system are provided free of charge, the coverage by the social security and private insurance is highly variable. Most DNA testing in the private sector is sent to private labs abroad. Predictive testing is rarely performed.

The survey by the Ministry of Health mentioned above (Alba et al., 2009) uncovered the fact that hospital-based genetic services have little or no connection with primary and secondary level health services. Indeed, there are no easy referral channels from primary and secondary level health facilities (where the bulk of patients are seen) to specialized centers, which leads to deficiencies in diagnostic and remedial capability of the system as a whole. The result of this inefficiency is that many patients with genetic conditions are not diagnosed nor properly cared for and followed up, and their families are not counseled appropriately. In order to address this problem, a two-prong strategy of (a) training in the basics of clinical genetics (suspicion of genetic conditions and appropriate referral to specialized centers) to physicians and other health professionals providing primary care, and (b) improving the referral channels to specialized centers, was developed and carried out under diverse auspices. The first program to be launched was part of a European Union–funded project to develop tools to train primary care health personnel in clinical genetics, of which Argentina was a demonstration site. This project (CAPABILITY, 2009) was directed in Argentina by Dr. Cristina Barreiro, based in the Garrahan National Pediatric Hospital of Buenos Aires, the main pediatric hospital of the country. In three years, hundreds of health professionals in several remote areas were trained to suspect and refer patients with possible genetic conditions to genetic centers in major cities (Capability Argentina, 2009). A second, similar program has been developed more recently by the National Medical Genetics Center of the Ministry of Health. These programs are being followed now by efforts to regionalize genetic services throughout the country, under the stewardship of the Ministry of Health. It is expected that these initiatives will improve the reach of medical genetics and provide better services to populations that have been underserved until now.

The Ministry of Health has recently shown interest in improving access to, and in the performance of the existing clinical genetics services in the public system throughout the country. A National Advisory Commission on Genetics and Health, appointed in 2005, conducted countrywide survey of genetic services and in 2009 submitted a proposal to the Ministry for their improvement.(Alba et al., 2009). The proposal calls for increasing the number of positions of clinical geneticists in public hospitals, and for modernization and expansion of the lab equipment for cytogenetics and molecular genetics in key hospitals. The Commission also proposed the organization of a network of genetic services to maximize their efficiency, avoid duplication of services, and channel referrals in a regionalized manner.

Voluntary abortion is illegal in Argentina, except to protect the life or health of the mother, or in cases of rape. Notwithstanding, it is estimated that 600,000 voluntary abortions occur annually in the country (Zamberlin, 2007), which is close to the number of annual live-born deliveries (750,000). The main effect of criminalization of abortion has been to keep maternal mortality high, as 31% of it is due to unsafe induced abortions. Furthermore, it has prevented the development of prenatal diagnosis services within the public and social security health systems. Although prenatal ultrasound is widely available, couples carrying pregnancies with fetal abnormalities do not have the option of legal termination of pregnancy (exception for anencephaly, thanks to a recent ruling of the Supreme Court in a landmark case). In contrast, prenatal diagnosis is widely practiced in the private sector for middle- and upper class couples, who do not suffer from any legal consequences (Gadow et al., 2006).

NEWBORN SCREENING

In 1986 Congress enacted a law mandating newborn screening for PKU, to which congenital hypothyroidism (CH) was added in 1990 and cystic fibrosis (CF) in 1995. Implementation of screening, however, took much longer as funds were not appropriated fast enough. Coverage has varied markedly by jurisdiction and health subsector, with the wealthiest cities and provinces and the private sector having better coverage and follow-up than in poorer areas and in the public system. In 2007, a new law was passed by Congress mandating newborn screening in all the territories of the country for the following conditions: PKU, CH, CF, galactosemia, congenital adrenal hyperplasia, biotinidase deficiency, retinopathy of prematurity, congenital deafness, Chagas disease, and syphilis (Argentina, 2008).

The law states that the public sector and social security insurance must cover the expenses of screening and treatment for these disorders.

There is no national system of laboratories for newborn screening, nor a quality control system. The tests are performed in several laboratories, either in public hospitals or outsourced to the private sector. Data from the Ministry of Health (Ministerio de Salud,,2008) show that by 2007, 22 of 28 provinces were testing for PKU and CH. Of those, 40% were also screening for CF and 10% for galactosemia. Population coverage is correlated with resources, political will, and degree of development and organization of health services. In one-third of provinces providing newborn screening, coverage was less than 50%. The highest coverage (close to 100%) occurs in the city of Buenos Aires (with only 30,000 births per year of a total of 750,000 for the country. The prevalence at birth among 300,000 babies tested between 2000 and 2010 in the city of Buenos Aires was as follows: PKU (1/27,120), congenital hypothyroidism (1/2000), congenital adrenal hyperplasia (1/16,650), cystic fibrosis (1/5,878), Chagas disease (1/2046), syphilis (1/153) (Buenos Aires, 2011). There are no data on follow-up or long-term management of affected infants, which presumably is very variable.

MEDICAL GENETICS EDUCATION

Physicians and allied health professions graduate virtually without formal courses on clinical genetics, genetic counseling, or public health genetics/genomics. There are 2–3 postgraduate training positions available annually for clinical genetics and cytogenetics. In contrast, graduates in biological sciences generally have a good fund of knowledge in molecular genetics, and many proceed to get postgraduate training in its medical applications. Some of these molecular geneticists go on to work in molecular diagnostic labs or in genetics/genomics research.

RESEARCH IN MEDICAL GENETICS

An epidemiologic study of factors associated with congenital malformations has been in operation for 40 years in Argentina and South America: the Latin American Collaborative Study on Congenital Malformations (ECLAMC; Castilla and Orioli, 2004). This project provides baselines for the prevalence of congenital malformations in the region, and its data are utilized to monitor environmental effects on the prevalence at birth of congenital defects. Research in clinical genetics is concentrated in dysmorphology/cytogenetics, cancer, and selected single gene disorders such as skeletal dysplasias, storage diseases, muscular dystrophies, fragile-X syndrome, cystic fibrosis, thalassemia, and congenital deafness.

Biomedical research in Argentina has a long tradition in academic centers. A National Council for Scientific and Technological research (CONICET) has existed for 50 years (www.conicet.gov.ar). In addition, the Ministry of Health funds public health research projects, including in genetics (www.msal.gov.ar/htm/Site/salud_investiga/index.html). Genetics research is conducted in several centers, such as the Department of Molecular Biology of the University of Buenos Aires (www.fcen.uba.ar), the Instituto de Investigaciones de Ingeniería Genética y Biología Molecular (www.conicet.gov.ar/php/datos_inst.php?n=02650), the Fundación Leloir. (www.leloir.org.ar), the Instituto Multidisciplinario de Biología Celular (IMBICE; www.imbice.org.ar), and the Instituto de Biología Molecular y Celular of the Universidad Nacional de Rosario (www.ibr.gov.ar). Areas of genetics/genomics research include basic molecular genetics, immunogenetics, molecular population genetics, forensic genetics, genome sequencing (the *T. cruzi* genome was sequenced by a multinational team in which Argentine geneticists played a key role), gene therapy, and stem cell research. In 2008, a new Ministry of Science and Technology (www.mincyt.gov.ar) has engaged in a concerted effort with the Ministry of Health and the private sector to apply genomic approaches in biotechnology and biomedical research.

FORENSIC GENETICS

Fueled by the need to solve problems of identity of several hundred children of political dissidents, abducted and appropriated illegally by the military during the dictatorship of 1976–1983, forensic genetics has had a speedy development in Argentina under the auspices of the government after the restoration of democracy in 1984. The reality faced by Argentine society in 1984 was that the military that oppressed the country during 1976–1983 had resorted to egregious violations of human rights to enforce their rule, including kidnappings, torture, and assassinations of 30,000 citizens, most of whom still remain disappeared. Among the disappeared there were several hundred women that were pregnant at the time of their abduction. These women were allowed to carry on their pregnancies and deliver their babies in clandestine detention centers, only to be assassinated shortly thereafter and their babies "distributed" to childless couples within the security apparatus, who then raised them in captivity under a false identity. As soon as democracy was restored in 1984, a National Genetic Data Bank was created to store DNA samples of living relatives and perform DNA testing to families who were searching for their disappeared children and grandchildren, as well as individuals presumed to be children of the disappeared (Penchaszadeh, 1997). This quest continues today and is credited to having led to the recovery of the true identity of 100 children and young adults over 25 years of work (Abuelas de Plaza de Mayo, 2009). A related field of research linking forensic genetics with human rights has been the research onto the identity of remains of 800 persons exhumed from mass graves by the Argentine Forensic Anthropology Team after democracy was restored (www.eaaf.org). This organization is currently engaged in a large population project, whereby public appeals are made through the media to elicit DNA samples from anyone who may think he or she has a relative disappeared by the dictatorship. The objective is to identify remains by matching their DNA, recovered from the bones, to the DNA of living relatives. So far, 5000 people have contributed DNA, which has enabled authorities to establish the identity of the remains of 80 persons (EEAF, 2009).

PREVENTION OF CONGENITAL AND GENETIC CONDITIONS

Folic acid fortification of flour has been mandated by law in 2003, although there is no good evidence of its implementation nor of its influence on the prevalence at birth of neural tube defects.

Screening carriers of prevalent recessive conditions is not an established policy. While beta-thalassemia is relatively prevalent in the country because of Italian descent, there are no systematic carrier detection programs (Feli-Rorres et al., 2002). Similarly, and in spite of the sizable Jewish community in Argentina, carrier testing for genetic diseases common in this group is not part of medical practice. It may be that the lack of legal reproductive options for carrier couples is keeping these programs from developing. The political pressure of the Catholic church has made prenatal genetic diagnosis unavailable in the public health system. It is hoped that scientific progress and pressures from health professionals and patient organizations will make these services more available soon.

CONCLUSION

The unmet needs of population-based genetics/genomics services in Argentina, and existing barriers for their development, are similar to those already reported for Latin America as a whole (Penchaszadeh and Beiguelman, 1988; Penchaszadeh, 2004):

a. Lack of universal access to healthcare, with a fragmented, inefficient, and inequitable health system
b. Load of unmet needs in infectious diseases, malnutrition, prenatal and newborn care
c. Deficient genetics knowledge by health professionals and health officials
d. Misperception that genetic services are too expensive and deal only with rare diseases
e. Erroneous association of genetic services with abortion
f. Discrimination practices by for-profit health insurance leading to poor coverage for genetic conditions
g. Lack of explicit public policies or programs in genetic healthcare and lack of governmental regulation of genetic testing

Recommendations to overcome existing barriers for genetics/genomics in population health in Argentina should be the same as those put forth by the World Health Organization for developing countries (Penchaszadeh et al., 1999) and by recent meetings of experts in medical genetics of the Pan American Health Organization (Kofman-Alfaro and Penchaszadeh, 2004; PAHO, 2007b) and of the World Health Organization (WHO, 2011).

a. Increase government funding, improve efficiency, and ensure equity in the access to the public healthcare system
b. Coordination of genetic services in the public system on a regionalized basis, avoiding duplications and deficiencies
c. Interactions between clinical geneticists, public health officers, and parent/patient organizations on the use of genetics in public health
d. Research on the prevalence and types of genetic disorders and genetic predispositions to common diseases
e. Education in genetics of health professionals, the public, and decision makers
f. Education of geneticists in community health and public health genetics
g. Interactions between medical educators, public health officers, medical professionals, researchers, and the industry, in the application of genetic/genomic technologies to public health
h. Proper state regulation of the application of genetic/genomic technologies to public health

REFERENCES

Abuelas de Plaza de Mayo. (2009). The Grandmothers and Genetics (Las Abuelas y la Genética, Spanish). Available at www.abuelas.org.ar.

Alba A, Barbero P, Barreiro C, Chertkoff L, Dain L, Ferreiro V, Francipane L, Frechtel G, Gallego M, Liascovich R, Meroni ME, Rozental S. (2009). Diseño y organización de una Red Nacional de Genética Médica (Design and organization of a national network of medical genetics.) Spanish. Unpublished (available from the author).

Argentina (2008). Newborn screening Law (Spanish). Available at http://test.e-legis-ar.msal.gov.ar/leisref/public/showAct.php?id=11449&word=pesquisa%20neonatal#'.

Argentina (2011). Indicadores Básicos. Ministerio de Salud. Available at: http://www.msal.gov.ar/index.php/component/content/article/49-publicaciones/173-argentina-salud-epidemis-global.

Argentina's National Commission on Disappeared People. (1986). *Nunca Más (Never Again): Report by Argentina's National Commission on Disappeared People*. London: Faber.

Barreiro C. (2007). Collaborative Study on Clinical Genetics. Argentina. Ministerio de Salud. Comisión Salud Investiga. Spanish. Available at: http://www.saludinvestiga.org.ar/.

Buenos Aires City. (2011). Newborn screening data (Spanish). Available at: http://buenosaires.gov.ar/areas/salud/redes/ppn/datos_2008.php?menu_id=30513.

CAPABILITY. (2009). Capacity Building for the Transfer of Genetic Knowledge into Practice and Prevention. www.capabilitynet.eu/Index.html.

Capability Argentina. (2009). Genetics for the practice and prevention of birth defects. www.Capabilityargentina.com.ar.

Castilla EE, Adams J. (1996). Genealogical information and the structure of rural Latin-American populations: reality and fantasy. *Hum Hered* 46(5):241–55.

Castilla EE, Orioli IM. (2004). ECLAMC: the Latin-American collaborative study of congenital malformations. *Community Genet* 7(2–3):76–94.

de Céspedes C, Saborío M, Trejos M, Abarca A, Sánchez A, Rojas L. (2004). Evolution and Innovations of the National Neonatal and High Risk Screening Program in Costa Rica. *Rev Biol Trop* 52(3):451–466 [Epub 2004 Dic 15].

Dodelson de Kremer R, Depetris de Boldini C, et al. (1987). Estimation of heterozygote frequency of Sandhoff disease in a high-risk Argentinian population. Predictive assignment of the genotype through statistical analysis. *Medicina (B Aires)* 47(5):455–463.

EAAF [Argentine Team of Forensic Anthropology] (2007). Available at: http://desaparecidos.org/notas/2007/11/arg-eaaf-solicita-muestras-de.html.

Gadow E, Petracchi F, Igarzabal L, Gadow A, Quadrelli R, Krupitzki H. (2006). Awareness and attitude toward prenatal diagnosis of chromosomal abnormalities in patients with no access to legal termination of pregnancy. *Prenat Diagn*10:885–891.

Heredero-Baute L. (2004). Community-based program for the diagnosis and prevention of genetic disorders in Cuba. *Community Genet* 7:130–136.

Horovitz DDG, Cabral de Almeida Cardoso MH, Llerena JC, Mattos RA. (2006). Birth defects in Brazil and health care: proposals for public policies in clinical genetics. *Cad. Saúde Pública, Rio de Janeiro* 22(12):2599–2609.

Kofman-Alfaro S, Penchaszadeh VB. (2004). Community Genetic Services in Latin America and Regional Network of Genetic Services. *Community Genet* 7:157–159, 2004.

Liascovich R, Rittler M, Castilla EE. (2001). Consanguinity in South America: demographic aspects. *Hum Hered* 51(1–2):27–34.

Marques-de-Faria AP, Faria-Ferraz VE, Acosta AX, Brunoni D. (2004). Clinical genetics in Brazil. *Community Genet* 7:95–105.

Martínez-Marignac VL, Bertoni B, Parra EJ, Bianchi NO. (2004). Characterization of admixture in an urban sample from Buenos Aires, Argentina, using uniparentally and biparentally inherited genetic markers. *Hum Biol* 76:543–557.

Ministerio de Salud. (2008). Program for the Early Detection of Congenital Defects. Accessible at: http://msal.gov.ar/htm/Site/promin/UCMISALUD/index.htm.

PAHO. (2007a). Health in the Americas: Argentina. Washington, DC. pp. 27–48.

PAHO. (2007b). Recommendations for the development of genetics in public health in Latin America. Report of a meeting of experts. Rio de Janeiro, June 21, 2007. Unpublished document. Available from the author, in Spanish.

Penchaszadeh VB. (1997). Genetic identification of children of the disappeared in Argentina. *JAMA* 52(1):16–27, 1997.

Penchaszadeh VB. (2004). Genetic Services in Latin America. *Community Genet* 7:65–69.

Penchaszadeh, VB. (2008). Argentina. Public health genomics. *Public Health Genomics* 12:59–65.

Penchaszadeh VB. (2011). Forced disappearance and suppression of identity of children in Argentina: Experiences in genetic identification. In: Gibbon S, Ventura-Santos R, Sans M. (eds.), *Genetic Admixture, Identity and Health in South America; Dialogues Between Anthropology and Genetics* (pp. 213–243). New York, Palgrave Macmillan.

Penchaszadeh VB, Beiguelman B. (1998): Medical genetics services in Latin America: Report of a meeting of experts. *Pan Am J Public Health* 3(6):409–420.

Penchaszadeh VB, Christianson AL, Giugliani R, Boulyjenkov V, Katz M. (1999). Services for the prevention and management of genetic disorders and birth defects in developing countries. *Community Genet* 2:196–201.

Penchaszadeh VB, Leone F, Rovere M. (2010). The health system in Argentina: an unequal struggle between equity and the market. *Ital J Public Health* 7(4):350–358.

WHO (2011). *Community Genetics Services in Low- and Middle-Income Countries. Report of a WHO Consultation.* Geneva, Switzerland, World Health Organization, 2010. ISBN 978 92 4 150114 9.

World Bank. (2011). Available at: http://web.worldbank.org/WBSITE/EXTERNAL/COUNTRIES/LACEXT/ARGENTINAEXTN/0,,contentMDK:22248065~pagePK:1497618~piPK:217854~theSitePK:316024,00.html.

Zamberlin, N. (2007). Abortion in Argentina. Buenos Aires; CEDES/FEIM/IPPF. Spanish. Available at: www.cedes.org.

106.

HEREDITARY ATAXIAS IN CUBA

Luis Velázquez-Pérez, José Miguel Laffita-Mesa, and Roberto Rodríguez-Labrada

INTRODUCTION

Hereditary ataxias (HA) are a group of chronic inherited neurodegenerative disorders with diverse inheritance patterns and variable clinical presentation. The most common type of HA are inherited in autosomal dominant manner. The autosomal dominant cerebellar ataxias (ADCA) are a clinically, pathologically, and genetically heterogeneous group of neurodegenerative disorders caused by degeneration of cerebellum and its afferent and efferent connections. The degenerative process may additionally involve the pontomedullar systems, pyramidal tracts, basal ganglia, cerebral cortex, peripheral nerves (ADCA I), and the retina (ADCA II), or can be limited to the cerebellum (ADCA III; Harding, 1993).

The most common of these dominantly inherited autosomal ataxias, ADCA I, includes many spinocerebellar ataxia (SCA) subtypes, some of which are caused by pathological CAG trinucleotide repeat expansion in the coding region on the mutated gene. Such is the case in SCA1, SCA2, SCA3/MJD, SCA17 and dentatorubral-pallidoluysian atrophy (DRPLA; Matilla et al., 2006). Among the almost 30 SCAs, SCA2 is the second most prevalent subtype worldwide, only surpassed by SCA3 (Schöls et al., 2004; Matilla et al., 2006). The disorder was first recognized in India in 1971 by Wadia and Swami, who were intrigued by the early and marked slowing of saccade movements associated to the cerebellar syndrome (Wadia and Swami, 1971). Contemporarily, in Cuba some neurologists were describing many families coming from the northeast region of the country with the distinct clinical picture of SCA2 (Vallés et al., 1978). Subsequent epidemiological surveys in this Cuban region allowed researchers to confirm the highest prevalence rate of SCA2 worldwide in the Holguin province, likely due to a founder effect (Orozco et al., 1989; Velázquez-Pérez et al., 2001, 2009a).

EPIDEMIOLOGY

The collective worldwide prevalence of SCAs is estimated at about 5–7 cases per 100,000 people, although higher figures have been reported in particular populations due to founder effects (Schöls et al., 2004). In the case of SCA2, the global prevalence is unknown because the majority of the few existing epidemiological studies have been performed in isolated geographical regions with families not large enough for linkage analysis. Nevertheless, large SCA2 families have been found in India, Martinique, Australia, Tunisia, Germany, Italy, Mexico, and especially in Cuba (Klockgether, 2007; Velázquez-Pérez et al., 2009a).

A recent epidemiological survey performed in the whole Cuban population disclosed the existence of 578 SCA2 patients encompassed in 163 families and distributed in 11 out 14 provinces, which represent 87% of the patients with SCAs and render a national prevalence rate of 6.57 cases per 10^5 inhabitants. Also, 7173 asymptomatic at-risk individuals in the SCA2 families were identified, 2060 of which (75.90%) were first-degree relatives, accounting for a national prevalence of SCA2 mutation near to 28.51 cases per 10^5 inhabitants. The highest concentration of SCA2 patients was observed in the Holguin province, with 413 SCA2 patients, representing 71.45% of all Cuban SCA2 patients and a prevalence rate of 40.18 per 10^5 inhabitants. Nevertheless, there are regions in this province where the prevalence reaches higher values, with remarkable figures in Baguanos municipality (141.7 per 10^5 inhabitants; see Figure 106-1). Also, this study found 1384 asymptomatic first-degree relatives in Holguin, which settled the prevalence rate for the mutation at 107.5 cases per 10^5 inhabitants (Velázquez-Pérez et al., 2009a).

Despite the dissemination of Cuban families through the island, the SCA2 prevalence in Holguin is different from other regions of the country. Cultural and environmental processes have long-lasting effects on allelic diversity, restricting mutations to selected populations (Takano

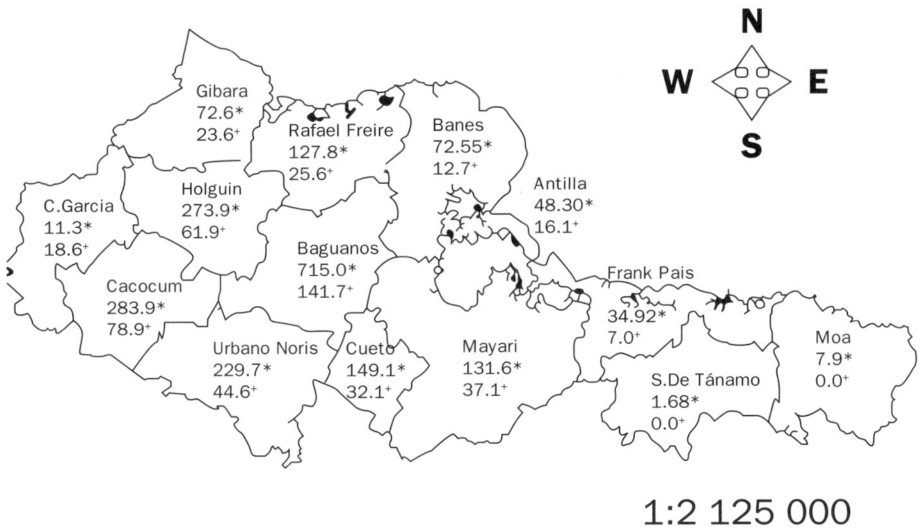

Figure 106-1 Prevalence rates of SCA2 patients (+) and SCA2 mutation (*) in Holguin province, Cuba (2006–2007).

et al., 1998). The endogamous and closed-cycle marriages of Holguin people together with putative restricted environmental unbalancing, limited SCA2 to the Holguin region (Velázquez-Pérez et al., 2009a). A similar explanatory mechanism has been proposed in SCA3 founders in Azoreans (Parduey, 2004).

CLINICAL FEATURES

The clinical picture of SCA2 includes a cerebellar syndrome in all patients, which is characterized by ataxic gait, cerebellar dysarthria, dysmetria, and dysdiadochokinesia (Orozco et al., 1989). Patients also exhibit abnormal tandem stance (95%), slow saccadic eye movements (91%), limited voluntary ocular movements (88%), loss of vibration sense (73%), areflexia or hyporeflexia (77%), and abnormal swallowing (76%). Autonomic abnormalities (urinary dysfunction, hypohidrosis, constipation, and sexual dysfunction) are presented in 78% of the cases and are accentuated in late stages of the disease (Velázquez-Pérez et al., 2009a). Additional SCA2 clinical features are sleep disturbances and cognitive disorders. Among sleep disturbances are included restless legs syndrome (Schöls et al., 1998, Abele et al., 2001, Irazno et al., 2007), muscle cramps, and insomnia, whereas among the cognitive disorders the frontal-executive dysfunctions and verbal short-term memory deficits stand out. as well as reduction of attention and concentration. Patients with dementia are rare (Storey E et al., 1999; Reynaldo-Arminan et al., 2002; Burk et al., 2003).

In the Cuban SCA2 population, around 80% of genetic transmissions exhibit anticipation. The age at onset varies from 3 to 79 years, with a mean of 33 years. There was no significant difference in the age at onset between males and females. The commonest first symptom of the disease is gait ataxia (97%), followed by cerebellar dysarthria (3%). Clinical features develop progressively with an increase in cerebellar syndrome, saccade slowing, and other features. The patients die approximately within 15–20 years after initial symptoms, but patients with larger CAG repeats have early age at onset and a faster progression to death (Velázquez-Pérez et al., 2009a). The main cause of death in SCA2 patients is bronchial pneumonia (63%), followed by bronchial aspiration and cardiovascular diseases, among others.

Rare symptoms, such as retinitis pigmentosa and myoclonus epilepsy, may be exhibited in SCA2 patients with very large CAG expansions (Babovic-Vuksanovic et al., 1998; Rufa et al., 2002; Tan et al., 2004.), whereas some patients with low-range CAG expansion may develop Parkinsonian signs (Lu et al., 2004; Charles et al., 2007).

MOLECULAR BIOLOGY

SCA2 is caused by a dynamic mutation consistent in the unstable expansion of the trinucleotide CAG repeat in a gene located in 12q24.1, which encodes for a polyglutamine tract in the protein ataxin 2 (Pulst et al., 1996; Imbert et al., 1996; Sanpei et al., 1996). In normal individuals, alleles contain between 13 and 31 CAG repeats. Of them, the alleles with 22 CAG repeats are the most frequent (76%) and usually have 2 CAA interruptions $[(CAG)_8$ CAA $(CAG)_4$ CAA $(CAG)_8]$ Five percent of the remaining normal alleles range between 13 and 21 CAG units, and 19% from 23 to 31. Indeed, the prevalence of these large normal alleles in Cuba is the highest compared to other regions with high or moderated prevalence of SCA2.

The pathological alleles have more than 32 CAG repeats, but for the range 32–36 they have incomplete penetrance (Pulst et al., 1996; Cancel et al., 1997; Geschwind et al., 1997). The most common size of the abnormal allele is 37 (72%). The expanded alleles lose one or both interrupting CAA-triplets, which promote the length instability. Expansions occur in 89%, and contractions in 11% of the offspring of affected patients. In paternal transmission there is higher variability in repeat lengths compared with maternal transmission. (Velázquez-Pérez et al., 2009a).

A global survey in the Cuban SCA2 population aimed at delineating the haplotype around the SCA2 mutation disclosed that the vast majority of Cuban families share the same haplotype 3-4-11 at D12S1333-D12S1672-D121332 loci. Slight variability is evident in branches emanating from them, showing the haplotype (6-6-5). This region encompasses 3cM around the CAG expansion. Neither these European founders nor the more recent studies of the same nature in SCA2 kindreds from Brazil, India, and mainly from Portugal, share the haplotypes at these loci (Ramos et al., 2010). Our studies of haplotypes disclosed that large normal alleles are devoid of the CAA interruption from its 5′ or 3′ CAG tracts (Figure 106-2).

The ataxin-2 protein (atxn2) is a polypeptide of 1312 amino acids encoded by 25 coding exons of the SCA2/ATXN2 gene, with at least five human isoforms. (Lastres-Becker et al., 2008a). Currently, the physiological function of atxn 2 is not well understood, but several findings suggest that atxn2 is involved in RNA metabolism. Bioinformatics predictions point to 4 essential motifs (Lsm, LsmAD, PABC, A2D) related overall with RNA metabolism. Also, this fact is supported by the intracellular location in the rough endoplasmic reticulum, and polyribosomes and their interaction with proteins involved in the RNA metabolism, such as A2BP1 and PABP (Satterfield and Pallanck 2006).

At the cellular level atxn2 mainly targets Purkinje cells in the cerebellum. It has been postulated that the polyQ expansion disrupts physiological function through a gain of toxic function, but knock-out models support that it could be a lack of function as well (Kiehl et al., 2006; Lastres-Becker et al., 2008b). Lately, it has been claimed that the function of atxn2 might be related to the development of obesity.

Like other polyglutamine diseases, the age at onset and severity of SCA2 symptoms inversely correlate with the length of the CAG repeat, accounting for ~80%, whereas the remaining variability suggests the existence of modifier genes, genetic polymorphisms, epigenetic factors, and unknown environmental determinants influencing age of onset in addition to the pathogenic allele itself (Velázquez-Pérez et al., 2009a). Supporting the abovementioned, long normal CAG repeats in the CACNA1A (Pulst et al., 2005) and RAI1 genes (Hayes et al., 2000), as well as the 10398G polymorphism in the mitochondrial complex I gene (Simon et al., 2007), are associated to earlier age at onset in Cuban SCA2 population.

ANIMAL MODELS

Animal models have been a useful tool to study the polyglutamine repeat expansion diseases because of their valuable insights into the understanding of these disorders. Specifically in SCA2, the use of RNA interference in the *Caenorhabditis elegans* model demonstrated the essential role of atx-2 gene in early embryonic development (Kiehl et al., 2000). Also, in *Drosophila melanogaster* the role of this gene was suggested in the dosage-sensitive regulation of cytoskeletal actin filament formation (Satterfield et al., 2002).

Regarding the murine models, two transgenic mouse models for ataxin-2 have been generated to date. The first one was produced by Huynh et al. (2000), who reported the use of the murine PcP2 (L7) promoter to direct the expression of the human SCA2 gene with an expanded allele of 58 CAG repeats. Using a rotarod test, they found

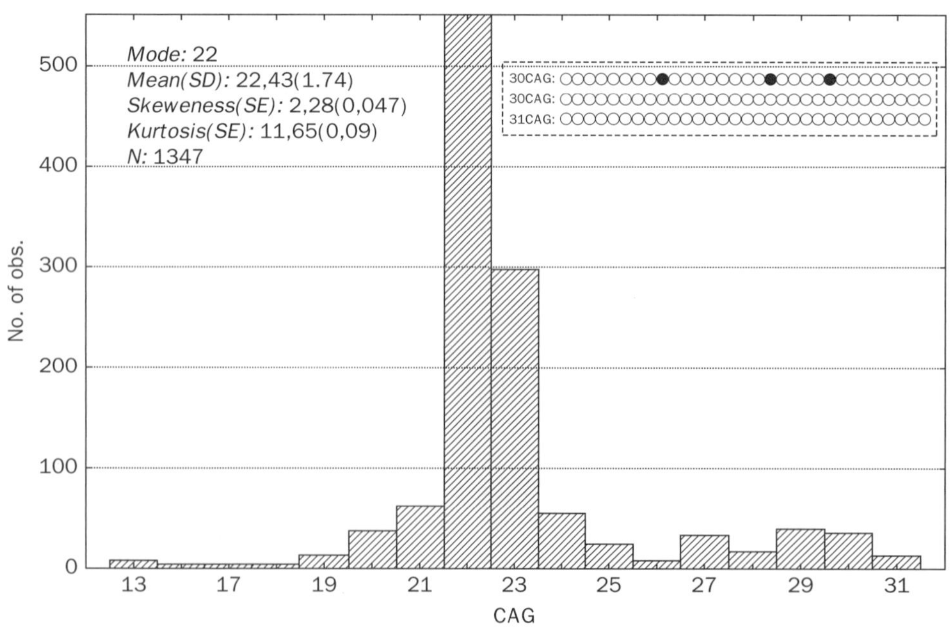

Figure 106-2 Distribution of unexpanded CAG repeat (13-31) at SCA2 locus in a large sample of Cuban population. White circles represent CAG triplets and black circles represent CAA triplets

that the animals became ataxic at 26 and 16 weeks for the heterozygous and homozygous transgenic mice, respectively. Also, they described progressive incoordination and morphological alterations of Purkinje cells in this animal model. In 2005, Aguiar and coworkers (Aguiar et al., 2006) generated transgenic mouse lines expressing a full-length human SCA2 gene, with 75 CAG units under the control of the human SCA2 self promoter. These animals developed the neurological phenotype in 12 weeks for heterozygous, whereas the homozygous mice did it in 6 weeks, suggesting that the promoter elements of the human SCA2 gene are more efficient in directing the expression of the transgene. Northern blot analysis revealed the ubiquitous expression of the transgene, but immunohistochemistry techniques demonstrated the specific degeneration of Purkinje neurons in the cerebellar cortex. The cell loss was associated with apoptosis, as was demonstrated by the immunodetection of the activated form of caspase-3 and *Bax* protein overexpression. Additionally, electron microscopy revealed some morphological markers of apoptosis such as chromatin condensation, loss of mitochondria crests, and dilatation of endoplasmic reticulum and Golgi complex. Both models are compatible with a gain-of-function or toxic gain-of-function of mutant SCA2 gene, but they cannot help us to understand the effects of in vivo ataxin-2 function. With this aim, two Sca2 knockout mice have been generated. Kiehl et al. (2006) reported that these animals were viable but their segregation studies indicated reduced frequency of homozygotes among offspring, especially of female pups. Macroscopic and microscopic analyses showed no histological alterations in the cerebellum in spite of a slight rotarod deficit. These authors were intrigued by the adult obesity that the mice developed on a fat-enriched diet. This latter feature was confirmed by Lastres-Becker et al. (2008b) in another Sca2 knock-out mouse model, which also leads to hepatosteatosis, dyslipidemia, and reduced fertility. This model showed decreased insulin receptor (Insr) protein in concordance to elevated Insr mRNA levels, indicating the posttranscriptional effect of ataxin-2 on the insulin receptor status. At the cerebellar level, increased concentration of gangliosides and sulfatides were observed, in addition to impaired cholesterol metabolism and disruption of sphingomyelin cycle, which suggests a dysfunction of cellular membranes and the second-messenger systems, leading to alteration of neuronal signaling.

NEUROIMAGING

Magnetic resonance imaging (MRI) shows early cerebellar and brainstem atrophy (Figure 106-3) with marked involvement of cerebellar cortex and pons/olive regions, supporting the presence of olivopontocerebellar atrophy. Also, frontotemporal atrophy is observed in advanced disease (Burk et al., 1996; Giuffrida et al., 1999; Figure 106-3). Voxel-based morphometry studies have revealed the atrophy of the cerebellar and brainstem white matter, as well as the symmetric loss of gray matter in the cerebellar vermis (Della Nave et al., 2008). Positron emission tomography (PET) studies showed a reduced regional glucose metabolism in the cerebellum, brainstem, and parietal cortex, which may occur years before the clinical onset of SCA2 (Inagaki et al., 2005). PET also revealed the loss of striatal dopamine transporter function with nigrostriatal atrophy, similar to the pattern observed in idiopathic Parkinson's disease (Boesch et al., 2004; Wullner et al., 2005; Inagaki et al., 2005). Imaging study by proton magnetic resonance spectroscopic demonstrated the loss of choline-containing compounds in SCA2 cerebella, suggesting the decreased production and/or the loss of cell membranes, as well as the reduced synthesis of precursors of acetylcholine. The same study demonstrated an increase of lactate levels in the cerebellum, suggesting the impairment of glycolysis and mitochondrial function (Boesch et al., 2001).

NEUROPATHOLOGY

The macroscopic examination of nervous structures in postmortem samples of SCA2 patients shows a significant

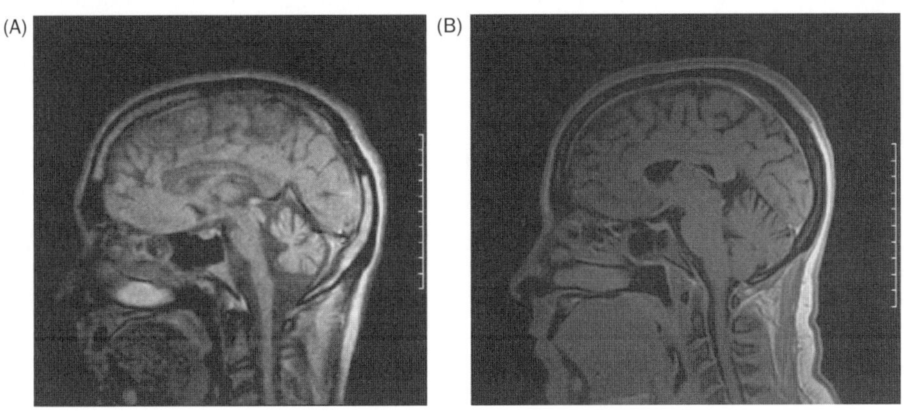

Figure 106-3 (A): MRI scans of a SCA2 patient (age 40 years, disease duration 8 years, CAG repeat size 39 units, SARA score 19) and (B): a healthy control (age 40 years, SARA score 0). Note the severe atrophy of cerebellum and brainstem in the SCA2 subject.

atrophy of the cerebellum, brainstem, and frontal lobe, as well as whiteness of the midbrain substantia nigra and reduction of the cerebral and cerebellar white matter. Microscopically, the cerebellum is characterized by an early and marked neuronal loss in the Purkinje cell layer, with reduction in the number of dendritic arborizations, and torpedo-like deformations of their axons. The number of granular neurons is diminished, usually toward late stages of the disease, whereas the dentate nucleus is relatively spared. Parallel fibers are sparse, and no climbing fibers are observed in the Purkinje cell dendritic trunks. In the brainstem, the most noteworthy microscopic finding is the marked loss of inferior olivary neurons, in addition to the degeneration of pontine and other precerebellar brainstem nuclei (Estrada et al., 1999). The neuropathological evaluation of cranial nerves shows that oculomotor, somatomotor, somatosensory, auditory, vestibular, and autonomic nuclei are notably affected by neuronal loss, and their associated fibers are atrophied and undergo demyelination (Gierga et al., 2005). Another important neuropathological marker of SCA2 is the notable reduction of neurons of the substantia nigra in the mesencephalon and the extensive degeneration of several thalamic nuclei, such as the reticular, fasciculosus, ventral anterior, lateral geniculate body, and the anterior nuclei (Rub et al., 2003, 2005). In the spinal cord is observed an early and progressive neuronal loss of dorsal ganglia root and reduction of motor neurons, usually in the cervical level and the Clarke's column (Rub et al., 2007).

NEUROCHEMICAL STUDIES

The most important finding in the biochemical characterization of SCA2 patients is the significant reduction of zinc, iron, and copper levels in CSF and serum. Zinc depletion seems to be caused by environmental deficit, and unknown metabolism factors depending on CAG repeats. The reduction of zinc levels could be associated with phenotypical features such as nerve conduction slowing, cognitive dysfunction, and inmunodepression at final stages of the disease, and could accentuate the dysfunction of cerebellar circuits, based in the important role of this element in control of synapses in the cerebellum (Gonzalez et al., 2005). On other hand, most biomarkers of antioxidant/pro-oxidant balance are significantly modified in SCA2 patients, with an increase in lipid peroxidation and oxidative damage to protein and DNA, as well as a significant reduction of the soluble antioxidant GSH. Also, the enzymes GST, SOD and CAT are modified in these patients, with a disruption of the balance of CAT/SOD (Almaguer et al., 2005). Another important finding is the decrease of erythropoietin levels in the CSF with a compensatory increase of this molecule in the serum, suggesting the existence of reduced capabilities of neuroprotection in the nervous system. We thought that these biochemical features might contribute to the high phenotypical variability of SCA2, and could be promising therapeutic targets to design future clinical trials.

NEUROPHYSIOLOGICAL FEATURES

NERVE CONDUCTION AND ELECTROMYOGRAPHY STUDIES

The most common electrophysiological finding in SCA2 patients is a predominantly sensory axonal neuropathy, expressed by the early and progressive reduction of sensory amplitudes, suggestive of dorsal root ganglionopathy. These alterations are associated with slowing of nerve conduction as a sign of demyelination, and secondary cause of axonal disorder. The progression rate of sensory axonal neuropathy is notably accentuated in patients with large CAG expansion sizes. Motor nerve conduction parameters are usually normal, but in patients with 10–15 years of disease duration it is possible to observe reduction of motor amplitudes (Velázquez-Pérez et al., 2007, 2010). Electromyographical findings reveal motor unit potentials (MUP) with light signs of polyphasism, increased amplitudes, and isolated contraction pattern as a result of decrease of cortical excitability in the corticospinal pathway. In advanced stages of the disease, signs can appear of denervation (fibrillations and fasciculations) and the contraction patterns become simple oscillations, indicating the loss of motorneurons in the anterior horn of the spinal cord (Velázquez-Pérez et al., 2009b).

SOMATOSENSORY EVOKED POTENTIALS (SSEP)

Tibial nerve SSEPs are characterized by a marked prolongation of the P40 component and central conduction time latencies. In the median nerve SSEP, there is a latency prolongation of N20 and N13 components in addition to reduction of amplitude of *Erb* potentials. In almost cases, the SSEPs show abnormal morphology and reduced reproducibility. These alterations get worse quickly in patients with a larger CAG repeat number, and can be detected in presymptomatic subjects (Velázquez-Pérez et al., 2007; Velázquez-Pérez, 2008). The SSEP alterations result from disturbed conduction at the posterior columns of the spinal cord, supported by neuropathological findings showing neuronal loss and demyelination of the posterior columns in addition to severe neurodegeneration in the cuneate and gracile nuclei, as well as neuronal loss in the ventral posterior nuclei in the thalamus.

BRAIN STEM AUDITORY EVOKED POTENTIALS (BSAEP)

BSAEPs have poor reproducibility and unstable morphology in 95% of the patients, in addition to the increase of latency of the waves III and V and the prolongation of the I–III interpeak interval. These abnormalities are common in patients with disease duration greater than 10 years, but the abnormal reproducibility and morphology can be detected as early as the preclinical stage (Velázquez-Pérez et al., 2007). These findings are in agreement with neuropathological observations in cranial nerves and nuclei involved in the auditory pathway.

VISUAL EVOKED POTENTIALS (VEP)

VEPs are frequently normal in SCA2 patients, but some patients in advances stages of the disease have prolonged P100 latencies with normal amplitudes. These findings reflect the integrity of the visual pathway in Cuban SCA2 patients, which allows us to distinguish SCA2 from the other spinocerebellar ataxias such as SCA1, SCA3 and SCA7. (Velázquez-Pérez et al., 2007; Velázquez-Pérez, 2008).

MOTOR EVOKED POTENTIALS

The study of the corticospinal tract by transcranial magnetic stimulation in SCA2 patients discloses an increase of central motor conduction time and motor threshold. Also, intracortical facilitation may be reduced and the induced cortical silent period prolonged. The progression of these abnormalities is dependent on the disease duration and ataxia severity, and they reflect the reduced excitability of the motor cortex, disturbed conduction along the pyramidal tract, and the loss of facilitatory influences of the cerebellum on the primary motor cortex (Restivo et al., 2000, 2004; Schwenkreis et al., 2002)

ELECTROCULOGRAPHY

The main oculomotor abnormality in SCA2 is the slowing of horizontal saccadic movements, which is the result of early pontine brainstem degeneration. This feature is electroculographically detected in 99% of the patients and in several presymptomatic subjects. The maximal saccade velocity is negatively correlated with the polyglutamine expansion and the ataxia score, but it is not significantly influenced by the disease duration. Other oculomotor abnormalities include hypometric saccades to 60^0 predictable amplitude and prolonged saccade latencies, reflecting the cerebellar and cortical/subcortical involvement, respectively (Velázquez-Pérez et al., 2004, 2009c; Velázquez-Pérez, 2008). These findings identify the saccade velocity as an objective and quantitative physiological parameter that is under strong genetic control, and appears to be a promising research tool to study the modulation of polyglutamine toxicity by modifier genes. In other hand, gain measurements in smooth pursuit movements and horizontal optokinetic nystagmus are slightly reduced in SCA2 patients, whereas vestibule-ocular reflex is normal (Buttner et al., 1998).

VIDEOPOLYSOMNOGRAPHY

Sleep disorders are common complaints of SCA2 patients, fundamentally toward the final stages of the disease. Clinically, the most prominent disorders are restless legs syndrome and muscle cramps, which appear in 45% of the cases. Patients with REM sleep behavior disorder, bruxism, and excessive daytime sleepiness are sparse. Polysomnographical evaluation reveals a reduction of REM sleep with decreased rapid eye movements, with a density in 70% of patients. These REM sleep abnormalities appear before the disease onset, and their progression rates depend on ataxia severity and disease duration. The REM sleep pathology in SCA2 is likely the result of the degeneration of REM-on neurons in the pons. Periodic leg movements (PLMs) are also frequent in 38% of SCA2 patients (Figure 106-4). They are directly associated with the clinical severity of the disease and their progression rate is notable. These movements suggest the dysfunctions in brain dopaminergic systems, and therefore they resolve the contradiction reported by Orozco et al., in 1989, which noted significant nigro-striatal impairment without notable clinical correlates. PLMs are sensible to dopaminergic treatments that could improve sleep quality and slow disease progression. Other, less prominent sleep abnormalities are the decrease of sleep efficiency, increase of arousal index and central apnea index, as well as REM sleep without atonia (Boesch et al., 2006; Tuin et al., 2006).

ELECTROENCEPHALOGRAPHY (EEG)

The study of conventional EEG in SCA2 patients shows a predominantly diffuse theta activity with reduced reactivity to eye opening in 72% of the cases. In the brain electrical activity mapping is observed a significant increase of absolute power of theta band with reduction of absolute power for alpha band (Figure 106-5). Paroxystic activity is observed in some patients, especially those who suffer from PLMs. These abnormalities are indicative of pathological effect of SCA2 mutation in the cerebral cortex, where the protein ataxin 2 is

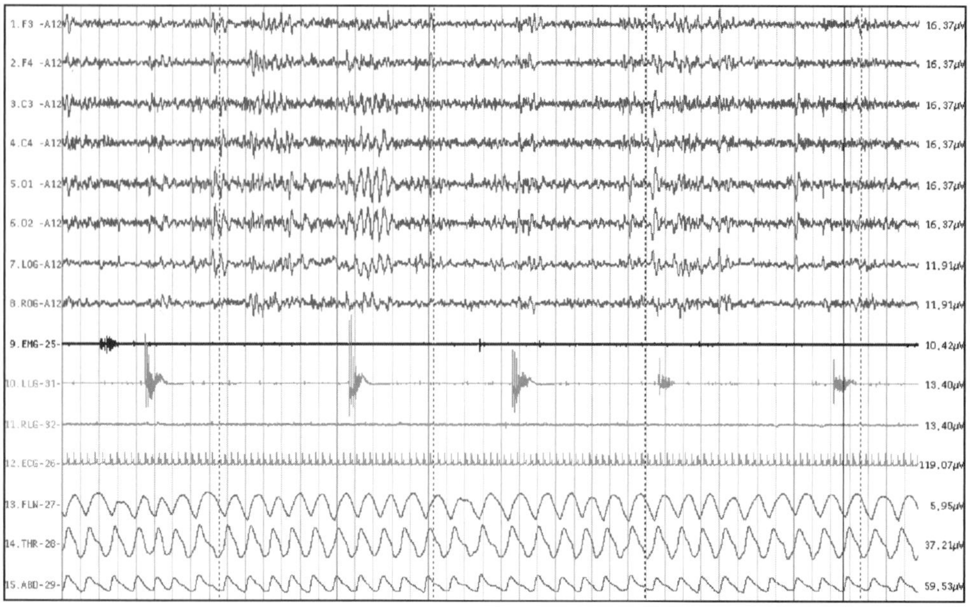

Figure 106-4 Two-minute epoch of stage 2 sleep showing periodic leg movements in a SCA2 patient 44 years old, with 12 years of disease duration, 39 CAG repeats in the SCA2 gene, and ataxia score of 15 units.

significantly expressed, and also they represent electrophysiological markers supporting the cognitive dysfunctions in SCA2.

Heart rate variability. In SCA2 patients, time domain HRV is characterized by increase of mean heart rate and reduction in the standard deviation of RR interval in approximately 70% of the cases. Frequency domain HRV studies reveal significant reductions for absolute spectral power in the 0.04–0.4 Hz range, with relevant increments of the low to high frequency spectral power bands ratio (LF/HF). These abnormalities are incremented during the Valsalva maneuver and deep breathing test (Pradhan et al., 2008; Julio et al., 2010). The decrease of heart rate variability is thought to be due to the loss of parasympathetic modulation, whereas frequency-domain analysis demonstrates sympathetic and parasympathetic dysfunction with predominant parasympathetic involvement.

Olfactory impairment. SCA2 patients show a significant impairment in olfactory threshold, identification, and discrimination. The UPSIT score is significantly reduced and correlates positively with the MMSE and ataxia score, but it is not influenced by age, age at onset, disease duration, and CAG repeats (Velazquez et al., 2006). These findings indicate the extension of neurodegenerative changes to the olfactory system at central and peripheral levels that sustain the cerebellar involvement in olfaction.

Visuomotor learning. The prism adaptation task let us identify the main visuomotor alterations in SCA2 patients, who have impaired adaptation decrement but, surprisingly, a normal aftereffect. The prism adaptation

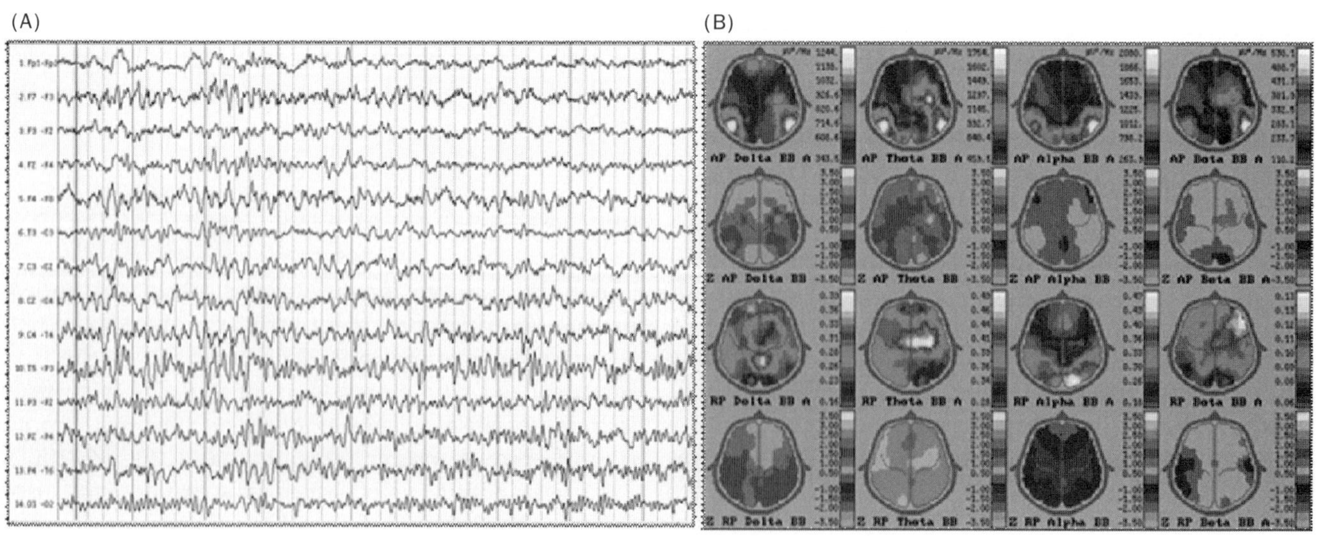

Figure 106-5 Conventional (A) and quantitative (B) EEG from a SCA2 patient 40 years old, with 10 years of disease duration, 40 CAG repeats in the SCA2 gene, and ataxia score of 17 units. Refer color figure.

capabilities are accentuated in patients with larger numbers of CAG repetitions. Also, the deterioration in the adaptation correlates with the motor performance and saccade velocity, suggesting that structures that degenerate in this disease may contribute to both adaptation and motor performance. In asymptomatic subjects, motor performance deficits are present in spite of the fact that motor learning capabilities are intact (Fernandez-Ruiz et al., 2007; Velázquez-Pérez et al., 2009d).

EARLY PRECLINICAL SIGNS

The study of the large and homogeneous population of asymptomatic first-degree relatives of SCA2 patients in Cuba has allowed us to characterize the preclinical stage of the disease and to identify electrophysiological biomarkers useful to predicting the disease onset and evaluating the effects of therapeutic options. The earliest subclinical sign appears in the central nervous system, up to 15 years before the ataxia onset. It is characterized by the slowing of horizontal saccades at 60^0 of target displacement, with amplitudes and latencies normal. This electrophysiological abnormality is accentuated in subjects with larger CAG repeats, and reflects the early dysfunction or degeneration of paramedian pontine reticular formation (Velázquez-Pérez et al., 2009c). This alteration is followed by the reduction of REM sleep percentage with decreased density of rapid eye movements, which precede the disease onset by approximately 10 years, although it progression during this stage is insidious. The REM sleep pathology in presymptomatic subjects suggests the early degeneration of REM-on neurons in the pons. Peripheral nerve involvement in SCA2 presymptomatics is shown by the decrease of sensory amplitudes in median and sural nerves, which may be found 8 to 5 years prior disease onset and in the last 4 years of the preclinical stage, respectively, and result in early loss of ganglion cell in the dorsal root ganglia (Velázquez-Pérez et al., 2010). SSEPs are characterized by abnormal morphology and reproducibility as well as increased P40, central conduction time, and N20 and N13 latencies (Velázquez-Pérez et al., 2007). Other subclinical signs of SCA2 are: T-mentonian reflex with prolonged latencies (Medrano et al., 2009), EMG with slight signs of polyphasism (Velázquez-Pérez et al., 2009b), motor performance deficits with normal motor learning capabilities, shown by the prism adaptation task (Velázquez-Pérez et al., 2009d), and reduced capabilities to identify specific odors in a sensible smell identification test (UPSIT). Finally, is important to make clear that the order and time of appearance of these abnormalities are not a rule, because the phenotypical and endophenotypical expression of SCA2 mutation could be appreciably modified by some factors such as the penetrance degree of the mutation, the existence of modifier genes and haplotypes, and environmental determinants.

THERAPEUTIC OPTIONS

To date, there is no specific treatment for SCA2. Physical and cognitive neurorehabilitation have palliative effects on motor and cognitive disorders. Cuban SCA2 patients have received a specialized neurorehabilitative program (Pérez-Ávila et al., 2004) since 1998, which has been applied to more than 300 patients and has allowed the recovery of motor, cognitive, and antioxidant functions in almost 75% of rehabilitated patients. Therefore, neurorehabilitation is considered to be an effective and essential therapeutic alternative, to be used together with any future pharmacological options in the treatment of the disease.

The number of clinical trials for SCA2 is small, and they have encompassed few patients, almost all with advanced disease evolution, limiting the accurate evaluation of the therapies' effects due to the significant extension of the neurodegeneration present in the subjects. Also, a majority of these studies have been designed to treat relatively infrequent symptoms, but their effect on improving the life quality of the patients is notable. For example, muscle cramps are successfully treated with magnesium, whereas deep brain stimulation was effective in treatment of coarse postural tremor in one case (Freund et al., 2007). Levodopa treatment alleviates the Parkinsonian signs in SCA2 patients.

A double-blinded and placebo-controlled clinical trial with 50 mg zinc sulphate in 36 Cuban SCA2 patients supported the increase of zinc levels in the serum and CSF of treated subjects, and it demonstrated the effectiveness of this treatment on cerebellar syndrome, peripheral neuropathy, and the restoration of antioxidant functions.

In Cuban SCA2 patients, the PLMs were significantly reduced and subjective sleep complaints improved as a result of a treatment protocol with 0.2 mg lisuride for one month.

The use of high doses of B-complex vitamins over three months improved sensory peripheral neuropathy, resulting in increased sensory amplitudes and nerve conduction velocity. Nevertheless, the central abnormalities, measured by somatosensory evoked potentials, remained unchanged.

REFERENCES

Abele M, Burk K, Laccone F, Dichgans J, Klockgether T. (2001). Restless legs syndrome in spinocerebellar ataxia types 1, 2, and 3. *J Neurol* 248(4):311–314.

Aguiar J, Fernandez J, Aguilar A et al. (2006) Ubiquitous expression of human SCA2 gene under the regulation of the SCA2 self

promoter cause specific Purkinje cell degeneration in transgenic mice. *Neurosci Lett* 392:202–206.
Almaguer L, Almaguer D, Gonzáles Y, Martinez E, Valcárcel P. (2005). Capacidad antioxidante total de en pacientes cubanos con ataxia Espinocerebelosa tipo 2. *Rev Mex Neuroci* 6(3):201–206.
Babovic-Vuksanovic D, Snow K, Patterson MC, Michels VV. (1998). Spinocerebellar ataxia type 2 (SCA 2) in an infant with extreme CAG repeat expansion. *Am J Med Genet* 79:383–387.
Boesch SM, Donnemiller E, Muller J, et al. (2004). Abnormalities of dopaminergic neurotransmission in SCA2: a combined ^{123} I-ßCIT and ^{123} I-IBZM SPECT study. *Mov Disord* 19:1320–1325.
Boesch SM, Frauscher B, Brandauer E, Wenning GK, Hogl B, Poewe W. (2006). Disturbance of rapid eye movement sleep in spinocerebellar ataxia type 2. *Mov Disord* 21(10)1751–1754.
Boesch SM, Schocke M, Burk K, et al. (2001). Proton Magnetic Resonance Spectroscopic Imaging Reveals Differences in Spinocerebellar Ataxia Types 2 and 6. *J Magn Res Imag* 13:553–559.
Burk K, Abele M, Fetter M, et al. (1996). Autosomal dominant cerebellar ataxia type I clinical features and MRI in families with SCA1, SCA2 and SCA3. *Brain* 119 (Pt 5):1497–1505
Burk K, Globas C, Bosch S, et al. (2003). Cognitive deficits in spinocerebellar ataxia type 1, 2, and 3. *J Neurol* 250:207–211.
Buttner N, Geschwind D, Jen JC, Perlman S, Pulst SM, Baloh RW. (1998). Oculomotor phenotypes in autosomal dominant ataxias. *Arch Neurol* 55(10):1353–1357.
Cancel G, Durr A, Didierjean O, et al. (1997) Molecular and clinical correlations in spinocerebellar ataxia 2: A study of 32 families. *Hum Mol Genet* 6:709–715.
Charles P, Camuzat a, Benammar N, et al. (2007). Are interrupted SCA2 CAG repeat expansions responsible for parkinsonism?. *Neurology* 69:1970–1975.
Della Nave R, Ginestroni A, Tessa C, et al. (2008). Structural Damage in Spinocerebellar Ataxia Type 2. A Voxel-Based Morphometry Study. *Mov Disord* 23(6):899–903.
Estrada R, Galarraga J, Orozco G, et al. (1999). Spinocerebellar ataxia 2 (SCA2): morphometric analyses in 11 autopsies characterize it as an olivo-ponto-cerebellar atrophy (OPCA) plus. *Acta Neuropathol* 97:306 –310.
Fernandez-Ruiz J, Velasquez-Perez L, Diaz R, et al. (2007). Prism adaptation in spinocerebellar ataxia type 2. *Neuropsychologia* 45:2692–2698.
Freund H-J, Barnikol UB, Nolte D, et al. (2007). Subthalamic-thalamic DBS in a case with spinocerebellar ataxia type 2 and severe tremor – an unusual clinical benefit. *Mov Disord* 22:732–735.
Geschwind DH, Perlman S, Figueroa CP, Treiman LJ, Pulst SM. (1997). The prevalence and wide clinical spectrum of the spinocerebellar ataxia type 2 trinucleotide repeat in patients with autosomal dominant cerebellar ataxia. *Am J Hum Genet* 60:842–850.
Gierga K, Burk K, Bauer M, et al. (2005). Involvement of the cranial nerves and their nuclei in spinocerebellar ataxia type 2 (SCA2). *Acta Neuropathol (Berl)* 109:617–631.
Giuffrida S, Saponara R, Restivo DA, et al. (1999). Supratentorial atrophy in spinocerebellar ataxia type 2: MRI study of 20 patients. *J Neurol* 246:383–378.
Gonzalez C, Sánchez G, Gonzalez Quevedo A, et al. (2005). Serum and cerebrospinal fluid levels of copper, iron and zinc in patients with Ataxia type SCA-2 from the province of Holguin in Cuba. Therap. Basic. *Dialog Clin Neurosci* 13 (4):12–16.
Harding AE. (1993). Clinical features and classification of inherited ataxias. In: Harding AE, Deufel T, eds. *Inherited Ataxias*, Vol 61, pp. 1–14. New York: Raven.
Hayes S, Turecki G, Brisebois K, et al. (2000). CAG repeat length in RAI1 is associated with age at onset variability in spinocerebellar ataxia type 2 (SCA2). *Hum Mol Genet* 9(12):1753–1758.
Huynh DP, Figueroa K, Hoang N, Pulst SM. (2000). Nuclear localization or inclusion body formation of ataxin-2 are not necessary for SCA2 pathogenesis in mouse or human. *Nat Genet* 26:44–50.
Imbert G, Saudou F, Yvert G, et al. (1996). Cloning of the gene for spinocerebellar ataxia 2 reveals a locus with high sensitivity to expand CAG/ glutamine repeats. *Nat Genet* 14:285–291.
Inagaki A, Iida A, Matsubara M, Inagaki H. (2005). Positron emission tomography and magnetic resonance imaging in spinocerebellar ataxia type 2: a study of symptomatic and asymptomatic individuals. *Eur J Neurol* 12:725–728.
Irazno A, Comella CL, Santamaria J. Oertel W. (2007). Restless legs syndrome in Parkinson's disease and other neurodegenerative diseases of the central nervous system. *Mov Disord* 22:S424–S430.
Julio MB, Gilberto MB, Andrés MG, Mario FB, Luis VP. (2010). Heart rate variability in type 2 spinocerebellar ataxia. *Acta Neurol Scand* 122(5):329–335.
Kiehl TR, Nechiporuk A, Figueroa KP, et al. (2006). Generation and characterization of Sca2 (ataxin-2) knockout mice. *Biochem Biophys Res Commun* 339:17–24.
Kiehl TR, Shibata H, Pulst SM. (2000). The ortholog of human ataxin-2 is essential for early embryonic patterning in C. elegans, *J Mol Neurosci* 15:231–241.
Klockgether T. Ataxias. In: Goetz CG, ed. (2007). *Textbook of Clinical Neurology*, Vol 1, pp. 741–757. WB Saunders, Philadelphia.
Lastres-Becker I, Brodesser S, Lutjohann D, et al. (2008b). Insulin receptor and lipid metabolism pathology in ataxin-2 knock-out mice. *Hum Mol Genet* 17:1465–1481.
Lastres-Becker I, Rüb U, Auburger G. (2008a). Spinocerebellar ataxia (SCA2). *Cerebellum* 7(2):115–124.
Lu CS, Wu Chou YH, Kuo PC, et al. (2004). The Parkinsonian phenotype of spinocerebellar ataxia type 2. *Arch Neurol* 61:35–38.
Matilla A, Goold R, Giunti P. (2006). Molecular pathogenesis of spinocerebellar ataxias. *Brain* 129:1357–1370.
Medrano J, Velazquez L, Canales N, Rodríguez R, Gonzalez Y. (2009). Estudio electrofisiológico de pares craneales en enfermos portadores asintomáticos de la SCA2. *Rev Neurol* 49:278–279.
Orozco DG, Estrada R, Perry T, Araña J, Fernández R. (1989). Dominantly inherited olivopontocerebellar atrophy from eastern Cuba. Clinical, neuropathological and biochemical findings. *J Neurol Sci* 93:37–50.
Parduey M. (2004). The pathogenesis of Machado Joseph disease: a high manganese/low magnesium initiated CAG expansion mutation in susceptible genotypes? *J Am Coll Nutr* 23:715S–729S.
Pérez-Ávila I, Fernández-Vieitez JA, Martínez-Góngora E, Ochoa-Mastrapa R, Velázquez-Manresa MG. (2004). Efectos de un programa de ejercicios físicos sobre variables neurológicas cuantitativas en pacientes con ataxia espinocerebelosa tipo 2 en estadio leve. *Rev Neurol* 39 (10):907–910.
Pradhan C, Yashavantha BS, Pal PK, Sathyaprabha TN. (2008). Spinocerebellar ataxias type 1, 2 and 3: a study of heart rate variability. *Acta Neurol Scand* 117:337–342
Pulst MS, Nechiporuk A, Nechiporuk T, et al. (1996). Moderate expansion of a normally biallelic trinucleotide repeat in spinocerebellar ataxia type 2. *Nature Genet* 14:269–276.
Pulst SM, Santos N, Wang D, et al. (2005). Spinocerebellar ataxia type 2: PolyQ repeat variation in the CACNA1A channel modifies age of onset. *Brain* 128:2297–2303.
Ramos EM, Martins S, Alonso I, et al. (2010). Common origin of pure and interrupted repeat expansions in spinocerebellar ataxia type 2 (SCA2). *Am J Med Genet Part B* 153B:524–531.
Restivo DA, Giuffrida S, Rapisarda G, et al. (2000). Central motor conduction to lower limb after transcranial magnetic stimulation in spinocerebellar ataxia type 2 (SCA2). *Clin Neurophys* 111:630–635.
Restivo DA, Lanza S, Giuffrida S, et al. (2004). Cortical silent period prolongation in spinocerebellar ataxia type 2 (SCA2). *Funct Neurol* 19:37–41.
Reynaldo-Arminan RD, Reynaldo-Hernandez R, Paneque-Herrera M, Prieto-Avila L, Perez-Ruiz E. (2002). Mental disorders in patients with spinocerebellar ataxia type 2 in Cuba. *Rev Neurol* 35:818–821.

Rub U, Del Turco D, Burk K, et al. (2005). Extended pathoanatomical studies point to a consistent affection of the thalamus in spinocerebellar ataxia type 2. *Neuropathol Appl Neurobiol* 31:127–140.

Rub U, Del Turco D, Del Tredici K, et al. (2003). Thalamic involvement in a spinocerebellar ataxia type 2 (SCA2) and spinocerebellar type 3 (SCA3) patient and its clinical relevance. *Brain* 126:1–16.

Rub U, Seidel K, Ozerden I, et al. (2007). Consistent affection of the central somatosensory system in spinocerebellar ataxia type 2 and type 3 and its significance for clinical symptoms and rehabilitative therapy. *Brain Res Rev* 53:235–249.

Rufa A, Dotti MT, Galli L, Orrico A, Sicurelli F, Federico A. (2002). Spinocerebellar ataxia type 2 (SCA2) associated with retinal pigmentary degeneration. *Eur Neurol* 47:128–129.

Sanpei K, Takano H, Igarashi S, et al. (1996). Identification of the spinocerebellar ataxia type 2 gene using a direct identification of repeat expansion and cloning technique, DIRECT. *Nature Genet* 14:277–284.

Satterfield TF, Jackson SM, Pallanck LJ. (2002). A Drosophila homolog of the polyglutamine disease gene SCA2 is a dosage-sensitive regulator of actin filament formation, *Genetics* 162:1687–1702.

Satterfield TF, Pallanck LJ. (2006). Ataxin-2 and its Drosophila homolog, ATX2, physically assemble with polyribosomes. *Hum Mol Genet* 15:2523–2532.

Schöls L, Haan J, Riess O, Amoiridis G, Przuntek H. (1998). Sleep disturbance in spinocerebellar ataxias: is the SCA 3 mutation a cause of restless legs syndrome? *Neurology* 51:1603–1607.

Schöls L, Bauer P, Schmidt T, Schulte T, Riess O. (2004). Autosomal dominant cerebellar ataxias: clinical features, genetics, and pathogenesis. *Lancet* 3:291–304.

Schwenkreis P, Tegenthoff M, Witscher K, et al. (2002). Motor cortex activation by transcranial magnetic stimulation in ataxia patients depends on the genetic defect. *Brain* 125:301–309.

Simon DK, Zheng K, Velazquez L, et al. (2007). Mitochondrial complex I gene variant associated with early age of onset in SCA2. *Arch Neurol* 64:1042–1044.

Storey E, Forrest SM, Shaw JH, Mitchell P, Gardner RJ. (1999). Spinocerebellar ataxia type 2: clinical features of a pedigree displaying prominent frontal-executive dysfunction. *Arch Neurol* 56:43–50.

Takano H, Cancel G, Ikeuchi T, et al. (1998). Close associations between prevalences of dominantly inherited spinocerebellar ataxias with CAG-repeat expansions frequencies of large normal CAG alleles in Japanese and Caucasian populations. *Am J Hum Genet* 63:1060–1066.

Tan NC, Zhou Y, Tan AS, Chong SS, Lee WL. (2004). Spinocerebellar ataxia type 2 with focal epilepsy–an unusual association. *Ann Acad Med Singapore* 33:103–106.

Tuin I, Voss U, Kang JS, et al. (2006). Stages of sleep pathology in spinocerebellar ataxia type 2 (SCA2). *Neurology* 67(11):1966–1972

Vallés L, Estrada GL, Bastecherrea SL. (1978). Algunas formas de heredoataxia en una región de Cuba. *Rev Neurol (Cubana)* 27:163–176.

Velázquez-Pérez L, Sánchez Cruz G, Santos Falcón N, et al. (2009a). Molecular epidemiology of spinocerebellar ataxias in Cuba: Insights into SCA2 founder effect in Holguin. *Neurosci Lett* 454:157–160.

Velazquez PL, Fernandez-Ruiz J, Diaz R, et al. (2006). Spinocerebellar ataxia type 2 olfactory impairment shows a pattern similar to other major neurodegenerative diseases. *J Neurol* 253(9):1165–1169.

Velázquez-Perez L, Díaz R, Perez R, et al. (2009d). Motor decline in clinically presymptomatic spinocerebellar ataxia type 2 gene carriers. *PLOS One* 4(4):5398–5402.

Velázquez-Perez L, Rodríguez-Labrada R, Canales-Ochoa N, et al. (2010). Progression markers of Spinocerebellar Ataxia 2. A twenty years neurophysiological follow up study. *J Neurol Sci* 290:22–26.

Velázquez-Pérez L, Rodríguez-Labrada R, Medrano-Montero J, et al. (2009b). Patrón electromiográfico en enfermos y portadores asintomáticos de la mutación SCA2. *Rev Neurol* (49)1:55–56.

Velázquez-Pérez L, Sanchez Cruz G, Canales Ochoa N, et al. (2007). Electrophysiological features in patients and presymptomatic relatives with spinocerebellar ataxia type 2. *J Neurol Sci* 263(1–2):158–164.

Velázquez-Pérez L, Santos FN, García R, Paneque HM, Hechavarría PR. (2001). Epidemiología de la Ataxia Cubana. *Rev Neurol* 32 (7):606–611.

Velázquez-Pérez L, Seifried C, Santos-Falcon N, et al. (2004). Saccade velocity is controlled by polyglutamine size in spinocerebellar ataxia 2. *Ann Neurol* 56(3):444–447.

Velázquez-Pérez L, Seifried C, Abele M, et al. (2009c). Saccade velocity is reduced in presymptomatic spinocerebellar ataxia type 2. *Clin Neurophys* 120:632–635.

Velázquez-Pérez L. (2008). Nerve conduction and evoked potentials studies as preclinical marker of SCA2 mutation. In: Gonzalez L, ed. *Spinocerebellar ataxia type 2. Main neurophysiological aspects into the diagnosis, prognosis and disease evolution.* 2nd ed. Holguín: Ediciones Holguín. p. 120–137. [Book in Spanish]

Wadia NH, Swami RK. (1971). A new form of heredo-familial spinocerebellar degeneration with slow eye movements (nine families). Brain 94:359–374.

Wüllner U, Reimold M, Abele M, et al. (2005). Dopamine transporter positron emission tomography in spinocerebellar ataxias type 1, 2, 3, and 6. *Arch Neurol* 62:1280–1285.

107.

LATIN AMERICAN SCHOOL OF HUMAN AND MEDICAL GENETICS AND LATIN AMERICAN NETWORK OF HUMAN GENETICS: PROMOTING EDUCATION, INTERACTION AND NETWORKING ACROSS LATIN AMERICA

Roberto Giugliani, Guilherme Baldo and Ursula Matte

The Latin American Network of Human Genetics (RELAGH) is a virtual society of Human Genetics. Nowadays it represents Latin America at the International Federation of Human Genetics Societies (IFGHS), and is building a set of tools to enhance cooperation and integration of persons, groups and societies in the continent, a growing need since ALAG - Latin American Association of Genetics – interrupted its operations in the mid-90s. Together with the "Genetics for All Institute" (*Instituto Genética para Todos* – IGPT) it organizes the Latin American School of Human and medical Genetics (ELAG) an annual course that aims to promote a "state of the art" course in Medical Genetics, exposing during one week a selected number of top-grade Latin American students with first rank researchers in the field, aiming to create new links among Latin American groups

THE LATIN AMERICAN NETWORK OF HUMAN GENETICS

A BRIEF HISTORY

The idea of forming a Latin American Network of Human Genetics was raised at the IX International Congress of Human Genetics in Rio de Janeiro, Brazil, in 1996, when the IFGHS was created. The IFGHS full members are representatives of continental societies of human genetics, like the ones of North America, Europe and Australasia. As various Latin-American Societies of Human, Medical or Clinical Genetics have asked for joining the IFGHS, it has been suggested that Latin America could constitute a similar group.

In 1997 a meeting took place in San Juan del Rio, Mexico, attended by representatives of Human Genetics from a dozen of Latin-American countries and Dr. Marcus Pembrey, representing IFHGS. It was then decided to organize RELAGH, establishing that it would have as members not only Societies dedicated to Human or Medical Genetics but also individuals from countries where there are no such societies.

A meeting in 1999 in Buenos Aires, Argentina, attended most by members of ECLAMC (Estudio Colaboractivo Latinoamericano de Malformaciones Congenitas), resulted in the nomination of an organizing committee formed by representatives from Brazil, Argentina and Mexico. After overcoming initial difficulties in the following years, RELAGH was formally constituted during the X International Congress of Human Genetics in Vienna, Austria in 2001. On the IFHGS meeting held at this Congress, RELAGH was accepted as a full member and part of its executive committee. Representatives of 5 member societies were present: Human Genetics Branch of the Argentinean Genetics Society, Brazilian Society of Clinical Genetics, Colombian Human Genetics Association, Mexican Human Genetics Association and Panamanian Human Genetics Society. José María Cantú (from Mexico) was nominated President for the first 2 years, and it was decided to organize a meeting and elect a Directory Board in two years time. Roberto Giugliani (from Brazil) housed the RELAGH website at Universidade Federal do Rio Grande do Sul, Brazil.

In 2003, during the Human Genome Meeting in Cancun, Mexico, the Directory Board was elected by representatives of the member societies. Roberto Giugliani was elected the new President for the period 2003-2005, being followed by Alejandro Giraldo (Colombia, 2005–2008) and Gerardo Jimenez-Arce (Costa-Rica, 2008-current)..

Presently, the following societies have joined RELAGH: Mexican Association of Human Genetics, Colombian Association of Human Genetics, Human Genetics Branch of the Argentinean Genetics Society, Brazilian Society of Clinical Genetics, Panamanian Society of Human Genetics and Human Genetics Branch of the Chilean Society of Genetics, but this number shall continue to grow in the next years(1).

SCOPE

More than just representing Latin America at the IFGHS, RELAGH aims to provide a link among Latin American Human and Medical Geneticists, fostering scientific and technical cooperation between its members, service centers and research groups. In order to that, a form for registration of individuals and groups is available at the website (www.ufrgs.br/relagh), in addition to the one for societies. In this way, a directory of geneticists is being organized and will be available soon. Also, a list of laboratories and research centers will be created, sharing information about research projects and availability of genetic tests or services.

Another important goal of RELAGH is to improve exchange of skills among Latin American countries by supporting educational activities. This support is being done through direct organization of activities. Initiatives linked to this goal were the first Latin American Congress of Human Genetics (CLAGH) held at Cartagena de las Indias, Colombia, in 2008, the proposal of an International Postgraduate Course in Human Genetics (under consideration) and the Latin American School of Medical Genetics, organized annually since 2005.

THE LATIN AMERICAN SCHOOL OF HUMAN AND MEDICAL GENETICS

Short courses of medical and human genetics to young students and professionals are already a tradition in North America and Europe, such as the Annual Short Course on Medical and Experimental Mammalian Genetics, held at Bar Harbor, USA, and the European School of Genetic Medicine, held in Italy. To prepare young researchers and professionals of Latin America to deal with the growing challenge of the genomic medicine seems to be a major challenge after the genomic revolution. Based on that, since 2005 RELAGH, in collaboration with CLADEGEN, organizes the Latin American School of Human and Medical Genetics (LASG) a one-week long course. The course receives 75 students from almost all Latin American countries and 30 faculty members from Latin America, North America and Europe.

The course aims to promote a "state of the art" course on Medical Genetics, exposing a selected number of top-grade students with first rank researchers in the field. In addition to attend to a strong teaching program, the students have the unique opportunity of discussing projects, techniques, clinical cases and scientific issues with the faculty members. The participation of students and faculty members from several Latin American countries aims to stimulate further cooperation among groups inside the continent. In addition, the organization provides several ways of formal and informal interaction between students and faculty members, such as the "Lunch with the Professor", and a social program.

ELAG PROGRAM

Traditionally, the course is structured by symposia in the morning and short-courses in the afternoon (Figure 107-2). The symposia main themes include topics as "Human Genetic Variability", "Genomic Medicine", "Therapy for Genetic Diseases" and "Ethical and Public Health Implications of the Genomic Revolution". Each symposia consists of 4 lectures related to the main theme. In addition, short courses with practical activities include topics as "Genetic Counseling", "Genetic Epidemiology", "Novel Therapeutic Strategies" and "New Tools for Genetic Analysis". Other activities include a debate on specific topics and a workshop on how to write a paper and how to apply for grants, aiming to stimulate students to publish their results and to obtain funding for future projects. Small changes in the course program are made every year based on suggestions from the former students and faculty members or according to specific needs noticed by the organizing committee. All activities are planned aiming to focus on the Latin American context.

PREVIOUS EDITIONS AND PRELIMINARY RESULTS

The first seven editions of the course were held in Brazil, at Caxias do Sul (2005, 2006, 2008, 2009, 2010 and 2011) and Angra do Reis (2007). During these years, a total of 533 students from 14 different Latin American countries attended the course (figure 107-1), almost all of them with scholarships to cover for meals and accommodation costs provided by the Organizing committee. The funding is obtained from Brazilian governmental agencies and from private companies, as well from the inscription fee. Faculty members have all their expenses paid but do not receive any financial compensation to participate in the course.

The course maintains its basic structure since the first edition, evolving with some minor adaptations. Improvements on the dynamic of ELAG were incorporated to enhance contact between students-faculty members, as marked seats during lunches and social events, including a multicultural evening and a soccer match.

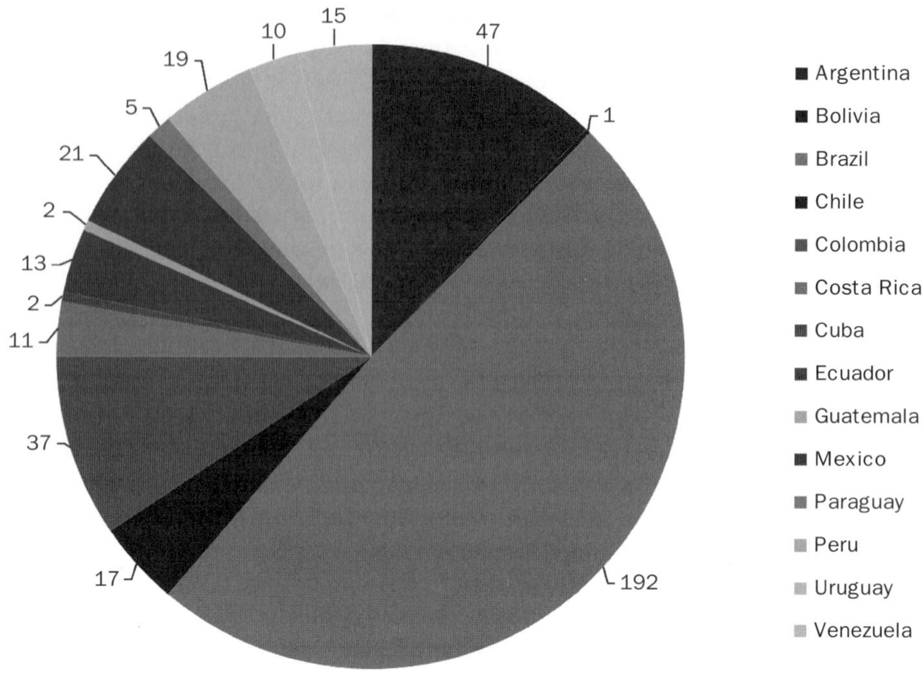

Figure 107-1 Number of students from different countries that have attended the ELAG course on its first 7 editions. Refer color figure.

Latin American School of Human And Medical Genetics
Day One: Welcome And Opening Lecture: 19:30 h

	Day 02	Day 03	Day 04	Day 05	Day 06	Day 07
Morning	***Symposium. 1: From Genetics to Genomics and beyond***	***Symposium 2: Human Genetic Variability***	***Symposium 3:Genomic Medicine I***	***Symposium 4: Genomic Medicine II***	***Symposium 5: Therapy for Genetic Disorders***	***Check-out, Farewell and Departures***
	From Genetics to Genomics and Beyond	*Genetics of Latin American Populations*	*Genetics of Development*	*Cancer Genetics*	*Enzyme Therapy*	*Meeting- Scientific and Organizing Committees*
	The new world of RNAs	*Genome wide admixture*	*Neurogenetics*	*Functional Genomics for Drug Discovery*	*Cell Therapy*	
	Coffee break	*Coffee break*	*Coffee break*	*Coffee break*	*Coffee break*	
	Proteomics	*Population Medical Genetics*	*Farmacogenomics*	*Prenatal Diagnosis*	*Gene Therapy*	
	Bioinformatics	*Epigenetics*	*New technologies of genetic analyses*	*Population Screening*	*New Therapies*	
12:30	*LUNCH*	*LUNCH*	*LUNCH*	*LUNCH*	*LUNCH*	
Afternoon	*Short courses 1: Genetic Epidemiology & 2: New Technologies in Genetic Research*	*Short courses 1: Genetic Epidemiology & 2: New Technologies in Genetic Research*	*Afternoon tour*	*Short courses 1 Genetic Epidemiology & 2: New Technologies in Genetic Research*	***Workshop "On papers and grants"***	
	Coffee break	*Coffee break*		*Coffee break*	*Coffee break*	
	Short courses 3: Genetic Counseling & 4: Novel Therapeutic Strategies	*Short courses 3: Genetic Counseling & 4: Novel Therapeutic Strategies*		*Short courses 3: Genetic Counseling & 4 Novel Therapeutic Strategies*	***Symposium 6: Public Health Implications***	
					Bringing New Therapies to Health Services	
					Pre-symptomatic Diagnosis and Predictive Medicine	
					Bioethics	
Evening	*Multicultural Evening*	*Dinner Debate: Personalized medicine*	*Free evening/ Dinner*	*Soccer Game Dinner*	*Barbecue*	

Figure 107-2 ELAG simplified program. Modified from the material provided to students in ELAG 2010 by the organizing committee.

Evaluating the impact of the course on a short-term basis is difficult, but anecdotic episodes shows that the initiative is accomplishing its aims. For example, there have been theoretical papers published by students who attended the school based on a simple hypothesis raised after informal conversations with some ELAG faculty members (2), and collaborative projects and students' exchange among groups from different countries have also been reported.

CONCLUSION

Although not replacing the role of national societies, RELAGH has the potential to improve integration and cooperation among medical and human geneticists from all Latin American countries, and of service centers and research groups as well, being a useful tool for the development of this field in the continent. The Latin American School of Human and Medical Genetics serves as a physical manifestation of the activities of RELAGH, offering a unique opportunity to young Latin American scientists to get in touch with the main researchers from the continent on specific subjects.

REFERENCES

Giugliani R, Matte U. The Latin-American network of human genetics. Community Genet 2004, 7(2–3):74–75.

Veit TD, Chies JA. Tolerance versus immune response – microRNAs as important elements in the regulation of the HLA-G gene expression. Transpl Immunol 2009, 20(4):229–231.

RUSSIA, CENTRAL ASIA, AND EUROPE

108.

GENETIC LEGACY OF POPULATIONS IN EURASIA

I.A. Kutuev, B.B. Yunusbayev, and Elza K. Khusnutdinova

INTRODUCTION

Studies of genetic variation within and between present-day human populations provide unique information about the past human history. Phylogeographic reconstructions based on the analysis of uniparentally inherited mitochondrial DNA (mtDNA) and Y-chromosome markers became very popular in recent years, owing to their ability to provide information about the place and the time of demographic events that can be associated with prehistoric events in human history. Because inferences about past population history are based on current phylogeographic patterns, it is essential to study present-day genetic diversity using as many population samples as possible, in order to cover all possible geographic regions inhabited by humans.

Numerous studies in the recent years have yielded considerable progress in understanding the phylogeography of Y-chromosome and mtDNA lineages, mainly in the western part of Eurasia. In contrast, eastern regions like central Asia, the Volga-Ural region, and the Caucasus have not been investigated systematically to the level currently available for the western Eurasian populations. Absence of detailed information on the phylogeography of maternal and paternal lineages in some eastern populations and populations from other regions hinders overall progress toward understanding genetic history in Eurasia. This chapter is based on a number of field expeditions in different populations of Eurasia, including the Volga-Ural, central Asia, and the Caucasus.

It is common to treat central Asia and the Volga-Ural region as a crossroads area that witnessed in the past numerous waves of population migrations. Until the Early Medieval Age, the southern Urals, according to archeological and paleoanthropological data, was inhabited by a Europeoid population that can be associated with the currently living representatives of different divisions of the Europeoid major race (Pontic type of the Indo-Mediterranean race and Suburalic type, which prevails among Volga basin Finnic populations; Akimova, 1974). The ancient population of the Volga basin could be linked to the present-day Baltic and Suburalic types of the Europeoid major race (Akimova, 1974). Alexeev and Ghokhman (1984) reviewed paleoanthropological studies and demonstrated that the ancient population of the central Asia region, between Neolithic times and up to the Iron Age, was predominantly Europeoid (Alexeev and Ghokhman, 1984). Turkic-speaking nomadic people from southern Siberia, and later, Tataro-Mongol invaders, are considered to be relatively recent newcomers into this area (Alexeev, 1974; Kuzeev, 1992).

POPULATION SAMPLES

A total of 2733 individuals were sampled from three regions during field expeditions in 1993–2003 conducted by authors' units (see details in Table 108-1). Informed consent was obtained from all donors. Ethnic origin of sampled individuals was ascertained up to three generations, as determined through oral interviews. Genomic DNA was extracted from peripheral blood lymphocytes using a standard phenol-chloroform method (Sambrook et al., 1989). The population data on Yakuts, Altays, Shors, Nganasans, Hants, Mansi, Selkups, Tuvinians, Dolgans, Turks, Azerbaijanians, and Estonians was included for comparison (Khusnutdinova et al., 2002).

The fragment 16024 -16400 of mtDNA hypervariable segment I (HVS I) was sequenced in a Perkin-Elmer ABI 377 DNA sequencer with a DYEnamic ET kit (Amersham Pharmacia Biotech). The nucleotide sequences obtained were compared with the revised Cambridge reference sequence (rCRS; Andrews et al., 1999). Haplogroups were identified by restriction fragment length polymorphism (RFLP) analysis of 39 polymorphic sites of the mtDNA coding region as described elsewhere (Finnila et al., 2001; Herrnstadt et al., 2002; Kivisild T. et al., 2002; Macaulay et al., 1999; Richards et al., 2000; Torroni et al., 1994; Yao et al., 2004).

Forty-two binary polymorphisms (M9, M89, YAP (M1), M174, M40, M35, M130, M48, 12f2, M267, M62, M172, M12, M201, M285, M342, P20, P15, P16, M286,

TABLE 108-1 GEOGRAPHIC AND LINGUISTIC DESCRIPTION OF STUDIED POPULATIONS

Population	Country, region and district	N	Linguistic affiliation
VOLGA-URAL REGION			
Bashkirs	Northwestern, northeastern, southwestern and southeastern ethnogeographic groups of Bashkortostan	258	Turkic branch of Altaic family
Gaininsk Bashkirs	Perm region	60	
Tatars	Al'met'evskii and Elabuzhskii districts of Tatarstan	228	
Chuvashis	Morgaushskii district of Chuvashia	55	
Komi-Permyaks	Komi-Permyak autonomous district	74	Finno-Ugric subfamily of Uralic family
Komi-Zyryans	Sysol'skii district of the Komi Republic	62	
Mari	Zvenigovskii district of Marii El	136	
Mordvinians	Staro-Shaiginskii district of Mordovia	102	
Udmurts	Malo-Purginskii district of Udmurtia and the Tatyshlinskii district of Bashkortostan	101	
CENTRAL ASIA			
Uzbeks	Samarkandskaya, Khorezmskaya, Tashkentskaya provinces of Uzbekistan	103	Turkic branch of Altaic family
Kazakhs	Kazakhstan	331	
Uygurs	Kazakhstan	121	
CAUCASUS			
Nogays	Karachay-Cherkessya	110	Turkic branch of Altaic family
Karanogays	Dagestan	102	
Karachays	Karachay-Cherkessya	106	
Kumyks	Dagestan	107	
Avars	Dagestan	42	Nakh-Dagestanian of Caucasus
Tabassarans	Dagestan	43	
Lezgis	Dagestan	31	
Ingushes	Dagestan	105	
Adyghes	Adyghea	154	Abkhaz-Adyghean of Caucasus
Cherkessians	Karachay-Cherkessia	126	
Kabardines	Kabardino-Balkariya	140	
Megrels	Georgia	76	Kartvelian

M406, M287, M170, M253, P37, M223, M52, M231, Tat (M46), P43, M128, M175, M20, M70, 92R7, M207, M242, M173, SRY 1532, M73, M269 and M124) were genotyped using either RFLP analysis or direct sequencing, following the hierarchy of the Y-chromosome phylogeny (Jobling et al., 2003; Karafet et al., 2008; YCC, 2002). The principal component analysis based on the haplogroup frequencies was performed using Statistica v. 5.5 (Statsoft 2001).

MTDNA VARIATION

The analysis of mtDNA variation among the Volga-Ural (Tatars, Bashkirs, Chuvashis, Mari, Mordvins, Udmurts, Komi), central Asia (Uzbeks, Kazakhs, Uighurs) and Caucasus (Nogays and Karanogays) populations revealed that the majority of maternal lineages are those distributed among Western European and Eastern European populations. Relatively high frequencies of East Asian (those specific to East Asian populations) G, D, C, Z, and F haplogroups among both Turkic-speaking population (Bashkirs, Nogays, Karanogays, Uzbeks, Kazakhs, Uighurs) and Finno-Ugric speaking populations (Udmurts and Komi-Permyaks) indicate significant genetic contribution of eastern Eurasian populations. Comparable high frequencies of East Asian and European lineages are consistent with the intermediate position of the Volga-Ural region. However, some inconsistent features in the distribution of frequencies of mtDNA lineages become apparent, keeping in perspective the linguistic affiliation of individual populations.

The interesting fact revealed is the high frequency of western Eurasian lineages in Uighurs (~45%). It is noticeable that still this population is surrounded by Mongol, Chinese, Kirgizis and Altays, in which frequency of western Eurasian lineages are less than 15% (Comas et al., 2004; Kolman et al., 1996; Yao et al., 2002, 2004; Yao and Zhang, 2002). Detailed analysis revealed high frequency of typical Anatolian, Iranian, and South Caucasus lineages, which led us to suggest that at least part of the genetic pool of modern Uighurs are common to Indo-Iranian nomads of Neolithic times.

It is quite obvious now that the so-called protoaltaic or protouralic genetic substrate doesn't exist. Even low-level resolution based on haplogroup frequency data demonstrate great differences between the most western (Gagauz) and eastern (Dolgans) Turkic speakers (our unpublished data). In an analysis of all the populations inhabiting Eurasia (belonging to other language families) the genetic landscape doesn't change drastically. This means that the modern genetic landscape is formed generally due to demographic processes in populations inhabiting this vast area.

Altaic and Uralic speakers inhabiting the European part of the continent are characterized by high frequency of western Eurasian lineages; those who inhabit the Asian part are characterized by high frequency of eastern Eurasian lineages. The only exceptions are Kalmyks and Nogays living in north Caucasus (Bakunin, 1995; Kereitov, 1999).

The low levels of eastern Eurasian mtDNA lineages in Gagauzes, Turks, Azeris, and Kumyks supports the hypothesis regarding recent expansion of Turkic languages to the west. At the same time, it's possible that a genetic pool of prototurkic people consisted mostly of western Eurasian lineages, and subsequent admixture with populations rich of eastern Eurasian lineages led to a drastic increase of the latter in Turkic-speaking populations living in Asian parts of Eurasia.

There are a couple of exceptions within the existing west–east gradient of mtDNA haplogroups frequency change. One of these exceptions is the Nogays inhabiting the northern part of Dagestan and Kabardino-Balkariya. The frequency of eastern Eurasian haplogroups in Nogays is up to 40%, but the neighboring populations (Kumyks, Karachays, and Balkars) have those haplogroups at a frequency lower than 7% (Bermisheva et al., 2004). This fact can be explained by the history of Nogays, which are remnants of the Nogay Horde that compared Turkic, Ugric and Mongol tribes. Nogays formed as an ethnicity quite recently (fourteenth to fifteenth centuries; Kereitov, 1999). The other exception is Kalmyks, people who came to North Caucasus some three centuries ago (Bakunin, 1995).

The interesting fact that mtDNA lineages revealed in Caucasus populations that don't belong to the A clade is that its frequency in Nogays and Kalmyks are up to 6% (Bermisheva et al., 2004). This fact demonstrates that eastern Eurasian lineages penetrated into the European part of the continent due to mass migrations of Mongols. At the same time, the admixture of nomadic Mongols with autochthonous populations of the Caucasus was insignificant.

It is generally believed that the first Turkic-speaking people were native to a region spanning from central Asia and across, throughout Siberia (Alexeev and Ghokhman, 1984). Assuming that a westward expansion of Turkic tribes involved the spread of mainly East Asian maternal lineages, we studied the distribution of mtDNA haplogroups among 18 Turkic-speaking populations scattered throughout Eurasia. Frequencies of East Asian mtDNA lineages ranged from 1% in Gagauz to 99% in Yakuts, with an increasing cline from west to northeast (Figure 108-1). While this frequency cline is consistent with a westward direction of migration, the proportion of East Asian mtDNA lineages within individual populations is too small to suggest significant genetic input associated with south Siberian populations. For example, Turkic-speaking populations residing in the Volga-Ural region (Tatars and Chuvashis) are more closely related genetically to their geographic neighbors than to their linguistic neighbors elsewhere. It looks as if indigenous populations of this region adopted Turkic languages without much genetic input, for example the Azerbaijanians in Caucasus (Nasidze et al., 2003) and Turks in Anatolia (Cinnioglu et al., 2004). The limited presence of East Asian mtDNA lineages in these populations most likely reflects consequences of language replacements, possibly via elite dominance. In general, this finding is consistent with our previous suggestion that geographic proximity rather than linguistic relatedness best explains genetic similarities between populations (Khusnutdinova et al., 2002). It should be noted that our earlier postulated mtDNA pool of south Siberia, and the conclusion that we have drawn based on it, is a result of apparent oversimplification of real demographic processes. A more realistic explanation could be offered if we take into account paleoanthropological data from the south Siberian and central Asian area. First of all, the prevalence of East Asian mtDNA lineages among hypothetical ancient Turkic people in south Siberia is not a solid assumption, because there is no direct evidence regarding the anthropological traits of that people. In this regard, it is worth noting the results of some paleoanthropological discoveries in the neighborhood of the postulated source region. For example, excavations at burial sites belonging to Turkic people of 700–800 AD in Tuva (south Siberian region in Russia) demonstrated detectable Europeoid traits in Mongoloid human skeletal remains (Alexeev and Ghokhman, 1984). Moreover, Alexeev and Ghokhman

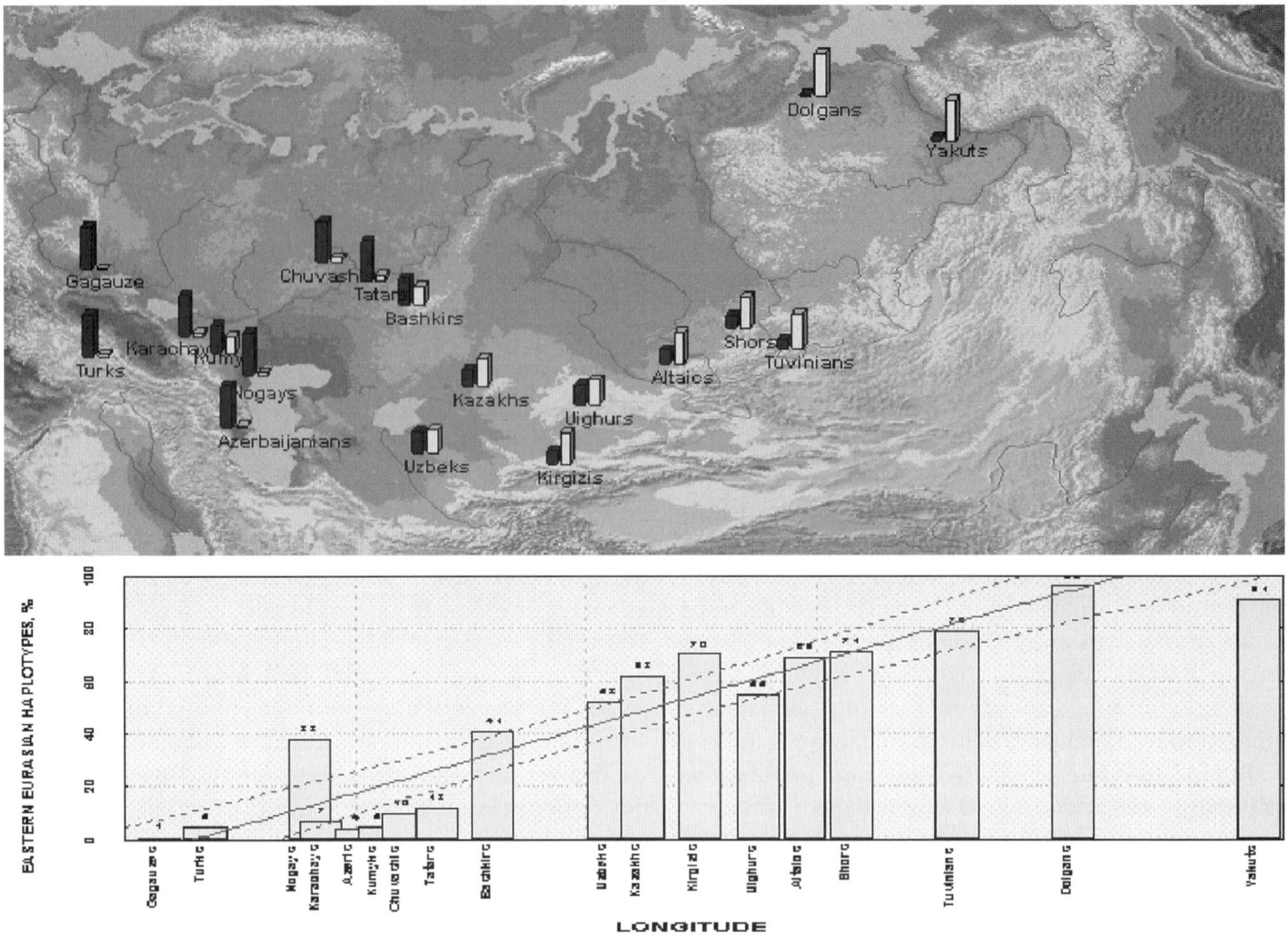

Figure 108-1 West to east frequency gradient of East Asian mtDNA lineages among 18 Turkic-speaking populations of Eurasia. Refer color figure.

(1984), based on their review of paleoanthropological studies, showed that a westward expansion of Turkic people from south Siberia through central Asia was accompanied by gradual admixture with indigenous Europeoid populations (Alexeev and Ghokhman, 1984). Thus, the limited presence of East Asian mtDNA lineages in the Volga basin (Tatars, Chuvashis), as well as in Caucasus (Azerbaijanians) and Anatolia (Turks; (Cinnioglu et al., 2004; Nasidze et al., 2003), can be explained by the arrival of already admixed Turkic-speaking populations in these regions. Paleoanthropological evidence from the central Asia region, therefore, makes it problematic or even impossible to estimate the proportion of gene flow associated with the arrival of Turkic-speaking people.

Geographic frequency distribution of Asian mtDNA lineages among 17 Finno-Ugric populations, with some exceptions, forms a similar west to east increasing cline.

Like in Turkic-speaking groups Asian mtDNA lineages dominate in northeastern Siberia, reaching maximum frequency in Nganasans (80%), and decrease toward the west: for example, in Estonians it is almost absent. A high proportion (60%–70%) of typical West European mtDNA lineages among Khanti, Mansi, and Selkups living in southern Siberia does not fit to the opposite west–east frequency gradient of European lineages. These populations are characterized by high frequency of U4 and low frequency of haplogroup W, which is typical of Finno-Ugric populations of the Volga-Ural region. This may be result of west to east migration from the Volga-Ural region (Bermisheva et al., 2004, 2002; Tambets et al., 2003; Villems et al., 2002).

So far we have analyzed mtDNA haplogroup diversity by dividing mtDNA variants into two broad groupings (European and Asian). In order to test whether frequency distribution of lineages correlate with our previous findings, we summarized mtDNA haplogroup frequencies using principal component analysis (PCA). Figure 108-2 demonstrates the projection of Volga-Ural, Caucasus, and central Asian populations in the plot of two principal components of mtDNA variability.

The frequency in the population of Volga-Ural region, north Caucasus, and central Asia explains 52.8% of haplogroups' frequency variability. The results obtained correspond with the east–west gradient of mtDNA haplogroups frequency alongside the Eurasian steppe belt. It is evident from the plot that studied populations are clustered

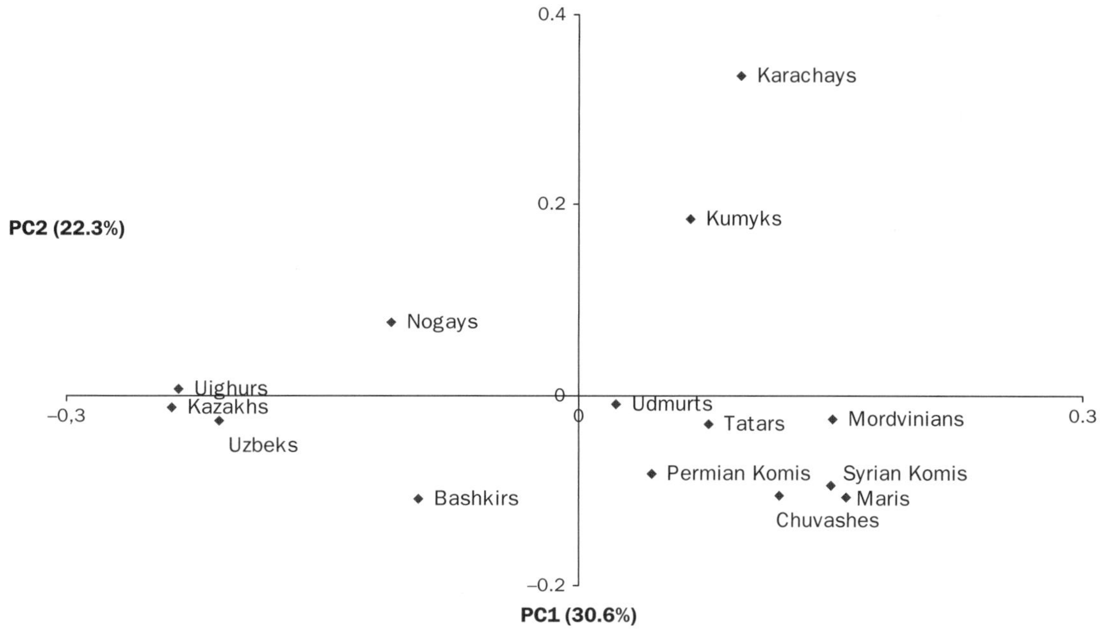

Fig. 108-2 Principal component analysis (2 dimensions) of mtDNA haplogroup frequencies in populations of Volga-Ural, central Asia, and the Caucasus.

according to their geographic positions. Some deviations, for example the close position of Nogays on the plot to central Asian populations, can be explained by the high proportion of typical East Asian mtDNA lineages in this population.

Populations on the plot are clustered together according to geography but not to linguistic affiliation of the investigated populations. Close proximity of Nogays and Bashkirs on the plot could be explained by a high percentage of eastern Eurasian lineages in both populations. High frequency of eastern Eurasian lineages in Nogays is not surprising, since they moved and settled in Caucasus quite recently (Bermisheva et al., 2004).

Analysis of three principal components (64.9%) hasn't revealed major changes (Figure. 108-3). Udmurts due to new dimension in the projection of the 3rd component are located quite far from other populations. This could be explained by relatively high frequency of haplogroup T in this population (0.238). Our preliminary summary of obtained results is that predominantly Europeoid population of Central Asia and Volga-Ural region experienced significant gene flow from eastern mongoloid populations that eventually reshaped genetic (mtDNA) landscape of Volga-Ural region and Central Asia (Fig. 108-3).

Y-CHROMOSOME VARIATION

A total of 25 distinct haplogroups were defined in the combined Y-chromosome pool of Volga-Ural, central Asian, and Caucasus populations. Most paternal lineages of the Volga-Ural and the central Asian populations belong to western Eurasian Y-chromosome haplogroups. Sub-clades of widely distributed R haplogroup, together with circum-Artic spread haplogroups N, predominate in all studied populations, constituting about 25% of paternal lineages in Kazakhs and up to 91% in Bashkirs. Consistent with mtDNA data, the Y-chromosome pool of haplogroups share considerable portions of eastern European lineages except in Kazakhs, carrying C (25%) and J (18%) lineages in relatively high frequencies. Haplogroup N3, the most frequent paternal lineage among populations inhabiting northern Siberia and Europe (Rootsi et al., 2007; Tambets et al., 2004; Zerjal et al., 2001) according to recent phylogeographic reconstruction, has dispersed throughout northern Eurasia, either from northern China or southern Siberia (Rootsi et al., 2007). High incidence of this lineage among Udmurts (55%) and other Volga-Ural Finno-Ugric populations (Mari, Komi-Ziryans, and Komi-Permyaks) is consistent with the scenario, according to which N3-bearing ancient populations witnessed numerous bottlenecks while migrating from Siberia to eastern Europe. Turkic-speaking populations in the Volga-Ural region (Chuvashis, Tatars, and Bashkirs) also share this haplogroup but in much lower frequencies. The high frequency of this lineage in the Volga-Ural region is likely associated with ancient Finno-Ugric tribes, and somewhat lower frequency in neighboring Turkic-speaking groups is either a "signature" of Finno-Ugric genetic input or the legacy of a common ancestral population for these two linguistically distinct groups. Indeed, taking into account that Karanogays from northeastern Caucasus (N3 – 3%), Kyrgyz and Dungans from central Asia (Zerjal et al., 2002), as well as 11 Altaic-speaking southern Siberian

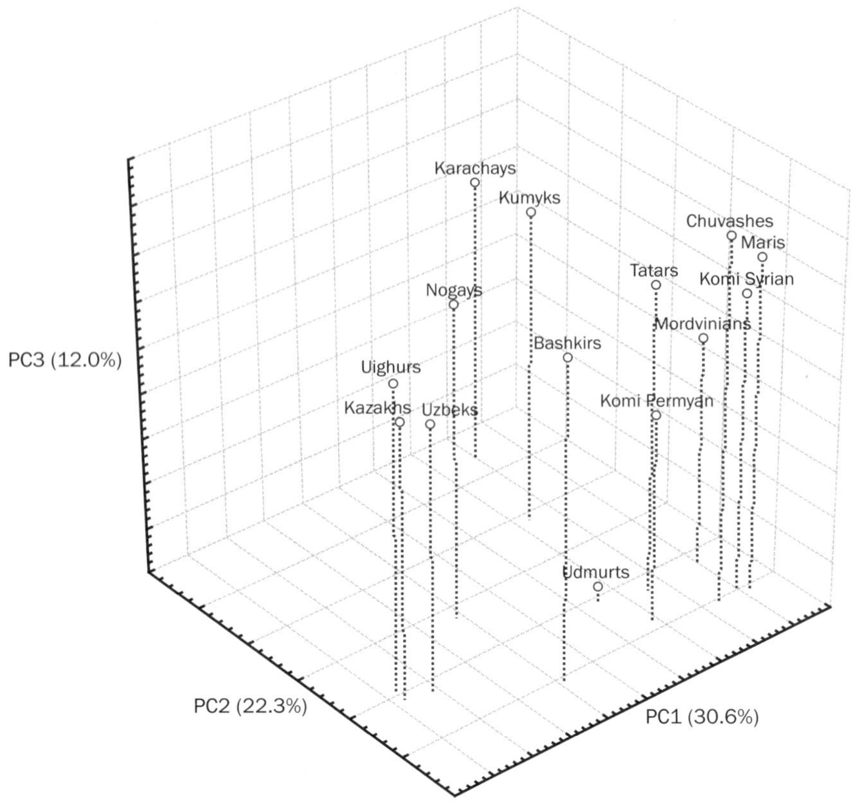

Fig. 108-3 Principal component analysis (3 dimensions) of mtDNA haplogroup frequencies in populations of Volga-Ural, central Asia and the Caucasus.

populations carry this haplogroup (frequency ranges from 2% in Shors to 25% in Tofalars; Derenko et al., 2006), it is possible that the N3 lineage in the past represented a common Siberian background of Finno-Ugric and Turkic tribes. Overall, haplogroups N subclades (N2, N3) regardless of they were brought by the ancestors of Turkic or Finno-Ugric populations, represent a significant amount of southern Siberian genetic legacy.

R subclades (R1b3, R1a1), which are widespread in Europe and the Middle East, are distributed unevenly in central Asia and the Volga-Ural region, ranging from 1% in Kazakhs to 82% in Trans-Ural Bashkirs. It was suggested that distinctive European geography of two main R lineages, R1a1 and R1b3, is associated with population expansion events during the last Ice Age (approximately 18 thousand years ago; Pericic et al., 2005; Semino et al., 2000), but the place and the time of origin of these two ancient lineages are not well established yet (Kivisild et al., 2003). There is increasing evidence that Y-chromosome haplogroup R1b3 is not specific to west Europe, but rather represents a genetic legacy of an ancient, widely distributed population of Eurasia. For example, recently Al-Zahery et al. (2003) and Cinnioglu et al. (2004) demonstrated that R1b3 Y-chromosomes in Middle Eastern populations are associated with the Taq I ht35 restriction profile (haplotype) of the complex 49a,f RFLP polymorphic system, while West European populations are characterized by the ht15 haplotype, which is restricted mainly to Europe and has a west–east decreasing gradient (Al-Zahery et al., 2003; Cinnioglu et al., 2004). Al-Zahery et al. (2003) suggested that haplotype 35, which has uniform distribution across west Eurasia, has been dispersed in west Eurasia earlier, and that haplotype 15 could have derived from it by a single mutation event. That mutation probably occurred inside Europe, and its extremely high incidence in western Europe could be the result of population bottlenecks during the last Glacial Maximum (Al-Zahery et al., 2003). These findings are very important because they demonstrate subdivision of the R1b3 lineage into as yet undefined lineages, one of which (ht15) has distinct geographic distribution in Europe. In this regard, decreasing frequency cline of haplogroup R1b3 from western Europe toward eastern Europe and unexpectedly high frequency of this lineage in the easternmost fringe, i.e., the southern Urals (>82%) and in the central Asian highlands among Tajiks, Kyrgyz, and Altai (>50%; Kivisild et al., 2003) imply that R1b3 Y-chromosomes in these two eastern regions may not be associated with the west–east population expansion from Europe. This proposal is plausible in the light of earlier presence of Europeoid populations in both regions. However, further genetic evidence is necessary in order to test the abovementioned proposal.

The principal component analysis of Y-chromosome haplogroups frequencies in populations of the three investigated regions reveal clustering of peoples in the Caucasus region (Figure 108- 4). It's quite interesting that different autochthonous Caucasus populations are scattered at the plot. Populations speaking Nakh-Dagestanian languages

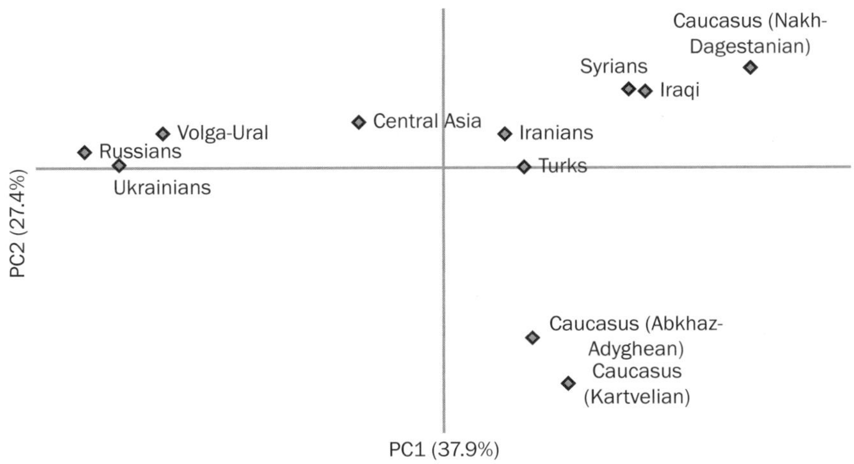

Figure 108-4 Principal component analysis of Y-chromosome haplogroup frequencies in populations of Volga-Ural region, central Asia, and the Caucasus.

(inhabiting Dagestan, Chechnya, and Ingushetia) are located quite close to Near Eastern populations (Syrians, Iraqis, Iranians, and Turks) due to their similar haplogroup frequency patterns, characterized mostly by high frequencies of haplogroups J1* and R1b1b2. However, the haplogroup frequencies vary in quite a wide range between Nakh-Dagestanian populations, in some cases demonstrating genetic drift in isolated populations. On the other hand, Abkhaz-Adyghean and Kartvelian speakers lie apart from other populations and form a close cluster characterized by a high frequency of haplogroup G2, and relatively high frequencies of R1a1, J2a*, and J2a2*. Eastern European populations (Volga-Ural populations, Russians, and Ukrainians) also form a close cluster due to similar haplogroup frequency distribution patterns.

CONCLUSIONS

Populations located in the Volga-Ural region, central Asia, and the Caucasus exhibit a complex mtDNA and Y-chromosome lineage composition, consisting mainly of western Eurasian lineages, with a somewhat lower but comparable contribution of eastern Asian lineages. Central Asians exhibit higher frequencies of East Asian lineages compared with their northern Volga-Ural neighbors. Overall, the genetic landscape in these three regions points to a stronger genetic influence of Turkic people on central Asian populations compared with their northern neighbors. Finally, genetic relationships between studied populations can be best explained by geographic proximity rather than linguistic affiliation.

ACKNOWLEDGMENTS

This work was supported by a Grant of the President of Russian Federation (t MK-2035.2008.4) to I. Kutuev, and a grant by the Russian Foundation for Basic Research (07-04-01016-a) to E. Khusnutdinova.

REFERENCES

Akimova MS. (1974). Anthropological investigations in Bashkiria. M.: Nauka.

Al-Zahery N, Semino O, Benuzzi G, et al. (2003). Y-chromosome and mtDNA polymorphisms in Iraq, a crossroad of the early human dispersal and of post-Neolithic migrations. *Mol Phylogenet Evol* 28 (3):458–472.

Alexeev VP. (1974). *Geography of human races*. Moscow: Nauka.

Alexeev VP, Ghokhman II. (1984). *Anthropology of the Asian Part of the USSR*. Moscow: Nauka.

Andrews RM, Kubacka I, Chinnery PF, Lightowlers RN, Turnbull DM, Howell N. (1999). Reanalysis and revision of the Cambridge reference sequence for human mitochondrial DNA. *Nat Genet* 23 (2):147.

Bakunin VM. (1995). *Description of Kalmyk Peoples, Especially Torgouts, their Habits, Khans and Masters*. Elista: Kalmyk Publishing House.

Bermisheva MA, Kutuev IA, Korshunova T Yu, Dubova NA, Villems R, Khusnutdinova EK. (2004). Phylogeographic analysis of mitochondrial DNA in the Nogays: A strong mixture of maternal lineages from eastern and western Eurasia. *Mol Biol (Mosk)* 38 (4):516–523.

Bermisheva M, Tambets K, Villems R, Khusnutdinova E. (2002). Diversity of mitochondrial DNA haplotypes in ethnic populations of the Volga-Ural region of Russia. *Mol Biol (Mosk)* 36 (6):990–1001.

Cinnioglu C, King R, Kivisild T, et al. (2004). Excavating Y-chromosome haplotype strata in Anatolia. *Hum Genet* 114 (2):127–148.

Comas D, Plaza S, Wells RS, Yuldaseva N, Lao O, Calafell F, Bertranpetit J. (2004). Admixture, migrations, and dispersals in Central Asia: evidence from maternal DNA lineages. *Eur J Hum Genet* 12 (6):495–504.

Derenko M, Malyarchuk B, Denisova GA, et al. (2006). Contrasting patterns of Y-chromosome variation in south Siberian populations from Baikal and Altai-Sayan regions. *Human Genetics* 118 (5):591–604.

Finnila S, Lehtonen MS, Majamaa K. (2001). Phylogenetic network for European mtDNA. *Am J Hum Genet* 68 (6):1475–1484.

Herrnstadt C, Elson JL, Fahy E, et al. (2002). Reduced-median-network analysis of complete mitochondrial DNA coding-region sequences for the major African, Asian, and European haplogroups. *Am J Hum Genet* 70 (5):1152–1171.

Jobling MA, Tyler Smith C. (2003). The human Y chromosome: an evolutionary marker comes of age. *Nat Rev Genet* 4 (8):598–612.

Karafet TM, Mendez FL, Meilerman MB, Underhill PA, Zegura SL, Hammer MF. (2008). New binary polymorphisms reshape and increase resolution of the human Y chromosomal haplogroup tree. *Genome Res* 18 (5):830–838.

Kereitov R Kh. (1999). *Ethnical history of Nogays*. Stavropol.

Khusnutdinova E, Bermisheva M, Malyarchuk B, et al. (2002). Towards a comprehensive understanding of the East European mtDNA heritage in its phylogeographic context. *Human Origins and Disease.* Cold Spring Harbor Laboratory, New York.

Kivisild T, Rootsi S, Metspalu M, et al. (2003). The genetic heritage of the earliest settlers persists both in Indian tribal and caste populations. *Am J Hum Genet* 72:313–332.

Kivisild T, Helle Tolk V, Parik J, et al. (2002). The emerging limbs and twigs of the East Asian mtDNA tree. *Mol Biol Evol* 19 (10):1737–1751 (erratum 1720:1162).

Kolman C, Sambuughin N, Bermingham E. (1996). Mitochondrial DNA analysis of Mongolian populations and implications for the origin of New World founders. *Genetics* 142 (4):1321–1334.

Kuzeev RG. (1992). *The People of the Central Volga Region and Southern Ural.* Moscow: Nauka.

Macaulay VA, Richards MB, Hickey E, et al. (1999). The emerging tree of West Eurasian mtDNAs: a synthesis of control-region sequences and RFLPs. *Am J Hum Genet* 64 (1):232–249.

Nasidze I, Sarkisian T, Kerimov A, Stoneking M. (2003). Testing hypotheses of language replacement in the Caucasus: evidence from the Y-chromosome. *Hum Genet* 112 (3):255–261.

Pericic M, Lauc LB, Klaric IM, et al. (2005). High-resolution phylogenetic analysis of Southeastern Europe (SEE) traces major episodes of paternal gene flow among Slavic populations. *Mol Biol Evol* 22 (10):1964–1975.

Richards M, Macaulay V, Hickey E, et al. (2000). Tracing European founder lineages in the Near Eastern mtDNA pool. *Am J Hum Genet* 67 (5):1251–1276.

Rootsi S, Zhivotovsky LA, Baldovic M, et al. (2007). A counter-clockwise northern route of the Y-chromosome haplogroup N from Southeast Asia toward Europe. *Eur J Hum Genet* 15 (2):204–211.

Sambrook J, Fritsch EF, Maniatis T. (1989). *Molecular cloning: a laboratory manual.* Cold Spring Harbor, NY: Cold Spring Harbor Laboratory Press.

Semino O, Passarino G, Oefner PJ, et al. (2000). The genetic legacy of Paleolithic Homo sapiens sapiens in extant Europeans: a Y chromosome perspective. *Science* 290 (5494):1155–1159.

Tambets K, Rootsi S, Kivisild T, et al. (2004). The Western and Eastern roots of the Saami—the story of genetic "outliers" told by mitochondrial DNA and Y chromosomes. *Am J Hum Genet* 74 (4):661–682.

Tambets K, Tolk HV, Kivisild T, et al. (2003). Complex signals for population expansions in Europe and beyond. In: Bellwood, P, Renfrew C, eds. *Examining the farming/language dispersal hypothesis.* Cambridge: Cambridge University Press, pp. 449–458.

Torroni A, Lott MT, Cabell MF, Chen YS, Lavergne L, Wallace DC. (1994). mtDNA and the origin of Caucasians: identification of ancient Caucasian- specific haplogroups, one of which is prone to a recurrent somatic duplication in the D-loop region. *Am J Hum Genet* 55 (4):760–776.

Villems R, Rootsi S, Tambets K, et al. (2002). Archaeogenetics of Finno-Ugric speaking populations. In: Julku K, ed. *The Roots of Peoples and Languages of Northern Eurasia IV.* Oulu: Societas Historiae Fenno-Ugricae, pp. 271–284.

Yao YG, Kong QP, Bandelt HJ, Kivisild T, Zhang YP. (2002). Phylogeographic differentiation of mitochondrial DNA in Han Chinese. *Am J Hum Genet* 70 (3):635–651.

Yao YG, Kong QP, Wang CY, Zhu C L, Zhang YP. (2004). Different matrilineal contributions to genetic structure of ethnic groups in the silk road region in China. *Mol Biol Evol* 21 (12):2265–2280.

Yao YG, Zhang YP. (2002). Phylogeographic analysis of mtDNA variation in four ethnic populations from Yunnan Province: new data and a reappraisal. *J Hum Genet* 47:311–318.

YCC. (2002). A nomenclature system for the tree of human Y-chromosomal binary haplogroups. *Genome Res* 12 (2):339–348.

Zerjal T, Beckman L, Beckman G, et al. (2001). Geographical, linguistic, and cultural influences on genetic diversity: Y-chromosomal distribution in Northern European populations. *Mol Biol Evol* 18 (6):1077–1087.

Zerjal T, Wells RS, Yuldasheva N, Ruzibakiev R, Tyler Smith C. (2002). A genetic landscape reshaped by recent events: y-chromosomal insights into central Asia. *Am J Hum Genet* 71 (3):466–482.

109.

MTDNA AND Y-CHROMOSOMAL VARIATION IN POPULATIONS OF SAKHA (YAKUTIA)

Sardana A. Fedorova, Elza K. Khusnutdinova, and Richard Villems

INTRODUCTION

The autonomous Sakha Republic (Yakutia) comprises 18% of Russia's total land area and occupies most of the northeastern part of the Eurasian continent. In spite of its huge territory, Yakutia is less populated due to severe natural and climatic conditions. Fewer than 1 million people inhabit the vast area, 80% of which includes mostly taiga, with a narrow strip of tundra in the north. The total number of indigenous people (Yukaghirs, Evenks, Evens, Yakuts, Dolgans, and Chukchi) account for up to 49% of the whole population of the Sakha. The indigenous population is heterogeneous for its anthropological types, linguistic affiliations, cultures, traditions, and types of economy (Table 109-1).

Archaeological records attest to the fact that the territory of northeastern Eurasia has been inhabited since the Upper Paleolithic time. The most ancient Paleolithic sites discovered on the Aldan River valley have been dated to ~32,000–33,000 years ago (Mochanov, 1977), although the dates are not universally accepted, and another site found recently on the Yana River above the Arctic circle at 71^{0}N has been dated to around 30,000 years ago (Pitulko et al., 2004).

The consecutive archaeological cultures in the territory of Sakha hint at multiple waves of migrations from southern territories surrounding the upper reaches of Yenisey, Lake Baikal, and the Amur River (Mochanov et al., 1983; Alekseev, 1996). Ancient tribes inhabiting this area since Neolithic times are regarded as presumable ancestors of different contemporary circumpolar ethnic groups speaking Paleoasiatic and Uralic languages (Okladnikov, 1955; Fedoseeva, 1980; Alekseev, 1996). Yukaghirs are, presumably, direct descendants of these Neolithic tribes (Okladnikov, 1955; Fedoseeva, 1980; Alekseev, 1996). Later on, Tungusic-speaking tribes, at first pedestrian, but who subsequently mastered reindeer breeding for transport, spread all over Siberia, including Yakutia (Vasilevich, 1969; Tugolukov, 1980). The mixture of Evenks with Yukaghirs and Koryaks caused the formation of Evens, a new Tungus-speaking ethnic group (Tugolukov, 1980).

The Turkic-speaking tribes of Sakha (Yakuts) migrated from the Baikal region (Okladnikov, 1955; Ksenofontov, 1992) to the basin of the middle Lena River relatively recently. Archaeological evidence points to the occurrence of remains of cattle and horse breeding in Central Yakutia since the thirteenth century (Gogolev, 1993), whereas Turkic runic inscriptions and some archaeological findings testify to the appearance of Turkic-speaking people in the Lena River basin since at least the sixth century (Alekseev, 1996). By the time of incorporation of Yakutia into the Russian Empire in the late seventeenth century, the total number of population paying tribute (*yasak*) accounted for 45,700, including 28,500 Sakha, 12,700 Tungus, and 4500–5000 Yukaghirs (Dolgikh, 1960). Dolgans are the youngest ethnic group of Siberia, formed in the eighteenth–nineteenth centuries, having absorbed the lineages of Tungus, Yakuts, Russians, Entsy, and Nentsy (Dolgikh, 1963).

The first genetic studies of the population of Sakha on the bases of mtDNA and Y-chromosomal analyses focused primarily on ancient human migrations in the territory of Eurasia (Zerjal et al., 1997; Torroni et al., 1998; Karafet et al., 2002) and inhabiting of the Americas (Torroni et al., 1993; Karafet et al., 1999). As a by-product, a very strong bottleneck in the Yakuts′ male lineage was determined (Zerjal et al., 1997), and later studies (Pakendorf et al., 2002, 2006; Khar'kov et al., 2008) have confirmed this finding. Data obtained by examination of the first hypervariable segment (HVSI), and a few coding region markers of mtDNA, have shown close genetic affinity of the Yakuts to the south Siberian population, and inappreciable Paleoasian contribution to their gene pool (Pakendorf et al., 2003, 2006; Fedorova et al., 2003; Tarskaya and Melton, 2006; Zlojutro et al., 2008). A recent study of arctic Siberian populations by Volodko et al. (2008) has shown the genetic discontinuity between the

TABLE 109-1 SAMPLE COMPOSITION, LINGUISTIC AFFILIATION AND CENSUS SIZES OF STUDIED POPULATIONS

Population	Linguistic affiliation	Census size in Russia, 2002[1]	Census size in Yakutia, 2002[1]	Sample size	
				mtDNA	NRY
Central Yakuts	Altaic (Turkic)	443,852	432,290	164	92
Vilyuy Yakuts				111	58
Northern Yakuts				148	66
Evenks	Altaic (Tungusic)	35,527	18,232	122	57
Evens	Altaic (Tungusic)	19,071	11,657	101	24
Yukaghirs	Isolate (Yukaghir)	1509	1097	22	11
Dolgans	Altaic (Turkic)	7261	1272	26	10

[1]data from http://www.perepis2002.ru/index.html?id=17

Yukaghir, the oldest population in Sakha, and adjoining Chukchi, the descendants of the latest inhabitants of Beringia. The core of the Yukaghir mtDNA pool, harboring haplogroups C4a, C4b, C5, D4 and D5, hints at the southern belt of Siberia as the original region of settlement of their ancestors. Nonetheless, remnants of the Yukaghir tribes form only a small part of the indigenous population of Yakutia.

In this paper, we present our data on mtDNA and Y-chromosomal variation in seven populations of Yakutia, at high phylogenetic resolution. We discuss these results in comparison with data available from other Siberian populations to identify possible source areas of haplogroups that are present in Yakutia.

MTDNA HAPLOGROUPS IN SAKHA AND THEIR POSSIBLE SOURCES

We analyzed mtDNA of 694 Native Yakutians from seven populations of Sakha by combining first hypervariable segment (HVSI) (nps 16024-16390) sequencing with screening of characteristic mutations in the coding region. A phylogenetic tree of studied mtDNA markers in Yakutian populations is shown in Figure 109-1, and haplogroups frequencies are presented in Table 109-2.

The population of Yakutia belongs to a common genetic continuum, covering northeast Eurasia (south and west Siberia, Mongolia, Russian Far East, Chukotka, and Kamchatka) characterized by a predominance of eastern Eurasian lineages and a high content of Asian mtDNA haplogroups C and D (Derenko et al., 2007).

<insert Fig. 109-1>

The haplogroup C accounted for 36%–55% of the mtDNA gene pool of populations under study. The diverse set of haplotypes fall into two major branches, C4 and C5, with the only exception, C7, found in one Yakut and two Evens. Subclades of C4 have some distinctive features in distribution: C4b3 has a high frequency in Evenks, whereas C4b4 seems to be specific for Yukaghirs. Distribution of C5 lineages has well-pronounced regional features: C5b has high frequencies in Evens and Yukaghirs, whereas C5c has contributed more to the Yakuts' gene pool.

Haplogroup D, the second major cluster, is represented by two predominant branches, D4 and D5. In D4, subhaplogroup D4i is represented more in Yakuts, while D4l is found more in Evenks, Evens, Yukaghirs, and Dolgans. D5 is represented by a few lineages of subclade D5a2, the only exception being one lineage of subclade D5b1, found in Northern Yakuts. Haplogroup D5a2 has spread at a low frequency in China, Siberia, and Japan, obtaining its highest frequency in Yakut populations (Fedorova et al., 2003; Pakendorf et al., 2006).

The higher diversity and frequency of D5a2 lineages in Yakuts hints at the possibility that this lineage had been brought to the Lena/Amga/Vilyuy valleys by progenitors of the Yakut people. This hypothesis is supported by the fact that 6 of 11 skeletons in 300–400-year-old Yakut burial site have HVSI sequence with transitions in nps 16092, 16172, 16189, 16223, 16266 and 16362 (Ricaut et al., 2004, 2006; Fedorova et al., 2008), a characteristic of clade D5a2. The star-like topology of the D5a2-cluster (Figure 109-1) coincides with a hypothesis that the predecessors of the Yakuts migrated to the Lena/Amga region in the recent past as a limited group, followed by rapid in situ expansion of founder lineages (Gogolev, 1993). The founder haplotype originates in the regions eastward of Lake Baikal, rather than west of Lake Baikal and central Asia, as the defining HVSI motif of clade D5a2 has been found in Mongols, Mongolic-speaking Daurs, and northern Han (Kolman et al., 1996; Tsai et al., 2001; Yao et al., 2002; Kong et al., 2003).

The contribution of the other 15 haplogroups in the gene pool of Yakutian populations, with rare exceptions, was less than 5%–10%. The results of phylogeographic analysis of lineages spread in Yakutia indicate their southern origin. The presence of mtDNA haplogroups typical for regions of southeast Asia and southern Siberia (A4,

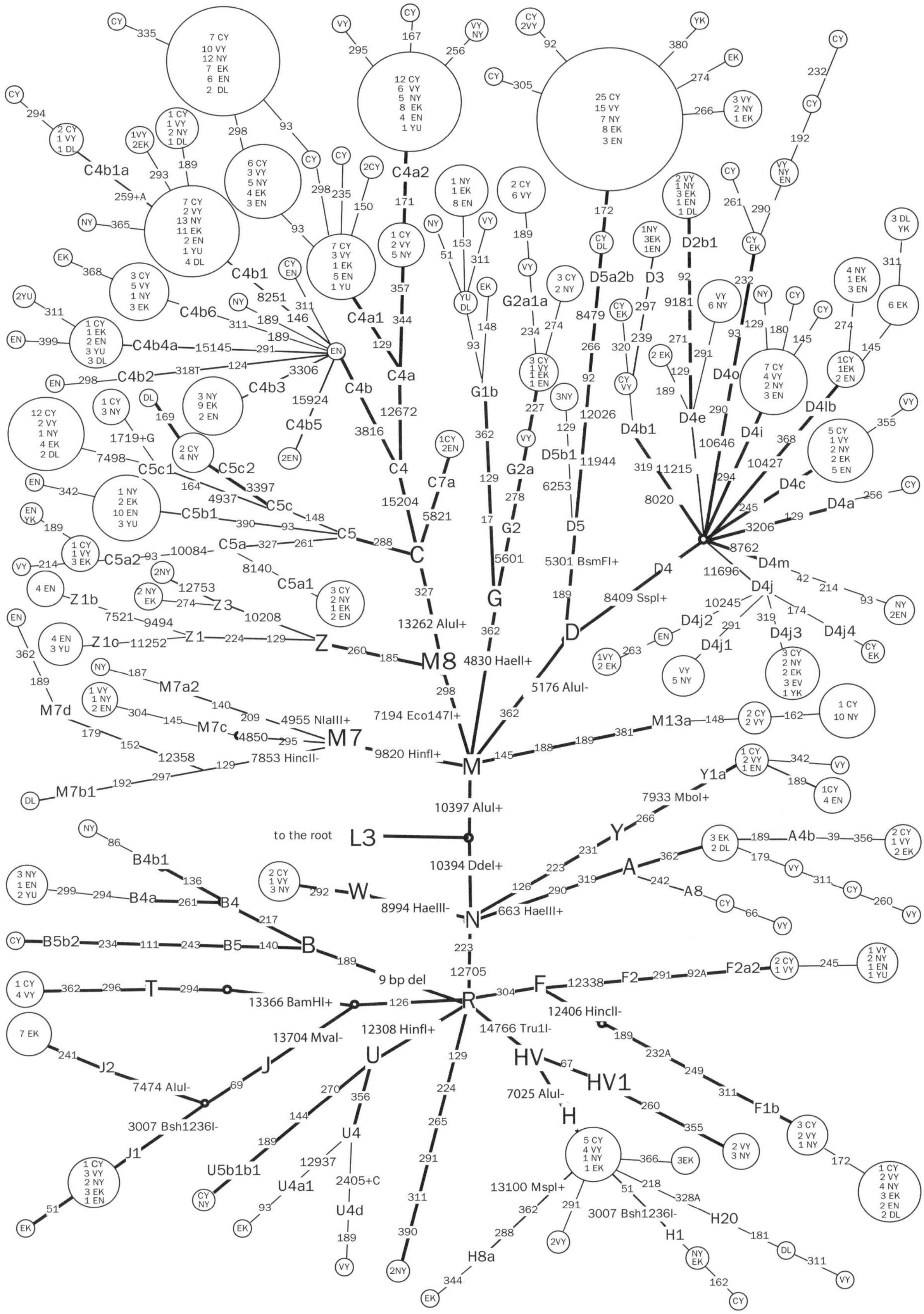

Figure 109-1 Phylogeny of mtDNA lineages observed in seven populations of Yakutia. Mutations relative to the revised reference sequence (Andrews et al., 1999) and RFLP positions are indicated on the branches. Populations are designated as follows: CY=Central Yakuts, VY=Vilyuy Yakuts, NY=Northern Yakuts, EK=Evenks, EN=Evens, YK=Yukaghirs, DL=Dolgans. Phylogeny construction was performed by hand, following a parsimony approach, and was confirmed by use of the program Network 4.5.1.0 (Fluxus Engineering website).

TABLE 109-2. DISTRIBUTION OF MTDNA HAPLOGROUPS IN POPULATIONS OF SAKHA (YAKUTIA)

	Frequency %						
	CY	VY	NY	EK	EN	YK	DL
Hg	n=164	n=111	n=148	n=122	n=101	n=22	n=26
A:	2.4	3.6	0	4.1	0	0	7.7
B:	0.6	0	2.7	0	1.0	9.1	0
C:	45.7	36.0	40.5	46.7	45.5	54.5	53.8
C4	33.5	32.4	33.1	38.5	29.7	36.4	42.3
C4a	23.8	23.4	18.9	16.4	17.8	9.1	7.7
C4a1	15.2	14.4	11.5	9.8	13.9	4.5	0
C4a2	8.5	9.0	7.4	6.6	4.0	4.5	7.7
C4b	9.8	9.0	14.2	22.1	11.9	27.3	34.6
C4b*	2.4	4.5	1.4	3.3	2.0	0	0
C4b1	6.7	4.5	10.8	10.7	2.0	4.5	23.1
C4b2	0	0	0	0	1.0	0	0
C4b3	0	0	2.0	7.4	2.0	0	0
C4b4	0.6	0	0	0.8	3.0	22.7	11.5
C4b5	0	0	0	0	2.0	0	0
C5	11.6	3.6	7.4	8.2	13.9	18.2	11.5
C5a	2.4	1.8	1.4	3.3	3.0	4.5	0
C5b	0	0	0.7	1.6	10.9	13.6	0
C5c	9.1	1.8	5.4	3.3	0	0	11.5
C7	0.6	0	0	0	2.0	0	0
D:	32.9	29.7	25.7	28.7	24.8	13.6	19.2
D4	15.9	11.7	17.6	20.5	21.8	9.1	15.4
D4a	0.6	0	0	0	0	0	0
D4b	1.2	0.9	0.7	3.3	1.0	0	0
D4c	3.0	1.8	1.4	1.6	5.0	0	0
D4e	0	2.7	4.7	4.1	1.0	0	3.8
D2b	0	1.8	0.7	2.5	1.0	0	3.8
D4i	5.5	3.6	2.0	0	3.0	0	0
D4j	2.4	1.8	4.7	4.1	4.0	4.5	0
D4l	0.6	0	2.7	6.6	5.0	4.5	11.5
D4m	0	0	0.7	0	2.0	0	0
D4o	2.4	0.9	0.7	0.8	1.0	0	0
D5a	17.1	18.0	8.1	8.2	3.0	4.5	3.8
D5a2	17.1	18.0	6.1	8.2	3.0	4.5	3.8
D5b1	0	0	2.0	0	0	0	0
F:	3.7	5.4	4.7	2.5	3.0	4.5	7.7
G:	4.9	3.6	6.8	2.5	8.9	4.5	3.8
G1b	0	0.9	1.4	1.6	7.9	4.5	3.8
G2a	4.9	2.7	5.4	0.8	1.0	0	0
H	3.7	6.3	1.4	4.9	0	0	3.8
HV1	0	1.8	2.0	0	0	0	0
J	0.6	2.7	1.4	9.0	1.0	0	0
M7	0	0.9	1.4	0	3.0	0	3.8
M13a	1.8	1.8	6.8	0	0	0	0
R1b	0	0	1.4	0	0	0	0
T	0.6	3.6	0	0	0	0	0
U	0.6	0.9	0.7	0.8	0	0	0
W	1.2	0.9	2.0	0	0	0	0
Y1a	1.2	2.7	0	0	5.0	0	0
Z:	0	0	2.7	0.8	7.9	13.6	0
Z1	0	0	0	0	7.9	13.6	0
Z3	0	0	2.7	0.8	0	0	0

Note: Populations are designated by the following letter codes: CY - Central Yakuts, VY - Vilyuy Yakuts, NY - Northern Yakuts, EK - Evenks, EN - Evens, YK - Yukaghirs, DL - Dolgans. Sequence classification into mtDNA haplogroups was based on generally accepted nomenclatures (Kivisild et al., 2002; Kong et al., 2006; Torroni et al., 1996; Richards et al., 1998; Macaulay et al., 1999; Loogväli et al., 2004; Achilli et al., 2005; Roostalu et al., 2007; Malyarchuk et al., 2008; Behar et al., 2008).

B4, B5, F1b, F2a, M7, M13; Kivisild et al., 2002; Yao et al., 2004; Derenko et al., 2007), the lower Amur River region (Y; Schurr et al., 1999), and central Asia (G2a; Comas et al., 2004) is, evidently, connected with ancient human migrations from south to north.

The phylogeography of some minor clades (D4e3 and Z1) found in the Yakutian populations indicates ancient human migration from Siberia to eastern Europe. Gene flow by a counter-clockwise northern route from inner Asia/southern Siberia to Europe is supported also by mtDNA hg D5 (Tambets et al., 2004) and Y-chromosomal haplogroup N1c data (Rootsi et al., 2007).

Study of mtDNA variation shows sharp genetic differences between the populations of Yakutia and those in the area of former Beringia. Lineages of haplogroups A2a, A2b, and D2a, specific for Chukchi and Eskimos (Volodko et al., 2008), and Z1d, found in Koryaks and Itelmen (Schurr et al., 1999), were not observed in Yakutia. The presence of haplogroups G1b and Y1a1 in Yakutia hints at gene flow from the Lower Amur/Southern Okhotsk region and Kamchatka, where these haplogroups are common (Schurr et al., 1999; Starikovskaya et al., 2005).

Western Eurasian mtDNA lineages have been detected in Evenks, Yakuts, Evens, and Dolgans, while they are absent in Yukaghirs. These lineages might have been carried to Yakutia by Russians, or they may have a more ancient origin in southern Siberian populations. It is known from historical records that the Russian population in Yakutia during an epoch of conquest comprised mainly men (Fisher, 1774), and resettlement of Yakutia by families of peasants and employees began at the end of the eighteenth century (Safronov, 1987). According to the first general census of the Russian Empire in 1897, Russians constituted 11.4% of the total population in Yakutia (Romanov, 1998). At the same time, detection of haplogroup J in remains from Yakut burials of the beginning of the seventeenth century (Ricaut et al., 2006) hints at the possibility that western Eurasian lineages might have an ancient origin. Haplogroups H8 and H20, found in Yakuts, Evenks, and Dolgans, are spread mainly in the Caucasus, the Middle and Near East, as well as among Altaians, Buryats, and Khamnigans in southern Siberia (Derenko et al., 2007; Roostalu et al., 2007). Three HVSI haplotypes identified in this study differ from founding haplotypes of haplogroups H8 and H20 (16288-16362 and 16218-16328A respectively) by one/two mutation(s), while they match those in Buryats exactly (Derenko et al., 2007), thus suggesting gene flow to Yakutia mediated by their common ancestors. Five Yakut mtDNAs belong to haplogroup HV1a1a, spread mainly in the Caucasus, Middle East, and Near East (Behar et al., 2008). Yakut lineage is defined by two additional transitions, A9303G and C16260T, from which at least the second mutation is shared by Buryat mtDNA (Derenko et al., 2007). These facts, altogether, hint at the possibility that haplogroups H8, H20, and HV1a1a in Yakutia originate in southern Siberia, to where they might have arrived from the Caucasus/Middle East. A U4d2 lineage, found in Vilyuy Yakut, is characterized by HVSI motif 16189-16356, which has been described earlier in Mansis in western Siberia (Derbeneva et al., 2002), and transition C10692T, shared by recently published Nganasan complete mtDNA sequence (designated as U4c in Volodko et al., 2008). So, it is likely that the U4d2 lineage was carried to Yakutia from western Siberia.

THE Y-CHROMOSOME PROFILES IN POPULATIONS OF SAKHA

Y-chromosome haplogroups were identified for 318 male individuals by analysis of 24 biallelic and six microsatellite markers (DYS19, DYS388, DYS390, DYS391, DYS392, DYS393) of NRY. The haplogroup distribution in the Yakutian populations is shown in 110-3 and STR-haplotypes in Table 109-4.

Ethnic Yakuts differ from other Siberian populations of Sakha by the striking prevalence of hg N1c (from 71% to 93% in different subgroups). In other populations of Yakutia, the frequency of N1c is significantly lower and varies in a range of 27%–50%. Hg N1c arose probably in the region of present-day China, and subsequently experienced serial bottlenecks in Siberia and secondary expansions in eastern Europe (Rootsi et al., 2007).

TABLE 109-3 DISTRIBUTION OF Y-CHROMOSOME HAPLOGROUPS IN POPULATIONS OF SAKHA (YAKUTIA).

	Frequency %						
	CY	VY	NY	EK	EN	YK	DL
Hg	n=92	n=58	n=66	n=57	n=24	n=11	n=10
E1b1b1a	0	0	0	0	0	0	40
C3	1.1	0	7.6	5.3	0	27.3	0
C3c	0	3.4	6.1	26.3	41.7	9.1	10
J	0	0	0	0	4.2	0	0
I*	2.2	1.8	0	0	0	0	0
I2a	0	0	0	1.8	0	9.1	0
O	0	0	0	0	0	9.1	0
N1b	2.2	0	10.6	8.8	8.3	0	0
N1c	85.9	93.0	71.2	50.9	41.7	27.3	50.0
R1a	6.5	1.8	4.5	7.0	0	18.2	0
R1b1b2	2.2	0	0	0	4.2	0	0

Note: Population codes are the same as those used in Table 109-2.

Source: Y-chromosomal haplogroups were named according to the most recent YCC nomenclature (Karafet et al., 2008).

The median-joining network of N1c lineages has a very strongly pronounced star-like phylogeny in Yakuts, characteristic for the founder effect for the whole paternal gene pool of this nation (Figure 109-2) and concordant with data obtained earlier (Zerjal et al., 1997; Pakendorf et al., 2002, 2006; Khar'kov et al., 2008). Hg N1b, most frequent in Uralic-speaking populations of western Siberia, is revealed with moderate frequencies in northern latitudes of Yakutia (in northern Yakuts, Evenks, and Evens). NRY haplogroup C has a particularly high frequency among Evenks, Evens, Yukaghirs, and Dolgans (Table 109-3). The majority of haplogroup C lineages consists of subcluster C3c defined by the M48 marker. Phylogenetic reconstructions do not determine the exact geographic origin of the branch C3c in Asia. Nevertheless, it should be noted that the high frequencies of C3c lineages are typical for Tungus-speaking (Evenks, Oroqens, Udegey, Ulchi, Negidals) and Mongol-speaking populations (Mongols, Khalkh, Uriankhai, Zakhchin, Khoton), and Kazakhs (Wells et al., 2001; Karafet et al., 2002; Lell et al., 2002; Katoh et al., 2005). C3c haplotypes are characterized by a wide spectrum and high frequency in the gene pool of Evenks and Evens. C3* lineages are represented more in the gene pool of Yukaghirs.

It was a surprising finding that the C3* lineages of Evenks coincide with haplotypes of male descendants of Genghis-Khan on the 5 STR level (Zerjal et al., 2003). Additionally, STR markers DYS389I, DYS389II, DYS437, DYS438, and DYS439 were typed in all C3* haplotypes, and three exact matches of Evenks C3* lineages by 10 overlapping markers from 14 STRs were found with those

TABLE 109-4 Y-CHROMOSOMAL STR HAPLOTYPES IN POPULATIONS OF YAKUTIA

		DYSs						Populations						
No.	Hg	19	388	390	391	392	393	CY	VY	NY	EK	EN	YK	DL
1	N1c	14	12	23	11	16	14	63	42	32	19	5		4
2	N1c	14	12	24	11	16	14	1	1	5	3			
3	N1c	14	12	24	11	15	14					1		
4	N1c	14	13	23	11	16	14		1					
5	N1c	14	12	23	11	15	14	11	4	2	4	1		
6	N1c	14	12	23	10	15	14				1			
7	N1c	14	12	23	11	14	14		1	1			3	
8	N1c	14	12	23	10	14	14			1	1	1		
9	N1c	14	12	22	10	14	14	1						
10	N1c	14	12	23	11	15	15					1		
11	N1c	14	12	23	11	14	13				1	1		
12	N1c	15	12	23	11	16	14	2		1				
13	N1c	14	12	23	12	16	14			1				
14	N1c	14	12	23	11	16	13		1					
15	N1c	14	12	23	10	16	14	1	3	1				1
16	N1c	14	12	23	10	16	13			3				
17	C3	16	13	24	9	11	14						3	
18	C3	16	14	25	10	11	13				3			
19	C3	17	12	23	10	11	13			1				
20	C3	15	12	23	10	11	14	1		2				
21	C3	15	13	24	10	11	13			2				
22	C3c	15	13	24	9	11	13		2		9	1		
23	C3c	15	13	24	9	10	13				2			
24	C3c	16	13	24	9	11	13				1	1		
25	C3c	17	13	24	9	11	13				2			1
26	C3c	17	13	23	9	11	13			4	1	8	1	
27	R1a	15	12	25	10	11	13	1			1			
28	R1a	15	12	25	11	11	13	2						
29	R1a	16	12	25	10	11	13	2		1				
30	R1a	16	12	25	11	11	13	1	1	2	1		1	
31	R1a	16	12	24	11	11	13				1		1	
32	R1a	17	12	25	12	11	13				1			
33	R1b	14	12	24	10	13	12	1						
34	R1b	14	12	24	11	13	12	1						
35	R1b	14	12	25	11	13	12					1		
36	N1b	14	12	23	10	14	13			4	5	2		
37	N1b	14	12	23	10	15	13	2		2				
38	N1b	14	12	23	11	14	13			1				
39	O	12	14	25	10	13	13						1	

(Continued)

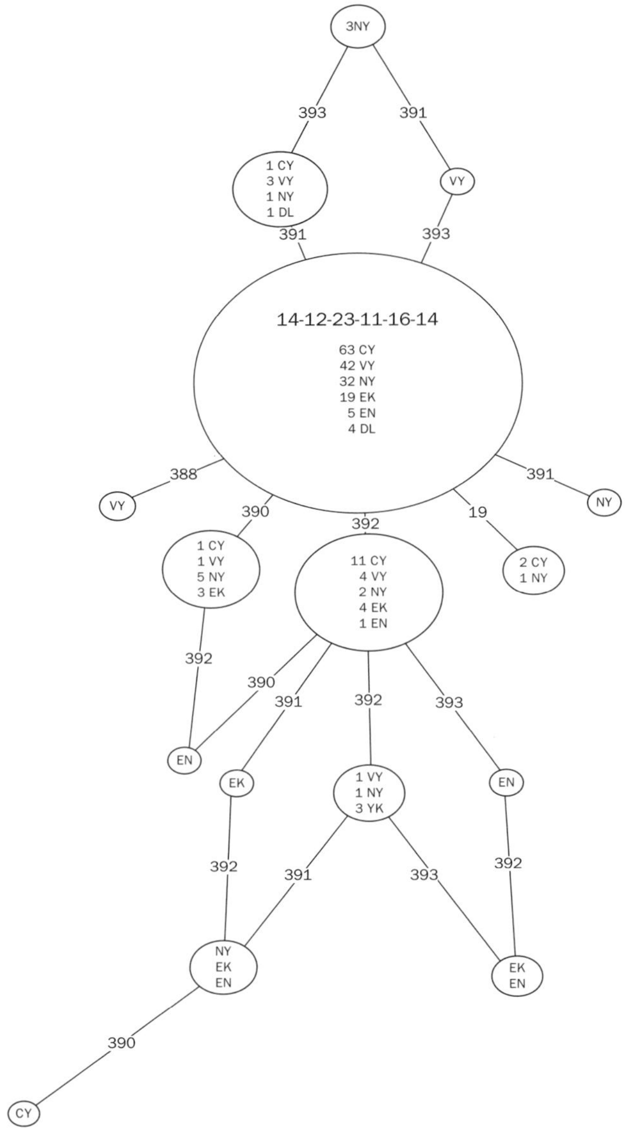

Figure 109-2 Phylogenetic network of Y-chromosomal hg N1c. Median joining network of hg N1c haplotypes was constructed based on data of 6 STRs in 226 individuals by using the program Network 4.5.1.0. The area of the circles is proportional to the haplotype frequency. STR markers are represented on lines. Populations are labeled as in Figure 110-1.

of the Genghis-Khan haplotype (Zerjal et al., 2003). This finding indicates a relatively recent male gene flow from Mongolia to Yakutia, and strongly supports those hypotheses that propose the spread of Tungus reindeer-breeding tribes from the Baikal region to other regions of northern Asia from the end of the first millennium (Tugolukov, 1980; Alekseev, 1996).

The presence of low frequency lineages of haplogroups R1, I, E1b1b1a, and J, typical for populations of Europe and the Near East in Yakutia, may point to recent admixture with Russians and other eastern Europeans of the former Russian Empire and Soviet Union, who started to arrive in eastern Siberia from the seventeenth century onward. As for the R1 and J lineages in the gene pool of the populations studied, their more ancient introgression cannot be excluded, because these plogroups are well presented in native south Siberian populations (Karafet et al, 2002; Derenko et al., 2006), sharing a variety of NRY haplogroups with populations inhabiting Yakutia.

Haplogroup Q is well represented in some west Siberian populations, as well as among easternmost Siberian Eskimos, and spread also in Evens and Yukaghirs (Karafet et al., 2002). Another rarity for populations of Yakutia, haplogroup O, common for Southeast Asian populations, was found in the Yukaghir gene pool in low frequency.

The differences in the Y-chromosome haplogroup spectrum between Yakutia, Chukotka, and Kamchatka are not so obvious, as in the case of mtDNA data that could be explained by recent extensive male gene flow between Yakutia and former Beringia, or by the forming of the male gene pool of Yakutia, Chukotka, and Kamchatka as a result of multiple migrations from the same source area. The proposition of recent male expansion from Yakutia to Chukotka and Kamchatka is not supported by phylogenetic analysis of STR lineages: Yakut N1c-haplotypes do not match those of Chukchi, Eskimos, Koryaks, and Evens, but form a distinct branch in the hg N1c network, differing from other populations (Derenko et al., 2007; Rootsi et al., 2007) except Evenks (Khar'kov et al., 2008).

So, the hypothesis of multiple migration waves is more preferable. Recent Y-chromosomes flowing from Chukotka and Kamchatka to Yakutia could be valid for northeastern populations of Yukaghirs and Evens, the immediate neighbors of Chukchi and Koryaks. Indeed, some haplotypes (7 and 8 in Table 109-4) are phylogenetically close to those of Chukchi and Eskimos, while others seem to be specific for Evens only (3 and 10 in Table 109-4). The coalescence time of N1c lineages for Yakuts is 1960 ± 950 years (calculated according to Zhivotovsky et al., 2004). If we exclude from calculation haplotypes 7 and 8 as probably derived from another source, then the time of coalescence decreases to 1550 ± 580.

While considering how the possible ancient Paleolithic migrations have affected the populations of Yakutia, the Y-chromosome haplogroups Q and C3*, which are omnipresent in Native Americans (Zegura et al., 2004), presumably should be older in Siberia, as they had to be present there already at the time of colonization of the Americas. The highest frequency of these two lineages is observed in Yukaghirs (27% C3* (our data) and 9% Q (Karafet et al., 2002). If we take into account the low content of these clusters in the gene pool of other modern indigenous populations of Yakutia, then they appear to hold a very small part of the Y-chromosomal lineages of ancient Paleolithic populations.

TABLE 109-4 Y-(CONTINUED)

No.	Hg	DYSs 19	388	390	391	392	393	Populations CY	VY	NY	EK	EN	YK	DL
40	E1b	13	12	24	10	11	13							4
41	I1a	15	14	22	10	11	13	1						
42	I1a	14	14	23	10	11	13	1						
43	I1a	14	14	22	11	11	13		1					
44	I2a	16	13	24	11	11	13				1			
45	I2a	15	13	24	11	11	13						1	
46	J	15	15	24	10	11	12					1		
								92	57	66	57	24	11	10

Note: Population codes are the same as those used in Table 109-2.

TABLE 109-5 GENETIC DIVERSITY OF POPULATIONS STUDIED

Population	mtDNA N	h^a (SE)	K (%)	NRY N	h^b (SE)	h^c (SE)
Central Yakuts	164	0.955 (0.008)	57 (35%)	92	0.26 (0.06)	0.52 (0.06)
Vilyuy Yakuts	111	0.965 (0.008)	50 (45%)	58	0.13 (0.06)	0.45 (0.08)
Northern Yakuts	148	0.962 (0.006)	46 (31%)	66	0.47 (0.07)	0.75 (0.06)
Evenks	122	0.952 (0.010)	42 (34%)	57	0.64 (0.04)	0.86 (0.04)
Evens	101	0.959 (0.007)	34 (34%)	24	0.67 (0.06)	0.86 (0.06)
Yukaghirs	22	0.952 (0.026)	14 (64%)	11	0.82 (0.08)	0.89 (0.07)
Dolgans	26	0.951 (0.022)	15 (58%)	10	0.64 (0.10)	0.73 (0.10)

N=sample size; h^a=haplotype diversity, calculated considering the sequence variation observed between nps 16024 and 16390; h^b=haplogroup diversity; h^c=STR haplotype diversity; K (%)=number of different haplotypes and percentage of different haplotypes in parentheses.

TABLE 109-6 AMOVA RESULTS OF MTDNA AND Y-CHROMOSOME HAPLOGROUPS IN YAKUTIA

Source of variance	Percentage of variation mtDNA haplogroups	Y-chromosome haplogroups
No. groups		
Among populations	1.85 (0.0000)	14.78 (0.0000)
Within populations	98.15 (0.0000)	85.22 (0.0000)
Yakuts and Evenks vs. Evens, Yukaghirs and Dolgans		
Among groups	1.61 (0.0276)	11.43 (0.1161)
Among populations within groups	1.17 (0.0000)	10.17 (0.0000)
Within populations	97.22 (0.0000)	78.41 (0.0000)
Yakuts, Tungus (Evenks+Evens), Yukaghirs, Dolgans		
Among groups	0.87 (0.0950)	19.52 (0.0049)
Among populations within groups	1.29 (0.0000)	2.12 (0.0194)
Within populations	97.84 (0.0000)	78.36 (0.0000)

P values are given in parentheses.

GENETIC DIVERSITY INDICES AND PC ANALYSIS

Genetic variation in the populations studied here was assessed by calculating haplogroup and haplotype diversity values and a number of different haplotypes (Table 109-5). All seven populations demonstrate relatively similar mtDNA haplotype diversity values, with Dolgans showing the lowest (0.951) and Vilyuy Yakuts the highest (0.965).

MtDNA haplotype diversity in populations of Sakha was slightly lower than in southern Siberian populations (Derenko et al., 2007). Y-chromosome haplogroup and STR haplotype diversity, differently from mtDNA data, varied between populations. Yakuts exhibited remarkably lower diversity compared with other populations, which is explained easily by high prevalence of one haplogroup (N1c).

The mtDNA and Y-chromosomal haplogroup frequencies were used as input vectors to perform a PC analysis. Yakut populations and Evenks presenting a similar spectrum of mtDNA haplogroups were clustered together, while Yukaghirs, Evens, and Dolgans were outliers (Figure 109-3a). AMOVA confirmed that the pattern

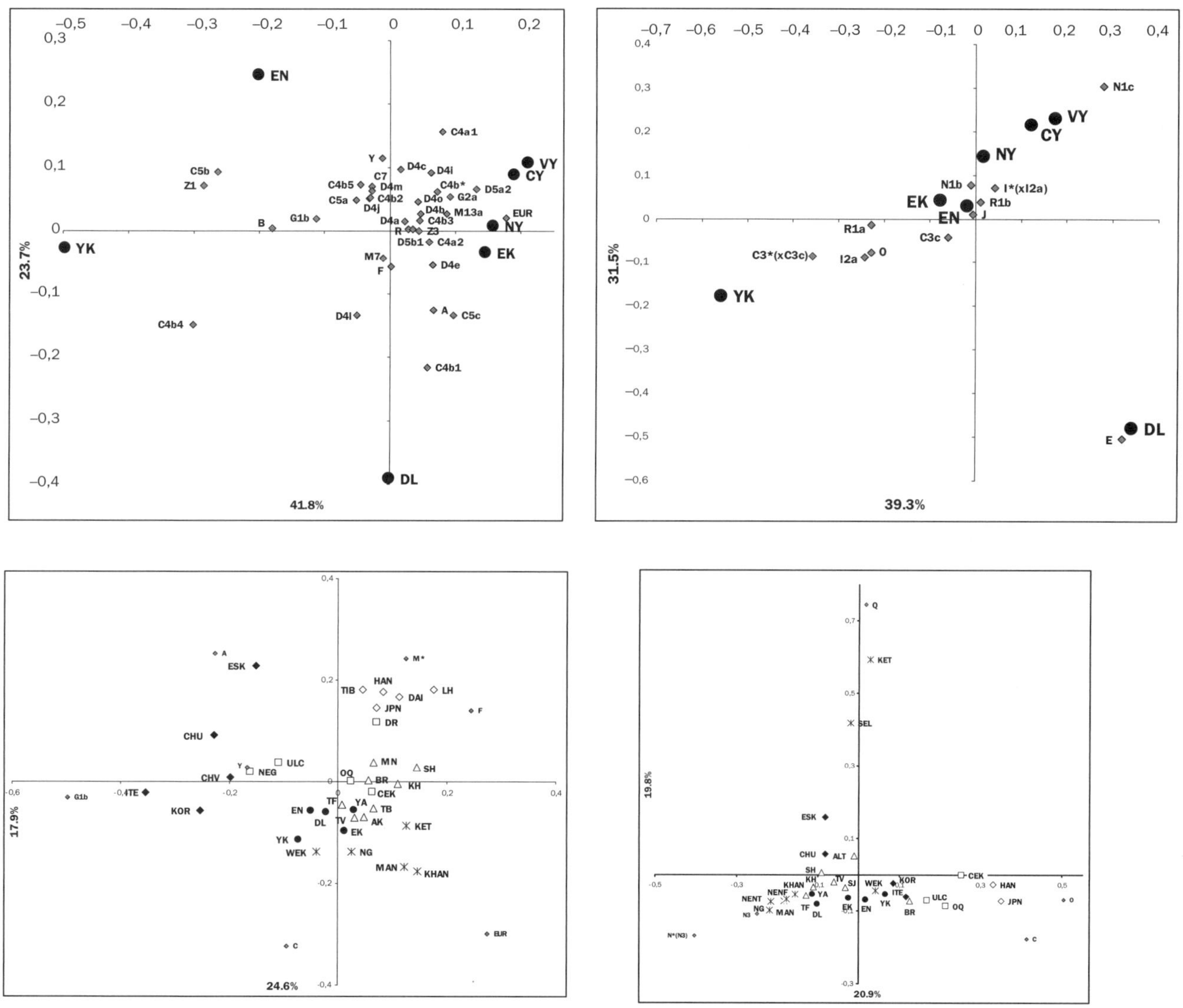

Figure 109-3 (A–D) Principal component analysis (PCA) based on mtDNA (a, c) and Y-chromosomal (b, d) haplogroup frequencies for population samples from Sakha (a, b) and from Northeast Eurasia (c, d). Abbreviations used are: CY=Central Yakuts, VY=Vilyuy Yakuts, NY=Northern Yakuts, EK=Evenks, EN=Evens, YK=Yukaghirs, DL=Dolgans; AK=Altaians-Kizhi, AL=Altaians, BR=Buryats, CHU=Chukchi, CEK=Evenks of China, CHV=Chuvantsi, DAI=Dai, DR=Daurs, ENT=Entsy, ESK=Eskimos, HAN=Chinese Han, ITE=Itelmen, KET=Kets, KH=Khakassians, KHAN=Khants, KHM=Khamnigans, KOR=Koryaks, LH=Lahu, MAN=Mansi, MN=Mongols, NEG=Negidals, NENF=Nentsy forest, NENT=Nentsy tundra, NG=Nganasans, OQ=Oroqens, SEL=Selkups, SH=Shors, SJ=Sojots, TB=Tubalars, TF=Tofalars, TIB = Tibetans, TL = Telenghits, TLT = Teleuts, TV = Tuvinians, ULC = Ulchi, WEK = Western Evenks, YA = Yakuts. Geographic areas are indicated as follows: Yakutia = filled circles, Chukotka and Kamchatka = diamonds, southern Siberia = triangles, western and Middle Siberia = stars, Japan = squares. Gray diamonds represent haplogroups. Principal component analysis (PCA) on haplogroup frequencies was performed using the software POPSTR, kindly provided by H. Harpending.

seen in the PC is significant (Table 109-6). When the populations were split into two groups (Yakuts and Evenks; Yukaghirs, Evens, and Dolgans), the difference among the groups was 1.61% of the total variation (P=0.0276). As in the case of mtDNA data, spectra of Y-chromosome haplogroups in Yukaghirs and Dolgans have reliable distinctions when comparing with other populations studied (Figure 109-3b), due to higher frequency of haplogroup with presumably ancient Paleolithic origin—C3*(xC3c), and haplogroup E, respectively. Common to the cluster including Evens and Evenks is high frequency of C3c lineages. The Yakut populations were separated from the others due to high frequency of hg N1c. AMOVA confirmed that the pattern seen in the PC is significant (Table 109-6).

Figure 109-3c shows the PC analysis of mtDNA variation of the Yakutian populations in the Siberian context. All Yakutian populations were clustered together with southern Siberian populations, while neighboring Koryaks and Itelmens, as well as Chukchi and Eskimos, were separated due to high frequency of haplogroup G1b and A, respectively. Tibetans, Dai, Lahu, Daurs, Japanese, and Han Chinese formed the separate group. PC analysis of Y-chromosomal haplogroups (Figure 109-3d) clustered the populations of Yakutia together with western Evenks and the populations of southern Siberia and Kamchatka. The populations of the Amur River region, Japanese, and Han Chinese clustered together. They were distanced from other populations due to higher frequency of haplogroups C and O.

CONCLUSIONS

Extended analysis of mtDNA and Y-chromosome variation confirms that Yakutia was colonized from the regions west and east of Lake Baikal, with minor gene flows from the Lower Amur/Southern Okhotsk region and/or Kamchatka, and western Siberia. Hence, those historical conceptions connecting ethnic groups of Yakutia with ancient people of southern Siberia are genetically confirmed. Traces of gene flow to Yakutia from the Caucasus/Middle East area through southern Siberia was shown on the basis of mtDNA haplogroups HV1, H20, and H8. The phylogeographic pattern of mtDNA haplogroups Z1a and D4e3, and Y-chromosomal haplogroup N1c, indicates deep common ancestry of populations of the circumpolar zone of Eurasia from Yakutia up to Scandinavia.

ACKNOWLEDGMENTS

We are indebted to the native people of Yakutia who participated in this study. We would like to thank M. Reidla, S. Rootsi, E. Metspalu, J. Parik, I.A. Kutuev, M.I. Tomsky, V.A. Argunov, F.A. Platonov, V.I. Kirillina, V.K. Pavlov, A.N. Nogovitsina and S.K. Kononova for help in this study. This work is supported by the republican program, "Development of Human Gene Diagnostics in the Sakha Republic," as part of the collaboration agreement between the Yakut Research Center of RAMS and Estonian Biocentre, the European Union European Regional Development Fund, through the Centre of Excellence in Genomics, Estonian Biocentre and Tartu University, and Estonian Basic Research Grant (SF 0270177s08).

REFERENCES

Achilli A, Perego UA, Bravi CM, et al. (2005). The phylogeny of the four pan-American mtDNA haplogroups: Implications for evolutionary and disease studies. *PLoS ONE* 3:e1764.

Alekseev AN. (1996). *Ancient Yakutia: The Iron Age and Medieval Epoch (in Russian).* Izdatel'stvo Instituta Arkheologii i Etnografii SO RAN, Novosibirsk.

Andrews RM, Kubacka I, Chinnery PF, Lightowlers RN, Turnbull DM, Howell N. (1999). Reanalysis and revision of the Cambridge reference sequence for human mitochondrial DNA. *Nat Genet* 23:147.

Behar DM, Villems R, Soodyall H, et al., Genographic Consortium. (2008). The dawn of human matrilineal diversity. *Am J Hum Genet* 82:1130–1140.

Comas D, Plaza S, Wells RS, et al. (2004). Admixture, migrations, and dispersals in Central Asia: Evidence from maternal DNA lineages. *Eur J Hum Genet* 12:495–504.

Derbeneva OA, Starikovskaya EB, Wallace DC, Sukernik RI. (2002). Traces of Early Eurasians in the Mansi of Northwest Siberia revealed by mitochondrial DNA analysis. *Am J Hum Genet* 70:1009–1014.

Derenko M, Malyarchuk B, Grzybowski T, et al. (2007). Phylogeographic analysis of mitochondrial DNA in Northern Asian populations. *Am J Hum Genet* 81:1025–1041.

Derenko M, Malyarchuk B, Denisova GA, et al. (2006). Contrasting patterns of Y-chromosome variation in South Siberian populations from Baikal and Altai-Sayan regions. *Hum Genet* 118:591–604.

Dolgikh BO. (1960). *The tribal composition of the peoples of Siberia in the 17th century (in Russian).* Izdatel'stvo Akademii Nauk SSSR, Moscow.

Dolgikh BO. (1963). The origin of Dolgans (in Russian). *Sibirsky etnografithesky sbornik.* Moscow: Izdatel'stvo Akademii Nauk SSSR.

Fedorova SA, Bermisheva MA, Villems R, Maksimova NR, Khusnutdinova EK. (2003). Analysis of mitochondrial DNA lineages in Yakuts. *Mol Biol (Mosk)* 37:544–553.

Fedorova SA, Stepanov AD, Adojaan M, et al. (2008). Phylogenetic analysis of ancient mitochondrial DNA lineages in Yakutia. *Mol Biol (Mosk)* 42:445–453.

Fedoseeva SA. (1980). *Ymiyakhtakhskaya culture of North-Eastern Asia* (in Russian). Izdatel'stvo Nauka, Novosibirsk.

Fisher I. (1774). *The History of Siberia (in Russian).* Izdatel'stvo Imperskoi Akademii Nauk, Saint-Petersburg.

Gogolev AI. (1993). *The Yakuts (in Russian).* Izdatel'stvo YaGU, Yakutsk.

Karafet TM, Osipova LP, Gubina MA, Posukh OL, Zegura SL, Hammer MF. (2002). High levels of Y-chromosome differentiation among native Siberian populations and the genetic signature of a boreal hunter-gatherer way of life. *Hum Biol* 74:761–789.

Karafet TM, Zegura SL, Posukh O, et al. (1999). Ancestral Asian source(s) of new world Y-chromosome founder haplotypes. *Am J Hum Genet* 64:817–831.

Karafet TM, Mendez FL, Meilerman MB, Underhill PA, Zegura SL, Hammer MF. (2008). New binary polymorphisms reshape and increase resolution of the human Y chromosomal haplogroup tree. *Genome Res* 18:830–838.

Katoh T, Munkhbat B, Tounai K, et al. (2005). Genetic features of Mongolian ethnic groups revealed by Y-chromosomal analysis. *Gene* 346:63–70.

Khar'kov VN, Stepanov VA, Medvedeva OF, et al. (2008). The origin of Yakuts: Analysis of Y chromosome haplotypes. *Mol Biol (Mosk)* 42:226–237.

Kivisild T, Tolk HV, Parik J, et al. (2002). The emerging limbs and twigs of the east Asian mtDNA tree. *Mol Biol Evol* 19:1737–1751.

Kolman CJ, Sambuughin N, Bermingham E. (1996). Mitochondrial DNA analysis of Mongolian populations and implications for the origin of New World founders. *Genetics* 142:1321–1334.

Kong Q-P, Yao Y.-G, Liu M, et al. (2003). Mitochondrial DNA sequence polymorphisms of five ethnic populations from northern China. *Hum Genet* 113:391–405.

Kong Q-P, Bandelt H.-J, Sun C, et al. (2006). Updating the East Asian phylogeny: A prerequisite for the identification of pathogenic mutations. *Hum Mol Genet* 1:2076–2086.

Ksenofontov GV. (1992). *Uraankghai-Sakhalar: Ancient history of Yakuts (in Russian).* Yakutsk: Natsionalnoe Izdatel'stvo Respubliki Sakha.

Lell JT, Sukernik RI, Starikovskaya YB, et al. (2002). The dual origin and Siberian affinities of Native American Y chromosomes. *Am J Hum Genet* 70:192–206.

Loogväli EL, Roostalu U, Malyarchuk BA, et al. (2004). Disuniting uniformity: A pied cladistic canvas of mtDNA haplogroup H in Eurasia. *Mol Biol Evol* 21:2012–2021.

Macaulay VA, Richards MB, Hickey E, et al. (1999). The emerging tree of West Eurasian mtDNAs: A synthesis of control-region sequences and RFLPs. *Am J Hum Genet* 64:232–249.

Malyarchuk B, Grzybowski T, Derenko M, et al. (2008). Mitochondrial DNA phylogeny in Eastern and Western Slavs. *Mol Biol Evol.* 25:1651–1658.

Mochanov YA. (1977). *Early stages of human occupation in Northeast Asia (in Russian).* Novosibirsk: Izdatel'stvo Nauka.

Mochanov YA, Fedoseeva SA, Alekseev AN, Kozlov VI, Kothmar NN, Sherbakova NM. (1983). *Archaeological sites in Yakutia (in Russian).* Novosibirsk: Izdatel'stvo Nauka.

Okladnikov AP. (1955). *The history of the Yakut ASSR: Yakutia before its incorporation into the Russian state (in Russian).* Moscow: Izdatel'stvo Akademii Nauk SSSR.

Pakendorf B, Morar B, Tarskaia L, et al. (2002). Y-chromosomal evidence for a strong reduction in male population size of Yakuts. *Hum Genet* 110:198–200.

Pakendorf B, Novgorodov IN, Osakovskij VL, Danilova AP, Protod'jakonov AP, Stoneking M. (2006). Investigating the effects of prehistoric migrations in Siberia: Genetic variation and the origins of Yakuts. *Hum Genet* 120:334–353.

Pakendorf B, Wiebe V, Tarskaia LA, et al. (2003). Mitochondrial evidence for admixed origins of Central Siberian populations. *Am J Phys Anthropol* 120:211–224.

Pitulko VV, Nikolsky PA, Girya EY, et al. (2004). The Yana RHS site: Humans in the arctic before the Last Glacial Maximum. *Science* 303:52–56.

Ricaut F-X, Kolodesnikov S, Keyser-Tracqui C, Alekseev AN, Crubezy E, Ludes B. (2004). Genetic analysis of human remains found in two eighteenth century Yakut graves at At-Dabaan. *Int J Legal Med* 118:24–31.

Ricaut F-X, Kolodesnikov S, Keyser-Tracqui C, Alekseev AN, Crubezy E, Ludes B. (2006). Molecular genetic analysis of 400-year old human remains found in two Yakut burial sites. *Am J Phys Anthropol* 129:55–63.

Richards MB, Macaulay VA, Bandelt HJ, Sykes BC. (1998). Phylogeography of mitochondrial DNA in western Europe. *Ann Hum Genet* 62:241–260.

Romanov IG. (1998). *Forming of Russian population in Yakutia (1917–1941) (in Russian).* Yakutsk: Izdatel'stvo Instituta Gumanitarnikh Issledovanii.

Roostalu U, Kutuev I, Loogväli EL, et al. (2007). Origin and expansion of haplogroup H, the dominant human mitochondrial DNA lineage in West Eurasia: The Near Eastern and Caucasian perspective. *Mol Biol Evol* 24:436–448.

Rootsi S, Zhivotovsky LA, Baldovic M, et al. (2007). A counter clockwise northern route of the Y-chromosome haplogroup N from Southeast Asia toward Europe. *Eur J Hum Genet* 15:204–211.

Safronov FG. (1987). *Spreading of agriculture in North-Eastern Siberia in XVII-XX cc. Historical relations of people of Yakutia with Russians (in Russian).*

Starikovskaya EB, Sukernik RI, Derbeneva OA, et al. (2005). Mitochondrial DNA diversity in indigenous populations of the southern extent of Siberia, and the origins of Native American haplogroups. *Ann Hum Genet* 69:67–89.

Tambets K, Rootsi S, Kivisild T, et al. (2004). The western and eastern roots of the Saami – the story of genetic "outliers" told by mitochondrial DNA and Y-chromosome. *Am J Hum Genet* 74, 661–682.

Tarskaya LA, Melton P. (2006). Comparative analysis of mitochondrial DNA of Yakuts and other Asian populations. *Genetika (Mosk)* 42:1703–1711. [the reference in the text is corrected]

Torroni A, Sukernik RI, Schurr TG, et al. (1993). MtDNA variation of aboriginal Siberians reveals distinct genetic affinities with Native Americans. *Am J Hum Genet* 53:591–608.

Torroni A, Huoponen K, Francalacci P, et al. (1996). Classification of European mtDNAs from an analysis of three European populations. *Genetics* 144:1835–1850.

Torroni A, Bandelt H-J, D'Urbano L, et al. (1998). MtDNA analysis reveals a major late Paleolithic population expansion from southwestern to northeastern Europe. *Am J Hum Genet* 62:1137–1152.

Tsai LC, Lin CY, Lee JCI, Chang JG, Linacre A, Goodwin W. (2001). Sequence polymorphism of mitochondrial D-loop DNA in the Taiwanese Han population. *Forensic Sci Int* 119:239–247.

Tugolukov VA. (1980). *Ethnic roots of Tungus (in Russian).* Moscow: Izdatel'stvo Nauka.

Vasilevich GM. (1969). *Evenki (in Russian).* Leningrad: Izdatel'stvo Nauka.

Volodko NV, Starikovskaya EV, Mazunin IO, et al. (2008). Mitochondrial genome diversity in arctic Siberians with particular reference to the evolutionary history of Beringia and Pleistocenic peopling of the Americas. *Am J Hum Genet* 82:1084–1100.

Wells RS, Yuldasheva N, Ruzibakiev R, et al. (2001). The Eurasian heartland: A continental perspective on Y-chromosome diversity. *Proc Natl Acad Sci USA* 98:10244–10249.

Safronov F.G. (1987) *Spreading of agriculture in North-Eastern Siberia in XVII-XX cc. Historical relations of people of Yakutia with Russians (in Russian).* Yakutskoe Knijnoe Izdatel'stvo, Yakutsk.

Schurr TG, Sukernik RI, Starikovskaya YB, Wallace DC. (1999). Mitochondrial DNA variation in Koryaks and Itel'men: Population replacement in the Okhotsk Sea – Bering Sea region during the Neolithic. *Am J Phys Anthropol* 108:1–39.

Yao Y-G, Kong Q-P, Bandelt HJ, Kivisild T, Zhang Y-P. (2002). Phylogeographic differentiation of mitochondrial DNA in Han Chinese. *Am J Hum Genet* 70:635–651.

Yao Y-G, Kong Q-P, Wang C-Y, Zhu C-L, Zhang Y-P. (2004). Different matrilineal contributions to genetic structure of ethnic groups in the Silk Road region in China. *Mol Biol Evol* 21:2265–2280.

Zegura SL, Karafet TM, Zhivotovsky LA, Hammer MF. (2004). High resolution SNPs and microsatellite haplotypes point to a single, recent entry of Native American Y chromosomes into the Americas. *Mol Biol Evol* 21:164–175.

Zerjal T, Xue Y, Bertorelle G, et al. (2003). The genetic legacy of the Mongols. *Am J Hum Genet* 72:717–721.

Zerjal T, Dashnyam B, Pandya A, et al. (1997). Genetic relationships of Asians and Northern Europeans, revealed by Y-chromosomal DNA analysis. *Am J Hum Genet* 60:1174–1183.

Zhivotovsky LA, Underhill PA, Cinnioglu C, et al. (2004). The effective mutation rate at Y chromosome short tandem repeats, with application to human population-divergence time. *Am J Hum Genet* 74:50–61.

Zlojutro M, Tarskaia LA, Sorensen M, Snodgrass JJ, Leonard WR, Crawford MH. (2008). The origins of the Yakut people: Evidence from mitochondrial DNA diversity. *Int J Hum Genet* 8:119–130.

110.

EPIDEMIOLOGY OF HEREDITARY DISEASES IN THE EUROPEAN SECTOR OF RUSSIA

E.K. Ginter and R.A. Zinchenko

INTRODUCTION

Genetic differentiation of human populations and its mechanisms remain the key problem of ethnogenetics and population genetics and a challenge for solving the intensity and direction of microevolution in different ethnic populations.

Investigation of the load of hereditary disorders (HDs) in human populations and its characteristics, such as size, structure (proportion of autosomal dominant, autosomal recessive and X-linked disorders), secular, and geographical dynamics, may be effective only if there is a close interaction between population and medical genetics. In the last case, it is possible to understand what kind of influence on a population can diminish the pressure of the load of HDs in the population, and how to optimize the organization of medical genetic service in the population. The information on the frequency of genetic diseases is also essential for estimating the load of HDs from mutagenesis by indirect or direct methods.

Unfortunately, our knowledge about the load of HDs and its peculiarities in different countries is too limited. In most cases we use data about the load that were received ad hoc; probably they are not very accurate, and may be distorted in any way, but mostly on the side of overestimation (Matsunaga, 1982). From our point of view, shared by some other authors, the most comprehensive and systematic investigation of the load of HDs and congenital anomalies is being conducted in British Columbia during the past 40 years, within the Register of Congenital and Hereditary Pathology in this Canadian province (Baird et al., 1988). The data from the British Columbia Health Surveillance Registry have been quoted in most of the reports from UNSCEAR (1977, 1982) as a baseline for estimation of the possible genetic risks to the human population from exposure to ionizing radiation (www.UNSCEAR.org). According to the last publication, in the population of British Columbia (with a size of approximately 3 million people) the frequency of autosomal dominant disorders is 1.4/1000, autosomal recessive disorders is 1.7/1000, and X-linked recessive disorders is 0.5/1000 newborns. But from these data it is unclear what is a proportion of sporadic cases in the loads of hereditary disorders, how the hereditary pathology is distributed in the province, and whether there are subpopulations with accumulation of some HDs, and some with others. The influence of some factors of population dynamics on the load of hereditary diseases is well known for populations of Finns (Norio, 1981), Ashkenazi Jews (Goodman, 1979), the Amish of the United States (McKusick, 1980), French Canadians (De Braekeleer & Dao, 1994a,b), and some others. It was postulated that a founder-drift mechanism determined the accumulation of different HDs, especially autosomal recessives, in those populations. On the contrary, in our study of central Asia populations of the former USSR, it was shown that the main factor determining the load of HDs there is the inbreeding of the population as a whole, and especially its nonrandom component due to marital genealogical assortativity (Ginter, 1986). This suggestion about the role of nonrandom inbreeding in the accumulation of autosomal recessives in central Asia was proved later by the study of some Arabic populations, which have approximately the same population structure as the central Asian ones.

In this paper we shall summarize some results of medical and population genetic studies carried out in some regions/ethnic groups of Russia (Figure 110-1): Russians from six populations (Krasnodar Province, Rostov, Kostroma, Kirov, Tver, and Bryansk regions), Adygean (Adygea Republic), Maris (Mari El Republic), Chuvashes (Chuvash Republic), Udmurt (Udmurt Republic), and Bashkirs (Bashkortostan Republic).

THE SOURCE OF THE DATA

The data discussed in this chapter is based on populationwide surveys covering 11 regions of Russia; the

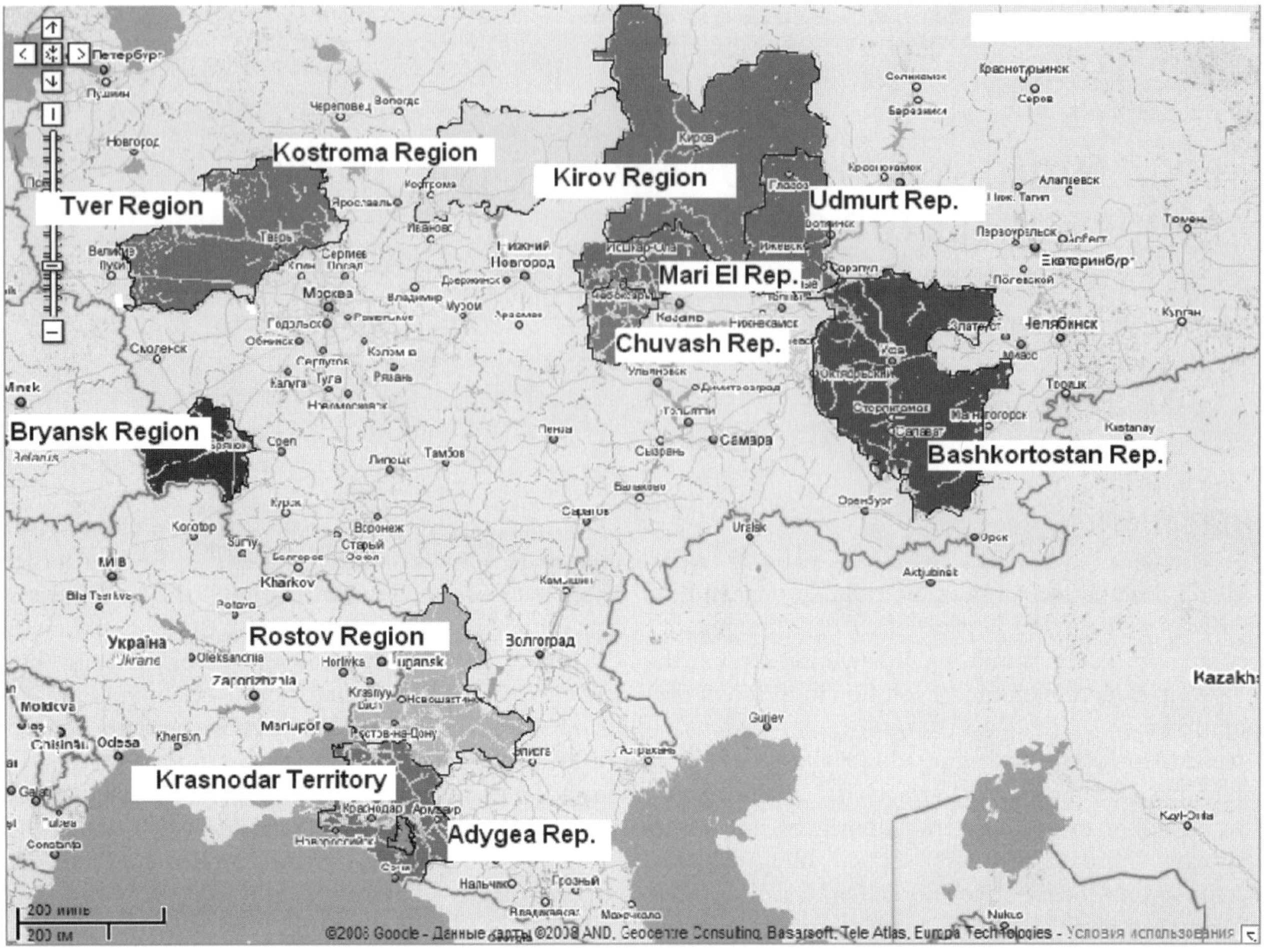

Figure 110-1 The geographical location of investigated populations

total size of population studied exceeds 2.5 x 10^6 people (Table 110-1). The analyzed sample consisted of six ethnic groups: Maris, Chuvashes, Adygeans, Udmurts, Bashkirs, and Russians. We analyzed a total of 1,660,000 Russians from the Kirov, Kostroma, Tver, Rostov and Bryansk regions, and Krasnodar Territory. The samples of Maris, Chuvashes, Bashkirs, and Udmurts were about 170,000–180,100 people each; the sample of Adygeans was 65,000 people. Maris and Udmurts are Finno-Ugric ethnic groups. The ethnogenesis of Chuvashes has been mainly contributed by local Finno-Ugrians and Turkic-speaking Bulgarians from the Volga-Kama region, and Bashkirs are Turkic-speaking. Adygeans are an autochthonous population of the northwestern Caucasus and, in anthropological terms, belong to the Balkan-Caucasian race. The Adygeans' population was bottlenecked about five generations ago, when only 4 out of 20 tribes remained in the territory of Russia. Study of the history of the formation, spread, and population dynamics of Maris, Chuvashes, Bashkirs, and Udmurts has shown that the sizes of these ethnic groups have considerably increased during the past three centuries. This rapid growth of small populations, retaining their traditional structure, favors various genetic processes, including the founder effect and

TABLE 110-1 THE SIZES OF INVESTIGATED POPULATIONS

Population	Number of districts	Population size	Ethnic group
Bryansk Region	1	88,200	Russian
Tver Region	2	76,000	Russian
Krasnodar Territory	6	426,700	Russian
Kirov Region	9	264,700	Russian
Kostroma Region and the Kostroma city	10	444,476	Russian
Rostov Region	8	320,925	Russian
Adygea Republic	4	65,000 36,800	Adygeans Russian
Mari El Republic	7	171,571 88,714	Maris Russian and other ethnic group
Chuvash Republic	6	178,722 85,768	Chuvashes Russian and other ethnic group
Udmurt Republic	6	155,356 112,299	Udmurts Russian and other ethnic group
Bashkortostan Republic	8	173,765 76,345	Bashkirs Russian and other ethnic group
Total	67	2,781,856	

genetic drift. The Kostroma and Kirov regions are two regions located in the northeast of European Russia. The Krasnodar territory and Rostov region are located in the southeast of European Russia. The Bryansk and Tver regions are located in the west of European Russia. Mari El and Chuvashia Republics are located along the Volga River; Bashkiria is located between the Volga River and the Ural Mountains, and Udmurt Republic is located between the Kama and Vyatka Rivers.

We used the protocol of genetic epidemiological studies that was developed at the laboratory of genetic epidemiology of the research center for medical genetics, of the Russian Academy of Medical Sciences, and has been used for more than 30 years. The unified methodology allows us to compare different populations.

The protocol consists of the population genetic and medical genetic sections, and is intended for comprehensive population analysis based on the data on different genetic systems, including the genes of HDs, DNA, and sometimes protein polymorphisms, demographic data including surnames frequencies (Ginter and Zinchenko, 2006; Ginter, 2002, 1997).

Part of this protocol is a medical genetic study proper, intended for detecting more than 2500 HDs out of a total of 5000–6000 known to date, and estimating the prevalence and spectrum of HD evaluation in various populations and ethnic groups. The medical genetic study of distribution of hereditary diseases in each population was made in three main steps: at the first step, special register cards (information cards) were distributed among local medical assistants and physicians, including the so-called district specialists (medical assistant, ophthalmologist, neurologist, dermatologist, and so on). On the card, symptoms of rather a great number of hereditary conditions were listed (Box 110-1).

BOX. 110-1. INFORMATION CARD

INFORMATION CARD

Dear Colleague!
The Research Center for Medical Genetics of the Russian Academy of Medical Sciences performs now an epidemiological study of hereditary disorders in your District. Here are some symptoms of hereditary disorders:

- hypotonia or hypertonia and seizures of newborns;
- high degree of mental retardation;
- congenital deafness, deaf-mutism;
- blindness, microphthalmos, congenital cataract, congenital glaucoma, coloboma, aniridia, nystagmus, ptosis of eyelids, constriction of visual fields, and night blindness;
- nanism, congenital limb deformities and reduction, vertebral defects, cranium and thorax anomalies, peculiar faces, cleft lip with or without cleft palate, polydactyly, sindactyly, combinations of different skeletal anomalies, joints dislocation;
- muscular atrophy or hypertrophy, joint limitation or /and contractures, muscular weakness, palsies, seizures, disturbances of gate, ataxia;
- altered skin pigmentation, thick or ichthyotic skin, hyperkeratosis palmar and plantar, hemangiomata and telangiectasia, multiple skin tumors, loose, redundant skin, epidermolysis, nail hypoplasia or dysplasia, alopecia, anodontia or hypodontia;
- congenital cardiac anomalies combined with other congenital anomalies;
- bleeding disorders;
- hypogenitalism, cryptorchism, hypospadia.

We ask you to present data on the patients with these symptoms living in your district to your local hospital. If there is more than one patient in the family, please show this on your card. Please show the first and second name of the patient's parents.

Yours sincerely

The data collected by the local medical personnel were combined with the data collected from other medical sources about handicapped patients, including specialized institutional schools for deaf and blind children, and were investigated further in detail. With this aim, at the second step all the families were visited by clinical geneticists, and the families with clearly nonhereditary pathology—traumas, infections, and so on—were excluded from the study. At the last step, the clinical investigations were performed by the same qualified specialists from leading research institutes of the Russian Academy of Medical Sciences and Russian Ministry of Health (a geneticist, a neurologist, an ophthalmologist, a dermatologist, a pediatrician, an otolaryngologist, and an orthopedist), which ensured unification of diagnostic criteria. In some cases, molecular, biochemical, X-ray, electromyographic, and other methods were used. The protocol of investigation was approved by the ethical committee of the Research Center for Medical Genetics. In all cases, the patients and their relatives gave their informed written consent to undergo molecular genetic analysis.

Since many HDs are heterogeneous, genetic analysis of the entire material was performed in order to test whether families with HDs had been correctly classified with respect to the mode of inheritance (autosomal dominant [AD], autosomal recessive [AR], or X-linked) and to determine the absolute and relative numbers of sporadic cases.

The collected data were subjected to segregation analysis. For this purpose all families, irrespective of clinical diagnosis, were classified into two groups according to the phenotype of the proband's parents. In sibships with one affected parent, the hypothesis of autosomal dominant inheritance was tested. The families ascertained through affected parents were tested for autosomal dominant inheritance using the χ^2 test. For the families ascertained through affected offspring segregation, frequency was estimated by the method of maximum likelihood scoring, using the ascertainment probability π (Cavalli-Sforza and Bodmer, 1971). The families with normal parents were tested for autosomal recessive inheritance. The complex segregation analysis was used in this case, with simultaneous maximum likelihood scoring of x (proportion of sporadic cases) and p (segregation frequency; Cavalli-Svorza and Bodmer, 1971; Morton and Rao, 1983). The sporadic cases were excluded from the sample of families with autosomal recessive disorders, but the cases with new dominant mutations were attributed to the sample with autosomal dominant disorders. Then, the prevalence rate was calculated and the spectrum of HDs was compiled.

The prevalence rates of AD and AR diseases were calculated as the number of cases per 1000 people examined, and the prevalence rate of X-linked diseases as the number of patients per 1000 men examined.

To study the genetic relationships between different populations/ethnic groups of Russia, we performed cluster analysis by the median-joining method. The "Statistica 6" software was used for cluster analysis. The analysis was performed separately for AD and AR diseases (199 and 168 diseases, respectively). Preliminarily, the matrix of genetic distances based on of the prevalence rates of HDs in 11 populations (Kostroma, Kirov, Bryansk, Tver, Rostov region and Krasnodar Territory), Chuvashes (Chuvashia), Adygeans (Adygea), Maris (Mari El), Udmurts (Udmurtia), and Bashkirs (Bashkortostan) were calculated.

To detect the accumulation of HDs in individual populations and ethnic groups, we calculated their prevalence rates in each population and compared them with the mean prevalence rate in all populations studied. Locally high or locally low frequencies of HD genes were detected using the F distribution according to Zhivotovsky (1991). Accumulation of individual diseases in districts or in ethnic groups was estimated using the F distribution at a significance level of $a< 0.001$.

The accumulation coefficient was calculated as the ratio of the number of diseases with locally high prevalence rates in a population to the population size (Zinchenko et al., 2001 a, b).

Another part of the protocol was the study of the genetic structure of each population with the use of population statistics (F_{ST}, endogamy index, and some other characteristics). To estimate F_{ST}, voting lists were used (total surnames lists in 9 regions, except for children under 18 years, were analyzed). Analysis of the marriage migration was used to calculate the endogamy index (more than 85,000 marriage records were collected). Random inbreeding F_{ST} was calculated by the "GenePop" (ver.3.3.) software (Russet, 2001) for Bashkirs in seven districts on the basis of frequencies of 53 alleles of 10 DNA markers (*STR/THOI, STR/FABP, STR/IVS6a, VNTR/PAH, VNTR/ApoB, VNTR/DAT1, VNTR/eNOS).*

The correlation analysis between the F_{ST} values and endogamy index, and the load of AD and AR pathology in populations, was estimated using the "Statistica 6" software with significance level of $p \leq 0.05$.

ANALYSIS OF THE DATA

Medical genetic surveys performed in 11 regions yielded initial information about a total of 49,400 patients and their family members. After examining all the patients, a diagnosis of monogenous hereditary pathology was made in 4650 families (6950 patients), for whom medical records were drawn up to contain full descriptions of the clinical and genetic status, case history, and the family tree. All families were subjected to the segregation analysis depending on the tentative pattern of inheritance.

SEGREGATION ANALYSIS

The results of the segregation analysis for the group of families presenting with AR pathology along different populations are given in Table 110-2.

Virtually in all the populations studied, the segregation frequencies did not differ from the expected segregation frequency for recessive diseases (0.25), with the proportions of sporadic cases being small.

For the families wherein the hypothesis of the AD pattern of inheritance of diseases was tested, we carried out two independent analyses (using the χ^2 test and Weinberg's proband method), depending on the method of the family ascertainment. The segregation analysis performed in the families with complete ascertainment showed that the ratio of the affected and apparently healthy people corresponded to the hypothesis of AD inheritance of diseases (50%).

The second group was composed of the families wherein ascertainment was carried out through the affected children. The values of the segregation frequencies for the group of families presenting with

TABLE 110-2 SEGREGATION FREQUENCIES, ASCERTAINMENT PROBABILITY AND THE NUMBER OF SPORADIC CASES IN FAMILIES WITH AUTOSOMAL RECESSIVE DISORDERS

Population	Ascertainment probability π	Segregation frequency	The number of sporadic cases
Bryansk Region	0.86	0.22±0.05	0.09±0.01
Kirov Region	0.77	0.26±0.03	0.13±0.08
Kostroma Region	0.81	0.22±0.01	0.13±0.08
Kostroma city	0.62	0.34±0.09	0.44±0.15
Krasnodar Territory	0.72	0.24±0.04	0.09±0.02
Adygea Republic	0.78	0.23±0.04	0.08±0.01
Mari El Republic	0.73	0.25±0.02	0.13±0.09
Chuvash Republic	0.86	0.24±0.09	0.11±0.09
Udmurt Republic	0.84	0.31±0.05	0.07±0.09
Bashkortostan Republic	0.85	0.24±0.05	0.13±0.09
Tver Region	0.77	0.20±0.07	0.13±0.09
Rostov Region	0.73	0.26±0.05	0.12±0.08

presumably autosomal dominant pathology are shown in Table 110-3.

Analyzing the segregation frequencies, either for the families presenting with presumably dominant pathology or for the recessive families, revealed no statistically significant differences from the expected value.

After the segregation analysis performed for the autosomal dominant and autosomal recessive families, we proceeded to the next stage of our work, i.e., calculation of the prevalence rates of, and an estimation of diversity of the populations presenting by autosomal dominant, autosomal recessive, and X-linked diseases.

TABLE 110-3 SEGREGATION FREQUENCIES AND ASCERTAINMENT PROBABILITY IN FAMILIES WITH AUTOSOMAL DOMINANT INHERITANCE

Population	Ascertainment probability π	Segregation frequency
Bryansk Region	0.73	0.55±0.08
Kirov Region	0.59	0.49±0.05
Kostroma Region	0.67	0.52±0.05
Kostroma city	0.63	0.61±0.05
Krasnodar Territory	0.74	0.51±0.09
Adygea Republic	0.72	0.53±0.08
Mari El Republic	0.60	0.56±0.06
Chuvash Republic	0.88	0.52±0.07
Udmurt Republic	0.76	0.54±0.07
Bashkortostan Republic	0.83	0.42±0.09
Tver Region	0.69	0.52±0.02
Rostov Region	0.61	0.48±0.05

LOAD OF HEREDITARY DISEASES IN POPULATIONS OF RUSSIA

Table 110-4 shows the prevalence rates of monogenic HDs in separate populations/ethnic groups.

As can be seen from Table 110-4, the load in all the populations examined is highest in those presenting with AD diseases, and lowest in those with X-linked diseases. There are no significant differences between the populations in their load of X-linked diseases. At least partially, this may be explained by the relatively low absolute values of the prevalence rate for this group of diseases. On the other hand, the differences between the populations by the load of AD and AR diseases appear statistically significant. First, the load of both AD and AR diseases is always higher in rural populations as compared with that in urban populations. The load of dominant diseases in the sample examined appeared to be 1.5 times higher in the urban populations than in the urban populations (1.79 ± 0.04 and 1.02 ± 0.04, respectively), with that of recessive diseases having proved to be 2 times higher (1.36 ± 0.03 and 0.75 ± 0.05, respectively). Secondly, there are significant differences observed within the groups of the rural and urban populations in the load of both dominant and recessive diseases. The differences are also seen while comparing the load of the autosomal hereditary pathology in various ethnic groups. For example, the load of AD and AR pathology among the Maris, Chuvashs, Bashkirs, and Udmurts is higher than that among the Russians and Adygeans (both urban and rural populations: $\chi^2_1 = 50.75$; $\chi^2_1 = 11.14$; $\chi^2_1 = 12.18$; $\chi^2_1 = 4.77$; $p < 0.05$). It is important to point out that there is differentiation in the load of autosomal diseases between the populations of different hierarchical levels. The lower the hierarchical level of the population concerned, the higher the differences observed between the populations. The greatest differences are revealed between the rural soviets of a particular region. For instance, in the Sharkansk region of the Republic of Udmurtia, the load of hereditary diseases in various rural soviets of approximately equal population (500–1000 people) varied from 0 to 16.34; in the Kugarchinsk region, from 0 to 18 per 1000 people. Less pronounced are differences in the load among separate regions of one district or republic (e. g., a variation from 1.95 to 5.77 per 1000 people between the loads in the rural population of two neighboring regions of the Republic of Udmurtia). And finally, the lowest differences were revealed between the examined subjects of the Russian Federation (Table 110-4).

TABLE 110-4 LOAD (PER 1000 PERSONS) OF HEREDITARY DISEASES IN SOME POPULATIONS OF RUSSIA

Populations \ Load	AD	AR	X-linked	Common
ETHNIC GROUP				
Bashkirs Bashkortostan Republic (rur)	3.13±0.12	1.60±0.09	0.53±0.09	5.26±0.16
Bashkirs Bashkortostan Republic (urb)	1.70±0.23	0.99±0.18	0.12±0.09	2.81±0.29
Udmurts Udmurt Republic (rur)	3.23±0.09	1.65±0.09	0.96±0.12	6.00±0.29
Udmurts Udmurt Republic (urb)	1.38±0.07	0.70±0.08	0.34±0.09	2.42±0.24
Chuvashes Chuvash Republic (rur)	1.79±0.06	1.25±0.06	0.56±0.08	3.60±0.16
Chuvashes Chuvash Republic (urb)	0.88±0.03	1.01±0.06	0.55±0.06	2.44±0.24
Maris Mari El Republic (rur)	2.23±0.13	1.43±0.10	0.33±0.08	3.99±0.17
Maris Mari El Republic (urb)	1.14±0.17	0.81±0.15	0.81±0.21	2.76±0.24
Adygeans Adygea Republic (rur)	1.07±0.13	1.41±0.15	0.50±0.13	2.98±2.14
Weighted mean for rural (in ethnic group)	2.30±0.06	1.47±0.05	0.56±0.05	4.32±0.09
Weighted mean for urban (in ethnic group)	1.20±0.08	0.85±0.07	0.42±0.06	2.47±0.12
RUSSIAN:				
Bryansk Region (rur.)	1.01±0.06	0.81±0.07	0.39±0.05	2.21±0.08
Bryansk Region (urb)	0.91±0.05	0.54±0.06	0.22±0.05	1.67±0.07
Kirov Region (rur)	1.56±0.12	1.28±0.11	0.36±0.07	3.20±0.08
Kirov Region (urb)	0.65±0.04	0.57±0.04	0.37±0.05	1.51±0.08
Tver Region (rur)	0.96±0.13	0.72±0.10	0.62±0.09	2.30±0.20
TverRegion (urb)	0.79±0.09	0.68±0.07	0.36±0.05	1.83±0.31
Kostroma Region (rur)	1.25±0.12	0.90±0.10	0.42±0.09	2.45±0.19
Kostroma Region and Kostroma city (urb)	0.75±0.05	0.42±0.04	0.42±0.01	1.26±0.07
Krasnodar Territory (urb)	1.01±0.08	0.64±0.07	0.35±0.06	2.00±0.06
Rostov Region (rur)	2.00±0.01	1.30±0.01	0.44±0.01	3.78±0.11
Rostov Region (urb)	1.34±0.05	1.03±0.06	0.19±0.04	2.56±0.15
Weighted mean for rural (Russian)	1.29±0.06	1.26±0.05	0.74±0.04	3.29±0.15
Weighted mean for urban (Russian)	0.84±0.05	0.65±0.07	0.32±0.03	1.81±0.09
Weighted mean for rural (all populations)	1.79±0.04	1.36±0.03	0.65±0.03	3.80±0.06
Weighted mean for urban(all populations)	1.02±0.04	0.75±0.05	0.37±0.03	2.14±0.04

Note: urb. – people from city and town, rur. – people from villages

Hence, analyzing the load of Mendelian disorders in the examined populations of Russia demonstrated that there exists a clear-cut differentiation between various populations, both within each of the 11 regions examined and between them, by the index concerned (there are strong grounds for believing that they also differ by the frequencies of genes of hereditary diseases). Mention should be made that the obtained estimations of the load, or prevalence of hereditary diseases, especially in the rural populations of the Kirov region, Republic of Mari El, and Chuvashia, by their absolute values are very close to the estimations of the hereditary diseases prevalence rate according to the data of the Register of Congenital and Hereditary Pathology in British Columbia (Canada; Baird et al., 1988).

EFFECT OF THE POPULATION DYNAMICS ON THE GENETIC LOAD

RELATIONSHIP BETWEEN THE LEVEL OF HEREDITARY PATHOLOGY LOAD AND THE VALUES OF RANDOM INBREEDING F_{ST}.

In order to elucidate what population genetic mechanisms are responsible for genetic differentiation of the populations

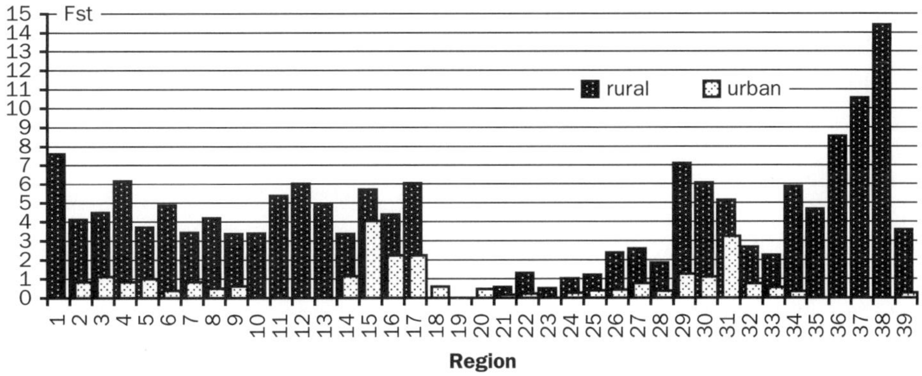

Figure 110-2 Random inbreeding (FST) 10^{-3} in populations of Russia
NOTE: 1–9 – Districts of Kirov Region, 10–17 – Districts of Kostroma Region, 18 – Kostroma City, 19–20 – Districts of Krasnodar Territory, 21–25 – Districts of Rostov Region; 26–27 – Districts of Tver Region, 29–33 – Districts of Mari El Republic; 34 – District of Bryansk Region, 35–38 – Districts of Adygea Republic, 39 – District of Chuvash Republic.

of Russia by the load of hereditary diseases in all the examined populations, the medical genetic study was simultaneously accompanied by the population genetic study proper. It should be emphasized that the unit measure of the survey in our medical genetic studies was the population of the region. That is the reason why the genetic structure of the populations being surveyed was frequently described by means of the *isonomic* method, allowing the obtaining of virtually unbiased values of random inbreeding (F_{st}) for the population of the region concerned.

We analyzed the dependence of hereditary diseases load on the genetic structure of the populations of the regions along nine regions of Russia (Krasnodar Territory, Kirov region, Kostroma region, Tver region, Bryansk region, Republic of Mari El, Chuvashia, and Adygea), wherein the values of F_{st} were obtained. The F_{st} values varied within extremely wide ranges, from 0 in some populations of the Krasnodar Territory to 14.9×10^{3} in the Teuchezhskii region of the Republic of Adygea (Figure 110-2). The weighted average value of random inbreeding for the urban populations amounted to 0.57×10^{-3}, whereas that for the rural populations equaled 4.36×10^{-3}. At the same time, both the rural and the urban populations showed variability of the random inbreeding values.

To carry out the analysis, all the elementary populations studied (a total of 74 populations) were subdivided into ten groups depending on the random inbreeding level (F_{st}). For each group, we calculated the average weighted values of the load of autosomal dominant and autosomal recessive pathology, and the values of random inbreeding (F_{st}). The Spearman's rank correlation ratios RsAD and RsAR amounted to 0.7596 and 0.8822, respectively. The ratio of correlation between the load of dominant and recessive pathologies amounted to 0.8735. Since the values of F_{st} constitute a measure of genetic drift, high values of correlation between random inbreeding and the load of AD and AR hereditary diseases make it possible to suggest that genetic drift is probably one of the leading factors determining differentiation of the populations of Russia by the load of hereditary diseases.

RELATIONSHIP BETWEEN THE LOAD OF HEREDITARY PATHOLOGY AND THE LEVEL OF ENDOGAMY OF THE POPULATIONS

In our population genetics studies, parallel to assessing random inbreeding for virtually all the populations studied, we also estimated the endogamy index, defined as the proportion of marriages contracted within a population to all marriages contracted in the population concerned. The total number of the analyzed marriages contracted during the period from 1970 to 2000, in all the populations, exceeded 85,000. The endogamy index gives an estimate of the intensity of pressure of migrations, on the one hand, and of the degree of population isolation on the other. The values of the endogamy index in the studied populations of Russia appeared to be as variable as the values of random inbreeding (Figure 110-3).

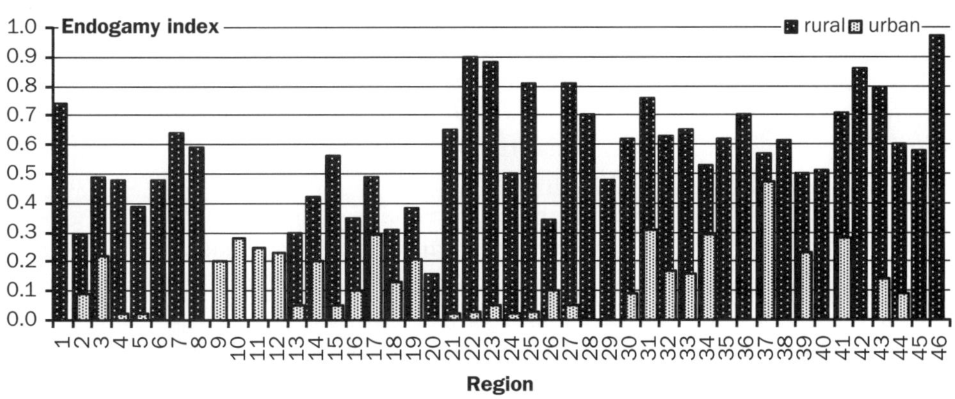

Figure 110-3 Index of endogamy in populations of Russia
NOTE: 1–5 – Districts of Kirov Region, 6–8 – Districts of Kostroma Region, 9 – Kostroma city, 10–12 – Districts of Krasnodar Territory, 13–14 – Districts of Tver Region, 15 – Districts of Bryansk Region, 16–20 – Districts of Rostov Region, 21–27 – Districts of Mari El Republic, 28 – Districts of Adygea Republic, 29–34 – Districts of Udmurt Republic, 35–40 – Districts of Chuvash Republic, 41–46 – Districts of Bashkortostan Republic.

Overall, they turned out to be higher in the rural populations, whereas variability of the endogamy index proved to be high both in the urban and in the rural populations. The endogamy index in the populations varies within extremely wide ranges, i.e., from 0 in the city of Kozmodemyansk to 0.97 in the Baltachevsk region of the Republic of Bashkortostan. The whole material was subdivided into 7 groups depending on the level of endogamy (a total of 59 populations with the estimated index of endogamy): Group One – endogamy index from 0 to 0.10; Group Two – from 0.11 to 0.29; Group Three – from 0.30 to 0.39; Group Four – from 0.40 to 0.49; Group Five – from 0.50 to 0.60; Group Six – from 0.61 to 0.70; and Group Seven – from 0.71 and more. For each group considered, we calculated the average weighted values of the endogamy index and the burden of AD and AR pathology, followed by analyzing the regression dependence. The Spearman's rank correlation coefficients rAD and rAR amounted to 0.9787 and 0.8583, respectively.

ANALYSIS OF THE EFFECT OF THE GENETIC STRUCTURE ASSESSED THROUGH THE CONDITIONALLY NEUTRAL DNA POLYMORPHISM ON THE LEVEL OF LOAD OF THE POPULATION WITH HEREDITARY PATHOLOGY

Yet another argument counting in favor of the fact that genetic drift is one of the leading factors determining the origin of genetic differentiation of populations by the load of hereditary diseases is the analysis of correlations between the autosomal hereditary pathology load and the F_{st} values evaluated via DNA polymorphic loci. The analysis is exemplified by the population of the Republic of Bashkortostan.

Table 110-5 shows the values of the load of AD and AR pathology for the rural Bashkirian population and the values of random inbreeding F_{st} along the regions, as assessed by the frequencies of alleles of ten polymorphic DNA loci of the nuclear genome.

We analyzed the linear dependence between the load of AD and AR pathology and the values of random inbreeding F_{st} along the regions. The correlation coefficients between the level of the load of the Bashkirian population with AD and AR pathology and the values of random inbreeding F_{st} were r = 0.8809 and r = 0.9304, respectively (level of significance $p \leq 0.05$).

DIVERSITY OF GENETIC DISEASES IN THE RUSSIAN POPULATIONS

In the 11 Russian regions surveyed, we revealed a total of 413 clinically differing hereditary diseases (199 AD, 168 AR, and 46 X-linked recessive forms). The number of genetic nosological entities might be substantially greater, since in many cases it was impossible to differentiate clinical variants of genetically heterogeneous HDs, such as, for example, pigment retinitis, congenital cataract, hereditary motor-sensory neuropathy, nonsyndromal sensorineural hearing impairment, etc.

The whole nosological spectrum of autosomal dominant, autosomal recessive, and X-linked diseases was subdivided into 5 groups according to the prevalence rate of separate entities in the whole sample (Tables 110-6, 110-7, and 110-8).

The analysis of the prevalence rate of HDs along the groups is shown in Table 110-9. In the whole population surveyed, we revealed a total of 29 frequently occurring diseases (encountered with an incidence rate of 1:50,000 and more). In essence, this group comprises the maximal number of patients (58.23%). However, mention should be made that in each population/ethnic group, the number and spectrum of frequently encountered HDs do differ. For example, Bashkirs had 49 hereditary diseases encountered with the prevalence of 1:50,000; Udmurts, 41; Chuvashs, 36; and Adygeans, 38:50,000. The qualitative and quantitative differences were also revealed for the spectrum of frequent HDs in the Russian populations, although they were less pronounced. In the Krasnodar Territory, we registered 21 frequent commonly encountered hereditary diseases, and in the Kirov Region, 29; in the Kostroma region, 21; in the Rostov region, 40; and in the Bryansk region, 32.

The highest number of nosological entities with the lowest number of patients was registered in the groups of especially uncommon diseases, whose prevalence rate is below than 1:200,001. The last two groups account for 82.8% of all nosological entities revealed.

Mention should be made of fundamental similarity of the spectrum of commonly occurring autosomal

TABLE 110-5 THE LOAD OF HEREDITARY DISEASES (PER 1000 PERSONS) AND FST (CALCULATED FOR BASHKIRS IN SEVEN DISTRICTS ON THE BASIS OF FREQUENCIES OF ALLELES OF 10 DNA MARKERS

	The load (per 1000 persons) of HDs		
District	AD	AR	F_{ST}
Baimaksky	2.93	1.65	0.0021
Burzyansky	3.34	2.12	0.0063
Abzelilovsky	2.45	1.33	0.0003
Arkhangelsky	2.40	1.31	0.0028
Salavatsky	2.59	1.38	0.0014
Askinsky	3.58	1.86	0.0029
Kugarchinsky	5.61	1.83	0.0075

TABLE 110-6 SPECTRUM OF AD DISEASES AND THEIR PREVALENCE RATES ($X10^{-5}$) IN THE POPULATIONS STUDIED

No.	Diagnosis	Bashkirs of Bashkortostan Republic	Udmurts of Udmurt Republic	Maris of Mari El Republic	Chuvashes of Chuvash Republic	Adygeans of Adygea Republic	Russian of Rostov Region	Russian of Kostroma Region	Russian of Krasnodar Territory	Russian of Kirov Region	Russian of Tver Region	Russian of Bryansk Region	Average prevalence rate
PREVALENCE RATE 1:50000 AND HIGHER													
1	CHARCOT-MARIE-TOOTH DISEASE	4.60±1.63	21.24±3.70	2.91±1.30	6.15±1.86		8.72±1.65	0.22±0.22	5.39±1.12	13.60±2.0	10.53±3.7	10.20±3.4	6.85±0.54
2	MENTAL RETARDATION	17.84±3.20	18.67±3.47	3.50±1.43	4.48±1.58			1.35±0.55			2.63±1.86		3.47±0.38
3	DYSTROPHIA MYOTONICA 1	21.29±3.50		0.58±0.58			2.49±0.88	0.45±0.32	0.23±0.23	0.76±0.53			2.16±0.30
4	NEUROFIBROMATOSIS. TYPE I	5.75±1.82	5.79±1.93	9.91±2.40	11.19±2.50	12.31±4.35	5.61±1.32	2.25±0.71	3.52±0.91	4.91±1.36		1.13±1.13	5.12±0.47
5	PTOSIS. HEREDITARY CONGENITAL	8.63±2.23	5.15±1.82	4.66±1.65	1.12±0.79	10.77±4.07	4.36±1.18	0.90±0.45	4.45±1.42	3.78±1.20	2.63±1.86	1.13±1.13	3.80±0.40
6	RETINITIS PIGMENTOSA	5.18±1.73		1.17±0.82	1.12±0.79		7.79±1.56	4.50±1.01	3.05±0.85	0.76±0.53		2.27±1.60	3.17±0.37
7	CONGENITAL HEREDITARY CATARACT	5.18±1.73	14.80±3.09	4.08±1.54	3.36±1.37		13.40±2.04	7.42±1.29	2.11±0.70	1.89±0.85		4.53±2.27	5.88±0.50
8	HYPOCHONDROPLASIA	7.48±2.08	0.64±0.64	7.58±2.10	3.92±1.48	16.92±5.10	1.87±0.76	5.17±1.08	4.45±1.42	6.04±1.51		3.40±1.96	4.73±0.45
9	POLYDACTYLY. POSTAXIAL. TYPE A1	12.09±2.64	5.15±1.82	8.74±2.26	2.24±1.12	9.23±7.37	4.67±1.21	0.90±0.45	0.70±0.41	2.64±1.00		2.27±1.60	3.59±0.39
10	OSTEOGENESIS IMPERFECTA	4.03±1.52	4.51±1.70	1.75±1.01		1.54±1.54	3.12±0.99	1.12±0.50	1.87±0.67	2.27±0.93	3.95±2.30	1.13±1.13	2.16±0.30
11	LIPOMATOSIS. MULTIPLE	3.45±1.41	4.51±1.70		2.24±1.12	6.15±3.08	3.43±1.03	0.45±0.32	6.09±1.20	2.27±0.93	1.32±1.32	4.53±2.27	3.00±0.36
12	ICHTHYOSIS VULGARIS	18.99±3.31	32.83±4.60	64.70±6.14	45.88±5.07	9.23±7.37	22.44±2.64	11.92±1.6	11.72±1.7	23.80±2.9	7.89±3.22	1.13±1.13	22.32±1.0
13	KERATOSIS PALMOPLANTARIS	31.08±4.23	13.52±2.95	8.74±2.26	10.63±2.44	4.62±2.67	5.30±1.29	1.35±0.55	2.34±0.74	0.38±0.38	2.63±1.86	3.40±1.96	6.38±0.52
14	MARFAN SYNDROME	5.75±1.82	9.91±2.40	3.50±1.43	3.92±1.48	1.54±1.54	4.99±1.25	0.67±0.39	1.41±0.57	0.76±0.53	3.95±2.30		3.00±0.36
15	EHLERS-DANLOS SYNDROME	17.84±3.20	9.66±2.14	1.75±1.01	8.39±2.17		11.84±1.92	0.45±0.32	5.86±1.20	7.56±0.17	3.95±2.30		6.43±0.52
16	DEAFNESS. AUTOSOMAL DOMINANT	2.30±1.15	5.79±1.93	4.08±1.54	1.68±0.97		1.56±0.70	2.92±0.81	3.28±0.88	3.78±1.20	2.63±1.86	27.21±5.6	3.85±0.40
PREVALENCE RATE 1:50001-1:100000													
17	HUNTINGTON DISEASE	1.73±0.99	1.93±1.12	1.17±0.82	2.24±1.12	1.54±1.54		0.67±0.39	1.41±0.57	1.51±0.76	2.63±1.86	3.40±1.96	1.31±0.23
18	FACIOSCAPULOHUMERAL MUSCULAR DYSTROPHY 1A (Landouzy–Dejerine muscular dystrophy)	3.45±1.41					0.31±0.31	1.35±0.55	1.41±0.57	0.38±0.38		4.53±2.27	1.01±0.21
19	NYSTAGMUS. CONGENITAL		3.22±1.44	5.83±1.84	2.24±1.12		4.05±1.17	0.22±0.22		0.38±0.38			1.44±0.25
20	COLOBOMA. OCULAR	2.88±1.29	3.22±1.44	0.58±0.58	1.12±0.79		2.80±0.95	0.22±0.22	0.70±0.41	1.13±0.65		1.13±1.13	1.27±0.23
21	SYNDACTYLY. TYPE I	2.88±1.29	1.93±1.12	3.50±1.43		3.08±2.18	3.74±1.08	0.90±0.45	0.23±0.23	3.02±1.07			1.73±0.27
22	SYNPOLYDACTYLY 1	0.58±0.58	1.93±1.12	2.91±1.30	2.80±1.25		2.80±0.95	0.45±0.32	0.94±0.47			3.40±1.96	1.35±0.24
23	ACHONDROPLASIA	5.75±1.82	1.29±0.91	1.17±0.82	1.68±0.97	1.54±1.54	2.80±0.95	1.57±0.60	0.94±0.47	1.89±0.85			1.82±0.28
24	EXOSTOSES. MULTIPLE. TYPE I	0.58±0.58			3.36±1.37		1.56±0.70	0.45±0.32	2.81±0.81	1.13±0.65	3.95±2.30		1.35±0.24

(Continued)

TABLE 110-6 (CONTINUED)

No.	Diagnosis	Bashkirs of Bashkorto-stan Republic	Udmurts of Udmurt Republic	Maris of Mari El Republic	Chuvashes of Chuvash Republic	Adygeans of Adygea Republic	Russian of Rostov Region	Russian of Kostroma Region	Russian of Krasnodar Territory	Russian of Kirov Region	Russian of Tver Region	Russian of Bryansk Region	Average prevalence rate
25	SCOLIOSIS. IDIOPATHIC	1.73±0.99	5.15±1.82				1.87±0.76	2.02±0.68			1.32±1.32		1.14±0.22
26	POLYDACTYLY. PREAXIAL I		2.57±1.29		0.56±0.56	6.15±3.08	1.56±0.70	2.02±0.68		1.51±0.76	1.32±1.32		1.18±0.22
27	SPLIT-HAND/FOOT MALFORMATION 1	2.88±1.29	1.93±1.12	1.75±1.01	1.68±0.97	3.08±2.18	1.56±0.70	0.67±0.39	0.23±0.23				1.06±0.21
28	NOONAN SYNDROME 1	3.45±1.41	7.08±2.14	1.75±1.01			3.74±1.08		0.70±0.41	1.51±0.76		1.13±1.13	1.69±0.27
PREVALENCE RATE 1:100001-1:200000													
29	SPASTIC PARAPLEGIA. AUTOSOMAL DOMINANT	4.60±1.63							0.70±0.41	3.02±1.07	1.32±1.32		0.85±0.19
30	TREMOR. HEREDITARY ESSENTIAL				1.12±0.79			0.45±0.32		3.78±1.20			0.59±0.16
31	MYOTONIA CONGENITA. AUTOSOMAL DOMINANT						2.49±0.88	0.67±0.39	1.64±0.62			1.13±1.13	0.80±0.18
32	TUBEROUS SCLEROSIS	4.60±1.63	1.29±0.91				0.62±0.44			0.38±0.38	1.32±1.32		0.59±0.16
33	MUSCULAR DYSTROPHY. LIMB-GIRDLE		1.93±1.12					1.35±0.55	0.70±0.41			3.40±1.96	0.63±0.16
34	MICROCEPHALY. AUTOSOMAL DOMINANT	4.03±1.52	1.93±1.12	3.50±1.43								6.80±2.78	0.93±0.19
35	STURGE-WEBER SYNDROME	0.58±0.58	2.57±1.29	1.17±0.82		1.54±1.54	1.56±0.70	0.22±0.22	0.47±0.33	0.38±0.38			0.72±0.17
36	OCULOPHARYNGEAL MUSCULAR DYSTROPHY						0.93±0.54		1.87±0.67	0.38±0.38			0.51±0.15
37	BLEPHAROPHIMOSIS. PTOSIS	7.48±2.08		0.58±0.58	1.12±0.79			0.45±0.32					0.76±0.18
38	RETINAL CONE DYSTROPHY		0.64±0.64					0.45±0.32	1.64±0.62	0.38±0.38		1.13±1.13	0.51±0.15
39	VITREORETINAL DEGENERATION	4.03±1.52	0.64±0.64					0.45±0.32		0.76±0.53			0.51±0.15
40	STARGARDT DISEASE		3.22±1.44				1.56±0.70		0.70±0.41	0.76±0.53		2.27±1.60	0.72±0.17
41	OPTIC ATROPHY 1	0.58±0.58	2.57±1.29	0.58±0.58	0.56±0.56		0.31±0.31	0.67±0.39	0.94±0.47	0.76±0.53		4.53±2.27	0.89±0.19
42	ANIRIDIA			0.58±0.58		1.54±1.54		2.25±0.71	0.70±0.41			1.13±1.13	0.68±0.17
43	CORNEAL DYSTROPHY			2.91±1.30					0.94±0.47	1.13±0.65			0.51±0.15
44	MICROPHTHALMIA. ISOLATED. WITH CATARACT	2.88±1.29	1.93±1.12	0.58±0.58				0.22±0.22		0.76±0.53		1.13±1.13	0.55±0.15
45	ACROKERATOSIS VERRUCIFORMIS		0.64±0.64						4.22±0.99				0.80±0.18
46	ECTODERMAL DYSPLASIA. HIDROTIC		1.93±1.12	4.08±1.54	0.56±0.56				0.70±0.41	0.76±0.53		2.27±1.60	0.76±0.18
47	BRACHYDACTYLY. TYPE B1	1.15±0.81			6.71±1.94								0.59±0.16
48	FRONTONASAL DYSPLASIA	4.60±1.63	1.29±0.91		1.12±0.79		0.93±0.54					2.27±1.60	0.72±0.17

49	ALBINOIDISM. OCULOCUTANEOUS. AUTOSOMAL DOMINANT		4.51±1.70		1.12±0.79				1.41±0.57			4.53±2.27	0.80±0.18
50	AXENFELD-RIEGER SYNDROME. TYPE 1	4.03±1.52					0.62±0.44					4.53±2.27	0.55±0.15
51	ECTRODACTYLY. ECTODERMAL DYSPLASIA. AND CLEFT LIP/PALATE SYNDROME			4.66±1.65	1.12±0.79		1.56±0.70		0.23±0.23	0.38±0.38			0.72±0.17
52	WAARDENBURG SYNDROME	1.73±0.99		1.75±1.01			1.56±0.70	1.57±0.60					0.76±0.18
53	GOLDENHAR SYNDROME	1.15±0.81	1.93±1.12	1.17±0.82	0.56±0.56		1.87±0.76	1.12±0.50	0.23±0.23	0.38±0.38	1.32±1.32		0.93±0.19
54	KLIPPEL-FEIL SYNDROME. AUTOSOMAL DOMINANT	0.58±0.58	1.29±0.91	0.58±0.58	1.68±0.97		2.49±0.88		0.23±0.23		1.32±1.32		0.72±0.17
55	POLAND SYNDROME	1.73±0.99	1.29±0.91	0.58±0.58	2.80±1.25			0.22±0.22	0.47±0.33	0.38±0.38			0.63±0.16
56	PRADER-WILLI SYNDROME	0.58±0.58	2.57±1.29	1.75±1.01	0.56±0.56	3.08±2.18	0.93±0.54	0.22±0.22	0.23±0.23	1.89±0.85		1.13±1.13	0.93±0.19
57	RUBINSTEIN-TAYBI SYNDROME	1.73±0.99	0.64±0.64			1.54±1.54	0.31±0.31		0.70±0.41	0.76±0.53		1.13±1.13	0.51±0.15
58	HOLT-ORAM SYNDROME	1.15±0.81	0.64±0.64	2.33±1.17	0.56±0.56	1.54±1.54	0.62±0.44	0.67±0.39	0.94±0.47	0.76±0.53	2.63±1.86		0.93±0.19
59	TRICHORHINOPHALANGEAL SYNDROME. TYPE I	1.15±0.81		0.58±0.58	0.56±0.56	12.31±4.35	0.31±0.31	0.45±0.32	0.23±0.23	1.13±0.65		3.40±1.96	0.93±0.19
60	ACRODENTAL DYSOSTOSIS OF WEYERS CURRY-HALL SYNDROME	6.91±1.99											0.51±0.15
61	ISOLATED GROWTH HORMONE DEFICIENCY	3.45±1.41	3.86±1.58										0.51±0.15
62	OTOSCLEROSIS			1.75±1.01			1.87±0.76				3.95±2.30		0.51±0.15
PREVALENCE RATE 1:200001–1:500000													
63	SPINOCEREBELLAR ATAXIA 1	0.58±0.58	2.57±1.29					0.22±0.22	0.47±0.33				0.34±0.12
64	MUSCULAR DYSTROPHY. CONGENITAL		0.64±0.64						1.64±0.62				0.34±0.12
65	DYSTONIA. DOPA-RESPONSIVE	1.15±0.81			2.24±1.12						2.63±1.86		0.34±0.12
66	VITREORETINOCHOROIDOPATHY		0.64±0.64				2.80±0.95						0.42±0.13
67	GLAUCOMA 3. PRIMARY CONGENITAL	3.45±1.41					0.93±0.54		0.47±0.33				0.47±0.13
68	BRACHYDACTYLY. TYPE E		0.64±0.64	2.33±1.17	2.24±1.12					0.76±0.53			0.47±0.13
69	EPIPHYSEAL DYSPLASIA. MULTIPLE			2.33±1.17					0.47±0.33		2.63±1.86		0.34±0.12
70	METAPHYSEAL CHONDRODYSPLASIA. SCHMID TYPE						1.25±0.62		0.94±0.47				0.34±0.12
71	RADIOULNAR SYNOSTOSIS		1.29±0.91	1.17±0.82	0.56±0.56				0.47±0.33		1.32±1.32		0.34±0.12
72	COSTOVERTEBRAL SEGMENTATION ANOMALIES			1.17±0.82	1.12±0.79		0.31±0.31		0.47±0.33		1.32±1.32		0.34±0.12
73	NEVUS ANEMICUS		0.64±0.64		2.24±1.12						5.26±2.63		0.38±0.13
74	ALOPECIA	0.58±0.58			1.12±0.79		0.62±0.44	0.45±0.32		0.76±0.53	1.32±1.32		0.42±0.13

(Continued)

TABLE 110-6 (CONTINUED)

No.	Diagnosis	Bashkirs of Bashkortostan Republic	Udmurts of Udmurt Republic	Maris of Mari El Republic	Chuvashes of Chuvash Republic	Adygeans of Adygea Republic	Russian of Rostov Region	Russian of Kostroma Region	Russian of Krasnodar Territory	Russian of Kirov Region	Russian of Tver Region	Russian of Bryansk Region	Average prevalence rate
75	ALPORT SYNDROME			1.75±1.01			0.31±0.31	0.90±0.45		0.76±0.53			0.42±0.13
76	WILLIAMS SYNDROME	1.73±0.99		0.58±0.58	0.56±0.56		0.62±0.44		0.47±0.33		1.32±1.32		0.42±0.13
77	LEOPARD SYNDROME		0.64±0.64				1.87±0.76		0.23±0.23				0.34±0.12
78	CROUZON SYNDROME		1.29±0.91		0.56±0.56			0.22±0.22	0.47±0.33	0.76±0.53			0.34±0.12
79	CORNELIA DE LANGE SYNDROME	1.73±0.99		0.58±0.58		3.08±2.18	0.31±0.31		0.23±0.23	0.38±0.38			0.38±0.13
80	TELANGIECTASIA. HEREDITARY HEMORRHAGIC. OF RENDU. OSLER		1.93±1.12					0.90±0.45	0.94±0.47				0.47±0.13
81	SILVER-RUSSELL SYNDROME	1.15±0.81			0.56±0.56	1.54±1.54	0.62±0.44			0.76±0.53			0.34±0.12
82	TREACHER COLLINS-FRANCESCHETTI SYNDROME	0.58±0.58	2.57±1.29		0.56±0.56			0.90±0.45					0.42±0.13
83	SPRENGEL DEFORMITY	1.15±0.81	1.29±0.91				0.93±0.54		0.70±0.41				0.42±0.13
84	ANGELMAN SYNDROME				2.24±1.12		0.93±0.54				1.32±1.32		0.34±0.12
85	OPITZ OCULOGENITOLARYNGEAL SYNDROME. TYPE II						2.49±0.88						0.34±0.12
86	CONTRACTURAL ARACHNODACTYLY. CONGENITAL (BEALS SYNDROME)				3.92±1.48			0.45±0.32					0.38±0.13
87	ALBINISM. OCULOCUTANEOUS. PARTIAL HYPOPIGMENTATION			1.17±0.82				0.45±0.32		0.76±0.53			0.25±0.10
88	ECTOPIA LENTIS. ISOLATED			0.58±0.58	0.56±0.56				0.47±0.33	0.76±0.53			0.25±0.10
89	SYNDACTYLY. TYPE V			1.17±0.82	1.68±0.97							2.27±1.60	0.30±0.11
90	ACROMESOMELIC DYSPLASIA						2.18±0.82						0.30±0.11
91	SPONDYLOMETAPHYSEAL DYSPLASIA			1.17±0.82	0.56±0.56		0.31±0.31				2.63±1.86		0.25±0.10
92	PSEUDOACHONDROPLASIA			0.58±0.58	0.56±0.56		0.62±0.44	0.45±0.32	0.23±0.23				0.30±0.11
93	CLEIDOCRANIAL DYSPLASIA	0.58±0.58			0.56±0.56		0.31±0.31	0.90±0.45					0.30±0.11
94	AMELOGENESIS IMPERFECTA								0.70±0.41	1.13±0.65			0.25±0.10
95	BLOUNT DISEASE						1.87±0.76						0.25±0.10
96	MONILETHRIX						1.87±0.76						0.25±0.10
97	DERCUM DISEASE								1.64±0.62				0.30±0.11
98	GILBERT SYNDROME	1.73±0.99					0.62±0.44				1.32±1.32		0.25±0.10

99	GREIG CEPHALOPOLYSYNDACTYLY SYNDROME	1.15±0.81					1.25±0.62						0.25±0.10
100	NAGER ACROFACIAL DYSOSTOSIS		4.51±1.70										0.30±0.11
101	PORPHYRIA	1.73±0.99									1.32±1.32	2.27±1.60	0.25±0.10
102	EPISTAXIS. HEREDITARY			1.75±1.01				0.67±0.39					0.25±0.10
PREVALENCE RATE 1:500001 AND LESS OFTEN													
103	SCAPULOPERONEAL MYOPATHY			1.17±0.82						0.76±0.53		1.13±1.13	0.21±0.09
104	MESODERMAL DYSGENESIS OF THE IRIS								1.17±0.52				0.21±0.09
105	ONYCHODYSPLASIA. ISOLATED CONGENITAL					1.54±1.54		0.90±0.45					0.21±0.09
106	PEMPHIGUS. BENIGN FAMILIAL (HAILEY-HAILEY DISEASE)			2.33±1.17								1.13±1.13	0.21±0.09
107	KLIPPEL-TRENAUNAY-WEBER SYNDROME				1.12±0.79				0.70±0.41				0.21±0.09
108	DARIER-WHITE DISEASE	2.88±1.29											0.21±0.09
109	DEAFNESS. CONDUCTIVE. WITH MALFORMED EXTERNAL EAR	0.58±0.58		1.17±0.82							2.63±1.86		0.21±0.09
110	NAIL-PATELLA SYNDROME						0.62±0.44		0.70±0.41				0.21±0.09
111	LYMPHEDEMA. HEREDITARY						0.31±0.31		0.47±0.33				0.13±0.07
112	CEREBELLAR ATAXIA						0.31±0.31		0.23±0.23				0.09±0.06
113	EPISODIC KINESIGENIC DYSKINESIA			0.58±0.58	0.56±0.56			0.22±0.22					0.13±0.07
114	OPHTHALMOPLEGIA. FAMILIAL STATIC							0.22±0.22	0.47±0.33				0.13±0.07
115	CHOREA. BENIGN HEREDITARY									0.76±0.53			0.09±0.06
116	MYOPATHY. SPHEROID BODY								0.23±0.23				0.04±0.04
117	ATAXIA. SPASTIC. 1											1.13±1.13	0.04±0.04
118	EMERY-DREIFUSS MUSCULAR DYSTROPHY	0.58±0.58					0.31±0.31						0.09±0.06
119	MYOPATHY. DISTAL 1						0.31±0.31						0.04±0.04
120	MOEBIUS SYNDROME			1.75±1.01									0.13±0.07
121	BEST VITELLIFORM MACULAR DYSTROPHY	0.58±0.58										1.13±1.13	0.09±0.06
122	OPTIC ATROPHY. CATARACT									0.76±0.53			0.09±0.06
123	LEBER CONGENITAL AMAUROSIS	1.73±0.99											0.13±0.07
124	COLOBOMA OF OPTIC NERVE			0.58±0.58									0.04±0.04
125	MICROPHTHALMIA. OPTIC ATROPHY							0.22±0.22					0.04±0.04
126	OPHTHALMOPLEGIA. FAMILIAL STATIC											1.13±1.13	0.04±0.04

(Continued)

TABLE 110-6 (CONTINUED)

No.	Diagnosis	Bashkirs of Bashkorto-stan Republic	Udmurts of Udmurt Republic	Maris of Mari El Republic	Chuvashes of Chuvash Republic	Adygeans of Adygea Republic	Russian of Rostov Region	Russian of Kostroma Region	Russian of Krasnodar Territory	Russian of Kirov Region	Russian of Tver Region	Russian of Bryansk Region	Average prevalence rate
127	RETINAL CONE-ROD DYSTROPHY	1.15±0.81											0.09±0.06
128	RETINOBLASTOMA	1.15±0.81	1.29±0.91	0.58±0.58	0.56±0.56								0.25±0.10
129	OPTIC ATROPHY 1 AND DEAFNESS	0.58±0.58											0.04±0.04
130	MICROPHTHALMIA. ISOLATED. WITH COLOBOM	1.15±0.81											0.09±0.06
131	KERATOCONUS	1.15±0.81											0.09±0.06
132	ACRODYSOSTOSIS						0.62±0.44		0.47±0.33				0.17±0.09
133	SPONDYLOEPIPHYSEAL DYSPLASIA CONGENITA		0.64±0.64										0.04±0.04
134	SPONDYLOEPIMETAPHYSEAL DYSPLASIA	1.73±0.99	0.64±0.64										0.17±0.09
135	BRACHYDACTYLY. TYPE C										1.32±1.32		0.04±0.04
136	BRACHYDACTYLY. TYPE A1	0.58±0.58			0.56±0.56		0.62±0.44						0.17±0.09
137	BRACHYDACTYLY. TYPE D						0.31±0.31						0.04±0.04
138	SYNDACTYLY. TYPE III		1.93±1.12										0.13±0.07
139	CHONDRODYSPLASIA PUNCTATA								0.23±0.23				0.04±0.04
140	SYNDACTYLY. TYPE IV							0.45±0.32					0.09±0.06
141	ICHTHYOSIS. CYCLIC. WITH EPIDERMOLYTIC HYPERKERATOSIS							0.22±0.22					0.04±0.04
142	PACHYDERMOPERIOSTOSIS. AUTOSOMAL DOMINANT	0.58±0.58											0.04±0.04
143	ARTHROGRYPOSIS. DISTAL	1.15±0.81											0.09±0.06
144	PSEUDOMONILETHRIX				1.68±0.97								0.13±0.07
145	EPIDERMOLYSIS BULLOSA HERPETIFORMIS. DOWLING-MEARA TYPE	2.30±1.15											0.17±0.09
146	EPIDERMOLYSIS BULLOSA DYSTROPHICA. PASINI TYPE							0.22±0.22					0.04±0.04
147	EPIDERMOLYSIS BULLOSA SIMPLEX. KOEBNER TYPE							0.22±0.22					0.04±0.04
148	EPIDERMOLYSIS BULLOSA SIMPLEX. WEBER-COCKAYNE TYPE	1.15±0.81						0.22±0.22	0.23±0.23				0.17±0.09

149	EPIDERMOLYTIC HYPERKERATOSIS						0.62±0.44						0.09±0.06
150	VATER ASSOCIATION						0.62±0.44	0.22±0.22	0.23±0.23				0.17±0.09
151	APERT SYNDROME							0.22±0.22					0.04±0.04
152	BLEPHARONASOFACIAL MALFORMATION SYNDROME	0.58±0.58			0.56±0.56								0.09±0.06
153	BECKWITH-WIEDEMANN SYNDROME						0.93±0.54			0.38±0.38			0.17±0.09
154	VAN DER WOUDE SYNDROME									1.13±0.65			0.13±0.07
155	VOLF PARKINSON WITTE SYNDROME					4.62±2.67							0.13±0.07
156	VON HIPPEL-LINDAU SYNDROME						0.31±0.31						0.04±0.04
157	BLEPHAROPHIMOSIS WITH OVARIAN FAILURE		2.57±1.29										0.17±0.09
158	LYMPHEDEMA-DISTICHIASIS SYNDROME			1.17±0.82									0.09±0.06
159	DEAFNESS. CURLY HAIR								0.47±0.33				0.09±0.06
160	KABUKI SYNDROME		0.64±0.64										0.04±0.04
161	COFFIN-SIRIS SYNDROME			0.58±0.58			0.62±0.44						0.13±0.07
162	LANGER-GIEDION SYNDROME						0.31±0.31			0.38±0.38			0.09±0.06
163	MARSHALL SYNDROME								0.47±0.33				0.09±0.06
164	BASAL CELL NEVUS SYNDROME		0.64±0.64										0.04±0.04
165	MARCUS GUNN PHENOMENON									0.38±0.38			0.04±0.04
166	OROFACIODIGITAL SYNDROME									0.76±0.53			0.09±0.06
167	OROFACIAL CLEFT SYNDROME	1.73±0.99											0.13±0.07
168	OTOFACIOCERVICAL SYNDROME	1.15±0.81											0.09±0.06
169	PROTEUS SYNDROME				0.56±0.56								0.04±0.04
170	PFEIFFER SYNDROME		0.64±0.64				0.62±0.44		0.23±0.23				0.17±0.09
171	MYOCLONUS AND ATAXIA (RAMSAY HUNT SYNDROME)								0.23±0.23				0.04±0.04
172	AXENFELD-RIEGER SYNDROME. TYPE 2	1.73±0.99										1.13±1.13	0.17±0.09
173	SOTOS SYNDROME						0.31±0.31		0.23±0.23				0.09±0.06
174	SAETHRE-CHOTZEN SYNDROME									0.38±0.38			0.04±0.04
175	GILLES DE LA TOURETTE SYNDROME						1.25±0.62						0.17±0.09
176	KLEEBLATTSCHAEDEL SYNDROME			0.58±0.58									0.04±0.04
177	PULMONARY FIBROSIS (HAMMAN-RICH DISEASE)	0.58±0.58											0.04±0.04
178	FREEMAN-SHELDON SYNDROME						0.31±0.31						0.04±0.04

(Continued)

TABLE 110-6 (CONTINUED)

No.	Diagnosis	Bashkirs of Bashkortostan Republic	Udmurts of Udmurt Republic	Maris of Mari El Republic	Chuvashes of Chuvash Republic	Adygeans of Adygea Republic	Russian of Rostov Region	Russian of Kostroma Region	Russian of Krasnodar Territory	Russian of Kirov Region	Russian of Tver Region	Russian of Bryansk Region	Average prevalence rate
179	CHONDRODYSPLASIA PUNCTATA. (Conradi's disease)						0.31±0.31			0.76±0.53			0.13±0.07
180	ALBINISM. OCULAR. WITH SENSORINEURAL DEAFNESS							0.45±0.32					0.09±0.06
181	PATELLA APLASIA-HYPOPLASIA						0.31±0.31						0.04±0.04
182	POPLITEAL PTERYGIUM SYNDROME						0.31±0.31						0.04±0.04
183	NEURAL TUBE DEFECTS	0.58±0.58					0.31±0.31						0.09±0.06
184	NEUROFIBROMATOSIS-NOONAN SYNDROME						0.31±0.31						0.04±0.04
185	FEMORAL-FACIAL SYNDROME			0.58±0.58								1.13±1.13	0.09±0.06
186	SPLIT-FOOT DEFORMITY WITH MANDIBULOFACIAL DYSOSTOSIS				0.56±0.56								0.04±0.04
187	CAMPTODACTYLY. TALL STATURE. AND HEARING LOSS SYNDROME				0.56±0.56								0.04±0.04
188	ACANTHOCYTOSIS			1.75±1.01									0.13±0.07
189	OCULODENTODIGITAL DYSPLASIA								0.23±0.23				0.04±0.04
190	ACROCEPHALOPOLYSYNDACTYLY TYPE III (SAKATI-NYHAN SYNDROME)								0.23±0.23				0.04±0.04
191	METATROPIC DYSPLASIA						0.31±0.31						0.04±0.04
192	HYPERTELORISM WITH ESOPHAGEAL ABNORMALITY AND HYPOSPADIAS (OPITZ-FRIAS SYNDROME)						0.31±0.31						0.04±0.04
193	ALBRIGHT HEREDITARY OSTEODYSTROPHY WITH MULTIPLE HORMONE RESISTANCE						1.25±0.62						0.17±0.09
194	HEMIFACIAL MICROSOMIA WITH RADIAL DEFECTS				0.56±0.56								0.04±0.04
195	DEAFNESS. CONGENITAL. WITH KERATOPACHYDERMIA AND CONSTRICTIONS OF FINGERS AND TOES				1.12±0.79								0.09±0.06
196	TREFOIL FACTOR			0.58±0.58									0.04±0.04
197	MENTAL RETARDATION. POSTAXIAL POLYDACTYLY	1.73±0.99											0.13±0.07
198	MICROCEPHALY-DEAFNESS SYNDROME	0.58±0.58											0.04±0.04
199	VON WILLEBRAND DISEASE					1.54±1.54					1.32±1.32		0.09±0.06

TABLE 110-7 SPECTRUM OF AR DISEASES AND THEIR PREVALENCE RATES ($X10^{-5}$) IN THE POPULATIONS STUDIED

No.	Diagnosis	Bashkirs of Bashkorto-stan Republic	Udmurts of Udmurt Republic	Maris of Mari El Republic	Chuvashes of Chuvash Republic	Adygeans of Adygea Republic	Russian of Rostov Region	Russian of Kostroma Region	Russian of Krasnodar Territory	Russian of Kirov Region	Russian of Tver Region	Russian of Bryansk Region	Average preva-lence rate
PREVALENCE RATE 1:50000 AND HIGHER													
1	MUSCULAR DYSTROPHY. LIMB-GIRDLE	1.15±0.81	5.79±1.93	2.33±1.17	1.68±0.97	4.62±2.67	1.87±0.76	2.92±0.81	2.58±0.78	4.91±1.36	1.32±1.32	2.27±1.60	2.79±0.34
2	MICROCEPHALY. PRIMARY AUTOSOMAL RECESSIVE	7.48±2.08	8.37±2.32	9.91±2.40	6.15±1.86		11.53±1.90	2.02±0.68	2.81±0.81	7.93±1.73	1.32±1.32	3.40±1.96	5.79±0.50
3	MENTAL RETARDATION	40.28±4.81	18.02±3.41	14.57±2.91	7.27±2.02	4.62±2.67	9.97±1.76	2.47±0.75	2.58±0.78	5.29±1.41	7.89±3.22	4.53±2.27	9.39±0.63
4	RETINITIS PIGMENTOSA	5.18±1.73	3.22±1.44	3.50±1.43	5.60±1.77		2.18±0.82	6.07±1.17	3.05±0.85	6.04±1.51	5.26±2.63		4.06±0.41
5	CONGENITAL HEREDITARY CATARACT	5.18±1.73	2.57±1.29	1.75±1.01	7.83±2.10		2.49±0.88	2.25±0.71	3.52±0.91	4.16±1.25	2.63±1.86	4.53±2.27	3.51±0.39
6	ICHTHYOSIFORM ERYTHRODERMA. CONGENITAL		0.64±0.64	4.66±1.65	3.36±1.37	3.08±2.18	1.56±0.70	3.60±0.90	2.81±0.81	3.78±1.20	2.63±1.86	3.40±1.96	2.75±0.34
7	GROWTH HORMONE DEFICIENCY	5.18±1.73	2.57±1.29	2.33±1.17	5.04±1.68		1.56±0.70	1.12±0.50	1.17±0.52	1.89±0.85		7.94±3.00	2.24±0.31
8	DEAFNESS. AUTOSOMAL RECESSIVE	33.95±4.42	32.18±4.55	12.82±2.73	25.74±3.79	43.08±8.14	37.90±3.30	2.70±0.78	7.73±1.35	8.31±1.79	13.16±4.2	7.94±3.00	16.95±0.9
PREVALENCE RATE 1:50001-1:100000													
9	SPASTIC PARAPLEGIA. AUTOSOMAL RECESSIVE		1.29±0.91	1.17±0.82	1.12±0.79		0.93±0.54	0.90±0.45	0.94±0.47	2.27±0.93	1.32±1.32		1.01±0.21
10	SPINAL MUSCULAR ATROPHY. TYPE I-III	1.73±0.99	5.15±1.82	1.75±1.01	1.68±0.97	1.54±1.54	0.31±0.31	1.12±0.50	2.58±0.78	2.64±1.00		2.27±1.60	1.82±0.28
11	MICROCEPHALY WITH SPASTIC QUADRIPLEGIA						4.67±1.21	0.67±0.39	0.94±0.47	2.27±0.93		3.40±1.96	1.31±0.24
12	ALBINISM. OCULOCUTANEOUS	1.15±0.81	5.79±1.93	4.08±1.54	1.68±0.97		1.25±0.62	0.45±0.32	0.23±0.23	1.89±0.85			1.40±0.24
13	USHER SYNDROME	0.58±0.58	1.29±0.91	1.75±1.01	1.12±0.79	1.54±1.54	2.49±0.88	1.80±0.64	1.64±0.62	1.51±0.76	1.32±1.32		1.56±0.26
14	THYROID HORMONOGENESIS. GENETIC DEFECT	7.48±2.08	1.93±1.12	2.33±1.17	4.48±1.58	7.69±3.44	1.25±0.62		1.41±0.57			3.40±1.96	1.94±0.29
15	PHENYLKETONURIA	1.15±0.81	1.29±0.91		0.56±0.56		2.80±0.95	1.35±0.55	2.34±0.74	1.51±0.76	3.95±2.30	4.53±2.27	1.73±0.27
16	HYPOTRICHOSIS. TOTAL. MARI TYPE			6.99±2.02	11.75								1.40±0.24
PREVALENCE RATE 1:100001-1:200000													
17	FRIEDREICH ATAXIA		1.29±0.91		0.56±0.56	3.08±2.18		1.57±0.60		3.02±1.07	1.32±1.32		0.89±0.19
18	METACHROMATIC LEUKODYSTROPHY	1.15±0.81			0.56±0.56	3.08±2.18	1.25±0.62	0.45±0.32		0.38±0.38	1.32±1.32		0.55±0.15
19	LEBER CONGENITAL AMAUROSIS	2.30±1.15	3.86±1.58	1.17±0.82		1.54±1.54	0.93±0.54	0.45±0.32	0.47±0.33	0.38±0.38			0.89±0.19
20	RETINOPATHY. PERICENTRAL PIGMENTARY. AUTOSOMAL RECESSIVE						0.31±0.31	2.02±0.68	1.41±0.57	0.76±0.53			0.76±0.18
21	GLAUCOMA. PRIMARY CONGENITAL	0.58±0.58	0.64±0.64	1.17±0.82	0.56±0.56		1.56±0.70			1.89±0.85		1.13±1.13	0.68±0.17
22	MICROPHTHALMIA. ISOLATED	1.15±0.81	0.64±0.64	0.58±0.58	2.24±1.12	1.54±1.54		0.22±0.22		0.76±0.53			0.51±0.15
23	SPONDYLOCOSTAL DYSOSTOSIS. AUTOSOMAL RECESSIVE	1.73±0.99	1.29±0.91	1.17±0.82	1.12±0.79			0.45±0.32		0.38±0.38			0.51±0.15

(Continued)

TABLE 110-7 (CONTINUED)

No.	Diagnosis	Bashkirs of Bashkortostan Republic	Udmurts of Udmurt Republic	Maris of Mari El Republic	Chuvashes of Chuvash Republic	Adygeans of Adygea Republic	Russian of Rostov Region	Russian of Kostroma Region	Russian of Krasnodar Territory	Russian of Kirov Region	Russian of Tver Region	Russian of Bryansk Region	Average prevalence rate
24	ALBINISM				1.68±0.97	9.23±7.37		0.67±0.39		1.89±0.85	1.32±1.32	4.53±2.27	0.93±0.19
25	EHLERS-DANLOS SYNDROME	2.88±1.29	0.64±0.64		1.12±0.79		1.56±0.70		0.23±0.23	1.51±0.76			0.76±0.18
26	CONSTRICTING BANDS. CONGENITAL ADAM COMPLEX. INCLUDED	1.73±0.99	1.29±0.91	0.58±0.58	1.12±0.79	1.54±1.54	1.56±0.70		0.70±0.41	0.38±0.38			0.76±0.18
27	MICROTIA WITH MEATAL ATRESIA AND CONDUCTIVE DEAFNESS	1.15±0.81	1.93±1.12		1.12±0.79		2.49±0.88	0.22±0.22					0.68±0.17
PREVALENCE RATE 1:200001–1:500000													
28	SPINOCEREBELLAR ATAXIA. AUTOSOMAL RECESSIVE			0.58±0.58	0.56±0.56	1.54±1.54	0.31±0.31	0.45±0.32	0.23±0.23		2.63±1.86		0.38±0.13
29	HYDROCEPHALUS			0.58±0.58		1.54±1.54	0.62±0.44			2.27±0.93			0.42±0.13
30	OPTIC ATROPHY		1.29±0.91		1.12±0.79				0.94±0.47				0.34±0.12
31	NYSTAGMUS. CONGENITAL MOTOR. AUTOSOMAL RECESSIVE			2.33±1.17	2.80±1.25								0.38±0.13
32	SPONDYLOEPIPHYSEAL DYSPLASIA TARDA. AUTOSOMAL RECESSIVE	0.58±0.58	0.64±0.64		0.56±0.56		1.56±0.70		0.23±0.23	0.38±0.38			0.42±0.13
33	ECTODERMAL DYSPLASIA. HYPOHIDROTIC. AUTOSOMAL RECESSIVE			1.75±1.01	0.56±0.56			0.45±0.32		0.76±0.53			0.34±0.12
34	DUBOWITZ SYNDROME		1.29±0.91	1.17±0.82	0.56±0.56		1.25±0.62	0.22±0.22					0.42±0.13
35	MUCOPOLYSACCHARIDOSIS			0.58±0.58		1.54±1.54	0.31±0.31	0.67±0.39	0.23±0.23	0.76±0.53			0.38±0.13
36	HALLERMANN-STREIFF SYNDROME	0.58±0.58			2.24±1.12	1.54±1.54	0.31±0.31	0.67±0.39	0.23±0.23	0.76±0.53			0.55±0.15
37	HOMOCYSTINURIA	0.58±0.58		0.58±0.58	1.12±0.79		0.62±0.44	0.22±0.22			2.63±1.86		0.38±0.13
38	OLIVOPONTOCEREBELLAR ATROPHY II. AUTOSOMAL RECESSIVE						0.62±0.44			1.51±0.76			0.25±0.10
39	CHARCOT-MARIE-TOOTH DISEASE							0.67±0.39		1.51±0.76			0.30±0.11
40	COCKAYNE SYNDROME	1.73±0.99		0.58±0.58					0.47±0.33				0.25±0.10
41	MICROPHTHALMIA. ISOLATED. WITH COLOBOMA		1.29±0.91	0.58±0.58			0.31±0.31		0.47±0.33				0.25±0.10
42	ICHTHYOSIS. LAMELLAR	1.15±0.81		0.58±0.58	1.68±0.97						1.32±1.32		0.30±0.11
43	LARSEN SYNDROME			1.17±0.82	1.68±0.97		0.31±0.31			0.38±0.38			0.30±0.11
44	HALLERVORDEN-SPATZ DISEASE			1.17±0.82		3.08±2.18		0.22±0.22					0.21±0.09
45	COHEN SYNDROME						0.93±0.54	0.22±0.22	0.23±0.23				0.21±0.09
46	LAURENCE-MOON SYNDROME		0.64±0.64				0.62±0.44				1.32±1.32	1.13±1.13	0.21±0.09
47	SECKEL SYNDROME		0.64±0.64		0.56±0.56		0.93±0.54						0.21±0.09
48	OSTEOGENESIS IMPERFECTA						0.31±0.31	0.45±0.32	0.47±0.33				0.21±0.09

PREVALENCE RATE 1:500001 AND LESS OFTEN													
49	SPASTIC ATAXIA. PARAPLEGIA						0.31±0.31						0.04±0.04
50	SPINAL MUSCULAR ATROPHY. TYPE IV		0.64±0.64										0.04±0.04
51	SPINAL MUSCULAR ATROPHY. SCAPULOPERONEAL		1.29±0.91										0.09±0.06
52	MYOCLONIC EPILEPSY OF UNVERRICHT AND LUNDBORG					4.62±2.67				0.38±0.38			0.17±0.09
53	LIPOFUSCIN STORAGE DISEASE	0.58±0.58					0.31±0.31						0.09±0.06
54	MYOPATHY. CONGENITAL. WITH FIBER-TYPE DISPROPORTION			0.58±0.58									0.04±0.04
55	MUSCULAR DYSTROPHY. DISTAL. LATE-ONSET. AUTOSOMAL RECESSIVE (MIYOSHI MYOPATHY)			1.17±0.82									0.09±0.06
56	SPINOCEREBELLAR DEGENERATION WITH SLOWER PROGRESS							0.45±0.32					0.09±0.06
57	CHONDRODYSTROPHIC MYOTONIA			0.58±0.58				0.22±0.22					0.09±0.06
58	CEREBELLAR HYPOPLASIA					1.54±1.54		0.22±0.22	0.23±0.23				0.13±0.07
59	ATAXIC DIPLEGIA WITH DEFECTIVE CELLULAR IMMUNITY							0.22±0.22					0.04±0.04
60	CEREBELLAR HYPOPLASIA AND MENTAL RETARDATION WITH OR WITHOUT QUADRUPEDAL LOCOMOTION								0.23±0.23				0.04±0.04
61	PARKINSON DISEASE. AUTOSOMAL RECESSIVE JUVENILE				1.12±0.79								0.09±0.06
62	MICROPHTHALMIA. SYNDROMIC									0.38±0.38			0.04±0.04
63	MICROPHTHALMIA. ISOLATED. WITH CATARACT	0.58±0.58			0.56±0.56								0.09±0.06
64	STARGARDT DISEASE 1	0.58±0.58											0.04±0.04
65	PTOSIS. HEREDITARY CONGENITAL				1.12±0.79								0.09±0.06
66	RETINAL DEGENERATION WITH NANOPHTHALMOS. CYSTIC MACULAR DEGENERATION						0.31±0.31		0.23±0.23				0.09±0.06
67	CONE-ROD DYSTROPHY						0.31±0.31		0.23±0.23		1.32±1.32		0.13±0.07
68	MACULAR DEGENERATION							0.22±0.22					0.04±0.04
69	EPIDERMOLYSIS BULLOSA DYSTROPHICA. HALLOPEAU-SIEMENS TYPE									1.13±0.65			0.13±0.07
70	APLASIA CUTIS CONGENITA OF LIMBS. RECESSIVE			0.58±0.58									0.04±0.04

(Continued)

TABLE 110-7 (CONTINUED)

No.	Diagnosis	Bashkirs of Bashkorto-stan Republic	Udmurts of Udmurt Republic	Maris of Mari El Republic	Chuvashes of Chuvash Republic	Adygeans of Adygea Republic	Russian of Rostov Region	Russian of Kostroma Region	Russian of Krasnodar Territory	Russian of Kirov Region	Russian of Tver Region	Russian of Bryansk Region	Average preva-lence rate
71	OSTEOCHONDROSIS DEFORMANS TIBIAE. FAMILIAL INFANTILE TYPE (BLOUNT DISEASE)						0.31±0.31						0.04±0.04
72	TUMORAL CALCINOSIS. HYPERPHOSPHATEMIC										2.63±1.86		0.09±0.06
73	ECTRODACTYLY			1.17±0.82									0.09±0.06
74	ANONYCHIA CONGENITA							0.22±0.22	0.23±0.23				0.09±0.06
75	HURLER SYNDROME			0.58±0.58									0.04±0.04
76	PSEUDORHEUMATOID DYSPLASIA. PROGRESSIVE. WITH HYPOPLASTIC TOES (CZECH DYSPLASIA. METATARSAL TYPE)								0.23±0.23				0.04±0.04
77	SPONDYLOEPIPHYSEAL DYSPLASIA TARDA WITH MENTAL RETARDATION						1.25±0.62						0.17±0.09
78	SPONDYLOEPIPHYSEAL DYSPLASIA TARDA						0.93±0.54		0.23±0.23				0.17±0.09
79	SPONDYLOEPIPHYSEAL DYSPLASIA TARDA WITH CHARACTERISTIC FACIES						0.31±0.31						0.04±0.04
80	LANGER MESOMELIC DYSPLASIA						0.93±0.54						0.13±0.07
81	DIASTROPHIC DYSPLASIA						0.31±0.31						0.04±0.04
82	TEL HASHOMER CAMPTODACTYLY SYNDROME						0.31±0.31						0.04±0.04
83	SPONDYLOENCHONDRODYSPLASIA							0.45±0.32			1.32±1.32		0.13±0.07
84	POLYDACTYLY. POSTAXIAL	1.73±0.99											0.13±0.07
85	CHONDRODYSPLASIA	0.58±0.58											0.04±0.04
86	EPIPHYSEAL DYSPLASIA. MULTIPLE							0.45±0.32					0.09±0.06
87	METAPHYSEAL DYSPLASIA (PYLE DISEASE)						0.31±0.31						0.04±0.04
88	METAPHYSEAL DYSPLASIA		0.64±0.64										0.04±0.04
89	METAPHYSEAL CHONDRODYSPLASIA				0.56±0.56		0.93±0.54						0.17±0.09
90	SPONDYLOMETAPHYSEAL DYSPLASIA	0.58±0.58	1.29±0.91										0.13±0.07
91	ECTROMELIA								0.23±0.23				0.04±0.04
92	ACROCALLOSAL SYNDROME				0.56±0.56								0.04±0.04
93	JOB SYNDROME										1.32±1.32		0.04±0.04
94	FEMUR-FIBULA-ULNA SYNDROME						0.31±0.31						0.04±0.04
95	ALSTROM SYNDROME			0.58±0.58									0.04±0.04

96	BARDET-BIEDL SYNDROME						0.31±0.31						0.04±0.04
97	WEILL-MARCHESANI SYNDROME			0.58±0.58					0.23±0.23				0.09±0.06
98	FACIOTHORACOSKELETAL SYNDROME (CAMPTODACTYLY SYNDROME. GUADALAJARA)				0.56±0.56								0.04±0.04
99	DYGGVE-MELCHIOR-CLAUSEN DISEASE							0.22±0.22	0.23±0.23	0.38±0.38			0.13±0.07
100	OCULODENTODIGITAL DYSPLASIA. AUTOSOMAL RECESSIVE			0.58±0.58									0.04±0.04
101	GORLIN-CHAUDHRY-MOSS SYNDROME								0.47±0.33				0.09±0.06
102	GAUCHER DISEASE								0.23±0.23				0.04±0.04
103	RHIZOMELIC CHONDRODYSPLASIA PUNCTATA (Conradi's disease)					1.54±1.54							0.04±0.04
104	CARPENTER SYNDROME								0.23±0.23				0.04±0.04
105	KARTAGENER SYNDROME			1.17±0.82	0.56±0.56								0.13±0.07
106	KLIPPEL-FEIL SYNDROME. AUTOSOMAL RECESSIVE	1.15±0.81											0.09±0.06
107	LARON SYNDROME						0.31±0.31						0.04±0.04
108	MARINESCO-SJOGREN SYNDROME					1.54±1.54							0.04±0.04
109	MARDEN-WALKER SYNDROME						0.31±0.31						0.04±0.04
110	POSTAXIAL POLYDACTYLY AND MENTAL RETARDATION (OLIVER SYNDROME)	0.58±0.58											0.04±0.04
111	OPITZ TRIGONOCEPHALY SYNDROME			0.58±0.58			0.62±0.44						0.13±0.07
112	PURETIC SYNDROME			0.58±0.58									0.04±0.04
113	ROBERTS SYNDROME		1.29±0.91							0.38±0.38			0.13±0.07
114	ROBINOW SYNDROME	1.15±0.81		1.17±0.82									0.17±0.09
115	PLASMA THROMBOPLASTIN ANTECEDENT DEFICIENCY (ROSENTHAL SYNDROME)								0.23±0.23				0.04±0.04
116	MAYER-ROKITANSKY-KUSTER-HAUSER SYNDROME				0.56±0.56								0.04±0.04
117	SMITH-LEMLI-OPITZ SYNDROME						0.31±0.31		0.23±0.23				0.09±0.06
118	TORIELLO-CAREY SYNDROME				0.56±0.56								0.04±0.04
119	FRASER SYNDROME				0.56±0.56								0.04±0.04
120	OROCRANIODIGITAL SYNDROME (JUBERG-HAYWARD SYNDROME)						0.31±0.31						0.04±0.04
121	ELLIS-VAN CREVELD SYNDROME						0.31±0.31		0.23±0.23				0.09±0.06
122	ADRENAL HYPERPLASIA. CONGENITAL					1.54±1.54					1.32±1.32		0.09±0.06
123	BRACHIOSKELETOGENITAL SYNDROME						0.31±0.31						0.04±0.04

(Continued)

TABLE 110-7 (CONTINUED)

No.	Diagnosis	Bashkirs of Bashkorto-stan Republic	Udmurts of Udmurt Republic	Maris of Mari El Republic	Chuvashes of Chuvash Republic	Adygeans of Adygea Republic	Russian of Rostov Region	Russian of Kostroma Region	Russian of Krasnodar Territory	Russian of Kirov Region	Russian of Tver Region	Russian of Bryansk Region	Average preva-lence rate
124	DIAMOND-BLACKFAN ANEMIA						0.31±0.31						0.04±0.04
125	CEREBROOCULAR DYSPLASIA-MUSCULAR DYSTROPHY SYNDROME						0.62±0.44						0.09±0.06
126	KAUFMAN OCULOCEREBROFACIAL SYNDROME	0.58±0.58					0.31±0.31						0.09±0.06
127	HEMIHYPERPLASIA. ISOLATED								0.23±0.23				0.04±0.04
128	ORAL-FACIAL-DIGITAL SYNDROME. TYPE IV						0.31±0.31		0.23±0.23				0.09±0.06
129	DWARFISM. LOW-BIRTH-WEIGHT TYPE. WITH UNRESPONSIVENESS TO GROWTH HORMONE						0.31±0.31						0.04±0.04
130	CARDIOMYOPATHY. CONGESTIVE. WITH HYPERGONADOTROPIC HYPOGONADISM						0.31±0.31						0.04±0.04
131	ALOPECIA-CONTRACTURES-DWARFISM MENTAL RETARDATION SYNDROME									1.13±0.65			0.13±0.07
132	MICROTIA. HEARING IMPAIRMENT. AND CLEFT PALATE								0.47±0.33				0.09±0.06
133	OPTIC ATROPHY AND DEAFNESS WITH MENTAL RETARDARION (WOLFRAM SYNDROME)								0.70±0.41				0.13±0.07
134	ACROMELIC FRONTONASAL DYSOSTOSIS							0.22±0.22					0.04±0.04
135	MENTAL RETARDATION. HYPOGONADISM				1.12±0.79								0.09±0.06
136	MICROPHTHALMIA WITH LIMB ANOMALIES								0.23±0.23				0.04±0.04
137	MULTICORE MYOPATHY WITH MENTAL RETARDATION. SHORT STATURE. AND HYPOGONADISM							0.45±0.32					0.09±0.06
138	MENTAL RETARDATION WITH OPTIC ATROPHY. FACIAL DYSMORPHISM				1.12±0.79								0.09±0.06
139	ICHTHYOSIS. MENTAL RETARDATION. DWARFISM. AND RENAL IMPAIRMENT				0.56±0.56								0.04±0.04
140	MICROTIA. HEARING IMPAIRMENT. AND CLEFT PALATE										1.32±1.32		0.04±0.04
141	MENTAL RETARDATION. FACIAL DYSMORPHISM										1.32±1.32		0.04±0.04
142	MENTAL RETARDATION. WHITE HAIR. HYPOGONADISM										1.32±1.32		0.04±0.04
143	RIGID SPINE MUSCULAR DYSTROPHY		0.64±0.64										0.04±0.04
144	CEREBELLAR ATAXIA. INFANTILE. WITH PROGRESSIVE EXTERNAL OPHTHALMOPLEGIA								0.23±0.23				0.04±0.04
145	DEAFNESS. ONYCHODYSTROPHY		0.64±0.64										0.04±0.04

146	MICROCEPHALY WITH MENTAL RETARDATION AND DIGITAL ANOMALIES		1.29±0.91										0.09±0.06
147	DEAFNESS. COCHLEAR. WITH MYOPIA AND INTELLECTUAL IMPAIRMENT		0.64±0.64										0.04±0.04
148	MENTAL RETARDATION. SHORT STATURE. BRACHYDACTYLY		0.64±0.64										0.04±0.04
149	DEAFNESS. SENSORINEURAL. WITH PITUITARY DWARFISM		0.64±0.64										0.04±0.04
150	HIRSCHSPRUNG DISEASE-MENTAL RETARDATION SYNDROME (MOWAT-WILSON SYNDROME)				1.12±0.79								0.09±0.06
151	SANDHOFF DISEASE					1.54±1.54							0.04±0.04
152	THALASSEMIAS							0.22±0.22					0.04±0.04
153	HYPOMELANOSIS				0.56±0.56								0.04±0.04
154	HISTIDINEMIA				0.56±0.56			0.45±0.32					0.13±0.07
155	IMINOGLYCINURIA							0.22±0.22					0.04±0.04
156	MUCOLIPIDOSIS									0.76±0.53			0.09±0.06
157	GLYCOGEN STORAGE DISEASE V (MCARDLE DISEASE)	0.58±0.58								0.38±0.38			0.09±0.06
158	CYSTIC FIBROSIS	0.58±0.58			0.56±0.56		0.31±0.31						0.13±0.07
159	OSTEOPETROSIS. AUTOSOMAL RECESSIVE				1.12±0.79								0.09±0.06
160	GANGLIOSIDOSIS		0.64±0.64										0.04±0.04
161	ACRODERMATITIS ENTEROPATHICA		0.64±0.64				0.31±0.31						0.09±0.06
162	MASTOCYTOSIS. CUTANEOUS. WITH SHORT STATURE. CONDUCTIVE HEARING LOSS AND MICROTIA		0.64±0.64	0.58±0.58	0.56±0.56								0.13±0.07
163	CHORIORETINAL ATROPHY. PROGRESSIVE BIFOCAL				0.56±0.56								0.04±0.04
164	MYOPATHY. CONGENITAL				1.12±0.79					0.38±0.38	1.32±1.32		0.17±0.09
165	MICROCEPHALIC PRIMORDIAL DWARFISM. TORIELLO TYPE	1.15±0.81										2.27±1.60	0.17±0.09
166	CLEFT LIP/PALATE WITH ABNORMAL THUMBS AND MICROCEPHALY	1.15±0.81											0.09±0.06
167	MICROCEPHALIC OSTEODYSPLASTIC PRIMORDIAL DWARFISM. TYPE II	0.58±0.58											0.04±0.04
168	GROWTH RETARDATION. DEAFNESS. FEMORAL EPIPHYSEAL DYSPLASIA. AND LACRIMAL DUCT OBSTRUCTION	0.58±0.58											0.04±0.04

TABLE 110-8 SPECTRUM OF X-LINKED DISEASES AND THEIR PREVALENCE RATES ($X10^{-5}$) IN THE POPULATIONS STUDIED

No.	Diagnosis	Bashkirs of Bashkorto-stan Republic	Udmurts of Udmurt Republic	Maris of Mari El Republic	Chuvashes of Chuvash Republic	Adygeans of Adygea Republic	Russian of Rostov Region	Russian of Kostroma Region	Russian of Krasnodar Territory	Russian of Kirov Region	Russian of Tver Region	Russian of Bryansk Region	Average preva-lence rate
PREVALENCE RATE 1:50000 AND AND HIGHER													
1	MUSCULAR DYSTROPHY. DUCHENNE TYPE		9.01±3.41	5.83±2.61			4.99±1.76	1.35±0.78	2.81±1.15	3.02±1.51	5.26±3.72		2.96±0.50
2	MENTAL RETARDATION. X-LINKED	19.57±4.75	18.03±4.82	9.33±3.30	22.38±5.00		9.35±2.41	2.70±1.10	0.94±0.66		21.05±7.44		7.61±0.80
3	NYSTAGMUS. CONGENITAL. X-LINKED		19.31±4.99	5.83±2.61	3.36±1.94	9.23±5.33	0.62±0.62		1.41±0.81	5.29±2.00			3.13±0.51
4	ICHTHYOSIS. X-LINKED	12.66±3.82	2.58±1.82	1.17±1.17	4.48±2.24	6.15±4.35	8.72±2.33	10.35±3.16	6.09±1.69	6.04±2.14			6.60±0.75
5	HEMOPHILIA A	3.45±1.99	10.30	3.50±2.02	6.71±2.74	12.31±6.15	0.62±0.62	4.50±1.42	3.75±1.33	7.56±2.39	2.63±2.63	2.27±2.27	4.65±0.63
PREVALENCE RATE 1:50000-1:100000													
6	MUSCULAR DYSTROPHY. BECKER TYPE		1.26±1.29				3.12±1.65	0.45±0.45	1.41±0.81		2.63±2.63	2.27±2.27	1.01±0.29
PREVALENCE RATE 1:100001-1:200000													
7	CHARCOT-MARIE-TOOTH DISEASE. X-LINKED	1.15±1.15				3.08±308			2.34±1.09				0.59±0.22
8	FACIOGENITAL DYSPLASIA	2.30±1.63		2.33±1.65	1.12±1.12	3.08±308	1.87±1.08						0.76±0.25
9	SPASTIC PARAPLEGIA. X-LINKED	4.60±2.30								2.27±1.31			0.59±0.22
10	HYDROCEPHALUS DUE TO CONGENITAL			2.33±1.65			0.62±0.62		1.87±0.94	1.51±1.07		2.27±2.27	0.85±0.27
11	RETINITIS PIGMENTOSA	1.15±1.15							1.87±0.94	1.51±1.07			0.59±0.22
12	CHOROIDEREMIA	3.45±1.99			3.36±1.94					0.76±0.76			0.59±0.22
13	OTOPALATODIGITAL SYNDROME			1.17±1.17					0.94±0.66	2.27±1.31			0.51±0.21
14	COFFIN-LOWRY SYNDROME						1.87±1.08		1.41±0.81				0.51±0.21
PREVALENCE RATE 1:200001-1:500000													
15	ALPORT SYNDROME. X-LINKED	4.60±2.30											0.34±0.17
16	HYPOPHOSPHATEMIC RICKETS	3.45±1.99							0.94±0.66				0.42±0.19
17	MUSCULAR DYSTROPHY. DUCHENNE TYPE WITH ICHTHYOSIS							1.35±0.78					0.25±0.15
18	MACULAR DYSTROPHY. X-LINKED								1.41±0.81				0.25±0.15
19	ECTODERMAL DYSPLASIA. HYPOHIDROTIC. X-LINKED (CHRIST-SIEMENS-TOURAINE SYNDROME)						1.25±0.88		0.47±0.47				0.25±0.15
20	RHIZOMELIC DYSPLASIA. SCOLIOSIS. AND RETINITIS PIGMENTOSA									2.27±1.31			0.25±0.15
21	RETT SYNDROME		1.26±1.29				0.62±0.62		0.47±0.47				0.25±0.15
22	MENTAL RETARDATION WITH PRIMORDIAL DWARFISM (RENPENNING SYNDROME)	1.15±1.15			2.24±1.58								0.25±0.15

23	HEMOPHILIA B	1.15±1.15						0.45±0.45	0.47±0.47				0.25±0.15
24	DEAFNESS. X-LINKED									2.27±1.31			0.25±0.15
PREVALENCE RATE 1:500001 AND LESS OFTEN													
25	EMERY-DREIFUSS MUSCULAR DYSTROPHY. X-LINKED								0.47±0.47				0.08±0.08
26	MYOPATHY. CONGENITAL			1.17±1.17									0.08±0.08
27	SPINOCEREBELLAR ATAXIA. X-LINKED								0.47±0.47				0.08±0.08
28	DYSTONIA. TORSION. X-LINKED								0.94±0.66				0.17±0.12
29	SPINOCEREBELLAR ATAXIA. X-LINKED 5		2.58±1.82										0.17±0.12
30	MICROPHTHALMIA. ISOLATED			1.17±1.17									0.08±0.08
31	ALBINISM. OCULAR		1.26±1.29						0.47±0.47				0.17±0.12
32	MICROPHTHALMIA. ISOLATED. WITH CATARACT								0.47±0.47				0.08±0.08
33	OPTIC ATROPHY					6.15±4.35							0.17±0.12
34	SPLIT-HAND/FOOT MALFORMATION				1.12±1.12								0.08±0.08
35	DANDY-WALKER MALFORMATION						0.62±0.62						0.08±0.08
36	SPINAL AND BULBAR MUSCULAR ATROPHY	1.15±1.15					0.62±0.62						0.17±0.12
37	PAINE SYNDROME									0.76±0.76			0.08±0.08
38	CORPUS CALLOSUM. AGENESIS OF. WITH ABNORMAL GENITALIA (PROUD SYNDROME)						1.25±0.88						0.17±0.12
39	MENTAL RETARDATION. X-LINKED. WITH GROWTH RETARDATION. DEAFNESS. AND MICROGENITALISM									1.51±1.07			0.17±0.12
40	HEMIHYPERPLASIA. ISOLATED							0.90±0.64					0.17±0.12
41	MENTAL RETARDATION. X-LINKED. WITH ICHTHYOSIS							0.90±0.64					0.17±0.12
42	MENTAL RETARDATION. X-LINKED. WITH MARFANOID HABITUS (LUJAN-FRYNS SYNDROME)						0.62±0.62						0.08±0.08
43	MUCOPOLYSACCHARIDOSIS TYPE II						0.62±0.62						0.08±0.08
44	MENTAL RETARDATION. X-LINKED. WITH GROWTH RETARDATION. DEAFNESS. AND MICROGENITALISM											2.27±2.27	0.08±0.08
45	POLYDACTYLY NEPHROTIC SYNDROME	2.30±1.63											0.17±0.12
46	TRIGONOCEPHALY WITH SHORT STATURE AND DEVELOPMENTAL DELAY (SAY AND MEYER SYNDROM)	1.15±1.15											0.08±0.08

TABLE 110-9 PREVALENCE RATES OF SOME AD, AR AND X-LINKED DISEASES IN THE 11 POPULATIONS STUDIED OF RUSSIA

No	Prevalence rate	Number (%) of affected				Number (%) of diseases			
		AD	AR	X-link.	Σ	AD	AR	X-link.	Σ
1	1:50000 and higher	2005 (57.65%)	1123 (57.03%)	295 (68.45%)	3423 (58.23%)	16 (8.04%)	8 (4.76%)	5 (10.87%)	29 (7.02%)
2	1:50001-1:100000	387 (11.13%)	288 (14.63%)	12 (2.78%)	687 (11.69%)	12 (6.03%)	8 (4.76%)	1 (1.17%)	21 (5.08%)
3	1:100001-1:200000	557 (16.01%)	187 (9.50%)	59 (13.69%)	803 (13.66%)	34 (17.09%)	11 (6.55%)	8 (17.39%)	53 (12.83%)
4	1:200001-1:500000	366 (10.52%)	156 (14.03%)	33 (7.66%)	555 (10.33%)	49 (24.62%)	21 (39.71%)	10 (21.74%)	80 (27.12%)
5	1:500001 and less often	163 (4.69%)	215 (10.92%)	32 (7.42%)	410 (6.98%)	88 (44.22%)	120 (71.43%)	22 (47.83%)	230 (55.69%)
		3478	1969	431	5878	199	168	46	413

and X-linked hereditary diseases, revealed in our investigations with the spectra of frequently encountered hereditary diseases in Europe (Carter, 1977) and British Columbia (Baird et al., 1988).

Many families with various hereditary diseases, in the process of medical genetic epidemiological studies, were subjected to DNA diagnosis (mucoviscidosis, phenylketonuria, Duchenne/Becker muscular dystrophy, hereditary isolated sensorineural hearing impairment, achondroplasia, brachydactyly type B, and so on). Resulting from the analysis performed, we supposed that there are peculiarities of ethnic distribution of genetic variants of separate hereditary diseases. For example, the Udmurts, Bashkirs, and Chuvashs affected by hereditary isolated sensorineural hearing impairment (HISHI) were found to have low rates of the 35delG mutation in the *GJB2* gene as compared with those in the populations of the Russians. In order check this finding, we analyzed the frequency of the 35delG mutation of the *GJB2* gene in four ethnic groups (Russians, Chuvashs, Bashkirs, and Udmurts). In each ethnic group, we studied DNA of apparently healthy individuals for the heterozygous carrier state of the respective mutation (not less than 800 chromosomes in each ethnic group). The incidence rate of the 35delG mutations in the *GJB2* gene was as follows: in Russians – 1.44%, in Chuvashs – 0.48%, in Udmurts – 0.25%, and in Bashkirs – 0.25%. Similar results were obtained on the same sample of blood specimens from apparently healthy individuals in relation to the F508del mutation in the *CFTR* gene (mucoviscidosis): in Russians, the incidence of this mutation amounted to 0.53%; in the Udmurts, to 0.11%; in the Chuvahs, 0.19%; with no mutation in the Maris and Bashkirs revealed. As for the R408W mutation in the *PAH* gene (phenylketonuria), the following findings were obtained: in the sample of Russians, the incidence rate of this mutation amounted to 1.06%, in the sample of Udmurts, 0.12%; in the sample of Chuvashs, 0.17%; in the sample of Bashkirs, 0.73%; and in the sample of Maris, 0.

CLUSTER ANALYSIS

In order to determine the geographic location of various populations and ethnic groups in a variety of the studied populations of Russia by the prevalence rate (10^{-3}) of AD and AR diseases, we performed the cluster analysis with the use of the average linkage method. The analysis comprised 199 diseases with the autosomal dominant pattern of inheritance, and 168 autosomal recessive diseases registered in all the populations studied, for whom we had preliminarily calculated the matrices of genetic distances (not shown). The respective dendrograms are presented in Figure 110-4 (for autosomal dominant diseases) and Figure 110-5 (for autosomal recessive diseases).

Analyzing the dendrograms, one cannot but notice fundamental similarity in the formation of the clusters in the two figures. First, all Russian populations are united into a single cluster, followed by the addition of the cluster of the Finno-Ugric population and the Chuvasian population, with the contribution of the Finno-Ugric component being quite considerable. The greatest genetic remoteness by genes of hereditary diseases was detected for the Bashkirs and the Adygeans. The coefficient of the linear correlation between the obtained correlation matrices sized 11 × 11 amounted to 0.73, which confirms similarity in the behavior of AD and AR diseases. It should be mentioned that the obtained pattern of genetic interrelationships between the studied populations/ethnic groups is largely similar to that obtained while using conditionally neutral polymorphic genetic systems.

ANALYSIS OF UNIFORMITY OF TERRITORIAL PREVALENCE OF HEREDITARY DISEASES

We performed the analysis of the whole nosological spectrum of hereditary diseases for uniformity of distribution

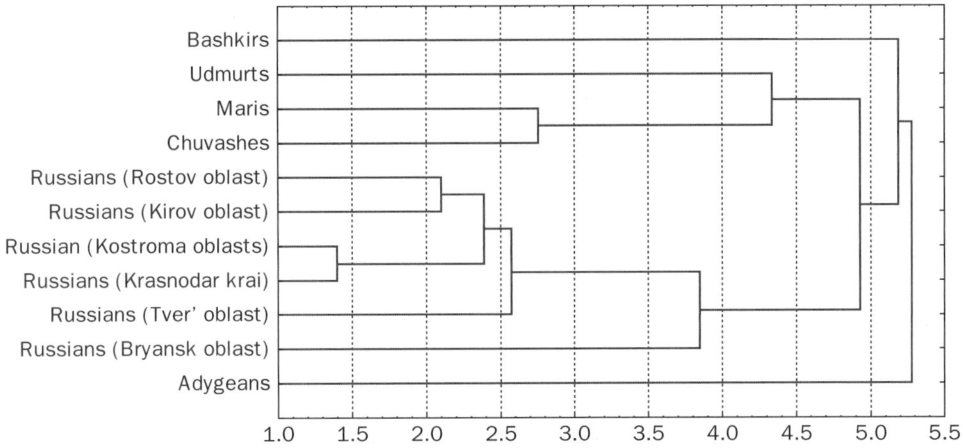

Figure 110-4 The genetic relationships among different populations/ethnic groups of Russia (AD diseases).

of separate nosological entities in various ethnic groups/populations considered in our sample.

In order to reveal the foci of local accumulation of hereditary diseases, we compared the prevalence rate of a particular hereditary disease in a particular population with the prevalence rate thereof in the whole population surveyed. In so doing, we used the F-distribution suitable for comparing rarely occurring events (Zhivotovsky, 1991). Table 110-10 shows the diseases, having displayed region-specific accumulation (marked with an asterisk *).

As is seen from Table 110-10, all the 11 populations studied were found to have accumulated one or another of nosological entities, although to varying degrees. The most prominent example of local accumulation of hereditary diseases is accumulation of at least three recessive diseases in the Republic of Chuvashia and Mari El. High prevalence of recessive erythrocytosis in Chuvashia had been known prior to our genetic epidemiological study in this republic (Sergeyeva et al., 1997). In 2004, Ang, with co-authors (Ang et al., 2004a, b) determined that this disease is caused by homozygosity of all sufferers for the same mutation 598C-T in the *VHL* gene. We demonstrated that recessive erythrocytosis is frequently encountered not only in Chuvashia but also in Mariy El and Udmurtia, with the gene frequency equaling 1.84%, 0.87%, and 0.47%, respectively. Our expeditionary studies revealed two more commonly occurring recessive diseases in Chuvashia and Mariy El. One of them was recessive lethal osteopetrosis (Ginter et al., 2001) caused by a c.807+G>A mutation in the *TCIRG* gene (Bliznetz et al., 2009). This mutation was found in all patients in the homozygous state and in their parents in the heterozygous state. The incidence rate of the mutant gene in Chuvashs amounted to 1.68%, and in Maris to 0.84%. The second disease we detected was hereditary hypotrichosis, presented in the OMIM Morbid Map and defined as *total hypotrichosis, Mari type* (60433790; Rogaev et al., 1999). We determined that hypotrichosis was caused in Chuvashs and Maris by homozygous deletion of exon 4 and its flanking intron sequences in the *LIPH* gene. A Δex4 mutation was found in all patients in the homozygous state (Kazantseva et al., 2006). The prevalence of the mutant gene in Chuvashia amounted to 2.72%, and in Mari El to 1.74%. This mutation was found also in the Udmurts and the Bashkirs (with the incidence rate equaling 0.44% and 0.31%, respectively).

To elucidate possible causes of locally high prevalence rates of separate nosological entities, we performed a correlation analysis of the dependence of hereditary diseases

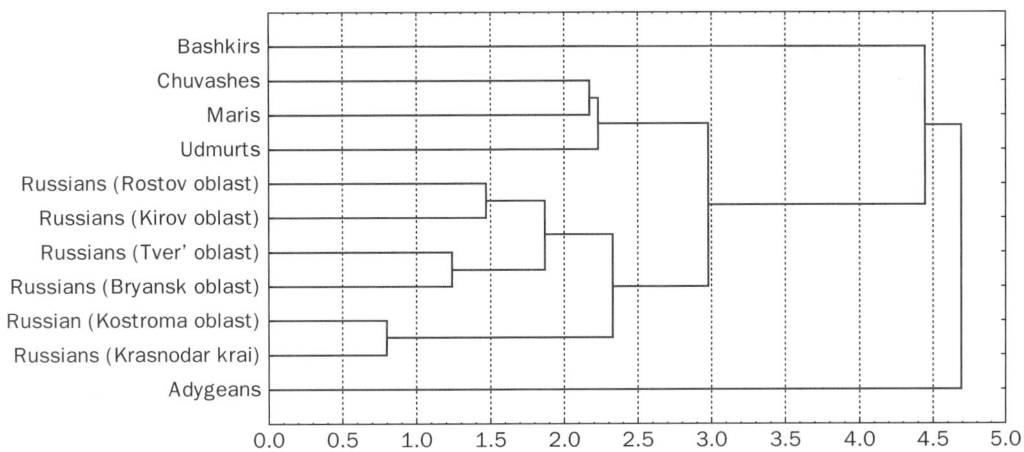

Figure 110-5 The genetic relationships among different populations/ethnic groups of Russia (AR diseases).

TABLE 110-10 THE SPECTRUM OF HEREDITARY DISEASES ACCUMULATED IN THE POPULATIONS STUDIED

	DIAGNOSIS	B	U	M	Ch	A	R	Ks	Kr	Kir	Tv	Br
AD DISEASES												
1	Charcot-Marie-Tooth Disease		*							*		
2	Spinocerebellar Ataxia 1		*									
3	Oculopharyngeal Muscular Dystrophy								*			
4	Muscular Dystrophy, Limb-Girdle											*
5	Facioscapulohumeral Muscular Dystrophy 1a	*										*
6	Muscular Dystrophy, Congenital								*			
7	Neurofibromatosis, Type I			*	*	*						
8	Spastic Paraplegia, Autosomal Dominant	*								*		
9	Tremor, Hereditary Essential									*		
10	Dystrophia Myotonica 1	*										
11	Myotonia Congenita, Autosomal Dominant						*					
12	Dystonia, Dopa-Responsive				*						*	
13	Tuberous Sclerosis	*										
14	Microcephaly, Autosomal Dominant	*		*								*
15	Mental Retardation	*	*									
16	Vitreoretinochoroidopathy						*					
17	Stargardt Disease		*									
18	Retinitis Pigmentosa						*					
19	Retinal Cone Dystrophy								*			
20	Vitreoretinal Degeneration	*										
21	Corneal Dystrophy, Posterior Polymorphous			*								
22	Optic Atrophy 1											*
23	Congenital Hereditary Cataract		*				*	*				
24	Glaucoma 3, Primary Congenital	*										
25	Mesodermal Dysgenesis of the Iris								*			
26	Nystagmus, Congenital			*			*					
27	Ptosis, Hereditary Congenital	*				*						
28	Blepharophimosis, Ptosis	*										
29	Aniridia							*				
30	Microphthalmia, Isolated, with Cataract		*									
31	Microphthalmia, Isolated, with Colobom	*										
32	Leber Congenital Amaurosis	*										
33	Achondroplasia	*										
34	Hypochondroplasia					*						
35	Brachydactyly, Type B1				*							
36	Brachydactyly, Type E			*	*							
37	Polydactyly, Postaxial, Type A1	*				*						
38	Polydactyly, Preaxial I					*						
39	Syndactyly, Type III		*									
40	Syndactyly, Type V				*							*

(Continued)

TABLE 110-10 (CONTINUED)

	DIAGNOSIS	B	U	M	Ch	A	R	Ks	Kr	Kir	Tv	Br
41	Split-Hand/Foot Malformation 1	*										
42	Exostoses, Multiple, Type I				*							
43	Scoliosis, Idiopathic		*									
44	Frontonasal Dysplasia	*										
45	Spondylometaphyseal Dysplasia										*	
46	Spondyloepimetaphyseal Dysplasia	*										
47	Epiphyseal Dysplasia, Multiple			*							*	
48	Acromesomelic Dysplasia						*					
49	Blount Disease						*					
50	Lipomatosis, Multiple								*			
51	Ichthyosis Vulgaris		*	*	*							
52	Keratosis Palmoplantaris	*	*	*	*							
53	Monilethrix						*					
54	Pseudomonilethrix		*									
55	Ectodermal Dysplasia, Hidrotic			*								
56	Epidermolysis Bullosa Herpetiformis, Dowling-Meara Type	*										
57	Acrokeratosis Verruciformis								*			
58	Telangiectasia, Hereditary Hemorrhagic, of Rendu, Osler		*									
59	Dercum Disease								*			
60	Albinoidism, Oculocutaneous, Autosomal Dominant		*									*
61	Nevus Anemicus				*						*	
62	Pemphigus, Benign Familial (Hailey-Hailey Disease)			*								
63	Acrodental Dysostosis of Weyers Curry-Hall Syndrome	*										
64	Albright Hereditary Osteodystrophy with Multiple Hormone Resistance						*					
65	Angelman Syndrome				*							
66	Axenfeld-Rieger Syndrome, Type 1	*										*
67	Axenfeld-Rieger Syndrome, Type 2	*										
68	Blepharophimosis with Ovarian Failure		*									
69	Contractural Arachnodactyly, Congenital (Beals Syndrome)				*							
70	Deafness, Conductive, with Malformed External Ear	*										
71	Ectrodactyly, Ectodermal Dysplasia, and Cleft Lip/Palate Syndrome			*								
72	Ehlers-Danlos Syndrome	*					*					
73	Hypertelorism with Esophageal Abnormality and Hypospadias (Opitz-Frias Syndrome)						*					
74	Klippel-Feil Syndrome						*					
75	Leopard Syndrome						*					
76	Lymphedema-Distichiasis Syndrome			*								
77	Marfan Syndrome		*									
78	Mental Retardation, Postaxial Polydactyly	*										
79	Moebius Syndrome			*								
80	Nager Acrofacial Dysostosis		*									

(Continued)

TABLE 110-10 (CONTINUED)

	DIAGNOSIS	B	U	M	Ch	A	R	Ks	Kr	Kir	Tv	Br
81	Nail-Patella Syndrome										*	
82	Noonan Syndrome 1		*									
83	Orofacial Cleft Syndrome	*										
84	Poland Syndrome				*							
85	Sotos Syndrome		*									
86	Sturge-Weber Syndrome		*									
87	Treacher Collins-Franceschetti Syndrome		*									
88	Trichorhinophalangeal Syndrome, Type I					*						*
89	Volf Parkinson Witte Syndrome					*						
90	Williams Syndrome	*										
91	Isolated Growth Hormone Deficiency	*	*									
92	Otosclerosis			*			*				*	
93	Porphyria	*										*
94	Epistaxis, Hereditary			*								
AR DISEASES												
95	Charcot-Marie-Tooth Disease									*		
96	Friedreich Ataxia									*		
97	Hydrocephalus									*		
98	Mental Retardation	*	*	*								
99	Metachromatic Leukodystrophy					*						
100	Microcephaly with Spastic Quadriplegia						*					
101	Microcephaly, Primary Autosomal Recessive			*			*					
102	Muscular Dystrophy, Distal, Late-Onset, Autosomal Recessive (Miyoshi Myopathy)			*								
103	Muscular Dystrophy, Limb-Girdle		*									
104	Myoclonic Epilepsy of Unverricht and Lundborg					*						
105	Olivopontocerebellar Atrophy II, Autosomal Recessive									*		
106	Spastic Paraplegia, Autosomal Recessive									*		
107	Spinal Muscular Atrophy, Scapuloperoneal		*									
108	Spinal Muscular Atrophy, Type I–III		*									
109	Spinocerebellar Ataxia, Autosomal Recessive										*	
110	Congenital Hereditary Cataract				*							
111	Leber Congenital Amaurosis		*									
112	Macular Degeneration					*						
113	Microphthalmia, Isolated				*							
114	Nystagmus, Congenital Motor, Autosomal Recessive			*	*							
115	Retinopathy, Pericentral Pigmentary, Autosomal Recessive							*				
116	Ectrodactyly			*								
117	Polydactyly, Postaxial	*										
118	Spondyloepiphyseal Dysplasia Tarda, Autosomal Recessive						*					

(Continued)

TABLE 110-10 (CONTINUED)

	DIAGNOSIS	B	U	M	Ch	A	R	Ks	Kr	Kir	Tv	Br
119	Spondylometaphyseal Dysplasia		*									
120	Tumoral Calcinosis, Hyperphosphatemic										*	
121	Albinism					*						*
122	Albinism, Oculocutaneous		*	*								
123	Epidermolysis Bullosa Dystrophica, Hallopeau-Siemens Type									*		
124	Ichthyosis, Lamellar				*							
125	Ectodermal Dysplasia, Hypohidrotic, Autosomal Recessive			*								
126	Hypotrichosis, Total, Mari Type			*	*							
127	Alopecia-Contractures-Dwarfism Mental Retardation Syndrome									*		
128	Cockayne Syndrome	*										
129	Ehlers-Danlos Syndrome	*										
130	Hallermann-Streiff Syndrome				*							
131	Hallervorden-Spatz Disease					*						
132	Kartagener Syndrome			*								
133	Larsen Syndrome				*							
134	Microcephalic Osteodysplastic Primordial Dwarfism, Type II	*										
135	Microcephalic Primordial Dwarfism, Toriello Type											*
136	Microtia with Meatal Atresia and Conductive Deafness						*					
137	Morquio Syndrome A			*								
138	Roberts Syndrome		*									
139	Robinow Syndrome	*										
140	Tel Hashomer Camptodactyly Syndrome	*										
141	Deafness, Autosomal Recessive		*		*	*	*					
142	Growth Hormone Deficiency	*			*							*
143	Homocystinuria										*	
144	Osteopetrosis, Autosomal Recessive				*							
145	Thyroid Hormonogenesis, Genetic Defect	*			*	*						
X-LINKED DISEASES												
146	Muscular Dystrophy, Duchenne Type		*									
147	Muscular Dystrophy, Becker Type						*					
148	Spinocerebellar Ataxia, X-Linked 5		*									
149	Spastic Paraplegia, X-Linked	*										
150	Mental Retardation, X-Linked	*	*		*						*	
151	Optic Atrophy					*						
152	Choroideremia	*			*							
153	Nystagmus, Congenital, X-Linked		*									
154	Faciogenital Dysplasia						*					

(Continued)

TABLE 110-10 (CONTINUED)

	DIAGNOSIS	B	U	M	Ch	A	R	Ks	Kr	Kir	Tv	Br
155	Rhizomelic Dysplasia, Scoliosis, and Retinitis Pigmentosa									*		
156	Hypophosphatemic Rickets	*										
157	Alport Syndrome, X-Linked	*										
158	Deafness, X-Linked									*		

Note: Populations: B-Bashkirs from the Republic of Bashkiria; U-Udmurts from the Republic of Udmurtia; M-Maris from the Republic of Mari El; Ch-Chuvashes from the Republic of Chuvashia; A-Adygeans from the Republic of Adygea; R-Russians from the Rostov Region; Ks-Russians from the Kostroma Region; Kr-Russians from the Krasnodar Territory; Kir-Russians from the Kirov Region; Tv-Russians from the Tver Region; Br-Russians from the Bryansk Region.

TABLE 110-11 AR AND AD DISEASES EXHIBITING LOCAL ACCUMULATION IN THE POPULATIONS STUDIED

Population	FST (10-3)	Endogamy	AD		AR	
			number of forms	coefficient of accumulation	number of forms	coefficient of accumulation
Bashkirs		0.75	29	0.17	9	0.05
Udmurts		0.61	21	0.14	9	0.06
Chuvashes		0.6	12	0.07	11	0.06
Maris	3.6	0.69	15	0.09	10	0.06
Adygeans	14.16	0.7	7	0.11	7	0.11
RUSSIANS:						
Rosrov Region	0.76	0.38	14	0.04	5	0.02
Kostroma Region	1.92	0.58	2	0.00	1	0.00
Krasnodar Territory	0.46	0.31	7	0.02	0	0.00
Kirov Region	2.92	0.52	3	0.01	7	0.03
Tver Region	1.54	0.34	6	0.08	3	0.04
Bryansk Region	1.96	0.56	9	0.10	3	0.03

accumulation on the genetic structure of the populations studied. Since the probability of accumulation of diseases in a population may well depend upon the size of this population, we normalized the number of entities having exhibited accumulation per the size of the respective population, thus having obtained the accumulation coefficients. As a measure of inbreeding of the populations examined, we used the average values of random inbreeding (F_{ST}) obtained by the isonomic method. The findings of this analysis are shown in Table 110-11. For the fullness of the picture, the analysis also comprised the data on the endogamy index reflecting the degree of isolation of the populations.

The coefficients of correlation between the coefficients of accumulation of AD and AR hereditary diseases and random inbreeding (F_{st}) amounted to 0.9705 and 0.8992, respectively. The coefficients of correlation between the coefficients of accumulation and the values of endogamy amounted to 0.5958 for AD pathology and 0.6295 for AP.

CONCLUSION

Based on a genetic epidemiological study, the prevalence rates of autosomal dominant, autosomal recessive, and X-linked recessive hereditary disorders (HDs) were determined in 11 regions of Russia, including six populations (Krasnodar Territory, Rostov, Kostroma, Kirov, Tver and Bryansk regions), Adygean (Adygea Republic), Maris (Mari El Republic), Chuvashes (Chuvash Republic), Udmurt (Udmurt Republic), and Bashkirs (Bashkortostan Republic). The size of the investigated population was more than 2.5 million people. Genetic differentiation between populations of different hierarchical levels was established using the HD data. The prevalence rate of all Mendelian disorders varied in the investigated populations from 1.26 to 6.00 per 1000 persons. The correlation analysis between the index endogamy and random inbreeding (F_{ST}) in the Russian populations, and the load of AD and AR pathology, has shown positive and significant correlations. It is likely that the genetic drift is the factor that determined genetic differentiation of populations by the prevalence of autosomal AD and AR. In the investigated populations, 413 HDs were revealed (6800 affected): 199 AD, 168 AR, and 46 X-linked recessive. It is established that 58.23% of patients with all forms of inheritance pattern represent common forms of hereditary disorders, however only 6% of all nosological entities. These 29 disorders meet with high frequencies in most investigated populations (16 with AD, 8 with AR, and 5 with X-linked inheritance). However, besides these 29 diseases, whose frequencies also varied between regions, specific common diseases are revealed in each population. It is proposed that genetic drift is also responsible for local accumulation of HDs.

Hence, population-wide surveys have made it possible for the first time to determine the size of the burden and spectrum of Mendelian hereditary diseases characteristic of the populations of the European part of Russia. This chapter also demonstrates the main factors of the population dynamics determining differentiation of the populations of Russia by the prevalence of Mendelian hereditary pathology, as well as determining that local accumulation of hereditary diseases in the populations studied is due to genetic drift.

REFERENCES

Ang SO, Chen H, Gordeuk VR, Sergueeva AI, et al. (2002a). Endemic polycythemia in Russia: Mutation in the VHL gene. *Blood Cells Molec Dis* 28:57–62.

Ang SO, Chen H, Hirota K, Gordeuk VR, et al. (2002b). Disruption of oxygen homeostasis underlies congenital Chuvash polycythemia. *Nature Genet* 32:614–621.

Baird PA, Anderson TW, Newcombe HB, Brian Lowry B. (1988). Genetic disorders in children and young adults: A population study. *Am J Hum Genet* 42:677–694.

Bliznetz EA, Tverskaya SM, Zinchenko RA et al. (2009). Genetic analysis of autosomal recessive osteopetrosis in Chuvashia: The unique splice mutation in *TCIRG1* gene spread by the founder effect. *Eur J Hum Genet* 17:664–672.

Carter CO. (1977). Monogenic disorders. *J Med Genet* 14:316–320.

Cavalli-Sforza LL, Bodmer WF. (1971). In: Kennedy D, Park RB, eds. *The Genetics of Human populations.* San Francisco: W.H. Freeman and Company.

De Braekeleer M, Ngoc Dao NT. (1994 a). Hereditary disorders in the French Canadian population of Quebec. I. In search of founders. *Hum Biol* 66:205–224.

De Braekeleer M, Ngoc Dao NT. (1994 b). Hereditary disorders in the French Canadian population of Quebec. II. Contribution of Perche. *Hum Biol* 66:225–250.

Ginter EK, Zinchenko RA (2006). *Genetic Structure and Hereditary Diseases in the Chuvash Population.* Cheboksary, Russia: Pegas (In Russian).

Ginter EK, Kirillov AG, Rogaev EI. (2001). Autosomal-recessive osteopetrosis in Chuvashia. *Genetika* (In Russian) 37:1152–1155.

Ginter EK. (1986). Medical genetic survey of Soviet Middle Asian Populations. In: Fogel F, Sperling K, eds. *Human Genetics. Proceedings of the 7th International Congress, Berlin 1986.* Berlin: Springer-Verlag.

Ginter EK (1997). *Medical Genetic Description of the Adygea Population.* Adygea, Maikop, Russia (In Russian).

Ginter EK, ed. (2002). *Hereditary Diseases in Human Populations.* Meditsina, Moscow, Russia (In Russian).

Goodman RM. (1979). *Genetic Disorders Among the Jewish People.* Baltimore: The John Hopkins Univ. Press.

Kazantseva AV, Goltsov A, Zinchenko RA, et al. (2006). Human hair growth deficiency is linked to genetic defect in the phospholipase gene *LIPH. Science* 314(5801):982–985.

Matsunaga EE. (1982). Perspectives in mutation epidemiology. 1. Incidence and prevalence of genetic disease in human populations. *Mut Research* 99:95–128.

McKusick, VA (1980). Medical genetic studies of the Amish, with comparison to other populations. In: Eriksson AW, ed. *Population Structure and Genetic Disorders.* London-New York-Toronto: Academic Press.

Morton NE, Rao DC. (1983). *Methods in Generic Epidemiology.* Basel: Karder.

Norio R. (1981). Diseases of Finland and Scandinavia. In: *Bio Cultural Aspects of Disease.* New York: Academic Press.

Rogaev EI, Zinchenko RA et al. (1999). Total hypotrichosis: Genetic form of alopecia not linked to hairless gene. *Lancet* 354:1097–1098.

Russet F. (2001). Inferences from spatial population genetics. In: Balding DJ, Bishop M, Cannings C, eds. *Handbook of Statistical Genetics.* New York: John Wiley & Sons.

Sergeyeva A, Gordeuk VR, Tokarev YN, Sokol L, Prchal JF, Prchal JT. (1997). Congenital polycythemia in Chuvashia. *Blood* 89:2148–2154.

UNSCEAR (1977). Sources and effects of ionizing radiation. In: *United Nations Scientific Committee on the Effects of Atomic Radiation.* New York: Report to the General Assembly with annexes, United Nations.

UNSCEAR. (1982). Genetic and somatic effects of ionizing radiation. In: *United Nations Scientific Committee on the Effects of Atomic Radiation.* New York, USA: Report to the General Assembly with annexes, United Nations.

Zhivotovsky LA. (1991). *Population Biometry.* Moscow, Russia: Nauka.

Zinchenko RA, El'chinova GI, Gavrilina SG, Ginter EK. (2001b). Analysis of Diversity of Autosomal Recessive Diseases in Russian Populations. *Russ J Genet* 37(11):1312–1322.

Zinchenko RA, El'chinova GI, Nurbaev SD, Ginter EK. (2001a). Diversity of autosomal dominant disorders in the populations of Russia. *Russ J Genet* 37(3):290–301.

111.

HEREDITARY DISEASES AMONG THE YAKUTS

Nadezda R. Maksimova and Anna N. Nogovitsina

INTRODUCTION

The Yakuts are Asians and live in the northeastern part of Siberia in the Republic of Sakha (Yakutia) of the Russian Federation. They emigrated from southern Siberia to a northern area in the thirteenth/fourteenth century AD and grew rapidly to more than 42,000 people in a restricted area. The total population of the Republic Sakha (Yakutia) is about 1 million people. About 80 different nationalities (the Yakuts, the Russians, the Evens, the Evenks, and others) live in the Republic.

The genetic counseling for the population of Republic of Sakha (Yakutia) was organized beginning in 1989, in a regional medical genetic consultation of Republican Hospital N1—National Center of Medicine of the Republic of Sakha (Yakutia). The main fields of interests of medical genetic consultation are: practical genetic counseling of families and patients with hereditary diseases and birth defects, neonatal screening for 5 inborn metabolic diseases, DNA-diagnostics for common hereditary diseases in the Republic, prenatal diagnosis for chromosomal and hereditary diseases, and monitoring of birth defects. The Republican Genetic Registry of the Hereditary and Inborn Pathology has been working in medical genetic consultation since 2001. There are over 3000 families with hereditary and birth defects in this registry (Maksimova et al., 2007a,b).

Medical genetic investigations in Siberia started in the 1980s (Puzyrev, 1984) and are in progress today in the territory of Tomsk oblast (the Russians and the Tatars), Tyva, Buryatia, Gornyi Altai, Khakassia, and Yakutia (Nazarenko et al., 2002a,b; Nogovitsina et al., 1999; Puzyrev et al., 1999; Puzyrev & Nazarenko, 2000). The medical genetic examination of the population of Yakutia is the longest in duration and the most intensive one. It has also proved to be the most informative, complete, and comprehensive. The clear formulation of the tasks by the Republican Hospital N1—National Center of Medicine of the Ministry of Health of the Republic of Sakha (Yakutia) and Yakut Research Center of the Siberian branch of the Russian Academy of Medical Sciences, as well as cooperation with scientists from other scientific institutions of Russia and with foreign colleagues (Goldfarb et al., 1996; Lunkes et al., 1994; Maksimova et al., 2007a,b) has also contributed to the conducting of these studies. According to genetic epidemiological investigations in the Yakut population, the load of autosomal dominant (AD) pathology is 1.7×10^{-3}, autosomal recessive (AR) is 0.7×10^{-3}, and X-linked is 0.4×10^{-3} (Nogovitsina, 2001).

The aim of this chapter is to present a survey of medical genetic investigations in Yakutia, primarily the clinical and molecular genetic descriptions of several hereditary diseases characterized by a high prevalence among the Yakuts as compared to their global incidence in the world. Several forms of pathologies, referred to as *Yakut hereditary diseases*, have been distinguished on the basis of the results of genetic epidemiological studies of Mendelian diseases in the population of the Republic of Sakha (Yakutia): spinocerebellar ataxia type I, myotonic dystrophy, oculopharyngeal muscular dystrophy, hereditary enzymopenic methemoglobinemia, 3-M syndrome, Friedreich's ataxia, and SOPH (Short stature, optic atrophy, Pelger-Huet anomaly) syndrome (Maksimova, 2009), autosomal recessive deafness 1A (Barashkov et al., 2011). Data on the molecular nature of mutations in genes responsible for these hereditary diseases are presented in Table 111-1.

HEREDITARY DISEASES IN THE YAKUTS

The aboriginal Yakut population has been studied for a wide spectrum of genetic markers, including those of the systems of immunological (erythrocytic antigens of blood groups) and biochemical (blood serum proteins and enzymes, blood erythrocytic enzymes) polymorphisms (Rychkov Yu, 2000), autosomal microsatellites (STR), Alu repeats (Stepanov, 2002), mitochondrial DNA lines, the Y-chromosome haplotype analysis (Puzyrev et al., 2003), candidate genes of widely distributed diseases (Puzyrev et al., 2007), and, finally, polymorphisms and mutations of hereditary diseases genes (Fedorova et al., 2005; Galeeva

TABLE 111-1 COMPARATIVE DATA FOR FIVE HEREDITARY DISEASES IN THE YAKUTS AND IN THE WORLD

Name of disease (OMIM *)	Type of inheritance**	Gene	Mutation in the Yakuts	Disease incidence per 100,000 individuals		Reference
				in the Yakuts	in the world	
Spinocerebellar ataxia type 1 (164400)	AD	*ATXN1*	Expansion of CAG repeats	36.8	1.0–2.0	Illarioshkin et al., 2006; Platonov et al., 2004
Myotonic dystrophy (160900)	AD	*DMPK*	Expansion of CTG repeats	21.3	4.0–5.0	Harper et al., 2004; Sukhomyasova, 2005
Oculopharyngeal muscular dystrophy (164300)	AD	*PABPN1*	Expansion of GCG repeats	11.1	1.0	Bouchard et al., 1997; Maksimova et al., 2007a
Hereditary enzymopenic methemoglobinemia (250800)	AR	*DIA1*	Pro269Leu	14.9	1.0	Galeeva et al., 2006; Tarskaya et al., 2004
3-M syndrome (273750)	AR	*CUL7*	4582insT	12.7	Unknown, about 60 cases	Cantu et al., 1981; Maksimova et al., 2007b; Miller et al., 1975; Temtamy et al., 2006
Friedreich's ataxia (229300)	AR	*FRDA*	Expansion of GAA repeats	2.8	2.0–5.0	Palau & Espinos, 2006
SOPH	AR	*NBAS*	R1914H	9.95	Unknown	Maksimova et al., 2010
Autosomal recessive deafness 1A (220290)	AR	*GJB2*	IVS1+1G4A	16.2	Rare	Barashkov et al., 2011

Notes: *OMIM, the number in the electron version of McKusick`s catalog «Mendelian Inheritance in Man» (MIM): www.ncbi.nlm.nih.gov/omim/

**AD and AR, autosomal dominant and autosomal recessive types of inheritance

et al., 2006; Goldfarb et al., 1996; Illarioshkin et al., 2006; Maksimova et al., 2007a,b; Popova et al., 2002). These studies, supplemented with demographic and vital statistics data, indicate that (1) there is a certain subdivision in the aboriginal Yakut population, but its level is low; (2) up to 98% of individuals under 20 are local residents, and the migration inflow is limited by the territory of Yakutia; (3) the level of isolation is rather high (a low frequency of international marriages); (4) the growth of the population is continued due to a high birth rate (the Yakut population has increased 14-fold in the past 250 years; Kucher, 2002). These factors determine a high degree of genetic homogeneity of the population and a large length of LD (linkage disequilibrium) blocks. This permitted the Yakut population to be regarded as an ideal one for identifying ethnospecific mutations of genes of Mendelian diseases, and for mapping genes of widely spread diseases (Peltonen et al., 2006; Puzyrev, 2003). Medical genetic investigations in this population, in parallel with successful molecular genetic studies of individual phenotypes, confirmed the above statements. The table below presents Mendelian phenotypes termed by us as Yakut hereditary diseases (Table 111-1).

SPINOCEREBELLAR ATAXIA, TYPE I (SCA1; OMIM 164400)

This is a hereditary autosomal dominant disease. The specific features of its clinical picture are late onset (by the age of 20 to 40) and high polymorphism of the combination of progressive cerebellum and pyramid symptoms. Until the mid-1970s this disease in Yakutia was considered to be the cerebellar form of Vilyui encephalomyelitis (Vladimirtsev, 1964; Zubri, 1975). The high incidence of the disease in Yakutia (38.6 per 100,000 individuals), compared to 1–2 per 100,000 in the world population, was evaluated as the Siberian focus of the disease and the largest in the world (Goldfarb et al., 1996; Illarioshkin et al., 2006; Lunkes et al., 1994). In the Republican Genetic Registry of the Hereditary and Inborn Pathology there are 147 patients from 33 Yakut families (Maksimova et al., 2008). The molecular basis of the disease is an increase in the number of trinucleotide CAG repeats, up to 39–71, as compared to normally 19–36 in the *SCA1* gene *(ATXN1)* localized in the 6p22–p23 region (Lunkes et al., 1994). Among different forms of hereditary ataxia in Yakutia, SCA1 occurs in 88.1% of all families according to molecular-genetic study (Platonov et al., 2004). The study of polymorphism of CAG repeats of the SCA1 gene in Yakut patients revealed a low level of genetic diversity compared to the one in Eurasian populations, and 10 allelic variants (from 26 to 40 CAG-repeats; Popova et al., 2002). DNA diagnostics of CAG repeats in the *ATXN1* gene is possible in regional medical genetic consultation. There is no specific therapy.

MYOTONIC DYSTROPHY (MD) (OMIM 160900)

Myotonic dystrophy is a hereditary autosomal dominant disease expressed as myotonia, myopathy, cardiovascular disturbances, endocrine vegetative disturbances, cataract, and deterioration of intellect. It can start in childhood, but the onset is more typical in the second or third decade of life. The gene of myotonic dystrophy (*DM*) is located in the 19q13.2 q13.3 region and consists of 15 exons. The molecular basis of the disease is the expansion of CTG repeat copies in the 3' untranslated region of the *DM* gene. The number of CTG repeats is normally from 5 to 37, with a permutation from 37 to 50, and in the case of pathology it is over 50 up to several thousands. The study of polymorphism of CTG repeats of the myotonin protein kinase gene in Yakut patients belonging to different ethnic groups (the Yakuts, Evenks, Evens, Yukagirs, Dolgans) revealed a low level of genetic diversity compared to that in Eurasian populations and a low level of unstable alleles (Fedorova et al., 2005; Popova et al., 2002). The MD frequency in different populations of the world is in the range from 4.0 to 5.0 per 100,000 individuals (Harper et al., 2004). The incidence of MD in Yakutia is considerably higher and constitutes 21.3 per 100,000 individuals (Sukhomyasova, 2005). These data, based on the number of patients (n= 92) registered in the Republican Genetic Register (Yakutsk), seem to be more real than those mentioned previously (Tarskaya et al., 2004). DNA diagnostics of CTG repeats in *DMPK* gene is possible in regional medical genetic consultation. There is no specific therapy.

OCULOPHARYNGEAL MUSCULAR DYSTROPHY (OPMD; OMIM 164300)

This is a hereditary disease of the nervous system with the autosomal dominant type of inheritance. It usually manifests itself in the fifth or sixth decade of life. The clinical picture of the disease is represented by progressive myasthenia and atrophy of proximal parts of limbs, dysphagia and dysphonation, ptosis, and facial muscle weakness. In Yakutia, the disease was known under the name *betyugen* as a hereditary form of bulbar paralysis (Zak, 1932). Later on, new cases of the disease were described already as OPMD (Korotov, 1994). Oculopharyngeal myodystrophy has been proved to have a high incidence predominantly in the Yakut population, with accumulation in the Ust-Aldan district. The OPMD incidence in the Yakut population is 1 per 11,000 or 11.1 per 100,000 individuals, which is ten- to twentyfold higher than in Europe (Maksimova et al., 2007a). The *OPMD* gene is located in the 14q11.2–13 region and responsible for the synthesis of poly-(A)-binding protein 2 (*PABPN1*) that serves as a factor of mRNA polyadenylation; a mutation identified in the gene is an increase in the number of copies of trinucleotide GCG repeats in the first exon (Braise et al., 1998). Normally the gene contains six tandem copies of GCG repeats, while patients with OPMD have 8–13 copies. In some populations the expansion of trinucleotide repeats results from simple addition of GCG repeats (Becher et al., 2001; Blumen et al., 2000); in other populations, a GCA insertion occurs along with the expansion of GCG repeats (Rodriguez et al., 2005). Molecular genetic examinations of 40 patients have been carried out for the first time in Yakutia, and a mutation has been identified in the first exon of the *PABPN1* gene consisting in an increase of GCG repeats up to 10 (Maksimova et al., 2007a). Direct sequencing of the gene in 17 families revealed the type of the mutation: insertion of four GCG repeats. DNA diagnostics of expanded GCG repeats in *PABPN1* gene is possible in regional medical genetic consultation. There is no specific therapy, but surgery can correct ptosis and dysphagia, at least temporarily.

HEREDITARY ENZYMOPENIC METHEMOGLOBINEMIA (HEM TYPE I; OMIM 250800)

This is a hereditary autosomal recessive disease that manifests itself in dermal and mucosal cyanosis, dyspnea, headache, mental retardation, increased level of methemoglobin, and deficiency of methemoglobin reductase in blood. Two forms of methemoglobin reductase deficiency are distinguished. In the case of form I, only symptoms of easily endurable methemoglobinemia are observed, and the deficiency of the enzyme is related to erythrocytic soluble cytochrome- b5-reductase (diaphorase). Form II is associated with mental retardation and a generalized enzyme deficiency involving both soluble and microsomal cytochrome- b5-reductase in erythrocytes and leucocytes. With a deficiency of NADH-cytochrome-b5-reductase, the produced methemoglobin is not transformed back in hemoglobin, and the level of methemoglobin in the blood of patients with the homozygous form averages about 18% (2% in heterozygotes, below 1% normally). HEM is a result of mutations in the diaphorase gene (*DIA1*) localized in the terminal part of the long arm of chromosome 22 (q13qter region). About 30 mutations in the *DIA1* gene are known at present, of which 19 lead to the development of HEM type I (Galeeva et al., 2006). In the world, HEM is often found among Eskimos in Alaska (Scott, 1960); single cases were described in other populations. The largest HEM focus in Yakutia was revealed in the 1970s (Tokarev et al., 1983). The incidence of HEM in Yakutia today is estimated at 1:5677 (Tarskaya et al., 2004). The first attempt of molecular genetic analysis of *DIA1* gene mutations among Yakut patients with HEM type I turned out to be unsuccessful (Nazarenko et al., 2002a,b). None of the known mutations in the *DIA1* gene (Len72Pro, Vall105Met, Arg57Gln) discovered in the nearest neighbors of the Yakuts (Japanese and Chinese) was identified. Later on, 16 unrelated Yakut patients from the same sample and their 8 healthy relatives were examined for the disequilibrium of the linkage of the disease with four markers flanking the *DIA1* gene, and then sequencing of all nine exons of the gene was carried out (Galeeva et al., 2006). A novel missense mutation was discovered in exon 9 leading to the substitution of Pro269Leu in the NADH-cytochrome-b5-reductase enzyme. DNA diagnostics of mutation Pro269Leu s in *DIA1* gene is possible in regional medical genetic consultation.

FRIEDREICH'S ATAXIA (OMIM 229300)

This is an autosomal recessive disorder. Gait ataxia is present from childhood. Areflexia of lower limbs, dysarthria, muscle weakness, and sensory disturbances complete the neurological presentation. Sensory nerve conduction is absent or severely reduced. Other features include pes cavus, scoliosis, cardiomyopathy, and diabetes mellitus. In the regional medical consultation we observed 11 Yakut patients with Friedreich's ataxia, with one patient being Russian. The frequency in the Yakuts is 2.8 per 100,000 Yakut individuals. It is caused by a dynamic mutation in the first exon of the frataxin gene (*FRDA*). Normal individuals have 7–36 GAA repeats; in contrast, FRDA patients have 90 to 1700 GAA repeats. All Yakut patients have GAA expansion in the *FRDA* gene. There is no specific therapy. DNA diagnostics of GAA repeats in the FRDA gene is available in regional medical genetic consultation.

However, genetically proved FA has never been reported in East Asians (Palau & Espinos, 2006). The reason is that expanded GAA alleles may be generated from a specific allele, LN, which is well observed in Caucasians and has never been observed in East Asians. The results indicate that origin of the Yakuts is different from East Asians, and it may be interesting to see the frequency of the specific allele, LN allele, which is thought to be the origin of the expanded allele.

AUTOSOMAL RECESSIVE DEAFNESS 1A (DFNB1A) (OMIM 220290)

Hereditary forms of hearing impairment caused by GJB2 (Cx26) mutations are the frequent sensory disorders registered among newborns in various human populations. In all, 15–20 cases of congenital/early onset HI in 14 000–15 000 live births (900–1:750 newborns) are detected per year in the Sakha Republic (Yakutia). According to the epidemiological data, high rate of congenital hearing impairment is caused by some hereditary forms of deafness spread in indigenous populations of Sakha Republic. Mutational analysis of GJB2 gene in 86 unrelated Yakut patients with congenital hearing impairment without other clinical features has been performed (Barashkov et al., 2011). In this study, authors registered a large cohort of Yakut patients homozygous for the IVS1+1G4A mutation (70 unrelated deaf subjects in total). Detailed audiological analysis of 40 deaf subjects with genotype IVS1+1G4A/IVS1+1G4A revealed significant association of this genotype with mostly symmetrical bilateral severe to profound hearing impairment. The highest among six investigated Eastern Siberian populations carrier frequency of the IVS1+1G4A mutation (11.7%) has been found in Yakut population.

YAKUT SHORT STATURE SYNDROME OR 3-M SYNDROME (OMIM 273750)

Under this name, the syndrome described in 2007 (Maksimova et al., 2007b) was included in the OMIM gene catalog as an alternative of the known 3-M syndrome (OMIM 273750). The reason for registering these two syndromes under the same catalog number is one and the same *CUL7* (Cullin 7) gene, only with different mutations.

Clinical features of 43 patients with Yakut short stature. Length of gestation varied from 35 to 42 weeks (median 37 weeks); five of 35 babies (14%) were delivered by caesarean section. The average of birth length was 42.0 cm (- 6.2 standard deviation score (SDS) (range -12.2 to -1.8)), and that of weight was 2330 kg. Average SDS in each age was from -4.23 to -6.57, and there was no statistically significant correlation between age and SDSs (r=0.13, P=0.07). The average of occipitofrontal head circumfluence was 36.3 cm (range 32 to 37), which was comparable for gestational age.

Eighteen (41.9%) infants had a severe condition at birth because of asphyxia and respiratory distress and 11 (25.6%) patients of them required ventilator assistance. In 5 families (11.6%) the newborns, who were diagnosed as having the same disease from clinical findings, died immediately after birth with unknown etiology. This respiratory failure was not due to respiratory distress syndrome (RDS) caused by a lack of lung surfactant or a lung infection. The clinical features of the patients are summarized in Table 111-2.

In radiological findings, slender tubular bones and ribs were observed only in one patient and high vertebral bodies with reduced anterior-posterior and transverse diameter in the lumbar region were observed in four patients (9.3%). Flared metaphyses was observed in 11 patients (25.6%) and spina bifida occulta was observed in nine patients (20.9%). Bone age of the patient's hands delays in the first decade, then catch up with normal bone age and outstrip the normal after the age of 15 years old. After the age of 16 years, the entire patient begins to have problems with the bones, such as arthrosis, osteochondroses, and osteoporoses.

TABLE 111-2 CLINICAL FEATURES AT THE FIRST TIME OF DIAGNOSIS IN 43 YAKUT PATIENTS WITH 3-M SYNDROME

	Frequency (percentages and numbers of patients who had each clinical feature.)				
	Age at first examination				
Feature	0-5 years	6-10 years	11-15 years	> = 16 years	total
Number of patients	25	8	5	5	43
HEAD AND FACE % (NO.)					
Hydrocephaloid skull	100 (25)	87.5 (7)	60.0 (3)	0	81.4 (35)
Characteristic face*	96.0 (24)	87.7(7)	100 (5)	100(5)	95.3 (41)
Hypertelorism	36.0 (9)	50.0 (4)	20.0 (1)	100 (5)	44.1 (19)
Epicanthus	56.0 (14)	12.5 (1)	40.0 (2)	0	39.9 (17)
Rich and wide eyebrows	72.0 (18)	50.0 (4)	80.0 (4)	40.0 (2)	65.1 (28)
Dysplasia of ears	40.0 (11)	25.0 (4)	100 (5)	0	46.5(20)
BODY % (NO.)					
Short neck	92.0 (23)	100 (8)	100 (5)	100 (5)	95.3 (41)
Short and wide thorax	100 (25)	100 (8)	100 (5)	100 (5)	100 (43)
Deformed sternum	96.0 (24)	37.5 (3)	60.0 (3)	0	69.8 (30)
Accentuated lumbar lordosis	100 (25)	87.5 (7)	100 (5)	0	86.0 (37)
Muscular hypotonia	100 (25)	87.5 (7)	100 (5)	0	86.0 (37)
Brachydactyly	100 (25)	100 (8)	100 (5)	100 (5)	100 (43)
Micromelia of hands and feet	100 (25)	100 (8)	100 (5)	100 (5)	100 (43)
Simian crease	56 (14)	75.0 (6)	40.0 (2)	0	51.2 (22)
Clinidactility of the 5th fingers	12.0 (3)	0	20.0 (1)	0	9.3 (4)
Huge abdomen	100 (25)	0	0	0	58.1 (25)
Prominent heels	100 (25)	100(8)	100 (5)	100 (5)	100 (43)
Birth heart defects	16 (4)	0	0	0	9.3 (4)
Hepatomegaly	20 (5)	0	0	0	11.6 (5)
Platypodia	36 (9)	37.5 (3)	20.0 (1)	0	30.2 (13)
CONSTITUTION % (NO.)					
Short proportional stature	100 (25)	100 (8)	100 (5)	100 (5)	100 (43)
Postnatal growth and weight failure	100 (25)	100 (8)	100 (5)	100 (5)	100 (43)
RADIOLOGICAL FINDINGS % (NO.)					
Scoliosis	4.0 (1)	37.5 (3)	60.0 (3)	0	16.3 (7)
Slender tubular bones and ribs	0	0	20.0 (1)	0	2.3 (1)
Flared metaphyses	8.0 (2)	12.5 (1)	60.0 (3)	100 (5)	25.6 (11)
High vertebral bones with reduced anterior-posterior and transverse diameter in lumbar region	8.0 (2)	0	40.0 (2)	0	9.3 (4)
Spina bifida occulta	8.0 (2)	12.5 (1)	60.0 (3)	60.0 (3)	20.9 (9)
Bone age of the hands correspondence with chronological age	0	12.5 (1)	60.0 (3)	0	9.3 (4)

Notes: * triangular face, hypoplastic midface, broad, frontal bossing, low nasal bridge, depressed nasal root, long philtrum, full lips, prominent mouth and lips, pointed chin.

Genome-wide genotyping analysis was carried out in 37 Yakut families (Maksimova et al., 2007b). It showed genetic linkage with *D6S282* locus (LOD score 24.6). A candidate region of the gene was established between *D6S1552* and *D6S271*markers in the Cullin 7 (*CUL7*) gene responsible for the development of 3-M syndrome, and mapped relatively recently by a team of scientists from several countries (Huber et al., 2005). Since the clinical data for Yakut patients were similar to the clinical picture of the rare 3-M syndrome, the authors concluded that *CUL7* was a good candidate gene for the Yakut variant of nanism, and performed direct sequencing of all 25 coding exons of the *CUL7* gene. In all 43 patients, a novel mutation,4582insT, was identified in the *CUL7* gene in the homozygous state. This permitted the correct diagnosis of the disease in the patients; their 34 relatives were found to be heterozygous carriers and five were healthy. 3-M syndrome (OMIM 273750) belongs to the group of hereditary syndromes with a pathological short stature. The disease was named by the initial letters of the surnames of three authors (Miller, McKusick, Malvaux) who first described this syndrome in 1975 (Miller et al., 1975). This rare hereditary disease with the autosomal recessive type of inheritance is characterized by prenatal and postnatal growth retardation, facial dysmorphias, normal intellect, and roentgenological changes in bones (slender tubular bones and long vertebrae). The frequency of 3-M syndrome in the world is unknown, since only about 60 clinical cases of the disease in different ethnic groups have been described in 30 years (Cantu et al., 1981; Temtamy et al., 2006). The *CUL7* (*KIAA0076,* Cullin 7) gene causing this disease was mapped and identified in 2005 in chromosome 6p21.1 (Huber et al., 2005). It consists of 26 exons and encodes the CUL7 protein that belongs to the family of cullins. This family comprises structurally associated proteins involved in the processes of cell regulation, signal transduction, oxygen regulation, and DNA repair. Huber et al. (2005) described 25 mutations for this gene among patients with 3-M syndrome in 29 families from Tunis, Morocco, France, Algeria, Syria, Portugal, Germany, Sri Lanka, Turkey, Austria, Italy, Surinam, India, and Brazil. Nineteen of these mutations are predictable premature termination of translation and six are missense mutations. In 22 families, the patients were consanguineous relatives. Thus, in Yakut patients the 3-M syndrome is caused by a single, previously nondescribed, mutation 4582insT in the *CUL7* gene. The sole mutation in the *CUL7*gene identified in all patients, and the identified haplotype, point to a founder effect. The incidence of the Yakut syndrome of short stature is 12.72 per 100,000 individuals (Maksimova et al., 2008). DNA diagnostics of mutation 4582insT in the *CUL7* is possible in regional medical genetic consultation.

SOPH (SHORT STATURE WITH OPTIC NERVE ATROPHY AND PELGER-HUET ANOMALY) SYNDROME

In the regional department of medical genetic consultation of the Republican Hospital No. 1—National Medical Centre in Yakutsk (Russia), we enrolled 31 Yakut families— 22 (64.7%) female and 12 (35.3%) male patients—in whom the following clinical criteria were fulfilled: short stature, autosomal recessive inheritance, and optic nerve atrophy (Maksimova et al., 2010). The average SDS of height at birth was −0.94 ± 0.29 in females, 0.31 ± 0.42 in males, and −0.65 ± 0.26 in total (mean ± SD). Although the SDS of height was within 1 SDS at birth, after 1 year of age, the average SDS of height was −4.44 ± 1.19 in females, −3.16 ± 1.06 in males, and −4.01 ± 1.29 in total (mean ± SD).

The clinical characteristics of the 34 patients are presented in Table 111-3. Ninety-seven per cent (33 cases) of the patients had normal intelligence. Eighty-eight per cent of the patients had a brachycephalic skull with hypoplasia of frontal and parietal tubers and a flat occipital area of the head. The patients had a characteristic face with straight nose, prominent glabella, small orbit, slight bilateral exophthalmos, hypoplastic cheekbone, narrowforehead, long philtrum, and thin lips. The patients' tissue turgor and skin elasticity were decreased in some patients. Size of the head was proportional for the body length, whereas length of the four limbs was relatively short for the body length. Their hands and feet were small (micromelia). Their feet were wide and small, accompanied by a high arch and sandal chink (about 90%). Brachydactyly was observed in all patients (100%). Male hypogonadism was not observed. Size of the uterus and ovaries of the female patients was normal, except in four individuals. One affected woman has a son of normal height. Brain magnetic resonance images (MRIs) were obtained for 24 patients and showed slight cerebellar atrophy (3 cases), Dandy-Walker malformation (1 case), cyst at internal capsule (1 case), and empty selae (1 case). Growth hormone provocation test with insulin was normal in all examined patients (13 cases). Serum free thyroxine was normal in all examined patients (28 cases).

One of the patients' initial symptoms was visual loss. The onset of visual loss was 4.3 ± 1.4 years of age (mean ± SD; n = 34). All patients had bilateral optic nerve atrophy (100%) and non-progressive impairment of visual acuity (0.23 ± 0.21 [mean ± SD; n = 25]). Disc paleness with different degree of and expanded excavation was observed on the fundi. Borders of the optic disc were clear, and arteries and veins were normal. None of the patients had visual field defect, central scotoma, or nystagmus. All examined patients11 had great difficulty reading any of the Ishihara pseudoisochromatic plates (24 plate edition, 1964). We also performed the Farnsworth panel D-15 test on these 11 patients and found that one patient made a tritan error. We obtained electric sensitivity threshold and labiality from 30 patients. The values were 245.00 ± 101.34 mkA for electric sensitivity threshold and 25.20 ± 3.49 Hz for labiality, indicating an impairment of functions of the retina and the optic nerve. Myopia (12 cases), strabismus (nine cases), and hypermetropia (three cases) were also observed.

The incidence of the syndrome is 9.95 per 100,000 Yakut individuals (Maksimova et al., 2009). On the basis

TABLE 111-3 CLINICAL FEATURES AT FIRST DIAGNOSIS IN 34 YAKUT PATIENTS WITH SHORT STATURE, OPTIC ATROPHY, PELGER-HUET ANOMALY SYNDROME

	Frequency; numbers of patients who had each clinical feature (percentage)		
	Sex		
	Women	Men	Total
Number of patients	22	12	34
CONSTITUTION			
Normal length at birth	11 (50)	9 (75)	20 (58)
Postnatal growth failure	22 (100)	12 (100)	34 (100)
CRANIOFACIAL FEATURES			
Brachycephalic skull	20 (90.9)	10 (83.3)	30 (88.2)
Hypoplasia of frontal and parietal tubers	20 (90.9)	11 (92)	31 (91.2)
Narrow forehead	20 (90.9)	11(91.7)	31 (91.2)
Long senile face	21 (95.4)	12 (100)	33 (97.1)
Small features of face	19 (86.4)	10 (83.3)	29 (85.3)
Facial asymmetry	19 (86.4)	5 (41.7)	24 (70.6)
Straight nose with prominent glabella	19 (86.4)	12 (100)	31 (91.2)
Thick/bushy eyebrows	15 (68.2)	9 (75)	24 (70.6)
Small orbit	22 (100)	11 (91.7)	33 (97.1)
Bilateral exophthalmos	20 (90.9)	11 (91.7)	31 (91.2)
Hypertelorism	3 (13.6)	2 (16.7)	5 (14.7)
Epicanthus	16 (72.7)	6 (50)	22 (64.7)
Hypoplastic cheekbone	21 (95.4)	12 (100)	33 (97.1)
Long philtrum	19 (86.4)	9 (75.0)	28 (82.4)
Thin lips	20 (90.9)	8 (66.7)	28 (82.4)
BODY AND EXTREMITY FEATURES			
Short neck	19 (86.4)	11 (91.7)	30 (88.2)
Loose and senile skin	22 (100)	12 (100)	34 (100)
Depressed turgor of tissue	22 (100)	12 (100)	34 (100)
Fine hair	17 (77.3)	8 (66.7)	25 (73.5)
Hypermobility of small joints	20 (90.9)	10 (83.3)	30 (88.2)
Muscular hypotonia	18 (81.8)	10 (83.3)	28 (82.4)
Micromelia	22 (100)	12 (100)	34 (100)
Brachydactyly	22 (100)	12 (100)	34 (100)
Syndactyly	1 (4.6)	1 (8.3)	2 (5.9)
Simian crease	3 (13.6)	2 (16.7)	5 (14.7)
Wide feet with high arch	19 (86.4)	11 (91.7)	30 (88.2)
Sandal chink	18 (81.8)	8 (66.7)	26 (76.5)
Wide big toe	18 (81.8)	8 (66.7)	26 (76.5)
OPHTHALMOLOGICAL FINDINGS			
Bilateral optic nerve atrophy	22 (100)	12 (100)	34 (100)
Strabismus	6 (27.2)	3 (25.0)	9 (26.5)
Pigmented nevus	1 (4.6)	1 (8.3)	2 (5.9)

(Continued)

TABLE 111-3 (CONTINUED)

	Frequency; numbers of patients who had each clinical feature (percentage)		
	Sex		
	Women	Men	Total
Myopia	9 (40.9)	3 (25.0)	12 (35.3)
Hypermetropia	1 (4.6)	2 (16.7)	3 (8.8)
RADIOLOGICAL FINDINGS			
Delay of chronological age	7 (31.8)	7 (58.3)	14 (41.2)
NEUROLOGICAL FINDINGS			
Normal intellectual function	21 (95.5)	12 (100)	33 (97.1)
PATHOLOGY OF OTHER SYSTEMS			
High voice with harsh timber	21 (95.5)	11 (91.7)	32 (94.1)
Hypoplasia of uterus	4 (18.2)	0	4 (11.8)
Insulin-dependent diabetes	2 (9.1)	0	2 (5.9)
Hypoplasia of thyroid gland	2 (9.1)	1 (8.3)	3 (8.8)

of these distinct clinical features, this new type named of syndrome as "short stature syndrome with optic atrophy and Pelger-Huët anomaly" (SOPH) (Maksimova et al., 2010). Genomewide homozygosity mapping was conducted in 33 patients in 30 families. The disease locus was mapped to the 1.1 Mb region on chromosome 2p24.3, including the neuroblastoma amplified sequence (*NBAS*) gene. Subsequently, 33 of 34 patients were identified with SOPH syndrome and had a 5741G/A nucleotide substitution (resulting in the amino acid substitution R1914H) in the *NBAS* gene in the homozygous state. None of the 203 normal Yakuts individuals had this substitution in the homozygous state. Immunohistochemical analysis revealed that the NBAS protein is well expressed in retinal ganglion cells, epidermal skin cells, and leukocyte cytoplasm in controls as well as a patient with SOPH syndrome.

CONCLUSION

Eight Yakut hereditary diseases (YHD) have been distinguished by two criteria. First, their rate of occurrence is significantly higher than in the world population, and five diseases—spinocerebellar ataxia type I, 3-M syndrome, SOPH syndrome, autosomal recessive deafness 1A type and enzymopenic methemoglobinemia—have been assigned to the world foci of accumulation of hereditary diseases. Second, specific features of the molecular nature of each of them have been established, which distinguish these diseases from similar phenotypes in other populations where they occur much more rarely.

Theoretical importance of molecular genetic investigations of ethnospecific and hereditary human pathologies, along with traditional research areas (genetic heterogeneity and clinical polymorphism of diseases, genetic diversity of human populations, precise establishment of population dynamics factors determining HD accumulation), also lies in the fact that such investigations provide a tool for systematization of the great diversity of hereditary pathologies, i.e., for specifying the available classifications of diseases and creating new ones, so necessary to clinicians. It is pertinent to note that approaching exact knowledge of the mechanisms of manifestation of a mutant gene in the phenotype of a Mendelian disease suggests more effective ways for identifying candidate genes of multifactorial diseases, since most of our knowledge in the field of molecular mechanisms underlying widely distributed diseases is the knowledge obtained in studying rare hereditary forms of pathology (Peltonen et al., 2006).

REFERENCES

Barashkov N, Dzhemileva L, Fedorova S, et al. (2011). Autosomal recessive deafness 1A (DFNB1A) in Yakut population isolate in Eastern Siberia: extensive accumulation of the splice site mutation IVS1+1G4A in GJB2 gene as a result of founder effect. *J Hum Genet* 56(9):631–639.

Becher MW, Morrison L, Davis LE, et al. (2001). Oculopharyngeal muscular dystrophy in Hispanic New Mexicans. *JAMA* 286:2437–2440.

Blumen SC, Korczyn AD, Lavoie H, et al. (2000). Oculopharyngeal MD among Bukhara Jews is due to a founder (GCG)9 mutation in the *PABP2* gene. *Neurology* 1:1268–1270.

Bouchard J-P, Brais B, Brunet D, et al. (1997). Recent studies on oculopharyngeal muscular dystrophy in Quebec. *Neuromusc Disord* 7:S22–S29.

Braise B, Bouchard J-P, Xie Ya-G, et al. (1998). Short GCG expansions in the *PABP2* gene cause oculopharyngeal muscular dystrophy. *Nat Genet*18:164–167.

Cantu JM, Garcia-Cruz D, Sanchez-Corona J, et al. (1981). 3-M slender-boned nanism: an intrauterine growth retardation syndrome. *Am J Dis Child* 135:905–908.

Fedorova SA, Khusainova RI, Kutuev IA, et al. (2005). Polymorphism of CTG-repeats in the *DMPK* gene in populations of Yakutia and Central Asia. *Mol Biol(Moscow)* 39:385–393.

Galeeva NM, Nazarenko LP, Nazarenko SA. (2006). Molecular genetic cause of hereditary methemoglobinemia type 1 in Yakutia. *Med Genet*5:15–20.

Goldfarb LG, Vasconselos O, Platonov FA, et al. (1996). Unstable triplet and phenotypic variability of spinocerebellar ataxia type 1. *Ann Neurol* 39:500–506.

Harper PS, Van Engelen B, Eymard B, Wilcox DE. (2004). *Myotonic Dystrophy: Present Management, Future Therapy.* New York: Oxford University Press.

Huber C, Dias-Santagata D, Glaser A, et al. (2005). Identification of mutations in *CUL7* in 3-M syndrome. *NatGenet* 37:1119–1124.

Illarioshkin SN, Rudenskaya GE, Ivanova-Smolenskaya IA, et al. (2006). *Nasledstvennye ataksii i paraplegii* [Hereditary Ataxias and Paraplegias] MEDpress-inform, Moscow.

Korotov MN, Kuz'mina ZM. (1994). Prevalence of hereditary degenerative nervous system diseases in Republic of Sakha. In *Aktual'nye Voprosy Zdorov'ya Naseleniya Respubliki Sakha* [Current Issues of Population Health in Republic of Sakha], Yakutsk, pp. 132–134.

Kucher AN. (2002). The population structure of indigenous and alien population of Siberian region. *Med Genet*1:262–270.

Lunkes A, Goldfarb LG, Platonov FA, et al. (1994). Autosomal dominant spinocerebellar ataxia (SCA) in Siberian ataxia type 1. *Exp Neurol* 126:310–312.

Maksimova NR, Korotov MN, Nikolaeva IA, et al. (2007a). Clinical and molecular genetic aspects of oculopharyngeal muscular dystrophy in Republic of Sakha (Yakutiya), *Genetika i patologiya* [Genetics and Pathology], Puzyrev VP, ed., Pechatnaya manufaktura, Tomsk, 8:160–161.

Maksimova N, Hara K, Miyashita A, et al. (2007b). Clinical, molecular and histopathological features of short stature syndrome with novel *CUL7* mutation in Yakuts: new population isolate in Asia. *J Med Genet* 44:772–778.

Maksimova NR, Sukhomyasova AL, Gurinova EE, et al. (2008). Ethno specific hereditary pathology in Yakutia: genetical epidemiological and social economic aspects. *Med Genetics* 10:35–43.

Maksimova NR. (2009). Clinical –genealogical and molecular-genetical pathology in Extended *Abstract of Doct. Sci. (Med.) Dissertation*, Tomsk.

Maksimova N, Hara K, Nikolaeva I, et al. (2010) Neuroblastoma amplified sequence gene is associated with a novel short stature syndrome characterised by optic nerve atrophy and PelgereHue¨t anomaly. *J Med Genet* 47:538–548.

Miller JD, McKusick V, Malvaux P, et al. (1975). The 3-M syndrome: a heritable low birthweight dwarfism. *Birth Defects Orig Artic Ser* 11:39–47.

Nazarenko LP, Nogovitsina AN, Salyukova OA, Filimonova MN. (2002a). Occurrence of monogenic hereditary diseases among rural population in Republic of Sakha (Yakutia), *Mediko-geneticheskoe issledovanie naseleniya Respubliki Sakha(Yakutiya).*[Medical Genetic Study of the Population of Republic of Sakha (Yakutia)], Collection Sci. Proc., Puzyrev VP, ed., Izd. YaF SO RAN, Yakutsk. pp. 92–114.

Nazarenko LP, Sazhenova EA, Nazarenko SA. (2002b). Mutation of Enzymopenic Hereditary Methemoglobinemia Gene in the Yakut Population, *Mediko-geneticheskoe issledovanie naseleniya Respubliki Sakha (Yakutiya).*[Medical Genetic Study of the Population of Republic of Sakha (Yakutia)], Collection Sci. Proc., Puzyrev VP, ed., Izd. *Yakut Filial, Ross Akad Nauk* Yakutsk: pp. 151–159.

Nogovitsina AN, Maksimova NR, Khandy MV, Alekseeva SP. (1999). Hereditary pathology in the families, asked for medical health at genetic counseling department of the National Medical Center Republic of Sakha (Yakutia) in 1990 to 1998. *Dal'nevost Med Zh*1:26–30.

Nogovitsina AN. (2001). Hereditary pathology load in population of republic of sakha (Yakutiya) and the analysis of regional medical genetic polyclinic activity. *Cand Sci (Med.) Dissertation*, Tomsk, 164 p.

Palau F, Espinos C. (2006). Autosomal recessive cerebellar ataxias. *Orphanet J Rare Diseases*1:47.

Peltonen L, Perola M, Naukkarien J, Palotie A. (2006). Lessons from studying monogenic disease for common disease. *Hum Mol Genet*15:R67–R74.

Platonov FA, Illarioshkin SN, Kononova SK, et al. (2004). Spinocerebellar ataxia type 1 in Yakutia: prevalence and clinical genetic comparisons. *Med Genet* 5:242–248.

Popova SN, Slominskii PA, Galushkin SN, et al. (2002). Allelic polymorphism of triplet repeats (CT6) and (CA6) in loci *DM, DRPLA*, and *SCA1* in several populations of Russia. *Russ J Genet*38:1549–1553.

Puzyrev VP. (1984). Characteristics of the incidence of hereditary diseases in western Siberia. *Vestn Ross Akad Med Nauk*7:76–78.

Puzyrev VP, Erdynieva LS, Kucher AN, Nazarenko LP. (1999). *Genetiko-epidemiologicheskoe issledovanie naseleniya Tuvy* (Genetic Epidemiologic Study of the Tuva Population) STT, Tomsk.

Puzyrev VP, Nazarenko LP. (2000). *Genetiko-epidemiologicheskoe issledovanie nasledstvennoi patologii v Zapadnoi Sibiri* (Genetic Epidemiological Study of Hereditary Disorders in Western Siberia) STT, Tomsk.

Puzyrev VP. (2003). Genetics of multifactorial diseases: between past and future. *Med Genet* 2:498–508.

Puzyrev VP, Stepanov VA, Golubenko MV, et al. (2003). mtDNA Lineages and Y-chromosomes in the Yakut populations. *Russ J Genet* 39:816–822.

Puzyrev VP, Freidin MB, Kucher AN. (2007). *Geneticheskoe raznoobrazie narodonaseleniya i bolezni cheloveka* (Genetic Diversity of Population and Human Diseases) Pechatnaya manufaktura, Tomsk.

Rodriguez M, Camejo C, Bertoni B. (2005). $(GCG)_{11}$ Founder Mutation in the *PABPN1* gene of OPMD Uruguayan families. *Neuromusc Disord* 15:185–190.

Rychkov Yu G (2000). *Genofond i genogeografiya narodonaseleniya* [Gene Pool and Genogeography of Population] Vol. 1: *Genofond naseleniya Rossii i sopredel'nykh stran* [Gene Pool of Population of Russia and Contiguous Countries], Nauka, St. Petersburg.

Scott EM. (1960). The relation of diaphorase of human erythrocytes to inheritance of methemoglobinemia. *J Clin Invest* 39:1176–1179.

Stepanov VA. (2002). *Etnogenomika naseleniya Severnoi Evrazii* (Population Ethnogenomics of Northern Eurasia). Pechatnaya manufaktura, Tomsk.

Sukhomyasova AL. (2005). Autosomal Dominant Myotonic Dystrophy in Republic of Sakha (Yakutiya), *Extended Abstract of Cand. Sci. (Med.) Dissertation* Tomsk, p. 22.

Tarskaya LA, Zinchenko RA, El'chinova GI, et al. (2004). The Structure and Diversity of Hereditary Pathology in Republic of Sakha (Yakutia). *Russ J Genet* 40:1530–1539.

Temtamy SA, Aglan MS, Ashour AM, et al. (2006). 3-M Syndrome: A Report of Three Egyptian Cases with Review of the Literature. *Clin. Dysmorphol* 15:55–64.

Tokarev YuN, Hollan SR, Korral Almonte HFM (1983). *Nasledstvennye anemii i gemoglobinopatii* (Hereditary Anemias and Hemoglobinopathies), Meditsina, Moscow.

Vladimirtsev AI. (1964). *Khronicheskii Yakutskii (Vilyuiskii) entsefalit za 12 let po materialam nevrologicheskogo otdeleniya yakutskoi respublikanskoi bol'nitsy* (Chronic Yakutsk (Vilyuisk) Encephalitis: 12-Year Study by the Neurological Department of the Republic Hospital), Yakutsk.

Zak M. (1932). Betyugen - Original Hereditary Form of Bulbar Palsy among Yakuts. *Sov Nevropatol. Psikhogigiena* 1:814–817.

Zubri GL, Goldfarb LG. (1975). Hereditary Cerebellar Ataxia in Yakutia, in *Pervaya Vsesoyuznaya konferenzia pomeditsinskoi genetike* (First All-Union Conference on Medical Genetics) Izd. AMN, Moscow:

112.

HEREDITARY DISEASES IN SIBERIAN POPULATIONS

Ludmila P. Nazarenko and V.P. Puzyrev

INTRODUCTION

A prominent place in human population genetics, in which the object of investigation is a population as an independent and relatively isolated unit, historically formed on a certain territory and self-reproduced within its area—is occupied by medical genetic examinations. An important achievement in genetics of the twenty-first century is deciphering of the human DNA nucleotide sequence. It provided medical geneticists with their own substrate for investigation, the genome, analogous to those of cardiologists (heart and vessels) and neuropathologists (nervous system; McKusick, 1997). Thus, molecular genetics has become a tool for studying population processes on a new level and for describing the molecular structure of genes that constitute gene pools of human populations and are the genetic basis of human diseases.

Genetic epidemiological investigations of hereditary diseases (HDs) and the load of hereditary pathologies in different ethnic populations in Russia have their own history. In 1968, this scientific direction was theoretically substantiated by N.P. Bochkov and N.V. Timofeeff-Ressovsky. They noted that the maintenance of mutations leading to pathologies and their accumulation in certain ethnic groups is favored by the specificity of ecological factors and traditional ways of life, especially reproductive behavior, and that investigations in this area may play a significant part in discovering new gene alleles (Bochkov and Timofeeff-Ressovsky, 1968). The first Russian publication by N.P. Bochkov et al. (1971) opened work on systematic examinations of isolated population groups in the Soviet Union (Bochkov et al., 1971). The Research Center for Medical Genetics of the Russian Academy of Medical Sciences has collected and analyzed data on the HD load among numerous ethnic groups in Central Asia and in the European part of Russia (Ginter, 2001, 2002; Zinchenko et al., 2007).

Medical genetic investigations in Siberia were started in the 1980s (Puzyrev, 1984) and are in progress today on the territory of the Tomsk region (Russians), Tyva, Buryatia, Gornyi Altai, Khakassia, and Yakutia (Puzyrev et al., 1999; Nogovitsyna et al., 1999; Puzyrev & Nazarenko, 2000; Nazarenko et al., 2002).

The aim of this chapter is to represent a survey of medical genetic investigations in Siberia.

SHORT EXCURSUS IN THE PROBLEM OF "ETHNIC" DISEASES

The problem of ethnic features of HD occurrence in the world population has a long history, and the term *features* was first referred to the HD load in individual populations. Two pioneer works, started in the 1950s and being continued at present in British Columbia (Canada) and in Northern Ireland (Baird et al., 1988; Stevenson, 1959), are considered to be the most demonstrative investigations in this aspect. The British Columbia (a province of Canada) register includes about 60 registration sources, with data about patients with hereditary pathology from hospitals, private practice physicians, welfare offices, etc., and statistical data about population size, movement, birth rate, and death rate. Analysis of the obtained data permit evaluation of the frequencies of some HDs. Summary data about frequency of birth defects and hereditary pathology in British Columbia is based on monitoring more than 1 million neonates. Another observation method of obtaining data concerning the patients with suppositional genetic pathology is used in Northern Ireland. Per saltum, the data concerning patients from all practicing physicians was obtained and it allowed the estimation of prevalence of autosomal dominant (AD) and autosomal-recessive pathology (AP) in Northern Ireland.

Another matter of interest regards populations preserving a traditional way of life and existing for a long time in relative isolation. In such populations a new allele is either introduced from the outside by the founder of a population (founder effect) or has occurred de novo, and it is problematic to distinguish these two situations. It is diseases accumulated in such populations for which the following terms are applied: *ethnically specific* hereditary diseases, *ethnically assigned* diseases, and diseases with *local accumulation*. Such diseases often receive the names of populations in which they occur more frequently than in neighboring populations or in the world population as a whole, and have specific mutations at their base ("Jewish" or "Finnish" diseases).

The first Jewish hereditary diseases were described when studying skeletal dysplasia and dwarfism in the Amish in Lancaster (Pennsylvania, United States) in the 1960s. The study was started with a visit of V. McKusick to the community of short-stature people (McKusick, 2006). Accumulation of two recessive forms of dwarfism was discovered. One form is the Ellis van Creveld syndrome, known for a long time but first recorded in very many cases within a single ethnic group. The second form (metaphyseal chondrodysplasia, type McKusick) was discovered in 77 patients in 53 Amish families and described for the first time. In addition, diseases associated with disturbances in lipid metabolism were found to be widely distributed in the Amish: Tay-Sachs disease, Niemann-Pick disease, and Gaucher's disease (adult form, type I). A review for these and other hereditary diseases in the Amish is given in McKusick, 1978. The results of molecular genetic studies of Jewish hereditary diseases have been published recently (Francomano and Biesecker, 2003).

Several conditions promoted the discovery of Finnish hereditary diseases: traditional population structure preserved within limited geographical territories of the country, high level of medical care, and accurate records of family histories in the parish books made throughout the last ten generations. Thirty-two hereditary diseases have been mapped in the Finnish population. They occur rather frequently in this population, and very rarely or never at all outside the country (Peltonen et al., 1999; Kere, 2001). Major mutations were detected in 70% of all mapped genes. It should be noted that the centers of origin of these diseases are located in just some geographical regions of the country, but the diseases are practically absent in the populations of large cities. Rare (compared to other populations) Finnish recessive diseases are congenital nephrotic syndrome (Finnish form), aspartylglucosaminuria, early childhood ceroid lipofuscinosis, congenital applanation, Melbry's dwarfism, and other diseases (Norio, 2003).

It should be mentioned in conclusion that the present situation in respect of Mendelian diseases is much the same as the situation with the classification of nonchromosomal congenital developmental defects based on phenotypic manifestations. The phenotypic description of hundreds of syndromal developmental defects had an important influence on the tactics of clinical attendance of patients and on genetic counseling of their families, and it is actively supplemented with findings from molecular biology, due to which new groups of diseases are distinguished. This pathway of systematization of hereditary diseases was defined in the 1960s by McKusick as "lumping and splitting" on the basis of the data on molecular genetics of phenotypes, specifying their attribution to one or another "genetic nosology" (McKusick, 1969). In this sense, the data on molecular genetics of Mendelian phenotypes, by which the spectra of hereditary diseases in different ethnic populations are compared, limit the conclusiveness of advanced statements about the peculiarities of population processes and regularities of formation of a load of hereditary diseases. It has been correctly noted that molecular genetic tests are qualitative, precise, and unambiguous as compared to clinical symptoms that are described quantitatively, sometimes being hardly perceptible and indistinguishable, and are clearly subjective (Biesecker, 1998).

APPROACHES TO MEDICAL GENETIC RESEARCH OF THE SIBERIAN POPULATIONS

It is possible to allocate for the territories of Siberia some groups of the population that differ according to their population structure (isolated and panmixed populations), degrees of urbanization (rural and urban population), and adaptation to climate and geographical conditions of inhabitancy (the migrated Russian population and natives—Khanty, Mancy, Nentsy, Selkups, Altaians, Ketas, Khakases, Tyvinians, and Yakuts). Our research on prevalence of hereditary pathology among the populations of Siberia concerned both aboriginal populations (the Tyvinians, the Altaians, the Khakases, the Yakuts), and Russian populations (Tomsk area, Republics of Altai and Khakassia). The organization of research remained strictly constant for all populations, and provided a number of sources of obtaining the data: questioning of medical workers with the purpose of revealing the possible hereditary pathology, data obtained by medical commissions of experts and specialized medical institutions (school for deaf and blind children, boarding schools for children with mental retardation). The data collected about the patients with presumed HDs in the population (usually an observation unit was a district) made up an initial database. The taped interviews with patients and members of their families were exposed to double medical examination. The first provided revealing of patients with suppositional hereditary pathology, and the purpose of the second was to find an exact diagnosis. The family material collected in such a way was subjected to complex segregation analysis for revealing a share of sporadic cases of diseases, with the analysis of loading in the population with Mendelian hereditary pathology. During study of the selected material, we used various methods of genetic and statistical analysis described in detail in works by E.K. Ginter and co-workers (Ginter et al., 1997) and in our published works (Nazarenko and Salyukova, 1997).

EPIDEMIOLOGY OF HEREDITARY DISEASES AMONG RUSSIAN POPULATIONS OF THE TOMSK REGION

The Tomsk region is a member of the Russian Federation and is a part of the Siberian Federal District. The administrative center is the city of Tomsk. There are 16 districts in Tomsk region. The region shares borders with Krasnoyarsk Krai, Tyumen, Omsk, Novosibirsk, and Kemerovo Oblasts. The area from the north to the south is about 600 km, and from the west to the east is 780 km. Most of the region's 316,900 km^2 territory is inaccessible because it is covered with taiga woods and swamps, in particular the Vasyugan bog is the largest in the world. Major ethnic groups living in the region are Russians (90.8%), Ukrainian-Belorussian (2.1%), Siberian Tatars (1.9%) and Volga German (1.3%).

PREVALENCE RATE OF HEREDITARY DISEASES AMONG THE RUSSIANS

Evaluation of hereditary diseases among the population of the Tomsk region are calculated by a hereditary pathology of the population of rural districts and small cities of the region, taking into account the number of rural and urban populations minus a share of the persons without children. So, according to the data received in the laboratory of population genetics of the Scientific Research Institute of Medical Genetics, Siberian Branch of the Russian Academy of Medical Science for the population of the Tomsk region, about 9% of married women of reproductive age are without children. Additionally, according to the results of the population census of 2002, about 3% of women and men of post-reproductive age were never married. In view of the above, in the present research the estimated share of persons who are childless is 12%.

In rural districts of the area, the estimation of the hereditary pathology had appeared variable enough for diseases with various modes of inheritance. So, the value of AD diseases varied from 0.25±0.12 up to 1.52±0.33 per 1000 persons, and the average estimation is $(0.82\pm0.05) \times 10^{-3}$. Thus, differences in value of some districts of the area have appeared highly significant ($\chi^2 = 70.89$; d.f. = 15; $P < 0.001$).

Estimations of AD diseases among all rural districts of Tomsk region can be conditionally divided by pathology into three groups, with high $(1.17–1.52) \times 10^{-3}$, average $(0.50–0.74) \times 10^{-3}$, and low $(0.25–0.37) \times 10^{-3}$ values. In these groups the differences in levels of HD load with the given mode of inheritance was not revealed, while between groups they were highly significant ($P < 0.001$).

The reason of such differentiation among rural districts of the area concerning AD pathology is difficult to establish. It is known that AD diseases depend on parameters of genetic structure of populations somewhat. Difference of the specified districts in the intensity of mutation processes is difficult enough to estimate, though not impossible. It is possible to assume only that the observable differentiation is the result of casual fluctuations of frequencies of "pathological" genes against a background of high intensity of migratory processes. It proves to be true also that a high proportion (70%) of families with AD diseases are those in which the affected parent was born outside the Tomsk area. Thus, it is possible to conclude that the basic share of the AD pathology in Tomsk area is "brought in" from various regions and their territorial distribution and, hence, estimations of the load are defined by casual distribution of carriers of mutant genes in the defined districts.

The load of X-linked pathology among rural populations also varied over a wide range: from zero value up to 0.93±0.17 per 1000 men (heterogeneity between districts is - $\chi^2 = 47.99$; d.f. = 15; $P < 0.001$). As mentioned above for AD diseases, it is noteworthy that in the X-linked pathologies, 66.7% of heterozygous mothers also were born outside the districts.

Levels of load of AR diseases among rural populations in the districts varied from 0.11±0.09 up to 1.56±0.14 per 1000 population (average – $0.74\pm0.05) \times 10^{-3}$. The obtained estimate of load of AR diseases in some districts appears to be highly significant ($\chi^2 = 110.61$; d.f. = 15; $P < 0.001$).

Three districts are sharply distinct from others. Tomsk and Teguldetsky districts have the highest levels of AR diseases at $(1.56\pm0.14) \times 10^{-3}$ and $(1.33\pm0.40) \times 10^{-3}$ respectively; the Asinovsky district appears to have the lowest AR diseases load, which is $(0.11\pm0.09) \times 10^{-3}$. Differences in HD load for the given districts, in comparison with the others, can be appreciably caused by random factors. However, for the Tomsk district the high level of AR diseases can be determined by intraregional migration of families having afflicted children, owing to the territorial relationship with the regional center, and hence greater availability of qualified medical aid. So, it has been shown that in at least in 11% of families with AR pathology registered in the Tomsk district, both parents of the patient are natives of other rural districts in the area. For the other districts of the Tomsk region, the values of HD load varied within much narrower limits, from 0.31±0.12 up to 0.61±0.20 per 1000 population ($\chi2 = 3.13$; d.f. = 12; $P > 0.05$).

It should be noted that in 64% of families with AR pathology, the parents were born outside the Tomsk region. The geographic distribution of their birthplaces was extensive, including Uzbekistan, Bashkiria, Mordovia, Stavropol Territory, Altai, Chuvashiya, etc. Thus, as in cases of the autosomal dominant and X-linked diseases, it is possible to speak about an appreciable share of "brought in" AR pathologies in the Tomsk region. Even when parents of the patients were born within the territory of the region, often it appeared that their grandfathers and grandmothers on both paternal and maternal lines were not local natives. A typical illustration of the situation is in the case of rare AR diseases, for example, Fanconi anemia registered in the Krivosheinsky district in a family with two patients. Parents of the patients were born in the Krivosheinsky district of the Tomsk area. However, the father of the proband is the descendant of immigrants from Belarus and Ukraine, and the mother is the descendant of immigrants of the last century from Lithuania. In turn, for this purpose, geographically closed areas are marked with a certain "concentration" of Fanconi's anemia gene, indicating that the this AR disease is an example of a genetic pathology "delivered" (prevalent) in the Tomsk region.

Highly variable for rural districts of the region, there were total estimations of hereditary pathology from 0.72×10^{-3} in the Zyrjansky district, up to 3.81×10^{-3} in the Tomsk district, and the average value was 2.04×10^{-3}. The structure of hereditary diseases for rural populations of the region was characterized by close values of the population load for AD and AR hereditary pathology $(0.82 \pm 0.05) \times 10^{-3}$ and $(0.74 \pm 0.05) \times 10^{-3}$ respectively ($P > 0.05$), and a considerably smaller share of contribution from the X-linked forms $(0.48 \pm 0.05) \times 10^{-3}$. The structure of hereditary diseases in the Siberian populations differs from Russian populations of the European part of Russia in that in the latter, there is statistically significant prevalence of AD forms above AR forms (Ginter et al., 1992).

Average estimation of the diseases with a varied mode of inheritance for the population of small cities appeared statistically significantly lower than for country populations. For inhabitants of small cities they were $(0.21 \pm 0.05) \times 10^{-3}$, $(0.35 \pm 0.06) \times 10^{-3}$, and $(0.03 \pm 0.05) \times 10^{-3}$ for autosomal dominant, autosomal recessive, and X-linked forms respectively, and the average value of total hereditary diseases was 0.60×10^{-3} (in comparison with rural districts for all estimations $P < 0.001$). There was also a different structure of HDs for the population of small cities, unlike the population of rural districts, which was characterized by insignificant prevalence of AR diseases above diseases with a dominant mode of inheritance, and practically did not differ from zero in load values from X-linked pathology.

There were low burden of AR diseases in the populations of small cities. The probable reason was higher intensity of migratory processes and lower parameters of subdivision in comparison with rural districts. The incomplete identification of patients was typical for AD and the X-linked diseases in the populations of small cities.

The structure of HDs in rural populations of the region was characterized by close values of AD and AR diseases—$(0.82 \pm 0.05) \times 10^{-3}$ and $(0.74 \pm 0.05) \times 10^{-3}$ respectively ($P > 0.05$)—with a contribution of X-linked forms at 0.48×10^{-3}. Total load of HDs of the rural population of the Tomsk region was 2.04×10^{-3}.

For the urban population of Tomsk, the load of AD and AR diseases also has appeared close, at 1.48×10^{-3} and 1.35×10^{-3} respectively ($P > 0.05$), and the contribution of X-linked forms is 0.66×10^{-3}. In Tomsk, the total load of HDs is 3.49×10^{-3}. Thus, the genetic load on the Tomsk population with AD and AR diseases is approximately identical; however, the city population has a higher level of load with HDs compared with the rural area. This difference is probably due to varied frequencies of mutant genes determined by various genetic structures of the populations. It is specified in the data of population-genetic research, which were collated in parallel with the medical genetic surveys.

DIVERSITY OF HEREDITARY DISEASES AMONG RUSSIANS

In this chapter, the results of epidemiological research of hereditary diseases with various modes of inheritance (autosomal dominant, autosomal recessive, X-linked, recessive and dominant)—i.e., that group of diseases growing out of the inherited genetic variations distributed in families according to Mendelian laws (Mendelian inheritance)—are presented.

The spectrum of AD diseases comprised 83 clinical forms (Table 112-1). The group with the hereditary syndromes that have included 36 nosological forms was the most varied. Among the syndromal pathologies were Marfan syndrome (21 families, 27 patients), Ehlers-Danlos syndrome (13 families, 24 patients), Fanconi renotubular syndrome (12 families, 22 patients), Gilbert syndrome (9 families, 17 patients), Crouzon syndrome (8 families, 17 patients), and Marie Unna hereditary hypotrichosis (6 families, 17 patients). It is necessary to note that among all families with the syndromal pathology, among 41% of the patients there was registered only one child, and the parents were healthy. From the formal genetic point of view, such families should be carried in a group of families with autosomal recessive pathology. But the clinico-genealogical analysis, using corresponding catalogues on hereditary diseases, has allowed us to classify the given diseases as autosomal dominant in inheritance. Incomplete penetrance and mutation *de novo* may be also the cause of the pathology in families.

Groups of diseases with the autosomal dominant mode of inheritance included skeletal and connective tissue diseases, including 13 nosological forms, the majority of which have arisen in two or more families, and the most widespread have appeared to be osteogenesis imperfecta (16 families, 36 patients), polydactyly (7 families, 20 patients), and chondrodysplasia (10 families, 16 patients). In the surveyed Russian population of the Tomsk region only an insignificant number of various clinical forms of diseases with the autosomal dominant mode of inheritance

TABLE 112-1 THE SPECTRUM AND PREVALENCES OF AD DISEASES AMONG RUSSIANS IN THE TOMSK REGION

No.	Disease	OMIM	Number of families	Number of patients	Prevalence rate, 1 case per:
DISEASES OF THE NERVOUS SYSTEM					
1.	Neurofibromatosis, type I	162200	27	45	19,712
2.	Charcot-Marie-Tooth disease, demyelinating	118200	16	23	38,567
3.	Spinocerebellar ataxia, type 1	164400	4	14	63,360
4.	Episodic kinesigenic dyskinesia	128200	9	9	98,560
5.	Episodic ataxia	108500	5	9	98,560
6.	Epilepsy, benign neonatal	121200	1	9	98,560
7.	Huntington disease	143100	5	7	126,720
8.	Tremor, hereditary essential	190300	2	4	221,760
9.	Facioscapulohumeral muscular dystrophy	158900	2	3	295,680
10.	Thomsen disease	160800	1	1	887,040
SKIN DISEASES					
11.	Vitiligo	193200	15	26	34,117
12.	Psoriasis	177900	8	23	38,567
13.	Alopecia areata	104000	8	16	55,440
14.	Nevus anemicus	163050	4	11	80,640
15.	Epidermolysis bullosa, Weber-Cockayne type	131800	3	5	177,408
16.	Pachyonychia congenita, type 1	167200	1	5	177,408
17.	Palmoplantar keratoderma, epidermolytic	144200	1	3	295,680
18.	Koilonychia	149300	1	1	887,040
SKELETAL AND CONNECTIVE TISSUE DISEASES					
19.	Osteogenesis imperfecta	166200	16	36	24,640
20.	Polydactyly, postaxial	174200	7	20	44,352
21.	Chondrodysplasia punctata	118650	10	16	55,440
22.	Spondyloepiphyseal dysplasia congenita	183900	4	9	98,560
23.	Achondroplasia	100800	5	8	110,880
24.	Hypochondroplasia	146000	5	8	110,880
25.	Metaphyseal chondrodysplasia, Schmid type	156500	4	8	110,880
26.	Split-hand/foot malformation	183600	4	6	147,840
27.	Contractural arachnodactyly	121050	3	6	147,840
28.	Exostoses, multiple	133700	3	5	177,408
29.	Epiphyseal dysplasia, multiple	132400	3	4	221,760
30.	Camptobrachydactyly	114150	1	3	295,680
31.	Osteochondromatosis, Ollier disease	166000	2	2	443,520
EYE DISEASES					
32.	Myopia	160700	12	28	31,680
33.	Retinitis pigmentosa	180100	14	21	42,240
34.	Ptosis, hereditary congenital	178300	6	13	68,234
35.	Cataract, total congenital	116700	5	13	68,234

(Continued)

TABLE 112-1 (CONTINUED)

No.	Disease	OMIM	Number of families	Number of patients	Prevalence rate, 1 case per:
35.	Optic atrophy	165500	5	12	73,920
36.	Heterochromia iridis	142500	5	8	110,880
37.	Nystagmus, congenital	164100	4	8	110,880
38.	Microphthalmia, isolated, with corectopia	156900	3	7	126,720
39.	Aniridia	106210	2	7	126,720
40.	Corneal dystrophy, lattice	122200	2	6	147,840
41.	Stargardt disease	600110	4	5	177,408
42.	Retinoblastoma	180200	3	5	177,408
BLOOD DISEASES					
43.	Spherocytosis	182900	20	35	25,344
44.	Von Willebrand disease	193400	4	6	147,840
45.	Erythrocytosis, familial	133100	1	2	443,520
46.	Cyclic hematopoiesis	162800	1	1	887,040
HEREDITARY SYNDROMES					
47.	Marfan	154700	21	27	32,853
48.	Ehlers-Danlos	130000	13	24	36,960
48.	Fanconi renotubular	134600	12	22	40,320
50.	Gilbert	143500	9	17	52,179
51.	Crouzon	123500	8	17	52,179
52.	Marie Unna hereditary hypotrichosis	146550	6	17	52,179
53.	Osler-Rendu-Weber disease	187300	6	12	73,920
54.	Sturge-Weber	185300	6	10	88,704
55.	Silver-Russell	180860	6	6	147,840
56.	Stickler	108300	5	6	147,840
57.	Klippel-Trenaunay-Weber	149000	5	6	147,840
58.	Noonan	163950	5	5	177,408
59.	Waardenburg, type I	193500	3	5	177,408
60.	Trichorhinophalangeal, type I	190350	2	5	177,408
61.	Treacher Collins-Franceschetti	154500	2	5	177,408
62.	Williams-Beuren	194050	4	4	221,760
63.	Waardenburg, type II	103470	1	4	221,760
64.	Prader-Willi	176270	3	3	295,680
65.	Rubinstein-Taybi	180849	3	3	295,680
66.	Poland	173800	3	3	295,680
67.	Digeorge	188400	2	3	295,680
68.	Aarskog	100050	2	3	295,680
99.	Leopard	151100	2	3	295,680
70.	Trichorhinophalangeal, type II	150230	2	2	443,520
71.	Hirschsprung disease	142623	2	2	443,520
72.	Duane-radial ray	607323	2	2	443,520

(Continued)

TABLE 112-1 (CONTINUED)

No.	Disease	OMIM	Number of families	Number of patients	Prevalence rate, 1 case per:
73.	Marshall	154780	1	2	443,520
74.	Hanhart	103300	1	1	887,040
75.	Robinow	180700	1	1	887,040
76.	EEC1	129900	1	1	887,040
77.	Kabuki	147920	1	1	887,040
78.	Opitz GBBB	145410	1	1	887,040
79.	Scheuermann disease	181440	1	1	887,040
80.	Riley-Smith	153480	1	1	887,040
81.	Nonne-Milroy lymphedema	153100	1	1	887,040
82.	Hemifacial microsomia	164210	1	1	887,040
	Total:		416	746	

have presented in individual families (30 diseases, i.e. 53.6% of the total number of clinical forms), and in 26 cases (46.3%) the pathology is registered in two and more families. Thus, "frequent" forms of the autosomal dominant pathology (1:55440) with the majority of affected is 413 persons or 55.4% of the total number of patients.

It is necessary to note, however, that in all cases families with the identical form of diseases have been territorially separated, and detailed genealogical analysis has not shown connections between them. Also, diseases as Marfan syndrome and Ehlers-Danlos, neurofibromatosis, Charcot-Marie-Tooth disease, and polydactyly, have shown rather high frequency of occurrence in Russian populations (Zinchenko et al., 2009), and that in itself can explain their observable accumulation.

The nosological spectrum of diseases with autosomal recessive mode of inheritance among the populations of the Tomsk region presents with 72 forms (Table 112-2). As in the case of autosomal dominant diseases, the most varied appears to be the group of hereditary syndromes, including 24 nosological forms, in which the majority have presented in individual families (17 syndromes). Muscular dystrophy (14 families, 19 patients), Charcot-Marie-Tooth disease (13 families, 19 patients), and spinal muscular atrophy, type I (12 families, 16 patients) were frequent AR diseases of the nervous system. Among the most "frequent" forms of AR metabolic diseases were phenylketonuria, hypothyroidism, and celiac disease. It is necessary to emphasize that all cases of these metabolic pathologies have been confirmed biochemically.

The "rare" nosological forms of eye diseases, which have presented only in individual families, on the other hand, were more often observed. However, the overwhelming share of families and patients with eye pathologies was a "frequent" nosology—a pigmental retinitis (30 families, 46 patients).

Only individual families had presented with 4 clinically different forms of blood diseases, among which were such "rare" diseases as coagulation factor XI deficiency, autoimmune hemolytic anemia, Fanconi anemia, and thrombasthenia of Glanzmann and Naegeli. In all cases the hereditary character of blood diseases was not in doubt, because in all families there were two or more patients.

Skeletal and connective tissue diseases include four nosological forms, two of which (chondrodysplasia calcificans metaphysealis and osteogenesis imperfecta) have presented in 4 families.

As a whole, for diseases with the autosomal recessive mode of inheritance, a ratio of "rare" and "frequent" forms of hereditary diseases differed from those for the autosomal dominant pathology. In favor of "rare" forms, corresponding shares for "rare" and "frequent" diseases have amounted to 54.1% and 45.9% respectively.

Autosomal recessive diseases are distributed in regular intervals throughout the territory of the region. In the majority of the inspected districts, there were 4–6 families with genetic diseases with a specific mode of inheritance. Thus, in 12 of 19 populations of the region the number of the identified families was equal to the number of observable nosological forms.

Among AR diseases in the territory of one district, simultaneously in two and more families there were 14 nosological forms, or 56% of all "frequent" nosologies. The majority of them were in the Tomsk district. Formally, such a situation could be considered as evidence of local accumulation of AR diseases in separate populations of the region. However, as well as in the case of AD diseases, families with the identical form of the disease have been territorially separated and in no one case was it possible to register connections between them. It is necessary to consider also that, as a whole, more than 60% of AR diseases

TABLE 112-2 THE SPECTRUM AND PREVALENCES OF AR DISEASES AMONG RUSSIANS IN THE TOMSK REGION

No.	Disease	OMIM	Number of families	Number of patients	Prevalence rate, 1 case per:
DISEASES OF THE NERVOUS SYSTEM					
1.	Muscular dystrophy, limb-girdle, type 2a	253600	14	19	46,686
2.	Charcot-Marie-Tooth disease, demyelinating	214400	13	19	46,686
3.	Spinal muscular atrophy, type I	253300	12	16	55,440
4.	Friedreich ataxia 1	229300	8	12	73,920
5.	Spastic paraplegia 5a	270800	6	10	88,704
6.	Hydrocephalus	236600	1	2	443,520
7.	Microcephaly, primary	251200	1	2	443,520
8.	Muscular dystrophy, limb-girdle, type 2c	253700	1	1	887,040
SKIN DISEASES					
9.	Albinism, oculocutaneous, type II	203200	4	7	126,720
10.	Ichthyosis congenita, harlequin fetus type	242500	3	5	177,408
11.	Melanosis, neurocutaneous	249400	2	3	295,680
SKELETAL AND CONNECTIVE TISSUE DISEASES					
12.	Chondrodysplasia calcificans metaphysealis	215050	4	8	110,880
13.	Osteogenesis imperfecta	259420	4	6	147,840
14.	Arthrogryposis multiplex congenita, neurogenic type	208100	3	3	295,680
15.	Osteochondrosis deformans tibiae, familial infantile type	259200	1	1	887,040
EYE DISEASES					
16.	Retinitis pigmentosa	268000	30	46	19,283
17.	Albinism, oculocutaneous	203200	4	7	126,720
18.	Cataract, congenital or juvenile	212500	3	7	126,720
19.	Glaucoma 3, primary congenital	231300	3	6	147,840
20.	Microphthalmia, syndromic 3	206900	1	1	887,040
BLOOD DISEASES					
21.	Coagulation factor XI deficiency	264900	2	4	221,760
22.	Anemia, autoimmune hemolytic	205700	1	3	295,680
23.	Fanconi anemia	227650	1	2	443,520
24.	Thrombasthenia of Glanzmann and Naegeli	273800	1	2	443,520
METABOLIC DISORDERS					
25.	Cystic fibrosis	219700	31	44	20,160
26.	Phenylketonuria	261600	23	26	34,117
27.	Hypothyroidism	218700	21	21	42,240
28.	Celiac disease	212750	15	19	46,686
29.	Gaucher disease	230800	4	5	177,408
30.	Wilson disease	277900	3	4	221,760
31.	MPS IS, Scheie syndrome	607016	3	4	221,760
32.	Tay-Sachs disease	272800	3	3	295,680
33.	MPS IVA, Morquio syndrome A	253000	2	2	443,520

(Continued)

TABLE 112-2 (CONTINUED)

No.	Disease	OMIM	Number of families	Number of patients	Prevalence rate, 1 case per:
34.	Ceroid lipofuscinosis, neuronal	204200	1	2	443,520
35.	GSD II, Pompe disease	232300	1	2	443,520
36.	GM1-gangliosidosis, type I	230500	1	2	443,520
37.	MPS IH, Hurler syndrome	607014	1	1	887,040
38.	Alkaptonuria	203500	1	1	887,040
39.	Hydroxyprolinemia	237000	1	1	887,040
40.	Fructose intolerance, hereditary	229600	1	1	887,040
41.	Sucrase-isomaltase deficiency, congenital	222900	1	1	887,040
HEREDITARY SYNDROMES					
42.	Bardet-Biedl	209900	6	10	88,704
43.	Klippel-Feil	214300	6	6	147,810
44.	Hydroxykynureninuria	236800	3	6	147,810
45.	Alport	203780	5	5	177,408
46.	Adrenal hyperplasia, congenital	201910	4	5	177,408
47.	Endocardial fibroelastosis and coarctation of abdominal aorta	226100	4	5	177,408
48.	Biliary malformation with renal tubular insufficiency	210550	2	5	177,408
49.	Meckel, type 1	249000	2	4	221,760
50.	Miller-Dieker lissencephaly	247200	2	3	295,680
51.	Smith-Lemli-Opitz	270400	2	3	295,680
52.	Pierre Robin	261800	1	3	295,680
53.	Bare lymphocyte	209920	1	3	295,680
54.	Usher, type I	276900	2	2	443,520
55.	COFS	214150	2	2	443,520
56.	Ciliary dyskinesia, primary	244400	2	2	443,520
57.	Faciocardiorenal	227280	1	1	887,040
58.	Ellis-Van Creveld	225500	1	1	887,040
59.	Shwachman-Diamond	260400	1	1	887,040
60.	Bowen-Conradi	211180	1	1	887,040
61.	Asplenia with cardiovascular anomalies	208530	1	1	887,040
62.	Bartter, antenatal	241200	1	1	887,040
63.	Chondrodysplasia, grebe type	200700	1	1	887,040
64.	Larsen	245600	1	1	887,040
65.	Alstrom	203800	1	1	887,040
OTHER HEREDITARY DISEASES					
66.	Deafness, neural, congenital moderate	221500	35	86	10,314
67.	Mental retardation	249500	16	38	23,343
68.	Deafness	220700	9	19	46,686
69.	Deafness, cochlear, with myopia and intellectual impairment	221200	3	7	126,720
70.	Deafness, neural, with atypical atopic dermatitis	221700	2	3	295,680
71.	Deafness, onychodystrophy, osteodystrophy, and mental retardation syndrome	220500	1	3	295,680
72.	Biliary atresia, extrahepatic	210500	1	2	443,520
	Total:		356	562	

have appeared to be "delivered" to the territory of the Tomsk region from various other regions of the country.

The X-linked hereditary diseases among the population of rural districts and small cities of the Tomsk area are represented by 15 clinical forms (Table 112-3). X-linked dominant hypophosphatemic rickets is identified in seven rural districts of the region, and was found among 35 patients of 24 families. Among the X-linked recessive diseases, most often there was colorblindness (26 families, 48 patients), hemophilia A (25 families, 30 patients), Duchenne muscular dystrophy (12 families, 17 patients), and fragile X-mental retardation syndrome (4 families, 11 patients). At the same time, seven nosological forms (58% of the total number of nosologies) are registered only in individual families.

From the point of view of the territorial distribution of the registered forms of X-linked pathology that we shall specify, the maximal number of families and patients, and also the greatest variety of the X-linked nosological forms, as well AD and AR diseases, is in the Tomsk district. On the other hand, the given mode of disease inheritance among inhabitants of 7 rural districts and 2 small cities of the region is not identified.

As a whole, in the structure of hereditary pathology among the population of the Tomsk region an appreciable variety of nosological forms of hereditary syndromes (42%) was also observed. The least were the groups of skin and blood diseases, where the share was only 6% of the total number of nosological forms. The contribution to the level of variety of the other inspected groups with hereditary diseases has appeared to be approximately similar, and varied from 12% up to 16%.

EPIDEMIOLOGY OF HEREDITARY DISEASES IN THE TYVA REPUBLIC

The Tyva Republic is a federal subject of Russia. The Republic is situated in the far south of Siberia, with the capital city of Kyzyl located at the geographic "center of Asia." It borders on Mongolia in the east and the south, on the Krasnoyarsk region in the north, Khakassia in northwest, Buryatiya in northeast, and the Altai Republic in the west. The eastern part of the republic is forested and elevated, and the west is a drier lowland. According to the 2002 Census, Tuvans, a Turkic people, make up 77.0% of the republic's population. Other groups include Russians (20.1%), Komi (0.5%), and a host of smaller groups, each accounting for less than 0.5% of the total population.

PREVALENCE RATE OF HEREDITARY DISEASES AMONG ABORIGINAL POPULATIONS

The total number of the inspected Tyva population was 198.4 thousand persons. In rural districts of the republic

TABLE 112-3 THE SPECTRUM AND PREVALENCES OF X-LINKED DISEASES AMONG RUSSIANS IN THE TOMSK REGION

No.	Disease	OMIM	Number of families	Number of patients	Prevalence rate in the male population, 1 case per:
1	Colorblindness, partial	303800	26	48	10,500
2	Hypophosphatemic rickets, X-linked dominant	307800	24	35	14,400
3	Hemophilia A	306700	25	30	16,800
4	Muscular dystrophy, Duchenne type	310200	12	17	29,647
5	Fragile X-mental retardation syndrome	300624	4	11	45,819
6	Ichthyosis, X-linked	308100	5	8	63,000
7	Retinitis pigmentosa	312600	5	6	84,000
8	Lowe oculocerebrorenal syndrome	309000	4	5	100,800
9	Microphthalmia, syndromic	309800	4	5	100,800
10	Ectodermal dysplasia, hypohidrotic	305100	2	4	126,000
11	Kallmann syndrome	308700	2	2	252,000
12	Faciogenital dysplasia	305400	2	2	252,000
13	Spondyloepiphyseal dysplasia tarda, X-linked	313400	1	2	252,000
14	Bloch-Sulzberger syndrome	308300	1	1	504,000
15	Opitz GBBB syndrome, X-linked	300000	1	1	504,000
	Total:		118	177	

the estimation of load of HDs is variable enough for diseases with a different mode of inheritance. So, for AD diseases, the value of the hereditary load ranged from 0.2±0.17 up to 1.85±0.45 per 1000 persons, and the average hereditary load of Tyvinians with AD diseases was 0.62±0.06 per 1000 persons (Table 112-4). Differences in load values of the population with AD pathology of different regions of the republic have appeared to be statistically significant (χ^2=26.25; d.f.=16; p<0.05). According to the level of the population load with AD pathology, it can be divided into regions with high (0.90×10^{-3} – 1.85×10^{-3}), average (0.50×10^{-3} – 0.73×10^{-3}), and low (0.20×10^{-3} – 0.42×10^{-3}) load. At the same time, the AD pathology load of the urban population (Kysyl City) appeared to be one of the lowest at 0.31×10^{-3}. The reason for such differentiation of the rural regions concerning AD pathology is impossible to determine.

The levels of the Tyva rural population load with AR diseases within the regions have been varied, from $(3.70\pm0.93) \times 10^{-3}$ up to $(0.20\pm0.16) \times 10^{-3}$. The average hereditary load of the republic rural population was 0.86×10^{-3}, and in Kysyl it was $(1.13\pm0.2) \times 10^{-3}$. The obtained data of the AR pathology load of the Tyva populations are considerably different according to the districts (χ^2 =34.49; d.f.=16; P<0.01). There are 3 districts with high levels of load of both dominant and recessive forms of the hereditary pathology: Todzhinsky, Ovyursky, and Ersinsky. The highest load of population with AR and X-linked diseases has been noted in Todzhinsky district: AR at $(3.7\pm0.93) \times 10^{-3}$ and X-linked at $(1.65\pm0.78) \times 10^{-3}$.

TABLE 112-4 THE SPECTRUM AND PREVALENCES OF AD DISEASES AMONG ABORIGINAL POPULATION IN THE REPUBLIC OF TUVA

No.	Disease	OMIM	Number of families	Number of patients	Prevalence rate, 1 case per:
DISEASES OF THE NERVOUS SYSTEM					
1.	Huntington disease	143100	8	16	12,400
2.	Neurofibromatosis, type I	162200	3	5	33,066
3.	Spinocerebellar ataxia, type 1	164400	1	2	99,200
SKIN DISEASES					
4.	Albinism, ocular, with sensorineural deafness	103470	2	5	19,840
5.	Nevus anemicus	163050	2	2	99,200
6.	Dyschromatosis symmetrica hereditaria	127400	1	2	99,200
7.	Vitiligo	193200	1	2	99,200
8.	Blue rubber bleb nevus	112200	1	1	198,400
SKELETAL AND CONNECTIVE TISSUE DISEASES					
9.	Osteogenesis imperfecta	166200	6	9	22,044
10.	Achondroplasia	100800	6	8	24,800
11.	Polydactyly, preaxial	174500	5	7	28,342
12.	Brachydactyly	112500	3	6	33,066
13.	Cleidocranial dysplasia	119600	3	4	49,600
14.	Split-hand/foot malformation	183600	2	4	49,600
15.	Acrodysostosis	101800	1	4	49,600
16.	Spondyloepiphyseal dysplasia congenita	183900	2	3	66,133
17.	Contractural arachnodactyly	121050	1	3	66,133
18.	Frontonasal dysplasia	136760	1	1	198,400
EYE DISEASES					
19.	Ptosis, hereditary congenital	178300	5	10	19,840
20.	Cataract, total congenital	116700	4	8	24,800
21.	Retinitis pigmentosa	180100	2	2	99,200
22.	Retinoblastoma	180200	1	1	198,400

(Continued)

TABLE 112-4 (CONTINUED)

No.	Disease	OMIM	Number of families	Number of patients	Prevalence rate, 1 case per:
HEREDITARY SYNDROMES					
23.	Marfan	154700	2	2	99,200
24.	Waardenburg, type I	193500	1	2	99,200
25.	Silver-Russell	180860	1	1	198,400
26.	Mccune-Albright	174800	1	1	198,400
27.	Klippel-Trenaunay-Weber	149000	1	1	198,400
28.	Treacher Collins-Franceschetti	154500	1	1	198,400
29.	Cornelia De Lange	122470	1	1	198,400
30.	Noonan	163950	1	1	198,400
31.	Holt-Oram	142900	1	1	198,400
32.	Rubinstein-Taybi	180849	1	1	198,400
33.	Opitz GBBB	145410	1	1	198,400
34.	Beckwith-Wiedemann	130650	1	1	198,400
35.	Popliteal pterygium	119500	1	1	198,400
36.	Oculodentodigital dysplasia	164200	1	1	198,400
	Total:		66	121	

The level of spreading AD diseases in this district is also high enough to increase the average republic value approximately two times. The high load of the population of this district with Mendelian HDs is connected with a large number of families having inborn forms of ophthalmological diseases (cataract, retinoblastoma, ptosis, anophtalmia). The load of the republic population with X-linked pathology varied widely, from zero values up to 1.65 ± 0.78 and 1.52 ± 0.36 for 1000 men. In Kysyl City, the level of the population load with AR and X-linked pathology became higher than in rural districts, and with dominant pathology it became lower. This is possibly connected with the number of undiagnosed families and patients with AD diseases in the urban population.

As a whole the average values of the Tyva population load were: with AD diseases $(0.58\pm0.05) \times 10^{-3}$, with AR diseases $(0.92\pm0.07) \times 10^{-3}$ and with X-linked diseases $(0.49\pm0.07) \times 10^{-3}$. The total value for the hereditary pathology load was 1.99×10^{-3} in the republic. The exact differentiation according to the degree of hereditary Mendelian diseases of rural and urban population was shown. The load of native Tyvinian population with AR diseases became statistically higher that with AD pathology. This effect is connected with the gene accumulation of recessive diseases in consanguineous families, and can be easily found with clinical geneological analysis. The accumulation of recessive mutant genes in the Tyvinian population can be explained by the "founder effect." Thus, high levels of accumulating mutant genes of some syndromes, in particular, microtia with meatal atresia and conductive deafness microtia, have been noted (OMIM 251800). This rare pathology among other populations has been found in 18 patients of 14 Tyvinian families living in six different districts of the republic.

DIVERSITY OF HEREDITARY DISEASES AMONG ABORIGINAL POPULATIONS

The spectrum of AD diseases is represented by 36 forms of Mendelian diseases including 14 syndromes, 10 forms of skeletal and connective tissue diseases, 5 forms of skin disorders, 4 forms of ophthalmologic pathology, and 3 forms of neurologic diseases. The most varied according to structure has appeared in a group of the hereditary syndromes. All syndromes were encountered in individual families with one patient in every category; only Waardenburg syndrome was registered with 2 patients in Kyzyl. The next group among diseases with AD mode of inheritance was that of skeletal and connective tissue diseases. Among cases with this systemic pathology osteogenesis imperfecta (6 families of 9 patients) has been seen more often. In Ovjursky district there were 2 families, one family with 3 patients, and the other with 2 patients, and in Erzinsky district there were 2 related families with 2 patients. Achondroplasia has been diagnosed in four families, and 3 families with 4 patients were relatives. Polydactyly, which has been seen in the Dzun-Hemchinsky district, also was diagnosed in three consanguineous families. Cleidocranial dysplasia was diagnosed in three consanguineous families in the Erzinsky district. Acrodysostosis was registered in the Tandinsky district in one family, but with four affected.

A variety of AR pathologies among the diagnosed native population of the republic is presented with 33 different clinically pathological forms of hereditary pathology (Table 112-5).

The group of neurologic diseases (14 families, 19 patients) is presented by such "frequent" forms as hydrocephalus and microcephaly, which have presented in several consanguineous families.

Dermatologic pathology also was presented in the examined population by 4 nosological forms (18 families, 26 patients), and the greater part of them were registered

TABLE 112-5 THE SPECTRUM AND PREVALENCES OF AR DISEASES AMONG ABORIGINAL POPULATION IN THE REPUBLIC OF TUVA

No.	Disease	OMIM	Number of families	Number of patients	Prevalence rate, 1 case per:
DISEASES OF THE NERVOUS SYSTEM					
1.	Hydrocephalus	236600	6	7	28,343
2.	Microcephaly, primary	251200	4	4	49,600
3.	Spastic paraplegia 5a	270800	1	3	66,133
4.	Friedreich ataxia 1	229300	1	2	99,200
5.	Spinal muscular atrophy, type I	253300	1	2	99,200
6.	Charcot-Marie-Tooth disease, demyelinating	214400	1	1	198,400
SKIN DISEASES					
7.	Albinism, oculocutaneous	203200	6	12	16,533
8.	Ichthyosis, lamellar	242300	9	10	19,840
9.	Ichthyosis congenita, harlequin fetus type	242500	2	3	66,133
10.	Melanosis, neurocutaneous	249400	1	1	198,400
SKELETAL AND CONNECTIVE TISSUE DISEASES					
11.	Osteogenesis imperfecta	259420	2	4	49,600
12.	Chondrodysplasia calcificans metaphysealis	215050	1	3	66,133
13.	Arthrogryposis multiplex congenita	208100	2	2	99,200
14.	Endocardial fibroelastosis	226000	1	1	198,400
EYE DISEASES					
15.	Cataract, congenital or juvenile	212500	13	13	15,261
16.	Microphthalmia, isolated 1	251600	2	2	99,200
17.	Microphthalmia, syndromic 3	206900	1	2	99,200
18.	Retinitis pigmentosa	268000	1	2	99,200
BLOOD DISEASES					
19.	Coagulation factor XI deficiency	264900	1	2	99,200
METABOLIC DISORDERS					
20.	Hypothyroidism	218700	13	13	15,261
21.	Cystic fibrosis	219700	4	6	33,067
22.	MPS IH, Hurler syndrome	607014	2	2	99,200
23.	MPS IVA, Morquio syndrome A	253000	1	1	198,400
HEREDITARY SYNDROMES					
24.	Laurence-Moon	245800	2	2	99,200
25.	Adrenal hyperplasia, congenital	201910	1	2	99,200

(Continued)

TABLE 112-5 (CONTINUED)

No.	Disease	OMIM	Number of families	Number of patients	Prevalence rate, 1 case per:
26.	Klippel-Feil	214300	1	2	99,200
27.	Seckel	210600	1	1	198,400
28.	Pierre Robin	261800	1	1	198,400
OTHER HEREDITARY DISEASES					
29.	Microtia with meatal atresia and conductive deafness	251800	14	18	11,022
30.	Deafness	220700	4	7	28,343
31.	Muscular dystrophy, limb-girdle	253600	2	2	99,200
32.	Deafness, neural, congenital moderate	221500	1	2	99,200
33.	Hemihyperplasia, isolated	235000	1	1	198,400
	Total:		104	136	

as "frequent" forms. Congenital ichthyosis was found in 9 of examined families and oculocutaneous albinism in 6 consanguineous families. They are the most widespread dermatologic pathology.

There was an interesting group of hereditary diseases of the skeletal and connective tissue, including 4 nosologies in which the majority have presented in two families. Osteogenesis imperfecta was seen in 2 families living in different districts but connected by relationship.

Ophthalmologic pathology was presented by such monogenic diseases as congenital cataract (13 families, 13 patients) and microphthalmia (2 families, 2 patients). Pigmental retinitis and anophthalmia were presented in individual families, but each family always had 2 patients. In all cases the hereditary character of the diseases was not in doubt.

Inborn disturbances of metabolism were registered in 20 families of 22 patients. More often the Tuva population had hypothyroidism (13 families, 13 patients), cystic fibrosis (4 families, 6 patients). In 3 families the diagnosis of MPS was confirmed biochemically. In one of the inspected families with MPS type I, there was a consanguineous marriage (between aunt and nephew).

All syndromes have presented in individual families, except for a case of Laurence-Moon syndrome, and all these forms were "rare" for the given population.

In a group of the patients suffering with the hearing disturbance, the most widespread pathology was microtia with meatal atresia and conductive deafness. The given pathology was observed in families living in several districts of the republic. Of 14 diagnosed families it was possible to determine that 6 of them were relatives.

A typical characteristic of the Tuva population was the absence of phenylketonuria.

Mendelian pathology among the X-linked recessive HDs were represented by 8 clinical forms (Table 112-6. Of

TABLE 112-6 THE SPECTRUM AND PREVALENCES OF X-LINKED DISEASES AMONG ABORIGINAL POPULATION IN THE REPUBLIC OF TUVA

No.	Disease	OMIM	Number of families	Number of patients	Prevalence rate in the male population, 1 case per:
1.	Ichthyosis, X-linked	308100	10	15	6613
2.	Fragile X-mental retardation syndrome	300624	4	9	11,022
3.	Spondyloepiphyseal dysplasia tarda, X-linked	313400	6	7	14,171
4.	Muscular dystrophy, Duchenne type	310200	2	5	19,840
5.	Hypophosphatemic rickets, X-linked dominant	307800	2	3	33,066
6.	Aarskog-Scott syndrome	305400	2	2	49,600
7.	Hemophilia A	306700	1	2	49,600
8.	Microphthalmia, syndromic	309800	1	1	99,200
	Total:		28	44	

the presented forms, the greatest frequency was in ichthyosis (10 families, 15 patients), fragile X-mental retardation syndrome (4 families, 9 patients), and spondyloepiphyseal dysplasia (6 families, 7 patients). Duchenne muscular dystrophy, hypophosphatemic rickets, and Aarskog-Scott syndrome were registered in two families, and only hemophilia A and microphthalmia were seen in individual families.

The received data allows judgment about a variety of hereditary diseases with various modes of inheritance among indigenous populations of the republic.

EPIDEMIOLOGY OF HEREDITARY DISEASES IN THE ALTAI REPUBLIC

The Altai Republic is a Federal subject of Russia. The capital is Gorno-Altaisk. The republic is situated in the very center of Asia at the junction of the Siberian taiga, the steppes of Kazakhstan, and the semi-deserts of Mongolia. Forests cover about 25% of the republic's territory. As per the 2002 Census, ethnic Russians make up 57.4% of the Altai Republic's population, with the ethnic Altay people numbering only 30.6%. Other groups include Kazakhs (6.0%), Telengits (1.2%), Tubalars (0.8%), Ukrainians (1437, 0.7%), and a host of smaller groups, each accounting for less than 0.5% of the total population.

PREVALENCE RATE OF HEREDITARY DISEASES AMONG THE REPUBLIC POPULATION

The total number of the examined population of the republic was 193.3 thousand persons. In the republic territory live three basic ethnic groups that have been examined (Russians, Altaians, and Kazakhs). About a quarter of the population live in the capital.

In the population of the Gorny Altai Republic, the estimation of hereditary load for native populations for diseases with different modes of inheritance appeared to be variable. So, for the AD diseases these estimations range from 0.54×10^{-3} up to 9.39×10^{-3} in various districts. For AR diseases, the range was from 0.41×10^{-3} up to 10.7×10^{-3}. The hereditary load of the population with X-linked pathology varied within narrow limits. The districts with high hereditary load were distinct from the others. Gorno-Altaisk, Mayminsk, and Chemalsky districts had the greatest estimations of load for AD and AR pathologies.

The values of hereditary load with AD pathology in Russian populations ranged from 0.13×10^{-3} up to 1.5×10^{-3}; for AR diseases the values ranged from 0.15×10^{-3} up to 0.9×10^{-3}, and the range of estimations of X-linked pathology load was not appreciable. The total load of X-linked pathology in the Russian populations of the republic was 0.12±0.05 per 1000 persons.

Thus, the load for AD pathology as a whole for the republic among Altaians was 1.14 per 1000 persons, and for Russians 0.54 per 1000 persons (χ^2=23.28; P=0.0001). The load for AR pathology was 2.57 and 0.53 per 1000 persons (χ^2=101.48; P=0.0001) respectively. Comparing the for hereditary pathology of urban and rural populations among Russians (χ^2=29.84; P=0.0001) and Altaians (χ^2=114.99; P=0.0001), statistically significant differences were seen. Estimating the X-linked-recessive pathology, nonsignificant differences were observed (P=0.731), with the load for Altaians being 0.13, and for Russians 0.12 per 1000 persons.

The hereditary pathology load for the Kazakh population in the Altai Republic was 0.34±0.2 for AD and 1.16±0.2 for AR per 1000 persons.

In summarizing the results of studying the hereditary pathology load of the population of the Altai Republic, it would be desirable to note the accumulation of patients with microtia with meatal atresia and conductive deafness in the republic. A lot of patients with that syndrome were observed in Tyva, too. Probably the "founder effect" took place in this case. Special population-genetic research for the resolution of this question would be necessary.

DIVERSITY OF HEREDITARY DISEASES AMONG VARIOUS ETHNIC GROUPS OF THE REPUBLIC

In the process of medical genetic research in the Altai Republic we had examined 234 families and 288 patients with a Mendelian hereditary pathology, including 133 families and 166 patients among Altaians, 93 families and 113 patients among Russians, and 8 families and 9 patients among Kazakhs. Considering the varied ethnohistorical parentage of Russians and Altaians, these two groups were presented separately. In addition, the group of Kazakhs in the Kosh-Agachsky district was emphasized because they were a major part of that district's population.

In all cases a mode of inheritance was diagnosed and resolved, according to which all Russian and Altaian families were distributed as follows: AD – 95 families and 122 patients; AR – 124 families and 148 patients; and X-linked mode of inheritance – 7 families and 9 patients.

Table 112-7 shows the spectrum of AD diseases in two ethnic groups: Russians and Altaians. Altaians, of all dominant forms, the most varied had a group of hereditary syndromes (18 forms). Among AD diseases the most often seen were branchiootic syndrome, amelogenesis imperfecta, and neurofibromatosis type I. Branchiootic syndrome was registered in 4 Altaian families and 10 patients (1:5900). Amelogenesis imperfecta was identified in three consanguineous families of Altaians, and was not seen in Russians. Prevalence of this disease among Altaians in the republic was 1 case per 8429 persons.

Among the Russians, by composition (13 forms) and by quantity of patients (28 persons), there was a group

TABLE 112-7 THE SPECTRUM AND PREVALENCES OF AD DISEASES IN THE REPUBLIC OF ALTAI

			Number of				Prevalence rate, 1 case per:	
			families		Patients			
No.	Disease	OMIM	Russians	Altaians	Russians	Altaians	Russians	Altaians
DISEASES OF THE NERVOUS SYSTEM								
1.	Neurofibromatosis, type I	162200	3	3	4	7	30,675	8,428
2.	Thomsen disease	160800	3	3	4	5	30,675	11,800
3.	Spinocerebellar ataxia, type 1	164400	4	-	4	-	30,675	-
4.	Tuberous sclerosis	191100	1	2	1	2	122,700	29,500
SKIN DISEASES								
5.	Vitiligo	193200	1	2	1	2	122,700	29,500
6.	Palmoplantar keratoderma, epidermolytic	144200	1	-	1	-	122,700	-
7.	Blue rubber bleb nevus	112200	-	1	-	1	-	59,000
SKELETAL AND CONNECTIVE TISSUE DISEASES								
8.	Osteogenesis imperfecta	166200	4	1	9	1	13,633	59,000
9.	Hypochondroplasia	146000	1	6	1	6	122,700	20,450
10.	Amelogenesis imperfecta	130900	-	3	-	7	-	8429
11.	Arthrogryposis multiplex congenita	108110	2	2	2	2	61,350	29,500
12.	Frontonasal dysplasia	136760	2	-	2	-	61,350	-
13.	Exostoses, multiple	133700	-	1	-	1	-	59,000
EYE DISEASES								
14.	Coloboma, ocular	120200	-	1	-	1	-	59,000
BLOOD DISEASES								
15.	Spherocytosis	182900	-	1	-	1	-	59,000
HEREDITARY SYNDROMES								
16.	Meniere disease	156000	5	1	7	1	17,528	59,000
17.	Ehlers-Danlos	130000	5	1	6	1	20,450	59,000
18.	Marfan	154700	4	2	4	2	30,675	29,500
19.	Rubinstein-Taybi	180849	2	2	2	3	61,350	19,666
20.	Silver-Russell	180860	2	-	2	-	61,350	-
21.	Treacher Collins-Franceschetti	154500	1	1	1	1	122,700	59,000
22.	Noonan	163950	1	1	1	1	122,700	59,000
23.	Sturge-Weber	185300	1	1	1	1	122,700	59,000
24.	Williams-Beuren	194050	1	-	1	-	122,700	-
25.	Beckwith-Wiedemann	130560	1	-	1	-	122,700	-
26.	Leopard	151100	1	-	1	-	122,700	-
27.	Poland	173800	1	-	1	-	122,700	-
28.	Sotos	117550	1	-	1	-	122,700	-
29.	Hirschsprung disease	142623	-	1	-	1	-	59,000
30.	Klippel-Trenaunay-Weber	149000	-	1	-	1	-	59,000
31.	Currarino	176450	-	1	-	1	-	59,000
32.	Parry-Romberg	141300	-	1	-	1	-	59,000
33.	EEC1	129900	-	1	-	1	-	59,000
OTHER HEREDITARY DISEASES								
34.	Branchiootic syndrome 1	602588	-	4	-	10	-	5900
35.	Oculoauriculovertebral dysplasia	164210	1	1	1	1	122,700	59,000
36.	Polycystic kidneys	173900	1	-	1	-	122,700	-
	Total:		50	45	60	62		

with hereditary syndromes. The most frequent among them there were Meniere's disease (7 patients) and Ehlers-Danlos syndrome (6 patients). In second place there was a group with skeletal and connective tissue pathology. The quantity of patients in this group was 14 persons, 9 of them suffering with osteogenesis imperfecta.

While inspecting the Kazakh population, 2 cases of disease with an AD mode of inheritance have been identified: Thomsen disease (1 family, 1 patient) and metaphyseal chondrodysplasia, Schmid type (1 family, 1 patient).

A variety of diseases with an AR mode of inheritance were presented, with 21 forms (Table 112-8). According to the frequency of occurrence, in first place among the Altaians there was a group with ear pathology in 3 clinical forms: microtia with meatal atresia and conductive deafness (26 patients), neural deafness (13 patients), and deafness, autosomal recessive (11 patients). And the greatest variety presented in the group of metabolic diseases in 8 clinical forms. Cystic fibrosis, phenylketonuria and hypothyroidism were seen more often.

Among the Russians the most widespread, according to the frequency of AR clinical forms, were groups of metabolic diseases, deafness, and lamellar ichthyosis.

AR pathologies among the Kazakhs included microtia with meatal atresia and conductive deafness (2 families, 3 patients), hydrocephalus, cystic fibrosis, congenital hypothyroidism, and Pierre Robin syndrome.

TABLE 112-8 THE SPECTRUM AND PREVALENCES OF AR DISEASES IN THE REPUBLIC OF ALTAI

			Number of				Prevalence rate, 1 case per:	
			families		patients			
No.	Disease	OMIM	Russians	Altaians	Russians	Altaians	Russians	Altaians
DISEASES OF THE NERVOUS SYSTEM								
1.	Charcot-Marie-Tooth disease, demyelinating	214400	1	1	1	1	122,700	59,000
2.	Muscular dystrophy, limb-girdle, type 2a	253600	1	-	1	-	122,700	-
3.	Microcephaly, primary	251200	1	-	1	-	122,700	-
4.	Spinal muscular atrophy, type I	253300	1	-	1	-	122,700	-
SKIN DISEASES								
5.	Albinism, oculocutaneous	203200	1	9	2	11	61,350	5,363
6.	Ichthyosis, lamellar	242300	3	-	5	-	24,540	-
METABOLIC DISORDERS								
7.	Cystic fibrosis	219700	9	9	9	12	13,633	4917
8.	Phenylketonuria	261600	-	10	-	10	-	5900
9.	Hypothyroidism	218700	1	7	1	7	122,700	8429
10.	Celiac disease	212750	2	1	2	2	61,350	29,500
11.	MPS IH, Hurler syndrome	607014	-	2	-	2	-	29,500
12.	MPS IVA, Morquio syndrome A	253000	-	1	-	1	-	59,000
13.	Wilson disease	277900	-	1	-	1	-	59,000
14.	Adrenal hyperplasia, congenital	201910	1	-	1	-	122,700	-
HEREDITARY SYNDROMES								
15.	Pierre Robin	261800	1	1	1	1	122,700	59,000
16.	Pendred	274600	3	-	4	-	30,675	-
OTHER HEREDITARY DISEASES								
17.	Microtia with meatal atresia and conductive deafness	251800	5	22	5	26	24,540	2269
18.	Deafness, neural, congenital moderate	221500	5	9	8	13	15,337	4538
19.	Deafness	220700	3	10	4	11	30,675	5363
20.	Deafness, onychodystrophy, osteodystrophy, and mental retardation syndrome	220500	-	1	-	2	-	29,500
21.	Pituitary dwarfism	262600	-	2	-	2	-	29,500
	Total:		38	86	46	102		

Phenotypic expression of microtia with meatal atresia was characterized by various deformations of auricles, from little change in sizes up to anotia. In most cases auricles were deformed. There were incidences of cartilaginous tissue, preauricular excrescences, and fistulas. Meatal atresia is more often unilateral. Almost all patients had conductive hearing impairment. Deafness was caused by a pathology of the middle ear, including deformation of hammer bone and anvils, absence of an anvil, or abnormal osteal prominences.

The X-linked recessive diseases were represented by 2 clinical forms among Altaians and 3 forms among Russians (Table 112-9). Among the Kazakh population, X-linked pathology was not registered. Russians had hypohidrotic ectodermal dysplasia more often. In all families, both Russian and Altai, the same mutation was found in 5 exon of the *EDA* gene, representing a deletion, presumably the size of 36 bp.

EPIDEMIOLOGY OF HEREDITARY DISEASES IN KHAKASSIA REPUBLIC

Republic of Khakassia is a federal subject of Russia located in south central Siberia. Abakan is the administrative center of Khakassia. It borders on the Kemerovo area, Krasnoyarsk region, Tyva Republic, and Altai Republic. According to the 2002 Russian Census, ethnic Russians make up 80.3% of the republic's population, while ethnic Khakas are only 12.0%. Other groups include ethnic Germans (1.7%), Ukrainians (1.5%), Tatars (0.7%), and a host of smaller groups, each accounting for less than 0.5% of the total population.

PREVALENCE RATE OF HEREDITARY DISEASES AMONG THE REPUBLIC POPULATION

Sizes of the HD load with various modes of inheritance were analyzed in two ethnic groups: the Khakases, with a total number of 56.8 thousand persons, and Russians, with a total number of 385.8 thousand persons. The analysis among the urban and rural populations of the was made separately, in two designated groups. These estimations were calculated taking into account the total number of districts and cities minus a share of persons without children.

The AD common load with pathology among Khakases $(1.06\pm0.16) \times 10^{-3}$ and the Russians $(0.55\pm0.07) \times 10^{-3}$ differ from each other, and was higher among the aboriginal population. Statistically significant differences were identified by comparing the average values of the common load with AD pathology among Khakases and Russians ($\chi^2 = 14.65$; P=0.0001). Unlike the common hereditary load, the analysis of AD load of the pathology among the Khakases and the Russians showed insignificant differentiation between the urban and rural populations, and statistically was not significant. Among the Russians living in the city, loading with AD diseases was a little below that among rural populations (0.51×10^{-3} and 0.65×10^{-3}, respectively). Statistically significant differences were not revealed ($\chi^2 = 3.3$; P=0.07). At the same time, among the Khakases living in city, loading with AD pathology (1.22×10^{-3}) was higher than among the inhabitants of villages (0.97×10^{-3}), but statistically significant differences were not revealed ($\chi^2 = 0.004$; P=0.95).

The loading with AR pathology of two ethnic groups of the Khakases and the Russians in the republic had statistically significant differences at comparison of the average values (χ^2= 44.69; P=0.0001). Loading with AR diseases of the Russian urban and rural populations did not differ from each other ($\chi^2 = 0.97$; P=0.32). At comparison of loading with AR pathology in the two ethnic groups, significant differences of loading among the Khakases urban and rural populations, and the Russians living in the same places, were perceived (χ^2=17.00; P<0.0001; $\chi^2 = 14.14$; P=0.0002).

At comparison of loading with X-linked HDs among the urban and rural populations, values among the rural Khakass population (0.65×10^{-3}) and among the rural Russians (0.04×10^{-3}) were insignificant. Among the urban Khakases it no X-linked pathology was registered. These values were higher among the Russian urban population (0.13×10^{-3}). The values among rural Khakases

TABLE 112-9 THE SPECTRUM AND PREVALENCES OF X-LINKED DISEASES IN THE REPUBLIC OF ALTAI

			Number of				Prevalence rate in the male population, 1 case per:	
			families		patients			
No.	Disease	OMIM	Russians	Altaians	Russians	Altaians	Russians	Altaians
1.	Ectodermal dysplasia, hypohidrotic	305100	2	1	4	1	15,337	29,500
2.	Fragile X-mental retardation syndrome	300624	2	-	2	-	30,675	-
3.	Muscular dystrophy, Duchenne type	310200	1	-	1	-	61,350	-
4.	Hemophilia A	306700	-	1	-	1	-	29,500
	Total:		5	2	7	2		

and the Russians were statistically significant (χ^2=5.84; P=0.01) and differed in loading with X-linked hereditary diseases, and among the Khakases they were higher. In total values of loading with X-linked diseases there were no perceived statistically significant differences among the Khakases (0.4×10^{-3}) and the Russians (0.1×10^{-3}; $\chi^2 = 2.39$; P=0.12). Among the urban population of the Khakases, the absence of registered X-linked pathology was apparently due to undiagnosed families and patients or because the ratio of urban and rural populations was different and the size of the Khakases rural population was higher.

In the two surveyed ethnic groups, the absence of differentiation according to the degree of loading with hereditary Mendelian diseases among the rural and urban populations was shown. At the same time, general values of loading with all types of hereditary pathology were higher among the Khakases compared to the Russians. The values of the general loading with hereditary pathology in the republic were as follows: AD diseases $(1.61\pm0.07) \times 10^{-3}$; AR diseases $(1.04\pm0.51) \times 10^{-3}$, and X-linked diseases $(0.51\pm0.3) \times 10^{-3}$.

DIVERSITY OF HEREDITARY DISEASES AMONG VARIOUS ETHNIC GROUPS OF THE REPUBLIC

As a result of the medical genetic examination in the Republic of Khakassia, there were 255 families diagnosed, and 352 patients with Mendelian hereditary pathology. Among them there were found 182 families with 247 patients: the Russians, 63 families with 93 patients; the Khakases, the remaining 10 families with 12 patients were representatives of other nationalities (the Armenians, the Germans, the Moldavians, the Tadjiks). Considering the varied ethnohistorical parentage of the Russians and the Khakases, these two groups were analyzed separately. In all cases they had been diagnosed, the type of inheritance according to which all families were distributed as follows: AD in 161 families, 230 patients; AR in 73 families, 90 patients; and X-linked in 11 families, 20 patients.

Among the urban and rural populations of the Republic, registered 21 nosological clinical Mendelian forms with AD type of inheritance were registered among the Khakases.

On studying a spectrum of autosomal dominant diseases among the Khakases (Table 112-10), the greatest variety of nosological forms was observed in two groups—diseases of skeletal and connective tissue (9 forms), and hereditary syndromes (9 forms). In the third place, both by quantity of forms and by quantity of families and patients, was neurological pathology with 3 forms, 7 families, and 12 patients. In this group of diseases there was also the most "frequent" AD disease among the Khakass population, namely, neurofibromatosis type I.

"Frequent" AD diseases among the Khakass population of the republic, with prevalence of 1 case per 50,000 persons, include 10 forms: neurofibromatosis type I; Alport syndrome; achondroplasia; multiple exostoses; Ehlers-Danlos syndrome; Charcot-Marie-Tooth disease; arthrogryposis; aplasia cutis; spondylometaphyseal dysplasia Kozlowski type; and Noonan syndrome. Among the examined population some HDs, such as the following, were diagnosed: Leopard syndrome, and a family case of multiple exostoses, which were "rare" and did not present among other examined populations.

Among the Russian population of Khakassia were AD diseases with the greatest variety in the group of hereditary syndromes—19 forms (Table 112-10), but patients had higher value of skeletal and connective tissue diseases, with 11 families and 16 patients. This grouping of the pathology included 13 forms.

Hereditary syndromes were registered in 39 families. The most "frequent" form was Marfan syndrome (1 case per 29,677).

In the group of diseases of skeletal and connective tissue, most often there were achondroplasia (11 families, 16 patients) and contractural arachnodactyly (5 families, 7 patients).

One more difference in the spectrum of autosomal dominant diseases among the Russians and aboriginal population was the spectrum of dermatologic and neurologic pathology.

Unlike the Khakases, among whom there was registered only one form, among the Russians there were registered 6 forms, and epidermolysis bullosa dystrophica was seen in 7 families (12 patients).

Neurologic diseases were registered in 30 families, among them Charcot-Marie-Tooth disease (13 families, 21 patients), neurofibromatosis, type I (6 families, 9 patients), facioscapulohumeral muscular dystrophy (6 families, 8 patients), and spastic paraplegia (5 families, 8 patients).

Among the Russian population there were found 8 "frequent" nosological forms: Charcot-Marie-Tooth disease (1:18,371); Marfan syndrome (1:29,677); achondroplasia (1:24,112); epidermolysis bullosa dystrophica (1:32,150); neurofibromatosis type I (1:42,867); facioscapulohumeral muscular dystrophy (1:48,225); spastic paraplegia (1:48,225); and contractural arachnodactyly (1:55,114).

Many AD diseases diagnosed in the two ethnic groups had only one patient, i.e., they were single events. Such situations can be explained due to either mutation de novo, or they can represent phenocopies of known AD diseases. Having analyzed the material, the share of single events to the ratio of all patients with AD pathology was, for the Khakases, 46.9% and for the Russians 39.8%. Some part of single cases could be explained by mutation processes, and the other part actually sporadic cases.

TABLE 112-10 THE SPECTRUM AND PREVALENCES OF AD DISEASES IN THE REPUBLIC OF KHAKASSIA

			Number of				Prevalence rate, 1 case per:	
			families		patients			
No.	Disease	OMIM	Russians	Khakases	Russians	Khakases	Russians	Khakases
DISEASES OF THE NERVOUS SYSTEM								
1.	Charcot-Marie-Tooth disease, demyelinating	118200	13	2	21	3	18,371	18,933
2.	Neurofibromatosis, type I	162200	6	4	9	8	42,867	7100
3.	Facioscapulohumeral muscular dystrophy	158900	6	1	8	1	48,225	56,800
4.	Spastic paraplegia	182600	5	-	8	-	48,225	-
SKIN DISEASES								
5.	Epidermolysis bullosa dystrophica	131750	7	-	12	-	32,150	-
6.	Ichthyosis vulgaris	146700	3	-	3	-	128,600	-
7.	Aplasia cutis congenita, nonsyndromic	107600	2	2	2	2	192,900	28,400
8.	Alopecia, familial focal	104110	1	-	2	-	192,900	-
9.	Epidermolytic hyperkeratosis	113800	1	-	2	-	192,900	-
10.	Pachyonychia congenita	167200	1	-	2	-	192,900	-
11.	Achondroplasia	100800	11	3	16	4	24,112	14,200
12.	Contractural arachnodactyly	121050	5	-	7	-	55,114	-
13.	Exostoses, multiple	133700	2	1	3	4	128,600	14,200
14.	Hypochondroplasia	146000	5	-	6	-	64,300	-
15.	Osteogenesis imperfecta	166200	2	1	4	1	96,450	56,800
16.	Arthrogryposis multiplex	108110	2	3	2	3	192,900	18,933
17.	Polydactyly, postaxial	174200	1	-	3	-	128,600	-
18.	Spondyloepiphyseal dysplasia	183900	1	1	1	2	385,800	28,400
19.	Acrodysostosis	101800	-	1	-	1	-	56,800
20.	Frontonasal dysplasia	136760	1	1	2	1	192,900	56,800
21.	Radioulnar synostosis	179300	1	1	1	1	385,800	56,800
22.	Spondylometaphyseal dysplasia, Kozlowski type	184252	1	1	1	1	385,800	56,800
23.	Orofacial cleft	119530	1	-	2	-	192,900	-
24.	Sprengel deformity	184400	1	-	1	-	385,800	-
EYE DISEASES								
25.	Ptosis, hereditary congenital	178300	2	-	2	-	192,900	-
26.	Sacral defect with anterior meningocele	600145	1	-	1	-	385,800	-
BLOOD DISEASES								
27.	Diamond-Blackfan anemia	105650	1	-	1	-	385,800	-
HEREDITARY SYNDROMES								
28.	Marfan	154700	8	-	13	-	29,677	-
29.	Ehlers-Danlos	130000	4	2	6	4	64,300	14,200

(Continued)

TABLE 112-10 (CONTINUED)

No.	Disease	OMIM	Number of families: Russians	Number of families: Khakases	Number of patients: Russians	Number of patients: Khakases	Prevalence rate, 1 case per: Russians	Prevalence rate, 1 case per: Khakases
30.	Alport	104200	-	4	-	6	-	9467
31.	Waardenburg, type I	193500	3	-	4	-	96,450	-
32.	Noonan	163950	2	2	2	2	192,900	28,400
33.	Holt-Oram	142900	2	1	3	1	128,600	56,800
34.	Prader-Willi	176270	3	-	3	-	128,600	-
35.	Silver-Russell	180860	3	-	3	-	128,600	-
36.	Trichorhinophalangeal, type I	190350	2	-	3	-	128,600	-
37.	Crouzon	123500	1	1	2	1	192,900	56,800
38.	Sotos	117550	2	-	2	-	192,900	-
39.	Gilbert	143500	1	-	2	-	192,900	-
40.	Treacher Collins-Franceschetti	154500	1	-	2	-	192,900	-
41.	Klippel-Trenaunay-Weber	149000	1	-	1	-	385,800	-
42.	Beckwith-Wiedemann	130650	1	-	1	-	385,800	-
43.	Poland	173800	1	-	1	-	385,800	-
44.	Rubinstein-Taybi	180849	1	-	1	-	385,800	-
45.	Saethre-Chotzen	101400	1	-	1	-	385,800	-
46.	Fanconi renotubular	134600	1	-	1	-	385,800	-
47.	Leopard	151100	-	1	-	1	-	56,800
OTHER HEREDITARY DISEASES								
48.	Prune belly	100100	1	-	1	-	385,800	-
49.	Mast cell disease	154800	-	1	-	1	-	56,800
50.	Oculodentodigital dysplasia	164200	-	1	-	1	-	56,800
51.	Polycystic kidneys	173900	2	-	4	-	96,450	-
52.	Hypercholesterolemia, autosomal dominant	143890	1	-	2	-	192,900	-
	Total:		126	35	181	49		

As a result of the analysis of the two ethnic groups a high frequency of occurrence for four AD diseases was found. These were Charcot-Marie-Tooth disease, neurofibromatosis type I, achondroplasia and Ehlers-Danlos.

The nosological spectrum of AR diseases found during medical genetic study among the Khakases had 17 nosological forms (Table 112-11). According to the frequency of occurrence, in the first place among the Khakases urban and rural populations there was a group of hereditary syndromes (6 forms, 7 patients); then, neurologic pathology (3 forms, 6 families, 10 patients), and a group of ear pathologies presenting in 2 clinical forms: microtia with meatal atresia and conductive deafness (4 patients), and autosomal recessive deafness (4 patients).

Among the Russian population of the Republic of Khakassia, AR diseases were presented by 22 nosological forms (Table 112-11).

In the Russian population the most widespread, according to the frequency of clinical forms, were groups of hereditary syndromes (6 forms), metabolic diseases (5 forms), and neurologic pathology (4 forms).

In the first place in the group of metabolic disorders was phenylketonuria (7 patients). The given disease had not presented among the Khakass population.

Analyzing the spectrum of AR pathology, it would be desirable to pay attention to the fact that some diseases were registered only among the Khakass population (ataxia-telangiectasia and Werner syndrome (1:28,400)).

TABLE 112-11 THE SPECTRUM AND PREVALENCES OF AR DISEASES IN THE REPUBLIC OF KHAKASSIA

			Number of				Prevalence rate, 1 case per:	
			families		patients			
No.	Disease	OMIM	Russians	Khakases	Russians	Khakases	Russians	Khakases
DISEASES OF THE NERVOUS SYSTEM								
1.	Spinal muscular atrophy, type III	253400	2	-	2	-	192,900	-
2.	Myopathy unspecified	AR	2	4	2	7	192,900	8114
3.	Friedreich ataxia 1	229300	1	-	1	-	385,800	-
4.	Dystonia, torsion	224500	1	1	1	1	385,800	56,800
5.	Ataxia-telangiectasia	208900	-	1	-	2	-	28,400
SKIN DISEASES								
6.	Epidermolysis bullosa, Herlitz type	226700	2	-	2	-	192,900	-
7.	Ichthyosis, lamellar	242300	1	-	2	-	192,900	-
8.	Albinism, oculocutaneous	203100	-	1	-	2	-	28,400
9.	Ichthyosiform erythroderma, congenital, nonbullous	242100	-	1	-	1	-	56,800
SKELETAL AND CONNECTIVE TISSUE DISEASES								
10.	Osteogenesis imperfecta	259420	4	2	4	2	96,450	28,400
11.	Achondrogenesis, type II	200610	1	-	1	-	385,800	-
EYE DISEASES								
12.	Microphthalmia, isolated 1	251600	2	-	3	-	128,600	-
13.	Cataract, congenital or juvenile	212500	-	1	-	1	-	56,800
METABOLIC DISORDERS								
14.	Phenylketonuria	261600	7	-	7	-	55,114	-
15.	Hypothyroidism	218700	3	-	3	-	128,600	-
16.	Celiac disease	212750	2	-	3	-	128,600	-
17.	Vitamin D-dependent rickets, type I	264700	1	-	2	-	192,900	-
18.	Cystic fibrosis	219700	1	1	1	1	385,800	56,800
HEREDITARY SYNDROMES								
18.	Pierre Robin	261800	5	-	5	-	77,160	-
19.	Adrenal hyperplasia, congenital	201910	2	-	3	-	128,600	-
20.	Roberts	268300	1	1	1	1	385,800	56,800
21.	Kartagener	244400	1	1	1	1	385,800	56,800
22.	Multiple pterygium, Escobar variant	265000	1	-	1	-	385,800	-
23.	Short rib-polydactyly syndrome, type II	263520	1	-	1	-	385,800	-
24.	Werner	277700	-	1	-	2	-	28,400
25.	Mayer-Rokitansky-Kuster-Hauser	277000	-	1	-	1	-	56,800
26.	Shwachman-Diamond	260400	-	1	-	1	-	56,800
27.	C syndrome	211750	-	1	-	1	-	56,800
OTHER HEREDITARY DISEASES								
28.	Microtia with meatal atresia and conductive deafness	251800	3	3	3	4	128,600	14,200
29.	Deafness	220700	-	1	-	4	-	14,200
30.	Biliary atresia, extrahepatic	210500	1	-	1	-	385,800	-
31.	Hyperbilirubinemia, Rotor type	237450	-	1	-	1	-	56,800
	Total:		48	25	54	36		

We could not find "frequent" AR diseases (with prevalence 1:50,000 and more often) among the Russians. At the same time, among the Khakass populations five diseases could be said to be "frequent": unspecified myopathy; microtia with meatal atresia and conductive deafness; autosomal recessive deafness; ataxia-telangiectasia; oculocutaneous albinism; osteogenesis imperfecta; and Werner syndrome.

The X-linked diseases were presented among aboriginal populations by 2 clinical forms (Table 112-12). The diseases of this type (insipidus diabetes and Duchenne muscular dystrophy) had not presented among the urban population.

Among the Russian populations (urban and rural) the X-linked diseases had greater variety than among the Khakass populations, and were represented by 7 forms (Table 112-12). Among the Russians in Khakassia, as well as among the Russians of Altai, hypohidrotic ectodermal dysplasia was observed more often.

The structure of Mendelian hereditary pathology showed essential differences among the aboriginal and Russian populations: namely, absence of phenylketonuria among the Khakases, while at the same time there was a higher prevalence of hereditary syndromes among the Russians.

As a special peculiarity it is necessary to note that in the Republic of Khakassia there were diagnosed such "rare" syndromes as Lesch-Nyhan syndrome, Werner syndrome, anemia Diamond-Blackfan, Bloch-Sulzberger syndrome, and Leopard syndrome. They were absent in the territories of the previously surveyed regions, the Republics of Altai and Tyva.

REPUBLIC YAKUTIA (SAKHA): THE "YAKUT" HEREDITARY DISEASES

The Sakha (Yakutia) Republic is a Federal subject of Russia. At half the size of the Far Eastern Federal District, it is the largest subnational governing body by area in the world at 3,100,000 km^2, yet it has a population of less than one million (950,700) inhabitants, 64.2% in urban areas. There are 33 districts and 13 cities. Its capital is Yakutsk. According to the 2002 Census, the national composition is 45.54% Yakuts, 41.15% Russians, 3.65% Ukrainians, and 3.15% Evenks and Evens, etc.

PREVALENCE RATE AND DIVERSITY OF HEREDITARY DISEASES AMONG THE REPUBLIC POPULATIONS

During genetic epidemiological inspection the data concerning the loading with HDs among the aboriginal populations of four native villages (ulus) of Yakutia had been obtained. The comparative analysis has shown that between separate subpopulations, there was no difference in estimations of loading with HDs. At the same time, it was found that the loading of the population of the republic with AD diseases (1.7×10^{-3}) differed to a statistically significant extent from the loading with AR diseases (0.7×10^{-3}), with obvious prevalence of AD forms ($\chi^2 = 30.7$; P=0.00001).

There is a high concentration of mutant genes of separate diseases among Yakuts.

First of all, AD diseases with late onset—in particular spinocerebellar ataxia type I (SCA1), oculopharyngeal muscular dystrophy (OPMD), and myotonic dystrophy (MD)—were recorded. The molecular basis of the diseases is an increase in the number of trinucleotide repeats. Achondroplasia was another "frequent" form of AD pathology. Among the AR diseases, there occurred hereditary enzymopenic methemoglobinemia (HEM) type I, Charcot-Marie-Tooth disease, and ichthyosis. In

TABLE 112-12 THE SPECTRUM AND PREVALENCES OF X-LINKED DISEASES IN THE REPUBLIC OF KHAKASSIA

			Number of				Prevalence rate in the male population, 1 case per:	
			families		patients			
No.	Disease	OMIM	Russians	Khakases	Russians	Khakases	Russians	Khakases
1.	Ectodermal dysplasia, hypohidrotic	305100	2	-	4	-	48,225	-
2.	Muscular dystrophy, Duchenne type	310200	1	1	2	1	96,450	28,400
3.	Bloch-Sulzberger syndrome	308300	1	-	2	-	96,450	-
4.	Lesch-Nyhan syndrome	300322	1	-	1	-	192,900	
5.	Hypophosphatemic rickets, X-linked dominant	307800	1	-	1	-	192,900	-
6.	Hemophilia A	306700	1	-	1	-	192,900	-
7.	Aarskog-Scott syndrome	305400	1	-	1	-	192,900	-
8.	Diabetes insipidus, nephrogenic, X-linked	304800	-	2	-	7	-	4057
	Total:		8	3	12	8		

the group of X-linked diseases there occurred, with high frequency, hemophilia A and ichthyosis.

The description of the load of hereditary diseases among Yakuts in details can be found in Chapter 111 of this book.

CONCLUSION

Uniformity of the methodical principles used for the examined populations allows us to compare the loading with hereditary pathology in the studied ethnic groups. The obtained data on the loading of aboriginal populations of Siberia with HDs are presented in Table 112-13. Comparing the data obtained during medical genetic examination of aboriginal populations of Siberia, there is an obvious differentiation of the studied populations (the Tyvinians, the Altaians, the Khakases, and the Yakuts) according to the values of loading with AR and AD diseases. Studying the reasons for such differentiation is the basic object of researching population genetics.

The load of hereditary diseases in the studied Turkic populations varies in sufficiently wide limits: from 0.1×10^{-3} (among the Altaians in the X-linked pathologies) up to 1.7×10^{-3} (among theYakuts in AD pathologies). However for the majority of the steady, generated populations, the degree of loading with HDs amounts to approximately 2.0–3.8 per 1,000 persons. As for loading with HDs in all surveyed populations, the share of AD pathology is typically large; however, in the populations of the republics of Tyva and Altai, the higher prevalence of AR diseases is noted in the general level of loading the population with HDs. The comparative analysis of loading with hereditary pathology of the aboriginal (Turkic) populations of Siberia has shown that the loading with AD pathology is, in all four studied populations, approximately at one level, except for the Republic of Tyva. Loading with AR pathology is at approximately one level except for the Republic of Altai, where the loading is 2.5 times higher, and that is connected with the population structure (in particular subdivisions of native populations; Puzyrev and Nazarenko, 2000). The casual inbreeding (Puzyrev et al., 1999) is a measure of the genetic drift caused by casual fluctuation of frequencies of genes in sectioned and small populations in Tyvinians. The major factor of population dynamics influencing frequency of mutant genes among the Russian population of the Tomsk region is migration (Puzyrev and Nazarenko, 2000). While carrying out the examination in the Republics of Yakutia, Altai, Tyva, and Khakassia, there were registered patients with microtia with meatal atresia and conductive deafness.

Epidemiological study of Mendelian HDs among the populations of Siberia have allowed us to make some conclusions. First of all, HDs are widespread in different populations and ethnic groups extremely non-uniformly; that is, caused by features of coarse population-genetic and demographic processes, and also connected with their ethnic history. Thus, the most adequate parameter for planning and organization of medical genetic services is not the size of the population, but features of its structure, size of HD load, and its qualitative characteristics. It assumes that a differentiated approach must be applied to the organization of medical genetic services in territories of such a big and multinational country as Russia. Thus, population-epidemiological research of the laws of formation of HD loads in certain populations and ethnic groups testify not only to a different level of loaded populations with HDs, but also their specificity. This fact highlights the importance of studying regional features of prevalence of hereditary diseases and creations of preventive registrations of HDs in certain regions of the country.

On the basis of the results of genetic epidemiological investigations, and the information collected in the Genetic Register of Siberia, several more candidates for "Yakut hereditary diseases" can be distinguished: nanism syndrome with optic nerve subatrophy and Pelger anomaly of leukocytes, Friedreich`s disease, and microtia with meatal atresia and conductive deafness.

The investigation of Mendelian hereditary pathology in different ethnic groups in Siberia is an important component for determining a strategy to address regional issues of genetic health. Monitoring of the hereditary diseases on the individual and population levels became the basis for the formation of the medical genetic service.

TABLE 112-13 THE LOAD OF HEREDITARY DISEASES AMONG ABORIGINAL POPULATION IN SIBERIA

Population	Number of persons (in thousand)	Load 1×10^{-3}			
		AD diseases	AR diseases	X-linked pathology	Total
Yakutia (Sakha) Republic	102.4	1.7	0.7	0.4	2.8
Tyva Republic	198.4	0.6	0.9	0.5	2.0
Altai Republic	59.0	1.1	2.6	0.1	3.8
Khakassia Republic	56.8	1.1	0.8	0.4	2.3

REFERENCES

Baird PA, Anderson NW, Newcombe NB, Lowry RB. (1988). Genetic disorders in children and young adults: a population study. *Am J Hum Genet* 42:677–694.

Biesecker LG. (1998). Lumping and splitting: molecular biology in genetics clinic. *Clin Genet* 53:3–7.

Bochkov NP, Timofeeff-Ressovsky NV. (1968). Geographic environment and human genetic pathology In: *Meditsinskaya geografya* [Medical Geography] (pp. 13–18). Nauchn. Tr. Obninskogo otdela GO SSSR (Sci. Proc. Obninsk Branch GS Soviet Union).

Bochkov NP, Anfalova TV, Garikavtseva RF, et al. (1971). Medical genetic study of the population of the western Pamir: Sex ratio, hereditary neuropsychiatric diseases, and general anthropogenetic traits. *Genetika (Moscow)* 7(5):149–159.

Fedorova SA, Khusainova RI, Kutuev IA, et al. (2005). Polymorphism of CTG-Repeats in the DMPK Gene in Populations of Yakutia and Central Asia. *Mol Biol* (Moscow) 39(3):385–393.

Francomano CA, Biesecker LG. (2003). Medical genetic studies in the Amish. *Am J Med Genet* 121:1–83.

Galeeva NM, Nazarenko LP, Nazarenko SA. (2006). Molecular genetic cause of hereditary methemoglobinemia type 1 in Yakutia. *Med Genet* 5:15–20.

Ginter EK, Golubcov VI, Petrin AN, et al. (1997). *Medical genetic examination of Adygei population.* Majkop.

Ginter EK, Revasov AA, Petrin AN. (1992). *Vlijanie urbanizacii na gruz nasledstvennyh boleznej v populjacijah.* [Man's heredity and environment.] Moscow: Nauka.

Ginter EK. (2001). Population genetics and medicine. *Vestn RAMN* 10:25–31.

Ginter EK. (2002). *Hereditary Diseases in Human Populations.* Moscow: Meditsina.

Goldfarb LG, Vasconselos O, Platonov FA, et al. (1996). Unstable triplet and phenotypic variability of spinocerebellar ataxia type 1. *Ann Neurol* 39:500–506.

Kere J. (2001). Human population genetics: lessons from Finland. *Ann Rev Genomics Hum Genet* 2: 103–128.

Kucher AN. (2002). The population structure of indigenous and alien populations of Siberian region. *Med Genet* 1(6):262–270.

Lunkes A, Goldfarb LG, Platonov FA, et al. (1994) Autosomal dominant spinocerebellar ataxia (SCA) in Siberian ataxia type 1. *Exp Neurol* 126:310–312.

Maksimova NR, Korotov MN, Nikolaeva IA, et al. (2007a). Clinical and molecular genetic aspects of oculopharyngeal muscular dystrophy in Republic of Sakha (Yakutiya). In: Puzyrev VP (ed.), *Human Genetics and Pathology* (pp. 160–161). Tomsk: Pechatnaya Manufaktura.

Maksimova NR, Nogovitsyna AN, Puzyrev VP. (2007b). Rare 3-M Syndrome among Yakuts: clinical and molecular genetic characteristic. In: Puzyrev VP (ed.), *Human Genetics and Pathology* (pp. 162–167). Tomsk: Pechatnaya Manufaktura.

Maksimova N, Hara K, Miyashita A, et al. (2007c). Clinical, molecular and histopathological features of short stature syndrome with novel CUL7 mutation in Yakuts: new population isolate in Asia. *J Med Genet* 44:772–778.

McKusick VA. (1969). On Lumpers and Splitters, or the Nosology of Genetic Disease. *Perspect Biol Med* 12:298–312.

McKusick VA. (1978). *Medical Genetic Studies of the Amish.* Baltimore: John Hopkins University Press.

McKusick VA. (1997). Genomics: Structural and Functional Studies of Genomes. *Genomics* 45:244–249.

McKusick VA. (2006). A 60-Year Tale of Spots, Maps and Genes. *Annu Res Genomics Hum Genet* 7:1–27.

Nazarenko LP, Nogovitsyna AN, Salyukova OA, Filimonova MN. (2002). Occurrence of Monogenic Hereditary Diseases among Rural Population in Republic of Sakha (Yakutia). In: *Medical Genetic Study of the Population of Republic of Sakha (Yakutia).* Yakutsk: YaF SO RAN.

Nazarenko LP, Salyukova OA. (1997). Monogenic hereditary diseases of Tomsk area. In: Puzyrev VP (ed.), *Human Genetics and Pathology* (pp. 65–86). Tomsk: Pechatnaya Manufaktura.

Nogovitsyna AN, Maksimova NR, Khandy MV, Alekseeva SP. (1999). Hereditary pathology in the families who asked for treatment at the genetic counseling department of the National Medical Center, Republic of Sakha (Yakutia) in 1990 to 1998. *Dalinevost Med Zh* 1:26–30.

Norio R. (2003). Finnish disease heritage. *Hum Genet* 112:441–526.

Peltonen L, Jalanko A, Varilo T. (1999). Molecular genetics of the Finnish disease heritage. *Hum Mol Genet* 8(10):1913–1923.

Platonov FA, Illarioshkin SN, Kononova SK, et al. (2004). Spinocerebellar ataxia type 1 in Yakutia: Prevalence and clinical genetic comparisons. *Med Genet* 5:242–248.

Puzyrev VP. (1984). Characteristics of the Incidence of Hereditary Diseases in Western Siberia. *Vestn Ross Akad Med Nauk* 7:76–78.

Puzyrev VP, Erdynieva LS, Kucher AN, Nazarenko LP. (1999). *Genetic Epidemiologic Study of the Tuva Population.* Tomsk: STT.

Puzyrev VP, Freidin MB, Kucher AN. (2007). *Genetic Diversity of Population and Human Diseases.* Tomsk: Pechatnaya Manufaktura.

Puzyrev VP, Maksimova NR. (2008). Hereditary diseases among Yakuts. *Russ J Genet* 44(10):1141–1147.

Puzyrev VP, Nazarenko LP. (2000). *Genetic Epidemiological Study of Hereditary Disorders in Western Siberia.* Tomsk: STT.

Puzyrev VP, Stepanov VA, Golubenko MV, et al. (2003). mtDNA Lineages and Y-Chromosomes in the Yakut Populations. *Russ J Genet* 39 (7):816–822.

Stevenson AC. (1959). The load of hereditary defects in human populations. *Radiat Res* 1:306–325.

Sukhomyasova AL. (2005). *Autosomal Dominant Myotonic Dystrophy in Republic of Sakha (Yakutiya).* Extended Abstract of Cand. Sci. (Med.) Dissertation, Tomsk.

Tarskaya LA, Zinchenko RA, Elichinova GI, et al. (2004). The structure and diversity of hereditary pathology in Republic of Sakha (Yakutia). *Russ J Genet* 40(11):1530–1539.

Zinchenko RA, Amelina SS, Shokarev RA, et al. (2009). Epidemiology of monogenic hereditary diseases in Rostov Oblast: Population dynamic factors determining the differentiation of the load of hereditary diseases in eight districts. *Russ J Genet* 45(4):469–477.

Zinchenko RA, Elichinova GI, Galkina VA, et al. (2007). Genetic Differentiation of ethnic groups of Russia evaluated by genes of hereditary disorders. *Med Genet* 6:2(36):29–37.

113.

HEREDITARY DISEASES IN THE VOLGA-URAL REGION OF RUSSIA

Irina M. Khidiyatova, Irina R. Gilyazova, Vita L. Akhmetova and Elza K. Khusnutdinova

INTRODUCTION

The main feature of modern medical genetics, analyzing hereditary diseases, is active use of methods of molecular biology, which allow us to detect structural disturbances in the human genome—in fact, to discover the first cause in pathogenesis of diseases, and to develop highly informative methods of diagnostics and effective methods of medical treatment on that basis. Many hereditary diseases are characterized by clinical polymorphism and genetic heterogeneity; that's why the decision of the most importance in solving the problems of medical genetics is the determination of population singularity and structural features of genes in determining the development of these diseases. That means it is necessary to develop optimal methods of DNA diagnostics for each particular region, on the one hand, and on the other hand, to broaden our knowledge about the structure of human genome and its functioning and make contributions to gene pool analysis in different populations. Here we report the main results of molecular analysis of genes responsible for phenylketonuria, Wilson disease, Duchenne muscular dystrophy, hereditary motor and sensory neuropathies, nonsyndromic hereditary deafness, myotonic dystrophy, and Huntington disease development in the Volga-Ural region. This region is at the boundary between Europe and Asia; its populations are each characterized by complex ethnogenesis, specific gene pool, and specific population structure, and combine both Caucasian and Mongoloid components in varying proportion. Linguistically, populations of the region belong to the Turkic (Tatars, Bashkirs, Chuvash), Finno-Ugric (Mari, Mordvinias, Udmurts, Komi), and Slavonic (Russians, Ukrainians) stocks (Khusnutdinova, 1999).

Genomic analysis of monogenic hereditary disorders was initially carried out in the Bashkortostan Republic. The spectrum and frequency of mutations in these genes, typical for the main ethnic groups of the Bashkortostan Republic (Bashkirs, Tatars, and Russians) were shown. These findings became a base for an optimal scheme of DNA-diagnostics development for these hereditary diseases in the Volga-Ural region and ascertainment of the mutations origin leading to the diseases.

PHENYLKETONURIA

Phenylketonuria is the most common inborn error of amino acid metabolism in Europeans. It is caused by an autosomal recessive deficiency of the hepatic enzyme phenylalanine hydroxylase (PAH). Failure to convert phenylalanine to tyrosine leads to an increase of phenylalanine in the body fluids and severe mental retardation unless phenylalanine intake is restricted. The average frequency of phenylketonuria in Europe is approximately 1:100,000 newborns (Woo et al., 1993). The disease is caused by mutations in phenylalanine hydroxylase gene (PAH), located on 12q22–24 chromosome (Lidsky et al., 1984). The *PAH* gene consists of 13 exons; its length is about 90,000 bp. More than 450 mutations in *PAH* gene are described, with a spectrum and frequency of mutations that are characterized by considerable population differences (http:www.mcgill.ca/pahdb). The most frequent mutation in Europe is *R408W (p.Arg408Trp)*.

We analyzed structural features of the *PAH* gene in 59 patients with phenylketonuria, and 126 members of their families, from Bashkortostan (Akhmetova et al., 2003). The frequency of *p.Arg408Trp* mutation in common samples of patients was 53%, which corresponds to the frequency of the mutation in European populations. *p.Arg408Trp* mutations appeared to be the most frequent among patients of Russian (60%) and Tatar (23%) ethnic origins.

In the *PAH* gene we have identified six mutations in patients from Bashkortostan: *p.Arg408Trp* (53%), *p.Arg261Gln* (10%), *p.Arg252Trp* (3%), *p.Pro281Leu* (3%), *p.ArgR252Pro* (1%) and 1315+del4 (1%). The last two mutations were described for the first time. The majority of examined patients were compound heterozygous, carrying the most widespread mutation *p.Arg408Trp* on the one chromosome, and either one of the mutations described above or some unknown mutation on the other. Direct DNA diagnostics in families with phenylketonuria was informative in 91.1% of cases. We have analyzed haplotype association of four intragenic polymorphic markers—VNTR (3'-terminal site), MspI (a) (site of intron 8, containing polymorphic MspI-site), STR (site of intron 3), PvuII (a) (site of intron 2, containing PvuII site)—with mutations in *PAH* gene for methods of indirect DNA diagnostics development. It is established that *p.Arg408Trp* mutation is associated with haplotype 380-a-240-A2 (linkage disequilibrium ΔSt = 0.66). According to Giannattasio et al. (1997), *p.Arg408Trp* mutation is also associated with the same haplotype, designated as RLP-haplotype2, in Baltic and Slovenian populations. *p.Arg261Gln* mutation is associated with the haplotype mentioned above (ΔSt=0.79). Haplotype 380-a-240-A2 (ΔSt=0.66) is determined on 48% of mutant chromosomes and only on 4% of normal chromosomes. Thus, these data may be evidence of a Baltic-Slavic origin of phenylketonuria in Bashkortostan.

WILSON DISEASE

Wilson disease (WD) is a heavy autosomal recessive disorder characterized by copper and protein metabolism disturbances, resulting in combined liver and brain disruptions owing to the toxic effects of copper accumulating in human organs. The frequency of Wilson disease in world populations is approximately 1:30,000–40,000 (Scheinberg & Sternlieb, 1984). The gene for Wilson disease (*ATP7B*), coding copper-transporting ATPase, is located on the13q14.3 chromosome and consists of 80,000 bp and 21 exons (Frydman et al., 1985; Bull et al., 1993). More than 370 mutations are identified and described in the ATP7B gene. The mutation spectrum is heterogeneous in various populations (Wilson Disease Mutation Database, http://www.wilsondisease.med.ualberta.ca; Human Gene Mutation Database, http://www.hgmd.cf.ac.uk/ac/index.php). The *p.His1069Gln* mutation is the most widespread in European populations at 10%–70% (Stapelbroek et al., 2004).

Molecular genetic analysis of Wilson disease was performed in 44 WD patients and 104 members of their families (33 families: 6 families of Russian ethnic origin, 9 Tatar, 4 Bashkir, 2 Chuvash; probands in 12 families were from mixed marriages; Karunas et al., 2009). We have screened exons 3 through 20 of the *ATP7B* gene. Using SSCP analysis followed by direct sequencing, we have characterized the molecular defect in 81.82% of WD chromosomes and identified 9 different mutations: *c.2152 G>C* (*p.Ala718Pro*) and *c.2304dupC* (*p.Met769HisfsX26*) – in the 8th exon; *c.3170T>C* (*p.Leu1057Pro*), *c.3190G>A* (*p.Glu1064Lys*), and *c.3207C>A* (*p.His1069Gln*) in 14th exon; *c.3402delC* (*p.Ala1135GlnfsX13*) in the 15th exon; *c.3556+1G>T* (*p.IVS16+1G>T*) in the 16th intron; *c.3914T>C* (*p.Leu1305Pro*) and *3943delCA; 3947delG* (*p.Lys1315_Arg1316delinsGlu*) in 19th exon; and 3 polymorphic variants, (*c.1366G>C* (*p.Val.456Leu*), *c.3419T>C* (*p.Val1140Ala*), and *c.3903+6T>C*). Four of the mutations revealed—*p.Ala718Pro, p.Leu1057Pro, p.IVS16+1G>T,* and *p.Lys1315_Arg1316delinsGlu*—had not been described before.

The common *p.His1069Gln* mutation was represented in 43.94% of WD chromosomes; its frequency rate in Bashkirs was 44.44%, in Russians 47. 62%, and in Tatars 36.36%. Mutation *p.Lys1315_Arg1316delinsGlu* was detected in 18.18% of Tatar chromosomes and 22.22% of Bashkir chromosomes.

Alleles and haplotypes, statistically significantly associated with mutant and normal chromosomes, were detected using polymorphism of microsatellite loci *D13S316*, *D13S133*, and *D13S228* analysis in families with Wilson disease and control DNA samples (Karunas et al., 2000). The most widespread haplotype (40%) on mutant chromosomes was 6–18-3, which was not revealed on normal chromosomes. Haplotypes 7–18-2 (19%) and 7–18-4 (9%) were the most widespread on normal chromosomes; their frequency on mutant chromosomes was 7%. Haplotypes association analysis of polymorphic markers with mutations has shown that 90% of chromosomes with a *p.His1069Gln* mutation had the most widespread haplotype of 6–18-3, revealed also on chromosomes with these mutations in other investigated populations. Taking this fact into consideration, we assumed that *p.His1069Gln* mutation appeared long ago and had a common source of origin. Three chromosomes bearing *3402delC* mutation had identical haplotype 4–7-4, and one chromosome with *p.Glu1064Lys* mutation had 6–7-2 haplotype, which was revealed neither on other mutant chromosomes in families with Wilson disease nor on normal chromosomes. The mutation *p.IVS16+1G>T*, revealed for the first time, had the same haplotype 7–18-4, which had not appeared on other mutant chromosomes and was detected in three WD patients of Tatar ethnic origin, confirming its Turkish origin.

DUCHENNE MUSCULAR DYSTROPHY

Duchenne muscular dystrophy (DMD) is the most frequent X-linked recessive hereditary disorder of the neuromuscular system, caused by mutations in the dystrophin gene (Goodfellow et al., 1985). The frequency of the disease is 1: 3500 boys (Emery, 1991). The dystrophin gene (*DMD*), the largest of all known human genes, is located on the X (p21) chromosome, consists of 2500 kb, and contains 79 exons. The most widespread mutations (25%–70%) in the *DMD* gene are deletions, occupying from one to several exons and located usually at two "hot" points -in 5' -terminal (exons 6–19) and 3'-terminal (exons 40–53) (Beggs et al., 1990).

We analyzed the deletions spectrum and deletion points of disruption distribution in the *DMD* gene in 70 patients from 63 families living in the Bashkortostan Republic, using multiplex PCR analysis. Families suffering from Duchenne muscular dystrophy had various ethnic origins: Russians - 20; Tatars – 23; Bashkirs – 12; Chuvashes – 3; Ukrainians – 2; Maris – 2; Kabardinians – 1. Deletions from 1 to 9 exons long were detected in 20 out of 63 unrelated patients. Four of them were patients of Russian ethnic origin, 7 Tatars, 2 Bashkir, 1 Maris, 1 Ukrainian, and 1 Kabardinians, making up 31.8% of the total number of patients. Thirteen deletions (65.0%) were located in the 3'-terminal of the gene, and seven (35.0%) in its 5'- "hot" area. Thus, the deletions ratio in distal and proximal areas of "hot" points in patients from Bashkortostan was 1.86:1.

The family character of the disease was revealed in five of seven proximal deletions; it could not be excluded in 2 cases that confirm frequent proximal region deletions in the disease with family character, and in distal region deletions in sporadic cases. The distribution of deletions in the *DMD* gene in various ethnic groups appeared interesting. Although separate ethnic groups of patients extracted from the whole sample were very small, we found that in three ethnic groups—Russians, Tatars and Bashkirs—the most numerous of all distribution of *DMD* gene deletions was heterogeneous; namely, 20.0% in Russians, (4 patients of 20), 47.82% in Tatars (11 patients of 23) and 16.6% in Bashkirs (2 patients of 12), but the differences between separate ethnic groups were not statistically significant. Disruption points localization was specific. There were 2 "hot" areas of the gene: the greatest number was fixed between in 5'-region between exons 3–19 (30.0%) and exons 19–42 (12.5%); in 3'-region between exons 50–51 (15.0%), 44–45 (10.0%), 45–46 (10.0%), and 51–52 (7.5%). One disruption point (2.5%) was revealed between exons 46–47, 47–48, 48–49, 49–50, 52–53 and 53–60.

SSCP analysis and consequent sequencing of DNA samples without large deletions revealed two new point mutations not described before, and one polymorphic variant. In 3 patients from 2 families of Russian ethnic origin we found *c.401_404delCCAA* (*p.Thr134ThrfsX7*) in the 6th exon, with a frequency of 3.17% among unrelated patients. Mutation *c.6626delA* (*p.Lys2210ArgfsX11*) (1.6%) was revealed in the 46th exon in one patient of Tatar ethnic origin. We found mutation *c.7728T>C* (*p.Asn2575Asn*) in the 53th exon, the functional significance of which is unknown. Thus, a specific mutation spectrum of *DMD* gene was found in patients from Bashkortostan Republic that is very important for analysis of structural and functional features of the gene, and for of the development of optimal DNA diagnostics approaches.

HEREDITARY MOTOR AND SENSORY NEUROPATHY

Hereditary motor and sensory neuropathies (HMSN; Charcot–Marie–Tooth disease, CMT) represent a genetically heterogeneous group of diseases affecting the peripheral nervous system. Typical symptoms of most HMSN cases are progressive weakness and atrophy of the distal leg muscles and the hand muscles in latter stages, depression of tendon reflex, foot deformities of the pes cavus type, and sensory disturbance of the polyneuritic type (Harding & Thomas, 1980). The disease mechanics include degenerative changes of myelin sheath, or of the axons of motor and sensitive fibers of peripheral nerves or spinal roots. Accordingly, two main forms of the disease are distinguished, including the demyelinating form, HMSN I, and the axonal form, HMSN II. Clinical differentiation between these two types is possible only using electroneuromyographic (ENMG) investigation, which provides evaluation of median nerve conduction velocity (MNCV). Within the two HMSN types, the autosomal dominant, autosomal recessive, as well as X-linked forms were all described (Dyck et al., 1993). To date, more than 25 HMSN loci have been mapped and 22 genes identified (http://www.neuro.wustl.edu/neuromuscular/time/hmsn.html). Prevalence of different HMSN forms in world populations varies, ranging from 10 to 40 in 100,000 individuals (Skre, 1974; Emery, 1991; Ionasescu, 1995). In the European countries and United States, the most prevalent form is HMSN type 1A (HMSN 1A), which is caused by duplication of the gene for peripheral myelin protein (*PMP22*; Timmerman et al., 1990; Patel et al., 1990; Lebo et al., 1992). The second most frequent form is HMSN type 1X, a dominant X-linked form, which is found in 10% to 20% of all HMSN cases (Ionasescu et al., 1993; Nelis et al., 1996). The genetic causes of this disease form are the mutations in the *GJB1*gene (gap junction B1 type), mapped to the long arm of X-chromosome region Xq131 (Bergoffen et al., 1993). At present, more than 300 different mutations, involving all regions of the protein molecule, have been described (http://www.molgen.ua.ac.be/CMTMMutations). The functional importance of the mutations is different, depending on their type and localization. Consequently, clinical characteristics of HMSN type 1X are variable. Furthermore, the HMSN 1X phenotype is not always clearly differentiated from HMSN type II, demonstrating the features of both myelinopathy and axonal degeneration (Ionasescu et al., 1996).

Mutations of the *MPZ* (major peripheral myelin protein zero) gene are also a frequent cause of HMSN. Gene *MPZ* is mapped to the region 1q21.3 – q23, consists of 6 exons, and more than 120 point mutations are known to result in HMSN (www.molgen.ua.ac.be/CMTMutations). Most of them refer to HMSN type I/HMSN IB (34 mutations) and other variants of HMSN I (41 mutations). The analysis of the *MPZ* gene in Dejerine-Sottas syndrome (DSS) showed sixteen mutations; both sporadic (Hayasaka et al., 1993) and recessive variants of the disease are known (Ikegami et al., 1996; Warner et al., 1996). Some *MPZ* gene mutations result in congenital dominant hypomyelinating neuropathy (CHN, CMT4E; Warner et al., 1996; Kochanski et al., 2004). Thirteen various mutations in the *MPZ* gene were found in HMSN type II (CMT2I, CMT2J; Chapon et al., 1999; Misu et al., 2000; Senderek et al., 2000). The frequency of the *MPZ* gene mutations among HMSN patients of different ethnic origin varies. The highest frequency of mutations is demonstrated in HMSN patients from Spain at 10.6% (Bort et al., 1997), and the lowest in Finland at 4.8% (Silander et al., 1998). The frequency of *MPZ* gene mutations in Russia is estimated at 4.6% (Mersiyanova et al., 2000).

In addition, the mutations in the early growth response 2 (*EGR2*) gene (10q21.1-q22.1), segregating as either dominant or recessive alleles, have been demonstrated in HMSNI and congenital hypomyelination (Warner et al., 1998). The coding part of the gene consists of 4300 bp and is represented by 2 exons. Although only 9 mutations and 1 polymorphism are known today in this gene, it is being intensively investigated in molecular genetic analysis of HMSN in populations.

In recent years, a survey on the prevalence and molecular bases of HMSN has been carried out in Bashkortostan. The HMSN frequency in Bashkortostan Republic is 10.3:100,000, and HMSN type I is prevalent in this region (Krupina et al., 2006).

We examined CMT1A duplication of 17p11.2-p12, point mutations of *PMP22, MPZ (P0), GJB1 (Cx32)* and *EGR2* genes in 173 patients from131 unrelated families (42 Russian, 37 Tatar, 27 Bashkir, 3 Ukrainian families, and one patient of Chuvash, Mari, and Azerbaijanian origin; 19 probands originated from interethnic marriages).

The frequency of the *PMP22* gene duplication, which is the cause of HMSNIA, was determined. It constituted 26.72% of all HMSN types, which was somewhat lower compared to other regions of Russia. *PMP22* gene duplication was found in Tatars more often (37.14%) than in Russians (31.43%) and Bashkirs (14.78%). Deletion of the *PMP22* gene, resulting in development of hereditary neuropathy with liability to pressure palsies, was found in 2 patients (1.53%).

SSCP analysis, followed by direct sequencing, allowed determination of only one nucleotide change, *c.309–29 G>A* in *PMP22* gene, which hadn't been described previously. In the *MPZ* gene we detected mutation *p.Ser88Ley* in one patient (0.8%) *(c.263C>G). This mutation was described earlier by* Fabrizi et al. (2000). The nucleotide change *c.714C>T* (*p.Ser238Sser*) was found in two patients. In the *EGR2* gene in HMSN patients from Bashkortostan Republic, no changes were revealed.

In the *GJB1* gene, 4 missense mutations were detected (Khidiyatova et al., 2008): *p.Pro87Ala* (*c.259C>G*), with frequency of 10% among HMSN patients, *p.Arg22Gln* (*c.65G>A*) (2.98%); *p.Arg15Gln* (*c.44G>A*) (1.5%), and *p.Thr86Ile* (*c.257C>T*) (0.8%); the last mutation had not been described previously. It was demonstrated that the frequency of the *p.Pro87Ala* mutation in the HMSN patients from Bashkortostan was high, but the distribution of this mutation in different ethnic groups was heterogeneous. The highest mutation frequency was observed among Bashkirs, where it was the reason for HMSN in 33% of the cases. The mutation was rare among Russians, and was nearly absent among Tatars. According to the literature data, the mutation *p.Pro87Ala* was earlier described in Belgian patients (Nelis et al., 1997). Taken together, the *GJB1*mutations were detected in 18 out of 131 unrelated HMSN patients from Bashkortostan, which constituted 13.7% for this sample. These data are consistent with those published elsewhere, and according to which in European populations, HMSN 1X constitutes 10% to 20% of all HMSN cases (Nelis et al., 1996).

At present we continue to study HMSN in the Bashkortostan Republic, analyzing genes responsible for other rare disease forms.

NONSYNDROMIC HEREDITARY DEAFNESS

Congenital deafness is a relatively common human disorder, occurring on average in 1/1000 newborns (Chalestori et. al., 2007). Nonsyndromic autosomal recessive deafness (DFNB1) is the most common form of inherited hearing loss; it is clinically polymorphic and genetically heterogeneous. This is a neurosensory disorder with distorted acoustic sense. Mutations in the *GJB2* and *GJB6* genes for DFNB1 (13q12) are the most frequent cause of monogenic hearing loss, and they are responsible for about half of all cases of autosomal recessive prelingual hearing loss (Morton et al., 2006; Petersen & Willems, 2006).

In many Caucasian populations, approximately 50% of recessive nonsyndromic deafness can be attributed to mutations in the connexin 26 (*GJB2*) gene (Denoyelle et al., 1999). More than 110 different mutations in *GJB2* have been identified in patients with nonsyndromic deafness, and a significant difference in the frequency and distribution of the mutations has been observed in different populations (http://www.iro.es/cx26deaf.html). Most interestingly, a single *GJB2* gene mutation, *c.35delG*, accounts for up to 70% of northern and southern European, as well as

American Caucasian populations, with a carrier frequency ranging from 1.3% to 2.8%. Another *GJB2* gene mutation, *c.167delT*, accounts for 40% of the pathologic alleles in the Jewish deaf population and has 4% carrier frequency among Ashkenazi Jews (Morell et al., 1998). Both *c.35delG* and *c.167delT* mutations are absent or exceptionally low in some Asian populations, in which another mutation, *c.235delC*, is the most prevalent; this mutation accounts for up to 80% of pathogenic *GJB2* alleles among Japanese (Kudo et al., 2000), and Chinese (Liu et al., 2002), with carrier rates ranging from 1.0% to 1.3%. Interestingly, *c.235delC* has not been detected in South Asian populations from India, Pakistan, Bangladesh, and Sri Lanka, in which the prevalent *GJB2* mutations are *Trp24X* and *Trp77X* (Liu et al., 2002).

We report the spectrum of *GJB2* and *GJB6* mutations among 235 patients from 100 unrelated families of different ethnic affiliation (38 families of Russian ethnic origin, 32 Tatar, 5 Bashkir, and 25 mixed and other ethnicity) with nonsyndromic sensorineural hearing loss in Bashkortostan (Khidiyatova et al., 2002, Dzhemileva et al., 2002, Khusnutdinova & Dzhemileva, 2005). Thirteen of 53 unrelated patients (24.5%) had at least one *c.35delG* allele and 40 (75%) were *c.35delG* homozygous. Most often this mutation arises on chromosomes of patients of Tatar and Russian origin, with a frequency of 40.9% and 55.2%, respectively.

The *c.312del14* mutation of the *GJB2* gene accounts for up to 4% of pathogenic *GJB2* alleles among patients with nonsyndromic hearing impairment from Bashkortostan. Also, the *c.312del14* mutation in *GJB2* is the second most frequent cause of nonsyndromic hearing impairment in Bashkortostan. In addition to the *c.35delG* mutation, five different GJB2 mutations, including three novel mutations, were identified in our samples.

One child carried the *c.35delG* mutation and a new, *c.254C>A* mutation (*p.Ser86Tyr*) that impaired the transmembrane portion (M2) of the protein. The second mutation was *c.235delC*. This deletion resulted in a frameshift converting codon 78 from leucine to cysteine, followed by a stop codon at position 79. The third and fourth (novel) mutations were *c.313/314delAA* and *c.314delA*. These mutations cause a frameshift at codon 111, resulting in a stop codon at positions 112 and 117, respectively. The mutation *c.360delG* was found in only one family. This deletion caused a frame shift that resulted in chain termination after addition of novel amino acids. Furthermore, we also identified a novel mutation, *c.557A>T (p.Ile186Phe)*, in the GJB3 gene. This transversion impaired the transmembrane portion (M3) of the protein. It was detected in 2.5% of all mutant chromosomes with sensorineural hearing loss. Interestingly, the audiogram of these patients showed a bilateral mild sensorineural hearing loss at all frequencies, but mainly at high frequencies.

In parallel, we have performed molecular genetic analysis of 700 DNA samples from 7 Volga-Ural populations and established the frequency of the *GJB2* variants in Volga-Ural population. This study has revealed prevalence of the *GJB2* sequence variations among the Volga-Ural region population. The *GJB2* gene contribution in deafness among deaf patients from Bashkortostan (46% of all patients' alleles) is mainly defined by the recessive mutation *c.35dellG*. Its frequencies in Mordvinians (1/16) and in Russians (1/46) were similar to those established for the populations of southern Europe, including Spain (3%; Estivill et al., 1998), Greece (3.5%; Denoyelle et al., 1999), and Italy, and for the Estonian population, which has the highest known frequency of *c.35delG* mutation (Gasparini et al., 2000). Other ethnic groups of the Volga–Ural region were similar in *c.35delG* frequency (0.02) to the north and central European populations and to American Caucasians (Gasparini et al., 2000).

MYOTONIC DYSTROPHY

Myotonic dystrophy (MD) is a severe autosomal dominant neuromuscular disorder with a great variety of clinical signs that suggest multisystemic lesion, including myotonia, progressive muscular weakness, cardiomyopathy, and cataracts. The incidence of the disease is 1:8000 in the global population and 1:7500 in Europe (Harper, 1989). A cause of MD is expansion of the CTG repeat in the 3'-untranslated region of the myotonic dystrophy gene (*DMPK*), which is located on chromosome 19q13.2–13.3 and codes for myotonin protein kinase acting as serine/threonine protein kinase (Mahadevan et al., 1992). The CTG repeat number varies from 5 to 35 in healthy individuals and amounts to several thousand in MD patients (Harris et al., 1996). It has been shown that the age at onset is in inverse proportion to the severity of MD, and in direct proportion to CTG repeat expansion. MD families are characterized by anticipation; i.e., a decline in the age of onset and an increase in severity in subsequent generations (Harper, 1989). We have analyzed the frequency distribution of the normal *DMPK* alleles in 11 Volga-Ural region populations (Fatkhlislamova et al., 1999) and in 72 MD patients and 54 members of their families, living in the Bashkortostan Republic (Khidiyatova et al., 2000). Population-specific features and heterogeneity of $(CTG)_n$ repeat distribution in the Volga-Ural region were revealed. The most frequent alleles among 25 revealed in the Volga-Ural region, corresponding to 5–34 trinucleotides, were $(CTG)_5$ and $(CTG)_{11-14}$. Alleles with more repeats (n>30) were detected in populations of Tatars and Bashkirs of the Abzelil district of Bashkortostan. Comparative analysis showed that in terms of $(CTG)_n$ repeat alleles distribution in the DMPK gene, the Volga-Ural region occupies an intermediate position between populations of Europe and Asia, gravitating toward Asian populations. Factual heterogeneity of this locus was 81%. All these data suggest that this polymorphic DNA locus can be used as a highly informative population marker.

Myotonic dystrophy incidence in Bashkortostan is heterogeneous and varies from 1.05 to 17.7 per 100,000 of population. More than one-third of MD families were descended from submountainous and mountainous areas of the South Urals (Magzhanov, 1985), where gene *DMPK* alleles with more than 30 CTG-triplets are encountered. A CTG repeat expansion in the *DMPK* gene was detected in 67 of 72 MD patients (91.9%) and in 12 healthy members of their families. The size of the expanded repeat in MD patients ranged from 400 bp (approximately 130 CTG units) to 5000 bp (approximately 1500–1600 CTG units). Analyzing the inheritance of $(CTG)_n$ repeat expansion we detected both an increase (67.7%) of CTG repeat number and anticipation, and conservation of initial repeat number, as well as its reduction (17.6%) in next generations. Changes on expansion did not depend on paternal or maternal CTG repeat transmission.

HUNTINGTON'S DISEASE

Huntington's disease is one of the most severe progressive neurodegenerative disorders, characterized by extended choreiform movements, progressive cognitive impairment resulting in deep dementia, and personality changes. The prevalence of HD in different populations varies from 4–10:100,000 (Hayden, 1981). The anticipation phenomenon is also typical for Huntington's disease. The disease is associated with the trinucleotide repeat $(CAG)_n$ expansion in exon 1 of the *IT15* (*HD*) gene. The *IT15* gene was mapped to the short arm of chromosome 4 in the region 4p16.3. It is demonstrated that normal alleles carried from 9 to 35 $(CAG)_n$ repeats, pre-mutation alleles $(CAG)_{36-39}$ associated with incomplete penetrance, and clinical symptoms of the disease inevitably develop in the presence of the $(CAG)_{\geq 40}$ allele (Andrew et al., 1993). The *IT15* gene includes a number of polymorphic sites linked with $(CAG)_n$ repeats, including $(CCG)_n$ repeats located close to the $(CAG)_n$ repeat in the 1st exon of the gene. In some cases, neutral del2642 polymorphism in the *IT15* gene is revealed in exon 58, about 150 kb downstream of the polymorphic $(CAG)_n$ repeats.

To examine the nature of genetic instability of the trinucleotide repeat $(CAG)_n$ expansion in *IT15* gene, and the origin of the repeats in Huntington's disease patients in the Volga-Ural region, we have analyzed $(CAG)_n$-, $(CCG)_n$- repeats polymorphism and *IT15* gene del2642 in 11 populations of the Volga-Ural region and in 59 patients and 127 members of their families (Kutuev et al., 2003, 2004). The allele and genotype $(CAG)_n$- repeats frequencies distribution in the *IT15* gene in different populations is strongly heterogeneous. Allelic diversity (Rs) appeared to range from 9.7 in southeast Bashkirs to 18.00 in Chuvashes, and in the Volga-Ural populations it was found to be 13.72±2.12. The alleles frequency distribution of $(CAG)_n$ repeats corresponds to normal (P<0.0001) single-mode, with a peak in $(CAG)_{17}$ that displays similarities with those of European populations. The allele frequencies of $(CAG)_{29}$ were 0.06, 0.01, 0.02, and 0.01 in populations of Russians, Chuvashes, Mordva, and Udmurts, respectively. $(CAG)_{29-35}$ alleles in Bashkirs, Tatars, Maris, and Komis have not been found. It was demonstrated that the Volga-Ural region populations are quite similar, with the most frequent alleles being $(CCG)_7$ (0.333 on average) and $(CCG)_{10}$ (0.333 on average). The regression analysis showed a strong negative correlation between the age of onset and $(CAC)_n$ repeat length on mutant chromosome. The analysis of the deletion del2642 frequency in the *IT15* gene in the Volga-Ural region populations demonstrated low polymorphism of the gene site (area) (PIC=0.161) with a high prevalence of allele without deletion (del-). The average deletion frequency (del+) was 0.10±0.03, which was similar to most European populations. With the help of logit regression analysis it is established that the probability of the mutation appearing becomes higher in the presence of $(CAG)_{\geq 20}$, which confirms deletion origin on chromosomes with the number of (CAG) repeats more than 20 (Illarioshkin et al., 1996).

The analysis of Huntington's disease prevalence showed that its frequency in the Bashkir population was 0.89:100,000, while in populations of Russians and Tatars it was 4.02 and 5.14:100,000, respectively. The $(CAG)_n$ repeat expansion was analyzed in 59 patients with clinical diagnoses of Huntington disease. Direct DNA analysis has revealed the $(CAG)_n$ repeat expansion in the first exon of the *IT15* gene in 57 of these patients, belonging to 46 families. The cohort of 127 at-risk individuals was also examined. The $(CAG)_n$ repeat expansion was detected in 27 out of 127 asymptomatic at-risk individuals. In patients from Bashkortostan, almost complete linkage disequilibrium between mutant alleles and $(CCG)_7$ and del2642 mutation was revealed. Three main haplotypes found in the mutant chromosomes—$(CCG)_7$/del-, $(CCG)_7$/del+, and $(CCG)_{10}$/del- — pointed to the existence of at least three sources of Huntington's disease in the region. The haplotype frequency distribution pattern observed displayed similarities with those in European populations. Detected linkage disequilibrium indicated not only the existence of different sources of the disease, but also the fact that the $(CCG)_n$ track played the role of *cis*-acting factor, modifying stability of the $(CAG)_n$ repeats.

CONCLUSION

Taking into consideration all the results received, it is possible to conclude that according to the structure of the genes responsible for hereditary diseases development, the Volga-Ural region is heterogeneous, and is most similar to European populations. At the same time, in patients of different ethnic origin and various hereditary disorders we identified mutations and polymorphic variants that had not been described in the world before, and that determine the genetic singularity of the Volga-Ural region populations. On the basis of the received data, DNA diagnostics of analyzed hereditary diseases was developed and introduced to genetic consulting.

REFERENCES

Akhmetova VL, Viktorova TV, Murzabaeva SSH, et al. (2003). Analysis of mutations of the phenylalanine hydroxylase gene in phenylketonuria patients from Bashkortostan. *Russian Journal of Medical Genetics 4,* 182–87.

Andrew SE, Goldberg YP, Kremer B, et al. (1993). The relationship between trinucleotide (CAG) repeat length and clinical features of Huntington's disease. *Nature Genet 4*, 398–403.

Beggs AH, Koenig M, Boyce FM, et al. (1990). Detection of 98% of DMD/BMD gene by polymerase chain reaction. *Human Genet 86*, 45–48.

Bergoffen J, Scherer SS, Wang S, et al. (1993) Connexin mutations in X-linked Charcot-Marie-Tooth Disease. *Science 24*, 2039–2042.

Bort S, Nelis E, Timmerman V, et al. (1997). Mutational analysis of the MPZ, PMP22 and Cx32 genes in patients of Spanish ancestry with Charcot-Marie-Tooth disease and hereditary with liability to pressure palsies. *Hum Genet 99*, 746–754.

Bull PC, Thomas GR, Rommens JM, Forbes JR, Cox DW. (1993). The Wilson disease gene is a putative copper transporting P-type ATPase similar to the Menkes gene. *Nature Genet 5,* 327–337.

Chalestori H, Farthud D, Patton A, et al. (2007). Familial and Sporadic GJB2-Related Deafness in Iran: Review of gene Mutations. *Iranian J Publ Health 36,* 1–14.

Chapon F, Latour P, Diraison P, et al. (1999). Axonal phenotype of Charcot-Marie-Tooth disease associated with a mutation in the myelin protein zero gene. *J Neurol Neurosurg Psychiat 66,* 779–782.

Denoyelle F, Marlin S, Weil L, et al. (1999). Clinical features of the prevalent form of childhood deafness, DFNB1, due to a connexin 26 defect: implications for genetics counseling. *Lancet 353,* 1298–1303.

Dyck PJ, Chance P, Lebo R, Carney JA. (1993) Hereditary Motor and Sensory Neuropathies. *Peripheral Neuropathy 2,* 1094–1136.

Dzhemileva LU, Khidiyatova IM, Khabibulin RM, et al. (2002). Novel mutations and haplotype analysis of genomic polymorphisms of GJB2 and GJB3 genes associated with profound and moderately severe hearing loss in patients from Bashkortostan. *Balkan J Med Genet 5,* 41–46.

Emery AEH. (1991). Population frequencies of inherited neuromuscular diseases – a world survey. *Neuromusc Disord 1,* 19–29.

Estivill X, Fortina P, Surrey S, et al. (1998). Connexin-26 mutations in sporadic and inherited sensorineural deafness. *Lancet 351,* 394–398.

Fabrizi GM, Taioli F, Cavallaro T, Rigatelli F, Simonati A, Mariani G, Perrone P, Rizzuto N. (2000). Focally folded myelin in Charcot-Marie-Tooth neuropathy type 1B with Ser49Leu in the myelin protein zero. *Acta Neuropathol 100*(3), 299–304.

Fatkhlislamova RI, Khidiyatova IM, Khusnutdinova EK, et al. (1999). Analysis of the polymorphism of the CTG repeat of the myotonic dystrophy gene in ethnic populations of the Volga-Ural region. *Russian Journal of Genetics 35,* 988–993.

Frydman M, Bonne-Tamir B, Farrer LA, et al. (1985). Assignment of the gene for Wilson disease to chromosome 13: linkage to the ersterase D locus. *Proc Natl Acad Sci USA 82,* 1819–1821.

Gasparini P, Rabionet R, Barbujani G, et al. (2000). High carrier frequency of the 35delG deafness mutation in European populations. Genetic Analysis Consortium of GJB2 35delG. *Eur J Hum Genet 8*(1), 19–23.

Giannattasio S, Jurgelevicius V, Lattanzio P, et al. (1997). Phenilketonuria mutations and linked haplotypes in the Lithuanian population: origin of the most common R408W mutation. *Hum Hered 47,* 155–160.

Goodfellow PN, Davies KE, Ropers HH. (1985). Report of the committee on the genetic constitution of the X and Y chromosomes. *Cytogenet Cell Genet 40*, 296–352.

Harding AE, Thomas, PK. (1980) The clinical features of hereditary motor and sensory neuropathy types I and II. *Brain 103,* 259–280.

Harper PS. (1989). Myotonic Dystrophy. Philadelphia: Saunders.

Harris S, Moncrieff C, Johnson K. (1996). Myotonic dystrophy: will the real gene please step forward. *Hum Mol Genet 5,* 1417–1423.

Hayasaka K, Himoro M, Sato W, et al. (1993). Charcot-Marie-Tooth neuropathy type 1B is associated with mutations of the myelin P(0) gene. *Nature Genet 5,* 31–34.

Hayden MR. (1981) *Huntington's Chorea.* Berlin-Heidelberg-New York: Springer-Verlag.

Ikegami T, Nicholson G, Ikeda H, et al. (1996) A novel homozygous mutation of the myelin P(0) gene producing Dejerine-Sottas disease (hereditary motor and sensory neuropathy type III). *Biochem Biophys Res Commun 222,* 107–110.

Illarioshkin SN, Ivanova-Smolenskaya IA, Markova ED, et al. (1996). Trinucleotide repeat expansion as a new mutation mechanism in Huntington's disease: theoretical and applied aspects. *Russian Journal of Genetics 32*, 103–109.

Ionasescu V, Lonasescu R, Searby C. (1993) Screening for dominantly inherited Charcot-Marie-Tooth neuropathies. *Muscle Nerve 16,* 1232–1238.

Ionasescu V, Searby C, Sheffield V, et al. (1996) Autosomal dominant Charcot-Marie-Tooth axonal neuropathy mapped on chromosome 7p (CMT2D). *Hum Mol Genet 5*, 1373–1375.

Ionasescu V. (1995) Charcot-Marie-Tooth neuropathies: from clinical description to molecular genetics. *Muscle Nerve 18,* 267–275.

Karunas AS, Magzhanova AR, Magzhanov RV, et al. (2009). Molecular-genetic study of Wilson disease in Bashkortostan Republic. *Russian Journal of Medical Genetics 8*, 16–22.

Karunas AS, Mersiyanova IV, Polyakov AV, et al. (2000). Analysis of mutations and haplotypes of polymorphic markers in patients with Wilson-Konovalov disease from Bashkortostan. *Russian Journal of Genetics 36*, 972–979.

Khidiyatova IM, Bagautdinova EG, Galieva DV, et al. (2008). spectrum and frequency of mutations in the connexin 32 gene (*GJB1*) in hereditary and sensory neuropathy type 1X patients from Bashkortostan. *Russian Journal of Genetics 44*, 1385–1391.

Khidiyatova IM, Dzhemileva LU, Khabibullin RM, et al. (2002). Frequency of the 35delG mutation of the connexin 26 gene (GJB2) in patients with non-syndromic recessive deafness from Bashkortostan and in ethnic groups of the Volga-Ural region. *Russian Journal of Molecular Biology 36*, 438–441.

Khidiyatova IM, Fatkhlislamova RI, Magzhanov RV, et al. (2000). A study of the expansion and mutation frequency of the CTG repeat of the myotonic dystrophy gene. *Russian Journal of Genetics 36*, 1410–1413.

Khusnutdinova EK. (1999). *Molecular ethnogenetics of populations of the Volga-Ural region.* Ufa: Gilem.

Khusnutdinova EK, Dzhemileva LU. (2005). Molecular-genetic analysis of non-syndromic autosomal recessive deafness in patients and Volga-Ural populations. *Vestnik of Biotechnology and physic-chemical biology of Yu A 1,* 24–31.

Kochanski A, Kabzinska D, Nowakowski A. (2004). An axonal form of Charcot-Marie-Tooth disease with a novel missense mutation in the myelin protein zero gene. *J Peripher Nerv Syst 9*(1), 1–2.

Krupina NB, Khidiyatova IM, Latypova EG, et al. (2006). Clinical epidemiological study of inherited motor and sensorial neuropathy and the screening for *PMP22* Gene duplications among patients in Republic of Bashkortostan. *Med Genet 11,* 126–128.

Kudo T, Ikeda K, Kure S, et al. (2000). Novel mutation in the connexin 26 gene (GJB2) responsible for childhood deafness in the Japanese population. *Am J Med Genet 90,* 141–145.

Kutuev IA, Khusainova RI, Khidiyatova IM, et al. (2003). Analysis of polymorphism of the Huntington disease gene in ethnic populations of the Volga-Ural region *Russian Journal of Molecular Biology 37,* 818–826.

Kutuev IA, Khusainova RI, Khidiyatova IM, et al. (2004). IT 15 gene analysis in Huntington's disease families. *Russian Journal of Genetics 40,* 1123–1130.

Lebo RV, Lynch ED, Bird TD. (1992). Multicolor in situ hybridization and linkage analysis order Charcot-Marie-Tooth Type I (*CMT 1A*) gene–region markers. *Am J Hum Genet 50,* 42–55.

Lidsky AS, Robson KJH, Thirumalachary C, et al. (1984). The PKU locus in man is on chromosome12. *Am J Hum Genet 36,* 527–533.

Liu XZ, Xia XJ, Ke XM, et al. (2002). The prevalence of connexin 26 (GJB2) mutations in the Chinese population. *Hum Genet 111,* 394–397.

Magzhanov RV. (1985). Hereditary diseases of nervous system. In: *Epidemiology of nervous system disorders in Bashkiria.* Bashkir State Medical University, Russia.

Mahadevan M, Tsilfidis C, Sabourin L, et al. (1992). Myotonic dystrophy mutation: an unstable CTG repeat in the 3'untranslated region of the gene. *Science 255,* 1253–1255.

Mersiyanova IV, Perepelov AV, Polyakov AV, et al. (2000). A new variant of Charcot- Marie-Tooth disease type 2 is probably the result of a mutation in the neurofilament-light gene. *Am J Hum Genet 67,* 37–46.

Misu K, Yoshihara T, Shikama Y, et al. (2000). An axonal form of Charcot-Marie-Tooth disease showing distinctive features in association with mutations in the peripheral myelin protein zero gene (thr124met or asp75val). *J Neurol Neurosurg Psychiat 69,* 806–811.

Morell R J, Kim HJ, Hood LJ, et al. (1998). Mutations in the connexin 26 gene (GJB2) among Ashkenazi Jews with non-syndromic recessive deafness. *Nat Engl J Med 339,* 1500–1505.

Morton C, Walter E, Nance MD. (2006). Newborn hearing screening – a silent revolution. *The New England Journal of Medicine 354,* 2151–2164.

Nelis E, Simokovic S, Timmerman V, et al. (1997). Mutation analysis of the connexin 32 (*Cx32*) gene in Charcot-Marie-Tooth Neuropathy Type I: Identification of five new mutations. *Hum Mutat 9,* 47–52.

Nelis E, Van Broeckhoven C, Timmerman V, et al. (1996) Estimation of the mutation frequencies in Charcot-Marie-Tooth Disease Type 1 and hereditary neuropathy with liability to pressure palsies: a European collaborative study. *Eur J Hum Genet 4,* 25–33.

Patel PI, Franco B, Garcia C. (1990). Genetic mapping of autosomal dominant Charcot-Marie-Tooth Disease in a large French-Acadian kindred: identification of new linked markers on chromosome 17. *Am J Hum Genet 46,* 801–809.

Petersen M, Willems P. (2006). Non-syndromic, autosomal-recessive deafness. *Clin Genet 69,* 371–392.

Scheinberg IH, Sternlieb I. (1984). Wilson's disease. In: Lloyd H, Smith J, eds. *Major Problems of Internal Medicine.* Philadelphia: Saunders.

Senderek J, Hermanns B, Lehmann U, et al. (2000). Charcot-marie-Tooth neuropathy type 2 and Po point mutations: two novel amino acid substitutions (Asp61Gly; Tyr119Cys) and a possible "hotspot" on Thr124Met. *Brain Path 10,* 235–248.

Silander K, Juvonen V, Ignatius J, et al. (1998). Spectrum of mutations in Finnish patients with Charcot-Marie-Tooth disease and related neuropathies. *Hum Mutat 12,* 59–68.

Skre H. (1974) Genetic and clinical aspects of Charcot-Marie-Tooth disease. *Clin Genet 6,* 98–118.

Stapelbroek JM, Bollen CW, van Amstel JK, van Erpecum KJ, van Hattum J, van den Berg LH, Klomp LW, Houwen RH, et al. (2004). The H1069Q mutation in ATP7B is associated with late and neurologic presentation in Wilson disease: results of a meta-analysis *J Hepatol 41,* 758–763.

Timmerman V, Raeymaekers P, De Jonghe P. (1990) Assignment of the Charcot-Marie-Tooth neuropathy type I (*CMT 1A*) gene to 17p11.2-p12. *Am J Hum Genet 47,* 680–685.

Warner LE, Hilz MJ, Appel SH, et al. (1996). Clinical phenotypes of different MPZ(P0) mutations may include Charcot-Marie-Tooth type 1B, Dejerine-Sottas, and congenital hypomyelination. *Neuron 17,* 451–460.

Warner LE, Mancias P, Butler IJ, et al. (1998). Mutations in the early growth response 2 (EGR2) gene are associated with hereditary myelinopathies. *Nature Genet 18,* 382–384.

Woo SLC, Lidsky AS, Guttler F, et al. (1993). Cloned human phenylalanine hydroxilase gene permits prenatal diagnosis and carrier detection of classical phenilketonuria. *Nature 306,* 151–155.

114.

INHERITED NEUROPSYCHIATRIC DISORDERS IN RUSSIA

D. Gaysina, A. Zainullina, and Elza K. Khusnutdinova

INTRODUCTION

The brainstem raphe serotonin (5-hydroxytryptamine, 5-HT) system is the most widely distributed neurotransmitter system in the brain; serotonergic raphe neurons diffusely project to a variety of brain regions (e.g., cortex, amygdala, hippocampus; Hensler et al., 1994). In addition to its neurotransmitter role, 5-HT is an important regulator of morphogenetic activities during early brain development, as well as during adult neurogenesis and plasticity, indicating cell proliferation, migration, differentiation, and synaptogenesis (Azmitia, 1999; Gould, 1999). In the brain of adult humans, nonhuman primates, and other mammals, 5-HT neurotransmission is a major modulator of emotional behavior and integrates cognition, sensory processing, motor activity, and circadian rhythms including food intake, sleep and reproductive activity. The regulation of a range of physiological functions is due to interaction with several other neurotransmitter systems (Benes et al., 2000).

The abundance of evidence suggesting that the serotonergic system is altered in different psychiatric conditions has prompted investigators to scrutinize the potential role of the serotonin system genes in the pathophysiology of schizophrenia, unipolar depression, and suicidal behavior.

This review describes serotonergic system genes, namely serotonin transporter (*SLC6A4*) and serotonin receptors (*HTR1A*, *HTR1B*, and *HTR2A*), and polymorphic features that have been identified. We review the progress made in functional analyses to elucidate the effect of polymorphisms on expression level and other transport properties. We review association studies attempting to establish relationships between this genetic diversity and the presence and severity of psychiatric disorders, including data of our research group.

SEROTONIN TRANSPORTER

The 5-HT transporter (5-HTT) is believed to be primarily responsible for the termination of action of 5-HT after it is released from the nerve terminal into the synapse. It is located on the presynaptic neuron and takes up one 5-HT molecule concurrently with one Na+ ion, decreasing extracellular fluid concentrations of 5-HT to levels where postsynaptic receptor activation ceases. The 5-HTT thus plays a critical role in regulating serotonergic neurotransmission in numerous projection fields throughout brain, including regions crucial to emotional behavior, such as motor behavior, emotional experience, and memory (Lesch et al., 1996). Consistent with this view, brain 5-HTT is the initial site of action of widely used 5-HT reuptake inhibitor antidepressants and drugs of abuse.

SEROTONIN TRANSPORTER GENE AND POLYMORPHISMS

The human serotonin transporter protein is encoded by a single gene (*5-HTT* or *SLC6A4*, LocusLink ID: 6532), mapped to chromosome 17q11.1–q12 (Ramamoorthy et al., 1993). It spans ~31 kilobase pairs (kbp) and contains 14 exons (Lesch and Mossner, 1998).

Expression of the *SLC6A4* is regulated by a combination of positive and negative elements acting through a basal promoter unit that is defined by a variable number of tandem repeat units consisting of 22 bp. This polymorphic repetitive element, *5-HTT* gene-linked polymorphic region (*5-HTTLPR*), is unique to humans and simian primates. In humans, there are two common alleles—termed "S" (Short) and "L" (Long) allele—that correspond to the presence of 14- or 16-repeat units of a 20–23 bp incomplete repeat. Additional low frequency alleles have been identified including 15-, 18-, 19-, 20- and 22-repeat units.

The effect of the *5-HTTLPR* length variability on 5-HTT function is determined by studying the relationship between *5-HTTLPR* genotype, *SLC6A4* transcription, and 5-HT uptake activity in human lymphoblastoid cell lines. The homozygous L/L polymorphism has been shown to have higher transcriptional activity for the *SLC6A4* in vitro than genotypes containing one or two S alleles (Heils et al., 1996). Additionally, the uptake of 5-HT is approximately twofold higher in cells containing the homozygous L/L form of the *5-HTTLPR* than either the L/S or S/S forms (Lesch and Mossner, 1998). Other factors, such as seasonal variation, also interact with the *5-HTTLPR* genotype to influence both the uptake potential and expression of the 5-HTT (Hanna et al., 1998; Willeit et al., 2003). Furthermore, the L/L genotype exhibits greater 5-HTT expression and transport activity in cultured pulmonary artery smooth muscle cells (PA-SMC) (Eddahibi et al., 2001).

Another polymorphism, a 17 bp variable number of tandem repeats (termed *STin2* VNTR), located in intron 2, involves two major alleles (termed "*STin2*.10" and "*STin2*.12") that correspond to 10- or 12-repeat units of the 17 bp VNTR (Lesch et al., 1994). Despite its location outside of the coding or promoter regions of the 5-HTT gene, expression in transgenic embryos and embryonic stem cells suggests that the *Stin2* VNTR polymorphism acts as a strong regulator of transcription (MacKenzie and Quinn, 1999; Fiskerstrand et al., 1999). The *STin2*.12 demonstrated greater transcriptional activation, which could be due to an increase in the number of transcriptional elements present in the larger 12-repeat construct, or the size difference that potentially allows a more favorable conformation of DNA for binding transcriptional factors. These findings support that intron 2 of the 5-HTT gene plays a role in transcriptional regulation, and the polymorphic region may direct different levels of expression. As a variant with potential effects on transcriptional activity, the *Stin2* VNTR has promise for association with clinical conditions.

A small number of SNPs were identified in the course of cloning of the 5-HTT cDNA and gene. The impact of these amino acid variants on expression or transport function remains unexplored.

POPULATION DIFFERENCES IN THE *SLC6A4* GENOTYPE AND ALLELE DISTRIBUTION

It is known that ethnic differences can affect allele and genotype distributions (Engen et al., 2006). Differences in population genetic structure and substructure between cases and controls can lead to false positive or false negative association tests. Ancestry differences corresponding to ethnic groups may be important in determining disease risk factors and optimizing treatment (Tian et al., 2008).

Examination of different ethnic groups revealed wide variation in frequencies; the *STin2*.10 allele ranged from a frequency of 0.02 in the Japanese to 0.47 in European Americans. The *STin2*.10 was present at 0.26 in African Americans. This is worth noting, as this difference in frequencies between the European and African Americans is greater than the size of effects commonly observed in association studies, in which these two ethnically diverse populations are often combined. A rare *STin2*.9 allele with a frequency of 0.01 and *STin2*.11 allele with a frequency of 0.13 in a Mbuti population have also been identified. (Gelernter et al., 1999).

Several studies revealed large variation in allele frequencies of the *5-HTTLPR* among a variety of ethnic populations. The allele frequencies between Caucasian and Asian populations are different, since S allele is found in 42% of Caucasians and in 79% of Asians. Rare 15-, 19-, 20-, and 22-repeat alleles are often limited to a particular ethnic ancestry (Gelernter et al., 1999).

In contrast to other Caucasian populations, populations from Russia comprise different ethnic groups, which could contribute to the different allele frequencies of the *SLC6A4*. Therefore, we investigated ethnic differences in the polymorphism of the *SLC6A4* across several European populations. The total sample of 1804 healthy individuals from four European populations (Croats, Russians, Tatars, and Bashkirs) was analyzed for the *5-HTTLPR* genotype and allele frequencies (Noskova et al., 2008).

The genotype and allele frequencies of the *5-HTTLPR* among all four populations were significantly different ($p < 0.001$) when men and women were analyzed separately (Table 114-1). There were significant differences between Croats, Russians, Tatars, and Bashkirs in the genotype ($p < 0.001$) and allele ($p < 0.001$) frequencies in men, and in the genotype ($p = 0.007$) and allele ($p = 0.002$) frequencies in women.

Among men, significant differences in *5-HTTLPR* genotype and allele frequencies were found between Croats and Tatars ($p < 0.001$ for genotypes and alleles), and Croats and Bashkirs ($p < 0.001$ for genotypes and alleles). Russian men had significantly different genotype and allele frequencies compared to Bashkir men ($p = 0.005$ and $p < 0.001$), and similar genotype and allele frequencies compared to men of Croat and Tatar origin ($p > 0.05$). Tatar and Bashkir men had similar genotype and allele frequencies of the *5-HTTLPR* ($p > 0.05$). Among women, the only significant differences were between Croats and

TABLE 114-1 FREQUENCIES OF THE *5-HTTLPR* GENOTYPES AND ALLELES IN MEN AND WOMEN OF FOUR EUROPEAN POPULATIONS—CROATIAN, RUSSIAN, TATAR, AND BASHKIR

	Croats		Russians		Tatars		Bashkirs	
	Men	Women	Men	Women	Men	Women	Men	Women
N	461	204	238	260	184	196	11	110
L/L	0.38	0.40	0.34	0.28	0.29	0.27	0.21	0.34
L/S	0.49	0.45	0.45	0.52	0.45	0.45	0.46	0.46
S/S	0.13	0.15	0.21	0.19	0.26	0.29	0.33	0.20
L	0.62	0.62	0.57	0.55	0.52	0.49	0.44	0.57
S	0.38	0.38	0.43	0.45	0.48	0.51	0.56	0.43

Tatars in the genotype and allele frequencies ($p < 0.001$). Other populations did not differ significantly from one another.

To elucidate which genotype contributed to these ethnic differences, male and female subjects from the four populations were analyzed using a recessive model for S allele (S/S homozygotes versus L allele carriers). Significant differences in genotypes were found between populations in men ($p < 0.001$), but not in women ($p = 0.01$, NS due to Bonferroni correction). Croat and Tatar men ($p < 0.001$) and women ($p < 0.001$), and Croat and Bashkir men ($p < 0.001$) differed significantly, while the other studied groups had similar frequency (because of lower power of performed tests (pc < 0.800) and/or Bonferroni correction).

The frequencies of the *5-HTTLPR* genotypes and alleles were significantly different between the four populations when data for men and women were analyzed together ($p < 0.001$ for both genotypes and alleles; Table 114-1). The largest difference was found in the frequency of the S/S genotype: 14% in Croats, 20% in Russians, and 27% in Tatars and Bashkirs. Genotype frequencies were significantly different between Croats and Bashkirs ($p < 0.001$), and Croats and Tatars ($p < 0.001$), while Russians did not differ significantly in the genotype frequency of the *5-HTTLPR* from Croats and Tatars, ($p = 0.01$ and 0.033, NS after Bonferroni correction), and Bashkirs did not differ significantly in the genotype frequency of the *5-HTTLPR* from Russians and Tatars.

Allele frequencies were significantly different between Croats and other populations: Russians ($p = 0.003$), Bashkirs ($p < 0.001$) and Tatars ($p < 0.001$). Russians did not differ significantly in the allele frequency of the *5-HTTLPR* from Tatars and Bashkirs ($p = 0.026$ and 0.027, respectively; NS after Bonferroni correction). Bashkirs and Tatars did not differ significantly in the allele frequency, either.

Our results confirmed that ethnic differences in the *5-HTTLPR*, previously found between Caucasian and Asian populations (Serretti et al., 2006; Kunugi et al., 1997; Gelernter et al., 1999), exist between Caucasian subjects from the four European populations. The different distribution of the alleles and genotypes of the *5-HTTLPR* from individuals living in Europe (Willeit et al., 2003; Szekely et al., 2004; Surtees et al., 2006; Noskova et al., 2008; Lang et al., 2004; Gutierrez et al., 1998), China (Li and He, 2007), Korea (Kim et al., 2000), and Japan (Katsuragi et al., 1999) is shown in Table 114-2.

These ethnic differences might be responsible for the inconsistency (i.e., nonreplication) of the associations of *5-HTTLPR* with suicidal behavior, anxiety, depression, and schizophrenia (Li and He, 2007; Lesch and Gutknecht, 2005). Our findings emphasize the importance of taking ethnicity into account when studying genetic basis of psychiatric disorders.

SCHIZOPHRENIA

There are a large number of association studies between the two *SLC6A4* polymorphisms and schizophrenia. However, neither polymorphism has yet been confirmed as a susceptibility locus for schizophrenia, since neither association has been consistently confirmed in replication studies. While none of the previous association studies showed statistically significant evidence of association or linkage disequilibrium (LD) between the *S* or *L* allele of the *5–HTTLPR* polymorphism and schizophrenia, two previous studies found association of *L/L* genotype with intensity of hallucinations or schizoaffective subtype

TABLE 114-2 POPULATION DIFFERENCES OF THE *5-HTTLPR* GENOTYPE AND ALLELE FREQUENCY (%) DISTRIBUTION

	Genotypes			Alleles	
Country	L/L	L/S	S/S	L	S
Croatia (Noskova et al, 2008)	38.2	47.5	14.3	62.0	38.0
Russia: Russians (Noskova et al, 2008)	31.3	48.8	19.9	55.7	44.3
Russia: Tatars (Noskova et al, 2008)	27.9	44.7	27.4	50.3	49.7
Russia: Bashkirs (Noskova et al, 2008)	26.4	46.4	27.2	49.6	50.4
Germany (Lang et al., 2004)	37.3	44.7	18.0	59.6	40.4
Austria (Willeit et al., 2003)	34.9	44.5	20.6	57.2	42.8
UK (Surtees et al., 2006)	33.4	48.7	17.9	57.7	42.3
Hungary (Szekely et al., 2004)	35.1	45.7	19.2	58.0	42.0
Spain (Gutierrez et al., 1998)	34.9	44.6	20.5	57.2	42.8
Italy (Nonnis et al., 2008)	14.0	57.3	28.7	42.7	57.3
PR China (Li et al., 2007)	13.0	31.0	56.0	28.0	72.0

of schizophrenia. On the other hand, three population-based association studies and one family-based association study have demonstrated the *STin2* 12 allele to be significantly associated with an increased vulnerability to schizophrenia. While a majority of the previous association studies failed to replicate such previous positive findings, most of them reported nonsignificant overrepresentation of the *STin2* 12 allele in schizophrenic patients (Fan and Sklar, 2005).

Ambiguous association study findings are not unusual for genetic analyses of complex genetic diseases (Ioannidis et al., 2001). Many factors can contribute to variability in association study findings. Among them, small sample size leads to low statistical power of individual studies with a concomitant increased rate of both false positive and false negative results, and subsequent difficulty replicating or refuting previous findings (Hirschhorn and Altshuler, 2002). To deal with these ambiguities, meta-analysis of individual studies can be performed, thereby increasing power and allowing for a more global interpretation of all available data (Munafo and Flint, 2004).

One meta-analyses showed no statistically significant evidence for the association between the *S* allele of the *5-HTTLPR* polymorphism and schizophrenia (OR = 0.99, 95% CI 0.92–1.07, Z 0.23, P = 0.82) from 19 population-based association studies consisting of 2990 case and 3875 control subjects (Fan and Sklar, 2005) However, highly significant evidence for association between the *STin2* 12 allele (OR = 1.24, 95% CI 1.11–1.38, Z 3.82, P = 0.00014) and schizophrenia was found from 12 population-based association studies consisting of 2177 cases and 2369 control subjects.

Our research group investigated both *5-HTTLPR* and *Stin2* VNTR for association with schizophrenia in Russian and Tatar populations (Zainullina et al., 2003, 2005). The results of our study did not confirm the association of *SLC6A4* with complex phenotype of schizophrenia in total sample, or in three different ethnic groups (Tables 114-3 and 115-4). These results have been included in the above mentioned review and meta-analysis (Fan and Sklar, 2005).

However, further analysis in our sample showed population-specific differences for the type of disease progression, continuous versus shift-like (Tables 114-5 and 114-6).

Russians with continuous schizophrenia had significantly higher frequencies of L/L homozygotes ($p = 0.048$) and L allele ($p = 0.017$) than the corresponding subgroup of Tatars. (Zainullina et al., 2003). In Tatars, frequency of *Stin2* 12/12 genotype was significantly higher in patients with continuous schizophrenia than in the control group ($p = 0.047$), while in Russians, patients with continuous schizophrenia did not significantly differ in *Stin2* genotype frequency from controls. Thus, the two ethnic groups proved to differ in genetic predisposition to continuous schizophrenia.

TABLE 114-3 ALLELE AND GENOTYPE DISTRIBUTION OF THE *5-HTTLPR* IN THREE POPULATIONS OF THE SCHIZOPHRENIA CASE-CONTROL STUDY

		Genotypes			Alleles	
Groups	N	L/L	L/S	S/S	L	S
Control	394	0.26	0.47	0.27	0.49	0.51
– Bashkirs	132	0.20	0.48	0.33	0.44	0.56
– Tatars	136	0.27	0.47	0.26	0.51	0.49
– Russians	126	0.30	0.47	0.23	0.54	0.46
SZ	334	0.25	0.49	0.26	0.49	0.51
– Bashkirs	107	0.20	0.53	0.27	0.46	0.55
– Tatars	105	0.23	0.44	0.33	0.45	0.55
– Russians	122	0.32	0.48	0.20	0.56	0.44

TABLE 114-4 ALLELE AND GENOTYPE DISTRIBUTION OF THE *STIN2* VNTR IN THREE POPULATIONS OF THE SCHIZOPHRENIA CASE-CONTROL STUDY

		Genotypes				Alleles		
Groups	N	12/12	12/10	10/10	Rare	12	10	9
Control	454	0.45	0.39	0.14	0.02	0.65	0.34	0.01
– Bashkirs	142	0.49	0.33	0.16	0.02	0.66	0.33	0.01
– Tatars	168	0.46	0.42	0.10	0.02	0.68	0.31	0.01
– Russians	144	0.38	0.40	0.17	0.05	0.59	0.39	0.02
SZ	339	0.48	0.35	0.15	0.02	0.67	0.32	0.01
– Bashkirs	108	0.53	0.28	0.17	0.02	0.68	0.31	0.01
– Tatars	107	0.48	0.37	0.15	0.00	0.66	0.34	0.00
– Russians	124	0.45	0.40	0.13	0.02	0.66	0.33	0.01

UNIPOLAR DEPRESSION

Polymorphisms of the *SLC6A4* (*5-HTTLPR* and *Stin2* VNTR) have been widely studied for association with unipolar depression and depressive symptoms. The

TABLE 114-5 ALLELE AND GENOTYPE DISTRIBUTION OF THE *5-HTTLPR* IN CONTINUOUS AND SHIFT-LIKE TYPES OF SCHIZOPHRENIA

		Genotypes			Alleles	
Groups	N	L/L	L/S	S/S	L	S
SZ continuous: Total	183	0.24	0.47	0.29	0.48	0.52
–Bashkirs	61	0.26	0.54	0.30	0.43	0.57
–Tatars	57	0.21	0.40	0.39	0.41	0.59
–Russians	65	0.34	0.46	0.20	0.57	0.43
SZ shift-like: Total	151	0.26	0.50	0.23	0.52	0.48
–Bashkirs	46	0.23	0.49	0.28	0.50	0.50
–Tatars	48	0.25	0.48	0.27	0.49	0.51
–Russians	57	0.30	0.51	0.19	0.55	0.45

TABLE 114-6 ALLELE AND GENOTYPE DISTRIBUTION OF THE *STIN2* VNTR IN CONTINUOUS AND SHIFT-LIKE TYPES OF SCHIZOPHRENIA

Groups	N	Genotypes				Alleles		
		12/12	12/10	10/10	Rare	12	10	9
SZ continuous: Total	187	0.53	0.30	0.14	0.02	0.70	0.29	0.01
–Bashkirs	62	0.56	0.24	0.16	0.03	0.70	0.28	0.02
–Tatars	58	0.57	0.29	0.14	0.00	0.72	0.28	0
–Russians	67	0.48	0.37	0.12	0.03	0.68	0.31	0.01
SZ shift-like: Total	152	0.42	0.41	0.16	0.01	0.63	0.37	0.003
–Bashkirs	46	0.48	0.33	0.17	0.02	0.65	0.34	0.0
–Tatars	49	0.37	0.47	0.16	0	0.60	0.40	0
–Russians	57	0.42	0.44	0.14	0	0.64	0.36	0

results of association studies examining *5-HTTLPR* in unipolar depression (major depressive disorder, MDD) have been mixed. Few studies have shown the significant association between *S* allele of the *5-HTTLPR* and unipolar depression, while the majority has not. Although Collier et al. (1996) failed to find the association of this marker with MDD in each of different European populations, they showed significant association in total sample (OR = 1.23, 95% CI 1.02–1.49, $p = 0.03$; Collier et al., 1996). To precisely ascertain small associations, two meta-analyses have been recently performed involving several thousand subjects (Lotrich and Pollock, 2004; Lasky-Su et al., 2005). One of the meta-analyses, comprising 1961 cases and 3402 controls, showed that the combined OR estimate was not significantly different than 1.0 (OR = 1.05, 95%CI 0.96–1.14, $p = 0.28$; Lasky-Su et al., 2005). According to the results of another recent meta-analysis, conducted by Lotrich and Pollock (2004) using published data till 2002 and taking into account ethnic origin of investigated samples (Asian populations were analyzed separately from European ones), the *S/S* genotype was associated with slightly higher risk of MDD (OR = 1.16, 95% CI 1.03–1.31, $P < 0.05$).

Our case-control association study of unipolar depression did not show significant association of the *5-HTTLPR* in either Tatars or Russians. (Table 114-7).

Efforts to identify an association of the *Stin2* VNTR polymorphism with affective disorder have yielded mixed results. One group has found an increase of the rare *Stin2 9* allele in unipolar depression in Scottish populations, supporting its contribution to affective disorder in only a small subset of patients (Ogilvie et al., 1996; Harmar et al., 1996). Furthermore, others have failed to replicate this finding in MDD (Hanna et al., 1998; Goldberg et al., 2009; Stober et al., 1996; Hoehe et al., 1998). An association of the *Stin2 12* allele with depression has also been reported (Kunugi et al., 1997). Both abovementioned meta-analyses tested results of nine (Lasky-Su et al., 2005) and eleven (Lotrich and Pollock, 2004) research studies of the *Stin2* VNTR polymorphism in affective disorders. No consistent effect of this polymorphism for MDD was revealed for either European populations or Asian ones, or in mixed groups.

Our comparative analysis of the genotype frequencies of the *Stin2* VNTR in the Tatar ethnic group showed significant differences between healthy donors and the UD patients ($p = 0.01$). This difference was caused by the lower frequency of the 12/10 genotype in the UD group (31%) compared to controls (49%) (OR = 0.62, 95%CI 0.44–0.88). There was no difference in allele frequency distribution between the UD patients and the healthy individuals (Table 114-8).

TABLE 114-7 ALLELE AND GENOTYPE DISTRIBUTION OF THE *5-HTTLPR* IN TWO POPULATIONS OF THE UNIPOLAR DEPRESSION CASE-CONTROL STUDY

Groups		N	Genotypes			Alleles	
			L/L	L/S	S/S	L	S
Total	UD	164	0.26	0.49	0.25	0.51	0.49
	Control	331	0.32	0.47	0.21	0.55	0.45
Tatars	UD	99	0.25	0.46	0.29	0.48	0.52
	Control	170	0.27	0.44	0.29	0.49	0.51
Russians	UD	75	0.34	0.53	0.13	0.61	0.39
	Control	161	0.31	0.52	0.17	0.57	0.43

TABLE 114-8 ALLELE AND GENOTYPE DISTRIBUTION OF THE *STIN2* VNTR IN TWO POPULATIONS OF THE UNIPOLAR DEPRESSION CASE-CONTROL STUDY

Groups		N	Genotypes				Alleles		
			12/12	12/10	10/10	Rare	12	10	9
Total	UD	168	0.39	0.35	0.20	0.06	0.58	0.39	0.03
	Control	328	0.38	0.51	0.08	0.03	0.64	0.34	0.02
Tatars	UD	99	0.44	0.31	0.16	0.09	0.62	0.34	0.04
	Control	170	0.38	0.49	0.10	0.02	0.63	0.36	0.01
Russians	UD	75	0.36	0.36	0.24	0.04	0.55	0.43	0.02
	Control	158	0.34	0.48	0.11	0.07	0.61	0.36	0.03

In Russians, genotype frequency distribution was significantly different between healthy individuals and UD patients ($p = 0.04$). Similarly to what was demonstrated in Tatars, 12/10 genotype was associated with lower risk for UD (OR = 0.64, 95CI 10.43–0.95), and 10/10 genotype with higher risk for UD (OR = 1.84, 95%CI 1.26–2.72) in Russians (Noskova et al., 2009).

SUICIDAL BEHAVIOR

The sample of 187 suicide attempters and 268 healthy control subjects was genotyped for the *5-HTTLPR* and *STin2* VNTR polymorphisms of the *SLC6A4* gene. We showed significant association of *5-HTTLPR L/L* genotype with suicidal behavior in Russians (OR = 1.65, 95%CI 0.99–2.75), but not in Tatars (Table 114-9).

Similarly, the frequency of *Stin2* VNTR 12/10 was higher in suicide attempters of Russian descent (OR = 1.79, 95%CI 1.01–3.16), but not in those of Tatar descent (Table 114-10).

We performed further analysis of the *SLC6A4* polymorphisms in the Russian group, separately in men and women. Cases were 147 suicide attempters (52 men, 98 women) with suicide attempts classified as violent (18%) or nonviolent–drug overdose (81%). The control group consisted of 244 volunteers (132 men and 112 women). Our findings indicate the contribution of the *SLC6A4* gene to susceptibility for suicidal behavior in women, but not in men. The L/L genotype (OR = 2.09, 95%CI 1.11–3.95) and L10 haplotype (OR = 1.77, 95%CI 1.01–3.17) were associated with suicide in Russian women only.

TABLE 114-9 ALLELE AND GENOTYPE DISTRIBUTION OF THE *5-HTTLPR* IN TWO POPULATIONS OF THE SUICIDAL BEHAVIOR CASE-CONTROL STUDY

Groups		N	Genotypes			Alleles	
			L/L	L/S	S/S	L	S
Total	SB	187	0.33	0.43	0.24	0.55	0.45
	Control	268	0.28	0.48	0.24	0.52	0.48
Tatars	SB	85	0.26	0.45	0.29	0.49	0.51
	Control	147	0.27	0.47	0.26	0.51	0.49
Russians	SB	102	0.39	0.42	0.19	0.60	0.40
	Control	132	0.30	0.48	0.22	0.53	0.47

Over the past few years, numerous studies on the association between the *5-HTTLPR* polymorphism and suicidal behavior have been published, however with conflicting results (Mann et al., 2001; Bondy et al., 2006; Lin and Tsai, 2004). The meta-analysis based on results of 18 studies provided evidence for association of the *5-HTTLPR* S allele with suicidal behavior in the psychiatric population, and also with violent suicide (Lin and Tsai, 2004). Perhaps our study was unable to find an effect of the S allele because the sample was composed of patients with suicidal behavior that was not violent (in 81% of cases) or lethal. Moreover, our suicide patients were recruited from the general population, with ICD-10 diagnostic information available only for a proportion of subjects. Therefore, we performed the analysis without controlling for psychiatric diagnosis. Additionally, our study was conducted in a Russian population, which has a higher frequency of the S allele than other Caucasian populations. Therefore, any effect of the *5-HTTLPR* S allele on suicidal behavior would require larger sample sizes than would be needed in other Caucasian populations. We also should emphasize that association of the L allele with suicide was true for Russian women only, but not for men. Unfortunately, gender was not included as a possible effect modifier in the majority of molecular genetic studies of suicide. Biological explanation of the L allele overrepresentation in suicidal women can involve an interaction between the *5-HTTLPR* polymorphism and low hormonal level. One study showed that in female attempters, the proportion of women carriers of the L allele in the menses or menopause was significantly higher than expected in the population, and almost significantly higher than in female attempters who were S allele carriers (Baca-Garcia et al., 2003).

Lopez de Lara and colleagues found a significant association of suicide completion with having at least one copy of the *STin2* 10 allele (Lopez et al., 2006). In our study,

TABLE 114-10 ALLELE AND GENOTYPE DISTRIBUTION OF THE *STIN2* VNTR IN TWO POPULATIONS OF THE SUICIDAL BEHAVIOR CASE-CONTROL STUDY

Groups		N	Genotypes				Alleles		
			12/12	12/10	10/10	Rare	12	10	9
Total	SB	169	0.50	0.35	0.12	0.03	0.61	0.38	0.01
	Control	301	0.39	0.44	0.14	0.03	0.64	0.34	0.02
Tatars	SB	79	0.45	0.40	0.12	0.03	0.63	0.35	0.02
	Control	157	0.38	0.50	0.11	0.01	0.69	0.30	0.01
Russians	SB	90	0.54	0.30	0.14	0.02	0.58	0.40	0.02
	Control	144	0.40	0.38	0.17	0.05	0.59	0.39	0.02

we didn't show the significant differences in genotype and allele frequencies of the *STin2* polymorphism between control and suicide groups While we found that L10 haplotype is associated with suicidal behavior in Russian females, the association seems to be not significant after Bonferroni correction. At the same time, the S12 haplotype was shown to be associated significantly with lower risk of suicide in the Russian female population (Gaysina et al, 2006). So, our results are in agreement with the report by Hranilovic et al. (2003) on the association of haplotype L10 with completed suicides in a Croatian population (Hranilovic et al., 2003). It is likely that the *SLC6A4* gene interacts with other genes and environmental factors, and that this interaction might have different phenotypic consequences ().

SEROTONIN RECEPTORS

GENES AND POLYMORPHISMS

Serotonin 1A receptor is involved in pathophysiology of psychiatric disorders. It also serves as a target for antidepressants such as selective serotonin reuptake in inhibitors. Functional mononuleotide exchange C-1019G (rs6295) in the promoter region of the serotonin receptor gene (*HTR1A*) (5q11.2-q13) is associated with changes in its transcription level (Lemonde et al., 2003; Strobel et al., 2003; Yu et al., 2006). It was shown previously that this polymorphic marker can affect serotonin receptor type A1 density, as well as the presynaptic membranes of raphe nucleus neurons, but not the postsynaptic membranes. This may produce downregulation of the serotonin activity (Lemonde et al., 2003). Association of UD with C–1019G was observed in some earlier studies. The associations of C–1019G with suicide attempts (Huang et al., 2004), and with anxiety traits in Germans have also been reported (Strobel et al., 2003).

Serotonin 1B receptor is highly important for regulation of serotonin neurotransmission since, along with serotonin receptor 1A, it serves both as an autoreceptor on serotonin containing neurons of raphe nuclei and as a heteroreceptor at other neurons lacking serotonin. It was suggested that polymorphism G861C (rs6296) of *HTR1B* could affect the expression of the gene and, accordingly, the receptor density (Huang et al., 2003). Huang et al. (1999) have demonstrated decreased serotonin receptor 1B density in the brain of suicide victims carrying G allele. As the G to C substitution at 861 position is synonymous, this locus does not modify amino acid structure of the receptor 1B protein. Therefore, its association with the decreased level of receptor activity can be caused by indirect effect on transcription or linkage disequilibrium with some functionally significant polymorphism.

Serotonin 2A receptor gene (*HTR2A*) (13q14-q21) is often considered as one of the major candidate genes for psychopathology. There is evidence that depressive patients have increased density of serotonin 2A receptor in the prefrontal cortex (Myers et al. 2007). A-1438G (rs6311) of *HTR2A* has been reported to be associated with various psychiatric disorders, such as unipolar depression, obsessive-compulsive disorder, food behavior disorder, and schizophrenia (Myers et al., 2007). Myers at al. (2007) have found that two *HTR2A* SNPs (A–1438G and A783G), assumed to be in strong linkage disequilibrium, could significantly modify promoter activity both in vivo and in vitro, and thus affect the level of gene expression and serotonin receptor A2 density in the brain. Thus, AG haplotype of the two SNPs (A783G – A-1438G) is associated with increased gene expression.

SCHIZOPHRENIA

HTR2A

We investigated association of the *HTR2A* A1438G polymorphism (rs6311) with schizophrenia in a sample of 344 patients and 393 controls of Russian, Tatar, and Bashkir descent (Table 114-11.)

When analyzing the total sample, significant differences were shown in genotype distribution between

TABLE 114-11 ALLELE AND GENOTYPE DISTRIBUTION OF THE *HTR2A* A1438G IN THREE POPULATIONS OF THE SCHIZOPHRENIA CASE-CONTROL STUDY

Groups	N	Genotypes			Alleles	
		A/A	A/G	G/G	A	G
Control	393	0.15	0.54	0.31	0.42	0.58
– Bashkirs	115	0.16	0.51	0.33	0.41	0.59
– Tatars	146	0.16	0.49	0.35	0.40	0.60
– Russians	132	0.13	0.61	0.26	0.44	0.56
SZ	344	0.18	0.44	0.37	0.41	0.59
– Bashkirs	108	0.16	0.41	0.43	0.36	0.64
– Tatars	109	0.22	0.47	0.31	0.45	0.55
– Russians	127	0.17	0.46	0.37	0.40	0.60

schizophrenia patients and controls ($P = 0.031$). There was a higher frequency of the A/G genotype in controls compared to that in the schizophrenia group (OR = 0.68; CI 95% 0.51–0.92). When taking ethnicity into account, the genotype frequency was significantly different between cases and controls among Russians only ($P = 0.045$), but not among Bashkirs or Tatars. The frequency of A/G genotype was increased in Russian controls (61%) compared to that in the Russian schizophrenia group (46%) (OR = 0.53; CI 95% 0.31 – 0.89).

Kishi et al. (2009a) detected no evidence of genetic association between 4 markers near *HTR2A* and mood disorders in the Japanese population (Kishi et al., 2009a). A recent study in a large European ancestry sample did not produce nominally significant evidence to support previous evidence for association of *HTR2A* with schizophrenia (Sanders et al., 2008). Thus, it is unlikely that common SNPs in these genes account for a substantial proportion of the genetic risk for schizophrenia, although small effects cannot be ruled out.

UNIPOLAR DEPRESSION

In our unipolar depression case-control study, three SNPs of the serotonin receptor genes were analyzed in total sample as well as in Russian and Tatar subgroups (Table 114-12).

TABLE 114-12 ALLELE AND GENOTYPE DISTRIBUTION OF THE *HTR1A*, *HTR1B* AND *HTR2A* POLYMORPHISMS IN TWO POPULATIONS OF THE UNIPOLAR DEPRESSION CASE-CONTROL STUDY

Groups			Genotypes			Alleles	
HTR1A (C1019G)		N	C/C	C/G	G/G	C	G
Total	UD	108	0.35	0.45	0.20	0.57	0.43
	Control	331	0.24	0.57	0.19	0.53	0.47
Tatars	UD	99	0.33	0.55	0.12	0.60	0.40
	Control	170	0.29	0.52	0.19	0.56	0.44
Russians	UD	75	0.20	0.55	0.25	0.48	0.52
	Control	161	0.22	0.60	0.18	0.51	0.49
HTR1B (G861C)			G/G	G/C	C/C	G	C
Total	UD	167	0.46	0.44	0.10	0.67	0.33
	Control	313	0.49	0.42	0.09	0.70	0.30
Tatars	UD	99	0.34	0.51	0.15	0.60	0.40
	Control	159	0.44	0.45	0.11	0.66	0.34
Russians	UD	75	0.52	0.41	0.07	0.73	0.27
	Control	154	0.53	0.42	0.05	0.74	0.26
HTR2A (A1438G)			A/A	A/G	G/G	A	G
Total	UD	167	0.08	0.46	0.47	0.31	0.69
	Control	328	0.15	0.50	0.35	0.40	0.60
Tatars	UD	99	0.09	0.48	0.43	0.33	0.67
	Control	168	0.15	0.42	0.43	0.36	0.64
Russians	UD	75	0.11	0.37	0.52	0.29	0.71
	Control	160	0.08	0.62	0.30	0.39	0.61

HTR1A

The association analysis of the *HTR1A* C-1019G (rs6295) in both Tatars and Russians did not show significant differences between UD patients and controls in genotype (P = 0.33 for Tatars; P = 0.43 for Russians) or allele frequency (P = 0.34 for Tatars; P = 0.55 for Russians; Noskova et al., 2009).

Several genetic studies have shown an association between the serotonin 1A (5-HT1A) receptor gene (*HTR1A*) and MDD; however, results have been rather inconsistent. A recent meta-analysis showed no association between *HTR1A* and MDD (Lopez-Leon et al., 2008). However, two very recent studies reported that rs6295 (C-1019G) in the promoter region of *HTR1A*, which regulates *HTR1A* transcription,(Le Francois et al., 2008; Lemonde et al., 2003) was associated with MDD in the Chinese population (Wu et al. 2008; Zhang et al. 2009), but not in the Japanese (Kishi et al., 2009b). Kishi et al. (2009b) conducted an updated meta-analysis of rs6295, which has been intensively investigated in other studies. They showed that rs6295 was associated with Asian MDD patients after correction for multiple testing ($p(Z)=0.0176$), but not with Caucasian MDD patients ($p(Z)=0.138$).

HTR1B

In our sample of unipolar depression case-control study, Russians and Tatars were analyzed separately for association with *HTR1B* rs6296 (G861C). In the Russian sample the frequency of most common genotype (G/G) was similar in the UD group (52%) and controls (53%). Minor allele frequency (C allele) was 27% and 26%, respectively. No significant differences were found in the allele ($p = 0.76$) and genotype ($p = 0.89$) distribution between UD patients and controls.

In the Tatar sample the most common genotype was G/C with a frequency of 51% in UD patients and 45% in controls. Minor allele frequency (C allele) was 40% and 34% in these groups, respectively. There were no significant differences between UD patients and controls in the allele ($p = 0.46$) and genotype ($p = 0.46$) distribution (Noskova et al., 2009).

Published findings for association of *HTR1B* G861C with unipolar depression are contradictory. Fehr et al. (2000), demonstrated an association of *HTR1B* G allele with alcohol addiction but not with UD, anxiety, and narcolepsy. Huang et al. (2004) have revealed the association of the *HTR1B* C allele with UD and with substance abuse in their USA sample. *HTR1B* C allele was also associated with antisocial behavior in ethnic Finns with alcohol addiction (Lappalainen et al., 1998) as well as with obsessive-compulsive disorder in Canadians (Mundo et al., 2000).

HTR2A

The study of A–1438G (rs6311) of the *HTR2A* gene have shown that heterozygous genotype *A/G* (52%) was the most frequent in the group of Russian controls, while homozygous genotype *G/G* (52%) was observed more frequently in the UD patient sample. The frequency of minor allele C was 29% in the control group and 39% in the UD patients. The analysis of the genotype ($p = 0.02$) and allele ($p = 0.02$) frequency distribution of A–1438G polymorphic locus in Russians revealed the presence of statistically significant differences between examined groups. In the individuals of Russian ethnicity, genotype G/G (OR = 1.68, 95%CI 1.16–2.44) and allele G (OR = 143, 95% CI 1.06–1.92) of *HTR2A* might be considered as a risk factor. In ethnic Tatars, no significant differences were observed between the UD patients and controls in allele ($p = 0.46$) and genotype ($p = 0.36$) distribution (Noskova et al., 2009).

SUICIDAL BEHAVIOR

HTR2A

We investigated association of the *HTR2A* A1438G polymorphism (rs6311) with suicidal behavior in a sample of 188 patients with attempts of suicide, and 272 normal controls, of Russian and Tatar descent (Table 114-13). For the A1438G, significant differences were found in total sample between patients and controls in the distribution of allele frequencies ($p = 0.028$) and genotype frequencies. The higher G/G genotype frequency was observed in controls (OR = 0.60; 95% CI = 0.38–0.95).

TABLE 114-13 ALLELE AND GENOTYPE DISTRIBUTION OF THE *HTR1B* AND *HTR2A* POLYMORPHISMS IN SUICIDAL BEHAVIOR CASE-CONTROL STUDY

Groups			Genotypes			Alleles	
HTR1B (G861C)		N	G/G	G/C	C/C	G	C
Total	SB	186	0.49	0.44	0.07	0.71	0.29
	Control	230	0.34	0.54	0.12	0.61	0.39
Tatars	SB	82	0.38	0.56	0.06	0.66	0.34
	Control	86	0.37	0.44	0.19	0.59	0.41
Russians	SB	104	0.59	0.34	0.07	0.75	0.25
	Control	144	0.32	0.66	0.08	0.62	0.38
HTR2A (A1438G)			G/G	G/A	A/A	G	A
Total	SB	188	0.20	0.63	0.17	0.52	0.48
	Control	272	0.31	0.56	0.13	0.59	0.41
Tatars	SB	85	0.20	0.62	0.18	0.51	0.49
	Control	140	0.36	0.50	0.13	0.62	0.38
Russians	SB	103	0.20	0.63	0.17	0.52	0.48
	Control	132	0.26	0.61	0.13	0.56	0.44

When performing analysis separately for Russians and Tatars, the higher prevalence of G/G genotypes was shown in the control group (OR = 0.44; 95% CI = 0.22–0.86). A allele was associated with higher risk of suicidal behavior in Tatars only (OR = 1.54; 95% CI = 1.03–2.31). In Russians no significant differences in genotype and allele frequencies between cases and controls were found. Thus, according to our data, A allele of the A1438G polymorphism of the *HTR2A* was associated with suicide attempts in Tatars, but not in Russians (Gaysina et al., 2003).

Recently, Li et al. (2006), using meta-analytic techniques, suggest a significant association between the A-1438G polymorphism and suicidal behavior [A-allele carriers vs. GG homozygotes: OR=0.67 (0.5–0.89, P=0.007)]. Saiz et al. (2007) found an excess of the -1438A allele in nonimpulsive suicide attempts, as compared with impulsive suicide attempts and the control group [v2(2) =11.92, corrected $p = 0.021$]. These results are in agreement with our findings in the Tatar population.

HTR1B

We investigated the effect of the *HTR1B* G861C polymorphism on suicidal behavior in two populations, Russian and Tatar (Table 114-13). Because of significant differences in genotype frequencies between Russians and Tatars in controls ($p = 0.01$) and in those with suicidal behavior ($p = 0.01$), we performed association analysis in Russian and Tatars separately. In Russian patients, the G/G genotype (OR = 2.83; 95% CI = 1.63–4.91) and the G allele (OR = 1.88; 95% CI = 1.24–2.84) were associated with risk of suicidal behavior. In Tatars, the significant differences between controls and patients with suicidal attempts were shown for the genotype frequencies ($p = 0.04$). Significantly higher frequency of the C/C genotype was observed in the control group (OR = 0.28; 95% CI = 0.09–0.89). According to our data, the *HTR1B* might be involved in biological susceptibility to suicidal behavior (Gaysina et al., 2003).

Our data is in agreement with the report of New et al. (2001), but contradict those by Turecki et al. (2003) and Nishiguchi et al. (2001), who reported no associations between this polymorphism and completed suicides, and to findings by Rujescu et al. (2003) of the lack of association with suicidal attempts.

CONCLUSIONS

Variants at the genes encoding for the serotonin transporter (*SLC6A4*) and receptors (*HTR1A*, *HTR1B*, *HTR2A*) have been investigated for association with many psychiatric disorders such as schizophrenia, mood disorders, and suicide. These genetic association studies report conflicting results. When positive, they showed very small impact of the genetic variants as risk factors. Conflicting and negative results could be due to a real marginal role of these genetic variants, or it could be caused by a lack of gene coverage of investigated SNPs and/or a lack of power in the most studies.

The Russian data indicate that ethnic differences in the serotonin transporter genes, previously found between Caucasian and Asian populations, also exist between Caucasian subjects from four European populations. Ethnic differences might be responsible for the inconsistency (i.e., nonreplication) of the associations of *SLC6A4* with suicidal behavior, anxiety, depression, and schizophrenia.

A wider investigation of the serotoninergic system genes may help in better understanding of their role in psychiatric disorders, preferably complemented with the use of proteomic or metabolomic approaches.

ACKNOWLEDGMENTS

This work was supported partly by grant of the Russian Foundation for Humanities (11-06-00554a).

REFERENCES

Azmitia EC. (1999). Serotonin neurons, neuroplasticity, and homeostasis of neural tissue. *Neuropsychopharmacology* 21 (2 Suppl):33S-45S.

Baca-Garcia E et al. (2003). A pilot study on a gene-hormone interaction in female suicide attempts. *Eur Arch Psychiatry Clin Neurosci* 253 (6):281–285.

Benes FM, Taylor JB, Cunningham MC. (2000). Convergence and plasticity of monoaminergic systems in the medial prefrontal cortex during the postnatal period: implications for the development of psychopathology. *Cereb Cortex* 10 (10):1014–1027.

Bondy B, Buettner A, Zill P. (2006). Genetics of suicide. *Mol Psychiatry* 11 (4):336–351.

Collier DA et al. (1996) A novel functional polymorphism within the promoter of the serotonin transporter gene: possible role in susceptibility to affective disorders. *Mol Psychiatry* 1(6):453–460.

Eddahibi S et al. (2001). Serotonin transporter overexpression is responsible for pulmonary artery smooth muscle hyperplasia in primary pulmonary hypertension. *J Clin Invest* 108(8):1141–1150.

Engen RM et al. (2006). Ethnic differences in pharmacogenetically relevant genes. *Curr Drug Targets* 7(12):1641–1648.

Fan JB, Sklar P. (2005). Meta-analysis reveals association between serotonin transporter gene STin2 VNTR polymorphism and schizophrenia. *Mol Psychiatry* 10(10):891, 928–938.

Fehr C et al. (2000). The HTR1B 861G>C receptor polymorphism among patients suffering from alcoholism, major depression, anxiety disorders and narcolepsy. *Psychiatry Res* 97(1):1–10.

Fiskerstrand CE, Lovejoy EA, Quinn JP. (1999). An intronic polymorphic domain often associated with susceptibility to affective disorders has allele dependent differential enhancer activity in embryonic stem cells. *FEBS Lett* 458(2):171–174.

Gaysina D, et al. (2003). Association study of the serotonin 2A and 1B receptor genes with suicidal behavior in two different populations from Russia. *Balkan J Med Genet* 6(1&2):51–56.

Gaysina D, Zainullina A, Gabdulhakov R, Khusnutdinova E. (2006). The serotonin transporter gene: polymorphism and haplotype analysis in Russian suicide attempters. *Neuropsychobiology* 54(1):70–74.

Gelernter J et al. (1999). Population studies of polymorphisms of the serotonin transporter protein gene. *Am J Med Genet* 88(1):61–66.

Goldberg TE et al. (2009). The serotonin transporter gene and disease modification in psychosis: evidence for systematic differences in allelic directionality at the 5-HTTLPR locus. *Schizophr Res* 111(1–3):103–108.

Gould E. (1999). Serotonin and hippocampal neurogenesis. *Neuropsychopharmacology* 21(2 Suppl):46S-51S.

Gutierrez B et al. (1998). Variability in the serotonin transporter gene and increased risk for major depression with melancholia. *Hum Genet* 103(3):319–322.

Hanna GL et al. (1998). Serotonin transporter and seasonal variation in blood serotonin in families with obsessive-compulsive disorder. *Neuropsychopharmacology* 18(2):102–111.

Harmar AJ et al. (1996). The serotonin transporter gene and affective disorder. *Cold Spring Harb Symp Quant Biol* 61:791–795.

Heils A et al. (1996). Allelic variation of human serotonin transporter gene expression. *J Neurochem* 66(6):2621–2624.

Hensler JG et al. (1994). Quantitative autoradiography of the serotonin transporter to assess the distribution of serotonergic projections from the dorsal raphe nucleus. *Synapse* 17(1):1–15.

Hirschhorn JN, Altshuler D. (2002). Once and again-issues surrounding replication in genetic association studies. *J Clin Endocrinol Metab* 87(10):4438–4441.

Hoehe MR et al. (1998). Serotonin transporter (5-HTT) gene polymorphisms are not associated with susceptibility to mood disorders. *Am J Med Genet* 81 (1):1–3.

Hranilovic D et al. (2003). Serotonin transporter gene promoter (5-HTTLPR) and intron 2 (VNTR) polymorphisms in Croatian suicide victims. *Biol Psychiatry* 54(9):884–889.

Huang YY et al. (2004). Human 5-HT1A receptor C(-1019)G polymorphism and psychopathology. *Int J Neuropsychopharmacol* 7(4):441–451.

Huang YY et al. (1999). Relationship of psychopathology to the human serotonin1B genotype and receptor binding kinetics in postmortem brain tissue. *Neuropsychopharmacology* 21(2):238–246.

Huang YY et al. (2003). Substance abuse disorder and major depression are associated with the human 5-HT1B receptor gene

(HTR1B) G861C polymorphism. *Neuropsychopharmacology* 28(1):163–169.

Ioannidis JP et al. (2001). Replication validity of genetic association studies. *Nat Genet* 29(3):306–309.

Katsuragi S et al. (1999). Association between serotonin transporter gene polymorphism and anxiety-related traits. *Biol Psychiatry* 45(3):368–370.

Kim DK et al. (2000). Serotonin transporter gene polymorphism and antidepressant response. *Neuroreport* 11(1):215–219.

Kishi T et al. (2009a). Genetic association analysis of serotonin 2A receptor gene (HTR2A) with bipolar disorder and major depressive disorder in the Japanese population. *Neurosci Res* 64(2):231–234.

Kishi T et al. (2009b). Serotonin 1A receptor gene and major depressive disorder: an association study and meta-analysis. *J Hum Genet* 54(11):629–633.

Kunugi H et al. (1997). Serotonin transporter gene polymorphisms: ethnic difference and possible association with bipolar affective disorder. *Mol Psychiatry* 2(6):457–462.

Lang UE et al. (2004). No association of a functional polymorphism in the serotonin transporter gene promoter and anxiety-related personality traits. *Neuropsychobiology* 49(4):182–184.

Lappalainen J et al. (1998). Linkage of antisocial alcoholism to the serotonin 5-HT1B receptor gene in 2 populations. *Arch Gen Psychiatry* 55(11):989–994.

Lasky-Su JA et al. (2005). Meta-analysis of the association between two polymorphisms in the serotonin transporter gene and affective disorders. *Am J Med Genet B Neuropsychiatr Genet* 133B(1):110–115.

Le Francois B et al. (2008). Transcriptional regulation at a HTR1A polymorphism associated with mental illness. *Neuropharmacology* 55(6):977–985.

Lemonde S et al. (2003). Impaired repression at a 5-hydroxytryptamine 1A receptor gene polymorphism associated with major depression and suicide. *J Neurosci* 23(25):8788–8799.

Lesch KP et al. (1994). Organization of the human serotonin transporter gene. *J Neural Transm Gen Sect* 95(2):157–162.

Lesch KP et al. (1996). Association of anxiety-related traits with a polymorphism in the serotonin transporter gene regulatory region. *Science* 274(5292):1527–1531.

Lesch KP, Gutknecht L (2005) Pharmacogenetics of the serotonin transporter. *Prog Neuropsychopharmacol Biol Psychiatry* 29(6):1062–1073.

Lesch KP, Mossner R. (1998). Genetically driven variation in serotonin uptake: is there a link to affective spectrum, neurodevelopmental, and neurodegenerative disorders? *Biol Psychiatry* 44(3):179–192.

Li D, Duan Y, He L. (2006). Association study of serotonin 2A receptor (5-HT2A) gene with schizophrenia and suicidal behavior using systematic meta-analysis. *Biochem Biophys Res Commun* 340(3):1006–1015.

Li D, He L. (2007) Meta-analysis supports association between serotonin transporter (5-HTT) and suicidal behavior. *Mol Psychiatry* 12(1):47–54.

Li J et al. (2007). Association between polymorphisms in serotonin transporter gene and attention deficit hyperactivity disorder in Chinese Han subjects. *Am J Med Genet B Neuropsychiatr Genet* 144B(1):14–19.

Lin PY, Tsai G. (2004). Association between serotonin transporter gene promoter polymorphism and suicide: results of a meta-analysis. *Biol Psychiatry* 55(10):1023–1030.

Lopez dL et al. (2006). STin2 variant and family history of suicide as significant predictors of suicide completion in major depression. *Biol Psychiatry* 59(2):114–120.

Lopez-Leon S et al. (2008). Meta-analyses of genetic studies on major depressive disorder. *Mol Psychiatry* 13(8):772–785.

Lotrich FE, Pollock BG. (2004). Meta-analysis of serotonin transporter polymorphisms and affective disorders. *Psychiatr Genet* 14(3):121–129.

MacKenzie A, Quinn J. (1999). A serotonin transporter gene intron 2 polymorphic region, correlated with affective disorders, has allele-dependent differential enhancer-like properties in the mouse embryo. *Proc Natl Acad Sci U S A* 96(26):15251–15255.

Mann JJ, Brent DA, Arango V. (2001). The neurobiology and genetics of suicide and attempted suicide: a focus on the serotonergic system. *Neuropsychopharmacology* 24(5):467–477.

Munafo MR, Flint J. (2004). Meta-analysis of genetic association studies. *Trends Genet* 20(9):439–444.

Mundo E et al. (2000). Is the 5-HT(1Dbeta) receptor gene implicated in the pathogenesis of obsessive-compulsive disorder? *Am J Psychiatry* 157(7):1160–1161.

Myers RL et al. (2007). Polymorphisms in the regulatory region of the human serotonin 5-HT2A receptor gene (HTR2A) influence gene expression. *Biol Psychiatry* 61(2):167–173.

New AS et al. (2001). Suicide, impulsive aggression, and HTR1B genotype. *Biol Psychiatry* 50(1):62–65.

Nishiguchi N et al. (2001). No evidence of an association between 5HT1B receptor gene polymorphism and suicide victims in a Japanese population. *Am J Med Genet* 105(4):343–345.

Nonnis MF et al. (2008). Genes regulating the serotonin metabolic pathway in the brain stem and their role in the etiopathogenesis of the sudden infant death syndrome. *Genomics* 91(6):485–491.

Noskova T et al. (2008). Ethnic differences in the serotonin transporter polymorphism (5-HTTLPR) in several European populations. *Prog Neuropsychopharmacol Biol Psychiatry* 32(7):1735–1739.

Noskova TG et al. (2009). Association of several polymorphic loci of serotoninergic genes with unipolar depression. *Genetika* 45 (6):842–848.

Ogilvie AD et al. (1996) Polymorphism in serotonin transporter gene associated with susceptibility to major depression. *Lancet* 347(9003):731–733.

Ramamoorthy S et al. (1993). Antidepressant- and cocaine-sensitive human serotonin transporter: molecular cloning, expression, and chromosomal localization. *Proc Natl Acad Sci U S A* 90(6):2542–2546.

Rujescu D et al. (2003). Lack of association between serotonin 5-HT1B receptor gene polymorphism and suicidal behavior. *Am J Med Genet B Neuropsychiatr Genet* 116B(1):69–71.

Saiz PA et al. (2007). Association study of serotonin 2A receptor (5-HT2A) and serotonin transporter (5-HTT) gene polymorphisms with schizophrenia. *Prog Neuropsychopharmacol Biol Psychiatry* 31(3):741–745.

Sanders AR et al. (2008). No significant association of 14 candidate genes with schizophrenia in a large European ancestry sample: implications for psychiatric genetics. *Am J Psychiatry* 165(4):497–506.

Serretti A et al. (2006). Serotonin transporter gene variants and behavior: a comprehensive review. *Curr Drug Targets* 7(12):1659–1669.

Stober G, Heils A, Lesch KP. (1996). Serotonin transporter gene polymorphism and affective disorder. *Lancet* 347(9011):1340–1341.

Strobel A et al. (2003). Allelic variation in 5-HT1A receptor expression is associated with anxiety- and depression-related personality traits. *J Neural Transm* 110(12):1445–1453.

Surtees PG et al. (2006) Social adversity, the serotonin transporter (5-HTTLPR) polymorphism and major depressive disorder. *Biol Psychiatry* 59(3):224–229.

Szekely A et al. (2004). Human personality dimensions of persistence and harm avoidance associated with DRD4 and 5-HTTLPR polymorphisms. *Am J Med Genet B Neuropsychiatr Genet* 126B(1):106–110.

Tian C et al. (2008). Analysis and application of European genetic substructure using 300 K SNP information. *PLoS Genet* 4(1):e4.

Turecki G et al. (2003). Suicide and serotonin: study of variation at seven serotonin receptor genes in suicide completers. *Am J Med Genet B Neuropsychiatr Genet* 118B(1):36–40.

Willeit M et al. (2003). A polymorphism (5-HTTLPR) in the serotonin transporter promoter gene is associated with DSM-IV

depression subtypes in seasonal affective disorder. *Mol Psychiatry* 8(11):942–946.
Wu Y et al. (2008). Association between the serotonin 1A receptor C(-1019)G polymorphism and major depressive disorder in the northern Han ethnic group in China. *Chin Med J (Engl)* 121(10):874–876.
Yu YW et al. (2006). Association study of two serotonin 1A receptor gene polymorphisms and fluoxetine treatment response in Chinese major depressive disorders. *Eur Neuropsychopharmacol* 16(7):498–503.
Zainullina A et al. (2005). Association analysis of polymorphic DNA loci of serotonin neurotransmitter system with risk of schizophrenia. *Medical Genetics (in Russian)* 4(3):123–127.
Zainullina AG et al. (2003). Association of the hSERT and SLC6A4 polymorphic markersof the serotonin transporter gene with schizophrenia in different ethnic groups. *Mol Biol (Mosk)* 37(4):601–606.
Zhang K et al. (2009). The combined effects of the 5-HTTLPR and 5-HTR1A genes modulates the relationship between negative life events and major depressive disorder in a Chinese population. *J Affect Disord* 114(1–3):224–231.

115.

GENETICS AND GENOMICS PERSPECTIVES OF MEDICAL AND HEALTH EDUCATION IN HUNGARY

Béla Melegh

INTRODUCTION

In Hungary, there were 36,088 registered medical doctors in December of 2008 serving a population of around 10 million citizens. There are specialist board examinations including one in clinical genetics. There are four medical universities in Hungary (Budapest, Debrencen, Pécs, and Szeged) producing approximately 800 MDs from the state-funded program each year. In addition, about 1000 English-speaking and 600 German-speaking new students enroll for medical training annually. The universities and colleges for health education are independent, but are affiliated to the medical universities. While the Bologna system was not introduced into the medical training, the BSc and MSc modules are being set up in numerous fields of health sciences. Every year approximately 6500 students enroll in health sciences training projects. Despite the broad spectrum of specialist training, there is no health specialist training program for any field of genetics.

In the early stages, national specialty board recognition was not accorded to medical genetics in Hungary. Although there were well characterized, proper training schemes for almost all of the major medical subspecialties, the clinical genetics subspecialty was introduced only in 1978. The medically qualified candidates with a general medical degree, with a minimum postgraduate training period of four years, were considered along with their curriculum vitae, list of publications, accompanied by a letter of recommendation from the head of the employing department to the Board of the Postgraduate Medical University, the legal authority of managing the postgraduate training programs. The selection is based on credentials assessed by a selection committee. Currently there are only 25 medically qualified human geneticists in Hungary; 16 of them also possess other qualifications (PhD or DSc degrees).

The bases of the currently available formal procedure for medical doctors was introduced in 1999; in Hungary, human genetics is a secondary medical specialty that can be achieved after completion of a two-year regular training followed by a successful board exam.

SCOPE AND LIMITATIONS OF GENOME-BASED BIOTECHNOLOGIES IN THE DEVELOPING WORLD

The genetic and genomic technologies are concentrated in the universities. The budget available for genome-based diagnostic tests is governmentally limited. A unique feature in the financing of the genetic tests is the method-based compensation: for the genetic testing of diseases, the fee of the required methods (PCRs, sequencings, etc.) are paid in Hungary.

CONSANGUINITY AND INTERBREEDING—GENETIC AND HEALTH IMPLICATIONS

In Hungary the total population is about 10 million; a significant number, about 600,000–800,000 are the Roma (Gypsy) people. According to data published in 1988, the occurrence of consanguinity is high in this population; the frequency of cousin marriage ranges between 2%–20%, while the average for the general population is less than 0.3% (Czeizel, 1988a; Assal et al., 1991). There are some typical genetically closed Roma populations in northeast Hungary. Although no detailed studies are available, the estimated negative effect of consanguinity is more pronounced in these regions.

DEVELOPING BIOTECHNOLOGY INFRASTRUCTURE FOR SERVICE AND RESEARCH

For diagnostic purposes the routine available laboratory methods include basic cytogenetic techniques, chromosome banding, different types of FISH methods, and

PCR-based molecular genetic techniques (sequencing/fragment analysis). For research purposes almost all methods, including different mass spectrometry facilities or the recent deep sequencing apparatuses (Applied Biosystems and Solexa) are available.

GENETIC TESTING AND SCREENING

Prior to requesting a genetic test, it is obligatory to offer genetic counseling in an officially registered center. In 2009 a total of 22 genetic units were available in Hungary, located in medical universities and all major medical centers. Genetic tests are available for almost all major inherited diseases. There is a virtual network of medical centers, identified through blocks of diseases representing a large spectrum of domestically available tests. Some laboratories are partners in international networks (EUROGENETEST, ORPHANET, Gendia, etc). In Hungary, there is a neonatal mass screening program for phenylketonuria, galactosemia, congenital hypothyroidism, and biotinidase deficiency; the regular mass screening with mass spectrometry was introduced in 2007.

MEDICAL GENETIC SERVICE PROVISION IN THE DEVELOPING WORLD

Hungarian authorities already had declared their interest in dealing with rare diseases at many forums. In 2009, the National Center for Rare Diseases Research was established at the University of Pécs. The goal of this center is to coordinate the development of existing and future networking of all centers dealing with diagnosis and treatment of rare inherited diseases.

The Hungarian Congenital Abnormality Registry, established in 1962, is very unique in Europe. It is obligatory to report all congenital abnormalities between 0–1 years of age. This initiative was appointed by the World Health Organization as a WHO Collaborating Center for the Community Control of Hereditary Diseases in 1968. The congenital anomaly registry has been a full member of the EUROCAT for seven years (Czeizel, 1978, 1988b; Dolk, 2005).

Following the genetic training initiative led by the European Society for Human Genetics, Hungary declared its official interest to reform the medical genetics training examination based on the four-year pan-European model. However, there is no clear plan for individuals with nonmedical qualifications (BSc, MSc, PhD) to pursue a career in genetic laboratory diagnostics. Selected health or biology MSc holders may enroll in a 4-year program in laboratory techniques, making them eligible to work in general diagnostic laboratories with a 2-year upgrade module for molecular genetics.

Hungary has professional representation for genetics, the Hungarian Society of Human Genetics (established in 1968) and the Clinical Genetics Advisory Board of Ministry of Health. Both organizations offer consultations to a number of government and nongovernmental authorities.

ETHICAL, LEGAL AND SOCIAL ISSUES [ELSI] OF GENETICS SERVICES AND GENOMICS RESEARCH

Law XX/2009 on "Regulation of genetic tests, genetic research, and biobanking" has been operative since July of 2009. This bill regulates methodology and procedures of genetic testing; the most important issue is the obligatory genetic counseling before the genetic test, and the results should be reported to the patients also in a genetic counseling unit by a qualified medical geneticist. The bill regulates the basic elements of the operation of biobanks and coding/storage of pathological specimens (Sándor, 2009, 2010; Sándor & Bárd, 2009).

REFERENCES

Assal S, Susanszky E, Czeizel A. (1991). High consanguinity rate in Hungarian gipsy communities. *Acta Pediatr Hung* 31:299–304.

Czeizel A. (1978). The baseline data of the Hungarian Congenital Malformation Register, 1970–1976. *Acta Paediatr Hung* 19:149–156.

Czeizel A. (1988a). The activities of the Hungarian Centre for Congenital Anomaly Control. *World Health Stat Q* 41:219–227.

Czeizel A. (1988b). Medical genetics in Hungary. *J Med Genet* 25:2–8.

Dolk H. (2005). EUROCAT: 25 years of European surveillance of congenital anomalies. *Arch Dis Child Fetal Neonat* 90:F355–358.

Sándor J. (2010). Biobanks: A successful fiction in the service of science? (Hungarian) *JURA* 16:116–126.

Sándor J, Bárd P. (2009). CELAB Paper Series No. 1. The Legal Regulation of Biobanks; National Report: Hungary, Budapest: CEU, CELAB, 2009.

Sándor J. (2009). From Private to Public? Legal Concepts of the Right to Privacy and Ownership in the Regulation of Biobanks In: Kris Dierickx, Pascal Borry (eds). *New Challenges for Biobanks: Ethics, Law and Governance*, Intersentia, Antwerp, pp. 123–136.

116.

HUMAN AND MEDICAL GENETICS IN HUNGARY

Béla Melegh

INTRODUCTION

Hungary represents a special nation in Europe, founded around 1100 years ago by the Magyars from the Carpathian Basin, originally from the eastern side of the Urals. The population genetic profile of the majority of Hungarian peoples is similar to that of the neighboring European nations. The Hungarian population includes several minority population groups, most notably Gypsy (Roma), representing about 8% of the total. Genetic disorders are represented by chromosomal abnormalities, congenital anomalies and malformation syndromes, classic common and rare Mendelian diseases, and polygenic/multifactorial contributions to communicable and noncommunicable diseases. Apart from classic genetic disorders, the present chapter also discusses the development of Hungary's genetic research and service infrastructure, including pharmacogenomics and the ethical and legal policies regulating the practice of human genetics and biobanking in the country.

THE POPULATION STRUCTURE AND VARIATIONS

The founders of the original Hungarian state are the Magyars, the ancient Hungarians, who settled in the Carpathian Basin at the end of the ninth century after two millennia of migration from East of the mountains of Urals. The Carpathian Basin region had been settled for thousands of years before the Magyars' arrival, by Avars, Slavs, Dacians, Romans, Sarmatians, Goths, Huns, and others; it is known that at the time of the Hungarian conquest the majority of the indigenous population was Slavic. The newly arrived Hungarians were estimated as 10% to 50% of the total population (Cavalli-Sforza et al., 1994). Several known settlements are documented in the national history: some groups came by invitation, others were assimilated gradually, while still others came as conquerors.

The first population genetic study was performed in the Hungarian population based on 26 genetic markers (including blood groups, serum proteins, and isoenzymes of red cells) in a collaboration between the German and Hungarian Academy of Sciences in the 1980s. Later, with modern molecular approaches in hand, mitochondrial DNA sequences of 27 ancient samples (10th–11th centuries), 101 contemporary Hungarian, and 76 Hungarian-speaking Sekler samples from Transylvania were analyzed (Tomory et al., 2007). The origin of the Hungarian-speaking Seklers, currently an isolated minority in Romanian Transylvania, is still matter of debate. As privileged people who guarded an exposed border region, Seklers lived in comparative isolation in the mountains in Transylvania from, at the latest, the twelfth century until the nineteenth. Mitochondrial DNA sequence data were compared with sequences of 7752 individuals representing 57 European and Asian populations. Only 2 of 27 ancient Hungarian samples were unambiguously of Asian origin; the rest belonged to one of the known Eurasian haplogroups with some Asian affinities. Strong differences were revealed when the ancient Hungarian samples were analyzed according to their apparent social status, as judged by grave goods. Commoners with poor archeological remains showed a predominance of mtDNA haplotypes and haplogroups common in west Eurasia (H, R, T), while individuals with high status, presumably conquering Hungarians that had rich grave goods—a horse's skull, arrowheads or spearheads, braid ornaments, earrings, etc.—showed a more heterogeneous haplogroup distribution, with haplogroups (N1a, X), which are present at very low frequencies in modern worldwide populations and are absent in the contemporary Hungarian and Sekler populations.

These findings suggest that significant genetic differences exist between the ancient and recent Hungarian-speaking populations, and no genetic continuity can be seen. The results also suggest that the ancient population from the tenth and eleventh centuries was genetically heterogeneous, and some Asian genetic influence is seen in the Hungarian conqueror population.

The analysis of maternal lineage also shows that the linguistic isolation of Hungarian-speaking populations in the Carpathian Basin had not been associated with significant genetic isolation. Gene flow from neighboring population migrations likely affected the Hungarian gene pool. Maternal lineages in the modern Hungarian gene pool bear the imprints of populations who have been living in the region for centuries. In contemporary Hungary, likely there is a dominating effect of populations already living in the Carpathian Basin, with influence from the Balkans and West Eurasia, while in the Seklers, the genetic effect of eastern and southern Europeans is more visible.

In another mtDNA-based approach, using 55 random DNA samples collected from average Hungarians, only 3 samples (5% of total) was found to have the M haplogroup, which has an unequivocally Asian origin (Nadasi et al., 2007). Furthermore, in another study on a pedigree with A7445G mutation-associated deafness, the U4b Asian haplogroup could be identified; this rare type had not been observed people living in Hungary, raising the possibility of an ancient Magyar origin of this variant (Maasz et al., 2008c).

Paternal lineages, which usually give a higher geographical resolution than the maternal, have also been studied. Total of 22 biallelic polymorphisms were typed from the non-recombining region of the human Y-chromosome in 100 men from modern Hungary and 97 Seklers. The results were compared with data from other European populations studied (Semino et al., 2000a), and the phylogeographic context of the Y-chromosome pool of the populations were analyzed (Csanyi et al., 2008). One specific Y-chromosomal base substitution (Tat, T→C), a relatively recent event (95% confidence interval 3140–6200 years), is a valuable marker in Finno-Ugric population studies (Zerjal et al., 1997). The C allele of the Tat polymorphism is widespread in all Uralic-speaking populations, except that it is absent or extremely rare among modern Hungarian-speaking populations (Lahermo et al., 1999; Rootsi et al., 2000; Semino et al., 2000a, 2000b; Tambets et al., 2001, 2004; Zerjal et al., 1997). Among the studied modern individuals, only one Sekler carries the Tat C allele, whereas out of four skeletal remains from the time of the Hungarian conquest, two possess the allele. The latter finding, even allowing for the low ancient sample number, appears to indicate a Siberian lineage of the invading Hungarians that later largely disappeared.

In the two more recent Hungarian-speaking populations, the Y-chromosomal binary markers share similar components described for other European populations, except for the presence of the haplogroup P*(xM173) in Sekler samples, which may reflect a Central Asian connection, and high frequency of haplogroup J in both Seklers and Hungarians. MDS analysis based on haplogroup frequency values confirms that modern Hungarian and Sekler populations are genetically closely related, and similar to populations from Central Europe and the Balkans. Another study analyzing a total of 6501 Y-chromosomes of 81 populations, including samples collected from 106 random Hungarian males, there was no statistical difference in the prevalence of the E-M78 and E-V13 haplogroups between Hungarians and the surrounding populations (Cruciani et al., 2007).

The lactase non-persistence *hypolactasia* is an autosomal recessive disorder (Sahi et al., 1973). The prevalence of the adult-type hypolactasia varies among populations. Lactase persistence most likely originated as a mutation in pastoralist populations who used milk as important nutriment for adult feeding (Sahi, 1994; Suarez et al., 1997). From the two phenotypes, the lactase non-persistent phenotype is the ancestral form. The prevalence of adult-type lactase non-persistence is 3%–70% among Caucasian populations in Europe, rare in Northern Europe, but increases in the south and west. It reaches approximately 100% in Asians (Gudmand-Hoyer et al., 1996; Suarez et al., 1997).

Recently a T/C SNP variant upstream of the *LCT* gene proved to be associated with lactase non-persistence. The functional role of C/T_{-13910} polymorphism in regulation of lactase gene expression was characterized, and it was found that it functions as a strong transcriptional enhancer (Olds et al., 2003; Troelsen et al., 2003). Both variants have enhancer activity, but the dominant lactase-persistent T_{-13910} variant has stronger enhancer effect than the non-persistent C_{-13910} allele (Olds et al., 2003; Troelsen et al., 2003). In individuals with the T/T_{-13910} and C/T_{-13910}, or with the C/C_{-13910} variant, lactase persistence or non-persistence may develop, respectively. The prevalence of different C/T_{-13910} lactase genotypes was evaluated also in Hungarian population samples (Nagy et al., 2009). Allele frequencies associated with lactase persistence (T_{-13910}) and non-persistence (C_{-13910}) were 37.8% and 62.2%, respectively. The 62.2% frequency of the lactase persistence in the Hungarian population is lower than that in Sweden and Finland (81.5%), higher than in France (43.1%) and Northern Italy (35.7%), and similar to Portugal (62%; Ingram et al., 2009). Moreover, the frequency of lactase non-persistence among those populations living in close proximity to the Magyars in the Siberian homeland significantly differs from the current Hungarian population: 71% in Northern Mansi, 78% in Nenets, 50% in Komi-Permiak, and 59% in Udmurtian.

More recently, in a study with analysis of more than 270,000 SNPs, genotyped with the Illumina 318K/370CNV chips and studying 3112 individuals across 16 European countries including 50 Hungarians, no unique genotype could be revealed in the Hungarian samples; they exhibited a high degree of similarity with the neighboring nations (Nelis et al., 2009).

Several subpopulations, like Seklers, Matyo, Kun, Sokac, and other Hungarian ethnic subpopulations living in Hungary were, to a greater or lesser extent, isolated populations even just decades ago. Their population genetic structure has not been investigated with modern molecular methods at all.

Roma (Gypsy/Romany) people originate from the Northern part of India; they began their migration about 1000 years ago (Gresham et al., 2001; Hajioff et al., 2000; Kalaydjieva et al., 2001a, 2001b; Mastana et al., 1992; McKee, 1997, Morar et al., 2004). The early history of the Romany people is still not clear, and the reason for their diaspora still remains largely an enigma in absence of a written history their origins. Contemporary populations sometimes suggested to have a close relationship with the Roma are the Dom people of Central Asia and the Banjara of India. Strong genetic evidence also suggests the Indian origin of Roma; 47.3% of Romany males carry the Y-chromosome haplogroup of H-M82, which rarely occurs outside of the Indian subcontinent. Mitochondrial haplogroup M, most common in Indian subjects and rare outside southern Asia, accounts for nearly 30% of Romany people (Kalaydjieva et al., 2001b). A more detailed study of Polish Roma shows this as an M5 lineage, which is specific to India (Malyarchuk et al., 2006).

Roma populations live in many countries throughout the world and are well known for their preserved traditions. Currently, the total Roma population size is estimated to be about 12 million in the world, and 8–10 million gypsies live in Europe today (Gresham et al., 2001; Morar et al., 2004). Hungary ranks fourth in Europe in terms of the estimated size of its Roma population, which is thought to be about 700,000 people (Gresham et al., 2001). According to the 2008 governmental survey, the number of Roma is between 600,000 and 800,000 in Hungary; therefore, about 6%–8% of the inhabitants of Hungary belong to this group of minorities (www.ksh.hu). Even nowadays, numerous Roma groups remain isolated; they marry traditionally only within their own community. The occurrence of consanguinity is still high in the Roma population, with a frequency of cousin marriage between 2%–20%, while the average for the general population is less than 0.3%. Besides the Romas, the Hungarian population also includes Slovaks, Germans, Croatians, Jews, and Serbs, whose major population genetic features have not been investigated at all.

After World War I, the postwar period for Hungary was quite dramatic. The nation had suffered its largest trauma with the Treaty of Trianon in 1921; Hungary lost two-thirds of its original territory, lost half of its citizens, the majority of its economy and intellectual power, and all that had a severely negative effect. After the Second World War, Hungary belonged to the East European bloc, was under the influence of the Soviet Union, and the borders were even worse than before. This rearrangement of borders resulted in a dictatorial change of the nationality of millions of people in the Carpathian Basin.

EPIDEMIOLOGY OF HUMAN DISEASE WITH EMPHASIS ON GENETIC CONDITIONS

In Hungary there are well-organized systems for the registration of different human diseases by the Hungarian National Health Insurance Fund (Országos Egészségbiztosítási Pénztár; OEP); the bilingual (Hungarian and English) Statistical Yearbook is available for a fee (www.oep.hu). Also, much useful related information is publicly available at the site of www.gyogyinfok.hu; other useful sites in this context can be found in the following parts of this chapter. However, records of genetic diseases are poor and obsolete generally. This is in part due to the officially used disease classification system, the International Statistical Classification of Diseases and Related Health Problems, which has been in use since 1993 and is therefore outmoded following tracking diseases in light of the current state of knowledge.

The current review focuses on common or peculiar aspects of genetic diseases, disease genetics, and accomplishments achieved in Hungary; the genetics of non-germ line mutations, belonging mainly to the field of pathology, are not systematically included.

CHROMOSOMAL DISORDERS

Basic chromosomal methods were introduced in Hungary in the second half of the 1960s under the supervision of Károly Méhes and Dezső Schuler. Today, all the medical universities are well equipped with microscopes and image analysis systems suitable for performing routine FISH methodologies. Besides the universities, there are also four county hospitals operating basic chromosome laboratories affiliated to genetic counseling units. In 2008 and 2009, 5050 and 5744 classic karyotype-based chromosome examinations were reported to and financed by the OEP, respectively; the number of bandings were 5130 and 5630, while the number of the reported and covered FISH were 4058 and 4806, respectively. These numbers do not include research tests.

In the research field, the description of premature centromere division as a manifestation of chromosome instability, and the possible role of this phenomenon in certain chromosome abnormalities, was a prominent and often cited observation of Károly Méhes, who was in addition a worldwide recognized clinical geneticist (Mehes et al., 1995). Following the primary observation, the principle of nonrandom centromere separation was established in various aneuploid cases (Bajnoczky et al., 1988; Mehes, 1978) and in association with chromosomal abnormalities (Mehes et al., 1982, 1990a), and tumors as well (Mehes et al., 2002). The other groups of clinical observations made during anthropometric, dysmorphologic, and cytogenetic investigations by him led to the recognition of the significance of mild errors of morphogenesis, also called "minor anomalies" (Meggyessy et al., 1987). The principles described in his book. *Informative Morphogenetic Variants in the Newborn Infant,* are now recognized worldwide (Mehes, 1988).

More theoretical findings were usable in disease prevention, e.g., studies concerning the mutagenic effect of drugs on cancer risk. Chromosome mutations and the number of sister chromatid exchanges induced by immunosuppressive treatment were investigated in certain types of renal diseases (Schuler et al., 1979). Chromosomal analysis of standard population and of persons occupationally exposed to mutagens were performed in order to detect a possible harmful effect, and showed that the frequency of chromosomal aberrations in people exposed to ionizing radiation showed a two- to sixfold increase of aberrant cells, and a two- to fourfold increase in cases of chemical mutagens (Gundy, 1989). A study confirmed the association between level of chromosomal aberrations and cancer risk, and provided novel information on the type of aberrations more strongly predictive of cancer risk (Boffetta et al., 2007). A study resulted in the identification of an aneugenic mutagenic chemical, trichlorfon, which caused Down syndrome in cluster of cases (Czeizel et al., 1993). Another study demonstrated the presence of Y-chromosomal sequences in Turner syndrome patients that were suspected as a risk factor for the development of gonadal tumors (Vlasak et al., 1999).

In some cases, observations made during research led to improved protocols of diagnostic investigations and screening, for example, in prenatal examinations in high-risk pregnancies (Beke et al., 2009; Papp et al., 1987). Among the various reports, the observations on ring chromosomes should be mentioned (Kosztolanyi, 1987). A ring chromosome 15 case with no significant deletion in the breakpoint area, showing normal growth and development, large hyperpigmented skin changes, multiple café au lait spots, and hypopigmentation of scalp hair, was published (Morava et al., 2003b). The first case for successive ring chromosome formation was also identified (Kellermayer et al., 2005b).

Analyses of the phenotype–genotype correlation provided useful information about mutations of different chromosomal regions. Among the translocation cases, a girl was described with submucous cleft palate, resolving mild hydrocephalus, severe hypotonia, and joint contractures. Detailed characterization of the translocation breakpoints was given, using molecular cytogenetic methods (Morava et al., 2003a). As a further interesting observation, exclusion of partial trisomy was performed by FISH in a pregnancy after determining the maternal pericentric inversion of chromosome 21 and the unbalanced condition in her affected son with Down syndrome (Tardy et al., 1997). In another report of a child who had short stature, mental retardation, coarctation of the aorta, and early developmental delay, characteristic clinical signs were described in association with a special karyotype of 46,XX,dup(13)(q12-14) (Hermann et al., 2000). Observations on a rare 45,X patient (Kellermayer et al., 2005a), and siblings with cryptic unbalanced translocation resulting in del 13q and dup 15q, were also published (Teszas et al., 2008). The clinical findings of a boy with a 2q33.3 deletion included minor anomalies of face and distal limbs, intrauterine and postnatal growth retardation, microcephaly, and moderate developmental delay (Riegel et al., 2001).

In a special case with immunological defects, a girl with characteristic pattern of minor anomalies, severe generalized muscular hypotonia, psychomotor retardation, juvenile rheumatoid arthritis, and chronic diarrhea was studied. An imbalanced form of a familiar translocation (46,XX,der(18),t(18;20)(p11.1;p11.1)pat) was revealed underlying these clinical features (Czako et al., 2002). In another interesting case with severe dystrophy, presenting also with cow's milk allergy, nasal regurgitation, and facial dimorphism, DiGeorge syndrome was finally diagnosed using FISH (Rozsai et al., 2009).

Cases with skeletal abnormalities were also found. These include a boy presenting with trigonocephaly, cleft palate, multiple minor anomalies, flexion deformities of elbows, cryptorchidism, and severe muscular hypotonia, who had an unbalanced karyotype with duplication of the distal 17q and deletion of 2p derived from a reciprocal familiar translocation, 46,XY,t(2;17)(p25;q24) (Czako et al., 2004). In another male infant with dysmorphic features, partial coronal craniosynostosis, congenital heart defect, muscle hypotonia, and hyperextensible joints, a microdeletion of the region 22q11.2 was revealed, as a novel chromosomal disease (Karteszi et al., 2004).

Chromosomal mosaicism was detected in a case with severe thoracolumbar torsion scoliosis, muscle hypotonia, hypermobility of all joints, developmental delay, and dysmorphic features (Morava et al., 2002). A young woman with Ullrich-Turner syndrome with a karyotype 45,X/46,X,idic(Y)(q11) had age-appropriate genitalia without virilization, in spite of the presence of the Y-derived marker chromosome in 70% of her lymphocytes (Morava et al., 2000). Chromosomal studies of an infant with all the characteristic features of Wiedemann-Rautenstrauch syndrome, but without progeroid signs, at the age of 2.5 years detected a mosaic triploidy/tetraploidy in fibroblasts, in 60% and 14% of the cells, respectively (Karteszi et al., 2006).

Still in the field of clinical chromosome research, as a pioneering approach for further therapy use, an artificial chromosome expression system has been developed (Duncan et al., 2007). This system is based on a platform satellite DNA-based artificial chromosome, which contains multiple integration acceptor sites that allow for the lambda integrase-mediated site-specific insertion of multiple copies of DNA sequences. The system has successfully entered into the pharmaceutical cellular protein production industry (Kennard et al., 2009), and proved its potential in the very first preclinical model experiments for cell replacement therapy (Katona et al., 2008). This combined artificial chromosome/stem cell model showed the potential of ex vivo artificial chromosome stem cell therapies in the treatment of reversible genetic diseases and cancers (Macnab et al., 2009).

CONGENITAL ANOMALIES AND MALFORMATION SYNDROMES

Károly Méhes, as a pediatric dysmorphologist, underlined the impact of minor morphological anomalies in a variety of disorders like mental retardation, isolated urinary tract malformations, diabetes, metabolic diseases, and malignancies (Mehes, 1979, 1983; Mehes et al., 1990b). He suggested that, although the existence of a supernumerary nipple in a renal field defect could be questioned (Meggyessy et al., 1987), an association of polythelia with cancer of the urogenital organs cannot be excluded (Mehes, 1996). In a series of studies, he observed a significant excess of minor malformations and extreme variants in children with acute lymphoblastic leukemia and also in their siblings (Mehes et al., 1998), and suggested that these anomalies might be interpreted as indicators of genetic instability at the macroscopic level, even if an increased prevalence of anomalies in sibs of children with leukemia cannot be regarded as a sign of predisposition to malignancy (Mehes et al., 1985). From a practical point of view, he emphasized the importance of recording any mild errors of morphogenesis in routine medical examinations, even in the era of genomic medicine—a skill that should be included in medical education, postgraduate courses, and clinical practice.

The Hungarian Congenital Abnormality Registry was established in 1962 as the first national-based registry of congenital anomalies in the world, and is still unique in Europe: all congenital abnormalities are reported between age 0–1 year. Under the direction of Andrew Czeizel, it became WHO's Collaborating Center for the Community Control of Hereditary Diseases (1984–1998). The registry is now a full member of the EUROCAT, since 2003. The Hungarian Case-Control Surveillance of Congenital Abnormalities was established in 1980, based on the comparison of cases from the Congenital Abnormaility Registry and 2–3 matched control newborns without any defect from the National Birth Registry of the Central Statistical Office. The exposures studied are based on prospective and medically recorded data in prenatal maternity logbooks, and at present this dataset, between 1980 and 2002, contains the highest number of cases (29,992) and controls (52,599) in the world, and thus is used in several international collaborations. To date, 156 papers have been published in the international periodicals based on this material.

As an example, Table 116-1 shows the key figures reported in the Hungarian Congenital Abnormality Registry during 2007. Much important information is available in Hungarian at the official site (www.oszmk.hu).

SELECTED COMMON MENDELIAN DISEASES

Compared with the surrounding nations, no special features among the common Mendelian diseases can be observed in Hungary; some aspects of the field are discussed in different parts of this chapter. By contrast, numerous unique features, case reports, and original clinical observations in the rare disease categories are available in relatively large numbers; examples are discussed below.

In the early decades of the last century of the previous millennium, when the human or medical genetics had yet become a distinct and separate discipline, the concept of hereditary degenerative diseases, called "heredodegeneration," was introduced in 1902 by a Hungarian neurologist, Ernő Jendrassik (1858–1921). Another neurologist, Károly Schaffer (1864–1935), described the pathomorphological features of neurodegeneration in Tay-Sachs disease in 1905; Schaffer suspected the changes of having a "lecithioid" nature (Schaffer 1905); some authors therefore referred to the disease as Tay-Sachs-Schaffer disease. Károly Csörsz (1892–1935) dealt with neurological diseases from their hereditary aspects, and also organized the first Hungarian twin registry. He made observations in Huntington's disease, described large pedigrees of deafness (Csorsz 1933), and his original report from 1928 is still referenced by the OMIM for the description of X-linked ichthyosis (OMIM #308100; Csorsz, 1928).

During and between the World Wars, science was repressed in Hungary; after the Second Word War, Hungary belonged to the Soviet bloc and genetics was centrally controlled by Lisenko's dogmas; several fields of the science, similar to Hungary's national feelings, were repressed by the Soviet methods. The Revolution of 1956 made a hole the fabric of communism, the fall of which was again initiated in Hungary in 1989 by the opening its national borders, representing the iron curtain, for the East German people.

After 1956, a relatively slow but growing accumulation of genetic discoveries could be seen in Hungary. At present, 62 suggestions are given for a search using the keyword of "Hungarian" in the OMIM—essentially the majority of these citations are, again, related to diseases belonging to the rare disease category. It is still not clear whether these are actually typical Hungarian types of rare

TABLE 116-1 DISTRIBUTION OF THE REGISTERED CONGENITAL ANOMALIES IN 2007

Anomalies	Births		Prenatal disorder recognition leading to abortion		Prenatal disorder recognition not leading to abortion		Total
	number	%	number	%	number	%	number
Acrania/Anencephaly	2	13.3	11	73.3	1	6.7	14
Encephalocele	2	50.0	2	50.0	0	0.0	4
Holoprosencephaly	1	25.0	3	0.75	0	0.0	4
Microcephaly	14	100	0	0.0	0	0.0	14
Hydrocephaly	8	27.6	16	55.2	5	17.2	29
Spina Bifida/Spinal Dysraphism	9	21.4	32	76.2	1	2.4	42
Other Neurological Anomalies	44	47.6	13	22.0	2	3.4	59
Anophthalmia/Microphthalmia	3	100	0	0.0	0	0.0	3
Congenital Cataracts	9	100	0	0.0	0	0.0	9
Other Eye Anomalies	40	100	0	0.0	0	0.0	40
Ear Anomalies	20	100	0	0.0	0	0.0	20
Branchial arc anomalies	11	39.3	16	57.1	1	3.6	28
Craniofacial Anomalies	33	100	0	0.0	0	0.0	33
Heart and Aortic Anomalies	1325	98.0	18	1.3	9	0.7	1352
Respiratory System Anomalies	47	94.0	1	2.0	1	2.0	49
Cleft Palate	31	100	0	0.0	0	0.0	31
Cleft Lip	66	95.7	1	1.4	2	2.9	69
Oesophageal atresia	12	100	0	0.0	0	0.0	12
Pyloric stenosis	37	100	0	0.0	0	0.0	37
Intestineal atresia / stenosis	15	93.8	0	0.0	1	6.3	16
Anal atresia/stenosis	18	100	0	0.0	0	0.0	18
Hirschsprung disease	12	100	0	0.0	0	0.0	12
Other gastrointestinal anomalies	90	93.8	3	3.1	2	2.1	95
Cryptorchid testes	199	100	0	0.0	0	0.0	199
Genital anomalies	48	94.1	0	0.0	3	5.9	51
Renal agenesis	30	88.2	3	8.8	1	2.9	34
Potter Syndrome	4	44.4	4	44.4	0	0.0	8
Cystic Renal Disease	19	65.5	2	6.9	8	27.6	29
Urogenital atresia anomalies	372	96.6	1	0.3	12	3.1	385
Renal extrophy	5	100	0	0.0	0	0.0	5
Hypospadias	257	100	0	0.0	0	0.0	257
Other urinary anomalies	68	97.1	1	1.4	1	1.4	70
Hip dislocation	174	100	0	0.0	0	0.0	174
Club foot	116	98.3	0	0.0	2	1.7	118
Polydactyly	77	100	0	0.0	0	0.0	77
Syndactyly	84	89.4	9	9.6	1	1.1	94
Limb Absence	16	64.0	7	28.0	2	8.0	25
Other limb anomalies	84	96.6	3	3.4	0	0.0	87
Skeletal Anomalies	21	91.3	2	8.7	0	0.0	23

(Continued)

TABLE 116-1 (CONTINUED)

Anomalies	Births		Prenatal disorder recognition leading to abortion		Prenatal disorder recognition not leading to abortion		Total
Torticollis	39	100	0	0.0	0	0.0	39
Absence of Diaphragm	21	95.5	0	0.0	1	4.5	22
Gastroschisis	8	40.0	12	60.0	0	0.0	20
Muscular anomalies	14	82.4	3	17.6	0	0.0	17
Skin anomalies	27	100	0	0.0	0	0.0	27
Asplenia	2	100	0	0.0	0	0.0	2
Endocrine anomalies	4	100	0	0.0	0	0.0	4
Congenital Tumors	23	100	0	0.0	0	0.0	23
Situs inversus	4	100	0	0.0	0	0.0	4
Down Syndrome	80	46.5	89	51.2	1	0.6	170
Chromosomal anomalies	19	19.4	76	77.6	3	3.1	98
Hydrops Foetalis	3	12.0	21	84.0	1	4.0	25
Conjoined twins	1	33.3	2	66.7	0	0.0	3
Minor anomalies	824	99.8	2	0.0	0	0.0	826
Random Combination	119	93.7	6	4.7	2	1.6	127
Multiplex Association	11	78.6	2	14.3	0	0.0	13
Robin Sequence	2	100	0	0.0	0	0.0	2
Multiplex Congenital Anomalies	104	87.4	13	10.9	2	1.7	119
Other Anomalies	4	36.4	7	63.6	0	0.0	11
Total	4732	91.2	381	7.3	65	1.3	5178

Still birth (2) and miscarriage (5) not included.

diseases, or if they just represent the site of their original description. Thus, in a family with meningocerebrovascular amyloidosis, a novel transthyretin gene missense mutation was observed (Asp18Gly). Based on the clinical features, pathologic alterations, and molecular findings, this disease represented a novel type of systemic familial amyloidosis with disease manifestation clinically restricted to the CNS, and was called "Hungarian type" since it was found to be manifested, for the first time, in a Hungarian patient (Garzuly et al., 1996).

With the introduction and development of new methods, observations could be used both in diagnostics and in research. Thus, for example, the original research method for Huntington chorea (Toth et al., 1997a, 1997b), Ivemark syndrome (Toth et al., 1998), steroid 21-hydroxylase deficiency (Theodoropoulou et al., 2001), hemophilia A and B (Klein et al., 2001), and epidermolysis bullosa simplex (Glasz-Bona et al., 2009), was later introduced into the genetic diagnostic procedure.

A number of selected novel mutations have been found related to hematologic, immunodeficiency, dermatological, and other diseases (Table 116-2).

In pharmacogenetics, which is a new field in Hungary, a small number of observations have been made. The comparative pharmacogenetic study of the VKORC1 haplotype profile in healthy Hungarian and Roma population samples revealed that the VKORC1 genetic profile in Hungarians was more similar to European Caucasians and Americans with European descent than to Roma samples (Sipeky et al., 2009a). Polymorphisms of the CYP2C9 gene in the two largest populations in Hungary, healthy Hungarians and Roma were investigated. A significant difference was found between the Hungarian and Roma populations regarding the *1/*1, *1/*3 and the *2/*3 genotypes (Sipeky et al., 2009b). Also, pharmacogenetic analyses showed a significant role of the high-activity Val allele of catechol-O-methyltransferase (COMT) Val158Met polymorphism in a Hungarian ADHD population (Kereszturi et al., 2008).

Besides the observations done in Hungarian average populations detailed above, some studies have also been done in the Roma population. Due to previous international efforts, some special hereditary disorders of the Roma population are already known. Thus, hereditary motor and sensory neuropathy, congenital cataracts, facial dysmorphism neuropathy, congenital myasthenia syndrome, limb-girdle muscular dystrophy, galactokinase deficiency, and polycystic kidney disease are known to accumulate in different European Romany populations.

As special domestic observations, the accumulation of recessively inherited autosomal multiplex malformation syndromes was observed in some Hungarian gypsy colonies. A special form is accepted now as orofaciodigital type VI, and as Váradi-Papp syndrome in the McKusick catalog (OMIM: 277170). Holoprosencephaly and severe facial dysmorphia are always present in the syndrome (Varadi et al., 1980).

Also, high incidence of short rib–polydactyly syndrome type VI was recognized in a Hungarian Roma subpopulation (Kovacs et al., 2006). A Roma patient with bigenic connexin mutations associated with hidrotic ectodermal dysplasia was also observed. The patient harbored a novel sporadic mutation (V41L) in GJA1 (Cx43), as well as a heterozygous coding variant (R127H) of GJB2 (Cx26) (Kellermayer et al., 2005c).

Molecular genetic analysis of the SLC22A5 gene in two nonconsanguineous Hungarian Roma children with carnitine-responsive cardiomyopathy, and in postmortem tissue specimens from sudden infant death cases, revealed a homozygous deletion of 17081C of the SLC22A5 gene (Melegh et al., 2004). Genomic DNA sequencing revealed novel mutation in the signal transducer and activator of transcription 3 gene (STAT3; H332Y) in a gypsy family (Jiao et al., 2008). Finally, a homozygous K76N mutation was also detected in two sisters from a consanguineous gypsy family (Miltenberger-Miltenyi et al., 2005).

COMPLEX NONCOMMUNICABLE DISEASES

A recent survey performed under the auspices of the Hungarian Prime Minister's Office and the Hungarian Academy of Sciences, with the purpose of summarizing the distribution of genetic mutations with major public health effect in Hungarian population, found that genetic research studies between 1995 and 2009 were conducted mainly on cardiovascular (26%), gastrointestinal (23%), and psychiatric (22%) diseases, while there were somewhat fewer on cancer (15%) and diabetes mellitus (12%) (Fiatal et al., 2009); here, we discuss the topic in this order.

CARDIOVASCULAR DISEASES

In 2003 a study was conducted to analyze the distribution of the platelet glycoprotein IIIa subunit A2 allele in different vascular subtypes of stroke disease in Hungary (Szolnoki et al., 2003). In the small vessel and mixed vascular pathology groups, the PLA2 allele frequency was similar to that in the controls, while it was found approximately twofold higher in patients with large vessel pathology (Szolnoki et al., 2003). In another study, the methylenetetrahydrofolate reductase (MTHFR) C677TT genotype, combined with the eNOS G894T or eNOS 894TT genotypes, occurred significantly more frequently in the subjects with ischemic stroke. The co-occurrence of the angiotensin-converting enzyme (ACE) D/D genotype and eNOS G894T or eNOS 894TT was calculated to be also more frequent in the ischemic stroke group (Szolnoki et al., 2005a). Two lymphotoxin-alpha (LTA) variants, namely the intron 1 252A-->G (252G) transition, which naturally coexists with an exon 3 804C-->A (804A) single-nucleotide polymorphism (SNP), have been investigated in ischemic stroke patients (Szolnoki et al., 2005b). The homozygous LTA allele with the 252G and 804C SNPs occurred more frequently in stroke patients. Specific sub-classification of the patients revealed an accumulation of these SNPs in large vessel pathology–associated cerebral infarction (Szolnoki et al., 2005b). A functional SNP of angiotensin II/angiotensin II type-1 receptor (AT1R) 1166C did not pose risk of stroke in Hungarian patients; however, in hypertensive smokers it was associated with an increased risk of ischemic stroke. Subgroup analysis revealed the same association for both small vessel and large vessel infarction (Szolnoki et al., 2006a). The angiotensin-converting enzyme (ACE) polymorphism ACE I/D can lead to an enhanced activity of the angiotensin II-AT1R axis; however, in a recent study the ACE D allele combined with the AT1R 1166C allele did not yield a risk of ischemic stroke. However, the co-occurrence of the homozygous ACE D/D and at least one AT1R 1166C allele was more frequent in the ischemic stroke group, and after specific subgroup analysis, this synergistic association was even stronger for small vessel ischemic stroke (Szolnoki et al., 2006b). Apolipoprotein (APO) A5 gene promoter region T-1131C polymorphism was associated with an increased level of triglycerides in large vessel–associated, small vessel–associated, and mixed group of ischemic stroke (Havasi et al., 2006). In another study the triglyceride levels of subjects carrying C56G allele were elevated. The serum total cholesterol levels did not differ between subjects with C or G alleles in each group. An accumulation of APOA5 56G allele was observed in the large vessel–associated stroke group compared to the healthy controls (Maasz et al., 2008b). In a recent study the IVS3+476A allele, but not the T1259C variant, showed significant accumulation in all stroke subgroups, too (Maasz et al., 2008a). The combination of the homozygous MTHFR 677TT genotype and at least one AT1R 1166C allele occurred frequently in the ischemic stroke patients. Specific sub-classification of the patients revealed an accumulation of this combination in small vessel–associated ischemic stroke (Szolnoki et al., 2007b). The galectin-2-encoding gene LGALS2 3279TT homozygote variant in combination with the LTA 252GG homozygote was significantly less frequent in ischemic stroke (Szolnoki et al., 2009).

The combination of the MTHFR 677 T and ACE D polymorphisms have been described as susceptibility variants for leukoaraiosis (Szolnoki et al., 2001, 2004). The KLC 1 56836 CC genetic variant was also described to increase the risk of the disease (Szolnoki et al., 2007a).

The presence of FXIII–A Leu34 in homozygous or heterozygous form did not change the risk of coronary artery disease (CAD); however, for patients with fibrinogen levels in the upper quartile, a statistically significant protection against myocardial infarction was conferred by carrying the Leu34 allele (Bereczky et al., 2007). The frequency of the homozygous carriers of the complement C4B*Q0 allele was significantly higher in CAD patients, and the frequencies of the TNFα -308A + C4A*Q0 allelic combinations and that of the TNFα -238A + C4A*6 were also significantly higher (Szalai et al., 2002). The prevalence of the MCP-1 -2518 G/G homozygotes was significantly higher among CAD patients, also (Szalai et al., 2002).

Neither of the ε4s of apolipoprotein E, the D allele of ACE, Leiden mutation of blood coagulation factor V, C677T mutation of MTHFR, PLA2 allele of platelet glycoprotein IIIa, nor F allele of C3 complement component gene variants were found to increase risk of coronary heart disease (Csaszar et al., 2001).

In preterm infants the homozygous carrier state of angiotensin II type 1 receptor A1166C SNP is protective against patent ductus Botall (Treszl et al., 2003), and the ACE polymorphism associated with an increased ACE activity is protective against heart failure (Nobilis et al., 2006). In acute renal failure, an SNP of heat shock protein 72 gene with lower HSP-70 levels increases the risk of disease (Fekete et al., 2003), while vascular endothelial growth factor SNPs can influence the risk of some congenital heart disease (Vannay et al., 2006).

GASTROINTESTINAL DISEASES

The possible effects of interleukin-23 receptor (IL23R) SNPs were investigated in an inflammatory bowel disease (IBD), namely Crohn's disease (CD). The prevalence of the homozygous rs10889677 AA and homozygous rs2201841 CC genotypes was increased in CD compared to a healthy control population, while the rs1884444 variant was found neutral. The association of three common NOD2/CARD15 susceptibility variants (Arg702Trp, Gly908Arg, and 1007finsC changes) and two functional polymorphisms of OCTN cation transporter genes (SLC22A4 1672C/T and SLC22A5 -207G/C mutations) was examined in pediatric and adult CD patients in 2006 (Bene et al., 2006). At least one CARD15 mutation was present in 52.6% of the children and in 34.5% of the adults compared to 14.3% in controls. Surprisingly, a strongly different mutation profile was detected in the pediatric versus adult patients. While the G908R and 1007finsC variants were 18.4% and 21.1% in the pediatric group, they were 1.82% and 11.8% in the adults, and were 1.02% and 3.06% in the controls, respectively. The R702W allele was increased approximately twofold in the adult subjects, while in the pediatric group it was only approximately 64% of the controls. No accumulation of the OCTN variants was observed in any patient group (Bene et al., 2006). Another study analyzed the possible influence of the previously mentioned functional variants of genes of OCTN cation transporters on the carnitine ester profile of patients with CD, with the result that the fasting propionyl-, butyryl- and isovalerylcarnitine levels were decreased; while the level of octenoyl-, myristoleyl-, palmitoyl- and oleylcarnitine were increased. After sorting the patients into SLC22A genotype-specific subgroups, no significant differences could be observed between them (Bene et al., 2007).

The prevalence of the +49A/G variant of the cytotoxic T-lymphocyte antigen 4 gene (CTLA4) was analyzed in patients with CD and ulcerative colitis, another form of IBD. The results showed that carriage of the +49G SNP in heterozygous or in homozygous form does not confer risk for CD (Magyari et al., 2007b). There was also an investigation recently on the possible effect of the novel IGR2096a_1 (rs12521868) and IGR2198a_1 (rs11739135) susceptibility loci on the development of IBD, in which the minor allele frequencies of IGR2096a_1 and IGR2198a_1 were increased in CD (Lakner et al., 2009). A study was conducted to verify the prevalence rate of the haplotype called TC, determined by combination of two functional alleles of OCTN cation transporter genes (SLC22A4 1672T and SLC22A5 -207C combination variants) in ulcerative colitis patients. There was no accumulation of the haplotype (Magyari et al., 2007a). Similar to CD, the carriage of the +49G CTLA4 SNP doesn't seem to confer risk for UC (Magyari et al., 2007b).

The frequencies of Arg64 allele in 3-BAR, -3826 A - G in UCP-1, -866G – A, and the 45 bp insertion of exon 8 in UCP-2 polymorphisms were not different in obese and healthy children (Bokor et al., 2007). The CD36 gene variations were associated with obesity and percentage of body fat in two independent populations of adolescents (Bokor et al., 2009).

AUTOIMMUNE DISEASES

In rheumatoid arthritis (RA) peptidylarginine deiminase type 4 (PADI4) gene variants (four exonic padi4_89: 163G/A, padi4_90: 245T/C, padi4_92: 335C/G, padi4_104: 349T/C and two intronic padi4_94: 17535226C/T and padi4_102: 17546809C/T) were examined in patients characterized for the presence of anti-CCP and rheumatoid factor. None of the PADI4 SNPs or haplotypes were accumulated (Farago et al., 2007). Three polymorphisms of IL23R were examined; an increased prevalence of the homozygous rs10889677 AA and homozygous rs2201841 CC genotypes was observed in RA. The rs1884444 mutation was found neutral (Farago et al., 2008). The prevalence of the C6607T SNP mapping to intron 1 of the SLC22A4 gene encoding the OCTN1 protein and the G24658C transversion in intron 6 of the gene encoding the RUNX1 transcription factor was also investigated. No statistically significant difference was found (Komlosi et al., 2008).

Mutations of the IL23R gene were examined in several other autoimmune diseases, also. A significant increase in the carriage of the T allele of rs11805303 and the A allele of rs1004189 in ankylosing spondylitis patients was observed, while for the rs10889677 variant, the prevalence of the AA genotype and the rs2201841 CC genotype showed a more than twofold increase in the patients. By contrast, the GA heterozygous genotype of rs11209026 variant showed a significant decrease in ankylosing spondylitis. Haplotype analysis revealed association of four IL23R haplotypes with the disease (Safrany et al., 2009b). The possible association of IL23R variants with Sjögren syndrome was analyzed in Hungarian patients for the first time in literature; however, no association was demonstrated (Safrany et al., 2009b). Similarly, when the same SNPs were examined in a group of systemic lupus erythematosus patients (Safrany et al., 2009a), no association was found. The rs10889677 A2370A genotype was significantly overrepresented among patients with relapsing/remitting multiple sclerosis (MS; Illes et al., 2008). The CT60 and +49A/G polymorphisms of the shared autoimmunity gene CTLA4 have been also tested in MS patients, but no association of these polymorphisms or respective haplotypes has been found (Greve et al., 2008). Both the GG genotype and the guanine allele (G) of TNF-alpha -376 polymorphism were detected significantly more often among patients with primary progressive MS compared with healthy controls (Losonczi et al., 2009).

In children, an association was detected between carrying G at –2518 of the CC chemokine MCP-1 gene regulatory region and the presence of asthma, and between asthma severity and homozygosity for the G allele (Szalai et al., 2001). Among the FcεRI-β Ile181Leu, a TNFα -308A, CCR2-64I, RANTES -403G, -28G, CCR5Δ32 polymorphisms, none associated with asthma (Nagy et al., 2002). Children infected with chlamydia pneumoniae–carrying variant mannose binding lectin alleles had a higher risk of developing asthma than infected children with normal mannose binding lectin genotype. This risk was especially high in children with chronic or recurrent infection (Nagy et al., 2003; Nagy et al., 2007).

In infants with necrotizing enterocolitis, the prevalence of an SNP with increased anti-inflammatory action (interleukin-4 receptor) was low (Heninger et al., 2002). Interleukin-18 SNPs may alter the severity of necrotizing enterocolitis; common cytokine and toll-like receptor SNPs have no role (Szebeni et al., 2006).

PSYCHIATRIC DISEASES

In major depressive disorder, a significantly higher frequency of the STin2.10/Stin2.10 homozygous genotype in the intron 2 of the serotonin transporter gene (termed STin2) was found (Sarosi et al., 2008). The brain-derived neurotrophic factor gene C270T polymorphism C/T genotype and the T allele were found significantly higher in schizophrenia patients (Szekeres et al., 2003).

The T/C mutation in intron 2 of CYP46 gene did not increase the susceptibility to late-onset sporadic Alzheimer's disease (Juhasz et al., 2005). Other results indicate that the PON2*S and ApoE4 alleles have interactive effect on the development of Alzheimer's disease (Janka et al., 2002).

CANCER

The first report on double heterozygote BRCA1 and BRCA2 mutations in a patient with breast and ovarian cancer was carried out on a Hungarian patient (Ramus et al., 1997). A high frequency of germline BRCA2 mutations among Hungarian male breast cancer patients was observed in 1999 by Csokay et al. Eighteen germline mutations were identified in the MLH1 and MSH2 genes in hereditary nonpolyposis colorectal cancer (HNPCC), of which sixteen were considered pathogenic; two were non-conservative missense alterations (Papp et al., 2007). Large germline deletions encompassing the last exons of the TACSTD1 gene co-segregated with HNPCC, also (Kovacs et al., 2009a).

Genetic analysis suggests that the P48T mutation of p16 is a strong melanoma-predisposing factor (Szell et al., 2007).

In children with acute lymphoid leukemia, the distribution of individual alleles and genotypes for MDR1 C3435T, and G2677T/A and the BCRP G34G and C421A, did not differ between patients and controls; however, the 2677G-3435T haplotype in the MDR1 gene occurred significantly more frequently, and the TC haplotype less frequently in patients (Semsei et al., 2008).

DIABETES MELLITUS AND METABOLIC DISEASES

Metabolic syndrome is known to mean a major risk for several cardiovascular and cerebrovascular diseases, such as coronary heart disease, acute myocardial infarction, and stroke. In Hungarian metabolic syndrome subjects, the triglyceride levels of patients with -1131C allele were significantly increased compared to the subjects with -1131T allele. In patients, the prevalence of the ApoA5 -1131C variant was increased compared to the healthy controls (Maasz et al., 2007). The IVS3+476A allele frequency was also increased in patients, whereas the 1259C allele frequency did not differ between the groups (Kisfali et al., 2008).

The insertion/deletion polymorphism of ACE was found to be associated with metabolic syndrome.

In the case of the CC allelic variant of the fructosamine-3-kinase gene rs1056534 polymorphism, type 2 diabetes was diagnosed at a later age than in cases of GC or GG variants (Mohas et al., 2009).

In children and adolescents with type 1 diabetes, the presence of the inflammatory allele of TNF-α gene is related to 24-hour ambulatory blood pressure values (Krikovszky et al., 2002). Investigation of G protein beta 3 subunit SNPs revealed no association between membrane transport and genotype (Deak et al., 2003). A study in diabetic children with different genotypes of lymphoid tyrosine phosphatase 22 gene (PTPN22) demonstrated the significance of PTPN22/Lyp 620Trp in altered lymphocyte activation characteristics (Aarnisalo et al., 2008).

GENETIC FACTORS IN COMMUNICABLE DISEASES

The Hungarian National Public Health and Medical Officer Service (Állami Népegészségügyi és Tisztiorvosi Szolgálat; ÁNTSZ) is a national institution financed by the public budget, which is responsible for controlling, coordinating, and supervising activities concerning public health, including environmental, settlement, food hygiene and nutrition, child and youth health, radiobiology and radiohygiene, chemical safety, epidemiology, health protection, and health services administration; it also controls several elements of the health service. The head of the service is the Chief Medical Officer of Hungary, who is directly under the direct control of the Minister of Health.

This system National Center for Epidemiology (Országos Epidomológiai Központ) collects the data weekly, and the published data are available in Hungarian (www.oek.hu/oekfile.pl?fid=2613). The year 2009 is shown in Table 116-3.

Scientific research is also highly centralized to assigned reference laboratories of particular pathogens. The majority of reference laboratories are within the National Center for Epidemiology (NCE) located in Budapest. Additional reference laboratories are found in selected regional institutes of the public health service network and university hospitals/health centers. Research interest in these laboratories mainly focuses on occurrence, epidemiologic monitoring, and molecular typing of human pathogens, although scientific interest in other areas (e.g., clinical and basic research) is also known (Bakos et al., 2007; Banyai et al., 2009; Deterding et al., 2008; Ferenczi et al., 2008; Jakab et al.. 2004; Vajo et al., 2009). Areas where strong records of publication can be seen include molecular epidemiological investigation of selected pathogens of man. It is a fortunate situation that many of the recently generated molecular epidemiological data have attracted international attention, and often resulted in invitations to research groups or principal investigators to participate in EU-funded projects and networks. Examples are monitoring of foodborne pathogens and postvaccination surveillance of childhood rotavirus infections (Banyai et al., 2007; Iturriza-Gomara et al., 2009; Kroneman et al., 2008; Lopman et al., 2004). Many of the communicable diseases are known to be of zoonotic origin (pathogens that jump from an animal species to human) and there are initiatives in Hungary to investigate related issues jointly, at the interface of human and veterinary medicine. Examples include detection and molecular characterization of hepatitis E virus (Reuter et al., 2009), rotavirus (Banyai et al., 2003; Matthijnssens et al., 2009), hantavirus (Jakab et al., 2007a, 2007b), tick-borne encephalitis virus (Holzmann et al., 1992), West Nile virus (Bakonyi et al., 2006), methicillin resistant *Staphylococcus aureus* (Juhasz-Kaszanyitzky et al., 2007), and *Escherichia coli* (Mag et al., 2009; Toth et al., 2009) from clinical specimens of humans and animals.

THE HEALTH DELIVERY SYSTEM AND PROVISION FOR GENETIC DISORDERS

As mentioned earlier, the National Health Insurance Fund (OEP) is the primary authority for covering healthcare expenditures in Hungary. During the past ten years, it received funding from three main sources: direct contributions from wages (from employers and from employees), general taxation, mainly to cover costs of people who are eligible but cannot contribute to the Fund, and assets that the central government made available from the central budget (approximately 25% of total). Hungary used 5.4% of its GDP (USD 18,754,000) in 2008; number of insured persons were 3.857 million. There were 34,793 doctors; 28,196 of them had a medical board exam certification, and a total of 27 persons had a human genetics qualifying exam certification in 2009—the majority of them had the old version of the certificate, and 17 of them had the new type.

In Hungary the financing of genetic tests is a method-based compensation: for the test of the diseases, fees for the genetic methods needed (PCRs, sequencings, etc.) are paid by the national insurance fund. The government's budget available for genome-based diagnostic tests is limited, and undergoes modifications yearly. As an example, in 2008 and 2009 the total number of PCR tests paid was 54,888 and 46,141, respectively.

Hungary has a professional society, the Hungarian Society of Human Genetics (established in 1968), and also

TABLE 116-2 SELECTED NOVEL MUTATIONS ASSOCIATED WITH MENDELIAN DISORDERS IN HUNGARIAN PATIENTS

	Disorder	Gene	Mutation	Family/patients	Reference
Hematologic disorders	FX deficiency	*FX*	G730A (G204R)	a one-year-old boy with subdural hematoma and bruising	(Bereczky et al., 2008)
	Factor V deficiency	*FV*	2952delT, 5493insG	male infant with subdural hematoma presented at birth	(Ajzner et al., 2002)
	Factor V deficiency	*FV*	G1651A (G493R)	a 34-year old female who had surgery-related remarkable blood loss at several occasions	(Kovacs et al., 2009)
	Glanzmann thrombasthenia	*GPIIb*	1619delC	a six-month-old boy homozygous	(Losonczy et al., 2005)
	Glanzmann thrombasthenia type II	*GPIIb*	C440G (L116V), 1772_1773insG (D560GfsX16), C2438A (H782N)	52-year-old patient with recurrent hemorrhagic symptoms	(Losonczy et al., 2007)
	Hemophilia A	*FVIII*	Large deletions involving intron 22	two patients	(Andrikovics et al., 2003)
	Hereditary angioedema	*C1INH*	G638A, C2238T, 2534_2535delCT, 2579_2620del42, G2533A, G2695A, 2696_2697insT, C4467T, A14224T, 14107delA, 16749_;16775dup, T16810A, C16885G	16 families	(Kalmar et al., 2003)
Immunodeficiency disorders	X-linked agammaglobulinemia	*BTK*	F30fsX52, Y39N, I355N, Q363X, R544T, L616V, I651S, and ivs8-1G>A	ten patients	(Toth et al., 2009b)
	Autoimmune polyendocrinopathy syndrome type I	*AIRE*	C449fsX479	one family	(Toth et al., 2009a)
	Autosomal recessive hyper-IgM syndrome	*AICDA*	E58K	one family	(Erdos et al., 2008a)
	Warts-hypogammaglobulinemia-infection-myelokathexis syndrome	*CXCR4*	S338X	one patient	(Alapi et al., 2007)
	Shwachman-Diamond syndrome	*SBDS*	N121T, R175W	one patient	(Erdos et al., 2006)
	X-linked lymphoproliferative syndrome	*SH2D1A*	G16D, C42Y	two families	(Erdos et al., 2005)
Dermatological diseases	Pseudoxanthoma elasticum	*ABCC6*	exon 24-25 deletion	one patient	(Katona et al., 2005)
	Hailey-Hailey disease	*ATP2C1*	1085insA, Q506X	two patients	(Racz et al., 2005b)
	Darier's disease	*ATP2A2*	559insT, 1288-6 A/G, 1320delT, 1821delC, A161D, R164S, T181R, Q790P, C909X	nine patients	(Racz et al., 2004, Racz et al., 2005a)
	Epidermolysis bullosa simplex	*KRT14*	N123K, R125G, V133L	three families	(Csikos et al., 2004)
	Epidemolytic palmoplantar keratoderma	*KRT9*	N160I	one family	(Csikos et al., 2003)
	Congenital ichthyosiform erythroderma	*TGM1*	Y503X, S669X	one patient	(Becker et al., 2003)
	Conradi-Hünermann-Happle syndrome	*EBP*	R147G	one patient	(Becker et al., 2001)

(*Continued*)

TABLE 116-2 (CONTINUED)

	Disorder	Gene	Mutation	Family/patients	Reference
Other Mendelian disorders	Cerebrotendinous xanthomatosis	*CYP27A1*	c.11_20dup	2 siblings with early-onset cataract	(Teszas et al., 2006)
	Duchenne/Becker muscular dystrophy	*DMD*	ex01ex30del, ex04ex09del, ex06ex12del, ex08ex30del, ex10ex12del, ex22ex41del, ex61ex64del	eight patients	(Piko et al., 2009)
	Fabry disease	*GLA*	A121fsX129, D165H, I289S, A291T	nine families	(Erdos et al., 2008b)
	Cartilage-hair hypoplasia syndrome	*RMRP*	236A>G, dupTACTACTCTGTGAA-12	one patient	(Erdos et al., 2008c)
	Gaucher disease	*GBA*	R48Q, V294fs	two patients	(Erdos et al., 2007)
	Biotinidase deficiency	*BTD*	E46X, T152P, R157C, N195S		(Milankovics et al., 2007)
	Juvenile hemochromatosis	*HJV*	G66X	one patient	(Janosi et al., 2005)
	Wilson's disease	*ATP7B*	N676I, S693Y, Y715H, M769L, W939C, P1273S, G1281D, G1341V	nine patients	(Folhoffer et al., 2007)
	Type I glycogen storage disease	*G6PC*	E110Q, D38V	one patient	(Parvari et al., 1997)

there is another professional board, called the Clinical Genetics Advisory Board of the Ministry of Health, headed by György Kosztolányi; this board is also available for consultations about questions that arise from different authorities, public or governmental.

THE PRACTICE OF GENETIC COUNSELING

In Hungary there are 20 genetic counseling outpatient clinics, which have contracts with the Hungarian National Health Insurance Fund, These units are located at or affiliated with the medical faculties of one of the four universities or major medical centers in Hungary; an additional five private units have all the necessary authorization to practice. In 2008 and 2009, a total of 22,260 and 23,026 visits at these outpatient clinics were recorded nationwide.

At the beginning, the national board recognition of clinical genetics (also medical or human genetics; these terms are used almost as synonyms) in Hungary was quite unique. Although there were well characterized, proper training schemes for almost all of the major medical subspecialties, the clinical genetics subspecialty was introduced only in 1978, and the candidates with general medical degrees, with a minimum postgraduate training period of four years, had to submit their curriculum vitae and list of publications, accompanied by a letter of recommendation from the head of the employing department, to the Board of the Postgraduate Medical University, the legal authority for the management of postgraduate training. Then, based on these credentials a committee decided the appropriateness for the qualification of the candidate. In 1985 there were only 25 registered human geneticists in Hungary; 16 of them had also other qualifications (PhD or DSc degrees).

The bases of the currently available formal procedure for medical doctors were introduced in 1999. Following the initiative by the European Society for Human Genetics in 2009, Hungary declared its official interest to reform the medical genetics training exam according to the pan-European model, with four years of training. Still, a special question is the future participation of the individuals with BSc-MSc-PhD qualifications in patient care and in laboratory diagnostics. There is a special 4-year training project for selected health or biology MSc holders that provides a training that qualifies eligible candidates to work in diagnostic laboratories.

The basic medical principle at the genetic outpatient clinics is the nondirective type of counseling. The popular professional and public view of nondirective genetic counseling limits the role of counselor to providing information to clients and assisting family members in making decisions in a fairly neutral fashion. This description of nondirective genetic counseling is generally believed to be incomplete. The recent, better understanding of what it means to respect autonomy shows that merely respecting client choices does not exhaust the counselor's duty. The relationship of the genetic counselor with the patient should also be governed by the counselor's commitment to the principle of beneficence. Thus, when nondirective counseling is reexamined in the mirror of both these principles, it becomes clear that there are cases in which counselors should attempt to persuade clients to reconsider their decisions. Such attempts are still consistent in the Hungarian practice with nondirective counseling, because while respecting the clients' decision-making authority, they insure that clients act with full knowledge of the moral consequences of their decisions.

In Hungary, anyone can apply for an appointment at the genetic counseling units for examination; the costs of the interviews and laboratory genetic tests are paid in full by the Hungarian National Health Insurance Fund to insured individuals. In Hungary there is no insurance-covered preimplantation genetics laboratory test available. The Down syndrome cytogenetic screening, from amnio or chorionic biopsy, or tested by any other method, is covered by insurance in women over 35 years age.

CURRENT GENETIC/GENOMIC RESEARCH INFRASTRUCTURE AND STRATEGIES

The "Genomic Consortium for Human Health" was formed in 2002 as an EU6-funded project with the goal to establish a well-functioning consortium, to collect and disseminate information of human genetic and genomic research test projects, and other programs existing, available, and planned in Hungary, on a newly established website (www.humangenom.hu), and to provide situation analysis and define strategic initiatives for human genomic research in Hungary. All available information related to clinical genetic testing, human genome projects, and the technical and methodological potential in Hungary were collected. The Consortium has recognized the immediate need for a web-based virtual net of Hungarian biobanks, and launched the www.biobank.hu homepage. The consortium also made a preliminary list of genomic facilities and activities (Figure 116-1). The genetic and genomic research infrastructure is concentrated mainly in the four universities having medical faculties (Figure 116-1), or in national institutions, which usually also have some kind of affiliation to one or more of these universities. For disease diagnostic purposes, the routine available methods consist of the different basic chromosome techniques, bandings, and different types of FISH methods; at the molecular level, PCR-based techniques and sequencing/fragment analysis are the basic methodologies. For research purposes almost all methods, including some sophisticated mass spectrometry facilities, or the recent deep sequencing apparatuses, like the systems developed by Applied Biosystems and Solexa, are also available.

Hungary is an official partner in the European Strategy Forum on Research Infrastructures (ESFRI), which is a known central strategic instrument to develop the scientific

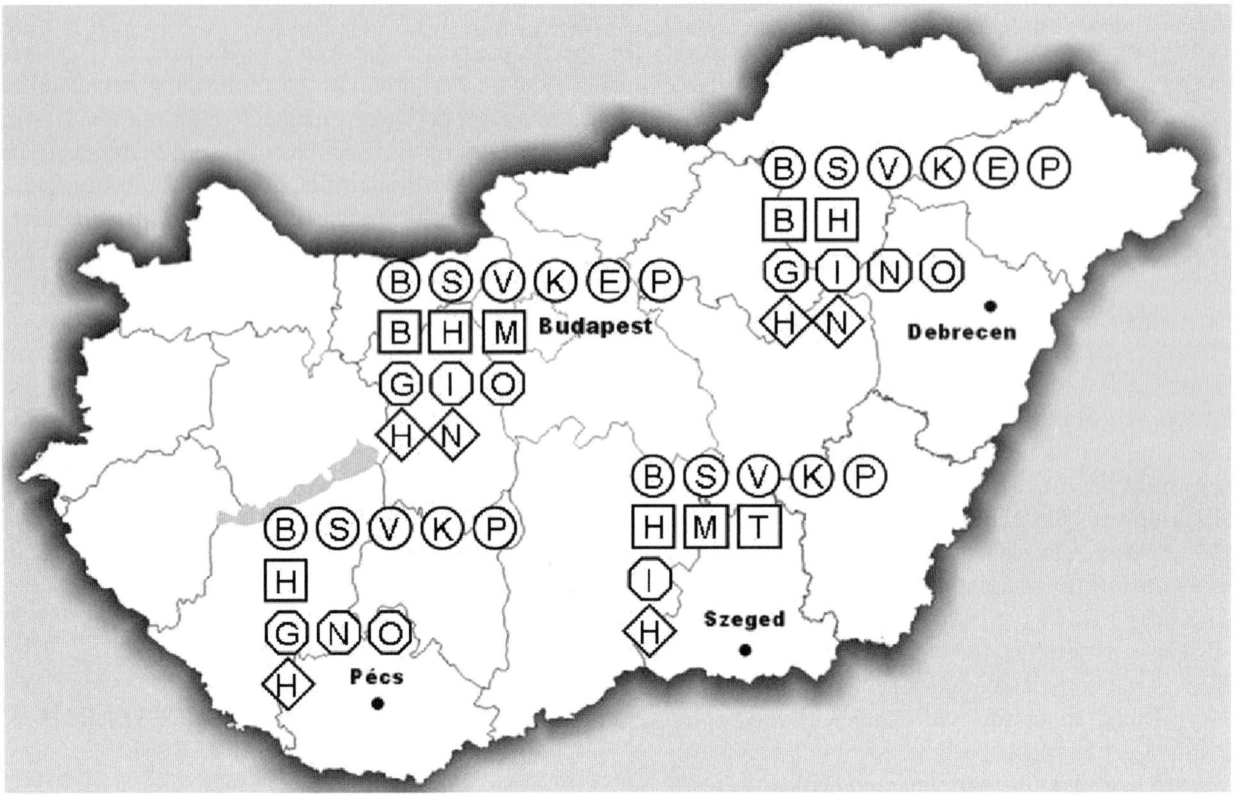

(B) -biobanks (S) - sequencing capacity (V) - analysis of genome variability (K) - gene expression analysis

(E) - epigenomics (P) - proteomics [B] - bioinformatics [H] - human genomics and postgenomics

[M] - genomic and postgenomic research on model organisms [T] - development of genomic and postgenomic technological platforms

(G) - pharmacogenomics (I) - immunogenomics (N) - public health genomics (O) - oncogenomics

◇H - genomics training part of the gradual and postgradual curriculum ◇N - international training courses or conferences in genomics and postgenomics

Figure 116-1 Genomic research and development activities in Hungary

integration of Europe and to strengthen its international outreach. The major goal of the ESFRI is to achieve competitive and open access to high-quality research infrastructure that supports and benchmarks the quality of the activities of European scientists, and attracts the best researchers from around the world. However, contrary to other European nations, Hungary agreed within the mission of ESFRI to support a coherent and strategy-led approach to policymaking on research infrastructure in Europe, and to facilitate multilateral initiatives leading to the better use and development of research infrastructure, at EU and international levels. Domestically an attempt was made to achieve this goal by the Government Decree of 1066/2007 (VIII. 29.) for the period of 2007 to 2010, with foresight mapping of the full and correct Hungarian research infrastructure of strategic importance. This goal is planned to be achieved by the NEKIFUT project, which is supported in part by the National Office for Research and Technology (NKTH; www.nkth.hu), the central and major Hungarian research and fund-handling authority. Another recent formation is the Genetic National Technology Platform (GNTP), which is also funded by the National Office for Research and Technology, which was established with the long-term objective to focus on support of research organizations, biotech, the healthcare and pharmaceutical industry, agriculture, technological development, and IT technology, to optimize the utilization of the results of genome technology and to facilitate what all these branches could contribute to the development of genome technology. Possibilities related to technological platforms and optimal and proportional development can be utilized according to the interest and expectations of all interested parties (http://www.genomika.net/gntp/fooldal.html).

Hungarian authorities already had declared their interest in the task of rare diseases at many forums, and in 2009 the National Research Center for Rare Diseases was established at the University of Pécs. The goal of this center is to coordinate the development of the existing and the future networking of excellence centers with rare disease diagnostic and treatment activities.

THE PUBLIC AND POLITICAL PERCEPTION OF HUMAN GENETIC/GENOMIC DISEASE

Generally, the exponential growth of human genetics and genomics ab ovo creates a gap between science and the public or political perception or understanding—the question is the size of those gaps. At this moment Hungary does not have an institutionalized, well established, deliberative mechanism to involve the broader segments of society into a potentially lengthy education and discussion of issues related to the current status of research, or topics of interest to contemporary genetics, the developing areas of current human genetics or genomic technologies, including even ethics and jurisdiction issues. As a consequence, there is a clear disconnect between the current state of the art and the current public perception of genetic diseases, and in public understanding of technologies, methods, and new areas of postgenomic trends, including the ethical and social implications. These are still discussed mainly in depths in professional groups, and the spread the information toward the public is often much poorer than would be appropriate.

Since the medical applications of genomics will reflect several new approaches to patient care, and knowing that the level of knowledge among the public about the principles of genetics is limited, it is important to know how these issues are viewed by different communities. Over the last 10 years several surveys have been carried out to obtain information of this kind. *Eurobarometer*, published in June 2005 (European Commission. Directorate General Press and Communication: Social values, Science and Technology. Special Eurobarometer 225, June 2005) presented the results of a survey on public attitudes toward social values, science, and technology. Question 11 of this survey asked whether the decisions about science and technology should be based primarily (1) on an analysis of the risks and benefits involved, or (2) on the moral and ethical issues involved. In Greece, most citizens (70%) believed that the analysis of the risks and benefits involved is most important; it was followed by Hungary with 67%. New EU member states can be found among the countries with the lowest number of citizens who choose answer 1—Poland (41%), Malta (41%), Lithuania (41%), Latvia (42%), and Estonia (43%). There were also some older member states among these, such as Ireland (41%), Luxembourg (45%) and Finland (46%). The answer to question (16.1) reflected an even more extreme picture. Here, EU citizens are asked whether we have the right to exploit nature for the sake of human wellbeing. Countries with the most citizens who agreed with this statement were Slovakia (77%), Poland (70%), and France (67%); countries with the lowest number of citizens who agreed were Hungary (10%), Austria (12%), and Germany (12%). When asked whether the next generation will enjoy a better quality of life than we do now (Question 7.2), an overwhelming majority of citizens in Lithuania (87%), Estonia (82%), Latvia (78%), Hungary (75%), Poland (75%), and the Slovak Republic (74%) said yes; while most older member states, like France (34%), the Netherlands (35%), Luxemburg (37%), and Sweden (42%) were more pessimistic.

As can be seen, the results of these surveys showed a marked diversity of response among different populations. However, perhaps looking at the trends over time there appears to be both an increase in knowledge but, at the same time, an increase in concern about these activities. Unfortunately, the global monetary crisis in 2009 had a severe effect on the Hungarian economy and health system; restrictions and regressions were seen in many areas of education and healthcare, none of which had a positive effect on the broader nationwide recognition of the importance of genetics.

As an example to illustrate the difficulty and complexity of the process, the process of adopting Parliamentary Act XXI of 2008—on the protection of human genetic data and the regulation of human genetic tests and research, and the regulation of biobanks (hereinafter referred to as "Genetic Act")—took more than five years. In this period of time the debate did not enrich the preliminary draft with more thoughts, and the text did not become more detailed; instead, several problematic issues were simply cut out from the text (Sandor et al., 2009). As a result, instead of making a much more comprehensive law on the use and processing of human genetic data in various fields, the law adopted by the parliament focused on major points of the diagnostic and research procedures, and on the diagnostic and possible research use of samples deposited in biobanks. In the earlier course of the legislative debate, lawyers and data protection activists were also mobilized and advocated for newer and stronger guarantees for the protection of genetic data, even if a wide majority of the public even did not know what these really meant. False fears of genetic discrimination, exploitation, or trafficking in data and materials to foreign countries, were also major concerns in the political discussions, so it seems that this law served as a kind of learning exercise for the wider public on the contemporary practical, legislative, societal, and ethical issues in genetics.

A part of this Genetic Act focuses on the regulation of the biobanks (Sandor et al., 2009). Since the original Genetic Act concept was proposed in 2003, several biobanks had already been initiated and had started to operate, and the scientists started to worry about the legal vacuum. Biobanks were established mainly attached to teaching hospitals or departments at universities and medical centers, systematically collecting disease-specific biological samples. As a result of this new biobanking trend observed in Hungary, the scope of the law was slightly shifted from the issues of genetic research toward issues related to the biobanks. The question of data protection was so dominant in the debate that some other, broader human rights aspects were entirely left out of the final version of the law. By focusing on data protection and creating stronger protections for the genetic data, some other elements of the ethical/legal framework were sacrificed, such as the prohibition of discrimination based on genetic characteristics. Instead, this was referred to general law and with relatively little public consultations conducted on biobanks in Hungary.

Several preliminary variants of the bills failed during the legislative process starting in 2003, before the current version—on the protection of personal genetic data, genetic diagnoses, and genetic research—was finally adopted (Sandor et al., 2009). At the beginning of the legislative process in 2003, the Minister for Health launched an invitation for public discussion, and published the proposal on the Ministry of Health website; principally the terms of discussion were based on the Oviedo Convention. It should be noted that this one was the first public consultations in the field of biotechnology or genetics and law in Hungary. During the procedure several open hearing days were organized in the Hungarian Parliament around real, burning, or virtual social issues, including questions of possible health insurance reform, for instance. Later, a Hungarian medical journal, the *Lege Artis Medicinæ* also published the first draft of the legislation and called on experts and the public for consultation. The proposal was also made available to anyone as an open announcement on the website of the Ministry of Health. Some public organizations and civil liberties organizations, such as the Hungarian Civil Liberties Union, participated in the debate that resulted in the amended version of the early drafts. In addition to the abovementioned actors, the Parliamentary Commissioner on Data Protection, the Health Science Council, members from the Hungarian Academy of Sciences, leading scientific experts, and medical and other professional organizations participated in the debate and submitted their written concerns or opinions with proposed amendments. Uncertainties about the various kinds of biobanks, and a fear of opening a debate also on preimplantation genetic testing, postponed the approval of the bill. Some opponents were against this law because they saw it as a sign of genetic exceptionalism, while the Ministry for Justice repeatedly raised concerns for even stronger protection of genetic data. Finally, the parliament started the discussion of the bill in March 2008, and after many years of debate outside of the parliament, a very short and simplified version of the original concept was adopted in April, 2008. This law ensures the strong protection of genetic data and, as such, follows the tradition of the data protection norms in Hungary (Sandor et al., 2009; Sandor, 2004, 2009).

DEVELOPMENTS IN GENETIC/GENOMIC BIOTECHNOLOGY FOR HEALTH IMPROVEMENT

Domestic and international governmental or private grants were used for biotechnology research purposes, and a few large research centers were built using these sources. These centers are affiliated to the universities, and involved in basic and translational research projects related to different areas of the health sciences. In Hungary there is a centralized biobank network, which is a member of the preparatory phase if the Biobanking and Biomolecular Resources Infrastructure (BBMRI; www.bbmri.eu).

Further, due to successfully implemented science funding, some technical innovations have also been introduced. Enzymes, kits, and small equipment are produced mainly by some SMEs (Small and medium enterprises)

ETHICAL, LEGAL AND SOCIETAL ISSUES

The Parliamentary Act XXI of 2008 on the protection of human genetic data and the regulation of human genetic studies, research, and biobanks entered into force on July1, 2008 (Sandor et al., 2009). One may state that it has become shorter and simpler than in the original policy paper in a desperate effort to avoid sensitive issues. Thus, the law addresses the use of genetic information only in the very narrow biomedical sector: namely, the fields of genetic testing, screening, and research. The law restricts the use of genetic data only in this biomedical context, so even in the lack of regulation of the broader use of genetic data based on the Act, genetic data processed for diagnostic or research purposes cannot be disseminated for the purposes of insurance. In Article 1 of the Genetic Act the purpose of the law is stated as "to lay down rules on human genetic tests and screening, and human genetic research, the conditions and purposes of the treatment of genetic data and rules on biobanks." The Act applies to genetic sampling procedures for diagnostic human genetic tests, and human genetic research performed under the Act in the territory of the Republic of Hungary, and the processing of genetic data irrespective of the place of sampling (Sandor et al., 2009; Sandor 2004, 2009).

According to Article 30 of the Act, all collections of cells, tissue samples, or other human biological materials established for genetic research or genetic diagnostics prior to the entry into force of the Act, excluding tissue banks for transplantation purposes and gamete banks for gamete donation and deposit, shall comply with the rules on biobanks of the Act by October1, 2009. Genetic samples and data anonymized in accordance with the requirements of the Genetic Act, even before its entry into force, may be treated in line with the Act for human genetic research and study. The Genetic Act is certainly not the last step in the Hungarian regulation of biobanks. Article 31 of the Act appoints the Minister of Health, who is empowered to lay down in Ministerial Decrees (a) detailed rules of genetic tests and screenings; (b) minimum conditions of the operation of laboratories engaged in diagnostic genetic tests and screenings; (c) accreditation and quality control requirements in laboratories engaged in genetic testing and research; (d) detailed rules of the storage, treatment, and processing of genetic samples and genetic data and the encoding of genetic data; (e) professional minimum conditions of the operation of the biobanks, and the aspects of authorization; and (f) formal requirements for the declarations of different consents, including consent before genetic tests and for biobanking of samples (Sandor et al., 2009).

It looks clear that the enumeration of domestic legal instruments other than the Genetic Act will follow the hierarchy of laws in the Hungarian legal system. At the moment, there are also further Acts applicable also to the Genetic Act: Act XLVII of 1997 on the Processing and Protection of Health Care Data and Associated Personal Data (hereinafter referred to as "Health Care Data Act") being more specific, and Act LXIII of 1992 on the Protection of Personal Data and Public Access to Data of Public Interest (hereinafter referred to as "Data Protection Act"). As background legislation, the Act CLIV of 1997 on Health Care (hereinafter referred to as "Health Care Act") and the Act IV of 1959 on the Civil Code of the Republic of Hungary (hereinafter referred to as "Civil Code") also provided guidance. In reverse chronological order, the following decrees are related to the Genetic Act: Ministerial Decree 3/2005. (II. 10.) of the Ministry of Health on Standards of Quality and Safety for the Collection, Testing, Processing, Storage and Distribution of Human Blood and Blood Components, and Certain Technical Requirements; Ministerial Decree 76/2004 (VIII. 19.) of the Ministry of Health, Welfare and Family on the Detailed Rules of the Determination, Collection and Processing of the Scope of Certain Specialized (Health, Professional) Data Incapable of Identification; Ministerial Decree 23/2002 (V. 9.) of the Ministry of Health on the Medical Research Conducted on Humans; and Ministerial Decree 62/1997(XII. 21) of the Ministry of Welfare on the Processing of Health and Related Personal Data (Sandor et al., 2009; Sandor 2004, 2009).

Hungary has special organizations to elaborate ethical guidelines and to review research programs. Each research protocol intending to involve human beings has usually to be reviewed by two levels of ethics committees. In this process, two different levels of ethics committees are involved: the national and the regional levels. At the national level, organized under the Health Science Council (ETT; founded in 1951), there are three different National Research Ethics Commissions: the Scientific and Research Ethics Commission (TUKEB), the Clinical Pharmacology Ethics Commission (KFEB), and the Human Reproduction Commission (HRB). In addition, the Health Ministry decree 23/2002 (V/9) on biomedical research involving human beings established 12 regional research ethics boards (REKEB) throughout the country. It depends on the nature of the research whether the protocol has to be reviewed on a national or just on a regional level; but all protocols that need national permission must have permission from the regional board prior to submission at the national level. Thus, according to the act, the review of research protocols carried out by the different levels of ethics commissions includes an evaluation both from a scientific and from an ethical point of view. The composition of the ethics commissions has to be multidisciplinary: in addition to medical professionals, representatives of non-healthcare professions (ethicist, lawyer, clergyman, and psychologist) are also members of the commissions. Hospitals also have local ethics committees; their competency is restricted to providing practical advice to directors about the local research activity; however, their permission is not enough itself to conduct research (Sandor et al., 2009; Sandor 2004, 2009).

REGULATION OF HUMAN GENETICS AND GENOMICS IN THE REGION

In the previous sections the Genetic Act was introduced. Here, some special aspects are discussed.

According to Article 6 of the Genetic Act, a crucial point is that the subjects have to be informed in genetic counseling units prior to sampling, by specialized medical experts, about the purpose of the sample collection, advantages and risks of participating or not participating in the study, possible consequences to the person concerned about the outcome, methods of the storage of the genetic sample and data, possible identification of genetic samples and data stored, and about the available possible formats of storage. When the genetic sample is used for research, the person concerned shall be informed about the provisions of Article 159 of the Health Care Act before signing their consent, on methods of the storage of the genetic sample and data, possible identification methods of genetic samples and data stored in different formats, and the inclusion into an archived collection of the genetic sample in the absence of another declaration by the person concerned (Sandor et al., 2009).

The person concerned shall be informed about the human genetic study in a consultation that is specifically adjusted to his or her needs. During the consultation the person concerned shall be assisted in the long-term processing of the possible consequences of the result and the choice of optimum treatment opportunities. Moreover, the "right not to know" is also respected by the law, which says that the person concerned may withdraw from knowing their genetic data by a declaration. The declaration is subject to withdrawal any time without any restriction or consequence. The person concerned shall be informed about this right, too. According to Article 8 Section 1, before a genetic sampling it is necessary to obtain the written consent of the person entitled to disposal, based on detailed information, irrespective of the purpose of the treatment of the genetic data. The Hungarian Genetic Act is built on the respect of self-determination of the individual concerned; Articles 8 to 11 deal with this right in detail. In order to be valid the consent has to include the following elements:

a. the consent of the person concerned to the following:
   - aa. taking a genetic sample from him or her,
   - ab. storing the genetic sample and the derived genetic data in a biobank and freely forwarding it to another biobank,
   - ac. storing the genetic sample and the derived genetic data in an archived collection;

b. declaration of the person concerned to the effect that they give their consent to:
   - ba. the use for the primary purpose of sampling,
   - bb. the use for any purpose under the Biobank Act apart from the provisions of point ba), and
   - bc. the use for research of the genetic sample and data only;

c. the declaration of the person concerned according to which he or she gives their consent to the storage of the genetic sample and data together with personal identification data, or the storage of the same in an encoded or, in the case under subpoint bc) of point b), in a pseudonym or anonymized format;

d. the declaration of the person concerned of having received and understood the information on the purpose of the sampling, advantages and risks of participating or not participating in the study, possible consequences to the person concerned and their close relatives of any outcome, methods of the storage of the genetic sample and data, possible identification of genetic samples and data stored in different formats; furthermore, the information under Article 159 Section (3) of the Health Care Act, and finally information about the details as to who has access to his or her genetic data (Sandor et al., 2009).

During the storage of the encoded genetic sample or data, the person concerned may be identified only for the interest of the person concerned or his or her close relative according to Article 8 of the Genetic Act. In the absence of a declaration when the person concerned does not expressly refuse it, the genetic samples and data may be placed in an archived collection in anonymized format. With Article 9 in line with genetic sampling from a deceased person, the study of a genetic sample stored together with personal identification data, or a sample which has been encoded, their use for human genetic research or the use of derived genetic data is possible only when the deceased person did not make a declaration objecting to it in their lifetime. In case of verbal consent, the declaration shall be documented in writing with the signature of two witnesses proving the existence of the oral declaration. An oral declaration made with the medical practitioner, signed by two witnesses, or a declaration at his or her disposal otherwise, shall be recorded in writing in the health documentation by the medical practitioner or shall be attached to the latter (Sandor et al., 2009).

According to Article 10 Section (1) the person concerned may withdraw his or her consent to the use of their genetic data being stored together with his or her personal identification data, code, or pseudonym at any time. For a declaration of withdrawal, the person concerned may request that the genetic sample and all the derived genetic data be destroyed. According to Article 10 Section (2), the genetic sample and data shall be destroyed immediately, but within 8 days at the latest for sample and data used for human genetic research purposes only, and following the 30th day after the submission of the request for samples and data stored for genetic research, within 45 days following the request. Following the receipt of the request, genetic samples and data may be released in the interest of a close relative only (Sandor J et al., 2009).

According to Article 7, when necessary for the prevention of an illness of a close relative, the knowledge of the nature of their illness, their therapy, or the assessment of the risk of illnesses for their descendants, the genetic data may be communicated to a close relative as well. For this purpose, the medical geneticist must call to the attention of the close relative known by them the necessity of knowing genetic data in the framework of a genetic counseling, or shall initiate the involvement of the direct relative into the genetic counseling with the person concerned.

Samples can be taken from persons under 14 years, i.e., persons without legal competency, or from persons between 14 and 18 years of age, who, in the Hungarian system are persons with limited legal competency. Similarly, samples can also be taken from persons with mental problems. In all these cases the guardian(s), for children typically the parents, give consent (Sandor et al., 2009; Sandor 2004, 2009).

TABLE 116-3 MONTHLY REGISTERED INFECTIOUS DISEASES IN HUNGARY (2008)

Disease	I.	II.	III.	IV.	V.	VI.	VII.	VIII.	IX.	X.	XI.	XII.	Total
Botulism	-	-	3	-	-	-	-	-	-	-	-	-	3
Typhus abdominalis*	-	-	-	-	-	-	-	-	-	-	1	-	1
Paratyphus*	-	-	-	-	-	-	-	1	1	-	-	-	2
Salmonellosis	210	335	273	369	757	840	927	826	945	534	843	307	7166
Shigellosis	-	2	2	-	3	49	4	4	3	5	6	-	78
Amoebiasis	4	2	2	3	5	-	1	1	3	-	6	2	29
Dyspepsia coli	3	1	4	1	2	4	7	-	3	6	3	3	37
Campylobacteriosis	280	326	283	336	557	632	646	547	583	508	522	343	5563
Yersiniosis	3	4	7	5	1	1	3	-	3	4	6	3	40
Cryptosporidiosis	1	-	-	-	-	1	2	1	2	1	2	-	10
Giardiasis	8	20	14	14	15	19	12	8	8	4	11	5	138
Enteritis infectiosa	2657	3927	4404	3731	2744	2306	2328	2165	2599	2718	3189	2832	35,600
Hepatitis A	13	4	8	11	5	7	18	25	16	18	26	17	168
Hepatitis B	6	10	11	11	8	6	6	6	7	4	7	6	88
Hepatitis C	5	-	3	5	4	3	4	5	1	2	1	1	34
Hepatitis E	2	5	2	3	-	2	5	3	1	3	4	2	32
Hepatitis infectiosa	-	-	1	-	1	1	2	-	1	2	-	2	10
AIDS	2	3	2	4	1	1	2	3	1	-	1	3	23
Pertussis	3	-	1	1	10	6	6	3	1	2	-	-	33
Scarlatina	546	4488	454	363	287	239	65	41	108	185	229	309	3274
Parotitis epidemica	8	3	1	1	-	1	-	-	-	-	-	-	14
Varicella	4161	3849	5012	4869	5107	3414	866	240	489	1653	2147	4605	36,412
Mononucleosis infectiosa	82	87	86	102	93	94	93	93	106	101	91	82	1110
Keratoconjunctivitis epid.	-	1	-	-	-	-	-	-	-	4	1	-	6
Legionellosis	3	1	1	1	2	3	4	1	5	2	1	3	27
Staphylococcosis	-	-	-	-	-	-	-	-	-	3	2	1	6
Meningitis epidemica	7	-	5	5	2	3	-	1	1	8	2	-	34
Hemophilus meningitis	1	-	-	-	2	-	1	-	1	-	-	1	6
Pneumococcus meningitis	7	6	5	9	6	4	1	4	8	4	6	5	65
Meningitis purulenta	11	13	11	8	11	9	7	15	10	11	3	6	115
Meningitis serosa	4	1	4	7	8	11	15	9	6	10	6	3	84
Encephalitis info	5	1	4	3	9	5	5	5	6	8	1	4	56
Tick-borne encephalitis	-	1	-	2	24	23	12	7	2	5	1	-	17
West Nile fever	-	-	-	-	-	-	-	9	8	2	-	-	19
Creutzfeldt–Jakob disease	7	4	1	3	2	1	-	4	-	2	1	-	25
Lyme disease	23	11	21	63	256	499	419	227	132	81	59	20	1811
Listeriosis	3	-	-	1	5	-	2	1	2	2	2	1	19
Leptospirosis	-	1	-	1	-	1	3	7	1	1	-	-	15
Tularemia	1	1	1	1	3	4	3	4	-	4	3	-	25
Tetanus	-	1	-	-	-	2	-	-	-	1	-	-	4

(Continued)

TABLE 116-3 (CONTINUED)

Disease	I.	II.	III.	IV.	V.	VI.	VII.	VIII.	IX.	X.	XI.	XII.	Total
Ornithosis	-	-	-	1	-	1	-	1	1	2	3	-	9
Q-fever	2	1	3	1	1	-	1	1	1	-	-	-	11
Hemorrhagic fever virus	1	1	-	-	1	1	-	2	-	2	-	1	9
Malaria*	-	-	-	1	-	-	-	2	-	1	-	1	5
Toxoplasmosis	7	9	9	8	7	7	6	3	8	10	7	3	84
Echinococcosis	-	-	-	-	1	1	2	1	1	1	-	-	7
Taeniasis	-	-	-	-	-	-	1	-	-	1	-	1	3
Trichinellosis	-	1	-	1	1	2	-	-	-	-	-	-	5
Ancylostomasis	-	-	-	-	1	-	-	-	-		-	-	1
Stronqyloidosis	-	-	-	1	-	-	1	-	1	-	-	-	3

* imported cases

ACKNOWLEDGMENTS

The author is indebted to Róza Ádány, Krisztián Bányai, Judit Bene, Zsuzsanna Bereczky, Judit Béres, Márta Czakó, Andrew Czeizel, Éva Erhardt, András Falus, György Fekete, Gyula Hadlaczky, Kinga Hadzsiev, Zsolt Illés, Veronika Karcagi, Sarolta Kárpáti, Richárd Kellermayer, Péter Kisfali, György Kosztolányi, Anita Maász, László Maródi, Dénes Molnár, László Muszbek, Edit Oláh, Zoltán Papp, Noémi Polgár, István Raskó, Mária Sasvári-Székely, Enikő Sáfrány, Judit Sándor, Andor Sebestyén, Csilla Sipeky, Csaba Szalai, Márta Széll, Zoltán Szolnoki, László Szőnyi, Attila Tordai, Beáta Tóth, Tivadar Tulassay, Zsolt Tulassay, and Barna Vásárhelyi for their help provided in the manuscript preparation.

Dedication: To the memory of Károly Méhes (1936–2007). We shall continue in his spirit.

REFERENCES

Aarnisalo J, Treszl A, Svec P, et al. (2008). Reduced CD4+T cell activation in children with type 1 diabetes carrying the PTPN22/Lyp 620Trp variant. *J Autoimmun* 31(1):13–21.

Ajzner EE, Balogh I, Szabo T, Marosi A, Haramura G, Muszbek L. (2002). Severe coagulation factor V deficiency caused by 2 novel frameshift mutations: 2952delT in exon 13 and 5493insG in exon 16 of factor 5 gene. *Blood* 99(2):702–705.

Alapi K, Erdos M, Kovacs G, Marodi L. (2007). Recurrent CXCR4 sequence variation in a girl with WHIM syndrome. *Eur J Haematol* 78(1):86–88.

Andrikovics H, Klein I, Bors A, et al. (2003). Analysis of large structural changes of the factor VIII gene, involving intron 1 and 22, in severe hemophilia A. *Haematologica* 88(7):778–784.

Bajnoczky K, Mehes K. (1988). Parental centromere separation sequence and aneuploidy in the offspring. *Hum Genet* 78(3):286–288.

Bakonyi T, Ivanics E, Erdelyi K, et al. (2006). Lineage 1 and 2 strains of encephalitic West Nile virus, central Europe. *Emerg Infect Dis* 12(4):618–623.

Bakos A, Banati F, Koroknai A, et al. (2007). High-resolution analysis of CpG methylation and in vivo protein-DNA interactions at the alternative Epstein-Barr virus latency promoters Qp and Cp in the nasopharyngeal carcinoma cell line C666-1. *Virus Genes* 35(2):195–202.

Banyai K, Bogdan A, Domonkos G, et al. (2009). Genetic diversity and zoonotic potential of human rotavirus strains, 2003–2006, Hungary. *J Med Virol* 81(2):362–370.

Banyai K, Deak J, Gray J, et al. (2007). [EuroRotaNet—European rotavirus strain surveillance network established with Hungarian participation]. *Orv Hetil* 148(43):2043–2045.

Banyai K, Gentsch JR, Glass RI, Szucs G. (2003). Detection of human rotavirus serotype G6 in Hungary. *Epidemiol Infect* 130(1):107–112.

Becker K, Csikos M, Horvath A, Karpati S. (2001). Identification of a novel mutation in 3beta-hydroxysteroid-Delta8-Delta7-isomerase in a case of Conradi-Hunermann-Happle syndrome. *Exp Dermatol* 10(4):286–289.

Becker K, Csikos M, Sardy M, Szalai ZS, Horvath A, Karpati S. (2003). Identification of two novel nonsense mutations in the transglutaminase 1 gene in a Hungarian patient with congenital ichthyosiform erythroderma. *Exp Dermatol* 12(3):324–329.

Beke A, Joo JG, Csaba A, et al. (2009). Incidence of chromosomal abnormalities in the presence of fetal subcutaneous oedema, such as nuchal oedema, cystic hygroma and non-immune hydrops. *Fetal Diagn Ther* 25(1):83–92.

Bene J, Komlosi K, Magyari L, et al. (2007). Plasma carnitine ester profiles in Crohn's disease patients characterized for SLC22A4 C1672T and SLC22A5 G-207C genotypes. *Br J Nutr* 98(2):345–350.

Bene J, Magyari L, Talian G, et al. (2006). Prevalence of SLC22A4, SLC22A5 and CARD15 gene mutations in Hungarian pediatric patients with Crohn's disease. *World J Gastroenterol* 12(34):5550–5553.

Bereczky Z, Balogh E, Katona E, et al. (2007). Modulation of the risk of coronary sclerosis/myocardial infarction by the interaction between factor XIII subunit A Val34Leu polymorphism and fibrinogen concentration in the high risk Hungarian population. *Thromb Res* 120(4):567–573.

Bereczky Z, Bardos H, Komaromi I, et al. (2008). Factor XDebrecen: Gly204Arg mutation in factor X causes the synthesis of a non-secretable protein and severe factor X deficiency. *Haematologica* 93(2):299–302.

Boffetta P, van der Hel O, Norppa H, et al. (2007). Chromosomal aberrations and cancer risk: Results of a cohort study from Central Europe. *Am J Epidemiol* 165(1):36–43.

Bokor S, Csernus K, Erhardt E, Burus I, Molnar D, Decsi T. (2007). Association of n-6 long-chain polyunsaturated fatty acids to -866

G/A genotypes of the human uncoupling protein 2 gene in obese children. *Acta Paediatr* 96(9):1350–1354.

Bokor S, Legry V, Meirhaeghe A, et al. (2009). Single-nucleotide Polymorphism of CD36 Locus and Obesity in European Adolescents. *Obesity (Silver Spring).*

Cavalli-Sforza LL, Menozzi P, Piazza A. (1995).. Genetics and the origin of European languages. *Proc Natl Acad Sci U.S.A.* 92(13):5836–5840.

Cruciani F, La Fratta R, Trombetta B, et al. (2007). Tracing past human male movements in northern/eastern Africa and western Eurasia: New clues from Y-chromosomal haplogroups E-M78 and J-M12. *Mol Biol Evol* 24(6):1300–1311.

Csanyi B, Bogacsi-Szabo E, Tomory G, et al. (2008). Y-chromosome analysis of ancient Hungarian and two modern Hungarian-speaking populations from the Carpathian Basin. *Ann Hum Genet* 72(Pt 4):519–534.

Csaszar A, Duba J, Melegh B, et al. (2001). Increased frequency of the C3*F allele and the Leiden mutation of coagulation factor V in patients with severe coronary heart disease who survived myocardial infarction. *Exp Clin Immunogenet* 18(4):206–212.

Csikos M, Hollo P, Becker K, Racz E, Horvath A, Karpati S. (2003). Novel N160I mutation of keratin 9 in a large pedigree from Hungary with epidermolytic palmoplantar keratoderma. *Acta Derm Venereol* 83(4):303–305.

Csikos M, Szalai Z, Becker K, et al. (2004). Novel keratin 14 gene mutations in patients from Hungary with epidermolysis bullosa simplex. *Exp Dermatol* 13(3):185–191.

Csokay B, Udvarhelyi N, Sulyok Z, et al. (1999). High frequency of germ-line BRCA2 mutations among Hungarian male breast cancer patients without family history. *Cancer Res* 59(5):995–998.

Csorsz K. (1928). Ichthyosis (X-linked). *Monatsschrift für Unfallheilkunde* 2:180.

Csorsz K. (1933). Taubstumme Familie. In: Benedek, L. (ed.), *Clinic for Nervous and Mental Diseases* (pp. 20–25), Hughling Jackson Memorial Volume. Debrecen.

Czako M, Riegel M, Morava E, Bajnoczky K, Kosztolanyi G. (2004). Opitz "C" trigonocephaly-like syndrome in a patient with terminal deletion of 2p and partial duplication of 17q. *Am J Med Genet A* 131(3):310–312.

Czako M, Riegel M, Morava E, Schinzel A, Kosztolanyi G. (2002). Patient with rheumatoid arthritis and MCA/MR syndrome due to unbalanced der(18) transmission of a paternal translocation t(18;20)(p11.1;p11.1). *Am J Med Genet* 108(3):226–228.

Czeizel AE, Elek C, Gundy S, et al. (1993). Environmental trichlorfon and cluster of congenital abnormalities. *Lancet* 341(8844):539–542.

Deak B, Dobos M, Kocsis I, et al. (2003). HbA1c levels and erythrocyte transport functions in complication-free type 1 diabetic children and adolescents. *Acta Diabetol* 40(1):9–13.

Deterding K, Constantinescu I, Nedelcu FD, et al. (2008). Prevalence of HBV genotypes in Central and Eastern Europe. *J Med Virol* 80(10):1707–1711.

Duncan A, Hadlaczky G. (2007). Chromosomal engineering. *Curr Opin Biotechnol* 18(5):420–424.

Erdos M, Alapi K, Balogh I, et al. (2006). Severe Shwachman-Diamond syndrome phenotype caused by compound heterozygous missense mutations in the SBDS gene. *Exp Hematol* 34(11):1517–1521.

Erdos M, Hodanova K, Tasko S, et al.(2007). Genetic and clinical features of patients with Gaucher disease in Hungary. *Blood Cells Mol Dis* 39(1):119–123.

Erdos M, Lakos G, Derfalvi B, Notarangelo LD, Durandy A, Marodi L. (2008a). Molecular genetic analysis of Hungarian patients with the hyper-immunoglobulin M syndrome. *Mol Immunol* 45(1):278–282.

Erdos M, Nemeth K, Toth B, et al. (2008b). Novel sequence variants of the alpha-galactosidase A gene in patients with Fabry disease. *Mol Genet Metab* 95(4):224–228.

Erdos M, Toth B, Almassy Z, Timar L, Marodi L. (2008c). [Cartilage-hair hypoplasia]. *Orv Hetil* 149(5):209–217.

Erdos M, Uzvolgyi E, Nemes Z, et al. (2005). Characterization of a new disease-causing mutation of SH2D1A in a family with X-linked lymphoproliferative disease. *Hum Mutat* 25(5):506.

Farago B, Magyari L, Safrany E, et al. (2008). Functional variants of interleukin-23 receptor gene confer risk for rheumatoid arthritis but not for systemic sclerosis. *Ann Rheum Dis* 67(2):248–250.

Farago B, Talian GC, Maasz A, et al. (2007). Prevalence of functional haplotypes of the peptidylarginine deiminase citrullinating enzyme gene in patients with rheumatoid arthritis: No influence of the presence of anti-citrullinated peptide antibodies. *Clin Exp Rheumatol* 25(4):523–528.

Fekete A, Treszl A, Toth-Heyn P, et al. (2003). Association between heat shock protein 72 gene polymorphism and acute renal failure in premature neonates. *Pediatr Res* 54(4):452–455.

Ferenczi E, Ban E, Abraham A, et al. (2008). Severe tick-borne encephalitis in a patient previously infected by West Nile virus. *Scand J Infect Dis* 40(9):759–761.

Fiatal S, Adany R. (2009). Frequency of genetic mutations representing increased susceptibility to chronic non-communicable diseases with high public health importance in the Hungarian population. *Nepegeszsegugy* 87(3):185–194.

Folhoffer A, Ferenci P, Csak T, et al. (2007). Novel mutations of the ATP7B gene among 109 Hungarian patients with Wilson's disease. *Eur J Gastroenterol Hepatol* 19(2):105–111.

Garzuly F, Vidal R, Wisniewski T, Brittig F, Budka H. (1996). Familial meningocerebrovascular amyloidosis, Hungarian type, with mutant transthyretin (TTR Asp18Gly). *Neurology* 47(6):1562–1567.

Glasz-Bona A, Medvecz M, Sajo R, et al. (2009). Easy method for keratin 14 gene amplification to exclude pseudogene sequences: New keratin 5 and 14 mutations in epidermolysis bullosa simplex. *J Invest Dermatol* 129(1):229–231.

Gresham D, Morar B, Underhill PA, et al. (2001). Origins and divergence of the Roma (gypsies). *Am J Hum Genet* 69(6):1314–1331.

Greve B, Simonenko R, Illes Z, et al. (2008). Multiple sclerosis and the CTLA4 autoimmunity polymorphism CT60: No association in patients from Germany, Hungary and Poland. *Mult Scler* 14(2):153–158.

Gudmand-Hoyer E, Skovbjerg H. (1996). Disaccharide digestion and maldigestion. *Scand J Gastroenterol Suppl* 216:111–121.

Gundy S. (1989). Cytogenetical studies on a large control population and on persons occupationally exposed to radiation and/or to chemicals. *Ann Ist Super Sanita* 25(4):549–555.

Hajioff S, McKee M. (2000). The health of the Roma people: A review of the published literature. *J Epidemiol Community Health* 54(11):864–869.

Havasi V, Szolnoki Z, Talian G, et al. (2006). Apolipoprotein A5 gene promoter region T-1131C polymorphism associates with elevated circulating triglyceride levels and confers susceptibility for development of ischemic stroke. *J Mol Neurosci* 29(2):177–183.

Heninger E, Treszl A, Kocsis I, Derfalvi B, Tulassay T, Vasarhelyi B. (2002). Genetic variants of the interleukin-18 promoter region (-607) influence the course of necrotising enterocolitis in very low birth weight neonates. *Eur J Pediatr* 161(7):410–411.

Hermann R, Soltesz G, Morava E, Kosztolanyi G, Czako M. (2000). De novo partial duplication of long arm of chromosome 13: dup(13) (q12-->q14). *Am J Med Genet* 92(4):296–297.

Holzmann H, Vorobyova MS, Ladyzhenskaya IP, et al. (1992). Molecular epidemiology of tick-borne encephalitis virus: Cross-protection between European and Far Eastern subtypes. *Vaccine* 10(5):345–349.

Illes Z, Safrany E, Peterfalvi A, et al. (2008). 3'UTR C2370A allele of the IL-23 receptor gene is associated with relapsing-remitting multiple sclerosis. *Neurosci Lett* 431(1):36–38.

Ingram CJ, Mulcare CA, Itan Y, Thomas MG, Swallow DM. (2009). Lactose digestion and the evolutionary genetics of lactase persistence. *Hum Genet* 124(6):579–591.

Iturriza-Gomara M, Dallman T, Banyai K, et al. (2009). Rotavirus surveillance in europe, 2005–2008: Web-enabled reporting and real-time analysis of genotyping and epidemiological data. *J Infect Dis* 200 Suppl 1:S215-S221.

Jakab F, Horvath G, Ferenczi E, Sebok J, Varecza Z, Szucs G. (2007a). Detection of Dobrava hantaviruses in Apodemus agrarius mice in the Transdanubian region of Hungary. *Virus Res* 128(1–2):149–152.

Jakab F, Meleg E, Banyai K, Melegh B, Timar L, Peterfai J, Szucs G. (2004). One-year survey of astrovirus infection in children with gastroenteritis in a large hospital in Hungary: Occurrence and genetic analysis of astroviruses. *J Med Virol* 74(1):71–77.

Jakab F, Sebok J, Ferenczi E, Horvath G, Szucs G. (2007b). First detection of Dobrava hantavirus from a patient with severe haemorrhagic fever with renal syndrome by SYBR Green-based real time RT-PCR. *Scand J Infect Dis* 39(10):902–906.

Janka Z, Juhasz A, Rimanoczy AA, Boda K, Marki-Zay J, Kalman J. (2002). Codon 311 (Cys --> Ser) polymorphism of paraoxonase-2 gene is associated with apolipoprotein E4 allele in both Alzheimer's and vascular dementias. *Mol Psychiatry* 7(1):110–112.

Janosi A, Andrikovics H, Vas K, et al. (2005). Homozygosity for a novel nonsense mutation (G66X) of the HJV gene causes severe juvenile hemochromatosis with fatal cardiomyopathy. *Blood* 105(1):432.

Jendrassik E. (1902). Beiträge zur Kenntniss der hereditären Krankenheiten. *Deutsche Zeitschrift für Nervenheilkunde* 22:444.

Jiao H, Toth B, Erdos M., et al. (2008). Novel and recurrent STAT3 mutations in hyper-IgE syndrome patients from different ethnic groups. *Mol Immunol* 46(1):202–206.

Juhasz A, Rimanoczy A, Boda K, et al. (2005). CYP46 T/C polymorphism is not associated with Alzheimer's dementia in a population from Hungary. *Neurochem Res* 30(8):943–948.

Juhasz-Kaszanyitzky E, Janosi S, Somogyi P, et al. (2007). MRSA transmission between cows and humans. *Emerg Infect Dis* 13(4):630–632.

Kalaydjieva L, Calafell F, Jobling MA, et al. (2001a). Patterns of inter- and intra-group genetic diversity in the Vlax Roma as revealed by Y chromosome and mitochondrial DNA lineages. *Eur J Hum Genet* 9(2):97–104.

Kalaydjieva L, Gresham D, Calafell F. (2001b). Genetic studies of the Roma (Gypsies): A review. *BMC Med Genet* 2:5.

Kalmar L, Bors A, Farkas H, et al. (2003). Mutation screening of the C1 inhibitor gene among Hungarian patients with hereditary angioedema. *Hum Mutat* 22(6):498.

Karteszi J, Kosztolanyi G, Czako M, Hadzsiev K, Morava E. (2006). Transient progeroid phenotype and lipodystrophy in mosaic polyploidy. *Clin Dysmorphol* 15(1):29–31.

Karteszi J, Kress W, Szasz M, et al. (2004). Partial craniosynostosis in a patient with deletion 22q11. *Genet Couns* 15(4):481–483.

Katona E, Aslanidis C, Remenyik E, et al. (2005). Identification of a novel deletion in the ABCC6 gene leading to Pseudoxanthoma elasticum. *J Dermatol Sci* 40(2):115–121.

Katona RL, I. Sinko, G. Hollo, et al. (2008). A combined artificial chromosome-stem cell therapy method in a model experiment aimed at the treatment of Krabbe's disease in the Twitcher mouse. *Cell Mol Life Sci* 65(23):3830–3838.

Kellermayer R, Czako M, Kiss-Laszlo Z, et al. (2005a). Alpha-thalassemia/mental retardation syndrome in a 45,X male. *Am J Med Genet A* 132(4):431–433.

Kellermayer R, Gyarmati J, Czako M, et al. (2005b). Mos 46,XX,r(18).ish r(18)(18ptel-,18qtel-)/46,XX.ish del(18)(18ptel-): An example for successive ring chromosome formation. *Am J Med Genet A* 139(3):234–235.

Kellermayer R, Keller M, Ratajczak P, et al. (2005c). Bigenic connexin mutations in a patient with hidrotic ectodermal dysplasia. *Eur J Dermatol* 15(2):75–79.

Kennard ML, Goosney DL, Monteith D, et al. (2009). The generation of stable, high MAb expressing CHO cell lines based on the artificial chromosome expression (ACE) technology. *Biotechnol Bioeng* 104(3):540–553.

Kereszturi E, Tarnok Z, Bognar E, et al. (2008). Catechol-O-methyltransferase Val158Met polymorphism is associated with methylphenidate response in ADHD children. *Am J Med Genet B Neuropsychiatr Genet* 147B(8):1431–1435.

Kisfali P, Mohas M, Maasz A, et al. (2008). Apolipoprotein A5 IVS3+476A allelic variant associates with increased trigliceride levels and confers risk for development of metabolic syndrome in Hungarians. *Circ J* 72(1):40–43.

Klein I, Andrikovics H, Bors A, Nemes L, Tordai A, Varadi A. (2001). A haemophilia A and B molecular genetic diagnostic programme in Hungary: A highly informative and cost-effective strategy. *Haemophilia* 7(3):306–312.

Komlosi K, Talian GC, Farago B, et al. (2008). No influence of SLC22A4 C6607T and RUNX1 G24658C genotypic variants on the circulating carnitine ester profile in patients with rheumatoid arthritis. *Clin Exp Rheumatol* 26(1):61–66.

Kosztolanyi G. (1987). Does "ring syndrome" exist? An analysis of 207 case reports on patients with a ring autosome. *Hum Genet* 75(2):174–179.

Kovacs KB, Tisza B, Komaromi I, Muszbek L, Bereczky Z. (2009b). Inherited factor V deficiency associated with a novel heterozygous missense mutation (p.G493R) in a patient with excessive surgical bleeding. *Thromb Haemost* 102(4):787–789.

Kovacs ME, Papp J, Szentirmay Z, Otto S, Olah E. (2009a). Deletions removing the last exon of TACSTD1 constitute a distinct class of mutations predisposing to Lynch syndrome. *Hum Mutat* 30(2):197–203.

Kovacs N, Sarkany I, Mohay G, et al. (2006). High incidence of short rib-polydactyly syndrome type IV in a Hungarian Roma subpopulation. *Am J Med Genet A* 140(24):2816–2818.

Krikovszky D, Vasarhelyi B, Toth-Heyn P, Korner A, Tulassay T, Madacsy L. (2002). Association between G-308A polymorphism of the tumor necrosis factor-alpha gene and 24-hour ambulatory blood pressure values in type 1 diabetic adolescents. *Clin Genet* 62(6):474–477.

Kroneman A, Harris J, Vennema H, et al. (2008). Data quality of 5 years of central norovirus outbreak reporting in the European Network for food-borne viruses. *J Public Health (Oxf)* 30(1):82–90.

Lahermo P, Savontaus ML, Sistonen P, et al. (1999). Y chromosomal polymorphisms reveal founding lineages in the Finns and the Saami. *Eur J Hum Genet* 7(4):447–458.

Lakner L, Csongei V, Sarlos P, et al. (2009). IGR2096a_1 T and IGR2198a_1 C alleles on IBD5 locus of chromosome 5q31 region confer risk for Crohn's disease in Hungarian patients. *Int J Colorectal Dis* 24(5):503–507.

Lopman B, Vennema H, Kohli E, et al. (2004). Increase in viral gastroenteritis outbreaks in Europe and epidemic spread of new norovirus variant. *Lancet* 363(9410):682–688.

Losonczi E, Bencsik K, Nagy ZF, et al. (2009). Tumour necrosis factor alpha gene (TNF-alpha) -376 polymorphism in Hungarian patients with primary progressive multiple sclerosis. *J Neuroimmunol* 208(1–2):115–118.

Losonczy G, Rosenberg N, Boda Z, et al. (2007). Three novel mutations in the glycoprotein IIb gene in a patient with type II Glanzmann thrombasthenia. *Haematologica* 92(5):698–701.

Losonczy G, Rosenberg N, Kiss C, et al. (2005). A novel homozygous mutation (1619delC) in GPIIb gene associated with Glanzmann thrombasthenia, the decay of GPIIb-mRNA and the synthesis of a truncated GPIIb unable to form complex with GPIIIa. *Thromb Haemost* 93(5):904–909.

M. Riegel, Morava E, Czako M, Kosztolanyi G, Schinzel A. (2001). Distal deletion, del(2)(q33.3q33.3), in a patient with severe growth deficiency and minor anomalies. *Am J Med Genet* 102(3):227–230.

Maasz A, Kisfali P, Horvatovich K, et al. (2007). Apolipoprotein A5 T-1131C variant confers risk for metabolic syndrome. *Pathol Oncol Res* 13(3):243–247.

Maasz A, Kisfali P, Jaromi L, et al. (2008a). Apolipoprotein A5 gene IVS3+G476A allelic variant confers susceptibility for development of ischemic stroke. *Circ J* 72(7):1065–1070.

Maasz A, Kisfali P, Szolnoki Z, Hadarits F, Melegh B. (2008b). Apolipoprotein A5 gene C56G variant confers risk for the development of large-vessel associated ischemic stroke. *J Neurol* 255(5):649–654.

Maasz A, Komlosi K, Hadzsiev K, et al. (2008c). Phenotypic variants of the deafness-associated mitochondrial DNA A7445G mutation. *Curr Med Chem* 15(13):1257–1262.

Macnab S, Whitehouse A. (2009). Progress and prospects: Human artificial chromosomes. *Gene Ther* 16(10):1180–1188.

Mag T, Nogrady N, Herpay M, Toth I, Rozgonyi F. (2010). Characterisation of verotoxin-producing Escherichia coli strains isolated from human patients in Hungary over a 7-year period. *Eur J Clin Microbiol Infect Dis.* 29:249–252.

Magyari L, Bene J, Komlosi K, et al. (2007a). Prevalence of SLC22A4 1672T and SLC22A5 -207C combination defined TC haplotype in Hungarian ulcerative colitis patients. *Pathol Oncol Res* 13(1):53–56.

Magyari L, Farago B, Bene J, et al. (2007b). No association of the cytotoxic T-lymphocyte associated gene CTLA4 +49A/G polymorphisms with Crohn's disease and ulcerative colitis in Hungarian population samples. *World J Gastroenterol* 13(15):2205–2208.

Malyarchuk BA, Grzybowski T, Derenko MV, Czarny J, Miscicka-Sliwka D. (2006). Mitochondrial DNA diversity in the Polish Roma. *Ann Hum Genet* 70(Pt 2):195–206.

Mastana SS, Papiha SS. (1992). Origin of the Romany gypsies—genetic evidence. *Z Morphol Anthropol* 79(1):43–51.

Matthijnssens J, Potgieter CA, Ciarlet M, et al. (2009). Are human P[14] rotavirus strains the result of interspecies transmissions from sheep or other ungulates that belong to the mammalian order Artiodactyla? *J Virol* 83(7):2917–2929.

McKee M. (1997). The health of gypsies. *BMJ* 315(7117):1172–1173.

Meggyessy V, Mehes K. (1987). Association of supernumerary nipples with renal anomalies. *J Pediatr* 111(3):412–413.

Mehes K, Buhler EM. (1995). Premature centromere division: A possible manifestation of chromosome instability. *Am J Med Genet* 56(1):76–79.

Mehes K, Kajtar P, Kosztolanyi G. (2002). Association of nonsyndromic Wilms tumor with premature centromere division (PCD). *Am J Med Genet* 112(2):215–216.

Mehes K, Kajtar P, Sandor G, Scheel-Walter M, Niethammer D. (1998). Excess of mild errors of morphogenesis in childhood lymphoblastic leukemia. *Am J Med Genet* 75(1):22–27.

Mehes K, Kosztolanyi G. (1990a). Premature centromere division of a translocation-carrier autosome. *Hum Genet* 85(3):379–380.

Mehes K, Pinter A. (1990b). Minor morphological aberrations in children with isolated urinary tract malformations. *Eur J Pediatr* 149(6):399–402.

Mehes K, Signer E, Pluss HJ, Muller HJ, Stalder G. (1985). Increased prevalence of minor anomalies in childhood malignancy. *Eur J Pediatr* 144(3):243–254.

Mehes K. (1978). Non-random centromere division: A mechanism of non-disjunction causing aneuploidy? *Hum Hered* 28(4):255–260.

Mehes K. (1979). Association of supernumerary nipples with other anomalies. *J Pediatr* 95(2):274–275.

Mehes K. (1983). Association of supernumerary nipples with other anomalies. *J Pediatr* 102(1):161.

Mehes K. (1988). Informative Morphogenetic Variants in the Newborn Infant. Akadémiai Kiadó Budapest, 1988.

Mehes K. (1996). Familial association of supernumerary nipple with renal cancer. *Cancer Genet Cytogenet* 86(2):129–130.

Mehes K. Bajnoczky K. (1982). Unusually early dividing chromosomes 13–15 in a child with retinoblastoma and 13q deletion. *Hum Genet* 61(1):78.

Melegh B, Bene J, Mogyorosy G, et al. (2004). Phenotypic manifestations of the OCTN2 V295X mutation: Sudden infant death and carnitine-responsive cardiomyopathy in Roma families. *Am J Med Genet A* 131(2):121–126.

Milankovics I, Kamory E, Csokay B, Fodor F, Somogyi C, Schuler A. (2007). Mutations causing biotinidase deficiency in children ascertained by newborn screening in Western Hungary. *Mol Genet Metab* 90(3):345–348.

Miltenberger-Miltenyi G, Szonyi L, Balogh L, Utermann G, Janecke AR. (2005). Mutation spectrum of type I glycogen storage disease in Hungary. *J Inherit Metab Dis* 28(6):939–944.

Mohas M, Kisfali P, Baricza E, et al. (2009). A Polymorphism within the Fructosamine-3-kinase Gene is Associated with HbA1c Levels and the Onset of Type 2 Diabetes Mellitus. *Exp Clin Endocrinol Diabetes* 118:209–212.

Morar B, Gresham D, Angelicheva D, et al. (2004). Mutation history of the roma/gypsies. *Am J Hum Genet* 75(4):596–609.

Morava E, Bartsch O, Czako M, et al. (2003a). Small inherited terminal duplication of 7q with hydrocephalus, cleft palate, joint contractures, and severe hypotonia. *Clin Dysmorphol* 12(2):123–127.

Morava E, Bartsch O, Czako M, Frensel A, Karteszi J, Kosztolanyi GY. (2003b). A girl with cutaneous hyperpigmentation, cafe au lait spots and ring chromosome 15 without significant deletion. *Genet Couns* 14(3):337–342.

Morava E, Czako M, Aszmann M, Illes T, Kosztolanyi GY. (2002). Tissue specific mosaicism of trisomy 9 in a patient with severe torsion scoliosis. *Genet Couns* 13(4):455–457.

Morava E, Hermann R, Czako M, Soltesz G, Kosztolanyi G. (2000). Isodicentric Y chromosome in an Ullrich-Turner patient without virilization. *Am J Med Genet* 91(2):99–101.

Nadasi E, Gyurus P, Czako M, et al. (2007). Comparison of mtDNA haplogroups in Hungarians with four other European populations: A small incidence of descents with Asian origin. *Acta Biol Hung* 58(2):245–256.

Nagy A, Keszei M, Kis Z, et al. (2007). Chlamydophila pneumoniae infection status is dependent on the subtypes of asthma and allergy. *Allergy Asthma Proc* 28(1):58–63.

Nagy A, Kozma GT, Bojszko A, Krikovszky D, Falus A, Szalai C. (2002). No association between asthma or allergy and the CCR5Delta 32 mutation. *Arch Dis Child* 86(6):426.

Nagy A, Kozma GT, Keszei M, Treszl A, Falus A, Szalai C. (2003). The development of asthma in children infected with Chlamydia pneumoniae is dependent on the modifying effect of mannose-binding lectin. *J Allergy Clin Immunol* 112(4):729–734.

Nagy D, Bogacsi-Szabo E, Varkonyi A, et al. (2009). Prevalence of adult-type hypolactasia as diagnosed with genetic and lactose hydrogen breath tests in Hungarians. *Eur J Clin Nutr* 63(7):909–912.

Nelis M, Esko T, Magi R, et al. (2009). Genetic structure of Europeans: A view from the North-East. *PLoS One* 4(5):e5472.

Nobilis A, Szabo M, Kocsis I, Sulyok E, Tulassay T, Vasarhelyi B. (2006). Angiotensin-converting enzyme DD genotype is preventive against circulatory failure in very-low-birthweight neonates. *Acta Paediatr* 95(6):747–750.

Olds LC, Sibley E. (2003). Lactase persistence DNA variant enhances lactase promoter activity in vitro: Functional role as a cis regulatory element. *Hum Mol Genet* 12(18):2333–2340.

Papp J, Kovacs ME, Olah E. (2007). Germline MLH1 and MSH2 mutational spectrum including frequent large genomic aberrations in Hungarian hereditary non-polyposis colorectal cancer families: Implications for genetic testing. *World J Gastroenterol* 13(19):2727–2732.

Papp Z, Toth Z, Torok O, Szabo M. (1987). Prenatal diagnosis policy without routine amniocentesis in pregnancies with a positive family history for neural tube defects. *Am J Med Genet* 26(1):103–110.

Parvari R, Lei KJ, Szonyi L, Narkis G, Moses S, and Chou JY. (1997). Two new mutations in the glucose-6-phosphatase gene cause glycogen storage disease in Hungarian patients. *Eur J Hum Genet* 5(4):191–195.

Piko H, Vancso V, Nagy B, Ban Z, Herczegfalvi A, and Karcagi V. (2009). Dystrophin gene analysis in Hungarian Duchenne/Becker muscular dystrophy families – detection of carrier status in symptomatic and asymptomatic female relatives. *Neuromuscul Disord* 19(2):108–112.

Racz E, Csikos M, and S. Karpati (2005b). Novel mutations in the ATP2C1 gene in two patients with Hailey-Hailey disease. *Clin Exp Dermatol* 30(5):575–577.

Racz E, Csikos M, Benko R, Kornsee Z, and S. Karpati (2005a). Three novel mutations in the ATP2A2 gene in Hungarian families with Darier's disease, including a novel splice site generating intronic nucleotide change. *J Dermatol Sci* 38(3):231–234.

Racz E, Csikos M, Kornsee Z, Horvath A, and S. Karpati (2004). Identification of mutations in the ATP2A2 gene in patients with Darier's disease from Hungary. *Exp Dermatol* 13(6):396–399.

Ramus SJ, Kote-Jarai Z, Friedman LS, et al. (1997). Analysis of BRCA1 and BRCA2 mutations in Hungarian families with breast or breast-ovarian cancer. *Am J Hum Genet* 60(5):1242–1246.

Reuter G, Fodor D, Forgach P, Katai A, Szucs G. (2009). Characterization and zoonotic potential of endemic hepatitis E virus (HEV) strains in humans and animals in Hungary. *J Clin Virol* 44(4):277–281.

Rootsi S, Kivisild T, Tambets K, Adojaan M, Parik J, et al. (2000). On the phylogeographic context of sex-specific genetic markers of Finno-Ugric populations. In: *The roots of peoples and languages of Northern Eurasia II and III*. pp. 148–164. (ed: A. Kiinnap) University of Tartu, Tartu,

Rozsai B, Kiss A, Csabi G, Czako M, Decsi T. (2009). Severe dystrophy in DiGeorge syndrome. *World J Gastroenterol* 15(11):1391–1393.

Safrany E, Pazar B, Csongei V, et al. (2009b). Variants of the IL23R gene are associated with ankylosing spondylitis but not with Sjogren syndrome in Hungarian population samples. *Scand J Immunol* 70(1):68–74.

Safrany E, R. Hobor, L. Jakab, et al. (2009a). Interleukin-23 receptor gene variants in Hungarian systemic lupus erythematosus patients. *Inflamm Res.* 59:159–164.

Sahi T, Isokoski M, Jussila J, Launiala K, Pyorala K. (1973). Recessive inheritance of adult-type lactose malabsorption. *Lancet* 2(7833):823–826.

Sahi T. (1994). Genetics and epidemiology of adult-type hypolactasia. *Scand J Gastroenterol Suppl* 202:7–20.

Sandor J, Bard P. (2009). The legal regulation of biobanks; national report: Hungary.

Sandor J. (2004). Society and genetic information: Codes and laws in the genetic era. Central European University Press, Budapest.

Sandor J. (2009). From private to public? Legal concepts of the right to privacy and ownership in the regulation of biobanks. In: *New challenges for biobanks: Ethics, law and governance.* pp. 123–136. (ed: Dierickx, K., P. Borry), Inersentia, Antwerp, 2009.

Sarosi A, Gonda X, Balogh G, et al. (2008). Association of the STin2 polymorphism of the serotonin transporter gene with a neurocognitive endophenotype in major depressive disorder. *Prog Neuropsychopharmacol Biol Psychiatry* 32(7):1667–1672.

Schaffer K. (1905). Zur Pathogenese der Tay-Sachs'schen amaurotischen Idiotie. *Neurologisches Centralblatt* 9/10:68.

Schuler D, Dobos M, Fekete G, Miltenyi M, Kalmar L. (1979). Chromosome mutations and chromosome stability in children treated with different regimes of immunosuppressive drugs. *Hum Hered* 29(2):100–105.

Semino O, Passarino G, Oefner PJ, et al. (2000a). The genetic legacy of Paleolithic Homo sapiens sapiens in extant Europeans: A Y chromosome perspective. *Science* 290(5494):1155–1159.

Semino O, Passarino G, Quintana-Murci L, et al. (2000b). MtDNA and Y chromosome polymorphisms in Hungary: Inferences from the palaeolithic, neolithic and Uralic influences on the modern Hungarian gene pool. *Eur J Hum Genet* 8(5):339–346.

Semsei AF, Erdelyi DJ, Ungvari I, et al. (2008). Association of some rare haplotypes and genotype combinations in the MDR1 gene with childhood acute lymphoblastic leukaemia. *Leuk Res* 32(8):1214–1220.

Sipeky C, Csongei V, Jaromi L, et al. (2009a). Vitamin K epoxide reductase complex 1 (VKORC1) haplotypes in healthy Hungarian and Roma population samples. *Pharmacogenomics* 10(6):1025–1032.

Sipeky C, Lakner L, Szabo M, et al. (2009b). Interethnic differences of CYP2C9 alleles in healthy Hungarian and Roma population samples: Relationship to worldwide allelic frequencies. *Blood Cells Mol Dis* 43(3):239–242.

Suarez FL, Savaiano D. (1997). Diet, genetics, and lactose intolerance. *Food Technology* 51:74–76.

Szalai C, Fust G, Duba J, et al. (2002). Association of polymorphisms and allelic combinations in the tumour necrosis factor-alpha-complement MHC region with coronary artery disease. *J Med Genet* 39(1):46–51.

Szalai C, Kozma GT, Nagy A, et al.(2001). Polymorphism in the gene regulatory region of MCP-1 is associated with asthma susceptibility and severity. *J Allergy Clin Immunol* 108(3):375–381.

Szebeni B, Szekeres R, Rusai K, et al. (2006). Genetic polymorphisms of CD14, toll-like receptor 4, and caspase-recruitment domain 15 are not associated with necrotizing enterocolitis in very low birth weight infants. *J Pediatr Gastroenterol Nutr* 42(1):27–31.

Szekeres G, Juhasz A, Rimanoczy A, Keri S, Janka Z. (2003). The C270T polymorphism of the brain-derived neurotrophic factor gene is associated with schizophrenia. *Schizophr Res* 65(1):15–18.

Szell M, Balogh K, Dobozy A, Kemeny L, Olah J. (2007). First detection of the melanoma-predisposing proline-48-threonine mutation of p16 in Hungarians: Was there a common founder either in Italy or in Hungary? *Melanoma Res* 17(4):251–254.

Szolnoki Z, Havasi V, Bene J, et al. (2005a). Endothelial nitric oxide synthase gene interactions and the risk of ischaemic stroke. *Acta Neurol Scand* 111(1):29–33.

Szolnoki Z, Havasi V, Talian G, et al. (2005b). Lymphotoxin-alpha gene 252G allelic variant is a risk factor for large-vessel-associated ischemic stroke. *J Mol Neurosci* 27(2):205–211.

Szolnoki Z, Havasi V, Talian G, et al. (2006a). Angiotensin II type-1 receptor A1166C polymorphism is associated with increased risk of ischemic stroke in hypertensive smokers. *J Mol Neurosci* 28(3):285–290.

Szolnoki Z, Kondacs A, Mandi Y, Somogyvari F. (2007a). Evaluation of the roles of the A185C and C406T kinesin light-chain 1 variants in the development of leukoaraiosis. *Neurosci Lett* 429(2–3):101–104.

Szolnoki Z, Maasz A, Magyari L, et al. (2006b). Coexistence of angiotensin II type-1 receptor A1166C and angiotensin-converting enzyme D/D polymorphism suggests susceptibility for small-vessel-associated ischemic stroke. *Neuromolecular Med* 8(3):353–360.

Szolnoki Z, Maasz A, Magyari L, et al. (2007b). The combination of homozygous MTHFR 677T and angiotensin II type-1 receptor 1166C variants confers the risk of small-vessel-associated ischemic stroke. *J Mol Neurosci* 31(3):201–207.

Szolnoki Z, Maasz A, Magyari L, et al. (2009). Galectin-2 3279TT variant protects against the lymphotoxin-alpha 252GG genotype associated ischaemic stroke. *Clin Neurol Neurosurg* 111(3):227–230.

Szolnoki Z, Somogyvari F, Kondacs A, et al. (2003). Increased prevalence of platelet glycoprotein IIb/IIIa PLA2 allele in ischaemic stroke associated with large vessel pathology. *Thromb Res* 109(5–6):265–269.

Szolnoki Z, Somogyvari F, Kondacs A, et al. (2004). Specific APO E genotypes in combination with the ACE D/D or MTHFR 677TT mutation yield an independent genetic risk of leukoaraiosis. *Acta Neurol Scand* 109(3):222–227.

Szolnoki Z, Somogyvari F, Kondacs A, Szabo M, Fodor L. (2001). Evaluation of the roles of common genetic mutations in leukoaraiosis. *Acta Neurol Scand* 104(5):281–287.

Tambets K, Rootsi S, Kivisild T, et al. (2004). The western and eastern roots of the Saami – the story of genetic "outliers" told by mitochondrial DNA and Y chromosomes. *Am J Hum Genet* 74(4):661–682.

Tambets K, Rootsi S, Kivisild T, Villems R. (2001). The concepts of Richard Indreko about the origin of the Finno-Ugric speakers and the population genetics of the extant North-East European populations. *TRAMES 55/50* 1:59–74.

Tardy EP, Toth A, Kosztolanyi G. (1997). Prenatal exclusion of segmental trisomy in familial chromosome 21 pericentric inversion by fluorescence in situ hybridization. *Prenat Diagn* 17(9):871–873.

Teszas A, Moller RS, Kellermayer R, et al. (2008). A cryptic unbalanced translocation resulting in del 13q and dup 15q. *Am J Med Genet A* 146A(19):2570–2573.

Teszas A, Pfund Z, Morava E, et al. (2006). Presenile cataract: Consider cholestanol. *Arch Ophthalmol* 124(10):1490–1492.

Theodoropoulou M, Barta C, Szoke M, et al. (2001). Prenatal diagnosis of steroid 21-hydroxylase deficiency by allele-specific amplification. *Fetal Diagn Ther* 16(4):237–240.

Tomory G, Csanyi B, Bogacsi-Szabo E, et al. (2007). Comparison of maternal lineage and biogeographic analyses of ancient and modern Hungarian populations. *Am J Phys Anthropol* 134(3):354–368.

Toth B, Boe Wolff AS, Halasz Z, et al. (2010). Novel sequence variation of AIRE and detection of interferon-omega antibodies in early infancy. *Clin Endocrinol (Oxf).* 72(5):641–647.

Toth B, Volokha A, Mihas A, et al. (2009). Genetic and demographic features of X-linked agammaglobulinemia in Eastern and Central Europe: A cohort study. *Mol Immunol* 46(10):2140–2146.

Toth I, Schmidt H, Kardos G, et al. (2009). Virulence genes and molecular typing of different groups of Escherichia coli O157 strains in cattle. *Appl Environ Microbiol* 75(19):6282–6291.

Toth T, Findlay I, Nagy B, Papp Z. (1997a). Accurate sizing of (CAG)n repeats causing Huntington disease by fluorescent PCR. *Clin Chem* 43(12):2422–2423.

Toth T, Hajdu J, Marton T, Nagy B, Papp Z. (1998). Connexin43 gene mutations and heterotaxy. *Circulation* 97(1):117–118.

Toth T, Nemeti M, Papp Z. (1997b). Detection of CAG repeats using silver staining in patients with Huntington disease in Hungary. *Am J Med Genet* 70(4):448–449.

Treszl A, Szabo M, Dunai G, et al. (2003). Angiotensin II type 1 receptor A1166C polymorphism and prophylactic indomethacin treatment induced ductus arteriosus closure in very low birth weight neonates. *Pediatr Res* 54(5):753–755.

Troelsen JT, Olsen J, Moller J, Sjostrom H. (2003). An upstream polymorphism associated with lactase persistence has increased enhancer activity. *Gastroenterology* 125(6):1686–1694.

Vajo Z, Tamas F, Sinka L, Jankovics I. (2010). Safety and immunogenicity of a 2009 pandemic influenza A H1N1 vaccine when administered alone or simultaneously with the seasonal influenza vaccine for the 2009–2010 influenza season: A multicentre, randomised controlled trial. *Lancet* 375:49–55.

Vannay A, Vasarhelyi B, Kornyei M, et al. (2006). Single-nucleotide polymorphisms of VEGF gene are associated with risk of congenital valvuloseptal heart defects. *Am Heart J* 151(4):878–881.

Varadi V, Szabo L, and Papp Z. (1980). Syndrome of polydactyly, cleft lip/palate or lingual lump, and psychomotor retardation in endogamic gypsies. *J Med Genet* 17(2):119–122.

Vlasak I, Plochl E, Kronberger G, et al. (1999). Screening of patients with Turner syndrome for "hidden" Y-mosaicism. *Klin Padiatr* 211(1):30–34.

Zerjal T, Dashnyam B, Pandya A, et al. (1997). Genetic relationships of Asians and Northern Europeans, revealed by Y-chromosomal DNA analysis. *Am J Hum Genet* 60(5):1174–1183.

117.

THE GENETICS OF ALCOHOL DEPENDENCE IN RUSSIA

A. Kazantseva, G. Faskhutdinova, and Elza K. Khusnutdinova

INTRODUCTION

Alcohol dependence (AD) is a widespread psychiatric disorder, affecting 5.4% of the general population. In addition to causing numerous serious medical disorders (e.g., liver and heart disease), alcohol dependence is associated with costly and adverse social consequences, such as disruption of families, crime, traumatic accidents, and lost productivity. Various types of research, including studies of identical and fraternal twins, adoptees, and families with multiple alcoholic members have provided evidence for a strong genetic component in the development of alcohol dependence, estimated approximately at 64% (Heath et al., 1997). Moreover, it is likely that some types of alcoholism are more genetically determined, such as the "male-limited" types of alcoholism proposed by Cloninger (1987) and severe forms of alcohol dependence (Walters, 2002). Numerous environmental risk factors for alcoholism have been identified, including biological, psychological, familial, and sociocultural factors (Gunzerath and Goldman, 2003). The complexity of alcoholism research as a heterogeneous disease includes differences in phenotypic characteristics (age of onset, drinking history, comorbid disorders, and the onset of withdrawal symptoms) which are probably confounded by the individual's success in socialization and sensitivity to stress and adverse conditions.

Numerous studies directed to detect genetic polymorphisms affecting the risk for alcoholism, or individual response to treatment, highlighted the presence of discrepancy caused by ethnic differences in the studied subjects, the presence of various comorbid symptoms, or investigation of different alcoholism types. Although DSM-IV and ICD-10 inventories have been reported to be reliable in alcoholism diagnostics, many studies were based on the typology proposed by Cloninger (1987) distinguishing two variants of alcohol dependence. Type I alcoholism was characterized by few antisocial behaviors, low impulsivity, decreased novelty seeking and a relatively late onset. By contrast, Type II alcohol dependence was defined as multisubstance dependence with antisocial behaviors, aggression, impulsivity, novelty seeking, and early onset, and was assumed to have a genetic component (Cloninger, 1987). Hence, the typology of alcoholism is therefore important and critical in both research and clinical practice.

NEUROTRANSMITTER SYSTEMS IN ALCOHOLISM

Alcoholism (mainly with antisocial impulsive features) was hypothesized to be caused by dysfunction of the serotonergic system: namely, decreased cerebrospinal fluid concentration of 5-hydroxyindoleacetic acid (5-HIAA), a marker of decreased serotonergic activity, and increased uptake of serotonin were reported in some alcoholic patients. Decreased serotonergic activity following chronic ethanol administration can lead to behavioral disinhibition, including impulsive aggression (Higley et al., 1996). These results prompted different scientific groups to examine the genetic association between alcoholism and certain serotonergic system genes.

It is likely that genetically mediated variability of the serotonin transporter (5-HTT) gene functioning can contribute to alcohol sensitivity. Moreover, it has been reported that a quantitative trait locus (QTL) related to alcohol preference in mice is located near the *5-HTT* gene (Melo et al., 1996), and serotonin reuptake inhibitors have shown some effectiveness for the treatment of alcoholism (Johnson, 2008a).

In addition to several regulatory domains controlling selective expression in serotonergic neurons, transcriptional activity of the *5-HTT* gene is modulated by the number of 20–23 bp repeats in the upstream regulatory region of the gene (*5-HTT* linked polymorphic region, *5-HTTLPR*). Since the more frequent alleles with 14 and 16 repeats were designated as S-allele and L-allele respectively, this polymorphism was also designated as 44 bp insertion/deletion polymorphism. Functional studies of the *5-HTT* promoter activity in cell lines, mRNA concentrations in the raphe complex of the human postmortem brain, platelet serotonin uptake, and content confirmed that S-allele was associated with lower levels of the *5-HTT* gene expression and function (Nakamura et al., 2000; Philibert et al., 2008). The correlation between *5-HTTLPR* and alcoholism had been supported (Kweon et al., 2005; Marques et al., 2006), while other studies failed to replicate these associations (Edenberg et al., 1998; Choi et al., 2006).

Polymorphism of variable number of 17 bp repeats located in intron 2 (*STin2* VNTR) was found to be in linkage disequilibrium with *5-HTTLPR* in different populations (Ebstein, 2006). In addition, the *STin2* polymorphism has been implicated in gene expression in embryonic stem cells (Fiskerstrand et al., 1999), and together with *5-HTTLPR* there was an observed combination effect on the rate of *5-HTT* mRNA transcription in lymphoblasts (Hranilovic et al., 2004).

Terminal serotonin autoreceptor (*HTR1B*) has attracted much attention due to the interesting results with an animal model of alcohol withdrawal: *HTR1B*-knockout mice demonstrated more aggressive behavior toward an intruder (Saudou et al., 1994) and showed increased spontaneous alcohol consumption (Crabbe et al., 1996). Moreover, an alcohol preference region has been mapped on mouse chromosome 9 very close to the location of *HTR1B* gene encoding protein specifically implicated in contributing to the reinforcing properties of alcohol (Phillips et al., 1998).

An involvement of serotonin 2A receptor (*HTR2A*) in alcohol dependence was suggested in several studies: *HTR2A* antagonists have been shown to reduce alcohol intake (Overstreet et al., 1997), while lower *HTR2A* receptor binding in the prefrontal cortex of alcohol-dependent patients with a positive family history of alcoholism was also indicated (Underwood et al., 2008). These results influenced active investigation of *-1438A/G* polymorphism in the *HTR2A* gene in reward-related diseases, and alcohol dependence in particular (Preuss et al., 2001, Nomura and Nomura, 2006).

The dopaminergic system has been primarily reported to be related to alcohol sensitivity. Behavioral reinforcement and locomotor stimulation evoked by alcohol was shown to be associated with the release of dopamine from mesolimbic dopaminergic terminals located in the nucleus accumbens (Koob, 1998). Moreover, dysfunction in dopaminergic transmission has been associated with craving for ethanol, and the increase in the desirability and reward of drinking was proposed to be mediated by changes in the reactivity of dopamine neurons (Kiefer and Mann, 2005). It has been shown that the density of dopamine receptors, the amount of dopamine released into the synapse, and the rapidity of its transport back into the cell by the dopamine transporter might affect the sensitivity and reactivity of dopaminergic system (Wise, 2002).

Animal and human studies suggested that alcohol can stimulate dopaminergic neurons in the ventral tegmental area, resulting in the lower density of dopamine D2 receptors (DRD2) in the limbic system in alcohol-preferring subjects (Hietala et al. 1994). Accordingly, the dopamine D2 receptor gene (*DRD2*) is one of the candidates of reward-related psychiatric disorders, alcoholism in particular. The most frequently examined polymorphism linked to this gene is the *Taq1A* polymorphism (rs1800497 or *32806C>T*) located in the 3'-untranslated region (3'-UTR) of *DRD2* (10 kb downstream of the last exon). Brain imaging studies have shown that individuals with *DRD2 Taq1A* A1-allele have a reduced number of dopamine D2 receptors (Pohjalainen et al., 1998; Laine et al., 2001; Ritchie and Noble, 2003); however, contradictory findings were also observed (Laruelle et al., 1998). Moreover, overexpression of the D2 dopamine receptor in rats, on the other hand, decreased voluntary alcohol intake significantly (Thanos et al., 2005), and high levels of D2 dopamine receptors have been reported to be a protective factor for alcohol dependence (Volkow et al., 2006). Since dopaminergic deficit prevails in A1-allele carriers, it could be hypothesized that individuals bearing the A1-allele might have more tendencies for reward-seeking (Comings and Blum, 2000). Hence, several molecular genetic studies involved *DRD2 Taq1A* polymorphism with respect to "reward-related psychiatric disorders," including addiction (Noble, 2003; Munafò et al., 2007; Kraschewski et al., 2009; Pinto et al., 2009).

However, recently this SNP has been shown not to reside in the *DRD2* gene, but in a neighboring ankyrin repeat and kinase domain containing 1 gene (*ANKK1;* Neville et al., 2004). This gene contains a single serine/threonine kinase domain, and is a member of an extensive family of proteins involved in signal transduction pathways. The rs1800497 causes an amino acid substitution within the 11th ankyrin repeat of *ANKK1* (*Glu713Lys*), which is unlikely to affect structural integrity, but rather affects substrate-binding specificity of the gene product (Neville et al., 2004). It has been hypothesized that the *ANKK1* gene may be involved in the dopaminergic reward pathway through signal transduction or other cellular response (Neville et al., 2004), since genes of related function are sometimes found clustered together.

Genetically mediated variability in the dopamine transporter gene (*SLC6A3*) functioning can also contribute to alcoholism predisposition. *SLC6A3* plays a critical role in the regulation of dopaminergic transmission by mediating active reuptake of dopamine from the synapse into the presynaptic terminal. Higher dopamine transporter densities among severe type II alcoholics were reported when compared with healthy controls (Tiihonen et al., 1995). Animal studies suggested that *SLC6A3*-knockout mice have decreased alcohol preference and consumption when compared with wild-type mice (Giros et al., 1996). VNTR polymorphism in 3'-UTR of *SLC6A3* gene was shown to be functional; however, increased gene expression was controversially associated either with 10-repeat allele (Fuke et al., 2001; VanNess et al., 2005) or 9-repeat allele (Miller and Madras, 2002). Several studies related dopamine transporter gene 9-repeat allele to severe alcohol withdrawal and alcohol dependence (Köhnke et al., 2005; Samochowiec et al., 2006). Single nucleotide polymorphism - rs27072 (*2319G>A*), located in 3'-UTR of *SLC6A3,* was observed in a few studies in relation to alcoholism (Ueno et al., 1999) and smoking (Ling et al., 2004), a symptom comorbid to alcoholism. Despite the absence of data indicating the functional significance of rs27072 polymorphism located in 3'-UTR of *SLC6A3* gene, there is some evidence pointing to the fact that sequence variation in the 3'-UTR of the gene could influence transcription, subcellular localization of mRNA, translation regulation, and/or maintenance of mRNA stability (Mignone et al., 2002).

Scientific interest in the dopamine D4 receptor (*DRD4*) gene with respect to alcohol dependence was evoked by a linkage study reporting QTL for alcoholism near the marker D11S1984 on the short arm of chromosome 11 in close proximity to the *DRD4* gene (Long et al., 1998). Moreover, animal studies suggested the role of the *DRD4* gene in novelty seeking behavior: *DRD4*-knockout mice showed a significantly reduced response to novel stimuli compared to wild-type mice (Dulawa et al., 1999), and their supersensitivity to the effects of alcohol were also revealed (Rubinstein et al., 1997).

The most intriguing results have been established while studying 48 bp VNTR polymorphism in exon III of *DRD4* gene. The presence of 2–11 repeat alleles has been reported, while the most frequent of those in Caucasians consisted of four (4R) and seven (7R) repeats (Ding et al., 2002). It has been reported that the four-repeat allele significantly slowed the blunted response to medication treatment, compared with the two-repeat and the seven-repeat 7R variants (Asghari et al., 1995). Primary findings indicated a positive correlation between the long-repeat alleles and novelty seeking scores (Ebstein et al., 1996; Benjamin et al., 1996), a trait usually assumed to be an endophenotype of alcohol dependence when increased. Subsequently, Skowronek et al. (2006) and Namkoong et al. (2008) argued that the long form of the *DRD4* gene is more prevalent in individuals with a high frequency of drug and alcohol use than in controls.

Some of the association studies of personality traits and disorders have focused on the role of polymorphisms in the promoter region of the *DRD4* gene. One of them is 120 bp VNTR polymorphism located 1.24 kb upstream of the initiation codon that occurs commonly as one or two repeats (Seaman et al., 1999). The 120-bp duplicated sequence was found to contain several transcription factor binding sites by in silico analysis (Seaman et al., 1999), suggesting that the duplication of this region might result in decreased promoter activity (D'Souza et al., 2004). Supporting this finding, Rogers et al. (2004) demonstrated that higher scores of novelty seeking were influenced by *DRD4* S/S genotype of 120 bp VNTR in the groups of bipolar and alcoholic families (Rogers et al., 2004).

Two further enzymes play a major role in dopamine and serotonin metabolism: catechol-O-methyltransferase (COMT) and monoamine oxidase A (MAO-A). Animal studies demonstrated that *MAOA*-knockout mice exhibited a modest increase in dopamine, and dramatic enhancement of aggressive traits (Shih and Chen, 1999). A functional 30 bp repeat polymorphism in the promoter region of the *MAOA* gene (*MAOA-LPR*), located approximately 1.2 kb upstream of the ATG codon, may alter transcriptional efficiency; the allele with 3 copies of the repeat sequence (L-allele) was transcribed about 2 times less efficiently than the allele with 4 copies of the repeat motif (H-allele)(Sabol et al., 2006; Balciuniene et al., 2002). The 3- and 4-repeat are the more common alleles, but others have also been found, including the 2-, 3.5-, 5-, and 6-repeat (Sabol et al., 2006). Molecular genetic studies revealed association of the low-activity 3-repeat allele of the *MAOA-LPR* with increased susceptibility to alcohol dependence (Schmidt et al., 2000; Contini et al., 2006).

The important role of *COMT* in the dopamine pathway has further been demonstrated by genetic and pharmacological studies in rodents. For example, *COMT*-knockout mice showed a number of behavioral and neurological effects, including alterations in anxiety-like and aggressive behavior (Gogos et al., 1998), and the more aggressive inbred strains had lower *COMT* expression levels in the hippocampus relative to the less aggressive strains (Fernandes et al., 2004). A common G to A polymorphism (rs4680) in the *COMT* gene at codon 158 results in the replacement of the amino acid methionine (Met) to valine (Val) in exon 4. This amino acid substitution results in threefold to fourfold variation in enzyme activity (Chen et al., 2004). The Val158 allele is commonly referred to as the high (H) activity allele, while the Met158 allele is termed the low (L) activity allele. The *COMT* activity distribution is trimodal, corresponding to the three *COMT* genotypes indicating additive effects on activity (Lachman et al., 1996). Published studies with respect to *COMT Val158Met* polymorphism are contradictory, indicating both association of high-activity allele (Serý et al., 2006; Enoch et al., 2006) and low-activity variant (Hallikainen et al., 2000; Tiihonen et al., 1999; Wang et al., 2001) and alcohol dependence.

Because there are many routes to alcoholism vulnerability, and many pathways to recovery, it is anticipated that many genetic loci will play a role. Hence, a gene–gene interaction approach might be more relevant in the study of alcoholism genetics. Moreover, serotonergic system genes are known to be involved in dopaminergic activity modulation, and vice versa (Ebstein, 2006). We therefore hypothesized that genes involved in serotonin and dopamine metabolic pathways might interact to increase susceptibility to alcohol dependence. In the present study we aimed to examine whether the main haplotype, or interaction effect between neurotransmitter system genes, is related to alcohol dependence. We also aimed to observe the best models accounting for higher risk in alcoholism predisposition. To diminish population and gender stratification, while analyzing our sample consisting of Russians, Tatars, and Bashkirs, only male subjects were analyzed, controlling for ethnicity.

THE RUSSIAN STUDY

The data used in this chapter is based on a study carried out on 302 subjects, aged 17–62 (M±SD = 40.8±11.0), identified with alcohol dependence and/or addiction. They originated from the Volga-Ural region of Russia, belonging to Russian (N = 113)(Slavic group of the Indo-European language family), Tatar (N = 91), and Bashkir populations (N = 98)(Turkic group of the Altaic language family). All enrolled individuals were hospitalized in the Republic Drug Abuse Hospital and Psychiatric Hospital (Ufa, Russia) with a diagnosis of acute alcoholic psychosis, made according to the ICD-10 criteria.

The control group consisted of 377 male volunteers with corresponding age and ethnicity: 130 Russians, 125 Tatars, 122 Bashkirs, aged 15–74 (M±SD = 38.1±14.3) without a personal or familial (first-degree) history of neuropsychiatric disorders and suicidal behavior. The study was approved by the Biological Ethics Committee of Bashkortostan Republic, and written informed consent was obtained from all the participants after they were acquainted with all the procedures. All participants were informed about the voluntary and confidential nature of their participation.

GENETIC EVALUATION

Genomic DNA was isolated from the peripheral blood using a standard phenol-chloroform technique. The PCR reaction was performed in a total volume of 15 μl containing 20–50 ng of genomic DNA, 200 μM of each dNTP, 0.5 μM of each primer (Syntol), 10× buffer (67 mM Tris-HCl, pH 8.8, 6.7 mM $MgCl_2$, 16.6 mM $(NH_4)_2SO_4$, 0.01% Tween-20 (Silex), and 0.5U of Taq polymerase (Silex). Thermal PCR conditions for 13 polymorphisms in *5-HTT (5-HTTLPR, STin2), HTR1B (861G/C), HTR2A (-1438A/G), MAO-A (MAOA-LPR, EcoRV), DRD2 (Taq1A, NcoI), DRD4* (120 bp VNTR in 5'-UTR, 48 bp VNTR in exon 3), *SLC6A3* (VNTR, *2319G/A), COMT (Val158Met)* genes, as well as primer sequences and restriction conditions, are demonstrated in Table 117-1. Digestion of PCR products was performed by overnight incubation with 5 U of corresponding restriction endonuclease (Fermentas) according to manufacturer's recommendations. Cleaved PCR products were resolved in 7% PAAG and visualized under ultraviolet light after staining with ethidium bromide. Each gel contained one lane of 100 bp ladder (2-Log DNA Ladder, New England Biolabs) to identify the fragments.

STATISTICAL ANALYSIS

Genotypes of all polymorphisms were evaluated for standard summary measures including genotyping rates, allele and genotype frequencies, and Hardy-Weinberg equilibrium, using Microsoft Excel macro PHARE version 2.1 (http://bioinformatics.org/macroshack/programs/PHARE/description.html).

For association analysis of individual genotypes and alleles, Fisher's exact test was carried out using the RxC program (http://www.marksgeneticsoftware.net/). Odds ratios (OR) with 95% confidence intervals (CI) were calculated using Woolf's method. For pairwise linkage disequilibrium and maximum likelihood analysis of haplotype distribution, the Haploview program was used.

We used logistic regression analyses implemented in STATA v.9.1 to investigate additive and interaction effects between SNPs identified in our best-fitting multi-locus models, and ICD-10 alcohol dependence. Separate logistic regression models were run for the Russian, Tatar, and Bashkir subjects. In the regression model, the phenotype (case-control status) served as the dependent variable, and covariates including all investigated polymorphisms of neurotransmitter system genes. Only those factors which contributed most to the difference between cases and controls, estimated through a likelihood ratio test (LR test) carried out on the variation of the log likelihood between two consecutive steps, would be retained in the final regression step. Dummy coding was used for *COMT Val158Met* genotypes, since additive effect of genotypes on *COMT* gene expression was established (Lachman et al., 1996), while ordinal coding was used for other polymorphic markers with the largest coding (1) indicated in Tables 117-4, 117-5, and 117-6.

Logistic regression analysis allows figuring out regression coefficients and OR (95%CI), permitting observation of the quantitative interrelation between dependent and independent variables while adjusting for the other variables in the model—the highest and the lowest scoring groups, for instance—without reducing the power (that occurs while performing multivariable analysis of variance with subsequent stratification of the sample). Significant positive or negative interrelation between independent factor and personality trait was established according to the sign of regression coefficient (β) and corresponding OR, with coding indicated in Tables 117-4, 117-5, 117-6. In the

TABLE 117-1 PCR CONDITIONS FOR INVESTIGATED POLYMORPHIC MARKERS

Gene	Polymorphism	Primers	Thermal PCR conditions	Restriction endonuclease	Allele (size, bp)	Reference
5-HTT	5-HTTLPR	5'-tcctccgctttggcgcctcttcc-3' 5'-tgggggttgcaggggagatcctg-3'	94°C - 4 min; 35 cycles: 94°C–30 sec; 66°C–1 min 20 sec; 72°C–1 min; extension: 72°C–10 min	-	*S (469) *L (512)	Wendland et al., 2006
	STin2	5'-gtcagtatcacaggctgcgag-3' 5'-tgttcctagtcttacgccagtg-3'	95°C - 3 min; 34 cycles: 95°C–30 sec; 55°C–30 sec; 72°C–1 min; extension: 72°C–10 min	-	*STin2.9 (249) *STin2.10 (266) *STin2.12 (299)	Fiskerstrand et al., 1999
HTR1B	861G>C	5'-gaaacagacgcccaacaggac-3' 5'-ccagaaaccgcgaaagaagaagat-3'	94°C - 5 min; 30 cycles: 94°C–20 sec; 58°C–1 min; 72°C–30 sec; extension: 72°C–10 min	HindII	*G (452 + 96) *C (310 + 142 + 96)	Hasegawa et al., 2002
HTR2A	–1438A >G (rs6311)	5'-aagctgcaaggtagcaacagc-3' 5'-aaccaacttatttcctaccac-3'	94°C - 4 min; 35 cycles: 94°C–20 sec; 60°C–1 min; 72°C–30 sec; extension: 72°C–10 min	MspI	*A (469) *G (243 + 226)	Nakamura et al., 1999
MAO-A	MAOA-LPR	5'-cccaggctgctccagaaac-3', 5'-ggacctgggcagttgtgc-3'	95°C - 3 min; 35 cycles: 94°C–40 sec; 60°C–30 sec; 72°C–50 sec; extension: 72°C–10 min	-	*3R (209) *4R (239) *5R (269)	Sabol et al., 2006
	EcoRV	5'-ttaaatggtctcgggaagg-3' 5'-gcccaatgacacagccttt-3'	94°C - 3 min; 30 cycles: 94°C–1 min; 57°C–2 min; 72°C–3 min; extension: 72°C–10 min	EcoRV	*T (456 + 32) *C (488)	Hotamisligil and Breakefield, 1991
DRD2	Taq1A (32806C >T) (rs1800497)	5'-acggctggccaagttgtcta-3' 5'-ccttcctgagtgtcatca-3'	94°C - 5 min; 35 cycles: 94°C–1 min; 50°C–1 min; 72°C–1 min; extension: 72°C–10 min	TaqI	*A1 (310) *A2 (180 + 130)	Chen et al., 1997
	NcoI (rs6275)	5'-cttgtccggctttacc-3' 5'-atcctgcagccatgg-3'	94°C - 3 min; 34 cycles: 94°C–30 sec; 59°C–30 sec; 72°C–1 min; extension: 72°C–10 min	Bsp19I	*N1 (454) *N2 (274 + 180)	Chen et al., 1997
DRD4	120 bp VNTR	5'-gaactctaagccgaccagag-3' 5'-gtgagcctcacacaggacaag-3'	94°C - 3 min; 35 cycles: 94°C–30 sec; 60°C–30 sec; 72°C–1 min; extension: 72°C–10 min	-	*S (274) *L (394)	Primer3[a]
	48 bp VNTR	5'-gcgactacgtggtctactcg-3' 5'-aggaccctcatggccttg-3'	96°C - 4 min; 35 cycles: 94°C–30 sec; 55°C–1 min; 72°C–1 min 30 sec; extension: 72°C–10 min	-	*2R (379) *3R (427) *4R (475) *5R (523) *6R (571) *7R (619) *8R (667)	Ding et al., 2002

(Continued)

TABLE 117-1 (CONTINUED)

Gene	Polymorphism	Primers	Thermal PCR conditions	Restriction endonuclease	Allele (size, bp)	Reference
SLC6A3	VNTR in 3'-UTR	5'-tgtggtgtagggaacggcctgag-3' 5'-cttcctggaggtcacggctcaagg-3'	94°C - 4 min; 31 cycles: 94°C–30 sec; 68°C–30 sec; 72°C–1 min 30 sec; extension: 72°C–10 min	-	*11R (523) *10R (483) *9R (443) *8R (403) *7R (363)	Ling et al., 2004
	2319G>A (rs27072)	5'-ccgtgtcttgtgttgctgta-3' 5'-acggggattctcagcaggtg-3'	95°C - 5 min; 30 cycles: 95°C–30 sec; 62°C–30 sec; 72°C–30 sec; extension: 72°C–10 min	MspI	*A (217) *G (135 + 82)	Ueno et al., 1999
COMT	Val158Met (rs4680)	5'-gagctgggggcctactgt-3' 5'-catgcacaccttgtccttga-3'	94°C - 4 min; 35 cycles: 94°C–30 sec; 62°C–30 sec; 72°C–30 sec; extension: 72°C–10 min	BstACI	*A (Met) (152) *G (Val) (131 + 22)	Primer3[a]

[a]–Primers designed under Primer 3 software (http://fokker.wi.mit.edu/primer3/input.htm)

case of OR >1, the presence of the highest coding allele was assumed to be the risk factor for alcohol dependence, while in the case of OR <1 the opposite correlation occurred. The critical level of model significance was established with p-value < 0.05. A correction for multiple testing was not performed, because only a single F-test assessing whether the independent variables (as a group) were associated with the outcome was employed (Katz, 2006).

THE OUTCOME

Genotype and allele frequencies of all investigated markers in three populations (Russian, Tatars, Bashkirs) are summarized in Table 117-2. To test Hardy-Weinberg equilibrium (HWE), the likelihood ratio test was used. Genotype and allele frequencies in alcohol-dependent and control subjects were not significantly different from those predicted under the HWE in all groups (P > 0.05).

Since enrolled individuals belong to Russian (Slavic group of the Indo-European language family), Tatar, and Bashkir populations (Turkic group of the Altaic language family), to differentiate the possible population stratification effects all statistical tests were performed in three populations from the Volga-Ural region of Russia separately. Allele frequencies of all investigated gene polymorphisms in alcohol-dependent (AD) patients and healthy controls are demonstrated in Table 117-2.

SINGLE GENE EFFECTS

Tests of single gene effects conducted between alcohol-dependent individuals and controls from the Russian population revealed an overrepresentation of the *HTR2A -1438A/G* G-allele ($\chi^2 = 6.37$, df = 1, p = 0.012, OR = 1.72, 95%CI 1.12–2.63) and G/G-genotype ($\chi^2 = 5.26$, df = 1, p = 0.022, OR = 2.06, 95%CI 1.10–3.86) in AD males compared to controls. Statistically significant differences in genotype distribution between AD patients and healthy donors were also observed with respect to *DRD4* gene polymorphisms (48 bp VNTR in exon 3 and 120 bp VNTR in 5'-UTR). Thus, individuals bearing *DRD4* 4R-allele of 48 bp VNTR polymorphism demonstrated protective effect for alcoholism predisposition ($\chi^2 = 4.81$, df = 1, p = 0.029, OR = 0.39, 95% CI 0.16–0.91) compared to those with alcohol addiction. Moreover, *DRD4* S-allele of 120 bp VNTR polymorphism was revealed to be the risk marker for alcoholism ($\chi^2 = 4.26$, df = 1, p = 0.039, OR = 2.07, 95%CI 1.03–4.14) when analyzed between AD and healthy males from the Russian population.

Analysis of allele and genotype frequencies distribution between AD and control individuals from the Tatar population showed the involvement of both *DRD2* and *MAOA* genes in alcoholism predisposition. Accordingly, *DRD2 NcoI* N1-allele was overrepresented in individuals with alcoholism ($\chi^2 = 3.88$, df = 1, p = 0.049, OR = 1.62, 95%CI 1.01–2.61) compared to control, and was concluded to be the risk factor for alcohol dependence in Tatar males. Additionally, the frequencies of *MAOA EcoRV* C-allele and *MAOA-LPR* H-allele have been observed to be significantly increased in AD males ($\chi^2 = 4.35$, df = 1, p = 0.037, OR = 1.58, 95%CI 1.03–2.43 and $\chi^2 = 9.75$, df = 1, p = 0.003, OR = 1.97, 95%CI 1.27–3.04, respectively) compared to control groups.

Subsequently performed analysis in individuals of Bashkir ethnicity revealed the increased frequency of

TABLE 117-2 GENOTYPED POLYMORPHISMS AND THEIR ALLELE AND GENOTYPE FREQUENCIES IN CASES AND CONTROLS FROM THREE POPULATIONS

			MAF					
			Russians		Tatars		Bashkirs	
Gene	SNP (rs no.)	Allele	Case (N = 113)	Control (N = 130)	Case (N = 91)	Control (N = 125)	Case (N = 98)	Control (N = 122)
5-HTT	5-HTTLPR	S	0.389	0.450	0.455	0.488	0.227	0.431
		L	0.611	0.550	0.545	0.512	0.773	0.569
	STin2	9R	0.027	0.017	0.007	0.010	0.036	0.007
		10R	0.393	0.383	0.359	0.402	0.393	0.330
		12R	0.580	0.600	0.634	0.588	0.571	0.663
HTR1B	861G>C	C	0.319	0.318	0.397	0.394	0.395	0.367
		G	0.681	0.682	0.603	0.606	0.605	0.633
HTR2A	-1438A>G	A	0.308	0.434	0.346	0.421	0.269	0.412
		G	0.692	0.566	0.654	0.579	0.731	0.588
MAOA	EcoRV	T	0.596	0.684	0.430	0.598	0.521	0.537
		C	0.404	0.316	0.570	0.402	0.479	0.463
	MAOA-LPR	3R	0.395	0.302	0.559	0.402	0.458	0.453
		4R	0.596	0.688	0.430	0.598	0.521	0.537
		5R	0.009	0.010	0.011	0.000	0.021	0.011
SLC6A3	VNTR	7-9R	0.241	0.167	0.233	0.210	0.219	0.148
		10-11R	0.759	0.833	0.767	0.79	0.781	0.852
	2319G>A (rs27072)	A	0.252	0.321	0.231	0.274	0.295	0.300
		G	0.748	0.679	0.769	0.726	0.705	0.700
DRD2	Taq1A (rs1800497)	A1	0.196	0.245	0.254	0.274	0.154	0.221
		A2	0.804	0.755	0.746	0.726	0.846	0.779
	NcoI (rs6275)	N1	0.570	0.624	0.392	0.511	0.500	0.562
		N2	0.430	0.376	0.608	0.489	0.500	0.438
DRD4	120 bp VNTR	S	0.238	0.131	0.221	0.183	0.188	0.240
		L	0.762	0.869	0.779	0.817	0.812	0.760
	48 bp VNTR	2R	0.106	0.069	0.083	0.135	0.061	0.099
		3R	0.048	0.049	0.060	0.063	0.061	0.116
		4R	0.846	0.882	0.839	0.739	0.867	0.736
COMT	Val158Met	L (Met)	0.514	0.585	0.448	0.493	0.403	0.453
		H (Val)	0.486	0.415	0.552	0.507	0.597	0.547

DRD4 4R-allele of 48bp VNTR marker ($\chi^2 = 4.58$, df = 1, p = 0.033, OR = 1.94, 95%CI 1.05–3.58) in AD individuals compared to healthy donors. Moreover, genes encoding enzymes involved in neurotransmitter degradation have been reported to be involved in alcohol dependence. Thus, *MAOA-LPR* H-allele and *COMT Val158Met* H-allele were shown to be the risk markers for alcoholism vulnerability ($\chi^2 = 8.67$, df = 1, p = 0.004, OR = 1.88, 95%CI 1.23–2.88 and $\chi^2 = 4.16$, df = 1, p = 0.041, OR = 2.85, 95%CI 1.04–8.10, respectively) in Bashkir males.

HAPLOTYPE BASED ANALYSIS

Since it has been established that gene-based haplotype—i.e., a combination of alleles located within a gene unit, independent of genetic heritability—potentially could be the most precise marker possible for a given gene, as it would contain all the variations in the gene (Hoehe, 2003), we conducted maximum-likelihood analysis of haplotype distributions in a pooled sample of cases and controls from each population separately. As a result of this analysis, the presence of pairwise linkage disequilibrium between polymorphic loci in *5-HTT (5-HTTLPR, STin2), MAOA (EcoRV, MAOA-LPR), DRD2 (Taq1A, NcoI), DRD4* (48 bp VNTR, 120 bp VNTR), and *SLC6A3* (VNTR, *2319A/G*) genes has been demonstrated. Lewontin coefficients (D') as a measure of linkage disequilibrium in Russians, Tatars, and Bashkirs between *5-HTTLPR* and *STin2* loci were 0.34, 0.38, and 0.34, correspondingly; between *MAOA-LPR* and *EcoRV* loci in *MAOA* gene were 0.29, 0.38 and 0.55; between *DRD2 Taq1A* and *NcoI* were 0.63, 0.52, and 0.61; between *DRD4* 48 bp VNTR and 120 bp VNTR – 0.30, 0.27, and 0.29; and between *SLC6A3* VNTR and *2319G/A* were 0.35, 0.65, and 0.45. Haplotype frequencies as well as haplotype association tests in three populations are shown in Table 117-3.

Analyses of the distribution of the estimated haplotype frequencies performed in the Russian males group revealed association of *SLC6A3* 9R*G haplotype (consisting of VNTR and *2319G/A* polymorphisms, respectively) with alcohol dependence ($\chi^2 = 4.92$, df = 1, p = 0.027, OR = 1.84, 95%CI 1.07–3.19). At the same time, the difference in haplotype distribution of *DRD4* between AD patients and control group have been demonstrated: *DRD4* 4R*L haplotype (consisting of 48 bp VNTR and 120 bp VNTR markers, respectively) was shown to be statistically significantly overrepresented in healthy individuals compared to those with alcoholism ($\chi^2 = 4.49$, df = 1, p = 0.034, OR = 0.62, 95%CI 0.39–0.97). These findings are congruent with results of single marker association tests reported in the Russian male group (indicated above); moreover, haplotype

TABLE 117-3 HAPLOTYPE FREQUENCIES OF *5-HTT, MAOA, DRD2, DRD4, SLC6A3* GENES IN ALCOHOLIC PATIENTS AND HEALTHY CONTROLS FROM THREE POPULATIONS

		Russian			Tatars			Bashkirs		
		Frequency			Frequency			Frequency		
Gene (Markers)	Haplotype	Case (N = 113)	Control (N = 130)	χ2 (P-value)	Case (N = 91)	Control (N = 125)	χ2 (P-value)	Case (N = 98)	Control (N = 122)	χ2 (P-value)
5-HTT (5-HTTLPR, STin2)	L*12R	0.314	0.279	0.716 (0.397)	0.278	0.296	0.153 (0.696)	0.084	0.237	3.47 (0.062)
	S*12R	0.266	0.317	1.512 (0.219)	0.355	0.382	0.278 (0.598)	0.487	0.422	0.453 (0.501)
	L*10R1	0.292	0.271	0.290 (0.590)	0.271	0.216	1.584 (0.208)	0.208	0.195	0.026 (0.873)
	S*10R	0.128	0.133	0.040 (0.842)	0.096	0.106	0.106 (0.745)	0.221	0.146	1.124 (0.289)
MAOA (VNTR, EcoRV)	L*T	0.386	0.379	0.022 (0.883)	0.258	0.467	16.652 **(<0.001)**	0.397	0.453	1.217 (0.270)
	H*C	0.228	0.179	1.53 (0.216)	0.366	0.293	2.176 (0.140)	0.383	0.263	6.04 **(0.015)**
	L*C	0.175	0.137	1.161 (0.282)	0.204	0.109	5.694 **(0.018)**	0.103	0.200	6.13 **(0.014)**
	H*T	0.211	0.305	4.427 **(0.036)**	0.172	0.131	1.247 (0.264)	0.117	0.084	1.16 (0.281)
DRD2 (Taq1A, NcoI)	A2*N2	0.403	0.337	1.855 (0.173)	0.370	0.424	1.026 (0.311)	0.487	0.514	0.071 (0.790)
	A2*N1	0.401	0.418	0.113 (0.737)	0.376	0.304	1.997 (0.158)	0.359	0.266	1.061 (0.302)
	A1*N1	0.169	0.207	0.970 (0.325)	0.232	0.195	0.740 (0.390)	0.141	0.177	0.213 (0.644)
	A1*N2	0.027	0.038	0.378 (0.538)	0.022	0.079	3.794 **(0.050)**	0.013	0.043	0.573 (0.449)
DRD4 (VNTR in exon 3, VNTR in 5'-UTR)	4R2*L	0.702	0.792	4.493 **(0.034)**	0.701	0.660	0.703 (0.402)	0.722	0.634	3.52 (0.061)
	4R*S	0.119	0.094	0.789 (0.374)	0.150	0.126	0.438 (0.508)	0.123	0.126	0.006 (0.941)
	2R*L	0.105	0.075	1.257 (0.262)	0.095	0.151	2.659 (0.103)	0.080	0.128	2.38 (0.123)
	2R2*S	0.074	0.039	2.741 (0.098)	0.054	0.063	0.116 (0.733)	0.075	0.112	1.712 (0.191)

(Continued)

TABLE 117-3 (CONTINUED)

Gene (Markers)	Haplotype	Russian Frequency Case (N = 113)	Control (N = 130)	χ2 (P-value)	Tatars Frequency Case (N = 91)	Control (N = 125)	χ2 (P-value)	Bashkirs Frequency Case (N = 98)	Control (N = 122)	χ2 (P-value)
SLC6A3 (VNTR, 2319G/A)	10R*G	0.545	0.564	0.184 (0.668)	0.560	0.538	0.212 (0.645)	0.545	0.562	0.130 (0.718)
	10R*A	0.215	0.267	1.673 (0.196)	0.210	0.252	1.063 (0.303)	0.289	0.288	0.0 (0.991)
	9R*G	0.203	0.120	4.921 **(0.027)**	0.207	0.195	0.112 (0.737)	0.160	0.145	0.194 (0.660)
	9R*A	0.037	0.049	0.364 (0.547)	0.023	0.015	0.300 (0.584)	0.006	0.005	0.0 (1.00)

[1] 9R- and 10R-alleles were included together, since the frequency of 9R-allele was low.

[2] 2R-, 3R-, 5R-, 6R-, 7R-alleles were included together versus 4R-allele, due to their functional significance (Asghari et al., 1995)

Statistically significant P-values are indicated in bold.

analysis indicated a distinct protective haplotype. We also investigated possible association of *MAOA* gene haplotypes (consisting of *MAOA-LPR* and *EcoRV* polymorphisms, correspondingly) with alcohol dependence. Thus, a protective role of *MAOA* H*T haplotype in alcohol dependence has been observed ($\chi^2 = 4.43$, df = 1, p = 0.036, OR = 0.61, 95%CI 0.38–0.97), proving the higher ability of haplotype construction in association with the trait of interest.

Subsequently, we performed analysis aimed to distinguish differences in haplotype frequencies distribution between AD patients and controls of Tatar descent. As a result of this analysis, marginal decrease in *DRD2* A1*N2-haplotype frequency (consisting of *Taq1A* and *NcoI* polymorphisms, correspondingly) in AD Tatar males was demonstrated ($\chi^2 = 3.79$, df = 1, p = 0.050, OR = 0.27, 95%CI 0.06–0.99). Moreover, the involvement of distinct *MAOA* gene haplotypes (consisting of *MAOA-LPR* and *EcoRV* polymorphisms, correspondingly) in alcohol dependence was revealed: L*C and L*T-haplotypes were prevalent in AD patients ($\chi^2 = 5.94$, df = 1, p = 0.018, OR = 2.11, 95%CI 1.13–3.94) and control group ($\chi^2 = 16.65$, df = 1, p < 0.001, OR = 2.52, 95%CI 1.59–4.00) respectively, indicating the highest impact of *EcoRV* polymorphism compared to *MAOA-LPR* in alcohol dependence.

Haplotype analysis conducted in individuals of Bashkir origin demonstrated significant increase in the frequency of *MAOA* H*C-haplotype consisting of *MAOA-LPR* and *EcoRV* polymorphisms, correspondingly ($\chi^2 = 6.04$, df = 1, p = 0.015, OR = 1.75, 95%CI 1.11–2.76), and decrease in L*C-haplotype frequency ($\chi^2 = 6.13$, df = 1, p = 0.014, OR = 0.47, 95%CI 0.25–0.86) among AD males compared to controls.

LOGISTIC REGRESSION MODELS

The present study also included a logistic regression to test the cause–effect relationship between genotypes (alleles) interaction, as well as single marker effects and alcohol dependence. Since some genotypes (alleles) have been chosen as a reference, the models estimate the relative risk of being alcohol-dependent while bearing another allele (genotype). As a result of logistic regression, differences occurred in covariates included in the best-fit models between three investigated populations.

For instance, logistic regression performed in Russians revealed 4 models evaluating the relative risk (or protective effect) of becoming alcohol-dependent. Model 1 in Table 117-4 indicates that in males, the presence of A-allele of *SLC6A3 2319G/A* polymorphism might play a protective role for alcoholism (P = 0.021, OR = 0.486) compared to those possessing G/G-genotype. Model 2 provides estimation of the effects of *DRD2*SLC6A3* gene polymorphisms interaction potentially affecting the risk of being alcohol dependent, demonstrating that carriers of *SLC6A3* A-allele together with *DRD2 Taq1A* A1-allele have about four times lower probability of becoming alcohol dependent (P = 0.029, OR = 0.237) compared to those with *SLC6A3* G/G- and *DRD2* A2/A2-genotypes. Surprisingly, once controlled for the *DRD2*SLC6A3* interaction, the relative risk associated to the main effect of *SLC6A3* G/G-genotype (Model 1) becomes no longer statistically significant. Inclusion of the *5-HTTLPR* marker in one logistic model (Model 3) revealed an even stronger role of *DRD2*SLC6A3* interaction in alcohol dependence. Accordingly, male carriers of *5-HTTLPR* L-allele with the constant other factors showed 2.58 times higher risk for being alcohol dependent compared to S/S-carriers (P = 0.023, OR = 2.586). Finally, Model 4 tests whether such genetic impact on the risk of becoming alcohol addicted may be mediated by the *HTR2A -1438A/G* polymorphism. The results clearly prove that once controlled for the other factors, possessing the *HTR2A* G/G-genotype increases the risk of alcoholism 2.6 times (P = 0.008, OR = 0.383) in Russian males. Inclusion of all these covariates in the

TABLE 117-4 LOGISTIC REGRESSION ANALYSIS OF POLYMORPHISMS IN NEUROTRANSMITTER SYSTEM GENES IN CASES AND CONTROLS FROM RUSSIAN POPULATION

Variables	Model (1)	Model (2)	Model (3)	Model (4)
SLC6A3 (2319A/G)				
A-allele	**0.486 (0.26-0.89) (P=0.021)**	0.840 (0.38-1.84) (P=0.664)	0.877 (0.39-1.95) (P=0.747)	0.981 (0.43-2.22) (P=0.964)
G/G	Referent	Referent	Referent	Referent
SLC6A3 (2319A/G)*DRD2 (TAQIA)				
A*A1		**0.237 (0.06-0.86) (P=0.029)**	**0.195 (0.05-0.74) (P=0.016)**	**0.166 (0.04-0.66) (P=0.010)**
G/G*A2/A2		Referent	Referent	Referent
5-HTTLPR				
L-allele			**2.586 (1.14-5.87) (P=0.023)**	**2.563 (1.11-5.90) (P=0.027)**
S/S			Referent	Referent
HTR2A (-1438A>G)				
A-allele				0.383 (0.19-0.78) (P=0.008)
G/G				Referent
Number of observation	243	243	243	243
Log likelihood	–115.66	–112.18	–109.52	–105.80
LR chi^2	5.45	12.41	17.71	25.17
P-value	**0.019**	**0.006**	**0.001**	**<0.001**
R^2	0.023	0.052	0.075	0.106
LR test	–	6.96	5.30	7.46
P-value	–	**0.031**	**0.021**	**0.006**

In bold are OR (95%CI) statistically significant at the 5% level.

final model (Model 4) significantly improves the fit of the model (the log likelihood decreases, passing from -109.52 to -105.8; LR test = 7.46; P-value = 0.006) and results in a 10.6% prediction of alcohol dependence.

With respect to AD patients and control males of Tatar origin, logistic regression reported the presence of three models having impacts on alcoholism predisposition. The logistic regression analysis results for Tatar individuals are summarized in Table 117-5. Model 1 demonstrated statistically significant involvement of *HTR2A*MAOA* gene polymorphisms interaction in alcohol dependence: male carriers of *HTR2A -1438A/G* A-allele together with *MAOA EcoRV* T-allele have 10.28 times higher risk of alcoholism compared to those bearing *HTR2A* G/G-genotype and *MAOA* C-allele (P = 0.002, OR = 10.277). Additionally, logistic regression analysis that also took *5-HTTLPR* into consideration (Model 2) confirmed the previously demonstrated association and improved Model 1 (the log likelihood decreases, passing from -90.4 to -76.3; LR test = 7.10; P-value = 0.029). Model 2 indicated that in those controlled for *HTR2A*MAOA* genes interaction, the presence of the *5-HTTLPR* L-allele significantly increased the probability of becoming alcohol dependent (P = 0.043, OR = 2.601). Independently from the first two models, Model 3 revealed the involvement of *DRD2*COMT* genes interaction in alcohol dependence: to be more precise, the protective effect for alcoholism predisposition was detected in males bearing N1-allele of *DRD2 NcoI* polymorphism together with *COMT Val158Met* H/L-genotype (P = 0.008, OR = 0.018) compared to those carrying *DRD2* N1-allele and *COMT* H/H-genotype. Hence, in Tatar males, two models of alcohol dependence including different genes were established, with total effect explained by Model 2 equal to 12.2% and by Model 3 equal to 10.6%.

Estimations of the effects of neurotransmitter system genes as covariates potentially affecting the risk of being alcohol addicted in Bashkir males (Table 117-6) clearly prove the significant impact of the *MAOA EcoRV* locus and *COMT Val158Met* marker independently: possessing either *MAOA* T-allele (Model 1; P = 0.024,

TABLE 117-5 LOGISTIC REGRESSION ANALYSIS OF POLYMORPHISMS IN NEUROTRANSMITTER SYSTEM GENES IN CASES AND CONTROLS FROM TATAR POPULATION

Variables	Model (1)	Model (2)	Model (3)
HTR2A (-1438A>G)*MAOA (ECORV)			
A*T	**10.277 (2.42–43.67) (P=0.002)**	**14.508 (2.89–72.84) (P=0.001)**	
G/G*C	Referent	Referent	
5-HTTLPR			
L-allele		**2.601 (1.03–6.57) (P=0.043)**	
S/S		Referent	
DRD2 (NCOI)*COMT (VAL158MET)			
N1*H/H			Referent
N1*H/L			**0.018 (0.001–0.344) (P=0.008)**
N1*L/L			0.083 (0.002–4.41) (P=0.220)
Number of observation	216	216	216
Log likelihood	–90.4	–76.3	–49.1
LR chi^2	13.80	21.17	11.58
P-value	**0.0032**	**0.0008**	**0.041**
R^2	0.071	0.122	0.106
LR test	–	7.10	–
P-value	–	0.029	–

In bold are OR (95%CI) statistically significant at the 5% level.

OR = 1.94) or *COMT* H-allele (Model 2; P = 0.028, OR = 2.85) significantly increased the probability of alcoholism. In order to detect the effect of genes interaction, *COMT Val158Met* and *MAOA-LPR* polymorphisms were included as an interaction in the subsequent model (Model 3), and were reported to improve significantly the model fit (the log likelihood decreases, passing from -117.98 to -101.70; LR test = 8.89; P-value = 0.031) compared to Model 2, although insignificant effect of such interaction was observed. However, while controlled for *MAOA EcoRV* and *COMT Val158Met* markers, the tendency to increase the risk of alcoholism in *MAOA-LPR* H-allele carriers compared to males with L/L-genotype was revealed (P = 0.063, OR = 10.520) in

TABLE 117-6 LOGISTIC REGRESSION ANALYSIS OF POLYMORPHISMS IN NEUROTRANSMITTER SYSTEM GENES IN CASES AND CONTROLS FROM BASHKIR POPULATION

Variables	Model (1)	Model (2)	Model (3)
MAOA (ECORV)			
T-allele	**1.94 (1.09–3.44) (P=0.024)**		**2.815 (1.23–6.44) (P=0.014)**
C-allele	Referent		Referent
COMT (VAL158MET)			
Val (H-allele)		**2.85 (1.11–7.27) (P=0.028)**	**9.616 (1.06–86.74) (P=0.044)**
Met/Met(L/L)		Referent	Referent

(Continued)

TABLE 117-6 (CONTINUED)

Variables	Model (1)	Model (2)	Model (3)
MAOA-LPR			
H-allele			10.520 (0.88–126.15) (P=0.063)
L-allele			Referent
COMT (VAL158MET)*MAOA-LPR			
H*H			0.135 (0.01–1.68) (P = 0.120)
L*L			Referent
Number of observation	220	220	220
Log likelihood	–133.93	–117.98	–101.70
LR chi^2	5.18	5.23	11.94
P-value	0.022	0.022	0.018
R^2	0.019	0.022	0.056
LR test	–	–	8.89
P-value	–	–	0.031

In bold are OR (95%CI) statistically significant at the 5% level.

Model 3, with the summarized effect of including all covariates being 5.6%.

DISCUSSION

In the present study we investigated single gene and haplotype effects, as well as effects of regression models involving additive and interaction effects of thirteen neurotransmitter system gene polymorphisms in alcohol dependence in healthy individuals of Tatar, Russian, and Bashkir descent, from the Volga-Ural region of Russia.

It is known that stratification effects may occur as a result of population diversity or other unknown causes. Since we recruited individuals differing in ethnicity (Tatars, Russians, and Bashkirs) supposed to be genetically close to Caucasians, the possibility of population stratification might occur. One of the sources includes differences in genotype frequencies of studied markers between individuals with different ethnicity.

Allele frequencies for 5-*HTTLPR* and *STin2* polymorphisms in control groups from all the studied populations did not significantly differ from each other, and were in accordance with those reported for Caucasians (Hranilovic et al., 2004; Lang et al., 2004; Vormfelde et al., 2006). With respect to *HTR1B 861G/C* polymorphism, significant difference in allele frequencies was observed between Tatars and Bashkirs involved in the present study, and individuals of German and Croatian origin (Stefulj et al., 2004), due to the increased frequency of the minor allele in our samples; however, allele frequencies of *861G/C* locus in the group of Russian males were congruent with those demonstrated in European populations (Stefulj et al., 2004). No significant differences were observed in *HTR2A -1438A/G* allele frequencies distribution between the three populations from the Volga-Ural region of Russia and Caucasians (Preuss et al., 2001; Rybakowski et al., 2006). The frequency of *MAOA-LPR* alleles in all investigated populations was congruent with that reported for individuals of Portuguese descent (Contini et al., 2006) and of German descent (Samochowiec et al., 1999), while *MAOA EcoRV* allele frequency was similar to that reported in Japanese subjects (Nakatome et al., 1998). *DRD2 TaqIA* allele frequencies were observed to be slightly increased in our control subjects compared to Caucasians (Reuter et al., 2006); however, the differences were not significant. With respect to *DRD2 NcoI* polymorphism, allele frequencies in Tatars and Bashkirs were similar to those reported for Chinese populations (Mo et al., 2007) and differed from those of individuals from Sweden (Rosmond et al., 2001) and Russia (Moscow)(Monakhov et al., 2008), while Russian males involved in the present study demonstrated the opposite relation. Comparison of allele frequencies in the *SLC6A3 VNTR* marker revealed a similar pattern of distribution between our samples of three ethnicities and individuals from Polish and Australian populations (Jorm et al., 2000; Samochowiec et al., 2001). However, significant differences in allele frequencies of *SLC6A3 2319G/A* marker were observed between our control groups and healthy individuals from both Caucasian (Feng et al., 2005) and Asian origin (Ling et al., 2004), due to increased frequency of the minor alleles in our sample. The comparative analysis of allele frequencies distribution of *DRD4* 120 bp VNTR polymorphism showed that healthy Bashkir males were close to a Chinese population (Li et al., 2004), while Russians and Tatars reported similarity to Caucasians (Szantai et al., 2005). These results could be explained by previously published data concluding that Bashkirs have 39.6% of east Eurasian mtDNA lineages, while Tatars have only 12% (Khusnutdinova et al., 2008). Interestingly, the subjects involved in the current study reported a significantly different pattern of alleles distribution with respect to *DRD4* 48 bp VNTR: Russians were characterized by overrepresentation of 4R allele, but the frequencies of 2R and 3R alleles were close to those reported in Caucasians (Mill et al., 2002); 4R-allele frequency in Tatars and Bashkirs was similar to that reported in a Chinese population (Li et al., 2000), 2R-allele frequency was at odds with that of Europeans (Mill et al., 2002); however, 3R-allele was significantly overrepresented. Moreover, it should be noted that the 7R-allele was very rare in all the studied ethnicities, as in Asian populations (Li et al., 2000) with overrepresentation of 4R-allele in Russians, and of 3R-allele in Tatars and Bashkirs. However, no differences in allele frequencies of *COMT Val158Met* polymorphism were demonstrated between Russians, Tatars, Bashkirs, and Caucasians (Ott et al., 2005; Golimbet et al., 2007). Assuming the demonstrated allele frequencies distributions, individuals of Tatar, Russian, and Bashkir descent might be concluded to be highly related to Caucasians with respect to the majority of the studied polymorphisms, although individual patterns of distributions were established that point to the necessity of a separate analysis of the three populations. Since other sources of population stratification involve differences in traditions, social norms, and religion (Leung et al. 1992)(orthodox in Russians, Muslim in Tatars and Bashkirs), it should be noted that differences in religious relationship to alcohol consumption probably modulates differences in lifestyle between the studied populations.

The confounding effect of gender was observed with respect to alcohol vulnerability (Cloninger, 1987). The known gender variations in alcohol dependence probably are caused by hormonal and social influences on the individual during various phases of development (Costa et al., 2000). Since both animal and human studies showed the involvement of sexual hormones in modulation of serotoninergic (Mann et al., 2001; Lu et al., 2003) and dopaminergic activity (Becker, 1999; Disshon et al., 1998; Thompson and Certain, 2005), the present study included only male subjects.

THE ROLE OF THE SEROTONIN TRANSPORTER GENE (*5-HTT*) IN ALCOHOL DEPENDENCE

Despite the majority of studies determining the role of dopaminergic system genes in alcohol addiction, published findings also determined the role of the serotoninergic system in this disorder—however, with the comorbidity of impulsive symptoms. Several case-control association studies and even two meta-analyses (Gorwood et al., 2004; Feinn et al., 2005) suggest an association between the low activity variant (S-allele) and alcohol dependence (Nellissery et al., 2003; Konishi et al., 2004) or the risk of relapse in abstinent alcohol-dependent patients (Pinto et al., 2008). Nevertheless, other studies have described an association of the long allele of this polymorphism with alcohol dependence (Hu et al., 2005; Kweon et al., 2005; Bleich et al., 2007; Gokturk et al., 2008; Johnson et al., 2008b). Our results, indicating the involvement of the *5-HTTLPR* L-allele (while controlled for the other factors in regression models) in increased predisposition to alcohol dependence, both in Tatars and Russian males, are congruent with data revealed in Korean populations (Kweon et al., 2005). Since serotonin reuptake inhibitors citalopram and fluvoxamine have been reported to have a positive effect on craving and relapse (Angelone et al., 1998), it is theoretically possible that subjects with the L-allele (associated with the higher levels of serotonin reuptake) are more vulnerable to alcohol dependence. Furthermore, the long allele has been associated with obsessive-compulsive disorder (Bengel et al., 1999), which shares various psychological features with alcohol craving. Another study investigating expression in alcohol-dependent individuals (Philibert et al., 2008) is congruent with our findings, since demonstrates that subjects with a lifetime history of alcohol dependence had higher levels of *5-HTT* mRNA. Moreover, it has been reported that L/L-genotype carriers might develop acute tolerance to alcohol effects more rapidly compared to individuals bearing S-allele (Corbin et al., 2006).

In contrast, association of the *5-HTTLPR* S-allele was seen among individuals with alcohol dependence complicated by either comorbid psychiatric condition, an early-onset alcohol dependence subtype, or antisocial behavior (Hallikainen et al., 1999; Feinn et al., 2005). Although the present study involved individuals with a severe alcohol dependence (acute alcoholic psychosis) considered to be genetically determined, the revealed association seems to be opposite to meta-analysis results confirming the S-allele to be an alcoholism predictor (Feinn et al., 2005). However, our preliminary study (Gorbunova et al., 2002) demonstrating the decreased frequency of S/S genotype in Russian alcoholic males with age of onset over 35 years—which suggested selection aimed at elimination of short allele homozygotes among patients with this disease—confirms that there are probably different genetic prerequisites for early-onset and late-onset alcohol dependence. Some studies have indicated that association of a short allele of the *5-HTT* gene with alcoholism might be observed, when compared to case-control subjects with antisocial history (Wu et al., 2008). Despite the similar mechanisms assumed to predispose an individual to addictive disorders, Saiz et al. (2009) demonstrated the differences in *5-HTTLPR* allele frequencies between heroin-dependent and alcohol-dependent patients: there was an excess of L-allele carriers in alcoholic patients in comparison to the heroin-dependent group (Saiz et al., 2009).

One of the major possibilities for the inconsistent results in the abovementioned gene–personality association studies might be due to bias in determining the subtype of alcoholism. For instance, Cloninger (1987) proposed two distinct types of alcohol use disorders: type I alcoholism characterized by late-onset drinking behavior, high harm avoidance, and low novelty seeking, and type II alcohol use disorders characterized by an early-onset dependence, more behavioral disturbances including antisocial behavior, low harm avoidance, and high novelty seeking. Wiesbeck et al. (2004) reported that higher harm avoidance was associated with the S-allele in primary alcohol dependence. Accordingly, predisposition to alcohol dependence might be explained by an indirect influence of S-allele on behavioral disinhibition (Twitchell et al., 2001) and on anxiety-related personality traits (Samochowiec et al., 2001) rather than by the presence of the S-allele itself.

Despite the functional significance of the *STin2* polymorphism (MacKenzie and Quinn, 1999), no single or haplotype effect of this marker on alcohol dependence was reported in the present study. However, previously reported findings revealed an excess of STin2.10 allele in alcohol-dependent males compared to healthy controls (Mokrović et al., 2008), while controversial results were also demonstrated—L*10 haplotype carriers were characterized by better alcohol treatment outcome (Florez et al., 2008).

THE ROLE OF THE SEROTONIN 2A RECEPTOR GENE (*HTR2A*) IN ALCOHOL DEPENDENCE

Published findings have indicated that the A-allele of *HTR2A -1438A/G* polymorphism is associated with impulsive traits in alcohol-dependent individuals (Preuss et al., 2001, Nomura et al., 2006). These data support the hypothesis that serotonergic deficit might be involved in alcoholism predisposition, via the involvement of the *HTR2A* high-activity A-allele (Parsons et al., 2004) in impulsive behavior, often assumed to be an important aspect of risk assessment in a variety of clinical situations such as pathological gambling, eating disorders, and many other uncontrollable behaviors. Since alcohol and nicotine abuse and dependence are multifactorial traits that frequently co-occur, the study conducted by Polina et al. (2009) seems to be in agreement with those previously mentioned: a higher frequency of A-allele was observed in a group of smokers than in nonsmoking controls. Accordingly, the revealed high risk of alcoholism in *HTR2A* A-allele male carriers from the Tatar population in the present study is congruent with the studies mentioned above (Preuss et al., 2001, Nomura et al., 2006).

On the other hand, we observed a single effect of the *HTR2A* G/G-genotype, as well as its involvement in the risk model of alcoholism in Russian males. This finding might be explained via the study of Nakamura et al. (1999) suggesting an association between the *HTR2A* G-allele and alcoholism mediated by the presence of inactive aldehyde dehydrogenase-2 (ALDH2). Unfortunately, we have not screened our samples for ALDH2 forms. Assuming a lower *HTR2A* receptor binding in the prefrontal cortex of alcohol-dependent patients with a positive family history of alcoholism (Underwood et al., 2008), and lower *HTR2A* receptor binding associated with the *HTR2A* G-allele (Turecki et al., 1999), the involvement of the G-allele in alcoholism risk demonstrated in Russian males is possible.

THE ROLE OF THE SEROTONIN 1B RECEPTOR GENE (*HTR1B*) IN ALCOHOL DEPENDENCE

Although *HTR1B* was considered to be the candidate gene for alcohol dependence due to its location at the site of several mouse alcohol QTLs on chromosome 9 (Phillips et al., 1998) and reported alcohol preference in *HTR1B*-knockout mice (Crabbe et al., 1996), the present study failed to reveal any differences in allele/genotype frequencies distribution of *HTR1B -861G/C* polymorphism between AD patients and controls. Despite several studies suggesting that the *HTR1B* -861C allele was implicated in alcohol dependence with antisocial behavior (Lappalainen et al., 1998), others indicated that the *HTR1B* -861G allele was more frequent in addicted males (Fehr et al., 2000). The possible role of *HTR1B -861G/C* polymorphism in alcoholism might be confounded by interaction with other genes of small effect. Namely, the study conducted in a Japanese population showed overrepresentation of the C-allele of the *861G/C* marker only in alcoholics with inactive ALDH2, compared to control group (Hasegawa et al., 2002).

A few studies revealed no association of this marker in the *HTR1B* gene in substance-dependent or alcohol-dependent patients (Gorwood et al., 2002; Kranzler et al., 2002; Sinha et al., 2003). Moreover, contrary to the previous data (Crabbe et al., 1996), subsequent findings indicated no difference in ethanol intake under free choice conditions (i.e., ethanol solution vs. water) between *HTR1B*-knockout mice and wild-type controls (Gorwood et al., 2002), which might explain the failure to detect an association in case-control studies. Another possible explanation of the discrepancy between previous studies (Lappalainen et al., 1998; Fehr et al., 2000) and ours is that the partial association was observed due to the absence of comorbid antisocial personality disorder and aggressive behavior in our AD patients.

THE ROLE OF THE DOPAMINE D2 RECEPTOR GENE (*DRD2*) IN ALCOHOL DEPENDENCE

The dopaminergic system was traditionally associated with the rewarding and reinforcing effects of drugs of abuse via increasing the concentration of dopamine in the extracellular spaces of several brain regions. However, positron emission tomography (PET) imaging studies have consistently shown long-lasting decreases in dopamine D2-like receptors on the surfaces of membranes in the brains of drug-addicted subjects, including alcoholics (Ritchie and Noble, 2003; Laine et al. 2001). Hence, dopaminergic system gene variants related to decrease in dopaminergic activity might be hypothesized to result in alcohol sensitivity. The majority of case-control studies as well as meta-analysis including AD patients from Asian or European populations reported an association of the *DRD2 TaqIA* A1-allele (demonstrated to be involved in downregulation of dopamine receptor activity) and alcoholism, mainly with antisocial comorbidity (Conner et al., 2005; Berggren et al., 2006; Munafò et al., 2007; Namkoong et al., 2008; Wu et al., 2008; Pinto et al., 2009). Contrary to these findings, we demonstrated the role of the *DRD2* A2/A2-genotype, while controlled for *SLC6A3* rs27072 polymorphism, in higher risk of alcoholism in Russians, supported by haplotype analysis carried out in Tatar males revealing the protective role of *DRD2* A1*N2-haplotype in alcohol dependence. Contradiction between our results and published studies might be not only due to ethnic differences, but due to the sensitivity of A1-allele carriers to stress (Kreek and Koob, 1998) and disinhibitory behavior (Lu et al., 2001) —factors known to mediate vulnerability to distinct alcoholism type. Thus, Ponce et al. (2008) supposed that A1-allele overrepresentation in AD patients was observed only in those with comorbid dissocial personality disorder; however, the subjects involved in the present study have not demonstrated any comorbidity. Additive effect of *DRD2* and *MAOA* genes is probably involved in alcoholism with anxiety symptoms: it was presented that A1/A1-genotype carriers might have a higher risk of alcoholism while possessing the *MAOA-LPR* low-activity allele (Huang et al., 2007). However, opposite findings indicated that the presence of A1/A1-genotype might predispose to alcoholism in individuals with antisocial personality disorder only in the case of presence of the *MAOA-LPR* high-activity allele (Wang et al., 2007).

Despite the similarity with respect to *DRD2 TaqIA* involvement in alcoholism in reported data, two studies (Hallikainen et al., 2003; Munafò et al., 2005) established that A1-allele carriers were found to have less alcohol intake, which is opposite to many studies associated this allele, with a higher risk for alcohol dependence, and this was confirmed by our findings.

Recent studies were devoted to assessing the role of distinct haplotype of the *DRD2* gene rather than single marker effect. To be more precise, Preuss et al. (2007) revealed that a distinct haplotype consisting of the *DRD2* A1-allele was significantly more common in heavy-drinking individuals, while that comprising A2-allele was overrepresented in individuals with smoking. Another study demonstrated that haplotypes containing either A1-allele or A2-allele occurred with a higher frequency in alcoholics with different comorbidity (Kraschewski et al., 2009), which suggests that not only the presence/absence of the A1-allele has a role in alcoholism predisposition, but also that the other polymorphic markers in linkage disequilibrium with *TaqIA* or distinct haplotypes are involved.

As a result of haplotype analysis, we demonstrated the protective role of *DRD2* A1*N2 haplotype in alcoholic addiction in Tatar males. Our data to some extent are congruent with results reported by Chen et al. (1997) and Luo et al. (2005) establishing an overrepresentation of *DRD2* A1*N1 haplotype in Han Chinese and Mexican-American patients, respectively; however, the differences in ethnicity might cause variations in risk haplotype between these studies.

THE ROLE OF THE DOPAMINE D4 RECEPTOR GENE (*DRD4*) IN ALCOHOL DEPENDENCE

Pioneer studies conducted with respect to *DRD4* 48 bp VNTR (Ebstein et al., 1996, Benjamin et al., 1996) suggested the role of longer repeat alleles in personality traits characterized by a tendency to novel stimuli, and were confirmed in several international publications (Strobel et al., 1999; Tomitaka et al., 1999). These findings catalyzed other researchers to conduct studies aimed at detecting possible associations between *DRD4* 48 bp VNTR long alleles and alcoholism, since increased novelty-seeking personality trait was found to be common in distinct type of alcoholics ("type 2 alcoholics") according to the Cloninger's classification (1987). However, subsequent studies revealed inconsistent results by using different statistical approaches in samples of different ethnicity (Bau et al., 2001, Hutchison et al., 2002, Soyka et al., 2002). Other researchers indicated that association of the DRD4 VNTR polymorphism with personality traits such as novelty seeking, possibly predisposing to alcoholism, may be of relevance (Ray et al., 2009).

In the current study we reported higher alcoholism risk in Russian males that are carriers of the *DRD4* 48 bp VNTR longer-repeat allele, while the *DRD4* 4R*L haplotype (consisting of 48 bp VNTR and 120 bp VNTR, respectively) was shown to have a protective role in predisposition to this disease. Our findings are at odds with those reporting an overrepresentation of *DRD4* 4R/4R-genotype in children of parents with no alcoholism history, compared to those with alcohol-dependent parents (Namkoong et al., 2008). Moreover, a few studies reported association of the *DRD4* seven-repeat allele with opioid dependence in Sephardic Jews and Israeli Arabs (Kotler et al., 1997) and methamphetamine abuse in a Chinese population (Chen et al., 2004), that are congruent with our data.

There are only a few findings estimating the involvement of *DRD4* 120bp VNTR in personality traits in alcohol-dependent patients. Accordingly, Rogers et al. (2004) demonstrated that *DRD4* 120 bp VNTR S/S carriers had higher novelty seeking, in a study involving individuals from alcoholic families. We also observed a higher frequency of S-allele of 120 bp VNTR in Russian AD individuals assumed to possess a higher tendency to novel stimuli. Since higher *DRD4* expression was associated with the S-allele of 120 bp VNTR (D'Souza et al., 2004; Keresztúri et al., 2007) and with the C-allele of *-521C/T* polymorphism (Okuyama et al., 1999), the study of Szilagyi et al. (2005) seems to be congruent with our findings: association of higher-activity genotype (-521C/C) with substance dependence was demonstrated.

Although Li et al. (2004) suggested the logistic regression model involving *COMT* and *DRD4* genes, we were unable to observe interaction between these genes in any ethnicity; however, a logistic model also including dopaminergic system genes interaction (*DRD2*COMT*) demonstrated in Tatar males was shown to be significantly involved in alcoholism risk.

THE ROLE OF THE DOPAMINE TRANSPORTER GENE (*SLC6A3*) IN ALCOHOL DEPENDENCE

Genetically mediated variability in the dopamine transporter gene (*SLC6A3*) (5p15.3) functioning can also contribute to individual sensitivity to alcohol dependence. Several published studies have reported a higher frequency of the *SLC6A3* 9R-allele of VNTR marker in 3'-UTR in alcoholics with severe alcohol withdrawal symptoms, delirium tremens, alcohol dependence, or more visual hallucinations during withdrawal compared to controls (Gorwood et al., 2003; Limosin et al., 2004; Kohnke et al., 2005; Samochowiec et al., 2006). These data were confirmed by our results to some extent, which revealed the increased risk of alcohol dependence in Russian males *SLC6A3* 9R*G-haplotype carriers. However, opposite to the present findings, Wernicke et al. (2002) reported higher prevalence of the A/A genotype of *SLC6A3 2319G/A* marker in patients with a history of alcohol withdrawal symptoms among individuals with *SLC6A3* VNTR 10R/10R genotype. Inconsistent findings also took place, distinguishing an association between the 7R-allele of the same VNTR and alcoholism in aldehyde dehydrogenase II-insufficient Japanese alcoholics (Muramatsu and Higuchi, 1995).

Despite the findings of some studies establishing an association between the *SLC6A3* rs27072 A-allele, and both smoking in Chinese (Ling et al. 2004) and alcoholism in Japanese men (Ueno et al. 1999), we detected the *SLC6A3* 9R*G haplotype to be a risk for alcoholism in Russians. These inconsistencies might be caused by different ethnic heterogeneity and the influence of differences in traditions and norms between individuals from Russian and East Asian populations. Obviously, our data confirm results demonstrated by Le Strat et al. (2008) in Caucasians; namely, overrepresentation of the *2319G*-allele in individuals with alcohol withdrawal seizures. Contradictory results reported by different research groups in regard to *SLC6A3 2319G/A* polymorphism could suggest the involvement of another polymorphic marker linked to *2319G/A* in addictive disorders. However, its location within the transcribed 3'-UTR is interesting, since these regions play an important role in the regulation of transcription efficiency, mRNA stability, mRNA subcellular localization (Mignone et al., 2002), and gene inactivation or degradation by target-specific binding of microRNAs (Presutti et al., 2006).

THE ROLE OF THE MONOAMINE OXIDASE A GENE (*MAOA*) IN ALCOHOL DEPENDENCE

It should be noted that our findings, demonstrating a high-activity *MAOA* allele (H-allele) of the *MAOA-LPR* polymorphism to be the risk marker for alcohol dependence, were observed in both populations belonging to a Turkic group of the Altaic language family (Tatars and Bashkirs), and a negative result was found in Russians (belonging to Slavic group of the Indo-European language family). These data point to the differences in ethnogenesis between the studied populations. However, published results conducted in ethnically different individuals seem to be in agreement with each other to some extent. Thus, a low-activity 3-repeat allele of the *MAOA-LPR* polymorphism was shown to be associated with antisocial alcoholism among German men (Schmidt et al., 2000), and with increased susceptibility to alcohol dependence and antisocial behaviors in a Brazilian sample (Contini et al., 2006). Saito et al. (2002) proposed a minimal association between the *MAOA* low-activity alleles and alcoholism, regardless of the presence or absence of antisocial behavior. However, our findings of the involvement of the high-activity *MAOA* allele in alcohol dependence are opposite to those indicated above (Schmidt et al., 2000; Contini et al., 2006). This discrepancy might be explained by different types of alcoholism between our study and previous studies, with the majority of them involving AD patients with antisocial behavior. This explanation is congruent with that proposed by Gerra et al. (2004) that the association with drug dependence might be a product of the influence of the *MAOA-LPR* on antisocial behavior and aggressiveness. In addition, higher probability for development of antisocial problems later in life was detected in maltreated males with *MAOA* low-activity alleles compared to those with the high-activity allele (Caspi et al., 2002). Similar *MAOA-LPR* polymorphism moderation of the impact of childhood trauma on adult psychopathology, and alcoholism in particular, was observed in sexually abused women (Ducci et al., 2008). Moreover, some researchers indicated the interaction effect of *MAOA-LPR* and *DRD2 Taq1A* or *ALDH2* gene polymorphisms in alcoholism with antisocial behavior (Huang et al., 2007; Lee et al., 2009). Assuming that the high-activity *MAOA* allele should predispose to increased dopamine metabolism and, hence, to decreased dopaminergic activity level observed in alcoholic-dependent patients (Hietala et al., 1994; Noble et al., 2003), our findings seem to be theoretically possible.

Despite the association of the *MAOA* T-allele of *EcoRV* polymorphism and alcohol dependence in Tatar and Bashkir males reported in the present study, published findings seem to be discrepant to ours, indicating no association of a single *EcoRV* markers or *MAOA* haplotypes (Lu et al., 2002).

THE ROLE OF THE CATECHOL-O-METHYLTRANSFERASE GENE (*COMT*) IN ALCOHOL DEPENDENCE

The genetic effect of the *COMT* gene in alcohol dependence was established in the current study while investigating Tatar and Bashkir males, both reporting higher risk for this disorder in the presence of the *COMT* high-activity allele (H-allele) of *Val158Met* polymorphism in the presence of either *DRD2* N1-allele (in Tatars) or alone (in Bashkirs). Previously published studies also indicated the association of H-allele (158Val-allele) with alcoholism in a Czech male population (Serý et al., 2006) and in Plains American Indian populations characterized by severe smoking (Enoch et al., 2006) a habit that is known to increase the risk for alcoholism. Another study based on haplotype analysis revealed protective and risk haplotypes for becoming nicotine dependent in European American males (Beuten et al., 2006). According to their results, a haplotype consisting of the low activity Met-allele (L-allele) of the *COMT* gene is protective to becoming nicotine dependent. Recently, Colilla et al. (2005) found overtransmission of the Met-allele in female ex-smokers, suggesting that the Met-allele is correlated with successful quitting. Our results suggest that individuals with high-activity *COMT* allele, due to their dopamine deficit, would possibly crave to increase the duration of reward derived from alcohol consumption (which stimulates the release of dopamine from neurons) and the risk of becoming alcohol dependent. In agreement with our suggestion, Köhnke et al. (2002) revealed that mean plasma homovanillic acid concentrations (known to be an indicator of central dopaminergic activity) were significantly lower in the studied abstinent alcoholic patients compared to healthy controls.

In contrast, several research groups observed *COMT* L-allele (158Met-allele) to be prevalent in individuals with late-onset (Hallikainen et al., 2000; Tiihonen et al., 1999) or early-onset alcoholism (Wang et al., 2001). The *COMT* L-allele has also been associated with elevated weekly alcohol consumption in male social drinkers (Kauhanen et al., 2000). Moreover, the L-allele has been associated with a more anxious, cautious personality (Olsson et al., 2005; Hettema et al., 2008) and with greater activation in response to negative emotional stimuli in the limbic system and connected prefrontal areas (Smolka et al., 2005). These findings allow to suppose that anxiety-related personality traits may be mediating for alcoholism: a more cautious temperament may be a vulnerability factor for "self-medication" maintenance drinking (Prescott et al., 2004). Thus, it is theoretically possible that the 158Met-allele could be both protective in one population and a risk factor in another population.

CONCLUSIONS

To date, enormous but inconsistent studies have focused on unraveling the role of genetic polymorphisms in alcohol dependence. Therefore, several possibilities for contradictory results might be suggested: the first relates to the study of alcoholism as a highly heterogeneous disorder; the second to inclusion of noncongruent populations (stratification bias); the third to performing statistical analysis with single marker association only, etc. The first problem has led the present work to focus on a more severe and genetically determined phenotype associated with alcohol dependence, such as acute alcoholic psychosis. With respect to the second point, we conducted the study controlled for the possible population stratification issues, enrolling only male subjects.

The most speculative step in the study refers to the set of markers included in the study, since gene–gene interactions are assumed to be a crucial issue in the study of such complex psychiatric disorders as alcohol dependence. Under a polygenic model, the simultaneous inheritance of diverse genetic variants is necessary for expression of the disease. Accordingly, functionally significant polymorphisms related to serotonergic and dopaminergic systems have been investigated, with subsequent logistic regression analyses resulting in identification of statistically significant models predicting alcohol dependence. Considering the results of the present study, the first hypothesis, suggesting that serotonergic and dopaminergic system genes might interact to increase susceptibility to alcohol dependence, was confirmed. Moreover, in agreement with the published data, diminished dopaminergic and serotonergic activity was demonstrated to be associated with increased risk of alcohol dependence in the present study. It should be noted that the maximal additive and interaction effect of covariates included into regression models was shown to comprise 10.6%, 12.2%, and 5.6% in individuals of Russian, Tatar, and Bashkir descent, respectively. In addition, the best models accounting for the higher risk in alcoholism predisposition were designed.

Future research is necessary to integrate genetic variations ever more closely, with effects on intermediate neurobiological processes such as reward, behavioral control, and anxiety response that are now understood to underlie addiction vulnerability and recovery.

ACKNOWLEDGMENTS

This work was supported partly by grants from the Russian Foundation for Humanities (08-06-00579a and 09-06-95601a/E) and the Russian Foundation for Basic Research (11-04-97032-R).

REFERENCES

Angelone SM, Bellini L, Di Bella D, Catalano M. (1998). Effects of fluvoxamine and citalopram in maintaining abstinence in a sample of Italian detoxified alcoholics. *Alcohol* 33(2):151–156.

Asghari V, Sanyal S, Buchwaldt S, et al. (1995). Modulation of intracellular cyclic AMP levels by different human dopamine D4 receptor variants. *J Neurochem* 65(3):1157–1165.

Balciuniene J, Emilsson L, Oreland L, et al. (2002). Investigation of the functional effect of monoamine oxidase polymorphisms in human brain. *Hum Genet* 110(1):1–7.

Bau CH, Almeida S, Costa FT, et al. (2001). DRD4 and DAT1 as modifying genes in alcoholism: Interaction with novelty seeking on level of alcohol consumption. *Mol Psychiatry* 6(1):7–9.

Becker JB. (1999). Gender differences in dopaminergic function in striatum and nucleus accumbens. *Pharmacol. Biochem. Behav.* 64(4): 803–812.

Bengel D, Greenberg BD, Corá-Locatelli G, et al. (1999). Association of the serotonin transporter promoter regulatory region polymorphism and obsessive-compulsive disorder. *Mol Psychiatry* 4(5):463–466.

Benjamin J, Li L, Patterson C, Greenberg BD, et al. (1996). Population and familial association between the D4 dopamine receptor gene and measures of Novelty Seeking. *Nat Genet* 12(1):81–84.

Berggren U, Fahlke C, Aronsson E, et al. (2006). The taqI DRD2 A1 allele is associated with alcohol-dependence although its effect size is small. *Alcohol Alcohol* 41(5):479–485.

Beuten J, Payne TJ, Ma JZ, Li MD. (2006). Significant association of catechol-O-methyltransferase (COMT) haplotypes with nicotine dependence in male and female smokers of two ethnic populations. *Neuropsychopharmacology* 31(3):675–684.

Bleich S, Bönsch D, Rauh J, et al. (2007). Association of the long allele of the 5-HTTLPR polymorphism with compulsive craving in alcohol dependence. *Alcohol Alcohol* 42(6):509–512.

Caspi A, McClay J, Moffitt TE, et al. (2002). Role of genotype in the cycle of violence in maltreated children. *Science* 297(5582):851–854.

Chen WJ, Lu ML, Hsu YP, et al. (1997). Dopamine D2 receptor gene and alcoholism among four aboriginal groups and Han in Taiwan. *Am J Med Genet* 74(2):129–136.

Chen J, Lipska BK, Halim N, et al. (2004). Functional analysis of genetic variation in catechol-O-methyltransferase (COMT): Effects on mRNA, protein, and enzyme activity in postmortem human brain. *Am J Hum Genet* 75(5):807–821.

Choi IG, Kee BS, Son HG, et al. (2006). Genetic polymorphisms of alcohol and aldehyde dehydrogenase, dopamine and serotonin transporters in familial and non-familial alcoholism. *Eur Neuropsychopharmacol* 16(2):123–128.

Cloninger CR. (1987). Neurogenetic adaptive mechanisms in alcoholism. *Science* 236(4800):410–416.

Colilla S, Lerman C, Shields PG, et al. (2005). Association of catechol-O-methyltransferase with smoking cessation in two independent studies of women. *Pharmacogenet. Genomics* 15(6):393–398.

Comings DE, Blum K. (2000). Reward deficiency syndrome: Genetic aspects of behavioral disorders. *Prog Brain Res* 126:325–341.

Conner BT, Noble EP, Berman SM, et al. (2005). DRD2 genotypes and substance use in adolescent children of alcoholics. *Drug Alcohol Depend* 79(3):379–387.

Contini V, Marques FZ, Garcia CE, et al. (2006). MAOA-uVNTR polymorphism in a Brazilian sample: Further support for the association with impulsive behaviors and alcohol dependence *Am J Med Genet B Neuropsychiatr Genet* 141B(3):305–308.

Corbin WR, Fromme K, Bergeson SE. (2006). Preliminary data on the association among the serotonin transporter polymorphism, subjective alcohol experiences, and drinking behavior. *J Stud Alcohol* 67(1):5–13.

Costa PT Jr, Herbst JH, McCrae RR, Siegler IC. (2000). Personality at midlife: Stability, intrinsic maturation, and response to life events. *Assessment* 7:365–378.

Crabbe JC, Phillips TJ, Feller DJ, et al. (1996). Elevated alcohol consumption in null mutant mice lacking 5-HT1B serotonin receptors. *Nat Genet* 14(1):98–101.

Ding YC, Chi HC, Grady DL, Morishima A, et al. (2002). Evidence of positive selection acting at the human dopamine receptor D4 gene locus. *Proc Natl Acad Science U S A* 99(1):309–314.

Disshon KA, Boja JW, Dluzen DE. (1998). Inhibition of striatal dopamine transporter activity by 17beta-estradiol. *Eur J Pharmacol* 345(2):207–211.

D'Souza UM, Russ C, Tahir E, et al. (2004). Functional effects of a tandem duplication polymorphism in the 5'flanking region of the DRD4 gene. *Biol Psychiatry* 56(9):691–697.

Ducci F, Enoch MA, Hodgkinson C, et al. (2008). Interaction between a functional MAOA locus and childhood sexual abuse predicts alcoholism and antisocial personality disorder in adult women. *Mol Psychiatry* 13(3):334–347.

Dulawa SC, Grandy DK, Low MJ, et al. (1999). Dopamine D4 receptor-knock-out mice exhibit reduced exploration of novel stimuli. *J Neurosci* 19(21):9550–9556.

Ebstein RP, Novick O, Umansky R, et al. (1996). Dopamine D4 receptor (D4DR) exon III polymorphism associated with the human personality trait of Novelty Seeking. *Nat Genet* 12(1):78–80.

Ebstein RP. (2006). The molecular genetic architecture of human personality: Beyond self-report questionnaires. *Mol Psychiatry* 11(5):427–445.

Edenberg HJ, Reynolds J, Koller DL, et al. (1998). A family-based analysis of whether the functional promoter alleles of the serotonin transporter gene HTT affect the risk for alcohol dependence. *Alcohol Clin Exp Res* 22(5):1080–1085.

Enoch MA, Waheed JF, Harris CR, et al. (2006). Sex differences in the influence of COMT Val158Met on alcoholism and smoking in plains American Indians. *Alcohol Clin Exp Res* 30(3):399–406.

Fehr C, Grintschuk N, Szegedi A, et al. (2000). The HTR1B 861G>C receptor polymorphism among patients suffering from alcoholism, major depression, anxiety disorders and narcolepsy. *Psychiatry Res* 97(1):1–10.

Feinn R, Nellissery M, Kranzler HR. (2005). Meta-analysis of the association of a functional serotonin transporter promoter polymorphism with alcohol dependence. *Am J Med Genet B Neuropsychiatr Genet* 133B(1):79–84.

Feng Y, Wigg KG, Makkar R, et al. (2005). Sequence variation in the 3'-untranslated region of the dopamine transporter gene and attention-deficit hyperactivity disorder (ADHD). *Am J Med Genet B Neuropsychiatr Genet* 139(1):1–6.

Fernandes C, Paya-Cano JL, Sluyter F, et al. (2004). Hippocampal gene expression profiling across eight mouse inbred strains: Toward understanding the molecular basis for behavior. *Eur J Neurosci* 19(9):2576–2582.

Fiskerstrand CE, Lovejoy EA, Quinn JP. (1999). An intronic polymorphic domain often associated with susceptibility to affective disorders has allele dependent differential enhancer activity in embryonic stem cells. *FEBS Lett* 458(2):171–174.

Florez G, Saiz P, Garcia-Portilla P, et al. (2008). Association between the Stin2 VNTR polymorphism of the serotonin transporter gene and treatment outcome in alcohol-dependent patients. *Alcohol Alcohol* 43(5):516–522.

Fuke S, Suo S, Takahashi N, et al. (2001). The VNTR polymorphism of the human dopamine transporter (DAT1) gene affects gene expression. *Pharmacogenomics J* 1(2):152–156.

Gerra G, Garofano L, Bosari S, et al. (2004). Analysis of monoamine oxidase A (MAO-A) promoter polymorphism in male heroin-dependent subjects: Behavioural and personality correlates. *J Neural Transm* 111(5):611–621.

Giros B, Jaber M, Jones SR, et al. (1996). Hyperlocomotion and indifference to cocaine and amphetamine in mice lacking the dopamine receptor. *Nature* 370:606–612.

Gogos JA, Morgan M, Luine V, et al. (1998). Catechol-O-methyltransferase-deficient mice exhibit sexually dimorphic changes in catecholamine levels and behavior. *Proc Natl Acad Science U S A* 95(17):9991–9996.

Gokturk C, Schultze S, Nilsson KW, et al. (2008). Serotonin transporter (5-HTTLPR) and monoamine oxidase (MAOA) promoter polymorphisms in women with severe alcoholism. *Arch Womens Ment Health* 11(5–6):347–355.

Golimbet VE, Alfimova MV, Gritsenko IK, Ebstein RP. (2007). Relationship between dopamine system genes and extraversion and novelty seeking. *Neuroscience Behav Physiol* 37(6):601–606.

Gorbunova EV, Galeeva AR, Khusnutdinova EK. (2002). Analysis of insertion-deletion polymorphism in serotonin transporter gene in men of different ethnic appurtenance with acute alcoholic psychosis. *Mol Gen Mikrobiol Virusol* 4:35–38.

Gorwood P, Aissi F, Batel P, et al. (2002). Reappraisal of the serotonin 5-HT(1B) receptor gene in alcoholism: Of mice and men. *Brain Res Bull* 57(1):103–107.

Gorwood P, Limosin F, Batel P, et al. (2003). The A9 allele of the dopamine transporter gene is associated with delirium tremens and alcohol-withdrawal seizure. *Biol Psychiatry* 53(1):85–92.

Gorwood P, Lanfumey L, Hamon M. (2004). Alcohol dependence and polymorphisms of serotonin-related genes. *Med Sci (Paris)* 20(12):1132–1138.

Gunzerath L, Goldman D. (2003). G x E: A NIAAA workshop on gene-environment interactions *Alcohol Clin Exp Res* 27(3):540–562.

Hallikainen T, Saito T, Lachman HM, et al. (1999). Association between low activity serotonin transporter promoter genotype and early onset alcoholism with habitual impulsive violent behavior. *Mol Psychiatry* 4(4):385–388.

Hallikainen T, Lachman H, Saito T, et al. (2000). Lack of association between the functional variant of the catechol-o-methyltransferase (COMT) gene and early-onset alcoholism associated with severe antisocial behavior. *Am J Med Genet* 96(3):348–352.

Hallikainen T, Hietala J, Kauhanen J, et al. (2003). Ethanol consumption and DRD2 gene TaqI a polymorphism among socially drinking males *Am J Med Genet A*. 119A(2):152–155.

Hasegawa Y, Higuchi S, Matsushita S, Miyaoka H. (2002). Association of a polymorphism of the serotonin 1B receptor gene and alcohol dependence with inactive aldehyde dehydrogenase-2. *J Neural Transm* 109(4):513–521.

Heath AC, Bucholz KK, Madden PA, et al. (1997). Genetic and environmental contributions to alcohol dependence risk in a national twin sample: Consistency of findings in women and men. *Psychol Med* 27:1381–1396.

Hettema JM, An SS, Bukszar J, et al. (2008). Catechol-O-methyltransferase contributes to genetic susceptibility shared among anxiety spectrum phenotypes. *Biol Psychiatry* 64(4):302–310.

Hietala J, West C, Syvälahti E, et al. (1994). Striatal D2 dopamine receptor binding characteristics in vivo in patients with alcohol dependence. *Psychopharmacology (Berl.)* 116(3):285–290.

Higley JD, Suomi SJ, Linnoila MA. (1996). Nonhuman primate model of type II excessive alcohol consumption? Part 1. Low cerebrospinal fluid 5-hydroxyindoleacetic acid concentrations and diminished social competence correlate with excessive alcohol consumption. *Alcohol Clin Exp Res* 20(4):629–642.

Hoehe MR. (2003). Haplotypes and the systematic analysis of genetic variation in genes and genomes. *Pharmacogenomics.* 4(5):547–570.

Hotamisligil GS, Breakefield XO. (1991). Human monoamine oxidase A gene determines levels of enzyme activity. *Am J Hum Genet* 49(2):383–392.

Hranilovic D, Stefulj J, Schwab S, et al. (2004). Serotonin transporter promoter and intron 2 polymorphisms: Relationship between allelic variants and gene expression. *Biol Psychiatry* 55(11):1090–1094.

Hu X, Oroszi G, Chun J, et al. (2005). An expanded evaluation of the relationship of four alleles to the level of response to alcohol and the alcoholism risk. *Alcohol Clin Exp Res* 29(1):8–16.

Huang SY, Lin WW, Wan FJ, et al. (2007). Monoamine oxidase-A polymorphisms might modify the association between the dopamine D2 receptor gene and alcohol dependence. *J Psychiatry Neurosci* 32(3):185–192.

Hutchison KE, McGeary J, Smolen A, et al. (2002). The DRD4 VNTR polymorphism moderates craving after alcohol consumption. *Health Psychol* 21(2):139–146.

Johnson BA. (2008a). Update on neuropharmacological treatments for alcoholism: Scientific basis and clinical findings. *Biochem Pharmacol* 75(1):34–56.

Johnson BA, Javors MA, Roache JD, et al. (2008b). Can serotonin transporter genotype predict serotonergic function, chronicity, and severity of drinking? *Prog Neuropsychopharmacol Biol Psychiatry* 32(1):209–216.

Jorm AF, Henderson AS, Jacomb PA, et al. (2000). Association of smoking and personality with a polymorphism of the dopamine transporter gene: Results from a community survey. *Am J Med Genet* 96(3):331–334.

Katz M. (2006). *Multivariable Analysis. A Practical Guide for Clinicians.* New York: Cambridge University Press.

Kauhanen J, Hallikainen T, Tuomainen TP, et al. (2000). Association between the functional polymorphism of catechol-O-methyltransferase gene and alcohol consumption among social drinkers. *Alcohol Clin Exp Res* 24(2):135–139.

Kereszturi E, Kiraly O, Csapo Z, et al. (2007). Association between the 120-bp duplication of the dopamine D4 receptor gene and attention deficit hyperactivity disorder: Genetic and molecular analyses. *Am J Med Genet B Neuropsychiatr Genet* 144B(2):231–236.

Khusnutdinova EK, Bermisheva MA, Kutuev IA, Yunusbayev BB, Villems R. (2008). Genetic landscape of the Central Asia and Volga-Ural region. *In* Nikolay Dobretsov, Nikolay Kolchanov, Alexey Rozanov, Georgy Zavarzin (Eds.), *Biosphere Origin and*

Evolution (pp. 373–381). Springer Science & Business Media, New York, USA.

Kiefer F, Mann K. (2005). New achievements and pharmacotherapeutic approaches in the treatment of alcohol dependence. *Eur J Pharmacol.* 526(1–3):163–171.

Köhnke MD, Zabetian CP, Anderson GM, et al. (2002). A genotype-controlled analysis of plasma dopamine beta-hydroxylase in healthy and alcoholic subjects: Evidence for alcohol-related differences in noradrenergic function. *Biol Psychiatry* 52(12):1151–1158.

Köhnke MD, Batra A, Kolb W, et al. (2005). Association of the dopamine transporter gene with alcoholism. *Alcohol Alcohol* 40(5):339–342.

Konishi T, Luo HR, Calvillo M, et al. (2004). ADH1B*1, ADH1C*2, DRD2 (-141C Ins), and 5-HTTLPR are associated with alcoholism in Mexican American men living in Los Angeles. *Alcohol Clin Exp Res* 28(8):1145–1152.

Koob GF. (1998). Circuits, drugs, and drug addiction. *Adv Pharmacol* 42:978–982.

Kotler M, Cohen H, Segman R, et al. (1997). Excess dopamine D4 receptor (D4DR) exon III seven repeat allele in opioid-dependent subjects. *Mol Psychiatry* 2(3):251–254.

Kranzler HR, Hernandez-Avila CA, Gelernter J. (2002). Polymorphism of the 5-HT1B receptor gene (HTR1B): Strong within-locus linkage disequilibrium without association to antisocial substance dependence. *Neuropsychopharmacology* 26(1):115–122.

Kraschewski A, Reese J, Anghelescu I, et al. (2009). Association of the dopamine D2 receptor gene with alcohol dependence: Haplotypes and subgroups of alcoholics as key factors for understanding receptor function. *Pharmacogenet Genomics* 19(7):513–527.

Kreek MJ, Koob GF. (1998). Drug dependence: Stress and dysregulation of brain reward pathways. *Drug Alcohol Depend* 51(1–2):23–47.

Kweon YS, Lee HK, Lee CT, et al. (2005). Association of the serotonin transporter gene polymorphism with Korean male alcoholics *J Psychiatr Res* 39(4):371–376.

Lachman HM, Papolos DF, Saito T, et al. (1996). Human catechol-O-methyltransferase pharmacogenetics: Description of a functional polymorphism and its potential application to neuropsychiatric disorders. *Pharmacogenetics* 6(3):243–250.

Laine TP, Ahonen A, Räsänen P, et al. (2001). The A1 allele of the D2 dopamine receptor gene is associated with high dopamine transporter density in detoxified alcoholics. *Alcohol* 36(3):262–265.

Lang UE, Bajbouj M, Wernicke C, et al. (2004). No association of a functional polymorphism in the serotonin transporter gene promoter and anxiety-related personality traits. *Neuropsychobiology* 49(4):182–184.

Lappalainen J, Long JC, Eggert M, et al. (1998). Linkage of antisocial alcoholism to the serotonin 5-HT1B receptor gene in 2 populations. *Arch Gen Psychiatry* 55(11):989–994.

Laruelle M, Gelernter J, Innis RB. (1998). D2 receptors binding potential is not affected by Taql polymorphism at the D2 receptor gene. *Mol Psychiatry* 3(3):261–265.

Le Strat Y, Ramoz N, Pickering P, et al. (2008). The 3' part of the dopamine transporter gene DAT1/SLC6A3 is associated with withdrawal seizures in patients with alcohol dependence. *Alcohol Clin Exp Res* 32(1):27–35.

Lee SY, Hahn CY, Lee JF, et al. (2009). MAOA-uVNTR polymorphism may modify the protective effect of ALDH2 gene against alcohol dependence in antisocial personality disorder. *Alcohol Clin Exp Res* 33(6):985–990.

Leung K, Au Y, Fernandez-Dols JM, Iwawaki S. (1992). Preference for methods of conflict processing in two collectivist cultures *Int J Psychol* 27:195–209.

Li T, Zhu ZH, Liu X, et al. (2000). Association analysis of polymorphisms in the DRD4 gene and heroin abuse in Chinese subjects. *Am J Med Genet* 96(5):616–621.

Li T, Chen CK, Hu X, et al. (2004). Association analysis of the DRD4 and COMT genes in methamphetamine abuse. *Am J Med Genet B Neuropsychiatr Genet* 129B(1):120–124.

Limosin F, Loze JY, Boni C, et al. (2004). The A9 allele of the dopamine transporter gene increases the risk of visual hallucinations during alcohol withdrawal in alcohol-dependent women. *Neuroscience Lett* 362(2):91–94.

Ling D, Niu T, Feng Y, Xing H, Xu X. (2004). Association between polymorphism of the dopamine transporter gene and early smoking onset: An interaction risk on nicotine dependence *J Hum Genet* 49(1):35–39.

Long JC, Knowler WC, Hanson RL, et al. (1998). Evidence for genetic linkage to alcohol dependence on chromosomes 4 and 11 from an autosome-wide scan in an American Indian population. *Am J Med Genet* 81(3):216–221.

Lu RB, Lee JF, Ko HC, Lin WW. (2001). Dopamine D2 receptor gene (DRD2) is associated with alcoholism with conduct disorder. *Alcohol Clin Exp Res* 25(2):177–184.

Lu RB, Lee JF, Ko HC, et al. (2002). No association of the MAOA gene with alcoholism among Han Chinese males in Taiwan. *Prog Neuropsychopharmacol Biol Psychiatry* 26(3):457–461.

Lu NZ, Eshleman AJ, Janowsky A, Bethea CL. (2003). Ovarian steroid regulation of serotonin reuptake transporter (SERT) binding, distribution, and function in female macaques. *Mol Psychiatry* 8(3):353–360.

Luo HR, Hou ZF, Wu J, et al. (2005). Evolution of the DRD2 gene haplotype and its association with alcoholism in Mexican Americans. *Alcohol* 36(2):117–125.

MacKenzie A, Quinn J. (1999). A serotonin transporter gene intron 2 polymorphic region, correlated with affective disorders, has allele-dependent differential enhancer-like properties in the mouse embryo. *Proc Natl Acad Science U S A* 96(26):15251–15252.

Mann JJ, Brent DA, Arango V. (2001). The neurobiology and genetics of suicide and attempted suicide: A focus on the serotonergic system. *Neuropsychopharmacology* 24(5):467–477.

Marques FZ, Hutz MH, Bau CH. (2006). Influence of the serotonin transporter gene on comorbid disorders among alcohol-dependent individuals. *Psychiatr Genet* 16(3):125–131.

Melo JA, Shendure J, Pociask K, Silver LM. (1996). Identification of sex-specific quantitative trait loci controlling alcohol preference in C57BL/6 mice. *Nat Genet* 13(2):147–153.

Mignone F, Gissi C, Liuni S, Pesole G. (2002). Untranslated regions of mRNAs. *Genome Biol* 3(3): review S0004.

Mill JS, Caspi A, McClay J, et al. (2002). The dopamine D4 receptor and the hyperactivity phenotype: A developmental-epidemiological study. *Mol Psychiatry* 7(4):383–391.

Miller GM, Madras BK. (2002). Polymorphisms in the 3'-untranslated region of human and monkey dopamine transporter genes affect reporter gene expression. *Mol Psychiatry* 7(1):44–55.

Mo GH, Lai IC, Wang YC, et al. (2007). Support for an association of the C939T polymorphism in the human DRD2 gene with tardive dyskinesia in schizophrenia. *Schizophr Res* 97(1–3):302–304.

Mokrović G, Matosić A, Hranilović D, et al. (2008). Alcohol dependence and polymorphisms of serotonin-related genes: Association studies. *Coll Antropol* 32 (Suppl 1):127–131.

Monakhov M, Golimbet V, Abramova L, et al. (2008). Association study of three polymorphisms in the dopamine D2 receptor gene and schizophrenia in the Russian population. *Schizophr Res* 100(1–3):302–307.

Munafò MR, Johnstone EC, Welsh KI, Walton RT. (2005). Association between the DRD2 gene Taq1A (C32806T) polymorphism and alcohol consumption in social drinkers. *Pharmacogenomics J* 5(2):96–101.

Munafò MR, Matheson IJ, Flint J. (2007). Association of the DRD2 gene Taq1A polymorphism and alcoholism: A meta-analysis of case-control studies and evidence of publication bias. *Mol Psychiatry* 12(5):454–461.

Muramatsu T, Higuchi S. (1995). Dopamine transporter gene polymorphism and alcoholism. *Biochem Biophys Res Commun* 211(1):28–32.

Nakamura T, Matsushita S, Nishiguchi N, et al. (1999). Association of a polymorphism of the 5HT2A receptor gene promoter region with alcohol dependence. *Mol Psychiatry* 4(1):85–88.

Nakamura M, Ueno S, Sano A, Tanabe H. (2000). The human serotonin transporter gene linked polymorphism (5-HTTLPR) shows ten novel allelic variants. *Mol Psychiatry* 5(1):32–38.
Nakatome M, Tun Z, Shimada S, Honda K. (1998). Detection and analysis of four polymorphic markers at the human monoamine oxidase (MAO) gene in Japanese controls and patients with Parkinson's disease. *Biochem Biophys Res Commun* 247(2):452–456.
Namkoong K, Cheon KA, Kim JW, et al. (2008). Association study of dopamine D2, D4 receptor gene, GABAA receptor beta subunit gene, serotonin transporter gene polymorphism with children of alcoholics in Korea: A preliminary study. *Alcohol* 42(2):77–81.
Nellissery M, Feinn RS, Covault J, et al. (2003). Alleles of a functional serotonin transporter promoter polymorphism are associated with major depression in alcoholics. *Alcohol Clin Exp Res* 27(9):1402–1408.
Neville MJ, Johnstone EC, Walton RT. (2004). Identification and characterization of ANKK1: A novel kinase gene closely linked to DRD2 on chromosome band 11q23.1. *Hum Mutat* 23(6):540–545.
Noble EP. (2003). D2 dopamine receptor gene in psychiatric and neurologic disorders and its phenotypes. *Am J Med Genet B Neuropsychiatr Genet* 116B(1):103–125.
Nomura M, Nomura Y. (2006). Psychological, neuroimaging, and biochemical studies on functional association between impulsive behavior and the 5-HT2A receptor gene polymorphism in humans. *Ann N Y Acad Science* 1086:134–143.
Okuyama Y, Ishiguro H, Toru M, Arinami T. (1999). A genetic polymorphism in the promoter region of DRD4 associated with expression and schizophrenia. *Biochem Biophys Res Commun* 258(2):292–295.
Olsson CA, Anney RJ, Lotfi-Miri M, et al. (2005). Association between the COMT Val158Met polymorphism and propensity to anxiety in an Australian population-based longitudinal study of adolescent health. *Psychiatr Genet* 15(2):109–115.
Ott U, Reuter M, Hennig J, Vaitl D. (2005). Evidence for a common biological basis of the Absorption trait, hallucinogen effects, and positive symptoms: Epistasis between 5-HT2a and COMT polymorphisms. *Am J Med Genet B Neuropsychiatr Genet* 137B(1):29–32.
Overstreet DH, McArthur RA, Rezvani AH, Post C. (1997). Selective inhibition of alcohol intake in diverse alcohol-preferring rat strains by the 5-HT2A antagonists amperozide and FG 5974. *Alcohol Clin Exp Res* 21(8):1448–1454.
Parsons MJ, D'Souza UM, Arranz MJ, et al. (2004). The -1438A/G polymorphism in the 5-hydroxytryptamine type 2A receptor gene affects promoter activity. *Biol Psychiatry* 56(6):406–410.
Philibert RA, Sandhu H, Hollenbeck N, et al. (2008). The relationship of 5HTT (SLC6A4) methylation and genotype on mRNA expression and liability to major depression and alcohol dependence in subjects from the Iowa Adoption Studies. *Am J Med Genet B Neuropsychiatr Genet* 147B(5):543–549.
Phillips TJ, Belknap JK, Buck KJ, Cunningham CL. (1998). Genes on mouse chromosomes 2 and 9 determine variation in ethanol consumption. *Mamm Genome* 9:936–941.
Pinto E, Reggers J, Gorwood P, et al. (2008). The short allele of the serotonin transporter promoter polymorphism influences relapse in alcohol dependence. *Alcohol Alcohol* 43(4):398–400.
Pinto E, Reggers J, Gorwood P, et al. (2009). The TaqI A DRD2 polymorphism in type II alcohol dependence: A marker of age at onset or of a familial disease? *Alcohol* 43(4):271–275.
Pohjalainen T, Rinne JO, Nagren K, et al. (1998). The A1 allele of the human D2 dopamine receptor gene predicts low D2 receptor availability in healthy volunteers. *Mol Psychiatry* 3(3):256–260.
Polina ER, Contini V, Hutz MH, Bau CH. (2009). The serotonin 2A receptor gene in alcohol dependence and tobacco smoking. *Drug Alcohol Depend* 101(1–2):128–131.
Ponce G, Hoenicka J, Jiménez-Arriero MA, et al. (2008). DRD2 and ANKK1 genotype in alcohol-dependent patients with psychopathic traits: Association and interaction study. *Br J Psychiatry* 193(2):121–125.
Prescott CA, Cross RJ, Kuhn JW, et al. (2004). Is risk for alcoholism mediated by individual differences in drinking motivations? *Alcohol Clin Exp Res* 28(1):29–39.
Presutti C, Rosati J, Vincenti S, Nasi S. (2006) Non coding RNA and brain. *BMC Neurosci* 7 (Suppl 1):S5.
Preuss UW, Koller G, Bondy B, et al. (2001). Impulsive traits and 5-HT2A receptor promoter polymorphism in alcohol dependents: Possible association but no influence of personality disorders. *Neuropsychobiology* 43(3):186–191.
Preuss UW, Zill P, Koller G, et al. (2007). D2 dopamine receptor gene haplotypes and their influence on alcohol and tobacco consumption magnitude in alcohol-dependent individuals. *Alcohol Alcohol* 42(3):258–266.
Ray LA, Bryan A, Mackillop J, et al. (2009). The dopamine D Receptor (DRD4) gene exon III polymorphism, problematic alcohol use and novelty seeking: Direct and mediated genetic effects. *Addict Biol* 14(2):238–244.
Reuter M, Schmitz A, Corr P, Hennig J. (2006). Molecular genetics support Gray's personality theory: The interaction of COMT and DRD2 polymorphisms predicts the behavioural approach system. Int. J. *Neuropsychopharmacol* 9(2):155–166.
Ritchie T, Noble EP. (2003). Association of seven polymorphisms of the D2 dopamine receptor gene with brain receptor-binding characteristics. *Neurochem Res* 28(1):73–82.
Rogers G, Joyce P, Mulder R, et al. (2004). Association of a duplicated repeat polymorphism in the 5'-untranslated region of the DRD4 gene with novelty seeking. *Am J Med Genet B Neuropsychiatr Genet* 126B(1):95–98.
Rosmond R, Rankinen T, Chagnon M, et al. (2001). Polymorphism in exon 6 of the dopamine D(2) receptor gene (DRD2) is associated with elevated blood pressure and personality disorders in men *J Hum Hypertens* 15(8):553–558.
Rubinstein M, Phillips TJ, Bunzow JR, et al. (1997). Mice lacking dopamine D4 receptors are supersensitive to ethanol, cocaine, and methamphetamine. *Cell* 90(6):991–1001.
Rybakowski F, Slopien A, Dmitrzak-Weglarz M, et al. (2006). The 5-HT2A -1438 A/G and 5-HTTLPR polymorphisms and personality dimensions in adolescent anorexia nervosa: Association study. *Neuropsychobiology* 53(1):33–39.
Sabol SZ, Hu S, Hamer D. (2006). A functional polymorphism in the monoamine oxidase A gene promoter. *Am J Med Genet B Neuropsychiatr Genet* 141B(3):305–308.
Saito T, Lachman HM, Diaz L, et al. J. (2002). Analysis of monoamine oxidase A (MAOA) promoter polymorphism in Finnish male alcoholics. *Psychiatry Res* 109(2):113–119.
Saiz PA, Garcia-Portilla MP, Florez G, et al. (2009). Differential role of serotonergic polymorphisms in alcohol and heroin dependence. *Prog Neuropsychopharmacol Biol Psychiatry* 33(4):695–700.
Samochowiec J, Lesch KP, Rottmann M, et al. (1999). Association of a regulatory polymorphism in the promoter region of the monoamine oxidase A gene with antisocial alcoholism. *Psychiatry Res* 86(1):67–72.
Samochowiec J, Rybakowski F, Czerski P, et al. (2001). Polymorphisms in the dopamine, serotonin, and norepinephrine transporter genes and their relationship to temperamental dimensions measured by the Temperament and Character Inventory in healthy volunteers. *Neuropsychobiology* 43(4):248–253.
Samochowiec J, Kucharska-Mazur J, Grzywacz A, et al. (2006). Family-based and case-control study of DRD2, DAT, 5HTT, COMT genes polymorphisms in alcohol dependence. *Neuroscience Lett* 410(1):1–5.
Saudou F, Amara DA, Dierich A, et al. (1994). Enhanced aggressive behavior in mice lacking 5-HT1B receptor. *Science* 265(5180):1875–1878.
Seaman MI, Fisher JB, Chang F, Kidd KK. (1999). Tandem duplication polymorphism upstream of the dopamine D4 receptor gene (DRD4). *Am J Med Genet* 88(6):705–709.
Schmidt LG, Sander T, Kuhn S, et al. (2000). Different allele distribution of a regulatory MAOA gene promoter polymorphism in

antisocial and anxious-depressive alcoholics. *J Neural Transm* 107(6):681–689.
Serý O, Didden W, Mikes V, et al. (2006). The association between high-activity COMT allele and alcoholism. *Neuro Endocrinol Lett* 27(1–2):231–235.
Shih JC, Chen K. (1999). MAO-A and -B gene knock-out mice exhibit distinctly different behavior. *Neurobiology (Bp)* 7(2):235–246.
Sinha R, Cloninger CR, Parsian A. (2003). Linkage disequilibrium and haplotype analysis between serotonin receptor 1B gene variations and subtypes of alcoholism. *Am J Med Genet B Neuropsychiatr Genet* 121B(1):83–88.
Skowronek MH, Laucht M, Hohm E, Becker K, Schmidt MH. (2006). Interaction between the dopamine D4 receptor and the serotonin transporter promoter polymorphisms in alcohol and tobacco use among 15-year-olds. *Neurogenetics* 7(4):239–246.
Smolka MN, Schumann G, Wrase J, et al. (2005). Catechol-O-methyltransferase val158met genotype affects processing of emotional stimuli in the amygdala and prefrontal cortex. *J Neuroscience* 25(4):836–842.
Soyka M, Preuss UW, Koller G, et al. (2002). Dopamine D 4 receptor gene polymorphism and extraversion revisited: Results from the Munich gene bank project for alcoholism. *J Psychiatr Res* 36(6):429–435.
Stefulj J, Büttner A, Skavic J, et al. (2004). Serotonin 1B (5HT-1B) receptor polymorphism (G861C) in suicide victims: Association studies in German and Slavic population. *Am J Med Genet B Neuropsychiatr Genet* 127B(1):48–50.
Strobel A, Wehr A, Michel A, Brocke B. (1999). Association between the dopamine D4 receptor (DRD4) exon III polymorphism and measures of Novelty Seeking in a German population. *Mol Psychiatry* 4(4):378–384.
Szantai E, Kiraly O, Nemoda Z, et al. (2005). Linkage analysis and molecular haplotyping of the dopamine D4 receptor gene promoter region. *Psychiatr Genet* 15(4):259–270.
Szilagyi A, Boór K, Székely A, et al. (2005). Combined effect of promoter polymorphisms in the dopamine D4 receptor and the serotonin transporter genes in heroin dependence. *Neuropsychopharmacol Hung* 7(1):28–33.
Thanos PK, Rivera SN, Weaver K, et al. (2005). Dopamine D2R DNA transfer in dopamine D2 receptor-deficient mice: Effects on ethanol drinking. *Life Sci* 77(2):130–139.
Thompson TL, Certain ME. (2005). Estrogen mediated inhibition of dopamine transport in the striatum: Regulation by G alpha i/o. *Eur J Pharmacol* 511(2–3):121–126.
Tiihonen J, Kuikka J, Bergström K, et al. (1995). Altered striatal dopamine re-uptake site densities in habitually violent and non-violent alcoholics. *Nat Med* 1(7):654–657.
Tiihonen J, Hallikainen T, Lachman H, et al. (1999). Association between the functional variant of the catechol-O-methyltransferase (COMT) gene and type 1 alcoholism. *Mol Psychiatry* 4(3):286–289.
Tomitaka M, Tomitaka S, Otuka Y, et al. (1999). Association between novelty seeking and dopamine receptor D4 (DRD4) exon III polymorphism in Japanese subjects. *Am J Med Genet* 88(5):469–471.
Turecki G, Brière R, Dewar K, et al. (1999). Prediction of level of serotonin 2A receptor binding by serotonin receptor 2A genetic variation in postmortem brain samples from subjects who did or did not commit suicide. *Am J Psychiatry* 156(9):1456–1458.
Twitchell GR, Hanna GL, Cook EH, et al. (2001). Serotonin transporter promoter polymorphism genotype is associated with behavioral disinhibition and negative affect in children of alcoholics. *Alcohol Clin Exp Res* 25(7):953–959.
Ueno S, Nakamura M, Mikami M, et al. (1999). Identification of a novel polymorphism of the human dopamine transporter (DAT1) gene and the significant association with alcoholism. *Mol Psychiatry* 4(6):552–557.
Underwood MD, Mann JJ, Huang YY, et al. (2008). Family history of alcoholism is associated with lower 5-HT2A receptor binding in the prefrontal cortex. *Alcohol Clin Exp Res* 32(4):593–599.
VanNess SH, Owens MJ, Kilts CD. (2005). The variable number of tandem repeats element in DAT1 regulates in vitro dopamine transporter density. *BMC Genet* 6:55.
Volkow ND, Wang GJ, Begleiter H, et al. (2006). High levels of dopamine D2 receptors in unaffected members of alcoholic families: Possible protective factors. *Arch Gen Psychiatry* 63(9):999–1008.
Vormfelde SV, Hoell I, Tzvetkov M, et al. (2006). Anxiety- and novelty seeking-related personality traits and serotonin transporter gene polymorphisms *J Psychiatr Res* 40(6):568–576.
Walters GD. (2002). The heritability of alcohol abuse and dependence: A meta-analysis of behavior genetic research. *Am J Drug Alcohol Abuse* 28(3):557–584.
Wang T, Franke P, Neidt H, et al. (2001). Association study of the low-activity allele of catechol-O-methyltransferase and alcoholism using a family-based approach. *Mol Psychiatry* 6(1):109–111.
Wang TJ, Huang SY, Lin WW, et al. (2007). Possible interaction between MAOA and DRD2 genes associated with antisocial alcoholism among Han Chinese men in Taiwan. *Prog Neuropsychopharmacol Biol Psychiatry* 31(1):108–114.
Wendland JR, Martin BJ, Kruse MR, et al. (2006). Simultaneous genotyping of four functional loci of human SLC6A4, with a reappraisal of 5-HTTLPR and rs25531. *Mol Psychiatry* 11(3):224–226.
Wernicke C, Smolka M, Gallinat J, et al. (2002). Evidence for the importance of the human dopamine transporter gene for withdrawal symptomatology of alcoholics in a German population. *Neuroscience Lett* 333(1):45–48.
Wiesbeck GA, Weijers HG, Wodarz N, et al. (2004). Serotonin transporter gene polymorphism and personality traits in primary alcohol dependence. *World J Biol Psychiatry* 5(1):45–48.
Wise RA. (2002). Brain reward circuitry: Insights from unsensed incentives. *Neuron* 36(2):229–240.
Wu CY, Wu YS, Lee JF, et al. (2008). The association between DRD2/ANKK1, 5-HTTLPR gene, and specific personality trait on antisocial alcoholism among Han Chinese in Taiwan. *Am J Med Genet B Neuropsychiatr Genet* 147B(4):447–453.

118.

MEDICAL GENETICS TEACHING AND TRAINING IN RUSSIA

S.I. Kozlova and E.K. Ginter

INTRODUCTION

As a teaching subject, medical genetics integrates aspects of biomedical and clinical science, thus creating novel bases for thought processes of practicing physicians. Because of that, every clinician of the twenty-first century will [would] understand the principles of genetics and apply them in his or her everyday practice, including understanding the pathogenesis, diagnostic features of the diseases, treatment, and strategies for prevention. The training of future physicians will include absorption of updated knowledge encompassing the subjects of human genetics and medical genetics (Young, 2005; Ginter, 2003; Harper, 2004; Bochkov, 2010; Kozlova, Demikova, 2007).

Speaking generally, medical education in the area of genetics includes three broad fields: formal knowledge, skills, and attitude.

Knowledge. Modern clinical practice implies recognition of the role of genetic factors in the maintenance of health and in the development of diseases. This, in turn, depends on knowledge concerning the structure and the function of relevant human genes, the mode of their transmission, and understanding of the importance of both gene–gene and gene–environment interactions.

Skills. Physicians should be able to derive a conclusion from the facts, to use available information in order to develop an adequate plan for establishing a diagnosis of hereditary or inborn disorder, and to help the patient and his or her family. Physicians should be taught the skill of communication. Information should be communicated to the patients as clearly as possible, not in directive form, without any hint of degrading attitudes toward people of differing levels of education or religious beliefs. The minimal list of other skills should include the charting of the pedigree; physical exam, including anthropometric evaluation; use of the reference literature, including materials available through the Internet; proper selection of the diagnostic test, including genetic and molecular assays; calculation of the genetic risk and risk of transmission; selecting of an adequate way toward prevention of the disease in the particular family; defining the psychological status of the family and possible deviations of behavioral response by the patient in response to the report delivered by the medical geneticist.

Attitude. Physicians should be taught to develop compassion and to get rid of judgmental attitudes toward the patient. A physician should recognize his own limitations and support motivation for self-improvement. A physician should strive to correctly estimate the seriousness of the condition and ways to prevent its development or occurrence; to respect any decision made by the patient, even if this decision differs from the doctor's own principles; to honestly, accurately, and nonjudgmentally present all possibilities; to maintain confidentiality (Walker, 1998; Baker et al., 1998).

POSTGRADUATE TRAINING IN MEDICAL GENETICS TO PHYSICIANS IN RUSSIA

There are three types of postgraduate medical education (PME) courses in medical genetics available in Russia:

1. teaching genetics to practicing physicians with an aim to improve their qualifications in the context of their actual specialization (thematic improvement in the area of medical genetics, or TI);
2. teaching and matriculating medical geneticists as a cadre for the positions open in the medical genetic counseling units and other practices in the need of this kind of specialists (specialization in a field of medical genetics); and
3. teaching practicing physicians and nurses participating in various genetic screening programs to collect and analyze the relevant data.

The contents of medical genetics courses aimed at these three categories of students may differ. However, the general strategy of the teaching remains the same, as in all cases it builds upon the following basics:

As medical genetics includes both an area of biomedical science and an area of the clinical specialization, it should be taught as a holistic concept with no regard to whether students are already practicing or still are in training;

Many aspects of medical genetics are overlapping with other fields of biomedical science, but specific questions of medical genetics are prevention or delaying the onset of hereditary diseases, application of the methods of DNA diagnostics to clinical practice, and genetic counseling.

Thematic Improvement in the Area of Medical Genetics. or TI[1].

In the last 20 years, medical genetics has penetrated all other clinical specialties. The contents of this kind of courses in medical genetics for medical students cannot be identical all over the world, as teaching materials are usually developed in accordance with the general level of medical practices in the region, while taking into account national priorities, number of practicing physicians, medical needs of the population, and many other variables. However, the teaching principles remain the same. So far, the United States, Canada, and United Kingdom have the most detailed medical genetics programs aimed at the topical improvement of medical students (http://www.kumc.edu/gec/prof/genecour.html; http://www.bshg.org.uk/documents/official_docs/undergrad.doc; http://www.pediatrics.uottawa.ca/eng/pgme_intapp_prog_genetics.html). Subsequently, these programs have influenced teaching the subject of medical genetics in other countries. In Russia also there is an obligatory program of the teaching of medical genetics at medical universities. During the same time, the preparation in medical genetics for physicians of different specialties in the different countries has had no strict regulation.

In Russia, TI in the area of medical genetics is offered by Departments of Medical Genetics both to physicians (mostly to pediatricians and OB/GYN) and biologists. Courses are organized in bi-weekly cycles of 72 hours each, including 60 hours of lectures and 12 hours of seminars. TI classes are not required by any professional standard, and physicians are assigned to these classes fairly randomly. Some Departments of Medical Genetics offer annual continuing education programs (144 hours, 1 month) by the term, and tenure-line faculties of the relevant departments in the Russian medical colleges. After completion of the courses, students receive a Certificate of TI credit in the area of genetics. To maintain their qualification, the term and tenure-line faculties of Russian medical colleges will undergo this type of TI certification every five years.

1. This is a subtitle (Auth.)

PRACTICING PHYSICIANS SPECIALIZING IN THE AREA OF MEDICAL GENETICS

In many countries, teaching and matriculation of physicians specializing in the area of genetic counseling take place in the medical graduate schools. In the United States, responsibility of genetic counselor certification lies on shoulders of the American Board of Genetic Counselors (http://www.abgc.net/ABGC/AmericanBoardofGeneticCounselors.asp). This Board was established in 1993 to certify individual physicians to practice as genetic counselors. Certification in other specialties, such as clinical geneticist, clinical biochemical geneticist, clinical cytogeneticist, or clinical molecular geneticist, is offered by the American Board of Medical Genetics (http://www.abmg.org/). In Great Britain, the same responsibilities are shared between the Joint Committee on Medical Genetics, whose parent bodies are the Royal College of Physicians, the British Society of Human Genetics, and the Royal College of Pathologists, and the General Medical Council's (GMC's) Education Committee. In Canada, medical geneticists are certified by examination held in the Canadian College of Medical Geneticists (http://www.ccmg-ccgm.org/cred_info.html). Practically everywhere, the preparation for sitting for the Board takes 3–4 years of postgraduate studies, which are in addition to a three-year graduate program and 2-year experience in teaching medical genetics.

In Russia, first efforts in matriculation of medical geneticists were made in 1980 with the introduction of a month-long TI course in genetics. Since 1988, when "physician-geneticist" and "physician-laboratory geneticist" became state-accredited specialties, these TI courses began to incorporate issuance of state certification in above-mentioned areas. Later, these specialties were converted to "Genetics" and "Laboratory Genetics." This type of specialization was made available in the context of the primary specialization (576 hours) for the physicians, and general continuing education (500 hours) for the biologists.

In 2007, according to an Order of the Ministry of Health (Russian Federation), specialization in the area of genetics has been named as one of basic clinical specializations that can be achieved in the context of internship (1 year) and residency (2-year) programs, after subsequent examination and certification. The only exclusion from this rule was made for already practicing physicians with 5 or more years of experience in the field of genetics, who did not have to go through primary specialization. These physician-geneticists can receive a certificate after training in a course lasting 576 hours (4 months). Additionally, there are programs aimed at general improvement of these physicians (144 hours TI). These programs take one month to complete and are required every 5 years in order to receive an endorsement of previously obtained certification in genetics. All certifications listed above require a passing grade on the final exam.

As opposed to specialization in the area of genetics, certification in the area of laboratory genetics now requires additional postgraduate training. This training could be obtained only by physicians already certified with an MD in Genetics after a 4-month retraining program (576 hours) or during a residency. After delivery of a certified examination, doctors receive a certificate in laboratory genetics. Biologists are trained via a general improvement course of 500 hours (3.5 months) and receive a certificate of general improvement in laboratory genetics. Biologists may be certified in laboratory genetics after a 3.5-month program (500 hours).

In 1988, the Ministry of Health (Russian Federation) adopted both a unified study program in medical genetics and a set of professional standards for the physicians in the area of medical genetics. According to these standards, the position of Physician-Geneticist could be occupied only by a certified physician with graduate degree who either certified in genetics or acquired specialization in genetics in the course of postgraduate studies. Physician-Geneticists should possess both theoretical knowledge and practical skills, recognize indications for application of specific genetic tests, be well versed in the adjacent clinical areas and modern diagnostic techniques, as well as methods of disease prevention and treatment of patients with hereditary disorders, be able to conduct medical counseling in different situations, strive to improve both his skills and the skills of nurses and physician assistants, advance understanding of genetics in the general population and other medical professionals, strictly adhere to ethical codex, and demonstrate good bedside manners. Furthermore, they should pass through cycles of TI on medical genetics not less often than once every five years.

The position of Physician-Laboratory Geneticist can be occupied only by a certified physician or a biologist with a graduate degree, who is either certified in genetics in the course of postgraduate studies or has acquired specialization in laboratory genetics in course of graduate studies. Physician-Laboratory Geneticists should know general principles of lab service organization, possess both theoretical knowledge and practical skills, be well versed in the modern diagnostics of hereditary and inborn pathologies, accurately maintain lab notes concerning his own studies, know how to prepare expert conclusions based on the lab findings, strive to improve both his skills and the skills of nurses and physician assistants, as well as strictly adhere to ethical codex. Every five years, a Physician-Laboratory Geneticist should pass cycles of TI in the area of laboratory genetics.

To ensure inclusion of recent discoveries and advances in clinical, biochemical, and molecular genetics, as well as in cytogenetics, standards of TI programs in genetics and laboratory genetics are revamped every 5 years. The unified education program in genetics consists of 5 chapters:

- Genetics and society and organization of genetic services in populations (15 hours).
- Human genetics (117 hours).
- Clinical genetics, general characteristics of the hereditary diseases (257 hours).
- Laboratory diagnostics of hereditary disease (68 hours).
- Prevention of hereditary diseases (107 hours).
- For a total of 576 hours, including 209 hours of lectures, 365 hours of seminars and laboratory practice, and 12 hours of exams.

Internship in genetics (1 year) approximates the same structure, but encompasses 1728 hours (48 weeks), including 405 lecture hours, 762 hours of hospital practice, and 173 hours of electives.

The first year of residency in genetics (out of 2 years) concentrates primarily on the theory of genetics with a focus on medical genetics. The second year of residency in genetics stresses the practical aspects, and minimally supervised experience in diagnostics of hereditary diseases and in genetic counseling.

A comparison of domestic educational programs in genetics to those implemented in other countries reveals much similarity both in structure and duration. However, Russian programs lack a number of chapters integral to other programs, particularly programs adopted in the United States, Canada, United Kingdom, Australia, Sweden, and Israel. Among these chapters are clinical sociology and psychology, philosophy of death and loss, and interpersonal communications. Additionally, some countries, for example, the United Kingdom, Canada, and Israel, include a mandatory research component. In Russia, this component is highly acclaimed but not required. All together, Russian programs are relatively deficient in the component that we have called "attitude."

EDUCATING THE PERSONNEL CONDUCTING GENETIC SCREENING IN NEONATES

Worldwide, all the programs of training in neonatal genetics screening include a mandatory educational chapter (WHO, 2000). The duration and content of this chapter depend on the details of the particular screening program. During recent years the genetic education of physicians was also promoted in the context of monitoring programs for congenital defects and neonatal screenings. Below, we give an example of the training program for physicians conducting neonatal screening (72 hours for each phase).

Phase I – collection of neonatal blood samples at 3rd–7th day of life, immediate transport of the sample to the lab, and laboratory tests (nurses, pediatricians/neonatologists, physicians/laboratory geneticists)

Phase II – confirming laboratory tests in preliminary positive samples collected during primary screening, including molecular diagnostics (physicians/laboratory geneticists, molecular diagnostics specialists)

Phase III – early treatment (no later than 1 month after birth), laboratory monitoring of the treatment efficiency (physicians–geneticists, physicians/laboratory geneticists, and treating physicians)

Phase IV – genetic counseling of the family members (physicians–geneticists) The following literature should be used during these trainings:

- Guidelines for the specialists participating in each of the phases of the screening program (primary screening, confirming laboratory testing, treatment, and genetic counseling)
- General recommendations concerning the prevention for each of the screened hereditary disorders and the control of the program
- Education materials describing various aspects of the neonatal screening for nonprofessionals

The training program "Neonatal Screening" serves as standalone chapter added to the educational requirements for physicians matriculating with a specialization in genetics. This educational program includes both basic and advanced training in neonatal screening, and allows the students to absorb and understand theory as well as acquire the skills for future individual practice. Lectures offered to the students include general principles of the development and implementation of the screening programs, etiology, pathogenesis, clinical course, genetics, molecular genetics and treatment of the screened disorders, genetic counseling of the families with affected, and efficiency of screening programs in different countries. Seminars and laboratory practices cover methods of primary screening, as well as principles and techniques of confirming laboratory diagnostics.

All three types of the postgraduate medical education (PME) courses in medical genetics are offered by each of three departments of medical genetics: Russian Medical Academy for Postgraduate Education (RMAPE), Moscow; Saint Petersburg Academy for Postgraduate Education; and Tomsk State Medical University. The department of medical genetics in RMAPE remains the leader in Russia. This department was founded in 1965, during the rebirth of genetics in Russia. At first, this department was known as the "Department of Biology and Genetics"; in 1985 it was renamed to "Department of Medical Genetics" and assumed a leadership position in the matriculation of geneticists, including physician-geneticists, cytogeneticists, biochemistry geneticists and specialists in DNA-based diagnostics. The alumni of the Department of Medical Genetics in RMAPE work as genetic counselors all over the country. At the present time, approximately 120 genetic counseling units function in all large cities and townships of Russia. Certification in genetics is required for cadre hired to these services.

Every year, the Department of Medical Genetics in RMAPE offers following courses:

1. Recent advances in the field of medical genetics. For professors of the medical schools – 1 month.
2. Genetics (the primary specialization). Certificate program for practicing physicians working in genetic service institutes more than 5 years – 4 months.
3. Laboratory Genetics (primary specialization). For biologists and physicians, with subsequent general improvement certification in laboratory genetics – 3.5 months.
4. Genetics and Laboratory Genetics. For physician-geneticists and laboratory geneticists with TI certificates seeking career advancement in the context of the area of their specialization – 1 month.
5. Recent advances in the field of medical genetics. For pediatricians and OB/GYN practitioners – 2 weeks.
6. DNA diagnostics of hereditary disease. For physician-laboratory geneticists – 2 weeks.
7. Recent advances in the field of medical genetics. For physicians practicing in hospitals and polyclinics – 1 month, taught by professors traveling to localities.
8. Internship in Genetics. For individuals matriculated by medical schools –1 year
9. Residency in Genetics – 2 years.
10. Residency in Laboratory Genetics – 2 years.

The main goal of the departments of medical genetics is to induct the students in three most important aspects of medical genetics – relevant formal knowledge, skills, and attitude. To achieve teaching efficiency, professors of the department developed computerized self-preparation tests and exams in genetics and laboratory genetics. These tests correspond to the Professional Standards in Medical Genetics, and the content of the Unified Study Program in medical genetics. Tests include 1000 questions covering various areas of medical genetics. The first few tests allow evaluation of the current state of students' understanding of genetics, while subsequent modules aim at advancement of the students' knowledge, through a process of individual learning sessions in computerized classrooms. The described tests and learning modules are mostly oriented toward individuals seeking certification as Physician-Geneticist or Physician-Laboratory Geneticist. In addition to computerized tests, the department employs other modern teaching techniques. For example, to teach physicians the clinical diagnostics of hereditary disorders by "portrait" method, professors rely on the video materials offered by the software POSSUM, and Oxford medical databases that include about 5000 hereditary syndromes with clinical descriptions, mode of inheritance, chromosome loci, and patient phenotypes. Teaching of laboratory genetics (cytogenetics) involves use of the software "Video-Test-Karyo." This utility allows a computer screen display of individual karyotypes. Additionally, the teaching process includes workshops allowing learning from direct exchange of the experience of different genetic counseling services all over the country. Workshops are indispensable for the proper digestion of the material, and for future employment in the organizations serving the families of patients with inherited disorders. This is especially important in light of the fact that the primary responsibility of the department is to prepare the cadre for genetic counseling services. Study plans also include site visits to the leading genetics research institutes, where students are acquainted with the latest discoveries in the field of medical genetics. The last step toward graduation is a final exam. This exam consists of three sequential parts that require three days of sitting. Exam parts are (1) the comprehensive oral part that includes solving of genetics problems; (2) a computerized test (200 questions); and (3) a test of diagnostic skills for the 150 most common hereditary diseases (computerized test) and a test of the laboratory skills to diagnose chromosomal diseases and inborn errors of metabolism.

CONCLUSION

During the last 20 years, the matriculation of the cadre for genetic counseling services in Russia has been performed by three Departments of Medical Genetics, all of which function as continued medical education institutions. The documents defining the goals of education in the area of medical genetics were developed by the Department of Education of the Ministry of Health, and affirmed by the Deputy of the Minister of Health. A clear tendency for an increase in the level of understanding of genetics, and the relevant skills acquired after TI Certification, has been seen recently. Another acknowledged tendency is for an increase in the number of both federal and private clinics with a funded position for a physician certified in genetics. The third tendency is toward fulfilling the need of more frequent updates of the knowledge base of physician-geneticists, and to an even larger extent, of laboratory geneticists. The latter tendency is explained by a burst of novel discoveries in the area of genetics and the fast-track transition of the obtained novel knowledge from bench to bedside. All observations mentioned above indicate the necessity of constant improvement to the current system of continued medical education programs in medical genetics.

REFERENCES

Walker AP. (1998), The practice of genetic counseling. In: Baker DL, Schuette JL, Uhlmann WR, eds. *A Guide to Genetic Counseling* (pp. 1–20). New York: Wiley-Liss, Inc.

Baker DL, Schuette JL, Uhlmann WR. (eds.) (1998). Appendix 1.1 Practice-Based Competencies In: *A Guide to Genetic Counseling* (pp. 21–26). New York: Wiley-Liss, Inc.

Bochkov NP. (2010). Clinical Genetics. (In Russian). Geotar-Media.

Ginter EK. (2003) Medical Genetics (in Russian), Medicine.

Harper PS. (2004). Practical Genetic Counselling (6th ed.), London: Arnold.

Kozlova SI, Demikova NS (2007) *Hereditary Syndromes and Genetic Counseling* (3-rd ed.). Author Academy.

WHO [World Health Organization]. Primary health care approaches for prevention and control of congenital disorders (2000), World Health Organization. Human Genetics Programme, Geneve.

Young ID. (2005) *Medical Genetics.* New York: Oxford University Press.

URL'S USED IN THE TEXT

http://www.kumc.edu/gec/prof/genecour.html;

http://www.bshg.org.uk/documents/official_docs/undergrad.doc;

http://www.pediatrics.uottawa.ca/eng/pgme_intapp_prog_genetics.html;

http://www.abgc.net/english/view.asp?x=1;

http://www.abgc.net/ABGC/AmericanBoardofGeneticCounselors.asp;

http://www.abmg.org;

http://www.ccmg-ccgm.org/cred_info.html.

119.

ETHICAL AND LEGAL ASPECTS OF MEDICAL GENETICS IN RUSSIA

V.L. Izhevskaya and V.I. Ivanov

INTRODUCTION

There are now a number of genetic tests available, and the possibility of conducting genetic testing has introduced radical changes in the approaches to diagnosis, prevention, and treatment of a vast variety of human diseases. At the same time, medical application of genetic testing poses a series of ethical, legal, and social issues, urgently requiring not only thorough consideration, but also a need to develop appropriate solutions in order to obtain the maximal return from application of novel methods in public healthcare.

Studies of ethical problems relating to the application of genomic technologies in public healthcare in Russia commenced in the late 1980s in conjunction with the Russian "Human Genome" program, and also within the framework of the international study, "Genetics and Ethics," headed by D. Wertz (The Shriver Center for Mental Retardation, Waltham, MA, USA) and J. Fletcher (University of Virginia Medical Center, USA).

We have carried out several questionnaire-mediated surveys among medical genetic clinic managers and medical geneticists in order to determine the ethical problems apparent in their practical work, and to obtain the specialists' opinions on the issues involved.

In the experience of the medical genetic clinic managers (survey conducted in 2003), the most frequently encountered problems in Russia were:

- poor availability and accessibility of medical genetic services for families needing genetic counseling (due to great distances between the cities with medical genetic clinics and the financial situation of individual patients)
- the attitude to abortion following prenatal diagnosis (both to termination of gestation once fetal pathology was confirmed, and to maintenance of pregnancy with an affected fetus)
- insufficient availability of genetic information on the patient and his/her relatives, spouses, and third parties
- problems arising in connection with testing for late-onset diseases

In practice, less commonly occurring problems involved the use of results obtained in genetic tests for nonmedical purposes, advising on incidental findings of testing, and refusal by the family of medical care apparently detrimental to the fetus or child.

In order to elucidate attitudes to the most controversial problems concerning genetic counseling in Russia, we carried out two (in 1992 and 2002) anonymous questionnaire-mediated surveys among geneticists who possessed vast experience in practical genetic counseling. The 1992 survey was carried out as part of the international study "Genetics and Ethics" under the leadership of D.C. Wertz and J.C. Fletcher. The 1992 study included a total of 45 participating medical geneticists, while that from 2002 included a total of 68.

These studies showed that the majority of Russian medical geneticists were inclined to be too directional, especially in cases of genetic counseling concerning prenatal detection of fetal pathology. More frequently, "negative" advice was given to the family (preferably offered by 18% to 87%, depending upon the disease) and was aimed at termination of a pregnancy with an affected fetus. For a substantial percentage of the physicians, the degree of severity and the age of onset of the disease, or the presence and severity of mental retardation were of little importance when suggesting that pregnancy be terminated. In the 10-year span between the two surveys there was a decrease in the percentage of geneticists offering negative advice for a variety of diseases. However, this decrease was accompanied and followed by a considerable increase in the proportion of those preferring to provide the family with objective information about the disease.

In both surveys, a substantial number of the specialists (approximately 20% in 1992 and 10% in 2002) were inclined, in some way or other, to exert pressure on the woman's choice of options (i.e., they attempted to impose or convey a feeling that the actions taken were incorrect; to persuade the woman not to have children; or to recommend sterilization).

Based on our surveys, Russian specialists appeared reluctant to reliably protect the patient's confidentiality. As few as 7% (in hemophilia and schizophrenia) to 13% (in familial hypercholesterolemia) of Russian geneticists in 1992 would have taken into consideration the patient's desire for confidentiality, and would not have advised the relatives on the genetic risks involved. The 2002 survey showed a significant increase in the percentage of physicians inclined to keep confidentiality with hypercholesterolemia and schizophrenia, with a similar increase observed for Huntington's chorea and hemophilia. A considerable percentage of geneticists in Russia (about 60%) strongly believed that bank-stored DNA samples should be readily available to blood relatives of the patient, without the prior consent of the latter. In 1992 in Russia, the number of those advocating or providing information to the spouse without the patient's consent was relatively high, varying widely from 30% to 60%, depending on the pathology involved. The 2002 survey revealed an increase in the percentage of physicians who considered the patient's consent to be an obligatory condition if they were to provide personal information to their partner, with a significant decrease in the percentage of physicians believing that such information could be given to the spouse without the patient's consent.

The proportion of Russian physicians prepared to reliably protect confidentiality of the patient from third parties in the 1990s ranged widely from 10% to 50%, depending on the disease detected. In the decade from the first survey, these figures have changed considerably.

Hence, the majority of Russian geneticists were inclined to both paternalistic decisions and direction. This was largely determined by insufficient regulation of the genetic services' activity in Russia, the peculiarities of the organization of the medical genetic services, and poor availability and accessibility to the facilities engaged in providing proper treatment of hereditary diseases, as well as medical and social rehabilitation.

Based on the findings, it was suggested that studying ethical issues related to medical genetics should be included in the professional training programs, with special emphasis on the documents on bioethics prepared by the UN, UNESCO, the European Council and its commissions, WHO, WMA, HUGO, the European Society of Human Genetics, the Forum of Ethics Committees of the CIS countries, and Russian legislation.

LEGAL ASPECTS OF GENETIC TESTING IN RUSSIA

Formation of a legislative basis for medical genetics in Russia that is adequate for the present-day development of this branch of knowledge is only just beginning. Juridical regulation in Russia is governed by general and special legislations, as well as by the appropriate by-laws. The basic provisions on protection of human rights are laid down in the Constitution of the Russian Federation, and secure the following:

- the right to health protection and medical care (Article 41);
- voluntary participation in medical, scientific, and other experiments (Article 21);
- personal privacy, as well as personal and family confidentiality (Article 23);
- debarment of collection, storage, use, and dissemination of information on a person's private life without his or her consent (Article 24).

The Russian Federation legislation currently in force firmly acknowledges not only the right to have access to medical and genetic information about oneself, but also the right not to have access. Article 32 of the Fundamentals of the Legislation of the Russian Federation on Health Protection defines these rights as follows: "Every citizen shall have the right to obtain the available information about his or her health status in a comprehensible form. The information concerning the citizen's health status shall be provided to him or her directly, and as far as persons less than 15 years of age are concerned, or citizens legislatively recognized as incapable – to their legal representatives. The information on health status shall not be given to the citizen against his or her will." Realization of the patient's right to information within the context of genetic testing is a guarantee that he/she will be able to freely choose the methods of examination and parenthood planning. The citizen's voluntary informed consent is an obligatory preliminary prerequisite of any medical intervention, also according to Article 32 of the Fundamentals of the Legislation of the Russian Federation on Heath Protection.

Public concern over genetic testing is, to a certain degree, based on anxiety that misuse of genetic data may occur, and that such data could be made available for unauthorized persons. Confidentiality of all personal data, including those obtained from genetic testing, pertains to fundamental human rights. The individual right to freely determine the boundaries of personal confidentiality comprises access to personal information and data, as well as to access to tissue samples not unreasonably considered genetic-information carriers. Provision of legal protection in this field is a factor of ultimate importance.

Confidentiality of information obtained from the patient during genetic studies and genetic counseling, as well as of the findings of genetic diagnosis, is ensured primarily through instituting medical confidentiality as envisaged by the Fundamentals of the Legislation on Health Protection, as well as instituting personal and family confidentiality as envisaged by Article 23 of the RF Constitution.

Article 61 of the Fundamentals of Legislation on Health Protection states that every citizen shall fully enjoy duly guaranteed confidentiality of information on the fact of his or her having sought medical attention, on his/her health status, diagnosis of the disease, as well as any data obtained during his/her examination and treatment. The data referred to above may be disclosed to other citizens, including officials, in the interests of examination and treatment of the patient, for carrying out research trials, for being published in scientific literature, or for use in an educational processes, only with the consent of the patient or his/her legal representative.

Disclosure of medical information protected by confidentiality shall be prosecuted by disciplinary, administrative, and criminal liability. This guarantee on the confidentiality of data constituting medical confidentiality is of such paramount importance that in the Decree of the RF president as of March 6, 1997 № 188, entitled "On the Approval of the List of Confidential Information List" it ranks first among the data related to professional activity.

Highly debatable ethically is the problem of handling information on the patient's genetic status when such information may be of interest to family members of the patient's spouse-to-be. The legislation currently in force does not cover interest in genetic information by the patient's blood relatives, nor regulates the physician's steps in the event of conflict between the patient and his/her relatives. Informing relatives that the patient is undergoing genetic testing is regulated by the Fundamentals of the Legislation, through the generally practiced procedure that prohibits providing the patient's relatives with information concerning the diagnosis without his/her consent.

Provision of information to the spouse-to-be prior to marriage is regulated by The Family Code of the Russian Federation. Article 15 of this document envisages, exceptionally, voluntary counseling on medical genetic problems of the persons contracting marriage, stating further on that the findings of the examination are to be subject to medical confidentiality, and may only be disclosed to the spouse-to-be with the consent of the person examined.

Obviously, the legal norms of the documents referred to above are not enough to reliably ensure an appropriate order of use and protection of genetic information. Sources of information on genetic status may be personalized registers of families with children suffering from congenital and hereditary diseases, as well as developmental malformations.

Yet another source of information on the genetic traits of a citizen that requires special protection from unauthorized access and possible improper use could be data obtained in the course of forensic medical molecular-genetic studies, as well as data obtained in realization of the provisions envisaged in the Federal Law № 242-FZ dated December 3, 2008 and entitled "On State Genome Registration in the Russian Federation," which is carried out for the purpose of personal identification.

Critical to realization of a solution to the problems stated above are the provisions laid down in the Federal Law of the Russian Federation "On Information, Informatization and Protection of Information" as of February 20, 1995 № 24-FZ, envisaging confidentiality of personal data, as well as the Federal Law of the Russian Federation as of July 27, 2006 № 152-FZ "On Personal Data," guaranteeing protection thereof in various methods of data processing.

From the evidence presented above, it can be seen that genetic testing and screening are increasingly applied in medical practice, touching the interests of a constantly growing number of people. Understanding not only the perceivable benefits to the patient and society, but also the potential hazards, widely discussing the problems involved, and making use of the international experience gained worldwide in solving ethical problems arising from application of genetic testing and screening, appear to be vitally important for the well-being of both the patients and society. However, in solving juridical and ethical issues concerning genetic testing we currently lag behind most of the socially, economically, and industrially advanced countries. The problems actively discussed in other countries (use of genetic tests for life and health insurance; testing of the under-aged for carriers of mutations definitely liable to lead to diseases with late onset, and for which there is no efficient treatment or cure; direct provision of genetic tests to the customer, etc.) have in Russia only been discussed by a narrow circle of specialists in medical genetics and bioethics, and have not yet become a subject of broad public discussion or of interest to legislators.

REFERENCES

Constitution of the Russian Federations (December 12, 1993).

Federal Law "Family Code" № 223-FZ as of December 29, 1993.

Federal Law "On Information, Informatization and Protection of Information" № 24-FZ as of February 20, 1995.

Federal Law "On State Genomic Registration in the Russian Federation" № 242-FZ as of December 3, 2008.

Federal Law of the Russian Federation as of July 27, 2006 № 152-FZ "On Personal Data."

Fundamentals of the Legislation of the Russian Federation on Health Protection as of July 22, 1993, № 5487–1.

120.

BIOETHICAL ASPECTS OF GENETICS AND GENOMICS IN YAKUT (SIBERIA)

S.K. Kononova, S.A. Fedorova, and Elza K. Khusnutdinova

INTRODUCTION

A number of population-specific reasons have led to an accumulation of several forms of hereditary diseases in the Yakutian population (Fedorova & Khusnutdinova, 2009). The most frequently encountered in Sakha (Yakutia) are neurodegenerative and neuromuscular diseases with the expansion of trinucleotide repeats and late manifestation: spinocerebellar ataxia of type 1 (SCA1), myotonic dystrophy (DM), and oculopharyngeal muscular dystrophy (Maximova et al., 2008). Before the year 2000, specialists studied the clinical and epidemiological characteristics of these diseases without the help of molecular-genetic methods of research, mostly due to the absence of a technical basis that would allow their use. Since 2000 well-equipped laboratories of the medical-genetic service of the Sakha (Yakutia) have begun to be introduced, along with trained specialists in molecular genetics. It seemed that with the use of DNA testing as a method of regular clinical analysis, many issues of genetic counseling, which could not be used before the introduction of molecular-genetic diagnosis, would be solved. In this case, however, the diagnostic capabilities exceeded the therapeutic ones and, due to this, DNA testing of monogenetic hereditary diseases has led to the appearance of a whole range of new issues, thus justifying the need for accepting the modern bioethical principles of voluntariness, autonomy of an individual, and confidentiality.

The relevance of bioethical research in Sakha (Yakutia) was reinforced by the need to regulate medical activities in new conditions of using genome technologies. We have outlined a number of bioethical issues of genetic counseling in the sphere of using DNA testing in everyday medical practice (Box 120-1).

BOX 120-1

- The form of a relationship between a doctor and a patient when directing the patient to DNA testing
- Informed consent for the procedure of DNA testing and giving blood to the DNA Bank of Hereditary Pathologies and Populations of the Peoples of the Sakha (Yakutia)
- Confidentiality of genetic information
- Presymptomatic DNA testing of diseases with late manifestation
- DNA testing of children
- Ethical issues of prenatal DNA testing for monogenetic diseases
- Issue of availability of DNA testing for everyone
- Question on the activities of the DNA Bank
- Genetic enlightenment of the population

INFORMED CONSENT

In our work, we practice the obtainment of the oral consent of a person to be tested. However, it is quite possible that at the moment of consenting, the patient does not fully understand the far-reaching negative effects of DNA diagnosis that might affect him. Moreover, it has been noted that the patients themselves are not ready to sign various documents, and do so very reluctantly. This has been shown in the results of our questionnaire, when to the question of what form would be the most desirable for obtaining the consent to DNA testing the following results were obtained: in written form signed by both the doctor and the patient – 44%, in written form signed by the patient only – 22%, orally – 25%, and didn't answer – 9% (Kononova et al., 2003). We think that when developing a form for the informed consent of a patient for presymptomatic SCA1 DNA testing, it is necessary to provide alternative choices (Box 120-2).

BOX 120-2

- Give blood and know the results of DNA testing;
- Give blood to the DNA Bank and don't know about the results of testing until a certain time;
- Do not give blood to the DNA Bank and refuse testing.

CONFIDENTIALITY

In our practice we adhere to the principle of full confidentiality. Encryption of DNA samples ensures that no third parties may identify the tested person. The results of DNA testing are revealed only during a personal talk between the geneticist and the tested person. The right to disclose genetic information to relatives is given only to the person who has been tested. However, as has been shown in practice, asymptomatic individuals do not hide the fact of carrying mutated SCA1 from their relatives. In families where cerebellar ataxia was passed on from generation to generation, a certain attitude is formed toward this disease.

Possibly, in the near future, in our society the issue of keeping confidentiality of genetic information after a presymptomatic DNA testing of late-manifestation diseases will become more acute. For example, in Yakutian villages the people are usually well aware of cases of ataxia in specific families, and that is why it is very hard to speak of information confidentiality within the region where the affected families live. This leads to a certain self-isolation of these families, and refusal of testing and supportive treatment; these families are generally more depressed and pessimistic. When the new carrier families and diseased people who have not even suspected the presence of cerebellar ataxia in their heritage are identified, the realization of this fact is very shocking for them.

DNA TESTING OF CHILDREN

The issue of DNA testing of children for illnesses with late manifestation is a very difficult and a highly discussed one. Firstly, a child cannot give an informed consent for testing: his or her parents or legal guardians must do that for him or her. Secondly, if the parents or the guardians learn about the positive result of testing, this can lead to a change in attitude toward the child, which is not really justified.

In our practice there were several situations connected with the issue of DNA testing of children. In the first case, the mother of a girl insisted on getting the results of DNA testing of her daughter with an aim of refusing a university education for her if the results of the analysis came out positive. The request of the mother was denied, as this case was viewed by us as genetic discrimination within a single family. In the second case, the parents asked for the results of their son who had been practicing boxing, so that he could be transferred to a more acceptable kind of sport if the presence of a mutation was confirmed. In this case we accepted the request of the parents. In another family the parents did their best to help their son with an early clinical manifestation of the disease to get a higher university education. In many cases the parents, learning that their family carries the mutation, naturally want to test their children. After the explanation of the reasons of why such testing was refused, they understand the situation and do not insist on carrying out the test. During expedition missions, biomaterial is collected from children for testing if the parents are willing, with an express condition that the results of DNA testing will be revealed confidentially, if needed, when the child reaches the necessary age (legal age). Undoubtedly, each family consultation is an individual case and requires an individual approach, and certain exemptions from the rules are possible, as quite a lot depends on the moral and psychological climate within the family, as well as the financial state and the level of education of the parents.

PRENATAL DIAGNOSTICS

Introduction of DNA testing as a new means of diagnosing hereditary diseases into the practice of genetic counseling in Sahka (Yakutia) gave reason to believe that there would be a demand for prenatal diagnostics among the patients from families with known hereditary diseases. The greatest expectations were connected with the use of prenatal SCA1 testing as the main instrument for preventing new births of children who would be carriers of the SCA1 mutation. That is why, in the early years of using the new method in genetic counseling, geneticists voiced their views on the advisability of a directive approach to DNA testing and prenatal SCA1 testing, taking into account the high social impact of the disease in Sakha (Yakutia). However, as a result of many discussions it was decided that it is necessary to strictly follow the international ethical principles (nondirective consultations, respect for any decision of a patient, voluntary testing, and confidentiality) relating to prenatal testing for diseases with late manifestation.

We have identified differences in approaches to genetic counseling of families carrying the diseases with dynamic mutations; namely, SCA1 and DM. There is a difference in the educational levels of women coming for prenatal testing: as a general rule, the women carriers of SCA1 mutation have higher or secondary professional education and are socially active and adapted, while the women with DM, for the most part, had secondary or lower level education. In most cases, at the moment of asking for medical-genetic help, the patients with DM were already ill, and thus were less adapted socially, were more passive and immature, and were mostly accompanied by healthy members of their families. When making morally difficult reproductive decisions (agreeing to or declining prenatal testing, agreeing to or declining abortion) patients with SCA1 have for the most part taken decisions by themselves and in a considered way; the number of motivated refusals for prenatal diagnostics was considerably higher, as well as the number of agreements for abortion if the results of DNA testing were positive. In contrast to the above, the women with DM had a lower number of refusals for prenatal testing, while the number of refusals for abortion in case of a positive DNA testing result was higher (most of the time, the refusals were unmotivated). In such cases the doctors could not affect the situation, and thus in 6 years, two children with congenital DM were born. In cases of DM, involvement of relatives in making vital decisions becomes inevitable, and, it seems, morally acceptable if the aim of prenatal testing is the birth of a healthy child to a woman affected by DM. However, the remaining close family ties among the Sakha people should considerably ease the genetic counseling of people carrying DM. Therefore, the main difference of genetic counseling for DM, as opposed to SCA1, is the inevitability of involving relatives of pregnant women in the process of informing and making difficult decisions for the whole family. The principle of confidentiality thus loses its meaning. This fact should be considered to be an acceptable ethical norm for people with DM who have asked for medical-genetic help.

ETHNIC FEATURES IN THE SPHERE OF DISCUSSED ISSUES

The mentality of the indigenous peoples of Sakha (Yakutia) has been forming in the process of a constant battle for survival in the harsh conditions of the North. It is possible to point out the unusual endurance for the unfavorable environmental conditions, hunger, cold etc., already described by ethnographers in the nineteenth century. Today, psychiatrists identify outward restraint in expressing emotional experiences as a national character trait of the Sakha people (Ivanov, 2009). The Sakha and the people of the north have a special attitude toward diseases—it is the ability to not dramatize the different manifestations of a disease. Emphasizing or ridiculing physical disabilities or diseases is viewed by the Sakha people very negatively. Compassion is the national character trait. According to many researchers the Sakha people have demonstrated a special, extraordinary compassion and care toward their sick or those who have physical disabilities, children, and have looked after them as best as they could. They were always very compassionate toward the disabled and the elderly. Another specific trait of the national character is the manifestation of the "Eastern fatalism." It is widely believed that the Yakut people highly appreciate the natural abilities of the body, believe in its huge potential for survival; therefore, when sick, the Yakut often chose to wait, and thus went to see healers and physicians late, preferring to leave it all to fate, which was considered to be a manifestation of the 'eastern fatalism' (Tyrylgin, 2002).. It seems that the idea of not resisting the fate is still very much alive in the psychology of the Sakha, even in our dynamic times.

Considering the above, during the process of genetic counseling we have noted several ethnic-specific peculiarities, which have a considerable impact on the process of genetic counseling and which had to be taken into account (Box 120-3).

BOX 120-3:

- The ability not to dramatize specific manifestations of any disease
- Adequate reaction to the results of DNA testing in families where the disease is hereditary
- The issues of disclosure or access to genetic information for spouses and relatives are not very acute due to the traditionally preserved family relations
- Impossibility of keeping genetic information confidential in rural areas of Sakha (Yakutia)
- Manifestation of the "eastern fatalism" and nonresistance to fate in the psychology of the indigenous peoples of Sakha (Yakutia).

SOCIAL AND LEGAL ASPECTS

Based on the experience of foreign countries, the practical side of genetic counseling in Sakha (Yakutia) is carried out in accordance with the accepted international bioethical standards for medical practice concerning the issues of genetic testing (WHO, 1997). In particular, we have developed the Rules of Operation of the DNA Bank, the form for informed consent on giving DNA to the DNA Bank, bioethical rules, and the form of informed consent on presymptomatic DNA testing. We have taken into account several ethnic-specific features, including research into psychological aspects of genetic counseling (Gurinova et al., 2006). Such ethical principles as the confidentiality of genetic information, nondirectivity of genetic counseling, and availability of DNA testing for monogenic diseases for families with hereditary pathologies, are codified in the new version of the Law of the Sakha (Yakutia) of 16.09.2003, "On Prevention of Some Human Hereditary Diseases and Congenital Development Disorders." In order to regulate the statutory powers for the prevention of some human hereditary diseases and congenital development disorders, the Yakutia Parliament adopted the article "On Prevention of Some Human Hereditary Diseases and Congenital Development Disorders," as per the Law of the Sakha (Yakutia) "On the Introduction of Amendments and Additions to the Law of the Sakha (Yakutia) of 16.09.2003 70-3 No 143-III.,. The above law, taking into account the requirements of the federal legislation, introduced amendments to legal norms that confirmed the state guarantees and rights of citizens in the sphere of hereditary disease research. It also introduced amendments to the ways in which prevention of the above diseases is organized in the territory of the republic (Law of the Sakha, 2003).

In 2005 the bioethical issues connected with genetic technologies attracted the attention of the UNESCO international organization. By our initiative and with the support of the National Committee of the Sakha (Yakutia) for UNESCO, the UNESCO office in Moscow, and the Department for the Ethics of Science and Technology in UNESCO Headquarters in Paris, a pilot program for the development of bioethics in the Sakha (Yakutia) was prepared. The main expert recommendation of the International Bioethics Seminar held within the framework of the scientific and practical conference, "Genetic Aspects of Human Pathology: Issues of Preserving the Gene Pool of the Indigenous Peoples of the North," with the participation of UNESCO experts, was to establish a network of acting ethical committees in the Sakha (Yakutia). In 2006 the first Committee on Biomedical Ethics was established in the republic at the Yakut Scientific Centre for Complex Medical Problems of the Siberian Branch of the Russian Academy of Medical Sciences, with an aim of developing an ethical expertise in protocols of scientific research with human participation. In 2007 a "Bioethical Issues in Modern Practical Health Care" roundtable was held within the framework of the 15th Assembly of Medical Workers and Society of the Sakha (Yakutia). The decisions of the rountable, in particular, underlined a recommendation to the Government of the Sakha (Yakutia), respective ministries and medical institutions, and public organizations to provide extensive support for the initiatives concerning the development of educational programs on ethics, as well as establishment and work of committees on ethics, taking into account the existing national and international experience. In accordance with the adopted resolution, the Healthcare Ministry of the Sakha (Yakutia) issued an order No 01-8/4-441 of 28.04.2009 on establishing a network of acting hospital ethical committees with the aim of ethical and legal regulation of medical activities in the Sakha (Yakutia) (Kononova et al., 2009).

CONCLUSION

Consequently, under the new conditions of using genetic technologies in Sakha (Yakutia), there were not only changes in the way doctors/geneticists of the medical genetic service of the republic worked, but also a realization by us of a number of important facts about, firstly, the changes in traditionally established relations between a doctor and a patient—for example, directive consultation being replaced with a nondirective one. Secondly, DNA testing is not a regular clinical analysis: identification of a mutation for the most part sounds like a sentence to a patient, and the results of DNA testing affect not only the tested person, but his family and relatives. Moreover, considering the fact that the benefits of presymptomatic DNA testing can be minimal, while the moral and psychological impact on the person, if a positive result for the existence of a mutation is confirmed, can be extensive.

Introduction of DNA testing into practical medicine requires a development of differentiated ethical approaches depending on the type of the hereditary disease, age of manifestation, seriousness of phenogenotypical manifestations, and clinical polymorphism of each hereditary disease.

The work of the UNESCO international organization in Sakha (Yakutia) gives us a real opportunity to develop bioethics in the republic in the right way, using the experience of other countries and taking into account ethnic traditions, to train our own specialists in bioethics, and to develop international cooperation in this sphere.

REFERENCES

Fedorova SA, Khusnutdinova EK. (2009). Ethnic genomics of the population of the Sakha (Yakutia). *Yakut Medical Journal* 2(26):34–37.

Gurinova EE, Sukhomyasova AL, Maximova NR, et al. (2006). Estimate of Socio-Demographic Indicators at the Presence of Some Hereditary Diseases. Bulletin of the Siberian Branch of the Russian Academy of Sciences (annex), pp. 67–68. Theses of the 13th International Congress of Near-Polar Medicine – Novosibirsk, 2006.

Ivanov IA. (2009). Behaviour of people in life threatening situations and results of acute reactions to stress in emergency conditions. *Yakut Medical Journal* 3(27):134–137.

Kononova SK, Sidorova OG, Fedorova SA, et al. (2009). The experience of researching ethical, legal, and social issues of using genetic technologies in Sakha (Yakutia). Theses of the 2nd Scientific and Practical Conference on the Relevant Issues of Medical Genetics in the Far North, Yakutsk, 4–5 June, 2009. *Yakut Medical Journal* 2(26):86–89.

Kononova SK, Sukhomyasova AL, Nogovitsyna AN, Maximova NR, Khusnutdinova EK. (2003). Bioethical issues of using modern genetic methods of diagnosing hereditary illnesses in the practice of medical-genetic consultation with the analysis of questionnaires of the indigenous peoples of Sakha (Yakutia). *Medical Genetics* 5:218–222.

Law of the Sakha (Yakutia). (2003). On Prevention of Some Human Hereditary Diseases and Congenital Development Disorders. The Law of the Sakha (Yakutia) 70–3 No 143-III of 16.09.2003.

Maximova NR, et al. (2008). Genetic-epidemiological and social-economic aspects of hereditary ethno-specific pathology in Sakha (Yakutia). *Medical Genetics* 10(76):35–43.

Tyrylgin MA. (2002). *Roots of Phenomenal Stamina of the Sakha People*. Yakutsk, 304.

WHO. (1997). Proposed International Guidelines on Ethical Issues in Medical Genetics and Genetic Services. Report of WHO Meeting on Ethical Issues in Medical Genetics. Geneva, December 15–16, 1997. WHO Human Genetic Program.

121.

INHERITED METABOLIC DISEASES AMONG MIGRANT ETHNIC POPULATIONS IN THE UNITED KINGDOM

Anupam Chakrapani and Maureen Cleary

INTRODUCTION

In the United Kingdom (UK), there are several regions in which the local population includes a large proportion of families who have migrated to the UK from the Indian subcontinent. In Manchester, Birmingham, and London metabolic clinics, therefore, there is a wide range of experience of the specific inherited metabolic disorders that are more common in this population. In this chapter we will describe the disorders seen in the UK population of South Asian background.

There are very few published reports detailing incidence of the inborn errors of metabolism (IEMs) among ethnic groups in the United Kingdom. However, one report from Birmingham found that the most common inborn errors seen in the Pakistani population living around the West Midlands were tyrosinemia type I, mucopolysaccharidosis type I, nonketotic hyperglycinemia, and hypertriglyceridemia (Hutchesson et al., 1998). The incidence of recorded IEMs was tenfold higher in the Pakistani population than the white population.

We discuss these disorders and other inborn errors by dividing the topic into disorders of intermediary metabolism, lysosomal, mitochondrial, peroxisomal, and transport disorders. Within these subdivisions we discuss only those metabolic disorders that we consider to be relevant for the families of South Asian origin. For most of these disorders, further information can be found in the relevant chapters of *Inborn Metabolic Diseases* (Fernandes et al., 2000) and *The Metabolic and Molecular Bases of Inherited Disease* (Scriver et al., 2001).

Two points apply to inborn errors of metabolism as a group: most disorders show autosomal recessive inheritance, making them more likely in children with consanguineous parents; and prenatal diagnosis is possible, using biochemical or molecular techniques, for the majority of inborn errors, apart from most mitochondrial disorders.

DISORDERS OF INTERMEDIARY METABOLISM

AMINO ACID DISORDERS

The amino acid disorders that occur in the South Asian population are maple syrup urine disease (MSUD], Nonketotic hyperglycinemia (NKH], tyrosinemia, and the urea cycle defects (UCD].

Maple Syrup Urine Disease

Maple syrup urine disease (MSUD] is an inborn error of branched-chain amino acid metabolism which arises due to the deficiency of branched-chain keto-acid dehydrogenase. It presents usually in the newborn period with an encephalopathy due to the accumulation of branched chain ketoacids (such as α-ketoisocaproic acid) as well as essential amino acid deficiency in the brain (Zinnanti 2009). In the acute presentation, the typical picture is of a full-term infant who, after a short period of well-being, develops poor feeding, lethargy, and progressively deteriorates to coma. In the comatose state, hypertonia is frequently seen and the EEG may show a burst-suppression pattern. There are no major biochemical disturbances. Affected patients are not severely dehydrated and have no metabolic acidosis. Ammonia levels are normal or mildly elevated, and blood lactate is normal. Plasma leucine, isoleucine, and valine are, however, markedly elevated. The acutely sick infant frequently needs exogenous toxin removal by hemodialysis or hemofiltration. Protein feeds should be stopped and anabolism encouraged by the provision of a high-energy intake using carbohydrate. In addition, insulin can be added to promote anabolism.

The objective of longer-term dietary management is to maintain plasma leucine levels at near normal concentrations using a branched-chain restricted protein diet, supplemented with an amino acid substitute, which specifically excludes leucine, isoleucine, and valine. The outcome depends on the severity of the acute presentation; intellectual performance is inversely related to the length of time of pretreatment, and is dependent on the quality of long-term metabolic control (Pollitt, 1997). Liver transplantation may be curative and should be considered a treatment option for severe cases (Shellmar et al., 2010). A minority of children with MSUD have a form that is responsive to thiamine, the cofactor for the branched-chain dehydrogenases. There is also an intermediate form of MSUD in which children have normal plasma leucine between episodes of metabolic decompensation. Both these groups have a better prognosis. In our experience, the acute severe form is more frequently seen in the South Asian population, but the intermediate form has also been observed.

Tyrosinemia Type 1

Tyrosinemia type I is an autosomal recessive disorder due to a defect of fumarylacetoacetase (FAA], the last step in tyrosine degradation. The defect results in the accumulation of intermediate metabolites maleyl and fumarylacetoacetate, highly reactive compounds that are toxic within the liver. These metabolites produce succinylacetone, which inhibits porphobilinogen synthase, accounting for the porphyria-like neurological crises seen in this condition.

In early infancy, tyrosinemia usually presents as acute liver failure with mild jaundice, coagulopathy, encephalopathy, and ascites. Hypoglycemia is a common finding. In older infants, failure to thrive, coagulopathy, hepatosplenomegaly, hypotonia, and rickets are frequently observed. Older children may present with chronic liver disease, cardiomyopathy, renal failure, or a porphyria-like syndrome with self-mutilation. Neurological symptoms include muscle weakness, particularly respiratory muscle weakness. Renal tubular dysfunction and hypophosphatemic rickets may occur at any age. The biochemical hallmark of this condition is the presence of succinylacetone in the urine. Other typical findings, in addition to deranged liver function, are markedly elevated alpha-fetoprotein levels, raised tyrosine, proximal tubular dysfunction with phosphaturia, and aminoaciduria and increased urinary delta-aminolevulinic acid with radiological evidence of rickets. The diagnosis is confirmed by measuring FAA activity in fibroblasts or lymphocytes, or by mutation analysis in informative families.

Treatment with NTBC (2-2-nitro-4-trifluoro-methylbezoyl-1, 3-cyclohexanedione or Nitisinone), in addition to dietary restriction, leads to the rapid clinical improvement and reduction of toxic metabolites within days, with disappearance within one month. Biochemical parameters of liver and renal function improve. Dietary restriction of phenylalanine and tyrosine is necessary to prevent ocular and skin complications associated with hypertyrosnaemia, which invariable occurs on treatment with NTBC. The long-term outcome for children with tyrosinemia type I treated with NTBC, and particularly the risk of hepatocellular carcinoma, is as yet unknown (Grompe et al., 1995).

The disease is found worldwide but is particularly common in Canada (incidence 1:500) (De Braekeleer & Larochelle, 1990) and in parts of Pakistan and northern Europe. In Birmingham, where newborn screening for tyrosinemia occurs, the incidence is 1:20,000, reflecting the mixed ethnic population, many of whom are of South Asian background (Hutchesson et al., 1996). Molecular analysis has demonstrated a "Pakistani" mutation (Rootwelt et al., 1996).

Urea Cycle Defects

The urea cycle is the main pathway of disposal of waste nitrogen. The sequence of reactions converts toxic ammonia molecules into the nontoxic product, urea, which is excreted in the urine. There are genetic defects of each of the enzyme steps with resulting hyperammonemia. These defective enzymes are:

- N-acetylglutamate synthase (NAGS);
- carbamoyl phosphate synthetase (CPS];
- ornithine transcarbamylase (OTC];
- arginosuccinic acid synthetase (citrullinemia];
- argininosuccinate lyase (ASA].

These disorders are autosomal recessive apart from OTC, which is X-linked. In the South Asian population in the UK, the more common disorders seen are citrullinemia and ASA. The urea cycle defects usually present due to the toxic effect on the brain of an elevated ammonia level. Patients may present at any age, but recognized patterns include neonatal onset, onset in childhood, and onset in adulthood. The typical neonate with a urea cycle defect is born healthy, but after a short symptom-free period develops poor feeding, vomiting, lethargy, and irritability due to hyperammonemia. Tachypnea, respiratory alkalosis with apnea, and seizures progressing to coma are seen. In childhood, the symptoms are generally less acute and nonspecific and may include anorexia, recurrent vomiting, and slow developmental progress. Older children and adults may present with an acute encephalopathy or have learning disability or fits. A history of natural avoidance of high-protein foods is common. Biochemically, in addition to hyperammonemia, there are typical derangements of amino acids: plasma citrulline is elevated in citrullinemia and ASA, and low in OTC and CPS; orotic aciduria is present in OTC, ASA, and citrullinemia. The diagnosis can be confirmed by mutation analysis of the appropriate gene or by measuring specific enzyme activity in the appropriate tissue: liver biopsy for OTC, CPS and either liver or skin fibroblasts for ASA and citrullinemia.

Treatment can be divided into the acute management of hyperammonemia and maintenance treatment. The initial aim of emergency treatment is to reduce ammonia as quickly as possible. Dietary protein should be withdrawn and intravenous dextrose, in combination with insulin, provided to minimize catabolism. Phenylbutyrate and sodium benzoate, which form conjugates with glutamine and glycine, respectively, are excreted in the urine and provide an alternative route for elimination of nitrogenous compounds. Arginine is given to supplement the arginine deficiency resulting from the block in the urea cycle. Pharmacological therapies alone may be insufficient, especially in the newborn, to reduce ammonia, and dialysis should then be commenced if hyperammonemia is severe.

Maintenance treatment consists of dietary restriction of protein, supplemented, if necessary, with essential amino acids, and a high-calorie diet sufficient to allow normal growth. In addition, oral sodium benzoate and/or phenylbutyrate are given. Frequent nutritional assessment and monitoring of plasma ammonia and amino acids are required. Children remain at lifelong risk of hyperammonemic coma, either during intercurrent illness or following dietary indiscretion. The outcome is generally poor: in neonatal onset cases, mortality is 50–80% with 70–80% survivors suffering severe neurodisability. In later onset cases, mortality rates of 20–40% are reported with 40–50% prevalence of disability (Nassogne 2005, Bachmann 2003). In view of the poor outcome, early liver transplantation (before the onset of severe brain damage) should be considered especially in severe cases (McKiernan & Leonard 2004). The most important determinant of outcome is the duration of neonatal coma—coma lasting longer than 48–72 hours is almost always associated with severe neurological abnormality (Chakrapani et al., 2000).

Citrullinemia type II

This is a disorder caused by autosomal recessively inherited defects of the gene SLC25A3, which codes for the aspartate-glutamate carrier in the inner mitochondrial membrane, resulting in secondary abnormalities in energy metabolism and the urea cycle (Saheki et al 2010). The disorder is common in Japan and South-East Asia. The diagnosis is suspected from typical amino acid patterns and confirmed by mutation analysis. Two clinical phenotypes result: neonatal intrahepatic cholestasis caused by citrin deficiency (NICCD) which manifests as neonatal cholestasis, and the later-onset citrullinaemia type II (CTLN2), which manifests as neuropsychiatric symptoms, encephalopathy and hyperammonaemia in adulthood. Patients typically self-select a high protein, low carbohydrate diet that is rich in nuts and soy protein. NICCD resolves spontaneously whereas CTLN2 is life-threatening and is treated with liver transplantation.

Nonketotic Hyperglycinemia

This is an autosomal recessive disorder of glycine catabolism. It is a severe neurological disorder with onset in the newborn period. In this condition, there is a defect in the glycine-cleavage system which results in an accumulation of glycine in the body. The pathological effects are seen from the effect on cerebral function. Affected infants become floppy and have depressed conscious level, usually within 48 hours of birth. Some infants show intrauterine damage manifest as agenesis of the corpus callosum or gyral malformations. Seizures are common, and conscious level decreases further with eventual apnea. The EEG frequently shows burst-suppression in the first few weeks of life changing to hypsarrhythmia in those who survive. Many infants die in the newborn period. In those who survive, severe psychomotor retardation is seen, most infants making little developmental progress. Later onset and more chronic forms of this disorder are also seen.

The diagnosis is made by measuring blood and CSF glycine levels. The CSF glycine is raised and the ratio of CSF to blood is >0.09, while normally it is <0.04. Confirmation can be made by enzyme analysis on liver biopsy or by mutation analysis of the genes coding for the subunits of the glycine cleavage enzyme (Applegarth & Toone 2001). Treatment is generally ineffective. Various modalities have been attempted without appreciably altering the prognosis of the disease, including sodium benzoate, ketamine and NMDA antagonists. A combination of sodium benzoate and dextromethorphan has allowed some children to survive the neonatal period and reduced the incidence of seizures, but not improved psychomotor development.

FATTY ACID OXIDATION DISORDERS

Fatty acid oxidation provides an important source of energy during fasting, especially in childhood when glycogen stores are limited. Hepatic fatty acid oxidation produces ketone bodies, which are an important secondary energy source for many tissues. These disorders are characterized by hypoketotic hypoglycemia, which may become evident on weaning or in association with the increased metabolic needs of intercurrent infection. Other clinical features include hypotonia, cardiomyopathy, rhabdomyolysis, metabolic acidosis, maternal fatty liver of pregnancy, or HELLP (hemolysis, elevated liver enzymes, and low platelets). The most dramatic presentation may be with a Reyes-like syndrome or an apparent sudden infant death syndrome, but patients may be clinically and biochemically normal between episodes. Ideally, the blood acylcarnitine profile and urinary organic analysis should be collected at the time of decompensation. Biochemical investigations show elevated aminotransferases, low plasma carnitine, abnormal acyl carnitine profile, especially during decompensation or following a controlled fast, elevated ratio of free fatty acids, ketones, increased urinary excretion of dicarboxylic acids, mild to moderate hyperammonemia, metabolic acidosis, and elevated plasma creatinine kinase. These disorders may be detected on newborn screening programmes that measure acylcarnitines using tandem mass spectrometry. As IEM screening expands in South Asia there will be better understanding of their incidence in this population.

Although overall the commonest disorder of FAO, occurring at a rate of approximately 1 in 10,000 of the UK population, medium chain acyl CoA dehydrogenase deficiency (MCAD] is uncommon in the South Asian population. Trifunctional protein deficiency, very long chain acyl CoA dehydrogenase deficiency (VLCAD], long-chain hydroxyacyl CoA dehydrogenase deficiency (LCHAD], carnitine transporter deficiency, and carnitine palmitoyl transferase I have all been observed in this population. Of these, carnitine transporter defects may be relatively more common in the South East Asian population (Lee et al., 2010). This group of disorders can cause maternal ill-health during pregnancy in the form of HELLP syndrome. This association was initially thought to occur only in association with LCHAD deficiency in the fetus (Iljst et al., 1996), but it is now recognized that this complication can occur in several other fatty acid oxidation defects.

The primary aim of treatment is to avoid excess fasting and suppress lipolysis. This may require the use of overnight nasogastric tube feeds or nocturnal uncooked cornstarch in the more severe disorders. In some disorders of long-chain fat oxidation, dietary long-chain fat should be restricted and medium chain triglycerides supplemented. Routine carnitine supplementation (100mg/kg/day) in FAO disorders remains controversial except in carntine transporter defect where supplementation at 100 mg/kg/day is the main treatment.

The prognosis for infants with a neonatal presentation of FAO is poor, with overall mortality in the first episode as high as 60%. In patients who survive to diagnosis, the prognosis is good for MCAD but variable for LCHAD and VLCAD (Morris et al., 1997). Carnitine transporter deficiency has a good prognosis once carnitine treatment is established. Intellectual development is usually normal as long as acute crises can be avoided.

ORGANIC ACID DISORDERS

Organic acids are carboxylic acids of low molecular weight, and are metabolites of amino acids, carbohydrates, and fats. Organic acid disorders are due to defects in the catabolism of the branched-chain amino acids isoleucine, isoleucine, and valine, and metabolism of propionate. The commonest disorders are: methylmalonic acidemia (MMA: methylmalonyl-CoA mutase deficiency], propionic acidemia (PA: propionyl-CoA carboxylase deficiency] and isovaleric acidemia (IVA: isovaleryl CoA dehydrogenase deficiency] and multiple carboxylase deficiency [MCD];; and their combined incidence is in the order of 1:30,000–40,000. A study undertaken in an Indian metabolic center found organic acid and respiratory chain disorders to be the most common inborn errors in their referral population (Muranjan and Kondurkar, 2001). This ethnicity is reflected in the immigrant South Asian population in the UK, where we have noticed the incidence of organic acidemias to be higher in the South Asian population than in the White population. In those of *South East* Asian background, MMA and Multiple Carboxylase deficiency appear to be the more common organic acid disorders. The presentation of the severe form of organic acidurias is typically of the severe form, with metabolic acidosis in the newborn. In contrast to the White population with these conditions, our experience is that hyperammonemia is a very common finding at presentation and, in some cases, outweighs the severity of the acidosis.

All disorders have a wide clinical spectrum of severity. The most severely affected present in the newborn period with encephalopathy, hypotonia, hepatomegaly, and subsequent coma. Some children present later with recurrent illness characterized by bouts of lethargy, abnormal behavior and altered consciousness, or occasionally with a slowly progressive form with failure to thrive and developmental delay. In general, patients are more susceptible to infection. Multiple carboxylase deficiency is notable for causing eczematous skin eruptions and alopecia.

Laboratory investigations demonstrate metabolic acidosis and ketosis, neutropenia, thrombocytopenia, and hyperammonemia. Plasma amino acids show hyperglycinemia, and urinary organic acids analysis are usually diagnostic showing a characteristic pattern of excretion. In MCD, however, there are typically metabolites of PA, IVA and 3-methylcrotonylglycine as several carboxylase pathways are affected. This pattern cannot be distinguished from biotinidase deficiency. Whilst in the white population biotinidase deficiency seems more common, MCD seems more likely in South Asians. The definitive diagnosis is achieved by demonstrating the relevant enzyme defect in cultured fibroblasts.

Initial management of the acute crisis involves intravenous 10% dextrose with sodium bicarbonate and dietary protein restriction. Insulin may be useful, especially if hyperglycemia occurs. Hyperammonemia requires specific treatment with sodium benzoate or dialysis (see "urea cycle disorders" above). In MMA, higher fluid intakes are necessary and all patients should have a trial of pharmacological doses of vitamin B_{12}. MCD patients usually respond well to oral biotin. Maintenance treatment of this group of disorders, who do not respond to vitamins, consists of an individually titrated low-protein, high-calorie diet with overnight nasogastric tube feeding. Carnitine to detoxify and metronidazole to decrease gut production of PA are prescribed, and sodium benzoate is useful in those with a recurrent hyperammonemia (Nyhan et al., 1991). In the vitamin B_{12}-responsive organic acidurias (MCD and MMA), treatment with vitamin B_{12} or biotin should be lifelong, with little need for dietary restriction.

With the exception of those responsive patients, affected individuals remain at risk of recurrent metabolic decompensation, often in association with intercurrent illness. The outlook for patients with neonatal presentation is poor. Neurological abnormalities are common, with severe hypotonia, progressive neurodevelopmental delay, and learning difficulties. Basal ganglia damage and stroke-like symptoms are common in patients who present early, but all affected patients are vulnerable to neurological damage (Nicolaides et al., 1998). Nutritional progress is difficult and systemic complications such as pancreatitis, cardiomyopathy, and osteoporosis occur. Tubulointerstitial nephritis with progressive renal impairment occurs in MMA, and may lead to renal failure in adolescence.

In view of this poor outlook, liver transplantation may be considered. Liver transplantation results in useful partial correction of PA. Diet is normalized, with decreased metabolite excretion and nutritional and developmental catch-up. There is complete protection against recurrent crises (Schlenzig et al., 1995). However, these benefits must be weighed against the risks of liver transplant. The effects of transplant are not as clear in severe MMA. Although transplantation also results in a useful functional correction of the metabolic defect, this may be less than in PA. In addition, transplantation does not appear to prevent vulnerability to neurological damage, particularly the development of metabolic stroke (Nyhan et al., 2002).

DISORDERS OF CARBOHYDRATE METABOLISM

GLYCOGEN STORAGE DISORDERS (GSD]

Glycogen is the main storage form of carbohydrate in animals and is most abundant in liver and muscle. Liver glycogen controls the export of glucose, which maintains blood glucose concentrations between meals, and is almost completely depleted within a few hours of fasting. Muscle glycogen provides muscles with readily available energy via glycolysis. The glycogen storage diseases are due to defects of glycogen synthesis or breakdown, each caused by a specific enzyme defect, and result in abnormal storage and/or deficient mobilization of glycogen (Wolfsdorf et al., 1999). Some enzyme defects are confined to the liver and are associated with hepatomegaly and hypoglycemia, whereas others affect only muscle glycogen metabolism and result in muscle cramps, weakness, and myopathy.

GSD 1a

Glycogen storage disease type I (GSD I] is due to defective breakdown of glucose-6-phosphate, resulting in decreased hepatic production of glucose and accumulation of glycogen in liver, kidney and intestine. Two subtypes of GSD 1 are currently recognized: GSD 1a, due to glucose-6-phosphatase deficiency, and GSD 1non-a due to defects of the glucose-6-phosphatase transporter.

Patients typically present in the neonatal period or early infancy with hepatomegaly, hypoglycemia, and tachypnea secondary to metabolic acidosis. Older children tend to have a doll-like face, thin extremities, short stature, and a protuberant abdomen due to massive hepatomegaly. The kidneys are also enlarged, but there is no cardiac involvement or splenomegaly. Children have a short tolerance to fasting and typically become hypoglycemic if a feed is delayed or if intake is reduced during intercurrent illness. Symptoms of hypoglycemia are usually accompanied by tachypnea secondary to lactic acidosis. Biochemical abnormalities are characterized by hypoglycemia, hyperlacticacidemia, hyperlipidemia and hyperuricemia. The diagnosis of GSD 1a can be confirmed by mutation analysis of the *G6PC* gene. The main aim of therapy is to prevent hypoglycemia and suppress the secondary metabolic derangements. This is achieved by providing a continuous supply of exogenous glucose, which maintains normal blood glucose concentrations and inhibits counter-regulatory responses. In infants this is best achieved by frequent daytime feeding. Continuous nocturnal enteral glucose feeds are usually required initially. In older children and adults, use of oral uncooked cornstarch, which is hydrolyzed in the gut to release glucose slowly over hours, may reduce the need for continuous or nocturnal feeding. With increasing age, tendency toward hypoglycemia becomes less, perhaps as a result of the natural decrease in metabolic rate (Smit, 1993).

Adequate dietary treatment can result in correction of many of the metabolic abnormalities and result in greatly reduced morbidity. There are many long-term complications associated with GSD. These include poor growth, delayed onset of puberty, polycystic ovaries (Lee et al., 1995a), osteopenia (Lee et al., 1995b), kidney disease, and liver adenomas. Early glomerular hyperfiltration slowly progresses to microalbuminuria, proteinuria, and decreasing glomerular filtration rate. Declining renal function and hypertension eventually occur, and renal dialysis and transplantation may be necessary. It is unclear whether dietary management can prevent long-term renal dysfunction and osteopenia. Liver adenomas are commonly present by the second or third decades of life and have the potential for malignant transformation. Though hepatic adenomas have been shown to regress following intensive dietary treatment, more recent long-term studies indicate that they may occur despite early continuous glucose therapy (Wolfsdorf & Crigler 1999). Liver transplantation is rarely undertaken but may be indicated for symptomatic multiple hepatic adenomata, to prevent the development of hepatocellular carcinoma and/or poor metabolic control (Matern et al., 1999). Successful liver transplantation restores normal metabolic balance, allows catch-up growth, and improves quality of life (Matern et al., 1999). However, it may not prevent the development of renal dysfunction with focal segmental glomerulosclerosis (Faivre et al., 1997).

Glycogen Storage Disease Type 1 Non-A (previously GSD 1B, 1C, and 1D)

Patients with GSD 1 non-a produce glucose-6-phosphatase but cannot transport its substrate (glucose-6-phosphate) to its site of action at the inner wall of the endoplasmic reticulum (ER]. The term *GSD 1 non-a* encompasses those disorders previously known as GSD types 1b, c and d. GSD 1 non-a is much less common than type 1a, the relative prevalence of GSD 1a: GSD 1 non-a being estimated to be around 5–10:1 (Fernandes & Smit, 2000). We have noticed a frequently occurring mutation among the South Asian families for GSD type 1 non-a (Gray, personal communication).

The clinical presentation, metabolic derangements, and complications are the same as in GSD 1a, with the additional finding of neutropenia and impaired neutrophil function. Neutropenia becomes apparent in infancy or early childhood and is usually intermittent (Visser et al., 2000). Inflammatory bowel disease resembling Crohn's disease may also occur. Biochemical abnormalities and liver histology are indistinguishable from GSD 1a. The diagnosis may be confirmed by mutation analysis of the *G6PT* gene. Failing that, it requires demonstration of deficient glucose-6-phosphatase activity in fresh liver biopsy tissue in which hepatocytes and microsomes are intact. Dietatic management of GSD-b patients is the same as in GSD 1a. However, dietary metabolic control has no effect on neutropenia and treatment with G-CSF (granulocyte colony stimulating factor) is recommended. Indications for liver transplantation are the same as those for GSD1a. For severe life-threatening neutropenia, bone marrow transplantation may be considered (Pierre et al 2008).

Glycogen Storage Disease Type III

GSD type III is caused by deficiency of the debrancher enzyme amylo-1,6-glucosidase, resulting in accumulation of partially broken down glycogen molecules (limit dextrin). Patients with liver and muscle involvement (GSD IIIa] have a generalized debrancher deficiency that affects liver, muscle, fibroblasts, cardiac muscle and erythrocytes (Chen, 2001), whereas patients with GSD IIIb have debrancher deficiency confined to the liver. Type IIIa is more common, occurring in about 80% of patients with GSDIII.

Presentation may be indistinguishable from GSDI but milder. The main clinical features are hypoglycemia, hepatomegaly, short stature, skeletal myopathy, hyperlipidemia, and cardiomyopathy. In contrast to GSDI, however, biochemical liver function abnormalities are common and renal enlargement is not present. Patients who have muscle involvement often develop slowly progressive skeletal myopathy and wasting, progressing from minimal signs in childhood to severe muscle weakness by the third or fourth decade of life (Coleman et al., 1992b). Left ventricular hypertrophy is common in patients with muscular involvement, and may lead to significant cardiac dysfunction in the long term (Kiechl et al., 1999). Diagnosis is suggested by the clinical picture and confirmed on enzyme assay either on leucocytes or liver biopsy or by mutation analysis of the *GDE* gene.

General principles of treatment and prevention of hypoglycemia are the same as for GSDI. Adequate dietary management is associated with catch-up growth, decreased liver size, and improved liver function. A high protein intake may help in improving glycemic control, as protein can be used for gluconeogenesis, a pathway that is intact in GSDIII (Slonim et al., 1983). However, progressive liver dysfunction and liver failure can occur. Hepatic adenomas have been reported in up to 25% of patients, but malignant transformation is rare. liver transplantation is indicated for carcinoma or progressive liver failure. The long-term outlook is favorable for patients without muscle involvement (GSD IIIb). For those with GSD IIIa, the prognosis depends on the severity of neuromuscular and cardiac disease. At present, there appears to be no satisfactory way of preventing progressive myopathy (Chen, 2001).

Deficiencies of the Liver Phosphorylase System: Glycogen Storage Disease Types VI and IX

Defects of the phosphorylase system are either due to deficient phosphorylase enzymes or to defects of the phosphorylase activating system (phosphorylase kinase). These systems are enzymatically distinct in the liver and skeletal muscles. Of the liver phosphorylase system defects, phosphorylase kinase defects (GSD IX] are much more common than liver phosphorylase deficiency (GSD VI].

The genetics of the phosphorylase kinase system is very complex, as the enzyme consists of 4 subunits encoded on different genes (X-chromosome as well as autosomes), which are differentially expressed in different tissues. The clinical expression of individual enzyme deficiencies and isoforms is variable and depends on both the severity and distribution of the enzyme defect.

Patients with GSD VI usually present with hepatomegaly and growth failure in early childhood. Symptoms of hypoglycemia, hyperlipidemia, and hyperketosis are mild if present. Plasma lactate and urate concentrations are usually normal. There is no cardiac or skeletal muscle involvement, and the condition has a benign course with reduction in hepatomegaly after puberty.

Phosphorylase kinase deficiency (GSD IX] is clinically and genetically more heterogeneous. Seventy-five percent of all cases have X-linked liver phosphorylase deficiency (type IXa), which manifests between the ages of 1 and 5 years with hepatomegaly, growth retardation, mild hypoglycemia, and mild elevation of hepatic transaminases, cholesterol, and triglycerides. Hepatomegaly and growth retardation usually resolve after puberty.

In the autosomal recessive forms of phosphorylase kinase deficiency (GSD IXb, c], however, which we have observed in the South Asian population, children present with more severe liver disease, which may progress to cirrhosis with or without skeletal myopathy (Burwinkel et al., 1998). Definitive diagnosis of phosphorylase and phosphorylase kinase deficiencies rely on specific enzyme assays in affected tissues, i.e., liver or muscle or mutation analysis. Enzymes can be measured in leukocytes and erythrocytes, but the presence of different isoenzymes can make interpretation difficult. Treatment for these conditions is symptomatic. Frequent feeds and overnight feeding, with or without corn starch (Nakai et al., 1994), may be necessary for more severely affected patients, but many do not require any specific intervention. The tendency to hypoglycemia diminishes with age, and catch-up growth usually occurs without any specific treatment. Rarer variants of GSD IX with isolated skeletal muscle involvement are known.

FRUCTOSE 1,6-BISPHOSPHATASE DEFICIENCY

Fructose-1,6-bisphosphatase deficiency results in impaired gluconeogenesis from all precursors including fructose. Maintenance of normal glucose concentrations therefore depends on dietary glucose and galactose and on hepatic glycogenolysis. Hypoglycemia results when these sources of glucose are exhausted. Accumulation of the gluconeogenic precursors lactate, alanine, and glycerol occurs. It is a rare disorder and incidence is unknown. Fructose-1,6-bisphosphatase deficiency is a life-threatening condition; about half of affected patients become symptomatic in the newborn period with lactic acidosis and hypoglycemia. Presenting symptoms include hyperventilation secondary to profound lactic acidosis, irritability, hypotonia, somnolence, apneic spells, coma, convulsions and hepatomegaly. Other patients may present in infancy or early childhood with hypoglycemia and acidosis triggered by a febrile illness. Subsequent attacks may be triggered by intercurrent illnesses, though patients remain very well between attacks. Ingestion of large quantities of fructose or sucrose are known to precipitate acute decompensation, though in contrast with hereditary fructose intolerance, children do not avoid sweet foods. Diagnosis is best achieved by mutation analysis of the *FBP1* gene; or alternatively by demonstrating the enzyme deficiency in liver, renal cortex, or jejunum. The enzyme is also expressed at low levels in leukocytes, and diagnosis can be attempted in these cells; deficient fructose-1,6-bisphosphatase activity is diagnostic, but normal activity does not rule out liver enzyme deficiency (Besley et al., 1994). The basic principle of long-term management is avoidance of fasting, particularly during intercurrent illnesses; overnight gastric drip-feeding may be required in very young infants. Dietary fructose and sucrose may have to be restricted, particularly during febrile illnesses. After diagnosis and institution of adequate management, the condition follows a relatively benign course. Fasting tolerance improves with age, and may be normal in adults (Moses et al., 1991).

DISORDERS OF TRIGLYCERIDE METABOLISM

The majority of the patients we have seen with type I hypertriglyceridemia due to lipoprotein lipase deficiency have been from South Asia. The diagnosis is suspected in asymptomatic children either when an incidental blood sample is noticed to have a creamy layer due to hyperchylomicronemia, or when hepatosplenomegaly or eruptive xanthomas are found. If this does not occur, then the child usually presents with abdominal pain due to pancreatitis. Treatment is with a low in fat with medium chain triglyceride oil supplementation. The outlook is good provided the diet is adhered to. There are over 30 mutations described in the LPL gene for this rare disorder. A similar picture can arise from Apo-II C deficiency, the cofactor for LPL. It is even rarer than LPL deficiency.

LYSOSOMAL STORAGE

The lysosomes are intracellular organelles containing a large number of different enzymes at acid pH, whose main function is the degradation of macromolecules. The lysosomal storage disorders are each due to a specific enzyme deficiency resulting in abnormal storage of partially degraded macromolecules in the lysosomes.

The clinical spectrum of the storage disorders is wide, ranging from prenatal hydrops fetalis to mild disease in adulthood. Suggestive signs may include coarsening of facial features, neurological deterioration, and hepatosplenomegaly. Patients with storage disorders often have a characteristic skeletal dysplasia (dysostosis multiplex) with a large skull, spinal deformities, and short, thick tubular bones. The liver and spleen are important sites for abnormal lysosomal storage, hence hepatosplenomegaly is a frequent finding, but the clinical picture is often dominated by neurodevelopmental regression.

SPHINGOLIPIDOSES

Sphingolipids are complex membrane lipids composed of sphingosine and a long-chain fatty acid. Various polar substances are attached to the hydroxyl group, giving rise to a range of different sphingolipids. The most basic sphingolipid, from which all others are derived, is ceramide. Since these complex molecules form an integral part of cerebral membranes, the sphingolipidoses, in which the catabolism is deranged, are often associated with relentless neurodegeneration.

GM2 GANGLIOSIDOSIS

There are four subdivisions of GM2 gangliosidosis: Tay-Sachs disease, Sandhoff's disease, and two variants where the enzyme is present but there is altered substrate specificity or there is activator protein deficiency. These are severe neurodegenerative disorders. A cherry-red spot of the eye is typically found. This feature, although characteristic, is not specific to Tay-Sachs, being present in Niemann-Pick and GM1 gangliosidosis. Tay-Sachs first presents usually in infancy, with an exaggerated startle response. Early milestones such as reaching out and sitting are achieved but, in general, are lost by the first birthday. Loss of visual attentiveness is also seen early. After the first year, progressive neurological deterioration occurs with spasticity, feeding difficulties, and seizures. Macrocephaly is common and death usually ensues around the age of 2–4 years. Sandhoff's disease follows a similar pattern. Both disorders have a later-onset variety in which ataxia and dementia begin in the second year of life, and the disease follows a more attenuated form. In the adult-onset form, the neurological picture differs; individuals may have psychosis, dystonia, or cerebellar ataxia. In these later-onset types, cherry-red spot is less common. In the adult form supranuclear ophthalmoplegia is seen.

There is no established treatment for this group of diseases. However, research is underway using "substrate deprivation therapy" in which the toxic metabolites are reduced by introducing another block earlier in the synthetic pathway (Butters et al., 2003).

METACHROMATIC LEUKODYSTROPHY

It is increasingly recognized that in nearly all the subgroups of sphingolipidoses there is a wide range of severity and hence different presentations are seen. This is certainly true in MLD, where presentations in infancy, childhood, adolescence, and adulthood are recognized. The classical picture of MLD is a previously apparently normal child developing problems around the time they learn to walk. The gait progressively deteriorates with spasticity, and reflexes are lost due to the peripheral neuropathy. MRI of the brain shows central demyelination and, frequently, cerebellar atrophy. In the juvenile form, ataxia and spasticity begin usually after the age of seven years. In these children, changes in bulbar function lead to speech difficulties as well as feeding problems. In the adult with MLD, psychiatric symptoms predominate.

The disease arises due to accumulation of sulfatide, a constituent of normal myelin. The diagnosis is made from measurement of the enzyme arylsulfatase. There is a pseudo-deficiency state recognized for this gene, in which low levels of arylsulfatase activity is seen with the artificial substrate in vitro, but in vivo this form allows sufficient activity to avoid the clinical picture of MLD. No effective treatment exists but there is some evidence to suggest that bone marrow transplant (BMT] can prevent deterioration if performed early enough in the course of the disease (Krivit et al., 1990), especially in the late onset forms.

KRABBE DISEASE

This is also a severe neurodegenerative disorder with infantile, childhood, and late-onset forms. The early-onset form develops symptoms in the first few months with irritability and poor feeding, and progresses quickly to spasticity and a vegetative state. There is no treatment but, once again, recent data suggest that in carefully chosen patients diagnosed very early in the course of their disease, BMT may have a role (Krivit et al., 1998), especially the late onset variants.

GM1 GANGLIOSIDOSIS

Infantile GM1 gangliosidosis causes symptoms early in life with hypotonia from birth. By the age of six months they have developmental delay and are developing a coarse physical appearance with puffy skin, maxillary hyperplasia, hypertrophied gums, and macroglossia. They have dysostosis multiplex. About half of the affected children have cherry-red spot. Rapid neurological deterioration is usual, and by the second year of life, individuals have seizures and swallowing difficulties. Death usually occurs by age two years. There is no curative treatment. Juvenile and adult forms are recognized in which there are no physical changes but a picture of neurological deterioration is observed.

NIEMANN-PICK DISEASE

Disorders in which there is storage of sphingomyelin were previously grouped together under the term Niemann-Pick disease, and three types were recognized: A, B, and C. However, only types A and B are due to sphingomyelinase deficiency. Type C is quite different and is due to a disorder of lipid trafficking. Nevertheless, these terms are still used.

Niemann-Pick Disease Types A and B (Sphingomyelinase Deficiency)

Niemann-Pick (NP] type A presents in infancy. It is a severe neurodegenerative disease. The early symptoms are nonspecific, including failure to thrive, feeding difficulties, and frequent respiratory infections. Hepatomegaly is prominent and, in contrast to infantile Gaucher, is more marked than splenomegaly. Neurological degeneration becomes apparent around 18 months of age with loss of visual attentiveness, and floppiness evolving to spasticity. Death usually occurs by around three years of age. Cherry-red spots lead on to retinal ganglion cell degeneration and optic atrophy; there is minimal corneal opacification.

Niemann-Pick disease type B presents after infancy with hepatosplenomegaly. There is usually no neurological involvement. Clinical problems may be hypersplenism arising from massive splenic enlargement, cirrhosis of the liver, and pulmonary infiltration. Occasional patients have had ataxia but normal intelligence.

Both NP types A and B are due to sphingomyelinase deficiency, and the different severity is due to a milder deficiency in type B. They have a higher prevalence in the Ashkenazi Jewish population. There is no recognized treatment for Type A, but Type B may be amenable to bone marrow transplantation (Vellodi et al., 1987) and, more recently, enzyme replacement therapy has shown positive results on the mouse model with human trials underway (Miranda et al., 2000). The characteristic finding on bone marrow examination of lipid-laden histiocytes may be helpful in diagnosis.

Niemann-Pick Disease Type C

This disorder is biochemically quite different from A and B, although there is sphingomyelin stored in the spleen and liver. Types C Niemann-Pick disease is due to impaired intracellular transport of lipids (including cholesterol, sphingosine and complex sphingolipids) due to defects in the NPC1 or NPC2 genes which play a central role in intracellular lipid trafficking. The vast majority of cases (~95%) are caused by defects in the NPC1 gene; the clinical presentation and course for NPC2 defects is identical to NPC1 defects... Type C Niemann-Pick disease has widely varying modes of presentation: neonatal cholestatic liver disease which spontaneously resolves, neonatal liver disease which is rapidly progressive resulting in death; childhood detection learning difficulties and neurological problems such as cataplexy and ataxia; or adolescent/adult presentation with psychiatric manifestations (Vanier 2010). Early onset supranuclear vertical gaze palsy is a characteristic symptom. Neurological progression occurs with the development of a seizure disorder, dystonia, cerebellar ataxia and dementia. Occasional cases are diagnosed with isolated splenomegaly at any age. Rate of progression is variable and depends on the age at onset of neurological symptoms. Early (neonatal) onset patients rarely survive beyond the age of 5 years, childhood onset patients usually survive upto the 2nd decade and patients with the adult onsetforms may survive into the 4th or 5th decades or beyond.

Diagnosis is based upon demonstration of intracellular lipid trafficking defect by fillipin staining or by mutation analysis. Treatment with substrate deprivation therapy (Miglustat) may help stabilize the course of the disease in NPC1 (Wraith et al 2009). Bone marrow transplantation can be attempted in NPC2 disease (Bonney JIMD 2011).

MUCOPOLYSACCHARIDOSES

The mucopolysaccharidoses (MPS] are a group of progressive multisystem disorders affecting the bones and joints, brain, liver and spleen, upper airways, and eyes. They can be grouped according to the underlying enzyme deficiency (Table121-1].

Of the different forms, San Fillipo syndrome or type III is the most common, followed by Hurler syndrome (type I).. MPS IV (Morquio syndrome) is common is some communities of Pakistani origin. Hunter's syndrome (X-linked) is rare in the South Asian population and such a diagnosis deserves further consideration to exclude multiple sulfatase deficiency (autosomal recessive) in which the deficiency of several enzymes causes features of Hunter's syndrome in addition to neurological signs of metachromatic leukodystrophy and ichthyosis of the skin.

TABLE 121-1 CLASSIFICATION OF MUCOPOLYSACCHARIDOSES (MPS]

DISORDER	ENZYME DEFECT	CLINICAL FEATURES	HEPATOSPLENOMEGALY (-, absent; +, mild; ++, moderate; +++, marked	OUTCOME
MPS I (Hurler or Scheie)	α-iduronidase	Hurler: psychomotor retardation, coarse facial features, growth retardation, dysostosis multiplex, corneal clouding, visceromegaly. Milder features in Scheie syndrome	Hurler: ++ Scheie: - to +	Hurler: death by 8–10 yrs. Scheie: normal lifespan; orthopedic and ophthalmological problems common
MPS II (Hunter)	Iduronate sulphatase	Symptoms similar to Hurler syndrome, no corneal clouding. Rare milder variant with no mental retardation	Severe: ++ Mild: - to +	Death by mid-teens Milder variant: normal life span
MPS III (San Fillipo) Type A Type B Type C Type D	Heparan-N-sulphatase α-acetyl-glucosaminidase AcetylCoA: glucosaminide-N-acetyltransferase N-acetyl-glucosamine 6-sulphatase	Commonest MPS disorder in UK, identical phenotype in all types. Marked mental retardation, severe behavioural and sleep disturbance. Mild somatic involvement, mild coarse features, no corneal clouding.	All forms: - to +	Survival into late teens or adulthood with severe mental retardation
MPS IV (Morquio) Type A Type B	N-acetyl-glucosamine 6-sulphatase β-galactosidase	Both types phenotypically similar. No mental retardation, severe skeletal deformities and growth retardation, cervical myelopathy a potentially fatal hazard, mild corneal clouding	Both forms: +	Survival into adulthood common if death does not occur earlier due to cervical myelopathy. Cardiopulmonary compromise later due to thoracic deformity
MPS VI (Maroteaux-Lamy)	Aryl-sulphatase B	Skeletal deformities similar to Hurler syndrome, but no mental retardation. Variable cardiac involvement, mild corneal clouding	+	Severe forms – survive into late teens Mild form – normal life span
MPS VII (Sly)	β-glucuronidase	Variable phenotype ranging from hydrops fetalis to mild adult type similar to MPS I	++	Variable depending on severity

The widespread nature of mucopolysaccharides within the connective tissue of the body results in a range of clinical effects. MPS effects include recurrent ENT infections, obstructive sleep apnea, deposits on the cardiac valves and infiltration of the cardiac muscle, inguinal and umbilical hernia, spinal cord compression, and pain and stiffness from the skeletal dysplasia. Therefore, management involves the input from a multidisciplinary team consisting of cardiology, anesthesia, orthopedics, ENT, neurosurgery, physiotherapy, audiology, speech therapy, and, of course, ophthalmology. Their care should ideally be coordinated by their pediatrician.

Anesthesia should be undertaken with care in MPS disorders, more specifically in types I, II, IV and VI as these children are difficult to intubate due to MPS deposition around the airway (Walker et al., 1994). Intubation is further compromised by the vulnerability of the cervical spinal cord in these subtypes.

The mainstay of management is symptomatic care. Symptomatic care is offered to types II, III and IV. Bone marrow transplant in carefully chosen children with Hurler syndrome (MPS1). Many features of the disease may show improvement: neurological decline is halted, the face remodels, liver and spleen return to normal, and ENT problems are improved. Corneal clouding is improved, with some patients showing complete clearing (Vellodi et al., 1997). The donor enzyme does not penetrate bone sufficiently, and the residual problems are orthopedic complications of the condition.

Enzyme replacement therapy (ERT] is now available for MPS type I, II & III (Wraith 2009). It is aimed primarily at the less severe forms of the condition to improve cardiorespiratory function and mobility. ERT is currently being researched for MPS VI.

MITOCHONDRIAL RESPIRATORY CHAIN DISORDERS

The mitochondria play a central role in cell metabolism by providing energy in the form of ATP. This is achieved by oxidative phosphorylation, which takes place on the respiratory chain located on the inner mitochondrial membrane. The mitochondria are unique among intracellular organelles, in that they contain several copies of their own DNA (mtDNA], although many components of the respiratory chain are encoded on the nuclear genome. Thus, abnormal mitochondrial oxidative phosphorylation can result from mutations of either the nuclear or the mitochondrial genomes. The inheritance of mitochondrial disorders can be maternal, sporadic, autosomal dominant, and autosomal recessive.

Disordered mitochondrial function can result in a wide variety of pathological and clinical manifestations. The clinical spectrum of mitochondrial respiratory chain disorders is very wide, and presentation may range from prenatal manifestation as hydrops fetalis to mild myopathy in adulthood. A number of mutations of mtDNA have been identified and are known to be associated with specific clinical syndromes.

It should be noted that a given mutation may result in clinical features that overlap several different mitochondrial syndromes, and that any one syndrome can be caused by a variety of mutations of the mitochondrial or the nuclear genome (Leonard and Schapira, 2000). Some patients can be diagnosed as having a recognized clinical syndrome, such as MELAS, MERFF or NARP (Table 121-2]; others have a combination of clinical signs and symptoms very suggestive of mitochondrial disease, but cannot be assigned to a particular syndrome. We have observed several clinical patterns of mitochondrial disease in the South Asian population.

Persistently elevated blood lactate levels are an important pointer to respiratory chain disease. However, lactic acidosis is not specific for respiratory chain defects and is commonly present in any sick neonate or infant, especially in the presence of significant liver disease. CSF lactate levels are less prone to fluctuation, and elevated levels are often found in mitochondrial disease with central nervous system involvement. Urine organic acid analysis may reveal abnormal excretion of lactate, fumarate, malate, succinate, 3-methylglutarate, and/or 3-methylglutaconate. Respiratory chain enzyme deficiency is often present in many tissues and, if demonstrable, confirms the diagnosis. Additional useful information is obtained from histological (steatosis, ragged red fibers), histochemical (staining for cytochrome oxidase and succinate dehydrogenase), and electron microscopic (number and morphology of mitochondria) studies on muscle tissue. Normal respiratory chain enzyme studies do not rule out mitochondrial disease. Mutation analysis is useful when the clinical picture suggests a syndrome known to be associated with particular mitochondrial or nuclear DNA mutations. This is rarely the case in the *paediatric* South Asian population but LHON due to the G11778A mutation which typically has onset from 15 to 30 years of age appears to be common in South East Asia (Yu-Wai-Man et al., 2008). As many cases of mitochondrial disease are due to nuclear DNA defects that are yet unknown, it is rarely possible to consider antenatal diagnosis. However, it has been reported that if respiratory chain defects are expressed in skin fibroblasts from the index case, reliable biochemical diagnosis may be possible on chorion villous cells (Wanders et al., 1996). Most mtDNA deletions are sporadic. In maternally inherited disorders antenatal diagnosis by foetal sampling is limited to mutations where the mutant loads correlates well with clinical outcome. Instead some centres offer oocyte sampling to test mutant load (Poulton et al., 2009). Any attempt at antenatal diagnosis of mitochondrial disease must be

TABLE 121-2 SOME RECOGNIZED MITOCHONDRIAL SYNDROMES

Syndrome	Genetics/Pathogenesis	Clinical features
Leigh (subacute necrotizing encephalomyelopathy)	Complex I or IV deficiency, mtDNA T8993G, T8993C, A3444G, Disorders of pyruvate metabolism	Hypotonia, lactic acidosis, psychomotor regression, brain stem dysfunction in childhood, with characteristic symmetrical brain stem and basal ganglia changes on MRI scan
MELAS (mitochondrial encephalomyelopathy with ragged red fibers)	mtDNA A3243G, T3721C	Recurrent stroke-like episodes not corresponding to vascular territories, ataxia, seizures, dementia, deafness, diabetes, myopathy with ragged red fibers
MERRF (myoclonic epilepsy with ragged red fibers)	mtDNA A8344G	Progressive myoclonic epilepsy, myopathy, ataxia, dementia, deafness, optic atrophy
NARP (neurogenic muscle weakness, ataxia, retinitis pigmentosa)	mtDNA T8993G, T8993C	Developmental delay, ataxia, myopathy, pigmentary retinopathy, epilepsy, dementia. Clinical & radiological features of Leigh's syndrome may be seen
LHON (Leber hereditary optic neuropathy)	mtDNA G11778A, G3460A, T14484C, other point mutations	Painless acute or subacute loss of vision, onset 18–30 years. Marked male predominance. 70% patients in UK have the G11778A mutation. Neurological signs in some genotypes, particularly MTND4160
CPEO (chronic progressive external ophthalmoplegia)	mtDNA duplications/deletions	Chronic progressive weakness of the external ocular muscles, occasional proximal limb weakness
KSS (Kearns-Sayre syndrome)	mtDNA duplications/deletions	Onset < 20 years, CPEO, pigmentary retinopathy and one or more of cardiac conduction defects, ataxia or CSF protein >100mg/dl
Pearson syndrome	MtDNA deletions	Sideroblastic anemia, exocrine pancreatic insufficiency, Fanconi syndrome, liver failure
MNGIE (mitochondrial neurogastrointestinal/ encephalomyopathy)		Gastrointestinal dysmotility, peripheral neuropathy, myopathy, ophthalmoplegia
mtDNA depletion syndrome	Quantitative reduction of mtDNA	liver failure, hypotonia, developmental delay, Fanconi syndrome

offered with careful genetic counselling (Thorburn and Dahl, 2001).

A number of different approaches have been tried in the treatment of respiratory chain disease, including cofactor therapy with riboflavin, artificial electron acceptors such as menadione and vitamin C, and a variety of free radical scavengers such as Coenzyme Q_{10}, vitamin E, ibedenone, carnitine, and methylene blue (Morris & Leonard, 1996). Experience with these modalities of treatment is largely anecdotal and the results variable (Horvath et al., 2008).

PEROXISOMAL DISORDERS

Peroxisomes are small membrane-bound intracellular organelles that contain about 40 different anabolic and catabolic enzymes. They serve many functions, including β-oxidation of very long-chain fatty acids (VLCFA], β-oxidation of phytanic acids, and β-oxidation of di- and trihydroxycholestanoic acids to chenodeoxycholic acid and cholic acid, which are bile acids precursors.

Peroxisomal disorders are classified into 2 main groups: (1) multiple enzyme deficiencies (such as the "Zellweger spectrum" disorders and rhizomelic chondrodysplasia punctata), which arise from defective peroxisome synthesis, assembly, and enzyme import, and (2) genetic deficiency of a single peroxisomal enzyme (such as adrenoleukodystrophy, classic Refsum's disease, and hyperoxaluria type I). Over 20 different disorders have been described. The overall prevalence has been estimated to be 1:25,000 (Roth, 1999). We have observed both single and multiple enzyme deficiencies in the South Asian population now resident in the UK. The disorders of peroxisome biogenesis are genetically heterogeneous and belong to 11 different complementation groups without correlation between complementation group and phenotype (Moser et al., 1995). Mutations in the genes responsible for peroxisomal assembly (PEX genes] are known to be associated with ZS, NALD, IRD and RCDP (Gartner, 2000).

DISORDERS WITH MULTIPLE ENZYME DEFICIENCIES

Patients who have a disorder of peroxisomal biogenesis lack normal peroxisomes. Four conditions are recognized: Zellweger syndrome (ZS], neonatal adrenoleukodystrophy (ALD], infantile Refsum's disease (RD] and rhizomelic chondrodysplasia punctata (CDP]. These are all associated with severely deranged peroxisomal assembly, loss of multiple enzyme activities, and multisystem involvement. Peroxisomes are absent or greatly reduced in number in skin fibroblasts and liver biopsy specimens. Zellweger syndrome is the most severe, infantile Refsum's disease less so, and neonatal ALD is between the two (Raymond & Moser, 1997).

ZELLWEGER SYNDROME (ZS]

Infants with ZS present in the neonatal period with characteristic dysmorphic features (prominent forehead, large anterior fontanelle, broad nasal bridge, epicanthal folds, high arched palate, micrognathia, redundant neck skin folds, clinodactyly, and talipes equinovarus). Neurological abnormalities are prominent, including severe hypotonia, areflexia, poor suck reflex, and seizures. Neuronal migration defects may be observed on neuroimaging. Other features such as corneal clouding, cataracts, pigmentary retinopathy, polycystic kidneys, cryptorchidism, dislocated hips, and stippled epiphyses on radiographs (chondrodysplasia punctata) may be present. Liver disease is common and includes hepatomegaly, conjugated hyperbilirubinemia, progression to cirrhosis, and liver failure in the first few months of life, but hepatic involvement is overshadowed by the neurological symptoms. Failure of psychomotor development is evident in early infancy, and survival beyond 1 year is rare. Patients with NALD and IRD have the same features as ZS but a milder phenotype. Infants have hepatomegaly and neonatal cholestasis; progressive liver disease is only significant in children who survive the first decade (Gould et al., 2001).

"Classic" rhizomelic chondrodysplasia punctata (RCDP] presents with dysmorphism and psychomotor retardation. Affected children have shortened proximal limbs, cataracts, and coronal clefts of the vertebral bodies. There may be ichthyosis.

DISORDERS DUE TO SINGLE PEROXISOMAL ENZYME DEFICIENCIES

There are many isolated peroxisomal enzyme deficiencies. Intact peroxisomes are found in liver biopsy and skin fibroblast specimens, and the biochemical and clinical abnormalities relate to the individual pathway. Although we have observed these phenotypes in the South Asian community, they are by no means common.

The diagnosis is made by measuring very long-chain fatty acids (VLCFA], DHPAT (blood), phytanate/pristanate (plasma), and plasmalogens (plasma). Morphological studies of liver/skin for fibroblasts may show complete absence/reduced or abnormal structure of peroxisomes. The diagnosis is confirmed by specific enzyme analysis in skin fibroblasts.

There is no treatment for multiple enzyme dysfunction, but supportive care with anticonvulsants, dietary supplements for the liver disease, and muscle relaxants is essential. Bile acid supplements have reduced cholestasis in a child with ZS (Setchell et al., 1992), while docosahexaenoic acid (DHA] has led to some clinical improvement in the "Zellweger spectrum" (Raymond & Moser 1997). Therapeutic options are available for some of the single-enzyme disorders.

TRANSPORT DISORDERS

GYRATE ATROPHY

This disorder arises due to deficiency of ornithine aminotransferase (OAT], a mitochondrial enzyme which, beyond infancy, catalyzes excess ornithine. Interestingly, early in life the OAT enzyme works in the reverse direction to allow ornithine production from which arginine is synthesized. In the newborn, therefore, ornithine levels are low, but following enzyme reversal, plasma ornithine levels are extremely high. The effects of this condition are in the eye and appear in adolescence. Initially, night vision is lost and then progressive daytime blindness follows. The condition is also associated with cataracts. There is considerable heterogeneity in this disorder even within families. The OAT enzyme is dependent on pyridoxal-phosphate as a cofactor, and in some patients administration of the cofactor has decreased plasma ornithine. Otherwise, treatment is dietary with a low-arginine diet. It has been shown that reduction of plasma ornithine levels can allay disease progression, although it will not reverse existing eye damage. This diet is very unpalatable and long-term compliance is difficult. Other suggested treatments, whose effects are not yet proven, include lysine and proline. There exists a secondary cerebral creatine deficiency in some patients and creatine supplementation has also been suggested in the minority of affected patients who also show some intellectual disability (Valayannopoulos et al., 2009). Diet should not be started until ornithine levels are high, since hyperammonemia is a risk from ornithine, and hence arginine, deprivation during this time.

SUMMARY

The South Asian population in the United Kingdom have largely settled in or near the main industrial cities of Manchester, Birmingham, and London. The experience of these metabolic centers of the pattern of inborn errors associated with that background is therefore expanding. For many families who have already experienced the misfortune of living with metabolic disease, it is important to realize that new treatments are being rapidly developed, and therefore the prognosis may have changed markedly over recent years. In view of the rarity of these disorders, referral to a specialist center should be offered to all families.

REFERENCES

Applegarth DA, Toone JR. (2001). Nonketotic hyperglycinemia (glycine encephalopathy): laboratory diagnosis. *Mol Genet Metab* 74(1–2):139–146.

Bachmann C. (2003). Outcome and survival of 88 patients with urea cycle disorders: a retrospective evaluation. *Eur J Pediatr* 162(6):410–416.

Bonney DK, O'Meara A, Shabani A, Imrie J, Bigger BW, Jones S, Wraith JE, Wynn RF. (2010). Successful allogeneic bone marrow transplant for Niemann-Pick disease type C2 is likely to be associated with a severe 'graft versus substrate' effect. *J Inherit Metab Dis* [Epub ahead of print].

Butters TD, Mellor HR, Narita K, Dwek RA, Platt FM. (2003). Small-molecule therapeutics for the treatment of glycolipid lysosomal storage disorders. *Philo Trans R Soc Lond B Biol Sci* 358:927–945.

Besley GTN, Walter JH, Lewis MA, et al. (1994). Fructose-1,6-bisphosphatase deficiency: severe phenotype with normal enzyme activity. *Journal of Inherited Metabolic Disease* 17:333–335.

Burwinkel B, Shiomi S, Al Zaben A. (1998). Liver glycogenosis due to phosphorylase kinase deficiency: PHKG2 gene structure and mutations associated with cirrhosis. *Human Molecular Genetics* 7:149–154.

Chakrapani A, Olpin S, Cleary M, Walter JH, Wraith JE, Besley GT. (2000). Trifunctional protein deficiency: three families with significant maternal hepatic dysfunction in pregnancy not associated with E474Q mutation. *J Inherit Metab Dis* 23, 826–834.

Chen Y-T. (2001). Glycogen Storage Diseases. In: Scriver CR, Beaudet AL, Sly WS, et al., eds. *The Metabolic and Molecular Bases of Inherited Disease*, 8th edition, pp. 1521–1552. New York: McGraw-Hill.

Coleman RA, Winter HS, Wolf B, et al. (1992b). Glycogen storage disease type III (glycogen debranching enzyme deficiency): correlation of biochemical defects with myopathy and cardiomyopathy. *Annals of Internal Medicine* 116:896–900.

de Braekeleer M,. Larochelle J. (1990) Genetic epidemiology of hereditary tyrosinaemia in Quebec and in Saguenay-Lac-St-Jean. *American Journal of Human Genetics* 47:302–307.

Faivre L, Houssin D, Valayer J, et al. (1997). Long-term outcome of liver transplantation in patients with glycogen storage disease. *Journal of Pediatric Gastroenterology and Nutrition* 24:485.

Fernandes J, Smit GPA. (2000). The glycogen-storage diseases. In: Fernandes J, Saudubray J-M, van den Berghe G. (eds.), *Inborn Metabolic Diseases: Diagnosis and Treatment* (3rd ed., pp. 87–102). Berlin–Heidelberg–New York: Springer-Verlag.

Fernandes J, Koster J, Grose W, et al. (1974). Hepatic phosphorylase deficiency: Its differentiation from other glycogenoses. *Archives of Disease in Childhood* 49:186–191.

Fernandes J, Saudubray JM, van den Berghe G, eds. *Inborn Metabolic Diseases*, 3rd Edition. (2000) Springer-Verlag, Heidelberg.

Gartner J. (2000) Disorders related to peroxisomal membranes. *Journal of Inherited Metabolic Disease* 23(3):264–272.

Grompe M, Lindstedt S, al-Dhalimy M, et al. (1995). Pharmacological correction of neonatal lethal hepatic dysfunction in a murine model of hereditary tyrosinaemia type 1. *Nature Genetics* 10(4):453–460.

Gould SJ, Raymond GV, Valle D. (2001). The peroxisome biogenesis disorders. In: Scriver CR, Beaudet AL, Sly WS, et al., eds. *The Metabolic and Molecular Bases of Inherited Disease*, 8th edition, pp. 3181–3218. New York: McGraw-Hill.

Horvath R, Gorman G, Chinnery PF. (2008). How can we treat mitochondrial encephalomyopathies? Approaches to therapy. *Neurotherapeutics* 5(4):558–568. Review.

Hutchesson ACJ, Hall SK, Preece MA, Green A. (1996). Screening for tyrosinaemia type I. *Archives of Disease in Childhood* 74:F191–F194.

Hutchesson AC, Bundet S, Preece MA, Hall SK, Green A. (1998). A comparison of disease and gene frequencies of inborn errors of metabolism among different ethnic groups in the West Midlands, UK. *J Med Genet* 35(5): 366–370.

Iljst L, Ruiter JP, Hoovers JM, Jakobs ME, Wanders RJ. (1996). Common missense mutation G1528C in long-chain 3-hydroxyacyl-CoA dehydrogenase deficiency. Characterization and expression of the mutant protein, mutation analysis on genomic DNA and chromosomal localization of the mitochondrial trifunctional protein alpha-subunit gene. *Journal of Clinical Investigation* 98(4):1028–1033.

Imrie J, Wraith JE. (2001). Isolated splenomegaly as the presenting feature of Niemann-Pick disease type C. *Arch Dis Child* 2001 84(5):427–429.

Kiechl S, Kohlendorfer U, Thaler C, et al. (1999). Different clinical aspects of debrancher deficiency myopathy. *Journal of Neurology Neurosurgery and Psychiatry* 67:364–368.

Krivit W, Shapiro E, Kennedy W, et al. (1990). Treatment of late infantile metachromatic leucodystrophy by bone marrow transplantation. *N Engl J Med* 322:28–32.

Krivit W, Shapiro EG, Peters C, et al. (1998). Hematopoietic stem-cell transplantation in globoid-cell leukodystrophy. *N Engl J Med* 338:1119–1126.

Lee NC, Tang NL, Chien YH, Chen CA, Lin SJ, Chiu PC, Huang AC, Hwu WL. (2010). Diagnoses of newborns and mothers with carnitine uptake defects through newborn Screening. *Mol Genet Metab.* 100(1):46–50 [Epub 2009 Dec 28].

Lee PJ, Patel A, Hindmarsh PC, et al. (1995a). The prevalence of polycystic ovaries in the hepatic glycogen storage diseases: its association with hyperinsulinism. *Clinical Endocrinology* (Oxford) 42:601–606.

Lee PJ, Patel JS, Fewtrell M, et al. (1995b). Bone mineralisation in type 1 glycogen storage disease. *European Journal of Pediatrics* 154:483–487.

Leonard JV, McKiernan PJ. (2004). The role of liver transplantation in urea cycle disorders *Mol Genet Metab* 81(Suppl 1):S74–S78.

Leonard JV, Schapiro AH. (2000). Mitochondrial respiratory chain disorders I: mitochondrial defects. *Lancet* 355:299–304.

Matern D, Starzl TE, Arnaout W, et al. (1999). Liver transplantation for glycogen storage disease types I, III, and IV. *European Journal of Pediatrics* 158 Suppl2:S43–48.

Miranda SR, He X, Simonaro CM, Gatt S, Dagan A, Desnick RJ, Schuchman EH. (2000). Infusion of recombinant human acid sphingomyelinase into Niemann-Pick disease mice leads to visceral, but not neurological correction of the pathophysiology. *FASEB J.* 14(13):1988–1995.

Morris AA, Clayton PT, Surtees RA, Leonard JV. (1997). Clinical outcomes in long-chain 3-hydroxyacyl-coenzyme A dehydrogenase deficiency. *Journal of Pediatrics* 131(6):938.

Morris AAM, Leonard JV. (1996). The treatment of congenital lactic acidosis. *Journal of Inherited Metabolic Disease* 19:573–580.

Moser AB, Rasmussen M, Naidu S, et al. (1995). Phenotype of patients with peroxisomal disorders subdivided into sixteen complementation groups. *Journal of Pediatrics* 127(1):13–22.

Moses SW, Bashan N, Flasterstein BF, et al. (1991). Fructose-1,6-bisphosphatase deficiency in Israel. *Israel Journal of Medical Sciences* 27:1–4.

Muranjan M, Kondurkar P. (2001). Clinical features of organic acidaemias: experience at a tertiary care center in Mumbai. *Indian Pediatrics* 38: 518–524.

Nakai A, Shigematsu Y, Takano T, et al. (1994). Uncooked starch treatment for hepatic phosphorylase kinase deficiency. *European Journal of Pediatrics* 153:581–583.

Nassogne MC, Héron B, Touati G, Rabier D, Saudubray JM. (2005). Urea cycle defects: management and outcome. *J Inherit Metab Dis.* 28(3):407–414.

Nicolaides P, Leonard J, Surtees R. (1998). Neurological outcome of methylmalonic acidaemia. *Arch Dis Child* 78, 508–512.

Nyhan WL, Rice-Asaro M, Acosta P. (1991). Advances in the treatment of amino acid and organic acid disorders. In: Desnick RJ, ed. *Treatment of Genetic Diseases*, pp. 45–67. New York: Churchill Livingstone.

Nyhan WL, Gargus JJ, Boyle K, Selby R, Koch R. (2002). Progressive neurologic disability in methylmalonic acidemia despite transplantation of the liver. *Eur J Pediatr* 161, 377–379.

Pierre G, Chakupurakal G, McKiernan P, Hendriksz C, Lawson S, Chakrapani A. (2008). Bone marrow transplantation in glycogen storage disease type 1b. *J Pediatr* 152(2):286–288.

Pollitt R. (1997). Neonatal screening for inborn errors of metabolism: cost, yield and outcome: a review. *Health Technology Assessment* 1(7):1–4.

Poulton J, Kennedy S, Oakeshott P, Wells D. (2009). Preventing transmission of maternally inherited mitochondrial DNA diseases. *BMJ* 30:338.

Raymond GV, Moser HW. (1997). Clinical diagnosis and therapy of peroxisomal diseases. In: Applegarth DA, Dimmick JE, Hall JG, eds. *Organelle Diseases: Clinical features, Diagnosis, pathogenesis and management*, pp. 169–192. London: Chapman & Hall.

Rootwelt H, Hoie K, Berger R, Kvittingen EA. (1996). Fumarylacetoacetase mutations in tyrosinaemia type I. *Human Mutation* 7, 239–243.

Roth KS. (1999). Peroxisomal disease—common ground for pediatrician, cell biologist, biochemist, pathologist, and neurologist. *Clinical Pediatrics* (Philadelphia) 38(2):73–75.

Saheki T, Inoue K, Tushima A, Mutoh K, Kobayashi K. (2010). Citrin deficiency and current treatment concepts. *Mol Genet Metab* 100(Suppl 1):S59–S64.

Schlenzig JS, Poggi-Travert F, Laurent J, et al. (1995). Liver transplantation in two cases of propionic acidaemia. *Journal of Inherited Metabolic Disease* 18: 448–461.

Scriver CR, Beudet AL, Sly WS, Valle D, eds. (2001). *The metabolic and molecular bases of inherited disease*, 7th Edition. New York: McGraw-Hill.

Setchell KD, Bragetti P, Zimmer-Nechemias L, et al. (1992). Oral bile acid treatment and the patient with Zellweger syndrome. *Hepatology* 15(2):198–207.

Shellmer DA, Devito Dabbs A, Dew MA, Noll RB, Feldman H, Strauss KA, Morton DH, Vockley J, Mazariegos GV. (2010). Cognitive and adaptive functioning after liver transplantation for maple syrup urine disease: A case series. *Pediatr Transplant* [Epub ahead of print].

Slonim AE, Coleman RA, Moses S, et al. (1983). Amino acid disturbances in type III glycogenosis: differences from type I glycogenosis. *Metabolism* 32:70–74.

Smit JP. (1993). The long-term outcome of patients with glycogen storage disease type 1a. *European Journal of Pediatrics* 152 (Suppl1):S52–55.

Thorburn DR, Dahl HH. (2001). Mitochondrial disorders: genetics, counseling, prenatal diagnosis and reproductive options. *Am J Med Genet.* 106(1):102–114.

Valayannopoulos V, Boddaert N, Mention K, Touati G, Barbier V, Chabli A, Sedel F, Kaplan J, Dufier JL, Seidenwurm D, Rabier D, Saudubray JM, de Lonlay P. (2009). Secondary creatine deficiency in ornithine delta-aminotransferase deficiency. *Mol Genet Metab.* 97(2):109–113.

Vanier MT. (2010). Niemann-Pick Disease type C. *Orphanet J Rare Dis* 5:16.

Vellodi A, Hobbs JR, O'Donnell NM, Coulter BS, Hugh-Jones K. (1987). Treatment of Niemann–Pick disease type B by allogenic bone marrow transplantation. *Br Med J* 295: 1375–1376.

Vellodi A, Young EP, Cooper A, et al. (1997). Bone marrow transplantation for mucopolysaccharidosis type I: experience of two British centres. *Arch Dis Child* 76(2):92–99.

Visser G, Rake JP, Fernandes J, et al. (2000). Neutropenia, neutrophil dysfunction, and inflammatory bowel disease in glycogen storage disease type Ib: results of the European Study on Glycogen Storage Disease type I. *Journal of Pediatrics* 137:187–191.

Walker RW, Darowski M, Morris P, Wraith JE.(1994). Anaesthesia and mucopolysaccharidoses. A review of airway problems in children. Anaesthesia. 49(12): 1078–1084.

Wanders RJ, Ruiter JP, Wijburg FA. (1996). Prenatal diagnosis of systemic disorders of the respiratory chain in cultured chorionic villus fibroblasts by study of ATP-synthesis in digitonin-permeabilized cells. *Journal of Inherited Metabolic Disease* 19:133–136.

Wolfsdorf JI, Crigler JF Jr. (1999). Effect of continuous glucose therapy begun in infancy on the long-term clinical course of patients with type I glycogen storage disease. *Journal of Pediatric Gastroenterology and Nutrition* 29:136–143.

Wolfsdorf JI, Holm IA, Weinstein DA. (1999). Glycogen storage diseases. Phenotypic, genetic, and biochemical characteristics, and therapy. *Endocrinology Metabolism Clinics of North America* 28:801–823.

Wraith JE. (2009). Enzyme replacement therapy for the management of the mucopolysaccharidoses. *Int J Clin Pharmacol Ther* 47(Suppl 1):S63–S65.

Wraith JE, Vecchio D, Jacklin E, Abel L, Chadha-Boreham H, Luzy C, Giorgino R, Patterson MC. (2010). Miglustat in adult and juvenile patients with Niemann-Pick disease type C: long-term data from a clinical trial. *Mol Genet Metab* 99(4):351–357.

Yu-Wai-Man P, Griffiths PG, Hudson G, Chinnery PF. (2009). Inherited mitochondrial optic neuropathies. *J Med Genet* 46(3):145–158.

Zinnanti WJ, Lazovic J, Griffin K, Skvorak KJ, Paul HS, Homanics GE, Bewley MC, Cheng KC, Lanoue KF, Flanagan JM. (2009). Dual mechanism of brain injury and novel treatment strategy in maple syrup urine disease. *Brain* 132(Pt 4):903–918.

122.

VIEWS OF INHERITED ILLNESS AMONG BRITISH PAKISTANI USERS OF GENETIC SERVICES

Alison Shaw and Jane Hurst

INTRODUCTION

Genetic counseling aims to provide information to help individuals referred for a known or suspected genetic condition (in the individual, a pregnancy, child, or relative) make informed reproductive decisions, by establishing a diagnosis, recurrence risks, and risk-management options. To be effective, the information provided needs to be assimilated within patients' prior understandings of illness causality, inheritance, and genetic risk (Richards and Ponder, 1996; Richards, 1998; Chapple et al., 1995). These understandings may be profoundly influenced by education, culture, and ethnic background. Individuals with graduate or postgraduate level education may readily accept a genetic explanation for their child's problem and benefit most from genetic counseling (Weil, 1991; Kelly et al., 2004). Within the American general public educated up to secondary/high school, basic misconceptions about genetics, such as the location of genes in the body, may be widespread, despite conversational familiarity with genetic terms (Lanie et al., 2004). Among 15–16 year old British school students taking compulsory science education, there is confusion over the nature of genetic information and mechanisms of Mendelian inheritance (Wood-Robinson et al., 2000; Lewis et al., 2000). Beliefs in the environmental or spiritual causes of illness may offer more satisfactory explanations for a child's problem than genetic theory, and may be patterned in culturally specific ways, with implications for genetic counseling across cross-cultural contexts (Panter-Brick, 1991; Meiser et al., 2001).

Patients' prior theories of illness causality and inheritance may influence their assimilation of genetic information and understandings of risk in clinically significant ways (Weil, 1991; Shiloh and Berkenstadt, 1992). Alternative theories of causality (Santos and Bizzo, 2005) or "personal" theories of inheritance (McAllister, 2003), such as "it only affects men," "it is on the mother's side," or "it skips generations" are often based upon observed patterns in the family history of inherited characteristics or medical problems (Hunt et al., 2002). "Lay" or "everyday" explanatory models can be at odds with genetic theory, such that genetic risk is overlooked or misunderstood (Richards, 1996, 1998; Davison, 1996). Cross-cultural differences in understandings of inheritance may also affect risk perception and the assimilation of genetic information (Richards, 1996; Meiser et al., 2001). Patrilineal kinship, such as is found in South and South East Asia, may provide a blueprint for conceptualizing biological inheritance as occurring through males that is "largely incompatible" with Mendelian inheritance, which recognizes an equal genetic contribution from both parents (Meiser et al., 2001; Richards, 1996). The belief that a causative gene is inherited from the father alone could mean that risk for recessive conditions in the wider family is underestimated, and carrier testing therefore considered irrelevant, but there is a dearth of data on the potential significance of this for genetic counseling (Meiser et al., 2001). Consanguineous marriage, common in the Middle East, South Asia, North Africa, and parts of southern Europe, also presents challenges to the assimilation of genetic information (Raz et al., 2003). The practice is associated with an elevated risk of a wide range of mostly rare recessive conditions (Bittles, 2001), but has important social and cultural correlates (Shaw, 2001; Shaw and Charsley, 2006). In this context, a genetic diagnosis, as something "in the family," is potentially stigmatizing not just of the individual but of the entire family. Individuals or couples may be reluctant to accept or discuss a diagnosis that may adversely affect their own or other family members' marriages or marriage prospects, and may negatively affect women's marriage prospects in particular (Raz et al., 2003). Patterns of authority and gendered hierarchy may also significantly affect communication about genetic risk within the family (Raz et al., 2003; Shaw, 2003).

Educational and cultural factors influencing access to and assimilation of genetic information may converge for members of minority ethnic groups in Western Europe, America, and Australia. Inequalities in accessing service may reflect miscommunication arising from patients' unfamiliarity with genetic and medical jargon, combined with problems with translation, where they are dependent on family members or interpreters for access to clinical information (Browner et al., 2003). Consanguineous marriage, and the issues it raises, is now increasingly relevant to genetic counseling referrals in Western Europe, North America, and Australia, given widespread migration from South and Southeast Asia and Africa (Port et al., 2005). There is, however, a tendency for medical practitioners to stereotype religious and cultural differences in the acceptability of prenatal testing in ways that adversely affect minority access to prenatal services; for instance, in the belief that Latinos in America or Muslims in Britain will not accept prenatal diagnosis for religious reasons (Modell et al., 2000; Hunt and de Voogd, 2005). Yet, individual beliefs and knowledge may play a larger role than "ethnicity" in the uptake of amniocentesis (Saucier et al., 2005), and religion cannot be taken as proxy for Pakistani women's attitudes to termination of pregnancy (Ahmed et al., 2006).

In this chapter, we focus on British Pakistani patients' understandings of genetics, illness causality, and inheritance, and the implications of these understandings for the assimilation of genetic risk information. Epidemiological research indicates that British Pakistanis have an elevated risk of recessive conditions associated with the practice of consanguineous marriage. In the general U.K. population, the risk of having a child with a serious genetic condition is about 3%, but this increases to about 6% with first-cousin marriage and to higher than this where there is a family history of complex consanguinity (Bundey and Alam, 1993). Theoretically, a proportion of this elevated risk could be managed by genetic testing.

THE EVIDENCE

This chapter is based on a fieldwork-based study of the impact of genetic risk information on families of Pakistani origin referred to a U.K. genetics clinic. The study had ethical approval from the National Health Service local research ethics committee. We used qualitative, flexible, patient-centered, and open ended research methods that enabled a focus on clients' perceptions, experiences, and views. The genetics clinic was based in a hospital in a district which, by comparison of data from U.K. 2001 Census enumeration districts, served the second largest Pakistani population in southern England outside London. Muslims number 10,527, or 6.5% of the district's population, and Pakistani Muslims number 9703 (6%). The Pakistani population of the town selected shared the broad demographic and socioeconomic features of the British Pakistani population generally (see Shaw, 2001; Harriss and Shaw, 2006).

Potential research participants were identified by the second author (JH), from all the genetics clinic referrals, their ethnicity inferred from their names, and invited to join the study in a letter of invitation sent out with clinical appointments or given to patients at clinical consultations. Potential participants were informed that the researcher (the first author) is a social anthropologist who speaks Urdu and English, and were reassured about confidentiality. Interested participants received further information about the project and a consent form in Urdu and English. All but two of those invited to participate consented to be part of the study. Over the study period (2001–2005), 61 families were recruited to the project via the genetics clinic. Most referrals concerned a problem in a couple's pregnancy, an infant, or child. Five additional families with an adult or older child with medical or intellectual problems—and who, it emerged, had previously received genetics advice—were recruited by the first author (AS) through local community contacts.

The majority of research participants reported family origins in villages, towns, or cities in Mirpur district in Azad Kashmir (16/36); northern Punjab (12/36); central Punjab (6/36); Sindh (1/36); and the North West Frontier (1/36). Most (61/66) couples were consanguineously related—46 as first cousins, 15 more distantly—and five were unrelated. In a majority of marriages, one partner was U.K.-raised or "second generation" and married to a spouse raised in Pakistan. The "second generation" adults had, in most cases, received some or all of their education in the U.K. to General Certificate of Secondary Education (GCSE) level or equivalent; three were graduates of U.K. colleges or universities, and one was a graduate from an American college. Usually the Pakistan-raised spouses chose to talk with the first author in Urdu, and in two couples both partners were educated in Pakistan and spoke little English.

The conditions observed in the study population reflect the background risk of genetic conditions shared with the general population, plus the elevated risk of recessive conditions associated with consanguinity. Cases of thalassemia were underrepresented, as these are seen by the thalassemia service. Overall, 69 different conditions were observed in 66 families (three families were affected by more than one condition). The majority of the conditions were recessives (44/69) of widely varying types (dysmorphic syndromes, lethal dysmorphic syndromes, and metabolic, neurological, hematological, skeletal, dermatological, and renal conditions). Nine conditions were autosomal dominant, four were chromosomal, nine were of unknown cause, and three were non-genetic.

The fieldwork employed several techniques of qualitative research: interviews, observations of clinical consultations, and participant observation. Pre-clinic interviews with participants were supplemented by observations of their clinical consultations, which identified potential disjunctures between clients' and clinicians' understandings. Emergent themes and issues, which could not have been anticipated at the outset, were then explored further in subsequent interviews. Participant observation enabled the first author to build longer-term relationships with some of the families being seen in the genetics clinics, and thus to develop a more complete picture of their understanding of the diagnosis and its implications than would have been possible from an interview alone. The first author sometimes accompanied people to other appointments for their child, and helped them access and understand clinical information, including through linguistic translation. Social science recognizes that the presence of a researcher inevitably influences the processes they are observing; in this case, the first author's participant observation was sometimes a necessary prerequisite to the research goal of understanding participants' responses to genetic risk information and its implications. It also yielded valuable insights into the difficulties people can experience in accessing and understanding genetic information (Shaw and Ahmed, 2004). The first author's role in facilitating access to clinical information was recognized by the hospital through an honorary clinical contract; if research participants sought clarification of clinical information, the first author would first check such information with the second author.

Interviews usually lasted about 2 hours and were conducted in participants' homes before or, more often, after the first clinical appointment. Initial interviews elicited adults' understandings of their own or their child or relative's problem, its nature and causality, and relevant family history, and were supplemented by participant observation at the clinic and followed by a second or third interview to discuss emerging themes. Most interviews were recorded in note form and written up in full as soon as possible afterward. A smaller proportion were fully tape-recorded and then transcribed. Data from participant observation were recorded in a fieldwork diary. Data were collected on a wide range of issues relating to the understanding of genetic risk in the context of family experience of particular conditions, and analysis was concurrent with data collection. Regular reading and re-reading of initial interviews and field observations identified themes for exploration in subsequent observation and interviews. Data were analyzed by identifying themes related to the topics of genetic risk, illness causality, and inheritance. As themes emerged, data were re-analyzed to develop and refine the findings. Several overarching themes were identified through this process, providing a framework through which we present our findings.

PERCEPTIONS OF GENETIC RISK IN COUSIN MARRIAGES

CONCERNS ABOUT HEALTH PROFESSIONAL PREJUDICE

Almost all respondents seemed to feel that mainstream medical and public opinion disapproved of cousin marriages as "causing" medical or intellectual disabilities among Pakistanis. The uncle of a child with a problem linked with a de novo mutation with a dominant effect reported that, prior to their genetics referral, a non-genetics health professional had informed the family that their child's problem was caused by their cousin marriage. Some respondents had postponed seeking a genetics referral because they believed cousin marriage had caused the problem and feared being blamed for it. Mrs. Z, married to a first cousin, delayed attending the genetics clinic until after her first child was born, for these reasons, despite relevant family history:

"We thought it was because we are married in the family that my niece is like that. When I came to the clinic where I met you, I felt 'these people are going to think badly of me because I am married to my cousin.' Even when my niece said, 'go for it. Go for the carrier test,' I felt they would think badly of me. They shouldn't, though, because a lot of the time the girls have no choice about who they marry, and about 90% of people who marry cousins do not have children with handicaps. But it is true, even the young girls in our community, they go round thinking we have abnormalities because we marry in the family. It makes you think twice about going to the genetics clinic, because you think you are going to be blamed."

When asked what he understood as "genetics," Mr. B, father of a boy with learning difficulties, said "genetics is what happens when you marry a cousin." Asked why he thought this, he said he was told by a South Asian support worker conducting a home visit that cousin marriage had caused his son's problems. He did not want to contradict her because his wife, who speaks no English, needed her help, but as the theory made no sense to him, he postponed requesting a genetics referral until his son was about to leave school. Mrs. C, when asked what she understood as "genetics," replied, "Genetics? It's about the genes. But I am only married to a cousin one way. My father and his mother are brother and sister, but my mother is from outside so it's not very much genetics in my family. If my mother and his father were sister and brother it would be more genetics."

Time after time, respondents expressed skepticism that cousin marriage causes birth defects. The following comment from a young father was typical: "I don't agree with this cousin marriages and birth defects argument. There are many people in our community who marry cousins, and in Pakistan, but you don't see them all having children with medical problems; and English people don't marry cousins, and they sometimes have children with medical problems." While most parents were consanguineously married, they were often skeptical of its relevance for the particular problem in their case. Following a genetics consultation at which Mrs. K learned she is a carrier of two recessive conditions, she said, "We were told that it was a genetic problem, but there was nothing in my family before…". The first author responded by asking, 'Do you know why they said it is genetic? Mrs. K replied, "Our Dads are brothers. Our Mums are sisters. You know how it's like in Pakistan…The closer you are, the more likely you marry in the family. If they did not know we were cousins, or if we weren't cousins, they would not say it is genetic."

Respondents were generally happy to give the genealogical details of their own and other consanguineous marriages in the family when reporting their family history during a genetics consultation. During clinical consultations, questions about parental consanguinity were asked in a nonjudgmental way. On one occasion, after the second author carefully explained the reasons for these questions, one father responded reassuringly, "Don't worry, we keep an open mind." Nonetheless, respondents sometimes commented afterward that they thought they would not have been asked about consanguinity if they were English. During a home-based interview, Mr. G, who is a carrier of a balanced translocation and whose brother has a chromosomal condition, volunteered emphatically that the condition in his family "is nothing to do with cousin marriages but most nurses and social workers think it is."

UNDERSTANDINGS OF THE CAUSES OF ILLNESS

ENVIRONMENTAL CAUSES

Parents drew from a range of everyday or common-sense explanations for a problem in their pregnancy, an infant, or a child. These included: accidents or infections, usually during the child's infancy; medical errors and medical negligence prenatally or at birth; and the behavior of the mother or, less often, the father, usually during pregnancy.

One case of deafness was attributed to a childhood accident in Pakistan. A skeletal problem in a child was attributed to a fever in infancy, during a visit to Pakistan, for which the infant received injections. The father of children with a recessive condition thought a snake bite he had received in Pakistan might have introduced the mutation he carried into his bloodstream, and asked if a blood transfusion might remove the gene from his body. Two cases of developmental delay were attributed to infant immunizations performed in the U.K. A grandmother attributed her grandson's dysmorphic syndrome to a botched circumcision operation performed locally, at which the infant had lost a lot of blood.

A few women suspected or blamed negligent prenatal services for not detecting their babies' problems early enough. Mrs. E, whose first baby died a few hours after its birth, complained that "they were not looking properly on the [prenatal] scan and they did not see anything." Mrs. C, whose child has a recessive condition with limb defects, suspected that her doctors told her they had not noticed the limb abnormalities prenatally so as to divert her concern from the fact that she had taken medication in pregnancy that, she felt, might have caused the problems: "How did they not know? They can tell a baby's sex, so why didn't they see this?…They said it's genetic, don't even think it is the medicine you took. Why did they say that? That's the question." Mrs. L, who has two children with a recessive condition, speculated that there were more birth problems among babies born to Pakistani mothers because these women were less likely than White women to be offered prenatal tests and thus the option of ending affected pregnancies. Mrs. K thought that early prenatal detection of her child's problems might have enabled a therapeutic treatment: "Do you remember, the doctor in genetics, she said they did not have his notes. Why? I want to know why…they should have had them. Is it true that it was apparent on the scan and they should have noticed it? This may be clutching at straws, but I think they could have found out and done something." Mrs. E thought an early Caesarian might have saved her second baby who was stillborn: "If they had got her out, before she died, could she have been okay?"

Mothers also worried that other things they had done, or that had happened to them in pregnancy, may have caused their child's problem. These included traveling in Pakistan, drinking insufficient water when traveling in Pakistan (in the case of renal problems), and physical abuse from a husband (in an infant with café-au-lait marks). Several women reported that relatives or neighbors blamed problems in a pregnancy or child on the mother's actions during pregnancy. These included walking out one evening to visit a brother without a mother-in-law's approval, and probably on the way annoying a malicious spirit (*jinn*), who might then have harmed the fetus (see Shaw, 2000: 209–212).

Eating a particular type of fish was mentioned in two cases of café-au-lait marks (diagnosed as neurofibromatosis type 1). Falling asleep with arms and legs curled up during an eclipse of the moon or sun (in Panjabi: *chānd garīn or sooraj garīn*) was mentioned as a possible cause of serious limb abnormalities (Roberts syndrome) and also, in the case of a cleft lip, where the mother was thought to have had a knife in her hand while she slept—the eclipse imprinting the knife image mark on the child's lip.

Mrs. W thought her son's café au-lait marks (in his case associated with neurofibromatosis) might have been caused by an eclipse which occurred when she was pregnant and when her husband was outside fixing telephone wires.

Almost all families reported no known relevant family history in their large extended kindreds spanning the U.K. and Pakistan; in almost all cases, they considered the condition had appeared in their family in the U.K. for the first time. Mr. B, grandfather of a boy with a metabolic syndrome, said, "This is not in our family in the village [in Pakistan]." Some parents and other relatives thought the U.K. climate and diet might have caused the child's problem. Mr. B commented, "there is more risk here than in Pakistan." In four cases, two involving learning difficulties and two also involving obesity, the parents thought the Pakistan climate might cure the child's problem, specifically through sweating, which might make the causative gene exit the body. Three parents asked whether the gene they carried might eventually leave their body, as a result of exercise, dietary change, ending a drug addiction, or blood treatments (including transfusion). One mother enquired, after being given information about recessive inheritance, whether she might have acquired the mutation associated with her second child's problems sometime after the birth of her first, unaffected child. Significantly, several respondents, speaking in Urdu, used the words "gene" and *jerasīm* interchangeably; the Urdu/Punjabi word *jerasīm* usually refers to a "germ" and implies an infectious agent; it may also be used to refer to semen.

RELIGIOUS EXPLANATIONS

Regardless of whether or not they acknowledged a problem as having a cause in a changed gene, chromosome, or some other factor, almost all parents saw the condition as an expression of God's will. Mrs. Y, who understood how the mechanism of recessive inheritance had caused her child's condition, said she did not understand "why" the condition had occurred, adding "only God knows why this has happened." For many parents, since the condition or problem represented the will of God, it was something they should endure. Several parents perceived their child's problem as a test from God of the parents' ability to cope with adversity. Accepting a genetic explanation was compatible with this view. As Mr. D, the father of a teenager with undiagnosed learning difficulties, put it: "It might give my mind a bit of rest to know. It might be easier to live with, rather than wondering about the injection. But a genetic explanation would not change anything. I have not seen any artificial limb that will be better than a limb God has made. Maybe it is the will of Allah that his brain is like that. No one can change it except God. You can try to find out why, but God knows best. Maybe God has done this to make me more responsible."

Mrs. U understood and accepted a genetic explanation for her teenage son's metabolic condition, and also considered her son as entrusted to her from God: "Science says it is from genetics. We went to the genetic clinic. They made a family tree, and they showed how it could have come down in the family because my husband's grandfather (father's father) and my mother's grandfather (father's father) were first cousins. I don't doubt that. I think this is true. But God has a purpose for everything… My God says that people like [my son] are higher than angels. Each child that is handicapped is a special gift from God for a special person, a person that God has chosen. God chooses special people for these special gifts. God is very kind. You understand *amānat* (safekeeping). A child or person like [my son] is *Khudā kī amānat*; God has given him to you for a special time." [Interview in Urdu]. Such beliefs were more clearly articulated among parents of older children with long-term problems, who had presumably had longer to reflect on the meaning of their predicament, and seemed to help counter feelings of responsibility or blame.

Acceptance of the problem as expressing "God's will" or a test from God was compatible with seeking treatments or cures (*ilāj*) by medical, spiritual, or other traditional means. As Mr. Y put it, "Allah says there is a cure for every disease, whether the doctor finds it or whether the spiritualist guide finds it. So you must try all routes." Those offered mutation research were happy to consent to enrolling their and their child's blood samples, hoping that "finding the gene" would lead to treatments or cures for their affected child. Additionally, most respondents considered prayer an important means to a possible cure, if God willed it. Some parents explained that certain verses of the Qur'an are believed to have healing powers, and the location of relevant verses is known to religious specialists in the U.K. and in Pakistan; parents may then take their child to see a religious specialist, or request them to send the relevant verses to be worn as an amulet (*tāvīs*). One father planned to take his deaf children to a particular *pīr* (saint) in Pakistan, if he could afford the trip. In three different cases, boys with syndromes involving learning difficulties and obesity had been taken to Pakistan to visit living *pīrs* or to visit tombs of deceased *pīrs* believed to exercise curative power. One woman had a tongue made out of silver that she donated to one of these shrines in gratitude, after her son, who has speech problems, began to talk, in fulfillment of the vow she had made when she took her son to the shrine.

Several participants also mentioned following dietary advice from *hakīms* (practitioners of traditional Islamic medicine) as sources of treatment. Just one mother, a college graduate educated in the United States whose son's metabolic condition is managed through medical dietary therapy, thought that while prayer is always important, the use of alternative remedies was not relevant for managing her son's problems—but she acknowledged that had she been living in Pakistan, she might well have consulted a *hakīm* or used herbal remedies.

LOCATING GENETIC SUBSTANCE

Most respondents considered blood to be where genetic problems are located; thus, as we have seen, alternative remedies often focused on removal of the infective agent or gene via treatment or alteration of the blood. However, in one case, a gene was thought to be located on the spine. Not long after hearing that a research group had located the causative gene on chromosome 9, the brother of a man with a recessive metabolic condition said: "I took him to a *hakīm*, and he found this gene in his spine. He uses a drill to numb the spine, and put the medication in. After an hour the medication brings something out of the spine. But the *hakīm* said there is a risk it goes wrong, and I am not going to do it." The first author asked, "How do you know the gene is on the spine?" This respondent replied: "Dr. N said it's on the ninth or tenth line. There are three genes and the middle one is missing from the twelfth line."

UNDERSTANDINGS OF BIOLOGICAL REPRODUCTION AND INHERITANCE

Although sexual intercourse was recognized as the mechanism by which a child is produced, people usually spoke of children as being given by God or the result of God's will rather than of human action. At a home visit prior to their first clinic appointment, one mother asked the first author, "What is this genetics, anyway?" The first author's answer involved an account of the fusion of gametes at conception that elicited embarrassed laughter, and a comment that "in our religion, it is not like this."

Parents' understandings of biological reproduction usually specified blood as the substance of intergenerational transmission. As one mother, speaking in Urdu, put it, "a baby is made from the blood of the father and this mixes with blood from the mother." Some respondents with previous genetics clinic experience described biological reproduction as involving the mixing of chromosomes or a mixing of genes. Mrs. G, whose husband's brother has a chromosome condition, said, "the chromosomes from the mother and father mix together and the baby starts." Those with experience of single gene disorders usually described conception as involving the "mixing" of genes. Fetal abnormalities or pregnancy loss were often described as the result of blood "not mixing," or resulting from "when the genes did not match." Younger adults, educated to GCSE level or above or its equivalent in Pakistan, sometimes mentioned the terms sperm and egg; older or Pakistan-raised participants rarely used either technical English or Urdu, or colloquial Panjabi terms, for the male and female gametes.

Sperm, occasionally described by the word *jerasīm*, was generally understood as concentrated blood and was thus seen to represent a "strong" genetic contribution to a child. When asked during a clinical consultation, "Are you and your wife related?" one couple replied, "Yes, but only through our mothers—our mothers are sisters—but not though our fathers," implying a weaker genetic link than if their fathers were brothers. After accompanying the parents of a child with a recessive condition to the clinic, the first author was asked by the mother to repeat the consultant's advice about the inheritance of the condition. The mother then interrupted the exposition by asking (in Urdu), "Are you saying the material from the mother and father is equal? Does the father not give more? (*bāp kī taraf se ziāda nāhīn milta*)?" A woman fluent in English and Urdu with whom the first author was discussing the translation of a leaflet on recessive inheritance asked her, "So the baby gets genetic material from both sides? So it has to come from the father and the mother? I thought it just came from the father. I had never known that." In the context of a discussion of marriage arrangements, a woman who had lost three children in infancy to a recessive disorder, and understood her one in four recurrence risk, told the first author she thought it would be safer to marry her daughter to a maternal rather than to a paternal relative, saying (in Urdu), "if the fathers of the boy are brothers, there is the most danger."

Generally, people were hazy about what the mother's generative contribution might be. Mrs. W said, "to speak about such things is shameful, especially before marriage... Even after marriage you don't find out what the words are. It is the same for a girl and for a boy, but here in England, for our children it is different; now they learn from school and they know all the words." [Interview in Urdu].

Nevertheless, all respondents considered the biological link between a mother and child in pregnancy to be very important. One British-raised mother thought this created a stronger genetic link between the mother and child than between father and child. One young Pakistan-raised mother, speaking in Urdu but using the English term "gametes," asserted: "The gametes from which a child is made come equally from the mother and father [but] the mother's contribution is greater... because for 9 months the child grows in the mother's body... the mother does not have periods and [that] blood goes into the baby. The mother's health and the food she eats in pregnancy are also very important for how the baby develops."

EVERYDAY THEORIES ABOUT INHERITED CHARACTERISTICS AND INHERITED RISK

Observations about shared family characteristics recognized a range of possible inheritance patterns. Sometimes people remarked on same-sex intergenerational resemblances—a daughter being like her mother, a son like his father; more often parents described characteristics as being transferred from either parent to a child, regardless of sex. One woman, speaking in Urdu, described her son as "dark like my husband, [but] he talks and eats like me" and her daughter as "fair like me, but [she[has her Dad's eyes... [and] her father's sister's hair." Characteristics such as having blue-gray eyes, pale skin, or only giving birth to boys were described as "skipping" a generation, showing up in grandchildren, on the basis of observed family resemblances. People also sometimes talked about characteristics as being on their "mother's side" or their "father's side" of the family, despite complex consanguinity.

While peoples' everyday understandings of illness causality, inheritance, and risk were elicited in some of the interviews, they rarely emerged within the genetics consultation itself. During clinical consultations, patients rarely indicated that clinical information did not match their own understandings (see also Featherstone et al., 2006). Occasionally, though, it emerged that, despite genetic counseling, patients' understandings of the inheritance of recessive conditions tended to stress "the mother's side" or "the father's side," in keeping with observed history of the condition in the family. "I thought it was only in my mother's family," said Miss S, whose parents are first cousins; her brother and her mother's brother's daughter have a recessive dermatological condition. She understood the condition "was only in my mother's family because that's where the problems are." At a second genetics consultation, after discussion of how both parents must be carriers of a recessive condition for a child to be affected, Miss S asked, with some astonishment, "Does this mean that my Dad's brothers and sisters could be carriers?" Previously, she had only considered her mother's siblings and their children to be at risk.

In another case, that of Mrs. N, it emerged that the wife and her siblings had attributed the genetic condition to he husband's 'side' of the family. Mrs. N said: "We thought my husband's family were carriers... through the men." In this case, Mr. and Mrs. N, who are first cousins, were referred for genetic advice because Mrs. N's eldest sister is married to Mr. N's eldest brother and has children with a recessive metabolic condition. The couple received a theoretical risk of having an affected child (1/16) and agreed to carrier testing, which would rule out the risk of having an affected child should only one partner be a carrier. The result showed Mr. N is a carrier and his wife is not. Interviewed 18 months later, Mrs. N reported that one of her younger sisters had recently given birth to a child with the same condition. Mrs. N and the baby's mother were devastated: "We thought my husband's family were carriers... through the men. My older sister is married to my husband's older brother. Two of her children have [the condition]. And my husband's two younger brothers each have children with a congenital problem, so it was in my husband's family because this is where the problems were. This is why I thought any children I have with my husband might also be at risk... It was not even in our dreams that she [the younger sister] or her husband could be carriers. He [her husband] is [my] mother's sister's son. My sister [the one with the new affected baby] always said it was in our husbands' family [i.e., Mrs. N's husband and his brothers] not on our side of the family. We thought it was only our uncle [Mrs. N's father-in-law, who is also her mother's brother] who had passed it on to his sons and one of them had passed it on to my niece and nephew. We thought it went through the men, in the male line. Now I know that this is wrong. We did not realize there has to be a wife who is a carrier for the kids to have got it... It was such a shock. The stress has been incredible—realizing it is in our side of the family too, not just in my husband's side. My sister, the one with the affected baby, is gutted. She has not even told my Mum that there is something wrong."

OVERVIEW

This chapter highlights a complex array of understandings of the causes of medical and intellectual problems. Some of these understandings concern environmental factors affecting a child postnatally, such as infections, accidents, and childhood immunizations. Others invoked prenatal causes in accidental or morally reprehensible actions performed by the mother or, less often, by the father, during pregnancy, such as a mother's insufficient intake of water as causes of a kidney problem, or physical abuse from a husband as causing marks of distinctive skin pigment. A tendency to blame the mother, or for mothers to feel themselves to blame for a problem in a pregnancy or child, is not restricted to British Pakistanis. However, some understandings (such as a belief in the capricious power of *jinn* or the potential for an eclipse to cause pregnancy loss or birth defects) were expressed in an idiom that reflects the cultural background of migrants with origins in rural Panjab or Mirpur, with parallels noted among other British Muslims of South Asian origin (see, e.g., Rozario, 2007). These were also sometimes challenged by the second generation. On the other hand, in cases of postnatally detected problems and unexpected infant deaths, some mothers blamed medical services for not detecting abnormalities early, and suspected medical malpractice or negligence. Their complaints can be viewed as expressions of women's covert resistance to being blamed or blaming themselves for their child's problem (see, e.g., Wilce, 1998), and to accepting genetic explanations, but also indicate a need for counseling to address patients' understandings of the problem in their pregnancy, infant, or child.

Many of the families in this study had little knowledge of genetics prior to their genetics referral, other than their awareness of the discourse of genetic risk associated with cousin marriage. Patients sometimes incorporated genetic information obtained during clinical consultations into their prior everyday understandings of illness causality in ways that did not reflect scientific genetics. Genes or chromosomes were generally understood to be located "in the blood," or as a characteristic of blood, or as infective agents rather than as located in every cell of the body and arising from the gametes at conception. Such misconceptions about the location of genes or chromosomes, despite "conversational familiarity" with these terms, are not restricted to British Pakistanis (Lanie et al., 2004). In an American telephone-interview study, only 34% of American adults with high school biology thought that genes were located in every cell; 24% thought they were located in the brain and/or the mind, and the rest did not know or gave some other explanation (Lanie et al., 2004).

Among our study participants, the common equation of blood, genes, and chromosomes sometimes led to confusion over genetic clinic requests for genetic blood tests after previous blood test results showing "the chromosomes are okay." The location of "the gene" as "in the blood" rather than located in every cell of the body was also associated with beliefs in the potential power of therapies aimed at removing the gene from the body, through cleansing the body or through blood transfusion. In one case, a gene was thought to be located in the spine, because chromosome 9 was understood as the ninth vertebra, with potentially serious therapeutic implications.

Regardless of their place and length of education, or their understandings of *how* a condition may have arisen, almost all respondents believed that ultimately only God knows *why* a particular condition occurred, for God decides whether someone is healthy or unwell, lives or dies. This belief was nonetheless compatible with the pursuit of scientific knowledge. Acceptance of a pragmatic or scientific explanation while simultaneously attributing a misfortune to forces beyond human understanding is not unique to British Pakistanis or to Muslims (Panter-Brick, 1991; Foster, 1976; Evans-Pritchard, 1976). Genetic explanations may pose particular cognitive challenges in comparison with other causal explanations by depending on notions of probability or chance, which leave a void that appeals to God's will serve to fill. Katie Featherstone and colleagues report that people of White British heritage attending a genetics clinic in South Wales often attributed being affected by genetic conditions to fate or destiny, as if in a kind of biological predestination (Featherstone et al., 2006).

For British Pakistanis, acceptance of a genetic explanation is further complicated by its association with a dominant discourse of disapproval of consanguineous marriage in contemporary Britain. Indeed, this issue seemed to overshadow many people's engagements with genetics. Rather than being perceived as a statement about elevated risk of a possible future event, the consanguinity thesis is often understood as a causal explanation for disability that fails to match the observed distribution of medical and learning problems in the Pakistani population in Britain and in Pakistan. It is thus experienced as stigmatizing British Pakistanis on account of this marriage preference. During this research, comparisons of family trees elicited in the clinic, with kin connections described in conversations with respondents at home, sometimes indicated an underreporting of complex consanguinity in the clinic. This might reflect a desire to minimize clinical disapproval, the difficulties clinicians sometimes have with reporting complex consanguinity in standard medical notes, or lay or personal theories of inheritance in which some consanguineous links (such as additional links through women) may be viewed as irrelevant. As a result, some risks may be underestimated by clinicians.

When explaining a condition's inheritance and associated risks, clinicians do not usually emphasize that genetic information is contained in every cell in the body and inherited in equal amounts from both parents, but they do usually consider or mention patients' everyday ideas about inheritance, such as their awareness that particular characteristics such as hair or eye color are passed on across the generations. We found patients' everyday knowledge of the inheritance of personal characteristics revealed a range of understandings of inheritance patterns: a characteristic might be inherited from either parent, regardless of sex; it might be inherited only by same-sex relatives across the generations; it might "skip" a generation, or it might be associated with the "mother's side" or the "father's side" of the family, references to "sides" of the family being particularly striking, given the cross-cutting ties of consanguinity in many families. This range of understandings of inheritance patterns implicitly recognizes a male and a female genetic contribution, and is remarkably similar to that noted among White Anglo-British families, whose everyday theories of inheritance recognize a "bilaterality" frequently modified by gender, generation, or birth order—for instance, in the idea that a condition only affects boys, or girls, or a first-born, or "skips" a generation, or that a trait runs through the "mother's side" or the "father's side" of the family (Davison, 1996; Richards and Ponder, 1996; Richards, 1996). As among White British families in South Wales, such models may offer templates for thinking about risk that single out particular kin as susceptible in ways that conflict with biogenetic theory (Davison, 1996; Featherstone et al., 2006). This research has identified an additional cultural blueprint that appears to prioritize inheritance through men. Mrs. N, for example, understood the recessive condition in her family to be inherited through men, in the male line... "We did not realize there has to be a wife who is a carrier to for the kids to have got it." Such thinking may be linked with a cultural emphasis on the male genetic contribution to a child, via blood or sperm. However, these ideas may only be drawn upon when they conform to the perceived family history of a condition (see also Hunt et al., 2002), and comprise only one set among a range of understandings of biological inheritance and illness causality.

Regardless of ethnic background, recessive inheritance is particularly difficult to understand: how can a condition be inherited when it has never before been seen in the family? (Chapple et al., 1995). The lack of a known family history for many rare recessive conditions is not unique to British Pakistanis (Pembrey, 1996). Nonetheless, high infant mortality, no routine investigation of the cause of infant deaths in Pakistan, and poor background knowledge of genetics are likely to contribute to misunderstandings associated with inheritance and genetic risk among British Pakistanis. British Pakistani parents and other relatives of children with thalassemia had insufficient knowledge to appreciate their risk of carrying a gene for this condition, which is relatively common among British South Asians (Ahmed et al., 2002). Our study has identified a range of areas in which British Pakistani clients' knowledge about genetics, risk, and inheritance in relation to a wide range of mostly recessive conditions can be improved in order to facilitate informed decision making.

IMPLICATIONS FOR GENETIC COUNSELING PRACTICE AND TRAINING

Patients negotiate their use of genetic information in managing reproductive and familiar risk in relation to a range of social, emotional, and cultural factors (see, e.g., Shaw and Hurst, 2008). Information provided during genetic counseling needs to be assimilated within patients' prior understandings of illness causality and inheritance if it is to be effective (Richards and Ponder, 1996; Richards 1998; Chapple et al., 1995). This chapter shows that patients' prior understandings of the causes of a clinically identified genetic condition may include a range of environmental or spiritual factors, and reflect poor prior knowledge of genetics. The chapter also shows that genetic information can be "assimilated" within patients' prior understandings, but in ways that may conflict with genetic theory, with potentially serious clinical consequences. Both observations are not unique to British Pakistani patients. Some specific components of these understandings, such as the idea of a gendered parental contribution to genetic inheritance or the notion that a genetic problem ultimately reflects God's will, may occur more frequently among particular patient groups, but these components of patients' understandings of genetics are not limited to any one population and cannot be assumed to be relevant in any given individual case.

In theory, genetic counseling includes first establishing patients' prior knowledge of illness causality, inheritance, and risk. In practice, patients' prior knowledge is rarely assessed in any systematic way. The findings reported here indicate that, when counseling individual patients, regardless of their ethnic background, it may be valuable for genetic counselors to elicit patients' prior understandings of the causality and inheritance of the condition they have come to discuss. This might be done by providing patients with opportunities to describe their own perception of the condition and their understanding of its inheritance or etiology. Genetic counselors would also need to be prepared to challenge mistaken beliefs, because mistaken beliefs may influence an individual or couple's reproductive decision making and risk communication in the family in clinically significant ways. Further research is necessary to investigate the most appropriate methods for eliciting patients' understandings, and working with these to enable patients to make more fully informed and effective use of genetic services. The need for clinical follow-up, as well as the timing of such an intervention, would also need to be considered, because patients may not fully appreciate the significance of genetic information, especially for other family members, when they first receive it.

In the training of genetic counselors, it would be important to incorporate awareness of the range of possible sources of misconception about genetics and the causes of a genetic problem that all patients may bring to a genetics consultation. Genetic counselors need to be aware of this range of ideas when eliciting patients' understandings, and when presenting genetic diagnostic and risk information, without assuming that any particular idea or set of ideas will be relevant in any individual case. In fostering what has variously been described as cultural sensitivity, cultural awareness, or cultural competence in delivering medical services to socially, ethnically, and culturally diverse populations, it is important to instill generic skills and to avoid stereotyping patients on the basis of assumed beliefs and practices (see, e.g., Kai, 2003; Papadopoulos, 2006); the same general point is relevant within genetic counseling practice.

ACKNOWLEDGMENTS

This research was supported by a grant awarded to Alison Shaw from the Wellcome Trust UK (GR063078MA) and was prospectively reviewed and approved by the relevant NHS local research ethics committee. We are grateful to colleagues at the Regional Hospital for issuing Alison Shaw with an honorary clinical contract for her research role as a medical anthropologist. We thank our colleagues in the Regional Genetics Service for their support of this research. We thank the editor and two anonymous reviewers for the Journal of Genetic Counseling for their comments and suggestions for revising the article on which this chapter is based (Shaw & Hurst, 2008). Above all, we thank all of the families who participated in this study.

REFERENCES

Ahmed S, Bekker H, Hewison J H, Kinsey S. (2002). Thalassaemia carrier testing in Pakistani adults: Behaviour, knowledge and attitudes. *Community Genetics 5*, 120–127.

Ahmed S, Green JM, Hewison J. (2006). Attitudes towards prenatal diagnosis and termination of pregnancy for thalassaemia in pregnant Pakistani women in the North of English. *Prenatal Diagnosis 26*, 248–257.

Bittles AH. (2001). Consanguinity and its relevance to clinical genetics. *Clinical Genetics 60*, 89–98.

Browner CH, Preloran HM, Casado MC, Bass HN, Walker AP. (2003). Genetic counseling gone awry: Miscommunication between prenatal genetic service providers and Mexican-origin clients. *Social Science and Medicine 56*, 1933–1946.

Bundey S, Alam H. (1993). A five-year prospective study of the health of children in different ethnic groups, with particular reference to the effect of inbreeding. *European Journal of Human Genetics 1*, 206–219.

Chapple A, May C, Campion P. (1995). Lay understanding of genetic disease: A British study of families attending a genetic counseling service. *J Genetic Counsel 2*, 281–300.

Davison C. (1996). Predictive genetics: The cultural implications of supplying probable futures. In T. Marteau, & M. Richards (Eds.), *The troubled helix: Social and psychological implications of the new human genetics* (pp. 317–330). Cambridge: Cambridge University Press.

Evans-Pritchard EE. (1976). *Witchcraft oracles and magic among the Azande*. Oxford: Clarendon.

Featherstone K, Atkinson P, Bharadwaj A, Clarke A. (2006). *Risky relations: Family kinship and the new genetics*. Oxford: Berg.

Foster, GM. (1976). Disease etiologies in non-Western medical systems. *American Anthropologist 78*, 773–782.

Harriss K, Shaw A. (2006). Family care and transnational kinship: British Pakistani experiences. In F. Ebtehaj, B. Lindley, & M. Richards (Eds.), *Kinship matters* (pp. 259–274). Cambridge: Hart Publishers.

Hunt K, Davison C, Emslie C, Ford G. (2002). Are perceptions of a family history of heart disease related to health-related attitudes and behaviour? *Health Education Research 15*, 131–143.

Hunt LM, de Voogd KB. (2005). Clinical myths of the cultural "other": Implications for Latino patient care. *Academic Medicine 80*, 918–924.

Kai, J. (Ed.) (2003). Ethnicity, health and primary care. Oxford: Oxford University Press.

Kelly K, Leventhal H, Marvin M, Toppmeyer D, Baran J Schwalb M. (2004). Cancer genetics knowledge and beliefs and receipt of results in Ashkenazi Jewish individuals receiving counseling for BRACA1/2 mutations. *Cancer Control 11*(4), 1–9.

Lanie AD, Jayaratne TE, Sheldon JP, et al. (2004). Exploring the public understanding of basic genetic concepts. *Journal of Genetic Counseling 13*, 305–320.

Lewis J, Leach J, Wood-Robinson C. (2000). All in the genes?—young people's understanding of the nature of genes. *Journal of Biological Education 34*(2), 74–79.

Meiser B, Eisenbruch M, Barlow-Stewart K, Tucker K, Steel Z, Goldstein D. (2001). Cultural aspects of cancer genetics: Setting a research agenda. *Journal of Medical Genetics 38*, 425–429.

McAllister M. (2003). Personal theories of inheritance, coping strategies, risk perception and engagement in hereditary nonpolyposis colon cancer families offered genetic testing. *Clinical Genetics 64*, 179–189.

Modell B, Harris R, Lane B, et al. (2000). during pregnancy: Audit from a national confidential enquiry. *British Medical Journal 320*, 337–341.

Panter-Brick C. (1991). Parental responses to consanguinity and genetic disease in Saudi Arabia. *Social Science and Medicine 33*, 1295–1302.

Papadopoulos I. (2006). *Transcultural Health and Social Care: The development of culturally competent practitioners*. Oxford: Elsevier.

Pembrey M. (1996). The new genetics: A user's guide. In T. Marteau, & M. Richards (Eds.), *The troubled helix: Social and psychological implications of the new human genetics* (pp. 63–81). Cambridge: Cambridge University Press.

Port KE, Mountain H, Nelson J, Bittles AH. (2005). Changing profile of couples seeking genetic counseling for consanguinity in Australia. *American Journal of Medical Genetics 132*, 159–163.

Raz AE, Atar M, Rodnay M, Shoham-Vardi I, Carmi R. (2003). Between acculturation and ambivalence: Knowledge of genetics and attitudes towards genetic testing in a consanguineous Bedouin community. *Community Genetics 6*, 88–95.

Richards M. (1996). Families, kinship and genetics. In T. Marteau, & M. Richards (Eds.), *The troubled helix: Social and psychological implications of the new human genetics* (pp. 249–273). Cambridge: Cambridge University Press.

Richards M. (1998). Annotation: Genetic research, family life, and clinical practice. *Journal of Child Psychiatry 39*, 291–305.

Richards, MPM, Ponder M. (1996). Lay understanding of genetics: A test of an hypothesis. *Journal of Medical Genetics 33*, 1032–1036.

Rozario S. (2007). Genetics, Religion and Identity: A study of British Bangladeshis (2004–7). Working Paper 93. Cardiff: Cardiff University School of Social Sciences.

Santos S, Bizzo N. (2005). From 'new' genetics to everyday knowledge: Ideas about how genetic diseases are transmitted in two large Brazilian families. *Science Education 89*(4), 564–576.

Saucier JB, Johnson D, Wicklund CA, Robbins-Furman P, Hecht JT, Monga M. (2005). Racial-ethnic differences in genetic amniocentesis uptake. *Journal of Genetic Counseling 14*, 189–195.

Shaw A. (2000). *Kinship and continuity: Pakistani families in Britain*. London: Routledge/Harwood.

Shaw A. (2001). Kinship, Cultural Preference and immigration: Consanguineous marriage among British Pakistanis. *Journal of the Royal Anthropological Institute 7*, 315–334.

Shaw A. (2003). *Genetic counseling for Muslim families of Pakistani and Bangladeshi origin in Britain. The encyclopedia of the human genome* (pp. 762–766). Nature Publishing Group.

Shaw A, Ahmed M. (2004). Translating genetic information into languages other than English: Lessons from an assessment of Urdu leaflets. *Journal of Genetic Counseling 13*, 321–342.

Shaw A, Charsley K. (2006). Rishtas: Adding emotion to strategy in understanding British Pakistani transnational marriages. *Global Networks 6*, 405–421.

Shaw A, Hurst J. (2008). "What is this genetics, anyway?": Understandings of genetics, illness causality and inheritance among British Pakistani users of genetic services. *J. Genetic Counseling 17*(4), 373–382.

Shaw A, Hurst J. (2009). "I don't see any point in telling them": Attitudes to sharing genetic information in the family and carrier testing of relatives among British Pakistani adults referred to a genetics clinic. *Ethnicity and Health, 14*(2), 205–224.

Shiloh S, Berkenstadt M. (1992). Lay conceptions of genetic disorders. *Birth Defects: Original Articles 28*, 191–200.

Weil J. (1991). Mothers' postcounseling beliefs about the causes of their children's genetic disorders. *American Journal of Human Genetics 48*, 145–153.

Wilce JE. (1998). *Eloquence in trouble: The poetics and politics of complaint in rural Bangladesh*. Oxford: Oxford University Press.

Wood-Robinson C, Lewis J, Leach J. (2000). Young people's understanding of the nature of genetic information in the cells of an organism. *Journal of Biological Education 35*, 29–36.

123.

SOCIAL, CULTURAL AND RELIGIOUS ISSUES RELATED TO GENETIC DISEASES AMONG THE BANGLADESHIS IN THE UK

Santi Rozario and Sophie Gilliat-Ray

In order to make sense of how British Bangladeshis respond to and cope with genetic disorder and associated medical illness, we need to understand Bangladeshi social, cultural, and religious life in its British context. In this interconnected global village, it is not only science, technologies, or commodities that travel across the many national boundaries, but also ideas, values, and religion. Thus to talk of one culture being clearly distinct from another culture is naïve, to say the least (Clarke & Parsons 1997:1).

The cultural and religious values that influence British Bangladeshis in relation to genetics are therefore fluid and complex. Minimally, one needs to bear in mind the transnational nature of British Bangladeshi families and the strong familial ties they maintain with their relatives in Sylhet, where most of the Bangladeshi migrants come from. Through these ongoing ties, Bangladesh impacts directly on their own beliefs, values, and culture in general.

The chapter provides an overview on the British Bangladeshis, followed by their ideas about illness including beliefs about spirit causation (*upri, bhut, nazar*). We then discuss the British Bangladeshis and the Islamic position on genetics, followed by the various misunderstandings and confusion about genetics on the part of our interviewees.

BACKGROUND: BRITISH BANGLADESHIS

Most of the British Bangladeshis, and all of our samples of patients and most of the *imams*, originated from Sylhet, a region in northwestern Bangladesh which is the place of origin of the bulk of the British Bangladeshi population.

From about the 1960s, Sylheti migration to Britain became more widespread. Many members of the Bangladeshi community have limited English language skills and they are said to lag behind most other ethnic groups in their access and utilization of educational opportunities in the United Kingdom. (Choudhury & Drake, 2001:20). Choudhury and Drake's comments, referring to the 1970s and 1980s, still apply to many of the families we worked with:

> It was and is doubly difficult for the Bangladeshi children of caterers to fully benefit from an English education. The father's hours of work meant disturbed sleep in the house and made it difficult for the children to be at school punctually. Many Bangladeshi boys, when they reached secondary school age, began part-time work in a relative's restaurant. The temptation of earning regular wages meant that they often left school at the earliest opportunity, then married young and started a family, so institutionalising a cycle of low pay and few career opportunities (Choudhury and Drake, 2001:20).

This is precisely the situation that we found during our research. Fathers usually do not return home from their restaurant or their taxi driving till early morning. There are times when they might even have a visitor drop in at that hour. Children would often be still up.[1] It is still the case that young boys often end up working in the restaurants. Some young teenagers with whom we discussed their future aspirations also told us they will probably work in a restaurant. It seemed to be an easy choice. However, a number of young men with small children voiced their concern about this scenario for the Bangladeshi boys, and told me they will make sure their sons become doctors, solicitors, or engineers. Thus perhaps this situation will change in the future. On the other hand, we found the girls were hoping to go further than year ten in their education, although often they were married at the age of

1. Working hours in restaurants interfere with many aspects of people's lives in the daytime, especially in tasks that requires men's participation. Thus, Hawthorne et al. (2003) found that Bangladeshi patients wanted to have appointments only at particular times.

eighteen or nineteen. If they were working before marriage, their husbands would often be opposed to their continuing with their jobs.

While the bulk of the Bangladeshi population, including all the families in our sample, come from the relatively disadvantaged background described above, it is worth noting that there are other components to the British Bangladeshi community with higher educational levels and more middle-class lifestyles (Rozario & Gilliat-Ray, 2007). The overall disadvantageous position of our interviewees, however, has, as we will see, a bearing on their health status.

FAMILY, SOCIAL STRUCTURE, PATTERNS OF MARRIAGE, AND RESIDENCE

In this section we look at the structure of families among British Bangladeshis. This material is important for two main reasons. First, how a family responds to and attempts to cope with the difficult situation of having a child with a serious genetic disorder depends greatly on the existing relationships within the immediate and wider family. Family members are in most cases the most immediate people one asks for help, but they also understand and respond to the situation in various ways that can assist or worsen the situation, and that are in many ways characteristic of a particular social and cultural context. For genetic counselors, biomedical staff, and others attempting to assist these families, it is important to understand ways in which the situation is viewed and made sense of (cf. Featherstone et al, 2006). For example, among Bangladeshi and North Indian families, women are generally seen as "outsiders" marrying into a male descent line, and when things go wrong they are the obvious targets for blame. We will describe some examples of this process later in the chapter.

Second, an important issue for the prevention of genetic disorders is whether and how far family members pass on information about genetic disorders to the extended family, particularly in relation to marriage between people who are already closely related genetically. In cases where genetic disorders may be inherited, biomedical practitioners, including genetic consultants and counselors, aim to advise the families concerned of the possible risks. One intention is that families whose members are carriers for particular disorders will convey this information to *fu*ture possible marriage partners, and will use genetic testing where possible so as to avoid contracting marriages with a high risk of genetic illness. Families with genetic illness in the general U.K. population do not necessarily follow biomedical advice in such matters, for a variety of reasons. These may include the extent to which they understand or accept the biomedical model of the causation of genetic illness, and this is also likely to relate to their own "folk understandings" of inheritance (Featherstone et al., 2006). Reasons may also include social and cultural barriers to the passing on of information.

Both of these factors apply, too, in the case of U.K. Bangladeshis, and investigating them was a major focus of the research.

Here, again, understanding the nature of Bangladeshi marriage, and in particular the process by which marriages are arranged and the reasons for the choice of particular marriage partners, is very important for understanding why information about possible genetic risk may not be passed on. Marriages between close kin are often the result of debts and obligations within the extended family, and in such cases the obligation to undertake the marriage may override any considerations about possible risk of genetic disorders.

The residential patterns of Bangladeshi marriages in the United Kingdom are also of significance in relation to this study. British Bangladeshi families are normally patrilocal, as in Bangladesh itself, meaning that on marriage a woman comes to live with her husband's parents. In many ways, Sylhetis have reproduced the village community of Bangladesh as much as they could in the United Kingdom. As we noted, the early migrants joined their relatives, neighbors, or fellow-villagers from Sylhet, and settled with them either in the same house or as close as possible nearby. Immediate relatives, e.g., two brothers with their wives, children, and parents continued to share the same household until the families became much larger, when they bought additional houses, often in the same street or at most a few minutes' walk from each other. In our sample we came across a row of three houses adjacent to each other belonging to members of the same family: an elderly couple with the young unmarried children in one house, a married son with his wife and children in another, and a married daughter with husband and children in another.

This pattern of close residence provides young families with small children with ready-made help and support from parents. Families with a sick child can also receive ready support from other family members. Even where married daughters and sons have moved away from the area and are living elsewhere as a nuclear family, the whole extended family gets together regularly at least on a weekly basis, often on Fridays, but also during the week. However, there were some families in our sample who were relatively isolated from the remainder of their families for various reasons.

On the whole, marriage continues to be arranged by families, although not necessarily without the consent of the prospective marriage partners, whose opinion will usually be asked. Young men generally have more influence over decisions regarding their marriages than young women. Most arranged marriages are between British-born women and men born and brought up in Sylhet.

There are also marriages between men born in Britain and women born and brought up in Bangladesh. By no means are all of these marriages between first cousins, though many are. Although there is a well-known Islamic preference for a man to be married to his father's brother's daughter, we found marriages to parallel cousins on either side (i.e., father's brother's daughter or mother's sister's daughter) or to cross-cousins (father's sister's daughter, mother's brother's daughter).

According to one *imam* we interviewed, some guardians come to ask for a *tabiz* (amulet) when they are trying to arrange the marriage of their daughter or son and s/he re*fus*es to oblige. "I then give them an appropriate verse from the Qur'an to change the feelings of the son or the daughter." As this suggests, there can be considerable pressure on daughters or sons to marry into a particular family. We found plenty of evidence of this in our own fieldwork. This is especially the case involving marriage with relatives from Bangladesh.

A number of female interviewees told me that British-born women feel that Bangladeshi boys born and brought up in the United Kingdom are no good. British-born men express similar opinions about British-born Bangladeshi women. Nevertheless, we found and were also told that the younger generation of British-born Bangladeshis often prefer to marry those also born in Britain. Frequently, too, they will seek an "educated" bride or groom. This may be because they have seen too many problems in marriages when one partner is born in the U.K. and the other in Bangladesh.

A high proportion of our sample in fact consisted of marriages between British-born men or women (most often women) and partners from Bangladesh, and it is worth examining the specific tensions involved in these families in more detail, because this again has considerable bearing on what happens when the family has a child with a genetic disorder. We have heard stories from several *imams* that we interviewed, as well as from the women themselves, of marital problems among such couples.

A British-born Bangladeshi woman, Sharifa, told us that when a boy born here marries a girl from Bangladesh she usually becomes a docile housewife, while the husband does whatever he likes, including womanizing outside the house. But the marriage does not usually break up, because the docile Bangladeshi wife puts up with whatever condition she finds herself in. Sharifa continued that when a girl born and brought up here marries a boy from Bangladesh, she may be less tolerant of his behavior. This is not necessarily a matter of women brought up in Bangladesh being more submissive. If such women want to object to their husbands' behavior, they are in a weak position to do so. They are likely to be totally dependent on their husbands for their British visas, to have poor English language skills, little knowledge of their rights under English law, and little or no access to assistance outside the husband's family. Sometimes these Bangladeshi wives have no relatives in the United Kingdom other than their husband's family, and even in the case of a cousin marriage her primary identity for them is as the wife of the household, not as their niece or cousin. A British-born wife of a Bangladeshi husband is ahead of her husband at least with her English skills. In addition, her own family is in the U.K. and she can turn to them for support if things go wrong.

Sharifa was in agreement with the *imams* that a major source of conflict concerned the imported Bangladeshi husband being committed to sending much of his income back home to his own family. "As soon as a man comes to this country, the remainder of his family members become disabled, and they expect a continuous supply of money from the U.K." This is a story we heard from many women, and we know of at least one marriage that broke up for precisely this reason. Another reason for marital problems, according to Sharifa, is when a British-born man insists on living with his parents after marriage, leading to conflicts between the wife and her in-laws. A third source of tension that she mentioned is over women working outside the house: "Men do not like their wives working." Sharifa herself was prevented from working by her Bangladeshi cousin-husband. She also commented that "men from Bangladesh have a bad image of girls born in this country." Her own husband would check up on her when she was out with her girlfriends. "Men don't like women to question. She should be blind, deaf and dumb, just cooking, cleaning and washing in the house."

Although Sharifa's own marriage is working well, this is the result of continuous negotiation. On the whole, it seems true that British-born Bangladeshi women like Sharifa are in a stronger position in relation to their husbands. Even so, their situation is not easy and things do not always work out. If things go really badly, the wife does not only have to deal with the trauma of her own marriage break-up, but also has to deal with her husband's family members, who may be her own relatives. She may be seen as causing dishonor for her own family in relation to the extended family, and her remarriage may be difficult.

Because marriage is a must for Bengali women, and mature women have no social position outside of marriage (Rozario, 2001, 2002a), it is understandable that most women will try to make their marriages work, even if it means a great deal of compromise on their part. In addition to marriage itself, it is important that a wife becomes a mother soon after marriage and gra*dua*lly gives birth to many more children. Large families (four to seven children) are the norm with Bangladeshi families. One- or two-child families are not common unless there are health problems within the family. One woman with two children told us how other women often asked her why she does not have more children, and whether they are having some marital problems.

The story of a British-born woman, married to her first cousin, who did not have any children during the first twelve years of her marriage is revealing of what it means to be a childless woman in Bangladeshi culture. She was perceived as inauspicious and not made welcome in other people's wedding rituals; her husband was told to marry again; her husband, his relatives, and other Bangladeshis felt she had some problems, not the husband. Her mother kept taking her to numerous *maulanas* (learned religious specialists) in an effort to cure her of any problems she might have had. In desperation they tried IVF, which was unsuccess*ful*, but at the end they had twins without any medical intervention. In cases where a couple fails to have children, it is automatically assumed that the problem must be on the woman's side. Similarly, when a child is born with genetic disorder, this too is usually blamed on the woman (Rozario, 2009).

IDEAS ABOUT ILLNESS: *UPRI* AND *NAZOR* (MALIGNANT SPIRITS)

With many health problems, British Bangladeshis will assume from the nature of the problem that *upri* ("supernatural" factors) may be involved. In Islamic terms, this involves the action of *jinns* (spirits) and families will look to a specialist in *jinn* (who may be a Muslim cleric, *imam*, or an independent specialist) for assistance. Although *upri* problems can generally be treated only by *imams* or other spiritual practitioners, and medical problems by doctors, in practice there is no sharp dividing line between these, and both kinds of treatment may be pursued for the same disorder.

We came across references to *upri* in relation to problems of genetic disorders very early on during our field research with British Bangladeshi families. Initially we did not know what people meant by *upri*, even though Rozario's research in rural Bangladesh had made her very familiar with people's routine association of particular health problems with *bhut* (i.e., evil spirits; Rozario 1998, 2002b; Spiro 2005; Wilce 1997, 2004; Blanchet 1984; Islam 1980). Very soon we realized that *upri* is parallel to *bhut*, and to another term used in rural Bangladesh, *kharap batash* ("bad air"), and like them, referred to illnesses and problems caused by malevolent spirits. *Upri* was used particularly by the British Bangladeshis, who came largely from the Sylhet region of Bangladesh. Rather than *bhut*, the standard Bangla term for evil spirits, Sylhetis tended to use the Arabic-derived term *jinn*, but their ideas about these spirits were very much the same as those Rozario was familiar with from rural Bangladesh in relation to *bhut*.

In rural Bangladesh, people associate such problems as children's diarrhea, young women's white discharge, children getting scared, excessive crying of babies, or babies not breastfeeding, with *bhut* and sometimes with *nazor* (evil eye; Bhopal, 1986). This was very much the case with British Bangladeshis as well. However, in addition to all this, many of our interviewees also considered *upri*, i.e., interference of *jinns*, perhaps through the blood or something, as possibly underlying their children's illnesses—although these had in most cases already been diagnosed as genetic disorders by medical personnel.

Bangladeshis generally believe that illnesses caused by *bhut* or *upri* cannot be cured by medical doctors, but have to be dealt with by traditional folk healers or religious healers *(imams)*. However, we found that families did not choose between medical treatment, religious, or other folk treatment. Rather, they usually sought medical treatment side by side with treatment from *imams*, and sometimes from other sources as well. Such behavior is quite a common occurrence in Bangladesh itself, as well as South Asia in general and in other such situations of medical pluralism (Nichter & Nichter, 1996; Connor & Samuel, 2001; Rozario and Samuel, 2002; Rozario, 1998). However, the point to emphasize here is that to our interviewees there was no real conflict between religious forms of healing and medical (*daktari*) treatment. That is, there is no sharp dividing line between *daktari* and *upri* problems, and families will use a combination of ways of understanding and attempting to remedy their children's conditions. In her article, Rozario (2009) discusses at length the way British Bangladeshis with children suffering from genetic disorders sought assistance from local and foreign *maulanas* for their children. Below, we share two short case studies; the first concerns Ketan and Mukul, with their young son suffering from Cockayne syndrome visiting a Bangladeshi *pir* based in the U.K. Ketan and Mukul had already visited this *pir* a number of times. Rozario expressed an interest to go to visit this *pir* and they took her to see him. The text below is excerpts from Rozario's field notes:

> When I asked Ketan why people call him pir, he said, "He who is called a pir, he has some miraculous power, that is why he is called a pir." It seems he is known to have cured some people. Ketan added that this *pir* doesn't take money, but his speech is rough. But it seems that Ketan thinks he tells the truth, since other people say this about him.
>
> When we went in, no one else was there. Apparently there is often a queue of people waiting to see him. We were lucky and got to see him immediately. He was sitting on the floor of his living room: a newspaper was spread on the floor and over it a wooden chopping board. He was chopping onions and potatoes as he talked. When asked by Ketan if he would take some food if they brought it, he said he would not.
>
> Apparently he cooks for himself, usually very simple vegetarian food; he has fish or meat once in a while. He said that meat is not good, and that one should boil meat before cooking and throw

out the water. Apparently Mukul does this too. He said that if meat is not boiled people get cancer, particularly lung cancer. When I asked Ketan later why they need to boil meat, he said Mukul boils it to get rid of fat.

The *pir's* wife left him a long time ago. He has a grown-up son in Bangladesh, and a daughter in the U.K. Mukul later told me that the *pir* did not like sleeping with his wife, he slept downstairs and she slept upstairs. The wife got fed up and left.

He does not charge people any fees for his advice, so he is very popular. He also does not give any *tabiz* (amulets). He came to the U.K. forty-five years ago. For twenty years, he used to take Qur'an classes twice a day.

Ketan started to talk about their son Abdul, saying he is not getting better, there is something wrong with his brain. The *pir* said that something is wrong with his kidney, too. When Ketan asked why Abdul's condition had occurred, the *pir* said it was because his parents had done something wrong. Either the parents made love when Mukul was menstruating, or Ketan put pressure on Mukul when pregnant and they were making love. He said that maybe the mother ate something very hot when she was pregnant. Ketan and Mukul both objected, saying that they did everything right, and followed all the rules. I talked to Mukul about this afterwards, and she said that this was true.

Ketan said that the doctors have said that Abdul will not live long. The *pir* said something like, "That's good, he will return for his parents." I asked Ketan later what he meant. Ketan told me, "When a small child like Abdul dies, they leave this world for the other world. There is a special place in the other world to which these children go. There is a prophet there and he teaches them everything, better than he would have been taught in this world. On the last day of judgment, the child goes and stands by his parents and asks Allah to take them to heaven. Allah listens to these children."

At the end, Ketan asked the pir, "Won't he get better?" The *pir* said that he will get better, but it will take time.

I asked the *pir* about *jinn* (spirits). He said, "I am the father of people, and also the father of *jinns*. Recently I converted four *jinns* to Islam. They live in my house, like a bodyguard." There are apparently three types of *jinn*. These three types are found in mosques, graveyards, and toilets. He said that the graveyard and the toilet *jinns* are dangerous.

When Ketan said that they were thinking about going to *hajj*, the *pir* dismissed the significance of *hajj*. He said that if people can say their *namaz* regularly, especially the early morning *namaz*, which most people find difficult to get up for, this is as good as going to *hajj*.

The *pir* was very critical about TV and radio shows, comparing these to a toilet. Like some of my other informants, he linked many of the health problems in this country to the fact that people do not sweat here because of constant cold weather. It is believed that when you sweat, a lot of illness comes out of the body.

At one stage he told Ketan and Mukul to say, "*Subhan Allah.*" I asked Ketan about this afterwards and he said, "This is a form of *dua*, a praise of Allah. When you see a nice thing you praise Allah. Thus, although Abdul is clearly sick, he has special gift, and Allah is testing you."

As treatment, the *pir* gave an *ayat* [a set of Qur'anic verses]. He recited them and Ketan wrote them down on a small piece of paper he had given them. The *ayat* was to be recited over some water and blown over. Then the water would be pani pora (consecrated water) and would be fed to Abdul. This arrangement was partly because Ketan and Mukul did not bring bottles for the *pir* to supply his own *pani pora,* and partly because the *pir's* own *fu* (blowing over the water) would be too strong, according to the *pir* himself. "If I do the *fu* [ritual], it will be strong, the child's brain will be burnt out."

He said about people unable to have children, that those who are greedy for money, they do not have children. When I asked him why people have genetic conditions (what I said was "were born with some ongoing health problems"), he answered. "*shongodushey hoy*" (improper method of love-making). He was referring to not following rules when making love. At one point he said "*Quran boro daktar*" (Qur'an is the big doctor).

The treatment provided by this *pir* and various other *imams* and *maulanas* for this child was not effective and their son eventually died.

Next we share another case study of a couple, Shemol and Hema, who had taken their sick son Mohammad to Bangladesh for a holiday. During their time there, a divination ritual was performed by an *imam* to find out cause of their son's ongoing illness, which the doctors referred to as xeroderma pigmentosum (XP) and declared was incurable. The ritual was performed some days before Rozario visited them in Bangladesh, so the following information is based on accounts given to her by several family members and by the girl who was used as a medium.

The *imam* got a young unmarried girl of about fifteen or sixteen years of age to be a medium between him and the *jinns*. The only ones present during the ritual apart from the *imam* and the girl were Hema, Mohammad, Shemol, and Shemol's mother; all other relatives were told to leave the room.

The *imam* told the girl to sit on a *namaz* (prayer) mat. She was made to hold tazbi (prayer beads) on both hands and she was blindfolded. He then did some Islamic prayers, and asked the girl to say what she could see. He asked her if she could see some *maulanas* (holy men), and how many, to which she said she could see six. Then he asked the girl if she could see any sweets, she said yes, they had six boxes of sweets in their hands. She was asked if she could see a park, and she replied that she could. Shemol wanted to know if there was a name of the park. The *imam* apparently said, "It is not necessary to ask about the name, there may be problems," but Shemol nevertheless asked the girl to see if she could see the name of a particular park in the city where he lived in the U.K. She replied that she could see this name.

His treatment included doing *fu* (ritual blowing over) for both Hema and Mohammad, offering six kilograms of sweets to the *jinns*, two *tabiz* (amulets) for Mohammad to wear on his neck and hand, one *tabiz* for Hema to wear on her body, *tel pora* (consecrated oil) to massage Mohammad with, and pani pora (consecrated water) for Mohammad, Hema, and Shemol to drink for fourteen days.

In addition he gave them fourteen pieces of paper with *dua* (Islamic prayers) written on them. They were instructed to wrap one piece of paper per day for fourteen days in cloth, dip it in the *tel pora*, burn it in fire, and let Mohammad and Hema inhale the smoke from these. When I visited them they were still using the water and inhaling the smoke. The *imam* was paid Tk.1500 (£15.00) altogether for sweets, candles etc. The *imam* was going to buy and take the sweets for the *jinns* to a particular spot in the village.

As a result of this ritual, the *imam* decided that some six *jinns* from the park had possessed Hema when she took a walk in the park near her house in the U.K. and that there have been problems ever since as a result. He said that it was from the mother that the son was also possessed by these *jinns*.

Evil spirits (*bhut* or *jinns*) are said to be able to take any form or shape to come near a victim, and usually reside in jungles, in big trees, and dirty places. In different parts of Bangladesh people refer to *batash* (bad wind), *batash lagse* (being attacked by bad wind), *upardosh* (being attacked from above), *upri*, (from above) *bhutey dhorchey* (being possessed by evil spirit), *jinn* (bad Islamic spirits) and poris (female spirits as causes of various diseases or illness; Gardner, 1992). In Rozario's work, as well as in the work of Thérèse Blanchet, it is clear that all these terms refer to the same category of dangers, i.e., evil spirits. Blanchet (1984:117) found in her work in a rural area of northern Bangladesh, that "good Muslims are not supposed to believe in *bhut*. Moreover, I have heard that people are sometimes frightened to mention *bhut* lest they retaliate." As Blanchet suggests, the use of more impersonal words such as *batash*, upordosh, *upri*, circumvents this problem. This might explain why most of interviewees used the term *upri*, *jinns* or *bhut* referring to evil spirits. They also use the terms *batash* and *batash lagche*.

It is not a surprise that the evil *jinns* who were supposed to have attacked Hema got hold of her from a park, *fu*ll of big trees. While in theory evil spirits (or *jinns*) can attack men or women alike, in practice women are thought to be much more vulnerable to their attacks than men. A British Bangladeshi *imam* told us how in order to avoid attack of evil spirits appropriate behavior is important. Some of his examples included how women should dress appropriately and cover their hair and watch where they go. He said *jinns* reside in "bad, dirty, messy places, e.g., you find them in jungles. *Jinn* will never take shelter where one says his *namaz*, keeps his fast." Such statements were also made by a number of families whom I interviewed. They imply that being attacked by *jinn* is not morally neutral; it is usually to some extent the fault of the victim or, in the case of a small child, of its guardians. As in Hema's case, a child's mother is the person most immediately at risk of being blamed.

ISLAM, BRITISH BANGLADESHIS, AND GENETICS

The British Bangladeshis have come into direct contact with the new purist Islam through various Islamic movements. We were struck by the much stronger outward expression of Islamic religiosity by British Bangladeshis than Rozario had observed in Bangladeshi Dhaka villages in the 1980s. Not surprisingly, Islam played a major part in the lives of our interviewees as they coped and dealt with the problems of the genetic illness of their children. For any mother or father of a child afflicted by a genetic disorder, the question of why it is their child who has been affected is difficult to escape. Genetics may give an explanation of the biological process, but even if people understand the genetic explanation, which is by no means certain, it does little to answer the question of the personal meaning of what has happened. For Muslims, as believers in a personal God, Allah, the question of Allah's role in what has happened is a natural one.

In an article some years ago discussing Muslim perspectives on health and suffering, Abdulaziz Sachedina pointed out the impossibility of thinking of Islamic views on any theological matter in monolithic terms.

> In fact, plurality in belief and practice is inherent to Islam which, like Rabbinical Judaism, invests the power of interpretation and decision-making not in an institution like a church, but in the person of experts in religious matters, the *ulama*. (Sachedina, 1999: 65).

At one level, Islam would seem to give a straightforward answer to the question of Divine responsibility. As Alison Shaw comments,

> For Muslims, health, illness and misfortune are ultimately the consequence of Allah's will, a belief which lends itself to a fatalism which may itself be a source of comfort in cases of terminal and incurable illness (Shaw 2000: 98).

This assertion is founded on well-known verses of the Qur'an, such as *Sura* 64, verse 11: "No affliction befalls, except it be by the leave of God" (quoted in Sachedina, 1999: 67). The corollary to the idea of affliction visited by a loving God is that those who have patience in the face of misfortune will receive their reward:

> Surely we will try you with something of fear and hunger, and loss of wealth and possessions, death, and the loss of fruits of your toil. Yet, give glad tidings to those who are patient who, when they are visited by an affliction, say, 'Surely we belong to God, and to Him we return.' Upon those rest blessings and mercy from their Lord, and those are the truly guided. (*Sura* 2, vv.155-57, cited in Sachedina, 1999: 69).

As Sachedina notes (1999: 66), there is a cryptic character to the Qur'anic reference, which leaves considerable room for divergence as to how the presence of misfortune and evil in the world is to be understood and how active or passive one should be in response to it. The *hadith*, the traditional accounts of the sayings of Muhammad, which, along with the Qur'an, form the basic source of Muslim law, are nevertheless on the whole very positive about the appropriateness of medical care:

> The *hadith* literature overwhelmingly regards illness as an affliction that needs to be cured by every possible legitimate means. In fact, in these traditions, the search for cure is founded upon unusual confidence generated by the divine promise reported in one of the early traditions that 'There is no disease that God has created, except that He has also created its treatment.' Hence, the purpose of medicine is to search for cure and provide necessary care to those afflicted with diseases. [. . . Thus] a physician's role as a healer is regarded as spiritually and morally commendable and a collective duty among the religious obligations, in that medicine is a religious necessity for society. (Sachedina, 1999: 71).

Yet the parents of a child with a serious genetic disorder may be faced with the fact that, according to the medical authorities who are advising them, there is no available cure for the condition. This brings them back up against the question of why this misfortune has happened to them, and is also likely to bring into action another kind of possible explanation for the disorder, based on the idea that it has arisen as the result of the action of malevolent spirits (*jinn*) or other related forces such as envy (*nazor*, the "evil eye"). This provides a *fu*rther set of responses, in that the action of *jinn* and *nazor* may be countered in various ways, but also raises *fu*rther questions, since *jinn* and *nazor* are only thought to be able to afflict a child as a result of negligence of some kind on the part of those who are caring for it, in particular the child's mother. In addition, this is an area where modernist versions of Islam are of relevance, since they often contest the Islamic legitimacy of the various countermeasures against spirit-affliction.

The various responses of the parents and family members in my sample include elements of all these ideas. Thus we were told many times that having a child with a genetic disorder should be understood as a gift from Allah: "Allah has given this for a good reason" or "Allah has given this for our own good." One of our interviewees, Khaleda, put it like this: "Allah gave this to test us. People say that Allah gives [suffering] to those who can bear it."

One of our male interviewees told us too, much in the words of the *hadith* we cited above, that Allah has created the medicine before he has given the illness.

The opposition of many of our interviewees to testing during pregnancy (via amniocentesis) and to the termination of pregnancy in the case of an affected fetus can also be seen in terms of the idea that one should not oppose Allah's will. We were often told by parents that if they were to have more children with the same problem, then, "Let it be. I will accept it." Thus, Shahanara told me that it was against their religion to terminate a pregnancy. "Allah's gift, whatever Allah does. Have iman in Allah, He will be kind." Another time Shahanara said, "We are Muslims, we don't want to destroy babies. It's Allah's wish if there is a miscarriage, but we never want to destroy it ourselves."

In fact, Islamic legal authorities are not unequivocally opposed to the termination of pregnancy, though the conditions in which it is seen as appropriate may be framed fairly narrowly, and there is no unified position on the legitimacy of termination in the case of a fetus affected by a serious genetic disorder. Instead, there is a variety of juridical opinions delivered by experts in Islamic law, which people may be informed about to varying degrees. Often these opinions (*fatwa*) are delivered as responses to a specific question. Here we return to Sachedina's point about plurality of belief and practice.

Vardit Rispler-Chaim, in a study based mainly on *fatwa* by Egyptian and Saudi authorities (muftis), looks at the question of a child's "right not to be born" in a situation such as that of genetic disorder where it cannot be assured of a reasonable quality of life (Rispler-Chaim 2003):

> The general Islamic attitude toward the pursuit of knowledge and "scientific research" is very favorable, so Muslims need not object to the prenatal tests themselves. Abdel Rahim Omran [a US-based scholar whose *Family Planning in the Legacy of Islam* studies the juridical literature on these issues] even states that "Muslim children have the right to be born with no actual or potential genetic disorders," emphasizing the awareness that genetic disorders can be identified during pregnancy, and sometimes even prevented. The real problem arises when the tests suggest that an abortion is in order.[2]
>
> Abortion is a complex subject in Islamic law and constitutes an action that is legally judged "reprehensible." This means that it is permitted only when a "good" reason can be *fu*rnished, and only up to 120 days into the pregnancy, which is the stage before ensoulment. Only when the continuation of pregnancy endangers the mother's life is abortion permitted later than 120 days of pregnancy. The muftis debate extensively about what is a "good" reason and what is not. Genetic fetal disorders are part of this debate. Should an impaired fetus be aborted? (Rispler-Chaim 2003: 84–85).

This 120-day period is derived from the *hadith* rather than the Qur'an, but it has become an established part of Shari'a law, which considers that after 120 days the fetus becomes, in effect, a person with rights of its own, and abortion is a crime against the person (Kyriakides-Yeldham, 2005: 216). In practice, Rispler-Chaim notes, the muftis are more willing to encourage contraception to prevent impaired fetuses from being conceived than to counsel abortion once they are conceived. Contraception is fairly generally permitted by these Egyptian and Saudi authorities, although most of the families in my sample were under the impression that it was un-Islamic, and a minority of muftis permit the sterilization of carriers of genetic diseases (2003: 86–87). Rozario discusses some other sources on these issues in an article from 2005; these present a similar picture (Rozario, 2005: 193–1944).

Imams at Bangladeshi mosques in the United Kingdom are, in our experience, Bangladeshis who have been trained in Bangladesh or elsewhere in the Muslim world and have come to the U.K. as adults. Among our interviewees, three *imams* were well educated and learned in Islam. Others did not appear to be very learned. One of the three educated *imams* told us that most of the Bangladeshi *imams* in this country were not well qualified. After their *madrassa* training, they do a short three-month course and then they become *imams*. They often do this so as to be qualified to come to the U.K.

This might explain the reservations of one of our informants about many Bangladeshi *imams* in the U.K. Shahanara said that she had heard a lot of stories about *imams*: "Some *imams* aren't good, they say incorrect[2] things in Islam's name: this is Islamic or this is not Islamic." She gave the example of many *imams* who are opposed in the name of Islam to women going outside the house. She said that this was not correct: "Our Prophet said women can go out with *parda* [i.e., wearing the *burqa*], but some *imams* are saying that women can't go out at all, can't work. Good, really big *imams*, they want women to study and to work."

None of the *imams* whom we interviewed had appeared to understand anything about of the nature of genetic illness, and none of them had been approached for advice regarding genetic illness.[3] While the families did go to see *imams*, it was mainly to ask them for amulets and other consecrated items that would help protect their child from attack by spirits (*jinn*). According to the *imams* themselves, people mostly went to see them for minor health problems, and for various social and marital problems. People might also go to see certain special *imams* or *maulanas* for divination rituals when they wanted to find out if bad *jinns* were involved in a problem, as we have seen above.

When we asked the *imams* nevertheless for their opinions on issues of prenatal testing, abortion, and contraception, they generally supported prenatal testing but differed on the question of termination of pregnancy. One said that it was permissible up to 120 days; others regarded it as not allowed under any circumstances, or allowed only if there was a risk to the mother's health. They mostly allowed contraception, though they tended to be more restrictive than the current Islamic literature in this area.[3]

We also asked them for their views on cousin marriages. They all felt that it was allowed, though one stated that Islam neither objected to it nor encouraged it. They were clearly unaware of any possible genetic problems with cousin marriages.

2. The attitude towards antenatal testing is fast changing.

3. One man, considering termination of a fetus if found to be injured or damaged, apparently consulted an *imam* about the matter.

MISUNDERSTANDING AND CONFUSION ABOUT GENETICS

In discussing the families' understanding of genetic illness, we need to bear in mind that communication between medical professionals and the families can often be quite problematic, as understanding the meaning and implications of genetic illness is far from simple for people without a technical knowledge of medical genetics, however carefully it is explained.

The complexity of the diagnosis and of the genetic disorder meant that people often had very little understanding of the real nature of their child's problem, or of what this would mean for the child in the long run. Their understanding of the implications if they planned to have more children in the future was also generally quite limited, as we will see. The term "genetic" meant very little to most people, and they kept finding other causes for their children's ongoing illness. Often, they assumed that the problem was the negligence of the medical system. Below we discuss two main areas of confusion on the part of our interviewees in relation to genetics: being affected and being a carrier, and risks in relation to recessive disorders.

BEING AFFECTED AND BEING A CARRIER

There was a general confusion about the distinction between being "affected" by a particular genetic disorder and being a "carrier" of that disorder. These terms appeared to mean very little to people. We give two short case studies below to illustrate people's confusion, but also to show how the complexity of the disorders means people do not bother sharing information with most family members.

The first case concerns Khaleda, a relatively young woman in her mid-thirties, and her husband Faruq, in his mid-sixties. Neither had much education; Khaleda is Faruq's niece, twice removed; they married after Faruq's first wife had died. Out of Faruq and Khaleda's five children, their fourth child, a daughter, was affected with thalassemia, which was diagnosed when she was eight months old.

In this case, they had had an interpreter during their consultation. Khaleda told me "Thalassemia is a bad thing, this eats up the blood, the blood degenerates, the blood does not increase." Faruq commented, "I too have thalassemia, she [referring to his wife sitting in the room] too has it. This is why the child got it. The doctor said there is no problem if the man has it, but if a woman has it, then the child gets it." I can imagine that they were told that if only one of them was a carrier, then the child would not be affected and that Faruq misinterpreted this comment.

However, later on Khaleda said, "We both have problems, it is both our problems that got to our child." It would seem that Khaleda understood the situation a little better than Faruq did. Khaleda kept talking about how thalassemia has four groups. She did not know which one affected her stepdaughter, who was a woman about of her own age and married with three children). Unfortunately, despite my perseverance, I was not able to interview Faruq's daughter by his first wife. Apparently this daughter is given saline or blood transfusions on a regular basis. I was not clear which, for Khaleda mentioned she received blood transfusions while Faruq said she received saline.

During one visit, Khaleda and Faruq talked about some operation that could be performed for her daughter's thalassemia. I never heard about this possibility for thalassemia before, and I am not sure where this information came from. However, they said they were not sure if the operation would cure their daughter.

It is clear that there were various misunderstandings about the cause of thalassemia for their daughter, and about whether and how it might be cured. Khaleda told me that she did not know before that marriage between relatives can lead to having sick children, but she has been informed of this by some medical staff.

Still on the issue of misunderstanding the concept of carrier, I move on to another thalassemia family. Afsana and Rabi are a young couple with only one daughter, who is aged two. Their mothers are sisters, so that they are first cousins. Afsana was born and brought up in the United Kingdom, and finished her schooling. Rabi had also completed 12 years of education in Bangladesh. According to their files, Afsana is a carrier of Hb J. Meerut trait, and Rabi is a carrier of beta-thalassemia trait; their daughter Asma is a carrier of both Hb J and b-thalassemia trait, but is not herself affected. However, there is 50% chance of Asma passing these traits to each of her future children. She will need genetic counseling when she is about 16 years of age, and it is important that her future husband is also tested.

Apparently Rabi was initially not tested for anything, but he voluntarily had some tests done, including a blood test, just before they took off for a long holiday in Bangladesh. While they were in Bangladesh, a letter came to their address in the U.K. and was opened by Afsana's sister. She phoned Afsana and Rabi to tell them that Rabi had thalassemia. I asked Afsana if her sister actually read out the letter to them. She said that she did, but the phone line was bad, and all they could pick up was that Rabi had thalassemia. When Rabi heard this, he did not know what thalassemia was, but he thought there was something terribly wrong with him and that he was going to die, and he apparently fainted at the news. They went to numerous doctors in Bangladesh trying to find out about Rabi's thalassemia and what could be done. More tests were carried out in Bangladesh, but the doctors found nothing. Rabi said that the doctors did not seem to know much about thalassemia anyway, and that the news caused them much headache during their stay in Bangladesh.

It was only after they returned to their home in the United Kingdom and went to see their genetic counselor, who had sent them a letter regarding counseling, that they found out there was nothing wrong with Rabi. He was a carrier of the beta-thalassemia trait, but is not affected.

However, it appears that they are still con*fu*sed about what being a "carrier" meant. When I asked Afsana if she was worried about her daughter, she said that she used to be worried, but now that she sees that Asma is doing everything normally, she is not worrying much. Later on she said, "She seems to be OK intellectually. She must be OK, mustn't she?" Then she asked, "What would the problem be called when a child of two is quiet, does not talk, or walk, and sits quietly in a corner?" She seemed to be comparing her daughter with another child she knew. Her daughter seemed to have been doing much better and she seemed happy. In this case, Afsana did not really have to worry about Asma for she was not affected, but I found that it was common for mothers of sick children to constantly compare the condition of their children with the condition of other sick children.

RISKS AND TESTS IN CASES OF RECESSIVE DISORDERS

We came across a number of families where there were children affected with recessive disorders, who were planning for *fu*rther marriages with first cousins. Not only this, but people were rejecting the idea of having blood tests or other genetic tests of possible marriage partners so as to avoid marriage with another carrier of the same gene.

As we have seen, the idea of a "carrier" was often poorly understood. There are other issues, however, involved in these decisions. Thus in one case, a woman (Fazeela) reported that her brother was a carrier of beta-thalassemia, as was their father. When discussing the risk factor of her brother getting married to someone who might also be a carrier, Fazeela seemed very keen that her brother's fiancée should be tested before marriage. However, I heard later that her brother had gone to Bangladesh and that they would not be asking the fiancée to be tested. Fazeela explained that in order to ask a *fu*ture bride to be tested, his brother would have to reveal to them he was himself a "carrier" of beta-thalassemia. As people do not understand the difference between someone being a "carrier" or being "affected," this revelation might be interpreted as him being sick with beta-thalassemia. This, of course, would also have implications for the whole family, and the marriageability of other siblings might also become an issue, as such ongoing family disorders are stigmatized in Bangladeshi culture (Rozario, 2007).

In another case, a family where the parents were first cousins and two children had been affected by a recessive genetic disorder were planning another consanguineous marriage. The wife's younger sister (Hasina) was to marry the husband's younger brother. Hasina was initially offered a blood test by a visiting genetic counselor, well before her plan to get married. She told the genetic counselor she would think about it, but later when I asked her she kept putting it off. Just before her marriage when I asked her again she said, "Is it really necessary to have the blood test?" When I nodded, she said, "Perhaps my husband and I could have the test done together after we got married."

In a sense, Hasina did not have much to gain from having the test done. The fact is, she was getting married to her first cousin anyway. It had all been arranged and there was no turning back. Hasina was taking a risk by marrying someone who might be a carrier, when she could also be a carrier. But I did not get the feeling that she was at all concerned about this, nor were her parents, or her married sister with two affected children. The genetic disorder in this case was not life-limiting, and can be reasonably well managed. I am not clear whether this was a conscious part of their decision, or whether they simply did not believe that the condition was genetic. For example, when we discussed genetic testing during pregnancy, I was told by both the sisters that they were opposed to it as it was not Islamic. I was also told that, in any case, it was really Allah's wish whether a child would be affected or not.

In another case, that of a girl affected with beta-thalassemia, we were discussing whether the parents were concerned about the marital prospects of their daughter. Khaleda said they were worried about Leila's marriage prospects. I asked Khaleda whether, when the time came for Leila to marry, they would tell the potential groom and his family about her thalassemia. She said, "We have to tell, otherwise there will be problems later." However, she went on, "Let her grow up, we will see, perhaps we will marry her with a relative or something." When I asked if the *fu*ture groom will be tested, she said that he will have to be tested. I asked whether, if she marries someone from Bangladesh, there were facilities there for such tests? She said that this illness is also there in Bangladesh, implying by this that the tests must also be available.

In this case, Khaleda was concerned that her daughter should not marry anyone who might be a carrier and felt that it was important that the *fu*ture groom should know of her status before they get married. However, at the same time, reflecting on the fact that there may be problem in finding someone to marry Leila under the circumstances, she thought that marrying her to a relative would be the answer. Thus she was prepared to take the risk of marrying her to a relative, despite the possibility that this might increase the chances of genetic disorders among their children.

People's willingness to take genetic risks in terms of marriage, their opting out of tests, and so forth, might be related to the fact that, as we have seen, they do not necessarily accept that consanguineous marriage has anything to do with genetic disorders. It might also reflect the fact that they have a duty to their extended families in Bangladesh, and that this gets priority over concerns there might be about possible "sickly" children. After all, the risk may be seen as not that high. Many of these disorders, although permanent, are not life-limiting or severe, and can be managed pretty well with the help of drugs, blood trans*fu*sions, and regular medical attention. Many of the children affected with genetic disorders probably would not have lived beyond a few months or a few years if they were in Bangladesh, but the British Bangladeshis are fortunate in that they have access to the excellent health-care facilities to look after the sick.

Another important reason, as we have seen with Fazeela's brother, is people's unwillingness to reveal their "faulty" genetic identity. This might also be an unwillingness to acknowledge that there might be something not quite right, which is genetic and which is "in the family" from way back in the past. Thus a common refrain from many of my interviewees was, "We have never had anything like this in our family before." Another common refrain was that "it is Allah who decides" who is born with a genetic condition and who is not. Together these imply that the family's descent line as such is not at fault.

Here, Armstrong et al's, (1998) notion of "revealed identity," referring to a genetic identity that a patient is provided with in the process of genetic counseling, is use*fu*l in understanding Bangladeshi people's behavior. Families will avoid if at all possible telling others that their child's problem is familial, hereditary, or genetic. Genetic disorders implicate the whole family, both the older generation and younger generation. So, by saying, "We have never had anything like this in our family before," people are perhaps trying to protect the unspoiled identity of their family and to treat the child's disease as a one-off thing. It is something given to them by Allah to test them, or because He loves them.[4]

On the other hand, if they accept the genetic counselors' suggestion, and ask for a potential bride or groom to be tested before marriage, then they are drawing attention to their own spoiled identity, and in the process revealing that it has already been spoiled in the past.

CONCLUSION

The above has, we hope, helped the reader to appreciate something of the nature of British Bangladeshi responses to genetic illness, and the various considerations that families might have in their minds when responding to information received from a genetic consultant or genetic counselor, or when making decisions as how to best deal with a seriously ill child.

By and large our interviewees usually sought medical help diligently, even if not accepting all their suggestions or options. However, in most cases they chose the Islamic option when they believed that it conflicted with the medical opinion of their British biomedical advisors, e.g., in relation to amniocentesis, contraception, termination of fetus, and so forth. As Rozario has suggested elsewhere, the reasons for this can be complex, and are in part tied up with the difficult and conflicted nature of Muslim identity politics in contemporary Britain (2009). If biomedical staff are to communicate effectively with British Bangladeshis, however, and to help achieve the best outcome for the families and children involved, it is important to understand the kinds of concerns that are significant for the family members involved. We hope that this chapter has contributed a little to such an understanding.

4. There is a widespread idea that a disabled child is a special gift from Allah.

ACKNOWLEDGMENT

This chapter is based on an ESRC-*fu*nded project on genetics, religion, and identity among British Bangladeshis from 2005–2007.[5] In our study we hoped to understand how Bangladeshi families in the United Kingdom made sense of genetic disorders, and how they used both Islamic and biomedical resources in coping with their situation. Some 20 families were involved in the study, with a wide variety of genetic disorders, many of them severe. Family members included both British-born Bangladeshis and recent migrants.

REFERENCES

Armstrong D, Michie S, Marteau T. (1998). Revealed identity: a study of the process of genetic counselling. *Social Science and Medicine* 47(2):1653–1658.

Bhopal RS. (1986). The inter-relationship of folk, traditional and western medicine within an Asian community in Britain. *Social Science and Medicine* 22(1):99–105.

Blanchet T. (1984). *Women, Pollution and Marginality: Meanings and Rituals of Birth in Rural Bangladesh.* Dhaka: International Centre for Diarrhoeal Disease Research, Bangladesh.

Choudhury Y, Drake P. (2001). *From Bangladesh to Birmingham: The History of Bangladeshis in Birmingham,* Birmingham: Birmingham Library Services.

Clarke A, Parsons E, eds. (1997) *Culture, Kinship and Genes: Towards Cross-cultural Genetics.* New York: Palgrave Macmillan.

5. The ESRC grant was awarded to Dr. Sophie Gilliat-Ray (Principal Investigator), Professor Angus Clarke, Professor Stephen Pattison, and Dr. Joanna Latimer. Santi Rozario was the senior research fellow and undertook all the research.

Connor L, Samuel G, eds. (2001). *Healing Powers and Modernity: Traditional Medicine, Shamanism, and Science in Asian Societies.* Westport & London: Bergin & Garvey.

Featherstone K, Atkinson P, Bharadwaj A, Clarke A. (2006). *Risky Relations: Family, Kinship and the New Genetics.* Oxford & New York: Berg.

Gardner, Katy (1992). *Songs at the River's Edge: Stories from a Bangladeshi Village.* Calcutta, Delhi: Rupa & Co.

Islam M. (1980). *Folk Medicine and Rural Women in Bangladesh.* Dhaka: Women for Women.

Kyriakides-Yeldham A. (2005). *Islamic medical ethics and the straight path of God. Islam and Christian-Muslim Relations* 16(3):213–225.

Nichter M, Nichter M. (1996). *Anthropology and International Health: Asian Case Studies.* Australia, UK, India: Gordon and Breach Publishers.

Rispler-Chaim V. (2003). The right not to be born: abortion of the disadvantaged fetus in contemporary *fatwas*. In Jonathan E. Brockopp (ed.) *Islamic Ethics of Life: Abortion, War, and Euthanasia,* pp. 81–95. Columbia: University of South Carolina Press.

Rozario S. (1998). The Dai and the Doctor: Discourses on Women's Reproductive Health in Rural Bangladesh. In Kalpana Ram and Margaret Jolly (eds.), *Modernities and Maternities: Colonial and Postcolonial Experiences in Asia and the Pacific*, pp. 144–176. Cambridge: Cambridge University Press.

Rozario S. (2001). *Purity and Communal Boundaries: Women, and Social Change in a Bangladeshi Village.* 2nd ed. with new intro. Dhaka: University Press Limited.

Rozario S. (2002a). Poor and 'Dark,' What is My *Future*?: Identity Construction and Adolescent Women in Bangladesh. In Lenore Manderson and Pranee Liamputtong Rice (eds.), *Coming of Age in South and Southeast Asia: Youth, Courtship and Sexuality,* pp. 42–57. Richmond, Surrey: Curzon Press.

Rozario S. (2002b). The Healer on the Margins: The Dai in Rural Bangladesh. In Santi Rozario and Geoffrey Samuel (eds.) *The Daughters of Hariti,* pp. 130–146. London and New York: Routledge.

Rozario S, Samuel G, eds. (2002). *The Daughters of Hariti.* London and New York: Routledge.

Rozario S. (2005). Genetics, religion and identity among British Bangladeshis: some initial findings. *Diversity in Health and Social Care* 2(3):187–196.

Rozario S. (2007). Growing up and living with neurofibromatosis 1 (NF1): A British Bangladeshi case study. *Journal of Genetic Counseling* 16(5): 551–559.

Rozario S. (2009). Allah is the scientist of the scientists: modern medicine and religious healing among British Bangladeshis. *Culture and Religion* 10(2):177–199.

Rozario S. (2007). (with S. Gilliat-Ray) Genetics, Religion and Identity: A Study of British Bangladeshis (2004–7). Cardiff University School of Social Sciences Working Paper No. 93. 105pp. ISBN 978-1-904815-64-8. Available at http://www.cardiff.ac.uk/socsi/research/publications/workingpapers/papers-91-100.html.

Sachedina A. (1999). Can God inflict unrequited pain on his creatures? Muslim perspectives on health and suffering." In John Hinnells & Roy Porter (eds.) *Religion, Health and Suffering,* pp. 65–84. London & New York: Kegan Paul International.

Shaw A. (2000). Conflicting Models of Risk: Clinical Genetics and British Pakistanis. In Pat Caplan (ed.) *Risk Revisited,* pp. 85–107. London: Pluto Press.

Spiro A. (2005). *Najar* or *Bhut* – evil eye or ghost affliction: Gujrati views about illness causation. *Anthropology & Medicine* 12(1):61–73.

Wilce JM. (1997). Discourse, power, and the diagnosis of weakness: encountering practitioners in Bangladesh. *Medical Anthropology Quarterly* 11(3):352:374.

Wilce JM. (2004). Madness, fear, and control in Bangladesh: clashing bodies of power/knowledge. *Medical Anthropology Quarterly* 18(3):357–375.

GLOSSARY: SELECTED TERMS AND PHRASES

Compiled and Edited by Dhavendra Kumar

Accession number: a unique number assigned to a nucleotide, protein, structure, or genome record by a sequence database builder.

Acrocentric: a chromosome having the centromere close to one end.

Algorithm: a step-by-step method for solving a computational problem.

Allele: an alternative form of a gene at the same chromosomal locus.

Allelic heterogeneity: different alleles for one gene.

Alternative splicing: a regulatory mechanism by which variations in the incorporation of coding regions (*see exon*) of the gene into messenger RNA (mRNA) lead to the production of more than one related protein, or isoform.

***Alu* repeat (or sequence):** one of a family of about 750,000 interspersed sequences in the human genome that are thought to have originated from the 7SL RNA gene; ***Alu*-PCR**: a PCR reaction that uses an oligonucleotide primer with a sequence derived from the *Alu* repeat.

Amino acid: a chemical subunit of a protein. Amino acids polymerize to form linear chains linked by peptide bonds called *polypeptides*. All proteins are made from twenty naturally occurring amino acids.

Amplification refractory mutation system (ARMS): an allele-specific PCR amplification reaction.

Annealing: the association of a complementary DNA (or RNA) strand to form the double-stranded structure.

Anonymous DNA: DNA not known to have a coding function.

Annotation: the descriptive text that accompanies a sequence in a database method.

Antibody: a protein produced by the immune system in response to an antigen (*see antigen*). Antibodies bind to their target antigen to help the immune system destroy the foreign entity.

Anticipation: a phenomenon in which the age of onset of a disorder is reduced and/or severity of the phenotype is increased in successive generations.

Apoptosis: programmed cell death.

Antigen: a molecule that is perceived by the immune system to be foreign.

Autosome: any chromosome other than a sex chromosome (X or Y) and the mitochondrial chromosome.

Autozygosity: in an inbred person, homozygosity for alleles identical by descent; ***autozygosity mapping***: a form of genetic mapping for autosomal recessive disorders in which affected individuals are expected to have two identical disease alleles by descent.

Bacterial artificial chromosome (BAC): DNA vectors into which large DNA fragments can be inserted and cloned in a bacterial host.

Bioinformatics: an applied computational system that includes development and utilization of facilities to store, analyze, and interpret biological data.

Biotechnology: the industrial application of biological processes; for example, recombinant DNA technology and genetic engineering.

BLAST (Basic Local Alignment Search Tool): a fast database similarity search tool used by the NCBI/PubMed that allows the world to search a number of genetic/ genomic databases; for example, GeneBank and OMIM.

Blastocyst: the mammalian embryo at the stage at which it is implanted into the wall of the uterus.

Bottleneck: a severe reduction in the number of individuals in a population, leading to a reduction in the genetic diversity of that population in later generations.

Candidate gene: any gene which. by virtue of a known property (function, expression pattern, chromosomal location, structural motif, etc.), is considered as a possible locus for a given disease.

Carrier: a person who carries an allele for a recessive disease (*see heterozygote*) without the disease phenotype, but can pass it on to the next generation; also used to denote a female carrying the mutation on one of the two X-chromosomes for an X-linked recessive disorder.

Carrier testing: carried out to determine whether an individual carries one copy of an altered gene for a particular recessive disease.

Cell cycle: series of tightly regulated steps that a cell goes through from its creation to division to form two daughter cells.

"Central Dogma": a term proposed by Francis Crick in 1957—*DNA is transcribed into RNA, which is translated into protein.*

cDNA (complementary DNA): a piece of DNA copied in vitro from mRNA by a reverse transcription enzyme.

Centimorgan (cM): a unit of genetic distance equivalent to 1% probability of recombination during meiosis; one centimorgan is equivalent, on average, to a physical distance of approximately 1 megabase in the human genome.

Centromere: the constricted region near the center of a chromosome that has critical role in cell division.

Chimera: a hybrid, particularly a synthetic DNA molecule that is the result of ligation of DNA fragments that come from different organisms, or an organism derived from more than one zygote.

Chromosome: subcellular structures that contain and convey the genetic material of an organism.

Chromosome painting: fluorescent labeling of whole chromosomes by a FISH procedure in which labeled probes consist of a complex mixture of different DNA sequences from a single chromosome.

Clinical proteomics: translational application of new protein-based technologies in clinical medicine.

Clinical sensitivity: the proportion of persons with a disease phenotype who test positive.

Clinical specificity: the proportion of persons without a disease phenotype who test negative.

Clone: a line of cells derived from a single cell and therefore carrying identical genetic material.

Cloning vector: a DNA construct such as a plasmid, modified viral genome (bacteriophage or phage), or artificial chromosome that can be used to carry a gene or fragment of DNA for purposes of cloning (for example, a bacterial, yeast, or mammalian cell).

Coding DNA (sequence): the portion of a gene that is transcribed into mRNA.

Codon: a three-base sequence of DNA or RNA that specifies a single amino acid.

Comparative genomics: the comparison of genome structure and function across different species in order to further an understanding of biological mechanisms and evolutionary processes.

Comparative genome hybridization (CGH): use of competitive fluorescence in situ hybridization to detect chromosomal regions that are duplicated or deleted; use of arrays of such regions (**arrayCGH**) is now increasingly applied in clinical genetics.

Complementary DNA (cDNA): DNA generated from an expressed messenger RNA through a process known as *reverse transcription.*

Complex diseases: diseases characterized by risk to relatives of an affected individual that is greater than the incidence of the disorder in the population

Complex trait: one which is not strictly Mendelian (dominant, recessive, or sex-linked) and may involve the interaction of two or more genes to produce a phenotype, or may involve *gene–environment* interactions.

Congenital: any trait, condition, or disorder that exists from birth.

Consanguinity: Marriage between two individuals having common ancestral parents, commonly between first cousins; an approved practice in some communities who share social, cultural, and religious beliefs. In genetic terms, two such individuals could be heterozygous by descent for an allele expressed as *coefficient of relationship*, and any offspring could be therefore homozygous by descent for the same allele expressed as *coefficient of inbreeding.*

Conserved sequence: a base sequence in a DNA molecule (or an amino acid sequence in a protein) that has remained essentially unchanged throughout evolution.

Copy-number: the number of different copies of a particular DNA sequence in a genome.

Cytoplasm: the internal matrix of a cell. The cytoplasm is the area between the outer periphery of a cell (the cell membrane) and the nucleus (in a eukaryotic cell).

Denaturation: dissociation of complementary strands to give single-stranded DNA and/or RNA.

Demographic transition: the change in the society from extreme poverty to a stronger economy, often associated by a transition in the pattern of diseases from malnutrition and infection to the intractable conditions of middle and old age; for example obesity, diabetes, and cancer.

Determinism (genetic): philosophical doctrine that human action is not free but determined by genetic factors.

Diploid: a genome (the total DNA content contained in each cell) that consists of two homologous copies of each chromosome.

Disease: a fluid concept influenced by societal and cultural attitudes that change with time and in response to new scientific and medical discoveries. The human genome sequence will dramatically alter how we define, prevent, and treat disease. Similar collections of symptoms and signs (*phenotype*) may have very different underlying genetic constitutions (*genotype*). As genetic capabilities increase, additional tools will become available to subdivide disease designations that are clinically identical (*see taxonomy of disease*).

Disease etiology: any factor or series of related events directly or indirectly causing a disease. For example, the genomics revolution has improved our understanding of disease determinants and provided a deeper understanding of molecular mechanisms and biological processes.

Disease expression: when a disease genotype is manifested in the *phenotype.*

Disease management: a continuous, coordinated healthcare process that seeks to manage and improve the health status of a patient over the entire course of a disease. The term may also apply to a patient population. Disease management services include disease prevention efforts as well as patient management.

Diversity, genomic: the number of base differences between two genomes divided by the genome size.

DNA (deoxyribonucleic acid): the chemical that comprises the genetic material of all cellular organisms.

DNA cloning: replication of DNA sequences ligated into a suitable vector in an appropriate host organism (*see cloning vector*).

DNA fingerprinting: use of hypervariable mini-satellite probe (usually those developed by Sir Alec Jeffreys) on a Southern blot to produce an individual-specific series of bands for identification of individuals or relationships. The technique is now extended to employ genomic polymorphisms (copy-number variation and single nucleotide polymorphisms) using PCR and next generation sequencing methods.

DNA library: a collection of cell clones containing different recombinant DNA clones.

DNA sequencing: technologies through which the order of base pairs in a DNA molecule can be determined; the *Sanger method* is commonly used, new methods include the *next generation sequencing techniques.*

Domain: a discrete portion of a protein with its own function. The combination of domains in a single protein determines its overall function.

Dominant: an allele (or the trait encoded by that allele) which produces its characteristic phenotype when present in the heterozygous form.

Dominant negative mutation: a mutation in *heterozygous* form that results in a mutant gene product inhibiting the function of the wild-type gene product.

Dosage effect: variation in the number of copies can result in aberrant gene expression or can be associated with disease phenotype.

Embryonic stem cells (ES cells): a cell line derived from undifferentiated, pluripotent cells from the embryo.

ENTREZ: an online search and retrieval system that integrates information from databases at NCBI. These databases include nucleotide sequences, protein sequences, macromolecular structures, whole genomes, OMIM, and MEDLINE, through PubMed (*see also BLAST*).

Environmental factors: may include chemical, dietary factors, infectious agents, physical and social factors.

Enzyme: a protein which acts as a biological catalyst that controls the rate of a biochemical reaction within a cell.

Epigenetic: a term describing non-mutational phenomena, such as methylation and histone modification that modify the expression of a gene; ***epigenomics*** is also used in the same context, but refers to several genomic regions having similar role or function.

Euchromatin: the fraction of the nuclear genome that contains transcriptionally active DNA and which, unlike *heterochromatin*, adopts a relatively extended conformation.

Eukaryote: an organism whose cells show internal compartmentalization in the form of membrane-bounded organelles, such as a nucleus (includes multicellular organisms—animals, plants, fungi, and algae).

Exon: the sections of a gene that code for its functional product. Eukaryotic genes may contain many *exons* interspersed with noncoding *introns.* An *exon* is represented in the mature mRNA product—the portions of an mRNA molecule that are left after all *introns* are spliced out— that serves as a template for protein synthesis.

Expression sequences tag (EST): partial or full complement DNA sequences, which can serve as markers for regions of the genome that encode expressed products.

Family history: an essential tool in clinical genetics. Interpreting the family history can be complicated by many factors, including small families, incomplete or erroneous family histories, consanguinity, variable penetrance, and the current lack of real understanding of the multiple genes involved in polygenic (complex) diseases.

Fluorescence in situ hybridization (FISH): a form of chromosome in situ hybridization in which a nucleic acid probe is labeled by incorporation of a ***fluorophore***, a chemical group that fluoresces when exposed to UV irradiation.

Founder effect: changes in allelic frequencies that occur when a small group is separated from a large population and establishes in a new location.

Frameshift mutation: the addition (duplication) or loss (deletion) of a number of DNA bases that is not a multiple of three, thus causing a shift in the reading frame of the gene. This shift leads to a change in the reading frame of all parts of a gene that are downstream from the mutation, leading to a premature stop codon and thus to a truncated protein product.

Functional genomics: the development and implementation of technologies to characterize the mechanisms through which genes and their products function and interact with each other and with the environment.

Gain-of-function mutation: a mutation that produces a protein that takes on a new or enhanced function.

Gene: the fundamental unit of heredity; in molecular terms, a gene comprises a length of DNA that encodes a functional product, which may be a polypeptide (a whole or constituent part of a protein or an enzyme) or a ribonucleic acid. It includes regions that precede and follow the coding region, as well as introns and exons. The exact boundaries of a gene are often ill-defined, since many promoter and enhancer regions dispersed over many kilobases may influence transcription.

Gene-based therapy: refers to all treatment regimens that employ or target genetic material. This includes (i) *transfection* (introducing cells whose genetic make-up is modified) (ii) *antisense* therapy, and (iii) *naked DNA* vaccination. *(see gene therapy)*

Gene expression: the process through which a gene is activated at a particular time and place so that its functional product is produced, such as transcription into mRNA followed by translation into protein.

Gene expression profile: the pattern of changes in the expression of a specific set of genes that is relevant to a disease or treatment. The detection of this pattern depends upon the use of specific gene expression measurement techniques.

Gene family: a group of closely related genes that make similar protein products.

Gene knockouts: a commonly used technique to demonstrate the phenotypic effects and/or variation related to a particular gene in a model organism, for example in mouse (*see knock-out*); absence of many genes may have no apparent effect upon phenotypes (though stress situations may reveal specific susceptibilities). Other single knockouts may have a catastrophic effect upon the organism, or be lethal so that the organism cannot develop at all.

Gene regulatory network: a functional map of the relationships between a number of different genes and gene products (proteins), regulatory molecules, etc., that define the regulatory response of a cell with respect to a particular physiological function.

Gene therapy: a therapeutic medical procedure that involves either replacing/manipulating or supplementing nonfunctional genes with healthy genes. Gene therapy can be targeted to somatic (body) or germ (egg and sperm) cells. In **somatic gene therapy** the recipient's genome is changed, but the change is not passed along to the next generation. In **germ-line gene therapy**, the parent's egg or sperm cells are changed with the goal of passing on the changes to their offspring.

Genetics: refers to the study of heredity, genes, and genetic material. In contrast to *genomics*, genetics is traditionally related to lower-throughput, smaller-scale emphasis on single genes, rather than on studying structure, organization, and function of the whole genome.

Genetic architecture: refers to the full range of genetic effects on a trait. Genetic architecture is a moving target that changes according to gene and genotype frequencies, distributions of environmental factors, and such biological properties as age and sex.

Genetic code: the relationship between the order of nucleotide bases in the coding region of a gene and the order of amino acids in the polypeptide product. It is universal, triplet, non-overlapping code, such that each set of three bases (termed a *codon*) specifies which of the 20 amino acids is present in the polypeptide chain product of a particular position.

Genetic counseling: an important process for individuals and families who have a genetic disease or who are at risk for such a disease. Genetic counseling provides patients and other family members information about their condition and helps them make informed decisions.

Genetic determinism: the unsubstantiated theory that genetic factors determine a person's health, behavior, intelligence, or other complex attributes.

Genetic discrimination: unfavorable discrimination directed toward an individual, a family, community, or an ethnic group on the basis of genetic information. Discrimination may include societal segregation, political persecution, limits on opportunities for education and training, lack of or restricted employment prospects, and restricted access to adequate personal financial planning, for example life insurance and mortgage financing.

Genetic drift: commonly used in population genetics, refers to a change in the prevalence of a hereditary condition or trait associated with population movements over a long period.

Genetic engineering: the use of molecular biology techniques such as restriction enzymes, ligation, and cloning to transfer genes among organisms (*also known as recombinant DNA cloning*).

Genetic enhancement: the use of genetic methodologies to improve functional capacities of an organism rather than to treat disease.

Genetic epidemiology: a field of research in which correlations are sought between phenotypic trends and genetic variation across population groups.

Genomic instability: an increased tendency of the genome to acquire mutations when various processes involved in maintaining and replicating the genome are dysfunctional.

Genetic screening: testing a population group to identify a subset of individuals at high risk for having or transmitting a specific genetic disorder

Genetic susceptibility: predisposition to a particular disease due to the presence of a specific allele or combination of alleles in an individual's genome.

Genetic test: an analysis performed on human DNA, RNA, genes, and/or chromosomes to detect heritable or acquired genotypes. A genetic test also is the analysis of human proteins and certain metabolites, which are predominantly used to detect heritable or acquired genotypes, mutations, or phenotypes.

Genetic testing: strictly refers to testing for a specific chromosomal abnormality or a DNA (nuclear or mitochondrial) mutation already known to exist in a family member. This includes diagnostic testing (postnatal or prenatal), presymptomatic or predictive genetic testing, or to establish carrier status. The individual concerned should have been offered full information on all aspects of the genetic test through the process of *nonjudgmental and nondirective* genetic counseling. Most laboratories require a formal, fully informed, signed consent before carrying out a test. Genetic testing commonly involves DNA/RNA-based tests for single gene variants, complex genotypes, acquired mutations, and measures of gene expression. Epidemiologic studies are needed to establish clinical validity of each method to establish sensitivity, specificity, and predictive value.

Genome: the complete set of chromosomal and extra-chromosomal DNA/RNA of an organism, a cell, an organelle, or a virus.

Genome annotation: the process through which landmarks in a genomic sequence are characterized using computational and other means; for example, genes are identified, predictions made as to the function of their products, their regulatory regions defined, and intergenic regions characterized *(see annotation)*.

Genome project: the research and technology development effort aimed at mapping and sequencing the entire genome of human beings and other organisms.

Genomics: the study of the structure and function of whole genome of an organism. The term is commonly used to refer large-scale, high-throughput molecular analyses of multiple genes, gene products, or regions of genetic material (DNA and RNA). The term also includes the comparative aspect of genomes of various species, their evolution, and how they relate to each other (*see comparative genomics*).

Genotype: the genetic constitution of an organism; commonly used in reference to a specific disease or trait.

Genomic profiling: complete genomic sequence of an individual including the expression profile; this may be targeted to specific requirements, for example most common complex diseases (diabetes, hypertension, and coronary heart disease).

Germ-line cells: a cell with a haploid chromosome content (also referred to as a *gamete*); in animals, sperm or egg, and in plants, pollen or ovum.

Germline mosaic (germinal mosaic, gonadal mosaic, gonosomal mosaic): an individual who has a subset of germline cells carrying a mutation that is not found in other germline cells.

Germline mutation: a gene change in the body's reproductive cells (egg or sperm) that becomes incorporated into the DNA of every cell in the body of offspring; germline mutations are passed on from parents to offspring, also called ***hereditary mutation***.

Haploid: describing a cell (typically a gamete) that has only a single copy of each chromosome (i.e., 23 in man).

Haplotype: a series of closely linked loci on a particular chromosome, which tend to be inherited together as a block.

Heterozygote: refers to a particular allele of a gene at a defined chromosome locus. A heterozygote has a different allelic form of the gene at each of the two homologous chromosomes.

Heterozygosity: the presence of different alleles of a gene in one individual or in a population—a measure of genetic diversity.

Homology: similarity between two sequences due to their evolution from a common ancestor, often referred to as ***homologs***.

Homozygote: refers to same allelic form of a gene on each of the two homologous chromosomes.

Human Genome Project: a program to determine the sequence of the entire three billion bases of the human genome.

Human Genetics Commission: established in December 1999 in order to provide strategic advice to the U.K. government on how new developments in human genetics will affect people and healthcare. It has a particular remit to advise on the social, ethical, and legal implications of these developments. [http://www.hgc.gov.uk/]

Identity by descent (IBD): alleles in an individual or in two people that are identical because they have been inherited from the same common ancestor, as opposed to ***identity by state*** **(IBS)**, which is coincidental possession of similar alleles in unrelated individuals (*see consanguinity*).

Immunogenomics: refers to the study of the organization, function, and evolution of vertebrate defense genes, particularly those encoded by the *major histocompatibility complex* (MHC) and the *leukocyte receptor complex* (LRC). Both complexes form integral parts of the immune system.

Intron: a noncoding sequence within eukaryotic genes which separates the exons (coding regions). Introns are spliced out of the messenger RNA molecule created from a gene after transcription, prior to protein translation (protein synthesis).

Isoforms/isozymes: alternative forms of protein/enzyme.

In situ hybridization: hybridization of a labeled nucleic acid to a target nucleic acid that is typically immobilized on a microscopic slide, such as DNA of denatured metaphase chromosomes (as in *fluorescent in situ hybridization* [FISH]) or the RNA in a section of tissue (as in *tissue in situ hybridization* [TISH]).

In vitro: (Latin) literally "in glass," meaning outside of the organism in the laboratory; usually a tissue culture.

In vivo: (Latin) literally "in life," meaning within a living organism.

Knock-out: a technique used primarily in mouse genetics to inactivate a particular gene in order to define its function.

Library: a collection of genomic or complementary DNA sequences from a particular organism that have been cloned in a vector and grown in an appropriate host organism (e.g., bacteria or yeast).

Ligase: an enzyme that can use ATP to create phosphate bonds between the ends of two DNA fragments, effectively joining two DNA molecules into one.

Linkage: the phenomenon whereby pairs of genes that are located in close proximity on the same chromosome tend to be co-inherited; **Linkage analysis**: a process of locating genes on the chromosome by measuring recombination rates between phenotypic and genetic markers (*see Lod score and positional cloning*).

Linkage disequilibrium: the nonrandom association in a population of alleles at nearby loci.

Locus: The specific site on a chromosome at which a particular gene or other DNA landmark is located.

LOD score: a statistical term, "log of the odd"; a measure of the likelihood of genetic linkage between loci; an LOD score greater than +3 is acceptable as evidence of linkage; one that is less than −2 is often taken as evidence against linkage.

Loss of function mutation: a mutation that decreases the production or function (or both) of the gene product.

Loss of heterozygosity (LOH): loss of alleles on one chromosome, detected by assaying for markers for which an individual is constitutionally heterozygous.

Lyonization: the process of random X-chromosome inactivation in mammals.

Meiosis: reductive cell division occurring exclusively in testis and ovary, and resulting in the production of haploid cells, including sperm cells and egg cells.

Mendelian genetics: classical genetics, focuses on monogenic genes with high penetrance. The Mendelian genetics is a true paradigm, and is used in discussing the mode of inheritance.

Messenger RNA (mRNA): RNA molecules that are synthesized from a DNA template in the nucleus (a *gene*) and transported to ribosomes in the cytoplasm, where they serve as a template for the synthesis of protein (*translation*).

Microarray diagnostics: a rapidly developing tool increasingly used in pharmaceutical and genomics research, and has the potential for applications in high-throughput diagnostic devices. Microarrays can be made of DNA sequences with known gene mutations, polymorphisms, and selected protein molecules.

Microsatellite DNA: small array (often less than 0.1 kb) of short tandemly repeated DNA sequences; **Mini-satellite DNA**: an intermediate size array (often 0.1 to 20 kb long) of short tandemly repeated DNA sequences. **Hypervariable mini-satellite** DNA is the basis of DNA fingerprinting and many VNTR markers.

Missense mutation: substitution of a single DNA base that results in a codon that specifies an alternative amino acid.

Mitochondria: cellular organelles present in eukaryotic organisms that enable aerobic respiration and generate the energy to drive cellular processes. Each mitochondria contains a small amount of circular DNA encoding a small number of genes (approximately 50).

Mitosis: cell division in somatic cells.

Molecular genetic testing: molecular genetic testing for use in patient diagnosis, management, and genetic counseling; this is increasingly used in presymptomatic (predictive) genetic testing of "at-risk" family members using a previously known disease-causing mutation in the family.

Molecular genetic screening: screening a section of the population known to be at a higher risk to be heterozygous for one of the mutations in the gene for a common autosomal recessive disease; for example, screening for cystic fibrosis in northern European populations and for beta-thalassemia in Mediterranean and Middle East population groups.

Mosaic: a genetic mosaic is an individual who has two or more genetically different cell lines derived from a single zygote.

Multifactorial disease: any disease or disorder caused by interaction of multiple genetic (polygenic) and environmental factors.

Multigene family: a set of evolutionarily related loci within a genome, at least one of which can encode a functional product.

Mutation: a heritable alteration in the DNA sequence.

Natural selection: the process whereby some of the inherited genetic variation within a population will affect the ability of individuals to survive to reproduce (*fitness*).

Neutral mutation: a change or alteration in DNA sequence that has no phenotypic effect (or has no effect on fitness).

Newborn screening: performed in newborns in state public health programs to detect certain genetic diseases for which early diagnosis and treatment are available.

Noncoding sequence: a region of DNA that is not translated into protein. Some noncoding sequences are regulatory portions of genes; others may serve structural purposes (telomeres, centromeres), while others may not have any function.

Nonsense mutation: substitution of a single DNA base resulting in a stop codon, thus leading to the truncation of a protein.

Nucleotide: a subunit of the DNA or RNA molecule. A nucleotide is a base molecule (adenine, cytosine, guanine, and thymine in the case of DNA) linked to a sugar molecule (deoxyribose or ribose) and phosphate groups.

Oncogene: an acquired mutant form of a gene that acts to transform a normal cell into a cancerous one.

OMIM: acronym for McKusick's ***O***n-line ***M***endelian ***I***nheritance in ***M***an, a regularly updated electronic catalog of inherited human disorders and phenotypic traits accessible on the NCBI network. Each entry is designated by a number (*OMIM number*).

Penetrance: the likelihood that a person carrying a particular mutant gene will have an altered phenotype (*see phenotype*).

PFGE (pulse field gel electrophoresis): a form of gel electrophoresis that permits size fractionation of large DNA molecules.

Pharmacogenomics: a broad term (*see also pharmacogenetics*), now increasingly applied to the identification of genes along with all regulatory sequences that could lead to new drug discovery, drug development, and assessing individuals' variations in terms of the efficacy or toxicity of a drug.

Phenotype: the clinical and/or any other manifestation or expression, such as a biochemical immunological alteration, of a specific gene or genes, environmental factors, or both.

Point mutation: the substitution of a single DNA base in the normal DNA sequence.

Polygenic trait or character: a character or trait determined by the combined action of a number of loci, each with a small effect.

Polymerase chain reaction (PCR): a molecular biology technique developed in the mid-1980s through which specific DNA segments may be amplified selectively.

Polymorphism: the stable existence of two or more variant allelic forms of a gene within a particular population, or among different populations.

Positional cloning: the technique through which candidate genes are located in the genome through their co-inheritance with linked markers. It allows genes to be identified that lack information regarding the biochemical actions of their functional product.

Predictive testing: determines the probability that a healthy individual with or without a family history of a certain disease might develop that disease.

Predisposition, genetic: increased susceptibility to a particular disease due to the presence of one or more gene mutations, and/or a combination of alleles (haplotype), not necessarily abnormal, that is associated with an increased risk for the disease, and/or a family history that indicates an increased risk for the disease; **Predisposition test:** a test for a genetic predisposition (incompletely penetrant conditions). Not all people with a positive test result will manifest the disease during their lifetimes

Preimplantation genetic diagnosis (PIGD): used following in vitro fertilization to diagnose a genetic disease or condition in a preimplantation embryo.

Prenatal diagnosis: used to diagnose a genetic disease or condition in a developing fetus.

Presymptomatic test: predictive testing of individuals with a family history. Historically, the term has been used when testing for diseases or conditions such as Huntington's disease, where the likelihood of developing the condition (known as *penetrance*) is very high in people with a positive test result.

Primer: a short nucleic acid sequence, often a synthetic oligonucleotide, which binds specifically to a single strand of a target nucleic acid sequence and initiates synthesis, using a suitable polymerase, of a complementary strand.

Probe: a DNA or RNA fragment that has been labeled in some way, and used in a *molecular hybridization* assay to identify closely related DNA or RNA sequences.

Prokaryote: an organism or cell lacking a nucleus and other membrane bounded organelles. Bacteria are prokaryotic organisms.

Promoter: a combination of short sequence elements to which RNA polymerase binds in order to initiate transcription of a gene.

Protein: a protein is the biological effector molecule encoded by sequences of a gene. A protein molecule consists of one or more polypeptide chains of amino acid subunits. The functional action of a protein depends on its three-dimensional structure, which is determined by its amino acid composition.

Protein truncation test: a method of screening for chain-terminating mutations by artificially expressing a mutant allele in a coupled transcription-translation system.

Proteome: all of the proteins present in a cell or organism.

Proteomics: the development and application of techniques to investigate the protein products of the genome and how they interact to determine biological functions.

Recessive: an allele that has no phenotypic effect in the heterozygous state.

Recombinant DNA technology: the use of molecular biology techniques, such as restriction enzymes, ligation, and cloning, to transfer genes among organisms (*see genetic engineering*).

Regulatory sequence: a DNA sequence to which specific proteins bind to activate or repress the expression of a gene.

Repeat sequences: a stretch of DNA bases that occurs in the genome in multiple identical or closely related copies.

Replication: a process by which a new DNA strand is synthesized by copying an existing strand, using it as a template for the addition of a complementary bases, catalyzed by a DNA polymerase enzyme.

Reproductive cloning: techniques aimed at the generation of an organism with an identical genome to an existing organism.

Restriction enzymes: a family of enzymes derived from bacteria that cut DNA at specific sequences of bases.

Restriction fragment length polymorphism (RFLP): a polymorphism due to difference in size of allelic restriction fragments as a result of restriction site polymorphism.

Ribonucleic acid (RNA): a single stranded nucleic acid molecule comprising a linear chain made up from four nucleotide subunits (A, C, G, and U). There are three types of RNA: messenger, transfer, and ribosomal.

Risk communication: an important aspect of genetic counseling that involves pedigree analysis, interpretation of the inheritance pattern, genetic risk assessment, and explanation to the family member (or the family).

Screening: carrying out of a test or tests, examination(s), or procedure(s) in order to expose undetected abnormalities, unrecognized (incipient) diseases, or defects. Examples are early diagnosis of cancer using mass X-ray mammography for breast cancer, and cervical smears for cancer of the cervix.

Segregation: the separation of chromosomes (and the alleles they carry) during meiosis; alleles on different chromosomes segregate randomly among the gametes (and the progeny).

Sensitivity (of a screening test): extent (usually expressed as a percentage) to which a method gives results that are free from false negatives; the fewer the false negatives, the greater the sensitivity. Quantitatively, sensitivity is the proportion of truly diseased persons in the screened population who are identified as diseased by the screening test.

Sex chromosome: the pair of chromosomes that determines the sex (gender) of an organism. In man, one X and one Y chromosome constitute a male, compared to two X chromosomes in a female.

Sex selection: preferential selection of the unborn child on the basis of gender for social and cultural purposes. However, this may be acceptable for medical reasons; for example, to prevent the birth of a male assessed to be at risk for an X-linked recessive disease.

Shotgun sequencing: a cloning method in which total genomic DNA is randomly sheared and the fragments ligated into a cloning vector, also referred to as "shotgun" cloning; one of the techniques employed in the human genome project.

Single nucleotide polymorphism (SNP): a common variant in the genome sequence; the human genome contains about 10 million SNPs.

Somatic: all of the cells in the body that are not gametes (germ-line).

Southern blot hybridization: a form of molecular hybridization in which the target nucleic acid consists of DNA molecules that have been size fractioned by gel electrophoresis and subsequently transferred to a nitrocellulose or nylon membrane.

Splicing: a process, prior to transcription by mRNA, by which *introns* are removed and the *exons* adjoined.

Stem cell: a cell that has the potential to differentiate into a variety of different cell types depending on the environmental stimuli it receives.

Stop codon: a codon that leads to the termination of a protein rather than to the addition of an amino acid. The three stop codons are TGA, TAA, and TAG.

Synteny: a large group of genes that appear in the same order on the chromosomes of two different species.

Systems biology: a bioinformatics tool; refers to simultaneous measurement of thousands of molecular components (such as transcripts, proteins, and metabolites) and integrate these disparate data sets with clinical end points, in a biologically relevant manner; this model can be applied in understanding the etiology and pathogenesis of disease.

Telomere: the natural end of the chromosome.

Transcription: the process through which a gene is expressed to generate a complementary RNA molecule on a DNA template using RNA polymerase.

Transcription factor: a protein that binds DNA at specific sequences and regulates the transcription of specific genes.

Transgene: a gene from one source that has been incorporated into the genome of another organism; T**ransgenic animal/plant:** a fertile animal or plant that carries an introduced gene(s) in its germ-line.

Translation: a process through which a polypeptide chain of amino acid molecules is generated as directed by the sequence of a particular messenger RNA sequence.

Tumor suppressor gene: a gene that serves to protect cells from entering a cancerous state; according to Knudson's "two-hit" hypothesis, both alleles of a particular tumor suppressor gene must acquire a mutation before the cell will enter a transformed cancerous state.

Unequal crossing over: recombination between non-allelic sequences on non-sister chromatids of homologous chromosomes.

X-chromosome inactivation: random inactivation of one of the two X-chromosomes in mammals by a specialized form of genetic imprinting (*see Lyonization*).

INDEX

D

E

H